RECENT HANDBOOK PUBLICATIONS

HANDBOOK OF ANTIBIOTIC COMPOUNDS
János Bérdy
 Senior Research Fellow
 Institute of Drug Research
 Budapest, Hungary

HANDBOOK OF CHROMATOGRAPHY:
 AMINO ACIDS AND AMINES
Stanley Blackburn
 Research Scientist
 Wool Industries Research Association
 Leeds, England

HANDBOOK OF CLINICAL CHEMISTRY
Mario Werner
 Professor of Pathology
 (Laboratory Medicine)
 The George Washington University Medical
 Center
 Washington, D.C.

HANDBOOK OF ENVIRONMENTAL
 RADIATION
Alfred W. Klement, Jr.
 Environmental Scientist (deceased)
 Private Consultant
 Kensington, Maryland

HANDBOOK OF GEOPHYSICAL
 EXPLORATION AT SEA
Richard A. Geyer
 Professor Emeritus
 Department of Oceanography
 Texas A & M University
 College Station, Texas

HANDBOOK OF IRRIGATION
 TECHNOLOGY
Herman J. Finkel
 Director
 Finkel & Finkel, Ltd.
 Consulting Engineers
 Haifa, Israel

HANDBOOK OF LASER SCIENCE AND
 TECHNOLOGY
Marvin J. Weber
 Laser Program
 Lawrence Livermore National Laboratory
 University of California
 Livermore, California

HANDBOOK OF LETHALITY GUIDES FOR
 LOW-ACID CANNED FOODS
C. R. Stumbo   **D. A. Evans**
K. S. Purohit   **F. J. Francis**
T. V. Ramakrishnan
 Department of Food Science and Nutrition
 University of Massachusetts
 Amherst, Massachusetts

HANDBOOK OF LUBRICATION
E. R. Booser
 Manager
 Systems Engineering Subsection
 Turbine Technology Laboratory
 General Electric Company
 Schenectady, New York

HANDBOOK OF PROCESSING AND
 UTILIZATION IN AGRICULTURE
Ivan A. Wolff
 Director (retired)
 Eastern Regional Research Center
 Agricultural Research Service
 U.S. Department of Agriculture
 Philadelphia, Pennsylvania

HANDBOOK OF TERPENOIDS:
 MONOTERPENOIDS
Sukh Dev
 Multi-Chem Research Centre
 Nandesari, Vadodara, India
J. S. Yadav
 National Chemical Laboratory
 Division of Organic Chemistry
 Poona, India
Anubhav Narula
 Lederle Laboratory
 American Cyanamid Co.
 Medical Research Div.
 Pearl River, New York

CRC Handbook
of
Chemistry and Physics

A Ready-Reference Book of Chemical and Physical Data

HANDBOOK OF CHEMISTRY AND PHYSICS

1983—1984

64th

EDITION

CRC PRESS, INC.

Editor-in-Chief

Robert C. Weast, Ph.D.

Associate Editors

Melvin J. Astle, Ph.D.
William H. Beyer, Ph.D.

In collaboration with a large number of professional chemists and physicists whose assistance is acknowledged in the list of general collaborators and in connection with the particular tables or sections involved.

CRC

CRC Press, Inc.
Boca Raton, Florida

PREFACE

The *Handbook of Chemistry and Physics,* continuing the policy of the past, is being revised at frequent intervals.

The general features and scheme of arrangement, which have received extensive endorsement in former editions have been retained. The aim throughout has been to present in condensed form as large amount of accurate, reliable and up-to-date information on the fields of chemistry and physics as was consistent with convenience in form and possibility of wide distribution and utility. A very large portion of the tables have been compiled especially for the *Handbook* from various authoritative collections of data from current journals and bulletins. Credit and acknowledgement are cited in such instances.

Since the beginning, special consideration has been given to requests and suggestions of those who have used former editions. In this way it has been hoped to develop the book along the lines most acceptable to those interested in a volume of this type. Suggestions and contributions are received each year from eminent chemists and physicists including those in the teaching profession and those engaged in research in industry. We believe this to have been of great value in the growth and development of the *Handbook.*

An attempt has been made to include material on all branches of chemistry, physics, and the closely allied sciences, which would be likely to find any extended use. However, in order to retain the convenience of moderate dimensions and of a single volume and at the same time allow for the natural growth due to the extension of knowledge in these sciences, and logical additions along the lines already developed, there has been the necessity to exclude types of material for use only in highly specialized lines of work.

Chemistry and physics, always closely related sciences, have been brought into more intimate relations by the more recent research results and discoveries. To an increasing extent the student of either science should have knowledge of the other. It would seem there should be a large field for a single volume containing constants, formulas, conversion tables, and physical data adequate for accurate computations. The generous response which the previous 63 editions have met indicates that the volumes have been found useful and it is with the hope of more nearly completely meeting the needs of chemists and physicists that this edition and succeeding editions are offered.

This current edition contains an extensive table on the Optical Properties of Metals in addition to tables on the Resistivities of Semi-Conducting Minerals and Minerals Listed in Order of Increasing Vickers Hardness. There has been extensive revision and expansion of the tables: Solubility Product Constants and the Electrochemical Series. Further revisions include the section dealing with the description of the Elements, the tables of Strengths of Chemical Bonds, Elements in the Earth's Crust, Elements in Sea Water, and Particle Properties with the latter including properties for stable particles, Mesons and Baryons. The nomenclature of the sugars which appear in the Organic Table has been updated and the structures of such sugars have been redrawn. The *Handbook* continues to list the publications of the National Standard Reference Data System Publications, and those of Fachinformation have been added.

There have been some errors in the book which have been called to the attention of the editor. All of these have been corrected in this edition. As always, suggestions for further improvement of the book are welcome and appreciated.

R. C. Weast
March, 1983

COLLABORATORS AND CONTRIBUTORS

L. Alberts, Ph.D.
Director-General
National Institute for Metallurgy
Randburg, South Africa

C. J. Allen
Illuminating Engineer
General Electric Company
Nela Park
Cleveland, Ohio

W. A. Anderson, Ph.D.
Analytical Instrument Research
Varian Associates
611 Hansen Way
Palo Alto, California

W. J. Armento
Oak Ridge National Laboratory
P. O. Box X
Oak Ridge, Tennessee

J. Askill, Ph.D.
Chairman, Physics Department
Millikin University
Decatur, Illinois

J. C. Bailar, Jr., Ph.D.
Professor of Inorganic Chemistry
University of Illinois
Urbana, Illinois

J. A. Bearden, Ph.D.
Department of Physics
Johns Hopkins University
Baltimore, Maryland

A. H. Benade, Ph.D.
Department of Physics
Case Western Reserve University
Cleveland, Ohio

F. F. Bentley
Air Force Materials Laboratory
Wright-Patterson Air Force Base,
  Ohio

William H. Beyer, Ph.D.
Professor of Mathematics and
  Statistics
Head of the Department of
  Mathematics and Statistics
University of Akron
Akron, Ohio

J. H. Billman, Ph.D.
Professor of Chemistry
Indiana University
Bloomington, Indiana

A. L. Bloom, Ph.D.
Spectra-Physics, Inc.
1255 Terra Bella Avenue
Mountain View, California

J. A. Bradley, Ph.D.
Dean, Newark College of
  Engineering
323 High Street
Newark, New Jersey

B. H. Brown, Ph.D.
Dartmouth College
Hanover, New Hampshire

J. E. Brown
Eastman Kodak Company
Rochester, New York

A. B. Burg, Ph.D.
Department of Chemistry
University of Southern California
Los Angeles, California

G. P. Burns, Ph.D.
Department of Physics
Mary Washington College
Fredericksburg, Virginia

A. F. Burr, Ph.D.
Department of Physics
New Mexico State University
Las Cruces, New Mexico

COLLABORATORS AND CONTRIBUTORS (Continued)

J. P. Catchpole, Ph.D.
   Admiralty Materials Laboratory
   Holton Heath
   Dorset, England

James C. Chang, Ph.D.
   Department of Chemistry
   University of Northern Iowa
   Cedar Falls, Iowa

Roger Clay
   Department of Physics
   The University of Adelaide
   Adelaide, South Australia 5001
   Australia

E. Richard Cohen, Ph.D.
   Associate Director
   Science Center/Aerospace and
      Systems Group
   North American Rockwell
      Corporation
   Thousand Oaks, California

Charles H. Corliss
   Spectroscopy Section
   Optical Physics Division
   National Bureau of Standards
   Washington, D. C.

M. Davies, Ph.D
   Edward Davies Chemical
      Laboratories
   University of Wales
   Aberystwyth, Wales

J. DeMent, D.Sc.
   Dement Laboratories
   4847 S.E. Division Street
   Portland Oregon

H. G. Deming, Ph.D
   2316 Tuttle Terrace
   Sarasota, Florida

E. DiCyan, Ph.D.
   Consulting Chemist
   420 Lexington Ave.
   New York, New York

Hans Dolezalek
   Department of the Navy
   Office of Naval Research
   Arlington, Virginia

A. P. Dunlop, Ph.D.
   Director of Chemical Research and
      Development
   John Stuart Research Laboratories
   The Quaker Oats Company
   Barrington, Illinois

L. M. Foster, Ph.D.
   Thomas J. Watson Research Center
   International Business Machines
      Corp.
   Yorktown Heights, New York

J. L. Franklin, Ph.D.
   Welch Professor of Chemistry
   William Marsh Rice University
   Houston, Texas

G. Fulford, Ph.D.
   Assistant Professor of Chemical
      Engineering
   University of Waterloo
   Waterloo, Ontario,
   Canada

Gladys H. Fuller
   NBS Institute of Science and
      Technology
   National Bureau of Standards
   Washington, D.C.

E. F. Furtsch, Ph.D.
   Department of Chemistry
   Virginia Poytechnic Institute
   Blacksburg, Virginia

L. A. Gillette, Ph.D.
   Manager, Product Development
   Pennsalt Chemical Corporation
   Philadelphia, Pennsylvania

H. Gilman, Ph.D.
   Department of Chemistry
   Iowa State University
   Ames, Iowa

B. Girling, M.Sc., F.I.M.A.
Department of Mathematics
The City University
London E.C.1, England

E. C. Gregg, Ph.D.
School of Medicine
Case Western Reserve University
Cleveland, Ohio

A. G. Gregory
Department of Physics
The University of Adelaide
Adelaide, South Australia 5001
Australia

R. R. Gupta, M.Sc., Ph.D.
Department of Chemistry
University of Rajasthan
Jaipur-4, India

C. R. Hammond
17 Greystone
West Hartford, Connecticut

W. E. Harris
Professor of Chemistry
University of Alberta
Edmonton, Alberta, Canada

H. J. Harwood, Ph.D.
Head, Durkee Famous Foods
Organic Chemistry Division
Glidden Company
Chicago, Illinois

R. L. Heath
Atomic Energy Division
Phillips Petroleum Co.
Idaho Falls, Idaho

R. W. Hoffman, Ph.D.
Department of Physics
Case Western Reserve University
Cleveland, Ohio

Jesse F. Hunsberger
RD #1
East Cedarville Road
Pottstown, Pennsylvania

C. D. Hurd, Ph.D.
Chemical Laboratory
Northwestern University
Evanston, Illinois

H. D. B. Jenkins, Ph.D.
Department of Molecular Sciences
University of Warwick
Coventry CV4 7AL
England

J. L. Kassner, Ph.D.
Department of Physics
The University of Missouri-Rolla
Rolla, Missouri

G. Lang, Dipl. Ing.
A-4963
St. Peter am Hart
Austria

Sidney O. Kastner, Ph.D.
National Aeronautics and Space
  Administration
Goddard Space Flight Center
Greenbelt, Maryland

Olga Kennard, Ph.D.
University Chemical Laboratory
Cambridge, England

T. G. Kennard, Ph.D.
20747 E. Palm Drive
Glendora, California

J. A. Kerr, Ph.D.
Chemistry Department
The University of Birmingham
Haworth Building
Birmingham 15, England

A. L. King, Ph.D.
Department of Physics
Dartmouth College
Hanover, New Hampshire

C. R. Kinney, Ph.D.
1318 27th Street
Des Moines, Iowa

R. Kretz, Ph.D.
Department of Science and
 Engineering
University of Ottawa
Ottawa, Ontario, Canada K1N 6N5

D. F. Lawden, Sc.D.
Department of Mathematics
University of Aston
Birmingham, England

K. Lee, Ph.D.
IBM Research Laboratory
San Jose, California

A. P. Levitt
75 Lovett Road
Newton Center, Massachusetts

M. M. MacMasters, Ph.D.
Department of Grain Science and
 Industry
Kansas State University
Manhattan, Kansas

W. Mahlig
Assistant Sales Manager
Laboratory Equipment Division
The W. S. Tyler Company
Cleveland, Ohio

C. J. Major, Ph.D.
Department of Chemical Engineering
University of Akron
Akron, Ohio

Georgia A. Martin, Ph.D.
Atomic and Plasma Radiation
 Division
National Bureau of Standards
Washington, D.C.

J. C. McGowan, Ph.D., D.Sc.
"Quantock"
13 Moreton Avenue
Harpenden, Herts AL5 2EU,
 England

L. Meites, Ph.D.
Head, Department of Chemistry
Clarkson College of Technology
Potsdam, New York

M. G. Mellon, Ph.D.
Professor Emeritus, Analytical
 Chemistry
Purdue University
West Lafayette, Indiana

H. B. Michaelson
IBM Journal of Research and
 Development
Armonk, New York

Karl Z. Morgan, Ph.D.
School of Nuclear Engineering
Georgia Institute of Technology
Atlanta, Georgia

R. R. Nimmo, Ph.D.
Professor of Physics
University of Otago
Dunedin, New Zealand

F. J. Norton, Ph.D.
General Electric Company
1133 Eastern Avenue
Schenectady, New York

F. M. Page
Department of Chemistry
The University of Aston
Birmingham B4 7ET, England

B. R. Pamplin, Ph.D.
School of Physics
University of Bath
Claverton Down
Bath BA2 7AY, England

W. Parker, Ph.D.
Deparment of Physics
University of California
Irvine, California

M. J. Parsonage
Chemistry Department
The University of Birmingham
Birmingham 15, England

Saul Patai, Ph.D.
Department of Organic Chemistry
Hebrew University of Jerusalem
Jerusalem, Israel

COLLABORATORS AND CONTRIBUTORS (Continued)

I. A. Pearl, Ph.D.
The Institute of Paper Chemistry
Appleton, Wisconsin

R. F. Peart, Ph.D.
The Plessey Co. Ltd.
Allen Clark Research Centre
Northampton, England

A. C. Peed, Jr.
Eastman Kodak Company
243 State Street
Rochester, New York

R. Pepinsky, Ph.D.
Department of Physics and
Astronomy
University of Florida
Gainesville, Florida

Daniel D. Pollock
Department of Mechanical and
Aerospace Engineering
State University of New York at
Buffalo
Buffalo, New York

H. A. Poul, Ph.D.
Professor of Physics
Oklahoma State University
Stillwater, Oklahoma

Richard L. Pratt
Staff Analyst
Data Corporation
Dayton, Ohio

H. J. Prebluda, Ph.D.
Consulting Biochemist
3 Belmont Circle
Trenton, New Jersey

I. B. Prettyman, M.S.
The Firestone Tire and Rubber Co.
1200 Firestone Parkway
Akron, Ohio

Zvi Rappoport, Ph.D.
Department of Organic Chemistry
Hebrew University of Jerusalem
Jerusalem, Israel

E. H. Ratcliffe
Institution of Electrical Engineers
Stenvenage
Herts SG1 1HQ
England

Joseph Reader, Ph.D.
Atomic and Plasma Radiation
Division
National Bureau of Standards
Washington, D.C.

M. C. Reed, Ph.D.
1368 Wood Valley Road
Mountainside, New Jersey

B. W. Roberts, Ph.D.
General Electric Research
Laboratory
Schenectady, New York

R. C. Roberts, Ph.D.
Professor Emeritus of Chemistry
Colgate University
Hamilton, New York

R. A. Robinson
School of Chemistry
The University
Newcastle-upon-Tyne NE1 7RU,
England

R. J. Rosen
Consulting Chemist
9301 Parkhill Drive
Los Angeles, California

A. H. Rosenfeld, Ph. D.
Lawrence Radiation Laboratory
University of California
Berkeley, California

Gordon D. Rowe
Specialist Lighting of GE Properties
General Electric Company
Cleveland, Ohio

A. L. Rozek
Velsicol Chemical Corporation
Chicago, Illinois

S. I. Salem, Ph.D.
Professor, Department of Physics
  and Astronomy
California State University
Long Beach, California

Calvin E. Scheldknecht
R.D. 8
Box 398
Gettysburg, Pennsylvania

G. T. Seaborg, Ph.D.
Lawrence Berkeley Laboratory
University of California
Berkeley, California

R. Shaw
Chemical Physicist
Physical Organic Program
Stanford Research Institute
Menlo Park, California

J. R. Shelton, Ph.D.
Department of Chemistry
Case Western Reserve University
Cleveland, Ohio

G. W. Smith, Ph.D.
General Motors Corporation
Research Laboratories
Warren, Michigan

J. M. Smith, B.S.E.E.
Product Planning Large Lamp
  Department
General Electric Company
Cleveland, Ohio

L. D. Smithson
Air Force Materials Laboratory
Wright-Patterson Air Force Base,
  Ohio

F. H. Spedding
Director, Ames Laboratory
Iowa State University
Ames, Iowa

R. H. Stokes, Ph.D.
Department of Chemistry
The University of New England
Armidale, N.S.W., Australia

Donald F. Swinehart, Ph.D.
Department of Chemistry
University of Oregon
Eugene, Oregon

A. Tarpinian
Army Materials and Mechanics
  Research Center
Arsenal Street
Watertown, Massachusetts

B. N. Taylor, Ph.D.
A-247-Metrology
National Bureau of Standards
Washington, D.C.

D. H. Tomlin, Ph.D.
Department of Physics
University of Reading
Reading, Berkshire, England

A. F. Trotman-Dickenson, Ph.D.
Institute of Technology
Cardiff, England

Petr Vanýsek, Ph.D.
Department of Chemistry
University of North
  Carolina
Chapel Hill, North Carolina

H. B. Vickery, Ph.D.
Connecticut Agricultural
  Experimental Station
112 Huntington Street
New Haven, Connecticut

John Weaver
Department of Chemical Engineering
  and Materials Science
University of Minnesota
Minneapolis, Minnesota

W. W. Wendlandt, Ph.D.
Professor of Chemistry
University of Houston
Houston, Texas

N. R. Whetten, Ph.D.
Research and Development Center
General Electric Company
P. O. Box 8
Schenectady, New York

COLLABORATORS AND CONTRIBUTORS (Continued)

Wolfgang L. Wiese, Ph.D.
Atomic Plasma and Radiation
  Division
National Bureau of Standards
Washington, D.C.

J. H. Yoe, Ph.D.
Professor Emeritus of Chemistry
University of Virginia
Charlottesville, Virginia

G. R. Yohe, Ph.D.
Illinois State Geological Survey
University of Illinois Campus
Urbana, Illinois

T. F. Young, Ph.D.
Division of Chemical Engineering
Argonne National Laboratory
Argonne, Illinois

J. Zabicky, Ph.D.
Department of Biophysics
The Weizmann Institute of Science
Rehovoth, Israel

S. Zuffanti, A.M.
Professor of Chemistry
Northeastern University
Boston, Massachusetts

The Publishers and Editors will be grateful to readers of this Handbook who will call their attention to errors that may be discovered. Suggestions for improvement are also welcome.

TABLE OF CONTENTS

MISCELLANEOUS MATHEMATICAL CONSTANTS

π CONSTANTS

$$\pi = 3.14159\ 26535\ 89793\ 23846\ 26433\ 83279\ 50288\ 41971\ 69399\ 37511$$
$$1/\pi = 0.31830\ 98861\ 83790\ 67153\ 77675\ 26745\ 02872\ 40689\ 19291\ 48091$$
$$\pi^2 = 9.86960\ 44010\ 89358\ 61883\ 44909\ 99876\ 15113\ 53136\ 99407\ 24079$$
$$\log_e \pi = 1.14472\ 98858\ 49400\ 17414\ 34273\ 51353\ 05871\ 16472\ 94812\ 91531$$
$$\log_{10} \pi = 0.49714\ 98726\ 94133\ 85435\ 12682\ 88290\ 89887\ 36516\ 78324\ 38044$$
$$\log_{10} \sqrt{2\pi} = 0.39908\ 99341\ 79057\ 52478\ 25035\ 91507\ 69595\ 02099\ 34102\ 92128$$

CONSTANTS INVOLVING e

$$e = 2.71828\ 18284\ 59045\ 23536\ 02874\ 71352\ 66249\ 77572\ 47093\ 69996$$
$$1/e = 0.36787\ 94411\ 71442\ 32159\ 55237\ 70161\ 46086\ 74458\ 11131\ 03177$$
$$e^2 = 7.38905\ 60989\ 30650\ 22723\ 04274\ 60575\ 00781\ 31803\ 15570\ 55185$$
$$M = \log_{10} e = 0.43429\ 44819\ 03251\ 82765\ 11289\ 18916\ 60508\ 22943\ 97005\ 80367$$
$$1/M = \log_e 10 = 2.30258\ 50929\ 94045\ 68401\ 79914\ 54684\ 36420\ 76011\ 01488\ 62877$$
$$\log_{10} M = 9.63778\ 43113\ 00536\ 78912\ 29674\ 98645\ -10$$

π^e AND e^π CONSTANTS

$$\pi^e = 22.45915\ 77183\ 61045\ 47342\ 71522$$
$$e^\pi = 23.14069\ 26327\ 79269\ 00572\ 90864$$
$$e^{-\pi} = 0.04321\ 39182\ 63772\ 24977\ 44177$$
$$e^{\frac{1}{2}\pi} = 4.81047\ 73809\ 65351\ 65547\ 30357$$
$$i^i = e^{-\frac{1}{2}\pi} = 0.20787\ 95763\ 50761\ 90854\ 69556$$

NUMERICAL CONSTANTS

$$\sqrt{2} = 1.41421\ 35623\ 73095\ 04880\ 16887\ 24209\ 69807\ 85696\ 71875\ 37695$$
$$\sqrt[3]{2} = 1.25992\ 10498\ 94873\ 16476\ 72106\ 07278\ 22835\ 05702\ 51464\ 70151$$
$$\log_e 2 = 0.69314\ 71805\ 59945\ 30941\ 72321\ 21458\ 17656\ 80755\ 00134\ 36026$$
$$\log_{10} 2 = 0.30102\ 99956\ 63981\ 19521\ 37388\ 94724\ 49302\ 67881\ 89881\ 46211$$
$$\sqrt{3} = 1.73205\ 08075\ 68877\ 29352\ 74463\ 41505\ 87236\ 69428\ 05253\ 81039$$
$$\sqrt[3]{3} = 1.44224\ 95703\ 07408\ 38232\ 16383\ 10780\ 10958\ 83918\ 69253\ 49935$$
$$\log_e 3 = 1.09861\ 22886\ 68109\ 69139\ 52452\ 36922\ 52570\ 46474\ 90557\ 82275$$
$$\log_{10} 3 = 0.47712\ 12547\ 19662\ 43729\ 50279\ 03255\ 11530\ 92001\ 28864\ 19070$$

OTHER CONSTANTS

$$\text{Euler's Constant } \gamma = 0.57721\ 56649\ 01532\ 86061$$
$$\log_e \gamma = -0.54953\ 93129\ 81644\ 82234$$
$$\text{Golden Ratio } \phi = 1.61803\ 39887\ 49894\ 84820\ 45868\ 34365\ 63811\ 77203\ 09180$$

EXPONENTIAL FUNCTIONS

x	e^x	$Log_{10}(e^x)$	e^{-x}	x	e^x	$Log_{10}(e^x)$	e^{-x}
0.00	1.0000	0.00000	1.000000	**0.50**	1.6487	0.21715	0.606531
0.01	1.0101	.00434	0.990050	0.51	1.6653	.22149	.600496
0.02	1.0202	.00869	.980199	0.52	1.6820	.22583	.594521
0.03	1.0305	.01303	.970446	0.53	1.6989	.23018	.588605
0.04	1.0408	.01737	.960789	0.54	1.7160	.23452	.582748
0.05	1.0513	0.02171	0.951229	**0.55**	1.7333	0.23886	0.576950
0.06	1.0618	.02606	.941765	0.56	1.7507	.24320	.571209
0.07	1.0725	.03040	.932394	0.57	1.7683	.24755	.565525
0.08	1.0833	.03474	.923116	0.58	1.7860	.25189	.559898
0.09	1.0942	.03909	.913931	0.59	1.8040	.25623	.554327
0.10	1.1052	0.04343	0.904837	**0.60**	1.8221	0.26058	0.548812
0.11	1.1163	.04777	.895834	0.61	1.8404	.26492	.543351
0.12	1.1275	.05212	.886920	0.62	1.8589	.26926	.537944
0.13	1.1388	.05646	.878095	0.63	1.8776	.27361	.532592
0.14	1.1503	.06080	.869358	0.64	1.8965	.27795	.527292
0.15	1.1618	0.06514	0.860708	**0.65**	1.9155	0.28229	0.522046
0.16	1.1735	.06949	.852144	0.66	1.9348	.28663	.516851
0.17	1.1853	.07383	.843665	0.67	1.9542	.29098	.511709
0.18	1.1972	.07817	.835270	0.68	1.9739	.29532	.506617
0.19	1.2092	.08252	.826959	0.69	1.9937	.29966	.501576
0.20	1.2214	0.08686	0.818731	**0.70**	2.0138	0.30401	0.496585
0.21	1.2337	.09120	.810584	0.71	2.0340	.30835	.491644
0.22	1.2461	.09554	.802519	0.72	2.0544	.31269	.486752
0.23	1.2586	.09989	.794534	0.73	2.0751	.31703	.481909
0.24	1.2712	.10423	.786628	0.74	2.0959	.32138	.477114
0.25	1.2840	0.10857	0.778801	**0.75**	2.1170	0.32572	0.472367
0.26	1.2969	.11292	.771052	0.76	2.1383	.33006	.467666
0.27	1.3100	.11726	.763379	0.77	2.1598	.33441	.463013
0.28	1.3231	.12160	.755784	0.78	2.1815	.33875	.458406
0.29	1.3364	.12595	.748264	0.79	2.2034	.34309	.453845
0.30	1.3499	0.13029	0.740818	**0.80**	2.2255	0.34744	0.449329
0.31	1.3634	.13463	.733447	0.81	2.2479	.35178	.444858
0.32	1.3771	.13897	.726149	0.82	2.2705	.35612	.440432
0.33	1.3910	.14332	.718924	0.83	2.2933	.36046	.436049
0.34	1.4049	.14766	.711770	0.84	2.3164	.36481	.431711
0.35	1.4191	0.15200	0.704688	**0.85**	2.3396	0.36915	0.427415
0.36	1.4333	.15635	.697676	0.86	2.3632	.37349	.423162
0.37	1.4477	.16069	.690734	0.87	2.3869	.37784	.418952
0.38	1.4623	.16503	.683861	0.88	2.4109	.38218	.414783
0.39	1.4770	.16937	.677057	0.89	2.4351	.38652	.410656
0.40	1.4918	0.17372	0.670320	**0.90**	2.4596	0.39087	0.406570
0.41	1.5068	.17806	.663650	0.91	2.4843	.39521	.402524
0.42	1.5220	.18240	.657047	0.92	2.5093	.39955	.398519
0.43	1.5373	.18675	.650509	0.93	2.5345	.40389	.394554
0.44	1.5527	.19109	.644036	0.94	2.5600	.40824	.390628
0.45	1.5683	0.19543	0.637628	**0.95**	2.5857	0.41258	0.386741
0.46	1.5841	.19978	.631284	0.96	2.6117	.41692	.382893
0.47	1.6000	.20412	.625002	0.97	2.6379	.42127	.379083
0.48	1.6161	.20846	.618783	0.98	2.6645	.42561	.375311
0.49	1.6323	.21280	.612626	0.99	2.6912	.42995	.371577
0.50	1.6487	0.21715	0.606531	**1.00**	2.7183	0.43429	0.367879

x	e^x	$\text{Log}_{10}(e^x)$	e^{-x}	x	e^x	$\text{Log}_{10}(e^x)$	e^{-x}
1.00	2.7183	0.43429	0.367879	1.50	4.4817	0.65144	0.223130
1.01	2.7456	.43864	.364219	1.51	4.5267	.65578	.220910
1.02	2.7732	.44298	.360595	1.52	4.5722	.66013	.218712
1.03	2.8011	.44732	.357007	1.53	4.6182	.66447	.216536
1.04	2.8292	.45167	.353455	1.54	4.6646	.66881	.214381
1.05	2.8577	0.45601	0.349938	1.55	4.7115	0.67316	0.212248
1.06	2.8864	.46035	.346456	1.56	4.7588	.67750	.210136
1.07	2.9154	.46470	.343009	1.57	4.8066	.68184	.208045
1.08	2.9447	.46904	.339596	1.58	4.8550	.68619	.205975
1.09	2.9743	.47338	.336216	1.59	4.9037	.69053	.203926
1.10	3.0042	0.47772	0.332871	1.60	4.9530	0.69487	0.201897
1.11	3.0344	.48207	.329559	1.61	5.0028	.69921	.199888
1.12	3.0649	.48641	.326280	1.62	5.0531	.70356	.197899
1.13	3.0957	.49075	.323033	1.63	5.1039	.70790	.195930
1.14	3.1268	.49510	.319819	1.64	5.1552	.71224	.193980
1.15	3.1582	0.49944	0.316637	1.65	5.2070	0.71659	0.192050
1.16	3.1899	.50378	.313486	1.66	5.2593	.72093	.190139
1.17	3.2220	.50812	.310367	1.67	5.3122	.72527	.188247
1.18	3.2544	.51247	.307279	1.68	5.3656	.72961	.186374
1.19	3.2871	.51681	.304221	1.69	5.4195	.73396	.184520
1.20	3.3201	0.52115	0.301194	1.70	5.4739	0.73830	0.182684
1.21	3.3535	.52550	.298197	1.71	5.5290	.74264	.180866
1.22	3.3872	.52984	.295230	1.72	5.5845	.74699	.179066
1.23	3.4212	.53418	.292293	1.73	5.6407	.75133	.177284
1.24	3.4556	.53853	.289384	1.74	5.6973	.75567	.175520
1.25	3.4903	0.54287	0.286505	1.75	5.7546	0.76002	0.173774
1.26	3.5254	.54721	.283654	1.76	5.8124	.76436	.172045
1.27	3.5609	.55155	.280832	1.77	5.8709	.76870	.170333
1.28	3.5966	.55590	.278037	1.78	5.9299	.77304	.168638
1.29	3.6328	.56024	.275271	1.79	5.9895	.77739	.166960
1.30	3.6693	0.56458	0.272532	1.80	6.0496	0.78173	0.165299
1.31	3.7062	.56893	.269820	1.81	6.1104	.78607	.163654
1.32	3.7434	.57327	.267135	1.82	6.1719	.79042	.162026
1.33	3.7810	.57761	.264477	1.83	6.2339	.79476	.160414
1.34	3.8190	.58195	.261846	1.84	6.2965	.79910	.158817
1.35	3.8574	0.58630	0.259240	1.85	6.3598	0.80344	0.157237
1.36	3.8962	.59064	.256661	1.86	6.4237	.80779	.155673
1.37	3.9354	.59498	.254107	1.87	6.4883	.81213	.154124
1.38	3.9749	.59933	.251579	1.88	6.5535	.81647	.152590
1.39	4.0149	.60367	.249075	1.89	6.6194	.82082	.151072
1.40	4.0552	0.60801	0.246597	1.90	6.6859	0.82516	0.149569
1.41	4.0960	.61236	.244143	1.91	6.7531	.82950	.148080
1.42	4.1371	.61670	.241714	1.92	6.8210	.83385	.146607
1.43	4.1787	.62104	.239309	1.93	6.8895	.83819	.145148
1.44	4.2207	.62538	.236928	1.94	6.9588	.84253	.143704
1.45	4.2631	0.62973	0.234570	1.95	7.0287	0.84687	0.142274
1.46	4.3060	.63407	.232236	1.96	7.0993	.85122	.140858
1.47	4.3492	.63841	.229925	1.97	7.1707	.85556	.139457
1.48	4.3929	.64276	.227638	1.98	7.2427	.85990	.138069
1.49	4.4371	.64710	.225373	1.99	7.3155	.86425	.136695
1.50	4.4817	0.65144	0.223130	2.00	7.3891	0.86859	0.135335

x	e^x	$\text{Log}_{10}(e^x)$	e^{-x}	x	e^x	$\text{Log}_{10}(e^x)$	e^{-x}
2.00	7.3891	0.86859	0.135335	**2.50**	12.182	1.08574	0.082085
2.01	7.4633	.87293	.133989	2.51	12.305	1.09008	.081268
2.02	7.5383	.87727	.132655	2.52	12.429	1.09442	.080460
2.03	7.6141	.88162	.131336	2.53	12.554	1.09877	.079659
2.04	7.6906	.88596	.130029	2.54	12.680	1.10311	.078866
2.05	7.7679	0.89030	0.128735	**2.55**	12.807	1.10745	0.078082
2.06	7.8460	.89465	.127454	2.56	12.936	1.11179	.077305
2.07	7.9248	.89899	.126186	2.57	13.066	1.11614	.076536
2.08	8.0045	.90333	.124930	2.58	13.197	1.12048	.075774
2.09	8.0849	.90768	.123687	2.59	13.330	1.12482	.075020
2.10	8.1662	0.91202	0.122456	**2.60**	13.464	1.12917	0.074274
2.11	8.2482	.91636	.121238	2.61	13.599	1.13351	.073535
2.12	8.3311	.92070	.120032	2.62	13.736	1.13785	.072803
2.13	8.4149	.92505	.118837	2.63	13.874	1.14219	.072078
2.14	8.4994	.92939	.117655	2.64	14.013	1.14654	.071361
2.15	8.5849	0.93373	0.116484	**2.65**	14.154	1.15088	0.070651
2.16	8.6711	.93808	.115325	2.66	14.296	1.15522	.069948
2.17	8.7583	.94242	.114178	2.67	14.440	1.15957	.069252
2.18	8.8463	.94676	.113042	2.68	14.585	1.16391	.068563
2.19	8.9352	.95110	.111917	2.69	14.732	1.16825	.067881
2.20	9.0250	0.95545	0.110803	**2.70**	14.880	1.17260	0.067206
2.21	9.1157	.95979	.109701	2.71	15.029	1.17694	.066537
2.22	9.2073	.96413	.108609	2.72	15.180	1.18128	.065875
2.23	9.2999	.96848	.107528	2.73	15.333	1.18562	.065219
2.24	9.3933	.97282	.106459	2.74	15.487	1.18997	.064570
2.25	9.4877	0.97716	0.105399	**2.75**	15.643	1.19431	0.063928
2.26	9.5831	.98151	.104350	2.76	15.800	1.19865	.063292
2.27	9.6794	.98585	.103312	2.77	15.959	1.20300	.062662
2.28	9.7767	.99019	.102284	2.78	16.119	1.20734	.062039
2.29	9.8749	.99453	.101266	2.79	16.281	1.21168	.061421
2.30	9.9742	0.99888	0.100259	**2.80**	16.445	1.21602	0.060810
2.31	10.074	1.00322	.099261	2.81	16.610	1.22037	.060205
2.32	10.176	1.00756	.098274	2.82	16.777	1.22471	.059606
2.33	10.278	1.01191	.097296	2.83	16.945	1.22905	.059013
2.34	10.381	1.01625	.096328	2.84	17.116	1.23340	.058426
2.35	10.486	1.02059	0.095369	**2.85**	17.288	1.23774	0.057844
2.36	10.591	1.02493	.094420	2.86	17.462	1.24208	.057269
2.37	10.697	1.02928	.093481	2.87	17.637	1.24643	.056699
2.38	10.805	1.03362	.092551	2.88	17.814	1.25077	.056135
2.39	10.913	1.03796	.091630	2.89	17.993	1.25511	.055576
2.40	11.023	1.04231	0.090718	**2.90**	18.174	1.25945	0.055023
2.41	11.134	1.04665	.089815	2.91	18.357	1.26380	.054476
2.42	11.246	1.05099	.088922	2.92	18.541	1.26814	.053934
2.43	11.359	1.05534	.088037	2.93	18.728	1.27248	.053397
2.44	11.473	1.05968	.087161	2.94	18.916	1.27683	.052866
2.45	11.588	1.06402	0.086294	**2.95**	19.106	1.28117	0.052340
2.46	11.705	1.06836	.085435	2.96	19.298	1.28551	.051819
2.47	11.822	1.07271	.084585	2.97	19.492	1.28985	.051303
2.48	11.941	1.07705	.083743	2.98	19.688	1.29420	.050793
2.49	12.061	1.08139	.082910	2.99	19.886	1.29854	.050287
2.50	12.182	1.08574	0.082085	**3.00**	20.086	1.30288	0.049787

x	e^x	$Log_{10}(e^x)$	e^{-x}	x	e^x	$Log_{10}(e^x)$	e^{-x}
3.00	20.086	1.30288	0.049787	**3.50**	33.115	1.52003	0.030197
3.01	20.287	1.30723	.049292	3.51	33.448	1.52437	.029897
3.02	20.491	1.31157	.048801	3.52	33.784	1.52872	.029599
3.03	20.697	1.31591	.048316	3.53	34.124	1.53306	.029305
3.04	20.905	1.32026	.047835	3.54	34.467	1.53740	.029013
3.05	21.115	1.32460	0.047359	**3.55**	34.813	1.54175	0.028725
3.06	21.328	1.32894	.046888	3.56	35.163	1.54609	.028439
3.07	21.542	1.33328	.046421	3.57	35.517	1.55043	.028156
3.08	21.758	1.33763	.045959	3.58	35.874	1.55477	.027876
3.09	21.977	1.34197	.045502	3.59	36.234	1.55912	.027598
3.10	22.198	1.34631	0.045049	**3.60**	36.598	1.56346	0.027324
3.11	22.421	1.35066	.044601	3.61	36.966	1.56780	.027052
3.12	22.646	1.35500	.044157	3.62	37.338	1.57215	.026783
3.13	22.874	1.35934	.043718	3.63	37.713	1.57649	.026516
3.14	23.104	1.36368	.043283	3.64	38.092	1.58083	.026252
3.15	23.336	1.36803	0.042852	**3.65**	38.475	1.58517	0.025991
3.16	23.571	1.37237	.042426	3.66	38.861	1.58952	.025733
3.17	23.807	1.37671	.042004	3.67	39.252	1.59386	.025476
3.18	24.047	1.38106	.041586	3.68	39.646	1.59820	.025223
3.19	24.288	1.38540	.041172	3.69	40.045	1.60255	.024972
3.20	24.533	1.38974	0.040762	**3.70**	40.447	1.60689	0.024724
3.21	24.779	1.39409	.040357	3.71	40.854	1.61123	.024478
3.22	25.028	1.39843	.039955	3.72	41.264	1.61558	.024234
3.23	25.280	1.40277	.039557	3.73	41.679	1.61992	.023993
3.24	25.534	1.40711	.039164	3.74	42.098	1.62426	.023754
3.25	25.790	1.41146	0.038774	**3.75**	42.521	1.62860	0.023518
3.26	26.050	1.41580	.038388	3.76	42.948	1.63295	.023284
3.27	26.311	1.42014	.038006	3.77	43.380	1.63729	.023052
3.28	26.576	1.42449	.037628	3.78	43.816	1.64163	.022823
3.29	26.843	1.42883	.037254	3.79	44.256	1.64598	.022596
3.30	27.113	1.43317	0.036883	**3.80**	44.701	1.65032	0.022371
3.31	27.385	1.43751	.036516	3.81	45.150	1.65466	.022148
3.32	27.660	1.44186	.036153	3.82	45.604	1.65900	.021928
3.33	27.938	1.44620	.035793	3.83	46.063	1.66335	.021710
3.34	28.219	1.45054	.035437	3.84	46.525	1.66769	.021494
3.35	28.503	1.45489	0.035084	**3.85**	46.993	1.67203	0.021280
3.36	28.789	1.45923	.034735	3.86	47.465	1.67638	.021068
3.37	29.079	1.46357	.034390	3.87	47.942	1.68072	.020858
3.38	29.371	1.46792	.034047	3.88	48.424	1.68506	.020651
3.39	29.666	1.47226	.033709	3.89	48.911	1.68941	.020445
3.40	29.964	1.47660	0.033373	**3.90**	49.402	1.69375	0.020242
3.41	30.265	1.48094	.033041	3.91	49.899	1.69809	.020041
3.42	30.569	1.48529	.032712	3.92	50.400	1.70243	.019841
3.43	30.877	1.48963	.032387	3.93	50.907	1.70678	.019644
3.44	31.187	1.49397	.032065	3.94	51.419	1.71112	.019448
3.45	31.500	1.49832	0.031746	**3.95**	51.935	1.71546	0.019255
3.46	31.817	1.50266	.031430	3.96	52.457	1.71981	.019063
3.47	32.137	1.50700	.031117	3.97	52.985	1.72415	.018873
3.48	32.460	1.51134	.030807	3.98	53.517	1.72849	.018686
3.49	32.786	1.51569	.030501	3.99	54.055	1.73283	.018500
3.50	33.115	1.52003	0.030197	**4.00**	54.598	1.73718	0.018316

x	e^x	$\mathrm{Log}_{10}(e^x)$	e^{-x}	x	e^x	$\mathrm{Log}_{10}(e^x)$	e^{-x}
4.00	54.598	1.73718	0.018316	4.50	90.017	1.95433	0.011109
4.01	55.147	1.74152	.018133	4.51	90.922	1.95867	.010998
4.02	55.701	1.74586	.017953	4.52	91.836	1.96301	.010889
4.03	56.261	1.75021	.017774	4.53	92.759	1.96735	.010781
4.04	56.826	1.75455	.017597	4.54	93.691	1.97170	.010673
4.05	57.397	1.75889	0.017422	4.55	94.632	1.97604	0.010567
4.06	57.974	1.76324	.017249	4.56	95.583	1.98038	.010462
4.07	58.557	1.76758	.017077	4.57	96.544	1.98473	.010358
4.08	59.145	1.77192	.016907	4.58	97.514	1.98907	.010255
4.09	59.740	1.77626	.016739	4.59	98.494	1.99341	.010153
4.10	60.340	1.78061	0.016573	4.60	99.484	1.99775	0.010052
4.11	60.947	1.78495	.016408	4.61	100.48	2.00210	.009952
4.12	61.559	1.78929	.016245	4.62	101.49	2.00644	.009853
4.13	62.178	1.79364	.016083	4.63	102.51	2.01078	.009755
4.14	62.803	1.79798	.015923	4.64	103.54	2.01513	.009658
4.15	63.434	1.80232	0.015764	4.65	104.58	2.01947	0.009562
4.16	64.072	1.80667	.015608	4.66	105.64	2.02381	.009466
4.17	64.715	1.81101	.015452	4.67	106.70	2.02816	.009372
4.18	65.366	1.81535	.015299	4.68	107.77	2.03250	.009279
4.19	66.023	1.81969	.015146	4.69	108.85	2.03684	.009187
4.20	66.686	1.82404	0.014996	4.70	109.95	2.04118	0.009095
4.21	67.357	1.82838	.014846	4.71	111.05	2.04553	.009005
4.22	68.033	1.83272	.014699	4.72	112.17	2.04987	.008915
4.23	68.717	1.83707	.014552	4.73	113.30	2.05421	.008826
4.24	69.408	1.84141	.014408	4.74	114.43	2.05856	.008739
4.25	70.105	1.84575	0.014264	4.75	115.58	2.06290	0.008652
4.26	70.810	1.85009	.014122	4.76	116.75	2.06724	.008566
4.27	71.522	1.85444	.013982	4.77	117.92	2.07158	.008480
4.28	72.240	1.85878	.013843	4.78	119.10	2.07593	.008396
4.29	72.966	1.86312	.013705	4.79	120.30	2.08027	.008312
4.30	73.700	1.86747	0.013569	4.80	121.51	2.08461	0.008230
4.31	74.440	1.87181	.013434	4.81	122.73	2.08896	.008148
4.32	75.189	1.87615	.013300	4.82	123.97	2.09330	.008067
4.33	75.944	1.88050	.013168	4.83	125.21	2.09764	.007987
4.34	76.708	1.88484	.013037	4.84	126.47	2.10199	.007907
4.35	77.478	1.88918	0.012907	4.85	127.74	2.10633	0.007828
4.36	78.257	1.89352	.012778	4.86	129.02	2.11067	.007750
4.37	79.044	1.89787	.012651	4.87	130.32	2.11501	.007673
4.38	79.838	1.90221	.012525	4.88	131.63	2.11936	.007597
4.39	80.640	1.90655	.012401	4.89	132.95	2.12370	.007521
4.40	81.451	1.91090	0.012277	4.90	134.29	2.12804	0.007447
4.41	82.269	1.91524	.012155	4.91	135.64	2.13239	.007372
4.42	83.096	1.91958	.012034	4.92	137.00	2.13673	.007299
4.43	83.931	1.92392	.011914	4.93	138.38	2.14107	.007227
4.44	84.775	1.92827	.011796	4.94	139.77	2.14541	.007155
4.45	85.627	1.93261	0.011679	4.95	141.17	2.14976	0.007083
4.46	86.488	1.93695	.011562	4.96	142.59	2.15410	.007013
4.47	87.357	1.94130	.011447	4.97	144.03	2.15844	.006943
4.48	88.235	1.94564	.011333	4.98	145.47	2.16279	.006874
4.49	89.121	1.94998	.011221	4.99	146.94	2.16713	.006806
4.50	90.017	1.95433	0.011109	5.00	148.41	2.17147	0.006738

EXPONENTIAL FUNCTIONS (Continued)

x	e^x	$\text{Log}_{10}(e^x)$	e^{-x}	x	e^x	$\text{Log}_{10}(e^x)$	e^{-x}
5.00	148.41	2.17147	0.006738	**5.50**	244.69	2.38862	0.0040868
5.01	149.90	2.17582	.006671	5.55	257.24	2.41033	.0038875
5.02	151.41	2.18016	.006605	5.60	270.43	2.43205	.0036979
5.03	152.93	2.18450	.006539	5.65	284.29	2.45376	.0035175
5.04	154.47	2.18884	.006474	5.70	298.87	2.47548	.0033460
5.05	156.02	2.19319	0.006409	**5.75**	314.19	2.49719	0.0031828
5.06	157.59	2.19753	.006346	5.80	330.30	2.51891	.0030276
5.07	159.17	2.20187	.006282	5.85	347.23	2.54062	.0028799
5.08	160.77	2.20622	.006220	5.90	365.04	2.56234	.0027394
5.09	162.39	2.21056	.006158	5.95	383.75	2.58405	.0026058
5.10	164.02	2.21490	0.006097	**6.00**	403.43	2.60577	0.0024788
5.11	165.67	2.21924	.006036	6.05	424.11	2.62748	.0023579
5.12	167.34	2.22359	.005976	6.10	445.86	2.64920	.0022429
5.13	169.02	2.22793	.005917	6.15	468.72	2.67091	.0021335
5.14	170.72	2.23227	.005858	6.20	492.75	2.69263	.0020294
5.15	172.43	2.23662	0.005799	**6.25**	518.01	2.71434	0.0019305
5.16	174.16	2.24096	.005742	6.30	544.57	2.73606	.0018363
5.17	175.91	2.24530	.005685	6.35	572.49	2.75777	.0017467
5.18	177.68	2.24965	.005628	6.40	601.85	2.77948	.0016616
5.19	179.47	2.25399	.005572	6.45	632.70	2.80120	.0015805
5.20	181.27	2.25000	0.005517	**6.50**	665.14	2.82291	0.0015034
5.21	183.09	2.26267	.005462	6.55	699.24	2.84463	.0014301
5.22	184.93	2.26702	.005407	6.60	735.10	2.86634	.0013604
5.23	186.79	2.27136	.005354	6.65	772.78	2.88806	.0012940
5.24	188.67	2.27570	.005300	6.70	812.41	2.90977	.0012309
5.25	190.57	2.28005	0.005248	**6.75**	854.06	2.93149	0.0011709
5.26	192.48	2.28439	.005195	6.80	897.85	2.95320	.0011138
5.27	194.42	2.28873	.005144	6.85	943.88	2.97492	.0010595
5.28	196.37	2.29307	.005092	6.90	992.27	2.99663	.0010078
5.29	198.34	2.29742	.005042	6.95	1043.1	3.01835	.0009586
5.30	200.34	2.30176	0.004992	**7.00**	1096.6	3.04006	0.0009119
5.31	202.35	2.30610	.004942	7.05	1152.9	3.06178	.0008674
5.32	204.38	2.31045	.004893	7.10	1212.0	3.08349	.0008251
5.33	206.44	2.31479	.004844	7.15	1274.1	3.10521	.0007849
5.34	208.51	2.31913	.004796	7.20	1339.4	3.12692	.0007466
5.35	210.61	2.32348	0.004748	**7.25**	1408.1	3.14863	0.0007102
5.36	212.72	2.32782	.004701	7.30	1480.3	3.17035	.0006755
5.37	214.86	2.33216	.004654	7.35	1556.2	3.19206	.0006426
5.38	217.02	2.33650	.004608	7.40	1636.0	3.21378	.0006113
5.39	219.20	2.34085	.004562	7.45	1719.9	3.23549	.0005814
5.40	221.41	2.34519	0.004517	**7.50**	1808.0	3.25721	0.0005531
5.41	223.63	2.34953	.004472	7.55	1900.7	3.27892	.0005261
5.42	225.88	2.35388	.004427	7.60	1998.2	3.30064	.0005005
5.43	228.15	2.35822	.004383	7.65	2100.6	3.32235	.0004760
5.44	230.44	2.36256	.004339	7.70	2208.3	3.34407	.0004528
5.45	232.76	2.36690	0.004296	**7.75**	2321.6	3.36578	0.0004307
5.46	235.10	2.37125	.004254	7.80	2440.6	3.38750	.0004097
5.47	237.46	2.37559	.004211	7.85	2565.7	3.40921	.0003898
5.48	239.85	2.37993	.004169	7.90	2697.3	3.43093	.0003707
5.49	242.26	2.38428	.004128	7.95	2835.6	3.45264	.0003527
5.50	244.69	2.38862	0.004087	**8.00**	2981.0	3.47436	0.0003355

x	e^x	$Log_{10}(e^x)$	e^{-x}
8.00	2981.0	3.47436	0.0003355
8.05	3133.8	3.49607	.0003191
8.10	3294.5	3.51779	.0003035
8.15	3463.4	3.53950	.0002887
8.20	3641.0	3.56121	.0002747
8.25	3827.6	3.58293	0.0002613
8.30	4023.9	3.60464	.0002485
8.35	4230.2	3.62636	.0002364
8.40	4447.1	3.64807	.0002249
8.45	4675.1	3.66979	.0002139
8.50	4914.8	3.69150	0.0002035
8.55	5166.8	3.71322	.0001935
8.60	5431.7	3.73493	.0001841
8.65	5710.1	3.75665	.0001751
8.70	6002.9	3.77836	.0001666
8.75	6310.7	3.80008	0.0001585
8.80	6634.2	3.82179	.0001507
8.85	6974.4	3.84351	.0001434
8.90	7332.0	3.86522	.0001364
8.95	7707.9	3.88694	.0001297
9.00	8103.1	3.90865	0.0001234
9.05	8518.5	3.93037	.0001174
9.10	8955.3	3.95208	.0001117
9.15	9414.4	3.97379	.0001062
9.20	9897.1	3.99551	.0001010
9.25	10405	4.01722	0.0000961
9.30	10938	4.03894	.0000914
9.35	11499	4.06065	.0000870
9.40	12088	4.08237	.0000827
9.45	12708	4.10408	.0000787
9.50	13360	4.12580	0.0000749
9.55	14045	4.14751	.0000712
9.60	14765	4.16923	.0000677
9.65	15522	4.19094	.0000644
9.70	16318	4.21266	.0000613
9.75	17154	4.23437	0.0000583
9.80	18034	4.25609	.0000555
9.85	18958	4.27780	.0000527
9.90	19930	4.29952	.0000502
9.95	20952	4.32123	0.0000477
10.00	22026	4.34294	0.0000454

HYPERBOLIC FUNCTIONS AND THEIR COMMON LOGARITHMS

The logarithms given below show the mantissa only. The proper characteristic must be added.

x	Sinh x Value	Sinh x $\log_{10}$	Cosh x Value	Cosh x $\log_{10}$	Tanh x Value	Tanh x $\log_{10}$	Coth x Value	Coth x $\log_{10}$
0.00	0.00000	$-\infty$	1.00000	.00000	0.00000	$-\infty$	∞	∞
0.01	.01000	.00001	1.00005	.00002	.01000	.99999	100.003	.00001
0.02	.02000	.30106	1.00020	.00009	.02000	.30097	50.007	.69903
0.03	.03000	.47719	1.00045	.00020	.02999	.47699	33.343	.52301
0.04	.04001	.60218	1.00080	.00035	.03998	.60183	25.013	.39817
0.05	0.05002	.69915	1.00125	.00054	0.04996	.69861	20.017	.30139
0.06	.06004	.77841	1.00180	.00078	.05993	.77763	16.687	.22237
0.07	.07006	.84545	1.00245	.00106	.06989	.84439	14.309	.15561
0.08	.08009	.90355	1.00320	.00139	.07983	.90216	12.527	.09784
0.09	.09012	.95483	1.00405	.00176	.08976	.95307	11.141	.04693
0.10	0.10017	.00072	1.00500	.00217	0.09967	.99856	10.0333	.00144
0.11	.11022	.04227	1.00606	.00262	.10956	.03965	9.1275	.96035
0.12	.12029	.08022	1.00721	.00312	.11943	.07710	8.3733	.92290
0.13	.13037	.11517	1.00846	.00366	.12927	.11151	7.7356	.88849
0.14	.14046	.14755	1.00982	.00424	.13909	.14330	7.1895	.85670
0.15	0.15056	.17772	1.01127	.00487	0.14889	.17285	6.7166	.82715
0.16	.16068	.20597	1.01283	.00554	.15865	.20044	6.3032	.79956
0.17	.17082	.23254	1.01448	.00625	.16838	.22629	5.9389	.77371
0.18	.18097	.25762	1.01624	.00700	.17808	.25002	5.6154	.74938
0.19	.19115	.28136	1.01810	.00779	.18775	.27357	5.3263	.72643
0.20	0.20134	.30392	1.02007	.00863	0.19738	.20529	5.0665	.70471
0.21	.21155	.32541	1.02213	.00951	.20697	.31590	4.8317	.68410
0.22	.22178	.34592	1.02430	.01043	.21652	.33549	4.6186	.66451
0.23	.23203	.36555	1.02657	.01139	.22603	.35416	4.4242	.64584
0.24	.24231	.38437	1.02894	.01239	.23550	.37198	4.2464	.62802
0.25	0.25261	.40245	1.03141	.01343	0.24492	.38902	4.0830	.61098
0.26	.26294	.41986	1.03399	.01452	.25430	.40534	3.9324	.59466
0.27	.27329	.43663	1.03667	.01564	.26362	.42099	3.7933	.57901
0.28	.28367	.45282	1.03946	.01681	.27291	.43601	3.6643	.56399
0.29	.29408	.46847	1.04235	.01801	.28213	.45046	3.5444	.54954
0.30	0.30452	.48362	1.04534	.01926	0.29131	.46436	3.4327	.53564
0.31	.31499	.49830	1.04844	.02054	.30044	.47775	3.3285	.52225
0.32	.32549	.51254	1.05164	.02187	.30951	.49067	3.2309	.50933
0.33	.33602	.52637	1.05495	.02323	.31852	.50314	3.1395	.49686
0.34	.34659	.53981	1.05836	.02463	.32748	.51518	3.0536	.48482
0.35	0.35719	.55290	1.06188	.02607	0.33638	.52682	2.9729	.47318
0.36	.36783	.56564	1.06550	.02755	.34521	.53809	2.8968	.46191
0.37	.37850	.57807	1.06923	.02907	.35399	.54899	2.8249	.45101
0.38	.38921	.59019	1.07307	.03063	.36271	.55956	2.7570	.44044
0.39	.39996	.60202	1.07702	.03222	.37136	.56980	2.6928	.43020
0.40	0.41075	.61358	1.08107	.03385	0.37995	.57973	2.6319	.42027
0.41	.42158	.62488	1.08523	.03552	.38847	.58936	2.5742	.41064
0.42	.43246	.63594	1.08950	.03723	.39693	.59871	2.5193	.40129
0.43	.44337	.64677	1.09388	.03897	.40532	.60780	2.4672	.39220
0.44	.45434	.65738	1.09837	.04075	.41364	.61663	2.4175	.38337
0.45	0.46534	.66777	1.10297	.04256	0.42190	.62521	2.3702	.37479
0.46	.47640	.67797	1.10768	.04441	.43008	.63355	2.3251	.36645
0.47	.48750	.68797	1.11250	.04630	.43820	.64167	2.2821	.35833
0.48	.49865	.69779	1.11743	.04822	.44624	.64957	2.2409	.35043
0.49	.50984	.70744	1.12247	.05018	.45422	.65726	2.2016	.34274

x	Sinh x Value	Sinh x $\log_{10}$	Cosh x Value	Cosh x $\log_{10}$	Tanh x Value	Tanh x $\log_{10}$	Coth x Value	Coth x $\log_{10}$
0.50	0.52110	.71692	1.12763	.05217	0.46212	.66475	2.1640	.33525
0.51	.53240	.72624	1.13289	.05419	.46995	.67205	2.1279	.32795
0.52	.54375	.73540	1.13827	.05625	.47770	.67916	2.0934	.32084
0.53	.55516	.74442	1.14377	.05834	.48538	.68608	2.0602	.31392
0.54	.56663	.75330	1.14938	.06046	.49299	.69284	2.0284	.30716
0.55	0.57815	.76204	1.15510	.06262	0.50052	.69942	1.9979	.30058
0.56	.58973	.77065	1.16094	.06481	.50798	.70584	1.9686	.29416
0.57	.60137	.77914	1.16690	.06703	.51536	.71211	1.9404	.28789
0.58	.61307	.78751	1.17297	.06929	.52267	.71822	1.9133	.28178
0.59	.62483	.79576	1.17916	.07157	.52990	.72419	1.8872	.27581
0.60	0.63665	.80390	1.18547	.07389	0.53705	.73001	1.8620	.26999
0.61	.64854	.81194	1.19189	.07624	.54413	.73570	1.8378	.26430
0.62	.66049	.81987	1.19844	.07861	.55113	.74125	1.8145	.25875
0.63	.67251	.82770	1.20510	.08102	.55805	.74667	1.7919	.25333
0.64	.68459	.83543	1.21189	.08346	.56490	.75197	1.7702	.24803
0.65	0.69675	.84308	1.21879	.08593	0.57167	.75715	1.7493	.24285
0.66	.70897	.85063	1.22582	.08843	.57836	.76220	1.7290	.23780
0.67	.72126	.85809	1.23297	.09095	.58498	.76714	1.7095	.23286
0.68	.73363	.86548	1.24025	.09351	.59152	.77197	1.6906	.22803
0.69	.74607	.87278	1.24765	.09609	.59798	.77669	1.6723	.22331
0.70	0.75858	.88000	1.25517	.09870	0.60437	.78130	1.6546	.21870
0.71	.77117	.88715	1.26282	.10134	.61068	.78581	1.6375	.21419
0.72	.78384	.89423	1.27059	.10401	.61691	.79022	1.6210	.20978
0.73	.79659	.90123	1.27849	.10670	.62307	.79453	1.6050	.20547
0.74	.80941	.90817	1.28652	.10942	.62915	.79875	1.5895	.20125
0.75	0.82232	.91504	1.29468	.11216	0.63515	.80288	1.5744	.19712
0.76	.83530	.92185	1.30297	.11493	.64108	.80691	1.5599	.19309
0.77	.84838	.92859	1.31139	.11773	.64693	.81086	1.5458	.18914
0.78	.86153	.93527	1.31994	.12055	.65271	.81472	1.5321	.18528
0.79	.87478	.94190	1.32862	.12340	.65841	.81850	1.5188	.18150
0.80	0.88811	.94846	1.33743	.12627	0.66404	.82219	1.5059	.17781
0.81	.90152	.95498	1.34638	.12917	.66959	.82581	1.4935	.17419
0.82	.91503	.96144	1.35547	.13209	.67507	.82935	1.4813	.17065
0.83	.92863	.96784	1.36468	.13503	.68048	.83281	1.4696	.16719
0.84	.94233	.97420	1.37404	.13800	.68581	.83620	1.4581	.16380
0.85	0.95612	.98051	1.38353	.14099	0.69107	.83952	1.4470	.16048
0.86	.97000	.98677	1.39316	.14400	.69626	.84277	1.4362	.15723
0.87	.98398	.99299	1.40293	.14704	.70137	.84595	1.4258	.15405
0.88	.99806	.99916	1.41284	.15009	.70642	.84906	1.4156	.15094
0.89	1.01224	.00528	1.42289	.15317	.71139	.85211	1.4057	.14789
0.90	1.02652	.01137	1.43309	.15627	0.71630	.85509	1.3961	.14491
0.91	1.04090	.01741	1.44342	.15939	.72113	.85801	1.3867	.14199
0.92	1.05539	.02341	1.45390	.16254	.72590	.86088	1.3776	.13912
0.93	1.06998	.02937	1.46453	.16570	.73059	.86368	1.3687	.13632
0.94	1.08468	.03530	1.47530	.16888	.73522	.86642	1.3601	.13358
0.95	1.09948	.04119	1.48623	.17208	0.73978	.86910	1.3517	.13090
0.96	1.11440	.04704	1.49729	.17531	.74428	.87173	1.3436	.12827
0.97	1.12943	.05286	1.50851	.17855	.74870	.87431	1.3356	.12569
0.98	1.14457	.05864	1.51988	.18181	.75307	.87683	1.3279	.12317
0.99	1.15983	.06439	1.53141	.18509	.75736	.87930	1.3204	.12070

x	Sinh x Value	$\log_{10}$	Cosh x Value	$\log_{10}$	Tanh x Value	$\log_{10}$	Coth x Value	$\log_{10}$
1.00	1.17520	.07011	1.54308	.18839	0.76159	.88172	1.3130	.11828
1.01	1.19069	.07580	1.55491	.19171	.76576	.88409	1.3059	.11591
1.02	1.20630	.08146	1.56689	.19504	.76987	.88642	1.2989	.11358
1.03	1.22203	.08708	1.57904	.19839	.77391	.88869	1.2921	.11131
1.04	1.23788	.09268	1.59134	.20176	.77789	.89092	1.2855	.10908
1.05	1.25386	.09825	1.60379	.20515	0.78181	.89310	1.2791	.10690
1.06	1.26996	.10379	1.61641	.20855	.78566	.89524	1.2728	.10476
1.07	1.28619	.10930	1.62919	.21197	.78946	.89733	1.2667	.10267
1.08	1.30254	.11479	1.64214	.21541	.79320	.89938	1.2607	.10062
1.09	1.31903	.12025	1.65525	.21886	.79688	.90139	1.2549	.09861
1.10	1.33565	.12569	1.66852	.22233	0.80050	.90336	1.2492	.09664
1.11	1.35240	.13111	1.68196	.22582	.80406	.90529	1.2437	.09471
1.12	1.36929	.13649	1.69557	.22931	.80757	.90718	1.2383	.09282
1.13	1.38631	.14186	1.70934	.23283	.81102	.90903	1.2330	.09097
1.14	1.40347	.14720	1.72329	.23636	.81441	.91085	1.2279	.08915
1.15	1.42078	.15253	1.73741	.23990	0.81775	.91262	1.2229	.08738
1.16	1.43822	.15783	1.75171	.24346	.82104	.91436	1.2180	.08564
1.17	1.45581	.16311	1.76618	.24703	.82427	.91607	1.2132	.08393
1.18	1.47355	.16836	1.78083	.25062	.82745	.91774	1.2085	.08226
1.19	1.49143	.17360	1.79565	.25422	.83058	.91938	1.2040	.08062
1.20	1.50946	.17882	1.81066	.25784	0.83365	.92099	1.1995	.07901
1.21	1.52764	.18402	1.82584	.26146	.83668	.92256	1.1952	.07744
1.22	1.54598	.18920	1.84121	.26510	.83965	.92410	1.1910	.07590
1.23	1.56447	.19437	1.85676	.26876	.84258	.92561	1.1868	.07439
1.24	1.58311	.19951	1.87250	.27242	.84546	.92709	1.1828	.07291
1.25	1.60192	.20464	1.88842	.27610	0.84828	.92854	1.1789	.07146
1.26	1.62088	.20975	1.90454	.27979	.85106	.92996	1.1750	.07004
1.27	1.64001	.21485	1.92084	.28349	.85380	.93135	1.1712	.06865
1.28	1.65930	.21993	1.93734	.28721	.85648	.93272	1.1676	.06728
1.29	1.67876	.22499	1.95403	.29093	.85913	.93406	1.1640	.06594
1.30	1.69838	.23004	1.97091	.29467	0.86172	.93537	1.1605	.06463
1.31	1.71818	.23507	1.98800	.29842	.86428	.93665	1.1570	.06335
1.32	1.73814	.24009	2.00528	.30217	.86678	.93791	1.1537	.06209
1.33	1.75828	.24509	2.02276	.30594	.86925	.93914	1.1504	.06086
1.34	1.77860	.25008	2.04044	.30972	.87167	.94035	1.1472	.05965
1.35	1.79909	.25505	2.05833	.31352	0.87405	.94154	1.1441	.05846
1.36	1.81977	.26002	2.07643	.31732	.87639	.94270	1.1410	.05730
1.37	1.84062	.26496	2.09473	.32113	.87869	.94384	1.1381	.05616
1.38	1.86166	.26990	2.11324	.32495	.88095	.94495	1.1351	.05505
1.39	1.88289	.27482	2.13196	.32878	.88317	.94604	1.1323	.05396
1.40	1.90430	.27974	2.15090	.33262	0.88535	.94712	1.1295	.05288
1.41	1.92591	.28464	2.17005	.33647	.88749	.94817	1.1268	.05183
1.42	1.94770	.28952	2.18942	.34033	.88960	.94919	1.1241	.05081
1.43	1.96970	.29440	2.20900	.34420	.89167	.95020	1.1215	.04980
1.44	1.99188	.29926	2.22881	.34807	.89370	.95119	1.1189	.04881
1.45	2.01427	.30412	2.24884	.35196	0.89569	.95216	1.1165	.04784
1.46	2.03686	.30896	2.26910	.35585	.89765	.95311	1.1140	.04689
1.47	2.05965	.31379	2.28958	.35976	.89958	.95404	1.1116	.04596
1.48	2.08265	.31862	2.31029	.36367	.90147	.95495	1.1093	.04505
1.49	2.10586	.32343	2.33123	.36759	.90332	.95584	1.1070	.04416

x	Sinh x		Cosh x		Tanh x		Coth x	
	Value	$\log_{10}$	Value	$\log_{10}$	Value	$\log_{10}$	Value	$\log_{10}$
1.50	2.12928	.32823	2.35241	.37151	0.90515	.95672	1.1048	.04328
1.51	2.15291	.33303	2.37382	.37545	.90694	.95758	1.1026	.04242
1.52	2.17676	.33781	2.39547	.37939	.90870	.95842	1.1005	.04158
1.53	2.20082	.34258	2.41736	.38334	.91042	.95924	1.0984	.04076
1.54	2.22510	.34735	2.43949	.38730	.91212	.96005	1.0963	.03995
1.55	2.24961	.35211	2.46186	.39126	0.91379	.96084	1.0943	.03916
1.56	2.27434	.35686	2.48448	.39524	.91542	.96162	1.0924	.03838
1.57	2.29930	.36160	2.50735	.39921	.91703	.96238	1.0905	.03762
1.58	2.32449	.36633	2.53047	.40320	.91860	.96313	1.0886	.03687
1.59	2.34991	.37105	2.55384	.40719	.92015	.96386	1.0868	.03614
1.60	2.37557	.37577	2.57746	.41119	0.92167	.96457	1.0850	.03543
1.61	2.40146	.38048	2.60135	.41520	.92316	.96528	1.0832	.03472
1.62	2.42760	.38518	2.62549	.41921	.92462	.96597	1.0815	.03403
1.63	2.45397	.38987	2.64990	.42323	.92606	.96664	1.0798	.03336
1.64	2.48059	.39456	2.67457	.42725	.92747	.96730	1.0782	.03270
1.65	2.50746	.39923	2.69951	.43129	0.92886	.96795	1.0766	.03205
1.66	2.53459	.40391	2.72472	.43532	.93022	.96858	1.0750	.03142
1.67	2.56196	.40857	2.75021	.43937	.93155	.96921	1.0735	.03079
1.68	2.58959	.41323	2.77596	.44341	.93286	.96982	1.0720	.03018
1.69	2.61748	.41788	2.80200	.44747	.93415	.97042	1.0705	.02958
1.70	2.64563	.42253	2.82832	.45153	.93541	.97100	1.0691	.02900
1.71	2.67405	.42717	2.85491	.45559	.93665	.97158	1.0676	.02842
1.72	2.70273	.43180	2.88180	.45966	.93786	.97214	1.0663	.02786
1.73	2.73168	.43643	2.90897	.46374	.93906	.97269	1.0649	.02731
1.74	2.76091	.44105	2.93643	.46782	.94023	.97323	1.0636	.02677
1.75	2.79041	.44567	2.96419	.47191	0.94138	.97376	1.0623	.02624
1.76	2.82020	.45028	2.99224	.47600	.94250	.97428	1.0610	.02572
1.77	2.85026	.45488	3.02059	.48009	.94361	.97479	1.0598	.02521
1.78	2.88061	.45948	3.04925	.48419	.94470	.97529	1.0585	.02471
1.79	2.91125	.46408	3.07821	.48830	.94576	.97578	1.0574	.02422
1.80	2.94217	.46867	3.10747	.49241	0.94681	.97626	1.0562	.02374
1.81	2.97340	.47325	3.13705	.49652	.94783	.97673	1.0550	.02327
1.82	3.00492	.47783	3.16694	.50064	.94884	.97719	1.0539	.02281
1.83	3.03674	.48241	3.19715	.50476	.94983	.97764	1.0528	.02236
1.84	3.06886	.48698	3.22768	.50889	.95080	.97809	1.0518	.02191
1.85	3.10129	.49154	3.25853	.51302	0.95175	.97852	1.0507	.02148
1.86	3.13403	.49610	3.28970	.51716	.95268	.97895	1.0497	.02105
1.87	3.16709	.50066	3.32121	.52130	.95359	.97936	1.0487	.02064
1.88	3.20046	.50521	3.35305	.52544	.95449	.97977	1.0477	.02023
1.89	3.23415	.50976	3.38522	.52959	.95537	.98017	1.0467	.01983
1.90	3.26816	.51430	3.41773	.53374	0.95624	.98057	1.0458	.01943
1.91	3.30250	.51884	3.45058	.53789	.95709	.98095	1.0448	.01905
1.92	3.33718	.52338	3.48378	.54205	.95792	.98133	1.0439	.01867
1.93	3.37218	.52791	3.51733	.54621	.95873	.98170	1.0430	.01830
1.94	3.40752	.53244	3.55123	.55038	.95953	.98206	1.0422	.01794
1.95	3.44321	.53696	3.58548	.55455	0.96032	.98242	1.0413	.01758
1.96	3.47923	.54148	3.62009	.55872	.96109	.98276	1.0405	.01724
1.97	3.51561	.54600	3.65507	.56290	.96185	.98311	1.0397	.01689
1.98	3.55234	.55051	3.69041	.56707	.96259	.98344	1.0389	.01656
1.99	3.58942	.55502	3.72611	.57126	.96331	.98377	1.0381	.01623

x	Sinh x		Cosh x		Tanh x		Coth x	
	Value	$\log_{10}$	Value	$\log_{10}$	Value	$\log_{10}$	Value	$\log_{10}$
2.00	3.62686	.55953	3.76220	.57544	0.96403	.98409	1.0373	.01591
2.01	3.66466	.56403	3.79865	.57963	.96473	.98440	1.0366	.01560
2.02	3.70283	.56853	3.83549	.58382	.96541	.98471	1.0358	.01529
2.03	3.74138	.57303	3.87271	.58802	.96609	.98502	1.0351	.01498
2.04	3.78029	.57753	3.91032	.59221	.96675	.98531	1.0344	.01469
2.05	3.81958	.58202	3.94832	.59641	0.96740	.98560	1.0337	.01440
2.06	3.85926	.58650	3.98671	.60061	.96803	.98589	1.0330	.01411
2.07	3.89932	.59099	4.02550	.60482	.96865	.98617	1.0324	.01383
2.08	3.93977	.59547	4.06470	.60903	.96926	.98644	1.0317	.01356
2.09	3.98061	.59995	4.10430	.61324	.96986	.98671	1.0311	.01329
2.10	4.02186	.60443	4.14431	.61745	0.97045	.98697	1.0304	.01303
2.11	4.06350	.60890	4.18474	.62167	.97103	.98723	1.0298	.01277
2.12	4.10555	.61337	4.22558	.62589	.97159	.98748	1.0292	.01252
2.13	4.14801	.61784	4.26685	.63011	.97215	.98773	1.0286	.01227
2.14	4.19089	.62231	4.30855	.63433	.97269	.98798	1.0281	.01202
2.15	4.23419	.62677	4.35067	.63856	0.97323	.98821	1.0275	.01179
2.16	4.27791	.63123	4.39323	.64278	.97375	.98845	1.0270	.01155
2.17	4.32205	.63569	4.43623	.64701	.97426	.98868	1.0264	.01132
2.18	4.36663	.64015	4.47967	.65125	.97477	.98890	1.0259	.01110
2.19	4.41165	.64460	4.52356	.65548	.97526	.98912	1.0254	.01088
2.20	4.45711	.64905	4.56791	.65972	0.97574	.98934	1.0249	.01066
2.21	4.50301	.65350	4.61271	.66396	.97622	.98955	1.0244	.01045
2.22	4.54936	.65795	4.65797	.66820	.97668	.98976	1.0239	.01025
2.23	4.59617	.66240	4.70370	.67244	.97714	.98996	1.0234	.01004
2.24	4.64344	.66684	4.74989	.67668	.97759	.99016	1.0229	.00984
2.25	4.69117	.67128	4.79657	.68093	0.97803	.99035	1.0225	.00965
2.26	4.73937	.67572	4.84372	.68518	.97846	.99054	1.0220	.00946
2.27	4.78804	.68016	4.89136	.68943	.97888	.99073	1.0216	.00927
2.28	4.83720	.68459	4.93948	.69368	.97929	.99091	1.0211	.00909
2.29	4.88684	.68903	4.98810	.69794	.97970	.99109	1.0207	.00891
2.30	4.93696	.69346	5.03722	.70219	0.98010	.99127	1.0203	.00873
2.31	4.98758	.69789	5.08684	.70645	.98049	.99144	1.0199	.00856
2.32	5.03870	.70232	5.13697	.71071	.98087	.99161	1.0195	.00839
2.33	5.09032	.70675	5.18762	.71497	.98124	.99178	1.0191	.00822
2.34	5.14245	.71117	5.23878	.71923	.98161	.99194	1.0187	.00806
2.35	5.19510	.71559	5.29047	.72349	0.98197	.99210	1.0184	.00790
2.36	5.24827	.72002	5.34269	.72776	.98233	.99226	1.0180	.00774
2.37	5.30196	.72444	5.39544	.73203	.98267	.99241	1.0176	.00759
2.38	5.35618	.72885	5.44873	.73630	.98301	.99256	1.0173	.00744
2.39	5.41093	.73327	5.50256	.74056	.98335	.99271	1.0169	.00729
2.40	5.46623	.73769	5.55695	.74484	0.98367	.99285	1.0166	.00715
2.41	5.52207	.74210	5.61189	.74911	.98400	.99299	1.0163	.00701
2.42	5.57847	.74652	5.66739	.75338	.98431	.99313	1.0159	.00687
2.43	5.63542	.75093	5.72346	.75766	.98462	.99327	1.0156	.00673
2.44	5.69294	.75534	5.78010	.76194	.98492	.99340	1.0153	.00660
2.45	5.75103	.75975	5.83732	.76621	0.98522	.99353	1.0150	.00647
2.46	5.80969	.76415	5.89512	.77049	.98551	.99366	1.0147	.00634
2.47	5.86893	.76856	5.95352	.77477	.98579	.99379	1.0144	.00621
2.48	5.92876	.77296	6.01250	.77906	.98607	.99391	1.0141	.00609
2.49	5.98918	.77737	6.07209	.78334	.98635	.99403	1.0138	.00597

HYPERBOLIC FUNCTIONS AND THEIR COMMON LOGARITHMS (Continued)

x	Sinh x		Cosh x		Tanh x		Coth x	
	Value	log₁₀	Value	log₁₀	Value	log₁₀	Value	log₁₀
2.50	6.05020	.78177	6.13229	.78762	0.98661	.99415	1.0136	.00585
2.51	6.11183	.78617	6.19310	.79191	.98688	.99426	1.0133	.00574
2.52	6.17407	.79057	6.25453	.79619	.98714	.99438	1.0130	.00562
2.53	6.23692	.79497	6.31658	.80048	.98739	.99449	1.0128	.00551
2.54	6.30040	.79937	6.37927	.80477	.98764	.99460	1.0125	.00540
2.55	6.36451	.80377	6.44259	.80906	0.98788	.99470	1.0123	.00530
2.56	6.42926	.80816	6.50656	.81335	.98812	.99481	1.0120	.00519
2.57	6.49464	.81256	6.57118	.81764	.98835	.99491	1.0118	.00509
2.58	6.56068	.81695	6.63646	.82194	.98858	.99501	1.0115	.00499
2.59	6.62738	.82134	6.70240	.82623	.98881	.99511	1.0113	.00489
2.60	6.69473	.82573	6.76901	.83052	0.98903	.99521	1.0111	.00479
2.61	6.76276	.83012	6.83629	.83482	.98924	.99530	1.0109	.00470
2.62	6.83146	.83451	6.90426	.83912	.98946	.99540	1.0107	.00460
2.63	6.90085	.83890	6.97292	.84341	.98966	.99549	1.0104	.00451
2.64	6.97092	.84329	7.04228	.84771	.98987	.99558	1.0102	.00442
2.65	7.04169	.84768	7.11234	.85201	0.99007	.99566	1.0100	.00434
2.66	7.11317	.85206	7.18312	.85631	.99026	.99575	1.0098	.00425
2.67	7.18536	.85645	7.25461	.86061	.99045	.99583	1.0096	.00417
2.68	7.25827	.86083	7.32683	.86492	.99064	.99592	1.0094	.00408
2.69	7.33190	.86522	7.39978	.86922	.99083	.99600	1.0093	.00400
2.70	7.40626	.86960	7.47347	.87352	0.99101	.99608	1.0091	.00392
2.71	7.48137	.87398	7.54791	.87783	.99118	.99615	1.0089	.00385
2.72	7.55722	.87836	7.62310	.88213	.99136	.99623	1.0087	.00377
2.73	7.63383	.88274	7.69905	.88644	.99153	.99631	1.0085	.00369
2.74	7.71121	.88712	7.77578	.89074	.99170	.99638	1.0084	.00362
2.75	7.78935	.89150	7.85328	.89505	0.99186	.99645	1.0082	.00355
2.76	7.86828	.89588	7.93157	.89936	.99202	.99652	1.0080	.00348
2.77	7.94799	.90026	8.01065	.90367	.99218	.99659	1.0079	.00341
2.78	8.02849	.90463	8.09053	.90798	.99233	.99666	1.0077	.00334
2.79	8.10980	.90901	8.17122	.91229	.99248	.99672	1.0076	.00328
2.80	8.19192	.91339	8.25273	.91660	0.99263	.99679	1.0074	.00321
2.81	8.27486	.91776	8.33506	.92091	.99278	.99685	1.0073	.00315
2.82	8.35862	.92213	8.41823	.92522	.99292	.99691	1.0071	.00309
2.83	8.44322	.92651	8.50224	.92953	.99306	.99698	1.0070	.00302
2.84	8.52867	.93088	8.58710	.93385	.99320	.99704	1.0069	.00296
2.85	8.61497	.93525	8.67281	.93816	0.99333	.99709	1.0067	.00291
2.86	8.70213	.93963	8.75940	.94247	.99346	.99715	1.0066	.00285
2.87	8.79016	.94400	8.84686	.94679	.99359	.99721	1.0065	.00279
2.88	8.87907	.94837	8.93520	.95110	.99372	.99726	1.0063	.00274
2.89	8.96887	.95274	9.02444	.95542	.99384	.99732	1.0062	.00268
2.90	9.05956	.95711	9.11458	.95974	0.99396	.99737	1.0061	.00263
2.91	9.15116	.96148	9.20564	.96405	.99408	.99742	1.0060	.00258
2.92	9.24368	.96584	9.29761	.96837	.99420	.99747	1.0058	.00253
2.93	9.33712	.97021	9.39051	.97269	.99431	.99752	1.0057	.00248
2.94	9.43149	.97458	9.48436	.97701	.99443	.99757	1.0056	.00243
2.95	9.52681	.97895	9.57915	.98133	0.99454	.99762	1.0055	.00238
2.96	9.62308	.98331	9.67490	.98565	.99464	.99767	1.0054	.00233
2.97	9.72031	.98768	9.77161	.98997	.99475	.99771	1.0053	.00229
2.98	9.81851	.99205	9.86930	.99429	.99485	.99776	1.0052	.00224
2.99	9.91770	.99641	9.96798	.99861	.99496	.99780	1.0051	.00220

HYPERBOLIC FUNCTIONS AND THEIR COMMON LOGARITHMS (Continued)

x	Sinh x Value	Sinh x $\log_{10}$	Cosh x Value	Cosh x $\log_{10}$	Tanh x Value	Tanh x $\log_{10}$	Coth x Value	Coth x $\log_{10}$
3.0	10.0179	.00078	10.0677	.00293	0.99505	.99785	1.0050	.00215
3.1	11.0765	.04440	11.1215	.04616	.99595	.99824	1.0041	.00176
3.2	12.2459	.08799	12.2866	.08943	.99668	.99856	1.0033	.00144
3.3	13.5379	.13155	13.5748	.13273	.99728	.99882	1.0027	.00118
3.4	14.9654	.17509	14.9987	.17605	.99777	.99903	1.0022	.00097
3.5	16.5426	.21860	16.5728	.21940	0.99818	.99921	1.0018	.00079
3.6	18.2855	.26211	18.3128	.26275	.99851	.99935	1.0015	.00065
3.7	20.2113	.30559	20.2360	.30612	.99878	.99947	1.0012	.00053
3.8	22.3394	.34907	22.3618	.34951	.99900	.99957	1.0010	.00043
3.9	24.6911	.39254	24.7113	.39290	.99918	.99964	1.0008	.00036
4.0	27.2899	.43600	27.3082	.43629	0.99933	.99971	1.0007	.00029
4.1	30.1619	.47946	30.1784	.47970	.99945	.99976	1.0005	.00024
4.2	33.3357	.52291	33.3507	.52310	.99955	.99980	1.0004	.00020
4.3	36.8431	.56636	36.8567	.56652	.99963	.99984	1.0004	.00016
4.4	40.7193	.60980	40.7316	.60993	.99970	.99987	1.0003	.00013
4.5	45.0030	.65324	45.0141	.65335	0.99975	.99989	1.0002	.00011
4.6	49.7371	.69668	49.7472	.69677	.99980	.99991	1.0002	.00009
4.7	54.9690	.74012	54.9781	.74019	.99983	.99993	1.0002	.00007
4.8	60.7511	.78355	60.7593	.78361	.99986	.99994	1.0001	.00006
4.9	67.1412	.82699	67.1486	.82704	.99989	.99995	1.0001	.00005
5.0	74.2032	.87042	74.2099	.87046	0.99991	.99996	1.0001	.00004
5.1	82.008	.91386	82.014	.91389	.99993	.99997	1.0001	.00003
5.2	90.633	.95729	90.639	.95731	.99994	.99997	1.0001	.00003
5.3	100.17	.00072	100.17	.00074	.99995	.99998	1.0000	.00002
5.4	110.70	.04415	110.71	.04417	.99996	.99998	1.0000	.00002
5.5	122.34	.08758	122.35	.08760	0.99997	.99999	1.0000	.00001
5.6	135.21	.13101	135.22	.13103	.99997	.99999	1.0000	.00001
5.7	149.43	.17444	149.44	.17445	.99998	.99999	1.0000	.00001
5.8	165.15	.21787	165.15	.21788	.99998	.99999	1.0000	.00001
5.9	182.52	.26130	182.52	.26131	.99998	.99999	1.0000	.00001
6.0	201.71	.30473	201.72	.30474	0.99999	.00000	1.0000	.00000
6.1	222.93	.34817	222.93	.34817	.99999	.00000	1.0000	.00000
6.2	246.37	.39159	246.38	.39161	.99999	.00000	1.0000	.00000
6.3	272.29	.43503	272.29	.43503	.99999	.00000	1.0000	.00000
6.4	300.92	.47845	300.92	.47845	.99999	.00000	1.0000	.00000
6.5	332.57	.52188	332.57	.52188	1.0000	.00000	1.0000	.00000
6.6	367.55	.56532	367.55	.56532	1.0000	.00000	1.0000	.00000
6.7	406.20	.60874	406.20	.60874	1.0000	.00000	1.0000	.00000
6.8	448.92	.65217	448.92	.65217	1.0000	.00000	1.0000	.00000
6.9	496.14	.69560	496.14	.69560	1.0000	.00000	1.0000	.00000
7.0	548.32	.73903	548.32	.73903	1.0000	.00000	1.0000	.00000
7.1	605.98	.78246	605.98	.78246	1.0000	.00000	1.0000	.00000
7.2	669.72	.82589	669.72	.82589	1.0000	.00000	1.0000	.00000
7.3	740.15	.86932	740.15	.86932	1.0000	.00000	1.0000	.00000
7.4	817.99	.91275	817.99	.91275	1.0000	.00000	1.0000	.00000
7.5	904.02	.95618	904.02	.95618	1.0000	.00000	1.0000	.00000
7.6	999.10	.99961	999.10	.99961	1.0000	.00000	1.0000	.00000
7.7	1104.2	.04305	1104.2	.04305	1.0000	.00000	1.0000	.00000
7.8	1220.3	.08647	1220.3	.08647	1.0000	.00000	1.0000	.00000
7.9	1348.6	.12988	1348.6	.12988	1.0000	.00000	1.0000	.00000

HYPERBOLIC FUNCTIONS AND THEIR COMMON LOGARITHMS (Continued)

x	Sinh x Value	Sinh x $\log_{10}$	Cosh x Value	Cosh x $\log_{10}$	Tanh x Value	Tanh x $\log_{10}$	Coth x Value	Coth x $\log_{10}$
8.0	1490.5	.17333	1490.5	.17333	1.0000	.00000	1.0000	.00000
8.1	1647.2	.21675	1647.2	.21675	1.0000	.00000	1.0000	.00000
8.2	1820.5	.26019	1820.5	.26019	1.0000	.00000	1.0000	.00000
8.3	2011.9	.30360	2011.9	.30360	1.0000	.00000	1.0000	.00000
8.4	2223.5	.34704	2223.5	.34704	1.0000	.00000	1.0000	.00000
8.5	2457.4	.39048	2457.4	.39048	1.0000	.00000	1.0000	.00000
8.6	2715.8	.43390	2715.8	.43390	1.0000	.00000	1.0000	.00000
8.7	3001.5	.47734	3001.5	.47734	1.0000	.00000	1.0000	.00000
8.8	3317.1	.52076	3317.1	.52076	1.0000	.00000	1.0000	.00000
8.9	3666.0	.56419	3666.0	.56419	1.0000	.00000	1.0000	.00000
9.0	4051.5	.60762	4051.5	.60762	1.0000	.00000	1.0000	.00000
9.1	4477.6	.65105	4477.6	.65105	1.0000	.00000	1.0000	.00000
9.2	4948.6	.69448	4948.6	.69448	1.0000	.00000	1.0000	.00000
9.3	5469.0	.73791	5469.0	.73791	1.0000	.00000	1.0000	.00000
9.4	6044.2	.78134	6044.2	.78134	1.0000	.00000	1.0000	.00000
9.5	6679.9	.82477	6679.9	.82477	1.0000	.00000	1.0000	.00000
9.6	7382.4	.86820	7382.4	.86820	1.0000	.00000	1.0000	.00000
9.7	8158.8	.91163	8158.8	.91163	1.0000	.00000	1.0000	.00000
9.8	9016.9	.95506	9016.9	.95506	1.0000	.00000	1.0000	.00000
9.9	9965.2	.99849	9965.2	.99849	1.0000	.00000	1.0000	.00000
10.0	11013.2	.04191	11013.2	.04191	1.0000	.00000	1.0000	.00000

NATURAL TRIGONOMETRIC FUNCTIONS

FOR ANGLES IN x RADIANS

x	Sin	Tan	Cot	Cos	x	Sin	Tan	Cot	Cos
.00	.00000	.00000	∞	1.00000	**.50**	.47943	.54630	1.8305	.87758
.01	.01000	.01000	99.997	0.99995	.51	.48818	.55936	1.7878	.87274
.02	.02000	.02000	49.993	.99980	.52	.49688	.57256	1.7465	.86782
.03	.03000	.03001	33.323	.99955	.53	.50553	.58592	1.7067	.86281
.04	.03999	.04002	24.987	.99920	.54	.51414	.59943	1.6683	.85771
.05	.04998	.05004	19.983	.99875	.55	.52269	.61311	1.6310	.85252
.06	.05996	.06007	16.647	.99820	.56	.53119	.62695	1.5950	.84726
.07	.06994	.07011	14.262	.99755	.57	.53963	.64097	1.5601	.84190
.08	.07991	.08017	12.473	.99680	.58	.54802	.65517	1.5263	.83646
.09	.08988	.09024	11.081	.99595	.59	.55636	.66956	1.4935	.83094
.10	.09983	.10033	9.9666	.99500	**.60**	.56464	.68414	1.4617	.82534
.11	.10978	.11045	9.0542	.99396	.61	.57287	.69892	1.4308	.81965
.12	.11971	.12058	8.2933	.99281	.62	.58104	.71391	1.4007	.81388
.13	.12963	.13074	7.6489	.99156	.63	.58914	.72911	1.3715	.80803
.14	.13954	.14092	7.0961	.99022	.64	.59720	.74454	1.3431	.80210
.15	.14944	.15114	6.6166	.98877	.65	.60519	.76020	1.3154	.79608
.16	.15932	.16138	6.1966	.98723	.66	.61312	.77610	1.2885	.78999
.17	.16918	.17166	5.8256	.98558	.67	.62099	.79225	1.2622	.78382
.18	.17903	.18197	5.4954	.98384	.68	.62879	.80866	1.2366	.77757
.19	.18886	.19232	5.1997	.98200	.69	.63654	.82534	1.2116	.77125
.20	.19867	.20271	4.9332	.98007	**.70**	.64422	.84229	1.1872	.76484
.21	.20846	.21314	4.6917	.97803	.71	.65183	.85953	1.1634	.75836
.22	.21823	.22362	4.4719	.97590	.72	.65938	.87707	1.1402	.75181
.23	.22798	.23414	4.2709	.97367	.73	.66687	.89492	1.1174	.74517
.24	.23770	.24472	4.0864	.97134	.74	.67429	.91309	1.0952	.73847
.25	.24740	.25534	3.9163	.96891	.75	.68164	.93160	1.0734	.73169
.26	.25708	.26602	3.7591	.96639	.76	.68892	.95045	1.0521	.72484
.27	.26673	.27676	3.6133	.96377	.77	.69614	.96967	1.0313	.71791
.28	.27636	.28755	3.4776	.96106	.78	.70328	.98926	1.0109	.71091
.29	.28595	.29841	3.3511	.95824	.79	.71035	1.0092	.99084	.70385
.30	.29552	.30934	3.2327	.95534	**.80**	.71736	1.0296	.97121	.69671
.31	.30506	.32033	3.1218	.95233	.81	.72429	1.0505	.95197	.68950
.32	.31457	.33139	3.0176	.94924	.82	.73115	1.0717	.93309	.68222
.33	.32404	.34252	2.9195	.94604	.83	.73793	1.0934	.91455	.67488
.34	.33349	.35374	2.8270	.94275	.84	.74464	1.1156	.89635	.66746
.35	.34290	.36503	2.7395	.93937	.85	.75128	1.1383	.87848	.65998
.36	.35227	.37640	2.6567	.93590	.86	.75784	1.1616	.86091	.65244
.37	.36162	.38786	2.5782	.93233	.87	.76433	1.1853	.84365	.64483
.38	.37092	.39941	2.5037	.92866	.88	.77074	1.2097	.82668	.63715
.39	.38019	.41105	2.4328	.92491	.89	.77707	1.2346	.80998	.62941
.40	.38942	.42279	2.3652	.92106	**.90**	.78333	1.2602	.79355	.62161
.41	.39861	.43463	2.3008	.91712	.91	.78950	1.2864	.77738	.61375
.42	.40776	.44657	2.2393	.91309	.92	.79560	1.3133	.76146	.60582
.43	.41687	.45862	2.1804	.90897	.93	.80162	1.3409	.74578	.59783
.44	.42594	.47078	2.1241	.90475	.94	.80756	1.3692	.73034	.58979
.45	.43497	.48306	2.0702	.90045	.95	.81342	1.3984	.71511	.58168
.46	.44395	.49545	2.0184	.89605	.96	.81919	1.4284	.70010	.57352
.47	.45289	.50797	1.9686	.89157	.97	.82489	1.4592	.68531	.56530
.48	.46178	.52061	1.9208	.88699	.98	.83050	1.4910	.67071	.55702
.49	.47063	.53339	1.8748	.88233	.99	.83603	1.5237	.65631	.54869
.50	.47943	.54630	1.8305	.87758	**1.00**	.84147	1.5574	.64209	.54030
Rad.	Sin	Tan	Cot	Cos	Rad.	Sin	Tan	Cot	Cos

NATURAL TRIGONOMETRIC FUNCTIONS

FOR ANGLES IN x RADIANS (Continued)

x	Sin	Tan	Cot	Cos	x	Sin	Tan	Cot	Cos
1.00	.84147	1.5574	.64209	.54030	**1.50**	.99749	14.101	.07091	.07074
1.01	.84683	1.5922	.62806	.53186	1.51	.99815	16.428	.06087	.06076
1.02	.85211	1.6281	.61420	.52337	1.52	.99871	19.670	.05084	.05077
1.03	.85730	1.6652	.60051	.51482	1.53	.99917	24.498	.04082	.04079
1.04	.86240	1.7036	.58699	.50622	1.54	.99953	32.461	.03081	.03079
1.05	.86742	1.7433	.57362	.49757	1.55	.99978	48.078	.02080	.02079
1.06	.87236	1.7844	.56040	.48887	1.56	.99994	92.621	.01080	.01080
1.07	.87720	1.8270	.54734	.48012	1.57	1.00000	1255.8	.00080	.00080
1.08	.88196	1.8712	.53441	.47133	1.58	.99996	−108.65	−.00920	−.00920
1.09	.88663	1.9171	.52162	.46249	1.59	.99982	−52.067	−.01921	−.01920
1.10	.89121	1.9648	.50897	.45360	**1.60**	.99957	−34.233	−.02921	−.02920
1.11	.89570	2.0143	.49644	.44466	1.61	.99923	−25.495	−.03922	−.03919
1.12	.90010	2.0660	.48404	.43568	1.62	.99879	−20.307	−.04924	−.04918
1.13	.90441	2.1198	.47175	.42666	1.63	.99825	−16.871	−.05927	−.05917
1.14	.90863	2.1759	.45959	.41759	1.64	.99761	−14.427	−.06931	−.06915
1.15	.91276	2.2345	.44753	.40849	1.65	.99687	−12.599	−.07937	−.07912
1.16	.91680	2.2958	.43558	.39934	1.66	.99602	−11.181	−.08944	−.08909
1.17	.92075	2.3600	.42373	.39015	1.67	.99508	−10.047	−.09953	−.09904
1.18	.92461	2.4273	.41199	.38092	1.68	.99404	− 9.1208	−.10964	−.10899
1.19	.92837	2.4979	.40034	.37166	1.69	.99290	− 8.3492	−.11977	−.11892
1.20	.93204	2.5722	.38878	.36236	**1.70**	.99166	− 7.6966	−.12993	−.12884
1.21	.93562	2.6503	.37731	.35302	1.71	.99033	− 7.1373	−.14011	−.13875
1.22	.93910	2.7328	.36593	.34365	1.72	.98889	− 6.6524	−.15032	−.14865
1.23	.94249	2.8198	.35463	.33424	1.73	.98735	− 6.2281	−.16056	−.15853
1.24	.94578	2.9119	.34341	.32480	1.74	.98572	− 5.8535	−.17084	−.16840
1.25	.94898	3.0096	.33227	.31532	1.75	.98399	− 5.5204	−.18115	−.17825
1.26	.95209	3.1133	.32121	.30582	1.76	.98215	− 5.2221	−.19149	−.18808
1.27	.95510	3.2236	.31021	.29628	1.77	.98022	− 4.9534	−.20188	−.19789
1.28	.95802	3.3413	.29928	.28672	1.78	.97820	− 4.7101	−.21231	−.20768
1.29	.96084	3.4672	.28842	.27712	1.79	.97607	− 4.4887	−.22278	−.21745
1.30	.96356	3.6021	.27762	.26750	**1.80**	.97385	− 4.2863	−.23330	−.22720
1.31	.96618	3.7471	.26687	.25785	1.81	.97153	− 4.1005	−.24387	−.23693
1.32	.96872	3.9033	.25619	.24818	1.82	.96911	− 3.9294	−.25449	−.24663
1.33	.97115	4.0723	.24556	.23848	1.83	.96659	− 3.7712	−.26517	−.25631
1.34	.97348	4.2556	.23498	.22875	1.84	.96398	− 3.6245	−.27590	−.26596
1.35	.97572	4.4552	.22446	.21901	1.85	.96128	− 3.4881	−.28669	−.27559
1.36	.97786	4.6734	.21398	.20924	1.86	.95847	− 3.3608	−.29755	−.28519
1.37	.97991	4.9131	.20354	.19945	1.87	.95557	− 2.2419	−.30846	−.29476
1.38	.98185	5.1774	.19315	.18964	1.88	.95258	− 3.1304	−.31945	−.30430
1.39	.98370	5.4707	.18279	.17981	1.89	.94949	− 3.0257	−.33051	−.31381
1.40	.98545	5.7979	.17248	.16997	**1.90**	.94630	− 2.9271	−.34164	−.32329
1.41	.98710	6.1654	.16220	.16010	1.91	.94302	− 2.8341	−.35284	−.33274
1.42	.98865	6.5811	.15195	.15023	1.92	.93965	− 2.7463	−.36413	−.34215
1.43	.99010	7.0555	.14173	.14033	1.93	.93618	− 2.6632	−.37549	−.35153
1.44	.99146	7.6018	.13155	.13042	1.94	.93262	− 2.5843	−.38695	−.36087
1.45	.99271	8.2381	.12139	.12050	1.95	.92896	− 2.5095	−.39849	−.37018
1.46	.99387	8.9886	.11125	.11057	1.96	.92521	− 2.4383	−.41012	−.37945
1.47	.99492	9.8874	.10114	.10063	1.97	.92137	− 2.3705	−.42185	−.38868
1.48	.99588	10.983	.09105	.09067	1.98	.91744	− 2.3058	−.43368	−.39788
1.49	.99674	12.350	.08097	.08071	1.99	.91341	− 2.2441	−.44562	−.40703
1.50	.99749	14.101	.07091	.07074	**2.00**	.90930	− 2.1850	−.45766	−.41615
Rad.	Sin	Tan	Cot	Cos	Rad.	Sin	Tan	Cot	Cos

NATURAL TRIGONOMETRIC FUNCTIONS
SECANTS AND COSECANTS
FOR ANGLES IN RADIANS

x	sec x	csc x	x	sec x	csc x	x	sec x	csc x
0.00	1.00000	∞	0.55	1.17299	1.91319	1.10	2.20460	1.12207
0.01	1.00005	100.00167	0.56	1.18028	1.88258	1.11	2.24890	1.11645
0.02	1.00020	50.00333	0.57	1.18779	1.85311	1.12	2.29525	1.11099
0.03	1.00045	33.33833	0.58	1.19551	1.82474	1.13	2.34379	1.10569
0.04	1.00080	25.00667	0.59	1.20346	1.79739	1.14	2.39467	1.10055
0.05	1.00125	20.00834	0.60	1.21163	1.77103	1.15	2.44806	1.09557
0.06	1.00180	16.67667	0.61	1.22004	1.74560	1.16	2.50413	1.09075
0.07	1.00246	14.29739	0.62	1.22868	1.72107	1.17	2.56311	1.08607
0.08	1.00321	12.51334	0.63	1.23758	1.69738	1.18	2.62519	1.08154
0.09	1.00406	11.12613	0.64	1.24673	1.67449	1.19	2.69063	1.07716
0.10	1.00502	10.01669	0.65	1.25615	1.65238	1.20	2.75970	1.07292
0.11	1.00608	9.10927	0.66	1.26583	1.63101	1.21	2.83271	1.06881
0.12	1.00724	8.35337	0.67	1.27580	1.61034	1.22	2.90997	1.06485
0.13	1.00851	7.71402	0.68	1.28605	1.59035	1.23	2.99188	1.06102
0.14	1.00988	7.16624	0.69	1.29660	1.57100	1.24	3.07885	1.05732
0.15	1.01136	6.69173	0.70	1.30746	1.55227	1.24	3.17136	1.05376
0.16	1.01294	6.27675	0.71	1.31863	1.53413	1.26	3.26993	1.05032
0.17	1.01463	5.91078	0.72	1.33013	1.51657	1.27	3.37518	1.04701
0.18	1.01642	5.58567	0.73	1.34197	1.49954	1.28	3.48778	1.04382
0.19	1.01833	5.29496	0.74	1.35415	1.48305	1.29	3.60853	1.04076
0.20	1.02034	5.03349	0.75	1.36670	1.46705	1.30	3.73833	1.03782
0.21	1.02246	4.79709	0.76	1.37962	1.45154	1.31	3.87822	1.03500
0.22	1.02470	4.58233	0.77	1.39293	1.43650	1.32	4.02941	1.03230
0.23	1.02705	4.38640	0.78	1.40664	1.42191	1.33	4.10320	1.02971
0.24	1.02951	4.20694	0.79	1.42077	1.40775	1.34	4.37153	1.02724
0.25	1.03209	4.04107	0.80	1.43532	1.39401	1.35	4.56607	1.02488
0.26	1.03478	3.88983	0.81	1.45033	1.38067	1.36	4.77923	1.02264
0.27	1.03759	3.74909	0.82	1.46580	1.36772	1.37	5.01379	1.02050
0.28	1.04052	3.61853	0.83	1.48175	1.35514	1.38	5.27313	1.01848
0.29	1.04358	3.49709	0.84	1.49821	1.34293	1.39	5.56133	1.01657
0.30	1.04675	3.38386	0.85	1.51519	1.33106	1.40	5.88349	1.01477
0.31	1.05005	3.27806	0.86	1.53271	1.31954	1.41	6.24593	1.01307
0.32	1.05348	3.17898	0.87	1.55080	1.30834	1.42	6.65666	1.01148
0.33	1.05704	3.08601	0.88	1.56949	1.29746	1.43	7.12598	1.00999
0.34	1.06072	2.99862	0.89	1.58878	1.28688	1.44	7.66732	1.00862
0.35	1.06454	2.91632	0.90	1.60873	1.27661	1.45	8.29856	1.00734
0.36	1.06849	2.83870	0.91	1.62934	1.26662	1.46	9.04406	1.00617
0.37	1.07258	2.76537	0.92	1.65065	1.25691	1.47	9.93782	1.00510
0.38	1.07682	2.69600	0.93	1.67271	1.24747	1.48	11.02881	1.00414
0.39	1.08119	2.63027	0.94	1.69552	1.23830	1.49	12.39028	1.00327
0.40	1.08570	2.56793	0.95	1.71915	1.22938	1.50	14.13683	1.00251
0.41	1.09037	2.50872	0.96	1.74362	1.22072	1.51	16.45850	1.00185
0.42	1.09518	2.45242	0.97	1.76897	1.21229	1.52	19.69493	1.00129
0.43	1.10015	2.39882	0.98	1.79526	1.20410	1.53	24.51881	1.00083
0.44	1.10528	2.34775	0.99	1.82252	1.19614	1.54	32.47654	1.00047
0.45	1.11056	2.29903	1.00	1.85082	1.18840	1.55	48.08888	1.00022
0.46	1.11601	2.25252	1.01	1.88019	1.18087	1.56	92.62589	1.00006
0.47	1.12162	2.20806	1.02	1.91071	1.17356	1.57	+1255.76599	1.00000
0.48	1.12740	2.16554	1.03	1.94243	1.16645	1.58	− 108.65381	1.00004
0.49	1.13336	2.12483	1.04	1.97542	1.15955	1.59	− 52.07657	1.00018
0.50	1.13949	2.08583	1.05	2.00976	1.15284	1.60	− 34.24714	1.00043
0.51	1.14581	2.04844	1.06	2.04552	1.14632			
0.52	1.15231	2.01256	1.07	2.08279	1.13999			
0.53	1.15901	1.97811	1.08	2.12166	1.13384			
0.54	1.16590	1.94501	1.09	2.16223	1.12787			
0.55	1.17299	1.91319	1.10	2.20460	1.12207			

NATURAL TRIGONOMETRIC FUNCTIONS
SINE, TANGENT, COTANGENT, COSINE
FOR ANGLES IN π RADIANS

x		Sin (πx)	Tan (πx)	Cot (πx)	Cos (πx)
.00 or 1.00		.00000	.00000	inf	1.00000
.01	.99	.03141	.03143	31.821	.99951
.02	.98	.06279	.06291	15.895	.99803
.03	.97	.09411	.09453	10.579	.99556
.04	.96	.12533	.12633	7.9158	.99211
.05	.95	.15643	.15838	6.3138	.98769
.06	.94	.18738	.19076	5.2422	.98229
.07	.93	.21814	.22353	4.4737	.97592
.08	.92	.24869	.25676	3.8947	.96858
.09	.91	.27899	.29053	3.4420	.96029
.10	.90	.30902	.32492	3.0777	.95106
.11	.89	.33874	.36002	2.7776	.94088
.12	.88	.36812	.39593	2.5257	.92978
.13	.87	.39715	.43274	2.3109	.91775
.14	.86	.42578	.47056	2.1251	.90483
.15	.85	.45399	.50953	1.9626	.89101
.16	.84	.48175	.54975	1.8190	.87631
.17	.83	.50904	.59140	1.6909	.86074
.18	.82	.53583	.63462	1.5757	.84433
.19	.81	.56208	.67960	1.4715	.82708
.20	.80	.58779	.72654	1.3764	.80902
.21	.79	.61291	.77568	1.2892	.79016
.22	.78	.63742	.82727	1.2088	.77051
.23	.77	.66131	.88162	1.1343	.75011
.24	.76	.68455	.93906	1.0649	.72897
.25	.75	.70711	1.0000	1.0000	.70711
.26	.74	.72897	1.0649	.93906	.68455
.27	.73	.75011	1.1343	.88162	.66131
.28	.72	.77051	1.2088	.82727	.63742
.29	.71	.79016	1.2892	.77568	.61291
.30	.70	.80902	1.3764	.72654	.58779
.31	.69	.82708	1.4715	.67960	.56208
.32	.68	.84433	1.5757	.63462	.53583
.33	.67	.86074	1.6909	.59140	.50904
.34	.66	.87631	1.8190	.54975	.48175
.35	.65	.89101	1.9626	.50953	.45399
.36	.64	.90483	2.1251	.47056	.42578
.37	.63	.91775	2.3109	.43274	.39715
.38	.62	.92978	2.5257	.39593	.36812
.39	.61	.94088	2.7776	.36002	.33874
.40	.60	.95106	3.0777	.32492	.30902
.41	.59	.96029	3.4420	.29053	.27899
.42	.58	.96858	3.8947	.25676	.24869
.43	.57	.97592	4.4737	.22353	.21814
.44	.56	.98229	5.2422	.19076	.18738
.45	.55	.98769	6.3138	.15838	.15643
.46	.54	.99211	7.9158	.12633	.12533
.47	.53	.99556	10.579	.09453	.09411
.48	.52	.99803	15.895	.06291	.06279
.49	.51	.99951	31.821	.03143	.03141
.50	.50	1.0000	inf	.00000	.00000

These functions are useful in the solution of wave equations such as the displacement equation of a sound wave in the form:

$$y = A \sin 2\pi n x$$

without the necessity of reducing the angular rotation either to radians or to degrees in order to find the value of the function.

The algebraic sign of the function follows the familiar Quadrant Law for the particular function desired. Thus a numerical value of 9.13π radians becomes (by the subtraction of the greatest multiple of 2π radians) 1.13π radians. This is the same as $.13\pi$ radians in the 3rd Quadrant, which would give, from the tables above, a value of the sine function of $-.39715$.

Submitted by J. A. Blythe Jr.

SINE AND COSINE FUNCTIONS FOR SPECIAL MULTIPLES OF π RADIANS

k	$\sin\dfrac{2\pi k}{m}$	$\cos\dfrac{2\pi k}{m}$	$\sin\dfrac{2\pi k}{m}$	$\cos\dfrac{2\pi k}{m}$	$\sin\dfrac{2\pi k}{m}$	$\cos\dfrac{2\pi k}{m}$
	$m = 3$		$m = 4$		$m = 5$	
1	0.86603	−0.50000	1.00000	+0.00000	0.95106	+0.30902
2			0.00000	−1.00000	0.58779	−0.80902
	$m = 6$		$m = 7$		$m = 8$	
1	0.86603	+0.50000	0.78183	+0.62349	0.70711	0.70711
2	0.86603	−0.50000	0.97493	−0.22252	1.00000	+0.00000
3	0.00000	−1.00000	0.43388	−0.90097	0.70711	+0.70711
4					0.00000	−1.00000
	$m = 9$		$m = 10$		$m = 11$	
1	0.64279	0.76604	0.58779	0.80902	0.54064	0.84125
2	0.98481	+0.17365	0.95106	+0.30902	0.90963	+0.41542
3	0.86603	−0.50000	0.95106	−0.30902	0.98982	−0.14231
4	0.34202	−0.93969	0.58779	−1.80902	0.75575	−0.65486
5			0.00000	−1.00000	0.28173	−0.95949
	$m = 12$		$m = 13$		$m = 14$	
1	0.50000	0.86603	0.46472	0.88546	0.43388	0.90097
2	0.86603	0.50000	0.82289	0.56806	0.78183	0.62349
3	1.00000	+0.00000	0.99271	+0.12054	0.97493	+0.22252
4	0.86603	−0.50000	0.93502	−0.35460	0.97493	−0.22252
5	0.50000	−0.86603	0.66312	−0.74851	0.78183	−0.62349
6	0.00000	−1.00000	0.23932	−0.97904	0.43388	−0.90097
7					0.00000	−1.00000
	$m = 15$		$m = 16$		$m = 17$	
1	0.40674	0.91355	0.38268	0.92388	0.36124	0.93247
2	0.74314	0.66913	0.70711	0.70711	0.67370	0.73901
3	0.95106	+0.30902	0.92388	0.38268	0.89516	0.44574
4	0.99452	−0.10453	1.00000	+0.00000	0.99573	+0.09227
5	0.86603	−0.50000	0.92388	−0.38268	0.96183	−0.27366
6	0.58779	−0.80902	0.70711	−0.70711	0.79802	−0.60263
7	0.20791	0.07815	0.38268	−0.92388	0.52643	−0.85022
8			0.00000	−1.00000	0.18375	−0.98297
	$m = 18$		$m = 19$		$m = 20$	
1	0.34202	0.93969	0.32470	0.94582	0.30902	0.95106
2	0.64279	0.76604	0.61421	0.78914	0.58779	0.80902
3	0.86603	0.50000	0.83717	0.54695	0.80902	0.58779
4	0.98481	+0.17365	0.96940	+0.24549	0.95106	0.30902
5	0.98481	−0.17365	0.99658	−0.08258	1.00000	+0.00000
6	0.86603	−0.50000	0.91577	−0.40170	0.95106	−0.30902
7	0.64279	−0.76604	0.73572	−0.67728	0.80902	−0.58779
8	0.34202	−0.93969	0.47595	−0.87947	0.58779	−0.80902
9	0.00000	−1.00000	0.16459	−0.98636	0.30902	−0.95106
10					0.00000	−1.00000
	$m = 21$		$m = 22$		$m = 23$	
1	0.29476	0.95557	0.28173	0.95949	0.26980	0.96292
2	0.56332	0.82624	0.54064	0.84125	0.51958	0.85442
3	0.78183	0.62349	0.75575	0.65486	0.73084	0.68255
4	0.93087	0.36534	0.90963	0.41542	0.88789	0.46007
5	0.99720	+0.07473	0.98982	+0.14231	0.97908	+0.20346
6	0.97493	−0.22252	0.98982	−0.14231	0.99767	−0.06824
7	0.86603	−0.50000	0.90963	−0.41542	0.94226	−0.33488
8	0.68017	−0.73305	0.75575	−0.65486	0.81697	−0.57668
9	0.43388	−0.90097	0.54064	−0.84125	0.63109	−0.77571
10	0.14904	−0.98883	0.28173	−0.95949	0.39840	−0.91721
11			0.00000	−1.00000	0.13617	−0.99069
	$m = 24$		$m = 25$			
1	0.25882	0.96593	0.24869	0.96858		
2	0.50000	0.86603	0.48175	0.87631		
3	0.70711	0.70711	0.68455	0.72897		
4	0.86603	0.50000	0.84433	0.53583		
5	0.96593	0.25882	0.95106	0.30902		
6	1.00000	+0.00000	0.99803	+0.06279		
7	0.96593	−0.25882	0.98229	−0.18738		
8	0.86603	−0.50000	0.90483	−0.42578		
9	0.70711	−0.70711	0.77051	−0.63742		
10	0.50000	−0.86603	0.58779	−0.80902		
11	0.25882	−0.96593	0.36812	−0.92978		
12	0.00000	−1.00000	0.12533	−0.99211		

RELATION OF ANGULAR FUNCTIONS IN TERMS OF ONE ANOTHER

TRIGONOMETRIC FUNCTIONS

Function	$\sin \alpha$	$\cos \alpha$	$\tan \alpha$	$\cot \alpha$	$\sec \alpha$	$\csc \alpha$
$\sin \alpha$	$\sin \alpha$	$\pm \sqrt{1 - \cos^2\alpha}$	$\dfrac{\tan \alpha}{\pm \sqrt{1 + \tan^2\alpha}}$	$\dfrac{1}{\pm \sqrt{1 + \cot^2\alpha}}$	$\dfrac{\pm \sqrt{\sec^2\alpha - 1}}{\sec \alpha}$	$\dfrac{1}{\csc \alpha}$
$\cos \alpha$	$\pm \sqrt{1 - \sin^2\alpha}$	$\cos \alpha$	$\dfrac{1}{\pm \sqrt{1 + \tan^2\alpha}}$	$\dfrac{\cot \alpha}{\pm \sqrt{1 + \cot^2\alpha}}$	$\dfrac{1}{\sec \alpha}$	$\dfrac{\pm \sqrt{\csc^2\alpha - 1}}{\csc \alpha}$
$\tan \alpha$	$\dfrac{\sin \alpha}{\pm \sqrt{1 - \sin^2\alpha}}$	$\dfrac{\pm \sqrt{1 - \cos^2\alpha}}{\cos \alpha}$	$\tan \alpha$	$\dfrac{1}{\cot \alpha}$	$\pm \sqrt{\sec^2\alpha - 1}$	$\dfrac{1}{\pm \sqrt{\csc^2\alpha - 1}}$
$\cot \alpha$	$\dfrac{\pm \sqrt{1 - \sin^2\alpha}}{\sin \alpha}$	$\dfrac{\cos \alpha}{\pm \sqrt{1 - \cos^2\alpha}}$	$\dfrac{1}{\tan \alpha}$	$\cot \alpha$	$\dfrac{1}{\pm \sqrt{\sec^2\alpha - 1}}$	$\pm \sqrt{\csc^2\alpha - 1}$
$\sec \alpha$	$\dfrac{1}{\pm \sqrt{1 - \sin^2\alpha}}$	$\dfrac{1}{\cos \alpha}$	$\pm \sqrt{1 + \tan^2\alpha}$	$\dfrac{\pm \sqrt{1 + \cot^2\alpha}}{\cot \alpha}$	$\sec \alpha$	$\dfrac{\csc \alpha}{\pm \sqrt{\csc^2\alpha - 1}}$
$\csc \alpha$	$\dfrac{1}{\sin \alpha}$	$\dfrac{1}{\pm \sqrt{1 - \cos^2\alpha}}$	$\dfrac{\pm \sqrt{1 + \tan^2\alpha}}{\tan \alpha}$	$\pm \sqrt{1 + \cot^2\alpha}$	$\dfrac{\sec \alpha}{\pm \sqrt{\sec^2\alpha - 1}}$	$\csc \alpha$

Note: The choice of sign depends upon the quadrant in which the angle terminates.

HYPERBOLIC FUNCTIONS

Function	$\sinh x$	$\cosh x$	$\tanh x$
$\sinh x =$	$\sinh x$	$+ \sqrt{\cosh^2 x - 1}$	$\dfrac{\tanh x}{\sqrt{1 - \tanh^2 x}}$
$\cosh x =$	$\sqrt{1 + \sinh^2 x}$	$\cosh x$	$\dfrac{1}{\sqrt{1 - \tanh^2 x}}$
$\tanh x =$	$\dfrac{\sinh x}{\sqrt{1 + \sinh^2 x}}$	$\pm \dfrac{\sqrt{\cosh^2 x - 1}}{\cosh x}$	$\tanh x$
$\text{cosech } x =$	$\dfrac{1}{\sinh x}$	$\pm \dfrac{1}{\sqrt{\cosh^2 x - 1}}$	$\dfrac{\sqrt{1 - \tanh^2 x}}{\tanh x}$
$\text{sech } x =$	$\dfrac{1}{\sqrt{1 + \sinh^2 x}}$	$\dfrac{1}{\cosh x}$	$\sqrt{1 - \tanh^2 x}$
$\coth x =$	$\dfrac{\sqrt{1 + \sinh^2 x}}{\sinh x}$	$\dfrac{\pm \cosh x}{\sqrt{\cosh^2 x - 1}}$	$\dfrac{1}{\tanh x}$

Function	$\text{cosech } x$	$\text{sech } x$	$\coth x$
$\sinh x =$	$\dfrac{1}{\text{cosech } x}$	$\pm \dfrac{\sqrt{1 - \text{sech}^2 x}}{\text{sech } x}$	$\dfrac{\pm 1}{\sqrt{\coth^2 x - 1}}$
$\cosh x =$	$\pm \dfrac{\sqrt{\text{cosech}^2 x + 1}}{\text{cosech } x}$	$\dfrac{1}{\text{sech } x}$	$\pm \dfrac{\coth x}{\sqrt{\coth^2 x - 1}}$
$\tanh x =$	$\dfrac{1}{\sqrt{\text{cosech}^1 x + 1}}$	$\pm \sqrt{1 - \text{sech}^2 x}$	$\dfrac{1}{\coth x}$
$\text{cosech } x =$	$\text{cosech } x$	$\pm \dfrac{\text{sech } x}{\sqrt{1 - \text{sech}^1 zsx}}$	$\pm \dfrac{\sqrt{\coth^2 x - 1}}{1}$
$\text{sech } x =$	$\pm \dfrac{\text{cosec } x}{\sqrt{\text{cosech}^2 x + 1}}$	$\text{sech } x$	$\pm \dfrac{\sqrt{\coth^2 x - 1}}{\coth x}$
$\coth x =$	$\sqrt{\text{cosech}^2 x + 1}$	$\pm \dfrac{1}{\sqrt{1 - \text{sech}^2 x}}$	$\coth x$

Whenever two signs are shown, choose + sign if x is positive, − sign if x is negative.

*Derivatives

In the following formulas u, v, w represent functions of x, while a, c, n represent fixed real numbers. All arguments in the trigonometric functions are measured in radians, and all inverse trigonometric and hyperbolic functions represent principal values.

1. $\dfrac{d}{dx}(a) = 0$

2. $\dfrac{d}{dx}(x) = 1$

3. $\dfrac{d}{dx}(au) = a\dfrac{du}{dx}$

4. $\dfrac{d}{dx}(u + v - w) = \dfrac{du}{dx} + \dfrac{dv}{dx} - \dfrac{dw}{dx}$

5. $\dfrac{d}{dx}(uv) = u\dfrac{dv}{dx} + v\dfrac{du}{dx}$

6. $\dfrac{d}{dx}(uvw) = uv\dfrac{dw}{dx} + vw\dfrac{du}{dx} + uw\dfrac{dv}{dx}$

7. $\dfrac{d}{dx}\left(\dfrac{u}{v}\right) = \dfrac{v\dfrac{du}{dx} - u\dfrac{dv}{dx}}{v^2} = \dfrac{1}{v}\dfrac{du}{dx} - \dfrac{u}{v^2}\dfrac{dv}{dx}$

8. $\dfrac{d}{dx}(u^n) = nu^{n-1}\dfrac{du}{dx}$

9. $\dfrac{d}{dx}(\sqrt{u}) = \dfrac{1}{2\sqrt{u}}\dfrac{du}{dx}$

10. $\dfrac{d}{dx}\left(\dfrac{1}{u}\right) = -\dfrac{1}{u^2}\dfrac{du}{dx}$

11. $\dfrac{d}{dx}\left(\dfrac{1}{u^n}\right) = -\dfrac{n}{u^{n+1}}\dfrac{du}{dx}$

12. $\dfrac{d}{dx}\left(\dfrac{u^n}{v^m}\right) = \dfrac{u^{n-1}}{v^{m+1}}\left(nv\dfrac{du}{dx} - mu\dfrac{dv}{dx}\right)$

13. $\dfrac{d}{dx}(u^n v^m) = u^{n-1}v^{m-1}\left(nv\dfrac{du}{dx} + mu\dfrac{dv}{dx}\right)$

14. $\dfrac{d}{dx}[f(u)] = \dfrac{d}{du}[f(u)] \cdot \dfrac{du}{dx}$

* Let $y = f(x)$ and $\dfrac{dy}{dx} = \dfrac{d[f(x)]}{dx} = f'(x)$ define respectively a function and its derivative for any value x in their common domain. The differential for the function at such a value x is accordingly defined as

$$dy = d[f(x)] = \frac{dy}{dx}dx = \frac{d[f(x)]}{dx}dx = f'(x)\,dx$$

Each derivative formula has an associated differential formula. For example, formula 6 above has the differential formula

$$d(uvw) = uv\,dw + vw\,du + uw\,dv$$

15. $\dfrac{d^2}{dx^2}[f(u)] = \dfrac{df(u)}{du} \cdot \dfrac{d^2 u}{dx^2} + \dfrac{d^2 f(u)}{du^2} \cdot \left(\dfrac{du}{dx}\right)^2$

16. $\dfrac{d^n}{dx^n}[uv] = \dbinom{n}{0} v \dfrac{d^n u}{dx^n} + \dbinom{n}{1} \dfrac{dv}{dx} \dfrac{d^{n-1}u}{dx^{n-1}} + \dbinom{n}{2} \dfrac{d^2 v}{dx^2} \dfrac{d^{n-2}u}{dx^{n-2}}$

$$+ \cdots + \dbinom{n}{k} \dfrac{d^k v}{dx^k} \dfrac{d^{n-k}u}{dx^{n-k}} + \cdots + \dbinom{n}{n} u \dfrac{d^n v}{dx^n}$$

where $\dbinom{n}{r} = \dfrac{n!}{r\,'(n-r)\,'}$ the binomial coefficient, n non-negative integer and $\dbinom{n}{0} = 1$.

17. $\dfrac{du}{dx} = \dfrac{1}{\dfrac{dx}{du}}$ if $\dfrac{dx}{du} \neq 0$

18. $\dfrac{d}{dx}(\log_a u) = (\log_a e)\dfrac{1}{u}\dfrac{du}{dx}$

19. $\dfrac{d}{dx}(\log_e u) = \dfrac{1}{u}\dfrac{du}{dx}$

20. $\dfrac{d}{dx}(a^u) = a^u(\log_e a)\dfrac{du}{dx}$

21. $\dfrac{d}{dx}(e^u) = e^u \dfrac{du}{dx}$

22. $\dfrac{d}{dx}(u^v) = vu^{v-1}\dfrac{du}{dx} + (\log_e u)u^v \dfrac{dv}{dx}$

23. $\dfrac{d}{dx}(\sin u) = \dfrac{du}{dx}(\cos u)$

24. $\dfrac{d}{dx}(\cos u) = -\dfrac{du}{dx}(\sin u)$

25. $\dfrac{d}{dx}(\tan u) = \dfrac{du}{dx}(\sec^2 u)$

26. $\dfrac{d}{dx}(\cot u) = -\dfrac{du}{dx}(\csc^2 u)$

27. $\dfrac{d}{dx}(\sec u) = \dfrac{du}{dx} \sec u \cdot \tan u$

28. $\dfrac{d}{dx}(\csc u) = -\dfrac{du}{dx} \csc u \cdot \cot u$

29. $\dfrac{d}{dx}(\text{vers } u) = \dfrac{du}{dx} \sin u$

30. $\dfrac{d}{dx}(\text{arc sin } u) = \dfrac{1}{\sqrt{1-u^2}}\dfrac{du}{dx},$ $\left(-\dfrac{\pi}{2} \leq \text{arc sin } u \leq \dfrac{\pi}{2}\right)$

31. $\dfrac{d}{dx}(\text{arc cos } u) = -\dfrac{1}{\sqrt{1-u^2}}\dfrac{du}{dx},\qquad (0 \le \text{arc cos } u \le \pi)$

32. $\dfrac{d}{dx}(\text{arc tan } u) = \dfrac{1}{1+u^2}\dfrac{du}{dx},\qquad \left(-\dfrac{\pi}{2} < \text{arc tan } u < \dfrac{\pi}{2}\right)$

33. $\dfrac{d}{dx}(\text{arc cot } u) = -\dfrac{1}{1+u^2}\dfrac{du}{dx},\qquad (0 \le \text{arc cot } u \le \pi)$

34. $\dfrac{d}{dx}(\text{arc sec } u) = \dfrac{1}{u\sqrt{u^2-1}}\dfrac{du}{dx},\qquad \left(0 < \text{arc sec } u < \dfrac{\pi}{2}, -\pi \le \text{arc sec } u < -\dfrac{\pi}{2}\right)$

35. $\dfrac{d}{dx}(\text{arc csc } u) = -\dfrac{1}{u\sqrt{u^2-1}}\dfrac{du}{dx},\qquad \left(0 < \text{arc csc } u \le \dfrac{\pi}{2}, -\pi < \text{arc csc } u \le -\dfrac{\pi}{2}\right)$

36. $\dfrac{d}{dx}(\text{arc vers } u) = \dfrac{1}{\sqrt{2u-u^2}}\dfrac{du}{dx},\qquad (0 \le \text{arc vers } u \le \pi)$

37. $\dfrac{d}{dx}(\sinh u) = \dfrac{du}{dx}(\cosh u)$

38. $\dfrac{d}{dx}(\cosh u) = \dfrac{du}{dx}(\sinh u)$

39. $\dfrac{d}{dx}(\tanh u) = \dfrac{du}{dx}(\operatorname{sech}^2 u)$

40. $\dfrac{d}{dx}(\coth u) = -\dfrac{du}{dx}(\operatorname{csch}^2 u)$

41. $\dfrac{d}{dx}(\operatorname{sech} u) = -\dfrac{du}{dx}(\operatorname{sech} u \cdot \tanh u)$

42. $\dfrac{d}{dx}(\operatorname{csch} u) = -\dfrac{du}{dx}(\operatorname{csch} u \cdot \coth u)$

43. $\dfrac{d}{dx}(\sinh^{-1} u) = \dfrac{d}{dx}[\log(u + \sqrt{u^2+1})] = \dfrac{1}{\sqrt{u^2+1}}\dfrac{du}{dx}$

44. $\dfrac{d}{dx}(\cosh^{-1} u) = \dfrac{d}{dx}[\log(u + \sqrt{u^2-1})] = \dfrac{1}{\sqrt{u^2-1}}\dfrac{du}{dx},\qquad (u>1, \cosh^{-1} u > 0)$

45. $\dfrac{d}{dx}(\tanh^{-1} u) = \dfrac{d}{dx}\left[\dfrac{1}{2}\log\dfrac{1+u}{1-u}\right] = \dfrac{1}{1-u^2}\dfrac{du}{dx},\qquad (u^2 < 1)$

46. $\dfrac{d}{dx}(\coth^{-1} u) = \dfrac{d}{dx}\left[\dfrac{1}{2}\log\dfrac{u+1}{u-1}\right] = \dfrac{1}{1-u^2}\dfrac{du}{dx},\qquad (u^2 > 1)$

47. $\dfrac{d}{dx}(\operatorname{sech}^{-1} u) = \dfrac{d}{dx}\left[\log\dfrac{1+\sqrt{1-u^2}}{u}\right] = -\dfrac{1}{u\sqrt{1-u^2}}\dfrac{du}{dx},\qquad (0 < u < 1, \operatorname{sech}^{-1} u > 0)$

48. $\dfrac{d}{dx}(\operatorname{csch}^{-1} u) = \dfrac{d}{dx}\left[\log\dfrac{1+\sqrt{1+u^2}}{u}\right] = -\dfrac{1}{|u|\sqrt{1+u^2}}\dfrac{du}{dx}$

49. $\dfrac{d}{dq} \displaystyle\int_p^q f(x)\,dx = f(q),$ [p constant]

50. $\dfrac{d}{dp} \displaystyle\int_p^q f(x)\,dx = -f(p),$ [q constant]

51. $\dfrac{d}{da} \displaystyle\int_p^q f(x, a)\,dx = \int_p^q \dfrac{\partial}{\partial a}[f(x, a)]\,dx + f(q, a)\dfrac{dq}{da} - f(p, a)\dfrac{dp}{da}$

INTEGRATION

The following is a brief discussion of some integration techniques. A more complete discussion can be found in a number of good text books. However, the purpose of this introduction is simply to discuss a few of the important techniques which may be used, in conjunction with the integral table which follows, to integrate particular functions.

No matter how extensive the integral table, it is a fairly uncommon occurrence to find in the table the exact integral desired. Usually some form of transformation will have to be made. The simplest type of transformation, and yet the most general, is substitution. Simple forms of substitution, such as $y = ax$, are employed almost unconsciously by experienced users of integral tables. Other substitutions may require more thought. In some sections of the tables, appropriate substitutions are suggested for integrals which are similar to, but not exactly like, integrals in the table. Finding the right substitution is largely a matter of intuition and experience.

Several precautions must be observed when using substitutions:

1. Be sure to make the substitution in the dx term, as well as everywhere else in the integral.
2. Be sure that the function substituted is one-to-one and continuous. If this is not the case, the integral must be restricted in such a way as to make it true. See the example following.
3. With definite integrals, the limits should also be expressed in terms of the new dependent variable. With indefinite integrals, it is necessary to perform the reverse substitution to obtain the answer in terms of the original independent variable. This may also be done for definite integrals, but it is usually easier to change the limits.

Example:

$$\int \frac{x^4}{\sqrt{a^2 - x^2}}\,dx$$

Here we make the substitution $x = |a| \sin \theta$. Then $dx = |a| \cos \theta\,d\theta$, and

$$\sqrt{a^2 - x^2} = \sqrt{a^2 - a^2 \sin^2 \theta} = |a|\sqrt{1 - \sin^2 \theta} = |a \cos \theta|$$

Notice the absolute value signs. It is very important to keep in mind that a square root radical always denotes the positive square root, and to assure the sign is always kept positive. Thus $\sqrt{x^2} = |x|$. Failure to observe this is a common cause of errors in integration.

Notice also that the indicated substitution is not a one-to-one function, that is, it does not have a unique inverse. Thus we must restrict the range of θ in such a way as to make the function one-to-one. Fortunately, this is easily done by solving for θ

$$\theta = \sin^{-1} \frac{x}{|a|}$$

and restricting the inverse sine to the principal values, $-\dfrac{\pi}{2} \le \theta \le \dfrac{\pi}{2}$.

Thus the integral becomes

$$\int \frac{a^4 \sin^4 \theta |a| \cos \theta \, d\theta}{|a| \, |\cos \theta|}$$

Now, however, in the range of values chosen for θ, $\cos \theta$ is always positive. Thus we may remove the absolute value signs from $\cos \theta$ in the denominator. (This is one of the reasons that the principal values of the inverse trigonometric functions are defined as they are.)

Then the $\cos \theta$ terms cancel, and the integral becomes

$$a^4 \int \sin^4 \theta \, d\theta$$

By application of integral formulas 299 and 296, we integrate this to

$$-a^4 \frac{\sin^3 \theta \cos \theta}{4} - \frac{3a^4}{8} \cos \theta \sin \theta + \frac{3a^4}{8} \theta + C$$

We now must perform the inverse substitution to get the result in terms of x. We have

$$\theta = \sin^{-1} \frac{x}{|a|}$$

$$\sin \theta = \frac{x}{|a|}$$

Then

$$\cos \theta = +\sqrt{1 - \sin^2 \theta} = \pm\sqrt{1 - \frac{x^2}{a^2}} = \pm \frac{\sqrt{a^2 - x^2}}{|a|}.$$

Because of the previously mentioned fact that $\cos \theta$ is positive, we may omit the $+$ sign. The reverse substitution then produces the final answer

$$\int \frac{x^4}{\sqrt{a^2 - x^2}} \, dx = -\tfrac{1}{4}x^3 \sqrt{a^2 - x^2} - \tfrac{3}{8}a^2 x \sqrt{a^2 - x^2} + \frac{3a^4}{8} \sin^{-1} \frac{x}{|a|} + C.$$

Any rational function of x may be integrated, if the denominator is factored into linear and irreducible quadratic factors. The function may then be broken into partial fractions, and the individual partial fractions integrated by use of the appropriate formula from the integral table. See the section on partial fractions for further information.

Many integrals may be reduced to rational functions by proper substitutions. For example,

$$z = \tan \frac{x}{2}$$

will reduce any rational function of the six trigonometric functions of x to a rational function of z. (Frequently there are other substitutions which are simpler to use, but this one will always work. See integral formula number 484.)

Any rational function of x and $\sqrt{ax + b}$ may be reduced to a rational function of z by making the substitution

$$z = \sqrt{ax + b}.$$

Other likely substitutions will be suggested by looking at the form of the integrand.

The other main method of transforming integrals is integration by parts. This involves applying formula number 5 or 6 in the accompanying integral table. The critical factor in this method is the choice of the functions u and v. In order for the method to be successful, $v = \int dv$ and $\int v \, du$ must be easier to integrate than the original integral. Again, this choice is largely a matter of intuition and experience.

Example:

$$\int x \sin x \, dx$$

Two obvious choices are $u = x$, $dv = \sin x \, dx$, or $u = \sin x$, $dv = x \, dx$. Since a preliminary mental calculation indicates that $\int v \, du$ in the second choice would be more, rather than less, complicated than the original integral (it would contain x^2), we use the first choice.

$$u = x \qquad\qquad du = dx$$
$$dv = \sin x \, dx \qquad\qquad v = -\cos x$$

$$\int x \sin x \, dx = \int u \, dv = uv - \int v \, du = -x \cos x + \int \cos x \, dx$$
$$= \sin x - x \cos x$$

Of course, this result could have been obtained directly from the integral table, but it provides a simple example of the method. In more complicated examples the choice of u and v may not be so obvious, and several different choices may have to be tried. Of course, there is no guarantee that any of them will work.

Integration by parts may be applied more than once, or combined with substitution. A fairly common case is illustrated by the following example.

Example:

$$\int e^x \sin x \, dx$$

Let

$$u = e^x \qquad \text{Then} \quad du = e^x \, dx$$
$$dv = \sin x \, dx \qquad\qquad v = -\cos x$$

$$\int e^x \sin x \, dx = \int u \, dv = uv - \int v \, du = -e^x \cos x + \int e^x \cos x \, dx$$

In this latter integral,

$$\text{let} \quad u = e^x \qquad \text{Then} \quad du = e^x \, dx$$
$$dv = \cos x \, dx \qquad\qquad v = \sin x$$

$$\int e^x \sin x \, dx = -e^x \cos x + \int e^x \cos x \, dx = -e^x \cos x + \int u \, dv$$

$$= -e^x \cos x + uv - \int v \, du$$

$$= -e^x \cos x + e^x \sin x - \int e^x \sin x \, dx$$

This looks as if a circular transformation has taken place, since we are back at the same integral we started from. However, the above equation can be solved algebraically for the required integral:

$$\int e^x \sin x \, dx = \tfrac{1}{2}(e^x \sin x - e^x \cos x)$$

In the second integration by parts, if the parts had been chosen as $u = \cos x$, $dv = e^x \, dx$, we would indeed have made a circular transformation, and returned to the starting place.

In general, when doing repeated integration by parts, one should never choose the function u at any stage to be the same as the function v at the previous stage, or a constant times the previous v.

The following rule is called the extended rule for integration by parts. It is the result of $n + 1$ successive applications of integration by parts.

If

$$g_1(x) = \int g(x)\,dx, \qquad g_2(x) = \int g_1(x)\,dx,$$

$$g_3(x) = \int g_2(x)\,dx, \ldots, g_m(x) = \int g_{m-1}(x)\,dx, \ldots,$$

then

$$\int f(x) \cdot g(x)\,dx = f(x) \cdot g_1(x) - f'(x) \cdot g_2(x) + f''(x) \cdot g_3(x) - + \cdots$$

$$+ (-1)^n f^{(n)}(x) g_{n+1}(x) + (-1)^{n+1} \int f^{(n+1)}(x) g_{n+1}(x)\,dx.$$

A useful special case of the above rule is when $f(x)$ is a polynomial of degree n. Then $f^{(n+1)}(x) = 0$, and

$$\int f(x) \cdot g(x)\,dx = f(x) \cdot g_1(x) - f'(x) \cdot g_2(x) + f''(x) \cdot g_3(x) - + \cdots + (-1)^n f^{(n)}(x) g_{n+1}(x) + C$$

Example:

If $f(x) = x^2, g(x) = \sin x$

$$\int x^2 \sin x\,dx = -x^2 \cos x + 2x \sin x + 2 \cos x + C$$

Another application of this formula occurs if

$$f''(x) = af(x) \quad \text{and} \quad g''(x) = bg(x),$$

where a and b are unequal constants. In this case, by a process similar to that used in the above example for $\int e^x \sin x\,dx$, we get the formula

$$\int f(x)g(x)\,dx = \frac{f(x) \cdot g'(x) - f'(x) \cdot g(x)}{b - a} + C$$

This formula could have been used in the example mentioned. Here is another example.

Example:

If $f(x) = e^{2x}, g(x) = \sin 3x$, then $a = 4, b = -9$, and

$$\int e^{2x} \sin 3x\,dx = \frac{3\,e^{2x} \cos 3x - 2\,e^{2x} \sin 3x}{-9 - 4} + C = \frac{e^{2x}}{13}(2 \sin 3x - 3 \cos 3x) + C$$

The following additional points should be observed when using this table.

1. A constant of integration is to be supplied with the answers for indefinite integrals.
2. Logarithmic expressions are to base $e = 2.71828\ldots$, unless otherwise specified, and are to be evaluated for the absolute value of the arguments involved therein.
3. All angles are measured in radians, and inverse trigonometric and hyperbolic functions represent principal values, unless otherwise indicated.
4. If the application of a formula produces either a zero denominator or the square root of a negative number in the result, there is usually available another form of the answer which

avoids this difficulty. In many of the results, the excluded values are specified, but when such are omitted it is presumed that one can tell what these should be, especially when difficulties of the type herein mentioned are obtained.

5. When inverse trigonometric functions occur in the integrals, be sure that any replacements made for them are strictly in accordance with the rules for such functions. This causes little difficulty when the argument of the inverse trigonometric function is positive, since then all angles involved are in the first quadrant. However, if the argument is negative, special care must be used. Thus if $u > 0$,

$$\sin^{-1} u = \cos^{-1}\sqrt{1 - u^2} = \csc^{-1}\frac{1}{u}, \text{ etc.}$$

However, if $u < 0$,

$$\sin^{-1} u = -\cos^{-1}\sqrt{1 - u^2} = -\pi - \csc^{-1}\frac{1}{u}, \text{ etc.}$$

See the section on inverse trigonometric functions for a full treatment of the allowable substitutions.

6. In integrals 340–345 and some others, the right side includes expressions of the form

$$A \tan^{-1}[B + C \tan f(x)].$$

In these formulas, the $\tan^{-1}$ does not necessarily represent the principal value. Instead of always employing the principal branch of the inverse tangent function, one must instead use that branch of the inverse tangent function upon which $f(x)$ lies for any particular choice of x.

Example:

$$\int_0^{4\pi} \frac{dx}{2 + \sin x} = \frac{2}{\sqrt{3}}\tan^{-1}\frac{2\tan\frac{x}{2} + 1}{\sqrt{3}}\Bigg]_0^{4\pi}$$

$$= \frac{2}{\sqrt{3}}\left[\tan^{-1}\frac{2\tan 2\pi + 1}{\sqrt{3}} - \tan^{-1}\frac{2\tan 0 + 1}{\sqrt{3}}\right]$$

$$= \frac{2}{\sqrt{3}}\left[\frac{13\pi}{6} - \frac{\pi}{6}\right] = \frac{4\pi}{\sqrt{3}} = \frac{4\sqrt{3}\pi}{3}$$

Here

$$\tan^{-1}\frac{2\tan 2\pi + 1}{\sqrt{3}} = \tan^{-1}\frac{1}{\sqrt{3}} = \frac{13\pi}{6},$$

since $f(x) = 2\pi$; and

$$\tan^{-1}\frac{2\tan 0 + 1}{\sqrt{3}} = \tan^{-1}\frac{1}{\sqrt{3}} = \frac{\pi}{6},$$

since $f(x) = 0$.

7. B_n and E_n where used in Integrals represents the Bernoulli and Euler numbers as defined in tables of Bernoulli and Euler polynomials contained in certain mathematics reference and handbooks, as for example, Beyer, W. H., *Handbook of Mathematical Sciences*, 5th ed., CRC Press, Inc., West Palm Beach 1978, 577–583.

ELEMENTARY FORMS

1. $\displaystyle\int a\,dx = ax$

2. $\displaystyle\int a\cdot f(x)\,dx = a\int f(x)\,dx$

3. $\displaystyle\int \phi(y)\,dx = \int \frac{\phi(y)}{y'}\,dy,$ where $y' = \dfrac{dy}{dx}$

4. $\displaystyle\int (u+v)\,dx = \int u\,dx + \int v\,dx,$ where u and v are any functions of x

5. $\displaystyle\int u\,dv = u\int dv - \int v\,du = uv - \int v\,du$

6. $\displaystyle\int u\frac{dv}{dx}\,dx = uv - \int v\frac{du}{dx}\,dx$

7. $\displaystyle\int x^n\,dx = \frac{x^{n+1}}{n+1},$ except $n = -1$

8. $\displaystyle\int \frac{f'(x)\,dx}{f(x)} = \log f(x),$ $(df(x) = f'(x)\,dx)$

9. $\displaystyle\int \frac{dx}{x} = \log x$

10. $\displaystyle\int \frac{f'(x)\,dx}{2\sqrt{f(x)}} = \sqrt{f(x)},$ $(df(x) = f'(x)\,dx)$

11. $\displaystyle\int e^x\,dx = e^x$

12. $\displaystyle\int e^{ax}\,dx = e^{ax}/a$

13. $\displaystyle\int b^{ax}\,dx = \frac{b^{ax}}{a\log b},$ $(b > 0)$

14. $\displaystyle\int \log x\,dx = x\log x - x$

15. $\displaystyle\int a^x \log a\,dx = a^x,$ $(a > 0)$

16. $\displaystyle\int \frac{dx}{a^2 + x^2} = \frac{1}{a}\tan^{-1}\frac{x}{a}$

17. $\displaystyle\int \frac{dx}{a^2 - x^2} = \begin{cases} \dfrac{1}{a}\tanh^{-1}\dfrac{x}{a} \\ \quad\text{or} \\ \dfrac{1}{2a}\log\dfrac{a+x}{a-x}, \quad (a^2 > x^2) \end{cases}$

18. $\displaystyle\int \frac{dx}{x^2 - a^2} = \begin{cases} -\dfrac{1}{a}\coth^{-1}\dfrac{x}{a} \\ \quad\text{or} \\ \dfrac{1}{2a}\log\dfrac{x-a}{x+a}, \quad (x^2 > a^2) \end{cases}$

19. $\int \dfrac{dx}{\sqrt{a^2 - x^2}} = \begin{cases} \sin^{-1} \dfrac{x}{|a|} \\ \quad\text{or} \\ -\cos^{-1} \dfrac{x}{|a|}, \quad (a^2 > x^2) \end{cases}$

20. $\int \dfrac{dx}{\sqrt{x^2 \pm a^2}} = \log(x + \sqrt{x^2 \pm a^2})$

21. $\int \dfrac{dx}{x\sqrt{x^2 - a^2}} = \dfrac{1}{|a|} \sec^{-1} \dfrac{x}{a}$

22. $\int \dfrac{dx}{x\sqrt{a^2 \pm x^2}} = -\dfrac{1}{a} \log\left(\dfrac{a + \sqrt{a^2 \pm x^2}}{x} \right)$

FORMS CONTAINING $(a + bx)$

For forms containing $a + bx$, but not listed in the table, the substitution $u = \dfrac{a + bx}{x}$ may prove helpful.

23. $\int (a + bx)^n \, dx = \dfrac{(a + bx)^{n+1}}{(n+1)b}, \qquad (n \neq -1)$

24. $\int x(a + bx)^n \, dx$

$$= \dfrac{1}{b^2(n+2)}(a + bx)^{n+2} - \dfrac{a}{b^2(n+1)}(a + bx)^{n+1}, \qquad (n \neq -1, -2)$$

25. $\int x^2(a + bx)^n \, dx = \dfrac{1}{b^3}\left[\dfrac{(a + bx)^{n+3}}{n+3} - 2a\dfrac{(a + bx)^{n+2}}{n+2} + a^2\dfrac{(a + bx)^{n+1}}{n+1} \right]$

26. $\int x^m(a + bx)^n \, dx = \begin{cases} \dfrac{x^{m+1}(a + bx)^n}{m + n + 1} + \dfrac{an}{m + n + 1}\displaystyle\int x^m(a + bx)^{n-1} \, dx \\ \quad\text{or} \\ \dfrac{1}{a(n+1)}\left[-x^{m+1}(a + bx)^{n+1} \right. \\ \qquad\qquad\qquad \left. + (m + n + 2)\displaystyle\int x^m(a + bx)^{n+1} \, dx \right] \\ \quad\text{or} \\ \dfrac{1}{b(m + n + 1)}\left[x^m(a + bx)^{n+1} - ma\displaystyle\int x^{m-1}(a + bx)^n \, dx \right] \end{cases}$

27. $\int \dfrac{dx}{a + bx} = \dfrac{1}{b} \log(a + bx)$

28. $\int \dfrac{dx}{(a + bx)^2} = -\dfrac{1}{b(a + bx)}$

29. $\int \dfrac{dx}{(a + bx)^3} = -\dfrac{1}{2b(a + bx)^2}$

30. $\int \dfrac{x \, dx}{a + bx} = \begin{cases} \dfrac{1}{b^2}[a + bx - a \log(a + bx)] \\ \quad\text{or} \\ \dfrac{x}{b} - \dfrac{a}{b^2} \log(a + bx) \end{cases}$

31. $\displaystyle\int \frac{x\,dx}{(a+bx)^2} = \frac{1}{b^2}\left[\log(a+bx) + \frac{a}{a+bx}\right]$

32. $\displaystyle\int \frac{x\,dx}{(a+bx)^n} = \frac{1}{b^2}\left[\frac{-1}{(n-2)(a+bx)^{n-2}} + \frac{a}{(n-1)(a+bx)^{n-1}}\right], \qquad n \neq 1, 2$

33. $\displaystyle\int \frac{x^2\,dx}{a+bx} = \frac{1}{b^3}\left[\frac{1}{2}(a+bx)^2 - 2a(a+bx) + a^2\log(a+bx)\right]$

34. $\displaystyle\int \frac{x^2\,dx}{(a+bx)^2} = \frac{1}{b^3}\left[a+bx - 2a\log(a+bx) - \frac{a^2}{a+bx}\right]$

35. $\displaystyle\int \frac{x^2\,dx}{(a+bx)^3} = \frac{1}{b^3}\left[\log(a+bx) + \frac{2a}{a+bx} - \frac{a^2}{2(a+bx)^2}\right]$

36. $\displaystyle\int \frac{x^2\,dx}{(a+bx)^n} = \frac{1}{b^3}\left[\frac{-1}{(n-3)(a+bx)^{n-3}}\right.$

$$\left. + \frac{2a}{(n-2)(a+bx)^{n-2}} - \frac{a^2}{(n-1)(a+bx)^{n-1}}\right], \qquad n \neq 1, 2, 3$$

37. $\displaystyle\int \frac{dx}{x(a+bx)} = -\frac{1}{a}\log\frac{a+bx}{x}$

38. $\displaystyle\int \frac{dx}{x(a+bx)^2} = \frac{1}{a(a+bx)} - \frac{1}{a^2}\log\frac{a+bx}{x}$

39. $\displaystyle\int \frac{dx}{x(a+bx)^3} = \frac{1}{a^3}\left[\frac{1}{2}\left(\frac{2a+bx}{a+bx}\right)^2 + \log\frac{x}{a+bx}\right]$

40. $\displaystyle\int \frac{dx}{x^2(a+bx)} = -\frac{1}{ax} + \frac{b}{a^2}\log\frac{a+bx}{x}$

41. $\displaystyle\int \frac{dx}{x^3(a+bx)} = \frac{2bx-a}{2a^2x^2} + \frac{b^2}{a^3}\log\frac{x}{a+bx}$

42. $\displaystyle\int \frac{dx}{x^2(a+bx)^2} = -\frac{a+2bx}{a^2x(a+bx)} + \frac{2b}{a^3}\log\frac{a+bx}{x}$

FORMS CONTAINING $c^2 \pm x^2,\ x^2 - c^2$

43. $\displaystyle\int \frac{dx}{c^2+x^2} = \frac{1}{c}\tan^{-1}\frac{x}{c}$

44. $\displaystyle\int \frac{dx}{c^2-x^2} = \frac{1}{2c}\log\frac{c+x}{c-x}, \qquad (c^2 > x^2)$

45. $\displaystyle\int \frac{dx}{x^2-c^2} = \frac{1}{2c}\log\frac{x-c}{x+c}, \qquad (x^2 > c^2)$

46. $\displaystyle\int \frac{x\,dx}{c^2 \pm x^2} = \pm\frac{1}{2}\log(c^2 \pm x^2)$

47. $\displaystyle\int \frac{x\,dx}{(c^2 \pm x^2)^{n+1}} = \mp\frac{1}{2n(c^2 \pm x^2)^n}$

48. $\displaystyle\int \frac{dx}{(c^2 \pm x^2)^n} = \frac{1}{2c^2(n-1)}\left[\frac{x}{(c^2 \pm x^2)^{n-1}} + (2n-3)\int \frac{dx}{(c^2 \pm x^2)^{n-1}}\right]$

49. $\displaystyle\int \frac{dx}{(x^2-c^2)^n} = \frac{1}{2c^2(n-1)}\left[-\frac{x}{(x^2-c^2)^{n-1}} - (2n-3)\int \frac{dx}{(x^2-c^2)^{n-1}}\right]$

50. $\displaystyle\int \frac{x\,dx}{x^2-c^2} = \frac{1}{2}\log(x^2-c^2)$

51. $\int \dfrac{x \, dx}{(x^2 - c^2)^{n+1}} = -\dfrac{1}{2n(x^2 - c^2)^n}$

FORMS CONTAINING $a + bx$ and $c + dx$

$$u = a + bx, \qquad v = c + dx, \qquad k = ad - bc$$

If $k = 0$, then $v = \dfrac{c}{a} u$

52. $\int \dfrac{dx}{u \cdot v} = \dfrac{1}{k} \cdot \log \left(\dfrac{v}{u} \right)$

53. $\int \dfrac{x \, dx}{u \cdot v} = \dfrac{1}{k} \left[\dfrac{a}{b} \log (u) - \dfrac{c}{d} \log (v) \right]$

54. $\int \dfrac{dx}{u^2 \cdot v} = \dfrac{1}{k} \left(\dfrac{1}{u} + \dfrac{d}{k} \log \dfrac{v}{u} \right)$

55. $\int \dfrac{x \, dx}{u^2 \cdot v} = \dfrac{-a}{bku} - \dfrac{c}{k^2} \log \dfrac{v}{u}$

56. $\int \dfrac{x^2 \, dx}{u^2 \cdot v} = \dfrac{a^2}{b^2 ku} + \dfrac{1}{k^2} \left[\dfrac{c^2}{d} \log (v) + \dfrac{a(k - bc)}{b^2} \log (u) \right]$

57. $\int \dfrac{dx}{u^n \cdot v^m} = \dfrac{1}{k(m-1)} \left[\dfrac{-1}{u^{n-1} \cdot v^{m-1}} - (m + n - 2)b \int \dfrac{dx}{u^n \cdot v^{m-1}} \right]$

58. $\int \dfrac{u}{v} dx = \dfrac{bx}{d} + \dfrac{k}{d^2} \log (v)$

59. $\int \dfrac{u^m \, dx}{v^n} = \begin{cases} \dfrac{-1}{k(n-1)} \left[\dfrac{u^{m+1}}{v^{n-1}} + b(n - m - 2) \int \dfrac{u^m}{v^{n-1}} dx \right] \\ \qquad\qquad \text{or} \\ \dfrac{-1}{d(n-m-1)} \left[\dfrac{u^m}{v^{n-1}} + mk \int \dfrac{u^{m-1}}{v^n} dx \right] \\ \qquad\qquad \text{or} \\ \dfrac{-1}{d(n-1)} \left[\dfrac{u^m}{v^{n-1}} - mb \int \dfrac{u^{m-1}}{v^{n-1}} dx \right] \end{cases}$

FORMS CONTAINING $(a + bx^n)$

60. $\int \dfrac{dx}{a + bx^2} = \dfrac{1}{\sqrt{ab}} \tan^{-1} \dfrac{x\sqrt{ab}}{a}, \qquad (ab > 0)$

61. $\int \dfrac{dx}{a + bx^2} = \begin{cases} \dfrac{1}{2\sqrt{-ab}} \log \dfrac{a + x\sqrt{-ab}}{a - x\sqrt{-ab}}, \qquad (ab < 0) \\ \qquad\qquad \text{or} \\ \dfrac{1}{\sqrt{-ab}} \tanh^{-1} \dfrac{x\sqrt{-ab}}{a}, \qquad (ab < 0) \end{cases}$

62. $\int \dfrac{dx}{a^2 + b^2 x^2} = \dfrac{1}{ab} \tan^{-1} \dfrac{bx}{a}$

63. $\int \dfrac{x \, dx}{a + bx^2} = \dfrac{1}{2b} \log (a + bx^2)$

64. $\int \dfrac{x^2 \, dx}{a + bx^2} = \dfrac{x}{b} - \dfrac{a}{b} \int \dfrac{dx}{a + bx^2}$

65. $\int \dfrac{dx}{(a + bx^2)^2} = \dfrac{x}{2a(a + bx^2)} + \dfrac{1}{2a} \int \dfrac{dx}{a + bx^2}$

66. $\displaystyle \int \frac{dx}{a^2 - b^2 x^2} = \frac{1}{2ab} \log \frac{a + bx}{a - bx}$

67. $\displaystyle \int \frac{dx}{(a + bx^2)^{m+1}} = \begin{cases} \dfrac{1}{2ma} \dfrac{x}{(a + bx^2)^m} + \dfrac{2m - 1}{2ma} \displaystyle\int \dfrac{dx}{(a + bx^2)^m} \\[2mm] \text{or} \\[2mm] \dfrac{(2m)!}{(m!)^2} \left[\dfrac{x}{2a} \displaystyle\sum_{r=1}^{m} \dfrac{r!(r - 1)!}{(4a)^{m-r}(2r)!(a + bx^2)^r} + \dfrac{1}{(4a)^m} \displaystyle\int \dfrac{dx}{a + bx^2} \right] \end{cases}$

68. $\displaystyle \int \frac{x\,dx}{(a + bx^2)^{m+1}} = -\frac{1}{2bm(a + bx^2)^m}$

69. $\displaystyle \int \frac{x^2\,dx}{(a + bx^2)^{m+1}} = \frac{-x}{2mb(a + bx^2)^m} + \frac{1}{2mb} \int \frac{dx}{(a + bx^2)^m}$

70. $\displaystyle \int \frac{dx}{x(a + bx^2)} = \frac{1}{2a} \log \frac{x^2}{a + bx^2}$

71. $\displaystyle \int \frac{dx}{x^2(a + bx^2)} = -\frac{1}{ax} - \frac{b}{a} \int \frac{dx}{a + bx^2}$

72. $\displaystyle \int \frac{dx}{x(a + bx^2)^{m+1}} = \begin{cases} \dfrac{1}{2am(a + bx^2)^m} + \dfrac{1}{a} \displaystyle\int \dfrac{dx}{x(a + bx^2)^m} \\[2mm] \text{or} \\[2mm] \dfrac{1}{2a^{m+1}} \left[\displaystyle\sum_{r=1}^{m} \dfrac{a^r}{r(a + bx^2)^r} + \log \dfrac{x^2}{a + bx^2} \right] \end{cases}$

73. $\displaystyle \int \frac{dx}{x^2(a + bx^2)^{m+1}} = \frac{1}{a} \int \frac{dx}{x^2(a + bx^2)^m} - \frac{b}{a} \int \frac{dx}{(a + bx^2)^{m+1}}$

74. $\displaystyle \int \frac{dx}{a + bx^3} = \frac{k}{3a} \left[\frac{1}{2} \log \frac{(k + x)^3}{a + bx^3} + \sqrt{3} \tan^{-1} \frac{2x - k}{k\sqrt{3}} \right], \quad \left(k = \sqrt[3]{\frac{a}{b}} \right)$

75. $\displaystyle \int \frac{x\,dx}{a + bx^3} = \frac{1}{3bk} \left[\frac{1}{2} \log \frac{a + bx^3}{(k + x)^3} + \sqrt{3} \tan^{-1} \frac{2x - k}{k\sqrt{3}} \right], \quad \left(k = \sqrt[3]{\frac{a}{b}} \right)$

76. $\displaystyle \int \frac{x^2\,dx}{a + bx^3} = \frac{1}{3b} \log(a + bx^3)$

77. $\displaystyle \int \frac{dx}{a + bx^4} = \frac{k}{2a} \left[\frac{1}{2} \log \frac{x^2 + 2kx + 2k^2}{x^2 - 2kx + 2k^2} + \tan^{-1} \frac{2kx}{2k^2 - x^2} \right],$
$$\left(ab > 0, k = \sqrt[4]{\frac{a}{4b}} \right)$$

78. $\displaystyle \int \frac{dx}{a + bx^4} = \frac{k}{2a} \left[\frac{1}{2} \log \frac{x + k}{x - k} + \tan^{-1} \frac{x}{k} \right], \quad \left(ab < 0, k = \sqrt[4]{-\frac{a}{b}} \right)$

79. $\displaystyle \int \frac{x\,dx}{a + bx^4} = \frac{1}{2bk} \tan^{-1} \frac{x^2}{k}, \quad \left(ab > 0, k = \sqrt{\frac{a}{b}} \right)$

80. $\displaystyle \int \frac{x\,dx}{a + bx^4} = \frac{1}{4bk} \log \frac{x^2 - k}{x^2 + k}, \quad \left(ab < 0, k = \sqrt{-\frac{a}{b}} \right)$

81. $\displaystyle \int \frac{x^2\,dx}{a + bx^4} = \frac{1}{4bk} \left[\frac{1}{2} \log \frac{x^2 - 2kx + 2k^2}{x^2 + 2kx + 2k^2} + \tan^{-1} \frac{2kx}{2k^2 - x^2} \right],$
$$\left(ab > 0, k = \sqrt[4]{\frac{a}{4b}} \right)$$

82. $\displaystyle\int \frac{x^2\, dx}{a + bx^4} = \frac{1}{4bk}\left[\log\frac{x-k}{x+k} + 2\tan^{-1}\frac{x}{k}\right], \qquad \left(ab < 0, k = \sqrt[4]{-\frac{a}{b}}\,\right)$

83. $\displaystyle\int \frac{x^3\, dx}{a + bx^4} = \frac{1}{4b}\log(a + bx^4)$

84. $\displaystyle\int \frac{dx}{x(a + bx^n)} = \frac{1}{an}\log\frac{x^n}{a + bx^n}$

85. $\displaystyle\int \frac{dx}{(a + bx^n)^{m+1}} = \frac{1}{a}\int \frac{dx}{(a + bx^n)^m} - \frac{b}{a}\int \frac{x^n\, dx}{(a + bx^n)^{m+1}}$

86. $\displaystyle\int \frac{x^m\, dx}{(a + bx^n)^{p+1}} = \frac{1}{b}\int \frac{x^{m-n}\, dx}{(a + bx^n)^p} - \frac{a}{b}\int \frac{x^{m-n}\, dx}{(a + bx^n)^{p+1}}$

87. $\displaystyle\int \frac{dx}{x^m(a + bx^n)^{p+1}} = \frac{1}{a}\int \frac{dx}{x^m(a + bx^n)^p} - \frac{b}{a}\int \frac{dx}{x^{m-n}(a + bx^n)^{p+1}}$

88. $\displaystyle\int x^m(a + bx^n)^p\, dx = \begin{cases}
\dfrac{1}{b(np + m + 1)}\left[x^{m-n+1}(a + bx^n)^{p+1}\right. \\
\qquad\qquad\qquad \left. - a(m - n + 1)\displaystyle\int x^{m-n}(a + bx^n)^p\, dx\right] \\[2pt]
\text{or} \\
\dfrac{1}{np + m + 1}\left[x^{m+1}(a + bx^n)^p\right. \\
\qquad\qquad\qquad \left. + anp\displaystyle\int x^m(a + bx^n)^{p-1}\, dx\right] \\[2pt]
\text{or} \\
\dfrac{1}{a(m+1)}\left[x^{m+1}(a + bx^n)^{p+1}\right. \\
\qquad\qquad\qquad \left. - (m + 1 + np + n)b\displaystyle\int x^{m+n}(a + bx^n)^p\, dx\right] \\[2pt]
\text{or} \\
\dfrac{1}{an(p+1)}\left[-x^{m+1}(a + bx^n)^{p+1}\right. \\
\qquad\qquad\qquad \left. + (m + 1 + np + n)\displaystyle\int x^m(a + bx^n)^{p+1}\, dx\right]
\end{cases}$

FORMS CONTAINING $c^3 \pm x^3$

89. $\displaystyle\int \frac{dx}{c^3 \pm x^3} = \pm\frac{1}{6c^2}\log\frac{(c \pm x)^3}{c^3 \pm x^3} + \frac{1}{c^2\sqrt{3}}\tan^{-1}\frac{2x \mp c}{c\sqrt{3}}$

90. $\displaystyle\int \frac{dx}{(c^3 \pm x^3)^2} = \frac{x}{3c^3(c^3 \pm x^3)} + \frac{2}{3c^3}\int \frac{dx}{c^3 \pm x^3}$

91. $\displaystyle\int \frac{dx}{(c^3 \pm x^3)^{n+1}} = \frac{1}{3nc^3}\left[\frac{x}{(c^3 \pm x^3)^n} + (3n - 1)\int \frac{dx}{(c^3 \pm x^3)^n}\right]$

92. $\displaystyle\int \frac{x\, dx}{c^3 \pm x^3} = \frac{1}{6c}\log\frac{c^3 \pm x^3}{(c \pm x)^3} \pm \frac{1}{c\sqrt{3}}\tan^{-1}\frac{2x \mp c}{c\sqrt{3}}$

93. $\displaystyle\int \frac{x\, dx}{(c^3 \pm x^3)^2} = \frac{x^2}{3c^3(c^3 \pm x^3)} + \frac{1}{3c^3}\int \frac{x\, dx}{c^3 \pm x^3}$

94. $\int \dfrac{x\,dx}{(c^3 \pm x^3)^{n+1}} = \dfrac{1}{3nc^3}\left[\dfrac{x^2}{(c^3 \pm x^3)^n} + (3n-2)\int \dfrac{x\,dx}{(c^3 \pm x^3)^n}\right]$

95. $\int \dfrac{x^2\,dx}{c^3 \pm x^3} = \pm\dfrac{1}{3}\log(c^3 \pm x^3)$

96. $\int \dfrac{x^2\,dx}{(c^3 \pm x^3)^{n+1}} = \mp\dfrac{1}{3n(c^3 \pm x^3)^n}$

97. $\int \dfrac{dx}{x(c^3 \pm x^3)} = \dfrac{1}{3c^3}\log\dfrac{x^3}{c^3 \pm x^3}$

98. $\int \dfrac{dx}{x(c^3 + x^3)^2} = \dfrac{1}{3c^3(c^3 \pm x^3)} + \dfrac{1}{3c^6}\log\dfrac{x^3}{c^3 \pm x^3}$

99. $\int \dfrac{dx}{x(c^3 \pm x^3)^{n+1}} = \dfrac{1}{3nc^3(c^3 \pm x^3)^n} + \dfrac{1}{c^3}\int \dfrac{dx}{x(c^3 \pm x^3)^n}$

100. $\int \dfrac{dx}{x^2(c^3 \pm x^3)} = -\dfrac{1}{c^3 x} \mp \dfrac{1}{c^3}\int \dfrac{x\,dx}{c^3 \pm x^3}$

101. $\int \dfrac{dx}{x^2(c^3 \pm x^3)^{n+1}} = \dfrac{1}{c^3}\int \dfrac{dx}{x^2(c^3 \pm x^3)^n} \mp \dfrac{1}{c^3}\int \dfrac{x\,dx}{(c^3 \pm x^3)^{n+1}}$

FORMS CONTAINING $c^4 \pm x^4$

102. $\int \dfrac{dx}{c^4 + x^4} = \dfrac{1}{2c^3\sqrt{2}}\left[\dfrac{1}{2}\log\dfrac{x^2 + cx\sqrt{2} + c^2}{x^2 - cx\sqrt{2} + c^2} + \tan^{-1}\dfrac{cx\sqrt{2}}{c^2 - x^2}\right]$

103. $\int \dfrac{dx}{c^4 - x^4} = \dfrac{1}{2c^3}\left[\dfrac{1}{2}\log\dfrac{c + x}{c - x} + \tan^{-1}\dfrac{x}{c}\right]$

104. $\int \dfrac{x\,dx}{c^4 + x^4} = \dfrac{1}{2c^2}\tan^{-1}\dfrac{x^2}{c^2}$

105. $\int \dfrac{x\,dx}{c^4 - x^4} = \dfrac{1}{4c^2}\log\dfrac{c^2 + x^2}{c^2 - x^2}$

106. $\int \dfrac{x^2\,dx}{c^4 + x^4} = \dfrac{1}{2c\sqrt{2}}\left[\dfrac{1}{2}\log\dfrac{x^2 - cx\sqrt{2} + c^2}{x^2 + cx\sqrt{2} + c^2} + \tan^{-1}\dfrac{cx\sqrt{2}}{c^2 - x^2}\right]$

107. $\int \dfrac{x^2\,dx}{c^4 - x^4} = \dfrac{1}{2c}\left[\dfrac{1}{2}\log\dfrac{c + x}{c - x} - \tan^{-1}\dfrac{x}{c}\right]$

108. $\int \dfrac{x^3\,dx}{c^4 \pm x^4} = \pm\dfrac{1}{4}\log(c^4 \pm x^4)$

FORMS CONTAINING $(a + bx + cx^2)$

$$X = a + bx + cx^2 \text{ and } q = 4ac - b^2$$

If $q = 0$, then $X = c\left(x + \dfrac{b}{2c}\right)^2$, and formulas starting with 23 should be used in place of these.

109. $\int \dfrac{dx}{X} = \dfrac{2}{\sqrt{q}}\tan^{-1}\dfrac{2cx + b}{\sqrt{q}}, \qquad (q > 0)$

110. $\int \dfrac{dx}{X} = \begin{cases} \dfrac{-2}{\sqrt{-q}}\tanh^{-1}\dfrac{2cx + b}{\sqrt{-q}} \\[2mm] \qquad \text{or} \\[2mm] \dfrac{1}{\sqrt{-q}}\log\dfrac{2cx + b - \sqrt{-q}}{2cx + b + \sqrt{-q}}, \qquad (q < 0) \end{cases}$

111. $\int \dfrac{dx}{X^2} = \dfrac{2cx + b}{qX} + \dfrac{2c}{q}\int \dfrac{dx}{X}$

112. $\displaystyle \int \frac{dx}{X^3} = \frac{2cx + b}{q}\left(\frac{1}{2X^2} + \frac{3c}{qX}\right) + \frac{6c^2}{q^2}\int \frac{dx}{X}$

113. $\displaystyle \int \frac{dx}{X^{n+1}} = \begin{cases} \dfrac{2cx + b}{nqX^n} + \dfrac{2(2n-1)c}{qn}\displaystyle\int \frac{dx}{X^n} \\[2mm] \quad\text{or} \\[2mm] \dfrac{(2n)!}{(n!)^2}\left(\dfrac{c}{q}\right)^n\left[\dfrac{2cx + b}{q}\displaystyle\sum_{r=1}^{n}\left(\dfrac{q}{cX}\right)^r\left(\dfrac{(r-1)!\,r!}{(2r)!}\right) + \int \frac{dx}{X}\right] \end{cases}$

114. $\displaystyle \int \frac{x\,dx}{X} = \frac{1}{2c}\log X - \frac{b}{2c}\int \frac{dx}{X}$

115. $\displaystyle \int \frac{x\,dx}{X^2} = -\frac{bx + 2a}{qX} - \frac{b}{q}\int \frac{dx}{X}$

116. $\displaystyle \int \frac{x\,dx}{X^{n+1}} = -\frac{2a + bx}{nqX^n} - \frac{b(2n-1)}{nq}\int \frac{dx}{X^n}$

117. $\displaystyle \int \frac{x^2}{X}\,dx = \frac{x}{c} - \frac{b}{2c^2}\log X + \frac{b^2 - 2ac}{2c^2}\int \frac{dx}{X}$

118. $\displaystyle \int \frac{x^2}{X^2}\,dx = \frac{(b^2 - 2ac)x + ab}{cqX} + \frac{2a}{q}\int \frac{dx}{X}$

119. $\displaystyle \int \frac{x^m\,dx}{X^{n+1}} = -\frac{x^{m-1}}{(2n - m + 1)cX^n} - \frac{n - m + 1}{2n - m + 1}\cdot\frac{b}{c}\int \frac{x^{m-1}\,dx}{X^{n+1}}$
$$+ \frac{m-1}{2n - m + 1}\cdot\frac{a}{c}\int \frac{x^{m-2}\,dx}{X^{n+1}}$$

120. $\displaystyle \int \frac{dx}{xX} = \frac{1}{2a}\log \frac{x^2}{X} - \frac{b}{2a}\int \frac{dx}{X}$

121. $\displaystyle \int \frac{dx}{x^2X} = \frac{b}{2a^2}\log \frac{X}{x^2} - \frac{1}{ax} + \left(\frac{b^2}{2a^2} - \frac{c}{a}\right)\int \frac{dx}{X}$

122. $\displaystyle \int \frac{dx}{xX^n} = \frac{1}{2a(n-1)X^{n-1}} - \frac{b}{2a}\int \frac{dx}{X^n} + \frac{1}{a}\int \frac{dx}{xX^{n-1}}$

123. $\displaystyle \int \frac{dx}{x^mX^{n+1}} = -\frac{1}{(m-1)ax^{m-1}X^n} - \frac{n + m - 1}{m - 1}\cdot\frac{b}{a}\int \frac{dx}{x^{m-1}X^{n+1}}$
$$- \frac{2n + m - 1}{m - 1}\cdot\frac{c}{a}\int \frac{dx}{x^{m-2}X^{n+1}}$$

FORMS CONTAINING $\sqrt{a + bx}$

124. $\displaystyle \int \sqrt{a + bx}\,dx = \frac{2}{3b}\sqrt{(a + bx)^3}$

125. $\displaystyle \int x\sqrt{a + bx}\,dx = -\frac{2(2a - 3bx)\sqrt{(a + bx)^3}}{15b^2}$

126. $\displaystyle \int x^2\sqrt{a + bx}\,dx = \frac{2(8a^2 - 12abx + 15b^2x^2)\sqrt{(a + bx)^3}}{105b^3}$

127. $\displaystyle \int x^m\sqrt{a + bx}\,dx = \begin{cases} \dfrac{2}{b(2m + 3)}\left[x^m\sqrt{(a + bx)^3} - ma\displaystyle\int x^{m-1}\sqrt{a + bx}\,dx\right] \\[2mm] \quad\text{or} \\[2mm] \dfrac{2}{b^{m+1}}\sqrt{a + bx}\displaystyle\sum_{r=0}^{m}\frac{m!(-a)^{m-r}}{r!(m-r)!(2r+3)}(a + bx)^{r+1} \end{cases}$

128. $\displaystyle\int \frac{\sqrt{a+bx}}{x}\,dx = 2\sqrt{a+bx} + a\int \frac{dx}{x\sqrt{a+bx}}$

129. $\displaystyle\int \frac{\sqrt{a+bx}}{x^2}\,dx = -\frac{\sqrt{a+bx}}{x} + \frac{b}{2}\int \frac{dx}{x\sqrt{a+bx}}$

130. $\displaystyle\int \frac{\sqrt{a+bx}}{x^m}\,dx = -\frac{1}{(m-1)a}\left[\frac{\sqrt{(a+bx)^3}}{x^{m-1}} + \frac{(2m-5)b}{2}\int \frac{\sqrt{a+bx}}{x^{m-1}}\,dx\right]$

131. $\displaystyle\int \frac{dx}{\sqrt{a+bx}} = \frac{2\sqrt{a+bx}}{b}$

132. $\displaystyle\int \frac{x\,dx}{\sqrt{a+bx}} = -\frac{2(2a-bx)}{3b^2}\sqrt{a+bx}$

133. $\displaystyle\int \frac{x^2\,dx}{\sqrt{a+bx}} = \frac{2(8a^2 - 4abx + 3b^2x^2)}{15b^3}\sqrt{a+bx}$

134. $\displaystyle\int \frac{x^m\,dx}{\sqrt{a+bx}} = \begin{cases} \dfrac{2}{(2m+1)b}\left[x^m\sqrt{a+bx} - ma\displaystyle\int \frac{x^{m-1}\,dx}{\sqrt{a+bx}}\right] \\[2mm] \text{or} \\[2mm] \dfrac{2(-a)^m\sqrt{a+bx}}{b^{m+1}}\displaystyle\sum_{r=0}^{m} \frac{(-1)^r m!(a+bx)^r}{(2r+1)r!(m-r)!a^r} \end{cases}$

135. $\displaystyle\int \frac{dx}{x\sqrt{a+bx}} = \frac{1}{\sqrt{a}}\log\left(\frac{\sqrt{a+bx} - \sqrt{a}}{\sqrt{a+bx} + \sqrt{a}}\right), \qquad (a > 0)$

136. $\displaystyle\int \frac{dx}{x\sqrt{a+bx}} = \frac{2}{\sqrt{-a}}\tan^{-1}\sqrt{\frac{a+bx}{-a}}, \qquad (a < 0)$

137. $\displaystyle\int \frac{dx}{x^2\sqrt{a+bx}} = -\frac{\sqrt{a+bx}}{ax} - \frac{b}{2a}\int \frac{dx}{x\sqrt{a+bx}}$

138. $\displaystyle\int \frac{dx}{x^n\sqrt{a+bx}} = \begin{cases} -\dfrac{\sqrt{a+bx}}{(n-1)ax^{n-1}} - \dfrac{(2n-3)b}{(2n-2)a}\displaystyle\int \frac{dx}{x^{n-1}\sqrt{a+bx}} \\[2mm] \text{or} \\[2mm] \dfrac{(2n-2)!}{[(n-1)!]^2}\left[-\dfrac{\sqrt{a+bx}}{a}\displaystyle\sum_{r=1}^{n-1}\frac{r!(r-1)!}{x^r(2r)!}\left(-\frac{b}{4a}\right)^{n-r-1}\right. \\[4mm] \left.\qquad\qquad + \left(-\dfrac{b}{4a}\right)^{n-1}\displaystyle\int \frac{dx}{x\sqrt{a+bx}}\right] \end{cases}$

139. $\displaystyle\int (a+bx)^{\pm\frac{n}{2}}\,dx = \frac{2(a+bx)^{\frac{2\pm n}{2}}}{b(2\pm n)}$

140. $\displaystyle\int x(a+bx)^{\pm\frac{n}{2}}\,dx = \frac{2}{b^2}\left[\frac{(a+bx)^{\frac{4\pm n}{2}}}{4\pm n} - \frac{a(a+bx)^{\frac{2\pm n}{2}}}{2\pm n}\right]$

141. $\displaystyle\int \frac{dx}{x(a+bx)^{\frac{m}{2}}} = \frac{1}{a}\int \frac{dx}{x(a+bx)^{\frac{m-2}{2}}} - \frac{b}{a}\int \frac{dx}{(a+bx)^{\frac{m}{2}}}$

142. $\displaystyle \int \frac{(a+bx)^{\frac{n}{2}}\,dx}{x} = b\int (a+bx)^{\frac{n-2}{2}}\,dx + a\int \frac{(a+bx)^{\frac{n-2}{2}}}{x}\,dx$

143. $\displaystyle \int f(x,\sqrt{a+bx})\,dx = \frac{2}{b}\int f\left(\frac{z^2-a}{b},z\right)z\,dz, \qquad (z=\sqrt{a+bx})$

FORMS CONTAINING $\sqrt{a+bx}$ and $\sqrt{c+dx}$

$$u = a + bx \qquad v = c + dx \qquad k = ad - bc$$

If $k=0$, then $v=\dfrac{c}{a}u$, and formulas starting with 124 should be used in place of these.

144. $\displaystyle \int \frac{dx}{\sqrt{uv}} = \begin{cases} \dfrac{2}{\sqrt{bd}}\tanh^{-1}\dfrac{\sqrt{bduv}}{bv}, \quad bd>o,\ k<o \\[2mm] \text{or} \\[2mm] \dfrac{2}{\sqrt{bd}}\tanh^{-1}\dfrac{\sqrt{bduv}}{du}, \quad bd>o,\ k>o. \\[2mm] \text{or} \\[2mm] \dfrac{1}{\sqrt{bd}}\log\dfrac{(bv+\sqrt{bduv})^2}{v}, \qquad (bd>0) \end{cases}$

145. $\displaystyle \int \frac{dx}{\sqrt{uv}} = \begin{cases} \dfrac{2}{\sqrt{-bd}}\tan^{-1}\dfrac{\sqrt{-bduv}}{bv} \\[2mm] \text{or} \\[2mm] -\dfrac{1}{\sqrt{-bd}}\sin^{-1}\left(\dfrac{2bdx+ad+bc}{|k|}\right), \qquad (bd<0) \end{cases}$

146. $\displaystyle \int \sqrt{uv}\,dx = \frac{k+2bv}{4bd}\sqrt{uv} - \frac{k^2}{8bd}\int \frac{dx}{\sqrt{uv}}$

147. $\displaystyle \int \frac{dx}{v\sqrt{u}} = \begin{cases} \dfrac{1}{\sqrt{kd}}\log\dfrac{d\sqrt{u}-\sqrt{kd}}{d\sqrt{u}+\sqrt{kd}} \\[2mm] \text{or} \\[2mm] \dfrac{1}{\sqrt{kd}}\log\dfrac{(d\sqrt{u}-\sqrt{kd})^2}{v}, \qquad (kd>0) \end{cases}$

148. $\displaystyle \int \frac{dx}{v\sqrt{u}} = \frac{2}{\sqrt{-kd}}\tan^{-1}\frac{d\sqrt{u}}{\sqrt{-kd}}, \qquad (kd<0)$

149. $\displaystyle \int \frac{x\,dx}{\sqrt{uv}} = \frac{\sqrt{uv}}{bd} - \frac{ad+bc}{2bd}\int \frac{dx}{\sqrt{uv}}$

150. $\displaystyle \int \frac{dx}{v\sqrt{uv}} = \frac{-2\sqrt{uv}}{kv}$

151. $\displaystyle \int \frac{v\,dx}{\sqrt{uv}} = \frac{\sqrt{uv}}{b} - \frac{k}{2b}\int \frac{dx}{\sqrt{uv}}$

152. $\displaystyle \int \sqrt{\frac{v}{u}}\,dx = \frac{v}{|v|}\int \frac{v\,dx}{\sqrt{uv}}$

153. $\displaystyle \int v^m\sqrt{u}\,dx = \frac{1}{(2m+3)d}\left(2v^{m+1}\sqrt{u} + k\int \frac{v^m\,dx}{\sqrt{u}}\right)$

154. $\int \dfrac{dx}{v^m \sqrt{u}} = -\dfrac{1}{(m-1)k}\left(\dfrac{\sqrt{u}}{v^{m-1}} + \left(m - \dfrac{3}{2}\right)b\int \dfrac{dx}{v^{m-1}\sqrt{u}}\right)$

155. $\int \dfrac{v^m \, dx}{\sqrt{u}} = \begin{cases} \dfrac{2}{b(2m+1)}\left[v^m\sqrt{u} - mk\int \dfrac{v^{m-1}}{\sqrt{u}}\,dx\right] \\ \quad\text{or} \\ \dfrac{2(m!)^2\sqrt{u}}{b(2m+1)!}\displaystyle\sum_{r=0}^{m}\left(-\dfrac{4k}{b}\right)^{m-r}\dfrac{(2r)!}{(r!)^2}v^r \end{cases}$

FORMS CONTAINING $\sqrt{x^2 \pm a^2}$

156. $\int \sqrt{x^2 \pm a^2}\,dx = \tfrac{1}{2}[x\sqrt{x^2 \pm a^2} \pm a^2\log(x + \sqrt{x^2 \pm a^2})]$

157. $\int \dfrac{dx}{\sqrt{x^2 \pm a^2}} = \log(x + \sqrt{x^2 \pm a^2})$

158. $\int \dfrac{dx}{x\sqrt{x^2 - a^2}} = \dfrac{1}{|a|}\sec^{-1}\dfrac{x}{a}$

159. $\int \dfrac{dx}{x\sqrt{x^2 + a^2}} = -\dfrac{1}{a}\log\left(\dfrac{a + \sqrt{x^2 + a^2}}{x}\right)$

160. $\int \dfrac{\sqrt{x^2 + a^2}}{x}\,dx = \sqrt{x^2 + a^2} - a\log\left(\dfrac{a + \sqrt{x^2 + a^2}}{x}\right)$

161. $\int \dfrac{\sqrt{x^2 - a^2}}{x}\,dx = \sqrt{x^2 - a^2} - |a|\sec^{-1}\dfrac{x}{a}$

162. $\int \dfrac{x\,dx}{\sqrt{x^2 \pm a^2}} = \sqrt{x^2 \pm a^2}$

163. $\int x\sqrt{x^2 \pm a^2}\,dx = \tfrac{1}{3}\sqrt{(x^2 \pm a^2)^3}$

164. $\int \sqrt{(x^2 \pm a^2)^3}\,dx = \dfrac{1}{4}\left[x\sqrt{(x^2 \pm a^2)^3} \pm \dfrac{3a^2 x}{2}\sqrt{x^2 \pm a^2}\right.$
$$\left. + \dfrac{3a^4}{2}\log(x + \sqrt{x^2 \pm a^2})\right]$$

165. $\int \dfrac{dx}{\sqrt{(x^2 \pm a^2)^3}} = \dfrac{\pm x}{a^2\sqrt{x^2 \pm a^2}}$

166. $\int \dfrac{x\,dx}{\sqrt{(x^2 \pm a^2)^3}} = \dfrac{-1}{\sqrt{x^2 \pm a^2}}$

167. $\int x\sqrt{(x^2 \pm a^2)^3}\,dx = \tfrac{1}{5}\sqrt{(x^2 \pm a^2)^5}$

168. $\int x^2\sqrt{x^2 \pm a^2}\,dx = \dfrac{x}{4}\sqrt{(x^2 \pm a^2)^3} \mp \dfrac{a^2}{8}x\sqrt{x^2 \pm a^2} - \dfrac{a^4}{8}\log(x + \sqrt{x^2 \pm a^2})$

169. $\int x^3\sqrt{x^2 + a^2}\,dx = (\tfrac{1}{5}x^2 - \tfrac{2}{15}a^2)\sqrt{(a^2 + x^2)^3}$

170. $\int x^3\sqrt{x^2 - a^2}\,dx = \dfrac{1}{5}\sqrt{(x^2 - a^2)^5} + \dfrac{a^2}{3}\sqrt{(x^2 - a^2)^3}$

171. $\int \dfrac{x^2\,dx}{\sqrt{x^2 \pm a^2}} = \dfrac{x}{2}\sqrt{x^2 \pm a^2} \mp \dfrac{a^2}{2}\log(x + \sqrt{x^2 \pm a^2})$

172. $\displaystyle\int \frac{x^3\,dx}{\sqrt{x^2 \pm a^2}} = \frac{1}{3}\sqrt{(x^2 \pm a^2)^3} \mp a^2\sqrt{x^2 \pm a^2}$

173. $\displaystyle\int \frac{dx}{x^2\sqrt{x^2 \pm a^2}} = \mp\frac{\sqrt{x^2 \pm a^2}}{a^2 x}$

174. $\displaystyle\int \frac{dx}{x^3\sqrt{x^2 + a^2}} = -\frac{\sqrt{x^2 + a^2}}{2a^2 x^2} + \frac{1}{2a^3}\log\frac{a + \sqrt{x^2 + a^2}}{x}$

175. $\displaystyle\int \frac{dx}{x^3\sqrt{x^2 - a^2}} = \frac{\sqrt{x^2 - a^2}}{2a^2 x^2} + \frac{1}{2|a^3|}\sec^{-1}\frac{x}{a}$

176. $\displaystyle\int x^2\sqrt{(x^2 \pm a^2)^3}\,dx = \frac{x}{6}\sqrt{(x^2 \pm a^2)^5} \mp \frac{a^2 x}{24}\sqrt{(x^2 \pm a^2)^3} - \frac{a^4 x}{16}\sqrt{x^2 \pm a^2}$

$$\mp \frac{a^6}{16}\log(x + \sqrt{x^2 \pm a^2})$$

177. $\displaystyle\int x^3\sqrt{(x^2 \pm a^2)^3}\,dx = \frac{1}{7}\sqrt{(x^2 \pm a^2)^7} \mp \frac{a^2}{5}\sqrt{(x^2 \pm a^2)^5}$

178. $\displaystyle\int \frac{\sqrt{x^2 \pm a^2}\,dx}{x^2} = -\frac{\sqrt{x^2 \pm a^2}}{x} + \log(x + \sqrt{x^2 \pm a^2})$

179. $\displaystyle\int \frac{\sqrt{x^2 + a^2}}{x^3}\,dx = -\frac{\sqrt{x^2 + a^2}}{2x^2} - \frac{1}{2a}\log\frac{a + \sqrt{x^2 + a^2}}{x}$

180. $\displaystyle\int \frac{\sqrt{x^2 - a^2}}{x^3}\,dx = -\frac{\sqrt{x^2 - a^2}}{2x^2} + \frac{1}{2|a|}\sec^{-1}\frac{x}{a}$

181. $\displaystyle\int \frac{\sqrt{x^2 \pm a^2}}{x^4}\,dx = \mp\frac{\sqrt{(x^2 \pm a^2)^3}}{3a^2 x^3}$

182. $\displaystyle\int \frac{x^2\,dx}{\sqrt{(x^2 \pm a^2)^3}} = \frac{-x}{\sqrt{x^2 \pm a^2}} + \log(x + \sqrt{x^2 \pm a^2})$

183. $\displaystyle\int \frac{x^3\,dx}{\sqrt{(x^2 \pm a^2)^3}} = \sqrt{x^2 \pm a^2} \pm \frac{a^2}{\sqrt{x^2 \pm a^2}}$

184. $\displaystyle\int \frac{dx}{x\sqrt{(x^2 + a^2)^3}} = \frac{1}{a^2\sqrt{x^2 + a^2}} - \frac{1}{a^3}\log\frac{a + \sqrt{x^2 + a^2}}{x}$

185. $\displaystyle\int \frac{dx}{x\sqrt{(x^2 - a^2)^3}} = -\frac{1}{a^2\sqrt{x^2 - a^2}} - \frac{1}{|a^3|}\sec^{-1}\frac{x}{a}$

186. $\displaystyle\int \frac{dx}{x^2\sqrt{(x^2 \pm a^2)^3}} = -\frac{1}{a^4}\left[\frac{\sqrt{x^2 \pm a^2}}{x} + \frac{x}{\sqrt{x^2 \pm a^2}}\right]$

187. $\displaystyle\int \frac{dx}{x^3\sqrt{(x^2 + a^2)^3}} = -\frac{1}{2a^2 x^2\sqrt{x^2 + a^2}} - \frac{3}{2a^4\sqrt{x^2 + a^2}}$

$$+ \frac{3}{2a^5}\log\frac{a + \sqrt{x^2 + a^2}}{x}$$

188. $\displaystyle\int \frac{dx}{x^3\sqrt{(x^2 - a^2)^3}} = \frac{1}{2a^2 x^2\sqrt{x^2 - a^2}} - \frac{3}{2a^4\sqrt{x^2 - a^2}} - \frac{3}{2|a^5|}\sec^{-1}\frac{x}{a}$

189. $\displaystyle\int \frac{x^m}{\sqrt{x^2 \pm a^2}}\,dx = \frac{1}{m}x^{m-1}\sqrt{x^2 \pm a^2} \mp \frac{m-1}{m}a^2\int \frac{x^{m-2}}{\sqrt{x^2 \pm a^2}}\,dx$

190. $\displaystyle \int \frac{x^{2m}}{\sqrt{x^2 \pm a^2}}\,dx = \frac{(2m)!}{2^{2m}(m!)^2}\left[\sqrt{x^2 \pm a^2}\,\sum_{r=1}^{m}\frac{r!(r-1)!}{(2r)!}(\mp a^2)^{m-r}(2x)^{2r-1} \right.$

$$\left. + (\mp a^2)^m \log(x + \sqrt{x^2 \pm a^2}) \right]$$

191. $\displaystyle \int \frac{x^{2m+1}}{\sqrt{x^2 \pm a^2}}\,dx = \sqrt{x^2 \pm a^2}\,\sum_{r=0}^{m}\frac{(2r)!(m!)^2}{(2m+1)!(r!)^2}(\mp 4a^2)^{m-r}x^{2r}$

192. $\displaystyle \int \frac{dx}{x^m\sqrt{x^2 \pm a^2}} = \mp\frac{\sqrt{x^2 \pm a^2}}{(m-1)a^2 x^{m-1}} \mp \frac{(m-2)}{(m-1)a^2}\int \frac{dx}{x^{m-2}\sqrt{x^2 \pm a^2}}$

193. $\displaystyle \int \frac{dx}{x^{2m}\sqrt{x^2 \pm a^2}} = \sqrt{x^2 \pm a^2}\,\sum_{r=0}^{m-1}\frac{(m-1)!m!(2r)!2^{2m-2r-1}}{(r!)^2(2m)!(\mp a^2)^{m-r}x^{2r+1}}$

194. $\displaystyle \int \frac{dx}{x^{2m+1}\sqrt{x^2 + a^2}} = \frac{(2m)!}{(m!)^2}\left[\frac{\sqrt{x^2 + a^2}}{a^2}\sum_{r=1}^{m}(-1)^{m-r+1}\frac{r!(r-1)!}{2(2r)!(4a^2)^{m-r}x^{2r}} \right.$

$$\left. + \frac{(-1)^{m+1}}{2^{2m}a^{2m+1}}\log\frac{\sqrt{x^2 + a^2} + a}{x} \right]$$

195. $\displaystyle \int \frac{dx}{x^{2m+1}\sqrt{x^2 - a^2}} = \frac{(2m)!}{(m!)^2}\left[\frac{\sqrt{x^2 - a^2}}{a^2}\sum_{r=1}^{m}\frac{r!(r-1)!}{2(2r)!(4a^2)^{m-r}x^{2r}} \right.$

$$\left. + \frac{1}{2^{2m}|a|^{2m+1}}\sec^{-1}\frac{x}{a} \right]$$

196. $\displaystyle \int \frac{dx}{(x-a)\sqrt{x^2 - a^2}} = -\frac{\sqrt{x^2 - a^2}}{a(x-a)}$

197. $\displaystyle \int \frac{dx}{(x+a)\sqrt{x^2 - a^2}} = \frac{\sqrt{x^2 - a^2}}{a(x+a)}$

198. $\displaystyle \int f(x, \sqrt{x^2 + a^2})\,dx = a\int f(a\tan u, a\sec u)\sec^2 u\,du, \qquad \left(u = \tan^{-1}\frac{x}{a}, a > 0 \right)$

199. $\displaystyle \int f(x, \sqrt{x^2 - a^2})\,dx = a\int f(a\sec u, a\tan u)\sec u\tan u\,du, \qquad \left(u = \sec^{-1}\frac{x}{a} \right.$

$$\left. a > 0 \right)$$

FORMS CONTAINING $\sqrt{a^2 - x^2}$

200. $\displaystyle \int \sqrt{a^2 - x^2}\,dx = \frac{1}{2}\left[x\sqrt{a^2 - x^2} + a^2\sin^{-1}\frac{x}{|a|} \right]$

201. $\displaystyle \int \frac{dx}{\sqrt{a^2 - x^2}} = \begin{cases} \sin^{-1}\dfrac{x}{|a|} \\ \text{or} \\ -\cos^{-1}\dfrac{x}{|a|} \end{cases}$

202. $\displaystyle \int \frac{dx}{x\sqrt{a^2 - x^2}} = -\frac{1}{a}\log\left(\frac{a + \sqrt{a^2 - x^2}}{x} \right)$

203. $\displaystyle \int \frac{\sqrt{a^2 - x^2}}{x}\,dx = \sqrt{a^2 - x^2} - a\log\left(\frac{a + \sqrt{a^2 - x^2}}{x} \right)$

204. $\displaystyle \int \frac{x\,dx}{\sqrt{a^2 - x^2}} = -\sqrt{a^2 - x^2}$

205. $\displaystyle \int x\sqrt{a^2 - x^2}\,dx = -\frac{1}{3}\sqrt{(a^2 - x^2)^3}$

206. $\int \sqrt{(a^2 - x^2)^3}\, dx = \frac{1}{4}\left[x\sqrt{(a^2 - x^2)^3} + \frac{3a^2 x}{2}\sqrt{a^2 - x^2} + \frac{3a^4}{2}\sin^{-1}\frac{x}{|a|} \right]$

207. $\int \dfrac{dx}{\sqrt{(a^2 - x^2)^3}} = \dfrac{x}{a^2\sqrt{a^2 - x^2}}$

208. $\int \dfrac{x\, dx}{\sqrt{(a^2 - x^2)^3}} = \dfrac{1}{\sqrt{a^2 - x^2}}$

209. $\int x\sqrt{(a^2 - x^2)^3}\, dx = -\frac{1}{5}\sqrt{(a^2 - x^2)^5}$

210. $\int x^2\sqrt{a^2 - x^2}\, dx = -\frac{x}{4}\sqrt{(a^2 - x^2)^3} + \frac{a^2}{8}\left(x\sqrt{a^2 - x^2} + a^2\sin^{-1}\frac{x}{|a|} \right)$

211. $\int x^3\sqrt{a^2 - x^2}\, dx = \left(-\frac{1}{5}x^2 - \frac{2}{15}a^2 \right)\sqrt{(a^2 - x^2)^3}$

212. $\int x^2\sqrt{(a^2 - x^2)^3}\, dx = -\frac{1}{6}x\sqrt{(a^2 - x^2)^5} + \frac{a^2 x}{24}\sqrt{(a^2 - x^2)^3}$

$$+ \frac{a^4 x}{16}\sqrt{a^2 - x^2} + \frac{a^6}{16}\sin^{-1}\frac{x}{|a|}$$

213. $\int x^3\sqrt{(a^2 - x^2)^3}\, dx = \frac{1}{7}\sqrt{(a^2 - x^2)^7} - \frac{a^2}{5}\sqrt{(a^2 - x^2)^5}$

214. $\int \dfrac{x^2\, dx}{\sqrt{a^2 - x^2}} = -\frac{x}{2}\sqrt{a^2 - x^2} + \frac{a^2}{2}\sin^{-1}\frac{x}{|a|}$

215. $\int \dfrac{dx}{x^2\sqrt{a^2 - x^2}} = -\dfrac{\sqrt{a^2 - x^2}}{a^2 x}$

216. $\int \dfrac{\sqrt{a^2 - x^2}}{x^2}\, dx = -\dfrac{\sqrt{a^2 - x^2}}{x} - \sin^{-1}\frac{x}{|a|}$

217. $\int \dfrac{\sqrt{a^2 - x^2}}{x^3}\, dx = -\dfrac{\sqrt{a^2 - x^2}}{2x^2} + \frac{1}{2a}\log\frac{a + \sqrt{a^2 - x^2}}{x}$

218. $\int \dfrac{\sqrt{a^2 - x^2}}{x^4}\, dx = -\dfrac{\sqrt{(a^2 - x^2)^3}}{3a^2 x^3}$

219. $\int \dfrac{x^2\, dx}{\sqrt{(a^2 - x^2)^3}} = \dfrac{x}{\sqrt{a^2 - x^2}} - \sin^{-1}\frac{x}{|a|}$

220. $\int \dfrac{x^3\, dx}{\sqrt{a^2 - x^2}} = -\frac{2}{3}(a^2 - x^2)^{\frac{3}{2}} - x^2(a^2 - x^2)^{\frac{1}{2}} = -\frac{1}{3}\sqrt{a^2 - x^2}(x^2 + 2a^2)$

221. $\int \dfrac{x^3\, dx}{\sqrt{(a^2 - x^2)^3}} = 2(a^2 - x^2)^{\frac{1}{2}} + \dfrac{x^2}{(a^2 - x^2)^{\frac{1}{2}}} = -\dfrac{a^2}{\sqrt{a^2 - x^2}} + \sqrt{a^2 - x^2}$

222. $\int \dfrac{dx}{x^3\sqrt{a^2 - x^2}} = -\dfrac{\sqrt{a^2 - x^2}}{2a^2 x^2} - \frac{1}{2a^3}\log\frac{a + \sqrt{a^2 - x^2}}{x}$

223. $\int \dfrac{dx}{x\sqrt{(a^2 - x^2)^3}} = \dfrac{1}{a^2\sqrt{a^2 - x^2}} - \frac{1}{a^3}\log\frac{a + \sqrt{a^2 - x^2}}{x}$

224. $\int \dfrac{dx}{x^2\sqrt{(a^2 - x^2)^3}} = \dfrac{1}{a^4}\left[-\dfrac{\sqrt{a^2 - x^2}}{x} + \dfrac{x}{\sqrt{a^2 - x^2}} \right]$

225. $\int \dfrac{dx}{x^3\sqrt{(a^2 - x^2)^3}} = -\dfrac{1}{2a^2 x^2\sqrt{a^2 - x^2}} + \dfrac{3}{2a^4\sqrt{a^2 - x^2}}$

$$-\frac{3}{2a^5}\log\frac{a + \sqrt{a^2 - x^2}}{x}$$

226. $\displaystyle\int \frac{x^m}{\sqrt{a^2 - x^2}}\,dx = -\frac{x^{m-1}\sqrt{a^2 - x^2}}{m} + \frac{(m-1)a^2}{m}\int \frac{x^{m-2}}{\sqrt{a^2 - x^2}}\,dx$

227. $\displaystyle\int \frac{x^{2m}}{\sqrt{a^2 - x^2}}\,dx = \frac{(2m)!}{(m!)^2}\left[-\sqrt{a^2 - x^2}\sum_{r=1}^{m}\frac{r!(r-1)!}{2^{2m-2r+1}(2r)!}a^{2m-2r}x^{2r-1} \right.$

$$\left. + \frac{a^{2m}}{2^{2m}}\sin^{-1}\frac{x}{|a|} \right]$$

228. $\displaystyle\int \frac{x^{2m+1}}{\sqrt{a^2 - x^2}}\,dx = -\sqrt{a^2 - x^2}\sum_{r=0}^{m}\frac{(2r)!(m!)^2}{(2m+1)!(r!)^2}(4a^2)^{m-r}x^{2r}$

229. $\displaystyle\int \frac{dx}{x^m\sqrt{a^2 - x^2}} = -\frac{\sqrt{a^2 - x^2}}{(m-1)a^2 x^{m-1}} + \frac{m-2}{(m-1)a^2}\int \frac{dx}{x^{m-2}\sqrt{a^2 - x^2}}$

230. $\displaystyle\int \frac{ax}{x^{2m}\sqrt{a^2 - x^2}} = -\sqrt{a^2 - x^2}\sum_{r=0}^{m-1}\frac{(m-1)!m!(2r)!2^{2m-2r-1}}{(r!)^2(2m)!a^{2m-2r}x^{2r+1}}$

231. $\displaystyle\int \frac{dx}{x^{2m+1}\sqrt{a^2 - x^2}} = \frac{(2m)!}{(m!)^2}\left[-\frac{\sqrt{a^2 - x^2}}{a^2}\sum_{r=1}^{m}\frac{r!(r-1)!}{2(2r)!(4a^2)^{m-r}x^{2r}} \right.$

$$\left. + \frac{1}{2^{2m}a^{2m+1}}\log\frac{a - \sqrt{a^2 - x^2}}{x} \right]$$

232. $\displaystyle\int \frac{dx}{(b^2 - x^2)\sqrt{a^2 - x^2}} = \frac{1}{2b\sqrt{a^2 - b^2}}\log\frac{(b\sqrt{a^2 - x^2} + x\sqrt{a^2 - b^2})^2}{b^2 - x^2},$

$$(a^2 > b^2)$$

233. $\displaystyle\int \frac{dx}{(b^2 - x^2)\sqrt{a^2 - x^2}} = \frac{1}{b\sqrt{b^2 - a^2}}\tan^{-1}\frac{x\sqrt{b^2 - a^2}}{b\sqrt{a^2 - x^2}}, \quad (b^2 > a^2)$

234. $\displaystyle\int \frac{dx}{(b^2 + x^2)\sqrt{a^2 - x^2}} = \frac{1}{b\sqrt{a^2 + b^2}}\tan^{-1}\frac{x\sqrt{a^2 + b^2}}{b\sqrt{a^2 - x^2}}$

235. $\displaystyle\int \frac{\sqrt{a^2 - x^2}}{b^2 + x^2}\,dx = \frac{\sqrt{a^2 + b^2}}{|b|}\sin^{-1}\frac{x\sqrt{a^2 + b^2}}{|a|\sqrt{x^2 + b^2}} - \sin^{-1}\frac{x}{|a|}$

236. $\displaystyle\int f(x, \sqrt{a^2 - x^2})\,dx = a\int f(a\sin u, a\cos u)\cos u\,du, \quad \left(u = \sin^{-1}\frac{x}{a},\ a > 0 \right)$

FORMS CONTAINING $\sqrt{a + bx + cx^2}$

$$X = a + bx + cx^2,\ q = 4ac - b^2, \text{and } k = \frac{4c}{q}$$

If $q = 0$, then $\sqrt{X} = \sqrt{c}\left| x + \dfrac{b}{2c} \right|$

237. $\displaystyle\int \frac{dx}{\sqrt{X}} = \begin{cases} \dfrac{1}{\sqrt{c}}\log(2\sqrt{cX} + 2cx + b) \\[2mm] \quad\text{or} \\[2mm] \dfrac{1}{\sqrt{c}}\sinh^{-1}\dfrac{2cx + b}{\sqrt{q}}, \quad (c > 0) \end{cases}$

238. $\displaystyle\int \frac{dx}{\sqrt{X}} = -\frac{1}{\sqrt{-c}}\sin^{-1}\frac{2cx + b}{\sqrt{-q}}, \quad (c < 0)$

239. $\displaystyle\int \frac{dx}{X\sqrt{X}} = \frac{2(2cx + b)}{q\sqrt{X}}$

240. $\int \dfrac{dx}{X^2 \sqrt{X}} = \dfrac{2(2cx + b)}{3q\sqrt{X}} \left(\dfrac{1}{X} + 2k \right)$

241. $\int \dfrac{dx}{X^n \sqrt{X}} = \begin{cases} \dfrac{2(2cx + b)\sqrt{X}}{(2n - 1)qX^n} + \dfrac{2k(n - 1)}{2n - 1} \int \dfrac{dx}{X^{n-1}\sqrt{X}} \\ \text{or} \\ \dfrac{(2cx + b)(n!)(n - 1)!4^n k^{n-1}}{q[(2n)!]\sqrt{X}} \sum_{r=0}^{n-1} \dfrac{(2r)!}{(4kX)^r (r!)^2} \end{cases}$

242. $\int \sqrt{X}\, dx = \dfrac{(2cx + b)\sqrt{X}}{4c} + \dfrac{1}{2k} \int \dfrac{dx}{\sqrt{X}}$

243. $\int X\sqrt{X}\, dx = \dfrac{(2cx + b)\sqrt{X}}{8c} \left(X + \dfrac{3}{2k} \right) + \dfrac{3}{8k^2} \int \dfrac{dx}{\sqrt{X}}$

244. $\int X^2\sqrt{X}\, dx = \dfrac{(2cx + b)\sqrt{X}}{12c} \left(X^2 + \dfrac{5X}{4k} + \dfrac{15}{8k^2} \right) + \dfrac{5}{16k^3} \int \dfrac{dx}{\sqrt{X}}$

245. $\int X^n\sqrt{X}\, dx = \begin{cases} \dfrac{(2cx + b)X^n\sqrt{X}}{4(n + 1)c} + \dfrac{2n + 1}{2(n + 1)k} \int X^{n-1}\sqrt{X}\, dx \\ \text{or} \\ \dfrac{(2n + 2)!}{[(n + 1)!]^2(4k)^{n+1}} \left[\dfrac{k(2cx + b)\sqrt{X}}{c} \sum_{r=0}^{n} \dfrac{r!(r + 1)!(4kX)^r}{(2r + 2)!} \right. \\ \left. \hspace{6cm} + \int \dfrac{dx}{\sqrt{X}} \right] \end{cases}$

246. $\int \dfrac{x\, dx}{\sqrt{X}} = \dfrac{\sqrt{X}}{c} - \dfrac{b}{2c} \int \dfrac{dx}{\sqrt{X}}$

247. $\int \dfrac{x\, dx}{X\sqrt{X}} = - \dfrac{2(bx + 2a)}{q\sqrt{X}}$

248. $\int \dfrac{x\, dx}{X^n \sqrt{X}} = - \dfrac{\sqrt{X}}{(2n - 1)cX^n} - \dfrac{b}{2c} \int \dfrac{dx}{X^n \sqrt{X}}$

249. $\int \dfrac{x^2\, dx}{\sqrt{X}} = \left(\dfrac{x}{2c} - \dfrac{3b}{4c^2} \right) \sqrt{X} + \dfrac{3b^2 - 4ac}{8c^2} \int \dfrac{dx}{\sqrt{X}}$

250. $\int \dfrac{x^2\, dx}{X\sqrt{X}} = \dfrac{(2b^2 - 4ac)x + 2ab}{cq\sqrt{X}} + \dfrac{1}{c} \int \dfrac{dx}{\sqrt{X}}$

251. $\int \dfrac{x^2\, dx}{X^n \sqrt{X}} = \dfrac{(2b^2 - 4ac)x + 2ab}{(2n - 1)cqX^{n-1}\sqrt{X}} + \dfrac{4ac + (2n - 3)b^2}{(2n - 1)cq} \int \dfrac{dx}{X^{n-1}\sqrt{X}}$

252. $\int \dfrac{x^3\, dx}{\sqrt{X}} = \left(\dfrac{x^2}{3c} - \dfrac{5bx}{12c^2} + \dfrac{5b^2}{8c^3} - \dfrac{2a}{3c^2} \right) \sqrt{X} + \left(\dfrac{3ab}{4c^2} - \dfrac{5b^3}{16c^3} \right) \int \dfrac{dx}{\sqrt{X}}$

253. $\int \dfrac{x^n\, dx}{\sqrt{X}} = \dfrac{1}{nc} x^{n-1} \sqrt{X} - \dfrac{(2n - 1)b}{2nc} \int \dfrac{x^{n-1}\, dx}{\sqrt{X}} - \dfrac{(n - 1)a}{nc} \int \dfrac{x^{n-2}\, dx}{\sqrt{X}}$

254. $\int x\sqrt{X}\, dx = \dfrac{X\sqrt{X}}{3c} - \dfrac{b(2cx + b)}{8c^2} \sqrt{X} - \dfrac{b}{4ck} \int \dfrac{dx}{\sqrt{X}}$

255. $\int xX\sqrt{X}\, dx = \dfrac{X^2\sqrt{X}}{5c} - \dfrac{b}{2c} \int X\sqrt{X}\, dx$

256. $\displaystyle\int x X^n \sqrt{X}\, dx = \frac{X^{n+1}\sqrt{X}}{(2n+3)c} - \frac{b}{2c}\int X^n \sqrt{X}\, dx$

257. $\displaystyle\int x^2 \sqrt{X}\, dx = \left(x - \frac{5b}{6c}\right)\frac{X\sqrt{X}}{4c} + \frac{5b^2 - 4ac}{16c^2}\int \sqrt{X}\, dx$

258. $\displaystyle\int \frac{dx}{x\sqrt{X}} = -\frac{1}{\sqrt{a}}\log\frac{2\sqrt{aX} + bx + 2a}{x}, \qquad (a > 0)$

259. $\displaystyle\int \frac{dx}{x\sqrt{X}} = \frac{1}{\sqrt{-a}}\sin^{-1}\left(\frac{bx + 2a}{|x|\sqrt{-q}}\right), \qquad (a < 0)$

260. $\displaystyle\int \frac{dx}{x\sqrt{X}} = -\frac{2\sqrt{X}}{bx}, \qquad (a = 0)$

261. $\displaystyle\int \frac{dx}{x^2\sqrt{X}} = -\frac{\sqrt{X}}{ax} - \frac{b}{2a}\int \frac{dx}{x\sqrt{X}}$

262. $\displaystyle\int \frac{\sqrt{X}\, dx}{x} = \sqrt{X} + \frac{b}{2}\int \frac{dx}{\sqrt{X}} + a\int \frac{dx}{x\sqrt{X}}$

263. $\displaystyle\int \frac{\sqrt{X}\, dx}{x^2} = -\frac{\sqrt{X}}{x} + \frac{b}{2}\int \frac{dx}{x\sqrt{X}} + c\int \frac{dx}{\sqrt{X}}$

FORMS INVOLVING $\sqrt{2ax - x^2}$

264. $\displaystyle\int \sqrt{2ax - x^2}\, dx = \frac{1}{2}\left[(x - a)\sqrt{2ax - x^2} + a^2\sin^{-1}\frac{x - a}{|a|}\right]$

265. $\displaystyle\int \frac{dx}{\sqrt{2ax - x^2}} = \begin{cases} \cos^{-1}\dfrac{a - x}{|a|} \\[2mm] \text{or} \\[2mm] \sin^{-1}\dfrac{x - a}{|a|} \end{cases}$

266. $\displaystyle\int x^n \sqrt{2ax - x^2}\, dx = \begin{cases} -\dfrac{x^{n-1}(2ax - x^2)^{\frac{3}{2}}}{n + 2} + \dfrac{(2n + 1)a}{n + 2}\displaystyle\int x^{n-1}\sqrt{2ax - x^2}\, dx \\[3mm] \text{or} \\[3mm] \sqrt{2ax - x^2}\left[\dfrac{x^{n+1}}{n + 2} - \displaystyle\sum_{r=0}^{n}\dfrac{(2n + 1)!(r!)^2 a^{n-r+1}}{2^{n-r}(2r + 1)!(n + 2)!n!}x^r\right] \\[3mm] \qquad\qquad + \dfrac{(2n + 1)!a^{n+2}}{2^n n!(n + 2)!}\sin^{-1}\dfrac{x - a}{|a|} \end{cases}$

267. $\displaystyle\int \frac{\sqrt{2ax - x^2}}{x^n}\, dx = \frac{(2ax - x^2)^{\frac{3}{2}}}{(3 - 2n)ax^n} + \frac{n - 3}{(2n - 3)a}\int \frac{\sqrt{2ax - x^2}}{x^{n-1}}\, dx$

268. $\displaystyle\int \frac{x^n\, dx}{\sqrt{2ax - x^2}} = \begin{cases} \dfrac{-x^{n-1}\sqrt{2ax - x^2}}{n} + \dfrac{a(2n - 1)}{n}\displaystyle\int \frac{x^{n-1}}{\sqrt{2ax - x^2}}\, dx \\[3mm] \text{or} \\[3mm] -\sqrt{2ax - x^2}\displaystyle\sum_{r=1}^{n}\dfrac{(2n)!r!(r - 1)!a^{n-r}}{2^{n-r}(2r)!(n!)^2}x^{r-1} \\[3mm] \qquad\qquad + \dfrac{(2n)!a^n}{2^n(n!)^2}\sin^{-1}\dfrac{x - a}{|a|} \end{cases}$

269. $\displaystyle\int \frac{dx}{x^n\sqrt{2ax-x^2}} = \begin{cases} \dfrac{\sqrt{2ax-x^2}}{a(1-2n)x^n} + \dfrac{n-1}{(2n-1)a}\displaystyle\int \dfrac{dx}{x^{n-1}\sqrt{2ax-x^2}} \\[2mm] \quad\text{or} \\[2mm] -\sqrt{2ax-x^2}\displaystyle\sum_{r=0}^{n-1}\dfrac{2^{n-r}(n-1)!n!(2r)!}{(2n)!(r!)^2 a^{n-r}x^{r+1}} \end{cases}$

270. $\displaystyle\int \frac{dx}{(2ax-x^2)^{\frac{3}{2}}} = \frac{x-a}{a^2\sqrt{2ax-x^2}}$

271. $\displaystyle\int \frac{x\,dx}{(2ax-x^2)^{\frac{3}{2}}} = \frac{x}{a\sqrt{2ax-x^2}}$

MISCELLANEOUS ALGEBRAIC FORMS

272. $\displaystyle\int \frac{dx}{\sqrt{2ax+x^2}} = \log(x+a+\sqrt{2ax+x^2})$

273. $\displaystyle\int \sqrt{ax^2+c}\,dx = \frac{x}{2}\sqrt{ax^2+c} + \frac{c}{2\sqrt{a}}\log(x\sqrt{a}+\sqrt{ax^2+c}), \quad (a>0)$

274. $\displaystyle\int \sqrt{ax^2+c}\,dx = \frac{x}{2}\sqrt{ax^2+c} + \frac{c}{2\sqrt{-a}}\sin^{-1}\left(x\sqrt{-\frac{a}{c}}\right), \quad (a<0)$

275. $\displaystyle\int \sqrt{\frac{1+x}{1-x}}\,dx = \sin^{-1}x - \sqrt{1-x^2}$

276. $\displaystyle\int \frac{dx}{x\sqrt{ax^n+c}} = \begin{cases} \dfrac{1}{n\sqrt{c}}\log\dfrac{\sqrt{ax^n+c}-\sqrt{c}}{\sqrt{ax^n+c}+\sqrt{c}} \\[3mm] \quad\text{or} \\[3mm] \dfrac{2}{n\sqrt{c}}\log\dfrac{\sqrt{ax^n+c}-\sqrt{c}}{\sqrt{x^n}}, \quad (c>0) \end{cases}$

277. $\displaystyle\int \frac{dx}{x\sqrt{ax^n+c}} = \frac{2}{n\sqrt{-c}}\sec^{-1}\sqrt{-\frac{ax^n}{c}}, \quad (c<0)$

278. $\displaystyle\int \frac{dx}{\sqrt{ax^2+c}} = \frac{1}{\sqrt{a}}\log(x\sqrt{a}+\sqrt{ax^2+c}), \quad (a>0)$

279. $\displaystyle\int \frac{dx}{\sqrt{ax^2+c}} = \frac{1}{\sqrt{-a}}\sin^{-1}\left(x\sqrt{-\frac{a}{c}}\right), \quad (a<0)$

280. $\displaystyle\int (ax^2+c)^{m+\frac{1}{2}}\,dx = \begin{cases} \dfrac{x(ax^2+c)^{m+\frac{1}{2}}}{2(m+1)} + \dfrac{(2m+1)c}{2(m+1)}\displaystyle\int (ax^2+c)^{m-\frac{1}{2}}\,dx \\[3mm] \quad\text{or} \\[3mm] x\sqrt{ax^2+c}\displaystyle\sum_{r=0}^{m}\dfrac{(2m+1)!(r!)^2 c^{m-r}}{2^{2m-2r+1}m!(m+1)!(2r+1)!}(ax^2+c)^r \\[3mm] \quad + \dfrac{(2m+1)!c^{m+1}}{2^{2m+1}m!(m+1)!}\displaystyle\int \dfrac{dx}{\sqrt{ax^2+c}} \end{cases}$

281. $\displaystyle\int x(ax^2+c)^{m+\frac{1}{2}}\,dx = \frac{(ax^2+c)^{m+\frac{3}{2}}}{(2m+3)a}$

282. $\int \dfrac{(ax^2 + c)^{m + \frac{1}{2}}}{x}\, dx = \begin{cases} \dfrac{(ax^2 + c)^{m + \frac{1}{2}}}{2m + 1} + c\int \dfrac{(ax^2 + c)^{m - \frac{1}{2}}}{x}\, dx \\ \text{or} \\ \sqrt{ax^2 + c}\, \displaystyle\sum_{r=0}^{m} \dfrac{c^{m-r}(ax^2 + c)^r}{2r + 1} + c^{m+1}\int \dfrac{dx}{x\sqrt{ax^2 + c}} \end{cases}$

283. $\int \dfrac{dx}{(ax^2 + c)^{m + \frac{1}{2}}} = \begin{cases} \dfrac{x}{(2m - 1)c(ax^2 + c)^{m - \frac{1}{2}}} + \dfrac{2m - 2}{(2m - 1)c}\int \dfrac{dx}{(ax^2 + c)^{m - \frac{1}{2}}} \\ \text{or} \\ \dfrac{x}{\sqrt{ax^2 + c}}\, \displaystyle\sum_{r=0}^{m-1} \dfrac{2^{2m - 2r - 1}(m - 1)!\, m!\,(2r)!}{(2m)!\,(r!)^2 c^{m-r}(ax^2 + c)^r} \end{cases}$

284. $\int \dfrac{dx}{x^m \sqrt{ax^2 + c}} = -\dfrac{\sqrt{ax^2 + c}}{(m - 1)cx^{m-1}} - \dfrac{(m - 2)a}{(m - 1)c}\int \dfrac{dx}{x^{m-2}\sqrt{ax^2 + c}}$

285. $\int \dfrac{1 + x^2}{(1 - x^2)\sqrt{1 + x^4}}\, dx = \dfrac{1}{\sqrt{2}} \log \dfrac{x\sqrt{2} + \sqrt{1 + x^4}}{1 - x^2}$

286. $\int \dfrac{1 - x^2}{(1 + x^2)\sqrt{1 + x^4}}\, dx = \dfrac{1}{\sqrt{2}} \tan^{-1} \dfrac{x\sqrt{2}}{\sqrt{1 + x^4}}$

287. $\int \dfrac{dx}{x\sqrt{x^n + a^2}} = -\dfrac{2}{na} \log \dfrac{a + \sqrt{x^n + a^2}}{\sqrt{x^n}}$

288. $\int \dfrac{dx}{x\sqrt{x^n - a^2}} = -\dfrac{2}{na} \sin^{-1} \dfrac{a}{\sqrt{x^n}}$

289. $\int \sqrt{\dfrac{x}{a^3 - x^3}}\, dx = \dfrac{2}{3} \sin^{-1} \left(\dfrac{x}{a}\right)^{\frac{3}{2}}$

FORMS INVOLVING TRIGONOMETRIC FUNCTIONS

290. $\int (\sin ax)\, dx = -\dfrac{1}{a} \cos ax$

291. $\int (\cos ax)\, dx = \dfrac{1}{a} \sin ax$

292. $\int (\tan ax)\, dx = -\dfrac{1}{a} \log \cos ax = \dfrac{1}{a} \log \sec ax$

293. $\int (\cot ax)\, dx = \dfrac{1}{a} \log \sin ax = -\dfrac{1}{a} \log \csc ax$

294. $\int (\sec ax)\, dx = \dfrac{1}{a} \log (\sec ax + \tan ax) = \dfrac{1}{a} \log \tan \left(\dfrac{\pi}{4} + \dfrac{ax}{2}\right)$

295. $\int (\csc ax)\, dx = \dfrac{1}{a} \log (\csc ax - \cot ax) = \dfrac{1}{a} \log \tan \dfrac{ax}{2}$

296. $\int (\sin^2 ax)\, dx = -\dfrac{1}{2a} \cos ax \sin ax + \dfrac{1}{2}x = \dfrac{1}{2}x - \dfrac{1}{4a} \sin 2ax$

297. $\int (\sin^3 ax)\, dx = -\dfrac{1}{3a}(\cos ax)(\sin^2 ax + 2)$

298. $\int (\sin^4 ax)\, dx = \dfrac{3x}{8} - \dfrac{\sin 2ax}{4a} + \dfrac{\sin 4ax}{32a}$

299. $\int (\sin^n ax)\, dx = -\dfrac{\sin^{n-1} ax \cos ax}{na} + \dfrac{n - 1}{n} \int (\sin^{n-2} ax)\, dx$

300. $\int (\sin^{2m} ax)\,dx = -\dfrac{\cos ax}{a} \sum\limits_{r=0}^{m-1} \dfrac{(2m)!(r!)^2}{2^{2m-2r}(2r+1)!(m!)^2} \sin^{2r+1} ax + \dfrac{(2m)!}{2^{2m}(m!)^2} x$

301. $\int (\sin^{2m+1} ax)\,dx = -\dfrac{\cos ax}{a} \sum\limits_{r=0}^{m} \dfrac{2^{2m-2r}(m!)^2(2r)!}{(2m+1)!(r!)^2} \sin^{2r} ax$

302. $\int (\cos^2 ax)\,dx = \dfrac{1}{2a} \sin ax \cos ax + \dfrac{1}{2}x = \dfrac{1}{2}x + \dfrac{1}{4a} \sin 2ax$

303. $\int (\cos^3 ax)\,dx = \dfrac{1}{3a}(\sin ax)(\cos^2 ax + 2)$

304. $\int (\cos^4 ax)\,dx = \dfrac{3x}{8} + \dfrac{\sin 2ax}{4a} + \dfrac{\sin 4ax}{32a}$

305. $\int (\cos^n ax)\,dx = \dfrac{1}{na} \cos^{n-1} ax \sin ax + \dfrac{n-1}{n} \int (\cos^{n-2} ax)\,dx$

306. $\int (\cos^{2m} ax)\,dx = \dfrac{\sin ax}{a} \sum\limits_{r=0}^{m-1} \dfrac{(2m)!(r!)^2}{2^{2m-2r}(2r+1)!(m!)^2} \cos^{2r+1} ax + \dfrac{(2m)!}{2^{2m}(m!)^2} x$

307. $\int (\cos^{2m+1} ax)\,dx = \dfrac{\sin ax}{a} \sum\limits_{r=0}^{m} \dfrac{2^{2m-2r}(m!)^2(2r)!}{(2m+1)!(r!)^2} \cos^{2r} ax$

308. $\int \dfrac{dx}{\sin^2 ax} = \int (\csc^2 ax)\,dx = -\dfrac{1}{a} \cot ax$

309. $\int \dfrac{dx}{\sin^m ax} = \int (\csc^m ax)\,dx = -\dfrac{1}{(m-1)a} \cdot \dfrac{\cos ax}{\sin^{m-1} ax} + \dfrac{m-2}{m-1} \int \dfrac{dx}{\sin^{m-2} ax}$

310. $\int \dfrac{dx}{\sin^{2m} ax} = \int (\csc^{2m} ax)\,dx = -\dfrac{1}{a} \cos ax \sum\limits_{r=0}^{m-1} \dfrac{2^{2m-2r-1}(m-1)!\,m!\,(2r)!}{(2m)!(r!)^2 \sin^{2r+1} ax}$

311. $\int \dfrac{dx}{\sin^{2m+1} ax} = \int (\csc^{2m+1} ax)\,dx =$

$\qquad -\dfrac{1}{a} \cos ax \sum\limits_{r=0}^{m-1} \dfrac{(2m)!(r!)^2}{2^{2m-2r}(m!)^2(2r+1)! \sin^{2r+2} ax} + \dfrac{1}{a} \cdot \dfrac{(2m)!}{2^{2m}(m!)^2} \log \tan \dfrac{ax}{2}$

312. $\int \dfrac{dx}{\cos^2 ax} = \int (\sec^2 ax)\,dx = \dfrac{1}{a} \tan ax$

313. $\int \dfrac{dx}{\cos^n ax} = \int (\sec^n ax)\,dx = \dfrac{1}{(n-1)a} \cdot \dfrac{\sin ax}{\cos^{n-1} ax} + \dfrac{n-2}{n-1} \int \dfrac{dx}{\cos^{n-2} ax}$

314. $\int \dfrac{dx}{\cos^{2m} ax} = \int (\sec^{2m} ax)\,dx = \dfrac{1}{a} \sin ax \sum\limits_{r=0}^{m-1} \dfrac{2^{2m-2r-1}(m-1)!\,m!\,(2r)!}{(2m)!(r!)^2 \cos^{2r+1} ax}$

315. $\int \dfrac{dx}{\cos^{2m+1} ax} = \int (\sec^{2m+1} ax)\,dx =$

$\qquad \dfrac{1}{a} \sin ax \sum\limits_{r=0}^{m-1} \dfrac{(2m)!(r!)^2}{2^{2m-2r}(m!)^2(2r+1)! \cos^{2r+2} ax}$

$\qquad\qquad\qquad + \dfrac{1}{a} \cdot \dfrac{(2m)!}{2^{2m}(m!)^2} \log (\sec ax + \tan ax)$

316. $\int (\sin mx)(\sin nx)\,dx = \dfrac{\sin(m-n)x}{2(m-n)} - \dfrac{\sin(m+n)x}{2(m+n)}, \qquad (m^2 \neq n^2)$

317. $\int (\cos mx)(\cos nx)\,dx = \dfrac{\sin(m-n)x}{2(m-n)} + \dfrac{\sin(m+n)x}{2(m+n)}, \qquad (m^2 \neq n^2)$

318. $\int (\sin ax)(\cos ax)\,dx = \dfrac{1}{2a} \sin^2 ax$

319. $\int (\sin mx)(\cos nx)\,dx = -\dfrac{\cos(m - n)x}{2(m - n)} - \dfrac{\cos(m + n)x}{2(m + n)}, \qquad (m^2 \neq n^2)$

320. $\int (\sin^2 ax)(\cos^2 ax)\,dx = -\dfrac{1}{32a}\sin 4ax + \dfrac{x}{8}$

321. $\int (\sin ax)(\cos^m ax)\,dx = -\dfrac{\cos^{m+1} ax}{(m + 1)a}$

322. $\int (\sin^m ax)(\cos ax)\,dx = \dfrac{\sin^{m+1} ax}{(m + 1)a}$

323. $\int (\cos^m ax)(\sin^n ax)\,dx = \begin{cases} \dfrac{\cos^{m-1} ax \sin^{n+1} ax}{(m + n)a} \\[2ex] \qquad + \dfrac{m - 1}{m + n}\int (\cos^{m-2} ax)(\sin^n ax)\,dx \\[2ex] \text{or} \\[2ex] -\dfrac{\sin^{n-1} ax \cos^{m+1} ax}{(m + n)a} \\[2ex] \qquad + \dfrac{n - 1}{m + n}\int (\cos^m ax)(\sin^{n-2} ax)\,dx \end{cases}$

324. $\int \dfrac{\cos^m ax}{\sin^n ax}\,dx = \begin{cases} -\dfrac{\cos^{m+1} ax}{(n - 1)a \sin^{n-1} ax} - \dfrac{m - n + 2}{n - 1}\int \dfrac{\cos^m ax}{\sin^{n-2} ax}\,dx \\[2ex] \text{or} \\[2ex] \dfrac{\cos^{m-1} ax}{a(m - n) \sin^{n-1} ax} + \dfrac{m - 1}{m - n}\int \dfrac{\cos^{m-2} ax}{\sin^n ax}\,dx \end{cases}$

325. $\int \dfrac{\sin^m ax}{\cos^n ax}\,dx = \begin{cases} \dfrac{\sin^{m+1} ax}{a(n - 1) \cos^{n-1} ax} - \dfrac{m - n + 2}{n - 1}\int \dfrac{\sin^m ax}{\cos^{n-2} ax}\,dx \\[2ex] \text{or} \\[2ex] -\dfrac{\sin^{m-1} ax}{a(m - n) \cos^{n-1} ax} + \dfrac{m - 1}{m - n}\int \dfrac{\sin^{m-2} ax}{\cos^n ax}\,dx \end{cases}$

326. $\int \dfrac{\sin ax}{\cos^2 ax}\,dx = \dfrac{1}{a \cos ax} = \dfrac{\sec ax}{a}$

327. $\int \dfrac{\sin^2 ax}{\cos ax}\,dx = -\dfrac{1}{a}\sin ax + \dfrac{1}{a}\log \tan\left(\dfrac{\pi}{4} + \dfrac{ax}{2}\right)$

328. $\int \dfrac{\cos ax}{\sin^2 ax}\,dx = -\dfrac{1}{a \sin ax} = -\dfrac{\csc ax}{a}$

329. $\int \dfrac{dx}{(\sin ax)(\cos ax)} = \dfrac{1}{a}\log \tan ax$

330. $\int \dfrac{dx}{(\sin ax)(\cos^2 ax)} = \dfrac{1}{a}\left(\sec ax + \log \tan \dfrac{ax}{2}\right)$

331. $\int \dfrac{dx}{(\sin ax)(\cos^n ax)} = \dfrac{1}{a(n - 1)\cos^{n-1} ax} + \int \dfrac{dx}{(\sin ax)(\cos^{n-2} ax)}$

332. $\int \dfrac{dx}{(\sin^2 ax)(\cos ax)} = -\dfrac{1}{a}\csc ax + \dfrac{1}{a}\log \tan\left(\dfrac{\pi}{4} + \dfrac{ax}{2}\right)$

333. $\int \dfrac{dx}{(\sin^2 ax)(\cos^2 ax)} = -\dfrac{2}{a}\cot 2ax$

334. $\displaystyle\int \frac{dx}{\sin^m ax \cos^n ax} =$
$$\begin{cases} -\dfrac{1}{a(m-1)(\sin^{m-1} ax)(\cos^{n-1} ax)} \\[2mm] \qquad +\dfrac{m+n-2}{m-1}\displaystyle\int \dfrac{dx}{(\sin^{m-2} ax)(\cos^n ax)} \\[2mm] \text{or} \\[2mm] \dfrac{1}{a(n-1)\sin^{m-1} ax \cos^{n-1} ax} \\[2mm] \qquad -\dfrac{m+n-2}{n-1}\displaystyle\int \dfrac{dx}{\sin^m ax \cos^{n-2} ax} \end{cases}$$

335. $\displaystyle\int \sin(a+bx)\,dx = -\frac{1}{b}\cos(a+bx)$

336. $\displaystyle\int \cos(a+bx)\,dx = \frac{1}{b}\sin(a+bx)$

337. $\displaystyle\int \frac{dx}{1 \pm \sin ax} = \mp \frac{1}{a}\tan\left(\frac{\pi}{4} \mp \frac{ax}{2}\right)$

338. $\displaystyle\int \frac{dx}{1 + \cos ax} = \frac{1}{a}\tan\frac{ax}{2}$

339. $\displaystyle\int \frac{dx}{1 - \cos ax} = -\frac{1}{a}\cot\frac{ax}{2}$

*340. $\displaystyle\int \frac{dx}{a + b\sin x} =$
$$\begin{cases} \dfrac{2}{\sqrt{a^2-b^2}}\tan^{-1}\dfrac{a\tan\frac{x}{2}+b}{\sqrt{a^2-b^2}} \\[4mm] \text{or} \\[2mm] \dfrac{1}{\sqrt{b^2-a^2}}\log\dfrac{a\tan\frac{x}{2}+b-\sqrt{b^2-a^2}}{a\tan\frac{x}{2}+b+\sqrt{b^2-a^2}} \end{cases}$$

*341. $\displaystyle\int \frac{dx}{a + b\cos x} =$
$$\begin{cases} \dfrac{2}{\sqrt{a^2-b^2}}\tan^{-1}\dfrac{\sqrt{a^2-b^2}\tan\frac{x}{2}}{a+b} \\[4mm] \text{or} \\[2mm] \dfrac{1}{\sqrt{b^2-a^2}}\log\left(-\dfrac{\sqrt{b^2-a^2}\tan\frac{x}{2}+a+b}{\sqrt{b^2-a^2}\tan\frac{x}{2}-a-b}\right) \end{cases}$$

*342. $\displaystyle\int \frac{dx}{a + b\sin x + c\cos x}$
$$= \begin{cases} \dfrac{1}{\sqrt{b^2+c^2-a^2}}\log\dfrac{b-\sqrt{b^2+c^2-a^2}+(a-c)\tan\frac{x}{2}}{b+\sqrt{b^2+c^2-a^2}+(a-c)\tan\frac{x}{2}}, & \text{if } a^2 < b^2+c^2,\, a\neq c \\[4mm] \text{or} \\[2mm] \dfrac{2}{\sqrt{a^2-b^2-c^2}}\tan^{-1}\dfrac{b+(a-c)\tan\frac{x}{2}}{\sqrt{a^2-b^2-c^2}}, & \text{if } a^2 > b^2+c^2 \\[4mm] \text{or} \\[2mm] \dfrac{1}{a}\left[\dfrac{a-(b+c)\cos x-(b-c)\sin x}{a-(b-c)\cos x+(b+c)\sin x}\right], & \text{if } a^2 = b^2+c^2,\, a\neq c. \end{cases}$$

*See note 6 on page A-40.

*343. $\displaystyle\int \frac{\sin^2 x \, dx}{a + b\cos^2 x} = \frac{1}{b}\sqrt{\frac{a+b}{a}}\tan^{-1}\left(\sqrt{\frac{a}{a+b}}\tan x\right) - \frac{x}{b}$, $\qquad (ab > 0, \text{ or } |a| > |b|)$

*344. $\displaystyle\int \frac{dx}{a^2\cos^2 x + b^2\sin^2 x} = \frac{1}{ab}\tan^{-1}\left(\frac{b\tan x}{a}\right)$

*345. $\displaystyle\int \frac{\cos^2 cx}{a^2 + b^2\sin^2 cx}\, dx = \frac{\sqrt{a^2+b^2}}{ab^2c}\tan^{-1}\frac{\sqrt{a^2+b^2}\tan cx}{a} - \frac{x}{b^2}$

346. $\displaystyle\int \frac{\sin cx \cos cx}{a\cos^2 cx + b\sin^2 cx}\, dx = \frac{1}{2c(b-a)}\log(a\cos^2 cx + b\sin^2 cx)$

347. $\displaystyle\int \frac{\cos cx}{a\cos cx + b\sin cx}\, dx = \int \frac{dx}{a + b\tan cx} =$

$$\frac{1}{c(a^2+b^2)}[acx + b\log(a\cos cx + b\sin$$

348. $\displaystyle\int \frac{\sin cx}{a\sin cx + b\cos cx}\, dx = \int \frac{dx}{a + b\cot cx} =$

$$\frac{1}{c(a^2+b^2)}[acx - b\log(a\sin cx + b\cos$$

*349. $\displaystyle\int \frac{dx}{a\cos^2 x + 2b\cos x \sin x + c\sin^2 x} =$

$$\begin{cases} \dfrac{1}{2\sqrt{b^2-ac}}\log\dfrac{c\tan x + b - \sqrt{b^2}}{c\tan x + b + \sqrt{b^2}} \\ \qquad\qquad\qquad\qquad (b^2 \\ \text{or} \\ \dfrac{1}{\sqrt{ac-b^2}}\tan^{-1}\dfrac{c\tan x + b}{\sqrt{ac-b^2}}, \quad (b^2 \\ \text{or} \\ -\dfrac{1}{c\tan x + b}, \qquad (b^2 = ac) \end{cases}$$

350. $\displaystyle\int \frac{\sin ax}{1 \pm \sin ax}\, dx = \pm x + \frac{1}{a}\tan\left(\frac{\pi}{4} \mp \frac{ax}{2}\right)$

351. $\displaystyle\int \frac{dx}{(\sin ax)(1 \pm \sin ax)} = \frac{1}{a}\tan\left(\frac{\pi}{4} \mp \frac{ax}{2}\right) + \frac{1}{a}\log\tan\frac{ax}{2}$

352. $\displaystyle\int \frac{dx}{(1 + \sin ax)^2} = -\frac{1}{2a}\tan\left(\frac{\pi}{4} - \frac{ax}{2}\right) - \frac{1}{6a}\tan^3\left(\frac{\pi}{4} - \frac{ax}{2}\right)$

353. $\displaystyle\int \frac{dx}{(1 - \sin ax)^2} = \frac{1}{2a}\cot\left(\frac{\pi}{4} - \frac{ax}{2}\right) + \frac{1}{6a}\cot^3\left(\frac{\pi}{4} - \frac{ax}{2}\right)$

354. $\displaystyle\int \frac{\sin ax}{(1 + \sin ax)^2}\, dx = -\frac{1}{2a}\tan\left(\frac{\pi}{4} - \frac{ax}{2}\right) + \frac{1}{6a}\tan^3\left(\frac{\pi}{4} - \frac{ax}{2}\right)$

355. $\displaystyle\int \frac{\sin ax}{(1 - \sin ax)^2}\, dx = -\frac{1}{2a}\cot\left(\frac{\pi}{4} - \frac{ax}{2}\right) + \frac{1}{6a}\cot^3\left(\frac{\pi}{4} - \frac{ax}{2}\right)$

356. $\displaystyle\int \frac{\sin x \, dx}{a + b\sin x} = \frac{x}{b} - \frac{a}{b}\int \frac{dx}{a + b\sin x}$

357. $\displaystyle\int \frac{dx}{(\sin x)(a + b\sin x)} = \frac{1}{a}\log\tan\frac{x}{2} - \frac{b}{a}\int \frac{dx}{a + b\sin x}$

358. $\displaystyle\int \frac{dx}{(a + b\sin x)^2} = \frac{b\cos x}{(a^2 - b^2)(a + b\sin x)} + \frac{a}{a^2 - b^2}\int \frac{dx}{a + b\sin x}$

*See note 6 on page A-30.

359. $\displaystyle\int \frac{\sin x \, dx}{(a + b \sin x)^2} = \frac{a \cos x}{(b^2 - a^2)(a + b \sin x)} + \frac{b}{b^2 - a^2}\int \frac{dx}{a + b \sin x}$

*360. $\displaystyle\int \frac{dx}{a^2 + b^2 \sin^2 cx} = \frac{1}{ac\sqrt{a^2 + b^2}} \tan^{-1} \frac{\sqrt{a^2 + b^2}\, \tan cx}{a}$

*361. $\displaystyle\int \frac{dx}{a^2 - b^2 \sin^2 cx} = \begin{cases} \dfrac{1}{ac\sqrt{a^2 - b^2}} \tan^{-1} \dfrac{\sqrt{a^2 - b^2}\,\tan cx}{a}, \quad (a^2 > b^2) \\[2ex] \text{or} \\[2ex] \dfrac{1}{2ac\sqrt{b^2 - a^2}} \log \dfrac{\sqrt{b^2 - a^2}\,\tan cx + a}{\sqrt{b^2 - a^2}\,\tan cx - a}, \quad (a^2 < b^2) \end{cases}$

362. $\displaystyle\int \frac{\cos ax}{1 + \cos ax}\, dx = x - \frac{1}{a}\tan\frac{ax}{2}$

363. $\displaystyle\int \frac{\cos ax}{1 - \cos ax}\, dx = -x - \frac{1}{a}\cot\frac{ax}{2}$

364. $\displaystyle\int \frac{dx}{(\cos ax)(1 + \cos ax)} = \frac{1}{a}\log\tan\left(\frac{\pi}{4} + \frac{ax}{2}\right) - \frac{1}{a}\tan\frac{ax}{2}$

365. $\displaystyle\int \frac{dx}{(\cos ax)(1 - \cos ax)} = \frac{1}{a}\log\tan\left(\frac{\pi}{4} + \frac{ax}{2}\right) - \frac{1}{a}\cot\frac{ax}{2}$

366. $\displaystyle\int \frac{dx}{(1 + \cos ax)^2} = \frac{1}{2a}\tan\frac{ax}{2} + \frac{1}{6a}\tan^3\frac{ax}{2}$

367. $\displaystyle\int \frac{dx}{(1 - \cos ax)^2} = -\frac{1}{2a}\cot\frac{ax}{2} - \frac{1}{6a}\cot^3\frac{ax}{2}$

368. $\displaystyle\int \frac{\cos ax}{(1 + \cos ax)^2}\, dx = \frac{1}{2a}\tan\frac{ax}{2} - \frac{1}{6a}\tan^3\frac{ax}{2}$

369. $\displaystyle\int \frac{\cos ax}{(1 - \cos ax)^2}\, dx = \frac{1}{2a}\cot\frac{ax}{2} - \frac{1}{6a}\cot^3\frac{ax}{2}$

370. $\displaystyle\int \frac{\cos x \, dx}{a + b \cos x} = \frac{x}{b} - \frac{a}{b}\int \frac{dx}{a + b \cos x}$

371. $\displaystyle\int \frac{dx}{(\cos x)(a + b \cos x)} = \frac{1}{a}\log\tan\left(\frac{x}{2} + \frac{\pi}{4}\right) - \frac{b}{a}\int \frac{dx}{a + b \cos x}$

372. $\displaystyle\int \frac{dx}{(a + b \cos x)^2} = \frac{b \sin x}{(b^2 - a^2)(a + b \cos x)} - \frac{a}{b^2 - a^2}\int \frac{dx}{a + b \cos x}$

373. $\displaystyle\int \frac{\cos x}{(a + b \cos x)^2}\, dx = \frac{a \sin x}{(a^2 - b^2)(a + b \cos x)} - \frac{b}{a^2 - b^2}\int \frac{dx}{a + b \cos x}$

*374. $\displaystyle\int \frac{dx}{a^2 + b^2 - 2ab \cos cx} = \frac{2}{c(a^2 - b^2)}\tan^{-1}\left(\frac{a + b}{a - b}\tan\frac{cx}{2}\right)$

*375. $\displaystyle\int \frac{dx}{a^2 + b^2 \cos^2 cx} = \frac{1}{ac\sqrt{a^2 + b^2}}\tan^{-1}\frac{a \tan cx}{\sqrt{a^2 + b^2}}$

*376. $\displaystyle\int \frac{dx}{a^2 - b^2 \cos^2 cx} = \begin{cases} \dfrac{1}{ac\sqrt{a^2 - b^2}}\tan^{-1}\dfrac{a \tan cx}{\sqrt{a^2 - b^2}}, \quad (a^2 > b^2) \\[2ex] \text{or} \\[2ex] \dfrac{1}{2ac\sqrt{b^2 - a^2}}\log\dfrac{a \tan cx - \sqrt{b^2 - a^2}}{a \tan cx + \sqrt{b^2 - a^2}}, \quad (b^2 > a^2) \end{cases}$

377. $\displaystyle\int \frac{\sin ax}{1 \pm \cos ax}\, dx = \mp \frac{1}{a}\log(1 \pm \cos ax)$

*See note 6 on page A-30.

378. $\displaystyle\int \frac{\cos ax}{1 \pm \sin ax}\, dx = \pm \frac{1}{a}\log\left(1 \pm \sin ax\right)$

379. $\displaystyle\int \frac{dx}{(\sin ax)(1 \pm \cos ax)} = \pm \frac{1}{2a(1 \pm \cos ax)} + \frac{1}{2a}\log\tan\frac{ax}{2}$

380. $\displaystyle\int \frac{dx}{(\cos ax)(1 \pm \sin ax)} = \mp \frac{1}{2a(1 \pm \sin ax)} + \frac{1}{2a}\log\tan\left(\frac{\pi}{4} + \frac{ax}{2}\right)$

381. $\displaystyle\int \frac{\sin ax}{(\cos ax)(1 \pm \cos ax)}\, dx = \frac{1}{a}\log\left(\sec ax \pm 1\right)$

382. $\displaystyle\int \frac{\cos ax}{(\sin ax)(1 \pm \sin ax)}\, dx = -\frac{1}{a}\log\left(\csc ax \pm 1\right)$

383. $\displaystyle\int \frac{\sin ax}{(\cos ax)(1 \pm \sin ax)}\, dx = \frac{1}{2a(1 \pm \sin ax)} \pm \frac{1}{2a}\log\tan\left(\frac{\pi}{4} + \frac{ax}{2}\right)$

384. $\displaystyle\int \frac{\cos ax}{(\sin ax)(1 \pm \cos ax)}\, dx = -\frac{1}{2a(1 \pm \cos ax)} \pm \frac{1}{2a}\log\tan\frac{ax}{2}$

385. $\displaystyle\int \frac{dx}{\sin ax \pm \cos ax} = \frac{1}{a\sqrt{2}}\log\tan\left(\frac{ax}{2} \pm \frac{\pi}{8}\right)$

386. $\displaystyle\int \frac{dx}{(\sin ax \pm \cos ax)^2} = \frac{1}{2a}\tan\left(ax \mp \frac{\pi}{4}\right)$

387. $\displaystyle\int \frac{dx}{1 + \cos ax \pm \sin ax} = \pm \frac{1}{a}\log\left(1 \pm \tan\frac{ax}{2}\right)$

388. $\displaystyle\int \frac{dx}{a^2\cos^2 cx - b^2\sin^2 cx} = \frac{1}{2abc}\log\frac{b\tan cx + a}{b\tan cx - a}$

389. $\displaystyle\int x(\sin ax)\, dx = \frac{1}{a^2}\sin ax - \frac{x}{a}\cos ax$

390. $\displaystyle\int x^2(\sin ax)\, dx = \frac{2x}{a^2}\sin ax - \frac{a^2x^2 - 2}{a^3}\cos ax$

391. $\displaystyle\int x^3(\sin ax)\, dx = \frac{3a^2x^2 - 6}{a^4}\sin ax - \frac{a^2x^3 - 6x}{a^3}\cos ax$

392. $\displaystyle\int x^m \sin ax\, dx = \begin{cases} -\dfrac{1}{a}x^m\cos ax + \dfrac{m}{a}\displaystyle\int x^{m-1}\cos ax\, dx \\[2mm] \text{or} \\[2mm] \cos ax \displaystyle\sum_{r=0}^{\left[\frac{m}{2}\right]} (-1)^{r+1}\,\dfrac{m!}{(m-2r)!}\cdot\dfrac{x^{m-2r}}{a^{2r+1}} \\[4mm] +\sin ax\displaystyle\sum_{r=0}^{\left[\frac{m-1}{2}\right]}(-1)^r\,\dfrac{m!}{(m-2r-1)!}\cdot\dfrac{x^{m-2r-1}}{a^{2r+2}} \end{cases}$

Note: $[s]$ means greatest integer $\leq s$; $[3\frac{1}{2}] = 3$, $[\frac{1}{2}] = 0$, etc.

393. $\displaystyle\int x(\cos ax)\, dx = \frac{1}{a^2}\cos ax + \frac{x}{a}\sin ax$

394. $\displaystyle\int x^2(\cos ax)\, dx = \frac{2x\cos ax}{a^2} + \frac{a^2x^2 - 2}{a^3}\sin ax$

395. $\displaystyle\int x^3(\cos ax)\, dx = \frac{3a^2x^2 - 6}{a^4}\cos ax + \frac{a^2x^3 - 6x}{a^3}\sin ax$

396. $\displaystyle\int x^m(\cos ax)\,dx = \begin{cases} \dfrac{x^m \sin ax}{a} - \dfrac{m}{a}\displaystyle\int x^{m-1}\sin ax\,dx \\ \qquad\text{or} \\ \sin ax \displaystyle\sum_{r=0}^{\left\lfloor\frac{m}{2}\right\rfloor}(-1)^r\,\dfrac{m!}{(m-2r)!}\cdot\dfrac{x^{m-2r}}{a^{2r+1}} \\ \qquad + \cos ax\displaystyle\sum_{r=0}^{\left\lfloor\frac{m-1}{2}\right\rfloor}(-1)^r\,\dfrac{m!}{(m-2r-1)!}\cdot\dfrac{x^{m-2r-1}}{a^{2r+2}}\end{cases}$

See note integral 392.

397. $\displaystyle\int \frac{\sin ax}{x}\,dx = \sum_{n=0}^{r}(-1)^n\,\frac{(ax)^{2n+1}}{(2n+1)(2n+1)!}$

398. $\displaystyle\int \frac{\cos ax}{x}\,dx = \log x + \sum_{n=1}^{r}(-1)^n\,\frac{(ax)^{2n}}{2n(2n)!}$

399. $\displaystyle\int x(\sin^2 ax)\,dx = \frac{x^2}{4} - \frac{x\sin 2ax}{4a} - \frac{\cos 2ax}{8a^2}$

400. $\displaystyle\int x^2(\sin^2 ax)\,dx = \frac{x^3}{6} - \left(\frac{x^2}{4a}-\frac{1}{8a^3}\right)\sin 2ax - \frac{x\cos 2ax}{4a^2}$

401. $\displaystyle\int x(\sin^3 ax)\,dx = \frac{x\cos 3ax}{12a} - \frac{\sin 3ax}{36a^2} - \frac{3x\cos ax}{4a} + \frac{3\sin ax}{4a^2}$

402. $\displaystyle\int x(\cos^2 ax)\,dx = \frac{x^2}{4} + \frac{x\sin 2ax}{4a} + \frac{\cos 2ax}{8a^2}$

403. $\displaystyle\int x^2(\cos^2 ax)\,dx = \frac{x^3}{6} + \left(\frac{x^2}{4a}-\frac{1}{8a^3}\right)\sin 2ax + \frac{x\cos 2ax}{4a^2}$

404. $\displaystyle\int x(\cos^3 ax)\,dx = \frac{x\sin 3ax}{12a} + \frac{\cos 3ax}{36a^2} + \frac{3x\sin ax}{4a} + \frac{3\cos ax}{4a^2}$

405. $\displaystyle\int \frac{\sin ax}{x^m}\,dx = -\frac{\sin ax}{(m-1)x^{m-1}} + \frac{a}{m-1}\int \frac{\cos ax}{x^{m-1}}\,dx$

406. $\displaystyle\int \frac{\cos ax}{x^m}\,dx = -\frac{\cos ax}{(m-1)x^{m-1}} - \frac{a}{m-1}\int \frac{\sin ax}{x^{m-1}}\,dx$

407. $\displaystyle\int \frac{x}{1 \pm \sin ax}\,dx = \mp\frac{x\cos ax}{a(1\pm\sin ax)} + \frac{1}{a^2}\log(1\pm\sin ax)$

408. $\displaystyle\int \frac{x}{1 + \cos ax}\,dx = \frac{x}{a}\tan\frac{ax}{2} + \frac{2}{a^2}\log\cos\frac{ax}{2}$

409. $\displaystyle\int \frac{x}{1 - \cos ax}\,dx = -\frac{x}{a}\cot\frac{ax}{2} + \frac{2}{a^2}\log\sin\frac{ax}{2}$

410. $\displaystyle\int \frac{x + \sin x}{1 + \cos x}\,dx = x\tan\frac{x}{2}$

411. $\displaystyle\int \frac{x - \sin x}{1 - \cos x}\,dx = -x\cot\frac{x}{2}$

412. $\displaystyle\int \sqrt{1 - \cos ax}\,dx = -\frac{2\sin ax}{a\sqrt{1-\cos ax}} = -\frac{2\sqrt{2}}{a}\cos\left(\frac{ax}{2}\right)$

413. $\displaystyle\int \sqrt{1 + \cos ax}\,dx = \frac{2\sin ax}{a\sqrt{1+\cos ax}} = \frac{2\sqrt{2}}{a}\sin\left(\frac{ax}{2}\right)$

414. $\displaystyle\int \sqrt{1 + \sin x}\,dx = \pm 2\left(\sin\frac{x}{2} - \cos\frac{x}{2}\right),$

$\left[\text{use } + \text{ if }(8k-1)\dfrac{\pi}{2} < x \le (8k+3)\dfrac{\pi}{2},\text{ otherwise } -;\ k \text{ an integer}\right]$

415. $\int \sqrt{1 - \sin x}\, dx = \pm 2\left(\sin\frac{x}{2} + \cos\frac{x}{2}\right)$,

[use $+$ if $(8k - 3)\dfrac{\pi}{2} < x \leq (8k + 1)\dfrac{\pi}{2}$, otherwise $-$; k an integer]

416. $\int \dfrac{dx}{\sqrt{1 - \cos x}} = \pm \sqrt{2}\, \log \tan\dfrac{x}{4}$,

[use $+$ if $4k\pi < x < (4k + 2)\pi$, otherwise $-$; k an integer]

417. $\int \dfrac{dx}{\sqrt{1 + \cos x}} = \pm \sqrt{2}\, \log \tan\left(\dfrac{x + \pi}{4}\right)$,

[use $+$ if $(4k - 1)\pi < x < (4k + 1)\pi$, otherwise $-$; k an integer]

418. $\int \dfrac{dx}{\sqrt{1 - \sin x}} = \pm \sqrt{2}\, \log \tan\left(\dfrac{x}{4} - \dfrac{\pi}{8}\right)$,

[use $+$ if $(8k + 1)\dfrac{\pi}{2} < x < (8k + 5)\dfrac{\pi}{2}$, otherwise $-$; k an integer]

419. $\int \dfrac{dx}{\sqrt{1 + \sin x}} = \pm \sqrt{2}\, \log \tan\left(\dfrac{x}{4} + \dfrac{\pi}{8}\right)$,

[use $+$ if $(8k - 1)\dfrac{\pi}{2} < x < (8k + 3)\dfrac{\pi}{2}$, otherwise $-$; k an integer]

420. $\int (\tan^2 ax)\, dx = \dfrac{1}{a}\tan ax - x$

421. $\int (\tan^3 ax)\, dx = \dfrac{1}{2a}\tan^2 ax + \dfrac{1}{a}\log \cos ax$

422. $\int (\tan^4 ax)\, dx = \dfrac{\tan^3 ax}{3a} - \dfrac{1}{a}\tan x + x$

423. $\int (\tan^n ax)\, dx = \dfrac{\tan^{n-1} ax}{a(n - 1)} - \int (\tan^{n-2} ax)\, dx$

424. $\int (\cot^2 ax)\, dx = -\dfrac{1}{a}\cot ax - x$

425. $\int (\cot^3 ax)\, dx = -\dfrac{1}{2a}\cot^2 ax - \dfrac{1}{a}\log \sin ax$

426. $\int (\cot^4 ax)\, dx = -\dfrac{1}{3a}\cot^3 ax + \dfrac{1}{a}\cot ax + x$

427. $\int (\cot^n ax)\, dx = -\dfrac{\cot^{n-1} ax}{a(n - 1)} - \int (\cot^{n-2} ax)\, dx$

428. $\int \dfrac{x}{\sin^2 ax}\, dx = \int x(\csc^2 ax)\, dx = -\dfrac{x \cot ax}{a} + \dfrac{1}{a^2}\log \sin ax$

429. $\int \dfrac{x}{\sin^n ax}\, dx = \int x(\csc^n ax)\, dx = -\dfrac{x \cos ax}{a(n - 1)\sin^{n-1} ax}$

$$-\dfrac{1}{a^2(n - 1)(n - 2)\sin^{n-2} ax} + \dfrac{(n - 2)}{(n - 1)}\int \dfrac{x}{\sin^{n-2} ax}\, dx$$

430. $\int \dfrac{x}{\cos^2 ax}\, dx = \int x(\sec^2 ax)\, dx = \dfrac{1}{a}x \tan ax + \dfrac{1}{a^2}\log \cos ax$

431. $\int \dfrac{x}{\cos^n ax}\, dx = \int x(\sec^n ax)\, dx = \dfrac{x \sin ax}{a(n - 1)\cos^{n-1} ax}$

$$-\dfrac{1}{a^2(n - 1)(n - 2)\cos^{n-2} ax} + \dfrac{n - 2}{n - 1}\int \dfrac{x}{\cos^{n-2} ax}\, dx$$

432. $\displaystyle\int \frac{\sin ax}{\sqrt{1 + b^2 \sin^2 ax}}\, dx = -\frac{1}{ab}\sin^{-1}\frac{b\cos ax}{\sqrt{1 + b^2}}$

433. $\displaystyle\int \frac{\sin ax}{\sqrt{1 - b^2 \sin^2 ax}}\, dx = -\frac{1}{ab}\log\left(b\cos ax + \sqrt{1 - b^2 \sin^2 ax}\right)$

434. $\displaystyle\int (\sin ax)\sqrt{1 + b^2 \sin^2 ax}\; dx = -\frac{\cos ax}{2a}\sqrt{1 + b^2 \sin^2 ax}$
$$-\frac{1 + b^2}{2ab}\sin^{-1}\frac{b\cos ax}{\sqrt{1 + b^2}}$$

435. $\displaystyle\int (\sin ax)\sqrt{1 - b^2 \sin^2 ax}\; dx = -\frac{\cos ax}{2a}\sqrt{1 - b^2 \sin^2 ax}$
$$-\frac{1 - b^2}{2ab}\log\left(b\cos ax + \sqrt{1 - b^2 \sin^2 ax}\right)$$

436. $\displaystyle\int \frac{\cos ax}{\sqrt{1 + b^2 \sin^2 ax}}\, dx = \frac{1}{ab}\log\left(b\sin ax + \sqrt{1 + b^2 \sin^2 ax}\right)$

437. $\displaystyle\int \frac{\cos ax}{\sqrt{1 - b^2 \sin^2 ax}}\, dx = \frac{1}{ab}\sin^{-1}(b\sin ax)$

438. $\displaystyle\int (\cos ax)\sqrt{1 + b^2 \sin^2 ax}\; dx = \frac{\sin ax}{2a}\sqrt{1 + b^2 \sin^2 ax}$
$$+\frac{1}{2ab}\log\left(b\sin ax + \sqrt{1 + b^2 \sin^2 ax}\right)$$

439. $\displaystyle\int (\cos ax)\sqrt{1 - b^2 \sin^2 ax}\; dx = \frac{\sin ax}{2a}\sqrt{1 - b^2 \sin^2 ax} + \frac{1}{2ab}\sin^{-1}(b\sin ax)$

440. $\displaystyle\int \frac{dx}{\sqrt{a + b\tan^2 cx}} = \frac{\pm 1}{c\sqrt{a - b}}\sin^{-1}\left(\sqrt{\frac{a - b}{a}}\sin cx\right), \qquad (a > |b|)$

$$\left[\text{use } + \text{ if } (2k - 1)\frac{\pi}{2} < x \le (2k + 1)\frac{\pi}{2}, \text{ otherwise } -; k \text{ an integer}\right]$$

FORMS INVOLVING INVERSE TRIGONOMETRIC FUNCTIONS

441. $\displaystyle\int (\sin^{-1} ax)\, dx = x\sin^{-1} ax + \frac{\sqrt{1 - a^2 x^2}}{a}$

442. $\displaystyle\int (\cos^{-1} ax)\, dx = x\cos^{-1} ax - \frac{\sqrt{1 - a^2 x^2}}{a}$

443. $\displaystyle\int (\tan^{-1} ax)\, dx = x\tan^{-1} ax - \frac{1}{2a}\log(1 + a^2 x^2)$

444. $\displaystyle\int (\cot^{-1} ax)\, dx = x\cot^{-1} ax + \frac{1}{2a}\log(1 + a^2 x^2)$

445. $\displaystyle\int (\sec^{-1} ax)\, dx = x\sec^{-1} ax - \frac{1}{a}\log(ax + \sqrt{a^2 x^2 - 1})$

446. $\displaystyle\int (\csc^{-1} ax)\, dx = x\csc^{-1} ax + \frac{1}{a}\log(ax + \sqrt{a^2 x^2 - 1})$

447. $\displaystyle\int \left(\sin^{-1} \frac{x}{a}\right) dx = x\sin^{-1}\frac{x}{a} + \sqrt{a^2 - x^2}, \qquad (a > 0)$

448. $\displaystyle\int \left(\cos^{-1} \frac{x}{a}\right) dx = x\cos^{-1}\frac{x}{a} - \sqrt{a^2 - x^2}, \qquad (a > 0)$

449. $\int \left(\tan^{-1} \dfrac{x}{a} \right) dx = x \tan^{-1} \dfrac{x}{a} - \dfrac{a}{2} \log (a^2 + x^2)$

450. $\int \left(\cot^{-1} \dfrac{x}{a} \right) dx = x \cot^{-1} \dfrac{x}{a} + \dfrac{a}{2} \log (a^2 + x^2)$

451. $\int x[\sin^{-1}(ax)] \, dx = \dfrac{1}{4a^2}[(2a^2x^2 - 1)\sin^{-1}(ax) + ax\sqrt{1 - a^2x^2}]$

452. $\int x[\cos^{-1}(ax)] \, dx = \dfrac{1}{4a^2}[(2a^2x^2 - 1)\cos^{-1}(ax) - ax\sqrt{1 - a^2x^2}]$

453. $\int x^n[\sin^{-1}(ax)] \, dx = \dfrac{x^{n+1}}{n+1}\sin^{-1}(ax) - \dfrac{a}{n+1}\int \dfrac{x^{n+1} \, dx}{\sqrt{1 - a^2x^2}}, \qquad (n \neq -1)$

454. $\int x^n[\cos^{-1}(ax)] \, dx = \dfrac{x^{n+1}}{n+1}\cos^{-1}(ax) + \dfrac{a}{n+1}\int \dfrac{x^{n+1} \, dx}{\sqrt{1 - a^2x^2}}, \qquad (n \neq -1)$

455. $\int x(\tan^{-1} ax) \, dx = \dfrac{1 + a^2x^2}{2a^2}\tan^{-1} ax - \dfrac{x}{2a}$

456. $\int x^n(\tan^{-1} ax) \, dx = \dfrac{x^{n+1}}{n+1}\tan^{-1} ax - \dfrac{a}{n+1}\int \dfrac{x^{n+1}}{1 + a^2x^2} \, dx$

457. $\int x(\cot^{-1} ax) \, dx = \dfrac{1 + a^2x^2}{2a^2}\cot^{-1} ax + \dfrac{x}{2a}$

458. $\int x^n(\cot^{-1} ax) \, dx = \dfrac{x^{n+1}}{n+1}\cot^{-1} ax + \dfrac{a}{n+1}\int \dfrac{x^{n+1}}{1 + a^2x^2} \, dx$

459. $\int \dfrac{\sin^{-1}(ax)}{x^2} \, dx = a \log \left(\dfrac{1 - \sqrt{1 - a^2x^2}}{x} \right) - \dfrac{\sin^{-1}(ax)}{x}$

460. $\int \dfrac{\cos^{-1}(ax) \, dx}{x^2} = -\dfrac{1}{x}\cos^{-1}(ax) + a \log \dfrac{1 + \sqrt{1 - a^2x^2}}{x}$

461. $\int \dfrac{\tan^{-1}(ax) \, dx}{x^2} = -\dfrac{1}{x}\tan^{-1}(ax) - \dfrac{a}{2} \log \dfrac{1 + a^2x^2}{x^2}$

462. $\int \dfrac{\cot^{-1} ax}{x^2} \, dx = -\dfrac{1}{x}\cot^{-1} ax - \dfrac{a}{2} \log \dfrac{x^2}{a^2x^2 + 1}$

463. $\int (\sin^{-1} ax)^2 \, dx = x(\sin^{-1} ax)^2 - 2x + \dfrac{2\sqrt{1 - a^2x^2}}{a}\sin^{-1} ax$

464. $\int (\cos^{-1} ax)^2 \, dx = x(\cos^{-1} ax)^2 - 2x - \dfrac{2\sqrt{1 - a^2x^2}}{a}\cos^{-1} ax$

465. $\int (\sin^{-1} ax)^n \, dx = \begin{cases} x(\sin^{-1} ax)^n + \dfrac{n\sqrt{1 - a^2x^2}}{a}(\sin^{-1} ax)^{n-1} \\ \qquad\qquad\qquad - n(n-1)\displaystyle\int (\sin^{-1} ax)^{n-2} \, dx \\ \text{or} \\ \displaystyle\sum_{r=0}^{\left[\frac{n}{2}\right]} (-1)^r \dfrac{n!}{(n-2r)!} x(\sin^{-1} ax)^{n-2r} \\ \qquad + \displaystyle\sum_{r=0}^{\left[\frac{n-1}{2}\right]} (-1)^r \dfrac{n!\sqrt{1 - a^2x^2}}{(n-2r-1)!a}(\sin^{-1} ax)^{n-2r-1} \end{cases}$

Note: $[s]$ means greatest integer $\leq s$. Thus $[3.5]$ means 3; $[5] = 5$, $[\frac{1}{2}] = 0$.

466. $\displaystyle\int (\cos^{-1} ax)^n \, dx = \begin{cases} x(\cos^{-1} ax)^n - \dfrac{n\sqrt{1-a^2x^2}}{a}(\cos^{-1} ax)^{n-1} \\ \qquad\qquad\qquad\qquad - n(n-1)\displaystyle\int (\cos^{-1} ax)^{n-2}\, dx \\ \qquad\qquad \text{or} \\ \displaystyle\sum_{r=0}^{\left[\frac{n}{2}\right]} (-1)^r \dfrac{n!}{(n-2r)!}\, x(\cos^{-1} ax)^{n-2r} \\ \qquad - \displaystyle\sum_{r=0}^{\left[\frac{n-1}{2}\right]} (-1)^r \dfrac{n!\sqrt{1-a^2x^2}}{(n-2r-1)!a}(\cos^{-1} ax)^{n-2r-1} \end{cases}$

467. $\displaystyle\int \frac{1}{\sqrt{1-a^2x^2}}(\sin^{-1} ax)\, dx = \frac{1}{2a}(\sin^{-1} ax)^2$

468. $\displaystyle\int \frac{x^n}{\sqrt{1-a^2x^2}}(\sin^{-1} ax)\, dx = -\frac{x^{n-1}}{na^2}\sqrt{1-a^2x^2}\,\sin^{-1} ax + \frac{x^n}{n^2a}$
$$+ \frac{n-1}{na^2}\int \frac{x^{n-2}}{\sqrt{1-a^2x^2}}\sin^{-1} ax\, dx$$

469. $\displaystyle\int \frac{1}{\sqrt{1-a^2x^2}}(\cos^{-1} ax)\, dx = -\frac{1}{2a}(\cos^{-1} ax)^2$

470. $\displaystyle\int \frac{x^n}{\sqrt{1-a^2x^2}}(\cos^{-1} ax)\, dx = -\frac{x^{n-1}}{na^2}\sqrt{1-a^2x^2}\,\cos^{-1} ax - \frac{x^n}{n^2a}$
$$+ \frac{n-1}{na^2}\int \frac{x^{n-2}}{\sqrt{1-a^2x^2}}\cos^{-1} ax\, dx$$

471. $\displaystyle\int \frac{\tan^{-1} ax}{a^2x^2+1}\, dx = \frac{1}{2a}(\tan^{-1} ax)^2$

472. $\displaystyle\int \frac{\cot^{-1} ax}{a^2x^2+1}\, dx = -\frac{1}{2a}(\cot^{-1} ax)^2$

473. $\displaystyle\int x\sec^{-1} ax\, dx = \frac{x^2}{2}\sec^{-1} ax - \frac{1}{2a^2}\sqrt{a^2x^2-1}$

474. $\displaystyle\int x^n\sec^{-1} ax\, dx = \frac{x^{n+1}}{n+1}\sec^{-1} ax - \frac{1}{n+1}\int \frac{x^n\, dx}{\sqrt{a^2x^2-1}}$

475. $\displaystyle\int \frac{\sec^{-1} ax}{x^2}\, dx = -\frac{\sec^{-1} ax}{x} + \frac{\sqrt{a^2x^2-1}}{x}$

476. $\displaystyle\int x\csc^{-1} ax\, dx = \frac{x^2}{2}\csc^{-1} ax + \frac{1}{2a^2}\sqrt{a^2x^2-1}$

477. $\displaystyle\int x^n\csc^{-1} ax\, dx = \frac{x^{n+1}}{n+1}\csc^{-1} ax + \frac{1}{n+1}\int \frac{x^n\, dx}{\sqrt{a^2x^2-1}}$

478. $\displaystyle\int \frac{\csc^{-1} ax}{x^2}\, dx = -\frac{\csc^{-1} ax}{x} - \frac{\sqrt{a^2x^2-1}}{x}$

FORMS INVOLVING TRIGONOMETRIC SUBSTITUTIONS

479. $\displaystyle\int f(\sin x)\, dx = 2\int f\!\left(\frac{2z}{1+z^2}\right)\frac{dz}{1+z^2}, \qquad \left(z = \tan\frac{x}{2}\right)$

480. $\displaystyle\int f(\cos x)\, dx = 2\int f\!\left(\frac{1-z^2}{1+z^2}\right)\frac{dz}{1+z^2}, \qquad \left(z = \tan\frac{x}{2}\right)$

*481. $\int f(\sin x)\,dx = \int f(u)\dfrac{du}{\sqrt{1-u^2}}, \qquad (u = \sin x)$

*482. $\int f(\cos x)\,dx = -\int f(u)\dfrac{du}{\sqrt{1-u^2}}, \qquad (u = \cos x)$

*483. $\int f(\sin x, \cos x)\,dx = \int f(u, \sqrt{1-u^2})\dfrac{du}{\sqrt{1-u^2}}, \qquad (u = \sin x)$

484. $\int f(\sin x, \cos x)\,dx = 2\int f\left(\dfrac{2z}{1+z^2}, \dfrac{1-z^2}{1+z^2}\right)\dfrac{dz}{1+z^2}, \qquad \left(z = \tan\dfrac{x}{2}\right)$

LOGARITHMIC FORMS

485. $\int (\log x)\,dx = x\log x - x$

486. $\int x(\log x)\,dx = \dfrac{x^2}{2}\log x - \dfrac{x^2}{4}$

487. $\int x^2(\log x)\,dx = \dfrac{x^3}{3}\log x - \dfrac{x^3}{9}$

488. $\int x^n(\log ax)\,dx = \dfrac{x^{n+1}}{n+1}\log ax - \dfrac{x^{n+1}}{(n+1)^2}$

489. $\int (\log x)^2\,dx = x(\log x)^2 - 2x\log x + 2x$

490. $\int (\log x)^n\,dx = \begin{cases} x(\log x)^n - n\int (\log x)^{n-1}\,dx & (n \neq -1) \\ \text{or} \\ (-1)^n n!\, x\displaystyle\sum_{r=0}^{n}\dfrac{(-\log x)^r}{r!} \end{cases}$

491. $\int \dfrac{(\log x)^n}{x}\,dx = \dfrac{1}{n+1}(\log x)^{n+1}$

492. $\int \dfrac{dx}{\log x} = \log(\log x) + \log x + \dfrac{(\log x)^2}{2\cdot 2!} + \dfrac{(\log x)^3}{3\cdot 3!} + \cdots$

493. $\int \dfrac{dx}{x\log x} = \log(\log x)$

494. $\int \dfrac{dx}{x(\log x)^n} = -\dfrac{1}{(n-1)(\log x)^{n-1}}$

495. $\int \dfrac{x^m\,dx}{(\log x)^n} = -\dfrac{x^{m+1}}{(n-1)(\log x)^{n-1}} + \dfrac{m+1}{n-1}\int \dfrac{x^m\,dx}{(\log x)^{n-1}}$

496. $\int x^m(\log x)^n\,dx = \begin{cases} \dfrac{x^{m+1}(\log x)^n}{m+1} - \dfrac{n}{m+1}\int x^m(\log x)^{n-1}\,dx \\ \text{or} \\ (-1)^n\dfrac{n!}{m+1}x^{m+1}\displaystyle\sum_{r=0}^{n}\dfrac{(-\log x)^r}{r!(m+1)^{n-r}} \end{cases}$

497. $\int x^p\cos(b\,\ln x)\,dx = \dfrac{x^{p+1}}{(p+1)^2 + b^2}\cdot[b\sin(b\,\ln x) + (p+1)\cos(b\,\ln x)] + c$

498. $\int x^p\sin(b\,\ln x)\,dx = \dfrac{x^{p+1}}{(p+1)^2 + b^2}\cdot[(p+1)\sin(b\,\ln x) - b\cos(b\,\ln x)] + c$

499. $\int [\log(ax+b)]\,dx = \dfrac{ax+b}{a}\log(ax+b) - x$

---

*The square roots appearing in these formulas may be plus or minus, depending on the quadrant of x. Care must be used to give them the proper sign.

500. $\displaystyle\int \frac{\log(ax + b)}{x^2}\, dx = \frac{a}{b}\log x - \frac{ax + b}{bx}\log(ax + b)$

501. $\displaystyle\int x^m[\log(ax + b)]\, dx = \frac{1}{m + 1}\left[x^{m+1} - \left(-\frac{b}{a}\right)^{m+1}\right]\log(ax + b)$

$$- \frac{1}{m + 1}\left(-\frac{b}{a}\right)^{m+1}\sum_{r=1}^{m+1}\frac{1}{r}\left(-\frac{ax}{b}\right)^r$$

502. $\displaystyle\int \frac{\log(ax + b)}{x^m}\, dx = -\frac{1}{m - 1}\frac{\log(ax + b)}{x^{m-1}} + \frac{1}{m - 1}\left(-\frac{a}{b}\right)^{m-1}\log\frac{ax + b}{x}$

$$+ \frac{1}{m - 1}\left(-\frac{a}{b}\right)^{m-1}\sum_{r=1}^{m-2}\frac{1}{r}\left(-\frac{b}{ax}\right)^r, (m > 2)$$

503. $\displaystyle\int \left[\log\frac{x + a}{x - a}\right]dx = (x + a)\log(x + a) - (x - a)\log(x - a)$

504. $\displaystyle\int x^m\left[\log\frac{x + a}{x - a}\right]dx = \frac{x^{m+1} - (-a)^{m+1}}{m + 1}\log(x + a) - \frac{x^{m+1} - a^{m+1}}{m + 1}\log(x - a)$

$$+ \frac{2a^{m+1}}{m + 1}\sum_{r=1}^{\left[\frac{m+1}{2}\right]}\frac{1}{m - 2r + 2}\left(\frac{x}{a}\right)^{m-2r+2}$$

See note integral 392.

505. $\displaystyle\int \frac{1}{x^2}\left[\log\frac{x + a}{x - a}\right]dx = \frac{1}{x}\log\frac{x - a}{x + a} - \frac{1}{a}\log\frac{x^2 - a^2}{x^2}$

506. $\displaystyle\int (\log X)\, dx =$
$$\begin{cases}\left(x + \dfrac{b}{2c}\right)\log X - 2x + \dfrac{\sqrt{4ac - b^2}}{c}\tan^{-1}\dfrac{2cx + b}{\sqrt{4ac - b^2}}, \\ \qquad\qquad\qquad (b^2 - 4ac < 0) \\ \text{or} \\ \left(x + \dfrac{b}{2c}\right)\log X - 2x + \dfrac{\sqrt{b^2 - 4ac}}{c}\tanh^{-1}\dfrac{2cx + b}{\sqrt{b^2 - 4ac}}, \\ \qquad\qquad\qquad (b^2 - 4ac > 0) \\ \text{where} \\ X = a + bx + cx^2 \end{cases}$$

507. $\displaystyle\int x^n(\log X)\, dx = \frac{x^{n+1}}{n + 1}\log X - \frac{2c}{n + 1}\int \frac{x^{n+2}}{X}\, dx - \frac{b}{n + 1}\int \frac{x^{n+1}}{X}\, dx$

$$\text{where } X = a + bx + cx^2$$

508. $\displaystyle\int [\log(x^2 + a^2)]\, dx = x\log(x^2 + a^2) - 2x + 2a\tan^{-1}\frac{x}{a}$

509. $\displaystyle\int [\log(x^2 - a^2)]\, dx = x\log(x^2 - a^2) - 2x + a\log\frac{x + a}{x - a}$

510. $\displaystyle\int x[\log(x^2 \pm a^2)]\, dx = \tfrac{1}{2}(x^2 \pm a^2)\log(x^2 \pm a^2) - \tfrac{1}{2}x^2$

511. $\displaystyle\int [\log(x + \sqrt{x^2 \pm a^2})]\, dx = x\log(x + \sqrt{x^2 \pm a^2}) - \sqrt{x^2 \pm a^2}$

512. $\displaystyle\int x[\log(x + \sqrt{x^2 \pm a^2})]\, dx = \left(\frac{x^2}{2} \pm \frac{a^2}{4}\right)\log(x + \sqrt{x^2 \pm a^2}) - \frac{x\sqrt{x^2 \pm a^2}}{4}$

513. $\displaystyle\int x^m[\log(x + \sqrt{x^2 \pm a^2})]\, dx = \frac{x^{m+1}}{m + 1}\log(x + \sqrt{x^2 \pm a^2})$

$$- \frac{1}{m + 1}\int \frac{x^{m+1}}{\sqrt{x^2 \pm a^2}}\, dx$$

514. $\displaystyle\int \frac{\log(x + \sqrt{x^2 + a^2})}{x^2}\, dx = -\frac{\log(x + \sqrt{x^2 + a^2})}{x} - \frac{1}{a}\log\frac{a + \sqrt{x^2 + a^2}}{x}$

515. $\displaystyle\int \frac{\log(x + \sqrt{x^2 - a^2})}{x^2}\, dx = -\frac{\log(x + \sqrt{x^2 - a^2})}{x} + \frac{1}{|a|}\sec^{-1}\frac{x}{a}$

516. $\displaystyle\int x^n \log(x^2 - a^2)\, dx = \frac{1}{n+1}\bigg[x^{n+1}\log(x^2 - a^2) - a^{n+1}\log(x - a)$

See note integral 392. $\displaystyle \qquad -(-a)^{n+1}\log(x + a) - 2\sum_{r=0}^{\left[\frac{n}{2}\right]}\frac{a^{2r}x^{n-2r+1}}{n - 2r + 1}\bigg]$

EXPONENTIAL FORMS

517. $\displaystyle\int e^x\, dx = e^x$

518. $\displaystyle\int e^{-x}\, dx = -e^{-x}$

519. $\displaystyle\int e^{ax}\, dx = \frac{e^{ax}}{a}$

520. $\displaystyle\int x\, e^{ax}\, dx = \frac{e^{ax}}{a^2}(ax - 1)$

521. $\displaystyle\int x^m e^{ax}\, dx = \begin{cases} \dfrac{x^m e^{ax}}{a} - \dfrac{m}{a}\displaystyle\int x^{m-1} e^{ax}\, dx \\[2mm] \qquad\text{or} \\[2mm] e^{ax}\displaystyle\sum_{r=0}^{m}(-1)^r\dfrac{m!\,x^{m-r}}{(m - r)!\,a^{r+1}} \end{cases}$

522. $\displaystyle\int \frac{e^{ax}\, dx}{x} = \log x + \frac{ax}{1!} + \frac{a^2 x^2}{2\cdot 2!} + \frac{a^3 x^3}{3\cdot 3!} + \cdots$

523. $\displaystyle\int \frac{e^{ax}}{x^m}\, dx = -\frac{1}{m-1}\frac{e^{ax}}{x^{m-1}} + \frac{a}{m-1}\int \frac{e^{ax}}{x^{m-1}}\, dx$

524. $\displaystyle\int e^{ax}\log x\, dx = \frac{e^{ax}\log x}{a} - \frac{1}{a}\int \frac{e^{ax}}{x}\, dx$

525. $\displaystyle\int \frac{dx}{1 + e^x} = x - \log(1 + e^x) = \log\frac{e^x}{1 + e^x}$

526. $\displaystyle\int \frac{dx}{a + be^{px}} = \frac{x}{a} - \frac{1}{ap}\log(a + be^{px})$

527. $\displaystyle\int \frac{dx}{ae^{mx} + be^{-mx}} = \frac{1}{m\sqrt{ab}}\tan^{-1}\left(e^{mx}\sqrt{\frac{a}{b}}\right), \quad (a > 0, b > 0)$

528. $\displaystyle\int \frac{dx}{ae^{mx} - be^{-mx}} = \begin{cases} \dfrac{1}{2m\sqrt{ab}}\log\dfrac{\sqrt{a}\,e^{mx} - \sqrt{b}}{\sqrt{a}\,e^{mx} + \sqrt{b}} \\[2mm] \qquad\text{or} \\[2mm] \dfrac{-1}{m\sqrt{ab}}\tanh^{-1}\left(\sqrt{\dfrac{a}{b}}\,e^{mx}\right), \quad (a > 0, b > 0) \end{cases}$

529. $\displaystyle\int (a^x - a^{-x})\, dx = \frac{a^x + a^{-x}}{\log a}$

530. $\displaystyle\int \frac{e^{ax}}{b + ce^{ax}}\, dx = \frac{1}{ac}\log(b + ce^{ax})$

531. $\displaystyle\int \frac{x\, e^{ax}}{(1 + ax)^2}\, dx = \frac{e^{ax}}{a^2(1 + ax)}$

532. $\int x\, e^{-x^2}\, dx = -\frac{1}{2}\, e^{-x^2}$

533. $\int e^{ax} [\sin (bx)]\, dx = \dfrac{e^{ax}[a \sin (bx) - b \cos (bx)]}{a^2 + b^2}$

534. $\int e^{ax} [\sin (bx)][\sin (cx)]\, dx = \dfrac{e^{ax}[(b - c) \sin (b - c)x + a \cos (b - c)x]}{2[a^2 + (b - c)^2]}$

$$- \dfrac{e^{ax}[(b + c) \sin (b + c)x + a \cos (b + c)x]}{2[a^2 + (b + c)^2]}$$

535. $\int e^{ax}[\sin (bx)][\cos (cx)]\, dx = \begin{cases} \dfrac{e^{ax}[a \sin (b - c)x - (b - c) \cos (b - c)x]}{2[a^2 + (b - c)^2]} \\[2ex] + \dfrac{e^{ax}[a \sin (b + c)x - (b + c) \cos (b + c)x]}{2[a^2 + (b + c)^2]} \\[2ex] \qquad\qquad\qquad \text{or} \\[1ex] \dfrac{e^{ax}}{\rho}[(a \sin bx - b \cos bx)[\cos (cx - \alpha)] \\[1ex] \qquad\qquad\qquad - c(\sin bx) \sin (cx - \alpha)] \\[1ex] \text{where} \\[1ex] \rho = \sqrt{(a^2 + b^2 - c^2)^2 + 4a^2 c^2}, \\[1ex] \rho \cos \alpha = a^2 + b^2 - c^2, \qquad \rho \sin \alpha = 2ac \end{cases}$

536. $\int e^{ax}[\sin (bx)][\sin (bx + c)]\, dx$

$$= \dfrac{e^{ax} \cos c}{2a} - \dfrac{e^{ax}[a \cos (2bx + c) + 2b \sin (2bx + c)]}{2(a^2 + 4b^2)}$$

537. $\int e^{ax}[\sin (bx)][\cos (bx + c)]\, dx$

$$= \dfrac{-e^{ax} \sin c}{2a} + \dfrac{e^{ax}[a \sin (2bx + c) - 2b \cos (2bx + c)]}{2(a^2 + 4b^2)}$$

538. $\int e^{ax}[\cos (bx)]\, dx = \dfrac{e^{ax}}{a^2 + b^2}[a \cos (bx) + b \sin (bx)]$

539. $\int e^{ax}[\cos (bx)][\cos (cx)]\, dx = \dfrac{e^{ax}[(b - c) \sin (b - c)x + a \cos (b - c)x]}{2[a^2 + (b - c)^2]}$

$$+ \dfrac{e^{ax}[(b + c) \sin (b + c)x + a \cos (b + c)x]}{2[a^2 + (b + c)^2]}$$

540. $\int e^{ax}[\cos (bx)][\cos (bx + c)]\, dx$

$$= \dfrac{e^{ax} \cos c}{2a} + \dfrac{e^{ax}[a \cos (2bx + c) + 2b \sin (2bx + c)]}{2(a^2 + 4b^2)}$$

541. $\int e^{ax}[\cos (bx)][\sin (bx + c)]\, dx$

$$= \dfrac{e^{ax} \sin c}{2a} + \dfrac{e^{ax}[a \sin (2bx + c) - 2b \cos (2bx + c)]}{2(a^2 + 4b^2)}$$

542. $\int e^{ax}[\sin^n bx]\, dx = \dfrac{1}{a^2 + n^2 b^2}\Bigg[(a \sin bx - nb \cos bx)\, e^{ax} \sin^{n-1} bx$

$$+ n(n - 1)b^2 \int e^{ax}[\sin^{n-2} bx]\, dx \Bigg]$$

543. $\int e^{ax}[\cos^n bx]\, dx = \dfrac{1}{a^2 + n^2 b^2}\Bigg[(a \cos bx + nb \sin bx)\, e^{ax} \cos^{n-1} bx$

$$+ n(n - 1)b^2 \int e^{ax}[\cos^{n-2} bx]\, dx \Bigg]$$

544. $\int x^m e^x \sin x \, dx = \dfrac{1}{2} x^m e^x (\sin x - \cos x) - \dfrac{m}{2} \int x^{m-1} e^x \sin x \, dx$

$$+ \dfrac{m}{2} \int x^{m-1} e^x \cos x \, dx$$

545. $\int x^m e^{ax} [\sin bx] \, dx = \begin{cases} x^m e^{ax} \dfrac{a \sin bx - b \cos bx}{a^2 + b^2} \\ \qquad - \dfrac{m}{a^2 + b^2} \int x^{m-1} e^{ax} (a \sin bx - b \cos bx) \, dx \\ \text{or} \\ e^{ax} \displaystyle\sum_{r=0}^{m} \dfrac{(-1)^r m! x^{m-r}}{\rho^{r+1}(m-r)!} \sin \left[bx - (r+1)\alpha \right] \\ \qquad \text{where} \\ \rho = \sqrt{a^2 + b^2}, \qquad \rho \cos \alpha = a, \qquad \rho \sin \alpha = b \end{cases}$

546. $\int x^m e^x \cos x \, dx = \dfrac{1}{2} x^m e^x (\sin x + \cos x)$

$$- \dfrac{m}{2} \int x^{m-1} e^x \sin x \, dx - \dfrac{m}{2} \int x^{m-1} e^x \cos x \, dx$$

547. $\int x^m e^{ax} \cos bx \, dx = \begin{cases} x^m e^{ax} \dfrac{a \cos bx + b \sin bx}{a^2 + b^2} \\ \qquad - \dfrac{m}{a^2 + b^2} \int x^{m-1} e^{ax} (a \cos bx + b \sin bx) \, dx \\ \text{or} \\ e^{ax} \displaystyle\sum_{r=0}^{m} \dfrac{(-1)^r m! x^{m-r}}{\rho^{r+1}(m-r)!} \cos \left[bx - (r+1)\alpha \right] \\ \qquad \text{where} \\ \rho = \sqrt{a^2 + b^2}, \qquad \rho \cos \alpha = a, \qquad \rho \sin \alpha = b \end{cases}$

548. $\displaystyle\int e^{ax}(\cos^m x)(\sin^n x)\,dx =$
$$\begin{cases}
\dfrac{e^{ax}\cos^{m-1}x\,\sin^n x\,[a\cos x + (m+n)\sin x]}{(m+n)^2 + a^2} \\[2mm]
\quad -\dfrac{na}{(m+n)^2 + a^2}\int e^{ax}(\cos^{m-1}x)(\sin^{n-1}x)\,dx \\[2mm]
\quad +\dfrac{(m-1)(m+n)}{(m+n)^2 + a^2}\int e^{ax}(\cos^{m-2}x)(\sin^n x)\,dx \\[2mm]
\text{or} \\[2mm]
\dfrac{e^{ax}\cos^m x\,\sin^{n-1}x\,[a\sin x - (m+n)\cos x]}{(m+n)^2 + a^2} \\[2mm]
\quad +\dfrac{ma}{(m+n)^2 + a^2}\int e^{ax}(\cos^{m-1}x)(\sin^{n-1}x)\,dx \\[2mm]
\quad +\dfrac{(n-1)(m+n)}{(m+n)^2 + a^2}\int e^{ax}(\cos^m x)(\sin^{n-2}x)\,dx \\[2mm]
\text{or} \\[2mm]
\dfrac{e^{ax}(\cos^{m-1}x)(\sin^{n-1}x)(a\sin x\cos x + m\sin^2 x - n\cos^2 x)}{(m+n)^2 + a^2} \\[2mm]
\quad +\dfrac{m(m-1)}{(m+n)^2 + a^2}\int e^{ax}(\cos^{m-2}x)(\sin^n x)\,dx \\[2mm]
\quad +\dfrac{n(n-1)}{(m+n)^2 + a^2}\int e^{ax}(\cos^m x)(\sin^{n-2}x)\,dx \\[2mm]
\text{or} \\[2mm]
\dfrac{e^{ax}(\cos^{m-1}x)(\sin^{n-1}x)(a\cos x\sin x + m\sin^2 x - n\cos^2 x)}{(m+n)^2 + a^2} \\[2mm]
\quad +\dfrac{m(m-1)}{(m+n)^2 + a^2}\int e^{ax}(\cos^{m-2}x)(\sin^{n-2}x)\,dx \\[2mm]
\quad +\dfrac{(n-m)(n+m-1)}{(m+n)^2 + a^2}\int e^{ax}(\cos^m x)(\sin^{n-2}x)\,dx
\end{cases}$$

549. $\displaystyle\int x\,e^{ax}(\sin bx)\,dx = \dfrac{x\,e^{ax}}{a^2 + b^2}(a\sin bx - b\cos bx)$
$$-\dfrac{e^{ax}}{(a^2 + b^2)^2}[(a^2 - b^2)\sin bx - 2ab\cos bx]$$

550. $\displaystyle\int x\,e^{ax}(\cos bx)\,dx = \dfrac{x\,e^{ax}}{a^2 + b^2}(a\cos bx + b\sin bx)$
$$-\dfrac{e^{ax}}{(a^2 + b^2)^2}[(a^2 - b^2)\cos bx + 2ab\sin bx]$$

551. $\displaystyle\int \dfrac{e^{ax}}{\sin^n x}\,dx = -\dfrac{e^{ax}[a\sin x + (n-2)\cos x]}{(n-1)(n-2)\sin^{n-1}x} + \dfrac{a^2 + (n-2)^2}{(n-1)(n-2)}\int \dfrac{e^{ax}}{\sin^{n-2}x}\,dx$

552. $\displaystyle\int \dfrac{e^{ax}}{\cos^n x}\,dx = -\dfrac{e^{ax}[a\cos x - (n-2)\sin x]}{(n-1)(n-2)\cos^{n-1}x} + \dfrac{a^2 + (n-2)^2}{(n-1)(n-2)}\int \dfrac{e^{ax}}{\cos^{n-2}x}\,dx$

553. $\displaystyle\int e^{ax}\tan^n x\,dx = e^{ax}\dfrac{\tan^{n-1}x}{n-1} - \dfrac{a}{n-1}\int e^{ax}\tan^{n-1}x\,dx - \int e^{ax}\tan^{n-2}x\,dx$

HYPERBOLIC FORMS

554. $\displaystyle\int (\sinh x)\,dx = \cosh x$

555. $\displaystyle\int (\cosh x)\,dx = \sinh x$

556. $\displaystyle\int (\tanh x)\, dx = \log \cosh x$

557. $\displaystyle\int (\coth x)\, dx = \log \sinh x$

558. $\displaystyle\int (\operatorname{sech} x)\, dx = \tan^{-1}(\sinh x)$

559. $\displaystyle\int \operatorname{csch} x\, dx = \log \tanh \left(\dfrac{x}{2}\right)$

560. $\displaystyle\int x(\sinh x)\, dx = x \cosh x - \sinh x$

561. $\displaystyle\int x^n(\sinh x)\, dx = x^n \cosh x - n \int x^{n-1}(\cosh x)\, dx$

562. $\displaystyle\int x(\cosh x)\, dx = x \sinh x - \cosh x$

563. $\displaystyle\int x^n(\cosh x)\, dx = x^n \sinh x - n \int x^{n-1}(\sinh x)\, dx$

564. $\displaystyle\int (\operatorname{sech} x)(\tanh x)\, dx = -\operatorname{sech} x$

565. $\displaystyle\int (\operatorname{csch} x)(\coth x)\, dx = -\operatorname{csch} x$

566. $\displaystyle\int (\sinh^2 x)\, dx = \dfrac{\sinh 2x}{4} - \dfrac{x}{2}$

567. $\displaystyle\int (\sinh^m x)(\cosh^n x)\, dx = \begin{cases} \dfrac{1}{m+n}(\sinh^{m+1} x)(\cosh^{n-1} x) \\[2mm] \qquad + \dfrac{n-1}{m+n}\displaystyle\int (\sinh^m x)(\cosh^{n-2} x)\, dx \\[2mm] \qquad\text{or} \\[2mm] \dfrac{1}{m+n}\sinh^{m-1} x \cosh^{n+1} x \\[2mm] \qquad - \dfrac{m-1}{m+n}\displaystyle\int (\sinh^{m-2} x)(\cosh^n x)\, dx, \quad (m+n \neq 0) \end{cases}$

568. $\displaystyle\int \dfrac{dx}{(\sinh^m x)(\cosh^n x)} = \begin{cases} -\dfrac{1}{(m-1)(\sinh^{m-1} x)(\cosh^{n-1} x)} \\[2mm] \qquad - \dfrac{m+n-2}{m-1}\displaystyle\int \dfrac{dx}{(\sinh^{m-2} x)(\cosh^n x)}, \quad (m \neq 1) \\[2mm] \qquad\text{or} \\[2mm] \dfrac{1}{(n-1)\sinh^{m-1} x \cosh^{n-1} x} \\[2mm] \qquad + \dfrac{m+n-2}{n-1}\displaystyle\int \dfrac{dx}{(\sinh^m x)(\cosh^{n-2} x)}, \quad (n \neq 1) \end{cases}$

569. $\displaystyle\int (\tanh^2 x)\, dx = x - \tanh x$

570. $\displaystyle\int (\tanh^n x)\, dx = -\dfrac{\tanh^{n-1} x}{n-1} + \int (\tanh^{n-2} x)\, dx, \quad (n \neq 1)$

571. $\displaystyle\int (\operatorname{sech}^2 x)\, dx = \tanh x$

572. $\displaystyle\int (\cosh^2 x)\, dx = \dfrac{\sinh 2x}{4} + \dfrac{x}{2}$

573. $\int (\coth^2 x)\,dx = x - \coth x$

574. $\int (\coth^n x)\,dx = -\dfrac{\coth^{n-1} x}{n-1} + \int \coth^{n-2} x\,dx, \quad (n \neq 1)$

575. $\int (\operatorname{csch}^2 x)\,dx = -\operatorname{ctnh} x$

576. $\int (\sinh mx)(\sinh nx)\,dx = \dfrac{\sinh (m+n)x}{2(m+n)} - \dfrac{\sinh (m-n)x}{2(m-n)}, \quad (m^2 \neq n^2)$

577. $\int (\cosh mx)(\cosh nx)\,dx = \dfrac{\sinh (m+n)x}{2(m+n)} + \dfrac{\sinh (m-n)x}{2(m-n)}, \quad (m^2 \neq n^2)$

578. $\int (\sinh mx)(\cosh nx)\,dx = \dfrac{\cosh (m+n)x}{2(m+n)} + \dfrac{\cosh (m-n)x}{2(m-n)}, \quad (m^2 \neq n^2)$

579. $\int \left(\sinh^{-1} \dfrac{x}{a}\right) dx = x \sinh^{-1} \dfrac{x}{a} - \sqrt{x^2 + a^2}, \quad (a > 0)$

580. $\int x\left(\sinh^{-1} \dfrac{x}{a}\right) dx = \left(\dfrac{x^2}{2} + \dfrac{a^2}{4}\right) \sinh^{-1} \dfrac{x}{a} - \dfrac{x}{4}\sqrt{x^2 + a^2}, \quad (a > 0)$

581. $\int x^n(\sinh^{-1} x)\,dx = \dfrac{x^{n+1}}{n+1} \sinh^{-1} x - \dfrac{1}{n+1} \int \dfrac{x^{n+1}}{(1+x^2)^{\frac{1}{2}}}\,dx, \quad (n \neq -1)$

582. $\int \left(\cosh^{-1} \dfrac{x}{a}\right) dx = \begin{cases} x \cosh^{-1} \dfrac{x}{a} - \sqrt{x^2 - a^2}, & \left(\cosh^{-1} \dfrac{x}{a} > 0\right) \\ \quad\text{or} \\ x \cosh^{-1} \dfrac{x}{a} + \sqrt{x^2 - a^2}, & \left(\cosh^{-1} \dfrac{x}{a} < 0\right), \quad (a > 0) \end{cases}$

583. $\int x\left(\cosh^{-1} \dfrac{x}{a}\right) dx = \dfrac{2x^2 - a^2}{4} \cosh^{-1} \dfrac{x}{a} - \dfrac{x}{4}(x^2 - a^2)^{\frac{1}{2}}$

584. $\int x^n(\cosh^{-1} x)\,dx = \dfrac{x^{n+1}}{n+1} \cosh^{-1} x - \dfrac{1}{n+1} \int \dfrac{x^{n+1}}{(x^2-1)^{\frac{1}{2}}}\,dx, \quad (n \neq -1)$

585. $\int \left(\tanh^{-1} \dfrac{x}{a}\right) dx = x \tanh^{-1} \dfrac{x}{a} + \dfrac{a}{2} \log (a^2 - x^2), \quad \left(\left|\dfrac{x}{a}\right| < 1\right)$

586. $\int \left(\coth^{-1} \dfrac{x}{a}\right) dx = x \coth^{-1} \dfrac{x}{a} + \dfrac{a}{2} \log (x^2 - a^2), \quad \left(\left|\dfrac{x}{a}\right| > 1\right)$

587. $\int x\left(\tanh^{-1} \dfrac{x}{a}\right) dx = \dfrac{x^2 - a^2}{2} \tanh^{-1} \dfrac{x}{a} + \dfrac{ax}{2}, \quad \left(\left|\dfrac{x}{a}\right| < 1\right)$

588. $\int x^n\left(\tanh^{-1} x\right) dx = \dfrac{x^{n+1}}{n+1} \tanh^{-1} x - \dfrac{1}{n+1} \int \dfrac{x^{n+1}}{1 - x^2}\,dx, \quad (n \neq -1)$

589. $\int x\left(\coth^{-1} \dfrac{x}{a}\right) dx = \dfrac{x^2 - a^2}{2} \coth^{-1} \dfrac{x}{a} + \dfrac{ax}{2}, \quad \left(\left|\dfrac{x}{a}\right| > 1\right)$

590. $\int x^n(\coth^{-1} x)\,dx = \dfrac{x^{n+1}}{n+1} \coth^{-1} x + \dfrac{1}{n+1} \int \dfrac{x^{n+1}}{x^2 - 1}\,dx, \quad (n \neq -1)$

591. $\int (\operatorname{sech}^{-1} x)\,dx = x \operatorname{sech}^{-1} x + \sin^{-1} x$

592. $\int x \operatorname{sech}^{-1} x\,dx = \dfrac{x^2}{2} \operatorname{sech}^{-1} x - \dfrac{1}{2}\sqrt{1 - x^2}$

593. $\int x^n \operatorname{sech}^{-1} x\,dx = \dfrac{x^{n+1}}{n+1} \operatorname{sech}^{-1} x + \dfrac{1}{n+1} \int \dfrac{x^n}{(1-x^2)^{\frac{1}{2}}}\,dx, \quad (n \neq -1)$

594. $\int \operatorname{csch}^{-1} x\,dx = x \operatorname{csch}^{-1} x + \dfrac{x}{|x|} \sinh^{-1} x$

595. $\int x \operatorname{csch}^{-1} x \, dx = \dfrac{x^2}{2} \operatorname{csch}^{-1} x + \dfrac{1}{2} \dfrac{x}{|x|} \sqrt{1 + x^2}$

596. $\int x^n \operatorname{csch}^{-1} x \, dx = \dfrac{x^{n+1}}{n+1} \operatorname{csch}^{-1} x + \dfrac{1}{n+1} \dfrac{x}{|x|} \int \dfrac{x^n}{(x^2 + 1)^{\frac{1}{2}}} \, dx, \qquad (n \neq -1)$

DEFINITE INTEGRALS

597. $\displaystyle\int_0^\infty x^{n-1} e^{-x} \, dx = \int_0^1 \left(\log \dfrac{1}{x} \right)^{n-1} dx = \dfrac{1}{n} \prod_{m=1}^\infty \dfrac{\left(1 + \dfrac{1}{m}\right)^n}{1 + \dfrac{n}{m}}$

$$= \Gamma(n), n \neq 0, -1, -2, -3, \ldots \quad \text{(Gamma Function)}$$

598. $\displaystyle\int_0^\infty t^n p^{-t} \, dt = \dfrac{n!}{(\log p)^{n+1}}, \qquad (n = 0, 1, 2, 3, \ldots \text{ and } p > 0)$

599. $\displaystyle\int_0^\infty t^{n-1} e^{-(a+1)t} \, dt = \dfrac{\Gamma(n)}{(a+1)^n}, \qquad (n > 0, a > -1)$

600. $\displaystyle\int_0^1 x^m \left(\log \dfrac{1}{x} \right)^n dx = \dfrac{\Gamma(n+1)}{(m+1)^{n+1}}, \qquad (m > -1, n > -1)$

601. $\Gamma(n)$ is finite if $n > 0$, $\Gamma(n+1) = n\Gamma(n)$

602. $\Gamma(n) \cdot \Gamma(1 - n) = \dfrac{\pi}{\sin n\pi}$

603. $\Gamma(n) = (n-1)!$ if $n = $ integer > 0

604. $\Gamma(\frac{1}{2}) = 2 \displaystyle\int_0^\infty e^{-t^2} \, dt = \sqrt{\pi} = 1.7724538509 \cdots = (-\frac{1}{2})!$

605. $\Gamma(n + \frac{1}{2}) = \dfrac{1 \cdot 3 \cdot 5 \cdots (2n - 1)}{2^n} \sqrt{\pi} \qquad n = 1, 2, 3, \ldots$

606. $\Gamma(-n + \frac{1}{2}) = \dfrac{(-1)^n 2^n \sqrt{\pi}}{1 \cdot 3 \cdot 5 \ldots (2n - 1)} \qquad n = 1, 2, 3, \ldots$

607. $\displaystyle\int_0^1 x^{m-1} (1 - x)^{n-1} \, dx = \int_0^\infty \dfrac{x^{m-1}}{(1 + x)^{m+n}} \, dx = \dfrac{\Gamma(m)\Gamma(n)}{\Gamma(m+n)} = B(m, n)$

(Beta function)

608. $B(m, n) = B(n, m) = \dfrac{\Gamma(m)\Gamma(n)}{\Gamma(m+n)}$, where m and n are any positive real numbers.

609. $\displaystyle\int_a^b (x - a)^m (b - x)^n \, dx = (b - a)^{m+n+1} \dfrac{\Gamma(m+1) \cdot \Gamma(n+1)}{\Gamma(m+n+2)}$,

$$(m > -1, n > -1, b > a)$$

610. $\displaystyle\int_1^\infty \dfrac{dx}{x^m} = \dfrac{1}{m-1}, \qquad [m > 1]$

611. $\displaystyle\int_0^\infty \dfrac{dx}{(1 + x)x^p} = \pi \csc p\pi, \qquad [p < 1]$

612. $\displaystyle\int_0^\infty \dfrac{dx}{(1 - x)x^p} = -\pi \cot p\pi, \qquad [p < 1]$

613. $\displaystyle\int_0^\infty \dfrac{x^{p-1} \, dx}{1 + x} = \dfrac{\pi}{\sin p\pi}$

$$= B(p, 1 - p) = \Gamma(p)\Gamma(1 - p), \qquad [0 < p < 1]$$

614. $\displaystyle\int_0^\infty \dfrac{x^{m-1} \, dx}{1 + x^n} = \dfrac{\pi}{n \sin \dfrac{m\pi}{n}}, \qquad [0 < m < n]$

615. $\displaystyle\int_0^\infty \frac{x^a\,dx}{(m + x^b)^c} = \frac{m^{\frac{a+1-bc}{b}}}{b}\left[\frac{\Gamma\left(\dfrac{a+1}{b}\right)\Gamma\left(c - \dfrac{a+1}{b}\right)}{\Gamma(c)}\right]$

$$\left(a > -1,\, b > 0,\, m > 0,\, c > \frac{a+1}{b}\right)$$

616. $\displaystyle\int_0^\infty \frac{dx}{(1 + x)\sqrt{x}} = \pi$

617. $\displaystyle\int_0^\infty \frac{a\,dx}{a^2 + x^2} = \frac{\pi}{2}$, if $a > 0$; 0, if $a = 0$; $-\dfrac{\pi}{2}$, if $a < 0$

618. $\displaystyle\int_0^a (a^2 - x^2)^{\frac{n}{2}}\,dx = \frac{1}{2}\int_{-a}^a (a^2 - x^2)^{\frac{n}{2}}\,dx = \frac{1 \cdot 3 \cdot 5 \ldots n}{2 \cdot 4 \cdot 6 \ldots (n + 1)} \cdot \frac{\pi}{2} \cdot a^{n+1}$ $\quad (n\text{ odd})$

619. $\displaystyle\int_0^a x^m(a^2 - x^2)^{\frac{n}{2}}\,dx = \begin{cases} \dfrac{1}{2}a^{m+n+1}B\left(\dfrac{m+1}{2}, \dfrac{n+2}{2}\right) \\ \qquad\text{or} \\ \dfrac{1}{2}a^{m+n+1}\dfrac{\Gamma\left(\dfrac{m+1}{2}\right)\Gamma\left(\dfrac{n+2}{2}\right)}{\Gamma\left(\dfrac{m+n+3}{2}\right)} \end{cases}$

620. $\displaystyle\int_0^{\pi/2}(\sin^n x)\,dx = \begin{cases} \displaystyle\int_0^{\pi/2}(\cos^n x)\,dx \\ \qquad\text{or} \\ \dfrac{1 \cdot 3 \cdot 5 \cdot 7 \ldots (n - 1)}{2 \cdot 4 \cdot 6 \cdot 8 \ldots (n)}\dfrac{\pi}{2}, \quad (n\text{ an even integer, } n \neq 0) \\ \qquad\text{or} \\ \dfrac{2 \cdot 4 \cdot 6 \cdot 8 \ldots (n - 1)}{1 \cdot 3 \cdot 5 \cdot 7 \ldots (n)}, \quad (n\text{ an odd integer, } n \neq 1) \\ \qquad\text{or} \\ \dfrac{\sqrt{\pi}}{2}\dfrac{\Gamma\left(\dfrac{n+1}{2}\right)}{\Gamma\left(\dfrac{n}{2} + 1\right)}, \quad (n > -1) \end{cases}$

621. $\displaystyle\int_0^\infty \frac{\sin mx\,dx}{x} = \frac{\pi}{2}$, if $m > 0$; 0, if $m = 0$; $-\dfrac{\pi}{2}$, if $m < 0$

622. $\displaystyle\int_0^\infty \frac{\cos x\,dx}{x} = \infty$

623. $\displaystyle\int_0^\infty \frac{\tan x\,dx}{x} = \frac{\pi}{2}$

624. $\displaystyle\int_0^\pi \sin ax \cdot \sin bx\,dx = \int_0^\pi \cos ax \cdot \cos bx\,dx = 0$, $\quad (a \neq b;\, a, b\text{ integers})$

625. $\displaystyle\int_0^{\pi/a}[\sin(ax)][\cos(ax)]\,dx = \int_0^\pi [\sin(ax)][\cos(ax)]\,dx = 0$

626. $\displaystyle\int_0^\pi [\sin(ax)][\cos(bx)]\,dx = \frac{2a}{a^2 - b^2}$, if $a - b$ is odd, or 0 if $a - b$ is even

627. $\displaystyle\int_0^\infty \frac{\sin x \cos mx\,dx}{x}$

$$= 0,\text{ if } m < -1\text{ or } m > 1;\, \frac{\pi}{4},\text{ if } m = \pm 1;\, \frac{\pi}{2},\text{ if } m^2 < 1$$

628. $\displaystyle\int_0^\infty \frac{\sin ax \sin bx}{x^2}\,dx = \frac{\pi a}{2}, \qquad (a \le b)$

629. $\displaystyle\int_0^\pi \sin^2 mx\,dx = \int_0^\pi \cos^2 mx\,dx = \frac{\pi}{2}$

630. $\displaystyle\int_0^\infty \frac{\sin^2 (px)}{x^2}\,dx = \frac{\pi p}{2}$

631. $\displaystyle\int_0^\infty \frac{\sin x}{x^p}\,dx = \frac{\pi}{2\Gamma(p)\sin(p\pi/2)}, \qquad 0 < p < 1$

632. $\displaystyle\int_0^\infty \frac{\cos x}{x^p}\,dx = \frac{\pi}{2\Gamma(p)\cos(p\pi/2)}, \qquad 0 < p < 1$

633. $\displaystyle\int_0^\infty \frac{1 - \cos px}{x^2}\,dx = \frac{\pi p}{2}$

634. $\displaystyle\int_0^\infty \frac{\sin px \cos qx}{x}\,dx = \left\{ 0, \quad q > p > 0; \quad \frac{\pi}{2}, \quad p > q > 0; \quad \frac{\pi}{4}, \quad p = q > 0 \right\}$

635. $\displaystyle\int_0^\infty \frac{\cos (mx)}{x^2 + a^2}\,dx = \frac{\pi}{2|a|}\, e^{-|ma|}$

636. $\displaystyle\int_0^\infty \cos (x^2)\,dx = \int_0^\infty \sin (x^2)\,dx = \frac{1}{2}\sqrt{\frac{\pi}{2}}$

637. $\displaystyle\int_0^\infty \sin ax^n\,dx = \frac{1}{na^{1/n}}\,\Gamma(1/n)\sin\frac{\pi}{2n}, \qquad n > 1$

638. $\displaystyle\int_0^\infty \cos ax^n\,dx = \frac{1}{na^{1/n}}\,\Gamma(1/n)\cos\frac{\pi}{2n}, \qquad n > 1$

639. $\displaystyle\int_0^\infty \frac{\sin x}{\sqrt{x}}\,dx = \int_0^\infty \frac{\cos x}{\sqrt{x}}\,dx = \sqrt{\frac{\pi}{2}}$

640. (a) $\displaystyle\int_0^\infty \frac{\sin^3 x}{x}\,dx = \frac{\pi}{4}$ (b) $\displaystyle\int_0^\infty \frac{\sin^3 x}{x^2}\,dx\, \frac{3}{4}\log 3$

641. $\displaystyle\int_0^\infty \frac{\sin^3 x}{x^3}\,dx = \frac{3\pi}{8}$

642. $\displaystyle\int_0^\infty \frac{\sin^4 x}{x^4}\,dx = \frac{\pi}{3}$

643. $\displaystyle\int_0^{\pi/2} \frac{dx}{1 + a\cos x} = \frac{\cos^{-1} a}{\sqrt{1 - a^2}}, \qquad (a < 1)$

644. $\displaystyle\int_0^\pi \frac{dx}{a + b\cos x} = \frac{\pi}{\sqrt{a^2 - b^2}}, \qquad (a > b \ge 0)$

645. $\displaystyle\int_0^{2\pi} \frac{dx}{1 + a\cos x} = \frac{2\pi}{\sqrt{1 - a^2}}, \qquad (a^2 < 1)$

646. $\displaystyle\int_0^\infty \frac{\cos ax - \cos bx}{x}\,dx = \log\frac{b}{a}$

647. $\displaystyle\int_0^{\pi/2} \frac{dx}{a^2 \sin^2 x + b^2 \cos^2 x} = \frac{\pi}{2ab}$

648. $\displaystyle\int_0^{\pi/2} \frac{dx}{(a^2 \sin^2 x + b^2 \cos^2 x)^2} = \frac{\pi(a^2 + b^2)}{4a^3 b^3}, \qquad (a, b > 0)$

649. $\displaystyle\int_0^{\pi/2} \sin^{n-1} x \cos^{m-1} x\,dx = \frac{1}{2}B\left(\frac{n}{2}, \frac{m}{2}\right), \qquad m \text{ and } n \text{ positive integers}$

650. $\int_0^{\pi/2} (\sin^{2n+1} \theta) \, d\theta = \dfrac{2 \cdot 4 \cdot 6 \ldots (2n)}{1 \cdot 3 \cdot 5 \ldots (2n+1)}, \qquad (n = 1, 2, 3 \ldots)$

651. $\int_0^{\pi/2} (\sin^{2n} \theta) \, d\theta = \dfrac{1 \cdot 3 \cdot 5 \ldots (2n-1)}{2 \cdot 4 \ldots (2n)} \left(\dfrac{\pi}{2} \right), \qquad (n = 1, 2, 3 \ldots)$

652. $\int_0^{\pi/2} \dfrac{x}{\sin x} \, dx = 2 \left\{ \dfrac{1}{1^2} - \dfrac{1}{3^2} + \dfrac{1}{5^2} - \dfrac{1}{7^2} + \cdots \right\}$

653. $\int_0^{\pi/2} \dfrac{dx}{1 + \tan^m x} = \dfrac{\pi}{4}$

654. $\int_0^{\pi/2} \sqrt{\cos \theta} \, d\theta = \dfrac{(2\pi)^{\frac{3}{2}}}{[\Gamma(\frac{1}{4})]^2}$

655. $\int_0^{\pi/2} (\tan^h \theta) \, d\theta = \dfrac{\pi}{2 \cos \left(\dfrac{h\pi}{2} \right)}, \qquad (0 < h < 1)$

656. $\int_0^{\infty} \dfrac{\tan^{-1}(ax) - \tan^{-1}(bx)}{x} \, dx = \dfrac{\pi}{2} \log \dfrac{a}{b}, \qquad (a, b > 0)$

657. The area enclosed by a curve defined through the equation $x^{\frac{b}{c}} + y^{\frac{b}{c}} = a^{\frac{b}{c}}$ where $a > 0$, c a positive odd integer and b a positive even integer is given by

$$\dfrac{\left[\Gamma\left(\dfrac{c}{b} \right) \right]^2}{\Gamma\left(\dfrac{2c}{b} \right)} \left(\dfrac{2ca^2}{b} \right)$$

658. $I = \displaystyle\iiint\limits_R x^{h-1} y^{m-1} z^{n-1} \, dv$, where R denotes the region of space bounded by

the co-ordinate planes and that portion of the surface $\left(\dfrac{x}{a} \right)^p + \left(\dfrac{y}{b} \right)^q + \left(\dfrac{z}{c} \right)^k = 1$,

which lies in the first octant and where $h, m, n, p, q, k, a, b, c$, denote positive real numbers is given by

$$\int_0^a x^{h-1} \, dx \int_0^{b\left[1 - \left(\frac{x}{a}\right)^p\right]^{\frac{1}{q}}} y^m \, dy \int_0^{c\left[1 - \left(\frac{x}{a}\right)^p - \left(\frac{y}{b}\right)^q\right]^{\frac{1}{k}}} z^{n-1} \, dz$$

$$= \dfrac{a^h b^m c^n}{pqk} \dfrac{\Gamma\left(\dfrac{h}{p} \right) \Gamma\left(\dfrac{m}{q} \right) \Gamma\left(\dfrac{n}{k} \right)}{\Gamma\left(\dfrac{h}{p} + \dfrac{m}{q} + \dfrac{n}{k} + 1 \right)}$$

659. $\int_0^{\infty} e^{-ax} \, dx = \dfrac{1}{a}, \qquad (a > 0)$

660. $\int_0^{\infty} \dfrac{e^{-ax} - e^{-bx}}{x} \, dx = \log \dfrac{b}{a}, \qquad (a, b > 0)$

661. $\int_0^{\infty} x^n e^{-ax} \, dx = \begin{cases} \dfrac{\Gamma(n+1)}{a^{n+1}}, & (n > -1, a > 0) \\ \qquad\text{or} \\ \dfrac{n!}{a^{n+1}}, & (a > 0, n \text{ positive integer}) \end{cases}$

662. $\int_0^{\infty} x^n \exp(-ax^p) \, dx = \dfrac{\Gamma(k)}{pa^k}, \qquad \left(n > -1, p > 0, a > 0, k = \dfrac{n+1}{p} \right)$

663. $\int_0^{\infty} e^{-a^2 x^2} \, dx = \dfrac{1}{2a} \sqrt{\pi} = \dfrac{1}{2a} \Gamma\left(\dfrac{1}{2} \right), \qquad (a > 0)$

664. $\int_0^{\infty} x e^{-x^2} \, dx = \dfrac{1}{2}$

665. $\displaystyle\int_0^\infty x^2 e^{-x^2}\,dx = \frac{\sqrt{\pi}}{4}$

666. $\displaystyle\int_0^\infty x^{2n} e^{-ax^2}\,dx = \frac{1\cdot 3\cdot 5\ldots(2n-1)}{2^{n+1}a^n}\sqrt{\frac{\pi}{a}}$

667. $\displaystyle\int_0^\infty x^{2n+1} e^{-ax^2}\,dx = \frac{n!}{2a^{n+1}},\qquad (a>0)$

668. $\displaystyle\int_0^1 x^m e^{-ax}\,dx = \frac{m!}{a^{m+1}}\left[1 - e^{-a}\sum_{r=0}^m \frac{a^r}{r!}\right]$

669. $\displaystyle\int_0^\infty e^{\left(-x^2+\frac{a^2}{x^2}\right)}\,dx = \frac{e^{-2a}\sqrt{\pi}}{2},\qquad (a>0)$

670. $\displaystyle\int_0^\infty e^{-nx}\sqrt{x}\,dx = \frac{1}{2n}\sqrt{\frac{\pi}{n}}$

671. $\displaystyle\int_0^\infty \frac{e^{-nx}}{\sqrt{x}}\,dx = \sqrt{\frac{\pi}{n}}$

672. $\displaystyle\int_0^\infty e^{-ax}(\cos mx)\,dx = \frac{a}{a^2+m^2},\qquad (a>0)$

673. $\displaystyle\int_0^\infty e^{-ax}(\sin mx)\,dx = \frac{m}{a^2+m^2},\qquad (a>0)$

674. $\displaystyle\int_0^\infty x\,e^{-ax}[\sin(bx)]\,dx = \frac{2ab}{(a^2+b^2)^2},\qquad (a>0)$

675. $\displaystyle\int_0^\infty x\,e^{-ax}[\cos(bx)]\,dx = \frac{a^2-b^2}{(a^2+b^2)^2},\qquad (a>0)$

676. $\displaystyle\int_0^\infty x^n e^{-ax}[\sin(bx)]\,dx = \frac{n![(a+ib)^{n+1}-(a-ib)^{n+1}]}{2i(a^2+b^2)^{n+1}},\qquad (i^2=-1, a>0)$

677. $\displaystyle\int_0^\infty x^n e^{-ax}[\cos(bx)]\,dx = \frac{n![(a-ib)^{n+1}+(a+ib)^{n+1}]}{2(a^2+b^2)^{n+1}},\qquad (i^2=-1, a>0)$

678. $\displaystyle\int_0^\infty \frac{e^{-ax}\sin x}{x}\,dx = \cot^{-1}a,\qquad (a>0)$

679. $\displaystyle\int_0^\infty e^{-a^2x^2}\cos bx\,dx = \frac{\sqrt{\pi}}{2a}\exp\left(-\frac{b^2}{4a^2}\right),\qquad (ab\neq 0)$

680. $\displaystyle\int_0^\infty e^{-t\cos\phi}\,t^{b-1}\sin(t\sin\phi)\,dt = [\Gamma(b)]\sin(b\phi),\qquad \left(b>0, -\frac{\pi}{2}<\phi<\frac{\pi}{2}\right)$

681. $\displaystyle\int_0^\infty e^{-t\cos\phi}\,t^{b-1}[\cos(t\sin\phi)]\,dt = [\Gamma(b)]\cos(b\phi),\qquad \left(b>0, -\frac{\pi}{2}<\phi<\frac{\pi}{2}\right)$

682. $\displaystyle\int_0^\infty t^{b-1}\cos t\,dt = [\Gamma(b)]\cos\left(\frac{b\pi}{2}\right),\qquad (0<b<1)$

683. $\displaystyle\int_0^\infty t^{b-1}(\sin t)\,dt = [\Gamma(b)]\sin\left(\frac{b\pi}{2}\right),\qquad (0<b<1)$

684. $\displaystyle\int_0^1 (\log x)^n\,dx = (-1)^n\cdot n!$

685. $\displaystyle\int_0^1 \left(\log\frac{1}{x}\right)^{\frac{1}{2}}\,dx = \frac{\sqrt{\pi}}{2}$

686. $\displaystyle\int_0^1 \left(\log\frac{1}{x}\right)^{-\frac{1}{2}}\,dx = \sqrt{\pi}$

687. $\int_0^1 \left(\log \frac{1}{x}\right)^n dx = n!$

688. $\int_0^1 x \log(1-x)\,dx = -\frac{3}{4}$

689. $\int_0^1 x \log(1+x)\,dx = \frac{1}{4}$

690. $\int_0^1 x^m (\log x)^n\,dx = \frac{(-1)^n n!}{(m+1)^{n+1}}, \qquad m > -1, n = 0, 1, 2, \ldots$

If $n \neq 0, 1, 2, \ldots$ replace $n!$ by $\Gamma(n+1)$.

691. $\int_0^1 \frac{\log x}{1+x}\,dx = -\frac{\pi^2}{12}$

692. $\int_0^1 \frac{\log x}{1-x}\,dx = -\frac{\pi^2}{6}$

693. $\int_0^1 \frac{\log(1+x)}{x}\,dx = \frac{\pi^2}{12}$

694. $\int_0^1 \frac{\log(1-x)}{x}\,dx = -\frac{\pi^2}{6}$

695. $\int_0^1 (\log x)[\log(1+x)]\,dx = 2 - 2\log 2 - \frac{\pi^2}{12}$

696. $\int_0^1 (\log x)[\log(1-x)]\,dx = 2 - \frac{\pi^2}{6}$

697. $\int_0^1 \frac{\log x}{1-x^2}\,dx = -\frac{\pi^2}{8}$

698. $\int_0^1 \log\left(\frac{1+x}{1-x}\right) \cdot \frac{dx}{x} = \frac{\pi^2}{4}$

699. $\int_0^1 \frac{\log x\,dx}{\sqrt{1-x^2}} = -\frac{\pi}{2}\log 2$

700. $\int_0^1 x^m \left[\log\left(\frac{1}{x}\right)\right]^n dx = \frac{\Gamma(n+1)}{(m+1)^{n+1}}, \qquad \text{if } m+1 > 0, n+1 > 0$

701. $\int_0^1 \frac{(x^p - x^q)\,dx}{\log x} = \log\left(\frac{p+1}{q+1}\right), \qquad (p+1 > 0, q+1 > 0)$

702. $\int_0^1 \frac{dx}{\sqrt{\log\left(\frac{1}{x}\right)}} = \sqrt{\pi}$, (same as integral 686)

703. $\int_0^\infty \log\left(\frac{e^x + 1}{e^x - 1}\right) dx = \frac{\pi^2}{4}$

704. $\int_0^{\pi/2} (\log \sin x)\,dx = \int_0^{\pi/2} \log \cos x\,dx = -\frac{\pi}{2}\log 2$

705. $\int_0^{\pi/2} (\log \sec x)\,dx = \int_0^{\pi/2} \log \csc x\,dx = \frac{\pi}{2}\log 2$

706. $\int_0^\pi x(\log \sin x)\,dx = -\frac{\pi^2}{2}\log 2$

707. $\int_0^{\pi/2} (\sin x)(\log \sin x)\,dx = \log 2 - 1$

708. $\displaystyle\int_0^{\pi/2} (\log \tan x)\,dx = 0$

709. $\displaystyle\int_0^{\pi} \log(a \pm b \cos x)\,dx = \pi \log\left(\frac{a + \sqrt{a^2 - b^2}}{2}\right), \qquad (a \geq b)$

710. $\displaystyle\int_0^{\pi} \log(a^2 - 2ab\cos x + b^2)\,dx = \begin{cases} 2\pi \log a, & a \geq b > 0 \\ 2\pi \log b, & b \geq a > 0 \end{cases}$

711. $\displaystyle\int_0^{\infty} \frac{\sin ax}{\sinh bx}\,dx = \frac{\pi}{2b} \tanh \frac{a\pi}{2b}$

712. $\displaystyle\int_0^{\infty} \frac{\cos ax}{\cosh bx}\,dx = \frac{\pi}{2b} \operatorname{sech} \frac{a\pi}{2b}$

713. $\displaystyle\int_0^{\infty} \frac{dx}{\cosh ax} = \frac{\pi}{2a}$

714. $\displaystyle\int_0^{\infty} \frac{x\,dx}{\sinh ax} = \frac{\pi^2}{4a^2}$

715. $\displaystyle\int_0^{\infty} e^{-ax}(\cosh bx)\,dx = \frac{a}{a^2 - b^2}, \qquad (0 \leq |b| < a)$

716. $\displaystyle\int_0^{\infty} e^{-ax}(\sinh bx)\,dx = \frac{b}{a^2 - b^2}, \qquad (0 \leq |b| < a)$

717. $\displaystyle\int_0^{\infty} \frac{\sinh ax}{e^{bx} + 1}\,dx = \frac{\pi}{2b} \csc \frac{a\pi}{b} - \frac{1}{2a}$

718. $\displaystyle\int_0^{\infty} \frac{\sinh ax}{e^{bx} - 1}\,dx = \frac{1}{2a} - \frac{\pi}{2b} \cot \frac{a\pi}{b}$

719. $\displaystyle\int_0^{\pi/2} \frac{dx}{\sqrt{1 - k^2 \sin^2 x}} = \frac{\pi}{2}\left[1 + \left(\frac{1}{2}\right)^2 k^2 + \left(\frac{1 \cdot 3}{2 \cdot 4}\right)^2 k^4\right.$

$$\left. + \left(\frac{1 \cdot 3 \cdot 5}{2 \cdot 4 \cdot 6}\right)^2 k^6 + \cdots\right], \text{ if } k^2 < 1$$

720. $\displaystyle\int_0^{\pi/2} \sqrt{1 - k^2 \sin^2 x}\,dx = \frac{\pi}{2}\left[1 - \left(\frac{1}{2}\right)^2 k^2 - \left(\frac{1 \cdot 3}{2 \cdot 4}\right)^2 \frac{k^4}{3}\right.$

$$\left. - \left(\frac{1 \cdot 3 \cdot 5}{2 \cdot 4 \cdot 6}\right)^2 \frac{k^6}{5} - \cdots\right], \text{ if } k^2 < 1$$

721. $\displaystyle\int_0^{\infty} e^{-x} \log x\,dx = -\gamma = -0.5772157\ldots$

722. $\displaystyle\int_0^{\infty} e^{-x^2} \log x\,dx = -\frac{\sqrt{\pi}}{4}(\gamma + 2\log 2)$

723. $\displaystyle\int_0^{\infty} \left(\frac{1}{1 - e^{-x}} - \frac{1}{x}\right) e^{-x}\,dx = \gamma = 0.5772157\ldots \qquad [\text{Euler's Constant}]$

724. $\displaystyle\int_0^{\infty} \frac{1}{x}\left(\frac{1}{1 + x} - e^{-x}\right) dx = \gamma = 0.5772157\ldots$

For n even:

725. $\displaystyle\int \cos^n x\,dx = \frac{1}{2^{n-1}} \sum_{k=0}^{\frac{n}{2}-1} \binom{n}{k} \frac{\sin(n - 2k)x}{(n - 2k)} + \frac{1}{2^n}\binom{n}{\frac{n}{2}} x$

726. $\int \sin^n x\, dx = \dfrac{1}{2^{n-1}} \displaystyle\sum_{k=0}^{\frac{n}{2}-1} \binom{n}{k} \dfrac{\sin\left[(n-2k)\left(\frac{\pi}{2}-x\right)\right]}{2k-n} + \dfrac{1}{2^n}\binom{n}{\frac{n}{2}} x$

For n odd:

727. $\int \cos^n x\, dx = \dfrac{1}{2^{n-1}} \displaystyle\sum_{k=0}^{\frac{n-1}{2}} \binom{n}{k} \dfrac{\sin(n-2k)x}{(n-2k)}$

728. $\int \sin^n x\, dx = \dfrac{1}{2^{n-1}} \displaystyle\sum_{k=0}^{\frac{n-1}{2}} \binom{n}{k} \dfrac{\sin\left[n-2k\right)\left(\frac{\pi}{2}-x\right)]}{2k-n}$

DIFFERENTIAL EQUATIONS

SPECIAL FORMULAS

Certain types of differential equations occur sufficiently often to justify the use of formulas for the corresponding particular solutions. The following set of tables I to XIV covers all first, second and nth order ordinary linear differential equations with constant coefficients for which the right members are of the form $P(x)e^{rx}\sin sx$ or $P(x)e^{rx}\cos sx$, where r and s are constants and $P(x)$ is a polynomial of degree n.

When the right member of a reducible linear partial differential equation with constant coefficients is not zero, particular solutions for certain types of right members are contained in tables XV to XXI. In these tables both F and P are used to denote polynomials, and it is assumed that no denominator is zero. In any formula the roles of x and y may be reversed throughout, changing a formula in which x dominates to one in which y dominates. Tables XIX, XX, XXI are applicable whether the equations are reducible or not.

The symbol $\binom{m}{n}$ stands for $\dfrac{m!}{(m-n)!n!}$ and is the $n+1$ st coefficient in the expansion of $(a+b)^m$. Also $0! = 1$ by definition.

The tables as herewith given are those contained in the text *Differential Equations* by Ginn and Company (1955) and are published with their kind permission and that of the author, Professor Frederick H. Steen.

Solution of Linear Differential Equations with Constant Coefficients

Any linear differential equation with constant coefficients may be written in the form

$$p(D)y = R(x),$$

where D is the differential operation

$$Dy = \frac{dy}{dx},$$

$p(D)$ is a polynomial in D,
y is the dependent variable,
x is the independent variable,
$R(x)$ is an arbitrary function of x.

A power of D represents repeated differentiation, that is

$$D^n y = \frac{d^n y}{dx^n}.$$

For such an equation, the general solution may be written in the form

$$y = y_c + y_p,$$

where y_p is any particular solution, and y_c is called the *complementary function*. This complementary function is defined as the general solution of the *homogeneous equation*, which is the original differential equation with the right side replaced by zero, i.e.

$$p(D)y = 0$$

The complementary function y_c may be determined as follows:

1. Factor the polynomial $p(D)$ into real and complex linear factors, just as if D were a variable instead of an operator.

DIFFERENTIAL EQUATIONS (Continued)

2. For each non-repeated linear factor of the form $(D - a)$, where a is real, write down a term of the form

$$ce^{ax},$$

where c is an arbitrary constant.

3. For each repeated real linear factor of the form $(D - a)^n$, write down n terms of the form

$$c_1 e^{ax} + c_2 x e^{ax} + c_3 x^2 e^{ax} + \cdots + c_n x^{n-1} e^{ax},$$

where the c_i's are arbitrary constants.

4. For each non-repeated conjugate complex pair of factors of the form $(D - a + ib)(D - a - ib)$, write down 2 terms of the form

$$c_1 e^{ax} \cos bx + c_2 e^{ax} \sin bx$$

5. For each repeated conjugate complex pair of factors of the form $(D - a + ib)^n (D - a - ib)^n$, write down $2n$ terms of the form

$$c_1 e^{ax} \cos bx + c_2 e^{ax} \sin bx + c_3 x e^{ax} \cos bx + c_4 x e^{ax} \sin bx + \cdots$$
$$+ c_{2n-1} x^{n-1} e^{ax} \cos bx + c_{2n} x^{n-1} e^{ax} \sin bx$$

6. The sum of all the terms thus written down is the complementary function y_c.

To find the particular solution y_p, use the following tables, as shown in the examples. For cases not shown in the tables, there are various methods of finding y_p. The most general method is called *variation of parameters*. The following example illustrates the method:

Find y_p for $(D^2 - 4)y = e^x$.

This example can be solved most easily by use of equation 63 in the tables following. However it is given here as an example of the method of variation of parameters.

The complementary function is

$$y_c = c_1 e^{2x} + c_2 e^{-2x}$$

To find y_p, replace the constants in the complementary function with unknown functions,

$$y_p = u e^{2x} + v e^{-2x}.$$

We now prepare to substitute this assumed solution into the original equation. We begin by taking all the necessary derivatives:

$$y_p = u e^{2x} + v e^{-2x}$$
$$y_p' = 2u e^{2x} - 2v e^{-2x} + u' e^{2x} + v' e^{-2x}$$

For each derivative of y_p except the highest, we set the sum of all the terms containing u' and v' to 0. Thus the above equation becomes

$$u' e^{2x} + v' e^{-2x} = 0 \quad \text{and} \quad y_p' = 2u e^{2x} - 2v e^{-2x}$$

Continuing to differentiate, we have

$$y_p'' = 4u e^{2x} + 4v e^{-2x} + 2u' e^{2x} - 2v' e^{-2x}$$

When we substitute into the original equation, all the terms not containing u' or v' cancel out. This is a consequence of the method by which y_p was set up.

Thus all that is necessary is to write down the terms containing u' or v' in the highest order derivative of y_p, multiply by the constant coefficient of the highest power of D in $p(D)$, and set it equal to $R(x)$. Together with the previous terms in u' and v' which were set equal to 0, this gives us as many linear equations in the first derivatives of the unknown functions as there are unknown functions. The first derivatives may then

DIFFERENTIAL EQUATIONS (Continued)

be solved for by algebra, and the unknown functions found by integration. In the present example, this becomes

$$u'e^{2x} + v'e^{-2x} = 0$$
$$2u'e^{2x} - 2v'e^{-2x} = e^x.$$

We eliminate v' and u' separately, getting

$$4u'e^{2x} = e^x$$
$$4v'e^{-2x} = -e^x.$$

Thus

$$u' = \tfrac{1}{4}e^{-x}$$
$$v' = -\tfrac{1}{4}e^{3x}.$$

Therefore, by integrating

$$u = -\tfrac{1}{4}e^{-x}$$
$$v = -\tfrac{1}{12}e^{3x}.$$

A constant of integration is not needed, since we need only one particular solution. Thus

$$y_p = ue^{2x} + ve^{-2x} = -\tfrac{1}{4}e^{-x}e^{2x} - \tfrac{1}{12}e^{3x}e^{-2x}$$
$$= -\tfrac{1}{4}e^x - \tfrac{1}{12}e^x = -\tfrac{1}{3}e^x,$$

and the general solution is

$$y = y_c + y_p = c_1e^{2x} + c_2e^{-2x} - \tfrac{1}{3}e^x.$$

The following examples illustrate the use of the tables.

Example 1. Solve $(D^2 - 4)y = \sin 3x$.

Substitution of $q = -4$, $s = 3$ in formula 24 gives

$$y_p = \frac{\sin 3x}{-9 - 4},$$

wherefore the general solution is

$$y = c_1e^{2x} + c_2e^{-2x} - \frac{\sin 3x}{13}.$$

Example 2. Obtain a particular solution of $(D^2 - 4D + 5)y = x^2e^{3x}\sin x$.

Applying formula 40 with $a = 2$, $b = 1$, $r = 3$, $s = 1$, $P(x) = x^2$, $s + b = 2$, $s - b = 0$, $a - r = -1$, $(a - r)^2 + (s + b)^2 = 5$, $(a - r)^2 + (s - b)^2 = 1$, we have

$$y_p = \frac{e^{3x}\sin x}{2}\left[\left(\frac{2}{5} - \frac{0}{1}\right)x^2 + \left(\frac{2(-1)2}{25} - \frac{2(-1)0}{1}\right)2x\right.$$

$$\left. + \left(\frac{3\cdot1\cdot2 - 2^3}{125} - \frac{3\cdot1\cdot0 - 0}{1}\right)2\right]$$

$$- \frac{e^{3x}\cos x}{2}\left[\left(\frac{-1}{5} - \frac{-1}{1}\right)x^2 + \left(\frac{1-4}{25} - \frac{1-0}{1}\right)2x\right.$$

$$\left. + \left(\frac{-1 - 3(-1)4}{125} - \frac{-1 - 3(-1)0}{1}\right)2\right]$$

$$= \left(\tfrac{1}{5}x^2 - \tfrac{4}{25}x - \tfrac{2}{125}\right)e^{3x}\sin x + \left(-\tfrac{2}{5}x^2 + \tfrac{28}{25}x - \tfrac{136}{125}\right)e^{3x}\cos x.$$

The special formulas effect a very considerable saving of time in problems of this type.

A-79

DIFFERENTIAL EQUATIONS (Continued)

Example 3.   Obtain a particular solution of $(D^2 - 4D + 5)y = x^2 e^{2x} \cos x$.   (Compare with Example 2.)

Formula 40 is not applicable here since for this equation $r = a$, $s = b$, wherefore the denominator $(a - r)^2 + (s - b)^2 = 0$.   We turn instead to formula 44.   Substituting $a = 2, b = 1, P(x) = x^2$ and replacing sin by cos, cos by $-$sin, we obtain

$$y_p = \frac{e^{2x} \cos x}{4}(x^2 - \tfrac{2}{4}) + \frac{e^{2x} \sin x}{2}\int (x^2 - \tfrac{1}{2})\,dx$$

$$= \left(\frac{x^2}{4} - \frac{1}{8}\right)e^{2x}\cos x + \left(\frac{x^3}{6} - \frac{x}{4}\right)e^{2x}\sin x,$$

which is the required solution.

Example 4.   Find z_p for $(D_x - 3D_y)z = \ln(y + 3x)$.

Referring to Table XV we note that formula 69 (not 68) is applicable. This gives

$$z_p = x \ln(y + 3x).$$

It is easily seen that $-\dfrac{y}{3}\ln(y + 3x)$ would serve equally well.

Example 5.   Solve $(D_x + 2D_y - 4)z = y \cos(y - 2x)$.

Since R in formula 76 contains a polynomial in x, not y, we rewrite the given equation in the form

$$(D_y + \tfrac{1}{2}D_x - 2)z = \tfrac{1}{2}y\cos(y - 2x).$$

Then

$$z_c = e^{2y}F(x - \tfrac{1}{2}y) = e^{2y}f(2x - y),$$

and by the formula

$$z_p = -\tfrac{1}{2}\cos(y - 2x)\cdot\left(\frac{y}{2} + \frac{\tfrac{1}{2}}{2}\right)$$

$$= -\tfrac{1}{8}(2y + 1)\cos(y - 2x).$$

Example 6.   Find z_p for $(D_x + 4D_y)^3 z = (2x - y)^2$.

Using formula 79, we obtain

$$z_p = \frac{\int\int\int u^2\,du^3}{[2 + 4(-1)]^3} = \frac{u^5}{5\cdot4\cdot3\cdot(-8)} = -\frac{(2x - y)^5}{480}.$$

Example 7.   Find z_p for $(D_x^3 + 5D_x^2 D_y - 7D_x + 4)z = e^{2x+3y}$.
By formula 87

$$z_p = \frac{e^{2x+3y}}{2^3 + 5\cdot2^2\cdot3 - 7\cdot2 + 4} = \frac{e^{2x+3y}}{58}.$$

Example 8.   Find z_p for

$$(D_x^4 + 6D_x^3 D_y + D_x D_y + D_y^2 + 9)z = \sin(3x + 4y).$$

Since every term in the left member is of *even* degree in the two operators D_x and D_y, formula 90 is applicable.
It gives

$$z_p = \frac{\sin(3x + 4y)}{(-9)^2 + 6(-9)(-12) + (-12) + (-16) + 9}$$

$$= \frac{\sin(3x + 4y)}{710}.$$

DIFFERENTIAL EQUATIONS

TABLE I: $(D - a)y = R$

R y_p

1. e^{rx} $\dfrac{e^{rx}}{r - a}$

2. $\sin sx^*$ $-\dfrac{a \sin sx + s \cos sx}{a^2 + s^2} = -\dfrac{1}{\sqrt{a^2 + s^2}} \sin\left(sx + \tan^{-1}\dfrac{s}{a}\right)$

3. $P(x)$ $-\dfrac{1}{a}\left[P(x) + \dfrac{P'(x)}{a} + \dfrac{P''(x)}{a^2} + \cdots + \dfrac{P^{(n)}(x)}{a^n}\right]$

4. $e^{rx}\sin sx^*$ Replace a by $a - r$ in formula 2 and multiply by e^{rx}.

5. $P(x)e^{rx}$ Replace a by $a - r$ in formula 3 and multiply by e^{rx}.

6. $P(x)\sin sx^*$ $-\sin sx\left[\dfrac{a}{a^2 + s^2}P(x) + \dfrac{a^2 - s^2}{(a^2 + s^2)^2}P'(x) + \dfrac{a^3 - 3as^2}{(a^2 + s^2)^3}P''(x) + \cdots\right.$

$$+ \left.\dfrac{a^k - \binom{k}{2}a^{k-2}s^2 + \binom{k}{4}a^{k-4}s^4 - \cdots}{(a^2 + s^2)^k}P^{(k-1)}(x) + \cdots\right]$$

$$-\cos sx\left[\dfrac{s}{a^2 + s^2}P(x) + \dfrac{2as}{(a^2 + s^2)^2}P'(x) + \dfrac{3a^2s - s^3}{(a^2 + s^2)^3}P''(x) + \cdots\right.$$

$$+ \left.\dfrac{\binom{k}{1}a^{k-1}s - \binom{k}{3}a^{k-3}s^3 + \cdots}{(a^2 + s^2)^k}P^{(k-1)}(x) + \cdots\right]$$

7. $P(x)e^{rx}\sin sx^*$ Replace a by $a - r$ in formula 6 and multiply by e^{rx}.

8. e^{ax} xe^{ax}

9. $e^{ax}\sin sx^*$ $-\dfrac{e^{ax}\cos sx}{s}$

10. $P(x)e^{ax}$ $e^{ax}\displaystyle\int P(x)\,dx$

11. $P(x)e^{ax}\sin sx^*$ $\dfrac{e^{ax}\sin sx}{s}\left[\dfrac{P'(x)}{s^3} - \dfrac{P'''(x)}{s^3} + \dfrac{P^v(x)}{s^5} - \cdots\right] - \dfrac{e^{ax}\cos sx}{s}\left[P(x) - \dfrac{P''(x)}{s^2} + \dfrac{P^{iv}(x)}{s^4} - \cdots\right]$

     * For $\cos sx$ in R replace "sin" by "cos" and "cos" by "$-$ sin" in y_p.

$$D^n = \dfrac{d^n}{dx^n} \qquad \binom{m}{n} = \dfrac{m!}{(m - n)!n!} \qquad 0! = 1$$

TABLE II: $(D - a)^2 y = R$

R y_p

12. e^{rx} $\dfrac{e^{rx}}{(r - a)^2}$

13. $\sin sx^*$ $\dfrac{1}{(a^2 + s^2)^2}[(a^2 - s^2)\sin sx + 2as\cos sx] = \dfrac{1}{a^2 + s^2}\sin\left(sx + \tan^{-1}\dfrac{2as}{a^2 - s^2}\right)$

14. $P(x)$ $\dfrac{1}{a^2}\left[P(x) + \dfrac{2P'(x)}{a} + \dfrac{3P''(x)}{a^2} + \cdots + \dfrac{(n + 1)P^{(n)}(x)}{a^n}\right]$

15. $e^{rx}\sin sx^*$ Replace a by $a - r$ in formula 13 and multiply by e^{rx}.

16. $P(x)e^{rx}$ Replace a by $a - r$ in formula 14 and multiply by e^{rx}.

17. $P(x)\sin sx^*$ $\sin sx\left[\dfrac{a^2 - s^2}{(a^2 + s^2)^2}P(x) + 2\dfrac{a^3 - 3as^2}{(a^2 + s^2)^3}P'(x) + 3\dfrac{a^4 - 6a^2s^2 + s^4}{(a^2 + s^2)^4}P''(x) + \cdots\right.$

$$+ \left.(k - 1)\dfrac{a^k - \binom{k}{2}a^{k-2}s^2 + \binom{k}{4}a^{k-4}s^4 - \cdots}{(a^2 + s^2)^k}P^{(k-2)}(x) + \cdots\right]$$

$$+ \cos sx\left[\dfrac{2as}{(a^2 + s^2)^2}P(x) + 2\dfrac{3a^2s - s^3}{(a^2 + s^2)^3}P'(x) + 3\dfrac{4a^3s - 4as^3}{(a^2 + s^2)^4}P''(x) + \cdots\right.$$

$$+ \left.(k - 1)\dfrac{\binom{k}{1}a^{k-1}s - \binom{k}{3}a^{k-3}s^3 + \cdots}{(a^2 + s^2)^k}P^{(k-2)}(x) + \cdots\right]$$

18. $P(x)e^{rx}\sin sx^*$ Replace a by $a - r$ in formula 17 and multiply by e^{rx}.

19. e^{ax} $\tfrac{1}{2}x^2 e^{ax}$

20. $e^{ax}\sin sx^*$ $-\dfrac{e^{ax}\sin sx}{s^2}$

21. $P(x)e^{ax}$ $e^{ax}\displaystyle\iint P(x)\,dx\,dx$

22. $P(x)e^{ax}\sin sx^*$ $-\dfrac{e^{ax}\sin sx}{s^2}\left[P(x) - \dfrac{3P''(x)}{s^2} + \dfrac{5P^{iv}(x)}{s^4} - \dfrac{7P^{vi}(x)}{s^6} + \cdots\right]$

$$-\dfrac{e^{ax}\cos sx}{s^2}\left[\dfrac{2P'(x)}{s} - \dfrac{4P'''(x)}{s^3} + \dfrac{6P^v(x)}{s^5} - \cdots\right]$$

     * For $\cos sx$ in R replace "sin" by "cos" and "cos" by "$-$ sin" in y_p.

TABLE III: $(D^2 + q)y = R$

R	y_p
23. e^{rx}	$\dfrac{e^{rx}}{r^2 + q}$
24. $\sin sx^*$	$\dfrac{\sin sx}{-s^2 + q}$
25. $P(x)$	$\dfrac{1}{q}\left[P(x) - \dfrac{P''(x)}{q} + \dfrac{P^{iv}(x)}{q^2} - \cdots + (-1)^k \dfrac{P^{(2k)}(x)}{q^k} \cdots\right]$
26. $e^{rx}\sin sx^*$	$\dfrac{(r^2 - s^2 + q)e^{rx}\sin sx - 2\,rse^{rx}\cos sx}{(r^2 - s^2 + q)^2 + (2rs)^2} = \dfrac{e^{rx}}{\sqrt{(r^2 - s^2 + q)^2 + (2rs)^2}}\sin\left[sx - \tan^{-1}\dfrac{2rs}{r^2 - s^2 + q}\right]$

27. $P(x)e^{rx}$

$$\dfrac{e^{rx}}{r^2 + q}\left[P(x) - \dfrac{2r}{r^2 + q}P'(x) + \dfrac{3r^2 - q}{(r^2 + q)^2}P''(x) - \dfrac{4r^3 - 4qr}{(r^2 + q)^3}P'''(x)\right.$$
$$\left. + \cdots + (-1)^{k-1}\dfrac{\binom{k}{1}r^{k-1} - \binom{k}{3}r^{k-3}q + \binom{k}{5}r^{k-5}q^2 - \cdots}{(r^2 + q)^{k-1}}P^{(k-1)}(x) + \cdots\right]$$

28. $P(x)\sin sx^*$

$$\dfrac{\sin sx}{(-s^2 + q)}\left[P(x) - \dfrac{3s^2 + q}{(-s^2 + q)^2}P''(x) + \dfrac{5s^4 + 10s^2q + q^2}{(-s^2 + q)^4}P^{iv}(x) + \cdots\right.$$
$$\left. + (-1)^k \dfrac{\binom{2k+1}{1}s^{2k} + \binom{2k+1}{3}s^{2k-2}q + \binom{2k+1}{5}s^{2k-4}q^2 + \cdots}{(-s^2 + q)^{2k}}P^{(2k)}(x) + \cdots\right]$$
$$- \dfrac{s\cos sx}{(-s^2 + q)}\left[\dfrac{2P'(x)}{(-s^2 + q)} - \dfrac{4s^2 + 4q}{(-s^2 + q)^3}P'''(x) + \cdots\right.$$
$$\left. + (-1)^{k+1}\dfrac{\binom{2k}{1}s^{2k-2} + \binom{2k}{3}s^{2k-4}q + \cdots}{(-s^2 + q)^{2k-1}}P^{(2k-1)}(x) + \cdots\right]$$

TABLE IV: $(D^2 + b^2)y = R$

R	y_p
29. $\sin bx^*$	$-\dfrac{x\cos bx}{2b}$
30. $P(x)\sin bx^*$	$\dfrac{\sin bx}{(2b)^2}\left[P(x) - \dfrac{P''(x)}{(2b)^2} + \dfrac{P^{iv}(x)}{(2b)^4} - \cdots\right] - \dfrac{\cos bx}{2b}\displaystyle\int\left[P(x) - \dfrac{P''(x)}{(2b)^2} + \cdots\right]dx$

* For $\cos sx$ in R replace "sin" by "cos" and "cos" by "$-$ sin" in y_p.

TABLE V: $(D^2 + pD + q)y = R$

R	y_p
31. e^{rx}	$\dfrac{e^{rx}}{r^2 + pr + q}$
32. $\sin sx^*$	$\dfrac{(q - s^2)\sin sx - ps\cos sx}{(q - s^2)^2 + (ps)^2} = \dfrac{1}{\sqrt{(q - s^2)^2 + (ps)^2}}\sin\left(sx - \tan^{-1}\dfrac{ps}{q - s^2}\right)$
33. $P(x)$	$\dfrac{1}{q}\left[P(x) - \dfrac{p}{q}P'(x) + \dfrac{p^2 - q}{q^2}P''(x) - \dfrac{p^3 - 2pq}{q^3}P'''(x) + \cdots\right.$

$$+ (-1)^n \dfrac{p^n - \binom{n-1}{1}p^{n-2}q + \binom{n-2}{2}p^{n-4}q^2 - \cdots}{q^n}P^{(n)}(x)\Big]$$

34. $e^{rx}\sin sx^*$ Replace p by $p + 2r$, q by $q + pr + r^2$ in formula 32 and multiply by e^{rx}.

35. $P(x)e^{rx}$ Replace p by $p + 2r$, q by $q + pr + r^2$ in formula 33 and multiply by e^{rx}.

TABLE VI: $(D - b)(D - a)y = R$

36. $P(x)\sin sx^*$

$$\dfrac{\sin sx}{b - a}\left[\left(\dfrac{a}{a^2 + s^2} - \dfrac{b}{b^2 + s^2}\right)P(x) + \left(\dfrac{a^2 - s^2}{(a^2 + s^2)^2} - \dfrac{b^2 - s^2}{(b^2 + s^2)^2}\right)P'(x)\right.$$
$$\left. + \left(\dfrac{a^3 - 3as^2}{(a^2 + s^2)^3} - \dfrac{b^3 - 3bs^2}{(b^2 + s^2)^3}\right)P''(x) + \cdots\right]$$
$$+ \dfrac{\cos sx}{b - a}\left[\left(\dfrac{s}{a^2 + s^2} - \dfrac{s}{b^2 + s^2}\right)P(x) + \left(\dfrac{2as}{(a^2 + s^2)^2} - \dfrac{2bs}{(b^2 + s^2)^2}\right)P'(x)\right.$$
$$\left. + \left(\dfrac{3a^2s - s^3}{(a^2 + s^2)^3} - \dfrac{3b^2s - s^3}{(b^2 + s^2)^3}\right)P''(x) + \cdots\right]^\dagger$$

37. $P(x)e^{rx}\sin sx^*$ Replace a by $a - r$, b by $b - r$ in formula 36 and multiply by e^{rx}.

38. $P(x)e^{ax}$

$$\dfrac{e^{ax}}{a - b}\left[\int P(x)dx + \dfrac{P(x)}{(b - a)} + \dfrac{P'(x)}{(b - a)^2} + \dfrac{P''(x)}{(b - a)^3} + \cdots + \dfrac{P^{(n)}(x)}{(b - a)^{n+1}}\right]$$

* For $\cos sx$ in R replace "sin" by "cos" and "cos" by "$-$ sin" in y_p.
† For additional terms, compare with formula 6.

TABLE VII: $(D^2 - 2aD + a^2 + b^2)y = R$

R	y_p

39. $P(x) \sin sx$*

$$\frac{\sin sx}{2b} \left[\left(\frac{s+b}{a^2 + (s+b)^2} - \frac{s-b}{a^2 + (s-b)^2} \right) P(x) + \left(\frac{2a(s+b)}{[a^2 + (s+b)^2]^2} - \frac{2a(s-b)}{[a^2 + (s-b)^2]^2} \right) P'(x) \right.$$
$$\left. + \left(\frac{3a^2(s+b) - (s+b)^3}{[a^2 + (s+b)^2]^3} - \frac{3a^2(s-b) - (s-b)^3}{[a^2 + (s-b)^2]^3} \right) P''(x) + \cdots \right]$$
$$- \frac{\cos sx}{2b} \left[\left(\frac{a}{a^2 + (s+b)^2} - \frac{a}{a^2 + (s-b)^2} \right) P(x) + \left(\frac{a^2 - (s+b)^2}{[a^2 + (s+b)^2]^2} - \frac{a^2 - (s-b)^2}{[a^2 + (s-b)^2]^2} \right) P'(x) \right.$$
$$\left. + \left(\frac{a^3 - 3a(s+b)^2}{[a^2 + (s+b)^2]^3} - \frac{a^3 - 3a(s-b)^2}{[a^2 + (s-b)^2]^3} \right) P''(x) + \cdots \right]^\dagger$$

40. $P(x)e^{rx} \sin sx$*   Replace a by $a - r$ in formula 39 and multiply by e^{rx}.

41. $P(x)e^{ax}$ $\dfrac{e^{ax}}{b^2} \left[P(x) - \dfrac{P''(x)}{b^2} + \dfrac{P^{iv}(x)}{b^4} - \cdots \right]$

42. $e^{ax} \sin sx$* $\dfrac{e^{ax} \sin sx}{-s^2 + b^2}$

43. $e^{ax} \sin bx$* $-\dfrac{xe^{ax} \cos bx}{2b}$

44. $P(x)e^{ax} \sin bx$* $\dfrac{e^{ax} \sin bx}{(2b)^2} \left[P(x) - \dfrac{P''(x)}{(2b)^2} + \dfrac{P^{iv}(x)}{(2b)^4} - \cdots \right] - \dfrac{e^{ax} \cos bx}{2b} \int \left[P(x) - \dfrac{P''(x)}{(2b)^2} + \dfrac{P^{iv}(x)}{(2b)^4} - \cdots \right] dx$

* For cos sx in R replace "sin" by "cos" and "cos" by "$-$ sin" in y_p.
$\dagger$ For additional terms, compare with formula 6.

TABLE VIII: $f(D)y = [D^n + a_{n-1}D^{n-1} + \cdots + a_1D + a_0]y = R$

R	y_p

45. e^{rx} $\dfrac{e^{rx}}{f(r)}$

46. $\sin sx$* $\dfrac{[a_0 - a_2 s^2 + a_4 s^4 - \cdots] \sin sx - [a_1 s - a_3 s^3 + a_5 s^5 + \cdots] \cos sx}{[a_0 - a_2 s^2 + a_4 s^4 - \cdots]^2 + [a_1 s - a_3 s^3 + a_5 s^5 - \cdots]^2}$

TABLE IX: $f(D^2)y = R$

47. $\sin sx$* $\dfrac{\sin sx}{f(-s^2)} = \dfrac{\sin sx}{a_0 - a_2 s^2 + \cdots \pm s^{2n}}$

TABLE X: $(D - a)^n y = R$

48. e^{rx} $\dfrac{e^{rx}}{(r - a)^n}$

49. $\sin sx$* $\dfrac{(-1)^n}{(a^2 + s^2)^n} \left\{ \left[a^n - \binom{n}{2} a^{n-2} s^2 + \binom{n}{4} a^{n-4} s^4 - \cdots \right] \sin sx + \left[\binom{n}{1} a^{n-1} s - \binom{n}{3} a^{n-3} s^3 + \cdots \right] \cos sx \right\}$

50. $P(x)$ $\dfrac{(-1)^n}{a^n} \left[P(x) + \binom{n}{1} \dfrac{P'(x)}{a} + \binom{n+1}{2} \dfrac{P''(x)}{a^2} + \binom{n+2}{3} \dfrac{P'''(x)}{a^3} + \cdots \right]$

51. $e^{rx} \sin sx$*   Replace a by $a - r$ in formula 49 and multiply by e^{rx}.

52. $e^{rx}P(x)$ Replace a by $a - r$ in formula 50 and multiply by e^{rx}.

53. $P(x) \sin sx$* $(-1)^n \sin sx [A_n P(x) + \binom{n}{1} A_{n+1} P'(x) + \binom{n+1}{2} A_{n+2} P''(x) + \binom{n+2}{3} A_{n+3} P'''(x) + \cdots]$
$$+ (-1)^n \cos sx [B_n P(x) + \binom{n}{1} B_{n+1} P'(x) + \binom{n+1}{2} B_{n+2} P''(x) + \binom{n+2}{3} B_{n+3} P'''(x) + \cdots]$$
$$A_1 = \frac{a}{a^2 + s^2}, \ A_2 = \frac{a^2 - s^2}{(a^2 + s^2)^2}, \ \cdots, \ A_k = \frac{a^k - \binom{k}{2} a^{k-2} s^2 + \binom{k}{4} a^{k-4} s^4 - \cdots}{(a^2 + s^2)^k}.$$
$$B_1 = \frac{s}{a^2 + s^2}, \ B_2 = \frac{2as}{(a^2 + s^2)^2}, \ \cdots, \ B_k = \frac{\binom{k}{1} a^{k-1} s - \binom{k}{3} a^{k-3} s^3 + \cdots}{(a^2 + s^2)^k}.$$

54. $P(x)e^{rx} \sin sx$*   Replace a by $a - r$ in formula 53 and multiply by e^{rx}.

55. $e^{ax}P(x)$ $\qquad$ $e^{ax}\displaystyle\int\int\cdots\int P(x)dx^n$

56. $P(x)e^{ax}\sin sx$* $\qquad$ $\dfrac{(-1)^{\frac{n-1}{2}}e^{ax}\sin sx}{s^n}\left[\binom{n}{n-1}\dfrac{P'(x)}{s}-\binom{n+2}{n-1}\dfrac{P'''(x)}{s^3}+\binom{n+4}{n-1}\dfrac{P^v(x)}{s^5}-\cdots\right]$

$$+\dfrac{(-1)^{\frac{n+1}{2}}e^{ax}\cos sx}{s^n}\left[\binom{n-1}{n-1}P(x)-\binom{n+1}{n-1}\dfrac{P''(x)}{s^2}+\binom{n+3}{n-1}\dfrac{P^{iv}(x)}{s^4}-\cdots\right]\quad(n\text{ odd})$$

$$\dfrac{(-1)^{\frac{n}{2}}e^{ax}\sin sx}{s^n}\left[\binom{n-1}{n-1}P(x)-\binom{n+1}{n-1}\dfrac{P''(x)}{s^2}+\binom{n+3}{n-1}\dfrac{P^{iv}(x)}{s^4}-\cdots\right]$$

$$+\dfrac{(-1)^{\frac{n}{2}}e^{ax}\cos sx}{s^n}\left[\binom{n}{n-1}\dfrac{P'(x)}{s}-\binom{n+2}{n-1}\dfrac{P'''(x)}{s^3}+\binom{n+4}{n-1}\dfrac{P^v(x)}{s^5}-\cdots\right]\quad(n\text{ even})$$

* For cos sx in R replace "sin" by "cos" and "cos" by "$-$ sin" in y_p.

TABLE XI: $(D-a)^nf(D)y=R$

57. e^{ax} $\qquad$ $\dfrac{x^n}{n!}\cdot\dfrac{e^{ax}}{f(a)}$

* For cos sx in R replace "sin" by "cos" and "cos" by "$-$ sin" in y_p.

TABLE XII: $(D^2+q)^ny=R$

R	y_p
58. e^{rx}	$e^{rx}/(r^2+q)^n$
59. $\sin sx$*	$\sin sx/(q-s^2)^n$
60. $P(x)$	$\dfrac{1}{q^n}\left[P(x)-\binom{n}{1}\dfrac{P''(x)}{q}+\binom{n+1}{2}\dfrac{P^{iv}(x)}{q^2}-\binom{n+2}{3}\dfrac{P^{vi}(x)}{q^3}+\cdots\right]$
61. $e^{rx}\sin sx$*	$\dfrac{e^{rx}}{(A^2+B^2)^n}\{[A^n-\binom{n}{2}A^{n-2}B^2+\binom{n}{4}A^{n-4}B^4-\cdots]\sin sx-[\binom{n}{1}A^{n-1}B-\binom{n}{3}A^{n-3}B^3+\cdots]\cos sx\}$

$$A=r^2-s^2+q,\qquad B=2rs$$

TABLE XIII: $(D^2+b^2)^ny=R$

62. $\sin bx$* $\qquad$ $(-1)^{\frac{n+1}{2}}\dfrac{x^n\cos bx}{n!(2b)^n}$ $(n\text{ odd})$, $\qquad\qquad (-1)^{\frac{n}{2}}\dfrac{x^n\sin bx}{n!(2b)^n}$ $(n\text{ even})$

TABLE XIV: $(D^n-q)y=R$

63. e^{rx} $\qquad$ $e^{rx}/(r^n-q)$

64. $P(x)$ $\qquad$ $-\dfrac{1}{q}\left[P(x)\dfrac{P^{(n)}(x)}{q}+\dfrac{P^{(2n)}(x)}{q^2}+\cdots\right]$

65. $\sin sx$* $\qquad$ $-\dfrac{q\sin sx+(-1)^{\frac{n-1}{2}}s^n\cos sx}{q^2+s^{2n}}$ $(n\text{ odd})$, $\qquad \dfrac{\sin sx}{(-s^2)^{n/2}-q}$ $(n\text{ even})$

66. $e^{rx}\sin sx$* $\qquad$ $\dfrac{Ae^{rx}\sin sr-Be^{rx}\cos sx}{A^2+B^2}=\dfrac{e^{rx}}{\sqrt{A^2+B^2}}\sin\left(sx-\tan^{-1}\dfrac{B}{A}\right)$

$$A=[r^n-\binom{n}{2}r^{n-2}s^2+\binom{n}{4}r^{n-4}s^4-\cdots]-q,\qquad B=[\binom{n}{1}r^{n-1}s-\binom{n}{3}r^{n-3}s^3+\cdots]$$

* For cos sx in R replace "sin" by "cos" and "cos" by "$-$ sin" in y_p.

TABLE XV: $(D_x+mD_y)z=R$

R	z_p
67 e^{ax+by}	$\dfrac{e^{ax+by}}{a+mb}$
68. $f(ax+by)$	$\dfrac{\int f(u)du}{a+mb}$, $u=ax+by$
69. $f(y-mx)$	$xf(y-mx)$
70. $\phi(x,y)f(y-mx)$	$f(y-mx)\int\phi(x,a+mx)dx$ $(a=y-mx$ after integration)

TABLE XVI: $(D_x + mD_y - k)z = R$

71. e^{ax+by} $\dfrac{e^{ax+by}}{a + mb - k}$

72. $\sin(ax + by)$* $-\dfrac{(a + bm)\cos(ax + by) + k\sin(ax + by)}{(a + bm)^2 + k^2}$

73. $e^{\alpha x + \beta y}\sin(ax + by)$*   Replace k in 72 by $k - \alpha - m\beta$ and multiply by $e^{\alpha x + \beta y}$

74. $e^{zx}f(ax + by)$ $\dfrac{e^{kx}\int f(u)\,du}{a + mb}$, $u = ax + by$

75. $f(y - mx)$ $-\dfrac{f(y - mx)}{k}$

76. $P(x)f(y - mx)$ $-\dfrac{1}{k}f(y - mx)\left[P(x) + \dfrac{P'(x)}{k} + \dfrac{P''(x)}{k^2} + \cdots + \dfrac{P^{(n)}(x)}{k^n}\right]$

77. $e^{kx}f(y - mx)$ $xe^{kx}f(y - mx)$

* For $\cos(ax + by)$ replace "sin" by "cos," and "cos" by "$-$ sin" in z_p.

$$D_z = \frac{\partial}{\partial x}; \qquad D_y = \frac{\partial}{\partial y}; \qquad D_x{}^k D_y{}^r = \frac{\partial^{k+r}}{\partial x^k \partial y^r}$$

TABLE XVII: $(D_x + mD_y)^n z = R$

R	z_p
78. e^{ax+by}	$\dfrac{e^{ax+by}}{(a + mb)^n}$
79. $f(ax + by)$	$\dfrac{\int\int \cdots \int\int f(u)\,du^n}{(a + mb)^n}$, $u = ax + by$
80. $f(y - mx)$	$\dfrac{x^n}{n!}f(y - mx)$
81. $\phi(x, y)f(y + mx)$	$f(y - mx)\int\int \cdots \int \phi(x, a + mx)\,dx^n$ ($a = y - mx$ after integration)

TABLE XVIII: $(D_x + mD_y - k)^n z = R$

82. e^{ax+by} $\dfrac{e^{ax+by}}{(a + mb - k)^n}$

83. $f(y - mx)$ $\dfrac{(-1)^n f(y - mx)}{k^n}$

84. $P(x)f(y - mx)$ $\dfrac{(-1)^n}{k^n}f(y - mx)\left[P(x) + \binom{n}{1}\dfrac{P'(x)}{k} + \binom{n+1}{2}\dfrac{P''(x)}{k^2} + \binom{n+2}{3}\dfrac{P'''(x)}{k^3} + \cdots\right]$

85. $e^{kx}f(ax + by)$ $\dfrac{e^{kx}\int\int \cdots \int\int f(u)\,du^n}{(a + mb)^n}$, $u = ax + by$

86. $e^{kx}f(y - mx)$ $\dfrac{x^n}{n!}e^{kx}f(y - mx)$

TABLE XIX: $[D_x{}^n + a_1 D_x{}^{n-1}D_y + a_2 D_x{}^{n-2}D_y{}^2 + \cdots + a^n D_y{}^n]z = R$

87. e^{ax+by} $\dfrac{e^{ax+by}}{a + a_1 a^{n-1}b + a_2 a^{n-2}b^2 + \cdots + a_n b^n}$

88. $f(ax + by)$ $\dfrac{\int\int \cdots \int f(u)\,du^n}{a^n + a_1 a^{n-1}b + a_2 a^{n-2}b^2 + \cdots + a^n b^n}$, $(u = ax + by)$

TABLE XX: $F(D_x, D_y)z = R$

89. e^{ax+by} $\dfrac{e^{ax+by}}{F(a, b)}$

TABLE XXI: $F(D_x{}^2, D_x D_y, D_y{}^2)z = R$

90. $\sin(ax + by)$* $\dfrac{\sin(ax + by)}{F(-a^2, -ab, -b^2)}$

* For $\cos(ax + by)$ replace "sin" by "cos", and "cos" by "$-$sin" in z_p.

DIFFERENTIAL EQUATIONS

Differential equation	Method of solution
Separation of variables $f_1(x) g_1(y) dx + f_2(x) g_2(y) dy = 0$	$$\int \frac{f_1(x)}{f_2(x)} dx + \int \frac{g_2(y)}{g_1(y)} dy = c$$
Exact equation $M(x,y)dx + N(x,y)dy = 0$ where $\partial M/\partial y = \partial N/\partial x$	$$\int M\,\partial x + \int \left(N - \frac{\partial}{\partial y} \int M\,\partial x \right) dy = c$$ where ∂x indicates that the integration is to be performed with respect to x keeping y constant.
Linear first order equation $\frac{dy}{dx} + P(x)y = Q(x)$	$$ye^{\int Pdx} = \int Qe^{\int Pdx} dx + c$$
Bernoulli's equation $\frac{dy}{dx} + P(x)y = Q(x)y^n$	$$ve^{(1-n)\int Pdx} = (1-n) \int Qe^{(1-n)\int Pdx} dx + c$$ where $v = y^{1-n}$. If $n = 1$, the solution is $$\ln y = \int (Q - P) \, dx + c$$
Homogeneous equation $\frac{dy}{dx} = F\left(\frac{y}{x}\right)$	$$\ln x = \int \frac{dv}{F(v) - v} + c$$ where $v = y/x$. If $F(v) = v$, the solution is $y = cx$
Reducible to homogeneous $(a_1 x + b_1 y + c_1) dx + (a_2 x + b_2 y + c_2) dy = 0$ $\frac{a_1}{a_2} \neq \frac{b_1}{b_2}$	Set $u = a_1 x + b_1 y + c_1$, $v = a_2 x + b_2 y + c_2$ Eliminate x and y and the equation becomes homogenous
Reducible to separable $(a_1 x + b_1 y + c_1) dx + (a_2 x + b_2 y + c_2) dy = 0$ $\frac{a_1}{a_2} = \frac{b_1}{b_2}$	Set $u = a_1 x + b_1 y$ Eliminate x or y and equation becomes separable

DIFFERENTIAL EQUATIONS (Continued)

$y\,F(xy)\,dx + x\,G(xy)\,dy = 0$	$\ln x = \displaystyle\int \frac{G(v)\,dv}{v\,\{G(v) - F(v)\}} + c$ where $v = xy$. If $G(v) = F(v)$, the solution is $xy = c$.
Linear, homogeneous second order equation $\dfrac{d^2 y}{dx^2} + b\,\dfrac{dy}{dx} + cy = 0$ b,c are real constants	Let m_1, m_2 be the roots of $m^2 + bm + c = 0$. Then there are 3 cases: Case 1. m_1, m_2 real and distinct: $\qquad y = c_1 e^{m_1 x} + c_2 e^{m_2 x}$ Case 2. m_1, m_2 real and equal: $\qquad y = c_1 e^{m_1 x} + c_2 x e^{m_1 x}$ Case 3. $m_1 = p + qi,\ m_2 = p - qi$: $\qquad y = e^{px}(c_1 \cos qx + c_2 \sin qx)$ where $p = -b/2,\ q = \sqrt{4c - b^2}/2$
Linear, nonhomogeneous second order equation $\dfrac{d^2 y}{dx^2} + b\,\dfrac{dy}{dx} + cy = R(x)$ b, c are real constants	There are 3 cases corresponding to those immediately above: Case 1. $y = c_1 e^{m_1 x} + c_2 e^{m_2 x}$ $\quad + \dfrac{e^{m_1 x}}{m_1 - m_2}\displaystyle\int e^{-m_1 x} R(x)\,dx$ $\quad + \dfrac{e^{m_2 x}}{m_2 - m_1}\displaystyle\int e^{-m_2 x} R(x)\,dx$ Case 2. $y = c_1 e^{m_1 x} + c_2 x e^{m_1 x}$ $\quad + x e^{m_1 x}\displaystyle\int e^{-m_1 x} R(x)\,dx$ $\quad - e^{m_1 x}\displaystyle\int x e^{-m_1 x} R(x)\,dx$ Case 3. $y = e^{px}(c_1 \cos qx + c_2 \sin qx)$ $\quad + \dfrac{e^{px}\sin qx}{q}\displaystyle\int e^{-px} R(x)\cos qx\,dx$ $\quad - \dfrac{e^{px}\cos qx}{q}\displaystyle\int e^{-px} R(x)\sin qx\,dx$

Euler or Cauchy equation	Putting $x = e^t$, the equation becomes
$x^2 \dfrac{d^2 y}{dx^2} + bx \dfrac{dy}{dx} + cy = S(x)$	$\dfrac{d^2 y}{dt^2} + (b-1) \dfrac{dy}{dt} + cy = S(e^t)$ and can then be solved as a linear second order equation.
Bessel's equation $x^2 \dfrac{d^2 y}{dx^2} + x \dfrac{dy}{dx} + (\lambda^2 x^2 - n^2)y = 0$	$y = c_1 J_n(\lambda x) + c_2 Y_n(\lambda x)$
Transformed Bessel's equation $x^2 \dfrac{d^2 y}{dx^2} + (2p+1)x \dfrac{dy}{dx} + (\alpha^2 x^{2r} + \beta^2)y = 0$	$y = x^{-p} \left\{ c_1 J_{q/r}\left(\dfrac{\alpha}{r} x^r\right) + c_2 Y_{q/r}\left(\dfrac{\alpha}{r} x^r\right) \right\}$ where $q = \sqrt{p^2 - \beta^2}$.
Legendre's equation $(1 - x^2)\dfrac{d^2 y}{dx^2} - 2x \dfrac{dy}{dx} + n(n+1)y = 0$	$y = c_1 P_n(x) + c_2 Q_n(x)$

FOURIER SERIES

1. If $f(x)$ is a bounded periodic function of period 2L (i.e. $f x + 2L) = f(x)$, and satisfies the *Dirichlet conditions:*
   A. In an period $f(x)$ is continuous, except possibly for a finite number of jump discontinuities.
   B. In any period $f(x)$ has only a finite number of maxima and minima.

Then $f(x)$ may be represented by the *Fourier series*

$$\frac{a_0}{2} + \sum_{n=1}^{\infty} \left(a_n \cos \frac{n\pi x}{L} + b_n \sin \frac{n\pi x}{L} \right),$$

where a_n and b_n are as determined below. This series will converge to $f(x)$ at every point where $f(x)$ is continuous, and to

$$\frac{f(x^+) + f(x^-)}{2}$$

(i.e. the average of the left-hand and right-hand limits) at every point where $f(x)$ has a jump discontinuity.

$$a_n = \frac{1}{L} \int_{-L}^{L} f(x) \cos \frac{n\pi x}{L} \, dx, \quad n = 0, 1, 2, 3, \ldots;$$

$$b_n = \frac{1}{L} \int_{-L}^{L} f(x) \sin \frac{n\pi x}{L} \, dx, \quad n = 1, 2, 3, \ldots$$

We may also write

$$a_n = \frac{1}{L} \int_{\alpha}^{\alpha + 2L} f(x) \cos \frac{n\pi x}{L} \, dx \text{ and } b_n = \frac{1}{L} \int_{\alpha}^{\alpha + 2L} f(x) \sin \frac{n\pi x}{L} \, dx,$$

where α is any real number. Thus if $\alpha = 0$,

$$a_n = \frac{1}{L} \int_0^{2L} f(x) \cos \frac{n\pi x}{L} \, dx, \quad n = 0, 1, 2, 3, \ldots;$$

$$b_n = \frac{1}{L} \int_0^{2L} f(x) \sin \frac{n\pi x}{L} \, dx, \quad n = 1, 2, 3, \ldots$$

2. If in addition to the above restrictions, $f(x)$ is even (i.e. $f(-x) = f(x)$), the Fourier series reduces to

$$\frac{a_0}{2} + \sum_{n=1}^{\infty} a_n \cos \frac{n\pi x}{L}.$$

That is, $b_n = 0$. In this case, a simpler formula for a_n is

$$a_n = \frac{2}{L} \int_0^L f(x) \cos \frac{n\pi x}{L} \, dx, \quad n = 0, 1, 2, 3, \ldots$$

3. If in addition to the restrictions in (1), $f(x)$ is an odd function (i.e. $f(-x) = -f(x)$), then the Fourier series reduces to

$$\sum_{n=1}^{\infty} b_n \sin \frac{n\pi x}{L}.$$

That is, $a_n = 0$. In this case, a simpler formula for the b_n is

$$b_n = \frac{2}{L} \int_0^L f(x) \sin \frac{n\pi x}{L} \, dx, \quad n = 1, 2, 3, \ldots$$

4. If in addition to the restrictions in (2) above, $f(x) = -f(L - x)$, then a_n will be 0 for all even values of n, including $n = 0$. Thus in this case, the expansion reduces to

$$\sum_{m=1}^{\infty} a_{2m-1} \cos \frac{(2m - 1)\pi x}{L}.$$

5. If in addition to the restrictions in (3) above, $f(x) = f(L - x)$, then b_n will be 0 for all even values of n. Thus in this case, the expansion reduces to

$$\sum_{m=1}^{\infty} b_{2m-1} \sin \frac{(2m - 1)\pi x}{L}.$$

(The series in (4) and (5) are known as *odd-harmonic series*, since only the odd harmonics appear. Similar rules may be stated for even-harmonic series, but when a series appears in the even-harmonic form, it means that $2L$ has not been taken as the smallest period of $f(x)$. Since any integral multiple of a period is also a period, series obtained in this way will also work, but in general computation is simplified if $2L$ is taken to be the smallest period.)

6. If we write the Euler definitions for $\cos \theta$ and $\sin \theta$, we obtain the complex form of the Fourier Series known either as the "Complex Fourier Series" or the "Exponential Fourier Series" of $f(x)$. It is represented as

$$f(x) = \frac{1}{2} \sum_{n=-\infty}^{n=+\infty} c_n e^{i\omega_n x},$$

where

$$c_n = \frac{1}{L} \int_{-L}^{L} f(x) e^{-i\omega_n x} dx, \quad n = 0, \pm 1, \pm 2, \pm 3, \ldots$$

with $\omega_n = \dfrac{n\pi}{L}, \quad n = 0, \pm 1, \pm 2, \ldots$

The set of coefficients $\{c_n\}$ is often referred to as the Fourier spectrum.

7. If both sine and cosine terms are present and if $f(x)$ is of period $2L$ and expandable by a Fourier series, it can be represented as

$$f(x) = \frac{a_0}{2} + \sum_{n=1}^{\infty} c_n \sin\left(\frac{n\pi x}{L} + \phi_n\right), \text{ where } a_n = c_n \sin \phi_n,$$

$$b_n = c_n \cos \phi_n, \quad c_n = \sqrt{a_n^2 + b_n^2}, \quad \phi_n = \arctan\left(\frac{a_n}{b_n}\right)$$

It can also be represented as

$$f(x) = \frac{a_0}{2} + \sum_{n=1}^{\infty} c_n \cos\left(\frac{n\pi x}{L} + \phi_n\right), \text{ where } a_n = c_n \cos \phi_n,$$

$$b_n = -c_n \sin \phi_n, \quad c_n = \sqrt{a_n^2 + b_n^2}, \quad \phi_n = \arctan\left(-\frac{b_n}{a_n}\right)$$

where ϕ_n is chosen so as to make a_n, b_n, and c_n hold.

8. The following table of trigonometric identities should be helpful for developing Fourier Series.

	n	n even	n odd	$n/2$ odd	$n/2$ even
$\sin n\pi$	0	0	0	0	0
$\cos n\pi$	$(-1)^n$	$+1$	-1	$+1$	$+1$
*$\sin \dfrac{n\pi}{2}$		0	$(-1)^{(n-1)/2}$	0	0
*$\cos \dfrac{n\pi}{2}$		$(-1)^{n/2}$	0	-1	$+1$
$\sin \dfrac{n\pi}{4}$			$\dfrac{\sqrt{2}}{2}(-1)^{(n^2+4n+11)/8}$	$(-1)^{(n-2)/4}$	0

*A useful formula for $\sin \dfrac{n\pi}{2}$ and $\cos \dfrac{n\pi}{2}$ is given by

$$\sin \frac{n\pi}{2} = \frac{(i)^{n+1}}{2}[(-1)^n - 1] \text{ and } \cos \frac{n\pi}{2} = \frac{(i)^n}{2}[(-1)^n + 1], \text{ where } i^2 = -1.$$

AUXILIARY FORMULAS FOR FOURIER SERIES

$$1 = \frac{4}{\pi}\left[\sin\frac{\pi x}{k} + \frac{1}{3}\sin\frac{3\pi x}{k} + \frac{1}{5}\sin\frac{5\pi x}{k} + \cdots\right] \qquad [0 < x < k]$$

$$x = \frac{2k}{\pi}\left[\sin\frac{\pi x}{k} - \frac{1}{2}\sin\frac{2\pi x}{k} + \frac{1}{3}\sin\frac{3\pi x}{k} - \cdots\right] \qquad [-k < x < k]$$

$$x = \frac{k}{2} - \frac{4k}{\pi^2}\left[\cos\frac{\pi x}{k} + \frac{1}{3^2}\cos\frac{3\pi x}{k} + \frac{1}{5^2}\cos\frac{5\pi x}{k} + \cdots\right] \qquad [0 < x < k]$$

$$x^2 = \frac{2k^2}{\pi^3}\left[\left(\frac{\pi^2}{1} - \frac{4}{1}\right)\sin\frac{\pi x}{k} - \frac{\pi^2}{2}\sin\frac{2\pi x}{k} + \left(\frac{\pi^2}{3} - \frac{4}{3^3}\right)\sin\frac{3\pi x}{k}\right.$$

$$\left. - \frac{\pi^2}{4}\sin\frac{4\pi x}{k} + \left(\frac{\pi^2}{5} - \frac{4}{5^3}\right)\sin\frac{5\pi x}{k} + \cdots\right] \; [0 < x < k]$$

$$x^2 = \frac{k^2}{3} - \frac{4k^2}{\pi^2}\left[\cos\frac{\pi x}{k} - \frac{1}{2^2}\cos\frac{2\pi x}{k} + \frac{1}{3^2}\cos\frac{3\pi x}{k} - \frac{1}{4^2}\cos\frac{4\pi x}{k} + \cdots\right]$$

$$[-k < x < k]$$

$$1 - \frac{1}{3} + \frac{1}{5} - \frac{1}{7} + \cdots = \frac{\pi}{4}$$

$$1 + \frac{1}{2^2} + \frac{1}{3^2} + \frac{1}{4^2} + \cdots = \frac{\pi^2}{6}$$

$$1 - \frac{1}{2^2} + \frac{1}{3^2} - \frac{1}{4^2} + \cdots = \frac{\pi^2}{12}$$

$$1 + \frac{1}{3^2} + \frac{1}{5^2} + \frac{1}{7^2} + \cdots = \frac{\pi^2}{8}$$

$$\frac{1}{2^2} + \frac{1}{4^2} + \frac{1}{6^2} + \frac{1}{8^2} + \cdots = \frac{\pi^2}{24}$$

FOURIER EXPANSIONS FOR BASIC PERIODIC FUNCTIONS

$$f(x) = \frac{4}{\pi} \sum_{n=1,3,5\ldots} \frac{1}{n} \sin \frac{n\pi x}{L}$$

$$f(x) = \frac{2}{\pi} \sum_{n=1}^{\infty} \frac{(-1)^n}{n} \left(\cos \frac{n\pi c}{L} - 1 \right) \sin \frac{n\pi x}{L}$$

$$f(x) = \frac{c}{L} + \frac{2}{\pi} \sum_{n=1}^{\infty} \frac{(-1)^n}{n} \sin \frac{n\pi c}{L} \cos \frac{n\pi x}{L}$$

$$f(x) = \frac{2}{L} \sum_{n=1}^{\infty} \sin \frac{n\pi}{2} \frac{\sin \left(\frac{1}{2} n\pi c/L \right)}{\frac{1}{2} n\pi c/L} \sin \frac{n\pi x}{L}$$

$$f(x) = \frac{2}{\pi} \sum_{n=1}^{\infty} \frac{(-1)^{n+1}}{n} \sin \frac{n\pi x}{L}$$

$$f(x) = \frac{1}{2} - \frac{4}{\pi^2} \sum_{n=1,3,5\ldots} \frac{1}{n^2} \cos \frac{n\pi x}{L}$$

$$f(x) = \frac{8}{\pi^2} \sum_{n=1,3,5\ldots} \frac{(-1)^{(n-1)/2}}{n^2} \sin \frac{n\pi x}{L}$$

$$f(x) = \frac{1}{2} - \frac{1}{\pi} \sum_{n=1}^{\infty} \frac{1}{n} \sin \frac{n\pi x}{L}$$

$$f(x) = \tfrac{1}{2}(1 + a) + \frac{2}{\pi^2 (1-a)} \sum_{n=1}^{\infty} \frac{1}{n^2} \left[(-1)^n \cos n\pi a - 1 \right] \cos \frac{n\pi x}{L} \; ; \; \left(a = \frac{c}{2L} \right)$$

$$f(x) = \frac{2}{\pi} \sum_{n=1}^{\infty} \frac{(-1)^{n-1}}{n} \left[1 + \frac{\sin n\pi a}{n\pi (1-a)} \right] \sin \frac{n\pi x}{L} \; ; \; \left(a = \frac{c}{2L} \right)$$

$$f(x) = \frac{1}{2} - \frac{4}{\pi^2 (1-2a)} \sum_{n=1,3,5\ldots} \frac{1}{n^2} \cos n\pi a \cos \frac{n\pi x}{L} \; ; \; \left(a = \frac{c}{2L} \right)$$

$$f(x) = \frac{2}{\pi} \sum_{n=1}^{\infty} \frac{(-1)^n}{n} \left[1 + \frac{1+(-1)^n}{n\pi(1-2a)} \sin n\pi a \right] \sin \frac{n\pi x}{L} \; ; \; \left(a = \frac{c}{2L} \right)$$

$$f(x) = \frac{4}{\pi} \sum_{n=1}^{\infty} \frac{1}{n} \sin \frac{n\pi}{4} \sin n\pi a \sin \frac{n\pi x}{L} \; ; \; \left(a = \frac{c}{2L} \right)$$

$$f(x) = \frac{9}{\pi^2} \sum_{n=1}^{\infty} \frac{1}{n^2} \sin \frac{n\pi}{3} \sin \frac{n\pi x}{L} \; ; \; \left(a = \frac{c}{2L} \right)$$

$$f(x) = \frac{32}{3\pi^2} \sum_{n=1}^{\infty} \frac{1}{n^2} \sin \frac{n\pi}{4} \sin \frac{n\pi x}{L} \; ; \; \left(a = \frac{c}{2L} \right)$$

$$f(x) = \frac{1}{\pi} + \frac{1}{2} \sin \omega t - \frac{2}{\pi} \sum_{n=2,4,6\ldots} \frac{1}{n^2-1} \cos n\omega t$$

Extracted from graphs and formulas, pages 372, 373, Differential Equations in Engineering Problems, Salvadori and Schwarz, published by Prentice-Hall, Inc., 1954.

THE FOURIER TRANSFORMS*

Dr. R. E. Gaskell

For a piecewise continuous function $F(x)$ over a finite interval $0 \leq x \leq \pi$, the *finite Fourier cosine transform* of $F(x)$ is

$$f_c(n) = \int_0^\pi F(x) \cos nx \, dx \quad (n = 0, 1, 2, \ldots). \tag{4}$$

If x ranges over the interval $0 \leq x \leq L$, the substitution $x' = \pi x / L$ allows the use of this definition, also. The inverse transform is written

$$\bar{F}(x) = \frac{1}{\pi} f_c(0) - \frac{2}{\pi} \sum_{n=1}^{x} f_c(n) \cos nx \quad (0 < x < \pi) \tag{5}$$

where $\bar{F}(x) = \dfrac{[F(x + 0) + F(x - 0)]}{2}$. We observe that $\bar{F}(x) = F(x)$ at points of continuity. The formula

$$\begin{aligned} f_c^{(2)}(n) &= \int_0^\pi F''(x) \cos nx \, dx \\ &= -n^2 f_c(n) - F'(0) + (-1)^n F'(\pi) \end{aligned} \tag{6}$$

makes the finite Fourier cosine transform useful in certain boundary value problems.

Analogously, the *finite Fourier sine transform* of $F(x)$ is

$$f_s(n) = \int_0^\pi F(x) \sin nx \, dx \quad (n = 1, 2, 3, \ldots) \tag{7}$$

and

$$\bar{F}(x) = \frac{2}{\pi} \sum_{n=1}^{\infty} f_s(n) \sin nx \quad (0 < x < \pi) \tag{8}$$

Corresponding to (6) we have

$$\begin{aligned} f_s^{(2)}(n) &= \int_0^\pi F''(x) \sin nx \, dx \\ &= -n^2 f_s(n) - nF(0) - n(-1)^n F(\pi). \end{aligned} \tag{9}$$

FOURIER TRANSFORMS

If $F(x)$ is defined for $x \geq 0$ and is piecewise continuous over any finite interval, and if

$$\int_0^\infty F(x) \, dx$$

is absolutely convergent, then

$$f_c(\alpha) = \sqrt{\frac{2}{\pi}} \int_0^\infty F(x) \cos (\alpha x) \, dx \tag{10}$$

is the *Fourier cosine transform* of $F(x)$. Furthermore,

$$\bar{F}(x) = \sqrt{\frac{2}{\pi}} \int_0^\infty f_c(\alpha) \cos (\alpha x) \, d\alpha. \tag{11}$$

* From Beyer, W. H., Ed., *CRC Handbook of Mathematical Sciences*, 5th ed., CRC Press, Boca Raton, 1978, 592—598. With permission.

If $\lim\limits_{x \to \infty} \dfrac{d^n F}{dx^n} = 0$, an important property of the Fourier cosine transform

$$f_c^{(2r)}(\alpha) = \sqrt{\frac{2}{\pi}} \int_0^{\infty} \left(\frac{d^{2r}F}{dx^{2r}}\right) \cos(\alpha x)\, dx$$

$$= -\sqrt{\frac{2}{\pi}} \sum_{n=0}^{r-1} (-1)^n a_{2r-2n-1} \alpha^{2n} + (-1)^r \alpha^{2r} f_c(\alpha) \qquad (12)$$

where $\lim\limits_{x \to 0} \dfrac{d^r F}{dx^r} = a_r$, makes it useful in the solution of many problems.

Under the same conditions,

$$f_s(\alpha) = \sqrt{\frac{2}{\pi}} \int_0^{\infty} F(x) \sin(\alpha x)\, dx \qquad (13)$$

defines the *Fourier sine transform* of $F(x)$, and

$$\bar{F}(x) = \sqrt{\frac{2}{\pi}} \int_0^{\infty} f_s(\alpha) \sin(\alpha x)\, d\alpha. \qquad (14)$$

Corresponding to (12) we have

$$f_s^{(2r)}(\alpha) = \sqrt{\frac{2}{\pi}} \int_0^{\infty} \frac{d^{2r}F}{dx^{2r}} \sin(\alpha x)\, dx$$

$$= -\sqrt{\frac{2}{\pi}} \sum_{n=1}^{i} (-1)^n \alpha^{2n-1} a_{2r-2n} + (-1)^{r-1} \alpha^{2r} f_s(\alpha). \qquad (15)$$

Similarly, if $F(x)$ is defined for $-\infty < x < \infty$, and if $\displaystyle\int_{-\infty}^{\infty} F(x)\, dx$ is absolutely convergent, then

$$f(\alpha) = \frac{1}{\sqrt{2\pi}} \int_{-\infty}^{\infty} F(x) e^{i\alpha x}\, dx \qquad (16)$$

is the *Fourier transform* of $F(x)$, and

$$\bar{F}(x) = \frac{1}{\sqrt{2\pi}} \int_{-\infty}^{\infty} f(\alpha) e^{-i\alpha x}\, d\alpha. \qquad (17)$$

Also, if

$$\lim_{|x| \to \infty} \left|\frac{d^n F}{dx^n}\right| = 0 \quad (n = 1, 2, \ldots, r-1),$$

then

$$f^{(r)}(\alpha) = \frac{1}{\sqrt{2\pi}} \int_{-\infty}^{\infty} F^{(r)}(x) e^{i\alpha x}\, dx = (-i\alpha)^r f(\alpha). \qquad (18)$$

	$f_s(n)$	$F(x)$
1	$f_s(n) = \displaystyle\int_0^\pi F(x) \sin nx \, dx \ (n = 1, 2, \cdots)$	$F(x)$
2	$(-1)^{n+1} f_s(n)$	$F(\pi - x)$
3	$\dfrac{1}{n}$	$\dfrac{\pi - x}{\pi}$
4	$\dfrac{(-1)^{n+1}}{n}$	$\dfrac{x}{\pi}$
5	$\dfrac{1 - (-1)^n}{n}$	1
6	$\dfrac{2}{n^2} \sin \dfrac{n\pi}{2}$	$\begin{cases} x & \text{when } 0 < x < \pi/2 \\ \pi - x & \text{when } \pi/2 < x < \pi \end{cases}$
7	$\dfrac{(-1)^{n+1}}{n^3}$	$\dfrac{x(\pi^2 - x^2)}{6\pi}$
8	$\dfrac{1 - (-1)^n}{n^3}$	$\dfrac{x(\pi - x)}{2}$
9	$\dfrac{\pi^2(-1)^{n-1}}{n} - \dfrac{2[1 - (-1)^n]}{n^3}$	x^2
10	$\pi(-1)^n \left(\dfrac{6}{n^3} - \dfrac{\pi^2}{n} \right)$	x^3
11	$\dfrac{n}{n^2 + c^2} [1 - (-1)^n e^{c\pi}]$	e^{cx}
12	$\dfrac{n}{n^2 + c^2}$	$\dfrac{\sinh c(\pi - x)}{\sinh c\pi}$
13	$\dfrac{n}{n^2 - k^2} \ (k \neq 0, 1, 2, \cdots)$	$\dfrac{\sin k(\pi - x)}{\sin k\pi}$
14	$\begin{cases} \dfrac{\pi}{2} & \text{when } n = m \\ 0 & \text{when } n \neq m \end{cases} (m = 1, 2, \cdots)$	$\sin mx$
15	$\dfrac{n}{n^2 - k^2} [1 - (-1)^n \cos k\pi]$ $(k \neq 1, 2, \cdots)$	$\cos kx$
16	$\begin{cases} \dfrac{n}{n^2 - m^2} [1 - (-1)^{n+m}] \\ \quad \text{when } n \neq m = 1, 2, \cdots \\ 0 \quad \text{when } n = m \end{cases}$	$\cos mx$
17	$\dfrac{n}{(n^2 - k^2)^2} \ (k \neq 0, 1, 2, \cdots)$	$\dfrac{\pi \sin kx}{2k \sin^2 k\pi} - \dfrac{x \cos k(\pi - x)}{2k \sin k\pi}$
18	$\dfrac{b^n}{n} \ (\lvert b \rvert \leq 1)$	$\dfrac{2}{\pi} \arctan \dfrac{b \sin x}{1 - b \cos x}$
19	$\dfrac{1 - (-1)^n}{n} b^n \ (\lvert b \rvert \leq 1)$	$\dfrac{2}{\pi} \arctan \dfrac{2b \sin x}{1 - b^2}$

FINITE COSINE TRANSFORMS

	$f_c(n)$	$F(x)$
1	$f_c(n) = \displaystyle\int_0^\pi F(x) \cos nx \, dx \quad (n = 0, 1, 2, \cdots)$	$F(x)$
2	$(-1)^n f_c(n)$	$F(\pi - x)$
3	0 when $n = 1, 2, \cdots$; $f_c(0) = \pi$	1
4	$\dfrac{2}{n} \sin \dfrac{n\pi}{2}$; $f_c(0) = 0$	$\begin{cases} 1 \text{ when } 0 < x < \pi/2 \\ -1 \text{ when } \pi/2 < x < \pi \end{cases}$
5	$-\dfrac{1 - (-1)^n}{n^2}$; $f_c(0) = \dfrac{\pi^2}{2}$	x
6	$\dfrac{(-1)^n}{n^2}$; $f_c(0) = \dfrac{\pi^2}{6}$	$\dfrac{x^2}{2\pi}$
7	$\dfrac{1}{n^2}$; $f_c(0) = 0$	$\dfrac{(\pi - x)^2}{2\pi} - \dfrac{\pi}{6}$
8	$3\pi^2 \dfrac{(-1)^n}{n^2} - 6 \dfrac{1 - (-1)^n}{n^4}$; $f_c(0) = \dfrac{\pi^4}{4}$	x^3
9	$\dfrac{(-1)^n e^c \pi - 1}{n^2 + c^2}$	$\dfrac{1}{c} e^{cx}$
10	$\dfrac{1}{n^2 + c^2}$	$\dfrac{\cosh c(\pi - x)}{c \sinh c\pi}$
11	$\dfrac{k}{n^2 - k^2} [(-1)^n \cos \pi k - 1]$ $(k \neq 0, 1, 2, \cdots)$	$\sin kx$
12	$\dfrac{(-1)^{n+m} - 1}{n^2 - m^2}$; $f_c(m) = 0 \quad (m = 1, 2, \cdots)$	$\dfrac{1}{m} \sin mx$
13	$\dfrac{1}{n^2 - k^2} \quad (k \neq 0, 1, 2, \cdots)$	$-\dfrac{\cos k(\pi - x)}{k \sin k\pi}$
14	0 when $n = 1, 2, \cdots$; $f_c(m) = \dfrac{\pi}{2} \quad (m = 1, 2, \cdots)$	$\cos mx$

FOURIER SINE TRANSFORMS**

	$F(x)$	$f_s(\alpha)$
1	$\begin{cases} 1 & (0 < x < a) \\ 0 & (x > a) \end{cases}$	$\sqrt{\dfrac{2}{\pi}}\left[\dfrac{1 - \cos\alpha}{\alpha}\right]$
2	$x^{p-1}\ (0 < p < 1)$	$\sqrt{\dfrac{2}{\pi}}\,\dfrac{\Gamma(p)}{\alpha^p}\sin\dfrac{p\pi}{2}$
3	$\begin{cases} \sin x & (0 < x < a) \\ 0 & (x > a) \end{cases}$	$\dfrac{1}{\sqrt{2\pi}}\left[\dfrac{\sin[a(1 - \alpha)]}{1 - \alpha} - \dfrac{\sin[a(1 + \alpha)]}{1 + \alpha}\right]$
4	e^{-x}	$\sqrt{\dfrac{2}{\pi}}\left[\dfrac{\alpha}{1 + \alpha^2}\right]$
5	$xe^{-x^2/2}$	$\alpha e^{-\alpha^2/2}$
6	$\cos\dfrac{x^2}{2}$	$\sqrt{2}\left[\sin\dfrac{\alpha^2}{2}\,C\!\left(\dfrac{\alpha^2}{2}\right) - \cos\dfrac{\alpha^2}{2}\,S\!\left(\dfrac{\alpha^2}{2}\right)\right]^{*}$
7	$\sin\dfrac{x^2}{2}$	$\sqrt{2}\left[\cos\dfrac{\alpha^2}{2}\,C\!\left(\dfrac{\alpha^2}{2}\right) + \sin\dfrac{\alpha^2}{2}\,S\!\left(\dfrac{\alpha^2}{2}\right)\right]^{*}$

$*C(y)$ and $S(y)$ are the Fresnel integrals

$$C(y) = \frac{1}{\sqrt{2\pi}}\int_0^y \frac{1}{\sqrt{t}}\cos t\,dt,$$

$$S(y) = \frac{1}{\sqrt{2\pi}}\int_0^y \frac{1}{\sqrt{t}}\sin t\,dt.$$

**More extensive tables of the Fourier sine and cosine transforms can be found in Fritz Oberhettinger, "Tabellen zur-Fourier Transformation," Springer (1957).

FOURIER COSINE TRANSFORMS

	$F(x)$	$f_c(\alpha)$
1	$\begin{cases} 1 & (0 < x < a) \\ 0 & (x > a) \end{cases}$	$\sqrt{\dfrac{2}{\pi}}\,\dfrac{\sin a\alpha}{\alpha}$
2	$x^{p-1}\ (0 < p < 1)$	$\sqrt{\dfrac{2}{\pi}}\,\dfrac{\Gamma(p)}{\alpha^p}\cos\dfrac{p\pi}{2}$
3	$\begin{cases} \cos x & (0 < x < a) \\ 0 & (x > a) \end{cases}$	$\dfrac{1}{\sqrt{2\pi}}\left[\dfrac{\sin[a(1 - \alpha)]}{1 - \alpha} + \dfrac{\sin[a(1 + \alpha)]}{1 + \alpha}\right]$
4	e^{-x}	$\sqrt{\dfrac{2}{\pi}}\left(\dfrac{1}{1 + \alpha^2}\right)$
5	$e^{-x^2/2}$	$e^{-\alpha^2/2}$
6	$\cos\dfrac{x^2}{2}$	$\cos\left(\dfrac{\alpha^2}{2} - \dfrac{\pi}{4}\right)$
7	$\sin\dfrac{x^2}{2}$	$\cos\left(\dfrac{\alpha^2}{2} + \dfrac{\pi}{4}\right)$

FOURIER TRANSFORMS*

	$F(x)$	$f(\alpha)$
1	$\dfrac{\sin ax}{x}$	$\begin{cases} \sqrt{\dfrac{\pi}{2}} & \lvert \alpha \rvert < a \\ 0 & \lvert \alpha \rvert > a \end{cases}$
2	$\begin{cases} e^{iwx} & (p < x < q) \\ 0 & (x < p,\ x > q) \end{cases}$	$\dfrac{i}{\sqrt{2\pi}}\dfrac{e^{ip(w+\alpha)} - e^{iq(w+\alpha)}}{(w + \alpha)}$
3	$\begin{cases} e^{-cx+iwx} & (x > 0) \\ 0 & (x < 0) \end{cases} \quad (c > 0)$	$\dfrac{i}{\sqrt{2\pi}(w + \alpha + ic)}$
4	$e^{-px^2} \quad R(p) > 0$	$\dfrac{1}{\sqrt{2p}}\, e^{-\alpha^2/4p}$
5	$\cos px^2$	$\dfrac{1}{\sqrt{2p}} \cos\left[\dfrac{\alpha^2}{4p} - \dfrac{\pi}{4}\right]$
6	$\sin px^2$	$\dfrac{1}{\sqrt{2p}} \cos\left[\dfrac{\alpha^2}{4p} + \dfrac{\pi}{4}\right]$
7	$\lvert x \rvert^{-p} \quad (0 < p < 1)$	$\sqrt{\dfrac{2}{\pi}}\,\dfrac{\Gamma(1 - p)\sin\dfrac{p\pi}{2}}{\lvert \alpha \rvert^{(1-p)}}$
8	$\dfrac{e^{-a\lvert x \rvert}}{\sqrt{\lvert x \rvert}}$	$\dfrac{\sqrt{\sqrt{(a^2 + \alpha^2)} + a}}{\sqrt{a^2 + \alpha^2}}$
9	$\dfrac{\cosh ax}{\cosh \pi x} \quad (-\pi < a < \pi)$	$\sqrt{\dfrac{2}{\pi}}\,\dfrac{\cos\dfrac{a}{2}\cosh\dfrac{\alpha}{2}}{\cosh \alpha + \cos a}$
10	$\dfrac{\sinh ax}{\sinh \pi x} \quad (-\pi < a < \pi)$	$\dfrac{1}{\sqrt{2\pi}}\,\dfrac{\sin a}{\cosh \alpha + \cos a}$
11	$\begin{cases} \dfrac{1}{\sqrt{a^2 - x^2}} & (\lvert x \rvert < a) \\ 0 & (\lvert x \rvert > a) \end{cases}$	$\sqrt{\dfrac{\pi}{2}}\, J_0(a\alpha)$
12	$\dfrac{\sin[b\sqrt{a^2 + x^2}]}{\sqrt{a^2 + x^2}}$	$\begin{cases} 0 & (\lvert \alpha \rvert > b) \\ \sqrt{\dfrac{\pi}{2}}\, J_0(a\sqrt{b^2 - \alpha^2}) & (\lvert \alpha \rvert < b) \end{cases}$
13	$\begin{cases} P_n(x) & (\lvert x \rvert < 1) \\ 0 & (\lvert x \rvert > 1) \end{cases}$	$\dfrac{i^n}{\sqrt{\alpha}}\, J_{n+\frac{1}{2}}(\alpha)$
14	$\begin{cases} \dfrac{\cos[b\sqrt{a^2 - x^2}]}{\sqrt{a^2 - x^2}} & (\lvert x \rvert < a) \\ 0 & (\lvert x \rvert > a) \end{cases}$	$\sqrt{\dfrac{\pi}{2}}\, J_0(a\sqrt{a^2 + b^2})$
15	$\begin{cases} \dfrac{\cosh[b\sqrt{a^2 - x^2}]}{\sqrt{a^2 - x^2}} & (\lvert x \rvert < a) \\ 0 & (\lvert x \rvert > a) \end{cases}$	$\sqrt{\dfrac{\pi}{2}}\, J_0(a\sqrt{\alpha^2 - b^2})$

*More extensive tables of Fourier transforms can be found in W. Magnus and F. Oberhettinger, "Formulas and Theorems of the Special Functions of Mathematical Physics," pp. 116–120. Chelsea (1949).

The following functions appear among the entries of the tables on transforms.

Function	Definition	Name
$Ei(x)$	$\int_{-x}^{x} \dfrac{e^{v}}{v}\, dv$; or sometimes defined as $$-Ei(-x) = \int_{x}^{\infty} \dfrac{e^{-v}}{v}\, dv$$	Sine, Cosine, and Exponential Integral tables pages 548–556
$Si(x)$	$\int_{0}^{x} \dfrac{\sin v}{v}\, dv$	Sine, Cosine, and Exponential Integral tables pages 548–556
$Ci(x)$	$\int_{x}^{\infty} \dfrac{\cos v}{v}\, dv$; or sometimes defined as negative of this integral	Sine, Cosine, and Exponential Integral tables pages 548–546
$erf(x)$	$\dfrac{2}{\sqrt{\pi}} \int_{0}^{x} e^{-v^{2}}\, dv$	Error function
$erfc(x)$	$1 - erf(x) = \dfrac{2}{\sqrt{\pi}} \int_{x}^{\infty} e^{-v^{2}}\, dv$	Complementary function to error function
$L_{n}(x)$	$\dfrac{e^{x}}{n!} \dfrac{d^{n}}{dx^{n}} (x^{n} e^{-x})$, $n = 0, 1, \cdots$	Laguerre polynomial of degree n

BESSEL FUNCTIONS *

1. **Bessel's differential equation for a real variable x is**

$$x^2 \frac{d^2 y}{dx^2} + x \frac{dy}{dx} + (x^2 - n^2) y = 0$$

2. **When n is not an integer, two independent solutions of the equation are $J_n(x)$ and $J_{-n}(x)$, where**

$$J_n(x) = \sum_{k=0}^{\infty} \frac{(-1)^k}{k! \, \Gamma(n + k + 1)} \left(\frac{x}{2}\right)^{n+2k}$$

3. **If n is an integer $J_{-n}(x) = (-1)^n J_n(x)$, where**

$$J_n(x) = \frac{x^n}{2^n n!} \left\{ 1 - \frac{x^2}{2^2 \cdot 1!(n+1)} + \frac{x^4}{2^4 \cdot 2!(n+1)(n+2)} \right.$$
$$\left. - \frac{x^6}{2^6 \cdot 3!(n+1)(n+2)(n+3)} + \cdots \right\}$$

4. **For $n = 0$ and $n = 1$, this formula becomes**

$$J_0(x) = 1 - \frac{x^2}{2^2(1!)^2} + \frac{x^4}{2^4(2!)^2} - \frac{x^6}{2^6(3!)^2} + \frac{x^8}{2^8(4!)^2} - \cdots$$

$$J_1(x) = \frac{x}{2} - \frac{x^3}{2^3 \cdot 1! \, 2!} + \frac{x^5}{2^5 \cdot 2! \, 3!} - \frac{x^7}{2^7 \cdot 3! \, 4!} + \frac{x^9}{2^9 \cdot 4! \, 5!} - \cdots$$

5. **When x is large and positive, the following asymptotic series may be used**

$$J_0(x) = \left(\frac{2}{\pi x}\right)^{\frac{1}{2}} \left\{ P_0(x) \cos\left(x - \frac{\pi}{4}\right) - Q_0(x) \sin\left(x - \frac{\pi}{4}\right) \right\}$$

$$J_1(x) = \left(\frac{2}{\pi x}\right)^{\frac{1}{2}} \left\{ P_1(x) \cos\left(x - \frac{3\pi}{4}\right) - Q_1(x) \sin\left(x - \frac{3\pi}{4}\right) \right\},$$

where

$$P_0(x) \sim 1 - \frac{1^2 \cdot 3^2}{2!(8x)^2} + \frac{1^2 \cdot 3^2 \cdot 5^2 \cdot 7^2}{4!(8x)^4} - \frac{1^2 \cdot 3^2 \cdot 5^2 \cdot 7^2 \cdot 9^2 \cdot 11^2}{6!(8x)^6} + \cdots$$

$$Q_0(x) \sim -\frac{1^2}{1!8x} + \frac{1^2 \cdot 3^2 \cdot 5^2}{3!(8x)^3} - \frac{1^2 \cdot 3^2 \cdot 5^2 \cdot 7^2 \cdot 9^2}{5!(8x)^5} + - \cdots$$

$$P_1(x) \sim 1 + \frac{1^2 \cdot 3 \cdot 5}{2!(8x)^2} - \frac{1^2 \cdot 3^2 \cdot 5^2 \cdot 7 \cdot 9}{4!(8x)^4} + \frac{1^2 \cdot 3^2 \cdot 5^2 \cdot 7^2 \cdot 9^2 \cdot 11 \cdot 13}{6!(8x)^6} - + \cdots$$

$$Q_1(x) \sim \frac{1 \cdot 3}{1!8x} - \frac{1^2 \cdot 3^2 \cdot 5 \cdot 7}{3!(8x)^3} + \frac{1^2 \cdot 3^2 \cdot 5^2 \cdot 7^2 \cdot 9 \cdot 11}{5!(8x)^5} - \cdots$$

[In $P_1(x)$ the signs alternate from + to − after the first term]

* From Beyer, W. H., Ed., *CRC Handbook of Mathematical Sciences*, 5th ed., CRC Press, Boca Raton, 1978, 500—503. With permission.

6. If $x > 25$, it is convenient to use the formulas

$$J_0(x) = A_0(x) \sin x + B_0(x) \cos x$$
$$J_1(x) = B_\tau(x) \sin x - A_1(x) \cos x,$$

where

$$A_0(x) = \frac{P_0(x) - Q_0(x)}{(\pi x)^{\frac{1}{2}}} \quad \text{and} \quad A_1(x) = \frac{P_1(x) - Q_1(x)}{(\pi x)^{\frac{1}{2}}}$$

$$B_0(x) = \frac{P_0(x) + Q_0(x)}{(\pi x)^{\frac{1}{2}}} \quad \text{and} \quad B_1(x) = \frac{P_1(x) + Q_1(x)}{(\pi x)^{\frac{1}{2}}}$$

7. The zeros of $J_0(x)$ and $J_1(x)$

If $j_{0,s}$ and $j_{1,s}$ are the s'th zeros of $J_0(x)$ and $J_1(x)$ respectively, and if $a = 4s - 1$, $b = 4s + 1$

$$j_{0,s} \sim \frac{1}{4} \pi a \left\{ 1 + \frac{2}{\pi^2 a^2} - \frac{62}{3\pi^4 a^4} + \frac{15,116}{15\pi^6 a^6} - \frac{12,554,474}{105\pi^8 a^8} + \frac{8,368,654,292}{315\pi^{10} a^{10}} - + \cdots \right\}$$

$$j_{1,s} \sim \frac{1}{4} \pi b \left\{ 1 - \frac{6}{\pi^2 b^2} + \frac{6}{\pi^4 b^4} - \frac{4716}{5\pi^6 b^6} + \frac{3,902,418}{35\pi^8 b^8} - \frac{895,167,324}{35\pi^{10} b^{10}} + \cdots \right\}$$

$$J_1(j_{0,s}) \sim \frac{(-1)^{s+1} 2^{\frac{3}{2}}}{\pi a^{\frac{1}{2}}} \left\{ 1 - \frac{56}{3\pi^4 a^4} + \frac{9664}{5\pi^6 a^6} - \frac{7,381,280}{21\pi^8 a^8} + \cdots \right\}$$

$$J_0(j_{1,s}) \sim \frac{(-1)^s 2^{\frac{3}{2}}}{\pi b^{\frac{1}{2}}} \left\{ 1 + \frac{24}{\pi^4 b^4} - \frac{19,584}{10\pi^6 b^6} + \frac{2,466,720}{7\pi^8 b^8} - \cdots \right\}$$

8. Table of zeros for $J_0(x)$ and $J_1(x)$

$$J_1(\alpha_n) = 0 \qquad J_0(\beta_n) = 0$$

Roots α_n	$J_1(\alpha_n)$	Roots β_n	$J_0(\beta_n)$
2.4048	0.5191	0.0000	1.0000
5.5201	−0.3403	3.8317	−0.4028
8.6537	0.2715	7.0156	0.3001
11.7915	−0.2325	10.1735	−0.2497
14.9309	0.2065	13.3237	0.2184
18.0711	−0.1877	16.4706	−0.1965
21.2116	0.1733	19.6159	0.1801

9. Recurrence formulas

$$J_{n-1}(x) + J_{n+1}(x) = \frac{2n}{x} J_n(x) \qquad n J_n(x) + x J_n'(x) = x J_{n-1}(x)$$

$$J_{n-1}(x) - J_{n+1}(x) = 2 J_n'(x) \qquad n J_n(x) - x J_n'(x) = x J_{n+1}(x)$$

10. If J_n is written for $J_n(x)$ and $J_n^{(k)}$ is written for $\dfrac{d^k}{dx^k}\{J_n(x)\}$, then the following derivative relationships are important

$$J_0^{(r)} = -J_1^{(r-1)}$$

$$J_0^{(2)} = -J_0 + \frac{1}{x} J_1 = \frac{1}{2} (J_2 - J_0)$$

$$J_0^{(3)} = \frac{1}{x} J_0 + \left(1 - \frac{2}{x^2}\right) J_1 = \frac{1}{4} (-J_3 + 3J_1)$$

$$J_0^{(4)} = \left(1 - \frac{3}{x^2}\right) J_0 - \left(\frac{2}{x} - \frac{6}{x^3}\right) J_1 = \frac{1}{8} (J_4 - 4J_2 + 3J_0), \text{ etc.}$$

11. Half order Bessel functions

$$J_{\frac{1}{2}}(x) = \sqrt{\frac{2}{\pi x}} \sin x$$

$$J_{-\frac{1}{2}}(x) = \sqrt{\frac{2}{\pi x}} \cos x$$

$$J_{n+\frac{1}{2}}(x) = -x^{n+\frac{1}{2}} \frac{d}{dx} \{x^{-(n+\frac{1}{2})} J_{n+\frac{1}{2}}(x)\}$$

$$J_{n-\frac{1}{2}}(x) = x^{-(n+\frac{1}{2})} \frac{d}{dx} \{x^{n+\frac{1}{2}} J_{n+\frac{1}{2}}(x)\}$$

n	$\left(\frac{\pi x}{2}\right)^{\frac{1}{2}} J_{n+\frac{1}{2}}(x)$	$\left(\frac{\pi x}{2}\right)^{\frac{1}{2}} J_{-(n+\frac{1}{2})}(x)$
0	$\sin x$	$\cos x$
1	$\dfrac{\sin x}{x} - \cos x$	$-\dfrac{\cos x}{x} - \sin x$
2	$\left(\dfrac{3}{x^2} - 1\right) \sin x - \dfrac{3}{x} \cos x$	$\left(\dfrac{3}{x^2} - 1\right) \cos x + \dfrac{3}{x} \sin x$
3	$\left(\dfrac{15}{x^3} - \dfrac{6}{x}\right) \sin x - \left(\dfrac{15}{x^2} - 1\right) \cos x$	$-\left(\dfrac{15}{x^3} - \dfrac{6}{x}\right) \cos x - \left(\dfrac{15}{x^2} - 1\right) \sin x$
	etc.	

12. Additional solutions to Bessel's equation are

$Y_n(x)$ (also called Weber's function, and sometimes denoted by $N_n(x)$)

$H_n^{(1)}(x)$ and $H_n^{(2)}(x)$ (also called Hankel functions)

These solutions are defined as follows

$$Y_n(x) = \begin{cases} \dfrac{J_n(x) \cos (n\pi) - J_{-n}(x)}{\sin (n\pi)} & n \text{ not an integer} \\[3mm] \lim_{v \to n} \dfrac{J_v(x) \cos (v\pi) - J_{-v}(x)}{\sin (v\pi)} & n \text{ an integer} \end{cases}$$

$$H_n^{(1)}(x) = J_n(x) + i Y_n(x)$$
$$H_n^{(2)}(x) = J_n(x) - i Y_n(x)$$

The additional properties of these functions may all be derived from the above relations and the known properties of $J_n(x)$.

13. Complete solutions to Bessel's equation may be written as

$$c_1 J_n(x) + c_2 J_{-n}(x) \qquad \text{if } n \text{ is not an integer,}$$

or

or

$$\left.\begin{array}{l} c_1 J_n(x) + c_2 Y_n(x) \\[2mm] c_1 H_n^{(1)}(x) + c_2 H_n^{(2)}(x) \end{array}\right\} \text{for any value of } n$$

14. The modified (or hyperbolic) Bessel's differential equation is

$$x^2 \frac{d^2 y}{dx^2} + x \frac{dy}{dx} - (x^2 + n^2)y = 0$$

15. When n is not an integer, two independent solutions of the equation are $I_n(x)$ and $I_{-n}(x)$, where

$$I_n(x) = \sum_{k=0}^{\infty} \frac{1}{k!\,\Gamma(n + k + 1)} \left(\frac{x}{2}\right)^{n+2k}$$

16. If n is an integer,

$$I_n(x) = I_{-n}(x) = \frac{x^n}{2^n n!} \left\{ 1 + \frac{x^2}{2^2 \cdot 1!(n + 1)} + \frac{x^4}{2^4 \cdot 2!(n + 1)(n + 2)} \right.$$
$$\left. + \frac{x^6}{2^6 \cdot 3!(n + 1)(n + 2)(n + 3)} + \cdots \right\}$$

17. For $n = 0$ and $n = 1$, this formula becomes

$$I_0(x) = 1 + \frac{x^2}{2^2(1!)^2} + \frac{x^4}{2^4(2!)^2} + \frac{x^6}{2^6(3!)^2} + \frac{x^8}{2^8(4!)^2} + \cdots$$

$$I_1(x) = \frac{x}{2} + \frac{x^3}{2^3 \cdot 1!2!} + \frac{x^5}{2^5 \cdot 2!3!} + \frac{x^7}{2^7 \cdot 3!4!} + \frac{x^9}{2^9 \cdot 4!5!} + \cdots$$

18. Another solution to the modified Bessel's equation is

$$K_n(x) = \begin{cases} \dfrac{1}{2}\,\pi\, \dfrac{I_{-n}(x) - I_n(x)}{\sin(n\pi)} & n \text{ not an integer} \\[4mm] \lim_{v \to n} \dfrac{1}{2}\,\pi\, \dfrac{I_{-v}(x) - I_v(x)}{\sin(v\pi)} & n \text{ an integer} \end{cases}$$

This function is linearly independent of $I_n(x)$ for all values of n. Thus the **complete** solution to the modified Bessel's equation may be written as

$$c_1 I_n(x) + c_2 I_{-n}(x) \qquad n \text{ not an integer}$$

or

$$c_1 I_n(x) + c_2 K_n(x) \qquad \text{any } n$$

19. The following relations hold among the various Bessel functions:

$$I_n(z) = i^{-m} J_m(iz)$$
$$Y_n(iz) = (i)^{n+1} I_n(z) - \frac{2}{\pi} i^{-n} K_n(z)$$

Most of the properties of the modified Bessel function may be deduced from the known properties of $J_n(x)$ by use of these relations and those previously given.

20. Recurrence formulas

$$I_{n-1}(x) - I_{n+1}(x) = \frac{2n}{x} I_n(x) \qquad I_{n-1}(x) + I_{n+1}(x) = 2I_n'(x)$$

$$I_{n-1}(x) - \frac{n}{x} I_n(x) = I_n'(x) \qquad\qquad I_n'(x) = I_{n+1}(x) + \frac{n}{x} I_n(z)$$

The Gamma Function *

Definition: $\Gamma(n) = \displaystyle\int_0^\infty t^{n-1} e^{-t}\, dt \quad n > 0$

Recursion Formula: $\Gamma(n + 1) = n\Gamma(n)$

$\Gamma(n + 1) = n!$ if $n = 0,1,2,\ldots$ where $0! = 1$

For $n < 0$ the gamma function can be defined by using

$$\Gamma(n) = \frac{\Gamma(n + 1)}{n}$$

Graph:

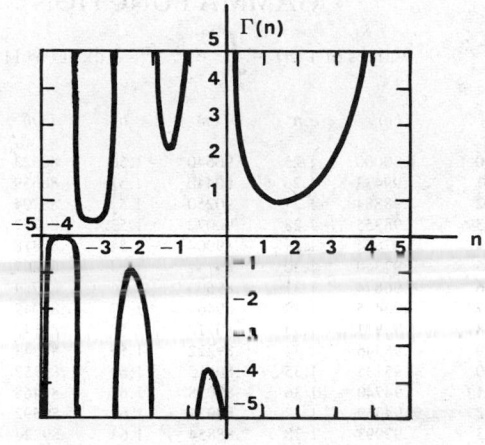

Special Values: $\qquad\qquad \Gamma(\tfrac{1}{2}) = \sqrt{\pi}$

$$\Gamma(m + \tfrac{1}{2}) = \frac{1 \cdot 3 \cdot 5 \cdots (2m - 1)}{2^m}\sqrt{\pi} \qquad m = 1,2,3,\ldots$$

$$\Gamma(-m + \tfrac{1}{2}) = \frac{(-1)^m 2^m \sqrt{\pi}}{1 \cdot 3 \cdot 5 \cdots (2m - 1)} \qquad m = 1,2,3,\ldots$$

Definition: $\quad \Gamma(x + 1) = \displaystyle\lim_{k \to \infty} \frac{1 \cdot 2 \cdot 3 \cdots k}{(x + 1)(x + 2)\cdots(x + k)} k^x$

$$\frac{1}{\Gamma(x)} = xe^{\gamma x} \prod_{m = 1}^{\infty} \left\{ \left(1 + \frac{x}{m}\right) e^{-x/m} \right\}$$

This is an infinite product representation for the gamma function where γ is Euler's constant.

* From Beyer, W. H., Ed., *CRC Handbook of Mathematical Sciences*, 5th ed., CRC Press, Boca Raton, 1978, 484—485. With permission.

Properties:

$$\Gamma'(1) = \int_0^\infty e^{-x} \ln x \, dx = -\gamma$$

$$\frac{\Gamma'(x)}{\Gamma(x)} = -\gamma + \left(\frac{1}{1} - \frac{1}{x}\right) + \left(\frac{1}{2} - \frac{1}{x+1}\right) + \ldots + \left(\frac{1}{n} - \frac{1}{x+n-1}\right) + \ldots$$

$$\Gamma(x+1) = \sqrt{2\pi x}\; x^x e^{-x} \left\{ 1 + \frac{1}{12x} + \frac{1}{288x^2} - \frac{139}{51,840x^3} + \ldots \right\}$$

This is called *Stirling's asymptotic series.*

If we let $x = n$ a positive integer, then a useful approximation for $n!$ where n is large [e.g. $n > 10$] is given by *Stirling's formula*

$$n! \approx \sqrt{2\pi n}\; n^n e^{-n}$$

GAMMA FUNCTION*

Values of $\Gamma(n) = \int_0^\infty e^{-x} x^{n-1}\, dx$; $\Gamma(n+1) = n\Gamma(n)$

n	$\Gamma(n)$	n	$\Gamma(n)$	n	$\Gamma(n)$	n	$\Gamma(n)$
1.00	1.00000	1.25	.90640	1.50	.88623	1.75	.91906
1.01	.99433	1.26	.90440	1.51	.88659	1.76	.92137
1.02	.98884	1.27	.90250	1.52	.88704	1.77	.92376
1.03	.98355	1.28	.90072	1.53	.88757	1.78	.92623
1.04	.97844	1.29	.89904	1.54	.88818	1.79	.92877
1.05	.97350	1.30	.89747	1.55	.88887	1.80	.93138
1.06	.96874	1.31	.89600	1.56	.88964	1.81	.93408
1.07	.96415	1.32	.89464	1.57	.89049	1.82	.93685
1.08	.95973	1.33	.89338	1.58	.89142	1.83	.93969
1.09	.95546	1.34	.89222	1.59	.89243	1.84	.94261
1.10	.95135	1.35	.89115	1.60	.89352	1.85	.94561
1.11	.94740	1.36	.89018	1.61	.89468	1.86	.94869
1.12	.94359	1.37	.88931	1.62	.89592	1.87	.95184
1.13	.93993	1.38	.88854	1.63	.89724	1.88	.95507
1.14	.93642	1.39	.88785	1.64	.89864	1.89	.95838
1.15	.93304	1.40	.88726	1.65	.90012	1.90	.96177
1.16	.92980	1.41	.88676	1.66	.90167	1.91	.96523
1.17	.92670	1.42	.88636	1.67	.90330	1.92	.96877
1.18	.92373	1.43	.88604	1.68	.90500	1.93	.97240
1.19	.92089	1.44	.88581	1.69	.90678	1.94	.97610
1.20	.91817	1.45	.88566	1.70	.90864	1.95	.97988
1.21	.91558	1.46	.88560	1.71	.91057	1.96	.98374
1.22	.91311	1.47	.88563	1.72	.91258	1.97	.98768
1.23	.91075	1.48	.88575	1.73	.91466	1.98	.99171
1.24	.90852	1.49	.88595	1.74	.91683	1.99	.99581
						2.00	1.00000

* For large positive values of x, $\Gamma(x)$ approximates Stirling's asymptotic series

$$x^x e^{-x} \sqrt{\frac{2\pi}{x}} \left[1 + \frac{1}{12x} + \frac{1}{288x^2} - \frac{139}{51840x^3} - \frac{571}{2488320x^4} + \ldots \right]$$

The Beta Function *

Definition: $B(m,n) = \int_0^1 t^{m-1}(1-t)^{m-1}\,dt \quad m > 0, n > 0$

Relationship with Gamma Function
$$B(m,n) = \frac{\Gamma(m)\Gamma(n)}{\Gamma(m+n)}$$

Properties:

$$B(m,n) = B(n,m)$$

$$B(m,n) = 2\int_0^{\pi/2} \sin^{2m-1}\theta \cos^{2n-1}\theta\,d\theta$$

$$B(m,n) = \int_0^\infty \frac{t^{m-1}}{(1+t)^{m+n}}\,dt$$

$$B(m,n) = r^n(r+1)^m \int_0^1 \frac{t^{m-1}(1-t)^{n-1}}{(r+t)^{m+n}}\,dt$$

The Error Function

Definition: $\operatorname{erf} x = \dfrac{2}{\sqrt{\pi}}\displaystyle\int_0^x e^{-t^2}\,dt$

Series: $\operatorname{erf} x = \dfrac{2}{\sqrt{\pi}}\left(x - \dfrac{x^3}{3} + \dfrac{1}{2!}\dfrac{x^5}{5} - \dfrac{1}{3!}\dfrac{x^7}{7} + \ldots\right)$

Property: $\operatorname{erf} x = -\operatorname{erf}(-x)$

Relationship with Normal Probability Function f(t): $\displaystyle\int_0^x f(t)\,dt = \dfrac{1}{2}\operatorname{erf}\left(\dfrac{x}{\sqrt{2}}\right)$

To evaluate $\operatorname{erf}(2.3)$, one proceeds as follows: Since $\dfrac{x}{\sqrt{2}} = 2.3$, one finds $x = (2.3)(\sqrt{2}) = 3.25$. In the normal probability function table (page 753), one finds the entry 0.4994 opposite the value 3.25. Thus $\operatorname{erf}(2.3) = 2(0.4994) = 0.9988$.

$$\operatorname{erfc} z = 1 - \operatorname{erf} z = \dfrac{2}{\sqrt{\pi}}\int_z^\infty e^{-t^2}\,dt$$

is known as the complimentary error function.

* From Beyer, W. H., Ed., *CRC Handbook of Mathematical Sciences,* 5th ed., CRC Press, Boca Raton, 1978, 499. With permission.

ORTHOGONAL POLYNOMIALS *

I

Name: Legendre   *Symbol*: $P_n(x)$ *Interval*: $[-1, 1]$

Differential Equation: $(1 - x^2)y'' - 2xy' + n(n + 1)y = 0$
$$y = P_n(x)$$

Explicit Expression: $P_n(x) = \dfrac{1}{2^n} \sum_{m=0}^{[n/2]} (-1)^m \dbinom{n}{m} \dbinom{2n - 2m}{n} x^{n - 2m}$

Recurrence Relation: $(n + 1)P_{n+1}(x) = (2n + 1)xP_n(x) - nP_{n-1}(x)$

Weight: 1   *Standardization*: $P_n(1) = 1$

Norm: $\displaystyle\int_{-1}^{+1} [P_n(x)]^2 \, dx = \dfrac{2}{2n + 1}$

Rodrigues' Formula: $P_n(x) = \dfrac{(-1)^n}{2^n n!} \dfrac{d^n}{dx^n} \{(1 - x^2)^n\}$

Generating Function: $R^{-1} = \displaystyle\sum_{n=0}^{\infty} P_n(x)z^n; \; -1 < x < 1, \; |z| < 1,$
$$R = \sqrt{1 - 2xz + z^2}.$$

Inequality: $|P_n(x)| \le 1, \; -1 \le x \le 1.$

II

Name: Tschebysheff, First Kind   *Symbol*: $T_n(x)$ *Interval*: $[-1, 1]$

Differential Equation: $(1 - x^2)y'' - xy' + n^2 y = 0$
$$y = T_n(x)$$

Explicit Expression: $\dfrac{n}{2} \displaystyle\sum_{m=0}^{[n/2]} (-1)^m \dfrac{(n - m - 1)!}{m!(n - 2m)!} (2x)^{n - 2m} = \cos(n \arccos x) = T_n(x)$

Recurrence Relation: $T_{n+1}(x) = 2xT_n(x) - T_{n-1}(x)$

Weight: $(1 - x^2)^{-1/2}$ *Standardization*: $T_n(1) = 1$

Norm: $\displaystyle\int_{-1}^{+1} (1 - x^2)^{-1/2}[T_n(x)]^2 \, dx = \begin{cases} \pi/2, & n \ne 0 \\ \pi, & n = 0 \end{cases}$

Rodrigues' Formula: $\dfrac{(-1)^n(1 - x^2)^{1/2}\sqrt{\pi}}{2^{n+1}\Gamma(n + \frac{1}{2})} \dfrac{d^n}{dx^n} \{(1 - x^2)^{n - (1/2)}\} = T_n(x)$

Generating Function: $\dfrac{1 - xz}{1 - 2xz + z^2} = \displaystyle\sum_{n=0}^{\infty} T_n(x)z^n, \; -1 < x < 1, \; |z| < 1.$

Inequality: $|T_n(x)| \le 1, \; -1 \le x \le 1.$

* From Beyer, W. H., Ed., *CRC Handbook of Mathematical Sciences*, 5th ed., CRC Press, Boca Raton, 1978, 557—560. With permission.

III

Name: Tschebysheff, Second Kind   *Symbol*: $U_n(x)$ *Interval*: $[-1, 1]$

Differential Equation: $(1 - x^2)y'' - 3xy' + n(n + 2)y = 0$

$$y = U_n(x)$$

Explicit Expression:
$$U_n(x) = \sum_{m=0}^{[n/2]} (-1)^m \frac{(m - n)!}{m!(n - 2m)!} (2x)^{n-2m}$$

$$U_n(\cos \theta) = \frac{\sin[(n + 1)\theta]}{\sin \theta}$$

Recurrence Relation: $U_{n+1}(x) = 2xU_n(x) - U_{n-1}(x)$

Weight: $(1 - x^2)^{1/2}$ *Standardization*: $U_n(1) = n + 1$

Norm: $\displaystyle\int_{-1}^{+1} (1 - x^2)^{1/2}[U_n(x)]^2 \, dx = \frac{\pi}{2}$

Rodrigues' Formula:
$$U_n(x) = \frac{(-1)^n(n + 1)\sqrt{\pi}}{(1 - x^2)^{1/2}2^{n+1}\Gamma(n + \frac{3}{2})} \frac{d^n}{dx^n}\{(1 - x^2)^{n+(1/2)}\}$$

Generating Function:
$$\frac{1}{1 - 2xz + z^2} = \sum_{n=0}^{\infty} U_n(x)z^n, \quad -1 < x < 1, \quad |z| < 1.$$

Inequality: $|U_n(x)| \leq n + 1, \quad -1 \leq x \leq 1.$

IV

Name: Jacobi   *Symbol*: $P_n^{(\alpha,\beta)}(x)$ *Interval*: $[-1, 1]$

Differential Equation:

$$(1 - x^2)y'' + [\beta - \alpha - (\alpha + \beta + 2)x]y' + n(n + \alpha + \beta + 1)y = 0$$

$$y = P_n^{(\alpha,\beta)}(x)$$

Explicit Expression:
$$P_n^{(\alpha,\beta)}(x) = \frac{1}{2^n} \sum_{m=0}^{n} \binom{n + \alpha}{m}\binom{n + \beta}{n - m}(x - 1)^{n-m}(x + 1)^m$$

Recurrence Relation:
$$\begin{aligned}
2(n + 1)(n + \alpha + \beta + 1)&(2n + \alpha + \beta) P_{n+1}^{(\alpha,\beta)}(x) \\
&= (2n + \alpha + \beta + 1)[(\alpha^2 - \beta^2) + (2n + \alpha + \beta + 2) \\
&\quad \times (2n + \alpha + \beta)x] P_n^{(\alpha,\beta)}(x) \\
&\quad - 2(n + \alpha)(n + \beta)(2n + \alpha + \beta + 2) P_{n-1}^{(\alpha,\beta)}(x)
\end{aligned}$$

Weight: $(1 - x)^\alpha(1 + x)^\beta; \quad \alpha, \beta > 1$ *Standardization*: $P_n^{(\alpha,\beta)}(x) = \dbinom{n + \alpha}{n}$

Norm:
$$\int_{-1}^{+1} (1 - x)^\alpha(1 + x)^\beta[P_n^{(\alpha,\beta)}(x)]^2 \, dx = \frac{2^{\alpha+\beta+1}\Gamma(n + \alpha + 1)\Gamma(n + \beta + 1)}{(2n + \alpha + \beta + 1)n!\,\Gamma(n + \alpha + \beta + 1)}$$

Rodrigues' Formula:
$$P_n^{(\alpha,\beta)}(x) = \frac{(-1)^n}{2^n n!(1 - x)^\alpha(1 + x)^\beta} \frac{d^n}{dx^n}\{(1 - x)^{n+\alpha}(1 + x)^{n+\beta}\}$$

IV (Continued)

Generating Function: $R^{-1}(1 - z + R)^{-\alpha}(1 + z + R)^{-\beta} = \sum_{n=0}^{\infty} 2^{-\alpha-\beta} P_n^{(\alpha,\beta)}(x) z^n,$

$$R = \sqrt{1 - 2xz + z^2}, \ |z| < 1$$

Inequality: $\max_{-1 \le x \le 1} | P_n^{(\alpha,\beta)}(x) | = \begin{cases} \binom{n + q}{n} \sim n^q \text{ if } q = \max(\alpha, \beta) \ge -\frac{1}{2} \\[2mm] | P_n^{(\alpha,\beta)}(x') | \sim n^{-1/2} \text{ if } q < -\frac{1}{2} \\ x' \text{ is one of the two maximum points nearest} \\[2mm] \dfrac{\beta - \alpha}{\alpha + \beta + 1} \end{cases}$

V

Name: Generalized Laguerre   *Symbol*: $L_n^{(\alpha)}(x)$ *Interval*: $[0, \infty]$

Differential Equation: $xy'' + (\alpha + 1 - x)y' + ny = 0$

$$y = L_n^{(\alpha)}(x)$$

Explicit Expression: $L_n^{(\alpha)}(x) = \sum_{m=0}^{n} (-1)^m \binom{n + \alpha}{n - m} \frac{1}{m!} x^m$

Recurrence Relation: $(n + 1)L_{n+1}^{(\alpha)}(x) = [(2n + \alpha + 1) - x]L_n^{(\alpha)}(x) - (n + \alpha)L_{n-1}^{(\alpha)}(x)$

Weight: $x^\alpha e^{-x}, \ \alpha > -1$ *Standardization*: $L_n^{(\alpha)}(x) = \dfrac{(-1)^n}{n!} x^n + \cdots$

Norm: $\displaystyle\int_0^{\infty} x^\alpha e^{-x} [L_n^{(\alpha)}(x)]^2 \, dx = \dfrac{\Gamma(n + \alpha + 1)}{n!}$

Rodrigues' Formula: $L_n^{(\alpha)}(x) = \dfrac{1}{n! \, x^\alpha e^{-x}} \dfrac{d^n}{dx^n} \{x^{n+\alpha} e^{-x}\}$

Generating Function: $(1 - z)^{-\alpha-1} \exp\left(\dfrac{xz}{z - 1}\right) = \sum_{n=0}^{\infty} L_n^{(\alpha)}(x) z^n$

Inequality: $| L_n^{(\alpha)}(x) | \le \dfrac{\Gamma(n + \alpha + 1)}{n! \, \Gamma(\alpha + 1)} e^{x/2}; \quad \begin{matrix} x \ge 0 \\ \alpha > 0 \end{matrix}$

$$| L_n^{(\alpha)}(x) | \le \left[2 - \dfrac{\Gamma(\alpha + n + 1)}{n! \, \Gamma(\alpha + 1)} \right] e^{x/2}; \quad \begin{matrix} x \ge 0 \\ -1 < \alpha < 0 \end{matrix}$$

Name: Hermite *Symbol:* $H_n(x)$ *Interval:* $[-\infty, \infty]$

Differential Equation: $y'' - 2xy' + 2ny = 0$

Explicit Expression: $H_n(x) = \displaystyle\sum_{m=0}^{[n/2]} \frac{(-1)^m \, n! \, (2x)^{n-2m}}{m! \, (n-2m)!}$

Recurrence Relation: $H_{n+1}(x) = 2 \, x \, H_n(x) - 2n H_{n-1}(x)$

Weight: e^{-x^2} *Standardization:* $H_n(1) = 2^n x^n + \ldots$

Norm: $\displaystyle\int_{-\infty}^{\infty} e^{-x^2} [H_n(x)]^2 \, dx = 2^n \, n! \, \sqrt{\pi}$

Rodriques' Formula: $H_n(x) = (-1)^n \, e^{x^2} \dfrac{d^n}{dx^n} (e^{-x^2})$

Generating Function: $e^{-t^2 + 2tx} = \displaystyle\sum_{n=0}^{\infty} H_n(x) \, \frac{t^n}{n!}$

Inequality: $|H_n(x)| < e^{\frac{x^2}{2}} \, k \, 2^{n/2} \, \sqrt{n!}$ $k \approx 1.086435$

NORMAL PROBABILITY FUNCTION

Areas under the Standard Normal Curve from 0 to z

z	0	1	2	3	4	5	6	7	8	9
0.0	.0000	.0040	.0080	.0120	.0160	.0199	.0239	.0279	.0319	.0359
0.1	.0398	.0438	.0478	.0517	.0557	.0596	.0636	.0675	.0714	.0754
0.2	.0793	.0832	.0871	.0910	.0948	.0987	.1026	.1064	.1103	.1141
0.3	.1179	.1217	.1255	.1293	.1331	.1368	.1406	.1443	.1480	.1517
0.4	.1554	.1591	.1628	.1664	.1700	.1736	.1772	.1808	.1844	.1879
0.5	.1915	.1950	.1985	.2019	.2054	.2088	.2123	.2157	.2190	.2224
0.6	.2258	.2291	.2324	.2357	.2389	.2422	.2454	.2486	.2518	.2549
0.7	.2580	.2612	.2652	.2673	.2704	.2734	.2764	.2794	.2823	.2852
0.8	.2881	.2910	.2939	.2967	.2996	.3023	.3051	.3078	.3106	.3133
0.9	.3159	.3186	.3212	.3238	.3264	.3289	.3315	.3340	.3365	.3389
1.0	.3413	.3438	.3461	.3485	.3508	.3531	.3554	.3577	.3599	.3621
1.1	.3643	.3665	.3686	.3708	.3729	.3749	.3770	.3790	.3810	.3830
1.2	.3849	.3869	.3888	.3907	.3925	.3944	.3962	.3980	.3997	.4015
1.3	.4032	.4049	.4066	.4082	.4099	.4115	.4131	.4147	.4162	.4177
1.4	.4192	.4207	.4222	.4236	.4251	.4265	.4279	.4292	.4306	.4319
1.5	.4332	.4345	.4357	.4370	.4382	.4394	.4406	.4418	.4429	.4441
1.6	.4452	.4463	.4474	.4484	.4495	.4505	.4515	.4525	.4535	.4545
1.7	.4554	.4564	.4573	.4582	.4591	.4599	.4608	.4616	.4625	.4633
1.8	.4641	.4649	.4656	.4664	.4671	.4678	.4686	.4693	.4699	.4706
1.9	.4713	.4719	.4726	.4732	.4738	.4744	.4750	.4756	.4761	.4767
2.0	.4772	.4778	.4783	.4788	.4793	.4798	.4803	.4808	.4812	.4817
2.1	.4821	.4826	.4830	.4834	.4838	.4842	.4846	.4850	.4854	.4857
2.2	.4861	.4864	.4868	.4871	.4875	.4878	.4881	.4884	.4887	.4890
2.3	.4893	.4896	.4898	.4901	.4904	.4906	.4909	.4911	.4913	.4916
2.4	.4918	.4920	.4922	.4925	.4927	.4929	.4931	.4932	.4934	.4936
2.5	.4938	.4940	.4941	.4943	.4945	.4946	.4948	.4949	.4951	.4952
2.6	.4953	.4955	.4956	.4957	.4959	.4960	.4961	.4962	.4963	.4964
2.7	.4965	.4966	.4967	.4968	.4969	.4970	.4971	.4972	.4973	.4974
2.8	.4974	.4975	.4976	.4977	.4977	.4978	.4979	.4979	.4980	.4981
2.9	.4981	.4982	.4982	.4983	.4984	.4984	.4985	.4985	.4986	.4986
3.0	.4987	.4987	.4987	.4988	.4988	.4989	.4989	.4989	.4990	.4990
3.1	.4990	.4991	.4991	.4991	.4992	.4992	.4992	.4992	.4993	.4993
3.2	.4993	.4993	.4994	.4994	.4994	.4994	.4994	.4995	.4995	.4995
3.3	.4995	.4995	.4995	.4996	.4996	.4996	.4996	.4996	.4996	.4997
3.4	.4997	.4997	.4997	.4997	.4997	.4997	.4997	.4997	.4997	.4998
3.5	.4998	.4998	.4998	.4998	.4998	.4998	.4998	.4998	.4998	.4998
3.6	.4998	.4998	.4999	.4999	.4999	.4999	.4999	.4999	.4999	.4999
3.7	.4999	.4999	.4999	.4999	.4999	.4999	.4999	.4999	.4999	.4999
3.8	.4999	.4999	.4999	.4999	.4999	.4999	.4999	.4999	.4999	.4999
3.9	.5000	.5000	.5000	.5000	.5000	.5000	.5000	.5000	.5000	.5000

F(z) below refers to area under Standard Normal Curve from $-\infty$ to z

z	1.282	1.645	1.960	2.326	2.576	3.090
F(z)	.90	.95	.975	.99	.995	.999
2[1 − F(z)]	.20	.10	.05	.02	.01	.002

PERCENTAGE POINTS, STUDENT'S t-DISTRIBUTION

This table gives values of t such that

$$F(t) = \int_{-\infty}^{t} \frac{\Gamma\left(\frac{n+1}{2}\right)}{\sqrt{n\pi}\,\Gamma\left(\frac{n}{2}\right)} \left(1 + \frac{x^2}{n}\right)^{-\frac{n+1}{2}} dx$$

for n, the number of degrees of freedom, equal to 1, 2, . . ., 30, 40, 60, 120, ∞; and for $F(t) = 0.60, 0.75, 0.90, 0.95, 0.975, 0.99, 0.995,$ and 0.9995. The t-distribution is symmetrical, so that $F(-t) = 1 - F(t)$

F n	.60	.75	.90	.95	.975	.99	.995	.9995
1	.325	1.000	3.078	6.314	12.706	31.821	63.657	636.619
2	.289	.816	1.886	2.920	4.303	6.965	9.925	31.598
3	.277	.765	1.638	2.353	3.182	4.541	5.841	12.924
4	.271	.741	1.533	2.132	2.776	3.747	4.604	8.610
5	.267	.727	1.476	2.015	2.571	3.365	4.032	6.869
6	.265	.718	1.440	1.943	2.447	3.143	3.707	5.959
7	.263	.711	1.415	1.895	2.365	2.998	3.499	5.408
8	.262	.706	1.397	1.860	2.306	2.896	3.355	5.041
9	.261	.703	1.383	1.833	2.262	2.821	3.250	4.781
10	.260	.700	1.372	1.812	2.228	2.764	3.169	4.587
11	.260	.697	1.363	1.796	2.201	2.718	3.106	4.437
12	.259	.695	1.356	1.782	2.179	2.681	3.055	4.318
13	.259	.694	1.350	1.771	2.160	2.650	3.012	4.221
14	.258	.692	1.345	1.761	2.145	2.624	2.977	4.140
15	.258	.691	1.341	1.753	2.131	2.602	2.947	4.073
16	.258	.690	1.337	1.746	2.120	2.583	2.921	4.015
17	.257	.689	1.333	1.740	2.110	2.567	2.898	3.965
18	.257	.688	1.330	1.734	2.101	2.552	2.878	3.922
19	.257	.688	1.328	1.729	2.093	2.539	2.861	3.883
20	.257	.687	1.325	1.725	2.086	2.528	2.845	3.850
21	.257	.686	1.323	1.721	2.080	2.518	2.831	3.819
22	.256	.686	1.321	1.717	2.074	2.508	2.819	3.792
23	.256	.685	1.319	1.714	2.069	2.500	2.807	3.767
24	.256	.685	1.318	1.711	2.064	2.492	2.797	3.745
25	.256	.684	1.316	1.708	2.060	2.485	2.787	3.725
26	.256	.684	1.315	1.706	2.056	2.479	2.779	3.707
27	.256	.684	1.314	1.703	2.052	2.473	2.771	3.690
28	.256	.683	1.313	1.701	2.048	2.467	2.763	3.674
29	.256	.683	1.311	1.699	2.045	2.462	2.756	3.659
30	.256	.683	1.310	1.697	2.042	2.457	2.750	3.646
40	.255	.681	1.303	1.684	2.021	2.423	2.704	3.551
60	.254	.679	1.296	1.671	2.000	2.390	2.660	3.460
120	.254	.677	1.289	1.658	1.980	2.358	2.617	3.373
∞	.253	.674	1.282	1.645	1.960	2.326	2.576	3.291

* This table is abridged from the "Statistical Tables" of R. A. Fisher and Frank Yates published by Oliver & Boyd. Ltd., Edinburgh and London, 1938. It is here published with the kind permission of the authors and their publishers.

ELECTRONIC CONFIGURATION OF THE ELEMENTS

By Laurence S. Foster

References: F. H. Spedding and A. H. Daane, Editors, *The Rare Earths*, John Wiley and Sons, Inc. Publishers, New York, 1961. R. F. Gould, Editor, *Lanthanide–Actinide Chemistry*, Advances in Chemistry Series, No. 71, American Chemical Society, Washington, D.C., 1967: Paper No. 14, Mark Fred, *Electronic Structure of the Actinide Elements.*

Atomic No.	Element	K (1) s	L (2) s p	M (3) s p d	N (4) s p d f	O (5) s p d f	P (6) s p d f	Q (7) s p d f
1	H	1						
2	He	2						
3	Li	2	1					
4	Be	2	2					
5	B	2	2 1					
6	C	2	2 2					
7	N	2	2 3					
8	O	2	2 4					
9	F	2	2 5					
10	Ne	2	2 6					
11	Na	2	2 6	1				
12	Mg	2	2 6	2				
13	Al	2	2 6	2 1				
14	Si	2	2 6	2 2				
15	P	2	2 6	2 3				
16	S	2	2 6	2 4				
17	Cl	2	2 6	2 5				
18	Ar	2	2 6	2 6				
19	K	2	2 6	2 6 ..	1			
20	Ca	2	2 6	2 6 ..	2			
21	Sc	2	2 6	2 6 1	2			
22	Ti	2	2 6	2 6 2	2			
23	V	2	2 6	2 6 3	2			
24	Cr	2	2 6	2 6 5*	1			
25	Mn	2	2 6	2 6 5	2			
26	Fe	2	2 6	2 6 6	2			
27	Co	2	2 6	2 6 7	2			
28	Ni	2	2 6	2 6 8	2			
29	Cu	2	2 6	2 6 10*	1			
30	Zn	2	2 6	2 6 10	2			
31	Ga	2	2 6	2 6 10	2 1			
32	Ge	2	2 6	2 6 10	2 2			
33	As	2	2 6	2 6 10	2 3			
34	Se	2	2 6	2 6 10	2 4			
35	Br	2	2 6	2 6 10	2 5			
36	Kr	2	2 6	2 6 10	2 6			
37	Rb	2	2 6	2 6 10	2 6 ..	1		
38	Sr	2	2 6	2 6 10	2 6 ..	2		
39	Y	2	2 6	2 6 10	2 6 1	2		
40	Zr	2	2 6	2 6 10	2 6 2 ..	2		
41	Nb	2	2 6	2 6 10	2 6 4*..	1		
42	Mo	2	2 6	2 6 10	2 6 5 ..	1		
43	Tc	2	2 6	2 6 10	2 6 6 ..	1		
44	Ru	2	2 6	2 6 10	2 6 7 ..	1		
45	Rh	2	2 6	2 6 10	2 6 8 ..	1		
46	Pd	2	2 6	2 6 10	2 6 10*..	0		
47	Ag	2	2 6	2 6 10	2 6 10 ..	1		
48	Cd	2	2 6	2 6 10	2 6 10 ..	2		
49	In	2	2 6	2 6 10	2 6 10 ..	2 1		
50	Sn	2	2 6	2 6 10	2 6 10 ..	2 2		
51	Sb	2	2 6	2 6 10	2 6 10 ..	2 3		
52	Te	2	2 6	2 6 10	2 6 10 ..	2 4		
53	I	2	2 6	2 6 10	2 6 10 ..	2 5		
54	Xe	2	2 6	2 6 10	2 6 10 ..	2 6		

Atomic No	Element	K (1) s	L (2) s p	M (3) s p d	N (4) s p d f	O (5) s p d f	P (6) s p d f	Q (7) s p d f
55	Cs	2	2 6	2 6 10	2 6 10 ..	2 6 ...	1	
56	Ba	2	2 6	2 6 10	2 6 10 ..	2 6 ...	2	
57	La	2	2 6	2 6 10	2 6 10 ..	2 6 1 ..	2	
58	Ce	2	2 6	2 6 10	2 6 10 2*	2 6	2	
59	Pr	2	2 6	2 6 10	2 6 10 3	2 6	2	
60	Nd	2	2 6	2 6 10	2 6 10 4	2 6	2	
61	Pm	2	2 6	2 6 10	2 6 10 5	2 6	2	
62	Sm	2	2 6	2 6 10	2 6 10 6	2 6	2	
63	Eu	2	2 6	2 6 10	2 6 10 7	2 6	2	
64	Gd	2	2 6	2 6 10	2 6 10 7	2 6 1	2	
65	Tb	2	2 6	2 6 10	2 6 10 9*	2 6	2	
66	Dy	2	2 6	2 6 10	2 6 10 10	2 6	2	
67	Ho	2	2 6	2 6 10	2 6 10 11	2 6 1	2	
68	Er	2	2 6	2 6 10	2 6 10 12	2 6	2	
69	Tm	2	2 6	2 6 10	2 6 10 13	2 6	2	
70	Yb	2	2 6	2 6 10	2 6 10 14	2 6	2	
71	Lu	2	2 6	2 6 10	2 6 10 14	2 6 1	2	
72	Hf	2	2 6	2 6 10	2 6 10 14	2 6 2 ..	2	
73	Ta	2	2 6	2 6 10	2 6 10 14	2 6 3 ..	2	
74	W	2	2 6	2 6 10	2 6 10 14	2 6 4 ..	2	
75	Re	2	2 6	2 6 10	2 6 10 14	2 6 5 ..	2	
76	Os	2	2 6	2 6 10	2 6 10 14	2 6 6 ..	2	
77	Ir	2	2 6	2 6 10	2 6 10 14	2 6 7 ..	2	
78	Pt	2	2 6	2 6 10	2 6 10 14	2 6 9 ..	1	
79	Au	2	2 6	2 6 10	2 6 10 14	2 6 10 ..	1	
80	Hg	2	2 6	2 6 10	2 6 10 14	2 6 10 ..	2	
81	Tl	2	2 6	2 6 10	2 6 10 14	2 6 10 ..	2 1	
82	Pb	2	2 6	2 6 10	2 6 10 14	2 6 10 ..	2 2	
83	Bi	2	2 6	2 6 10	2 6 10 14	2 6 10 ..	2 3	
84	Po	2	2 6	2 6 10	2 6 10 14	2 6 10 ..	2 4	
85	At	2	2 6	2 6 10	2 6 10 14	2 6 10 ..	2 5	
86	Rn	2	2 6	2 6 10	2 6 10 14	2 6 10 ..	2 6	
87	Fr	2	2 6	2 6 10	2 6 10 14	2 6 10 ..	2 6 ...	1
88	Ra	2	2 6	2 6 10	2 6 10 14	2 6 10 ..	2 6 ...	2
89	Ac	2	2 6	2 6 10	2 6 10 14	2 6 10 ..	2 6 1 ..	2
90	Th	2	2 6	2 6 10	2 6 10 14	2 6 10 ..	2 6 2 ..	2
91	Pa	2	2 6	2 6 10	2 6 10 14	2 6 10 2*	2 6 1 ..	2
92	U	2	2 6	2 6 10	2 6 10 14	2 6 10 3	2 6 1 ..	2
93	Np	2	2 6	2 6 10	2 6 10 14	2 6 10 4	2 6 ..	2
94	Pu	2	2 6	2 6 10	2 6 10 14	2 6 10 6	2 6 ..	2
95	Am	2	2 6	2 6 10	2 6 10 14	2 6 10 7	2 6 ..	2
96	Cm	2	2 6	2 6 10	2 6 10 14	2 6 10 7	2 6 1 ..	2
97	Bk	2	2 6	2 6 10	2 6 10 14	2 6 10 9*	2 6 ..	2
98	Cf	2	2 6	2 6 10	2 6 10 14	2 6 10 10	2 6 ..	2
99	Es	2	2 6	2 6 10	2 6 10 14	2 6 10 11	2 6 ..	2
100	Fm	2	2 6	2 6 10	2 6 10 14	2 6 10 12	2 6 ..	2
101	Md	2	2 6	2 6 10	2 6 10 14	2 6 10 13	2 6 ..	2
102	No	2	2 6	2 6 10	2 6 10 14	2 6 10 14	2 6 ..	2
103	Lr	2	2 6	2 6 10	2 6 10 14	2 6 10 14	2 6 1 ..	2
104	—	2	2 6	2 6 10	2 6 10 14	2 6 10 14	2 6 2 ..	2

* Note irregularity.

THE ELEMENTS
C. R. Hammond

One of the most striking facts about the elements is their unequal distribution and occurrence in nature. Present knowledge of the chemical composition of the universe, obtained from the study of the spectra of stars and nebulae, indicates that hydrogen is by far the most abundant element and may account for more than 90% of the atoms or about 75% of the mass of the universe. Helium atoms make up most of the remainder. All of the other elements together contribute only slightly to the total mass.

The chemical composition of the universe is undergoing continuous change. Hydrogen is being converted into helium, and helium is being changed into heavier elements. As time goes on, the ratio of heavier elements increases relative to hydrogen. Presumably, the process is not reversible.

Burbidge, Burbidge, Fowler, and Hoyle, and more recently, Peebles, Penzias, and others have studied the synthesis of elements in stars. To explain all of the features of the nuclear abundance curve — obtained by studies of the composition of the earth, meteorites, stars, etc. — it is necessary to postulate that the elements were originally formed by at least eight different processes: (1) hydrogen burning, (2) helium burning, (3) χ process, (4) e process, (5) s process, (6) r process, (7) p process, and (8) the X process. The X process is thought to account for the existence of light nuclei such as D, Li, Be, and B. Common metals such as Fe, Cr, Ni, Cu, Ti, Zn, etc. were likely produced early in the historyof our galaxy. It is also probable that most of the heavy elements on earth and elsewhere in the universe were originally formed in supernovae, or in the hot interior of stars.

Studies of the solar spectrum have led to the identification of 67 elements in the sun's atmosphere; however, all elements cannot be identified with the same degree of certainty. Other elements may be present in the sun, although they have not yet been detected spectroscopically. The element helium was discovered on the sun before it was found on earth. Some elements such as scandium are relatively more plentiful in the sun and stars than here on earth.

Minerals in lunar rocks brought back from the moon on the Apollo missions consist predominantly of *plagioclase* {(Ca, Na)(Al,Si)O_4O_8} and *pyroxene* {(Ca,Mg,Fe)$_2$ Si$_2$O$_6$} — two minerals common in terrestrial volcanic rock. No new elements have been found on the moon that cannot be accounted for on earth; however, three minerals, *armalcolite* {(Fe,Mg)Ti$_2$O$_5$}, *pyroxferroite* {CaFe$_6$ (SiO$_3$) $_7$}, and *tranquillityite* {Fe$_8$(Zr,Y)Ti$_3$Si$_3$O$_2$}, are new. The oldest known terrestrial rocks are about 3.75 billion years old. One rock, known as the "Genesis Rock," brought back from the Apollo 15 Mission, is about 4.15 billion years old. This is only about one-half billion years younger than the supposed age of the moon and solar system. Lunar rocks appear to be relatively enriched in refractory elements such as chromium, titanium, zirconium, and the rare earths, and impoverished in volatile elements such as the alkali metals, in chlorine, and in noble metals such as nickel, platinum, and gold.

Even older than the "Genesis Rock" are *carbonaceous chondrites,* a type of meteorite that has fallen to earth and has been studied. These are some of the most primitive objects of the solar system yet found. The grains making up these objects probably condensed directly out of the gaseous nebula from which the sun and planets were born. Most of the condensation of the grains probably was completed within 50,000 years of the time the disk of the nebula was first formed — about 4.6 billion years ago. It is now thought that this type of meteorite may contain a small percentage of presolar dust grains. The relative abundances of the elements in these meteorites are about the same as the abundances found in the solar chromosphere.

The X-ray fluorescent spectrometer sent with the Viking I spacecraft to Mars shows that the Martian soil contains about 12 to 16% iron, 14 to 15% silicon, 3 to 8% calcium, 2 to 7% aluminum, and one half to 2% titanium. The gas chromatograph — mass spectrometer on Viking II found no trace of organic compounds that should be present if life ever existed there.

F. W. Clarke and others have carefully studied the composition of rocks making up the crust of the earth. Oxygen accounts for about 47% of the crust, by weight, while silicon comprises about 28% and aluminum about 8%. These elements, plus iron, calcium, sodium, potassium, and magnesium, account for about 99% of the composition of the crust.

Many elements such as tin, copper, zinc, lead, mercury, silver, platinum, antimony, arsenic, and gold, which are so essential to our needs and civilization, are among some of the rarest elements in the earth's crust. These are made available to us only by the processes of concentration in ore bodies. Some of the so-called *rare-earth* elements have been found to be much more plentiful than originally thought and are about as abundant as uranium, mercury, lead, or bismuth. The least abundant rare-earth or *lanthanide* element, thulium, is now believed to be more plentiful on earth than silver, cadmium, gold, or iodine, for example. Rubidium, the 16th most abundant element, is more plentiful than chlorine while its compounds are little known in chemistry and commerce.

It is now thought that at least 24 elements are essential to living matter. The four most abundant in the human body are hydrogen, oxygen, carbon, and nitrogen. The seven next most common, in order of abundance, are calcium, phosphorus, chlorine, potassium, sulfur, sodium, and magnesium. Iron, copper, zinc, silicon, iodine, cobalt, manganese, molybdenum, fluorine, tin, chromium, selenium, and vanadium are needed and play a role in living matter. Boron is also thought essential for some plants, and it is possible that aluminum, nickel, and germanium may turn out to be necessary.

Ninety-one elements occur naturally on earth. Minute traces of plutonium-244 have been discovered in rocks mined in Southern California. This discovery supports the theory that heavy elements were produced during creation of the solar system. While technetium and promethium have not yet been found naturally on earth, they have been found to be present in stars. Technetium has been identified in the spectra of certain "late" type stars, and promethium lines have been identified in the spectra of a faintly visible star HR465 in Andromeda. Promethium must have been made very recently near the star's surface for no known isotope of this element has a half-life longer than 17.7 years.

It has been suggested that californium is present in certain stellar explosions known as supernovae; however, this has not been proved. At present no elements are found elsewhere in the universe that cannot be accounted for here on earth.

All atomic mass numbers from 1 to 238 are found naturally on earth except for masses 5 and 8. About 280 stable and 67

THE ELEMENTS (continued)

naturally radioactive isotopes occur on earth totalling 347. In addition, the neutron, technetium, promethium, and the transuranic elements (lying beyond uranium) up to Element 107 have been produced artificially. West German scientists in late 1982 confirmed the existance of Element 107 made by Soviet scientists and announced synthesis of Element 109. Laboratory processes have now extended the radioactive mass numbers beyond 238 to 266. Each element from atomic number 1 to 107 is known to have at least one radioactive isotope. About 1700 different nuclides (the name given to different kinds of nuclei, whether they are of the same or different elements) are now recognized. Many stable and radioactive isotopes are now produced and distributed by the Oak Ridge National Laboratory, Oak Ridge, Tenn., U.S.A., to customers licensed by the U.S. Atomic Energy Commission.

The nucleus of an atom is characterized by the number of protons it contains, usually denoted by Z, and by the number of neutrons, N. Isotopes of an element have the same value of Z, but different values of N. The *mass number* A, is the sum of Z and N. For example, Uranium-238 has a mass number of 238, and would contain 92 protons and 146 neutrons.

There is evidence that the definition of chemical elements must be broadened to include the electron. Several compounds known as *electrides*, have recently been made of alkaline metal elements and electrons. A relatively stable combination of a positron and electron, known as *positronium* has also been studied.

In addition to the proton, neutron, and electron, there are considerably more than 100 other fundamental particles which have been discovered or hypothesized. The majority of these fall into one of two classes, *leptons* or *hadrons.* The leptons comprise just four known particles, the electron, the *muon* (μ meson), and two kinds of *neutrinos.* The muon is essentially similar to the electron and has a charge of -1, but it is 200 times heavier. The neutrino is either of two stable particles of small (probably zero) rest mass, carrying no charge. Also there are four *antileptons,* identical to the corresponding leptons in some respects, such as mass, but they have properties exactly opposite those of the leptons. The *positron,* for example, is an antilepton, with a charge of $+1$. Leptons cannot be broken into smaller units and are considered to be elementary. On the other hand, hadrons are complex and thought to have internal structure. Protons and neutrons, which make up atomic nuclei, are hadrons.

Elementary particle physics is not yet clearly understood, but groupings and arrangements of these particles have been made resembling the periodic table of chemical elements. This has led to the speculation that hadrons are composed of three (or possibly more simpler components called *quarks.* Quarks are presumed to be elementary particles. There is presently no evidence that quarks exist in isolation. Many physicists now hold that all the matter and energy in the universe is controlled by four fundamental natural forces: the electromagnetic force, gravity, a weak nuclear force, and a strong nuclear force. Each of these natural forces is passed back and forth among the basic particles of matter by unique force-carrying particles. The electromagnetic force is carried by the *photon,* the weak nuclear force by the intermediate vector *boson,* and gravity by the *gravitation.* There is now evidence of the existence of a new particle, known as the *gluon,* that binds quarks together by carrying the strong nuclear force.

The available evidence leads to the conclusion that elements 89 (actinum) through 103 (lawrencium) are chemically similar to the rare-earth or lanthanide elements (elements 57 to 71, inclusive). These elements therefore have been named *actinides* after the first member of this series. Those elements beyond uranium that have been produced artificially have the following names and symbols: neptunium, 93 (Np); plutonium, 94 (Pu); americium, 95 (Am); curium, 96 (Cm); berkelium, 97 (Bk); californium, 98 (Cf); einsteinium, 99 (Es); fermium, 100 (Fm); mendelevium, 101 (Md); nobelium, 102 (No); and lawrencium 103 (Lr). It is now claimed that Elements 104, 105, 106, 107, and 109 have been produced and identified. Names and chemical symbols have been suggested for Elements 104 and 105, but have not been officially adopted. Names for Elements 106, 107, and 109 have not yet been suggested.

Element 104 is expected to have chemical properties similar to those of hafnium and would not be a member of the actinide series. Element 105 probably would have chemical properties similar to those of tantalum, Element 106 similar to tungsten, and Element 107 similar to rhenium, and Element 109 similar to iridium.

There is still thought to be a possibility of producing elements beyond Element 109 either by bombardment of heavy isotopic targets with heavy ions, or by the irradiation of uranium or transuranic elements with the instantaneous high flux of neutrons produced by underground nuclear explosions. The limit will be set by the yields of the nuclear reactions and by the half-lives of radioactive decay. It has been suggested that Elements 102 and 103 have abnormally short lives only because they are in a pocket of instability, and that this region of instability might begin to "heal" around Element 105. If so, it may be possible to produce heavier isotopes with longer half-lives. It has also been suggested that Element 114, with a mass number of 298, and Element 126, with a mass number of 310, may be sufficiently stable to make discovery and identification possible. Calculations indicate that Element 110, a homolog of platinum, may have a half-life of as long as 100 million years. Searches have already been made by workers in a number of laboratories for Element 110 and its neighboring elements in naturally occurring platinum. Recent studies of the xenon component ($Xe^{131-136}$) of certain carbonaceous chronditic meteorites suggest that Elements 113, 114, or 115 may have been its progenitor.

There are many claims in the literature of the existence of various allotropic modifications of the elements, some of which are based on doubtful or incomplete evidence. Also, the physical properties of an element may change drastically by the presence of small amounts of impurities. With new methods of purification, which are now able to produce elements with 99.9999% purity, it has been necessary to restudy the properties of the elements. For example, the melting point of thorium changes by several hundred degrees by the presence of a small percentage of ThO_2 as an impurity. Ordinary commercial tungsten is brittle and can be worked only with difficulty. Pure tungsten, however, can be cut with a hacksaw, forged, spun, drawn, or extruded. In general, the value of physical properties given here applies to the pure element, when it is known.

Many of the chemical elements and their compounds are toxic and should be handled with due respect and care. In recent years there has been a greatly increased knowledge and awareness of the health hazards associated with chemicals, radioactive

B-3

THE ELEMENTS (continued)

materials, and other agents. Anyone working with the elements and certain of their compounds should become thoroughly familiar with the proper safeguards to be taken. Reference should be made to publications such as the following:

1. *Code of Federal Regulations,* Title 29, Labor, chapter XVII, section 1910.93 of subpart G, redesignated as 1910.1000 at 40 FR (Federal Register) 23072. May 28, 1975; amended at 41 FR 11505, March 19, 1976; 4l FR 35184, August 20, 1976; FR 46784, October 22, 1976; 42 FR 3304, January 18, 1977 (corrections) and additional amendments and corrections as issued, U.S. Government Printing Office, Supt. of Documents, Washington, D.C.
2. *Code of Federal Regulations,* Title 10, Energy, Chapter 1, Nuclear Regulatory Commission, section 20.103 — 20.108; 20.201 — 207; 20.301 — 305; 20.401 — 409; 20.501 — 2; 20.601; appendices, correstions, and amendments.
3. *Occupational Safety and Health Reporter* (latest edition with amendments and corrections), Bureau of National Affairs, Washington, D.C.
4. *Atomic Energy Law Reporter,* Commerce Clearing House, Chicago, IL.
5. *Nuclear Regulation Reporter,* Commerce Clearing House, Chicago, IL.
6. *Maximum Permissible Body Burdens and Maximum Permissible Concentrations of Radionuclides in Air and in Water for Occupational Exposure,* with addenda, U.S. Department of Commerce, N.B.S, Handbook No. 69, (NCRP Report No. 22), latest edition, National Council on Radiation Protection and Measurements (NCRP), Bethesda, MD.; also refer to Permissible Quarterly Intakes of Radionuclides. Handbook of Chemistry and Physics, 64th Edition.
7. *TLVs® Threshold Limit Values for Chemical Substances and Physical Agents in Workroom Environment with Intended Changes,* latest edition, American Conference of Governmental Industrial Hygienists, Cincinnati, Ohio.

Actinium — (Gr. *aktis, aktinos,* beam or ray), Ac; at. wt. (227); at no. 89; m.p. 1050°C; b.p. 3200± 300°C (est.); sp. gr. 10.07 (calc.). Discovered by Andre Debierne in 1899 and independently by F. Giesel in 1902. Occurs naturally in association with uranium minerals. Actinium-227, a decay product of uranium-235, is a beta emitter with a 2l.6-year half-life. Its principal decay products are thorium-227 (18.5-day half-life), radium-223 (11.4-day half-life), and a number of short-lived products including radon, bismuth, polonium, and lead isotopes. In equilibrium with its decay products, it is a powerful source of alpha rays. Actinium metal has been prepared by the reduction of actinium fluoride with lithium vapor at about 1100 to 1300°C. The chemical behavior of actinium is similar to that of the rare earths, particularly lanthanum. Purified actinium comes into equilibrium with its decay products at the end of 185 days, and then decays according to its 21.6-year half-life. It is about 150 times as active as radium, making it of value in the production of neutrons.

Aluminum — (L. *alumen, alum*), Al; at. wt. 26.98154; at. no. 13; m.p. 660.37°C; b.p. 2467°C; sp. gr. 2.6989 (20°C); valence 3. The ancient Greeks and Romans used *alum* in medicine as an astringent, and as a mordant in dyeing. In 1761 de Morveau proposed the name *alumine* for the base in alum, and Lavoisier, in 1787, thought this to be the oxide of a still undiscovered metal. Wohler is generally credited with having isolated the metal in 1827, although an impure form was prepared by Oersted two years earlier. In 1807, Davy proposed the name *alumium* for the metal, undiscovered at that time, and later agreed to change it to *aluminum.* Shortly thereafter, the name *aluminium* was adopted to conform with the "ium" ending of most elements, and this spelling is now in use elsewhere in the world. *Aluminium* was also the accepted spelling in the U.S. until 1925, at which time the American Chemical Society officially decided to use the name *aluminum* thereafter in their publications. The method of obtaining aluminum metal by the electrolysis of alumina dissolved in *cryolite* was discovered in 1886 by Hall in the U.S. at about the same time by Heroult in France. Cryolite, a natural ore found in Greenland, is no longer widely used in commercial production, but has been replaced by an artifical mixture of sodium, aluminum, and calcium fluorides. *Bauxite,* an impure hydrated oxide ore, is found in large deposits in Jamaica, Australia, Surinam, Guyana, Arkansas, and elsewhere. The Bayer process is most commonly used today to refine bauxite so it can be accommodated in the Hall-Heroult refining process, used to produce most aluminum. Aluminum can now be produced from clay, but the process is not economically feasible at present. Aluminum is the most abundant metal to be found in the earth's crust (8.1%), but is never found free in nature. In addition to the minerals mentioned above, it is found in feldspars, granite, and in many other common minerals. Pure aluminum, a silvery-white metal, possesses many desirable characteristics. It is light, nontoxic, has a pleasing appearance, can easily be formed, machined, or cast, has a high thermal conductivity, and has excellent corrosion resistance. It is nonmagnetic and nonsparking, stands second among metals in the scale of malleability, and sixth in ductility. It is extensively used for kitchen utensils, outside building decoration, and in thousands of industrial applications where a strong, light, easily constructed material is needed. Although its electrical conductivity is only about 60% that of copper per area of cross section, it is used in electrical transmission lines because of its light weight. Pure aluminum is soft and lacks strength, but it can be alloyed with small amounts of copper, magnesium, silicon, manganese, and other elements to impart a variety of useful properties. These alloys are of vital importance in the construction of modern aircraft and rockets. Aluminum, evaporated in a vacuum, forms a highly reflective coating for both visible light and radiant heat. These coatings soon form a thin layer of the protective oxide and do not deteriorate as do silver coatings. They have found application in coatings for telescope mirrors, in making decorative paper, packages, toys, and in many other uses. The compounds of greatest importance are aluminum oxide, the sulfate, and the soluble sulfate with potassium (alum). The oxide, alumina, occurs naturally as ruby, sapphire, corundum, and emery, and is used in glassmaking and refractories. Synthetic ruby and sapphire have found application in the construction of lasers for producing coherent light. In 1852, the price of aluminum was about $545/lb, and just before Hall's discovery in 1886, about $11.00. The price rapidly dropped to 30c and has been as low as 15c/lb.

Americium — (the Americas), Am; at. wt. 243; at. no. 95; m.p. 994± 4°C; b.p. 2607°C; sp. gr. 13.67 (20°C); valence 2, 3, 4, 5, or 6. Americium was the fourth transuranium element to be discovered; the isotope Am^{241} was identified by Seaborg, James, Morgan, and Ghiorso late in 1944 at the wartime Metallurgical Laboratory (now the Argonne National Laboratory) of the University of Chicago as the result of successive neutron capture reactions by plutonium isotopes in a nuclear reactor:

$$Pu^{239}\,(n\gamma)Pu^{240}\,(n\gamma)Pu^{241} \xrightarrow{\beta} Am^{241}$$

Since the isotope Am^{241} can be prepared in relatively pure form by extraction as a decay product over a period of years from strongly neutron-bombarded plutonium Pu^{241} this isotope is used for much of the chemical investigation of this element. Better suited is the isotope Am^{243} due to its longer half-life (8.8× 10^3 years as compared to 470 years for Am^{241}). A mixture of the isotopes Am^{241}, Am^{242}, and Am^{243} can be prepared by intense neutron irradiation of Am^{241} according to the reactions Am^{241} (n, γ) Am^{242} (n, γ) Am^{243}. Nearly isotopically pure Am^{243} can be prepared by a sequence of neutron bombardments and chemical separations as follows: neutron bombardment of Am^{241} yields Pu^{242} by the reactions Am^{241} (n, γ) Am^{242} → Pu^{242}, after chemical separation the Pu^{242} can be transformed to Am^{243} via the reactions Pu^{242} (n, γ) Pu^{243} → Am^{243}, and the Am^{243} can be chemically separated. Fairly pure Pu^{242} can be prepared more simply by very intense neutron irradiation of Pu^{239} as the result of successive neutron-capture reactions. Americium metal has been prepared by reducing the trifluoride with barium vapor at 1000 to 1200°C or the dioxide by lanthanum metal. The luster of freshly prepared americium metal is whiter and more silvery than plutonium or neptunium prepared in the same manner. It appears to be more malleable than uranium or neptunium and tarnishes slowly in dry air at room temperature. Americium is thought to exist in two forms: an alpha form which has a double hexagonal close-packed structure and a loose-packed cubic beta form. Americium must be handled with great care to avoid personal contamination. As little as 0.03 μg of Am^{241} is the maximum permissible total body burden. The alpha activity from Am^{241} is about three times that of radium. When gram quantities of Am^{241} are handled, the intense gamma activity makes exposure a serious problem. Americium dioxide, AmO$_2$, is the most important oxide. AmF$_3$, AmF$_4$, AmCl$_3$, AmBr$_3$, AmI$_3$, and other compounds have been prepared. The isotope Am^{241} has been used as a portable source for gamma radiography. It has also been used as a radioactive glass thickness gage for the flat glass industry, and as a source of ionization for smoke detectors. Americium-241 is available from the A.E.C. at a cost of $150/g and Americium-243 at a cost of $100/mg.

Antimony — (Gr. anti plus monos — a metal not found alone), Sb, at. wt. 121.75; at. no. 51; m.p. 630.74°C; b.p. 1950°C; sp. gr. 6.691 (20°C); valence 0, −3, +3, or +5. Antimony was recognized in compounds by the ancients and was known as a metal at the beginning of the 17th century and possibly much earlier. It is not abundant, but is found in over 100 mineral species. It is sometimes found native, but more frequently as the sulfide, *stibnite* (Sb$_2$S$_3$); it is also found as antimonides of the heavy metals, and as oxides. It is extracted from the sulfide by roasting to the oxide, which is reduced by salt and scrap iron; from its oxides it is also prepared by reduction with carbon. Two allotropic forms of antimony exist: the normal stable, metallic form, and the amorphous gray form. The so-called explosive antimony is an ill-defined material always containing an appreciable amount of halogen; therefore, it no longer warrants consideration as a separate allotrope. The yellow form, obtained by oxidation of *stibine,* SbH$_3$, is probably impure, and is not a distinct form. Metallic antimony is an extremely brittle metal of a flaky, crystalline texture. It is bluish white and has a metallic luster. It is not acted on by air at room temperature, but burns brilliantly when heated with the formation of white fumes of Sb$_2$O$_3$. It is a poor conductor of heat and electricity, and has a hardness of 3 to 3.5. Antimony, available commercially with a purity of 99.999 + %, is finding use in semiconductor technology for making infrared detectors, diodes, and Hall-effect devices. Commercial-grade antimony is widely used in alloys with percentages ranging from 1 to 20. It greatly increases the hardness and mechanical strength of lead. Batteries, antifriction alloys, type metal, small arms and tracer bullets, cable sheathing, and minor products use about half the metal produced. Compounds taking up the other half are oxides, sulfides, sodium antimonate, and antimony trichloride. These are used in manufacturing flame- proofing compounds, paints, ceramic enamels, glass, and pottery. Tartar emetic (hydrated potassium antimonyltartate) has been used in medicine. Antimony and many of its compounds are toxic. Exposure to antimony and its compounds should not exceed 0.5 Mg/M^3 (8-hr time weighted average 40-hr work week).

Argon — (Gr. *argos,* inactive), Ar; at. wt. 39.948; at. no. 18; freezing pt. −189.2°C; b.p. −185.7°C; density 1.7837 g/l. Its presence in air was suspected by Cavendish in 1785, discovered by Lord Rayleigh and Sir William Ramsay in 1894. The gas is prepared by fractionation of liquid air, the atmosphere containing 0.94% argon. The atmosphere of Mars contains 1.6% of Ar^{40} and 5 p.p.m. of Ar^{36}. Argon is two and one half times as soluble in water as nitrogen, having about the same solubility as oxygen. It is recognized by the characteristic lines in the red end of the spectrum. It is used in electric light bulbs and in fluorescent tubes at a pressure of about 3 mm, and in filling photo tubes, glow tubes, etc. Argon is also used as an inert gas shield for arc welding and cutting, as a blanket for the production of titanium and other reactive elements, and as a protective atmosphere for growing silicon and germanium crystals. Argon is colorless and odorless, both as a gas and liquid. It is available in high-purity form. Commercial argon is available at a cost of about 3c per cubic foot. Argon is considered to be a very inert gas and is not known to form true chemical compounds, as do krypton, xenon, and radon. However, it does form a hydrate having a dissociation pressure of 105 atm at 0°C. Ion molecules such as (ArKr)$^+$, (ArXe)$^+$, (NeAr)$^+$ have been observed spectroscopically. Argon also forms a clathrate with β hydroquinone. This clathrate is stable and can be stored for a considerable time, but a true chemical bond does not exist. Van der Waals' forces act to hold the argon. Naturally occurring argon is a mixture of three isotopes. Five other radioactive isotopes are now known to exist.

Arsenic — (L. *arsenicum,* Gr. *arsenikon,* yellow orpiment, identified with *arsenikos,* male, from the belief that metals were different sexes; Arab, *az-zernikh,* the orpiment from Persian *zerni-zar,* gold), As; at. wt. 74.9216; at. no. 33; valence −3, 0, +3 or +5. Elemental arsenic occurs in two solid modifications: yellow, and gray or metallic, with specific gravities of

THE ELEMENTS (continued)

1.97, and 5.73, respectively. Gray arsenic, the ordinary stable form, has a m.p. of 817°C (28 atm) and sublimes at 613°C. Several other allotropic forms of arsenic are reported in the literature. It is believed that Albertus Magnus obtained the element in 1250 A.D. In 1649 Schroeder published two methods of preparing the element. It is found native, in the sulfides *realgar* and *orpiment,* as arsenides and sulfarsenides of heavy metals, as the oxide, and as arsenates. *Mispickel* or arsenopyrite (FeSAs) is the most common mineral, from which on heating the arsenic sublimes leaving ferrous sulfide. The element is a steel gray, very brittle, crystalline, semimetallic solid; it tarnishes in air, and when heated is rapidly oxidized to arsenous oxide (As_2O_3) with the odor of garlic. Arsenic and its compounds are poisonous. Exposure to arsenic and its compounds (as/As) should not exceed 0.5 mg/M^3 (8-hr time-weighted average-40 hr work week.) These values, however, are being studied, and may be lowered. Arsenic is also used in bronzing, pyrotechny, and for hardening and improving the sphericity of shot. The most important compounds are white arsenic (As_2O_3), the sulfide, Paris green $3Cu(AsO_2)_2 \cdot Cu(C_2H_3O_2)_2$, calcium arsenate, and lead arsenate; the last three have been used as agricultural insecticides and poisons. Marsh's test makes use of the formation and ready decomposition of arsine (AsH_3). Arsenic is available in high-purity form. It is finding increasing uses as a doping agent in solid-state devices such as transistors. Gallium arsenide is used as a laser material to convert electricity directly into coherent light.

Astatine — (Gr. *astatos,* unstable), At; at. wt. ~210; at. no. 85; m.p.302°C; b.p. 337°C (est.); valence probably 1, 3, 5, or 7. Synthesized in 1940 by D.R. Corson, K. R. MacKenzie, and E. Segre at the University of California by bombarding bismuth with alpha particles. The longest-lived isotope, At^{210}, has a half-life of only 8.3 hr. Twenty isotopes are now known. Minute quantities of At^{215}, At^{218}, amd ^{219} exist in equilibrium in nature with naturally occurring uranium and thorium isotopes, and traces of At^{217} are in equilibrium with U^{233} and Np^{239} resulting from interaction of thorium and uranium with naturally produced neutrons. The total amount of astatine present in the earth's crust, however, totals less than 1 oz. Astatine can be produced by bombarding bismuth with energetic alpha particles to obtain the relatively long-lived $At^{209-211}$, which can be distilled from the target by heating it in air. Only about 0.05 µg of astatine have been prepared to date. The "time of flight" mass spectrometer has been used to confirm that this highly radioactive halogen behaves chemically very much like other halogens, particularly iodine. The interhalogen compounds AtI, AtBr, and AtCl are known to form, but it is not yet known if astatine forms diatomic astatine molecules. HAt and CH_3At (methyl astatide) have been detected. Astatine is said to be more metallic than iodine, and, like iodine, it probably accumulates in the thyroid gland. Workers at the Brookhaven National Laboratory have recently used reactive scattering in crossed molecular beams to identify and measure elementary reactions involving astatine.

Barium — (Gr. *barys,* heavy), Ba; at. wt. 137.33 ±0.01, at. no. 56; m.p. 725°C; b.p.1640°C; sp. gr. 3.5 (20°C); valence 2. Baryta was distinguished from lime by Scheele in 1774; the element was discovered by Sir Humphrey Davy in 1808. It is found only in combination with other elements, chiefly in *barite* or *heavy spar* (sulfate) and *witherite* (carbonate) and is prepared by electrolysis of the chloride. Barium is a metallic element, soft, and when pure is silvery white like lead; it belongs to the alkaline earth group, resembling calcium chemically. The metal oxidizes very easily and should be kept under petroleum or other suitable oxygen-free liquids to exclude air. It is decomposed by water or alcohol. The metal is used as a "getter" in vacuum tubes. The most important compounds are the peroxide (BaO_2), chloride, sulfate, carbonate, nitrate, and chlorate. Lithopone, a pigment containing barium sulfate and zinc sulfide, has good covering power, and does not darken in the presence of sulfides. The sulfate, as permanent white or *blanc fixe,* is also used in paint, in X-ray diagnostic work, and in glassmaking. *Barite* is extensively used as a weighting agent in oilwell drilling fluids, and also in making rubber. The carbonate has been used as a rat poison, while the nitrate and chlorate give colors in pyrotechny. The impure sulfide phosphoresces after exposure to the light. The compounds and the metal are not expensive. Barium metal (99.5 + % pure) costs about $20.00/ lb. All barium compounds that are water or acid soluble are poisonous. Naturally occurring barium is a mixture of seven stable isotopes. Thirteen other radioactive isotopes are known to exist.

Berkelium — (*Berkeley,* home of the University of California), Bk; at. wt. 247; at. no. 97; valence 3 or 4; sp. gr. 14 (est.). Berkelium, the eighth member of the actinide transition series, was discovered in December 1949 by Thompson, Ghiorso, and Seaborg, and was the fifth transuranium element synthesized. It was produced by cyclotron bombardment of milligram amounts of Am^{241} with helium ions at Berkeley, California. The first isotope produced had a mass number of 243 and decayed with a half-life of 4.6 hr. Eight isotopes are now known and have been synthesized. The existence of Bk^{249}, with a half-life of 314 days, make it feasible to isolate berkelium in weighable amounts so that its properties can be investigated with macroscopic quantities. One of the first visible amounts of a pure berkelium compound, berkelium chloride, was produced in 1962. It weighed 3 billionth of a gram. Berkelium has not yet been prepared in elemental form, but it is expected to be a silvery metal, easily soluble in dilute mineral acids, and readily oxidized by air or oxygen at elevated temperatures to form the oxide. X-ray diffraction methods have been used to identify the following compounds: BkO_2, Bk_2O_3, BkF_3, $BkCl_3$, and BkOCl. As with other actinide elements, berkelium tends to accumulate in the skeletal system. The maximum permissible body burden of Bk^{249} in the human skeleton is about 0.0004 µ. Because of its rarity, berkelium presently has no commercial or technological use.

Beryllium — (Gr. *beryllos, beryl;* also called Glucinium or Glucinum, Gr. *glykys,* sweet), Be; at. wt. 9.01218; at. no. 4; m.p. 127±5°C; b.p. 2970°C; sp. gr. 1.848 (20° C); valence 2. Discovered as the oxide by Vauquelin in beryl and in emeralds in 1798. The metal was isolated in 1828 by Wohler and by Bussy independently by the action of potassium on beryllium chloride. Beryllium is found in some 30 mineral species, the most important of which are *bertrandite, beryl, chrysoberyl,* and *phenacite. Aquamarine* and emerald are precious forms of *beryl. Beryl* ($3BeO \cdot Al_2O_3 \cdot 6SiO_2$) and *bertrandite* ($4BeO \cdot 2SiO_2 \cdot H_2O$) are the most important commercial sources of the element and its compounds. Most of the metal is now prepared by reducing beryllium fluoride with magnesium metal. Beryllium metal did not become readily available to industry until 1957. The metal, steel gray in color, has many desirable properties. It is one of the lightest of all metals, and has one of the highest melting points of the light metals. Its modulus of elasticity is about one third greater than that of steel. It resists attack by concentrated nitric acid, has excellent thermal conductivity, and is nonmagnetic. It has a high permeability to X-rays, and when bombarded by alpha particles, as from radium or polonium, neutrons are produced in the ratio of about 30 neutrons/million alpha particles. At ordinary temperatures beryllium resists oxidation in air, although its ability to scratch glass is probably due to the formation of a thin layer of the oxide. Beryllium is used as an alloying agent in producing beryllium copper; which is extensively used for springs, electrical contacts, spot-welding electrodes, and nonsparking tools. It has found application as a structural material for high-speed aircraft, missiles, and spacecraft, and communication satellites. It is being used in the windshield frame, brake discs, support beams, and other structural components of the space shuttle. Because beryllium is relatively transparent to X-rays, ultra-thin Be-foil is finding use in X-ray lithography for reproduction of micro-miniature integrated circuits.

Beryllium is used in nuclear reactors as a reflector or moderator for it has a low thermal neutron absorption cross section. It is used in gyroscopes, computer parts, and instruments where lightness, stiffness, and dimensional stability are required. The oxide has a very high melting point and is also used in nuclear work and ceramic applications. Beryllium and its salts are toxic and should be handled with the greatest of care. Beryllium and its compounds should not be tasted to verify the sweetish nature of beryllium (as did early experimenters). The metal, its alloys, and its salts can be handled safely if certain work codes are observed, but no attempt should be made to work with beryllium before becoming familiar with proper safeguards. Exposure to beryllium dust in air should be limited to 2μ g/M³ (8-hr time-weighted average — 40-hr week), with a ceiling concentration of 5μ g/M³. A maximum peak above the acceptable ceiling concentration for an 8-hr shift is 25 μ g/M³ for a maximum duration of 30 min. These values are being reviewed and studied.

Beryllium metal in vacuum cast billet form is priced roughly at \$150/lb. Fabricated forms are more expensive.

Bismuth — (Ger. *Weisse Masse,* white mass; later *Wisuth* and *Bisemutum*), Bi; at. wt. 208.9808; at. no. 83; m.p. 271.3°C; b.p. 1560± 5°C; sp. gr. 9.747 (20°C); valence 3 or 5. In early times bismuth was confused with tin and lead. Claude Geoffroy the Younger showed it to be distinct from lead in 1753. It is a white, crystalline, brittle metal with a pinkish tinge. It occurs native. The most important ores are *bismuthinite* or bismuth glance (Bi_2S_3) and *bismite* (Bi_2O_3). Peru, Japan, Mexico, Bolivia, and Canada are major bismuth producers. Much of the bismuth produced in the U.S. is obtained as a by-roduct in refining lead, copper, tin, silver, and gold ores. Bismuth is the most diamagnetic of all metals, and the thermal conductivity is lower than any metal, except mercury. It has a high electrical resistance, and has the highest Hall effect of any metal (i.e., greatest increase in electrical resistance when placed in a magnetic field). "Di manol" is a permanent magnet of high coercive force, made of MnBi, by the U.S. Naval Surface Weapons Center. Bismuth expands 3.32% on solidification. This property makes bismuth alloys particularly suited to the making of sharp castings of objects subject to damage by high temperatures. With other metals such as tin, cadmium, etc. bismuth forms low melting alloys which are extensively used for safety devices used in fire detection and extinguishing systems. Bismuth is used in producingmalleable irons and is finding use as a catalyst for making acrylic fibers. When bismuth is heated in air it burns with a blue flame, forming yellow fumes of the oxide. The metal is also used as a thermocouple material (has highest negativy known), and has found application as a carrier for U^{235} or U^{233} fuel in atomic reactors. Its soluble salts are characterized by forming insoluble basic salts on the addition of water, a property sometimes used in detection work. Bismuth oxychloride is used extensively in cosmetics. Bismuth subnitrate and subcarbonate are used in medicine. Bismuth metal costs about \$8/lb in large quantities.

Boron — (Ar. *Buraq,* Pers. *Burah*), B; at. wt. 10.81; at. no. 5; m.p. 2079°C; b.p. sublimes 2550°C; sp. gr. of crystals 2.34, of amorphous variety 2.37; valence 3. Boron compounds have been known for thousands of years, but the element was not discovered until 1808 by Sir Humphry Davy and by Gay-Lussac and Thenard. The element is not found free in nature, but occurs as orthoboric acid usually in certain volcanic spring waters and as borates in *borax* and *colemanite. Ulexite,* another foron mineral, is interesting as it is nature's own version of "fiber optics." By far the most important source of boron is the mineral *rasorite,* also known as *kernite,* found in the Mojave dersert of California. Extensive *borax* deposits are also found in Turkey. Boron exists naturally as 19.78%₅ B^{10} isotope and 80.22%₅B^{11} isotope. High-purity crystalline boron may be prepared by the vapor phase reduction of boron trichloride or tribromide with hydrogen on electrically heated filaments. The impure, or amorphous, boron, a brownish-black powder, can be obtained by heating the trioxide with magnesium powder. Boron of 99.9999% purity has been produced and is available commercially. Elemental boron has an energy band gap of 1.50 to 1.56 eV, which is higher than that of either silicon or germanium. It has interesting optical characteristics, transmitting portions of the infrared, and is a poor conductor of electricity at room temperature, but a good conductor at high temperature. Amorphous boron is used in pyrotechnic flares to provide a distinctive green color, and in rockets as an igniter. The most important compounds of boron are boric, or boracic, acid, widely used as a mild antiseptic, and borax ($Na_2B_4O_7 \cdot 10H_2O$), which serves as a cleansing flux in welding and as a water softener in washing powders. Boron compounds are used in production of enamels for covering steel of refrigerators, washing machines, and like products. Boron compounds are also extensively used in the manufacture of borosilicate glasses. Other boron compounds show promise in treating arthritis. The isotope boron 10 is used as a control for nuclear reactors, as a shield for nuclear radiation, and in instruments used for detecting neutrons. Boron nitride has remarkable properties and can be used to make a material as hard as diamond. The nitride also behaves like an electrical insulator but conducts heat like a metal. It also has lubricating properties similar to graphite. The hydrides are easily oxidized with considerable energy liberation, and have been studied for use as rocket fuels. Demand is increasing for boron filaments, a high-strength, lightweight material chiefly employed for advanced aerospace structures. Boron is similar to carbon in that it has a capacity to form stable covalently bonded molecular networks. Carboranes, metalloboranes, phosphacarboranes, and other families comprise thousands of compounds. Crystalline boron (99%) costs about \$5/g. Amorphous boron costs about \$2/g. Elemental boron is not considered to be a poison, but assimilation of its compounds has a cumulative poisonous effect. Care must be taken in handling these.

THE ELEMENTS (continued)

Bromine — (Gr. *bromos,* stench), Br; at. wt. 79.904; at. no. 35; m.p. −7.2°C; b.p. 58.78°C; density of gas 7.59 g/l, liquid 3.12 (20°C); valence 1, 3, 5, or 7. Discovered by Balard in 1826, but not prepared in quantity until 1860. A member of the halogen group of elements, it is obtained from natural brines from wells in Michigan and Arkansas. Little bromine is extracted today from seawater, which contains only about 85 ppm. Bromine is the only liquid nonmetallic element. It is a heavy, mobile, reddish-brown liquid, volatilizing readily at room temperature to a red vapor with a strong disagreeable odor, resembling chlorine, and having a very irritating effect on the eyes and throat; it is readily soluble in water or carbon disulfide, forming a red solution; it is less active than chlorine but more so than iodine; it unites readily with many elements and has a bleaching action; when spilled on the skin it produces painful sores. It presents a serious health hazard, and maximum safety precautions should be taken when handling it.Much of the bromine output in the U.S. is used in the production of ethylene dibromide, a lead scavenger used in making gasoline antiknock compounds. Lead in gasoline, however, is presently being drastically reduced, due to environmental considerations. This will greatly affect future production of bromine. Bromine is also used in making fumigants, flameproofing agents, water purification compounds, dyes, medicinals, sanitizers, inorganic bromides for photography, etc. Organic bromides are also important.

Cadmium — (L. *cadmia;* Gr. *kadmeia* — ancient name for calamine, zinc carbonate), Cd; at. wt. 112.41 ± 0.01; at. no. 48; m.p. 320.9°C; b.p. 765°C; sp. gr. 8.65 (20°C); valence 2. Discovered by Stromeyer in 1817 from an impurity in zinc carbonate. Cadmium most often occurs in small quantities associated with zinc ores, such as *sphalerite* (ZnS). *Greenockite* (CdS) is the only mineral of any consequence bearing cadmium. Almost all cadmium is obtained as a by-product in the treatment of zinc, copper, and lead ores. It is a soft, bluish-white metal which is easily cut with a knife. It is similar in many respects to zinc. It is a component of some of the lowest melting alloys; it is used in bearing alloys with low coefficients of friction and great resistance to fatigue; it is used extensively in electroplating, which accounts for about 60% of its use. It is also used in many types of solder, for standard E.M.F. cells, for Ni-Cd batteries, and as a barrier to control atomic fission. Cadmium compounds are used in black and white television phosphors and in blue and green phosphors for color TV tubes. It forms a number of salts, of which the sulfate is the most common; the sulfide is used as a yellow pigment. Cadmium and solutions of its compounds are toxic. Failure to appreciate the toxic properties of cadmium may cause workers to be unwittingly exposed to dangerous fumes. Silver solder, for example, which contains cadmium, should be handled with care. Serious toxicity problems have been found from long-term exposure and work with cadmium plating baths. Exposure to cadmium dust should not exceed 0.05 mg/M³ (8-hr time-weighted average, 40-hr week). The ceiling concentration (maximum), for a period of 15 min, should not exceed 0.15 mg/M³. Cadmium oxide fume exposure (8-hr, 40-hr week) should not exceed 0.05 mg/M³, and the maximum concentration should not exceed 0.05 mg/M³. These values are presently being restudied and recommendations have been made to reduce the exposure.The current price of cadmium is about $12/lb. It is available in high purity form.

Calcium — (L. *calx,* lime), Ca; at. wt. 40.08; at. no. 20; m.p. 839 ± 2°C; b.p. 1484°C; sp. gr. 1.55 (20°C); valence 2. Though lime was prepared by the Romans in the first century under the name calx, the metal was not discovered until 1808. After learning that Berzelius and Pontin prepared calcium amalgam by electrolyzing lime in mercury, Davy was able to isolate the impure metal. Calcium is a metallic element, fifth in abundance in the earth's crust, of which it forms more than 3%. It is an essential constituent of leaves, bones, teeth, and shells. Never found in nature uncombined, it occurs abundantly as *limestone* (CaCO₃), *gypsum* (CaSO₄·2H₂O), and *fluorite* (CaF₂); *apatite* is the fluophosphate or chlorophosphate of calcium. The metal has a silvery color, is rather hard, and is prepared by electrolysis of the fused chloride to which calcium fluoride is added to lower the melting point. Chemically it is one of the alkaline earth elements; it readily forms a white coating of nitride in air, reacts with water, burns with a yellow-red flame, forming largely the nitride. The metal is used as a reducing agent in preparing other metals such as thorium, uranium, zirconium, etc., and is used as a deoxidizer, desulfurizer, or decarburizer for various ferrous and nonferrous alloys. It is also used as an alloying agent for aluminum, beryllium, copper, lead, and magnesium alloys, and serves as a "getter" for residual gases in vacuum tubes, etc. Its natural and prepared compounds are widely used. Quicklime (CaO), made by heating limestone and changed into slaked lime by the careful addition of water, is the great cheap base of chemical industry with countless uses. Mixed with sand it hardens as mortar and plaster by taking up carbon dioxide from the air. Calcium from limestone is an important element in Portland cement. The solubility of the carbonate in water containing carbon dioxide causes the formation of caves with stalactites and stalagmites and hardness in water. Other important compounds are the carbide (CaC₂), chloride (CaCl₂), cyanamide (CaCN₂), hypochlorite (Ca(OCl)₂), nitrate (Ca(NO₃)₂), and sulfide (CaS).

Californium — (State and University of California), Cf; at. wt. 251; at. no. 98. Californium, the sixth transuranium element to be discovered, was produced by Thompson, Street, Ghiorso, and Seaborg in 1950 by bombarding microgram quantities of Cm²⁴² with 35 MeV helium ions in the Berkeley 60-in. cyclotron. Californium (III) is the only ion stable in aqueous solutions, all attempts to reduce or oxidize californium (III) having failed. The isotope Cf²⁴⁹ results from the beta decay of Bk²⁴⁹ while the heavier isotopes are produced by intense neutron irradiation by the reactions:

$$Bk^{249} (n\gamma)Bk^{250} \xrightarrow{\beta} Cf^{250} \text{ and } Cf^{249}(n\gamma)Cf^{250}$$

followed by

$$Cf^{250}(n\gamma)Cf^{251}(n\gamma)Cf^{252}$$

The existence of the isotopes Cf²⁴⁹, Cf²⁵⁰, Cf²⁵¹, and Cf²⁵² makes it feasible to isolate californium in weighable amounts so that its properties can be investigated with macroscopic quantities. Californium-252 is a very strong neutron emitter. One microgram releases 170 million neutrons per minute, which presents biological hazards. Proper safeguards should be used in handling californium. In 1960 a few tenths of a microgram of californium trichloride CfCl₃, californium oxychloride, CfOCl, and californium oxide, Cf_2O_3, were first prepared. Reduction of californium to its metallic state has not yet been accomplished. Because californium is a very efficient source of neutrons, many new uses are expected for it. It has already found use in neutron moisture gages and in well-logging (the determination of water and oil-bearing layers.) It is also being used as a portable neutron source for discovery of metals such as gold or silver by on-the-spot activation analysis. Cf^{252} is now being offered for sale by the A.E.C. at a cost of $10/0.1µg. As of May 1975, more than 63 mg have been produced and sold. It has been suggested that californium may be produced in certain stellar explosions, called *supernovae*, for the radioactive decay of Cf^{254} (55-day half-life) agrees with the characteristics of the light curves of such explosions observed through telescopes. This suggestion, however, is questioned.

Carbon — (L. *carbo,* charcoal), C; at. wt. 12 exactly (C^{12}); at. wt. (natural carbon) 12.011; at. no. 6; m.p. $\sim$ 3550°C; graphite sublimes at 3367 ± 25°C; triple point: (graphite-liquid-gas), 3627 ± 50° at a pressure of 10.1 MPa and (graphite-diamond-liquid), 3830—3930° at a pressure of 12—13 GPa; sp. gr. amorphous 1.8 to 2.1, graphite 1.9 to 2.3, diamond 3.15 to 3.53 (depending on variety); gem diamond 3.513 (25°C); valence 2, 3, or 4. Carbon, an element of prehistoric discovery, is very widely distributed in nature. It is found in abundance in the sun, stars, comets, and atmospheres of most planets. Carbon in the form of microscopic diamonds is found in some meteorites. Natural diamonds are found in *kimberlite* of ancient volcanic "pipes," such as found in South Africa, Arkansas, and elsewhere. Diamonds are now also being recovered from the ocean floor off the Cape of Good Hope. About 30% of all industrial diamonds used in the U.S. are now made synthetically. The energy of the sun and stars can be attributed at least in part to the well-known carbon-nitrogen cycle. Carbon is found free in nature in three allotropic forms: amorphous, graphite, and diamond. A fourth form, known as "white" carbon, is now thought to exist. Graphite is one of the softest known materials while diamond is one of the hardest. Graphite exists in two forms: alpha and beta. These have identical physical properties, except for their crystal structure. Naturally occurring graphites are reported to contain as much as 30% of the rhombohedral (beta) form, whereas synthetic materials contain only the alpha form. The hexagonal alpha type can be converted to the beta by mechanical treatment, and the beta form reverts to the alpha on heating it above 1000°C. In 1969 a new allotropic form of carbon was produced during the sublimation of pyrolytic graphite at low pressures. Under free-vaporization conditions above $\sim$2550 K, "white" carbon forms as small transparent crystals on the edges of the basal planes of graphite. The interplanar spacings of "white" carbon are identical to those of a carbon form noted in the graphitic gneiss from the Ries (meteoritic) Crater of Germany. "White" carbon is a transparent birefringent material. Little information is presently available about this allotrope. In combination, carbon is found as carbon dioxide in the atmosphere of the earth and dissolved in all natural waters. It is a component of great rock masses in the form of carbonates of calcium (limestone), magnesium, and iron. Coal, petroleum, and natural gas are chiefly hydrocarbons. Carbon is unique among the elements in the vast number of variety of compounds it can form. With hydrogen, oxygen, and nitrogen, and other elements, it forms a very large number of compounds, carbon atom often being linked to carbon atom. There are upwards of a million or more known carbon compounds, many thousands of which are vital to organic and life processes. Without carbon, the basis for life would be impossible. While it has been thought that silicon might take the place of carbon in forming a host of similar compounds, it is now not possible to form stable compounds with very long chains of silicon atoms. The atmosphere of Mars contains 96.2% CO_2. Some of the most important compounds of carbon are carbon dioxide (CO_2), carbon monoxide (CO), carbon disulfide (CS_2), chloroform ($CHCl_3$), carbon tetrachloride (CCl_4), methane (CH_4), ethylene (C_2H_4), acetylene (C_2H_2), benzene (C_6H_6), ethyl alcohol (C_2H_5OH), acetic acid (CH_3COOH), and their derivatives. Carbon has seven isotopes. In 1961 the International Union of Pure and Applied Chemistry adopted the isotope carbon-12 as the basis for atomic weights. Carbon-14, an isotope with a half-life of 5730 years, has been widely used to date such materials as wood, archeological speciments, etc. Carbon-13 is now commercially available at a cost of $700/g.

Cerium — (named for ts *Ceres*, which was discovered in 1801 only 2 years before the element), Ce; at. wt. 140.12; at. no. 58; m.p. 799°C; b.p. 3426°C; sp. gr. 6.657 (25°C); valence 3 or 4. Discovered in 1803 by Klaproth and by Berzelius and Hisinger; metal prepared by Hillebadadorton n187. Cerium is the mostaudant of themetals of the so-called rar ats it is found in a number of minerals including *allanite* (also known as *orthite*), *monazite, bastnasite, cerite,* and *samarskite*. Monazite and bastnasite are presently the two most important sources of cerium. Large deposits of monazite found on the beaches of Travancore, India, in river sands in Brazil, and deposits of *allanite* in the western United States, and *bastnasite* in Southern California will supply cerium, thorium, and the other rare-earth metals for many years to come. Metallic cerium is prepared by metallothermic reduction techniques, such as by reducing cerous fluoride with calcium, or by electrolysis of molten cerous chloride or other cerous halides. The metallothermic technique is used to produce high-purity cerium. Crspecially interesting because of its variable electronic structure. The energy of the inner 4f level is nearly the same as that of the outer or valence electrons, and only small amounts of energy are required to change the relative occupancy of these electronic levels. This gives rise to dual valency states. For example, a volume change of about 10% occurs when cerium is subjected to high pressures or low temperatures. It appears that the valence changes from about 3 to 4 when it is cooled or compressed. The low temperature behavior of cerium is complex. Four allotropic modifications are thought to exist: cerium at room temperature and at atmospheric pressure is known as γ cerium. Upon cooling to −23°C, γ cerium changes to β cerium. The remaining γ cerium starts to change to α cerium when cooled to −158°C, and the transformation is complete at −196°C. α Cerium has a density of 8.24; δ cerium exists above 726°C. At atmospheric pressure, liquid cerium is more dense than its solid form at the melting point. Cerium is an iron-gray lustrous metal. It is malleable, and oxidizes very readily at room temperature, especially in moist air. Except for europium, cerium is the most reactive of the "rare-earth" metals. It slowly decomposes in cold water, and rapidly in hot water. Alkali solutions and dilute and concentrated acids attack the metal rapidly. The pure metal is likely to ignite if scratched with a knife. Ceric salts are orange red or yellowish; cerous salts are usually white. Cerium is a component of misch metal, which is extensively used in the manufacture of pyrophoric alloys for cigarette lighters, etc. While cerium is not radioactive, the impure commercial grade may contain traces of thorium, which is radioactive. The oxide is an important constituent of incandescent gas mantles and it is emerging as a hydrocarbon catalyst in ''self-cleaning'' ovens. In this application it can be incorporated into oven walls to prevent the collection of cooking residues. As ceric sulfate it finds extensive use as a volumetric oxidizing agent in quantitative analysis. Cerium compounds are used in the manufacture of glass, both as a component and as a decolorizer. The oxide is finding increased use as a glass polishing agent instead of rouge, for it is much faster than rouge in polishing glass surfaces. Cerium, with other rare earths, is used in carbon-arc lighting, especially in the motion picture industry. It is also finding use as an important catalyst in petroleum refining and in metallurgical and nuclear applications. In small lots, 99.9% cerium costs about $2/g.

Cesium — (L. *caesius*, sky blue), Cs; at. wt. 132.9054; at. no. 55; m.p. 28.40± 0.01°C; b.p. 669.3°C; sp. gr. 1.873 (20°C); valence 1. Cesium was discovered spectroscopically by Bunsen and Kirchhoff in 1860 in mineral water from Durkheim. Cesium, an alkali metal, occurs in *lepidolite, pollucite* (a hydrated silicate of aluminum and cesium), and in other sources. One of the world's richest sources of cesium is located at Bernic Lake, Manitoba. The deposits are estimated to contain 300,000 tons of pollucite, averaging 20% cesium. It can be isolated by electrolysis of the fused cyanide and by a number of other methods. Very pure, gas-free cesium can be prepared by thermal decomposition of cesium azide. The metal is characterized by a spectrum containing two bright lines in the blue along with several others in the red, yellow, and green. It is silvery white, soft, and ductile. It is the most electropositive and most alkaline element. Cesium, gallium, and mercury are the only three metals that are liquid at room temperature. Cesium reacts explosively with cold water, and reacts with ice at temperatures above −116°C. Cesium hydroxide, the strongest base known, attacks glass. Because of its great affinity for oxygen the metal is used as a ''getter'' in electron tubes. It is also used in photoelectric cells, as well as a catalyst in the hydrogenation of certain organic compounds. The metal has recently found application in ion propulsion systems. Although these are not usable in the earth's atmosphere, 1 lb of cesium in outer space theoretically will propel a vehicle 140 times as far as the burning of the same amount of any known liquid or solid. Cesium is used in atomic clocks, which are accurate to 5 sec in 300 years. Its chief compounds are the chloride and the nitrate. The present price of cesium is about $25/g.

Chlorine — (Gr. *chloros*, greenish yellow), Cl; at. wt. 35.453; at. no. 17; m.p. −100.98°C; b.p.−34.6°C; density 3.214 g/l; sp. gr. 1.56(−33.6°C); valence 1, 3, 5, or 7. Discovered in 1774 by Scheele, who thought it contained oxygen; named in 1810 by Davy, who insisted it was an element. In nature it is found in the combined state only, chiefly with sodium as common salt (NaCl), *carnallite* ($KMgCl_3 \cdot 6H_2O$), and *sylvite* (KCl). It is a member of the halogen (salt-forming) group of elements and is obtained from chlorides by the action of oxidizing agents and more often by electrolysis; it is a greenish-yellow gas, combining directly with nearly all elements. At 10°C one volume of water dissolves 3.10 volumes of chlorine, at 30°C only 1.77 volumes. Chlorine is widely used in making many everyday products. It is used for producing safe drinking water the world over. Even the smallest water supplies are now usually chlorinated. It is also extensively used in the production of paper products, dyestuffs, textiles, petroleum products, medicines, antiseptics, insecticides, foodstuffs, solvents, paints, plastics, and many other consumer products. Most of the chlorine produced is used in the manufacture of chlorinated compounds of sanitation, pulp bleaching, disinfectants, and textile processing. Further use is in the manufacture of chlorates, chloroform, carbon tetrachloride, and in the extraction of bromine. Organic chemistry demands much from chlorine, both as an oxidizing agent and in substitution, since it often brings desired properties in an organic compound when substituted for hydrogen, as in one from of synthetic rubber. Chlorine is a respiratory irritant. The gas irritates the mucous membranes and the liquid burns the skin. As little as 3.5 ppm can be detected as an odor, and 1000 ppm is likely to be fatal after a few deep breaths. It was used as a war gas in 1915. Exposure to chlorine should not exceed 1 ppm(8-hr time-weighted average — 40 hr week.)

Chromium — (Gr. *chroma*, color), Cr; at. wt. 51.996; at. no. 24; m.p. 1857 ± 20°C; b.p. 2672°C; sp. gr. 7.18 to 7.20 (20°C); valence chiefly 2, 3, or 6. Discovered in 1797 by Vauquelin, who prepared the metal the next year. Chromium is a steel-gray, lustrous, hard metal that takes a high polish. The principal ore is *chromite* ($FeCr_2O_4$), which is found in Southern Rhodesia, U.S.S.R., Transvaal, Turkey, Iran, Albania, Finland, Democratic Republic of Madagascar, and the Philippines. The metal is usually produced by reducing the oxide with aluminum. Chromium is used to harden steel, to manufacture stainless steel, and to form many useful alloys. Much is used in plating to produce a hard, beautiful surface and to prevent corrosion. Chromium is used to give glass an emerald green color. It finds wide use as a catalyst. All compounds of chromium are colored; the most important are the chromates of sodium and potassium (K_2CrO_4) and the dichromates ($K_2Cr_2O_7$) and the potassium and ammonium chrome alums, as $KCr(SO_4)_2 \cdot 12H_2O$. The dichromates are used as oxidizing agents in quantitative analysis, also in tanning leather. Other compounds are of industrial value; lead chromate is chrome yellow, a valued pigment. Chromium compounds are used in the textile industry as mordants, and by the aircraft and other industries for anodizing aluminum. The refractory industry has found chromite useful for forming bricks and shapes, as it has a high melting point, moderate thermal expansion, and stability of crystalline structure. Chromium compounds are toxic and should be handled with proper safeguards.

Cobalt — (*Kobald*, from the German, goblin or evil spirit, *cobalos*, Greek, mine), Co; at. wt. 58.9332; at. no. 27; m.p. 1495°C; b.p. 2870°C; sp. gr. 8.9 (20°C); valence 2 or 3. Discovered by Brandt about 1735. Cobalt occurs in the minerals *cobaltite, smaltite,* and *erythrite,* and is often associated with nickel, silver, lead, copper, and iron ores, from which it is most frequently obtained as a by-product. It is also present in meteorites. Important ore deposits are found in Zaire, Morocco, and Canada. The U.S. Geological Survey has announced that the bottom of the north central Pacific Ocean may have cobalt-rich deposits at relatively shallow depths in waters close to the Hawaiian Islands and other U.S. Pacific territories. Cobalt is a brittle, hard metal, closely resembling iron and nickel in appearance. It has a magnetic permeability of about two thirds that of iron. Cobalt tends to exist as a mixture of two allotropes over a wide temperature range; the β-form predominates below 400°C, and the α above that temperature. The transformation is sluggish and accounts in part for the wide variation

THE ELEMENTS (continued)

in reported data on physical properties of cobalt. It is alloyed with iron, nickel and other metals to make Alnico, an alloy of unusual magnetic strength with many important uses. Stellite ® alloys, containing cobalt, chromium, and tungsten, are used for high-speed, heavy-duty, high-temperature cutting tools, and for dies. Cobalt is also used in other magnet steels and stainless steels, and in alloys used in jet turbines and gas turbine generators. The metal is used in electroplating because of its appearance, hardness, and resistance to oxidation. The salts have been used for centuries for the production of brilliant and permanent blue colors in porcelain, glass, pottery, tiles, and enamels. It is the principal ingredient in Sevre's and Thenard's blue. A solution of the chloride ($CoCl_2 \cdot 6H_2O$) is used as sympathetic ink. The cobalt ammines are of interest; the oxide and the nitrate are important. Cobalt carefully used in the form of the chloride, sulfate, acetate, or nitrate has been found effective in correcting a certain mineral deficiency disease in animals. Soils should contain 0.13 to 0.30 ppm of cobalt for proper animal nutrition. Cobalt-60, an artifical isotope, is an important gamma ray source, and is extensively used as a tracer and a radiotherapeutic agent. Single compact sources of Cobalt-60 are readily available and have an equivalent gamma ray output equal to thousands of grams of radium. The cost of Cobalt-60 varies from about $1 to $10/curie, depending on quantity and specific activity. Exposure to cobalt (metal fumes and dust) should be limited to 0.05 mg/M³ (8-hr time-weighted average, 40 hr week).

Columbium — see Niobium.

Copper — (L. *cuprum*, from the island of Cyprus), Cu; at. wt. 63.546; at. no. 29; m.p. 1083.4 ± 0.2°C; b.p. 2567 °C; sp. gr. 8.96 (20°C); valence 1 or 2. The discovery of copper dates from prehistoric times; it is said to have been mined for more than 5000 years. It is one of man's most important metals. Copper is reddish colored, takes on a bright metallic luster, and is malleable, ductile, and a good conductor of heat and electricity (second only to silver in electrical conductivity). The electrical industry is one of the greatest users of copper. Copper occasionally occurs native, and is found in many minerals such as *cuprite, malachite, azurite, chalcopyrite,* and *bornite.* Large copper ore deposits are found in the U.S., Chile, Zambia, Zaire, Peru, and Canada. The most important copper ores are the sulfides, oxides, and carbonates. From these, copper is obtained by smelting, leaching, and by electrolysis. Its alloys, brass and bronze, long used, are still very important; all American coins are now copper alloys; monel and gun metals also contain copper. The most important compounds are the oxide and the sulfate, blue vitriol; the latter has wide use as an agricultural poison and as an algicide in water purification. Copper compounds such as Fehling's solution are widely used in analytical chemistry in tests for sugar. High-purity copper (99.999 + %) is available commercially.

Curium — (Pierre and Marie Curie), Cm; at. wt. 247; at. no. 96; m.p. 1340 ± 40°C; sp. gr. 13.51 (calc.); valence 3 and 4. Although curium follows americium in the periodic system, it was actually known before americium and was the third transuranium element to be discovered. It was identified by Seaborg, James, and Ghiorso in 1944 at the wartime Metallurgical Laboratory in Chicago as a result of helium-ion bombardment of Pu^{239} in the Berkeley, California, 60-in. cyclotron. Visible amounts (30 µg) of Cm^{242}, in the form of the hydroxide, were first isolated by Werner and Perlman of the University of California in 1947. In 1950, Crane, Wallmann, and Cunningham found that the magnetic susceptibility of microgram samples of CmF_3 was of the same magnitude as that of GdF_3. This provided direct experimental evidence for assigning an electronic configuration to Cm^{+3}. In 1951, the same workers prepared curium in its elemental form for the first time. Thirteen isotopes of curium are now known. The most stable, Cm^{247}, with a half-life of 16 million years, is so short compared to the earth's age that any primordial curium must have disappeared long ago from the natural scene. Minute amounts of curium probably exist in natural deposits of uranium, as a result of a sequence of neutron captures and β decays sustained by the very low flux of neutrons naturally present in uranium ores. The presence of natural curium, however, has never been detected. Cm^{242} and Cm^{244} are available in multigram quantities. Cm^{248} has been produced only in milligram amounts. Curium is similar in some regards to gadolinium, its rare-earth homolog, but it has a more complex crystal structure. Curium is silver in color, is chemically reactive, and is more electropositive than aluminum. CmO_2, Cm_2O_3, CmF_3, CmF_4, $CmCl_3$, $CmBr_3$, and CmI_3 have been prepared. Most compounds of trivalent curium are faintly yellow in color. The A.E.C. is attempting to produce several kilograms of Cm^{244}, an isotope with a 17.6-year half-life, by neutron irradiation of plutonium in a nuclear reactor. Ultimately, it is possible that it may be produced in ton quantities by converting a number of plutonium production reactors to the manufacture of Cm^{244}. Cm^{242} generates about three thermal watts of energy per gram. This compares to one-half thermal watt per gram of Pu^{238}. This suggests use for curium as an isotope power source. Cm^{244} is now offered for sale by the A.E.C. at $100/ mg. Curium absorbed into the body accumulates in the bones, and is therefore very toxic as its radiation destroys the red-cell forming mechanism. The maximum permissible total body burden of Cm^{244}(soluble) in a human being is 0.3 µCi (microcurie).

Deuterium, an isotope of hydrogen — see Hydrogen.

Dysprosium — (Gr. *dysprositos*, hard to get at), Dy; at. wt. 162.50; at. no. 66; m.p. 1412°C; b.p.2562°C; sp. gr. 8.550 (25°C); valence 3. Dysprosium was discovered in 1886 by Lecoq de Boisbaudran, but not isolated. Neither the oxide nor the metal was available in relatively pure form until the development of ion-exchange separation and metallographic reduction techniques by Spedding and associates about 1950. Dysprosium occurs along with other so-called rare-earth or lanthanide elements in a variety of minerals such as *xenotime, fergusonite, gadolinite, euxenite, polycrase,* and *blomstrandine.* The most important sources, however, are from *monazite* and *bastnasite.* Dysprosium can be prepared by reduction of the trifluoride with calcium. The element has a metallic, bright silver luster. It is relatively stable in air at room temperature, and is readily attacked and dissolved, with the evolution of hydrogen, by dilute and concentrated mineral acids. The metal is soft enough to be cut with a knife and can be machined without sparking if overheating is avoided. Small amounts of impurities can greatly affect its physical properties. While dysprosium has not yet found many applications, its thermal neutron absorption cross-section and high melting point suggest metallurgical uses in nuclear control applications for alloying with special stainless steels. A dysprosium oxide-nickel cermet has found use in cooling nuclear reactor control rods. This cermet absorbs neutrons readily without swelling or contracting under prolonged neutron bombardment. In combination with vanadium and other earth, dysprosium has been used in making laser materials. Dysprosium-cadmium calcogenides, as sources of infrared radiation, have been used for studying chemical reactions. The cost of dysprosium metal has dropped in recent years since the development of ion- exchange and solvent extraction techniques, and the discovery of large ore bodies. The metal is still expensive, however, and costs about $3/g in purities of 99 + %.

Einsteinium — (Albert Einstein), Es; at wt. 254; at no. 99. Einsteinium, the seventh transuranic element of the actinide series to be discovered, was identified by Ghiorso and co-workers at Berkeley in December 1952 in debris from the first large thermonuclear or "hydrogen" bomb explosion, which took place in the Pacific in November 1952. The isotope produced was the 20-day Es^{253} isotope. In 1961, a sufficient amount of einsteinium was produced to permit separation of a macroscopic amount of Es^{253}. This sample weighed about 0.01 μg. A special magnetic-type balance was used in making this determination. Es^{253} so produced was used to produce mendelevium (Element 101). About 3 μg of einsteinium has been produced at Oak Ridge National Laboratores by irradiating for several years kilogram quantities of Pu^{239} in a reactor to produce Pu^{242}. This was then fabricated into pellets of plutonium oxide and aluminum powder, and loaded into target rods for an initial 1-year irradiation at the A.E.C.'s Savannah River Plant, followed by irradiation in a HFIR (High Flux Isotopic Reactor). After 4 months in the HFIR the targets were removed for chemical separation of the einsteinium from californium. Eleven isotopes of einsteinium are now recognized. Es^{254} has the longest half-life (276 days). Tracer studies using Es^{253} show that einsteinium has chemical properties typical of a heavy trivalent, actinide element.

Element 104 — In 1964, workers of the Joint Nuclear Research Institute at Dubna (U.S.S.R.) bombarded plutonium with accelerated 113 to 115 MeV neon ions. By measuring fission tracks in a special glass with a microscope, they detected an isotope that decays by spontaneous fission. They suggested that this isotope, which had a half-life of 0.3 ± 0.1 sec might be 104^{260}, produced by the following reaction:

$$_{94}Pu^{242} + {}_{10}Ne^{22} \longrightarrow 104^{260} + 4n$$

Element 104, the first *transactinide* element, is expected to have chemical properties similar to those of hafnium. It would, for example, form a relatively volatile compound with chlorine (a tetrachloride). The Soviet scientists have performed experiments aimed at chemical identification, and have attempted to show that the 0.3-sec activity is more volatile than that of the relatively nonvolatile actinide trichlorides. This experiment does not fulfill the test of chemically separating the new element from all others, but it provides important evidence for evaluation. New data, reportedly issued by Soviet scientists, have reduced the half-life of the isotope they worked with from 0.3 to 0.15 sec. The Dubna scientists suggest the name *kurchatovium* and symbol *Ku* for Element 104, in honor of Igor Vasilevich Kurchatov (1903-1960), late Head of Soviet Nuclear Research. In 1969, Ghiorso, Nurmia, Harris, K.A.Y. Eskola, and P.L. Eskola of the University of California at Berkeley reported they had positively identified two, and possibly three, isotopes of Element 104. The group also indicated that after repeated attempts so far they have been unable to produce isotope 104^{260} reported by the Dubna group in 1964. The discoveries at Berkeley were made by bombarding a target of Cf^{249} with C^{12} nuclei of 71 MeV, and C^{13} nuclei of 69 MeV. The combination of C^{12} with Cf^{249} followed by instant emission of four neutrons produced Element 104^{257}. This isotope has a half-life of 4 to 5 sec, decaying by emitting an alpha particle into No^{253}, with a half-life of 105 sec. The same reaction, except with the emission of three neutrons, was thought to have produced 104^{258}. with a half-life of about 1/100 sec. Element 104^{259} is formed by the merging of a C^{13} nuclei with Cf^{249}, followed by emission of three neutrons. This isotope has a half-life of 3 to 4 sec, and decays by emitting an alpha particle into No 255, which has a half-life of 185 sec. Thousands of atoms of 104^{257} and 104^{259} have been detected. The Berkeley group believe their identification of 104^{258} is correct, but they do not attach the same degree of confidence to this work as to their work on 104^{257} and 104^{259}. The Berkeley group proposes for the new element the name *rutherfordium* (symbol Rf), in honor of Ernest R. Rutherford, New Zealand physicist. The claims for discovery and the naming of Element 104 are still in question.

Element 105 — In 1967 G.N. Flerov reported that a Soviet team working at the Joint Institute for Nuclear Research at Dubna may have produced a few atoms of 105^{260} and 105^{261} by bombarding Am^{243} with Ne^{22}. Their evidence was based on time-coincidence measurements of alpha energies. More recently, it was reported that early in 1970 Dubna scientists synthesized Element 105 and that by the end of April 1970 "had investigated all the types of decay of the new element and had determined its chemical properties." The Soviet group has not proposed a name for Element 105. In late April 1970, it was announced that Ghiorso, Nurmia, Harris, K.A.Y. Eskola, and P.L.Eskola, working at the University of California at Berkeley, had positively identified Element 105. The discovery was made by bombarding a target of Cf^{249} with a beam of 84 MeV nitrogen nuclei in the Heavy Ion Linear Accelerator (HILAC). When a N^{15} nuclei is absorbed by a Cf^{249} nucleus, four neutrons are emitted and a new atom of 105^{260} with a half-life of 1.6 sec is formed. While the first atoms of Element 105 are said to have been detected conclusively on March 5, 1970, there is evidence that Element 105 had been formed in Berkeley experiments a year earlier by the method described. Ghiorso and his associates have attempted to confirm Soviet findings by more sophisticated methods without success. The Berkeley Group proposes the name *hahnium*, after the late German scientist Otto Hahn (1879-1968), and *Ha* for the chemical symbol.

More recently, in October 1971, it was announced that two new isotopes of Element 105 were synthesized with the heavy ion linear accelerator by A. Ghiorso and co-workers at Berkeley. Element 105^{261} was produced both by bombarding Cl^{250} with N^{15} and by bombarding Bk^{249} with O^{16}. The isotope emits 8.93-MeV α particles and decays to Lr^{257} with a half-life of about 1.8 sec. Element 105^{262} was produced by bombarding Bk^{249} with O^{18}. It emits 8.45 MeV α particles and decays to Lr^{258} with a half-life of about 40 sec.

Element 106 — In June 1974, members of the Joint Institute for Nuclear Research in Dubna, U.S.S.R., reported their discovery of Element 106, which they claim to have synthesized. In September 1974, workers of the Lawrence Berkeley and

THE ELEMENTS (continued)

Livermore Laboratories also reported creation of Element 106 "without any scientific doubt." The LBL and LLL Group used the SuperHILAC to accelerate O^{18} ions onto a Cf^{249} target. Element 106 was created by the reaction $Cf^{249}(O^{18}\ 4n)_{106}X^{263}$, which decayed by α emission to rutherfordium, and then by α emission to nobelium, which in turn further decayed by α emission. An elaborate detection system not only looked for correlations between the new element and its daughter, but also between daughter and granddaughter. The element so identified had α energies of 9.06 and 9.25 MeV with a half-life of 0.9 ± 0.2 sec. At Dubna, 280-MeV ions of Cr^{54} from the 310-cm cyclotron were used to strike targets of Pb^{206}, Pb^{207}, and Pb^{208}, in separate runs. Foils exposed to a rotating target disc were used to detect spontaneous fission activities, the foils being etched and examined microscopically to detect the number of fission tracts and the half-life of the fission activity. Other experiments were made to aid in confirmation of the discovery. Neither the Dubna team nor the Berkeley- Livermore Group has proposed a name as yet for Element 106.

Element l07 — In 1976 Soviet scientists at Dubna announced they had synthesized Element 107 by bombarding Bi^{204} with heavy nuclei of Cr^{54}. It is reported that earlier experiments in 1975 had allowed scientists "to glimpse" the new element for 2/1000 sec. A rapidly rotating cylinder, coated with a thin layer of the bismuth metal, was used as the target. This was bombarded by a stream of Cr^{54} ions fired tangentially. The existence of Element 107 was confirmed by a team of West German physicists at the Heavy Ion Research Laboratory at Darmstadt, who created and identified six nuclei of Element 107.

Element 109 — On August 29, 1982, Element 109 was made and discovered by physicists of the Heavy Ion Research Laboratory, Darmstadt, West Germany, by bombing a target of Bi-209 with accelerated nuclei of Fe-58. If the combined energy of two nuclei is sufficiently high, the repulsive forces between the nuclei can be overcome. In this experiment it took a week of target bombardment to produce a single fused nucleus. The team confirmed the existence of Element 109 by four independent measurements. The newly formed atom recoiled from the target at a predicted velocity and was separated from smaller, faster nuclei by a newly developed velocity filter. The time of flight to the detector and the striking energy were measured and found to match predicted values. The nucleus of $_{109}X^{266}$ started to decay 5 msec after striking the detector. A high-energy α particle was emitted, producing $_{107}X^{262}$. This in turn emitted an α particle, becoming $_{105}Ha^{358}$, which in turn captured an electron and became $_{104}Rf^{258}$. This in turn decayed into other nuclides. This experiment demonstrated the feasibility of using fusion techniques as a method of making new, heavy nuclei. West German scientists as well as those at Berkeley will soon attempt to use cold fusion to make Element 116 by bombarding curium-248 with calcium-48 nuclei. The fission product is expected to be Element 116, which should decay through a whole series of hitherto unknown nuclides. Names have not yet been proposed for Elements 106, 107, or 109.

Erbium — (Ytterby, a town in Sweden), Er; at. wt. 167.26; at. no. 68; m.p. 159°C; b.p. 2863°C; sp. gr. 9.066 (25°C); valence 3. Erbium, one of the so-called rare-earth elements of the lanthanide series, is found in the minerals mentioned under dysprosium above. In 1842 Mosander separated "yttria," found in the mineral gadolinite, into three fractions which he called yttria, erbia, and terbia. The names erbia and terbia became confused in this early period. After 1860, Mosander's terbia was known as erbia, and after 1877, the earlier known erbia became terbia. The erbia of this period was later shown to consist of five oxides, now known as erbia, scandia, holmia, thulia and ytterbia. By 1905 Urbain and James independently succeeded in isolating fairly pure Er_2O_3. Klemm and Bommer first produced reasonably pure erbium metal in 1934 by reducing the anhydrous chloride with potassium vapor. The pure metal is soft and malleable and has a bright, silvery, metallic luster. As with other rare-earth metals, its properties depend to a certain extent on the impurities present. The metal is fairly stable in air and does not oxidize as rapidly as some of the other rare-earth metals. Naturally occurring erbium is a mixture of six isotopes, all of which are stable. Nine radioactive isotopes of erbium are also recognized. Recent production techniques, using ion-exchange reactions, have resulted in much lower prices of the rare-earth metals and their compounds in recent years. The cost of 99 + % erbium metal is about $3/g in small quantities. Erbium is finding nuclear and metallurgical uses. Added to vanadium, for example, erbium lowers the hardness and improves workability. Most of the rare-earth oxides have sharp absorption bands in the visible, ultraviolet, and near infrared. This property, associated with the electronic structure, gives beautiful pastel colors to many of the rare-earth salts. Erbium oxide gives a pink color and has been used as a colorant in glasses and porcelain enamel glazes.

Europium — (Europe), Eu; at. wt. 151.96; at. no. 63; m.p. 822°C; b.p. 1597°C; sp. gr. 5.243 (25°C); valence 2 or 3. In 1890 Boisbaudran obtained basic fractions from samarium-gadolinium concentrates which had spark spectral lines not accounted for by samarium or gadolinium. These lines subsequently have been shown to belong to europium. The discovery of europium is generally credited to Demarcay, who separated the earth in reasonably pure form in 1901. The pure metal was not isolated until recent years. Europium is now prepared by mixture Eu_2O_3 with a 10%-excess of lanthanum metal and heating the mixture in a tantalum crucible under high vacuum. The element is collected as a silvery-white metallic deposit on the walls of the crucible. As with other rare-earth metals, except for lanthanum, europium ignites in air at about 150 to 180°C. Europium is about as hard as lead and is quite ductile. It is the most reactive of the rare-earth metals, quickly oxidizing in air. It resembles calcium in its reaction with water. Bastnasite and monazite are the principal ores containing europium. Europium has been identified spectroscopically in the sun and certain stars. Seventeen isotopes are now recognized. Europium isotopes are good neutron absorbers and are being studied for use in nuclear control applications. Europium oxide is now widely used as a phosphor activator and europium-activated yttrium vanadate is in commercial use as the red phosphor in color TV tubes. Europium-doped plastic has been used as a laser material. With the development of ion-exchange techniques and special processes, the cost of the metal has been greatly reduced in recent years. Europium is one of the rarest and most costly of the rare-earth metals. It is priced at about $50/g.

Fermium — (Enrico Fermi), Fm; at. wt. 257; at. no. 100. Fermium, the eighth transuranium element of the actinide series to be discovered, was identified by Ghiorso and co-workers in 1952 in the debris from a thermonuclear explosion in the

Pacific in work involving the University of California Radiation Laboratory, the Argonne National Laboratory, and the Los Alamos Scientific Laboratory. The isotope produced was the 20-hr Fm^{255}. During 1953 and early 1954, while discovery of elements 99 and 100 was withheld from publication for security reasons, a group from the Nobel Institute of Physics in Stockholm bombarded U^{238} with O^{16} ions, and isolated a 30-min α-emitter, which they ascribed to 100^{250}, without claiming discovery of the element. This isotope has since been identified positively, and the 30-min half-life confirmed. The chemical properties of fermium have been studied solely with tracer amounts, and in normal aqueous media only the (III) oxidation state appears to exist. The isotope Fm^{254} and heavier isotopes can be produced by intense neutron irradiation of lower elements such as plutonium by a process of successive neutron capture interspersed with beta decays until these mass numbers and atomic numbers are reached. Ten isotopes of fermium are known to exist. Fm^{257}, with a half-life of about 80 days, is the longest lived. Fm^{250}, with a half-life of 30 min, has been shown to be a product of decay of Element 102^{254}. It was by chemical identification of Fm^{250} that it was certain that Element 102 (nobelium) had been produced.

Fluorine — (L. and F. *fluere*, flow, or flux), F; at. wt. 18.998403 ± 0.000001; at. no. 9; m.p. −219.62°C (1 atm.); b.p. −188.14°C (1 atm); density 1.696 g/l (O°C, 1 atm.); sp. gr. of liquid 1.108 at b.p.; valence 1. In 1529, Georgius Agricola described the use of fluorspar as a flux, and as early as 1670 Schwandhard found that glass was etched when exposed to fluorspar treated with acid. Scheele and many later investigators, including Davy, Gay-Lussac, Lavoisier, and Thenard, experimented with hydrofluoric acid, some experiments ending in tragedy. The element was finally isolated in 1886 by Moisson after nearly 74 years of continuous effort. Fluorine occurs chiefly in *fluorspar* (CaF_2) and *cryolite* (Na_2AlF_6), but is rather widely distributed in other minerals. It is a member of the halogen family of elements, and obtained by electrolyzing a solution of potassium hydrogen fluoride in anhydrous hydrogen fluoride in a vessel of metal or transparent fluorspar. Modern commercial production methods are essentially variations on the procedures first used by Moisson. Fluorine is the most electronegative and reactive of all elements. It is a pale yellow, corrosive gas, which reacts with practically all organic and inorganic substances. Finely divided metals, glass, ceramics, carbon, and even water burn in fluorine with a bright flame. Until World War II, there was no commercial production of elemental fluorine. The atom bomb project and nuclear energy applications, however, made it necessary to produce large quantities. Safe handling techniques have now been developed and it is possible at present to transport liquid fluorine by the ton. Fluorine and its compounds are used in producing uranium (from the hexafluoride) and more than 100 commercial fluorochemicals, including many well-known high-temperature plastics. Hydrofluoric acid is extensively used for etching the glass of light bulbs, etc. Fluorochloro hydrocarbons are extensively used in air conditioning and refrigeration. It has been suggested that fluorine can be substituted for hydrogen wherever it occurs in organic compounds, which could lead to an astronomical number of new fluorine compounds. The presence of fluorine as a soluble fluoride in drinking water to the extent of 2 ppm may cause mottled enamel in teeth, when used by children acquiring permanent teeth; in smaller amounts, however, fluorides are said to be beneficial and used in water supplies to prevent dental cavities. Elemental fluorine is being studied as a rocket propellant as it has an exceptionally high specific impulse value. Compounds of fluorine with rare gases have now been confirmed. Fluorides of xenon, radon, and krypton are among those reported. Elemental fluorine and the fluoride ion are highly toxic. The free element has a characteristic pungent odor, detectable in concentrations as low as 20 ppb, which is below the safe working level. The recommended maximum allowable concentration for a daily 8-hr time-weighted exposure is 0.1 ppm.

Francium — (France), Fr; at. no. 87; at. wt. 223; m.p. 27°C; b.p. 677°C; valence 1. Discovered in 1939 by Mlle. Marguerite Perey of the Curie Institute, Paris. Francium, the heaviest known member of the alkali metal series, occurs as a result of an alpha disintegration of actinium. It can also be made artificially by bombarding thorium with protons. While it occurs naturally in uranium minerals, there is probably less than an ounce of francium at any time in the total crust of the earth. It has the highest equivalent weight of any element, and is the most unstable of the first 101 elements of the periodic system. Twenty isotopes of francium are recognized. The longest lived, Fr^{223}(AcK), a daughter of Ac^{227}, has a half-life of 22 min. This is the only isotope of francium occurring in nature. Because all known isotopes of francium are highly unstable, knowledge of the chemical properties of this element comes from radiochemical techniques. No weighable quantity of the element has been prepared or isolated. The chemical properties of francium most closely resemble cesium.

Gadolinium — (*gadolinite,* a mineral named for Gadolin, a Finnish chemist), Gd; at. wt. 157.25; at. no. 64; m.p. 1313±1°C; b.p. 3266°C; sp. gr. 7.9004 (25°C); valence 3. Gadolinia, the oxide of gadolinium, was separated by Marignac in 1880 and Lecoq de Boisbaudran independently isolated the element from Mosander's "yttria" in 1886. The element was named for the mineral *gadolinite* from which this rare earth was originally obtained. Gadolinium is found in several other minerals, including *monazite* and *bastnasite,* which are of commercial importance. The element has been isolated only in recent years. With the development of ion-exchange and solvent extraction techniques, the availability and price of gadolinium and the other rare-earth metals have greatly improved. Seventeen isotopes of gadolinium are now recognized; seven occur naturally. The metal can be prepared by the reduction of the anhydrous fluoride with metallic calcium. As with other related rare-earth metals, it is silvery white, has a metallic luster, and is malleable and ductile. At room temperature, gadolinium crystallizes in the hexagonal, close-packed α form. Upon heating to 1262°C, α gadolinium transforms into the β form, which has a body-centered cubic structure. The metal is relatively stable in dry air, but in moist air it tarnishes with the formation of a loosely adhering oxide film which spalls off and exposes more surface to oxidation. The metal reacts slowly with water and is soluble in dilute acid. Gadolinium has the highest thermal neutron capture cross-section of any known element (49,000 barns). Natural gadolinium is a mixture of seven isotopes. Two of these, Gd^{155} and Gd^{157}, have excellent capture characteristics, but they are present naturally in low concentrations. As a result, gadolinium has a very fast burnout rate and has limited use as a nuclear control rod material. It has been used in making gadolinium yttrium garnets, which have microwave applications. Compounds of gadolinium are used in making phosphors for color TV tubes. The metal has unusual superconductive properties. As little as 1% gadolinium has been found to improve the workability and resistance of iron, chromium, and related alloys to high

THE ELEMENTS (continued)

temperatures and oxidation. Gadolinium ethyl sulfate has extremely low noise characteristics and may find use in duplicating the performance of h.f. amplifiers, such as the maser. The metal is ferromagnetic. Gadolinium is unique for its high magnetic moment and for its special Curie temperature (above which ferromagnetism vanishes) lying just at room temperature. This suggests uses as a magnetic component that senses hot and cold. The price of the metal is $4/g.

Gallium — (L. *Gallia*, France; also from Latin, *gallus*, a translation of Lecoq, a cock); Ga; at. wt. 69.737 ± 0.006; at. no. 3l; m.p. 29.78°C; b.p.2403°C; sp. gr. 5.904 (29.6°C) solid; sp. gr. 6.095 (29.8°C) liquid; valence 2 or 3. Predicted and described by Mendeleev as ekaaluminum, and discovered spectroscopically by Lecoq de Boisbaudran in 1875, who in the same year obtained the free metal by electrolysis of a solution of the hydroxide in KOH. Gallium is often found as a trace element in *diaspore, sphalerite, germanite, bauxite,* and *coal.* Some flue dusts from burning coal have been shown to contain as much as 1.5% gallium. It is the only metal, except for mercury, cesium, and rubidium, which can be liquid near room temperatures; this makes possible its use in high-temperature thermometers. It has one of the longest liquid ranges of any metal and has a low vapor pressure even at high temperatures. There is a strong tendency for gallium to supercool below its freezing point. Therefore, seeding may be necessary to initiate solidification. Ultra-pure gallium has a beautiful, silvery appearance, and the solid metal exhibits a conchoidal fracture similar to glass. The metal expands 3.1% on solidifying; therefore, it should not be stored in glass or metal containers, as they may break as the metal solidifies. Gallium wets glass or porcelain, and forms a brilliant mirror when it is painted on glass. It has found recent use in doping semiconductors and producing solid-state devices such as transistors. High-purity gallium is attacked only slowly by mineral acids. Magnesium gallate containing divalent impurities such as Mn^{+2} is finding use in commercial ultraviolet activated powder phosphors. Gallium arsenide is capable of converting electricity directly into coherent light. Gallium readily alloys with most metals, and has been used as a component in low-melting alloys. Its toxicity appears to be of a low order, but should be handled with care until more data are forthcoming. The metal can be supplied in ultrapure form (99.99999 + %). The cost is about $3/g.

Germanium —(L. *Germania*, Germany), Ge; at. wt. 72.59; at. no. 32; m.p. 937.4°C; b.p. 2,830°C; sp. gr. 5.323 (25°C); valence 2 and 4. Predicted by Mendeleev in 1871 as ekasilicon, and discovered by Winkler in 1886. The metal is found in *argyrodite,* a sulfide of germanium and silver; in *germanite,* which contains 8% of the element; in zinc ores; in coal; and in other minerals. The element is frequently obtained commercially from flue dusts of smelters processing zinc ores, and has been recovered from the by-products of combustion of certain coals. Its presence in coal insures a large reserve of the element in the years to come. Germanium can be separated from other metals by fractional distillation of its volatile tetrachloride. The tetrachloride may then be hydrolyzed to give GeO_2; then the dioxide can be reduced with hydrogen to give the metal. Recently developed zone-refining techniques permit the production of germanium of ultra-high purity. The element is a gray white metalloid, and in its pure state is crystalline and brittle, retaining its luster in air at room temperature. It is a very important semiconductor material. Zone-refining techniques have led to production of crystalline germanium for semiconductor use with an impurity of only one part in 10^{10}. Doped with arsenic, gallium, or other elements, it is used as a transistor element in thousands of electronic applications. Its application as a semiconductor element now provides the largest use for germanium. Germanium is also finding many other applications including use as an alloying agent, as a phosphor in fluorescent lamps, and as a catalyst. Germanium and germanium oxide are transparent to the infrared and are used in infrared spectroscopes and other optical equipment, including extremely sensitive infrared detectors. Germanium oxide's high index of refraction and dispersion has made it useful as a component of glasses used in wide-angle camera lenses and microscope objectives. The field of organogermanium chemistry is becoming increasingly important. Certain germanium compounds have a low mammalian toxicity, but a marked activity against certain bacteria, which makes them of interest as chemotherapeutic agents. The cost of germanium is about $300/lb.

Gold — (Sanskrit *Jval;* Anglo-Saxon *gold),* Au (L. *aurum,* shining dawn); at. wt. 196.9665; at. no. 79; m.p. 1064.43°C; b.p. 3080°C; sp. gr. 18.88 (20°C); valence 1 or 3. Known and highly valued from earliest times, gold is found in nature as the free metal and in tellurides; it is very widely distributed and is almost always associated with quartz or pyrite. It occurs in veins and alluvial deposits, and is often separated from rocks and other minerals by sluicing or panning operations. About two thirds of the world's gold output now comes from South Africa, and about two thirds of the total U.S. production comes from South Dakota and Nevada. The metal is recovered from its ores by cyaniding, amalgamating, and smelting processes. Refining is also frequently done by electrolysis. Gold occurs in sea water to the extent of 0.1 to 2 mg/ton, depending on the location where the sample is taken. As yet, no method has been found for recovering gold from sea water profitably. It is estimated that all the gold in the world, so far refined, could be placed in a single cube 60 ft. on a side. Of all the elements, gold in its pure state is undoubtedly the most beautiful. It is metallic, having a yellow color when in a mass, but when finely divided it may be black, ruby, or purple. The Purple of Cassius is a delicate test for auric gold. It is the most malleable and ductile metal; 1 oz of gold can be beaten out to 300 ft². It is a soft metal and is usually alloyed to give it more strength. It is a good conductor of heat and electricity, and is unaffected by air and most reagents. It is used in coinage and is a standard for monetary systems in many countries. It is also extensively used for jewelry, decoration, dental work, and for plating. It is used for coating certain space satellites, as it is a good reflector of infrared and is inert. Gold, like other precious metals, is measured in troy weight; when alloyed with other metals, the term *carat* is used to express the amount of gold present, 24 carats being pure gold. For many years the value of gold was set by the U.S. at $20.67 /troy ounce; in 1934 this value was fixed by law at $35.00/ troy ounce, 9/10th fine. On March 17, 1968, because of a gold crisis, a two-tiered pricing system was established whereby gold was still used to settle international accounts at the old $35.00/troy ounce price while the price of gold on the private market would be allowed to fluctuate. Since this time, the price of gold on the free market has fluctuated widely. On March 19, 1968, President Johnson signed into law a bill removing the last statutory requirement for a gold backing against U.S. currency. On August 15, 1971, President Nixon announced an embargo on U.S. gold to settle international accounts, and on May 18, 1972, U.S. monetary gold was revalued at $38/ troy ounce. In February

THE ELEMENTS (continued)

1973 the U.S. in effect devalued the dollar by another 10%, decreasing the dollar in relation to gold from $38.00 to $42.22 / troy ounce. On September 2l, 1973, President Nixon signed a bill ratifying the action taken earlier in February; the bill also restored to private U.S. citizens the right to own gold after December 31, 1973 (private possession, except for gold in the form of jewelry, certain coins, etc., had been prohibited since 1933 when the U.S. went off the gold standard). The final version of the bill gave the President discretion to lift the ban when he determined that private ownership would not impair the monetary position during this period of instability. President Ford signed final legislation on August 14, 1974, lifting the 41-year ban, to be effective after December 31, 1974. The price of gold on the free market reached a price of $620/tr. oz. in January 1980. The most common gold compounds are auric chloride ($AuCl_3$) and chlorauric acid ($HAuCl_4$), the latter being used in photography for toning the silver image. Gold has 18 isotopes; Au^{198}, with a half-life of 2.7 days, is used for treating cancer and other diseases. Disodium aurothiomalate is administered intramuscularly as a treatment for arthritis. A mixture of one part nitric acid with three of hydrochloric acid is called *aqua regia* (because it dissolved gold, the King of Metals). Gold is available commercially with a purity of 99.999 + %. For many years the temperature assigned to the freezing point of gold has been 1063.0°C; this has served as a calibration point for the International Temperature Scales (ITS-27 and ITS-48) and the International Practical Temperature Scale (ITPS-48). In 1968, a new International Practical Temperature Scale (ITPS-68) was adopted, which demands that the freezing point of gold be changed to 1064.43°C. Although workers in precision temperature measurement should adopt IPTS-68 immediately, many of the scale changes are of minor significance to the routine user. IPTS-68 has defined several other fixed temperature points, among which are the boiling points of hydrogen, neon, oxygen, and sulfur, and the freezing points of zinc, silver, tin, lead, antimony, and aluminum. As of December 31, 1982, gold was priced at about $450/tr. oz.

Hafnium — (*Hafnia,* Latin name for Copenhagen), Hf; at. wt. 178.49; at. no. 72; m.p. 2227 ± 20°C; b.p. 4602°C; sp. gr. 13.31 (?) 20°C; valence 4. Hafnium was thought to be present in various minerals and concentrations many years prior to its discovery, in 1923, credited to D. Coster and G. von Hevesey. On the basis of the Bohr theory, the new element was expected to be associated with zirconium. It was finally identified in *zircon* from Norway, by means of X-ray spectroscope analysis. It was named in honor of the city in which the discovery was made. Most zirconium minerals contain 1 to 5% hafnium. It was originally separated from zirconium by repeated recrystallization of the double ammonium or potassium fluorides by von Hevesey and Jantzen. Metallic hafnium was first prepared by van Arkel and deBoer by passing the vapor of the tetraiodide over a heated tungsten filament. Almost all hafnium metal now produced is made by reducing the tetrachloride with magnesium or with sodium (Kroll Process). Hafnium is a ductile metal with a brilliant silver luster. Its properties are considerably influenced by the impurities of zirconium present. Of all the elements, zirconium and hafnium are two of the most difficult to separate. Their chemistry is almost identical, however, the density of zirconium is about half that of hafnium. Very pure hafnium has been produced, with zirconium being the major impurity. Because hafnium has a good absorption cross section for thermal neutrons (almost 600 times that of zirconium), has excellent mechanical properties, and is extremely corrosion resistant, it is used for reactor control rods. Such rods are used in nuclear submarines. Hafnium has been successfully alloyed with iron, titanium, niobium, tantalum, and other metals. Hafnium carbide is the most refractory binary composition known, and the nitride is the most refractory of all known metal nitrides (m.p. 3310° C). Hafnium is used in gas-filled and incandescent lamps, and is an efficient "getter" for scavenging oxygen and nitrogen. Finely divided hafnium is pyrophoric and can ignite spontaneously in air. Care should be taken when machining the metal or when handling hot sponge hafnium. At 700°C hafnium rapidly absorbs hydrogen to form the composition $HfH_{1.86}$. Hafnium is resistant to concentrated alkalis, but at elevated temperatures reacts with oxygen, nitrogen, carbon, boron, sulfur, and silicon. Halogens react directly to form tetrahalides. Exposure to hafnium should not exceed 0.5 mg/M^3 (8-hr time-weighted average — 40-hr week). The price of the metal is in the broad range of $100 to $500/lb, depending on purity and quantity. The yearly demand for hafnium in the U.S. is now in excess of 175,000 lb.

Hahnium — see Element 105.

Helium — (Gr. *helios,* the sun), He; at. wt. 4.00260; at. no. 2; m.p. below −272.2°C (26atm); b.p. − 268.934°C; density 0.1785 g/1 (0°C, 1 atm); liquid density 7.62 lb/ft^3 at. b.p.; valence usually 0. Evidence of the existence of helium was first obtained by Janssen during the solar eclipse of 1868 when he detected a new line in the solar spectrum; Lockyer and Frankland suggested the name *helium* for the new element; in 1895, Ramsay discovered helium in the uranium mineral *clevite,* and it was independently discovered in clevite by the Swedish chemists Cleve and Langlet about the same time. Rutherford and Royds in 1907 demonstrated that α particles are helium nuclei. Except for hydrogen, helium is the most abundant element found throughout the universe. It has been detected spectroscopically in great abundance, especially in the hotter stars, and it is an important component in both the proton-proton reaction and the carbon cycle, which account for the energy of the sun and stars. The fusion of hydrogen into helium provides the energy of the hydrogen bomb. The helium content of the atmosphere is about 1 part in 200,000. While it is present in various radioactive minerals as a decay product, the bulk of the Free World's supply is obtained from wells in Texas, Oklahoma, and Kansas. The only helium plant in the Free World outside the U.S. is near Swift River, Saskatchewan. The cost of helium fell from $2500/$ft^3$ in 1915 to 1.5c/ft^3 in 1940. The U.S. Bureau of Mines has set the price of Grade A helium at $35/1000 ft^3. Helium has the lowest melting point of any element and has found wide use in cryogenic research, as its boiling point is close to absolute zero. Its use in the study of superconductivity is vital. Using liquid helium, Kurti and co-workers, and others, have succeeded in obtaining temperatures of a few microdegrees K by the adiabatic demagnetization of copper nuclei, starting from about 0.01 K. Five isotopes of helium are known. Liquid helium (He^4) exists in two forms: He^4 I and He^4 II, with a sharp transition point at 2.174 K (3.83 cm Hg). He^4I (above this temperature) is a normal liquid, but He^4II (below it) is unlike any other known substance. It expands on cooling; its conductivity for heat is enormous; and neither its heat conduction nor viscosity obeys normal rules. It has other peculiar properties. Helium is the only liquid that cannot be solidified by lowering the temperature. It remains liquid

THE ELEMENTS (continued)

down to absolute zero at ordinary pressures, but it can readily be solidified by increasing the pressure. Solid He³ and He⁴ are unusual in that both can readily be changed in volume by more than 30% by application of pressure. The specific heat of helium gas is unusually high. The density of helium vapor at the normal boiling point is also very high, with the vapor expanding greatly when heated to room temperature. Containers filled with helium gas at 5 to 10° K should be treated as though they contained liquid helium due to the large increase in pressure resulting from warming the gas to room temperature. While helium normally has a 0 valence, it seems to have a weak tendency to combine with certain other elements. Means of preparing helium difluoride have been studied, and species such as HeNe and the molecular ions He^+ and He^{++} have been investigated. Helium is widely used as an inert gas shield for arc welding; as a protective gas in growing silicon and germanium crystals, and in titanium and zirconium production; as a cooling medium for nuclear reactors, and as a gas for supersonic wind tunnels. A mixture of 80% helium and 20% oxygen is used as an artificial atmosphere for divers and others working under pressure. Helium is extensively used for filling balloons as it is a much safer gas than hydrogen. While its density is almost twice that of hydrogen, it has about 98% of the lifting power of hydrogen. At sea level, 1000 ft³ of helium lifts 68.5 lb. One of the recent largest uses for helium has been for pressuring liquid fuel rockets. A Saturn booster such as used on the Apollo lunar missions requires about 13 million cubic feet of helium for a firing, plus more for checkouts.

Holmium — (L. *Holmia*, for Stockholm), Ho; at. wt. 164.9304; at. no. 67; m.p. 1474°C; b.p.2695°C; sp. gr. 8.795 (25°C); valence + 3. The spectral absorption bands of holmium were noticed in 1878 by the Swiss chemists Delafontaine and Soret, who announced the existence of an "Element X." Cleve, of Sweden, later independently discovered the element while working on erbia earth. The element is named after Cleve's native city. Pure holmia, the yellow oxide, was prepared by Homberg in 1911. Holmium occurs in *gadolinite, monazite,* and in other rare-earth minerals. It is commercially obtained from *monazite,* occurring in that mineral to the extent of about 0.05%. It has been isolated in pure form only in recent years. It can be separated from other rare earths by ion-exchange and solvent extraction techniques, and isolated by the reduction of its anhydrous chloride or fluoride with calcium metal. Pure holmium has a metallic to bright silver luster. It is relatively soft and malleable, and is stable in dry air at room temperature, but rapidly oxidizes in moist air and at elevated temperatures. The metal has unusual magnetic properties. Few uses have yet been found for the element. The element, as with other rare earths, seems to have a low acute toxic rating. The price of 99 + % holmium metal is about $8/g.

Hydrogen — (Gr. *hydro*, water, and *genes*, forming), H; at. wt. (natural) 1.0079; at. wt. (H¹)1.007822; at. no. 1; m.p. −259.14°C; b.p. −252.87°C; density 0.08988 g/l; density (liquid) 70.8 g/l (−253°C); density (solid) 70.6 g/l (−262°C); valence 1. Hydrogen was prepared many years before it was recognized as a distinct substance by Cavendish in 1766. It was named by Lavoisier. Hydrogen is the most abundant of all elements in the universe, and it is thought that the heavier elements were, and still are, being built from hydrogen and helium. It has been estimated that hydrogen makes up more than 90% of all the atoms or three quarters of the mass of the universe. It is found in the sun and most stars, and plays an important part in the proton-proton reaction and carbon-nitrogen cycle, which accounts for the energy of the sun and stars. It is thought that hydrogen is a major component of the planet Jupiter and that at some depth in the planet's interior the pressure is so great that solid molecular hydrogen is converted into solid metallic hydrogen. In 1973, it was reported that a group of Russian experimenters may have produced metallic hydrogen at a pressure of 2.8 Mbar. At the transition the density changed from 1.08 to 1.3 g/cm³. Earlier, in 1972, a Livermore (California) group also reported on a similar experiment in which they observed a pressure-volume point centered at 2 Mbar. It has been predicted that metallic hydrogen may may be metastable; others have predicted it would be a superconductor at room temperature. On earth, hydrogen occurs chiefly in combination with oxygen in water, but it is also present in organic matter such as living plants, petroleum, coal, etc. It is present as the free element in the atmosphere, but only to the extent of less than 1 ppm, by volume. It is the lightest of all gases, and combines with other elements, sometimes explosively, to form compounds. Great quantities of hydrogen are required commercially for the fixation of nitrogen from the air in the Haber ammonia process and for the hydrogenation of fats and oils. It is also used in large quantities in methanol production, in hydrodealkylation, hydrocracking, and hydrodesulfurization. It is also used as a rocket fuel, for welding, for production of hydrochloric acid, for the reduction of metallic ores, and for filling balloons. The lifting power of 1 ft³ of hydrogen gas is about 0.076 lb at 0°C, 760 mm pressure. Production of hydrogen in the U.S. alone now amounts to about 3 billion cubic feet per year. It is prepared by the action of steam on heated carbon, by decomposition of certain hydrocarbons with heat, by the electrolysis of water, or by the displacement from acids by certain metals. It is also produced by the action of sodium or potassium hydroxide on aluminum. Liquid hydrogen is important in cryogenics and in the study of superconductivity as its melting point is only a few degrees above absolute zero. The ordinary isotope of hydrogen, ₁H¹, is known as *protium*. In 1932, Urey announced the preparation of a stable isotope, deuterium (₁H² or D) with an atomic weight of 2. Two years later an unstable isotope, tritium (₁H³), with an atomic weight of 3 was discovered. Tritium has a half-life of about 12.5 years. One atom of deuterium is found mixed in with about 6000 ordinary hydrogen atoms. Tritium atoms are also present but in much smaller proportion. Tritium is readily produced in nuclear reactors and is used in the production of the hydrogen bomb. It is also used as a radioactive agent in making luminous paints, and as a tracer. The current price of tritium, to authorized personnel, is about $2/Ci; deuterium gas is readily available, without permit, at about $1/l. Heavy water, deuterium oxide (D_2O), which is used as a moderator to slow down neutrons, is available without permit at a cost of 6c to $1/g, depending on quantity and purity. Quite apart from isotopes, it has been shown that hydrogen gas under ordinary conditions is a mixture of two kinds of molecules, known as *ortho*- and *para*-hydrogen, which differ from one another by the spins of their electrons and nuclei. Normal hydrogen at room temperature contains 25% of the *para* form and 75% of the *ortho* form. The *ortho* form cannot be prepared in the pure state. Since the two forms differ in energy, the physical properties also differ. The melting and boiling points of *para*-hydrogen are about 0.1°C lower than those of normal hydrogen. Consideration is being given to an entire economy based on solar- and nuclear-generated hydrogen. Located in remote regions, power plants would electrolyze sea water; the hydrogen produced would travel to distant cities by

THE ELEMENTS (continued)

pipelines. Pollution-free hydrogen could replace natural gas, gasoline, etc., and could serve as a reducing agent in metallurgy, chemical processing, refining, etc. It could also be used to hydrolyze trash into methane and ethylene. Public acceptance, high capital investment, and the high present cost of hydrogen with respect to present fuels are but a few of the problems facing establishment of such an economy.

Indium — (from the brilliant indigo line in its spectrum), In; at. wt. 114.82; at. no. 49; m.p. 156.61°C; b.p. 2080°C; sp. gr. 7.31 (20°C); valence 1, 2, or 3. Discovered by Reich and Richter, who later isolated the metal. Indium is most frequently associated with zinc minerals, and it is from these that most commercial indium is now obtained; however, it is also found in iron, lead, and copper ores. Until 1924, a gram or so constituted the world's supply of this element in isolated form. It is probably about as abundant as silver. About 4 million troy ounces of indium are now produced annually in the Free World. Canada is presently producing more than 1,000,000 troy ounces annually. The present cost of indium is about $1 to $5/g, depending on quantity and purity. It is available in ultrapure form. Indium is a very soft, silvery-white metal with a brilliant luster. The pure metal gives a high-pitched "cry" when bent. It wets glass, as does gallium. It has found application in making low-melting alloys; an alloy of 24% indium-76% gallium is liquid at room temperature. It is used in making bearing alloys, germanium transistors, rectifiers, thermistors, and photoconductors. It can be plated onto metal and evaporated onto glass, forming a mirror as good as that made with silver but with more resistance to atmospheric corrosion. There is evidence that indium has a low order of toxicity; however, care should be taken until further information is available.

Iodine — (Gr. *iodes,* violet), I; at. wt. 126.9045; at. no. 53; m.p. 113.5°C; b.p. 184.35°C; density of the gas 11.27 g/l; sp. gr. solid 4.93 (20°C); valence 1, 3, 5, or 7. Discovered by Courtois in 1811. Iodine, a halogen, occurs sparingly in the form of iodides in sea water from which it is assimilated by seaweeds, in Chilean saltpeter and nitrate-bearing earth, known as *caliche,* in brines from old sea deposits, and in brackish waters from oil and salt wells. Ultrapure iodine can be obtained from the reaction of potassium iodide with copper sulfate. Several other methods of isolating the element are known. Iodine is a bluish-black, lustrous solid, volatilizing at ordinary temperatures into a blue-violet gas with an irritating odor; it forms compounds with many elements, but is less active than the other halogens, which displace it from iodides. Iodine exhibits some metallic-like properties. It dissolves readily in chloroform, carbon tetrachloride, or carbon disulfide to form beautiful purple solutions. It is only slightly soluble in water. Iodine compounds are important in organic chemistry and very useful in medicine. Twenty-three isotopes are recognized. Only one stable isotope, I^{127}, is found in nature. The artificial radioisotope I^{131}, with a half-life of 8 days, has been used in treating the thyroid gland. The most common compounds are the iodides of sodium and potassium (KI) and the iodates (KIO_3). Lack of iodine is the cause of goiter. The iodide, and thyroxin which contains iodine, are used internally in medicine, and a solution of KI and iodine in alcohol is used for external wounds. Potassium iodide finds use in photography. The deep blue color with starch solution is characteristic of the free element. Care should be taken in handling and using iodine, as contact with the skin can cause lesions; iodine vapor is intensely irritating to the eyes and mucous membranes. The maximum allowable concentration of iodine in air should not exceed 1 mg/M³ (8-hr time-weighted average — 40-hr).

Iridium — (L. *iris,* rainbow), Ir; at. wt. 192.22; at. no. 77; m.p. 2410°C; b.p. 4130°C; sp. gr. 22.42 (17°C); valence 3 or 4. Discovered in 1803 by Tennant in the residue left when crude platinum is dissolved by aqua regia. The name iridium is appropriate, for its salts are highly colored. Iridium, a metal of the platinum family, is white, similar to platinum, but with a slight yellowish cast. It is very hard and brittle, making it very hard to machine, form, or work. It is the most corrosion-resistant metal known, and was used in making the standard meter bar of Paris, which is a 90% platinum-10% iridium alloy. This meter bar has since been replaced as a fundamental unit of length (see under Krypton). Iridium is not attacked by any of the acids nor by aqua regia, but is attacked by molten salts, such as NaCl and NaCN. Iridium occurs uncombined in nature with platinum and other metals of this family in alluvial deposits. It is recovered as a by-product from the nickel mining industry. Iridium has found use in making crucibles and apparatus for use at high temperatures. It is also used for electrical contacts. Its principal use is as a hardening agent for platinum. With osmium, it forms an alloy which is used for tipping pens and compass bearings. The specific gravity of iridium is only very slightly lower than that of osmium, which has been generally credited as being the heaviest known element. Calculations of the densities of iridium and osmium from the space lattices give values of 22.65 and 22.61 g/cm³, respectively. These values may be more reliable than actual physical measurements. At present, therefore, we know that either iridium or osmium is the densest known element, but the data do not yet allow selection between the two. Iridium costs about $300/ troy ounce.

Iron — (Anglo-Saxon, *iron*), Fe (L. *ferrum*); at. wt. 55.847; at. no. 26; m.p. 1535°C; b.p. 2750°C; sp. gr. 7.874 (20°C); valence 2, 3, 4, or 6. The use of iron is prehistoric. Genesis mentions that Tubal-Cain, seven generations from Adam, was "an instructor of every artificer in brass and iron." A remarkable iron pillar, dating to about A.D. 400, remains standing today in Delhi, India. This solid shaft of wrought iron is about 7¼ m high by 40 cm in diameter. Corrosion to the pillar has been minimal although it has been exposed to the weather since its erection. Iron is a relatively abundant element in the universe. It is found in the sun and many types of stars in considerable quantity. Its nuclei are very stable. Iron is found native as a principal component of a class of meteorites known as *siderites,* and is a minor constituent of the other two classes. The core of the earth, 2150 mi in radius, is thought to be largely composed of iron with about 10% occluded hydrogen. The metal is the fourth most abundant element, by weight, making up the crust of the earth. The most common ore is *hematite* (Fe_2O_3), from which the metal is obtained by reduction with carbon. Iron is found in other widely distributed minerals such as *magnetite,* which is frequently seen as *black sands* along beaches and banks of streams. *Taconite* is becoming increasingly important as a commercial ore. Common iron is a mixture of four isotopes. Six other isotopes are known to exist. Iron is a vital constituent of plant and animal life, and appears in hemoglobin. The pure metal is not often encountered in commerce, but is usually alloyed with carbon or other metals. The pure metal is very reactive chemically, and rapidly corrodes, especially in moist air or at elevated temperatures. It has four allotropic forms, or ferrites, known as α, β, γ, and δ, with transition points at 770, 928, and 1530°C. The α form is magnetic, but when transformed into the β form, the magnetism disappears although the lattice remains unchanged. The relations of these forms are peculiar. Pig iron is an alloy containing about 3% carbon with varying amounts of S, Si, Mn, and P. It is hard, brittle, fairly fusible, and is used to produce other alloys, including steel. Wrought iron contains only a few tenths of a percent of carbon, is tough, malleable, less fusible, and has usually a "fibrous" structure. Carbon steel is an alloy of iron with carbon, with small amounts of Mn, S, P, and Si. Alloy steels are carbon steels with other additives such as nickel, chromium, vanadium, etc. Iron is the cheapest and most abundant, useful, and important of all metals.

Krypton — (Gr. *kryptos,* hidden), Kr; at. wt. 83.80; at. no. 36; m.p. −156.6°C; b.p. −152.30± 0.10°C; density 3.733 g/l (O°C); valence usually 0. Discovered in 1898 by Ramsay and Travers in the residue left after liquid air had nearly boiled away. Krypton is present in the air to the extent of about 1 ppm. The atmosphere of Mars has been found to contain 0.3 ppm of krypton. It is one of the "noble" gases. It is characterized by its brilliant green and orange spectral lines. Naturally occurring krypton contains six stable isotopes. Fifteen other unstable isotopes are now recognized. The spectral lines of krypton are easily produced and some are very sharp. In 1960 it was internationally agreed that the fundamental unit of length, the meter, should be defined in terms of the orange-red spectral line of Kr⁸⁶, corresponding to the transition 5p[O₁/₂]₁ − 6d[O₁/₂]₁, as follows: *1 m = 1,650,763.73 wavelengths (in vacuo) of the orange-red line of Kr⁸⁶.* This replaces the standard meter of Paris, which was defined in terms of a bar made of a platinum-iridium alloy. Solid krypton is a white crystalline substance with a face-centered cubic structure which is common to all the "rare gases." While krypton is generally thought of as a rare gas that normally does not combine with other elements to form compounds, it now appears that the existence of some krypton compounds is established. Krypton difluoride has been prepared in gram quantities and can be made by several methods. A higher fluoride of krypton and a salt of an oxyacid of krypton also have been reported. Molecule ions of ArKr⁺ and KrH⁺ have been identified and investigated, and evidence is provided for the formation of KrXe or KrXe⁺. Krypton clathrates have been prepared with hydroquinone and phenol. Kr⁸⁵ has found recent application in chemical analysis. By imbedding the isotope in various solids, *kryptonates* are formed. The activity of these kryptonates is sensitive to chemical reactions at the surface. Estimates of the concentration of reactants are therefore made possible. Krypton is used commercially with argon as a low-pressure filling gas for fluorescent lights. It is used in certain photographic flash lamps for high-speed photography. Uses thus far have been limited because of its high cost. Krypton gas presently costs about $30/ℓ.

Kurchatovium — see Element 104.

Lanthanum — (Gr. *lanthanein,* to lie hidden), La; at. wt. 138.9055; at. no. 57; m.p. 921°C; b.p. 3457°C; sp. gr. 6.145 (25°C), valence 3. Mosander in 1839 extracted a new earth, *lanthana,* from impure cerium nitrate, and recognized the new element. Lanthanum is found in rare-earth minerals such as *cerite, monazite, allanite,* and *bastnasite.* Monazite and bastnasite are principal ores in which lanthanum occurs in percentages up to 25 and 38%, respectively. Mischmetal, used in making lighter flints, contains about 25% lanthanum. Lanthanum was isolated in relatively pure form in 1923. Ion-exchange and solvent extraction techniques have led to much easier isolation of the so-called "rare-earth" elements. The availability of lanthanum and other rare earths has improved greatly in recent years. The metal can be produced by reducing the anhydrous fluoride with calcium. Lanthanum is silvery white, malleable, ductile, and soft enough to be cut with a knife. It is one of the most reactive of the rare-earth metals. It oxidizes rapidly when exposed to air. Cold water attacks lanthanum slowly, and hot water attacks it much more rapidly. The metal reacts directly with elemental carbon, nitrogen, boron, selenium, silicon, phosphorus, sulfur, and with halogens. At 310°C, lanthanum changes from a hexagonal to a face-centered cubic structure, and at 865°C it again transforms into a body-centered cubic structure. Natural lanthanum is a mixture of two stable isotopes, La¹³⁸ and La¹³⁹. Seventeen other radioactive isotopes are recognized. Rare-earth compounds containing lanthanum are extensively used in carbon lighting applications, especially by the motion picture industry for studio lighting and projection. This application consumes about 25% of the rare-earth compounds produced. La₂O₃ improves the alkali resistance of glass, and is used in making special optical glasses. Small amounts of lanthanum, as an additive, can be used to produce nodular cast iron. There is current interest in hydrogen sponge alloys containing lanthanum. These alloys take up to 400 times their own volume of hydrogen gas, and the process is reversible. Heat energy is released every time they do so; therefore these alloys have possibilities in energy conservation systems. Lanthanum and its compounds have a low to moderate acute toxicity rating; therefore, care should be taken in handling them. The metal costs about $25/oz.

Lawrencium — (Ernest O. Lawrence, inventor of the cyclotron), Lr; at. no. 103; at. mass no. 257; valence +3(?). This member of the 5f transition elements (actinide series) was discovered in March 1961 by A. Ghiorso, T. Sikkeland, A.E. Larsh, and R.M. Latimer. A 3-μg californium target, consisting of a mixture of isotopes of mass number 249, 250, 251, and 252, was bombarded with either B¹⁰ or B¹¹. The electrically charged transmutation nuclei recoiled with an atmosphere of helium and were collected on a thin copper conveyor tape which was then moved to place collected atoms in front of a series of solid-state detectors. The isotope of element 103 produced in this way decayed by emitting an 8.6-MeV alpha particle with a half-life of 8 sec. In 1967, Flerov and associates of the Dubna Laboratory reported their inability to detect an alpha emitter with a half-life of 8 sec which was assigned by the Berkeley group to 103²⁵⁷. This assignment has been changed to Lr²⁵⁸ or Lr²⁵⁹. In 1965, the Dubna workers found a longer-lived lawrencium isotope, Lr²⁵⁶, with a half-life of 35 sec. In 1968, Ghiorso and associates at Berkeley were able to use a few atoms of this isotope to study the oxidation behavior of lawrencium. Using solvent extraction techniques and working very rapidly, they extracted lawrencium ions from a buffered aqueous solution into an organic solvent, completing each extraction in about 30 sec. It was found that lawrencium behaves differently from dipositive nobelium and more like the tripositive elements earlier in the actinide series.

Lead — (Anglo-Saxon *lead*), Pb (L. *plumbum*); at. wt. 207.2; at. no. 82; m.p. 327.502°C; b.p. 1740°C; sp.gr. 11.35 (20°C); valence 2 or 4. Long known, mentioned in Exodus. The alchemists believed lead to be the oldest metal and associated it with the planet Saturn. Native lead occurs in nature, but it is rare. Lead is obtained chiefly from *galena* (PbS) by a roasting process. *Anglesite* (PbSO$_4$), *cerussite* (PbCO$_3$),and *minim* (Pb$_3$O$_4$) are other common lead minerals. Lead is a bluish-white metal of bright luster, is very soft, highly malleable, ductile, and a poor conductor of electricity. It is very resistant to corrosion; lead pipes bearing the insignia of Roman emperors, used as drains from the baths, are still in service. It is used in containers for corrosive liquids (such as in sulfuric acid chambers) and may be toughened by the addition of a small percentage of antimony or other metals. Natural lead is a mixture of four stable isotopes: Pb^{204} (1.48%), Pb^{206} (23.6%), Pb^{207} (22.6%), and Pb ^{208} (52.3%). Lead isotopes are the end products of each of the three series of naturally occurring radioactive elements: Pb^{206} for the uranium series, Pb^{207} for the actinium series, and Pb^{208} for the thorium series. Seventeen other isotopes of lead, all of which are radioactive, are recognized. Its alloys include solder, type metal, and various antifriction metals. Great quantities of lead, both as the metal and as the dioxide, are used in storage batteries. Much metal also goes into cable covering, plumbing, ammunition, and in the manufacture of lead tetraethyl, used as an antiknock compound in gasoline. The metal is very effective as a sound absorber, is used as a radiation shield around X-ray equipment and nuclear reactors, and is used to absorb vibration. White lead, the basic carbonate, sublimed white lead (PbSO$_4$), chrome yellow (PbCrO$_4$), red lead (Pb$_3$O$_4$), and other lead compounds are used extensively in paints, although in recent years the use of lead in paints has been drastically curtailed to eliminate or reduce health hazards. Lead oxide is used in producing fine "crystal glass" and "flint glass" of a high index of refraction for achromatic lenses. The nitrate and the acetate are soluble salts. Lead salts such as lead arsenate have been used as insecticides, but their use in recent years has been practically eliminated in favor of less harmful organic compounds. Care must be used in handling lead as it is a cumulative poison. Environmental concern with lead poisoning has resulted in a national program to reduce the concentration of lead in gasoline.

Lithium — (Gr. *lithos*, stone), Li; at. wt. 6.941; at. no. 3; m.p. 180.54°C; b.p. 1342°C; sp. gr. 0.534 (20 °C); valence 1. Discovered by Arfvedson in 1817. Lithium is the lightest of all metals, with a density only about half that of water. It does not occur free in nature; combined it is found in small amounts in nearly all igneous rocks and in the waters of many mineral springs. *Lepidolite, spodumene, petalite,* and *amblygonite* are the more important minerals containing it. Lithium is presently being recovered from brines of Searles Lake, in California, and from those in Nevada. Large deposits of spodumene are found in North Carolina. The metal is produced electrolytically from the fused chloride. Lithium is silvery in appearance, much like Na and K, other members of the alkali metal series. It reacts with water, but not as vigorously as sodium. Lithium imparts a beautiful crimson color to a flame, but when the metal burns strongly the flame is a dazzling white. Since World War II, the production of lithium metal and its compounds has increased greatly. Because the metal has the highest specific heat of any solid element, it has found use in heat transfer applications; however, it is corrosive and requires special handling. The metal has been used as an alloying agent, is of interest in synthesis of organic compounds, and has nuclear applications. It ranks as a leading contender as a battery anode material as it has a high electrochemical potential. Lithium is used in special glasses and ceramics. The glass for the 200-in. telescope at Mt. Palomar contains lithium as a minor ingredient. Lithium chloride is one of the most hygroscopic materials known, and it, as well as lithium bromide, is used in air conditioning and industrial drying systems. Lithium stearate is used as an all-purpose and high-temperature lubricant. Other lithium compounds are used in dry cells and storage batteries. The metal is priced at about $20/lb.

Lutetium — (Lutetia, ancient name for Paris, sometimes called *cassiopeium* by the Germans), Lu; at. wt. 174.967 ± 0.003; at. no. 71; m.p. 1663°C; b.p. 3395°C; sp. gr. 9.840 (25°C); valence 3. In 1907, Urbain described a process by which Marignac's ytterbium (1879) could be separated into the two elements, ytterbium (neoytterbium) and lutetium. These elements were identical with "aldebaranium" and "cassiopeium," independently discovered by von Welsbach about the same time. Charles James of the University of New Hampshire also independently prepared the very pure oxide, *lutecia*, at this time. The spelling of the element was changed from *lutecium* to *lutetium* in 1949. Lutetium occurs in very small amounts in nearly all minerals containing yttrium, and is present in *monazite* to the extent of about 0.003%, which is a commercial source. The pure metal has been isolated only in recent years and is one of the most difficult to prepare. It can be prepared by the reduction of anhydrous LuCl$_3$ or LuF$_3$ by an alkali or alkaline earth metal. The metal is silvery white and relatively stable in air. While new techniques, including ion-exchange reactions, have been developed to separate the various rare-earth elements, lutetium is still the most costly of all naturally occurring rare earths. It is slightly more abundant than thulium. It is now priced at about $75/g or $30,000/lb. Lu^{176}occurs naturally (2.6%) with Lu^{175} (97.4%). It is radioactive with a half-life of about 3 × 10^{10} years. Stable lutetium nuclides, which emit pure beta radiation after thermal neutron activation, can be used as catalysts in cracking, alkylation, hydrogenation, and polymerization. Virtually no other commercial uses have been found yet for lutetium, as it is one of the most costly of the rare earth metals. While lutetium, like other rare-earth metals, is thought to have a low toxicity rating, it should be handled with care until more information is available.

Magnesium — (*Magnesia*, district in Thessaly) Mg; at. wt. 24.305; at. no. 12; m.p. 648.8±0.5°C; b.p. 1090°C; sp. gr. 1.738 (20°C); valence 2. Compounds of magnesium have long been known. Black recognized magnesium as an element in 1755. It was isolated by Davy in 1808, and prepared in coherent form by Bussy in 1831. Magnesium is the eighth most abundant element in the earth's crust. It does not occur uncombined, but is found in large deposits in the form of *magnesite, dolomite,* and other minerals. The metal is now principally obtained in the U.S. by electrolysis of fused magnesium chloride derived from brines, wells, and sea water. Magnesium is a light, silvery-white, and fairly tough metal. It tarnishes slightly in air, and finely divided magnesium readily ignites upon heating in air and burns with a dazzling white flame. It is used in flashlight photography, flares, and pyrotechnics, including incendiary bombs. It is one third lighter than aluminum, and in alloys is essential for airplane and missile construction. The metal improves the mechanical, fabrication, and welding characteristics of aluminum when used as an alloying agent. Magnesium is used in producing nodular graphite in cast iron, and is used as an additive to conventional propellants. It is also used as a reducing agent in the production of pure uranium and other metals from their salts. The hydroxide (*milk of magnesia*), chloride, sulfate (*Epsom salts),* and citrate are used in medicine. Dead-burned magnesite is employed for refractory purposes such as brick and liners in furnaces and converters.

THE ELEMENTS (continued)

Organic magnesium compounds (Grignard's reaction) are important. Magnesium is an important element in both plant and animal life. Chlorophylls are magnesium-centered porphyrins. The adult daily requirement of magnesium is about 300 mg/day, but this is affected by various factors. Great care should be taken in handling magnesium metal, especially in the finely divided state, as serious fires can occur. Water should not be used on burning magnesium or on magnesium fires.

Manganese — (L. *magnes,* magnet, from magnetic properties of pyrolusite; It. *manganese,* corrupt form of *magnesia*), Mn; at. wt. 54.9380; at. no. 25; m.p. 1244±3°C; b.p. 1962°C; sp. gr. 7.21 to 7.44, depending on allotropic form; valence 1, 2, 3, 4, 6, or 7. Recognized by Scheele, Bergman, and others as an element and isolated by Gahn in 1774 by reduction of the dioxide with carbon. Manganese minerals are widely distributed; oxides, silicates, and carbonates are the most common. The discovery of large quantities of manganese nodules on the floor of the oceans holds promise as a source of manganese. These nodules contain about 24% manganese together with many other elements in lesser abundance. Most manganese today is obtained from ores found in the U.S.S.R., Brazil, Australia, Republic of So. Africa, Gabon, and India. *Pyrolusite* (MnO_2) and *rhodochrosite* ($MnCO_3$) are among the most common manganese minerals. The metal is obtained by reduction of the oxide with sodium, magnesium, aluminum, or by electrolysis. It is gray-white, resembling iron, but is harder and very brittle. The metal is reactive chemically, and decomposes cold water slowly. Manganese is used to form many important alloys. In steel, manganese improves the rolling and forging qualities, strength, toughness, stiffness, wear resistance, hardness, and hardenability. With aluminum and antimony, especially with small amounts of copper, it forms highly ferromagnetic alloys. Manganese metal is ferromagnetic only after special treatment. The pure metal exists in four allotropic forms. The alpha form is stable at ordinary temperature; gamma manganese, which changes to alpha at ordinary temperatures, is said to be flexible, soft, easily cut, and capable of being bent. The dioxide (pyrolusite) is used as a depolarizer in dry cells, and is used to "decolorize" glass that is colored green by impurities of iron. Manganese by itself colors glass an amethyst color, and is responsible for the color of true amethyst. The dioxide is also used in the preparation of oxygen, chlorine, and in drying black paints. The permanganate is a powerful oxidizing agent and is used in quantitative analysis and in medicine. Manganese is widely distributed throughout the animal kingdom. It is an important trace element and may be essential for utilization of vitamin B_1. Exposure to manganese dusts, fume, and compounds (as Mn) should not exceed the ceiling value of 5 mg/M^3 for even short periods because of the toxicity of the element in larger quantity.

Mendelevium — (Dmitri Mendeleev), Md; at. wt. 257; at. no. 101; valence +2, +3. Mendelevium, the ninth transuranium element of the actinide series to be discovered, was first identified by Ghiorso, Harvey, Choppin, Thompson, and Seaborg early in 1955 as a result of the bombardment of the isotope Es^{253} with helium ions in the Berkeley 60 in. cyclotron. The isotope produced was Md^{256}, which has a half-life of 77 min. This first identification was notable in that Md^{256} was synthesized on a one-atom-at-a-time basis. Four isotopes are now recognized. Md^{258} has a half-life of 2 months. This isotope has been produced by the bombardment of an isotope of einsteinium with ions of helium. It now appears possible that eventually Md^{258} can be made so that some of its physical properties can be determined. Md^{256} has been used to elucidate some of the chemical properties of mendelevium in aqueous solution. Experiments seem to show that the element possesses a moderately stable dipositive (II) oxidation state in addition to the tripositive (III) oxidation state, which is characteristic of actinide elements.

Mercury — (Planet *Mercury*), Hg (*hydrargyrum,* liquid silver); at. wt. 200.59; at. no. 80; m.p. −38.842°C; b.p. 356.58°C; sp. gr. 13.546 (20°C); valence 1 or 2. Known to ancient Chinese and Hindus; found in Egyptian tombs of 1500 B.C. Mercury is the only common metal liquid at ordinary temperatures. It only rarely occurs free in nature. The chief ore is *cinnabar* (HgS). Spain and Italy produce about 50% of the world's supply of the metal. The commercial unit for handling mercury is the "flask," which weighs 76 lb. The metal is obtained by heating cinnabar in a current of air and by condensing the vapor. It is a heavy, silvery-white metal; a rather poor conductor of heat, as compared with other metals, and a fair conductor of electricity. It easily forms alloys with many metals, such as gold, silver, and tin, which are called *amalgams*. Its ease in amalgamating with gold is made use of in the recovery of gold from its ores. The metal is widely used in laboratory work for making thermometers, barometers, diffusion pumps, and many other instruments. It is used in making mercury-vapor lamps and advertising signs, etc. and is used in mercury switches and other electrical apparatus. Other uses are in making pesticides, mercury cells for caustic soda and chlorine production, dental preparations, antifouling paint, batteries, and catalysts. The most important salts are mercuric chloride $HgCl_2$ (corrosive sublimate — a violent poison), mercurous chloride Hg_2Cl_2 (calomel, occasionally still used in medicine), mercury fulminate ($Hg(ONC)_2$, a detonator widely used in explosives), and mercuric sulfide (HgS, vermillion, a high-grade paint pigment). Organic mercury compounds are important. It has been found that an electrical discharge causes mercury vapor to combine with neon, argon, krypton, and xenon. These products, held together with van der Waals' forces, correspond to HgNe, HgAr, HgKr, and HgXe. Mercury is a virulent poison and is readily absorbed through the respiratory tract, the gastrointestinal tract, or through unbroken skin. It acts as a cumulative poison since only small amounts of the element can be eliminated at a time by the human organism. Since mercury is a very volatile element, dangerous levels are readily attained in air. Air saturated with mercury vapor at 20°C contains a concentration that exceeds the toxic limit many times. The danger increases at higher temperatures. *It is therefore important that mercury be handled with care.* Containers of mercury should be securely covered and spillage should be avoided. If it is necessary to heat mercury or mercury compounds, it should be done in a well-ventilated hood. Methyl mercury is a dangerous pollutant and is now widely found in water and streams. The U.S. National Bureau of Standards has recently redetermined the triple point of mercury and found it to be −38.84168°C.

Molybdenum — (Gr. *molybdos,* lead), Mo; at. wt. 95.94 ± 0.01; at. no. 42; m.p. 2617°C; b.p. 4612°C; sp. gr. 10.22 (20°C); valence 2, 3, 4?, 5?, or 6. Before Scheele recognized molybdenite as a distinct ore of a new element in 1778, it was confused with graphite and lead ore. The metal was prepared in an impure form in 1782 by Hjelm. Molybdenum does not occur native, but is obtained principally from *molybdenite* (MoS_2). *Wulfenite* ($PbMoO_4$), and *powellite* ($Ca(MoW)O_4$) are also minor commercial ores. Molybdenum is also recovered as a by-product of copper and tungsten mining operations. The metal is prepared from the powder made by the hydrogen reduction of purified molybdic trioxide or ammonium molybdate. The metal is silvery white, very hard, but is softer and more ductile than tungsten. It has a high elastic modulus, and only tungsten and tantalum, of the more readily available metals, have higher melting points. It is a valuable alloying agent, as it contributes to the hardenability and toughness of quenched and tempered steels. It also improves the strength of steel at high temperatures. It is used in certain nickel-based alloys, such as the "Hastelloys,®" which are heat-resistant and corrosion-resistant to chemical solutions. Molybdenum oxidizes at elevated temperatures. The metal has found recent application as electrodes for electrically heated glass furnaces and forehearths. The metal is also used in nuclear energy applications and for missile and aircraft parts. Molybdenum is valuable as a catalyst in the refining of petroleum. It has found application as a filament material in electronic and electrical applications. Molybdenum is an essential trace element in plant nutrition. Some lands are barren for lack of this element in the soil. Molybdenum sulfide is useful as a lubricant, especially at high temperatures where oils would decompose. Almost all ultra-high strength steels with minimum yield points up to 300,000 psi(lb/in.^{2}) contain molybdenum in amounts from 0.25 to 8%.

Neodymium — (Gr. *neos*, new, and *didymos*, twin), Nd; at. wt. 144.24; at. no. 60; m.p. 1021°C; b.p.3068°C; sp. gr. 6.80 and 7.007, depending on allotropic form; valence 3. In 1841, Mosander, extracted from *cerite* a new rose-colored oxide, which he believed contained a new element. He named the element *didymium*, as it was "an inseparable twin brother of lanthanum." In 1885 von Welsbach separated didymium into two new elemental components, *neodymia* and *praseodymia*, by repeated fractionation of ammonium didymium nitrate. While the free metal is in *misch metal*, long known and used as a pyrophoric alloy for lighter flints, the element was not isolated in relatively pure form until 1925. Neodymium is present in misch metal to the extent of about 18%. It is present in the minerals *monazite* and *bastnasite*, which are principal sources of rare-earth metals. The element may be obtained by separating neodymium salts from other rare earths by ion-exchange or solvent extraction techniques, and by reducing anhydrous halides such as NdF$_3$ with calcium metal. Other separation techniques are possible. The metal has a bright silvery metallic luster. Neodymium is one of the more reactive rare-earth metals and quickly tarnishes in air, forming an oxide that spalls off and exposes metal to oxidation. The metal, therefore, should be kept under light mineral oil or sealed in a plastic material. Neodymium exists in two allotropic forms, with a transformation from a double hexagonal to a body-centered cubic structure taking place at 860°C. Natural neodymium is a mixture of seven stable isotopes. Seven other radioactive isotopes are recognized. Didymium, of which neodymium is a component, is used for coloring glass to make welder's goggles. By itself, neodymium colors glass delicate shades ranging from pure violet through wine-red and warm gray. Light transmitted through such glass shows unusually sharp absorption bands. The glass has been used in astronomical work to produce sharp bands by which spectral lines may be calibrated. Glass containing neodymium can be used as a laser material in place of ruby to produce coherent light. Neodymium salts are also used as a colorant for enamels. The price of the metal is about $2/g. Neodymium has a low-to-moderate acute toxic rating. As with other rare earths, neodymium should be handled with care.

Neon — (Gr. *neos*, new), Ne; at. wt. 20.179; at. no. 10; m.p. −248.67°C; b.p. −246.048°C (1 atm); density of gas 0.89990 g/l (1 atm, 0°C); density of liquid at b.p. 1.207 g/cm^3; valence 0. Discovered by Ramsay and Travers in 1898. Neon is a rare gaseous element present in the atmosphere to the extent of 1 part in 65,000 of air. It is obtained by liquefaction of air and separated from the other gases by fractional distillation. Natural neon is a mixture of three isotopes. Five other unstable isotopes are known. It is a very inert element; however, it is said to form a compound with fluorine. It is still questionable if true compounds of neon exist, but evidence is mounting in favor of their existence. The following ions are known from optical and mass spectrometric studies: Ne$^+$, (NeAr)$^+$, (NeH)$^+$, and (HeNe)$^+$. Neon also forms an unstable hydrate. In a vacuum discharge tube, neon glows reddish orange. Of all the rare gases, the discharge of neon is the most intense at ordinary voltages and currents. Neon is used in making the common neon advertising signs, which accounts for its largest use. It is also used to make high-voltage indicators, lightning arrestors, wave meter tubes, and TV tubes. Neon and helium are used in making gas lasers. Liquid neon is now commercially available and is finding important application as an economical cryogenic refrigerant. It has over 40 times more refrigerating capacity per unit volume than liquid helium and more than three times that of liquid hydrogen. It is compact, inert, and is less expensive than helium when it meets refrigeration requirements. Neon has costs about $1.50/l.

Neptunium — (Planet *Neptune*), Np; at. wt. 237.0482; at. no. 93; m.p. 640± 1°C; b.p. 3902°C (est.); sp. gr. 20.25 (20°C); valence 3, 4, 5, and 6. Neptunium was the first synthetic transuranium element of the actinide series discovered; the isotope Np^{239} was produced by McMillan and Abelson in 1940 at Berkeley, California, as the result of bombarding uranium with cyclotron-produced neutrons. The isotope Np^{237} (half-life of 2.14 × 10^6 years) is currently obtained in gram quantities as a by-product from nuclear reactors in the production of plutonium. Trace quantities of the element are actually found in nature due to transmutation reactions in uranium ores produced by the neutrons which are present. Neptunium is prepared by the reduction of NpF$_3$ with barium or lithium vapor at about 1200°C. Neptunium metal has a silvery appearance, is chemically reactive, and exists in at least three structural modifications: α-neptunium, orthorhombic, density 20.25 g/cm^3; β-neptunium (above 280°C), tetragonal, density (313° C) 19.36 g/cm^3; γ-neptunium (above 577°), cubic, density (600°C) 18.0 g/cm^3. Neptunium has four ionic oxidation states in solution: Np^{+3} (pale purple), analogous to the rare earth ion Pm^{+3}, Np^{+4} (yellow green); NpO$^+$ (green blue); and NpO^{++} (pale pink). These latter oxygenated species are in contrast to the rare earths which exhibit only simple ions of the (II), (III), and (IV) oxidation states in aqueous solution. The element forms tri- and tetrahalides such as NpF$_3$, NpF$_4$, NpCl$_4$, NpBr$_3$, NpI$_3$, and oxides of various compositions such as are found in the uranium-oxygen system, including Np$_3$O$_8$ and NpO$_2$. Thirteen isotopes of neptunium are now recognized. The A.E.C. has Np^{237} available for sale to its licensees and for export. This isotope can be used as a component in neutron detection instruments. It is offered at a price of $225/g or $0.25/mg for quantities less than 1g.

Nickel — (Ger, *Nickel,* Satan or "Old Nick's and from *kupfernickel,* Old Nick's copper), Ni; at. wt. 58.71; at. no. 28; m.p. 1453°C; b.p. 2732°C; sp. gr. 8.902 (25° C); valence 0,1,2,3. Discovered by Cronstedt in 1751 in kupfernickel(*niccolite*). Nickel is found as a constituent in most meteorites and often serves as one of the criteria for distinguishing a meteorite from other minerals. Iron meteorites, or *siderites,* may contain iron alloyed with from 5 to nearly 20% nickel. Nickel is obtained commercially from *pentlandite* and *pyrrhotite* of the Sudbury region of Ontario, a district that produces about 50% of the nickel for the Free World. Other deposits are found in New Caledonia, Australia, Cuba, Indonesia, and elsewhere. Nickel is silvery white and takes on a high polish. It is hard, malleable, ductile, somewhat ferromagnetic, and a fair conductor of heat and electricity. It belongs to the iron-cobalt group of metals and is chiefly valuable for the alloys it forms. It is extensively used for making stainless steel and other corrosion-resistant alloys such as Invar®, Monel®, Inconel®, and the Hastelloys®. Tubing made of a copper-nickel alloy is extensively used in making desalination plants for converting sea water into fresh water. Nickel is also now used extensively in coinage and in making nickel steel for armor plate and burglar-proof vaults, and is a component in Nichrome®, Permalloy®, and constantan. Nickel added to glass gives a green color. Nickel plating is often used to provide a protective coating for other metals, and finely divided nickel is a catalyst for hydrogenating vegetable oils. It is also used in ceramics, in the manufacture of Alnico magnets, and in the Edison® storage battery. The sulfate and the oxides are important compounds. Natural nickel is a mixture of five stable isotopes; seven other unstable isotopes are known.Exposure to nickel metal and soluble compounds (as Ni) should not exceed 1mg/M^3 (8-hr time-weighted average — 40-hr week). Nickel carbonyl exposure, however, should not exceed 0.007mg/M^3, and is considered to be a very toxic material. Nickel sulphide fume and dust is recognized as having carcinogenic potential.

Niobium — *(Niobe,* daughter of Tantalus), Nb; or **Columbium** (*Columbia,* name for America), Cb; at. wt. 92.9064; at. no. 41; m.p. 2468± 10°C; b.p. 4742°C; sp. gr. 8.57 (20°C); valence 2, 3, 4?, 5. Discovered in 1801 by Hatchett in an ore sent to England more than a century before by John Winthrop the Younger, first goveror of Connecticut. The metal was first prepared in 1864 by Blomstrand, who reduced the chloride by heating it in a hydrogen atmosphere. The name "niobium" was adopted by the International Union of Pure and Applied Chemistry in 1950 after 100 years of controversy. Many leading chemical societies and government organizations refer to it by this name. Most metallurgists, leading metal societies, and all but one of the leading U.S. commercial producers, however, still refer to the metal as "columbium." The element is found in *niobite* (or *columbite*), *niobite-tantalite, pyrochlore,* and *euxenite.* Large deposits of niobium have been found associated with *carbonatites* (carbon-silicate rocks), as a constituent of *pyrochlore.* Extensive ore reserves are found in Canada, Brazil, Nigeria, Zaire, and in the U S S R. The metal can be isolated from tantalum, and prepared in several ways. It is a shiny, white, soft, and ductile metal, and takes on a bluish cast when exposed to air at room temperatures for a long time. The metal starts to oxidize in air at 200°C, and when processed at even moderate temperatures must be placed in a protective atmosphere. It is used as an alloying agent in carbon and alloy steels and in nonferrous metals. These alloys have improved strength and other desirable properties, and have been used in pipeline construction. The metal has a low capture cross-section for thermal neutrons. It is used in arc welding rods for stabilized grades of stainless steel. Thousands of pounds of niobium have been used in advanced air frame systems such as were used in the Gemini space program. The element has superconductive properties; superconductive magnets have been made with Nb-Zr wire, which retains its superconductivity in strong magnetic fields. This type of application offers hope of direct large-scale generation of electric power. Sixteen isotopes of niobium are known. Niobium metal (99.5% pure) is priced at about $30/lb.

Nitrogen — (L. *nitrum,* Gr. *nitron,* native soda; genes, *forming*), N; at. wt. 14.0067; at. no. 7; m.p. −209.86 °C; b.p. −195.8°C; density l.2506 g/l; sp. gr. liquid 0.808 (−195.8°C), solid 1.026 (−252°C); valence 3 or 5. Discovered by Daniel Rutherford in 1772, but Scheele, Cavendish, Priestley, and others about the same time studied "burnt or dephlogisticated air," as air without oxygen was then called. Nitrogen makes up 78% of the air, by volume. The atmosphere of Mars, by comparison, is 2.6% nitrogen. The estimated amount of this element in our atmosphere is more than 4000 billion tons. From this inexhaustible source it can be obtained by liquefaction and fractional distillation. Nitrogen molecules give the orange-red, blue-green, blue-violet, and deep violet shades to the aurora. The element is so inert that Lavoisier named it *azote,* meaning without life, yet its compounds are so active as to be most important in foods, poisons, fertilizers, and explosives. Nitrogen can be also easily prepared by heating a water solution of ammonium nitrite. Nitrogen, as a gas, is colorless, odorless, and a generally inert element. As a liquid it is also colorless and odorless, and is similar in appearance to water. Two allotropic forms of solid nitrogen exist, with the transition from the α to the β form taking place at −237 °C. When nitrogen is heated, it combines directly with magnesium, lithium, or calcium; when mixed with oxygen and subjected to electric sparks, it forms first nitric oxide (NO) and then the dioxide (NO_2); when heated under pressure with a catalyst with hydrogen, ammonia is formed (Haber process). The ammonia thus formed is of the utmost importance as it is used in fertilizers, and it can be oxidized to nitric acid (Ostwald process). The ammonia industry is the largest consumer of nitrogen. Large amounts of the gas are also used by the electronics industry, which uses the gas as a blanketing medium during production of such components as transistors, diodes, etc. Large quantities of nitrogen are used in annealing stainless steel and other steel mill products. The drug industry also uses large quantities. Nitrogen is used as a refrigerant both for the immersion freezing of food products and for transportation of foods. Liquid nitrogen is also used in missile work as a purge for components, insulators for space chambers, etc., and by the oil industry to build up great pressures in wells to force crude oil upward. Sodium and potassium nitrates are formed by the decomposition of organic matter with compounds of the metals present. In certain dry areas of the world these saltpeters are found in quantity. Ammonia, nitric acid, the nitrates, the five oxides (N_2O, NO, N_2O_3, NO_2, and N_2O_5), TNT, the cyanides, etc. are but a few of the important compounds. Nitrogen gas prices vary from 2c to $2.75 per 100 ft^3, depending on purity, etc. Production of elemental nitrogen in the U.S. is more than 9 million short tons per year.

Nobelium — (Alfred Nobel, discoverer of dynamite), No; at. no. 102; valence +2, + 3. Nobelium was unambiguously discovered and identified in April 1958 at Berkeley by A. Ghiorso, T. Sikkeland, J. R. Walton, and G.T. Seaborg, who used a new double-recoil technique. A heavy-ion linear accelerator (HILAC) was used to bombard a thin target of curium (95% Cm^{244} and 4.5% Cm^{246}) with C^{12} ions to produce 102^{254} according to the Cm^{246} (C^{12} 4n) reaction. Earlier in 1957 workers of the U.S., Britain, and Sweden announced the discovery of an isotope of element 102 with a 10-min half-life at 8.5 MeV, as a result of bombarding Cm^{244} with C^{13} nuclei. On the basis of this experiment the name *nobelium* was assigned and accepted by the Commission on Atomic Weights of the International Union of Pure and Applied Chemistry. The acceptance of the name was premature, for both Russian and American efforts now completely rule out the possibility of any isotope of element 102 having a half-life of 10 min in the vicinity of 8.5 MeV. Early work in 1957 on the search for this element, in Russia at the Kurchatov Institute, was marred by the assignment of 8.9± 0.4 MeV alpha radiation with a half-life of 2 to 40 sec, which was too indefinite to support claim to discovery. Confirmatory experiments at Berkeley in 1966 have shown the existence of 102^{254} with a 55-sec half-life, 102^{252} with a 2.3-sec half-life, and 102^{257} with a 23-sec half-life. Four other isotopes are now recognized, one of which — 102^{255} — has a half-life of 3 min. In view of the discoverer's traditional right to name an element, the Berkeley group, in 1967, suggested that the hastily given name *nobelium,* along with the symbol No, be retained.

Osmium — (Gr. *osme,* a smell), Os; at. wt. 190.2; at. no. 76; m.p. 3045± 30°C; b.p. 5027 ± 100°C; sp. gr. 22.57; valence 0 to +8, more usually +3, +4, +6, and +8. Discovered in 1803 by Tennant in the residue left when crude platinum is dissolved by *aqua regia.* Osmium occurs in *iridosmine* and in platinum-bearing river sands of the Urals, North America, and South America. It is also found in the nickel- bearing ores of the Sudbury, Ontario region along with other platinum metals. While the quantity of platinum metals in these ores is very small, the large tonnages of nickel ores processed make commercial recovery possible. The metal is lustrous, bluish white, extremely hard, and brittle even at high temperatures. It has the highest melting point and lowest vapor pressure of the platinum group. The metal is very difficult to fabricate, but the powder can be sintered in a hydrogen atmosphere at a temperature of 2000°C. The solid metal is not affected by air at room temperature, but the powdered or spongy metal slowly gives off osmium tetroxide, which is a powerful oxidizing agent and has a strong smell. The tetroxide is highly toxic, and boils at 130° C (760 mm). Concentrations in air as low as 10^{-7}g/m³ can cause lung congestion, skin damage, or eye damage. Exposure to osmium tetroxide should not exceed 0.002 mg/M³ (8-hr time-weighted average — 40-hr work week). The tetroxide has been used to detect fingerprints and to stain fatty tissue for microscope slides. The metal is almost entirely used to produce very hard alloys, with other metals of the platinum group, for fountain pen tips, instrument pivots, phonograph needles, and electrical contacts. The price of 99% pure osmium powder — the form usually supplied commercially — is about $20/g or $1000/tr. oz., depending on quantity and supplier. The measured densities of iridium and osmium seem to indicate that osmium is slightly more dense than iridium, so osmium has generally been credited with being the heaviest known element. Calculations of the density from the space lattice, which may be more reliable for these elements than actual measurements, however, give a density of 22.65 for iridium compared to 22.61 for osmium. At present, therefore, we know either iridium or osmium is the heaviest element, but the data do not allow selection between the two.

Oxygen — (GR. *oxys,* sharp, acid, and *genes,* forming; acid former), O; at. wt. (natural) 15.99994; at. no. 8; m.p. −218.4°C; b.p. −182.962°C; density 1.429 g/l (O °C); sp. gr. liquid 1.14 (−182.96°C); valence 2. For many centuries, workers occasionally realized air was composed of more than one component. The behavior of oxygen and nitrogen as components of air led to the advancement of the phlogiston theory of combustion, which captured the minds of chemists for a century. Oxygen was prepared by several workers, including Bayen and Borch, but they did not know how to collect it, did not study its properties, and did not recognize it as an elementary substance. Priestley is generally credited with its discovery, although Scheele also discovered it independently. Oxygen is the third most abundant element found in the sun, and it plays a part in the carbon-nitrogen cycle, one process thought to give the sun and stars their energy. Oxygen under excited conditions is responsible for the bright red and yellow-green colors of the aurora. Oxygen, as a gaseous element, forms 21% of the atmosphere by volume from which it can be obtained by liquefaction and fractional distillation. The atmosphere of Mars contains about 0.15% oxygen. The element and its compounds make up 49.2 %, by weight, of the earth's crust. About two thirds of the human body and nine tenths of water is oxygen. In the laboratory it can be prepared by the electrolysis of water or by heating potassium chlorate with manganese dioxide as a catalyst. The gas is colorless, odorless, and tasteless. The liquid and solid forms are a pale blue color and are strongly paramagnetic. Ozone (O_3), a highly active allotropic form of oxygen, is formed by the action of an electrical discharge or ultraviolet light on oxygen. Ozone's presence in the atmosphere (amounting to the equivalent of a layer 3 mm thick at ordinary pressures and temperatures) is of vital importance in preventing harmful ultraviolet rays of the sun from reaching the earth's surface. There has been recent concern that aerosols in the atmosphere may have a detrimental effect on this ozone layer. Ozone is toxic and exposure should not exceed 0.2mg/M³ (8-hr time-weighted average — 40-hr work week.) Undiluted ozone has a bluish color. Liquid ozone is bluish black, and solid ozone is violet-black. Oxygen is very reactive and capable of combining with most elements. It is a component of hundreds of thousands of organic compounds. It is essential for respiration of all plants and animals and for practically all combustion. In hospitals it is frequently used to aid respiration of patients. Its atomic weight was used as a standard of comparison for each of the other elements until 1961 when the International Union of Pure and Applied Chemistry adopted carbon 12 as the new basis. Oxygen has eight isotopes. Natural oxygen is a mixture of three isotopes. Oxygen 18 occurs naturally, is stable, and is available commercially. Water (H_2O with 1.5% O^{18}) is also available. Commercial oxygen consumption in the U.S. is estimated to be 20 million short tons per year and the demand is expected to increase substantially in the next few years. Oxygen enrichment of steel blast furnaces accounts for the greatest use of the gas. Large quantities are also used in making synthesis gas for ammonia and methanol, ethylene oxide, and for oxy-acetylene welding. Air separation plants produce about 99% of the gas, electrolysis plants about 1%. The gas costs 5c/ ft³ in small quantities, and about $15/ton.

Palladium — (named after the asteroid *Pallas,* discovered about the same time; Gr. *Pallas,* goddess of wisdom), Pd; at.

wt. 106.4; at. no. 46; m.p. 1554°C; b.p. 2970°C; sp. gr. 12.02 (20°C); valence 2, 3, or 4. Discovered in 1803 by Wollaston. Palladium is found along with platinum and other metals of the platinum group in placer deposits of the U.S.S.R., South and North America, Ethiopia, and Australia. It is also found associated with the nickel-copper deposits of South Africa and Ontario. Its separation from the platinum metals depends upon the type of ore in which it is found. It is a steel-white metal, does not tarnish in air, and is the least dense and lowest melting of the platinum group of metals. When annealed, it is soft and ductile; cold working greatly increases its strength and hardness. Palladium is attacked by nitric and sulfuric acid. At room temperatures the metal has the unusual property of absorbing up to 900 times its own volume of hydrogen, possibly forming Pd_2H. It is not yet clear if this is a true compound. Hydrogen readily diffuses through heated palladium and this provides a means of purifying the gas. Finely divided palladium is a good catalyst and is used for hydrogenation and dehydrogenation reactions. It is alloyed and used in jewelry trades. White gold is an alloy of gold decolorized by the addition of palladium. Like gold, palladium can be beaten into leaf as thin as 1/250,000 in. The metal is used in dentistry, watchmaking, and in making surgical instruments and electrical contacts. The metal sells for about $100/tr. oz.

Phosphorus — (Gr. phosphoros, light bearing; ancient name for the planet Venus when appearing before sunrise), P; at. wt. 30.97376; at. no. 15; m.p. (white) 44.1°C; b.p. (white) 280°C; sp. gr. (white) 1.82, (red) 2.20, (black) 2.25 to 2.69; valence 3 or 5. Discovered in 1669 by Brand, who prepared it from urine. Phosphorus exists in four or more allotropic forms: white (or yellow), red, and black (or violet). White phosphorus has two modifications: α and β with a transition temperature at −3.8°C. Never found free in nature, it is widely distributed in combination with minerals. Phosphate rock, which contains the mineral apatite, an impure tri-calcium phosphate, is an important source of the element. Large deposits are found in the U.S.S.R., in Morocco, and in Florida, Tennessee, Utah, Idaho, and elsewhere. Phosphorus is an essential ingredient of all cell protoplasm, nervous tissue, and bones. Ordinary phosphorus is a waxy white solid; when pure it is colorless and transparent. It is insoluble in water, but soluble in carbon disulfide. It takes fire spontaneously in air, burning to the pentoxide. It is very poisonous, 50mg constituting an approximate fatal dose. Exposure to white phosphorus should not exceed $0.1mg/M^3$ (8-hr time-weighted average — 40-hr work week). White phosphorus should be kept under water, as it is dangerously reactive in air, and it should be handled with forceps, as contact with the skin may cause severe burns. When exposed to sunlight or when heated in its own vapor to 250°C, it is converted to the red variety, which does not phosphoresce in air as does the white variety. This form does not ignite spontaneously and it is not as dangerous as white phosphorus. It should, however, be handled with care as it does convert to the white form at some temperatures and it emits highly toxic fumes of the oxides of phosphorus when heated. The red modification is fairly stable, sublimes with a vapor pressure of 1 atm at 417°C, and is used in the manufacture of safety matches, pyrotechnics, pesticides, incendiary shells, smoke bombs, tracer bullets, etc. White phosphorus may be made by several methods. By one process, tri-calcium phosphate, the essential ingredient of phosphate rock, is heated in the presence of carbon and silica in an electric furnace or fuel-fired blast furnace. Elementary phosphorus is liberated as vapor and may be collected under water. If desired, the phosphorus vapor and carbon monoxide produced by the reaction can be oxidized at once in the presence of moisture or water to produce phosphoric acid, an important compound in making super-phosphate fertilizers. In recent years, concentrated phosphoric acids, which may contain as much as 70 to 75% P_2O_5 content, have become of great importance to agriculture and farm production. World-wide demand for fertilizers has caused record phosphate production in recent years. Phosphates are used in the production of special glasses, such as those used for sodium lamps. Bone-ash, calcium phosphate, is also used to produce fine chinaware and to produce mono-calcium phosphate used in baking powder. Phosphorus is also important in the production of steels, phosphor bronze, and many other products. Trisodium phosphate is important as a cleaning agent, as a water softener, and for preventing boiler scale and corrosion of pipes and boiler tubes. Organic compounds of phosphorus are important.

Platinum — (Sp. platina, silver), Pt; at. wt. 195.09; at. no. 78; m.p. 1772°C; b.p. 3827 ± 100°C; sp. gr. 21.45 (20°C); valence 1?, 2, 3, or 4. Discovered in South America by Ulloa in 1735 and by Wood in 1741. The metal was used by pre-Columbian Indians. Platinum occurs native, accompanied by small quantities of iridium, osmium, palladium, ruthenium, and rhodium, all belonging to the same group of metals. These are found in the alluvial deposits of the Ural mountains, of Columbia, and of certain western American states. Sperrylite ($PtAs_2$), occurring with the nickel-bearing deposits of Sudbury, Ontario, is the source of a considerable amount of the metal. The large production of nickel offsets there being only one part of the platinum metals in two million parts of ore. Platinum is a beautiful silvery-white metal, when pure, and is malleable and ductile. It has a coefficient of expansion almost equal to that of soda-lime-silica glass, and is therefore used to make sealed electrodes in glass systems. The metal does not oxidize in air at any temperature, but is corroded by halogens, cyanides, sulfur, and caustic alkalis. It is insoluble in hydrochloric and nitric acid, but dissolves when they are mixed as aqua regia, forming chloroplatinic acid (H_2PtCl_6), an important compound. The metal is extensively used in jewelry, wire, and vessels for laboratory use, and in many valuable instruments including thermocouple elements. It is also used for electrical contacts, corrosion-resistant apparatus, and in dentistry. Platinum-cobalt alloys have magnetic properties. One such alloy made of 76.7% Pt and 23.3% Co, by weight, is an extremely powerful magnet that offers a B-H (max) almost twice that of Alnico V. Platinum resistance wires are used for constructing high-temperature electric furnaces. The metal is used for coating missile nose cones, jet engine fuel nozzles, etc., which must perform reliably for long periods of time at high temperatures. The metal, like palladium, absorbs large volumes of hydrogen, retaining it at ordinary temperatures but giving it up at red heat. In the finely divided state platinum is an excellent catalyst, having long been used in the contact process for producing sulfuric acid. It is also used as a catalyst in cracking petroleum products. There is also much current interest in the use of platinum as a catalyst in fuel cells and in antipollution devices for automobiles. Platinum anodes are extensively used in cathodic protection systems for large ships and ocean-going vessels, pipelines, steel piers, etc. Fine platinum wire will glow red hot when placed in the vapor of methyl alcohol. It acts here as a catalyst, converting the alcohol to formaldehyde. This phenomenon has been used commercially to produce cigarette lighters and hand warmers. Hydrogen and oxygen explode in the presence of platinum. The price of platinum has varied widely; more than a century ago it was used to adulterate gold. It was nearly eight times as valuable as gold in 1920. The price in December 1982 was about $450/tr. oz.

Plutonium — (Planet *pluto*), Pu; at. no. 94; isotopic mass Pu^{239} 239.13 (physical scale); sp. gr. (α modification) 19.84 (25°C); m.p. 641°C; b.p. 3232°C; valence 3, 4, 5, or 6. Plutonium was the second transuranium element of the actinide series to be discovered. The isotope Pu^{238} was produced in 1940 by Seaborg, McMillan, Kennedy, and Wahl by deuteron bombardment of uranium in the 60-in. cyclotron at Berkeley, California. Plutonium also exists in trace quantities in naturally occurring uranium ores. It is formed in much the same manner as neptunium, by irradiation of natural uranium with the neutrons which are present. By far of greatest importance is the isotope Pu^{239}, with a half-life of 24,360 years, produced in extensive quantities in nuclear reactors from natural uranium:

$$U^{238}(n\gamma)U^{239} \xrightarrow{\beta} Np^{239} \xrightarrow{\beta} Pu^{239}$$

Fifteen isotopes of plutonium are known. Plutonium has assumed the position of dominant importance among the transuranium elements because of its successful use as an explosive ingredient in nuclear weapons and the place which it holds as a key material in the development of industrial use of nuclear power. One kilogram is equivalent to about 22 million kilowatt hours of heat energy. The complete detonation of a kilogram of plutonium produces an explosion equal to about 20,000 tons of chemical explosive. Its importance depends on the nuclear property of being readily fissionable with neutrons and its availability in quantity. The world's nuclear-power reactors are now producing about 20,000 kg of plutonium/yr. By 1982 it is estimated that about 300,000 kg will have been accumulated. The various nuclear applications of plutonium are well known. Pu^{238} has been used in the Apollo lunar missions to power seismic and other equipment on the lunar surface. As with neptunium and uranium, plutonium metal can be prepared by reduction of the trifluoride with alkaline-earth metals. The metal has a silvery appearance and takes on a yellow tarnish when slightly oxidized. It is chemically reactive. A relatively large piece of plutonium is warm to the touch because of the energy given off in alpha decay. Larger pieces will produce enough heat to boil water. The metal readily dissolves in concentrated hydrochloric acid, hydroiodic acid, or perchloric acid with formation of the Pu^{+3} ion. The metal exhibits six allotropic modifications having various crystalline structures. The densities of these vary from 16.00 to 19.86 g/cm³. Plutonium also exhibits four ionic valence states in aqueous solutions: Pu^{+3} (blue lavender), Pu^{+4} (yellow brown), PuO^{+} (pink?), and PuO^{+2} (pink orange). The ion PuO^{+} is unstable in aqueous solutions, disproportionating into Pu^{+4} and PuO^{+2} the Pu^{+4} thus formed, however, oxidizes the PuO^{+} into PuO^{+2}, itself being reduced to Pu^{+3}, giving finally Pu^{+3} and PuO^{+2}. Plutonium forms binary compounds with oxygen: PuO, PuO_2, and intermediate oxides of variable composition; with the halides: PuF_3, PuF_4, $PuCl_3$, $PuBr_3$, PuI_3; with carbon, nitrogen, and silicon: PuC, PuN, $PuSi_2$. Oxyhalides are also well known: $PuOCl$, $PuOBr$, $PuOI$. Because of the high rate of emission of alpha particles and the element being specifically absorbed by bone marrow, plutonium, as well as all of the other transuranium elements except neptunium, are radiological poisons and must be handled with very special equipment and precautions. Plutonium is a very dangerous radiological hazard. Scientists at Berkeley believe that linear catechoylamide carboxylate has potential for removing plutonium from living tissue and bone as well as from nuclear-reactor wastes. Precautions must also be taken to prevent the unintentional formation of a critical mass. Plutonium in liquid solution is more likely to become critical than solid plutonium. The shape of the mass must also be considered where criticality is concerned. Plutonium-238 is available from the A.E.C. at a cost of about $700/g (80 to 89% enriched.)

Polonium — (Poland, native country of Mme. Curie), Po; at. mass. (~209); at. no. 84; m.p. 254°C; b.p. 962°C; sp. gr. (alpha modification) 9.32; valence −2, 0, +2, +3 (?), +4, and +6. Polonium was the first element discovered by Mme. Curie in 1898, while seeking the cause of radioactivity of pitchblende from Joachimsthal, Bohemia. The electroscope showed it separating with bismuth. Polonium is also called Radium F. Polonium is a very rare natural element. Uranium ores contain only about 100μg of the element per ton. Its abundance is only about 0.2% of that of radium. In 1934 it was found that when natural bismuth (Bi^{209}) was bombarded by neutrons, Bi^{210}, the parent of polonium, was obtained. Milligram amounts of polonium may now be prepared this way, by using the high neutron fluxes of nuclear reactors. Polonium-210 is a low-melting, fairly volatile metal, 50% of which is vaporized in air in 45 hr at 55°C. It is an alpha emitter with a half-life of 138.39 days. A milligram emits as many alpha particles as 5 g of radium. The energy released by its decay is so large (27.5 cal /Ci/day or 140 W/g) that a capsule containing about half a gram reaches a temperature above 500°C. The capsule also presents a contact gamma-ray dose rate of 1.2R/hr. A few curies of polonium exhibit a blue glow, caused by excitation of the surrounding gas. Because almost all alpha radiation is stopped within the solid source and its container, giving up its energy, polonium has attracted attention for uses as a lightweight heat source for thermoelectric power in space satellites. Polonium has more isotopes than any other element. Twenty-seven isotopes of polonium are known, with atomic masses ranging from 192 to 218. Polonium-210 is the most readily available. Isotopes of mass 209 (half-life 103 years) and mass 208 (half-life 2.9 years) can be prepared by alpha, proton, or deuteron bombardment of lead or bismuth in a cyclotron, but these are expensive to produce. Metallic polonium has been prepared from polonium hydroxide and some other polonium compounds in the presence of concentrated aqueous or anhydrous liquid ammonia. Two allotropic modifications are known to exist. Polonium is readily dissolved in dilute acids, but is only slightly soluble in alkalis. Polonium salts of organic acids char rapidly; halide ammines are reduced to the metal. Polonium can be mixed or alloyed with beryllium to provide a source of neutrons. It has been used in devices for eliminating static charges in textile mills, etc.; however, beta sources are more commonly used and are less dangerous. It is also used on brushes for removing dust from photographic films. The polonium for these is carefully sealed and controlled, minimizing hazards to the user. Polonium-210 is very dangerous to handle in even milligram or microgram amounts, and special equipment and strict control is necessary. Damage arises from the complete absorption of the energy of the alpha particle into tissue. The maximum permissible body burden for ingested polonium is

THE ELEMENTS (continued)

only 0.03μCi, which represents a particle weighing only 6.8× 10^{-12}g. Weight for weight it is about 2.5 × 10^{11} times as toxic as hydrocyanic acid. The maximum allowable concentration for soluble polonium compounds in air is about 2× 10^{-11} μCi/cm³. Polonium is available commercially on special order with an A.E.C. permit from the Oak Ridge National Laboratory.

Potassium — (English, *potash* — pot ashes; L. *kalium,* Arab. *qali,* alkali), K; at. wt. 39.0983±0.0003; at. no. 19; m.p. 63.25°C; b.p. 760°C; sp. gr. 0.862 (20°C); valence 1. Discovered in 1807 by Davy, who obtained it from caustic potash (KOH); this was the first metal isolated by electrolysis. The metal is the seventh most abundant and makes up about 2.4% by weight of the earth's crust. Most potassium minerals are insoluble and the metal is obtained from them only with great difficulty. Certain minerals, however, such as *sylvite, carnallite, langbeinite,* and *polyhalite* are found in ancient lake and sea beds and form rather extensive deposits from which potassium and its salts can readily be obtained. Potash is mined in Germany, New Mexico, California, Utah, and elsewhere. Large deposits of potash, found at a depth of some 3000 ft in Saskatchewan, promise to be important in coming years. Potassium is also found in the ocean, but is present only in relatively small amounts, compared to sodium. The greatest demand for potash has been in its use for fertilizers. Potassium is an essential constituent for plant growth and it is found in most soils. Potassium is never found free in nature, but is obtained by electrolysis of the hydroxide, much in the same manner as prepared by Davy. Thermal methods also are commonly used to produce potassium (such as by reduction of potassium compounds with CaC_2, C, Si, or Na). It is one of the most reactive and electropositive of metals; except for lithium, it is the lightest known metal. It is soft, easily cut with a knife, and is silvery in appearance immediately after a fresh surface is exposed. It rapidly oxidizes in air and must be preserved in a mineral oil such as kerosene. As with other metals of the alkali group, it decomposes in water with the evolution of hydrogen. It catches fire spontaneously on water. Potassium and its salts impart a violet color to flames. Nine isotopes of potassium are known. Ordinary potassium is composed of three isotopes, one of which is K^{40} (0.0118%), a radioactive isotope with a half-life of 1.28 × 10^9 years. The radioactivity presents no appreciable hazard. An alloy of sodium and potassium (NaK) is used as a heat-transfer medium. Many potassium salts are of utmost importance, including the hydroxide, nitrate, carbonate, chloride, chlorate, bromide, iodide, cyanide, sulfate, chromate, and dichromate. Metallic potassium is available commercially for about $15/oz in small quantities.

Praseodymium — (Gr. *prasios,* green, and *didymos,* twin), Pr; at. wt. 140.9077; at. no. 59; m.p. 931°C; b.p. 3512 °C; sp. gr. (α) 6.773 (β) 6.64; valence 3 or 4. In 1841 Mosander extracted the rare earth *didymia* from *lanthana;* in 1879, Lecoq de Boisbaudran isolated a new earth, *samaria,* from didymia obtained from the mineral *samarskite.* Six years later, in 1885, von Welsbach separated didymia into two other earths, *praseodymia* and *neodymia,* which gave salts of different colors. As with other rare earths, compounds of these elements in solution have distinctive sharp spectral absorption bands or lines, some of which are only a few Angstroms wide. The element occurs along with other rare-earth elements in a variety of minerals. *Monazite* and *bastnasite* are the two principal commercial sources of the rare-earth metals. Ion-exchange and solvent extraction techniques have led to much easier isolation of the rare earths and the cost has dropped greatly in the past few years. Praseodymium can be prepared by several methods, such as by calcium reduction of the anhydrous chloride of fluoride. Misch metal, used in making cigarette lighters, contains about 5% praseodymium metal. Praseodymium is soft, silvery, malleable, and ductile. It was prepared in relatively pure form in 1931. It is somewhat more resistant to corrosion in air than europium, lanthanum, cerium, or neodymium, but it does develop a green oxide coating that spalls off when exposed to air. As with other rare-earth metals it should be kept under a light mineral oil or sealed in plastic. The rare-earth oxides, including Pr_2O_3, are among the most refractory substances known. Along with other rare earths, it is widely used as a core material for carbon arcs used by the motion picture industry for studio lighting and projection. Salts of praseodymium are used to color glasses and enamels; when mixed with certain other materials, praseodymium produces an intense and unusually clean yellow color in glass. Didymium glass, of which praseodymium is a component, is a colorant for welder's goggles. The metal (99 + % pure) is priced at about $1.00/g.

Promethium — (*Prometheus,* who, according to mythology, stole fire from heaven), Pm; at. no. 61; m.p. 1168± 6°C; b.p. 2460°C; sp. gr. 7.22±0.02 (25°C); valence 3. In 1902 Branner predicted the existence of an element between neodymium and samarium, and this was confirmed by Moseley in 1914. In 1941, workers at Ohio State University irradiated neodymium and praseodymium with neutrons, deuterons, and alpha particles, resp., and produced several new radioactivities, which most likely were those of element 61. Wu and Segre, and Bethe, in 1942, confirmed the formation; however, chemical proof of the production of element 61 was lacking because of the difficulty in separating the rare earths from each other at that time. In 1945, Marinsky, Glendenin, and Coryell made the first chemical identification by use of ion-exchange chromatography. Their work was done by fission of uranium and by neutron bombardment of neodymium. Searches for the element on earth have been fruitless, and it now appears that promethium is completely missing from the earth's crust. Promethium, however, has been identified in the spectrum of the star HR^{465} in Andromeda. This element is being formed recently near the star's surface, for no known isotope of promethium has a half-life longer than 17.7 years. Thirteen isotopes of promethium, with atomic masses from 141 to 154, are now known. Promethium — 147, with a half- life of 2.5 years, is the most generally useful. Promethium — 145 is the longest lived, and has a specific activity of 940 Ci /g. It is a soft beta emitter; although no gamma rays are emitted, X-radiation can be generated when beta particles impinge on elements of a high atomic number, and great care must be taken in handling it. Promethium salts luminesce in the dark with a pale blue or greenish glow, due to their high radioactivity. Ion-exchange methods led to the preparation of about 10 g of promethium from atomic reactor fuel processing wastes in early 1963. Little is yet generally known about the properties of metallic promethium. Two or more allotropic modifications are thought to exist. The element has applications as a beta source for thickness gages, and it can be absorbed by a phosphor to produce light. Light produced in this manner can be used for signs or signals that require dependable operation; it can be used as a nuclear-powered battery by capturing light in photocells which convert it into electric current. Such a battery, using Pm^{147}, would have a useful life of about 5 years. Promethium shows promise as a portable X-ray unit, and it may become useful as a heat source to provide auxiliary power for space probes and satellites. More than 30 promethium compounds have been prepared. Most are colored. Promethium—147 is available to A.E.C. licensees at a cost of about 50c/Ci.

Protactinium — (Gr. *protos,* first), Pa; at. wt. 231.0359; at. no. 91; m.p.< 1600°C; b.p. . . . ; sp. gr. 15.37 (calc.); valence 4 or 5. The first isotope of element 91 to be discovered was Pa^{234m}, also known as UX_2, a short-lived member of the naturally occurring U^{238} decay series. It was identified by K. Fajans and O. H. Gohring in 1913 and they named the new element *brevium.* When the longer-lived isotope Pa ^{231} was identified by Hahn and Meitner in 1918, the name protoactinium was adopted as being more consistent with the characteristics of the most abundant isotope. Soddy, Cranston, and Fleck were also active in this work. The name *protoactinium* was shortened to *protactinium* in 1949. In 1927, Grosse prepared 2 mg of a white powder, which was shown to be Pa_2O_5. Later, in 1934, from 0.1 g of pure Pa_2O_5 he isolated the element by two methods, one of which was by converting the oxide to an iodide and "cracking" it in a high vacuum by an electrically heated filament by the reaction

$$2PaI_5 \longrightarrow 2Pa + 5I_2$$

Protactinium has a bright metallic luster which it retains for some time in air. The element occurs in *pitchblende* to the extent of about 1 part Pa^{231} to 10 million of ore. Ores from Zaire have about 3 ppm. Protactinium has 13 isotopes, the most common of which is Pr^{231} with a half-life of 32,500 years. A number of protactinium compounds are known, some of which are colored. The element is superconductive below 1.4 K. An indirect measurement indicates that protactinium has a vapor pressure of 5.1×10^{-5} at 1927°C. The element is a dangerous toxic material and requires precautions similar to those used when handling plutonium. In 1959 and 1961, it was announced that the Great Britain Atomic Energy Authority extracted by a 12-stage process 125 g of 99.9 %protactinium, the world's only stock of the metal for many years to come. The extraction was made from 60 tons of waste material at a cost of about $500,000. Protactinium is one of the rarest and most expensive naturally occurring elements. It was reported that this stock was being distributed to laboratories around the world at a cost of about $2800/g. The element is an alpha emitter (5.0 MeV), and is a radiological hazard similar to polonium.

Radium — (L. *radius,* ray), Ra; at. wt. 226.0254; at. no. 88; m.p. 700°C; b.p. 1140°C; sp. gr. 5?; valence 2. Radium was discovered in 1898 by M. and Mme. Curie in the *pitchblende* or *uraninite* of North Bohemia, in which it occurs. There is about 1 g of radium in 7 tons of pitchblende. The element was isolated in 1911 by Mme. Curie and Debierne by the electrolysis of a solution of pure radium chloride, employing a mercury cathode; on distillation in an atmosphere of hydrogen this amalgam yielded the pure metal. Originally, radium was obtained from the rich pitchblende ore found at Joachimsthal, Bohemia. The *carnotite* sands of Colorado furnish some radium, but richer ores are found in the Democractic Republic of Zaire and the Great Bear Lake region of Canada. Radium is present in all uranium minerals, and could be extracted, if desired, from the extensive wastes of uranium processing. Large uranium deposits are located in Ontario, New Mexico, Utah, Australia, and elsewhere. Radium is obtained commercially as the bromide or chloride; it is doubtful if any appreciable stock of the isolated element now exists. The pure metal is brilliant white when freshly prepared, but blackens on exposure to air, probably due to formation of the nitride. It exhibits luminescence, as do its salts; it decomposes in water and is somewhat more volatile than barium. It is a member of the alkaline-earth group of metals. Radium imparts a carmine red color to a flame. Radium emits alpha, beta, and gamma rays and when mixed with beryllium produces neutrons. One gram of Ra^{226} undergoes 3.7×10^{10} disintegrations per sec. The *curie (Ci)* is defined as that amount of radioactivity which has the same disintegration rate as 1 g of Ra^{226}. Sixteen isotopes are now known; radium 226, the common isotope, has a half-life of 1620 years. One gram of radium produces about 0.0001 ml(stp) of emanation, or radon gas, per day. This is pumped from the radium and sealed in minute tubes, which are used in the treatment of cancer and other diseases. One gram of radium yields about 1000 cal per year. Radium is used in producing self-luminous paints, neutron sources, and in medicine for the treatment of disease. Some of the more recently discovered radioisotopes, such as Co^{60}, are now being used in place of radium. Some of these sources are much more powerful, and others are safer to use. Radium loses about 1% of its activity in 25 years, being transformed into elements of lower atomic weight. Lead is a final product of disintegration. The study of radium has greatly altered our ideas of the structure of the atom. Radium is a radiological hazard. (Stored radium should be ventilated to prevent build-up of radon.) Inhalation, injection, or body exposure to radium can cause cancer and other body disorders. The maximum permissible burden in the total body for Ra^{226} is $0.2\mu Ci$ (microcuries).

Radon — (from *radium;* called *niton* at first, L. *nitens,* shining), Rn; at. wt. ($\sim$ 222); at. no. 86; m.p.$-$ 71°C; b.p. -61.8°C; density of gas 9.73 g/l; sp. gr. liquid 4.4 at -62°C, solid 4; valence usually 0. The element was discovered in 1900 by Dorn, who called it *radium emanation.* In 1908 Ramsay and Gray, who named it *niton,* isolated the element and determined its density, finding it to be the heaviest known gas. It is essentially inert and occupies the last place in the zero group of gases in the Periodic Table. Since 1923, it has been called radon. Twenty isotopes are known. Radon—222, coming from radium, has a half-life of 3.823 days and is an alpha emitter; radon—220, emanating naturally from thorium and called *thoron,* has a half-life of 54.5 sec and is also an alpha emitter. Radon—219 emanates from actinium and is called *actinon.* It has a half-life of 3.92 sec and is also an alpha emitter. It is estimated that every square mile of soil to a depth of 6 in. contains about 1 g of radium, which releases radon in tiny amounts to the atmosphere. Radon is present in some spring waters, such as those at Hot Springs, Arkansas. On the average, one part of radon is present to 1 sextillion parts of air. At ordinary temperatures radon is a colorless gas; when cooled below the freezing point, radon exhibits a brilliant phosphorescence which becomes yellow as the temperature is lowered and orange-red at the temperature of liquid air. It has been reported that fluorine reacts with radon, forming radon fluoride. Radon clathrates have also been reported. Radon is still produced for therapeutic use by a few hospitals by pumping it from a radium source and sealing it in minute tubes, called seeds or needles, for application to patients. This practice has now been largely discontinued as hospitals can order the seeds directly from suppliers, who make up the seeds with the desired activity for the day of use. Radon is available at a cost of about \$4/mCi. Care must be taken in handling radon, as with other radioactive materials. The main hazard is from inhalation of the element and its solid daughters, which are collected on dust in the air. The maximum permissible concentration of Rn^{222} in air has been set at $3\times 10^{-8}\ \mu$ Ci/cc (lung) for an 8-hr day, 40-hr work week. Good ventilation should be provided where radium, thorium, or actinium is stored to prevent build-up of this element. Radon build-up is also a health consideration in uranium mines.

Rhenium — (L. *Rhenus*, Rhine), Re; at. wt. 186.2; at. no. 75; m.p. 3180°C; b.p. 5627°C (est.); sp. gr. 21.02 (20°C); valence −1, + 1, 2, 3, 4, 5, 6, 7. Discovery of rhenium is generally attributed to Noddack, Tacke, and Berg, who announced in 1925 they had detected the element in platinum ores and *columbite*. They also found the element in *gadolinite* and *molybdenite*. By working up 660 kg of molybdenite they were able in 1928 to extract 1 g of rhenium. The price in 1928 was \$ 10,000/g. Rhenium does not occur free in nature or as a compound in a distinct mineral species. It is, however, widely spread throughout the earth's crust to the extent of about 0.001 ppm. Commercial rhenium in the U.S. today is obtained from molybdenite roaster-flue dusts obtained from copper-sulfide ores mined in the vicinity of Miami, Arizona, and elsewhere in Arizona and Utah. Some molybdenites contain from 0.002 to 0.2% rhenium. More than 150,000 troy ounces of rhenium are now being produced yearly in the United States. The total estimated Free World reserve of rhenium metal is 3500 tons. Natural rhenium is a mixture of two stable isotopes. Sixteen other unstable isotopes are recognized. Rhenium metal is prepared by reducing ammonium perrhenate with hydrogen at elevated temperatures. The element is silvery white with a metallic luster; its density is exceeded only by that of platinum, iridium, and osmium, and its melting point is exceeded only by that of tungsten and carbon. It has other useful properties. The usual commercial form of the element is a powder, but it can be consolidated by pressing and resistance-sintering in a vacuum or hydrogen atmosphere. This produces a compact shape in excess of 90% of the density of the metal. Annealed rhenium is very ductile, and can be bent, coiled, or rolled. Rhenium is used as an additive to tungsten and molybdenum-based alloys to impart useful properties. It is widely used for filaments for mass spectrographs and ion gages. Rhenium-molybdenum alloys are superconductive at 10 K. Rhenium is also used as an electrical contact material as it has good wear resistance and withstands arc corrosion. Thermocouples made of Re-W are used for measuring temperatures up to 2200°C, and rhenium wire is used in photoflash lamps for photography. Rhenium catalysts are exceptionally resistant to poisoning from nitrogen, sulfur, and phosphorus, and are used for hydrogenation of fine chemicals, hydrocracking, reforming, and the disproportionation of olefins. Rhenium costs about \$50/troy ounce. Little is known of its toxicity; therefore, it should be handled with care until more data are available.

Rhodium — (Gr. *rhodon*, rose), Rh; at. wt. 102.9055; at. no. 45; m.p. 1966 ± 3°C; b.p. 3727 ±100°C; sp. gr. 12.41 (20°C); valence 2, 3, 4, 5, and 6. Wollaston discovered rhodium in 1803-4 in crude platinum ore he presumably obtained from South America. Rhodium occurs native with other platinum metals in river sands of the Urals and in North and South America. It is also found with other platinum metals in the copper-nickel sulfide ores of the Sudbury, Ontario region. Although the quantity occurring here is very small, the large tonnages of nickel processed make the recovery commercially feasible. The annual world production of rhodium is only 2 or 3 tons. The metal is silvery white and at red heat slowly changes in air to the sesquioxide. At higher temperatures it converts back to the element. Rhodium has a higher melting point and lower density than platinum. Its major use is as an alloying agent to harden platinum and palladium. Such alloys are used for furnace windings, thermocouple elements, bushings for glass fiber production, electrodes for aircraft spark plugs, and laboratory crucibles. It is useful as an electrical contact material as it has a low electrical resistance, a low and stable contact resistance, and is highly resistant to corrosion. Plated rhodium, produced by electroplating or evaporation, is exceptionally hard and is used for optical instruments. It has a high reflectance and is hard and durable. Rhodium is also used for jewelry, for decoration, and as a catalyst. Exposure to rhodium (metal fume and dust, as Rh) should not exceed 0.1mg/M³ (8-hr. time-weighted average, 40-hr wk.). Soluble salts should not exceed 0.001mg/M³. In December 1982 rhodium cost about \$800/tr. oz.

Rubidium — (L. *rubidius*, deepest red), Rb; at. wt. 85.4678; at. no. 37; m.p. 38.89°C; b.p. 686°C; sp. gr. (solid) 1.532(20°C), (liquid) 1.475 (39°C); valence 1, 2, 3, 4. Discovered in 1861 by Bunsen and Kirchoff in the mineral *lepidolite* by use of the spectroscope. The element is much more abundant than was thought several years ago. It is now considered to be the 16th most abundant element in the earth's crust. Rubidium occurs in *pollucite, carnallite, leucite,* and *zinnwaldite,* which contains traces up to 1%, in the form of the oxide. It is found in lepidolite to the extent of about 1.5%, and is recovered commercially from this source. Potassium minerals, such as those found at Searles Lake, California, and potassium chloride recovered from brines in Michigan also contain the element and are commercial sources. It is also found along with cesium in the extensive deposits of *pollucite* at Bernic Lake, Manitoba. Rubidium can be liquid at room temperature. It is a soft, silvery-white metallic element of the alkali group and is the second most electropositive and alkaline element. It ignites spontaneously in air and reacts violently in water, setting fire to the liberated hydrogen. As with other alkali metals, it forms amalgams with mercury and it alloys with gold, cesium, sodium, and potassium. It colors a flame yellowish violet. Rubidium metal can be prepared by reducing rubidium chloride with calcium, and by a number of other methods. It must be kept under a dry mineral oil or in a vacuum or inert atmosphere. Seventeen isotopes of rubidium are known. Naturally occurring rubidium is made of two isotopes, Rb^{85} and Rb^{87}. Rubidium—87 is present to the extent of 27.85% in natural rubidium and is a beta emitter with a half-life of 5×10^{11} years. Ordinary rubidium is sufficiently radioactive to expose a photographic film in about 30 to 60 days. Rubidium forms four oxides: Rb_2O, Rb_2O_2, Rb_2O_3, Rb_2O_4. Because rubidium can be easily ionized, it is being considered for use in "ion engines" for space vehicles; however, cesium is somethwat more efficient for this purpose. It is also proposed for use as a working fluid for vapor turbines and for use in a thermoelectric generator using the magnetohydrodynamic principle where rubidium ions are formed by heat at high temperature and passed through a magnetic field. These conduct electricity and act like an armature of a generator and cause electricity to be generated. Rubidium is used as a getter in vacuum tubes and as a photocell component. It has been used in making special glasses. $RbAg_4I_5$ is important, as it has the highest room conductivity of any known ionic crystal. At 20°C its conductivity is about the same as dilute sulfuric acid. This suggests use in thin film batteries and other applications. The present cost in small quanitites is about $20/g.

Ruthenium — (L. *Ruthenia,* Russia), Ru; at. wt. 101.07; at. no. 44, m.p. 2310°C; b.p. 3900°C; sp. gr. 12.41 (20°C); valence 0,1,2,3,4,5,6,7,8. Berzelius and Osann in 1827 examined the residues left after dissolving crude platinum from the Ural mountains in *aqua regia.* While Berzelius found no unusual metals, Osann thought he found three new metals, one of which he named ruthenium. In 1844 Klaus, generally recognized as the discoverer, showed that Osann's ruthenium oxide was very impure and that it contained a new metal. Klaus obtained 6 g of ruthenium from the portion of crude platinum that is insoluble in *aqua regia.* A member of the platinum group, ruthenium occurs native with other members of the group in ores found in the Ural mountains and in North and South America. It is also found along with other platinum metals in small but commercial quantities in *pentlandite* of the Sudbury, Ontario, nickel-mining region, and in *pyroxinite* deposits of South Africa. The metal is isolated commercially by a complex chemical process, the final stage of which is the hydrogen reduction of ammonium ruthenium chloride, which yields a powder. The powder is consolidated by powder metallurgy techniques or by argon-arc welding. Ruthenium is a hard, white metal and has four crystal modifications. It does not tarnish at room temperatures, but oxidizes in air at about 800°C. The metal is not attacked by hot or cold acids or *aqua regia,* but when potassium chlorate is added to the solution, it oxidizes explosively. It is attacked by halogens, hydroxides, etc. Ruthenium can be plated by electrodeposition or by thermal decomposition methods. The metal is one of the most effective hardeners for platinum and palladium, and is alloyed with these metals to make electrical contacts for severe wear resistance. A ruthenium-molybdenum alloy is said to be superconductive at 10.6 K. The corrosion resistance of titanium is improved a hundredfold by addition of 0.1% ruthenium. It is a versatile catalyst. Hydrogen sulfide can be split catalytically by light using an aqueous suspension of CdS particles loaded with ruthenium dioxide. It is thought this may have application to removal of H_2S from oil refining and other industrial processes. Compounds in at least eight oxidation states have been found, but of these, the $+2$, $+3$, and $+4$ states are the most common. Ruthenium tetroxide, like osmium tetroxide, is highly toxic. In addition, it may explode. Ruthenium compounds show a marked resemblance to those of osmium. The metal is priced at about $4/g or $60/troy ounce.

Rutherfordium — See Element 104.

Samarium — (*Samarskite,* a mineral), Sm; at. wt. 150.4; at. no. 62; m.p. 1077 ± 5°C; b.p. 1791°C; sp. gr. (α) 7.520 (β) 7.40; valence 2 or 3. Discovered spectroscopically by its sharp absorption lines in 1879 by Lecoq de Boisbaudran in the mineral *samarskite,* named in honor of a Russian mine official, Col. Samarski. Samarium is found along with other members of the rare-earth elements in many minerals, including *monazite* and *bastnasite,* which are commercial sources. It occurs in monazite to the extent of 2.8%. While *misch metal* containing about 1% of samarium metal, has long been used, samarium has not been isolated in relatively pure form until recent years. Ion-exchange and solvent extraction techniques have recently simplified separation of the rare earths from one another; more recently, electrochemical deposition, using an electrolytic solution of lithium citrate and a mercury electrode, is said to be a simple, fast, and highly specific way to separate the rare earths. Samarium metal can be produced by reducing the oxide with barium or lanthanum. Samarium has a bright silver luster and is reasonably stable in air. Two crystal modifications of the metal exist, with a transformation point at 917°C. The metal ignites in air at about 150°C. Sixteen isotopes of samarium exist. Natural samarium is a mixture of seven isotopes, three of which are unstable with long half-lives. Samarium, along with other rare earths, is used for carbon-arc lighting for the motion picture industry. The sulfide has excellent high-temperature stability and good thermoelectric efficiencies up to 1100°C. $SmCo_5$ has been used in making a new permanent magnet material with the highest resistance to demagnetization of any known material. It is said to have an intrinsic coercive force as high as 28,000 oersteds. Samarium oxide has been used in optical glass to absorb the infrared. Samarium is used to dope calcium fluoride crystals for use in optical masers or lasers. Compounds of the metal act as sensitizers for phosphors excited in the infrared; the oxide exhibits catalytic properties in the dehydration and dehydrogenation of ethyl alcohol. It is used in infrared absorbing glass and as a neutron absorber in nuclear reactors. The metal is priced at about $5/g. Little is known of the toxicity of samarium; therefore, it should be handled carefully.

Scandium — (L. *Scandia,* Scandinavia), Sc; at. wt. 44.9559; at. no. 21; m.p. 1541°C; b.p. 2831°C; sp. gr. 2.989 (25°C); valence 3. On the basis of the Periodic System, Mendeleev predicted the existence of *ekaboron,* which would have an atomic weight between 40 of calcium and 48 of titanium. The element was discovered by Nilson in 1876 in the minerals *euxenite* and *gadolinite,* which had not yet been found anywhere except in Scandinavia. By processing 10 kg of euxenite and other residues of rare-earth minerals, Nilson was able to prepare about 2 g of scandium oxide of high purity. Cleve later pointed out that Nilson's scandium was identical with Mendeleev's ekaboron. Scandium is apparently a much more abundant element in the sun and certain stars than here on earth. It is about the 23rd most abundant element in the sun, compared to the 50th most abundant on earth. It is widely distributed on earth, occurring in very minute quantities in over 800 mineral species. The blue color of beryl (aquamarine variety) is said to be due to scandium. It occurs as a principal component in the rare mineral *thortveitite,* found in Scandinavia and Malagasy. It is also found in the residues remaining after the extraction of tungsten from Zinnwald *wolframite,* and in *wiikite* and *bazzite.* Most scandium is presently being recovered from *thortveitite* or is extracted as a by-product from uranium mill tailings. Metallic scandium was first prepared in 1937 by Fischer, Brunger, and Grieneisen, who electrolyzed a eutectic melt of potassium, lithium, and scandium chlorides at 700 to 800°C. Tungsten wire and a pool of molten zinc served as the electrodes in a graphic crucible. Pure scandium is now produced by reducing scandium fluoride with calcium metal. The production of the first pound of 99% pure scandium metal was announced in 1960 as having been made under a U.S. Air Force contract. Scandium is a silvery-white metal which develops a slightly yellowish or pinkish cast upon exposure to air. It is relatively soft, and resembles yttrium and the rare-earth metals more than it resembles alumi-

THE ELEMENTS (continued)

num or titanium. It is a very light metal and has a higher melting point than aluminum, making it of interest to designers of space missiles. Scandium is not attacked by a 1:1 mixture of conc. HNO_3 and 48% HF. Scandium reacts rapidly with many acids. Eleven isotopes of scandium are recognized. The metal is still expensive, costing about $50/g with a purity of about 99.9%. Scandium oxide costs about :15/g. About 12 kg of scandium)as Sc_2O_3) are now being used yearly in the U.S. to produce high-intensity lights and the radioactive isotope Sc^{46} used as a tracing agent in refinery crackers for crude oil, etc. Scandium iodide added to mercury vapor lamps produces a highly efficient light source resembling sunlight, which is important for indoor or night-time color TV transmission. Little is yet known about the toxicity of scandium; therefore, it should be handled with care.

Selenium — (Gr. *Selene,* moon), Se; at. wt. 78.96; at. no. 34; m.p. (gray) 217°C; b.p. (gray) 684.9 ± 1.0°C; sp. gr. (gray) 4.79, (vitreous) 4.28; valence −2, +4, or +6. Discovered by Berzelius in 1817, who found it associated with tellurium, named for the earth. Selenium is found in a few rare minerals, such as *crooksite* and *clausthalite.* In years past it has been obtained from flue dusts remaining from processing copper sulfide ores, but the anode muds from electrolytic copper refineries now provide the source of most of the world's selenium. Selenium is recovered by roasting the muds with soda or sulfuric acid, or by smelting them with soda and niter. Selenium exists in several allotropic forms. Three are generally recognized, but as many as six have been claimed. Selenium can be prepared with either an amorphous or crystalline structure. The color of amorphous selenium is either red, in powder form, or black, in vitreous form. Crystalline monoclinic selenium is a deep red; crystalline hexagonal selenium, the most stable variety, is a metallic gray. Natural selenium contains six stable isotopes. Fourteen other nuclides and isomers have been characterized. The element is a member of the sulfur family and resembles sulfur both in its various forms and in its compounds. Selenium exhibits both photovoltaic action, where light is converted directly into electricity, and photoconductive action, where the electrical resistance decreases with increased illumination. These properties make selenium useful in the production of photocells and exposure meters for photographic use, as well as solar cells. Selenium is also able to convert a.c. electricity to d.c., and is extensively used in rectifiers. Below its melting point selenium is a p-type semiconductor and is finding many uses in electronic and solid-state applications. It is used in Xerography for reproducing and copying documents, letters, etc. It is used by the glass industry to decolorize glass and to make ruby-colored glasses and enamels. It is also used as a photographic toner, and as an additive to stainless steel. Elemental selenium has been said to be practically nontoxic and is considered to be an essential trace element; however, hydrogen selenide and other selenium compounds are extremely toxic, and resemble arsenic in their physiological reactions. Hydrogen selenide in a concentration of 1.5 ppm is intolerable to man. Selenium occurs in some soils in amounts sufficient to produce serious effects on animals feeding on plants, such as locoweed, grown in such soils. Exposure to selenium compounds (as Se) in air should not exceed 0.2 mg/M³ (8-hr time-weighted average — 40-hr week). Selenium is priced at about $20/lb. It is also available in high purity form at a somewhat higher cost.

Silicon — (L. *silex, silicis,* flint), Si; at. wt. 28.0855 ± 0.0003; at. no. 14; m.p. 1410°C; b.p. 2355°C; sp. gr. 2.33 (25°C); valence 4. Davy in 1800 thought silica to be a compound and not an element; later in 1811, Gay Lussac and Thenard probably prepared impure amorphous silicon by heating potassium with silicon tetrafluoride. Berzelius, generally credited with the discovery, in 1824 succeeded in preparing amorphous silicon by the same general method as used earlier, but he purified the product by removing the fluosilicates by repeated washings. Deville in 1854 first prepared crystalline silicon, the second allotropic form of the element. Silicon is present in the sun and stars and is a principal component of a class of meteorites known as *aerolites.* It is also a component of *tektites,* a natural glass of uncertain origin. Silicon makes up 25.7% of the earth's crust, by weight, and is the second most abundant element, being exceeded only by oxygen. Silicon is not found free in nature, but occurs chiefly as the oxide, and as silicates. *Sand, quartz, rock crystal, amethyst, agate, flint, jasper,* and *opal* are some of the forms in whch the oxide appears. *Granite, hornblende, asbestos, feldspar, clay, mica,* etc. are but a few of the numerous silicate minerals. Silicon is prepared commercially by heating silica and carbon in an electric furnace, using carbon electrodes. Several other methods can be used for preparing the element. Amorphous silicon can be prepared as a brown powder, which can be easily melted or vaporized. Crystalline silicon has a metallic luster and grayish color. The Czochralski process is commonly used to produce single crystals of silicon used for solid-state or semiconductor devices. Hyperpure silicon can be prepared by the thermal decomposition of ultra-pure trichlorosilane in a hydrogen atmosphere, and by a vacuum float zone process. This product can be doped with boron, gallium, phosphorus, or arsenic, etc. to produce silicon for use in transistors, solar cells, rectifiers, and other solid-state devices which are used extensively in the electronics and space-age industries. Hydrogenated amorphous silicon has shown promise in producing economical cells for converting solar energy into electricity. Silicon is a relatively inert element, but it is attacked by halogens and dilute alkali. Most acids, except hydrofluoric, do not affect it. Silicones are important products of silicon. They may be prepared by hydrolyzing a silicon organic chloride, such as dimethyl silicon chloride. Hydrolysis and condensation of various substituted chlorosilanes can be used to produce a very great number of polymeric products, or silicones, ranging from liquids to hard, glasslike solids with many useful properties. Elemental silicon transmits more than 95% of all wavelengths of infrared, from 1.3 to 6.7μm. Silicon is one of man's most useful elements. In the form of sand and clay it is used to make concrete and brick; it is a useful refractory material for high-temperature work, and in the form of silicates it is used in making enamels, pottery, etc. Silica, as sand, is a principal ingredient of glass, one of the most inexpensive of materials with excellent mechanical, optical, thermal, and electrical properties. Glass can be made in a very great variety of shapes, and is used as containers, window glass, insulators, and thousands of other uses. Silicon tetrachloride can be used to iridize glass. Silicon is important in plant and animal life. Diatoms in both fresh and salt water extract silica from the water to build up their cell walls. Silica is present in ashes of plants and in the human skeleton. Silicon is an important ingredient in steel; silicon carbide is one of the most important abrasives and has been used in lasers to produce coherent light of 4560 A. Regular grade silicon (97%) costs about

THE ELEMENTS (continued)

50c. Silicon 99.7% pure costs about $7/lb; hyperpure silicon may cost as much as $100/lb. Miners, stonecutters, and others engaged in work where siliceous dust is breathed in large quantities often develop a serious lung disease known as *silicosis*.

Silver — (Anglo-Saxon, *Seolfor siolfur*), Ag (L. argentum), at. wt. 107.86815; at. no. 47; m.p. 961.93°C; b.p. 2212°C; sp. gr. 10.50 (20°C); valence 1, 2, Silver has been known since ancient times. It is mentioned in Genesis. Slag dumps in Asia Minor and on islands in the Aegean Sea indicate that man learned to separate silver from lead as early as 3000 B.C. Silver occurs native and in ores such as *argentite* (Ag_2S) and *horn silver* (AgCl); lead, lead-zinc, copper, gold, and copper-nickel ores are principal sources. Mexico, Canada, Peru, and the U.S. are the principal silver producers in the western hemisphere. Silver is also recovered during electrolytic refining of copper. Commercial fine silver contains at least 99.9% silver. Purities of 99.999 + % are available commercially. Pure silver has a brilliant white metallic luster. It is a little harder than gold and is very ductile and malleable, being exceeded only by gold and perhaps palladium. Pure silver has the highest electrical and thermal conductivity of all metals, and possesses the lowest contact resistance. It is stable in pure air and water, but tarnishes when exposed to ozone, hydrogen sulfide, or air containing sulfur. The alloys of silver are important. Sterling silver is used for jewelry, silverware, etc. where appearance is paramount. This alloy contains 92.5% silver, the remainder being copper or some other metal. Silver is of utmost importance in photography, about 30% of the U.S. industrial consumption going into this application. It is used for dental alloys. Silver is used in making solder and brazing alloys, electrical contacts, and high capacity silver-zinc and silver-cadmium batteries. Silver paints are used for making printed circuits. It is used in mirror production and may be deposited on glass or metals by chemical deposition, electrodeposition, or by evaporation. When freshly deposited, it is the best reflector of visible light known, but it rapidly tarnishes and loses much of its reflectance. It is a poor reflector of ultraviolet. Silver fulminate ($Ag_2C_2N_2O_2$), a powerful explosive, is sometimes formed during the silvering process. Silver iodide is used in seeding clouds to produce rain. Silver chloride has interesting optical properties as it can be made transparent; it also is a cement for glass. Silver nitrate, or *lunar caustic,* the most important silver compound, is used extensively in photography. While silver itself is not considered to be toxic, most of its salts are poisonous due to the anions present. Exposure to silver (metal and soluble compounds, as Ag) in air should not exceed 0.01mg/ M^3, (8-hr time-weighted average — 40-hr week). Silver compounds can be absorbed in the circulatory system and reduced silver deposited in the various tissues of the body. A condition, known as *argyria*, results, with a greyish pigmentation of the skin and mucous membranes. Silver has germicidal effects and kills many lower organisms effectively without harm to higher animals. Silver for centuries has been used traditionally for coinage by many countries of the world. In recent times, however, consumption of silver has greatly exceeded the output. In 1939, the price of silver was fixed by the U.S. Treasury at 71c/troy ounce, and at 90.5c/ troy ounce in 1946. In November 1961 the U.S. Treasury suspended sales of nonmonetized silver, and the price stabilized for a time at about $1.29, the melt-down value of silver U.S. coins. The Coinage Act of 1965 authorized a change in the metallic composition of the three U.S. subsidiary denominations to clad or composite type coins. This was the first change in U.S. coinage since the monetary system was established in 1792. Clad dimes and quarters are made of an outer layer of 75% Cu and 25% Ni bonded to a central core of pure Cu. The composition of the one- and five-cent pieces remains unchanged. One-cent coins are 95% Cu and 5% Zn. Five-cent coins are 75% Cu and 25% Ni. Old silver dollars are 90% Ag and 10% Cu. Earlier subsidiary coins of 90% Ag and 10% Cu officially are to circulate alongside the clad coins; however, in practice they have largely disappeared (Gresham's Law), as the value of the silver is now greater than their exchange value. Silver coins of other countries have largely been replaced with coins made of other metals. On June 24, 1968, the U.S. Government ceased to redeem U.S. Silver Certificates with silver. Since that time, the price of silver has fluctuated widely. The U.S. Government discontinued selling silver to domestic users and foreign buyers on November 10, 1970. As of December 31, 1982, the price of silver was about $11/tr. oz.

Sodium — (English, *soda;* Medieval Latin, *sodanum,* headache remedy), Na (L. *natrium*); at. wt. 22.9898; at. no. 11; m.p. 97.81± 0.03°C; b.p. 882.9°C; sp. gr. 0.971 (20°C); valence 1. Long recognized in compounds, sodium was first isolated by Davy in 1807 by electrolysis of caustic soda. Sodium is present in fair abundance in the sun and stars. The D lines of sodium are among the most prominent in the solar spectrum. Sodium is the sixth most abundant element on earth, comprising about 2.6% of the earth's crust; it is the most abundant of the alkali group of metals of which it is a member. The most common compound is sodium chloride, but it occurs in many other minerals, such as *soda niter, cryolite, amphibole, zeolite, sodalite,* etc. It is a very reactive element and is never found free in nature. It is now obtained commercially by the electrolysis of absolutely dry fused sodium chloride. This method is much cheaper than that of electrolyzing sodium hydroxide, as was used several years ago. Sodium is a soft, bright, silvery metal which floats on water, decomposing it with the evolution of hydrogen and the formation of the hydroxide. It may or may not ignite spontaneously on water, depending on the amount of oxide and metal exposed to the water. It normally does not ignite in air at temperatures below 115°C. Sodium should be handled with respect, as it can be dangerous when improperly handled. Metallic sodium is vital in the manufacture of sodamide and sodium cyanide, sodium peroxide, and sodium hydride. It is used in preparing tetraethyl lead, in the reduction of organic esters, and in the preparation of organic compounds. The metal may be used to improve the structure of certain alloys, to descale metal, to purify molten metals, and as a heat transfer agent. An alloy of sodium with potassium, NaK, is also an important heat transfer agent. Sodium compounds are important to the paper, glass, soap, textile, petroleum, chemical, and metal industries. Soap is generally a sodium salt of certain fatty acids. The importance of common salt to animal nutrition has been recognized since prehistoric times. Among the many compounds that are of the greatest industrial importance are common salt (NaCl), soda ash (Na_2CO_3), baking soda ($NaHCO_3$), caustic soda (NaOH), Chile saltpeter ($NaNO_3$), di- and tri-sodium phosphates, sodium thiosulfate (hypo, $Na_2S_2O_3 \cdot 5H_2O$), and borax ($Na_2B_4O_7 \cdot 10H_2O$). Seven isotopes of sodium are recognized. Metallic sodium is priced at about 15 to 20c/lb in quantity. On a volume basis, it is the cheapest of all metals. Sodium metal should be handled with great care. It should be maintained in an inert atmosphere and contact with water and other substances with which sodium reacts should be avoided.

Strontium — (*Strontian,* town in Scotland), Sr; at. wt. 87.62; at. no. 38; m.p. 769°C; b.p. 1384°C; sp. gr. 2.54; valence 2. Isolated by Davy by electrolysis in 1808; however, Adair Crawford in 1790 recognized a new mineral (strontianite) as differing from other barium minerals (baryta). Strontium is found chiefly as *celestite* ($SrSO_4$) and *strontianite* ($SrCO_3$). The metal can be prepared by electrolysis of the fused chloride mixed with potassium chloride, or is made by reducing strontium oxide with aluminum in a vacuum at a temperature at which strontium distills off. Three allotropic forms of the metal exist, with transition points at 235 and 540°C. Strontium is softer than calcium and decomposes water more vigorously. It does not absorb nitrogen below 380°C. It should be kept under kerosene to prevent oxidation. Freshly cut strontium has a silvery appearance, but rapidly turns a yellowish color with the formation of the oxide. The finely divided metal ignites spontaneously in air. Volatile strontium salts impart a beautiful crimson color to flames, and these salts are used in pyrotechnics. Natural strontium is a mixture of four stable isotopes. Twelve other unstable isotopes are known to exist. Of greatest importance is Sr^{90} with a half-life of 28 years. It is a product of nuclear fallout and presents a health problem. This isotope is one of the best long-lived high-energy beta emitters known, and is used in SNAP (Systems for Nuclear Auxiliary Power) devices. These devices hold promise for use in space vehicles, remote weather stations, navigational buoys, etc., where a lightweight, long-lived, nuclear-electric power source is needed. Strontium hydroxide has been used in sugar refining; however, lime is replacing its use as it is cheaper. Strontium titanate is an interesting optical material as it has an extremely high refractive index and an optical dispersion greater than that of diamond. It has been used as a gemstone, but it is very soft. It does not occur naturally. The applications of strontium are similar to those of barium and calcium, but there are few advantages and the cost is much higher. Strontium metal costs about $6 to $8/lb.

Sulfur — (Sanskrit, *sulvere;* L. *sulphurium*), S; at. wt. 32.06; at. no. 16; m.p. (rhombic) 112.8°C, (monoclinic) 119.0°C; b.p. 444.674°C; sp. gr. (rhombic) 2.07, (monoclinic) 1.957 (20°C); valence 2, 4, or 6. Known to the ancients; referred to in Genesis as *brimstone.* Sulfur is found in meteorites. A dark area near the crater Aristarchus on the moon has been studied by R. W. Wood with ultraviolet light. This study suggests strongly that it is a sulfur deposit. Sulfur occurs native in the vicinity of volcanoes and hot springs. It is widely distributed in nature as *iron pyrites, galena, sphalerite, cinnabar, stibnite, gypsum, Epsom salts, celestite, barite,* etc. Sulfur is commercially recovered from wells sunk into the salt domes along the Gulf Coast of the U.S. It is obtained from these wells by the Frasch process, which forces heated water into the wells to melt the sulfur that is then brought to the surface. Sulfur also occurs in natural gas and petroleum crudes and must be removed from these products. Formerly this was done chemically, which wasted the sulfur. New processes now permit recovery, and these sources promise to be very important. Large amounts of sulfur are being recovered from Alberta gas fields. Sulfur is a pale yellow, odorless, brittle solid, which is insoluble in water but soluble in carbon disulfide. In every state, whether gas, liquid or solid, elemental sulfur occurs in more than one allotropic form or modification; these present a confusing multitude of forms whose relations are not yet fully understood. Amorphous or "plastic" sulfur is obtained by fast cooling of the crystalline form. X-ray studies indicate that amorphous sulfur may have a helical structure with eight atoms per spiral. Crystalline sulfur seems to be made of rings, each containing eight sulfur atoms, which fit together to give a normal X-ray pattern. Ten isotopes of sulfur exist. Four occur in natural sulfur, none of which is radioactive. A finely divided form of sulfur, known as *flowers of sulfur,* is obtained by sublimation. Sulfur readily forms sulfides with many elements. sulfur is a component of black gunpowder, and is used in the vulcanization of natural rubber and as a fungicide. It is also used extensively in making phosphatic fertilizers. A tremendous tonnage is used to produce sulfuric acid, the most important manufactured chemical. It is used in making sulfite paper and other papers, as a fumigant, and in the bleaching of dried fruits. The element is a good electrical insulator. Organic compounds containing sulfur are very important. Calcium sulfate, ammonium sulfate, carbon disulfide, sulfur dioxide, and hydrogen sulfide are but a few of the other many important compounds of sulfur. Sulfur is essential to life. It is a minor constituent of fats, body fluids, and skeletal minerals. Carbon disulfide, hydrogen sulfide, and sulfur dioxide should be handled carefully. Hydrogen sulfide in small concentrations can be metabolized, but in higher concentrations it quickly can cause death by respiratory paralysis. It is insidious in that it quickly deadens the sense of smell. Sulfur dioxide is a dangerous component in atmospheric air pollution. In 1975, University of Pennsylvania scientists reported synthesis of polymeric sulfur nitride, which has the properties of a metal, although it contains no metal atoms. The material has unusual optical and electrical properties. High-purity sulfur is commercially available in purities of 99.999 + %.

Tantalum — (Gr. *Tantalos,* mythological character, father of *Niobe*), Ta; at. wt. 180.9479; at. no. 73; m.p. 2996 °C; b.p. 5425 ± 100°C; sp. gr. 16.654; valence 2?,3,4?, or 5. Discovered in 1802 by Ekeberg, but many chemists thought niobium and tantalum were identical elements until Rose, in 1844, and Marignac, in 1866, indicated and showed that niobic and tantalic acids were two different acids. The early investigators only isolated the impure metal. The first relatively pure ductile tantalum was produced by von Bolton in 1903. Tantalum occurs principally in the mineral *columbite-tantalite* (Fe, Mn)(Nb,Ta)$_2$O$_6$. Tantalum ores are found in the Republic of Zaire, Brazil, Mozambique, Thailand, Portugal, Nigeria, and Canada. Mines at Bernic Lake, Manitoba, have reserves of 900,000 tons of ore averaging about 0.15% tantalum oxide. Separation of tantalum from niobium requires several complicated steps. Several methods are used to commercially produce the element, including electrolysis of molten potassium fluotantalate, reduction of potassium fluotantalate with sodium, or reacting tantalum carbide with tantalum oxide. Sixteen isotopes of tantalum are known to exist. Natural tantalum contains two isotopes; one of these, Ta^{180}, is present in very small quantity (0.0123%) and is unstable with a very long half-life of > 10^{13} years. Tantalum is a gray, heavy, and very hard metal. When pure, it is ductile and can be drawn into fine wire, which is used as a filament for evaporating metals such as aluminum. Tantalum is almost completely immune to chemical attack at temperatures below 150°C, and is attacked only by hydrofluoric acid, acidic solutions containing the fluoride ion, and free sulfur trioxide. Alkalis attack it only slowly. At high termperatures, tantalum becomes much more reactive. The element has a melting point exceeded only by tungsten and rhenium. Tantalum is used to make a variety of alloys with desirable properties

THE ELEMENTS (continued)

such as high melting point, high strength, good ductility, etc. Scientists at Los Alamos have produced a tantalum carbide graphite composite material, which is said to be one of the hardest materials ever made. The compound has a melting point of 6760°F. Tantalum has good "gettering" ability at high temperatures, and tantalum oxide films are stable and have good rectifying and dielectric properties. Tantalum is used to make electrolytic capacitors and vacuum furnace parts, which account for about 60% of its use. The metal is also widely used to fabricate chemical process equipment, nuclear reactors, and aircraft and missile parts. Tantalum is completely immune to body liquids and is a nonirritating metal. It has, therefore, found wide use in making surgical appliances. Tantalum oxide is used to make special glass with a high index of refraction for camera lenses. The metal has many other uses. In powdered form it costs about $40/lb. Sheet tantalum and fabricated forms are more expensive.

Technetium — (Gr. *technetos,* artifical), Tc; at. wt. 98.9062; at. no. 43; m.p. 2172°C; b.p. 4877°C; sp. gr. 11.50 (calc.); valence 0, +2, +4, +5, +6, and +7. Element 43 was predicted on the basis of the periodic table, and was erroneously reported as having been discovered in 1925, at which time it was named *masurium.* The element was actually discovered by Perrier and Segre in Italy in 1937. It was found in a sample of molybdenum, which was bombarded by deuterons in the Berkeley cyclotron, and which E. Lawrence sent to these investigators. Technetium was the first element to be produced artificially. Since its discovery, searches for the element in terrestrial materials have been made without success. If it does exist, the concentration must be very small. Technetium has been found in the spectrum of S-, M-, and N-type stars, and its presence in stellar matter is leading to new theories of the production of heavy elements in the stars. Sixteen isotopes of technetium, with atomic masses ranging from 92 to 107, are known. Tc^{97} has a half-life of 2.6×10^6 years. Tc^{98} has a half-life of 1.5×10^6 years. The isomeric isotope Tc^{95m}, with a half-life of 61 days, is useful for tracer work, as it produces energetic gamma rays. Technetium metal has been produced in kilogram quantities. The metal was first prepared by passing hydrogen gas at 1100°C over Tc_2S_7. It is now conveniently prepared by the reduction of ammonium pertechnetate with hydrogen. Technetium is a silvery-gray metal that tarnishes slowly in moist air. Until 1960, technetium was available only in small amounts and the price was as high as $2800/g. It is now offered commercially to holders of A.E.C. permits at a price of $90 to $100/g. The chemistry of technetium is said to be similar to that of rhenium. Technetium dissolves in nitric acid, aqua regia, and conc. sulfuric acid, but is not soluble in hydrochloric acid of any strength. The element is a remarkable corrosion inhibitor for steel. It is reported that mild carbon steels may be effectively protected by as little as 5 ppm of $KTcO_4$ in aerated distilled water at temperatures up to 250°C. This corrosion protection is limited to closed systems, since technetium is radioactive and must be confined. Tc^{99} has a specific activity of 6.2×10^8 disintegrations per sec/g. Activity of this level must not be allowed to spread. Tc^{99} is a contamination hazard and should be handled in a glove box. The metal is an excellent superconductor at 11°K and below.

Tellurium — (L. *tellus,* earth), Te; at. wt. 127.60; at. no. 52; m.p. 449.5 ± 0.3 °C; b.p. 989.8 ± 3.8°C; sp. gr. 6.24 (20°C); valence 2, 4, or 6. Discovered by Muller von Reichenstein in 1782; named by Klaproth, who isolated it in 1798. Tellurium is occasionally found native, but is more often found as the telluride of gold (*calaverite*), and combined with other metals. It is recovered commercially from the anode muds produced during the electrolytic refining of blister copper. The U.S., Canada, Peru, and Japan are the largest Free World producers of the element. Crystalline tellurium has a silvery-white appearance, and when pure exhibits a metallic luster. It is brittle and easily pulverized. Amorphous tellurium is formed by precipitating tellurium from a solution of telluric or tellurous acid. Whether this form is truly amorphous, or made of minute crystals, is open to question. Tellurium is a p-type semiconductor, and shows greater conductivity in certain directions, depending on alignment of the atoms. Its conductivity increases slightly with exposure to light. It can be doped with silver, copper, gold, tin, or other elements. In air, tellurium burns with a greenish-blue flame, forming the dioxide. Molten tellurium corrodes iron, copper, and stainless steel. Tellurium and its compounds are probably toxic and should be handled with care. Workmen exposed to as little as 0.01 mg/m³ of air, or less, develop "tellurium breath," which has a garlic-like odor. Twenty-one isotopes of tellurium are known, with atomic masses ranging from 115 to 135. Natural tellurium consists of eight isotopes, one of which, Te^{127}, is unstable. It is present to the extent of 0.87% and has a half-life of 1.2×10^{13} years. Tellurium improves the machinability of copper and stainless steel, and its addition to lead decreases the corrosive action of sulfuric acid to lead and improves its strength and hardness. Tellurium is used as a basic ingredient in blasting caps, and is added to cast iron for chill control. Tellurium is used in ceramics. Bismuth telluride has been used in thermoelectric devices. Tellurium costs about $20/lb, with a purity of about 99.5%.

Terbium — (*Ytterby,* village in Sweden), Tb; at. wt. 158.9254; at. no. 65; m.p. 1356°C; b.p. 3123°C; sp. gr. 8.229; valence 3, 4. Discovered by Mosander in 1843. Terbium is a member of the lanthanide or "rare earth" group of elements. It is found in *cerite, gadolinite,* and other minerals along with other rare earths. It is recovered commercially from *monazite* in which it is present to the extent of 0.03%, from *xenotime,* and from *euxenite,* a complex oxide containing 1% or more of terbia. Terbium has been isolated only in recent years with the development of ion-exchange techniques for separating the rare-earth elements. As with other rare earths, it can be produced by reducing the anhydrous chloride or fluoride with calcium metal in a tantalum crucible. Calcium and tantalum impurities can be removed by vacuum remelting. Other methods of isolation are possible. Terbium is reasonably stable in air. It is a silvery-gray metal, and is malleable, ductile, and soft enough to be cut with a knife. Two crystal modifications exist, with a transformation temperature of 1315°C. Nineteen isotopes with atomic masses ranging from 147 to 164 are recognized. The oxide is a chocolate or dark maroon color. Sodium terbium borate is used as a laser material and emits coherent light at 5460A. Terbium is used to dope calcium fluoride, calcium tungstate, and strontium molybdate, used in solid-state devices. The oxide has potential application as an activator for green phosphors used in color TV tubes. It can be used with ZrO_2 as a crystal stabilizer of fuel cells which operate at elevated temperature. Few other uses have been found. The element is priced at about $3/g or $900/lb. Little is known of the toxicity of terbium. It should be handled with care as with other lanthanide elements.

Thallium — (Gr. *thallos,* a green shoot or twig), Tl; at. wt. 204.37; at. no. 81; m.p. 303.5°C; b.p. 1457 ± 10°C; sp. gr. ll.85 (20°C); valence 1, or 3. Thallium was discovered spectroscopically in 1861 by Crookes. The element was named after the beautiful green spectral line, which identified the element. The metal was isolated both by Crookes and Lamy in 1862 about the same time. Thallium occurs in *crooksite, lorandite,* and *hutchinsonite.* It is also present in *pyrites* and is recovered from the roasting of this ore in connection with the production of sulfuric acid. It is also obtained from the smelting of lead and zinc ores. Extraction is somewhat complex and depends on the source of the thallium. Manganese nodules, found on the ocean floor, contain thallium. When freshly exposed to air, thallium exhibits a metallic luster, but soon develops a bluish-gray tinge, resembling lead in appearance. A heavy oxide builds up on thallium if left in air, and in the presence of water the hydroxide is formed. The metal is very soft and malleable. It can be cut with a knife. Twenty isotopic forms of thallium, with atomic masses ranging from 191 to 210, are recognized. Natural thallium is a mixture of two isotopes. The element and its compounds are toxic and should be handled carefully. Contact of the metal with the skin is dangerous, and when melting the metal adequate ventilation should be provided. Exposure to thallium (soluble compounds) — skin, as Tl, should not exceed 0.1mg/M³ (8-hr time-weighted average — 40-hr week). Thallium is suspect of carcinogenic potential for man. Thallium sulfate has been widely employed as a rodenticide and ant killer. It is odorless and tasteless, giving no warning of its presence. Its use, however, has been prohibited in the U.S. since 1975 as a household insecticide and rodenticide. Continued sales subjects dealers to civil and criminal penalties. The electrical conductivity of thallium sulfide changes with exposure to infrared light, and this compound is used in photocells. Thallium bromide-iodide crystals have been used as infrared detectors. Thallium has been used, with sulfur or selenium and arsenic, to produce low melting glasses which become fluid between 125 and 150°C. These glasses have properties at room temperatures similar to ordinary glasses and are said to be durable and insoluble in water. Thallium oxide has been used to produce glasses with a high index of refraction. Thallium has been used in treating ringworm and other skin infections; however, its use has been limited because of the narrow margin between toxicity and therapeutic benefits. A mercury-thallium alloy, which forms a eutectic at 8.5% thallium, is reported to freeze at −60°C, some 20° below the freezing point of mercury, Commercial thallium metal costs about $8/lb.

Thorium — (*Thor,* Scandinavian god of war), Th; at. wt. 232.0381; at. no. 90; m.p. 1750°C; b.p. ∼4790°C; sp. gr. 11.72; valence +2(?), +3(?), +4. Discovered by Berzelius in 1828. Thorium occurs in *thorite* (ThSiO₄) and in *thorianite* (ThO₂ + UO₂). Large deposits of thorium minerals have been reported in New England and elsewhere, but these have not yet been exploited. Thorium is now thought to be about three times as abundant as uranium and about as abundant as lead or molybdenum. The metal is a source of nuclear power. There is probably more energy available for use from thorium in the minerals of the earth's crust than from both uranium and fossil fuels. Any sizable demand for thorium as a nuclear fuel is still several years in the future. Work has been done in developing thorium cycle converter-reactor systems. Several prototypes, including the HTGR (high-temperature gas-cooled reactor) and MSTR (molten salt converter reactor experiment), have operated. While the HTGR reactors are efficient, they are not expected to become important commercially for 10 or more years because of certain operating difficulties. Thorium is recovered commercially from the mineral *monazite,* which contains from 3 to 9% ThO₂ along with most rare-earth minerals. Much of the internal heat the earth has been attributed to thorium and uranium. Several methods are available for producing thorium metal: it can be obtained by reducing thorium oxide with calcium, by electrolysis of anhydrous thorium chloride in a fused mixture of sodium and potassium chlorides, by calcium reduction of thorium tetrachloride mixed with anhydrous zinc chloride, and by reduction of thorium tetrachloride with an alkali metal. Thorium was originally assigned a position in Group IV of the periodic table. Because of its atomic weight, valence, etc., it is now considered to be the second member of the *actinide* series of elements. When pure, thorium is a silvery-white metal which is air-stable and retains its luster for several months. When contaminated with the oxide, thorium slowly tarnishes in air, becoming gray and finally black. The physical properties of thorium are greatly influenced by the degree of contamination with the oxide. The purest specimens often contain several tenths of a percent of the oxide. High-purity thorium has been made. Pure thorium is soft, very ductile, and can be cold-rolled, swaged, and drawn. Thorium is dimorphic, changing at 1400°C from a cubic to a body-centered cubic structure. Thorium oxide has a melting point of 3300°C, which is the highest of all oxides. Only a few elements, such as tungsten, and a few compounds, such as tantalum carbide, have higher melting points. Thorium is slowly attacked by water, but does not dissolve readily in most common acids, except hydrochloric. Powdered thorium metal is often pyrophoric and should be carefully handled. When heated in air, thorium turnings ignite and burn brilliantly with a white light. The principal use of thorium has been in the preparation of the Welsbach mantle, used for portable gas lights. These mantles, consisting of thorium oxide with about 1% cerium oxide and other ingredients, glow with a dazzling light when heated in a gas flame. Thorium is an important alloying element in magnesium, imparting high strength and creep resistance at elevated temperatures. Because thorium has a low work-function and high electron emission, it is used to coat tungsten wire used in electronic equipment. The oxide is also used to control the grain size of tungsten used for electric lamps; it is also used for high-temperature laboratory crucibles. Glasses containing thorium oxide have a high refractive index and low dispersion. Consequently, they find application in high quality lenses for cameras and scientific instruments. Thorium oxide has also found use as a catalyst in the conversion of ammonia to nitric acid, in petroleum cracking, and in producing sulfuric acid. Twelve isotopes of thorium are known with atomic masses ranging from 223 to 234. All are unstable. Th²³² occurs naturally and has a half-life of 1.41 × 10¹⁰ years. It is an alpha emitter. Th²³² goes through six alpha and four beta decay steps before becoming the stable isotope Pb²⁰⁸. Th²³² is sufficiently radioactive to expose a photographic plate in a few hours. Thorium disintegrates with the production of thoron (radon²²⁰), which is an alpha emitter and presents a radiation hazard. Good ventilation of areas where thorium is stored or handled is therefore essential. Thorium and is compounds are subject to licensing and control by the U.S. Atomic Energy Commission. Thorium metal (99.9%) costs about $10/lb.

Thulium — (*Thule,* the earliest name for Scandinavia), Tm; at. wt. 168.9342; at. no. 69; m.p. 1,545 ± l5° C; b.p. 1947°C; sp. gr. 9.321(25°); valence 2, 3. Discovered in 1879 by Cleve. Thulium occurs in small quantities along with other rare earths in a number of minerals. It is obtained commercially from *monazite,* which contains about 0.007% of the element. Thulium is the least abundant of the earth elements, but with new sources recently discovered, it is now considered to be about as rare as silver, gold, or cadmium. Ion-exchange and solvent extraction techniques have recently permitted much easier separation of the rare earths, with much lower costs. Only a few years ago, thulium metal was not obtainable at any cost; in 1950 the oxide sold for $450/g. Thulium metal now costs from $3 to $20/g, depending on the purity, quantity, and supplier. Thulium can be isolated by reduction of the oxide with lanthanum metal or by calcium reduction of the anhydrous fluoride. The pure metal has a bright, silvery luster. It is reasonably stable in air, but the metal should be protected from moisture in a closed container. The element is silver-gray, soft, malleable, and ductile, and can be cut with a knife. Sixteen isotopes are known, with atomic masses ranging from 161 to 176. Natural thulium, Tm^{169} is stable. Because of the relatively high price of the metal, thulium has not yet found many practical applications. Tm^{169} bombarded in a nuclear reactor can be used as a radiation source in portable X-ray equipment. Tm^{171} is potentially useful as an energy source. Natural thulium also has possible use in *ferrites* (ceramic magnetic materials) used in microwave equipment. As with other lanthanides, thulium has a low-to-moderate acute toxic rating. It should be handled with care.

Tin — (Anglo-Saxon, *tin*), Sn (L. *stannum*); at. wt. 118.69; at. no. 50; m.p. 231.9681°C; b.p. 2270°C; sp. gr. (gray) 5.75, (white) 7.31; valence 2, 4. Known to the ancients. Tin is found chiefly in *cassiterite* (SnO_2). Most of the world's supply comes from Malaya, Boliva, Indonesia, Zaire, Thailand, and Nigeria. The U.S. produces almost none, although occurrences have been found in Alaska and California. Tin is obtained by reducing the ore with coal in a reverberatory furnace. Ordinary tin is composed of nine stable isotopes; 13 unstable isotopes are also known. Ordinary tin is a silvery-white metal, is malleable, somewhat ductile, and has a highly crystalline structure. Due to the breaking of these crystals, a "tin cry" is heard when a bar is bent. The element has two or perhaps three allotropic forms. On warming, gray, or α tin, with a cubic structure, changes at 13.2°C into white, or β tin, the ordinary form of the metal. White tin has a tetragonal structure. Some authorities believe a γ form exists between 161°C and the melting point; however, other authorities discount its existence. When tin is cooled below 13.2°C, it changes slowly from white to gray. This change is affected by impurities such as aluminum and zinc, and can be prevented by small additions of antimony or bismuth. This change from the α to β form is called the tin pest. There are few if any uses for gray tin. Tin takes a high polish and is used to coat other metals to prevent corrosion or other chemical action. Such tin plate over steel is used in the so-called tin can for preserving food. Alloys of tin are very important. Soft solder, type metal, fusible metal, pewter, bronze, bell metal, Babbitt metal, White metal, die casting alloy, and phosphor bronze are some of the important alloys using tin. Tin resists distilled, sea, and soft tap water, but is attacked by strong acids, alkalis, and acid salts. Oxygen in solution accelerates the attack. When heated in air, tin forms SnO_2, which is feebly acid, forming stannate salts with basic oxides. The most important salt is the chloride ($SnCl_2 \cdot H_2O$), which is used as a reducing agent and as a mordant in calico printing. Tin salts sprayed onto glass are used to produce electrically conductive coatings. These have been used for panel lighting and for frost-free wind- shields. Most window glass is now made by floating molten glass on molten tin (float glass) to produce a flat surface (Pilkington process). Of recent interest is a crystalline tin-niobium alloy that is superconductive at very low temperatures. This promises to be important in the construction of super-conductive magnets that generate enormous field strengths but use practically no power. Such magnets, made of tin-niobium wire, weigh but a few pounds and produce magnetic fields that, when started with a small battery, are comparable to that of a 100 ton electromagnet operated continuously with a large power supply. The small amount of tin used in canned foods is quite harmless. The agreed limit of tin content in U.S. foods is 300 mg/kg. The trialkyl and triaryl tin compounds are used as biocides and must be handled carefully. Over the past 25 years the price of tin has varied from 50c/lb to its present price of $6/lb.

Titanium — (L. *Titans,* the first sons of the Earth, myth.), Ti; at. wt. 47.90; at. no. 22; m.p. 1660 ± 10°C; b.p. 3287°C; sp. gr. 4.54; valence 2, 3, or 4. Discovered by Gregor in 1791; named by Klaproth in 1795. Impure titanium was prepared by Nilson and Pettersson in 1887; however, the pure metal (99.9%) was not made until 1910 by Hunter by heating $TiCl_4$ with sodium in a steel bomb. Titanium is present in meteorites and in the sun. Rocks obtained during the Apollo 17 lunar mission showed presence of 12.1% TiO_2. Analyses of rocks obtained during earlier Apollo missions show lower percentages. Titanium oxide bands are prominent in the spectra of M-type stars. The element is the ninth most abundant in the crust of the earth. Titanium is almost always present in igneous rocks and in the sediments derived from them. It occurs in the minerals *rutile, ilmenite,* and *sphene,* and is present in titanates and in many iron ores. Titanium is present in the ash of coal, in plants, and in the human body. The metal was a laboratory curiosity until Kroll, in 1946, showed that titanium could be produced commercially by reducing titanium tetrachloride with magnesium. This method is largely used for producing the metal today. The metal can be purified by decomposing the iodide. Titanium, when pure, is a lustrous, white metal. It has a low density, good strength, is easily fabricated, and has excellent corrosion resistance. It is ductile only when it is free of oxygen. The metal burns in air and is the only element that burns in nitrogen. Titanium is resistant to dilute sulfuric and hydrochloric acid, most organic acids, moist chlorine gas, and chloride solutions. Natural titanium consists of five isotopes with atomic masses from 46 to 50. All are stable. Four other unstable isotopes are known. Natural titanium is reported to become very radioactive after bombardment with deuterons. The emitted radiations are mostly positrons and hard gamma rays. The metal is dimorphic. The hexagonal α form changes to the cubic β form very slowly at about 880°C. The metal combines with oxygen at red heat, and with chlorine at 550°C. Titanium is important as an alloying agent with aluminum, molybdenum, manganese, iron, and other metals. Alloys of titanium are principally used for aircraft and missiles where lightweight strength and ability to withstand extremes of temperature are important. Titanium is a strong as steel, but 45% lighter. It is 60% heavier than aluminum, but twice as strong. Titanium has potential use in desalination plants for converting sea water into fresh water. The metal has excellent resistance to sea water and is used for propeller shafts, rigging, and other parts of ships exposed to salt water. A titanium anode coated with platinum has been used to provide cathodic protection from corrosion by salt water. Titanium metal is considered to be physiologically inert. When pure, titanium dioxide is relatively clear and has an extremely high index of refraction with an optical dispersion higher than diamond. It is produced artificially for use as a gemstone, but it is relatively soft. Star sapphires and rubies exhibit their asterism as a result of the presence of TiO_2. Titanium dioxide is extensively used for both house paint and artist's paint, as it is permanent and has good covering power. Titanium oxide pigment accounts for the largest use of the element. Titanium paint is an excellent reflector of infrared, and is extensively used in solar observatories where heat causes poor seeing conditions. Titanium tetrachloride is used to iridize glass. This compound fumes strongly in air and has been used to produce smoke screens. The price of titanium metal powder (99.7%) is about $25/lb.

Tungsten — (Swedish, *tung sten,* heavy stone); also known as *wolfram*(from *wolframite,* said to be named from *wolf rahm* or *spumi lupi,* because the ore interfered with the smelting of tin and was supposed to devour the tin), W; at. wt. 183.85; at. no. 74; m.p. 3410 ± 20°C; b.p. 5660°C, sp. gr. 19.3 (20°C); valence 2, 3, 4, 5, or 6. In 1779 Peter Woulfe examined the mineral now known as *wolframite* and concluded it must contain a new substance. Scheele, in 1781, found that a new acid could be made from *tung sten* (a name first applied about 1758 to a mineral now known as *scheelite*). Scheele and Bergman suggested the possibility of obtaining a new metal by reducing this acid. The de Elhuyar brothers found an acid in *wolframite* in 1783 that was identical to the acid of *tungsten* (tungstic acid) of Scheele, and in that year they succeeded in obtaining the element by reduction of this acid with charcoal. Tungsten occurs in *wolframite,* $(Fe, Mn)WO_4$; *scheelite,* $CaWO_4$; *huebnerite,* $MnWO_4$; and *ferberite,* $FeWO_4$. Important deposits of tungsten occur in California, Colorado, South Korea, Bolivia, U.S.S.R., and Portugal. China is reported to have about 75% of the world's tungsten resources. Natural tungsten contains five stable isotopes. Twelve other unstable isotopes are recognized. The metal is obtained commercially by reducing tungsten oxide with hydrogen or carbon. Pure tungsten is a steel-gray to tin-white metal. Very pure tungsten can be cut with a hacksaw, and can be forged, spun, drawn, and extruded. The impure metal is brittle and can be worked only with difficulty. Tungsten has the highest melting point and lowest vapor pressure of all metals, and at temperatures over 1650°C has the highest tensile strength. The metal oxidizes in air and must be protected at elevated temperatures. It has excellent corrosion resistance and is attacked only slightly by most mineral acids. The thermal expansion is about the same as borosilicate glass, which makes the metal useful for glass-to-metal seals. Tungsten and its alloys are used extensively for filaments for electric lamps, electron and television tubes, and for metal evaporation work; for electrical contact points for automobile distributors, X-ray targets; windings and heating elements for electrical furnaces; and for numerous space missile and high-temperature applications. High-speed tool steels, Hastelloy ®, Stellite ®, and many other alloys contain tungsten. Tungsten carbide is of great importance to the metal-working, mining, and petroleum industries. Calcium and magnesium tungstates are widely used in fluorescent lighting; other salts of tungsten are used in the chemical and tanning industries. Tungsten disulfide is a dry, high-temperature lubricant, stable to 500°C. Tungsten bronzes and other tungsten compounds are used in paints. Tungsten powder costs about $15/lb.

Uranium — (Planet *Uranus*), U; at. wt. 238.029; at. no. 92; m.p. 1132.3 ± 0.8°C; b.p. 3818°C; sp. gr. ~ 18.95; valence 2, 3, 4, 5, or 6. Yellow-colored glass, containing more than 1% uranium oxide and dating back to 79 A.D., has been found near Naples, Italy. Klaproth recognized an unknown element in *pitchblende* and attempted to isolate the metal in 1789. The metal apparently was first isolated in 1841 by Peligot, who reduced the anhydrous chloride with potassium. Uranium is not as rare as it was once thought. It is now considered to be more plentiful than mercury, antimony, silver, or cadmium, and is about as abundant as molybdenum or arsenic. It occurs in numerous minerals such as *pitchblende, uraninite, carnotite, autunite, uranophane, davidite,* and *tobernite.* It is also found in *phosphate rock, lignite, monazite sands,* and can be recovered commercially from these sources. The A.E.C. purchases uranium in the form of acceptable U_3O_8 concentrates. This incentive program has greatly increased the known uranium reserves. Uranium can be prepared by reducing uranium halides with alkali or alkaline earth metals or by reducing uranium oxides by calcium, aluminum, or carbon at high temperatures. The metal can also be produced by electrolysis of KUF_5 or UF_4, dissolved in a molten mixture of $CaCl_2$ NaCl. High-purity uranium can be prepared by the thermal decomposition of uranium halides on a hot filament. Uranium exhibits three crystallographic modifications as follows:

$$\alpha \xrightarrow{667°C} \beta \xrightarrow{772°C} \gamma$$

Uranium is a heavy, silvery-white metal which is pyrophoric when finely divided. It is a little softer than steel, and is attacked by cold water in a finely divided state. It is malleable, ductile, and slightly paramagnetic. In air, the metal becomes coated with a layer of oxide. Acids dissolve the metal, but it is unaffected by alkalis. Uranium has fourteen isotopes, all of which are radioactive. Naturally occurring uranium nominally contains 99.2830% by weight U^{238}, 0.7110% U^{235}, and 0.0054% U^{234}. Studies show that the percentage weight of U^{235} in natural uranium varies by as much as 0.1%, depending on the source. The A.E.C. has adopted the value of 0.711 as being their "official" percentage of U^{235} in natural uranium. Natural uranium is sufficiently radioactive to expose a photographic plate in an hour or so. Much of the internal heat of the earth is thought to be attributable to the presence of uranium and thorium. U^{238} with a half-life of 4.51×10^9 years, has been used to estimate the age of igneous rocks. The origin of uranium, the highest member of the naturally occurring elements — except perhaps for traces of neptunium or plutonium — is not clearly understood, although it may be presumed that uranium is a decay product of elements of higher atomic weight, which may have once been present on earth or elsewhere in the universe. These original elements may have been created as a result of a primordial "creation," known as "the big bang," in a supernovae, or in some other stellar processes. Uranium is of great importance as a nuclear fuel. U^{238} can be converted into fissionable plutonium by the following reactions:

THE ELEMENTS (continued)

$$U^{238}\,(n\gamma)U^{239}\xrightarrow{\beta-}Np^{239}\xrightarrow{\beta-}Pu^{239}$$

This nuclear conversion can be brought about in "breeder" reactors where it is possible to produce more new fissionable material than the fissionable material used in maintaining the chain reaction. U^{235} is of even greater importance, for it is the key to the utilization of uranium. U^{235}, while occurring in natural uranium to the extent of only 0.71%, is so fissionable with slow neutrons that a self-sustaining fission chain reaction can be made to occur in a reactor constructed from natural uranium and a suitable moderator, such as heavy water or graphite, alone. U^{235} can be concentrated by gaseous diffusion and other physical processes, if desired, and used directly as a nuclear fuel, instead of natural uranium, or used as an explosive. Natural uranium, slightly enriched with U^{235} by a small percentage, is used to fuel nuclear power reactors for the generation of electricity. Natural thorium can be irradiated with neutrons as follows to produce the important isotope U^{233}.

$$Th^{232}\,(n\gamma)Th^{233}\xrightarrow{\beta-}Pa^{233}\xrightarrow{\beta-}U^{233}$$

While thorium itself is not fissionable, U^{233} is, and in this way may be used as a nuclear fuel. One pound of completely fissioned uranium has the fuel value of over 1500 tons of coal. The uses of nuclear fuels to generate electrical power, to make isotopes for peaceful purposes, and to make explosives are well known. The estimated world-wide capacity of the 189 nuclear power reactors in operation in 1979 amounted to about 110 million kilowatts. Uranium in the U.S.A. is controlled by the Atomic Energy Commission. New uses are being found for "depleted uranium, i.e., uranium with the percentage of U^{235} lowered to about 0.2%. It has found use in inertial guidance devices, gyro compasses, counterweights for aircraft control surfaces, as ballast for missile reentry vehicles, and as a shielding material. Uranium metal is used for X-ray targets for production of high-energy X-rays; the nitrate has been used as photographic toner, and the acetate is used in analytical chemistry. Crystals of uranium nitrate are triboluminescent. Uranium salts have also been used for producing yellow "vaseline" glass and glazes. Uranium and its compounds are highly toxic, both from a chemical and radiological standpoint. Finely divided uranium metal, being pyrophoric, presents a fire hazard. The maximum recommended allowable concentration of soluble uranium compound in air (based on chemical toxicity) is 0.05 mg/M^3 (8-hr time-weighted average — 40-hr week); for insoluble compounds the concentration is set at 0.25 mg/M^3 of air. The maximum permissible total body burden of natural uranium (based on radiotoxicity) is 0.2μCi for soluble compounds.

Vanadium — (Scandinavian goddess, *Vanadis*), V; at. wt. 50.9415; at. no. 23; m.p. 1890 ± 10°C; b.p. 3380°C; sp. gr. 6.ll (18.7°C); valence 2, 3, 4, or 5. Vanadium was first discovered by del Rio in 1801. Unfortunately, a French chemist incorrectly declared del Rio's new element was only impure chromium; del Rio thought himself to be mistaken and accepted the French chemist's statement. The element was rediscovered in 1830 by Sefstrom, who named the element in honor of the Scandinavian goddess *Vanadis* because of its beautiful multicolored compounds. It was isolated in nearly pure form by Roscoe, in 1867, who reduced the chloride with hydrogen. Vanadium of 99.3 to 99.8% purity was not produced until 1927. Vanadium is found in about 65 different minerals among which are *carnotite, roscoelite, vanadinite,* and *patronite* — important sources of the metal. Vanadium is also found in phosphate rock and certain iron ores, and is present in some crude oils in the form of organic complexes. It is also found in small percentages in meteorites. Commercial production from petroleum ash holds promise as an important source of the element. High-purity ductile vanadium can be obtained by reduction of vanadium trichloride with magnesium or with magnesium-sodium mixtures. Much of the vanadium metal being produced is now made by calcium reduction of V_2O_5 in a pressure vessel, an adaption of a process developed by McKechnie and Seybolt. Natural vanadium is a mixture of two isotopes, V^{50} (0.24%) and V^{51} (99.76%). V^{50} is slightly radioactive, having a half-life of 6 × 10^{15} years. Seven other unstable isotopes are recognized. Pure vanadium is a bright white metal, and is soft and ductile. It has good corrosion resistance to alkalis, sulfuric and hydrochloric acid, and salt waters, but the metal oxidizes readily above 660°C. The metal has good structural strength and a low-fission neutron cross section, making it useful in nuclear applications. Vanadium is used in producing rust-resistant, spring, and high-speed tool steels. It is an important carbide stabilizer in making steels. About 80% of the vanadium now produced is used as ferrovanadium or as a steel additive. Vanadium foil is used as a bonding agent in cladding titanium to steel. Vanadium pentoxide is used in ceramics and as a catalyst. It is also used as a mordant in dyeing and printing fabrics and in the manufacture of aniline black. Vanadium-gallium tape has been used in producing a superconductive magnet with a field of 175,000 gauss. Vanadium and its compounds are toxic and should be handled with care. Exposure to V_2O_2 dust (as V) should not exceed the ceiling value of 0.05mg/M^3, and exposure to V_2O_2 fume (as V) should not exceed 0.1mg/M^3 (8-hr time-weighted average — 40-hr week). Ductile vanadium is commercially available. Commercial vanadium metal, of about 95% purity, costs about $10/lb.

Wolfram — see Tungsten.

Xenon — (Gr. *xenon*, stranger), Xe; at. wt. 131.30; at. no. 54; m.p. −111.9°C; b.p. −107.1 ± 3°C; density (gas) 5.887 ± 0.009 g/l, sp. gr. (liquid) 3.52 (−109°C); valence usually O. Discovered by Ramsay and Travers in 1898 in the residue left after evaporating liquid air components. Xenon is a member of the so-called noble or "inert" gases. It is present in the atmosphere to the extent of about one part in twenty million. Xenon is present in the Martian atmosphere to the extent of 0.08 ppm. The element is found in the gases evolved from certain mineral springs, and is commercially obtained by extraction from liquid air. Natural xenon is composed of nine stable isotopes. In addition to these, 22 unstable nuclides and isomers have been characterized. Before 1962, it had generally been assumed that xenon and other noble gases were unable to form compounds. Evidence has been mounting in the past few years that xenon, as well as other members of the zero valence elements, do form compounds. Among the "compounds" of xenon now reported are xenon hydrate, sodium perxenate, xenon deuterate, difluoride, tetrafluoride, hexafluoride, and $XePtF_6$ and $XeRhF_6$. Xenon trioxide, which is highly explosive, has been prepared. More than 80 xenon compounds have been made with xenon chemically bonded to fluorine and oxygen. Some xenon compounds are colored. Metallic xenon has been produced, using several hundred kilobars of pressure. Xenon in a vacuum tube produces a beautiful blue glow when excited by an electrical discharge. The gas is used in making electron tubes, stroboscopic lamps, bactericidal lamps, and lamps used to excite ruby lasers for generating coherent light. Xenon is used in the atomic energy field in bubble chambers, probes, and other applications where its high molecular weight is of value. It is also potentially useful as a gas for ion engines. The perxenates are used in analytical chemistry as oxidizing agents. Xe^{133} and Xe^{135} are produced by neutron irradiation in aircooled nuclear reactors. Xe^{133} has useful applications as a radioisotope. The element is available in sealed glass containers for about $20/l of gas at standard pressure. Xenon is not toxic, but its compounds are highly toxic because of their strong oxidizing characteristics.

Ytterbium — (Ytterby, village in Sweden), Yb; at. wt. 173.04; at. no. 70; m.p. 819°C; b.p. 1194°C; sp. gr. (α) 6.965 (β) 6.54; valence 2, 3. Marignac in 1878 discovered a new component, which he called *ytterbia*, in the earth then known as *erbia*. In 1907, Urbain separated ytterbia into two components, which he called *neoytterbia* and *lutecia*. The elements in these earths are now known as *ytterbium* and *lutetium*, respectively. These elements are identical with *aldebaranium* and *cassiopeium*, discovered independently and at about the same time by von Welsbach. Ytterbium occurs along with other rare earths in a number of rare minerals. It is commercially recovered principally from *monazite sand*, which contains about 0.03%. Ion-exchange and solvent extraction techniques developed in recent years have greatly simplified the separation of the rare earths from one another. The element was first prepared by Klemm and Bonner in 1937 by reducing ytterbium trichloride with potassium. Their metal was mixed, however, with KCl. Daane, Dennison, and Spedding prepared a much purer form in 1953 from which the chemical and physical properties of the element could be determined. Ytterbium has a bright silvery luster, is soft, malleable, and quite ductile. While the element is fairly stable, it should be kept in closed containers to protect it from air and moisture. Ytterbium is readily attacked and dissolved by dilute and concentrated mineral acids and reacts slowly with water. Ytterbium normally has two allotropic forms with a transformation point at 798°C. The alpha form is a room-temperature, face-centered, cubic modification, while the high-temperature beta form is a body-centered cubic form. Another body-centered cubic phase has recently been found to be stable at high pressures at room temperatures. The alpha form ordinarily has metallic-type conductivity, but becomes a semiconductor when the pressure is increased above 16,000 atm. The electrical resistance increases tenfold as the pressure is increased to 39,000 atm and drops to about 80% of its standard temperature-pressure resistivity at a pressure of 40,000 atm. Natural ytterbium is a mixture of seven stable isotopes. Seven other unstable isotopes are known. Ytterbium metal has possible use in improving the grain refinement, strength, and other mechanical properties of stainless steel. One isotope is reported to have been used as a radiation source as a substitute for a portable X-ray machine where electricity is unavailable. Few other uses have been found. Ytterbium metal is commercially available with a purity of about 99 + % for about $1.00/g or $300/lb. Ytterbium has a low acute toxic rating.

Yttrium — (*Ytterby*, village in Sweden near Vauxholm), Y; at. wt. 88.9059; at. no. 39; m.p. 1522 ± 8°C; b.p. 3338°C; sp. gr. 4.469 (25°C); valence 3. *Yttria*, which is an earth containing yttrium, was discovered by Gadolin in 1794. *Ytterby* is the site of a quarry which yielded many unusual minerals containing rare earths and other elements. This small town, near Stockholm, bears the honor of giving names to *erbium, terbium,* and *ytterbium* as well as *yttrium*. In 1843 Mosander showed that yttria could be resolved into the earths of three elements. The name yttria was reserved for the most basic one; the others were named *erbia* and *terbia*. Yttrium occurs in nearly all of the rare-earth minerals. Analysis of lunar rock samples obtained during the Apollo missions show a relatively high yttrium content. It is recovered commercially from *monazite sand,* which contains about 3%, and from *bastnasite,* which contains about 0.2%. Wohler obtained the impure element in 1828 by reduction of the anhydrous chloride with potassium. The metal is now produced commercially by reduction of the fluoride with calcium metals. It can also be prepared by other techniques. Yttrium has a silvery-metallic luster and is relatively stable in air if their temperature exceeds 400°C, and finely divided yttrium is very unstable in air. Turnings of the metal, however, ignite in air. Yttrium oxide is one of the most important compounds of yttrium and accounts for the largest use. It is widely used in making YVO_4europium, and Y_2O_3europium phosphors to give the red color in color television tubes. Many hundreds of thousands of pounds are now used in this application. Yttrium oxide also is used to produce yttrium-iron-garnets, which are very effective microwave filters. Yttrium iron, aluminum, and gadolinium garnets, with formulas such as $Y_3Fe_5O_{12}$ and $Y_3Al_5O_{12}$, have interesting magnetic properties. Yttrium iron garnet is also exceptionally efficient as both a transmitter and transducer of acoustic energy. Yttrium aluminum garnet, with a hardness of 8.5, is also finding use as a gemstone (simulated diamond). Small amounts of yttrium (0.1 to 0.2%) can be used to reduce the grain size in chromium, molybdenum, zirconium, and titanium, and to increase strength of aluminum and magnesium alloys. Alloys with other useful properties can be obtained by using yttrium as an additive. The metal can be used as a deoxidizer for vanadium and other nonferrous metals. The metal has a low cross section for nuclear capture. Y^{90}, one of the isotopes of yttrium, exists in equilibrium with its parent Sr^{90}, a product of atomic explosions. Yttrium has been considered for use as a nodulizer for producing nodular cast iron, in which the graphite forms compact nodules instead of the usual flakes. Such iron has increased ductility. Yttrium is also finding application in laser systems and as a catalyst for ethylene polymerization. It has also potential use in ceramic and glass formulas as the oxide has a high melting point and imparts shock resistance and low expansion characteristics to glass. Natural yttrium contains but one isotope, Y^{89}. Twenty other unstable nuclides and isomers have been characterized. Yttrium metal of 99 + % purity is commercially available at a cost of about 60c/g or $150/lb.

Zinc — (Ger. *Zink*, of obscure origin), Zn; at. wt. 65.38; at. no. 30; m.p. 419.58°C; b.p. 907°C; sp. gr. 7.133 (25°C); valence 2. Centuries before zinc was recognized as a distinct element, zinc ores were used for making brass. Tubal-Cain, seven generations from Adam, is mentioned as being an "instructor in every artificer in brass and iron." An alloy containing 87% zinc has been found in prehistoric ruins in Transylvania. Metallic zinc was produced in the 13th century A.D. in India by reducing calamine with organic substances such as wool. The metal was rediscovered in Europe by Marggraf in 1746, who showed that it could be obtained by reducing *calamine* with charcoal. The principal ores of zinc are *sphalerite* or *blende* (sulfide), *smithsonite* (carbonate), *calamine* (silicate), and *franklinite* (zinc, manganese, iron oxide). Zinc can be obtained by roasting its ores to form the oxide and by reduction of the oxide with coal or carbon, with subsequent distillation of the metal. Other methods of extraction are possible. Naturally occurring zinc contains five stable isotopes. Ten other unstable nuclides and isomers are recognized. Zinc is a bluish- white, lustrous metal. It is brittle at ordinary temperatures but malleable at 100 to 150°C. It is a fair conductor of electricity, and burns in air at high red heat with evolution of white clouds of the oxide. The metal is employed to form numerous alloys with other metals. Brass, nickel silver, typewriter metal, commercial bronze, spring brass, German silver, soft solder, and aluminum solder are some of the more important alloys. Large quantities of zinc are used to produce die castings, used extensively by the automotive, electrical, and hardware industries. An alloy called *Prestal,* consisting of 78% zinc and 22% aluminum is reported to be almost as strong as steel but as easy to mold as plastic. It is said to be so plastic that it can be molded into form by relatively inexpensive die casts made of ceramics and cement. It exhibits superplasticity. Zinc is also extensively used to galvanize other metals such as iron to prevent corrosion. Neither zinc nor zirconium is ferromagnetic, but $ZrZn_2$ exhibits ferromagnetism at temperatures below 35 K. Zinc oxide is a unique and very useful material to modern civilization. It is widely used in the manufacture of paints, rubber products, cosmetics, pharmaceuticals, floor coverings, plastics, printing inks, soap, storage batteries, textiles, electrical equipment, and other products. It has unusual electrical, thermal, optical, and solid-state properties that have not yet been fully investigated.

Lithopone, a mixture of zinc sulfide and barium sulfate, is an important pigment. Zinc sulfide is used in making luminous dials, X-ray and TV screens, and fluorescent lights. The chloride and chromate are also important compounds. Zinc is an essential element in the growth of human beings and animals. Tests show that zinc-deficient animals require 50% more food to gain the same weight of an animal supplied with sufficient zinc. Zinc is not considered to be toxic, but when freshly formed ZnO is inhaled a disorder known as the *oxide shakes* or *zinc chills* sometimes occurs. It is recommended that where zinc oxide is encountered good ventilation be provided to avoid concentrations exceeding 5 mg/M³, (time-weighted over an 8-hr exposure, 40-hr work week). Zinc costs roughly 40c/lb

Zirconium — (Arabic *zargun,* gold color), Zr; at. wt. 91.22; at. no. 40; m.p. 1852 ± 2°C; b.p. 4377°C; sp. gr. 6.506 (20°C); valence +2, +3, and +4. The name *zircon* probably originated from the arabic word *zargun,* which describes the color of the gemstone now known as *zircon, jargon, hyacinth, jacinth,* or *ligure.* This mineral, or its variations, is mentioned in biblical writings. The mineral was not known to contain a new element until Klaproth, in 1789, analyzed a jargon from Ceylon and found a new earth, which Werner named zircon (*silex circonius*), and Klaproth called *Zirkonerde* (*zirconia*). The impure metal was first isolated by Berzelius in 1824 by heating a mixture of potassium and potassium zirconium fluoride in a small iron tube. Pure zirconium was first prepared in 1914. Very pure zirconium was first produced in 1925 by van Arkel and de Boer by an iodide decomposition process they developed. Zirconium is found in abundance in S-type stars, and has been identified in the sun and meteorites. Analyses of lunar rock samples obtained during the various Apollo missions to the moon show a surprisingly high zirconium oxide content, compared with terrestial rocks. Naturally occurring zirconium contains five isotopes, one of which, Zr^{96} (abundant to the extent of 2.80%,) is unstable with a very long half-life of $> 3.6 \times 10^{17}$ years. Fifteen other unstable nuclides and isomers of zirconium have been characterized. Zircon, $ZrSiO_4$, the principal ore, is found in deposits in Florida, South Carolina, Australia, and Brazil. *Baddeleyite,* found in Brazil, is an important zirconium mineral. It is principally pure ZrO_2 in crystalline form having a hafnium content of about 1%. Zirconium also occurs in some 30 other recognized mineral species. Zirconium is produced commercially by reduction of the chloride with magnesium (the Kroll Process), and by other methods. It is a grayish-white lustrous metal. When finely divided, the metal may ignite spontaneously in air, especially at elevated temperatures. The solid metal is much more difficult to ignite. The inherent toxicity of zirconium compounds is low. Hafnium is invariably found in zirconium ores, and the separation is difficult. Commercial-grade zirconium contains from 1 to 3% hafnium. Zirconium has a low absorption cross section for neutrons, and is therefore used for nuclear energy applications, such as for cladding fuel elements. Zirconium has been found to be extremely resistant to the corrosive environment inside atomic reactors, and it allows neutrons to pass through the internal zirconium construction material without appreciable absorption of energy. Commercial nuclear power generation now takes more than 90% of zirconium metal production. Reactors of the size now being made may use as much as a half-million lineal feet of zirconium alloy tubing. Reactor-grade zirconium is essentially free of hafnium. *Zircaloy* is an important alloy developed specifically for nuclear applications. Zirconium is exceptionally resistant to corrosion by many common acids and alkalis, by sea water, and by other agents. It is used extensively by the chemical industry where corrosive agents are employed. Zirconium is used as a getter in vacuum tubes, as an alloying agent in steel, in making surgical appliances, photoflash bulbs, explosive primers, rayon spinnerets, lamp filaments, etc. It is used in poison ivy lotions in the form of the carbonate as it combines with *urushiol.* With niobium, zirconium is superconductive at low temperatures and is used to make superconductive magnets, which offer hope of direct large-scale generation of electric power. Alloyed with zinc, zirconium becomes magnetic at temperatures below 35 K. Zirconium oxide (zircon) has a high index of refraction and is used as a gem material. The impure oxide, zirconia, is used for laboratory crucibles that will withstand heat shock, for linings of metallurgical furnaces, and by the glass and ceramic industries as a refractory material. Its use as a refractory material accounts for a large share of all zirconium consumed. Commercial-grade zirconium metal sponge is priced at about $7/lb. Fabricated zirconium parts are higher in cost.

NOMENCLATURE OF INORGANIC CHEMISTRY

AMERICAN VERSION*

By permission of the Committee on Publications of the International Union of Pure and Applied Chemistry.

INDEX TO IUPAC INORGANIC RULES

*This is the United States presentation of these IUPAC Inorganic Rules as published in J. Am. Chem. Soc., 82, 5525 (1960)

1. ELEMENTS

1.1. Names and Symbols of the Elements

1.11.—The elements should have the symbols given in the following table (Table I). It is desirable that the names should differ as little as possible among the different languages, but as complete uniformity is hard to achieve, separate lists have been drawn up in English and in French. The English list only is reproduced here.

TABLE I

ELEMENTS

Name	Symbol	Atomic number	Name	Symbol	Atomic number	Name	Symbol	Atomic number
Actinium	Ac	89	Gold (Aurum)	Au	79	Praseodymium	Pr	59
Aluminum	Al	13	Hafnium	Hf	72	Promethium	Pm	61
Americium	Am	95	Helium	He	2	Protactinium	Pa	91
Antimony	Sb	51	Holmium	Ho	67	Radium	Ra	88
Argon	Ar	18	Hydrogen	H	1	Radon	Rn	86
Arsenic	As	33	Indium	In	49	Rhenium	Re	75
Astatine	At	85	Iodine	I	53	Rhodium	Rh	45
Barium	Ba	56	Iridium	Ir	77	Rubidium	Rb	37
Berkelium	Bk	97	Iron (Ferrum)	Fe	26	Ruthenium	Ru	44
Beryllium	Be	4	Krypton	Kr	36	Samarium	Sm	62
Bismuth	Bi	83	Lanthanum	La	57	Scandium	Sc	21
Boron	B	5	Lead (Plumbum)	Pb	82	Selenium	Se	34
Bromine	Br	35	Lithium	Li	3	Silicon	Si	14
Cadmium	Cd	48	Lutetium	Lu	71	Silver (Argentum)	Ag	47
Calcium	Ca	20	Magnesium	Mg	12	Sodium	Na	11
Californium	Cf	98	Manganese	Mn	25	Strontium	Sr	38
Carbon	C	6	Mendelevium	Md	101	Sulfur	S	16
Cerium	Ce	58	Mercury	Hg	80	Tantalum	Ta	73
Cesium	Cs	55	Molybdenum	Mo	42	Technetium	Tc	43
Chlorine	Cl	17	Neodymium	Nd	60	Tellurium	Te	52
Chromium	Cr	24	Neon	Ne	10	Terbium	Tb	65
Cobalt	Co	27	Neptunium	Np	93	Thallium	Tl	81
Copper (Cuprum)	Cu	29	Nickel	Ni	28	Thorium	Th	90
Curium	Cm	96	Niobium	Nb	41	Thulium	Tm	69
Dysprosium	Dy	66	Nitrogen	N	7	Tin (Stannum)	Sn	50
Einsteinium	Es	99	Nobelium	No	102	Titanium	Ti	22
Erbium	Er	68	Osmium	Os	76	Tungsten (Wolfram)	W	74
Europium	Eu	63	Oxygen	O	8	Uranium	U	92
Fermium	Fm	100	Palladium	Pd	46	Vanadium	V	23
Fluorine	F	9	Phosphorus	P	15	Xenon	Xe	54
Francium	Fr	87	Platinum	Pt	78	Ytterbium	Yb	70
Gadolinium	Gd	64	Plutonium	Pu	94	Yttrium	Y	39
Gallium	Ga	31	Polonium	Po	84	Zinc	Zn	30
Germanium	Ge	32	Potassium	K	19	Zirconium	Zr	40

◆The committees reaffirm the name niobium for element 41 in spite of the fact that many in the United States, particularly outside of chemical circles, still retain the name columbium.

1.12.—The names placed in parentheses (after the trivial names) in the list in Table I shall always be used when forming names derived from those of the elements, *e.g.*, aurate, ferrate, wolframate and not goldate, ironate, tungstate.

For some compounds of sulfur, nitrogen, and antimony, derivatives of the Greek name θεῖον, the French name *azote*, and the Latin name *stibium*, respectively, are used.

Although the name nickel agrees with the chemical symbol, it is essentially a trivial name and is spelled so differently in various languages (niquel, nikkel, *etc.*) that it is recommended that names of derivatives be formed from the Latin name *niccolum*, *e.g.*, niccolate instead of nickelate. The name mercury should be used as the root name also in languages where the element has another name (mercurate, *not* hydrargyrate).

In the cases in which different names have been used, the Commission has selected one based upon considerations of prevailing usage and practicability. It should be emphasized that their selection carries no implication regarding priority of discovery.

◆Tungstate and nickelate are both so well established in American practice that the committees object to changing them to wolframate and niccolate.

1.13.—Any new metallic elements should be given names ending in -ium. Molybdenum and a few other elements have long been spelled without an "i" in most languages, and the Commission hesitates to insert it.

1.14.—All new elements shall have two-letter symbols.

1.15.—All isotopes of an element should have the same name. For hydrogen the isotope names protium, deuterium, and tritium may be retained, but it is undesirable to assign isotopic names instead of numbers to other elements. They should be designated by mass numbers as, for example, "oxygen-18."

◆The list in 3.21 implies that D is an acceptable symbol for deuterium, whereas ^{2}H is used in 1.32. It is recommended that D and T be allowed for deuterium and tritium, respectively. *Cf.* comments made at 1.31, 1.32.

NOMENCLATURE OF INORGANIC CHEMISTRY (Continued)

1.2. Names for Groups of Elements and their Subdivisions

1.21.—The use of the collective names: halogens (F, Cl, Br, I, and At), chalcogens (O, S, Se, Te, and Po), and halides and chalcogenides for their compounds, alkali metals (Li to Fr), alkaline earth metals (Ca to Ra), and inert gases may be continued. The name rare earth metals may be used for the elements Sc, Y, and La to Lu inclusive; the name lanthanum series for the elements no. 57–71 (La to Lu inclusive), and the name lanthanides for the elements 58–71 (Ce to Lu inclusive) are recommended. Elements no. 89 (Ac) to 103 form the actinium series, and the name actinides is reserved for the elements in which the 5f shell is being filled. The name transuranium elements is also approved for the elements following uranium.

◆The collective term halogenides used in the Rules has been replaced in this version by halides, which is almost universally used in English and is unambiguous.

The inclusion of Sc with the rare earths is questioned by some. No need is seen for the terms lanthanum series and actinium series, particularly since the latter term is used for a radioactive series. The use of a collective term for elements 58–71 is approved, although it is suggested that lanthanoid is preferable to lanthanide because of the use of -ide for binary compounds; similarly, actinoid is preferable to actinide. Definition by means of atomic numbers is recommended in both cases rather than on the basis of interpretation (e.g., filling of 5f shells).

1.22.—The word metalloid should not be used to denote nonmetals.

1.3. Indication of Mass, Charge, etc., on Atomic Symbols

1.31.—The mass number, atomic number, number of atoms, and ionic charge of an element may be indicated by means of four indices placed around the symbol. The positions are to be occupied thus

| left upper index | right lower index | mass number | number of atoms |
| left lower index | right upper index | atomic number | ionic charge |

Ionic charge should be indicated by A^{n+} rather than by A^{+n}.

Example: $^{32}_{16}S_2^{2+}$ represents a doubly ionized molecule containing two atoms of sulfur, each of which has the atomic number 16 and mass number 32.

The following is an example of an equation for a nuclear reaction

$$^{26}_{12}Mg + ^4_2He = ^{29}_{13}Al + ^1_1H$$

◆Although the practice of American chemists and physicists in general has been to put the mass number at the upper right of the symbol, the committees recognize the advantage of putting it at the upper left so that the upper right is available for the ionic charge as needed.

1.32.—Isotopically labeled compounds may be described by adding to the name of the compounds the symbol of the isotope in parentheses.

Examples:

$^{32}PCl_3$ phosphorus(^{32}P) trichloride (spoken: phosphorus-32 trichloride)

$H^{36}Cl$ hydrogen chloride ($^{35}Cl^d$) (spoken: hydrogen chloride-36)

$^{15}NH_3$ ammonia(^{15}N) (spoken: ammonia nitrogen-15)

The position of the labeled atom may be indicated by placing the isotope symbol immediately after the locant (name of the group concerned).

Example: $^2H_2{}^{35}SO_4$ sulfuric(^{35}S) acid(^2H)

If this method gives names which are ambiguous or difficult to pronounce, the whole group containing the labeled atom may be indicated.

Examples:

$HOSO_2{}^{35}SH$	thiosulfuric(^{35}SH) acid
$^{15}NO_2NH_2$	nitramide($^{15}NO_2$), not nitr(^{15}N)amide
$NO_2{}^{15}NH_2$	nitramide($^{15}NH_2$)
$HO_3S^{18}O{-}^{18}OSO_3H$	peroxo($^{18}O_2$)disulfuric acid

1.4. Allotropes

If systematic names for gaseous and liquid modifications are required, they should be based on the size of the molecule, which can be indicated by Greek numerical prefixes (listed in 2.251). If the number of atoms is large and unknown, the prefix poly may be used. To indicate ring and chain structures the prefixes cyclo and catena may be used.

Examples:

Symbol	Trivial Name	Systematic Name
H	atomic hydrogen	monohydrogen
O_2	(common) oxygen	dioxygen
O_3	ozone	trioxygen
P_4	white phosphorus (yellow phosphorus)	tetraphosphorus
S_8	λ-sulfur	cycloöctasulfur or octasulfur
S_n	μ-sulfur	catenapolysulfur or polysulfur

For the nomenclature of solid allotropic forms the rules in Section 8 may be applied.

◆The use of prefixes to indicate ring and chain structures is favored by the committees, but it should be pointed out that ino (not catena) has been used by mineralogists for indicating chain structures in silicates, along with cyclo and other prefixes (neso, phyllo, tecto, soro) denoting structure, all used without italics or hyphens (e.g., inosilicates). Cf. use of catena in 7.42 for chains of alternating, not self-linking, atoms.

NOMENCLATURE OF INORGANIC CHEMISTRY (Continued)

2. FORMULAS AND NAMES OF COMPOUNDS IN GENERAL

Many chemical compounds are essentially binary in nature and can be regarded as combinations of ions or radicals; others may be treated as such for the purpose of nomenclature.

Some chemists have expressed the opinion that the name of a compound should indicate whether it is ionic or covalent. Such a distinction is made in some languages (*e.g.*, in German: Natriumchlorid *but* Chlorwasserstoff), but it has not been made consistently, and indeed it seems impossible to introduce this distinction into a consistent system of nomenclature, because the line of demarcation between these two categories is not sharp. In these rules a system of nomenclature has been built on the basis of the endings -ide and -ate, and it should be emphasized that these are intended to be applied both to ionic and covalent compounds. If it is desired to avoid such endings for neutral molecules, names can be given as coördination compounds in accordance with **2.24** and Section **7**.

2.1. Formulas

2.11.—Formulas provide the simplest and clearest method of designating inorganic compounds. They are of particular importance in chemical equations and in descriptions of chemical procedure. However, their general use in text is not recommended, although in some cases a formula, on account of its compactness, may be preferable to a cumbersome and awkward name.

2.12.—The *empirical formula* is formed by juxtaposition of the atomic symbols to give the *simplest possible* formula expressing the stoichiometric composition of the compound in question. The empirical formula may be supplemented by indication of the crystal structure—see Section **8**.

2.13.—For compounds consisting of discrete molecules the *molecular formula*, *i.e.*, a formula corresponding with the correct molecular weight of the compound, should be used, *e.g.*, S_2Cl_2 and $H_4P_2O_6$ and not SCl and H_2PO_3. When the molecular weight varies with temperature, *etc.*, the simplest possible formula generally may be chosen, *e.g.*, S, P, and NO_2 instead of S_8, P_4, and N_2O_4, unless it is desirable to indicate the molecular complexity.

2.14.—In the *structural formula* the sequence and spatial arrangement of the atoms in a molecule are indicated.

2.15.—In formulas the *electropositive constituent* (cation) should always be placed first, *e.g.*, KCl, $CaSO_4$.

This also applies in Romance languages even though the electropositive constituent is placed last in the *name*, *e.g.*, KCl, chlorure de potassium.

If the compound contains more than one electropositive or more than one electronegative constituent, their sequence is determined by Rules **6.32** and **6.33**.

2.16.—In the case of binary compounds between nonmetals that constituent should be placed first which appears earlier in the sequence: B, Si, C, Sb, As, P, N, H, Te, Se, S, At, I, Br, Cl, O, F.

Examples: NH_3, H_2S, N_4S_4, S_2Cl_2, Cl_2O, OF_2

◆Because N_4S_4 is definitely a nitride, not a sulfide, the committees prefer to write the formula S_4N_4, name it sulfur nitride, and cite it as an exception rather than as an example.

2.161.—For compounds containing three or more elements, however, the sequence should in general follow the order in which the atoms are actually bound in the molecule or ion, *e.g.*, NCS^-, not CNS^-, HOCN (cyanic acid), and HONC (fulminic acid).

Although formulas such as HNO_3, $HClO_4$, H_2SO_4, do not agree with this rule and HNO_3 does not even follow the main rule in **2.16**, the Commission does not at this time wish to break the old custom of putting the central atom immediately after the hydrogen atom in such cases (*cf* Section **5**). The formula for hypochlorous acid may be written HOCl or HClO.

2.17.—In intermetallic compounds the constituents should be placed in the order

Fr, Cs, Rb, K, Na, Li	Pt, Ir, Os, Pd, Rh, Ru, Ni, Co, Fe
Ra, Ba, Sr, Ca, Mg, Be	Au, Ag, Cu
103, No, Md, Fm, Es, Cf, Bk, Cm, Am, Pu,	Hg, Cd, Zn
Np, U, Pa, Th, Ac, Lu–La, Y, Sc	Tl, In, Ga, Al
Hf, Zr, Ti	Pb, Sn, Ge
Ta, Nb, V	Bi, Sb
W, Mo, Cr	Po
Re, Tc, Mn	

Nonmetals (except Sb) in the order given in **2.16**.

Deviations from this order may be allowed, *e.g.*, when compounds with analogous structures are compared (AgZn and AgMg).

2.18.—The number of identical atoms or atomic groups in a formula is indicated by means of Arabic numerals, placed below and to the right of the symbol or symbols in parentheses ( ) or brackets [ ] to which they refer. Water of crystallization and similar loosely bound molecules, however, are designated by means of Arabic numerals before their formulas.

Examples: $CaCl_2$ not $CaCl^2$
$[Co(NH_3)_6]Cl_3$ not $[Co6NH_3]Cl_3$
$[Co(NH_3)_6]_2(SO_4)_3$
$Na_2SO_4.10H_2O$

2.19.—The prefixes *cis, trans, sym, asym* may be used in their usual senses. The prefixes may be connected with the formula by a hyphen and it is recommended that they be italicized.

Example: *cis*-$[PtCl_2(NH_3)_2]$

2.2. Systematic Names

Systematic names of compounds are formed by indicating the constituents and their proportions according to the following rules. (For the order of the constituents see also the later sections.)

2.21.—The name of the *electropositive constituent* (or that treated as such according to **2.16**) will not be modified (see, however, **2.2531**).

In Germanic languages the electropositive constituent is placed first, but in Romance languages it is customary to place the electronegative constituent first.

2.22.—If the *electronegative constituent* is monatomic its name is modified to end in -ide. For binary compounds of the nonmetals the name of the element standing later in the sequence in **2.16** is modified to end in -ide.

Examples: Sodium chloride, calcium sulfide, lithium nitride, arsenic selenide, calcium phosphides, nickel arsenide, aluminum borides, iron carbides, boron hydrides, phosphorus hydrides, hydrogen chloride, hydrogen sulfide, silicon carbide, carbon disulfide, sulfur hexafluoride, chlorine dioxide, oxygen difluoride.

Certain polyatomic groups are also given the ending -ide—see **3.22**.

In the Romance languages the endings -ure, -uro, and -eto are used instead of -ide. In some languages the word *oxyde* is used, whereas the ending -ide is used in the names of other binary compounds; it is recommended that the ending -ide be universally adopted in these languages.

◆Nitrogen sulfide has been taken out of the examples. *Cf.* comment at 2.16.

2.23.—If the electronegative constituent is polyatomic it should be designated by the termination -ate. In certain exceptional cases the terminations -ide and -ite are used—see **3.22**.

2.24.—In inorganic compounds it is generally possible in a polyatomic group to indicate a *characteristic atom* (as in ClO^-) or a *central atom* (as in ICl_4^-). Such a polyatomic group is designated a *complex*, and the atoms, radicals, or molecules bound to the characteristic or central atom are termed *ligands*.

In this case the name of a negatively charged complex should be formed from the name of the characteristic or central element (as indicated in **1.12**) modified to end in -ate.

Anionic ligands are indicated by the termination -o. Further details concerning the designation of ligands, the definition of "central atom," *etc.*, appear in Section 7.

Although the terms sulfate, phosphate, *etc.*, were originally the names of the anions of particular oxo acids, the names sulfate, phosphate, *etc.*, should now designate quite generally a negative group containing sulfur or phosphorus, respectively, as the central atom, irrespective of its oxidation state (the designation of the oxidation state is discussed in later rules) and the number and nature of the ligands. The complex is indicated by brackets [ ], but this is not always necessary.

Examples:

$Na_2[SO_4]$	sodium tetraoxosulfate	$Na_3[PS_4]$	sodium tetrathiophosphate
$Na_2[SO_3]$	sodium trioxosulfate	$Na[PCl_6]$	sodium hexachlorophosphate
$Na_2[S_2O_3]$	sodium trioxothiosulfate	$K[PO_2F_2]$	potassium dioxodifluorophosphate
$Na[SO_3F]$	sodium trioxofluorosulfate	$K[POCl_2(NH)]$	potassium oxodichloroimidophosphate
$Na_3[PO_4]$	sodium tetraoxophosphate		

In many cases these names may be abbreviated, *e.g.*, sodium sulfate, sodium thiosulfate (see **2.26**), and in other cases trivial names may be used (*cf.* **2.3**, **3.224**, and Section 5). It should be pointed out, however, that the principle is quite generally applicable, to compounds containing organic ligands also, and its use is recommended in all cases where trivial names do not exist.

The coördination principle applied in this rule may also be applied to complexes which are positive or neutral (*cf.* **3.1** and Section 7). However, neutral complexes which are as a rule considered as binary compounds are given names according to **2.22**, **2.16**. Thus, SO_3, sulfur trioxide, not trioxosulfur.

◆In the examples it would seem that full coördination-type names should be given, *e.g.*, either sodium tetraoxosulfate (VI) or disodium tetraoxosulfate. *Cf.* 7.32 and comment at 7.312.

2.25.—Indication of the Proportions of the Constituents.

2.251.—The *stoichiometric proportions* may be denoted by means of Greek numerical prefixes (mono, di, tri, tetra, penta, hexa, hepta, octa, ennea, deca, hendeca, and dodeca) preceding without hyphen the names of the elements to which they refer. It may be necessary in some languages to supplement these numerals with hemi ($1/2$) and the Latin sesqui ($3/2$).

The prefix mono may generally be omitted. Beyond 12, Greek prefixes are replaced by Arabic numerals (with or without hyphen according to the custom of the language), because they are more readily understood.

This sytem is applicable to all types of compounds and is especially suitable for binary compounds of the nonmetals.

When it is required to indicate the number of entire groups of atoms, particularly when the name includes a numerical prefix with a different significance, the multiplicative numerals (Latin bis, Greek tris, tetrakis, *etc.*) are used and the whole group to which they refer may be placed in parentheses if necessary.

Examples:

N_2O	dinitrogen oxide	Fe_3O_4	triiron tetraoxide
NO_2	nitrogen dioxide	U_3O_8	triuranium octaoxide
N_2O_4	dinitrogen tetraoxide	MnO_2	manganese dioxide
N_2S_5	dinitrogen pentasulfide	$Ca_3[PO_4]_2$	tricalcium diorthophosphate
S_2Cl_2	disulfur dichloride	$Ca[PCl_6]_2$	calcium bis(hexachlorophosphate)

In indexes it may be convenient to italicize a numerical prefix at the beginning of the name and connect it to the rest of the name with a hyphen, but this is not desirable in text, e.g., tri-Uranium octaoxide.

Since the degree of polymerization of many substances varies with temperature, state of aggregation, etc., the name to be used should normally be based upon the simplest possible formula of the substance except when it is required specifically to draw attention to the degree of polymerization.

Example: The name nitrogen dioxide may be used for the equilibrium mixture of NO_2 and N_2O_4. Dinitrogen tetraoxide means specifically N_2O_4.

◆In accordance with the organic nomenclature rules and well-established practice, it is recommended that "or Latin" be inserted between "Greek" and "numerical prefixes" in the first sentence and that "nona" replace "ennea," and "undeca" replace "hendeca."

Extreme caution is advised in the omission of numerical prefixes, including mono (cf. second set of examples in 5.23), because of the frequent use of names such as chloroplatinate (cf. 2.26 and the last sentence in 5.24).

2.252.—The proportions of the constituents also may be indicated indirectly by *Stock's system*, that is, by Roman numerals representing the oxidation number or stoichiometric valence of the element, placed in parentheses immediately following the name. For zero the Arabic 0 will be used. When used in conjunction with symbols the Roman numeral may be placed above and to the right.

The Stock notation can be applied to both cations and anions, but preferably should *not* be applied to compounds between nonmetals.

In employing the Stock notation, use of the Latin names of the elements (or Latin roots) is considered advantageous.

Examples:

$FeCl_2$	iron(II) chloride or ferrum(II) chloride
$FeCl_3$	iron(III) chloride or ferrum(III) chloride
MnO_2	manganese(IV) oxide
BaO_2	barium(II) peroxide
$Pb^{II}_2Pb^{IV}O_4$	dilead(II) lead(IV) oxide or trilead tetraoxide
$K_4[Ni(CN)_4]$	potassium tetracyanoniccollate (0)
$K_4[Fe(CN)_6]$	potassium hexacyanoferrate(II)
$Na_2[Fe(CO)_4]$	sodium tetracarbonylferrate(—II)

◆While the committees favor the extended use of the Stock notation, they suggest that in some cases the system of Ewens and Bassett (designation of the aggregate charge of a complex ion by an Arabic numeral in parentheses following the name, similar to the use as superior notations with formulas) is advantageous and should be allowed as an alternate (cf. 3.17 and comment at 7.323). Mixed use of the two systems, while not desirable in any one context, does not affect indexing and should not lead to confusion. See comment at 1.12.

2.253.—The following systems are in use but are not recommended:

2.2531.—The system of indicating valence by means of the suffixes *-ous* and *-ic* added to the root of the name of the cation may be retained for elements exhibiting not more than two valences.

2.2532.—"*Functional*" nomenclature (such as "nitric anhydride" for N_2O_5) is not recommended apart from the name *acid* to designate the acid function (Section 5).

◆Apparently there is no objection to acid anhydride as a class name (cf. 5.32). Other functional derivatives of acids are named as such in the Rules (5.3).

2.26.—In systematic names it is not always necessary to indicate stoichiometric proportions. In many instances it is permissible to omit the numbers of atoms, oxidation numbers, etc., when they are not required in the particular circumstances. For instance, these indications are not generally necessary with elements of essentially constant valence.

Examples:

> sodium sulfate instead of sodium tetraoxosulfate
> aluminum sulfate instead of aluminum(III) sulfate
> potassium chloroplatinate(IV) instead of potassium hexachloroplatinate(IV)
> potassium cyanoferrate(III) instead of potassium hexacyanoferrate(III)
> phosphorus pentaoxide instead of diphosphorus pentaoxide

2.3. Trivial Names

Certain well-established trivial names for oxo acids (Section 5) and for hydrogen compounds (water, ammonia, hydrazine) are still acceptable. For some other hydrogen compounds these names are approved:

B_2H_6	diboran	PH_3	phosphine	SbH_3	stibine	P_2H_4	diphosphine
SiH_4	silane	AsH_3	arsine	Si_2H_6	disilane, *etc.*	As_2H_4	diarsine

In some languages names of the type "Chlorwasserstoff" are in use and may be retained if national nomenclature committees so wish.

Purely trivial names, free from false scientific implications, such as soda, Chile saltpeter, quicklime, are harmless in industrial and popular literature; but old incorrect scientific names such as sulfate of magnesia, Natronhydrat, sodium muriate, carbonate of lime, should be avoided under all circumstances, and they should be eliminated from technical and patent literature.

◆For BH_3 (omitted in the Rules) borane rather than the previously used borine has been recommended by the Advisory Committee on the Nomenclature of Organic Boron Compounds of the ACS in a report not yet published.

Because soda is an ambiguous term, it is suggested that it be replaced by soda ash.

3. NAMES FOR IONS AND RADICALS

3.1. Cations

3.11.—Monatomic cations should be named like the corresponding element, without change or suffix, except as provided by **2.2531**.

Examples:

Cu^+ the copper(I) ion Cu^{2+} the copper(II) ion I^+ the iodine cation

◆For I^+, iodide(I) cation is more consistent with recommended practice. *Cf.* 3.21.

3.12.—The preceding principle should apply also to polyatomic cations corresponding to radicals for which special names are given in **3.32**, *i.e.*, these names should be used without change or suffix.

Examples: NO^+ the nitrosyl cation NO_2^+ the nitryl cation

◆Polyatomic here and in 3.13, 3.14, 3.223, 3.32, and 5.2 seems to be limited to more than one *kind* of atom, and hence heteratomic would be a more precise term. It is agreed that nitryl and not nitronium should be used in all cases (*cf.* 3.151).

3.13.—Polyatomic cations formed from monatomic cations by the addition of other ions or neutral atoms or molecules (ligands) will be regarded as complex and will be named according to the rules given in Section **7**.

Examples:

$[Al(H_2O)_6]^{3+}$ the hexaaquoaluminum ion $[CoCl(NH_3)_5]^{2+}$ the chloropentamminecobalt ion

For some important polyatomic cations which fall in this section, radical names given in **3.32** may be used alternatively, *e.g.*, for UO_2^{2+} the name uranyl(VI) ion in place of dioxouranium(VI) ion.

3.14.—Names for polyatomic cations derived by addition of protons to monatomic anions are formed by adding the ending -onium to the root of the name of the anion element.

Examples: phosphonium, arsonium, stibonium, oxonium, sulfonium, selenonium, telluronium, and iodonium ions.

Organic ions derived by substitution in these parent cations should be named as such, whether the parent itself is a known compound or not: for example $(CH_3)_4Sb^+$, the tetramethylstibonium ion.

The ion H_3O^+, which is in fact the monohydrated proton, is to be known as the oxonium ion when it is believed to have this constitution, as for example in $H_3O^+ClO_4^-$, oxonium perchlorate. The widely used term hydronium should be kept for the cases where it is wished to denote an indefinite degree of hydration of the proton, as, for example, in aqueous solution. If, however, the hydration is of no particular importance to the matter under consideration, the simpler term hydrogen ion may be used. The latter also may be used for the indefinitely solvated proton in nonaqueous solvents; but definite ions such as $CH_3OH_2^+$ and $(CH_3)_2OH^+$ should be named as derivatives of the oxonium ion, *i.e.*, as methyl- and dimethyloxonium ions, respectively.

◆The committees concur in oxonium for the ion H_3O^+, but see little reason for encouraging retention of the term hydronium ion because hydrogen ion adequately designates an indeterminate degree of hydration.

3.15.—Ions from Nitrogen Bases.

3.151.—The name ammonium for the ion NH_4^+ does not conform to **3.14**, but should be retained. This decision does *not* release the word nitronium for other uses: this would lead to inconsistencies when the rules are applied to other elements.

3.152.—Substituted ammonium ions derived from nitrogen bases with names ending in -amine will receive names formed by changing -amine to -ammonium. For example, $HONH_3^+$, the hydroxylammonium ion.

3.153.—When the nitrogen base is known by a name ending otherwise than in -amine, the cation name is to be formed by adding the ending -ium to the name of the base (if necessary omitting a final -e or other vowel).

Examples: hydrazinium, anilinium, glycinium, pyridinium, guanidinium, imidazolium.

The names uronium and thiouronium, though inconsistent with this rule, may be retained.

3.16.—Cations formed by adding protons to nonnitrogenous bases may also be given names formed by adding -ium to the name of the compound to which the proton is added.

Examples: dioxanium, acetonium.

In the case of cations formed by adding protons to acids, however, their names are to be formed by adding the word acidium to the name of the corresponding anion, and not that of the acid itself. For example, $H_2NO_3^+$, the nitrate acidium ion; $H_2NO_2^+$, the nitrite acidium ion; and $CH_3COOH_2^+$, the acetate acidium ion. Note, however, that when the anion of the acid is monatomic **3.14** will apply; for example, FH_2^+ is the fluoronium ion.

◆In accord with present practice, nitric acidium ion, *etc.*, are preferred to nitrate acidium ion, *etc.* $CH_3COOH_2^+$ is organic.

3.17.—Where more than one ion is derived from one base, as, for example, $N_2H_5^+$ and $N_2H_6^{2+}$, their charges may be indicated in their names as the hydrazinium(1+) and the hydrazinium(2+) ion, respectively.

◆*Cf.* comment at 2.252 and the use of Stock notation with ions or radicals in 3.13 and 3.32.

3.2. Anions

3.21.—The names for monatomic anions shall consist of the name (sometimes abbreviated) of the element, with the termination -ide. Thus

H^-	hydride ion	Br^-	bromide ion	Se^{2-}	selenide ion	As^{3-}	arsenide ion
D^-	deuteride ion	I^-	iodide ion	Te^{2-}	telluride ion	Sb^{3-}	antimonide ion
F^-	fluoride ion	O^{2-}	oxide ion	N^{3-}	nitride ion	C^{4-}	carbide ion
Cl^-	chloride ion	S^{2-}	sulfide ion	P^{3-}	phosphide ion	Si^{4-}	silicide ion
		B^{3-}	boride ion				

Expressions of the type "chlorine ion" are used particularly in connection with crystal structure work and spectroscopy; the Commission recommends that whenever the charge corresponds with that indicated above, the termination -ide should be used.

◆*Cf.* comments at 1.15 and 1.32 regarding ^2H and D.

3.22. Polyatomic Anions.

3.221.—Certain polyatomic anions have names ending in -ide. These are

OH^-	hydroxide ion	I_3^-	triiodide ion	$NHOH^-$	hydroxylamide ion
O_2^{2-}	peroxide ion	HF_2^-	hydrogen difluoride ion	$N_2H_3^-$	hydrazide ion
O_2^-	hyperoxide ion	N_3^-	azide ion	CN^-	cyanide ion
O_3^-	ozonide ion	NH^{2-}	imide ion	C_2^{2-}	acetylide ion
S_2^{2-}	disulfide ion	NH_2^-	amide ion		

Names for other polysulfide, polyhalide, *etc.*, ions may be formed in analogous manner. The OH^- ion should not be called the hydroxyl ion. The name hydroxyl is reserved for the OH group when neutral or positively charged, whether free or as a substituent (*cf.* **3.12** and **3.32**).

◆Superoxide is well established in English for O_2^- and no advantage is seen in changing to hyperoxide.

3.222.—Ions such as SH^- and O_2H^- will be called the hydrogen sulfide ion and the hydrogen peroxide ion, respectively. This agrees with **6.2**, and names such as hydrosulfide are not needed.

◆*Cf.* comment at 6.2. All "fused" names (as hydrogensulfide here and methylisocyanide in 5.33) in the original version have been written as two words in this version. For a rule on the written form of the names of compounds see *J. Chem. Education*, 8, 1336–8 (1931)

3.223.—The names for other polyatomic anions shall consist of the name of the central atom with the termination -ate, which is used quite generally for complex anions. Atoms and groups attached to the central atom shall generally be treated as ligands in a complex (*cf.* **2.24** and Section 7) as, for example, $[Sb(OH)_6]^-$, the hexahydroxoantimonate(V) ion.

This applies also when the exact composition of the anion is not known; *e.g.*, by solution of aluminum hydroxide or zinc hydroxide in sodium hydroxide, aluminate and zincate ions are formed.

3.224.—It is quite practicable to treat oxygen in the same manner as other ligands (**2.24**), but it has long been customary to ignore the name of this element altogether in anions and to indicate its presence and proportion by means of a series of prefixes (hypo-, per-, *etc.*, see Section 5) and sometimes also by the suffix -ite in place of -ate.

The termination -ite has been used to denote a lower oxidation state and may be retained in trivial names in these cases

NO_2^-	nitrite	$PH_2O_2^-$	hypophosphite	$S_2O_2^{2-}$	thiosulfite
$N_2O_2^{2-}$	hyponitrite	AsO_3^{3-}	arsenite	SeO_3^{2-}	selenite
NOO_2^-	peroxonitrite	SO_3^{2-}	sulfite	ClO_2^-	chlorite
PHO_3^{2-}	phosphite	$S_2O_5^{2-}$	disulfite (pyrosulfite)	ClO^-	hypochlorite
$P_2H_2O_5^{2-}$	diphosphite (pyrophosphite)	$S_2O_4^{2-}$	dithionite		

(and correspondingly for the other halogens)

The Commission does not recommend the use of any such names other than those listed. A number of other names ending in -ite have been used, *e.g.*, antimonite, tellurite, stannite, plumbite, ferrite, manganite, but in many cases such compounds are known in the solid state to be double oxides and are to be treated as such (*cf.* **6.5**), *e.g.*, $Cu(CrO_2)_2$ copper(II) chromium(III) oxide, not copper chromite. Where there is reason to believe that a definite salt with a discrete anion exists, the name is formed in accordance with **2.24**. By dissolving, for example, Sb_2O_3, SnO, or PbO in sodium hydroxide an antimonate(III), a stannate(II), or a plumbate(II) is formed in the solution.

Concerning the use of prefixes hypo-, per-, *etc.*, see the list of acids (table in **5.214**). For all new compounds and even for the less common ones listed in the table in **3.224** or derived from the acids listed in the table in **5.214**, it is preferable to use the system given in **2.24** and in Sections 5 and 7.

◆For phosphite names see comment at 6.2.

3.3. Radicals

3.31.—A radical is here regarded as a group of atoms which occurs repeatedly in a number of different compounds. Sometimes the same radical fulfils different functions in different cases, and accordingly different names often have been assigned to the same group. The Commission considers it desirable to reduce this diversity and recommends that formulas or systematic names be used to denote all new radicals, instead of introducing new trivial names. The list of names for ions and radicals on page B-169 gives an extensive selection of radical names at present in use in inorganic chemistry.

◆The list of names for ions and radicals following the Rules is very useful and can be made more so by additions (as of dithio, nitrilo, and azido in the last column) and by greater attempt at uniformity with organic usage. Some of the terms (as nitride and amide) listed as anions are used also of covalent compounds. A single atom may function like a radical as defined above and may be named similarly, as chloro and oxo.

3.32.—Certain radicals containing oxygen or other chalcogens have special names ending in -yl, and the Commission approves the provisional retention of

HO	hydroxyl	S_2O_5	pyrosulfuryl	PO	phosphoryl	ClO_2	chloryl
CO	carbonyl	SeO	seleninyl	VO	vanadyl	ClO_3	perchloryl
NO	nitrosyl	SeO_2	selenonyl	PuO_2	plutonyl		(and similarly
SO	sulfinyl	CrO_2	chromyl		(similarly for		for other
	(thionyl)	UO_2	uranyl		other actinide		halogens)
SO_2	sulfonyl	NpO_2	neptunyl		elements)		
	(sulfuryl)	NO_2	nitryl[1]	ClO	chlorosyl		

[1]The name nitroxyl should not be used for this group since the name nitroxylic acid has been used for H_2NO_2. Although the word nitryl is firmly established in English, nitroyl may be a better model for many other languages.

Names such as the above should be used only to designate compounds containing these discrete groups. The use of thionyl and sulfuryl should be restricted to the halides. Names such as bismuthyl and antimonyl are not approved because the compounds do not contain BiO and SbO groups, respectively; such compounds are to be designated as oxide halides (**6.4**).

Radicals analogous to the above containing other chalcogens in place of oxygen are named by adding the prefixes thio-, seleno-, *etc.*

Examples:

PS	thiophosphoryl	CSe selenocarbonyl

In cases where radicals may have different valences, the oxidation number of the characteristic element should be indicated by means of the Stock notation. For example, the uranyl group UO_2 may refer either to the ion UO_2^{2+} or to the ion UO_2^+; these can be distinguished as uranyl(VI) and uranyl(V), respectively. In like manner, VO may be vanadyl(V), vanadyl(IV), and vanadyl(III).

These polyatomic radicals always are treated as forming the positive part of the compound.

Examples:

$COCl_2$	carbonyl chloride	$NO_2HS_2O_7$	nitryl hydrogen disulfate
NOS	nitrosyl sulfide	S_2O_5ClF	pyrosulfuryl chloride fluoride
PON	phosphoryl nitride	$SO_2(N_3)_2$	sulfonyl azide
$PSCl_3$	thiophosphoryl chloride	SO_2NH	sulfonyl imide
POCl	phosphoryl(III) chloride	IO_2F	iodyl fluoride

By using the same radical names regardless of unknown or controversial polarity relationships, names can be formed without entering into any controversy. Thus, for example, the compounds NOCl and $NOClO_4$ are quite unambiguously denoted by the names nitrosyl chloride and nitrosyl perchlorate, respectively.

◆Caution is urged in the use of some of these radical names: Vanadyl, for example, has been used for VO_2 as well as for VO (*cf.* also the naming of $VOSO_4$ in 6.42). Most of these radical names (except hydroxyl and thionyl) can be regarded as derived from the names of acids which have lost all of their hydroxyls (analogous to -yl or -oyl organic acid radical names) by the use of -yl and -osyl for radicals from -ic and -ous acids, respectively; this is implied in the footnote above on nitroxyl. The use of the Stock notation in only one example (phosphoryl(III) chloride) seems confusing. It might be clearer to indicate stoichiometric proportions, *e.g.*, phosphoryl (mono)chloride, thiophosphoryl trichloride (*cf.* phosphoryl triamide in 5.34) or to use phosphorosyl for phosphoryl(III).

The restriction of the use of thionyl and sulfuryl to the halides was agreed upon at a joint meeting of the inorganic and organic nomenclature commissions of the IUPAC in 1951.

3.33.—It should be noted that the same radical may have different names in inorganic and organic chemistry. To draw attention to such differences the prefix names of radicals as substituents in organic compounds have been listed together with the inorganic names in the list of names printed at the end of the Rules. Names of purely organic compounds, of which many are important in the chemistry of coördination compounds (Section 7), should agree with the nomenclature of organic chemistry.

Organic chemical nomenclature is to a large extent based on the principle of substitution, *i.e.*, replacement of hydrogen atoms by other atoms or groups. Such "substitutive names" are extremely rare in inorganic chemistry; they are used, *e.g.*, in the following cases: NH_2Cl is called chloramine, and $NHCl_2$ dichloramine. These names may be retained in the absence of better terms. Other substitutive names (derived from "sulfonic acid" as a name for HSO_3H) are fluoro- and chlorosulfonic acid, aminosulfonic acid, iminodisulfonic acid, and nitrilotrisulfonic acid. These names should preferably be replaced by the following

FSO_3H	fluorosulfuric acid	$NH(SO_3H)_2$	imidodisulfuric acid
$ClSO_3H$	chlorosulfuric acid	$N(SO_3H)_3$	nitridotrisulfuric acid
NH_2SO_3H	amidosulfuric acid		

Names such as chlorosulfuric acid and amidosulfuric acid might be considered to be substitutive names derived by substitution of *hydroxyl* groups in sulfuric acid. From a more fundamental point of view, however (see **2.24**), such names are formed by adding hydroxyl, amide, imide, *etc.*, groups together with oxygen atoms to a sulfur atom, "sulfuric acid" in this connection standing as an abbreviation for "trioxosulfuric acid."

Another organic-chemical type of nomenclature, the formation of "conjunctive names," is also met in only a few cases in inorganic chemistry, *e.g.*, the hydrazine- and hydroxylaminesulfonic acids. According to the principles of inorganic chemical nomenclature these compounds should be called hydrazido- and hydroxylamidosulfuric acid.

◆These are not true "conjunctive names" since sulfonic acid is not a compound. For the naming of partial amides *cf.* also 5.34.

NOMENCLATURE OF INORGANIC CHEMISTRY (Continued)

4. CRYSTALLINE PHASES OF VARIABLE COMPOSITION

Isomorphous replacement, interstitial solutions, intermetallic compounds, and other nonstoichiometric compounds (berthollides)

4.1.—If an intermediate crystalline phase occurs in a two-component (or more complex) system, it may obey the law of constant composition very closely, as in the case of sodium chloride, or it may be capable of varying in composition over an appreciable range, as occurs for example with FeS. A substance showing such a variation is called a *berthollide*.

In connection with the berthollides the concept of a characteristic or ideal composition is frequently used. A unique definition of this concept seems to be lacking. In one case it may be necessary to use a definition based upon lattice geometry and in another to base it on the ratio of valence electrons to atoms. Sometimes one can state several characteristic compositions, and at other times it is impossible to say whether a phase corresponds to a characteristic composition or not.

In spite of these difficulties it seems that the concept of a characteristic composition can be used in its present undefined form for establishing a system of notation for phases of variable composition. It also seems possible to use the concept even if the characteristic composition is not included in the known homogeneity range of the phase.

4.2.—For the present, mainly formulas should be used for berthollides and solid solutions, since strictly logical names tend to become inconveniently cumbersome. The latter should be used only when unavoidable (*e.g.*, for indexing), and may be written in the style of iron(II) sulfide (iron-deficient); molybdenum dicarbide (excess carbon), or the like. Mineralogical names should be used only to designate actual minerals and not to define chemical composition; thus the name calcite refers to a particular mineral (contrasted with other minerals of similar composition) and is not a term for the chemical compound whose composition is properly expressed by the name calcium carbonate. (The mineral name may, however, be used to indicate the structure type—see **6.52.**)

4.3.—A general notation for the berthollides, which can be used even when the mechanism of the variation in composition is unknown, is to put the sign $\sim$ (read as *circa*) before the formula. (In special cases it may also be printed above the formula.)

$$\text{Examples:} \quad \sim\text{FeS}, \qquad \overset{\sim}{\text{CuZn}}$$

The direction of the deviation may be indicated when required:

$$\sim\text{FeS (iron-deficient)}; \qquad \sim\text{MoC}_2 \text{ (excess carbon)}$$

4.4.—For a phase where the variable composition is solely or partially caused by replacement, atoms or atomic groups which replace each other are separated by a comma and placed together between parentheses.

If possible the formula ought to be written so that the limits of the homogeneity range are represented when one or other of the two atoms or groups is lacking. For example the symbol (Ni,Cu) denotes the complete range from pure Ni to pure Cu; likewise K(Br,Cl) comprises the range from pure KBr to pure KCl. If only part of the homogeneity range is referred to, the major constituent should be placed first.

Substitution accompanied by the appearance of vacant positions (combination of substitutional and interstitial solution) receives an analogous notation. For example, $(\text{Li}_2,\text{Mg})\text{Cl}_2$ denotes the homogeneous phase from LiCl to MgCl$_2$ where the anion lattice structure remains the same but one vacant cation position appears for every substitution of 2Li$^+$ by Mg^{2+}.

The formula $(\text{Mg}_3,\text{Al}_2)\text{Al}_6\text{O}_{12}$ represents the homogeneous phase from the spinel MgAl$_2$O$_4$ ($= \text{Mg}_3\text{Al}_6\text{O}_{12}$) to the spinel form of Al$_2$O$_3$ ($= \text{Al}_2\text{Al}_6\text{O}_{12}$).

The solid solutions between CaF$_2$ and YF$_3$, where cation substitution is accompanied by interstitial addition of F$^-$, would be represented by the formula $(\text{Ca,YF})\text{F}_2$. It is important to note that this formula is based purely on considerations of composition, and it does not imply that YF^{2+} takes over the actual physical position of Ca^{2+}. On the same basis a notation for the plagioclases would be $(\text{NaSi,CaAl})\text{Si}_2\text{AlO}_8$.

4.5.—A still more complete notation, which should always be used in more complex cases, may be constructed by indicating in a formula the variables that define the composition. Thus, a phase involving simple substitution may be written $A_x B_{1-x}$.

$$\text{Examples:} \quad \text{Ni}_x\text{Cu}_{1-x} \text{ and } \text{KBr}_x\text{Cl}_{1-x}$$

This shows immediately that the total number of atoms in the lattice is constant. Combined substitutional and interstitial or subtractive solution can be shown in an analogous way. The commas and parentheses called for in **4.4** are not required in this case.

For example, the homogeneous phase between LiCl and MgCl$_2$ becomes $\text{Li}_{2x}\text{Mg}_{1-x}\text{Cl}_2$ and the phase between MgAl$_2$O$_4$ and Al$_2$O$_3$ can be written $\text{Mg}_{3x}\text{Al}_{2(1-x)}\text{Al}_6\text{O}_{12}$, which shows that it cannot contain more Mg than that corresponding to MgAl$_2$O$_4$ ($x = 1$). The other examples given in **4.4** will be given the formulas $\text{Ca}_x\text{Y}_{1-x}\text{F}_{3-x}$ and $\text{Na}_x\text{Ca}_{1-x}\text{Si}_{2+x}\text{Al}_{2-x}\text{O}_8$. In the case of the γ-phase of the Ag–Cd system, which has the characteristic formula Ag$_5$Cd$_8$, the Ag and Cd atoms can replace one another to some extent and the notation would be $\text{Ag}_{5\pm x}\text{Cd}_{8\mp x}$.

Further examples:

$$\text{Fe}_{1-x}\text{Sb} \quad \text{Fe}_{1-x}\text{O} \quad \text{Fe}_{1-x}\text{S} \quad \text{Cu}_{2-x}\text{O} \qquad \text{Na}_{1-x}\text{WO}_3 \text{ (sodium tungsten bronzes)}$$

For $x = 0$ each of these formulas corresponds to a characteristic composition. If it is desired to show that the variable denoted by x can attain only small values, this may be done by substituting ϵ for x.

Likewise a solid solution of hydrogen in palladium can be written as PdH$_x$, and a phase of the composition M which has dissolved a variable amount of water can be written M(H$_2$O)$_x$.

When this notation is used, a particular composition can be indicated by stating the actual value of the variable x. Probably the best way of doing this is to put the value in parentheses after the general formula. For example, $Li_{4-x}Fe_{3x}Ti_{2(1-x)}O_6$ ($x = 0.35$). If it is desired to introduce the value of x into the formula itself, the mechanism of solution is more clearly understood if one writes $Li_{4-0.35}Fe_{3\times0.35}Ti_{2(1-0.35)}O_6$ instead of $Li_{3.65}Fe_{1.05}Ti_{1.30}O_6$.

5. ACIDS

Many of the compounds which now according to some definitions are called acids do not fall into the classical province of acids. In other parts of inorganic chemistry functional names are disappearing and it would have been most satisfactory to abolish them also for those compounds generally called acids. Names for these acids may be derived from the names of the anions as in Section 2, e.g., hydrogen sulfate instead of sulfuric acid. The nomenclature of acids has, however, a long history of established custom, and it appears impossible to systematize acid names without drastic alteration of the accepted names of many important and well-known substances.

The present rules are aimed at preserving the more useful of the older names while attempting to guide further development along directions which should allow new compounds to be named in a more systematic manner.

5.1. Binary and Pseudobinary Acids

Acids giving rise to the -ide anions defined by **3.21** and **3.221** will be named as binary and pseudobinary compounds of hydrogen, e.g., hydrogen chloride, hydrogen sulfide, hydrogen cyanide.

For the compound HN_3 the name hydrogen azide is recommended in preference to hydrazoic acid.

5.2. Acids Derived from Polyatomic Anions

Acids giving rise to anions bearing names ending in -ate or in -ite may also be treated as in **5.1**, but names more in accordance with custom are formed by using the terminations -ic acid and -ous acid corresponding with the anion terminations -ate and -ite, respectively. Thus chloric acid corresponds to chlorate, sulfuric acid to sulfate, and phosphorous acid to phosphite.

This nomenclature may also be used for less common acids, e.g., hexacyanoferric acids correspond to hexacyanoferrate ions. In such cases, however, systematic names of the type hydrogen hexacyanoferrate are preferable.

Most of the common acids are oxo acids, i.e., they contain only oxygen atoms bound to the characteristic atom. It is a long established custom not to indicate these oxygen atoms. It is mainly for these acids that long-established names will have to be retained. Most other acids may be considered as coördination compounds and be named as such.

◆Polyatomic means heteroatomic here, as opposed to pseudobinary. *Cf.* comment at 3.12.

5.21. Oxo Acids.—For the oxo acids the ous–ic notation to distinguish between different oxidation states is applied in many cases. The -ous acid names are restricted to acids corresponding to the -ite anions listed in the table in **3.224**.

Further distinction between different acids with the same characteristic element is in some cases effected by means of prefixes. This notation should not be extended beyond the cases listed below.

5.211.—The prefix hypo- is used to denote a lower oxidation state, and may be retained in these cases

$H_4B_2O_4$	hypoboric acid	HPH_2O_2	hypophosphorous acid
$H_2N_2O_2$	hyponitrous acid	$HOCl$	hypochlorous acid (and similarly for the other halogens)
$H_4P_2O_6$	hypophosphoric acid		

5.212.—The prefix per- is used to designate a higher oxidation state and may be retained for $HClO_4$, perchloric acid, and similarly for the other elements in Group VII.

The prefix per- should not be confused with the prefix peroxo- (see **5.22**).

5.213.—The prefixes ortho- and meta- have been used to distinguish acids differing in "water content." These names are approved

H_3BO_3	orthoboric acid	H_5IO_6	orthoperiodic acid	$(H_2SiO_3)_n$	metasilicic acids
H_4SiO_4	orthosilicic acid	H_6TeO_6	orthotelluric acid	$(HPO_3)_n$	metaphosphoric acids
H_3PO_4	orthophosphoric acid	$(HBO_2)_n$	metaboric acids		

For the acids derived by removing water from orthoperiodic or orthotelluric acid, the systematic names should be used, e.g., HIO_4 tetraoxoiodic(VII) acid.

The prefix pyro- has been used to designate an acid formed from two molecules of an ortho acid minus one molecule of water. Such acids can now generally be regarded as the simplest cases of isopoly acids (*cf.* **7.5**). The prefix pyro- may be retained for pyrosulfurous and pyrosulfuric acids and for pyrophosphorous and pyrophosphoric acids, although in these cases also the prefix di- is preferable.

◆The use of orthoperiodic acid and orthotelluric acid is approved, but the question of names for HIO_4, etc., needs further study because of confusion in the literature.

5.214.—The accompanying Table II contains the accepted names of the oxo acids (whether known in the free state or not) and some of their thio and peroxo derivatives (**5.22** and **5.23**).

For the less common of these acids systematic names would seem preferable, for example

H_2MnO_4	manganic(VI) acid, to distinguish it from H_3MnO_4, manganic(V) acid
$HReO_4$	tetraoxorhenic(VII) acid, to distinguish it from H_3ReO_5, pentaoxorhenic(VII) acid
H_2ReO_4	tetraoxorhenic(VI) acid, to distinguish it from $HReO_3$, trioxorhenic(V) acid; H_3ReO_4, tetraoxorhenic(V) acid; and $H_4Re_2O_7$, heptaoxodirhenic(V) acid
H_2NO_2	dioxonitric(II) acid instead of nitroxylic acid.

Trivial names should not be given to such acids as HNO, $H_2N_2O_3$, $H_2N_2O_4$, of which salts have been described. These salts are to be designated systematically as oxonitrates(I), trioxodinitrates(II), tetraoxodinitrates(III), respectively.

The names germanic acid, stannic acid, antimonic acid, bismuthic acid, vanadic acid, niobic acid, tantalic acid, telluric acid, molybdic acid, wolframic acid, and uranic acid may be used for substances with indefinite "water content" and degree of polymerization.

◆Unless trivial names clash with good nomenclature practices or are ambiguous, retention of well-established ones or the use of formulas is urged (especially for HNO, $H_2N_2O_3$, etc.) until structures are known. Systematic coordination-type names in the case of these nitrogen acids, for example, imply a structure that is ruled out by our present state of knowledge.

If hexahydroxyantimonic acid is considered a trivial name, hexahydroxy- might be preferable (cf. the systematic name hexahydroxyantimonate(V) ion in 3.223). For the analogous use of peroxo and peroxy, see comment at 5.22.

TABLE II

NAMES FOR OXO ACIDS

H_3BO_3	orthoboric acid or (mono)boric acid	H_2SO_4	sulfuric acid
$(HBO_2)_n$	metaboric acids	$H_2S_2O_7$	disulfuric or pyrosulfuric acid
$(HBO_2)_3$	trimetaboric acid	H_2SO_5	peroxo(mono)sulfuric acid
$H_4B_2O_4$	hypoboric acid	$H_2S_2O_8$	peroxodisulfuric acid
H_2CO_3	carbonic acid	$H_2S_2O_3$	thiosulfuric acid
HOCN	cyanic acid	$H_2S_2O_6$	dithionic acid
HNCO	isocyanic acid	H_2SO_3	sulfurous acid
HONC	fulminic acid	$H_2S_2O_5$	disulfurous or pyrosulfurous acid
H_4SiO_4	orthosilicic acid	$H_2S_2O_2$	thiosulfurous acid
$(H_2SiO_3)_n$	metasilicic acids	$H_2S_2O_4$	dithionous acid
HNO_3	nitric acid	H_2SO_2	sulfoxylic acid
HNO_4	peroxonitric acid	$H_2S_xO_6$ ($x = 3,4...$)	polythionic acids
HNO_2	nitrous acid	H_2SeO_4	selenic acid
HOONO	peroxonitrous acid	H_2SeO_3	selenious acid
H_2NO_2	nitroxylic acid	H_6TeO_6	(ortho)telluric acid
$H_2N_2O_2$	hyponitrous acid	H_2CrO_4	chromic acid
H_3PO_4	(ortho)phosphoric acid	$H_2Cr_2O_7$	dichromic acid
$H_4P_2O_7$	diphosphoric or pyrophosphoric acid	$HClO_4$	perchloric acid
$H_5P_3O_{10}$	triphosphoric acid	$HClO_3$	chloric acid
$H_{n+2}P_nO_{3n+1}$	polyphosphoric acids	$HClO_2$	chlorous acid
$(HPO_3)_n$	metaphosphoric acids	HClO	hypochlorous acid
$(HPO_3)_3$	trimetaphosphoric acid	$HBrO_3$	bromic acid
$(HPO_3)_4$	tetrametaphosphoric acid	$HBrO_2$	bromous acid
H_3PO_5	peroxo(mono)phosphoric acid	HBrO	hypobromous acid
$H_4P_2O_8$	peroxodiphosphoric acid	H_5IO_6	(ortho)periodic acid
$(HO)_2OP-PO(OH)_2$	hypophosphoric acid	HIO_3	iodic acid
$(HO)_2P-O-PO(OH)_2$	diphosphoric(III,V) acid	HIO	hypoiodous acid
H_2PHO_3	phosphorous acid	$HMnO_4$	permanganic acid
$H_4P_2O_5$	diphosphorous or pyrophosphorous acid	H_2MnO_4	manganic acid
		$HTcO_4$	pertechnetic acid
HPH_2O_2	hypophosphorous acid	H_2TcO_4	technetic acid
H_3AsO_4	arsenic acid	$HReO_4$	perrhenic acid
H_3AsO_3	arsenious acid	H_2ReO_4	rhenic acid
$HSb(OH)_6$	hexahydroxyantimonic acid		

5.22. Peroxo Acids.—The prefix peroxo, when used in conjunction with the trivial names of acids, indicates substitution of –O– by –O–O– (cf. **7.312**).

Examples:
HNO_4 peroxonitric acid H_2SO_5 peroxosulfuric acid
H_3PO_5 peroxophosphoric acid $H_2S_2O_8$ peroxodisulfuric acid
$H_4P_2O_8$ peroxodiphosphoric acid

◆Peroxy, as recommended in the 1940 Rules (inorganic), is more acceptable than peroxo to organic chemists. It is not necessary that the use with trivial names conform with the use of peroxo denoting a coördinated ligand; e.g., peroxysulfuric acid or trioxoperoxosulfuric(VI) acid.

5.23. Thio Acids.—Acids derived from oxo acids by replacement of oxygen by sulfur are called *thio acids* (cf. **7.312**).

Examples: $H_2S_2O_2$ thiosulfurous acid $H_2S_2O_3$ thiosulfuric acid   HSCN thiocyanic acid

When more than one oxygen atom can be replaced by sulfur the number of sulfur atoms generally should be indicated

H_3PO_3S monothiophosphoric acid H_3AsS_3 trithioarsenious acid
$H_3PO_2S_2$ dithiophosphoric acid H_3AsS_4 tetrathioarsenic acid
H_2CS_3 trithiocarbonic acid

The prefixes seleno- and telluro- may be used in a similar manner.

5.24. Chloro Acids, etc.—Acids containing ligands other than oxygen and sulfur are generally designated according to the rules in Section 7.

Examples:

$HAuCl_4$	hydrogen tetrachloraurate(III) or tetrachloroauric(III) acid
H_2PtCl_4	hydrogen tetrachloroplatinate(II) or tetrachloroplatinic(II) acid
H_2PtCl_6	hydrogen hexachloroplatinate(IV) or hexachloroplatinic(IV) acid
$H_4Fe(CN)_6$	hydrogen hexacyanoferrate(II) or hexacyanoferric(II) acid
$H[PHO_2F]$	hydrogen hydridodioxofluorophosphate or hydridodioxofluorophosphoric acid
HPF_6	hydrogen hexafluorophosphate or hexafluorophosphoric acid
H_2SiF_6	hydrogen hexafluorosilicate or hexafluorosilicic acid
H_2SnCl_6	hydrogen hexachlorostannate(IV) or hexachlorostannic(IV) acid
HBF_4	hydrogen tetrafluoroborate or tetrafluoroboric acid
$H[B(OH)_2F_2]$	hydrogen dihydroxodifluoroborate or dihydroxodifluoroboric acid
$H[B(C_6H_5)_4]$	hydrogen tetraphenylborate or tetraphenylboric acid

It is preferable to use names of the type hydrogen tetrachloroaurate(III).

For some of the more important acids of this type abbreviated names may be used, *e.g.*, chloroplatinic acid, fluorosilicic acid.

◆For the use of hydrido in the fifth example see comment at 7.312.

5.3. Functional Derivatives of Acids

Functional derivatives of acids are compounds formed from acids by substitution of OH and sometimes also O by other groups. In this borderline between organic and inorganic chemistry organic-chemical nomenclature principles prevail.

◆The intention of the statement "organic-chemical nomenclature principles prevail" is not clear, since most of the examples given in the sections immediately following are not named according to organic practice. *Cf.* 3.33.

5.31. Acid Halides.—The names of acid halides are formed from the name of the corresponding acid radical if this has a special name, *e.g.*, sulfuryl chloride, phosphoryl chloride.

In other cases these compounds are named as oxide halides according to rule **6.41**, *e.g.*, MoO_2Cl_2, molybdenum dioxide dichloride.

5.32. Acid Anhydrides.—Anhydrides of inorganic acids generally should be given names as oxides, *e.g.*, N_2O_5 dinitrogen pentaoxide, *not* nitric anhydride or nitric acid anhydride.

5.33. Esters. Esters of inorganic acids are given names in the same way as the salts, *e.g.*, dimethyl sulfate, diethyl hydrogen phosphate.

If, however, it is desired to specify the constitution of the compound, a name based on the nomenclature for coördination compounds should be used.

Example:

$(CH_3)_4[Fe(CN)_6]$	tetramethyl hexacyanoferrate(II)
or	or
$[Fe(CN)_2(CH_3NC)_4]$	dicyanotetrakis(methyl isocyanide)iron(II)

◆According to common organic practice for esters (ethers, sulfides, *etc.*) and to the naming of inorganic salts (*e.g.*, sodium sulfate, not disodium sulfate), methyl sulfate would be used instead of dimethyl sulfate. However, no objection is seen to the more specific name. Such names as methyl sulfate are better for alphabetic listing, as in indexes.

5.34. Amides.—The names for amides may be derived from the names of acids by replacing acid by amide, or from the names of the acid radicals.

Examples:

$SO_2(NH_2)_2$	sulfuric diamide or sulfonyl diamide
$PO(NH_2)_3$	phosphoric triamide or phosphoryl triamide

If not all hydroxyl groups of the acid have been replaced by NH_2 groups, names ending in -amidic acid may be used: this is an alternative to naming the compounds as complexes.

Examples:

NH_2SO_3H	amidosulfuric acid or sulfamidic acid
$NH_2PO(OH)_2$	amidophosphoric acid or phosphoramidic acid
$(NH_2)_2PO(OH)$	diamidophosphoric acid or phosphorodiamidic acid

Abbreviated names (sulfamide, phosphamide, sulfamic acid) are often used but are not recommended.

◆The use of adjectives from names of inorganic acids may lead to confusion because an -ic or -ous adjective (as chromic) may refer to a higher- or lower-valent form of the element as well as to the acid (a possibility that does not arise, of course, with adjectives from names of organic acids).

Names of the type phosphoramidic acid are recommended in the report of the Advisory Committee on the Nomenclature of Organic Phosphorus Compounds of the Division of Organic Chemistry of the ACS published in 1952, but are not as acceptable to the inorganic nomenclature committees as a whole as the amido- or coördination-type names. *Cf.* 3.33.

The retention of sulfamic acid and sulfamide as trivial names is favored by the committees. An acceptable systematic name for NH_2SO_3H would be ammonia-sulfur trioxide, in keeping with its probable structure.

5.35. Nitriles.—The suffix -nitrile has been used in the names of a few inorganic compounds, *e.g.* $(PNCl_2)_3$, trimeric phosphonitrile chloride. According to **2.22** such compounds can be designated as nitrides, *e.g.*, phosphorus nitride dichloride. Accordingly there seems to be no reason for retention of the name nitrile (and nitrilo, *c.f.* 3.33) in inorganic chemistry.

◆Nitrilo is used in organic chemistry, though not given in the list at the end of the Rules.

6. SALTS AND SALT-LIKE COMPOUNDS

Among salts particularly there persist many old names which are bad and misleading, and the Commission wishes to emphasize that any which do not conform to these Rules should be discarded.

6.1. Simple Salts

Simple salts fall under the broad definition of binary compounds given in Section **2**, and their names are formed from those of the constituent ions (given in Section **3**) in the manner set out in Section **2**.

6.2. Salts Containing Acid Hydrogen ("Acid" Salts[1])

Names are formed by adding the word hydrogen, to denote the replaceable hydrogen present, immediately in front of the name of the anion.

The nonacidic hydrogen present, *e.g.*, in the phosphite ion, is included in the name of the anion and is not explicitly cited (*e.g.*, Na_2PHO_3, sodium phosphite).

Examples:

$NaHCO_3$	sodium hydrogen carbonate
NaH_2PO_4	sodium dihydrogen phosphate
$NaH[PHO_3]$	sodium hydrogen phosphite

◆The use of "fused" hydrogen names (as hydrogencarbonate in the original version) is not acceptable in English; the present practice of running hydrogen as a separate word is preferred and has been followed throughout this version. *Cf.* 3.222, 6.324, 6.333. If necessary for clarity, parentheses can be used, as in naming ligands, *e.g.*, (hydrogen carbonato). The use of hydro (as in hydrocarbonato) is not acceptable because of conflicts with organic usage, where hydro denotes addition of hydrogen to unsaturated compounds.

It seems safer to cite even nonacidic hydrogen present in an anion like PHO_3^{2-} (unless the ion has a specific name), because of current usage. (*Cf.* triethyl phosphite in 7.412, seventh example).

6.3. Double Salts, Triple Salts, etc.

6.31.—In formulas all the cations shall precede the anions; in names the principles embodied in Section 2 shall be applied. In those languages where cation names are placed after anion names the adjectives double, triple, *etc.* (their equivalents in the language concerned) may be added immediately after the anion name. The number so implied concerns the number of *kinds* of cation present and *not* the total number of such ions.

[1]For "basic" salts see 6.4.

6.32.—Cations.

6.321.—Cations shall be arranged in order of increasing valence (except hydrogen, *cf.* **6.2** and **6.324**).

6.322. The cations of each valence group shall be arranged in order of decreasing atomic number, with the polyatomic radical ions (*e.g.*, ammonium) at the end of their appropriate group.
◆Alphabetical order would be simpler here and even for 6.321.

6.323. Hydration of Cations.—Owing to the prevalence of hydrated cations, many of which are in reality complex, it seems unnecessary to disturb the cation order in order to allow for this; but if it is necessary to draw attention specifically to the presence of a particular hydrated cation this may be done by writing, for example, "hexaaquo" or "tetraaquo" before the name of the simple ion. Apart from this exception, however, all complex ions should be placed after simple ones in the appropriate valence group.

◆*Cf.* comment at 7.322.

6.324. Acidic Hydrogen.—When hydrogen is considered to be present as a cation its name shall be cited last among the cations. Actually acidic hydrogen will in most cases be bound to an anion and shall be cited together with this (**6.2**). If the salt contains only one anion, acidic hydrogen shall be cited in the same place whichever view is taken of the function of this hydrogen. Nonacidic hydrogen shall be either not explicitly cited (*cf.* **6.2**) or designated hydrido (*cf.* **5.24** and **7.311**). For salts with more than one anion see **6.333.**

Examples:

$KMgF_3$	potassium magnesium fluoride
$TlNa(NO_3)_2$	thallium(I) sodium nitrate or thallium sodium dinitrate
$KNaCO_3$	potassium sodium carbonate
$NH_4MgPO_4.6H_2O$	ammonium magnesium phosphate hexahydrate
$NaZn(UO_2)_3(C_2H_3O_2)_9.6H_2O$	sodium zinc triuranyl acetate hexahydrate
$Na[Zn(H_2O)_6](UO_2)_3(C_2H_3O_2)_9$	sodium hexaaquozinc triuranyl acetate
$NaNH_4HPO_4.4H_2O$	sodium ammonium hydrogen phosphate tetrahydrate

◆*Cf.* comments at 6.2, 7.312. In the fifth and sixth examples, either triuranyl(VI) or nonaacetate should be specified. *Cf.* 3.32, 6.34.

6.33.—Anions.

6.331.—Anions are to be cited in this group order

1. H^-
2. O_2^- and OH^- (in that order)
3. Simple (*i.e.*, one element only) inorganic anions, other than H^- and O^{2-}
4. Inorganic anions containing two or more elements, other than OH^-
5. Anions of organic acids and organic substances exerting an acid function

◆The committees consider it preferable to cite H^- last in accordance with usage.

6.332.—Within group 3 the ions shall be cited in the order given in **2.16**, the inclusion of O in that list being taken as referring to all oxygen anions apart from O^{2-} (*i.e.*, O_2^-, *etc.*).

Within group 4, anions containing the smallest number of atoms shall be cited first, and in the case of two ions containing the same number of atoms they shall be cited in order of decreasing atomic number of the central atoms. Thus CO_3^{2-} should precede CrO_4^{2-}, and the latter should precede SO_4^{2-}.

Within group 5 the anions shall be cited in alphabetical order.

◆Again alphabetical order would be simpler, within groups 3 and 4 as well as 5. *Cf.* comment at 7.251.

6.333.—Acidic hydrogen should be cited together with the anion to which it is attached. If it is not known to which anion the hydrogen is bound, it should be cited last among the cations.

◆*Cf.* comment at 6.2.

6.34.—The stoichiometric method is the most practicable for indicating the proportions of the constituents. It is not always essential to give the numbers of all the anions, provided the valences of all the cations are either known or indicated.

Examples:

$NaCl.NaF.2Na_2SO_4$ or $Na_6ClF(SO_4)_2$	(hexa)sodium chloride fluoride (bis)sulfate
$Ca_5F(PO_4)_3$	(penta)calcium fluoride (tris)phosphate

The parentheses in these cases mean that numerical prefixes may not be necessary. The multiplicative numerical prefixes bis, tris, *etc.*, should be used in connection with anions, because disulfate, triphosphate, *etc.*, designate isopoly anions.

6.4.   Oxide and Hydroxide Salts ("Basic" salts formerly oxy and hydroxy salts)

6.41.—For the purposes of nomenclature, these should be regarded as double salts containing O^{2-} and OH^- anions, and Section **6.3** may be applied in its entirety.

6.42.   Use of the Prefixes Oxy and Hydroxy.—In some languages the citation in full of all the separate anion names presents no trouble and is strongly recommended (*e.g.*, copper oxide chloride), to the exclusion of the oxy form wherever possible. In some other languages, however, such names as "oxyde et chlorure double de cuivre" are so far removed from current practice that the present system of using oxy- and hydroxy-, *e.g.*, oxychlorure de cuivre, may be retained in such cases.

Examples:

$Mg(OH)Cl$	magnesium hydroxide chloride
$BiOCl$	bismuth oxide chloride
$LaOF$	lanthanum oxide fluoride
$VOSO_4$	vanadium(IV) oxide sulfate
$CuCl_2.3Cu(OH)_2$ or $Cu_2(OH)_3Cl$	dicopper trihydroxide chloride
$ZrOCl_2.8H_2O$	zirconium oxide (di)chloride octahydrate

6.5.   Double Oxides and Hydroxides

The terms "mixed oxides" and "mixed hydroxides" are not recommended. Such substances preferably should be named double, triple, *etc.*, oxides or hydroxides as the case may be.

Many double oxides and hydroxides belong to several distinct groups, each having its own characteristic structure type, which is sometimes named after some well-known mineral of the same group (*e.g.*, perovskite, ilmenite, spinel). Thus, $NaNbO_3$, $CaTiO_3$, $CaCrO_3$, $CuSnO_3$, $YAlO_3$, $LaAlO_3$, and $LaGaO_3$ all have the same structure as perovskite, $CaTiO_3$. Names such as calcium titanate may convey false implications and it is preferable to name such compounds as double oxides and double hydroxides unless there is clear and generally accepted evidence of cations and oxo or hydroxo anions in the structure. This does not mean that names such as titanates or aluminates should always be abandoned, because such substances may exist in solution and in the solid state (*cf.* **3.223**).

◆"Multiple" has been used in English as a class term including double, triple, etc. (oxides or the like).

6.51.—In the double oxides and hydroxides the metals shall be named in the same order as for double salts (**6.32**).

6.52.—When required the structure type may be added in parentheses and in italics after the name, except that when the type name is also the mineral name of the substance itself then the italics should not be used (*cf.* **4.2**).

Examples:

$NaNbO_3$	sodium niobium trioxide (*perovskite* type)
$MgTiO_3$	magnesium titanium trioxide (*ilmenite* type)
$FeTiO_3$	iron(II) titanium trioxide (ilmenite)
$4CaO.Al_2O_3.nH_2O$ or $Ca_2Al(OH)_7.nH_2O$	dicalcium aluminum hydroxide hydrate
but $Ca_3[Al(OH)_6]_2$	(tri)calcium (bis)-[hexahydroxoaluminate]
$LiAl(OH)_4.2MnO_2$ or $LiAlMn^{IV}_2O_4(OH)_4$	lithium aluminum dimanganese(IV) tetraoxide tetrahydroxide

7. COÖRDINATION COMPOUNDS

7.1. Definitions

In its oldest sense the term *coördination compound* is taken as referring to molecules or ions in which there is an atom (A) to which are attached other atoms (B) or groups (C) to a number in excess of that corresponding to the oxidation number of the atom A. However, the system of nomenclature originally evolved for the compounds within this narrow definition has proved useful for a much wider class of compounds, and for the purposes of nomenclature the restriction "in excess . . . oxidation number" is to be omitted. Any compound formed by addition of one or several ions and/or molecules to one or more ions or/and molecules may be named according to the same system as strict coördination compounds.

The effect of this is to bring many simple and well-known compounds under the same nomenclature rules as accepted coördination compounds; the result is to reduce the diversity of names and avoid many controversial issues, because it should be understood that there is no intention of implying that any structural analogy necessarily exists between different compounds merely on account of a common system of nomenclature. The system extends also to many addition compounds.

In the rules which follow certain terms are used in the senses here indicated: the atom referred to above as (A) is known as the *central* or *nuclear* atom, and all other atoms which are directly attached to A are known as *coördinating atoms*. Atoms (B) and groups (C) are called *ligands*. A group containing more than one *potential* coördinating atom is termed a *multidentate* ligand, the number of potential coördinating atoms being indicated by the terms *unidentate, bidentate, etc.* A *chelate* ligand is a ligand *attached to one* central atom through *two or more* coördinating atoms, while a *bridging group* is attached to *more than one* atom. The whole assembly of one or more central atoms with their attached ligands is referred to as a *complex*, which may be an uncharged molecule or an ion of either polarity. A *polynuclear complex* is a complex which contains *more than one* nuclear atom, their number being designated by the terms *mononuclear, dinuclear, etc.*

◆Some dissatisfaction with the definition of coördination compounds was expressed, though the broad definition (last sentence of first paragraph) was generally approved.

Central atom or *center of coördination* is to be preferred to *nuclear* atom because of other senses of nucleus, especially of an atom. Possible replacements for polynuclear, etc., are polycentric, *etc.*, or bridged complex since bridging group is in common use.

In the United States the Greek-Latin hybrid terms polydentate and monodentate seem to be used more than the all-Latin multidentate and unidentate.

7.2. Formulas and Names for Complex Compounds in General

7.21. Central Atoms.—In *formulas* the symbol for the central atom(s) should be placed *first* (except in formulas which are primarily structural), the anionic and neutral, *etc.*, ligands following as prescribed in **7.25**, and the formula for the whole complex entity (ion or molecule) should be placed in brackets [ ].

In *names* the central atom(s) should be placed *after* the ligands.

◆It is considered preferable to place the whole complex in brackets, especially when more than one central atom is present, but not essential with only one central atom or with complex ions or nonionic complexes, because brackets may be needed sometimes for indicating concentrations.

7.22. Indication of Valence and Proportion of Constituents.—The oxidation number of the central atom is indicated by means of the Stock notation (**2.252**). Alternatively the proportion of constituents may be given by means of stoichiometric prefixes (**2.251**).

7.23.—Formulas and names may be supplemented with the prefixes *cis, trans, etc.* (**2.19**).

7.24. Terminations.—Complex anions shall be given the termination -ate (*cf.* **2.23**, **2.24** and **3.223**). Complex cations and neutral molecules are given no distinguishing termination. For further details concerning the names of ligands see **7.3**.

7.25. Order of Citation of Ligands in Complexes.—

first: anionic ligands
next: neutral and cationic ligands

7.251.—The anionic ligands shall be cited in the order

1. H^-
2. O^{2-}, OH^- (in that sequence)
3. Other simple anions (*i.e.*, one element only)
4. Polyatomic anions
5. Organic anions in alphabetical order

The sequence within categories 3 and 4 should be that given in **6.332**.

◆H^- is preferably named last, not first (*cf.* comment at 6.331). Alphabetical order is strongly recommended for simplicity at least within 3 and 4, where only rare uses are involved. The intention seems to be to include under 3 monatomic ions (*i.e.*, one *atom*—rather than one *element*—only), in other words to exclude N_3, I_3, etc. Under 4 the insertion of "inorganic" between "Polyatomic" and "anions" is recommended.

7.252.—Neutral and cationic ligands shall be cited in the order given

first: water, ammonia (in that sequence)
then: other inorganic ligands in the sequence in which their coördinating elements appear in the list given in **2.16**
last: organic ligands in alphabetical order.

◆For the inorganic ligands alphabetical order again is recommended. The use of parentheses wherever there is any possibility of ambiguity should be stressed, as illustrated in examples under 7.321: potassium trichloro(ethylene)platinate(II), where parentheses are given in the Rules, and tetra(pyridine)platinum(II) tetrachloroplatinate(II), where they have been added in this version. Parentheses might also be helpful with "thiocyanato" preceded by a numerical prefix (see last example in 7.311) and are definitely required with two-word names for ligands recommended instead of the fused names of the original version (6.2), as in the seventh example in 7.412: di-μ-carbonyl-bis{carbonyl(triethyl phosphite)cobalt}. Since brackets denote complexes, braces can be used in formulas or names where needed in order to avoid the use of two sets of parentheses (braces were so used in some but not all such cases in the original version).

7.3. Names for Ligands

7.31.—Anionic Ligands.

7.311.—The names for anionic ligands, whether inorganic or organic, end in -o (see, however, **7.324**). In general, if the anion name ends in -ide, -ite, or -ate, the final -e is replaced by -o, giving -ido, -ito, and -ato, respectively.

Examples:

Li[AlH$_4$]	lithium tetrahydridoaluminate
Na[BH$_4$]	sodium tetrahydridoborate
K$_2$[OsNCl$_5$]	potassium nitridopentachloroösmate(VI)
[Co(NH$_2$)$_2$(NH$_3$)$_4$]OC$_2$H$_5$	diamidotetraamminecobalt(III) ethanolate
[CoN$_3$(NH$_3$)$_5$]SO$_4$	azidopentaamminecobalt(III) sulfate
Na$_3$[Ag(S$_2$O$_3$)$_2$]	sodium bis(thiosulfato)argentate(I)
[Ru(HSO$_3$)$_2$(NH$_3$)$_4$]	bis(hydrogen sulfito)tetraammineruthenium
(II) NH$_4$[Cr(SCN)$_4$(NH$_3$)$_2$]	ammonium tetrathiocyanatodiamminechromate (III)

◆For "ethanolate" in the fourth example "ethoxide" may be preferred.

7.312.—These anions do not follow exactly the above rule; modified forms have become established:

F$^-$	fluoride	fluoro (*not* fluo)	O$_2^{2-}$	peroxide	peroxo
Cl$^-$	chloride	chloro	HS$^-$	hydrogen sulfide	thiolo
Br$^-$	bromide	bromo	S^{2-}	sulfide	thio[1] (sulfido)
I$^-$	iodide	iodo	(but:S$_2^{2-}$	disulfide	disulfido)
O^{2-}	oxide	oxo	CN$^-$	cyanide	cyano
OH$^-$	hydroxide	hydroxo			

[1]The name thio has long been used to denote the ligand S^{2-} when it can be regarded as replacing O^{2-} in an oxo acid or its anion. The general use of this name will prevent confusion between the two interpretations of disulfido as S$_2^{2-}$ or two S^{2-} ligands.

By analogy with hydroxo, CH$_3$O$^-$, *etc.*, are called methoxo, *etc.* For CH$_3$S$^-$, *etc.*, the systematic names methanethiolato, *etc.*, are used.

Examples:

K[AgF$_4$]	potassium tetrafluoroargentate(III)
K$_2$[NiF$_6$]	potassium hexafluoroniccolate(IV)
Ba[BrF$_4$]$_2$	barium tetrafluorobromate(III)
Na[AlCl$_4$]	sodium tetrachloroaluminate
Cs[ICl$_4$]	cesium tetrachloroiodate(III)
K[Au(OH)$_4$]	potassium tetrahydroxoaurate(III)
K[CrOF$_4$]	potassium oxotetrafluorochromate(V)
K$_2$[Cr(O)$_2$O$_2$(CN)$_2$(NH$_3$)]	potassium dioxoperoxodicyanoamminechromate(VI)
Na[BH(OCH$_3$)$_3$]	sodium hydridotrimethoxoborate
K$_2$[Fe$_2$S$_2$(NO)$_2$]	dipotassium dithiotetranitrosyldiferrate

◆It is strongly recommended on the basis of past and present usage, analogy with chloro, *etc.*, and euphony that hydro be added to the list of modified forms of names for anionic ligands and that hydrido be abandoned (*cf.* also examples in 7.311).

While the Subcommittee on Coördination Compounds recognizes the usefulness of the invariable -o ending for anionic ligands, some members of the general committees do not see a sharp enough distinction between such ligands and the same groups in organic compounds to justify a departure from organic usage by using hydroxo, methoxo, *etc.*, instead of hydroxy, methoxy, *etc.* These members therefore favor adding hydroxy, methoxy, *etc.*, to the hydrocarbon radicals excepted from the rule of -o endings for anions (7.324). For the use of peroxo, see comment at 5.22.

The -o of a negative ligand should not be elided before another vowel (chloroösmate, chloroiodo). This agrees with organic practice. *Cf.* comment at 7.322.

It should be pointed out here as well as in the list of names for ions and radicals at the end of the Rules that the approved organic name for unsubstituted HS is mercapto (not thiol, as also given in the list) and for the alkyl- and aryl-substituted radicals methylthio, *etc.*

7.313.—Ligands derived from organic compounds not normally called acids, but which function as such in complex formation by loss of a proton, should be treated as anionic and given the ending -ato. If, however, no proton is lost, the ligand must be treated as neutral—see **7.32**.

Examples:

[Ni(C$_4$H$_7$N$_2$O$_2$)$_2$] bis(dimethylglyoximato)nickel(II)   [Cu(C$_5$H$_7$O$_2$)$_2$] bis(acetylacetonato)copper(II)

bis(8-quinolinolato)silver(II)   bis(4-fluorosalicylaldehydato)copper(II)   *N,N'*-ethylenebis(salicylideneiminato)cobalt(II)

NOMENCLATURE OF INORGANIC CHEMISTRY (Continued)

◆Although according to 3.33 the name for the ligand in the second example should be derived from the systematic name 2,4-pentanedione instead of from acetylacetone, acetylacetonato perhaps conveys better the idea that it is the enol form that is involved. However, the coördination subcommittee questions the use of the termination -ato rather than plain -o especially in cases where the -ate terms (as dimethylglyoximate) are not accepted organic practice. Such -ato terms are especially misleading in the last two examples, where the -ato belongs with the hydroxyl part of the name, not the aldehyde or imine part to which it is attached.

7.32.—Neutral and Cationic Ligands.

7.321.—The name of the coördinated molecule or cation is to be used without change, except in the special cases provided for in **7.322**.

Examples:

[CoCl$_2$(C$_4$H$_8$N$_2$O$_2$)$_2$]	dichlorobis(dimethylglyoxime)cobalt(II) (*cf.* nickel derivative given in **7.313**)
cis-[PtCl$_2$(Et$_3$P)$_2$]	*cis*-dichlorobis(triethylphosphine)platinum(II)
[CuCl$_2$(CH$_3$NH$_2$)$_2$]	dichlorobis(methylamine)copper(II)
[Pt py$_4$] [PtCl$_4$]	tetra(pyridine)platinum(II) tetrachloroplatinate(II)
[Fe(dipy)$_3$]Cl$_2$	tris(dipyridyl)iron(II) chloride
[Co en$_3$]$_2$(SO$_4$)$_3$	tris(ethylenediamine)cobalt(III) sulfate
[Zn{NH$_2$CH$_2$CH(NH$_2$)CH$_2$NH$_2$}$_2$]I$_2$	bis(1,2,3-triaminopropane)zinc iodide
K[PtCl$_3$(C$_2$H$_4$)]	potassium trichloro(ethylene)platinate(II) or potassium trichloromonoethyleneplatinate(II)
[PtCl$_2${H$_2$NCH$_2$CH(NH$_2$)CH$_2$NH$_3$}]Cl	dichloro(2,3 - diaminopropylammonium)platinum(II) chloride
[Cr(C$_6$H$_5$NC)$_6$]	hexakis(phenyl isocyanide)chromium

◆In the fifth example, bipyridine is preferred to dipyridyl in organic practice, and in the seventh example 1,2,3-propanetriamine to 1,2,3-triaminopropane.

7.322.—Water and ammonia as neutral ligands in coördination complexes are called "aquo" and "ammine," respectively.

In the tentative rules it was proposed to change the old-established "aquo" to "aqua," thus keeping the -o termination consistently for anionic ligands alone. However, as the old form is so widely used, many regarded the change as too pedantic, and the Commission has decided to retain "aquo" as an exception.

Examples:

[Cr(H$_2$O)$_6$]Cl$_3$	hexaaquochromium(III) chloride or hexaaquochromium trichloride
[Al(OH)(H$_2$O)$_5$]$^{++}$	the hydroxopentaaquoaluminum ion
[Co(NH$_3$)$_6$]ClSO$_4$	hexaamminecobalt(III) chloride sulfate
[CoCl(NH$_3$)$_5$]Cl$_2$	chloropentaamminecobalt(III) chloride
[CoCl$_3$(NH$_3$)$_2${(CH$_3$)$_2$NH}]	trichlorodiammine(dimethylamine)cobalt(III)

◆Hexaquo, pentaquo, *etc.*, are used in the examples in the original version, but have been changed in this version to hexaaquo, *etc.*, for conformity with the latest approved organic practice. (*Cf.* hexaammonium, given with two a's separated by a hyphen in 7.6, second example, in the original version.)

7.323.—The groups NO, NS, CO, and CS, when linked directly to a metal atom, are to be called nitrosyl, thionitrosyl, carbonyl, and thiocarbonyl, respectively. In computing the oxidation number these radicals are treated as neutral.

Examples:

Na$_2$[Fe(CN)$_5$NO]	disodium pentacyanonitrosylferrate
K$_3$[Fe(CN)$_5$CO]	tripotassium pentacyanocarbonylferrate
K[Co(CN)(CO)$_2$(NO)]	potassium cyanodicarbonylnitrosylcobaltate(0)
HCo(CO)$_4$	hydrogen tetracarbonylcobaltate($-$I)
[Ni(CO)$_2$(Ph$_3$P)$_2$]	dicarbonylbis(triphenylphosphine)nickel(0)
[Fe en$_3$] [Fe(CO)$_4$]	tris(ethylenediamine)iron(II) tetracarbonylferrate ($-$II)
Mn$_2$(CO)$_{10}$ or [CO$_5$]Mn–Mn(CO)$_5$]	decacarbonyldimanganese(0) or bis(pentacarbonylmanganese)

◆The necessity of arbitrarily considering these groups as always neutral can be avoided by not using the oxidation number but instead the stoichiometric proportions (as in some of these examples) or the Ewens-Bassett system (*cf.* comment at 2.252). Thus the anion in the last example in 7.312 (where NO is known to be positive) would by this system be named dithiotetranitrosyldiferrate (2–).

7.324.—Anions derived from hydrocarbons are given radical names without -o, but are counted as negative when computing the oxidation number.

The consistent introduction of the ending -o would in this case lead to unfamiliar names, *e.g.*, phenylato or phenido for C$_6$H$_5{}^-$. On the other hand, if the radicals were counted as neutral ligands, the central atom would have to be given an unusual oxidation number, *e.g.*, $-$I for boron in K[B(C$_6$H$_5$)$_4$], instead of III.

Examples:

K[B(C$_6$H$_5$)$_4$]	potassium tetraphenylborate
K[SbCl$_5$C$_6$H$_5$]	potassium pentachloro(phenyl)antimonate(V)
K$_2$[Cu(C$_2$H)$_3$]	potassium triethynylcuprate(I)
K$_4$[Ni(C$_2$C$_6$H$_5$)$_4$]	potassium tetrakis(phenylethynyl)niccolate(0)
[Fe(CO)$_4$(C$_2$C$_6$H$_5$)$_2$]	tetracarbonylbis(phenylethynyl)iron(II)
Fe(C$_5$H$_5$)$_2$	bis(cyclopentadienyl)iron(II)
[Fe(C$_5$H$_5$)$_2$]Cl	bis(cyclopentadienyl)iron(III) chloride
[Ni(NO)(C$_5$H$_5$)]	nitrosylcyclopentadienylnickel

◆The use of radical names such as cyclopentadienyl in the examples does not seem consistent with the use of Stock notations; for the sixth example ("ferrocene"), iron(II) cyclopentadienide is preferred by some. This matter is presumably part of the whole organometallic problem to be dealt with by the new IUPAC joint organic-inorganic subcommittee.

For niccolate, see comment at 1.12.

7.33.—Alternative Modes of Linkage of Some Ligands.—Where ligands are capable of attachment by different atoms this may be denoted by adding the symbol for the atom by which attachment occurs at the end of the name of the ligand. Thus the dithioöxalato group may be attached through S or O, and these are distinguished as dithioöxalato-S,S' and dithioöxalato-O,O', respectively.

In some cases different names are already in use for alternative modes of attachment, as, for example, thiocyanato ($-SCN$) and isothiocyanato ($-NCS$), nitro ($-NO_2$), and nitrito ($-ONO$). In these cases existing custom may conveniently be retained.

Examples:

$$K_2\left[Ni\begin{pmatrix} S-CO \\ | \\ S-CO \end{pmatrix}_2 \right]$$

potassium bis(dithioöxalato-S,S')niccolate(II)

dichloro(2-N,N-dimethylaminoethyl 2-aminoethyl sulfide-N',S)platinum(II)

$K_2[Pt(NO_2)_4]$	potassium tetranitroplatinate(II)
$Na_3[Co(NO_2)_6]$	sodium hexanitrocobaltate(III)
$[Co(NO_2)_3(NH_3)_3]$	trinitrotriamminecobalt(III)
$[Co(ONO)(NH_3)_5]SO_4$	nitritopentaamminecobalt(III) sulfate
$[Co[NCS](NH_3)_5]Cl_2$	isothiocyanatopentaamminecobalt(III) chloride

◆Thioöxalato in the original version has been changed to dithioöxalato.

7.4. Di- and Polynuclear Compounds

7.41.—Bridging Groups.

7.411.—A bridging group shall be indicated by adding the Greek letter μ immediately before its name and separating this from the rest of the complex by a hyphen. Two or more bridging groups of the same kind are indicated by di-μ-, etc.

7.412.—If the number of central atoms bound by one bridging group exceeds two, the number shall be indicated by adding a subscript numeral to the μ.

This system of notation allows simply of distinction between, for example, μ-disulfido (one S_2 bridge) and di-μ-sulfido (two S bridges). It is also capable of extension to much more complex and unsymmetrical structures by use of the conventional prefixes cis, trans, asym, and sym where necessary.

Examples:

$[(NH_3)_5Cr-OH-Cr(NH_3)_5]Cl_5$
μ-hydroxo-bis{pentaamminechromium(III)}chloride

di-μ-chloro-dichlorobis(triethylarsine)diplatinum(II) (three possible isomers: asym, sym-cis, and sym-trans; the last is shown)

di-μ-thiocyanato-dithiocyanatobis(tripropylphosphine)diplatinum(II)

$[(CO)_3Fe(CO)_3Fe(CO)_3]$	tri-μ-carbonyl-bis(tricarbonyliron)
$[(CO)_3Fe(SEt)_2Fe(CO)_3]$	di-μ-ethanethiolato-bis(tricarbonyliron)
$[(C_5H_5)(CO)Fe(CO)_2Fe(CO)(C_5H_5)]$	di-μ-carbonyl-bis(carbonylcyclopentadienyliron)
$[(CO)\{P(OEt)_3\}Co(CO)_2Co(CO)\{P(OEt)_3\}]$	di-μ-carbonyl-bis{carbonyl(triethyl phosphite)-cobalt}
$[Au(CN)(C_3H_7)_2]_4$	cyclo-tetra-μ-cyano-tetrakis(dipropylgold)
$[CuI(Et_3As)]_4$	tetra-μ_3-iodo-tetrakis{triethylarsinecopper(I)}
$[Be_4O(CH_3COO)_6]$	μ_4-oxo-hexa-μ-acetato-tetraberyllium

7.42. Extended Structures.—Where bridging causes an indefinite extension of the structure it is best to name compounds primarily on the basis of their over-all composition; thus the compound having the composition represented by the formula $CsCuCl_3$ has an anion with the structure:

$$\left[\cdots\text{Cl}-\underset{\overset{|}{\text{Cl}}}{\overset{\overset{\text{Cl}}{|}}{\text{Cu}}}-\text{Cl}-\underset{\overset{|}{\text{Cl}}}{\overset{\overset{\text{Cl}}{|}}{\text{Cu}}}-\text{Cl}-\underset{\overset{|}{\text{Cl}}}{\overset{\overset{\text{Cl}}{|}}{\text{Cu}}}-\text{Cl}-\underset{\overset{|}{\text{Cl}}}{\overset{\overset{\text{Cl}}{|}}{\text{Cu}}}\cdots \right]^{n-}$$

This may be expressed in the formula $(Cs^+)_n$ $[(CuCl_3)_n]^{n-}$ which leads to the simple name cesium catena-μ-chloro-dichlorocuprate(II). If the structure were in doubt, however, the substance would be called cesium copper(II) chloride (as a double salt).

◆*Cf.* comment on *catena* at 1.4.

7.5. Isopoly Anions

The structure of many complicated isopoly anions has now been cleared up by X-ray work and it turns out that the indication of the several μ-oxo and oxo atoms in the name does not convey any clear picture of the structure and is therefore of little value.

For the time being it is sufficient to indicate the number of atoms by Greek prefixes, at least until isomers are found. When all atoms have their "normal" oxidation states (*e.g.*, W^{VI}), it is not necessary to give the numbers of the oxygen atoms, if all the others are indicated.

Examples:

$K_2S_2O_7$	dipotassium disulfate
$K_2S_3O_{10}$	dipotassium trisulfate
$Na_5P_3O_{10}$	pentasodium triphosphate
$K_2Cr_4O_{13}$	dipotassium tetrachromate
$Na_2B_4O_7$	disodium tetraborate
NaB_5O_8	sodium pentaborate
$Ca_3Mo_7O_{24}$	tricalcium heptamolybdate
$Na_7HNb_6O_{19}.15H_2O$	heptasodium monohydrogen hexaniobate-15-water
$K_2Mg_2V_{10}O_{28}.16H_2O$	dipotassium dimagnesium decavanadate-16-water

7.6. Heteropoly Anions

The central atom or atoms should be cited last in the name and first in the formula of the anion (*cf.* 7.21), *e.g.*, wolframophosphate, *not* phosphowolframate.

If the oxidation number has to be given, it may be necessary to place it immediately after the atom referred to and not after the ending -ate, in order to avoid ambiguity.

The method formerly recommended for naming iso- and heteropoly anions by giving the number of atoms in parentheses is not practicable in more complicated cases.

Examples:

$(NH_4)_3PW_{12}O_{40}$	triammonium dodecawolframophosphate
$(NH_4)_6TeMo_6O_{24}.7H_2O$	hexaammonium hexamolybdotellurate heptahydrate
$Li_3HSiW_{12}O_{40}.24H_2O$	trilithium (mono)hydrogen dodecawolframosilicate-24-water
$K_6Mn^{IV}Mo_9O_{32}$	hexapotassium enneamolybdomanganate(IV)
$Na_6P^V_2Mo_{18}O_{62}$	hexasodium 18-molybdodiphosphate(V)
$Na_4P^{III}_2Mo_{12}O_{41}$	tetrasodium dodecamolybdodiphosphate(III)
$K_7Co^{II}Co^{III}W_{12}O_{42}.16H_2O$	heptapotassiumdodecawolframocobalt(II)-cobalt(III)ate-16-water
$K_3PV_2Mo_{10}O_{39}$	tripotassium decamolybdodivanadophosphate

◆The coördination subcommittee would prefer not to have sections 7.5 and 7.6 included under 7 and recommends that they should be studied by a special subcommittee. Some cyclic isomers are already known, and isopoly cations also are being investigated.

Cf. comment at 1.12 for stand on wolframate and wolframo, and 2.251 for nona instead of ennea (fourth example).

Isopoly and heteropoly are not separate words in the original version.

7.7. Addition Compounds

The ending -ate is now the accepted ending for *anions* and should generally not be used for addition compounds. Alcoholates are the *salts* of alcohols and this name should not be used to indicate alcohol of crystallization. Analogously addition compounds containing ether, ammonia, *etc.*, should *not* be termed etherates, ammoniates, *etc.*

However, one exception has to be recognized. According to the commonly accepted meaning of the ending -ate, "hydrate" would be, and was formerly regarded as, the name for a *salt* of water, *i.e.*, what is now known as a hydroxide; the name hydrate has now a very firm position as the name of a compound containing water of crystallization and is allowed also in these Rules to designate water bound in an unspecified way; it is considered to be preferable even in this case to avoid the ending -ate by using the name "water" (or its equivalent in other languages) when possible.

NOMENCLATURE OF INORGANIC CHEMISTRY (Continued)

The names of addition compounds may be formed by connecting the names of individual compounds by hyphens (short dashes) and indicating the number of molecules by Arabic numerals. When the added molecules are organic, however, it is recommended to use multiplicative numeral prefixes (bis, tris, tetrakis, *etc.*) instead of Arabic figures to avoid confusion with the organic-chemical use of Arabic figures to indicate position of substituents.

Examples:

$CaCl_2.6H_2O$	calcium chloride–6-water (or calcium chloride hexahydrate)
$3CdSO_4.8H_2O$	3-cadmium sulfate–8-water
$Na_2CO_3.10H_2O$	sodium carbonate–10-water (or sodium carbonate decahydrate)
$AlCl_3.4C_2H_5OH$	aluminum chloride–4-ethanol or –tetrakisethanol
$BF_3.(C_2H_5)_2O$	boron trifluoride–diethyl ether
$BF_3.2CH_3OH$	boron trifluoride–bismethanol
$BF_3.H_3PO_4$	boron trifluoride–phosphoric acid
$BiCl_3.3PCl_5$	bismuth trichloride–3-(phosphorus pentachloride)
$TeCl_4.2PCl_5$	tellurium tetrachloride–2-(phosphorus pentachloride)
$(CH_3)_4NAsCl_4.2AsCl_3$	tetramethylammonium tetrachloroarsenate(III)–2-(arsenic trichloride)
$CaCl_2.8NH_3$	calcium chloride–8-ammonia
$8H_2S.46H_2O$	8-(hydrogen sulfide)–46-water
$8Kr.46H_2O$	8-krypton–46-water
$6Br_2.46H_2O$	6-dibromine–46-water
$8CHCl_3.16H_2S.136H_2O$	8-chloroform–16-(hydrogen sulfide)–136-water

These names are not very different from a pure verbal description which may in fact be used, *e.g.*, calcium chloride with 6 water, compound of aluminum chloride with 4 ethanol, *etc.*

If it needs to be shown that added molecules form part of a complex, the names are given according to 7.2 and 7.3.

Examples:

$FeSO_4.7H_2O$ or $[Fe(H_2O)_6]SO_4.H_2O$	iron(II) sulfate heptahydrate or hexaaquoiron(II) sulfate monohydrate
$PtCl_2.2PCl_3$ or $[PtCl_2(PCl_3)_2]$	platinum(II) chloride–2-(phosphorus trichloride) or dichlorobis(phosphorus trichloride)-platinum(II)
$AlCl_3.NOCl$ or $NO[AlCl_4]$	aluminum chloride–nitrosyl chloride or nitrosyl tetrachloroaluminate
$BF_3.Et_3N$ or $[BF_3(Et_3N)]$	boron trifluoride–triethylamine or trifluoro(triethylamine)boron

◆Only some so-called "addition compounds" are known to be coordination compounds, and the formulas and names can show such structures. Those that are lattice compounds and those of unknown structure do not really belong in 7.

The committees like hydrate and ammoniate and see no particular advantage in dropping them (*cf.* the use of hydrate terms in examples in 6.324; 6-hydrate, *etc.*, as well as hexahydrate, *etc.*, are considered acceptable).

No reason can be seen for deviating here from usual organic practice by using multiplicative prefixes with simple names: tetraethanol is just as clear as tetrakisethanol, and dimethanol as bismethanol. Parentheses can always be used if there is danger of any ambiguity.

The use of short dashes ("en" dashes) instead of hyphens between the names (as in the examples here, but not in the original version) makes for greater ease in reading.

It is considered preferable by some to place the electron donor first in both the formulas and name: $(C_2H_5)_2O.BF_3$, diethyl ether–boron trifluoride (note that organic usage favors ethyl ether over diethyl ether).

8. POLYMORPHISM

Minerals occurring in nature with similar compositions have different names according to their crystal structures; thus, zinc blende, wurtzite; quartz, tridymite, and cristobalite. Chemists and metallographers have designated polymorphic modifications with Greek letters or with Roman numerals (α-iron, ice-I, *etc.*). The method is similar to the use of trivial names, and is likely to continue to be of use in the future in cases where the existence of polymorphism is established, but not the structures underlying it. Regrettably there has been no consistent system, and some investigators have designated as α the form stable at ordinary temperatures, while others have used α for the form stable immediately below the melting point, and some have even changed an already established usage and renamed α-quartz β-quartz, thereby causing confusion. If the α–β nomenclature is used for two substances A and B, difficulties are encountered when the binary system A–B is studied.

NOMENCLATURE OF INORGANIC CHEMISTRY (Continued)
LIST OF NAMES FOR IONS AND RADICALS

In inorganic chemistry substitutive names seldom are used, but the organic-chemical names are shown to draw attention to certain differences between organic and inorganic nomenclature.

Atom or group	as neutral molecule	as cation or cationic radical[1]	as anion	as ligand	as prefix for substituent in organic compounds
H	monohydrogen	hydrogen	hydride	hydrido	
F	monofluorine		fluoride	fluoro	fluoro
Cl	monochlorine	chlorine	chloride	chloro	chloro
Br	monobromine	bromine	bromide	bromo	bromo
I	monoiodine	iodine	iodide	iodo	iodo
I_3			triiodide		
ClO		chlorosyl	hypochlorite	hypochlorito	
ClO_2	chlorine dioxide	chloryl	chlorite	chlorito	
ClO_3		perchloryl	chlorate	chlorato	
ClO_4			perchlorate		
IO		iodosyl	hypoiodite		iodoso
IO_2		iodyl			iodyl or iodoxy
O	monoöxygen		oxide	oxo	oxo or keto
O_2	dioxygen		$O_2{}^{2-}$: peroxide	peroxo	peroxy
			$O_2{}^-$: hyperoxide		
HO	hydroxyl		hydroxide	hydroxo	hydroxy
HO_2	(perhydroxyl)		hydrogen peroxide	hydrogen peroxo	hydroperoxy
S	monosulfur		sulfide	thio (sulfido)	thio
HS	(sulfhydryl)		hydrogen sulfide	thiolo	thiol or mercapto
S_2	disulfur		disulfide	disulfido	
SO	sulfur monoxide	sulfinyl (thionyl)			sulfinyl
SO_2	sulfur dioxide	sulfonyl (sulfuryl)	sulfoxylate		sulfonyl
SO_3	sulfur trioxide		sulfite	sulfito	
HSO_3			hydrogen sulfite	hydrogen sulfito	
S_2O_3			thiosulfate	thiosulfato	
SO_4			sulfate	sulfato	
Se	selenium		selenide	seleno	seleno
SeO		seleninyl			seleninyl
SeO_2		selenonyl			selenonyl
SeO_3	selenium trioxide		selenite	selenito	
SeO_4			selenate	selenato	
Te	tellurium		telluride	telluro	telluro
CrO_2		chromyl			
UO_2		uranyl			
NpO_2		neptunyl			
PuO_2		plutonyl			
AmO_2		americyl			
N	mononitrogen		nitride	nitrido	
N_3			azide	azido	
NH			imide	imido	imino
NH_2			amide	amido	amino
NHOH			hydroxylamide	hydroxylamido	hydroxylamino
N_2H_3			hydrazide	hydrazido	hydrazino
NO	nitrogen oxide	nitrosyl		nitrosyl	nitroso
NO_2	nitrogen dioxide	nitryl		nitro	nitro
ONO			nitrite	nitrito	
NS		thionitrosyl			
$(NS)_n$		thiazyl (e.g., trithiazyl)			
NO_3			nitrate	nitrato	
N_2O_3			hyponitrite	hyponitrito	
P	phosphorus		phosphide	phosphido	
PO		phosphoryl			phosphoroso
PS		thiophosphoryl			
PH_2O_3			hypophosphite	hypophosphito	
PHO_3			phosphite	phosphito	
PO_4			phosphate	phosphato	
AsO_4			arsenate	arsenato	
VO		vanadyl			
CO	carbon monoxide	carbonyl		carbonyl	carbonyl
CS		thiocarbonyl			
CH_3O	methoxyl		methanolate	methoxo	methoxy
C_2H_5O	ethoxyl		ethanolate	ethoxo	ethoxy
CH_3S			methanethiolate	methanethiolato	methylthio
C_2H_5S			ethanethiolate	ethanethiolato	ethylthio
CN		cyanogen	cyanide	cyano	cyano
OCN			cyanate	cyanato	cyanato
SCN			thiocyanate	thiocyanato and isothiocyanato	thiocyanato and isothiocyanato
SeCN			selenocyanate	selenocyanato	selenocyanato
TeCN			tellurocyanate	tellurocyanato	
CO_3			carbonate	carbonato	
HCO_3			hydrogen carbonate	hydrogen carbonato	
CH_3CO_2			acetate	acetato	acetoxy
CH_3CO	acetyl	acetyl			acetyl
C_2O_4			oxalate	oxalato	

[1]If necessary, oxidation state is to be given by Stock notation.

◆Although some additions might be made to this list, especially in the last column, and a few changes suggested, no attempt has been made to do so at this time. *Cf.* comments at such rules as 3.31, 3.32, 5.35, 6.2, and 7.312.

A rational system should be based upon crystal structure, and the designations α, β, γ, etc., should be regarded as provisional, or as trivial names. The designations should be as short and understandable as possible, and convey a maximum of information to the reader. The rules suggested here have been framed as a basis for future work, and it is hoped that experience in their use may enable more specific rules to be formulated at a later date.

8.1.—For chemical purposes (*i.e.*, when particular mineral occurrences are not under consideration) polymorphs should be indicated by adding the crystal system after the name or formula. For example, zinc sulfide(cub.) or ZnS(cub.) corresponds to zinc blende or sphalerite, and ZnS(hex.) to wurtzite. The Commission considers that these abbreviations might with advantage be standardized internationally:

cub.	=	cubic	hex.	=	hexagonal
c.	=	body-centered	trig.	=	trigonal
f.	=	face-centered	mon.	=	monoclinic
tetr.	=	tetragonal	tric.	=	triclinic
o-rh.	=	orthorhombic			

Slightly distorted lattices may be indicated by use of the *circa* sign, $\sim$. Thus, for example, a slightly distorted face-centered cubic lattice would be expressed as $\sim$f.cub.

8.2.—Crystallographers may find it valuable to add the space-group; it is doubtful whether this system would commend itself to chemists where **8.1** is sufficient.

8.3.—Simple well-known structures may also be designated by giving the type-compound in italics in parentheses; but this system often breaks down because many structures are not referable to a type in this way. Thus, AuCd above 70° may be written as AuCd(cub.) or as AuCd(*CsCl-type*); but at low temperature only as AuCd(o-rh.), as its structure cannot be referred to a type.

◆*Cf.* 6.5 and 6.52.

ABBREVIATIONS USED IN TABLE OF PHYSICAL CONSTANTS
OF INORGANIC COMPOUNDS

a	acid	fus	fused	prop	properties
abs	absolute	fxd	fixed	purp	purple
ac. a	acetic acid	gel., gelat	gelatinous	pyr	pyridine
acet	acetone	gl	glass	quad	quadrilateral
act	active	glac	glacial	quest	questioned
al	alcohol	glit	glittering	rect	rectangular
alk	alkali	glob	globular	redsh	reddish
amm	ammonium	glyc	glycerin	reg	regular
amor	amorphous	gran	granular	rhbdr	rhombohedral
anh	anhydrous	greas	greasy	rhomb	rhombic, ortho-rhombic
appr	approximately	grn	green		
aq	aqua, water	h	hot	s	soluble
aq. reg	aqua regia	hex	hexagonal	satd	saturated
asym	asymmetrical	ht	heat	sld	solid
atm	atmospheres	hyd	hydrolyzed	sensit	sensitive
bipyr	bipyramidal	hydx	hydroxides	sc	scales
bl	blue	hyg	hygroscopic	sec	secondary
blk	black	i	insoluble	silv	silver
boil	boiling	ign	ignites	sl	slightly
br., brn	brown	ind	indigo	sly	slowly
brnsh	brownish	indef	indefinite	sm	small
bz	benzene	infl., inflam	inflammable	sod	sodium
c	cold	infus	infusible	soln	solution
calc	calculated	irid	iridescent	solv	solvents
carb	carbon	leaf	leaflets	spont	spontaneous
caust	caustic	lem	lemon	st	steel
chl	chloroform	lgr	ligroin	stab	stable
choc	chocolate	lng	long	subl	sublimes
cit. a	citric acid	lq., liq	liquid	suffoc	suffocating
col	colorless	lt	light	sulfd	sulfides
coll	colloidal	lum	luminous	sulf	sulfur
com'l	commercial	lust	lustrous	sym	symmetrical
comp	compounds	me., meth	methyl	tabl	tablets
compl	completely	met	metal or metal-lic	tart. a	tartaric acid
conc	concentrated			tetr	tetragonal
const	constant	micr	microscopic	tetrah	tetrahedral
cont	contains	min	mineral	tol	toluene
corros	corrosive	misc	miscible	trac	trace, traces
cr	crystalline	mixt	mixture	trans	transparent
cub	cubic	mod	modifications	translu	translucent
d., dec	decomposes	monbas	monobasic	tri., trig	trigonal
deliq	deliquescent	mon-H	monohydrogen	tribas	tribasic
deriv	derivative	monocl	monoclinic	tricl	triclinic
dibas	dibasic	near	nearly	trim	trimetric
di-H	dihydrogen	need	needles	tr	transition point
dil	dilute	nit	nitrate	turp	turpentine
dimorph	dimorphous	oct	octahedral	unpleas	unpleasant
disg	disagreeable	odorl	odorless	unst	unstable
dk	dark	offen	offensive	v	very
doubt	doubtful	olv	olive	vac	vacuum
duct	ductile	opt	optical or optically	var	various
effl	efflorescent			viol	violent, violence
em	emerald	or	orange		
eth	ether	ord	ordinary	visc	viscous
ev	evolves	org	organic	vitr	vitreous
evln	evolution	oxal	oxalate or oxalic	vlt	violet
ex	excess			volt., volat	volatizes
exist	existence	pa	pale	wh	white
exp	explodes	pet	petroleum	wh. lt	white light
extr	extreme(ly)	pl	plates	yel	yellow
f., fr	from	pois	poisonous	yelsh	yellowish
feath	feathery	polymorph	polymorphous	∞	soluble in all proportions
fl	flakes	powd	powder		
floc	flocculent	ppt	precipitate	>	above
fluo, fluores	fluorescent	pr	prisms	<	below
form	formic	press	pressure		
fum	fuming	prob	probably		

No.	Name	Synonyms and Formulae	Mol. wt.	Crystalline form, properties and index of refraction	Density or spec. gravity	Melting point, °C	Boiling point, °C	Solubility, in grams per 100 cc		
								Cold water	Hot water	Other solvents
	Actinium									
a1	**Actinium**	Ac	227	silv wh met, cub		1050	3200 ± 300	d to Ac(OH)$_3$		
a2	bromide	AcBr$_3$	466.73	wh, hex	5.85	subl 800		s		
a3	chloride, tri-	AcCl$_3$	333.36	wh cr, hex	4.81	subl 960		s		
a4	fluoride, tri-	AcF$_3$	284	wh cr, hex	7.88			i	i	
a5	hydroxide	Ac(OH)$_3$	278.02	wh				i		
a6	iodide	AcI$_3$	607.71	wh		subl 700–800		s		
a7	oxalate	Ac$_2$(C$_2$O$_4$)$_3$	718.06					i		
a8	oxide, sesqui-	Ac$_2$O$_3$	502	wh cr, hex	9.19			l		
a9	sulfide, sesqui-	Ac$_2$S$_3$	550.19	dk, cub	6.75					
a10	**Aluminum**	Al	26.98154	silv wh duct met, cub	2.702	660.37	2467	i	i	s alk, HCl,H$_2$SO$_4$; i conc HNO$_3$, h ac a
a11	acetate, tri-	Al(C$_2$H$_3$O$_2$)$_3$	204.12	wh solid		d		v sl s	d	
a12	acetylacetonate	Al(C$_5$H$_7$O$_2$)$_3$	324.31	col, monocl	1.27	193, subl vac	314	i	i	v s al; s eth, bz
a13	orthoarsenate	AlAsO$_4$.8H$_2$O	310.02	wh powd	3.001	− H$_2$O		i	i	sl s a
a14	benzoate	Al(C$_7$H$_5$O$_2$)$_3$	390.33	wh cr powd				v sl s		
a15	benzyloxide	Aluminum benzylate. Al(C$_7$H$_7$O)$_3$	348.38			59–60	283–284$^{0.5}$			
a16	boride	AlB$_{12}$	156.71	dk red-blk, monocl	2.55$_4^{18}$			i		s hot HNO$_3$; i a, alk
a17	boride, di-	AlB$_2$	48.60	copper red, hex	3.19					
a18	bromate	Al(BrO$_3$)$_3$.9H$_2$O	572.84	wh cr, hygr		62.3	d 100	s	s	sl s a
a19	bromide	AlBr$_3$ (or Al$_2$Br$_6$)	266.71	col, rhomb pl, deliq	2.64^{10} (fused)	97.5	263.3^{747}	s with viol	d	s al, acet, CS$_2$
a20	bromide, hexa-hydrate	AlBr$_3$.6H$_2$O	374.80	col-yelsh need, deliq	2.54	93	d 135	s	d	s al, amyl al; sl s CS$_2$
a21	bromide, penta-decylhydrate	AlBr$_3$.15H$_2$O	536.91	col need		7.5	d 7	s	s	s al
a22	butoxide, tert-	Al(C$_4$H$_9$O)$_3$	246.33	wh cr	1.0251$_0^{20}$	subl 180, m.p >300, sealed tube.				v s org sol
a23	carbide	Al$_4$C$_3$	143.96	yel-grn, hex	2.36	stab to 1400	d 2200^{400}	d to CH$_4$	d	d dil a; i acet
a24	chlorate	Al(ClO$_3$)$_3$.6H$_2$O	385.43	col, rhbdr, deliq		d		vs	vs	s dil HCl
a25	perchlorate	Al(ClO$_4$)$_3$.6H$_2$O	433.43	col, hygr	2.020	82	−6H$_2$O, 178 d 262	s	s	
a26	chloride	AlCl$_3$ (or Al$_2$Cl$_6$)	133.34	wh to col, hex, odor HCl, v deliq	fus 2.44^{25} liq 1.31^{200}	190$^{2.5}$ atm	182.7^{752} subl. 177.8	69.9^{15} with viol	s d	100$^{12.5}$ abs al; 0.072^{25} chl; sCCl$_4$, eth sl s bz
a27	chloride, hexa-hydrate	AlCl$_3$.6H$_2$O	241.43	col, rhomb, deliq, 1.6	2.398	d 100		s	v s ev HCl	50 abs al; s eth; sl s HCl
a28	chloride, hexammine	AlCl$_3$.6NH$_3$	235.52	col cr, hygr	1.412$_4^{25}$	d		s		
a29	diethylmalonate deriv	Al(C$_7$H$_{11}$O$_4$)$_3$	504.47		1.084^{100}	98		i		s org solv
a30	ethoxide	Al(C$_2$H$_5$O)$_3$	162.14	wh cr	1.142$_0^{20}$	134	205^{14}	d	s	v sl s al, eth
a31	α-ethylacetoacetate deriv	Al(C$_6$H$_9$O$_3$)$_3$	414.39	wh cr	1.101^{80}	78–79	190–200^{11}	s		s lgr
a32	ferrocyanide	Al$_4$[Fe(CN)$_6$]$_3$.17H$_2$O	1050.05	br powd				sl s	sl s	s dil a
a33	fluoride	AlF$_3$	83.98	col, tricl	2.882$_4^{25}$	1291 subl^{760}		0.559^{25}	s	i a, alk, al, acet
a34	fluoride	AlF$_3$.3½H$_2$O	147.03	wh cr powd	1.914$_4^{25}$	−2H$_2$O 100	anhydr 250	i	sl s	d ac a
a35	fluoride, mono-hydrate	Nat. fluellite. AlF$_3$.H$_2$O.	101.99	col, rhomb, 1.473, 1.490, 1.511	2.17			sl s	sl s	
a36	fluosilicate	Nat topaz. 2AlFO.SiO$_2$	184.04	rhomb, 1.619, 1.620, 1.627	3.58			i	i	
a37	hydroxide	Nat. boehmite. AlO(OH)	59.99	wh, orthorhomb microcr	3.01	− H$_2$O, trans to γ-Al$_2$O$_3$		i	i	s h a, h alk
a38	hydroxide	Nat. diaspore. AlO(OH)	59.99	col, rhomb cr	3.3–3.5	− H$_2$O, trans to Al$_2$O$_3$		i	i	s h a, h alk
a39	hydroxide	Al(OH)$_3$	78.00	wh, monocl	2.42	− H$_2$O, 300		i	i	s a, alk; i al
a40	iodide	AlI$_3$ (or Al$_2$I$_6$)	407.69	br pl, cont free I$_2$, deliq	3.98^{25}	191	360	s d	s	s al, eth, CS$_2$, liq NH$_3$
a41	iodide, hexa-hydrate	AlI$_3$.6H$_2$O	515.79	wh-yel cr, hygr	2.63	d 185	d	v s	v s	s al, CS$_2$
a42	isopropoxide	Al(C$_3$H$_7$O)$_3$	204.25	wh cr	1.0346$_0^{20}$	118.5	140.5^8	d		s al, bz, chl
a43	lactate	Al(C$_3$H$_5$O$_3$)$_3$	294.20	wh-yelsh powd				v s		

No.	Name	Synonyms and Formulae	Mol. wt.	Crystalline form, properties and index of refraction	Density or spec. gravity	Melting point, °C	Boiling point, °C	Solubility, in grams per 100 cc		
								Cold water	Hot water	Other solvents
	Aluminum									
a44	nitrate	$Al(NO_3)_3.9H_2O$	375.13	col, rhomb, deliq, 1.54	....	73.5	d 150	63.7^{25}	v s d	100 al; s alk, acet, HNO_3. a
a45	nitride	AlN	40.99	wh cr, hex	3.26	>2200 (in N_2)	subl 2000	d (NH_3)	d	d a, alk
a46	oleate (com'l)	$Al(C_{18}H_{33}O_2)_3(?)$	871.37	wh powd, existence doubted except as basic salt				d	s	i al; v sl s bz
a47	oxalate	$Al_2(C_2O_4)_3.4H_2O$	390.08	wh powd				i	i	i al; s a
a48	oxide	Al_2O_3	101.96	col, hex, 1.768, 1.760	3.965^{25}	**2072**	2980	i	i	v sl s a, alk
a49	oxide	α-Alumina, nat. corundum. Al_2O_3	101.96	col, rhomb cr, 1.765	3.97	2015 ± 15	2980 ± 60	0.000098^{29}	i	v sl s a, alk
a50	oxide	γ-Alumina. Al_2O_3	101.96	wh micr cr, 1.7..	3.5–3.9	tr to α		i	i	sl s a, alk
a51	oxide, monohydrate	$Al_2O_3.H_2O$	119.98	col, rhomb, 1.624 ± 0.003	3.014			i	i	v sl s a, alk
a52	oxide, trihydrate	Nat. gibbsite, hydraargilite. $Al_2O_3.3H_2O$	156.01	wh monocl cr, 1.577, 1.577, 1.595	2.42	tr to $Al_2O_3.H_2O$ (Boehmite)		i	i	s h a, alk
a53	oxide, trihydrate	Nat. bayerite. $Al_2O_3.3H_2O$	156.01	wh micr cr, 1.583	2.53	tr to $Al_2O_3H_2O$ (Boehmite)		i	i	s hot a, alk
a54	metaphosphate	$Al(PO_3)_3$	263.90	col, tetr	2.779			i	i	i a
a55	palmitate, mono- (com'l)	$Al(OH)_2C_{16}H_{31}O_2$	316.41	wh	1.095	200		i		s alk, hydrocarb
a56	1-phenol-4-sulfonate	$Al(C_6H_5O_4S)_3$	546.49	redsh-wh powd				s		s al, glyc
a57	phenoxide	$Al(C_6H_5O)_3$	306.27	grayish-wh cr mass	1.23	d 265		d		s al, eth, chl
a58	orthophosphate	$AlPO_4$	121.95	wh rhomb pl, 1.546, 1.556, 1:578	2.566	>1500		i	i	s a, alk, al
a59	propoxide	$Al(C_3H_7O)_3$	204.25	wh cr	1.0578^{20}_0	106	248^{14}	d	d	s al
a60	salicylate	$Al(C_7H_5O_3)_3$	438.33	redsh-wh powd				i		i al; s alk
a61	selenide	Al_2Se_3	290.84	lt brn powd, unstable in air	3.437^{18}_4			d	d	d a
a62	silicate	Nat. sillimanite, andalusite, cyanite. $Al_2O_3.SiO_2$	162.04	wh, rhomb, 1.66..	3.247	1545 tr to $Al_2O_3.2SiO_2$	>1545	i	i	d HF; i HCl; s fus alk
a63	silicate	Nat. mullite. $3Al_2O_3.2SiO_2$	426.05	col, rhomb, 1.638, 1.642, 1.653	3.156	1920		i	i	i a, HF
a64	stearate, tri-	$Al(C_{18}H_{35}O_2)_3$	877.42	wh powd	1.010	103		i		s al, bz, turp, alk
a65	sulphate	$Al_2(SO_4)_3$	342.15	wh powd, 1.47..	2.71	d 770		31.3^0	98.1^{100}	s dil a; sl s al
a66	sulfate, hydrate	Nat. alunogenite. $Al_2(SO_4)_3.18H_2O$	666.43	col, monocl, 1.474, 1.467, 1.483	1.69^{17}	d 86.5		86.9^0	1104^{100}	i al
a67	sulfide	Al_2S_3	150.16	yel, hex, odor H_2S, d moist air	2.02^{13}	1100	subl 1500 (N_2)	d		s a; i acet
a68	thallium sulfate	Aluminum thallium alum. $AlTl(SO_4)_2.12H_2O$	639.66	col, oct, 1.50112..	2.325^{20}_4	91		4.84^0	65.19^{60}	
a69	**Americium**	Am	243.13	silvery, hex		**994 ± 4**	2607 (extrap)			s dil a
a70	bromide	$AmBr_3$	482.86	wh, orthorhomb		subl		s		
a71	chloride	$AmCl_3$	349.49	pink, hex	5.78	subl 850		s		
a72	fluoride	AmF_3	300.12	pink, hex	9.53			i		
a73	iodide	AmI_3	623.84	yel, orthoromb	6.9			s		
a74	oxide	Am_2O_3	534.26	redsh-brn, cub or tan, or hex						s min a
a75	oxide, di-	AmO_2	275.13	blk, cub	11.68					s min a
a76	**Ammonia**	NH_3	17.03	col gas; liq, 0.817^{-79}, $1.325^{16.5}$	0.7710 g/l 760 mm	−77.7	−33.35	89.9	7.4^{100}	13.20^{20} al; s eth, org solv
a77	Ammonia-d_3	Trideuterio ammonia. ND_3	20.05			−74	−30.9			
	Ammonium									
v78	acetate	$NH_4C_2H_3O_2$	77.08	wh cr, hygr	1.17^{20}_4	114	d	148^4	d	7.89^{15} MeOH; s al; sl s acet
a79	acetate, hydrogen	$(NH_4)H(C_2H_3O_2)_2$	137.14	col need, deliq		66		s		s al
a80	aluminum chloride	$NH_4Cl.AlCl_3$	186.83	wh cr		304		s		
a81	aluminum sulfate	$NH_4Al(SO_4)_2$	237.14	col, hex	2.45^{20}			s		s glyc; i al

No.	Name	Synonyms and Formulae	Mol. wt.	Crystalline form, properties and index of refraction	Density or spec. gravity	Melting point, °C	Boiling point, °C	Solubility, in grams per 100 cc		
								Cold water	Hot water	Other solvents
	Ammonium									
a82	aluminum sulfate, hydrate	Nat. tschernigite. $NH_4Al(SO_4)_2.12H_2O$	453.33	col, cub, 1.459	1.64	93.5	$-10H_2O$, 120	15^{20}	v s	s dil a; i al
a83	orthoarsenate	$(NH_4)_3AsO_4.3H_2O$	247.08	rhomb cr		d, $-NH_3$		sl s		
a84	orthoarsenate, di-H	$NH_4H_2AsO_4$	158.98	col, tetr, 1.577, 1.522	2.311^9	d, $-NH_3^{300}$		33.74^0	122.4^{90}	
a85	orthoarsenate mono-H	$(NH_4)_2HAsO_4$	176.00	col, monocl, odor NH_3	1.989	d		s	d	i al
a86	metaarsenite	NH_4AsO_2	124.96	col, rhomb pr, hygr				v s	d	i al, acet; sl s NH_4OH
a87	azide	NH_4N_3	60.06	col pl	1.346	160	subl 134 expl	20.16^{30}	27.04^{10}	1.06^{20} 80 % al; i eth, bz
a88	benzene sulfonate	$NH_4C_6H_5SO_3$	175.21	rhomb	1.342	d 271–275		98	320	19 c al; i eth, bz
a89	benzoate	$NH_4C_7H_5O_2$	139.16	col, rhomb	1.260	d 198	subl 160	$19.6^{14.5}$	83.3^{100}	1.63^{25} al; s glyc; i eth
a90	pentaborate	Aluminum decaborate. $NH_4B_5O_8.4H_2O$	272.15					7.03^{18}		
a91	peroxyborate	$NH_4BO_3.\frac{1}{2}H_2O$	85.86	wh cr		d		$1.55^{17.5}$		i al
a92	tetraborate	Ammonium biborate. $(NH_4)_2B_4O_7.4H_2O$	263.38	col, tetr		d		7.27^{18}	52.68^{90}	sl s acet; i al
a93	bromate	NH_4BrO_3	145.95	col, hex		expl		v s	vs	
a94	bromide	NH_4Br	97.95	cub, coll hygr, 1.712^{25}	2.429	subl 452	235 vac	97^{25}	145.6^{100}	10^{78} al; s acet, eth, NH_3
a95	dibromoiodide	NH_4IBr_2	304.75	met-grn pr, hygr		198		v s	s	s eth
a96	bromoplatinate	$(NH_4)_2PtBr_6$	710.62	red-brn cub	4.26^{24}	d 145		0.59^{20}	0.36^{100}	
a97	bromoselenate	$(NH_4)_2SeBr_6$	594.49	red, oct cr	3.326	d		d	d	sl s eth
a98	bromostannate	$(NH_4)_2SnBr_6$	634.22	col, cub	3.50	d		v s		
a99	cadmium chloride	$4NH_4Cl.CdCl_2$	397.27	col, rhomb, 1.6038	2.01			s		
a100	calcium arsenate	$NH_4CaAsO_4.6H_2O$	305.13	col, monocl	1.905^{15}	d 140		0.02	s	s NH_4Cl; i NH_4OH
a101	calcium phosphate	$NH_4CaPO_4.7H_2O$	279.20	1.561^{15}		d		i	d	s a
a102	carbamate	$NH_4NH_2CO_2$	78.07	col, rhomb		subl 60		v s	d	v s NH_4OH; sl s al; i acet
a103	carbamate acid carbonate	Sal volatile. $NH_4NH_2CO_2.NH_4HCO_3$	157.13	wh cr		subl		25^{11}	h /66	d al, s glyc; i acet
a104	carbonate	$(NH_4)_2CO_3.H_2O$	114.10	col, cub		d 58		100^{15}	d	i al, NH_3, CS_2; s dil MeOH
a105	carbonate, hydrogen	Ammonium bicarbonate. NH_4HCO_3	79.06	col, rhomb or monocl, 1.423, 1.536, 1.555	1.58	107.5 (d 36–60)	subl	11.9^0	d	i al, acet
a106	cerium nitrate(ic)	$(NH_4)_2Ce(NO_3)_6$	548.23	or, monocl				141^{25}	227^{80}	s HNO_3, al
a107	cerium nitrate(ous)	$2NH_4NO_3.Ce(NO_3)_3.4H_2O$	558.28	monocl		74		318.20	817.4^{65}	
a108	cerium sulfate(ous)	$(NH_4)_2SO_4.Ce_2(SO_4)_3.8H_2O$	844.69	monocl	2.523	$-6H_2O$, 100	$-8H_2O$,150	3.29^{49} (anhydr)		
a109	chlorate	NH_4ClO_3	101.49	col, monocl need.	1.80	102 expl		28.7^0	115^{75}	sl s al
a110	perchlorate	NH_4ClO_4	117.49	col, rhomb, 1.482	1.95	d		10.74^0	42.45^{85}	s acet; sl s al
a111	chloride	Sal ammoniac. NH_4Cl	53.49	col, cub, 1.642	1.527	subl 340	520	29.7^0	75.8^{100}	0.6^{19} al; s liq NH_3; i acet, eth
a112	chloroaurate	NH_4AuCl_4	356.82	yel, monocl or rhomb			520	s		sl s al
a113	chloroaurate, hydrate	$(NH_4AuCl_4)_4.5H_2O$	1517.35	yel, monocl		$-5H_2O$, 100		s		s al
a114	chlorogallate	NH_4GaCl_4	229.57	wh cr		275		v s	v s	s al; i pet eth
a115	chloroiridate	$(NH_4)_2IrCl_6$	441.00	red-blk, cub	2.856	d		0.69^{14}	4.38^{80}	i al; s HCl
a116	chloroiridite	$(NH_4)_3IrCl_6.1\frac{1}{2}H_2O$	486.06					s		
a117	chloroosmate	$(NH_4)_2OsCl_6$	439.00	cub	2.93					
a118	chloropalladate	$(NH_4)_2PdCl_6$	355.20	red-brn, cub	2.418	d		sl s		
a119	chloropalladite	$(NH_4)_2PdCl_4$	284.29	olive grn, tetr	2.17	d		s		i al
a120	hexachloroplatinate	$(NH_4)_2PtCl_6$	443.89	yel, cub, 1.8	3.065	d		0.7^{15}	1.25^{100}	0.005 al; i eth, conc HCl
a121	chloroplatinite	$(NH_4)_2PtCl_4$	372.98	red, rhomb (tetr)	2.936	d 140–150		s	s	i al
a122	chloroplumbate	$(NH_4)_2PbCl_6$	455.98	yel, cub	2.925	d 120		sl s	d	s a
a123	chlorostannate	$(NH_4)_2SnCl_6$	367.49	wh, cub	2.4	d		$33.^{14.5}$	v s	
a124	tetrachlorozincate	$ZnCl_2.2NH_4Cl$	243.26	wh pl, rhomb, hygr	1.879	d 150		v s		
a125	chromate	$(NH_4)_2CrO_4$	152.08	yel, monocl	1.91^{12}	d 180		40.5^{30}	d	i al; sl s NH_3, acet
a126	dichromate	$(NH_4)_2Cr_2O_7$	252.06	or, monocl	2.15^{25}	d 170		30.8^{15}	89^{30}	s al; i acet
a127	peroxychromate	$(NH_4)_3CrO_8$	234.11	red-brn, cub		d 40	expl 50	sl s		i al, eth, sl s NH_3; expl H_2SO_4
a129	chromium sulfate(ic)	$NH_4Cr(SO_4)_2.12H_2O$	478.34	grn or vlt, cub; 1.4842	1.72	94, $-9H_2O$, 100		21.2^{25}	32.8^{40}	s al, dil a
a130	citrate, di(sec.)	$(NH_4)_2HC_6H_5O_7$	226.19	wh gran or powd	1.48			100		sl s al
a131	citrate, tri-(tert.)	$(NH_4)_3C_6H_5O_7$	243.22	wh cr, deliq		d		v s		i al, eth, acet
a132	cobalt orthophosphate(ous)	$(NH_4)CoPO_4.H_2O$	189.96	vlt cr powd				i	d	s a

No.	Name	Synonyms and Formulae	Mol. wt.	Crystalline form, properties and index of refraction	Density or spec. gravity	Melting point, °C	Boiling point, °C	Solubility, in grams per 100 cc		
								Cold water	Hot water	Other solvents
	Ammonium									
a133	cobalt (II)-selenate	$(NH_4)_2SeO_4.CoSO_4.6H_2O$	489.02	ruby-red, monocl, 1.526, 1.532, 1.541	2.228_4^{20}					
a134	cobalt sulfate(ous)	$(NH_4)_2SO_4.CoSO_4.6H_2O.$	395.23	ruby-red, monocl, 1.490, 1.495, 1.503	1.902			20.5^{20}	45.4^{80}	i al
a135	copper chloride(ic)	$2NH_4Cl.CuCl_2.2H_2O$	277.46	blue, tetrag, 1.744, 1.724	1.993	d 110		33.8^0	99.3^{80}	s a, al; sl s NH_3
a136	copper iodide(ous)	$NH_4I.CuI.H_2O$	353.40	rhomb pl				d	d	s NH_4I
a137	cyanate	NH_4OCN	60.06	wh cr	1.342_4^{20}	d 60		v s	d	sl s al; i eth
a138	cyanide	NH_4CN	44.06	col, cub	1.02^{100}	d 36	subl 40	v s	d	v s al
a139	cyanaurate	$NH_4Au(CN)_4.H_2O$	337.09	col pl		d 200		v s		v s al; i eth
a140	cyanaurite	$NH_4Au(CN)_2$	267.04	col, cub		d 100		v s	v s	s al; i eth
a141	cyanoplatinite	$(NH_4)_2Pt(CN)_4.H_2O.$	**353.61**	yel cr				s		
a142	ethylsulfate	$NH_4C_2H_5SO_4.$	143.16			99		s		
a143	ferricyanide	$(NH_4)_3Fe(CN)_6$	266.06	red cr, rhomb		d		v s		
a144	ferrocyanide	$(NH_4)_4Fe(CN)_6.3H_2O.$	338.15	yel, monocl, turns bl in air		d		s	d	i al
a145	fluoantimonite	$(NH_4)_2SbF_5$	252.84	col, rhomb		d, subl		108		
a146	fluoborate	NH_4BF_4	104.84	wh, rhomb	1.871^{15}	subl		25^{16}	97^{100}	s NH_4OH
a147	fluogallate	$(NH_4)_3GaF_6$	237.83	wh oct cr		d >250– GaF_3			sl s	
a148	fluogermanate	$(NH_4)_2GeF_6$	222.66	col hex pr and bipyram, 1.428, 1.425	2.564_{25}^{25}			s		i al, MeOH
a149	fluophosphate, di-	$NH_4PO_2F_2$	119.01	col, rhomb		213		s	s	s al, acet
a150	fluophosphate, hexa-	NH_4PF_6	163.00	col, pl	2.180^{18}	d		s	s	s al, acet
a151	fluoride	NH_4F	37.04	col, hex, deliq	1.009^{25}	subl		100^0	d	s al; i NH_3
a152	fluoride, hydrogen	NH_4HF_2	57.04	rhomb or tetr, deliq, 1.390	1.50	125.6		v s	v s	sl s al
a153	fluorosulfonate	$NH_4FSO_3.$	117.10	wh need		d 245		v s	v s	sl s al; s MeOH
a154	fluosilicate	α-Nat. cryptohalite. $(NH_4)_2SiF_6$	178.14	α oct, β hex, col α 1.3696	α 2.011 β 2.152	d		18.6^{17}	55.5^{100}	sl s al; i acet
a155	fluosulfonate	$NH_4SO_3F.$	117.10	col, need		244.7		s	s	sl s al; s MeOH; d NH_4OH
a156	fluotitanate	$(NH_4)_2TiF_6$	197.97	hex pr		d		s	s	i al, eth
a157	fluozirconate	$(NH_4)_2ZrF_6$	241.29	rhomb, hex	1.154			s	s	
a158	fluozirconate	$(NH_4)_3ZrF_7$	278.32	col, cub	1.433			sl s		
a159	formate	$NH_4CHO_2.$	63.06	wh, monocl, deliq.	1.280	116	d 180	102^0	531^{80}	s al, eth, NH_3
a160	gallium sulfate	Ammonium gallium alum. $Ga(NH_4)(SO_4)_2.12H_2O$	496.07	cub, 1.468	1.777			30.84^{25}		0.00875^{25} 70% EtOH
a161	trihydrogen para-periodate	$(NH_4)_2H_3IO_6$	262.00	col, rhomb	2.85					
a162	hydroxide	NH_4OH	35.05	at ord temp in sol only		−77		s		
a163	iodate	NH_4IO_3	192.94	col, rhomb or monocl	3.309^{21}	d 150		2.06^{15}	14.5^{101}	
a164	iodide	NH_4I	144.94	col, cub, hygr, 1.7031	2.514^{25}	subl 551	220 vac	154.2^0	250.3^{100}	v s al, acet, NH_3; sl s eth
a165	triiodide	$NH_4I_3.$	398.75	dk br, rhomb	3.749	d 175		s d	d	
a166	iodoplatinate	$(NH_4)_2PtI_6$	992.59	blk, cub	4.61					i al
a167	iridium chloride (III)	$(NH_4)_3IrCl_6.H_2O$	477.05	grn-br-bl, cub		d 350		10.5^{19}		
a168	iridium sulfate	$NH_4Ir(SO_4)_2.12H_2O$	618.54	yel-red cr		106		s		
a169	iron (III) chloride	$2NH_4Cl.FeCl_3.H_2O$	287.20	ruby-red, rhomb, hygr, 1.78	1.99	234		v s	v s	
a170	iron (III) fluoride	$3NH_4F.FeF_3.$	223.95	col to lt yel, oct	1.96			sl s	sl s	
a171	iron (II) selenate	$(NH_4)_2SeO_4.FeSO_4.6H_2O$	485.93	lt grn, monocl, 1.5226, 1.5260, 1.5334	2.191_4^{20}					
a172	iron (III) sulfate	$(NH_4)_2SO_4.Fe_2(SO_4)_3.$	532.02	wh, hex	2.49^{22}	d 420		44.15^{25}		
a173	iron sulfate(ic)	$NH_4Fe(SO_4)_2.12H_2O$	482.19	vlt, cub oct, 1.4854	1.71	39–41	−12H_2O, 230	124.0^{25}	400^{100}	i al; s dil a
a174	iron sulfate(ous)	$(NH_4)_2SO_4.FeSO_4.6H_2O.$	392.14	grn, monocl, 1.487, 1.492, 1.499	1.864_4^{20}	d 100–110		26.9^{20}	73.0^{80}	i al
a175	lactate	$NH_4C_3H_5O_3.$	107.11	col-yelsh liq	$1.19-1.21^{15}$			∞		∞ al
a176	laurate, acid (mixt.)	$NH_4C_{12}H_{22}O_2.C_{12}H_{24}O_2.$	417.68	wh		75	d	s		4.8⁷ al; sl s eth, acet
a177	magnesium arsenate	$NH_4MgAsO_4.6H_2O.$	289.36	col, tetrg, 1.608	1.932^{15}	d		0.038^{20}	0.024^{80}	s a; i al
a178	magnesium carbonate	$(NH_4)_2CO_3.MgCO_3.4H_2O$	252.47	wh				s	v s	s a; i al

No.	Name	Synonyms and Formulae	Mol. wt.	Crystalline form, properties and index of refraction	Density or spec. gravity	Melting point, °C	Boiling point, °C	Solubility, in grams per 100 cc		
								Cold water	Hot water	Other solvents
	Ammonium									
a179	magnesium chloride	$NH_4Cl.MgCl_2.6H_2O$	256.80	col, rhomb, deliq	1.456	$-2H_2O$, 100 d		16.7		d al
a180	magnesium chromate	$(NH_4)_2CrO_4.MgCrO_4.6H_2O$	400.51	yel, monocl, 1.636, 1.637, 1.653	1.84	d		v s	v s	
a181	magnesium phosphate	Guanite, struvite $NH_4MgPO_4.6H_2O$	245.41	col, rhomb, 1.495, 1.496, 1.504	1.711	d		0.0231^0	0.0195^{80}	v s dil a; i al
a182	magnesium selenate	$(NH_4)_2SeO_4.MgSeO_4.6H_2O$	454.40	col monocl pr, 1.507, 1.509, 1.517	2.058^{20}_4			sl s		
a183	magnesium sulfate	Nat. boussingaulite. $(NH_4)_2SO_4.MgSO_4.6H_2O$	360.60	col, monocl, 1.472, 1.473, 1.479	1.723	>120	d 250	17.92^0	64.7^{100} (anhydr)	
a184	l-malate, hydrogen	$NH_4HC_4H_4O_5$	151.08	col, rhomb	1.5	161	d	$32.2^{15.7}$		
a185	permanganate	NH_4MnO_4	136.97	rhomb	2.208^{10}	d 110		7.9^{15}	d	
a186	manganese phosphate(ic)	$NH_4MnPO_4.H_2O$	185.96	wh cr				0.0031	0.05	i al, NH_4 salts
a187	manganese sulfate(ous)	$(NH_4)_2SO_4.MnSO_4.6H_2O$	391.23	pa red, monocl, 1.480, 1.484, 1.491	1.83			51.3^{25}	v s	
a188	molybdate	$(NH_4)_2MoO_4$	196.01	col, monocl pr	2.276^{25}_4	d		s, d	d	s a; i al, NH_3, acet
a189	paramolybdate	"Molybdic acid" com'l. $(NH_4)_6Mo_7O_{24}.4H_2O$	1235.86	col-yelsh, monocl	2.498	$-H_2O$, 90	d 190	43	d	i al; s a, alk
a190	permolybdate	$3(NH_4)_2O.5MoO_3.2MoO_4.6H_2O$	608.17	lt yel monocl pr	2.975	d 170		v s	v s	sl s al
a191	molybdotellurate	$(NH_4)_6TeMo_6O_{24}.7H_2O$	1321.56	col, rhomb	2.78	d 550		s	s	
a192	myristate, acid (mixt.)	$NH_4C_{14}H_{27}O_2.C_{14}H_{28}O_2$	473.78	wh solid		75–90	d	sl s	s	s al; i eth
a193	nickel chloride	$NH_4Cl.NiCl_2.6H_2O$	291.20	grn, monocl, deliq	1.654			v s	v s	
a194	nickel sulfate	Double nickel salt. $(NH_4)_2SO_4.NiSO_4.6H_2O$	395.00	dk bl-grn, monocl, 1.495, 1.501, 1.508	1.923			10.4^{20}	30^{80}	i al; s $(NH_4)_2SO_4$
a195	nitrate	NH_4NO_3	80.04	col, rhomb, (monocl >32.1°)	1.725^{25}	169.6	210^{11}	118.3^0	871^{100}	3.8^{20} al; 17.1^{20} MeOH s acet, NH_3; i eth
a196	nitratocerate	$(NH_4)_2Ce(NO_3)_6$	548.23	yel-red, monocl				142.6^{25}	232^{90}	s al; sl s HNO_3
a197	nitrite	NH_4NO_2	64.04	wh-yelsh cr	1.69	60–70 expl	30 subl vac	v s	s	s dil al; i eth
a198	oleate, acid (mixt.)	$NH_4C_{18}H_{33}O_2.C_{18}H_{34}O_2$	581.97	wh powd		d 78		s	s	80^{50} al; 13.3^{15} eth
a199	osmium chloride	$(NH_4)_2OsCl_6$	439.00	blk, oct	2.93^{25}	subl 170		d	d	s HCl
a200	oxalate	$(NH_4)_2C_2O_4.H_2O$	142.11	col, rhomb, 1.439, 1.546, 1.594	1.50	d		2.54^0	11.8^{50}	i NH_3
a201	oxalate, acid	Ammonium binoxalate. $NH_4HC_2O_4.H_2O$	125.08	col, rhomb	1.556	$-H_2O$, 170		s		i eth, bz
a202	oxalatoferrate (III)	$(NH_4)_3Fe(C_2O_4)_3.3H_2O$	428.07	grn, monocl	1.78	d 165		42.7^0	345^{100}	
a203	palladium (II)-chloride	$(NH_4)_2PdCl_4$	284.29	grnsh-yel, tetr	2.17	d		v s	v s	s dil al; i abs al
a204	palladium (IV)-chloride	$(NH_4)_2PdCl_6$	355.20	red oct cr	2.418	d $-Cl$		sl s	sl s	
a205	palmitate, (acid)- (mixt.)	$NH_4C_{16}H_{31}O_2.C_{16}H_{32}O_2$	529.90	yelsh soapy mass or yel powd		>100		sl s	s	8.8^{50} al; 0.23^{13} eth
a206	metaperiodate	NH_4IO_4	208.94	col, tetr	3.056^{18}	expl		2.7^{16}		
a207	hypophosphate	$(NH_4)_2H_2P_2O_6$	196.04	col cr		170		7^{25}	25^{100}	
a208	orthophosphate	$(NH_4)_3PO_4.3H_2O$	203.13	wh pr				26.1^{25}		sl s dil NH_4OH; i NH_3, acet
a209	orthophosphate, di-H	$NH_4H_2PO_4$	115.03	col, tetr, 1.525, 1.479	1.803^{19}	190		22.7^0	173.2^{100}	i acet
a210	orthophosphate-mono-H	$(NH_4)_2HPO_4$	132.05	col, monocl, 1.52	1.619	d 155	d	57.5^{10}	106.0^{70}	i al, acet, NH_3
a211	hypophosphite	$NH_4H_2PO_2$	83.03	rhomb tabl	1.634	200	d 240	s		s al, NH_3; i acet
a212	orthophosphite, di-H	$NH_4H_2PO_3$	99.03	col, monocl pr		123	d 145	171^0	260^{31}	i al
a213	phosphofluoride, hexa-	NH_4PF_6	163.00	col, cub	2.180^{18}_4	d		74.8^{20}		s al, acet; d h a
a214	phosphomolybdate	Ammonium molybdo phosphate. $(NH_4)_3P(Mo_3O_{10})_4$	1876.35	yel powd		d		sl s	sl s	s alk; i al, HNO_3
a215	phosphotungstate	$(NH_4)_3P(W_3O_{10})_4$	2931.27	wh				sl s	sl s	
a216	picramate	$NH_4C_6H_4N_3O_5$	216.15	redsh-brn cr powd				s		
a217	picrate	$NH_4C_6H_2N_3O_7$	246.14	red or yel, rhomb	1.719	d	expl 423	1.1^{20}	s	sl s al
a219	praseodymium sulfate	$(NH_4)_2SO_4.Pr_2(SO_4)_3.8H_2O$	846.26	cr	$2.531^{16.5}$	$-8H_2O$, 170		sl s		
a220	propionate	$NH_4C_3H_5O_2$	91.11	pr, deliq	1.108^{25}	45		v s		s al, ac a
a221	perrhenate	NH_4ReO_4	268.24	wh, hex pl	3.97	d	d	6.1^{20}	32.34^{80}	

No.	Name	Synonyms and Formulae	Mol. wt.	Crystalline form, properties and index of refraction	Density or spec. gravity	Melting point, °C	Boiling point, °C	Solubility, in grams per 100 cc		
								Cold water	Hot water	Other solvents
	Ammonium									
a222	rhodanid	NH_4NCS	76.12	monocl cr, to rhomb at 92, ~1.685	1.305	149.6	d 170	128[0]	347[60]	v s al; s MeOH, acet; i $CHCl_3$
a223	rhodium chloride	$(NH_4)_3RhCl_6.1H_2O$	387.75	dk red rhomb need		$-H_2O$, 140		s	s	sl s al; s dil NH_4Cl
a224	rhodium sulfate	Ammonium rhodium alum. $Rh(NH_4)(SO_4)_2.12H_2O$	529.25	orange, 1.5150		102				
a225	d-saccharate, hydrogen	$NH_4HC_6H_8O_8$	227.17	wh need or monocl pr				1.22[15]	24.35[100]	i c al; s h al
a226	salicylate	$NH_4C_7H_6O_3$	155.15	col, monocl			subl	111[25]	v s	28.8[25] al
a227	selenate	$(NH_4)_2SeO_4$	179.03	col, monocl, 1.561, 1.563, 1.585	2.194[20, 4]	d		117[7]	197[100]	i al, acet, NH_3
a228	selenate, hydrogen	NH_4HSeO_4	162.00	rhomb	2.162	d				
a229	selenide	$(NH_4)_2Se$	115.04	col, or wh cr		d		s		
a230	sodium phosphate, hydrate	Microsmic salt. $NH_4NaHPO_4.4H_2O$	209.07	col, monocl, 1.439, 1.442, 1.469	1.574	d 97				
a231	stearate, acid (mixt.)	$NH_4C_{18}H_{35}O_2.C_{18}H_{36}O_2$	586.00	wh cr		d 110		v s		0.3[25] al; 0.19[25] eth; 0.08[25] acet
a232	succinate	$(NH_4)_2.C_4H_4O_4$	152.15	col cr	1.37			s		sl s al
a233	sulfamate	$NH_4NH_2SO_3$	114.12	large, pl, deliq		125	d 160	166.6[10]	357[50]	i al
a234	sulfate	Nat. mascagnite. $(NH_4)_2SO_4$	132.14	col, rhomb, 1.521, 1.523, 1.533	1.769[50]	d 235		70.6[0]	103.8[100]	i al, acet, NH_3
a235	sulfate, hydrogen	Ammonium bisulfate. NH_4HSO_4	115.11	col, rhomb, 1.473	1.78	146.9	d	100	v s	sl s al; i acet
a236	peroxydisulfate	$(NH_4)_2S_2O_8$	228.18	col, monocl, 1.498, 1.502, 1.587	1.982	d 120		58.2[0]	v s	
a237	sulfide, hydro-	NH_4HS	51.11	wh rhomb, 1.74	1.17	118[150 atm]	88.4[19 atm]	128.1[0]	d	s al; NH_3
a238	sulfide, mono-	$(NH_4)_2S$	68.14	col yel cr (>-18), hygr		d		v s	d	s al; v s NH_3
a239	sulfide, penta-	$(NH_4)_2S_5$	196.40	yel pr		d 115		v s		s al, i eth, CS_2
a240	sulfite	$(NH_4)_2SO_3.H_2O$	134.15	col, monocl, 1.515	1.41[25]	d 60–70	subl 150	32.4[0]	60.4[100]	sl s al; i acet
a241	sulfite, hydrogen	Ammonium bisulfite. NH_4HSO_3	99.10	rhomb pr, deliq	2.03	subl 150 (in N_2)		71.8[0]	84.7[60]	
a242	dl-tartrate	$(NH_4)_2C_4H_4O_6$	184.15	col, monocl, d, α 1.55, β 1.581	1.601	d		58.01[15]	81.17[60]	sl s al
a243	dl-tartrate, hydrogen	$NH_4HC_4H_4O_6$	167.12	col, monocl pr, 1.561, 1.591	1.636	d		2.35[15]	3.24[25]	i al, s a, alk
a244	tellurate	$(NH_4)_2TeO_4$	227.67	wh powd	3.024[24.5]	d		s	s	i al; s dil a
a245	thallium chloride	$3NH_4Cl.TlCl_3.2H_2O$	507.23	col	2.39			s		
a246	thioantimonate	$(NH_4)_3SbS_4.4H_2O$	376.18	yel pr		d		71.2[0]	d	i al, eth
a247	thiocarbamate	$NH_4CS_2NH_2$	110.20	yel cr		d 50		v s		s al; sl s eth
a248	thiocarbonate, tri-	$(NH_4)_2CS_3$	144.28	yel cr, hygr		subl		v s		sl s al, eth
a249	thiocyanate	NH_4SCN	76.12	col, monocl, deliq	1.305	149.6	d 170	128[0]	v s	s al, acet, NH_3
a250	dithionate	$(NH_4)_2S_2O_6.\frac{1}{2}H_2O$	205.21	col, monocl	1.704	d 130		135[0]	v s	i al
a251	thiosulfate	$(NH_4)_2S_2O_3$	148.20	col, monocl, hygr.	1.679	d 150		v s	103.3[100]	i al; sl s acet
a252	titanium oxalate, basic	$(NH_4)_2TiO(C_2O_4)_2.H_2O$	294.03					v s		
a253	uranylcarbonate	$2(NH_4)_2CO_3.UO_2CO_3.2H_2O$	558.24	yel, monocl	2.773	d 100		5.8[19]	d	s($NH_4)_2CO_3$, aq.SO_2
a254	uranylfluoride, penta-	$(NH_4)_3UO_2F_5$	419.14	tetr cr, 1.495	3.186	subl		s		
a255	valerate	$NH_4C_5H_9O_2$	119.16	col or wh cr		d		s		s al, eth
a256	metavanadate	NH_4VO_3	116.98	wh-yelsh or col cr	2.326	d 200		0.52[15]	6.95[96], d	
a257	vanadium sulfate	$NH_4V(SO_4)_2.12H_2O$	477.28	red to bl	1.687	49		28.45[20]		
a258	zinc sulfate	$(NH_4)_2SO_4.ZnSO_4.6H_2O$	401.66	wh monocl, 1.489, 1.493, 1.499	1.931	d		7[0] (anhydr)	42[80] (anhydr)	
a259	**Antimony**	Sb	121.75	silv wh met, hex	6.684[25]	630.5	**1750**	i	i	s hot conc H_2SO_4, aq reg
a260	bromide, tri-	$SbBr_3$	361.48	col rhomb, 1.74	4.148[23]	96.6	280	d	d	s HCl, HBr, CS_2, NH_3, al, acet
a261	chloride, penta-	$SbCl_5$	299.02	wh liq or monocl, 1.601[14]	liq 2.336[20, 4]	2.8	79[22]	d	d	s HCl, tarta, $CHCl_3$
a262	chloride, tri-	Butter of antimony. $SbCl_3$	228.11	col, rhomb, deliq	3.140[25]	73.4	283	601.6[0]	∞[80]	s abs al, HCl, tart a, $CHCl_3$, bz, acet
a263	fluoride, penta-	SbF_5	216.74	col oily liq	2.99[23] liq	7	149.5	s		s KF
a264	fluoride, tri-	SbF_3	178.75	col, rhomb	4.379[20.9]	292	subl 319	384.7[0]	**563.6[30]**	s NH_3
a265	hydride	SbH_3	124.77	inflamm gas	gas 4.36[15]; liq 2.26[-25]	-88	$-17.1[751]$	0.41[0]		1500 ml A; 2500 ml CS_2
a266	iodide, penta-	SbI_5	756.27	br		79	400.6			
a267	iodide, tri-	SbI_3	502.46	ruby-red, hex, 2.78[Li], 2.36[Li]	4.917[17]	170	401	d	d	i al; s CS_2, bz, HI, HCl
a268	iodosulfide	SbSI	280.72	dk red		392	d	i	i	d conc HCl; i CS_2

No.	Name	Synonyms and Formulae	Mol. wt.	Crystalline form, properties and index of refraction	Density or spec. gravity	Melting point, °C	Boiling point, °C	Solubility, in grams per 100 cc Cold water	Hot water	Other solvents
	Antimony									
a269	mercaptoacetamide..	Antimony thiogly-colamide. $Sb(C_2H_4NOS)_3$	392.12	wh cr............		139		200		
a270	nitrate, basic......	$2Sb_2O_3.N_2O_5(?)$........	691.01	wh glossy cr....		d		d		d a; v sl s conc HNO_3
a271	nitride...........	SbN	135.76	or powd......		d		d		
a272	oxide, penta-......	Sb_2O_5 (or Sb_4O_{10}).....	323.50	yel powd........	3.80 (dep on temp)	— O, 380 — O_2, 930		v sl s	v sl s	v sl s KOH, HCl, HI
a273	oxide, tetra-......	Nat. cervantite. Sb_2O_4 (or $Sb_2O_3.Sb_2O_5$)	307.50	wh powd, 2.00....	5.82	— O, 930		v sl s	v sl s	v sl s KOH, HCl, HI
a274	oxide, tri-......	Nat. senarmontite. Sb_2O_3 (or Sb_4O_6)	291.50	wh, cub, 2.087....	5.2	656	1550 subl	v sl s	sl s	s KOH, HCl, tart, a, ac a
a275	oxide, tri-......	Nat. valentinite. Sb_2O_3(or Sb_4O_6)	291.50	col, rhomb, 2.18, 2.35, 2.35	5.67	656	1550	v sl s	sl s	**s KOH, HCl, tart, a, ac a**
a276	III oxychloride(ous)	SbOCl...........	173.20	wh monocl.......		d 170		i	d	s acet, HCl, CS_2; i al, eth, $CHCl_3$
a277	III oxychloride(ous).	$Sb_4O_5Cl_2$........	637.90	monocl......	5.01	d 320		i		s HCl; i al, eth
a278	oxyhydrate........	$H_4Sb_2O_7$......	359.13	wh, amorph......		— H_2O, 200		sl s	sl s	s alk
a279	III oxysulfate, di-(ous)	$Sb_2O_2SO_4$......	371.56	wh.............	4.89			d	d	s, H_2SO_4
a280	potassium tartrate...	Tartar emetic. $K(SbO)C_4H_4O_6.\frac{1}{2}H_2O$	333.93	col cr.........	2.6	— $\frac{1}{2}H_2O$, 100		8.3	33.3	i al; 6.7 glyc
a281	selenide.........	Sb_2Se_3......	480.38	gray cr........		611		v sl s		s conc HCl
a282	III sulfate......	$Sb_2(SO_4)_3$......	531.68	wh powd, deliq..	3.625⁴	d		i	d	s a
a283	sulfide, penta-.....	Sb_2S_5......	403.82	yel powd, prism.	4.120	d 75		i	i	i al; s HCl, alk, NH_4HS
a284	sulfide, tri-.....	Nat. stibnite. Sb_2S_3.....	339.69	blk, rhomb, 3.194, 4.064, 4.303	4.64	550	ca 1150	0.000175¹⁸		s al, NH_4HS, K_2S, HCl; i ac a
a285	sulfide, tri-.....	Sb_2S_3......	339.69	yel-red, amorph.	4.12	550	ca 1150	0.000173¹⁸		s al, NH_4SH, K_2S, HCl; i ac a
a286	d-tartrate.......	$Sb_2(C_4H_4O_6)_3.6H_2O$	795.84	wh cr powd......				s		
a287	telluride, tri-.....	Sb_2Te_3......	626.30	gray.........	6.50¹³	629				s HNO_3, aq reg
a288	**Argon**........	Ar.............	39.948	col inert gas..	1.784 g/l cr: 1.65⁻²³³, liq: 1.40⁻¹⁸⁶	— 189.2	185.7	5.6⁴ cm⁹	3.01⁵⁰ cm³	
a289	**Arsenic**.......	As.............	74.9216	gray met, hex-rhomb	5.727¹⁴	817 (28 atm.)	613 subl	i	i	s HNO_3
a290	**Arsenic**.......	As_4...........	299.69	yel, cub......	2.026¹⁸	d 358		i	i	s CS_2, bz
a291	**Arsenic acid, meta-**	$(HO)AsO_2$	123.93	wh, hygr.....		d	forms ortho-arsenic acid	d		
a292	ortho-.......	$H_3AsO_4.\frac{1}{2}H_2O$	150.95	wh translu hygr cr	2.0–2.5	35.5	— H_2O 160	302¹²·⁵	50	s al, alk, glyc
a293	pyro-........	$H_4As_2O_7$	265.87	col pr......		forms ortho-arsenic acid 206 d				
	Arsenic									
a294	bromide, tri-.....	$AsBr_3$........	314.65	col-yel hygr pr....	3.54²⁵	32.8	221	d	d	s HCl, HBr, CS_2
a295	chloride, tri-.......	$AsCl_3$........	181.28	oily liq or need, 1.621¹⁴ F	liq 2.163²⁰	— 8.5	**130.2**	d	d	s HBr, HCl, PCl_3, al, eth
a296	fluoride, penta-.....	AsF_5........	169.91	col gas........	7.71 g/l	— 80	— 53	s		s alk, al, eth, bz
a297	fluoride, tri-.....	AsF_3........	131.92	oily liq.......	liq 2.666	— 8.5	— 63⁷⁵²	d	d	s al, eth, bz, NH_4OH
a298	hydride........	Arsine. AsH_3	77.95	col gas.......	gas 2.695 liq 1.689⁸⁴·⁹	— 116.3	— 55 (d 300)	20 ml		s $CHCl_3$, bz
a299	iodide, di-.....	AsI_2........	328.73	red pr.......		d 136		d		s al, eth, $CHCl_3$, bz, CS_2
a300	iodide, penta-.....	AsI_5........	709.44		3.93	76				
a301	iodide, tri-.....	AsI_3........	455.64	red hex, ca. 2.59, ca. 2.23	4.39¹³	146	403	6²⁵	30 d	5.2 CS_2; s al, eth, bz, $CHCl_3$
a302	oxide, pent-.....	As_2O_5........	229.84	wh amor, deliq.	4.32	d 315		150¹⁶	76.7¹⁰⁰	s al, a, alk
a303	oxide, tri-.....	As_2O_3........	197.84	amor or vitreous	3.738	**312.3**		3.7²⁰	10.14¹⁰⁰	s alk, alk carb, HCl
a304	oxide, tri-.....	Nat. arsenolite. As_2O_3 (or As_4O_6)	197.84	col, cub or fibr, 1.755	3.865²⁵	subl 193		1.2²	11.46¹⁰⁰	s al, alk, HCl
a305	oxide, tri-.....	Nat. claudetite. As_2O_3 (or As_4O_6)	197.84	col, monocl, 1.871, 1.92, 2.01	4.15	193, subl 312.3	457.2	1.2²	11.46¹⁰⁰	s al, alk, HCl
a306	(III)oxychloride(ous)	AsOCl...........	126.37	brnsh......		d		d	d	
a307	phosphide, mono-....	AsP........	105.90	br-red powd....		subl, d		d	d	sl s CS_2; s H_2SO_4, HCl; i al, eth, $CHCl_3$
a308	selenide........	As_2Se_3........	386.72	br cr.......	4.75	ca. 360		i	d	s alk
a309	sulfide, di-.....	Nat. realgar. As_2S_2.....	213.97	red-br monocl, 2.46, 2.59, 2.61	α 3.506¹⁹ β 3.254¹⁹	α tr 267 β 307	565	i	i	s K_2S, $NaHCO_3$
a310	sulfide, penta-......	As_2S_5........	310.16	yel.............		subl, d 500		0.000136⁰	i	s alk, alk sulf, HNO_3

No.	Name	Synonyms and Formulae	Mol. wt.	Crystalline form, properties and index of refraction	Density or spec. gravity	Melting point, °C	Boiling point, °C	Solubility, in grams per 100 cc		
								Cold water	Hot water	Other solvents
	Arsenic									
a311	sulfide, tri-	Nat. orpiment. As_2S_3	246.04	yel or red, monocl, 2.4, 2.81, 3.02 (Li)	3.43	300	707	0.00005[18]	sl s	s al, alk, alk carb
	Auric or Aurous	*See* **Gold**								
b1	**Barium**	Ba	137.34	yel-silv met	3.51[20]	725	1640	d, ev H_2	d	s al; i bz
b2	acetate	$Ba(C_2H_3O_2)_2$	255.43	col cr	2.468			58.8[0]	75[100]	
b3	acetate, hydrate	$Ba(C_2H_3O_2)_2.H_2O$	273.45	col, tricl, 1.500, 1.517, 1.525	2.19	$-H_2O$, 150		76.4[26]	75[100]	sl s al
b4	amide	$Ba(NH_2)_2$	169.39	gray-wh cr		280		d	d	i liq NH_3
b5	*ortho*arsenate	$Ba_3(AsO_4)_2$	689.86		5.10	1605		0.055		s a, NH_4Cl
b6	*ortho*arsenate, mono-H	$BaHAsO_4.H_2O$	295.28	col, rhomb or monocl, 1.635	1.93[15]	$-H_2O$, 150		sl s	d	s HCl
b7	arsenide	Ba_3As_2	561.86	br	4.1[15]			d		d Cl_2, F_2, Br_2
b8	azide	$Ba(N_3)_2$	221.38	monocl pr	2.936	$-N_2$, 120	expl	17.3[17]		abs al 0.017[16]; i eth
b9	azide, hydrate	$Ba(N_3)_2.H_2O$	239.40	tricl, 1.7		expl		v s	v s	sl s al; i eth
b10	benzoate	$Ba(C_7H_5O_2)_2.2H_2O$	415.60	col, nacreous leaf		$-2H_2O$, 100		s		sl s al
b11	boride, hexa-	BaB_6	202.21	met-blk, cub	4.36[18]	2270		i	i	s HNO_3; i HCl
b12	bromate	$Ba(BrO_3)_2.H_2O$	411.16	col, monocl	3.99[18]	d 260		0.3[0]	5.67[100]	i al; s acet
b13	bromide	$BaBr_2$	297.16	col cr, 1.75	4.781[24]	847	d	104.1[20]	149[100]	v s al, MeOH
b14	bromide, dihydrate	$BaBr_2.2H_2O$	333.19	col, monocl, 1.713, 1.727, 1.744	3.58[24]	880, ($-H_2O$, 75)	$-2H_2O$, 120	151[20]	204[100]	v s MeOH, s al
b15	bromofluoride	$BaBr_2.BaF_2$	472.49	col pl	4.96[18]			d	d	i al; s conc HCl, conc HNO_3
b16	bromoplatinate	$BaPtBr_6.10H_2O$	992.04	monocl	3.71					
b17	butyrate	$Ba(C_4H_7O_2)_2.2H_2O$	347.57	col				37.42[0]	42.12[80]	
b18	carbide	BaC_2	161.36	gray, tetr	3.75			d to C_2H_2		d a
b19	carbonate(α)	$BaCO_3$	197.35	wh, hex	4.43	1740[90 atm]	d	0.002[20]	0.006[100]	s a, NH_4Cl; i al
b20	carbonate(β)	$BaCO_3$	197.35			tr to α, 982		0.0022[18]	0.0065[100]	s a, NH_4Cl; i al
b21	carbonate(γ)	Nat. witherite. $BaCO_3$	197.35	wh rhomb, 1.529, 1.676, 1.677	4.43	to β, 811	d 1450	0.0022[18]	0.0065[100]	s a, NH_4Cl; i al
b22	chlorate	$Ba(ClO_3)_2.H_2O$	322.26	col monocl, 1.5622, 1.577, 1.635	3.18	414 ($-H_2O$, 120)	$-O$, 250	27.4[15]	111.2[100]	sl s al, acet, HCl
b23	*per*chlorate	$Ba(ClO_4)_2$	336.24	col, hex	3.2	505		198.5[25]	562.3[100]	v s al
b24	*per*chlorate, hydrate	$Ba(ClO_4)_2.3H_2O$	390.29	col, hex, 1.533	2.74	d 400		198[33]	s	al 124[62]
b25	chloride α	$BaCl_2$	208.25	col, monocl; 1.7303, 1.7367, 1.7420	3.856[24]	tr to cub 962	1560	37.5[26]	59[100]	sl s HCl, HNO_3; v sl s al
b26	chloride β	$BaCl_2$	208.25	col, cub	3.917	963	1560			v sl s al
b27	chloride, hydrate	$BaCl_2.2H_2O$	244.28	col, monocl, 1.629, 1.642, 1.658 n_D^{26}	3.097[24]	$-2H_2O$, 113	35.7[20]		58.7[100]	sl s HCl, HNO_3; v sl s al
b28	*hypo*chlorite	$Ba(ClO)_2.2H_2O$	276.28	col cr		d				
b29	chlorofluoride	$BaCl_2.BaF_2$	383.58	col, tetr, 1.640	4.51[18]			d	d	i al; s conc HCl, HNO_3
b30	chloroplatinate	$BaPtCl_6.6H_2O$	653.24	or-yel, monocl	2.868	$-5H_2O$, 70	d	s		d a, al; i MeOH, eth
b31	chloroplatinite	$BaPtCl_4.3H_2O$	528.35	dk red pr	2.868	$-3H_2O$, 150		v s		s al
b32	chromate	$BaCrO_4$	253.33	yel, rhomb	4.498[15]			0.00034[16]	0.00044[38]	s min a
b33	*di*chromate	$BaCr_2O_7$	353.33	red, monocl				sl s		s h conc H_2SO_4
b34	*di*chromate, hydrate	$BaCr_2O_7.2H_2O$	389.36	bright red-yel need		$-2H_2O$, 120		d		s conc CrO_3 soln
b35	chromite	$BaO.4Cr_2O_3$	761.30	grn-blk, hex	5.4[15]			i	i	s a, fus carb
b26	citrate	$Ba_3(C_6H_5O_7)_2.7H_2O$	916.33	wh powd		$-7H_2O$, 150		0.0406[18]	0.0572[25]	sl s al; s HCl
b37	cyanide	$Ba(CN)_2$	189.38	wh cr powd				80[14]		18[14] 70 % al
b38	cyanoplatinite	$BaPt(CN)_4.4H_2O$	508.56	(a) monocl, yel, α 1.6704 (b) grn, rhomb	a) 2.076 b) 2.085	$-2H_2O$, 100		3[16]	25[100]	i al
b39	dithionate	$Ba(SO_3)_2.2H_2O$	333.50	col, rhomb or monocl, 1.586, 1.595, 1.607	4.536[18.5]	d 120		24.75[18]	90.9[100]	sl s al
b40	ethylsulfate	$Ba(C_2H_5SO_4)_2.2H_2O$	423.62	wh lust leaf				s		sl s al
b41	ferrocyanide	$Ba_2Fe(CN)_6.6H_2O$	594.73	yel, monocl	2.666	$-H_2O$, 40		0.17[16]	0.9[100]	i al
b42	fluogallate	$Ba_3(GaF_6)_2.H_2O$	797.46	wh cr	4.06	$-\frac{1}{2}H_2O$, 110; $-\frac{1}{2}H_2O$, 230		i		s HF
b43	fluoride	BaF_2	175.34	col, cub, 1.4741 n_D^{26}	4.89	1355	2137	0.12[25]	sl s	s a, NH_4Cl
b44	fluoroiodide	$BaF_2.BaI_2$	566.49	pl	5.21[18]			d	d	i al; s conc HCl, conc NHO_3

No.	Name	Synonyms and Formulae	Mol. wt.	Crystalline form, properties and index of refraction	Density or spec. gravity	Melting point, °C	Boiling point, °C	Solubility, in grams per 100 cc		
								Cold water	Hot water	Other solvents
	Barium									
b45	fluosilicate	BaSiF$_6$	279.42	rhomb need	4.29$_4^{21}$	d 300		0.026^{17}	0.09^{100}	i al; sl s a, NH$_4$Cl
b46	formate	Ba(CHO$_2$)$_2$	227.38	col, rhomb, 1.573, 1.597, 1.636	3.21	d		27.76^0	39.71^{90}	i al, eth
b47	gluconate	Ba(C$_6$H$_{11}$O$_7$)$_2$.3H$_2$O	581.69	pr or rhomb leaf		−3H$_2$O, 100; d 120		3.3$^{16.4}$		i al
b48	hydride	BaH$_2$	139.36	gray cr	4.21^0	d 675	1400(?)	d to Ba(OH)$_2$ + H$_2$		d a
b49	hydroxide	Ba(OH)$_2$.8H$_2$O	315.48	col, monocl, 1.471, 1.502, 1.50	2.18^{16}	78	−8H$_2$O, 78^{64}	5.6^{15}	94.7^{78}	sl s al; i acet
b50	hyponitrite	BaN$_2$O$_2$.4H$_2$O	269.41	wh cr powd	2.742^{25}			0.008^0		s HNO$_3$, HCl
b51	iodate	Ba(IO$_3$)$_2$	487.15	monocl	4.998	d		0.008^0	197^{100}	s HNO$_3$, HCl
b52	iodate, hydrate	Ba(IO$_3$)$_2$.H$_2$O	505.17	col, monocl	4.657^{18}	−H$_2$O, 200		v sl s	sl s	s HNO$_3$, HCl; i al, acet, H$_2$SO$_4$
b53	iodide	BaI$_2$	391.15	col cr	5.15$_4^{25}$	740		170^0		al 77^{20}
b54	iodide, hydrate	BaI$_2$.2H$_2$O	427.18	col rhomb, deliq	5.15	−H$_2$O, 98.9; −2H$_2$O, 539; d 740	200^{15}	269^{100}		1.07^{15} al; s acet
b55	iodide, hydrate	BaI$_2$.6H$_2$O	499.24	col, hex		25.7		410^0	v s	v s al
b56	laurate	Ba(C$_{12}$H$_{23}$O$_2$)$_2$	535.97	wh leaf cr		260		0.008$^{15.3}$	0.011^{60}	0.008^{25} al; 0.006^{25} eth
b57	l-malate	BaC$_4$H$_4$O$_5$	269.41					0.883^{20}	1.044^{90}	
b58	malonate	BaC$_3$H$_2$O$_4$.H$_2$O	257.40	col				0.143^0	0.326^{90}	
b59	manganate	BaMnO$_4$	256.28	gray-grn, hex	4.85			v sl s		s a
b60	per-manganate	Ba(MnO$_4$)$_2$	375.21	br-vlt cr	3.77	d 200		62.5^{11}	75.4^{25}	d al
b61	methylsulfate	Ba(CH$_3$SO$_4$)$_2$.2H$_2$O	395.56	col effl cr				s		s al
b62	molybdate	BaMoO$_4$	297.28	wh powd	4.65	1480		0.0058^{23}		sl s a
b63	myristate	Ba(C$_{14}$H$_{27}$O$_2$)$_2$	592.08					0.007^{25}	0.010^{60}	0.009^{25} al; 0.003^{25} eth 0.046^{15} MeOH
b64	nitrate	Nitrobarite. Ba(NO$_3$)$_2$	261.35	col cub, 1.572	3.24^{23}	592	d	8.7^{20}	34.2^{100}	i al, al s a
b65	nitride	Ba$_3$N$_2$	440.03	yel-br	4.783$_4^{0}$		1000 vac	d	d	
b66	nitrite	Ba(NO$_2$)$_2$	229.35	col, hex	3.23^{23}	d 217		67.5^{20}	300^{100}	sl s al
b67	nitrite, hydrate	Ba(NO$_2$)$_2$.H$_2$O	247.37	col-yelsh, hex	3.173^{20}	d 115		63^{20}	109.6^{80}	1.6 al; v s HCl; i acet
b68	oxalate	BaC$_2$O$_4$	225.36	cr	2.658	d 400		0.0093^{18}	0.0228^{100}	i al; s NH$_4$Cl, a
b69	oxide	BaO	153.34	col, cub, wh-yelsh powd, 1.98	5.72	1918	ca 2000	3.48^{20}	90.8^{100}	s dil a, al; i acet, NH$_3$
b70	oxide, per-	BaO$_2$	169.34	wh-gray powd	4.96	450	−O, 800	v sl s	d	s dil a; i acet
b71	oxide, per-, hydrate	BaO$_2$.8H$_2$O	313.46	col, hex	2.292	−8H$_2$O, 100		0.168	d	s dil a; i al, eth, acet
b72	palmitate	Ba(C$_{16}$H$_{31}$O$_2$)$_2$	648.19	wh cr powd		d		0.004^{15}	0.007^{60}	0.008$^{15.5}$ al; 0.001^{15} eth
b73	hypophosphate	BaPO$_3$	216.31	need				sl s		s al; v sl s ac a
b74	orthophosphate di-	BaHPO$_4$	233.32	wh, rhomb, 1.635, 1.617	4.165^{15}	d 410^{710}		0.01−0.02		s a, NH$_4$Cl
b75	orthophosphate, mono-	Ba(H$_2$PO$_4$)$_2$	331.31	tricl	2.9^4			d	d	s a
b76	orthophosphate, tri-	Ba$_3$(PO$_4$)$_2$	601.96	wh, cub	4.1^{16}			i	i	s a
b77	pyrophosphate	Ba$_2$P$_2$O$_7$	448.62	wh, rhomb	3.9^{20}			0.01	sl s	s a, NH$_4$ salts
b78	hypophosphite	Ba(H$_2$PO$_2$)$_2$.H$_2$O	285.33	wh, monocl	2.90$_4^{17}$	d 100−150		30^{15}	33^{100}	i al
b79	propionate	Ba(C$_3$H$_5$O$_2$)$_2$.H$_2$O	301.50	rhomb, β 1.518		d 300		48^0	67.9^{90}	0.08 al
b80	salycylate	Ba(C$_7$H$_5$O$_3$)$_2$.H$_2$O	437.65	wh need				s		
b81	selenate	BaSeO$_4$	280.30	wh, rhomb	4.75	d		0.0118	0.138^{100}	s HCl; i HNO$_3$
b82	selenide	BaSe	216.30	wh cub disc, n_D, 2.268	5.02			d	d	d HCl
b83	metasilicate	BaSiO$_3$	213.42	col, rhomb, 1.673, 1.674, 1.678	4.399	1604		i	d	s HCl
b84	metasilicate, hydrate	BaSiO$_3$.6H$_2$O	351.52	rhomb, 1.542, 1.548, 1.548	2.59			0.17	d	
b85	stearate	Ba(C$_{18}$H$_{35}$O$_2$)$_2$	704.13	wh powd				0.004^{15}	0.006^{60}	0.005$^{15.5}$ al, 0.008^{25} al, 0.001^{15} eth
b86	succinate	BaC$_4$H$_4$O$_4$	253.37	wh powd				0.421^0	0.237^{60}	sl s al
b87	sulfate	Nat. barite. prec. blanc fixe. BaSO$_4$	233.40	wh, rhomb (monocl), 1.637, 1.638, 1.649	4.50^{15}	1580	tr 1149 monocl	0.000222^{18} 0.000246^{25}	0.000336^{50} 0.000413^{100}	0.006 s 3 % HCl; sl s H$_2$SO$_4$
b88	peroxydisulfate	BaS$_2$O$_8$.4H$_2$O	401.52	wh, monocl		d		52.2^0	d	d al
b89	sulfide, hydro-	Ba(HS)$_2$.4H$_2$O	275.56	yel, rhomb		d 50		s		i al
b90	sulfide, mono-	BaS	169.40	col, cub, n_D 2.155	4.25^{15}	1200		d	d	i al
b91	sulfide, tetra-	BaS$_4$.H$_2$O	283.61	red or yel, rhomb	2.988	d 300		41^{15}	v s	i al, CS$_2$
b92	sulfide, tri-	BaS$_3$	233.53	yel-grn cr		d 554		s		i al

No.	Name	Synonyms and Formulae	Mol. wt.	Crystalline form, properties and index of refraction	Density or spec. gravity	Melting point, °C	Boiling point, °C	Solubility, in grams per 100 cc		
								Cold water	Hot water	Other solvents
	Barium									
b93	sulfite..........	$BaSO_3$......	217.40	col, cub hex......		d		0.02^{20}	0.002^{80}	v s HCl
b94	tartrate..........	$BaC_4H_4O_6.H_2O$......	303.52	wh cr..........	$2.980^{20.5}$			0.026^{18}	0.058^{90}	0.032^{18} al
b95	tellurate..........	$BaTeO_4.3H_2O$......	383.07	volum wh......	4.2^{200}	d >200		sl s	sl s	s HCl, NHO_3
b96	*pyrotellurate,* hydrogen	$Ba(HTe_2O_7)_2.H_2O$......	891.76	volum ppt; hot: yel; cold: wh				s	s	s a
b97	telluride..........	$BaTe$......	264.94	yel-wh, cub, disc, n_D 2.440	5.13					d a
b98	thiocarbonate......	$BaCS_3$......	245.54	yel, hex......		d		1.08^0	d	i al
b99	thiocyanate......	$Ba(SCN)_2.2H_2O$......	289.53	need, deliq......	2.286^{18}	$-H_2O$, 160		4.3^{20}	s	35^{20} al
b100	thiosulfate......	BaS_2O_3......	249.47	wh, rhomb......		d 220		0.2		
b101	thiosulfate, hydrate	$BaS_2O_3.H_2O$......	267.48	wh, rhomb......	3.5^{18}	d 100		0.208^{20}		i al
b102	titanate..........	$BaTiO_3$......	233.24	tetr and hex, 2.40	tetr 6.017, hex 5.806					
b103	tungstate..........	$BaWO_4$......	385.19	col, tetr......	5.04			sl s	sl s	d a
b104	*pyrovanadate*......	$Ba_2V_2O_7$......	488.56	wh cr......		863				
b105	**Beryllium**......	Glucinum. Be.	9.0122	grey met, hex.....	1.85^{20}	1278 ± 5	$2970^{(5mm)}$	i	sl s, d	s dil a, alk; i Hg
b106	acetate..........	$Be(C_2H_3O_2)_2$......	127.10	col pl..........		d 300		i		i al, eth, CCl_4
b107	acetate, basic......	$BeO(C_2H_3O_2)_6$......	406.32	oct..........	1.36^4	284	331	sl d	d	s chl, ac a; sl s al, eth
b108	acetate propionate, basic	$Be_4O(C_2H_3O_2)_3.$ $(C_3H_5O_2)_3$	448.40			127	330			
b109	aluminate..........	Nat. chrysoberyl. $BeAl_2O_4$	126.97	rhomb, 1.747, 1.748, 1.757	3.76	1870				i a
b110	aluminum silicate...	Nat. beryl. $Be_3Al_2(SiO_3)_6$	537.51	transp, hex, col, 1.580, 1.547	2.66	1410 ± 100				i a
b111	aluminum silicate...	Nat. euclase. $Be_2Al_2(SiO_4)_2.(OH)_2$	290.17	monocl, 1.652, 1.655, 1.671	3.1					
b112	benzenesulfonate....	$Be(C_6H_5O_3S)_2$......	323.35	monocl......				v s	v s	v s al, acet, ac a; i eth, bz, CS_2, CCl_4
b113	*ortho*borate, basic...	Nat. hambergite. $Be_2(OH)BO_3$	93.84	rhomb, 1.560, 1.591, 1.631	2.35					
b114	bromide..........	$BeBr_2$......	168.83	wh need, deliq...	3.465^{25}	490 ± 10 subl	520	s	v s	s al, eth; i bz; 18.56 pyr
b115	butyrate, basic......	$Be_4O(C_4H_7O_2)_6$......	574.64				239^{19}			
b116	carbide..........	Be_2C......	30.04	yel, hex..........	1.90^{15}	>2100 d		d	d	s a; d alk
b117	carbonate, basic......	$BeCO_3 + Be(OH)_2$......	112.05	wh powd......				i	d	s a, alk
b118	chloride..........	$BeCl_2$......	79.92	col need, deliq....	1.899^{25}	405	520 (488)	v s	v s, d	v s al, eth, pyr; sl s bz, chl, CS_2; i NH_3, acet
b119	fluoride..........	BeF_2......	47.01	col, amorph, <1.33	1.986^{25}	subl 800		∞	∞	sl s al; s H_2SO_4
b120	hydride..........	BeH_2......	11.03	wh cr..........		d 125		d	d	i eth, tol
b121	iodide..........	BeI_2......	262.82	col need..........	4.325^{25}	510 ± 10	590	d	d	s al, eth, CS_2
b122	nitrate..........	$Be(NO_3)_2.3H_2O$......	187.07	wh-yel cr, deliq...	1.557	60	142	v s	v s	v s al
b123	nitride..........	Be_3N_2......	55.05	col, cub..........		2200 ± 100	d 2240	d	d	d a, conc alk; i al
b124	oxalate..........	$BeC_2O_4.3H_2O$......	151.08	rhomb, β 1.487......		$-H_2O$, 100; $-3H_2O$, 220	d 350	38.22^{25}		
b125	oxide..........	Nat. bromellite. BeO...	25.01	wh hex, 1.719, 1.733	3.01	2530 ± 30	ca 3900	0.00002^{30}		s conc H_2SO_4, fus KOH
b126	oxide..........	$BeO.xH_2O$......		wh amorph powd or gel		d		i	i	s a, alk, $(NH_4)_2CO_3$
b127	2,4-pentanedione deriv	Beryllium acetyl-acetonate $Be(C_5H_7O_2)_2$	207.23	wh, monocl......	1.168^4	108	270	sl s	d	s al, eth, a
b128	*ortho*phosphate......	$Be_3(PO_4)_2.3H_2O$......	271.03			$-H_2O$, 100		s	s	s ac a
b129	propionate basic.....	$Be_4O(C_3H_5O_2)_6$......	490.43			120				
b130	selenate..........	$BeSeO_4.4H_2O$......	224.03	col, rhomb, 1.466, 1.501, 1.503	2.03	$-2H_2O$, 100; $-4H_2O$, 300		56.7^{25}	s	
b131	*di*-silicate..........	Nat. bertrandite. $Be_4Si_2O_7(OH)_2$	238.23	rhomb, 1.591, 1.605, 1.604	2.6					
b132	*ortho*silicate..........	Nat. phenazite. Be_2SiO_4	110.11	tricl, col, 1.654, 1.670	3.0					i a
b133	stearate (com'l)....	$Be(C_{18}H_{35}O_2)_2$......	575.97	wh, waxy..........		45		i	i	i al; s eth, CCl_4
b134	sulfate..........	$BeSO_4$......	105.07		2.443	d 550–600		i	d to $BeSO_4$ $4H_2O$	
b135	sulfate, hydrate.....	$BeSO_4.4H_2O$......	177.14	col, tetr, 1.472, 1.440	$1.713^{10.5}$	$-2H_2O$, 100; $-4H_2O$, 400		42.5^{25}	100^{100}	sl s conc H_2SO_4; i al
b136	sulfide..........	BeS......	41.08	reg..........	2.36			d	d	
b137	**Bismuth**..........	Bi.	208.9808	rhomb silver-wh or redsh met	9.80^{20}	271.3	1560 ± 5^{760}	i	i	s h H_2SO_4, HNO_3 aq, reg; sl s h HCl
b138	acetate..........	$Bi(C_2H_3O_2)_3$......	386.12	wh cr..........		d		i	i	s ac a

No.	Name	Synonyms and Formulae	Mol. wt.	Crystalline form, properties and index of refraction	Density or spec. gravtiy	Melting point, °C	Boiling point, °C	Solubility, in grams per 100 cc		
								Cold water	Hot water	Other solvents
	Bismuth									
b139	*ortho*arsenate	$BiAsO_4$	347.90	wh, monocl, 2.14, 2.15, 2.18	7.14	....	....	i	i	sl s h conc HNO_3
b140	benzoate	$Bi(C_7H_5O_2)_3$	572.33	wh powd	....	....	....			s a; i eth
b141	bromide, tri-	$BiBr_3$	448.71	yel cr powd, deliq	5.72_4^{25}	218	453	d to BiOBr	d	i al; s HCl, HBr, eth; v s liq NH_3
b142	carbonate, basic	Bismuth oxycarbonate. $Bi_2O_2CO_3$	509.97	wh powd	6.86	....	d	i	i	s a
b143	chloride, tetra-	$BiCl_4$	350.79	col cr	....	226				
b144	chloride, tri-	$BiCl_3$	315.34	wh cr, deliq	4.75_4^{25}	230–232	447	d to BiOCl	d	s a, al, eth, acet
b145	*di*chromate, basic	$(BiO)_2.Cr_2O_7$	665.94	yel-or-red cr	....	....	....	i	i	s a; i alk
b146	citrate	$BiC_6H_5O_7$	398.08	wh cr	3.458	d	....	sl s	sl s	sl s al; s NH_4OH
b147	fluoride, tri-	BiF_3	265.98	gray cr, cub, 1.74	5.32^{20}	727		i		s inorg a; i liq NH_3, al
b148	gallate, basic	Com'l dermatol. $Bi(OH)_2C_7H_5O_5$ (appr)	412.11	yel, amorph	d	....			i	i al, eth
b149	hydroxide	$Bi(OH)_3$ (appr)	260.00	wh amorph powd	4.36	$-H_2O$, 100; d 415	$-1\frac{1}{2}H_2O$, 400	0.00014	d	s a; i or sl s conc alk
b150	iodate	$Bi(IO_3)_3$	733.69	wh	....			i		i HNO_3
b151	iodide, di-	BiI_2	462.79	red need, rhomb	....	d 400	subl vac	s		s al, MeOH
b152	iodide, tri-	BiI_3	589.69	redsh, hex	5.778^{15}	408	*ca* 500	i	d	3.5^{20} al; i a; s NH_3; s HCl, HI
b153	lactate, *dl*	$Bi(C_3H_5O_3).7H_2O$	512.22	pr, need	....			14.4^{25}		i al
b154	molybdate	$Bi(MoO_4)_3$	897.78	yel-wh tetr need	6.07	643				v s a
b155	nitrate	$Bi(NO_3)_3.5H_2O$	485.07	col tricl, sl hygr	2.83	d 30	$-5H_2O$, 80	d	d	v s HNO_3; s a, glyc; i al 42^{19} acet
b156	nitrate, basic	$BiONO_3.H_2O$	305.00	hex leaf	4.928^{18}	$-H_2O$, 105	$-HNO_3$, 260	i	i	s a; i al
b157	oxalate	$Bi_2(C_2O_4)_3.7H_2O$	808.73	wh powd	....	$-6H_2O$, su 130		d	....	s inorg a; i al, eth
b158	oxide, mono-	BiO	224.97	dk-gray powd	7.15^{19}	d *ca* 180 (Bi_2O_3)		sl d	d	d dil a; s dil KOH
b159	oxide, pent-	Bi_2O_5	497.96	dk red or br	5.10	$-O$, 150	$-2O$, 357	i	i	s KOH
b160	oxide, tetr-	$Bi_2O_4.2H_2O$	517.99	br powd	5.6	$-H_2O$, 110	$-2H_2O$, 180	i	i	s a
b161	oxide, tri-	Bi_2O_3	465.96	yel, rhomb	8.9	825 ± 3	1890 (?)	i	i	s a
b162	oxide, tri-	Bi_2O_3	465.96	gray-blk, cub	8.20	tr 704		i	i	s a
b163	oxide, tri-	Bi_2O_3	465.96	wh-lt yel, rhomb, 1.91	8.55	860	....	i	i	sl s a
b164	oxybromide	BiOBr	304.89	col cr or wh powd	8.082^{15}	d red ht		i	i	i al; s a
b165	oxychloride	BiOCl	260.43	wh cr or powd, 2.15	7.72^{15}	red ht.		i	i	s a; i NH_3, tart a, acet
b166	oxyfluoride	BiOF	243.98	wh cr or powd	7.5_{20}^{20}	d red ht	....	i	i	s a
b167	oxyiodide	BiOI	351.88	red cr, tetr	7.922	d red ht		i	i	s a; i al, $CHCl_3$
b168	*ortho*phosphate	$BiPO_4$	303.95	wh, monocl	6.323^{15}	d		i	i	s HCl; i al, dil HNO_3
b169	propionate, basic	$BiOC_3H_5O_2$	297.97	wh powd	....				i	v s dil HCl; i al
b170	salicylate, basic	Bismuth subsalicylate. $Bi(C_7H_5O_3)_3.Bi_2O_3$ (appr)	1086.29	wh micr cr (variable comp)					i	s a, alk; i al, eth
b171	selenide, tri-	Nat. guanajuatite. Bi_2Se_3	654.84	blk, rhomb	6.82	710	d	i	i	i alk
b172	silicate	Nat. eulytite. $2Bi_2O_3.3SiO_2$	1112.17	yel, cub, 2.05	6.11			i	i	s, d HCl, HNO_3
b173	sulfate	$Bi_2(SO_4)_3$	706.14	wh need	5.08^{15}	d 405		d	d	s a
b174	mono-sulfide	BiS	241.04	dk gray powd	7.6–7.8	680 (in CO_2)	d	v sl s		....
b175	sulfide, tri-	Nat. bismuthinite, bismuthglance. Bi_2S_3	514.15	br-blk, rhomb, 1.340, 1.456, 1.459	7.39	d 685		0.000018^{18}		s HNO_3; i dil a
b176	tartrate	$Bi_2(C_4H_4O_6)_3.6H_2O$	970.27	wh powd	2.595^{25}	$-3H_2O$, 105		i	i	s a, alk; i al
b177	tellurate	Montanite. $Bi_2TeO_6.2H_2O$	677.59	biaxial, β: 2.09	3.79					
b178	telluride, tri-	Nat. tetradymite. Bi_2Te_3	800.76	gray, rhbdr	7.7_4^{20}	573				d HNO_3
b179	vanadate	Nat. pucherite. $Bi_2O_3.V_2O_5$	647.84	red-grn, rhomb, 2.41, 2.50, 2.51 (Li)	6.25^{25}					
b180	**Bismuthic acid**	$HBiO_3$	257.99	red	5.75	$-H_2O$, 120	$-2O$, 357	i	i	s a, KOH
b181	**Borazole**	$B_3N_3H_6$	80.50	col liq	1^{-45}, 0.8614^{06}	-58	55	sl d	d	....
b182	**Boric Acid**, meta-	HBO_2	43.82	wh cr, cub, 1.619	2.486	236 ± 1	....	v sl s	sl s	

No.	Name	Synonyms and Formulae	Mol. wt.	Crystalline form, properties and index of refraction	Density or spec. gravity	Melting point, °C	Boiling point, °C	Solubility, in grams per 100 cc		
								Cold water	Hot water	Other solvents
	Boric acid									
b183	ortho-.............	Boracic acid. H_3BO_3...	61.83	col, tricl, 1.337, 1.461, 1.462	1.435^{15}	169 ± 1 tr to HBO_2	$-1\frac{1}{2}H_2O$, 300	6.35^{30}	27.6^{100}	28^{30} glyc; 0.0078 eth; 5.56 al; 20. 20^{25} MeOH; 1.92^{25} liq NH_3; sl s acet
b184	tetra- (pyro-).....	$H_2B_4O_7$........	157.26	vitr or wh powd....				s	s	s al
b185	fluo-...............	HBF_4.............	87.81	col liq........			d 130	∞	∞	s al
b186	**Borinoaminoborine**	B_3H_7N........	42.68	col liq........		-66.5	76.2			s triborine tri-amine
b187	**Boron**.............	B................	10.811	yel monocl or br amorph powd	2.34, 2.37 amorph	2300	2550	i	i	v sl s HNO_3
b188	arsenate..........	$BAsO_4$..........	149.73	wh cr tetrag, 1.681, 1.690	3.64	subl ca 700	....	v sl s	1.4^{100}	i al; s inorg a
b189	bromide, tri-......	BBr_3............	250.54	col fum liq, $n_D^{16.3}$ 1.5312	$2.6431^{18.4}_4$	-46	91.3 ± 0.25	d		s al, CCl_4
b190	bromide, di-, iodide..	BBr_2I..........	297.53	col liq........			125	d	d	
b191	bromide, mono-, diiodide	$BBrI_2$..........	344.53	col liq........			180	d	d	
b192	(di-) bromide, mono-, pentahydride	B_2H_5Br........	106.67	col gas.......		-104	ca 10	hyd to HBO_2, $HBr +$ H_2		
b193	(tetra-) carbide....	B_4C.............	55.26	blk rhbdr........	2.52	2350	>3500	i	i	i a; s fus alk
b194	chloride, tri-......	BCl_3............	117.17	col fum liq, $1.4195^{5.7}$ α line H_2	1.349^{11}_4	-107.3	12.5		d to HCl $+ H_3BO_3$	d al
b195	fluoride, tri-......	BF_3.............	67.81	col gas.......	2.99 g/l	-126.7	-99.9	106 (762 mm)	d	d al; s conc H_2SO_4
b196	fluoride dihydrate...	$BF_3.2H_2O$......	103.84	col liq; n_{HE}^{20} 1.31498	1.6316^{20}_4	6	....	d	d	s eth, dioxan
b197	hydride...........	Diborane, boroethane. B_2H_6	27.67	col gas........	liq: 0.447^{-11} sol: 0.577^{-183}	-165.5	-92.5	sl s d to H_3BO_3 $+ H_2$		d 1.6^0 al; s NH_4OH, conc H_2SO_4
b198	hydride...........	Dihydrotetraborane, borobutane. B_4H_{10}	53.32	col gas, pois....	0.56^{-35}	-120.8	16	sl s, d		s bz; d al
b199	hydride...........	Pentaborane. B_5H_9	63.13	col liq........	0.66^0	-46.82	58.4	d		
b200	hydride...........	Hexaborane. B_6H_{10}	74.95	col liq........	0.69^0	-65	0^{72}	d		
b201	hydride...........	Decaborane. $B_{10}H_{14}$	122.22	wh cr.........	0.94^{25}	99.5	213	sl s	d	v s CS_2; s al, eth bz
b202	iodide, tri-.......	BI_3.............	391.52	col pl, hygr.....	3.35^{50}	49.9	210	d	d	d al; v s CS_2, bz, CCl_4
b203	nitride...........	BN...............	24.82	wh, hex.......	2.25	subl ca 3000	....	i	sl d	sl s h a
b204	oxide.............	B_2O_3..........	69.62	rhomb cr, 1,64, 1.61	2.46 ± 0.01	450 ± 2	ca 1860	sl s	s	....
b205	oxide glass........	B_2O_3..........	69.62	col, vitr 1.485..	1.812^{25}_4	ca 450	....	1.1^0	15.7^{100}	s al, a
b206	phosphide.........	BP...............	41.78	maroon powd....		ign 200		i	i	i all solv
b207	triselenide........	B_2Se_3..........	258.50	yel-gray powd....				d	d	....
b208	(hexa-) silicide.....	B_6Si............	92.95	blk cr.........	2.47			i		s HNO_3; d H_2SO_4; i KOH
b209	(tri-) silicide......	B_3Si............	60.52	blk rhomb......	2.52			i		sl s HNO_3; d H_2SO_4, KOH
b210	sulfide, penta-.....	B_2S_5...........	181.94	col, tetrag......	1.85	390		d	d	d al
b211	sulfide, tri-.......	B_2S_3...........	117.81	wh cr or vitr....	1.55	310		d		sl s PCl_3, SCl_2; d al
b212	**Borotungstic acid**...	$H_5BW_{12}O_{40}.30H_2O$.....	3402.49	tetr, cr........	3	$45-51$		s		s al, eth
b213	**Bromic acid**.......	$HBrO_3$..........	128.92	known in sol only, col or yelsh		d 100		v s	d	
b214	**Bromine**..........	Br_2.............	159.808	dk red liq, 1.661..	2.928^{59}_9 3.119^{20}	-7.2	58.78	4.17^0, 3.58^{30}	3.52^{50}	v s al, eth, chl, CS_2
b215	azide.............	Bromoazide. BrN_3	121.93	cr, red liq......		ca 45	exp			s eth, KI; sl s bz, ligr
b216	chloride..........	BrCl.............	115.36	red-col liq or gas..		ca -66 d 10	ca 5	s d		s eth, CS_2
b217	fluoride, mono-.....	BrF..............	98.91	red-br gas......		d -33	-20			
b218	fluoride, penta-.....	BrF_5............	174.90	col liq........	2.466^{25}	-61.3	40.5	d	d	
b219	fluoride, tri-......	BrF_3............	136.90	col-gray-yel liq..	2.49^{115}	(-2) 8.8	135	d viol. to O_2, HOBr, HF, $HBrO_3$		d alk
b220	hydrate...........	$Br_2.10H_2O$......	339.97	red oct........	1.49	d 6.8	....	s		
b221	oxide, di-.........	BrO_2............	111.91	lt yel.........		d 0				
b222	oxide, mon-.......	Br_2O............	175.82	dk br.........		-17 to -18				s, d CCl_4
b223	(tri-) oxide, oct-....	Br_2O_8(or Br_3O_8)ₙ.......	367.72	wh...........		stable at -40				

No.	Name	Synonyms and Formulae	Mol. wt.	Crystalline form, properties and index of refraction	Density or spec. gravity	Melting point, °C	Boiling point, °C	Solubility, in grams per 100 cc		
								Cold water	Hot water	Other solvents
	Bromauric acid									
b224	**Bromauric acid**	$HAuBr_4.5H_2O$	607.69	red-br cr		27		v s		s al
b225	**Bromous acid, hypo-**	$HBrO$	96.92	known only in soln, col-yel		40 (vac)		s	s d	s al, eth, chl
b226	**Bromoplatinic acid**	$H_2PtBr_6.9H_2O$	838.70	red monocl, deliq		d -100		v s	v s	
c1	**Cadmium**	Cd	112.41	hex silv-wh malleable met	8.642	320.9	**765**	i	i	s a, NH_4NO_3, h H_2SO_4
c2	acetate	$Cd(C_2H_3O_2)_2$	230.50	col	2.341	256	d	v s		s MeOH
c3	acetate, hydrate	$Cd(C_2H_3O_2)_2.2H_2O$	266.54	col monocl, odor ac a	2.01	$-H_2O$, 130		v s	v s	v s al
c4	amide	$Cd(NH_2)_2$	144.45		3.05^{75}	d 120				
c5	ammonium chloride	$CdCl_2.NH_4Cl$	236.79	col need, rhomb	2.93	289		33.45^{16}	$43.99^{63.8}$	s al, MeOH
c6	ammonium sulfate	$Cd(NH_4)_2(SO_4)_2.6H_2O$	448.69	col monocl pr	2.061^{20}_4	100 d $-H_2O$		s	s	
c7	arsenate, hydrogen	$CdHAsO_4.H_2O$	270.34		4.164^{15}_4	>120				
c8	arsenide	Cd_3As_2	487.04	dk gray cub	6.21^{15}_4	721		i	i	sl s HCl; s HNO_3; i aq reg
c9	benzoate	$Cd(C_7H_5O_2)_2.2H_2O$	390.66					3.34^{20}		sl s al
c10	borate	$Cd(BO_2)_2.H_2O$	248.03	wh rhomb	3.758	d		125^{17}		i al
c11	borotungstate	$Cd_5(BW_{12}O_{40}).18H_2O$	6600.25	yel tricl		75		1250^{19}	v s	
c12	bromide	$CdBr_2$	272.22	yel cr	5.192^{25}	567	863	57^{10}	162^{104}	26.6^{15} al; 0.4^{15} eth; s HCl; 1.6^{18} acet
c13	bromide, tetrahydrate	$CdBr_2.4H_2O$	344.28	sm wh need, effl		tr 36		121^{10}	v s	25 al; s acet; sl s eth
c14	carbonate	$CdCO_3$	172.41	wh, trig	4.258^4	d <500		i	i	s a, KCN, NH_4 salts; i NH_3
c15	chlorate	$Cd(ClO_3)_2.2H_2O$	315.33	col pr, deliq	2.28^{18}	80		298^0	487^{65}	s al, a, acet
c16	chloride	$CdCl_2$	183.32	col, hex	4.047^{25}	568	960	140^{20}	150^{100}	1.52^{15} al; $1.7^{15.5}$ MeOH; i acet, eth
c17	chloride	$CdCl_2.2\frac{1}{2}H_2O$	228.35	col monocl, 1.6513	3.327	tr 34		168^{20}	180^{100}	2.05^{15} MeOH; sl s al
c18	chloroacetate, di-	$Cd(C_2HCl_2O_2)_2.2H_2O$	386.29	need	2.132^{15}					
c19	chloroacetate mono-	$Cd(C_2H_2ClO_2)_2.6H_2O$	407.47		1.942^{25}					
c20	chloroacetate, tri-	$Cd(C_2Cl_3O_2)_2.1\frac{1}{2}H_2O$	464.18	rhomb	2.093^{25}					
c21	chloroplatinate	$CdPtCl_6.3H_2O$	574.25	yel trig need	2.882	$-H_2O$, 170, d		s	s	
c23	chromite	$CdCr_2O_4$	280.39	grn to blk, cub	5.79^{17}			i	i	i a
c24	cyanide	$Cd(CN)_2$	164.44	cr		dec >200		1.7^{15}	s	s a, KCN, NH_4OH; i al
c25	ferrocyanide	$Cd_2Fe(CN)_6.xH_2O$						i	i	s HCl
c26	fluogallate	$[Cd(H_2O)_6].[GaF_6H_2O]$	403.22	col cr, 1.45	2.79	$-5H_2O$, 110		v s		
c27	fluoride	CdF_2	150.40	wh cub, 1.56	6.64	1100	1758	4.35^{25}		s a, HF; i al, NH_3
c28	fluosilicate	$CdSiF_6.6H_2O$	362.57	col hex				s	s	s 50 % al
c29	formate	$Cd(CHO_2)_2.2H_2O$	238.47	monocl	2.44	d		v s		
c30	fumarate	$Cd(C_4H_2O_4)$	226.48					0.9^{30}		
c31	hydroxide	$Cd(OH)_2$	146.41	wh, trig or amorph	4.79^{15}	d 300		0.00026^{25}		s a, NH_4 salts; i alk
c32	iodate	$Cd(IO_3)_2$	462.21	wh cr	6.43	d		s		s NHO_3; $NH_4.OH$
c33	iodide	CdI_2	366.21	grn-yel powd	5.670^{30}_4	387	796	86.2^{25}	125^{100}	110.5^{20} al; 41^{25} acet; sl s NH_3; 206.7^{25} MeOH
c34	lactate	$Cd(C_3H_5O_3)_2$	290.54	need				10	12.5	i al
c35	maleate	$Cd(C_4H_2O_4).2H_2O$	262.49					0.66^{30}		
c36	permanganate	$Cd(MnO_4)_2.6H_2O$	458.36		2.81	d 95		v s	v s	s a, NH_4OH, KCN
c37	molybdate	$CdMoO_4$	272.34	yel pl	5.347			sl s		s a, NH_4OH, KCN
c38	nitrate	$Cd(NO_3)_2$	236.41	col		350		109^0	326^{60} 682^{100}	v s a; s et ac
c39	nitrate, tetrahydrate	$Cd(NO_3)_2.4H_2O$	308.47	wh pr, need, hygr	2.455^{17}_4	59.4	132	215		s al, NH_3; i HNO_3
c40	nitrocobaltate (III)	Cadium cobaltinitrite. $Cd[Co(NO_2)_6]_2$	1007.13	yel		d 175		sl s	v s	d a, alk, org solv
c41	oxalate	CdC_2O_4	200.42	col cr	3.32^{18}_4	d 340				s a; i al
c42	oxalate, trihydrate	$CdC_2O_4.3H_2O$	254.45	col cr		d		0.005^{18}	0.009	
c43	oxide	CdO	128.40	br, amorph	6.95	>1500	d 900–1000	i	i	s a, NH_4 salts; i alk
c44	oxide	CdO	128.40	br cub, 2.49 (Li)	8.15	>1500	subl 1559	i	i	s a, NH_4 salts; i alk
c45	orthophosphate	$Cd_3(PO_4)_2$	527.14	col, amorph		1500		i		s a, NH_4 salts

No.	Name	Synonyms and Formulae	Mol. wt.	Crystalline form, properties and index of refraction	Density or spec. gravity	Melting point, °C	Boiling point, °C	Solubility, in grams per 100 cc		
								Cold water	Hot water	Other solvents
	Cadmium									
c46	*pyro*phosphate	$Cd_2P_2O_7$	398.74	wh cr leaf	4.965^{15}	above red heat		sl s	s	s a, NH_3
c47	phosphate, dihydrogen	$Cd(H_2PO_4)_2.2H_2O$	342.41	col, tricl	2.74^{15}_4	d 100				i al, eth; s HCl
c48	phosphide	Cd_3P_2	399.15	grn, tetr need	5.60	700				s d HCl; s exp conc HNO_3
c49	potassium cyanide	$Cd(CN)_2.2KCN$	294.68	col, glossy, oct	1.847			33.3	100	i al
c50	potassium sulfate	$CdK_2(SO_4)_2.2H_2O$	322.69	tricl col tab	2.922^{15}			42.89^{15}	47.40^{40}	
c51	salycilate	$Cd(C_7H_5O_3)_2.H_2O$	404.65	wh need				sl s	s	s al, eth, glycerol a, NH_4OH
c52	selenate	$CdSeO_4.2H_2O$	291.39	rhomb	3.63	$-1H_2O$, 100; $-2H_2O$, 170		v s		
c53	selenide	$CdSe$	191.36	grn-br or red powd, hex	5.81^{15}_4	>1350		i		d a
c54	*meta*silicate	$CdSiO_3$	188.48	col, rhomb, 1.739	4.93	1242		v sl s		
c55	sulfate	$CdSO_4$	208.46	wh, rhomb	4.691^{20}_4	1000		75.5^0	60.8^{100}	i al, acet, NH_3
c56	sulfate, hydrate	$CdSO_4.H_2O$	226.48	monocl	3.79^{20}	tr 108		s	s	i al
c57	sulfate, hydrate	$CdSO_4.7H_2O$	334.57	col, monocl	2.48	tr 4		s		i al
c58	sulfate, hydrate	$3CdSO_4.8H_2O$	769.50	col, monocl, 1.565	3.09	tr 41.5		113^0	s	
c59	sulfide	Nat. greenockite. CdS.	144.46	yel-or, hex, 2.506, 2.529	4.82	1750^{100atm}	subl in N_2, 980	0.000013^{18}	colloid	s a; v sl s NH_4OH
c60	sulfite	$CdSO_3$	192.46	cr		d		sl s		i al; s a, NH_4OH
c61	tartrate	$CdC_4H_4O_6$	260.47	wh cr powd				sl s		s a, NH_4OH
c62	telluride	$CdTe$	240.00	blk, cub	6.20^{15}	1041		i		i a; d HNO_3
c63	tungstate	$CdWO_4$	360.25	yel cr				0.05		s NH_4OH
	Cadmium complexes									
c64	tetramminecadmium *per*rhenate	$[Cd(NH_3)_4](ReO_4)_2$	681.16		3.714^{25}_4					0.037 conc NH_4OH
c65	tetrapyridine cadmiumfluosilicate	$[Cd(C_5H_5N)_4]SiF_6$	570.89	wh, tricl	2.282					
c66	**Calcium**	Ca	40.08	silv wh soft met, cub	1.54	839±2	1484	d to H_2+ $Ca(OH)_2$	d	s a, liq NH_3; sl s al; i bz
c67	acetate	$Ca(C_2H_3O_2)_2$	158.17	col cr; 1.55, 1.56, 1.57		d		37.4^0	29.7^{100}	sl s al
c68	acetate, dihydrate	$Ca(C_2H_3O_2)_2.2H_2O$	194.21	col need		$-1H_2O$, 84		34.7^{20}	33.5^{80}	
c69	acetate, monohydrate	$Ca(C_2H_3O_2)_2.H_2O$	176.19	col need		d		43.6^0	34.3^{100}	sl s al
c70	aluminate	$CaAl_2O_4$ (or $CaO.Al_2O_3$)	158.04	wh monocl, tricl or rhomb; 1.643, 1.665, 1.663	2.981^{25}	1600		d		s HCl; i HNO_3, H_2SO_4
c71	(tri-)aluminate	$Ca_3Al_2O_6$ (or $3CaO.Al_2O_3$)	270.20	wh, cub, 1.710	3.038^{25}	d 1535		i		s a
c72	(tri-)aluminate hexahydrate	$3CaO.Al_2O_3.6H_2O$	378.29	col, oct, 1.603	2.52^{20}	d 700–800		d		
c73	aluminosilicate	$2Ca.Al_2O_3.SiO_2$	274.20	col, tetr, 1.669, 1.658	3.048	1590 ± 2				d a
c74	aluminosilicate	Nat. anorthite. $CaAl_2.Si_2O_8$ (or $CaO.Al_2O_3.2SiO_2$)	278.21	wh, tricl, 1.5832	2.765	1551				
c75	*ortho*arsenate	$Ca_3(AsO_4)_2$	398.08	col amorph powd	3.620	1.455		0.013^{25}		
c76	arsenate, trihydrate	Nat. haidingerite. $2CaO.As_2O_5.3H_2O$	396.04	col, rhomb, 1.590, 1.602, 1.638	2.967					
c77	arsenide	Ca_3As_2	270.08	red cr	3.031^{25}	d		d	d	d a; s h HNO_3
c78	azide	$Ca(N_3)_2$	124.12	col, rhomb, hyg		$-3H_2O$, 110; exp 144–156		38.1^0	45^{15}	0.211^{16} al; i eth
c79	benzoate	$Ca(C_7H_5O_2)_2.3H_2O$	336.36	col, rhomb	1.436	$-3H_2O$, 110		2.7^0	8.3^{80}	
c80	*meta*borate	$Ca(BO_2)_2$	125.70	col, flat rhomb pr, 1.550, 1.660, 1.680		1154		sl s		s a, NH_4 salts; sl s ac a
c81	*meta*borate, hexahydrate	$Ca(BO_2)_2.6H_2O$	233.79	col, tetr, 1.520, 1.502	1.88			0.25^{20}		
c82	*tetra*borate	CaB_4O_7	195.32	readily vitrified		986				
c83	boride	CaB_6	104.95	blk, cub	2.3^{20}	2235		i	i	s HNO_3; sl s conc H_2SO_4
c84	bromide	$CaBr_2$	199.90	col, rhomb need, deliq	3.353^{25}	sl d 730	806–812	142^{20}	312^{105}	s al, acet, a; sl s NH_3, MeOH
c85	bromate	$Ca(BrO_3)_2.H_2O$	313.91	monocl cr	3.329	$-H_2O$, 180		v s	v s	
c86	bromide, hexahydrate	$CaBr_2.6H_2O$	307.99	col, hex cr	2.295	38.2	149	594^0	1360^{25}	s al, acet, a

No.	Name	Synonyms and Formulae	Mol. wt.	Crystalline form, properties and index of refraction	Density or spec. gravity	Melting point, °C	Boiling point, °C	Solubility, in grams per 100 cc		
								Cold water	Hot water	Other solvents
	Calcium									
c87	butyrate	$Ca(C_4H_7O_2)_2.3H_2O$	268.32	col cr				s	sl s	
c88	carbide	CaC_2	64.10	col, tetr, 1.75	2.22	stab 25–447	2300	d	d	
c89	carbonate	Nat. aragonite. $CaCO_3$	100.09	col, rhomb, 1.530, 1.681, 1.685	2.930	tr to calcite 520	d 825	0.00153^{25}	0.00190^{75}	s a, NH_4Cl
c90	carbonate	Nat. calcite. $CaCO_3$	100.09	col, rhomb or hex, 1.6583, 1.4864	2.710^{18}	1339^{1025}	d 898.6	0.0014^{25}	0.0018^{75}	s a, NH_4Cl
c91	carbonate, hexahydrate	$CaCO_3.6H_2O$	208.18	col, monocl, 1.460, 1.535, 1.545	$1,771^0$					
c92	chlorate	$Ca(ClO_3)_2$	206.99	wh cr, hyg		340 ± 10 (−some O)		s	s	s al, acet
c93	chlorate, dihydrate	$Ca(ClO_3)_2.2H_2O$	243.01	wh-yelsh, rhomb, or monocl, deliq	2.711	$−H_2O$, 100		177.7^8	v s	s al, acet
c94	perchlorate	$(CaClO_4)_2$	238.98	col cr	2.651	d 270		188.6^{25}	v s	166.2^{25}_{25} al; 237.4 MeOH
c95	chloride	$CaCl_2$	110.99	col, cub, deliq 1.52	2.15^{25}_4	782	>1600	74.5^{20}	159^{100}	s al, acet, ace a
c96	chloride aluminate	$3CaO.Al_2O_3.CaCl_2.10H_2O$	561.33	col, monocl or hex, hex, 1.550, 1.535	1.892^{14}	$−H_2O$, 105	$−8H_2O$, 350	sl s	d	s a
c97	chloride, dihydrate	$CaCl_2.2H_2O$	147.02	col cr	0.835	stab	326^{60}	97.7^0	326^{60}	50^{60} al
c98	chloride, hexahydrate	$CaCl_2.6H_2O$	219.08	col, trig, deliq, 1.417, 1.393	1.71^{25}	29.92	$−4H_2O$, 30, $−6H_2O$, 200	279^0	536^{20}	s al
c99	chloride, monohydrate	$CaCl_2.H_2O$	129.00	col cr, deliq		260		76.8^0	249^{100}	s al; i acet
c100	chloride fluoride orthophosphate	$3Ca_3(PO_4)_2.CaClF$	1025.08	col cr, 1.674, 1.631	3.14	1270		v sl s		
c101	chlorite	$Ca(ClO_2)_2$	174.98	wh, cub	2.71			d	d	i al
c102	hypochlorite	$Ca(ClO)_2$	142.98	wh powd or flat pl, 1.54n, 1.60	2.35	d 100		s		i al
c103	chlorite, basic	$Ca(ClO_2)_2.2Ca(OH)_2$	257.16	wh, hex, 1.51, 1.585	2.10			sl s solns with 5–6 % avail Cl	d	d a
c104	hypochlorite, basic	Bleaching powder, chlorinated lime. $Ca(ClO)_2.CaCl_2.xCa(OH)_2.xH_2O$	comp varies	wh powd strong Cl odor		d		d evln Cl		d a
c105	hypochlorite, trihydrate	$Ca(ClO)_2.3H_2O$	197.03	tetr pl, 1.535, 1.63	2.1	$−3H_2O$, 60				
c106	chromate	$CaCrO_4.2H_2O$	192.09	yel, monocl pr		$−2H_2O$, 200		16.3^{20}	18.2^{45}	s a, al
c107	chromite	$CaCr_2O_4$	208.07	ol grn, cub need	4.8^{18}	2090		i	i	i a; s fus K_2CO_3
c108	cinnamate	$Ca(C_9H_7O_2)_2.3H_2O$	388.44	col cr				0.22^2	1.34^{100}	
c109	citrate	$Ca_3(C_6H_5O_7)_24H_2O$	570.51	wh need		$−4H_2O$, 120		0.85^{18}	0.096^{23}	0.0065^{18} al
c110	cyanamide	$CaCN_2$	80.10	col, hex, rhbdr		1300 subl >1150		d evl NH_3	d	
c111	cyanide	$Ca(CN)_2$	92.12	wh powd		d >350		d	d	
c112	cyanoplatinite	$CaPt(CN)_4.5H_2O$	429.31	yel-grn fluoresc, rhomb, 1.6226		$−5H_2O$, 100		s		
c113	ferricyanide	$Ca_3[Fe(CN)_6]_212H_2O$	760.42	red need, deliq				v s	v s	
c114	ferrite, mono-	$CaO.Fe_2O_3$	215.77	dk redsh r, rhomb, 2.58, 2.43 (Na)	5.08	1250		i	i	v sl s a
c115	ferrocyanide	$Ca_2Fe(CN)_611$ or $12H_2O$	490.28	yel tricl, 1.570, 1.582, 1.596	1.68	d		86.8^{25}	115^{65}	i al
c116	fluosilicate	$CaSiF_6$	182.16	col, tetr	2.66^{18}			sl s		s al, HF, HCl
c117	fluoride	Nat. fluorite. CaF_2	78.08	col, cub luminisc w heat, 1.434	3.180	1423	ca 2500	0.0016^{18}	0.0017^{26}	s NH_4 salts; sl s a; i acet
c118	fluosilicate, dihydrate	$CaSiF_6.2H_2O$	218.19	col, tetrag	2.254			sl s d		s HCl, HF; i al
c119	formate	$Ca(CHO_2)_2$	130.12	col, rhomb, 1.510, 1.514, 1.578	2.015	d		16.2^0	18.4^{100}	i al
c120	fumarate	$CaC_4H_2O_4.3H_2O$	208.18	col, rhomb				2.11^{30}		
c121	d-gluconate	$Ca(C_6H_{11}O_7)_2.H_2O$	448.40	wh cr powd, need		$−H_2O$, 120		3.3^{15}		v sl s al
c122	glycerophosphate	$CaC_3H_6(OH)_2PO_4$	210.16	wh cr powd, hyg		d 170		2^{25}	sl s	i al
c123	hydride	CaH_2	42.10	wh, rhomb cr	1.9	816 (in H_2) d ca 600		d H_2+ $Ca(OH)_2$		d a
c124	hydroxide	$Ca(OH)_2$	74.09	col, hex, 1.574, 1.545	2.24	$−H_2O$, 580	d	0.185^0	0.077^{100}	s NH_4 salts, a; i al
c125	hyponitrite	$CaN_2O_2.4H_2O$	172.15	wh cr	1.834	d 320				d dil a
c126	iodate	Nat. lautarite. $Ca(IO_3)_2$	389.89	col, monocl	4.519^{15}	d 540		0.20^{15}	0.67^{90}	s HNO_3; i al
c127	iodate, hexahydrate	$Ca(IO_3)_2.6H_2O$	497.98	col, rhomb		d 35		0.13^0	1.22^{100}	s HNO_3
c128	iodide	CaI_2	293.89	yelsh-wh, hex, deliq	3.956^{25}_4	784	ca 1100	209^{20}	426^{100}	126^{20} MeOH; s al, acet, a

No.	Name	Synonyms and Formulae	Mol. wt.	Crystalline form, properties and index of refraction	Density or spec. gravity	Melting point, °C	Boiling point, °C	Solubility, in grams per 100 cc		
								Cold water	Hot water	Other solvents
	Calcium									
c129	iodide, hexahydrate	$CaI_2.6H_2O$	401.98	yel, hex need	2.55	d 42	160	757^0	1680^{30}	s a, al, acet
c130	iron (III) aluminate	Calcium (tetra-) alumino-ferite, nat. celite. $4CaO.Fe_2O_3.Al_2O_3$	485.97	brn, rhomb, 1.98, 2.05, 2.08 all for λ	3.77	1418		3		
c131	isobutyrate	$Ca(C_4H_7O_2)_2.5H_2O$	304.35	col powd				20	sl s	
c132	lactate	$Ca(C_3H_5O_3)_2.5H_2O$	308.30	wh need, effl		$-3H_2O$, 100		3.1^0	7.9^{30}	sl s a; i al, eth
c133	laurate	$Ca(C_{12}H_{23}O_2)_2.H_2O$	456.73	wh need, effl		182–183		0.004^{15}	0.055^{100}	0.059^{15}, 1.72^{75} al
c134	linoleate	$Ca(C_{18}H_{31}O_2)_2$	598.97	wh amorph powd				i		s al, eth
c135	magnesium carbonate	Nat. dolomite. $CaCO_3.MgCO_3$	184.41	col, trig, 1.6817, 1.5026	2.872	d 730–760		0.032^{18}		
c136	magnesium *meta*silicate	Nat. diopside. $CaO.MgO.2SiO_2$	216.56	col, monocl, 1.665, 1.672, 1.695	3.275	1391		i	i	i HCl
c137	magnesium *ortho*silicate	Nat. merwinite. $3CaO.MgO.SiO_2$	328.72	col to pa grn, monocl, 1.708, 1.711, 1.718	3.150					
c138	dl-malate	$CaC_4H_4O_5.3H_2O$	226.20	col, rhomb, 1.545, 1.555, 1.575				0.321^0	$0.451^{27.5}$	i al
c139	l-malate	$CaC_4H_4O_5.2H_2O$	208.18	col				0.812^0	$1.224^{27.5}$	s al
c140	malate, dihydrogen	$Ca(HC_4H_4O_5)_2.6H_2O$	414.33	rhomb, or wh cr powd, 1.493, 1.507, 1.545				sl s		
c141	maleate	$CaC_4H_2O_4.H_2O$	172.15	col rhomb, 1.495, 1.575, 1.640				2.89^{25}	3.21^{40}	
c142	malonate	$CaC_3H_2O_4.4H_2O$	214.19					0.44^0	0.72^{100}	
c143	*per*manganate	$Ca(MnO_4)_2.5H_2O$	368.03	purp cr	2.4	d		331^{14}	338^{25}	s NH_4OH
c144	α-methylbutyrate	Calvium ethylmethyl-acetate. $Ca(C_5H_9O_2)_2$	242.34					24.24^0	25.65^{70}	
c145	molybdate	Nat. powellite $CaMoO_4$	200.01	col, tetr, 1.967, 1.978	4.38–4.53			i	d	s a i al, eth
c146	nitrate	$Ca(NO_3)_2$	164.09	col, cub, hyg	2.504^{18}	561		121.2^{18}	376^{100}	14^{15} al; s MeOH, liq NH_3, acet; i eth
c147	nitrate, tetrahydrate	$Ca(NO_3)_2.4H_2O$	236.15	col, monocl, deliq, 1.465, 1.498, 1.504	α 1.896, β 1.82	α 42.7, β 39.7	d 132	266^0	660^{30}	s al, acet
c148	nitrate, trihydrate	$Ca(NO_3)_2.3H_2O$	218.14	col, tricl		51.1				
c149	nitride	Ca_3N_2	148.25	brn cr, hex	2.63^{17}	1195		d	d	s dil a; d abs al
c150	nitrite	$Ca(NO_2)_2.H_2O$	150.11	col-yelsh, hex, deliq	2.23^{34}	$-H_2O$, 100		45.9^0	89.6^{31}	sl s al
c151	nitrite, tetrahydrate	$Ca(NO_2)_2.4H_2O$	204.15	col cr, tetr	1.674^0_9	$-2H_2O$, 44		74.9^0	106^{42}	s al
c152	oleate	$Ca(C_{18}H_{33}O_2)_2$	603.01	wh wax-like cr		83–84		0.04^{15}	0.03^{50}	sl s eth
c153	oxalate	CaC_2O_4	128.10	col, cub	2.2^4	d		0.00067^{13}	0.0014^{95}	s a; i ac a
c154	oxalate, hydrate	$CaC_2O_4.H_2O$	146.12	col	2.2	$-H_2O$, 200		i	i	s a; i ac a
c155	oxide	Lime, calcia. CaO	56.08	col, cub, 1.838	3.25–3.38	2614	2850	0.131^{10} d	0.07^{80} d	s a
c156	oxide, per-	CaO_2	72.08	wh, tetr, 1.895	2.92^{25}_4	d 275		sl s		s a
c157	oxide, per-octahydrate	$CaO_2.8H_2O$	216.20	wh, tetr, pearly	1.70	$-8H_2O$, 200	d 275 expl	sl s	d	s a, NH_4 salts; i al, eth
c158	palmitate	$Ca(C_{16}H_{31}O_2)_2$	550.93	wh or yelsh, wh fatty powd				0.003^{25}		v al s al; 0.008^{25} eth
c159	1-phenol-4 sulfonate(p-)	$Ca[C_6H_4(OH)SO_3]_2.H_2O$	404.43	wh to pinkish powd				s		s al
c160	phenoxide	$Ca(OC_6H_5)_2$	226.29	redsh powd				sl s		al s al
c161	*hypo*phosphate	$Ca_2P_2O_6.2H_2O$	274.13	gel				i		s HCl
c162	*meta*phosphate	$Ca(PO_3)_2$	198.02	col, 1.588, 1.595	2.82	975		i	i	i a
c163	*ortho*phosphate, di-(sec)	Nat. brushite. $CaHPO_4.2H_2O$	172.09	wh, tricl, 1.5576, 1.5457, 1.5392	2.306^{16}	$-H_2O$, 109		0.0316^{25}	0.075^{100}	i al, s a
c164	*ortho*phosphate, mono-(prim.)	$Ca(H_2PO_4)_2.H_2O$	252.07	col, tricl, deliq, 1.5292, 1.5176, 1.4392	2.220^{16}	$-H_2O$, 109	d 203	1.8^{30}	d	s a
c165	*ortho*phosphate, tri-(tert.)	Nat. whitlockite. $Ca_3(PO_4)_2$	310.18	wh amorph powd, 1.629, 1.626	3.14	1670		0.002	d	i al; s a
c166	*pyro*phosphate	$Ca_2P_2O_7$	254.10	col, biax, 1.585, 1.604	3.09	1230		i		s a
c167	*pyro*phosphate, pentahydrate	$Ca_2P_2O_7.5H_2O$	344.18	col, monocl, 1.539, 1.545, 1.551	2.25			sl s		s a; i NH_4Cl
c168	phosphide	Ca_3P_2	182.19	gray lumps	2.51	ca 1600		d ev PH_3		s a; i al, eth, bs
c169	*hypo*phosphite	$Ca(H_2PO_2)_2$	170.06	wh-gray, monocl		d		15.4^{25}	12.5^{100}	i al
c170	*ortho*phosphite, di-	$2CaHPO_3.3H_2O$	294.17	wh		d		sl s	d	s NH_4Cl
c171	*ortho*plumbate	Ca_2PbO_4	351.35	red-br cr	5.71	d		i	d	s a
c172	propionate	$Ca(C_3H_5O_2)_2.H_2O$	204.24	col, monocl tabl				49^0	55.8^{100}	i al
c173	l-quinate	$Ca(C_7H_{11}O_6)_2.10H_2O$	602.56	rhomb leaf		50, $-10H_2O$ 120		16^{15}		i al

No.	Name	Synonyms and Formulae	Mol. wt.	Crystalline form, properties and index of refraction	Density or spec. gravity	Melting point, °C	Boiling point, °C	Solubility, in grams per 100 cc		
								Cold water	Hot water	Other solvents
	Calcium									
c174	salicylate	$Ca(C_7H_5O_3)_2.2H_2O$	350.34	wh, oct		$-2H_2O$, 120		4^{25}	s	s al
c175	selenate	$CaSeO_4$	183.04	col	2.88			7.9^6	5.4^{67}	
c176	selenate, dihydrate	$CaSeO_4.2H_2O$	219.07	col, monocl	2.68					
c177	selenide	CaSe	119.04	cub, 2.274	3.57					
c178	metasilicate(α)	Nat. pseudowollastonite. $CaSiO_3$	116.16	col, monocl, 1.610, 1.611, 1.664	2.905	1540		0.0095^{17}		s HCl
c179	metasilicate(β)	Nat. wollastonite. $CaSiO_3$	116.16	col, monocl, 1.616, 1.629, 1.631	2.5	tr 1200				
c180	di-orthosilicate (I)	Ca_2SiO_4	172.24	col, monocl, 1.717, 1.705	3.27	2130				
c181	di-orthosilicate (II)	Ca_2SiO_4	172.24	col, rhomb, 1.717, 1.735	3.28	tr to (I) 1420				
c182	di-orthosilicate (III)	Ca_2SiO_4	172.24	col, monocl, 1.642, 1.645, 1.654	2.97	tr to 675				
c183	(tri-)silicate	Nat. alite. Ca_3SiO_5 or $(3CaO.SiO_2)$	228.32	col, monocl, α 1.718, β 1.724		**1900** (incogr)				
c184	silicide	$CaSi_2$	96.25		2.5			i	d	s a, alk
c185	stearate	$Ca(C_{18}H_{35}O_2)_2$	607.04	cr powd		179–180		0.004^{15}		i al, eth
c186	succinate	$CaC_4H_4O_4.3H_2O$	212.22	col, 1.460, 1.540, 1.610				0.193^{10}	0.89^{80}	
c187	sulfate	Nat. anhydrite. $CaSO_4$	136.14	col, rhomb, or monocl, 1.569, 1.575, 1.613	2.960	monocl 1450	rhomb tr to monocl 1193	0.209^{30}	0.1619^{100}	s a, NH_4 salts, $Na_2 S_2O_3$, glyc
c188	sulfate	Soluble anhydrite. $CaSO_4$	136.14	col, hex or tricl, 1.505, 1.548	2.61	tr to rhomb >200				
c189	sulfate half-hydrate	Plaster of Paris. $CaSO_4.\frac{1}{2}H_2O$	145.15	wh powd		$-\frac{1}{2}H_2O$, 163		0.3^{20}	sl s	s a, NH_4 salts, $Na_2S_2O_3$, glyc
c190	sulfate dihydrate	Nat. gypsum. $CaSO_4.2H_2O$	172.17	col, monocl, 1.521, 1.523, 1.530	2.32	$-1\frac{1}{2}H_2O$, 128	$-2H_2O$, 163	0.241	0.222^{100}	s a, NH_4 salts, $Na_2S_2O_3$, glyc
c191	sulfide	Nat. oldhamite. CaS	72.14	col, cub, 2.137	2.5	d		0.021^{18} d	0.048^{60} d	d a
c192	sulfide, hydro-	$Ca(HS)_2.6H_2O$	214.32	col pr		d 15–18		v s		s al,
c193	sulfite	$Ca(SO_3).\frac{1}{2}H_2O$	130.16	col, hex		$-\frac{1}{2}H_2O$, >250		0.0043^{18}	0.0011^{100}	s H_2SO_4
c194	sulfite, dihydrogen	$Ca(HSO_3)_2$	202.22	yelsh liq, strong SO_2 odor				s		s a
c195	d-tartrate	$CaC_4H_4O_6.4H_2O$	260.21	col, rhomb, 1.525, 1.535, 1.550		d		0.0266^0	$0.0689^{37.5}$	sl s al
c196	dl-tartrate	$CaC_4H_4O_6.4H_2O$	260.21	tricl, powd or need		$-4H_2O$, 200		0.0032^0	$0.0078^{37.5}$	s HCl; i ac, a
c197	mesotartrate	$CaC_4H_4O_6.3H_2O$	242.20	wh, monocl or tricl pr		$-3H_2O$ <170		i	0.16^{100}	0.28^{18}, 0.85^{100} ac a
c198	telluride	CaTe	167.68	cub, 2.51, 2.58	4.873					
c199	tellurite	$CaTeO_3$	215.68	wh fl		>960		sl s	s	s a
c200	thiocarbonate, tri-	$CaCS_3$	148.28	yel cr				s		i al
c201	thiocyanate	$Ca(SCN)_2.3H_2O$	210.29	wh cr, deliq				v s	v s	v s al
c202	di-thionate	$Ca(S_2O_6)_2.4H_2O$	272.27	col, trig, 1.5496	2.176			16^0	30^{30}	
c203	thiosulfate	$CaS_2O_3.6H_2O$	260.30	tricl	1.872	d		100^3	d	s al
c204	metatitanate	Nat. perovskite. $CaTiO_3$	135.98	col, cub, rhomb, β 2.34	4.10	1975				
c205	tungstate		287.93	wh, tetr, 1.9263, 1.9107	6.062^{20}			0.00064^{15}	0.00012^{100}	
c206	tungstate	Nat. scheelite. $CaWO_4$	287.93	col or w sc, tetr. 1.918, 1.934	6.06			0.2		i al, a; s NH_4Cl
c207	metatungstate	$Ca_2H_4[H_2(W_2O_7)_6].27H_2O$	3500.96	col, tric		$-7H_2O$, 105	$-10H_2O$, d			d a
c208	valerate	$Ca(C_5H_9O_2)_2$	242.33					8.28^0	7.39^{100}	
c209	metazirconate	$CaZrO_3$	179.30	col, monocl	4.78	2550				
c210	**Carbon**	Diamond. C	12.01	col, cub, 2.4173	3.51	Diamond trans. to graphite		i	i	i a, alk
c211	carbon	Graphite. C	12.01	blk, hex	2.25^{20}	subl 3652–97	in vac. at 1500°C	i	i	s liq Fe; i a, alk
c212	carbon, amorphous	C	12.01	amorph, blk	1.8–2.1	subl 3652–97		i	i	i a, alk
c213	(di)-bromide, hexa-	Hexabromomethane. C_2Br_6	503.48	rhomb pr, 1.740, 1.847, 1.863	3.823	148–149 d	210	i		s CS_2; v sl s al, eth
c214	bromide, tetra-	Tetrabromomethane. CBr_4	331.65	col, monocl or oct	3.42	tr to oct 48.4; m.p. 90.1	189.5	0.024^{30}		s al, eth, chl
c215	(di)-bromide, tetra-	Tetrabromoethylene. C_2Br_4	343.66			57.5	227			
c216	(di)-chloride, hexa-	Hexachloro ethane. C_2Cl_6	236.74	col, rhmb, tricl or cub	2.091	subl 187			i	s al, eth, oils
c217	chloride, tetra-	Tetrachloromethane. CCl_4	153.81	col liq, 1.4601	1.5867^{20}_{20}	-23	76.8		v sl s	s al, bz, chl, eth
c218	(di)-chloride, tetra-	Tetrachloroethylene. C_2Cl_4	165.83	col liq, eth odor, 1.5055	1.6311^{15}_4	-22.4	120.8			s al, eth

No.	Name	Synonyms and Formulae	Mol. wt.	Crystalline form, properties and index of refraction	Density or spec. gravity	Melting point, °C	Boiling point, °C	Solubility, in grams per 100 cc		
								Cold water	Hot water	Other solvents
	Carbon									
c219	fluoride, tetra-	Tetrafluoromethane. CF_4	87.99	col gas	1.96^{-184}	-184	-128	sl s		
c220	iodide, tetra-	Tetraiodomethane. CI_4	519.63	dk red, cub	4.34^{20}	d 171		i	d	s al, CS_2, eth, MeOH, bz
c221	oxide, di-	CO_2	44.01	col gas or col liq	1.977g/l, liq 1.101^{-37}, solid 1.56^{-79}	$-56.6^{5.2atm}$	-78.5 subl	171.3^0 cm³ 0.348^0g 0.145^{25}g	90.1^{20} cm³ 0.097^{40} g 0.058^{60}g	31^{15} cm³ al, s acet
c222	oxide, mon-	CO	28.01	col odorl pois gas	1.250g/l, liq 0.793	-199	-191.5	3.5^0 cm	2.32^{20} cm³	s al, bz, ac a, Cu_2Cl_2
c223	oxide, sub-	C_3O_2	68.03	col gas or liq, 1.4538	liq 1.114^0	-111.3	7	d		
c224	oxysulfide	COS	60.07	col gas, pois	gas 1.073 g/l⁰ liq 1.24^{-87}	-138.2	-50.2	54^{20} ml		s al; v s CS_2
c225	selenide, di-	CSe_2	169.93	golden yel liq, 1.845^{20}	2.6626^{25}_4	-45.5	125–126	i		d al; s CS_2, tol
c226	selenide, sulfide	$CSeS$	123.04	yel oily liq	1.9874	-85	84.5	i	i	sl s al; s CS_2
c227	sulfide, di-	CS_2	76.14	col liq, inflamm, 1.62950^{18}	1.261^{22}_{20}	-110.8	46.3	0.22^{22}	0.14^{50}	s al, eth
c228	sulfide, mono-	CS	44.08	red powd	1.66	d 200		i		i al; s eth, CS_2
c229	sulfide, sub-	C_3S_2	100.16	red liq	1.274	-0.5	d 90			
c230	sulfide telluride	$CSTe$	171.68	yel-red liq	2.9^{-60}	-54	d >-54			s CS_2, bz
c231	sulfochloride	Thiophosgene. $CSCl_2$	114.98	yel-red liq	1.509^{15}		73.5			
c232	**Carbonic acid**	H_2CO_3	62.03	exists in solution only				s		
c233	**Carbonyl bromide**	Carbon oxybromide. $COBr_2$	187.83	col liq			64.5			
c234	**Carbonyl chloride**	Phosgene, carbon oxychloride, $COCl_2$	98.92	col gas, pois	1.392	-104	8.3	d		d al, a; v s bz, tol s ac a
c235	**Carbonyl fluoride, di-**	COF_2	66.01	col gas, hyg	sol 1.388^{-190} liq 1.139^{-114}	-114	-83.1	d		
c236	**Carbonyl selenide**	$COSe$	106.97	col gas, very pois	liq $1.812^{4.1}$	-124.4	-21.7	d		s $COCl_2$
c237	**Cerium**	Ce	140.12	gray met, cub or hex	hex 6.657 cub 6.757	799	3426	sl d	d	s dil min a, i alk
c238	(III) acetate	$Ce(C_2H_3O_2)_3$	317.26	col		d 308		20^{15}	12^{75}	
c239	(III) acetate hydrate	$Ce(C_2H_3O_2)_3.1\frac{1}{2}H_2O$	344.28	wh-redsh cr powd		$-1\frac{1}{2}$ H_2O, 115	d	26.5^{15}	16.2^{75}	31^{15} al, s acet
c240	boride, hexa-	CeB_6	204.98	blue met, cub		2190	d	i	i	i HCl
c241	boride, tetra-	CeB_4	183.36	tetr	5.74					
c242	III bromate	$Ce(BrO_3)_3.9H_2O$	685.98	redsh-wh, hex		49		s		
c243	bromide	$CeBr_3.H_2O$	397.86	col need, deliq		d		v s	v s	v s al
c244	carbide	CeC_2	164.14	red, hex	5.23			d	d	s a
c245	carbonate	$Ce_2(CO_3)_3.5H_2O$	550.37	wh cr				i		s a; sl s $(NH_4)_2CO_3$
c246	carbonate fluoride	Nat. bastnaesite. $CeFCO_3$	219.13	hex, 1.717, 1.817	5					
c247	chloride	$CeCl_3$	246.48	col cr, deliq	3.92^0	848	1727	100	d	30 al; s acet
c248	citrate	$Ce(C_6H_5O_7).3\frac{1}{2}H_2O$	392.28	wh powd				i		s dil min a
c249	(III) cyanoplatinite	$Ce_2[Pt(CN)_4]_3.18H_2O$	1502.00	yel-bl lust, monocl	2.657	$-13\frac{1}{2}H_2O$ 100–110	d	s		
c250	(III) fluoride	CeF_3	197.12	wh, hex	6.16	1460	2300	i		
c251	(IV) fluoride	$CeF_4.H_2O$	234.13	col microcr, 1.614	4.77	ca 650	d	i		s a
c252	hydride	CeH_3	143.14	dk bl amorph powd		ign		d		
c253	(III) hydroxide	$Ce(OH)_3$	191.14	wh gelat ppt						s a, $(NH_4)_2CO_3$; i alk
c254	(IV) iodate	$Ce(IO_3)_4$	839.73	yel cr				0.015^{20}		
c255	(III) iodate	$Ce(IO_3)_3.2H_2O$	700.86	cr				0.16^{25}		s HNO_3
c256	(III) iodide	$CeI_3.9H_2O$	682.97	wh or redsh-wh cr		752	1397	v s		v s al
c257	(III) molybdate	$Ce_2(MoO_4)_3$	760.05	yel, tetr, 2.019, 2.007	4.83	973				
c258	(III) nitrate	$Ce(NO_3)_3.6H_2O$	434.23	col or redsh cr (trac La, Di), deliq		$-3H_2O$, 150	d 200	v s	v s	50 al; s acet
c259	(IV) nitrate, basic	$Ce(OH)(NO_3)_3.3H_2O$	397.07	long red need				s		
c260	(III) oxalate	$Ce_2(C_2O_4)_3.9H_2O$	706.44	yel-wh cr		d		v sl s		s H_2SO_4, HCl; i $H_2C_2O_4$, alk, eth, al
c261	(IV) oxide	$CeO_2.xH_2O$		yelsh gelat ppt						s a; sl s alk carb, i alk
c262	(III) oxide	Ce_2O_3	328.24	gray-grn, trig	6.86	1692, ign 200		i	i	s H_2SO_4, i HCl

No.	Name	Synonyms and Formulae	Mol. wt.	Crystalline form, properties and index of refraction	Density or spec. gravity	Melting point, °C	Boiling point, °C	Cold water	Hot water	Other solvents
								Solubility, in grams per 100 cc		
	Cerium									
c263	(IV) oxide(di-)	Ceria. CeO_2	172.12	brn-wh, cub	7.132^{23}	ca 2600		i	i	s H_2SO_4, HNO_3; i dil a
c264	oxychloride	CeOCl	191.57	purp leaf				i		s dil a
c265	(III) 2,4-pentanedione	Cerium acetylacetonate. $Ce(C_5H_7O_2)_3.3H_2O$	491.50	lt yel cr ppt		131–132		d		v s al
c266	(III) metophosphate	$Ce(PO_3)_3$	377.04	micr need	3.272					i a
c267	(III) orthophosphate	Nat. monazite. $CePO_4$	235.09	red, monocl or yel, rhomb, 1.795	5.22			i	i	s a; i al
c268	(III) salicylate	$Ce(C_7H_5O_3)_3$	551.47	wh-redsh wh powd				i		i al
c269	(III) selenate	$Ce_2(SeO_4)_3$	709.11	rhomb	4.450			30.55^0	2.513^{100}	
c270	silicide	$CeSi_2$	196.29		5.67^{17}			i		
c271	(IV) sulfate	$Ce(SO_4)_2$	332.24	deep yel cr	3.91^{18}	d 195		sl d, forms basic salts		
c272	(III) sulfate	$Ce_2(SO_4)_3$	568.42	col to grn, monocl or rhomb	3.912	$d\ 920^{746}$		10.1^0	2.25^{100}	
c273	(IV) sulfate, dihydrate	$Ce(SO_4)_2.4H_2O$	404.31	yel, rhomb				v s d		s dil H_2SO_4
c274	(III) sulfate, monohydrate	$Ce_2(SO_4)_3.9H_2O$	730.56	hex need	2.831			11.87^{15}	0.42^{30}	
c275	(III) sulfate, octahydrate	$Ce_2(SO_4)_3.8H_2O$	712.54	pink cr, tricl	2.886^{17}	$-8H_2O$, 630		12^{20}	6^{50}	
c276	(III) sulfate, pentahydrate	$Ce_2(SO_4)_3.5H_2O$	658.50	monocl	3.17			3.90^{50}	0.514^{100}	
c277	(III) sulfide	Ce_2S_3	376.43	red cr, br-dk powd, purp	5.020^{11}	d 2100 (vac)		i	d	s dil a
c278	(III) tungstate	$Ce_2(WO_4)_3$	1023.78	yel, tetr	6.77^{17}	1089				
	Cerium complexes									
c280	hexantipyrinecerium perchlorate	$[Ce(C_{11}H_{12}N_2O)_6].(ClO_4)_3$	1567.86	col, hex cr		d 295–300		1.08^{20}		
c281	hexantipyrinecerium iodide (III)	$[Ce(C_{11}H_{12}H_{12}O)_6].I_3$	1650.22	large yel cr		268–270		15.10^{20}		
c282	**Cesium**	Cs	132.9054	silv met cr hex	1.8785^{15}	**28.40±0.01**	669.3	d		s liq NH_3
c283	acetate	$CsC_2H_3O_2$	191.95	deliq		194		$945.1^{-2.5}$	$1345.5^{88.5}$	
c284	aluminum sulfate	$CsAl(SO_4)_2.12H_2O$	568.19	col, cub, 1.4587	1.97	117		0.34^0	42.54^{100}	s dil al
c285	amide	$CsNH_2$	148.93	wh need	3.44^{25}_4	262 ± 1		d		s liq NH_3
c286	azide	CsN_3	174.93	col need, deliq		310		224.2^0		1.037^{16} al; i eth
c287	benzoate	$CsC_7H_5O_2$	254.02					294.5^0	398.5^{100}	
c288	borofluoride	$CsBF_4$	219.71	rhomb, 1.350	3.20	550 d		1.6^{17}	ca 30^{100}	s dil NH_3
c289	borohydride	$CsBH_4$	147.75	wh, cub, 1.498	2.404			v s		sl s al; i eth, bz
c290	bromate	$CsBrO_3$	260.81	hex, ca 2.15	4.109^{15}	ca 420 d		3.66^{25}	5.32^{35}	
c291	bromide, mono-	CsBr	212.81	col, cub, 1.6984	4.44, liq 3.04^{700}	636	1300	124.3^{25}	v s	s a
c292	bromide, tri-	$CsBr_3$	372.63	rhomb		180				
c293	dibromochloride	$CsBr_2Cl$	328.18	yel-red, rhomb		191	150, $-Br_2$	s		d al, acet
c294	bromochloride iodide	CsIBrCl	375.17	yel-red, rhomb		235	d 290	s		s al
c295	bromoiodide di-	$CsIBr_2$	419.63	rhomb	4.25	248	d 320	4.61^{20}		s al
c296	carbonate	Cs_2CO_3	325.82	col cr, deliq		d 610		260.5^{15}	v s	11^{19} al; s eth
c297	carbonate, hydrogen	$CsHCO_3$	193.92	rhomb		$175-\frac{1}{2}H_2O$		209.3^{15}	v s	s al
c298	chlorate	$CsClO_3$	216.36	sm cr	3.57			$6.28^{19.8}$	76.5^{90}	
c299	perchlorate	$CsClO_4$	232.36	rhomb, at 219 cub, 1.4752, 1.4788, 1.4804	3.327^4	d 250		2.00^{25}	28.57^{99}	0.093^{25} al; 0.7878^{25} al; 0.150^{25} acet
c300	chlorobromide	$CsBrCl_2$	283.72	glossy-yel, rhomb		205		s		d al, eth
c301	chloride	CsCl	168.36	col, cub, deliq, 1.6418	3.988	645	1290	$162.22^{0.7}$	$259.56^{89.5}$	33.7^{25} MeOH; v s al; i acetone
c302	chloroiodide	$CsICl_2$	330.72	or, trig	3.86	230	d 290	s		s al
c303	chloroaurate	$CsAuCl_4$	471.68	yel, monocl				0.5^{10}	27.5^{100}	s al
c304	chlorobromide, di-	$CsBrCl_2$	283.72	glossy-yel, rhomb		205				
c305	chlorodibromide	$CsBr_2Cl$	328.18	yel		191				
c306	chloroiodide, di-	$CsICl_2$	330.72	or, trig	3.86	230	d 290	s		s al
c307	chloroplatinate	Cs_2PtCl_6	673.62	yel, cub	4.197 ± 0.004	d 570		0.024^0	0.377^{100}	i al
c309	chlorostannate	$CsSnCl_6$	597.22	wh, cub	3.33					
c310	chromate	Cs_2CrO_4	381.80	yel pr, rhomb	4.237			71.4^{13}	95.5^{90}	
c311	chromium sulfate	Cesium chromium alum. $Cs[Cr(H_2O)_6](SO_4)_2.6H_2O$	593.21	vlt cr	2.064	116		9.4^{25}		
c312	cyanide	CsCN	158.92	very sm wh cr	2.93			v s	v s	
c313	fluoride	CsF	151.90	cub, deliq, 1.478 ± 0.005^{15}	4.115	682	1251	367^{18}		191^{15} MeOH; i Diox, Pyr
c314	fluoride	CsF. $1\frac{1}{2}H_2O$	178.93		4.10	703		366.5^{18}		
c315	fluorogermanate	Cs_2GeF_6	452.39	isotrop cr, reg oct	4.10			sl s	v s	sl s a
c316	fluosilicate	Cs_2SiF_6	407.89	wh, cub	3.372^{17}			60^{17}		i al
c317	fluotellurite	Cs_2TeF_8	355.50	col need				d	d	s HF soln
c318	formate	$CsCHO_2$	172.92		1.0169^{21}_4				$2012^{26.4}$	

No.	Name	Synonyms and Formulae	Mol. wt.	Crystalline form, properties and index of refraction	Density or spec. gravity	Melting point, °C	Boiling point, °C	Cold water	Hot water	Other solvents
	Cesium									
c319	formate..........	$CsCHO_2.H_2O$	195.94	....	....	41, $-H_2O$		....		
c320	gallium selenate.....	$CsGa(SeO_4)_2.2H_2O$	704.72	col cr............	....			4.14^{25}		
c321	gallium sulfate.....	$CsGa(SO_4)_2.12H_2O$	610.93	col cub, 1.46495........	2.113			1.21^{25}	....	0.0035^{25} 75 % al
c322	hydride..........	CsH	133.91	wh cr, cub.......	3.41	d		d	d	d a; i org solv
c323	hydrofluoride.....	$CsF.HF$	171.91	need, deliq........		160		v s		v s a; i al
c324	hydrogencarbide ...	$CsHC_2$	157.94	trsp cr............		300		d		
c325	hydroxide..........	$CsOH$	149.91	lt yel, deliq......	3.675	272.3		395.5^{15}		s al
c326	iodate..........	$CsIO_3$	307.81	wh, monocl........	4.85	....		2.6^{24}		
c327	*meta*periodate......	$CsIO_4$	323.81	wh rhomb pl.....	4.259^{15}_4	....		2.15^{15}	s	
c328	iodide..........	CsI	259.81	rhomb, deliq, 1.7876	4.510^{25}_4	626	1280	44^0	160^{61}	s al
c329	iodide, penta-.....	CsI_5	767.43	bl, tricl.......		73				
c330	iodide, tri-..........	CsI_3	513.62	blk, rhomb......	4.47	207.5		sl s	sl s	s al
c331	iodotetrachloride....	$CsICl_4$	401.62	pale or needles...	3.374^{-10}	228	d	sl s	sl s	....
c332	iron (II) sulfate.....	$Cs_2SO_4.FeSO_4.6H_2O$	621.87	lt grn, monocl, 1.500, 1.504, 1.509	2.791^{20}_4	ca 70		101.1^{25} (anhyd)		
c333	iron (III) sulfate....	$CsFe(SO_4)_2.12H_2O$	597.06	pa-vlt cr, 1.484...	2.061^{20}	ca 90		s	s	
c334	magnesium sulfate...	$Cs_2SO_4.MgSO_4.6H_2O$	590.34	col, monocl, 1.486, 1.452	2.676^{20}_4					
c335	*per*manganate......	$CsMnO_4$	251.84	....	3.597	d 320		0.097^1	1.27^{50}	sl s al
c336	mercury bromide(ic).	$CsBr.2HgBr_2$	933.63	rhomb				0.807^{17}		sl s al
c337	mercury chloride(ic).	$CsCl.HgCl_2$	439.85	col, cub or rhomb, 1.792		....		1.44^{17}	....	i abs al
c338	nitrate..........	$CsNO_3$	194.91	col, hex or cub, 1.55, 1.56	3.685 liq 2.71^{500}	414	d	9.16^0	196.8^{100}	s acet, sl s al
c339	nitrate, hydrogen....	$CsNO_3.NHO_3$	257.92	oct........		100				
c340	nitrate, dihydrogen..	$CsNO_3.2HNO_3$	320.92	col pl.....		32–36				
c341	nitrate..........	$CsNO_2$	178.91	yel cr		....		v s	v s	
c342	oxalate..........	$Cs_2C_2O_4$	353.82	....	3.230^{15}			282.9^{25}		
c343	oxide..........	Cs_2O	281.81	or need.......	4.25	d 400; m.p. 490 (in N_2)		v s	d	s a
c344	oxide, per-.....	Cs_2O_2	297.81	pa yel need......	4.25	400	650. $-O_2$	s	d	s a
c345	oxide, tri-..........	Cesium oxide, sesqui- Cs_2O_3	313.81	choc br cr, cub....	4.25	400		d		s a
c346	phthalate, hydrogen.	$CsHC_8H_4O_4$	298.03	rhomb........	2.178					
c347	polonium chloride....	Cs_2PoCl_6	688.53	cub, 1.86........	3.82					
c348	rhodium sulfate....	$CsRh(SO_4)_2.12H_2O$	644.12	yel, oct........	2.238	110–111		sl s		
c349	rhodium sulfate....	$CsRh(SO_4)_2.12H_2O$	644.12	or cr........	2.22^{20}_4	111		sl s	s	
c350	salicylate..........	$CsC_7H_5O_3$	270.02	....				196.2^0	1522^{100}	
c351	selenate......	Cs_2SeO_4	408.77	rhomb, deliq, 1.5950, 1.5060, 1.5964	4.4528^{20}_4			244.8^{12}		
c352	sulfate..........	Cs_2SO_4	361.87	col rhomb, or hex, 1.560, 1.564, 1.566	4.243	1010	tr hex 600	167^0	220^{100}	i al, acet
c353	sulfate, hydrogen...	$CsHSO_4$	229.97	col rhomb pr....	3.352^{16}	d		s		
c354	sulfide..........	$Cs_2S.4H_2O$	369.94	wh cr, deliq......		....		v s	v s	
c355	sulfide, di-.....	CsS_2	329.94	dk red, amorph...		460	>800	hgr		
c356	sulfide, di-.....	$Cs_2S_2.H_2O$	347.95	tetr.......		....		s		
c357	sulfide, hexa-.....	Cs_2S_6	458.19	br red......		186				
c358	sulfide, penta-.....	Cs_2S_5	426.13	....	2.806^{15}	210				
c359	sulfide, tetra-.....	Cs_2S_4	394.06	yel.......		d 160				
c360	sulfide, tri-.....	Cs_2S_3	362.00	yel leaf......		217	780			
c361	tartrate, hydrogen...	$CsHC_4H_4O_6$	281.99	wh, rhomb cr.....		....		9.7^{25}	98^{100}	
c362	*l*-tartrate.........	$Cs_2C_4H_4O_6$	413.88	col, trig.......	3.03^{14}			v s d	v s	
c363	vanadium sulfate....	Cesium vanadium alum· $VCs(SO_4)_2.12H_2O$	592.15	red, cub, 1.4780	2.033^{20}_4	82	$-12H_2O$, 230; d 300	0.464^{10}	sl s	
c364	**Chloramine**, mono-..	NH_2Cl	51.48	yel liq........		−66		s		s al, eth; v sl s CCl_4, bz
c365	**Chloric acid**........	$HClO_3.7H_2O$	210.57	known only as col sol	$1.282^{14.2}$	<−20	d 40	v s	s	....
c366	**Chloric acid, per-**....	$HClO_4$	100.46	col liq unstable...	1.764^{22}	−112	39^{56}	∞		
c367	**Chloric acid, per-**....	Hydronium perchlorate. $HClO_4.H_2O$ or $(H_3O)^+(ClO_4)^-$	118.47	need, fairly stab	1.88, liq 1.776^{50}	50	exp 110	v s	v s	....
c368	per, dihydrate....	$HClO_4.2HO$	136.49	stab liq........	1.65	−17.8	200	v s	v s	s al

No.	Name	Synonyms and Formulae	Mol. wt.	Crystalline form, properties and index of refraction	Density or spec. gravity	Melting point, °C	Boiling point, °C	Solubility, in grams per 100 cc		
								Cold water	Hot water	Other solvents
	Chlorine									
c369	Chlorine	Cl_2	70.906	grnsh-yel gas, or liq, or rhomb cr; gas 1.000768, liq 1.367	3.214^0	−100.98	−34.6	310^{10} cm³ 1.46^0 g	177^{30} cm³ 0.57^{30} g	s alk
c370	azide	chlor(o)azide ClN_3	77.48	gas, expl				sl s		d alk
c371	fluoride, mono-	ClF	54.45	col gas	1.62^{-100}	−154±5	−100.8	d	d	
c372	fluoride, tri-	ClF_3	92.45	col gas	1.77^{13}	−83	11.3	d	d	
c373	hydrate	$Cl_2.8H_2O$	215.03	lt yel, rhomb	1.23	d 9.6		i		s alk
c374	oxide, di-	ClO_2	67.45	yel red gas, or red cr, expl	3.09^{11} g/l	−59.5	9.9^{731} exp	2000^4 cm³	d to $HClO_3$, Cl_2, O_2	s alk, H_2SO_4
c375	oxide, hept-	Cl_2O_7	182.90	col oil		−91.5	82	s d		s bz
c376	oxide, mono-	Cl_2O	86.91	yel-red gas, or red-br liq	3.89^0 g/l	−20	3.8^{766} exp	200 cm³	d to HOCl	s alk, H_2SO_4
c377	oxide, tetr-	ClO_4 or Cl_2O_8	99.45				d	s d		s bz
c378	chloroauric acid	$HAuCl_4.4H_2O$	411.85	brt yel need, deliq		d		s	v s	s al, eth
c379	chloroplatinic acid	$H_2PtCl_6.6H_2O$	517.92	red br pr, deliq	2.431	60		v s	v s	s al, eth
c380	chlorostannic acid	$H_2SnCl_6.6H_2O$	441.52	col leaf	1.93	9		s		
c381	Chlorosulfonic acid	$ClSO_3H$	116.52	col fum liq, 1.437^{14}	1.766^{18}	−80	158	d to H_2SO_4 + HCl		d al, a; i CS_2
c382	Chlorotetroxy fluoride	ClO_4F	118.45	col gas, v exp		−167.3	−15.9			
c383	Chloryl (per-)fluoride	ClO_2F	102.45	gas	1.392^{25}	−146	−46.8			
c384	Chromium	Cr	51.996	steel gray, cub v hard	7.20^{28}	1857±20	2672	i	i	s dil H_2SO_4, HCl; i HNO_3, aq reg
c385	(II) acetate	$Cr(C_2H_3O_2)_2$	170.09	red cr				sl s	s	sl s al
c386	(III) acetate	$Cr(C_2H_3O_2)_3.H_2O$	247.15	gray-grn powd or blsh-grn pasty mass				s		i al
c387	arsenide, mon-	CrAs	126.92	gray, hex	6.35^{16}			i	i	i a
c388	boride, mono-	CrB	62.81	silv cr, orthorhomb	6.17	2760(?)		i	i	v dil Na_2O_2
c389	(II) bromide	$CrBr_2$	211.81	wh cr	4.356	842		s	s	s al
c390	(III) bromide	$CrBr_3$	291.72	olv gr, hex	4.250^{25}_4	subl		i	s	v s al; d alk
c391	bromide, hexahydrate	$[CrBr_2(H_2O)_4]Br.2H_2O$	399.81	grn cr, deliq				s	s tr to vlt	s al; i eth
c392	bromide, hexahydrate	$[Cr(H_2O)_6]Br_3$	399.81	blsh gray to vlt	5.4^{17}			v s	v s	i al
c393	(tri-)carbide, di-	Cr_3C_2	180.02	gray, rhomb	6.68	1890	3800	i	i	
c394	carbonyl	$Cr(CO)_6$	220.06	col, orthorhomb	1.77	d 110	210 exp	i	i	i al, eth, ac a; sl s CHI_3, CCl_4
c395	(II) chloride	$CrCl_2$	122.90	wh need, deliq	2.878^{25}	824		v s	v s	i al, eth
c396	(III) chloride	$CrCl_3$	158.35	vlt, trig	2.76^{15}	ca 1150	subl 1300	i	sl s	i al, acet, MeOH, eth
c397	chloride, hexahydrate	$[Cr(H_2O)_4Cl_2].2H_2O$	266.45	vlt, monocl	1.76	83		58.5^{25}	s	s al; i eth; sl s acet
c398	(II) fluoride	CrF_2	89.99	grn, cr, monocl	4.11	1100	>1300	sl s		i al; s h HCl
c399	(III) fluoride	CrF_3	108.99	grn, rhomb	3.8	>1000	subl 1100−1200	i		i al, NH_3; sl s a; s HF
c400	(II) hydroxide	$Cr(OH)_2$	86.01	yel-br		v		d		s a
c401	iodate, hydrate	$[Cr(H_2O)_6]I_3.3H_2O$	594.85	dk vlt cr, hygro	4.915^{25}_4	41−HI		v s	v s	s al, acet; i CHI
c402	(II) iodide	CrI_2	305.80	graysh powd	5.196	856	subl vac 800	s		
c403	(III) iodide	CrI_3	432.71	shiny blk cr	4.915^{25}_4	>600	−I_2, vac 350			
c404	(III) nitrate	$Cr(NO_3)_3.7\frac{1}{2}H_2O$	373.13	br, monocl		100	d	s	s	
c405	(III) nitrate	$Cr(NO_3)_3.9H_2O$	400.15	purple, monocl		60	d 100	s	s	s a, alk, al, acet
c406	nitride, mono-	CrN	66.00	cub or amorph	5.9	d 1700		i	i	sl s aq reg
c407	(II) oxalate	$CrC_2O_4.H_2O$	158.03	yel cr powd	2.468			sl s	s	s dil a
c408	(III) oxalate	$Cr_2(C_2O_4)_3.6H_2O$	476.14	red, amorph, hyg		120, −H_2O tr to grn		s	s	v s (red) al, eth; i (grn) al
c409	oxide, di-	CrO_2	84.00	br-blk powd		300, −O		i	i	s HNO_3
c410	(II) oxide, mon-	CrO	68.00	blk powd				i	i	i dil HNO_3
c411	(III) oxide, sesqui-	Cr_2O_3	151.99	grn, hex, 2.551	5.21	2266±25	4000	i	i	i a, alk, al
c412	(III) oxide, sesqui-	$Cr_2O_3.xH_2O$	varies	vlt, amorph or bl-gray grn gel				i	i	s a, alk; sl s NH_4OH
c413	oxide, tri-	Chromic anhydride, "chromic acid", CrO_3	99.99	red, rhomb, deliq	2.70	196	d	61.7^0	67.45^{100}	s al, eth, H_2SO_4, HNO_3
c414	oxychloride	CrO_2Cl_2	154.90	dk red liq	1.911	−96.5	117	d	d	d al; s eth, ac a
c415	2,4-pentanedione	Chromium acetyl-acetonate. $Cr(C_5H_7O_2)_3$	349.33			216	340	i		s org solv; i lgr

No.	Name	Synonyms and Formulae	Mol. wt.	Crystalline form, properties and index of refraction	Density or spec. gravity	Melting point, °C	Boiling point, °C	Solubility, in grams per 100 cc		
								Cold water	Hot water	Other solvents
	Chromium									
c416	(III) *orthophosphate*	CrPO₄.2H₂O	183.00	vlt cr	2.42²².⁵			sl s		s a, alk; i ac a
c417	(III) *orthophosphate*	CrPO₄.6H₂O	255.06	vlt, tricl, 1.568, 1.591, 1.699	2.121¹⁴	100				
c418	*pyrophosphate*	Cr₄(P₂O₇)₃	729.81	pa grn, monocl	3.2			i	i	s alk
c419	phosphide, mono-	CrP	82.97	gray-blk cr	5.7¹⁵			i		s HNO₃, HF
c420	silicide	Cr₃Si₂	212.17	gray, tetr pr	5.5⁰			i	i	s HCl, HF; i H₂SO₄, HNO₃
c421	(II) sulfate	CrSO₄.7H₂O	274.17	bl cr				12.35⁰	d	s ls al; s NH₄OH
c422	(III) sulfate	Cr₂(SO₄)₃	392.18	vlt or red powd	3.012			i, s*		sl s al; i a
c423	(III) sulfate	Cr₂(SO₄)₃.15H₂O	662.41	vlt, amorph sc	1.867¹⁷	100	−10H₂O, 100	s	d 67	i al
c424	(III) sulfate	Cr₂(SO₄)₃.18H₂O	716.45	bl vlt, cub oct, 1.564	1.7²¹	−12H₂O, 100		120²⁰	s	s al
c425	(II) sulfide, mono-	CrS	84.06	blk powd, hex	4.85	1550		i		v s a
c426	(III) sulfide, sesqui-	Cr₂S₃	200.18	brn-blk powd	3.77¹⁹	−S, 1350		i, d	i	s HNO₃; d al
c427	(III) sulfite	Cr₂(SO₃)₃	344.18	grnsh-wh	2.2	d				
c428	(II) tartrate	CrC₄H₄O₆	200.07	bl powd	2.33			i		sl s a; i ac a
c429	**Chromium complexes**									
	hexammine chromium-(III) chloride	[Cr(NH₃)₆]Cl₃.H₂O	278.55	yel cr	1.585			s		
c430	hexaureachromium-(III) fluosilicate	[Cr(CON₂H₄)₆]₂.[SiF₆]₃.3H₂O	1304.94	lt grn leaf				0.522²⁰		i al
c431	hexaureachromium-(III) *perrhenate*	[C₂(CON₂H₄)₆].(ReO₄)₃	1162.92	grn need	2.652²⁵			1.786		0.667 al
c432	chloropentammine chromium chloride	[Cr(NH₃)₅Cl]Cl₂	243.51	red, oct	1.696			0.65¹⁶		i HCl
c433	**Cobalt**	Co	58.933	silv gray met, cub	8.9	1495	2870	i	i	s a
c434	(III) acetate	Co(C₂H₃O₂)₃	236.07	grn, oct		d 100		hydr readily		s a, glac ac a
c435	(II) acetate	Co(C₂H₃O₂)₂.4H₂O	249.08	red-vlt, monocl, deliq, 1.542	1.705¹⁹	−4H₂O, 140		s	s	s a, al
c436	aluminate	(approx) Thenard's blue. CoAl₂O₄	176.89	bl, cub				i	i	
c437	(II) *orthoarsenate*	Co₃(AsO₄)₂.8H₂O	598.75	vlt-red, monocl, 1.626, 1.661, 1.669	3.178¹⁵	d		i	i	s dil a, NH₄OH
c438	arsenic sulfide	Nat. cobaltite. CoAsS	156.92	gray-redsh	6.2–6.3	d				
c439	arsenide	Co₂As	192.79	cr powd	8.28	950		i	i	i HCl, H₂SO₄; s HNO₃, aq reg
c440	(II) benzoate	Co(C₇H₅O₂)₂.4H₂O	373.23	gray red leaf		−4H₂O, 115		v s		
c441	boride, mono-	CoB	69.74	pr	7.25¹⁸			d	d	s HNO₃, aq reg
c442	(II) bromate	Co(BrO₃)₂.6H₂O	422.84	red, oct				45.5¹⁷		s NH₄OH
c443	(II) bromide	CoBr₂	218.75	grn, hex, deliq	4.909²⁵	678 (in N₂)		66.7⁴⁹	68.1⁹⁷	77.1²⁰ al; 58.6⁵⁰ MeOH; s eth, acet
c444	(II) bromide hexahydrate	CoBr₂.6H₂O	326.84	red-vlt pr, deliq	2.46	47–48, −4H₂O 100	−6H₂O, 130	s red color	153.2⁹⁷	s blk color, al, a, eth
c445	bromoplatinate	CoPtBr₆.12H₂O	949.66	trig	2.762					
c446	carbonate	Nat. spherocobaltite. CoCO₃	118.94	red, trig, 1.855, 1.60	4.13	d		i	i	s a; i NH₃
c447	(II) carbonate, basic	2CoCO₃.Co(OH)₂.H₂O	534.74	vlt-red pr				i	d	s a, (NH₄)₂CO₃
c448	carbonyl tetra-	Dicobalt octacarbonyl. [Co(CO)₄]₂ or Co₂(CO)₈	341.95	or cr or dk br, microcr	1.73¹⁸	51	d 52	i	i	sl s al; s CS₂, eth
c449	carbonyl, tri-	Tetracobalt dodeca-carbonyl. [Co(CO)₃]₄ or Co₄(CO)₁₂	571.86	blk cr				sl s		s bz; d Br
c450	(II) chlorate	Co(ClO₃)₂.6H₂O	333.93	red, cub, deliq, 1.55	1.92	50	d 100	558.3⁰	v s	s al
c451	(II) perchlorate	Co(ClO₄)₂	257.83	red need 1.510, 1.490	3.327			100⁰	115⁴⁵	s al, acet
c452	(II) perchlorate	Co(ClO₄)₂.5H₂O	347.88	red, hex		143		100.13⁰	115.10⁴⁵	v s al, acet i CHCl₃
c453	perchlorate	Co(ClO₄)₂.6H₂O	365.93	red pr		d 1534	d	259¹⁵		s al, acet
c454	perchlorate	Co(ClO₄)₂.6H₂O	365.93	red, oct, deliq, 1.55		d 182		255¹⁵		s al, acet
c455	(II) chloride	CoCl₂	129.84	bl, hex, hygr	3.356²⁵	724 (in HCl gas)	1049	45⁷	105⁹⁶	54.4 al; 8.6 acet; 38.5 MeOH; sl s eth
c456	(III) chloride	CoCl₃	165.29	red cr or yel cr	2.94	subl		s		

* Several chromic salts exist in two forms, a soluble and insoluble modification.

No.	Name	Synonyms and Formulae	Mol. wt.	Crystalline form, properties and index of refraction	Density or spec. gravity	Melting point, °C	Boiling point, °C	Solubility, in grams per 100 cc		
								Cold water	Hot water	Other solvents
	Cobalt									
c457	(II) chloride, dihydrate	$CoCl_2.2H_2O$	165.87	red-vlt, monocl or tricl, 1.625, 1.671, 1.67	2.477_{24}^{25}			s	s	v sl s eth
c458	(II) chloride, hexahydrate	$CoCl_2.6H_2O$	237.93	red, monocl	1.924_{25}^{25}	86	$-6H_2O$, 110	76.7^0	190.7^{100}	v s (bl col) al; s acet; 0.29 eth
c459	chloroplatinate	$CoPtCl_6.6H_2O$	574.83	trig	2.699	d				
c460	chlorostannate	$CoSnCl_6.6H_2O$	498.43	rhomb or trig		d 100				
c461	(II) chromate	$CoCrO_4$	174.93	gray blk cr		d		i	d	s a, NH_4OH
c462	(II) citrate	$Co_3(C_6H_5O_7)_2.2H_2O$	591.04	rose-red		$-2H_2O$, 150		0.8		
c463	(II) cyanide, dihydrate	$Co(CN)_2.2H_2O$	147.00	buff anhyd bl-vlt powd	anhydr 1.872_{25}	$-2H_2O$, 280	d 300	0.00418^{18}		s KCN, HCl, NH_4OH
c464	(II) cyanide, trihydrate	$Co(CN)_2.3H_2O$	165.01	red-gray powd, amorph		$-3H_2O$, 250		i		s KCN
c465	(II) ferricyanide	$Co_3[Fe(CN)_6]_2$	600.71	red need				i		s NH_4OH; i HCl
c466	(II) ferrocyanide	$Co_2Fe(CN)_6.xH_2O$		gray-grn				i		s KCN; i HCl
c467	(II) fluogallate	$[Co(H_2O)_6][GaF_5.H_2O]$	349.75	pink cr, monocl (?), 1.45	2.35	$-5H_2O$, 110		sl s		d a
c468	(II) fluoride	CoF_2	96.93	pink monocl	4.46_4^{25}	ca 1200	1400	1.5^{25}	s	sl s a; i al, eth, bz
c469	(III) fluoride	CoF_3	115.93	br, hex	3.88	d to $Co(OH)_3$				i, al, eth, bz
c470	fluoride	$Co_2F_6.7H_2O$	357.96	grn powd	2.314_{25}			d		s H_2SO_4
c471	(II) fluoride, tetrahydrate	$CoF_2.4H_2O$	168.99	α: red, rhomb oct, β: rose cr powd	2.192_4^{25}	d 200		s	s	i al
c472	fluosilicate	$CoSiF_6.6H_2O$	309.10	pink trig, 1.382, 1.387	2.113^{19}			$118.1^{21.5}$		
c473	(II) formate	$Co(CHO_2)_2.2H_2O$	185.00	red cr	2.129^{22}	$-2H_2O$, 140	d 175	5.03^{20}		s a, NH_4 salts
c474	(II) hydroxide	$Co(OH)_2$	92.95	rose-red, rhomb	3.597^{15}	d		0.00032		s a, NH_4 salts; i alk
c475	(III) hydroxide	$Co_2O_3.3H_2O$	219.91	blk-brn powd	4.46	d	$-H_2O$, 100	0.00032		s a; i al
c476	(II) iodate	$Co(IO_3)_2$	408.74	bl-vlt need	5.000^{19}	d 200		0.45^{18}	1.33^{100}	s HCl, HNO_3. h H_2SO_4
c477	(II) iodate, hexahydrate	$Co(IO_3)_2.6H_2O$	516.83	red, oct	3.689^{21}	d 61	$-4H_2O$, 135	s		
c478¹	(II) iodide (α) stable	CoI_2	312.74	blk hex, hyg	5.68	515 (vac)	570 (vac)	159^0	420^{100}	v s al, acet
c478²	(II) iodide (β)	CoI_2	312.74	yel need, unstab	5.45_{25}	d 400		s		
c479	(II) iodide, dihydrate	$CoI_2.2H_2O$	348.77	grn, deliq		d 100		376.2^{45}	s	
c480	(II) iodide, hexahydrate	$CoI_2.6H_2O$	420.83	br-red hex, hygr	2.90	d 27, $-6H_2O$, 130		s	s	s al, eth, chl
c481	iodoplatinate	$CoPtI_6.9H_2O$	1177.59	trig	3.618					
c482	linoleate	$Co(C_{18}H_{31}O_2)_2$	617.83	br, amorph				i		s al, eth, acet
c483	(II) nitrate	$Co(NO_3)_2.6H_2O$	291.04	red, monocl, 1.52	1.87_4^{25}	55–56	$-3H_2O$, 55	133.8^0	217^{30}	$100.0^{11.5}$ al; s acet; sl s NH_3
c484	nitrosylcarbonyl	$Co(NO)(CO)_3$	172.97	cherryred liq		-1.05	48.6; d 55	i		s al, eth, acet, bz
c485	(II) oleate	$Co(C_{18}H_{33}O_2)_2$	621.86	br, amorph				i		s al, eth, oils, bz
c486	(II) oxalate	CoC_2O_4	146.95	wh or redsh	3.021^{25}	d 250		i		s a, NH_4OH
c487	oxalate, dihydrate	$CoC_2O_4.2H_2O$	182.98	pink cr		$-H_2O$, ca 190		v sl s	sl s	v sl s a; s NH_4OH
c488	(II) oxide	CoO	74.93	pink cub	6.45	1795 ± 20		i	i	s a; i al, NH_4OH
c489	(III) oxide	Co_2O_3	165.86	blk-gray, hex, or rhomb	5.18	d 895		i	i	s a; i al
c490	(II, III) oxide	Co_3O_4	240.80	blk, cub	6.07	tr to CoO 900–950		i	i	v sl s a; i aq reg
c491	palmitate	$Co(C_{16}H_{31}O_2)_2$	569.78			70.5				s pyr, hot CS_2, CCl_4; sl s eth; i MeOH, acet
c492	(II) orthophosphate	$Co_3(PO_4)_2$	366.74	redsh cr	2.587^{25}			i	i	s H_3PO_4, NH_4OH
c493	(II) orthophosphate, dihydrate	$Co_3(PO_4)_2.2H_2O$	402.77	pink powd				i		s H_3PO_4
c494	(II) orthophosphate, octahydrate	$Co_3(PO_4)_2.8H_2O$	510.87	redsh powd	2.769^{25}	$-8H_2O$, 200		sl s		s min a, H_3PO_4; i al
c495	phosphide	Co_2P	148.84	gray need	6.4^{15}	1386		i	i	s HNO_3, aq reg
c496	(II) propionate	$Co(C_3H_5O_2)_2.3H_2O$	259.12	dk-red cr		ca 250		anh 33.5^{11}		v s al
c497	(II) perrhenate	$Co(ReO_4)_2.5H_2O$	649.41	dk pink		d		d		
c498	(II) selenate, heptahydrate	$CoSeO_4.7H_2O$	328.00	monocl	2.135					
c499	selenate, hexahydrate	$CoSeO_4.6H_2O$	309.98	red, monocl, 1.5225	2.25^{17}			s	s	
c500	(II) selenate, pentahydrate	$CoSeO_4.5H_2O$	291.97	ruby red, tricl	2.512	d		v s		

No.	Name	Synonyms and Formulae	Mol. wt.	Crystalline form, properties and index of refraction	Density or spec. gravity	Melting point, °C	Boiling point, °C	Solubility, in grams per 100 cc		
								Cold water	Hot water	Other solvents
	Cobalt									
c501	selenide, mono-	CoSe	137.89	yel, hex	7.65	red heat				s HNO_3, aq reg; i alk
c502	(II) orthosilicate	Co_2SiO_4	209.95	vlt cr, rhomb	4.63	1345		i	i	s dil HCl
c503	silicide	CoSi	87.03	rhomb		1395				s HCl; i HNO_3, H_2SO_4
c504	silicide, di-	$CoSi_2$	115.11	rhomb	5.3	1277				
c505	(di-)silicide	Co_2Si	145.95	gray cr	7.28^0	1327				d a
c506	(II) orthostannate	Co_2SnO_4	300.55	grnsh-bl, cub	6.30^{18}					i H_2SO_4; s h HCl
c507	(II) sulfate	$CoSO_4$	155.00	dk blsh, cub	3.71^{25}_{25}	d 735		36.2^{20}	83^{100}	1.04^{18} MeOH; i NH_3
c508	(II) sulfate, hepta-hydrate	Nat. bieberite. $CoSO_4.7H_2O$	281.10	red-pink, monocl, 1.477, 1.483, 1.489	1.948^{25}_{25}	96.8	$-7H_2O$, 420	60.4^3	67^{70}	2.5^3 al; 54.5^{18} MeOH
c509	(II) sulfate, hexa-hydrate	$CoSO_4.6H_2O$	263.09	red, monocl, 1.531 1.549, 1.552	2.019^{15}_{15}	$-2H_2O$, 95				
c510	(II) sulfate, mono-hydrate	$CoSO_4.H_2O$	173.01	red cr, 1.603, 1.639, 1.683	3.075^{25}	d		s	s	
c511	(III) sulfate	$Co_2(SO_4)_3.18H_2O$	730.33	bl-grn		d 35		s d		s H_2SO_4; i pyr
c512	sulfide, di-	CoS_2	123.06	blk, cub	4.269			i		s HNO_3, aq reg
c513	sulfide, mono-	Nat. sycoporite. CoS	91.00	redsh, silv-wh, oct	5.45^{18}	>1116		0.00038^{18}		sl s a
c514	(III) sulfide, sesqui	Co_2S_3	214.06	blk cr	4.8					d a, aq reg
c515	(tri-) sulfide	Cobalt sulfide, tetra-(ous, ic) Nat. linneite. Co_3S_4	305.06	dk gray, cub	4.86	d 480				
c516	(II) sulfite	$CoSO_3.5H_2O$	229.07	red				i		s H_2SO_3
c517	tartrate	$CoC_4H_4O_6$	207.01	redsh, monocl				sl s		s dil a
c518	thiocyanate	$Co(SCN)_2.3H_2O$	229.14	vlt, rhomb		$-3H_2O$, 105		s		s al, MeOH, eth
c519	orthotitanate	Co_2TiO_4	229.76	grnsh-blk, cub	5.07–5.12					s conc HCl; sl s dil HCl
c520	(II) tungstate	$CoWO_4$	306.78	bl-grn, monocl	8.42			i		s h conc a; sl s c dil a
	Cobalt complexes									
c521	hexammine cobalt (II) bromide	$CoBr_2.6NH_3$	320.93	dk pink cr	1.871^{25}_4	d 258		d		
c522	diamminecobalt (II) chloride [α]	$CoCl_2.2NH_3$	163.90	rose cr	2.097	273				
c523	diamminecobalt (II) chloride (β)	$CoCl_2.2NH_3$	163.90	bl-vlt	2.073	tr to α, 210 (in NH_3)				
c524	hexamminecobalt (II) chloride	$[Co(NH_3)_6]Cl_2$	232.02	rose red, oct	1.497	d		d		s NH_4OH; i abs al
c525	hexamminecobalt (III) chloride	$Co(NH_3)_6Cl_3$	267.46	wine-red, monocl	1.710^{25}_4	$-1NH_3$, 215		5.9^{10}	$12.74^{46.6}$	s conc HCl; i al, NH_4OH
c526	hexamminecobalt (II) iodide	$CoI_2.6NH_3$	414.93	dk pink, cub	2.096^{25}_4	141^{100mm}				
c527	hexamminecobalt (III) nitrate	$Co(NH_3)_6.(NO_3)_3$	347.13	yel, tetr	1.804^{25}_4			1.7^{25}	v s	v sl s dil a
c528	hexamminecobalt (III) perrhenate	$[Co(NH_3)_6](ReO_4)_3.2H_2O$	947.74	or-yel pr	3.329^{25}			0.0469		
c529	hexamminecobalt (II) sulfate	$CoSO_4.6NH_3$	257.18	pink powd	1.654^{25}_4	d 116^{760}		d		v s dil NH_3
c530	hexamminecobalt (III) sulfate	$[Co(NH_3)_6]_2(SO_4)_3.5H_2O$	700.50	dk yel, monocl	1.797^{25} anhydr	$-4H_2O$, 100	$-5H_2O$, 150	$1.4^{17.4}$		
c531	ammonium tetra-nitrodiammine (III) cobaltate	Erdmann's salt. $NH_4[Co(NH_3)_2(NO_2)_4]$	295.12	redsh-pa brn, rhomb, 1.78, 1.78, 1.74	1.876^{25}					
c532	aquapentammine-cobalt (III) chloride (roseo)	$[Co(NH_3)_5.H_2O]Cl_3$	268.45	brick red cr	1.7^{25}	d 100		24.87^{25}		sl s HCl; i al
c533	aquapentammine-cobalt (III)-sulfate (roseo)	$[Co(NH_3)_5H_2O]_2(SO_4)_3.2H_2O$	638.34	red, tetr	1.854^{20}	$-3H_2O$, 99	d 110	$1^{17.2}$	1.72^{27}	s H_2SO_4
c534	cis-chloroaquo-tetramminecobalt (III) chloride	$[Co(NH_3)_4(H_2O)Cl]Cl_2$	251.42	vlt, rhomb	1.847	d		1.4^0		s a; i al
c535	chloropentammine cobalt-(III) chloride (purpureo)	$[Co(NH_3)_5Cl]Cl_2$	250.45	dk red-vlt, rhomb	1.819^{25}_{25}	d		0.4^{25}	$1.03^{46.6}$	s conc H_2SO_4; i al

No.	Name	Synonyms and Formulae	Mol. wt.	Crystalline form, properties and index of refraction	Density or spec. gravity	Melting point, °C	Boiling point, °C	Solubility, in grams per 100 cc		
								Cold water	Hot water	Other solvents
	Cobalt complexes									
c536	triethylenediaminecobalt-(III) chloride	$Co[C_2H_4(NH_2)_2]_3Cl_3.3H_2O$	399.64	br pr............	1.542^{17}	256; $-3H_2O$, 100	....	v s		
c537	trinitrotriamminecobalt	$Co(NH_3)_3(NO_2)_3$.......	248.04	yel, rhomb pl or leaf	1.992^{25}_4	d 158	exp 164	$0.177^{16.5}$	0.28^{25}	....
c538	trinitrotetramminecobalt-(III) nitrate	$[Co(NH_3)_4(NO_2)_2]NO_2$..	265.07	yel, rhomb....	1.922^{17}			3^{20}		
c539	potassium tetranitrodiamminecobaltate (III)	$K[Co(NH_3)_2(NO_2)_4]$....	316.12	yel, rhomb....	2.076^{15}			$1.758^{16.5}$		
	Colombium............	see Niobium.								
c540	Copper............	Cu............	**63.546**	redsh met, cub...	8.92	**1063.4±0.2**	**2567**	i	i	s HNO_3, h H_2SO_4; v s l s HCl, NH_4OH
c541	acetate, basic......	Blue verdigris. $Cu(C_2H_3O_2)_2.CuO.6H_2O$	369.26	grnsh-bl powd				sl s		s dil a, NH_4OH; sl s al
c542	(II) acetate........	Neutral verdigris. $Cu(C_2H_3O_2)_2.H_2O$	199.65	dk grn powd, 1.545, 1.550	1.882, anhydr-1.93	115	d 240	7.2	20	7.14 al; s eth
c543	(II) acetate meta-arsenate	Paris green. $Cu(C_2H_3O_2)_2.3Cu(AsO_2)_2$ (approx)	1013.77	em grn powd....				i		s a, NH_4OH; i al
c544	(III) acetylide.....	Cu_2C_2........	151.10	red, amorph, expl.		exp		v sl s		s a, KCN
c545	amine azide.......	$Cu(NH_3)_2(N_3)_2$.......	181.64	dk grn cr, exp....		d 100–105	exp 202	i	d	d a; i MeOH
c546	(II) diamminechloride, di-	$Cu(NH_3)_2Cl_2$.......	168.51	grn cr........	2.32^{25}_4	260–270	d 300	i		s NH_4OH; i abs a
c547	(II) hexamminechloride, di-	$Cu(NH_3)_6Cl_2$.......	236.63	bl, cub........	1.48^{25}_4	....		v s		
c548	tetrammine dithionate	$[Cu(NH_3)_4]S_2O_6$........	291.79	vlt-bl cr........		d 160		s	d	
c549	(II) tetrammine nitrate	$[Cu(NH_3)_4](NO_3)_2$........	155.67	dk-bl, oct......	1.91^{25}_4	d 210 exp		s		
c550	(II) amine nitrate...	$[Cu(NH_3)_4](NO_3)_2$.......	223.61	vlt-bl, tetr.....		$-2NH_3$ 97		v s		
c551	tetrammine sulfate...	Cuprum ammoniacale. $[Cu(NH_3)_4]SO_4.H_2O$	245.74	dk-bl, rhomb, unstab	1.79^{25}_4	$-NH_3.H_2O$, 30		$18.05^{21.5}$		
c552	(tri-)antimonide....	Cu_3Sb........	312.37	gray........	8.51	687				
c553	(II) orthoarsenate...	$Cu_3(AsO_4)_2.4H_2O$.....	540.52	blsh-grn......				i	i	s a, NH_4OH
c554	(II) orthoarsenate, di-H	$Cu_3H_2(AsO_4)_4.2H_2O$....	911.42	bl........				i		s a, NH_4OH
c555	arsenide...........	Cu_3As_2........	467.54	bl, oct........	7.56			i	i	s a, NH_4OH
c556	tri-arsenide.......	Nat. domeykite. Cu_3As.	265.54	hex........	8.0	830				
c557	(II) orthoarsenite, hydrogen(?)	Scheele's green. $CuHAsO_3(?)$	187.47	grn powd......		d		i	i	s a, NH_4OH; i al
c558	(I) azide........	CuN_3........	105.56	col cr, v exp....	3.26			0.00075^{20}		d conc H_2SO_4; s NH_4Cl
c559	(II) azide........	$Cu(N_3)_2$........	147.58	brn-red or brn-yel cr, exp	2.604	exp 215		0.008^{20}		v s dil a
c560	(II) benzoate.......	$Cu(C_7H_5O_2)_2.2H_2O$....	341.80	lt bl cr powd....		$-H_2O$, 110		sl s		s dil a; sl s al
c561	(II) metaborate....	$Cu(BO_2)_2$........	149.16	blsh grn cr powd.	3.859	....		s		
c562	boride..........	Cu_3B_2........	212.24	yel........	8.116					
c563	(II) bromate.......	$Cu(BrO_3)_2.6H_2O$.....	427.45	bl-grn, cub......	2.583	d 180	$-6H_2O$, 200	v s		s NH_4OH
c564	(I) bromide........	CuBr (or Cu_2Br_2)......	143.45	wh, cub, 2.116...	4.98	492	1345	v sl s	d	s HBr, HCl, HNO_3, NH_4OH; i acet
c565	(II) bromide.......	$CuBr_2$........	223.31	blk, monocl, deliq	4.77^{25}_4	498		v s		s al, acet, NH_3, pyr; i bz
c566	trioxybromide......	$CuBr_2.3Cu(OH)_2$......	516.02	em grn, rhomb...	4.00	$-H_2O$, 210–215	d 240–250	i	d	s dil min a, NH_4OH; v s ac a
c567	(II) butyrate.......	$Cu(C_4H_7O_2)_2.2H_2O$....	273.77	dk grn cr......				v sl s		s al, eth, NH_4OH, dil a
c568	(I) carbonate......	Cu_2CO_3........	187.09	yel........	4.40	d		i	i	s a, NH_4OH
c569	(II) carbonate, basic.	Nat. malachite. $CuCO_3.Cu(OH)_2$	221.11	dk grn, monocl, 1.655, 1.875, 1.909	4.0	d 200		i	d	0.026 aq CO_2; s a, NH_4OH, KCN; i al
c570	(II) carbonate, basic.	Nat. azurite, chessylite. $2CuCO_3.Cu(OH)_2$	344.65	bl, monocl, 1.730, 1.758, 1.838	3.88	d 220		i	d	s NH_4OH, h $NaHCO_3$
c571	(II) chlorate.......	$Cu(ClO_3)_2.6H_2O$.....	338.53	grn, cub, deliq...		65	d 100	207^0	v s	s al, acet
c572	(II) chlorate, basic..	$Cu(ClO_3)_2.3Cu(OH)_2$.....	523.11	grn cr or amorph.	3.55	d	....	i	i	s dil a

No.	Name	Synonyms and Formulae	Mol. wt.	Crystalline form, properties and index of refraction	Density or spec. gravity	Melting point, °C	Boiling point, °C	Solubility, in grams per 100 cc		
								Cold water	Hot water	Other solvents
	Copper									
c573	perchlorate	$Cu(ClO_4)_2$	262.43	monocl, 1.495, 1.505, 1.522	2.225^{23}	82.3		s	s	
c574	perchlorate, hexahydrate	$Cu(ClO_4)_2.6H_2O$	370.53	lt bl, tricl, deliq, 1.505	2.225^{25}_4	82	d 120	vs		s al, eth
c575	(I) chloride(ous)	Nat. nantokite. $CuCl$ (or Cu_2Cl_2)	98.99	wh, cub, 1.93	4.14	430	1490	0.0062		s HCl, NH_4OH, eth; i al
c576	(II) chloride	$CuCl_2$	134.44	br, yel powd, hygr	3.386^{25}_4	620	993 d to CuCl	70.6^0	107.9^{100}	53^{15} al; 68^{15} MeOH; s h H_2SO_4, acet
c577	(II) chloride, basic	$CuCl_2.Cu(OH)_2$	232.00	yel-grn, hex	3.78	$-H_2O$, 250	d red heat	d	d	
c578	(II) chloride, dihydrate	Nat. eriochalcite. $CuCl_2.2H_2O$	170.47	bl-grn, rhomb, deliq. 1.644, 1.683, 1.731	2.54	$-2H_2O$, 100	d	110.4^0	192.4^{100}	s al, NH_4OH
c579	chloride, thioureate	$CuCl.3[CS(NH_2)_2]$	327.35	col pr, 1.758, 1.17719	1.73	168		vs		
c580	(II) chromate, basic	$CuCrO_4.2CuO.2H_2O$	374.64	yel br		$-2H_2O$, 260		i		s dil a, NH_4OH; i al
c581	(II) dichromate	$CuCr_2O_7.2H_2O$	315.56	blk cr, deliq	2.283	$-2H_2O$, 100		vs	d	s a, NH_4OH, al
c582	(I) chromite	$Cu_2Cr_2O_4$	295.07	gray blk cub pl	5.24^{20}			i	i	s HNO_3
c583	(II) citrate	$2Cu_2C_6H_4O_7.5H_2O$	720.43	blsh grn powd		$-H_2O$, 100		i		s a, NH_4OH
c584	(I) cyanide	$CuCN$	89.56	wh, monocl pr	2.92	473 (in N_2)	d	i	i	s HCl, KCN, NH_4OH; sl s liq NH_3
c585	(II) cyanide	$Cu(CN)_2$	115.58	yel-grn powd		d			i	s a, alk, KCN, pyr
c586	ethylacetoacetate	$Cu(C_6H_9O_3)_2$	321.81	grn need		192–193	subl	i		v s al, eth; 10^{80} bz
c587	(I) ferricyanide	$Cu_3Fe(CN)_6$	402.57	br red				i		i HCl; s NH_4OH
c588	(II) ferricyanide	$Cu_3[Fe(CN)_6]_2.14H_2O$	866.74	yel-grn				i		i HCl; s NH_4OH
c589	(II) ferrocyanide	Hatchett's brown. $Cu_2Fe(CN)_6.xH_2O$		red brn				i	i	i a, NH_3; s NH_4OH
c590	(I) fluogallate	$[Cu(H_2O)_6][GaF_6.H_2O]$	354.36	pa bl, monocl(?), 1.45	2.20	$-5H_2O$, 110		sl s		s HF
c591	(I) fluoride	CuF (or Cu_2F_2)	82.54	red cr, (exist?)		908	subl, 1100	i		s HCl, HF; d HNO_3; i al
c592	(II) fluoride	CuF_2	101.54	wh, tricl	4.23	d 950		4.7^{20}	s	s dil min a; i al
c593	(II) fluoride dihydrate	$CuF_2.2H_2O$	137.57	bl, monocl	2.93^{25}_4	d		4.7^{20}	d	s HCl, HF, HNO_3, al; i acet, NH_3
c594	(I) fluosilicate	Cu_2SiF_6	269.16	red powd		d to SiF_4			d 100	
c595	(II) fluosilicate	$CuSiF_6.4H_2O$	277.68	monocl pr	2.158			42.8		
c596	(II) fluosilicate hexahydrate	$CuSiF_6.6H_2O$	313.71	bl, rhomb, deliq, 1.409, 1.408	2.207			232^{17}		0.16^{20} 92 % al
c597	(II) formate	$Cu(CHO_2)_2$	153.55	bl, monocl	1.831		d	12.5	d	0.25 al
c598	(II) formate tetrahydrate	$Cu(CHO_2)_2.4H_2O$	225.61	bl cr	1.81	$-H_2O$, 130		6.2		s alk; sl s al
c599	(II) glycerine deriv	$Cu(C_3H_4NO_2)_2.H_2O$	229.67	bl need		$-H_2O$, 130		0.57^{15}		s alk
c600	hydride	CuH (or Cu_2H_2)	64.55	red-brn, (exist?)	6.38	d sl 55–60		i	d 65	d HCl
c601	(II) hydroxide	$Cu(OH)_2$	97.56	bl gel cr powd	3.368	$-H_2O$, d		i	d	s a, NH_4OH, KCN
c602	(II) trihydroxychloride	γ: Paratacamite δ: atacamite $CuCl_2.3Cu(OH)_2$	427.11	γ: grn, hex, δ: grn rhomb; γ: 1.743, 1.849, δ:1.861, 1.861, 1.880, grn lt	(γ)3.75	$-H_2O$, 250		i	i	v s a
c603	(II) trihydroxynitrate	$Cu(NO_3)_2.3Cu(OH)_2$	480.27	dk grn, rhomb or moncl	rhomb, 3.41 monocl, 3.378	$-H_2O$ ~400		i	d	v s a
c604	(II) iodate	$Cu(IO_3)_2$	413.35	grn, moncl	5.241^{15}	d		0.1364^{15}	i	s dil HNO_3, dil H_2SO_4
c605	(II) iodate, basic	$Cu(OH)IO_3$	255.45	grn, rhomb	4.873	d 290		i		s dil H_2SO_4
c606	(II) iodate, monohydrate	Nat. bellingerite. $Cu(IO_3)_2.H_2O$	431.36	bl, tricl	4.872	$-H_2O$, 248	d 290	0.33^{15}	0.65^{100}	s dil H_2SO_4, NH_4OH; i al, dil HNO_3
c607	paraperiodate	Cu_2HIO_6	350.00	grn cr powd		d 110		i	i	s HNO_3, NH_4OH
c608	(I) iodide	Nat. marshite. CuI (or Cu_2I_2)	190.44	wh or brnsh-wh, cub, 2.346	5.62	605	1290	0.0008^{18}		s dil HCl, KI, KCN, conc H_2SO_4, liq NH_3
c609	(II) lactate	$Cu(C_3H_5O_3)_2.2H_2O$	277.71	dk bl, monocl				16.7	45^{100}	s NH_4OH; sl s al
c610	(II) laurate	$Cu(C_{12}H_{23}O_2)_2$	462.17	lt bl powd		111–113		sl s	sl s	
c611	mercury iodide (α)	Cu_2HgI_4	835.29	red, tetr	6.116	tr ca 67		i		
c612	mercury iodide (β)	Cu_2HgI_4	835.29	choc, cub	6.102			i		

No.	Name	Synonyms and Formulae	Mol. wt.	Crystalline form, properties and index of refraction	Density or spec. gravity	Melting point, °C	Boiling point, °C	Solubility, in grams per 100 cc		
								Cold water	Hot water	Other solvents
	Copper									
c613	(II) nitrate, hexahydrate	$Cu(NO_3)_2.6H_2O$	295.64	bl cr, deliq	2.074	$-3H_2O$, 26.4	....	243.7^0	∞	s al
c614	(II) nitrate, trihydrate	$Cu(NO_3)_2.3H_2O$	241.60	bl cr, deliq	2.32_4^{25}	114.5	$-HNO_3$, 170	137.8^0	1270^{100}	$100^{12.5}$ al; v s liq NH_3
c615	nitride	Cu_3N	204.63	dk grn powd	5.84_4^{25}	d 300	....	d	....	d a
c616	(II) nitrite, basic	$Cu(NO_2)_2.3Cu(OH)_2$	448.22	grn powd		d 120	....	i	d	v s dil a; sl s al; s NH_4OH
c617	(II) hyponitrite, basic	$Cu(NO)_2.Cu(OH)_2$	221.11	pea grn amorph, hygr	....	d<100	....	i		s dil a; v s NH_4OH; d $NaOH$
c618	(II) nitroprusside	$CuFe(CN)_5NO.2H_2O$	315.51	wh-grnsh powd			....	i	....	s alk; i al
c619	(II) oleate	$Cu(C_{18}H_{33}O_2)_2$	626.47	br powd or grn-bl mass, pois				i	i	s eth
c620	(II) oxalate	$CuC_2O_4.\frac{1}{2}H_2O$	160.57	bl wh				0.00253^{25}	....	s NH_4OH; i ac a
c621	(I) oxide	Nat. cuprite. Cu_2O	143.08	red, oct cub, 2.705	6.0	1235	$-O$, 1800	i	i	s HCl, NH_4Cl, NH_4OH; sl s HNO_3; i al
c622	(II) oxide	Nat. tenorite. CuO	79.54	blk, monocl, β 2.63	6.3–6.49	1326		i	i	s a, NH_4Cl, KCN
c623	oxide, per-	$CuO_2.H_2O$	113.55	br or brnsh-blk cr		d 60		i	....	i al, s d a
c624	oxide, sub-	Cu_3O	270.16	olv grn, (exist?)		d		i	....	d a
c625	(II) oxychloride	Nat. atacamite. $Cu_2(OH)_3Cl$ (or $CuCl_2.3Cu(OH)_2$)	213.56	grn, orthorhomb	3.76–3.78					....
c626	(II) oxychloride	Brunswick green. $CuCl_2.3CuO.4H_2O$(?)	445.13	grn powd, or em grn to grnsh-blk, rhomb		$-3H_2O$, 140			d 100	s a, NH_4OH
c627	(II) palmitate	$Cu(C_{16}H_{31}O_2)_2$	574.39	grn-bl powd	....	120		i	....	s h bz, CS_2, CCl_4; sl s al, eth; i MeOH, acet
c628	2,4-pentanedione	Copper acetylacetonate. $Cu(C_5H_7O_2)_2$	261.76	bl cr		>230	subl	i	....	sl s al; s chl
c629	(I) phenyl	C_6H_5Cu	140.65	col powd		d 80		d	d	i al, CS_2; s pyr
c630	(II) orthophosphate	$Cu_3(PO_4)_2.3H_2O$	434.61	bl, rhomb		d		i	sl s	s a, NH_4OH, H_3PO_4; i NH_3
c631	(tri-) phosphide	Cu_3P	221.59	gray-blk	6.4–6.8	d		i	....	s HNO_3; i HCl
c632	(di-) pyridine chloride(di)	$Cu(C_5H_5N)_2Cl_2$	292.65	grn-bluish, monocl, 1.60, 1.75	1.76	d 263		s	d	sl s c al, chl
c633	(II) salicylate	$Cu(C_7H_5O_3)_2.4H_2O$	409.83	bl-grn need				v s	....	v s al, NH_4OH
c634	(II) selenate	$CuSeO_4.5H_2O$	296.57	bl, tricl, 1.56	2.559	$-4H_2O$, 50–100	$-5H_2O$, 150	25.7^{15}	d	s a, NH_4OH; v sl s acet; i a
c635	(I) selenide	Cu_2Se	206.04	blk, cub	6.749_4^{30}	1113		i	....	d HCl
c636	(II) selenide	CuSe	142.50	grn-blk hex pl, unstab	5.99	d red heat		i	i	sl s HCl, NH_4OH; s h HNO_3
c637	selenite	$CuSeO_3.2H_2O$	226.53	bl-grn, rhomb	3.31_4^{25}	$-H_2O$, 100		i	i	....
c638	silicide	Cu_4Si	282.25	wh met	7.53	850		i	....	i HCl; d HNO_3
c639	(II) stearate	$Cu(C_{18}H_{35}O_2)_2$	630.50	lt grn-bl amorph powd	....	125		i	....	s eth, h bz, chl, turp; sl s pyr; i MeOH, acet
c640	(I) sulfate	Cu_2SO_4	223.14	gray powd, 1.724, 1.733, 1.739	3.605	$+O$, 200		d	....	s conc HCl, NH_3 glac ac a
c641	(II) sulfate	Nat. hydrocyanite. $CuSO_4$	159.60	grn, wh, rhomb, 1.733	3.603	sl d above 200	d 650 to CuO	14.3^0	75.4^{100}	1.04^{18} MeOH; i al
c642	(II) sulfate, basic	Nat. brochantite. $CuSO_4.3Cu(OH)_2$	452.27	grn, monocl, 1.728, 1.771, 1.800	3.78	d 300		i	....	s a, NH_4OH
c643	(II) sulfate, pentahydrate	Blue vitriol, nat. chalcanthite. $CuSO_4.5H_2O$	249.68	bl, tricl, 1.514, 1.537, 1.543	2.284	$-4H_2O$, 110	$-5H_2O$, 150	31.6^0	203.3^{100}	15.6^{18} MeOH; i al
c644	(I) sulfide	Nat. chalcocite. Cu_2S	159.14	blk, rhomb	5.6	1100		$\times 10^{-14}$	....	s HNO_3, NH_4OH; i acet
c645	(II) sulfide	Nat. covellite. CuS	95.60	blk, monocl or hex, 1.45	4.6	tr 103	d 220	0.000033^{18}	....	s HNO_3, KCN, h HCl, H_2SO_4; i al, alk
c646	(I) sulfite, monohydrate	$Cu_2SO_3.H_2O$	225.16	red pr	4.46^{15}			i	....	d dil a; s NH_4OH
c647	(I) sulfite, monohydrate	$Cu_2SO_3.H_2O$	225.16	wh, hex	3.83^{15}	d		sl s		s HCl, NH_4OH; i al, eth
c648	(I, II) sulfite, dihydrate	Chevreul's salt. $Cu_2SO_3.CuSO_3.2H_2O$	386.78	red cr	3.57	d 200		i	i	s HCl, NH_4OH; sl s HNO_3
c649	(II) tartrate	$CuC_4H_4O_6$	211.61	lt bl powd				v sl s		s a, alk
c650	(II) tartrate, trihydrate	$CuC_4H_4O_6.3H_2O$	265.66	lt gray-bl powd		d		0.02^{15}	0.14^{85}	s a, alk

No.	Name	Synonyms and Formulae	Mol. wt.	Crystalline form, properties and index of refraction	Density or spec. gravity	Melting point, °C	Boiling point, °C	Cold water	Hot water	Other solvents
	Copper									
c651	telluride	Cu_2Te	254.68	bl-blk, oct	7.27					s $Br + H_2O$; i HCl, H_2SO_4
c652	telluride	Nat. rickardite. Cu_4Te_3	636.96	purp, tetr	7.54					
c653	tellurite	$CuTeO_3$	239.10	blk glass				i	i	s conc HCl
c654	(I) thiocyanate	$CuSCN$	121.62	wh	2.843	1084		0.0005[18]		s NH_4OH; sl s ac a; i al; d min a
c655	(II) thiocyanate	$Cu(SCN)_2$	179.70	blk		d 100		d	d	s a, NH_4OH
c656	(II) tungstate	$CuWO_4.2H_2O$	347.42	lt-grn, oct		red heat		0.1[15]		s NH_4OH; sl s ac a; i al; d min a
c657	xanthate	Copper ethylxanthogenate. $Cu(C_3H_5OS_2)_2$	305.94	yel ppt		d		i		s NH_4OH; v sl s al; i CS_2
	Copper complexes									
c658	diamminecopper (II) acetate	$Cu(C_2H_3O_2)_2.2NH_3$	215.69	vlt bl cr		d ca 175		s d		s ac a, NH_4OH; i al
c659	tetrammine copper (II) sulfate	$[Cu(NH_3)_4]SO_4.H_2O$	245.74	bl, rhomb	1.81	d 150		18.5[21.5]	d	i al
c660	tetrapyridine copper-(II) fluosilicate	$[Cu(C_5H_5N)_4]SiF_6$	522.03	purp-bl, rhomb	2.108					
c661	tetrapyridine copper-(II) *perrhenate*	$[Cu(C_5H_5N)_4](ReO_4)_2$	880.34	bl cr, monocl	2.338			0.5555		
c662	**Cyanic acid** isocyanic acid	$HOCN$	43.03	liq	1.14_4^{20}			s d		s eth, bz, tol
c663	**Cyanoauric acid**	$HAu(CN)_4.3H_2O$	356.09	tab		50	d	s		s al, eth
c664	**Cobalticyanic acid**	$[H_3Co(CN)_6]_2.H_2O$	454.14	col need, deliq		d 100		s		s al, HCl, dil HNO_3, dil H_2SO_4
c665	**Cyanogen**	$(CN)_2$	52.04	col gas, pungent odor v pois	2.335 g/l, liq: $0.9577^{-21.17}$	-27.9	-20.7	450^{20} cm^3		230 cm^3 al, 500 cm^3 eth
c666	**Cyanogen compounds**	See organic tables								
d1	**Deuterium**	Heavy hydrogen. D_2	4.032	col gas	llg $0.169^{-250.9}$	-254.6	-249.7	sl s		
d2	deuterium chloride	DCl	37.47	gas		-114.8	-81.6			
d3	deuterium oxide	Heavy water. D_2O	20.031	col liq or hex cr, 1.33844[20]	1.105_4^{20}	3.82	101.42^{760}	24.1 cm^3 11.9[25] cm^3	8.4[60] cm^3 7.12[90] cm^3	
d4	**Dysprosium**	Dy	162.50	met, hex	8.5500	1412	2562	i	i	s a
d5	acetate	$Dy(C_2H_3O_2)_3.4H_2O$	411.64	yel need		d 120		s		v sl s al
d6	bromate	$Dy(BrO_3)_3.9H_2O$	708.36	yel hex need		78	$-6H_2O$, 110	v s		sl s al
d7	bromide	$DyBr_3$	402.23	col cr		881	1480	s	s	
d8	carbonate	$Dy_2(CO_3)_3.4H_2O$	577.06			$-3H_2O$, 15		i		
d9	chloride	$DyCl_3$	268.85	shining yel pl	3.67_4^0	718	1500	s	s	
d10	chromate	$Dy_2(CrO_4)_3.10H_2O$	853.13	yel cr		$-3\frac{1}{2}H_2O$, 150	d		1.002^{25}	
d11	fluoride	DyF_3	219.50	col cr		1360	>2200	i	i	i dil a
d12	iodide	DyI_3	543.21	yelsh grn cr		955	1320	s	s	
d13	nitrate	$Dy(NO_3)_3.5H_2O$	438.58	yel cr		88.6		s		
d14	oxalate	$Dy_2(C_2O_4)_3.10H_2O$	769.21	wh pr		$-H_2O$, 40		i	i	s dil a
d15	oxide	Dysprosia. Dy_2O_3	373.00	wh powd	7.81[27]	2340 ± 10				grn soln a
d16	*orthophosphate*	$DyPO_4.5H_2O$	347.55	yelsh-wh powd		$-5H_2O$ > 200				s dil a, ac a
d17	selenate	$Dy_2(SeO_4)_3.8H_2O$	898.00	yel need		$-8H_2O$, 200		v s		i al
d18	sulfate	$Dy_2(SO_4)_3.8H_2O$	757.31	bril yel cr		stab 110	$-8H_2O$, 360	5.072^{20}	3.34[40]	
e1	**Erbium**	Er	167.28	dk gray powd	9.006	1529	2863	i	i	s a
e2	acetate	$Er(C_2H_3O_2)_3.4H_2O$	416.48	wh cr, tricl	2.114					
e3	bromide	$ErBr_3.9H_2O$	569.15	vlt-rose cr		950	1460	s	s	
e4	chloride	$ErCl_3.6H_2O$	381.73	pink cr, deliq		774	1500	s		s al
e5	fluoride	ErF_3	224.28	rose cr		1350	2200	i	i	i dil a
e6	iodide	ErI_3	547.99	vlt-red cr		1020	1280	s	s	
e7	nitrate	$Er(NO_3)_3.5H_2O$	443.37	redsh cr		$-4H_2O$, 130		s		s al, eth, acet
e8	oxalate	$Er_2(C_2O_4)_3.10H_2O$	778.77	redsh micr powd	2.64(?)	d 575		i	i	i dil a
e9	oxide	Erbia. Er_2O_3	382.56	rose red powd, tr to cub at 1300	8.640	infus		0.00049[24]		sl s min a
e10	sulfate	$Er_2(SO_4)_3$	622.74	wh powd, hygr	3.678	d 630		43[0]		
e11	sulfate, octahydrate	$Er_2(SO_4)_3.8H_2O$	766.87	rose red, monocl	3.217	$-8H_2O$, 400		16[20]	6.53[40]	
e12	**Europium**	Eu	151.96	steel gray met, cub	5.2434	822	1597	i	i	
e13	(II) bromide	$EuBr_2$	311.78			677	1880	s		
e14	(III) bromide	$EuBr_3$	391.69			702	d	s		
e15	(II) chloride	$EuCl_2$	222.87	wh, amorph		727	>2000	s	s	s a
e16	(III) chloride	$EuCl_3$	258.32	yel need	4.89[20]	850				
e17	(II) fluoride	EuF_2	189.96	brt yel	6.495	1380	>2400	i	i	
e18	(III) fluoride	EuF_3	208.96	col		1390	2280	i	i	i dil a

No.	Name	Synonyms and Formulae	Mol. wt.	Crystalline form, properties and index of refraction	Density or spec. gravity	Melting point, °C	Boiling point, °C	Solubility, in grams per 100 cc		
								Cold water	Hot water	Other solvents
	Europium									
e19	(II) iodide	EuI$_2$	405.77	br to olv grn cr	5.50$_4^{25}$	527	1580	s	s	
e20	(III) iodide	EuI$_3$	532.68			877	d	s	s	
e21	(III) nitrate	Eu(NO$_3$)$_3$.6H$_2$O	446.07	col		85 (sealed tube)		v s	v s	
e22	oxide	Eu$_2$O$_3$	351.92	pa rose powd	7.42			i	i	i dil a
e23	(II) sulfate	EuSO$_4$	248.02	col, orthorhomb	4.989^{20}			i	i	i dil a
e24	(III) sulfate	Eu$_2$(SO$_4$)$_3$.8H$_2$O	736.23	pa rose cr	4.95 (anh)	−8H$_2$O, 375		2.563^{20}	1.93^{40}	
—	**Ferric or ferrous**	See *Iron*								
f1	**Ferricyanic acid**	H$_3$Fe(CN)$_6$	214.98	grn-brn need, deliq		d		s	s	s al
f2	**Ferrocyanic acid**	H$_4$Fe(CN)$_6$	215.99	wh need, bl in moist air		d		s	s	s al; i eth
f3	**Fluoboric acid**	HBF$_4$	87.81	col liq		d 130		∞	s	∞ al
f4	**Fluorophosphoric acid, di-**	HPO$_2$F$_2$	101.98	col fum liq	1.583$_4^{25}$	−75	116	s		
f5	**Fluorophosphoric acid, hexa-**	HPF$_6$	145.97	col fum liq	ca 1.65 (65%)	31 (6H$_2$O)				
f6	**Fluorophosphonic acid, mono-**	H$_2$PO$_3$F	99.99	col visc liq	1.818$_4^{25}$			∞		
f7	**Fluorine**	F	18.998403	grn yel gas, pois, 1.000195	1.69^{15}g/1 1.51^{-188}	−219.62	−188.14	HF+O$_3$	d	
f8	(di-)oxide	F$_2$O	54.00	col gas or yel brn liq	liq 1.90$^{-233.8}$	−223.8	−144.8	sl s d	i	sl s alk, a
f9	oxide, di-	Dioxygen fluoride. F$_2$O$_2$	70.00	brn gas, red liq, orange solid	sol 1.912^{-165} liq 1.45^{-57}	−163.5	−57			
f10	**Fluosilicic acid**	H$_2$SiF$_6$.xH$_2$O	hydr (not known)	col fum coros liq, 1.3465^{25} (60.97% soln)	1.4634^{25} (60.97% soln)		d	s	s	sl s alk
f11	dihydrate	H$_2$SiF$_6$.2H$_2$O	180.12	wh cr, fum, deliq		d		s	s	s alk
f12	**Fluosulfonic acid**	HSO$_3$F	100.07	col liq	1.743^{15}	−87.3	165.5	s		
g1	**Gadolinium**	Gd	157.25	col or lt yel met, hex	7.9004	1313	3266	i	i	s a
g2	acetate tetrahydrate	Gd(C$_2$H$_3$O$_2$)$_3$.4H$_2$O	406.45	col, tricl	1.611			11.6^{25}		
g3	acetylacetonate, trihydrate	Gd[CH(COCH$_3$)$_2$]$_3$.3H$_2$O	508.63			143.5–145		i	i	
g4	bromide, hexahydrate	GdBr$_3$.6H$_2$O	505.07	rhomb pl	2.844^{16}			s	s	s HBr
g5	chloride	GdCl$_3$	263.61	col monocl, pr	4.52^0	609		s	s	
g6	chloride, hexahydrate	GdCl$_3$.6H$_2$O	371.70	wh pr, deliq	2.424^0			s	s	
g7	iodide	GdI$_3$	537.96	citr yel		926	1340	s	s	
g8	fluoride	GdF$_3$	214.25					i		sl s h HF
g9	dimethylphosphate	Gd[(CH$_3$)$_2$PO$_4$]$_3$	532.37					23.0^{25}	6.7^{100}	
g10	nitrate, hexahydrate	Gd(NO$_3$)$_3$.6H$_2$O	451.36	tricl, deliq	2.332	91		v s	v s	s al
g11	nitrate, pentahydrate	Gd(NO$_3$)$_3$.5H$_2$O	433.34	pr	2.406^{15}	92		i	i	v sl s conc HNO$_3$
g12	oxalate	Gd$_2$(C$_2$O$_4$)$_3$.10H$_2$O	758.71	monocl		−6H$_2$O, 110		i		s HNO$_3$; v sl s H$_2$SO$_4$
g13	oxide	Gadolinia. Gd$_2$O$_3$	362.50	wh amorph powd, hygr	7.407^{15}	2330 ± 20		v sl s		s a
g14	selenate	Gd$_2$(SeO$_4$)$_3$.8H$_2$O	887.50	monocl, pearly	3.309	−8H$_2$O, 130		s	s	
g15	sulfate	Gd$_2$(SO$_4$)$_3$	602.68	col	4.139$^{14.5}$	d 500		3.98^0	2.26$^{34.4}$	
g16	sulfate, octahydrate	Gd$_2$(SO$_4$)$_3$.8H$_2$O	746.81	col, monocl	3.010$^{14.6}$			3.28^{20}	2.54^{40}	
g17	sulfide	Gd$_2$S$_3$	410.69	yel mass, hyg	3.8			d		d a
g18	**Gallium**	Ga	69.72	gray-blk ortho-rhomb, tendency to undercool	sol 5.904$^{29.6}$ liq 6.095$^{29.8}$	29.78	2403	i	i	s a, i alk
g19	acetate, basic	4Ga(C$_2$H$_3$O$_2$)$_3$. 2Ga$_2$O$_3$.5H$_2$O	1452.37	wh micro cr		d 160		s d	d	i ac a
g20	acetylacetonate	2,4-Pentanedione deriv. Ga(C$_5$H$_7$O$_2$)$_3$	367.05	monocl or pl, rhomb or pyram, rhomb pyram	1.42, 1.41	194–195	subl 140^{10}	s	s	s acet
g21	arsenide	GaAs	144.64	dk-gray cub cr		1238				
g22	bromide, tri-	GaBr$_3$	309.45	wh cr	3.69$_4^{25}$	121.5 ± 0.6	278.8	s	s	sl s NH$_3$
g23	bromide, tri-, hexammine	GaBr$_3$.6NH$_3$	411.63	wh powd				d	d	sl s NH$_3$
g24	bromide, tri-, monammine	GaBr$_3$.NH$_3$	326.48	wh powd	3.112^{25}	124		d	d	sl s NH$_3$
g25	perchlorate	Ga(ClO$_4$)$_3$.6H$_2$O	476.16			d 175		v s		v s al
g26	chloride, di-	GaCl$_2$	140.63	wh cr, deliq		164	535	d	d	s bz
g27	chloride, tri-	GaCl$_3$	176.03	wh need, deliq	2.47$_4^{25}$ liq 2.36$_{80}^{80}$	77.9 ± 02	201.3	v s	v s	s bz, CCl$_4$, CS$_2$

No.	Name	Synonyms and Formulae	Mol. wt.	Crystalline form, properties and index of refraction	Density or spec. gravity	Melting point, °C	Boiling point, °C	Solubility, in grams per 100 cc		
								Cold water	Hot water	Other solvents
	Gallium									
g28	chloride, tri-, hexammine	GaCl₃.6NH₃	278.26					d	d	s NH₃
g29	chloride, tri-, monoammine	GaCl₃.NH₃	193.11	wh powd	2.189²⁵	124		d	d	s NH₃
g30	ferrocyanide	Ga₄[Fe(CN)₆]₃	914.74		d	subl 800	ca 1000	d	d	i conc HCl
g31	fluoride, tri-	GaF₃	126.72	wh powd	4.47 ± 0.01	subl 800 (in N₂)	ca 1000	0.002	i	v sl s dil a; s HF
g32	fluoride, tri-, trihydrate	GaF₃.3H₂O	180.76	wh powd		−H₂O (vac) 140		i	sl s	sl s dil H₂F₂; v s dil HCl
g33	fluoride, tri-, triammine	GaF₃.3NH₃	177.81	wh powd		−NH₃, 100		d	d	
g34	hydride	Digallane, galloethane. Ga₂H₆	145.49	col liq		−21.4	139, d>130	d	d	d a, alk
g35	hydroxide	Ga(OH)₃	120.74	wh		d 440		i	i	s dil a
g36	hydroxyquinoline deriv.	Ga(C₉H₆NO)₃	502.18	grn-yel cr		>150	subl vac	0.0001	0.0012	s a, alk; sl s al
g37	iodide, tri-	GaI₃	450.43	lt yel cr	4.15²⁵₄	212 ± 1	subl 345		d	
g38	iodide, tri-, hexammine	GaI₃.6HN₃	552.62	wh powd				d	d	
g39	iodide, tri-, monoammine	GaI₃.NH₃	467.46	wh powd	3.635²⁵₄			d	d	
g40	nitrate	Ga(NO₃)₃ · X H₂O		wh cr, deliq		d 110	d to Ga₂O₃ 200	v s	v s	s abs al; i eth
g41	nitride	GaN	83.73	dk gray powd	6.1	subl 800		i	i	i dil a; sl s h conc H₂SO₄, h conc NaOH
g42	oxide, sesqui-(α)	Ga₂O₃	187.44	wh, hex, rhomb, 1.92, 1.95	6.44	1900;tr to β 600		i	i	s alk; v sl s h a
g43	oxide, sesqui-(β)	Ga₂O₃	187.44	monocl, rhomb	5.88	1795 ± 15		i		s alk; v sl s ha a
g44	oxide, sesqui-, monohydrate	Ga₂O₃.H₂O	205.45	wh micr cr, orthorhom, 1.84	5.2	−H₂O, 400 tr to Ga₂O₃		i	i	sl s a; s alk
g45	oxide, sub-	Ga₂O	155.44	blk brn powd	4.77²⁵₄	>660	subl >500	i	i	s a, alk
g46	oxalate	Ga₂(C₂O₄)₃.4H₂O	475.56	wh micro cr, hygr		−4H₂O, 180	d 200	0.4		
g47	oxychloride	6GaOCl.14H₂O	979.25	oct				i		v s KOH; i dil HNO₃; s acet
g48	selenate	Ga₂(SeO₄)₃.16H₂O	856.56	col, monocl or tricl cr				57.5²⁵	v s	
g49	selenide, mono-	GaSe	148.68	dk red-br greasy leaf	5.03²⁵₄	960 ± 10				
g50	selenide, sesqui-	Ga₂Se₃	376.32	rdsh-bl brittle, hard	4.92²⁵₄	>1020 ± 10				
g51	selenide, sub-	Ga₂Se	218.40	bl	5.02²⁵₄					
g52	sulfate	Ga₂(SO₄)₃	427.62	wh powd		diss 690⁷⁰⁰		v s	v s	s al; i eth
g53	sulfate, hydrate	Ga₂(SO₄)₃.18H₂O	751.90	oct cr				v s	v s	s 60 % al; i eth
g54	sulfide, mono-	GaS	101.78	yel cr	3.86²⁵₄	965 ± 10		i	d	s a, alk
g55	sulfide, sesqui-	Ga₂S₃	235.63	yel cr, or wh amorph	3.65²⁵	1255 ± 10		d	d	s a, alk
g56	sulfide, sub-	Ga₂S	171.50	dk gray	4.18²⁵	d vac 800		d	d	s a, alk
g57	telluride, mono-	GaTe	197.32	blk soft cr	5.44²⁵₄	824 ± 2				
g58	telluride, sesqui-	Ga₂Te₃	522.24	blk brittle cr	5.57²⁵₄	790 ± 2				
	Germane									
g59	bromo-	GeH₃Br	155.52	col liq	2.34²⁹·⁵	−32	52	d	d	i al; d alk
g60	chloro-	GeH₃Cl	111.07	col liq	1.75⁻²⁵	−52	28.0	d	d	i al; d alk
g61	chloro trifluoro-	GeF₃Cl	165.04	col gas		−66.2	−20.63	d	d	s abs al
g62	dibromo-	GeH₂Br₂	234.42		2.80⁰	−15.0	89.0	d	d	i al; d alk
g63	dichloro-	GeH₂Cl₂	145.51	col liq	1.90⁻⁶⁸	−68.0	69.5	d	d	i al; d alk
g64	dichlorodifluoro-	GeCl₂F₂	181.49	col gas		−51.8	−2.8	d	d	s abs al
g65	tribromo-	Germanium bromoform GeHBr₃	313.33	col liq		−24	d	d	d	d alk
g66	trichloro-	Germanium chloroform. GeHCl₃	179.96	col liq	1.93⁰	−71	75.2 d	d	d	d alk
g67	trichlorofluoro-	GeCl₃F	197.95	col liq		−49.8	37.5	d		s abs al
g68	**Germanium**	Ge	72.59	gray-wh met-cub	5.35²⁰₂₀	937.4	2830	i	i	s h H₂SO₄, aq reg; i alk

No.	Name	Synonyms and Formulae	Mol. wt.	Crystalline form, properties and index of refraction	Density or spec. gravtiy	Melting point, °C	Boiling point, °C	Solubility, in grams per 100 cc		
								Cold water	Hot water	Other solvents
	Germanium									
g69	bromide, di-	$GeBr_2$	232.41	col need or pl		122	d	d	d	s a, $GeBr_4$, al; i bz
g70	bromide, tetra-	$GeBr_4$	392.23	gray-wh oct, 1.6269	3.132^{29}_{29}	26.1	186.5	d	d	s abs al, eth, bz; i conc H_2SO_4
g71	chloride, di-	$GeCl_2$	143.50	wh powd		d to Ge+ $GeCl_4$		d	d	s $GeCl_4$; i al, chl
g72	chloride, tetra-	$GeCl_4$	214.41	col liq, 1.464	1.8443^{30}	−49.5	84	d	d	s al, eth; v s dil HCl; i conc HCl conc H_2SO_4
g73	fluoride, di-	GeF_2	110.59	wh cr, hygr		d>350	subl	s	v s	
g74	fluoride, tetra-	GeF_4	148.58	col gas or liq, not liq at atm press	$2.46^{-35.5}$	subl −37		d to GeO+ H_2GeF_6		
g75	fluoride, tetra-	$GeF_4.3H_2O$	202.63	wh cr, deliq		d		s	s	
g76	hydride	Digermane. Ge_2H_6	151.23	liq	1.98^{-100}	−109	29; d 215	d		s liq NH_3
g77	hydride	Trigermane. Ge_3H_8	225.83	col liq	2.2^{20}	−105.6	110.5; d 195	i	i	s CCl_4
g78	hydride, tetra-	Germane. GeH_4	76.62	col gas	1.523^{-142}	−165	−88.5; d 350	i	i	s liq NH_3, NaOCl; sl s h HCl
g79	imide	$Ge(NH)_2$	102.62	wh amorph powd		d 150			d to NH_3 +GeO_2	
g80	iodide, di-	GeI_2	326.40	or hex pl	5.37	d	subl vac 240	s	s d	s conc HI, dil a; sl s CCl_4, chl; i CS_2
g81	iodide, tetra-	GeI_4	580.21	red-or, cub	4.322^{25}_{25}	144	d 440	s d		d al, acet; s CS_2, CCl_4, bz, MeOH
g82	(tri-) nitride, di-	Ge_3N_2	245.78	blk cr			subl 650			
g83	(tri-) nitride, tetra-	Ge_3N_4	273.80	wh-lt brn powd	5.25^{25}	d 450		i	i	i a, alk
g84	oxide, d-(insoluble)	GeO_2	104.59	tetr	6.239	1086±5		i		sl s NaOH; i HCl
g85	oxide, di-(soluble)	GeO_2	104.59	col, hex, 1.650	4.228^{25}	1115.0±4		0.447^{25}	1.07^{100}	s a, alk; i HCl HF; one form NaOH, NH_4OH; one form
g86	oxide, mono-	GeO	88.59	blk cr powd, 1.607		subl 710		i	i	s Cl_2 water, H_2O_2+NH_4OH; i a, alk
g87	oxychloride	$GeOCl_2$	159.50	col liq		−56	d>20	d	d	i all solv
g88	selenide	$GeSe_2$	230.51	orange, rhomb(?)	4.56^{25}	707±3	d	i	i	v sl s a; sl s alk
g89	sulfide, di-	GeS_2	136.72	wh powd, or wh, orthorhomb	2.94^{14}	ca 800	subl >600	0.45 d	d to GeO_2 +H_2S	s alk, alk sulf; i al, eth, a; 3.112 liq NH_3
g90	sulfide, mono-	GeS	104.65	yel-red amorph, or rhomb bipyram, blk	amorph: 3.31 rhomb:4.01	530	subl 430	0.24	i	s HCl, alk or alk sulf; sl s NH_4OH; 0.0473 liq NH_3
	Glucinum	See **Beryllium**								
g91	**Gold**	Au	196.967	yel duct met, cub, coll blue-viol	18.88^{20}	1064.43	3080	i	i	s aq reg, KCN, h H_2SO_4; i a
g92	(I) bromide	AuBr	276.88	yel-gray mass, or cr powd	7.9	d 115		i	i	d a; s NaCN
g93	(III) bromide	$AuBr_3$	436.69	gray powd, or br cr		97.5-Br, 160		sl s		s eth, al
g94	(I) chloride	AuCl	232.42	yel cr	7.4	170 d to $AuCl_3$	d 289.5	v sl s d	d	s HCl, HBr
g95	(III) chloride	$AuCl_3$ or Au_2Cl_6	303.33	claret-red cr pr	3.9	d 254	subl 265	68	v s	s al, eth; sl s NH_3; i CS_2
g96	(I) cyanide	AuCN	222.98	lt yel cr powd	7.12^{25}	d		v sl s	v sl s	s KCN, NH_4OH; i eth, alk
g97	(III) cyanide	Cyanoauric acid. $Au(CN)_3.3H_2O$ or $HAu(CN)_4.3H_2O$	329.07	col pl, hygr		d 50		v s	d, v s	s al, eth
g98	(I) iodide	AuI	323.87	grnsh-yel powd	8.25	d 120		v sl s	sl s d	s KI
g99	(III) iodide	AuI_3	577.68	dk grn		d		i	d	s iodides
g100	(III) nitrate, hydrogen	Nitratoauric acid. $AuH(NO_3)_4.3H_2O$ or $H[Au(NO_3)_4].3H_2O$	500.04	yel, tricl, oct	2.84	d 72		s, d		s HNO_3
g101	(III) oxide	Au_2O_3	441.93			−O, 160	−3O, 250	i	i	s HCl, conc HNO_3, NaCN
g102	(III) oxide	$Au_2O_3.xH_2O$				−1½H_2O, 250		$5.7×10^{-11}$ 25		s HCl, NaCN, conc HNO_3
g103	phosphide	Au_2P_3	486.86	gray	6.67		d			i HCl, dil HNO_3
g104	selenide	Au_2Se_3	630.81		4.65^{22}					
g105	(I) sulfide	Au_2S	426.00	br blk powd		d 240		i fresh sol	ppt coll	s aq reg, KCN; i a
g106	(III) sulfide	Au_2S_3	490.13	br blk powd	8.754	d 197		i		i al, eth, s Na_2S

No.	Name	Synonyms and Formulae	Mol. wt.	Crystalline form, properties and index of refraction	Density or spec. gravity	Melting point, °C	Boiling point, °C	Solubility, in grams per 100 cc		
								Cold water	Hot water	Other solvents
	Gold									
g107	telluride, di-	Nat. krennerite. $AuTe_2$	452.16	1) rhomb, 2) monocl, 3) tricl yel earthy to massive	8.2–9.3	d 472	....	i	i	
h1	**Hafnium**	Hf	178.49	hex	13.31^{20}	2227 ± 20	4602	i	i	s HF
h2	bromide	$HfBr_4$	498.13	wh		subl 420				
h3	carbide	HfC	190.54		12.20	ca 3890	....		i	
h4	chloride	$HfCl_4$	320.30	wh		subl 319		d		s MeOH, acet
h5	fluoride	HfF_4	254.48	monocl, 1.56						
h6	iodide	HfI_4	686.11				400 subl vac			
h7	nitride	HfN	192.50	yel-brn, cub		3305				
h8	oxide	Hafnia. HfO_2	210.49	wh, cub	9.68^{20}	2758 ± 25	$\sim5400(?)$	i	i	
h9	oxychloride	$HfOCl_2.8H_2O$	409.52	col				s		
h10	**Helium**	He	4.0026	col gas, inert odorless	0.1785^0 g/l liq $0.147^{-270.3}$	$-272.2^{26/m}$	-268.9	0.94^0 cm^3 0.94^{25} cm^3	1.05^{50} cm^3 1.21^{75} cm^3	i al; absorbed by Pt
h11	**Holmium**	Ho	164.9304	met, hex	8.7947	1474	2695	i	i	
h12	bromide	$HoBr_3$	404.66	lt yel		914	1470	s	s	
h13	chloride	$HoCl_3$	271.29	lt yel		718	1500	s	s	
h14	iodide	HoI_3	545.64	lt yel		989	1300	s	s	
h15	fluoride	HoF_3	221.93	lt yel		1143	>2200	i	i	i dil a
h16	oxalate	$Ho_2(C_2O_4)_3.10H_2O$	774.10	pa tan		$-H_2O$, 40	d	i	i	i dil a
h17	oxide	Holmia. Ho_2O_3	377.86	tan				i	i	s a
h18	**Hydrazine**	NH_2NH_2	32.05	col liq or wh cr, 1.470^{22}	$1.004^{2\ 5}$	2.0	113.5	v s	....	s al
h19	azide	$N_2H_4.HN_3$	75.07	wh pr, deliq, 1.53, 1.76		75.4		v s	v s	1.2^{23} al; 6.1^{23} MeOH; i CS_2 bz
h20	fluogermanate	$2N_2H_4.H_2GeF_6$	252.80	monocl pr, 1.452, 1.460, 1.464	2.406^{25}_{25}			s		
h21	fluosilicate	$N_2H_4.H_2SiF_6$	176.14	cr		d 186		v s	sl s al	
h22	formate	$N_2H_4.2CH_2O_2$	124.10			128		s		
h23	hydrate	$N_2H_4.H_2O$	50.07	col fum liq or cub cr, 1.42842	1.03^{21}	-40	118.5^{740}	∞	∞	s al; i eth, chl
h24	hydrochloride, di-	$N_2H_4.2HCl$	104.97	col vitr, oct	1.42	198, $-HCl$	d 200	27.2^{22}	v s	sl s al
h25	hydrochloride, mono-	$N_2H_4.HCl$	68.51	wh need		89	d 240	v s	v s al; v s liq NH_3
h26	hydroiodide	Hydrazine monoiodide. $N_2H_4.HI$	159.98	col pr		124–126		s		
h27	nitrate, di-	$N_2H_4.2HNO_3$	158.07	col cr		104 (rapid heat); d 80 (slow heat)		v s	d	
h28	nitrate, mono-	$N_2H_4.HNO_3$	95.06	col dimorph need (α, β)		$\alpha70.71$; $\beta62.09$	subl 140	174.9^{10}	2127^{60}	sl s al
h29	oxalate	$2N_2H_4.H_2C_2O_4$	154.14	wh need		148		200^{35}		0.0003^{22} al; i eth
h30	perchlorate	$N_2H_4.HClO_4.\frac{1}{2}H_2O$	141.51	exp	1.939	137	d 145	d		s al; i eth, bz, chl, CS_2
h31	hypophosphate	$N_2H_4.2H_2PO_3$	194.03			152				
h32	orthophosphate	$N_2H_4.H_3PO_4$	130.05	cr, hygr		82		v s		
h33	orthophosphite	$N_2H_4.2H_3PO_3$	196.06			82				
h34	orthophosphite	$N_2H_4.H_3PO_3$	114.05			36		v s		
h35	picrate	$N_2H_4.HC_6H_2N_3O_7.\frac{1}{2}H_2O$	270.16			201.3		s	s	
h36	selenate	$N_2H_4.H_2SeO_4$	177.02	col cr powd, unstab		exp		v sl s	v s	
h37	sulfate	$N_2H_4H_2SO_4$	130.13	col, rhomb	1.37	254	d	3.415^{25}	14.39^{80}	i al
h38	sulfate	$(N_2H_4)_2.H_2SO_4$	162.18	col cr, hygr		85		202.2^{25}	554.4^{50}	i al
h39	tartrate	$(N_2H_4)C_4H_6O_6$	182.13	col cr; $[\alpha]^{20}_4 +22.5$		182–183		6.0^0		
h40	**Hydroazoic acid**	Azoimide. HN_3	43.03	col liq	1.09^{25}	-80	37	∞	∞	s al, alk, eth
h41	**Hydrogen**	H_2	2.0158	col gas, cub sol	gas 0.0899 g/l liq 0.070	-259.14	-252.8	2.14^0 cm^3 1.91^{25} cm^3	0.85^{50} cm^3 1.89^{60} cm^3	6.925^0 cm^3 al
h42	antimonide	Stibine H_3Sb	124.77	col gas, pois	gas 5.30^0 g/l liq 2.26^{-25}	-88.5	-17	20 cm^3	4 cm^3	1500 cm^3 al; 2500 cm^3 CS_2
h43	arsenide	Arsine H_3As	77.95	col gas, pois	3.484 g/l	-113.5	-55, d 230	20 cm^3	sl s	sl s al, alk
h44	arsenide (solid)	H_2As_2	151.86	br powd		d 200	....	i	d	i al, eth, CS_2, alk; s HNO_3
h45	bismuthide	Bismuthine. H_3Bi	212.00	liq, v unstab			22			
h46	bromide	Hydrobromic acid. HBr	80.92	col gas or pa yel liq, 1.325	gas 3.5^0 g/l liq 2.77^{-67}	-88.5	-67.0	221^0	130^{100}	s al
h47	bromide (const. boiling)	HBr(47 %) + H_2O		col liq	1.49	-11	126			
h48	bromide, dihydrate	$HBr.2H_2O$	116.95	wh cr, col liq	2.11^{-15}	-11	....	s	s	

No.	Name	Synonyms and Formulae	Mol. wt.	Crystalline form, properties and index of refraction	Density or spec. gravity	Melting point, °C	Boiling point, °C	Solubility, in grams per 100 cc		
								Cold water	Hot water	Other solvents
	Hydrogen									
h49	bromide, monohydrate	HBr.H₂O	98.93	col liq	1.78	stab	−3.6 to −15.5 between 1–2.5 atm			
h50	chloride	Hydrochloric acid. HCl	36.46	col gas or col liq, pois	1.187$^{-84.9}$ gas 1.00045 g/l	−114.8	−84.9	82.3^0	56.1^{60}	327 cm³ al; s eth, bz
h51	chloride (const. boiling)	HCl(20.24 %)+H₂O		col liq	1.097		110			
h52	chloride, dihydrate	HCl.2H₂O	72.49	col liq	1.4618.4	−17.7	d	d	∞	s al
h53	chloride, monohydrate	HCl.H₂O	54.48	col liq	1.48	−15.35		∞	∞	s al
h54	chloride, trihydrate	HCl.3H₂O	90.51	col liq		−24.4	d	∞		s al
h55	cyanide	Hydrocyanic acid. HCN	27.03	col liq or gas, pois, liq 1.2675^{10}	gas 0.901 g/l liq 0.699^{22}	−14	26	∞	∞	∞ al; s eth
h56	fluoride	Hydrofluoric acid. HF	20.01	col fum cor liq, or gas; liq 1.90	0.991$^{19.54}$ liq 0.987	−83.1	19.54	∞	v s	
h57	fluoride (const. boiling)	HF(35.35 %)+H₂O		col liq			120			
h58	iodide	Hydroiodic acid. HI	127.91	col gas, or pa yel liq, $n_D^{16.5}$ 1.466	gas 5.66^0 g/l liq 2.85$^{-4.7}$	−50.8	−35.38^4 atm	42.5^0 cm³	v s	s al
h59	iodide (const. boiling)	HI(57 %)+H₂O		col or pa yel fum liq	1.70^{15}		127^{774}			
h60	iodide, dihydrate	HI.2H₂O	163.94	col liq		−43		∞		
h61	iodide, tetrahydrate	HI.4H₂O	199.97	col liq		−36.5		∞		
h62	iodide, trihydrate	HI.3H₂O	181.96	col liq		−48		∞		
h63	oxide	Water. H₂O	18.0153	col liq or hex cr, liq 1.333, sol 1.309, 1.313	1.000$^4_4$	0.000	100.000			∞ al
h64	oxide, per-	H₂O₂	34.01	col liq; 1.414^{22}	1.4067^{25}	−0.41	150.2^{760}	∞		s al, eth; i pet eth
h65	phosphide	Phosphine. H₃P	34.04	col gas rather good odor or col liq, 1.317 liq	gas 1.529 g/l liq 0.746^{-90}	−133.5	−87.4	26^0 cm³	i	s al, eth, Cu₂Cl₂
h66	phosphide	H₄P₂	65.98	col liq	1.012	−90	57.5^{735}	i	i	s al, turp
h67	phosphide	(H₂P₄)₃	377.73	yel solid	1.83^{19}	ign 160	d	i	i	i al; s P, P₂H₄
h68	sulfide	H₂S	34.08	col gas, infl, liq 1.374	1.539^0 g/l	−85.5	−60.7	437^0 cm³	186^{40} cm³	9.54^{20} cm³ al; s CS₂
h69	sulfide, di-	H₂S₂	66.14	yel oil, 1.885	1.334^{20}	−89.6	70.7^{760}	d		s bz, eth, CS₂; i al
h70	sulfide, penta-	H₂S₅	162.34	clear yel oil	1.67^{16}	−50	50^4			s bz, eth, CS₂; i al
h71	sulfide, tetra-	H₂S₄	130.27	lt yel liq	1.588^{15}	−85				s bz, eth, CS₂; i al
h72	sulfide, tri-	H₂S₃	98.21	brt yel liq, 1.705^{15}	1.496^{15}	−52	d 90			s bz, eth, CS₂; i al
h73	telluride	H₂Te	129.62	col gas or yel need	gas 5.81 g/l liq 2.57^{-20}	−49	−2	s unstab		s al, alk
h74	**Hydroxylamine**	NH₂OH	33.03	wh need or col liq, deliq	1.204	33.05	56.5	s	d	s a, al, MeOH; v sl s eth, chl, bz, CS₂
h75	acetate	NH₂OH.CH₃CO₂	92.08	col cr		87	subl 90	v s		
h76	bromide	NH₂OH.HBr	113.95	wh, monocl	2.35$^4_4$			v s	v s	i eth
h77	fluogermanate	(NH₂OH)₂.H₂GeF₆.2H₂O	290.69	monocl pr, 1.418, 1.438, 1.443	2.229^{25}			s		s abs al
h78	fluosilicate	(NH₂OH)₂.H₂SiF₆.2H₂O	246.18	scales				v s		i al
h79	formate	NH₂OH.HCO₂	78.05	col need		76	d 80	v s	d	s h al; i eth
h80	hydrochloride	NH₂OH.HCl	69.49	col, monocl	1.67^{17}	151	d	83^{17}	v s	4.43^{20} al; 16.4^{20} MeOH; s gluc; i eth
h81	iodide	NH₂OH.HI	160.94	col need, hygr		83–84 exp		v s	d	v s MeOH; sl s eth
h82	nitrate	NH₂OH.HNO₃	96.04	wh		48	d <100	v s	d	v s al
h83	orthophosphate	(NH₂OH)₃.H₃PO₄	197.09			148 exp		1.9^{20}	16.8^{90}	
h84	sulfate	(NH₂OH)₂.H₂SO₄	164.14	col, monocl		d 170	d	32.9^0	68.5^{20}	sl s al; s eth
i1	**Indium**	In	114.82	soft silv wh met, tetr	7.30^{20}	**156.61**	**2080**	i	i	s a; v sl s NaOH
i2	antimonide	InSb	236.57	cr		535				
i3	arsenide	InAs	189.74	met cr		943				i a
i4	bromide, di-	InBr₂	274.64	pa yel solid	4.22^{25}	235	632 subl	d		s a
i5	bromide, mono-	InBr	194.73	red br solid	4.96^{25}	220	662 subl	d		s a
i6	bromide, tri-	InBr₃	354.55	wh to yel need, deliq	4.74^{25}	436 ± 2	subl	v s		
i7	perchlorate	In(ClO₄)₃.8H₂O	557.29	col cr, deliq		ca 800	d 200	v s	d	s abs al; sl s eth
i8	chloride, di-	InCl₂	185.73	wh rhomb, deliq	3.655^{25}	235	550–570	d	d	
i9	chloride, mono-	InCl	150.27	1) yel or 2) dk red, deliq	4.19^{25} yel 4.18^{25} red	225 ± 1	608	d	d	s a

No.	Name	Synonyms and Formulae	Mol. wt.	Crystalline form, properties and index of refraction	Density or spec. gravity	Melting point, °C	Boiling point, °C	Solubility, in grams per 100 cc		
								Cold water	Hot water	Other solvents
	Indium									
i10	chloride, tri-	$InCl_3$	221.18	wh pl, deliq	3.46_4^{25}	586 subl 300	volat 600	v s	v s	sl s al, eth
i11	cyanide	$In(CN)_3$	192.87	wh ppt					unstab	i dil a; s HCN; v sl s NaOH
i12	fluoride	InF_3	171.82	col	$4.39^{25} \pm 1$	1170 ± 10	>1200	0.040^{25}		s a; i al, eth
i13	fluoride, trihydrate	$InF_3.3H_2O$	225.86	cr		$-3H_2O$, 100		8.49^{22}		s a; i al, eth
i14	fluoride, nonahydrate	$InF_3.9H_2O$	333.95	wh need		d		sl s	d	s HCl, HNO_3; i al, eth
i15	hydroxide	$In(OH)_3$	165.84	wh ppt		$-H_2O < 150$		i		s a; v sl s NaOH; i NH_4OH
i16	iodate	$In(IO_3)_3$	639.53	wh cr			d	0.067^{20}		s dil HNO_3, dil H_2SO_4; s d HCl
i17	iodide, di-	InI_2	368.63		4.71^{25}	212				s a
i18	iodide, mono-	InI	241.72	br red solid	5.31	351	711–715		sl d	i al, eth, chl; s dil a
i19	iodide, tri-	InI_3	495.53	carmine red, or yel cr	4.69	210		v unstable s		s a, chl, bz yxl
i20	methylate	$In(CH_3)_3$	159.93	col cr	1.568_{19}^{19}			d		d al, MeOH; s liq NH_3, eth; v s acet, bz
i21	nitrate	$In(NO_3)_3.3H_2O$	354.88	pl, deliq		$-2H_2O$, 100	d	v s		s al
i22	nitrate	$In(NO_3)_3.4\frac{1}{2}H_2O$	381.90	need, deliq		$-4\frac{1}{2}H_2O$	d	v s		s al
i23	oxide, mon-	InO	130.81	wh gray				i		s a
i24	oxide, sesqui-	In_2O_3	277.64	red brn, (h) pa yel (c) amorph and trig	7.179		volat 850	i		amorph s a; cr i a
i25	oxide, sub-	In_2O	245.64	blk cr	6.99_4^{25}	subl vac 565–700				s HCl
i26	phosphide	InP	145.79	brittle mass met		1070				v sl s min a
i27	selenate	$In_2(SeO_4)_3.10H_2O$	838.67	cr, deliq				v s		
i28	selenide, sesqui-	In_2Se_3	466.52	blk cr, or soft dk scales	5.67_4^{25}	890 ± 10				s, d conc a
i29	sulfate	$In_2(SO_4)_3$	517.83	wh gray powd, monocl pr, hygr	3.438			s	v s	
i30	sulfate	$In_2(SO_4)_3.9H_2O$	679.96	wh powd, hygr	3.44	d 250		v s		
i31	sulfate, dihydrate	$In_2(SO_4)_3.H_2SO_4.7H_2O$	742.01	rhomb cr		$-7H_2O$, $-H_2SO_4$, ca 250				
i32	sulfide, mono-	InS	146.88	dk	5.18^{25}	692 ± 5	subl vac 850			s HCl, HNO_3
i33	sulfide, sesqui-	In_2S_3	325.83	red cr or yel ppt	4.90	1050	subl ca 850 in high vac	i		s a; sl s Na_2S
i34	sulfide, sub-	In_2S	261.70	yel or blk need	5.87^{25}	653 ± 5				
i35	sulfite, basic	$2In_2O_3.3SO_2.8H_2O$	891.59	wh cr		$-3H_2O$, 100	$-8H_2O$, 260	i		
i36	telluride, sesqui-	$InTe$	242.42	dk met, shiny	6.29_4^{25}	696 ± 2				i HCl, s HNO_3
i37	telluride	In_2Te_3	612.44	bl brittle cr	5.78	667				
i38	**Iodic acid**	HIO_3	175.91	col or pa yel cr powd, rhomb	4.629^0	d 110		286^0 310^{16}	473^{80} 576^{101}	v s 87% al; sl s HNO_3; i abs al, eth, chl
i39	metaper-	HIO_4	191.91	col		subl 110	d 138	v s, d		
i40	*ortho*paraper-	H_5IO_6 or $HIO_4.2H_2O$	227.94	wh monocl, deliq			d 140	113	v s	s al, eth
i41	**Iodine**	I_2	253.809	vlt blk met lust, rhomb, 3.34	4.93	**113.5**	184.35^{atm}	0.029^{20} 0.030^{25}	0.078^{50}	20.5^{15} al; 16.46^{25} bz 20.6^{17} eth; s chl glyc, KI; 24^{25} eth 23^{25} MeOH 20.15^{25} CS_2; 2.91^{25} CCl_4
i42	azide	Iod(o)azide. IN_3	168.92	yel, exp				s d		s $Na_2S_2O_3$
i43	bromide, mono-	IBr	206.81	dk gray cr	4.415^0	(42) subl 50	d 116	s d		s al, eth, chl, CS_2
i44	bromide, tri-	IBr_3	366.63	br liq				s		s al
i45	chloride, mono-(α)	ICl	162.36	dk-red need, cub, red br oily liq	3.1822^0	27.2	97.4	d to HIO_3 +Cl		s al, eth, CS_2, HCl
i46	chloride, mono-(β)	ICl	162.36	brn red, rhombic 6 sided pl	liq 3.24^{34}	13.92	97.4; d 100	d		s al, eth, HCl
i47	chloride, tri-	ICl_3	233.26	yel brn, rhomb, red liq	3.117^{15}	101^{16atm}	d 77	s d		s al, eth, CCl_4, ac a, bz
i48	cyanide	ICN	152.92	wh cr				sl s	sl s	s al, eth
i49	fluoride, hepta-	IF_7	259.89	col cr or liq	liq 2.8^6	5.5	4.5 subl	v s, d	d	d a, alk
i50	fluoride, penta-	IF_5	221.90	col liq	3.75	9.6	98	d	d	d a, ak
i51	oxide, di- (or tetra-)	IO_2 or I_2O_4	158.90	lem-yel cr	4.2_{10}^{10}	d slow 75 rap 130		d to $HIO_3 + I_2$		s H_2SO_4; sl s acet; i al, eth

No.	Name	Synonyms and Formulae	Mol. wt.	Crystalline form, properties and index of refraction	Density or spec. gravity	Melting point, °C	Boiling point, °C	Cold water	Hot water	Other solvents
	Iodine									
i52	oxide, pent-	I_2O_5	333.81	wh trim	4.799_4^{25}	d 300–350		187.4^{13}	v s	i abs al, eth, chl, CS_2; sl s dil a
i53	(tetra-) oxide non-	I_4O_9	651.61	yel powd, hygr		d 75				
i54	**Iodo platinic acid**	$H_2PtI_6.9H_2O$	1120.67	blk, monocl, deliq		>100		v s, d	d	
i55	**Iodous acid, hypo-**	Iodine hydroxide. HOI	143.91	only in sol, yel to grayish				d	d	
i56	**Iridium**	Ir	192.20	silv wh met, cub	22.421	2410	4130	i	i	sl s aq reg; i a, alk
i57	bromide, tetra-	$IrBr_4$	511.84	blk, deliq		d		s d		s al
i58	bromide, tri-	$IrBr_3.4H_2O$	503.99	olv-grn cr		$-3H_2O$, 100		v s		i al
i59	carbonyl	$Ir_2(CO)_8$	608.48	yel cr		subl 160 (in CO_2)				s CCl_4
i60	carbonyl	$Ir_4(CO)_{12}$	1104.93	yel cr		subl 250 (in CO_2)				i CCl_4
i61	carbonyl chloride	$Ir(CO)_2Cl_2$	319.13	col need		d 140		d		d HCl, KOH
i62	chloride, di-	Ir_2Cl_2	263.11	blk gray cr (exist ?)		d 773		s		i a, alk
i63	chloride, tetra-	$IrCl_4$	334.01	dk-brn, amorph, hygr		d		s	d	s al, dil HCl
i64	chloride, tri-	$IrCl_3$	298.56	olv-grn hex, or trig	5.30	d 763		i	i	i a, alk
i65	fluoride, hexa-	IrF_6	306.19	yel glass or tetr	6.0	44.4^{20}	53	d	d	
i66	iodide, tetra-	IrI_4	699.82	blk		d 100		i	i	i al, s KI
i67	iodide, tri-	IrI_3	572.91	grn		d		sl s	s	sl s al
i68	oxalic acid	$H_3[Ir(C_2O_4)_3.xH_2O$		pa yel cr				v s	v s	sl s al; i eth
i69	oxide, di-	IrO_2	224.20	blk tetr or bl cr	11.665	d 1100		0.0002^{20}	i	i a, alk
i70	oxide, di- hydrate	$IrO_2.2H_2O$ or $Ir(OH)_4$	260.23	indigo bl cr		$-2H_2O$, 350		i	i	s HCl
i71	oxide, sesqui-	Ir_2O_3	432.40	bl-blk (exist ?)		$-O$, 400		i		s H_2SO_4, h HCl; i alk
i72	oxide, sesqui-	$Ir_2O_3.xH_2O$		olive green		d		i		s a, alk
i73	phosphor chloride	IrP_2Cl_{12} or $IrCl_3.3PCl_5$	710.56	yel cr		d 250		sl s	d 100	sl s, d al
i74	selenide	$IrSe_2$	350.12	dk gray cr powd		d 600–700 (in CO_2)				sl s aq reg; i a
i75	sulfate	$Ir_2(SO_4)_3.xH_2O$		yel cr		d		s		
i76	sulfide, di-	IrS_2	256.33	br-blk	8.43_4^{25}	d 300		i		s aq reg; i a
i77	sulfide, hydro-	$Ir(HS)_4.2H_2O$	327.45	choc br		d		i		s HNO_3
i78	sulfide, mono-	IrS	224.26	bl-blk			d	i		s K_2S, i a
i79	sulfide, sesqui-	Ir_2S_3	480.59	br-blk	9.64_4^{25}	d		sl s		s K_2S, HNO_3
i80	telluride	$IrTe_2$	575.00	dk gray cr	9.5_4^{25}			i	i	i a; s h aq reg
	Iridium complexes									
i81	aquopentammine iridium-chloride	$[Ir(NH_3)_5H_3(H_2O)]Cl_3$	402.54	wh micro cr	2.474_4^{15}	$-H_2O$, 100		ca 75^{25}	d	i al, eth
i82	chloropentammine iridiumchloride	$[Ir(NH_3)_5Cl]Cl_2$	383.71	pa yel, rhomb	$2.681_4^{15.5}$	d		sl s	s	i HCl
i83	hexammine iridium chloride	$[Ir(NH_3)_6]Cl_3$	400.74	col, rhomb	$2.434_4^{15.5}$			20		
i84	hexammine iridium nitrate	$[Ir(NH_3)_6](NO_3)_3$	408.40	col, tetr micro cr	2.395_1^{15}			1.7^{14}		
i85	nitratopentammine iridium nitrate	$[Ir(NH_3)_5(NO_3)](NO_3)_2$	463.37	wh micro cr	2.515_4^{15}	heat exp		0.28	2.5^{100}	
i86	**Iron**	Fe	55.847	silv met, cub	7.86	1535	2750	i	i	s a; i alk, al, eth
i87	(II) acetate	$Fe(C_2H_3O_2)_2.4H_2O$	246.00	lt grn need, monocl		d		v s		
i88	(III) acetate, basic	$FeOH(C_2H_3O_2)_2$	190.94	br-red powd				i		s a, al
i89	(III) acetylacetonate	$Fe(C_5O_2H_7)_3$	353.18	rubyred, rhomb	1.33	184		sl s	sl s	s al, acet, bz, chl
i90	(II) orthoarsenate	$Fe_3(AsO_4)_2.6H_2O$	553.47	grn amorph powd		d		i		s dil HCl; sl s NH_4OH
i91	(III) orthoarsenate	Nat. scorodite. $FeAsO_4.2H_2O$	230.80	grn, rhomb, 1.765, 1.774, 1.797	3.18	d		i	i	s HCl; i HNO_3
i92	(III) orthoarsenite, basic	$2FeAsO_3.Fe_2O_3.5H_2O$	607.30	br-yel powd		d		sl s		s a, alk
i93	(II) pyroarsenite	$Fe_2As_2O_5$	341.53	grn-wh				i		s NH_4OH
i94	arsenide	FeAs	130.77	wh	7.83	1020		v sl s		
i95	arsenide, di-	Arsenoferrite. $FeAs_2$	205.69	silv gray, cub	7.4	990		i		sl s HNO_3; i HCl
i96	(III) benzoate	$Fe(C_7H_5O_2)_3$	419.20	br powd				i		s h al, h eth
i97	boride	FeB	66.66	gray cr	7.15^{18}			i		s HNO_3, h conc H_2SO_4

No.	Name	Synonyms and Formulae	Mol. wt.	Crystalline form, properties and index of refraction	Density or spec. gravity	Melting point, °C	Boiling point, °C	Solubility, in grams per 100 cc		
								Cold water	Hot water	Other solvents
	Iron									
i98	(II) bromide	$FeBr_2$	215.67	grn-yel, hex	4.636^{25}	d 684 (?)		109^{10}	170^{95}	s al; sl s bz
i99	(III) bromide	$FeBr_3$ or Fe_2Br_6	295.57	dk red-brn, rhomb (?) deliq		subl d		s	s	s al, eth, sl s NH_3
i100	(III) bromide-, hexahydrate	$FeBr_3.6H_2O$	403.68	dk grn		27		v s	v s	s al, eth
i101	(III) cacodylate	$Fe[(CH_3)_2AsO_2]_3$	466.82	yelsh amorph powd				6.67		sl s al
i102	carbide	Fe_3C	179.55	gray, cub	7.694	1837		i	i	s a
i103	(II) carbonate	Nat. siderite. $FeCO_3$	115.85	gray, trig, 1.875, 1.633	3.8	d		0.0067^{25}		s CO_2 sol
i104	(II) carbonate, hydrate	$FeCO_3.H_2O$	133.86	amorph		d		sl s		s a, CO_2 sol
i105	carbonyl, ennea-	$Fe_2(CO)_9$	363.79	yel met cr, hex	2.085^{18}	d 80		i	i	v sl s al, MeOH; d HNO_3; i a
i106	carbonyl, penta-	$Fe(CO)_5$	195.90	visc yel liq	liq 1.457^{21}	−21	102.8^{749}	i		s al, eth, bz, alk, conc H_2SO_4
i107	carbonyl, tetra-	$Fe(CO)_4$	167.89	dk grn lust cr, tetr	1.996^{18}	d 140−150		i		s org solv, conc HNO_3 h H_2SO_4
i108	(II) perchlorate	$Fe(ClO_4)_2$	254.75	wh or grnsh-wh need, hygr		d>100		v s		
i109	(II) perchlorate hexahydrate	$Fe(ClO_4)_2.6H_2O$	362.84	grn, 1.493, 1.478		d>100		97.8^0	116.1^{60}	86.5^{20} al; s $HClO_4$
i110	oxychloride	FeOCl	107.30	brn, rhomb	3.1	d 200				
i111	(II) chloride	Nat. lawrencite. $FeCl_2$	126.75	grn to yel, hex deliq, 1.567	3.16^{25}_4	670−674	subl	64.4^{10}	105.7^{100}	100 al; s acet i eth
i112	(II) chloride, dihydrate	$FeCl_2.2H_2O$	162.78	grn, monocl	2.358					
i113	(II) chloride, tetra-hydrate	$FeCl_2.4H_2O$	198.81	bl-grn, monocl, deliq	1.93			160.1^{10}	415.5^{100}	s al; sl s acet
i114	(III) chloride	Nat. molysite. $FeCl_3$ or Fe_2Cl_6	162.21	blk-brn hex	2.898^{25}_4	306	d 315	74.4^0	535.7^{100}	v s al, MeOH; eth; 63^{18} acet
i115	(III) chloride, hydrate	$FeCl_3.2\frac{1}{2}H_2O$	207.24	dk yel-red, rhomb, deliq		56		v s	v s	v s al, eth
i116	(III) chloride, hexa-hydrate	$FeCl_3.6H_2O$	270.30	br yel cr mass, v deliq		37	280−285	91.9^{20}	∞	s al, eth
i117	(II) chloride, hexammine	$FeCl_2.6NH_3$	228.94	wh powd	1.928^{25}_4					
i118	(II) chloroplatinate	$FePtCl_6.6H_2O$	571.75	yel, hex	2.714	d		v s	v s	
i119	(III) dichromate	$Fe_2(Cr_2O_7)_3$	759.66	red-brn gran				s		s a
i120	(II) chromite	$FeCr_2O_4$	223.84	brn-blk, cub	4.97^{20}			i	i	sl s a
i121	(II) citrate	$FeC_6H_6O_7.H_2O$	263.97	wh micr, rhomb		d 350 (in H_2)		sl s		s NH_4OH
i122	(III) citrate	$FeC_6H_5O_7.5H_2O$	335.03	red-brn scales				sl s	s	i al
i123	hydroxide	Nat. goethite. FeO(OH)	88.85	brn, blksh, rhomb, 2.260, 2.394, 2.400	4.28	−$\frac{1}{2}H_2O$, 136				s HCl
i124	(II) ferricyanide	$Fe_3[Fe(CN)_6]_2(?)$	591.45	deep bl		d		i		i al, dil a
i125	(III) ferricyanide	Berlin green. $Fe[Fe(CN)_6]$	267.80	cub						
i126	(II) ferrocyanide	$Fe_2[Fe(CN)_6]$	323.65	bl-wh, amorph	1.601^{25}_4	d 100	d 430 (vac)	i		
i127	(III) ferrocyanide	$Fe_4[Fe(CN)_6]_3$	859.25	dk bl cr		d		i	i	s HCl, H_2SO_4; i al, eth
i128	(II) fluoride	FeF_2	93.84	wh, rhomb	4.09^{25}_4	>1000 (?)		sl s		s a; i al, eth
i129	(II) fluoride, octa-hydrate	$FeF_2.8H_2O$	237.97	grn-bl	4.20 (anh)	−$8H_2O$, 100		sl s	s	s HF, a; i al, eth
i130	(II) fluoride, tetra-hydrate	$FeF_2.4H_2O$	165.91	wh, rhomb	2.095	d		v sl s		s a; sl s al, eth
i131	(III) fluoride	FeF_3 or Fe_2F_6	112.84	grn, rhomb	3.52	>1000		sl s	s	s a; i al, eth
i132	(III) fluoride, tetra-hydrate	$FeF_3.4\frac{1}{2}H_2O$	193.91	yel cr		−$3H_2O$, 100	d	sl s	s	i al
i133	(II) fluosilicate	$FeSiF_6.6H_2O$	306.01	col, trig, 1.361, 1.385	1.961			128.2		
i134	(III) fluosilicate	$Fe_2(SiF_6)_3$ (exist ?)	537.92	flesh col, gel				s	s, d	
i135	(III) formate	$Fe(CHO_2)_3.H_2O$	208.92	red cr or powd				s		v sl s al
i136	(III) glycerophos-phate	$Fe_2[C_3H_5(OH)_2.PO_4]_3$	**621.39**	yelsh-grn sc or powd				50^{25}		i al
i137	(II) hydrogen cyanide	$H_4[Fe(CN)_6]$	215.99	wh, rhomb	1.536^{25}_4	d 190		s	s	v s al; s a; i acet
i138	(III) hydrogen cyanide	$H_3[Fe(CN)_6]$	214.98	brn-yel need		d 50−60		s	s	v s al; i eth

No.	Name	Synonyms and Formulae	Mol. wt.	Crystalline form, properties and index of refraction	Density or spec. gravity	Melting point, °C	Boiling point, °C	Solubility, in grams per 100 cc		
								Cold water	Hot water	Other solvents
	Iron									
i139	(II) hydroxide	$Fe(OH)_2$	89.86	pa grn hex, or wh amorph	3.4	d		0.00015^{18}		s a, NH_4Cl; i alk
i140	(III) hydrosulfate	Iron(ic) tetrasulfate. Nat. rhomboklas. $Fe_2O_3.4SO_3.9H_2O$	642.08	wh to pink powd, 1.533, 1.550, 1.635	2.172	$-6H_2O$, 80		s		sl s abs al
i141	(III) iodate	$Fe(IO_3)_3$	580.55	grn yel powd	4.80^{20}_4	d 130		sl s		i dil HNO_3
i142	(II) iodide	FeI_2	309.66	gray hex, hygr	5.315	red heat		s	s	s al, acet
i143	(II) iodide, tetrahydrate	$FeI_2.4H_2O$	381.72	gray blk cr, deliq	2.873	d 90-98		v s	d	s al, eth
i144	(II) lactate	$Fe(C_3H_5O_3)_2.3H_2O$	287.96	grn-wh cr or powd		d		2.1^{10}	8.5^{100}	s alk citrate; v sl s al; i eth
i145	(III) lactate	$Fe(C_3H_5O_3)_3$	323.06	br, amorph, deliq				s	v s	i eth
i146	(III) malate	$Fe_2(C_4H_4O_5)_3$	507.91	br scales, hygr				s		s al
i147	methanoarsenate	$Fe(CH_3AsO_3)_3$	525.60	redsh br lustr scales				50		i al, eth
i148	(II) nitrate	$Fe(NO_3)_2.6H_2O$	287.95	grn, rhomb		60.5		83.5^{20}	166.7^{61}	
i149	(III) nitrate	$Fe(NO_3)_3.6H_2O$	349.95	cub		35	d	150^0	∞	
i150	(III) nitrate	$Fe(NO_3)_3.9H_2O$	404.02	col-pa vlt, monocl, deliq	1.684	47.2	d 125	s	s	s al, acet; sl s HNO_3
i151	nitride	Fe_4N	237.39		6.57(?)					
i152	nitride	Fe_2N or Fe_3N_2	125.70	gray	6.35	d 200		i		s HCl, H_2SO_4
i153	nitrosyl carbonyl	$Fe(NO)_2(CO)_2$	171.88	dk red cr	1.56	18.5	d 50, 110	i		s org solv
i154	(III) oleate	$Fe(C_{18}H_{33}O_2)_3$	900.23	br-red fatty lumps				i		s a, al, eth
i155	(II) oxalate	$FeC_2O_4.2H_2O$	179.90	pa yel, rhomb	2.28	d 190		0.022	0.026	s a
i156	(III) oxalate	$Fe_2(C_2O_4)_3.5H_2O$	465.83	yel micro cr powd		d 100		v s	v s	s a; i al
i157	(II) oxide	Nat. wuestite. FeO	71.85	blk, cub, 2.32	5.7	**1369 ± 1**		i		s a; i al, alk
i158	(III) oxide	Nat. hematite. Fe_2O_3	159.69	red-brn to blk, trig, 3.01, 2.94(Li)	5.24	**1565**		i	i	s HCl, H_2SO_4; sl s HNO_3
i159	oxide	Iron ferrosoferric, nat. magnetite. Fe_3O_4	231.54	blk, cub or red-blk powd, 2.42	5.18	1594 ± 5		i	i	s conc a; i al, eth
i160	(III) oxide, hydrate	$Fe_2O_3.xH_2O$		red-brn amorph powd or gelat	2.44-3.60	all H_2O, 350-400		i	i	s a; i al, eth
i161	(II) orthophosphate	Nat. vivianite. $Fe_3(PO_4)_2.8H_2O$	501.61	wh-bl, monocl, 1.579, 1.603, 1.633	2.58			i	i	s a; i ac a
i162	(III) orthophosphate	$FePO_4.2H_2O$	186.85	pink, monocl	2.74	d		v sl s	0.67^{100}	s HCl, H_2SO_4; i HNO_3
i163	(III) pyrophosphate	$Fe_4(P_2O_7)_3.9H_2O$	907.36	yel-wh powd				i		s a, alk citr
i164	phosphide, mono-	FeP	86.82	rhomb	$6.07 (5.2^{20})$					
i165	(di-)phosphide	Fe_2P	142.67	bl-gray cr or powd	6.56	1290		i	i	s aq reg, HNO_3+ HF; i dil a
i166	(tri-)phosphide	Fe_3P	198.51	gray	6.74	1100		i		
i167	(III) hypophosphite	$Fe(H_2PO_2)_3$	250.81	wh or gray-wh powd		d		0.043^{25}	0.083^{100}	s alk citr
i168	(II) metasilicate	Nat. grunerite. $FeSiO_3$	131.93	gray-grn, rhomb, 1.672, 1.697, 1.717	3.5	1146				
i169	orthosilicate	Nat. fayalite. Fe_2SiO_4	203.78	col, rhomb	4.34	1503 (?)		i	i	d HCl
i170	silicide	FeSi	83.93	yel-gray, oct	6.1			i	i	i aq reg
i171	(II) sulfate	Nat. szomolnikite. $FeSO_4.H_2O$	169.96	off-wh, monocl	2.970^{25}			sl s	s	
i172	(III) sulfate	$Fe_2(SO_4)_3$	399.87	yel rhomb, hygr, 1.814	3.097^{18}	d 480		sl s	d	i H_2SO_4, NH_3
i173	(II) sulfate, heptahydrate	Nat. melanterite. $FeSO_4.7H_2O$	278.05	bl-grn, monocl, 1.471, 1.478, 1.486	1.898	64 $-6H_2O$, 90	$-7H_2O$, 300	15.65	48.6^{50}	sl s al; s abs MeOH
i174	(II) sulfate, pentahydrate	Nat siderotil. $FeSO_4.5H_2O$	242.02	wh, tricl, 1.526, 1.536, 1.542	2.2	$-5H_2O$, 300		s	s	i al
i175	(II) sulfate, tetrahydrate	$FeSO_4.4H_2O$	224.01	grn monocl pr, 1.533, 1.535	2.23-2.29					
i176	(III) sulfate, enneahydrate	Nat. coquimbite. $Fe_2(SO_4)_3.9H_2O$	562.01	rhomb, deliq, 1.552, 1.558	2.1	$-7H_2O$, 175		440	d	s abs al
i177	sulfide, di-	Nat. pyrite. FeS_2	119.98	yel, cub	5.0	1171		0.00049		d HNO_3, dil a
i178	sulfide, di-	Nat. marcasite. FeS_2	119.98	yel, rhomb	4.87	tr 450	d	0.00049		d HNO_3, i dil a
i179	(II) sulfide	Nat. troilite. FeS	87.91	blk-brn, hex	4.74	1193-1199	d	0.00062^{18}		s d a; i NH_3
i180	(III) sulfide	Fe_2S_3	207.87	yel-grn	4.3	d		sl d	d FeS+S	d a
i181	(II) sulfite	$FeSO_3.3H_2O$	189.96	grnsh or wh cr		d 250		v sl s		s SO_2 sol; i al
i182	tantalate	Nat. tapiolite. $Fe(TaO_3)_2$	513.73	lt brn, tetr, 2.27, 2.42 (Li)	7.33					
i183	d-tartrate	$FeC_4H_4O_6$	203.92	wh cr				0.877^{16}	v sl s	v s min a; s NH_4OH
i184	(II) thiocyanate	$Fe(SCN)_2.3H_2O$	226.06	grn, rhomb		d		v s		s al, eth, acet
i185	(III) thiocyanate	$Fe(SCN)_3$ or $Fe_2(SCN)_6$	230.09	blk-red, cub, deliq			d	v s		s al, eth, acet

No.	Name	Synonyms and Formulae	Mol. wt.	Crystalline form, properties and index of refraction	Density or spec. gravity	Melting point, °C	Boiling point, °C	Cold water	Hot water	Other solvents
	Iron									
i186	(II) thiosulfate	$FeS_2O_3.5H_2O$	258.05	grn cr, deliq				v s	d	v s al
i187	tungstate	Nat. ferberite. $FeWO_4$	303.69	tetr, 2.40 (Li)	6.64					
i188	metavanadate	$Fe(VO_3)_2$	352.67	grayish-brn powd.						s a; i al
k1	**Krypton**	Kr	83.80	inert gas	gas 3.736g/l liq 2.155 at −152.9	−156.6	−152.30 ± 0.10	11.0^0 cm³ 6.0^{25} cm³	4.67^{50} cm³	
11	**Lanthanum**	La	138.91	wh met, tarnish in air, α: hex, β: cub above 350	α 6.1453, β 6.17	921	3457	d	d	s min a; i conc H_2SO_4
12	acetate	$La(C_2H_3O_2)_3.1\frac{1}{2}H_2O$	343.07					16.88^{25}		
13	boride, hexa-	LaB_6	203.78	purp met cub	2.61	2210	d	i	i	i HCl
14	bromate	$La(BrO_3)_3.9H_2O$	684.77	hex pr		37.5	−7H₂O, 100	28.5^{15}		i al
15	bromide	$LaBr_3.7H_2O$	504.74	col cr	5.057^{25} anh	783 ± 3 anh	1577	v s		v s al; i eth
16	carbide	LaC_2	162.93	yel cr	5.02			d	d	s H_2SO_4; i conc HNO_3
17	carbonate	$La_2(CO_3)_3.8H_2O$	601.97	wh	2.6–2.7			i	i	s dil a; sl s aq CO_2; i acet
18	chloride	$LaCl_3$	245.27	wh cr, deliq	3.842^{25}	860	>1000	v s	d	v s al, pyr; i eth, bz
19	chloride, hepta-hydrate	$LaCl_3.7H_2O$	371.38	wh, tricl, hygr		d 91		v s	v s	v s al
110	hexaantipyrin perchlorate	$[La(C_{11}H_{12}N_2O)].(ClO_4)_3$	1566.65	col, hex cr		d 290–295		1.50^{20}		
111	hydroxide	$La(OH)_3$	189.93	wh powd		d		i		s a
112	iodate	$La(IO_3)_3$	663.62	col, cr				1.7^{25}		
113	iodide	LaI_3	519.62	gray-wh, rhomb, hygr	5.63	772		v s		s acet
114	molybdate	$La_2(MoO_4)_3$	757.63	tetr	4.77^{16}	1181		0.00179^{25}	0.0033^{25}	
115	nitrate	$La(NO_3)_3.6H_2O$	433.02	col cr, deliq, tricl		40	d 126	151.1^{25}	v s	v s al; s acet
116	oxalate	$La_2(C_2O_4)_3.9H_2O$	704.02	wh cr		d		0.00008^{25}		s min a
117	oxide	Lanthana. La_2O_3	325.82	wh rhomb, or amorph	6.51^{15}	2307	4200	0.0004^{20}	d	s a, NH_4Cl; i acet
118	sulfate	$La_2(SO_4)_3$	566.00	wh powd, hygr	3.60^{15}	d 1150		3.0	0.69^{100}	sl s al; i acet
119	sulfate, hydrate	$La_2(SO_4)_3.9H_2O$	728.14	col hex, 1.564	2.821	d white heat		3.8^0	0.87^{100}	sl s HCl; i al
120	sulfide	La_2S_3	374.01	red-yel cr, hex	4.91^{11}	2100–2150 vac		d	d	s a
121	**Lead**	Pb	207.19	silv-blsh wh soft met, cub, 2.01	11.3437^{16} Ra-Pb 11.288^{20}_{20} UPb 11.2960^{16}	327.502	1740	i	i	s HNO_3, h conc H_2SO_4
122	abietate	$Pb(C_{20}H_{29}O_2)_2$	810.10	brn lumps or yelsh-wh powd				i		
123	acetate	$Pb(C_2H_3O_2)_2$	325.28	wh cr	3.25^{20}_4	280		44.3^{20}	221^{50}	s glyc; v sl s al
124	acetate, basic	$Pb_2(OH)(C_2H_3O_2)_3$	608.52	wh				v s		sl s al
125	acetate, basic	$Pb(C_2H_3O_2)_2.3PbO.H_2O$	1012.86	wh powd				i		
126	acetate, basic	$Pb(C_2H_3O_2)_2.Pb(OH)_2.H_2O$	584.50	wh, monocl				v s		v s al
127	acetate, decahydrate	$Pb(C_2H_3O_2)_2.10H_2O$	505.44	wh, rhomb cr	1.69	22		s	s	i al
128	acetate, trihydrate	Sugar of lead. $Pb(C_2H_3O_2)_2.3H_2O$	379.33	wh, monocl, β 1.567	2.55	−H₂O, 75	d 200	45.61^{15}	200^{100}	i al
129	acetate, tetra-	$Pb(C_2H_3O_2)_4$	443.37	col, monocl	2.228^{17}	175		d		d al; s chl, h ac a
130	diantimonate	$Pb_2Sb_2O_7$	769.88	dk yel powd	6.72			i	i	v sl s HCl
131	orthoantimonate	$Pb_3(SbO_4)_2$	993.07	or-yel powd	6.58^{20}_4			i	i	v sl s HCl
132	orthoantimonate	Nat. monimolite. $Pb_3(SbO_4)_2$	993.07	orange powd	6.58^{20}_4			i		i dil a
133	metaarsenate	$Pb(AsO_3)_2$	453.03	hex tabl	6.42^{15}			d	d	s HNO_3
134	orthoarsenate	$Pb_3(AsO_4)_2$	899.41	wh cr, v pois	7.80	1042, sl d 1000		v sl s		s HNO_3
135	orthoarsenate, di-	Nat. schultenite. $PbHAsO_4$	347.12	monocl leaf, α 1.90, γ 1.97	5.79	d 720	−H₂O, 220	i	sl s	s HNO_3, caust alk
136	orthoarsenate, mono-	$Pb(H_2AsO_4)_2$	489.06	tricl, 1.74, 1.82	4.46^{15}	d 140		d		s HNO_3
137	pyroarsenate	$Pb_2As_2O_7$	676.22	rhomb, β 2.03	6.85^{15}_{15}	802		i	d	s HCl, HNO_3; i ac a
138	metaarsenite	$Pb(AsO_2)_2$	421.03	wh powd	5.85			i		s HNO_3
139	orthoarsenite	$Pb_3(AsO_3)_2.xH_2O$	585.00	wh powd	5.85			i		s alk, HNO_3
140	azide	$Pb(N_3)_2$	291.23	col need, or powd			expl 350	0.023^{18}	0.09^{70}	v s ac a; i NH_4OH
141	metaborate	$Pb(BO_2)_2.H_2O$	310.82	wh cr powd	5.598, anhydr	−H₂O, 160		i	i	s a; i alk

No.	Name	Synonyms and Formulae	Mol. wt.	Crystalline form, properties and index of refraction	Density or spec. gravity	Melting point, °C	Boiling point, °C	Solubility, in grams per 100 cc		
								Cold water	Hot water	Other solvents
	Lead									
142	borofluoride	$Pb(BF_4)_2$	380.80	cr pr				d		d al
143	bromate	$Pb(BrO_3)_2.H_2O$	481.02	col, monocl	5.53	d 180		1.38^{20}	sl s	
144	bromide	$PbBr_2$	367.01	wh, rhomb	6.66	373	916	0.4554^0 0.8441^{20}	4.71^{100}	s a, KBr; sl s NH₃; i al
145	butyrate	$Pb(C_4H_7O_2)_2$	381.39	col scales, pois		90		i	i	s dil HNO₃
146	caprate	$Pb(C_{10}H_{18}O_2)_2$	549.71			103–104		i	i	0.0029^{20} eth
147	caproate	$Pb(C_6H_{11}O_2)_2$	437.50			73–74		i	i	1.09^{25} eth
148	caprylate	Lead octoate. $Pb(C_8H_{15}O_2)_2$	493.60	wh leaf		83.5–84.5		i	i	s al; 0.0938 eth
149	carbonate	Nat. cerussite. $PbCO_3$	267.20	col, rhomb, 1.804, 2.076, 2.078	6.6	d 315		0.000011^{20}	d	s a, alk; i NH₃, al
150	carbonate, basic	White lead, hydrocerussite. $2PbCO_3.Pb(OH)_2$	775.60	wh powd, or hex.	6.14	d 400		i	i	sl s aq CO₂; s HNO₃; i al
151	cerotate	$Pb(C_{26}H_{51}O_2)_2$	998.57	wh need		113		i		i al, eth; s bz
152	chlorate	$Pb(ClO_3)_2$	374.09	wh monocl, deliq	3.89	d 230		v s		s al
153	chlorate, hydrate	$Pb(ClO_3)_2.H_2O$	392.11	wh, monocl, deliq	4.037	d 110		151.3^{18}	171^{80}	s al
154	perchlorate	$Pb(ClO_4)_2.3H_2O$	460.14	wh, rhomb	2.6	d 100		499.7^{25}		s al
155	chloride	Nat. cotunite. $PbCl_2$	278.10	wh, rhomb, 2.199, 2.217, 2.260	5.85	501	950^{760}	0.99^{20}	3.34^{100}	sl s dil HCl, NH₃; i al; s NH₄ salts
156	chloride, tetra-	$PbCl_4$	349.00	yel oily liq	3.18^0	−15	expl 105	d (Cl₂)	d	s conc HCl
157	chloride, sulfide	$PbCl_2.3PbS$	995.86	red				i	d	d a, alk; i dil a
158	chlorite	$Pb(ClO_2)_2$	342.09	yel, monocl		expl 126		0.095^{20}	0.42^{100}	s KOH
159	chromate	Nat. crocoite, chrome yellow. $PbCrO_4$	323.18	yel, monocl, 2.31, 2.37(Li), 2.66	6.12^{15}	844	d	0.0000058^{25}	i	s a, alk; i ac a, NH₃
160	chromate, basic	Chrome red. $PbCrO_4.PbO$	546.37	red cr powd	6.63			i	i	s a, alk
161	chromate, basic	$Pb_2(OH)_2CrO_4$	564.39	red amorph or cr	6.63	920		i	i	s KOH
162	dichromate	$PbCr_2O_7$	423.18	red cr				d		s a, alk
163	citrate	$Pb_3(C_6H_5O_7)_2.3H_2O$	1053.82	wh cr powd				s		v sl s al
164	cyanate	$Pb(OCN)_2$	291.22	wh need		d		i	sl s	
165	cyanide	$Pb(CN)_2$	259.23	yelsh-wh powd, pois				sl s	s	s KCN
166	enanthate	$Pb(C_7H_{13}O_2)_2$	465.55	wh leaf		91.5		sl s		i al
167	ethylsulfate	$Pb(C_2H_5SO_4)_2.2H_2O$	493.57	col liq, pois				s		
168	ferricyanide	$Pb_3[Fe(CN)_6]_2.5$ (or 6) H_2O	1135.55	blk-brn to red, monocl pr		−H₂O, 110–120 d		sl s	s, d 100	s alk, HNO₃
169	ferrite	$PbFe_2O_4$	382.88	hex		1530 d, 725				
170	ferrocyanide	$Pb_2Fe(CN)_6.3H_2O$	680.38	yelsh-wh powd		−H₂O, 100		i		sl s H₂SO₄
171	fluoride	PbF_2	245.19	col, rhomb, pois	8.24	855	1290	0.064^{20}		s HNO₃; i acet, NH₃
172	fluorochloride	Nat. matlockite. $PbFCl$	261.64	wh, tetr, 2.145, 2.006	7.05	601		0.037^{25}	0.1081^{100}	
173	fluosilicate	$PbSiF_6.2H_2O$	385.30	col, monocl		d		s	v s	
174	fluosilicate, tetrahydrate	$PbSiF_6.4H_2O$	421.33	col, monocl		d<100				
175	formate	$Pb(CHO_2)_2$	297.23	wh, rhomb, lust, 1.789, 1.852, 1.877	4.63	d 190		1.6^{15}	20^{100}	i al
176	hydride, di-	PbH_2	209.21	gray powd		d				
177	hydroxide	$Pb(OH)_2$	241.20	wh, amorph		d 145		0.0155^{20}	sl s	s a, alk; i ac a
178	hydroxide	$Pb_2O(OH)_2$ or $2PbO.H_2O$	464.39	wh cub, or amorph powd, pois	7.592	d 145		0.014	sl s	s alk, ac a, HNO₃
179	iodate	$Pb(IO_3)_2$	557.00	wh	6.155^{20}	d 300		0.0012^2	0.003^{25}	sl s HNO₃; i NH₃
180	paraperiodate	$PbHIO_6$	415.10	wh cr		d 130		i	i	s dil HNO₃
181	paraperiodate, hydrate	$PbHIO_6.H_2O$	433.11	amorph		−H₂O, 110		i	i	sl s dil HNO₃
182	iodide, basic	$PbI_2.PbO.H_2O$	702.20	rhomb cr	6.83^{20}	d 100				
183	iodide, di-	PbI_2	461.00	yel hex powd, pois	6.16	402	954	0.044^0 0.063^{20}	0.41^{100}	s alk, KI; i al
184	iodide, mono-	PbI	334.09	pa yel		d 300		0.1		
185	isobutyrate	$Pb(C_4H_7O_2)_2$	381.39	wh pr		<100		9.1^{15}		
186	lactate	$Pb(C_3H_5O_3)_2$	385.33	wh cr powd				s		s h al
187	laurate	$Pb(C_{12}H_{23}O_2)_2$	605.82	chalky wh powd		104.7		0.009^{35}		0.008^{25} al; $0.007^{14.5}$ eth
188	lignocenate	$Pb(C_{24}H_{47}O_2)_2$	942.47	wh powd		117		i		v s h bz; sl s al; i eth
189	malate	$Pb(C_4H_4O_5).3H_2O$	393.31	wh powd				sl s		v sl s al
190	melissate	$Pb(C_{31}H_{61}O_2)_2$	1138.85	wh powd		115–116		i	i	s boil tol, ac a; sl s h bz, chl; i al, eth
191	molybdate	Nat. wulfenite. $PbMoO_4$	367.13	col-lt yel, tetr pl	6.92^{25}_4	1060–1070		i		d conc H₂SO₄; s a, KOH; i al
192	myristate	$Pb(C_{14}H_{27}O_2)_2$	661.93	wh powd		107		0.005^{35}	0.006^{50}	0.004^{25} al; $0.010^{14.5}$ eth

No.	Name	Synonyms and Formulae	Mol. wt.	Crystalline form, properties and index of refraction	Density or spec. gravity	Melting point, °C	Boiling point, °C	Solubility, in grams per 100 cc		
								Cold water	Hot water	Other solvents
	Lead									
193	2-naphthalene-sulfonate	$Pb(C_{10}H_7SO_3)_2$........	621.65	wh cr powd, pois.				i		s al
194	nitrate	$Pb(NO_3)_2$	331.20	col, cub or monocl, pois, 1.782	4.53^{20}	d 470		37.65^0 56.5^0	127^{100}	8.77^{22} 43 % al; s alk, NH_3
195	nitrate, basic	$Pb(OH)NO_3$	286.20	wh rhomb cr	5.93	d 180		19.4^{19}	s	s a
196	nitrite	$3PbO.N_2O_3.H_2O$	763.60	lt yel powd				v s		s dil HNO_3
197	oleate	$Pb(C_{18}H_{33}O_2)_2$	770.12					i		6.46^{20} eth; s pet eth; sl s al
198	oxalate	PbC_2O_4	295.21	wh powd	5.28	d 300		0.00016^{18}		s HNO_3; i al
199	oxide-, di-	Plattnerite. PbO_2	239.19	bz, tetr, ω 2.3(Li)	9.375	d 290	i	i		s dil HCl; sl s ao a; i al
1100	oxide, mono-	Litharge. PbO	223.19	yel, tetr	9.53	**886**		0.0017^{20}		s HNO_3. alk, Pb acet, NH_4Cl, $CaCl_2$, $SrCl_2$
1101	oxide, mono-	Massicot. PbO	223.19	yel, rhomb, 2.51, 2.61(Li), 2.71	8.0			0.0023^{22}	i	s alk
1102	oxide, red	Minium. Pb_3O_4	685.57	red cr sc, or amorph powd	9.1	d 500		i	i	s HCl, acet a; i al
1103	oxide, sesqui-	Pb_2O_3	462.38	or-yel powd, amorph		d 370		i	d	d a
1104	oxide, sub-	Pb_2O	430.38	blk, amorph	8.342	d		i	i	s a, alk
1105	oxychloride	$PbCl_2.3PbO$	947.66	yel				0.0056^{18}	0.07^{74}	
1106	oxychloride	Cassel yellow. $PbCl_2.7PbO$	1840.42	yel cr, or powd				i		
1107	oxychloride	Fiedlerite. $2PbCl_2.PbO.H_2O$	797.40	monocl, 1.816, 2.1023, 2.026	5.88^{20}	d 150				s HNO_3
1108	oxychloride	Nat. laurionite. $PbCl_2.Pb(OH)_2$	519.29	rhomb	6.24	d 142				
1109	oxychloride	Matlockite. $PbCl_2.PbO$	501.29						i	s alkalies, hot conc HCl
1110	oxychloride	Nat. matlockite. $PbCl_2.Pb(OH)_2$	519.29	wh, tetrag, 2.04, 2.15, 2.15	7.21	d 524		0.0095^{18}		s alk
1111	oxychloride	Nat. mendipite. $PbCl_2.2PbO$	724.47	yel, rhomb, 2.24, 2.27, 2.31	7.08	693		i	i	s alk
1112	oxychloride	Nat. paralaurionite. $PbCl_2.PbO.H_2O$	519.29	col to wh, monocl pr, 2.146	6.05^{15}	d 150				
1113	palmitate	$Pb(C_{16}H_{31}O_2)_2$	718.04	wh powd		112.3		0.005^{35}	0.007^{50}	s al; 0.148^{20} eth
1114	phenolate	Lead phenate, lead carbolate. $Pb(OH)OC_6H_5$	317.30	yelsh-wh powd				i		
1115	phenolsulfonate	Lead sulfocarbolate. $Pb[C_6H_4(OH)SO_3]_2.5H_2O$	643.60	wh lustr need				s		s al
1116	metaphosphate	$Pb(PO_3)_2$	365.13	col cr		800(?)		v sl s		
1117	orthophosphate	$Pb_3(PO_4)_2$	811.51	col or wh powd, hex, 1.970, 1.936	6.9–7.3	1014		0.000014^{20}	i	s HNO_3, alk; i ac a, al
1118	orthophosphate, di-	$PbHPO_4$	303.17	rhomb	5.661^{15}	d				s HNO_3, alk, NH_4Cl
1119	orthophosphate, mono-	$Pb(H_2PO_4)_2$	401.16	need						s alk, dil HNO_3, h conc HCl; i acet a
1120	phosphide	PbP_5	362.06	blk unstable, inflam		d 400 (vac)		d	d	d dil a
1122	orthophosphite	$PbHPO_3$	287.17	wh, powd		d		i		s HNO_3
1123	picrate	$Pb(C_6H_2N_3O_7)_2.H_2O$	681.45	yel, need	2.831^{20}	$-H_2O$, 130	expl	0.88^{15}		
1124	proprionate, tetra-	$Pb(C_3H_6O_2)_4$	499.48	solid		132				
1125	pyrophosphate	$Pb_2P_2O_7$	588.32	wh, rhomb	5.8^{20}	824		i	i	s HNO_3, KOH
1126	pyrophosphate, hydrate	$Pb_2P_2O_7.H_2O$	606.34	wh, rhomb		806 anhydr		i	d	s HNO_3, KOH, $Na_4P_2O_7$
1127	selenate	$PbSeO_4$	350.17	wh, rhomb	6.37^{20}_4	d		i	i	s conc a
1128	selenide	Nat. clausthalite. PbSe	286.15	gray, cub	8.10^{15}	1065		i		s HNO_3
1129	metasilicate	Nat. alamosite. $PbSiO_3$	283.27	col or wh, monocl	6,49	766		i		d a
1130	orthosilicate, di-	Nat. barysilite. $Pb_3Si_2O_7$	582.55	wh, trig, 2.070, 2.050	6.707			i		
1131	stearate	$Pb(C_{18}H_{35}O_2)_2$	774.15	wh powd		115.7		0.05^{35}	0.06^{50}	$0.005^{14.5}$ eth; i al
1132	sulfate	Nat. anglesite. $PbSO_4$	303.25	wh, monocl, or rhomb, 1.877, 1.822, 1.894	6.2	1170		0.00425^{25}	0.0056^{40}	s NH_4 salts; sl s conc H_2SO_4; i a
1133	sulfate, basic	Nat. lanarkite. $PbSO_4.PbO$	526.44	wh, monocl, 1.93, 1.99, 2.02	6.92	977		0.0044^0	v sl s	sl s H_2SO_4
1134	sulfate, hydrogen	$Pb(HSO_4)_2.H_2O$	419.34	wh cr		d		0.0001^{18} d		sl s H_2SO_4
1135	peroxydisulfate	$PbS_2O_8.3H_2O$	453.36	deliq				v s		
1136	sulfide	Nat. galena. PbS	239.25	bl met cub, 3.921	7.5	1114		0.000086^{18}		s a; i al, KOH
1137	sulfite	$PbSO_3$	287.25	wh powd		d		i	i	s HNO_3
1138	tartrate, dl-	$PbC_4H_4O_6$	355.26	wh cr powd	2.53^{19}			0.0025^{20}	0.0074^{100}	s HNO_3, KOH; i al, ac a, NH_4 ac

No.	Name	Synonyms and Formulae	Mol. wt.	Crystalline form, properties and index of refraction	Density or spec. gravity	Melting point, °C	Boiling point, °C	Solubility, in grams per 100 cc		
								Cold water	Hot water	Other solvents
	Lead									
1139	dithionate	$PbS_2O_6.4H_2O$	439.38	trig, 1.635, 1.653	3.22	d		115.0[20]		
1140	thiosulfate	PbS_2O_3	319.32	wh cr	5.18	d		0.03		s a, $Na_2S_2O_3$
1141	metatitanate	$PbTiO_3$	303.09	yel. rhomb-pyr	7.52			i	i	
1142	telluride	Nat. altaite. PbTe	334.79	wh, cub	8.164_4^{20}	917				i a
1143	thiocyanate	$Pb(SCN)_2$	323.35	wh, monocl	3.82	d 190		0.05[20]	0.2[100]	s KCNS, HNO_3
1144	tungstate	Nat. stolzite. $PbWO_4$	455.04	tetr, 2.269, 2.182	8.23			i		i HNO_3, s KOH
1145	tungstate	Nat. raspite. $PbWO_4$	455.04	col, monocl, 2.27, 2.27, 2.30	1123			0.03		d a; i al
1146	metavanadate	$Pb(VO_3)_2$	405.07	yel powd				sl s		d HCl; s dil HNO_3
1147	**Lithium**	Li	6.939	silver white, soft	0.534[20]	180.54	1342	d		
1148	acetate	$LiC_2H_3O_2.2H_2O$	102.01	wh, rhomb, α 1.40, β 1.50		70		300[15]	v s	21.5 al
1149	acetylsalicylate	$LiC_9H_7O_4$	186.09	wh powd hygr, d in moist air				100		25 al
1150	metaaluminate	$LiAlO_2$ (or $Li_2Al_2O_4$)	65.92	wh, rhomb, 1.604, 1.614	2.55_4^{25}	1900–2000		i		
1151	aluminum hydride	$LiAlH_4$	37.95	wh cr powd	0.917	d 125		d		ca 30 eth
1152	amide	$LiNH_2$	22.96	col need, cub	1.178[17.5]	380–400	d 750–200 subl	s	d	sl s liq NH_3, al; i eth, bz
1153	antimonide	Li_3Sb	142.57		3.2[17]	>950		d	d	d a
1154	orthoarsenate	Li_3AsO_4	159.74	wh powd, rhomb	3.07[15]			v sl s		s dil ac a; i pyr
1155	azide	LiN_3	48.96	col cr, hygr		d 115–298		66.41[16]		20.26[16] abs al; i eth
1156	benzoate	$LiC_7H_5O_2$	128.06	wh cr or powd				33[25]	40[100]	7.7[25] al, 10[78] al
1157	metaborate	$LiBO_2$	49.75	wh, tricl	1.397[41.7]	845		2.57[20]	11.83[80]	
1158	metaborate	$LiBO_2.8H_2O$	193.87	col, trig	1.38[14.9]	47				
1159	pentaborate	$LiB_5O_8.8H_2O$	522.10	wh	1.72	300–350 −8H_2O		36.3[45]	194[100]	3.9[30] al; 22[33] glycerine; i bz
1160	tetraborate	$Li_2B_4O_7$	169.11	wh cr		930		2.89[20]	5.45[100]	i org solv
1161	borohydride	$LiBH_4$	21.78	rhomb cr	0.66	d 279		s d		s eth
1162	borohydrate	$LiBH_4$	21.78	wh, orthorhomb	0.666	275 d		v sl s		d al, 2.5 eth
1163	bromide	LiBr	86.85	wh, cub, deliq, 1.784	3.464[25]	550	1265	145[4]	254[90]	73[40] al; 8 MeOH; s al, eth; sl s pyrid
1164	bromide, dihydrate	$LiBr.2H_2O$	122.28	wh cr		−1H_2O; 44		246.0[20]	v s	s al
1165	carbide	Li_2C_2	37.90	wh cr or powd	1.65[18]			d	d	s a
1166	carbonate	Li_2CO_3	73.89	wh, monocl, 1.428, 1.567, 1.572	2.11	723	d 1310[760]	1.54[0]	0.72[100]	i al; acet
1167	carbonate, acid	Lithium bicarbonate. $LiHCO_3$	67.96	wh				5.5[13]		
1168	chlorate	$LiClO_3$	90.39	col, rhomb need, deliq, α 1.63, γ 1.64	1.1190_4^{18} (18%Soln)	127.6	300 d	500[27]		v s al; 0.142[25] acetone
1169	chlorate	$LiClO_3.\frac{1}{2}H_2O$ (or $\frac{1}{3}H_2O$)	99.39	wh, tetr, deliq		65(?)	$-\frac{1}{2}H_2O$, 90 d 290	v s	v s	v s al
1170	perchlorate	$LiClO_4$	106.39	wh	2.428	236	430 d	60.0[25]	150[89]	152[25] al; 182[25] MeOH; 114[25] eth; 137[25] acetone
1171	perchlorate, trihydrate	$LiClO_4.3H_2O$	160.44	wh, hex	1.841	95 deliq 236 (anhydr)	d 100 −2H_2O	130[25]		72.9[25] al; 156[25] MeOH; 96.2[25] acetone; 0.096[25] eth
1172	chloride	LiCl	42.39	wh, cub, 1.662	2.068[25]	605	1325–1360	63.7[0]	130[95]	25.10[30] al; 42.36[25] MeOH; 4.11[25] acetone; 0.538[23.9] NH_4OH
1173	chloride, monohydrate	$LiCl.H_2O$	60.41	wh cr, hygr	1.78	−H_2O >98		86.2[30]	s	s HCl
1174	chloroplatinate	$Li_2PtCl_6.6H_2O$	529.78	or prism		−6H_2O, 180		v s	v s	v s al; i eth
1175	dichromate, dihydrate	$Li_2Cr_2O_7.2H_2O$	265.90	orange-red cr, deliq	2.34[20]	187 d	110 −2H_2O	187[30]	278[100]	s reacts al
1177	citrate	$Li_3C_6H_5O_7.4H_2O$	281.98	col cr or powd, deliq		−4H_2O, 105		74.5[25]	66.7[100]	sl s al, eth
1178	fluoride	LiF	25.94	wh, cub, 1.3915	2.635[20]	845	1676	0.27[18]		i al; s HF
1179	fluosilicate	$Li_2SiF_6.2H_2O$	191.99	wh, monocl, 1.300, 1.296	2.33[12]	−2H_2O, 100	d	73[17]		s al; i eth, acet
1180	fluosulfonate	$LiSO_3F$	106.00	wh powd		360		v s	s	v s al, eth, acet; i ligorin
1181	formate, monohydrate	$H.COOLi.H_2O$	69.97	wh, rhomb	1.46	−H_2O, 94	d 230	27.85[18]	57.05[98]	sl s al, acet; i bz
1182	gallium hydride	$LiGaH_4$	80.69	wh cr				d	d	s eth
1183	gallium nitride	$LiGaN_2$	118.55	lt gr powd	3.35	d 800		d	d	s a, alk
1184	metagermanate	Li_2GeO_3	134.47	monocl, 1.7	3.53[21]	1239		0.85[25]		s a

No.	Name	Synonyms and Formulae	Mol. wt.	Crystalline form, properties and index of refraction	Density or spec. gravity	Melting point, °C	Boiling point, °C	Solubility, in grams per 100 cc		
								Cold water	Hot water	Other solvents
	Lithium									
1185	hydride	LiH	7.95	wh cr	0.82	680	....	d	....	v sl s a
1186	hydroxide	LiOH	23.95	wh tetr, 1.464, 1.452	1.46	450	d 924	12.8²⁰	17.5¹⁰⁰	sl s al
1187	hydroxide, mono-hydrate	LiOH.H₂O	41.96	wh monocl, 1.460, 1.524	1.51			22.3¹⁰	26.8⁸⁰	sl s al; i eth
1188	iodate	LiIO₃	181.84	wh, hex, hygr	4.502³²			80.3¹⁸	....	i al
1189	iodide	LiI	133.84	wh, cub, 1.955 ± 0.003	4.076	449	1180 ± 10	165²⁰	433⁸⁰	250.8²⁰ al; 42.6¹⁸ acet 343.4²⁰ MeOH; v s NH₄OH
1190	iodide, trihydrate	LiI.3H₂O	187.89	col-yelsh, hex, hygr	3.48	73 – H₂O	−2H₂O, 80 −H₂O, 300	151⁰	201.2⁶⁰	s abs al, acet
1191	laurate	LiC₁₂H₂₃O₂	206.25	wh powd		229.2–229.8		0.154¹⁶·³	0.178²⁵	0.322²⁵ al; 0.008¹⁵·³ eth; 0.240²⁵ acet
1192	permanganate	LiMnO₄.3H₂O	179.92	cub	2.06	d 190		71.43¹⁶		d alk
1193	molybdate	Li₂MoO₄	173.82	wh trig, hygr	2.66	705		v s		
1194	myristate	LiC₁₄H₂₇O₂	234.31			223.6–224.2		0.027¹⁶·³ 0.036²⁵	0.062⁵⁰	0.010¹⁵·³ eth; 0.331¹⁵ acet; 0.155²⁰ al
1195	nitrate	LiNO₃	68.94	wh, trig, 1.735, 1.735	2.38	264	d 600	89.8²⁷·⁵⁵	234¹⁰⁰	s NH₄OH, al; 37.15 pyridine
1196	nitrate, trihydrate	LiNO₃.3H₂O	122.99	col need		−2½H₂O, 29.9	−3H₂O, 61.1	34.8⁰	57.48²⁹·⁶	s al, MeOH, acet
1197	nitride	Li₃N	34.82	red-brn amorph, or blk-gray cr, cub		tr 840–850 (in N₂)				
1198	nitrite	LiNO₂.H₂O	70.96	col flat need	1.615⁰	>100	d	125⁰	459⁵⁰	v s abs al
1199	oxalate	Li₂C₂O₄	101.90	col, rhomb, 1.465, 1.53, 1.696	2.121¹⁷·⁵	d		8¹⁹·⁵		i al, eth
1200	oxalate, acid	LiHC₂O₄.H₂O	113.99			d		8¹⁷		
1201	oxide	Li₂O	29.88	wh cr, cub, n_D 1.644	2.013²⁵·²	>1700	1200⁶⁰⁰	6.67⁰ d	10.02¹⁰⁰	
1202	palmitate	LiC₁₆H₃₁O₂	262.36	wh powd		224.5		0.01¹⁸	0.015²⁵	0.347¹⁵ acet; 0.077²⁰ al; 0.005¹⁵·⁸ eth
1203	metaphosphate	LiPO₃	85.91	col pl	2.461	red heat		i	i	s a
1204	orthophosphate	Li₃PO₄	115.79	col, rhomb	2.537¹⁷·⁵	837		0.039¹⁸		s a, NH₄OH; i acet
1205	orthophosphate	Li₃PO₄.½H₂O	124.80	wh cr powd	2.41	−½H₂O, 100		0.04²⁵		s a
1206	phosphate, di- H	LiH₂PO₄	103.93	col cr, hygr	2.461	>100				
1207	potassium sulfate	LiKSO₄	142.10	col, hex; n_D 1.4723, 1.4717	2.393²⁰			s	s	
1208	potassium dl-tartrate	LiKC₄H₄O₆.H₂O	212.13	col, monocl, β 1.523 (red)	1.610			s		
1209	salicylate	LiC₇H₅O₃	144.06	wh, powd, deliq		d		133.3		50 al
1210	selenide	Li₂Se.9H₂O	254.98	col, rhomb, deliq				d		
1211	metasilicate	Li₂SiO₃	89.96	col, rhomb; α 1.584, γ 1.604	2.52²⁵₄	1204		i	s d	s dil HCl
1212	orthosilicate	Li₄SiO₄	119.84	col, rhomb; α 1.594, γ 1.614	2.392²⁵₄	1256		i	d	d a
1213	silicide	Li₆Si₂	97.81	bl cr, hygr	ca. 1.12	d 600 (vac)		d	d	d a; i NH₃ turp
1214	sodium fluoaluminate	Li₃Na₃(AlF₆)₂	371.73	cub cr, 1.3395	2.774	710		0.074¹⁸		
1215	stearate	LiC₁₈H₃₅O₂	290.41	wh cr	220.5–221.5			0.010¹⁸		0.010²⁵ al; 0.040¹⁸ eth; 0.457¹⁵ acet
1216	sulfate	Li₂SO₄	109.94	α monocl; β hex or rhomb, γ cub 500°C; β 1.465	2.221		845	26.1⁰	23¹⁰⁰	i abs al, acet
1217	sulfate, hydrogen	LiHSO₄	104.01	col pr	2.123¹³	120		d		
1218	sulfate, mono-hydrate	Li₂SO₄.H₂O	127.95	col cr, monocl, 1.465, 1.477, 1.488	880			34.9²⁵	29.2¹⁰⁰	11.5³⁰ al + H₂O (23.9 % alco); i acet, pryidine
1219	sulfide	Li₂S	45.94	wh-yel, cub, deliq	1.66	900–975		v s	v s	v s al
1220	sulfide, hydro-	LiHS	40.01	wh powd, hygr				s	s	s al
1221	sulfite, monohydrate	Li₂SO₃.H₂O	111.96	wh need, α 1.53, γ 1.59		455 d	140 – H₂O	24.9³⁰	22⁸⁰	i org solv
1222	tartrate	Li₂C₄H₄O₆.H₂O	179.97	wh cr powd				s		
1223	thallium dl-tartrate	LiTlC₄H₄O₆.2H₂O	395.41	tricl	3.144					
1224	thiocyanate	LiSCN	65.02	wh cr, deliq, n_D 1.333				v s		s methylacet
1225	dithionate	Li₂S₂O₆.2H₂O	210.03	col, rhomb, 1.5602	2.158	d		v s		
1226	tungstate	Li₂WO₄	261.73	col, trig	3.71	742		v s	v s	d a; i al
1227	**Lutetium**	Cassiopeium. Lu	174.97	met, hex	9.8404	1663	3395			

PHYSICAL CONSTANTS OF INORGANIC COMPOUNDS (Continued)

No.	Name	Synonyms and Formulae	Mol. wt.	Crystalline form, properties and index of refraction	Density or spec. gravity	Melting point, °C	Boiling point, °C	Cold water	Hot water	Other solvents
	Lutetium									
l228	bromide	LuBr₃	414.70		1025	1400		s	s	
l229	chloride	LuCl₃	281.33	col cr	3.98	905	subl 750	s		
l230	fluoride	LuF₃	231.97			1182	2200	i	i	
l231	iodide	LuI₃	555.68			1050	1200	s	s	
l232	oxalate	Lu₂(C₂O₄)₃.6H₂O	722.09	wh cr		50 (−H₂O)		i	i	i dil a
l233	oxide	Lu₂O₃	397.94	cub cr	9.42			i	i	
l234	sulfate	Lu₂(SO₄)₃.8H₂O	782.25	col cr				42.27²⁰	16.93⁴⁰	
m1	**Magnesium**	Mg	24.312	silv wh met, hex	1.74⁵	**648.8**	1107	i	d to Mg(OH)₂	s min a, conc. HF, NH₄ salts, i CrO₃, alk.
m2	acetate	Mg(C₂H₃O₂)₂	142.40	wh cr	1.42	323 d		v s	v s	5.25¹⁵ MeOH
m3	acetate, tetrahydrate	Mg(C₂H₃O₂)₂.4H₂O	214.46	col, monocl deliq, β 1.491	1.454	80		120¹⁵	∞	v s al
m4	aluminate	Nat. spinel. MgAl₂O₄	142.27	col, cub, 1.723	3.6	2135				sl s H₂SO₄; v sl s dil HCl; i HNO₃
m5	amide	Mg(NH₂)₂	56.36	gray powd		d 350–400	d	d	d	v sl s liq NH₃; d al
m6	antimonate	MgO.Sb₂O₅.12H₂O	579.99	hex or tricl cr	2.60 (hex)	−12H₂O, 200		v sl s		d HCl
m7	antimonide	Mg₃Sb₂	316.44	met hex pl	4.088²⁵₄	961		i		
m8	orthoarsenate	Nat. hoernesite. Mg₃(AsO₄)₂.8H₂O	494.90	wh monocl	2.60–2.61					
m9	orthoarsenate	Mg₃(AsO₄)₂.22H₂O	747.11	wh cr	1.788	−17H₂O, 100	−21H₂O, 220	i	i	s a, NH₄Cl
m10	orthoarsenate, mono-H	Nat. roesslerite. MgHAsO₄.7H₂O	290.35	monocl	1.943¹⁵	−5H₂O, 100		d		
m11	arsenide	Mg₃As₂	222.78	brn-red, cub	3.148²⁵₄	800		d	d	d dil a, al
m12	orthoarsenite	Mg₃(AsO₃)₂	318.78	wh				s	v s	s a, NH₄Cl; i NH₄OH
m13	benzoate	Mg(C₇H₅O₂)₂.3H₂O	320.59	wh powd		−3H₂O, 110	d 200	6.16¹⁵	19.6¹⁰⁰	s al
m14	bismuthide	Mg₃Bi₂	490.90	met hex	5.945²⁵₁	823				
m15	bismuth nitrate	3Mg(NO₃)₂.2Bi(NO₃)₃.24H₂O	1543.31	col cr, deliq	2.32¹⁶₁₅	71		d	d	s HNO₃
m16	diborate	Nat. ascharite. Mg₂B₂O₄.H₂O	168.26	orthorhmb, 1.54	2.60–2.70					
m17	metaborate	Nat. pinnoite. Mg(BO₂)₂.3H₂O	163.98	yel, tetr, pyram, 1.565, 1.575	2.27–2.30					
m18	metaborate, octahydrate	Mg(BO₂)₂.8H₂O	254.05	col, tetr, 1.565, 1.575	2.30			i	v sl s	s a
m19	orthoborate	Mg₃(BO₃)₂	190.55	col, rhomb, 1.6527, 1.6537, 1.6548	2.99²¹			i	i	s min a; i ac a
m20	boride	MgB₄	89.18	bl		d 1200 (vac)		d		sl s a
m21	bromate	Mg(BrO₃)₂.6H₂O	388.22	col, cub, 1.514	2.29	−6H₂O, 200	d	42¹⁸	v s	i al
m22	bromide	MgBr₂	184.13	wh hex cr, deliq	3.72²⁵₄	700		101.50²⁰	125.6¹⁰⁰	6.9 al; 21.8²⁰ MeOH
m23	bromide, hexahydrate	MgBr₂.6H₂O	292.22	col, hex pr or need, hygr, fluo in x-rays	2.00	172.4		316⁰	v s	s al, acet; sl s NH₃
m24	bromoplatinate	MgPtBr₆.12H₂O	915.04	trig	2.802					
m25	carbonate	Nat. magnesite. MgCO₃	84.32	wh, trig, 1.717, 1.515	2.958	d 350	−CO₂, 900	0.0106		s a, aq+CO₂; i acet, NH₃
m26	carbonate, basic artinite	Nat. artinite. MgCO₃.Mg(OH)₂.3H₂O	196.69	wh, rhomb, 1.489, 1.534, 1.557	2.02²⁰					
m27	carbonate, basic	Nat. hydromagnesite. 3MgCO₃.Mg(OH)₂.3H₂O	365.34	wh, rhomb, 1.527, 1.530, 1.540	2.16	d		0.04	0.011	s a, NH₄ salts
m28	carbonate, pentahydrate	Nat. lansfordite. MgCO₃.5H₂O	174.40	wh, monocl, 1.456, 1.476, 1.502	1.73	d in air		0.176⁷	0.375²⁰	s HCl, MgSO₄ soln
m29	carbonate, trihydrate	Nat. nesquehonite. MgCO₃.3H₂O	138.37	col, rhomb need, 1.495, 1.501, 1.526	1.850	165		0.179¹⁶	d	s a; 1.4 aq+CO₂
m30	chlorate	Mg(ClO₃)₂.6H₂O	299.31	wh, rhomb need, deliq	1.80²⁵	35	d 120	128.6¹⁸	v s	s al
m31	perchlorate	Mg(ClO₄)₂	223.21	deliq	2.21¹⁸	d 251		99.3²⁵	v s	23.96²⁵ al
m32	perchlorate, hexahydrate	Mg(ClO₄)₂.6H₂O	331.31	wh, rhomb cr, 1.482, 1.458	1.98	185–190		v s	v s	
m33	perchlorate, hexammine	Mg(ClO₄)₂.6NH₃	325.40	wh, cub	1.41²⁰₄					s liq NH₃; s d al
m34	chloride	MgCl₂	95.22	wh, lustr hex cr, 1.675, 1.59	2.316–2.33	714	1412	54.25²⁰	72.7¹⁰⁰	7.40⁶⁰ al

No.	Name	Synonyms and Formulae	Mol. wt.	Crystalline form, properties and index of refraction	Density or spec. gravity	Melting point, °C	Boiling point, °C	Solubility, in grams per 100 cc		
								Cold water	Hot water	Other solvents
	Magnesium									
m35	chloride, hexahydrate	Nat. bischofite. $MgCl_2.6H_2O$	203.31	col, monocl, deliq, 1.495, 1.507 1.528	1.569	d 116–118	d	167	367	s al
m36	chloropalladate	$MgPdCl_6.6H_2O$	451.52	hex	2.12	d				
m37	chloroplatinate	$MgPtCl_6.6H_2O$	540.21	yel, trig	2.692	$-H_2O$, 180		s	s	
m38	chlorostannate	$MgSnCl_6.6H_2O$	463.81	tricl	2.08	d 100				
m39	chromate	$MgCrO_4.7H_2O$	266.41	yel, rhomb, 1.521, 1.550, 1.568	1.695		211.5^{18}	v s		
m40	chromite	$MgCr_2O_4$	192.30	dk-grn or red, cub	4.6^{20}			i	i	s conc H_2SO_4; i dil a, dil alk
m41	citrate, nono-H	$MgHC_6H_5O_7.5H_2O$	304.50	wh gran powd				20^{25}	s	s a; i al
m42	cyanide	$Mg(CN)_2$	76.35			d 300 to $MgCN_2$	d 600	s	d	
m43	cyanoplatinite	$MgPt(CN)_4.7H_2O$	449.58	red, pr	2.185^{16}	$-H_2O$, 45		v s	v s	i al, eth
m44	ferrite	$MgFeO_4$	200.00	blk, oct, 2.35	4.44–4.60	1750 ± 25				s conc HCl; i dil a, h HNO_3, al
m45	ferrocyanide	$Mg_2Fe(CN)_6.12H_2O$	476.76	pa yel cr		d ca 200		33		i al
m46	fluoride	Nat. sellaite. MgF_2	62.31	col, tetr, faint vlt, lumin, 1.378, 1.390; 3.14		1261	2239	0.0076^{18}	i	s HNO_3; sl s a; i al
m47	fluosilicate	$MgSiF_6$	166.39	wh, cr or powd				65		
m48	fluosilicate, hexahydrate	$MgSiF_6.6H_2O$	274.48	wh, trig	1.788	d 120		$64.8^{17.5}$		i al
m49	formate	$Mg(CHO_2)_2.2H_2O$	150.38	col, rhomb		$-2H_2O$, 100		14^0 (anh)	24^{100} (anh)	i al, eth
m50	orthogermanate	Mg_2GeO_4	185.21	wh ppt				0.0016^{25}		s a; i alk
m51	germanide	Mg_2Ge	121.21			1115				
m52	hydride	MgH_2	26.33	wh tetr cr or mass		d 280 (vac)		d viol		i eth
m53	hydroxide	Nat. brucite. $Mg(OH)_2$	58.33	col, hex pl, 1.559, 1.580	2.36	$-H_2O$, 350		0.0009^{18}	0.004^{100}	s a, NH_4 salts
m54	iodate	$Mg(IO_3)_2.4H_2O$	446.18	wh, monocl	$3.3^{18.5}$	$-4H_2O$, 210 d		10.2^{20}	19.3^{100}	
m55	iodide	MgI_2	278.12	wh, hex, deliq	4.43^{25}_4	d < 637		148^{18}	164.9^{110}	s al, eth, NH_3
m56	iodide, octahydrate	$MgI_2.8H_2O$	422.24	wh powd, deliq		d 41		81^{20}	90.3^{80}	s al, eth
m57	lactate	$Mg(C_3H_5O_3)_2.3H_2O$	256.50	wh cr powd, v bitter taste				3.3	16.7^{100}	i al, eth
m58	laurate	$Mg(C_{12}H_{23}O_2)_2.2H_2O$	458.97	wh lumps		150.4		0.007^{25}	0.041^{100}	0.415^{15} al; 0.012^{25} eth
m59	permanganate	$Mg(MnO_4)_2.6H_2O$	370.27	dk purp need, deliq	2.18(?)	d		v s	d	s MeOH, ac a
m60	molybdate	$MgMoO_4$	184.25	rhomb, tricl	2.208			13.7^{25}		
m61	myristate	$Mg(C_{14}H_{27}O_2)_2$	479.05	wh powd		131.6		$0,006^{15}$	0.014^{60}	0.189^{25} al; 0.007^{25} eth
m62	nitrate, dihydrate	$Mg(NO_3)_2.2H_2O$	184.35	col pr	2.0256^{25}	129		s	s	s al, liq NH_3; sl s conc HNO_3
m63	nitrate, hexahydrate	$Mg(NO_3)_2.6H_2O$	256.41	col, monocl, deliq	1.6363^{25}_4	89	d 330	125	v s	s al, liq NH_3
m64	nitride	Mg_3N_2	100.95	grn-yel, powd or mass	2.712^{25}_4	d 800	subl 700 (vac)	d	d	s a; i al
m65	nitrite, trihydrate	$Mg(NO_2)_2.3H_2O$	170.37	wh pr, hygr		d 100		s		s al
m66	oleate	$Mg(C_{18}H_{33}O_2)_2$	587.24	yel powd or mass				0.024^{25}		6.64^{20} al; s linseed oil; sl s eth
m67	oxalate	$MgC_2O_4.2H_2O$	148.36	wh powd	2.45	d 150		0.07^{16}	0.08^{100}	s alk, a, oxalate
m68	oxide	Nat. periclase. MgO	40.31	col, cub, 1.736	3.58^{25}	**2852**	3600	0.00062	0.0086^{30}	s a, NH_4 salts; i al
m69	oxide, per-	MgO_2	56.31	wh powd				i	i	s a
m70	palmitate	$Mg(C_{16}H_{31}O_2)_2$	535.16	wh cr need or lumps		121.5		0.008^{25}	0.009^{60}	0.047^{25} al; 0.003^{25} eth
m71	orthophosphate	$Mg_3(PO_4)_2$	262.88	rhomb pl, iridisc		1184		i	i	s NH_4 salts, i liq NH_3
m72	orthophosphate	$Mg_3(PO_4)_2.22H_2O$	659.22	col, monocl pr	1.640^{15}	$-18H_2O$, 100	d 200	v sl s		d a
m73	orthophosphate, mono-H	Nat. newberyite. $MgHPO_4.3H_2O$	174.34	wh, rhomb, 1.514, 1.518, 1.533	2.123^{15}	$-H_2O$, 205	d 550–650	sl s		s a
m74	orthophosphate, mono-H, heptahydrate	$MgHPO_4.7H_2O$	246.40	wh, monocl need	1.728^{15}	$-4H_2O$, 100	d 550–650	0.3	0.2	s a; i al
m75	orthophosphate, octahydrate.	Nat. bobierite. $Mg_3(PO_4)_2.8H_2O$	407.00	wh, monocl pl, 1.510, 1.520, 1.543	2.195^{15}	$-5H_2O$, 150	$-8H_2O$, 400	v sl s		s NH_4 citrate
m76	orthophosphate, tetrahydrate	$Mg_3(PO_4)_2.4H_2O$	334.97	monocl	1.64^{15}			0.0205		s a; i NH_4 salts
m77	phosphide	Mg_3P_2	134.88	yel-grn cub cr	2.055			d	d	d dil min a; sl d conc H_2SO_4

No.	Name	Synonyms and Formulae	Mol. wt.	Crystalline form, properties and index of refraction	Density or spec. gravity	Melting point, °C	Boiling point, °C	Solubility, in grams per 100 cc Cold water	Hot water	Other solvents
	Magnesium									
m78	*hypo*phosphite	$Mg(H_2PO_2)_2.6H_2O$	262.38	wh, ditetrag	$1.59_4^{12.5}$	$-5H_2O$, 100	$-6H_2O$, 180	20^{25}		i al, eth
m79	*ortho*phosphite	$MgHPO_3.3H_2O$	158.34						0.25	s a
m80	*pyro*phosphate	$Mg_2P_2O_7$	222.57	col, tab monocl, 1.602, 1.604, 1.615	2.559, (3.06)	1383		i	i	s a; i al
m81	platinocyanide	$MgPt(CN)_4.7H_2O$	449.58	red cr, 1.561	2.185^{16}	$-2H_2O$, 45		s	s	s al; i eth
m82	salicylate	$Mg(C_7H_5O_3)_2.4H_2O$	370.61	col or sl redsh cr powd, effl				s		s al
m83	selenate	$MgSeO_4.6H_2O$	275.36	col, monocl, 1.468, 1.489, 1.491	1.928			v s	v s	
m84	selenide	$MgSe$	103.27	lght gray powd or cr, 2.44	4.21			d	d	d a
m85	*meta*silicate	Nat. clinoenstatite. $MgSiO_3$	100.40	wh, monocl, α 1.651, γ 1.660	3.192_4^{25}	d 1557		i	i	v sl s HF
m86	*ortho*silicate	Nat. forstérite. Mg_2SiO_4	140.71	wh, orthorhmb, 1.65, 1.66, 1.67	3.21	1910		i		d h HCl
m87	(di-) silicide	Mg_2Si	76.71	blue cub	1.94	1102		i	d	s a, NH_4Cl, HCl
m88	silicofluoride	$MgSiF_6.6H_2O$	274.48	wh hex-rhomb, 1.3439, 1.3602	1.788	d 100		60^{25}	5	s dil a, v sl s HF; i al
m89	stannide	Mg_2Sn	167.31	blsh-wh met		778		s		s dil HCl
m90	stearate	$Mg(C_{18}H_{35}O_2)_2$	591.27	wh powd or lumps		**86-88**		0.0003^{15} 0.004^{25}	0.008^{50}	0.020^{25} al; 0.003^{25} eth
m91	sulfate	$MgSO_4$	120.37	col, rhomb cr, 1.56	2.66	d 1124		26^0	73.8^{100}	s al, glyc; 1.16^{18} eth; i acet
m92	sulfate, heptahydrate	Epsom salt, nat. epsomite. $MgSO_4.7H_2O$	246.48	col, rhomb or monocl, 1.433, 1.455, 1.461	1.68	$-6H_2O$, 150	$-7H_2O$, 200	71^{20}	91^{40}	sl s al, glyc
m93	sulfate, monohydrate	Nat. kieserite. $MgSO_4.H_2O$	138.39	col monocl pr, 1.523, 1.535, 1.586	2.445				68.4^{100}	
m94	sulfide	MgS	56.38	pa red-brn, cub, rhomb, 2.271	2.84	d >2000		d	d	s a, PCl_3
m95	sulfite	$MgSO_3.6H_2O$	212.47	wh, rhomb or hex, 1.511, 1.464 (hex)	1.725	$-6H_2O$, 200	d	1.25	s	i al, NH_3
m96	*d*-tartrate	$MgC_4H_4O_6.5H_2O$	262.46	wh, rhomb	1.67	$-4H_2O$, 100	$-5H_2O$, 200	0.8^{18}	1.44^{90}	s min a; i al, NH_3
m97	*d*-tartrate, hydrogen	$Mg(HC_4H_4O_6)_2.4H_2O$	394.54	wh, rhomb	1.72				1.893^{100}	
m98	telluride	$MgTe$	151.91	wh, hex cr	3.86			d		d a
m99	thiosulfate	$MgS_2O_3.6H_2O$	244.53	col, rhomb pr	1.818^{24}	$-3H_2O$, 170	d	v s	v s	i al
m100	thiotellurite	Mg_3TeS_4	360.86	pa yel cr mass				s	s	s al
m101	tungstate	$MgWO_4$	272.16	col, monocl	5.66			i		d a; i al
m102	**Manganese**	Mn	54.938	gray-pink met, cub or tetr	7.20	1244 ± 3	**1962**	d	d	s dil a
m103	(II) acetate	$Mn(C_2H_3O_2)_2$	173.02	brn cr	1.74			s, d		s al
m104	acetate, tetrahydrate	$Mn(C_2H_3O_2)_2.4H_2O$	245.08	pa red, monocl	1.589			s		s al
m105	arsenide, mono-	$MnAs$	129.86	blk, hex	6.17-6.20 (5.55)	d 400		i	i	s HCl, aq reg
m106	arsenide, di-	Mn_2As	184.80			1400		i	i	s aq reg
m107	arsenide, tri-	Mn_3As_2	314.66	magnetic, (exist ?)				i	i	s aq reg
m108	(II) benzoate	$Mn(C_7H_5O_2)_2.4H_2O$	369.23	pa red pr				6.55^{15}		
m109	boride, di-	MnB_2	76.56	gray-vlt cr	6.9			d	d	s a
m110	boride, mono-	MnB	65.75	cr powd	6.2^{15}					
m111	bromide, di-	$MnBr_2$	214.76	rose cr	4.385_4^{25}	d		127.3^0	228^{100}	i NH_3
m112	bromide, di-, tetrahydrate	$MnBr_2.4H_2O$	286.82	α stable, rose monocl, deliq β labile, col, rhomb		d 64.3		296.7^0		
m113	carbide	Mn_3C	176.83	tetr	6.89^{17}			d	d	s a
m114	(II) carbonate	Nat. rodochrosite. $MnCO_3$	114.95	rose, rhomb, lt brn in air	3.125	d		0.0065^{25}		s dil a, aq CO_2; i al, NH_3
m115	chloride, di-	Scacchite. $MnCl_2$	125.84	pink, cub cr, deliq	2.977_4^{25}	650	1190	72.3^{25}	123.8^{100}	s al; i eth, NH_3
m116	chloride, di-, tetrahydrate	$MnCl_2.4H_2O$	197.91	rose, monocl, deliq	2.01	58	$-H_2O$, 106; $-4H_2O$, 198	151^8	656^{100}	s al; i eth
m117	chloride, tri-	$MnCl_3$	161.30	brn cr or grnsh-blk		d sl				s abs al
m118	chloroplatinate	$MnPtCl_6.6H_2O$	570.84	trig	2.692	d		i		
m119	chromite	$MnCr_2O_4$	222.93	gray-blk, cub	4.97^{20}			i	i	i a
m120	(II) citrate	$Mn_3(C_6H_5O_7)_2$	543.02	wh-redsh powd				v sl s		s Na-citr sol
m121	(II) ferrocyanide	$Mn_2Fe(CN)_6.7H_2O$	447.94	grnsh-wh powd				i		s HCl; i NH_4 salts
m122	fluogallate	$[Mn(H_2O)_6][GaF_5.H_2O]$	345.76	pink, orthorhomb, 2.22, 1.45	2.22	d 230		v s		s HF
m123	fluosilicate	$MnSiF_6.6H_2O$	305.11	rose, hex pr, 1.357, 1.374	1.903	d		140^{18}	v s	s al

No.	Name	Synonyms and Formulae	Mol. wt.	Crystalline form, properties and index of refraction	Density or spec. gravity	Melting point, °C	Boiling point, °C	Solubility, in grams per 100 cc		
								Cold water	Hot water	Other solvents
	Manganese									
m124	fluoride, di-	MnF_2	92.93	red, tetr, or redsh powd	3.98	856		0.66^{40}	0.48^{100}	s a; i al, eth
m125	fluoride, tri-	MnF_3	111.93	red cr	3.54	d		d	d	s a
m126	formate	$Mn(CHO_2)_2.2H_2O$	181.00	rhomb	1.953	d		s	s	
m127	(II) glycerophosphate	$MnC_3H_7O_6P$	225.00	wh or sl red powd				sl s		s a, citr a; i al
m128	hydroxide	$MnO(OH)_2$	104.95	blk-brn, amorph (exist ?)	2.58			v sl s		
m129	(II) hydroxide	Nat. pyrochroite. $Mn(OH)_2$	88.95	wh-pink, trig 1.723, 1.681	3.258^{13}	d		0.0002^{18}		s a, NH_4 salts; i alk
m130	(III) hydroxide	Magnanite. $MnO(OH)$	87.94	br-blk, rhomb. 2.24, 2.24, 2.53 (Li)	4.2–4.4	d		i	i	s HCl, h H_2SO_4
m131	iodide, di-	MnI_2	308.75	pink, hex cr, deliq, br. in air	5.0^1	638 (vac) d ca 80	subl vac 500	s	s	0.02^{25} NH_3
m132	iodide, di-, tetrahydrate	$MnI_2.4H_2O$	380.81	rose, monocl, deliq		d		s	v s	
m133	*hexa*iodoplatinate	$MnPtI_6.9H_2O$	1173.59	trig	3.604^{20}_4	d				
m134	(II) nitrate	$Mn(NO_3)_2.4H_2O$	251.01	col, or rose, monocl	1.82	25.8	129.4	426.4^0	∞	v s al
m135	(II) lactate	$Mn(C_3H_5O_3)_2.3H_2O$	287.04	pa red, monocl		d		10	v s	s al
m136	(II) oxalate	MnC_2O_4	142.96	wh cr powd	$2.43^{21.7}$	d 150		i	i	s a, NH_4Cl
m137	(II) oxalate, dihydrate	$MnC_2O_4.2H_2O$	178.98	redsh-wh oct cr powd		$-2H_2O$, 100	d	0.0312^{25}	0.037^{36}	
m138	(II) oxalate, trihydrate	$MnC_2O_4.3H_2O$	197.00	pink, tricl		$-H_2O$, 25				
m139	(II, III) oxide	Nat. hausmannite. Mn_3O_4	228.81	blk, tetr (rhomb), 2.46, (Li) 2.15 (Li)	4.856	**1564**		i	i	s HCl
m140	oxide, di-	Nat. pyrolusite. MnO_2	86.94	blk, rhomb, or brn-blk powd	5.026	$-O$, 535		i	i	s HCl; i HNO_3, acet
m141	oxide, hept-	Mn_2O_7	221.87	dk red oil, hyg, exp	2.396^{20}_4	5.9	d 55, exp 95	v s	d	s H_2SO_4
m142	(II) oxide, mon-	Nat. manganosite, MnO	70.94	grn, cub, 2.16	5.43–5.46 (3.7–3.9)			i	i	s a, NH_4Cl
m143	(III) oxide, sesqui-	Nat. braunite. Mn_2O_3	157.87	blk, cub (tetr)	4.50	$-O$, 1080		i	i	s a; i ac a
m144	oxide, tri-	MnO_3	102.94	redsh, deliq (exist ?)		d		s	d	s alk, H_2SO_4
m145	(III) *meta*phosphate	$Mn_2(PO_3)_6.2H_2O$	619.74					sl s	s	
m146	(II) *ortho*phosphate	Nat. reddingite. $Mn_3(PO_4)_2.3H_2O$	408.80	rose or yelsh-wh rhomb, 1.651, 1.656, 1.683	3.102			i		
m147	(III) *ortho*phosphate	$MnPO_4.H_2O$	167.92	gray cr powd		$-H_2O$, 300	d	i		s h conc H_2SO_4, conc HCl, molten H_3PO_4
m148	(II) *ortho*phosphate, di-H	$Mn(H_2PO_4)_2.2H_2O$	284.94			$-H_2O$, >100		s		i al
m149	(II) *ortho*phosphate mono-H	$MnHPO_4.3H_2O$	204.97	red, rhomb or pink powd, 1.656				sl s	d	s a; i al
m150	(II) *pyro*phosphate	$Mn_2P_2O_7$	283.82	br-pink, monocl, 1.695, 1.704, 1.710	3.707^{25}	1196		i		s a
m151	(II) *pyro*phosphate, trihydrate	$Mn_2P_2O_7.3H_2O$	337.87	wh, amorph powd				i		s $K_2P_2O_7$ sol, H_2SO_4 i acet
m152	phosphide, mono-	MnP	85.91	dk gray	5.39^{21}	1190		i	i	sl s HNO_3
m153	(tri-)phosphide, di-	Mn_3P_2	226.76	dk gray	5.12^{18}	1095		i	i	sl s dil HNO_3
m154	(II) *hypo*phosphite	$Mn(H_2PO_2)_2.H_2O$	202.93	rose cr or powd		$-H_2O$, >150		12.5	16.7	i al
m155	(II) *ortho*phosphite	$MnHPO_3.H_2O$	152.93	redsh		$-H_2O$, 200		sl s		s $MnSO_4$, $MnCl_2$
m156	selenate, dihydrate	$MnSeO_4.2H_2O$	233.93	rhomb	2.95–3.01			s	s	
m157	selenate, pentahydrate	$MnSeO_4.5H_2O$	287.97	pa red, trig	2.33–2.39					
m158	selenide	MnSe	133.90	gray, cub	5.55^{15}			i		d dil a
m159	selenite	$MnSeO_3.2H_2O$	217.93	monocl cr				v sl s	v sl s	
m160	(II) *meta*silicate	Nat. rhodonite. $MnSiO_3$	131.02	red, tricl, 1.733, 1.740, 1.744	3.72^{25}	1323		i		i HCl
m161	silicide, di-	$MnSi_2$	111.11	gray, oct	5.24^{13}			i	i	s HF, alk; i HNO_3, H_2SO_3
m162	silicide, mono-	MnSi	83.02	tetrah	5.90^{15}	1280		i	i	s HF; v sl s a
m163	(di-)silicide,	Mn_2Si	137.96	quadr pr	6.20^{15}	1316		i	i	s HCl, NaOH; i HNO_3
m164	(II) sulfate	$MnSO_4$	151.00	redsh	3.25	700	d 850	52^5	70^{70}	s al; i eth
m165	(III) sulfate	$Mn_2(SO_4)_3$	398.06	grn cr, deliq, hex	3.24	d 160		d	d	s HCl, dil H_2SO_4; i conc. H_2SO_4, HNO_3

PHYSICAL CONSTANTS OF INORGANIC COMPOUNDS (Continued)

No.	Name	Synonyms and Formulae	Mol. wt.	Crystalline form, properties and index of refraction	Density or spec. gravity	Melting point, °C	Boiling point, °C	Cold water	Hot water	Other solvents
	Manganese									
m166	(II) sulfate, dihydrate	$MnSO_4.2H_2O$	187.03	(exist ?)	2.526[15]	stab 57–117		85.27[25]	106.8[55]	
m167	(II) sulfate, heptahydrate	$MnSO_4.7H_2O$	277.11	red monocl or rhomb	2.09	–7H$_2$O, 280; stab +9		172	118[13]	i al
m168	(II) sulfate, hexahydrate	$MnSO_4.6H_2O$	259.09	(exist?)		stab +5 to +8		147.4	1.345[28]	
m169	(II) sulfate, monohydrate	Nat. szmikite. $MnSO_4.H_2O$	169.01	pa pink monocl, 1.562, 1.595, 1.632	2.95	stab 57–117		98.47[48]	79.8[100]	
m170	(II) sulfate, pentahydrate	$MnSO_4.5H_2O$	241.08	rose, tricl, 1.495, 1.508, 1.514	2.103[15]	stab 9–26		124[0]	142[54]	
m171	(II) sulfate, tetrahydrate	Common form. $MnSO_4.4H_2O$	223.06	pink, monocl or rhomb effl, 1.508, 1.522	2.107	stab 26–27		105.3[0]	111.2[54]	i al
m172	(II) sulfate, tri-hydrate	$MnSO_4.3H_2O$	205.05	(exist?)	2.356[15]	stab 30–40		74.22[5]	99.31[57]	
m173	(II) sulfide	Nat. alabandite. MnS	87.00	grn cub or pink amorph, 2.70 (Li)	3.99	d		0.00047[18]		s dil a, al; i (NH$_4$)$_2$S
m174	(II) sulfide	3MnS.H$_2$O	279.05	gray-pink		d		0.0006	i	s dil a; i (NH$_4$)$_2$ S
m175	(IV) sulfide	Nat. hauerite. MnS$_2$	119.07	blk cub, 2.69 (Li)	3.463	d		i	i	d HCl
m176	(II) tantalate	$Mn(TaO_3)_2$	512.83	blk, rhomb, 2.22, 2.25, 2.29	7.03					
m177	(II) tartrate	$MnC_4H_4O_6$	203.01	wh powd				v sl s		s dil a
m178	(II) thiocyanate	$Mn(SCN)_2.3H_2O$	225.16	deliq		–3H$_2$O, 160–170		s	v s	v s al
m179	(II) dithionate	$Mn(SO_3)_2$	215.06	tricl cr	1.757			s	v s	
m180	(II) titanate	Nat. pyrophanite. MnTiO$_3$	150.84	yel, trig, 2.481, 2.210	4.54	1360				
m181	valerate	$Mn(C_5H_9O_2)_2.2H_2O$	293.22	br powd				s		
m182	**Manganic acid, per-**	HMnO$_4$	119.94					v s	d	
m183	**Manganocyanic acid.**	$H_4Mn(CN)_6$	215.08			d		i		v s al; i eth
m184	Mercury	Quicksilver. Hg	200.59	sil wh met, liq.	13.5000[0]	38.87	356.58	i	i	s HNO$_3$, i dil HCl, HBr, HI cold H$_2$SO$_4$
m185	(I) acetate	$Hg_2(C_2H_3O_2)_2$	519.27	micaceous plates		d		0.75[12]		s HNO$_3$, H$_2$SO$_4$; i al, eth
m186	(II) acetate	$Hg(C_2H_3O_2)_2$	318.76	wh, sc or powd	3.270	d		25[10]	100[100]	s al, ac a
m187	(II) acetylide	3HgC$_2$.H$_2$O	691.85	wh powd	5.3	expl		i	i	i al
m188	(II) orthoarsenate	$Hg_3(AsO_4)_2$	879.61	yel				v sl s		s HCl, HNO$_3$
m189	(I) orthoarsenate mono-H	Hg_2HAsO_4	541.11					i		s HNO$_3$; i ac a, NH$_4$OH
m190	(I) azide	$Hg_2(N_3)_2$	485.22	wh cr		expl, d by light		0.025		
m191	(II) benzoate	$Hg(C_7H_5O_2)_2.H_2O$	460.84	wh cr powd		165		1.2[15]	2.5[100]	s al, NaCl, NH$_4$Cl, bz
m192	(I) bromate	$Hg_2(BrO_3)_2$	656.99	cr		d		d		sl s HNO$_3$
m193	(II) bromate	$Hg(BrO_3)_2.2H_2O$	492.44	cr		d 130–140		0.15	1.6	s HCl, HNO$_3$, Hg(NO$_3$)$_2$
m194	(I) bromide	Hg_2Br_2	561.00	wh, yel, tetr	7.307	subl 345		0.0000004[25]		s a; i al, acet
m195	(II) bromide	HgBr$_2$	360.41	col rhomb	6.109[25] 5.12[240]	236	322	0.61[25]	4.0[100]	15[0] al; s MeOH; v sl s eth
m196	bromide iodide	HgBrI	407.40	yel, rhomb		229	360			s al, eth
m197	(I) carbonate	Hg$_2$CO$_3$	461.19	yel br cr		d 130		0.0000045	d	s NH$_4$Cl; i al
m198	(II) carbonate, basic.	HgCO$_3$.2HgO	693.78	br red				i		s NH$_4$Cl, aq CO$_2$
m199	(I) chlorate	$Hg_2(ClO_3)_2$	568.08	wh, rhomb	6.409	d 250		s	d	s al, ac a
m200	(II) chlorate	$Hg(ClO_3)_2$	367.49	wh need	4.998	d		25		
m201	(I) chloride	Calomel. Hg$_2$Cl$_2$	472.09	wh, tetr, 1.973, 2.656	7.150	subl 400		0.00020[25]	0.001[43]	s aq reg, Hg(NO$_3$)$_2$; sl s HCl, h HNO$_3$; i al, eth
m202	(II) chloride	Corrosive sublimate. HgCl$_2$	271.50	col, rhomb or wh powd pois, 1.859	5.44[25], liq 4.44[280]	276	302	6.9[20]	48[100]	33[25] al; 4 eth; s ac a, pyr
m203	(I) chromate	Hg$_2$CrO$_4$	517.17	red need, or powd		d		v sl s	sl s	s HCN, HNO$_3$; i al, ac
m204	(II) chromate	HgCrO$_4$	316.58	red, rhomb		d		sl s, d	d	s HN$_4$Cl; d a; i acet
m205	(II) cyanide	Hg(CN)$_2$	252.63	col, tetr, or wh powd pois, 1.645	3.996	d		9.3[14]	33[100]	25[19.5] MeOH; 8 al; s NH$_3$, glyc; i bz
m206	(I) fluoride	Hg$_2$F$_2$	439.18	yel, cub	8.73[15]	570	d	d to Hg$_2$O		
m207	(II) fluoride	HgF$_2$	238.59	col, cub	8.95[15]	d 645	650	d		s HF, dil HNO$_3$
m208	(I) fluosilicate	Hg$_2$SiF$_6$.2H$_2$O	579.29	col pr	2.134			sl s		i HCl

No.	Name	Synonyms and Formulae	Mol. wt.	Crystalline form, properties and index of refraction	Density or spec. gravity	Melting point, °C	Boiling point, °C	Solubility, in grams per 100 cc		
								Cold water	Hot water	Other solvents
	Mercury									
m209	(II) fluosilicate	$HgSiF_6.6H_2O$	450.76	col, rhomb, deliq		d easily		0.4^{17}	d	i al
m210	(I) formate	$Hg_2(CHO_2)_2$	491.22	glist scales		d				
m211	(II) fulminate	$Hg(CNO)_2$	284.62	wh, cub	4.42	expl		sl s	s	s al, NH_4OH
m212	(I) iodate	$Hg_2(IO_3)_2$	750.99	yelsh		d 250		i	i	s dil HCl, conc HNO_3
m213	(II) iodate	$Hg(IO_3)_2$	550.48	wh, amorph powd				i		s HCl, NH_4Cl, NaCl, KI; i HNO_3
m214	(I) iodide	Hg_2I_2	654.99	yel, tetr or amorph powd	7.70	subl 140	d 290	v sl s		s KI, NH_4OH; i al, eth
m215	(II) iodide (α)	HgI_2	454.90	red, tetr	6.36^{25}_4	tr 127		0.01^{25}		3.18^{25} acet; 2.23^{25} al; s chl; d NH_4OH
m216	(II) iodide (β)	HgI_2	454.90	yel, rhomb cr or powd	6.094^{127}_4	259	354	v sl s	sl s	s eth, KI, $Na_2S_2O_3$; v sl s al
m217	(I) nitrate	$Hg_2(NO_3)_2.2H_2O$	561.22	col, monocl, effl	4.79^4	70		d	s, d	s dil HNO_3; i NH_4OH
m218	(II) nitrate	$Hg(NO_3)_2.\frac{1}{2}H_2O$	333.61	wh-yelsh cr or powd, deliq	4.39	79	d	v s	d	s acet, HNO_3, NH_3; i al
m219	(II) nitrate	$Hg(NO_3)_2.H_2O$	342.61	col cr or wh powd, deliq	4.3			s		s HNO_3; i al
m220	(I) nitrite	$Hg_2(NO_2)_2$	493.19	yel	7.33	d 100		d		
m221	nitride	Hg_3N_2	629.78	br powd		expl		d		s NH_4OH, NH_4 salts; d a
m222	(I) oxalate	$Hg_2C_2O_4$	489.20					i	i	sl s HNO_3
m223	(II) oxalate	HgC_2O_4	288.61			d		0.0107^{20}		s HCl; sl s HNO_3
m224	(I) oxide	Hg_2O	417.18	blk or brnish-blk powd	9.8	d 100		i	i	s HNO_3
m225	(II) oxide	Nat. montroydite. HgO	216.59	yel or red, rhomb, 2.37, 2.5, 2.65	11.1^4	d 500		0.0053^{25}	0.0395^{100}	s a; i al, eth, acet, alk, NH_3
m226	(II) oxybromide	$HgBr_2.3HgO$	1010.17	yel cr				i	sl s	v s al
m227	(II) oxychloride	$HgCl_2.2HgO$	704.67	red hex, or blk monocl	red 8.16–8.43 blk 8.53					
m228	(II) oxychloride	$HgCl_2.3HgO$	921.26	yel, hex	7.93	d 260		i	d	
m229	(II) oxycyanide	$Hg(CN)_2.HgO$	469.22	wh need or cr powd	4.437^{19}	expl		1.25	s	
m230	(II) oxyfluoride	$HgF_2.HgO.H_2O$	473.19	yel cr		d 100		d		s dil HNO_3
m231	(II) oxyiodide	$HgI_2.3HgO$	1104.17	yel br				d	d	s HI
m232	(II) selenide	Nat. tiemannite. HgSe	279.55	gray plates	8.266	vac subl		i		s aq reg
m233	(I) sulfate	Hg_2SO_4	497.24	col monocl, wh-yelsh powd	7.56	d	d	0.06^{25}	0.09^{100}	s HNO_3, H_2SO_4
m234	(II) sulfate	$HgSO_4$	296.65	col rhomb or wh powd	6.47	d		d		s a, NaCl; i al, acet, NH_3
m235	(II) sulfate, basic	$HgSO_4.2HgO$	729.83	lem yel powd	6.44		volat	0.003^{16}	sl s	s a; i al
m236	(I) sulfide	Hg_2S	433.24	blk		d		i		i al, $(NH_4)_2$ S
m237	(II) sulfide (α)	Cinnabar, vermillion. HgS	232.65	red cr hex, or powd, 2.854, 3.201	8.10	subl 583.5		0.000001^{18}		s aq reg, Na_2S; i al, HNO_3
m238	(II) sulfide (β)	Metacinnabar. HgS	232.65	blk, cub or amorph powd	7.73	583.5		i		s aq reg, Na_2S, alk; i al, HNO_3
m239	(I) tartrate	$Hg_2C_4H_4O_6$	549.25	yelsh-wh cr powd				i		i a
m240	(I) orthotellurate	HgH_4TeO_6	428.22	trans, orthorhomb		d 20		slow d	rapid d	
m241	(II) orthotellurate	Hg_2TeO_6	825.37	amber, cub		unalt at 140		i	i	s HCl, HNO_3
m242	(I) thiocyanate	$Hg_2(SCN)_2$	517.34			d		i		s HCl, KCNS
m243	(II) thiocyanate	$Hg(SCN)_2$	316.75	wh powd, pois		d		0.07^{25}	s	s NH_4 salts, HCl, NH_3, KCN; sl s al, eth
m244	(I) tungstate	Hg_2WO_4	649.03	yel, amorph		d		i	i	d a; i al
m245	(II) tungstate	$HgWO_4$	448.44	yel		d		i	d	d a; i al
	Mercury nitrogen compounds									
m246	mercury (II) bromide, ammono-basic	$Hg(NH_2)Br$	296.52	wh powd		d		d		s NH_4OH; i al
m247	mercury (II) bromide, diammine	$Hg(NH_3)_2Br_2$	394.47	wh powd		180		d		s NH_4Cl, NH_4Br, NH_4I
m248	mercury (II) chloride ammonobasic	Infusible ppt. $Hg(NH_2)Cl$	252.07	wh powd or sm pr	5.70	infus		0.14	d 100	d a; i al
m249	mercury (II) chloride, aquobasic ammonobasic	Chloride of Millon's base. OHg_2NH_2Cl	468.66	pa yel or wh powd		d >120		sl s		s HCl, HNO_3
m250	mercury (II) chloride, diammine	Fusible white ppt. $Hg(NH_3)_2Cl_2$	305.56	rhombd		300		i	i	s a, KI

No.	Name	Synonyms and Formulae	Mol. wt.	Crystalline form, properties and index of refraction	Density or spec. gravity	Melting point, °C	Boiling point, °C	Solubility, in grams per 100 cc		
								Cold water	Hot water	Other solvents
	Mercury nitrogen compounds									
m251	mercury (II) iodide, ammonobasic	$Hg(NH_2)I$	353.52							i eth
m252	mercury (II) iodide aquobasic ammonobasic	Iodide of Millon's base. OHg_2NH_2I	560.11	yel to brn		>128	expl	i		s d, HCl, KI soln
m253	mercury (II) iodide, diammine	$Hg(NH_3)_2I_2$	488.48	col or pa yel powd or need				d		s NH_4OH
m254	**Millons's base**	$(HO)_2Hg_2NH_2OH$	468.22		4.083[18]					
m255	**Molybdenum**	Mo	95.94	silv-wh met, or gray-blk powd, cub	10.2	2610	5560	i	i	s h conc HNO_3, h conc H_2SO_4, aq reg; sl s HCl; i HF, NH_3
m256	boride, (di-)	MoB_2	117.56	rhomb	7.12					
m257	boride, (mono-)	MoB	106.75	tetr	8.65					
m258	(di-) boride	Mo_2B	202.69	tetr	9.26					
m259	bromide, di-	$MoBr_2$ (or Mo_2Br_4)	255.76	yel red, amorph	4.88[17.5]			i	i	s alk; i a, aq reg
m260	bromide, tetra-	$MoBr_4$	415.58	blk need, deliq	d		volat	v s		d alk
m261	bromide, tri-	$MoBr_3$	335.67	dk-grn need	d			i	i	d alk, NH_3; i a
m262	carbide, mono-	MoC	107.95	gray, hex	8.20[20]	2692		i	i	sl s HNO_3, HF, h H_2SO_4, HCl; i alk hydr
m263	carbide(di-)	Mo_2C	203.89	wh, hex pr	8.9	2687		i		sl s HNO_3, HF h H_2SO_4, aq reg, HCl; i alk
m264	carbonyl	$Mo(CO)_6$	264.00	wh cr, rhomb diamagnet	1.96	d 150, without meltg	156.4[766]	i	i	s bz; sl s eth
m265	carbonyl, tri-pyridine, tri-	$Mo(CO)_3(C_5H_5N)_3$	287.08	yel-brn cr	d					
m266	chloride, di-	$MoCl_2$ (or Mo_6Cl_6)	166.85	yel, amorph	3.714[25]	d		i	i	s HCl, H_2SO_4, alk, NH_4OH, al, acet
m267	chloride, penta-	$MoCl_5$	273.21	grn-blk cr, trig deliq	2.928	194	268	d	d	s conc min a, liq NH_3, CCl_4, chl; s d al, eth
m268	chloride, tetra-	$MoCl_4$	237.75	brn powd or cr, deliq	d		vol	d	d*	s conc min a; d al, eth
m269	chloride, tri-	$MoCl_3$	202.30	dk red need or powd	3.578[25]	d		i	sl d	s conc H_2SO_4, conc HNO_3; v sl s al, eth; i HCl; d alk
m270	fluoride, hexa-	MoF_6	209.93	col cr	liq 2.55[17.5]	17.5[406]	35[760]	s d	d	s NH_4OH, alk
m271	hydrotetrachloro-hydroxide, di-	$[Mo_2Cl_4(H_2O)_2](OH)_2.6H_2O$	607.77	lt yel cr		$-H_2O$ 35–300		i	i	i a, al
m272	hydroxide	$Mo(OH)_3$ (or $Mo_2O_3.3H_2O$)	146.96	blk powd	d			0.2		sl s H_2SO_4, HCl; s 30 % H_2O_2
m273	hydroxide	$MoO(OH)_3$ (or $Mo_2O_5.3H_2O$)	162.96	br to blk powd				0.2 (coll)		s a, alk carb; i alk hydr
m274	(VI) hydroxide	$MoO_3.2H_2O$	179.97	lt yel, monocl pr	3.124[15]			0.05[15]		s dil H_2O_2, alk hydr; sl s a
m275	hydroxytetra-bromide, di-	$Mo_2Br_4(OH)_2$	641.47	red powd						s alk
m276	hydroxytetra-bromide, diocta-hydrate	$Mo_2Br_4(OH)_2.8H_2O$	785.59	golden yel cr	d	d				s HCl; d alk, HNO_3
m277	hydroxytetra-chloride, di-	$Mo_2Cl_4(OH)_2.2H_2O$	499.68	pa yel, amorph					i	s conc a; i al
m278	iodide, di-	MoI_2	349.75	brn powd	5.278[25.4]			i	d	v sl s a
m279	iodide, tetra-	MoI_4	603.56	blk cr	d 100					
m280	oxide, di-	MoO_2	127.94	lead gray, tetr or monocl	6.47			i		sl s h conc H_2SO_4; i alk, HCl, HF
m281	oxide, pent-	Mo_2O_5	271.88	vlt-bl powd (exist?)						s h H_2SO_4, h HCl
m282	oxide, pent-	"Molybd. blue" $Mo_2O_5.xH_2O$ (variations in Mo and O)		dk blue coll or powd	3.6[18]	d		s		s a, MeOH; i acet, bz, chl
m283	oxide, sesqui-	Mo_2O_3	239.88	blk, opaque (exist?)				i	i	i a, alk, NH_4OH
m284	oxide, tri-	Molybdic anhydride. Nat. Molybdite. MoO_3	143.94	col or wh-yel, rhomb	4.692[21]	795	subl 1155[760]	0.1066[18]	2.055[70]	s a, alk sulf, NH_4OH
m285	oxydibromide, di-	MoO_2Br_2	287.76	yel-red tabl, deliq				s		

No.	Name	Synonyms and Formulae	Mol. wt.	Crystalline form, properties and index of refraction	Density or spec. gravity	Melting point, °C	Boiling point, °C	Solubility, in grams per 100 cc		
								Cold water	Hot water	Other solvents
	Molybdenum									
m286	oxy*tetra*chloride	MoOCl₄	253.75	grn cr, deliq		subl		s		
m287	oxy*tri*chloride	MoOCl₃	218.30	grn cr		subl 100		d		
m288	oxy*di*chloride, di-	MoO₂Cl₂	198.84	yelsh wh scaly cr	3.31¹⁷	subl		s	s	s al, eth
m289	oxy*di*chloride, dihydrate	MoO₂Cl₂.H₂O	216.86	pa yel cr		subl		s	s	sl s al, acet, eth
m290	oxy*hexa*chloride, tri-	Mo₂O₂Cl₆	452.60	rubyred or dk vlt cr		d		d		s eth
m291	oxy*penta*chloride, tri-	Mo₂O₃Cl₅	417.14	dk brn-blk cr, deliq		melts easily	subl	s	s	
m292	oxychloride acid	MoO(OH)₂Cl₂	216.63	wh need, deliq		d 160		v s		s al, eth, acet
m293	oxy*di*fluoride, di-	MoO₂F₂	165.94	wh cr, hygr	3.494²⁵	subl 270		v s	v s	s al, MeOH; i eth, chl, tol
m294	oxy*tetra*fluoride	MoOF₄	187.93	col-wh, deliq	3.001²⁵	98	180	s		s al, eth, CCl₄, s d H₂SO₄; v sl s bz
m295	*meta*phosphate	Mo(PO₃)₆	569.77	yel powd	3.28⁰			i	i	sl s h aq reg; i HCL, HNO₃, H₂SO₄
m296	phosphide	MoP (or Mo₂P₂)	126.93	gray-grn cr powd	6.167					s h HNO₃
m297	phosphide	MoP₂	157.89	blk powd	5.35²⁵					s HNO₃, h conc H₂SO₄, aq reg; i conc HCl
m298	silicide	MoSi₂	152.11	gray, met, tetr	6.31²⁰·⁵					i a, aq reg; v s HF+HNO₃
m299	sulfide, di-	Nat. molybdenite. MoS₂	160.07	blk luster, hex	4.80¹⁴	1185	subl 450, d in air	i	i	s h H₂SO₄, aq reg, HNO₃; i dil a, conc H₂SO₄
m300	sulfide, penta	Mo₂S₅.3H₂O	406.25	dk br powd		−H₂O, 135	d	i	i	s NH₄OH, alk, sulfides
m301	sulfide, sesqui-	Mo₂S₃	288.07	steel gray need	5.91¹⁵	d 1100	vol 1200			i conc HCl; d h HNO₃
m302	sulfide, tetra-	MoS₄	224.20	brn powd		d		i	i	i a; s h H₂SO₄, alk sulfide
m303	sulfide, tri-	MoS₃	192.13	blk pl		d	d	sl s	s	s alk sulf, conc KOH
m304	**Molybdic acid**	H₂MoO₄.H₂O (or MoO₃.2H₂O)	179.97	yel, monocl	3.124¹⁵	−H₂O, 70	d	0.133¹⁸	2.568⁷⁰	s alk hydr, alk carb; sl s a
m305	anhydrous	H₂MoO₄ (or MoO₃.H₂O)	161.95	wh or sl yelsh, hex	3.112	−H₂O, 70		sl s	sl s	s alk, NH₄OH, H₂SO₄; i NH₃
m306	**Molybdic arsenic acid**	As₂O₅.6MoO₃.18H₂O	1475.75	col, trig	2.493¹⁹·⁸	−15H₂O, 150		v s		s abs al; i chl, CS₂
m307	**Molybdic phosphoric acid**	H₇[P(Mo₂O₇)₆].28H₂O	2365.71	yel, oct	2.53	78		d		
m308	**Molybdic silicic acid**	H₈[Si(Mo₂O₇)₆].28H₂O	2363.83	yel, tetr		45	d 100	600¹⁴		s dil a; i bz, chl, CS₂
n1	**Neodymium**	Nd	144.24	silv-wh to yelsh met, hex to m.p. 868, cub from 868	hex 7.004 cub 6.80	1024	3027	d		
n2	acetate	Nd(C₂H₃O₂)₃.H₂O	339.39					26.2		
n3	acetylacetonate	Nd[CH(COCH₃)₂]₃	441.57	pink cr	1.618	150−152				
n4	*hexa*antipyrin perchlorate	[Nd(C₁₁H₁₂N₂O)₆].(ClO₄)₃	1571.99	rose, hex cr		d 285−289		0.99²⁰		
n5	bromate	Nd(BrO₃)₃.9H₂O	690.10	red, hex		66.7	−9H₂O, 150	151²⁵		
n6	bromide	NdBr₃	383.97	grn cr		684	1540	sl s		
n7	carbide	NdC₂	168.26	yel, hex leaf	5.15	d		d	d	s dil a; i conc HNO₃
n8	chloride	NdCl₃	250.60	rose-vlt pr	4.134²⁵	784	1600	96.7¹³	140¹⁰⁰	44.5 al; i eth, chl
n9	chloride, hexahydrate	NdCl₃.6H₂O	358.69	red, rhomb		124	−6H₂O, 160	246¹³	511¹⁰⁰	v s al
n10	chromate	Nd₂(CrO₄)₃.8H₂O	780.58	yel cr				0.027		
n11	fluoride	NdF₃	201.24	pa lilac		1410	2300	i	i	
n12	iodide	NdI₃	524.95	blk cr powd		775±3	1370	s	s	
n13	kojate	Nd(C₆H₄O₄)₃	567.60	lt choc		d 275		i		
n14	manganous nitrate	2Nd(NO₃)₃.3Mn(NO₃)₂.24H₂O	1629.72	vlt-red	2.114	77		77.4³⁰		
n15	magnesium nitrate	2Nd(NO₃)₃.3Mg(NO₃)₂.24H₂O	1537.84	vlt-red	2.020	109		69.5³⁰		
n16	dimethylphosphate	Nd[(CH₃)₂PO₄]₃	519.35	pa lilac, hex pl				56.1²⁵	22.3¹⁰⁰	
n17	molybdate	Nd₂(MoO₄)₃	768.29	tetr, 2.005	5.14¹⁸	1176				
n18	nickel nitrate	2Nd(NO₃)₃.3Ni(NO₃)₂.24H₂O	1641.04	blsh-grn	2.202	105.6		71.5³⁰		
n19	nitrate	Nd(NO₃)₃.6H₂O	438.35	tricl				152.9²⁵		s al, acet
n20	nitride	NdN	158.25	blk powd				d		
n21	oxalate	Nd₂(C₂O₄)₃.10H₂O	732.69	rose cr				0.000074²⁵		
n22	oxide	Neodymia. Nd₂O₃	336.48	lt bl powd, red fluores	7.24	~1900		0.00019²⁹	0.003⁷⁵	s a

No.	Name	Synonyms and Formulae	Mol. wt.	Crystalline form, properties and index of refraction	Density or spec. gravity	Melting point, °C	Boiling point, °C	Solubility, in grams per 100 cc		
								Cold water	Hot water	Other solvents
	Neodymium									
n23	sulfate	$Nd_2(SO_4)_3.8H_2O$	720.79	red, monocl, 1.41, 1.551, 1.562	2.85	1176	....	8^{20}	5.4^{40}	....
n24	sulfide	Nd_2S_3	384.67	oliv grn powd	5.179^{11}	d	....	i	d	s dil a
n25	Neon	Ne	20.183	inert gas col sol, cub	gas: 0.9002^0 g/l; liq: $1.204^{-245.9}$	−248.67	−245.9	1.47^{20} cm³	....	s liq O_2
n26	Neptunium	Np	237.00	α: orthorhomb silvery β: tetr (above 278) γ: cub (above 500)	α: 20.45 β: 19.36^{313} γ: 18.0^{600}	630±1 α 278±5 stab 278–570				s HCl
n27	bromide, tri-	$NpBr_3$	476.73	α: hex β: grn orthorhomb	α 6.92	subl ca 800	....	s		
n28	chloride, tetra-	$NpCl_4$	378.81	red-brn tetr	4.92	538		s		
n29	chloride, tri-	$NpCl_3$	343.36	wh, hex	5.38	ca 800	....	s		
n30	fluoride, hexa-	NpF_6	350.99	brn, orthorhomb		53	d	d		
n31	fluoride, tetra-	NpF_4	312.99	grn, monocl	6.8			i		i conc HNO_3
n32	fluoride, tri-	NpF_3	294.00	purple, hex	9.12			i		
n33	iodide, tri-	NpI_3	617.71	brn, orthorhomb	6.82			s		
n34	oxide, di-	NpO_2	269.00	apple grn, cub	11.11			i		s conc a
n35	(tri-) oxide, octa-	Np_3O_8	839.00	brn, cub		d 500	....			s HNO_3
n36	Nickel	Ni	58.71	silv met, cub	8.90	1455	2730	i	i	s dil HNO_3; sl s HCl, H_2SO_4; i NH_3
n37	acetate	$Ni(C_2H_3O_2)_2$	176.80	grn pr	1.798	d	16.6			i al
n38	acetate, tetrahydrate	$Ni(C_2H_3O_2)_2.4H_2O$	248.86	grn pr	1.744	d	16			s dil al
n39	antimonide	Nat. breithauptite. NiSb	180.46	lt copper red, hex	7.54	1158	d 1400			
n40	orthoarsenate, octahydrate	$Ni_3(AsO_4)_2.8H_2O$	598.09	yelsh-grn powd	4.98	....	....	i	....	s a
n41	arsenide	Nat. niccolite. NiAs	133.63	hex	7.57^0	968		i	i	s aq reg
n42	orthoarsenite, acid	$NiH_4(AsO_3)_4.H_2O$	691.87	grn-wh		d		i		s a, alk
n43	benzenesulfonate	$Ni(C_6H_5SO_3)_2.6H_2O$	481.11	grn, monocl	1.742^{15}	−8H₂O	d	14.0^{14}	51.5^{93}	5.9 al, 1.5 eth
n44	boride	NiB	69.52	pr	7.39^{15}		d	d	d	s aq reg, HNO_3
n45	bromate	$Ni(BrO_3)_2.6H_2O$	422.62	monocl	2.575	d		28		
n46	bromide	$NiBr_2$	218.53	yel brn, deliq	5.098^{27}	963		112.8^0	155.1^{100}	s al, eth, NH_4OH
n47	bromide, trihydrate	$NiBr_2.3H_2O$	272.57	yelsh-grn need, deliq		−3H₂O, 300		199^0	315.7^{100}	s al, eth, NH_4OH
n48	bromoplatinate	$NiPtBr_6.6H_2O$	841.35	trig	3.715					
n49	di-N-butyldithiocarbamate	(NBC).Ni[(C₄H₉)₂NCSS]₂	467.47	dk olive grn powd	1.29	89–90		i	i	sl s bz, pet comp; i al
n50	carbide	Ni_3C	188.14	dk gray powd	7.957^{25}			i	i	
n51	carbonate	$NiCO_3$	118.72	lt grn, rhomb		d		0.0093^{25}	i	s a
n52	carbonate, basic	$2NiCO_3.3Ni(OH)_2.4H_2O$	587.67	lt grn cr or brn powd		d		i	d	s a, NH_4 salts
n53	carbonate, basic	Zaratite. $NiCO_3.$ $2Ni(OH)_2.4H_2O(?)$	376.23	emerald grn, cub, 1.56–1.61	2.6	....		i	i	s h dil HCl, NH_4OH
n54	carbonyl	$Ni(CO)_4$	170.75	col, volat, inflamm, liq, or need	1.32^{17}	−25	43	$0.018^{9.8}$	....	s aq reg, al, eth, bz, HNO_3; i dil a, dil alk
n55	chlorate	$Ni(ClO_3)_2.6H_2O$	333.70	dk red	2.07	d 80		0.9^{27}		
n56	perchlorate	$Ni(ClO_4)_2.6H_2O$	365.70	grn, hex need, 1.518, 1.498	....	140		222.5^0	273.7^{45}	s al, acet, chl
n57	chloride	$NiCl_2$	129.62	yel sc, deliq	3.55	1001	subl 973	64.2^{20}	87.6^{100}	s al, NH_4OH; i NH_3
n58	chloride, hexahydrate	$NiCl_2.6H_2O$	237.70	grn, monocl, deliq ~1.57	....			254^{20}	599^{100}	v s al
n59	chloropalladate	$NiPdCl_6.6H_2O$	485.92	hex	2.353					
n60	chloroplatinate	$NiPtCl_6.6H_2O$	574.61	trig	2.798					
n61	cyanide	$Ni(CN)_2$	110.75	yel-brn				i	i	s KCN
n62	cyanide, tetrahydrate	$Ni(CN)_2.4H_2O$	182.81	lt grn pl or powd pois		−4H₂O, 200 d		i	i	s KCN, NH_4OH, alk; sl s dil a
n63	ferrocyanide	$Ni_2Fe(CN)_6.xH_2O$		grn-wh	1.892(?)			i	i	s KCN, NH_4OH; i HCl
n64	fluogallate	$[Ni(H_2O)_6][GaF_5.H_2O]$	349.53	pa grn, monocl(?), 1.45	2.45	−5H₂O, 110	....	sl s		s HF
n65	fluoride	NiF_2	96.71	grn, tetr	4.63	subl 1000 (in HF)		4^{25}		s a, alk, eth, NH_3
n66	fluosilicate	$NiSiF_6.6H_2O$	308.88	grn, trig, 1.391, 1.407	2.134	d				
n67	formate, dihydrate	$Ni(CHO_2)_2.2H_2O$	184.78	grn cr	2.154	d		s		
n68	(II) hydroxide	$Ni(OH)_2$ (or $NiO.xH_2O$)	92.72	grn cr, or amorph	4.15(3.65)	d 230		0.013		s a, NH_4OH
n69	iodate	$Ni(IO_3)_2$	408.52	yel need	5.07			1.1^{90}	1.0^{90}	
n70	iodate, tetrahydrate	$Ni(IO_3)_2.4H_2O$	480.59	hex		d 100		1.4^{30}	1.1^{90}	
n71	iodide	NiI_2	312.52	blk cr, deliq	5.834	797		124.2^0	188.2^{100}	s al

No.	Name	Synonyms and Formulae	Mol. wt.	Crystalline form, properties and index of refraction	Density or spec. gravity	Melting point, °C	Boiling point, °C	Solubility, in grams per 100 cc		
								Cold water	Hot water	Other solvents
	Nickel									
n72	dimethylglyoxime	$Ni(HC_4H_6N_2O_2)_2$	288.94	scarlet red cr		subl 250		i	i	s a, abs al; i a ac, NH₄OH
n73	nitrate, hexahydrate	$Ni(NO_3)_2.6H_2O$	290.81	grn, monocl, deliq	2.05	56.7	136.7	238.5⁹	v s	s al, NH₄OH
n74	oleate	$Ni(C_{18}H_{33}O_2)_2$	621.64	grn oil		18-20				
n75	oxalate	$NiC_2O_4.2H_2O$	182.76	lt grn powd				i		s a, NH₄ salts; v sl s oxal a
n76	oxide, mono-	Nat. bunsenite. NiO	74.71	grn-blk, cub, 2.1818(red)	6.67	1984		i	i	s a, NH₄OH
n77	*orthophosphate*	$Ni_3(PO_4)_2.8H_2O$	510.20	apple grn pl or emerald cr gran		d		i	i	s a, NH₄ salts; i me acet. et acet
n78	*pyrophosphate*	$Ni_2P_2O_7.xH_2O$		grn	3.93 (anhydr)					s a, NH₄OH
n79	(di-)phosphide	Ni_2P	148.39	gray cr	6.31¹⁵	1112		i		s HNO₃+HF; i a
n80	(penta)phosphide, (di-)	Ni_5P_2	355.50	need or tabl cr		1185				
n81	(tri-)phosphide, (di-)	Ni_2P_3	238.08	dk grn-blk	5.99			i	i	s HNO₃; i HCl
n82	*hypophosphite*	$Ni(H_2PO_2)_2.6H_2O$	296.78	grn	1.82¹⁹·⁵	d 100		s		
n83	selenate	$NiSeO_4.6H_2O$	309.76	grn, tetr, 1.5393	2.314			s		
n84	selenide	NiSe	137.67	wh or gray, cub	8.46	red heat		i		s aq reg, HNO₃; i a, HCl
n85	silicide	Ni_2Si	145.51		7.2¹⁷	1309		i		i a
n86	stearate	$Ni(C_{18}H_{35}O_2)_2$	625.67	grn powd		100		i		s CCl₄, pyr; sl s acet; i MeOH, eth
n87	sulfate	$NiSO_4$	154.78	yel, cub	3.68	d 848⁷⁶⁰		29.3⁰	83.7¹⁰⁰	i al, eth, acet
n88	sulfate, heptahydrate	Morenosite. $NiSO_4.7H_2O$	280.88	grn, rhomb, 1.467, 1.489, 1.492	1.948	99; −H₂O 31.5	−6H₂O, 103	75.6¹⁵·⁵	475.8¹⁰⁰	s al
n89	sulfate, hexahydrate	Single nickel salt. $NiSO_4.6H_2O$	262.86	α: bl, tetr β: grn, monocl, 1.511, 1.487	2.07	tr: 53.3	−6H₂O, 103	62.52⁰	340.7¹⁰⁰	12.5 MeOH; v s al, NH₄OH
n90	sulfide, mono-	Nat. millerite. NiS	90.77	blk; trig or amorph	5.3-5.65	797		0.00036¹⁸		s aq reg, HNO₃, KHS; sl s a
n91	sulfide, sub-	Heazlewoodite. Ni_3S_2	240.26	pa yelsh bronze met, lust	5.82	790		i		s HNO₃
n92	(II, III) sulfide	Poydomite. Ni_3S_4	304.39	gray-blk, cub	4.7			i		s HNO₃
n93	sulfite	$NiSO_3.6H_2O$	246.86	grn, tetrah				i		s HCl, H_2SO_4
n94	dithionate	$NiS_2O_6.6H_2O$	326.93	grn, tricl	1.908	d				
	Nickel Complexes									
n95	diaquotetrammine nickel (II) nitrate	$[Ni(NH_3)_4(H_2O)_2].(NO_3)_2$	286.87	grn cr				s		i al
n96	hexamminenickel (II) bromide	$[Ni(NH_3)_6]Br_2$	320.71	vlt powd	1.837			v s	d	
n97	hexamminenickel (II) chlorate	$[Ni(NH_3)_6](ClO_3)_2$	327.80		1.52	180		d to Ni(NH₃)₄		
n98	hexamminenickel (II) chloride	$[Ni(NH_3)_6]Cl_2$	231.80	blsh, cub	1.468²⁵			s	d	s NH₄OH; i al
n99	hexamminenickel (II) iodide	$[Ni(NH_3)_6]I_2$	414.70	pa bl, cub	2.101	d		d		s NH₄OH
n100	hexamminenickel (II) nitrate	$[Ni(NH_3)_6](NO_3)_2$	284.90	bl, oct or cub				4.46		
n101	tetrapyridinnickel (II) fluosilic	$[Ni(C_5H_5N)_4]SiF_6$	517.20	bl grn, rhomb	2.307					
n102	**Niobium**	Columbium. Nb	92.906	steel gray, lustr met cub, 1.80	8.57	2468 ± 10	4742	i	i	s fus alk; i HCl, HNO₃, aq reg
n103	boride	NbB_2	114.53	hex	6.97	2900(?)				
n104	bromide, penta-	$NbBr_5$	492.46	purp red		265.2	361.6	d		s al, ethyl bromide
n105	carbide	NbC	104.92	blk, cub or lavender-gray powd	7.6	3500		i		s HNO₃, HF
n106	chloride, penta-	$NbCl_5$	270.17	yel-wh, deliq	2.75	204.7	254	d		s al, HCl, CCl₄
n107	fluoride, penta-	NbF_5	187.90	col, monocl pr, hygr	3.293	72-73	236	d		s al; sl s chl, CS₂, H_2SO_4
n108	hydride	NbH	93.91	gray powd	6.6	infus				s HF, conc H_2SO_4; i HCl, alk, HNO₃
n109	nitride	NbN	106.91	blk, cub	8.4	2573				s HF+HNO₃; i HNO₃
n110	oxalate, hydrogen	$Nb(HC_2O_4)_5$	538.05	col, monocl				d	d	s $H_2C_2O_4$; d al
n111	oxide, di-	NbO_2	124.90	blk	5.9			i	i	sl s alk; i a
n112	oxide, mon- (or di-)	NbO (or Nb_2O_2)	108.91	blk, cub	7.30			i	i	s a, alk; i al, NHO₃
n113	oxide, pent-	Nb_2O_5	265.81	wh, rhomb	4.47	1485 ± 5		i	i	s HF, alk; i a
n114	oxide, pent-, hydrate	$Nb_2O_5.xH_2O$				d		i		s conc H_2SO_4, conc HCl, HF, alk; i NH₃
n115	oxide, tri-(sesqui)	Nb_2O_3	233.81	bl-blk		1780				

No.	Name	Synonyms and Formulae	Mol. wt.	Crystalline form, properties and index of refraction	Density or spec. gravity	Melting point, °C	Boiling point, °C	Solubility, in grams per 100 cc		
								Cold water	Hot water	Other solvents
	Niobium									
n116	oxybromide	$NbOBr_3$	348.63	yel cr		subl		d		s a
n117	oxychloride	$NbOCl_3$	215.26	col need		subl 400		s, d	d	s al, H_2SO_4; i HCl
n118	potassium fluoride	$NbOF_3.2KF.H_2O$	300.12	monocl leaf, lustr fatty				7.8	v s	
n119	**Nitric acid**	HNO_3	63.01	col liq, corr, pois, 1.5027_4^{25} $1.397^{16.4}$		-42	83	∞	∞	d al, viol; s eth
n120	const boil	68% $HNO_3 + 32\% H_2O$		col liq	1.41_4^{20}		120.5	∞	∞	
n121	**Nitrogen**	N_2	28.0134	col gas, col liq, sol cub cr	gas 1.2506 g/l; liq $0.8081^{-195.8}$ sol $1.026^{-252.5}$	-209.86	-195.8	2.33^0 cm^3	1.42^{40} cm^3	sl s al
n122	chloride, tri-	NCl_3	120.37	yel oil or rhomb cr	1.653	<-40	<71, expl 95	i	d	s chl, bz, CCl_4; CS_2, PCl_3
n123	fluoride, tri-	NF_3	71.00	col gas	liq: 1.537^{-129}	-206.60	-128.8	v sl s		
n124	iodide, tri-	NI_3	394.72	blk, expl		expl	subl vac	i	d	s KCNS, $Na_2S_2O_3$
n125	iodide, tri-, monoammine	$NI_3.NH_3$	411.75	dk red, rhomb	3.5	d>20	expl	i	d	s HCl, KCNS, $Na_2S_2O_3$; i abs al
n126	oxide(ic)	NO	30.01	col gas, bl liq, sol liq 1.330^{-90}	gas 1.3402 g/l liq; $1.269^{-150.2}$	-163.6	-151.8	7.34^0 cm^3	2.37^{60} cm^3	3.4 cm^3 H_2SO_4; 26.6 cm^3 al; s $FeSO_4$, CS_2
n127	oxide(ous)	N_2O	44.01	col gas or liq or cub cr, 1.0005_{760}^{20}	1.977_{760}^0 g/l	-90.8	-88.5	130^0 cm^3	56.7^{25} cm^3	s al, eth, H_2SO_4
n128	oxide, pent-	Nitric anhydride. N_2O_5	108.01	wh, rhomb or hex	1.642^{18}	30	d 47	s	d to HNO_3	s chl
n129	oxide, tri-	NO_3	62.00	blsh gas		d at ord temp				s eth
n130	peroxide	Nitrogen oxide, di-. NO_2	46.01	col sol, yel liq or brn gas, 1.40^{20}	1.4494_{20}^{20}	-11.20	21.2	s, d		s alk, CS_2, chl
n131	(di-) oxide(tri-)	Nitrous anhydride. N_2O_3	76.01	red-brn gas, bl sol or liq	1.447^2	-102	d 3.5	s	d	s eth, a, alk
n132	oxi (tri-) fluoride	NO_3F	81.06	col gas expl	liq; $1.507^{-45.9}$ sol: $1.951^{-193.2}$	-175	-45.9	d		s acet; expl al, eth
n133	sulfide, penta-	N_2S_5	188.33	gray cr		10–11	d	d	d	s CS_2, eth; i bz, al
n134	sulfide, tetra-	Tetranitrogen tetrasulfide sulfur-nitride. N_4S_4	184.28	yel cr, 2.046	2.24^{18}	d 178				s al, bz, CS_2
n135	**Nitrosyl bromide**	$NOBr$	109.92	br gas or dk br liq	>1.0	-55.5	-2	d	d	s alk
n136	perchlorate	$NOClO_4.H_2O$	147.47	rhomb, deliq	2.169	d 100		d		expl al, eth
n137	chloride	$NOCl$	65.46	yel gas or yel-red liq or cr	gas: 2.99 g/l liq: 1.417^{-12}	-64.5	-5.5	d	d	s fum H_2SO_4
n138	fluoborate	$NOBF_4$	116.81	col, rhomb cr, hygr	2.185_4^{25}	subl $250^{0.01}$		d		
n139	fluoride	NOF	49.00	col gas	2.176 g/l	-134	-56			d to $HNO_2 + HF$
n140	**Nitrosylsulfuric acid**	Chamber crystals. $NOHSO_4$	127.08	col, rhomb		d 73.5		d		s H_2SO_4
n141	**Nitrosylsulfuric anhydride**	$(NOSO_3)_2O$	236.14	tetr		217	360	d		s H_2SO_4
n142	**Nitrous acid**	HNO_2	47.01	only in sol (pa bl)				d		
n143	hypo-	$H_2N_2O_2$	62.03	wh, sol		expl		s		
n144	**Nitryl chloride**	NO_2Cl	81.46	pa yel-br gas	gas: 2.57 g/l liq: 1.32^{14}	<-31	5	d		
n145	fluoride	NO_2F	65.00	col gas, col sol	2.90 g/l	-139	-63.5	d		d al, eth, chl
o1	**Osmium**	Os	190.20	gray-blsh met, hex	22.48^{20}	2700	>5300	i	i	sl s aq reg, HNO_3; i NH_3
o2	carbonyl chloride	$Os(CO)_3Cl_2$	345.14	col pr	269–273	d 280		i	i	s NaOH; i a
o3	chloride, di-	$OsCl_2$	261.11	dk brn, deliq	d			i	sl d	s al, eth, HNO_3; sl s alk
o4	chloride, tetra-	$OsCl_4$	332.01	red br need		subl		sl s, d		i al
o5	chloride, tri-	$OsCl_3$	296.56	br, cub		d 500–600		v s		s alk, a, a; sl s eth
o6	chloride, tri-, trihydrate	$OsCl_3.3H_2O$	350.61	dk gr cr		d		v s		s al
o7	fluoride, hexa-	OsF_6	304.19	grn cr		32.1	45.9	d	d	
o8	fluoride, tetra-	OsF_4	266.19	br powd				d	d	
o9	iodide	OsI_4	697.82	vlt-blk, hygr met, lust				v s	d	s al

No.	Name	Synonyms and Formulae	Mol. wt.	Crystalline form, properties and index of refraction	Density or spec. gravity	Melting point, °C	Boiling point, °C	Solubility, in grams per 100 cc		
								Cold water	Hot water	Other solvents
	Osmium									
o10	oxide, di-, brown	OsO_2	222.20	brn cr	$11.37^{21.4}$	30 % tr to OsO_4, 500		i	i	i a
o11	oxide, di-, black	OsO_2	222.20	blk powd	7.71^{21}	tr to br 350–400		i	i	s dil HCl
o12	oxide, mon-	OsO	206.20	blk				i	i	
o13	oxide, sesqui-	Os_2O_3	428.40	dk brn	d			i	i	i a
o14	oxide, tetra-	OsO_4	254.10	a) col, monocl b) yel mass	4.906^{22}	a) 39.5 b) 41.0	130	5.70^{10}	6.23^{25}	250 ± 10^{20} CCl_4; s al, eth, NH_4OH, $POCl_3$
o15	sulfide, di-	OsS_2	254.33	blk, cub	9.47	d		i	i	s HNO_3; i alk
o16	sulfide, tetra-	OsS_4	318.46	br blk (exist?)		d		i		s dil HNO_3; i $(NH_4)_2$ S
o17	sulfite	$OsSO_3$	270.26	bl blk		d		i		s dil HCl, alk
o18	telluride	$OsTe_2$	445.40	gray-blk cr		ca 600				i a; d dil HNO_3
o19	**Oxygen**	O_2	31.9988	col gas, sol hex cr	gas: 1.429 g/l, liq: 1.149^{-183} sol: $1.426^{-252.5}$	−218.4	−182.962	4.89^0 cm^3 3.16^{25} cm^3	2.46^{50} cm^3 2.30^{100} cm^3	2.78^{25} al
o20	fluoride	OF_2	54.00	col gas, unst	liq: $1.90^{-223.8}$	−223.8	−144.8	sl s, d	i	sl s a, alk
o21	**Ozone**	O_3	47.9982	col gas, or dk bl liq, or bl-blk cr, liq: 1.2226	gas: 2.144^0 g/l liq: $1.614^{-195.4}$ g/l	-192.7 ± 2 1.0	−111.9	49^0 cm^3		s alk sol, oils
p1	**Palladium**	Pd	106.40	silv wh, met, cub	$12.02^{2\ 0}$ $11.40^{22.5}$	1554	2970	i	i	s aq reg, h HNO_3, H_2SO_4; sl s HCl
p2	bromide	$PbBr_2$	266.22	red br	5.173^{16}	d		i	i	s HBr
p3[1]	chloride	$PdCl_2$	177.31	dk red, cub need, deliq	4.0^{18}	d 500		s	s	s HBr; acet
p3[2]	chloride, dihydrate	$PdCl_2.2H_2O$	213.34	br pr, deliq	d			v s	v s	s HCl, acet
p4	cyanide	$Pd(CN)_2$	158.44	yelsh-wh		d		i	i	s KCN, NH_4OH; i dil a
p5	fluoride, di-	PdF_2	144.40	br, tetr	5.80	volat	d red heat	sl s, d	i	s HF
p6	fluoride, tri-	PdF_3	163.40	blk, rhomb	5.06	d	d	d	d	s HF
p7	hydride	Pd_2H (or Pd_4H_2)	213.81	silv met (exist?)	10.76	d				
p8	iodide	PdI_2	360.21	blk powd	6.003^{18}	d 350		i	i	s KI; i al, eth, dil HCl
p9	nitrate	$Pd(NO_3)_2$	230.41	br yel, rhomb, deliq		d		s, d		s HNO_3
p10	oxide, di-	$PdO_2.xH_2O$		dull red		d − H_2O, −O		i	i,d	s a, alk
p11	oxide, mon-	PdO	122.40	grnsh-bl or amber mass, or blk powd	8.70^{20}_4	870		i	i	i aq reg
p12	oxide, mon-, hydrate	$PdO.xH_2O$		yel to brn		d		i	i	s a, NH_3, NH_4Cl
p13	selenate	$PdSeO_4$	249.36	dk brn-red, rhomb, deliq	6.5	d red heat		v s	v s	i al, eth, alk; s NH_3
p14	selenide	PdSe	185.36	dk gray		<960		i		s aq reg
p15	selenide, di-	$PdSe_2$	264.32	olive gray, hex		<1000		i		v s aq reg; v sl s HNO_3; i alk
p16	silicide	PdSi	134.49	cr	7.31^{15}			i		
p17	sulfate	$PdSO_4.2H_2O$	238.50	red-br cr, deliq		d		v s	d	
p18	sulfide, di-	PdS_2	170.53	dk br cr	$4.7–4.8^{25}_4$	d		i	i	s aq reg, $(NH_4)_2S$
p19	sulfide, mono-	PdS	138.46	brn-blk tetr	6.6^{25}_4	d 950		i	i	sl s HNO_3, aq reg; i HCl, $(NH_4)_2S$
p20	sulfide, sub-	Pd_2S	244.86	grn gray (exist?)	7.303^{15}	d 800		i	i	sl s a, aq reg
p21	telluride, di-	$PdTe_2$	361.60	silvery cr, hex				i	i	v s aq reg; s HNO_3; i alk
	Palladium Complexes									
p22	diamminepalladium (II) hydroxide	$Pd(NH_3)_2(OH)_2$	174.48	yel micr cr		>105		v s	d	
p23	dichlorodiammine-palladium (II) *trans* (or *α*)	$Pd(NH_3)_2.Cl_2$	211.37	yel, tetr	2.5	d		0.304^{10}	s, d	s, d a; s NH_4OH; i chl, acet
p24	tetramminepalladium (II) chloride	$Pd(NH_3)_4Cl_2.H_2O$	263.44	col, tetr	1.91^{16}	d 120		v s		
p25	tetramminepalladium tetrachloropalladate	Vauquelin's salt. $Pd(NH_3)_4.PdCl_4$	422.73	pink powd or need	2.489^{21}	tr yel 184 d above 192		i	sl s	sl s dil HCl; s KOH
p26	**Phospham**	PN_2H	60.00	wh, amorph		infus		i	d	s conc H_2SO_4, alk; i a
p27	**Phosphomolybdic acid**	Molybdophosphoric acid $H_3[P(Mo_3O_{10})_4]$		yel, tetr		78–90		s	s	

No.	Name	Synonyms and Formulae	Mol. wt.	Crystalline form, properties and index of refraction	Density or spec. gravity	Melting point, °C	Boiling point, °C	Solubility, in grams per 100 cc		
								Cold water	Hot water	Other solvents
	Phosphonium									
	Phosphonium									
p28	bromide...........	PH_4Br	114.91	col, cub	gas: 2.464 g/l	subl ca 30	38.8[794]	d	d	....
p29	chloride...........	PH_4Cl	70.46	col, cub		28[46 atm]	subl	d		
p30	iodide...........	PH_4I	161.91	col, tetr, deliq	2.86	18.5, subl 61.8	80	d		d, s a, alk
p31	sulfate...........	$(PH_4)_2SO_4$	166.07					d		
p32	**Phosphoramide**	Phosphorylamide. $PO(NH_2)_3$	95.04	wh, cr		d		40.66[15]	i	sl, me
	Phosphoric acid									
p33	difluoro-...........	$H_2PO_2F_2$	102.99	col, fum liq	1.583_4^{25}	-96.5 ± 0.1	115.9 sl d			
p34	hypo-...........	$H_4P_2O_6.2H_2O$	198.01	col, rhomb deliq		70	d 100	d to $H_3PO_3+HPO_3$		
p35	meta-...........	HPO_3	79.98	col, vitrous, deliq	2.2–2.5	subl		d to H_3PO_4	d	s al; i liq. CO_2
p36	monofluo-...........	H_2PO_3F	99.99	oily, col liq	1.818	-80				
p37	ortho-...........	H_3PO_4	98.00	col, liq, or rhomb cr, deliq	1.834[18]	42.35	$-\frac{1}{2}H_2O$, 213	548	v s	s al
p38	ortho-...........	$2H_3PO_4.H_2O$	214.01	col, hex pr deliq		29.32	d	v s		
p39	pyro-...........	$H_4P_2O_7$	177.98	col, need or liq, hygr		61		709[23]	d to H_3PO_4	v s al, eth
p40	**Phosphorus**, black...	P_4	123.89504	blk, incombust...	2.70					i CS_2, conc H_2SO_4
p41	red...........	P_4	123.89504	redsh-brn, cub, or amorph powd, (mix of col and vlt?)	2.34	590[43 atm]	ign 200 280	v sl s	i	s abs al; i CS_2, eth, NH_3
p42	violet...........	P_4	123.89504	vlt, monocl.....	2.36	590				i org solv
p43	yellow...........	Phosphorus, white. P_4	123.89504	yel (or wh) cub or wax like solid, 0.144	1.82[20]	44.1	280	0.0003[15]	sl s	0.3 al; 880[10]CS_2; s bz, NH_3, alk, eth, chl, tol
p44	bromide, penta-...	PBr_5	430.52	yel, rhomb		d < 100	d 106	d		s CS_2, CCl_4, bz
p45	bromide, tri-...	PBr_3	270.70	col, fum liq, 1.697[26.6]	2.852[15]	-40	172.9	d		d al; s eth, chl, CS_2, CCl_4
p46	bromide(di-) chloride, tri-	PBr_2Cl_3	607.16	or cr		d 35		d		
p47	bromide(hepta-) chloride, di-	PBr_7Cl_2	661.24	pr			d			s PCl_3, PCl_5
p48	bromide(mono-) chloride, tetra-	$PBrCl_4$	252.69	yel cr			d			
p49	bromide(octa-) chloride, tri-	PBr_8Cl_3	776.60	brn need		25		d		
p50	bromide(di-) fluoride, tri-	PBr_2F_3	247.79	pa yel liq		-20	d 15	d		d glass
p51	bromide nitride.....	$(PNBr_2)_3$	614.40	col, rhomb		190	subl v 150	i		s eth; sl s chl, CS_2
p52	chloride, di-...........	PCl_2 (or P_2Cl_4 ?)...	101.88	col liq		-28	180	hydr		
p53	chloride, penta-...	PCl_5	208.24	yelsh-wh, tetr, fum	gas: 4.65[296] g/l	d 166.8 (press)	subl 162	d		d a; s CS_2, CCl_4
p54	chloride, tri-...	PCl_3	137.33	col, fum liq, 1.516[14]	1.574[21]	-112	75.5[749]	d	d	s eth, bz, chl, CS_2, CCl_4
p55	chloride(di-) fluoride, tri-	PCl_2F_3	158.88	col liq	5.4 g/l	-8	10			
p56	chloride(tri-)iodide, di-	PCl_3I_2	391.14	red, hex		d 259		d		s CS_2
p57	chloride(di-)nitride...	$(PNCl_2)_3$	347.66	rhomb	1.98	114	256.5	i	d	s al, eth, bz, chl, a ac, CS_2
p58	chloride(di-)nitride...	$(PNCl_2)_4$	463.55		2.18_{24}^{24}	123.5	328.5			
p59	chloride(di-)nitride...	$(PNCl_2)_5$	579.43			41	224[13], polym >250			
p60	chloride(di-)nitride...	$(PNCl_2)_6$	695.32			90	262[13], polym >250			
p61	cyanide...........	$P(CN)_3$	109.03	wh need		subl 130		d		v s eth; sl s h bz
p62	fluoride, penta-...	PF_5	125.97	col gas	5.805 g/l	-83	-75	d		
p63	fluoride, tri-...	PF_3	87.97	col gas	3.907 g/l	-151.5	-101.5	d		s al; d alk
p64	hydride, tri-...	Phosphine. PH_3 ...	34.00	col gas, pois		-133	-87.7	0.26 vol 20		
p65	iodide, di	P_2I_4	569.57	or, tricl		110		d	d	s CS_2
p66	iodide, tri-...........	PI_3	411.68	red, hex, deliq	4.18	61	d	d	d	v s CS_2
p67	oxide, pent-...........	Phosphoric anhydride. P_2O_5 (or P_4O_{10})	141.94	wh, monocl or powd, v deliq	2.39	580–585	subl 300	d to H_3PO_4	d	s H_2SO_4; i acet, NH_3
p68	oxide, sesqui-.....	Phosphorus trioxide. P_4O_6 (or P_2O_3)	219.89	col, or wh powd, monocl cr, deliq	2.135[21]	23.8	175.4	d to H_3PO_3		s chl, bz, eth, CS_2
p69	oxide, tetra-.....	P_2O_4	125.95	col, rhomb, deliq	2.54[21]	>100	180 vac	v s to H_3PO_3	d	
p70	oxide, tri-...........	P_2O_3 (or P_4O_6)...........	109.95	col, or wh powd, or monocl, deliq	2.135[21]	23.8	173.8 (in N_2)	d to H_3PO_3	d	s CS_2, eth, chl, bz

No.	Name	Synonyms and Formulae	Mol. wt.	Crystalline form, properties and index of refraction	Density or spec. gravity	Melting point, °C	Boiling point, °C	Solubility, in grams per 100 cc		
								Cold water	Hot water	Other solvents
	Phosphorus									
p71	oxybromide	$POBr_3$	286.70	col pl	2.822	56	189.5	d		s H_2SO_4, CS_2, eth, bz, chl
p72	oxydibromide chloride	$POBr_2Cl$	242.27	sol or liq	liq: 2.45^{50}	30	165	d		
p73	oxybromide chloride, di	$POBrCl_2$	197.79	tabl or liq	liq: 2.104^{14}	13	137.6	d		
p74	oxychloride	$POCl_3$	153.33	col, fum liq, $1.460^{25.1}$	1.675	2	105.3	d	d	d al, a
p75	oxychloride	$P_2O_3Cl_4$	251.76	col fum liq	liq: 1.58[7]	<-50	212	d		
p76	oxyfluoride	POF_3	103.97	col gas	4.69 g/l	-68	-39.8	d		d al
p77	oxynitride	PON	60.98	wh, amorph		red heat		i	i	i a, alk
p78	oxysulfide	$P_4O_6S_4$	348.15	wh, tetr, deliq		102	295	d		50 CS_2
p79	selenide, penta-	P_2Se_5	456.75	dk red-blk need		d			d	s CCl_4; i CS_2
p80	(tetra-)selenide, tri-	P_4Se_3	360.78	or red cr	1.31	242	360–400			
p81	(tetra-)sulfide, hepta-	P_4S_7	348.34	lt yel cr	2.19^{17}	310	523			sl s CS_2
p82	sulfide, penta-	P_2S_5 (or P_4S_{10})	222.27	gray-yel cr, deliq	2.03	286–90	514	i	d	0.22 CS_2; s alk
p83	sesquisulfide	Tetraphosphorus trisulfide. P_4S_3	220.09	yel, rhomb	2.03^{17}	174	408	i	d	100^{17} CS_2; 11.1^{80} bz
p84	thiobromide	$PSBr_3$	302.76	yel, oct	2.85^{17}	38	d 212	s		s eth, CS_2, PCl_3
p85	thiochloride	$PSCl_3$	169.40	col, fum liq	1.668	-35	125	sl d	d	s CS_2
p86	thiocyanate	$P(SCN)_3$	205.22	liq	1.625^{18}	$ca\ -4$	265	d		s al, eth, bz, CS_2
	Phosphorous acid									
p87	hypo-	$H(H_2PO_2)$	66.00	col oily liq or deliq cr	1.493^{19}	26.5	d 130	s	v s	v s al, eth
p88	meta-	HPO_2	63.98	feather like cr			d			
p89	ortho-	$H_2(HPO_3)$	82.00	col-yel cr, deliq	$1.651^{21.2}$	73.6	d 200	309[0]	694[40]	s al
p90	pyro-	$H_4P_2O_5$	145.98	need		38	d 120	d		
p91	**Phosphotungstic acid**	Tungstophosphoric acid $H_3[P(W_3O_{10})_4].14H_2O$	3132.39	yel-grn cr, tricl		d		s		s al, eth
p92	phosphotungstic acid.	Dodecatungstophosphoric acid. $H_3[P(W_3O_{10})_4].24H_2O$	3312.54	trig		89		s		
p93	**Platinic acid,** hexachloro	$H_2PtCl_6.6H_2O$	517.92	red, brn, deliq	2.431	60		v s	v s	s al, eth
p94	tetracyano	$H_2Pt(CN)_4$	301.18			d 100		v s	v s	v s al, eth, chl
p95	hexahydroxy-	$H_2Pt(OH)_6$	299.15	yel need		$-2H_2O$, 100	$-3H_2O$, 120	i	v sl s	s H_2SiF_6, dil a, alk
p96	hexaiodo-	$H_2PtI_6.9H_2O$	1120.67	blk-red, deliq				s		
p97	**Platinum**	Pt	195.09	silv met, cub	21.45^{20}	1772	3827 ± 100			s aq reg, fus alk
p98	arsenide	Nat. sperrylith. $PtAs_2$	344.93	gray, cub	11.8	d >800		sl d	sl d	v sl s a
p99	bromic acid	$H_2PtBr_6.9H_2O$	838.70	br-red, monocl, hygr				v s	v s	v s al, eth
p100	(II) bromide, di-	$PtBr_2$	354.91	br	6.65^{25}_4	d 250		i	i	i al; s HBr, KBr, aq Br
p101	(IV) bromide, tetra-	$PtBr_4$	514.73	dk cr	5.69^{25}_4	d 180		0.41^{20}	sl s	v s al, eth, HBr
p102	carbonyl bromide	$[Pt_2(CO)_2]Br_4$	765.84	lt red, need, hygr.	5.115^{25}_4	d 180		s d		s abs al, CCl_4, bz
p103	carbonyl chloride, di-	$Pt(CO)Cl_2$	294.00	yel need	4.2346^{25}_4	195, subl 240 in CO_2	d 300	d	d	s conc HCl, H_2SO_4, al
p104	dicarbonyl chloride, di-	$Pt(CO)_2Cl_2$	322.02	lt yel need	3.4882^{25}_4	142	$-CO$, 210	d	d	d HCl; s CCl_4
p105	diplatinum dicarbonyl tetrachloride	$Pt_2(CO)_2Cl_4$	588.01	or-yel need	4.235^{25}_4	195	subl 240 (in CO_2)	d		d HCl
p106	diplatinum tricarbonyl tetrachloride	$Pt_2(CO)_3Cl_4$	616.02	or-yel need		130	d 250	d		s h CCl_4, d al
p107	carbonyl diiodide	PtI_2CO	476.91	red cr	5.257	d 140–150		d		s d, al; s bz
p108	carbonyl sulfide	$Pt(CO)S$	255.16	br-blk		d 300–400		d		d alk, al
p109	chloric acid	$H_2PtCl_6.6H_2O$	517.92	brn-red cr, hygr	2.431	60	d >115	s	s	s abs al, acet; v s eth
p110	(II) chloride, di-	$PtCl_2$	266.00	olive grn, hex	6.05	d 581 (in Cl_2)		v sl s		i al, eth; s HCl, NH_4OH
p111	(IV) chloride, tetra-	$PtCl_4$	336.90	br-red cr	4.303^{25}_4	d 370 (in Cl_2)		58.7^{25}	v s	sl s al, NH_3; s acet; i eth
p112	(IV) chloride, tetra-, hydrate	$PtCl_4.5H_2O$	426.98	red, monocl	2.43	$-H_2O$, 100		v s	v s	s al, eth
p113	chloride, tri-	$PtCl_3$	301.45	grnsh-blk	5.256^{25}	435		sl s	s	s h HCl; v sl s conc HCl

No.	Name	Synonyms and Formulae	Mol. wt.	Crystalline form, properties and index of refraction	Density or spec. gravity	Melting point, °C	Boiling point, °C	Solubility, in grams per 100 cc		
								Cold water	Hot water	Other solvents
	Platinum									
p114	*dichloro*carbonyl, dichloride	Pt(COCl₂)Cl₂	364.91	yel cr		d		v s		sl s al; v sl s CCl₄
p115	(II) cyanide, di-	Pt(CN)₂	247.13	yel-br cr			i		i	i al, a, alk; s KCN
p116	(II) fluoride, di-	PtF₂	233.09	yelsh-grn			i		i	
p117	fluoride, hexa-	PtF₆	309.08	dk red solid, very unstable		57.6				
p118	(IV) fluoride, tetra-	PtF₄	271.08	deep red, fused, or yel-lt brn cr, deliq		d red heat		s d	v s	s a, alk
p119	(II) hydroxide	Pt(OH)₂	229.10	blk		d		i	i	s HCl, HBr, alk; i H₂SO₄, dil HNO₃
p120	(II) hydroxide, hydrate	Pt(OH)₂.2H₂O	265.14			−2H₂O, 100		i	i	s conc a
p121	*mono*hydroxy chloric acid	H₂[PtCl₅(OH)].H₂O	409.39	red-brn cr, hygr						
p122	(II) iodide, di-	PtI₂	448.90	blk powd	6.403²⁵₄	d 360		i	i	i eth, a; s HI; sl s Na₂SO₃
p123	(IV) iodide, tetra-	PtI₄	702.71	brn, amorph, or blk cr	6.064²⁵₄	d 130		s d		s al, acet, alk, HI, KI, liq NH₃
p124	iodide, tri-	PtI₃	575.80	blk, like graphite	7.414²⁵₄	d 270		i	i	i al, eth; s KI
p125	(II) oxide, mon-	PtO	211.09	vlt-blk	14.9¹⁵	d 550		i	i	s HCl; i a, aq reg
p126	(II) oxide, mon-, dihydrate	PtO.2H₂O	247.12			−2H₂O, 140–150		i	i	s conc HCl, conc H₂SO₄, conc HNO₃
p127	(IV) oxide, di-	PtO₂	227.03	blk	10.2	450		i	i	i a, aq reg
p128	(IV) oxide, di-dihydrate	Platinic hydroxide PtO₂.2H₂O or Pt(OH)₄	263.12			−2H₂O, 100		i	i	s HCl, aq reg, KOH
p129	(IV) oxide, di-, hydrate	PtO₂.H₂O	245.10					i	i	i aq reg, ac a, HCl, sl s NaOH
p130	(IV) oxide, di-trihydrate	PtO₂.3H₂O	281.13	ochre		d 300		i	i	i aq reg, HCl
p131	(IV) oxide, di-, tetrahydrate	Hydroxoplatinic acid. PtO₂.4H₂O or H₂Pt(OH)₆	299.15	yel need		−2H₂O, 100	−3H₂O, 120	i	i	s a, dil alk
p132	(II, IV) oxide,	Pt₃O₄	649.27			d		i	i	i a, aq reg
p133	oxide, sesqui-	Pt₂O₃.3H₂O	492.22			−H₂O, 100		i	i	s conc H₂SO₄, caust alk
p134	oxide, tri-	PtO₃	243.09	redsh-brn powd						s HCl, H₂SO₄; sl s HNO₃, H₂SO₄
p135	*pyro*phosphate	PtP₂O₇	369.03	grn-yel	4.85	d 600		v sl s		
p136	phosphide	PtP₂	257.04	met shine	9.01²⁵₄	ca 1500		i	i	i a; v sl s aq reg
p137	selenide, di-	PtSe₂	353.01	blk or gray cr or amorph	7.65	d when dry				s aq reg; sl s HNO₃, H₂SO₄
p138	selenide, tri-	PtSe₃	431.97	bl flakes	7.15	d 140		i	i	i conc a, CS₂; s aq reg
p139	sulfate	Pt(SO₄)₂.4H₂O	459.27	yel pl				s	d	s al, eth, a
p140	(IV) sulfide, di-	PtS₂	259.22	blk-brn powd	7.66²⁵₄	d 225–250		i	i	s HCl, HNO₃; i (NH₄)₂S
p141	(II) sulfide, mono-	PtS	227.15	blk, tetr	10.04²⁵₄	d		i	i	s (NH₄)₂ S; i a, alk
p142	sulfide, sesqui-	Pt₂S₃	486.37	gray (exist ?)	5.52	d		i	i	sl s aq reg; i a
p143	(III) sulfuric acid	H₂[Pt(SO₄)₄(H₂O)₂].9½H₂O	983.62	or, red, tricl		d 150		s	s	d alk
p144	telluride	PtTe₂	450.29	gray, hex		1200–1300				sl s Na₂S, (NH₄)₂S
	Platinum complexes									
p145	tetramine platinum (II) chloride	[Pt(NH₃)₄]Cl₂.H₂O	352.13	col, tetr, 1.672, 1.667	2.737	250, −H₂O, 100		i	i	
p146	tetrammineplatinum (II) chloroplatinate	Magnus, salt. [Pt(NH₃)₄]PtCl₄	600.11	grn or red, tetr	<4.1	d		sl s	sl s	
p147	tetrachlorodiammine platinum (IV), *trans-*	[Pt(NH₃)₂]Cl₄	370.96		3.3	200–216				
p148	tetrachlorodiammine platinum (IV), *cis-*	[Pt(NH₃)₂]Cl₄	370.96	or-yel, rhomb or hexag pl or need		240				
p149	**Plumbous, plumbic**	see Lead								
p150	**Plutonium** α	Pu	242.00	sil wh, monocl	19.84	**641**	**3232**			s HCl; i HNO₃, conc H₂SO₄

No.	Name	Synonyms and Formulae	Mol. wt.	Crystalline form, properties and index of refraction	Density or spec. gravity	Melting point, °C	Boiling point, °C	Solubility, in grams per 100 cc		
								Cold water	Hot water	Other solvents
	Plutonium									
p151	β	Pu	239.05	monocl	17.70	stab 122 ± 2 to 206 ± 3				
p152	γ	Pu	239.05	orthorhomb	17.14	stab 206 ± 3 to 319 ± 5				
p153	δ	Pu	239.05	cub	15.92	stab 319 ± 5 to 451 ± 4				
p154	δ'	Pu	239.05	tetrag	16.00	stab 451 ± 4 to 476 ± 5				
p155	ε	Pu	239.05	cub	16.51	stab 476 ± 5 to 639.5 ± 2				
p156	bromide, tri-	PuBr₃	481.73	grn, orthorhomb	6.69	681		s		
p157	chloride, tri-	PuCl₃	348.36	emerald grn, hex	5.70	760		s		s dil a
p158	fluoride, hexa-	PuF₆	355.99	redsh-brn, orthorhomb		50.75	62.3	d		
p159	fluoride, tetra-	PuF₄	317.99	pa brn, monocl	7.0 ± 0.2	1037				
p160	fluoride, tri-	PuF₃	299.00	purple, hex	9.32	1425(± 3)		i		
p161	iodide, tri-	PuI₃	622.71	bright grn, orthorhomb	6.92	777		s		
p162	nitride	PuN	256.01	blk, cub	14.25			hydrol		s HCl, H₂SO₂
p163	oxalate	Pu(C₂O₄)₂.6H₂O	526.13	yel-grn				i		
p164	oxide, di-	PuO₂	274.00	yelsh-grn, cub	11.46					sl s h conc H₂SO₄, HNO₃, HF
p165	sulfate	Pu(SO₄)₂	434.12	light pink						s dil min a
p166	sulfate, tetrahydrate	Pu(SO₄)₂.4H₂O	506.18	coral pink		d 280				s dil min a
p167	**Polonium**	Po	210.05	α-Po: simple cub; β-Po: rhbr	9.4 (for β-Po)	254	962	sl s		s dil min a; v sl s dil KOH
p168	ammonium chloride	(NH₄)₂PoCl₆	458.85		2.76					
p169	tetrabromide	PoBr₄	529.67	bright red, cub		330 (in Br atm)	360²⁰⁰			s al, acet; i bz, CCl₄
p170	dichloride	PoCl₂	280.96	ruby red, orthorhomb	6.50	subl 190				s dil HNO₃
p171	tetrachloride	PoCl₄	351.86	yel, monocl or tric		300 (in Cl atm)	390	s, d		s HCl; sl s al, acet
p172	tetraiodide	PoI₄	717.67	blk cr		200 (in N atm subl)				sl s al, acet; i bz, CCl₄
p173	dioxide	PoO₂	242.05	red, tetr		d 500				
p174	selenate	2PoO₂.SeO₃	611.06	wh powd		>400				s dil HCl
p175	sulfate, basic	2PoO₂.SO₃	564.16	wh powd		>400, d 550				s dil HCl
p176	disulfate	Po(SO₄)₂	402.17	purp		d 550				i al; v s dil HCl
p177	monosulfide	PoS	242.11	blk		d 275				i al; sl s dil HCl
p178	**Potassium**	Kalium. K	39.0983	cub silv met	0.86²⁰	63.25	760	d to KOH	d	d al; s a, Hg, NH₃
p179	acetate	KC₂H₃O₂	98.15	wh, lust powd, deliq	1.57²⁵	292		253²⁰	492⁶²	33 al; 24.24¹⁵ MeOH; s liq NH₃; i eth, acet
p180	acetate, acid	K₂C₂H₃O₂.HC₂H₃O₂	158.20	col, need or pl , hygr		148	d 200	d	d	s al, acet
p181	acetyl salicylate	KC₉H₇O₄.2H₂O	254.29			65				
p182	metaaluminate	K₂Al₂O₄.3H₂O	250.22	col cr				v s, d	v s, d	s alk; i al
p183	aluminosilicate	Nat. orthoclase. KAlSi₃O₈ (or K₂O.Al₂O₃.6SiO₂)	278.34	wh, monocl, 1.518, 1.524, 1.526	2.56	ca 1200				
p184	aluminosilicate	Nat. microcline. KAlSi₃O₈ (or K₂O.Al₂O₃.6SiO₂)	278.34	wh, tricl 1.522, 1.526, 1.530	2.54-2.57	1140-1300				
p185	aluminosilicate	Nat. muscovite, white mica. KAl₃Si₃O₁₀.(OH)₂ (or K₂O.3Al₂O₃.6SiO₂.2H₂O)	398.31	col, monocl, 1.551, 1.587, 1.581	2.76-2.80	d		i		
p186	aluminum metasilicate	Nat. leucite. KAlSi₂O₆	218.25	col cr, 1.508	2.47	1686 ± 5		i	i	d a
p187	aluminum orthosilicate	Nat. kaliophilite. KAlSiO₄	158.17	col, hex or rhomb (hex→rhomb 1540°) hex: 1.532, 1.572; rhomb: 1.528, 1.536	2.5	ca 1800 (rhomb)				
p188	aluminum sulfate	Nat. kalinite. KAl(SO₄)₂.12H₂O	474.39	col, cub, oct or monocl, cub: 1.454, 1.4564; hex: 1.456, 1.429	1.757²⁰₁₄	92.5 −9H₂O, 64.5	−12H₂O, 200	11.4²⁰	v s	i al, acet; s dil a
p189	amide	Potassamide. KNH₂	55.12	col-wh, or yel-grn, hygr		335	subl 400	d	d	d al; s liq NH₃
p190	peroxylammine sulfonate	(KSO₃)₂NO	268.39	yel cr, expl				0.62³, d	6.6²⁹, d	i al

No.	Name	Synonyms and Formulae	Mol. wt.	Crystalline form, properties and index of refraction	Density or spec. gravity	Melting point, °C	Boiling point, °C	Solubility, in grams per 100 cc		
								Cold water	Hot water	Other solvents
	Potassium									
p191	ammonium tartrate	$KNH_4C_4H_4O_6$	205.21	wh, cr powd				v s		
p192	antimonate, hydroxo-	"*Pyro*"-antimonate. $KSb(OH)_6.\frac{1}{2}H_2O$	271.90	wh gran or cr powd				2.82²⁰	s	
p193	antimonide	K_3Sb	239.06	yel-grn		812		d		d air
p194	antimony tartrate	$KSbC_4H_4O_7.\frac{1}{2}H_2O$	333.93	col, rhomb, 1.620, 1.636, 1.638	2.607	$-H_2O$, 100		5.26⁸·⁷	35.7¹⁰⁰	i al; 6.67²⁵ glyc
p195	*ortho*arsenate	K_3AsO_4	256.23	col need, deliq		1310		18.87	v s	4 al
p196	*ortho*arsenate, di-H	KH_2AsO_4	180.04	col, tetr, 1.567, 1.518	2.867	288		19⁶	v s	i al; s NH_3, a; 59.5 glyc
p197	*ortho*arsenate, mono H	K_2HAsO_4	218.13	col monocl pr		d 300		18.86⁶	s	i al
p198	*meta*arsenite	$KAsO_2$	146.02	wh powd, hygr				s	s	sl s al
p199	*ortho*arsenite	K_3AsO_3	240.23	col need				v s		s al
p200	*meta*arsenite, acid	$KH(AsO_2)_2.H_2O$	271.97					s		sl s al
p201	aurate	$KAuO_2.3H_2O$ (or $2H_2O$)	322.11	lt yel need		d		s	d	s al
p202	azide	KN_3	81.12	col, tetr	2.04	350 (vac)		49.6¹⁷	105.7¹⁰⁰	s al; i eth
p203	benzoate	$KC_7H_5O_2.3H_2O$	214.26	wh cr powd		$-3H_2O$, 110 d		52²⁵	112¹⁰⁰	s al
p204	*di*borane	Diboranidex. $K_2B_2H_6$	105.87	wh, cub cr, 1.493	1.18	subl 400, vac		d		
p205	*penta*borate	Pentaboranidex. $K_2B_5H_9$	141.32	wh powd		d <180		s, d	d	
p206	*di*borane, dihydroxy	$K_2B_2H_6O_2$	137.87	col, cub cr	1.39	$d \rightarrow K$, 400–500		s, d		s al; d a
p207	*meta*borate	KBO_2 (or $K_2B_2O_4$)	81.91	col, hex 1.526, 1.450		950		71³⁰	v s	i al, eth
p208	*penta*borate	$KB_5O_8.4H_2O$	293.21	col, rhomb		780			0.007°	
p209	*peroxy*borate	$KBO_3.\frac{1}{2}H_2O$	106.92	wh cr		$-O_2$, 100	d 150	1.22⁰		i al, eth
p210	*tetra*borate	$K_2B_4O_7.8H_2O$	377.57	col, monocl	1.74 (anhydr)	d		26.7³⁰	v s	
p011	*borohydride*	KBH_4	53.94	wh, cub, 1.494	1.178	d 500		19.3³⁰	v s	0.25 al; 0.56 MeOH; i eth
p213	*boroxalate*	$KHC_2O_4.HBO_2.2H_2O$	207.98			$-H_2O$, 110		v s	v s	i, d al
p214	borotartrate	Solution: cream of tartar. $K_2C_4H_4O_6$	213.99	wh cr powd	1.832			v s		i al, eth, chl
p215	bromate	$KBrO_3$	167.01	col, trig	3.27¹⁷·⁵	434 d 370		13.3⁴⁰	49.75¹⁰⁰	sl s al; i acet
p216	bromide	KBr	119.01	col, cub, sl hygr, 1.559	2.75²⁵	734	1435	53.48⁰	102¹⁰⁰	0.142²⁵ al; sl s eth; s glyc
p217	bromoaurate	$K[AuBr_4]$	555.71	red-brn, rhomb		d 120		sl s		s al
p218	bromoaurate, dihydrate	$K[AuBr_4].2H_2O$	591.74	vlt, monocl cr	4.08			19.5¹⁵	204⁶⁷	s al, KBr; d eth
p219	bromoiodide, di-	$KIBr_2$	325.82	red, rhomb		60	d 180	v s		
p220	*hexa*bromoplatinate	$K_2[PtBr_6]$	752.75	dk red-brn, cub	4.66²⁴	d>400		2.02³⁰	10¹⁰⁰	i al
p221	*tetra*bromoplatinite	$K_2[PtBr_4]$	592.93	br, rhomb				v s	v s	
p222	bromoplatinite, dihydrate	$K_2[PtBr_4].2H_2O$	628.96	blk, rhomb	3.747²⁵	$-H_2O$, vac		v s	v s	
p223	bromostannate	$K_2[SnBr_6]$	676.35	wh cr	3.783					
p224	cacodylate	$K[(CN_3)_2AsO_2].H_2O$	194.10	wh cr				s		sl s al; i eth
p225	cadmium cyanide	$K_2[Cd(CN)_4]$	294.68	col, oct	1.846	450		33.3	100¹⁰⁰	s al
p226	cadmium iodide	$2KI.CdI_2.2H_2O$	734.25	wh-yelsh cr powd, deliq	3.359			137¹⁵		s a, al, eth
p227	calcium chloride	Chlorocalcite. $KCl.CaCl_2$	185.54	col cub, β 1.52		754		s		
p228	calcium magnesium sulfate	Polyhalite. $K_2Ca_2Mg(SO_4)_4.2H_2O$	602.95	wh, trig, 1.548, 1.562, 1.567	2.775					
p228	calcium magnesium sulfate	Krugite. $K_2Ca_4Mg(SO_4)_6.2H_2O$	875.24	gray cr	2.801					
p229	calcium sulfate	Kaluszite, syngenite. $K_2Ca(SO_4)_2.H_2O$	328.42	col, monocl, 1.500, 1.517, 1.518	2.60	1004		0.25	d	i al; s a
p230	*d*-camphorate	$K_2C_{10}H_{14}O_4.5H_2O$	366.50	col, need cluster, hygr		$-5H_2O$, 110		260¹⁴		s al
p231	carbide	KHC_2	64.13	col, rhomb cr	1.37					
p232	carbonate	K_2CO_3	138.21	col, monocl, hygr, 1.531	2.428¹⁹	891	d	112²⁰	156¹⁰⁰	i al, acet
p233	*peroxy*carbonate	$K_2C_2O_6.H_2O$	216.22			200–300		s		
p234	carbonate, dihydrate	$K_2CO_3.2H_2O$	174.24	col, monocl, hygr, 1.380, 1.432, 1.573	2.043	$-H_2O$, 130		146.9	331¹⁰⁰	
p235	carbonate, hydrogen	$KHCO_3$	100.12	col, monocl, 1.482	2.17	d 100–200		22.4	60⁶⁰	i al
p236	carbonate, trihydrate	$2K_2CO_3.3H_2O$	330.47	col, monocl, 1.380, 1.482, 1.573	2.043			129.4	268.3¹⁰⁰	i al, conc HN_4OH
p237	carbonyl	$(KCO)_6$	402.68	gray-red		expl		expl		d al
p238	chlorate	$KClO_3$	122.55	col, monocl, 1.409, 1.517, 1.524	2.32	356	d 400	7.1³⁰	57¹⁰⁰	14.1¹⁰⁰ 50 % al; sl s glyc, liq NH_3; i acet; s alk
p239	*per*chlorate	$KClO_4$	138.55	col, rhomb, 1.4717, 1.4724, 1.476	2.52¹⁹	610 ± 10	d 400	0.75⁰	21.8¹⁰⁰	v sl s al; i eth

No.	Name	Synonyms and Formulae	Mol. wt.	Crystalline form, properties and index of refraction	Density or spec. gravity	Melting point, °C	Boiling point, °C	Solubility, in grams per 100 cc		
								Cold water	Hot water	Other solvents
	Potassium									
p240	chloride..........	Nat. sylvite. KCl.....	74.56	cub, col 1.490....	1.984	770	subl 1500	23.8^{20}	56.7^{100}	sl s al; s eth, glyc, alk
p241	*hypo*chlorite........	KClO........	90.55	in sol only......		d		v s	v s	....
p242	chloroaquoruthenate (III) penta-	$K_2[Ru(H_2O)Cl_5]$......	374.55	rose pr........		$-H_2O$, 200		s	s	sl s al
p243[1]	chloroaurate........	$K[AuCl_4]$.....	377.88	yel, monocl.....	3.75	d 357		61.8^{20}	80.2^{60}	25 al; s a
p243[2]	chloroaurate, dihydrate	$K[AuCl_4].2H_2O$.....	413.91	yel , rhomb pl....				s	s	s al, eth
p244	chlorochromate....	Peligot's salt. $KCrO_3Cl$.	174.55	red, monocl.....	2.497	d		s, d	....	s a
p245	chlorohydroxo- ruthenate	$K_2[Ru(OH)Cl_5]$.......	373.55	brn-red cr........		d		s, d	d	i al
p246	chloroiodate (III) ...	$KICl_4$........	307.82	yel, rhomb.......	1.76^{45}	d		d		d eth
p247	chloroiodide, di-...	$KICl_2$........	236.91	col, monocl.....		60	d 215	d		
p248	chloroiridate.....	$K_3[IrCl_6]$......	483.22	blk, cub.......	3.546	d		125^{19}	6.67	i al, KCl, HN_4OH
p249	chloronitrosyl- ruthenate (III) penta-	$K_2Ru(NO)Cl_5$......	386.55	dk red, rhomb....		d		12^{25}	80^{60}	i al
p250	chloroösmate (III)...	$K_3[OsCl_6].3H_2O$....	574.27	dk red cr......		$-3H_2O$, 150		v s		s a; i eth
p251	chloroösmate (IV)...	K_2OsCl_6.......	481.12	red, cub.......		d		sl s	s	i al; s dil HCl
p252	chloropalladate.....	$K_2[PdCl_6]$.......	397.32	red, cub.......	2.738	d		sl s, d		i al; sl s HCl
p253	chloropalladite.....	K_2PdCl_4......	326.42	red-brn, tetr (yel cub)	2.67	d 105		s	v s	s KCl, HN_4OH; i al
p254	*hexa*chloroplatinate..	$K_2[PtCl_6]$.....	486.01	yel, cub, ∼ 1.825	3.499^{24}	d 250		0.481^2	5.22^{100}	i al, eth
p255	*tetra*chloroplatinite...	$K_2[PtCl_4]$.....	415.11	red-brn tetr, 1.64, 1.67	3.38	d		0.93^{16}	5.3^{100}	i al
p256	chloroplumbate.....	K_2PbCl_6.....	498.11	lt yel, cub.....		d 190		d	....	s h HCl
p257	chlororhenate (IV)..	K_2ReCl_6......	477.12	yel-grn, oct.....	3.34			0.8	d	d alk; sl s HCl
p258	chlororhenate (V)...	K_2ReOCl_5.....	457.67	grn hex pl.....				s	d	s a; i al, eth
p259	chlororhodate, hexa-	$K_3RhCl_6.3H_2O$....	486.98	red, tricl........	3.291	d		d		sl s al, KCl
p260	chlororhodite, penta-	K_2RhCl_5......	358.37	red, rhomb....		d		sl s	d	i al
p261	chlororuthenate (IV)	K_2RuCl_6.....	391.99	blk, cub.......		d		s, d	....	i al
p262	chlorostannate....	K_2SnCl_6......	409.63	col, cub, 1.657..	2.71			s	s	i al
p263	chlorotellurate.....	K_2TeCl_6......	418.52	pale yel, octa- hedral		d		d	d	s HCl
p264	chromate..........	Nat. tarapacaite. K_2CrO_4	194.20	yel, rhomb, β 1.74	2.732^{18}	968.3		62.9^{20}	79.2^{100}	i al
p265	*di*chromate........	$K_2Cr_2O_7$.....	294.19	red, monocl or tricl, 1.738	2.676^{25}_4	tricl→ monocl 241.6 m.p. 398	d 500	4.9^0	102^{100}	i al
p266	*peroxy*chromate....	K_3CrO_8.....	297.30	brn-red, cub....		d 170		sl s		i a, al, eth
p267	chromium sulfate....	Potassium chromium alum. $K[Cr(SO_4)_2].12H_2O$	499.41	vlt-ruby red, cub, oct, 1.4814	1.826^{25}	89	$-10H_2O$, 100 $-12H_2O$, 400	24.39^{25}	50	i al; s dil a
p268	chromium chromate, basic	$K_2CrO_4.2[Cr(OH).CrO_4]$	564.19	vlt brn amorph powd	2.28^{14}	300		i	....	i al, acet a
p269	citrate...........	$K_3C_6H_5O_7.H_2O$.....	324.42	wh cr.......	1.98	d 230		167^{15}	199.7^{31}	sl s al; s glyc
p270	citrate, monobasic...	$KH_2C_6H_5O_7$.....	230.22	wh cr powd.....				s		
p271	cobalt carbonate, hydrogen(ous)	$KHCO_3.CoCO_3.4H_2O$..	291.12	rose need......				d		
p272	cobalt (II) cyanide..	$K_4[Co(CN)_6]$........	371.45	redsh-brn cr, deliq	2.039^{25}_4			v s	v s	d a; i al, eth, $CHCl_3$
p273	cobalt (III) cyanide..	$K_3[Co(CN)_6]$........	332.35	wh-yel, monocl pr	1.878^{25}_4			sl s	sl s	s dil HCl, dil HNO_3; sl s al
p274	cobaltinitrite.......	Fischer's salt. $K_3[Co(NO_2)_6]$	452.27	yel pr, cub......				0.9^{17}	d	i al
p275	cobaltinitrite, hydrate	$K_3[Co(NO_2)_6].H_2O$.....	470.29	yel cr powd....				i	s, d	s min a; sl s ac a; i al, eth
p276	cobaltinitrite, hydrate	$K_3[Co(NO_2)_6].1\frac{1}{2}H_2O$....	479.30	yel, tetr.......		d 200		0.089^{17}	sl s	i al, meth
p277	cobaltmalonate (II)..	$K_2[Co(C_3H_2O_4)_2]$.....	341.23	....	2.234					
p278	cobalt sulfate (II)...	$K_2SO_4.CoSO_4.6H_2O$....	437.36	red pr, monocl, 1.481, 1.487, 1.500	2.218			25.5^0	108.4^{40}	....
p279	copperchloride......	$KCl.CuCl_2$......	209.00	red need........	2.86			s	s	i al
p280	cyanate...........	KOCN........	81.12	col, tetrag.....	2.056^{20}	d 700–900		75^{25}	s	i al
p281	cyanide...........	KCN........	65.12	col, cub, wh gran, deliq, very pois, 1.410	1.52^{16}	634.5		50	100	$0.88^{19.5}$ al; $4.91^{19.5}$ MeOH; s glyc
p282	cyanoargentate (I)..	Potassium argento- cyanide. $K[Ag(CN)_2]$	199.01	cub, 1.625	2.36			25^{20}	100	4.85 % al; i a
p283	cyanoaurate........	$K[Au(CN)_2]$..........	288.10	col, rhomb.......	3.45			14.3	200	sl s al; i eth
p284	cyanoaurate (III) ...	$K[Au(CN)_4].1\frac{1}{2}H_2O$....	367.16	col tabl........		d 200		s	v s	s al

No.	Name	Synonyms and Formulae	Mol. wt.	Crystalline form, properties and index of refraction	Density or spec. gravity	Melting point, °C	Boiling point, °C	Cold water	Hot water	Other solvents
	Potassium									
p285	cyanocadmate	$K_2[Cd(CN)_4]$	294.68	col, cub	1.85			33	100[100]	sl s al
p286	cyanochromate (III)	$K_3[Cr(CN)_6]$	325.41	yel, rhomb	1.71			30.9[20]		i al
p287	cyanocobaltate (II)	$K_4[Co(CN)_6]$	371.42					s	s	i al, eth
p288	cyanocobaltate (III)	$K_3[Co(CN)_6]$	332.32	yel, monocl	1.906	d		s	s	i al; sl s NH_3
p289	cyanocuprate (I)	$K_3[Cu(CN)_4]$	284.92	col, rhbdr		d		v s		
p290	cyanomanganate (II)	$K_4[Mn(CN)_6].3H_2O$	421.50	deep bl, tetr				s	d	
p291	cyanomanganate (III)	$K_3[Mn(CN)_6]$	328.35	red, rhomb, 1.553, 1.555, 1.571 (Li)				s		
p292	cyanomercurate	$K_2[Hg(CN)_4]$	382.87	col, cr pois				s		s al
p293	cyanomolybdate	$K_4[Mo(CN)_8].2H_2O$	496.52	yel, rhomb	2.337_4^{25} (anhyd)	$-H_2O$, 105–110		v s	v s	i eth; 0.0017[20] al
p294	cyanonickelate (II)	$K_2[Ni(CN)_4].H_2O$	259.00	red-yel, monocl cr or powd	1.875[11]	$-H_2O$, 100		s		d a
p295	cyanoosmite	$K_4[Os(CN)_6].6H_2O$	556.76	col, yel, monocl, β 1.607		d		sl s	s	i al, eth
p296	cyanoplatinite	$K_2[Pt(CN)_4].3H_2O$	431.41	col, yel, rhomb, blue fluor, deliq	2.455[16]	$-3H_2O$, 100	d 400–600	sl s	s	sl s al, eth, H_2SO_4
p297	cyanotungstate (IV)	$K_4[W(CN)_8].2H_2O$	584.43	lt yel- grn cr powd	1.989_4^{25} (anhydr)	$-2H_2O$, 115		130[18]	s	i al, eth
p298	ethylsulfate	$KC_2H_5SO_4$	164.23	wh, monocl	1.843			s		s al
p299	ferricyanide	$K_3Fe(CN)_6$	329.26	red, monocl, 1.566, 1.569, 1.583	1.85[25]	d		33[4]	77.5[100]	i al; s acet
p300	ferrocyanide	Yellow prussate of potash. $K_4Fe(CN)_6.3H_2O$	422.41	lem yel, monocl, β 1.577	1.85[17]	$-3H_2O$, 70	d	27.8[12] anhydr 14.5[0]	90.6[96.3] anhydr 74[96]	s acet; i al, eth. NH_3
p301	fluoberyllate	K_2BeF_4	163.21	col, rhomb		red heat		2[20]	5.26[100]	
p302	fluoborate	Nat. avogadrite. KBF_4	125.91	col, rhomb or cub, 1.324, 1.320, 1.325	2.498[10]	d 350	d	0.44[20]	6.27[100]	sl s al, eth; i alk
p303	fluogermanate	K_2GeF_6	264.78	wh, hex		730	ca 835	0.542[18]	2.58[100]	
p304	fluomanganate (IV)	K_2MnF_6	247.13	yel, hex, tab		d	d	d	d	s conc HCl
p305	fluoniobate, penta-	Potassium oxyniobate. $K_2NbOF_5.H_2O$	300.12	col, monocl pl or leaf				7.69	s	s d conc HF
p306	fluorescein deriv	$K_2C_{20}H_{10}O_5$	408.50	yelsh-red powd				s		
p307	fluoride	KF	58.10	col, cub deliq, 1.363	2.48	858	1505	92.3[18]	v s	s HF, NH_3; i al
p308	fluoride, acid	KHF_2	78.11	col, cub, deliq	2.37	d ca 225	d	41[21]	v s	s $KC_2H_3O_2$; i al
p309	fluoride, dihydrate	$KF.2H_2O$	94.13	col, monocl pr, deliq, 1.352	2.454	41	156	349.3[18]	v s	s HF; i al
p310	hexafluorophosphate	KPF_6	184.07			ca 575	d	9.3[25]	20.6[50]	
p311	fluorotungstate	$2KF.WO_2F_2.H_2O$	388.06	monocl		$-H_2O$, red heat		6[17]		
p312	fluosilicate	Nat, hieratite. K_2SiF_6	220.25	col, cub or hex 1.3991	hex 3.08 cub 2.665[17]	d		0.12[17.5] 6.9[19]	0.954[100]	s HCl; i NH_3; v sl s al
p313	fluostannate	$K_2SnF_6.H_2O$	328.90	monocl pr	3.053			3.7[18]	33.3[100]	i al, NH_3
p314	fluosulfonate	$KFSO_3$	138.16	short, thick pr		311		6.9[19]		
p315	fluotantalate	K_2TaF_7	392.14	col, rhomb	4.56; 5.24			sl s, d		sl s HF
p316	fluotellurate, di-	$K_2TeO_2F_2.3H_2O$	345.84	micros oct, monocl		d		sl s	sl s	s HF
p317	fluothorate	$K_2ThF_6.4H_2O$	496.29	col		d		2.15	6.6	d a; i al
p318	fluotitanate	$K_2TiF_6.H_2O$	258.11	col, monocl lust leaf		$-H_2O$, 32 m.p. 780	d	0.556[0]	1.27[21]	sl s min a; i NH_3
p319	fluozirconate	K_2ZrF_6	283.41	col, monocl, 1.466, 1.455	3.48			0.781[2]	25[100]	i NH_3
p320	formate	$KCHO_2$	84.12	col, rhomb deliq	1.91	167.5	d	331[18]	657[80]	s al; i eth
p321	gadolinium sulfate	$K_2SO_4.Gd_2(SO_4)_3.2H_2O$	812.98	cr	3.503[16]			s	s	s K_2SO_4
p322	gallium sulfate	$KGa(SO_4)_2.12H_2O$	517.13	col cr	1.895			s		
p323	digermanate	$K_2Ge_2O_5$	303.38	wh cr	4.31[21.5]	>83		s		s a
p324	metagermanate	K_2GeO_3	198.79	wh cr	3.40[21.5]	823		s		s a
p325	tetragermanate	$K_2Ge_4O_9$	512.55	wh cr	4.12[21.5]	1083		s		s a
p326	glycerophosphate	$K_2C_3H_7PO_6$	248.26	col-sl yelsh mass, hygr				v s	v s	s al
p327	hydride	KH	40.11	wh need 1.453	1.47	d		d	d	i CS_2, eth, bz
p328	hydroxide	KOH	56.11	wh, rhomb deliq	2.044	360.4 ± 0.7	1320–1324	107[15]	178[100]	v s al; i eth, NH_3
p329	(tri-)hydroxyl-ammine trisulfonate	$(KSO_3)_3.NO.1\frac{1}{2}H_2O$	414.52	col, monocl pr		$-H_2O$, 100–200		4[15]	sl d	
p330	hexahydroxyplatinate	$K_2[Pt(OH)_6]$	375.34	yel, rhomb	5.18	d 160		s	s	i al
p331	imidolsulfonate	$(KSO_3)_2NH$	253.34	col, monocl	2.515	d 170–180	d 360–440, vac	1.3[22]	d	i HNO_3
p332	iodate	KIO_3	214.00	col, monocl	3.93_4^{22}	560	d>100	4.74[0]	32.3[100]	s KI; i al, NH_3
p333	iodate, acid	$KIO_3.HIO_3$	389.92	col, monocl				1.33[15]		i al

No.	Name	Synonyms and Formulae	Mol. wt.	Crystalline form, properties and index of refraction	Density or spec. gravity	Melting point, °C	Boiling point, °C	Solubility, in grams per 100 cc		
								Cold water	Hot water	Other solvents
	Potassium									
p334	iodate, acid	$KIO_3.2HIO_3$	565.83	col, tricl				4.15		
p335	*meta*periodate	KIO_4	230.00	col, tetr, 1.6205	3.618_4^{15}	582	$-O, 300$	0.66^{13}	s	v sl s KOH
p336	iodide	KI	166.01	col or wh, cub or gran, 1.677	3.13	681	1330	127.5^0	208^{100}	1.88^{25} al; 1.31^{25} acet; sl s eth; s NH_3
p337	iodide, tri-	$KI_3.\frac{1}{2}H_2O$	428.82	dk bl, monocl, deliq	3.498	31	d 225	v s		s al, KI
p338	iodoaurate	$KAuI_4$	743.69	blk lust cr		d 150		s, d		s dil KI sol
p339	iodoiridite	K_3IrI_6	1070.93	gr cr		d		i	i	i al
p340	iodomercurate (II) tetra-	K_2HgI_4 (or $KI.HgI_2$)	786.41	yel cr, deliq				v s		i al
p341	iodomercurate (II) tri-	Potassium mercury iodide. $KHgI_3$ (or $KI.HgI_2$)	620.47	yel pr, deliq		105		v s		341^{34} al; s KI sol, ac a, eth
p342	iodoplatinate	K_2PtI_6	1034.72	blk, rect	4.96_4^{25}			s	s, d	i al
p343	iridium chloride	Potassium hexachloro iridate. K_2IrCl_6	483.12	redsh-blk, cub	3.549	d		1.12^{20}	s	i al
p344	iridium oxalate	$K_3[Ir(C_2O_4)_3].4H_2O$	645.63	orange, tricl cr	2.510^{19}	$-H_2O, 120$	d 160	i	v s	i al, eth
p345	iron chloride, (III)	Nat. erythrosiderite. $2KCl.FeCl_3.H_2O$	329.33	red, orthorhomb	2.372					
p346	iron chloride (II)	Rinneite. $3KCl.NaCl.FeCl_2$	408.86	rhbdr, 1.589, 1.590	2.3					
p347	iron (III) oxalate	$K_3Fe(C_2O_4)_3.3H_2O$	491.26	emerald grn, monocl, 1.5019, 1.5558, 1.5960	2.133_4^{20}	$-3H_2O, 100$	d 230	4.7^0	118^{100}	i al
p348	iron sulfate (III)	$KFe(SO_4)_2.12H_2O$	599.32	vlt, cub oct, 1.452	1.83	33		$20^{12.5}$	v s	i al
p349	iron sulfate (III)	Nat. krausite. $K_2SO_4.Fe_2(SO_4)_3.24H_2O$	1006.51	pa yel-grn, monocl, 1.482	1.806	28	d 33			
p350	iron sulfate (II)	$K_2SO_4.FeSO_4.6H_2O$	434.27	grn pr, monocl, 1.476, 1.482, 1.497	2.169	d		s	s	
p351	iron sulfide	$KFeS_2$	159.08	purp, hex	2.563			d		
p352	lactate	$KC_3H_5O_3.xH_2O$						s		s al; i eth
p353	laurate	$KC_{12}H_{23}O_2$	238.42	amorph						4.5^{15} al
p354	laurate, acid (mixt)	$KC_{12}H_{23}O_2.C_{12}H_{24}O_2$	438.74	wh, wax like sol		160				$0.904^{13.5}$ al
p355	lead chloride	Nat. pseudocotunnite. $2KCl.PbCl_2$	427.21	yel		490		s		
p356	magnesium carbonate, hydrogen	$KHCO_3.MgCO_3.4H_2O$	256.50	col, tricl or rhomb	2.98			d		
p357	magnesium chloride	Nat. carnallite. $KCl.MgCl_2.6H_2O$	277.86	col, rhomb, deliq, 1.466, 1.475, 1.494	1.61	265		64.5^{19}	d	d al
p358	magnesium chloride sulfate	Nat. kainite. $KCl.MgSO_4.3H_2O$	248.98	col, monocl	2.131			79.56^{18}		i al, eth
p359	magnesium chromate	$K_2CrO_4.MgCrO_4.2H_2O$	370.53	tricl	2.59			d		
p360	magnesium phosphate, hexahydrate	$KMgPO_4.6H_2O$	266.48	wh, rhomb cr		$-5H_2O, 110$		d		
p361	magnesium selenate	$K_2SeO_4.MgSeO_4.6H_2O$	496.52	col, monocl, 1.497, 1.499, 1.514	2.3645_4^{20}	$-2H_2O, 33$		s	s	
p362	magnesium sulfate	Nat. langbeinite. $K_2SO_4.2MgSO_4$	415.01	tetrah	2.829	927				
p363	magnesium sulfate	Nat. leonite. $K_2SO_4.MgSO_4.4H_2O$	366.70	col, monocl, 1.483, 1.487, 1.490	2.201^{20}			v s		
p364	magnesium sulfate	$K_2SO_4.MgSO_4.6H_2O$	402.73	col, monocl, 1.461, 1.463, 1.476	2.15	d 72		19.26^0 25^{20}	59.8^{75}	
p365	malate	$K_2C_4H_4O_5$	210.28	col, viscid mass				s		
p366	manganate	K_2MnO_4	197.14	grn, rhomb		d 190		d	d	s KOH
p367	*per*manganate	$KMnO_4$	158.04	purple rhomb, 1.59	2.703	d < 240		6.38^{20}	25^{65}	d al; v s MeOH, acet; s H_2SO_4
p368	manganese chloride(ous)	Chloromanganokalite. $4KCl.MnCl_2$	424.06	trig. 1.50	2.31			s	s	
p369	manganese sulfate(ic)	$KMn(SO_4)_2.12H_2O$	502.35	vlt, cub, (oct)				d		
p370	manganese sulfate(ous)	Manganolongbeinite. $K_2SO_4.2MnSO_4$	476.20	rose, tetrah. 1.572	3.02	850				
p371	mercury tartrate(ous)	$KHgC_4O_6$	387.76	wh cr powd				i		i al
p372	methionate	Potassium methane disulfonate. $K_2CH_2(SO_3)_2$	252.36	monocl, β 1.539	2.376			s		
p373	methylsulfate	$2KCH_3SO_4.H_2O$	318.41	wh cr				s		s al

PHYSICAL CONSTANTS OF INORGANIC COMPOUNDS (Continued)

No.	Name	Synonyms and Formulae	Mol. wt.	Crystalline form, properties and index of refraction	Density or spec. gravity	Melting point, °C	Boiling point, °C	Cold water	Hot water	Other solvents
	Potassium									
p374	molybdate	K_2MoO_4	238.14	wh powd or 4-sid pr, deliq	2.91^{18}	919	d 1400	184.6^{25}	v s	i al
p375	permolybdate	$K_2O.3MoO_4.3H_2O$	596.06	lt yel cr, monocl		d 180		sl s	s	v sl s al
p376	trimolybdate	$K_2O.3MoO_3.3H_2O$	580.06	wh need		571 (anhydr)		0.22^{15}	s	
p377	molybdenum cyanate	$K_4[Mo(CN)_8].2H_2O$	496.52	yel, rhomb	2.337^{25}_4 (anhydr)	$-2H_2O$, 105–110		v s	v s	0.0017²⁰ abs al; i eth; d HCl, H_2SO_4
p378	myristate, acid (mixt)	$KC_{14}H_{27}O_2.C_{14}H_{28}O_2$	494.85	wh, waxlike sol		153				$0.453^{13.5}$ al
p379	naphthalene–1,5-disulfonate	$K_2C_{10}H_6(SO_3)_2.2H_2O$	400.53	monocl, 1.485, 1.669, 1.697	1.797			s		
p380	nickelsulfate	$K_2SO_4.NiSO_4.6H_2O$	437.13	bl, monocl, 1.484, 1.492, 1.505	2.124	d < 100		7^0	60.8^{75}	
p381	nitrate	Saltpeter. KNO_3	101.11	col, rhomb or trig, 1.335, 1.5056, 1.5064	2.109^{16}	tr-trig 129 m.p. 334	d 400	13.3^0	247^{100}	i dil al, eth; s liq NH_3, glyc
p382	nitride	K_2N	131.31	grnsh-blk		d		d		
p383	nitrite	KNO_2	85.11	wh-yelsh pr, deliq	1.915	440	d	281^0	413^{100}	s hot al; v s liq NH_3
p384	m-nitrophenoxide	$KOC_6H_4NO_2.2H_2O$	213.24	flat or need	1.691^{20}	$-H_2O$, 130	d	16.3^{15}		s al
p385	p-nitrophenoxide	$KOC_6H_4NO_2.2H_2O$	213.24	yel leaf	1.652^{20}	$-2H_2O$, 130	d	7.5^{15}		sl s al
p386	nitroplatinite	$K_2Pt(NO_2)_4$	457.32	col, monocl		d		3.8^{15}	s	
p387	nitroprusside	$K_2[Fe(NO)(CN)_5].2H_2O$	330.18	red, monocl, hygr				100^{16}		s al
p388	nitrososulfate	$K_2SO_3(NO)_2$	218.28	col need		d 127, expl		$12^{14.5}$, d		i al
p389	oleate	$KC_{18}H_{33}O_2$	320.56	yelsh or brnsh soft mass or cr, 1.452				25	s	4.315¹³·⁵ al; 100⁵⁰ al; 3.5⁵⁵ eth
p390	oleate, acid (mixt)	$KC_{18}H_{33}O_2.C_{18}H_{34}O_2$	603.03	wh, wax-like solid		95		s	s	$5.2^{13.5}$ al
p391	osmate	$K_2OsO_4.2H_2O$	368.43	vlt, cub, hygr		$-H_2O$, >100		sl s	s, d	i al, eth
p392	osmiumchloride	K_2OsCl_6	481.17	blk oct	3.42^{17}	d 000		s	s	i al; s HCl
p393	osmylchloride	$K_2OsO_2Cl_4$	442.21	red, tetr	3.42	d 200 (in H atm)		s	s	
p394	osmyloxalate	$K_2[OsO_2(C_2O_4)_2].2H_2O$	512.47	brn need, tricl		$-H_2O$, 80	d 180	0.75^{15}	3.0^{15}	
p395	oxalate	$K_2C_2O_4.H_2O$	184.24	wh, monocl, 1.440, 1.485 1.550	$2.127^{1.9}$	$-H_2O$, 100			33^{16}	
p396	oxalate, hydrogen	KHC_2O_4	128.11	col, monocl, 1.382, 1.553, 1.573	2.044	d		2.5	16.7^{100}	i al, eth
p396	oxalate, hydrogen. monohydrate	$KHC_2O_4.H_2O$	146.13	rhomb	$2.044^{1.9}$					
p398	oxalate, tetra-	$KHC_2O_4.H_2C_2O_4.2H_2O$	254.20	col, tricl, 1.415, 1.536, 1.560	1.836	d		1.8^{13}		d al
p399	oxaloferrate (II)	$K_2[Fe(C_2O_4)_2].2H_2O$	346.12	gold need		d		s	s	
p400	oxaloferrate (II)	$K[Fe(C_2O_4)_2].2\frac{1}{2}H_2O$	316.03	br cr		d		92^{21}	d	i al
p401	oxaloferrate (III)	$K_3[Fe(C_2O_4)_3].3H_2O$	491.25	grn, monocl		$-3H_2O$, 100	d 230	4.7^0	117.7^{100}	i al, NH_3; s acet
p402	oxalatoplatinate	$K_2[Pt(C_2O_4)_2].2H_2O$	485.36	col, monocl pr	3.04^{12}	$-H_2O$, 100		sl s	s	
p403	oxalatouranate (IV)	$K_4[U(C_2O_4)_4].5H_2O$	836.59	yel, monocl	2.563					
p404	oxide, mon-	K_2O	94.20	col, cub, hygr	2.32^0	d 350		v s	v s	s al, eth
p405	peroxide	K_2O_2	110.20	wh, amorph, deliq		490	d			
p406	oxide, super-	KO_2	71.10	yel, cub leaf	2.14	380	d	v s, d		d al
p407	oxide, tri- (sesqui-)	K_2O_3	126.20	red		430		ev O_2		d dil H_2SO_4
p408	palladium chloride	$K_2(PdCl_4)$	326.42	yel-grnsh-br cr, tetr, 1.710, 1.523	2.67	524		s	s	sl s⁸⁰ al
p409	palladium chloride	$K_2(PdCl_6)$	397.32	lt red, oct	2.738	d 170		v sl s	sl s	s HCl; i al
p410	palladium oxalate	$K_2[Pd(C_2O_4)_2].4H_2O$	432.71	yel need		dec in air, $-4H_2O$, 80		0.833^{27}	$9.98^{82.9}$	
p411	palmitate, acid (mixt)	$KC_{16}H_{31}O_2.C_{16}H_{32}O_2$	550.96	wh, fatty sol		138				0.198^{13} al
p412	1-phenol-2-sulfonate(o-)	$KC_6H_4(OH)SO_3.H_2O$	230.29	rhomb, 1.527, 1.568, 1.647	1.87	400		s		s al
p413	1-phenol-4-sulfonate(p-)	$KC_6H_4(OH)SO_3$	212.27	rhomb, 1.571, 1.608, 1.694	1.87	>260				
p414	phenylsulfate	$KC_6H_5SO_4$	212.27	phenyl leaf		d 150–160	d	14^{15}		v sl s al
p415	metaphosphate, hexa-	$(KPO_3)_6$	708.44	col mass, hygr	2.107	810	1320	s	s	i al
p416	metaphosphate, tetra-	$(KPO_3)_4.2H_2O$	508.33	col cr		$-2H_2O$, 100		100^{15}	s	
p417	orthophosphate	K_3PO_4	212.28	col, rhomb, deliq	2.564^{17}	1340		90^{20}		i al
p418	orthophosphate, di-H	KH_2PO_4	136.09	col, tetr, deliq, 1.510, 1.4864	2.338	252.6		33^{25}	83.5^{90}	i al
p419	orthophosphate, mono-H	K_2HPO_4	174.18	wh, amorph, deliq		d		167^{20}	v s	v s al
p420	pyrophosphate	$K_4P_2O_7.3H_2O$	384.40	col, deliq	2.33	$-2H_2O$, 180	$-3H_2O$, 300	s	v s	i al
p421	subphosphate	$K_2PO_3.4H_2O$	229.24	col, rhomb		40	$-4H_2O$, 150	v s	v s	

PHYSICAL CONSTANTS OF INORGANIC COMPOUNDS (Continued)

No.	Name	Synonyms and Formulae	Mol. wt.	Crystalline form, properties and index of refraction	Density or spec. gravity	Melting point, °C	Boiling point, °C	Cold water	Hot water	Other solvents
	Potassium									
p422	*hypo*phosphite	KH_2PO_2	104.09	wh, hex, deliq		d		200^{25}	330	v sl s abs al, NH_3; i eth; 11.1^{25} chl
p423	*ortho*phosphite, di-H	KH_2PO_3	120.09	wh cr, deliq		d		220^{20}	v s	i al
p424	phthalate, hydrogen	$KHC_8H_4O_4$	204.23	col, rhomb	1.636			10^{25}	33^{100}	
p425	picrate	$KC_6H_2N_3O_7$	267.20	yel-redsh or grnsh, rhomb, 1.527, 1.903, 1.952	1.852	expl 310		0.5^{15}	25^{100}	0.184^{25} al
p426	piperate	$KC_{12}H_9O_4$	256.31	lt yel cr powd				sl s	v s	
p427	platinate, hydroxo-	$K_2Pt(OH)_6$	375.34	yel, rhomb		d 160		s		i al
p428	platinorhodanide	$K_2[Pt(CNS)_6].2H_2O$	657.82	red, rhomb				s	8^{60}	s h al
p429	platinum iodide, hexa-	K_2PtI_6	1034.72	blk, cub	4.963^{25}_4			s		sl s al
p430	plumbate, hydroxo-	$K_2Pb(OH)_6$	387.44	col, rhomb		d	d			s KOH
p431	praseodymium sulfate	$3K_2SO_4.Pr_2(SO_4)_3.2H_2O$	1110.81	cr	3.275^{16}			sl s		s HCl, HNO_3
p432	propionate	$KC_3H_5O_2.H_2O$	130.19	wh cr, hygr, or wh leaf, deliq		$-H_2O$, 120		207^{16}	359	22.2^{13} 95 % al
p433	propyl sulfate	$KC_3H_7SO_4$	178.25	wh cr powd				v s		
p434	*per*rhenate	$KReO_4$	289.30	wh, tetr, 1.643	4.887	550	1360–1370	1.21^{20}	14.0^{100}	v sl s al
p435	rhenium (IV) chloride	$K_2[ReCl_6]$	477.12	yel-grn cr, oct	3.34	d		s		s a; i conc H_2SO_4; d h H_2SO_4
p437	rhenium (V) oxychloride	$K_2[ReOCl_5]$	457.67	yel-grn cr, rhomb or monocl, 1.52		d		s		sl s HCl; s H_2SO_4; i al, eth
p438	rhenium oxycyanide	$K_3[ReO_2(CN)_4]$	439.58	red cr, monocl	2.70^{25}_4	d 300–400, vac		s	s	v sl s al; i alk
p439	rhodium cyanide	$K_3[Rh(CN)_6]$	376.32	pa yel, monocl, 1.5498, 1.5513, 1.5634				v s	v s	
p440	rhodium oxalate	$K_3[Rh(C_2O_4)_3].4\frac{1}{2}H_2O$	565.34	col, tricl	2.171^{20}_4	$-4\frac{1}{2}H_2O$, 190		v s	v s	
p441	rhodium sulfate	$KRh(SO_4)_2.12H_2O$	646.38	yel, cub	2.23			s		
p442	ruthenate	$K_2RuO_4.H_2O$	261.29	blk, tetr		$-H_2O$, 200	d 400 vac	v s	d	d a, al
p443	*per*ruthenate	$KRuO_4$	204.17	blk, tetr		d 44		sl s	s, d	
p444	*d*-saccharate, acid	$KHC_6H_9O_8$	248.24	rhomb need				1.1^6	s	
p445	salicylate	$KC_7H_5O_3$	176.22	wh powd				s		s al
p446	santoninate	$KC_{15}H_{19}O_4$	302.42	wh cr powd, deliq				s		s al
p447	selenate	K_2SeO_4	221.16	col, rhomb, hygr, 1.535, 1.539, 1.545	3.066			110.5^0	122.2^{100}	
p448	selenide	K_2Se	157.16	wh cub, reddens in air, hygr	2.851^{15}			s d	s	
p449	selenite	K_2SeO_3	205.16	wh, deliq		d 875		s		sl s al
p450	selenocyanate	$KSeCN$	144.08	need, deliq	2.347	d 100		s		d a; s al
p451	selenocyanoplatinate	$K_2Pt(SeCN)_6$	903.16	rhomb	$3.378^{12.5}$	d 80		s		
p452	selenothionate	$K_2SeS_2O_6$	317.29	col, monocl pr		d 250		s, d		
p453	*di*silicate	$K_2Si_2O_5$	214.37	col, rhomb, 1.502, 1.513	2.456^{25}_4	1015 ± 10		s	s	i al
p454	*meta*silicate	K_2SiO_3	154.29	col, rhomb (?), 1.520, 1.528		976		s	s	i al
p455	*tetra*silicate	$K_2Si_4O_9.H_2O$	352.56	wh, rhomb, α 1.495, β 1.535	2.417	d 400		s	s	i al
p456	*di*silicate, hydrogen	$KHSi_2O_5$	176.28	wh, rhomb, 1.480, 1.530	2.417^{15}_4	515				d HCl
p457	silicotungstate	$K_4SiW_{12}O_{40}.18H_2O$	3354.95	col, hex		$-17H_2O$, 100		33.3^{20}	v s	v s acet; s MeOH; sl s al; eth, bz
p458	silver carbonate	$KAgCO_3$	206.98	rect pl	3.769	d	d	d	d	
p459	silver nitrate	$KNO_3.AgNO_3$	270.98	monocl	3.219	125		v s	v s	
p460	sodium antimony tartrate	$KNaSbC_4H_2O_7$	346.97	wh, scales or powd				s		
p461	sodium carbonate	$KNaCO_3.6H_2O$	230.19	monocl, hygr, effl	1.61–1.63^{14}	$-6H_2O$, 100		185.2^{15}		
p462	sodium ferricyanide	$K_2Na[Fe(CN)_6]$	313.15	or-red, monocl		d		50^{25}	80^{60}	
p463	sodium nitrocobaltate (III)	$K_2Na[Co(NO_2)_6].H_2O$	454.18	yel cr	1.633	135		0.07^{25}		i al
p464	sodium sulfate	$3K_2SO_4.Na_2SO_4$	664.84	col, rhbdr	2.7			s	s	
p465	sodium tartrate	Rochelle salt, seignette salt. $KNaC_4H_4O_6.4H_2O$	282.23	col, rhomb, 1.492, 1.493, 1.496	1.790	70–80	$-4H_2O$, 215	26^0	66^{26}	v sl s al
p466	sorbate	$KC_6H_7O_2$	150.22	col cr	1.363^{25}_{20}	d 270		58.5^{25}		sl s MeOH

No.	Name	Synonyms and Formulae	Mol. wt.	Crystalline form, properties and index of refraction	Density or spec. gravity	Melting point, °C	Boiling point, °C	Solubility, in grams per 100 cc		
								Cold water	Hot water	Other solvents
	Potassium									
p467	stannate, hydroxo-	$K_2Sn(OH)_6$	298.94	col, trig	3.197			85^{10}	110.5^{20}	sl s KOH; i al, acet
p468	stearate	$KC_{18}H_{35}O_2$	322.58	wh cr powd					s	$0.145^{13.5}$ al; i eth, chl, CS_2
p469	stearate, acid (mixt)	$KC_{18}H_{35}O_2.C_{18}H_{36}O_2$	607.07	wh powd		153		s	s	$0.091^{13.5}$ al
p470	strontium chromium oxalate(ic)	$KSrCr(C_2O_4)_3.6H_2O$	550.87	grnsh blk cr	2.155^{13}					
p471	styphnate	$KC_6H_2N_3O_8.H_2O$	301.22	yel, monocl pr		$-H_2O, 120$ expl		1.54^{20}		v sl s al
p472	succinate	$K_2C_4H_4O_4.3H_2O$	248.32	rhomb, hygr	1.564			s		
p473	succinate, hydrogen	$KHC_4H_4O_4$	156.18	monocl	1.767	d 242				
p474	succinate, hydrogen, dihydrate	$KHC_4H_4O_4.2H_2O$	192.21	rhomb, 1.417, 1.530, 1.533	1.616			s		s al
p475	succinate, hydrogen	$KHC_4H_4O_4.C_4H_6O_4$	274.27	monocl	1.56	162				
p476	sulfate	Nat. arcanite. K_2SO_4	174.27	col, rhomb or hex, 1.494, 1.495, 1.497	2.662	tr 588 m. p. 1069	1689	12^{25}	24.1^{100}	i al, acet, CS_2
p477	peroxydisulfate	$K_2S_2O_8$	270.33	col, tricl, 1.461, 1.467, 1.566	2.477	d<100		1.75^0	5.3^{20}	i al
p478	pyrosulfate	$K_2S_2O_7$	254.33	col, need	2.512^{25}_4	>300	d	s	d	
p479	sulfate, hydrogen	Nat. mercallite, misenite. $KHSO_4$	136.17	col, rhomb, deliq 1.480	2.322	214	d	36.3^0	121.6^{100}	i al, acet
p480	sulfide, di-	K_2S_2	142.33	red-yel cr		470		s	d	s al
p481	sulfide, hydro-	KHS	72.17	yel, rhomb deliq	1.68-1.70	455		d	d	s al
p482	sulfide, mono-	K_2S	110.27	yel-br, cub, deliq	1.805^{14}	840		s	v s	s al, glyc; i eth
p483	sulfide, mono-, pentahydrate	$K_2S.5H_2O$	200.34	col, rhomb		60	$-3H_2O, 150$	s		s al, glyc; i eth
p484	sulfide, penta-	K_2S_5	238.52	or cr, hygr		206	d 300	v s	v s	sl s al
p485	sulfide, tetra-	K_2S_4	206.46	red-brn cr		145	d 850	s		s al
p486	sulfide, tetra-, dihydrate	$K_2S_4.2H_2O$	242.49	yel				s	s	sl s al
p487	sulfide, tri-	K_2S_3	174.40	br-yel cr		252		s	d	s al
p488	sulfide, di-, trihydrate	$K_2S_2.3H_2O$	196.38	yel				v s	v s	s al
p489	sulfite	$K_2SO_3.2H_2O$	194.30	wh-yelsh, hex		d		100	<100	sl s al, l NH₃, d dil a
p490	pyrosulfite	Potassium metabisulphite. $K_2S_2O_5$	222.33	col, monocl pl	2.34	d 190		sl s		sl s al; i eth
p491	sulfite, hydrogen	$KHSO_3$	120.17	col cr		d 190		s	s	i al
p492	d-tartrate	$K_2C_4H_4O_6.\frac{1}{2}H_2O$	235.28	col, monocl, β 1.526	1.98^{20}_4	$-H_2O, 155$	d 200-220	150^{14}	278^{100}	sl s al
p493	dl-tartrate	$K_2C_4H_4O_6.2H_2O$	262.31	col, monocl	1.984	$-2H_2O, 100$		100^{25}		
p494	d-tartrate, hydrogen	$KHC_4H_4O_6$	188.18	col, rhomb, 1.511, 1.550, 1.590	1.984^{18}			0.37	6.1^{100}	s a, alk; i al, ac a
p495	dl-tartrate, hydrogen	$KHC_4H_4O_6$	188.18	col, monocl	1.954			0.42^{15}	7.0^{100}	i al; s min a
p496	metatellurate	K_2TeO_4	269.80	soft glutinous mass		d 200		d		
p497	orthotellurate	$K_2H_4TeO_6.3H_2O$	359.88	col, rhomb, deliq		$-H_2O$	$-O, 300$	sl s	s	i al; sl s KOH
p498	telluride	K_2Te	205.80	col, cub, hygr	2.51			s, d	s	
p499	tellurite	K_2TeO_3	253.80	wh cr, deliq		d 460-470		v s	v s	s h K_2CO_3, KOH
p500	tellurium chloride	K_2TeCl_6	418.52	yel, oct, hygr	2.645			s d,	s, d al; s dil HCl
p501	thioantimonate	$2K_3SbS_4.4\frac{1}{2}H_2O$	815.68	yel cr, deliq				300^0	400^{80}	i al
p502	thioarsenate	K_3AsS_4	320.48	wh cr, deliq			d	v s		i al
p503	thioarsenite	K_3AsS_3	288.42			d		s		i al
p504	thiocarbonate, tri-	K_2CS_3	186.41	yel-red-brn cr, deliq		d		v s	s	s NH₃; sl s al; i eth
p505	thiocyanate	$KNCS$	97.18	col, rhomb pr, deliq	1.886^{14}	173.2	d 500	177.2^0	217^{20}	s al; 20.75^{22} acet; 0.18^{12} amyl al; v s liq NH₃
p506	dithionate	$K_2S_2O_6$	238.32	col, trig, 1.455, 1.515	2.278	d		6	66^{100}	i al
p507	pentathionate	$K_2S_5O_6.1\frac{1}{2}H_2O$	361.55	col, rhomb, —1.63	2.112	d		s	d	i al
p508	tetrathionate	$K_2S_4O_6$	302.46	col, monocl, 1.6057	2.296	d		v s		i al
p509	trithionate	$K_2S_3O_6$	270.39	col, rhomb, 1.475, 1.480, 1.487	2.304	d 30-40		s	d	i al
p510	thioplatinate	$K_2Pt_2S_3$	1050.95	bl-gray cr	6.44^{15}	d, ign		i		d HCl
p511	thiosulfate, penta-hydrate	$3K_2S_2O_3.5H_2O$	661.07	col, rhomb		d		$150.2^{17.2}$		
p512	metathiostannate	$K_2SnS_3.3H_2O$	347.13	dk brn oil	1.847^{18}	$-3H_2O, 100$		s		i al
p513	thiosulfate	$K_2S_2O_3\frac{1}{2}.H_2O$	196.34	col, monocl, deliq	2.590 anhydr 2.23	$-H_2O, 200$	d	96.1^0	312^{90}	i al
p514	tungstate	$K_2WO_4.2H_2O$	362.08	col, monocl deliq	3.113	tr 388, 921		51.5	151.5	d a; i al
p515	metatungstate	$K_6[H_2W_{12}O_{40}].18H_2O$	3407.08	hex		ca 930		s	v s	d a
p516	metaauranate	K_2UO_4	380.23	or-yel, rhomb				i		v s a
p517	uranyl acetate	$KUO_2(C_2H_3O_2)_3.H_2O$	504.28	tetr	3.296^{15} ($\frac{1}{2}H_2O$)	$-H_2O, 275$		s		
p518	uranyl carbonate	$2K_2CO_3.UO_2CO_3$	606.46	yel, hex		$-CO_2$		7.4^{15}	d	s K_2CO_3 sol; i a

No.	Name	Synonyms and Formulae	Mol. wt.	Crystalline form, properties and index of refraction	Density or spec. gravity	Melting point, °C	Boiling point, °C	Solubility, in grams per 100 cc		
								Cold water	Hot water	Other solvents
	Potassium									
p519	uranyl sulfate......	$K_2SO_4.UO_2SO_4.2H_2O$...	576.39	yel, monocl	$3.363^{19.1}$	$-2H_2O$, 120	....	s		
p520	urate, acid.........	$KHC_5H_2N_4O_3$...	206.21	wh powd				sl s		
p521	*meta*vanadate....	KVO_3	138.04	col cr				sl s	s	sl s KOH; i al
p522	vanadium sulfate....	Potassium vanadium alum. $KV(SO_4)_2.12H_2O$	498.35	violet cubic	1.783^{20}_4	20	$-12H_2O$, 230	1984^{10}		
p523	*ethyl*xanthate......	KC_2H_5OCSS	160.30	wh to pa yel cr or cr powd	$1.558^{21.5}$	d>200		v s	d	s al; i eth
p524	**Praseodymium** (α form)	Pr..............	140.907	pa yel, met, hex up to 798	**6.773**	**931**	**3512**	d		s a
p525	(β form)	Pr..............	140.907	cub	6.64	935	3127	d		s a
p526	acetate......	$Pr(C_2H_3O_2)_3.3H_2O$...	372.09	grn need				v s		
p527	acetylacetonate.....	$Pr(C_5H_7O_2)_3$	438.24	cr ppt		146				s CS_2
p528	bromate......	$Pr(BrO_3)_3.9H_2O$...	686.77	grn, hex		56.5	$-7H_2O$, 170	196^{25}		
p529	bromide......	$PrBr_3$	380.63	grn cr powd		691	1547	d, sl s		
p530	carbide.......	PrC_2	164.93	yel cr	5.10	d		d	d	s dil a
p531	carbonate........	$Pr_2(CO_3)_3.8H_2O$...	605.96	grn silky pl		$-6H_2O$, 100		i		s a
p532	chloride.......	$PrCl_3$	247.27	bl grn need	4.02^{25}	786	1700	103.9^{13}	∞ 100	v s al; 2.4 pyr; i chl, eth
p533	chloride, hepta-hydrate	$PrCl_3.7H_2O$	373.37	grn, tricl	2.25^{17}	115		334^{13}	∞ 100	s al, HCl
p534	hexantipyrine *perchlorate*	$[Pr(C_{11}H_{12}N_2O)_6].(ClO_4)_3$	1568.65	grn, hex leaf		d 286–291				
p535	iodide.........	PrI_3	521.62	gr cr, hygr		737		v s		
p536	molybdate......	$Pr_2(MoO_4)_3$	761.63	grass-green tetr	4.84	1030				
p537	oxalate.......	$Pr_2(C_2O_4)_3.10H_2O$	726.03	lt grn cr				i		s a
p538	oxide, di-.....	PrO_2	172.91	br-bl powd	6.82	>350 tr to Pr_6O_{10}				
p539	oxide, sesqui-...	Preseodymia. Pr_2O_3	329.81	yel-grn, amorph	7.07	d		0.0000020^{29}		s a
p540	selenate..........	$Pr_2(SeO_4)_3$	710.69		4.30^{15}			36^0	3^{92}	
p541	sulfate........	$Pr_2(SO_4)_3$	570.00	lt grn powd	3.72^{16}			23.7^0 17.7^{20}	1.02^{95}	
p542	sulfate, octahydrate	$Pr_2(SO_4)_3.8H_2O$...	714.12	grn, monocl, 1.540, 1.549, 1.561	$2.827^{18.1}$			17.4^{20}	sl s	
p543	sulfate, pentahydrate	$Pr_2(SO_4)_3.5H_2O$	660.08	monocl pr	3.176^{16}				1.85^{85}	
p544	sulfide........	Pr_2S_3	378.01	br powd	5.042^{11}	d		i	d	s dil a
p545	**Protactinium**........	Pa	231.10	gray met, tetrag	15.37	**<1600**				
p546	chloride........	$PaCl_4$	372.91	yel-grn, tetrag		subl 400 in vac		s		s dil HCl
p547	fluoride........	PaF_4	307.09	monocl				i		
p548	oxide, di-.....	PaO_2	263.10	blk, cub						
p549	oxide, pent-.......	Pa_2O_5	542.20	wh, cub						s dil HF
r1	**Radium**...........	Ra..............	**226.0254**	silver wh met	5(?)	700	**<1140**	d, ev H_2		d a
r2	bromide......	$RaBr_2$	385.82	col-yelsh, monocl	5.79	728	subl 900	s	s	s al
r3	bromide dihydrate	$RaBr_2.2H_2O$	421.85	wh, monocl		$-2H_2O$, 100		s	s	
r4	carbonate.........	$RaCO_3$	286.01	wh, or sl brnsh, monocl				i		d a
r5	chloride.........	$RaCl_2$	296.91	col-yelsh, monocl	4.91	1000		s	s	s al
r6	chloride, dihydrate	$RaCl_2.2H_2O$	332.94	wh, monocl, discol		$-2H_2O$, 100		s	s	s HCl
r7	iodate........	$Ra(IO_3)_2$	575.81					0.0175^0	0.170^{100}	
r8	nitrate........	$Ra(NO_3)_2$	350.01	cr				13.9^{20}		
r9	sulfate........	$RaSO_4$	322.06	col, rhomb				0.00000002^{29}	0.000005^{44}	i a
r10	**Radon**..........	Niton, Radium emanation. Rn	222.00	col gas, opaque cr	gas 9.73 g/l liq 4.4^{-62} sol 4.0	-71	-61.8	51.0^0 cm^3 22.4^{25} cm^3	13.0^{60} cm^3	sl s al, org liqu
r11	**Rhenium**...........	Re	186.2	met lust, hex	20.53	3180	**5627** (est)	i	i	s dil HNO_3, H_2O_2; sl s H_2SO_4; i HCl
r12	bromide........	$ReBr_3$	425.93	grn-blk cr		subl 500, vac				s dil H_2SO_4, HBr, liq NH_3
r13	carbonyl, penta-.....	$[Re(CO)_5]_2$	652.51	col, cub cr		d 250				v sl s org solv
r14	chloride, penta-.....	$ReCl_5$	363.47	dk grn to blk	4.9	d	d	d	d	s HCl, alk
r15	chloride, tetra-.....	$ReCl_4$	328.01	blk (exist. ?)			500	s, d	s, d	s HCl
r16	chloride, tri-......	$ReCl_3$	292.56	dk red, hex		>550		s	s	s a, alk, liq NH_3 al; sl s eth
r17	fluoride........	ReF_4	262.19	dk grn	5.383^{26}	124.5	d 500	d		s a
r18	fluoride, hexa-.....	ReF_6	300.19	pa yel, v hygr	liq 6.1573, sld $3.616^{18.5}$	18.8	47.6	s, d	s, d	d HNO_3, H_2SO_4
r19	iodide pentacarbonyl	$ReI.5CO$	453.16	yel, rhombd		200	d 400, subl vac 90	i	....	s bz
r20	oxide, di-.....	ReO_2	218.20	blk	11.4^{25}_4	d 1000		i	i	s conc HCl, H_2O_2
r21	oxide, hept-........	Re_2O_7	484.40	yel pl or hex or powd, hygr	6.103	ca 297	subl 250	v s	v s	v s al; s alk, a
r22	oxide, per-.........	$Re_2O_5(?)$	500.40	wh	8.4	145	....	v s	v s	s alk; sl s eth

No.	Name	Synonyms and Formulae	Mol. wt.	Crystalline form, properties and index of refraction	Density or spec. gravity	Melting point, °C	Boiling point, °C	Cold water	Hot water	Other solvents
	Rhenium									
r23	oxide, sesqui-	Dirhenium trioxide $Re_2O_3.xH_2O$		unstable				d, ev H_2		
r24	oxide, tri-	ReO_3	234.20	red, blue, cub	6.9–7.4		d 400	i	i	s H_2O_2, HNO_3
r25	oxybromide, tri-	ReO_3Br	314.11	wh		39.5	163			
r26	oxychloride, tri-	ReO_3Cl	269.65	col liq	3.867_4^{20}	4.5	131^{760}	d	d	s CCl_4
r27	oxytetrachloride	$ReOCl_4$	344.01	or need		29.3	223.00	d	d	
r28	oxytetrafluoride	$ReOF_4$	278.19	wh	lq 3.717 sol 4.032	39.7	62.7			
r29	oxydifluoride, di-	ReO_2F_2	256.20	col		156				
r30	sulfide, di-	ReS_2	250.33	blk tr leaf	7.506_4^{20}		d	i	i	s HNO_3; i al, alk, HCl
r31	sulfide, hepta-	Re_2S_7	596.85	blk powd	4.866		d	i	i	s HNO_3, H_2O_2; alk; i HCl
r32	**Rhodium**	Rh	102.905	gray-wh, cub	12.4	1966 ± 3	3727 ± 100	i	i	s H_2SO_4+HCl, conc H_2SO_4; sl s a, aq reg
r33	amminechloride, hexa-	$[Rh(NH_3)_6]Cl_3$	311.45	rhomb pl	2.008_4^{25}	$-NH_3$ 210, d		12.5^8	s	
r34	carbonylchloride, basic	$RhCl_3.RhO.3CO$	376.75	ruby red need		subl 125.5		sl s	d	s CCl_4, ac a, bz
r35	chloride, tri-,	$RhCl_3$	209.26	br-red powd, deliq		d 450–500	subl 800	i	i	i a, aq reg
r36	chloride, tri-	$RhCl_3.xH_2O$		dk red, deliq		d 100		v s		s al, HCl; i eth
r37	fluoride, tri-	RhF_3	159.90	red, rhomb	5.38	>600 subl		i	i	i a, alk
r38	iodide, tri-	RhI_3	483.62	blk						
r39	nitrate	$Rh(NO_3)_3.2H_2O$	324.93	red, deliq				s	s	i al
r40	oxide, di-	RhO_2	134.90	br				i	i	i a, alk
r41	oxide, di-, dihydrate	$RhO_2.2H_2O$	170.93	olive grn		d		i		s HCl, acet a, alk
r42	oxide, sesqui-	Rh_2O_3	253.81	gray cr or amorph	8.20	d 1100–1150		i	i	i a, aq reg, KOH
r43	oxide, sesqui-, pentahydrate	$Rh_2O_3.5H_2O$	343.88	yel ppt		d		i	s	s a
r44	sulfate	$Rh_2(SO_4)_3.4H_2O$	566.05	red		d		s	s	
r45	sulfate	$Rh_2(SO_4)_3.12H_2O$	710.18	lt yel cr		d		v s	d	i al
r46	sulfate	$Rh_2(SO_4)_3.15H_2O$	764.22	pa yel cr		d		v s	d	i al, eth
r47	sulfide, hydro-	$Rh(HS)_3$	202.12	blk		d		i	d	s aq reg, aq Br; i Na_2S
r48	sulfide, mono-	RhS	134.97	gray-blk cr		d		i	i	i a, aq reg
r49	sulfide, sesqui-	Rh_2S_3	302.00	blk	6.40_4^{25}	d		i	i	i a, aq reg, aq Br
r50	sulfite	$Rh_2(SO_3)_3.6H_2O$	554.09	yel cr		d		s		i al
r51	**Rubidium**	Rb	85.47	soft, silver wh met	1.532 liq: $1.475^{38.5}$	38.89	686	d	d	d al; s a
r52	acetate	$RbC_2H_3O_2$	144.52	col, nacreous leaf, hygr		246		$86^{44.7}$	$89.3^{99.4}$	
r53	aluminium sulfate	$RbAl(SO_4)_2.12H_2O$	520.76	col, cub, oct, 1.457, 1.45232, 1.46618	1.867^0	99		2.59^{20}	43.25^{80}	
r54	azide	RbN_3	127.49	col need or plates	2.7876	d ca 310		107.1^{16}		0.182^{16} al; i eth
r55	borofluoride	$RbBF_4$	172.27	very sm rhomb cr, 1.333	2.820^{20}	590	d 500	0.6^{17}	10^{100}	
r56	borohydride	$RbBH_4$	100.31	white, cubic	1.920	1.487		v s		sl s al, i ether, bz
r57	bromate	$RbBrO_3$	213.38	cub	3.68	430		2.93^{25}	5.08^{40}	
r58	bromide	RbBr	165.38	col, cub, 1.5530	3.35 liq: 2.79^{730}	693	1340	98^8	$205.2^{113.5}$	i al, sl s acet
r59	bromide, tri-	$RbBr_3$	325.20	red, rhomb		d 140				
r60	bromochloroiodide	RbIBrCl	327.74	rhomb		d 200				s al, d eth
r61	bromoiodide, di-	$RbIBr_2$	372.13	red, rhomb	3.84	225	d 265	s		s al, d eth
r62	carbonate	Rb_2CO_3	230.95	col cr, deliq		837	d 740	450^{20}	s	0.7 abs al
r63	carbonate, acid	$RbHCO_3$	146.49	wh, rhomb		d 175		53.73^{20}		2.0 al
r64	chlorate	$RbClO_3$	168.92	trim	3.19			5.0^{19}	62.8^{100}	
r65	perchlorate	$RbClO_4$	184.92	rhomb, 1.4701	2.80	fus	d	0.5^0	18^{100}	0.009^{25} al, 0.06^{25} MeOH
r66	chloride	RbCl	120.92	col, cub, 1.493^{25}	2.80: liq: 2.088^{760}	718	1390	77^0	138.9^{100}	0.08^{25} al, 1.41^{25} MeOH; v sl s NH_3
r67	chlorobromide, di-	$RbBrCl_2$	236.29	rhomb		d 110				
r68	chlorodibromide	$RbBr_2Cl$	280.74	rhomb		76				
r69	chloroiodide, di-	$RbICl_2$	283.28	dk orange, rhomb		180–200	d 265	v s	v s	
r70	chloroplatinate	Rb_2PtCl_6	578.75	yel, cub	$3.94^{17.5}$	d		0.184^0	0.634^{100}	i al
r71	chloroplatinate, hexa-	$Rb_2[PtCl_6]$	578.75	yel, cub	$3.94^{17.5}$	d		0.014^0	0.33^{100}	i al
r72	chromate	Rb_2CrO_4	286.93	yel, rhomb, ~1.71	3.518			62^0	95.7^{90}	
r73	dichromate	$Rb_2Cr_2O_7$	386.93	tricl or monocl, >1.95, 1.70	3.02 monocl 3.125 tricl		tricl: monocl:	4.96^{18} 5.42^{18}	27.3^{80} 28.1^{80}	
r74	chromiumsulfate	$RbCr(SO_4)_2.12H_2O$	545.77	vlt cub, 1.482	1.946	107		43.4^{25}		

No.	Name	Synonyms and Formulae	Mol. wt.	Crystalline form, properties and index of refraction	Density or spec. gravity	Melting point, °C	Boiling point, °C	Solubility, in grams per 100 cc		
								Cold water	Hot water	Other solvents
	Rubidium									
r75	cobalt (II) sulfate...	$RbSO_4.CoSO_4.6H_2O$....	449.22	rubyred, monocl, 1.486, 1.491, 1.501	2.56^{15}			9.3^{25}	s	
r76	coppersulfate......	$Rb_2SO_4.CuSO_4.6H_2O$...	534.70	monocl, 1.489, 1.491, 1.504	2.57			10.28^{25}		
r77	cyanide.........	$RbCN$	111.49	col cr powder....	2.32			s	s	i al, eth
r78	galliumsulfate......	$RbGa(SO_4)_2.12H_2O$...	563.50	col cr, 1.46579	1.962					
r79	fluoride.........	RbF	104.47	col, cubic, 1.398..	3.557	795	1410	130.6^{18}		s dil HF; i al, eth, NH_3
r80	fluorogermanate.....	Rb_2GeF_6	357.52	wh cr..........				sl s	v s	
r81	fluosilicate........	Rb_2SiF_6	313.02	cub, oct........	3.332^{20}			0.16^{20}	1.35^{100}	s a, i al
r82	fluosulfonate......	$RbFSO_3$	184.53	need.........		304				
r83	iodate.........	$RbIO_3$	260.37	monocl or cub....	$4.33^{19.5}$	d		2.1^{23}		v s HCl
r84	*metaperiodate*.....	$RbIO_4$	276.37	tetr.........	3.918^{16}			0.65^{13}		
r85	iodide.........	RbI	212.37	col, cub, 1.6474..	3.55; liq: 2.87^{825}	647	1300	152^{17}	163^{25}	s liq NH_3; 0.674^{25} acet
r86	iodide, tri-......	RbI_3	466.18	blk, rhomb......	4.03^{22}	190		s		
r87	iodide, cmpd. with SO_2	$RbI.4SO_2$	468.63	lemon yel.......		13.5				
r88	iron (II) selenate...	$Rb_2SeO_4.FeSeO_4.6H_2O$..	620.79	blue-grn monocl, prism, 1.513, 1.520, 1.532	2.819					
r89	iron (III) selenate...	$RbFe(SeO_4)_2.12H_2O$...	643.42	cub, 1.507^{18}....	2.31^{11}	45	$-12H_2O$, 100			
r90	iron (II) sulfate....	$Rb_2SO_4.FeSO_4.6H_2O$...	527.00	gr monocl, prism, 1.4815, 1.4874, 1.4977	2.516	d 60		24.2^{25} (anhydr)		
r91	iron (III) sulfate....	$RbFe(SO_4)_2.12H_2O$...	549.62	cub, 1.4823......	1.91–1.95	48–53	$4.55^{6.6}$	52.6^{90}		
r92	hydride.........	RbH	86.48	col need.......	2.60	d 300		d	d	d a
r93	hydroxide.......	$RbOH$	102.48	gray-wh, deliq....	3.203^{11}	301 ± 0.9		180^{15}	v s	s al
r94	neodymium nitrate..	$2(?)RbNO_3.Nd(NO_3)_3.4H_2O$	697.26	redsh-vlt pl.....	2.56	47	$-4H_2O$, 60			
r95	nitrate.........	$RbNO_3$	147.47	col, hex cub, rhomb or tricl, hygr, 1.51, 1.52, 1.524	3.11; liq: 2.395^{100}	tr cub 161.4 m.p. 310	tricl rhom 219	44.28^{16}	452^{100}	sl s acet; v s HNO_3
r96	nitrate, hydrogen....	$RbNO_3.HNO_3$	210.49	tetr......		62				
r97	nitrate, hydrogen....	$RbNO_3.2HNO_3$	273.50	col need.......		45				
r98	magnesium sulfate...	$Rb_2SO_4.MgSO_4.6H_2O$...	495.47	col, monocl, 1.467, 1.469, 1.478	2.386^{20}_4			20.2^{25} (anhydr)		
r99	permanganate......	$RbMnO_4$	204.41	cr.........	$3.235^{10.4}$	d 295		0.5^0	4.7^{60}	
r100	oxide, mon-.......	Rb_2O	186.94	col-yel, cub.....	3.72	d 400		s d	s d	s liq NH_3
r101	oxide, per-........	Rb_2O_2	202.94	yel, cub.......	3.65^0	570	d 1011^{760} mm	dec to RbOH $+H_2$		
r102	oxide, super......	RbO_2, unstable.......	117.47	yel plates......	3.80	432	d 1157[1] atm			
r103	oxide (tetr-)......	Rb_2O_4	234.94	dk orange cr, deliq		dec 500 vac				
r104	oxide, tri- (sesqui)...	Rb_2O_3 (or Rb_4O_6).....	218.94	blk cub......	3.53^0	489		s d		
r105	praseodymium nitrate	$2RbNO_3.Pr(NO_3)_3.4H_2O$	693.93	grnsh, monocl, need hygr	2.50	63.5	$-4H_2O$, 60			
r106	selenate........	Rb_2SeO_4	313.90	col, rhomb, 1.5515, 1.5537, 1.5582	3.90			159^{12}		
r107	silicofluoride......	Rb_2SiF_6	313.02	col, oct or hex...	3.3383^{20}_4			sl s	s	s a, i al
r108	sulfate.........	Rb_2SO_4	267.00	col, rhomb hex, 1.513, 1.513, 1.514	3.613^{20}_4 liq 2.53^{1100}	1060 trig 653	*ca* 1700	42.4^{10}	81.8^{100}	i acet; sl s NH_3
r109	sulfate, hydrogen....	$RbHSO_4$	182.54	rhomb, 1.473.....	2.892^{16}	< red heat				
r110	sulfide, di-.......	Rb_2S_2	235.07	dk red......		420	volat >850			
r111	sulfide, hexa-......	Rb_2S_6	363.32	brn-red......		201				
r112	sulfide, mono-.....	Rb_2S	203.00	wh-pale yel.....	2.912	530 d vac		v s	v s	
r113	sulfide, mono-tetrahydrate	$Rb_2S.4H_2O$	275.07	cr, deliq......				v s	v s	
r114	sulfide, penta-.....	Rb_2S_5	331.26	red, rhomb, deliq..	2.618^{15}	225		d		s 70 % al; i eth, chl
r115	sulfide, tri-.......	Rb_2S_3	267.13	redsh yel......		213				
r116	tartrate, d & l.....	$Rb_2C_4H_4O_6$	319.01	trig.........	2.658^{20}_4			200^{25}		i toluol
r117	*dl*-tartrate, hydrogen.	$RbHC_4H_4O_6$	234.55	trim pr.......	2.282	201 d		1.18^{25}	11.7^{100}	
r118	vanadium sulfate....	Rubidium vanadium alum. $RbV(SO_4)_2.12H_2O$	544.72	yellow, cubic, 1.4689	1.915^{20}_4	64	230 $-12H_2O$, 300 dec	2.56^{10}		

No.	Name	Synonyms and Formulae	Mol. wt.	Crystalline form, properties and index of refraction	Density or spec. gravity	Melting point, °C	Boiling point, °C	Solubility, in grams per 100 cc		
								Cold water	Hot water	Other solvents
	Ruthenium									
r119	**Ruthenium**	Ru	101.07	gray-wh or silv brittle met, hex	12.30	**2310**	3900	i	i	i aq reg, a, al; s fus alk
r120	carbonyl, penta-	Ru(CO)$_5$	241.12	col liq		−22				s al, bz
r121	chloride, tetra-	RuCl$_4$.5H$_2$O	332.96	rdsh-br cr, hygr		d		s		s al
r122	chloride, tri-	RuCl$_3$	207.43	br cr, deliq	3.11	d >500		i	d	sl s al; s HCl; i CS$_2$
r123	fluoride, penta-	RuF$_5$	196.06	dk grn cr	2.963[16.5]	101	250	d	d	
r124	hydroxide	Ru(OH)$_3$	152.09	blk powd				v sl s		s a; i alk
r125	oxide, di-	RuO$_2$	133.07	dk bl, tetr	6.97	d		i	i	i a; s fus alk
r126	oxide, tetr-	RuO$_4$	165.07	yel, rhomb need	3.29[21]	25.5	d 108	2.033[20]	2.249[74]	s al, a, alk, CCl$_4$
r127	oxychloride Ruthenium red. ammoniated	Ru$_2$(OH)$_2$Cl$_4$.7NH$_3$.3H$_2$O	551.19	brn-red powd				s		
r128	silicide	RuSi	129.16	met pr	5.40[4]			i	i	s HNO$_3$+HF
r129	sulfide	Nat. laurite. RuS$_2$	165.20	gray-blk, cub	6.99	d 1000		i	i	i a; s fus alk
s1	**Samarium**	Sm	150.35	wh-gray met, hex	7.520	1077	1791	i		s a
s2	acetate	Sm(C$_2$H$_3$O$_2$)$_3$.3H$_2$O	381.53		1.94			15[25]		
s3	acetylacetonate	Sm(C$_5$H$_7$O$_2$)$_3$	447.68	cr mass		146–147		i		
s4	benzylacetonate	Sm(C$_{10}$H$_9$O$_2$)$_3$.2H$_2$O	669.92	straw color		103–105				s org solv
s5	bromate	Sm(BrO$_3$)$_3$.9H$_2$O	696.21	yel, hex		75	−9H$_2$O, 150	114[25]		v sl s al
s6	(II) bromide	SmBr$_2$	310.17	dk brn	5.1	508	1880	d		
s7	(III) bromide	SmBr$_3$.6H$_2$O	498.17	yel cr, deliq	2.971[22]	640				
s8	carbide	SmC$_2$	174.37	yel, hex	5.86			d	d	s, d a
s9	(II) chloride	SmCl$_2$	221.26	red-brn cr	4.56[25]	740		s, d		i al, CS$_2$
s10	(III) chloride	SmCl$_3$	256.71	yelsh-wh cr, hygr	4.46[18]	678±2	d	92.4[10]	99.9[50]	v s al; 6.4[25] pyr
s11	(III) chloride, hexahydrate	SmCl$_3$.6H$_2$O	364.80	grn-yel, tricl, hygr	2.383	−5H$_2$O, 110				
s12	chromate	Sm$_2$(CrO$_4$)$_3$.8H$_2$O	792.80	yel				0.043[25]		
s13	(II) fluoride	SmF$_2$	188.35			1306	>2400	i	i	
s14	(III) fluoride	SmF$_3$	207.35			1306	2300	i	i	
s15	hydroxide	Sm(OH)$_3$	201.37	pa yel powd				i		s a; i alk
s16	(II) iodide	SmI$_2$	404.16	dk brn		527	1580	d		
s17	(III) iodide	SmI$_3$	531.06	or-yel cr		850				
s18	iodate	Sm(IO$_3$)$_3$	672.98			d 075		i		
s19	(III) methyl-phosphate, di	Sm[(CH$_3$)$_2$PO$_4$]$_3$	525.47	cream col, hex pr				35.2[25]	10.8[95]	
s20	(III) molybdate	Sm$_2$(MoO$_4$)$_3$	780.51	vlt, rhomb oct	5.36					
s21	(III) nitrate	Sm(NO$_3$)$_3$.6H$_2$O	444.46	pa yel, tricl	2.375	78–79		v s		
s22	(III) oxalate	Sm$_2$(C$_2$O$_4$)$_3$.10H$_2$O	744.91	wh cr				0.000054		s H$_2$SO$_4$
s23	oxide, sesqui-	Samaria. Sm$_2$O$_3$	348.70	wh-yelsh powd	8.347			i		v s a
s24	(III) sulfate	Sm$_2$(SO$_4$)$_3$.8H$_2$O	733.01	lt yel, monocl, 1.543, 1.552, 1.563	2.930	−8H$_2$O, 450		2.67[20] 4.4[25]	1.99[40]	
s25	(III) sulfide	Sm$_2$S$_3$	396.89	yelsh-pink	5.729	1900			d	d dil a
s26	sulfate, basic	Sm$_2$O$_3$SO$_4$	428.76	yel powd		d 1100		i		i dil H$_2$SO$_4$
s27	**Scandium**	Sc	44.956	silv met, cubic or hex	2.9890	1541	2831	d, ev H$_2$		
s28	acetylacetonate	Sc(C$_5$H$_7$O$_2$)$_3$	342.29	col pl		187.5	subl 210–215			s al, bz, chl
s29	bromide	ScBr$_3$	284.68		3.914	subl >1000				
s30	chloride	ScCl$_3$	151.32	col cr	2.39[25]	939	subl 800–850	v s	v s	i abs al
s31	hydroxide	Sc(OH)$_3$	95.98	col amorph				i		s dil a
s32	nitrate	Sc(NO$_3$)$_3$	230.97	col, deliq		150		s		s al
s33	nitrate, tetrahydrate	Sc(NO$_3$)$_3$.4H$_2$O	303.03	col, pr, deliq		−4H$_2$O, 100		v s		
s34	oxalate	Sc$_2$(C$_2$O$_4$)$_3$.5H$_2$O	444.05	cr		−4H$_2$O, 140				
s35	oxide	Scandia. Sc$_2$O$_3$	137.91	wh powd	3.864			i	i	s h a
s36	sulfate	Sc$_2$(SO$_4$)$_3$	378.10	col cr	2.579	d		10.3[25]	v s	
s37	sulfate, hexahydrate	Sc$_2$(SO$_4$)$_3$.6H$_2$O	486.19			−4H$_2$O, 100 −6H$_2$O, 250		v s		
s38	sulfate, pentahydrate	Sc$_2$(SO$_4$)$_3$.5H$_2$O	468.17		2.519	−3H$_2$O, 100		54.6[25]		
s39	**Selenic acid**	H$_2$SeO$_4$	144.97	wh, hex prism, hygr	3.004[15/4]	58 eas undercools	d 260	1300[30]	∞[60]	s H$_2$SO$_4$; d al; i NH$_3$
s40	monohydrate	H$_2$SeO$_4$.H$_2$O	162.99	wh, need	2.627[15/4] liq 2.3564[15/4]	26 eas undercools	205	v s	v s	
s41	tetrahydrate	H$_2$SeO$_4$.4H$_2$O	217.03	col liq		51.7 eas undercools	−H$_2$O, 172[85]	∞		s H$_2$SO$_4$; d org solv
s42	**Selenium**	Se	78.96	blsh-gray, met hex	4.81[20/4]	217	684.9 ± 1.0	i	i	s H$_2$SO$_4$, CHCl$_3$; i al; v sl s CS$_2$
s43	"	Se	78.96	red, monocl prism	4.50	170–180 trsf to hex	684.8	i	i	0.14[6.5] CS$_2$; s H$_2$SO$_4$, HNO$_3$
s44	"	Se	78.96	red amorph, blk vitr	red 4.26 blk 4.28	tr to hex, 60–80	684.8	i	i	s H$_2$SO$_4$, CS$_2$, bz

No.	Name	Synonyms and Formulae	Mol. wt.	Crystalline form, properties and index of refraction	Density or spec. gravity	Melting point, °C	Boiling point, °C	Solubility, in grams per 100 cc		
								Cold water	Hot water	Other solvents
	Selenium									
s45	bromide, "mono-"	Diselenium dibromide. Se_2Br_2	317.74	dk red liq	3.604^{15}		227 d	d	d	d al; s CS_2, $CHCl_3$, C_2H_5Br
s46	bromide, tetra-	$SeBr_4$	398.60	or-red-brn cr		d 75		d	d	s CS_2, chl, C_2H_5Br, HCl
s47	bromide (mono-) chloride, tri-,	$SeBrCl_3$	265.23	yel br cr		190				i CS_2
s48	bromide (tri-) chloride	$SeBr_3Cl$	354.14	or cr, hygr		d				v sl s CS_2
s49	carbide	SeC_2	102.98	yel liq, 1.845	2.682^{20}_4	45.5	$125-126^{760}$	i		s CS_2, eth, CCl_4, bz, al
s50	chloride "mono"-	Diselenium dichloride. Se_2Cl_2	228.83	br red liq, 1.596	2.77^{25}_4	−85	d 130	d	d	d al, eth; s CS_2, chl, CCl_4, bz
s51	chloride, tetra-	$SeCl_4$	220.77	wh-yel, cub, deliq. 1.807	$3.78-3.85^{360}$	305, subl 170−196	d 288	d	d	i al, eth, CS_2; s $POCl_3$; d a, alk
s52	fluoride, hexa-	SeF_6	192.95	col gas, 1.895	3.25^{-28} g/l	−39, subl −46.6	−34.5	s d		
s53	fluoride, tetra-	SeF_4	154.95	col liq or wh cr		m.p. −13.8 frz −90	>100	d	d	
s54	hydride	H_2Se	80.98	col gas, pois	gas 3.664^{760} air; liq: $2.004^{-41.5}$	60.4	−41.5	3.77^4	$270^{22.5}$	s CS_2, $COCl_2$
s55	iodide, "mono"-	Diselenium diiodide. Se_2I_2	411.73	steelgray cr (exist ?)		68−70	d 100	d	d	
s56	nitride	Se_4N_4	371.87	amorph, or yel-brickred, hygr		expl 160−200	d	i	sl d	i al, eth; v sl s acet ac, bz
s57	oxide, di-	SeO_2	110.96	wh, monocl, col, tetr, pois, >1.76	3.95^{18}_{15}	340−350 subl 315−317		38.4^{14}	82.5^{65}	6.67^{14} al; $4.35^{15.3}$ acet; $1.11^{13.9}$ ac a; s bz
s58	oxide, tri-	SeO_3	126.96	pa yel cub or fibre, deliq	3.6	118	d 180	d, v s	d, v s	s al, conc H_2SO_4; i eth, bz, chl, CCl_4
s59	oxybromide	Selenyl bromide. $SeOBr_2$	254.78	red yel cr	liq 3.38^{50}	41.6	217^{710} d	d		s CS_2, CCl_4, chl, H_2SO_4, bz
s60	oxychloride	$SeOCl_2$	165.87	col-yel, liq 1.651^{20}	2.42^{22}	8.5	176.4	d		s CS_2, CCl_4, chl, bz
s61	sulfur oxy*tetra*chloride	$SeSO_2Cl_4$	300.83	hex pr		165	183	d		
s62	oxyfluoride	Selenyl fluoride. $SeOF_2$	132.96	col liq	2.67	4.6	124	d		s al, CCl_4
s63	sulfide, di-	SeS_2	143.09	br red-yel		<100	d	i		d sq reg, HNO_3; s $(NH_4)_2S$
s64	sulfide, "mono"-	SeS	111.02	or-yel tabl or powd	3.056^0	d 118−119		i	i	s CS_2; i eth; d al
s65	sulfur oxide	$SeSO_3$	159.02	grn pr or yel powd		−SO_2, 40		d, 118		s H_2SO_4; i SO_3
s67	**Selenious acid**	H_2SeO_3	128.97	col, hex, deliq	3.004^{15}_4	d 70	−H_2O	167^{20}	v s	v s al; i NH_3
s68	**Silane, bromo-**	SiH_3Br	111.02	col gas, expl in air	1.72^{-80} 1.533^0	−94	1.9			
s69	bromotrichloro-	$SiBrCl_3$	214.35	col liq	1.826	−62	80.3	d	d	
s70	chloro-	SiH_3Cl	66.56	col gas	gas: 3.033 g/l liq: 1.145^{-112}	−118.1	−30.4			
s71	dibromo-	SiH_2Br_2	189.92	col liq, inflam	2.17^0	−70.1	66	d		d alk
s72	dibromo-	Silicobromoform. $SiHBr_3$	268.82	col liq, inflam	2.7^{17}	−73	109	d		d NH_3
s73	dibromodichloro-	$SiBr_2Cl_2$	258.81	col liq	2.172^{25}_4	−45.5	104	d	d	
s74	dichloro-	SiH_2Cl_2	101.01	gas	gas 4.599 g/l liq 1.42^{-122}	−122	8.3	d	d	
s75	dichloro difluoro-	$SiCl_2F_2$	136.99	gas	6.2784 g/l	−144±2	−31.7±0.2	d	d	
s76	(hexa-)hexaoxocyclo-	Siloxane. $Si_6O_9H_6$	222.56	wh, pl	1.32^{20}	d 140, inflam		sl d	d	
s77	monochloro trifluoro-	$SiClF_3$	120.53	gas	5.455 g/l	−138.0±2	−70.0±0.2	d	d	
s78	monoiodo-	SiH_3I	158.01	col liq	$2.035^{14.8}$	−57.0	45.5			
s79	(tri-)nitrilo-	Silicylamine, tri-$(SiH_3)_3N$	107.34	col inflam liq	0.895^{-106}	−105.6				
s81	tribromochloro-	$SiBr_3Cl$	303.27	col liq	2.497^{25}_4	−20.8±1	126−128	d	d	
s82	trichloro-	Silicochloroform. $SiHCl_3$	135.45	col liq	1.34	−126.5	33^{768} mm	d	d	s CS_2, CCl_4, chl, bz
s83	trichloroiodo-	$SiCl_3I$	261.35	col liq		>−60	113.5	d	d	
s84	trifluoro-	Silicofluoroform. $SiHF_3$	86.09	col gas	3.86^0 g/l	−131.4	ca −95			d al, eth, alk; s tol
s85	triiodo-	Silicoiodoform. $SiHI_3$	409.81	red liq	3.314^{20}	8	220	d	d	s CS_2, bz

No.	Name	Synonyms and Formulae	Mol. wt.	Crystalline form, properties and index of refraction	Density or spec. gravity	Melting point, °C	Boiling point, °C	Solubility, in grams per 100 cc		
								Cold water	Hot water	Other solvents
	Silicane cyanate									
s86	Silicane cyanate....	Si(OCN)₄..........	196.16	sol or liq........	1.414$^{20}_4$	34.5 ± 0.5	247.2 ± 0.5 0.5^{760}	d	....	....
s87	diimide..........	Si(NH)₂..........	58.12	wh powd..........		d 900				
s88	isocyanate..........	Si(NCO)₄..........	196.16	sol or liq........	1.434$^{25}_4$	26.0 ± 0.5	185.6 ± 0.3^{760}	d		i acet; s bz, CCl₄, CS₂
s89	**Silicic acid, di-**.....	H₂SiO₄..........	138.18	col cr..........		d 150		i	i	s NH₃, HF
s90	meta-...........	H₂SiO₃..........	78.10	col, amorph........		d room temp		i	i	s NH₃, HF, h alk
s91	**Silicon**..........	Si..........	28.0855	steel gray, large to micr cr, cub	2.32–2.34	1410	2355	i	i	s HF + HNO₃; i HF
s92	acetate, tetra-.....	Si(C₂H₃O₂)₄.....	264.27	col cr, hygr.....		subl 110 d 160–170	148^6 mm			d al; sl s acet, bz
s93	bromide, tetra-.....	Tetrabromosilane. SiBr₄	347.72	col fum liq, sol cub	liq: 2.7715$^{25}_4$ sol: 3.292^{-79}	5.4	154	d	d	d H₂SO₄
s94	(di-)bromide, hexa-..	Si₂Br₆..........	535.62	wh, rhomb..........		95	240	d	d	s CS₂; d KOH
s95	bromide(di-) sulfide,.	SiSBr₂..........	219.97	col pl..........		93	150$^{18.3}$ mm	d	d	s bz, CS₂
s96	carbide..........	SiC..........	40.10	col-blk, hex or cub, 2.654, 2.697	3.217	~2700, subl, d		i	i	i a; s fus KOH
s97	chloride, tetra-.....	Tetrachlorosilane. SiCl₄	169.90	col fum liq......	liq: 1.483^{20} sol: 1.90^{-97} gas: 7.59 g/l	−70	57.57	d	d	d al
s98	(di-)chloride, hexa-..	Hexachlorodisilane. Si₂Cl₆	268.89	col liq, 1.4748^{18}	1.58^0	−1	145^{769}	d	d	d al
s99	chloride(di-) sulfide.	SiSCl₂..........	131.06	col pr..........		75	92$^{22.5}$	d	d	s CCl₄, CS₂ bz
s100	chloride(tri-) sulfide,. hydro	SiCl₂HS..........	167.52	col liq..........	1.45	....	96–100	d	d	d al
s101	fluoride, tetra-.....	Tetrafluorosilane. SiF₄	104.08	col gas..........	gas: 4.69 g/l^{760} liq: 1.66^{-95}	−90.2	−86	d	d	s abs al, HF; i eth
s102	(di-)fluoride, hexa.	Hexafluorodisilane Si₂F₆	170.16	gas..........	7.709 g/l	−18.7	10.5	d	d	....
s103	hydride..........	Silane, silicane. SiH₄	32.12	col gas..........	liq: 0.68^{-85} gas: 1.44 g/l	−185	−111.8^{760} mm	i	....	d KOH
s104	hydride..........	Disilane, disilicane. Si₂H₆	62.22	col gas..........	gas: 2.865 g/l liq: 0.686^{-20}	−132.5	−14.5	sl d		s al, bz, CS₂
s105	hydride..........	Trisilane, trisilanepropane. Si₃H₈	92.32	col liq..........	liq: 0.743^0; gas: 4.15 g/l$^0_{760}$	−117.4	52.9	d		d CCl₄
s106	hydride..........	Tetrasilane, tetrasilane butane. Si₄H₁₀	122.42	col liq..........	liq: 0.79^0 gas 5.48 g/l$^0_{760}$	−108	84.3	d		
s107	iodide, tetra-.....	Tetraiodosilane. SiI₄...	535.70	col, cub..........	4.198	120.5	287.5	d		2.2^{27} CS₂
s108	(di-)iodide, hexa-..	Hexaiodosilane. Si₂I₆..	817.60	col, hex..........		d 250	d	d	d	19^{19} CS₂
s109	nitride..........	Si₃N₄..........	140.28	gray-wh amorph powd	3.44	1900 press	....	i	i	s HF
s110	oxide, di-..........	Nat. cristobalite. SiO₂.	60.08	col, cub or tetr, 1.487, 1.484	2.32	1723 ± 5	2230 (2590)	i	i	s HF; v sl s alk
s111	oxide, di-..........	Nat. lechatelierite. SiO₂	60.08	col, amorph, vitr, 1.4588	2.19	....	2230 (2590)	i	i	s HF; v sl s alk
s112	oxide, di-..........	Nat. opal. SiO₂.xH₂O...		col, amorph 1.41–1.46	2.17–2.20	>1600	....	i	i	s HF; v sl s alk
s113	oxide, di-..........	Nat. tridymite. SiO₂...	60.08	col, rhomb, 1.469, 1.470, 1.471	2.26$^{25}_4$	1703	2230 (2590)	i	i	s HF; v sl s alk
s114	oxide, di-..........	Nat. quartz. SiO₂.....	60.08	col, hex, 1.544, 1.553	2.635–2.660	1610	2230 (2590)	i	i	s HF; v sl s alk
s115	oxide, mon-........	SiO..........	44.09	wh, cub..........	2.13	>1702	1880	i	i	s dil HF + HNO₃
s116	oxychloride.........	Chlorosiloxane. Si₂OCl₆.	284.89	col liq..........		28.1 ± 0.2	137	d		∞ CS₂, CCl₄
s117	oxyfluoride.........	Si₂OF₆.........	186.16	col gas..........	1.358 liq	−47.8 ± 0.5	−23.3	d	d	d alk
s118	sulfide, di-..........	SiS₂..........	92.21	wh need, rhomb..	2.02	subl 1090	white heat	d		d al, liq NH₃; s dil alk; i bz
s119	sulfide, mono-.....	SiS..........	60.15	yel need..........	1.853^{15}	subl 940^{20}	....	d	d	d al, alk
s120	thiocyanate.........	Si(CNS)₄..........	260.41	wh, rhomb need..	1.409$^{20}_4$	143.8	314.2	d		d al, a, alk; i eth, CS₂, CHCl₃
s121	**Silicotungstic acid.**	H₄[Si(W₃O₁₀)₄]·26H₂O	3346.47	wh-sl yel cr, deliq	....	....	....	v s	v s	v s al
s122	**Silicyl oxide**........	Disiloxane. (SiH₃)₂O ...	78.22	col gas..........	gas: 3.491 g/l liq: 0.881^{-80}	−144	−15.2	v sl s	sl d	....

No.	Name	Synonyms and Formulae	Mol. wt.	Crystalline form, properties and index of refraction	Density or spec. gravity	Melting point, °C	Boiling point, °C	Solubility, in grams per 100 cc		
								Cold water	Hot water	Other solvents
	Siloxane									
s123	**Siloxane, (di-), oxide.**	$[H(O)Si]_2.O$	106.19	wh volum subst		expl ca 300		sl s		s, d HF; d al
s124	**Silver**	Ag	107.868	wh met, cub 0.54	10.5^{20}	961.93	2212	i	i	s HNO_3, h H_2SO_4, KCN; i alk
s125	acetate	$AgC_2H_3O_2$	166.92	wh pl	3.259^{15}	d		1.02^{20}	2.52^{80}	s dil HNO_3
s126	acetylide	Ag_2C_2	239.76	wh ppt		expl				s a; sl s al
s127	$ortho$arsenate	Ag_3AsO_4	462.53	dk red, cub	6.657^{25}	d		0.00085^{20}		s NH_4OH, ac a
s128	$ortho$arsenite	Ag_3AsO_3	446.53	yel, powd		d 150		0.00115^{20}	i	s ac a, NH_4OH, HNO_3; i al
s129	azide	AgN_3	149.89	wh rhomb pr, expl		252	297	i	0.01^{100}	s KCN, dil HNO_3; sl s NH_4OH
s130	benzoate	$AgC_7H_5O_2$	228.99	wh powd				0.262^{25}	s	0.017 al
s131	$tetra$borate	$Ag_2B_4O_7.2H_2O$	407.01	wh cr				sl s		s a
s132	bromate	$AgBrO_3$	235.78	col, tetr, 1.874, 1.920	5.206	d		0.196^{25}	1.33^{90}	s NH_4OH; sl s HNO_3
s133	bromide	Bromyrite: AgBr	187.78	pa yel, 2.253	6.473^{25}	432	d>1300	8.4×10^{-6}	0.00037^{100}	s KCN, $Na_2S_2O_3$, NaCl sol; sl s NH_4OH; i al
s134	carbonate	Ag_2CO_3	275.75	yel powd	6.077	d 218		0.0032^{20}	0.05^{100}	s NH_4OH, $Na_2S_2O_3$; i al
s135	chlorate	$AgClO_3$	191.32	wh, tetr	4.430^{20}_4	230	d 270	10^{15}	50^{80}	sl s al
s136	perchlorate	$AgClO_4$	207.32	wh, cr, deliq	2.806^{25}	d 486		557^{25}	s	s al; 101 tol; 5.28 bz
s137	chloride	Nat. cerargyrite. AgCl	143.32	wh, cub, 2.071	5.56	455	1550	0.000089^{10}	0.0021^{100}	s NH_4OH, $Na_2S_2O_3$, KCN
s138	chlorite	$AgClO_2$	175.32	yel cr		105 expl		0.45^{25}	2.13^{100}	
s139	chromate	Ag_2CrO_4	331.73	red, monocl	5.625			0.0014^0	0.008^{70}	s NH_4OH, KCN
s140	dichromate	$Ag_2Cr_2O_7$	431.72	red, tricl	4.770	d		0.0083^{15}	d	s a, NH_4OH, KCN
s141	citrate	$Ag_3C_6H_5O_7$	512.71	wh need		d		0.028^{18}	sl s	s a, NH_4OH, KCN, $Na_2S_2O_3$
s142	cyanate	AgOCN	149.89	col	4.00	d		sl s	s	s KCN, HNO_3, NH_4OH
s143	cyanide	AgCN	133.84	wh, hex	3.95	d 320		0.000023^{20}		s HNO_3, NH_4OH, KCN, $Na_2S_2O_3$
s144	ferricyanide	$Ag_3Fe(CN)_6$	535.56					0.000066^{20}		i a; s NH_4OH, h $(NH_4)_2CO_3$
s145	ferrocyanide	$Ag_4Fe(CN)_6.H_2O$	661.45	wh				i	i	s KCN; i a, NH_4 salts, NH_4OH
s146	fluogallate	$Ag_3[GaF_6].10H_2O$	687.47	col, orthorhomb cr, 1.493	2.90			v s		i al
s147	fluoride	AgF	126.87	wh, cub, deliq	$5.852^{15.5}$	435	ca 1159	$182^{15.5}$	205^{108}	sl s NH_4OH
s148	fluoride, di-	AgF_2	145.87	brn, rhomb	4.57–4.58	690	d 700	d	d	
s149	(di-)fluoride	Ag_2F	234.74	yel, hex	8.57	d 90		d		
s150	fluosilicate	$Ag_2SiF_6.4H_2O$	429.88	wh powd or col cr, deliq		>100	d	v s		
s151	fulminate	$Ag_2C_2N_2O_2$	299.77	need		expl		0.075^{13}	s	i NHO_3; s NH_4OH
s152	iodate	$AgIO_3$	282.77	col, rhomb	$5.525^{16.5}$	>200	d	0.003^{10}	0.019^{60}	s HNO_3, NH_4OH, KI
s153	periodate	$AgIO_4$	298.77	or yel, tetrag	5.57	d 180		d		s HNO_3
s154	iodide(α)	Nat. iodyrite. AgI	234.77	yel, hex 2.21, 2.22	5.683^{30}_4	tr 146 to β		$2.8 \times 10^{-7.25}$	$2.5 \times 10^{-6.60}$	s KCN, $Na_2S_2O_3$, KI; sl s NH_4OH
s155	iodide (β)	AgI	234.77	or, cub	$6.010^{14.6}_4$	558	1506			
s156	iodomercurate (α)	Ag_2HgI_4	923.94	yel, tetrag	6.02	tr to β 50.7		i		s KI, KCN; i dil a
s157	iodomercurate (β)	Ag_2HgI_4	923.94	red, cub	5.90	d 158		i		s KI, KCN; i dil a
s158	hydrogen(tri-) $para$periodate	$Ag_2H_3IO_6$	441.69	yel, rhomb	5.68^{25}	60 d		1.68^{25}		s HNO_3
s159	hyponitrite	$Ag_2N_2O_2$	275.75	yel	5.75^{30}	d 110		v sl s		d HNO_3, H_2SO_4
s160	lactate	$AgC_3H_5O_3.H_2O$	214.96	wh or sl gray cr, powd				ca 7.7		
s161	laurate	$AgC_{12}H_{23}O_2$	307.19	wh, greasy powd		212.5				0.007^{25} al; 0.008^{15} eth
s162	levunilate	$AgC_5H_7O_3$	222.98	leaf				0.67^{17}	d	
s163	permanganate	$AgMnO_4$	226.81	dk vlt, monocl	4.27^{25}	d		0.55^0	$1.69^{28.5}$	d al
s164	mercury iodide (α)	Ag_2HgI_4	923.94	yel, tetrag	6.02	trst 50.7		i		
s165	mercury iodide (β)	Ag_2HgI_4	923.94	red, cub	5.90	158 d		i		
s166	myristate	$AgC_{14}H_{27}O_2$	335.24			211		0.007^{25}		0.006^{25} al; 0.007^{15} eth
s167	nitrate	$AgNO_3$	169.87	col, rhomb, 1.729, 1.744, 1.788	4.352^{19}	212	d 444	122^0	952^{190}	s eth, glyc; v sl s abs al

No.	Name	Synonyms and Formulae	Mol. wt.	Crystalline form, properties and index of refraction	Density or spec. gravity	Melting point, °C	Boiling point, °C	Solubility, in grams per 100 cc		
								Cold water	Hot water	Other solvents
	Silver									
s168	nitrite	$AgNO_2$	153.88	wh, rhomb	4.453^{26}	d 140		0.155^0	1.363^{60}	s ac a, NH_4OH; i al
s169	nitroplatinite	$Ag_2[Pt(NO_2)_4]$	594.85	yel-brn monocl pr		d 100		sl s	s	
s170	nitroprusside	$Ag_2[FeNO(CN)_5]$	431.68	lt pink				i		s NH_4OH; i al, HNO_3
s171	oxalate	$Ag_2C_2O_4$	303.76	col cr	5.029^4	expl 140		0.00339^{18}		s KCN, NH_4OH, a
s172	oxide	Ag_2O	231.74	br-blk, cub	$7.143^{16.6}$	**d 230**		0.0013^{20}	0.0053^{80}	s a, KCN, NH_4OH, al
s173	oxide, per	Ag_3O_2 (or AgO)	247.74	gray-blk, cub	7.44	d>100		i		s H_2SO_4, HNO_3, NH_4OH
s174	palmitate	$AgC_{16}H_{31}O_2$	363.29	wh, greasy powd		209		0.0012^{20}	0.006^{25}	0.007^{15} eth; 0.006^{25} al
s175	metaphosphate	$AgPO_3$	186.84	wh, amorph	6.37	ca 482		i		s HNO_3, NH_4OH
s176	orthophosphate	Ag_3PO_4	418.58	yel, cub	6.370^{25}	849		$0.00065^{19.5}$		s a, KCN, NH_4OH, NH_3
s177	orthophosphate, mono-H	Ag_2HPO_4	311.75	wh, trig	1.8036	d 110				
s178	pyrophosphate	$Ag_4P_2O_7$	605.42	wh	$5.306^{7.5}$	585		i	i	s a, NH_4OH, KCN, ac a
s179	propionate	$AgC_3H_5O_2$	180.94	wh leaf or need	2.687^{25}_4			0.842^{20}	2.04^{80}	
s180	perrhenate	$AgReO_4$	358.07	wh cr, tetrag or rhomb	7.05	430		0.32^{20}		
s181	salicylate	$AgC_7H_5O_3$	244.99	wh to redsh-wh cr				sl s		s al
s182	selenate	Ag_2SeO_4	358.73	wh, orthorhomb cr	5.72			0.118^{20}		
s183	selenide	Ag_2Se	294.70	thin gray pl, cub	8.0	880	d	i		s NH_4OH, h HNO_3
s184	stearate	$AgC_{18}H_{35}O_2$	391.35	wh powd amorph		205		0.006^{20}		0.006^{25} al; 0.006^{25} eth
s185	sulfate	Ag_2SO_4	311.80	wh, rhomb, 1.7583, 1.7719, 1.7852	$5.45^{29.2}$	652	d 1085	0.57^0	1.41^{100}	s a, NH_4OH; i al
s186	sulfide	Nat. acanthite. Ag_2S	247.80	gray-blk, rhomb	7.326	tr 175	d		v sl s	s KCN, conc H_2SO_4, HNO_3
s187	sulfide	Nat. argentite. Ag_2S	247.80	blk, cub	7.317	825	d	8.4×10^{-1} ^{5}		s KCN, a
s188	sulfite	Ag_2SO_3	295.80	wh cr		d 100		v sl s		s a, NH_4OH, KCN; i HNO_3
s189	d-tartrate	$Ag_2C_4H_4O_6$	363.81	wh, scales	3.423^{15}	d		0.2^{18}	0.203^{25}	s a, KCN, NH_4OH
s190	orthotellurate, tetra-H	$Ag_2H_4TeO_6$	443.40	straw yel, rhomb bipyr		d>200		i		s KCN, NH_4OH
s191	telluride	Nat. hessite, Ag_2Te	343.34	gray, cub	8.5	955		i	i	s KCN, NH_4OH
s192	tellurite	Ag_2TeO_3	391.36	yel-wh ppt		250-bl 450-pa yel		i		s KCN, NH_3
s193[1]	thioantimonite	Nat. pyrargyrite. Ag_3SbS_3	541.55	red, trig, 3.084 2.881 (Li)	5.76	486		i	i	s HNO_3
s193[2]	thioarsenite	Nat. proustite. Ag_3AsS_3	494.72	scarlet red, trig, 3.088, 2.792	5.49	490		i	i	s HNO_3
s194	thiocyanate	$AgSCN$	165.95	col cr		d		0.0000021^{25}	0.00064^{100}	s NH_4OH; i a
s195	di-thionate	$Ag_2S_2O_6.2H_2O$	411.90	rhomb cr, ~1.662	3.61					
s196	thiosulfate	$Ag_2S_2O_3$	327.87	wh cr		d		sl s		s $Na_2S_2O_3$, NH_4OH
s197	tungstate	Ag_2WO_4	463.59	pa yel cr				0.05^{15}		s KCN, NH_4OH, HNO_3
	Silver complex									
s198	diamminesilver perrhenate	$[Ag(NH_3)_2]ReO_4$	392.13	col monocl cr	3.901					1.618 conc NH_4OH
s199	**Sodium**	Na	22.9898	silv, met cub, 4.22	0.97	**97.81 ± 0.03**	**882.9**	d to $NaOH+H_2$		d al; i eth, bz
s200	acetate	$NaC_2H_3O_2$	82.03	wh gr powd, monocl, 1.464	1.528	324		119^0	170.15^{100}	sl s al
s201	acetate trihydrate	$NaC_2H_3O_2.3H_2O$	136.08	col, monocl pr, effl, β 1.464	1.45	58	123, $-3H_2O$, 120	76.2^0	138.8^{50}	2.1^{18} al; s eth
s202	alumina trisilicate	Nat. albite. $NaAlSi_3O_8$ (or $Na_2O.Al_2O_3.6SiO_2$)	262.22	col, tricl, 1.525, 1.529, 1.536	2.61	1100			sl d	s HCl; d dil al
s203	metaaluminate	$NaAlO_2$	81.97	wh amorph powd, hygr, 1.566, 1.595, 1.580		1800		s	v s	i al
s204	aluminum chloride	$NaCl.AlCl_3$	191.78	wh-yelsh cr powd, hygr		185		s	s	
s205	aluminum meta-silicate	Nat. jadeite. $Na_2O.Al_2O_3.4SiO_2$	404.28	col, monocl	3.3	1000–1060		i	i	d HCl

No.	Name	Synonyms and Formulae	Mol. wt.	Crystalline form, properties and index of refraction	Density or spec. gravity	Melting point, °C	Boiling point, °C	Solubility, in grams per 100 cc		
								Cold water	Hot water	Other solvents
	Sodium									
s206	aluminum *ortho*-silicate	Nat. nephelite. $Na_2O.Al_2O_3.2SiO_2$	284.11	col, hex, 1.537 ± 0.002	2.619[21]	1526	....	i	d	d a
s207	aluminum sulfate	$NaAl(SO_4)_2.12H_2O$	458.28	col, cub oct, 1.4388	1.6754[20]	61	....	110[15] (anhydr)	146[30] (anhydr)	....
s208	amide	Sodamide. $NaNH_2$	39.01	wh, conchoid fract	....	210	400	d	d	d hot al; 0.1 liq NH_3
s209	ammonium phosphate	Microcosmic salt, stercorite. $NaNH_4HPO_4.4H_2O$	209.07	col, monoc, 1.439, 1.441, 1.469	1.554	d 79	....	16.7	100	i al, acet
s210	ammonium sulfate	$NaNH_4SO_4.2H_2O$ ...	173.12	wh, rhomb	1.63[15]	d 80	....	s	s	....
s211	ammonium tartrate	$NaNH_4C_4H_4O_6.4H_2O$	261.16	wh, rhomb	1.590		....	21.09[0]		s Na_2S sol
s212	*meta*antimonate	Leuconine. $NaSbO_3$	192.74	wh powd	....		....	i	s	s Na_2S sol
s213	antimonate, hydroxy	"Pyroantimonate". $NaSb(OH)_6$	246.78	pseudo cub	....		....	0.03[12.3]	0.3[100]	sl s al
s214	*pyro*antimonate, dihydro-	$Na_2H_2Sb_2O_7.6H_2O(?)$	511.58	wh, tetrag	....	d 280	....	i	0.28[100], d	
s215	antimonide	Na_3Sb	190.72	blk powd or bl cr, inflamm	....	856	....	d		sl s NH_3
s216	*meta*antimonite	$NaSbO_2.3H_2O$	230.78	col, rhomb	2.864	d	....	d		
s217	*meta*arsenate	$NaAsO_3$	145.91	rhomb, effl, 1.479, 1.502, 1.527	2.301	615	....	v s		
s218	orthoarsenate	$Na_3AsO_4.12H_2O$	423.93	col, trig or hex prism, 1.457, 1.466	1.752–1.804	86.3	....	38.9[15.5]		1.67 al; 50[15] glyc
s219	orthoarsenate, di-H	$NaH_2AsO_4.H_2O$	181.94	col, rhomb or monocl, 1.583, 1.553, 1.507	2.53	130, $-H_2O$, 100	d 200–280	s		
s220	orthoarsenate, mono-H	$Na_2HAsO_4.7H_2O$	312.01	col, monocl, pois, 1.462, 1.466, 1.478	1.88	130, $-5H_2O$, 50	d 180	5.46[0]	100[100]	s glyc; sl s al
s221	orthoarsenate, mono-H	$Na_2HAsO_4.12H_2O$	402.09	col, monocl, effl, 1.445, 1.466, 1.451	1.736	28	$-12H_2O$, 100	56[14]	140.7[30]	sl s al; i liq Cl
s222	*pyro*arsenate	$Na_4As_2O_7$	353.79	wh cr	2.205	850	d 1000	v s		....
s223	arsenate fluoride	$2Na_3AsO_4.NaF.19H_2O$	800.06	wh, cub, 1.4657, 1.4693, 1.4726	2.849[25]		....	10[75]		
s224	arsenite	Sodium metaarsenite (?) (com'l) $NaAsO_2$ (or mixt with Na_3AsO_3)	129.91	gray-wh powd, pois	1.87		....	v s	v s	sl s al
s225	arsenotartrate	$Na(AsO)C_4H_4O_6.2\frac{1}{2}H_2O$	307.02	shiny cr, pois	....	$-2\frac{1}{2}H_2O$, 275	d 275	6.5[19]	....	i al
s226	azide	NaN_3	65.01	col, hex	1.846[20]	d Na+N	d in vac	41.7[17]		0.314[16] al; s liq NH_3, i eth
s227	barbital	$NaC_8H_{11}N_2O_3$	206.18	wh powd	....		....	20[25]	40[100]	sl s al; i eth
s228	benzenesulfonate	$NaC_6H_5SO_3$	180.16	wh cr	....		....	35.8[30]	v s	
s229	benzoate	$NaC_7H_5O_2$	144.11	col cr, or wh amorph, or gran powd	....		....	66[20]	74.2[100]	1.64[25] al
s230	*meta*bismuthate	$NaBiO_3$	279.97	yel-brn powd (com'l), yel (pure)	....		....	i	d	d a
s231	*meta*borate	$NaBO_2$	65.80	col, hex pr	2.464	966	1434	26[20]	36[35]	
s232	*meta*borate, tetra-hydrate	$NaBO_2.4H_2O$	137.86	tricl, coll	....	57	$-H_2O$, 120	v s	v s	
s233	*meta*borate, peroxy-hydrate	Sodium perborate (com'l). $NaBO_2.H_2O_2.3H_2O$	153.86	col, monocl	....	63.0	$-H_2O$, 130–150	2.55[15]	3.75[32]	s a, al, glyc
s234	*tetra*borate	$Na_2B_4O_7$	201.22	cr, 1.5010	2.367	741	d 1575	1.06[0]	8.79[40]	i al
s235	*tetra*borate, deca-hydrate	Borax. $Na_2B_4O_7.10H_2O$	381.37	col, monocl, effl, 1.447, 1.469, 1.472	1.73	75, $-8H_2O$, 60	$-10H_2O$, 320	2.01[0]	170[100]	v sl s al; s glyc; i a
s236	*tetra*borate, penta-hydrate	$Na_2B_4O_7.5H_2O$	291.30	col, cub or hex, deliq, 1.461	1.815	$-H_2O$, 120	....	22.65[55] (anhydr)	52.3[100]	
s237	borohydride	$NaBH_4$	37.83	white cub, 1.542..	1.074	400 dec	....	55[25]	v s	4 al; 16.4 MeOH; s pyr; i eth
s238	bromate	$NaBrO_3$	150.90	col, cub, 1.594..	3.339[17.5]	381	....	27.5[0]	90.9[100]	i al
s239	bromide	$NaBr$	102.90	col, cub, hygr, 1.6412	3.203[$\frac{25}{4}$]	747	1390	116.0[50]	121[100]	sl s al
s240	bromide, dihydrate	$NaBr.2H_2O$	138.93	col, monocl pr	2.176	$-2H_2O$, 51	....	79.5[0]	118.6[80.5]	2.31[25] al; s liq NH_3; 17.42[15] MeOH
s241	bromoaurate	$NaAuBr_4.2H_2O$	575.62	br-blk cr	....		....	s		
s242	bromoiridite	$Na_2IrBr_6.12H_2O$	956.81	dk grn, rhomb, effl	....	100	$-H_2O$, 150	....		s NH_4OH

No.	Name	Synonyms and Formulae	Mol. wt.	Crystalline form, properties and index of refraction	Density or spec. gravity	Melting point, °C	Boiling point, °C	Cold water	Hot water	Other solvents
	Sodium									
s243	bromoplatinate	$Na_2PtBr_6.6H_2O$	828.62	dk red, tricl	3.323	d 150	....	v s	v s	v s al
s244	cacodylate	$Na[(CH_3)_2AsO_2].3H_2O$	214.03	wh		ca 60	$-H_2O$, 120	200^{15-20}	....	40^{25} al; 100^{15-20} 90 % al
s245	calcium sulfate	$Na_2Ca(SO_4)_2.2H_2O$	314.21	col, monocl need.	2.64	$-2H_2O$, 80	....	d	d	
s246	d-camphorate	$Na_2C_{10}H_{14}O_4.3H_2O$	298.25	wh need, hygr		$-3H_2O$, 100	....	122^{14}		s al
s247	carbide	Na_2C_2	70.00	wh powd	1.575^{15}	ca 700		d	d	s a; d al
s248	carbonate	Na_2CO_3	105.99	wh powd, hygr, 1.535	2.532	851	d	7.1^0	45.5^{100}	sl s abs al; i acet
s249	carbonate, decahydrate	Washing soda. $Na_2CO_3.10H_2O$	286.14	wh, monocl, 1.405, 1.425, 1.440	1.44^{15}	32.5-34.5	$-H_2O$, 33.5	21.52^0	421^{104}	i al
s250	carbonate, heptahydrate	$Na_2CO_3.7H_2O$	232.10	rhomb bipyr, effl	1.51	$-H_2O$, 32		16.90	33.9^{35}	....
s251	carbonate, monohydrate	Crystal carbonate, thermonatrite. $Na_2CO_3.H_2O$	124.00	col, rhomb, deliq, 1.506, 1.509	2.25	$-H_2O$, 100		33	52.08	14^{25} glyc; i al, eth
s252	carbonate, sesqui-	$Na_2CO_3.NaHCO_3.2H_2O$	226.03	col, monocl, 1.5073	2.112	d		13^0	42^{100}	
s253	carbonate hydrogen	$NaHCO_3$	84.00	wh, monocl pr, 1.500	2.159	$-CO_2$,270		6.9^0	16.4^{60}	sl s al
s254	chlorate	$NaClO_3$	106.44	col, cub or trig, 1.513	2.490^{15}	248-261	d	79^0	230^{100}	s al, liq NH_3, glyc
s255	perchlorate	$NaClO_4$	122.44	wh, rhomb, deliq, 1.4606, 1.4617, 1.4731		d 482	d	s	v s	s al
s256	perchlorate, hydrate	$NaClO_4.H_2O$	140.46	col rhbdr, deliq	2.02	130	d 482	209^{15}	284^{50}	s al
s257	chloride	Common salt, nat. halite. NaCl	58.44	col, cub, 1.5442	2.165^{25}_4	801	1413	35.7^0	39.12^{100}	sl s al, liq, NH_3; s glyc; i HCl
s258	chlorite	$NaClO_2$	90.44	wh, cr, hygr		d 180-200		39^{17}	55^{60}	....
s259	chlorite, pentahydrate	$NaOCl.5H_2O$	164.52	col		24.5		29.3^5	94.2^{33}	
s260	hypochlorite	NaOCl	74.44	in solution only						
s261	hypochlorite, dihydrate	$NaOCl.2\frac{1}{2}H_2O$	119.48	col, hygr		57.5		v s		
s262	chloroaurate	$NaAuCl_4.2H_2O$	397.80	yel, rhomb, ω 1.545 ϵ>1.75		d 100		150^{10}	990^{60}	v s al, eth
s263	chloroiridate	$Na_2IrCl_6.6H_2O$	558.99	dull red-blk, tricl		d 600		v s	v s	sl s al
s264	chloroiridite	$Na_3IrCl_6.12H_2O$	690.07	dk grn cr		$-H_2O$, 50		31.46^{15}	307.26^{85}	
s265	chloroosmate	$Na_2OsCl_6.2H_2O$	484.93	or-red, rhomb pr				v s		s al
s266	chloropalladite	$Na_2PdCl_4.3H_2O$	348.24	br-red cr, deliq				v s		s al
s267	chloroplatinate	Na_2PtCl_6	453.79	or-yel powd, hygr		tr 150-160		s	v s	s al
s268	hexachloroplatinate	$Na_2PtCl_6.6H_2O$	561.88	or-red, tricl	2.500	$-6H_2O$, 100		66^{15}	v s	11.9 al, MeOH; i eth
s269	chloroplatinite	$Na_2PtCl_4.4H_2O$	454.98	red pr		100	$-H_2O$, 150	s		s al
s270	chlororhodite, hexa-	Na_3RhCl_6	384.59	red, tricl		d>550		v s		
s271	chlororhodite, hexa-, hydrate	$Na_3RhCl_6.18H_2O$	708.87	garnet red, oct, effl		d 904, effl		v s		i al
s272	chromate	Na_2CrO_4	161.97	yel, rhomb bipyram	2.710-2.736			87.3^{30}		sl s al; s MeOH
s273	chromate decahydrate	$Na_2CrO_4.10H_2O$	342.13	yel, monocl, deliq	1.483	19.92		50^{10}	126^{100}	sl s al; i ac a
s274	dichromate	$Na_2Cr_2O_7.2H_2O$	298.00	red, monocl pr, deliq, 1.661, 1.699, 1.751	2.52^{13}	$-2H_2O$, 100 356.7 (anhydr)	d 400 (anhydr)	238^0 (anhydr) 180^{20}	508^{80} (anhydr) 433^{98}	i al
s275	peroxychromate	Na_3CrO_8	248.96	or pl		d 115		sl s		i al, eth
s276	cinnamate	$NaC_9H_7O_2$	170.14	wh cr powd				9.1	5^{100}	0.625 90 % al; s glyc
s277	citrate, dihydrate	$Na_3C_6H_5O_7.2H_2O$	294.10	wh cr, gran or powd		$-2H_2O$, 150		72^{25}	167^{100}	0.625 90 % al; s glyc
s278	citrate, pentahydrate	$Na_3C_6H_5O_7.5(or 5\frac{1}{2})H_2O$	348.15	wh, rhomb	$1.857^{33.5}$	$-5H_2O$, 150	d	92.6^{25}	250^{100}	sl s al
s279	cobaltinitrite	$Na_3Co(NO_2)_6$	403.94	yelsh-brnsh cr powd				v s, d		sl s al; d min a; i dil ac a
s280	cyanamide, mono-	$NaHCN_2$	64.02	wh cr powd, hygr				v s		
s281	cyanate	NaOCN	65.01	col need	1.937^{20}	d 700 vac		s	s	v sl s eth, bz
s282	cyanide	NaCN	49.01	col, cub, deliq, pois, 1.452		563.7	1496	48^{10}	82^{35}	sl s al; s NH_3
s283	cyanoaurite	Sodium aurocyanide. $NaAu(CN)_2$	271.99					s		
s284	cyanocuprate (I)	$NaCu(CN)_2$	138.57		1.013^{20}	d 100		s		
s285	cyanoplatinite	$Na_2[Pt(CN)_4].3H_2O$	399.19	col, tricl	2.646	$-3H_2O$, 120-125		s	s	s al
s286	enanthate	Sodium heptanoate. $NaC_7H_{13}O_2$	152.17	wh cr powd or leaf		240-350		s		s al
s287	ethyl acetoacetate	$NaC_6H_9O_3$	152.13	need		d		d		s eth
s288	ethyl sulfate	$NaC_2H_5SO_4.H_2O$	166.13	wh, hex pl, deliq		d		164^{17}		d alk, H_2SO_4; 142 al

Sodium

No.	Name	Synonyms and Formulae	Mol. wt.	Crystalline form, properties and index of refraction	Density or spec. gravity	Melting point, °C	Boiling point, °C	Cold water	Hot water	Other solvents
s289	ferrate (III)	Ferrite. $Na_2Fe_2O_4$	221.67	br, hex pl or need	4.05			d		v s dil HCl
s290	ferricyanide	$Na_3Fe(CN)_6.H_2O$	298.92	red cr, deliq				18.9^0	67^{100}	i al
s291	ferrocyanide	Yellow prussiate of soda. $Na_4Fe(CN)_6.10H_2O$ Sodium hexacyanoferrate (II).	484.04	pa yel, monocl, 1.519, 1.530 1.544	1.458			31.85^{20}	156.5^{98}	i al
s292	fluoaluminate	Na_3AlF_6	209.94	col, monocl, β 1.364	2.90	1000		sl s		i HCl,; d alk
s293	fluoantimonate	$NaSbF_6$	258.73	rhomb	3.375^{18}	<1360		128.6^{20}		s al, acet
s294	fluoberyllate	Na_2BeF_4	130.99	wh, rhomb or monocl	d			1.47^{18}	2.94^{100}	
s295	fluoborate	$NaBF_4$	109.79	wh, rhomb	2.47^{20}	sl d 384	d	108^{26}	210^{100}	sl s al; d H_2SO_4
s296	fluoride	Nat. villiaumite. NaF	41.99	col, cub or tetr, 1.336	2.558^{41}	993	1695	4.22^{18}		s HF; v sl s al
s297	fluoride, hydrogen	$NaF.HF$	61.99	col, or wh cr powd, rhdr	2.08			s	s	
s298	fluoride orthophosphate	$NaF.Na_3PO_4.12H_2O$	422.11		2.2165			12^{25}	57.5^{50}	
s299	fluoroacetate, mono-	$NaC_2H_2FO_2$	100.02	wh powd		200		111^{25}		1.4^{25} al; 5^{25} MeOH; 0.04^{25} acet; 0.0049^{25} CCl_4
s300	fluorophosphate, hexa-	$NaPF_6.H_2O$	185.97		2.369^{19}			103.2^0		
s301	fluorophosphate, mono-	Na_2PO_3F	143.95	col		ca 625		25		
s302	fluosilicate	Na_2SiF_6	188.06	col, hex, 1.312, 1.309	2.679	d		0.0652^{17}	2.46^{100}	i al
s303	fluosulfonate	$NaSO_3F$	122.05	shiny leaf, hygr		d red heat		s		s al, acet; i eth
s304	formaldehyde-sulfoxylate	$NaHSO_2.CH_2O.2H_2O$	154.12	rhomb pr, hygr		64	d	v s		d a; s al, alk
s305	formate	$NaCHO_2$	68.01	col, monocl, deliq	1.92^{20}	253	d	97.2^{20}	160^{100}	sl s al; i eth
s306	2-furanacrylate	$NaC_7H_5O_3$	160.10	lt brn powd	1.919	d		s	s	sl s al; i eth
s307	metagermanate	Na_2GeO_3	166.57	wh, monocl, deliq, 1.59	3.31^{22}	1083			d	s a
s308	metagermanate, heptahydrate	$Na_2GeO_3.7H_2O$	292.68	col, rhomb		83		24.6^0 45.5^{25}		s a
s309	(mono-) d-glutamate	$NaC_5H_8NO_4$	169.11	wh cr		d		v s		sl s al
s310	glycerophosphate, monohydrate	$Na_2C_3H_7O_6P.H_2O$	234.05	yelsh visc liq ; wh cr or powd				s		s al
s311	glycerophosphate, pentahydrate	$Na_2C_3H_7O_6P.5\frac{1}{2}H_2O$	315.12	wh pl, sc or powd		>130		v s		i al
s312	gold sulfide	$NaAuS.4H_2O$	324.08	col, monocl		d		s		s al
s313	hydride	NaH	24.00	silver need, 1.470	0.92	d 800		d	d	s molten Na; i CS_2, CCl_4, NH_3, bz
s314	hydroxide	$NaOH$	40.00	wh, deliq, 1.3576	2.130	318.4	1390	42^0	347^{100}	v s al, glyc; i acet, eth
s315	iodate	$NaIO_3$	197.89	wh, rhomb	$4.277^{17.5}$	d		9^{20}	34^{100}	i al; s ac a
s316	metaperiodate	$NaIO_4$	213.89	col, tetr	**4.174**	d 300		14.44^{25}	$38.9^{81.5}$	s H_2SO_4, HNO_3, ac a
s317	metaperiodate, trihydrate	$NaIO_4.3H_2O$	267.94	col, rhombdsh, effl	3.219^{18}_4	d 175		18.78^{25}	$36.4^{34.5}$	
s318	paraperiodate	Na_5IO_6	337.85	wh		800 d		d		
s319	(tri-)paraperiodate	$Na_3H_2IO_6$	293.88	col, hexag				sl s		s con NaOH sol
s320	iodide	NaI	149.89	col, cub, 1.7745	3.667^{25}_4	661	1304	184^{25}	302^{100}	42.57^{25} al; 39.9^{25} acet; s glyc
s321	iodide, dihydrate	$NaI.2H_2O$	185.92	col, monocl	$2.448^{20.8}$	752		317.9^0	1550^{100}	v s NH_3
s322	iodoplatinate	$Na_2PtI_6.6H_2O$	1110.59	brn, monocl	3.707			v s		s al
s323	iridium chloride	Sodium hexachloroiridate. $Na_3IrCl_6.12H_2O$	690.07	olive cr, rhomb or trig-rhomb		50		s		i al
s324	iron (III) nitrosopentacyanide	$Na_2[Fe(CN)_5NO].2H_2O$	297.95	ruby red, rhomb, 1.605, 1.575, 1.56	1.687^{25}	$-H_2O$, 100	d 160	40^{16}	s	
s325	iron (III) oxalate	$Na_3[Fe(C_2O_4)_3].5\frac{1}{2}H_2O$	487.96	grn, monocl	$1.973^{17.5}$	$-4H_2O$	d 300	32^0	182^{100}	
s326	iron (III) sulfate	$3Na_2SO_4.Fe_2(SO_4)_3.6H_2O$	934.09	wh, trig, 1.558, 1.613	2.5	$-6H_2O$, 100		d v sl		i al
s327	lactate	$NaC_3H_5O_3$	112.06	col or yelsh liq, very hygr		17	d 140	v s		s al; i eth
s328	lithium sulfate	$Na_2Li(SO_4)_2.6H_2O$	376.12	col, ditrig	2.009	$-6H_2O$, 50		s	s	
s329	magnesium carbonate	$Na_2CO_3.MgCO_3$	190.31	wh, rhomb	2.729^{15}	677 CO_2 1240/atm		d	d	
s330	magnesium sulfate	Nat. bloedite. $Na_2SO_4.MgSO_4.4H_2O$	334.48	col, monocl, 1.486, 1.488, 1.489	2.23			s		
s331	magnesium tartrate	$Na_2Mg(C_4H_4O_6)_2.10H_2O$	546.59	wh, monocl pr or powd				s	s	

No.	Name	Synonyms and Formulae	Mol. wt.	Crystalline form, properties and index of refraction	Density or spec. gravity	Melting point, °C	Boiling point, °C	Solubility, in grams per 100 cc		
								Cold water	Hot water	Other solvents
	Sodium									
s332	manganate	$Na_2MnO_4.10H_2O$	345.07	grn, monocl		17		s	d	
s333	permanganate	$NaMnO_4$	141.93	red cr. deliq		d		v s	v s	
s334	permanganate, trihydrate	$NaMnO_4.3H_2O$	195.97	purp cr, deliq	2.47	d 170		v s	v s	s NH_3; d alk
s335	methanearsenate	$Na_2CH_3AsO_3.6H_2O$	292.03	wh cr powd		130–140		ca 100		sl s al; i bz, eth, oils
s336	methoxide	$CH_3ONa.2CH_3OH$	118.11	wh powd		d, $-CH_3OH$		s, d		s CH_3OH
s337	methylsulfate	$NaCH_3SO_4.H_2O$	152.10	col cr, hygr				s		s al
s338	molybdate	Na_2MoO_4	205.92	opaque wh	3.28^{18}	687		s 44.3	84^{100}	
s339	molybdate, dihydrate	$Na_2MoO_4.2H_2O$	241.95	wh, rhbdr	3.28(?)	$-2H_2O$, 100		56.2^0	115.5^{100}	i meth acet
s340	decamolybdate	$Na_2Mo_{10}O_{31}.21H_2O$	1879.68	wh, monocl pr				sl s	0.842^{100}	
s341	dimolybdate	$Na_2Mo_2O_7$	349.86	wh need		612		sl s	sl s	
s342	octamolybdate	$Na_2Mo_4O_{2.5}.17H_2O$	1519.75	monocl cr		$-H_2O$, 20		v s	v s	
s343	paramolybdate	$Na_6Mo_7O_{24}.22H_2O$	1589.84	col, monocl, effl		$700 - H_2O$, 100–120		117.9^{30} (anhydr)	s	
s344	tetramolybdate	$Na_2Mo_4O_{13}.6H_2O$	745.82	yel need				39.8^{21}	v s	
s345	trimolybdate	$Na_2Mo_3O_{10}.7H_2O$	619.96	need acicular		$528 - 6H_2O$, 100–120		3.878^{20}	13.7^{100}	
s346	metaniobate	$Na_2Nb_2O_6.7H_2O$	453.90	pseudo-cub	4.512–4.559	$-H_2O$, 100		s		
s347	nitrate	Soda niter. $NaNO_3$	84.99	col, trig or rhbdr, 1.587, 1.336	2.261	306.8	d 380	92.1^{25}	180^{100}	s al, MeOH; v s NH_3; v sl s acet; sl s glyc
s348	nitride	Na_3N	82.98	dk gray		d 300		d		
s349	nitrite	$NaNO_2$	69.00	col-yel, rhomb pr, hygr	2.168^0	271	d 320	81.5^{15}	163^{100}	0.3^{20} eth; 4.4^{20} MeOH; 3 abs al; v s NH_3
s350	hyponitrite	$Na_2N_2O_2$	105.99		1.728^{25}	d 300		d		i al
s351	p-nitrophenoxide	$NaOC_6H_4NO_2.4H_2O$	233.15	yel, monocl pr		$-2H_2O$, 36 $-4H_2O$, 120	d	5.97^{25}		sl s al
s352	nitroplatinite	$Na_2Pt(NO_2)_4$	425.09	pa yel rhomb or monocl pr, effl				s	s	
s353	nitroprusside	$Na_2[Fe(NO)(CN)_5].2H_2O$	297.95	red, rhomb	1.72			40^{16}		s al
s354	oleate	$NaC_{18}H_{33}O_2$	304.45	wh cr, or yel amorph gran		232–235		10^{12}		s al; sl s eth
s355	oxalate	$Na_2C_2O_4$	134.00	col cr, or wh powd	2.34	d 250–270		3.7^{20}	6.33^{100}	i al, eth
s356	oxalate, hydrogen	$NaHC_2O_4.H_2O$	130.03	wh, monocl		$-H_2O$, 100	d 200	1.7^{15}	21^{100}	
s357	oxalatoferrate (III)	$Na_3Fe(C_2O_4)_3.xH_2O$	$388.88 + xH_2O$	grn, monocl cr	$1.973^{17.5}$	$-H_2O$, 100–120		32.5	182^{100}	
s358	oxide, mon-	Na_2O	61.98	wh-gray, deliq	2.27	subl 1275		d	d	d al
s359	oxide, per-	Na_2O_2	77.98	yel-wh powd	2.805	d 460	d 657	s	d	d al, NH_3; s dil a; i alk
s360	oxide, per-, octahydrate	$Na_2O_2.8H_2O$	222.10	wh, hex		d 30	d	s	d	i al
s361	palmitate	$NaC_{16}H_{31}O_2$	278.41	wh cr		270				
s362	pentobarbital	$NaC_{11}H_{17}N_2O_3$	248.26	wh				s	s, d	s al
s363	phenobarbital	$NaC_{12}H_{11}N_2O_3$	254.22	wh				v s		s al; i eth, chl
s364	1-phenol-4-sulfonate(p-)	$NaC_6H_4(OH)SO_3.2H_2O$	232.19	col, monocl or gran, sl effl		d		23.8^{25}	125^{100}	0.75^{25} al; 20^{25} glyc
s365	phenoxide	$NaOC_6H_5$	116.10	wh cr need, deliq				v s		s al, acet; d a
s366	phenylcarbonate	$NaC_7H_5O_3$	160.11	col powd		d 120		d		d acet
s367	hypophosphate	$Na_4P_2O_6.10H_2O$	430.06	col, monocl, 1.477, 1.482, 1.504	1.823	d		1.49^{25}	5.46^{50}	
s368	hypophosphate, di-H	$Na_2H_2P_2O_6.6H_2O$	314.03	col, monocl, 1.468, 1.490, 1.504	1.849	250 (anhydr)	$-6H_2O$, 100	2.35	25	s dil H_2SO_4, NH_4OH; i al
s369	metaphosphate, hexa-	Graham's salt. $(NaPO_3)_6$	611.17	col glass, 1.482 ± 0.002				v s		
s370	metaphosphate, tri-, hexahydrate	Knorre's salt. $(NaPO_3)_3.6H_2O$	413.98	col, tricl, effl, 1.433, 1.442, 1.446		53; $-6H_2O$, 50		s		
s371	orthophosphate	$Na_3PO_4.10H_2O$	344.09	col, oct	$2.536^{17.5}$ (anhydr)	100		8.8 (anhydr)		
s372	orthophosphate	$Na_3PO_4.12H_2O$	380.12	col, trig, 1.446, 1.452	1.62^{20}	d 73.3–76.7	$-12H_2O$, 100	1.5^0	157^{70}	i CS_2, al
s373	orthophosphate, di-H	$NaH_2PO_4.H_2O$	137.99	col, rhomb, 1.456, 1.458, 1.487	2.040	$-H_2O$, 100	d 204	59.9^0	427^{100}	v sl s eth, chl, tol; i al
s374	orthophosphate, di-H	$NaH_2PO_4.2H_2O$	156.01	col, rhomb, 1.4629	1.91	60		v s	v s	—
s375	orthophosphate, mono-H	Sörensen's sodium phosphate. $Na_2HPO_4.2H_2O$	177.99	rhomb bispheroidal, 1.463	2.066^{16}	$-2H_2O$, 95		100^{50}	117^{80}	i al
s376	orthophosphate, mono-H	$Na_2HPO_4.7H_2O$	268.07	col, monocl pr, 1.442	1.679	$-5H_2O$, 48.1		104^{40}		i al
s377	orthophosphate, mono-H	$Na_2HPO_4.12H_2O$	358.14	col, rhomb or monocl, eff, wh powd, 1.432, 1.436, 1.437	1.52	$-5H_2O$, 35.1	$-12H_2O$, 100	4.15	87.4^{34}	i al

No.	Name	Synonyms and Formulae	Mol. wt.	Crystalline form, properties and index of refraction	Density or spec. gravity	Melting point, °C	Boiling point, °C	Cold water	Hot water	Other solvents
	Sodium									
s378	*pyro*phosphate	$Na_4P_2O_7$	265.90	wh cr, 1.425	2.534	880		3.16^0	40.26^{100}	
s379	*pyro*phosphate	$Na_4P_2O_7.10H_2O$	446.06	col, monocl, 1.450, 1.453, 1.460	1.815–1.836	$-H_2O$, 93.8 m.p. 880		5.41^0	93.11^{100}	i al, NH_3
s380	*pyro*phosphate, di-H	$Na_2H_2P_2O_7.6H_2O$	330.03	monocl, 1.4599, 1.4645, 1.4649	1.85	$-H_2O$, 220		6.9^0	35^{40}	
s381	phosphide	Na_3P	99.94	red		d		d, PH_3		
s382	*hypo*phosphite	$NaH_2PO_2.H_2O$	105.99	col, monocl, deliq		d viol		100^{25}	667^{100}	v s al; s glyc; sl s NH_3, NH_4OH
s383	*ortho*phosphite, di-H	$NaH_2PO_3.2\frac{1}{2}H_2O$	149.01	col, monocl, 1.419, 1.431, 1.449		42	$-2\frac{1}{2}H_2O$, 100	56^0	193^{42}	
s384	*ortho*phosphite, mono-H	$Na_2HPO_3.5H_2O$	216.04	wh, rhomb deliq, β, 1.443	53	d 200–250		s	v s	i al, NH_4OH
s385	triphosphate	Sodium tripoly-phosphate. $Na_5P_3O_{10}$	367.86	powd and gran				14.5^{25}	32.5^{100}	
s386	phthalate	$Na_2C_8H_4O_4$	210.10	wh powd or pearly pl		$-H_2O$, 150				
s387	platinate, hydroxo-	$Na_2Pt(OH)_6$	343.11	yel or red-brn, hex		$-3H_2O$, 150–170	d	s		i al; sl s HCl
s388	platinum cyanide	$Na_2[Pt(CN)_4].3H_2O$	399.19	col, tricl	2.646	$-H_2O$, 120–125		s		s al
s389	plumbate, hydroxo-	$Na_2Pb(OH)_6$	355.21	yel-wh lumps, hygr				d to PbO_2		d a; s alk
s390	potassium(*dl*)-tartrate	$NaKC_4H_4O_6$	210.14	col, tricl		90–100	d 200	47.4^6 (anhydr)	v s	
s391	propionate	$NaC_3H_5O_2$	96.06	wh, gran powd				s		s al
s392	perrhenate	$NaReO_4$	273.19	col, hex pl, hygr	5.39	300 (in O_2) d 440 (vac)		100^{20}		s al
s393	*pyrohypo*rhenate	$Na_4Re_2O_7.2H_2O$	594.37	sandy yel cr				0.004		
s394	rhodiumchloride	$Na_3RhCl_6.12H_2O$	600.78	dk-red cr, monocl pr		$-12H_2O$, 120		v s	v s	i al
s395	rhodiumnitrite	$Na_3[Rh(NO_2)_6]$	447.91	wh cr		d 360		40^{17}	s	i al; d a
s396	perruthenate	$NaRuO_4.H_2O$	206.07	blk cr, lamellar		d 440 vac		v s	d	
s397	salicylate	$NaC_7H_5O_3$	160.11	wh cr powd				111^{15}	125^{25}	17^{15} al; 25 glyc
s398	selenate	Na_2SeO_4	188.94	col, rhomb	$3.213^{17.4}$			84^{35}	72.8^{100}	
s399	selenate, decahydrate	$Na_2SeO_4.10H_2O$	369.09	col, monocl	1.603–1.620	ca 32 trans		43.5^{20}	340^{100}	
s400	selenide	Na_2Se	124.94	wh to red, cr, deliq	2.625^{10}	>875		d		i NH_3
s401	selenite	$Na_2SeO_3.5H_2O$	263.01	wh cr, tetrag				s	s	i al
s402	silicate	Waterglass. $Na_2O.xSiO_2(x=3-5)$		col, amorph, deliq				s	s	i al, K and Na salts
s403	*di*silicate	$Na_2Si_2O_5$	182.15	rhomb pearly luster 1.500, 1.510		874		s	s	
s404	*meta*silicate	Na_2SiO_3	122.06	col, monocl, α 1.518, γ 1.527	2.4	1088		s	s, d	i al, K and Na salts
s405	*meta*silicate	$NaSiO_3.9H_2O$	284.20	col, rhomb bi-pyramid, effl		40–48	$-6H_2O$, 100	v s	v s	s dil NaOH; i al, a
s406	*ortho*silicate	Na_4SiO_4	184.04	col, hex, 1.530		1018		s	s	
s407	silicotungstate, dodeca-	$Na_4[Si(W_3O_{10})_4].20H_2O$ **Sodium tungstosilicate**	3326.53	col, tricl		$-7H_2O$, 100	d	v s	v s	s a, sl s al
s408	stannate, hydroxo-	$Na_2Sn(OH)_6$	266.71	col, hex or wh powd or lumps		$-3H_2O$, 140		$61.3^{15.5}$	50^{100}	i al, acet
s409	stearate	$NaC_{18}H_{35}O_2$	306.47	wh fatty powd				s	s	s h al
s410	succinate	$Na_2C_4H_4O_4.6H_2O$	270.15	wh, gran or powd		$-6H_2O$, 120		21.45^0	86.63^{75}	v sl s al
s411	succinate, tetra-hydroxy-	Sodium dihydroxy tartrate. $Na_2C_4H_8O_8.3H_2O$	280.10			d		0.032^0	d	d min a; i al, eth
s412	sulfanilate	$NaC_6H_4(NH_2)SO_3$	195.17	wh lust cr leaf				s		
s413	sulfate, anhydr	Na_2SO_4	142.04	monocl (between ca 160–185), 1.480		884; tr to hex ca 241		s	$42-5^{100}$	s HI
s414	sulfate, anhydr	Nat. thenardite. Na_2SO_4	142.04	orthorhomb,1.484, 1.477, 1.471	2.68			4.76^0	42.7^{100}	s glyc; i al
s415	sulfate, decahydrate	Glauber's salt, mirabilite. $Na_2SO_4.10H_2O$	322.19	col, monocl, effl, 1.394, 1.396, 1.398	1.464	32.38	$-10H_2O$, 100	11^0	92.7^{30}	i al
s416	sulfate, heptahydrate	$Na_2SO_4.7H_2O$	268.15	wh, rhomb or tetrag		tr to anhydr 24.4		19.5^0	44^{20}	i al
s417	*pyro*sulfite	Sodium metabisulfite. $Na_2S_2O_5$	190.10	wh powd or cr (+7H_2O)	1.4	>d 150		54^{20}	81.7^{100}	sl s al; s glyc
s418	*pyro*sulfate	$Na_2S_2O_7$	222.16	wh, transluc cr, deliq	2.658^{25}	400.9	d 460	s		s fum H_2SO_4
s419	sulfate hydrogen	$NaHSO_4$	120.06	col, tricl	2.435^{13}	>315	d	28.6^{25}	100^{100}	sl s al; i NH_3
s420	sulfate hydrogen, monohydrate	$NaHSO_4.H_2O$	138.07	col, monocl, deliq ~1.46	$2.103^{12.5}_4$	58.54 ± 0.5		ca 67, d	d	d al
s421	sulfide, hydro-	$NaHS$	56.06	col, rhomb or wh gran cr, deliq		350		vs		s al

No.	Name	Synonyms and Formulae	Mol. wt.	Crystalline form, properties and index of refraction	Density or spec. gravity	Melting point, °C	Boiling point, °C	Solubility, in grams per 100 cc		
								Cold water	Hot water	Other solvents
	Sodium									
s422	sulfide, hydrodi-hydrate	NaHS.2H₂O	92.09	col need, deliq		d		s	s	d a; s al
s423	sulfide, mono-	Na₂S	78.04	wh cr, deliq	1.856¹⁴	1180		15.4¹⁰	57.2⁹⁰	d a; sl s al; i eth
s424	sulfide, mono-hydrate	Na₂S.9H₂O	240.18	col, tetr, deliq	1.427⁴¹⁶	d 920		47.5¹⁰	96.7¹⁰	d, sl s al
s425	sulfide, penta-	Na₂S₅	206.30	yel (exist ?)		251.8		s	s	s al
s426	sulfide, tetra-	Na₂S₄	174.24	yel, cub, hygr		275	d	s		s al
s427	sulfide, hydro-trihydrate	NaHS.3H₂O	110.11	col, lust rhomb cr		22	d	s	s	s al
s428	sulfite	Na₂SO₃	126.04	wh powd or hex, prism 1.565, 1.515	2.633¹⁵·⁴	d red heat	d	12.54⁰	28.3⁸⁰	sl s al; i liq Cl₂, NH₃
s429	sulfite hydrate	Na₂SO₃.7H₂O	252.15	col, monocl, effl	1.539¹⁵	−7H₂O, 150	d	32.8⁰	196⁴⁰	sl s al
s430	*hydro*sulfite	Dithionite, hyposulfite Na₂S₂O₄.2H₂O	210.14	col, monocl(?) cr, or yel-wh powd		d 52		25.4²⁰	d	d a; s alk; i al
s431	sulfite, hydrogen	NaHSO₃	104.06	wh, monocl, yel in sol, 1.526	1.48	d		v s	v s	sl s al
s432	d(& l)-tartrate	Na₂C₄H₄O₆.2H₂O	230.08	col, rhomb, 1.545, 1.49	1.818	−2H₂O, 150		29⁶	66⁴³	i al
s433	d-tartrate, hydrogen	NaHC₄H₄O₆.H₂O	190.09	wh cr powd, rhomb, 1.53, 1.54, 1.60		−H₂O, 100	d 234	6.7¹⁸	9.2³⁰	
s434	dl-tartrate hydrogen	NaHC₄H₄O₆.H₂O	190.09	col, monocl or tricl, 1.53, 1.54, 1.60		−H₂O, 100	d 219	8.9¹⁹		
s435	*ortho*tellurate, tetra-H	Na₂H₄TeO₆	273.61	col, hex pl		d	d	0.77¹⁸	2¹⁰⁰	s h dil HNO₃; i NaOH
s436	telluride	Na₂Te	173.58	wh cr powd very hygr, d in air	2.90	953		v s, d	v s, d	
s437	tellurite	Na₂TeO₃	221.58	wh, rhomb pr				sl s	s	
s438	thioantimonate	Schlippe's salt. Na₃SbS₄.9H₂O	481.11	pa yel, cub	1.806	87	d 234	20.15⁰	100¹⁰⁰	i al, eth
s439	thioarsenate	Na₃AsS₄.8H₂O	416.27	yel, monocl, β 1.6802		d		v s	d	i al
s440	thiocarbonate, tri	Na₂CS₃.H₂O	172.20	yel need, deliq		d 75		s	d	s al; i eth, bz
s441	thiocyanate	NaSCN	81.07	col, rhomb deliq pois, ∼1.625		287		139.31²¹·³	225¹⁰⁰	v s al, acet
s442	*di*thionate	Na₂S₂O₆.2H₂O	242.13	col, rhomb, 1.482, 1.495, 1.519	2.189	−2H₂O, 110	−SO₂, 267	47.6¹⁶	90.9¹⁰⁰	s HCl; i al
s443	thiosulfate	Na₂S₂O₃	158.11	col, monocl	1.667			50	231¹⁰⁰	i al
s444	thiosulfate, pentahydrate	"Hypo", sodium hyposulfite. Na₂S₂O₃.5H₂O	248.18	col, monocl, effle, 1.489, 1.508, 1.536	1.729¹⁷	40–45 d 48	−5H₂O, 100	79.4⁰	291.1⁴⁵	s NH₃; i al
s445	thiosulfoaurate (I)	Na₃[Au(S₂O₃)₂].2H₂O	526.22	wh cr, monocl	3.09	−H₂O, 150	d	50		i al
s446	*tri*titanate	Na₂Ti₃O₇	301.68	wh need, monocl	3.35–3.50	1128		i		s h HCl
s447	tungstate	Na₂WO₄	293.83	wh, rhomb	4.179	698		57.5⁰ 73.2²¹	96.9¹⁰⁰	
s448	tungstate, dihydrate	Na₂WO₄.2H₂O	329.86	col pl, rhomb, 1.5533	3.23–3.25	−2H₂O, 100 anhydr 698		41⁰	123.5¹⁰⁰	sl s NH₃; i al, a
s449	*meta*tungstate	Na₂O.4WO₃.10H₂O	1169.53	col, oct		706.6		s	v s	i a
s450	*para*tungstate	Na₁₀W₁₂O₄₁.16H₂O	2097.12	col, tricl	3.987	−12H₂O, 100; −16H₂O, 300		8	d	
s451	*meta*uranate	Na₂UO₄	348.01	gr yel or red pl, rhomb pr				i	i	s dil a, alk carb
s452	uranyl acetate	NaUO₂(C₂H₃O₂)₃	470.15	yel, tetr pr, 1.501	2.56					
s453	uranyl carbonate	2Na₂CO₃.UO₂CO₃	542.02	yel cr		d 400		sl s		i al
s454	urate	Na₂C₅H₂N₄O₃.H₂O	230.09	wh gran powd or hard cr nodules					1.3¹⁰⁰	v sl s 90 % al
s455	urate, acid	NaHC₅H₂N₄O₃	190.09	wh gran powd				0.083	0.8¹⁰⁰	
s456	valerate	NaC₅H₉O₂	124.12	wh cr or mass, hygr		140		s		s al
s457	*meta*vanadate	NaVO₃	121.93	col, monocl pr		630		21.1²⁵	38.8⁷⁵	
s458	*ortho*vanadate	Na₃VO₄	183.94	col, hex pr		850–866		s		i al
s459	*ortho*vanadate, decahydrate	Na₃VO₄.10H₂O	364.06	wh, cub or hex cr, 1.5305, 1.5398, 1.5475				s	s	
s460	*ortho*vanadate, hexadecylhydrate	Na₃VO₄.16H₂O	472.15	col need		866 (anhydr)		v s	d	i al
s461	*pyro*vanadate	Na₄V₂O₇	305.84	col, hex		632–654		s	s	i al
s462	ethylxanthate	NaC₂H₅OCSS	144.19	yelsh powd				s	s	s al
s463	zinc uranyl acetate	NaZn(UO₂)₃(C₂H₃O₂)₉.9H₂O	1591.98	monocl cr, α 1.475, γ 1.480				i		s al
	Stannous	See under tin								
	Stannic	See under tin								

No.	Name	Synonyms and Formulae	Mol. wt.	Crystalline form, properties and index of refraction	Density or spec. gravity	Melting point, °C	Boiling point, °C	Solubility, in grams per 100 cc		
								Cold water	Hot water	Other solvents
	Strontium									
s464	**Strontium**	Sr	87.62	silv wh to pa yel met	2.6^{20}	769	1384	d	d	s a, al, liq NH_3
s465	acetate	$Sr(C_2H_3O_2)_2$	205.71	wh cr	2.099	d		36.9	36.4^{97}	0.26^{15} MeOH
s466	acetate	$Sr(C_2H_3O_2)_2.\frac{1}{2}H_2O$	214.72	wh cr powd		$-\frac{1}{2}H_2O$, 150		s	sl s al	
s467	*ortho*arsenate, acid	$SrHAsO_4.H_2O$	245.56	rhomb, need	3.606^{15}; 4.035 (anhydr)	$-H_2O$, 125		$0.284^{15.5}$	d	s a
s468	*ortho*arsenite	$Sr_3(AsO_3)_2.4H_2O$	580.76	cr, or wh powd				sl s		s a; sl s al
s469	borate, *tetra-*	$SrB_4O_7.4H_2O$	314.92						77^{100}	s HNO_3; NH_4 salts
s470	boride, *hexa-*	SrB_6	152.49	blk, cub	3.39^{15}	2235		i	i	s HNO_3; i HCl
s471	bromate	$Sr(BrO_3)_2.H_2O$	361.45	col yelsh, monocl, hygr	3.773	$-H_2O$, 120	d 240	33^{16}		
s472	bromide	$SrBr_2$	247.44	wh, hex need, hygr, 1.575	4.216^{24}	643	d	100^{20}	222.5^{100}	s al, amyl al
s473	bromide, hexahydrate	$SrBr_2.6H_2O$	355.53	col, hex, hygr	2.386^{25}_4	tr to $2H_2O$, 88.6	$-6H_2O$, >180	204.2^0	∞	63.9^{20} al; 113.4^{30} MeOH; 0.6^{20} acet; i eth
s474	carbide	SrC_2	111.64	blk, tetr	3.2	>1700		d	d	d a
s475	carbonate	Nat. strontianite. $SrCO_3$	147.63	col, rhomb, or wh powd trfrs to $-$hex at 926, 1.516, 1.664, 1.666	3.70	1497^{69atm}	$-CO_2$, 1340	0.0011^{18}	0.065^{100}	0.12 aq CO_2; s a, NH_4 salts
s476	chlorate	$Sr(ClO_3)_2$	254.52	col, rhomb, or wh powd, 1.516, 1.605, 1.626	3.152	d 120		174.9^{18}	v s	s dil al; i abs al
s477	*perchlorate*	$Sr(ClO_4)_2$	286.52	col cr, hygr				310^{25}	v s	212 MeOH; 181 al; i eth
s478	chloride	$SrCl_2$	158.53	col, cub 1.650^{25}	3.052	875	1250	53.8^{20}	100.8^{100}	v sl s abs al, acet; i NH_3
s479	chloride, dihydrate	$SrCl_2.2H_2O$	194.56	transp leaf, 1.594, 1.595, 1.617	2.6715^{25}					
s480	chloride, fluoride	$SrCl_2.SrF_2$	284.14	col, tetr, 1.651, 1.627	4.18	962		d	d	s conc HNO_3, conc HCl, i al
s481	chloride, hexahydrate	$SrCl_2.6H_2O$	266.62	col, trig, 1.536, 1.487	1.93	115, $-4H_2O$, 60	$-6H_2O$, 100	106.2^0	205.8^{40}	3.8^6 al
s482	chromate	$SrCrO_4$	203.61	yel, monocl	3.895^{15}			0.12^{15}	3^{100}	s HCl, HNO_3, ac a, NH_4 salts
s483	cyanide	$Sr(CN)_2.4H_2O$	211.72	wh, rhomb, deliq		d		v s		
s484	cyanoplatinite	$Sr[Pt(CN)_4].5H_2O$	476.86	col, monocl pr, 1.696		$-5H_2O$, 150				s abs al
s485	glycerophosphate	$SrC_3H_7O_6P$	257.68	wh powd				sl s		i al
s486	ferrocyanide	$Sr_2Fe(CN)_6.15H_2O$	657.42	yel, monocl				50	100	
s487	fluoride	SrF_2	125.62	col, cub or wh powd, 1.442	4.24	1473	2489^{760}	0.011^0	0.012^{27}	s hot HCl; i HF, al, acet
s488	fluosilicate	$SrSiF_6.2H_2O$	265.73	monocl	$2.99^{17.5}$	d		3.2^{15}	v s	s HCl; 0.065^{15} 50 % al
s489	formate	$Sr(CHO_2)_2$	177.66	col, rhomb, 1.559, 1.547, 1.598	2.693	71.9		9.1^0	34.4^{100}	
s490	formate, dihydrate	$Sr(CHO_2)_2.2H_2O$	213.69	col, rhomb, 1.484, 1.521, 1.538	2.25	d, $-2H_2O$, 100		$11.62^{28.6}$	26.57^{100}	i al, eth
s491	hydride	$SrH_2(?)$	89.64	wh, rhomb, hygr	3.72	d 675	subl 1000 (in H_2)	d	d	d al
s492	hydroxide	$Sr(OH)_2$	121.63	wh, deliq	3.625	375 (in H_2)	$-H_2O$, 710	0.41^0	21.83^{100}	s a, NH_4Cl
s493	hydroxide, octahydrate	$Sr(OH)_2.8H_2O$	265.76	col, tetr, deliq, 1.499, 1.476	1.90	$-8H_2O$, 100		0.90^0	47.71^{100}	s a, NH_4Cl; i acet
s494	iodate	$Sr(IO_3)_2$	437.43	tricl	5.045^{15}			0.03^{15}	0.8^{100}	
s495	iodide	SrI_2	341.43	col pl	4.549^{25}_4	515	d	165.3^0	383^{100}	4.5^{30} al; 0.31^0 NH_4OH; s MeOH
s496	iodide, hexahydrate	$SrI_2.6H_2O$	449.52	col-yelsh, hex, deliq	2.672^{25}	d 90		448.9^0	∞	s al; i eth
s497	lactate	$Sr(C_3H_5O_3)_2.3H_2O$	319.81	wh cr or gran powd		$-3H_2O$, 120		25	200^{100}	sl s al
s498	*permanganate*	$Sr(MnO_4)_2.3H_2O$	379.54	purpl, cub	2.75	d 175		270^0	291^{18}	
s499	molybdate	$SrMoO_4$	247.56	col, tetr ~1.91	4.54^{25}_{35}	d		0.0104^{17}		s a
s500	nitrate	$Sr(NO_3)_2$	211.63	col, cub	2.986	570		70.9^{18}	100^{90}	0.012 abs al; v s NH_3; sl s acet
s501	nitrate, tetrahydrate	$Sr(NO_3)_2.4H_2O$	283.69	col, monocl	2.2	$-4H_2O$, 100	1100 tr SrO	60.43^0	206.5^{100}	s liq NH_3; v sl s abs al, acet; i HNO_3

No.	Name	Synonyms and Formulae	Mol. wt.	Crystalline form, properties and index of refraction	Density or spec. gravity	Melting point, °C	Boiling point, °C	Solubility, in grams per 100 cc		
								Cold water	Hot water	Other solvents
	Strontium									
s502	nitride	Sr_3N_2	290.87					d	d	s HCl
s503	nitrite	$Sr(NO_2)_2.H_2O$	197.65	col, hex, 1.588	2.408_0^0	$-H_2O>100$	d 240	58.9^0	182^{100}	0.42^{20} 90 % al
s504	*hyponitrite*	$SrN_2O_2.5H_2O$	237.71	wh need	2.173_4^{25}			v sl s	sl s	v sl s NH_3
s505	oxalate	$SrC_2O_4.H_2O$	193.64	col cr		$-H_2O$, 150		0.0051^{18}	0.15^{100}	s HCl, HNO_3
s506	oxide	SrO	103.62	gray-wh, cub, 1.810	4.7	2430	~3000	0.69^{20}	22.85^{100}	30 fus KOH; sl s al; i eth, acet
s507	oxide, per-	SrO_2	119.62	wh powd	4.56	d 215^{780}		0.018^{20}	d	v s al, NH_4Cl; i acet
s508	oxide, per-, octahydrate	$SrO_2.8H_2O$	263.74	col cr	1.951	$-8H_2O$, 100	d	0.018^{20}	d	s NH_4Cl; i al, acet, NH_4OH
s509	orthophosphate, di-	$SrHPO_4$	183.60	col, rhomb	3.544^{15}	1.62		i	i	s a, NH_4 salts
s510	salicylate	$Sr(C_7H_5O_3)_2.2H_2O$	397.88	col cr		d		5.6^{25}	28.6^{100}	1.5^{25}, 9.5^{78} al
s511	selenate	$SrSeO_4$	230.58	col, rhomb	4.23			i	i	s hot HCl; i HNO_3
s512	selenide	SrSe	166.58	wh, cub, 2.220	4.38			d	d	s HCl
s513	*metasilicate*	$SrSiO_3$	163.70	col, pr monocl, 1.599, 1.637	3.65	1580		i	i	
s514	*orthosilicate*	$SrSiO_4$	179.70	monocl, 1.728, 1.732, 1.758	3.84	>1750				
s515	sulfate	Nat. celestite. $SrSO_4$	183.68	col, rhomb, 1.622, 1.624, 1.631	3.96	1605		0.0113^0	0.014^{30}	sl s a; i al, dil H_2SO_4
s516	sulfate, hydrogen	$Sr(HSO_4)_2$	281.76	col		d		d	d	14^{70} H_2SO_4
s517	sulfide, hydro	$Sr(HS)_2$	153.76	col, cub need, 2.107		d		s	d	
s518	sulfide, mono	SrS	119.68	col, lt gray, cub, 2.107	3.70^{15}	>2000		i	d	d a
s519	sulfide, tetra-	$SrS_4.6H_2O$	323.97	redsh cr, hygr		25	$-4H_2O$, 100	s	s	s al
s520	sulfite	$SrSO_3$	167.68	col cr		d		0.0033^{17}		v s H_2SO_4; s a, HCl
s521	tartrate	$SrC_4H_4O_6.4H_2O$	307.75	wh, monocl	1.966			0.112^0	0.755^{85}	s dil HCl, dil HNO_3
s522	telluride	SrTe	215.22	wh, cub, 2.198	1.00					
s523	thiocyanate	$Sr(SCN)_2.3H_2O$	257.83	deliq		$-3H_2O$, 100	d 160–170	v s		v s al
s524	thiosulfate	$SrS_2O_3.5H_2O$	289.82	monocl need	2.17^{17}	$-4H_2O$, 100		2.5^{13}	57^{100}	i al
s525	*dithionate*	$SrS_2O_6.4H_2O$	319.81	trig, 1.530, 1.525	2.373	$-4H_2O$, 78		22^{16}	67^{100}	i al
s526	tungstate	$SrWO_4$	335.47	col, tetr	6.187	d		0.14^{15}		id a; i al
s527	**Sulfamic acid**	Amidosulfuric, amino-sulfonic acid. NH_2SO_3H	97.09	col, rhomb	2.126^{25}	200 d	d	14.68	47.08^{80}	v sl s al, eth, acet; i CS_2, CCl_4
s528	**Sulfamide**	Sulfuryl amide. $SO_2(NH_2)_2$	96.11	rhomb pl	1.611	91.5	d 250	s		s al
s529	**Sulfur(α)**	S_8	256.512	yel, rhomb, 1.957	2.07^{20}	112.8 95.5 (revers.) 444.6	444.674	i	i	23^0 CS_2; sl s tol, al, bz, eth, liq NH_3; s CCl_4
s530	(β)	S_8	256.512	pa yel, monocl	1.96	119.0	444.674	i	i	70 CS_2; s al, bz
s531	(γ)	S_8	256.512	pa yel, amorph	1.92	ca 120	444.6	i	i	i CS_2
s532	bromide, mono-	S_2Br_2	223.95	red liq, 1.730	2.63	-40	$54^{0.2}$	d	d	s CS_2
s533	chloride, di-	SCl_2	102.97	dk red liq, 1.557[11]	1.621_{15}^{15}	-78	d 59	d	d	s CCl_4, bz; d al, eth
s534	chloride, mono-	S_2Cl_2	135.03	yel-red liq, 1.666[14]	1.678	-80	135.6	d	d	s bz, eth, CS_2
s535	chloride, tetra-	SCl_4	173.88	yel-br liq		-30	d -15	d	d	
s536	fluoride, hexa-	SF_6	146.05	col gas	gas 6.602 g/l liq $1.88^{-50.5}$	-50.5	-63.8 (subl)	sl s	sl s	s al, KOH
s537	fluoride, mono-	S_2F_2	102.12	col gas	liq 1.5^{-100}	-120.5	-38.4	d	d	d KOH
s538	fluoride, tetra-	SF_4	108.06	gas (exist ?)		-124	-40	d	d	
s539	(di) fluoride, deca-	S_2F_{10}	254.11	col liq	2.08_4^0	-92	29			d fus caust
s540	(tetra-) nitride, di-	S_4N_2	156.27	red liq or gray solid	1.901^{18}	23	d 100 expl	i		s eth; sl s al, CS_2
s541	(tetra-) nitride, tetra-	S_4N_4	184.28	or red, monocl	2.22^{15}	subl 179	expl 160	d		s CS_2, chl, bz, NH_3; sl s al, eth
s542	(tri-)*di*nitrogen dioxide	$S_3N_2O_2$	156.20	pa yel cr		100.7	d	i		s al, bz
s543	oxide, di-	SO_2	64.06	col gas or liq suffoc odor	gas 2.927 g/l liq 1.434	-72.7	-10	22.8^0	0.58^{90}	s al, ac a, H_2SO_4
s544	oxide, hept-	Sulfur oxide, per- S_2O_7	176.12	visc liq, or need		0	subl 10	d	d	s H_2SO_4
s545	oxide, mono-	SO (or S_2O_2)	48.06	col gas		d	d		d	
s546	oxide, sesqui-	S_2O_3	112.13	bl-grn cr		d 70–95			d	s al, eth, fum H_2SO_4
s547	oxide, tetra-	Sulfurperoxide. SO_4	96.06	wh		d 0–3		s, d		d dil H_2SO_4
s548	oxide, tri-(α)	SO_3	80.06	silky fibr need, stable modific	1.97^{20}	16.83	44.8	d	d	forms fum H_2SO_4
s549	oxide, tri-(β)	$(SO_3)_2$	160.12	asbestos like fibre, metastable		62.4	50 (subl)	d	d	forms fum H_2SO_4

No.	Name	Synonyms and Formulae	Mol. wt.	Crystalline form, properties and index of retraction	Density or spec. gravity	Melting point, °C	Boiling point, °C	Cold water	Hot water	Other solvents

Sulfur(α)

No.	Name	Synonyms and Formulae	Mol. wt.	Crystalline form, properties and index of retraction	Density or spec. gravity	Melting point, °C	Boiling point, °C	Cold water	Hot water	Other solvents
s550	oxide, tri-(γ)	SO_3	80.06	vitreous, orthorhomb, metastable	liq 1.920_4^{20} sld 2.29^{-10}	16.8	44.8	d	d	forms fum H_2SO_4
s551	oxytetrachloride, mono-	S_2OCl_4	221.94	dk red liq	1.656^0		60	d	d	d al
s552	oxytetrachloride, tri-	$S_2O_3Cl_4$	253.94	wh, rhomb need or pl		d 57		d	d	d al
s553	trithiazyl chloride	S_4N_3Cl	205.73	pa yel cr		d 170 (vac)		d	d	
s554	**Sulfuric acid**	H_2SO_4	98.08	col liq	1.841 (96–98 %)	10.36 (100 %) 3.0 (98 %)	330±0.5 (100%)	∞ ev heat	∞	d al
s555	dihydrate	$H_2SO_4.2H_2O$	134.11	col liq, 1.405	1.650^0	−38.9	167	∞	∞	d al, eth
s556	hexahydrate	$H_2SO_4.6H_2O$	206.17	liq		−54		v s	v s	
s557	monohydrate	$H_2SO_4.H_2O$	116.09	col liq or monocl cr, 1.438	1.788	8.62	290	∞	∞	d al
s558	octahydrate	$H_2SO_4.8H_2O$	242.20	liq		−62		v s	v s	
s559	peroxidi-	Per(di-)sulfuric acid. $H_2S_2O_8$	194.14	hygr cr		d 65	d	d	d	s al, eth, H_2SO_4
s560	peroximono-	Permonosulfuric acid. Caro's acid H_2SO_6	114.08			d 45		d	d	s H_3PO_4
s561	pyro-	$H_2S_2O_7$	178.14	col cr, hygr	1.9^{20}	35	d	d	d	d al
s562	tetrahydrate	$H_2SO_4.4H_2O$	170.14			−27		∞	∞	d al, eth
s563	**Sulfurous acid**	H_2SO_3	82.08	in sol only	ca 1.03			s		s al, eth, ac a
s564	**Sulfuryl chloride**	SO_2Cl_2	134.97	col liq, 1.444	1.6674_4^{20}	−54.1	69.1	d	d	s bz, ac a
s565	chloride fluoride	SO_2ClF	118.52	col gas	1.623^0 g/l	−124.7	7.1	d		
s566	fluoride	SO_2F_2	102.06	col gas	gas 3.72 g/l liq 1.7	−136.7	−55.4	10^9		s al, CCl_4; sl s alk
s567	pyro-, chloride	$S_2O_5Cl_2$	215.03	col liq, 1.937^{20}	gas 9.6 g/l liq 1.818_4^{11}	−39 to −37	152.5	d	d	d a
t1	**Tantalum**	Ta	180.948	gray black hard metal, cub or powd	met 16.6 powd 14.401	2996	5425 ± 100	i	i	s HF, fus alk; i a
t2	boride, di-	TaB_2	202.57		11.15	3000(?)				
t3	bromide	$TaBr_5$	580.49	yel cr	4.67	265	348.8	d		s abs al, eth
t4	carbide	TaC	192.96	blk, cub	13.9	3880	5500	i	i	sl s H_2SO_4. HF
t5	chloride, penta-	$TaCl_5$	358.21	lt yel, vitr cr powd	3.68^{27}	216	242	d		s abs al, H_2SO_4
t6	fluoride	TaF_5	275.94	col, tetrag, deliq	4.74	96.8	229.5	s		s HF, eth
t7	nitride	TaN	194.95	br bronze or blk, hex	16.30	3360 ± 50		i	i	sl s aq reg, HF, HNO_3
t8	oxide, pent-	Ta_2O_5	441.89	col, rhomb	8.2	1872 ± 10		i	i	s fus $KHSO_4$, HF; i a
t9	oxide, pent- hydrate	Tantalic acid $Ta_2O_5xH_2O$		col gel				s		s alk, exc conc HNO_3; i a
t10	oxide, tetr-	Ta_2O_4 (or TaO_2)	425.89	dk gray powd		oxidizes		i	i	i a
t11	sulfide	Ta_2S_4 (or TaS_2)	490.15	blk powd or cr		>1300		i	i	sl s HF+HNO_3; i HCl
t12	**Telluric acid, ortho-**	$Te(OH)_6$ or $H_2TeO_4.2H_2O$	229.64	wh, monocl pr	3.071	136		s		sl s dil a, HNO_3; i abs al, acet, eth
t13	**Telluric acid,**	$Te(OH)_6$ or H_6TeO_6	229.64	wh, cub	3.158	136		s		sl s dil a, HNO_3; i abs al, acet, eth
t14	**Tellurium**	Te	127.60	br blk, amorph, 1.0025	6.00	449.5 ± 0.3	989.8 ± 3.8	i	i	s H_2SO_4, HNO_3, aq reg, KCN, KOH; i HCl, CS_2
t15	**Tellurium**	Te	127.60	rhomb silv wh met, 1.0025	6.25	452	1390	i	i	s H_2SO_4, HNO_3, aq reg, KCN KOH; i HCl, CS_2
t16	bromide, di-	$TeBr_2$	287.42	brn to gray grn, need, unstable		210	339	d		s eth; sl s a; d NaOH
t17	bromide, tetra-	$TeBr_4$	447.27	or cr	4.31_4^{15}	380 ± 6	d 421	sl s	d	s eth, a, tart a, NaOH
t18	chloride, di-	$TeCl_2$	198.50	blk cr or amorph, unstable	7.05	209 ± 5	327	d	d	s min a, tart a; d NaOH
t19	chloride, tetra-	$TeCl_4$	269.41	wh to yel cr, deliq	3.26^{18} 2.559^{223}	224	380^{760}	s d	s d	s HCl, bz, al, chl, CCl_4; i CS_2
t20	ethoxide	$Te(OC_2H_5)_4$	307.85			20	$107-107.5^{5.5}$			
t21	fluoride, hexa-	TeF_6	241.59	col gas unpleas odor	sol 4.006^{-191} liq 2.56^{-35}	−36	+35.5	d		d a, alk
t22	fluoride, tetra-	TeF_4	203.59	wh cr hygr		subl	>97	d	d	
t23	hydride	H_2Te	129.62	col gas pois	4.49	−48.9	-2.2^{760}	v s	s	d al

No.	Name	Synonyms and Formulae	Mol. wt.	Crystalline form, properties and index of refraction	Density or spec. gravity	Melting point, °C	Boiling point, °C	Solubility, in grams per 100 cc		
								Cold water	Hot water	Other solvents
	Tellurium									
t24	iodide, di-	TeI_2	381.41	blk cr (exist ?)		subl		i	i	
t25	iodide, tetra,-	TeI_4	635.22	blk cr	5.403_4^{15}	280	d	sl s	d	s alk, aq NH_3, HI
t26	methoxide	$Te(OCH_3)_4$	251.74	solid		123–124				
t27	oxide, di-	Tellurite. TeO_2	159.60	wh, tetr or rhomb, 2.00, 2.18(Li), 2.35	tetr 5.67^{15} rhomb 5.91^0	733	1245	i	i	s HCl, hot HNO_3, alk; i NH_4OH
t28	oxide, mon-	TeO	143.60	blk, amorph (exist ?)	5.682	d 370 (in CO_2)	d	i	i	s dil a, H_2SO_4, KOH
t29	oxide, tri-	TeO_3	175.60	α yel amorph β gray cr	α 5.075_4^{105} β 6.21	d 395		i	i	d conc HCl; s hot KOH; i a, al
t30	sulfide	TeS_2	191.72	red-blk powd amorph (exist ?)				i		i a; s alk sulf
t31	sulfoxide	$TeSO_3$	207.66	deep red amorph		d 30	d	d	d	s H_2SO_4
t32	**Tellurous acid**	H_2TeO_3(?)	177.61	wh flocks, indef not isolated	3.05	d 40		0.00067	d	s a, NaOH; sl s NH_4OH; i al
t33	**Terbium**	Tb	158.924	silv-gray met, hex	8.2294	1360±4	3123	i	i	s a
t34	bromide	$TbBr_3$	398.65			827	1490	s	s	
t35	chloride hexa-hydrate	$TbCl_3.6H_2O$	373.38	col pr cr, deliq	4.35 (anhydr)	588 (anhydr)	$-H_2O$, 180–200 (in HCl gas)	v s		
t36	fluoride	TbF_3	215.92			1172	2280(?)	i	i	i dil a
t37	iodide	TbI_3	539.64			946	>1300	s	s	
t38	dimethylphosphate	$Tb[(CH_3)_2PO_4]_3$	534.05					12.6^{25}	8.07^{40}	
t39	nitrate	$Tb_2(NO_3)_3.6H_2O$	453.03	col, monocl cr		893		s		
t40	oxalate	$Tb_2(C_2O_4)_3.10H_2O$	762.06	wh cr	2.60	$-H_2O$, 40		i	i	i dil a
t41	oxide	Terbia. Tb_2O_3	365.85	wh solid				i		s dil a
t42	oxide, per-	Tb_4O_7	747.69	dk-brn or blk solid		$-O_2$		i	i	s hot conc a
t43	sulfate	$Tb_2(SO_4)_3.8H_2O$	750.16	wh cr		$-8H_2O$, 360		3.561^{20}	2.51^{40}	
t44	**Thallium**	Tl	204.37	bl-wh met, tetr	11.85	303.5	1457±10	i	i	s HNO_3. H_2SO_4; sl s HCl
t45	acetate	$TlC_2H_3O_2$	263.42	silk wh cr, deliq	3.765^{17}	131		v s		v s al, $CHCl_3$; i acet
t46	aluminium sulfate	$TlAl(SO_4)_2.12H_2O$	639.66	oct, 1.488	2.306^{20}	91		11.78^{25}		
t47	azide	TlN_3	246.39	yel, tetr		330 (vac)		0.1712^0	0.31^6	i al, eth
t48	bromate	$TlBrO_3$	332.28	col, need				0.35^{20}	s	s dil al
t49	bromide, di-	Bromothallate(ous). Tl_2Br_4 or $Tl_3^I[Tl^{III}Br_4]$	728.38	yel need		d	d	d	d	
t50	bromide, mono-	$TlBr$	284.28	yel-wh, cub, 2.4–2.8	$7.557^{17.1}$	480	815	0.05^{25}	0.25^{68}	s al; i HB_2, acet
t51	bromide, tri-	$TlBr_3$	444.10	yel, deliq, unstable		d		s	v s	v s al
t52	carbonate	Tl_2CO_3	468.75	col, monocl	7.11	273		$4.03^{15.5}$	27.2^{100}	i abs al, eth, acet
t53	chlorate	$TlClO_3$	287.82	need (rhomb ?)	5.047^9			20^0	57.31^{100}	
t54	perchlorate	$TlClO_4$	303.82	col, rhomb	4.89	501	d	20.5^{30}	167^{100}	sl s al
t55	chloride	$TlCl$	239.82	wh reg discol in air, 2.247	7.004_4^{30}	430	720	$0.29^{15.6}$	$2.41^{99.35}$	i al, acet; d a
t56	chloride, tri-	$TlCl_3$	310.73	hex pl, hygr		25	d	v s		s al, eth
t57	chloride, tri-	$TlCl_3.H_2O$	328.74	col, need		$-H_2O$, 60	d 100	v s	d	v s al, eth
t58	chloride, tri-	$TlCl_3.4H_2O$	382.79	col, need		37	$-4H_2O$, 100	86.2^{17}	d	s al, eth
t59	chloroplatinate	Tl_2PtCl_6	816.55	pale or cr	5.76^{17}			0.0064^{15}	0.05^{100}	
t60	chromate	Tl_2CrO_4	524.73	yel				0.03^{60}	0.21^{00}	sl s a, alk; i ac a
t61	dichromate	$Tl_2Cr_2O_7$	624.73	red				i		d a
t62	chromium sulfate	Thallium chromium alum. $Tl[Cr(H_2O)_6](SO_4)_2.6H_2O$	664.67	vlt cr	2.394	92		163.8^{25}		
t63	cyanate	$TlCNO$	246.39	col, need	5.487_4^{20}			s	v s	sl s al
t64	cyanide	$TlCN$	230.39	tabl	6.523	d		$16.8^{28.5}$		s a
t65	ethoxide	$(TlOC_2H_5)_4$	997.73	col liq	3.522	-3	d 80	s d		9.11^{25} al; s bz; i liq NH_3
t66	ethylate	$TlOC_2H_5$	249.43	liq, 1.6714^{20}	3.493_4^{20}	-3	d 130	s d		sl s al; s eth
t67	ferrocyanide	$Tl_4Fe(CN)_6.2H_2O$	1065.46	yel, tricl	4.641			0.37^{18}	3.93^{101}	
t68	fluogallate	$Tl_3(GaF_6H_2O)$	591.47	col, orthorhomb	6.44			s		
t69	fluoride, mono-	TlF	223.37	col, cub, oct	8.23^4	327	655	78.6^{15}	d	sl s al
t70	fluoride, tri-	TlF_3	261.37	olive grn	8.36^{25}	d 550		d		i conc HCl
t71	fluosilicate	$Tl_2SiF_6.2H_2O$	586.85	hex pl	5.72			v s		
t72	formate	$TlHCO_2$	249.39	col, need, hygr	4.967^{104}	101		500^{10}		v s MeOH, sl s al; i $CHCl_3$

Thallium

No.	Name	Synonyms and Formulae	Mol. wt.	Crystalline form, properties and index of refraction	Density or spec. gravity	Melting point, °C	Boiling point, °C	Cold water	Hot water	Other solvents
t73	(I) hydroxide	TlOH	221.38	pa yel, need		d 139		25.9^0	52^{40}	s al
t74	iodate	TlIO$_3$	379.27	wh need		tr to (β) 170		0.058^{20}	sl s	sl s HNO$_3$
t75	iodide (α)	TlI	331.27	yel, rhomb	7.29	tr to (β) 170		0.0006^{20}	0.12^{100}	s liq NH$_3$
t76	iodide (β)	TlI	331.27	red, cub	$7.098^{14.7}$	440	823	i	i	i al
t77	iodide, tri-	TlI$_3$	585.08	blk, lust rhomb		d		s		
t78	iron (III) sulfate	TlFe(SO$_4$)$_2$.12H$_2$O	668.52	pink, oct, n_D^{17} 1.524	2.351^{15}	$-$H$_2$O, ca 100		36.15^{25} (anhydr)		
t79	magnesium sulfate	Tl$_2$SO$_4$.MgSO$_4$.6H$_2$O	733.26	wh dull cr, 1.5660, 1.5836, 1.5900	3.573_4^{20}	$-$6H$_2$O, 40		d 0		
t80	methoxide	TlOCH$_3$	235.40	wh cr powd		d>120		s d		1.70^{25} CH$_3$OH; 3.16^{25} bz
t81	molybdate	Tl$_2$MoO$_4$	568.68	wh powd or cr		vol red heat		i	v sl s	i al; s alk carb, conc NH$_4$OH, HF
t82	myristate	TlC$_{14}$H$_{27}$O$_2$	431.74	wh powd		120–3				0.52^{25} 50 % al
t83	(I) nitrate (α)	TlNO$_3$	266.37	cubic		206	430	9.55^{20}	4.13^{100}	i al, s acet
t84	(I) nitrate (β)	TlNO$_3$	266.37	trig		tr 145 to (α)				
t85	(I) nitrate (γ)	TlNO$_3$	266.37	rhomb, α 1.817	$5.556^{21.4}$	tr 75 to (β)		3.91^0	414^{100}	i al; s acet
t86	(III) nitrate	Tl(NO$_3$)$_3$	390.38	cr				s		
t87	(III) nitrate	Tl(NO$_3$)$_3$.3H$_2$O	444.43	col, rhomb, deliq		182	s 100	d	d	i a
t88	nitrite	TlNO$_2$	250.38	yel micro cr		182		32.10^{25}	95.78^{98}	i a
t89	oleate	TlC$_{18}$H$_{33}$O$_2$	485.83	wh cr clusters		131–2		0.05^{15}	0.38^{60}	3.0^{25} al
t90	oxalate	Tl$_2$C$_2$O$_4$	496.76	monocl pr	6.31			1.48^{15}	9.02^{100}	
t91	oxalate, tetra-	TlH$_3$(C$_2$O$_4$)$_2$.2H$_2$O	419.46	tricl, leaf, 1.5097, 1.6319, 1.6538	2.992^{17}	d 100		76.9^{23}	v s	i cold al; s hot al
t92	(I) oxide	Tl$_2$O	424.74	blk, deliq	9.52^{16}	300	1080^{760}–O, 1865	v s d to TlOH		s a, al
t93	(III) oxide	Tl$_2$O$_3$	456.74	col, amorph pr, hex	hex 10.19^{22}, am 9.65^{21}	717 ± 5	$-$2O, 875	i	i	s a; i alk
t94	palmitate	TlC$_{16}$H$_{31}$O$_2$	459.79	cr need		115–117		0.01^{15}	0.07^{60}	1.04^{45} al
t95	phenoxide	TlOC$_6$H$_5$	297.48	wh cr		233–5		d		s hot bz; sl s lgr
t96	orthophosphate	Tl$_3$PO$_4$	708.02	col, need	6.89^{10}			0.5^{15}	0.67^{100}	i al, s NH$_4$ salts
t97	orthophosphate, (di)-β	TlH$_2$PO$_4$	301.36	monocl	4.726	ca 190		sl s	sl s	i al
t98	pyrophosphate	Tl$_4$P$_2$O$_7$	991.42	monocl pr	6.786^{20}	>120		40		
t99	picrate	TlC$_6$H$_2$N$_3$O$_7$	432.47	red, monocl or yel tricl	red 3.164^{17}, yel 2.993^{17}	expl 723–725		0.135^0	2.43^{70}	0.40 CH$_3$OH
t100	rhodanide	TlCNS	262.45	glossy leaflets, rhomb, tetr cr	4.954_4^{20}	d low temp		0.393^{15}		0.024^0 liq SO$_2$; i acet; s MeOH
t101	selenate	Tl$_2$SeO$_4$	551.70	rhomb need, 1.949, 1.959, 1.964		>400		2.13^{10}	8.5^{50}	i al, eth
t102	selenide	Tl$_2$Se	487.70	gray leaf	9.05_4^{25}	340		i		s a; i acet a
t103	silver nitrate	TlNO$_3$.AgNO$_3$	436.25	wh cr powd		75		s		
t104	stearate	TlC$_{18}$H$_{35}$O$_2$	487.85	need		119		0.005^{15}	0.095^{75}	0.18^{15} al, 0.60^{60} al
t105	(I) sulfate	Tl$_2$SO$_4$	504.80	col, rhomb, 1.860, 1.867, 1.885	6.77	632	d	4.87^{20}	19.14^{100}	
t106	(I) sulfate hydrogen	TlHSO$_4$	301.44	pr need		120 d				v sl s dil H$_2$SO$_4$
t107	(III) sulfate	Tl$_2$(SO$_4$)$_3$.7H$_2$O	823.03	col leaf		$-$6H$_2$O, 220		d	d	s dil H$_2$SO$_4$
t108	(I) sulfide	Tl$_2$S	440.80	bl-blk tetr	8.46	448.5	d	0.02^{20}	sl s	s a; i alk, acet
t109	(III) sulfide	Tl$_2$S$_3$	504.93	blk, amorph		260 (in N$_2$)	d	i	i	s hot H$_2$SO$_4$
t110	sulfite	Tl$_2$SO$_3$	488.80	wh cr	6.427			3.34^{15}	v s	i al
t111	tartrate(dl)	Tl$_2$C$_4$H$_4$O$_6$	556.81	monocl	4.659	d 165		13.3^{15}		
t112	metatellurate	Tl$_2$TeO$_4$	600.34	heavy wh ppt	$6.760^{17.6}$	red heat		sl s	sl s	
t113	thiocyanate	TlSCN	262.45	col, tetr	4.956^{20}			0.315^{20}	0.727^{40}	i al
t114	dithionate	Tl$_2$S$_2$O$_6$	568.86	monocl	5.57_4^{20}	d		41.8^{19}		
t115	thiosulfate	Tl$_2$S$_2$O$_3$	520.87	wh rhomb cr		d 130		sl s	v s	
t116	metavanadate	TlVO$_3$	303.41	gray cr	6.09^{17}	424		0.87^{11}	0.21^{100}	
t117	pyrovanadate	Tl$_4$V$_2$O$_7$	1031.36	light yel	8.21^{19}	454		0.2^{14}	0.26^{100}	
t118	**Thiocarbonyl chloride**	Thiophosgene. CSCl$_2$	114.98	red yel liq, 1.5442	1.509^{15}		73.5	d		d al; s eth
t119	**Thiocarbonyl chloride, tetra-**	CSCl$_4$	185.89	yel	1.712^{13}	146–147			d	
t120	**Thiocyanic acid (iso)**	HSCN (HNCS)	59.09	col mass or gas		>-110	polym to solid -90	v s		v s al, eth, bz
t121	**Thiocyanogen**	(SCN)$_2$	116.16	liq, or yel solid		-2 to -3		d		s al, eth, CS$_2$, CCl$_4$

No.	Name	Synonyms and Formulae	Mol. wt.	Crystalline form, properties and index of refraction	Density or spec. gravtiy	Melting point, °C	Boiling point, °C	Solubility, in grams per 100 cc		
								Cold water	Hot water	Other solvents
	Thionyl bromide									
t122	**Thionyl** bromide	SOBr₂	207.88	or, yel liq	2.68¹⁸	−52	138⁷⁷², 68⁴⁰	d	d	s bz, chl, CS₂, CCl₄
t123	chloride	SOCl₂	118.97	col, or yel liq; 1.527¹⁰	1.655¹⁰₄	−105	78.8⁷⁴⁶ d 140	d	d	d a, al, alk; s bz, chl
s124	chloride fluoride	SOClF	102.51	gas		−139.5	12.2			
t125	fluoride	SOF₂	86.06	col, gas	gas 2.93 g/l liq 1.780⁻¹⁰⁰	−110.5	−43.8	d	d	s eth, bz, chl, acet, AsCl₃
t126	**Thiophosphoramide**	Thiophosphorylamide. PS(NH₂)₃	111.11	yel wh cr	1.7¹⁴	d 200		14.3⁷⁵	d	sl, me
t127	**Thiophosphoryl** bromide	PSBr₃	302.76	yel, cub	2.85¹⁷	37.8	125²⁵	d		s eth, CS₂, PCl₃
t128	thiophosphoryl bromide, hydrate	PSBr₃.H₂O	320.78	yel cr	2.794¹⁵	35				
t129	thiophosphoryl bromide (mono-) chloride, di-	PSBrCl₂	213.85	yel liq	2.12⁰	−30	d 150	d		
t130	thiophosphoryl bromide (di-) chloride	PSBr₂Cl	258.31	pa grn fum liq	2.48⁰	−60	95⁵⁰	d		
t131	thiophosphoryl chloride	PSCl₃	169.40	col liq, 1.563 (c)	1.635	−35	125	d		s bz, CS₂, CCl₄
t132	thiophosphoryl fluoride	PSF₃	120.03	gas	3.8⁷·⁵ᵃᵗᵐ	d		sl s d		s eth; i bz, CS₂
t133	**Thiosulfuric acid**	H₂S₂O₃	114.14	in sol only				s		
t134	**Thorium**	Th	232.038	gray, cub radioactive	11.7			i	i	s HCl, H₂SO₄, aq reg; sl s HNO₃
t135	boride, hexa-	ThB₆	300.02	dk viol blk met, cub	6.4¹⁵	2195		i	i	s HNO₃; i H₂SO₄, HCl, HF, aq alk
t136	boride, tetra-	ThB₄	275.28	tetr pr	7.5¹⁶			i	i	s HNO₃ HCl, hot H₂SO₄
t137	bromide	ThBr₄	551.67	col cr, hygr	5.67	subl 610	725	s	s	
t138	carbide	ThC₂	256.06	yel, tetr	8.96¹⁸	2655 ± 25	ca 5000 (?)	d		v sl s conc a
t139	carbonate	Th(CO₃)₂	352.06	exist?				i	d	s conc Na₂CO₃
t140	chloride	ThCl₄	373.85	wh, rhomb, deliq	4.59	770 ± 2 subl 820	d 928	v s	v s	s al, a, KCl; sl s eth
t142	tetracyanoplatinate	Th[Pt(CN)₄]₂.16H₂O	1118.61	yel-grn, rhomb	2.460			sl s	s	
t143	fluoride	ThF₄	308.03	wh cub powd	6.32²⁴	>900				sl d dil H₂SO₄, HCl; i conc H₂SO₄
t144	fluoride	ThF₄.4H₂O	380.09	cr		−H₂O, 100	−2H₂O, 140−200	0.017²⁵		i HF
t145	hydroxide	Th(OH)₄	300.02	wh gelat		d		i	i	s a; i alk, HF
t146	iodate	Th(IO₃)₄	931.65					i	i d	s dil H₂SO₄; i dil HNO₃
t147	iodide, tetra-	ThI₄	739.66	yel		566	839	s		
t148	nitrate	Th(NO₃)₄	480.06	plates, deliq		d 500		v s		s al
t149	nitrate	Th(NO₃)₄.4H₂O	552.12	col cr		swells		v s		v s al; sl s acet; 36.9 eth
t150	nitrate	Th(NO₃)₄.12H₂O	696.24	col leaf, deliq		d		v s		v s al, a
t151	nitride	Th₃N₄	752.14	dk brn powd, or blk cr				sl d	d	s HCl
t152	oxalate	Th(C₂O₄)₂	408.08	wh cr	4.637¹⁶	d		0.0017¹⁷	0.0017⁶⁰	s h aq (NH₄)₂C₂O₄; sl s a
t153	oxalate	Th(C₂O₄)₂.6H₂O	516.17	wh amorph powd				i		s Na₂CO₃, (NH₄)₂C₂O₄ sol; i HNO₃
t154	oxide, di-	Thorianite. ThO₂	264.04	wh cub, 2.20 (liq)	9.86	3220 ± 50	4400	i	i	s hot H₂SO₄; i dil a, alk
t155	oxysulfide	ThOS	280.10	yel cr	6.44⁰	d		i		s aq reg; sl s HNO₃
t156	2,4-pentanedione	Thorium acetylacetonate. Th(C₅H₇O₂)₄	628.48	col cr		171 subl 160¹⁰	260−270¹⁰	sl s		v s al, chl; s eth
t157	hypophosphate	ThP₂O₆.11H₂O	588.15	wh amorph ppt		−11H₂O, 160		i		i a, alk
t158	metaphosphate	Th(PO₃)₄	547.93	col rhomb pr	4.08¹⁶·⁴					
t159	orthophosphate	Th₃(PO₄)₄.4H₂O	1148.06	wh gelat				i	i	s 30° HCl; i a
t160	picrate	Th(C₆H₂N₃O₇)₄.10H₂O	1324.59					0.305²⁵		
t161	selenate	Th(SeO₄)₂.9H₂O	680.09	col, monocl	3.026	−8H₂O, 200	d 1500	0.5⁰	2.0¹⁰⁰	
t162	orthosilicate	Thorite ThSiO₄	324.12	col, tetr 1.80, 1.81	6.82¹⁶			v sl s		i a
t163	silicide	ThSi₂	288.21	blk, tetr	7.96¹⁶					s hot HCl; sl s H₂SO₄
t164	sulfate	Th(SO₄)₂	424.16	wh cr, hygr	4.225¹⁷			s	s	i a; v s NH₄C₂H₅O₂
t165	sulfate	Th(SO₄)₂.4H₂O	496.22	wh need, or cr powd		−4H₂O, 400		9.41¹⁷ (anhydr)	2.54⁶⁰ (anhydr)	i a

No.	Name	Synonyms and Formulae	Mol. wt.	Crystalline form, properties and index of refraction	Density or spec. gravity	Melting point, °C	Boiling point, °C	Solubility, in grams per 100 cc		
								Cold water	Hot water	Other solvents
	Thorium									
t166	sulfate	$Th(SO_4)_2.6H_2O$	532.26					1.63^{15}	6.64^{60}	i a
t167	sulfate	$Th(SO_4)_2.8H_2O$	568.29	monocl, prism, 1.5168		$-4H_2O, 42$		1.88^{25}	3.71^{44}	i a
t168	sulfate	$Th(SO_4)_2.9H_2O$	586.31	wh monocl	2.77	$-9H_2O, 400$		1.57^{20}	6.67^{55}	i a
t169	sulfide	ThS_2	296.17	dk brn-blk cr	7.30_4^{75}	1925 ± 50 (vac)		i	d 200	s hot aq reg; sl s a
t170	*pyro*vanadate	$ThV_2O_7.6H_2O$	554.01	yel				i	i	s conc a
t171	**Thulium**	Tm	168.934	silv wh met, hex.	9.3208	1545	1947	i	i	
t172	bromide	$TmBr_3$	408.66			952	1440	s	s	
t173	chloride	$TmCl_3.7H_2O$	401.40	grn cr, deliq		824	1440	v s		v s al
t174	fluoride	TmF_3	225.93			1158	>2200	i	i	s dil a
t175	iodide	TmI_3	549.65	brt yel cr		1015	1260	s	s	
t176	oxalate	$Tm_2(C_2O_4)_3.6H_2O$	710.02	grn-wh ppt		$-H_2O, 50$		i		s alk oxal sol; i dil a
t177	oxide	Thulia. Tm_2O_3	385.87	grn-wh powd						sl s min a
t178	**Tin gray**	Sn	118.69	gray, cub	5.75	**231.9681**	2270	i	i	s HCl, H_2SO_4, aq reg, alk; sl s dil HNO_3
t179	**Tin white**	Sn	118.69	wh met, tetr	7.28	231.88 stable 13.2−161	2260	i	i	s HCl, H_2SO_4, aq reg, alk; sl s dil HNO_3
t180	**Tin brittle**	Sn	118.69	wh, rhomb	6.52−56	231.89 stable >161	2260	i	i	s HCl, H_2SO_4, aq reg, alk; sl s dil HNO_3
t181	(II) acetate	$Sn(C_2H_3O_2)_2$	236.78	yelsh powd		182	240	d		s dil HCl
t182	*pyro*arsenate	$Sn_2As_2O_7$	499.22	flocculent ppt		$d\ As_2O_3 + SnO_2$		i	i	i conc ac a
t183	(II) bromide	$SnBr_2$	278.51	pa yel, rhomb	5.117^{17}	215.5	620	85.2^0	222.5^{100}	s al, eth, acet
t184	(IV) bromide	$SnBr_4$	438.33	col, rhomb pyr, deliq	liq 3.34^{35}	31	202^{734}	s d	d	s acet, PCl_3, $AsBr_3$
t185	bromide chloride(tri-)	$SnBrCl_3$	304.96	col liq	2.51^{13}	-31	50^{30}			
t186	bromide(di-) chloride(di-)	$SnBr_2Cl_2$	349.41		2.82^{13}	-20	65^{30} d 191	d	d	
t187	bromide(tri-) chloride	$SnBr_3Cl$	393.87	liq	3.12^{13}	1	73^{30}			
t188	bromide(di-) iodode(di-)	$SnBr_2I_2$	532.32	or-red, hex pl	3.631^{15}	50	225	s	d<80	
t189	(II) chloride	$SnCl_2$	189.60	wh, rhomb	3.95_4^{25}	246	652	83.9^0	269.8^{15} d	s al, eth, acet, et acet, me acet, pyr
t190	(II) chloride dihydrate	$SnCl_2.2H_2O$	225.63	wh, monocl	$2.710^{15.5}$	37.7	d	d	d	s al, eth, acet, glac ac a
t191	(IV) chloride	$SnCl_4$	260.50	col liq, solid cub, 1.512	liq 2.226	-33	114.1	s	d	s eth
t192	(IV) chloride penta-hydrate	$SnCl_4.5H_2O$	350.58	monocl cr		stable 19−56		s		
t193	(IV) chloride tetra-hydrate	$SnCl_4.4H_2O$	332.56	opaque		stable 56−83	s			
t194	(IV) chloride tri-hydrate	$SnCl_4.3H_2O$	314.55	col, monocl cr		80	stable 64−83	s		
t195	(IV) chloride *di*ammine	$SnCl_4.2NH_3$	294.56	cr				s		d HCl
t196	chloride(tri-) bromide	$SnCl_3Br$	304.96	col liq	2.51^{13}	-31	50^{30}			
t197	chloride (di-) iodide(di-)	$SnCl_2I_2$	443.40	red mobile liq	3.287^{15}		297	s	d	s chl, bz CS_2
t198	(IV) chloride nitrosyl-chloride	$SnCl_4.2NOCl$	391.42	pa yel, oct cr	2.60	180		d	d	
t199	(IV) chromate	$Sn(CrO_4)_2$	350.68	br yel cr powd		d		s		
t200	(II) ferricyanide	$Sn_3[Fe(CN)_6]_2$	779.98	wh		d		i	i	s HCl
t201	(II) ferrocyanide	$Sn_2Fe(CN)_6$	449.33	wh gel				i	i	d HCl
t202	(IV) ferrocyanide	$SnFe(CN)_6$	330.64					i	i	d h HCl
t203	(II) fluoride	Fluoristan. SnF_2	156.69	wh, monocl cr				s		
t204	(IV) fluoride	SnF_4	194.68	wh, monocl cr, hygr	4.780^{19}	705 subl		v s	d	
t205	hydride	Stannane. SnH_4	122.72	gas		$d-150$	-52			s $AgNO_3$, $HgCl_2$, conc alk, conc H_2SO_4
t206	iodide	SnI_2	372.50	yelsh-red to red monocl need	5.285	320	717	0.98^{20}	4.03^{100}	v s NH_4OH, HI soln
t207	(IV) iodide	SnI_4	626.31	or red cub, 2.106	4.473^0	144.5	364.5	s	d	141.1^{25} CS_2; 6.03^{15} CCl_4; 17.88^{25} bz

No.	Name	Synonyms and Formulae	Mol. wt.	Crystalline form, properties and index of refraction	Density or spec. gravtiy	Melting point, °C	Boiling point, °C	Cold water	Hot water	Other solvents
	Tin									
t208	(II) nitrate	$Sn(NO_3)_2.20H_2O$	603.07	col leaf		−20		d	d	d HNO_3
t209	(II) nitrate, basic	$SnO.Sn(NO_3)_2$	377.39	wh cr mass		d >100 expl		d	d	
t210	(IV) nitrate	$Sn(NO_3)_4$	366.71	silky need		d 50		d		
t211	(II) oxide, mon-	SnO	134.69	blk, cub (tetr)	6.446^0	d 1080^{600}		i	i	s a, alk; sl s NH_4Cl
t212	oxide, mon-hydrate	$SnO.xH_2O$		wh powd or yellow-brn cr					d to SnO	d a; alk; s alk carb; i NH_4OH
t213	(IV) oxide, di	Nat. cassiterite. SnO_2	150.60	wh, tetr, (also hex or rhomb), 1.997, 2.093	6.95	1630	subl 1800–1900	i	i	d KOH, NaOH; i aq reg
t214	oxide, di-hydrate	α-Stannic acid or "ordinary" stannic acid. $SnO_2.xH_2O$		amorph or gel				i	i	s a, alk, K_2CO_3
t215	oxide, di-hydrate	β-Stannic acid or "meta" stannic acid. $SnO_2.xH_2O$		wh, amorph or gel				i	i	i a, K_2CO_3; sol alk
t216	(II) metaphosphate	$Sn(PO_3)_2$	276.63	amorph mass	$3.380^{22.8}$			i	i	
t217	(II) orthophosphate	$Sn_3(PO_4)_2$	546.01	wh, amorph	3.823^{17}			i	i	d a, alk
t218	(II) orthophosphate, di-H	$Sn(H_2PO_4)_2$	312.66	wh, rhomb cr	$3.167^{22.8}$	d	d		d	
t219	(II) orthophosphate, mono-H	$SnHPO_4$	214.67	cr	$3.476^{15.5}$	stabl >100	d	i	i	s dil min a
t220	(II) pyrophosphate	$Sn_2P_2O_7$	411.32	amorph powd	$4.009^{16.4}$					s conc a
t221	phosphide, mono-	SnP	149.66	silv wh	6.56	d	d	i	i	s HCl; i HNO_3
t222	phosphide, tri-	SnP_3	211.61	cr	4.10^0	<415 d to Sn_4P_3		i	i	d HNO_3; i HCl
t223	tetraphosphide, tri-	Sn_4P_3	567.68	wh cr	5.181	d <480		i	i	d fixed alk hydr, HCl
t224	phosphorus chloride	$SnCl_4.PCl_5$	468.74	col cr		subl 200		d	d	
t225	(II) selenide	$SnSe$	197.65	steelgray cr	6.179^0	861		i	i	d HCl, HNO_3, aq reg, alk sulf
t226	(II) sulfate	$SnSO_4$	214.75	wh yelsh or powd		>360 (SO_3)		33^{15}		s H_2SO_4
t227	(IV) sulfate	$Sn(SO_4)_2.2H_2O$	346.84	wh, hex pr, deliq				v s	d	s eth, dil H_2SO_4, HCl
t228	(II) sulfide	SnS	150.75	gray-blk cub, monocl	5.22^{25}	882	1230	0.00000218		d HCl, alk, $(NH_4)_2S$
t229	(IV) sulfide	Mosaic gold. SnS_2	182.82	gold yel, hex	4.5	d 600		0.0002^{18}		d alk sulf, aq reg, alk hydr, PCl_5, $SnCl_2$; i a
t230	(IV) sulfur chloride	$SnCl_4.2SCl_4$	608.25	yel cr		37	d <40	d	d	s eth, bz, CS_2, ethyl acet; d HNO_3
t231	tartrate	$SnC_4H_4O_6$	266.76	heavy wh powd				s		v s dil HCl
t232	(II) telluride	$SnTe$	246.29	gray cr	6.48	780	d	i	i	d alk sulf
t233	(IV) telluride	$SnTe_2$	373.89	blk, flocc ppt				i	i	d dil a, alk
t234	**Titanic acid, ortho-**	α Titanic acid. H_2TiO_4	113.91	wh		d		v sl s d		s dil HCl, dil H_2SO_4, conc alk
t235	**Titanium**	Ti	47.90	α hex, tr β cub 838, silv gray	4.5^{20}	**1660±10**	**3287**	i	i	s dil a
t236	boride, di-	TiB_2	69.52	hex	4.50	2900				
t237	bromide, di-	$TiBr_2$	207.72	blk powd	4.31	d >500		s ev H_2		
t238	bromide, tetra-	$TiBr_4$	367.54	or yel, deliq	2.6	39	230	d		s abs al, abs eth
t239	bromide, tri-	$TiBr_3.6H_2O$	395.72	redsh-viol or dk blue cr, deliq		115	d 400	v s		v s al, acet
t240	carbide	TiC	59.91	gr met, cub	4.93	3140±90	4820	i	i	s aq reg, HNO_3
t241	chloride, di-	$TiCl_2$	118.81	lt br-blk, hex, deliq	3.13	subl H_2	d 475 vac	d		s al, i eth, chl, CS_2
t242	chloride, tetra-	$TiCl_4$	189.71	lt yel liq, $1.61^{10.5}$, sol 2.06^{-79}	liq 1.726	−25	136.4	s	d	s dil HCl, al
t243	chloride, tri-	$TiCl_3$	154.26	dk viol, deliq	2.64	d 440	660^{106}	s	s	v s al; s HCl; i eth
t244	fluoride, tetra-	TiF_4	123.89	wh powd, hygr	$2.798^{20.5}$	>400 (pressure)	284 (subl.)	s d		s H_2SO_4, al, C_6H_5N; i eth
t245	fluoride, tri-	TiF_3	104.90	purp-red or vlt	3.40	1200	1400	red s vlt i		
t246	hydride	TiH_2	49.92	gray powd	3.9^{12}	d 400				
t247	iodide, di-	TiI_2	301.71	blk, hygr	4.99	600	1000	d		d alk; s conc HF, conc HCl
t248	iodide, tetra-	TiI_4	555.52	red, cub	4.3	150	377.1	v s	d	
t249	nitride	TiN	61.91	yel-bronze, cub	5.22	2930		i	i	sl s hot aq reg + HF
t250	oxalate	$Ti_2(C_2O_4)_3.10H_2O$	540.01	yel pr				s	s	i al, eth
t251	oxide, di-	Nat. brookite. TiO_2	79.90	wh, rhomb, 2.583, 2.586, 2.741	4.17	1825		i	i	s H_2SO_4, alk; i a
t252	oxide, di-	Nat. octahedrite. anatase. TiO_2	79.90	br-blk, tetr, 2.554, 2.493	3.84			i	i	s H_2SO_4, alk; i a

No.	Name	Synonyms and Formulae	Mol. wt.	Crystalline form, properties and index of refraction	Density or spec. gravity	Melting point, °C	Boiling point, °C	Solubility, in grams per 100 cc		
								Cold water	Hot water	Other solvents
	Titanium									
t253	oxide, di-	Nat. rutile. TiO_2	79.90	col, tetr, 2.616, 2.903	4.26	1830–1850	2500–3000	i	i	s H_2SO_4, alk; i a
t254	oxide, mon-	TiO	63.90	yel blk, pr	4.93	1750	>3000			s dil H_2SO_4; i HNO_3
t255	oxide, sesqui-	Ti_2O_3	143.80	vlt blk, trig	4.6	2130 d		i	i	s H_2SO_4; i HCl, HNO_3
t256	phosphide	TiP	78.87	gray, met	3.95^{25}			i	i	i a
t257	sulfate	$Ti_2(SO_4)_3$	383.98	green powd				i	i	s dil a; i al, eth, conc H_2SO_4
t258	sulfate, basic	$TiOSO_4$	159.96	wh or sl yelsh powd, 1.80–1.89				d		
t259	sulfide, di-	TiS_2	112.03	yel sc	3.22^{20}			hyd sl	d in steam	d HCl; s dil HNO_3, H_2SO_4
t260	sulfide, mono-	TiS	79.96	redsh solid	4.12				i	s conc H_2SO_4; i HCl, HF, dil H_2SO_4
t261	sulfide, sesqui	Ti_2S_3	191.99	grayish-blk cr	3.584			i	i	s conc H_2SO_4, conc HNO_3; i dil H_2SO_4, dil HCl
t262	**Tungsten**	Wolfram. W	183.85	gray-blk, cub	19.35^{30}_4	3410 ± 20	5660	i	i	v sl s HNO_3, H_2SO_4, aq reg; s HNO_3+HF, fus NaOH+ $NaNO_3$; i HF, KOH
t263	arsenide	WAs_2	333.69	blk cr	6.9^{18}	d red heat		i		d hot HNO_3, hot H_2SO_4
t264	boride, di-	WB_2	205.47	silvery, oct	10.77	ca 2900		i	i	s aq reg
t265	bromide, di-	WBr_2	343.67	bl-blk need		d 400		d		
t266	bromide, penta-	WBr_5	583.40	vlt-brn need, hygr		276	333	d		s abs al, chl, eth, alk
t267	bromide, hexa-	WBr_6	663.30	bl-blk, need	6.9	232		i	d	s abs a, eth, CS_2, NH_4OH
t268	carbide	WC	195.86	blk, hex	15.63^{18}	2870 ± 50	6000	i		s HNO_3+HF, aq reg
t269	(di-)carbide	W_2C	379.71	blk, hex	17.15	2860	6000	i		s HNO_3+HCl
t270	carbonyl	$W(CO)_6$	351.91	col, rhomb cr	2.65	d ~150	175^{766}	i	i	s fum HNO_3; v sl s al, eth, bz
t271	chloride, di-	WCl_2	254.76	gray, amorph	5.436			d		
t272	chloride, hexa-	WCl_6	396.57	dk bl, cub	3.52^{25}_4	275	346.7		d^{60}	s al, eth, bz, CCl_4; v s CS_2, POCl
t273	chloride, penta-	WCl_5	361.12	blk. deliq	3.875^{25}_4	248	275.6		d to W_2O_5	v sl s CS_2
t274	chloride, tetra-	WCl_4	325.66	gray, deliq	4.624^{25}_4	d		d		
t275	fluoride, hexa-	WF_6	297.84	col gas, or lt yel liq	liq 3.44 gas 12.9 g/l	$2.5^{4.20}$	17.5	d	d	s alk
t276	iodide, di-	WI_2	437.66	br-gr, amorph	6.799^{25}_4	d		i	d	s alk; i al CS_2
t277	iodide, tetra-	WI_4	691.47	blk, cr	5.2^{18}	d		i	d	s abs al; i eth, chl, turp
t278	nitride, di-	WN_2	211.86	brn, cub		above 400 (vac)		d	d	
t279	oxide, di-	WO_2	215.85	br, cub	12.11	1500–1600 (in N_2)	ca 1430 subl 800	i	i	s a, KOH
t280	oxide, pent-	Mineral blue. W_2O_5 or W_4O_{11}	447.70 or 911.39	blue-vlt, triicl		subl 800–900	ca 1530 d 2000	i	i	i a
t281	oxide, tri-	WO_3	231.85	yel, rhomb, or yel-or powd	7.16	1473		i	i	s hot alk; sl s HF; i a
t282	oxydibromide, di-	WO_2Br_2	375.67	red, prism		d				
t283	oxytetrabromide	$WOBr_4$	519.49	blk, deliq		277	327	d	d	
t284	oxytetrachloride	$WOCl_4$	341.66	red, need		211	227.5	d	d	s CS_2, S_2Cl_2, bz
t285	oxydichloride, di-	WO_2Cl_2	286.75	lt yel tabl		266		s	d	i al; s NH_4OH, alk
t286	oxytetrafluoride-	WOF_4	75.84	col pl, hygr		110	187.5			sl s CS_2; i CCl_4
t287	phosphide	WP	214.82	gray, prism	8.5			i		s HNO_3+HF; i alk, HCl
t288	phosphide	WP_2	245.80	blk cr	5.8	d		i	i	s HNO_3+HF, aq reg; i al, eth
t289	phosphide	W_2P	398.67	dk gray prism	5.21	d				s fus Na_2CO_3+ $NaNO_3$; i a, aq reg

No.	Name	Synonyms and Formulae	Mol. wt.	Crystalline form, properties and index of refraction	Density or spec. gravity	Melting point, °C	Boiling point, °C	Solubility, in grams per 100 cc		
								Cold water	Hot water	Other solvents
	Tungsten									
t290	silicide.........	WSi_2............	240.02	blue, gray, tetrag.	9.4	above 900	....	i	i	s HNO_3+HF; i aq reg
t291	sulfide, di-........	Nat. tungstenite. WS_2	247.98	dk gray, hexag.	7.5^{10}	d 1250	....	i		s HNO_3+HF, fus alk; i al
t292	sulfide, tri-.......	WS_3...........	280.04	choc brn powd...		d 50	....	sl s	s	s alk
t293	**Tungstic acid, meta-**	$H_2W_4O_{13}.9H_2O$........	1107.55	col, tetrag.......	3.93	d 50		88.57^{22}	$111.87^{41.5}$	$110.76^{24.3}$ eth; s al
t294	Tungstic acid, ortho-	H_2WO_4........	249.86	yel powd, 2.24...	5.5	− H_2O, 100	1473	i	sl s	s alk, HF, NH_3; i most a
t295	Tungstic acid, ortho-	$H_2WO_4.H_2O$.......	267.88	wh........		$H_2W_2O_7$ at 100		sl s		s alk
u1	**Uranic acid meta-**	Uranyl hydroxide. H_2UO_4(or $UO_2(OH)_2$)	304.04	yel, rhomb, or powd	5.926	− H_2O 250−300		i	i	s a, alk carb
u2	**Uranium**.........	U........	238.03	silvery, cubic, radioactive	19.05 ± 0.02^{25}	1132.3 ± 0.8	3818	i	i	s a; i alk, al
u3	boride, di-........	UB_2........	259.65	hex........	12.70	2365				
u4	bromide tetra-.....	UBr_4........	557.67	br leaf, deliq	5.35	516	792^{760}	v s	v s	s d al, MeOH; i bz; s liq NH_3
u5	bromide tri-......	UBr_3........	477.76	dk brn need, hygr	6.53	730	volat	s	....	d al
u6	dicarbide........	UC_2........	262.05	met cr........	11.28^{16}	2350−2400	4370^{760}	d	d	i al; d dil inorg a
u7	chloride, penta-....	UCl_5........	415.30	dk green, gray need, red by trans light, hydr	3.81(?)	d 300		d		s abs als, a acet, NH_4Cl; d ac a; i bz, eth
u8	chloride, tetra-.....	UCl_4........	379.84	dk grn met, cub oct, hygr	4.87	590 ± 1	792^{760}	v s	s	s al, acet, ac a; i eth, $CHCl_3$
u9	chloride, tri-......	UCl_3........	344.39	dk red need, hygr	5.44^{25}_{4}	842 ± 5		s	s	s MeOH, acet, glac acet a; i eth
u10	fluoride, hexa-.....	UF_6........	352.02	col cr, deliq, monocl	4.68^{21}	64.5−64.8	56.2^{765}	d	d	d al, eth; s CCl_4, chl; i CS_2
u11	fluoride, tetra-.....	UF_4........	314.02	green, tricl need.	6.70 ± 0.10	960 ± 5		v sl s	....	i dil a, alk; s conc a, conc alk
u12	fluoride, tri-......	UF_3........	295.03	blk cr or fused...		d above 1000	....	sl d	....	v sl s dil inorg a
u13	hydride........	UH_3........	241.05	blk-brn powd....	10.95	....		i	i	i al, acet, liq NH_3; sl s dil HCl; d HNO_3
u14	hydride........	UH_3........	241.05	blk powd, cub...	11.4					
u15	iodide, tetra-......	UI_4........	745.65	blk, need........	5.6^{18}	506	759	s	s d	
u16	nitride, mono-.....	UN........	252.04	br powd........	14.31	ca 2630 ± 50				i HCl, H_2SO_4
u17	oxide, di-........	UO_2........	270.03	br-blk rhomb, or cub	10.96	2878 ± 20		i	i	s HNO_3, conc H_2SO_4
u18	oxide, per-.......	$UO_4.2H_2O$........	338.06	pa yel cr, hygr...		d 115		0.0006^{20}	0.008^{90}	d HCl
u19	oxide, tri-.......	Uranyl oxide. UO_3...	286.03	yel-red powd.....	7.29	d		i	i	s HNO_3, HCl
u20	tri-oxide, oct-......	U_3O_8........	842.09	olive green-blk...	8.30	d 1300 to UO_2		i	i	s HNO_3, H_2SO_4
u21	(IV) sulfate......	$U(SO_4)_2.4H_2O$.......	502.21	grn, rhomb......		−$4H_2O$, 300		23^{11}	9^{63} (anhydr)	s dil a
u22	(IV) sulfate......	$U(SO_4)_2.8H_2O$.......	574.28	....		d 90		11.3^{18}	58.2^{62}	i al; s dil a
u23	(IV) sulfate......	$U(SO_4)_2.9H_2O$.......	592.29	grnsh, monocl.....		−$7H_2O$, 230	−$9H_2O$ red heat			s dil H_2SO_4
u24	sulfide, di-......	US_2........	302.16	gray-blk, tetr....	7.96^{25}	>1100	oxidizes	sl d	....	s conc HCl; d HNO_3 v s al
u25	sulfide, mono-.....	US........	270.09	blk amorph powd.	10.87	above 2000		i	i	i HCl, HNO_3
u26	sulfide, sesqui-....	U_2S_3........	572.25	gray blk, rhomb need		ign				s +O aq reg conc HNO_3; i dil a
u27	**Uranyl acetate**......	$UO_2(C_2H_3O_2)_2.2H_2O$..	422.19	yel, rhomb.....	2.893^{15}	−$2H_2O$, 110	d 275	7.694^{15}	d	v s al
u28	benzoate........	$UO_2(C_7H_5O_2)_2$.....	512.26	yel powd.........				sl s	sl s al	sl s al
u29	bromide........	UO_2Br_2........	429.85	grn-yel need, hygr				s d		s al, eth
u30	perchlorate........	$UO_2(ClO_4)_2.6H_2O$...	577.02	yel cr, deliq, rhomb		90 d 110				
u31	chloride........	UO_2Cl_2........	340.93	yel, deliq........		578	d	320^{18}	v s	s al, amyl al, eth
u32	formate........	$UO_2(CHO_2)_2.H_2O$...	378.08	yel, oct........	3.695^{19}	−H_2O, 110		7.2^{15}		sl s form a; 0.74^{15} MeOH, 2.37^{15} acet
u33	iodate.........	$UO_2(IO_3)_2$........	619.83	yel, rhomb......	5.2	d 250		s		i HNO_3
u34	iodate.........	$UO_2(IO_3)_2.H_2O$...	637.85	α prismatic, stable, β pyramidal	α 5.220^{18} β 5.052^{18}			α 0.1049^{18} β 0.1214^{18}		
u35	iodide........	UO_2I_2........	523.84	red, deliq........		d in air		s al, eth, bz		s al, eth, bz
u36	nitrate.........	$UO_2(NO_3)_2.6H_2O$...	502.13	yel, rhomb, deliq, 1.4967	2.807^{13}	60.2 d 100	118	....	∞ 60	v s al, eth, ac a, acet, MeOH
u37	oxalate.........	$UO_2C_2O_4.3H_2O$...	412.09	yel cr........		−H_2O, 110		0.8^{14}	3.3^{100}	s inorg a, alk, oxal a
u38	phosphate, mono-H	$UO_2HPO_4.4H_2O$...	438.07	yel pl, tetr........				i	i	s HNO_3, aq Na_2CO_3; i ac a
u39	potassium carbonate.	$UO_2CO_3.2K_2CO_3$...	606.46	yel cr........		−CO_2, 300		7.4^{15}	d	i al

No.	Name	Synonyms and Formulae	Mol. wt.	Crystalline form, properties and index of refraction	Density or spec. gravity	Melting point, °C	Boiling point, °C	Solubility, in grams per 100 cc		
								Cold water	Hot water	Other solvents
	Uranyl									
u40	sodium carbonate...	$UO_2CO_3.2Na_2CO_3$...	542.02	yel cr........				sl s	....	i al
u41	sulfate............	$UO_2SO_4.3H_2O$...	420.14	yel-grn cr...	$3.28^{16.5}$	d 100	....	$20.5^{15.5}$	22.2^{100}	24.3^{13} conc H_2SO_4; 30^{13} conc HCl
u42	sulfate............	$2(UO_2SO_4).7H_2O$...	858.29	yel.........		anh 300		sl s		s H_2SO_3
u43	sulfide............	UO_2S............	302.09	brn-blk, tetr....		d 40-50		sl s	s d	s dil a, dil al, $(NH_4)_2CO_3$; i abs al
u44	sulfite............	$UO_2SO_3.4H_2O$...	422.15	pa-gr cr....						s H_2SO_3
v1	**Vanadic acid, meta.**	HVO_3..........	99.95	yel sc........				i		s a, alk; i NH_4OH
v2	tetra-........	$H_2V_4O_{11}$..........	381.78	br amorph....				i		s a, alk, NH_4OH
v3	**Vanadium**........	V..........	50.9415	lt gray met, cub, 3.03	5.96	1890±10	**3380**	i	i	s aq reg, HNO_3, H_2SO_4, HF; i HCl, alk
v4	boride, di-........	VB_2..........	72.56	hex........	5.10					
v5¹	bromide, tri-........	VBr_3..........	290.67	grn-blk, deliq...	4.00^{18}	d		s		s al, eth; i HBr
v5²	carbide............	VC..........	62.95	blk. cub.......	5.77	2810	3900	i		s HNO_3, fus KNO_3; i HCl, H_2SO_4
v6	chloride, di-........	VCl_2..........	121.85	grn, hex, deliq.	3.23^{18}			s d	s d	s al, eth
v7	chloride, tetra-........	VCl_4..........	192.75	red-br liq.......	1.816^{30}	$-28±2$	148.5^{755}	s d		s abs al, eth, chl, acet a
v8	chloride, tri-........	VCl_3..........	157.30	pink cr, deliq....	3.00^{18}	d		s d	s d	s abs al, eth
v9	fluoride, penta-........	VF_5..........	145.93	....	2.177^{19}		111.2^{758}			s al
v10	fluoride, tetra-........	VF_4..........	126.94	br yel.......	2.975^{22}	d 325		s		s acet; sl s al, chl
v11	fluoride, tri-........	VF_3..........	107.94	grn, rhomb.....	3.363^{19}	>800	subl	i		s al, chl, CS_2
v12	fluoride, tri-........	$VF_3.3H_2O$..........	161.98	dk gr, rhomb....		$-3H_2O, 100$		s	v s d	i abs al
v13	iodide, di-........	VI_2..........	304.71	vlt-rose, hex....	5.44	750-800 subl vac		s		i al, CCl_4, CS_2, bz
v14	iodide, tri-........	$VI_3.6H_2O$..........	539.75	gr cr, deliq.....		d		v s		s al
v15	nitride............	VN..........	64.95	blk, cub.......	6.13	2320		i		sl s aq reg
v16	**oxide**........	VO (or V_2O_2)......	66.94	lt gray cr........	5.758^{14}	ign		i	i	s a
v17	oxide, di- (or tetr)-...	VO_2 (or V_2O_4)........	82.94	bl cr........	4.339	1967		i	i	s a, alk
v18	oxide, pent-........	V_2O_5........	181.88	yel-red, rhomb, 1.46, 1.52, 1.76	3.357^{18}	690	d 1750	0.8^{20}		s a, alk; i abs al
v19	oxide, sesqui-........	Vanadium trioxide. V_2O_3	149.88	blk cr........	4.87^{18}_4	1970		sl s	s	s HNO_3, HF, alk
v20	oxybromide........	VOBr........	146.85	vlt, oct........	4.00^{18}	d 480		v sl s		s acet, anhydr eth, acet
v21	oxy di-bromide......	$VOBr_2$........	226.76	br powd, deliq....		d 180		s		
v22	oxytribromide......	$VOBr_3$........	306.67	red liq.......	$2.933^{14.5}$	d 180	130^{100}			
v23	oxychloride........	VOCl........	102.39	yel brn powd....	$2.824, 3.64^{20}$		127	i		v s HNO_3
v24	oxydichloride......	$VOCl_2$........	137.85	grn, deliq........	2.88^{13}			d		s dil HNO_3
v25	oxytrichloride......	$VOCl_3$........	173.30	yel liq.......	1.829	$-77±2$	126.7	s d		s al, eth, ac a
v26	oxydifluoride......	VOF_2........	104.94	yel........	3.396^{19}	d				sl s acet
v27	oxytrifluoride......	VOF_3........	123.94	yel-wh, hygr....	2.459^{19}	300	480			
v28	silicide, di-........	VSi_2........	107.11	met pr........	4.42			i	i	s HF; i al, eth, a
v29	(dl-)silicide........	V_2Si........	129.97	silv wh pr.......	5.48^{17}			i	i	s HF; i al, eth, a
v30	sulfate (hypovanadous)	$VSO_4.7H_2O$........	273.11	vlt, monocl.....		d in air				
v31	sulfide, mono- or (di-)	VS (or V_2S_2)........	83.01	blk pl (exist ?)..	4.20	d		i		s hot H_2SO_4, HNO_3; sl s KSH; i HCl, alk
v32	sulfide, penta-........	V_2S_5........	262.20	blk-grn powd....	3.0	d		i		s HNO_3, alk sulf, alk
v33	sulfide, sesqui- or (tri-)	V_2S_3........	198.08	grn-blk pl, or powd	4.72^{21}	d>600		i		s alk sulf; sl s, alk, HCl, HNO_3, H_2SO_4
v34	**Vanadyl sulfate**.....	$VOSO_4$........	163.00	bl.........				v s		
w1	**Water**........	H_2O........	18.01534	col liq, or col hex cr	liq 1.000^4 sld 0.9168^0	0.00	100.00			s al
w2	**Water heavy**........	Deuterium oxide. D_2O	20.03	col liq or hex cr, 1.33844^{20}	1.105^{20}_4	3.82	101.42	∞	∞	∞ al; sl s eth
w3	**Wolfram**........	See tungsten.								
x1	**Xenon**........	Xe........	131.30	col inert gas....	gas 5.887 g/l ±0.009 liq 3.52^{-109} solid 2.7^{-140}	-111.9	$-107.1±3$	24.1^0 cm³ 11.9^{25} cm³	8.4^{40}, 7.12^{90}	
y1	**Ytterbium**........	Yb........	173.04	cub........	6.9654 up to 789 6.54 above 789	819±5	1194	i		s a
y2	(III) acetate........	$Yb(C_2H_3O_2)_3.4H_2O$...	422.24	hex pl........	2.09	$-4H_2O, 100$		v s	v s	
y3	(II) bromide........	$YbBr_2$........	332.86	....	5.91^{25}_4	677	1800	s	s	s dil a

No.	Name	Synonyms and Formulae	Mol. wt.	Crystalline form, properties and index of refraction	Density or spec. gravity	Melting point, °C	Boiling point, °C	Solubility, in grams per 100 cc		
								Cold water	Hot water	Other solvents
	Ytterbium									
y4	(III) bromide	$YbBr_3$	412.77	col cr		956	d	s	s	
y5	(II) chloride	$YbCl_2$	243.95	grn-yel cr	5.08	702	1900	s	s	s dil a
y6	(III) chloride	$YbCl_3.6H_2O$	387.49	grn, rhomb cr, deliq	2.575	865 $-6H_2O$, 180		v s	v s	s abs al
y7	(II) fluoride	YbF_2	211.04			1052	2380	i	i	
y8	fluoride	YbF_3	230.04			1157	2200	i	i	i dil a
y9	(II) iodide	YbI_2	426.85	lt yel, hex cr	5.40_4^{25}	780 ± 4	1300 d(700) vac	s	s	s dil a
y10	(III) iodide	YbI_3	553.75	gold yel cr		d 700	d	s	s	s dil a
y11	(III) oxalate	$Yb_2(C_2O_4)_3.10H_2O$	790.29	col cr	2.644			0.00033^{25}		sl s dil a
y12	(III) oxide	Ytterbia. Yb_2O_3	394.08	col	9.17			i	i	s h dil a
y13	(III) selenate	$Yb_2(SeO_4)_3.8H_2O$	919.08	hex pl	3.30			s d	s	
y14	(III) selenite	$Yb_2(SeO_3)_3$	726.95					i		
y15	(III) sulfate	$Yb_2(SO_4)_3$	634.26	col cr	3.793	d 900		44.2^0	4.7^{100}	
y16	(III) sulfate, octohydrate	$Yb_2(SO_4)_3.8H_2O$	778.39	prism	3.286			35.9^{25}	21.1^{40}	
y17	**Yttrium**	Y	**88.905**	gray-blk met, hex	4.4689	1522	3338	sl d	d	v s dil a; s h KOH
y18	acetate	$Y(C_2H_3O_2)_3.4H_2O$	338.10	col, tricl					9.03^{25}	
y19	bromate	$Y(BrO_3)_3.9H_2O$	634.76	hex pr		74	$-6H_2O$, 100	168^{25}		sl s al; i eth
y20	bromide	YBr_3	328.63	deliq		904		v s		s al; i eth
y21	bromide hydrate	$YBr_3.9H_2O$	490.77	col tabl, deliq				v s		sl s al; i eth
y22	carbide	YC_2	112.93	yel., microcr	4.13^{18}			d		
y23	carbonate	$Y_2(CO_3)_3.3H_2O$	411.88	wh-redsh powd						s dil min a, $(NH_4)_2$ CO_3; sl s aq CO_2; i al, eth
y24	chloride	YCl_3	195.26	shiny wh leaf	2.67	721	1507	78^{10}	82^{50}	60.1^{15} al; 60.6^{15} pyr
y25	chloride, hexahydrate	$YCl_3.6H_2O$	303.36	redsh-wh, rhomb, deliq	2.18^{18}	$-5H_2O$, 100		217^{20}	235^{50}	s al; i eth
y26	chloride, monohydrate	$YCl_3.H_2O$	213.28	col cr		$-H_2O$, 160		v s		
y27	fluoride	YF_3	145.90	gelat	4.01	1387		i		v sl s dil a
y28	hydroxide	$Y(OH)_3$	139.93	wh-yel gelat or powd		d		i	i	s a, NH_4Cl; i alk
y29	iodide	YI_3	469.62	wh, cr, deliq		1004	$650-700^{0.02}$	v s		s al, acet; sl s eth
y30	molybdate	$Y_2(MoO_4)_3.4H_2O$	729.68	grayish or yelsh, tetr pl, 2.03	4.79_{16}^{16}	1347				
y31	nitrate, hexahydrate	$Y(NO_3)_3.6H_2O$	383.01	col, redsh cr, deliq	2.68	$-3H_2O$, 100		$134.7^{22.5}$		v s al, eth, HNO_3
y32	nitrate, tetrahydrate	$Y(NO_3)_3.4H_2O$	346.98	redsh-wh pr	2.682			s		s al, HNO_3
y33	oxalate	$Y_2(C_2O_4)_3.9H_2O$	604.01	wh cr powd		d		0.0001		sl s HCl
y34	oxide	Yttria. Y_2O_3	225.81	col-yelsh, cub or powd	5.01	2410		0.00018^{29}		s a; i alk
y35	sulfate	$Y_2(SO_4)_3$	465.99	wh powd	2.52	d 1000		5.38^{25}	s	s sat K_2SO_4 sol
y36	sulfate, octahydrate	$Y_2(SO_4)_3.8H_2O$	610.12	col-redsh, monocl, 1.543, 1.549, 1.576	2.558	$-8H_2O$, 120	d 700	7.47^{16} (anhydr)	1.99^{96} (anhydr)	i al, alk; s conc H_2SO_4
y37	sulfide	Y_2S_3	273.99	yel-gr powd						d a
y38	**Yttrium hexaanti-pyrine** perchlorate	$[Y(C_{11}H_{12}N_2O)_6](ClO_4)_3$	1516.60	col, hex cr		d 293-296		0.55^{20}		
y39	hexaantipyrine iodide	$[Y(C_{11}H_{12}N_2O)_6]I_3$	1598.96	col cr		280-282		4.65^{20}		
z1	**Zinc**	Zn	**65.38**	bluish-wh met, hex	7.14	**419.58**	907	i	i	s a, alk, ac a
z2	acetate	$Zn(C_2H_3O_2)_2$	183.46	col, monocl	1.84	d 200	subl vac	30^{20}	44.6^{100}	2.8^{25} al; 166.79^{79} al
z3	acetate, dihydrate	$Zn(C_2H_3O_2)_2.2H_2O$	219.49	col, monocl, β 1.494	1.735	237	$-2H_2O$, 100	31.1^{20}	66.6^{100}	2 al
z4	acetylacetonate	$Zn(C_5H_7O_2)_2$	263.59	need		138	subl	v s d		v s bz, acet; s al
z5	aluminate	Nat. gahnite. $ZnAl_2O_4$	183.33	cub, grn 1.78	4.58			i	i	i a; sl s alk
z6	amide	$Zn(NH_2)_2$	97.42	wh powd, amorph	2.13^{25}	d 200 vac		d	d	i al, eth
z7	antimonide	Zn_3Sb_2	439.61	silv wh, rhomb pr	6.33	570		d		
z8	orthoarsenate	Nat. koettigite. $Zn_3(AsO_4)_2.8H_2O$	618.08	monocl, 1.662, 1.683, 1.717	3.309^{15}	$-1H_2O$, 100				s HNO_3, H_3PO_4, alk
z9	orthoarsenate, basic	Nat. adamite. $Zn_2(AsO_4)_2.Zn(OH)_2$	573.34	col, rhomb	4.475^{15}	d 250				
z10	orthoarsenate, hydrogen	$ZnHAsO_4.4H_2O$	277.36	wh, rhomb		$-H_2O$, 327		d	d	
z11	arsenide	Zn_3As_2	345.95	met-gray, tetr	5.528	1015		i		d a
z12	benzoate	$Zn(C_7H_5O_2)_2$	307.60	wh powd				2.46^{20}	1.44^{20}	
z13	borate	$3ZnO.2B_2O_3$	383.35	wh tricl cr, or amorph powd	cr 4.22 powd 3.64	980		s		cr i HCl; amorph; s HCl
z14	bromate	$Zn(BrO_3)_2.6H_2O$	429.23	wh, cub, 1.5452	2.566	100	$-6H_2O$, 200	v s	∞	
z15	bromide	$ZnBr_2$	225.19	col, rhomb, hygr n_D^{18} 1.5452	4.201_4^{25}	394	650	447^{20}	675^{100}	v s al, eth, acet; s NH_4OH

No.	Name	Synonyms and Formulae	Mol. wt.	Crystalline form, properties and index of refraction	Density or spec. gravity	Melting point, °C	Boiling point, °C	Solubility, in grams per 100 cc		
								Cold water	Hot water	Other solvents
	Zinc									
z16	butyrate.........	$Zn(C_4H_7O_2)_2.2H_2O$	275.60	wh pr				10.7^{16}	d	
z17	caproate.........	$Zn(C_6H_{11}O_2)_2$	295.68					$1.03^{24.5}$		
z18	carbonate........	Nat. smithsonite. $ZnCO_3$	125.39	col, trig, 1.818, 1.618	4.398	$-CO_2$, 300		0.001^{15}		s a, alk, NH_4 salts; i NH_3, acet, pyr
z19	chlorate.........	$Zn(ClO_3)_2.4H_2O$	304.33	col yelsh, cub, deliq	2.15	d 60	d	262^{20}	v s	167 al; s acet, eth, glyc
z20	chlorate, per-....	$Zn(ClO_4)_2.6H_2O$	372.36	wh, rhomb, deliq. 1.508, 1.480	2.252 ± 0.01	105–107	d 200	s		s al
z21	chloride.........	$ZnCl_2$	136.28	wh, hex, deliq, 1.681, 1.713	2.91^{25}	283	732	432^{25}	615^{100}	$100^{12.5}$ al; v s eth; i NH_3
z22	chloroplatinate.....	$ZnPtCl_6.6H_2O$	581.27	yel, trig, hygr	2.717^{12}	d 160		v s	v s	v s al; d H_2SO_4
z23	chromate........	$ZnCrO_4$	181.36	lem-yel pr	3.40			i	d	s a, liq NH_3; i acet
z24	chromate........	$ZnCr_2O_4$	233.36	dk grn to black, cub	5.30^{15}					
z25	*di*chromate......	$ZnCr_2O_7.3H_2O$	335.40	redsh-brn cr, or or-yel powd, hygr				v s	d	i al, eth; s a
z26	citrate..........	$Zn_3(C_6H_5O_7)_2.2H_2O$	610.35					sl s		
z27	cyanide.........	$Zn(CN)_2$	117.41	col, rhomb	1.852	d 800		0.0005^{20}		s alk, KCN, NH_3; i al
z28	ferrate (III)......	Ferrite. $ZnFe_2O_4$	241.06	blk, oct	5.33^{20}	1590				s conc HCl; i dil a, alk
z29	ferrocyanide......	$Zn_2Fe(CN)_6$	342.69	wh powd	1.85^{25}_4			i		s excess alk; i dil a
z30	ferrocyanide, trihydrate	$Zn_2Fe(CN)_6.3H_2O$	396.74	wh powd		d		i	i	i al, HCl; d NaOH; s NH_4OH; v sl s NH_3
z31	fluoride.........	ZnF_2	103.37	col, monocl or tricl	4.95^{25}_4	872	ca 1500	1.62^{20}	s	s hot a, NH_4OH; i al, NH_3
z32	fluoride, tetrahydrate	$ZnF_2.4H_2O$	175.43	col, rhomb	2.255	$-4H_2O$, 100	tr to ZnO, 3000	1.6^{18}	s	s a, alk, NH_4OH
z33	fluosilicate......	$ZnSiF_6.6H_2O$	315.54	col, hex pr, 1.3824, 1.3956	2.104	d 100		v s		
z34	formaldehyde-sulfoxylate	$Zn(HSO_2.CH_2O)_2$	255.56	rhomb pr		d		v s	v s	d a; i al
z35	formaldehyde-sulfoxylate, basic	$Zn(OH)HSO_2.CH_2O$	177.47	rhomb pr		d		i	i	d a; i al
z36	formate.........	$Zn(CHO_2)_2$	155.41	col, cr	2.368	d		3.80	62^{100}	
z37	formate.........	$Zn(CHO_2)_2.2H_2O$	191.44	wh, monocl, 1.513, 1.526, 1.566	2.207^{20}	$-2H_2O$, 140	d	5.2^{20}	38^{100}	i al
z38	gallate..........	$ZnGa_2O_4$	268.81	wh fine cr, 1.74	6.15 calc	<800		i	i	i org solv; s dil a, NH_4OH
z39	glycerophosphate....	$ZnC_3H_7O_6P$	235.43	wh amorph powd				s		i al, eth
z40	hydroxide(ε)......	$Zn(OH)_2$	99.38	col, rhomb	3.053	d 125		v sl s		s a, alk
z41	iodate..........	$Zn(IO_3)_2$	415.18	wh, need	5.0632^{5}	d		0.87	1.31	s alk, HNO_3
z42	iodate, dihydrate....	$Zn(IO_3)_2.2H_2O$	451.21	wh, cr powd	4.223^{25}_4	$-H_2O$, 200		0.877	1.32	s HNO_3, NH_4OH
z43	iodide..........	ZnI_2	319.18	col, hexag	4.7364^{25}_4	446	d 624	432^{18}	511^{100}	s a, al, eth, NH_3, $(NH_4)_2CO_3$
z44	*d*-lactate........	$Zn(C_3H_5O_3)_2.2H_2O$	279.45					5.7^{15}	9^{33}	0.104 h 98 % al
z45	*dl*-lactate........	$Zn(C_3H_5O_3)_2.3H_2O$	297.47	wh, rhomb cr				1.67^{106}	16.7^{100}	v sl s al
z46	laurate..........	$Zn(C_{12}H_{23}O_2)_2$	464.00	wh powd		128		0.011^{15}	0.019^{100}	0.010^{15} al
z47	*per*manganate.....	$Zn(MnO_4)_2.6H_2O$	411.33	vlt-br or bl, deliq.	2.47	$-5H_2O$, 100		33.3	v s	d al, a
z48	nitrate, trihydrate...	$Zn(NO_3)_2.3H_2O$	243.43	col, need		45.5			327.3^{40}	
z49	nitrate, hexahydrate...	$Zn(NO_3)_2.6H_2O$	297.47	col, tetrag	2.065^{14}	36.4	$-6H_2O$, 105–131	184.3^{20}	∞	v s al
z50	nitride..........	Zn_3N_2	224.12	gray	6.22^{25}_4			d		s HCl
z51	oleate..........	$Zn(C_{18}H_{33}O_2)_2$	628.30	wax-like solid		70		i		s al, eth, bz, CS_2; sl s acet
z52	oxalate..........	$ZnC_2O_4.2H_2O$	189.42	wh powd	3.28^{25}_4	d 100		0.00079^{18}		s a, alk
z53	oxide...........	Nat. zincite. ZnO	81.37	wh, hex, 2.008, 2.029	5.606	1975		0.00016^{29}		s a, alk, NH_4Cl; i al, NH_3
z54	oxide, per-.......	$ZnO.\frac{1}{2}H_2O$	106.38	yelsh, powd	3.00 ± 0.08	$-O_2$, vac		sl d	d	d al, eth, acet
z55	1-phenol-4-sulfonate(p)	$Zn(C_6H_5SO_4)_2.8H_2O$	555.83	col cr or fine wh powd, effl		$-8H_2O$, 125		62.5	250^{100}	55.6^{25} al
z56	*ortho*phosphate.....	$Zn_3(PO_4)_2$	386.05	col, rhomb	3.998^{15}	900		i	i	s a, NH_4OH; i al
z57	*ortho*phosphate, dihydrogen	$Zn(H_2PO_4)_2.2H_2O$	295.38	tricl		d 100		d		
z58	*ortho*phosphate, octahydrate	$Zn_3(PO_4)_2.8H_2O$	530.18	rhomb pl	3.109^{15}			i		s alk

No.	Name	Synonyms and Formulae	Mol. wt.	Crystalline form, properties and index of refraction	Density or spec. gravity	Melting point, °C	Boiling point, °C	Cold water	Hot water	Other solvents
	Zinc									
z59	*ortho*phosphate, tetrahydrate	α—Hopeite. $Zn_3(PO_4)_2.4H_2O$	458.11	col, rhomb, 1.572, 1.591, 1.59	3.04	tr >105	. . . .	i	i	v s a, NH_4OH, NH_4 salts
z60	*ortho*phosphate tetrahydrate	β-Hopeite. $Zn_3(PO_4)_2.4H_2O$	458.11	col, rhomb, 1.574, 1.582, 1.582	3.03	tr >140	. . . .	i	i	v s a, NH_4OH, NH_4 salts
z61	*ortho*phosphate tetrahydrate	Parahopeite. $Zn_3(PO_4)_2.4H_2O$	458.11	col, tricl, 1.614, 1.625, 1.665	3.75	tr >163	. . . .	i	i	v s a, NH_4OH, NH_4 salts
z62	*pyro*phosphate	$Zn_2P_2O_7$	304.68	wh powd	3.75^{21}			i	i	s a, alk, NH_4OH
z63	phosphide	Zn_3P_2	258.06	dk gray, tetrag, pois	4.55^{13}	>420	1100; subl in H_2	d		d H_2SO_4 ev H_3P s HNO_3; s (viol) dil a; i al
z64	hypophosphite	$Zn(H_2PO_2)_2.H_2O$	213.36	col, cr powd, hygr	. . . .			s		s alk
z65	picrate	$Zn(C_6H_2N_3O_7)_2.8H_2O$	665.69	yel cr powd, expl	. . . .	expl		s		
z66	salicylate	$Zn(C_7H_5O_3)_2.3H_2O$	393.65	need	. . . .			5^{20}		s al
z67	selenate	$ZnSeO_4.5H_2O$	298.40	wh, tricl	2.591^{20}_4	d >50		s		
z68	selenide	$ZnSe$	144.33	yelsh to redsh, cub, 2.89	5.42^{15}_4	>1100		i		s a; d HNO_3
z69	silicate	Nat. hemimorphite. $2ZnO.SiO_2.H_2O$	240.84	rhomb, or trigon 1.614, 1.617, 1.636	3.45			i	i	
z70	*meta*silicate	$ZnSiO_3$	141.45	col, rhomb	3.42	1437		i		i a
z71	*ortho*silicate	Nat. willemite. Zn_2SiO_4	222.82	trig, 1.694, 1.723	4.103	1509		i	i	s acet a
z72	stearate	$Zn(C_{18}H_{35}O_2)_2$	632.33	light powd	. . . .	130		i		i al, eth
z73	sulfate	Nat. zinkosite. $ZnSO_4$	161.43	col, rhomb, 1.658, 1.669, 1.670	3.54^{25}_4	d 600		s	s	sl s al; s MeOH, glyc
z74	sulfate, heptahydrate	Nat. goslarite. $ZnSO_4.7H_2O$	287.54	col, rhomb, effl, 1.457, 1.480, 1.484	$1.957^{?}_{?}$	100	$-7H_2O$, 280	96.5^{20}	663.6^{100}	sl s al, glyc
z75	sulfate, hexahydrate	$ZnSO_4.6H_2O$	269.50	col, monocl or tetrag	2.072^{11}	$-5H_2O$, 70		s	117.3^{100}	
z76	sulfide, (α)	Nat. wurtzite. ZnS	97.43	col, hex, 2.356, 2.378	3.98	**1700 ± 20** **atm** 1185		0.00069^{18}		v s a; i ac a
z77	sulfide, (β)	Nat. sphalerite. ZnS	97.43	col, cub, 2.368	4.102^{25}	tr 1020		0.000065^{18}		v s a
z78	sulfide, monohydrate	$ZnS.H_2O$	115.45	yelsh-wh powd	3.98	1049		i		s a
z79	sulfite	$ZnSO_3.2H_2O$	181.46	wh, cr powd	. . . .	$-2H_2O$, 100	d 200	0.16	d	i al; s H_2SO_3
z80	tartrate	$ZnC_4H_4O_6.H_2O$ (or $2H_2O$)	231.46	wh powd	. . . .			0.055^{20}		s KOH, NaOH
z81	tellurate	Zn_3TeO_6	419.71	wh, gran ppt	. . . .			i	i	s a
z82	telluride	$ZnTe$	192.97	red, cub, 3.56	6.34^{15}	1238.5		d		s d a
z83	thiocyanate	$Zn(SCN)_2$	181.53	wh powd, deliq	. . . .			s		s al, NH_4OH
z84	valerate	$Zn(C_5H_9O_2)_2.2H_2O$	303.65	wh glist sc or powd	. . . .			2.6^{24-25}	s	ca 2.5 al; v sl s eth
	Zinc complexes									
z85	diamminezinc chloride	$[Zn(6H_3)_2]Cl_2$	170.34	col, rhomb, 1.625, 1.590	2.10	210.8	d 271	d		
z86	tetrammine perrhenate	$[Zn(NH_3)_4](ReO_4)_2$	633.89	wh, cub cr	3.608^{25}_4					0.1852 conc NH_4OH
z87	tetrapyridine fluosilicate	$[Zn(C_5H_5N)_4]SiF_6$	523.86	wh, rhomb	2.197					
z88	**Zirconium**	Zr	91.22	silver gray, met	6.49	**1852 ± 2**	**4377**	i	i	s HF, aq reg; sl s a
z89	bromide, di-	$ZrBr_2$	**112.84**	blk powd, ign in air	. . . .	d >350		d ev H_2		
z90	boride, di-	ZrB_2	**251.04**	hex	6.085	ca 3000		. . . .		
z93	bromide, tetra-	$ZrBr_4$	410.86	wh cr powd, deliq	. . . .	450 ± 1^{15atm}	357 subl	i d		s liq NH_3, acetone; i bz, CCl_4
z94	bromide, tri-	$ZrBr_3$	330.95	bl-blk powd	. . . .	d 350		d ev H_2		
z95	carbide	ZrC	103.23	gray met, cub	6.73	3540	5100	i		sl s conc H_2SO_4
z96	carbonate, basic	$3ZrO_2.CO_2.H_2O$	431.68	wh, amorph powd	. . . .			i		s a
z97	chloride, di-	$ZrCl_2$	162.13	blk	3.6^{18}	d 350		d ev H_2		
z98	chloride, tetra-	$ZrCl_4$	233.03	wh cr	2.803^{15}	437^{25atm}	subl 331	s	d	s al, eth, conc HCl
z99	chloride, tri-	$ZrCl_3$	197.58	br cr	3.00^{15}	d 350		d ev H_2		s $-H_2$ conc al; i org cpd
z100	fluoride	ZrF_4	167.21	wh hex, 1.59	4.43	subl ~ 600		1.388^{25}	d	sl s HF
z101	hydride	ZrH_2	93.24	gray-blk powd	. . . .			i		s dif HF, conc a
z102	hydroxide	$Zr(OH)_4$	159.25	wh amorp powd	3.25	$-2H_2O$, 500		0.02	i	s min a
z103	iodide	ZrI_4	598.84	wh need, hygr	6.3^{atm}	499 ± 2	d ~600	s d	s	d al; s eth; v sl s CS_2, bz; i liq NH_3
z104	nitrate	$Zr(NO_3)_4.5H_2O$	429.32	col cr, deliq, 1.60, 1.61	. . . .			v s	. . . .	s al

No.	Name	Synonyms and Formulae	Mol. wt.	Crystalline form, properties and index of refraction	Density or spec. gravity	Melting point, °C	Boiling point, °C	Solubility, in grams per 100 cc		
								Cold water	Hot water	Other solvents
	Zirconium									
z105	nitride	ZrN	105.23	yel-brn cr	7.09	2980 ± 50		i	i	sl s inorg ac; s conc H_2SO_4. HF, aq reg
z106	oxide	Nat. baddeleyite. ZrO_2	123.22	col-yel-brn, monocl, 2.13, 2.19, 2.20	5.89	ca 2700	ca 5000	i	i	s H_2SO_4, HF
z107	oxide	Zirconia. ZrO_2 $HfO_2 < 2\%$	123.22	wh, monocl below 1000°, cub, above	5.6	2715		i	i	s H_2SO_4, HF
z108	oxide	Zirconium hydroxide, zirconic acid. $ZrO_2.xH_2O$		gel or wh amorph powd	3.25	$-2H_2O$, 550		0.02		s acids; i al, alk
z109	phosphide	ZrP_2	153.17	gray, brittle	4.77^{25}_4			i		v s conc hot H_2SO_4
z110	selenate	$Zr(SeO_4)_2.4H_2O$	449.20	hex trsp cr				s		sl s al, conc a
z111	selenite	$Zr(SeO_3)_2$	345.14	wh sm cr	4.3	d $\sim$ 400		i	i	sl s H_2SO_4
z112	orthosilicate	Zircon, hyacinth. $ZrSiO_4$	183.30	tetr, var colors, 1.92–96; 1.97–2.02	4.56	2550		i	i	i a, aq reg, alk
z113	silicide	$ZrSi_2$	147.39	steel gray rhomb, lust met	4.88^{22}			i	i	s HF; i inorg a, aq reg
z114	sulfate	$Zr(SO_4)_2$	283.35	microcr powd, hygr	3.22^{16}	410 d		s	v s	
z115	sulfate	$Zr(SO_4)_2.4H_2O$	355.41	wh cr powd, rhomb	3.22^{16}	$-3H_2O$, 135–150		v s		i al
z116	sulfide	ZrS_2	155.35	steelgray cr, hexag	3.87	$\sim$ 1550		i	i	i a
z117	**Zirconyl bromide**	$ZrOBr_2.xH_2O$		brill need, deliq		$-H_2O$, 120		s		s hot conc HBr
z118	chloride	$ZrOCl_2.8H_2O$	322.25	wh, need, tetr, effl, 1.552, 1.563		$-6H_2O$, 150	$-8H_2O$, 210	s	d	s al, eth; sl s HCl
z119	iodide	$ZrOI_2.8H_2O$	505.15	col, need, hygr		d		v s	v s	s al; v s eth
z120	sulfide	Zirconium sulfoxide. $ZrOS$	139.28	yel powd	4.87	ign in air		i	i	

GRAVIMETRIC FACTORS AND THEIR LOGARITHMS
Rudolf Loebel

Compiled from International atomic weights of 1964. To facilitate use of this table the group of substances weighed under each element as well as the substance sought under each substance weighed are arranged in alphabetical order of their formulas.

Weighed	Sought	Factor	Log of Factor +10	Reciprocal of Factor	Log of Reciprocal of Factor +10
Aluminum					
Al	Al_2O_3	1.88946	10.27634	0.52925	9.72366
	$AlPO_4$	4.51987	10.65513	0.22125	9.34488
Al_4C_3	Al_2O_3	1.41653	10.15122	0.70595	9.84877
$Al(C_9H_6ON)_3$	Al	0.05873	8.76886	17.02811	11.23114
	Al_2O_3	0.11096	9.04516	9.01226	10.95483
$AlCl_3$	Al_2O_3	0.38233	9.58244	2.61554	10.41756
AlF_3	CaF_2	1.39464	10.14446	0.71703	9.85554
Al_2O_3	Al	0.52925	9.72366	1.88946	10.27634
	Al_4C_3	0.70595	9.84877	1.41653	10.15122
	$AlCl_3$	2.61552	10.41756	0.38233	9.58244
	$AlPO_4$	2.39214	10.37879	0.41804	9.62122
	$Al_2(SO_4)_3$	3.35567	10.52578	0.29800	9.47422
	$Al_2(SO_4)_3.18H_2O$	6.53605	10.81531	0.15300	9.18469
	$K_2SO_4.Al_2(SO_4)_3.24H_2O$	9.30532	10.96873	0.10747	9.03129
	$(NH_4)_2SO_4.Al_2(SO_4)_3.24H_2O$	8.89216	10.94901	0.11246	9.05100
$AlPO_4$	Al	0.22125	9.34488	4.51977	10.65512
	Al_2O_3	0.41804	9.62122	2.39211	10.37878
	P_2O_5	0.58196	9.76489	1.71833	10.23511
$Al_2(SO_4)_3$	Al_2O_3	0.29800	9.47422	3.35570	10.52578
$Al_2(SO_4)_3.18H_2O$	Al_2O_3	0.15300	9.18469	6.53595	10.81531
CaF_2	AlF_3	0.71703	9.85554	1.39464	10.14446
$K_2SO_4.Al_2(SO_4)_3.24H_2O$	Al_2O_3	0.10747	9.03129	9.30492	10.96872
$(NH_4)_2SO_4.Al_2(SO_4)_3.$					
24H_2O	Al_2O_3	0.11246	9.05100	8.89205	10.94900
P_2O_5	$AlPO_4$	1.71831	10.23510	0.58196	9.76489
Ammonium					
Ag	NH_4Br	0.90802	9.95810	1.10130	10.04191
	NH_4Cl	0.49589	9.69538	2.01657	10.30462
	NH_4I	1.34366	10.12829	0.74424	9.87171
$AgBr$	NH_4Br	0.52161	9.71735	1.91714	10.28265
	NH_4Cl	0.37323	9.57198	2.67931	10.42802
AgI	NH_4I	0.61737	9.79055	1.61977	10.20945
$BaSO_4$	$(NH_4)_2SO_4$	0.56615	9.75293	1.76632	10.24707
Br	NH_4Br	1.22574	10.08840	0.81583	9.91160
Cl	NH_4	0.50880	9.70655	1.96541	10.29345
	NH_4Cl	1.50881	10.17864	0.66277	9.82136
HCl	NH_4Cl	1.46710	10.16646	0.68162	9.83354
I	NH_4I	1.14214	10.05772	0.87555	9.94229
$MgNH_4PO_4.6H_2O$	NH_3	0.06941	8.84142	14.40714	11.15857
	NH_4	0.07352	8.86641	13.60174	11.13360
	$(NH_4)_2O$	0.10613	9.02584	9.42241	10.97416
N	NH_3	1.21589	10.08489	0.82244	9.91510
	NH_4	1.28785	10.10987	0.77649	9.89014
	NH_4Cl	3.81903	10.58195	0.26184	9.41804
	NH_4NO_3	5.71466	10.75699	0.17499	9.24302
	$(NH_4)_2O$	1.85899	10.26928	0.53793	9.73072
	$(NH_4)_2SO_4$	4.71699	10.67367	0.21200	9.32634
NH_3	$MgNH_4PO_4.6H_2O$	14.40648	11.15855	0.06941	8.84142
	N	0.82244	9.91510	1.21589	10.08489
	NH_4	1.05919	10.02498	0.94412	9.97503
	NH_4Cl	3.14093	10.49706	0.31837	9.50293
	$(NH_4)_2CO_3$	2.82099	10.45040	0.35448	9.54959
	NH_4HCO_3	4.64199	10.66671	0.21542	9.33329
	NH_4NO_3	4.69998	10.67210	0.21277	9.32791
	$(NH_4)_2O$	1.52891	10.18438	0.65406	9.81562
	NH_4OH	2.05783	10.31341	0.48595	9.68659
	$(NH_4)_2PtCl_6$	13.03213	11.11501	0.07673	8.88497
	$(NH_4)_2SO_4$	3.87945	10.58877	0.25777	9.41123
	N_2O_5	3.17106	10.50121	0.31535	9.49879
	Pt	5.72763	10.75797	0.17459	9.24202
	SO_3	2.35053	10.37117	0.42544	9.62884
NH_4	Cl	1.96540	10.29345	0.50880	9.70655
	$MgNH_4PO_4.6H_2O$	13.60144	11.13359	0.07352	8.86641
	N	0.77648	9.89013	1.28786	10.10986
	NH_3	0.94412	9.97503	1.05919	10.02498

Weighed	Sought	Factor	Log of Factor +10	Reciprocal of Factor	Log of Reciprocal of Factor +10
Ammonium (contd.)					
NH_3	NH_4Cl	2.96542	10.47208	0.33722	9.52792
	$(NH_4)_2PtCl_6$	12.30389	11.09005	0.08128	8.90998
	Pt	5.40757	10.73301	0.18493	9.26701
NH_4Br	Ag	1.10130	10.04191	0.90802	9.95810
	AgBr	1.91713	10.28265	0.52161	9.71735
	Br	0.81583	9.91160	1.22575	10.08840
NH_4Cl	Ag	2.01656	10.30461	0.49589	9.69538
	AgCl	2.67934	10.42802	0.37323	9.57198
	Cl	0.66277	9.82136	1.50882	10.17864
	HCl	0.68162	9.83354	1.46709	10.16646
	N	0.26185	9.41805	3.81898	10.58194
	NH_3	0.31838	9.50294	3.14090	10.49706
	NH_4	0.33722	9.52791	2.96542	10.47208
	$(NH_4)_2O$	0.48677	9.68732	2.05436	10.31268
	NH_4OH	0.65516	9.81635	1.52634	10.18365
	$(NH_4)_2PtCl_6$	4.14913	10.61795	0.24101	9.38204
	Pt	1.82354	10.26091	0.54838	9.73908
$(NH_4)_2CO_3$	NH_3	0.35448	9.54959	2.82103	10.45040
NH_4HCO_3	NH_3	0.21542	9.33329	4.64209	10.66672
NH_4I	Ag	0.74422	9.87170	1.34369	10.12830
	AgI	1.61977	10.20946	0.61737	9.79055
	I	0.87555	9.94228	1.14214	10.05772
NH_4NO_3	NH_3	0.21277	9.32791	4.69991	10.67209
	$(NH_4)_2PtCl_6$	2.77280	10.44292	0.36064	9.55709
	N_2O_5	0.67470	9.82911	1.48214	10.17089
	Pt	1.21865	10.08588	0.82058	9.91412
$(NH_4)_2O$	$MgNH_4PO_4.6H_2O$	9.42281	10.97418	0.10613	9.02584
	N	0.53733	9.73069	1.86105	10.26931
	NH_3	0.65407	9.81562	1.52889	10.18438
	NH_4Cl	2.05437	10.31268	0.48677	9.68732
	$(NH_4)_2PtCl_6$	8.52378	10.93063	0.11732	9.06937
	N_2O_5	2.07406	10.31682	0.48215	9.68319
	Pt	3.74621	10.57359	0.26694	9.42641
NH_4OH	N	0.39967	9.60170	2.50206	10.39830
	NH_3	0.48595	9.68660	2.05782	10.31341
	NH_4	0.51471	9.71156	1.94284	10.28843
	NH_4Cl	1.52633	10.18365	0.65517	9.81635
	$(NH_4)_2PtCl_6$	6.33297	10.80158	0.15790	9.19841
	Pt	2.78334	10.44456	0.35928	9.55544
$(NH_4)_2PtCl_6$	NH_3	0.07674	8.88502	13.03101	11.11497
	NH_4	0.08128	8.90998	12.30315	11.09002
	NH_4Cl	0.24102	9.38206	4.14903	10.61794
	NH_4NO_3	0.36065	9.55709	2.77277	10.44291
	$(NH_4)_2O$	0.11732	9.06937	8.52370	10.93063
	NH_4OH	0.15791	9.19841	6.33272	10.80159
	$(NH_4)_2SO_4$	0.29768	9.47375	3.35931	10.52625
$(NH_4)_2SO_4$	$BaSO_4$	1.76632	10.24709	0.56615	9.75291
	H_2SO_4	0.74223	9.87054	1.34729	10.12947
	N	0.21200	9.32634	4.71698	10.67367
	NH_3	0.25777	9.41123	3.87943	10.58877
	$(NH_4)_2PtCl_6$	3.35927	10.52625	0.29768	9.47375
	$(NH_4)_2SO_4$	1.47640	10.16921	0.67732	9.83079
	SO_3	0.60589	9.78239	1.65046	10.21761
N_2O_5	NH_3	0.31535	9.49879	3.17108	10.50121
	NH_4NO_3	1.48214	10.17089	0.67470	9.82911
	$(NH_4)_2O$	0.48214	9.68318	2.07409	10.31682
Pt	NH_3	0.17459	9.24202	5.72770	10.75798
	NH_4	0.18493	9.26701	5.40745	10.73300
	NH_4Cl	0.54838	9.73908	1.82355	10.26092
	NH_4NO_3	0.82058	9.91412	1.21865	10.08588
	$(NH_4)_2O$	0.26694	9.42641	3.74616	10.57364
	NH_4OH	0.35928	9.55544	2.78334	10.44456
	$(NH_4)_2SO_4$	0.67732	9.83079	1.47641	10.16921
SO_3	NH_3	0.42543	9.62883	2.35056	10.37118
	$(NH_4)_2SO_4$	1.65046	10.21760	0.60589	9.78239
Antimony					
$K(SbO)C_4H_4O_6.\frac{1}{2}H_2O$	Sb	0.36460	9.56182	2.74273	10.43819
	Sb_2O_3	0.43647	9.63995	2.29111	10.36005

Weighed	Sought	Factor	Log of Factor +10	Reciprocal of Factor	Log of Reciprocal of Factor +10
Antimony (contd.)					
$K(SbO)C_4H_4O_6.\frac{1}{2}H_2O$	Sb_2O_4	0.46043	9.66317	2.17188	10.33684
	Sb_2S_3	0.50862	9.70640	1.96610	10.29360
Sb	$K(SbO)C_4H_4O_6.\frac{1}{2}H_2O$	2.74275	10.43819	0.36460	9.56182
	Sb_2O_3	1.19713	10.07814	0.83533	9.92186
	Sb_2O_4	1.26283	10.10137	0.79187	9.89866
	Sb_2O_5	1.32854	10.12340	0.75271	9.87659
	Sb_2S_3	1.39503	10.14458	0.71683	9.85540
	Sb_2S_5	1.65840	10.21969	0.60299	9.78031
Sb_2O_3	$K(SbO)C_4H_4O_6.\frac{1}{2}H_2O$	2.29111	10.36005	0.43647	9.63995
	Sb	0.83533	9.92186	1.19713	10.07814
	Sb_2O_4	1.05489	10.02320	0.94797	9.97680
	Sb_2O_5	1.10978	10.04523	0.90108	9.95476
	Sb_2S_3	1.16532	10.06645	0.85813	9.93356
	Sb_2S_5	1.38532	10.14155	0.72185	9.85845
Sb_2O_4	$K(SbO)C_4H_4O_6.\frac{1}{2}H_2O$	2.17190	10.33684	0.46043	9.66317
	Sb	0.79187	9.89866	1.26283	10.10137
	Sb_2O_3	0.94797	9.97680	1.05489	10.02300
	Sb_2O_5	1.05203	10.02203	0.95054	9.97797
	Sb_2S_3	1.10468	10.04324	0.90523	9.95671
	Sb_2S_5	1.31324	10.11834	0.76148	9.88166
Sb_2O_5	Sb	0.75270	9.87663	1.32853	10.12337
	Sb_2O_3	0.90108	9.95477	1.10977	10.04523
	Sb_2O_4	0.95054	9.97798	1.05202	10.02202
	Sb_2S_5	1.24828	10.09632	0.80109	9.90368
Sb_2S_3	$K(SbO)C_4H_4O_6.\frac{1}{2}H_2O)$	1.96603	10.29360	0.50863	9.70640
	Sb	0.71681	9.85542	1.39503	10.14458
	Sb_2O_3	0.85814	9.93356	1.16532	10.06645
	Sb_2O_4	0.90524	9.95676	1.10469	10.04332
	Sb_2O_5	0.97294	9.97879	1.05000	10.02121
Sb_2S_5	Sb	0.60299	9.78031	1.65840	10.21969
	Sb_2O_3	0.72186	9.85834	1.38570	10.14167
	Sb_2O_4	0.76148	9.88166	1.31323	10.11834
	Sb_2O_5	0.80110	9.90369	1.24828	10.09631
Arsenic					
As	As_2O_3	1.32032	10.12068	0.75739	9.87932
	As_2O_5	1.53387	10.18579	0.65195	9.81422
	As_2S_3	1.64194	10.21535	0.60904	9.78465
	As_2S_5	2.06991	10.31597	0.48311	9.68405
	$BaSO_4$	4.67291	10.66959	0.21400	9.33041
	$Mg_2As_2O_7$	2.07191	10.31637	0.48265	9.68364
	$MgNH_4AsO_4.\frac{1}{2}H_2O$	2.53968	10.40478	0.39375	9.59523
AsO_3	$BaSO_4$	2.84821	10.45457	0.35110	9.54543
	$Mg_2As_2O_7$	1.26286	10.10135	0.79185	9.89865
	$MgNH_4AsO_4.\frac{1}{2}H_2O$	1.54797	10.18976	0.64601	9.81024
AsO_4	$BaSO_4$	2.52018	10.40143	0.39680	9.59857
	$Mg_2As_2O_7$	1.11742	10.04821	0.89492	9.95178
	$MgNH_4AsO_4.\frac{1}{2}H_2O$	1.36969	10.13662	0.73009	9.86337
As_2O_3	As	0.75739	9.87932	1.32032	10.12068
	As_2O_5	1.16173	10.06511	0.86079	9.93490
	As_2S_3	1.24359	10.09468	0.80412	9.90532
	As_2S_5	1.56773	10.19527	0.63781	9.80472
	$BaSO_4$	3.53922	10.54891	0.28255	9.45110
	$Mg_2As_2O_7$	1.56925	10.19569	0.63725	9.80431
	$MgNH_4AsO_4.\frac{1}{2}H_2O$	1.92353	10.28410	0.51988	9.71590
As_2O_5	As	0.65195	9.81422	1.53386	10.18579
	As_2O_3	0.86077	9.93489	1.16175	10.06511
	As_2S_3	1.07046	10.02957	0.93418	9.97043
	As_2S_5	1.34947	10.13016	0.74103	9.86984
	$BaSO_4$	3.04648	10.48380	0.32825	9.51621
	$Mg_2As_2O_7$	1.35077	10.13058	0.74032	9.86942
	$MgNH_4AsO_4.\frac{1}{2}H_2O$	1.65573	10.21899	0.60396	9.78101
As_2S_3	As	0.60903	9.78464	1.64196	10.21536
	As_2O_3	0.80412	9.90532	1.24360	10.09468
	As_2O_5	0.93418	9.97043	1.07046	10.02957
	As_2S_5	1.26065	10.10060	0.79324	9.89939
	$Mg_2As_2O_7$	1.26187	10.10102	0.79247	9.89898
As_2S_5	As	0.48311	9.68405	2.06992	10.31595
	As_2O_3	0.63786	9.80472	1.56774	10.19527

Weighed	Sought	Factor	Log of Factor +10	Reciprocal of Factor	Log of Reciprocal of Factor +10
Arsenic (contd.)					
As_2S_5	As_2O_5	0.74103	9.86984	1.34947	10.13016
	As_2S_3	0.79324	9.89940	1.26065	10.10060
$BaSO_4$	As	0.21399	9.33039	4.67312	10.66961
	AsO_3	0.35110	9.54543	2.84819	10.45457
	AsO_4	0.39679	9.59856	2.52022	10.40143
	As_2O_3	0.28255	9.45110	3.53920	10.54890
	As_2O_5	0.32825	9.51621	3.04646	10.48379
$Ca_2As_2O_7$	As	0.43814	9.64161	2.28238	10.35839
	As_2O_3	0.57849	9.76230	1.72864	10.23770
$Mg_2As_2O_7$	As	0.48264	9.68363	2.07194	10.31638
	AsO_3	0.79184	9.89864	1.26288	10.10136
	AsO_4	0.89491	9.95178	1.11743	10.04822
	As_2O_3	0.63725	9.80431	1.56924	10.19569
	As_2O_5	0.74031	9.86942	1.35079	10.13059
	As_2S_3	0.79248	9.89899	1.26186	10.10102
$MgNH_4AsO_4.\frac{1}{2}H_2O$	As	0.39375	9.59523	2.53968	10.40478
	AsO_3	0.64600	9.81023	1.54799	10.18977
	AsO_4	0.73009	9.86337	1.36969	10.13662
	As_2O_3	0.51988	9.71590	1.92352	10.28410
	As_2O_5	0.60396	9.78101	1.65574	10.21899
Barium					
Ba	$BaCO_3$	1.43694	10.15744	0.69592	9.84256
	$BaCrO_4$	1.84455	10.26589	0.54214	9.73411
	$BaSiF_6$	2.03444	10.30844	0.49154	9.69156
	$BaSO_4$	1.69943	10.23030	0.58843	9.76969
$BaCl_2$	$BaCO_3$	0.94766	9.97665	1.05523	10.02334
	$BaCrO_4$	1.21647	10.08510	0.82205	9.91490
	$BaSO_4$	1.12077	10.04952	0.89224	9.95048
$BaCl_2.2H_2O$	$BaSO_4$	0.95546	9.98021	1.04662	10.01979
$BaCO_3$	Ba	0.69592	9.84256	1.43695	10.15747
	$BaCl_2$	1.05523	10.02334	0.94766	9.97665
	$BaCrO_4$	1.28366	10.10845	0.77902	9.89155
	$Ba(HCO_3)_2$	1.31426	10.11869	0.76088	9.88132
	BaO	0.77700	9.89042	1.28700	10.10958
	$BaSO_4$	1.18267	10.07286	0.84554	9.92713
	CO_2	0.22300	9.34830	4.48430	10.65170
$BaCrO_4$	Ba	0.54214	9.73411	1.84455	10.26589
	$BaCl_2$	0.82205	9.91490	1.21647	10.08510
	$BaCO_3$	0.77902	9.89155	1.28366	10.10846
	BaO	0.60530	9.78197	1.65207	10.21803
BaF_2	$BaSiF_6$	1.59359	10.20238	0.62751	9.79762
$Ba(HCO_3)_2$	$BaCO_3$	0.76088	9.88132	1.31427	10.11869
$Ba(NO_3)_2$	$BaSO_4$	0.89306	9.95088	1.11975	10.04912
BaO	$BaCO_3$	1.28701	10.10958	0.77699	9.89042
	$BaCrO_4$	1.65208	10.21803	0.60530	9.78197
	$BaSiF_6$	1.82233	10.26061	0.54878	9.73934
	$BaSO_4$	1.52211	10.18244	0.65698	9.81756
	CO_2	0.28701	9.45790	3.48420	10.54210
BaO_2	$BaSO_4$	1.37829	10.13934	0.72554	9.86066
BaS	$BaSO_4$	1.37780	10.13919	0.72579	9.86081
$BaSiF_6$	Ba	0.49152	9.69154	2.03451	10.30846
	BaF_2	0.62751	9.79762	1.59360	10.20238
	BaO	0.54878	9.73939	1.82222	10.26060
$BaSO_4$	Ba	0.58843	9.76969	1.69944	10.23030
	$BaCl_2$	0.89225	9.95049	1.12076	10.04952
	$BaCl_2.2H_2O$	1.04662	10.01979	0.95546	9.98021
	$BaCO_3$	0.84554	9.92713	1.18268	10.07286
	$Ba(NO_3)_2$	1.11975	10.04912	0.89306	9.95088
	BaO	0.65698	9.81756	1.52212	10.18244
	BaO_2	0.72554	9.86066	1.37828	10.13934
	BaS	0.72579	9.86081	1.37781	10.13911
CO_2	$BaCO_3$	4.48421	10.65169	0.22300	9.34830
	BaO	3.48421	10.54211	0.28701	9.45790
Beryllium					
Be	BeO	2.77530	10.44331	0.36032	9.55668
$BeCl_2$	BeO	0.31297	9.49551	3.19519	10.50450
BeO	Be	0.36032	9.55668	2.77531	10.44331
	$BeCl_2$	3.19525	10.50451	0.31296	9.49549

Weighed	Sought	Factor	Log of Factor +10	Reciprocal of Factor	Log of Reciprocal of Factor +10
Beryllium (contd.)					
BeO	$BeSO_4.4H_2O$	7.08211	10.85017	0.14120	9.14983
$Be_2P_2O_7$	Be	0.09389	8.97262	10.65076	11.02738
$BeSO_4.4H_2O$	BeO	0.14120	9.14983	7.08215	10.85017
Bismuth					
Bi	$BiAsO_4$	1.66475	10.22135	0.60069	9.77865
	Bi_2O_3	1.11484	10.04721	0.80600	9.95270
	BiOCl	1.24620	10.09559	0.80244	9.90441
	Bi_2S_3	1.23014	10.08996	0.81292	9.91005
$BiAsO_4$	Bi	0.60069	9.77865	1.66475	10.22135
	Bi_2O_3	0.66968	9.82587	1.49325	10.17413
$Bi(NO_3)_3.5H_2O$	Bi_2O_3	0.48030	9.68151	2.08203	10.31849
	BiOCl	0.53689	9.72988	1.86258	10.27011
Bi_2O_3	Bi	0.89699	9.95279	1.11484	10.04722
	$BiAsO_4$	1.49326	10.17414	0.66968	9.82587
	$Bi(NO_3)_3.5H_2O$	2.08204	10.31849	0.48030	9.68151
	BiOCl	1.11783	10.04837	0.89459	9.95163
	$BiONO_3$	1.23180	10.09054	0.81182	9.90946
	Bi_2S_3	1.10343	10.04275	0.90627	9.95726
BiOCl	Bi	0.80244	9.90441	1.24619	10.09559
	$Bi(NO_3)_3.5H_2O$	1.86256	10.27011	0.53690	9.72989
	Bi_2O_3	0.89458	9.95162	1.11784	10.04838
	$BiONO_3$	1.10195	10.04215	0.90748	9.95784
$BiONO_3$	Bi_2O_3	0.81182	9.90946	1.23180	10.09054
	BiOCl	0.90748	9.95784	1.10195	10.04216
$BiPO_4$	Bi	0.68754	9.83730	1.45446	10.16270
	Bi_2O_3	0.76651	9.88452	1.30461	10.11548
Bi_2S_3	Bi	0.81291	9.91005	1.23015	10.08996
	Bi_2O_3	0.90627	9.95726	1.10342	10.04274
Boron					
B	B_2O_3	3.21987	10.50784	0.31057	9.49216
	KBF_4	11.64619	11.06618	0.08586	8.93379
BO_2	B_2O_3	0.81314	9.91016	1.22981	10.08984
BO_3	B_2O_3	0.59192	9.77226	1.68942	10.22774
B_2O_3	B	0.31057	9.49216	3.21990	10.50785
	BO_2	1.22982	10.08985	0.81313	9.91016
	BO_3	1.68943	10.22774	0.59192	9.77226
	B_4O_7	1.11491	10.04724	0.89693	9.95276
	H_3BO_3	1.77630	10.24952	0.56297	9.75049
	KBF_4	3.61678	10.55832	0.27647	9.44165
	$Na_2B_4O_7.10H_2O$	2.73896	10.43758	0.36510	9.56241
B_4O_7	B_2O_3	0.89693	9.95276	1.11491	10.04724
$C_{20}H_{16}N_4.HBF_4$	B	0.02702	8.43169	37.00962	11.56832
	B_2O_3	0.08698	8.93942	11.49690	11.06058
H_3BO_3	B_2O_3	0.56297	9.75049	1.77629	10.24951
	KBF_4	2.03624	10.30883	0.49110	9.69117
KBF_4	B	0.08587	8.93384	11.64551	11.06616
	B_2O_3	0.27648	9.44167	3.61690	10.55834
	H_3BO_3	0.49110	9.69117	2.03625	10.30885
	$Na_2B_4O_7.10H_2O$	0.75725	9.87924	1.32057	10.12076
$Na_2B_4O_7.10H_2O$	B_2O_3	0.36510	9.56241	2.73898	10.43759
	KBF_4	1.32056	10.12075	0.75725	9.87924
Bromine					
Ag	Br	0.74079	9.86969	1.34991	10.13030
	BrO_3	1.18575	10.07399	0.84335	9.92601
	HBr	0.75013	9.87514	1.33310	10.12487
AgBr	Br	0.42555	9.62895	2.34990	10.37105
	BrO_3	0.68116	9.83325	1.46808	10.16675
	HBr	0.43092	9.63440	2.32062	10.36561
AgCl	Br	0.55754	9.74627	1.79359	10.25372
Br	Ag	1.34991	10.13030	0.74079	9.86969
	AgBr	2.34991	10.37105	0.42555	9.62895
	AgCl	1.79358	10.25372	0.55754	9.74627
	O	0.10011	9.00047	9.98901	10.99952
BrO_3	Ag	0.84335	9.92601	1.18575	10.07399
	AgBr	1.46809	10.16676	0.68116	9.83325
O	Br	9.98899	10.99952	0.10011	9.00047
HBr	Ag	1.33309	10.12486	0.75014	9.87514
	AgBr	2.32064	10.36561	0.43092	9.63440

Weighed	Sought	Factor	Log of Factor +10	Reciprocal of Factor	Log of Reciprocal of Factor +10
Cadmium					
Cd	$CdCl_2$	1.63087	10.21242	0.61317	9.78758
	$Cd(NO_3)_2$	2.10329	10.32290	0.47545	9.67711
	CdO	1.14235	10.05780	0.87539	9.94221
	CdS	1.28523	10.10898	0.77807	9.89102
	$CdSO_4$	1.85463	10.26825	0.53919	9.73174
$Cd(C_{13}H_8O_2N)_2$	Cd	0.21095	9.32418	4.74046	10.67582
	CdO	0.24098	9.38198	4.14972	10.61802
$CdCl_2$	Cd	0.61317	9.78758	1.63087	10.21242
	CdO	0.70045	9.84538	1.42765	10.15462
	CdS	0.78807	9.89657	1.26892	10.10343
	$CdSO_4$	1.13720	10.05584	0.87935	9.94417
$Cd(C_9H_6NO)_2$	Cd	0.28050	9.44793	3.56506	10.55207
	CdO	0.32043	9.50573	3.12081	10.49426
CdMoO4	Cd	0.41272	9.61566	2.42295	10.38434
	CdO	0.47147	9.67345	2.12103	10.32655
$Cd(NO_3)_2$	Cd	0.47545	9.67711	2.10327	10.32290
	CdO	0.54312	9.73490	1.84121	10.26510
	CdS	0.61106	9.78608	1.63650	10.21392
	$CdSO_4$	0.88177	9.94536	1.13408	10.05464
CdO	Cd	0.87539	9.94221	1.14235	10.05781
	$CdCl_2$	1.42765	10.15462	0.70045	9.85438
	$Cd(NO_3)_2$	1.84120	10.26510	0.54312	9.73490
	CdS	1.12508	10.05118	0.88883	9.94882
	$CdSO_4$	1.62352	10.21046	0.61595	9.78955
CdS	Cd	0.77807	9.89102	1.28523	10.10898
	$CdCl_2$	1.26893	10.10343	0.78807	9.89657
	$Cd(NO_3)_2$	1.63651	10.21392	0.61106	9.78608
	CdO	0.88883	9.94882	1.12507	10.05118
	$CdSO_4$	1.44303	10.15928	0.69299	9.84072
$CdSO_4$	Cd	0.53919	9.73174	1.85463	10.26825
	$CdCl_2$	0.87935	9.94417	1.13720	10.05584
	$Cd(NO_3)_2$	1.13408	10.05464	0.88177	9.94536
	CdO	0.61595	9.78955	1.62351	10.21046
	CdS	0.69299	9.84072	1.44302	10.15928
Calcium					
$BaSO_4$	CaS	0.30908	9.49007	3.23541	10.50993
	$CaSO_4$	0.58329	9.76588	1.71441	10.23411
	$CaSO_4.2H_2O$	0.73766	9.86786	1.35564	10.13214
Ca	$CaCl_2$	2.76921	10.44235	0.36111	9.55764
	$CaCO_3$	2.49726	10.39746	0.40044	9.60253
	CaF_2	1.94810	10.28961	0.51332	9.71039
	CaO	1.39920	10.14588	0.71469	9.85411
	$CaSO_4$	3.39671	10.53106	0.29440	9.46894
	Cl	1.76911	10.24776	0.56526	9.75225
$Ca_3(AsO_4)_2$	$Mg_2As_2O_7$	0.77990	9.89204	1.28222	10.10798
$Ca_2As_2O_7$	Ca	0.23439	9.36994	4.26639	10.63006
	CaO	0.32795	9.51581	3.04925	10.48420
$CaC_2O_4.H_2O$	Ca	0.27430	9.43823	3.64564	10.56178
	CaO	0.38379	9.58410	2.60550	10.41589
$CaCl_2$	Ca	0.36111	9.55764	2.76924	10.44236
	$CaCO_3$	0.90179	9.95511	1.10891	10.04489
	CaO	0.50527	9.70352	1.97914	10.29648
	$CaSO_4$	1.22660	10.08870	0.81526	9.91130
	Cl	0.63885	9.80540	1.56531	10.19460
$CaCO_3$	Ca	0.40044	9.60253	2.49725	10.39751
	$CaCl_2$	1.10890	10.04489	0.90179	9.95511
	$Ca(HCO_3)_2$	1.61964	10.20942	0.61742	9.79058
	CaO	0.56030	9.74842	1.78476	10.25158
	$CaSO_4$	1.36018	10.13360	0.73520	9.86641
	$CaSO_4.2H_2O$	1.72015	10.23557	0.58134	9.76442
	CO_2	0.43970	9.64316	2.27428	10.35684
	HCl	0.72854	9.86245	1.37261	10.13755
CaF_2	Ca	0.51332	9.71039	1.94810	10.28961
	$CaSO_4$	1.74360	10.24145	0.57353	9.75855
$Ca(HCO_3)$	$CaCO_3$	0.61742	9.79058	1.61964	10.20942
	CaO	0.34594	9.53900	2.89067	10.46100
$Ca(IO_3)_2$	Ca	0.10280	9.01199	9.72763	10.98801
	CaO	0.14384	9.15788	6.95217	10.84212

Weighed	Sought	Factor	Log of Factor +10	Reciprocal of Factor	Log of Reciprocal of Factor +10
Calcium (contd.)					
$Ca(NO_3)_2$	N_2O_5	0.65824	9.81838	1.51920	10.18162
CaO	Ca	0.71470	9.85412	1.39919	10.14588
	$CaCl_2$	1.97914	10.29648	0.50527	9.70352
	$CaCO_3$	1.78477	10.25158	0.56030	9.74842
	CaF_2	1.39230	10.14373	0.71824	9.85627
	$Ca(HCO_3)_2$	2.89069	10.40100	0.34594	9.53900
	$Ca_3(PO_4)_2$	1.84368	10.26569	0.54239	9.73431
	$CaSO_4$	2.42760	10.38518	0.41193	9.61482
	$CaSO_4.2H_2O$	3.07008	10.48715	0.32572	9.51285
	Cl	1.26437	10.10187	0.79091	9.89813
	CO_2	0.78477	9.89474	1.27426	10.10527
	MgO	0.71879	9.85660	1.39123	10.14340
	SO_3	1.42760	10.15461	0.70048	9.84540
$Ca_3(PO_4)_2$	CaO	0.54239	9.73431	1.84369	10.26569
	$CaSO_4$	1.31672	10.11950	0.75946	9.88051
	$Mg_2P_2O_7$	0.71755	9.85585	1.39363	10.14415
	$(NH_4)_3PO_4.12MoO_3$	12.09843	11.08273	0.08266	8.91730
	P_2O_5	0.45761	9.66050	2.18527	10.33950
CaS	$BaSO_4$	3.23538	10.50992	0.30908	9.49007
$CaSO_4$	$BaSO_4$	1.71441	10.23411	0.58329	9.76581
	Ca	0.29440	9.46894	3.39674	10.53107
	$CaCl_2$	0.81526	9.91130	1.22660	10.08870
	$CaCO_3$	0.73520	9.86641	1.36017	10.13360
	CaF_2	0.57353	9.75855	1.74359	10.24145
	CaO	0.41193	9.61482	2.42760	10.38518
	$Ca_3(PO_4)_2$	0.75946	9.88051	1.31673	10.11950
	SO_3	0.58809	9.76944	1.70042	10.23056
$CaSO_4.2H_2O$	$BaSO_4$	1.35564	10.13213	0.73766	9.86786
	$CaCO_3$	0.58134	9.76443	1.72016	10.23557
	CaO	0.32572	9.51285	3.07012	10.48716
	SO_3	0.46502	9.66747	2.15045	10.33252
$CaWO_4$	WO_3	0.80523	9.90592	1.24188	10.09408
Cl	Ca	0.56526	9.75225	1.76910	10.24775
	$CaCl_2$	1.56531	10.19460	0.63885	9.80540
	CaO	0.79091	9.89813	1.26437	10.10188
CO_2	$CaCO_3$	2.27428	10.35684	0.43970	9.64316
	CaO	1.27427	10.10526	0.78476	9.89474
HCl	$CaCO_3$	1.37256	10.13753	0.72857	9.86247
$Mg_2As_2O_7$	$Ca_3(AsO_4)_2$	1.28221	10.10796	0.77990	9.89204
MgO	CaO	1.39118	10.14339	0.71881	9.85661
$Mg_2P_2O_7$	$Ca_3(PO_4)_2$	1.39365	10.14415	0.71754	9.85585
$(NH_4)_3PO_4.12MoO_3$	$Ca_3(PO_4)_2$	0.08265	8.91725	12.09892	11.08275
N_2O_5	$Ca(NO_3)_2$	1.51920	10.18162	0.65824	9.81838
P_2O_5	$Ca_3(PO_4)_2$	2.18521	10.33949	0.45762	9.66051
SO_3	CaO	0.70046	9.84539	1.42763	10.15462
	$CaSO_4$	1.70043	10.23056	0.58809	9.76944
	$CaSO_4.2H_2O$	2.15046	10.33253	0.46502	9.66747
WO_3	$CaWO_4$	1.24188	10.09408	0.80523	9.90592
Carbon					
Ag	CN	0.24120	9.38238	4.14594	10.61762
	HCN	0.25054	9.39888	3.99138	10.60112
	KCN	0.60369	9.78081	1.65648	10.21918
AgCN	CN	0.19433	9.28854	5.14589	10.71146
	HCN	0.20185	9.30503	4.95417	10.69497
	KCN	0.48638	9.68698	2.05601	10.31302
AgCNS	CNS	0.34999	9.54406	2.85722	10.45594
$Ag_4Fe(CN)_6$	C	0.11200	9.04922	8.92857	10.95078
AgOCN	OCN	0.28033	9.44767	3.56722	10.55233
$BaCO_3$	C	0.06086	8.78433	16.43115	11.21567
	CO_2	0.22300	9.34830	4.48431	10.65170
	CO_3	0.30408	9.48299	3.28861	10.51701
BaO	CO_2	0.28701	9.45790	3.48420	10.54210
	CO_2 (bicarbonate)	0.57402	9.75593	1.74210	10.24107
$BaSO_4$	CNS	0.24885	9.39594	4.01849	10.60406
C	$BaCO_3$	16.4305	11.21565	0.06086	8.78433
	CO_2	3.66409	10.56397	0.27292	9.43604
$CaCO_3$	CO_2	0.43970	9.64316	2.27428	10.35684
$Ca(HCO_3)_2$	CO_2	0.54295	9.73476	1.84179	10.26524

GRAVIMETRIC FACTORS AND THEIR LOGARITHMS (Continued)

Weighed	Sought	Factor	Log of Factor +10	Reciprocal of Factor	Log of Reciprocal of Factor +10
Carbon (contd.)					
CaO	CO_2	0.78477	9.89474	1.27426	10.10526
	CO_2 (bicarbonate)	1.56954	10.19577	0.63713	9.80423
CN	Ag	4.14599	10.61762	0.24120	9.38238
	AgCN	5.14599	10.71147	0.19433	9.28854
CNS	AgCNS	2.85720	10.45594	0.34999	9.54406
	$BaSO_4$	4.01846	10.60406	0.24885	9.39594
	CuCNS	2.09394	10.32097	0.47757	9.67904
CO_2	$BaCO_3$	4.48420	10.65169	0.22301	9.34832
	$Ba(HCO_3)_2$	2.94672	10.46934	0.33963	9.53066
	BaO	3.48421	10.54211	0.28701	9.45790
	C	0.27292	9.43603	3.66408	10.56397
	$CaCO_3$	2.27426	10.35684	0.43970	9.64316
	$Ca(HCO_3)_2$	1.84174	10.26523	0.54296	9.73477
	CaO	1.27426	10.10526	0.78477	9.89474
	CO_3	1.36354	10.13467	0.73339	9.86534
	Cs_2CO_3	7.40329	10.86943	0.13508	9.13059
	$CsHCO_3$	4.40632	10.64407	0.22695	9.35593
	$FeCO_3$	2.63249	10.42037	0.37987	9.57964
	$Fe(HCO_3)_2$	2.02092	10.30555	0.49482	9.69445
	K_2CO_3	3.14049	10.49700	0.31842	9.50300
	$KHCO_3$	2.27491	10.35696	0.43958	9.64304
	K_2O	2.14049	10.33050	0.46718	9.66948
	Li_2CO_3	1.67887	10.22502	0.59564	9.77498
	$LiHCO_3$	1.54410	10.18868	0.64763	9.81133
	Li_2O	0.67887	9.83179	1.47304	10.16821
	$MgCO_3$	1.91595	10.28239	0.52193	9.71761
	$Mg(HCO_3)_2$	1.66265	10.22080	0.60145	9.77920
	MgO	0.91595	9.96188	1.09176	10.03812
	$MnCO_3$	2.61185	10.41695	0.38287	9.58305
	$Mn(HCO_3)_2$	2.01059	10.30332	0.49737	9.69668
	MnO	1.61185	10.20733	0.62041	9.79268
	Na_2CO_3	2.40829	10.38171	0.41523	9.61829
	$NaHCO_3$	1.90882	10.28077	0.52388	9.71923
	Na_2O	1.40829	10.14869	0.71008	9.85131
	$(NH_4)_2CO_3$	2.18329	10.33911	0.45802	9.66088
	NH_4HCO_3	1.79632	10.25439	0.55669	9.74561
	$PbCO_3$	6.07135	10.78328	0.16471	9.21672
	Rb_2CO_3	5.24767	10.71996	0.19056	9.28003
	$RbHCO_3$	3.32856	10.52225	0.30043	9.47774
	Rb_2O	4.24767	10.62815	0.23542	9.37184
	$SrCO_3$	3.35446	10.52562	0.29811	9.47438
	$Sr(HCO_3)_2$	2.3818	10.37690	0.41985	9.62309
	SrO	2.35446	10.37189	0.42473	9.62811
CO_3	$BaCO_3$	3.2886	10.51701	0.30408	9.48299
	CO_2	0.73339	9.86533	1.36353	10.13466
Cs_2CO_3	CO_2	0.13507	9.13056	7.40357	10.86944
$CsHCO_3$	CO_2	0.22695	9.35593	4.40626	10.64407
CuCNS	CNS	0.47757	9.67904	2.09393	10.32097
$FeCO_3$	CO_2	0.37987	9.57964	2.63248	10.42037
$Fe(HCO_3)_2$	CO_2	0.49482	9.69445	2.02094	10.30556
HCN	Ag	3.99137	10.60112	0.25054	9.39888
	AgCN	4.95408	10.69497	0.20185	9.30503
KCN	Ag	1.65648	10.21918	0.60369	9.78081
	AgCN	2.05602	10.31302	0.48638	9.68698
K_2CO_3	CO_2	0.31842	9.50300	3.14051	10.49700
$KHCO_3$	CO_2	0.43958	9.64304	2.27490	10.35696
K_2O	CO_2	0.46718	9.66948	2.14050	10.33052
Li_2CO_3	CO_2	0.59564	9.77498	1.67887	10.22502
Li_2O	CO_2	1.47304	10.16821	0.67887	9.83179
$MgCO_3$	CO_2	0.52193	9.71761	1.91597	10.28239
$Mg(HCO_3)_2$	CO_2	0.60145	9.77920	1.66265	10.22080
MgO	CO_2	1.09176	10.03812	0.91595	9.96188
$MnCO_3$	CO_2	0.38287	9.58305	2.61185	10.41695
MnO	CO_2	0.62041	9.79268	1.61184	10.20733
Na_2CO_3	CO_2	0.41523	9.61829	2.40830	10.38171
$NaHCO_3$	CO_2	0.52388	9.71923	1.90883	10.28077
Na_2O	CO_2	0.71008	9.85131	1.40829	10.14869
$(NH_4)_2CO_3$	CO_2	0.45802	9.66088	2.18331	10.33911

Weighed	Sought	Factor	Log of Factor +10	Reciprocal of Factor	Log of Reciprocal of Factor +10
Carbon (contd.)					
NH_4HCO_3	CO_2	0.55669	9.74561	1.79633	10.25439
$PbCO_3$	CO_2	0.16471	9.21672	6.07128	10.78328
Rb_2CO_3	CO_2	0.19056	9.28003	5.24769	10.71996
$RbHCO_3$	CO_2	0.30043	9.47774	3.32856	10.52225
Rb_2O	CO_2	0.23542	9.37184	4.24773	10.62815
$SrCO_3$	CO_2	0.29811	9.47438	3.35447	10.52562
$Sr(HCO_3)_2$	CO_2	0.41984	9.62308	2.38186	10.37691
SrO	CO_2	0.42472	9.62810	2.35449	10.37190
Cerium					
Ce	$Ce_2(C_2O_4)_3.3H_2O$	2.13513	10.32943	0.46836	9.67058
	$Ce(NO_3)_4$	2.77005	10.44249	0.36100	9.55751
	$Ce(NO_3)_4(NH_4NO_3)_2.H_2O$	4.04111	10.60650	0.24746	9.39351
	CeO_2	1.22838	10.08933	0.81408	9.91067
	Ce_2O_3	1.17128	10.06866	0.85377	9.93134
	$Ce_2(SO_4)_3$	2.02833	10.30714	0.49302	9.69286
$Ce_2(C_2O_4)_3.3H_2O$	Ce	0.46835	9.67057	2.13516	10.32943
	$Ce_2(SO_4)_3$	0.95000	9.97772	1.05263	10.02228
$Ce(NO_3)_4$	Ce	0.36100	9.55751	2.77008	10.44249
	CeO_2	0.44345	9.64684	2.25505	10.35316
	Ce_2O_3	0.42284	9.62618	2.36496	10.37382
$Ce(NO_3)_4(NH_4NO_3)_2.$					
H_2O	Ce	0.24745	9.39349	4.04122	10.60651
	CeO_2	0.30396	9.48282	3.28991	10.51719
	Ce_2O_3	0.28984	9.46216	3.45018	10.53784
CeO_2	Ce	0.81408	9.91067	1.22838	10.08933
	$Ce(NO_3)_4$	2.25505	10.35316	0.44345	9.64685
	$Ce(NO_3)_4(NH_4NO_3)_2.H_2O$	3.28986	10.51718	0.30396	9.48281
	Ce_2O_3	0.95352	9.97933	1.04875	10.02068
Ce_2O_3	Ce	0.85377	9.93134	1.17128	10.06866
	$Ce(NO_3)_4$	2.36498	10.37383	0.42284	9.62618
	$Ce(NO_3)_4(NH_4NO_3)_2.H_2O$	3.45016	10.53784	0.28984	9.46216
	CeO_2	1.04874	10.02067	0.95353	9.97933
	$Ce_2(SO_4)_3$	1.73172	10.23848	0.57746	9.76152
$Ce_2(SO_4)_3$	Ce	0.49302	9.69286	2.02832	10.30714
	$Ce_2(C_2O_4)_3.3H_2O$	1.05265	10.02220	0.94998	9.97771
	Ce_2O_3	0.57746	9.76152	1.73172	10.23848
Cesium					
$AgCl$	$CsCl$	1.17468	10.06992	0.85130	9.93008
Cl	Cs	3.74877	10.57389	0.26675	9.42610
	$CsCl$	4.74877	10.67658	0.21058	9.32342
Cs	Cl	0.26675	9.42610	3.74883	10.57390
	$CsCl$	1.26675	10.10269	0.78942	9.89731
	Cs_2O_3	1.22576	10.08840	0.81582	9.91159
	Cs_2O	1.06019	10.02539	0.94323	9.97462
	Cs_2PtCl_6	2.53422	10.40385	0.39460	9.59616
	Cs_2SO_4	1.36139	10.13398	0.73454	9.86602
$CsB(C_6H_5)_4$	Cs	0.29394	9.46826	3.40205	10.53174
	Cs_2O	0.31164	9.49365	3.20883	10.50635
$CsCl$	$AgCl$	0.85130	9.93008	1.17467	10.06993
	Cl	0.21058	9.32342	4.74879	10.67658
	Cs	0.78942	9.89731	1.26675	10.10269
	Cs_2O	0.83693	9.92269	1.19484	10.07731
	Cs_2PtCl_6	2.00055	10.30114	0.49986	9.69885
	Cs_2SO_4	1.07470	10.03129	0.93049	9.96871
Cs_2CO_3	Cs	0.81582	9.91159	1.22576	10.08841
	Cs_2PtCl_6	2.06746	10.31544	0.48369	9.68457
	Cs_2SO_4	1.11065	10.04557	0.90037	9.95442
$CsClO_4$	Cs	0.57199	9.75739	1.74828	10.24261
	$CsCl$	0.72457	9.86008	1.38013	10.13992
	Cs_2O_3	0.70110	9.84578	1.42631	10.15421
	Cs_2O	0.60642	9.78277	1.64902	10.21723
	Cs_2SO_4	0.77870	9.89137	1.28419	10.10863
Cs_2O	Cs	0.94323	9.97462	1.06019	10.02539
	$CsCl$	1.19483	10.07731	0.83694	9.92269
	Cs_2PtCl_6	2.39034	10.37846	0.41835	9.62154
	Cs_2SO_4	1.28410	10.10860	0.77870	9.89140
	SO_3	0.28410	9.45347	3.51989	10.54653
Cs_2PtCl_6	Cs	0.39460	9.59616	2.53421	10.40385
	$CsCl$	0.49986	9.69885	2.00056	10.30115

Weighed	Sought	Factor	Log of Factor +10	Reciprocal of Factor	Log of Reciprocal of Factor +10
Cesium (contd.)					
Cs_2PtCl_6	Cs_2CO_3	0.48368	9.68456	2.06748	10.31544
	Cs_2O	0.41835	9.62154	2.39034	10.37846
Cs_2SO_4	Cs	0.73454	9.86602	1.36140	10.13399
	CsCl	0.93048	9.96871	1.07471	10.03129
	Cs_2CO_3	0.90037	9.95442	1.11065	10.04557
	Cs_2O	0.77876	9.89140	1.28409	10.10860
SO_3	Cs_2O	3.51987	10.54652	0.28410	9.45347
	Cs_2SO_4	4.51988	10.65513	0.22124	9.34486
Chlorine					
Ag	Cl	0.32866	9.51675	3.04266	10.48325
	HCl	0.33801	9.52893	2.95850	10.47107
AgCl	Cl	0.24736	9.39333	4.04269	10.60667
	ClO_3	0.58226	9.76512	1.71745	10.23488
	ClO_4	0.69389	9.84129	1.44115	10.15871
	HCl	0.25440	9.40552	3.93082	10.59448
$BaCrO_4$	Cl	0.27990	9.44700	3.57270	10.55299
Ca	Cl	1.76911	10.24776	0.56526	9.75225
Cl	Ag	3.04262	10.48325	0.32866	9.51675
	AgCl	4.04262	10.60666	0.24736	9.39333
	$BaCrO_4$	3.57276	10.55300	0.27990	9.44700
	Ca	0.56526	9.75225	1.76910	10.24775
	HCl	1.02843	10.01217	0.97236	9.98783
	K	1.10292	10.04255	0.90668	9.95745
	KCl	2.10292	10.32282	0.47553	9.67718
	Li	0.19572	9.29164	5.10934	10.70837
	Mg	0.34288	9.53514	2.91647	10.46486
	$MgCl_2$	1.34288	10.12804	0.74467	9.87196
	MnO_2	1.22609	10.08852	0.81560	9.91148
	Na	0.64846	9.81188	1.54212	10.18811
	NaCl	1.64846	10.21708	0.60663	9.21708
Cl	NH_4	0.50880	9.70655	1.96541	10.29346
	$PbCrO_4$	4.55787	10.65876	0.21940	9.34124
ClO_3	AgCl	1.71745	10.23488	0.58226	9.76512
	KCl	0.89340	9.95105	1.11932	10.04895
	NaCl	0.70032	9.84530	1.42792	10.15471
ClO_4	AgCl	1.44114	10.15870	0.69390	9.84130
	KCl	0.74967	9.87487	1.33392	10.12513
	NaCl	0.58765	9.76912	1.70169	10.23088
HCl	Ag	2.95850	10.47107	0.33801	9.52893
	AgCl	3.93086	10.59448	0.25440	9.40552
	NH_4Cl	1.46710	10.16646	0.68162	9.83354
	$(NH_4)_2SO_4$	1.81206	10.25817	0.55186	9.74183
K	Cl	0.90668	9.95745	1.10292	10.04255
KCl	Cl	0.47553	9.67718	2.10292	10.32282
	ClO_3	1.11932	10.04895	0.89340	9.95105
	ClO_4	1.33393	10.12513	0.74966	9.87486
Li	Cl	5.10924	10.70836	0.19572	9.29164
Mg	Cl	2.91650	10.46487	0.34288	9.53514
$MgCl_2$	Cl	0.74467	9.87196	1.34288	10.12804
MnO_2	Cl	0.81560	9.91148	1.22610	10.08853
Na	Cl	1.54212	10.18811	0.64896	9.81188
NaCl	Cl	0.60663	9.78293	1.64845	10.21708
	ClO_3	1.42791	10.15470	0.70032	9.84530
	ClO_4	1.70168	10.23088	0.58765	9.76912
NH_4	Cl	1.96539	10.29345	0.50880	9.70655
NH_4Cl	HCl	0.68162	9.83354	1.46709	10.16646
$(NH_4)_2SO_4$	HCl	0.55186	9.74184	1.81205	10.25817
$PbCrO_4$	Cl	0.21940	9.34124	4.55789	10.65876
Chromium					
Ag_2CrO_4	Cr	0.15674	9.19518	6.37999	10.80482
	Cr_2O_3	0.22895	9.35974	4.36777	10.64026
	CrO_3	0.30143	9.47919	3.31752	10.52082
	CrO_4	0.34966	9.54365	2.85992	10.45635
$BaCrO_4$	Cr	0.20525	9.31228	4.87211	10.68772
	CrO_3	0.39472	9.59629	2.53344	10.40371
	CrO_4	0.45788	9.66075	2.18398	10.33924
	Cr_2O_3	0.29998	9.47709	3.33356	10.52291
	$Cr_2(SO_4)_3.18H_2O$	1.41404	10.15046	0.70719	9.84954

Weighed	Sought	Factor	Log of Factor +10	Reciprocal of Factor	Log of Reciprocal of Factor +10
Chromium (contd.)					
$BaCrO_4$	$PbCrO_4$	1.27573	10.10576	0.78386	9.89424
Cr	$BaCrO_4$	4.87210	10.68772	0.20525	9.31228
	Cr_2O_3	1.46155	10.16482	0.68421	9.83519
	$PbCrO_4$	6.21547	10.79347	0.16089	9.20653
$Cr(C_9H_6NO)_3$	Cr	0.10733	9.03072	9.31706	10.96928
	Cr_2O_3	0.15677	9.19526	6.37877	10.80473
CrO_3	$BaCrO_4$	2.53345	10.40372	0.39472	9.59629
	Cr	0.51999	9.71600	1.92311	10.28401
	Cr_2O_3	0.75999	9.88081	1.31581	10.11920
	K_2CrO_4	1.94210	10.28827	0.51491	9.71173
	$K_2Cr_2O_7$	1.47105	10.16762	0.67979	9.83237
	$PbCrO_4$	3.23199	10.50947	0.30941	9.49053
CrO_4	$BaCrO_4$	2.18399	10.33925	0.45788	9.66075
	$PbCrO_4$	2.78618	10.44501	0.35891	9.55499
Cr_2O_3	$BaCrO_4$	3.33351	10.52291	0.29998	9.47709
	Cr	0.68420	9.83518	1.46156	10.16482
	CrO_3	1.31580	10.11919	0.75999	9.88081
	CrO_4	1.52634	10.18364	0.65516	9.81635
	$PbCrO_4$	4.25265	10.62866	0.23515	9.37135
$Cr_2(SO_4)_3.18H_2O$	$BaCrO_4$	0.70718	9.84953	1.41407	10.15047
	$PbCrO_4$	0.90217	9.95529	1.10844	10.04471
K_2CrO_4	CrO_3	0.51491	9.71173	1.94209	10.28827
	$PbCrO_4$	1.66418	10.22120	0.60090	9.77880
$K_2Cr_2O_7$	CrO_3	0.67979	9.83237	1.47104	10.16762
	$PbCrO_4$	2.10707	10.04104	0.45515	9.65815
$PbCrO_4$	Cr	0.16089	9.20652	6.21543	10.79347
	CrO_3	0.30941	9.49053	3.23196	10.50946
	CrO_4	0.35891	9.55499	2.78621	10.44501
	Cr_2O_3	0.23514	9.37133	4.25279	10.62868
	$Cr_2(SO_4)_3.18H_2O$	1.10843	10.04471	0.90218	9.95529
	K_2CrO_4	0.60090	9.77880	1.66417	10.22119
	$K_2Cr_2O_7$	0.45514	9.65815	2.19713	10.34186
Cobalt					
Co	$Co(NO_3)_2.6H_2O$	4.93731	10.69349	0.20254	9.30651
	$Co(NO_2)_3.(KNO_2)_3$	7.67433	10.88504	0.13030	9.11494
	CoO	1.27148	10.10431	0.78649	9.89569
	Co_3O_4	1.36197	10.13417	0.73423	9.86583
	$CoSO_4$	2.63001	10.41996	0.38023	9.58005
	$CoSO_4.7H_2O$	4.76984	10.67851	0.20965	9.32149
$Co(C_{10}H_6O_2N).2H_2O$	$(CoSO_4)_2.(K_2SO_4)_3$	7.06550	10.84914	0.14153	9.15085
	Co	0.09638	8.98399	10.37560	11.01602
	CoO	0.12255	9.08831	8.15993	10.91169
$(Co[C_5H_5N]_4).(SCN)_2$	Co	0.11990	9.07882	8.34028	10.92118
	CoO	0.15246	9.18316	6.55910	10.81684
$Co(C_9H_6ON)_2$	Co	0.16972	9.22973	5.89206	10.77027
	Co_2O_3	0.23883	9.37809	4.18708	10.62191
$Co(NO_3)_2.6H_2O$	Co	0.20254	9.30651	4.93730	10.69349
$Co(NO_2)_3.(KNO_2)_3$	Co	0.13030	9.11494	7.67460	10.88506
	CoO	0.16568	9.21927	6.03573	10.78073
CoO	Co	0.78648	9.89569	1.27149	10.10432
	$Co(NO_2)_3.(KNO_2)_3$	6.03571	10.78073	0.16568	9.21927
	Co_3O_4	1.07117	10.02986	0.93356	9.97014
	$CoSO_4$	2.06845	10.31564	0.48345	9.68435
	$(CoSO_4)_2.(K_2SO_4)_3$	5.55690	10.74483	0.17996	9.25518
Co_3O_4	Co	0.73423	9.86583	1.36197	10.13417
	CoO	0.93356	9.97014	1.07117	10.02986
$Co_2P_2O_7$	Co	0.40392	9.60630	2.47574	10.39371
	CoO	0.51357	9.71060	1.94715	10.28940
	Co_3O_4	0.55012	9.74046	1.81779	10.25953
	$CoSO_4$	1.06230	10.02624	0.94135	9.97375
	$CoSO_4.7H_2O$	1.92661	10.28479	0.51905	9.71521
$CoSO_4$	Co	0.38023	9.58005	2.62999	10.41996
	CoO	0.48345	9.68435	2.06847	10.31565
$CoSO_4.7H_2O$	Co	0.20965	9.32150	4.76985	10.67851
	CoO	0.26657	9.42581	3.75136	10.57419
$(CoSO_4)_2.(K_2SO_4)_3$	Co	0.14153	9.15085	7.06553	10.84914
	CoO	0.17996	9.25518	5.55679	10.74482
Columbium	See Niobium				

Weighed	Sought	Factor	Log of Factor +10	Reciprocal of Factor	Log of Reciprocal of Factor +10
Copper					
Cu	$Cu_2C_2H_3O_2.(AsO_2)_3$	3.98875	10.60084	0.25071	9.39917
	CuCNS	1.91407	10.28196	0.52245	9.71804
	CuO	1.25181	10.09754	0.79884	9.90246
	Cu_2O	1.12590	10.05150	0.88818	9.94850
	Cu_2S	1.25228	10.09770	0.79854	9.90230
	$CuSO_4.5H_2O$	3.92949	10.59433	0.25449	9.40567
$Cu_2C_2H_3O_2.(AsO_2)_3$	Cu	0.25071	9.39917	3.98867	10.60083
CuCNS	Cu	0.52245	9.71804	1.91406	10.28195
	CuO	0.65400	9.81558	1.52905	10.18443
$(Cu[C_5H_5N]_2.(SCN)_2$	Cu	0.18804	9.27425	5.31802	10.72575
	CuO	0.23539	9.37179	4.24827	10.62821
$Cu(C_7H_6NO_2)_2$	Cu	0.18922	9.27697	5.28485	10.72303
	CuO	0.23687	9.37451	4.22172	10.62549
$Cu(C_9H_6NO)_2$	Cu	0.18059	9.25669	5.53741	10.74330
	CuO	0.22606	9.35422	4.42360	10.64578
$Cu(C_{12}H_{10}ONS)_2$	Cu	0.12795	9.10704	7.81555	10.89296
	CuO	0.16017	9.20458	6.24337	10.79542
$Cu(NH_2C_6H_4CO_2)_2$	Cu	0.18922	9.27697	5.28485	10.72303
	CuO	0.23687	9.37451	4.22172	10.62549
CuO	Cu	0.79884	9.90246	1.25182	10.09754
	CuCNS	1.52904	10.18442	0.65401	9.81558
	Cu_2S	1.00038	10.00016	0.99962	9.99983
	$CuSO_4.5H_2O$	3.13905	10.49680	0.31857	9.50320
Cu_2O	Cu	0.88817	9.94850	1.12591	10.05150
	CuO	1.11183	10.04603	0.89942	9.95396
	Cu_2S	1.11224	10.04620	0.89909	9.95380
Cu_2S	Cu	0.79854	9.90230	1.25229	10.09770
	CuO	0.99962	9.99983	1.00004	10.00002
	Cu_2O	0.89908	9.95380	1.11225	10.04620
	$CuSO_4.5H_2O$	3.13787	10.49663	0.31869	9.50337
$CuSO_4.5H_2O$	Cu	0.25449	9.40567	3.92943	10.59433
	CuO	0.31857	9.50320	3.13903	10.49679
	Cu_2S	0.31869	9.50337	3.13785	10.49663
$Mg_2As_2O_7$	$Cu_2C_2H_3O_2.(AsO_2)_3$	1.08844	10.03681	0.91875	9.96320
Dysprosium					
Dy_2O_3	Dy	0.87131	9.94017	1.14770	10.05983
Erbium					
Er	Er_2O_3	1.14349	10.05824	0.87452	9.94177
Er_2O_3	Er	0.87452	9.94177	1.14349	10.05824
Europium					
Eu_2O_3	Eu	0.86361	9.93632	1.15793	10.06368
Fluorine					
BaF_2	$BaSiF_6$	1.59359	10.20238	0.62751	9.79762
$BaSiF_6$	BaF_2	0.62751	9.79762	1.59360	10.20238
	F	0.40795	9.61061	2.45128	10.38939
	HF	0.42960	9.63306	2.32775	10.36694
	H_2SiF_6	0.51568	9.71238	1.93919	10.28762
	SiF_4	0.37248	9.57110	2.68471	10.42889
	SiF_6	0.50847	9.70627	1.96668	10.29373
CaF_2	F	0.48664	9.68721	2.05491	10.31279
	6HF	0.51248	9.70968	1.95130	10.29033
	H_2SiF_6	0.61517	9.78900	1.62557	10.21100
	SiF_6	0.60657	9.78288	1.64861	10.21712
$CaSO_4$	F	0.27910	9.44576	3.58295	10.55424
	HF	0.29391	9.46821	3.40240	10.53178
F	$BaSiF_6$	2.45125	10.38939	0.40796	9.61062
	CaF_2	2.05491	10.31279	0.48664	9.68721
	$CaSO_4$	3.58293	10.55433	0.27910	9.44576
	H_2SiF_6	1.26407	10.10177	0.79110	9.89823
	K_2SiF_6	1.93244	10.28611	0.51748	9.71389
2HF	H_2SiF_6	3.60115	10.55644	0.27769	9.44356
6HF	H_2SiF_6	1.20038	10.07932	0.83307	9.92068
H_2SiF_6	$BaSiF_6$	1.93917	10.28762	0.51568	9.71237
	CaF_2	1.62555	10.21100	0.61518	9.78900
	F	0.79109	9.89823	1.26408	10.10178
	2HF	0.27769	9.44356	3.60114	10.55644
	6HF	0.83307	9.92068	1.20038	10.07932
	K_2SiF_6	1.52875	10.18434	0.65413	9.81566

Weighed	Sought	Factor	Log of Factor +10	Reciprocal of Factor	Log of Reciprocal of Factor +10
Fluorine (contd.)					
H_2SiF_6	SiF_4	0.72232	9.85873	1.38443	10.14129
	SiF_6	0.98601	9.99388	1.01419	10.00612
2KF	K_2SiF_6	1.89568	10.27777	0.52752	9.72224
K_2SiF_6	F	0.51748	9.71389	1.93244	10.28611
	6HF	0.54493	9.73634	1.83510	10.26366
	H_2SiF_6	0.65413	9.81566	1.52875	10.18434
	KF	0.52751	9.72223	1.89570	10.27777
	SiF_6	0.64498	9.80955	1.55044	10.19045
Na_2SiF_6	F	0.60615	9.78258	1.64976	10.21742
	6HF	0.63831	9.80503	1.56664	10.19498
	H_2SiF_6	0.76622	9.88435	1.30511	10.11565
	2NaF	0.44655	9.64987	2.23939	10.35013
	SiF_4	0.55344	9.74307	1.80688	10.25693
	SiF_6	0.75550	9.87823	1.32363	10.12177
PbClF	F	0.07261	8.86100	13.77221	11.13900
SiF_4	$BaSiF_6$	2.68467	10.42889	0.37249	9.57111
	H_2SiF_6	1.38444	10.14128	0.72231	9.85872
SiF_6	$BaSiF_6$	1.96666	10.29373	0.50848	9.70627
	CaF_2	1.64862	10.21712	0.60657	9.78288
	H_2SiF_6	1.01419	10.00612	0.98601	9.99388
	K_2SiF_6	1.55043	10.19045	0.64498	9.80955
Gallium					
Ga	Ga_2O_3	1.34423	10.12847	0.74392	9.87153
$Ga(C_9H_4NOBr_2)_3$	Ga	0.07146	8.85406	13.99384	11.14594
	Ga_2O_3	0.09606	8.98254	10.41016	11.01746
Ga_2O_3	Ga	0.74392	9.87153	1.34423	10.12847
Ga_2S_3	Ga	0.59178	9.77216	1.68982	10.22785
Germanium					
Ge	GeO_2	1.44083	10.15861	0.69404	9.84138
$(C_9H_6NOH)_4$.					
$(GeO_2.12MoO_3)$	Ge	0.03009	8.47842	33.23363	11.52158
	GeO_2	0.04335	8.63699	23.06805	11.36301
GeO_2	Ge	0.69404	9.84139	1.44084	10.15861
K_2GeF_6	Ge	0.27415	9.43799	3.64764	10.56201
Mg_2GeO_4	Ge	0.39193	9.59321	2.55148	10.40679
	GeO_2	0.56471	9.75183	1.77082	10.24817
Gold					
Au	$AuCl_3$	1.53998	10.18749	0.64936	9.81249
	$HAuCl_4.4H_2O$	2.09095	10.32034	0.47825	9.67966
	$KAu(CN)_4.H_2O$	1.81836	10.25968	0.54995	9.74032
$AuCl_3$	Au	0.64938	9.81250	1.53993	10.18750
C_6H_5SAu	Au	0.64339	9.80847	1.55427	10.19153
$HAuCl_4.4H_2O$	Au	0.47825	9.67966	2.09096	10.32034
$KAu(CN)_4.H_2O$	Au	0.54995	9.74032	1.81835	10.25967
Hafnium					
HfO_2	Hf	0.84797	9.92838	1.17929	10.07162
$Hf(C_6H_5CHOHCO_2)_4$	Hf	0.22794	9.35782	4.38712	10.64218
	HfO_2	0.26880	9.42943	3.72024	10.57057
Holmium					
Ho_2O_3	Ho	0.87297	9.94100	1.14551	10.05900
Hydrogen					
AgCNS	HCNS	0.35606	9.55152	2.80852	10.44848
$BaSO_4$	HCNS	0.25317	9.40341	3.94992	10.59659
CuCNS	HCNS	0.48586	9.68651	2.05821	10.31349
H	H_2O	8.93644	10.95116	0.11190	9.04883
	O	7.93644	10.89963	0.12600	9.10037
HCNS	AgCNS	2.80846	10.44847	0.35607	9.55154
	$BaSO_4$	3.94991	10.59659	0.25317	9.40341
	CuCNS	2.05822	10.31350	0.48586	9.68651
H_2O	H	0.11190	9.04883	8.93655	10.95117
O	H	0.12600	9.10037	7.93651	10.89963
Indium					
In	In_2O_3	1.20902	10.08244	0.92712	9.91757
	In_2S_3	1.41888	10.15194	0.70478	9.84805
$In(C_9H_6ON)_3$	In	0.20980	9.32181	4.76644	10.67819
	In_2O_3	0.25365	9.40423	3.94244	10.59577
In_2O_3	In	0.82711	9.91756	1.20903	10.08244
In_2S_3	In	0.70478	9.84805	1.41888	10.15194

Weighed	Sought	Factor	Log of Factor +10	Reciprocal of Factor	Log of Reciprocal of Factor +10
Iodine					
Ag	HI	1.18580	10.07401	0.84331	9.92599
	I	1.17646	10.07058	0.85001	9.92942
AgCl	I	0.88544	9.94716	1.12938	10.05284
AgI	HI	0.54483	9.73626	1.83543	10.26374
	I	0.54054	9.73283	1.85000	10.26717
	IO_3	0.74498	9.87214	1.34232	10.12786
	IO_4	0.81313	9.91016	1.22982	10.08985
	I_2O_5	0.71091	9.85181	1.40665	10.14819
	I_2O_7	0.77906	9.89157	1.28360	10.10843
HI	Ag	0.84331	9.92599	1.18580	10.07401
	AgI	1.83542	10.26374	0.54483	9.73626
	Pd	0.41590	9.61899	2.40442	10.38101
	PdI_2	1.40799	10.14860	0.71023	9.85140
	TlI	2.58981	10.41327	0.38613	9.58673
I	Ag	0.85000	9.92942	1.17647	10.07058
	AgCl	1.12937	10.05283	0.88545	9.94716
	AgI	1.85000	10.26717	0.54054	9.73283
	Pd	0.41921	9.62243	2.38544	10.37757
	PdI_2	1.41917	10.15203	0.70464	9.84797
	TlI	2.61039	10.41671	0.38308	9.58329
IO_3	AgI	1.34231	10.12785	0.74498	9.87214
	PdI_2	1.02971	10.01272	0.97115	9.98729
	TlI	1.89402	10.27738	0.52798	9.72262
IO_4	AgI	1.22981	10.08984	0.81313	9.91016
	PdI_2	0.94341	9.97470	1.05998	10.02530
	TlI	1.73528	10.23937	0.57628	9.76063
I_2O_5	AgI	1.40665	10.14819	0.71091	9.95181
	PdI_2	1.07907	10.03305	0.92672	9.96695
	TlI	1.98481	10.29772	0.50383	9.70228
I_2O_7	AgI	1.28360	10.10843	0.77906	9.89157
	PdI_2	0.98467	9.99329	1.01557	10.00671
	TlI	1.81120	10.25797	0.55212	9.74203
Pd	HI	2.40437	10.38100	0.41591	9.61900
	I	2.38542	10.37757	0.41921	9.62243
PdI_2	HI	0.71023	9.85140	1.40799	10.14860
	I	0.70463	9.70463	1.41918	10.15204
	IO_3	0.97114	9.98728	1.02972	10.01272
	IO_4	1.05998	10.02530	0.94341	9.97470
	I_2O_5	0.92672	9.96695	1.07907	10.03305
	I_2O_7	1.01556	10.00671	0.98468	9.99330
TlI	HI	0.38613	9.58673	2.58980	10.41327
	I	0.38308	9.58329	2.61042	10.41671
	IO_3	0.52798	9.72262	1.89401	10.27738
	IO_4	0.57627	9.76063	1.73530	10.23937
	I_2O_5	0.50382	9.70228	1.98484	10.29772
	I_2O_7	0.55212	9.74203	1.81120	10.25797
Iron					
Ag	$Fe_7(CN)_{18}$ (Prussian blue)	0.44253	9.64594	2.25973	10.35406
CN	$Fe_7(CN)_{18}$	1.83474	10.26358	0.54504	9.73643
CO_2	$FeCO_3$	2.63249	10.42037	0.37987	9.57964
	$Fe(HCO_3)_2$	2.02092	10.30555	0.49482	9.69445
	FeO	1.63249	10.21287	0.61256	9.78715
Fe	$Fe(HCO_3)_2$	3.18517	10.50313	0.31395	9.49686
	FeO	1.28648	10.10940	0.77731	9.89059
	Fe_2O_3	1.42973	10.15526	0.69943	9.84474
	$FePO_4$	2.70056	10.43145	0.37029	9.56854
	FeS	1.57414	10.19704	0.63527	9.80296
	$FeSO_4$	2.72009	10.43458	0.36763	9.56541
	$FeSO_4.7H_2O$	4.97817	10.69707	0.20088	9.30294
	$FeSO_4.(NH_4)_2SO_4.6H_2O$	7.02054	10.84637	0.14244	9.15363
$FeAsO_4$	$Mg_2As_2O_7$	0.79702	9.90147	1.25467	10.09853
$Fe(C_6H_5N[NO]O)_3$	Fe	0.11953	9.07748	8.36610	10.92252
	Fe_2O_3	0.17090	9.23274	5.85138	10.76726
$Fe(C_9H_6NO)_3$	Fe	0.11437	9.05821	8.74355	10.94169
	Fe_2O_3	0.16351	9.21354	6.11583	10.78646
$FeCl_3$	Fe_2O_3	0.49225	9.69219	2.03149	10.30781
$Fe_7(CN)_{18}$	Ag	2.25971	10.35405	0.44253	9.64594
	CN	0.54503	9.73642	1.83476	10.26358

Weighed	Sought	Factor	Log of Factor +10	Reciprocal of Factor	Log of Reciprocal of Factor +10
Iron (contd.)					
$FeCO_3$	CO_2	0.37986	9.57962	2.63255	10.42038
	FeO	0.62013	9.79248	1.61257	10.20752
	Fe_2O_3	0.68918	9.83833	1.45100	10.16167
$Fe(HCO_3)_2$	CO_2	0.49482	9.69445	2.02094	10.30555
	Fe	0.31396	9.49687	3.18512	10.50313
	FeO	0.40390	9.60627	2.47586	10.39373
	Fe_2O_3	0.44887	9.65212	2.22782	10.34788
FeO	CO_2	0.61256	9.78715	1.63249	10.21285
	Fe	0.77732	9.89060	1.28647	10.10940
	$FeCO_3$	1.61256	10.20751	0.62013	9.79248
	$Fe(HCO_3)_2$	2.47588	10.39373	0.40390	9.60627
	Fe_2O_3	1.11134	10.04584	0.89981	9.95415
	$FePO_4$	2.09918	10.32205	0.47638	9.67795
	FeS	1.22360	10.08764	0.81726	9.91236
	SO_3	1.11436	10.04703	0.89738	9.95298
Fe_2O_3	Fe	0.69943	9.84474	1.42974	10.15524
	$FeCl_3$	2.03149	10.30781	0.49225	9.69219
	$FeCO_3$	1.45099	10.16167	0.68918	9.83833
	$Fe(HCO_3)_2$	2.22781	10.34788	0.44887	9.65212
	$Fe(HCO_3)_3$	2.99200	10.47596	0.33422	9.52403
	FeO	0.89981	9.95415	1.11135	10.04584
	Fe_3O_4	0.96660	9.98525	1.03455	10.01475
	$FePO_4$	1.88886	10.27620	0.52942	9.72380
	FeS	1.10101	10.04113	0.90826	9.95821
	$FeSO_4$	1.90252	10.27933	0.52562	9.72067
	$FeSO_4.7H_2O$	3.48190	10.54182	0.28720	9.45818
	$Fe_2(SO_4)_3$	2.50406	10.39864	0.39935	9.60135
	$FeSO_4.(NH_4)_2SO_4.6H_2O$	4.91040	10.69112	0.20365	9.30888
Fe_3O_4	Fe_2O_3	1.03455	10.01475	0.96660	9.98525
$FePO_4$	Fe	0.37029	9.56854	2.70059	10.43145
	FeO	0.47638	9.67795	2.09916	10.32204
	Fe_2O_3	0.52942	9.72380	1.88886	10.27620
FeS	Fe	0.63527	9.80296	1.57413	10.19704
	FeO	0.81726	9.91236	1.22360	10.08764
	Fe_2O_3	0.90826	9.95821	1.10101	10.04179
$FeSO_4$	Fe	0.36763	9.56541	2.72013	10.43459
	Fe_2O_3	0.52562	9.72067	1.90252	10.27933
	SO_3	0.52704	9.72184	1.89739	10.27817
$FeSO_4.7H_2O$	Fe	0.20088	9.30294	4.97810	10.69706
	Fe_2O_3	0.28720	9.45818	3.48189	10.54182
$FeSO_4.(NH_4)_2SO_4.6H_2O$	Fe	0.14244	9.15363	7.02050	10.84637
	Fe_2O_3	0.20365	9.30888	4.91039	10.69112
$Fe_2(SO_4)_3$	Fe_2O_3	0.39935	9.60135	2.50407	10.39864
$Mg_2As_2O_7$	$FeAsO_4$	1.25468	10.09853	0.79702	9.90147
SO_3	FeO	0.89738	9.95298	1.11436	10.04703
	$FeSO_4$	1.89739	10.27816	0.52704	9.72184
Lanthanum					
La	La_2O_3	1.17277	10.06921	0.85268	9.93079
La_2O_3	La	0.85268	9.93079	1.17277	10.06921
Lead					
$BaSO_4$	$PbSO_4$	1.29927	10.11370	0.76966	9.88630
Pb	$PbCl_2$	1.34225	10.12783	0.74502	9.87217
	$PbCO_3$	1.28958	10.11044	0.77545	9.88955
	$(PbCO_3)_2.Pb(OH)_2$	1.24781	10.09615	0.80140	9.90385
	$PbCrO_4$	1.55982	10.19307	0.64110	9.80693
	PbO	1.07722	10.03231	0.92832	9.96770
	PbO_2	1.15445	10.06238	0.86621	9.93762
	$Pb(OH)_2$	1.16415	10.06601	0.85900	9.93399
	PbS	1.15474	10.06248	0.86600	9.93752
	$PbSO_4$	1.46363	10.16543	0.68323	9.83457
$Pb(C_2H_3O_2).3H_2O$	$PbCrO_4$	0.85198	9.93043	1.17374	10.06957
	$PbSO_4$	0.79944	9.90279	1.25088	10.09722
$Pb(C_7H_5NO_2)_2$	Pb	0.43396	9.63746	2.30436	10.36255
	PbO	0.46747	9.66975	2.13971	10.33024
$PbCl_2$	Pb	0.74502	9.87218	1.34225	10.12783
	PbO	0.80255	9.90448	1.24603	10.09553
$PbCO_3$	Pb	0.77541	9.88954	1.28964	10.11047
	PbO	0.83529	9.92184	1.19719	10.07816

Weighed	Sought	Factor	Log of Factor +10	Reciprocal of Factor	Log of Reciprocal of Factor +10
Lead (contd.)					
$PbCO_3$	$PbSO_4$	1.13492	10.05497	0.88112	9.94504
$(PbCO_3)_2.Pb(OH)_2$	Pb	0.80141	9.90386	1.24780	10.09614
	$PbCrO_4$	1.25007	10.09693	0.79996	9.90307
	$PbSO_4$	1.17298	10.06929	0.85253	9.93017
$PbCrO_4$	Pb	0.64110	9.80692	1.55982	10.19307
	$Pb(C_2H_3O_2)_2.3H_2O$	1.17371	10.06956	0.85200	9.93044
	$(PbCO_3)_2.Pb(OH)_2$	0.79996	9.90307	1.25006	10.09693
	PbO	0.69061	9.83922	1.44800	10.16077
	Pb_3O_4	0.70710	9.84947	1.41423	10.15052
	$PbSO_4$	0.93833	9.97235	1.06572	10.02765
$Pb(NO_3)_2$	PbO	0.67388	9.82859	1.48394	10.17141
	PbO_2	0.72219	9.85865	1.38468	10.14135
	$PbSO_4$	0.91561	9.96171	1.09217	10.03829
PbO	Pb	0.92831	9.96770	1.07723	10.03231
	$PbCl_2$	1.24602	10.09553	0.80256	9.90448
	$PbCO_3$	1.19718	10.07816	0.83530	9.92184
	$PbCrO_4$	1.44800	10.16077	0.69061	9.83923
	$Pb(NO_3)_2$	1.48393	10.17141	0.67389	9.82859
	PbO_2	1.07169	10.03007	0.93311	9.96993
	PbS	1.07196	10.03018	0.93287	9.96982
	$PbSO_4$	1.35871	10.13313	0.73599	9.86687
PbO_2	Pb	0.86622	9.93763	1.15444	10.06237
	$Pb(NO_3)_2$	1.38467	10.14135	0.72219	9.85865
	PbO	0.93311	9.96993	1.07169	10.03007
	$PbSO_4$	1.26782	10.10306	0.78876	9.89694
Pb_3O_4	$PbCrO_4$	1.41423	10.15052	0.70710	9.84948
	$PbSO_4$	1.32702	10.12288	0.75357	9.87712
$Pb(OH)_2$	Pb	0.85900	9.93399	1.16414	10.06601
$Pb_2P_2O_7$	Pb	0.70434	9.84778	1.41977	10.15222
	PbO	0.75874	9.88009	1.31797	10.11991
PbS	Pb	0.86600	9.93752	1.15473	10.06249
	PbO	0.93287	9.96982	1.07196	10.03017
	$PbSO_4$	1.26750	10.10296	0.78895	9.89705
$PbSO_4$	$BaSO_4$	0.76966	9.88630	1.29928	10.11370
	Pb	0.68323	9.83457	1.46364	10.16543
	$Pb(C_2H_3O_2)_2.3H_2O$	1.25088	10.09722	0.79944	9.90279
	$PbCO_3$	0.88112	9.94504	1.13492	10.05497
	$(PbCO_3)_2.Pb(OH)_2$	0.85253	9.93071	1.17298	10.06929
	$PbCrO_4$	1.06572	10.02765	0.93833	9.97236
	$Pb(NO_3)_2$	1.09216	10.03828	0.91562	9.96172
	PbO	0.73599	9.86687	1.35871	10.13313
	PbO_2	0.78876	9.89694	1.26781	10.10305
	Pb_3O_4	0.75357	9.87712	1.32702	10.12288
	PbS	0.78895	9.89705	1.26751	10.10295
Lithium					
CO_2	Li_2CO_3	1.67887	10.22504	0.59564	9.77498
	$LiHCO_3$	1.54410	10.18868	0.64763	9.81133
	Li_2O	0.67887	9.83179	1.47304	10.16821
Li	LiCl	6.10924	10.78599	0.16369	9.21402
	Li_2CO_3	5.32404	10.72624	0.18783	9.27377
	Li_2O	2.15283	10.33201	0.46450	9.66699
	Li_3PO_4	5.56218	10.74524	0.17979	9.25477
	Li_2SO_4	7.92189	10.89883	0.12623	9.10115
LiCl	Li	0.16369	9.21402	6.10911	10.78598
	Li_2CO_3	0.87147	9.94025	1.14749	10.05975
	Li_2O	0.35239	9.54702	2.83776	10.45297
	Li_3PO_4	0.91045	9.95926	1.09836	10.04074
	Li_2SO_4	1.29670	10.11284	0.77119	9.88716
Li_2CO_3	CO_2	0.59564	9.77498	1.67887	10.22502
	Li	0.18782	9.27374	5.32425	10.72626
	LiCl	1.14747	10.05974	0.87148	9.94026
	$LiHCO_3$	1.83963	10.26473	0.54359	9.73527
	Li_2O	0.40436	9.60677	2.47304	10.39323
	Li_3PO_4	1.04473	10.01901	0.95719	9.98100
$LiHCO_3$	CO_2	0.64762	9.81132	1.54412	10.18868
	Li_2CO_3	0.54363	9.73530	1.83949	10.26470
	Li_2O	0.21982	9.34207	4.54918	10.65794
	Li_3PO_4	0.56795	9.75431	1.76072	10.24569

Weighed	Sought	Factor	Log of Factor +10	Reciprocal of Factor	Log of Reciprocal of Factor +10
Lithium (contd.)					
Li_2O	CO_2	1.47304	10.16821	0.67887	9.83179
	Li	0.46449	9.66698	2.15290	10.33303
	LiCl	2.83767	10.45296	0.35240	9.54704
	Li_2CO_3	2.47304	10.39323	0.40436	9.60677
	$LiHCO_3$	4.54890	10.65791	0.21983	9.34209
	Li_3PO_4	2.58364	10.41223	0.38705	9.58777
	Li_2SO_4	3.67975	10.56582	0.27175	9.43417
	SO_3	2.67872	10.42809	0.37317	9.57191
Li_3PO_4	Li	0.17979	9.25477	5.56204	10.74523
	LiCl	1.09835	10.04074	0.91046	9.95926
	Li_2CO_3	0.95718	9.98099	1.04474	10.01901
	$LiHCO_3$	1.76070	10.24569	0.56796	9.75432
	Li_2O	0.38705	9.58777	2.58365	10.41223
	$Li_2.SO_4.H_2O$	1.65761	10.21949	0.60328	9.78052
Li_2SO_4	Li	0.12623	9.10116	7.92205	10.89883
	LiCl	0.77118	9.88716	1.29671	10.11284
	Li_2O	0.27175	9.43417	3.67975	10.56582
	Li_3PO_4	0.70213	9.84642	1.42424	10.15358
	SO_3	0.72823	9.86227	1.37319	10.13773
$Li_2SO_4.H_2O$	Li_3PO_4	0.60328	9.78052	1.65761	10.22948
SO_3	Li_2O	0.37317	9.57191	2.67974	10.42809
	Li_2SO_4	1.37318	10.13773	0.72824	9.86227
Lutetium					
Lu_2O_3	Lu	0.87938	9.94418	1.13716	10.05582
Magnesium					
$BaSO_4$	$MgSO_4$	0.51574	9.71243	1.93896	10.28757
	$MgSO_4.7H_2O$	1.05605	10.02368	0.94692	9.97631
Br	Mg	0.15212	9.18219	6.57376	10.81781
	$MgBr_2$	1.15212	10.06150	0.86797	9.93850
	$MgBr_2.6H_2O$	1.82807	10.26200	0.54703	9.73801
Cl	Mg	0.34288	9.53514	2.91647	10.46486
	$MgCl_2$	1.34288	10.12804	0.74467	9.87196
	$MgCl_2.6H_2O$	2.86643	10.45734	0.34887	9.54266
CO_2	$MgCO_3$	1.91595	10.28239	0.52193	9.71761
	MgO	0.91595	9.96187	1.09176	10.03812
I	Mg	0.09579	8.98132	10.43950	11.01868
	MgI_2	1.09578	10.03972	0.91259	9.96028
Mg	Br	6.57262	10.81780	0.15212	9.18219
	Cl	2.91650	10.46487	0.34288	9.53514
	I	10.43965	11.01869	0.09579	8.98132
	$MgCO_3$	3.46829	10.54011	0.28833	9.45989
	MgO	1.65807	10.21960	0.60311	9.78040
	$Mg_2P_2O_7$	4.57731	10.66061	0.21847	9.33939
	$MgSO_4$	4.95122	10.69471	0.20197	9.30529
$MgBr_2$	Br	0.86796	9.93850	1.15213	10.06150
$MgBr.6H_2O$	Br	0.54702	9.73800	1.82809	10.26200
MgC_2O_4	Mg	0.21643	9.33532	4.62043	10.66468
	MgO	0.35886	9.55493	2.78660	10.44508
$Mg(C_9H_6NO)_2$	Mg	0.07777	8.89081	12.85843	11.10919
	$MgBr_2$	0.58898	9.77010	1.69785	10.22990
	$MgBr_2.6H_2O$	0.93455	9.97060	1.07003	10.02939
	$MgCO_3$	0.26972	9.43091	3.70755	10.56909
	$MgCl_2$	0.30458	9.48370	3.28321	10.51630
	$MgCl_2.6H_2O$	0.65014	9.81301	1.53813	10.18699
	$Mg(HCO_3)_2$	0.46813	9.67037	2.13616	10.32963
	MgO	0.12895	9.11042	7.75494	10.88958
	$MgSO_4$	0.38505	9.58552	2.59707	10.41448
	$MgSO_4.7H_2O$	0.78842	9.89676	1.26834	10.10324
$MgCl_2$	Cl	0.74467	9.87196	1.34288	10.12804
	$Mg_2P_2O_7$	1.16872	10.06771	0.85564	9.93229
$MgCl_2.6H_2O$	Cl	0.34887	9.54266	2.86640	10.45723
	$Mg_2P_2O_7$	0.54753	9.73841	1.82638	10.26159
$MgCl_2.KCl.6H_2O$	$Mg_2P_2O_7$	0.40059	9.60270	2.49632	10.39730
$MgCO_3$	CO_2	0.52193	9.71761	1.91597	10.28239
	Mg	0.28833	9.45989	3.46825	10.54011
	$Mg(HCO_3)_2$	1.73559	10.23945	0.57617	9.76055
	MgO	0.47807	9.67949	2.09174	10.32051
	$Mg_2P_2O_7$	1.31977	10.12049	0.75771	9.87950

Weighed	Sought	Factor	Log of Factor +10	Reciprocal of Factor	Log of Reciprocal of Factor +10
Magnesium (contd.)					
$Mg(HCO_3)_2$	$MgCO_3$	0.57617	9.76055	1.73560	10.23945
	MgO	0.27545	9.44004	3.63042	10.55996
	$Mg_2P_2O_7$	0.76041	9.88105	1.31508	10.11895
MgI_2	I	0.91258	9.96027	1.09579	10.03973
MgO	CO_2	1.09176	10.03812	0.91595	9.96187
	Mg	0.60311	9.78040	1.65807	10.21960
	$MgCO_3$	2.09176	10.32051	0.47807	9.67949
	$Mg(HCO_3)_2$	3.63045	10.55996	0.27545	9.44004
	$Mg_2P_2O_7$	2.76064	10.44101	0.36223	9.55898
	$MgSO_4$	2.98613	10.47511	0.33488	9.52489
	SO_3	1.98611	10.29800	0.50350	9.70200
$Mg_2P_2O_7$	Mg	0.21847	9.33939	4.57729	10.66061
	$MgCl_2$	0.85563	9.93229	1.16873	10.06771
	$MgCl_2.6H_2O$	1.82638	10.26159	0.54753	9.73841
	$MgCl_2.KCl.6H_2O$	2.49633	10.39730	0.40059	9.60270
	$MgCO_3$	0.75771	9.87950	1.31977	10.12049
	$Mg(HCO_3)_2$	1.31508	10.11896	0.76041	9.88105
	MgO	0.36224	9.55900	2.76060	10.44101
	$MgSO_4$	1.08168	10.03410	0.92449	9.96590
	$MgSO_4.7H_2O$	2.21488	10.34535	0.45149	9.65465
$MgSO_4$	$BaSO_4$	1.93896	10.28757	0.51574	9.71243
	Mg	0.20197	9.30529	4.85123	10.69471
	MgO	0.33488	9.52489	2.98614	10.47511
	$Mg_2P_2O_7$	0.92449	9.96590	1.08168	10.03410
	SO_3	0.66511	9.82289	1.50351	10.17711
$MgSO_4.7H_2O$	$BaSO_4$	0.94693	9.97632	1.05604	10.02368
	$Mg_2P_2O_7$	0.45149	9.65465	2.21489	10.34535
	SO_3	0.32482	9.51164	3.07863	10.48836
SO_3	MgO	0.50350	9.70200	1.98610	10.29800
	$MgSO_4$	1.50351	10.17711	0.66511	9.82289
	$MgSO_4.7H_2O$	3.07863	10.48836	0.32482	9.51164
Manganese					
$BaSO_4$	$MnSO_4$	0.64696	9.81088	1.54569	10.18912
CO_2	$MnCO_3$	2.61184	10.41694	0.38287	9.58305
	MnO	1.61184	10.20733	0.62041	9.79268
Mn	$MnCO_3$	2.09230	10.32062	0.47794	9.67937
	MnO	1.29122	10.11100	0.77446	9.88900
	MnO_2	1.58246	10.19933	0.63193	9.80067
	Mn_2O_3	1.43684	10.15741	0.69597	9.84259
	Mn_3O_4	1.38830	10.14248	0.72031	9.85752
	$Mn_2P_2O_7$	2.58308	10.41213	0.38713	9.58786
MnC_2O_4	Mn	0.38429	9.58466	2.60220	10.41534
	Mn_3O_4	0.53352	9.72715	1.87434	10.27285
$Mn(C_9H_6NO)_2$	Mn	0.16005	9.20426	6.24805	10.79574
	Mn_3O_4	0.22220	9.34674	4.50045	10.65326
$MnCO_3$	CO_2	0.38287	9.58305	2.61185	10.41695
	Mn	0.47794	9.67937	2.09231	10.32063
	$Mn(HCO_3)_2$	1.53961	10.18741	0.64952	9.81259
	MnO	0.61713	9.79038	1.62040	10.20962
	Mn_3O_4	0.66353	9.82186	1.50709	10.17814
	$Mn_2P_2O_7$	1.23456	10.09152	0.81001	9.90849
	MnS	0.75689	9.87903	1.32120	10.12097
	$MnSO_4$	1.31365	10.11848	0.76124	9.88152
$Mn(HCO_3)_2$	$MnCO_3$	0.64952	9.81259	1.53960	10.18741
	MnO	0.40084	9.60297	2.49476	10.39703
	Mn_3O_4	0.43098	9.63446	2.32029	10.36555
MnO	CO_2	0.62041	9.79268	1.61184	10.20733
	Mn	0.77446	9.88900	1.29122	10.11099
	$MnCO_3$	1.62041	10.20963	0.61713	9.79038
	$Mn(HCO_3)_2$	2.49479	10.39703	0.40084	9.60297
	Mn_2O_3	1.11277	10.04641	0.89866	9.95360
	Mn_3O_4	1.07518	10.03148	0.93008	9.96852
	$Mn_2P_2O_7$	2.00049	10.30114	0.49988	9.69887
	MnS	1.22647	10.08865	0.81535	9.91134
	$MnSO_4$	2.12865	10.32811	0.46978	9.67189
	SO_3	1.12864	10.05255	0.88602	9.94744
MnO_2	Mn	0.63193	9.80067	1.58245	10.19933
	Mn_3O_4	0.87731	9.94315	1.13985	10.05684

Weighed	Sought	Factor	Log of Factor +10	Reciprocal of Factor	Log of Reciprocal of Factor +10
Manganese (contd.)					
MnO_2	$Mn_2P_2O_7$	1.63233	10.21281	0.61262	9.78719
Mn_2O_3	Mn	0.69597	9.84259	1.43684	10.15741
	MnO	0.89865	9.95359	1.11278	10.04641
	Mn_3O_4	0.96622	9.98508	1.03496	10.01492
Mn_3O_4	Mn	0.72030	9.85751	1.38831	10.14249
	$MnCO_3$	1.50709	10.17814	0.66353	9.82186
	$Mn(HCO_3)_2$	2.32032	10.36555	0.43098	9.63446
	MnO	0.93007	9.96852	1.07519	10.03149
	MnO_2	1.13984	10.05684	0.87732	9.94316
	Mn_2O_3	1.03496	10.01492	0.96622	9.98508
	$MnSO_4$	1.97978	10.29662	0.50511	9.70339
$Mn_2P_2O_7$	Mn	0.38713	9.58786	2.58311	10.41214
	$MnCO_3$	0.81000	9.90849	1.23457	10.09152
	MnO	0.49987	9.69886	2.00052	10.30114
	MnO_2	0.61262	9.78719	1.63233	10.21281
	$MnSO_4$	1.06405	10.02696	0.93981	9.97304
MnS	Mn	0.63146	9.80035	1.58363	10.19966
	$MnCO_3$	1.32120	10.12097	0.75689	9.87903
	MnO	0.81535	9.91134	1.22647	10.08865
	$MnSO_4$	1.73559	10.23945	0.57617	9.76055
$MnSO_4$	$BaSO_4$	1.54570	10.18913	0.64696	9.81088
	Mn	0.36383	9.56090	2.74854	10.43910
	MnO	0.46978	9.67189	2.12866	10.32811
	Mn_3O_4	0.50510	9.70338	1.97981	10.29663
	$Mn_2P_2O_7$	0.93980	9.97304	1.06406	10.02696
	MnS	0.57617	9.76055	1.73560	10.23945
	SO_3	0.53021	9.72445	1.88605	10.27555
SO_3	MnO	0.88603	9.94745	1.12863	10.05255
	$MnSO_4$	1.88604	10.27555	0.53021	9.72445
Mercury					
Hg	$HgCl$	1.17673	10.07068	0.84981	9.92932
	$HgCl_2$	1.35351	10.13146	0.73882	9.86854
	HgO	1.07976	10.03332	0.92613	9.96667
	HgS	1.15983	10.06440	0.86220	9.93561
$Hg_3(AsO_4)_2$	Hg	0.68413	9.83514	1.46171	10.16486
$Hg(C_{12}H_{10}ONS)_2$	Hg	0.31681	9.50080	3.15647	10.49920
$HgCl$	Hg	0.84981	9.92932	1.17673	10.07068
	$HgCl_2$	1.15023	10.06079	0.86939	9.93921
	$HgNO_3$	1.11248	10.04629	0.89889	9.95371
	HgO	0.91760	9.96265	1.08980	10.03735
	Hg_2O	0.88369	9.94630	1.13162	10.05370
	HgS	0.98564	9.99372	1.01457	10.00629
$HgCl_2$	Hg	0.73882	9.86854	1.35351	10.13146
	$HgCl$	0.86939	9.93921	1.15023	10.06079
	HgS	0.85691	9.93294	1.16698	10.06706
$Hg(CN)_2$	HgS	0.92091	9.96422	1.08588	10.03578
Hg_2CrO_4	Hg	0.77572	9.88971	1.28912	10.11029
$HgNO_3$	$HgCl$	0.89889	9.95371	1.11248	10.04629
	HgS	0.88598	9.94742	1.12869	10.05257
$Hg(NO_3)_2$	HgS	0.71673	9.85536	1.39523	10.14464
$Hg(NO_3)_2.H_2O$	HgS	0.67903	9.83189	1.47269	10.16812
HgO	Hg	0.92613	9.96667	1.07977	10.03333
	$HgCl$	1.08980	10.03735	0.91760	9.96265
	HgS	1.07415	10.03106	0.93097	9.96894
Hg_2O	$HgCl$	1.13160	10.05369	0.88370	9.94630
	HgS	1.11535	10.04741	0.89658	9.95259
HgS	Hg	0.86220	9.93561	1.15982	10.06440
	$HgCl$	1.01457	10.00629	0.98564	9.99372
	$HgCl_2$	1.16698	10.06706	0.85691	9.93294
	$Hg(CN)_2$	1.08588	10.03578	0.92091	9.96422
	$HgNO_3$	1.12869	10.05257	0.88598	9.94743
	$Hg(NO_3)_2$	1.39522	10.14464	0.71673	9.85536
	$Hg(NO_3)_2.H_2O$	1.47268	10.16811	0.67903	9.83189
	HgO	0.93097	9.96894	1.07415	10.03106
	Hg_2O	0.89656	9.95258	1.11537	10.04741
	$HgSO_4$	1.27509	10.10554	0.78426	9.89446
$HgSO_4$	HgS	0.78426	9.89446	1.27509	10.10554
Molybdenum					
Mo	MoO_3	1.50031	10.17618	0.66653	9.82382

Weighed	Sought	Factor	Log of Factor +10	Reciprocal of Factor	Log of Reciprocal of Factor +10
Molybdenum					
Mo	MoC	1.12518	10.05122	0.88875	9.94878
	MoS$_3$	2.00261	10.30159	0.49935	9.69841
	PbMoO$_4$	3.82666	10.58282	0.26132	9.41717
MoC	C	0.11127	9.04638	8.89715	10.95362
	Mo	0.88874	9.94877	1.12519	10.05122
MoO$_2$(C$_9$H$_6$NO)$_2$	Mo	0.23049	9.36265	4.33858	10.63735
	MoO$_3$	0.34580	9.53883	2.89184	10.46118
MoO$_3$	Mo	0.66653	9.82382	1.50031	10.17617
	MoS$_3$	1.33479	10.12541	0.74918	9.87459
	(NH$_4$)$_2$MoO$_4$	1.36175	10.13410	0.73435	9.86590
	(NH$_4$)$_3$PO$_4$.12MoO$_3$	1.08630	10.03595	0.92056	9.96405
	PbMoO$_4$	2.55058	10.40664	0.39207	9.59336
MoS$_3$	Mo	0.49935	9.69841	2.00260	10.30159
	MoO$_3$	0.74918	9.87459	1.33479	10.12541
	(NH$_4$)$_2$MoO$_4$	1.02019	10.00868	0.98021	9.99132
MoSi	Mo	0.77352	9.88847	1.29279	10.11152
	Si	0.22645	9.35497	4.41599	10.64503
(NH$_4$)$_2$MoO$_4$	MoO$_3$	0.73435	9.86590	1.36175	10.13410
	MoS$_3$	0.98021	9.99132	1.02019	10.00868
	(NH$_4$)$_3$PO$_4$.12MoO$_3$	0.79771	9.90185	1.25359	10.09816
	PbMoO$_4$	1.87302	10.27254	0.53390	9.72746
(NH$_4$)$_3$PO$_4$.12MoO$_3$	MoO$_3$	0.92054	9.96404	1.08632	10.03596
	(NH$_4$)$_2$MoO$_4$	1.25359	10.09816	0.79771	9.90185
PbMoO$_4$	Mo	0.26132	9.41717	3.82673	10.58283
	MoO$_3$	0.39207	9.59336	2.55056	10.40664
	(NH$_4$)$_2$MoO$_4$	0.53390	9.72746	1.87301	10.27254
Neodymium					
Nd	Nd$_2$O$_3$	1.16639	10.06684	0.85735	9.93315
Nd$_2$O$_3$	Nd	0.85735	9.93316	1.16638	10.06684
Nickel					
Ni	Ni(C$_4$H$_7$N$_2$O$_2$)$_2$	4.92148	10.69209	0.20319	9.30790
	Ni(NO$_3$)$_2$.6H$_2$O	4.95231	10.69481	0.20193	9.30520
	NiO	1.27254	10.10467	0.78584	9.89533
	NiSO$_4$	2.63618	10.42099	0.37934	9.57903
	NiSO$_4$.7H$_2$O	4.78419	10.67981	0.20902	9.32019
Ni(C$_4$H$_7$N$_2$O$_2$)$_2$	Ni	0.20319	9.30790	4.92150	10.69210
	NiO	0.25857	9.41258	3.86742	10.58742
(Ni[C$_5$H$_5$N]$_4$)(SCN)$_2$	Ni	0.11950	9.07737	8.36820	10.92263
	NiO	0.15207	9.18204	6.57592	10.81776
Ni(C$_7$H$_6$O$_2$N)$_2$	Ni	0.17736	9.24886	5.63825	10.75114
	NiO	0.22573	9.35359	4.43007	10.64641
Ni(C$_9$H$_6$NO)$_2$	Ni	0.16918	9.22835	5.91086	10.77165
	NiO	0.21529	9.33303	4.64490	10.66698
Ni(NO$_3$)$_2$.6H$_2$O	Ni	0.20193	9.30520	4.95221	10.69480
	NiO	0.25696	9.40987	3.89166	10.59013
	NiSO$_4$	0.53231	9.72616	1.87860	10.27383
NiO	Ni	0.78584	9.89533	1.27252	10.10467
	Ni(C$_4$H$_7$N$_2$O$_2$)$_2$	3.86749	10.58743	0.25857	9.41258
	Ni(NO$_3$)$_2$.6H$_2$O	3.89171	10.59014	0.25696	9.40987
	NiSO$_4$	2.07161	10.31631	0.48272	9.68370
	NiSO$_4$.7H$_2$O	3.75960	10.57514	0.26599	9.42487
NiSO$_4$	Ni	0.37934	9.57903	2.63616	10.42098
	Ni(NO$_3$)$_2$.6H$_2$O	1.87859	10.27384	0.53231	9.72616
	NiO	0.48272	9.68370	2.07159	10.31630
	NiSO$_4$.7H$_2$O	1.81482	10.25884	0.55102	9.74117
NiSO$_4$.7H$_2$O	Ni	0.20902	9.32019	4.78423	10.67981
	NiO	0.26599	9.42487	3.75954	10.57514
	NiSO$_4$	0.55102	9.74117	1.81482	10.25884
Niobium					
C	Nb	7.73497	10.88846	0.12928	9.11153
	NbC	8.73493	10.94126	0.11448	9.05873
	$\frac{1}{2}$Nb$_2$O$_5$	11.06508	11.04396	0.09038	8.95607
Nb	C	0.12928	9.11153	7.73515	10.88848
	Nb$_2$O$_5$	1.43053	10.15550	0.69904	9.84450
NbC	C	0.11448	9.05873	8.73515	10.94127
	Nb	0.88552	9.94720	1.12928	10.05280
Nb$_2$O$_5$	Nb	0.69904	9.84450	1.43053	10.15550
Nitrogen					
AgNO$_2$	HNO$_2$	0.30552	9.48504	3.27311	10.51496

Weighed	Sought	Factor	Log of Factor +10	Reciprocal of Factor	Log of Reciprocal of Factor +10
Nitrogen (contd.)					
$AgNO_2$	N_2O_3	0.24698	9.39266	4.04891	10.60733
$C_{20}H_{16}N_4 \cdot HNO_3$	N	0.37312	9.57185	2.68010	10.42815
	NO_3	0.16517	9.21793	6.05437	10.78207
HNO_2	$AgNO_2$	3.27310	10.51496	0.30552	9.48504
HNO_3	N	0.22228	9.34690	4.49883	10.65310
	NH_3	0.27027	9.13180	3.70000	10.50820
	NH_4Cl	0.84889	9.92885	1.17801	10.07115
	$(NH_4)_2PtCl_6$	3.52221	10.54682	0.28391	9.45318
	NO	0.47619	9.67778	2.10000	10.32222
	Pt	1.54801	10.18977	0.64599	9.81023
	SO_3	0.63528	9.80297	1.57411	10.19703
	N_2O_5	0.53413	9.72765	1.87220	10.27235
N	HNO_3	4.49877	10.65310	0.22229	9.34692
	$NaNO_3$	6.06816	10.78306	0.16479	9.21693
	NH_3	1.21589	10.08489	0.82244	9.91510
	NH_4Cl	3.81902	10.58195	0.26185	9.41805
	$(NH_4)_2PtCl_6$	15.84563	11.19991	0.06311	8.80010
	$(NH_4)_2SO_4$	4.71699	10.67367	0.21200	9.32634
	NO_2	3.28454	10.51648	0.30446	9.48353
	NO_3	4.42680	10.64609	0.22590	9.35392
	N_2O_3	2.71341	10.43352	0.36854	9.56648
	N_2O_5	3.85566	10.58610	0.25936	9.41390
	Pt	6.96416	10.84287	0.14359	9.15712
	SO_3	2.85798	10.45606	0.34990	9.54394
$NaNO_3$	N	0.16479	9.21603	6.06800	10.78307
	N_2O_5	0.63539	9.80304	1.57384	10.19696
NH_3	HNO_3	3.69998	10.56820	0.27027	9.43180
	N	0.82244	9.91510	1.21589	10.08489
	NO_4	3.04079	10.50119	0.27467	9.43881
	N_2O_5	3.17107	10.50121	0.31535	9.49879
$NH_4B(C_6H_5)_4$	N	0.04153	8.61836	24.07898	11.38164
	NH_3	0.05049	8.70320	19.80590	11.29679
	NH_4	0.05348	8.72819	18.69858	11.27181
NH_4Cl	HNO_3	1.17799	10.07115	0.84890	9.92886
	N	0.26185	9.41805	3.81898	10.58195
	NO_3	1.15914	10.06413	0.86271	9.93586
	N_2O_5	1.00959	10.00415	0.99050	9.99585
$(NH_4)_2PtCl_6$	HNO_3	0.28392	9.45320	3.52212	10.54680
	N	0.06311	8.80010	15.84535	11.19990
	NO_3	0.27938	9.44620	3.57935	10.55380
	N_2O_5	0.24333	9.38620	4.10965	10.61380
$(NH_4)_2SO_4$	N	0.21200	9.32634	4.71698	10.67367
	N_2O_5	0.81740	9.91243	1.22339	10.08757
NO	HNO_3	2.10000	10.32222	0.47619	9.67778
	NO_2	1.53320	10.18560	0.65223	9.81440
	NO_3	2.06641	10.31522	0.48393	9.68478
	N_2O_3	1.26660	10.10264	0.78952	9.89736
	N_2O_5	1.79980	10.25522	0.55562	9.74478
NO_2	N	0.30446	9.48353	3.28450	10.51647
	NO	0.65223	9.81440	1.53320	10.18560
NO_3	N	0.22590	9.35392	4.42674	10.64608
	NH_3	0.27467	9.43881	3.64073	10.56119
	NH_4Cl	0.86270	9.93586	1.15915	10.06413
	NO	0.48393	9.68478	2.06641	10.31522
	Pt	1.57318	10.19678	0.63566	9.80322
N_2O_3	$AgNO_2$	4.04875	10.60732	0.24699	9.39268
	N	0.36854	9.56648	2.71341	10.43352
	NO	0.78951	9.89736	1.26661	10.10264
N_2O_5	KNO_3	1.87217	10.27235	0.53414	9.72766
	N	0.25936	9.41390	3.85564	10.58610
	$NaNO_3$	1.57382	10.19695	0.63540	9.80305
	NH_3	0.31535	9.49879	3.17108	10.50121
	NH_4Cl	0.99049	9.99585	1.00960	10.00415
	$(NH_4)_2PtCl_6$	4.10965	10.61380	0.24333	9.38620
	$(NH_4)_2SO_4$	1.22339	10.08757	0.81740	9.91243
	NO	0.55562	9.74478	1.79980	10.25522
	Pt	1.80621	10.25677	0.55365	9.74324
	SO_3	0.74124	9.86996	1.34909	10.13004

Weighed	Sought	Factor	Log of Factor +10	Reciprocal of Factor	Log of Reciprocal of Factor +10
Nitrogen (contd.)					
Pt	HNO_3	0.64595	9.81020	1.54811	10.18980
	N	0.14358	9.15709	6.96476	10.84291
	NO_3	0.63562	9.80320	1.57327	10.19680
	N_2O_5	0.55364	9.74323	1.80623	10.25678
SO_3	HNO_3	1.57410	10.19703	0.63528	9.80297
	N	0.34990	9.54394	2.85796	10.45605
	N_2O_5	1.34908	10.13004	0.74125	9.86996
Osmium					
Os	OsO_4	1.33649	10.12597	0.74823	9.87404
OsO_4	Os	0.74823	9.87404	1.33649	10.12597
Palladium					
K_2PdCl_6	Pd	0.26781	9.42783	3.73399	10.57217
	$PdCl_2.2H_2O$	0.53687	9.72987	1.86265	10.27013
Pd	K_2PdCl_6	3.73402	10.57217	0.26781	9.42783
	$PdCl_2.2H_2O$	2.00460	10.30205	0.49883	9.69795
	PdI_2	3.38534	10.52960	0.29539	9.47040
	$Pd(NO_3)_2$	2.16541	10.33554	0.46181	9.66446
$Pd(C_4H_7N_2O_2)_2$	Pd	0.31610	9.49982	3.16356	10.50018
$PdCl_2.2H_2O$	K_2PdCl_6	1.86263	10.27012	0.53688	9.72988
	Pd	0.49883	9.69795	2.00469	10.30204
$PdCl_2.C_{12}H_8N_2$	Pd	0.29762	9.47366	3.35999	10.52634
PdI_2	Pd	0.29539	9.47040	3.38536	10.52960
$Pd(NO_3)_2$	Pd	0.46181	9.66446	2.16539	10.33554
Phosphorus					
Ag_3PO_4	P	0.07400	8.86923	13.51351	11.13077
	PO_4	0.22689	9.35582	4.40742	10.64418
	P_2O_5	0.16954	9.22927	5.89831	10.77072
$Ag_4P_2O_7$	P	0.10232	9.00996	9.77326	10.99004
	PO_4	0.31374	9.49657	3.18735	10.50343
	P_2O_5	0.23446	9.37007	4.26512	10.62993
Al_2O_3	P_2O_5	1.39214	10.14368	0.71832	9.85632
$AlPO_4$	PO_4	0.77875	9.89140	1.28411	10.10860
	P_2O_5	0.58196	9.76489	1.71833	10.23511
$Ca_3(PO_4)_2$	P_2O_5	0.45762	9.66051	2.18522	10.33949
$FePO_4$	PO_4	0.62971	9.79914	1.58803	10.20086
	P_2O_5	0.47058	9.67263	2.12504	10.32737
$Mg_2P_2O_7$	H_3PO_4	0.88059	9.94471	1.13560	10.05523
	Na_2HPO_4	1.27565	10.10573	0.78391	9.89427
	$Na_2HPO_4.12H_2O$	3.21828	10.50763	0.31073	9.49238
	$NaNH_4.HPO_4.4H_2O$	1.87870	10.27386	0.53228	9.72614
	P	0.27833	9.44456	3.59286	10.55544
	PO_4	0.85341	9.93116	1.17177	10.06884
	P_2O_5	0.63776	9.80466	1.56799	10.19535
Na_2HPO_4	$Mg_2P_2O_7$	0.78392	9.89427	1.27564	10.10573
	P_2O_5	0.49995	9.69893	2.00020	10.30105
$Na_2HPO_4.12H_2O$	$Mg_2P_2O_7$	0.31072	9.49237	3.21833	10.50763
	P_2O_5	0.19816	9.29702	5.04643	10.70299
$NaNH_4HPO_4.4H_2O$	$Mg_2P_2O_7$	0.53228	9.72614	1.87871	10.27386
	P_2O_5	0.33946	9.53079	2.94586	10.46921
$(NH_4)_3PO_4.12MoO_3$	P	0.01651	8.21775	60.56935	11.78223
	PO_4	0.05061	8.70424	19.75894	11.29577
	P_2O_5	0.03782	8.57772	26.44104	11.42227
P	Ag_3PO_4	13.51403	11.13078	0.07400	8.86923
	$Ag_4P_2O_7$	9.77314	10.99004	0.10232	9.00996
	$Mg_2P_2O_7$	3.59282	10.55544	0.27833	9.44456
	$(NH_4)_3PO_4.12MoO_3$	60.5786	11.78234	0.01651	8.21775
	P_2O_5	2.29136	10.36009	0.43642	9.63990
	$P_2O_5.24MoO_3$	58.0564	11.76385	0.01723	8.23616
	$U_2P_2O_{11}$	11.52587	11.06167	0.08676	8.93832
PO_4	Ag_3PO_4	4.40744	10.64418	0.22689	9.35582
	$Ag_4P_2O_7$	3.18739	10.50343	0.31374	9.49657
	$AlPO_4$	1.28410	10.10860	0.77876	9.89140
	$FePO_4$	1.58803	10.20086	0.62971	9.79914
	$Mg_2P_2O_7$	1.17175	10.06884	0.85342	9.93116
	$(NH_4)_3PO_4.12MoO_3$	19.7570	11.29572	0.05062	8.70432
	$P_2O_5.24MoO_3$	18.9344	11.27725	0.05281	8.72272
	$U_2P_2O_{11}$	3.75902	10.57507	0.26603	9.42493
P_2O_5	Ag_3PO_4	5.89780	10.77069	0.16955	9.22930

Weighed	Sought	Factor	Log of Factor +10	Reciprocal of Factor	Log of Reciprocal of Factor +10
Phosphorus (contd.)					
P_2O_5	$Ag_4P_2O_7$	4.26520	10.62994	0.23446	9.37007
	Al_2O_3	0.71831	9.85631	1.39216	10.14369
	$AlPO_4$	1.71831	10.23510	0.58197	9.76440
	$Ca_3(PO_4)_2$	2.18521	10.33949	0.45762	9.66051
	$FePO_4$	2.12502	10.32736	0.47058	9.67263
	$Mg_2P_2O_7$	1.56798	10.19534	0.63770	9.80466
	Na_2HPO_4	2.00020	10.30107	0.49995	9.69893
	$Na_2HPO_4.12H_2O$	5.04623	10.70297	0.19817	9.29704
	$NaNH_4HPO_4.4H_2O$	2.94578	10.46920	0.33947	9.53080
	$(NH_4)_3PO_4.12MoO_3$	26.4377	11.42222	0.03783	8.57784
	P	0.43642	9.63990	2.29137	10.36010
	$P_2O_5.24MoO_3$	25.3370	11.40376	0.03947	8.59622
	$U_2P_2O_{11}$	5.03013	10.70158	0.19880	9.29842
$P_2O_5.24MoO_3$	P	0.01722	8.23603	58.07201	11.76397
	PO_4	0.05281	8.72272	18.93581	11.27728
	P_2O_5	0.03947	8.59627	25.33570	11.40373
$U_2P_2O_{11}$	P	0.08676	8.93832	11.52605	11.06176
	PO_4	0.26603	9.42493	3.75897	10.57507
	P_2O_5	0.19880	9.29842	5.03018	10.70158
Platinum					
$H_2PtCl_6.6H_2O$	K_2PtCl_6	0.93852	9.97244	1.06551	10.02756
	Pt	0.37673	9.57603	2.65442	10.42397
K_2PtCl_6	$H_2PtCl_6.6H_2O$	1.06551	10.02756	0.93852	9.97244
	Pt	0.40141	9.60359	2.49122	10.39541
	$PtCl_4$	0.69320	9.84086	1.44259	10.15915
	$PtCl_4.5H_2O$	0.87854	9.94376	1.13825	10.05624
$(NH_4)_2PtCl_6$	Pt	0.43950	9.64296	2.27531	10.35704
	$PtCl_4$	0.75897	9.88022	1.31758	10.11978
	$PtCl_6$	0.91854	9.96310	1.08868	10.03690
Pt	$H_2PtCl_6.6H_2O$	2.65442	10.42397	0.37673	9.57603
	K_2O	0.48287	9.68383	2.07095	10.31617
	K_2PtCl_6	2.49121	10.39641	0.40141	9.60359
	$(NH_4)_2PtCl_6$	2.27531	10.35704	0.43950	9.64296
	$PtCl_4$	1.72690	10.23727	0.57907	9.76273
	$PtCl_4.5H_2O$	2.18863	10.34017	0.45691	9.65983
$PtCl_4$	K_2PtCl_6	1.44259	10.15915	0.69320	9.84086
	$(NH_4)_2PtCl_6$	1.31757	10.11978	0.75897	9.88022
	Pt	0.57907	9.76273	1.72691	10.23727
$PtCl_4.5H_2O$	K_2PtCl_6	1.13825	10.05624	0.87854	9.94376
	Pt	0.45691	9.65983	2.18861	10.34017
$PtCl_6$	K_2PtCl_6	1.19199	10.07628	0.83893	9.92373
	$(NH_4)_2PtCl_6$	1.08869	10.03691	0.91854	9.96310
Potassium					
Ag	KBr	1.10328	10.04269	0.90639	9.95732
	KCl	0.69116	9.83958	1.44684	10.16042
	$KClO_3$	1.13612	10.05543	0.88019	9.94458
	$KClO_4$	1.28444	10.10872	0.77855	9.89129
	KCN	0.60369	9.78081	1.65648	10.21919
	KI	1.53895	10.18723	0.64979	9.81277
$AgBr$	KBr	0.63378	9.80194	1.57783	10.19806
	$KBrO_3$	0.88939	9.94909	1.12437	10.05091
$AgCl$	KCl	0.52019	9.71616	1.92237	10.28384
	$KClO_3$	0.85508	9.93201	1.16948	10.06799
	$KClO_4$	0.96672	9.98530	1.03443	10.01470
$AgCN$	KCN	0.48638	9.68698	2.05601	10.31302
AgI	KI	0.70709	9.84947	1.41425	10.15053
	KIO_3	0.91154	9.95978	1.09704	10.04022
$BaCrO_4$	K_2CrO_4	0.76658	9.88456	1.30450	10.11544
	$K_2Cr_2O_7$	0.58064	9.76391	1.72224	10.23609
$BaSO_4$	$KHSO_4$	0.58343	9.76599	1.71400	10.23401
	K_2S	0.47244	9.67435	2.11667	10.32565
	K_2SO_4	0.74664	9.87311	1.33933	10.12689
Br	K	0.48933	9.68960	2.04361	10.31040
	KBr	1.48933	10.17299	0.67144	9.82701
CaF_2	$KF.2H_2O$	2.41114	10.38223	0.41474	9.61778
$CaSO_4$	$KF.2H_2O$	1.38286	10.14078	0.72314	9.85922
Cl	K	1.10292	10.04255	0.90668	9.95745
	KCl	2.10292	10.32282	0.47553	9.67718

Weighed	Sought	Factor	Log of Factor +10	Reciprocal of Factor	Log of Reciprocal of Factor +10
Potassium (contd.)					
Cl	$KClO_3$	3.45677	10.53867	0.28929	9.46133
	$KClO_4$	3.90808	10.59196	0.25588	9.40804
	K_2O	1.32856	10.12338	0.75269	9.87662
CO_2	$KHCO_3$	2.27491	10.35696	0.43958	9.64304
	K_2CO_3	3.14049	10.49700	0.31842	9.50300
	K_2O	2.14049	10.33051	0.46718	9.66948
I	KI	1.30811	10.11665	0.76446	9.88335
	KIO_3	1.68634	10.22695	0.59300	9.77306
K	Br	2.04360	10.31040	0.48933	9.68960
	Cl	0.90668	9.95745	1.10292	10.04256
	KBr	3.04360	10.48339	0.32856	9.51661
	KCl	1.90668	10.28028	0.52447	9.71972
	$KClO_3$	3.13419	10.49614	0.31906	9.50387
	$KClO_4$	3.54337	10.54941	0.28222	9.45059
	KI	4.24546	10.62793	0.23555	9.37208
	KNO_3	2.58572	10.41259	0.38674	9.58742
	K_2O	1.20458	10.08084	0.83016	9.91916
	K_2PtCl_6	6.21467	10.79342	0.16091	9.20658
	K_2SO_4	2.22835	10.34799	0.44876	9.65201
	Pt	2.49463	10.39701	0.40086	9.60299
K_3AsO_4	$Mg_2As_2O_7$	0.60584	9.78236	1.65060	10.21764
$K(B[C_6H_5]_4)$	K	0.10912	9.03790	9.16422	10.96210
	K_2O	0.13144	9.11873	7.60803	10.88127
KBr	Ag	0.90639	9.95732	1.10328	10.04269
	AgBr	1.57783	10.19806	0.63378	9.80194
	Br	0.67144	9.82701	1.48934	10.17299
	K	0.32856	9.51662	3.04358	10.48338
	K_2O	0.39580	9.59748	2.52653	10.40253
$KBrO_3$	AgBr	1.12436	10.05091	0.88939	9.94909
$KC_{12}H_4O_{12}N_7$	K	0.08192	8.91339	12.20703	11.08661
	K_2O	0.09868	8.99423	10.13377	11.00577
KCl	Ag	1.44685	10.16043	0.69116	9.83958
	AgCl	1.92238	10.28384	0.52019	9.71616
	Cl	0.47553	9.67718	2.10292	10.32283
	K	0.52447	9.71972	1.90669	10.28028
	$KClO_3$	1.64379	10.21585	0.60835	9.78415
	$KClO_4$	1.85840	10.26914	0.53810	9.73086
	K_2CO_3	0.92692	9.96704	1.07884	10.03296
	$K_2Cr_2O_7$	1.97299	10.29512	0.50684	9.70487
	$KHCO_3$	1.34289	10.12804	0.74466	9.87196
	KNO_3	1.35614	10.13230	0.73739	9.86770
	K_2O	0.63180	9.80058	1.58278	10.19942
	K_2PtCl_6	3.25941	10.51314	0.30680	9.48686
	K_2SO_4	1.16871	10.06770	0.85564	9.93229
	Pt	1.30836	10.11673	0.76432	9.88328
$KClO_3$	Ag	0.88019	9.94458	1.13612	10.05543
	AgCl	1.16948	10.06799	0.85508	9.93201
	Cl	0.28929	9.46133	3.45674	10.53867
	KCl	0.60835	9.78415	1.64379	10.21584
$KClO_4$	Ag	0.77855	9.89129	1.28444	10.10872
	AgCl	1.03443	10.01470	0.96672	9.98530
	Cl	0.25588	9.40804	3.90808	10.59196
	K	0.28222	9.45059	3.54333	10.54941
	KCl	0.53810	9.73086	1.85839	10.26914
	K_2O	0.33995	9.53142	2.94161	10.46858
KCN	Ag	1.65648	10.21918	0.60369	9.78081
	AgCN	2.05602	10.31302	0.48638	9.68698
K_2CO_3	CO_2	0.31842	9.50300	3.14051	10.49700
	KCl	1.07883	10.03295	0.92693	9.96705
	K_2O	0.68158	9.83352	1.46718	10.16648
	KOH	0.81191	9.90951	1.23166	10.09049
	K_2PtCl_6	3.51638	10.54610	0.28438	9.45390
	K_2SO_4	1.26085	10.10067	0.79312	9.89934
K_2CrO_4	$BaCrO_4$	1.30449	10.11544	0.76658	9.88456
$K_2Cr_2O_7$	$BaCrO_4$	1.72220	10.23608	0.58065	9.76391
	KCl	0.50685	9.70488	1.97297	10.29512
	K_2O	0.32021	9.50543	3.12295	10.49456
$KF.2H_2O$	CaF_2	0.41474	9.61778	2.41115	10.38222

Weighed	Sought	Factor	Log of Factor +10	Reciprocal of Factor	Log of Reciprocal of Factor +10
Potassium (contd.)					
$KF.2H_2O$	$CaSO_4$	0.72314	9.85922	1.38286	10.14078
K_2HAsO_4	$Mg_2As_2O_7$	0.71165	9.85227	1.40519	10.14774
$KHCO_3$	KCl	0.74466	9.87196	1.34289	10.12804
	K_2O	0.47046	9.67252	2.12558	10.32748
	K_2PtCl_6	2.42716	10.38510	0.41200	9.61490
	K_2SO_4	0.87029	9.93966	1.14904	10.06034
$KHSO_4$	$BaSO_4$	1.71401	10.23401	0.58343	9.76599
	K_2SO_4	0.63988	9.80610	1.56279	10.19389
KI	Ag	0.64980	9.81278	1.53894	10.18722
	AgI	1.41425	10.15053	0.70709	9.84947
	I	0.76446	9.88335	1.30811	10.11665
	K	0.23555	9.37208	4.24538	10.62792
	K_2O	0.28374	9.45292	3.52435	10.54708
KIO_3	AgI	1.09705	10.04023	0.91154	9.95978
	I	0.59300	9.77305	1.68634	10.22695
$KMnO_4$	Mn_2O_3	0.49948	9.69852	2.00208	10.30148
	MnS	0.55051	9.74007	1.81650	10.25924
K_2MnO_4	Mn_2O_3	0.40041	9.60250	2.49744	10.39749
	MnS	0.44132	9.64475	2.26593	10.35525
KNO_2	K_2SO_4	1.02380	10.01022	0.97675	9.98978
	N_2O_3	0.44656	9.64988	2.23934	10.35012
KNO_3	K	0.38674	9.58742	2.58572	10.41260
	KCl	0.73739	9.86770	1.35613	10.13230
	K_2O	0.46586	9.66826	2.14657	10.33174
	K_2PtCl_6	2.40344	10.38083	0.41607	9.61917
	K_2SO_4	0.86179	9.93540	1.16038	10.06460
	N	0.13853	9.14154	7.21865	10.85846
	NH_3	0.16841	9.22615	5.03683	10.77356
	NO	0.29678	9.47243	3.36950	10.52757
	N_2O_5	0.53414	9.72766	1.87217	10.27235
K_2O	Cl	0.75269	9.87662	1.32857	10.12338
	CO_2	0.46718	9.66948	2.14050	10.33052
	K	0.83016	9.91916	1.20459	10.08085
	KBr	2.52667	10.40255	0.39578	9.59745
	KCl	1.58284	10.19944	0.63178	9.80057
	$KClO_3$	2.60186	10.41529	0.38434	9.58472
	$KClO_4$	2.94155	10.46858	0.33996	9.53143
	K_2CO_3	1.46718	10.16648	0.68158	9.83352
	$K_2Cr_2O_7$	3.12294	10.49456	0.32021	9.50543
	$KHCO_3$	2.12558	10.32748	0.47046	9.67252
	KI	3.52439	10.54709	0.28374	9.45292
	KNO_3	2.14655	10.33174	0.46586	9.66826
	KOH	1.19122	10.07599	0.83948	9.92401
	K_2PtCl_6	5.15918	10.71258	0.19383	9.28742
	K_2SO_4	1.84990	10.26715	0.54057	9.73285
	N_2O_5	1.14657	10.05940	0.87217	9.94060
KOH	K_2CO_3	1.23164	10.09048	0.81193	9.90952
	K_2O	0.83946	9.92400	1.19124	10.07600
K_2PtCl_6	K	0.16091	9.20658	6.21465	10.79342
	KCl	0.30680	9.48686	3.25948	10.51315
	K_2CO_3	0.28438	9.45390	3.51642	10.54610
	$KHCO_3$	0.41200	9.61490	2.42718	10.38510
	KNO_3	0.41606	9.61916	2.40350	10.38084
	K_2O	0.19383	9.28742	5.15916	10.71258
	K_2SO_4	0.35856	9.55456	2.78893	10.44544
	$K_2SO_4.Al_2(SO_4)_3.24H_2O$	1.95218	10.29052	0.51225	9.70948
	$K_2SO_4.Cr_2(SO_4)_3.24H_2O$	2.05512	10.31283	0.48659	9.68716
K_2S	$BaSO_4$	2.11666	10.32565	0.47244	9.67435
	K_2SO_4	1.58039	10.19877	0.63276	9.80124
K_2SiO_3	SiO_2	0.38943	9.59043	2.56786	10.40958
K_2SO_4	$BaSO_4$	1.33933	10.12689	0.74664	9.87311
	K	0.44876	9.65201	2.22836	10.34799
	KCl	0.85565	9.93230	1.16870	10.06770
	K_2CO_3	0.79312	9.89934	1.26084	10.10066
	$KHCO_3$	1.14904	10.06034	0.87029	9.93966
	$KHSO_4$	1.56281	10.19390	0.63987	9.80609
	KNO_2	0.97676	9.98979	1.02379	10.01021
	KNO_3	1.16038	10.06460	0.86179	9.93540

Weighed	Sought	Factor	Log of Factor +10	Reciprocal of Factor	Log of Reciprocal of Factor +10
Potassium (contd.)					
K_2SO_4	K_2O	0.54057	9.73285	1.84990	10.26715
	K_2PtCl_6	2.78889	10.44543	0.35857	9.55457
	K_2S	0.63276	9.80124	1.58038	10.19876
	SO_3	0.45942	9.66221	2.17666	10.33779
$K_2SO_4.Al_2(SO_4)_3.24H_2O$	K_2PtCl_6	0.51224	9.70947	1.95221	10.29053
$K_2SO_4.Cr_2(SO_4)_3.24H_2O$	K_2PtCl_6	0.48658	9.68715	2.05516	10.31284
$Mg_2As_2O_7$	K_3AsO_4	1.65059	10.21764	0.60584	9.78236
	K_2HAsO_4	1.40519	10.14774	0.71165	9.85227
Mn_2O_3	$KMnO_4$	2.00207	10.30148	0.49948	9.69852
	K_2MnO_4	2.49731	10.39747	0.40043	9.60253
MnS	$KMnO_4$	1.81649	10.25923	0.55051	9.74077
	K_2MnO_4	2.26592	10.35524	0.44132	9.64475
N	KNO_3	7.21847	10.85845	0.13853	9.14154
NH_3	KNO_3	5.93678	10.77355	0.16844	9.22645
NO	KNO_3	3.36954	10.52757	0.29678	9.47243
N_2O_3	KNO_2	2.23934	10.35012	0.44656	9.64988
N_2O_5	KNO_3	1.87217	10.27235	0.53414	9.72765
	K_2O	0.87217	9.94060	1.14657	10.05940
Pt	K	0.40086	9.60299	2.49464	10.39701
	KCl	0.76431	9.88327	1.30837	10.11673
SiO_2	K_2SiO_3	2.56783	10.40957	0.38943	9.59043
SO_3	K_2SO_4	2.17664	10.33779	0.45942	9.66221
Praseodymium					
Pr	Pr_2O_3	1.17032	10.06832	0.85447	9.93169
Pr_2O_3	Pr	0.85447	9.93169	1.17032	10.06831
Pr_6O_{11}	Pr	0.82770	9.91787	1.20817	10.08213
Rhenium					
$C_{20}H_{16}N_4.HReO_4$	Re	0.33038	9.51901	3.02682	10.48098
$(C_6H_5)_4AsReO_4$	Re	0.29392	9.46823	3.40229	10.53177
Rhodium					
Na_3RhCl_6	Rh	0.26757	9.42744	3.73734	10.57256
Rh	Na_3RhCl_6	3.73735	10.57256	0.26757	9.42744
$RhCl_3$	Rh	0.49175	9.69174	2.03355	10.30826
Rubidium					
$AgCl$	Rb	0.59635	9.77550	1.67687	10.22450
	$RbCl$	0.84368	9.92618	1.18528	10.07382
Cl	Rb	2.41080	10.38216	0.41480	9.61784
	$RbCl$	3.41071	10.53284	0.29319	9.46715
Rb	$AgCl$	1.67688	10.22450	0.59635	9.77550
	Cl	0.41480	9.61784	2.41080	10.38216
	$RbCl$	1.41477	10.15069	0.70683	9.84932
	Rb_2CO_3	1.35106	10.13068	0.74016	9.86933
	Rb_2O	1.09360	10.03886	0.91441	9.96114
	Rb_2PtCl_6	3.38569	10.52965	0.29536	9.47035
	Rb_2SO_4	1.56195	10.19367	0.64023	9.80634
$RbB(C_6H_5)_4$	Rb	0.21119	9.32467	4.73507	10.67533
	Rb_2O	0.23096	9.36354	4.32975	10.63647
$RbCl$	$AgCl$	1.18527	10.07382	0.84369	9.92618
	Cl	0.29319	9.46715	3.41076	10.53285
	Rb	0.70683	9.84932	1.41477	10.15069
	Rb_2CO_3	0.95493	9.97997	1.04720	10.02003
	Rb_2O	0.77296	9.08816	1.29373	10.11184
	Rb_2PtCl_6	2.39301	10.37896	0.41788	9.62105
	Rb_2SO_4	1.10399	10.04297	0.90581	9.95704
Rb_2CO_3	Rb	0.74019	9.86934	1.35100	10.12066
	$RbCl$	1.04720	10.02003	0.45493	9.97997
	$RbHCO_3$	1.26864	10.10335	0.78825	9.89666
	Rb_2PtCl_6	2.50595	10.39897	0.39905	9.60103
	Rb_2SO_4	1.15609	10.06299	0.86498	9.93701
$RbClO_4$	Rb	0.46220	9.66483	2.16357	10.33517
	Rb_2CO_3	0.62446	9.79550	1.60138	10.20449
	$RbCl$	0.65390	9.81551	1.52929	10.18448
	$RbHCO_3$	0.79218	9.89883	1.26234	10.10118
	Rb_2O	0.50546	9.70369	1.97840	10.29632
	Rb_2SO_4	0.72193	9.85850	1.38518	10.14151
$RbHCO_3$	Rb_2CO_3	0.78831	9.89670	1.26854	10.10330
	Rb_2PtCl_6	1.97546	10.29567	0.50621	9.70433
	Rb_2SO_4	0.91136	9.95969	1.09726	10.04031

Weighed	Sought	Factor	Log of Factor +10	Reciprocal of Factor	Log of Reciprocal of Factor +10
Rubidium (contd.)					
Rb_2O	Rb	0.91441	9.96114	1.09360	10.03887
	RbCl	1.29368	10.11182	0.77299	9.88817
	Rb_2PtCl_6	3.09591	10.49079	0.32301	9.50922
	Rb_2SO_4	1.42827	10.15481	0.70015	9.84519
Rb_2PtCl_6	Rb	0.29537	9.47037	3.38558	10.52963
	RbCl	0.41787	9.62103	2.39309	10.37896
	Rb_2CO_3	0.39905	9.60103	2.50595	10.39898
	$RbHCO_3$	0.50624	9.70436	1.97535	10.29565
	Rb_2O	0.32301	9.50922	3.09588	10.49079
Rb_2SO_4	Rb	0.64022	9.80633	1.56196	10.19367
	RbCl	0.90577	9.95701	1.10403	10.04298
	Rb_2CO_3	0.86498	9.93701	1.15610	10.06300
	$RbHCO_3$	1.09730	10.04033	0.91133	9.95968
	Rb_2O	0.70015	9.84520	1.42827	10.15481
Samarium					
Sm_2O_3	Sm	0.86235	9.93568	1.15962	10.06431
Scandium					
Sc_2O_3	Sc	0.65196	9.81422	1.53384	10.18578
$Sc(C_9H_6NO)_3.$					
(C_9H_6NOH)	Sc	0.07221	8.85860	13.84850	11.14140
	Sc_2O_3	0.11076	9.04438	9.02853	10.95562
Selenium					
H_2SeO_3	Se	0.61224	9.78692	1.63335	10.21308
H_2SeO_4	Se	0.54466	9.73613	1.83601	10.26389
$PbSeO_4$	Se	0.22550	9.35315	4.43459	10.64685
	SeO_2	0.31689	9.50091	3.15567	10.49909
Se	H_2SeO_3	1.63330	10.21308	0.61223	9.78691
	H_2SeO_4	1.83599	10.26387	0.54467	9.73613
	SeO_2	1.40527	10.14776	0.71161	9.85224
	SeO_3	1.60790	10.20626	0.62193	9.79374
SeO_2	Se	0.71161	9.85224	1.40526	10.14776
SeO_3	Se	0.62193	9.79374	1.60790	10.20626
Silicon					
$BaSiF_6$	SiF_4	0.37249	9.57111	2.68464	10.42888
	SiO_2	0.21503	9.33250	4.65051	10.66750
H_2SiO_3	SiO_2	0.76933	9.88611	1.29983	10.11388
K_2SiF_6	SiF_4	0.47249	9.67439	2.11645	10.32561
	SiO_2	0.27277	9.43580	3.66609	10.56420
Si	SiO_2	2.13932	10.33027	0.46744	9.66973
SiC	C	0.29955	9.47647	3.33834	10.52353
	Si	0.70045	9.84538	1.42765	10.15462
SiF_4	$BaSiF_6$	0.26847	9.42890	3.72481	10.57110
	K_2SiF_6	2.11645	10.32561	0.47249	9.67439
SiF_4	SiO_2	0.57730	9.76140	1.73220	10.23860
SiO_2	$BaSiF_6$	4.65041	10.66749	0.21503	9.33250
	H_2SiO_3	1.29983	10.11388	0.76933	9.88611
	K_2SiF_6	3.66614	10.56421	0.27277	9.43580
	Si	0.46744	9.66973	2.13931	10.33027
	SiF_4	1.73221	10.23861	0.57730	9.76140
	SiO_3	1.26627	10.10252	0.78972	9.89747
	SiO_4	1.53256	10.18542	0.65250	9.81458
	Si_2O	0.60058	9.77857	1.66506	10.22182
	$Si(OH)_4$	1.59965	10.20403	0.62514	9.79597
$SiO_2.12MoO_3$	Si	0.01571	8.19618	63.65372	11.80382
	SiO_2	0.03362	8.52660	29.74420	11.47340
SiO_3	SiO_2	0.78972	9.89747	1.26627	10.10252
SiO_4	SiO_2	0.65250	9.81458	1.53257	10.18542
Si_2O	SiO_2	1.66505	10.22142	0.60058	9.77857
$Si(OH)_4$	SiO_2	0.62514	9.79598	1.59964	10.20402
Silver					
Ag	AgBr	1.74079	10.24075	0.57445	9.75925
	AgCl	1.32866	10.12341	0.75264	9.87659
	AgCN	1.24120	10.09384	0.80567	9.90616
	AgI	2.17645	10.33775	0.45946	9.66225
	$AgNO_3$	1.57481	10.19723	0.63500	9.80277
	Ag_2O	1.07416	10.03107	0.93096	9.96893
	Ag_3PO_4	1.29347	10.11176	0.77311	9.88824

Weighed	Sought	Factor	Log of Factor +10	Reciprocal of Factor	Log of Reciprocal of Factor +10
Silver (contd.)					
Ag	$Ag_4P_2O_7$	1.40313	10.14710	0.71269	9.85290
	Br	0.74079	9.86970	1.34991	10.13030
	Cl	0.32866	9.51675	3.04266	10.48325
	I	1.17646	10.07058	0.85001	9.92942
AgBr	Ag	0.57445	9.75925	1.74080	10.24075
	Br	0.42555	9.62895	2.34990	10.37105
AgCl	Ag	0.75264	9.87659	1.32866	10.12342
	$AgNO_3$	1.18526	10.07382	0.84370	9.92619
	Ag_2O	0.80845	9.90765	1.23693	10.09235
	Br	0.55754	9.74628	1.79359	10.25372
	Cl	0.24736	9.39333	4.04269	10.60667
AgCN	Ag	0.80567	9.90616	1.24120	10.09384
Ag_2CrO_4	Ag	0.65034	9.81314	1.53766	10.18686
AgI	Ag	0.45946	9.66225	2.17647	10.33775
	I	0.54054	9.73283	1.85000	10.26717
$AgNO_3$	Ag	0.63499	9.80277	1.57483	10.19723
	AgCl	0.84370	9.92619	1.18526	10.07381
Ag_2O	Ag	0.93096	9.96893	1.07416	10.03107
	AgCl	1.23693	10.09235	0.80845	9.90765
Ag_3PO_4	Ag	0.77311	9.88824	1.29348	10.11176
$Ag_4P_2O_7$	Ag	0.71269	9.85290	1.40313	10.14710
Br	Ag	1.34991	10.13030	0.74079	9.86970
	AgBr	2.34991	10.37105	0.42555	9.62895
	AgCl	1.79358	10.25372	0.55754	9.74628
Cl	Ag	3.04261	10.48325	0.32867	9.51676
	AgCl	4.04262	10.60666	0.24736	9.39333
I	Ag	0.85000	9.92942	1.17647	10.07058
	AgI	1.85000	10.26717	0.54054	9.73283
Sodium					
Ag	NaBr	0.95392	9.97951	1.04831	10.02049
	NaCl	0.54179	9.73383	1.84573	10.26617
	NaI	1.38958	10.14288	0.71964	9.85712
AgBr	NaBr	0.54798	9.73876	1.82488	10.26123
AgCl	NaCl	0.40777	9.61042	2.45236	10.38959
	$NaClO_3$	0.74267	9.87080	1.34649	10.21921
	$NaClO_4$	0.85429	9.93161	1.17056	10.06839
AgI	NaI	0.63846	9.80513	1.56627	10.19487
$BaSO_4$	$NaHSO_4$	0.51439	9.71129	1.94405	10.28871
	$NaHSO_4.H_2O$	0.59157	9.77201	1.69042	10.22799
	Na_2S	0.33438	9.52424	2.99061	10.47577
	Na_2SO_3	0.54002	9.73241	1.85178	10.26759
	$Na_2SO_3.7H_2O$	1.08032	10.03355	0.92565	9.96645
	Na_2SO_4	0.60857	9.78431	1.64320	10.21569
	$Na_2SO_4.10H_2O$	1.38043	10.14001	0.72441	9.85998
B_2O_3	$Na_2B_4O_7$	1.44511	10.15990	0.69199	9.84010
	$Na_2B_4O_7.10H_2O$	2.73895	10.43758	0.36510	9.56241
Br	Na	0.28770	9.45894	3.47584	10.54106
	NaBr	1.28770	10.10981	0.77658	9.89019
	Na_2O	0.38781	9.58862	2.57858	10.41138
$CaCl_2$	NaCl	1.05321	10.02252	0.94948	9.97749
$CaCO_3$	Na_2CO_3	1.05894	10.02488	0.94434	9.97513
CaF_2	NaF	1.07551	10.03161	0.92979	9.96838
CaO	Na_2CO_3	1.88996	10.27645	0.52911	9.72355
$CaSO_4$	Na_2CO_3	0.77853	9.89128	1.28447	10.10873
Cl	Na	0.64846	9.81188	1.54212	10.18812
	NaCl	1.64846	10.21708	0.60663	9.78292
	Na_2O	0.87410	9.94156	1.14403	10.05844
CO_2	Na_2CO_3	2.40829	10.38171	0.41523	9.61829
	Na_2O	1.40829	10.14869	0.71008	9.85131
H_3BO_3	$Na_2B_4O_7$	0.81356	9.91039	1.22917	10.08961
	$Na_2B_4O_7.10H_2O$	1.54195	10.18807	0.64853	9.81193
I	Na	0.18116	9.25806	5.51998	10.74194
	NaI	1.18115	10.07231	0.84663	9.92769
	Na_2O	0.24419	9.38773	4.09517	10.61227
KBF_4	$Na_2B_4O_7$	0.39954	9.60156	2.50288	10.39844
	$Na_2B_4O_7.10H_2O$	0.75725	9.87924	1.32057	10.12076
$Mg_2As_2O_7$	Na_2HAsO_3	1.09454	10.03923	0.91363	9.96077
	Na_2HAsO_4	1.19761	10.07832	0.83500	9.92169

Weighed	Sought	Factor	Log of Factor +10	Reciprocal of Factor	Log of Reciprocal of Factor +10
Sodium (contd.)					
$MgCl_2$	NaCl	1.22756	10.08904	0.81462	9.91096
$Mg_2P_2O_7$	Na_2HPO_4	1.27565	10.10573	0.78391	9.89427
	$Na_2HPO_4.12H_2O$	3.21828	10.50763	0.31072	9.49237
	$NaNH_4HPO_4.4H_2O$	1.87870	10.27386	0.53228	9.72614
	Na_3PO_4	1.47318	10.16825	0.67880	9.83174
	$Na_4P_2O_7.10H_2O$	2.00414	10.30193	0.49897	9.69807
N	$NaNO_3$	6.06815	10.78306	0.16479	9.21693
Na	Br	3.47585	10.54106	0.28770	9.45894
	Cl	1.54212	10.18811	0.64846	9.81188
	I	5.52003	10.74194	0.18116	9.25806
	NaBr	4.47585	10.65088	0.22342	9.34912
	NaCl	2.54213	10.40520	0.39337	9.59480
	Na_2CO_3	2.30513	10.36269	0.43382	9.63731
	$NaHCO_3$	3.65410	10.56278	0.27367	9.43723
	NaI	6.52003	10.81425	0.15337	9.18574
	Na_2O	1.34797	10.12968	0.74186	9.87032
	Na_2SO_4	3.08921	10.48985	0.32371	9.51016
$Na_2B_4O_7$	B_2O_3	0.69198	9.84009	1.44513	10.15991
	H_3BO_3	1.22916	10.08961	0.81356	9.91039
	KBF_4	2.50287	10.39844	0.39954	9.60156
$Na_2B_4O_7.10H_2O$	B_2O_3	0.36510	9.56241	2.73898	10.43759
	H_3BO_3	0.64853	9.81193	1.54195	10.18807
	KBF_4	1.32057	10.12076	0.75725	9.87924
NaBr	Ag	1.04831	10.02048	0.95392	9.97951
	AgBr	1.82489	10.26123	0.54798	9.73876
	Br	0.77658	9.89019	1.28770	10.10981
	Na	0.22342	9.34912	4.47588	10.65088
	Na_2O	0.30116	9.47880	3.32049	10.52120
$NaC_2H_3O_2.Mg(C_2H_3O_2)_2.$					
$3UO_2(C_2H_3O_2)_2.6\frac{1}{2}H_2O$	Na	0.01527	8.18375	65.50075	11.81625
	NaBr	0.06833	8.83464	14.63400	11.16536
	Na_2CO_3	0.03519	8.54646	28.41474	11.45354
	NaCl	0.03881	8.58895	25.76589	11.41104
	NaF	0.02788	8.44536	35.86286	11.55465
	$NaHCO_3$	0.05579	8.74654	17.92500	11.25346
	NaI	0.09954	8.99801	10.04590	11.00199
	$NaNO_3$	0.05644	8.75162	17.71667	11.24837
	Na_2O	0.02058	8.31345	48.59086	11.68655
	NaOH	0.02656	8.42426	37.64746	11.57574
	Na_2SO_4	0.04716	8.67361	21.20261	11.32638
$NaC_2H_3O_2.Zn(C_2H_3O_2)_2.$					
$3UO_2(C_2H_3O_2)_2.6\frac{1}{2}H_2O$	Na	0.01495	8.17461	66.89410	11.85239
	NaBr	0.06691	8.82549	14.94545	11.17450
	Na_2CO_3	0.03446	8.53730	29.01999	11.46270
	NaCl	0.03800	8.57981	26.31440	11.42019
	NaF	0.02733	8.43621	36.62601	11.56379
	$NaHCO_3$	0.05466	8.73739	18.30663	11.26261
	NaI	0.09747	8.98886	10.25977	11.01114
	$NaNO_3$	0.05527	8.74247	18.09398	11.25753
	NaOH	0.02601	8.41511	38.44970	11.58489
	Na_2O	0.02015	8.30430	49.62532	11.69570
	Na_2SO_4	0.04618	8.66446	21.65392	11.33554
Na_2CO_3	$CaCO_3$	0.94434	9.97513	1.05894	10.02488
	CaO	0.52911	9.72355	1.88997	10.27645
	$CaSO_4$	1.28447	10.10873	0.77853	9.89128
	CO_2	0.41523	9.61829	2.40830	10.38171
	Na	0.43381	9.63730	2.30516	10.36270
	NaCl	1.10281	10.04250	0.90677	9.95750
	$NaHCO_3$	1.58520	10.20008	0.63084	9.79992
	NaOH	0.75474	9.87780	1.32496	10.12220
	Na_2O	0.58477	9.76699	1.71007	10.23301
	Na_2SO_4	1.34015	10.12715	0.74619	9.87285
$Na_2CO_3.10H_2O$	Na_2SO_4	0.49639	9.69582	2.01455	10.30417
NaCl	Ag	1.84573	10.26617	0.54179	9.73383
	AgCl	2.45236	10.38958	0.40777	9.61042
	Cl	0.60663	9.78292	1.64845	10.21707
	Na	0.39337	9.59480	2.54214	10.40520
	$NaClO_3$	1.82129	10.26038	0.54906	9.73962

Weighed	Sought	Factor	Log of Factor +10	Reciprocal of Factor	Log of Reciprocal of Factor +10
Sodium (contd.)	$NaClO_4$	2.09503	10.32118	0.47732	9.67881
	Na_2CO_3	0.90677	9.95750	1.10282	10.04251
	$NaHCO_3$	1.43742	10.15759	0.69569	9.84242
	Na_2HPO_4	1.21451	10.08440	0.82338	9.91560
	Na_2O	0.53025	9.72448	1.88590	10.27552
$NaCl$	Na_2SO_4	1.21521	10.08465	0.82290	9.91535
$NaClO_3$	$AgCl$	1.34650	10.12921	0.74267	9.87080
	$NaCl$	0.54906	9.73961	1.82129	10.26038
$NaClO_4$	$AgCl$	1.17055	10.06839	0.85430	9.93161
	$NaCl$	0.47732	9.67881	2.09503	10.32119
NaF	CaF_2	0.92979	9.96838	1.07551	10.03161
Na_2HAsO_3	$Mg_2As_2O_7$	0.91362	9.96077	1.09455	10.03923
Na_2HAsO_4	$Mg_2As_2O_7$	0.83499	9.92168	1.19762	10.07832
$NaHCO_3$	Na	0.27366	9.43721	3.65417	10.56279
	$NaCl$	0.69569	9.84242	1.43742	10.15759
	Na_2CO_3	0.63083	9.79991	1.58521	10.20009
	Na_2O	0.36889	9.56690	2.71084	10.43310
Na_2HPO_4	$Mg_2P_2O_7$	0.78392	9.89427	1.27564	10.10573
	$NaCl$	0.82338	9.91560	1.21451	10.08440
	Na_2O	0.43660	9.64008	2.29043	10.35992
	$Na_4P_2O_7$	0.93654	9.97153	1.06776	10.02847
	P_2O_5	0.49994	9.69892	2.00024	10.30108
$Na_2HPO_4.12H_2O$	$Mg_2P_2O_7$	0.31072	9.49237	3.21833	10.50763
	$Na_4P_2O_7$	0.37122	9.56963	2.69382	10.43037
	P_2O_5	0.19816	9.29702	5.04643	10.70298
$NaHSO_3$	SO_2	0.61564	9.78933	1.62433	10.21076
$NaHSO_4$	$BaSO_4$	1.94404	10.28871	0.51439	9.71129
$NaHSO_4.H_2O$	$BaSO_4$	1.69039	10.22799	0.59158	9.77201
NaI	Ag	0.71964	9.85712	1.38958	10.14288
	AgI	1.56626	10.19486	0.63846	9.80513
	I	0.84662	9.92769	1.18117	10.07231
	Na	0.15337	9.18574	6.52018	10.81426
	Na_2O	0.20674	9.31542	4.83699	10.68458
$NaNH_4HPO_4.4H_2O$	$Mg_2P_2O_7$	0.53228	9.72614	1.87871	10.27386
	NH_3	0.08146	8.91094	12.27596	11.08905
	P_2O_5	0.33946	9.53079	2.94586	10.46921
$NaNO_3$	N	0.16479	9.21693	6.06833	10.78307
	Na_2O	0.36461	9.56183	2.74266	10.43818
	NH_3	0.20037	9.30183	4.99077	10.69817
	NO	0.35303	9.54781	2.83262	10.45219
	N_2O_5	0.63539	9.80304	1.57384	10.19696
Na_2O	Br	2.57854	10.41137	0.38782	9.58863
	Cl	1.14401	10.05843	0.87412	9.94157
	CO_2	0.71008	9.85131	1.40829	10.14869
	I	4.09501	10.61225	0.24420	9.38775
	Na	0.74185	9.87032	1.34798	10.12968
	$NaBr$	3.32039	10.52119	0.30117	9.47881
	$NaCl$	1.88587	10.27551	0.53026	9.72449
	Na_2CO_3	1.71008	10.23302	0.58477	9.76699
	$NaHCO_3$	2.71078	10.43310	0.36890	9.56691
	Na_2HPO_4	2.29044	10.35992	0.43660	9.64008
	NaI	4.83686	10.68457	0.20675	9.31545
	$NaNO_3$	2.74265	10.43817	0.36461	9.56183
	$NaOH$	1.29065	10.11081	0.77480	9.88919
	Na_2SO_4	2.29176	10.36017	0.43635	9.63984
	N_2O_5	1.74269	10.24122	0.57383	9.75878
	SO_3	1.29176	10.11118	0.77414	9.88882
$NaOH$	Na_2CO_3	1.32495	10.12220	0.75475	9.87780
	Na_2O	0.77479	9.88918	1.29067	10.11082
$Na_4P_2O_7$	Na_2HPO_4	1.06775	10.02847	0.93655	9.97153
	$Na_2HPO_4.12H_2O$	0.26938	9.43037	3.71223	10.56963
$Na_4P_2O_7.10H_2O$	$Mg_2P_2O_7$	0.49897	9.69807	2.00413	10.30193
Na_2S	$BaSO_4$	2.99062	10.47576	0.33438	9.52424
Na_2SO_3	$BaSO_4$	1.85176	10.26758	0.54003	9.73242
	SO_2	0.50827	9.70609	1.96746	10.29391
$Na_2SO_3.7H_2O$	$BaSO_4$	0.92564	9.96644	1.08033	10.03356
	SO_2	0.25407	9.40495	3.93592	10.59505
Na_2SO_4	$BaSO_4$	1.64319	10.21569	0.60857	9.78431
	Na	0.32370	9.51014	3.08928	10.48986

Weighed	Sought	Factor	Log of Factor +10	Reciprocal of Factor	Log of Reciprocal of Factor +10
Sodium (contd.)					
	NaCl	0.82289	9.91534	1.21523	10.08466
	Na_2CO_3	0.74619	9.87285	1.34014	10.21715
	$Na_2CO_3.10H_2O$	2.01451	10.30417	0.49640	9.69583
	Na_2O	0.43635	9.63984	2.29174	10.36016
$Na_2SO_4.10H_2O$	SO_3	0.56365	9.75101	1.77415	10.24899
	$BaSO_4$	0.72441	9.85998	1.38043	10.14001
$Na_2U_2O_7.?ZnU_2O_7$	Na	0.02369	8.37457	42.21190	11.62544
	Na_2O	0.03193	8.50420	31.31851	11.49579
NH_3	$NaNH_4HPO_4.4H_2O$	12.27607	11.08906	0.08146	8.91094
	$NaNO_3$	4.99070	10.69816	0.20037	9.30183
NO	$NaNO_3$	2.83258	10.45218	0.35304	9.54782
N_2O_5	$NaNO_3$	1.57382	10.19695	0.63540	9.80305
	Na_2O	0.57382	9.75878	1.74271	10.24123
P_2O_5	Na_2HPO_4	2.00020	10.30107	0.49995	9.69893
	$Na_2HPO_4.12H_2O$	5.04623	10.70297	0.19817	9.29704
	$NaNH_4HPO_4.4H_2O$	2.94578	10.46920	0.33947	9.53080
SO_2	$NaHSO_3$	1.62442	10.21070	0.61560	9.78930
	Na_2SO_3	1.96756	10.29393	0.50824	9.70607
SO_3	Na_2O	0.77414	9.88882	1.29176	10.11118
	Na_2SO_4	1.77414	10.24899	0.56365	9.75101
Strontium					
CO_2	$SrCO_3$	3.35446	10.52562	0.29811	9.47438
SO_3	SrO	1.29424	10.11202	0.77265	9.88798
	$SrSO_4$	2.29422	10.36063	0.43588	9.63937
Sr	$SrCO_3$	1.68489	10.22657	0.50351	9.77343
	$Sr(NO_3)_2$	2.41532	10.38298	0.41402	9.61702
	SrO	1.18261	10.07284	0.84559	9.92716
	$SrSO_4$	2.09633	10.32146	0.47702	9.67854
$SrCl_2$	$SrCO_3$	0.93124	9.96900	1.07384	10.03094
	SrO	0.65363	9.81533	1.52992	10.18467
	$SrSO_4$	1.15865	10.06395	0.86307	9.93605
$SrCO_3$	CO_2	0.29811	9.47438	3.35447	10.52562
	Sr	0.59351	9.77343	1.68489	10.22657
	$SrCl_2$	1.07383	10.03095	0.93125	9.96907
	$Sr(HCO_3)_2$	1.42010	10.15232	0.70418	9.84768
	$Sr(NO_3)_2$	1.43352	10.15640	0.69758	9.84359
	SrO	0.70189	9.84627	1.42472	10.15373
	$SrSO_4$	1.24419	10.09489	0.80374	9.90511
SrC_2O_4	Sr	0.49886	9.69798	2.00457	10.30202
	SrO	0.58996	9.77082	1.69503	10.22918
$Sr(HCO_3)_2$	$SrCO_3$	0.70417	9.84768	1.42011	10.15232
	SrO	0.49425	9.69395	2.02327	10.30602
$Sr(IO_3)_2$	Sr	0.20031	9.30170	4.99226	10.69829
	SrO	0.23688	9.37453	4.22155	10.69829
$Sr(NO_3)_2$	Sr	0.41402	9.61702	2.41434	10.38298
	$SrCO_3$	0.69759	9.84360	1.43351	10.15640
	SrO	0.48963	9.68987	2.04236	10.31014
	$SrSO_4$	0.86793	9.93849	1.15217	10.06151
SrO	SO_3	0.77265	9.88798	1.29425	10.11202
	Sr	0.84559	9.92716	1.18261	10.07284
	$SrCl_2$	1.52992	10.18467	0.65363	9.81533
	$SrCO_3$	1.42472	10.15373	0.70189	9.84627
	$Sr(HCO_3)_2$	2.02326	10.30605	0.49425	9.69395
	$Sr(NO_3)_2$	2.04237	10.31014	0.48963	9.68987
	$SrSO_4$	1.77263	10.24862	0.56413	9.75138
$SrSO_4$	SO_3	0.43588	9.63937	2.29421	10.36063
	Sr	0.47703	9.67855	2.09630	10.32145
	$SrCl_2$	0.86308	9.93605	1.15864	10.06395
	$SrCO_3$	0.80373	9.90511	1.24420	10.09489
	$Sr(NO_3)_2$	1.15217	10.06151	0.86793	9.93848
	SrO	0.56413	9.75138	1.77264	10.24862
Sulfur					
As_2S_3	H_2S	0.41555	9.61862	2.40645	10.38138
	S	0.39097	9.59214	2.55774	10.40786
$BaSO_4$	FeS_2	0.25702	9.40997	3.89077	10.59003
	H_2S	0.14602	9.16441	6.84838	10.83559
	H_2SO_3	0.35166	9.54612	2.84366	10.45388
	H_2SO_4	0.42021	9.62347	2.37976	10.37653

Weighed	Sought	Factor	Log of Factor +10	Reciprocal of Factor	Log of Reciprocal of Factor +10
Sulfur (contd.)					
	S	0.13738	9.13792	7.27908	10.86208
	SO$_2$	0.27448	9.43851	3.64325	10.56149
	SO$_3$	0.34302	9.53532	2.91528	10.46468
	SO$_4$	0.41158	9.61445	2.42966	10.38555
C$_{12}$H$_{12}$N$_2$.H$_2$SO$_4$	S	0.11357	9.05526	8.80514	10.94474
	SO$_4$	0.34026	9.53181	2.93893	10.46819
CdS	H$_2$S	0.23591	9.37275	4.23890	10.62725
	S	0.22196	9.34627	4.50532	10.65372
FeS$_2$	BaSO$_4$	3.89081	10.59004	0.25702	9.40997
H$_2$S	As$_2$S$_3$	2.40646	10.38138	0.41555	9.61862
	BaSO$_4$	6.84859	10.83560	0.14602	9.16441
	CdS	4.23885	10.62725	0.23591	9.37275
	SO$_3$	2.34924	10.37093	0.42567	9.62907
H$_2$SO$_3$	BaSO$_4$	2.84363	10.45287	0.35166	9.54612
H$_2$SO$_4$	BaSO$_4$	2.37974	10.37653	0.42021	9.62347
	(NH$_4$)$_2$SO$_4$	1.34728	10.12946	0.74224	9.87054
	SO$_3$	0.81631	9.91186	1.22502	10.08814
(NH$_4$)$_2$SO$_4$	H$_2$SO$_4$	0.74223	9.87054	1.34729	10.12946
	SO$_3$	0.60589	9.78239	1.65046	10.21760
S	As$_2$S$_3$	2.55775	10.40786	0.39097	9.59214
	BaSO$_4$	7.27919	10.86208	0.13738	9.13792
	CdS	4.50536	10.65373	0.22196	9.34627
SO$_2$	BaSO$_4$	3.64329	10.56149	0.27448	9.43851
SO$_2$	BaSO$_4$	2.91524	10.46468	0.34302	9.53532
	H$_2$S	0.42567	9.62907	2.34924	10.37093
	(NH$_4$)$_2$SO$_4$	1.65047	10.21761	0.60589	9.78239
SO$_4$	BaSO$_4$	2.42968	10.38555	0.41158	9.61445
Tantalum					
Ta	TaCl$_5$	1.97965	10.29659	0.50514	9.70341
	Ta$_2$O$_5$	1.22105	10.08674	0.81897	9.91327
TaC	C	0.06225	8.79413	16.06451	11.20587
	Ta	0.93775	9.97209	1.06638	10.02791
TaCl$_5$	Ta	0.50514	9.70341	1.97965	10.29659
	Ta$_2$O$_5$	0.61680	9.79041	1.62127	10.20985
Ta$_2$O$_4$	Ta$_2$O$_5$	1.03757	10.01602	0.96379	9.98398
Ta$_2$O$_5$	Ta	0.81897	9.91327	1.22105	10.08673
	TaCl$_5$	1.62126	10.20985	0.61680	9.79014
	Ta$_2$O$_4$	0.96379	9.98398	1.03757	10.01602
Tellurium					
H$_2$TeO$_4$	Te	0.65906	9.81893	1.51731	10.18108
H$_2$TeO$_4$.2H$_2$O	Te	0.55565	9.74480	1.79969	10.25520
Te	H$_2$TeO$_4$	1.51732	10.18108	0.65906	9.81893
	H$_2$TeO$_4$.2H$_2$O	1.79969	10.25520	0.55565	9.74480
	TeO$_2$	1.25078	10.09718	0.79950	9.90282
	TeO$_3$	1.37618	10.13868	0.72665	9.86133
	(TeO$_2$)$_2$SO$_3$	1.56450	10.19438	0.63918	9.80562
TeO$_2$	Te	0.79950	9.90282	1.25078	10.09718
TeO$_3$	Te	0.72665	9.86133	1.37618	10.13868
(TeO$_2$)$_2$SO$_3$	Te	0.63918	9.80562	1.56450	10.19438
Terbium					
Tb$_4$O$_7$	Tb	0.85021	9.92953	1.17618	10.07048
Thallium					
(C$_6$H$_5$)$_4$AsTlCl$_4$	Tl	0.28014	9.44738	3.56964	10.55263
Tl	TlCl	1.17346	10.06947	0.85218	9.93053
	Tl$_2$CO$_3$	1.14682	10.05949	0.87198	9.94051
	Tl$_2$CrO$_4$	1.28377	10.10849	0.77896	9.89151
	TlHSO$_4$	1.47497	10.16878	0.67798	9.83122
	TlI	1.62093	10.20976	0.61693	9.79024
	TlNO$_3$	1.30337	10.11507	0.76724	9.88493
	Tl$_2$O	1.03914	10.01668	0.96233	9.98332
	Tl$_2$PtCl$_6$	1.99772	10.30055	0.50057	9.69946
	Tl$_2$SO$_4$	1.23501	10.09167	0.80971	9.90833
TlCl	Tl	0.85218	9.93053	1.17346	10.06947
	Tl$_2$PtCl$_6$	1.70240	10.23106	0.58741	9.76894
Tl$_2$CO$_3$	Tl	0.87198	9.94051	1.14682	10.05951
	Tl$_2$PtCl$_6$	1.74197	10.24104	0.57406	9.75896
Tl$_2$CrO$_4$	Tl	0.77895	9.89151	1.28378	10.10850
TlHSO$_4$	Tl	0.67798	9.83122	1.47497	10.16878

Weighed	Sought	Factor	Log of Factor +10	Reciprocal of Factor	Log of Reciprocal of Factor +10
Thallium (contd.)					
TlI	Tl	0.61693	9.79024	1.62093	10.20976
	Tl_2PtCl_6	1.23244	10.09076	0.81140	9.90924
$TlNO_3$	Tl	0.76724	9.88493	1.30338	10.11507
	Tl_2PtCl_6	1.53272	10.18546	0.65243	9.81453
Tl_2O	Tl	0.96223	9.98332	1.03914	10.01668
	Tl_2PtCl_6	1.92247	10.28386	0.52016	9.71614
Tl_2PtCl_6	Tl	0.50057	9.69946	1.99772	10.30054
	TlCl	0.58741	9.76894	1.70239	10.23106
	Tl_2CO_3	0.57406	9.75896	1.74198	10.24105
	TlI	0.81140	9.90924	1.23244	10.09076
	$TlNO_3$	0.65244	9.81454	1.53271	10.18546
	Tl_2O	0.52016	9.71614	1.92249	10.28386
	Tl_2SO_4	0.61821	9.79114	1.61757	10.20886
Tl_2SO_4	Tl	0.80971	9.90833	1.23501	10.09167
	Tl_2PtCl_6	1.61757	10.20886	0.61821	9.79114
Thorium					
Th	ThO_2	1.13790	10.05610	0.87881	9.94390
$Th(C_9H_6NO)_4.$					
(C_9H_6NOH)	Th	0.24327	9.38609	4.11066	10.61391
	ThO_2	0.27682	9.44220	3.61246	10.55781
$ThCl_4$	ThO_2	0.70626	9.84896	1.41590	10.15103
$Th(NO_3)_4.6H_2O$	ThO_2	0.44898	9.65223	2.22727	10.34777
ThO_2	Th	0.87881	9.94390	1.13790	10.05610
	$ThCl_4$	1.41590	10.15103	0.70626	9.84896
	$Th(NO_3)_4.6H_2O$	2.22729	10.34778	0.44898	9.65223
Thulium					
Tm_2O_3	Tm	0.87561	9.94231	1.14206	10.05769
Tin					
Sn	$SnCl_2$	1.59744	10.20342	0.62600	9.79657
	$SnCl_2.2H_2O$	1.90100	10.27898	0.52604	9.72102
	$SnCl_4$	2.19479	10.34140	0.45562	9.65860
	$SnCl_4.(NH_4Cl)_2$	3.09622	10.49083	0.32297	9.50916
	SnO	1.13480	10.05492	0.88121	9.94508
	SnO_2	1.26961	10.10367	0.78764	9.89633
$SnCl_2$	Sn	0.62600	9.79657	1.59744	10.20342
	SnO_2	0.79478	9.90025	1.25821	10.09975
$SnCl_2.2H_2O$	Sn	0.52604	9.72102	1.90100	10.27898
	SnO_2	0.66786	9.82469	1.49732	10.17531
$SnCl_4$	Sn	0.45562	9.65860	2.19481	10.34139
	SnO_2	0.57846	9.76227	1.72873	10.23772
$SnCl_4.(NH_4Cl)_2$	Sn	0.32297	9.50916	3.09626	10.49084
	SnO_2	0.41005	9.61284	2.43873	10.38716
SnO	Sn	0.88121	9.94508	1.13480	10.05491
	SnO_2	1.11879	10.04875	0.89382	9.95125
SnO_2	Sn	0.78764	9.89633	1.26962	10.10367
	$SnCl_2$	1.25821	10.09976	0.79474	9.90023
	$SnCl_2.2H_2O$	1.49731	10.17531	0.66786	9.82469
	$SnCl_4$	1.72871	10.23772	0.57847	9.76228
	$SnCl_4.(NH_4Cl)_2$	2.43872	10.38716	0.41005	9.61284
	SnO	0.89382	9.95125	1.11879	10.04875
Titanium					
K_2TiF_6	F	0.47472	9.67644	2.10650	10.32356
	K	0.32573	9.51286	3.07003	10.48714
	Ti	0.19951	9.29996	5.01228	10.70004
	TiO_2	0.33279	9.52217	3.00490	10.47783
Ti	K_2TiF_6	5.01232	10.70004	0.19951	9.29996
	TiC	1.25073	10.09717	0.79953	9.90283
	TiO_2	1.66806	10.22222	0.59950	9.77779
TiC	C	0.20049	9.30210	4.98778	10.69791
	Ti	0.79953	9.90283	1.25073	10.09717
$TiO(C_9H_4NOCl_2)_2$	Ti	0.09776	8.99016	10.22913	11.00984
	TiO_2	0.16306	9.21235	6.13271	10.78765
$TiO(C_9H_6ON)_2$	Ti	0.13600	9.13354	7.35294	10.86646
	TiO_2	0.22685	9.35574	4.40820	10.64426
TiO_2	K_2TiF_6	3.00488	10.47782	0.33279	9.52217
	Ti	0.59950	9.77779	1.66806	10.22221
	TiC	0.74969	9.87488	1.33388	10.12512

Weighed	Sought	Factor	Log of Factor +10	Reciprocal of Factor	Log of Reciprocal of Factor +10
Tungsten					
$FeWO_4$	W	0.60539	9.78203	1.65183	10.21797
	WO_3	0.76344	9.88277	1.30986	10.11722
$MgWO_4$	W	0.67552	9.82964	1.48034	10.17036
	WO_3	0.85189	9.93038	1.17386	10.06962
$MnWO_4$	W	0.60719	0.78332	1.64693	10.21668
	WO_3	0.76571	9.88406	1.30598	10.11593
$PbWO_4$	W	0.40403	9.60641	2.47506	10.39359
	WO_3	0.50952	9.70716	1.96263	10.29284
W	W_2C	1.03266	10.01396	0.96837	9.98604
	WC	1.06532	10.02748	0.93869	9.97252
	WO_2	1.17405	10.06969	0.85175	9.93031
	WO_3	1.26108	10.10075	0.79297	9.89925
W_2C	C	0.03163	8.50010	31.61555	11.49990
	W	0.96837	9.98604	1.03266	10.01396
WC	C	0.06133	8.78767	16.30523	11.21233
	W	0.93868	9.97252	1.06533	10.02749
WO_2	W	0.85175	9.93031	1.17405	10.06969
$WO_2(C_9H_6ON)_2$	W	0.36467	9.56190	2.74221	10.43810
	WO_3	0.45987	9.66264	2.17453	10.33736
WO_3	W	0.79297	9.89926	1.26108	10.10074
Uranium					
U	UO_2	1.13444	10.05478	0.88149	9.94522
	U_3O_8	1.17925	10.07160	0.84800	9.92840
	$U_2P_2O_{11}$	1.49981	10.17604	0.66675	9.82396
UO_2	U	0.88149	9.94522	1.13444	10.05478
	U_3O_8	1.03950	10.01682	0.96200	9.98318
	$U_2P_2O_{11}$	1.32268	10.12145	0.75604	9.87854
$UO_2(C_9H_6ON)_2 \cdot (C_9H_7ON)$	U	0.33835	9.52937	2.95552	10.47063
	UO_2	0.38384	9.58415	2.60525	10.41585
U_3O_8	U	0.84799	9.92839	1.17926	10.07161
	UO_2	0.96199	9.98317	1.03951	10.01683
	$UO_2(NO_3)_2 \cdot 6H_2O$	1.78864	10.25252	0.55908	9.74747
$UO_2(NO_3)_2 \cdot 6H_2O$	U_3O_8	0.55909	9.74748	1.78862	10.25252
$U_2P_2O_{11}$	U	0.66675	9.82396	1.49981	10.17603
	UO_2	0.75639	9.87875	1.32207	10.12125
Vanadium					
V	VC	1.23578	10.09194	0.80921	9.90806
	V_2O_5	1.78518	10.25168	0.56017	9.74832
VC	C	0.19080	9.28058	5.24109	10.71942
	V	0.80921	9.90806	1.23577	10.09194
VO_4	V_2O_5	0.79120	9.89829	1.26390	10.10171
V_2O_5	V	0.56017	9.74832	1.78517	10.25168
	VC	0.69225	9.84026	1.44456	10.15974
	VO_4	1.26390	10.10171	0.79120	9.89829
Ytterbium					
Yb	YbO_3	1.13870	10.05641	0.87819	9.94359
Yb_2O_3	Yb	0.87820	9.94359	1.13869	10.05640
Yttrium					
Y	Y_2O_3	1.26994	10.10378	0.78744	9.89622
Y_2O_3	Y	0.78744	9.89622	1.26994	10.10378
Zinc					
$BaSO_4$	ZnS	0.41744	9.62059	2.39555	10.37941
	$ZnSO_4 \cdot 7H_2O$	1.23196	10.09059	0.81171	9.90940
Zn	$ZnNH_4PO_4$	2.72877	10.43596	0.36647	9.56404
	ZnO	1.24476	10.09508	0.80337	9.90492
	$Zn_2P_2O_7$	2.33043	10.36744	0.42911	9.63257
	ZnS	1.49044	10.17332	0.67094	9.82668
$Zn(C_9H_6NO)_2$	Zn	0.18483	9.26677	5.41038	10.73323
	ZnO	0.23007	9.36186	4.34650	10.63814
$ZnCl_2$	ZnO	0.59708	9.77603	1.67482	10.22397
$ZnCO_3$	ZnO	0.64899	9.81224	1.54086	10.18777
$(Zn[C_5H_5N]_2) \cdot (SCN)_2$	Zn	0.19241	9.28423	5.19724	10.71577
	ZnO	0.23951	9.37932	4.17519	10.62068
$ZnNH_4PO_4$	Zn	0.36646	9.56403	2.72881	10.43597
	ZnO	0.45616	9.65912	2.19221	10.34089
ZnO	Zn	0.80337	9.90492	1.24476	10.09509
	$ZnCl_2$	1.67482	10.22397	0.59708	9.77603

Weighed	Sought	Factor	Log of Factor +10	Reciprocal of Factor	Log of Reciprocal of Factor +10
Zinc (contd.)					
	$ZnCO_3$	1.54086	10.18776	0.64899	9.81224
	$ZnNH_4PO_4$	2.19221	10.34088	0.45616	9.65912
	$Zn_2P_2O_7$	1.87219	10.27235	0.53413	9.72765
	ZnS	1.19737	10.07823	0.83516	9.92177
	$ZnSO_4.7H_2O$	3.53373	10.54823	0.28299	9.45177
$Zn_2P_2O_7$	Zn	0.42911	9.63257	2.33040	10.36743
	ZnO	0.53413	9.72765	1.87220	10.27235
ZnS	$BaSO_4$	2.39556	10.37941	0.41744	9.62059
	Zn	0.67094	9.82668	1.49045	10.17332
	ZnO	0.83516	9.92177	1.19738	10.07822
	$ZnSO_4.7H_2O$	2.95125	10.47001	0.33884	9.52999
$ZnSO_4.7H_2O$	$BaSO_4$	0.81171	9.90940	1.23197	10.09060
	ZnO	0.28299	9.45177	3.53369	10.54823
	ZnS	0.33884	9.52999	2.95125	10.47001
Zirconium					
K_2ZrF_6	Zr	0.32187	9.50768	3.10684	10.49232
	ZrF_4	0.58999	9.77084	1.69494	10.22915
	ZrF_6	0.72407	9.85978	1.38108	10.14022
	ZrO_2	0.43478	9.63827	2.30001	10.36173
Zr	ZrO_2	1.35080	10.13059	0.74030	9.86941
	ZrC	1.13166	10.05372	0.88366	9.94629
$Zr(BrC_8H_6O_3)_4$	Zr	0.09020	8.95521	11.08647	11.04479
	ZrO_2	0.12184	9.08579	8.20749	10.91421
$Zr(C_8H_7O_3)_4$	Zr	0.13110	9.11760	7.62777	10.88240
	ZrO_2	0.17709	9.24819	5.64685	10.75101
ZrC	C	0.11635	9.06577	8.59476	10.93424
	Zr	0.88366	9.94629	1.13166	10.05372
ZrO_2	Zr	0.74030	9.86941	1.35080	10.13059
	ZrC	0.83777	9.92313	1.19365	10.07687
$Zr_2P_2O_7$	Zr	0.34402	9.53658	2.90681	10.46342
	ZrO_2	0.46470	9.66717	2.15193	10.33283

PHYSICAL CONSTANTS OF MINERALS

Compiled by Ralph Kretz

The following table presents data for many of the more common minerals.

In order to avoid duplication and save space, very few cross references are given in the body of the table. If the name sought is not found in the table, consult the **synonym index** given below.

Specific gravities are given at normal atmospheric temperatures, a more precise statement being valueless considering the large variations in natural minerals.

Hardness is given in terms of Mohs' scale. (See under Hardness.)

Indices of refraction for the sodium line, λ = 5893 Å, unless otherwise indicated. Li, λ = 6708 Å. Indices will invariably be given in the order ω, ϵ or α, β, γ. Uniaxial crystals are considered positive if $\epsilon > \omega$, negative if $\omega > \epsilon$. Biaxial crystals are considered positive if β is nearer α in value than it is γ and negative if β is nearer γ than α.

ABBREVIATIONS

Abbreviation	Meaning of abbreviation	Abbreviation	Meaning of abbreviation	Abbreviation	Meaning of abbreviation
bl	blue	grn	green	rhbdr	rhombohedral
blk	black	grnsh	greenish	rhomb	rhombic
blksh	blackish	hex	hexagonal	somet	sometimes
blsh	bluish	iridesc	iridescent	tarn	tarnishes
br	brown	monocl	monoclinic	tetr	tetragonal
brnsh	brownish	oft	often	tricl	triclinic
col	colorless	pa	pale	vlt	violet
cub	cubic	purp	purple	wh	white
dk	dark	(R)	radioactive	yel	yellow
Fe	Fe, ferrous iron	redsh	redish	yelsh	yellowish
Fe^{+3}	Fe, ferric iron				

SYNONYM INDEX

Compound sought	Listed	Compound sought	Listed
Acmite	Aegirine	Lead sulfate	Anglesite
Agate	Quartz (impure)	Lead sulfide	Galena
Aluminum hydroxide	Boehmite, Diaspore, Gibbsite	Limonite	Goethite (impure)
Amphibole	Actinolite, Anthophyllite, Cummingtonite, Glaucophane, Hornblende, Riebeckite, Tremolite	Lithiophyllite	Triphylite
		Lithium mica	Lepidolite
		Lodestone	Magnetite
Antimony oxide	Senarmontite, Valentinite	Magnesium carbonate	Magnesite
Antimony sulfide	Stibnite	Magnesium hydroxide	Brucite
Arsenic oxide	Arsenolite, Claudetite	Magnesium oxide	Periclase
Arsenic sulfide	Orpiment, Realgar	Magnesium sulfate	Kieserite
Barium carbonate	Witherite	Manganese carbonate	Rhodochrosite
Barium sulfate	Barite	Manganese hydroxide	Pyrochroite
Barytes	Barite	Manganese oxide	Hausmannite, Manganosite, Pyrolusite
Bauxite	Gibbsite, Boehmite, Diaspore		
Brimstone	Sulfur	Manganese sulfide	Alabandite
Bronzite	Orthopyroxene	Meerschaum	Serpentine
Cadmium sulfide	Greenockite	Mica	Muscovite, Paragonite, Phlogopite, Biotite, Lepidolite
Calamine	Hemimorphite		
Calcium carbonate	Aragonite, Calcite, Vaterite	Native copper	Copper
Calcium sulfate	Anhydrite, Gypsum	Native gold	Gold
Calcium sulfide	Oldhamite	Nickel oxide	Bunsenite
Carborundum	Moissanite	Nickel sulfide	Millerite
Chalcedony	Quartz (impure, fibrous)	Orthite	Allanite
Chinaclay	Kaolinite	Penninite	Chlorite
Chloanthite	Skutterodite	Peridote	Olivine
Chromespinel	Chromite	Pistacite	Epidote
Chrysolite	Serpentine	Pitchblende	Uraninite
Clinoptolite	Heulandite	Plagioclase	Albite, Oligoclase, Andesine, Anorthite
Clayminerals	Illite, Kaolinite, Montmorillonite		
Clinochlore	Chlorite	Potassium chloride	Sylvite
Cobaltbloom	Erythrite	Potassium sulfate	Arcanite
Copper chloride	Nantokite	Pyroxene	Diopsite, Angite, Aegirine, Jadeite, Pigeonite, Eustatite, Orthopyroxene
Copper oxide	Cuprite		
Copper sulfide	Chalcocite, Covellite, Digenite		
Emerald	Beryl	Rocksalt	Halite
Emery	Mixture of Corundum, Magnetite and other minerals	Ruby	Corundum
		Sapphire	Corundum
Epsom salt	Epsomite	Silica	Christobalite, Quartz, Tridymite
Feldspar	Orthoclase, Microcline, Anorthoclase, Albite, Oligoclase, Andesine, Anorthite	Silver chloride	Cerargyrite
		Silver iodide	Jodyrite, Miersite
		Silver sulfide	Acanthite, Argentite
Fibrolite	Sillimanite	Smalltite	Skutterotite
Flint	Quartz (impure)	Soapstone	Mixture of Talc and other minerals
Fluorapatite	Apatite	Sodium chloride	Halite
Fluorspar	Fluorite	Sodium sulfate	Thenardite
Garnet	Almandine, Pyrope, Spessartite, Andradite, Grossularite, Uvarovite, Hydrogrossularite	Strontium carbonate	Strontianite
		Strontium sulfate	Celestite
		Thorium oxide	Thorianite
Garnierite	Serpentine (Ni-bearing)	Tin oxide	Cassiterite
Glauber salt	Mirabilite	Titanite	Sphene
Hyacinth	Zircon	Titanium oxide	Anatase, Brookite, Rutile
Iceland spar	Calcite	Uranium oxide	Uraninite
Idocrase	Vesuvianite	Zeolite	Natrolite, Mesolite, Scolecite, Thomasonite, Harmatome, Eddingtonite, Heulandite, Stilbite, Phillipsite, Chabazite, Gmelinite, Levyn, Laumontite, Mordenite
Iron carbonate	Siderite		
Iron hydroxide	Goethite, Lepidocrocite		
Iron oxide	Hematite, Magnetite		
Iron spinel	Hercynite		
Iron sulfide	Marcasite, Pyrite, Pyrrhotite	Zincblende	Sphalerite
Lapis lazuli	Lazurite	Zinc carbonate	Smithsonite
Lead carbonate	Cerussite	Zinc oxide	Zincite
Lead chloride	Cotunnite	Zinc spinel	Gahnite
Lead chromate	Crocoite	Zinc sulfide	Sphalerite, Wurtzite
Lead oxide	Litharge, Minium	Zirconium oxide	Baddeleyite

Name	Formula	Sp. gr.	Hardness	Crystalline form and color	Index of refraction (Na) η; ω ϵ α β γ
Acanthite	AgS	7.2–7.3	2–2.5	rhomb.(?), iron-blk.	
Actinolite	$Ca_2((Mg,Fe)_5Si_8O_{22}(OH,F)_2$	3.02–3.44	5–6	monocl., pa. to dk. grn.	1.599–1.688, 1.612–1.697, 1.622–1.705
Aegirine	$NaFe^{+3}Si_2O_6$	3.55–3.60	6	monocl., dk. grn. to grnsh. blk.	1.750–1.776, 1.780–1.820, 1.800–1.836
Åkermanite	$Ca_2MgSi_2O_7$	2.944	5–6	tetr., col., gray-grn., br.	1.632, 1.640
Alabandite	MnS	4.050	3.5–4	cub., iron-blk., tarn., br.	
Albite	$NaAlSi_3O_8$	2.63	6–6.5	tricl., col., wh., somet. yel., pink, grn.	1.527, 1.531, 1.538
Allanite	$(Ca,Mn,Ce,La,Y,Th)_2(Fe,Fe^{+3},Ti)(Al,Fe^{+3})_2$ $Si_3O_{12}(OH)$	3.4–4.2	5–6.5	monocl., pa. br. to blk.	1.690–1.791, 1.700–1.815, 1.706–1.828
Allemontite	AsSb	5.8–6.2	3–4	hex., tin-wh. to redsh., gray, tarn. gray-brnsh. blk.	
Almandine	$Fe_3Al_2Si_3O_{12}$	4.318	6–7.5	cub., red, dk. red, blk.	1.830
Altaite	PbTe	8.15	3	cub., tin-wh., yelsh., tarn. bronze yel.	
Aluminite	$Al_2(SO_4)(OH)_4.7H_2O$	1.66–1.82	1	monocl.(?), wh.	1.459, 1.464, 1.470
Alunite	$(K,Na)Al_3(SO_4)_2(OH)_6$	2.6–2.9	3.5–4	rhbdr., wh., gray, yel., redsh., br.	1.572, 1.592
Alunogen	$Al_2(SO_4)_3.18H_2O$	1.77	1.5–2	tricl., col., wh., yelsh. wh., redsh. wh.	1.459–1.475, 1.461–1.478, 1.470–1.485
Amblygonite	$(Li,Na)Al(PO_4)(F,OH)$	3.0–3.1	5.5–6	tricl., wh., yelsh. wh., grnsh. wh., blsh. wh., gray	1.591, 1.604, 1.613
Analcite	$NaAlSi_2O_6.H_2O$	2.24–2.29	5.5	cub., wh., pink, gray	1.479–1.493
Anatase	TiO_2	3.90	5.5–6	tetr., br., yelsh. br., redsh. br., bl., blk., grn., gray	2.5612, 2.4880
Andalusite	Al_2OSiO_4	3.13–3.16	6.5–7.5	rhomb., pink, wh., red	1.629–1.640, 1.633–1.644, 1.638–1.650
Andesine	$([Na_{0.7–0.5}[CaAl]_{0.3–0.5})AlSi_2O_8$	2.65–2.68	6–6.5	tricl., wh., gray, grn.	1.544–1.555, 1.548–1.558, 1.551–1.563
Andorite	$PbAgSb_3S_6$	5.33–5.37	3–3.5	rhomb., dk. steel gray, somet. tarn. yel. or iridesc.	
Andradite	$Ca_3Fe_2^{+3}Si_3O_{12}$	3.859	6–7.5	cub., brnsh. red, blk., somet. yel., grn.	1.887
Anglesite	$PbSO_4$	6.37–6.39	2.5–3	rho.nb., col., wh., somet. gray, yelsh., grn. tinge	1.8771, 1.8826, 1.8937
Anhydrite	$CaSO_4$	2.98	3.5	rhomb., col., blsh. wh., vlt.	1.5698, 1.5754, 1.6136
Ankerite	$Ca(Fe,Mg,Mn)(CO_3)_2$	2.8–3.1	3.5–4	rhbdr., br., redsh. br., grnsh. br., pink	1.690–1.750, 1.510–1.548
Anorthite	$CaAl_2Si_2O_8$	2.76	6–6.5	tricl., wh., yel., grn., blk.	1.577, 1.585, 1.590
Anorthoclase	$(Na,K)AlSi_3O_8$	2.56–2.60	6	tricl., col., wh.	1.523, 1.528, 1.529
Anthophyllite	$(Mg,Fe)_7Si_8O_{22}(OH,F)_2$	2.85–3.57	5.5–6	rhomb., wh., gray, grn., br., yelsh. br., dk. br.	1.596–1.694, 1.605–1.710, 1.615–1.722
Antimony	Sb	6.61–6.72	3–3.5	hex., tin-wh.	
Apatite	$Ca_5(PO_4)_3(OH,F,Cl)$	3.1–3.35	5	hex., grn., wh., yel., br. red, bl.	1.629–1.667, 1.624–1.666
Apophyllite	$KFCa_4Si_8O_{20}.8H_2O$	2.33–2.37	4.5–5	tetr., col., wh., pink, pa. yel.	1.534–1.535, 1.535–1.537
Aragonite	$CaCO_3$	2.94–2.95	3.5–4	rhomb., col., wh.	1.530–1.531, 1.680–1.681, 1.685–1.686
Arcanite	K_2SO_4	2.663		rhom., col., wh.	1.4935, 1.4947, 1.4973
Argentite	Ag_2S	7.2–7.4	2–2.5	cub., blksh. lead gray	
Arsenic	As	5.63–5.78	3.5	hex., tin-wh., tarn. dk. gray	
Arsenolite	As_2O_3	3.86–3.88	1.5	cub., wh., somet. blsh., yelsh., redsh. tinge	1.755
Arsenopyrite	FeAsS	5.9–6.2	5.5–6	monocl., silver-wh., to steel gray	
Atacamite	$Cu_2(OH)_3Cl$	3.74–3.78	3–3.5	rhomb., grn., dk. grn., blksh. grn.	1.831, 1.861, 1.880
Augelite	$Al_2(PO_4)(OH)_3$	2.696	4–5	monocl., col., wh., yelsh. wh., rose	1.5736, 1.5759, 1.5877
Augite	$(Ca,Na)(Mg,Fe,Al)(Al,Si)_2O_6$	3.23–3.52	5.5–6	monocl., pa. br., br., purp. br., grn., blk.	1.671–1.735, 1.672–1.741, 1.703–1.761
Autunite	$Ca(UO_2)_2(PO_4)_2.10–12H_2O$	3.1–3.2	2–2.5	tetr., yel., somet. grnsh. yel. to pa. grn.	1.577, 1.553
Axinite	$(Ca,Mn,Fe)_3Al_2BSi_4O_{12}(OH)$	3.26–3.36	6.5–7	tricl., br., yelsh.	1.674–1.693, 1.681–1.701, 1.684–1.704
Azurite	$Cu_3(OH)_2(CO_3)_2$	3.77	3.5–4	monocl., azure bl., dk. bl., pa. bl.	1.730, 1.758, 1.838
Baddeleyite	ZrO_2	5.4–6.02	6.5	monocl., col., yel., gr., redsh. br., br., blk.	2.13, 2.19, 2.20
Barite	$BaSO_4$	4.50	3–3.5	rhomb., col., wh., somet. dk. br., gray	1.6362, 1.6373, 1.6482
Benitoite	$BaTi(SiO_3)_3$	3.65	6–6.5	rhbdr., bl., purp., col.	1.757, 1.804
Bertrandite	$Be_4Si_2O_7(OH)_2$	2.6	6	rhomb., col.	1.589, 1.602, 1.613
Beryl	$Be_3Al_2Si_6O_{18}$	2.66–2.83	7.5–8	hex., col., wh., blsh. grn., grnsh. yel., yel., bl.	1.565–1.590, 1.567–1.598
Beryllonite	$NaBe(PO_4)$	2.81	5.5–6	monocl., col., wh., pa. yel.	1.5520, 1.5579, 1.561
Biotite	$K(Mg,Fe)_3AlSi_3O_{10}(OH,F)_2$	2.7–3.3	2.5–3	monocl., blk., dk. br., redsh. br.	1.565–1.625–1.605–1.696, 1.605–1.696
Bismuth	Bi	9.70–9.83	2–2.5	rhbdr., silver-wh. to redsh wh.	
Bismuthinite	Bi_2S_3	6.75–6.81	2	rhomb., lead gray to tin-wh., tarn. yel. or iridesc.	
Bixbyite	$(Mn,Fe)_2O_3$	4.945	6–6.5	cub., blk.	
Bloedite	$Na_2Mg(SO_4)_2.4H_2O$	2.22–2.28	2.5–3	monocl., col., somet. blsh.-grn. or redsh.	1.483, 1.486, 1.487
Boehmite	$AlO(OH)$	3.01–3.06	3.5	rhomb., wh.	1.64–1.65, 1.65–1.66, 1.65–1.67
Boracite	$Mg_3B_7O_{13}Cl$	2.91–2.97	7–7.5	rhomb., col., wh., gray, yel., blsh.-grn., grn.	1.66, 1.66, 1.67
Borax	$Na_2B_4O_7.10H_2O$	1.715	2–2.5	monocl., col., wh., gray, blsh. or grnsh-wh.	1.4466, 1.4687, 1.4717
Bornite	Cu_5FeS_4	5.06–5.08	3	cub., copper red to pinchbeck br., tarn. purp., iridesc.	
Boulangerite	$Pb_5Sb_4S_{11}$	6.0–6.2	2.5–3	monocl., blsh. lead gray, oft. with yel. spots	
Bournonite	$PbCuSbS_3$	5.80–5.86	2.5–3	rhomb., steel gray to blk.	
Braggite	PtS	10.0		tetr., steel gray	
Braunite	$(Mn,Si)_2O_3$	4.72–4.83	6–6.5	tetr., brns. blk. to steel gray	
Bravoite	$(Ni,Fe)S_2$	4.62	5.5–6	cub., steel gray	
Breithauptite	NiSb	8.23	5.5	hex., pa. copper red to vlt., tarn.	
Brochantite	$Cu_4(SO_4)(OH)_6$	3.79	3.5–4	monocl., emerald-grn. to blksh. grn., pa. grn.	1.728, 1.771, 1.800
Bromyrite	AgBr	6.47	2.5	cub., col., gray, yelsh., grnsh.-br.	2.253
Brookite	TiO_2	4.08–4.20	5.5–6	rhomb., br., yelsh. br., redsh. br., blk.	2.5831, 2.5843, 2.7004
Brucite	$Mg(OH)_2$	2.38–3.40	2.5	hex., wh., pa. grn., gray, bl., yel., br.	1.560–1.590, 1.580–1.600
Bunsenite	NiO	6.898	5.5	cub., dk. pistachio-grn.	(Li) 2.37
Cacoxenite	$Fe_4(PO_4)_3(OH).12H_2O$	2.2–2.4	3–4	hex., yel. to brnsh.-yel., redsh. yel., somet. grnsh.	1.575–1.585, 1.635–1.656
Calcite	$CaCO_3$	2.715–2.94	3	rhbdr., col., wh., somet. gray, yel., pink, bl.	1.658–1.740, 1.486–1.550
Caledonite	$Cu_2Pb_5(SO_4)_3(CO_3)(OH)_6$	5.75–5.77	2.5–3	rhomb., dk. grn., blsh. grn.	1.815–1.831, 1.863–1.869, 1.906–1.912
Calomel	$HgCl$	7.15	1.5	tetr., col., wh., gray, yelsh. wh., br.	1.973, 2.656
Cancrinite	$(Na,Ca)_{7–8}Al_6Si_6O_{24}(CO_3SO_4Cl)_{1.5–2}.1–5H_2O$	2.51–2.42	5–6	hex., col., wh., pa. bl., pa. grn., yel., redsh.	1.528–1.507, 1.503–1.495
Carnallite	$KMgCl_3.6H_2O$	1.602	2.5	rhomb., col., wh., oft. redsh., somet. yel., bl.	1.466, 1.475, 1.494
Carnotite	$K_2(UO_2)_2(VO_4)_2.3H_2O$		1–2	rhomb. or monocl., bright yel., yel., grnsh. yel.	1.75, 1.92, 1.95
Cassiterite	SnO_2	6.99	6–7	tetr., yelsh. or redsh. br., brnsh.-blk.	2.006, 2.0972
Celestite	$SrSO_4$	3.96	3–3.5	rhomb., col., wh., pa. bl., redsh., grnsh., brnsh.	1.621–1.622, 1.623–1.624, 1.630–1.631
Celsian	$BaAl_2Si_2O_8$	3.10–3.39	6–6.5	monocl., col., wh.	1.579–1.587, 1.583–1.593, 1.588–1.600
Cervantite	$Sb_2O_4(?)$	6.64	4–5	rhomb.(?), yel., wh., somet. redsh.-wh.	
Cerargyrite	AgCl	5.55	2.5	cub., col., gray, grnsh.-br., tarn. purp., yelsh.	2.071
Cerussite	$PbCO_3$	6.53–6.57	3–3.5	rhomb., col., wh., gray, somet. bl., blk., grn., yelsh.	1.8036, 2.0765, 2.0786
Chabazite	$(Ca,Na)_2Al_2Si_4O_{12}.6H_2O$	2.05–2.10	4.5	rhbdr., redsh.-wh., wh., yelsh., grnsh.	1.470–1.494
Chalcocite	Cu_2S	5.5–5.8	2.5–3	rhomb., blksh., lead gray	
Chalcanthite	$CuSO_4.5H_2O$	2.28	2.5	tricl., dk. bl. to sky bl., somet. grnsh.	1.514, 1.537, 1.543
Chalcopyrite	$CuFeS_2$	4.1–4.3	3.5–4	tetr., brass-yel., tarn., iridisc.	
Chiolite	$Na_5Al_3F_{14}$	3.00	3.5–4	tetr., wh. to col.	1.349, 1.342
Chlorite	$(Mg,Al,Fe)_{12}(Si,Al)_8O_{20}(OH)_{16}$	2.6–3.3	2–3	monocl., grn., wh., yel., pink, br., red	1.57–1.66, 1.57–1.67, 1.57–1.67
Chloritoid	$(Fe,Mg,Mn)_2(AlFe^{+3})Al_3O_2SiO_4(OH)_4$	3.51–3.80	6.5	monocl., tricl., dk. grn.	1.713–1.730, 1.719–1.734, 1.723–1.740

Name	Formula	Sp. gr.	Hardness	Crystalline form and color	Index of refraction (Na) η; ω ϵ / α β γ
Chondrodite	$Mg(OH,F)_2.2Mg_2SiO_4$	3.16–3.26	6.5	monocl., yel., br., red	1.592–1.615, 1.602–1.627, 1.621–1.646
Chromite	$FeCr_2O_4$	4.5–5.1	5.5	cub., blk.	2.16
Chrysoberyl	$BeAl_2O_4$	3.65–3.85	8.5	rhomb., grn., yel.	1.746, 1.748, 1.756
Chrysocolla	$CuSiO_3.2H_2O$	~2.4	2	rhomb., (?)., grn., bl., br., blk.	1.575, 1.597, 1.598
Cinnabar	HgS	8.090	2–2.5	hex., red, brnsh. red, gray	(Li) 2.814, 3.143
Claudetite	As_2O_3	4.15	2.5	monocl., col. to wh.	1.87, 1.92, 2.01
Clinozoisite	$Ca_2Al_3Si_3O_{12}(OH)$	3.21–3.38	6.5	monocl., col., pa. yel., gray, grn.	1.670–1.715, 1.674–1.725, 1.690–1.734
Cobaltite	$CoAsS$	6.33	5.5	cub., silver wh., redsh., steel gray, blk.	
Colemanite	$Ca_2B_6O_{11}.5H_2O$	2.42–2.43	4.5	monocl., col., wh., yelsh. wh., gray	1.586, 1.592, 1.614
Columbite	$(Fe,Mn)(Cb,Ta)_2O_6$	5.15–5.25	6	rhomb., iron blk. to br. blk.	
Connellite	$Cu_{19}(SO_4)Cl_4(OH)_{32}.3H_2O(?)$	3.36	3	hex., azure bl.	1.724–1.738, 1.746–1.758
Copiapite	$(Fe,Mg)Fe_4^{+3}(SO_4)_6(OH)_2.20H_2O$	2.08–2.17	2.5–3	tricl., yel., grnsh. yel.	1.51–1.53, 1.53–1.55, 1.58–1.60
Copper	Cu	8.95	2.5–3	cub., red	
Coquimbite	$Fe_2(SO_4)_3.9H_2O$	2.10–2.12	2.5	hex., pa. vlt. to dk. amethystine, yelsh., grnsh.	1.53–1.55, 1.55–1.57
Cordierite	$Al_3(Mg,Fe)_2Si_5AlO_{18}$	2.53–2.78	7	rhomb., grayish-bl., bl., dk. bl.	1.522–1.558, 1.524–1.574, 1.527–1.578
Corundum	Al_2O_3	4.022	9	hex., col., bl., yel., purp., grn., pink, red	1.767–1.772, 1.759–1.763
Cotunnite	$PbCl_2$	5.80	2.5	rhomb., col. to wh., somet. yelsh., grnsh.	2.199, 2.217, 2.260
Covellite	CuS	4.6–4.76	1.5–2	hex., indigo bl., dk. bl., iridesc. brass yel. to red	
Cristobalite	SiO_2	2.33	6–7	tetr.(?)., col., wh., red.	1.487, 1.484
Crocoite	$PbCrO_4$	5.96–6.02	2.5–3	monocl., red, orange red, orange yel.	2.29, 2.36, 2.66
Cryolite	Na_3AlF_6	2.96–2.98	2.5	monocl., col. to wh., brnsh., redsh., blk.	1.338, 1.338, 1.339
Cryolithionite	$Na_3Li_3Al_2F_{12}$	2.77	2.5–3	cub., col. to wh.	1.3395
Cubanite	$CuFe_2S_3$	4.03–4.18	3.5	rhomb., brass to bronze yel.	
Cummingtonite	$(Mg,Fe)_7Si_8O_{22}(OH)_2$	3.2–3.5	5–6	monocl., dk. grn., br.	1.635–1.665, 1.644–1.675, 1.655–1.698
Cuprite	Cu_2O	6.14	3.5–4	cub., red, somet. blk.	
Danburite	$CaSi_2B_2O_8$	3.0	7	rhomb., pa. yel., col., dk. yel., yelsh. br.	1.63, 1.63–1.64, 1.63–1.64
Datolite	$CaBSiO_4(OH)$	2.96–3.00	5–5.5	monocl. col., wh., yelsh., grnsh., pinksh.	1.622–1.626, 1.649–1.654, 1.666–1.670
Daubreelite	Cr_2FeS_4	3.80–3.82	?	cub., blk.	
Derbylite	$Fe_6Ti_6Sb_2O_{23}(?)$	4.53	5	rhomb., pitch blk.	2.45, 2.45, 2.51
Diamond	C	3.50–3.53	10	cub., col., pa. yel. to dk. yel., pa. br. to dk. br., wh., blsh. wh.	2.4175
Diaspore	$AlO(OH)$	3.3–3.5	6.5–7	rhomb., wh., graysh. wh., col.	1.682–1.706, 1.705–1.725, 1.730–1.752
Digenite	Cu_2xS	5.546	2.5–3	cub., bl. to blk.	
Diopside	$CaMgSi_2O_6$	3.22–3.38	5.5–6.5	monocl., wh., pa. grn., dk. grn.	1.664–1.695, 1.672–1.701, 1.695–1.721
Dioptase	$CuSi_6O_{18}.6H_2O$	3.5	5	rhbdr., emerald grn.	1.64–1.66, 1.70–1.71
Dolomite	$CaMg(CO_3)_2$	2.86	3.5–4	rhbdr., wh., oft. yel. or br. tinge, col.	1.679, 1.500
Douglasite	$K_2FeCl_4.2H_2O(?)$	2.16		pa. grn., tarn. brnsh. red	1.485–1.491, 1.497–1.503
Dyscrasite	Ag_3Sb	9.67–9.81	3.5–4	rhomb., silver wh., tarn. gray, yelsh. or blksh.	
Eddingtonite	$BaAl_2Si_3O_{10}.4H_2O$	2.7–2.8		rhomb. or monocl., col., pink, br. wh.	1.541, 1.553, 1.557
Eglestonite	Hg_4OCl_2	8.4	2.5	cub., yel., orange-yel. to dk. brnsh., tarn. bl.	2.47–2.51
Emplectite	$CuBiS_2$	6.38	2	rhomb., gray to tin wh.	
Empressite	$AgTe$	7.510	3–3.5	pa. bronze	
Enargite	Cu_3AsS_4	4.4–4.5	3	rhomb., gray-blk. to iron-blk.	
Enstatite	$MgSiO_3$	3.209	5–6	rhomb., col., gray, grn., yel., brn.	1.650–1.662, 1.653–1.671, 1.658–1.680
Epidote	$Ca_2Fe^{+3}Al_2Si_3O_{12}(OH)$	3.38–3.49	6	monocl., grn., yel., gray	1.715–1.751, 1.725–1.784, 1.734–1.797
Epsomite	$MgSO_4.7H_2O$	1.675–1.679	2–2.5	rhomb., col., wh. pink, grn.	1.4325, 1.4554, 1.4609
Erythrite	$(Co,Ni)_3(AsO_4)_2.8H_2O$	3.06	1.5–2.5	monocl., crimson-red, red, pa. pink	1.626, 1.661, 1.699
Eucairite	$CuAgSe$	7.6–7.8	2.5	silver wh. to lead gray	
Euclasite	$BeAlSiO_4(OH)$	3.0–3.1	7.5	monocl., col., pa. grn., bl.	1.651, 1.655, 1.671
Eudialite	$(Na,Ca,Fe)_6ZrSi_6O_{18}(OH,Cl)(?)$	2.8–3.1	5–6	hex., pa. pink, red, br.	1.59–1.61, 1.59–1.61
Eulytite	$Bi_4Si_3O_{12}$	6.6	4.5	cub., br., yel., gray	2.05
Euxenite	$(Y,Ca,Ce,U,Th)(Cb,Ta,Ti)_2O_6$	5.0–5.9	5.5–6.5	rhomb., blk., grnsh. or brnsh. tint.	~2.2
Fayalite	Fe_2SiO_4	4.392	6.5	rhomb., grnsh., yelsh.	1.827, 1.869, 1.879
Ferberite	$FeWO_4$	7.51	4–4.5	monocl., br. to blk.	(Li)2.37–2.43
Fergussonite	$(Y,Er,Ce,Fe)(Cb,Ta,Ti)O_4$	5.6–5.8	5.5–6.5	tetr., gray, yel., br., dk. br.	2.1
Fluorite	CaF_2	3.18	4	cub., bl., purp., wh., col., yel., grn.	1.433–1.435
Forsterite	Mg_2SiO_4	3.222	7	rhomb., wh., grn., yelsh., yelsh.	1.635, 1.651, 1.670
Franklinite	$ZnFe_2^{+3}O_4$	5.07–5.34	5.5–6.5	Cub., blk. to br.-blk.	(Li) ~2.36
Gahnite	$ZnAl_2O_4$	4.62	7.5–8	cub., dk. bl.-grn., somet. yelsh. or brnsh.	1.79–1.81
Galena	PbS	7.57–7.59	2.5–2.75	cub. lead gray	
Galenabismuthite	$PbBi_2S_4$	7.04	2.5–3.5	rhomb., pa. gray to tin-wh., lead gray, somet. tarn., yel. or irid.	
Ganomalite	$(Ca,Pb)_{10}(OH,Cl)_2(Si_2O_7)_3$	5.4–5.7	3–4	hex., col., gray	1.910, 1.945
Gaylussite	$Na_2Ca(CO_3)_2.5H_2O$	1.991	2.5–3	monocl., col. to yelsh. wh., graysh. wh., wh.	1.4435, 1.5156, 1.5233
Gehlenite	$Ca_2Al_2SiO_7$	3.038	5–6	tetr., col., gray-grn., br.	1.669, 1.658
Geikielite	$MgTiO_3$	4.05	5–6	rhbdr., brnsh blk., blsh.	2.31, 1.95
Gibbsite	$Al(OH)_3$	2.38–2.42	2.5–3.5	monocl., wh., graysh., grnsh. or redsh.-wh.	1.56–1.58, 1.56–1.58, 1.58–1.60
Glauberite	$Na_2Ca(SO_4)_2$	2.75–2.85	2.5–3	monocl., gray, yelsh., somet. col., redsh.	1.515, 1.535, 1.536
Glauconite	$(K,Na,Ca)_{1.2-2}(Fe^{+3},Al,Fe,Mg)_4Si_{7-7.6}Al_{1-0.4}O_{20}(OH)_4.nH_2O$	2.4–2.95	2	monocl., col., ye.lsh. grn., grn., blsh. gray	1.592–1.610, 1.614–1.641, 1.614–1.641
Glaucophane	$Na_2Mg_3Al_2Si_8O_{22}(OH)_2$	3.08–3.30		monocl., gray, lavender bl.	1.606–1.661, 1.622–1.667, 1.627–1.670
Gmelinite	$(Ca,Na_2)Al_2Si_4O_{12}.6H_2O$	~2.1	4.5	rhbdr., wh., redsh.-wh., yelsh., grnsh.	1.476–1.494, 1.474–1.480
Goethite	$FeO(OH)$	3.3–4.3	5–5.5	rhomb., blksh-br., yelsh. or redsh.-br., yel.	2.260–2.275, 2.393–2.409, 2.398–2.515
Gold	Au	19.3	2.5–3	cub., yel.	
Goslarite	$ZnSO_4.7H_2O$	1.978	2–2.5	rhomb., col., wh., somet. br., grn., bl.	1.4568, 1.4801, 1.4844
Graphite	C	2.09–2.23	1–2	hex., iron-blk. to steel gray	
Greenockite	CdS	4.9	3–3.5	hex., yel. to orange	2.506, 2.529
Grossularite	$Ca_3Al_2Si_3O_{12}$	3.594	6.5–7.5	cub., wh., yel., br., red	1.734
Gummite (R)	$UO_3.H_2O$	3.9–6.4	2.5–5	yel., orange, redsh.-yel., red, br. blk.	
Gypsum	$CaSO_4.2H_2O$	2.30–2.37	2	monocl., col., wh., somet. gray, red, yel., br.	1.519–1.521, 1.523–1.526, 1.529–1.531
Halite	$NaCl$	2.16–2.17	2.5	cub., col., wh., orange, red	1.544
Hambergite	$Be_2(OH)(BO_3)$	2.36	7.5	rhomb., col. to gray, wh., yel.	1.56, 1.59, 1.63
Hanksite	$Na_{22}K(SO_4)_9(CO_3)_2Cl$	2.562	3–3.5	hex., col., somet. pa.-yelsh. or gray	1.481, 1.461
Harmotome	$BaAl_2Si_6O_{16}.6H_2O$	2.41–2.47	4.5	monocl., or rhomb., col., wh., pink, gray, yel.	1.503–1.508, 1.505–1.509, 1.508–1.514
Hausmannite	Mn_3O_4	4.83–4.85	5.5	tetr., brnsh.-blk.	(Li) 2.46, 2.15
Haüyne	$(Na,Ca)_{4-8}Al_6Si_6O_{24}(SO_4,S)_{1-2}$	2.44–2.50	5.5–6	cub., wh., gray, grn., bl.	1.496–1.505
Hedenbergite	$CaFeSi_2O_6$	3.50–3.56	6	monocl., brnsh.-grn., dk. grn., blk.	1.716–1.726, 1.723–1.730, 1.741–1.751
Helvite	$Mn_4Be_3Si_3O_{12}S$	3.20–3.44	6	cub., yel., br., redsh.-brn.	1.728–1.749
Hematite	Fe_2O_3	5.26	5–6	rhbdr., steel gray, dull red to bright red	3.22, 2.94
Hemimorphite	$Zn_4Si_2O_7(OH)_2.H_2O$	3.45	5	rhomb., col., wh., pa. bl., pa. grn., yel.	1.614, 1.617, 1.636
Hercynite	$FeAl_2O_4$	4.40	7.5–8	cub., blk.	1.835
Herderite	$CaBe(PO_4)(Fe,OH)$	2.95–3.01	5–5.5	monocl., col. to pa. yel. or grnsh.-wh.	1.592, 1.612, 1.621
Hessite	Ag_2Te	8.24–8.45	2–3	monocl., (<149.5°), cub. (>149.5°), gray	
Heulandite	$(Ca,Na_2)Al_2Si_7O_{18}.6H_2O$	2.1–2.2	3.5–4	pseudo-monocl., col., wh., yel., pink, red, gray, br.	1.491–1.505, 1.493–1.503, 1.500–1.512

Name	Formula	Sp. gr.	Hardness	Crystalline form and color	Index of refraction (Na) η; ω ϵ ω β γ
Hopeite	$Zn_3(PO_4)_2.4H_2O$	3.0–3.1	3.25	rhomb., col. to grayish-wh., pa. yel.	1.57–1.59, 1.58–1.60, 1.58–1.60
Hornblende	$(Ca,Na,K)_{2-3}(Mg,Fe,Fe^{+3}Al)_5Si_6(Si,Al)_2O_{22}(OH,F)_2$	3.02–3.45	5–6	monocl., grn., dk. grn., blk.	1.615–1.705, 1.618–1.714, 1.632–1.730
Huebnerite	$MnWO_4$	7.12	4–4.5	monocl., yel.-br. to red br., somet. br., blk.	2.17, 2.22, 2.32
Humite	$Mg(OH,F)_2.3Mg_2SiO_4$	3.2–3.32	6	rhomb., yel., orange	1.607–1.643, 1.619–1.653, 1.639–1.675
Huntite	$Mg_3Ca(CO_3)_4$	2.696		rhomb.(?)., wh.	
Hydrogrossularite	$Ca_3Al_2Si_2O_8(SiO_4)_{1-m}(OH)_{4m}$	3.594–3.13	6–7.5	cub., wh., buff, pa. grn., gray, pink	1.734–1.675
Hydromagnesite	$Mg_4(OH)_2(CO_3)_3.3H_2O$	2.236	3.5	monocl., col. to wh.	1.520–1.526, 1.524–1.530, 1.544–1.546
Illite	$K_{1-1.5}Al_4Si_{7-6.5}Al_{1-1.5}O_{20}(OH)_4$	2.6–2.9	1–2	monocl., wh.	1.54–1.57, 1.57–1.61, 1.57–1.61
Ilmenite	$FeTiO_3$	4.68–4.76	5–6	rhbdr., iron-blk.	
Iodyrite	AgI	5.69	1	hex., col. on exposure to light, yel., br.	2.21, 2.22
Jadeite	$NaAlSi_2O_6$	3.24–3.43	6	monocl., col., wh., grn., grnsh. bl.	1.640–1.658, 1.645–1.663, 1.652–1.673
Jamesonite	$Pb_4FeSb_6S_{14}$	5.63	2.5	monocl., gray-blk., somet. tarn. iridesc.	
Jarosite	$KFe_3(SO_4)_2(OH)_6$	2.91–3.26	2.5–3.5	rhbdr., ocherous, amber yel. to dk. br.	1.820, 1.715
Kainite	$KMg(SO_4)Cl.3H_2O$	2.15	2.5–3	monocl., col., gray bl., vlt., yelsh., redsh.	1.494, 1.505, 1.516
Kaliophyllite	$KAlSiO_4$	2.61	6	hex., col.	1.532, 1.537
Kaolinite	$Al_4Si_4O_{10}(OH)_8$	2.61–2.68	2–2.5	tricl. or monocl., wh., redsh.-wh., grnsh.-wh.	1.533–1.565, 1.559–1.569, 1.560–1.570
Kernite	$Na_2B_4O_7.4H_2O$	1.908	2.5	monocl., col., wh.	1.454, 1.472, 1.488
Kieserite	$MgSO_4.H_2O$	2.571	3.5	monocl., col., gray, wh., yelsh.	1.520, 1.533, 1.584
Kyanite	Al_2OSiO_4	3.53–3.65	5.5–7	tricl., bl., wh., gray, grn., yel., pink	1.712–1.718, 1.721–1.723, 1.727–1.734
Lanarkite	$Pb_2(SO_4)O$	6.92	2–2.5	monocl., gray to grnsh. wh., pa. yel.	1.925–1.931, 2.004–2.010, 2.033–2.039
Lanthanite	$(La,Ce)_2(CO_3)_3.8H_2O$	2.69–2.74	2.5–3	rhomb., col. to wh., pink, yelsh.	1.51–1.53, 1.584–1.590, 1.610–1.616
Laumontite	$CaAl_2Si_4O_{12}.4-3.5H_2O$	2.2–2.3	3.3.5	monocl., col., wh., red, yel., brn.	1.502–1.514, 1.512–1.522, 1.514–1.525
Laurionite	$Pb(OH)Cl$	6.24	3–3.5	rhomb., col. to wh.	2.08, 2.12, 2.16
Lawsonite	$CaAl_2(OH)_2Si_2O_7.H_2O$	3.05–3.10	6	rhomb., col., wh.	1.655, 1.674–1.675, 1.684–1.686
Lazulite	$(Mg,Fe)Al_2(PO_4)_2(OH)_2$	3.08–3.38	5.5–6	monocl., bl., blsh. wh., dk. bl., blsh. grn.	1.604–1.626, 1.626–1.654, 1.637–1.663
Lazurite	$Na_4SSi_3Al_3O_{12}$	2.38–2.45	5–5.5	cub., berlin bl., azure bl., grnsh. bl., vlt.	1.500
Leadhillite	$Pb_4(SO_4)(CO_3)_2(OH)_2$	6.55	2.5–3	monocl., col. to wh., gray, pa. grn., pa. bl., yelsh.	1.87, 2.00, 2.01
Lepidocrocite	$FeO(OH)$	4.05–4.31	5	rhomb., ruby-red to red-br.	1.94, 2.20, 2.51
Lepidolite	$K_2(Li,Al)_{5-6}Si_{6-7}Al_{2-1}O_{20}(OH,F)_4$	2.80–2.90	2.5–4	monocl., col., pa. pink, pa. purp.	1.525–1.548, 1.551–1.585, 1.554–1.587
Leucite	$KAlSi_2O_6$	2.47–2.50	5.5–6	tetr., (pseudo-cub.) wh., gray	1.508–1.511
Levyne	$(Ca,Na_2)Al_2Si_4O_{12}.6H_2O$	~2.1	4.5	rhbdr., wh., redsh. wh., grnsh.	1–496–1.505, 1.491–1.500
Litharge	PbO	9.14	2	tetr., red	(Li) 2.665, 2.535
Loellingite	$FeAs_2$	7.39–7.41	5–5.5	rhomb., silver wh. to steel-gray	
Magnesite	$MgCO_3$	2.98–3.44	3.5–4.5	rhbdr., wh., col., somet. yel., br.	1.700–1.782, 1.500, 1.500
Magnetite	Fe_3O_4	5.175	5.5–6.5	cub., blk. to br. blk.	2.42
Malachite	$Cu_2(OH)_2(CO_3)$	4.03–4.07	3.5–4	monocl., bright grn. to dk. grn., blksh. grn.	1.652–1.658, 1.872–1.878, 1.906–1.912
Manganite	$MnO(OH)$	4.32–4.43	4	monocl., dk. steel gray to iron-blk.	(Li) 2.25, 2.25, 2.53
Manganosite	MnO	5.364	5.5	cub., emerald grn., tarn. bl.	
Marcasite	FeS_2	4.887	6–6.5	rhomb., pa. bronze-yel., tin-wh.	
Marialite	$Na_4Al_3Si_9O_{24}Cl$	2.50–2.62	5.0	tetr., wh., pa. grnsh. yel., gray, br.	1.546–1.550, 1.540–1.541
Marshite	CuI	5.08	2.5	cub., col. to pa. yel., on exposure to light, red	2.346
Mascagnite	$(NH_4)_2SO_4$	1.768	2–2.5	rhomb., col., gray, wh.	1.5202, 1.5230, 1.5330
Matlockite	$PbFCl$	7.12	2.5–3	tetr., col. or yel. to pa. amber, grnsh.	2.145, 2.006
Meionite	$Ca_4Al_6Si_6O_{24}CO_3$	2.78	5–6	tetr., wh., pa. grnsh. yel., gray, grn.	1.590–1.600, 1.556–1.562
Melanterite	$FeSO_4.7H_2O$	1.898	2	monocl., grn., grnsh. bl., grnsh. wh.	1.47, 1.48, 1.49
Melilite	$(Ca,Na,K)_2(Mg,Fe,Fe^{+3},Al,Si)_3O_7$	2.95–3.05	5–6	tetr., yelsh., br., grn.-br.	1.624–1.666, 1.616–1.661
Mellite	$Al_2C_{12}O_{12}.18H_2O$	1.64	2–2.5	tetr., yel., redsh., brnsh., somet. wh.	1.5393, 1.5110
Mendipite	$Pb_3O_2Cl_2$	7.24	2.5	rhomb., col. to wh., gray, oft. yel., red, bl. tinge	2.22–2.26, 2.25–2.29, 2.29–2.33
Mesolite	$Na_2Ca_2(Al_2Si_3O_{10}).8H_2O$	~2.26	5	monocl., col., wh., gray, yel., pink, red	β = 1.504–1.508
Metacinnabar	HgS	7.65	3	cub., graysh.-blk.	
Microcline	$KAlSi_3O_8$	2.56–2.63	6–6.5	tricl., col., wh., pink, red, yel., grn.	1.514–1.529, 1.518–1.533, 1.521–1.539
Microlite	$(Na,Ca)_2Ta_2O_6(O,OH,F)$	4.2–6.4	5–5.5	cub., pa. yel. to br., somet. red, grn.	~2.0
Miersite	AgI	5.64–5.68	2.5	cub., canary-yel.	2.18–2.22
Millerite	NiS	5.3–5.7	3–3.5	hex., pa. brass-yel. to bronze-yel., gray, tarn. iridesc.	
Mimetite	$Pb_5(AsO_4,PO_4)_3Cl$	7.24	3.5–4	hex. pa. yel. to yelsh. br., orange-yel., wh.	2.147, 2.128
Minium	Pb_3O_4	8.9–9.2	2.5	scarlet red, bl. red, somet. yel. tint.	(Li) 2.40–2.44
Mirabilite	$Na_2SO_4.10H_2O$	1.490	1.5–2	monocl., col. to wh.	1.391–1.397, 1.393–1.399, 1.395–1.401
Moissanite	SiC	3.218	9.5	hex., grn. to blk., somet. blsh., red	2.647–2.649, 2.689–2.693
Molybdenite	MoS_2	4.62–4.73	1–1.5	hex., lead-gray	
Monazite	$(Ce,La,Th)PO_4$	5.0–5.3	5	monocl., yel., br., redsh. br.	1.774–1.800, 1.777–1.801, 1.828–1.851
Monetite	$CaH(PO_4)$	2.929	3.5	tricl., wh., pa. yelsh.-wh.	1.587, ~1.615, 1.640
Monticellite	$CaMgSiO_4$	3.08–3.27	5.5	rhomb., col.	1.639–1.654, 1.646–1.664, 1.653–1.674
Montmorillonite	$(0.5Ca,Na)_{0.7}(Al,Mh,Fe)_4(Si,Al)_8O_{20}(OH)_4.nH_2O$	2–3	1–2	monocl., wh., yel., grn.	1.48–1.61, 1.50–1.64, 1.50–1.64
Montroydite	HgO	11.23	2.5	rhomb., dk. red to brnsh. red, br.	(Li) 2.37, 2.5, 2.65
Mordenite	$(Na_2,K_2Ca)Al_2Si_{10}O_{24}.7H_2O$	2.12–2.15	3–4	rhomb., col., wh., red, yel.	1.472–1.483, 1.475–1.485, 1.477–1.487
Muscovite	$KAl_2Si_3AlO_{10}(OH,F)_2$	2.77–2.88	2.5–3	monocl., col., pa. grn., pa. red, pa. br.	1.552–1.574, 1.582–1.610, 1.587–1.616
Nantokite	$CuCl$	4.136	2.5	cub., col. to wh., grayish, grn.	1.925–1.935
Natrolite	$Na_2Al_2Si_3O_{10}.2H_2O$	2.20–2.26	5	rhomb., col., wh., gray, yel., pink, red	1.473–1.483, 1.476–1.486, 1.485–1.496
Nepheline	$Na_2KAl_4Si_4O_{16}$	2.56–2.665	5.5–6	hex., col., wh., gray	1.529–1.546, 1.526–1.542
Newberyite	$MgH(PO_4).3H_2O$	2.10	3.0–3.5	rhomb., col. to wh.	1.511–1.517, 1.514–1.520, 1.530–1.536
Niccolite	$NiAs$	7.784	5–5.5	hex., pa. copper-red, tarn. gray to blk.	
Nosean	$Na_8Al_6Si_6O_{24}SO_4$	2.30–2.40	5.5	cub., gray, bl., br.	1.495
Oldhamite	CaS	2.58	4	cub., pa. chestnut-br.	2.137
Oligoclase	$([NaSi]_{0.9-0.7}[CaAl]_{0.1-0.3})AlSi_2O_8$	2.63–2.65	6–6.5	tricl., col., wh., gray, pink	1.533–1.544, 1.537–1.548, 1.543–1.552
Olivenite	$Cu_2(AsO_4)(OH)$	3.9–4.5	3	rhomb., olive grn., grnsh.-br., br., gray	1.75–1.78, 1.79–1.82, 1.83–1.87
Olivine	$(Mg,Fe)_2SiO_4$	3.22–4.39	6.5–7	rhomb., olive grn., grayish grn. to yelsh. br.	1.63–1.83, 1.65–1.87, 1.67–1.88
Opal	$SiO_2.nH_2O$	1.73–2.16	~6	col., wh., yel., br., red, grn., bl., blk., amorp.	1.41–1.46
Orpiment	As_2S_3	3.49	1.5–2	monocl., yel., brnsh. yel.	(Li) 2.4, 2.81, 3.02
Orthoclase	$KAlSi_3O_8$	2.55–2.63	6–6.5	monocl., col., wh., pink, red, yel., grn.	1.518–1.529, 1.522–1.533, 1.522–1.539
Orthopyroxene	$(Mg,Fe)SiO_3$	3.209–3.96	5–6	rhomb., col., gray, grn., grn., dk. brn.	1.650–1.768, 1.653–1.770, 1.658–1.788
Paragonite	$NaAl_2Si_3AlO_{10}(OH)_2$	2.85	2.5	monocl., pa. yel.	1.564–1.580, 1.594–1.609, 1.600–1.609
Parisite	$(Ce,La,Na)FCO_2.CaCO_3$	4.42	4.5	hex., brnsh., col.	1.672, 1.771
Pectolite	$Ca_2NaH(SiO_3)_3$	2.86–2.90	4.5–5	tricl., col., wh.	1.595–1.610, 1.605–1.615, 1.632–1.645
Penfieldite	$Pb_2Cl_6(OH)_2$	6.6		hex., wh.	2.13, 2.21
Pentlandite	$(Fe,Ni)_9S_8$	4.6–5.0	3.5–4	cub., pa. bronze-yel.	
Percylite	$PbCuCl_2(OH)_2(?)$	?	2.5	cub(?), sky bl.	2.04–2.06
Periclase	MgO	3.55–3.68	5.5	cub., col. to gray-wh., yel., brnsh. yel., grn., bl.	1.7350
Pekovskite	$CaTiO_3$	3.97–4.26	5.5	pseudo cub., blk., gray-blk., brnsh. bl., redsh. br., br., yel.	2.30–2.38
Petalite	$LiAlSi_4O_{10}$	2.412–2.422	6.5	monocl., wh., gray, somet. pink, grn.	1.504–1.507, 1.510–1.513, 1.516–1.523

Name	Formula	Sp. gr.	Hard-ness	Crystalline form and color	Index of refraction (Na) η; ω ϵ α β γ
Pharmacosiderite	$Fe_3(AsO_4)_2(OH)_3.5H_2O$	2.797	2.5	cub., olive-grn. to yel., br., redsh.	1.676–1.704
Phenakite	Be_2SiO_4	2.98	7.5	rhbdr., col., rose, yel., br.	1.654, 1.670
Phillipsite	$(0.5Ca,Na,K)_3Al_5Si_5O_{16}.6H_2O$	2.2	4–4.5	monocl. or rhomb., col., wh., pink, gray, yel.	1.483–1.504, 1.484–1.509, 1.496–1.514
Phlogopite	$KMg_3AlSi_3O_{10}(OH,F)_2$	2.76–2.90	2–2.5	monocl., col., yelsh.-br., grn., redsh.-br., br.	1.530–1.590, 1.557–1.637, 1.558–1.637
Phosgenite	$Pb_2(CO_3)Cl_2$	6.133	2–3	tetr., yelsh. wh. to yelsh. br., br., somet. wh., rose, gray	2.1181, 2.1446
Piemontite	$Ca_2(Mn,Fe^{+3},Al)_2Si_3O_{12}(OH)$	3.45–3.52	6	monocl., redsh. brn., blk.	1.732–1.794, 1.750–1.807, 1.762–1.829
Pigeonite	$(Mg,Fe,Ca)(Mg,Fe)Si_2O_6$	3.30–3.46	6	monocl., br., grnsh. br., blk.	1.682–1.722, 1.684–1.722, 1.705–1.751
Platinum	Pt	14–19	4–4.5	cub., whitish, steel gray to dk. gray	
Pollucite	$CsAlSi_2O_6$	2.9	6.5	cub., (pseudo-cub.) col.	1.507–1.527
Polybasite	$(Ag,Cu)_{16}Sb_2S_{11}$	6.0–6.2	2–3	monocl., iron-blk.	
Powellite	$Ca(Mo,W)O_4$	4.21–4.25	3.5–4	tetr., straw-yel., br., grnsh., somet. gray, bl., blk.	1.959–1.982, 1.967–1.993
Prehnite	$Ca_2Al_2Si_3O_{10}(OH)_2$	2.90–2.95	6–6.5	rhomb., pa. grn., yel., gray, wh.	1.611–1.632, 1.615–1.642, 1.632–1.665
Proustite	Ag_3AsS_3	5.57	2–2.5	rhbdr., scarlet-vermillion	3.0877, 2.7924
Pseudobrookite	Fe_2TiO_5	4.33–4.39	6	rhomb., dk. red-br. to brnsh. blk. and blk.	(Li) 2.38, 2.39, 2.42
Psilomelane	$BaMn^{+2}Mn_8^{+4}O_{16}(OH)_4$	4.71	5–6	rhomb., iron-blk. to steel-gray	
Pumpellyite	$Ca_4(Mg,Fe,Mn)(Al,Fe^{+3},Ti)_5(OH)_3Si_6O_{23}.2H_2O$	3.18–3.23	6	monocl., grn., blsh. grn., br.	1.674–1.702, 1.675–1.715, 1.688–1.722
Pyrargyrite	Ag_3SbS_3	5.85	2.5	rhbdr., deep red	(Li) 3.084, 2.881
Pyrite	FeS_2	5.018	6–6.5	cub., pa. brass-yel., tarn. iridesc.	
Pyrochlore	$NaCaCb_2O_6F$	4.2–6.4	5–5.5	cub., br. to blk., yelsh., redsh. or blksh. br.	
Pyrochroite	$Mn(OH)_2$	3.23–3.27	2.5	hex., col. to pa. grn. or bl., tarn. br. to blk.	1.72, 1.68
Pyrolusite	MnO_2	5.04–5.08	6–6.5	tetr., pa. steel-gray, iron-gray, blk., blsh.	
Pyromorphite	$Pb_5(PO_4,AsO_4)_3Cl$	7.00–7.08	3.5–4	hex., grn., yel., br., orange, brnsh. red., gray	2.058, 2.048
Pyrope	$Mg_3Al_2Si_3O_{12}$	3.582	6–7.5	cub., red, pink	1.714
Pyrophyllite	$Al_2Si_4O_{10}(OH)_2$	2.65–2.90	1–2	monocl., wh., yel., pa. bl., gray-grn., brnsh.-grn.	1.534–1.556, 1.568–1.589, 1.596–1.601
Pyrrhotite	$Fe_{1-0.8}S$	4.58–4.65	3.5–4.5	hex., bronze-yel. to br., tarn., somet. iridesc.	
Quartz	SiO_2	2.65	7	rhbdr., col., wh., blk., purp., grn., bl., rose	1.544, 1.553
Rammelsbergite	$NiAs_2$	7.0–7.2	5.5–6	tin. wh., redsh. tinge	
Raspite	$PbWO_4$	8.46	2.5–3	monocl., yelsh. br., pa. yel., gray	1.25–1.29, 1.25–1.29, 1.28–1.32
Realgar	AsS	3.56	1.5–2	monocl., aurora-red to orange-yel.	2.538, 2.684, 2.704
Riebeckite	$Na_2Fe_3Fe_2^{+3}Si_8O_{22}(OH,F)_2$	3.02–3.42	5	monocl., dk. bl., bl.	1.654–1.701, 1.662–1.711, 1.668–1.717
Rhodochrosite	$MnCO_3$	3.70	3.5–4	rhbdr., pink, red, br., brnsh.-yel.	1.816, 1.597
Rhodonite	$(Mn,Fe,Ca)SiO_3$	3.57–3.76	5.5–6.5	tricl., pink to brnsh. red	1.711–1.738, 1.716–1.741, 1.724–1.751
Rutile	TiO_2	4.23–5.5	6–6.5	tetr., redsh. brn. to red, somet. yelsh., blsh.	2.605–2.613, 2.899–2.901
Safflorite	$(Co,Fe)As_2$	7.0–7.5	4.5–5	rhomb., tin-wh., tarn. dk. gray	
Samarskite	$(Y,Er,Ce,U,Ca,Fe,Pb,Th)(Cb,Ta,Ti,Sn)_2O_6$	5.69	5–6	rhomb., velvet blk., somet. brnsh. tint	~2.20
Sapphirine	$(Mg,Fe)_2Al_4O_6SiO_4$	3.40–3.58	7.5	monocl., pa. bl., pa. grn.	1.701–1.717, 1.703–1.720, 1.705–1.724
Scapolite	$(Na,Ca)_4Al_3(Al,Si)_3Si_6O_{24}(Cl,F,OH,CO_3,SO_4)$	2.50–2.78	5–6	tetr., col., wh., pa. grnsh. yel., gray, bl.	1.546–1.600, 1.540–1.562
Scheelite	$CaWO_4$	6.08–6.12	4.5–5	tetr., yelsh. wh., pa. yel., brnsh., col., wh., gray	1.920, 1.936
Scolecite	$CaAl_2Si_3O_{10}.3H_2O$	2.25–2.29	5	monocl., col., wh., gray, yel., pink, red	1.507–1.513, 1.516–1.520, 1.517–1.521
Scorodite	$Fe^{+3}(AsO_4).2H_2O$	3.28	3.5–4	rhomb., pa. grn., gray grn., br. somet. col., blsh., yel.	1.784, 1.795, 1.814
Sellaite	MgF_2	3.15	5	tetr., col. to wh.	1.378, 1.390
Senarmontite	Sb_2O_3	5.50	2–2.5	pseudo-cub., col., gray-wh.	2.087
Serpentine	$Mg_3Si_2O_5(OH)_4$	~2.55	2.5–3.5	monocl., wh., yel., gray, bl., blsh. grn.	1.53–1.57, 1.56, 1.54–1.57
Siderite	$FeCO_3$	3.96	3.5–4	rhbdr., yelsh. br., br., dk. br.	1.875, 1.635
Sillimanite	Al_2SiO_5	3.23–3.27	6.5–7.5	rhomb., col., wh., yelsh., br., grnsh.	1.654–1.661, 1.658–1.662, 1.637–1.683
Silver	Ag	10.1–11.1	2.5–3	cub., wh., tarn. gray or blk.	
Skutterudite	$(Co,Ni)As_3$	6.1–6.9	5.5–6	cub., between tin-wh. and silver-gray, tarn. gray or iridesc.	
Smithsonite	$ZnCO_3$	4.42–4.44	4–4.5	rhbdr., grayish wh. to dk. gray, grnsh., brnsh. wh.	1.848, 1.621
Sodalite	$Na_8Al_6Si_6O_{24}Cl_2$	2.27–2.33	5.5–6	cub., bl., grn., yel., gray, pink	1.483–1.487
Sperrylite	$PtAs_2$	10.58	6–7	cub., tin-wh.	
Spessartite	$Mn_3Al_2Si_3O_{12}$	4.190	6–7.5	cub., blk., dk. red, brnsh. red., bl., yelsh. orange	1.800
Sphalerite	ZnS	3.9–4.1	3.5–4	cub., br., blk., yel., red, wh.	2.369
Sphene	$CaTiSiO_4(O,OH,F)$	3.45–3.55	5	monocl., col., wh., yel., br., blk.	1.843–1.950, 1.870–2.034, 1.943–2.110
Spinel	$MgAl_2O_4$	3.55	7.5–8	cub., grn., red, bl., br. to col.	1.719
Spodumene	$LiAlSi_2O_6$	3.03–3.22	6.5–7	monocl., col., gray-wh., pa. bl., pa. grn., yelsh.	1.648–1.663, 1.655–1.669, 1.662–1.679
Stannite	Cu_2FeSn_4	4.3–4.5	4	tetr., steel gray to iron blk.	
Staurolite	$(Fe,Mg)_2(AlFe^{+3})_9O_6SiO_4(O,OH)_2$	3.74–3.83	7.5	monocl., brn., redsh., yelsh.	1.739–1.747, 1.745–1.753, 1.752–1.761
Stercorite	$Na(NH_4)H(PO_4).4H_2O$	1.615		tricl., wh., velsh., brnsh.	1.439, 1.442, 1.469
Stibiotantalite	$Sb(Ta,Cb)O_4$	5.7–7.5	5.5	rhomb., dk. br. to pa. yel.-br., red-br., grnsh.-yel.	2.38, 2.41, 2.46
Stibnite	Sb_2S_3	4.61–4.65	2	rhomb., lead-gray to steel-gray	
Stilbite	$(Ca,Na_2K_2)Al_2Si_7O_{18}.7H_2O$	2.1–2.2	3.5–4	monocl., col., wh., col., pink, red, gray, br.	1.484–1.500, 1.492–1.507, 1.494–1.513
Stilpnomelane	$(K,Na,Ca)_{0–1.4}(Fe^{+3}Fe,Mg,Al)_{6–8}Si_8O_{20}(OH)_4(O,OH,H_2O)_{4–8}$	2.59–2.96	3–4	monocl., br., dk. br., redsh. br., dk. grn.	1.543–1.634, 1.576–1.745, 1.576–1.745
Stolzite	$PbWO_4$	7.9–8.4	2.5–3	tetr., redsh. br., yelsh. gray, straw-yel., grnsh.	2.26–2.28, 2.18–2.20
Strengite	$Fe^{+3}(PO_4).2H_2O$	2.90	3.5–4.5	rhomb., red, carmine, vlt., near col.	1.707, 1.719, 1.741
Strontianite	$SrCO_3$	3.72	3.5	rhomb., col., wh., yel., grnsh., brnsh.	1.516–1.520, 1.664–1.667, 1.666–1.669
Struvite	$Mg(NH_4)(PO_4).6H_2O$	1.71	2	rhomb., col., somet. yelsh., brnsh.	1.495, 1.496, 1.504
Sulfur	S	2.07	1.5–2.5	rhomb., yel., brnsh., grnsh., redsh., gray	1.9579, 2.0377, 2.2452
Sylvanite	$(Ag,Au)Te_2$	8.161	1.5–2	monocl., steel-gray to silver-wh.	
Sylvite	KCl	1.99	2	cub., col., wh., somet. grayish, blsh., yelsh., red	1.49031
Talc	$Mg_3Si_4O_{10}(OH)_2$	2.58–2.83	1	monocl., col., wh., pa. grn., dk. grn., br.	1.539–1.550, 1.589–1.594, 1.589–1.600
Tantalite	$(Fe,Mn)(Ta,Cb)_2O_6$	7.90–8.00	6.5	rhomb., iron-bl. to br.-blk.	2.26, 2.32, 2.43
Tapiolite	$FeTa_2O_6$	7.9	6–6.5	tetr., blk.	(Li) 2.27, 2.42
Tellurobismuthite	Bi_2Te_3	7.800–7.830	1.5–2	rhbdr., pa. lead-gray	
Terlinguaite	Hg_2OCl	8.725	2.5	monocl., yel. to grnsh.-yel., somet. br.	(Li) 2.33–2.37, 2.62–2.66, 2.64–2.68
Tetrahedrite	$(Cu,Fe)_{12}Sb_4S_{13}$	4.6–5.1	3–4.5	cub., flint-gray to iron-blk. to dull-blk.	
Thenardite	Na_2SO_4	2.664	2.5–3	rhomb., col., grayish-wh., yelsh., yelsh. br., redsh.	1.464–1.471, 1.473–1.477, 1.481–1.485
Thermonatrite	$Na_2CO_3.H_2O$	2.255	1–1.5	rhomb., col. to wh., grayish, yelsh.	1.420, 1.506, 1.524
Thomsenolite	$NaCaAlF_6.H_2O$	2.981	2	monocl., col. to wh., somet. brnsh., redsh.	1.4072, 1.4136, 1.4150
Thomsonite	$NaCa_2([Al,Si]_5O_{10})_2.6H_2O$	2.10–2.39	5–5.5	rhomb., col., wh., pink, br.	1.497–1.530, 1.513–1.533, 1.518–1.544
Thorianite (R)	ThO_2	9.7	6.5	cub., dk. gray to brnsh.-blk., blk.	~2.20
Thorite (R)	$ThSiO_4$	5.2–5.4	4.5–5	tetr., orange-yel., brnsh. to blk.	~1.8
Topaz	$Al_2SiO_3(OH,F)_2$	3.49–3.57	8	rhomb., col., wh., yel., gray, grn., red, bl.	1.606–1.629, 1.609–1.631, 1.616–1.638

Name	Formula	Sp. gr.	Hardness	Crystalline form and color	Index of refraction (Na) η; ω ϵ / ω β γ
Torbernite (R)	$Cu(UO_2)_2(PO_4)_2.8\text{-}12H_2O$	3.22	2-2.5	tetr., various shades of grn.	1.592, 1.582
Tourmaline	$Na(Mg,Fe,Mn,Li,Al)_3Al_6Si_6O_{18}(BO_3)_3(OH,F)_4$	3.03-3.25	7	rhbdr., blk., bl., grn., yel., red, col., br.	1.635-1.675, 1.610-1.650
Tremolite	$Ca_2Mg_5Si_8O_{22}(OH,F)_2$	3.0	5-6	monocl., col., gray, wh.	1.599, 1.612, 1.622
Tridymite	SiO_2	2.27	7	rhomb., col., wh.	1.471-1.479, 1.472-1.480, 1.474-1.483
Triphyllite-Lithiophyllite	$Li(Fe,Mn)PO_4$	3.34-3.58	4-5	rhomb., blsh. or grnsh. gray to yelsh. br., br.	1.66-1.70, 1.67-1.70, 1.68-1.71
Troegerite (R)	$(UO_2)_3(AsO_4)_2.12H_2O$		2-3	tetr., lemon-yel.	1.58-1.59, 1.625-1.635
Trona	$Na_3H(CO_3)_2.2H_2O$	2.14	2.5-3	monocl., gray or yelsh. wh., col.	1.412, 1.492, 1.540
Turquois	$Cu(Al,Fe^{+3})_6(PO_4)_4(OH)_8.4H_2O$	2.6-3.2	4.5-6	tricl., bl., grn., grnsh.-gray	1.61-1.78, 1.62-1.84, 1.65-1.84
Ullmannite	$NiSbS$	6.61-6.69	5-5.5	cub., steel-gray to silver-wh.	
Uraninite (R)	UO_2	8.0-11	5-6	cub., steel-blk., brnsh.-blk., grayish, grn.	
Uvarovite	$Ca_3Cr_2Si_3O_{12}$	3.90	7.5	cub., emerald-grn.	1.86
Valentinite	Sb_2O_3	5.76	2.5-3	rhomb., col. to wh., somet. yelsh., redsh., gray, br.	2.18, 2.35, 2.35
Vanadinite	$Pb_5(VO_4)_3Cl$	6.5-7.1	2.75-3	hex., orange-red, red, brnsh.-red, br., brnsh.-yel., yel.	2.416, 2.350
Variscite-Strengite	$(Al,Fe^{+3})(PO_4).2H_2O$	2.57-2.87	3.5-4.5	rhomb., pa. grn., grn., blsh.-grn., red, vlt., col.	1.563-1.707, 1.588-1.719, 1.594-1.741
Vaterite	$CaCO_3$	2.645		hex., col.	1.550, 1.640-1.650
Vermiculite	$(Mg,Ca)_{0.7}(Mg,Fe^{+3}Al)_6(Al,Si)_8O_{20}(OH)_4.8H_2O$	~2.3	~1.5	monocl., col., yel., grn., br.	1.525-1.564, 1.545-1.583, 1.545-1.583
Vesuvianite	$Ca_{10}(Mg,Fe)_2Al_4(Si_2O_7)_2(SiO_4)_5(OH,F)_4$	3.33-3.43	6-7	tetr., yel., grn., br.	1.700-1.746, 1.703-1.752
Villiaumite	NaF	2.79	2-2.5	cub., carmine, (nat.), col. (artif.)	1.327
Vivianite	$Fe_3(PO_4)_2.8H_2O$	2.67-2.69	1.5-2	monocl., col., tarn. pa. bl., grnsh. bl., dk. bl., blsh. blk.	1.579-1.616, 1.602-1.656, 1.629-1.675
Wagnerite	$Mg_2(PO_4)F$	3.15	5-5.5	monocl., yel., gray, somet. red, grn.	1.568, 1.572, 1.582
Wavellite	$Al_3(OH)_3(PO_4)_2.5H_2O$	2.36	3.25-4	rhomb., grnsh. wh., grn. to yel., somet. br., bl., wh.	1.520-1.535, 1.526-1.543, 1.545-1.561
Whewellite	$Ca(C_2O_4).H_2O$	2.23	2.5-3	monocl., col., somet. yelsh., brnsh.	1.491, 1.554, 1.650
Willemite	Zn_2SiO_4	3.9-4.1	5.5	rhbdr., wh., yel., grn., red, gray, br.	1.691, 1.719
Witherite	$BaCO_3$	4.29-4.30	3.5	rhomb., col., wh., gray, yelsh. br.	1.529, 1.676, 1.677
Wolframite	$(Fe,Mn)WO_4$	7.12-7.51	4-4.5	monocl., dk. gray, brnsh. blk. to iron blk.	(Li) ~2.26, 2.32, 2.42
Wollastonite	$CaSiO_3$	2.87-3.09	4.5-5	tricl., wh., col., gray, pa. grn.	1.616-1.640, 1.628-1.650, 1.631-1.653
Wulfenite	$PbMoO_4$	6.5-7.0	2.75-3	tetr., orange-yel. to yel., gray, grn., br., red	2.403, 2.283
Wurtzite	ZnS	3.98	3.5-4	hex., brnsh. blk.	2.356, 2.378
Xenotime	$Y(PO_4)$	4.4-5.1	4-5	tetr., yelsh. br. to redsh. br., somet. gray, wh., pa. yel., grnsh.	1.721, 1.816
Zeunerite (R)	$Cu(UO_2)_2(AsO_4)_2.10\text{-}16H_2O$			tetr.	1.602-1.610
Zincite	ZnO	5.64-5.68	4	hex., orange-yel. to dk. red, somet. yel.	2.013, 2.029
Zircon	$ZrSiO_4$	4.6-4.7	7.5	tetr., redsh. br., yel., gray, grn., col.	1.923-1.960, 1.968-2.015
Zoisite	$Ca_2Al_3Si_3O_{12}(OH)$	3.15-3.365	6	rhomb., gray, grnsh., brnsh.	1.685-1.705, 1.688-1.710, 1.697-1.705

X-Ray Crystallographic Data, Molar Volumes, and Densities of Minerals and Related Substances

From U.S. Geological Survey Bulletin 1248 by
Richard A. Robie, Philip M. Bethke and Keith M. Beardsley

An extensive list of references and the bases for the calculations and the selection of data are given in the above referenced Bulletin. Bulletin 1248 may be obtained from the Superintendent of Documents, U.S. Government Printing Office, Washington, D.C., 20402.

Z; The number of gram formula weights per unit cell.

r; Indicates the data were obtained at an unspecified room temperature and may be taken as 25° ±5°C.

*; Indicates the measurements were made on a natural specimen which may have deviated slightly from the listed formula. Densities for these minerals were calculated using the formula weight for the stoichiometric phase.

hex-R; Rhombohedral symmetry. To distinguish from true hexagonal symmetry.

X-Ray Crystallographic Data of Minerals

Name and formula	Crystal system	Space group	Structure type	Z	a_o	b_o	c_o	α_o	β_o	γ_o	Cell volume 10^{-24} cm^3	Molar volume cm^3	Molar volume cal bar^{-1}	X-Ray density grams cm^{-3}	Temp. °C	
							Elements									
1 Silver Ag	cubic	Fm3m(225)	face-centered cubic	4	4.0862 ±.0002						68.227 ±.010	10.272 ±.002	.24556 ±.00008	10.501 ±.002	25	1
2 Arsenic As	hex-R	R3̄m(166)	arsenic	6	3.760 ±.001		10.555 ±.003				129.23 ±.08	12.972 ±.002	.31007 ±.00023	5.776 ±.004	26	2
3 Gold Au	cubic	Fm3m(225)	face-centered cubic	4	4.0786 ±.0002						67.847 ±.010	10.215 ±.002	.24420 ±.00008	19.282 ±.003	25	3
4 Bismuth Bi	hex-R	R3̄m(166)	arsenic	6	4.5459 ±.0010		11.8622 ±.0030				212.29 ±.11	21.309 ±.011	.50934 ±.00030	9.8071 ±.0050	26	4
5 Diamond C*	cubic	Fd3m(227)	diamond	8	3.5670 ±.0001						45.385 ±.004	3.4166 ±.0003	.08170 ±.00005	3.5155 ±.0003	25	5
6 Graphite C*	hex.	C6/mmc(194)	graphite	4	2.4612 ±.0001		6.7079 ±.0010				35.189 ±.006	5.2982 ±.0009	.12668 ±.00007	2.2670 ±.0004	15	6
7 Copper Cu	cubic	Fm3m(225)	face-centered cubic	4	3.6150 ±.0005						47.242 ±.020	7.1128 ±.0030	.17005 ±.00012	8.9331 ±.0037	25	7
8 α-Iron Fe	cubic	Im3m(229)	body-centered cubic	2	2.8664 ±.0005						23.551 ±.012	7.0918 ±.0037	.16954 ±.00013	7.8748 ±.0041	25	8
9 Nickel Ni	cubic	Fm3m(225)	face-centered cubic	4	3.5238 ±.0005						43.756 ±.019	6.5880 ±.0028	.15750 ±.00011	8.9117 ±.0038	25	9
10 Lead Pb	cubic	Fm3m(225)	face-centered cubic	4	4.9505 ±.0005						121.32 ±.04	18.267 ±.006	.43663 ±.00018	11.342 ±.003	25	10
11 Platinum Pt	cubic	Fm3m(225)	face-centered cubic	4	3.9231 ±.0005						60.379 ±.023	9.0909 ±.0035	.21732 ±.00013	21.460 ±.008	25	11
12 orthorhombic Sulfur S	orth.	Fddd(70)	S₈ ring molecules	128	10.4646 ±.0020	12.8660 ±.0020	24.4860 ±.0040				3296.73 ±.97	15.511 ±.005	.37078 ±.00015	2.0671 ±.0006	25	12
13 monoclinic Sulfur S	mon.	P2₁/c(14)	S₈ ring molecules	48	11.04 ±.03	10.98 ±.03	10.92 ±.03		96.73 ±.50		1314.6 ±6.4	16.49 ±.08	.3943 ±.0020	1.944 ±.009	103	13
14 rhombohedral Sulfur S	hex-R	R3̄(148)	S₆ ring molecules	18	10.818 ±.002		4.280 ±.001				433.78 ±.19	14.514 ±.006	.34693 ±.00020	2.2092 ±.0010	r	14
15 Antimony Sb	hex-R	R3̄m(166)	arsenic	6	4.310 ±.001		11.279 ±.003				181.45 ±.09	18.213 ±.010	.43535 ±.00028	6.685 ±.004	26	15
16 Selenium Se	hex.	P3₁21(152) P3₂21(154)		3	4.3642 ±.0008		4.9588 ±.0008				81.793 ±.033	16.420 ±.007	.39249 ±.00020	4.8088 ±.0019	26	16
17 Silicon Si	cubic	Fd3m(227)	diamond	8	5.4305 ±.0003						160.15 ±.03	12.056 ±.002	.28819 ±.00009	2.3296 ±.0004	25	17
18 β-Tin (white) Sn	tet.	14₁/amd(141)		4	5.8315 ±.0008		3.1813 ±.0006				108.18 ±.04	16.289 ±.005	.38935 ±.00017	7.2867 ±.0024	26	18
19 Tellurium Te	hex.	P3₁21(152) P3₂21(154)		3	4.4570 ±.0008		5.9290 ±.0010				102.00 ±.04	20.476 ±.008	.48944 ±.00024	6.2316 ±.0025	25	19

X-Ray Crystallographic Data of Minerals

No.	Name and formula	Crystal system	Space group	Structure type	Z	a_o	b_o	c_o	β_o	Cell volume 10^{-24} cm³	Molar volume cm³	Molar volume cal bar⁻¹	X-Ray density grams cm⁻³	Temp. °C
20	Zinc Zn	hex.	P6₃/mmc(194)	hexagonal close packed	2	2.665 ±.001		4.947 ±.001		30.428 ±.024	9.162 ±.007	.2190 ±.0002	7.134 ±.006	25
	Sulfides, arsenides, tellurides, selenides, and sulfosalts													
21	Shandite β-Ni₃Pb₂S₂*	hex-R	R3̄m(166)		3	5.576 ±.010		13.658 ±.010		367.76 ±1.35	73.83 ±.27	1.765 ±.007	8.867 ±.033	r
22	High-Argentite AgS I	cubic			4	6.269 ±.020				246.4 ±2.4	37.09 ±.36	.8866 ±.0085	6.680 ±.064	600
23	Argentite AgS II	cubic			2	4.870 ±.008				115.5 ±.6	34.78 ±.17	.8313 ±.0041	7.125 ±.035	189
24	Acanthite AgS III	mon.	P2₁/c(14)		4	4.228 ±.002	6.228 ±.005	7.862 ±.003	99.58 ±.30	227.08 ±.29	34.19 ±.04	.8172 ±.0011	7.248 ±.009	25
25	High-Naumanite AgSe	cubic			2	4.993 ±.016				124.48 ±1.20	37.48 ±.36	.8959 ±.0087	7.862 ±.076	170
26	AgTe I	cubic			2	5.29 ±.01				148.0 ±.8	44.58 ±.26	1.065 ±.006	7.702 ±.044	825
27	AgTe II	cubic			2	6.585 ±.010				285.54 ±1.30	42.99 ±.20	1.028 ±.005	7.986 ±.036	250
28	Hessite Ag₂Te III	mon.	P2₁/c(14)		4	8.09 ±.02	4.48 ±.03	8.96 ±.02	123.33 ±.30	271.33 ±1.43	40.85 ±.22	.9764 ±.0052	8.405 ±.044	r
29	Ag₁.₈₅Cu₀.₁₅S I	cubic			4	6.110 ±.010				228.10 ±1.12	34.34 ±.17	.8209 ±.0041	6.635 ±.033	300
30	Ag₁.₈₅Cu₀.₁₅S II	cubic			2	4.825 ±.005				112.33 ±.35	33.83 ±.11	.8085 ±.0026	6.736 ±.021	116
31	Jalpaite Ag₁.₅₅Cu₀.₄₅S III	tet.			16	8.673 ±.004		11.756 ±.006		884.30 ±.93	33.286 ±.035	.79559 ±.00088	6.8455 ±.0072	r
32	Ag₁.₅₅Cu₀.₄₅S I	cubic			4	5.961 ±.009				211.82 ±.96	31.89 ±.14	.7623 ±.0035	6.283 ±.029	196
33	Ag₁.₅₅Cu₀.₄₅S II	hex.			2	4.138 ±.004		7.105 ±.007		105.36 ±.23	31.73 ±.07	.7583 ±.0017	6.316 ±.014	100
34	Stromeyerite Ag₀.₉₃Cu₁.₀₅S III	orth.	Cmcm(63)		4	4.066 ±.002	6.6 ±.04	7.072 ±.007		214.84 ±.18	32.35 ±.03	.7732 ±.0007	6.194 ±.005	r
35	Eucairite AgCuSe	orth.	pseudo P4/nmm(129)		10	4.105 ±.010	20.3 ±.05	6.1 ±.01		527.12 ±1.62	31.75 ±.10	.7588 ±.0024	7.887 ±.024	r
36	Petzite AgAuTe₂*	cubic	I4₁32(214)		8	10.38 ±.02				1118.4 ±6.5	84.19 ±.49	2.012 ±.012	9.214 ±.053	r
37	Maldonite Au₂Bi	cubic	Fd3m(227)	Cu₂Mg	8	7.958 ±.002				503.98 ±.38	37.94 ±.03	.9068 ±.0007	15.891 ±.012	465
38	High-Digenite Cu₂S I	cubic			4	5.725 ±.010				187.64 ±.98	28.25 ±.15	.6753 ±.0036	5.633 ±.030	465
39	High-Chalcocite Cu₂S II	hex.			2	3.961 ±.004		6.722 ±.007		91.34 ±.21	27.50 ±.06	.6574 ±.0015	5.786 ±.013	152
40	Chalcocite Cu₂S III	orth.	Ab2m(39)		96	11.881 ±.004	27.32 ±.01	13.491 ±.004		4379.5 ±2.5	27.475 ±.016	.65671 ±.00043	5.7924 ±.0034	r
41	Digenite Cu₁.₇₆S (Cu rich side)	cubic		deformed fluorite	4	5.5695 ±.0010				172.76 ±.09	26.012 ±.014	.6217 ±.0004	5.605 ±.003	25
42	Digenite Cu₁.₇₇S (S rich side)	cubic		deformed fluorite	4	5.5542 ±.0010				171.34 ±.09	25.798 ±.014	.6166 ±.0004	5.602 ±.005	25
43	Berzelianite Cu₂Se	cubic			4	5.85 ±.01				200.2 ±1.0	30.14 ±.15	.7205 ±.0037	6.835 ±.035	170
44	High-Bornite Cu₅FeS₄*	cubic			1	5.50 ±.01				166.4 ±.9	100.2 ±.5	2.395 ±.013	5.008 ±.027	240

X-Ray Crystallographic Data of Minerals

Name and formula	Crystal system	Structure type	Space group	Z	a_o	b_o	c_o	α_o	β_o	γ_o	Cell volume 10^{-24} cm³	Molar volume cm³	Molar volume cal bar⁻¹	X-Ray density grams cm⁻³	Temp. °C	
Metastable Bornite Cu₅FeS₄	cubic			8	10.94 ±.02						1309.34 ±7.18	98.57 ±.54	2.356 ±.013	5.091 ±.028	r	45
Low-Bornite Cu₅FeS₄*	tet.		P̄42₁c(144)	16	10.94 ±.02		21.88 ±.04				2618.7 ±10.7	98.57 ±.54	2.356 ±.010	5.091 ±.021	r	46
Umangite Cu₃Se₂	tet.		P4/mmm(123)	2	6.402 ±.010		4.276 ±.010				175.25 ±.68	52.77 ±.21	1.261 ±.005	6.604 ±.026	r	47
Heazlewoodite Ni₃S₂	hex-R		R32(155)	3	5.746 ±.001		7.134 ±.002				203.98 ±.09	40.95 ±.02	.9788 ±.0005	5.867 ±.003	r	48
Maucherite Ni₁₁As₈	tet.		P4₁2₁2(92)	4	6.870 ±.001		21.81 ±.01				1029.36 ±.56	154.98 ±.08	3.7043 ±.0021	8.0343 ±.0044	r	49
Pentlandite Fe₉.₂₅Ni₈.₇₅S₈	cubic		Fm3m(225)	4	10.196 ±.010						1059.96 ±3.12	159.59 ±.47	3.8144 ±.0113	4.823 ±.014	r	50
Pentlandite Fe₉.₇₅Ni₈.₂₅S₈	cubic		Fm3m(225)	4	10.095 ±.010						1028.77 ±3.06	154.89 ±.46	3.702 ±.011	4.998 ±.015	r	51
Sternbergite AgFe₂S₃*	orth.		Cmmm(63)	8	11.60 ±.02	12.675 ±.020	6.63 ±.01				974.81 ±2.71	73.39 ±.20	1.754 ±.005	4.303 ±.012	r	52
Argentopyrite AgFe₂S₃*	orth.		Pmmm(47)	4	6.64 ±.01	11.47 ±.02	6.45 ±.02				491.2 ±1.9	73.96 ±.29	1.768 ±.007	4.269 ±.017	r	53
Realgar AsS*	mon.		P2₁/m(11)	16	9.29 ±.05	13.53 ±.05	6.57 ±.03		106.55 ±.30		791.6 ±6.4	29.80 ±.24	.7122 ±.0058	3.591 ±.029	r	54
Oldhamite CaS	cubic	rock salt	Fm3m(225)	4	5.689 ±.006						184.12 ±.58	27.722 ±.088	.6626 ±.0021	2.602 ±.0021	r	55
Greenockite CdS	hex.	zincite	P6₃mc(186)	2	4.1354 ±.0010		6.7120 ±.0010				99.407 ±.050	29.934 ±.015	.71549 ±.00041	4.8261 ±.0024	r	56
Hawleyite CdS	cubic	sphalerite	F43m(216)	4	5.833 ±.002						198.46 ±.20	29.88 ±.03	.7142 ±.0008	4.835 ±.005	r	57
(hypothetical) CdS	cubic	rock salt	Fm3m(225)	4	5.516 ±.002						167.83 ±.18	25.27 ±.03	.6040 ±.0007	5.717 ±.006	r	58
Cadmoselite CdSe	hex.	zincite	P6₃mc(186)	2	4.2977 ±.0010		7.0021 ±.0010				112.00 ±.05	33.727 ±.016	.80614 ±.00044	5.6738 ±.0028	r	59
CdTe	cubic	sphalerite	F43m(216)	4	6.4805 ±.0006						272.16 ±.08	40.977 ±.012	.97943 ±.00032	5.8569 ±.0016	25	60
(hypothetical) CoS	cubic	sphalerite	F43m(216)	4	5.339 ±.001						152.19 ±.09	22.91 ±.02	.5477 ±.0004	3.971 ±.002	r	61
Chalcopyrite (CuFeS₂) CuFeS₁.₉₀	tet.		I42d(122)	4	5.2988 ±.0010		10.434 ±.005				292.96 ±.18	44.109 ±.027	1.0543 ±.0007	4.0878 ±.0025	r	62
Cubanite CuFe₂S₃	orth.		Pcmn(62)	4	6.46 ±.01	11.12 ±.01	6.23 ±.01				447.53 ±1.08	67.38 ±.16	1.611 ±.004	4.026 ±.010	r	63
Covellite CuS	hex.		P6₃/mmc(194)	6	3.792 ±.001		16.34 ±.01				203.48 ±.16	20.42 ±.02	.4882 ±.0005	4.682 ±.001	r	64
Klockmannite CuSe	hex.	deformed covellite	P6₃/mmc(194)	78	14.206 ±.010		17.25 ±.05				3014.8 ±9.7	23.28 ±.08	.5564 ±.0018	6.122 ±.020	r	65
Troilite FeS	hex.	niccolite	P6₃/mmc(194)	2	3.446 ±.003		5.877 ±.001				60.439 ±.106	18.20 ±.03	.4350 ±.0008	4.830 ±.009	28	66
Pyrrhotite Fe.₈₈₀S	hex.	defect niccolite	P6₃/mmc(194)	2	3.446 ±.001		5.848 ±.002				60.14 ±.04	18.11 ±.01	.4329 ±.0003	4.793 ±.003	28	67
Pyrrhotite Fe.₈₈₈S	hex.	defect niccolite	P6₃/mmc(194)	2	3.440 ±.001		5.709 ±.003				58.507 ±.046	17.62 ±.02	.4211 ±.0004	4.625 ±.004	28	68
(hypothetical) FeS	cubic	sphalerite	F43m(216)	4	5.455 ±.001						162.32 ±.09	24.44 ±.01	.5842 ±.0004	3.597 ±.002	r	69
(hypothetical) FeS	hex.	zincite	P6₃mc(186)	2	3.872 ±.001		6.345 ±.002				82.38 ±.05	24.81 ±.02	.5930 ±.0004	3.544 ±.002	r	70

X-Ray Crystallographic Data of Minerals

No.	Name and formula	Crystal system	Space group	Structure type	Z	a_o	b_o	c_o	α_o	β_o	γ_o	Cell volume 10^{-24} cm³	Molar volume cm³	Molar volume cal bar⁻¹	X-Ray density grams cm⁻³	Temp. °C	
71	Cinnabar HgS	hex.	P3₁21(152) P3₂1(154)	cinnabar	3	4.149 ±.001		9.495 ±.002				141.55 ±.07	28.416 ±.015	.6792	8.187 ±.004	r	71
72	Metacinnabar HgS	cubic	F$\bar{4}$3m(216)	sphalerite	4	5.8517 ±.0010						200.38 ±.10	30.169 ±.016	.7211 ±.0004	7.712 ±.004	r	72
73	Tiemannite HgSe	cubic	F$\bar{4}$3m(216)	sphalerite	4	6.0853 ±.0050						225.34 ±.56	33.928 ±.084	.8110 ±.0020	8.239 ±.020	r	73
74	Coloradoite HgTe	cubic	F$\bar{4}$3m(216)	sphalerite	4	6.4600 ±.0006						269.59 ±.08	40.590 ±.011	.97016 ±.00032	8.0855 ±.0023	r	74
75	Alabandite MnS	cubic	Fm3m(225)	rock salt	4	5.2234 ±.0005						142.51 ±.04	21.457 ±.006	.51289 ±.00019	4.0546 ±.0012	r	75
76	(hypothetical) MnS	cubic	F$\bar{4}$3m(216)	sphalerite	4	5.611 ±.002						176.65 ±.19	26.60 ±.03	.6357 ±.0007	3.271 ±.004	r	76
77	(hypothetical) MnS	hex.	P6₃mc(186)	zincite	2	3.986 ±.001		6.465 ±.002				88.96 ±.05	26.79 ±.02	.6403 ±.0004	3.248 ±.002	r	77
78	Nicolite NiAs	hex.	P6₃/mmc(194)	niccolite	2	3.618 ±.001		5.034 ±.001				57.07 ±.03	17.18 ±.01	.4108 ±.0003	7.776 ±.005	r	78
79	Millerite NiS	hex-R	R3m(160)	niccolite	9	9.616 ±.001		3.152 ±.001				252.41 ±.10	16.891 ±.006	.40374 ±.00020	5.3743 ±.0020	r	79
80	Breithauptite NiSb	hex.	P6₃/mmc(194)	niccolite	2	3.942 ±.001		5.155 ±.001				69.37 ±.04	20.89 ±.01	.4994 ±.0004	8.639 ±.005	r	80
81	Galena PbS	cubic	Fm3m(225)	rock salt	4	5.9360 ±.0005						209.16 ±.05	31.492 ±.008	.75272 ±.00024	7.5973 ±.0019	26	81
82	Clausthalite PbSe	cubic	Fm3m(225)	rock salt	4	6.1255 ±.0005						229.84 ±.06	34.605 ±.009	.82713 ±.00025	8.2690 ±.0020	r	82
83	Teallite PbSnS₂	orth.	Pbnm(62)	GeS	2	4.266 ±.003	11.4 ±.06	4.090 ±.002				199.24 ±.21	59.996 ±.063	1.4340 ±.0016	6.501 ±.007	r	83
84	Altaite PbTe	cubic	Fm3m(225)	rock salt	4	6.4606 ±.0005						269.66 ±.06	40.601 ±.009	.97043 ±.00027	8.2459 ±.0019	r	84
85	Cooperite PtS	tet.	P4₂/mmc(131)		2	3.4699 ±.0006		6.1098 ±.0010				73.563 ±.028	22.152 ±.008	.5295 ±.0003	10.254 ±.004	r	85
86	Herzenbergite SnS	orth.	Pbnm(62)	GeS	4	4.328 ±.002	11.19 ±.004	3.978 ±.01				192.66 ±.12	29.01 ±.02	.6933 ±.0005	5.197 ±.003	r	86
87	Sphalerite ZnS	cubic	F$\bar{4}$3m(216)	sphalerite	4	5.4093 ±.0005						158.28 ±.04	23.831 ±.007	.56962 ±.00020	4.0885 ±.0011	r	87
88	Wurtzite ZnS	hex.	P6₃mc(186)	zincite	2	3.8230 ±.0010		6.2565 ±.0010				79.190 ±.043	23.846 ±.013	.56998 ±.00036	4.0859 ±.0022	r	88
89	Stilleite ZnSe	cubic	F$\bar{4}$3m(216)	sphalerite	4	5.6685 ±.0005						182.14 ±.05	27.424 ±.007	.65548 ±.00022	5.2630 ±.0014	r	89
90	ZnTe	cubic	F$\bar{4}$3m(216)	sphalerite	4	6.1020 ±.0006						227.20 ±.07	34.209 ±.010	.81765 ±.00029	5.6410 ±.0017	r	90
91	Orpiment As₂S₃*	mon.	P2₁/n(14)		4	11.49 ±.02	9.59 ±.02	4.2 ±.0		90.45 ±.30		468.3 ±1.7	70.51 ±.25	1.685 ±.006	3.490 ±.013	r	91
92	Bismuthinite Bi₂S₃	orth.	Pbnm(62)	stibnite	4	11.150 ±.004	11.300 ±.004	3.981 ±.001				501.59 ±.28	75.520 ±.043	1.8050 ±.0011	6.8081 ±.0038	26	92
93	Tellurobismuthite Bi₂Te₃	hex-R	R3m(166)	Bi₂Te₃S	3	4.3835 ±.0020		30.487 ±.03				507.33 ±.47	101.85	2.4342 ±.0023	7.862 ±.007	25	93
94	Stibnite Sb₂S₃	orth.	Pbnm(62)	stibnite	4	11.229 ±.004	11.310 ±.004	3.8389 ±.0010				487.54 ±.28	73.406 ±.042	1.7545 ±.0010	4.6276 ±.0026	25	94
95	Linnaeite Co₃S₄	cubic	Fd3m(227)	spinel	8	9.401 ±.001						830.85 ±.27	62.548	1.4950 ±.0005	4.8772 ±.0016	r	95
96	Greigite Fe₃S₄	cubic	Fd3m(227)	spinel	8	9.876 ±.002						963.26 ±.59	72.32 ±.04	1.733 ±.001	4.079 ±.003	r	96

X-Ray Crystallographic Data of Minerals

	Name and formula	Crystal system	Space group	Structure type	Z	a_o	b_o	c_o	α_o	β_o	γ_o	Cell volume 10^{-24} cm³	Molar volume cm³	Molar volume cal bar⁻¹	X-Ray density grams cm⁻³	Temp. °C	
97	Daubreelite $FeCr_2S_4$	cubic	Fd3m(227)	spinel	8	9.966 ±.005						989.83 ±1.49	74.52 ±.11	1.781 ±.003	3.866 ±.006	r	97
98	Violarite $FeNi_2S_4$	cubic	Fd3m(227)	spinel	8	9.464 ±.005						847.66 ±1.34	63.81 ±.10	1.525 ±.002	4.725 ±.008	r	98
99	Polymidite Ni_3S_4	cubic	Fd3m(227)	spinel	8	9.480 ±.001						851.97 ±.27	64.138 ±.020	1.5330 ±.0005	4.7458 ±.0015	r	99
100	Co-Safflorite $CoAs_2$	mon.		deformed marcasite	2	5.049 ±.002	5.872 ±.002	3.127 ±.001		90.45 ±.20		92.706 ±.057	27.92 ±.02	.6672 ±.0005	7.479 ±.005	26	100
101	Safflorite $(Co,Fe)As_2$	orth.	Pnnm(58)	marcasite	2	5.231 ±.002	5.953 ±.002	2.962 ±.002				92.237 ±.078	27.775 ±.024	.6639 ±.0006	7.461 ±.006	26	101
102	Cobaltite $CoAsS$*	cubic	P2₁3(198)	NiSbS	4	5.60 ±.05						175.62 ±4.70	26.44 ±.71	.6320 ±.0170	6.275 ±.168	r	102
103	Glaucodot $(Co,Fe)AsS$*	orth.	Cmmm(65)		24	6.64 ±.05	28.39 ±.10	5.64 ±.05				1063.2 ±12.9	26.68 ±.32	.6377 ±.0078	6.161 ±.075	r	103
104	Cattierite CoS_2	cubic	Pa3(205)	pyrite	4	5.5345 ±.0005						169.53 ±.05	25.524 ±.007	.61009 ±.00021	4.8213 ±.0013	r	104
105	Trogtalite $CoSe_2$	cubic	Pa3(205)	pyrite	4	5.8588 ±.0010						201.11 ±.10	30.279 ±.016	.72374 ±.00042	7.1618 ±.0037	r	105
106	Loellingite $FeAs_2$	orth.	Pnnm(58)	marcasite	2	5.300 ±.002	5.981 ±.002	2.882 ±.001				91.357 ±.056	27.51 ±.02	.6576 ±.0005	7.477 ±.005	26	106
107	Arsenopyrite $FeAsS$*	tri.	P1̄(2)		4	5.760 ±.010	5.690 ±.005	5.785 ±.005	90.00 ±.20	112.23 ±.20	90.00 ±.20	175.51 ±.44	26.42 ±.07	.6316 ±.0016	6.162 ±.015	r	107
108	Gudmundite $FeSbS$*	mon.	B2₁/d(14)		8	10.00 ±.05	5.93 ±.03	6.73 ±.03		90.00 ±.50		399.09 ±3.35	30.04 ±.25	.7181 ±.0061	6.978 ±.059	r	108
109	Pyrite FeS_2	cubic	Pa3(205)	pyrite	4	5.4175 ±.0005						159.00 ±.04	23.940 ±.007	.57221 ±.00020	5.0116 ±.0014	r	109
110	Marcasite FeS_2*	orth.	Pnnm(58)	marcasite	2	4.443 ±.002	5.423 ±.002	3.3876 ±.0015				81.622 ±.060	24.579 ±.018	.58749 ±.00047	4.8813 ±.0036	25	110
111	Ferroselite $FeSe_2$	orth.	Pnnm(58)	marcasite	2	4.801 ±.005	5.778 ±.005	3.587 ±.004				99.50 ±.17	29.96 ±.05	.7162 ±.0013	7.134 ±.013	r	111
112	Frohbergite $FeTe_2$	orth.	Pnnm(58)	marcasite	2	5.265 ±.005	6.265 ±.005	3.869 ±.002				127.62 ±.17	38.43 ±.05	.9185 ±.0013	8.094 ±.011	r	112
113	Hauerite MnS_2	cubic	Pa3(205)	pyrite	4	6.1014 ±.0006						227.14 ±.07	34.198 ±.010	.81741 ±.00029	3.4816 ±.0010	28	113
114	Molybdenite MoS_2	hex.	P6₃/mmc(194)	molybdenite	2	3.1604 ±.0010		12.295				106.35 ±.07	32.025 ±.021	.76547 ±.00055	4.9982 ±.0033	26	114
115	Rammelsbergite $NiAs_2$	orth.	Pnnm(58)	marcasite	2	4.757 ±.002	5.797 ±.004	3.542 ±.002				97.645 ±.096	29.41 ±.03	.7030 ±.0007	7.091 ±.007	26	115
116	Pararammelsbergite $NiAs_2$	orth.	Pbca(61)		8	5.75 ±.01	5.82 ±.01	11.428 ±.02				382.42 ±1.15	28.79 ±.09	.6882 ±.0021	7.244 ±.022	r	116
117	Gersdorffite $NiAsS$	cubic	P2₁3(198)		4	5.693 ±.001						184.51 ±.10	27.78 ±.01	.6640 ±.0004	5.964 ±.003	26	117
118	Vaesite NiS_2	cubic	Pa3(205)	pyrite	4	5.6873 ±.0005						183.96 ±.05	27.697 ±.007	.66203 ±.00022	4.4350 ±.0012	r	118
119	$NiSe_2$	cubic	Pa3(205)	pyrite	4	5.9604 ±.0010						211.75 ±.11	31.882 ±.016	.76204 ±.00043	6.7948 ±.0034	20	119
120	Melonite $NiTe_2$	hex.	P3̄m1(164)	cadmium iodide	1	3.869 ±.010		5.308 ±.010				68.81 ±.38	41.44 ±.23	.9905 ±.0055	7.575 ±.042	84	120
121	Sperrylite $PtAs_2$	cubic	Pa3(205)	pyrite	4	5.968 ±.005						212.56 ±.53	32.00 ±.08	.7650 ±.0020	10.778 ±.027	r	121
122	Laurite RuS_2	cubic	Pa3(205)	pyrite	4	5.60 ±.02						175.6 ±1.9	26.44 ±.28	.6320 ±.0068	6.248 ±.067	r	122

X-Ray Crystallographic Data of Minerals

	Name and formula	Crystal system	Space group	Structure type	Z	a_0	b_0	c_0	α_0	β_0	γ_0	Cell volume 10^{-24} cm³	Molar volume cm³	Molar volume cal bar⁻¹	X-Ray density grams cm⁻³	Temp. °C	
123	Tungstenite WS_2	hex.	$P6_3/mmc(194)$	molybdenite	2	3.154 ±.001		12.362 ±.004				106.50 ±.08	32.069 ±.023	.76652 ±.00059	7.7325 ±.0055	26	123
124	Co-Skutterudite $CoAs_{3-x}$ $CoAs_{2.95}$	cubic	$Im3(204)$		8	8.2060 ±.0010						552.58 ±.20	41.599 ±.015	.99428 ±.00041	6.7298 ±.0025	r	124
125	Fe-Skutterudite $FeAs_{3-x}$ $FeAs_{2.95}$	cubic	$Im3(204)$		8	8.1814 ±.0010						547.62 ±.20	41.226 ±.015	.98537 ±.00041	6.7158 ±.0025	r	125
126	Ni-Skutterudite $NiAs_{3-x}$ $NiAs_{2.95}$	cubic	$Im3(204)$		8	8.3300 ±.0010						578.01 ±.21	43.513 ±.016	1.0400 ±.0004	6.4286 ±.0023	r	126
127	Tennantite $Cu_{12}As_4S_{13}$	cubic	$I\bar{4}3m(217)$	tetrahedrite	2	10.190 ±.004						1058.09 ±1.25	318.62 ±.38	7.1652 ±.0090	4.642 ±.006	r	127
128	Tetrahedrite $Cu_{12}Sb_4S_{13}$	cubic	$I\bar{4}3m(217)$	tetrahedrite	2	10.327 ±.004						1101.3 ±1.3	331.64 ±.39	7.9266 ±.0094	5.024 ±.006	r	128
129	Enargite Cu_3AsS_4	orth.	$Pnm2(34)$		2	6.426 ±.005	7.42 ±.00?	6.14 ±.05				293.03 ±.38	88.24 ±.12	2.109 ±.003	4.463 ±.006	26	129
130	Luzonite Cu_3AsS_4*	tet.	$I\bar{4}2m(121)$		2	5.289 ±.005		10.40 ±.008				292.04 ±.60	87.94 ±.18	2.1019 ±.0043	4.478 ±.009	26	130
131	Famatinite Cu_3SbS_4*	tet.	$I\bar{4}m(121)$		2	5.384 ±.005		10.770 ±.008				312.19 ±.62	94.01 ±.19	2.2469 ±.0045	4.687 ±.009	26	131
132	Proustite Ag_3AsS_3	hex-R	$R3c(161)$		6	10.816 ±.001		8.648 ±.0013				880.89 ±.21	88.420 ±.19	2.1133 ±.0006	5.595 ±.009	26	132
133	Pyrargyrite Ag_3SbS_3	hex-R	$R3c(161)$		6	11.052 ±.002		8.777 ±.0420				922.18 ±.40	92.564 ±.040	2.2124 ±.0010	5.8506 ±.0025	26	133
134	Miargyrite $AgSbS_2$*	mon.	$Cc(9)$		8	12.862 ±.013	4.111 ±.004	13.220 ±.0.0		98.63 ±.15		691.10 ±1.14	52.027 ±.086	1.244 ±.002	5.646 ±.009	r	134

Oxides and hydroxides

	Name and formula	Crystal system	Space group	Structure type	Z	a_0	b_0	c_0	α_0	β_0	γ_0	Cell volume 10^{-24} cm³	Molar volume cm³	Molar volume cal bar⁻¹	X-Ray density grams cm⁻³	Temp. °C	
135	Corundum Al_2O_3	hex-R	$R\bar{3}c(167)$	corundum	6	4.7591 ±.0004		12.9894 ±.0030				254.78 ±.07	25.575 ±.007	.61128 ±.00022	3.9869 ±.0011	25	135
136	Boehmite $AlO(OH)$*	orth.	$Cmcm(63)$	lepidocrocite	4	2.868 ±.003	12.227 ±.003	3.70 ±.03				129.75 ±.17	19.535 ±.026	.46695 ±.00067	3.071 ±.004	26	136
137	Diaspore $AlO(OH)$*	orth.	$Pbnm(62)$		4	4.401 ±.005	9.421 ±.005	2.845 ±.02				117.96 ±.17	17.760 ±.026	.4245 ±.0007	3.378 ±.005	r	137
138	Gibbsite $Al(OH)_3$	mon.	$P2_1/n(14)$		8	9.719 ±.002	5.070? ±.001?	8.6412 ±.000?		94.57 ±.25		424.49 ±.20	31.956 ±.015	.7638 ±.0004	2.441 ±.001	25	138
139	Arsenolite As_2O_3	cubic	$Fd3m(227)$	diamond	16	11.074 ±.005						1358.0 ±1.8	51.118 ±.069	1.2218 ±.0017	3.870 ±.005	25	139
140	Claudetite As_2O_3	mon.	$P2_1/n(14)$		4	5.339 ±.002	12.984 ±.005	4.5405 ±.00.0		94.27 ±.10		313.88 ±.19	47.259 ±.028	1.1296 ±.0007	4.1863 ±.0025	25	140
141	Bromellite BeO	hex.	$P6_3mc(186)$	zincite	2	2.6979 ±.0005		4.3792 ±.0005				27.592 ±.011	8.3086 ±.0032	.19862 ±.00012	3.0104 ±.0012	26	141
142	Bismite α-Bi_2O_3	mon.	$P2_1/c(14)$	pseudo orthorhombic	8	8.166 ±.005	13.827 ±.010	5.850 ±.004		90.00 ±.20		660.53 ±.77	49.73 ±.06	1.1885 ±.0014	9.371 ±.011	25	142
143	Lime CaO	cubic	$Fm3m(225)$	rock salt	4	4.8108 ±.0005						111.34 ±.03	16.764 ±.005	.40071 ±.00017	3.3453 ±.0010	26	143
144	Portlandite $Ca(OH)_2$	hex.	$P\bar{3}m1(164)$	CdI_2	1	3.5933 ±.0005		4.9065 ±.002?				54.888 ±.027	33.056 ±.016	.79011 ±.00043	2.2415 ±.0011	26	144
145	Monteponite CdO	cubic	$Fm3m(225)$	rock salt	4	4.6953 ±.0010						103.51 ±.07	15.585 ±.010	.37254 ±.00028	8.2386 ±.0053	27	145
146	Cerianite CeO_2	cubic	$Fm3m(225)$	fluorite	4	5.4110 ±.0020						158.43 ±.18	23.853 ±.026	.57016 ±.00068	7.216 ±.008	26	146
147	CoO	cubic	$Fm3m(225)$	rock salt	4	4.260 ±.002						77.31 ±.11	11.64 ±.02	.2782 ±.0004	6.438 ±.009	26	147

X-Ray Crystallographic Data of Minerals

#	Name and formula	Crystal system	Space group	Structure type	Z	a_0	b_0	c_0	α_0	β_0	γ_0	Cell volume 10^{-24} cm³	Molar volume cm³	Molar volume cal bar⁻¹	X-Ray density grams cm⁻³	Temp. °C
148	Eskolaite Cr_2O_3	hex-R	R3̄c(167)	corundum	6	4.9607 ±.0020		13.599 ±.010				289.82 ±.32	29.090 ±.032	.6953 ±.0008	5.225 ±.006	r
149	Tenorite CuO	mon.	C2/c(15)		4	4.684 ±.005	3.425 ±.005	5.129 ±.005		99.47 ±.17		81.16 ±.17	12.22 ±.03	.2921 ±.0007	6.509 ±.014	26
150	Cuprite Cu_2O	cubic	Pn3m(224)		2	4.2696 ±.0010						77.833 ±.055	23.437 ±.016	.56021 ±.00044	6.1047 ±.0043	26
151	Wustite $Fe_{.933}O$	cubic	Fm3m(225)	defect rock salt	4	4.3088 ±.0003						79.996 ±.017	12.044 ±.003	.28791 ±.00011	5.7471 ±.0012	17
152	Hematite Fe_2O_3	hex-R	R3̄c(167)	corundum	6	5.0329 ±.0010		13.7492 ±.0010				301.61 ±.12	30.274 ±.012	.72361 ±.00034	5.2749 ±.0021	25
153	Magnetite Fe_3O_4	cubic	Fd3m(227)	spinel	8	8.3940 ±.0005						591.43 ±.11	44.524 ±.008	1.0642 ±.0002	5.2003 ±.0009	22
154	Goethite α-FeO(OH)*	orth.	Pbnm(62)		4	4.596 ±.005	9.957 ±.010	3.021 ±.003				138.2 ±.2	20.82 ±.04	.4975 ±.0009	4.269 ±.008	r
155	Lepidocrocite γ-FeO(OH)*	orth.	Amam(63)		4	3.868 ±.010	12.525 ±.010	3.066 ±.003				148.54 ±.43	22.364 ±.064	.5346 ±.0016	3.973 ±.011	r
156	α-Ga_2O_3	hex-R	R3̄c(167)	corundum	6	4.9793 ±.0010		13.429 ±.003				288.34 ±.13	28.943 ±.013	.69179 ±.00036	6.4762 ±.0030	24
157	Low-germania GeO_2	tet.	P4₂/mnm(136)	rutile	2	4.3963 ±.0010		2.8626 ±.0010				55.327 ±.032	16.660 ±.010	.39824 ±.00027	6.2777 ±.0036	25
158	High-germania GeO_2	hex.	P3₁21(152) P3₂21(154)	α-quartz	3	4.987 ±.002		5.652 ±.002				121.73 ±.11	24.438 ±.021	.58413 ±.0056	4.2797 ±.0038	26
159	Ice H_2O	hex.	P6₃/mmc(194)		4	4.5212 ±.0010		7.3666 ±.0010				130.41 ±.06	19.635 ±.009	.46932 ±.00026	.9175 ±.0004	0
160	Hafnia HfO_2	mon.	P2₁/c(14)	baddeleyite	4	5.1156 ±.0010	5.1722 ±.0010	5.2948 ±.0010		99.18 ±.08		138.30 ±.06	20.823 ±.008	.49772 ±.00025	10.108 ±.004	r
161	Montroydite HgO	orth.	Pnma(62)		4	6.608 ±.003	5.518 ±.003	3.519 ±.003				128.3 ±.1	19.32 ±.02	.4618 ±.0006	11.21 ±.01	25
162	Periclase MgO	cubic	Fm3m(225)	rock salt	4	4.2117 ±.0005						74.709 ±.027	11.248 ±.004	.26889 ±.00014	3.5837 ±.0013	25
163	Brucite $Mg(OH)_2$	hex.	P3̄m1(164)	CdI_2	1	3.147 ±.004		4.769 ±.004				40.90 ±.11	24.63 ±.07	.5888 ±.0016	2.368 ±.006	26
164	Manganosite MnO	cubic	Fm3m(225)	rock salt	4	4.4448 ±.0005						87.813 ±.030	13.221 ±.004	.31604 ±.00015	5.3653 ±.0018	26
165	Pyrolusite MnO_2	tet.	P4₂/mnm(136)	rutile	2	4.388 ±.003		2.865 ±.002				55.16 ±.08	16.61 ±.02	.3971 ±.0007	5.234 ±.008	r
166	Bixbyite Mn_2O_3	cubic	Ia3(206)	Tl_2O_3	16	9.411 ±.005						833.5 ±1.3	31.37 ±.05	.7499 ±.0012	5.032 ±.008	25
167	Hausmanite Mn_3O_4	tet.	I4₁/amd(141)		8	8.136 ±.005		9.422 ±.005				623.68 ±.84	46.95 ±.06	1.1222 ±.0016	4.873 ±.007	20
168	Molybdite MoO_3	orth.	Pbnm(62)		4	3.962 ±.002	13.858 ±.005	3.697 ±.004				202.98 ±.25	30.56 ±.04	.7305 ±.0010	4.710 ±.006	26
169	Bunsenite NiO	cubic	Fm3m(225)	rock salt	4	4.177 ±.002						72.88 ±.10	10.97 ±.02	.2623 ±.0004	6.809 ±.010	26
170	Litharge PbO red	tet.	P4/nmm(129)		2	3.9759 ±.0040		5.023 ±.004				79.40 ±.17	23.91 ±.05	.5715 ±.0013	9.334 ±.020	27
171	Massicot PbO yellow	orth.	Pb2a(32)		4	5.489 ±.003	4.755 ±.004	5.891 ±.004				153.8 ±.2	23.15 ±.03	.5533 ±.0007	9.641 ±.012	27
172	Minium Pb_3O_4	tet.	P4₂/mbc(135)		4	8.815 ±.005		6.565 ±.003				510.13 ±.62	76.81 ±.09	1.836 ±.002	8.926 ±.009	25
173	Senarmontite Sb_2O_3	cubic	Fm3m(225)	arsenic trioxide	16	11.152 ±.003						1386.9 ±1.1	52.206 ±.042	1.2478 ±.0011	5.5837 ±.0045	26

X-Ray Crystallographic Data of Minerals

#	Name and formula	Crystal system	Space group	Structure type	Z	a_o	b_o	c_o	β_o	Cell volume 10^{-24} cm³	Molar volume cm³	Molar volume cal bar⁻¹	X-Ray density grams cm⁻³	Temp. °C
174	Valentinite Sb₂O₃	orth.	Pccn(56)	antimony trioxide	4	4.914 ±.002	12.46 ±.00	5.21 ±.004		332.13 ±.31	50.007 ±.047	1.1952 ±.0012	5.8292 ±.0054	25
175	Cervantite Sb₂O₄	cubic	Fd3m(227)		8	10.305 ±.005				1094.3 ±1.6	82.38 ±.12	1.9690 ±.0029	3.733 ±.005	26
176	Selenolite SeO₂	tet.	P4₂/mbc(135) P4₃bc(106)		8	8.35 ±.01		5.68 ±.01		354.2 ±1.1	26.66 ±.08	.6373 ±.0020	4.161 ±.013	26
177	α-Quartz SiO₂*	hex.	P3₂21(152) P3₁21(154)		3	4.9136 ±.0001		5.4051 ±.0001		113.01 ±.01	22.688 ±.001	.54229 ±.00007	2.6483 ±.0001	25
178	β-Quartz SiO₂*	hex.	P6₂22(181) P6₄22(180)		3	4.999 ±.001		5.592 ±.020		118.15 ±.06	23.718 ±.013	.5669 ±.0004	2.533 ±.002	575
179	α-Cristobalite SiO₂	tet.	P4₁2₁2(92) P4₃2₁2(96)		4	4.971 ±.003		6.918 ±.003		170.95 ±.22	25.739 ±.033	.61521 ±.00083	2.3344 ±.0030	25
180	β-Cristobalite SiO₂	cubic	Fd3m(227)		8	7.1382 ±.0010				363.72 ±.15	27.381 ±.012	.65447 ±.00032	2.1944 ±.0009	405
181	Keatite SiO₂	tet.	P4₁2₁2(92) P4₃2₁2(96)		12	7.456 ±.003		8.604 ±.045		478.3 ±.5	24.01 ±.02	.5738 ±.0006	2.503 ±.003	r
182	β-Tridymite SiO₂	hex.	P6₂c(172) P6₃/mmc(194)		4	5.0463 ±.0020		8.263 ±.0030		182.08 ±.16	27.414 ±.024	.65527 ±.00062	2.1917 ±.0019	405
183	Coesite SiO₂*	mon.	B2/b(15)		16	7.152 ±.001	12.37 ±.002	7.12 ±.001	120.00 ±.17	548.37 ±.95	20.641 ±.036	.49338 ±.00090	2.9110 ±.0050	25
184	Stishovite SiO₂*	tet.	P4/mnm(136)	rutile	2	4.1790 ±.0010		2.6649 ±.0010		46.540 ±.028	14.014 ±.009	.33500 ±.00025	4.2874 ±.0026	r
185	Melanophlogite SiO₂*	cubic	Pm3n(223)	clathrate type	46	13.402 ±.004				2407.2 ±2.2	31.516 ±.028	.75325 ±.00072	1.9065 ±.0017	r
186	Cassiterite SnO₂	tet.	P4/mnm(136)	rutile	2	4.738 ±.003		3.188 ±.003		71.57 ±.11	21.55 ±.03	.5151 ±.0009	6.992 ±.011	26
187	Tellurite TeO₂*	orth.	Pbca(61)	tellurite	8	5.607 ±.003	12.034 ±.005	5.463 ±.003		368.61 ±.32	27.750 ±.024	.66328 ±.00062	5.7514 ±.0050	25
188	Paratellurite TeO₂*	tet.	P4₁2₁2(92) P4₃2₁2(96)		4	4.810 ±.002		7.613 ±.002		176.14 ±.15	26.52 ±.02	.6339 ±.0006	6.018 ±.005	25
189	Thorianite ThO₂	cubic	Fm3m(225)	fluorite	4	5.5952 ±.0005				175.16 ±.05	26.373 ±.007	.63038 ±.00021	10.012 ±.003	25
190	Rutile TiO₂	tet.	P4/mnm(136)		2	4.5937 ±.0005		2.9598 ±.0000		62.500 ±.025	18.820 ±.008	.44986 ±.00023	4.2453 ±.0017	25
191	Anatase TiO₂	tet.	I4₁/amd(141)		4	3.785 ±.002		9.51 ±.00		136.30 ±.17	20.522 ±.025	.4905 ±.0007	3.893 ±.005	r
192	Brookite TiO₂	orth.	Pcab(61)		8	5.456 ±.002	9.182 ±.005	5.14 ±.00		257.6 ±.2	19.40 ±.02	.4636 ±.0005	4.119 ±.004	r
193	Titanium sesquioxide Ti₂O₃	hex-R	R3̄c(167)	corundum	6	5.149 ±.002		13.645 ±.010		313.2 ±.3	31.44 ±.03	.7515 ±.0009	4.574 ±.005	r
194	Uraninite UO₂	cubic	Fm3m(225)	fluorite	4	5.4682 ±.0010				163.51 ±.09	24.618 ±.014	.58843 ±.00037	10.969 ±.006	26
195	Karelianite V₂O₃	hex-R	R3̄c(167)	corundum	6	4.952 ±.002		14.00 ±.03		297.36 ±.32	29.848 ±.032	.71342 ±.00081	5.0216 ±.0054	r
196	Zincite ZnO	hex.	P6₃mc(186)	zincite	2	3.2495 ±.0005		5.206 ±.000		47.615 ±.015	14.338 ±.005	.34273 ±.00016	5.6750 ±.0018	25
197	Baddeleyite ZrO₂	mon.	P2₁/c(14)	baddeleyite	4	5.1454 ±.0010	5.2075 ±.0010	5.31 ±.001	99.23 ±.08	140.46 ±.06	21.148 ±.009	.50548 ±.00025	5.8267 ±.0023	r

Multiple oxides

#	Name and formula	Crystal system	Space group	Structure type	Z	a_o	b_o	c_o	β_o	Cell volume 10^{-24} cm³	Molar volume cm³	Molar volume cal bar⁻¹	X-Ray density grams cm⁻³	Temp. °C
198	Spinel MgAl₂O₄	cubic	Fd3m(227)	spinel	8	8.080 ±.002				527.5 ±.4	39.71 ±.03	.9492 ±.0008	3.583 ±.003	26

X-Ray Crystallographic Data of Minerals

	Name and formula	Crystal system	Space group	Structure type	Z	a_0	b_0	c_0	α_0	β_0	γ_0	Cell volume 10^{-24} cm³	Molar volume cm³	Molar volume cal bar⁻¹	X-Ray density grams cm⁻³	Temp. °C	
199	Hercynite FeAl₂O₄	cubic	Fd3m(227)	spinel	8	8.150 ±.004						541.3 ±.6	40.75 ±.05	.9740 ±.0011	4.265 ±.005	25	199
200	Galaxite MnAl₂O₄	cubic	Fd3m(227)	spinel	8	8.258 ±.002						563.2 ±.4	42.39 ±.03	1.013 ±.001	4.078 ±.003	25	200
201	Gahnite ZnAl₂O₄	cubic	Fd3m(227)	spinel	8	8.0848 ±.0020						528.45 ±.39	39.783 ±.030	.95088 ±.00075	4.6083 ±.0034	26	201
202	Magnetite FeFe₂O₄	cubic	Fd3m(227)	spinel	8	8.3940 ±.0005						591.43 ±.11	44.524 ±.008	1.0642 ±.0002	5.2003 ±.0009	22	202
203	Jacobsite MnFe₂O₄	cubic	Fd3m(227)	spinel	8	8.499 ±.002						613.9 ±.4	46.22 ±.03	1.105 ±.001	4.990 ±.004	25	203
204	Trevorite NiFe₂O₄	cubic	Fd3m(227)	spinel	8	8.339 ±.003						579.9 ±.6	43.65 ±.05	1.043 ±.001	5.370 ±.006	25	204
205	Picrochromite MgCr₂O₄	cubic	Fd3m(227)	spinel	8	8.333 ±.003						578.6 ±.6	43.56 ±.05	1.041 ±.001	4.415 ±.005	26	205
206	Ilmenite FeTiO₃	hex-R	R3̄(148)	ilmenite	6	5.093 ±.005		14.055 ±.020				315.73 ±.75	31.69 ±.08	.7574 ±.0019	4.788 ±.012	r	206
207	Geikielite MgTiO₃	hex-R	R3̄(148)	ilmenite	6	5.054 ±.005		13.898 ±.010				307.44 ±.65	30.86 ±.07	.7376 ±.0016	3.896 ±.008	26	207
208	Pyrophanite MnTiO₃	hex-R	R3̄(148)	ilmenite	6	5.155 ±.005		14.18 ±.01				326.3 ±.7	32.76 ±.07	.7829 ±.0017	4.605 ±.010	r	208
209	Cobalt Titanate CoTiO₃	hex-R	R3̄(148)	ilmenite	6	5.066 ±.001		13.918 ±.005				309.34 ±.17	31.05 ±.02	.7422 ±.0004	4.986 ±.003	r	209
210	Perovskite CaTiO₃	orth.	Pcmn(62)	perovskite	4	5.3670 ±.0010	7.6438 ±.0010	5.4439 ±.0010				223.33 ±.07	33.626 ±.010	.80371 ±.00028	4.0439 ±.0012	r	210
211	Chrysoberyl BeAl₂O₄	orth.	Pmnb(62)	olivine	4	5.4756 ±.0020	9.4041 ±.0030	4.4267 ±.0020				227.94 ±.15	34.320 ±.023	.82031 ±.00059	3.6997 ±.0025	25	211
Halides																	
212	Halite NaCl	cubic	Fm3m(225)	rock salt	4	5.6402 ±.0002						179.43 ±.02	27.015 ±.003	.64571 ±.00011	2.1634 ±.0002	26	212
213	Sylvite KCl	cubic	Fm3m(225)	rock salt	4	6.2931 ±.0002						249.23 ±.02	37.524 ±.004	.89690 ±.00013	1.9868 ±.0002	25	213
214	Villiaumite NaF	cubic	Fm3m(225)	rock salt	4	4.6342 ±.0005						99.523 ±.032	14.984 ±.005	.35818 ±.00016	2.8021 ±.0009	25	214
215	Chlorargyrite AgCl	cubic	Fm3m(225)	rock salt	4	5.5491 ±.0005						170.87 ±.05	25.727 ±.007	.61493 ±.00021	5.5710 ±.0015	26	215
216	Bromargyrite AgBr	cubic	Fm3m(225)	rock salt	4	5.7745 ±.0005						192.55 ±.05	28.991 ±.008	.69294 ±.00022	6.4772 ±.0017	26	216
217	Nantockite CuCl	cubic	F4̄3m(216)	sphalerite	4	5.416 ±.003						158.87 ±.26	23.92 ±.04	.5717 ±.0010	4.139 ±.007	25	217
218	Marshite CuI	cubic	F4̄3m(216)	sphalerite	4	6.0507 ±.0010						221.52 ±.11	33.353 ±.017	.7972 ±.0004	5.710 ±.003	26	218
219	Miersite AgI	cubic	F4̄3m(216)	sphalerite	4	6.4963 ±.0010						274.16 ±.10	41.278 ±.020	.9866 ±.0004	5.688 ±.003	r	219
220	Iodargyrite AgI	hex.	P6₃mc(186)	zincite	2	4.5955 ±.0010		7.5005 ±.0033				137.18 ±.10	41.308 ±.030	.9873 ±.0009	5.683 ±.004	25	220
221	Calomel HgCl	tet.	14/mm(139)		4	4.478 ±.005		10.910 ±.005				218.77 ±.50	32.939 ±.075	.7873 ±.0018	7.166 ±.016	26	221
222	Fluorite CaF₂	cubic	Fm3m(225)	fluorite	4	5.4638 ±.0004						163.11 ±.04	24.558 ±.005	.58701 ±.00017	3.1792 ±.0007	25	222
223	Sellaite MgF₂	tet.	P4₂/mnm(136)	rutile	2	4.621 ±.001		3.050 ±.001				65.13 ±.04	19.61 ±.01	.4688 ±.0003	3.177 ±.002	18	223

X-Ray Crystallographic Data of Minerals

No.	Name and formula	Crystal system	Space group	Structure type	Z	a_o	b_o	c_o	α_o	β_o	γ_o	Cell volume 10^{-24} cm³	Molar volume cm³	Molar volume cal bar⁻¹	X-Ray density grams cm⁻³	Temp. °C
224	Chloromagnesite $MgCl_2$	hex-R	R3̄m(166)		3	3.632 ±.004		17.795 ±.016				203.29	40.81 ±.10	.9754 ±.0024	2.333 ±.006	r
225	Lawrencite $FeCl_2$	hex-R	R3̄m(166)		3	3.593 ±.003		17.58 ±.09				196.55 ±1.06	39.46 ±.21	.9431 ±.0051	3.212 ±.017	r
226	Scacchite $MnCl_2$	hex-R	R3̄m(166)		3	3.711 ±.002		17.59 ±.07				209.79 ±.86	42.11 ±.17	1.007 ±.004	2.988 ±.012	r
227	Cotunnite $PbCl_2$	orth.	Pnmb(62)		4	4.535 ±.005	7.32 ±.01	9.05 ±.01				312.74 ±.64	47.09 ±.10	1.1254 ±.0023	5.906 ±.012	26
228	Matlockite $PbFCl$	tet.	P4/nmm(129)		2	4.106 ±.005		7.23 ±.01				121.89 ±.34	36.70 ±.10	.8773 ±.0025	9.853 ±.028	26
229	Cryolite Na_3AlF_6*	mon.	P2₁/n(14)		2	5.40 ±.01	5.60 ±.01	7.776 ±.010		90.18 ±.25		235.1 ±.7	70.81 ±.20	1.692 ±.005	2.965 ±.009	r
230	Neighborite $NaMgF_3$	orth.	Pcmn(62)	perovskite	4	5.363 ±.001	7.676 ±.001	5.503 ±.001				226.54 ±.07	34.11 ±.01	.8152 ±.0003	3.058 ±.001	18

Carbonates and nitrates

No.	Name and formula	Crystal system	Space group	Structure type	Z	a_o	b_o	c_o	α_o	β_o	γ_o	Cell volume 10^{-24} cm³	Molar volume cm³	Molar volume cal bar⁻¹	X-Ray density grams cm⁻³	Temp. °C
231	Calcite $CaCO_3$	hex-R	R3̄c(167)	calcite	6	4.9899 ±.0010		17.064 ±.002				367.96 ±.15	36.934 ±.015	.88278 ±.00041	2.7100 ±.0011	26
232	Otavite $CdCO_3$	hex-R	R3̄c(167)	calcite	6	4.9204 ±.0010		16.298 ±.003				341.72 ±.15	34.300 ±.015	.81983 ±.00041	5.0265 ±.0022	26
233	Cobalticalcite $CoCO_3$	hex-R	R3̄c(167)	calcite	6	4.6581 ±.0010		14.958 ±.003				281.07 ±.13	28.213 ±.013	.67435 ±.00036	4.2159 ±.0020	26
234	Siderite $FeCO_3$	hex-R	R3̄c(167)	calcite	6	4.6887 ±.0010		15.373 ±.003				292.68 ±.14	29.378 ±.014	.70219 ±.00037	3.9436 ±.0018	26
235	Magnesite $MgCO_3$	hex-R	R3̄c(167)	calcite	6	4.6330 ±.0010		15.016 ±.003				279.13 ±.13	28.018 ±.013	.66969 ±.00036	3.0095 ±.0014	26
236	Rhodochrosite $MnCO_3$	hex-R	R3̄c(167)	calcite	6	4.7771 ±.0010		15.664 ±.003				309.57 ±.14	31.073 ±.014	.74272 ±.00039	3.6992 ±.0017	26
237	Nickelous Carbonate $NiCO_3$	hex-R	R3̄c(167)	calcite	6	4.5975 ±.0010		14.723 ±.002				269.51 ±.12	27.052 ±.012	.64660 ±.00034	4.3886 ±.0020	26
238	Smithsonite $ZnCO_3$	hex-R	R3̄c(167)	calcite	6	4.6528 ±.0010		15.025 ±.003				281.69 ±.13	28.275 ±.013	.67583 ±.00037	4.4343 ±.0021	26
239	Dolomite $CaMg(CO_3)_2$*	hex-R	R3̄(148)	calcite	3	4.8079 ±.0010		16.010 ±.003				320.50 ±.15	64.341 ±.029	1.5378 ±.0008	2.8661 ±.0013	26
240	Huntite $Mg_3Ca(CO_3)_4$*	hex-R	R32(155)	calcite	3	9.498 ±.003		7.816 ±.004				610.63 ±.44	122.58 ±.10	2.9299 ±.0024	2.880 ±.002	26
241	Norsethite $BaMg(CO_3)_2$*	hex-R	R32(155)	calcite	3	5.020 ±.005		16.75 ±.02				365.6 ±.8	73.39 ±.17	1.754 ±.004	3.838 ±.009	r
242	Vaterite $CaCO_3$	hex.			6	7.135 ±.005		8.524 ±.007				375.80 ±.61	37.72 ±.06	.9016 ±.0015	2.653 ±.004	r
243	Witherite $BaCO_3$	orth.	Pnam(62)	aragonite	4	6.430 ±.005	8.904 ±.005	5.314 ±.005				304.24 ±.41	45.81 ±.06	1.095 ±.002	4.308 ±.006	26
244	Aragonite $CaCO_3$	orth.	Pnam(62)	aragonite	4	5.741 ±.005	7.968 ±.005	4.959 ±.005				226.85 ±.33	34.15 ±.05	.8164 ±.0012	2.930 ±.004	26
245	Cerussite $PbCO_3$	orth.	Pnam(62)	aragonite	4	6.152 ±.005	8.436 ±.005	5.195 ±.005				269.61 ±.38	40.59 ±.06	.9702 ±.0014	6.582 ±.009	26
246	Strontianite $SrCO_3$	orth.	Pnam(62)	aragonite	4	6.029 ±.005	8.414 ±.005	5.107 ±.005				259.07 ±.37	39.01 ±.06	.9323 ±.0014	3.785 ±.005	26
247	Shortite $Na_2Ca_2(CO_3)_3$	orth.	Amm2(38)		2	4.961 ±.005	11.03 ±?	7.02 ±?				389.6 ±1.0	117.3 ±.3	2.804 ±.007	2.610 ±.007	r
248	Malachite $Cu_2(OH)_2CO_3$	mon.	P2₁/a(14)		4	9.502 ±.007	11.974 ±.007	3.240 ±.003		98.75 ±.25		364.35 ±.54	54.86 ±.08	1.311 ±.002	4.030 ±.006	25

X-Ray Crystallographic Data of Minerals

	Name and formula	Crystal system	Space group	Structure type	Z	a_o	b_o	c_o	α_o	β_o	γ_o	Cell volume 10^{-24} cm³	Molar volume cm³	Molar volume cal bar⁻¹	X-Ray density grams cm⁻³	Temp. °C	
249	Azurite $Cu_3(OH)_2(CO_3)_2$	mon.	P2₁/a(14)		2	5.008 ±.005	5.844 ±.005	10.336 ±.005		92.45 ±.25		302.22 ±.43	91.01 ±.13	2.1752 ±.0031	3.787 ±.005	25	249
250	Niter KNO_3	orth.	Pnam(62)	aragonite	4	6.431 ±.005	9.164 ±.005	5.414 ±.005				319.07 ±.42	48.04 ±.06	1.148 ±.002	2.105 ±.003	26	250
251	Soda Niter $NaNO_3$	hex-R	R3̄c(167)	calcite	6	5.0696 ±.0010		16.829 ±.005				374.57 ±.19	37.598 ±.019	.89866 ±.00049	2.2606 ±.0011	25	251
252	Gerhardite $Cu_2(NO_3)(OH)_3$	orth.	P2₁2₁2₁(19)		4	6.075 ±.004	13.812 ±.008	5.592 ±.004				469.21 ±.53	70.65 ±.08	1.689 ±.002	3.399 ±.004	r	252
	Sulfates and borates																
253	Barite $BaSO_4$	orth.	Pnma(62)	barite	4	8.878 ±.005	5.450 ±.005	7.152 ±.003				346.05 ±.40	52.10 ±.06	1.245 ±.002	4.480 ±.005	26	253
254	Anhydrite $CaSO_4$	orth.	Amma(63) Ccmm(63)	anhydrite	4	6.991 ±.005	6.996 ±.005	6.238 ±.005				305.09 ±.39	45.94 ±.06	1.098 ±.002	2.964 ±.004	26	254
255	Anglesite $PbSO_4$	orth.	Pnma(62)	barite	4	8.480 ±.005	5.398 ±.005	6.958 ±.005				318.50 ±.38	47.95 ±.06	1.146 ±.002	6.324 ±.008	25	255
256	Celestite $SrSO_4$	orth.	Pnma(62)	barite	4	8.359 ±.005	5.352 ±.005	6.866 ±.005				307.17 ±.41	46.25 ±.06	1.105 ±.002	3.972 ±.005	26	256
257	Zinkosite $ZnSO_4$	orth.	Pnma(62)	barite	4	8.588 ±.008	6.740 ±.006	4.770 ±.005				276.10 ±.46	41.57 ±.07	.9936 ±.0017	3.883 ±.006	25	257
258	Arcanite K_2SO_4	orth.	Pnma(62)	arcanite	4	5.772 ±.005	10.072 ±.005	7.483 ±.004				435.03 ±.49	65.50 ±.07	1.566 ±.002	2.661 ±.003	25	258
259	Mascagnite $(NH_4)_2SO_4$	orth.	Pnma(62)	arcanite	4	7.782 ±.005	5.993 ±.005	10.636 ±.005				496.04 ±.57	74.68 ±.09	1.7851 ±.0021	1.7693 ±.0020	25	259
260	Thenardite Na_2SO_4	orth.	Fddd(70)	thenardite	8	5.863 ±.005	12.304 ±.005	9.821 ±.005				708.47 ±.76	53.33 ±.06	1.275 ±.002	2.663 ±.003	25	260
261	Gypsum $CaSO_4 \cdot 2H_2O$*	mon.	C2/c(15)		4	5.68 ±.01	15.18 ±.01	6.29 ±.01		113.83 ±.22		496.1 ±1.5	74.69 ±.22	1.785 ±.005	2.305 ±.007	r	261
262	Epsomite $MgSO_4 \cdot 7H_2O$	orth.	P2₁2₁2₁(19)	epsomite	4	11.86 ±.01	11.99 ±.01	6.858 ±.007				975.22 ±1.53	146.83 ±.23	3.5094 ±.0055	1.679 ±.003	25	262
263	Goslarite $ZnSO_4 \cdot 7H_2O$	orth.	P2₁2₁2₁(19)	epsomite	4	11.779 ±.005	12.050 ±.005	6.822 ±.005				968.29 ±.72	145.79 ±.11	3.4845 ±.0026	1.9723 ±.0015	25	263
264	Mirabilite $Na_2SO_4 \cdot 10H_2O$	mon.	P2₁/c(14)		4	11.51 ±.01	10.38 ±.01	12.83 ±.01		107.75 ±.17		1459.9 ±2.6	219.8 ±.4	5.253 ±.009	1.466 ±.003	24	264
265	Chalcanthite $CuSO_4 \cdot 5H_2O$	tri.	P1̄(2)		2	6.1045 ±.0050	10.72 ±.01	5.949 ±.007	97.57 ±.17	107.28 ±.17	77.43 ±.17	361.88 ±.72	108.97 ±.22	2.6045 ±.0052	2.2912 ±.0046	r	265
266	Brochantite $Cu_4SO_4(OH)_6$*	mon.	P2₁/c(14)		4	13.066 ±.010	9.85 ±.01	6.022 ±.010		103.27 ±.25		754.3 ±1.8	113.6 ±.2	2.715 ±.006	3.982 ±.005	25	266
267	Syngenite $K_2Ca(SO_4)_2 \cdot H_2O$	mon.	P2₁/m(11)		2	9.775 ±.005	7.156 ±.005	6.251 ±.005		104.00 ±.25		424.27 ±.68	127.76 ±.20	3.0535 ±.0049	2.5707 ±.0041	r	267
268	Alunite $KAl_3(SO_4)_2(OH)_6$	hex-R	R3m(160)		3	6.982 ±.01		17.32 ±.01				731.2 ±1.1	146.8 ±.8	3.508 ±.02	2.822 ±.004	r	268
269	Natroalunite $NaAl_3(SO_4)_2(OH)_6$	hex-R	R3m(160)		3	6.974 ±.005		16.69 ±.01				702.99 ±1.09	141.1 ±.2	3.373 ±.005	2.821 ±.004	r	269
270	Hexahydrite $MgSO_4 \cdot 6H_2O$	mon.	C2/c(15)		8	10.110 ±.005	7.212 ±.004	24.41 ±.01		98.30 ±.10		1761.2 ±1.6	132.58 ±.12	3.1689 ±.0029	1.7232 ±.0015	25	270
271	Leonhardtite $MgSO_4 \cdot 4H_2O$	mon.	P2₁/n(14)		4	5.922 ±.006	13.604 ±.004	7.905 ±.005		90.85 ±.20		636.78 ±.78	95.88 ±.12	2.2915 ±.0029	2.0071 ±.0025	r	271
272	Melanterite $FeSO_4 \cdot 7H_2O$	mon.	P2₁/c(14)		4	14.072 ±.010	6.503 ±.007	11.041 ±.010		105.57 ±.15		973.29 ±1.69	146.54 ±.25	3.5025 ±.0061	1.8972 ±.0033	r	272
273	Vanthoffite $MgSO_4 \cdot 3Na_2SO_4$	mon.	P2₁/c(14)		2	9.797 ±.003	9.217 ±.003	8.199 ±.003		113.50 ±.10		678.96 ±.65	204.45 ±.20	4.8866 ±.0047	2.6730 ±.0025	r	273

X-Ray Crystallographic Data of Minerals

	Name and formula	Crystal system	Space group	Structure type	Z	a_0	b_0	c_0	α_0	β_0	γ_0	Cell volume 10^{-24} cm³	Molar volume cm³	Molar volume cal bar⁻¹	X-Ray density grams cm⁻³	Temp. °C	
274	Dolerophanite $Cu_2O(SO_4)$	mon.	C2/m(15)		4	9.355 ±.010	6.3.3 ±.005	7.628 ±.035		122.29 ±.10		380.77 ±.70	57.33 ±.11	1.3703 ±.0026	4.171 ±.008	r	274
275	Retgersite $NiSO_4.6H_2O$	tet.	P4₂2₁2(92) P4₂2₁(96)		4	6.782 ±.004		18.28 ±.1				840.80 ±.16	12E.59 ±.16	3.0257 ±.0040	2.076 ±.003	25	275
276	Colemanite $CaB_3O_4(OH)_3.H_2O$*	mon.	P2₁/a(14)		4	8.743 ±.004	11.254 ±.002	6.102 ±.003		110.12 ±.08		564.26 ±.49	84.957 ±.073	2.0306 ±.0018	2.4194 ±.0021	r	276
277	Borax $Na_2B_4O_7.10H_2O$	mon.	C2/c(15)		4	11.858 ±.005	10.674 ±.005	12.197 ±.005		106.68 ±.03		1478.8 ±1.1	222.66 ±.17	5.3217 ±.0041	1.7128 ±.0013	r	277
278	Kernite $Na_2B_4O_7.4H_2O$	mon.	P2₁/c(14)		4	7.022 ±.003	9.151 ±.004	15.676 ±.008		108.83 ±.25		953.40 ±1.61	143.55 ±.24	4.4309 ±.0058	1.9038 ±.0032	r	278
279	Hambergite $Be_2BO_3.(OH,F)$*	orth.	Pbca(61)		8	9.755 ±.001	12.201 ±.001	4.436 ±.001				526.79 ±.14	39.658 ±.011	.9479 ±.0003	2.3363 ±.0006	r	279

Phosphates, molybdates, and tungstates

	Name and formula	Crystal system	Space group	Structure type	Z	a_0	b_0	c_0	α_0	β_0	γ_0	Cell volume 10^{-24} cm³	Molar volume cm³	Molar volume cal bar⁻¹	X-Ray density grams cm⁻³	Temp. °C	
280	Berlinite $AlPO_4$	hex.	P3₁21(152) P3₂21(154)	a-quartz	3	4.942 ±.005		10.97 ±.047				232.03 ±.50	46.58 ±.10	1.113 ±.002	2.618 ±.006	25	280
281	Xenotime YPO_4	tet.	I4₁/amd(141)	zircon	4	6.885 ±.005		5.942 ±.045				283.57 ±.48	42.69 ±.07	1.020 ±.002	4.307 ±.008	26	281
282	Hydroxylapatite $Ca_5(PO_4)_3OH$	hex.	P6₃/m(176)	apatite	2	9.418 ±.003		6.883 ±.003				528.7 ±.5	159.2 ±2	3.805 ±.004	3.155 ±.004	r	282
283	Fluorapatite $Ca_5(PO_4)_3F$	hex.	P6₃/m(176)	apatite	2	9.3684 ±.0030		6.8841 ±.0030				523.25 ±.41	157.56 ±.12	3.7659 ±.0030	3.2007 ±.0025	25	283
284	Chlorapatite $Ca_5(PO_4)_3Cl$	hex.	P6₃/m(176)	apatite	2	9.629 ±.005		6.77 ±.003				544.16 ±.61	163.86 ±.19	3.916 ±.004	3.178 ±.004	r	284
285	Carbonate-apatite $Ca_{10}(PO_4)_6CO_3H_2O$	hex.	P6₃/m(176)	apatite	1	9.436 ±.010		6.883 ±.0				530.74 ±1.36	319.6 ±20	7.640 ±.020	3.281 ±.008	r	285
286	Turquois $CuAl_6(PO_4)_4(OH)_8.4H_2O$*	tri.	P1̄(2)		1	7.424 ±.008	7.629 ±.008	9.910 ±.0	68.61 ±.20	69.71 ±.20	65.08 ±.20	461.40 ±1.12	277.9 ±7	6.6416 ±.0162	2.927 ±.007	25	286
287	Powellite $CaMoO_4$	tet.	I4₁/a(100)	scheelite	4	5.226 ±.005		11.43 ±.63				312.17 ±.63	47.00 ±.09	1.1234 ±.0023	4.256 ±.009	25	287
288	Wulfenite $PbMoO_4$	tet.	I4₁/a(100)	scheelite	4	5.435 ±.005		12.10 ±.07				357.72 ±.69	53.859 ±.104	1.2873 ±.0025	6.816 ±.013	25	288
289	Scheelite $CaWO_4$	tet.	I4₁/a(100)	scheelite	4	5.242 ±.005		11.372 ±.05				312.49 ±.61	47.049 ±.092	1.1245 ±.0023	6.120 ±.012	25	289
290	Stolzite $PbWO_4$	tet.	I4₁/a(100)	scheelite	4	5.4616 ±.0030		12.046 ±.05				359.32 ±.42	54.100 ±.064	1.2931 ±.0016	8.4110 ±.0099	25	290
291	Ferberite $FeWO_4$	mon.	P2/c(13)	wolframite	2	4.732 ±.005	5.703 ±.005	4.965 ±.004		90.00 ±.05		134.11 ±.17	40.38 ±.05	.9652 ±.0013	7.520 ±.010	r	291
292	Huebnerite $MnWO_4$	mon.	P2/c(13)	wolframite	2	4.834 ±.004	5.758 ±.004	4.999 ±.010		91.18 ±.10		139.11 ±.20	41.89 ±.06	1.001 ±.010	7.228 ±.010	r	292
293	Wolframite $Fe_5Mn_5WO_4$	mon.	P2/c(13)	wolframite	2	4.782 ±.004	5.730 ±.004	4.982 ±.003		90.57 ±.10		136.53 ±.18	41.11 ±.06	.9826 ±.0014	7.376 ±.010	r	293
294	Sanmartinite $ZnWO_4$	mon.	P2/c(13)	wolframite	2	4.691 ±.003	5.720 ±.003	4.925 ±.003		89.36 ±.20		132.14 ±.14	39.79 ±.04	.9511 ±.0010	7.872 ±.008	25	294

Ortho and ring structure silicates

	Name and formula	Crystal system	Space group	Structure type	Z	a_0	b_0	c_0	α_0	β_0	γ_0	Cell volume 10^{-24} cm³	Molar volume cm³	Molar volume cal bar⁻¹	X-Ray density grams cm⁻³	Temp. °C	
295	Forsterite Mg_2SiO_4	orth.	Pbnm(62)	olivine	4	4.758 ±.002	10.21 ±.02	5.98 ±.02				290.81 ±.18	43.786 ±.027	1.0465 ±.0007	3.2136 ±.0020	25	295
296	Fayalite Fe_2SiO_4	orth.	Pbnm(62)	olivine	4	4.817 ±.005	10.477 ±.04	6.105 ±.01				308.11 ±.62	46.389 ±.093	1.1088 ±.0023	4.3928 ±.0088	r	296
297	Tephroite Mn_2SiO_4*	orth.	Pbnm(62)	olivine	4	4.871 ±.005	10.636 ±.04	6.238 ±.005				322.87 ±.45	48.612 ±.067	1.1619 ±.0017	4.1545 ±.0058	r	297

X-Ray Crystallographic Data of Minerals

Name and formula	Crystal system	Space group	Structure type	Z	a_0	b_0	c_0	α_0	β_0	γ_0	Cell volume 10^{-24} cm³	Molar volume cm³	Molar volume cal bar⁻¹	X-Ray density grams cm⁻³	Temp. °C	
Lime Olivine γ-Ca₂SiO₄	orth.	Pbnm(62)	olivine	4	5.091 ±.010	11.371 ±.020	6.782 ±.010				392.61 ±1.19	59.11 ±.18	1.4129 ±.0043	2.914 ±.009	r	298
Nickel Olivine Ni₂SiO₄	orth.	Pbnm(62)	olivine	4	4.727 ±.002	10.121 ±.005	5.915 ±.002				282.98 ±.21	42.61 ±.03	1.0184 ±.0008	4.917 ±.004	r	299
Cobalt Olivine Co₂SiO₄	orth.	Pbnm(62)	olivine	4	4.782 ±.002	10.301 ±.005	6.003 ±.002				295.70 ±.21	44.52 ±.03	1.0642 ±.0008	4.716 ±.003	r	300
Monticellite CaMgSiO₄	orth.	Pbnm(62)	olivine	4	4.827 ±.005	11.084 ±.005	6.376 ±.005				341.13 ±.47	51.362 ±.071	1.2276 ±.0017	3.046 ±.004	r	301
Kerschsteinite CaFeSiO₄	orth.	Pbnm(62)	olivine	4	4.886 ±.005	11.146 ±.005	6.434 ±.010				350.39 ±.67	52.756 ±.101	1.2609 ±.0025	3.564 ±.007	r	302
Knebelite MnFeSiO₄*	orth.	Pbnm(62)	olivine	4	4.854 ±.010	10.602 ±.010	6.162 ±.010				317.11 ±.88	47.74 ±.13	1.1412 ±.0032	4.249 ±.012	r	303
Glauchroite CaMnSiO₄	orth.	Pbnm(62)	olivine	4	4.944 ±.004	11.19 ±.01	6.529 ±.005				361.2 ±.9	54.38 ±.14	1.2997 ±.0032	3.441 ±.009	r	304
Fluor-Norbergite MgSiO₄,MgF₂*	orth.	Pnmb(62)		4	8.727 ±.005	10.271 ±.010	4.709 ±.002				422.09 ±.51	63.551 ±.077	1.5190 ±.0019	3.194 ±.004	25	305
Chondrodite 2MgSiO₄,MgF₂*	mon.	P2₁/c(14)		2	7.89 ±.03	4.743 ±.020	10.29 ±.03		109.03 ±.30		364.0 ±2.4	109.6 ±.7	2.620 ±.017	3.136 ±.004	r	306
Fluor-Humite 3MgSiO₄,MgF₂*	orth.	Pnma(62)		4	10.243 ±.005	20.72 ±.02	4.735 ±.002				1004.9 ±1.2	151.31 ±.18	3.6163 ±.0042	3.2017 ±.0037	25	307
Clinohumite 4MgSiO₄,MgF₂*	mon.	P2₁/c(14)		2	13.68 ±.04	4.75 ±.02	10.27 ±.02		100.83 ±.50		655.5 ±3.8	197.4 ±1.1	4.717 ±.027	3.167 ±.018	r	308
Grossularite Ca₃Al₂Si₃O₁₂	cubic	Ia3d(230)	garnet	8	11.851 ±.001						1664.43 ±.42	125.30 ±.03	2.9948 ±.0008	3.595 ±.001	25	309
Uvarovite Ca₃Cr₂Si₃O₁₂	cubic	Ia3d(230)	garnet	8	11.999 ±.002						1727.57 ±.86	130.05 ±.07	3.1084 ±.0016	3.848 ±.002	26	310
Andradite Ca₃Fe₂Si₃O₁₂	cubic	Ia3d(230)	garnet	8	12.048 ±.001						1748.82 ±.44	131.65 ±.03	3.1466 ±.0008	3.860 ±.001	25	311
Goldmanite Ca₃V₂Si₃O₁₂	cubic	Ia3d(230)	garnet	8	12.070 ±.005						1758.42 ±2.19	132.38 ±.16	3.1639 ±.0040	3.765 ±.005	r	312
Almandite Fe₃Al₂Si₃O₁₂	cubic	Ia3d(230)	garnet	8	11.526 ±.001						1531.21 ±.40	115.27 ±.04	2.7551 ±.0008	4.318 ±.001	25	313
Pyrope Mg₃Al₂Si₃O₁₂	cubic	Ia3d(230)	garnet	8	11.459 ±.001						1504.67 ±.39	113.27 ±.03	2.7074 ±.0008	3.559 ±.001	25	314
Spessartite Mn₃Al₂Si₃O₁₂	cubic	Ia3d(230)	garnet	8	11.621 ±.001						1569.39 ±.41	118.15 ±.03	2.8238 ±.0008	4.190 ±.001	25	315
Zircon ZrSiO₄*	tet.	I4/amd(141)	zircon	4	6.604 ±.005		5.979 ±.005				260.76 ±.45	39.261 ±.068	.9384 ±.0017	4.669 ±.008	25	316
Thorite ThSiO₄	tet.	I4/amd(141)	zircon	4	7.143 ±.004		6.327 ±.003				322.82 ±.39	48.60 ±.06	1.1617 ±.0015	6.668 ±.008	r	317
Coffinite USiO₄	tet.	I4/amd(141)	zircon	4	6.995 ±.004		6.263 ±.005				306.45 ±.43	46.140 ±.064	1.103 ±.002	7.155 ±.010	25	318
Kyanite Al₂SiO₅*	tri.	P1̄(2)		4	7.123 ±.001	7.848 ±.002	5.564 ±.008	89.92 ±.15	101.25 ±.08	105.97 ±.08	292.83 ±.45	44.09 ±.07	1.054 ±.002	3.675 ±.006	25	319
Andalusite Al₂SiO₅*	orth.	Pnnm(58)		4	7.7959 ±.0050	7.8983 ±.0020	5.5583 ±.0020				342.25 ±.27	51.530 ±.040	1.2316 ±.0010	3.145 ±.002	25	320
Sillimanite Al₂SiO₅*	orth.	Pbnm(62) Pnma(62)		4	7.4843 ±.0030	7.6730 ±.0030	5.7711 ±.0040				331.42 ±.30	49.899 ±.044	1.1927 ±.0011	3.248 ±.003	25	321
3.2 Mullite 3Al₂O₃,2SiO₂	orth.			3/4	7.557 ±.002	7.6876 ±.0020	2.8842 ±.0010				167.56 ±.07	134.55 ±.07	3.2159 ±.0016	3.166 ±.002	r	322
2.1 Mullite 2Al₂O₃,SiO₂	orth.	Pbam(55)		6/5	7.5788 ±.0020	7.6909 ±.0020	2.8883 ±.0010				168.35 ±.09	84.492 ±.043	2.0195 ±.0011	3.125 ±.002	r	323

X-Ray Crystallographic Data of Minerals

Name and formula	Crystal system	Space group	Structure type	Z	a_0	b_0	c_0	α_0	β_0	γ_0	Cell volume 10^{-24} cm³	Molar volume cm³	Molar volume cal bar⁻¹	X-Ray density grams cm⁻³	Temp. °C	
324 Staurolite $Fe_2Al_9Si_4O_{23}(OH)_2$*	mon.	C2/m(15)		2	7.90 ±.10	16.65 ±.15	5.63 ±.10		90.00 ±.25		740.5 ±17.5	223.0 ±5.5	5.330 ±.126	3.825 ±.090	r	324
325 Topaz $Al_2(SiO_4)(OH)_2$*	orth.	Pmnb(62)		4	8.394 ±.005	8.792 ±.007	4.645 ±.005				343.10 ±.41	51.66 ±.06	1.2347 ±.0015	3.563 ±.005	26	325
326 Phenacite Be_2SiO_4*	hex-R	R$\bar{3}$(148)	phenacite	18	12.472 ±.005		8.252 ±.005				1111.6 ±1.1	37.194 ±.037	.8890 ±.0009	2.960 ±.003	25	326
327 Willemite Zn_2SiO_4*	hex-R	R$\bar{3}$(148)	phenacite	18	13.94 ±.01		9.30 ±.01				1566.6 ±2.3	52.42 ±.08	1.253 ±.002	4.251 ±.006	25	327
328 Dioptase CuH_2SiO_4*	hex-R	R$\bar{3}$(148)	phenacite	18	14.61 ±.02		7.80 ±.01				1441.9 ±4.4	48.24 ±.15	1.153 ±.004	3.247 ±.010	r	328
329 Larnite β-Ca_2SiO_4*	mon.	P2₁/n(14)		4	5.48 ±.02	6.76 ±.02	9.28 ±.02		94.55 ±.33		342.7 ±1.8	51.60 ±.27	1.233 ±.006	3.338 ±.017	r	329
330 Akermanite $Ca_2MgSi_2O_7$	tet.	P$\bar{4}$2₁m(113)	melilite	2	7.8435 ±.0030		5.010 ±.0030				308.22 ±.30	92.812 ±.090	2.2183 ±.0022	2.9375 ±.0029	r	330
331 Gehlenite $Ca_2Al_2SiO_7$	tet.	P$\bar{4}$2₁m(113)	melilite	2	7.690 ±.003		5.0675 ±.0030				299.67 ±.29	90.239 ±.088	2.1568 ±.0022	3.0387 ±.0030	r	331
332 Fe-Gehlenite $Ca_2Fe_2SiO_7$	tet.	P$\bar{4}$2₁m(113)	melilite	2	7.54 ±.01		4.85 ±.01				276.01 ±.79	83.12 ±.24	1.9865 ±.0057	3.994 ±.011	r	332
333 Hardystonite $Ca_2ZnSi_2O_7$*	tet.	P$\bar{4}$2₁m(113)	melilite	2	7.87 ±.03		5.01 ±.02				310.3 ±2.7	93.44 ±.24	2.233 ±.019	3.357 ±.029	r	333
334 Sodium Melilite $NaCaAlSi_2O_7$	tet.	P$\bar{4}$2₁m(113)	melilite	2	8.511 ±.005		4.80 ±.02				348.35 ±.46	104.90 ±.90	2.507 ±.003	2.462 ±.003	r	334
335 Beryl $Be_3Al_2(Si_6O_{18})$*	hex.	P6/mmc(192)	beryl	2	9.215 ±.005		9.19 ±.010				675.98 ±.82	203.55 ±.25	4.8651 ±.0060	2.641 ±.003	25	335
336 Indialite high Cordierite $MgAl_3(AlSi_5O_{18})$	hex.	P6/mmc(192)	beryl	2	9.7698 ±.0030		9.357 ±.0030				773.02 ±.54	232.78 ±.3	5.5636 ±.0039	2.513 ±.002	25	336
337 Low Cordierite $MgAl_3(AlSi_5O_{18})$	orth.	Cccm(66)		4	9.721 ±.003	17.062 ±.006	9.33 ±.010				1548.96 ±.88	233.22 ±.3	5.5741 ±.0032	2.508 ±.001	25	337
338 Fe-Indialite $Fe_2Al_3(AlSi_5O_{18})$	hex.	P6/mmc(192)	beryl	2	9.860 ±.010		9.28 ±.010				781.75 ±1.80	235.40 ±.54	5.6264 ±.0130	2.753 ±.006	r	338
339 Fe-Cordierite $Fe_2Al_3(AlSi_5O_{18})$	orth.	Cccm(66)	cordierite	4	9.726 ±.010	17.065 ±.010	9.28 ±.010				1541.40 ±2.47	232.08 ±.37	5.5468 ±.0089	2.792 ±.005	r	339
340 Mn-Indialite $Mn_2Al_3(AlSi_5O_{18})$	hex.	P6/mmc(192)	beryl	2	9.925 ±.010		9.29 ±.010				793.11 ±.81	238.5 ±.8	5.708 ±.013	2.706 ±.006	r	340
341 Sapphirine $MgAl_2O_4SiO_4$*	mon.	P2₁/c(14)		8	11.26 ±.03	14.46 ±.03	9.95 ±.02		125.33 ±.50		1321.7 ±9.7	99.50 ±.73	2.378 ±.017	3.464 ±.025	r	341
342 Elbaite $NaLiAl_{1.8}B_3Si_6O_{27}(OH)_4$*	hex-R	R3m(160)	tourmaline	3	15.842 ±.010		7.005 ±.010				1523.4 ±2.9	305.82 ±.140	7.3093 ±.0140	3.271 ±.006	r	342
343 Schorl $NaFe_3Al_6B_3Si_6O_{27}(OH)_4$*	hex-R	R3m(160)	tourmaline	3	16.032 ±.010		7.149 ±.010				1591.3 ±3.0	319.45 ±.60	7.635 ±.014	3.297 ±.006	r	343
344 Dravite $NaMg_3Al_6B_3Si_6O_{27}(OH)_4$	hex-R	R3m(160)	tourmaline	3	15.942 ±.010		7.22 ±.010				1589.99 ±2.97	319.9 ±.60	7.629 ±.014	3.004 ±.006	r	344
345 Uvite $CaMg_4Al_5B_3Si_6O_{27}(OH)_4$	hex-R	R3m(160)	tourmaline	3	15.86 ±.01		7.19 ±.01				1566.3 ±2.9	314.4 ±.6	7.515 ±.014	3.095 ±.006	r	345
346 Sphene $CaTiSiO_5$*	mon.	A2/a(15)		4	7.07 ±.01	8.72 ±.01	6.56 ±.01		113.95 ±.25		369.61 ±1.13	55.65 ±.04	1.330 ±.004	3.523 ±.011	r	346
347 Datolite $CaBSiO_4(OH)$*	mon.	P2₁/c(14)		4	9.62 ±.03	7.60 ±.03	4.84 ±.02		90.15 ±.25		353.9 ±2.3	53.28 ±.35	1.273 ±.008	3.003 ±.020	r	347
348 Euclase $AlBeSiO_4(OH)$*	mon.	P2₁/a(14)		4	4.763 ±.005	14.29 ±.02	4.618 ±.001		100.25 ±.10		309.30 ±.64	46.57 ±.02	1.113 ±.002	3.116 ±.007	r	348
349 Chloritoid $H_2FeAl_2SiO_7$*	mon.	C2/c(15)		8	9.48 ±.01	5.48 ±.01	18.18 ±.01		101.77 ±.25		924.6 ±2.2	69.61 ±.6	1.664 ±.004	3.619 ±.008	r	349

X-Ray Crystallographic Data of Minerals

No.	Name and formula	Crystal system	Space group	Structure type	Z	a_0	b_0	c_0	α_0	β_0	γ_0	Cell volume 10^{-24} cm³	Molar volume cm³	Molar volume cal bar⁻¹	X-Ray density grams cm⁻³	Temp. °C
350	Hemimorphite $Zn_4(OH)_2Si_2O_7 \cdot H_2O$*	orth.	Imm2(35)		2	8.370 ±.005	10.719 ±.005	5.120 ±.005				459.36 ±.57	138.32 ±.17	3.306 ±.004	3.482 ±.004	25
351	Zoisite $Ca_2Al_3(SiO_4)_3OH$	orth.	Pnma(62)		4	16.15 ±.01	5.581 ±.005	10.06 ±.01				906.74 ±1.34	136.52 ±.20	3.263 ±.005	3.328 ±.005	r
352	Clinozoisite $Ca_2Al_3(SiO_4)_3OH$	mon.	P2₁/m(11)		2	8.887 ±.007	5.581 ±.005	10.14 ±.01		115.93 ±.33		452.30 ±1.45	136.20 ±.44	3.255 ±.010	3.336 ±.011	r
353	Epidote $Ca_2Al_{2.8}Fe_{0.2}(SiO_4)_3OH$*	mon.	P2₁/m(11)		2	8.89 ±.02	5.63 ±.01	10.19 ±.02		115.40 ±.30		460.72 ±1.97	138.7 ±.7	3.316 ±.014	3.587 ±.015	r
354	Piemontite $Ca_2Al_{2.5}Mn_{0.5}(SiO_4)_3OH$*	mon.	P2₁/m(11)		2	8.95 ±.02	5.70 ±.01	9.41 ±.02		115.70 ±.50		432.56 ±2.38	130.3 ±.7	3.113 ±.017	3.810 ±.021	r
355	Lawsonite $CaAl_2Si_2O_7(OH)_2 \cdot H_2O$	orth.	Cccm(63)		4	8.787 ±.005	5.836 ±.005	13.123 ±.008				672.96 ±.80	101.32 ±.12	2.4217 ±.0029	3.101 ±.004	r

Chain and band structure silicates

No.	Name and formula	Crystal system	Space group	Structure type	Z	a_0	b_0	c_0	α_0	β_0	γ_0	Cell volume 10^{-24} cm³	Molar volume cm³	Molar volume cal bar⁻¹	X-Ray density grams cm⁻³	Temp. °C
356	Enstatite $MgSiO_3$*	orth.	Pcab(61)		16	8.829 ±.010	18.22 ±.01	5.192 ±.005				835.21 ±1.32	31.44 ±.05	.7514 ±.0012	3.194 ±.005	r
357	Clinoenstatite $MgSiO_3$	mon.	P2₁/c(15)		8	9.620 ±.005	8.825 ±.005	5.188 ±.005		108.33 ±.17		418.10 ±.66	31.47 ±.05	.7523 ±.0012	3.190 ±.005	r
358	Protoenstatite $MgSiO_3$	orth.	Pbcn(60)		8	9.25 ±.01	8.74 ±.01	5.32 ±.01				430.10 ±1.05	32.38 ±.08	.7739 ±.0019	3.101 ±.008	r
359	High Clinoenstatite $MgSiO_3$	tri.			8	10.000 ±.005	8.934 ±.004	5.170 ±.003	88.27 ±.05	70.03 ±.04	91.01 ±.04	433.72 ±.40	32.65 ±.03	.7804 ±.0008	3.075 ±.003	r
360	Clinoferrosilite $FeSiO_3$	mon.	P2₁/c(14)		8	9.7085 ±.0010	9.0872 ±.0011	5.2284 ±.004		108.43 ±.05		437.60 ±.15	32.943 ±.011	.7874 ±.0003	4.005 ±.002	r
361	Orthoferrosilite $FeSiO_3$	orth.	Pcab(61)	enstatite	16	9.080 ±.002	18.431 ±.004	5.238 ±.001				876.6 ±.54	33.00 ±.02	.7887 ±.0008	3.998 ±.002	r
362	Diopside $CaMg(SiO_3)_2$	mon.	C2/c(15)	diopside	4	9.743 ±.005	8.923 ±.005	5.251 ±.003		105.93 ±.25		438.97 ±.69	66.09 ±.10	1.580 ±.003	3.277 ±.005	r
363	Hedenbergite $CaFe(SiO_3)_2$*	mon.	C2/c(15)	diopside	4	9.854 ±.010	9.024 ±.010	5.263 ±.010		104.23 ±.33		453.64 ±1.28	68.30 ±.16	1.632 ±.005	3.632 ±.005	r
364	Johannsenite $CaMn(SiO_3)_2$*	mon.	C2/c(15)	diopside	4	9.83 ±.03	9.04 ±.03	5.27 ±.02		105.00 ±.50		452.35 ±2.87	68.11 ±.43	1.628 ±.010	3.629 ±.023	r
365	Ureyite $NaCr(SiO_3)_2$	mon.	C2/c(15)	diopside	4	9.550 ±.016	8.712 ±.007	5.273 ±.008		107.44 ±.16		418.6 ±1.1	63.02 ±.16	1.506 ±.004	3.605 ±.009	r
366	Jadeite $NaAl(SiO_3)_2$*	mon.	C2/c(15)	diopside	4	9.409 ±.005	8.564 ±.005	5.220 ±.005		107.50 ±.20		401.15 ±.67	60.40 ±.10	1.444 ±.002	3.347 ±.006	r
367	Acmite (Aegirine) $NaFe(SiO_3)_2$	mon.	C2/c(15)	diopside	4	9.658 ±.005	8.795 ±.005	5.294 ±.005		107.42 ±.20		429.06 ±.70	64.60 ±.11	1.544 ±.003	4.411 ±.007	r
368	Ca Tschermak Molecule $CaAl_2SiO_6$	mon.	C2/c(15)	diopside	4	9.615 ±.005	8.661 ±.005	5.272 ±.003		106.12 ±.20		421.77 ±.59	63.50 ±.09	1.518 ±.002	3.435 ±.005	r
369	Spodumene $LiAl(SiO_3)_2$	mon.	C2/c(15)	diopside	4	9.451 ±.002	8.387 ±.002	5.208 ±.001		110.07 ±.03		387.7 ±1	58.37 ±.02	1.395 ±.001	3.188 ±.001	r
370	β-Spodumene $LiAl(SiO_3)_2$	tet.	P4₂2₁2(96) P4₂2₁2(92)		4	7.5332 ±.0008		9.1540 ±.0008				519.48 ±.12	78.215 ±.018	1.8694 ±.0005	2.379 ±.001	r
371	Pectolite $Ca_2NaH(SiO_3)_3$*	tri.	P1(2)		2	7.99 ±.01	7.04 ±.01	7.02 ±.01	90.05 ±.25	95.27 ±.25	102.47 ±.25	383.84 ±.99	115.58 ±.30	2.763 ±.007	2.876 ±.007	r
372	Wollastonite $CaSiO_3$*	tri.	P1(2)		6	7.94 ±.01	7.32 ±.01	7.07 ±.01	90.03 ±.25	95.37 ±.25	103.43 ±.25	397.82 ±1.03	39.93 ±.10	.9544 ±.0025	2.909 ±.008	r
373	Parawollastonite $CaSiO_3$*	mon.	P2(4)		12	15.417 ±.004	7.321 ±.002	7.066 ±.003		95.40 ±.10		793.98 ±.47	39.85 ±.02	.9524 ±.0006	2.915 ±.002	r
374	Pseudowollastonite $CaSiO_3$*	tri.	P2(4)		24	6.90 ±.02	11.78 ±.02	19.65 ±.02	90.00 ±.30	90.80 ±.30	90.00 ±.30	1597.0 ±5.6	40.08 ±.14	.9579 ±.0034	2.899 ±.010	r

X-Ray Crystallographic Data of Minerals

No.	Name and formula	Crystal system	Space group	Structure type	Z	a_o	b_o	c_o	α_o	β_o	γ_o	Cell volume 10^{-24} cm³	Molar volume cm³	Molar volume cal bar⁻¹	X-Ray density grams cm⁻³	Temp. °C
375	Rhodonite MnSiO3*	tri.	P1̄(2)		10	7.682 ±.002	11.818 ±.003	6.707 ±.002	92.36 ±.05	93.95 ±.05	105.66 ±.05	583.77 ±.31	35.158 ±.9	.8403 ±.0005	3.727 ±.002	r
376	Bustamite CaMn(SiO3)2*	tri.	A1̄(2)		6	7.736 ±.003	7.157 ±.003	13.824 ±.010	90.52 ±.25	94.58 ±.25	103.87 ±.25	740.38 ±1.08	74.32 ±.1	1.776 ±.003	3.326 ±.005	r
377	Pyroxmangite MnFe(SiO3)2*	tri.	P1̄(2)		7	7.56 ±.02	17.45 ±.05	6.67 ±.02	84.00 ±.30	94.30 ±.30	113.70 ±.30	800.77 ±4.29	68.90 ±.36	1.647 ±.009	3.817 ±.020	r
378	Tremolite Ca2Mg5[Si8O22](OH)2*	mon.	C2/m(12)	tremolite	2	9.840 ±.010	18.052 ±.020	5.275 ±.010		104.70 ±.25		906.34 ±2.43	272.92 ±.75	6.523 ±.018	2.977 ±.008	r
379	Fluor-tremolite Ca2Mg5[Si8O22]F2	mon.	C2/m(12)	tremolite	2	9.781 ±.007	18.01 ±.01	5.267 ±.005		104.52 ±.25		898.18 ±1.56	270.46 ±.45	6.464 ±.011	3.018 ±.005	20
380	Ferrotremolite Ca2Fe5[Si8O22](OH)2	mon.	C2/m(12)	tremolite	2	9.97 ±.01	18.34 ±.02	5.30 ±.01		104.50 ±.10		938.24 ±2.92	282.55 ±.69	6.753 ±.017	3.434 ±.008	r
381	Grunerite Fe7[Si8O22](OH)2	mon.	C2/m(12)	tremolite	2	9.572 ±.005	18.44 ±.01	5.342 ±.007		101.77 ±.25		923.08 ±1.63	277.96 ±.49	6.644 ±.012	3.603 ±.006	r
382	Cummingtonite (hypo.) Mg7[Si8O22](OH)2	mon.	C2/m(12)	tremolite	2	9.476 ±.010	17.935 ±.010	5.292 ±.005		102.23 ±.25		878.97 ±1.58	264.68 ±.45	6.326 ±.011	2.950 ±.005	r
383	Riebeckite Na2Fe3Fe2[Si8O22](OH)2	mon.	C2/m(12)	tremolite	2	9.729 ±.020	18.065 ±.020	5.334 ±.010		103.31 ±.25		912.29 ±2.37	274.71 ±.85	6.566 ±.021	3.407 ±.011	r
384	Magnesioriebeckite Na2Mg3Fe2[Si8O22](OH)2	mon.	C2/m(12)	tremolite	2	9.733 ±.010	17.946 ±.020	5.299 ±.010		103.30 ±.25		900.74 ±2.37	271.2 ±.71	6.483 ±.017	3.102 ±.008	r
385	Glaucophane I Na2Mg3Al2[Si8O22](OH)2	mon.	C2/m(12)	tremolite	2	9.748 ±.010	17.915 ±.010	5.273 ±.010		102.78 ±.25		898.04 ±2.35	270.42 ±.71	6.463 ±.017	2.898 ±.008	r
386	Glaucophane II Na2Mg3Al2[Si8O22](OH)2	mon.	C2/m(12)	tremolite	2	9.663 ±.010	17.696 ±.020	5.277 ±.010		103.67 ±.10		876.79 ±2.17	264.02 ±.65	6.310 ±.016	2.968 ±.007	r
387	Fluor-edenite NaCa2Mg5[AlSi7O22]F2	mon.	C2/m(12)	tremolite	2	9.847 ±.005	18.00 ±.01	5.282 ±.010		104.83 ±.25		905.03 ±1.51	272.55 ±.46	6.514 ±.011	3.076 ±.005	r
388	Fluor-richterite Na2CaMg5[Si8O22]F2	mon.	C2/m(12)	tremolite	2	9.823 ±.005	17.96 ±.01	5.268 ±.005		104.33 ±.25		900.47 ±1.48	271.11 ±.45	6.481 ±.011	3.033 ±.005	r
389	Anthophyllite Mg7[Si8O22](OH)2	orth.	Pnma(62)		4	18.61 ±.02	13.01 ±.01	5.24 ±.06				1756.3 ±7.0	264.4 ±1.1	6.320 ±.025	2.953 ±.012	r

Framework structure silicates

No.	Name and formula	Crystal system	Space group	Structure type	Z	a_o	b_o	c_o	α_o	β_o	γ_o	Cell volume 10^{-24} cm³	Molar volume cm³	Molar volume cal bar⁻¹	X-Ray density grams cm⁻³	Temp. °C
390	Microcline KAlSi3O8	tri.	C1̄(2)		4	8.582 ±.002	12.964 ±.002	7.222 ±.002	90.62 ±.10	115.92 ±.10	87.68 ±.10	722.06 ±.67	108.72 ±.10	2.5984 ±.0025	2.560 ±.002	r
391	High Sanidine KAlSi3O8	mon.	C2/m(12)		4	8.615 ±.002	13.031 ±.003	7.177 ±.002		115.98 ±.10		724.28 ±.69	109.05 ±.10	2.6064 ±.0025	2.552 ±.002	r
392	Orthoclase KAlSi3O8*	mon.	C2/m(12)		4	8.562 ±.003	12.996 ±.004	7.193 ±.003		116.02 ±.15		719.25 ±1.02	108.2 ±.15	2.5883 ±.0037	2.570 ±.004	r
393	Fe-Sanidine KFeSi3O8	mon.	C2/m(12)		4	8.689 ±.008	13.12 ±.01	7.319 ±.007		116.10 ±.30		749.28 ±2.24	112.81 ±.34	2.6964 ±.0081	2.723 ±.008	r
394	Fe-Microcline KFeSi3O8	tri.	C1̄(2)		4	8.68 ±.01	13.10 ±.01	7.340 ±.007	90.75 ±.25	116.05 ±.25	86.23 ±.25	748.09 ±1.92	112.63 ±.29	2.692 ±.007	2.727 ±.007	r
395	Low Albite NaAlSi3O8	tri.	C1̄(2)		4	8.139 ±.002	12.788 ±.003	7.160 ±.002	94.27 ±.10	116.57 ±.10	87.68 ±.10	664.65 ±.60	100.07 ±.09	2.3918 ±.0022	2.620 ±.002	26
396	High Albite (Analbite) NaAlSi3O8	tri.	C1̄(2)		4	8.160 ±.002	12.870 ±.003	7.106 ±.002	93.54 ±.10	116.36 ±.10	90.19 ±.10	667.00 ±.60	100.43 ±.09	2.4003 ±.0022	2.611 ±.002	r
397	Anorthite CaAl2Si2O8	tri.	P1̄(2)	primitive cell	8	8.177 ±.002	12.877 ±.003	14.169 ±.003	93.17 ±.02	115.85 ±.02	91.22 ±.02	1338.9 ±6	100.79 ±.04	2.4090 ±.0011	2.760 ±.001	r
398	Synthetic CaAl2Si2O8	hex.	P6₃/mcm(193)		2	5.10 ±.02		14.72 ±.02				331.57 ±2.64	99.85 ±.79	2.386 ±.019	2.786 ±.022	r
399	Synthetic CaAl2Si2O8	orth.	P2,2,2(18)		2	8.22 ±.02	8.60 ±.02	4.83 ±.01				341.44 ±1.35	102.85 ±.41	2.457 ±.010	2.706 ±.011	r

X-Ray Crystallographic Data of Minerals

	Name and formula	Crystal system	Space group	Structure type	Z	a_0	b_0	c_0	a_0	β_0	γ_0	Cell volume 10^{-24} cm³	Molar volume cm³	Molar volume cal bar⁻¹	X-Ray density grams cm⁻³	Temp. °C	
400	Celsian BaAl₂Si₂O₈*	mon.	I2/c(15)		8	8.627 ±.010	13.045 ±.010	14.408 ±.020		115.20 ±.25		1467.1 ±4.2	110.45 ±.31	2.640 ±.008	3.400 ±.010	r	400
401	Paracelsian BaAl₂Si₂O₈*	mon.	P2₁/a(14)		4	8.58 ±.02	9.583 ±.020	9.08 ±.02		90.00 ±.50		746.6 ±2.9	112.4 ±.4	2.687 ±.010	3.340 ±.013	r	401
402	Banalsite BaNa₂Al₄Si₄O₁₆*	orth.			4	8.50 ±.02	9.97 ±.02	16.72 ±.03				1416.9 ±5.1	213.3 ±.8	5.099 ±.018	3.092 ±.011	r	402
403	Danburite CaB₂Si₂O₈*	orth.	Pnam(62)		4	8.04 ±.02	8.77 ±.02	7.74 ±.02				545.8 ±2.3	82.17 ±.35	1.964 ±.008	2.992 ±.013	r	403
404	Low Nepheline NaAlSiO₄	hex.	C6₃(178)		8	9.986 ±.005		8.330 ±.004				719.38 ±.80	54.16 ±.06	1.294 ±.002	2.623 ±.003	r	404
405	High Carnegeite NaAlSiO₄	cubic			4	7.325 ±.004						393.03 ±.64	59.18 ±.10	1.414 ±.002	2.401 ±.004	750	405
406	Kaliophilite natural KAlSiO₄*	hex.	P6₃22(182)		54	26.930 ±.010		8.522 ±.004				5352.4 ±4.7	59.69 ±.05	1.427 ±.001	2.650 ±.002	r	406
407	Kaliophilite synthetic KAlSiO₄	hex.	P6₃(173) P6₃22(182)		2	5.180 ±.002		8.559 ±.004				198.89 ±.18	59.89 ±.05	1.431 ±.001	2.641 ±.002	r	407
408	Kalsilite KAlSiO₄	hex.	P6₃(173)		2	5.1597 ±.0020		8.7032 ±.0030				200.66 ±.17	60.424 ±.051	1.4442 ±.0031	2.618 ±.002	r	408
409	Leucite KAlSi₂O₆	tet.	I4₁/a(100)		16	13.074 ±.003		13.738 ±.003				2348.23 ±1.19	88.389 ±.045	2.1126 ±.0011	2.469 ±.001	25	409
410	High Leucite KAlSi₂O₆*	cubic	Ia3d(230)		16	13.43 ±.05						2422.3 ±27.1	91.18 ±1.02	2.179 ±.024	2.394 ±.027	625	410
411	Fe-Leucite KFeSi₂O₆	tet.	I4₁/a(100)		16	13.205 ±.002		13.970 ±.003				2435.98 ±.91	91.692 ±.034	2.1915 ±.0009	2.695 ±.001	25	411
412	Petalite LiAlSi₄O₁₀*	mon.	P2₁/n(14)		2	11.32 ±.03	5.14 ±.01	7.62 ±.02		105.90 ±20		426.41 ±1.57	128.4 ±.5	3.069 ±.01	2.385 ±.009	r	412
413	Marialite Na₄Al₃Si₉O₂₄Cl	tet.	I4/m(87) P4/m(83)		2	12.064 ±.008		7.514 ±.004				1093.6 ±1.6	329.3 ±.5	7.871 ±.011	2.566 ±.004	r	413
414	Meionite Ca₄Al₆Si₆O₂₄CO₃	tet.	I4/m(87) P4/m(83)		2	12.174 ±.008		7.652 ±.015				1134.07 ±2.68	341.5 ±.8	8.162 ±.019	2.737 ±.007	r	414
	Sheet structure silicates																
415	Muscovite KAl₂[AlSi₃O₁₀](OH)₂*	mon.	C2/c(15)	2M₂ mica	4	5.203 ±.005	8.995 ±.005	20.030 ±.005		94.47 ±.33		934.57 ±1.21	140.71 ±.18	3.363 ±.004	2.831 ±.004	r	415
416	Paragonite NaAl₂[AlSi₃O₁₀](OH)₂*	mon.	C2/c(15)	2M₁ mica	4	5.13 ±.03	8.89 ±.05	19.32 ±.10		95.17 ±.50		877.52 ±8.47	132.1 ±1.3	3.158 ±.031	2.893 ±.028	r	416
417	Lepidolite K₂Al₃Li₃[AlSi₃O₁₀](OH)₄*	mon.	C2/c(15)	2M₂ mica	4	9.2 ±.1	5.3 ±.1	20.0 ±.2		98.00 ±.50		965.7 ±23.2	290.8 ±7.0	6.950 ±.167	2.698 ±.065	r	417
418	Phlogopite KMg₃[AlSi₃O₁₀](OH)₂	mon.	Cm(8)	1M mica	2	5.326 ±.010	9.210 ±.010	10.311 ±.010		100.17 ±.10		497.83 ±1.19	149.91 ±.36	3.5830 ±.0086	2.784 ±.007	r	418
419	Fluor-phlogopite KMg₃[AlSi₃O₁₀]F₂	mon.	Cm(8)	1M mica	2	5.299 ±.005	9.188 ±.005	10.135 ±.005		99.92 ±.10		486.07 ±.60	146.37 ±.18	3.498 ±.004	2.878 ±.004	r	419
420	Annite KFe₃[AlSi₃O₁₀](OH)₂	mon.	Cm(8)	1M mica	2	5.391 ±.010	9.350 ±.005	10.313 ±.020		99.70 ±.25		512.40 ±1.45	154.30 ±.44	3.688 ±.010	3.318 ±.009	r	420
421	Ferriannite KFe₃[AlFeSi₃O₁₀](OH)₂	mon.	C2/m(12)	2M mica	2	5.430 ±.002	9.404 ±.003	10.341 ±.006		100.07 ±.20		519.92 ±.51	156.56 ±.15	3.7419 ±.0037	3.454 ±.003	r	421
422	Margarite CaAl₂[Al₂Si₂O₁₀](OH)₂*	mon.	C2/c(15)	2M mica	4	5.13 ±.02	8.92 ±.03	19.50 ±.05		95.00 ±.20		888.9 ±5.2	133.8 ±.8	3.199 ±.019	2.975 ±.017	r	422
423	Talc Mg₃Si₄O₁₀(OH)₂*	mon.	C2/c(15)	2M₁	4	5.287 ±.007	9.158 ±.010	18.95 ±.01		99.50 ±.20		904.94 ±1.71	136.25 ±.26	3.2565 ±.0062	2.784 ±.005	r	423
424	Pyrophyllite Al₂Si₄O₁₀(OH)₂*	mon.	C2/c(15)	2M₁	4	5.14 ±.02	8.90 ±.02	18.55 ±.03		99.92 ±.20		835.9 ±4.0	125.9 ±.6	3.008 ±.015	2.863 ±.014	r	424

X-Ray Crystallographic Data of Minerals

	Name and formula	Crystal system	Space group	Structure type	Z	a_0	b_0	c_0	α_0	β_0	γ_0	Cell volume 10^{-24} cm³	Molar volume cm³	Molar volume cal bar⁻¹	X-Ray density grams cm⁻³	Temp. °C	
425	Minnesotaite $Fe_3Si_4O_{10}(OH)_2$*	mon.	C2/c(15)		4	5.4 ±.1	9.2 ±.1	19.4		100.00 ±.50		971.8 ±19.2	146.3 ±2.9	3.497 ±.069	3.239 ±.064	r	425
426	Dickite $Al_2Si_2O_5(OH)_4$*	mon.	Cc(9)		4	5.150 ±.002	8.940 ±.006	14.736 ±.00?		103.58 ±.10		659.49 ±.49	99.30 ±.07	2.3733 ±.0018	2.600 ±.002	r	426
427	Kaolinite $Al_2Si_2O_5(OH)_4$*	tri.	P1(1)		2	5.155 ±.007	8.945 ±.01?	7.40? ±.00?	91.68 ±.35	104.87 ±.35	89.93 ±.35	330.48 ±.86	99.52 ±.26	2.3785 ±.0062	2.594 ±.007	r	427
428	Nacrite $Al_2Si_2O_5(OH)_4$*	mon.	Cc(9)		4	8.909 ±.010	5.14? ±.0?	15.697 ±.02?		113.70 ±.25		658.9 ±2.1	99.21 ±.32	2.3713 ±.0076	2.602 ±.008	r	428
	Zeolites																
429	Analcite $NaAlSi_2O_6 \cdot H_2O$	cubic	Ia3d(230)		16	13.733 ±.005						2589.98 ±2.83	97.49 ±.1	2.3301 ±.0026	2.258 ±.003	r	429
430	Natrolite $Na_8Al_8Si_{10}O_{10} \cdot 2H_2O$*	orth.	Fdd2(43)		8	18.30 ±.02	18.65 ±.0?	6.60 ±.01				2250.1 ±4.8	169.59 ±.57	4.049 ±.009	2.245 ±.005	r	430

HEAT CAPACITY OF ROCK FORMING MINERALS

The units of heat capacity at constant pressure, C_p, in the table are cal kg⁻¹ deg⁻¹. Values of these units are given for several temperatures with the temperatures being in degrees C.

Heat capacity at other temperatures may be calculated by use of the equation $C_p = a + bT - cT^{-2}$. The units of these constants are cal kg⁻¹ deg⁻¹

Mineral	Heat capacity at various temperatures						Constants for heat capacity equation		
	-200	0	200	400	800	1200	a	10³b	10⁶c
Albite	—	0.1695	0.236	0.26	0.286	—	1.018	0.187	0.268
Amphibole	—	0.177	0.246	0.27	0.296	—	1.067	1.183	0.281
Apatite	—	0.24	—	—	—	—	—	—	—
Arsenopyrite	—	0.103 at 55°C	—	—	—	—	—	—	—
Asbestos	—	0.195	—	—	—	—	—	—	—
Barite	0.047	0.1076	0.12	0.1315	0.1555	—	0.383	0.253	0
Cassiterite	—	0.0814	0.103	0.115	0.132	—	0.387	0.157	0.007
Chalcopyrite	—	0.129 at 50°C	—	—	—	—	—	—	—
Diamond	—	0.104	0.253	0.328	0.445	—	0.754	1.067	0.454
Dolomite	—	0.222 at 60°C	—	—	—	—	—	—	—
Fluorite	0.0525	0.203	0.213	0.222	0.24	—	0.798	0.204	0
Galena	0.034	0.0496	0.0528	0.0562	—	—	0.188	0.007	0
Garnet	—	0.177 at 58°C	—	—	—	0.263	—	—	—
β-Graphite	—	0.152	0.282	0.348	0.45	—	0.932	0.913	0.4077
Hematite	—	0.146	0.189	0.215	0.258	—	0.640	0.420	0.111
Ice	0.156	0.492	—	—	—	—	—	—	—
Kaolinite	—	0.222	0.244	—	—	—	0.806	0.463	0.0

Mineral	Heat capacity at various temperatures						Constants for heat capacity equation		
	-200	0	200	400	800	1200	a	10³b	10⁶c
Labradorite	—	0.196 at 60°C	—	—	—	—	—	—	—
Magnesite	0.0385	0.207	—	—	—	—	—	—	—
Magnetite	—	0.1435	0.1985	0.222	—	—	0.744	0.340	0.177
Mica (mono-crystal)	—	0.208	—	—	—	—	—	—	—
Microcline	—	0.163	0.227	0.248	0.342	—	0.988	0.166	0.263
Oligoclase	—	0.2035 at 60°C	—	—	—	—	—	—	—
Olivine	—	0.189 at 36°C	—	—	—	—	—	—	—
Orthoclase	—	0.146	0.226	0.251	0.347	—	0.043	0.124	0.351
Pyrite	0.0179	0.1195	0.142	0.165	—	—	0.373	0.466	0
Pyroxene	—	0.18	0.246	0.275	—	—	0.973	0.336	0.233
α-Quartz	0.0414	0.167	0.232	0.2695	—	—	0.7574	0.607	0.168
β-Quartz	—	—	—	—	0.28	0.3165	0.763	0.383	0
Serpentine	—	0.227	—	—	—	—	—	—	—
Siderite	0.056	0.163	—	—	—	—	—	—	—
Talc	—	0.208 at 59°C	—	—	—	—	—	—	—
Zircon	—	0.146 at 60°C	—	—	—	—	—	—	—

RESISTIVITIES OF SEMICONDUCTING MINERALS (ZERO FREQUENCY)

Native elements	ϱ (ohm-m)	Native elements	ϱ (ohm-m)
Diamond (C)	2.7	Gersdorffite, NiAsS	1 to 160×10^{-6}
Sulfides		Glaucodote, (Co, Fe)AsS	5 to 100×10^{-6}
Argentite, Ag_2S	1.5 to 2.0×10^{-3}	Antimonide	
Bismuthinite, Bi_2S_3	3 to 570	Dyscrasite, Ag_3Sb	0.12 to 1.2×10^{-6}
Bornite, $Fe_2S_3 \cdot nCu_2S$	1.6 to 6000×10^{-6}	Arsenides	
Chalcocite, Cu_2S	80 to 100×10^{-6}	Allemonite, $SbAs_2$	70 to 60,000
Chalcopyrite, $Fe_2S_3 \cdot Cu_2S$	150 to 9000×10^{-6}	Lollingite, $FeAs_2$	2 to 270×10^{-6}
Covellite, CuS	0.30 to 83×10^{-6}	Nicollite, NiAs	0.1 to 2×10^{-6}
Galena, PbS	6.8×10^{-6} to 9.0×10^{-2}	Skutterudite, $CoAs_3$	1 to 400×10^{-6}
Haverite, MnS_2	10 to 20	Smaltite, $CoAs_2$	1 to 12×10^{-6}
Marcasite, FeS_2	1 to 150×10^{-3}	Tellurides	
Metacinnabarite, 4HgS	2×10^{-6} to 1×10^{-3}	Altaite, PbTe	20 to 200×10^{-6}
Millerite, NiS	2 to 4×10^{-7}	Calavarite, $AuTe_2$	6 to 12×10^{-6}
Molybdenite, MoS_2	0.12 to 7.5	Coloradoite, HgTe	4 to 100×10^{-6}
Pentlandite, (Fe, Ni)$_9S_8$	1 to 11×10^{-6}	Hessite, Ag_2Te	4 to 100×10^{-6}
Pyrrhotite, Fe_7S_8	2 to 160×10^{-6}	Nagyagite, $Pb_6Au(S, Te)_{14}$	20 to 80×10^{-6}
Pyrite, FeS_2	1.2 to 600×10^{-3}	Sylvanite, $AgAuTe_4$	4 to 20×10^{-6}
Sphalerite, ZnS	2.7×10^{-3} to 1.2×10^4	Oxides	
Antimony-sulfur compounds		Braunite, Mn_2O_3	0.16 to 1.0
Berthierite, $FeSb_2S_4$	0.0083 to 2.0	Cassiterite, SnO_2	4.5×10^{-4} to 10,000
Boulangerite, $Pb_5Sb_4S_{11}$	2×10^3 to 4×10^4	Cuprite, Cu_2O	10 to 50
Cylindrite, $Pb_3Sn_4Sb_2S_{14}$	2.5 to 60	Hollandite, (Ba, Na, K)Mn_8O_{16}	2 to 100×10^{-3}
Franckeite, $Pb_5Sn_3Sb_2S_{14}$	1.2 to 4	Ilmenite, $FeTiO_3$	0.001 to 4
Hauchecornite, $Ni_9(Bi, Sb)_2S_8$	1 to 83×10^{-6}	Magnetite, Fe_3O_4	52×10^{-6}
Jamesonite, $Pb_4FeSb_6S_{14}$	0.020 to 0.15	Manganite, $MnO \cdot OH$	0.018 to 0.5
Tetrahedrite, Cu_3SbS_3	0.30 to 30,000	Melaconite, CuO	6000
Arsenic-sulfur compounds		Psilomelane, $KMnO \cdot MnO_2 \cdot nH_2O$	0.04 to 6000
Arsenopyrite, FeAsS	20 to 300×10^{-6}	Pyrolusite, MnO_2	0.007 to 30
Cobaltite, CoAsS	6.5 to 130×10^{-3}	Rutile, TiO_2	29 to 910
Enargite, Cu_3AsS_4	0.2 to 40×10^{-3}	Uraninite, UO	1.5 to 200

From Carmichael, R. S., ed., *Handbook of Physical Properties of Rocks*, Vol. I, CRC Press, 1982.

MINERALS ARRANGED IN ORDER OF INCREASING VICKERS HARDNESS NUMBERS

Mineral species	Mean	Range	Remarks	Mineral species	Mean	Range	Remarks
Graphite	12	12		Boulangerite	166	157—183	
Molybdenite	17	16—19	$\perp$ to cleavage	Dyscrasite	167	162—178	
	23	21—28	$\parallel$ to cleavage	Berthierite	171	155—185	
Bismuth	18	16—19		Zinkenite	178	162—207	
Tellurbismuth	21	20—21		Emplectite	191	168—213	$\parallel$ elongation
Argentite	24	20—30			222	197—238	$\perp$ to elongation
Hessite	33	28—41		Bournonite	192	185—199	
Orpiment	38	23—52		Chalcopyrite	194	186—219	
Electrum	40	34—44		Blende	198	186—209	
Stromeyerite	41	38—44		Cuprite	199	192—218	
Altaite	51	48—57		Stannite	210	197—221	
Gold	51	50—52		Cubanite	213	199—228	
Silver	53	48—63		Pentlandite	215	202—230	
Realgar	56	53—60		Tenorite	236	209—254	
Digenite	61	56—67		Millerite	236	225—256	Isotropic sections
Arsenic	63	57—69			254	235—280	$\parallel$ to elongation
Pyrargyrite	71	50—97	$\perp$ to cleavage		348	318—376	$\perp$ to elongation
	106	98—126	$\parallel$ to cleavage	Pyrrhotite	248	230—259	Anistropic sections
Covellite	72	69—78			303	280—318	Isotropic sections
Galena	76	71—84		Alabandite	251	240—266	
Pyrolusite	76	76	Average hardness $\perp$ to fibers	Coffinite	258	236—333	
	252	252	Average hardness $\parallel$ to fibers	Chalcostibite	276	264—285	
	279	256—346	Isotropic sections	Niccolite	336	328—348	Anistropic sections
	292	225—405	Microcrystalline		446	433—455	Isotropic sections
Stibnite	77	42—109		Tennantite	338	320—361	
Chalcophanite	81	71—85	$\perp$ to cleavage	Freibergite	345	317—375	
	124	103—165	$\parallel$ to cleavage	Scheelite	348	285—429	
	133	110—178	Isotropic sections	Tetrahedrite	351	328—367	
Chalcocite	84	68—98		Famatinite	363	333—397	
Antimony	89	83—99		Wolframite	373	357—394	
Jamesonite	99	96—105	Granular allotriomorphic sections	Manganite	410	367—459	
	113	105—121	Prismatic sections	Carrollite	463	351—566	
Bornite	103	97—105		Lollingite	486	421—556	$\perp$ to elongation
Bismuthinite	107	92—119			825	739—920	$\parallel$ elongation
Miargyrite	110	104—123		Siegenite	524	503—533	
Sylvanite	110	102—125		Ullmannite	525	498—542	
Kobellite	116	69—173		Betafite	525	503—560	
Proustite	123	109—135		Ilmenite	536	519—553	Possible differences in composition
Platinum	126	125—127			681	659—703	
Copper	134	120—143		Goethite	554	525—620	Microcrystalline
Naumannite	148	115—185			803	772—824	Coarsely crystalline
Zincite	154	150—157	$\perp$ to cleavage	Magnetite	560	530—599	
	304	295—318	$\parallel$ to cleavage	Breithauptite	563	542—584	
Pearceite	160	153—164		Psilomelane	572	503—627	
Enargite	160	133—185	$\perp$ to cleavage	Hausmannite	587	541—613	
	272	245—346	$\parallel$ to cleavage	Braunite	595	584—605	

Mineral species	Mean	Range	Remarks	Mineral species	Mean	Range	Remarks
Pyrochlore	613	572—665		Pararammelsbergite	772	762—803	
Hollandite	620	560—724		Coronadite	784	767—813	
Skutterudite	653	589—724		Columbite-tantalite	803	724—882	
Gersdorffite	698	665—743		Uraninite	808	782—839	
Maucherite	704	685—724		Maghemite	946	894—988	
Euxenite	707	599—782		Bixbyite	1,018	1,003—1,033	
Rammelsbergite	712	687—778		Thorianite	1,918	988—1,115	
Brannerite	720	710—730		Cassiterite	1,053	1,027—1,075	
Pitchblende	720	673—803	Fresh specimens, oxidation produces marked decrease in hardness	Arsenopyrite	1,094	1,048—1,127	
				Bravoite	1,097	1,003—1,288	
Lepidocrocite	724	690—782		Marcasite	1,113	941—1,288	
Jacobsite	734	724—745		Glaucodot	1,124	1,071—1,166	
Davidite	745	707—803		Rutile	1,139	1,074—1,210	
Hematite	755	739—822	Microcrystalline	Pyrite	1,165	1,027—1,240	
	1,000	920—1,062	Coarsely crystalline	Cobaltite	1,200	1,176—1,226	
				Chromite	1,206	1,195—1,210	

From Carmichael, R. S., ed., *Handbook of Physical Properties of Rocks,* Vol. I, CRC Press, 1982.

SOLUBILITY PRODUCT CONSTANTS

James C. Chang

The following solubility product constants are calculated from the free energies of formation of the substances as solids and those of the aqueous ions at their standard states of m = 1. Thus, for the reaction

$$M_m X_n(s) \leftrightarrows mM^{n+}(aq) + nX^{m-}(aq),$$

$$\Delta G^\circ = m\Delta G_f^\circ(M^{n+}, aq) + n\Delta G_f^\circ(X^{m-}, aq) - \Delta G^\circ(M_m X_n, s)$$

where $M_m X_n$ is the slightly soluble substance, M^{n+} and X^{m-} are the two ions produced in solution by the dissociation of $M_m X_n$. Then the solubility product constant, K_{sp}, is calculated by using the equation

$$\ln K_{sp} = -\frac{\Delta G^\circ}{RT}$$

The values in the following table are for K_{sp} at 25°C.

Substance	Formula	Solubility product	Substance	Formula	Solubility product
Barium carbonate	$BaCO_3$	2.60×10^{-9}	Lithium carbonate	Li_2CO_3	1.07×10^{-3}
Barium sulfate	$BaSO_4$	1.05×10^{-10}	Magnesium carbonate	$MgCO_3$	1.15×10^{-5}
Bismuth sulfide	Bi_2S_3	1.80×10^{-99}	Magnesium fluoride	MgF_2	8.43×10^{-8}
Cadmium oxalate 3-hydrate	$CdC_2O_4 \cdot 3H_2O$	1.44×10^{-8}	Magnesium hydroxide	$Mg(OH)_2$	5.66×10^{-12}
Cadmium hydroxide	$Cd(OH)_2$	5.33×10^{-15}	Manganese(II) carbonate	$MnCO_3$	2.23×10^{-11}
Cadmium sulfide	CdS	1.40×10^{-29}	Manganese(II) hydroxide	$Mn(OH)_2$	2.04×10^{-13}
Cadmium carbonate	$CdCO_3$	6.20×10^{-12}	Manganese(II) iodate	$Mn(IO_3)_2$	4.19×10^{-7}
Cadmium iodate	$Cd(IO_3)_2$	2.43×10^{-8}	Manganese(II) oxalate 2-hydrate	$MnC_2O_4 \cdot 2H_2O$	1.72×10^{-7}
Calcium carbonate	$CaCO_3$	4.95×10^{-9}	Manganese(II) sulfide	MnS	4.55×10^{-14}
Calcium fluoride	CaF_2	1.61×10^{-10}	Mercury(II) iodide	HgI_2	2.86×10^{-29}
Calcium hydroxide	$Ca(OH)_2$	7.88×10^{-6}	Mercury(II) sulfide	HgS	6.38×10^{-53}
Calcium oxalate l-hydrate	$CaC_2O_4 \cdot H_2O$	1.96×10^{-8}	Mercury(I) bromide	Hg_2Br_2	6.36×10^{-23}
Calcium phosphate	$Ca_3(PO_4)_2$	2.02×10^{-33}	Mercury(I) chloride	Hg_2Cl_2	1.45×10^{-18}
Calcium sulfate	$CaSO_4$	3.73×10^{-5}	Mercury(I) iodide	Hg_2I_2	5.31×10^{-29}
Cobalt(II) hydroxide	$Co(OH)_2$	1.09×10^{-15}	Mercury(I) sulfate	Hg_2SO_4	7.97×10^{-7}
Cobalt(II) sulfide	CoS	9.70×10^{-21}	Nickel(II) carbonate	$NiCO_3$	1.45×10^{-7}
Copper(II) iodate l-hydrate	$Cu(IO_3)_2 \cdot H_2O$	6.85×10^{-8}	Nickel(II) hydroxide	$Ni(OH)_2$	5.54×10^{-16}
Copper(II) oxalate	CuC_2O_4	4.42×10^{-10}	Nickel(II) iodate	$Ni(IO_3)_2)$	4.73×10^{-5}
Copper(II) sulfide	CuS	1.28×10^{-36}	Nickel(II) sulfide	NiS	1.08×10^{-21}
Copper(I) bromide	$CuBr$	6.16×10^{-9}	Silver(I) bromate	$AgBrO_3$	9.30×10^{-5}
Copper(I) chloride	$CuCl$	1.72×10^{-7}	Silver(I) bromide	$AgBr$	5.32×10^{-13}
Copper(I) iodide	CuI	1.29×10^{-12}	Silver(I) carbonate	Ag_2CO_3	8.58×10^{-12}
Copper(I) sulfide	Cu_2S	2.24×10^{-48}	Silver(I) chloride	$AgCl$	1.76×10^{-10}
Copper(I) thiocyanate	$CuSCN$	3.35×10^{-16}	Silver(I) chromate	Ag_2CrO_4	1.11×10^{-12}
Iron(III) hydroxide	$Fe(OH)_3$	2.67×10^{-39}	Silver(I) iodate	$AgIO_3$	3.15×10^{-8}
Iron(II) carbonate	$FeCO_3$	3.13×10^{-11}	Silver(I) iodide	AgI	8.49×10^{-17}
Iron(II) hydroxide	$Fe(OH)_2$	4.79×10^{-17}	Silver(I) oxalate	$Ag_2C_2O_4$	5.44×10^{-12}
Iron(II) sulfide	FeS	1.57×10^{-19}	Silver(I) phosphate	Ag_3PO_4	1.05×10^{-16}
Lead bromide	$PbBr_2$	6.56×10^{-6}	Silver(I) sulfate	Ag_2SO_4	1.19×10^{-5}
Lead carbonate	$PbCO_3$	1.48×10^{-13}	Silver(I) sulfide	Ag_2S	6.62×10^{-50}
Lead chloride	$PbCl_2$	1.70×10^{-5}	Silver(I) thiocyanate	$AgSCN$	1.03×10^{-12}
Lead fluoride	PbF_2	1.07×10^{-7}	Strontium carbonate	$SrCO_3$	5.60×10^{-10}
Lead hydroxide	$Pb(OH)_2$	1.40×10^{-20}	Strontium oxalate 1-hydrate	$SrC_2O_4 \cdot H_2O$	1.51×10^{-7}
Lead iodate	$Pb(IO_3)_2$	3.63×10^{-13}	Strontium sulfate	$SrSO_4$	3.42×10^{-7}
Lead iodide	PbI_2	8.49×10^{-9}	Tin(II) sulfide	SnS	3.23×10^{-28}
Lead oxalate	PbC_2O_4	8.55×10^{-10}	Zinc hydroxide	$Zn(OH)_2$	7.68×10^{-17}
Lead sulfate	$PbSO_4$	1.81×10^{-8}	Zinc oxalate 2-hydrate	$ZnC_2O_4 \cdot 2H_2O$	1.40×10^{-9}
Lead sulfide	PbS	8.81×10^{-29}	Zinc sulfide	ZnS	2.91×10^{-25}

PROPERTIES OF RARE EARTH METALS

F. H. Spedding

Symbol	M. P. °C[1]	B. P. °C[1]	Heat of Vaporization $\Delta H_{v,0}$ Kcal/g atm[2]	Density	Atomic Vol. (cm³/mole)	Metallic Radius A	Electrical Resistivity Polycrystalline Wire 298°K (ohm-cm x 10⁻⁶)[3]	Residual Resistivity Wire 4.2°K (ohm-cm x 10⁻⁶)[3]	Compressibili cm²/kg x 10⁻⁶
Sc	1541	2831	89.9	2.989	15.041	1.640	52	3	2.26
Y	1522	3338	101.3	4.469	19.894	1.801	59	2	2.68
La	921	3457	103.1	6.145	22.603	1.879	61-80*	SC	4.04
Ce	799	3426	101.1	γ=6.767 β=6.657	20.400 21.049	1.820	70-80	10	4.10
Pr	931	3512	85.3	6.773	20.805	1.828	68	1	3.21
Nd	1021	3068	78.5	7.007	20.585	1.821	65	7	3.0
Pm	1168	2700 est.							
Sm	1077	1791	49.2	7.520	20.001	1.804	91	7	3.34
Eu	822	1597	41.9 ΔH°298	5.243	28.981	1.984	91	1	8.29
Gd	1313	3266	95.3	7.900	19.904	1.801	127	1	2.56
Tb	1356	3123	93.4	8.229	19.312	1.783	114	4	2.45
Dy	1412	2562	70.0	8.550	19.006	1.774	100	5	2.55
Ho	1474	2695	72.3	8.795	18.753	1.766	88	3	2.47
Er	1529	2863	76.1	9.066	18.450	1.757	71	3	2.39
Tm	1545	1947	55.8	9.321	18.124	1.746	74	3	2.47
Yb	819	1194	36.5	6.965	24.843	1.939	28	2	7.39
Lu	1663	3395	102.2	9.840	17.781	1.735	60	2	2.38

* Crystal usually mixture of α hcp and fcc lattice. SC - Superconductor
** Best values in author's opinion. Many numbers are taken from P. W. Bridgman's publication.
[1] Corrected for new temperature scale. Best values Ames Laboratory 2/1/72.
[2] Values from Thermodynamic Properties of Metals and Alloys (Review) R. Hultgren, R. Orr and K. Kelley. Supplements 1967.
[3] Weighted average of original papers 1/1/72.

Symbol	Crystal Structure at Room Temperature 25°C		Allotropic Forms Transition Temperatures Expressed in °C	
Sc	Hex. (to 1335°)	a=3.3088A, c=5.2680A	bcc(above 1335°)	
Y	Hex. (to 1478°)	a=3.6482A, c=5.7318A	bcc(above 1478°)	a=4.08A
La	Hex. (to 310°) (usually contains fcc also)	a=3.7740A, c=12.171A	fcc(310°-865°) bcc(above 865°)	a=4.26A
Ce	fcc(~ 0° to 726°)	a=5.160A	fcc(below-157° on cooling) (up to -94° on heating) Hex. (below-23° on cooling) (up to 168° on heating)	a=4.85A a=3.68A c=11.92A
Pr	Hex. (to 795°)	a=3.6721A, c=11.832A	bcc(above 795°)	a=4.12A
Nd	Hex. (to 863°)	a=3.6583A, c=11.7966A	bcc(above 863°)	a=4.13A
Sm	*Rhom (to 926°)	a=8.9834A, α=23° 49.5′	bcc(above 926°)	a=4.07A
Eu	bcc	a=4.5827A		
Gd	Hex. (to 1235°)	a=3.6336A, c=5.7810A	bcc(above 1235°)	a=4.05A†
Tb	Hex. (to 1289°)	a=3.6055A, c=5.6966A	bcc(above 1289°)	a=4.02A †
Dy	Hex. (to 1381°)	a=3.5915A, c=5.6501A	bcc(above 1381°)	a=3.98A †
Ho	Hex.	a=3.5778A, c=5.6178A	Not present pure metals	
Er	Hex.	a=3.5592A, c=5.5850A	Not present pure metals	
Tm	Hex.	a=3.5375A, c=5.5540A	Not present pure metals	
Yb	Hex. (to 795°)	a=5.4848A	bcc(above 795°)	a=4.44A
Lu	Hex.	a=3.5052A, c=5.5494A	Not present pure metals	

† Extrapolated from magnesium alloy studies. Hex. refers to close packed hexagonal. La, Ce, Pr, and Nd have a stacking order ABAC, the other rare earths ABAB. fcc refers to close packed face centered cubic with a stacking order ABC, ABC, bcc refers to body centered cubic. These high temperature forms are very soft and deform very easily.
* The rhombic Sm cell can be expressed as hexagonal with a=3.6290A, c=26.207 and possess a stacking order ABABCBCAC

DENSITY OF LIQUID ELEMENTS

Gernot Lang

Data in this table consist of those published in various places in the literature. Generally, there has been excellent agreement among values obtained by different methods of measurement. Temperatures are in degrees C and densities in grams per cubic centimeter. Grams per cc × 62.427961 = pounds per cu ft.

Element:	Purity:	ρ_{mp}	ρ_{t_1}		ρ_{t_2}		ρ_{t_3}		Atm.:	Method:	Ref.
			t	ρ	t	ρ	t	ρ			
Ag	—	9.32	1100	9.17	1200	9.07	—	—	—	IBF	74
Ag	—	9.33	1100	9.20	1200	9.10	1300	9.00	—	DBF	33
Ag	99.9	—	1100	9.170	1200	9.070	1300	8.980	—	IBF	Q
Ag	—	9.285	1100	9.15	1200	9.07	1300	8.97	—	IBF	24
Ag	—	9.343	1100	9.20	—	—	—	—	H_2	Bubble pressure	46
Ag	—	9.33 ±0.01	1100	9.18	1200	9.08	1300	8.97	Ar	Bubble pressure	47
Ag	—	9.346	1100	9.22	1200	9.13	1300	9.04	—	DBF	39
Al	99.4	2.41	—	—	—	—	—	—	—	BF	62
Al	99.4	2.38₄	—	—	—	—	—	—	—	Pycnometer	15
Al	99.996	2.368	700	2.357	800	2.332	900	2.304	—	BF	25
Al	99.99	2.39	—	—	—	—	—	—	—	Pycnometer	60
Al	99.997	2.39	677	2.380	807	2.334	912	2.292	Ar	Bubble pressure	11
Au	99.96	17.28	1100	17.24	1200	17.12	1300	16.99	H_2	IBF	24
Au	—	17.361	1100	17.221	1200	17.099	1300	16.950	—	IBF	P
B	99.8	2.08 ±0.03	—	—	—	—	—	—	vac.	Weighing and vol. determin.	80
Ba	—	3.325		$\rho_t = 3.476 - (2.14 \cdot 10^{-4})t \ (t°C)$					Ar	DBF	1
Ba	—	3.320		$\rho_t = 3.847 - 5.26 \cdot 10^{-4} T \ (T°K)$					—	calculated	K
Be	—	—	1500	1.42 ±0.04	—	—	—	—	—	calculated	22
Bi	—	10.07	300	10.03	400	9.91	500	9.78	—	Manometer pressure	31
Bi	—	10.02	400	9.87	600	9.62	800	9.40	—	Dilatometer	7
Bi	—	10.07	400	9.90	500	9.78	600	9.66	—	Dilatometer	53
Bi	—	10.04	400	9.91	600	9.67	800	9.44	—	DBF	34
Bi	—	—	400	9.86	600	9.62	700	9.49	N_2	Bubble pressure	78
Bi	99.94	10.03	400	9.88	600	9.62	700	9.51	—	DBF	66
Bi	99.90	—	800	9.43	900	9.28	1000	9.13	H_2	Bubble pressure	57
Bi	—	10.02	300	9.98	400	9.85	500	9.73	—	Dilatometer	30
Bi	99.98	10.07	500	9.78	700	9.53	900	9.30	Ar	Bubble pressure	49
Bi	—	10.057	700	9.51 ±0.03	—	—	—	—	—	—	L
Ca	—	1.365		$\rho_t = 1.613 - 2.21 \cdot 10^{-4} T \ (T°K)$					—	calculated	K
Cd	—	8.02		$\rho_t = 8.02 - 0.00110 \ (t-320) \ (t°C)$					—	Manometer pressure	31
Cd	99.97	—	340	8.009	400	7.942	500	7.821	N_2	Bubble pressure	28
Cd	—	—	385	7.94	479	7.83	560	7.74	Ar	Bubble pressure	76
Co	—	—	1600	8.08 – 8.13	—	—	—	—	—	Drop volume	59
Co	—	—	1500	ca. 8.05	—	—	—	—	—	BF	26
Co	99.99	7.67	1500	7.66	1600	7.54	1700	7.42	Ar	Bubble pressure	50
Co	—	—	1500	7.70	—	—	—	—	—		81
Co	99.53	7.76		$\rho_t = 9.51 - 9.88 \cdot 10^{-4} T \ (T°K)$					vac.	Sessile drop (mp – 2200)	U
Cr	—	—	1950	6.00 ±0.13	—	—	—	—	vac.	Sessile drop	18
Cr	—	6.46	—	—	—	—	—	—	—	calculated	3
Cs	—	1.843	110	1.800	310	1.691	510	1.575	—	Pycnometer	H
Cs	—	1.851		$\rho_t = 1.853 - 5.71 \cdot 10^{-4} t \ (t°C)$					—	Pycnometer (100–750°C)	F
Cu	—	7.99	1100	7.97	1200	7.81	1300	7.66	—	IBF	6
Cu	—	7.96₂	1100	7.94	1200	7.77	1300	7.62	—	Dilatometer	85
Cu	—	7.92₄	1100	7.91	1300	7.76	1500	7.61	—	DBF	84
Cu	99.99	7.940	1100	7.924	1200	7.846	1300	7.764	—	IBF	Q
Cu	—	—	1100	8.10	—	—	—	—	—	Drop volume	43
Cu	—	7.87₅	1100	7.86	1300	7.70	1500	7.55	—	DBF	45
Cu	—	—	1100	7.90	—	—	—	—	—	Pycnometer	52
Cu	—	7.99₂	1100	7.98	1300	7.81	1500	7.66	Ar	DBF	8
Cu	—	—	1100	8.07	—	—	—	—	—	Bubble pressure	Beer in 49
Cu	—	8.090		$\rho_t = 9.370 - 9.442 \cdot 10^{-4} T \ (T°K)$					—	Drop volume	Mehairy in 49
Cu	—	8.03	1100	8.02	1300	7.86	1500	7.70	Ar	Bubble pressure	49
Fe	—	7.13	—	—	—	—	—	—	—	DBF	75
Fe	—	7.24	—	—	—	—	—	—	vac.	Drop volume	35
Fe	—	—	1550	7.01	—	—	—	—	air	Casting	79
Fe	—	7.15	—	—	—	—	—	—	—	DBF	42
Fe	—	6.99		$\rho_t = 8.523 - 8.358 \cdot 10^{-4} T \ (T°K)$					Ar	DBF	38
Fe	Carbonyl	—	1550	7.189	—	—	—	—	—		27
Fe	—	7.015		$\rho_t = 8.618 - 8.83 \cdot 10^{-4} T \ (T°K)$					Ar	DBF	C
Fe	—	—	1555	6.98	—	—	—	—	—	Bubble pressure	Beer in 49
Fe	99.96	7.03₅	1550	7.01	1600	6.94	1700	6.80	Ar	Bubble pressure	49
Fe	—	—	1550	7.13	—	—	—	—	—		70
Fe	99.9	7.02		$\rho_t = 8.50 - 8.17 \cdot 10^{-4} T \ (T°K)$					vac.	Sessile drop	U

Element:	Purity:	ρ_{mp}	ρ_{t_1}		ρ_{t_2}		ρ_{t_3}		Atm.:	Method:	Ref.
			t	ρ	t	ρ	t	ρ			
Ga	—	6.20 ± 0.01	—	—	—	—	—	—	—	—	E
Ge	—	5.52	1000	5.50	1100	5.44	1200	5.39	—	Volumetric measurement	58
Ge	—	5.57_5	1000	5.56	1100	5.51	1200	5.45	N_2	Pycnometer	40
Ge	99.990	5.49	1000	5.46	1200	5.36	1500	5.21	Ar	Bubble pressure	49
Hf	—	12.0	—	—	—	—	—	—	—	calculated	3
Hg	—	—	0	13.5951	20	13.5457	100	13.3514	—	—	—
In	—	—	231	6.99	302	6.93	421	6.84	Ar	Bubble pressure	76
Ir	—	20.0	—	—	—	—	—	—	—	Calculated	3
K	—	0.826	span $\rho_t = 0.826 - 0.000222\,(t - 62.4)$ (t°C)						—	—	72
K	—	0.819	$\rho_t = 0.819 - 0.238 \cdot 10^{-3}\,(t - 64)$ (t°C)						—	Pycnometer (64–1400°C)	D
K	—	0.828	110	0.817	310	0.772	510	0.724	—	Pycnometer	H
Li	—	0.515	$\rho_t = 0.515 - 0.101 \cdot 10^{-3}\,(t - 200)$ (t°C)						—	Pycnometer (200–1600°C)	D
Li	—	0.518	$\rho_t = 0.5368 - 1.021 \cdot 10^{-4}\,t$ (t°C)						—	Pycnometer (400–1125°C)	F
Mg	—	—	700	1.575	800	1.555	—	—	—	DBF	67
Mg	99.5	1.585	700	1.570	800	1.550	900	1.525	—	IBF	N
Mg	—	1.57	—	—	—	—	—	—	—	Estimated	17
Mg	—	1.590	$\rho_t = 1.834 - 2.67 \cdot 10^{-4}\,T$ (T°K)						Ar	DBF	B
Mn	—	—	1440	$5.84 \pm 2\%$	—	—	—	—	He	Volumetric method	M
Mn	—	6.43	—	—	—	—	—	—	—	—	71
Mo	99.7	9.33	—	—	—	—	—	—	vac.	Drop volume	64
Mo	—	9.35	—	—	—	—	—	—	—	calculated	3
Na	—	0.938_5	$\rho_t = 0.938_5 - 0.000260\,(t - 96.5)$ (t°C)						—	—	72
Na	—	—	$\rho_t = 0.938 - 1.9 \cdot 10^{-4}\,T$ (T°K)						—	Bubble pressure	82
Na	—	0.927	$\rho_t = 0.927 - 0.238 \cdot 10^{-3}\,(t - 100)$ (t°C)						—	Pycnometer	D
Na	—	0.928	410	0.854	610	0.806	810	0.754	—	Pycnometer	H
Nb, Cb	—	7.83	—	—	—	—	—	—	—	calculated	3
Ni	—	—	1500	ca. 8.04	—	—	—	—	—	BF	26
Ni	—	7.905	$\rho_t = 9.908 - 11.598 \cdot 10^{-4}\,T$ (T°K)						Ar	DBF	C
Ni	99.85	7.77	1500	7.71_5	1600	7.59_5	1700	7.48	Ar	Bubble pressure	49
Ni	99.99	7.78	$\rho_t = 7.78 - 0.006\,(t - 1453)$ (t°C)						vac.	Sessile drop	21
Ni	—	—	1500	7.78	—	—	—	—	—	—	81
Ni	99.95	7.91	$\rho_t = 9.51 - 10.00 \cdot 10^{-4}\,T$ (T°K)						vac.	Sessile drop	U
Os	—	20.1	—	—	—	—	—	—	—	calculated	3
Pb	—	10.71	$\rho_t = 10.71 - 0.00139\,(t - 327)$ (t°C)						—	Manometer pressure	31
Pb	99.98	—	340	10.57	400	10.49	440	10.43	N_2	Bubble pressure	28
Pb	—	—	568	10.380	640	10.294	720	10.201	—	DBF	O
Pb	—	—	400	10.56	500	10.43	700	10.17	N_2	Bubble pressure	78
Pb	—	—	437	10.52	623	10.28	705	10.16	Ar	Bubble pressure	76
Pb	99.9923	10.678	$\rho_t = 10.678 - 13.174 \cdot 10^{-4}\,T$ (T°K)						Ar	DBF	A
Pd	—	10.7	—	—	—	—	—	—	—	calculated	20
Pd	—	10.7	—	—	—	—	—	—	—	calculated	3
Pd	99.95	—	1600	10.43	1700	10.31	1800	10.18_5	Ar	Bubble pressure	49
Pt	99.84	19.7 ± 0.25	—	—	—	—	—	—	vac.	Drop volume	19
Pt	99.999	—	1800	18.82 ± 0.02	—	—	—	—	Ar	Sessile drop	44
Pt	—	18.91	—	—	—	—	—	—	Ar	Bubble pressure	48
Pt	—	—	1800	18.82	1850	18.67_5	1875	18.60_5	Ar	Bubble pressure	49
Pu	—	—	$\rho_t = 17.57 - 1.45 \cdot 10^{-3}\,t$ (t°C) ± 0.21						—	(655–P60°C)	T
Pu	99.95	16.623	699	16.548	824	16.370	950	16.185	—	Pycnometer	32
Rb	—	1.48	—	—	—	—	—	—	—	estimated	56
Rb	—	1.463	$\rho_t = 1.481 - 4.51 \cdot 10^{-4}\,t$ (t°C)						—	Pycnometer	F
Rb	—	—	45.5	1.4505	47.4	1.4498	49.3	1.4469	vac.	Pycnometer	G
Re	99.4	18.9	—	—	—	—	—	—	vac.	Drop volume	64
Re	—	18.7	—	—	—	—	—	—	—	calculated	3
Rh	—	10.65	—	—	—	—	—	—	—	calculated	20
Rh	—	11.1	—	—	—	—	—	—	—	calculated	3
Ru	—	10.9	—	—	—	—	—	—	—	calculated	3
S	—	1.819	120–160°C: $\rho_t = 1.901 - 8.00 \cdot 10^{-4}\,t$ (t°C)						N_2	Bubble pressure	61
Sb	—	6.49	700	6.45	800	6.39	1000	6.27	—	IBF	7
Sb	99.52	—	650	6.530	700	6.509	800	6.424	N_2	Bubble pressure	28
Sb	—	—	700	6.41	800	6.38	900	6.35	H_2	Bubble pressure	78

DENSITY OF LIQUID ELEMENTS (Continued)

Element:	Purity:	ρ_{mp}	ρ_{t_1}		ρ_{t_2}		ρ_{t_3}		Atm.:	Method:	Ref.
			t	ρ	t	ρ	t	ρ			
Sb	—	—	650	6.52	700	6.50	—	—	H_2	Bubble pressure	23
Sb	—	—	728	6.44	827	6.38	917	6.32	Ar	Bubble pressure	76
Sb	99.9	6.50	650	6.49	700	6.45			—	DBF	66
Sb	99.992	6.483	\multicolumn{6}{}{$\rho_t = 6.596 + 2.022 \cdot 10^{-4}\,T - 3.629 \cdot 10^{-7}\,T^2$ (T°K)}						DBF, Bubble pressure	37
Sb	99.6	6.46_5	700	6.42	900	6.30	1200	6.13	Ar	Bubble pressure	49
Se	—	3.987	\multicolumn{6}{}{$\rho_t = 3.987 - 16 \cdot 10^{-4}\,(T - 493)$ (T°K)}						—	—	13
Se	—	3.985	\multicolumn{6}{}{$\rho_t = 3.985 - 15.5 \cdot 10^{-4}\,(T - 490)$ (T°K)}						—	Pycnometer	10
Se	—	4.06	\multicolumn{6}{}{$\rho_t = 4.06 - 5 \cdot 10^{-4}\,(T - 491)$ (T°K)}						—	—	4
Se	—	3.984	—	—	—	—	—	—	—	—	1
Se	—	4.01	250	3.97	300	3.91	400	3.79	—	BF	51
Si	—	—	1550	2.54	—	—	—	—	—	—	14
Si	99.9	2.52_5	1450	2.51	1500	2.49	1600	2.46	Ar	Bubble pressure	49
Sn	—	6.988	300	6.94	500	6.81	800	6.64	—	DBF	12
Sn	—	6.98	—	—	—	—	—	—	—	DBF	63
Sn	—	7.01	\multicolumn{6}{}{$\rho_t = 7.01 - 0.00074\,(t - 232)$ (t°C)}						—	Manometer pressure	31
Sn	—	6.97	300	6.92	500	6.78	800	6.57	—	IBF	74
Sn	—	6.896	300	6.93	500	6.78	600	6.71	—	Dilatometer	53
Sn	—	6.966	300	6.92	600	6.72	1200	6.34	—	DBF	84
Sn	—	6.983	300	6.93	400	6.86	500	6.79	—	Pycnometer	73
Sn	—	7.00	300	6.95	500	6.80	600	6.74	N_2	Bubble pressure	65
Sn	—	—	625	6.67	—	—	—	—	N_2	Bubble pressure	78
Sn	—	6.99	300	6.94	400	6.87	—	—	H_2	Bubble pressure	23
Sn	—	6.993	\multicolumn{6}{}{$\rho_t = 6.98 - 0.00074\,(t - 250)$ (t°C)}						—	Manometer pressure	69
Sn	—	6.968 ± 0.005	300	6.93	400	6.87	500	6.79	—	Dilatometer	30
Sn	—	6.968	300	6.96	400	6.84	500	6.77	Ar	Bubble pressure	83
Sn	—	6.978 ± 0.022	300	6.93	600	6.70	1200	6.29	Ar	Bubble pressure	49
Ta	—	15.0	—	—	—	—	—	—	—	calculated	3
Te	—	5.75	460	5.75	500	5.74	600	5.70	—	Pycnometer	40
Te	—	—	460	5.58	600	5.53	—	—	N_2	Bubble pressure	77
Te	99.9999	5.797	460	5.79	600	5.72	700	5.66	Ar	Bubble pressure	49
Ti	98.7	4.11 ± 0.08	—	—	—	—	—	—	—	Capillary method	16
Ti	—	4.15	—	—	—	—	—	—	—	calculated	7
Tl	—	11.29	412	11.13	504	11.01	651	10.77	Ar	Bubble pressure	76
U	—	17.907	\multicolumn{6}{}{$\rho_t = 19.356 - 10.328 \cdot 10^{-4}\,T$ (T°K)}						Ar	DBF	29
V	—	5.55	—	—	—	—	—	—	—	calculated	3
W	—	17.6	—	—	—	—	—	—	—	estimated	9
W	99.8	17.7	—	—	—	—	—	—	vac.	Drop volume	64
W	—	17.5	—	—	—	—	—	—	—	calculated	3
Zn	—	6.59	\multicolumn{6}{}{$\rho_t = 6.59 - 0.00097\,(t - 419)$ (t°C)}						—	Manometer pressure	31
Zn	—	—	600	6.35	—	—	—	—	—	IBF	7
Zn	—	6.56_2	500	6.47	600	6.37	700	6.29	—	Dilatometer	53
Zn	—	6.55_1	500	6.47	600	6.38	700	6.29	—	Pycnometer	73
Zn	—	6.64_5	500	6.56	600	6.45	800	6.25	—	DBF	67
Zn	—	6.64_4	500	6.58	600	6.49	800	6.22	—	DBF	68
Zn	—	6.64	500	6.57	600	6.50	700	6.43	N_2	Bubble pressure	65
Zn	—	—	500	6.55	—	—	—	—	H_2	Bubble pressure	78
Zn	—	6.64_5	—	—	—	—	—	—	—	Pycnometer	41
Zn	99.995	6.562	500	6.417	600	6.370	700	6.259	—	IBF	25
Zn	—	6.57_5	500	6.49	600	6.37	700	6.26	Ar	Bubble pressure	83
Zn	99.999	$6.57_7 \pm 0.012$	500	6.48	600	6.37	700	6.27	Ar	Bubble pressure	49
Zr	—	5.80	—	—	—	—	—	—	—	calculated	3

DENSITY OF LIQUID ELEMENTS

SUPPLEMENTARY TABLE

Gernot Lang

Element	Purity	ρ_{mp}	ρt_1		ρt_2		ρt_3		Atm.	Method	Ref.
			t°C	ρ	t°C	ρ	t°C	ρ			
Ag		9.36	\multicolumn{6}{}{$\rho_t = 9.36 - 0.00114\,(t - t_{mp})$}					(t°C)	Bubble pressure	110
Ag	99.95		\multicolumn{6}{}{$\rho_t = 9.318 - 1.08 \cdot 10^{-3}\,(t - 1000) \pm 0.006$ (1000–1200°C)}					H_2	Bubble pressure	81
Ag	99.999	9.31	\multicolumn{6}{}{$\rho_t = 10.31 - 1.05 \cdot 10^{-3} \cdot t$ (t°C) (960–1150)}					Ar	Sessile drop	88
Al	99.99	2.39	\multicolumn{6}{}{$\rho_t = 2.60 - 3.2 \cdot 10^{-4}\,t$ (t°C) (900–1750°C)}					He	Drop volume	87
Al	99.999	2.365	\multicolumn{6}{}{$\rho_t = 2.576 - 3.2 \cdot 10^{-4}\,t$ (t°C)}					He	Sessile drop	108
Al	99.998	2.375	\multicolumn{6}{}{$\rho_t = 2.375 - 5.4 \cdot 10^{-4}\,(t - 660)$}					(t°C) Ar	Sessile drop	97

Element	Purity	ρ_{mp}	ρ_{t_1}		ρ_{t_2}		ρ_{t_3}		Atm.	Method	Ref.
			$t°C$	ρ	$t°C$	ρ	$t°C$	ρ			
Ba	99.5	3.32		$\rho_t = 3.59 - 2.74 \cdot 10^{-4}T$				(T°K)		Bubble pressure	90
Bi	99.9999	10.05		$\rho_t = 10.05 - 1.41 \cdot 10^{-3}\,(t - t_{mp})$				(t°C)	N_2	Volume determin.	96
Bi	99.98	10.04		$\rho_t = 10.04 - 1.44 \cdot 10^{-3}\,(t - t_{mp})$				(t°C)	N_2	Volume determin.	96
Bi	99.999	10.114		$\rho_t = 10.406 - 1.078 \cdot 10^{-3}t$ (271–414)				(t°C)	N_2	DBF	89
Bi	—	10.031		$\rho_t = 10.336 - 12.367 \cdot 10^{-4}t$				(t°C)	vac.	Pycnometer	92
Ca	—	1.365		$\rho_t = 1.613 - 2.21 \cdot 10^{-4}T$				(T°K)		Bubble pressure	90
Cd	99.999	7.996		$\rho_t = 8.388 - 12.205 \cdot 10^{-4}t$				(t°C)	Ar	Pycnometer	91
Co	99.53	7.78		$\rho_t = 9.5_7 - 10.1_7 \cdot 10^{-4}T$ (1780–2200)				(T°K)	Ar	Drop volume	103
Co	99.67	7.75		$\rho_t = 9.7_1 - 1.1_7 \cdot 10^{-3}T$				(T°K)	Ar	Bubble pressure	109
Cs	—	1.86		$\rho_t = 1.84/1 + 1.1755 \cdot 10^{-4}\,(t-82.4) + 7.656 \cdot 10^{-8}$ $(t-82.4)^2$ (t°F, 29–704°C)					Ar	Pycnometer	86
Cs	—	—	40	1.840	271	1.711	376	1.655	Ar	Dilatometer	106
Fe	—	7.06		$\rho_t = 7.05 - 7.3 \cdot 10^{-4}\,(t-1550)$				(t°C)	H_2, He	Sessile drop	101
Fe	99.9	7.03		$\rho_t = 8.5_7 - 8.5_3 \cdot 10^{-4}T$ (1800–2150)				(T°K)	Ar	Drop volume	103
Fe	99.98	7.05		$\rho_t = 8.7_8 - 0.95_8 \cdot 10^{-3}T$				(T°K)	Ar	Bubble pressure	109
Fr			100	2.29						calculated	100
Ga	—	6.1136		$\rho_t = 6.11564 - 7.37437 \cdot 10^{-4}t + 1.37767 \cdot 10^{-7}t^2$ (t°C) (50–600°C)						Pycnometer	98
Hg	—	—	0	13.5951	20	13.5457	100	13.3514		—	105
In	99.999	7.032		$\rho_t = 7.153 - 0759 \cdot 10^{-3}t$				(t°C)	N_2	DBF (160–532)	89
In	99.999	7.016		$\rho_t = 7.146 - 8.362 \cdot 10^{-4}t$				(t°C)	Ar	Pycnometer	91
Mg	99.5		959	1.52	1053	1.47	—	—		Bubble pressure	90
Mo		9.1							vac.	Pendant drop and drop weight	95
Nb, Cb	—	7.6							vac.	Pendant drop and drop weight	95
Ni	—	7.82		$\rho_t = 7.81 - 8.7 \cdot 10^{-4}\,(t-1460)$				(t°C)	H_2, He	Sessile drop	U
Ni	—	7.65							vac.	Pendant drop and drop weight	95
Ni	99.95	7.95		$\rho_t = 9.8_1 - 10.8_0 \cdot 10^{-4}T$ (1890–2150)				(T°K)	Ar	Drop volume	103
Pb	99.999	10.330		$\rho_t = 10.650 - 9.8 \cdot 10^{-4}t$ (1000–1600)				(t°C)	He	Sessile drop	108
Pb	99.997	10.660		$\rho_t = 11.060 - 12.220 \cdot 10^{-4}t$				(t°C)		Pycnometer	107
Pd	99.998	10.379		$\rho_t = 12.193 - 11.69 \cdot 10^{-4}t$ (MP–1700)				(t°C)	He	Sessile drop	108
Rb	99.4	1.385		$\rho_t = 1.55643 - 2.6511 \cdot 10^{-4}t - 6.26779/t$ (t°F, 66–1076°C)					Ar	Dilatometer	106
Rb	—	1.484		$\rho_t = 1.472/1 + 1.3309 \cdot 10^{-4}\,(t-102) + 5.2106 \cdot 10^{-8}\,(t-102)^2$ (t°F, 39–730°C)					Ar	Pycnometer	86
Sb	99.999	6.452		$\rho_t = 6.818 - 5.8 \cdot 10^{-4}t$ (1000–1600°C)				(t°C)	He	Sessile drop	108
Sb	99.999	6.535		$\rho_t = 6.962 - 0.673 \cdot 10^{-3}t$ (635–745°C)				(t°C)	N_2	DBF	89
Sb	—	6.493		$\rho_t = 6.902 - 6.486 \cdot 10^{-4}t$ (MP–746°C)				(t°C)	Ar	Pycnometer	93
Si	99.9999		1500	2.46					Ar		94
Sn	99.999		246	6.964	336	6.900	400	6.854	vac.	Pycnometer	111
Sn	99.999	7.01		$\rho_t = 7.16 - 6.3 \cdot 10^{-4}t$				(t°C)		Bubble pressure	104
Sn	99.999	6.981		$\rho_t = 7.135 - 0.663 \cdot 10^{-3}t$ (MP–438°C)				(t°C)	N_2	DBF	89
Sn	99.999	6.973		$\rho_t = 7.139 - 7.125 \cdot 10^{-4}t$				(t°C)		Pycnometer	107
Sr	99.5	2.375		$\rho_t = 2.648 - 2.62 \cdot 10^{-4}T$				(T°K)		Bubble pressure	90
Te	—	5.71		$\rho_t = 5.71 - 3.6 \cdot 10^{-4}\,(t-t_{mp})$				(t°C)		Bubble pressure	110
Te	99.7	5.86		$\rho_t = 5.86 - 0.73 \cdot 10^{-3}\,(t-t_{mp})$				(t°C)	N_2	Volume determin.	96
Ti	—	4.10							vac.	Pendant drop and drop weight	95
U		17.27		$\rho_t = 19.520 - 16.01 \cdot 10^{-4}T$				(T°K)	vac.	Pycnometer	102
V	—	5.3							vac.	Pendant drop and drop weight	95
Zr	—	5.60							vac.	Pendant drop and drop weight	95

DENSITY OF LIQUID ELEMENTS (Continued)

References

1. Addison and Pulham, J. Chem. Soc., 3873 (1962).
2. Allen, Trans, AIME, **227,** 1175 (1963).
3. Allen, Trans. AIME, **230,** 1537 (1964).
4. Astachov, Penin and Dobkina, Zh. Fiz. Chim., **20,** 403 (1946).
5. Beer (in Lucas), Mém. Sci. Rev. Mét., **61,** 1, 97 (1964).
6. Bornemann and Sauerwald, Z. Metallkunde, **14,** 145 (1922).
7. Bornemann and Sauerwald, Z. Metallkunde, **14,** 254 (1922).
8. Cahill and Kirshenbaum, J. Phys. Chem., **66,** 1080 (1962).
9. Calverley, Proc. Phys. Soc., **70,** 1040 (1957).
10. Campbell and Epstein, J. Amer. Chem. Soc., **64,** 2679 (1942).
11. Coy and Mateer, Trans. ASM, **58,** 99 (1965).
12. Day, Sosman and Hostetter, Am. J. Sci., **187,** 1 (1914).
13. Dobinsky and Weselowsky, Bull. Int. Acad. Polon., A, 446 (1936).
14. Dshemilev, Popel and Zarevski, Fiz. Met. i Met., **18,** 83 (1964).
15. Edwards and Moorman, Chem & Met. Eng., **24,** 61 (1921).
16. Eljutin and Maurach, Izv. A. N., OTN, **4,** 129 (1956).
17. Eremenko, Ukr. Chim. Zh., **28,** 427 (1962).
18. Eremenko and Naidich, Izv. A. N., OTN, **2,** 111 (1959).
19. Eremenko and Naidich, Izv. A. N., OTN, **6,** 129 (1959).
20. Eremenko and Naidich, Izv. A. N., OTN, **6,** 100 (1961).
21. Eremenko and Nishenko, Ukr. Chim. Zh., **30,** 125 (1964).
22. Eremenko, Nishenko and Taj-Shou-Wej, Izv. A. N., OTN, **3,** 116 (1960).
23. Fisher and Phillips, J. Metals, **6,** 1060 (1954).
24. Gebhardt and Becker, Z. Metallkunde, **42,** 111 (1951).
25. Gebhardt, Becker and Dorner, Aluminium, **31,** 315 (1955).
26. Geld and Vertman, Fiz. Met i Met., **10,** 793 (1960).
27. Gogiberidse and Kekelidse, Izv. A. N., **3,** 125 (1963).
28. Greenaway, J. Inst. Met., **74,** 133 (1947)
29. Grosse, Cahill and Kirshenbaum, J. Amer. Chem. Soc., **83,** 4665 (1961).
30. Herczynska, Naturwiss., **47,** 200 (1960).
31. Hogness, J. Amer. Chem. Soc., **43,** 1621 (1921).
32. Jones, Ofte, Rohr and Wittenberg, Trans. ASM, **55,** 819 (1962).
33. Jouniaux, Bull. Soc. Chim. France, **47,** 524 (1930).
34. Jouniaux, Bull. Soc. Chim. France, **51,** 677 (1932).
35. Kingery and Humenik, J. Phys. Chem., **57,** 359 (1953).
36. Kirshenbaum and Cahill, Trans. ASM, **55,** 844 (1962).
37. Kirshenbaum and Cahill, Trans. ASM, **55,** 849 (1962).
38. Kirshenbaum and Cahill, Trans. AME, **224,** 816 (1962).
39. Kirshenbaum, Cahill and Grosse, J. Inorg. Nucl. Chem., **24,** 333 (1962).
40. Klemm et al., Monatsh. Chem., **83,** 629 (1952).
41. Knappwost and Restle, Z. Elektrochemie, **58,** 112 (1954).
42. Königer and Nagel, GieBerei TWB, **13,** 57 (1960).
43. Kozakevitch, Châtel, Urbain and Sage, Rev. de Mét., **52,** 139 (1955).
44. Kozakevitch and Urbain, C. R., Paris, **253,** 2229 (1961).
45. Lang, WADC-Techn. Rep., 57–488 (1957).
46. Lauermann and Metzger, Z. Phys. Chem., **216,** 37 (1961).
47. Lucas, C. R., Paris, **250,** 1850 (1960).
48. Lucas, C. R., Paris, **253,** 2526 (1961).
49. Lucas, Mém. Sci. Rev. Mét., **61,** 1 (1964).
50. Lucas, Mém. Sci. Rev. Mét., **61,** 97 (1964).
51. Lucas and Urbain, C. R., Paris, **258,** 6403 (1964).
52. Malmberg, J. Inst. Met., **89,** 137 (1960).
53. Matuyama, Sci. Rep. RITU, **18,** 19 (1929).
54. Mehairy and Wood (in Lucas), Mém. Sci. Rev. Mét., **61,** 1 (1964).
55. Metals Handbook: Properties and Selection of Metals, 8th edit., **1,** (1961).
56. Metals Reference Book, C. J. Smithells, 4th edit., **1,** 688 (1967).
57. Metzger, Z. Phys. Chem., **211,** 1 (1959).
58. Mokrovski and Regel, J. Phys. Techn., **22,** 1281 (1952).
59. Monma and Suto, J. Jap. Inst. Met., **1,** 69 (1960).
60. Naidich and Eremenko, Fiz. Met. i Met., **6,** 62 (1961).
61. Ono and Matsushima, Sci. Rep. RITU, **9,** 309 (1957).
62. Pascal and Jouniaux, C. R., Paris, **158,** 414 (1914).
63. Pascal and Jouniaux, Z. Elektrochemie, **22,** 72 (1916).
64. Pekarev, Izv. Vyss. Utch. Saved., Tsvetn. Met., **6,** 111 (1963).
65. Pelzel, Berg-u. Hütt. Mon. Hefte, Leoben, **93,** 248 (1948).

DENSITY OF LIQUID ELEMENTS (Continued)

66. Pelzel, Z. Metallkunde, **50**, 392 (1959).
67. Pelzel and Sauerwald, Z. Metallkunde, **33**, 229 (1941).
68. Pelzel and Schneider, Z. Metallkunde, **35**, 121 (1943).
69. Pokrovski and Saidov, Zh. Fiz. Chim., **29**, 1601 (1955).
70. Popel, Smirnov, Zarevski, Dshemilev and Pastuchov, Izv. A. N., **1**, 62 (1965).
71. Popel, Zarevski and Dshemilev, Fiz. Met. i Met., **18**, 468 (1964).
72. Rink, C. R., Paris, **189**, 39 (1929).
73. Saeger and Ash, J. Res. Nat. Bur. Stand., **8**, 37 (1932).
74. Sauerwald, Z. Metallkunde, **14**, 457 (1922).
75. Sauerwald and Widawski, Z. anorg. allg. Chem., **155**. 1 (1926).
76. Schneider and Heymer, Z. anorg. allg. Chem., **286**, 118 (1956).
77. Smith and Spitzer, J. Phys. Chem., **66**, 946 (1962).
78. Stauffer, Thesis Göttingen (1953).
79. Stott and Rendall, J. Iron & Steel Inst., **175**, 374 (1953).
80. Tavadse, Bairamashvili, Chantadse and Zagareishvili, Doklady A. N., **150**, 544 (1963).
81. Tavadse, Bairamashvili and Chantadse, Doklady A. N., **162**, 67 (1965).
82. Taylor, J. Inst. Met., **83**, 143 (1954).
83. Übelacker and Lucas, C. R., Paris, **254**, 1622 (1962).
84. Widawski and Sauerwald, Z. anorg. allg. Chem., **192**, 145 (1930).
85. Zimmermann and Esser, Arch. f. Eisenh., **2**, 867 (1929).
A) Kirshenbaum, Cahill and Grosse, J. Inorg. Nucl. Chem., **22**, 33 (1961).
B) McGonigal, Kirshenbaum and Grosse, J. Phys. Chem., **66**, 737 (1962).
C) Grosse and Kirshenbaum, J. Inorg. Nucl. Chem., **25**, 331 (1963).
D) Golchova, Teplofiz. Vysok. Temp., **4**, 360 (1966).
E) Bosio, C. R., Paris, **259**, 4545 (1964).
F) Spilrajn and Jakimovich, Teplofiz. Vysok. Temp., **5**, 239 (1967).
G) Jakimovich and Saars, Teplofiz. Vysok. Temp., **5**, 532 (1967).
H) Stone, Ewing, Spann, Steinkuller, Williams and Miller, J. Chem. Engng. Data, **11**, 320 (1966).
I) Shirai, Hamada and Kobayashi, J. Chem. Soc. Japan, **84**, 968 (1963).
K) Grosse and McGonigal, J. Phys. Chem., **68**, 414 (1964).
L) Cubicciotti, J. Phys. Chem., **68**, 537 (1964).
M) Watolin and Esin, Fiz. Met. i Met., **16**, 936 (1963).
N) Gebhardt, Becker and Trägner, Z. Metallkunde, **46**, 90 (1955).
O) Kubaschewski and Hörnle, Z. Metallkunde, **42**, 129 (1951).
P) Gebhardt and Dorner, Z. Metallkunde, **42**, 353 (1951).
Q) Gebhardt and Wörwag, Z. Metallkunde, **42**, 358 (1951).
R) Been, Edwards, Teeter and Chalkins, NEPA-1585, US-AEC (1950).
S) da C. Andrade and Dobbs, Proc. Roy. Soc., **211**, 12 (1952).
T) Wilkinson: Extractive and Physical Metallurgy of Plutonium and its Alloys, New York-London, 1960.
U) Saito and Sakuma, J. Jap. Inst. Met., **31**, 1140 (1967).
86. Achener, HTLMHTTM., Vol. I, Oak Ridge, p. 5, November 1964.
87. Ayushina, Lewin and Geld, Zh. Fiz. Chim., **42**, 2799 (1968).
88. Bernard and Lupis, Met. Trans., **2**, 555 (1971).
89. Berthou and Tougas, Met. Trans., **1**, 2978 (1970).
90. Bohdansky and Schins, J. Inorg. Nucl. Chem., **30**, 2331 (1968).
91. Crawley, Trans. AIME, **242**, 2237 (1968).
92. Crawley and Kiff, Met. Trans., **2**, 609 (1971).
93. Crawley and Kiff, Met. Trans., **3**, 158 (1972).
94. Eljutin, Kostikow and Lewin, Izv. Vys. Uch. Sav., Tsvetn. Met., 2, 131 (1970).
95. Eljutin, Kostikow and Penkow, Poroshk. Met., 9, 46 (1970).
96. Keskar and Hruska, Met. Trans., **1**, 2357 (1970).
97. Körber and Löhberg, Gießereiforschung, **23**, 173 (1971).
98. Koster, Hensel and Franck, Ber. Bunsenges., **74**, 43 (1970).
99. Nagamori, Trans. AIME, **245**, 1897 (1969).
100. Osminin, Zh. Fiz. Chim., **43**, 2610 (1969).
101. Popel, Shergin and Zarewski, Zh. Fiz. Chim., **43**, 2365 (1969).
102. Rohr and Wittenberg, J. Phys. Chem., **74**, 1151 (1970).
103. Saito and Sakuma, Sci. Rep. RITU, **22**, 57 (1970).
104. Schwaneke and Folke, Us-Bur. Min., Invest. Rep. No. 7372 (1970).
105. Stoffhutte. Taschenbuch der Werkstoffkunde, 4th ed. p. 1059, 1967.
106. Tepper, Murchison, Zelenak and Roehlich, HTLMHTTM, Vol. I, Oak Ridge, p. 26, November 1964.
107. Thresh, Crawley and White, Trans. AIME, **242**, 819 (1968).
108. Watolin, Esin, Uchow and Dubinin, Trudy Inst. Met. Sverdlovsk, **18**, 73 (1969).
109. Watanabe, Trans. Jap. Inst. Met., **12**, 17 (1971).
110. Wobst and Rentzsch, Z. Phys. Chem., **240**, 36 (1969).
111. Crawley, Trans. AIME, **245**, 1655 (1969).

HEAT OF FUSION OF SOME INORGANIC COMPOUNDS

RUDOLF LOEBEL

Values in parentheses are of uncertain reliability

Compound	Formula	M.P.,°C	H_f cal/g mole	H_f cal/g	Compound	Formula	M.P.,°C	H_f cal/g mole	H_f cal/g
Actinium[227]	Ac	1050 ± 50	(3400)	(11.0)	Bismuth trichloride	$BiCl_3$	223.8	2600	8.2
Aluminum	Al	658.5	2550	94.5	Bismuth trifluoride	BiF_3	726.0	(6200)	(23.3)
Aluminum bromide	Al_2Br_6	87.4	5420	10.1	Bismuth trioxide	Bi_2O_3	815.8	6800	14.6
Aluminum chloride	Al_2Cl_6	192.4	19600	63.6	Boron	B	2300	(5300)	(490)
Aluminum iodide	Al_2I_6	190.9	7960	9.8	Boron tribromide	BBr_3	−48.8	(700)	(2.9)
Aluminum oxide	Al_2O_3	2045.0	(26000)	(256.0)	Boron trichloride	BCl_3	−107.8	(500)	(4.3)
Antimony	Sb	630	4770	39.1	Boron trifluoride	BF_3	−128.0	480	7.0
Antimony tribromide	$SbBr_3$	96.8	3510	9.7	Boron triiodide	BI_3	31.8	(1000)	(2.5)
Antimony trichloride	$SbCl_3$	73.3	3030	13.3	Boron trioxide	B_2O_3	448.8	5500	78.9
Antimony pentachloride	$SbCl_5$	4.0	2400	8.0	Bromine	Br_2	−7.2	2580	16.1
Antimony trioxide	Sb_4O_6	655.0	(26990)	(46.3)	Bromine pentafluoride	BrF_5	−61.4	1355	7.07
Antimony trisulfide	Sb_4S_6	546.0	11200	33.0	Cadmium	Cd	320.8	1460	12.9
Argon	Ar	−190.2	290	7.25	Cadmium bromide	$CdBr_2$	567.8	(5000)	(18.4)
Arsenic	As	816.8	(6620)	(22.0)	Cadmium chloride	$CdCl_2$	567.8	5300	28.8
Arsenic pentafluoride	AsF_5	80.1	3000	16.5	Cadmium fluoride	CdF_2	1110	(5400)	(35.9)
Arsenic tribromide	$AsBr_3$	30.0	2810	8.9	Cadmium iodide	CdI_2	386.8	3660	10.0
Arsenic trichloride	$AsCl_3$	−16.0	2420	13.3	Cadmium sulfate	$CdSO_4$	1000	4790	22.9
Arsenic trifluoride	AsF_3	−6.0	2486	18.9	Calcium	Ca	851	2230	55.7
Arsenic trioxide	As_4O_6	312.8	8000	22.2	Calcium bromide	$CaBr_2$	729.8	4180	20.9
Barium	Ba	725	1830	13.3	Calcium carbonate	$CaCO_3$	1282	(12700)	(126)
Barium bromide	$BaBr_2$	846.8	6000	21.9	Calcium chloride	$CaCl_2$	782	6100	55
Barium chloride	$BaCl_2$	959.8	5370	25.9	Calcium fluoride	CaF_2	1382	4100	52.5
Barium fluoride	BaF_2	1286.8	3000	17.1	Calcium nitrate	$Ca(NO_3)_2$	560.8	5120	31.2
Barium iodide	BaI_2	710.8	(6800)	(17.3)	Calcium oxide	CaO	2707	(12240)	(218.1)
Barium nitrate	$Ba(NO_3)_2$	594.8	(5900)	(22.6)	Calcium metasilicate	$CaSiO_3$	1512	13400	115.4
Barium oxide	BaO	1922.8	13800	93.2	Calcium sulfate	$CaSO_4$	1297	6700	49.2
Barium phosphate	$Ba_3(PO_4)_2$	1727	18600	30.9	Carbon dioxide	CO_2	−57.6	1900	43.2
Barium sulfate	$BaSO_4$	1350	9700	41.6	Carbon monoxide	CO	−205	199.7	7.13
Beryllium	Be	1278	—	260.0	Cyanogen	C_2N_2	−27.2	2060	39.6
Beryllium bromide	$BeBr_2$	487.8	(4500)	(26.6)	Cyanogen chloride	CNCl	−5.2	2240	36.4
Beryllium chloride	$BeCl_2$	404.8	(3000)	(30)	Cerium	Ce	775	2120	15.1
Beryllium fluoride	BeF_2	796.8	(6000)	(127.6)	Cerium (III) chloride	$CeCl_3$	820.8	(8000)	(32.4)
Beryllium iodide	BeI_2	479.8	(4500)	(17.1)	Cerium (IV) fluoride	CeF_4	976	(10000)	(46.3)
Beryllium oxide	BeO	2550.0	17000	679.7					
Bismuth	Bi	271	2505	12.0					
Bismuth tribromide	$BiBr_3$	216.8	(4000)	(8.9)					

B-227

Compound	Formula	M.P.,°C	H_f cal/g mole	H_f cal/g	Compound	Formula	M.P.,°C	H_f cal/g mole	H_f cal/g
Cesium	Cs	28.3	500	3.7	Hydrogen fluoride	HF	−83.11	1094	54.7
Cesium chloride	CsCl	641.8	3600	21.4	Hydrogen iodide	HI	−50.91	686.3	5.4
Cesium fluoride	CsF	704.8	(2450)	(16.1)	Hydrogen nitrate	HNO_3	−47.2	601	9.5
Cesium nitrate	$CsNO_3$	406.8	3250	16.6	Hydrogen oxide (water)	H_2O	0	1436	79.72
Chlorine	Cl_2	−103 ± 5	1531	22.8	Deuterium oxide	D_2O	3.78	1516	75.8
Chromium	Cr	1890	3660	62.1	Hydrogen peroxide	H_2O_2	−0.7	2920	8.58
Chromium (II) chloride	$CrCl_2$	814	7700	65.9	Hydrogen selenate	H_2SeO_4	57.8	3450	23.8
Chromium (III) sequioxide	Cr_2O_3	2279	4200	27.6	Hydrogen sulfate	H_2SO_4	10.4	2360	24.0
Chromium trioxide	CrO_3	197	3770	37.7	Hydrogen sulfide	H_2S	−85.6	5683	16.8
Cobalt	Co	1490	3640	62.1	Hydrogen sulfide (di-)	H_2S_2	−89.7	1805	27.3
Cobalt (II) chloride	$CoCl_2$	727	7390	56.9	Hydrogen telluride	H_2Te	−49.0	1670	12.9
Cobalt (II) fluoride	CoF_2	1201	(9000)	(92.9)	Indium	In	156.3	781	6.8
Copper	Cu	1083	3110	49.0	Iodine	I_2	112.9	3650	14.3
Copper (I) bromide	CuBr	487	(2300)	(16.03)	Iodine chloride (α)	ICl	17.1	2660	16.4
Copper (II) chloride	$CuCl_2$	430	4890	24.7	Iodine chloride (β)	ICl	13.8	2270	13.3
Copper (I) chloride	CuCl	429	2620	26.4	Iron	Fe	1530.0	3560	63.7
Copper (I) cyanide	$Cu_2(CN)_2$	473	(5400)	(30.1)	Iron carbide	Fe_3C	1226.8	12330	68.6
Copper (I) iodide	CuI	587	(2600)	(13.6)	Iron (III) chloride	Fe_2Cl_6	303.8	20500	63.2
Copper (II) oxide	CuO	1446	2820	35.4	Iron (II) chloride	$FeCl_2$	677	7800	61.5
Copper (I) oxide	Cu_2O	1230	(13400)	(93.6)	Iron (II) oxide	FeO	1380	(7700)	(107.2)
Copper (I) sulfide	Cu_2S	1129	5500	34.6	Iron oxide	Fe_3O_4	1596	33000	142.5
Dysprosium	Dy	1407	4100	25.2	Iron penta-carbonyl	$Fe(CO)_5$	−21.2	3250	16.5
Dysprosium chloride	$DyCl_3$	646	(7000)	(26.0)	Iron (II) sulfide	FeS	1195	5000	56.9
Erbium	Er	1496	4100	24.5	Lanthanum	La	920	2400	17.4
Erbium trichloride	$ErCl_3$	775	(7000)	(26.0)	Lanthanum chloride	$LaCl_3$	861	(9000)	(36.6)
Europium	Eu	826	2500	16.4	Lead	Pb	327.3	1224	5.9
Europium trichloride	$EuCl_3$	622	(8000)	(20.9)	Lead bromide	$PbBr_2$	487.8	4290	11.7
Fluorine	F_2	−219.6	244.0	6.4	Lead chloride	$PbCl_2$	497.8	5650	20.3
Gadolinium	Gd	1312	3700	23.8	Lead fluoride	PbF_2	823	1860	7.6
Gadolinium trichloride	$GdCl_3$	608	(7000)	(26.6)	Lead iodide	PbI_2	412	5970	17.9
Gallium	Ga	29	1336	19.1	Lead molybdate	$PbMoO_4$	1065	(25800)	70.8
Germanium	Ge	959	(8300)	(114.3)	Lead oxide	PbO	890	2820	12.6
Gold	Au	1063	3030	15.3	Lead sulfide	PbS	1114	4150	17.3
Hafnium	Hf	2214	(6000)	(34.1)	Lead sulfate	$PbSO_4$	1087	9600	31.6
Holmium	Ho	1461	4100	24.8	Lead tungstate	$PbWO_4$	1123	(15200)	(33.4)
Holmium trichloride	$HoCl_3$	717	(8000)	(29.5)	Lithium	Li	178.8	1100	158.5
Hydrogen	H_2	−259.25	28	13.8	Lithium borate, meta-	$LibO_2$	845	(5570)	(111.9)
Hydrogen bromide	HBr	−86.96	575.1	7.1	Lithium bromide	LiBr	552	2900	33.4
Hydrogen chloride	HCl	−114.3	476.0	13.0	Lithium chloride	LiCl	614	3200	75.5

HEAT OF FUSION OF SOME INORGANIC COMPOUNDS (Continued)

Compound	Formula	M.P.,°C	H_f cal/g mole	H_f cal/g	Compound	Formula	M.P.,°C	H_f cal/g mole	H_f cal/g
Lithium fluoride	LiF	896	(2360)	(91.1)	Neodymium trichloride	NdCl$_3$	758	(8000)	(31.9)
Lithium hydroxide	LiOH	462	2480	103.3	Neon	Ne	−248.6	77.4	3.83
Lithium iodide	LiI	440	(1420)	(10.6)	Nickel	Ni	1452	4200	71.5
Lithium metasilicate	Li$_2$SiO$_3$	1177	7210	80.2	Nickel chloride	NiCl$_2$	1030	18470	142.5
Lithium molybdate	Li$_2$MoO$_4$	705	4200	24.1	Nickel subsulfide	Ni$_3$S$_2$	790	5800	25.8
Lithium nitrate	LiNO$_3$	250	6060	87.8	Niobium	Nb	2496	(6500)	(68.9)
Lithium orthosilicate	Li$_4$SiO$_4$	1249	7430	60.5	Niobium pentachloride	NbCl$_5$	211	8400	30.8
Lithium sulfate	Li$_2$SO$_4$	857	3040	27.6	Niobium pentoxide	Nb$_2$O$_5$	1511	24200	91.0
Lithium tungstate	Li$_2$WO$_4$	742	(6700)	(25.6)	Nitrogen	N$_2$	−210	172.3	6.15
Lutetium	Lu	1651	4600	26.3	Nitrogen oxide (ic)	NO	−163.7	549.5	18.3
Lutetium chloride	LuCl$_3$	904	(9000)	(32)	Nitrogen oxide (ous)	N$_2$O	−90.9	1563	35.5
Magnesium	Mg	650	2160	88.9	Nitrogen tetroxide	N$_2$O$_4$	−13.2	5540	60.2
Magnesium bromide	MgBr$_2$	711	8300	45.0	Osmium	Os	2700	(7000)	(36.7)
Magnesium chloride	MgCl$_2$	712	8100	82.9	Osmium tetroxide (white)	OsO$_4$	41.8	2340	9.2
Magnesium fluoride	MgF$_2$	1221	5900	94.7	Osmium tetroxide (yellow)	OsO$_4$	55.8	4060	15.5
Magnesium oxide	MgO	2642	18500	459.0	Oxygen	O$_2$	−218.8	106.3	3.3
Magnesium phosphate	Mg$_3$(PO$_4$)$_2$	1184	(11300)	(42.9)	Palladium	Pd	1555	4120	38.6
Magnesium silicate	MgSiO$_3$	1524	14700	146.4	Phosphoric acid	H$_3$PO$_4$	42.3	2520	25.8
Magnesium sulfate	MgSO$_4$	1327	3500	28.9	Phosphoric acid, hypo-	H$_4$P$_2$O$_6$	54.8	8300	51.2
Manganese	Mn	1220	3450	62.7	Phosphorous acid, hypo-	H$_3$PO$_2$	17.3	2310	35.0
Manganese dichloride	MnCl$_2$	650	7340	58.4	Phosphorous acid, ortho-	H$_3$PO$_3$	73.8	3070	37.4
Manganese metasilicate	MnSiO$_3$	1274	(8200)	(62.6)	Phosphorus, yellow	P$_4$	44.1	600	4.8
Manganese (II) oxide	MnO	1784	13000	183.3	Phosphorus oxychloride	POCl$_3$	1.0	3110	20.3
Manganese oxide	Mn$_3$O$_4$	1590	(39000)	(170.4)	Phosphorus pentoxide	P$_4$O$_{10}$	569.0	17080	60.1
Mercury	Hg	−39	557.2	2.7	Phosphorus trioxide	P$_4$O$_6$	23.7	3360	15.3
Mercury bromide	HgBr$_2$	241	3960	10.9	Platinum	Pt	1770	4700	24.1
Mercury chloride	HgCl$_2$	276.8	4150	15.3	Potassium	K	63.4	574	14.6
Mercury iodide	HgI$_2$	250	4500	9.9	Potassium borate, meta-	KBO$_2$	947	(5660)	(69.1)
Mercury sulfate	HgSO$_4$	850	(1440)	(4.8)	Potassium bromide	KBr	742	5000	42.0
Molybdenum	Mo	2622	(6600)	(68.4)	Potassium carbonate	K$_2$CO$_3$	897	7800	56.4
Molybdenum dichloride	MoCl$_2$	726.8	6000	3.58	Potassium chloride	KCl	770	6410	85.9
Molybdenum hexafluoride	MoF$_6$	17	2500	11.9	Potassium chromate	K$_2$CrO$_4$	984	6920	35.6
Molybdenum trioxide	MoO$_3$	795	(2500)	(17.3)	Potassium cyanide	KCN	623	(3500)	(53.7)
Neodymium	Nd	1020	1700	11.8	Potassium dichromate	K$_2$Cr$_2$O$_7$	398	8770	29.8
					Potassium fluoride	KF	875	6500	111.9

HEAT OF FUSION OF SOME INORGANIC COMPOUNDS (Continued)

Compound	Formula	M.P.,°C	H_f cal/g mole	cal/g	Compound	Formula	M.P.,°C	H_f cal/g mole	cal/g
Potassium hydroxide	KOH	360	(1980)	(35.3)	Silver cyanide	AgCN	350	2750	20.5
Potassium iodide	KI	682	4100	24.7	Silver iodide	AgI	557	2250	9.5
Potassium nitrate	KNO_3	338	2840	28.1	Silver nitrate	$AgNO_3$	209	2755	16.2
Potassium peroxide	K_2O_2	490	6100	55.3	Silver sulfide	Ag_2S	841	3360	13.5
Potassium phosphate	K_3PO_4	1340	8900	41.9	Silver sulfate	Ag_2SO_4	657	(4280)	(13.7)
Potassium pyro-phosphate	$K_4P_2O_7$	1092	14000	42.4	Sodium	Na	97.8	630	27.4
Potassium sulfate	K_2SO_4	1074	8100	46.4	Sodium borate, meta-	$NaBO_2$	966	8660	134.6
Potassium sulfocyanide	KCNS	179	2250	23.1	Sodium bromide	NaBr	747	6140	59.7
Potassium tungstate	K_2WO_4	929	(4400)	(13.6)	Sodium carbonate	Na_2CO_3	854	7000	66.0
Praseodymium	Pr	931	2700	19.0	Sodium chloride	NaCl	800	7220	123.5
Praseodymium chloride	$PrCl_3$	786	(8000)	(32.3)	Sodium chlorate	$NaClO_3$	255	5290	49.7
Promethium	Pm	(1027)	(3000)	(21)	Sodium cyanide	NaCN	562	(4360)	(88.9)
Rhenium	Re	3167 ± 60	(7900)	(42.4)	Sodium fluoride	NaF	992	7000	166.7
Rhenium heptoxide	Re_2O_7	296	15340	30.1	Sodium hydroxide	NaOH	322	2000	50.0
Rhenium hexafluoride	ReF_6	19.0	5000	16.6	Sodium iodide	NaI	662	5340	35.1
Rubidium	Rb	38.9	525	6.1	Sodium molybdate	Na_2MoO_4	687	3600	17.5
Rubidium bromide	RbBr	677	3700	22.4	Sodium nitrate	$NaNO_3$	310	3760	44.2
Rubidium chloride	RbCl	717	4400	36.4	Sodium phosphate, meta-	$NaPO_3$	988	(4960)	(48.6)
Rubidium fluoride	RbF	833	4130	39.5	Sodium peroxide	Na_2O_2	460	5860	75.1
Rubidium iodide	RbI	638	2990	14.0	Sodium pyro-phosphate	$Na_4P_2O_7$	970	(13700)	(51.5)
Rubidium nitrate	$RbNO_3$	305	1340	9.1	Sodium silicate, meta-	Na_2SiO_3	1087	10300	84.4
Samarium	Sm	1072	2600	17.3	Sodium silicate, di-	$Na_2Si_2O_5$	884	8460	46.4
Samarium chloride	$SmCl_2$	562	(6000)	(27.1)	Sodium silicate, aluminum-	$NaAlSi_3O_8$	1107	13150	50.1
Scandium	Sc	1538	3800	84.4	Sodium sulfide	Na_2S	920	(1200)	15.4
Scandium trichloride	$ScCl_3$	940	(19000)	(125.4)	Sodium sulfate	Na_2SO_4	884	5830	41.0
Selenium	Se	217	1220	15.4	Sodium thiocyanate	NaCNS	323	4450	54.8
Selenium oxychloride	$SeOCL_3$	9.8	1010	6.1	Sodium tungstate	Na_2WO_4	702	5800	19.6
Silicon	Si	1427	9470	337.0	Strontium	Sr	757	2190	25.0
Silicondioxide (Cristobalite)	SiO_2	2100	2100	35.0	Strontium bromide	$SrBr_2$	643	4780	19.3
Silicondioxide (Quartz)	SiO_2	1470	3400	56.7	Strontium chloride	$SrCl_2$	872	4100	26.5
Disilane, hexafluoro-	Si_2F_6	−28.6	3900	22.9	Strontium fluoride	SrF_2	1400	4260	34.0
Silicon tetrachloride	$SiCl_4$	−67.7	1845	10.8					
Silver	Ag	961	2700	25.0					
Silver bromide	AgBr	430	2180	11.6					
Silver chloride	AgCl	455	3155	22.0					

HEAT OF FUSION OF SOME INORGANIC COMPOUNDS (Continued)

Compound	Formula	M.P.,°C	H_f cal/g mole	H_f cal/g	Compound	Formula	M.P.,°C	H_f cal/g mole	H_f cal/g
Strontium oxide	SrO	2430	16700	161.2	Tin chloride, ous-	SnCl₂	247	3050	16.0
Sulfur (monatomic)	S	119	295	9.2	Tin iodide, ic-	SnI₄	143.4	(4330)	(6.9)
Sulfur dioxide	SO₂	−73.2	2060	32.2	Tin oxide	SnO	1042	(6400)	(46.8)
Sulfur trioxide (α)	SO₃	16.8	2060	25.8	Titanium	Ti	1800	(5000)	(104.4)
Sulfur trioxide (β)	SO₃	32.3	2890	36.1	Titanium bromide, tetra-	TiBr₄	38	(2060)	(5.6)
Sulfur trioxide (γ)	SO₃	62.1	6310	79.0	Titanium chloride, tetra-	TiCl₄	−23.2	2240	11.9
Tantalum	Ta	2996 ± 50	(7500)	34.6 to 41.5	Titanium dioxide	TiO₂	1825	(11400)	(142.7)
Tantalum pentachloride	TaCl₅	206.8	9000	25.1	Titanium oxide	TiO	991	14000	219
Tantalum pentoxide	Ta₂O₅	1877	48000	108.6	Tungsten	W	3387	(8420)	(45.8)
Tellurium	Te	453	3230	25.3	Tungsten dioxide	WO₂	1270	13940	60.1
Terbium	Tb	1356	3900	24.6	Tungsten hexafluoride	WF₆	−0.5	1800	6.0
Thallium	Tl	302.4	1030	5.0	Tungsten tetrachloride	WCl₄	327	6000	18.4
Thallium bromide, mono-	TlBr	460	5990	21.0	Tungsten trioxide	WO₃	1470	13940	60.1
Thallium carbonate	Tl₂CO₃	273	4400	9.5	Uranium²³⁵	U	~1133	3700	20
Thallium chloride, mono-	TlCl	427	1260	17.7	Uranium tetrachloride	UCl₄	590	10300	27.1
Thallium iodide, mono-	TlI	440	3125	9.4	Vanadium	V	1917	(4200)	(70)
Thallium nitrate	TlNO₃	207	2290	8.6	Vanadium dichloride	VCl₂	1027	8000	65.6
Thallium sulfide	Tl₂S	449	3000	6.8	Vanadium oxide	VO	2077	15000	224.0
Thallium sulfate	Tl₂SO₄	632	5500	10.9	Vanadium pentoxide	V₂O₅	670	15560	85.5
Thorium	Th	1845	(<4600)	(<19.8)	Xenon	Xe	−111.6	740	5.6
Thorium dioxide	ThO₂	2952	291100	1102.0	Ytterbium	Yb	823	2200	12.7
Thorium chloride	ThCl₄	765	22500	61.6	Yttrium	Y	1504	4100	46.1
Thulium	Tm	1545	4400	26.0	Yttrium oxide	Y₂O₃	2227	25000	110.7
Thulium chloride	TmCl₃	821	(9000)	(32.6)	Yttrium trichloride	YCl₃	709	(9000)	(46)
Tin	Sn	231.7	1720	14.4	Zinc	Zn	419.4	1595	24.4
Tin bromide, ic-	SnBr₄	29.8	3000	6.8	Zinc chloride	ZnCl₂	283	(5540)	(40.6)
Tin bromide, ous-	SnBr₂	231.8	(1720)	(6.1)	Zinc oxide	ZnO	1975	4470	54.9
Tin chloride, ic-	SnCl₄	−33.3	2190	8.4	Zinc sulfide	ZnS	1700 ± 20	(9100)	(93.3)
					Zirconium	Zr	1857	(5500)	(60)
					Zirconium dichloride	ZrCl₂	727	7300	45.0
					Zirconium oxide	ZrO₂	2715	20800	168.8

Table of the Isotopes

Compiled by
Russell L. Heath
National Reactor Testing Station
Idaho Nuclear Corporation
Idaho Falls, Idaho

The following describes the information contained in each column of this table:

Column 1.　This column lists the isotopes with the atomic number as a subscript and the atomic weight as a superscript. Isomers are indicated by the addition of the letter m to the superscript. Although they are not isotopes, the elements and their corresponding atomic numbers are also included in this column.

Column 2.　% Natural Abundance. This column lists isotopic abundance in percent.

Column 3.　Atomic Mass. This column lists atomic mass of isotopes in the physical scale. The atomic masses are all relative to the mass of the Carbon 12 isotope which has arbitrarily been assigned 12.00000 atomic mass units as recommended by the International Union of Pure and Applied Chemistry and the International Commission of Atomic Weights.

Column 4.　Lifetime. This column lists the half-life of the radioactive isotopes. The notation used is: μs = microseconds, ms = milliseconds, s = seconds, m = minutes, d = days, and y = years.

Column 5.　This column lists the observed modes of decay for all radioactive isotopes. Symbols used are: β^- = negative beta emission, β^+ = positron emission, α = alpha particle decay, E.C. = orbital electron capture, I.T. = isomeric transition from upper to lower isomeric state, n = neutron emission, and S.F. = spontaneous fission.

Column 6.　Decay Energy. This lists the currently adopted values for the total disintegration energies (Q) in MeV. Where more than one mode of decay exists separate values are given.

Column 7.　Particle Energies. This column lists the end-point energies of beta particle transitions or discrete energies of alpha particles in MeV.

Column 8.　Particle Intensities. This column lists the intensities of beta groups or alpha particle transitions in percent.

Column 9.　Thermal Neutron Capture Cross Section. This column lists cross sections for thermal-neutron capture in barns (10^{-24} cm^2) or in millibarns (mb). Where neutron capture for a given nucleus may produce two isomers, the separate cross sections are listed if known. Fission cross sections are listed for the heavy elements and are designated by the symbol σ_f.

Column 10. I. This table lists the Nuclear Spin or Angular Momentum in units of h/2π.

Column 11. μ. This column lists values of the Nuclear Magnetic Moments of the isotope in nuclear magnetons, with diamagnetic correction.

Isotope	% nat. abundance	Atomic mass	Lifetime	Modes of decay	Decay energy (MeV)	Particle energies (MeV)	Particle intensities	Thermal neutron capture cross-section	I	μ
n^1		1.008665	12m	β^-	0.7825	0.7825	100%		1/2	−1.9131
H		1.00797								
$_1H$	99.985	1.007825						.322 ± 2b	1/2	+2.79278
$_1H^2$	0.015	2.0140						0.51 ± 0.01mb	1	+0.85742
$_1H^3$		3.01605	12.26y	β^-	0.01861	0.01861	100%	<6μb	1/2	
He		4.0026						≥7mb		
$_2He^3$	0.00013	3.01603						<0.1mb	1/2	−2.1275
$_2He^4$	99.99987	4.00260						≈0	0	
$_2He^5$		5.0123								
$_2He^6$		6.01888	0.81s	β^-	3.5098	3.5098	100%		0	<0.16
$_2He^8$		8.0375	0.1225s	β^- η	13	13	88% 12%			
Li		6.941								
$_3Li^5$		5.0125								
$_3Li^6$	7.5	6.01512						45 ± 10mb	1	+0.82202
$_3Li^7$	92.5	7.01600						37 ± 4mb	3/2	+3.2564
$_3Li^8$			0.855s	β^-, α	13	13			(2)	+1.6532
$_3Li^9$			0.175s	β^-, η	13.5	13.5 11	75% 25%			
Be		9.0122						9.2 ± 0.5mb		
$_4Be^6$		6.0197								
$_4Be^7$		7.0169	53.37d	EC	0.861					
$_4Be^8$		8.0053	2×10^{-16}s	α						
$_4Be^9$	100%	9.01218						9.2 ± 0.5mb	3/2	−1.1776
$_4Be^{10}$		10.0135	2.5×10^6y	β^-	0.55	0.55	100%	≤1mb		
$_4Be^{11}$		11.0216	13.6s	β^-	11.5	11.5 9.3 4.7 3.6	61% 29% 6% 4%			
B		10.811								
$_5B^8$		8.0246	0.77s	β^+	18	13.7				
$_5B^9$		9.0133	8×10^{-19}s	P						
$_5B^{10}$	20	10.0129						($\sigma[\eta, \alpha]$ 3836b) 0.5 ± 0.2b	3	+1.8007
$_5B^{11}$	80	11.00931						5 ± 3mb	3/2	+2.6885
$_5B^{12}$		12.0143	0.02s	β^-, α	13.4	13.4 9.0	~100% 1.3%		(1)	+1.003
$_5B^{13}$		13.0178	0.019s	β^-	13.4	13.4	93%			

Isotope	% nat. abundance	Atomic mass	Lifetime	Modes of decay	Decay energy (MeV)	Particle energies (MeV)	Particle intensities	Thermal neutron capture cross-section	I	μ
C		12.01115						3.4 ± 0.2mb		
$_6C^{10}$			19.45s	β^+	3.61	1.87				
$_6C^{11}$			20.3min	β^+, EC	1.98	0.98			3/2	± 1.03
$_6C^{12}$	98.89	12.0000						3.4 ± 0.3mb	0	
$_6C^{13}$	1.11	13.00335						0.9 ± 0.2mb	1/2	+ 0.7024
$_6C^{14}$			5730y	β^-	0.156	0.156		< 10^{-6}b	0	
$_6C^{15}$			2.4s	β^-	9.8	9.82 4.51	32% 68%			
$_6C^{16}$			0.74s	β^-, η						
N		14.0067								
$_7N^{12}$			0.011s	β^+	16.4	16.38				
$_7N^{13}$										
$_7N^{13}$			9.97m	β^+	1.19	1.19			1/2	± 0.3221
$_7N^{14}$	99.64	14.00307						75 ± 7.5mb	1	+ 0.4036
$_7N^{15}$	0.36	15.00011						24 ± 8μb	1/2	− 0.2831
$_7N^{16}$			7.2s	β^-	10.44	10.44 4.27	26% 68%			
$_7N^{17}$			4.16s	β^-, (η)	8.68	8.7 7.8 4.12 3.3 2.7	1.6% 2.6% 45% 45% 5%			
$_7N^{18}$			0.63s	β^-	13.9	9.4	100%			
O		15.9994								
$_8O^{13}$			0.0087s	β^+, (P)	17.8	6.40 (proton) 6.79 (proton)	100† 24			
$_8O^{14}$			71.0s	β^+	5.14	1.813 4.09	99.35% 0.65%			
$_8O^{15}$			124s	β^+	2.76	1.70	100%		1/2	± 0.7189
$_8O^{16}$	99.76	15.99491						0.178 ± 0.025mb	0	
$_8O^{17}$	0.04							235 ± 10mb	5/2	− 1.8937
$_8O^{18}$	0.20							0.21 ± 0.04mb	0	
$_8O^{19}$			29s	β^-	4.819	4.60 3.25	42% 58%		5/2	
$_8O^{20}$			14s	β^-	3.81	2.7				
F		18.9984						9.8 ± 0.7mb		
$_9F^{17}$			66s	β^+	2.76	1.73	100%		(5/2)	± 4.722
$_9F^{18}$			109.7m	β^+, EC	1.65	0.635 EC	97% 3%			
$_9F^{19}$	100%	18.99840						9.8 ± 0.7mb	1/2	+ 2.6288

Isotope	% nat. abundance	Atomic mass	Lifetime	Modes of decay	Decay energy (MeV)	Particle energies (MeV)	Particle intensities	Thermal neutron capture cross-section	I	μ
$_9F^{20}$			11.4s	β^-					(2)	+2.094
$_9F^{21}$			4.4s	β^-	5.68	5.4 4.0				
$_9F^{22}$			4.0s	β^-	12	11				
Ne		20.183						38 ± 10mb		
$_{10}Ne^{17}$			0.10s	β^+, (P)						
$_{10}Ne^{18}$			1.5s	β^+	4.45	3.42 2.38	93% 7%			
$_{10}Ne^{19}$			17.5s	β^+	3.24	2.24			1/2	−1.887
$_{10}Ne^{20}$	90.51	19.99244						38 ± 10mb	0	
$_{10}Ne^{21}$	0.27	20.99395							3/2	−0.6618
$_{10}Ne^{22}$	9.22	21.99138						36 ± 10mb	0	
$_{10}Ne^{23}$			37.6s	β^-	4.38	4.38 3.95 2.4	67% 32% 1.0%			
$_{10}Ne^{24}$			3.38m	β	2.47	1.98 1.10	92% 8%			
Na		22.9898						534 ± 5mb		
$_{11}Na^{20}$			0.39s	$\beta^+(\alpha)$	14					
$_{11}Na^{21}$			23s	β^+	3.54	2.51	98%		3/2	+2.386
$_{11}Na^{22}$			2.602y	β^+, EC				$40,000 \pm 5000$b	3	+1.746
$_{11}Na^{23}$	100%	22.9898						400 ± 30mb	3/2	+2.2175
$_{11}Na^{24}$			15.0h	β^-	5.51	4.17 1.389	0.003% 99 + %		4	+1.690
$_{11}Na^{25}$			60s	β^-	3.83	4.0 3.1 2.6	65% 25% 7%			
$_{11}Na^{26}$			1.0s	β^-	8.5	6.7	>80%			
Mg		24.312						64 ± 2mb		
$_{12}Mg^{20}$			0.6s	β^+	9					
$_{12}Mg^{21}$			0.12s	$\beta^+(P)$						
$_{12}Mg^{23}$			12.1s	β^+	4.06	3.10				
$_{12}Mg^{24}$	78.99	23.98504						52 ± 15mb	0	
$_{12}Mg^{25}$	10.00	24.98584						0.18 ± 0.05b	5/2	−0.8553
$_{12}Mg^{26}$	11.01	25.98259						0.03 ± 0.005b	0	
$_{12}Mg^{27}$			9.5min	β^-	2.61	1.75 1.59	58% 42%			

Isotope	% nat. abundance	Atomic mass	Lifetime	Modes of decay	Decay energy (MeV)	Particle energies (MeV)	Particle intensities	Thermal neutron capture cross-section	I	μ
$_{12}\text{Mg}^{28}$			21h	β^-	1.84	0.46				
Al		26.98153						232 ± 3mb		
$_{13}\text{Al}^{24}$			2.1s	$\beta^+(\alpha)$	14	8.5				
$_{13}\text{Al}^{25}$			7.2s	β^+	4.26	3.24	100%			
$_{13}\text{Al}^{26m}$			6.4s	β^+		3.21				
$_{13}\text{Al}^{26}$			7.4×10^5y	β^+, EC	4.003	1.16				
$_{13}\text{Al}^{27}$	100%	26.98153						232 ± 3mb	5/2	+3.6414
$_{13}\text{Al}^{28}$			2.31m	β^-	4.635	2.82	100%			
$_{13}\text{Al}^{29}$			6.6min	β^-	3.68	2.5 / 1.4	70% / 30%			
$_{13}\text{Al}^{30}$			3.3s	β^-	7.3	5.0				
Si		28.086						160 ± 20mb		
$_{14}\text{Si}^{25}$			0.23s	$\beta^+(P)$						
$_{14}\text{Si}^{26}$			2.1s	β^+	5.08	3.83 / 3.00	66† / 34			
$_{14}\text{Si}^{27}$			4.2s	β^+	4.810	3.85				
$_{14}\text{Si}^{28}$	92.23	27.97693						160 ± 40mb	0	
$_{14}\text{Si}^{29}$	4.67	28.97649						280 ± 90mb	1/2	−0.5553
$_{14}\text{Si}^{30}$	3.10	29.97376						100 ± 10mb	0	
$_{14}\text{Si}^{31}$			2.62h	β^-	1.48	1.49	99+%			
$_{14}\text{Si}^{32}$			~650y	β^-	0.21	0.21				
P		30.9738						190 ± 10mb		
$_{15}\text{P}^{28}$			0.28s	β^+	13.8	10.6	50%			
$_{15}\text{P}^{29}$			4.4s	β^+	4.95	3.96				
$_{15}\text{P}^{30}$			2.50m	β^+	4.24	3.27				
$_{15}\text{P}^{31}$	100%	30.97376						190 ± 10mb	1/2	+1.1317
$_{15}\text{P}^{32}$			14.3d	β^-	1.710	1.710	100%		1	−0.2523
$_{15}\text{P}^{33}$			25d	β^-	0.248	0.249	100%			
$_{15}\text{P}^{34}$			12.4s	β^-	5.1	5.1 / 3.2	75% / 25%			
S		32.064								
$_{16}\text{S}^{29}$			0.19s	$\beta^+(P)$						
$_{16}\text{S}^{30}$			1.4s	β^+	6.11	5.09 / 4.42	20† / 80			
$_{16}\text{S}^{31}$			2.7s	β^+	5.44	4.39				

Isotope	% nat. abundance	Atomic mass	Lifetime	Modes of decay	Decay energy (MeV)	Particle energies (MeV)	Particle intensities	Thermal neutron capture cross-section	I	μ
$_{16}S^{32}$	95.0%	31.97207							0	
$_{16}S^{33}$	0.76%	32.97146						15 ± 10mb	3/2	+0.6433
$_{16}S^{34}$	4.22%	33.96786						200 ± 100mb	0	
$_{16}S^{35}$			88d	β^-	0.1674	0.1674	100%		3/2	±1.0
$_{16}S^{36}$	0.02%	35.96709						140 ± 40mb	0	
$_{16}S^{37}$			5.06m	β	4.8	4.3 1.6	10% 90%			
$_{16}S^{38}$			2.87h	β^-	3.0	3.0 1.1	5% 95%			
Cl		35.453								
$_{17}Cl^{32}$			0.31s	$\beta^+(\alpha)$	13.2	9.5	~50%			
$_{17}Cl^{33}$			2.5s	β^+	5.57	4.51				
$_{17}Cl^{34m}$			32.0m	β^+, IT		2.5 1.3 4.5	26† 26 47			
$_{17}Cl^{34}$			1.56s	β^+	5.482	4.50				
$_{17}Cl^{35}$	75.77%	34.96885						44 ± 2h	3/2	+0.82183
$_{17}Cl^{36}$			3.1×10^5y	β^- β^+, EC	0.712 1.14	0.714 (β^-)	98.1% 1.9%	100 ± 300	2	+1.285
$_{17}Cl^{37}$	24.23%							430 ± 100mb	3/2	+0.68411
$_{17}Cl^{38m}$			0.74s	IT						
$_{17}Cl^{38}$			37.3m	β^-	4.91	4.81 2.77 1.11	53% 16% 31%			
$_{17}Cl^{39}$			55.5m	β^-	3.44	3.45 2.18 1.91	7% 8% 85%			
$_{17}Cl^{40}$			1.4m	β^-	7.5	~7.5 ~3.2				
Ar		39.948								
$_{18}Ar^{35}$			1.83s	β^+	5.96	4.93	93%		(3/2)	+0.632
$_{18}Ar^{36}$	0.34%	35.96755						6 ± 2b	0	
$_{18}Ar^{37}$			35d	EC	0.814	EC	100%		3/2	+0.95
$_{18}Ar^{38}$	0.07%	37.96272						0.8 ± 0.2b	0	
$_{18}Ar^{39}$			265y	β^-	3.44	0.565	100%	500 ± 200b	7/2	-1.3
$_{18}Ar^{40}$	99.59%							650 ± 30mb	0	
$_{18}Ar^{41}$			1.83h	β^-	2.491	1.198 2.49	99% 0.78%	0.5 ± 0.1b		
$_{18}Ar^{42}$			33y	β^-	0.60					
K		39.0983								
$_{19}K^{37}$			1.23s	β^+	6.16	~5.1	98%			

Isotope	% nat. abundance	Atomic mass	Lifetime	Modes of decay	Decay energy (MeV)	Particle energies (MeV)	Particle intensities	Thermal neutron capture cross-section	I	μ
$_{19}K^{38m}$			0.95s	β^+		5.0	100%			
$_{19}K^{38}$			7.71m	β^+, EC	5.93	2.68			3	+1.374
$_{19}K^{39}$	93.26	38.96371						2.2 ± 0.2b	3/2	0.3914
$_{19}K^{40}$	0.01 %	39.964	1.28 × 10⁹y	β^-, β^+, EC	1.35 1.505	β^- β^+	1.35% 0.49%	70 ± 20b	4	−1.298
$_{19}K^{41}$	6.73%							1.3 ± 0.2b	3/2	+0.2149
$_{19}K^{42}$			12.4h	β^-	0.60	3.52 1.97	82% 18%		2	−1.140
$_{19}K^{43}$			22.4h	β^-	1.82	1.81 1.24 0.84 0.46	1.3% 3.5% 87% 8%		3/2	±0.163
$_{19}K^{44}$			22.0m	β^-	5.2	5.3 4.0 2.6	~35% 9% 38%			
$_{19}K^{45}$			16m	β^-	4.2	4.0 2.1 1.1	15† 60 25		3/2	±0.173
Ca		40.08						0.44 ± 0.02b		
$_{20}Ca^{37}$			0.173s	β^+(P)	11.5					
$_{20}Ca^{38}$			0.66s	β^+						
$_{20}Ca^{39}$			0.87s	β^+	6.50	5.49				
$_{20}Ca^{40}$	96.941%	39.96259						430 ± 40mb	0	
$_{20}Ca^{41}$			8 × 10⁴y	EC	0.427				7/2	−1.595
$_{20}Ca^{42}$	0.647%	41.95863						700 ± 160mb		
$_{20}Ca^{43}$	0.135%	42.95878						6.2 ± 1.1b	7/2	−1.317
$_{20}Ca^{44}$	2.086%	43.95549						1.1 ± 0.3b		
$_{20}Ca^{45}$			165d	β^-	0.252	0.258	100%			
$_{20}Ca^{46}$	0.004%	45.9537						250 ± 100mb		
$_{20}Ca^{47}$			4.53d	β^-	1.979	1.979 1.48 0.67	16% 2% 82%			
$_{20}Ca^{48}$	0.187%	47.9524						1.1 ± 0.1b		
$_{20}Ca^{49}$			8.8m	β^-	5.26	1.95 0.9	88† 12			
$_{20}Ca^{50}$			9s	β^-	3.9					
Sc		44.95592						25 ± 2b		
$_{21}Sc^{40}$			0.179s	β^+	14.5	9.1				
$_{21}Sc^{41}$			0.60s	β^+	6.50	5.6	100%			
$_{21}Sc^{42m}$			61s	β^+		2.82				

Isotope	% nat. abundance	Atomic mass	Lifetime	Modes of decay	Decay energy (MeV)	Particle energies (MeV)	Particle intensities	Thermal neutron capture cross-section	I	μ
$_{21}Sc^{42}$			0.68s	β^+	6.431	5.3	100%			
$_{21}Sc^{43}$			3.92h	β^+, EC	2.2	1.20 0.82 0.50	79† 17 4		7/2	+4.62
$_{21}Sc^{44m}$			2.44d	IT EC		98.6% 1.4%			6	+3.88
$_{21}Sc^{44}$			3.92h	β^+	3.647	1.471	99%		2	+2.56
$_{21}Sc^{45}$	100%	44.95592						25 ± 2b 11 ± 4b (Sc^{46m})	7/2	+4.7564
$_{21}Sc^{46m}$			20s	IT	0.142					
$_{21}Sc^{46}$			83.80d	β^-	2.367	1.48 0.357	0.004% 100%	8.0 ± 1b	4	+3.03
$_{21}Sc^{47}$			3.43d	β^-	0.60	0.600 0.439	40% 60%		7/2	+5.34
$_{21}Sc^{48}$			1.83d	β^-	3.98	0.65 0.47	94% 6%		6	
$_{21}Sc^{49}$			57.5m	β^-	2.008	2.01	100%			
$_{21}Sc^{50m}$			0.35s	IT	0.258					
$_{21}Sc^{50}$			1.72m	β^-	6.1					
Ti		47.90						6.1 ± 0.2b		
$_{22}Ti^{43}$			0.56s	β^+	6.8	5.5	100%			
$_{22}Ti^{44}$			48y	EC	0.16					
$_{22}Ti^{45}$			3.09h	β^+, EC	2.059	1.02	99+%		7/2	±0.095
$_{22}Ti^{46}$	8.0%	45.95263						0.6 ± 0.2b		
$_{22}Ti^{47}$	7.5%	46.9518						1.7 ± 0.3b	5/2	−0.7883
$_{22}Ti^{48}$	73.7%							8.3 ± 0.6b		
$_{22}Ti^{49}$	5.5%	48.94787						1.9 ± 0.5b	7/2	−1.1039
$_{22}Ti^{50}$	5.3%	49.9448						140 ± 30mb		
$_{22}Ti^{51}$			5.8m	β^-	2.46	2.13 1.50	94% 6%			
V		50.942						5.06 ± 0.06b		
$_{23}V^{46}$			0.426s	β^+	7.054	6.0	100%			
$_{23}V^{47}$			33m	β^+, EC	2.92	1.89				
$_{23}V^{48}$			16.0d	β^+, EC	4.013	0.698				
$_{23}V^{49}$			330d	EC	0.601					
$_{23}V^{50}$	0.25%	49.9472	$6 \times 10^{15}y$	EC				100 ± 60b	6	+3.3470
$_{23}V^{51}$	99.75%	50.9440						4.8 ± 0.2b	7/2	+5.149

Isotope	% nat. abundance	Atomic mass	Lifetime	Modes of decay	Decay energy (MeV)	Particle energies (MeV)	Particle intensities	Thermal neutron capture cross-section	I	μ
$_{23}V^{52}$			3.76m	β^-	3.97	2.47				
$_{23}V^{53}$			2.0m	β^-	3.5	2.50	100%			
$_{23}V^{54}$			55s	β^-	7.3	3.3	100%			
Cr		51.996						3.1 ± 0.2b		
$_{24}Cr^{48}$			23h	EC	1.4					
$_{24}Cr^{49}$			41.9m	β^-	5.26	1.54 1.39 0.73	50† 35 15		5/2	±0.48
$_{24}Cr^{50}$	4.35%	49.9461						16.0 ± 0.5b		
$_{24}Cr^{51}$			27.8d	EC	0.752				7/2	±0.94
$_{24}Cr^{52}$	83.79%	51.9405						0.76 ± 0.06b		
$_{24}Cr^{53}$	9.50%	52.9407						18.2 ± 1.5b	3/2	−0.4744
$_{24}Cr^{54}$	2.36%	53.9389						380 ± 40mb		
$_{24}Cr^{55}$			3.5m	β^-	2.59	2.50	100%			
$_{24}Cr^{56}$			5.9m	β^-	1.6	1.5				
Mn								13.3 ± 0.1b		
$_{25}Mn^{50m}$		54.9380	2m	β^+, EC						
$_{25}Mn^{50}$			0.286s	β^+	7.631	6.6	100%			
$_{25}Mn^{51}$			45m	β^+, EC	3.19	2.17	97%		5/2	±3.60
$_{25}Mn^{52m}$			21m	IT, β^+, EC	0.383	2% 98%			2	±0.0077
$_{25}Mn^{52}$			5.7d	β^+, EC	4.708	0.575			6	±3.075
$_{25}Mn^{53}$			2×10^6y	EC	0.598		100%	~170b	7/2	±5.05
$_{25}Mn^{54}$			303d	EC	1.379			<10b	3	±3.30
$_{25}Mn^{55}$	100%	54.9380						13.3 ± 0.1b	5/2	±3.444
$_{25}Mn^{56}$			2.576h	β^-	3.702	2.84 1.03 0.72 0.30	47% 34% 18% 1%		3	+3.218
$_{25}Mn^{57}$			1.7m	β^-	2.7	2.55 1.10	82% 18%			
$_{25}Mn^{58}$			1.1m	β^-	6.5					
Fe		55.847						2.56 ± 0.05b		
$_{26}Fe^{52}$			8.2h	β^+, EC	2.37	0.80				
$_{26}Fe^{53}$			8.5m	β^+, EC	3.98	2.8 2.4 1.6	50† 38 12			
$_{26}Fe^{54}$	5.8%	53.9396						2.8 ± 0.4b		

Isotope	% nat. abundance	Atomic mass	Lifetime	Modes of decay	Decay energy (MeV)	Particle energies (MeV)	Particle intensities	Thermal neutron capture cross-section	I	μ
$_{26}Fe^{55}$			2.6y	EC	0.232					
$_{26}Fe^{56}$	91.8 %	55.9349						2.5 ± 0.3b		
$_{26}Fe^{57}$	2.1 %	56.9354						2.5 ± 0.2b	1/2	+0.0902
$_{26}Fe^{58}$	0.3%	57.9333						1.23 ± 0.05b		
$_{26}Fe^{59}$			45.1 ± 0.5d	β^-		1.573 0.475 0.273	0.3% 51% 48%		3/2	
$_{26}Fe^{60}$			3×10^5y	β^-	0.14					
$_{26}Fe^{61}$			6.0m	β^-	3.8	2.8				
Co		58.9332						37.5 ± 0.2b		
$_{27}Co^{54m}$			1.5m	β^+		4.3	100%			
$_{27}Co^{54}$			0.1945 s	β^+	8.252	>7.4				
$_{27}Co^{55}$			18.2h	β^+, EC	3.46	1.51 1.04 0.79	45† 33 2		(7/2)	±4.3
$_{27}Co^{56}$			77d	β^+, EC	4.57	1.50 0.44	96† 1		4	±3.85
$_{27}Co^{57}$			270d	EC	0.837				7/2	±4.65
$_{27}Co^{58m}$			9.0h	IT	0.025			1.4×10^5 $\pm 1 \times 10^4$b		
$_{27}Co^{58}$			71.3d	β^+, EC	2.309	1.3 0.474	0.0006% 83% EC 15% β^+	1700 ± 200b	2	±4.06
$_{27}Co^{59}$	100%	58.9332						19.9 ± 0.91b (Co^{60m}) 17 ± 2b (Co^{60})	7/2	+4.649
$_{27}Co^{60m}$			10.5m	IT β^-	0.05860	1.56	99% 1%	58 ± 8b		
$_{27}Co^{60}$			5.26 ± 0.01y		2.819	0.315 0.663 1.488	99.87% 0.008% 0.12%	2.0 ± 0.2b	5	+3.81
$_{27}Co^{61}$			1.650h ± 0.05	β^-	1.29	1.22 ± 0.04	100%			
$_{27}Co^{62m}$			1.6m	β^-						
$_{27}Co^{62}$			13.9m	β^-	5.22	2.88 0.88	75% 25%			
$_{27}Co^{63}$			52s	β^-	3.6	3.6	100%			
Ni		58.71						4.54 ± 0.1b		
$_{28}Ni^{56}$			6.10d	EC	2.11					
$_{28}Ni^{57}$			36.0h	β^+, EC	3.24	0.85 0.72 0.35	87† 11 2			
$_{28}Ni^{58}$	68.27%	57.9353						4.4 ± 0.4b		

Isotope	% nat. abundance	Atomic mass	Lifetime	Modes of decay	Decay energy (MeV)	Particle energies (MeV)	Particle intensities	Thermal neutron capture cross-section	I	μ
$_{28}Ni^{59}$			8×10^4y	EC	1.072		100%			
$_{28}Ni^{60}$	26.10%	59.9308						2.6 ± 0.2b		
$_{28}Ni^{61}$	1.13%	60.9310						2.5 ± 0.9b	3/2	−0.7487
$_{28}Ni^{62}$	3.59%	61.9283						15 ± 2b		
$_{28}Ni^{63}$			92y	β^-	0.067	0.067	100%			
$_{28}Ni^{64}$	0.91%	63.9280						1.52 ± 0.24b		
$_{28}Ni^{65}$			2.521h ± 0.005	β^-	2.139	0.412 0.513 0.655 1.022 2.139	0.7% 1.2% 30.2% 9.9% 58%	20 ± 2b		
$_{28}Ni^{66}$			54.6h ± 0.9	β^-	0.20	0.20	100%			
$_{28}Ni^{67}$			50s	β	4.1	~2 2.3 3.2 4.1	22% 24% 4% 50%			
Cu		63.54						3.8 ± 0.1b		
$_{29}Cu^{58}$			3.20s	β^+	8.569	8.2				
$_{29}Cu^{59}$			82.0s ± 0.4	β^+, EC	4.8	2.10 2.48 2.90 3.32 3.44 3.75	1.4% 10.9% 7.8% 4.1% 3.6% 70.8%			
$_{29}Cu^{60}$			23.0m ± 0.3	β^+, EC	6.12	0.52 0.55 0.61 0.77 1.02 1.08 1.71 1.83 1.91 2.00 2.48 3.00 3.92	0.1% 0.2% 0.2% 0.3% 1.0% 1.0% 0.8% 2.4% 11.8% 47.0% 4.2% 14.4% 8.4%		2	
$_{29}Cu^{61}$			3.41h	β^+, EC	2.242	0.307 0.560 0.933 1.148 1.216	0.03% 2.5% 6.3% 2.4% 52%		3/2	+2.13
$_{29}Cu^{62}$		61.9326	9.8m	β^+, EC	3.939	0.87 1.75 2.923	1.5% 1.8% 93.9%		1	
$_{29}Cu^{63}$	69.2%	62.9296						4.5 ± 0.2b	3/2	+2.226
$_{29}Cu^{64}$		63.9298	12.9h	β^- β^+, EC	0.573 1.677	0.573 0.654	39.6% 19.3%	<6000b	1	−0.216

Isotope	% nat. abundance	Atomic mass	Lifetime	Modes of decay	Decay energy (MeV)	Particle energies (MeV)	Particle intensities	Thermal neutron capture cross-section	I	μ
$_{29}Cu^{65}$	30.8 %	64.9278						$2.3 \pm 0.3b$	3/2	+ 2.385
$_{29}Cu^{66}$			5.10m ± 0.02	β^-	2.633	0.760 1.64 2.63	0.25 % 9 % 91 %	$130 \pm 30b$	1	± 0.283
$_{29}Cu^{67}$			61.88h ± 0.14	β^-	0.576	0.183 0.395 0.484 0.577	0.9 % 51 % 28 % 20 %			
$_{29}Cu^{68}$			30s	β^-	4.6	3.5 2.7 2.3	75 % 22 % 3 %			
Zn		65.37						$1.10 \pm 0.04b$		
$_{30}Zn^{60}$			2.1m	β^+, EC	4.16					
$_{30}Zn^{61}$			$88 \pm 1s$	β^+. EC	5.91	4.27 4.94 5.44 5.91	7.8 % 1.1 % 10.8 % 79.6 %			
$_{30}Zn^{62}$			$9.15 \pm 0.03h$	β^+, EC	1.679	0.66	15.2 %			
$_{30}Zn^{63}$			$38.40 \pm 0.04m$	β^+, EC	3.365	0.26 0.79 0.972 1.40 1.69 2.34	0.002 % 0.08 % 0.1 % 6.9 % 9.7 % 76 %		3/2	− 0.282
$_{30}Zn^{64}$	48.6 %	63.9291						$0.82 \pm 0.01b$	0	
$_{30}Zn^{65}$			243.6d ± 0.1	β^+, EC	1.353	0.325	1.54 %		5/2	+ 0.769
$_{30}Zn^{66}$	27.9 %	65.9260							0	
$_{30}Zn^{67}$	4.1 %	66.9721							5/2	+ 0.8755
$_{30}Zn^{68}$	18.8 %	67.9249						$0.09 \pm 0.01b$ (Zn^{69m}) $1.0 \pm 0.1b$ (Zn^{69})	0	
$_{30}Zn^{69m}$			13.9h	IT	0.4387					
$_{30}Zn^{69}$			$58 \pm 2m$	β^-	0.925	0.925	100 %			
$_{30}Zn^{70}$	0.6 %	69.9253						$9 \pm 1mb$ (Zn^{71m}) $90 \pm 10mb$ (Zn^{71})		
$_{30}Zn^{71m}$			4.0h	β^-		1.45				
$_{30}Zn^{71}$			2.4m	β^-	2.61	2.61 2.1	82 % 14 %			
$_{30}Zn^{72}$			46.5h	β^-	0.45	0.30 0.25	90 % 10 %			

Isotope	% nat. abundance	Atomic mass	Lifetime	Modes of decay	Decay energy (MeV)	Particle energies (MeV)	Particle intensities	Thermal neutron capture cross-section	I	μ
Ga		69.72						$3.1 \pm 0.3b$		
$_{31}Ga^{63}$			31s	β^+, EC	5.5	5.5	100%			
$_{31}Ga^{64}$			2.6m	β^+, EC	7.08	0.85 1.31 1.60 2.27 2.64 2.69 2.80 2.87	0.05% 0.04% 1.6% 5.3% 6.9% 31% 2.5% 31%			
$_{31}Ga^{65}$			15.2m	β^+, EC	3.26	0.266 0.769 0.895 1.190 1.327 1.370 1.468 2.030 2.122 2.183 2.237	0.003% 0.64% 0.8% 2.1% 0.4% 6.3% 1.5% 10.3% 53.1% 6.9% 7.2%			
$_{31}Ga^{66}$			9.4h	β^+, EC	5.175	0.367 0.720 0.770 0.935 1.373 1.781 4.153	1.1% 0.16% 0.8% 3.8% 0.1% 0.56% 48.6%		0	
$_{31}Ga^{67}$			78.1h	EC	1.0003				3/2	+1.850
$_{31}Ga^{68}$			68.3m	β^+, EC	2.919	0.82 1.895	1.2% 88%		1	±0.0117
$_{31}Ga^{69}$	60%	68.9257						$1.8 \pm 0.4b$	3/2	+2.016
$_{31}Ga^{70}$			21.1m	β^-	1.66	1.65 0.61 0.44	99+% 0.3% 0.5%		1	
$_{31}Ga^{71}$	40%	70.9249						$0.15 \pm 0.05b$	3/2	+2.562
$_{31}Ga^{72}$			14.10h	β^-	4.00	1.51 1.97 2.52 3.15	6% 4% 8% 8%		3	−0.1322
$_{31}Ga^{73}$			4.9h	β^-	1.55	0.4 1.19	5% 95%			
$_{31}Ga^{74}$			7.9m	β^-	5.60	2.6 4.0?				
$_{31}Ga^{75}$			2m	β^-	3.3	5.4	96%			
$_{31}Ga^{76}$			32s	β^-		~6				
Ge		72.59						2.3 ± 0.26		

Isotope	% nat. abundance	Atomic mass	Lifetime	Modes of decay	Decay energy (MeV)	Particle energies (MeV)	Particle intensities	Thermal neutron capture cross-section	I	μ
$_{32}Ge^{65}$			1.5m	β^+, EC	~6.5	3.8 4.8 5.5	2% 3% 94%			
$_{32}Ge^{66}$			9.4h	β^+, EC	3.0	1.2 1.44 1.60 1.69 1.74 1.87 1.93	2.7% 12.6% 43.1% 1.1% 4.0% 3.8% 16.5%			
$_{32}Ge^{67}$			19m	β^+, EC	4.43	0.68 1.24 1.60 1.96 2.33 2.58 3.08 3.24	0.08% 4.6% 11.4% 2.6% 7.7% 2.7% 13% 50.8%			
$_{32}Ge^{68}$			287d	EC	~0.7					
$_{32}Ge^{69}$			39h	β^+, EC	2.227	0.333 0.631 1.213	0.2% 2.3% 32.5%			
$_{32}Ge^{70}$	20.5%	69.9243						$0.28 \pm 0.07b$	0	
$_{32}Ge^{71m}$			20ms	IT	0.198					
$_{32}Ge^{71}$			11.4d	EC	0.235		100%		1/2	+0.546
$_{32}Ge^{72}$	27.4%	71.9217						$0.98 \pm 0.09b$	0	
$_{32}Ge^{73m}$			0.53s	IT	0.067		100%			
$_{32}Ge^{73}$	7.8%	72.9234						$14 \pm 1b$	9/2	−0.8792
$_{32}Ge^{74}$	36.5%	73.9219						$0.14 \pm 0.3b$	0	
$_{32}Ge^{75m}$			? 48.9s	IT	0.139		100%			
$_{32}Ge^{75}$			82.78m ±0.04	β^-	1.20	0.58 0.72 0.92 0.98 1.19	0.4% 0.3% 11% 1.4% 87%			
$_{32}Ge^{76}$	7.8%	75.9214						$0.09 \pm 0.02b$	0	
$_{32}Ge^{77m}$			54s	IT β^-	0.159 2.75	2.7 2.9	1† 10			
$_{32}Ge^{77}$			$11.3 \pm 0.01h$	β^-	2.75	0.71 1.38 2.20	23% 35% 42%			
$_{32}Ge^{78}$			1.47h	β^-	0.99	0.69 0.71	6% 94%			
As		74.9216						$4.30 \pm 0.10b$		
$_{33}As^{69}$			15m	β^+, EC	3.9	2.9				

Isotope	% nat. abundance	Atomic mass	Lifetime	Modes of decay	Decay energy (MeV)	Particle energies (MeV)	Particle intensities	Thermal neutron capture cross-section	I	μ
$_{33}As^{70}$			53.0m ±0.6	β^+, EC	6.222	1.44 2.14 2.89	10† 75 6			
$_{33}As^{71}$			62h	β^+, EC	2.01	0.25 0.81	30%			
$_{33}As^{72}$			26h	β^+, EC	4.36	0.62 1.027 1.33 1.45 1.84 2.50 3.34	4% 3% 3% 5% 2% 56% 17%			
$_{33}As^{73}$			80.3d	EC	0.37					
$_{33}As^{74m}$			8.0s	IT	0.283					
$_{33}As^{74}$			17.9d	β^-	1.36	0.72 1.35	14% 18%			
				β^+, EC		0.91 1.51	26% 3.5%			
$_{33}As^{75}$	100%	74.9216						4.30 ± 0.10b	3/2	+1.439
$_{33}As^{76}$			26.5h	β^-	2.97	0.35 1.20 1.75 2.40 2.96	3% 6% 6% 32% 53%		2	−0.905
$_{33}As^{77}$			38.83h ±0.05	β^-	0.686	0.165 0.437 0.477 0.684	0.68% 0.6% 1.6% 97%			
$_{33}As^{78}$			91m	β^-	4.3	1.4 4.1	30% 70%			
$_{33}As^{79}$			9.0m	β^-	2.2	1.25 1.43 1.70 1.80 2.15	1.5% 0.5% 2% 1.5% 95%			
$_{33}As^{80}$			15.3s	β^-	6.0	3.0 3.5 3.7 4.1 4.2 4.5 5.3 6.0	0.4% 0.5% 4.3% 3.6% 1.7% 1.4% 32% 56%			
$_{33}As^{81}$			33s	β^-	3.8	3.8				
Se	78.96							12.2 ± 0.6b		
$_{34}Se^{70}$			44m	β^+, EC						
$_{34}Se^{71}$			5m	β^+, EC	4.4	3.4				
$_{34}Se^{72}$			8.4d	EC	0.6		100%			

Isotope	% nat. abundance	Atomic mass	Lifetime	Modes of decay	Decay energy (MeV)	Particle energies (MeV)	Particle intensities	Thermal neutron capture cross-section	I	μ
$_{34}Se^{73}$			7.1h	β^+, EC	2.75	0.25 0.75 1.32 1.68	1† 10 88 1			
$_{34}Se^{73}$			42m	β^+, EC	1.8					
$_{34}Se^{74}$	0.9 %	73.9225						55 ± 5b	0	
$_{34}Se^{75}$			120.4d	β	0.865	0.293 0.465 0.625 0.601	0.1 % 94 % <2 % 3 %		5/2	
$_{34}Se^{76}$	9.0 %	75.9192						21 ± 2b	0	
$_{34}Se^{77m}$			17.5s	IT	0.161		100 %			
$_{34}Se^{77}$	7.6 %	76.9199						42 ± 4b	1/2	+0.534
$_{34}Se^{78}$	23.5 %	77.9173						0.33 ± 0.04b (Se^{79m}) 0.20 ± 0.106 (Se^{79})	0	
$_{34}Se^{79}$			6.5×10^4y	β^-	0.154	0.154	100 %		7/2	−1.02
$_{34}Se^{80}$	49.8 %	79.9165						0.08 ± 0.01b	0	
$_{34}Se^{81m}$			57m	IT	0.103		100 %			
$_{34}Se^{81}$			18.6m	β^-	1.576	0.74 1.01 1.51	? % 2 % ~93 %			
$_{34}Se^{82}$	9.2 %	81.9167						40 ± 10mb (Se^{83m}) 6 ± 1mb (Se^{83})	0	
$_{34}Se^{83m}$			70s	β^-		1.5 2.4 3.5	25 % 25 % 50 %			
$_{34}Se^{83}$			23m	β^-	3.6	0.83 0.87 0.91 0.96 1.55 2.52 3.25	1 % 13 % 29 % 37 % 5 % 10 % 5 %			
$_{34}Se^{84}$			3.3m	β^-						
$_{34}Se^{85}$			39s	β^-						
Br		79.909						6.8 ± 0.1b		
$_{35}Br^{74}$			36m	β^+, EC	~6.8	4.7				
$_{35}Br^{75}$			1.7h	β^+, EC	2.72	0.3 0.6 0.8 1.70	19† 15 20 46			
$_{35}Br^{76}$			16.1h	4.6	β^+, EC		1.2 1.7 2.4 3.1 3.6	~3 % 9 % 12 % 33 % ~8 %		1	±0.548

Isotope	% nat. abundance	Atomic mass	Lifetime	Modes of decay	Decay energy (MeV)	Particle energies (MeV)	Particle intensities	Thermal neutron capture cross-section	I	μ
$_{35}\mathrm{Br}^{77}$			57h	β^+, EC	1.365	0.361	0.9%		3/2	
$_{35}\mathrm{Br}^{78}$			6.4m	β^+, EC	3.573	1.937 2.52	11% 82%			
$_{35}\mathrm{Br}^{79}$	50.69%	78.9183						2.6 ± 0.2b	3/2	+2.106
$_{35}\mathrm{Br}^{80m}$			4.4h	IT	0.085				5	+1.317
$_{35}\mathrm{Br}^{80}$			17.6m	β^-	β^+, EC 1.871 β^- 2.01	0.866 0.70 0.76 1.38 2.05	β^+ 7.2% 0.23% β^- 1.1% 0.2% 0.004% 0.08%		1	±0.514
$_{35}\mathrm{Br}^{81}$	49.31%	80.9163						3.0 ± 0.2b (Br^{82m}) ^{c} 0.26b (Br^{82})	3/2	+2.270
$_{35}\mathrm{Br}^{82m}$			? 6.1m	IT β^-	0.046	1.659 2.357	0.04% 0.2%			
$_{35}\mathrm{Br}^{82}$			35.5h	β^-	3.092	0.257 0.440	~2% ~98%		5	±1.626
$_{35}\mathrm{Br}^{83}$			2.41h	β^-	0.97	0.395 0.925	1.4% 98.6%			
$_{35}\mathrm{Br}^{84}$			31.8m	β^-	4.8	0.89 1.45 2.9 3.9 4.8	19% 14% 15% 14% 32%			
$_{35}\mathrm{Br}^{85}$			3.0m	β^-	2.8	2.8	100%			
$_{35}\mathrm{Br}^{86}$			54s	β^-	7.1					
$_{35}\mathrm{Br}^{87}$			55s	β^-, η	6.1	8.0 2.6	30† 70			
$_{35}\mathrm{Br}^{88}$			16s	β^-, η						
$_{35}\mathrm{Br}^{89}$			45s	β^-, η						
$_{35}\mathrm{Br}^{90}$			1.6s	β^-, η						
Kr		83.80						24.5 ± 1.0b		
$_{36}\mathrm{Kr}^{74}$			20m	β^+, EC	4.1	3.1				
$_{36}\mathrm{Kr}^{75}$			5m	β^+, EC	~5.1					
$_{36}\mathrm{Kr}^{76}$			14.8h	EC	~1.2					
$_{36}\mathrm{Kr}^{77}$			1.19h	β^+, EC	2.99	1.64 1.68 1.85 1.86	2.3% 34.4% 40.7% 4.2%			
$_{36}\mathrm{Kr}^{78}$	0.35%	77.9204						2.0 ± 0.5b		
$_{36}\mathrm{Kr}^{79}$			34.9h	β^+, EC	1.628	0.209 0.299 0.325 0.604	0.03% 0.01% 0.2% 7.3%			

Isotope	% nat. abundance	Atomic mass	Lifetime	Modes of decay	Decay energy (MeV)	Particle energies (MeV)	Particle intensities	Thermal neutron capture cross-section	I	μ
$_{36}Kr^{80}$	2.25%	79.9164						$14 \pm 2b$		
$_{36}Kr^{81}$			$2.1 \times 10^5 y$	EC	0.29		100%			
$_{36}Kr^{82}$	11.6%	81.9135						$3 \pm 1b$ (Kr^{83m}) $32 \pm 15b$ (Kr^{83})	0	
$_{36}Kr^{83m}$			1.86h	IT						
$_{36}Kr^{83}$	11.5%							$180 \pm 40b$	9/2	-0.970
$_{36}Kr^{84}$	57.0%							$0.10 \pm 0.03b$	0	
$_{36}Kr^{85m}$			4.39hr	IT β^-	0.305	0.82				
$_{36}Kr^{85}$			10.76y	β^-	0.67				9/2	± 1.005
$_{36}Kr^{86}$	17.3%							$0.06 \pm 0.02b$	0	
$_{36}Kr^{87}$			76m	β^-	3.89	3.8 3.3 1.3	~70% ~5% 25%			
$_{36}Kr^{88}$			2.80h	β^-	1.?	0.?? 0.90 2.8	?0?% 12% 20%			
$_{36}Kr^{89}$			3.2m	β^-, η	4.6	4.0				
$_{36}Kr^{90}$			33s	β^-, η		~1 2.0 2.80	105 120 236			
$_{36}Kr^{91}$			10s	β^-		~3.6				
$_{36}Kr^{92}$			3.0s	β^-						
$_{36}Kr^{93}$			2.0s	β^-						
$_{36}Kr^{94}$			1.4s	β^-						
Rb		85.47						$0.5 \pm 0.1b$		
$_{37}Rb^{79}$			21m	β^+, EC	~3.6					
$_{37}Rb^{80}$			34s	β^+, EC	5.1	3.5 4.1	40% 60%			
$_{37}Rb^{81m}$			32m	IT	0.085				9/2	
$_{37}Rb^{81}$			4.7h	β^+, EC	2.26	0.575 1.05 1.4	2.3% 30% ~50%		3/2	$+2.05$
$_{37}Rb^{82m}$			6.4h	β^+		0.80	100%		5	$+1.5$
$_{37}Rb^{83}$			83d	EC	0.83				5/2	$+1.4$
$_{37}Rb^{84m}$			20m	EC, IT						
$_{37}Rb^{84}$			33.0d	β^+, EC	2.68	0.78 1.657	53† 47		2	-1.32

Isotope	% nat. abundance	Atomic mass	Lifetime	Modes of decay	Decay energy (MeV)	Particle energies (MeV)	Particle intensities	Thermal neutron capture cross-section	I	μ
$_{37}Rb^{85}$	72.17%	84.9117						$55 \pm 5mb$ (Rb^{86m}) $0.45 \pm 0.04b$ (Rb^{86})	5/2	+1.352
$_{37}Rb^{86m}$			1.04m	IT	0.56		100%			
$_{37}Rb^{86}$			18.66d	β^-	1.78	0.71 1.78	8.8% 91.2%		2	−1.691
$_{37}Rb^{87}$	27.83%		$5 \times 10^{11}y$	β^-	0.274	0.274	100%	$0.12 \pm 0.03b$	3/2	+2.750
$_{37}Rb^{88}$			17.8m	β^-	5.2	2.5 3.6 5.3	9% 13% 78%	$1.0 \pm 0.2b$	2	±0.51
$_{37}Rb^{89}$			15.4m	β^-	3.92	0.67 1.61 2.8 3.92	28% 53% 5% 7%			
$_{37}Rb^{90}$			2.9m	β^-	6.6	1.3 2.0 2.2 2.5 4.4 5.8 6.6	10% 5% 17% 23% 15% 10%			
$_{37}Rb^{91}$			1.2m	β^-	~5.5	4.6				
$_{37}Rb^{92}$			5.3s	β^-	~7.6					
$_{37}Rb^{93}$			5.6s	β^-						
$_{37}Rb^{94}$			2.9s	β^-						
$_{37}Rb^{95}$			< 2.5s	β^-						
Sr		87.62						$1.21 \pm 0.06b$		
$_{38}Sr^{80}$			1.7h	EC	2.0					
$_{38}Sr^{81}$			29m	β^+, EC	~3.8					
$_{38}Sr^{82}$			25d	EC	~0.4					
$_{38}Sr^{83}$			33h	β^+	2.21	0.44 0.81 1.15	1.3% 7.4% 11%			
$_{38}Sr^{84}$	0.5%	83.9134						0.57 ± 0.5 (Sr^{85m}) 0.31 ± 0.03 (Sr^{85})		
$_{38}Sr^{85m}$			70m	IT, EC	0.238					
$_{38}Sr^{85}$			64d	EC	1.11					
$_{38}Sr^{86}$	9.9%	85.9094						$0.8 \pm 0.1b$ (Sr^{87m})	0	
$_{38}Sr^{87m}$			2.83h	IT, EC			~100%			
$_{38}Sr^{87}$	7.0%	86.9089							9/2	−1.093
$_{38}Sr^{88}$	82.6%	87.9056						$5 \pm 1mb$	0	
$_{38}Sr^{89}$			52d	β^-	1.463	1.463 0.55	>99% <0.01%	$0.5 \pm 0.1b$		

Isotope	% nat. abundance	Atomic mass	Lifetime	Modes of decay	Decay energy (MeV)	Particle energies (MeV)	Particle intensities	Thermal neutron capture cross-section	I	μ
$_{38}Sr^{90}$			28.1y	β^-	0.546	0.546	100%	$0.9 \pm 0.5b$		
$_{38}Sr^{91}$			9.67h	β^-	2.67	0.62 1.09 1.36 2.03 2.67	7% 33% 29% 4% 26%			
$_{38}Sr^{92}$			2.71h	β^-	1.9	0.55 1.5	90% 10%			
$_{38}Sr^{93}$			8m	β^-	4.8	2.2 2.6 2.8	10% 25% 65%			
$_{38}Sr^{94}$						2.1				
$_{38}Sr^{95}$			0.8m	β^-						
Y		88.905						$1.3 \pm 0.1b$		
$_{39}Y^{82}$			12m	β^+, EC	~7.5					
$_{39}Y^{83}$			7.5m	β^+, EC						
$_{39}Y^{84}$			41m	β^+, EC	6.3	2.9 3.5				
$_{39}Y^{85m}$			2.7h	β^+, EC		1.54	50%			
$_{39}Y^{85}$			5.0h	β^+, EC	3.26	1.1 2.1 2.24	4% 10% 55%			
$_{39}Y^{86m}$			48m	IT						
$_{39}Y^{86}$			14.6h	β^+, EC	5.27	1.04 1.25 1.60 2.02 2.34 3.15	4% 11% 5% 4% 1.1% 0.5%			
$_{39}Y^{87m}$			2.83h	IT	0.388					
$_{39}Y^{87}$			80h	β^+, EC	1.7	0.7	~0.3%			
$_{39}Y^{88}$			106.6d ± 0.1	β^+, EC	3.621	0.76				
$_{39}Y^{89m}$			16s	IT	0.909					
$_{39}Y^{89}$	100%	88.9059						$1.3 \pm 0.1b$ (Y^{90}) $1.0 \pm 0.2mb$ (Y^{90m})	1/2	−0.1373
$_{39}Y^{90m}$			3.1h	IT β^-	0.685 2.95					
$_{39}Y^{90}$			64h	β^-	2.27	2.273	~100%		2	−1.63
$_{39}Y^{91m}$			50m	IT	0.555					
$_{39}Y^{91}$			58.8d	β^-	1.545	1.545 0.33	99.7% 0.3%	$1.4 \pm 0.3b$	1/2	±0.164

Isotope	% nat. abundance	Atomic mass	Lifetime	Modes of decay	Decay energy (MeV)	Particle energies (MeV)	Particle intensities	Thermal neutron capture cross-section	I	μ
$_{39}Y^{92}$			3.53h	β^-	3.63	1.80 2.14 2.25 2.70 3.63	0.8% 1.5% 2.3% 3.1% 86%			
$_{39}Y^{93}$			10.2h	β^-	2.89	0.71 1.47 1.95 2.62 2.89	1.8% 0.9% 3.0% 4% 90%			
$_{39}Y^{94}$			20.3m	β^-	5.0	5.0				
$_{39}Y^{95}$			10.9m	β^-	~4.2					
$_{39}Y^{96}$			2.3m	β^-	~6.9	3.5				
Zr		91.22						0.182 ± 0.005b		
$_{40}Zr^{81}$			~10m	β^+, EC						
$_{40}Zr^{82}$			10m	β^+, EC						
$_{40}Zr^{83}$			~7m	β^+, EC						
$_{40}Zr^{84}$			16m	β^+, EC						
$_{40}Zr^{85}$			1.4h	β^+, EC						
$_{40}Zr^{86}$			16.5h	β^+, EC	1.2					
$_{40}Zr^{87}$			1.6h	β^+, EC	3.50	2.10				
$_{40}Zr^{88}$			85d	EC	~0.5					
$_{40}Zr^{89m}$			4.18m	IT, β^+		2.4				
$_{40}Zr^{89}$			78.4h	β^+, EC	2.834					
$_{40}Zr^{90m}$			0.81s	IT	2318.7					
$_{40}Zr^{90}$	51.4%	89.9043						0.10 ± 0.07b		
$_{40}Zr^{91}$	11.2%	90.9053						1.58 ± 0.12b	5/2	−1.303
$_{40}Zr^{92}$	17.1%	91.9046						0.25 ± 0.12b		
$_{40}Zr^{93}$			1.5×10^6y	β^-	0.090	0.060	≥95%	<4b		
$_{40}Zr^{94}$	17.5%	93.9061						75 ± 8mb		
$_{40}Zr^{95}$			65d	β^-	1.121	0.360 0.396 0.890 1.130	43% 55% 2% 0.4%			
$_{40}Zr^{96}$	2.80%	95.9082	$> 3.6 \times 10^{17}$y					0.05 ± 0.01b		
$_{40}Zr^{97}$			17h	β^-	2.67	0.46 1.91				
$_{40}Zr^{98}$			1m	β^-	~1.5					
Nb		92.906						1.15 ± 0.05b		

Isotope	% nat. abundance	Atomic mass	Lifetime	Modes of decay	Decay energy (MeV)	Particle energies (MeV)	Particle intensities	Thermal neutron capture cross-section	I	μ
$_{41}Nb^{88}$			14m	β^+, EC	~7.2	3.2				
$_{41}Nb^{89m}$			42m	β^+, EC		3.1				
$_{41}Nb^{89}$			1.9h	β^+, EC	3.9	2.9				
$_{41}Nb^{90m}$			24s	IT						
$_{41}Nb^{90}$			14.6h	β^+, EC	6.11	0.55 0.86 1.50	4† 9 87			
$_{41}Nb^{91m}$			62d	EC IT						
$_{41}Nb^{91}$				EC	1.1					
$_{41}Nb^{92m}$			10.14 ± 0.03d	β^+, EC						
$_{41}Nb^{92}$			> 350y	γ						
$_{41}Nb^{93}$	100%	92.9060						0.15 ± 0.1b (Nb^{94m}) 1.15 ± 0.05b (Nb^{94})	9/2	+6.167
$_{41}Nb^{94m}$			6.26 + 0.01m	IT β^-	0.0415 2.1	1.15				
$_{41}Nb^{94}$			2.0 × 10^{4}y	β^-	2.06	0.60	100%	11 ± 2b		
$_{41}Nb^{95m}$				IT	0.2357		100%			
$_{41}Nb^{95}$			35.15 ± 0.03d	β^-	0.925	0.1597 0.924	~100% ≤0.07%	<7b		
$_{41}Nb^{96}$			23.4h	β^-	3.15	0.71 0.70 0.80	27% 68% 5%			
$_{41}Nb^{97m}$			1.0m	IT	0.743					
$_{41}Nb^{97}$			72m	β^-	1.93	0.97 1.27	2% 98%			
$_{41}Nb^{98m}$			< 2m	β^-						
$_{41}Nb^{98}$			51m	β^-	4.6	3.1				
$_{41}Nb^{99}$			10s	β^-						
$_{41}Nb^{99}$			2.4 m	β^-	3.2					
$_{41}Nb^{100}$			3.0m	β^-	~6.1					
$_{41}Nb^{100}$			11m	β^-	~6.1	3.1 3.5 42	~45% ~45% ~10%			
$_{41}Nb^{101}$			1.0m	β^-						
Mo		95.94						2.65 ± 0.05b		

Isotope	% nat. abundance	Atomic mass	Lifetime	Modes of decay	Decay energy (MeV)	Particle energies (MeV)	Particle intensities	Thermal neutron capture cross-section	I	μ
$_{42}Mo^{88}$			27m	β^+, EC						
$_{42}Mo^{89}$			7m	β^+, EC		4.0 4.9				
$_{42}Mo^{90}$			5.7h	β^+, EC	1.465	1.085	25%			
$_{42}Mo^{91m}$			66s	IT β^+, EC	0.658 6.1					
$_{42}Mo^{91}$			15.49m	IT, β^+	4.46	3.44				
$_{42}Mo^{92}$	14.8%	91.9063						<0.006b (Mo^{93m}) <0.3b (Mo^{93})	0	
$_{42}Mo^{93m}$			6.9h	IT	2.428					
$_{42}Mo^{93}$			>100y	EC						
$_{42}Mo^{94}$	9.3%	93.9047							0	
$_{42}Mo^{95}$	15.9%	94.90584						14.4 ± 0.5b	5/2	−0.9133
$_{42}Mo^{96}$	16.7%	95.9046						1.2 ± 0.6b	0	
$_{42}Mo^{97}$	9.6%	96.9058						2.2 ± 0.7b	5/2	−0.9325
$_{42}Mo^{98}$	24.1%	97.9055						0.14 ± 0.02b	0	
$_{42}Mo^{99}$			66.69 ± 0.06h	β^-	1.37	0.26 0.45 0.86 1.19	0.3% 17% 1% 82%			
$_{42}Mo^{100}$	9.6%	99.9076						0.20 ± 0.05b	0	
$_{42}Mo^{101}$			14.6m	β^-	2.82	0.84 1.23 1.61 2.23	38% 13% 25% 10%			
$_{42}Mo^{102}$			11m	β^-	1.2	1.2				
$_{42}Mo^{103}$			62s	β^-						
$_{42}Mo^{104}$			1.3m	β^-		2.2 4.8				
$_{42}Mo^{105}$			40s	β^-						
Tc		98.9062								
$_{43}Tc^{92}$			4.4m	β^+, EC	7.9	4.1	92%			
$_{43}Tc^{93m}$			43m	IT EC	0.39 ~3.6					
$_{43}Tc^{93}$			2.7h	β^+, EC	3.19	0.64 0.80	32% 11%			
$_{43}Tc^{94m}$			53m	β^+, EC	~4.3	2.47				
$_{43}Tc^{94}$			293m	β^+, EC	4.26	0.816	11%			
$_{43}Tc^{95m}$			61d	IT β^+, EC	0.0389 ~1.7	0.49 0.71	10† 25			

Isotope	% nat. abundance	Atomic mass	Lifetime	Modes of decay	Decay energy (MeV)	Particle energies (MeV)	Particle intensities	Thermal neutron capture cross-section	I	μ
$_{43}Tc^{95}$			20.0h	EC	1.66					
$_{43}Tc^{96m}$			52m	IT	0.0344					
$_{43}Tc^{96}$					2.9					
$_{43}Tc^{97m}$			90d	IT	0.0965					
$_{43}Tc^{97}$		96.906	2.6×10^6y	EC	~0.3					
$_{43}Tc^{98}$			1.5×10^6y	β^-	1.7	0.3				
$_{43}Tc^{99m}$		98.906	6.0h	IT	0.1427					
$_{43}Tc^{99}$			2.12×10^5y	β^-	0.292	0.292		22 ± 3b	9/2	$+5.68$
$_{43}Tc^{100}$			17s	β^-		2.2 2.88 3.38				
$_{43}Tc^{101}$			14m	β^-	1.63	1.07 1.32				
$_{43}Tc^{102}$			4.5m	β^-	4.4	~2				
$_{43}Tc^{102}$			5s	β^-	4.4	4.4				
$_{43}Tc^{103}$			50s	β^-	2.4	2.0 2.2				
$_{43}Tc^{104}$			18m	β^-	5.8	2.4 3.3 4.6 >5.3				
$_{43}Tc^{105}$			8m	β^-	3.4					
$_{43}Tc^{106}$			37s	β^-						
$_{43}Tc^{107}$			29 s	β^-						
Ru		101.07						3.0 ± 0.8b		
$_{44}Ru^{93}$			50s	β^+						
$_{44}Ru^{94}$			57m	EC						
$_{44}Ru^{95}$			1.7h	β^+, EC	2.03	0.7 1.01 1.33	1.3† 12 4			
$_{44}Ru^{96}$	5.5%	95.9076						0.21 ± 0.02b		
$_{44}Ru^{97}$			2.9d	EC	~1.2					
$_{44}Ru^{98}$	1.9%	97.9055						<8b		
$_{44}Ru^{99}$	12.7%	98.9061						10.6 ± 0.6b	5/2	-0.63
$_{44}Ru^{100}$	12.6%	99.9030						10.4 ± 0.7b		
$_{44}Ru^{101}$	17.1%							3.1 ± 0.9b	5/2	-0.69
$_{44}Ru^{102}$	31.6%	101.9037								
$_{44}Ru^{103}$			39.6d	β^-	0.74	0.90 0.203 0.382 0.445	7% 89% 1% 1%			

Isotope	% nat. abundance	Atomic mass	Lifetime	Modes of decay	Decay energy (MeV)	Particle energies (MeV)	Particle intensities	Thermal neutron capture cross-section	I	μ
						0.700	3%			
$_{44}Ru^{104}$	18.6%	103.9055						0.47 ± 0.2b		
$_{44}Ru^{105}$			4.44h	β^-	1.871	0.49 1.06 1.08 1.15 1.52 1.75 1.87	5% 14% 13% 44% ~2% ≤10% 11%	0.2 ± 0.02b		
$_{44}Ru^{106}$			367d	β^-	0.0394	0.0392	100%	146 ± 45mb		
$_{44}Ru^{107}$			4.2m	β^-	3.2	2.1 3.2				
$_{44}Ru^{108}$			4.5m	β^-	1.3	1.1 1.3	28% 72%			
Rh		102.905						150 ± 5b		
$_{45}Rh^{97}$			32m	β^+, EC	3.49	1.8 2.06 2.47				
$_{45}Rh^{98}$			8.7m	β^+, EC	4.2	2.5				
$_{45}Rh^{99m}$			4.7h	β^+, EC	~2.05	0.74				
$_{45}Rh^{99}$			16d	β^+, EC	~2.05	0.44 0.59 0.71 1.03				
$_{45}Rh^{100}$			20h	β^+, EC	3.64	0.20 0.58 0.8 1.30 1.6 2.12 2.68	~2% 3%			
$_{45}Rh^{101m}$			4.5d	IT EC	0.157 ~0.7		10%			
$_{45}Rh^{101}$			3.1y	EC	0.56					
$_{45}Rh^{102}$			2.9y	EC	2.3					
$_{45}Rh^{102}$			206d	β^+, EC β^-	2.32 1.15	β^- 0.59 1.15 β^+0.20 0.36 0.83 1.30	2% 20% 7% 3% 37% 29%			
$_{45}Rh^{103}$	100%	102.9048						11 ± 1b (Rh^{104m}) 139 ± 5b (Rh^{104})	1/2	−0.0883
$_{45}Rh^{104m}$			4.4m	IT β^-	0.129 ~2.5			800 ± 100b	5	

Isotope	% nat. abundance	Atomic mass	Lifetime	Modes of decay	Decay energy (MeV)	Particle energies (MeV)	Particle intensities	Thermal neutron capture cross-section	I	μ
$_{45}Rh^{104}$			43s	EC β^-	1.15 2.47	0.7 1.9 2.5	0.08% 1.4% 98.5%	40 ± 30b		
$_{45}Rh^{105m}$			45s	IT	0.129					
$_{45}Rh^{105}$			35.9h	β^-	0.565	0.249 0.262 0.568	20% 5% 75%	5700 ± 1200b (Rh^{106m}) 13,000 ± 2000b (Rh^{106})		
$_{45}Rh^{106}$			30s	β^-	3.54	2.0 2.4 3.1 3.53	3% 12% 11% 68%			
$_{45}Rh^{106}$			130m	β^-	~3.63	0.79 0.95 1.18 1.62	40% 38% 11% 10%			
$_{45}Rh^{107}$			22m	β^-	1.51	0.84 0.94 1.14 1.20				
$_{45}Rh^{108}$			17s	β^-	~4.5	3.5 4.0 4.5	22% 17% 51%			
$_{45}Rh^{111}$			<1h	β^-	~2.5					
$_{45}Rh^{110}$			5s	β^-	5.5	5.5				
Pd		106.4						6.0 ± 1.0b		
$_{46}Pd^{98}$			17m	EC β^+						
$_{46}Pd^{99}$			22m	β^+, EC	3.8	2.0				
$_{46}Pd^{100}$			4.0d	EC	~0.6					
$_{46}Pd^{101}$			8.4h	β^+, EC	1.99	0.49 0.78	17† 83			
$_{46}Pd^{102}$	1.0%	101.9049						4.8 ± 1.5b		
$_{46}Pd^{103}$			17d	EC	0.56					
$_{46}Pd^{104}$	11.0%	103.9036								
$_{46}Pd^{105}$	22.2%	104.9046							5/2	−0.642
$_{46}Pd^{106}$	27.3%	105.9032						13 ± 2mb (Pd^{107m}) 279 ± 29mb (Pd^{107})		
$_{46}Pd^{107m}$			22s	IT	0.21					
$_{46}Pd^{107}$			~7 × 10⁶y	β^-	0.035	0.035				
$_{46}Pd^{108}$	26.7%	107.90389						0.20 ± 0.04b (Pd^{109m}) 12 ± 2b (Pd^{109})	0	
$_{46}Pd^{109m}$			4.7m	IT	0.188					
$_{46}Pd^{109}$			13.47h	β^-	1.115	0.248 0.255 0.278	0.023% 0.002% 0.002%			

Isotope	% nat. abundance	Atomic mass	Lifetime	Modes of decay	Decay energy (MeV)	Particle energies (MeV)	Particle intensities	Thermal neutron capture cross-section	I	μ
						0.383	0.037%			
						0.394	0.022%			
						0.416	0.005%			
						0.703	0.010%			
						0.806	0.027%			
						1.030	99.87%			
$_{46}Pd^{110}$	11.8%							20 ± 15mb (Pd^{111m}) 0.4 ± 0.1b (Pd^{111})		
$_{46}Pd^{111m}$			5.5h	IT β^-	0.17		75% 25%			
$_{46}Pd^{111}$			22m	β^-	2.2	0.6 0.7 1.1 1.2 2.2				
$_{46}Pd^{112}$			21h	β^-	0.30	0.30	100%			
$_{46}Pd^{113}$			1.4m	β^-						
$_{46}Pd^{114}$			2.4m	β^-						
$_{46}Pd^{115}$			45s	β^-						
Ag		107.870						63.8 ± 0.6b		
$_{47}Ag^{102}$			15m	β^+, EC	~5.3					
$_{47}Ag^{103m}$			5.7s	IT	0.138					
$_{47}Ag^{103}$			66m	β^+,EC	2.6	1.3 1.6			7/2	
$_{47}Ag^{104m}$			30m	IT β^+, EC	0.02	2.7			2	+3.7
$_{47}Ag^{104}$			67m	β^+, EC	4.27				5	+4.0
$_{47}Ag^{105}$			40d	EC					1/2	±1.01
$_{47}Ag^{106m}$			8.4d	EC					6	
$_{47}Ag^{106}$			24m	β^+, EC	2.96	1.45 1.96	11† 89		1	
$_{47}Ag^{107m}$			44.3s	IT	0.093					
$_{47}Ag^{107}$	51.83%	106.90509						35 ± 5b	1/2	−0.1135
$_{47}Ag^{108m}$			>5y	IT EC	0.110		10% 90%			
$_{47}Ag^{108}$			2.42m	β^+, EC β^-	1.92 1.64	β^- 1.8 1.2 0.47 β^+ 0.88	96% 1.8% ~0.02%		1	+2.8
$_{47}Ag^{109m}$			40s	IT	0.0877				7/2	±4.3
$_{47}Ag^{109}$	48.17%	108.9047						4.2 ± 0.7b (Ag^{110m}) 89 ± 4b (Ag^{110})	1/2	−0.1305
$_{47}Ag^{110m}$			253d	β^- IT	~2.9 0.12	.0.087 0.529 1.5	61% 36% 0.6%	82 ± 11b	6	+3.604

Isotope	% nat. abundance	Atomic mass	Lifetime	Modes of decay	Decay energy (MeV)	Particle energies (MeV)	Particle intensities	Thermal neutron capture cross-section	I	μ
$_{47}Ag^{110}$			24.4s	β^-	~2.9	1.40 2.14 2.87	0.05% 12% 88%			
$_{47}Ag^{111m}$			74s	IT	0.07					
$_{47}Ag^{111}$			7.5d	β^-	1.05	0.69 0.79 1.04	6% 1% 93%		1/2	−0.146
$_{47}Ag^{112}$			3.2h	β^-	4.01	1.96 3.35 3.94	10% 22% 54%		2	±0.054
$_{47}Ag^{113m}$			1.2m	β^-						
$_{47}Ag^{113}$			5.3h	β^-	2.00	2.0			1/2	±0.159
$_{47}Ag^{114}$			4.5s	β^-	4.6	4.6				
$_{47}Ag^{114}$			2m	β^-	4.6					
$_{47}Ag^{115}$			20s	β^-						
$_{47}Ag^{115}$			20m	β^-	3.3	1.0 2.1 2.7 3.0 3.2				
$_{47}Ag^{116}$			2.5m	β^-	~6.1					
$_{47}Ag^{111}$			1.1m	μ						
Cd		112.40						2450 ± 20b		
$_{48}Cd^{103}$			10m	β^+, EC						
$_{48}Cd^{104}$			57m	EC	~1.2					
$_{48}Cd^{105}$			55m	β^+, EC	~2.8	0.8 1.69				
$_{48}Cd^{106}$	1.2%	105.907						1.0 ± 0.5b		
$_{48}Cd^{107}$			6.5h	β^+, EC	1.417	0.302			5/2	−0.6144
$_{48}Cd^{108}$	0.9%	107.9040						1.5 ± 0.5b	0	
$_{48}Cd^{109}$			450 ± 5d	EC	0.16			700 ± 100b	5/2	−0.8270
$_{48}Cd^{110}$	12.4%	109.9030						0.10 ± 0.03b (Cd^{111m})	0	
$_{48}Cd^{111m}$			48.6m	IT						
$_{48}Cd^{111}$	12.8%	110.9042							1/2	−0.5943
$_{48}Cd^{112}$	24.0%	111.9028						0.06 ± 0.02b (Cd^{113m})	0	
$_{48}Cd^{113m}$			14y	β^-	0.58	0.58	99+%		11/2	−1.087
$_{48}Cd^{113}$	12.3%	112.9046						20,000 ± 300b	1/2	−0.6217
$_{48}Cd^{114}$	28.8%	113.9036						0.36 ± 0.007b (Cd^{115m}) 0.30 ± 0.015b (Cd^{115})	0	

Isotope	% nat. abundance	Atomic mass	Lifetime	Modes of decay	Decay energy (MeV)	Particle energies (MeV)	Particle intensities	Thermal neutron capture cross-section	I	μ
$_{48}Cd^{115m}$			43d	IT	0.18	0.34			11/2	− 1.042
						0.68	1.6%			
				β^-	1.62	1.62	97%			
$_{48}Cd^{115}$			53.5h	β^-	1.45	0.58⎫	42%		1/2	− 0.648
						0.86⎭				
						1.11	58%			
$_{48}Cd^{116}$	7.6%	115.9050						27 ± 5mb (Cd^{117m})	0	
								50 ± 8mb (Cd^{117})		
$_{48}Cd^{117m}$			3.4h	β^-	2.65	0.41				
						0.70				
$_{48}Cd^{117}$			2.4h	β^-	2.52	0.65				
						1.79				
						2.23				
$_{48}Cd^{118}$			49m	β^-	~0.8					
$_{48}Cd^{119}$			2.7m	β^-	3.5	~3.5				
$_{48}Cd^{119}$			10m	β^-	3.5	~3.5				
In		114.82						194 ± 2b		
$_{49}In^{106}$			5.3m	β^+, EC	6.5	3.1				
						4.85				
$_{49}In^{107}$			32m	β^+, EC	3.5	2.2				
$_{49}In^{108}$			39m	β^+, EC		2.28				
						2.66				
						3.50				
$_{49}In^{108}$			56m	β^+, EC		1.29				
$_{49}In^{109}$			4.3h	β^+, EC	2.02				9/2	+ 5.53
$_{49}In^{110m}$			4.9h	EC β^+					7	± 10.5
$_{49}In^{110}$			67m	β^+, EC	3.93	2.25			2	
$_{49}In^{111}$			2.81d						9/2	+ 5.53
$_{49}In^{112m}$			20.7m	IT	0.156					
$_{49}In^{112}$			14m	β^- β^+, EC	0.66	β^- 0.66	44%			
						β^+ 1.0	6%			
						1.52	28%			
$_{49}In^{113m}$			100m	IT	0.3916				1/2	− 0.210
$_{49}In^{113}$	4.3%	112.9043						3.2 ± 0.7b (In^{114m_2})	9/2	+ 5.523
								4.5 ± 0.7b (In^{114m_1})		
								3 ± 1b (In^{114})		
$_{49}In^{114m_2}$			0.0425	IT	0.502					

Isotope	% nat. abundance	Atomic mass	Lifetime	Modes of decay	Decay energy (MeV)	Particle energies (MeV)	Particle intensities	Thermal neutron capture cross-section	I	μ
$_{49}\text{In}^{114m_1}$			50.0d	IT EC	0.191		96.5% 3.5%		5	+4.7
$_{49}\text{In}^{114}$			72s	β^+, EC β^-	1.44 1.986	β^- 1.989 β^+ 0.40	98+% 0.039%			+1.7
$_{49}\text{In}^{115m}$			4.50h	IT β^-	0.335 0.82		95% 5%		1/2	−0.244
$_{49}\text{In}^{115}$	95.7%	114.9041						85 + 10b (In^{116m_2}) 72 ± 4b (In^{116m_1}) 42 ± 4b (In^{116})	9/2	+5.534
$_{49}\text{In}^{116m_2}$			2.16s	IT	0.22					
$_{49}\text{In}^{116m_1}$			54m	β^-	3.39	0.60 0.87 1.00	21% 28% 51%		5	+4.3
$_{49}\text{In}^{116}$			14s	β^-	3.33	3.3	99%			
$_{49}\text{In}^{117m}$			1.93h	IT β^-	0.314 ~1.77	1.62 1.77	47% 16% 37%		1/2	±0.25
$_{49}\text{In}^{117}$			44m	β^-	1.47	0.74			9/2	
$_{49}\text{In}^{118}$			5s	β^-	4.2	4.2				
$_{49}\text{In}^{118}$			4.4m	β^-	4.2	1.3 2.0				
$_{49}\text{In}^{119m}$			18m	IT β^-	0.30 2.8	2.7	5%			
$_{49}\text{In}^{119}$			2.1m	β^-	2.5	1.6				
$_{49}\text{In}^{120}$			3.2s	β^-	~5.6	4.4 5.6	15% 85%			
$_{49}\text{In}^{120}$			46s	β^-	~5.3	2.2 3.1	~40% ~27%			
$_{49}\text{In}^{121}$			3.1m	β^-	~3.6	3.7				
$_{49}\text{In}^{121}$			30s	β^-						
$_{49}\text{In}^{122}$			8s	β^-	~6.7					
$_{49}\text{In}^{123}$			36s	β^-	~4.5					
$_{49}\text{In}^{123}$			10s	β^-						
$_{49}\text{In}^{124}$			4s	β^-	7.4	5				
Sn		118.69						0.63 ± 0.01b		
$_{50}\text{Sn}^{108}$			9m	EC						
$_{50}\text{Sn}^{109}$			18.1m	β^+, EC		~1.6				
$_{50}\text{Sn}^{110}$			4.0h	EC						
$_{50}\text{Sn}^{111}$			35m	β^+, EC	2.52	1.51	35.3%			
$_{50}\text{Sn}^{112}$	1.0%	111.9040						0.35 ± 0.08b (Sn^{113m}) 0.8 ± 0.2 (Sn^{113})		

Isotope	% nat. abundance	Atomic mass	Lifetime	Modes of decay	Decay energy (MeV)	Particle energies (MeV)	Particle intensities	Thermal neutron capture cross-section	I	μ
$_{50}Sn^{113m}$			20m	IT EC	0.079 1.1		91% 9%			
$_{50}Sn^{113}$			115d	EC	1.02					
$_{50}Sn^{114}$	0.7%	113.9030								
$_{50}Sn^{115}$	0.4%	114.9035							1/2	−0.918
$_{50}Sn^{116}$	14.7%	115.9021						$6 \pm 2mb$ (Sn^{117m})	0	
$_{50}Sn^{117m}$			14d	IT	0.317					
$_{50}Sn^{117}$	7.7%	116.9031							1/2	−1.000
$_{50}Sn^{118}$	24.3%	117.9018						$10 \pm 6mb$ (Sn^{119m})	0	
$_{50}Sn^{119m}$			250d	IT	0.089				(3/2)	+0.67
$_{50}Sn^{119}$	8.6%	118.9034							1/2	−1.046
$_{50}Sn^{120}$	32.4%							$1 \pm 1mb$ (Sn^{121m}) $140 \pm 30mb$ (Sn^{121})	0	
$_{50}Sn^{121m}$			76y	β^-	0.45	0.35	100%			
$_{50}Sn^{121}$			27h	β^-	0.383	0.42	100%		3/2	±0.70
$_{50}Sn^{122}$	4.6%	121.9034						1.0 ± 0.5 mb (Sn^{123m}) $0.15 \pm 0.02b$ (Sn^{123})		
$_{50}Sn^{123m}$			125d	β^-	1.42	1.42	100%			
$_{50}Sn^{123}$			42m	β^-	1.46	1.46				
$_{50}Sn^{124}$	5.6%	123.9052						$0.14 \pm 0.02b$ (Sn^{125m}) $4 \pm 2mb$ (Sn^{125})		
$_{50}Sn^{125m}$			9.7m	β^-	2.39	0.281 0.447 0.480 0.658 0.694 0.910 1.045 1.473 1.751 2.062	0.002% 0.2% 0.03% 0.9% 0.1% 0.3% <0.02% 0.2% <0.1% 98.3%			
$_{50}Sn^{125}$			9.4d	β^-	2.34	0.34 0.46 0.95 1.3 2.37	1.9% 1.9% <0.25% ~0.8% 95%			
$_{50}Sn^{126}$			$\sim 10^5 y$	β^-	~0.3					
$_{50}Sn^{127}$			2.1h	β^-		1.5				
$_{50}Sn^{127}$			4m	β^-	~3.1	2.7				

Isotope	% nat. abundance	Atomic mass	Lifetime	Modes of decay	Decay energy (MeV)	Particle energies (MeV)	Particle intensities	Thermal neutron capture cross-section	I	μ
$_{50}Sn^{128}$			59m	β^-	1.3	0.8				
Sb		121.75						$5 \pm 1b$		
$_{51}Sb^{112}$			0.9m	β^+, EC						
$_{51}Sb^{113}$			6.7m	β^+, EC	4.47	1.85 2.42				
$_{51}Sb^{114}$			3.3m	β^+, EC	6.3	2.7				
$_{51}Sb^{115}$			31m	β^+, EC	3.03	1.51				
$_{51}Sb^{116m}$			60m	β^+, EC		1.16				
$_{51}Sb^{116}$			15m	β^+, EC	4.6	1.5 2.4				
$_{51}Sb^{117}$			2.8h	β^+, EC	1.82	0.57			5/2	+2.67
$_{51}Sb^{118m_2}$			0.87s	IT						
$_{51}Sb^{118m_1}$			5.1h	β^+, EC	3.9					
$_{51}Sb^{118}$			3.5m	β^+, EC	3.7	1.5 2.7			1	+2.1
$_{51}Sb^{119}$			38hr	EC			100%		5/2	+3.45
$_{51}Sb^{120}$			15.9m	β^+, EC	2.69	1.70			1	+2.3
$_{51}Sb^{121}$	57.3%	120.9038						$55 \pm 10mb$ (Sb^{122m}) $6.2 \pm 0.3b$ (Sb^{122})	5/2	+3.359
$_{51}Sb^{122m}$			4.2m	IT	0.162					
$_{51}Sb^{122}$			2.8d	β^- β^+, EC	1.972 1.62	β^+ 0.56 β^- 0.74 1.40 1.97	4% 63% 30%		2	−1.90
$_{51}Sb^{123}$	42.7%	122.9041						$15 \pm 4mb$ (Sb^{124m_2}) $30 \pm 8mb$ (Sb^{124m_1}) $3.4 \pm 0.8b$ (Sb^{124})	7/2	+2.547
$_{51}Sb^{124m_2}$			21m	IT	0.035					
$_{51}Sb^{124m_1}$			93s	IT β^-	0.010					
$_{51}Sb^{124}$			60.3 $\pm$ 0.2d	β^-	2.916	0.06 0.23 0.621 0.950 1.01 1.59 1.67 2.317	2% 11% 50% 5% 1.5% 5% 3% 22%	$6.5 \pm 1.5b$	3	

Isotope	% nat. abundance	Atomic mass	Lifetime	Modes of decay	Decay energy (MeV)	Particle energies (MeV)	Particle intensities	Thermal neutron capture cross-section	I	μ
$_{51}Sb^{125}$			2.7y	β^-	0.764	0.1 0.619 0.45 0.299 0.240 0.125	2% 13% 6% 43% 1% 28%			
$_{51}Sb^{126m}$			19m	IT β^-		1.9				
$_{51}Sb^{126}$			12.5d	β^-	3.7	1.9				
$_{51}Sb^{127}$			93h	β^-	1.60	0.86 1.11 1.57	50% 20% 30%			
$_{51}Sb^{128}$			11m	β^-	4.3	2.5				
$_{51}Sb^{129}$			4.3h	β^-	2.5	1.87				
$_{51}Sb^{130}$			2.6m	β^-						
$_{51}Sb^{131}$			25m	β^-						
$_{51}Sb^{132}$			2.1m	β^-						
$_{51}Sb^{133}$			4.2m	β^-						
Te		127.60						4.7 ± 0.1b		
$_{52}Te^{115}$			6.0m	β^+, EC	4.5	2.8				
$_{52}Te^{116}$			2.50h	β^+, EC	1.6	0.44			0	
$_{52}Te^{117}$			61m	β^+, EC	3.50	1.7			1/2	
$_{52}Te^{118}$			6.00d	EC	~0.3					
$_{52}Te^{119m}$			4.7d	EC	~2.6				11/2	
$_{52}Te^{119}$			15.9h	β^+, EC	2.294	0.627	5%		1/2	±0.25
$_{52}Te^{120}$	0.1%	119.9045						0.34 ± 0.06b (Te^{121m}) 2.0 ± 0.3b (Te^{121})		
$_{52}Te^{121m}$			154d	IT EC, β^+	0.293 ~1.6		90%			
$_{52}Te^{121}$			17d	EC	1.29					
$_{52}Te^{122}$	2.5%	121.9030						1.1 ± 0.5b (Te^{123m}) 1.7 ± 0.8b (Te^{123})		
$_{52}Te^{123m}$			117d	IT	0.247					
$_{52}Te^{123}$	0.9%	122.9042	1.2×10^{13}y	EC	~0.06			410 ± 30b	1/2	-0.7359
$_{52}Te^{124}$	4.6%	123.9028						40 ± 25mb (Te^{125m}) 6.8 ± 1.3b (Te^{125})		
$_{52}Te^{125m}$			58d	IT					3/2	$\sim+0.7$
$_{52}Te^{125}$	7.0%	124.9044						1.56 ± 0.16b	1/2	-0.8871

Isotope	% nat. abundance	Atomic mass	Lifetime	Modes of decay	Decay energy (MeV)	Particle energies (MeV)	Particle intensities	Thermal neutron capture cross-section	I	μ
$_{52}Te^{126}$	18.7%	125.9032						$125 \pm 23mb$ (Te^{127m}) $0.90 \pm 0.15b$ (Te^{127})	0	
$_{52}Te^{127m}$			109d	IT β^-	0.0887 0.77		99.2%			
$_{52}Te^{127}$			9.4h	β^-	0.69	0.28 0.50 0.70	0.32% 0.005% 99.7%			
$_{52}Te^{128}$	31.7%	127.9047						$14 \pm 4mb$ (Te^{129m}) $155 \pm 40mb$ (Te^{129})	0	
$_{52}Te^{129m}$			34d	β^- IT	1.48 0.105	0.91 1.60				
$_{52}Te^{129}$			69m	β^-	1.48	0.39 0.67 1.00 1.45				
$_{52}Te^{130}$	34.5%	129.9067						$0.02 \pm 0.01b$ (Te^{131m}) $0.2 \pm 0.1b$ (Te^{131})	0	
$_{52}Te^{131m}$			30h	β^- IT	2.46 0.1817 (18%)	0.45 0.48 0.50 0.53 0.56 0.63 0.84 2.46	10% 11% 12% 3% 13% 7% 16% 5%			
$_{52}Te^{131}$			25m	β^-		0.80 0.86 0.88 1.32 1.80 2.14	1.4% 1.6% 1.5% 1.0% 20% 62%			
$_{52}Te^{132}$			78h	β^-	0.50	0.22	100%			
$_{52}Te^{133m}$			50m	IT β^-	0.334		13%			
$_{52}Te^{133}$			12.5m	β^-	2.7	2.4 1.3	~30% ~70%			
$_{52}Te^{134}$			42m	β^-						
$_{52}Te^{135}$			<2m	β^-						
I		126.9044						$6.2 \pm 0.2b$		
$_{53}I^{117}$			7m	β^+, EC						
$_{53}I^{118}$			14m	β^+, EC	7	5.5				
$_{53}I^{119}$			19m	β^+, EC						
$_{53}I^{120}$			1.3h	β^+, EC	5.6	2.1 4.0				

Isotope	% nat. abundance	Atomic mass	Lifetime	Modes of decay	Decay energy (MeV)	Particle energies (MeV)	Particle intensities	Thermal neutron capture cross-section	I	μ
$_{53}I^{121}$			2.1h	β^+, EC	2.36	1.2				
$_{53}I^{122}$			3.5m	β^+, EC	4.14	1.8 2.6 3.1				
$_{53}I^{123}$			13.3h	EC	~1.4				5/2	
$_{53}I^{124}$			4.2d	β^+, EC	3.17	0.79 1.53 2.13	8† 46 46		2	
$_{53}I^{125}$			60d	EC	0.149			900 ± 90b	5/2	+3.0
$_{53}I^{126}$			13d	β^+, EC β^-	2.150 1.251	β^+ 1.129 β^- 0.385 0.865 1.25	6† 29 12		2	
$_{53}I^{127}$	100%	126.9044						6.2 ± 0.2b	5/2	+2.808
$_{53}I^{128}$			25.08 ± 0.05m	β^+, EC β^-	1.27 2.14	β^- 1.13 1.67 2.12	1.6% 13% 79%		1	
$_{53}I^{129}$			1.7×10^7y	β^-	0.189	0.189	100%	19 ± 2b (I^{130m}) 9 ± 1b (I^{130})	7/2	+2.617
$_{53}I^{130m}$			8.82 ± 0.04m	IT						
$_{53}I^{130}$			12.3 ± 0.1h	β^-	2.99	0.62 1.04 1.7	52% 48% 0.4%	18 ± 3b	5	
$_{53}I^{131}$			8.070 ± 0.009d	β^-	0.970	0.257 0.333 0.487 0.606 0.806	1.6% 6.9% 0.5% 90.4% 0.6%	~0.7b	7/2	+2.74
$_{53}I^{132}$			2.3h	β^-	3.56	0.72 0.80 0.91 0.98 1.16 1.60 1.75 2.12	16% 8% 5% 18% 20% 9% 6% 18%		4	± 3.08
$_{53}I^{133}$			20.9 ± 0.1h	β^-	1.80	0.7 0.94 1.27			7/2	+2.84
$_{53}I^{134}$			52m	β^-	4.2	1.10 1.32 1.56 1.72 1.86 2.28 2.46	1.0% 23% 15% <4% 9% 12% 25%			
$_{53}I^{135}$			6.7h	β^-	~2.8	0.5 1.0 1.4	35% 40% 25%		7/2	
$_{53}I^{136}$			83s	β^-	7.0	2.7 4.3 5.6	9% 23% 16%			

Isotope	% nat. abundance	Atomic mass	Lifetime	Modes of decay	Decay energy (MeV)	Particle energies (MeV)	Particle intensities	Thermal neutron capture cross-section	I	μ
						7.0	~6%			
$_{53}I^{137}$			23s	β^-, η						
$_{53}I^{138}$			5.9s	β^-, η						
$_{53}I^{139}$			2s	β^-, η						
Xe		131.30						24.5 ± 1b		
$_{54}Xe^{118}$			6m	β^+, EC						
$_{54}Xe^{119}$			6m	β^+, EC						
$_{54}Xe^{120}$			40m	β^+, EC						
$_{54}Xe^{121}$			39m	β^+, EC	3.0	2.8				
$_{54}Xe^{122}$			20h	EC	~0.5					
$_{54}Xe^{123}$			2.1h	β^+, EC	2.8	1.6				
$_{54}Xe^{124}$	0.1%	123.9061						100 ± 20b		
$_{54}Xe^{125m}$			55s	IT						
$_{54}Xe^{125}$			17h	EC						
$_{54}Xe^{126}$	0.1%	125.9042						1.5 ± 1.0b		
$_{54}Xe^{127m}$			73s	IT						
$_{54}Xe^{127}$			36.4d	EC	0.44					
$_{54}Xe^{128}$	1.9%	127.9035						0.43 ± 0.1b (Xe^{129m}) <8b (Xe^{129})		
$_{54}Xe^{129m}$			8.0d	IT	0.236					
$_{54}Xe^{129}$	26.4%	128.9048						21 ± 7b	1/2	−0.7768
$_{54}Xe^{130}$	4.1%	129.9035						0.34 ± 0.08b (Xe^{131m}) <26b (Xe^{131})		
$_{54}Xe^{131m}$			11.8d	IT	0.163				3/2	+0.6908
$_{54}Xe^{131}$	21.2%	130.9051						110 ± 20b		
$_{54}Xe^{132}$	26.9%	131.9042						0.53 ± 0.10b (Xe^{133m}) 50 ± 20mb (Xe^{133})	0	
$_{54}Xe^{133m}$			2.26d	IT	0.233					
$_{54}Xe^{133}$			5.27d	β^-	0.427	0.267 0.347	0.7% 99.3%			
$_{54}Xe^{134}$	10.4%	133.9054						0.23 ± 0.02b	0	
$_{54}Xe^{135m}$			15.6m		0.526					

Isotope	% nat. abundance	Atomic mass	Lifetime	Modes of decay	Decay energy (MeV)	Particle energies (MeV)	Particle intensities	Thermal neutron capture cross-section	I	μ
$_{54}Xe^{135}$			9.2h	β^-	1.16	0.55 0.92	3% 97%	2.64×10^6b $\pm 0.1 \times 10^6$		
$_{54}Xe^{136}$	8.9%	135.9072						280 ± 28mb	0	
$_{54}Xe^{137}$			3.9m	β^-	4.0	3.6 4.1	33% 67%			
$_{54}Xe^{138}$			17m	β^-	2.8					
$_{54}Xe^{139}$			42s	β^-						
$_{54}Xe^{140}$			16s	β^-						
$_{54}Xe^{141}$			2s	β^-						
$_{54}Xe^{142}$			~1.5s	β^-						
Cs		132.905						30.0 ± 1.5b		
$_{55}Cs^{123}$			8m	β^+, EC						
$_{55}Cs^{125}$			45m	β^+, EC		2.05			1/2	+1.41
$_{55}Cs^{126}$			1.6m	β^+, EC		3.8				
$_{55}Cs^{127}$			6.2h	β^+, EC	2.1	1.02 0.677	2.5† 3		1/2	+1.46
$_{55}Cs^{128}$			3.8m	β^+, EC	3.93	1.30 1.90 2.44 2.88	8† 16 39 100			
$_{55}Cs^{129}$			32h	EC	~1.1				1/2	+1.479
$_{55}Cs^{130}$			30m	β^+, EC β^-	2.99 0.442	1.97 0.442	2%		1	±1.4
$_{55}Cs^{131}$			9.70d	EC	0.35				5/2	+3.54
$_{55}Cs^{132}$			6.5d	β^+, EC	2.09				2	+2.22
$_{55}Cs^{133}$	100%	132.9051						2.6 ± 0.2b (Cs^{134m}) 27.4 ± 1.5b (Cs^{134})	7/2	+2.579
$_{55}Cs^{134m}$			2.90 ±0.01h	IT	0.138				8	+1.096
$_{55}Cs^{134}$			2.05y	β^-	2.062	0.089 0.410 0.662	28% 1% 71%	134 ± 12b	4	+2.990
$_{55}Cs^{135}$			3×10^6y	β^-	0.210	0.210	100%	8.7 ± 0.5b	7/2	+2.729
$_{55}Cs^{136}$			13d	β^-	2.54	0.341 0.56 0.657	93% 2% 7%		5	±3.70
$_{55}Cs^{137}$			30.23 ±0.16y	β^-	1.176	0.511 1.176	94% 6%	110 ± 33 mb	7/2	+2.838

Isotope	% nat. abundance	Atomic mass	Lifetime	Modes of decay	Decay energy (MeV)	Particle energies (MeV)	Particle intensities	Thermal neutron capture cross-section	I	μ
$_{55}Cs^{138}$			32.2m	β^-	4.83	1.49 2.20 2.39 2.53 2.62 2.94 3.4	0.5% 10% 36% 5% 16% 12% 21%		3	±0.5
$_{55}Cs^{139}$			9.5m	β^-	4.0					
$_{55}Cs^{140}$			66s	β^-	6.1					
$_{55}Cs^{141}$			24s	β^-						
$_{55}Cs^{142}$			2.3s	β^-						
$_{55}Cs^{143}$			2.0s	β^-						
$_{55}Cs^{144}$				β^-						
Ba		137.34						$1.2 \pm 0.1b$		
$_{56}Ba^{123}$			2m	β^+, EC						
$_{56}Ba^{125}$			6m	β^+, EC						
$_{56}Ba^{126}$			97m	EC	~1.3					
$_{56}Ba^{127}$			10m	β^+, EC	~3.6					
$_{56}Ba^{128}$			2.4d	EC	~0.7					
$_{56}Ba^{129+129m}$			~2.5h	β^+, EC	2.45	1.0 1.24 1.43	6† 24 78			
$_{56}Ba^{130}$	0.1%	129.9062						$2.5 \pm 0.3b$ (Ba^{131m}) $11 \pm 3b$ (Ba^{131})		
$_{56}Ba^{131m}$			15m	IT	0.18					
$_{56}Ba^{131}$			12d	EC	1.16					
$_{56}Ba^{132}$	0.1%	131.9057						$8.5 \pm 1.0b$		
$_{56}Ba^{133m}$			38.9h	IT	0.288					
$_{56}Ba^{133}$				EC	0.488					
$_{56}Ba^{134}$	2.4%	133.9043						$0.16 \pm 0.02b$ (Ba^{135m}) $2 \pm 2b$ (Ba^{135})	0	
$_{56}Ba^{135m}$			28.7h	IT	0.268					
$_{56}Ba^{135}$	6.6%	134.9056						$5.8 \pm 0.9b$	3/2	$+0.8365$
$_{56}Ba^{136m}$			0.32s	IT	2.04					
$_{56}Ba^{136}$	7.9%	135.9044						$10 \pm 1mb$ (Ba^{137m}) $0.4 \pm 0.4b$ (Ba^{137})	0*	
$_{56}Ba^{137m}$			2.55m	IT	0.6616					
$_{56}Ba^{137}$	11.2%							$5.1 \pm 0.4b$	3/2	$+0.9357$

Isotope	% nat. abundance	Atomic mass	Lifetime	Modes of decay	Decay energy (MeV)	Particle energies (MeV)	Particle intensities	Thermal neutron capture cross-section	I	μ
$_{56}Ba^{138}$	71.7%	137.9050						$0.35 \pm 0.15b$	0	
$_{56}Ba^{139}$			82.9m	β^-	2.34	0.95 2.23 2.38	<1% 27% 72%	$4 \pm 1b$		
$_{56}Ba^{140}$			12.8d	β^-	1.05	0.47 0.58 0.89 1.01 1.02	34% 11% ~3% 33% 19%	$1.6 \pm 0.3b$		
$_{56}Ba^{141}$			18m	β^-	3.0	2.0 2.4 2.6 2.8 3.0				
$_{56}Ba^{142}$			11m	β^-	2.2	1.0 ~1.7				
$_{56}Ba^{143}$			12s	β^-						
La		138.91						$8.9 \pm 0.2b$		
$_{57}La^{126}$			1.0m	β^+, EC						
$_{57}La^{127}$			3.5m	β^+, EC						
$_{57}La^{128}$			4.4m	β^+, EC						
$_{57}La^{129}$			10m	β^+, EC	~4.0					
$_{57}La^{130}$			8.7m	β^+, EC	~5					
$_{57}La^{131}$			59m	β^+, EC	2.96	0.70 1.42 1.94	17† 56 27			
$_{57}La^{132}$			4.5h	β^+, EC	5.3	3.8				
$_{57}La^{133}$			4.0h	β^+, EC	2.2	~1.2				
$_{57}La^{134}$			6.7m	β^+, EC	3.77	2.7				
$_{57}La^{135}$			19.5h	EC	~1.2	0.81				
$_{57}La^{136}$			9.5m	β^+, EC	2.9	1.8				
$_{57}La^{137}$			6×10^4y	EC	~0.5					
$_{57}La^{138}$	0.09%	137.9068							5	+3.707
$_{57}La^{139}$	99.91%	138.9061						$8.8 \pm 0.7b$	7/2	
$_{57}La^{140}$			$40.22 \pm 0.1h$	β^-	3.769	1.25 1.36 1.42 1.69 2.17	15% 35% 4% 18% 27%	$2.8 \pm 0.3b$	3	
$_{57}La^{141}$			3.9h	β^-	2.43	0.9 2.43	~5% ~95%			
$_{57}La^{142}$			92m	β^-	4.51	0.86 1.06	12% 16%			

Isotope	% nat. abundance	Atomic mass	Lifetime	Modes of decay	Decay energy (MeV)	Particle energies (MeV)	Particle intensities	Thermal neutron capture cross-section	I	μ
						1.20	5%			
						1.79	11%			
						1.98	19%			
						2.11	24%			
						2.31	6%			
						2.33	1%			
						2.98	1.7%			
						3.85	2.4%			
						4.52	12%			
$_{57}La^{143}$			14m	β^-	3.3	3.3				
$_{57}La^{144}$			40s	β^-	5.6					
Ce		140.12						$0.73 \pm 0.08b$		
$_{58}Ce^{132}$			4.2h	EC	~1.2					
$_{58}Ce^{133}$			6.3h	β^+, EC	~2.8	1.3				
$_{58}Ce^{134}$			72h	EC	~0.16					
$_{58}Ce^{135}$			17.2h	β^+, EC	~2.1	0.81				
$_{58}Ce^{136}$	0.2%							$0.95 \pm 0.25b$ (Ce^{137m}) $6.3 \pm 1.5b$ (Ce^{137})		
$_{58}Ce^{137m}$			34.4h	IT EC	0.255				(11/2)	± 0.89
$_{58}Ce^{137}$			9.0h	β^+, EC						
$_{58}Ce^{138}$	0.3%	137.9057						$15 \pm 5mb$ (Ce^{139m}) $1.1 \pm 0.3b$ (Ce^{139})		
$_{58}Ce^{139m}$			55s	IT	0.746					
$_{58}Ce^{139}$			140d	EC	0.27				(3/2)	± 0.9
$_{58}Ce^{140}$	88.4%	139.9053						$0.58 \pm 0.06b$		
$_{58}Ce^{141}$			33d	β^-	0.581	0.444 0.582	60% 40%	$29 \pm 3b$	7/2	± 0.97
$_{58}Ce^{142}$	11.1%	141.9090						$0.95 \pm 0.05b$		
$_{58}Ce^{143}$			33h	β^-	1.44	0.28 0.50 0.72 1.09 1.38	1% 3% 16% 42% 38%	$6.0 \pm 0.7b$	3/2	
$_{58}Ce^{144}$			284.9 $\pm 0.8d$	β^-		0.175 0.24 0.309	24% 4.5% 76%	$1.0 \pm 0.1b$		
$_{58}Ce^{145}$			3.0m	β^-	~2.6	~2.0				
$_{58}Ce^{146}$			14m	β^-	~1.0	0.7				
$_{58}Ce^{147}$			65s	β^-						
$_{58}Ce^{148}$			43s	β^-						
Pr		140.907						$11.5 \pm 1.0b$		
$_{59}Pr^{134}$			17m	β^+ EC						

Isotope	% nat. abundance	Atomic mass	Lifetime	Modes of decay	Decay energy (MeV)	Particle energies (MeV)	Particle intensities	Thermal neutron capture cross-section	I	μ
$_{59}Pr^{135}$			22m	β^+, EC		2.5				
$_{59}Pr^{136}$			1.1h	β^+, EC		~2.0				
$_{59}Pr^{137}$			1.5h	β^+, EC	2.7	1.7				
$_{59}Pr^{138}$			2.1h	β^+, EC	4.79	1.65				
$_{59}Pr^{139}$			4.5h	β^+, EC	2.11	1.09				
$_{59}Pr^{140}$			3.39m	β^+, EC	3.34	2.32				
$_{59}Pr^{141}$	100%	140.9074						$3.9 \pm 0.5b$	5/2	+4.3
$_{59}Pr^{142}$			19.2h	β^-	2.16	0.64 2.15	4% 96%	$18 \pm 3b$	2	±0.26
$_{59}Pr^{143}$			13.7d	β^-	0.933	0.933	100%	$89 \pm 10b$	7/2	
$_{59}Pr^{144}$			17.3m	β^-	2.989	2.99 0.80 2.29	98% 1% 1.2%			
$_{59}Pr^{145}$			5.98h	β^-	1.80	1.80	95%			
$_{59}Pr^{146}$			24m	β^-	4.2	3.7 2.3	56% 44%			
$_{59}Pr^{147}$			12.0m	β^-	2.7	1.0 1.4 2.1 2.7	≤15% 30% 50% ≤ 5%			
$_{59}Pr^{148}$			2.0m	β^-	4.5					
Nd		144.24						$49 \pm 2b$		
$_{60}Nd^{138}$			22m	β^+	3.4	~2.4				
$_{60}Nd^{139m}$			5.5h	IT β^+	0.232					
$_{60}Nd^{139}$			5.2h	EC β^+	4.1	1.0 3.3	8† 100			
$_{60}Nd^{140}$			3.3d	EC						
$_{60}Nd^{141}$			2.5h	EC β^+	1.80	EC β^+ 0.78	95% 5%			
$_{60}Nd^{141m}$			64s	IT	0.755				3/2	
$_{60}Nd^{142}$	27.2%	141.9075						$18.8 \pm 0.7b$		
$_{60}Nd^{143}$	12.2%	142.9096						$330 \pm 10b$ $20 \pm 2mb$ (η, α)	7/2	−1.08
$_{60}Nd^{144}$	23.8%	143.9099	$\approx 5 \times 10^{15}$y	α		1.8		$4.0 \pm 0.5b$		
$_{60}Nd^{145}$	8.3%	144.9122						$50 \pm 4b$	7/2	−0.66
$_{69}Nd^{146}$	17.2%	145.9127						$1.4 \pm 0.2b$		
$_{60}Nd^{147}$			11.1d	β^-	0.91	0.82 0.38	77% 17%		5/2	±0.59

Isotope	% nat. abundance	Atomic mass	Lifetime	Modes of decay	Decay energy (MeV)	Particle energies (MeV)		Particle intensities	Thermal neutron capture cross-section	I	μ
$_{60}Nd^{148}$	5.7%	147.9165							2.5 ± 0.2b		
$_{60}Nd^{149}$			1.73h	β^-	1.66	1.55 1.45 1.40 1.12 1.01	1.01 1.12 1.40 1.45 1.55	17% 18% 18% 25% 11%		5/2	
$_{60}Nd^{150}$	5.6%	149.9207							1.3 ± 0.3b		
$_{60}Nd^{151}$			10m	β^-	2.31	2.14 2.06 1.99 1.78 1.47		17% 40% 9% 11% 16%			
Pm											
$_{61}Pm^{141}$			22m	β^+	3.6	2.6					
$_{61}Pm^{142}$			34s	β^+	4.82	β^+, 3.8 EC		69% 31%			
$_{61}Pm^{143}$			265d	EC	1.1	EC					
$_{61}Pm^{144}$			$\approx$400d	EC		EC					
$_{61}Pm^{145}$		144.913	17.7y	EC	0.14	EC					
$_{61}Pm^{146}$			$\approx$710d	EC β^-	1.72 1.53	EC β^-, 0.78		35%	8400 ± 1680b		
$_{61}Pm^{147}$		146.915	2.5y	β^-	0.225	0.225		100%	90 ± 30b (Pm^{110m}) 100 ± 30b (Pm^{148})	7/2	± 2.7
$_{61}Pm^{148m}$			42d	IT β^-	2.59	β^- 0.5 0.6 0.68 0.8 IT		37% 19% 8% 24% 6%	$25,000 \pm 4000$b		
$_{61}Pm^{148}$			5.39d	β^-	2.46	1.1 2.0 2.6		41% 14% 45%		1	+ 2.0
$_{61}Pm^{149}$			53.1h	β^-	1.06	0.19 0.47 0.78 1.06		0.6% 0.3% 10% 89%	1350 ± 400b	7/2	
$_{61}Pm^{150}$			2.7h	β^-	3.43	1.5 1.8 2.3 2.8 3.2					
$_{61}Pm^{151}$			28h	β^-	1.19	0.35 0.50 0.73 0.84 0.98 1.05 1.13 1.19		11% 6% 10% 43% 3% 11% 6% 10%	<700b	5/2	± 1.8
$_{61}Pm^{152}$			6m	β^-		2.2					
$_{61}Pm^{154}$			2.5m	β^-		2.5					
Sm		150.35							5820 ± 100b		

Isotope	% nat. abundance	Atomic mass	Lifetime	Modes of decay	Decay energy (MeV)	Particle energies (MeV)	Particle intensities	Thermal neutron capture cross-section	I	μ
$_{62}Sm^{142}$			72m	EC β^+		EC β^+ 1.03	90% 10%			
$_{62}Sm^{143m}$			64s	IT	0.748					
$_{62}Sm^{143}$			8.9m	EC β^+	3.3	EC 2.5	50% 50%			
$_{62}Sm^{144}$	3.1%	143.9117						≈ 0.7b		
$_{62}Sm^{145}$			340d	EC	0.65			≈ 110b		
$_{62}Sm^{146}$		146.9129	7×10^7y	α		2.55				
$_{62}Sm^{147}$	15.1%	146.9146	1.06×10^{11}y	α	2.314	2.314		75 ± 11b	7/2	−0.813
$_{62}Sm^{148}$	11.3%	147.9146	1.2×10^{13}y	α	2.001	2.04		2.4 ± 0.3b		
$_{62}Sm^{149}$	13.9%	148.9169	$\sim 4 \times 10^{14}$y	α	1.90	1.84		$41,000 \pm 2000$b 50 ± 10mb (η, α)	7/2	−0.66
$_{62}Sm^{150}$	7.4%	149.9170						102 ± 5b		
$_{62}Sm^{151}$			93 ± 8y	β^-	0.076	0.055 0.076	1.7% 98.3%	$15,000 \pm 1800$b		
$_{62}Sm^{152}$	26.6%	151.9195						210 ± 10b		
$_{62}Sm^{153}$			46.8 ± 0.1h	β^-	0.801	0.679 0.698 0.801	32% 48% 20%		3/2	−0.03
$_{62}Sm^{154}$	22.6%	153.9220						5.5 ± 1.1b		
$_{62}Sm^{155}$			22m	β^-	1.65	1.65 1.50	93% 7%		3/2	
$_{62}Sm^{156}$			9.4h	β^-	0.72	0.43 0.70	44% 51%			
$_{62}Sm^{157}$			0.5m	β^-						
Eu		151.96						4100 ± 100b		
$_{63}Eu^{144}$			10s	β^+, EC	6.32					
$_{63}Eu^{145}$			5.9d	β^+, EC						
$_{63}Eu^{146}$			4.6d	β^+, EC	3.86	0.80 1.47 2.1	0.4% 3.3% 0.14%			
$_{63}Eu^{147}$			22d	β^+, EC	1.8					
$_{63}Eu^{148}$			54d	β^+, EC	3.11	0.92	0.13%			
$_{63}Eu^{149}$			106d	EC	0.80	0.25 0.52 0.78 0.80	5% 4% 5% 85%			
$_{63}Eu^{150m}$			12.8h	β^+, EC β^-	2.25 1.01	β^- 1.01 β^+ 1.24				
$_{63}Eu^{150}$			5y	EC	2.25					

Isotope	% nat. abundance	Atomic mass	Lifetime	Modes of decay	Decay energy (MeV)	Particle energies (MeV)	Particle intensities	Thermal neutron capture cross-section	I	μ
$_{63}Eu^{151m}$			$58\mu s$	IT	0.196				7/2	+2.57
$_{63}Eu^{151}$	47.8%	150.9196						$3.8 \pm 1.9b$ (Eu^{152m_2}) $3000 \pm 200b$ (Eu^{152m_1}) $5000 \pm 300b$ (Eu^{152})	5/2	+3.464
$_{63}Eu^{152m_2}$			96m	IT	0.147					
$_{63}Eu^{152m_1}$			9.2h	β^- β^+, EC	1.82 1.857	β^- 0.55 1.88 β^+ 0.77 0.895	10† 0.004% 0.007%			
$_{63}Eu^{152}$			13y	β^+, EC β^-	1.86 1.82	β^- 0.22 0.36 0.68 1.05 1.46 β^+ 0.47 0.713	9† 13 51 6 21 0.005% 0.014%		3	±1.924
$_{63}Eu^{153}$	52.2%	152.9209						$450 \pm 100b$	5/2	+1.530
$_{63}Eu^{154}$			16y	β^-	1.97	β^- 0.27 0.59 0.89 1.86	20% 45% 23% 12%	$1500 \pm 400b$	3	±2.000
$_{63}Eu^{155}$			1.81y	β^-	0.240	0.10 0.15 0.19 0.25	34% 40% 10% 16%	$11,000 \pm 1000b$		
$_{63}Eu^{156}$			15d	β^-	2.45	0.25 0.26 0.27 0.43 0.49 1.08 1.21 1.28 2.45	8% 9% 2.2% 8% 33% 2.3% 1.3% 6% 31%			
$_{63}Eu^{157}$			15.2h	β^-	1.34	0.55 0.65 0.86 0.90 1.28 1.35	2% 10% 35% 15% 30% 3%			
$_{63}Eu^{158}$			46m	β^-	3.5	1.11 1.9 2.5				
$_{63}Eu^{159}$			18m	β^-	2.57	1.0 1.5 1.75 1.90 2.35 2.57	10% 11% 11% 21% 21% 25%			
$_{63}Eu^{160}$			$\approx 2.5m$	β^-	3.6					
Gd		157.25						$49,000 \pm 2000b$		

Isotope	% nat. abundance	Atomic mass	Lifetime	Modes of decay	Decay energy (MeV)	Particle energies (MeV)	Particle intensities	Thermal neutron capture cross-section	I	μ
$_{64}Gd^{145}$			25m	EC β^+	5.2	EC β^+ 2.4	55% 45%			
$_{64}Gd^{146}$			4.6d	EC	~1.1					
$_{64}Gd^{147}$			35h	β^+, EC	~2.2					
$_{64}Gd^{148}$		147.9177	≈130y	α	3.27	3.18				
$_{64}Gd^{149}$		148.9189	9d	EC, α	3.17	α 3.01	0.0005%			
$_{64}Gd^{150}$		149.9185	2.1×10^6y	α	2.80	2.73	100%			
$_{64}Gd^{151}$			120d	EC, α	~0.4 2.61	2.60	~8 × 10^{-7}%			
$_{64}Gd^{152}$	0.2%	151.9195	1.1×10^{14}y	α	2.24	2.15	100%	<125b		
$_{64}Gd^{153}$			242d	EC	0.243					
$_{64}Gd^{154}$	2.2%	153.9207						102 ± 7b		
$_{64}Gd^{155}$	14.8%	154.9226						$61,000 \pm 1000$b	3/2	−0.27
$_{64}Gd^{156}$	20.5%	155.9221								
$_{64}Gd^{157}$	15.7%	156.9339						$254,000 \pm 2000$b	3/2	−0.36
$_{64}Gd^{158}$	24.8%	157.9241						3.5 ± 1.0b		
$_{64}Gd^{159}$			18h	β^-	0.95	0.59 0.89 0.95	13% 24% 63%		3/2	
$_{64}Gd^{160}$	21.8%	159.9271						0.77 ± 0.01b		
$_{64}Gd^{161}$			3.7m	β^-	2.02	1.5 1.6		$(9.6 \pm 5) \times 10^4$b		
Tb		158.924						30 ± 10b		
$_{65}Tb^{147}$			24m	β^+, EC						
$_{65}Tb^{148}$			70m	β^+, EC	5.6	2.6 4.6				
$_{65}Tb^{149m}$			4.3m	EC α		3.99	0.025%			
$_{65}Tb^{149}$			4.1h	EC α	4.06	EC α 2.95	84% 16%			
$_{65}Tb^{150}$			3.1h	β^+, EC	4.79	1.5 2.8 3.6	5† 5 10			
$_{63}Tb^{151}$		150.9230	18h	EC α	~2.7 3.49	EC α 3.44	99 + % 0.0005%			
$_{65}Tb^{152m}$			4m	β^+, EC α		α	0.002%			
$_{65}Tb^{152}$			18h	β^+, EC	4.2	0.42 0.85 1.30 1.93 2.82	1† 4 14 45 100			
$_{65}Tb^{153}$			55h	EC	~1.6					
$_{65}Tb^{154}$			8.5h	EC						

Isotope	% nat. abundance	Atomic mass	Lifetime	Modes of decay	Decay energy (MeV)	Particle energies (MeV)	Particle intensities	Thermal neutron capture cross-section	I	μ
$_{65}Tb^{154}$			21h	β^+, EC		EC	99 + %			
$_{65}Tb^{155}$			5.6d	EC	0.9					
$_{65}Tb^{156m}$			5.5h	IT EC						
$_{65}Tb^{156}$			5.4d	EC					(3)	±1.4
$_{65}Tb^{157}$			150y	EC	0.06		100%			
$_{65}Tb^{158m}$			11s	IT	0.111					
$_{65}Tb^{158}$			1.2×10^3y	EC β^-	1.20 0.95	0.64 0.86	1% 13%		3	±1.74
$_{65}Tb^{159}$	100%	159.9250						30 ± 10b	3/2	+1.95
$_{65}Tb^{160}$			73d	β^-	1.72	0.32 0.36 0.43 0.46 0.68 0.95 1.55 1.74	5% 10% 4% 44% 7% 29% 0.12% 0.34%	525 ± 100b	3	
$_{65}Tb^{161}$			6.9d	β^-	0.58	0.04 0.17 0.21 0.45 0.57 0.38	0.07% 0.03% 0.11% 26% 64% 10%		3/2	
$_{65}Tb^{162}$			7.5m	β^-	~2.8					
$_{65}Tb^{163m}$			7m	β^-						
$_{65}Tb^{163}$			6.5h	β^-	1.68	1.1 1.4 1.5 1.65	15% 40% 15% 30%			
$_{65}Tb^{164}$			23h	β^-	~3.8					
Dy		162.50						930 ± 20b		
$_{66}Dy^{149}$			≈15m	EC						
$_{66}Dy^{150}$			7.2m	α β^+, EC		α 4.2	18% 82%			
$_{66}Dy^{151}$			18m	β^+, EC α	~2.9 4.1	4.06	6%			
$_{66}Dy^{152}$		151.9244	2.5h	β^+, EC α	0.4 3.75	α 3.7	0.05%			
$_{66}Dy^{153}$		152.9254	6.4h	β^+, EC α	~2.1 3.57	α 3.48				
$_{66}Dy^{154m}$			13h	α		3.37				
$_{66}Dy^{154}$		153.9248	≈10^6y	α	2.93	2.85				
$_{66}Dy^{155}$			10.2h	β^+, EC β^+, EC		0.85 1.08	2% 0.14%			
$_{66}Dy^{156}$	0.06 %	155.9238								
$_{66}Dy^{157}$			8.1h	EC						

Isotope	% nat. abundance	Atomic mass	Lifetime	Modes of decay	Decay energy (MeV)	Particle energies (MeV)	Particle intensities	Thermal neutron capture cross-section	I	μ
$_{66}$Dy^{158}	0.1%	157.9240						96 ± 20b		
$_{66}$Dy^{159}			144d	EC	0.38				3/2	
$_{66}$Dy^{160}	2.34%	159.9248						55 ± 9b		
$_{66}$Dy^{161}	18.9%	160.9266						585 ± 50b	5/2	−0.46
$_{66}$Dy^{162}	25.5%	161.9265						200 ± 50b		
$_{66}$Dy^{163}	24.9%	162.9284						140 ± 30b	5/2	+0.64
$_{66}$Dy^{164}	28.2%	163.9288						2100 ± 400b (Dy^{165m},4) 2600 ± 200b (Dy^{165m} + Dy^{165},4)		
$_{66}$Dy^{165m}			75s	IT β^-	0.108 1.30	β^- 0.89 1.0	87† 13			
$_{66}$Dy^{165}			2.3h	β^-	1.30	0.22 0.24 0.3 1.25 1.30	0.1% 0.03% 1.3% 15% 83%		7/2	
$_{66}$Dy^{166}			81.5h	β^-	0.482	0.05 0.40 0.482	1% 94% 5%			
$_{66}$Dy^{167}			4.4m	β^-						
Ho		164.930						65 ± 2b		
$_{67}$Ho^{151}			42s	α β^+, EC		4.60	30%			
$_{67}$Ho^{151}			36s	α β^+, EC		4.51	20%			
$_{67}$Ho^{152}			2.4m	α β^+, EC	4.50 6.4	4.38	30%			
$_{67}$Ho^{152}			52s	α β^+, EC	4.57	4.45	19%			
$_{67}$Ho^{153}			9m	α	4.02	3.92	0.3%			
$_{67}$Ho^{154}			7m	β^+, EC	~5.5					
$_{67}$Ho^{155}			50m	β^+, EC		2.1				
$_{67}$Ho^{156}			55m	β^+, EC	≈5	1.3 1.80 2.9	4† 14 1			
$_{67}$Ho^{157}			14m	β^+, EC						
$_{67}$Ho^{158m}			29m	IT β^+. EC	0.0673 4.1	1.32				
$_{67}$Ho^{158}			11m	β^+, EC	4.04					
$_{67}$Ho^{159m}			6.9s	IT	~2.0					
$_{67}$Ho^{159}			33m	EC	~1.8					
$_{67}$Ho^{160m}			5.0h	IT	3.36		66%			

Isotope	% nat. abundance	Atomic mass	Lifetime	Modes of decay	Decay energy (MeV)	Particle energies (MeV)	Particle intensities	Thermal neutron capture cross-section	I	μ
				β^+, EC		0.31 0.57 1.01	0.01% 0.2% 0.08%			
$_{67}Ho^{160}$			23m	β^+, EC	3.30	1.9	0.02%			
$_{67}Ho^{161m}$			6s	IT	0.211					
$_{67}Ho^{161}$			2.5h	EC	~0.8				7/2	
$_{67}Ho^{162m}$			68m	IT EC	~2.26		63%			
$_{67}Ho^{162}$			15m	β^+, EC	2.16	~1.06 ~1.14	$\sim56\%$ $\sim44\%$			
$_{67}Ho^{163m}$			1.1s	IT	0.305					
$_{67}Ho^{163}$			$>10^3$y	EC	~0.01					
$_{67}Ho^{164}$			37m	β^- EC	1.03 1.11	0.99	53% 47%			
$_{67}Ho^{165}$	100%	164.9303						3.45 ± 0.35b (Ho^{166m}) 61.2 ± 2.0b (Ho^{166})	7/2	+4.08
$_{67}Ho^{166}$			26.9h	β^-	1.847	0.18 0.38 1.76 1.84	0.3% 1% 47% 52%		0*	
$_{67}Ho^{166m}$			1.2×10^3y	β^-	1.847	≤0.067				
$_{67}Ho^{167}$			3.1h	β^-	1.0	0.28 0.96	50% 50%			
$_{67}Ho^{168}$			3.3m	β^-	3.3	≈2.2				
$_{67}Ho^{169}$			4.8m	β^-	2.1	1.20 1.95	75% 25%			
$_{67}Ho^{170}$			45s	β^-	4.2	≈3.1				
Er		167.26						160 ± 30b		
$_{68}Er^{158}$			2.5h	β^+, EC	0.8					
$_{68}Er^{159}$			36m	β^+, EC						
$_{68}Er^{160}$			29h	EC	~0.8					
$_{68}Er^{161}$			3.1h	β^+, EC	2.4	1.2				
$_{68}Er^{162}$	0.1%	161.9288						160 ± 30b		
$_{68}Er^{163}$			75m	β^+, EC	1.21	0.19	0.004%			
$_{68}Er^{164}$	1.6%	163.9293						13 ± 5b		
$_{68}Er^{165}$			10.3h	EC	0.37				5/2	±0.65
$_{68}Er^{166}$	33.4%	165.9304						10 ± 5b (Er^{167m}) 30 ± 5b ($Er^{167m} + Er^{167}$)		
$_{68}Er^{167m}$			2.3s	IT	0.208					

Isotope	% nat. abundance	Atomic mass	Lifetime	Modes of decay	Decay energy (MeV)	Particle energies (MeV)	Particle intensities	Thermal neutron capture cross-section	I	μ
$_{68}Er^{167}$	22.9%	166.9320						700 ± 50b	7/2	−0.564
$_{68}Er^{168}$	27.0%	167.9324						1.9 ± 0.2b		
$_{68}Er^{169}$			9.4d	β^-	0.34	0.332 0.340	42% 58%		1/2	+0.513
$_{68}Er^{170}$	15.0%	169.9355						6 ± 1b		
$_{68}Er^{171}$			7.52h	β^-	1.490	0.50 0.54 0.59 0.77 0.83 1.065 1.49	0.4% 0.5% 3.6% 0.4% 0.2% 91% 2.3%		5/2	±0.70
$_{68}Er^{172}$			49h	β^-	0.91	0.30 0.38 0.43 0.91	42% 44% 12% <10%			
Tm		168.934						115 ± 15b		
$_{69}Tm^{161}$			30m	EC	3.5					
$_{69}Tm^{162}$			77m	β^+, EC	4.89					
$_{69}Tm^{162}$			22m	β^+, EC	4.89	0.9 2.3 3.82	3† 12 10			
$_{69}Tm^{163}$			1.8h	β^+, EC	2.27	0.40 1.1	7† 100		1/2	
$_{69}Tm^{164}$			1.9m	β^+, EC	~3.8	1.3 2.94	4† 100			
$_{69}Tm^{165}$			30.1h	β^+, EC	1.57	0.1	~0.004%			
$_{69}Tm^{166}$			7.7h	β^+, EC	3.04	1.10 1.94	0.3% 1.5%		2	±0.05
$_{69}Tm^{167}$			9.6d	EC	1.16				1/2	
$_{69}Tm^{168}$			87d	EC	1.72					
$_{69}Tm^{169}$	100%	168.9344						115 ± 15b	1/2	−0.230
$_{69}Tm^{170}$			128.6 ± 0.3d	β^-	0.967	0.886 0.97	23% 77%	92 ± 4b	1	±0.25
$_{69}Tm^{171}$			1.92y	β^-	0.098	0.097 0.030	98% 2%	4.5 ± 0.2b	1/2	±0.23
$_{69}Tm^{172}$			63.6h	β^-	1.88	0.13 0.22 0.28 0.33 0.41 0.71 1.62 1.80 1.88	~0.05% 1% 10% 4% 14% 8% 1% 39% 23%			
$_{69}Tm^{173}$			8.2h	β^-	1.32	0.80 0.84 1.26	23% 75% 2%			

Isotope	% nat. abundance	Atomic mass	Lifetime	Modes of decay	Decay energy (MeV)	Particle energies (MeV)	Particle intensities	Thermal neutron capture cross-section	I	μ
$_{69}Tm^{174m}$			5.2m	β^-	3.0	0.7 1.2	~20% ~80%			
$_{69}Tm^{174}$			5.5m	β^-	2.5	2.5	100%			
$_{69}Tm^{175}$			20m	β^-	2.5	2.0				
$_{69}Tm^{176}$			1.5m	β^-	4.2	4.2				
Yb		173.04						$37 \pm 3b$		
$_{70}Yb^{164}$			76m	EC						
$_{70}Yb^{165}$			10m	β^+, EC	~2.7					
$_{70}Yb^{166}$			57.5h	EC	0.3					
$_{70}Yb^{167}$			18m	β^+, EC	1.96	0.65	0.4%			
$_{70}Yb^{168}$	0.1%	167.9339						$3200 \pm 400b$		
$_{70}Yb^{169}$			32d	EC	~1.2					
$_{70}Yb^{170}$	3.1%	169.9349						$9.4 \pm 0.9b$		
$_{70}Yb^{171}$	14.3%	170.9365						$50 \pm 5b$	1/2	+0.4919
$_{70}Yb^{172}$	21.9%	171.9366						$0.4 \pm 0.1b$		
$_{70}Yb^{173}$	16.2%	172.9383						$19 \pm 2b$	5/2	−0.6776
$_{70}Yb^{174}$	31.7%	173.9390						$16 \pm 4b$ (Yb^{175m}) $65 \pm 5b$ $(Yb^{175m} + Yb^{175})$		
$_{70}Yb^{175m}$			0.067s	IT	0.513					
$_{70}Yb^{175}$			101h	β^-	0.467	0.073 0.22 0.35 0.466	11% <0.1% 2.1% 87%		(7/2)	±0.13
$_{70}Yb^{176}$	12.7%	175.9427						$5.5 \pm 1.0b$		
$_{70}Yb^{177m}$			6.5s	IT	0.33					
$_{70}Yb^{177}$			1.9h	β^-	1.40	0.17 0.18 0.25 1.1 1.25 1.3 1.40	3% 6% 1% 2% 17% 9% 60%			
Lu		174.97						$75 \pm 2b$		
$_{71}Lu^{167}$			54m	β^+, EC	3.0	1.1 1.5	1† 1			
$_{71}Lu^{168}$			7.1m	β^+, EC	~4.6	1.20				
$_{71}Lu^{169m}$			2.7m	IT	0.02%					
$_{71}Lu^{169}$			34h	β^+, EC	2.26	0.4 1.2	17† 100			
$_{71}Lu^{170m}$			0.7s	IT	0.093					

Isotope	% nat. abundance	Atomic mass	Lifetime	Modes of decay	Decay energy (MeV)	Particle energies (MeV)	Particle intensities	Thermal neutron capture cross-section	I	μ
$_{71}Lu^{170}$			2.0d	β^+, EC	3.41	2.39	0.19%			
$_{71}Lu^{171m}$			76s	IT	0.071					
$_{71}Lu^{171}$			8.3d	β^+, EC	~1.6					
$_{71}Lu^{172m}$			3.7m	IT	0.041	γ Energies 41.86				
$_{71}Lu^{172}$			6.70d	EC	~2.7					
$_{71}Lu^{173}$			≈1.37y	EC	0.69					
$_{71}Lu^{174m}$			140d	IT EC	0.171 ~1.7					
$_{71}Lu^{174}$			3.6y	EC	1.5					
$_{71}Lu^{175}$	97.4%	174.9409						21 ± 3b	7/2	+2.23
$_{71}Lu^{176m}$			3.7h	β^-	1.3	1.22 1.31	~65% ~35%		1	+0.318
$_{71}Lu^{176}$	2.6%		3×10^{10}y	β^-	1.02	0.42	100%	7 ± 2b (Lu^{177m}) 2050 ± 50b (Lu^{177})	7	+3.18
$_{71}Lu^{177m}$			155d	IT β^-	0.97		22% 78%	gamma rays in equilibrium with Lu^{177} (* with IT)		
$_{71}Lu^{177}$			6.7d	β^-	0.497	0.17 0.38 0.497	6.7% 7% 86%		7/2	+2.24
$_{71}Lu^{178}$			20m	β^-		1.5				
$_{71}Lu^{178}$			30m	β^-	2.25	2.25				
$_{71}Lu^{179}$			4.6h	β^-	1.34	1.1 1.35	13% 87%			
$_{71}Lu^{180}$			2.5m	β^-	3.3	3.3				
Hf		178.49						103 ± 3b		
$_{72}Hf^{168}$			22m	β^+, EC		EC β^+ 1.7				
$_{72}Hf^{169}$			1.5h	β^+, EC	~1.3					
$_{72}Hf^{170}$			12.2h	β^+, EC						
$_{72}Hf^{171}$			11h	EC						
$_{72}Hf^{172}$			≈5y	EC	~1					
$_{72}Hf^{173}$			23.6h	EC	>2					
$_{72}Hf^{174}$	0.2%	173.9403	2.0×10^{15}y	α	2.55	≈2.55		390 ± 55b		
$_{72}Hf^{175}$			70d	EC	0.59					
$_{72}Hf^{176}$	5.2%	175.94165						15 ± 15b		
$_{72}Hf^{177}$	18.5%	176.9435						1.1 ± 0.1b (Hf^{178m})	7/2	+0.61

Isotope	% nat. abundance	Atomic mass	Lifetime	Modes of decay	Decay energy (MeV)	Particle energies (MeV)	Particle intensities	Thermal neutron capture cross-section	I	μ
								380 ± 30b (Hf^{178})		
$_{72}Hf^{178m_2}$			>10y	IT	2.5					
$_{72}Hf^{178m_1}$			5s	IT	1.148					
$_{72}Hf^{178}$	27.1%	177.9439						52 ± 6b (Hf^{179m}) 75 ± 10b ($Hf^{179m} + Hf^{179}$)	0	
$_{72}Hf^{179m}$			18.6s	IT	0.375					
$_{72}Hf^{179}$	13.8%	178.9460						0.34 ± 0.03b (Hf^{180m}) 65 ± 15b (Hf^{180})	9/2	−0.47
$_{72}Hf^{180m}$			5.5h	IT	1.143					
$_{72}Hf^{180}$	35.2%	179.9468						12.6 ± 0.7b	0	
$_{72}Hf^{181}$			42.4 ± 0.2d	β^-	1.023	0.4 0.408	7% 93%	$40 {+40 \atop -20}$b		
$_{72}Hf^{182}$			9×10^6y	β^-	~0.5	~0.5				
$_{72}Hf^{183}$			1.1h	β^-	2.2	0.7 1.2 1.6	25% 50% 25%			
Ta		180.948						22 ± 1b		
$_{73}Ta^{172}$			44m	β^+, EC	~5					
$_{73}Ta^{173}$			3.7h	β^+, EC						
$_{73}Ta^{174}$			1.3h	β^+, EC						
$_{73}Ta^{175}$			10.5h	EC	~2.3					
$_{73}Ta^{176}$			8.0h	EC	~3					
$_{73}Ta^{177}$			56.6h	EC	1.17					
$_{73}Ta^{178}$			9.4m	β^+, EC	1.9	0.80 0.89	7† 10			
$_{73}Ta^{178}$			2.1h	β^+, EC	1.9	1				
$_{73}Ta^{179}$			~600y	EC	0.115					
$_{73}Ta^{180m}$			8.1h	EC β^-	0.7 0.50	0.61 0.71	3% 10%			
$_{73}Ta^{180}$	0.012%	179.9415	$>10^{13}$y	EC, β^-						
$_{73}Ta^{181m}$			$6.8\mu s$	IT	0.0063					
$_{73}Ta^{181}$	99.988%	180.9480						10 ± 2mb (Ta^{182m}) 22 ± 1b (Ta^{182})	7/2	+2.36

Isotope	% nat. abundance	Atomic mass	Lifetime	Modes of decay	Decay energy (MeV)	Particle energies (MeV)	Particle intensities	Thermal neutron capture cross-section	I	μ
$_{73}\text{Ta}^{182m}$			16.5m	IT	0.503					
$_{73}\text{Ta}^{182}$			115d	β^-	1.811	0.17 0.23 0.35 0.37 0.43 0.50 1.48 1.71	29% 3% 2.2% ~2% 42% 0.9% 0.24% 0.05%	8200 ± 600b		
$_{73}\text{Ta}^{183}$			5.1d	β^-	1.068	0.48 0.62 0.66 0.76 0.78	0.7% 84% 6% 7% ~3%		7/2	
$_{73}\text{Ta}^{184}$			8.7h	β^-	2.75	1.19 1.45 1.76 2.22 2.64	93% 6% 0.9% 0.1% 0.15%			
$_{73}\text{Ta}^{185}$			50m	β^-	2.0	0.15 1.7	~30% ~70%			
$_{73}\text{Ta}^{186}$			10.5m	β^-	3.7	2.2				
W		183.85						18.5 ± 0.5b		
$_{74}\text{W}^{173}$			16m	EC						
$_{74}\text{W}^{174}$			31m	EC						
$_{74}\text{W}^{175}$			34m	EC						
$_{74}\text{W}^{176}$			2.5h	β^+, EC	≈ 1.5	~2				
$_{74}\text{W}^{177}$			135m	EC	≈ 2					
$_{74}\text{W}^{178}$			21.5d	EC	≈ 0.3					
$_{74}\text{W}^{179m}$			5.2m	IT	0.222					
$_{74}\text{W}^{179}$			38m	EC	≈ 1.2					
$_{74}\text{W}^{180m}$			0.005s	IT						
$_{74}\text{W}^{180}$	0.1%	179.9470						10 ± 10b		
$_{74}\text{W}^{181m}$			14μs	IT						
$_{74}\text{W}^{181}$			140d	EC	0.19					
$_{74}\text{W}^{182}$	26.3	181.9483						20.7 ± 0.5b	0	
$_{74}\text{W}^{183m}$			5.3s	IT	0.309					
$_{74}\text{W}^{183}$	14.3%	182.9503						10.2 ± 0.3b	1/2	+0.1172
$_{74}\text{W}^{184}$	30.7%	183.9510						2.4 ± 0.4mb (W^{185m}) 1.8 ± 0.2b (W^{185})	0	
$_{74}\text{W}^{185m}$			1.62m	IT	0.368					
$_{74}\text{W}^{185}$			75.8d	β^-	0.432	0.432	100%		3/2	
$_{74}\text{W}^{186}$	28.6%	185.9543						37 ± 2b	0	

Isotope	% nat. abundance	Atomic mass	Lifetime	Modes of decay	Decay energy (MeV)	Particle energies (MeV)	Particle intensities	Thermal neutron capture cross-section	I	μ
$_{74}W^{187}$			24h	β^-	1.315	0.43 0.47 0.55 0.63 ~0.71 1.316	0.5% ~1% 4.3% 59% 8% 19%	$90 \pm 40b$	3/2	
$_{74}W^{188}$			69d	β^-	0.349	0.059 0.285 0.349	0.9% 0.6% 98.6%			
$_{74}W^{189}$			11.5m	β^-	2.5	2.0 2.5				
Re		186.2						$85 \pm 5b$		
$_{75}Re^{177}$			17m	β^+, EC	≈ 3	>0.4				
$_{75}Re^{178}$			15m	β^+, EC	≥ 4.1	3.1				
$_{75}Re^{179}$			20m	EC	≈ 2.5					
$_{75}Re^{180}$			2.4m	β^+, EC		1.1				
$_{75}Re^{180}$			20h	β^+, EC	2.9	≈ 1.9				
$_{75}Re^{181}$			18h	EC	≈ 1.7					
$_{75}Re^{182}$			12.7h	β^+, EC	2.86	0.55 1.74	0.06% 0.2%			
$_{75}Re^{182}$			64h	EC						
$_{75}Re^{183}$			71d	EC	~0.9					
$_{75}Re^{184m}$			169d	IT	0.188					
$_{75}Re^{184}$			3.8d	EC	~1.6					
$_{75}Re^{185}$	37.4%	184.9530						$110 \pm 5b$	5/2	+3.172
$_{75}Re^{186}$			90h	β^-	1.071	0.30 0.933 1.071	0.1% 21% 74%		1	+1.73
$_{75}Re^{187}$	62.6%	186.9560	7×10^{10}y	β^-	<0.01	≤ 0.008		$2.0 \pm 0.5b$ (Re^{184m}) $75 \pm 4b$ (Re^{184})	5/2	+3.204
$_{75}Re^{188m}$			18.7m	IT	0.172					
$_{75}Re^{188}$			16.7h	β^-	2.116	0.21 0.23 0.30 0.37 0.41 0.56 0.71 0.715 0.77 1.09 1.54 1.96 2.116	0.05% 0.06% 0.02% 0.026% 0.14% 0.012% 0.35% 0.026% 0.025% 0.5% 1.2% 18% 80%	<2b	1	+1.78
$_{75}Re^{189}$			24h	β^-	1.00	0.72 0.80 1.02				

Isotope	% nat. abundance	Atomic mass	Lifetime	Modes of decay	Decay energy (MeV)	Particle energies (MeV)	Particle intensities	Thermal neutron capture cross-section	I	μ
$_{75}Re^{190}$			2.8m	β^-	3.1	1.7				
$_{75}Re^{190m}$			2.8h	IT						
$_{75}Re^{191}$			9.8m	β^-	1.8	1.8				
$_{75}Re^{192}$			6s	β^-		2.5				
Os		190.2						$15.3 \pm 0.7b$		
$_{76}Os^{181}$			23m	EC	≈ 2.8					
$_{76}Os^{182}$			22h	EC	≈ 1.1					
$_{76}Os^{183m}$			9.9h	IT EC	0.171 ≈ 2.2	IT EC	46% 54%			
$_{76}Os^{183}$			12h	EC	≈ 2.0					
$_{76}Os^{184}$	0.02%	183.9526						$< 200b$		
$_{76}Os^{185}$			94d	EC	0.982					
$_{76}Os^{186}$	1.58%	185.9539								
$_{76}Os^{187}$	1.6%	186.9560							1/2	+0.0643
$_{76}Os^{188}$	13.3%	187.9560								
$_{76}Os^{189m}$			5.7h	IT	0.03					
$_{76}Os^{189}$	16.1%	188.9586						$0.26 \pm 0.03mb$ (Os^{190m})	3/2	+0.6566
$_{76}Os^{190m}$			9.9m	IT	1.706					
$_{76}Os^{190}$	26.4%	189.9586						$12 \pm 6b$ (Os^{191m}) $4 \pm 2b$ (Os^{191})		
$_{76}Os^{191m}$			13h	IT	0.074					
$_{76}Os^{191}$			15d	β^-	0.310	0.13				
$_{76}Os^{192}$	41.0%									
$_{76}Os^{193}$			31h	β^-	1.132	0.57 0.67 0.85 0.99 1.059	4% 7% 1% 9% 6%			
$_{76}Os^{194}$			6.0y	β^-	0.097	0.009 0.054 0.097	2% 98% <2%			
$_{76}Os^{195}$			6.5m	β^-	2.0	~ 2				
Ir		192.2						$425 \pm 15b$		
$_{77}Ir^{182}$			15m	β^+, EC	≈ 5.5					

Isotope	% nat. abundance	Atomic mass	Lifetime	Modes of decay	Decay energy (MeV)	Particle energies (MeV)	Particle intensities	Thermal neutron capture cross-section	I	μ
$_{77}Ir^{183}$			0.9h	EC						
$_{77}Ir^{184}$			3.2h	β^+, EC	> 4.3					
$_{77}Ir^{185}$			14h	EC	≈ 2.5					
$_{77}Ir^{186}$			16h	β^+, EC	3.83	1.06 1.37 1.94	0.5% 1.9%			
$_{77}Ir^{186}$			1.7h	β^+, EC		0.8 1.30 1.93 2.6 3.4	22† 12 44 20 1			
$_{77}Ir^{187}$			10.5h	EC	≈ 1.6					
$_{77}Ir^{188}$			41h	β^+, EC		1.21 1.66				
$_{77}Ir^{189}$			13.3d	EC	≈ 0.6					
$_{77}Ir^{190m_2}$			3.2h	IT β^+, EC	0.175 ~2.3		6% 94%			
$_{77}Ir^{190m_1}$			1.2h	IT	0.026					
$_{77}Ir^{190}$			11d	EC	2.1					
$_{77}Ir^{191m}$			4.9s	IT	0.042				11/2	± 6
$_{77}Ir^{191}$	37.3%	190.9609						$0.4 \pm 0.2b$ (Ir^{192m_2}) $610 \pm 60b$ (Ir^{192m_1}) $300 \pm 30b$ (Ir^{192})	3/2	+ 0.18
$_{77}Ir^{192m_2}$			> 5y	IT	0.161					
$_{77}Ir^{192m_1}$			1.4m	IT β^-	0.058 1.50	0.9 1.2 1.5	0.0025% 0.008% 0.007%			
$_{77}Ir^{192}$			74d	β^- β^+, EC	1.453 1.2	β^-, 0.24 0.536 0.672 β^+, 0.24	8% 41% 46% 1.5×10^{-5}%	$1100 \pm 400b$ (Ir^{193m})	4	
$_{77}Ir^{193m}$			12d	IT	0.08			$0.05 \pm 0.02b$ (Ir^{194m_2})	[1/2?]	< 2
$_{77}Ir^{193}$	62.7%	192.9633						$110 \pm 15b$	3/2	+ 0.18
$_{77}Ir^{194}$			17.4h	β^-	2.24	0.98 1.62	2.0% 1.2%		1	
$_{77}Ir^{195}$			4.2h	β^-	1.0	0.6 1.0	≈ 25% ≈ 75%			
$_{77}Ir^{196m}$			84m	β^-	3.40	0.24 0.39 0.66 1.08	10% 4.7% 5.4% 80%			
$_{77}Ir^{196}$			52s	β^-	3.22	0.3 0.4	~1% ~1%			

Isotope	% nat. abundance	Atomic mass	Lifetime	Modes of decay	Decay energy (MeV)	Particle energies (MeV)	Particle intensities	Thermal neutron capture cross-section	I	μ
						0.82	~1%			
						1.09	16%			
						3.22	80%			
$_{77}\text{Ir}^{197}$			7m	β^-	2.0	1.5	~50%			
						2.0	~50%			
$_{77}\text{Ir}^{198}$			50s	β^-	4.4	~3.6				
Pt		195.09						9 ± 1b		
$_{78}\text{Pt}^{173}$			short	α	6.34	α 6.19				
$_{78}\text{Pt}^{174}$			0.7s	β^+, EC	6.17		20%			
				α		6.03	80%			
$_{78}\text{Pt}^{175}$			2.1s	α	6.09	5.95				
$_{78}\text{Pt}^{176}$			6s	α	5.87	5.74	1%			
				β^+, EC			99%			
$_{78}\text{Pt}^{177}$			6.6s	α	5.64	5.51	0.3%			
				β^+, EC			99 + %			
$_{78}\text{Pt}^{178}$			21s	α	5.56	5.28	0.07%			
				β^+, EC		5.44	1.3%			
$_{78}\text{Pt}^{179}$			33s	α	5.27	5.15	0.1%			
				β^+, EC						
$_{78}\text{Pt}^{180}$			50s	α	5.26	5.14	0.3%			
				β^+, EC						
$_{78}\text{Pt}^{181}$			51s	α	5.13	5.02	0.0006%			
				β^+, EC						
$_{78}\text{Pt}^{182}$			3.0m	α	4.95		≈0.02%			
				β^+, EC	≈3					
$_{78}\text{Pt}^{183}$			7m	β^+, EC	4.84	4.74				
				α						
$_{78}\text{Pt}^{184}$			20m	α	4.60	4.50				
				β^+, EC						
$_{78}\text{Pt}^{185}$			1.2h	EC						
$_{78}\text{Pt}^{186}$			3h	α	4.32	4.23				
				EC						
$_{78}\text{Pt}^{187}$			2h	EC						
$_{78}\text{Pt}^{188}$			10.2d	EC	0.51					
$_{78}\text{Pt}^{189}$			10.9h	EC						
$_{78}\text{Pt}^{190}$	0.01%	189.960	6×10^{11}y	α		3.18				
$_{78}\text{Pt}^{191}$			3.0d	EC	~2.0					
$_{78}\text{Pt}^{192}$	0.79%	191.9614	~10^{15}y	α		2.6		2 ± 1b (Pt^{193m})		
$_{78}\text{Pt}^{193m}$			4.3d	IT	0.148					
$_{78}\text{Pt}^{193}$			<500y	EC	0.045					
$_{78}\text{Pt}^{194}$	32.9%	193.9628						87 ± 13mb		

Isotope	% nat. abundance	Atomic mass	Lifetime	Modes of decay	Decay energy (MeV)	Particle energies (MeV)	Particle intensities	Thermal neutron capture cross-section	I	μ
								(Pt^{195m}) 1.2 $\pm$ 0.9b (Pt^{195})	0	
$_{78}$Pt^{195m}			4.1d	IT	0.2593					
$_{78}$Pt^{195}	33.8%	194.9648						27 $\pm$ 2b	1/2	+0.6060
$_{78}$Pt^{196}	25.3%	195.9650						0.06 $\pm$ 0.02b (Pt^{197m}) 0.9 $\pm$ 0.1b (Pt^{197})	0	
$_{78}$Pt^{197m}			80m	IT β^-	0.399 1.15		97%			
$_{78}$Pt^{197}			18h	β^-	0.75	0.48 0.67	10% 90%			
$_{78}$Pt^{198}	7.2%	197.9675						4.0 $\pm$ 0.5b		
$_{78}$Pt^{199m}			14.1s	IT	0.425					
$_{78}$Pt^{199}			30m	β^-	1.68	0.72 0.89 0.94 1.14 1.38 1.69	1% 10% 8% 14% 4% 63%	15 $\pm$ 10b		
$_{78}$Pt^{200}			11.5h	β^-	$\approx$0.7					
Au		196.967						98.8 $\pm$ 0.3b		
$_{79}$Au^{185}			7m	α	5.18	5.07				
$_{79}$Au^{186}			12m	EC						
$_{79}$Au^{187}			8m	EC						
$_{79}$Au^{188}			8m	β^+, EC	5.2					
$_{79}$Au^{189}			30m	EC						
$_{79}$Au^{190}			39m	β^+, EC	$\approx$4.4				1	$\pm$0.066
$_{79}$Au^{191}			3.2h	EC	$\approx$2.0				3/2	$\pm$0.137
$_{79}$Au^{192}			4.1h	β^+, EC	3.24	2.22	1%		1	$\pm$0.00785
$_{79}$Au^{193m}			3.9s	IT EC	0.290	I.T	99%			
$_{79}$Au^{193}			16h	EC	$\approx$1.1				3/2	$\pm$0.139
$_{79}$Au^{194}			39.5h	β^+, EC	2.51	1.21 1.55	1.3† 1.7		1	$\pm$0.074
$_{79}$Au^{195m}			31s	IT	0.318					
$_{79}$Au^{195}			183d	EC	0.227				3/2	$\pm$0.147
$_{79}$Au^{196}			6.18d	β^+, EC β^-	1.48 0.684	β^-, 0.259			2	$\pm$0.6
$_{79}$Au^{197m}			7.2s	IT	0.409				[1/2]†	+0.37†

Isotope	% nat. abundance	Atomic mass	Lifetime	Modes of decay	Decay energy (MeV)	Particle energies (MeV)	Particle intensities	Thermal neutron capture cross-section	I	μ
$_{79}Au^{197}$	100%	196.9666						$98.8 \pm 0.3b$	3/2	+0.14486
$_{79}Au^{198}$			2.693 ±0.005d	β^-	1.374	0.28 0.961 1.374	1.1% 98.9% 0.025%	$25,800 \pm 1200b$	2	+0.590
$_{79}Au^{199}$			3.15d	β^-	0.46	0.250 0.296 0.462	22.4% 71.6% 6.0%	$30 \pm 15b$	3/2	+0.270
$_{79}Au^{200}$			48.4m	β^-	2.21	0.61 1.84 2.21	24% 6% 70%			
$_{79}Au^{201}$			26m	β^-	1.5	1.0 1.5	5% 95%			
$_{79}Au^{203}$			5.5s	β^-	≈ 2.1	1.9				
Hg		200.59						$375 \pm 5b$		
$_{80}Hg^{185}$			50s	EC						
$_{80}Hg^{186}$			1.5m	EC						
$_{80}Hg^{187}$			3m	EC α		5.14?				
$_{80}Hg^{188}$			3.7m	EC						
$_{80}Hg^{189}$			10m	β^+, EC						
$_{80}Hg^{190}$			20m	EC						
$_{80}Hg^{191}$			55m	β^+, EC						
$_{80}Hg^{192}$			4.8h	EC	≈ 1.4					
$_{80}Hg^{193m}$			10h	IT β^+, EC	0.141	1.17	16%		13/2	−1.063
$_{80}Hg^{193}$			~6h	EC	2.34				3/2	−0.62
$_{80}Hg^{194}$			1.9y?	EC	≈ 0.4					
$_{80}Hg^{195m}$			40h	IT EC	0.176 ~1.7				13/2	−1.049
$_{80}Hg^{195}$			9.5h	EC	~1.5				1/2	+0.538
$_{80}Hg^{196}$	0.2%	195.9658						$120 \pm 15b$ (Hg^{197m}) $3000 \pm 100b$ (Hg^{197})		
$_{80}Hg^{197m}$			24h	IT EC	0.72 0.42	IT EC	94% 6%		13/2	−1.032
$_{80}Hg^{197}$			65h	EC	0.42				1/2	+0.5241
$_{80}Hg^{198}$	10.1%	197.9668						$0.02 \pm 0.01b$ (Hg^{199m})	0	
$_{80}Hg^{199m}$			43m	IT	0.533					
$_{80}Hg^{199}$	16.9%	198.9683						$2000 \pm 1000b$	1/2	+0.5027
$_{80}Hg^{200}$	23.1%	199.9683						<60b	0	
$_{80}Hg^{201}$	13.2%	200.9703						<60b	3/2	−0.5567

Isotope	% nat. abundance	Atomic mass	Lifetime	Modes of decay	Decay energy (MeV)	Particle energies (MeV)	Particle intensities	Thermal neutron capture cross-section	I	μ
$_{80}Hg^{202}$	29.7%	201.9706						$4.9 \pm 0.2b$	0	
$_{80}Hg^{203}$			$46.57 \pm 0.03d$	β^-	0.492	0.210 0.49	100% <0.004%		5/2	+0.84
$_{80}Hg^{204}$	6.8%	203.9735						$0.43 \pm 0.10b$	0	
$_{80}Hg^{205}$			5.5m	β^-	1.6	~1.4 ~1.6				
$_{80}Hg^{206}$			7.5m	β^-	1.31					
Tl		204.37						$3.4 \pm 0.5b$		
$_{81}Tl^{191}$			10m	β^+, EC						
$_{81}Tl^{192}$			11m	β^+, EC						
$_{81}Tl^{193m}$			2.1m	IT						
$_{81}Tl^{193}$			23m	EC						
$_{81}Tl^{194m}$			32.8m	EC						
$_{81}Tl^{194}$			33m	EC	≈5.4					
$_{81}Tl^{195m}$			3.5s	IT	0.482					
$_{81}Tl^{195}$			1.2h	β^+, EC		1.8			1/2	
$_{81}Tl^{196m}$			1.41h	IT EC	0.395	IT EC	4% 96%			
$_{81}Tl^{196}$			1.84h	EC	4.6					
$_{81}Tl^{197m}$			0.54s	IT	0.609					
$_{81}Tl^{197}$			2.8h	EC	2.2				1/2	
$_{81}Tl^{198m}$			1.87h	IT EC	0.543 ~4.0		55% 45%		7	
$_{81}Tl^{198}$			5.3h	β^+, EC	3.5	0.7 1.4 2.1 2.4	4† 10 12 14		2	± <0.002
$_{81}Tl^{199}$			7.4h	EC	1.1				1/2	+1.59
$_{81}Tl^{200}$			26.1h	β^+, EC	2.454	1.07 1.44	5.6† 1.0		2	± ≤0.15
$_{81}Tl^{201m}$			2.1ms	IT	0.93					
$_{81}Tl^{201}$			73h	EC	≈0.41				1/2	+1.60
$_{81}Tl^{202}$			12.0d	EC	1.1				2	± ≤0.15
$_{81}Tl^{203}$	29.50%	202.9723						$10 \pm 1b$	1/2	+1.6115
$_{81}Tl^{204}$			3.8y	β^- EC	0.763 0.34	0.7635 ± 0.0002 EC	97.9% 2%		2	±0.089
$_{81}Tl^{205}$	70.50%	204.9745						$0.5 \pm 0.2b$	1/2	+1.6274
$_{81}Tl^{206}$			4.19m	β^-	1.524	1.524	100%		0*	
$_{81}Tl^{207m}$			1.3s	IT	1.34					

Isotope	% nat. abundance	Atomic mass	Lifetime	Modes of decay	Decay energy (MeV)	Particle energies (MeV)	Particle intensities	Thermal neutron capture cross-section	I	μ
$_{81}Tl^{207}$			4.78m	β^-	1.44	0.53 1.44	0.16% 99+%			
$_{81}Tl^{208}$			3.10m	β^-	4.994	1.04 1.29 1.52 1.792 2.38	4% 24% 23% 49% 0.03%			
$_{81}Tl^{209}$			2.2m	β^-	3.98	1.99	100%			
$_{81}Tl^{210}$			1.3m	β^-	5.50	1.3 1.9 2.3	25% 56% 19%			
Pb		207.2						180 ± 10mb		
$_{82}Pb^{194}$			11m	EC						
$_{82}Pb^{195}$			17m	EC						
$_{82}Pb^{196}$			37m	EC	≈ 2.8					
$_{82}Pb^{197m}$			42m	IT EC	0.319 ≈ 4.5	IT EC	20% 80%			
$_{82}Pb^{197}$				EC	≈ 4.1					
$_{82}Pb^{198}$			2.4h	EC	≈ 1.8					
$_{82}Pb^{199m}$			12.2m	IT	0.424					
$_{82}Pb^{199}$			1.5h	β^+, EC	≈ 3.2	2.8	17%			
$_{82}Pb^{200}$			21.5h	EC	≈ 0.9					
$_{82}Pb^{201m}$			61s	IT	0.629					
$_{82}Pb^{201}$			9.4h	β^+, EC	≈ 2.0	0.55				
$_{82}Pb^{202m}$			3.62h	IT EC	2.17 2.17	IT EC	90% 10%			
$_{82}Pb^{202}$			$\sim 3 \times 10^5$y	EC	0.05					
$_{82}Pb^{203m}$			6.1s	IT	0.825					
$_{82}Pb^{203}$			52.1h	EC	0.96					
$_{82}Pb^{204m}$			66.9m	IT	2.186					
$_{82}Pb^{204}$	1.4%	203.973						0.66 ± 0.07b		
$_{82}Pb^{205m}$			4ms	IT	1.01					
$_{82}Pb^{205}$			3×10^7y	EC	≈ 0.035					
$_{82}Pb^{206}$	24.1%	205.9745						30 ± 1mb	0	
$_{82}Pb^{207m}$			0.8s	IT	1.633					
$_{82}Pb^{207}$	22.1%	206.9759						0.71 ± 0.01b	1/2	+0.5895
$_{82}Pb^{208}$	52.4%	207.9766						0.02 ± 0.01b	0	
$_{82}Pb^{209}$			3.30h	β^-	0.635	0.635	100%			

Isotope	% nat. abundance	Atomic mass	Lifetime	Modes of decay	Decay energy (MeV)	Particle energies (MeV)	Particle intensities	Thermal neutron capture cross-section	I	μ
$_{82}Pb^{210}$			21y	β^- α	0.061 3.72	β^-0.015 0.061 α 3.72	81% 19% 1.7×10^{-6}%			
$_{82}Pb^{211}$			36.1m	β^-	1.37	0.10 0.26 0.34 0.53 0.59 0.96 1.36	0.01% 0.7% 0.1% 6% 0.4% 1% 92%			
$_{82}Pb^{212}$			10.64h	β^-	0.58	0.17 0.35 0.589	5% 81% 14%			
$_{82}Pb^{213}$			10m	β^-	≈ 1.8					
$_{82}Pb^{214}$			26.8m	β^-	1.04	0.59 0.65 1.03	6%			
Bi		208.980						34 ± 2mb		
$_{83}Bi^{199}$			24m	EC α		 5.47	99+% 0.01%		9/2	
$_{83}Bi^{200}$			35m	EC	≈ 6.5				7	
$_{83}Bi^{201m}$			52m	EC α	≈ 4.2 ≈ 4.7	5.28				
$_{83}Bi^{201}$			1.8h	EC	≈ 4.2				9/2	
$_{83}Bi^{202}$			95m	EC	≈ 5.5				5	
$_{83}Bi^{203}$			11.8h	β^+, EC α	≈ 4.4 4.3	β^+ 0.74 1.35 α 4.85	33† 67		9/2	+4.59
$_{83}Bi^{204}$			11.2h	EC	≈ 4.4				6	+4.25
$_{83}Bi^{205}$			15.3d	β^+, EC	2.70	0.98	0.06%		9/2	+5.5
$_{83}Bi^{206}$			6.24d	β^+, EC	3.65	0.98	8×10^{-4}%		6	+4.56
$_{83}Bi^{207}$			30y	EC	2.40					
$_{83}Bi^{208}$			3.7×10^5y	EC	2.87					
$_{83}Bi^{209}$	100%	208.9804						19 ± 2mb	9/2	+4.080
$_{83}Bi^{210m}$			3×10^6y	α		4.43 4.57 4.92 4.96	0.4% 6% 36% 58%			
$_{83}Bi^{210}$			5.01d	β^- α	1.16 5.044	β^- 1.160 α 4.654 4.691	99+% 1.3×10^{-4}%	54 ± 5mb	1	±0.0442
$_{83}Bi^{211}$		210.9873	2.15m	α β^-	6.750 0.60	α 5.946 6.278 6.622 β^- 0.060	0.0037% 15.9% 84.1% 0.28%			
$_{83}Bi^{212}$			60.6m	β^- α	2.25 6.206	β^- 0.08 0.45 0.67	0.7% 0.1% 2.2%			

TABLE OF THE ISOTOPES (*Continued*)

Isotope	% nat. abundance	Atomic mass	Lifetime	Modes of decay	Decay energy (MeV)	Particle energies (MeV)	Particle intensities	Thermal neutron capture cross-section	I	μ
						0.93	1.4%			
						1.55	5%			
						2.27	54%			
						α 5.178	5×10^{-5} †			
							$\alpha_T = 36\%$			
						5.291	1.3×10^{-4}			
						5.334	1×10^{-3}			
						5.473	0.014			
						5.547	1.1			
						5.617	1.7			
						6.051	69.9			
						6.090	27.2			
$_{83}Bi^{213}$			47m	β^- α	1.42 5.98	β^- 0.96 ⎱ 1.39 ⎰ α 5.55 ⎱ 5.87 ⎰	97.8% 2.2%			
$_{83}Bi^{214}$			19.7m	β^- α	3.28 5.616	β^- 0.82 1.06 1.14 1.41 1.50 1.51 1.88 1.98 2.65 3.26 α 4.941 5.023 5.184 5.268 5.448 5.512	2% 8% 5% 8% 19% 19% 10% 2% 5% 19% 0.25† 0.21 0.6 5.8 53.9 39.2			
$_{83}Bi^{215}$			7m	β^-	2.2					
Po										
$_{84}Po^{192}$										
$_{84}Po^{193}$			Short	α	7.1	7.0				
$_{84}Po^{194}$			0.5s	α	6.97	6.85				
$_{84}Po^{195m}$			1.4s	α		6.72				
$_{84}Po^{195}$			3s	α	6.77	6.63				
$_{84}Po^{196}$			6s	α	6.67	6.53				
$_{84}Po^{197m}$			25s	α		6.39				
$_{84}Po^{197}$			54s	α	6.43	6.30				
$_{84}Po^{198}$			1.7m	α	6.29	6.16				
$_{84}Po^{199m}$			4.2m	EC α		EC α 6.047	74% 26%			
$_{84}Po^{199}$			5.0m	EC α	6.06	EC α 5.942	97% 3%			
$_{84}Po^{200}$			11m	EC α	≈3.6 est. 5.97	EC α 5.850	88% 12%			
$_{84}Po^{201m}$			9m	EC α		EC α 5.776	97% 3%			

Isotope	% nat. abundance	Atomic mass	Lifetime	Modes of decay	Decay energy (MeV)	Particle energies (MeV)	Particle intensities	Thermal neutron capture cross-section	I	μ
$_{84}Po^{201}$			15.1m	EC α	≈ 5.3 est. 5.79	EC α 5.677	99% 1.1%		3/2	
$_{84}Po^{202}$			45m	EC α	≈ 3.0 est. 5.69	EC α 5.581	98% 2%		0	
$_{84}Po^{203}$			42m	EC α	≈ 4.4 est. 5.59	EC α 5.49	99% 0.02%		5/2	
$_{84}Po^{204}$			3.6h	EC α	≈ 2.5 est. 5.48	EC α 5.376	99% 0.6%		0	
$_{84}Po^{205}$			1.8h	EC α	≈ 3.6 est. 5.34	EC α 5.25	99+% 0.07%		5/2	+0.26
$_{84}Po^{206}$		205.9805	8.8d	EC α	1.80 calc. 5.33	EC α 5.224	95% 5%		0	
$_{84}Po^{207m}$			2.8s	IT						
$_{84}Po^{207}$		206.9816	5.7h	EC β+ α	2.91 5.33	EC β+, 1.14, 0.80 α 5.11	99+% 0.5% ≈ 0.01%		5/2	≈ 0.27
$_{84}Po^{208}$		207.9813	2.93y	α EC	5.21 1.41 calc.	5.114 EC	99+% 0.006%			
$_{84}Po^{209}$		208.9825	103y	α EC	4.98 1.89 calc.	4.882 EC	99+% 0.5%		1/2	
$_{84}Po^{210}$		209.9829	138.4d	α	5.408	5.305		< 0.5mb (25s Po^{211}) < 30 mb (0.5s Po^{211})	0	
$_{84}Po^{211m}$			25s	α		8.88 8.31 8.00 7.28	7% 0.25 1.66 91%			
$_{84}Po^{211}$		210.9866	0.52s	α	7.592	7.448 6.891 6.551	99% 0.50 0.53			
$_{84}Po^{212m}$			45s	α		11.65 14.27 14.85	97% 1.00% 2.0%			
$_{84}Po^{212}$		211.9889	3.04×10^{-7}s	α		8.785 + long range α's				
$_{84}Po^{213}$		212.9928	4.2×10^{-6}s	α	8.54	8.377 7.61	99+% 0.003			
$_{84}Po^{214}$		213.9952	1.64×10^{-4}s	α	7.835	7.687 6.905 + long range α's	99+% 0.01			
$_{84}Po^{215}$		214.9995	1.78×10^{-3}s	α	7.524	7.384 6.954 6.948	100% 0.034 0.022			

Isotope	% nat. abundance	Atomic mass	Lifetime	Modes of decay	Decay energy (MeV)	Particle energies (MeV)	Particle intensities	Thermal neutron capture cross-section	I	μ
$_{84}Po^{216}$		216.0019	0.15s	α	6.906	6.777 5.984	100% 2.1×10^{-3}			
$_{84}Po^{217}$			< 10s	α β^-	6.67 ≈ 1.3 est.	6.55 β^-	80% < 20%			
$_{84}Po^{218}$		218.0089	3.05m	α β^-	6.111 0.278	α 6.002 5.181 β^-	99.98% 99.9+ 0.0011 0.019%			
At										
$_{85}At^{200}$			0.9m	EC α	6.55	6.47 6.42				
$_{85}At^{201}$			1.5m	EC α	-6.48	6.35				
$_{85}At^{202}$			3.0m	EC α	≈ 8.1 est. 6.359	EC α 6.234 6.120	88% 12% 36† 64			
$_{85}At^{203}$			7.4m	EC α	≈ 5.8 est. 6.211	EC α 6.092	86% 14%			
$_{85}At^{204}$			9.3m	EC α	7.1 est. 6.071	EC α 5.955	95.5% 4.52%			
$_{85}At^{205}$			26m	EC α	≈ 4.9 est. 6.018	EC α 5.902	82% 18%			
$_{85}At^{206}$			32m	EC α	≈ 6.0 est. 5.88	EC α 5.70	$\approx 12\%$ $\approx 88\%$			
$_{85}At^{207}$			1.79h	EC α	3.73 calc. 5.87	EC α 5.76	$\approx 90\%$ $\approx 10\%$			
$_{85}At^{208}$			1.6h	EC α	≈ 5.0 est. 5.77	EC α 5.65	99+% 0.5%			
$_{85}At^{209}$		208.986	5.5h	EC α	3.49 calc. 5.758	EC α 5.648	95% $\approx 5\%$			
$_{85}At^{210}$		209.987	8.3h	EC α	3.87 calc. 5.631	EC α 5.525 5.443 5.361	99+% 0.17% 32† 31 37			
$_{85}At^{211}$		210.9875	7.21h	EC α	0.79 calc. 5.981	EC α 5.868	59.1% 40.9%		9/2	
$_{85}At^{212m}$			0.12s	α		7.899 7.837	34% 66			
$_{85}At^{212}$			0.30s	α	7.81	7.678 7.616	80% 20			
$_{85}At^{213}$		212.993090	Short	α	9.4	9.06				

Isotope	% nat. abundance	Atomic mass	Lifetime	Modes of decay	Decay energy (MeV)	Particle energies (MeV)	Particle intensities	Thermal neutron capture cross-section	I	μ
$_{85}At^{214}$		213.9963	Short	α	8.94	8.78 8.44 8.23	100% <0.2 <0.1			
$_{85}At^{215}$		214.9987	$\approx 10^{-4}$ s	α	8.16	8.01 7.60	100% 0.05			
$_{85}At^{216}$		216.0024	$\approx 3 \times 10^{-4}$ s	α	7.95	7.81 7.71 7.60 7.58 7.49 7.40 7.33 7.25	97% 2.1 0.2 0.5 0.23 0.09 0.06			
$_{85}At^{217}$		217.0046	0.032s	α	7.199	7.066 6.820 6.623 6.486	100% 0.025 0.003 0.040			
$_{85}At^{218}$		218.0086	≈ 2 s	α β⁻	6.816 2.83 calc.	6.697 6.653	94% ≈ 6 0.1%			
$_{85}At^{219}$		219.0114	0.9m	α β⁻	6.39 1.7 calc.	α 6.28	97% $\approx 3\%$			
Rn										
$_{86}Rn^{204}$			75s	α	6.545	6.416				
$_{86}Rn^{205}$			1.8m	α	6.387	6.262				
$_{86}Rn^{206}$			6.5 m	α EC	6.382 ≈ 3.5 est.	α 6.258 EC	65% 35%			
$_{86}Rn^{207}$			11m	EC α	≈ 4.8 est. 6.23	EC α 6.15	96% 4%			
$_{86}Rn^{208}$			23m	EC α	≈ 3.0 est. 6.268	EC α 6.148	$\approx 80\%$ $\approx 20\%$			
$_{86}Rn^{209}$			30m	EC α	≈ 4.0 est. 6.161	EC α 6.044	83% 17%			
$_{86}Rn^{210}$		209.9897	2.42h	α EC	6.160 2.33 calc.	α EC 6.044	$\approx 96\%$ $\approx 4\%$			
$_{86}Rn^{211}$		210.9906	15h	EC α	2.89 calc. 5.966	EC α 5.853 5.785 5.619	74% 26% 33.5† 64.5 2.0			
$_{86}Rn^{212}$		211.9907	25m	α	6.39	6.271				
$_{86}Rn^{213}$			19ms	α	8.296	8.14				
$_{86}Rn^{215}$		214.9987	Short	α	8.78	8.6				
$_{86}Rn^{216}$		216.0002	45μs	α	8.201	8.05				
$_{86}Rn^{217}$		217.0039	540μs	α	7.888	7.742				
$_{86}Rn^{218}$		218.0056	0.035s	α	7.26	7.14 6.54	99.8% 0.16			

Isotope	% nat. abundance	Atomic mass	Lifetime	Modes of decay	Decay energy (MeV)	Particle energies (MeV)	Particle intensities	Thermal neutron capture cross-section	I	μ
$_{86}Rn^{219}$		219.0095	4.0s	α	6.944	6.817 6.551 6.527 6.423 6.310 6.222 6.157 6.146 6.101 5.999 5.946 5.785	81% 11.5 0.12 7.5 0.054 0.0026 0.017 0.0026 0.003 0.0044 0.0037 ≈ 0.001			
$_{86}Rn^{220}$		220.0114	55s	α	6.405	6.287 5.747	100% 0.07	$< 0.2b$		
$_{86}Rn^{221}$			25m	β^- α	≈ 1.0 est. 6.1	β^- 6.0 est.	$\approx 80\%$ $\approx 20\%$			
$_{86}Rn^{222}$		222.0175	3.823d	α	5.587	5.486 4.984 4.824	100% 8×10^{-2} 5×10^{-4}	$0.72 \pm 0.07b$		
$_{86}Rn^{223}$			43m	β^-						
$_{86}Rn^{224}$			1.9h	β^-						
Fr										
$_{87}Fr^{204}$			2.0s	α	7.17	7.03				
$_{87}Fr^{205}$			3.7s	α	7.05	6.92				
$_{87}Fr^{206}$			15.8s	α	6.93	6.80				
$_{87}Fr^{207}$			19s	α	6.91	6.78				
$_{87}Fr^{208}$			37s	α	6.79	6.66				
$_{87}Fr^{209}$			55s	α	6.79	6.66				
$_{87}Fr^{210}$			2.6m	α	6.68	6.56				
$_{87}Fr^{211}$			3.1m	α	6.68	6.56				
$_{87}Fr^{212}$		211.996	19m	EC α	≈ 5.2 est. 6.530	EC α 6.407 6.383 6.338 6.261 6.179 6.127 6.077 5.983 5.83	56% 44% 21† 23 12 40 2.1 1.0 0.4 0.1 0.05			
$_{87}Fr^{213}$			34s	α EC	6.91 2.10 calc.	6.78 EC	99.48% 0.52%			
$_{87}Fr^{214m}$			3.35ms	α	8.59	8.549 8.477	51% 49			
$_{87}Fr^{214}$			5.0ms	α	8.70	8.426 8.353	94.5% 5.5			
$_{87}Fr^{215}$			$\ll 1ms$	α	9.59	9.4				
$_{87}Fr^{217}$		217.0048	Short	α	8.5	8.31				
$_{87}Fr^{218}$		218.0095	Short	α	8.00	7.85 7.70	93% < 0.5			

Isotope	% nat. abundance	Atomic mass	Lifetime	Modes of decay	Decay energy (MeV)	Particle energies (MeV)	Particle intensities	Thermal neutron capture cross-section	I	μ
						7.55	5			
						7.52	1			
						7.36	<0.5			
$_{87}Fr^{219}$		219.0092	20ms	α	7.44	7.30	98.4%			
						7.14	0.3			
						6.95	0.8			
						6.72	0.2			
						6.68	0.3			
$_{87}Fr^{220}$		220.0123	27.5s	α	6.80	6.68	85%			
						6.64	13			
						6.58	1			
						6.53	0.5			
						6.48	<0.1			
						6.40	<0.1			
$_{87}Fr^{221}$		221.0142	4.8m	α	6.457	6.340	83.4%			
						6.242	1.34			
						6.126	15.1			
						6.075	0.15			
						6.036	0.003			
						5.979	0.49			
						5.965	0.08			
						5.938	0.17			
						5.925	0.03			
						5.813	0.004			
						5.782	0.005			
						5.775	0.06			
						5.696?	≈0.001			
						5.688	0.002			
$_{87}Fr^{222}$		222.018	14.8m	β⁻ n	2.03 calc. 5.87 calc.	β⁻ n	99+% 0.01–0.1%			
$_{87}Fr^{223}$		223.0198	22m	β⁻ α	1.15 ≈5.44 calc.	1.15 5.34	99.994% 0.006%			
$_{87}Fr^{224}$			≈2m	β⁻	≈2.9 est.					
Ra										
$_{88}Ra^{213}$			2.7m	α	6.89	6.74 6.61	50% 50			
$_{88}Ra^{214}$			2.6s	α	7.31	7.17				
$_{88}Ra^{215}$			1.6ms	α	8.90	8.73				
$_{88}Ra^{216}$			<1ms	α	9.48	9.3				
$_{88}Ra^{217}$			Short	α	9.18	9.0				
$_{88}Ra^{219}$		219.0100	Short	α	8.16	8.0				
$_{88}Ra^{220}$		220.0110	23ms	α	7.59	7.45 6.90	≈99% ≈1			
$_{88}Ra^{221}$		221.0139	30s	α	6.883	6.758	31%			
						6.665	21			
						6.610	35			
						6.588	8			
						6.578	3			
						6.46?	0.4			
						6.40?	0.3			
						6.25	0.7			
						6.16?	0.3			
$_{88}Ra^{222}$		222.0154	38s	α	6.68	6.56	96%			
						(6.23)	4			
						(5.36)	7×10^{-3}			
						(4.31)	6×10^{-3}			

Isotope	% nat. abundance	Atomic mass	Lifetime	Modes of decay	Decay energy (MeV)	Particle energies (MeV)	Particle intensities	Thermal neutron capture cross-section	I	μ
$_{88}Ra^{223}$		223.0186	11.43d	α	5.977	5.870	0.87%	σ_f		
						5.865?	<0.02	$1.0 \pm 0.5b$		
						5.856	0.32	$\sigma(\eta, \gamma)$		
						5.833?		$130 \pm 20b$		
						5.745	9.1			
						5.714	53.7			
						5.605	26			
						5.538	9.1			
						5.500	0.8			
						5.479	≈0.008			
						5.432	2.3			
						5.364	0.11			
						5.337	0.10			
						5.285	0.13			
						5.281	0.10			
						5.257	0.043			
						5.234	0.042			
						5.210	0.0054			
						5.171	0.026			
						5.150	0.021			
						5.133	≈0.0017			
						5.110	$≈6 \times 10^{-4}$			
						5.084	$≈6 \times 10^{-4}$			
						5.054	$≈2 \times 10^{-4}$			
						5.034	$≈4 \times 10^{-4}$			
						5.023	$≈6 \times 10^{-4}$			
						5.012	$≈4 \times 10^{-4}$			
$_{88}Ra^{224}$		224.0202	3.64d	α	5.787	5.684	94.5%	$12.0 \pm 0.5b$		
						5.447	5.5			
						5.159	0.007			
						5.049	0.007			
						5.032	0.003			
$_{88}Ra^{225}$			14.8d	β^-	0.39 calc.	0.32				
$_{88}Ra^{226}$		226.0254	1600y	α		4.781	94.5%	$20 \pm 3b$		
						4.598	5.5			
						4.340	6×10^{-3}			
						4.194	9×10^{-4}			
						4.160	2.7×10^{-4}			
$_{88}Ra^{227}$			41.2m	β^-	1.31	1.31				
$_{88}Ra^{228}$		228.031	$5.77 \pm 0.02y$	β^-	0.055	0.024	30%	$36 \pm 5b$		
						0.048	70			
$_{88}Ra^{230}$			1h	β^-	0.9 est.	1.2				
Ac										
$_{89}Ac^{221}$		221.0157	Short	α	7.75	7.63				
$_{89}Ac^{222}$		222.0178	5s	α	7.13	7.00	93†			
						6.96	6			
$_{89}Ac^{223}$		223.0191	2.2m	α	6.781	6.659	37.6%			
						6.648	42.1			
						6.56	13.3			
						6.52	3.8			
						6.47	3.2			
				EC	0.56	EC	1%			
$_{89}Ac^{224}$		224.0217	2.9h	EC	1.38	EC	90%			
				α	6.32	α	≈10%			
						6.210	20.4†			
						6.203	11.9			
						6.155	1.03			
						6.138	25.6			
						6.056	21.9			
						6.014	1.4			
						6.000	6.7			

Isotope	% nat. abundance	Atomic mass	Lifetime	Modes of decay	Decay energy (MeV)	Particle energies (MeV)	Particle intensities	Thermal neutron capture cross-section	I	μ
						5.968	0.2			
						5.958	~0.04			
						5.942	4.4			
						5.916	0.94			
						5.909	0.15			
						5.902	0.15			
						5.877	1.7			
						5.867	0.1			
						5.860	0.75			
						5.854	0.25			
						5.851	~0.07			
						5.841	0.55			
						5.837	0.26			
						5.803	~0.02			
						5.776	0.06			
						5.768	0.24			
						5.739	~0.05			
						5.718	0.12			
						5.709	0.12			
						5.641	0.06			
$_{89}Ac^{225}$		225.0231	10.0d	α	5.931	5.829	52%			
						5.793	29			
						5.732	9.5			
						5.724	2.6			
						5.683	1.0			
						5.638	4.0			
						5.610	0.8			
						5.58	0.8			
						5.55	0.07			
						5.49	≈0.02			
						5.45	0.1			
						5.40	≈0.01			
						5.37	≈0.01			
						5.33	0.07			
						5.30	0.1			
						5.23	≈0.02			
$_{89}Ac^{226}$			29h	β^-	1.12	β^-	77%			
						1.1	41†			
						0.89	59			
				EC	0.77	EC	23%			
				α	5.54	5.44	Weak			
$_{89}Ac^{227}$		227.0278	21.6y	β^-	0.043	β^- 0.044	98.6%	810 ± 20b	3/2	+1.1
				α	5.040	α	1.4%			
						4.950	47†			
						4.938	38			
						4.897	0.11			
						4.870	6.5			
						4.853	4.0			
						4.819	0.07			
						4.793	0.9			
						4.782	0.08			
						4.766	1.6			
						4.735	0.09			
						4.712	0.31			
						4.591	0.02			
						4.586	0.01			
						4.578	≈0.003			
						4.509	≈0.003			
						4.456	≈0.005			
						4.442	0.05			
						4.420	0.006			
						4.360	≈0.003			
$_{89}Ac^{228}$			6.13h	β^-	2.14	0.45	13%			
						0.62	6			
						1.18	53			
						1.70	6.7			

Isotope	% nat. abundance	Atomic mass	Lifetime	Modes of decay	Decay energy (MeV)	Particle energies (MeV)	Particle intensities	Thermal neutron capture cross-section	I	μ
						1.76	20			
						2.10	13			
$_{89}Ac^{229}$			66m	β^-	1.1	≈ 1.0				
$_{89}Ac^{230}$			<1m	β^-	≈ 2.9	≈ 2.2				
$_{89}Ac^{231}$			15m	β^-	2.1	2.1				
Th								$7.4 \pm 0.1b$		
$_{90}Th^{223}$		223.0209	0.9s	α	7.7	7.55				
$_{90}Th^{224}$		224.0214	1.05s	α	7.30	7.17	79%			
						6.99	19			
						6.77	1.2			
						6.70	0.4			
$_{90}Th^{225}$		225.0237	8.0m	α	6.92	α	$\geq 95\%$			
						6.797	9†			
						6.743	7			
						6.699	2			
						6.649	3			
						6.626	3			
						6.500	14			
						6.477	43			
						6.440	15			
						6.344	2			
						6.311	2			
				EC	0.68	EC	$\leq 5\%$			
$_{90}Th^{226}$		226.0249	30.9m	α	6.45	6.33	79%			
						6.22	19			
						6.10	1.7			
						6.03	0.6			
$_{90}Th^{227}$		227.0278	18.5d	α	6.145	6.037	24.5%	$\sigma_f = 1500 \pm 1000b$		
						6.007	2.9			
						5.988	0.002			
						5.976	23.4			
						5.958	3.0			
						5.914	0.8			
						5.908	0.17			
						5.865	2.4			
						5.806	1.3			
						5.794	0.31			
						5.761	0.23			
						5.755	20.3			
						5.726	0.034			
						5.712	4.89			
						5.707	8.20			
						5.669	3.63			
						5.691	1.50			
						5.673	0.057			
						5.666	2.06			
						5.639	0.018			
						5.620	0.0070			
						5.612	0.22			
						5.599	0.17			
						5.584	0.18			
						5.531	0.021			
						5.508	0.0166			
						5.479	0.0012			
						5.457	0.0027			
						5.407	0.0004			
						5.363	0.00066			
						5.334	0.0002			
						5.320	0.0002			
						5.263	0.0026			
						5.247	0.0032			
						5.227	0.0098			
						5.208	0.0070			

Isotope	% nat. abundance	Atomic mass	Lifetime	Modes of decay	Decay energy (MeV)	Particle energies (MeV)	Particle intensities	Thermal neutron capture cross-section	I	μ
						5.192	0.0038			
						5.179	0.0012			
						5.169	0.0017			
						5.145	0.004			
						5.127	0.00062			
						5.109	0.00028			
						5.082	0.00015			
						5.054	0.00023			
						5.032	0.00031			
$_{90}$Th^{228}		228.0287	1.913y	α	5.521	5.424	71%	120 ± 15b		
						5.342	28	σ_f = <0.3b		
						5.211	0.4			
						5.176	0.2			
						5.140	0.03			
$_{90}$Th^{229}		229.0316	7340y	α	5.167	5.054	6.7	σ_f = 32 ± 3b	5/2	+0.38
						5.034	≈0.2			
						5.009	≈0.1			
						4.977	3.4			
						4.967	6.0			
						4.930	0.25			
						4.899	10.7			
						4.842	58.2			
						4.811	11.4			
						4.793	1.0			
						4.756	1.5			
						4.683	0.4			
$_{90}$Th^{230}		230.0331	8.0 × 10^{4}y	α	4.767	4.684	76%	33 ± 10b		
						4.617	24			
						4.474	0.12			
						4.431	0.03			
$_{90}$Th^{231}			25.5h	β^-	0.381	0.299	39%			
						0.218	33			
						0.134	20			
						0.090	8			
$_{90}$Th^{232}		232.0382	1.41 × 10^{10}y	α	4.08	3.994	77%	7.4 ± 0.1b		
						3.935	23			
						3.809	0.2			
$_{90}$Th^{233}			22.2m	β^-	1.246	1.245	<87%	1500 ± 100b		
						1.158		σ_f = 15 ± 2b		
						1.073				
						0.88				
						0.79				
						0.58				
$_{90}$Th^{234}		234.0436	24.1d	β^-	0.263	0.191	65%	1.8 ± 0.5b		
						0.100	35			
Pa										
$_{91}$Pa^{225}			2.0s	α	≈7.5est.	7.25				
$_{91}$Pa^{226}		226.0278	1.8m	α	6.99	α	74%			
						6.86	52†			
						6.82	46			
						6.73	1			
				EC	2.78	EC	26%			
$_{91}$Pa^{227}		227.0289	38.3m	α	6.581	α	≈85%			
						6.465	50.7			
						6.423	11.8			
						6.415	15.2			
						6.401	9.6			
						6.376	2.6			
						6.356	8.0			
						6.336	0.7			

Isotope	% nat. abundance	Atomic mass	Lifetime	Modes of decay	Decay energy (MeV)	Particle energies (MeV)	Particle intensities	Thermal neutron capture cross-section	I	μ
						6.326	0.4			
						6.299	0.8			
				EC	1.00	EC	15%			
$_{91}Pa^{228}$		228.0310	26h	EC	2.10	EC	≈ 98%			
				α	6.227	α	≈ 2			
						6.118	10.5†			
						6.105	12.0			
						6.091	2.3			
						6.078	20.7			
						6.066	1.0			
						6.041	2.3			
						6.028	9.0			
						6.011	0.8			
						5.998	0.3			
						5.989	1.1			
						5.982	2.8			
						5.975	2.7			
						5.947	0.6			
						5.941	0.5			
						5.922	0.8			
						5.907	1.1			
						5.874	1.4			
						5.858	0.3			
						5.843	0.4			
						5.805	7.3			
						5.799	11.3			
						5.779	1.4			
						5.765	2.0			
						5.760	1.4			
						5.756	2.5			
						5.711	1.0			
$_{91}Pa^{229}$			1.5d	EC	0.22	EC	99 +%			
				α	5.837	α	0.25			
						5.74	0.5†			
						5.70	1.5			
						5.671	18.5			
						5.631	9.7			
						5.616	13.3			
						5.592	4.6			
						5.581	36.5			
						5.566	3.9			
						5.537	8.8			
						5.518	0.6			
						5.502	0.7			
						5.480	1.7			
						5.423	0.07			
						5.414	0.15			
						5.321	0.05			
$_{91}Pa^{230}$		230.035	17.7d	EC	1.30	EC	89.6%	$\sigma_f = 1500 \pm 250b$		
				β^-	0.56	β^- 0.41	10.4%			
				α	5.438	α	0.0032%			
						5.343	23†			
						5.338	15			
						5.325	18			
						5.310	13			
						5.299	17			
						5.286	3			
						5.274	3			
						5.267	3.5			
						5.215	0.5			
						5.181	0.5			
						5.151	0.4			
						5.117	0.6			
						5.082	0.7			
						5.058	0.4			
						4.971	0.7			
						4.932	0.4			
						4.796	≈ 0.03			
						4.764	0.2			

Isotope	% nat. abundance	Atomic mass	Lifetime	Modes of decay	Decay energy (MeV)	Particle energies (MeV)	Particle intensities	Thermal neutron capture cross-section	I	μ
$_{91}Pa^{231}$		231.0359	3.25×10^4y	α	5.148	5.051 5.024 5.022 5.017 5.005 4.977 4.967 4.943 4.926 4.892 4.845 4.787 4.729 4.705 4.673 4.635 4.624 4.591 4.558 4.500 4.14	11.0% $? \approx 2.5$ 22 ? 25.4 1.4 0.4 22.8 3.0 0.002 1.4 0.04 8.4 ≈ 1.0 1.5 0.1 0.1 0.015 0.008 0.003	200 ± 10b $\sigma_f = 10 \pm 5$mb	3/2	± 1.98
$_{91}Pa^{232}$			1.31d	β^-	1.34	0.32 1.19 1.30	98% 0.8 0.7	760 ± 100b $\sigma_f = 700 \pm 100$b		
$_{91}Pa^{233}$		233.040	27.0d	β^-	0.571	0.58 0.254 0.154	12% 56 32	22 ± 4b (Pa^{234m}) 21 ± 3b (Pa^{234})	3/2	+3.4
$_{91}Pa^{234m}$			1.17m	β^- IT		2.29 1.57 1.25	98 + % 0.13%			
$_{91}Pa^{234}$		234.043	6.75h	β^-	2.23	0.23 0.51 0.73 1.02 1.35	14% 66 11 7 ≤ 2			
$_{91}Pa^{235}$			23.7m	β^-	1.4	1.4				
$_{91}Pa^{236}$			12m	β^-	≈ 2.9	3.3				
$_{91}Pa^{237}$			39m	β^-	2.30	β^- 2.30 1.3 0.8	60% 30 10			
U								$\sigma(abs) = 7.595$ ± 0.07b $\sigma_f = 4.172$ ± 0.015b		
$_{92}U^{227}$		227.0309	1.3m	α	7.1	6.8				
$_{92}U^{228}$		228.0313	9.3m	α EC	6.80 0.36	α 6.68 6.59 6.44 6.40 EC	$\geq 95\%$ 70† 29 0.7 0.6 $\leq 5\%$			
$_{92}U^{229}$		229.0332	58m	EC α	1.32 6.473	EC α 6.362 6.334 6.299 6.262 6.225 6.187	$\approx 80\%$ $\approx 20\%$ 64† 20 11 1 3 1			

Isotope	% nat. abundance	Atomic mass	Lifetime	Modes of decay	Decay energy (MeV)	Particle energies (MeV)	Particle intensities	Thermal neutron capture cross-section	I	μ
$_{92}U^{230}$		230.0339	20.8d	α	5.991	5.887 5.816 5.666 5.661 5.585 5.542 5.532	67.5% 31.9 0.038 0.26 0.012 0.0005 0.0001	$\sigma_f = 25 \pm 10b$		
$_{92}U^{231}$			4.3d	EC α	0.36 5.55	EC α 5.46	99+% 0.0055%	$\sigma_f = 400 \pm 300b$		
$_{92}U^{232}$		232.0372	73.6y	α SF	5.414	5.320 5.263 5.137 4.997 4.946 4.929 4.51	68.6% 31.2 0.28 2.9×10^{-3} 1.7×10^{-4} 2.1×10^{-4} 2.4×10^{-5}	$75 \pm 10b$ $\sigma_f = 75 \pm 10b$ $\sigma(abs.) = 150 \pm 10b$		
$_{92}U^{233}$		233.0395	$1.62 \times 10^5 y$	α	4.909	4.824 4.804 4.796 4.783 4.758 4.754 4.751 4.729 4.701 4.687 4.681 4.664 4.656 4.641 4.634 4.626 4.615 4.611 4.590 4.572 4.565 4.538 4.513 4.507 4.503 4.483 4.465 4.457 4.411 4.404 4.309	84.4% 0.051 0.28 13.23 0.016 0.163 0.01 1.61 0.06 0.0028 0.01 0.042 0.005 0.003 0.01 ≤ 0.004 0.004 0.006 0.007 0.0023 0.0028 0.004 0.018 0.012 0.001 0.0014 0.003 0.0028 0.0004 0.0003 0.0009		5/2	+0.54
$_{92}U^{234}$	0.005	234.0409	$2.47 \times 10^5 y$	α	4.856	4.773 4.722 4.599 4.265 4.139	72% 28 0.35 4.5×10^{-5} 3×10^{-5}	$\sigma_{(abs)} = 95 \pm 7b$ $\sigma_f \leq 0.65b$		
$_{92}U^{235}$	0.720	235.0439	$7.1 \times 10^8 y$	α SF	4.681	4.597 4.556 4.502 4.445? 4.415 4.396 4.366 4.344? 4.323 4.266 4.216 4.157	4.6% 3.7 1.2 0.6 4 57 18 1.5 3 0.6 5.7 ≈ 0.5	$100.5 \pm 1.4b$ $\sigma_f = 579.5 \pm 2.0b$ $\sigma_{(abs)} = 679.9 \pm 2.3b$	7/2	-0.35
$_{92}U^{235m}$			26.1m	IT	<0.001					

Isotope	% nat. abundance	Atomic mass	Lifetime	Modes of decay	Decay energy (MeV)	Particle energies (MeV)	Particle intensities	Thermal neutron capture cross-section	I	μ
$_{92}U^{236}$		236.0457	2.39×10^7 y	α SF	4.573	4.493 4.443 4.330	74% 26 0.3	6.0 ± 0.5b $\sigma_f = 0$		
$_{92}U^{237}$			6.75d	β^-	0.517	0.248 ≈0.09	>80% ≈12	480 ± 160b $\sigma_f = 2$b		
$_{92}U^{238}$	99.275	238.0508	4.51×10^9 y	α SF	4.268	4.195 4.147 4.135	77% 23 0.23	2.720 ± 0.025b		
$_{92}U^{239}$			23.5m	β^-	1.28	1.29 1.21	20% 80	22 ± 5b $\sigma_f = 14 \pm 3$b		
$_{92}U^{240}$			14.1h	β^-	0.5	0.36 0.32	75% 25			
Np										
$_{93}Np^{229}$			4.0m	α		6.89				
$_{93}Np^{230}$			4.6m	α		6.66				
$_{93}Np^{231}$		231.0383	~50m	α	6.40	6.29				
$_{93}Np^{232}$			≈13m	EC	≈2.6					
$_{93}Np^{233}$		233.0406	35m	EC α	≈1.0 5.7	EC α 5.54	99+% 0.001%			
$_{93}Np^{234}$			4.40d	EC β^+	1.80	EC 0.8	99+% ≈0.05%	$\sigma_f = 900 \pm 300$b		
$_{93}Np^{235}$			410d	EC α	1.23 5.187	EC α 5.10 5.02 4.93 4.87	99+% 0.0016 3.8† 84 12 1			
$_{93}Np^{236m}$			>5000y	EC, β^-				$\sigma_f = 2800 \pm 800$b		
$_{93}Np^{236}$		236.0466	22h	β^- EC	0.52 0.91	0.51 0.36 EC	≈40% ≈10 ≈50%			
$_{93}Np^{237}$		237.0480	2.14×10^6 y	α	4.956	4.872 4.870 4.862 4.816 4.802 4.787 4.770 4.765 4.740 4.711 4.707 4.698? 4.693 4.663 4.658 4.638 4.598 4.594 4.580 4.573 4.514 4.385	0.4% 0.9 0.2 1.5 1.6 51 19 17 0.02 0.13 0.29 0.07 0.18 1.6 0.57 4.6 0.06 0.08 0.02 0.05 0.01 0.02	170 ± 5b $\sigma_f = 19 \pm 3$mb	5/2	+5
$_{93}Np^{238}$			2.1d	β^-	1.29	1.24 1.13?	38% 3	$\sigma(\eta, \gamma)$ 0 $\sigma_f = 2000 \pm 300$b	2	

Isotope	% nat. abundance	Atomic mass	Lifetime	Modes of decay	Decay energy (MeV)	Particle energies (MeV)	Particle intensities	Thermal neutron capture cross-section	I	μ
						0.28	20			
						0.25	31			
						0.20?	8			
$_{93}Np^{239}$			2.35d	β^-	0.723	0.713	7%	$35 \pm 10b$	5/2	
						0.654	4	(Np^{240m})		
						0.437	48	$25 \pm 15b$		
						0.393	13	(Np^{240})		
						0.332	28	$\sigma_f = <1b$		
$_{93}Np^{240m}$			7.3m	β^-	2.18	0.7	7%			
						1.30	10			
						1.60	31			
						2.18	52			
$_{93}Np^{240}$			63m	β^-	2.0	0.89				
$_{93}Np^{241}$			16m	β^-	1.4	1.4				
Pu										
$_{94}Pu^{232}$		232.0411	36m	EC α	1.0 6.70	EC α 6.70	$\leq 98\%$ $\geq 2\%$			
$_{94}Pu^{233}$			20m	EC α	≈ 2.0 6.42	EC 6.31	$99+\%$ 0.1%			
$_{94}Pu^{234}$		234.0433	9.0h	EC α	0.4 6.30	EC α 6.203 6.152 6.032	94% 6% 68† 32 0.4			
$_{94}Pu^{235}$			26m	EC α	1.1 5.96	EC α 5.86	$99+\%$ 0.003%			
$_{94}Pu^{236}$		236.0461	2.85y	α SF	5.868	5.769 5.722 5.616 5.21 5.08	69% 31 0.18 2.7×10^{-4} 6×10^{-4}	$\sigma_f = 170 \pm 35b$		
$_{94}Pu^{237m}$			0.18s	IT SF	0.145					
$_{94}Pu^{237}$		237.0483	45.6d	EC α SF	0.22 5.74	EC α 5.66 5.37	$99+\%$ $3.3 \times 10^{-3}\%$ 21† 79	$\sigma_f = 2500 \pm 500b$		
$_{94}Pu^{238}$		238.0495	86y	α SF	5.592	5.499 5.456 5.358 5.214 4.70	72% 28 0.09 0.005 1.1×10^{-4}	$560 \pm 25b$ $\sigma_f = 16.5 \pm 1.0b$		
$_{94}Pu^{239}$		239.0522	24,400y	α SF	5.243	5.157 5.145 5.112? 5.107 5.078 5.066 5.056 5.031 5.010 5.001 4.988 4.963 4.957 4.937 4.914	73.3% 15.1 ≤ 0.03 11.5 0.032 0.0009 0.021 0.005 0.008 0.0006 0.005 0.003 0.0005 0.003 0.0008	$265.7 \pm 3.7b$ $\sigma_f = 742.4 \pm 3.5b$ $\sigma_{(abs)} = 1008.1 \pm 4.9b$	1/2	+0.200

Isotope	% nat. abundance	Atomic mass	Lifetime	Modes of decay	Decay energy (MeV)	Particle energies (MeV)	Particle intensities	Thermal neutron capture cross-section	I	μ
						4.873	0.0007			
						4.863	0.0008			
						4.830	0.0015			
						4.801	0.0006			
						4.775	0.0006			
						4.754?	0.0004			
						4.743⎫ 4.739⎭	0.0026			
						4.695	0.0004			
						4.636	0.0002			
$_{94}$Pu^{240}		240.0540	6580y	α SF	5.255	5.168 5.123 5.020 4.856 4.49	76% 24 0.09 3×10^{-3} 2.1×10^{-5}	290 ± 15b		
$_{94}$Pu^{241}		241.057	13.2y	β^- α	0.0208 5.144	0.0208 α 5.051 5.041 4.998? 4.995 4.971 4.968? 4.896 4.853 4.798 4.73	99 + % 0.0023% 0.35† 1.0 0.36 1.1 83 12 1.2 0.03	360 ± 15b $\sigma_f = 1011 \pm 9$b $\sigma_{(abs)} = 1371 \pm 18$b	5/2	-0.73
$_{94}$Pu^{242}		242.0587	3.79×10^5y	α	4.98	4.903 4.863	76% 24	20 ± 3b		
$_{94}$Pu^{243}			4.98h	β^-	0.59	0.579 0.490	62% 38	75 ± 50b $\sigma_f = 196 \pm 16$b.		
$_{94}$Pu^{244}		244.064	8×10^7y	α SF	4.66	α		1.8 ± 0.3b		
$_{94}$Pu^{245}			10h	β^-	≈1.2			260 ± 150b		
$_{94}$Pu^{246}			10.9d	β^-	0.37	0.15 0.33	≈73% ≈27			
Am										
$_{95}$Am^{237}			≈1.3h	EC α	≈1.5 ≈6.2	EC 6.02	99+% 0.005%			
$_{95}$Am^{238}			1.9h	EC	≈2.3					
$_{95}$Am^{239}			12.1h	EC α	0.81 5.93	EC 5.78	99+% 0.005%			
$_{95}$Am^{240}			51h	EC	≈1.3					
$_{95}$Am^{241}		241.0567	458y	α	5.640	5.545 5.313 5.486 5.469? 5.443 5.417 5.389 5.322 5.279 5.244 5.223 5.194 5.182 5.178 5.156 5.114	0.25% 0.12 86 <0.04 12.7 ≈0.01 1.3 0.015 0.0005 0.0024 0.0013 0.0006 0.0009 0.0003 0.0007 0.0004	70 ± 5b (Am^{242m}) 670 ± 60b (Am^{242}) $\sigma_f = 3.15 \pm 0.1$b	5/2	$+1.59$

Isotope	% nat. abundance	Atomic mass	Lifetime	Modes of decay	Decay energy (MeV)	Particle energies (MeV)	Particle intensities	Thermal neutron capture cross-section	I	μ
$_{95}Am^{242m}$			152y	IT α	0.048 5.588	5.096 5.089 5.068 5.004 4.834 4.800 IT α 5.411 5.367 5.315 5.207 5.142 5.088 5.067	0.0004 0.0004 0.00014 0.001 0.0007 0.00009 99.52% 0.48 1.2† 1.5 0.8 89 6.1 0.3 0.2	$2400 \pm 500b$ $\sigma_f = 7200 \pm 300b$		
$_{95}Am^{242}$			16.0h	β^- EC	0.66 0.73	0.667 0.625 EC	$\approx 33\%$ ≈ 49 ≈ 18	$<200b$ $\sigma_f = 2100 \pm 200b$	1	± 0.382
$_{95}Am^{243}$		243.0614	7.37×10^3y	α	5.439	5.350 5.321 5.276 5.234 5.181 5.113 5.088 5.035 5.029 5.008 4.946 4.930 4.919 4.695	0.16% 0.12 87.9 10.6 1.1 5.4×10^{-3} 4×10^{-3} 2.2×10^{-3} 1.6×10^{-3} 3.4×10^{-4} 1.8×10^{-4} 8×10^{-5} 6×10^{-4}	$180 \pm 20b$	5/2	+1.4
$_{95}Am^{244m}$			26m	β^- EC		1.50 EC	$99 + \%$ 0.039%			
$_{95}Am^{244}$			10.1h	β^-	1.429	0.387	100%	$2300 \pm 300b$		
$_{95}Am^{245}$			2.1h	β^-	0.905	0.905 (0.709) (0.653)	78% ≈ 5 ≈ 17			
$_{95}Am^{246}$			25.0m	β^-	2.29	1.31 1.60 2.10	79% 14 7			
$_{95}Am^{247}$			20m	β^-		227 285				
Cm										
$_{96}Cm^{238}$		238.0530	2.5h	EC α	≈ 1.0 6.63	EC 6.51	$<90\%$ >10			
$_{96}Cm^{239}$			2.9h	EC	≈ 1.7	EC	100%			
$_{96}Cm^{240}$		240.0555	26.8d	α SF	6.441	6.292 6.249 6.150 5.992	71.1% 28.9 0.052 0.014			
$_{96}Cm^{241}$			35d	EC α	0.77 6.184	EC α 6.083 6.039 5.941 5.929 5.887 5.831	$\approx 99\%$ 1% 0.64% 0.10 71.5 16.3 11.5 2			

Isotope	% nat. abundance	Atomic mass	Lifetime	Modes of decay	Decay energy (MeV)	Particle energies (MeV)	Particle intensities	Thermal neutron capture cross-section	I	μ
						5.824	1.4			
						5.810	1.2			
						5.789	1.8			
						5.784	0.6			
						5.781	3.0			
						5.732	0.9			
$_{96}Cm^{242}$		242.0588	163d	α		6.115	74.2%	20 ± 10b	0	
						6.071	25.8	$\sigma_f = <5$b		
						5.974	0.036			
						5.819	4.6×10^{-3}			
						5.611	2×10^{-5}			
						5.516	2.5×10^{-4}			
						5.191	2.5×10^{-5}			
				SF		5.148	$\leq 5 \times 10^{-6}$			
$_{96}Cm^{243}$		243.0614	32y	α	6.168	α	99.7%	250 ± 50b		
						6.067	1.5	$\sigma_f = 700 \pm 50$b		
						6.057	4.7			
						6.010	1.1			
						5.993	5.6			
						5.907	0.1			
						5.876	0.7			
						5.786	73.5			
						5.742	10.6			
						5.686	1.6			
						5.682	0.2			
						5.646	0.03			
						5.639	0.14			
						5.622	0.06			
						5.612	≈ 0.04			
						5.609	0.01			
						5.607?				
						5.593	0.01			
						5.587	≈ 0.02			
						5.582	≈ 0.009			
						5.575	0.007			
						5.568	0.007			
						5.537	0.002			
						5.532	0.006			
						5.523	0.002			
						5.332	0.003			
						5.323	0.003			
						5.316	0.001			
						5.267	0.0015			
				EC	0.006	EC	0.3%			
$_{96}Cm^{244}$		244.0629	17.6y	α	5.902	5.806	76.4%	13 ± 5b		
						5.765	23.6	$\sigma_f = 1.0 \pm 0.5$b		
						6.664	0.02			
						5.513	3.4×10^{-3}			
						5.313	$\sim 4 \times 10^{-5}$			
						5.215	$\sim 1 \times 10^{-4}$			
						4.960	3×10^{-4}			
				SF		4.920	1.3×10^{-4}			
$_{96}Cm^{245}$		245.0653	9.3×10^3y	α	5.624	5.533	0.5%	250 ± 50b		
						5.492		$\sigma_f = 2000 \pm 80$b		
						5.468	2.7			
						5.464	2.0			
						5.359	87.6			
						5.305	4.5			
						5.255	0.3			
						5.245	0.7			
						5.240	0.5			
						5.231	0.6			
						5.199	0.35			
						5.175	0.25			
						5.158	0.3			
$_{96}Cm^{246}$		246.0674	5.5×10^3y	α	5.476	5.385	79%	8.4 ± 2.0b		
				SF		5.342	21			

Isotope	% nat. abundance	Atomic mass	Lifetime	Modes of decay	Decay energy (MeV)	Particle energies (MeV)	Particle intensities	Thermal neutron capture cross-section	I	μ
$_{96}Cm^{247}$		247.070	1.6×10^7y	α	≈ 5.3			180b $\sigma_f = 108 \pm 5b$		
$_{96}Cm^{248}$		248.072	4.7×10^5y	α SF	5.161	α 5.080 5.036 SF	89% 82† 18 11%	$6 \pm 2b$		
$_{96}Cm^{249}$		249.0758	64m	β^-	0.9	0.9		$1.6 \pm 1.0b$		
$_{96}Cm^{250}$			1.7×10^4y	SF		SF	100%			
Bk										
$_{97}Bk^{243}$			4.6h	EC α	1.51 6.871	EC α 6.758 6.718 6.666 6.605 6.574 6.542 6.502 6.446 6.394 6.210 6.182	99 + % 0.15% 15† 12 1.2 0.7 26 19 7 0.7 0.3 14 3.9			
$_{97}Bk^{244}$			4.4h	EC α	≈ 2.2 6.79	EC α 6.666 6.624	99 + % 0.006% $\approx 50†$ ≈ 50			
$_{97}Bk^{245}$			4.98d	EC α	0.82 6.464	EC α 6.358 6.317 6.265 6.200 6.153 6.124 6.087 6.038 5.985 5.889 5.858	99 + % 0.11% 16† 15 1.4 1.4 19 15 5.6 0.6 0.2 22 4.0			
$_{97}Bk^{246}$			1.8d	EC	≈ 1.5	EC	100%			
$_{97}Bk^{247}$		247.0702	1.4×10^3y	α	5.86	5.68 5.52 5.31	37% 58 5			
$_{97}Bk^{248}$			>9y	β^- EC	0.65 0.5	0.65 EC	70% 30%			
$_{97}Bk^{249}$		249.075	314d	β^- α SF	0.125 5.520	β^- 0.125 α 5.431 5.412 5.384 5.345 5.109 5.046	99 + % 0.0022% 6.7† 69 18 2.6 2.7 0.12	$1000 \pm 500b$		
$_{97}Bk^{250}$			3.22h	β^-	1.76	0.73 1.72 1.76	89% 6 5	$\sigma_f = 960 \pm 150b$		
Cf										
$_{98}Cf^{242}$			3.7m	α	7.49	7.37				

Isotope	% nat. abundance	Atomic mass	Lifetime	Modes of decay	Decay energy (MeV)	Particle energies (MeV)	Particle intensities	Thermal neutron capture cross-section	I	μ
$_{98}Cf^{244}$		244.0659	25m	α	7.59	7.18				
$_{98}Cf^{245}$			44m	EC α	1.52 7.23	EC 7.12	70% 30%			
$_{98}Cf^{246}$		246.0688	36h	α SF	6.87	6.760 6.718 6.63 6.47	78% 22 0.18			
$_{98}Cf^{247}$			2.5h	EC	≈0.8					
$_{98}Cf^{248}$		248.0724	360d	α SF	6.37	6.27 5.84	82% 18			
$_{98}Cf^{249}$		249.0748	360y	α SF	6.295	6.201 6.146 6.079 5.997? 5.948 5.905 5.848 5.812 5.784? 5.755 5.693	1.9% 1.1 0.4 0.08 3.3 3.0 1.2 84 0.5 4.4 0.4	300 ± 200b σ_f = 1735 ± 70b		
$_{98}Cf^{250}$		250.0766	13y	α SF	6.128	6.031 5.987 5.889	83% 17 0.22	1500 ± 500b σ_f = <350b		
$_{98}Cf^{251}$		251.080	≈800y	α	5.94	6.072 6.012 5.937 5.849 5.813 5.797 5.762 5.726 5.672 5.645 5.630 5.600 5.564 5.501	2.7% 12.3 0.3 27.8 4.1 2.8 3.9 1.1 34.7 2.8 4.8 0.5 1.8 0.3	2100 ± 1000b σ_f = 4000 ± 1000b		
$_{98}Cf^{252}$		252.082	2.65y	α SF	6.217	α 6.119 6.076 5.975 SF	97% 84.3† 15.5 0.28 3%	10 ± 5b		
$_{98}Cf^{253}$		253.0850	17.6d	β^- α	0.27 6.2	0.27 } 0.17?} α 5.979 5.921	99+% 0.31% 94.7† 5.3	≈165b		
$_{98}Cf^{254}$		254.087	60.5d	α SF	5.929	α 5.834 5.792 SF	≈0.2% 83† 17 99+%	<2b		
Es										
$_{99}Es^{246}$			7.3m	EC α	≈3.6 7.48	EC α 7.33	90% 10%			
$_{99}Es^{247}$			5.0m	EC α	≈2.2 ≈7.3	EC 7.33	≈93% ≈7%			

Isotope	% nat. abundance	Atomic mass	Lifetime	Modes of decay	Decay energy (MeV)	Particle energies (MeV)	Particle intensities	Thermal neutron capture cross-section	I	μ
$_{99}Es^{248}$			25m	EC α	≈ 2.8 6.99	EC 6.88	$99+\%$ $\approx 0.3\%$			
$_{99}Es^{249}$			2h	EC α	1.40 6.88	EC 6.77	$99+\%$ 0.13%			
$_{99}Es^{250}$			8h	EC	2.1	EC	100%			
$_{99}Es^{251}$			1.5d	EC α	≈ 0.4 6.59	EC α 6.48	$99+\%$ 0.53%			
$_{99}Es^{252}$		252.0829	$\approx 140d$	α	6.75	6.639 6.58 6.493 6.26 6.23 6.09 6.07? 6.02 5.99 5.94?	82† 13 2.3 0.8 0.33 0.6 0.11 1.1 0.07 0.09			
$_{99}Es^{253}$		253.0847	20.47d	α SF	6.747	6.640 6.631 6.601 6.597 6.559 6.547 6.504 6.486 6.436 6.256 6.216 6.165 6.03 5.93–6.04 5.91 5.73	90% 0.8 0.7 6.6 0.75 0.85 0.26 0.08 0.1 0.04 0.04 0.015 2×10^{-4} 1×10^{-4} 3×10^{-5} 8×10^{-5}	$150 \pm 5db$		
$_{99}Es^{254m}$			39.3h	β^- EC SF		β^- 1.127 0.48 EC	$99+\%$ 25† 75 0.08%	$\sigma_f = 1840 \pm 80b$		
$_{99}Es^{254}$		254.0881	276d	α	6.623	6.486 6.437 6.424 6.392 6.367 6.355 6.331 6.284 6.276 6.20 6.19 6.113 6.056	0.27% 93 1.7 0.13 2.9 0.74 0.05 0.16 0.22 0.05 0.05 0.33 0.16	$<40b$ $\sigma_f = 3060 \pm 180b$		
$_{99}Es^{255}$			38.3d	β^- α	≈ 0.5 ≈ 6.8	6.307	91.5% 8.5%	$\approx 40b$		
$_{99}Es^{256}$			Short	β^-						
Fm										
$_{100}Fm^{248}$		248.0772	0.6m	α						
$_{100}Fm^{249}$			$\approx 2.5m$	α	8.0	7.9				

Isotope	% nat. abundance	Atomic mass	Lifetime	Modes of decay	Decay energy (MeV)	Particle energies (MeV)	Particle intensities	Thermal neutron capture cross-section	I	μ
$_{100}Fm^{250}$		250.0795	30m	α	7.56	7.44	100%			
$_{100}Fm^{251}$			7h	EC α	≈ 1.6 ≈ 7.4	EC 6.90	$\approx 99\%$ $\approx 1\%$			
$_{100}Fm^{252}$		252.0827	23h	α	7.17	7.05	100%			
$_{100}Fm^{253}$			3d	EC α	0.4 7.3	EC α 6.96 6.91	89% 11% 4† 1			
$_{100}Fm^{254}$		254.0870	3.24h	α SF	7.310	7.200 7.158 7.061 SF	85% 14 0.9 0.055%			
$_{100}Fm^{255}$		255.090	20.1h	α	7.244	7.130 7.106 7.084 7.027 6.985 6.968 6.956 6.923 6.896 6.895 6.880 6.842 6.811 6.710 6.59 6.54 6.49 6.41	0.09% 0.10 0.43 93.4 0.11 5.3 0.024 0.017 0.01 0.60 0.013 0.002 0.12 0.03 0.020 0.015 0.0035 0.0003	26 ± 3b		
$_{100}Fm^{256}$			2.7h	SF α	6.96	SF 6.86	97% 3%			
$_{100}Fm^{257}$		257.095	80d	α	6.871	6.76? 6.703 6.526 6.450 6.36?	0.4% 3.2 94 2.2 0.5			
Md										
$_{101}Md^{255}$		255.0906	0.6h	EC α	≈ 0.6 7.46	EC 7.35	90% 10%			
$_{101}Md^{256}$			1.5h	EC α	1.4 7.28	EC 7.18	97% 3%			
$_{101}Md^{257}$			3h	EC α	7.18?	EC α 7.25? 7.08	$\approx 92\%$ $\approx 8\%$			
No										
$_{102}No^{251}$			1s	α	8.8	8.68? 8.58	$\approx 20\%$ ≈ 80			
$_{102}No^{252}$			2.1s	α SF	8.55	8.41 SF	$\approx 70\%$ 30%			
$_{102}No^{253}$			95s	α	8.2	8.01				
$_{102}No^{254}$			55s	α	8.23	8.10				
$_{102}No^{255}$			3m	α	8.24	8.11				

Isotope	% nat. abundance	Atomic mass	Lifetime	Modes of decay	Decay energy (MeV)	Particle energies (MeV)	Particle intensities	Thermal neutron capture cross-section	I	μ
$_{102}No^{256}$			2.7s	α SF	8.56	8.43 SF	99 + % 0.5%			
$_{102}No^{257}$			20s	α	8.4	8.27 8.23	$\approx 50\%$ ≈ 50			
Lr										
$_{103}Lr^{256}$			8s	α		8.6				
$_{103}Lr^{257}$										
Ku										
$_{104}Ku^{260}$			0.3s	SF						
$_{105}?^{260}$				α		9.4				
$_{105}?^{261}$			1.8s	α		8.93				
$_{105}?^{262}$			40s	α		8.45				

GAMMA ENERGIES AND INTENSITIES OF RADIONUCLIDES

Compiled by
Russell L. Heath

National Reactor Testing Station
Idaho Nuclear Corporation
Idaho Falls, Idaho

The following table lists the Gamma Energies and Intensities of Radionuclides for a large number of the radioactive isotopes. The values of the gamma ray energies are given in keV. The abbreviation Ann. Rad. refers to 511.006 keV photon associated with the annihilation of positrons in matter. The values for the gamma-ray intensities are given in percent of decays for a given gamma-ray transition and not the percent of transitions between two given energy levels in the daughter nucleus. Relative intensities for gamma rays are indicated by the symbol †.

Gamma energies (KeV)	Gamma intensities
$^{5}_{2}$He	
99	88%
$^{7}_{4}$Be	
477.575±0.02	
$^{11}_{4}$Be	
2120	32%
4640	2.1%
5060	2.4%
6790	4.4%
7970	1.7%
$^{12}_{5}$B	
1190	0.0001%
4430	1.3%
$^{13}_{5}$B	
3230	0.0005%
4430	1.3%
$^{13}_{5}$B	
3670	7%
$^{10}_{6}$C	
717	100%
1030	1.7%
$^{11}_{6}$C	
5299	68%
$^{13}_{7}$N	
4430	2.4%
$^{13}_{7}$N	
Ann. Rad.	
$^{16}_{7}$N	
6128.9±0.4	100†
7117.0±0.5	7
$^{17}_{7}$N	
870	7†
2190	1
$^{18}_{7}$N	
820	60†
1650	63
1980	100
2470	43
$^{14}_{8}$O	
2311	99%
Ann. Rad.	
$^{15}_{8}$O	
Ann. Rad.	
$^{19}_{8}$O	
112	2.7
200	97
1370	59
1440	2.7
$^{20}_{8}$O	
1070	
$^{17}_{9}$F	
Ann. Rad.	
$^{18}_{9}$F	
Ann. Rad.	
$^{21}_{9}$F	
345	100†

Gamma energies (KeV)	Gamma intensities
1380	13
$^{22}_{9}$F	
1274.52±0.07	15†
2060	10
$^{17}_{10}$Ne	
Ann. Rad.	
$^{18}_{10}$Ne	
Ann. Rad.	
1041	7%
$^{19}_{10}$Ne	
Ann. Rad.	
$^{23}_{10}$Ne	
440	1000†
1630	28
1960	0.4
2070	3.0
2270	0.2
2550	0.8
2870	0.1
2990	0.9
$^{24}_{10}$Ne	
472	100†
880	8
$^{20}_{11}$Na	
Ann. Rad.	
$^{21}_{11}$Na	
Ann. Rad.	
347	2.3%
$^{22}_{11}$Na	
Ann. Rad.	
1274.52±0.07	
$^{24}_{11}$Na	
1368.650±0.050	100%
2754.10±0.07	100%
3850	0.09%
4230	0.0015%
$^{25}_{11}$Na	
400	10†
580	9
980	10
1610	3
$^{26}_{11}$Na	
1820	
$^{20}_{12}$Mg	
Ann. Rad.	
$^{21}_{12}$Mg	
Ann. Rad.	
$^{23}_{12}$Mg	
Ann. Rad.	
440	9.1%
$^{27}_{12}$Mg	
80	0.7%
840	70%
1013	30%
$^{28}_{12}$Mg	
32	96†

Gamma energies (KeV)	Gamma intensities
391	31
950	29
350	70
$^{24}_{13}$Al	
Ann. Rad.	
1368	40†
2754	32
4200	15
5400	6
7100	7
$^{25}_{13}$Al	
Ann. Rad.	
$^{26m}_{13}$Al	
Ann. Rad.	
$^{26}_{13}$Al	
1830	99.7%
1120	3.7%
2960	0.3%
$^{27}_{13}$Al	
1780	100%
$^{28}_{13}$Al	
1280	85%
2430	15%
$^{30}_{13}$Al	
2240	10†
3520	6
$^{26}_{14}$Si	
Ann. Rad.	
820	
$^{27}_{14}$Si	
Ann. Rad.	
$^{31}_{14}$Si	
1260	<1%
$^{28}_{15}$P	
1790	75%
2600	
4440	10%
4900	
6100	
6700	
7000	
7600	5%
$^{29}_{15}$P	
1280	0.8%
2430	0.2%
$^{30}_{15}$P	
2160	0.5%
$^{34}_{15}$P	
2100	
4000	0.2%
$^{29}_{16}$S	
Ann. Rad.	
$^{30}_{16}$S	
Ann. Rad.	
$^{31}_{16}$S	
687	80%

Gamma energies (KeV)	Gamma intensities
$^{31}_{16}$S	
Ann. Rad.	
1270	1.1%
$^{37}_{16}$S	
3090	90%
$^{38}_{16}$S	
1880	0.5%
$^{33}_{17}$Cl	
2210	70%
3770	
4270	7%
4770	14%
$^{34}_{17}$Cl	
2900	0.3%
$^{34m}_{17}$Cl	
Ann. Rad.	
640	
770	
1176.9±1.0	32%†
2128.2±1.0	100%
3303.9±1.0	32%
4100	1%
146.5±0.3	IT (50%)
$^{34}_{17}$Cl	
Ann. Rad.	
$^{38m}_{17}$Cl	
660	
$^{38}_{17}$Cl	
1600	85†
2167.6±0.7	100
3760	0.05
$^{39}_{17}$Cl	
246	90†
1270	100
1520	85
$^{40}_{17}$Cl	
1460	
2830	
3100	
5800	
$^{35}_{18}$Ar	
Ann. Rad.	
1190	5%
1730	2%
$^{37}_{18}$Ar	
Cl X-ray	
$^{41}_{18}$Ar	
1293	99.2%
1660	0.05%
$^{37}_{19}$K	
Ann. Rad.	
2790	2%
$^{38m}_{19}$K	
Ann. Rad.	

Gamma energies (KeV)	Gamma intensities
$^{38}_{19}K$	
Ann. Rad.	
1650	weak
2167.6±0.7	99·†%
$^{40}_{19}K$	
Ann. Rad.	
1460	11%
$^{42}_{19}K$	
310	1.1†
600	0.1
900	0.1
1020	0.1
1524	100
1924	0.3
2440	0.2
$^{43}_{19}K$	
220.4±0.8	3%
372.6±1.0	85%
396.9±0.8	11%
593.6±1.0	13%
617.5±1.0	81%
990.2±2.0	
1021.9±2.0	~2%
$^{44}_{19}K$	
480	
630	
740	
900	1†
1060	1
1157.02±0.1	100
1499.63±0.25	13†
1740	
2080	
2170	
2550	12
3400	
3660	6
4400	
4600	
5000	0.6
$^{45}_{19}K$	
175	
500	
900	
1230	
1710	
1900	
2100	
2350	
2600	
3100	
$^{37}_{20}Ca$	
Ann. Rad.	
$^{38}_{20}Ca$	
Ann. Rad.	
3.5	
$^{39}_{20}Ca$	
Ann. Rad.	
$^{45}_{20}Ca$	
12.5	
$^{47}_{20}Ca$	
490	5%
520	0.05%
760	0.05%
810	5%
1290	71%
$^{49}_{20}Ca$	
3100	88%
4050	10%
4680	0.8%
$^{50}_{20}Ca$	
72	
258	
$^{40}_{21}Sc$	
Ann. Rad.	
3750	
$^{41}_{21}Sc$	
Ann. Rad.	
$^{42m}_{21}Sc$	
Ann. Rad.	
438	100†
1230	
1520	100

Gamma energies (KeV)	Gamma intensities
$^{42}_{21}Sc$	
Ann. Rad.	
$^{43}_{21}Sc$	
220	
370	
620	
960	
$^{44m}_{21}Sc$	
271.19±0.8 (IT)	98.6%
1001.99±0.1	
1126.16±0.15	
1157.02±0.1	
$^{44}_{21}Sc$	
Ann. Rad.	
1126.16±0.15	
1157.02±0.1	
1499.63±0.25	
$^{46}_{21}Sc$	
142	
889.25±0.07	100%
1120.50±0.07	100%
$^{47}_{21}Sc$	
159.38±0.08	40%
$^{48}_{21}Sc$	
175.4±0.2	6%
983.46±0.15	100%
1037.5±0.2	100%
1311.9±0.2	100%
$^{49}_{21}Sc$	
1780	0.03%
$^{50m}_{21}Sc$	
258	
$^{50}_{21}Sc$	
520	85†
1120	100
1570	100
$^{44}_{22}Ti$	
67.85±0.07	10†
78.40±0.10	14
$^{45}_{22}Ti$	
718	
1238	
1408	
1665	
$^{51}_{22}Ti$	
320.07±0.05	100†
605	1.5
928	4.4
$^{46}_{23}V$	
Ann. Rad.	
$^{47}_{23}V$	
1500	
1800	0.8%
2160	0.2%
$^{48}_{23}V$	
Ann. Rad.	
290.1±0.4	
928.3±0.3	
944.1±0.2	10†
983.46±0.15	100
1311.9±0.2	98
1437.6±0.4	
2240.1±0.2	2.7
$^{52}_{23}V$	
935	0.14%
1331	0.9%
1433	100%
1531	0.16%
$^{53}_{23}V$	
1000	100%
$^{54}_{23}V$	
834.795±0.04	100%
990	100%
2210	100%
$^{48}_{24}Cr$	
116	95
310	100%
$^{49}_{24}Cr$	
62.31±0.1	15†
90.7±0.1	30
152.85±0.1	14
610	

Gamma energies (KeV)	Gamma intensities
$^{51}_{24}Cr$	
320.080±0.013	9%
$^{56}_{24}Cr$	
26	
83	
$^{50m}_{25}Mn$	
Ann. Rad.	
660	<1†
790	3
1110	3
1280	<1
1450	1
$^{50}_{25}Mn$	
Ann. Rad.	
1560	
2030	
$^{52m}_{25}Mn$	
3.83	2%
Ann. Rad.	
$^{52}_{25}Mn$	
Ann. Rad.	
345.7	1%
630	
744	100%
846.8	100%
938.1	84%
1070	
1214	
1246	
1332	6%
1434	100%
1463	ω
2620	0.08%
$^{54}_{25}Mn$	
834.795±0.04	100%
$^{56}_{25}Mn$	
846.78±0.06	100†
1810.96±0.2	30
2113.2±0.2	15.3
2522.6±0.3	1.2
2657.24±0.2	0.7
2959.8±0.2	0.4
3367±1.0	0.21
$^{57}_{25}Mn$	
117	
134	
220	
350	
690	
$^{58}_{25}Mn$	
360	
410	
520	
570	
820	
1000	
1250	
1400	
1600	
2200	
2800	
$^{52}_{26}Fe$	
165	
Ann. Rad.	
383	
$^{53}_{26}Fe$	
Ann. Rad.	38%
377.5±0.5	
$^{55}_{26}Fe$	
Mn X-rays	
$^{59}_{26}Fe$	
142.45±0.08	0.8%
192.23±0.08	2.5%
334.81±0.10	0.3%
1099.27±0.08	56%
1291.58±0.10	44%
$^{61}_{26}Fe$	
130	11†
180	3
230	0.3
300	48
1030	98

Gamma energies (KeV)	Gamma intensities
1200	100
1650	1.4
1980	0.7
$^{54m}_{27}Co$	
Ann. Rad.	
410	1†
1140	1
1410	1
$^{54}_{27}Co$	
$^{55}_{27}Co$	
Ann. Rad.	
95	0.5%
245	<0.06%
390	
415	
480	12%
650	0.5%
750	0.2%
805	2.5%
892	2.4%
930	80%
985	0.9%
1060	0.4%
1210	1.0%
1320	6%
1370	4.2%
1410	13%
1580	0.09%
1800	0.06%
2190	0.11%
2310	
$^{56}_{27}Co$	
Ann. Rad.	
263	
788.0±0.2	0.36%
846.78±0.06	100%
977.4±0.1	1.5%
1037.85±0.06	14%
1175.08±0.07	1.6%
1140.18±0.1	
1175.1±0.1	
1238.29±0.04	64%
1360.219±0.04	4%
1771.33±0.06	14%
1810.46±0.07	
1963.64±0.1	0.68%
2015.33±0.07	2.6%
2034.90±0.07	6.6%
2112.95±0.1	0.56%
2212.8±0.1	0.60%
2598.52±0.1	14%
3009.56±0.1	0.6%
3202.18±0.1	2.9%
3253.61±0.1	7.2%
3273.16±0.1	1.6%
3451.68±0.1	0.72%
3547.5±0.2	0.2%
$^{57}_{27}Co$	
Fe X-rays	
14.41±0.020	8.4%
122.060±0.010	85%
136.471±0.010	11%
231	0.0005%
339.7	0.0048%
352.4	0.0037%
366.7	0.0007%
570.3	0.014%
692.1	0.26 0.16%
706.8	0.0067%
$^{58}_{27}Co$	
810.81±0.12	100†
864.02±0.17	1.38
1674.94±0.2	0.61
$^{60m}_{27}Co$	
58.60±0.010	
830	0.008%
1332.483±0.040	0.25%
2160	0.0008%
$^{60}_{27}Co$	
1173.226±0.040	99.88%
1332.483±0.46	100%

Gamma energies (KeV)	Gamma intensities
2158	0.001%
$^{61}_{27}$Co	
67.4±0.4	100%
$^{62}_{27}$Co	
1172	100†
1470	11
1740	11
2030	4
$^{56}_{28}$Ni	
163	99%
276	35%
472	35%
748	48%
812	87%
1560	14%
$^{57}_{28}$Ni	
Ann. Rad.	
129.3±1.0	14%
1377.6±1.0	86%
1757.8±1.0	6%
1783.0±1.5	1%
1882.4±1.0	8%
$^{58}_{28}$Ni	
Co X-rays	
$^{65}_{28}$Ni	
344	
366	4.8%
507	0.35%
610	0.15%
771	0.07%
855	0.07%
1115.51±0.07	15.2%
1480	
1624	0.76%
1725	0.53%
$^{57}_{28}$Ni	
890	24%
900	50%
1360	22%
$^{58}_{29}$Cu	
Ann. Rad.	
$^{59}_{29}$Cu	
Ann. Rad.	
340	3.6%
425	1.4%
465	4.1%
878	9.4%
1301	9.9%
1680	1.5%
$^{60}_{29}$Cu	
Ann. Rad.	
467	2.8%
826	19.2%
910	3.4%
1037	3.3%
1292	1.5%
1332.483±0.04	87.3%
1792	44.9%
1863	4.7%
1920	0.6%
1938	1.7%
2061±3	0.9%
2158±2	3.5%
2693	0.5%
2747	0.9%
2268	0.09%
2399	0.7%
2428	0.6%
3124	5.1%
3164±2	0.7%
3195±2	1.9%
3222	0.2%
3252	0.3%
3271±3	0.9%
4022±3	1.0%
4080	0.1%
4332	0.09%
4503	0.09%
4550	0.06%
4584	0.02%
$^{61}_{29}$Cu	
Ann. Rad.	
23.2±0.3	
44.8±0.5	
55.4±0.5	
67.6±0.2	4.9%
283.0±0.2	13%
373.6±0.3	1.9%
405.2±0.3	
529.3±0.3	0.3%
588.6±0.3	1.2%
656.01±0.2	9.6%
1074.5	0.03%
1185.0±0.3	3.6%
1446.7	0.04%
1543.6	0.03%
1611.1	0.025%
1662.9	0.04%
1730.4	0.04%
1997	0.004%
2122.4	0.04
$^{62}_{29}$Cu	
Ann. Rad.	
875.4	1%
1170	3%
$^{64}_{29}$Cu	
Ann. Rad.	
1348	0.6%
$^{66}_{29}$Cu	
833.6	0.25%
1039.0±1	9%
$^{67}_{29}$Cu	
91.26±0.02	6.2%
93.31±0.02	35%
184.6	45%
209.0	0.09%
300.2	0.6%
393.6	0.2%
$^{68}_{29}$Cu	
810	18†
1080	100
1340	3
188	5
$^{60}_{30}$Zn	
Ann. Rad.	
$^{61}_{30}$Zn	
Ann. Rad.	
470	10.9%
690	1.8%
966	2.9%
1640	6%
$^{62}_{30}$Zn	
Ann. Rad.	
40.88	
243.47±0.2	1.6%
246.99±0.2	1.0%
305.2±0.3	0.2%
349.7±0.3	0.3%
393.94±0.25	1.4%
507.35±0.18	12.8%
548.38±0.2	12.6%
596.76±0.2	20.3%
$^{63}_{30}$Zn	
Ann. Rad.	
669.8±0.4	11.2%
961.2±0.4	8.4%
1411.3±0.4	0.9%
1555	0.2%
$^{65}_{30}$Zn	
1115.51±0.07	50.6%
344	0.003%
771	0.003%
$^{69m}_{30}$Zn	
438.7±0.2	100%
$^{71m}_{30}$Zn	
130	10†
210	
385	100
495	73
609	68
760	6
880	
920	3
990	8
1120	4
1310	1
1370	0.7
1490	1.0
1640	0.11
1740	0.27
1850	0.08
1960	0.06
2400	0.08
$^{71}_{30}$Zn	
120	7†
385	10
510	100
680	2
920	24
1120	10
1630	1.1
$^{72}_{30}$Zn	
18	
46	
145	90%
190	10%
$^{63}_{31}$Ga	
Ann. Rad.	
$^{64}_{31}$Ga	
Ann. Rad.	
427.1	0.9%
756.7	1.6%
808.8	14%
991.6	46%
918.9	8.3%
1276.4	6.9%
1386.9	14%
1455.2	1.9%
1566.6	4.3%
1617.2	1.7%
1626.2	1.4%
1799.3	4.6%
1995.2	2.9%
2195.2	11%
2270.5	2.3%
2374.8	7.8%
2433.5	0.6%
2550	0.9%
2804.3	0.7%
3262.9	0.3%
3366.1	1.7%
3425.1	5.1%
4215	0.1%
4454.7	0.9%
4749.0	0.05%
$^{65}_{31}$Ga	
Ann. Rad.	
53.9±0.4	31%
61.2±0.4	15%
90.1±1	1%
115.2±0.4	55%
153.2±0.4	8.5%
207.1±0.4	1.8%
751.3±0.4	7.3%
769	1.2%
932.4±0.4	1.7%
1047	1.1%
1227	0.8%
1342	0.3%
1353	0.8%
1414	0.27%
1480	0.7%
1764	0.1%
1855	0.15%
1970	0.04%
$^{66}_{31}$Ga	
Ann. Rad.	
833.6	6%
1039.0	39%
1232.6	0.63%
1332	1.3%
1356.9	0.67%
1419.0	0.63%
1508.5	0.59%
1918.7	2.4%
2190.3	5.9%
2215.4	0.16%
2394	0.24%
2423.0	2.0%
2752.1	25%
2780.5	0.13%
2932	0.34%
3229.5	1.5%
3257	0.13%
3381.2	1.6%
3423.0	0.94%
3433.0	0.29%
3766	0.12%
3791.3	1.0%
4087.5	1.1%
4295.5	3.6%
4462.0	0.71%
4806.0	1.6%
$^{67}_{31}$Ga	
91.26	2.9%
93.31	70%
184.46±0.2	20.7%
200.96±0.2	2.3%
300.17±0.2	15%
353.0±0.4	
393.43±0.2	4.1%
494.4±0.3	0.08%
595.9±0.4	
703.6	0.012%
794.7	0.05%
887.9±0.4	0.12%
$^{68}_{31}$Ga	
578.3	0.04%
805.8	0.09%
1077.1±1	3.2%
1261.3	0.1%
1764.5	0.01%
1884.5±1	0.13%
2338	0.001%
$^{70}_{31}$Ga	
173	0.11%
1042	0.5%
1215	
$^{72}_{31}$Ga	
601	8†
630	24
690	
786	4
812	4
834	100
894	11
1050	7
1231	1.4
1263	1.3
1278	1.8
1377	0.4
1400	0.3
1465	5
1602	6
1685	1.0
1718	0.5
1866	7
2114	1.1
2201	35
2460	0.7
2491	11
2508	19
2620	0.2
2849	0.4
3050	
3050	0.01
3100	0.01
3340	0.007
3680	0.0005
$^{73}_{31}$Ga	
13.5	
53.39±0.2	9†
68.66±0.3	1
285.0±0.3	3
297.37±0.2	97
325.74±0.2	10
379.28±0.4	
739.37±0.25	
767.87±0.3	
833.3±0.3	

Gamma energies (KeV)	Gamma intensities	Gamma energies (KeV)	Gamma intensities	Gamma energies (KeV)	Gamma intensities	Gamma energies (KeV)	Gamma intensities
$^{71}_{31}$Ga		318.7±0.3	1.33%	1115.4±0.4	3.1±0.6	1.5121	0.34±0.20
53.92±0.4		532.8±0.3	0.2%	1125.7±0.4	3.0±0.6	1.5233	6.2
380	2†	553.3±0.4	0.61%	1151.8±0.4	6.2±1.4	1.5666	0.34±0.20
500	11	574.0±0.3	11.8%	1193.3±0.2	86.2±4.3	1.5879	0.54
595.6±0.2	87	587.8±0.4	0.27%	1202.7±0.6	2.0±1.0	1.7079	21.9
600	13	762.0±0.4	0.18%	1216.0±0.4	3.5±1.3	1.7813	4.8
720	2	788.0±0.4	0.32%	1242.4±0.3	10.4±2.8	1.8831	0.63
870	5	872.0±0.3	9.6%	1264.2±0.3	26.7±3.2	1.9450	0.10
980	4	1001.7±0.4		1280.8±0.4	3.4±0.9	1.9490	0.20
1110	5	1051.8±0.4	0.32%	1296.3±0.4	2.9±0.7	2.0077	3.6
1200	6	1106.5±0.3	26%	1309.5±0.5	16.1±2.9	2.0200	20.4
1330	5	1206.3±0.4	0.26%	1313.1±0.5	9.2±1.7	2.0647±0.0030	≪0.15
1460	6	1362.2±0.3	2.94%	1319.8±0.5	13.5±2.2	2.0955±0.0030	0.22±0.18
1560	2	1349.5±0.4	0.27%	1368.5±0.5	90.5±9.1	2.1576	0.45
1700	3	1487.8±1	0.06%	1452.6±0.7	5.9±1.4	2.2193	0.13±0.07
1760	3	1525.6±1	0.15%	1464.8±0.8	5.5±1.3	2.2561	0.16±0.08
1930	6	1572.3±1	0.13%	1477.1±0.6	11.1±1.9	2.3266	0.14±0.07
2350	45	1891.8±0.4	0.27%	1479.9±0.6	8.3±1.4	2.3334	0.10±0.07
2550	3	1924.0±0.5	0.08%	1495.1±0.5	18.6±3.0	2.4212	0.10
2730	3	2023.9±0.5	0.34%	1529.0±0.5	2.1±0.6	2.4250	0.10
2970	2	$^{71}_{32}$Ge		1539.4±0.5	4.9±1.4	2.4493	0.43
3230	3	23.0		1573.3±0.25	22.7±1.4	2.5195±0.0030	0.09
3410	2	175.0		1709.6±0.6	10.4±2.5	2.6372	0.37
$^{75}_{31}$Ga		$^{73m}_{32}$Ge		1719.7±0.6	14.7±2.9	2.7804±0.0030	0.09
360?	1%	13.5		1726.9±0.9	4.2±1.3	2.8523±0.0030	0.05
580	3%	54		1830.9±0.8	1.5±0.5	2.9649±0.0030	0.09
$^{76}_{31}$Ga		$^{75m}_{32}$Ge		1846.5±0.5	5.4±1.6	3.1256	0.08
560		139	~100%	1879.9±0.8	1.0±0.4	4.0906±0.0030	0.02
960		$^{75}_{32}$Ge		1928.8±0.8	0.7±0.3	4.3279±0.0030	0.02
1120		66	2.5†	2000.1±0.3	19.2±3.7	4.4345±0.0030	0.01
$^{65}_{32}$Ge		200	13	2038.5±0.5	1.2±0.3	$^{71}_{33}$As	
655		264	100	2077.4±0.5	8.0±2.3	Ann. Rad.	
1670		427	2.6	2089.7±0.5	7.3±2.2	23.0	
$^{66}_{32}$Ge		477	2.6	2126.5±0.4	4.8±1.5	174.70±0.2	
Ann. Rad.		628	1.3	2248.0±1.0	0.7±0.3	247.14±0.5	
43.83	43.8%	$^{77m}_{32}$Ge		2329.4±1.0	0.8±0.2	279.22±0.5	
65.11	7.8%	159 (IT)	55†	2341.5±0.4	16.4±3.2	326.53±0.3	
90.9	0.5%	215	100	$^{78}_{32}$Ge		349.87±0.4	
108.93	16.5%	$^{77}_{32}$Ge		277	94%	373.57±0.5	
120.05	0.4%	156.3±0.3	16.3±2.2†	294	6%	391.22±0.3	
147.87	1.0%	177.4±0.4	1.0±0.4	$^{69}_{33}$As		465.00±0.6	
154.85	0.6%	194.9±0.2	48.5±4.3	Ann. Rad.		499.85±0.17	
181.96	7.5%	211.03±0.04	998.0±90.0	86.8±1.0		526.57±0.2	
190.17	7.5%	215.51±0.04	871.0±78.0	146.0±0.5		572.15±0.3	
245.75	6.1%	255.0±0.25	2.6±2.1	232.8±0.4		615.26±0.3	
272.90	11.9%	264.45±0.025	1700.0±116.0	$^{70}_{33}$As		622.72±0.5	
302.51	2.4%	338.5±0.15	30.8±2.8	Ann. Rad.	1.98†	679.71±0.5	
338.01	9.9%	367.49±0.04	485.0±32.0	0.176±0.2	3.2	685.41±0.8	
381.85	29%	416.35±0.04	815.0±51.0	0.2405	0.26	707.0±0.3	
427.51	0.3%	439.5±0.23	10.3±5.1	0.2523	3.6	712.5±0.3	
492.76	0.4%	461.4±0.22	41.9±3.7	0.2942	0.23	747.0±0.3	
536.58	6.1%	475.5±0.15	35.9±3.3	0.2988	0.47	851.2±0.3	
706.04	3.8%	557.7±0.08	569.0±28.0	0.4480±0.0010	0.20	920.6±0.2	
$^{67}_{32}$Ge		582.5±0.18	25.5±3.9	0.4509±0.0010	0.13	1026.7±0.2	
Ann. Rad.		613.6±0.4	23.9±5.8	0.4922	1.2	1033.54±0.2	
166	89.7%	624.6±0.4	2.0±0.6	0.4970	3.1	1037.87±1.0	
166.5	24.2%	632.3±0.23	314.0±16.0	0.5952	20.0	1095.48±0.15	
360	5.3%	673.1±0.25	21.9±1.9	0.6076	4.8	1139.3±0.2	
558	0.09%	698.8±0.35	8.35±0.69	0.6530	0.73	1231.28±0.4	
661	0.26%	714.1±0.09	269.0±14.0	0.6684	25.9	1298.43±0.2	
720	0.14%	743.2±0.40	4.8±1.4	0.7448	25.5	1331.5±1.5	
728	3.9%	745.6±0.28	35.2±4.9	0.7602	0.30±0.15	$^{72}_{33}$As	
828	5.0%	749.9±0.28	32.6±4.6	0.8281	0.43	Ann. Rad.	
915	9.8%	766.8±0.25	26.0±2.6	0.8893	3.8	595.79±0.2	
981	2.4%	781.3±0.25	33.5±4.0	0.8931	2.4	600.84±0.17	
1082	2.3%	784.8±0.25	40.9±4.9	0.9019	1.7	629.88±0.15	10†
1162	1.7%	794.7±0.35	4.9±1.4	0.9057	14.9	634.6±0.4	
1283	0.35%	810.6±0.20	78.1±5.8	0.9421	1.7	756.7±0.3	
1450	0.9%	823.6±0.26	20.9±3.6	0.9538	0.58	765.7±0.3	
1477	8.5%	843.7±0.30	7.9±2.1	1.0400	100	786.42±0.17	
1644	2.1%	875.3±0.30	26.0±10.0	1.0993	5.4	834.03±0.13	100
1810	1.7%	907.2±0.35	22.3±4.1	1.1143	25.9	894.17±0.15	2
1837	0.4%	914.2±0.5	4.2±1.0	1.1181	3.9	1050.65±0.16	3
2004	0.8%	922.9±0.35	16.8±2.5	1.2183	0.22	1212.6±0.5	
2170	0.09%	925.5±0.30	26.2±3.9	1.2961	0.21±0.10	1215.3±0.5	
2559	0.13%	929.1±0.30	29.2±4.4	1.3322	0.75	1379.2±0.4	
2726	0.1%	939.6±0.35	6.5±1.6	1.3360	0.75	1390.3±0.3	
2991	0.1%	946.9±1.0	0.4±0.25	1.3394	10.9	1464.08±0.18	
3065	0.1%	959.1±0.8	2.6±0.7	1.3518	0.72	1475.8±0.2	
3157	0.1%	968.1±0.8	2.4±0.6	1.4125	10.5	1710.7±0.3	
3398		986.3±0.8	2.3±0.6	1.4183	0.61±0.20	1917.8±0.3	1
$^{68}_{32}$Ge		997.1±0.5	2.3±0.5	1.4961	1.9	1991.1±0.4	
Ann. Rad.		1062.1±1.5	1.4±0.5	1.5071	0.48±0.30	2071.7±0.4	
234.8±0.3	0.29%	1085.0±0.15	216.0±14.0			2090.6±0.5	

Gamma energies (KeV)	Gamma intensities
2105.2±0.5	
2109.6±0.5	
2201.57±0.1	4
2248.3±0.3	0.9
2507.7±0.16	
2523.4±0.4	
2528.8±0.4	
2620.9±0.4	
2780.0±0.5	0.6
3111.2±1.0	
3481.3±1.0	
3994.3±1.0	

$^{73}_{33}$As

Gamma energies (KeV)	Gamma intensities
13.3±0.2	044†
53.415±0.035	100

$^{74m}_{33}$As

Gamma energies (KeV)	Gamma intensities
283	100%

$^{74}_{33}$As

Gamma energies (KeV)	Gamma intensities
Ann. Rad.	
595.86±0.14	100†
608.4±0.2	0.2
634.73±0.15	25.6
887.2±0.7	0.048
993.6±0.8	0.021
1204.0±0.3	0.47
1604.0±1.0	0.013
2198.8±1.0	0.019

$^{76}_{33}$As

Gamma energies (KeV)	Gamma intensities
510	0.5†
559.1	100
562.8	1.0
657.04	15
665.4	
708	0.3
740	0.5
775	0.4
858	0.3
869	0.24
1213.3	
1216.25	10
1220	2
1228.63	2.3
1439.4	1.2
1453	0.54
1550	0.2
1789.8	0.77
1880	0.2
2096.6	1.3
2111.1	0.73
2434	0.05
2656	0.08

$^{77}_{33}$As

Gamma energies (KeV)	Gamma intensities
87.86±0.12	0.26%
161.86±0.12	0.34%
238.96±0.12	1.65%
270.74±0.15	0.008%
281.63±0.12	0.06%
520.63±0.15	0.61%

$^{78}_{33}$As

Gamma energies (KeV)	Gamma intensities
620	42†
700	15
830	8
990	2.5
1110	2.9
1210	4.2
1310	11
1490	2.1
1700	1.7
1820	1.4
1940	1.5
2050	1.0
2240	1.0
2650	1.4

$^{79}_{33}$As

Gamma energies (KeV)	Gamma intensities
96	100%
360	2%
430	2%
540	0.5%
730	0.5%
890	1%

$^{80}_{33}$As

Gamma energies (KeV)	Gamma intensities
666	
785 }	42%
812 }	1.4%

Gamma energies (KeV)	Gamma intensities
1220	3.6%
1640	4%
1770	1.7%
1840	0.5%
2300	0.3%
2350	0.4%

$^{70}_{34}$Se

Gamma energies (KeV)	Gamma intensities
Ann. Rad.	

$^{71}_{34}$Se

Gamma energies (KeV)	Gamma intensities
Ann. Rad.	
160	

$^{72}_{34}$Se

Gamma energies (KeV)	Gamma intensities
46.0	

$^{73}_{34}$Se

Gamma energies (KeV)	Gamma intensities
Ann. Rad.	
67.5±0.5	
360.7±0.5	
427.9±0.5	

$^{73}_{34}$Se

Gamma energies (KeV)	Gamma intensities
Ann. Rad.	
88	
251	
580	
1080	

$^{75}_{34}$Se

Gamma energies (KeV)	Gamma intensities
24.3	
66.05	1.6†
96.731±0.007	5.6
121.113±0.010	28
135.998±0.010	96
198.600±0.020	2.4
264.648±0.015	100
279.522±0.012	42
303.892±0.020	2.3
400.641±0.015	20

$^{77m}_{34}$Se

Gamma energies (KeV)	Gamma intensities
161	

$^{79m}_{34}$Se

Gamma energies (KeV)	Gamma intensities
103	

$^{81}_{34}$Se

Gamma energies (KeV)	Gamma intensities
270	0.3%
276	2.0%
290	1.3%
561	0.3%
565	1%
836	1%

$^{83m}_{34}$Se

Gamma energies (KeV)	Gamma intensities
356	15%
676	13%
989	14%
1031	13%
1063	4%
1664	2%
2054	10%
2147	0.2%

$^{83}_{34}$Se

Gamma energies (KeV)	Gamma intensities
226	31%
356	73%
357	3.5%
512	45%
554	3%
676	13%
720	22%
801	15%
837	14%
866	9%
1065	6%
1082	2%
1192	4%
1299	9%
1319	4%
1344	6%
1355	3%
1421	0.7%
1558	3%
1784	4%
1830	1.3%
1855	3%
1897	9%
2291	12%
2338	4%
2421	1%

$^{74}_{35}$Br

Gamma energies (KeV)	Gamma intensities
Ann. Rad.	
640	

$^{75}_{35}$Br

Gamma energies (KeV)	Gamma intensities
Ann. Rad.	
112.51±0.5	
141.50±0.25	
236.0±0.2	
286.70±0.18	
293.1±0.2	
316.0±0.4	
377.6±0.3	
428.2±0.3	
431.8±0.3	
566.1±0.3	
598.5±0.4	
608.8±0.3	
657.0±0.4	
771.0±0.6	
796.4±0.8	
860.0±1.0	
897.6±1.0	
912.1±0.6	
949.9±1.5	
952.8±1.0	
962.1±1.0	

$^{76}_{35}$Br

Gamma energies (KeV)	Gamma intensities
Ann. Rad.	
358.4±0.5	
399.31±0.5	
472.7±0.3	
490.1±0.5	
558.99±0.25 }	
563.1±0.6 }	95†
598.8±0.4	
604.2±0.5	
657.2±0.25	28
665.3±0.5	
681.6±0.5	
696.0±0.5	
727.1±0.5	
730.7±0.5	
789.2±0.5	
803.5±0.5	
836.6±0.5	
867.7±0.5	
882.2±0.5	
886.6±0.5	
901.0±0.5	
942.7±0.5	
981.2±0.5	
1029.7±0.5	
1032.2±0.5	
1129.9±0.4	
1212.9±0.5	
1216.2±0.3	
1228.7±0.5	
1369.0±0.4	
1380.4±0.4	
1428.8±0.6	
1439.2±0.6	
1453.7±0.5	
1470.9±0.5	
1488.5±0.4	
1560.1±0.6	
1568.2±0.5	
1578.7±0.6	
1787.6±0.5	
1853.1±0.25	
2096.2±0.4	
2110.8±0.4	
2135.0±0.4	
2281.2±0.4	
2329.2±0.5	
2334.0±0.7	
2348.7±0.5	
2390.50±0.3	
2439.0±0.6	
2510.2±0.5	
2600.7±0.6	
2617.6±0.7	
2792.15±0.25	8
2900.3±0.5	
2950.0±0.25	12
2996.8±0.3	
3023.4±0.7	
3092.3±0.7	
3603.5±0.5	3
3929.5±0.7	0.1
4436.7±1.0	0.07

$^{77}_{35}$Br

Gamma energies (KeV)	Gamma intensities
87.86±0.12	1.6%
161.86±0.12	1.9%
180.74±0.15	
187.34±0.2	
200.4±0.12	0.7%
231.3±0.2	
238.96±0.12	27%
243.3±0.2	
249.77±0.12	2.5%
270.74±0.15	1%
281.63±0.12	3%
297.15±0.13	8%
303.76±0.15	
384.98±0.18	0.6%
439.37±0.15	1.6%
484.52±0.18	3%
510.8±0.2	
Ann. Rad.	
520.63±0.15	23%
565.82±0.25	
567.94±0.2	
574.64±0.2	
578.85±0.18	7%
585.39±0.18	
662.5±0.5	
757.3±0.25	2%
817.8±0.25	3%
1005.2±0.3	2%

$^{78}_{35}$Br

Gamma energies (KeV)	Gamma intensities
Ann. Rad.	13%
614.1	

$^{80m}_{35}$Br

Gamma energies (KeV)	Gamma intensities
85	
37	

$^{80}_{35}$Br

Gamma energies (KeV)	Gamma intensities
	616†
2.8%	639
	666
0.2%	704
0.3%	812
6.8%	1257
83.6%	

$^{82m}_{35}$Br

Gamma energies (KeV)	Gamma intensities
46 (IT)	
698.4	
776.8	
1474.8	

$^{82}_{35}$Br

Gamma energies (KeV)	Gamma intensities
92.3±0.1	0.4%
100.9±0.1	
137.1±0.1	
221.28±0.05	2.3%
273.22±0.05	1.2%
295.5±0.1	
452.9±0.1	
554.24±0.05	73%
606.23±0.1	
619.02±0.05	43%
698.30±0.05	27%
776.45±0.05	83%
827.80±0.05	24%
1007.55±0.1	
1044.02±0.05	29%
1084	0.4%
1317.52±0.07	28%
1474.93±0.1	17%
1650.5±0.2	0.8%
1777.6±0.5	0.12%
1874	0.05%
1959	0.05%
2056	0.02%

$^{83}_{35}$Br

Gamma energies (KeV)	Gamma intensities
32	100%
93	100%
521	1.4%

Gamma energies (KeV)	Gamma intensities
⁸⁶₃₅Br	
270	1†
350	3
430	5
470	2
520	6
610	5
740	7
810	18
880	100
1010	20
1210	8
1470	4
1570	2
1740	4
1900	36
2050	4
2170	4
2470	16
2820	4
3030	8
3280	6
3930	25
⁸⁵₃₅Br	
305.0	23%
(Kr⁸⁵ᵐ)	
⁸⁷₃₅Br	
1440	100†
1850	18
2480	18
2640	16
2980	25
3180	16
3800	11
4000	7
4190	21
4800	17
5000	17
5200	12
⁸⁸₃₅Br	
760	
⁷⁴₃₆Kr	
Ann. Rad.	
⁷⁶₃₆Kr	
39	
73	
104	
135	
197	
267	
316	
360	
407	
452	
⁷⁷₃₆Kr	
24	4%
107.6	9%
131	84%
148	40%
279	3%
313	3%
607	
734	
⁷⁹₃₆Kr	
44	
136.0	0.7%
181	0.08%
208.5	0.7%
217.3	2%
261.3	11.5%
299.7	1.3%
306.7	2.5%
308	
345	0.1%
389.1	1.6%
397.4	10.1%
523	0.3%
526	0.2%
616	0.1%
726	0.05%
810	0.14%
833	1.7%
860	0.08%

Gamma energies (KeV)	Gamma intensities
935	0.1%
1026	0.13%
1072	0.09%
1115	0.4%
1165	0.09%
1332	0.5%
⁸¹₃₆Kr	
Br X-rays	
⁸³ᵐ₃₆Kr	
32	
93	
⁸⁵ᵐ₃₆Kr	
151.2	
305.0	
⁸⁵₃₆Kr	
514	
⁸⁷₃₆Kr	
402.4±0.2	100†
850	19
2050	~6
2570	42
⁸⁸₃₆Kr	
28	
166	20†
196.1±0.2	100
360	14
850	65
1369.6±0.2	
1550	40
2391.6±0.3	100
⁸⁹₃₆Kr	
230	85†
360	28
430	29
510	42
600	100
740	32
880	65
1120	45
1290	31
1510	88
1710	34
1930	10
2040	16
2230	10
2420	22
2570	10
2840	25
⁹⁰₃₆Kr	
105	15†
120	65
236	16
495	12
536	48
640	7
670	4
720	4
770	3
890	2
970	4
1110	48
1230	3
1340	4
1400	5
1540	17
1630	4
1700	5
1790	11
1940	2
2030	1
2120	2
2480	4
2580	2
2700	3
2940	2
3080	2
3170	1
3600	1
⁷⁹₃₇Rb	
Ann. Rad.	
⁸⁰₃₇Rb	
Ann. Rad.	
616	
704?	

Gamma energies (KeV)	Gamma intensities
⁸¹ᵐ₃₇Rb	
85	
⁸¹₃₇Rb	
190	100%
370	1.1%
450	29%
560	2.9%
820	1.1%
1040	0.9%
⁸²ᵐ₃₇Rb	
554.3	70%
619.1	41%
698.4	27%
776.8	83%
827.6	26%
1044.1	31%
1317.2	30%
1474.8	17%
⁸³₃₇Rb	
93	
520.43±0.2	46%
529.65±0.2	16%
552.63±0.2	16%
651	0.09%
682	0.03%
790.1±0.2	0.7%
801	0.3%
⁸⁴ᵐ₃₇Rb	
215.6±0.3	
248.3±0.4	
445.5	
463.6±0.4	
Ann. Rad.	
882.9±0.2	100†
1016	0.53
1892	1.1
⁸⁶ᵐ₃₇Rb	
560	~100%
⁸⁶₃₇Rb	
1078	8.8%
⁸⁸₃₇Rb	
898.0±0.2	63†
1390	6
1835.9±0.2	100
2110	4.5
2680	11
3010	1.4
3240	1.4
3520	1.1
3680	0.4
4780	1.4
⁸⁹₃₇Rb	
272.69±0.1	2.3†
287.97±0.1	1.0
657.76±0.05	16.9
947.72±0.05	16.4
1031.89±0.05	100
1248.07±0.05	75.4
1538.41±0.1	3.2
2008.21±0.1	3.4
2196.13±0.1	25.4
2569.96±0.1	16.6
2706.76±0.2	2.1
3508.43±0.5	2.3
⁹⁰₃₇Rb	
530	4%
720	4%
830	56%
860	6%
1030	5%
1110	7%
1240	2%
1400	5%
1700	3%
1820	3%
2200	2%
2510	2%
2710	1%
2840	2%
3070	5%
3340	15%
3540	5%
4130	11%

Gamma energies (KeV)	Gamma intensities
4340	13%
4370	4%
4600	5%
5100	2%
5200	4%
⁸⁰₃₈Sr	
580	
⁸¹₃₈Sr	
Ann. Rad.	
⁸²₃₈Sr	
Rb X-rays	
⁸³₃₈Sr	
41.9±0.5	
94.1±0.2	
159.7±0.25	
289.9±0.25	
381.56±0.15	
389.4±0.2	
418.4±0.2	
423.5±0.2	
438.2±0.2	
Ann. Rad.	
658.8±0.5	
736.9±1.0	
762.51±0.15	
778.35±0.2	
804.9±1.0	
818.92±0.3	
831.3±2.0	
818.3±2.0	
853.9±1.5	
890.1±2.0	
908.0±1.0	
944.6±1.5	
994.1±0.5	
1043.6±0.4	
1054.8±0.3	
1098.1±2.0	
1147.09±0.2	
1159.81±0.2	
1214.8±1.5	
1237.6±2.0	
1324.7±3.0	
1562.63±0.2	
1951.81±0.3	
2089.84±1.0	
2146.9±1.0	
⁸⁵ᵐ₃₈Sr	
7±0.1	
151.28±0.1	14% (EC)
231.69±0.1	85% (IT)
238.65±0.15	~4%(IT)
⁸⁵₃₈Sr	
360	0.002%
513.998±0.02	~100%
880	0.017%
⁸⁷₃₈Sr	
388.40±0.08	~100%
⁸⁹₃₈Sr	
910	<0.01%
⁹¹₃₈Sr	
118.31±0.1	0.21†
261.00±0.1	1.25
272.30±0.1	0.29
274.29±0.05	3.0
555.57±0.05	94.7
620.13±0.1	5.1
631.29±0.1	1.75
652.91±0.05	36.9
749.84±0.05	69.7
761.29±0.1	2.08
925.83±0.05	12.2
1024.29±0.05	100
1054.70±0.2	0.5
1281.09±0.1	2.75
1413.58±0.1	3.39
1473.83±0.2	0.47
1546.53±0.3	0.13
1723.63±0.3	0.40
⁹²₃₈Sr	
241.53±0.1	3.36†
430.45±0.1	4.64
953.32±0.1	4.30
1384.00±0.05	100

$^{93}_{38}$Sr

Gamma energies (KeV)	Gamma intensities
178	17†
210	6
255	13
360	7
400	5
450	9
530	7
600	100
710	35
880	73
1080	9
1180	19
1260	19
1660	16
1860	4
2190	4
2450	6
2580	5
3000	

$^{94}_{38}$Sr

Gamma energies (KeV)	Gamma intensities
1420	

$^{82}_{39}$Y

Ann. Rad.

$^{83}_{39}$Y

Ann. Rad.

$^{84}_{39}$Y

Gamma energies (KeV)	Gamma intensities
200	
453	
590	15†
795	100
982	90
1041	50
1270	9
1470	6

$^{85m}_{39}$Y

Gamma energies (KeV)	Gamma intensities
503	
700	
770	
920	

$^{85}_{39}$Y

Gamma energies (KeV)	Gamma intensities
Ann. Rad.	
230.86±0.2	
627.4±0.2	
699.06±0.2	
768.0±0.2	
1030.1±0.2	
1101.9±0.2	
1122.8±0.2	
1151.6±0.2	
1220.8±0.2	
1403.8±0.7	
1584.1±0.7	
1889.7±0.7	
2123.5±1.0	
2172.4±1.0	
2351.2±1.0	
2750	

$^{86m}_{39}$Y

Gamma energies (KeV)	Gamma intensities
10.2	
208.0	
152.2±0.5	
188.8±0.5	
191.8±0.5	
252.0±0.3	
265.2±0.3	
307.3±0.3	
331.4±0.3	
370.7±0.3	
443.2±0.25	
514±2	
580.6±0.3	
608.1±0.6	
627.7±0.15	
645.6±0.18	
703.3±0.2	
709.7±0.3	
767.7±0.4	
777.3±0.2	
826.2±0.5	
833.5±1.0	
896.6±2	
1024.0±0.25	
1076.8±0.2	

Gamma energies (KeV)	Gamma intensities
1153.1±0.3	
1253.6±0.8	
1349.5±0.5	
1407	
1854.1±0.3	
1920.4±0.3	
2567.2±0.6	
2610.7±1.0	
2830	
3060	
3340	
3660	
3860	

$^{87m}_{39}$Y

Gamma energies (KeV)	Gamma intensities
381.1±0.5	99%

$^{87}_{39}$Y

Gamma energies (KeV)	Gamma intensities
388.4±0.08	~100%
484.8±0.5	~100%
Ann. Rad.	

$^{88}_{39}$Y

Gamma energies (KeV)	Gamma intensities
898.00±0.03	92%
1836.075±0.05	~100%
2734.07±0.08	<1%

$^{89m}_{39}$Y

Gamma energies (KeV)	Gamma intensities
909.07±0.1	100%

$^{90m}_{39}$Y

Gamma energies (KeV)	Gamma intensities
202.4	100†
482	95.7
2320	0.4

$^{90}_{39}$Y

Gamma energies (KeV)	Gamma intensities
1.734	0.4%

$^{91m}_{39}$Y

Gamma energies (KeV)	Gamma intensities
555.59±0.07	~100%

$^{91}_{39}$Y

Gamma energies (KeV)	Gamma intensities
1.21	0.3%

$^{92}_{39}$Y

Gamma energies (KeV)	Gamma intensities
447.99±0.05	18.2†
492.17±0.1	7.8
560.81±0.05	18.5
844.12±0.1	9.1
912.66±0.1	4.8
934.44±0.05	100
972.35±0.2	0.6
1132.29±0.1	1.5
1405.44±0.05	32.3
1848.07±0.2	2.1
1885.9±0.5	0.2
2066	

$^{93}_{39}$Y

Gamma energies (KeV)	Gamma intensities
266.75±0.05	100†
428.31±0.1	0.1
478.15±0.1	0.2
658.21±0.1	1.8
680.27±0.05	9.7
714.42±0.1	0.3
743.51±0.1	1.9
947.12±0.05	29.6
1158.73±0.1	0.4
1182.5±0.1	0.5
1203.28±0.1	1.8
1425.39±0.1	3.6
1470.04±0.2	1.0
1827.15±0.1	1.4
1917.71±0.07	22.3
2184.2±0.2	1.4
2190.8±0.2	2.6
2472.5±0.3	0.2

$^{94}_{39}$Y

Gamma energies (KeV)	Gamma intensities
560	6%
920	43%
1130	5%
1420	
1650	2.4%
1900	1.6%
2130	2.4%
2570	1.5%
2840	0.7%
3060	1.3%
3530	1.1%

$^{96}_{39}$Y

Gamma energies (KeV)	Gamma intensities
700	
1000	
1500	

$^{81}_{40}$Zr

Ann. Rad.

$^{82}_{40}$Zr

Ann. Rad.

$^{83}_{40}$Zr

Ann. Rad.

$^{84}_{40}$Zr

Ann. Rad.

$^{85}_{40}$Zr

Ann. Rad.

$^{86}_{40}$Zr

Ann. Rad.

Gamma energies (KeV)	Gamma intensities
28	21†
243	100
612	5.4

$^{87}_{40}$Zr

Gamma energies (KeV)	Gamma intensities
1200	
2200	
Ann. Rad.	

$^{88}_{40}$Zr

Gamma energies (KeV)	Gamma intensities
392.6±0.3	100%

$^{89m}_{40}$Zr

Gamma energies (KeV)	Gamma intensities
588	94%
1506	6%

$^{89}_{40}$Zr

Gamma energies (KeV)	Gamma intensities
Ann. Rad.	
909.07±0.1	
1621.4±0.6	
1713.3±0.4	

$^{90m}_{40}$Zr

Gamma energies (KeV)	Gamma intensities
132.5±0.4	
2186.1±2.0	
2318.7±2.0	

$^{93}_{40}$Zr

Gamma energies (KeV)	Gamma intensities
235.7±0.2	
724.24±0.06	
756.87±0.07	

Gamma energies (KeV)	Gamma intensities
254.1±0.2	
355.6±0.1	
602.5±0.1	
703.8±0.1	
743.2±0.2	
971.5±0.4	
1147.9±0.2	
1276.1±0.2	
1362.7±0.2	
1712.6±0.4	
1750.6±0.3	
1852.0±0.3	

$^{88}_{41}$Nb

Gamma energies (KeV)	Gamma intensities
76	
141	
272	
399	
671	
1058	
1083	

$^{89m}_{41}$Nb

Gamma energies (KeV)	Gamma intensities
588	
(Zr^{89m})	
1506	

$^{89}_{41}$Nb

Gamma energies (KeV)	Gamma intensities
1510	
1626	
1713	
1832	
1911	
2220	
2572	
2574	
2961	
3018	
3093	
3282	
3333	
3513	
3577	
3838	
3908	

$^{90m}_{41}$Nb

Gamma energies (KeV)	Gamma intensities
3	
122.4	

$^{90}_{41}$Nb

Gamma energies (KeV)	Gamma intensities
132.5±0.3	5†
141.5±0.3	75
371.5±1.0	3
Ann. Rad.	
890.0±1.0	
1128.7±1.5	97
1269.5±2.0	
2186.27±1.0	14
2318.7±1.0	82

$^{91m}_{41}$Nb

Gamma energies (KeV)	Gamma intensities
104.5	
1210	

$^{92m}_{41}$Nb

Gamma energies (KeV)	Gamma intensities
Ann. Rad.	
912.66±0.1	1.8%
934.44±0.05	99%
1848.07±0.2	0.8%

$^{94m}_{41}$Nb

Gamma energies (KeV)	Gamma intensities
41.5	
703	0.002%
872	0.2%

$^{94m}_{41}$Nb

Gamma energies (KeV)	Gamma intensities
702.59±0.15	100%
871.16±0.15	100%

$^{95m}_{41}$Nb

Gamma energies (KeV)	Gamma intensities
235.7±0.2	100%

$^{95}_{41}$Nb

Gamma energies (KeV)	Gamma intensities
765.87±0.09	~100%

$^{96}_{41}$Nb

Gamma energies (KeV)	Gamma intensities
219.13±0.3	
241.46±0.1	
350.13±0.2	
372.20±0.1	
460.16±0.15	
480.43±0.15	
568.69±0.08	
719.50±0.08	
778.18±0.15	
810.53±0.15	
849.96±0.15	
1091.45±0.15	
1199.97±0.15	
1497.6±0.2	

$^{97m}_{41}$Nb

Gamma energies (KeV)	Gamma intensities
743.20±0.15	100%

$^{97}_{41}$Nb

Gamma energies (KeV)	Gamma intensities
658.0±0.2	98%
743.2±0.2	2%
1021.66±0.2	
1148.0±0.2	

$^{98}_{41}$Nb

Gamma energies (KeV)	Gamma intensities
330	9†
720	75
780	100
1160	30
1440	10
1520	4
1680	10
1880	4
1930	8
2240	
2440	
2700	

$^{99}_{41}$Nb

Gamma energies (KeV)	Gamma intensities
100	1†
260	1

$^{100}_{41}$Nb

Gamma energies (KeV)	Gamma intensities
140	10†
360	55
450	40
530	100
650	
2200	
2300	
2650	
2850	

$^{100}_{41}$Nb

Gamma energies (KeV)	Gamma intensities
530	100†
620	60
1040	10
1150	10
1470	5

Gamma energies (KeV)	Gamma intensities
$^{88}_{42}$Mo	
Ann. Rad.	
$^{89}_{42}$Mo	
Ann. Rad.	
$^{90}_{42}$Mo	
42.70	6†
122.4	36
162.9	0.43
203.1	0.26
257.34	14.2
323.2	0.08
421.0	0.001
445.4	0.036
472.2	0.005
489.8	0.003
941.5	0.004
990	0.001
1271	0.002
1387	0.001
1455	0.0008
1463	0.0004
1482	0.0004
$^{91m}_{42}$Mo	
658	57%
Ann. Rad.	
1210	16%
1530	22%
$^{91}_{42}$Mo	
Ann. Rad.	
$^{93m}_{42}$Mo	
264	100%
685	100%
1479	100%
$^{93}_{42}$Mo	
30.4	85%
$^{99}_{42}$Mo	
40.6	1†
140.3±0.2	
180.9±0.2	7
366.3±0.2	1.8
410	0.15
739.3±0.2	15
777.6±0.2	4
940	0.14
$^{101}_{42}$Mo	
80	3%
191	25%
193	2%
300	7%
400	2%
510	15%
590	21%
700	1%
704	11%
840	1%
890	15%
950	2%
1020	25%
1140	1%
1180	11%
1280	3%
1380	9%
1460	1%
1560	11%
1660	3%
2080	16%
$^{104}_{42}$Mo	
70	
$^{92}_{43}$Tc	
90	~20†
140	67
240	30
330	90
790	95
1540	100
$^{93m}_{43}$Tc	
390	63%
2660	18%
$^{93}_{43}$Tc	
860	3%
1350	65%
1490	33%
2030	0.4%

Gamma energies (KeV)	Gamma intensities
2440	0.4%
$^{94m}_{43}$Tc	
871.01±0.08	100±5
875.45±0.12	1.6±0.1
993.01±0.10	2.3±0.8
1195.8±0.4	0.9±0.2
1521.52±0.17	5.6±1.1
1868.50±0.25	6.2±0.4
2393±1	0.5±0.2
2739.7±0.3	4.5±0.5
3128.2±0.4	2.7±0.7
$^{94}_{43}$Tc	
449.07±0.06	2.6±1.0†
532.13±0.10	2.6±1.0
702.60±0.08	99.8±6
742.2±0.3	1.2±0.6
849.65±0.08	97.7±6
871.01±0.08	100±5
916.12±0.10	7.4±1.5
1362.7±0.6	0.9±0.4
1508.6±0.7	0.66±0.3
1591.7±0.3	2.4±1.0
$^{95m}_{43}$Tc	
Ann. Rad.	
38.9	
204.2	100†
584	55
763	1
784	2
788	15
823	12
837	39
1040	6
1630	0.1
$^{95}_{43}$Tc	
204	1%
680	2%
768	82%
840	11%
930	1.7%
1060	4%
$^{96m}_{43}$Tc	
34.4	
$^{96}_{43}$Tc	
312	0.5†
778.18±0.15	100
810.53±0.15	82
849.96±0.15	100
1119	17
$^{97m}_{43}$Tc	
96.5	100%
$^{97}_{43}$Tc	
Mo X-rays	
$^{98}_{43}$Tc	
669	100†
770	100
$^{99m}_{43}$Tc	
142.7	100%
$^{100}_{43}$Tc	
542	
600	
710	
810	
890	
1010	
1310	
1490	
1800	
$^{101}_{43}$Tc	
127	0.8†
130	2.6
183	0.4
186	2.3
235	0.8
307	100
380	1.8
410	0.4
545	8
640	1.0
720	1.2
850	0.36
940	0.24

Gamma energies (KeV)	Gamma intensities
$^{102}_{43}$Tc	
470	
$^{103}_{43}$Tc	
135	17†
210	10
350	
$^{104}_{43}$Tc	
360	
530	
630	
890	
1150	
1250	
1370	
1580	
1630	
1900	
2200	
2500	
2700	
3200	
3400	
3700	
4000	
4400	
4700	
$^{105}_{43}$Tc	
110	
$^{95}_{44}$Ru	
Ann. Rad.	
254.6±0.4	
290.3±0.3	
301.0±0.3	
336.5±0.2	
367.2±0.3	
511.006	
551.6±0.3	
591.1±0.3	
627.0±0.3	
653.0±0.5	
749.4±0.5	
807.6±0.3	
843.2±0.5	
1051.7±0.3	
1097.5±0.2	
1159.5±0.3	
1179.6±0.3	
1218.8±0.5	
1230.2±0.4	
1267.5±0.5	
1303.1±0.3	
1353.9±0.5	
1410.8±0.4	
1434.0±0.5	
1459.5±0.4	
2323.8±0.5	
$^{97}_{44}$Ru	
96.5	0.04†
215.4	100
324.4	9
565	<1
$^{103}_{44}$Ru	
39.8±0.1	0.072±0.008
53.3±0.1	0.30±0.03
65	<0.01
295.2±0.1	0.21±0.02
358	<0.03
443.0±0.1	0.31±0.03
497.1±0.1	100±10
557.1±0.1	0.76±0.07
610.3±0.1	4.8±0.4
$^{106}_{44}$Ru	
129.53	27%
149.04	2%
150	2%
188	2%
210	2%
262.84	9%
315.50	10%
326.1	3%
330.9	
393.36	3%

Gamma energies (KeV)	Gamma intensities
413.51	2%
469.38	19%
499.28	3%
575.19	1%
656.15	2%
676.32	6%
724.2	44%
875.8	2%
969.4	1.0%
1350	0.1%
1380	<0.1%
1580	<0.1%
1730	<0.1%
$^{107}_{44}$Ru	
195	14%
370	
480	
860	7%
930	4%
1030	4%
1290	4%
$^{108}_{44}$Ru	
165	28%
$^{95}_{45}$Rh	
Ann. Rad.	
190	
260	
430	
860	
1180	
1570	
1760	
1960	
2160	
2540	
$^{98}_{45}$Rh	
Ann. Rad.	
650	
$^{99m}_{45}$Ru	
Ann. Rad.	
89.3±0.6	
250.6±0.5	
276.6±0.5	
321.7±0.4	
340.3±0.25	
446.0±0.3	
558.2±0.4	
575.7±0.4	
921.0±0.7	
938.0±0.6	
1261.27±0.2	
$^{99}_{45}$Rh	
Ann. Rad.	
89.7±0.3†	100±4†
120.3±0.2	0.22±0.05
175.45±0.05	6.58±0.3
232.75±0.05	1.64±0.25
279.00±0.15	0.78±0.08
295.76±0.08	4.04±0.25
322.32±0.05	20.4±0.9
353.06±0.05	115±9
442.59±0.05	6.42±0.4
486.28±0.08	1.44±0.08
528.53±0.05	132±10
575.6±0.3	0.65±0.1
618.08±0.05	14.4±0.8
733.93±0.06	0.93±0.15
764.90±0.07	1.25±0.09
807.07±0.05	4.1±0.3
896.65±0.15	2.7±0.3
940.38±0.12	4.9±0.8
998.51±0.10	2.8±0.2
1009.6±0.2	0.19±0.05
1060.6±0.2	0.86±0.08
1088.88±0.15	1.0±0.2
1209.1±0.4	0.80±0.08
1293.34±0.15	0.95±0.05
1325.2±0.4	0.60±0.06
1482.8±0.3	0.42±0.05
1531.5±0.4	1.62±0.12
1571.9±0.6	0.74±0.07
1616.5±0.8	0.77±0.07
1660.9±0.4	0.20±0.05

Gamma energies (KeV)	Gamma intensities
1970.2±0.7	0.34±0.03
$^{100}_{45}$Rh	†
Ann. Rad.	†
302.2±0.3	
340.9±0.4	
370.7±1.5	
446.1±0.4	
539.5±0.3	
588.3±0.4	
591.0±1.0	
686.6±0.6	
822.5±0.4	
1033.7±0.6	
1107.6±0.4	
1341.4±0.4	
1362.2±0.3	
1553.7±0.3	
1865.5±0.5	
1930.2±0.4	
2376.2±0.3	
2530.3±0.5	
$^{101m}_{45}$Rh	
127.23±0.2	
157.35±0.2	1T
179.40±0.2	
183.91±0.2	
197.80±0.2	
233.22±0.2	
237.60±0.2	
306.52±0.2	
545.15±0.2	
595.73±0.3	
$^{101}_{45}$Rh	
127.23±0.01	117†
157.44±0.08	1T
179.64±0.06	
184.09±0.08	
197.02±0.04	(100)
233.72±0.08	
238.20±0.15	
306.87±0.03	
325.24±0.06	1
545.00±0.10	15
$^{102}_{45}$Rh	
345.89±0.12	0.87±0.10†
415.25±0.15	2.1±0.3
418.52±0.18	9.4±0.1
420.4±0.2	3.2±0.3
475.06±0.04	95±4
628.05±0.05	8.3±0.4
631.29±0.05	56±2
692.4±0.2	1.6±0.2
695.6±0.3	2.9±0.4
697.49±0.08	44±2
766.84±0.06	34±2
1046.59±0.07	34±2
1103.16±0.06	4.6±0.3
1112.84±0.07	19±1
1323.6±0.5	0.46±0.08
$^{102}_{45}$Rh	
Ann. Rad.	
415.25±0.15	0.03±0.02†
418.52±0.18	0.12±0.02
456.42±0.15	0.08±0.02
468.58±0.04	2.9±0.2
475.06±0.04	46±3
556.60±0.04	2.0±0.2
628.05±0.05	4.5±0.4
631.29±0.05	0.10±0.03
636.81±0.10	0.23±0.03
680.66±0.05	0.58±0.04
733.93±0.08	0.10±0.02
739.58±0.07	0.53±0.08
930.5±0.3	0.03±0.02
1046.59±0.07	0.43±0.03
1103.16±0.06	2.9±0.1
1105.7±0.3	0.39±0.03
1158.10±0.06	0.58±0.04
1362.08±0.20	0.39±0.05
1562.2±0.4	0.11±0.03
1568.7±0.6	0.01±0.01
1580.5±0.3	0.05±0.01
1786.4±0.4	0.01±0.01

Gamma energies (KeV)	Gamma intensities
2037.0±0.3	0.03±0.02
2261.3±0.4	0.02±0.02
$^{104m}_{45}$Rh	
31.86	
51.43 } IT	
77.55	
97.11	
560	
760	0.1%
770	0.1%
930	0.03%
1340	0.02%
1530	0.03%
1700	<0.03%
$^{104}_{45}$Rh	
555.5	2%
1240	0.12%
1800	<0.002%
$^{105m}_{45}$Rh	
129.4	100%
$^{105}_{45}$Rh	
38.8	
215.8	
280.54	0.17%
306.31	5%
319.24	19%
442.3	
497.0	
$^{106}_{45}$Rh	
511.9±0.2	100†
616.38±0.2	45
622.1±0.2	6
710	0.2
873.8	1.8
1050.4±0.2	6.8
1130	2.3
1490	0.16
1550	0.61
1770	0.19
1950	0.08
2090	0.13
2360	0.17
2640	0.03
$^{106}_{45}$Rh	
220	20†
407	20
450	35
511.9±0.2	100
620	33
735	47
820	40
1050.4±0.2	29
1140	14
1220	19
1560	2.1
1740	2.0
1860	2.0
2120	1.0
2280	0.5
$^{107}_{45}$Rh	
115	0.5%
285	3%
307	73%
570	1.8%
680	2.7%
$^{108}_{45}$Rh	
433.99±0.2	43%
614.29±0.2	22%
510	40%
1520	5%
$^{110}_{45}$Rh	
375	
$^{98}_{46}$Pd	
132	
Ann. Rad.	
$^{99}_{46}$Pd	
Ann. Rad.	
140	
275	
420	
670	
$^{100}_{46}$Pd	
32.7	

Gamma energies (KeV)	Gamma intensities
42.1	51†
55.8	3
74.8	70
84.0	100
126.1	33
139.7	1.2
151.7	2.5
158.8	8
$^{101}_{46}$Pd	
24	†
269.6±0.2	27
296.1±0.2	100
320.6±0.2	3.6
355.2±0.3	
428.2±0.4	
453.5±0.25	4.0
556.0±0.2	2.2
590.54±0.15	80
611.7±0.4	
723.8±0.4	16
748.6±0.4	3.9
854.0±1.0	
881.0±1.0	
992.8±0.4	5.5
1177.7±0.4	1.6
1201.8±0.4	7
1218.3±0.4	2.5
1288.7±0.4	10
1311.4±0.6	1.5
1638.0±2.5	
$^{103}_{46}$Pd	
53	†
66	0.001
298	0.009
324	0.003
362	0.06
498	0.013
$^{107m}_{46}$Pd	
210	
$^{109}_{46}$Pd	
188	
$^{109}_{46}$Pd	
44.8±0.6†	3.6
88.036±0.008	8900±800
103.6±0.7	2.2
134.7±0.7	3.2
145.9±0.7	2.7
311.5±0.5	(100)
390.9±0.8	2.5±0.5
413.5±1	49 {26^b
415.2±0.6	{23
424.7±0.8	1.8±0.4
448.2±0.8	2.6±0.6
551.3±0.8	1.5±0.5
557.8±0.5	6.2±0.8
602.4±0.5	21.5±2
636.1±0.5	27±3
647.3±0.6	65±5
701.8±0.8	9.2±1
707.3±0.8	4.5±0.5
736.7±0.7	5.0±0.6
781.8±0.7	33±3
(863)	<0.5
$^{111m}_{46}$Pd	
170	
others	
(see below)*	
$^{111}_{46}$Pd	
50	
70	
160	620†
280	71
290	
380	177
400	
450	36
500	32
560	68
630	100
750	39
830	10
900	9

Gamma energies (KeV)	Gamma intensities
960	11
*1080	18
1130	17
*1250	16
1380	9
1440	5
*1640	23
*1690	11
*1900	9
$^{112}_{46}$Pd	
18.5	20%
$^{102}_{47}$Ag	
Ann. Rad	
$^{103m}_{47}$Ag	
138	
$^{103}_{47}$Ag	
Ann. Rad.	
120	26†
150	23
240	10
270	34
380	
760	
920	
1010	12
1160	9
1280	13
$^{104m}_{47}$Ag	
Ann. Rad.	
555	
$^{104}_{47}$Ag	
Ann. Rad.	
167	
262	
355	
443	
478	
556	100†
750	20
767	57
854	15
938	35
1260	4
1340	10
1530	8
1620	9
1810	8
$^{105}_{47}$Ag	
64.0±0.1	
74.1	18†
89.9±0.2	
112.7±0.2	0.013
155.7±0.3	0.9
183.0±0.3	0.7
280.4±0.3	62
284.8±0.4	0.003
289.4±0.4	
306.3±0.3	1.6
311.8±0.4	
319.4±0.3	0.11
325.8±0.4	
329.0±0.4	
331.8±0.3	
344.7±0.3	69
360.8±0.3	
370.3±0.3	
329.8±0.3	
401.7±0.4	
415.0±0.4	
421.1±0.4	
437.4±0.4	
443.5±0.2	18
527.4	
560.9±0.4	
618.1±0.4	
644.8±0.3	16
650.9±0.4	
673.3±0.4	
681.4±0.5	
727.6±0.4	
743.4±0.4	
807.6±0.4	

Gamma energies (KeV)	Gamma intensities
962.8±0.5	
1088.2±0.3	4.2
$^{106m}_{47}$Ag	
194.9	
221.5	
228.5	
328.3	
309.9	
406.0	
429.5	
450.8	
474.2	
511.8	
600.9	
616.0	
680	
703.3	
716.2	
717.1	
748.2	
792.8	
803.9	
807.5	
824.5	
847.5	
1045.7	
1127.8	
1199	
1223	
$^{106}_{47}$Ag	
Ann. Rad.	
295	
450	
616.2±0.6	
621.4±0.6	
870.3±0.6	
1045.9±0.6	
1124.6±0.6	
1520	
1730	
1880	
2170	
$^{107m}_{47}$Ag	
93.1	100%
$^{108m}_{47}$Ag	
30.2	
79.49±0.18	
433.99±0.13	10†
614.29±0.15	10
722.81±0.18	10
$^{108}_{47}$Ag	
433.8	0.45%
510	
614.29±0.15	0.26%
633.2	1.7%
841	0.02%
1010	
$^{109m}_{47}$Ag	
88.036±0.008	
$^{110m}_{47}$Ag	
116.41 (1T)	
446.2	
620.10	
657.6	100†
677.5	10
686.8	7
706.6	10
744.2	
763.8	24
817.9	8
884.5	74
937.3	33
1384.3	22
1475.9	4
1505.2	11
1562.5	1.2
$^{110}_{47}$Ag	
657.6	12%
817.9	
1475.9	0.02%
$^{111m}_{47}$Ag	
70	100%

Gamma energies (KeV)	Gamma intensities
$^{111}_{47}$Ag	
95	
247	1%
342	6%
610	
$^{112}_{47}$Ag	
606.2±0.5	41%
617.0±0.4	
694.1±0.5	
716.8±0.8	
775.8±1.0	1%
850.8±0.6	2%
1103.4±1.0	1%
1252.2±1.0	0.5%
1311.5±1.0	1%
1386.7±0.5	4%
1539.0±1.0	1%
1612.9±0.7	2%
2055.8±0.8	
2105.7±0.6	2%
2211.0±1.5	
2359.8±2.5	0.5%
2506.4±1.0	
2685.0±1.0	0.6%
2827.2±1.0	0.7%
2749.5±2.0	0.6%
$^{113}_{47}$Ag	
97.2±0.3	
258.8±0.3	20
298.6±0.2	100†
316.2±0.3	10
333.1±0.3	5
339.3±0.3	5
364.6±0.3	1
382.0±0.3	1
583.7±0.3	5
672.4±0.3	9
680.9±0.3	9
884.0±0.3	4
988.7±0.3	5
1194.8±0.3	4
$^{114}_{47}$Ag	
570	
$^{115}_{47}$Ag	
110	1%
140	9%
170	2%
220	44%
240	7%
280	13%
360	10%
420	7%
450	1%
470	10%
620	1%
640	3%
1040	2%
1080	2%
1480	11%
1660	8%
1760	2%
1860	3%
1900	7%
2030	2%
2120	13%
2500	1%
$^{103}_{48}$Cd	
Ann. Rad	
220	
630	
850	
$^{104}_{48}$Cd	
66.7	
83.6	
123.6	
134.2	
$^{105}_{48}$Cd	
Ann. Rad.	
25.5	
27.7	
263	
292.5	

Gamma energies (KeV)	Gamma intensities
308	
312	
317	
321	
325	
336	
341	
347	
433	
607	
1901	
1960	
2000	
2050	
2280	
2320	
$^{107}_{48}$Cd	
Ann. Rad.	
32.6	
93.1(1T)	
300.7	
324.6	
422.6	
526.4	
596.8	
624.5	
719.8	
786.7	
796.0	
799.0	
818	
828.9	
897	
949.0	
1049	
1130	
1222.4	
$^{109}_{48}$Cd	
88.036±0.008	
$^{111m}_{48}$Cd	
149.6	30†
246	94
$^{113m}_{48}$Cd	
265	~0.1%
$^{115m}_{48}$Cd	
130	0.07†
162	1.3
210	0.15
292	0.2
485	16
935	100
1130	5
1240	45
1420	2
$^{115}_{48}$Cd	
35	0.5†
230	0.6
262	2.0
485	10
530	26
$^{117m}_{48}$Cd	
89	
161	
222	
273	
293	
314	
345	
366	6†
434	
462	
565	
631	2
702	3
715	5
748	3
762	
832	
862	3
880	
931	2
1029	4

Gamma energies (KeV)	Gamma intensities
1052	
1065	11
1117	5
1165	
1233	8
1248	3
1260	2
1338	9
1371	
1408	9
1433	12
1562	7
1652	
1659	
1682	3
1723	9
1939	
1998	17
2095	2
2311	
2319	4
2394	
2407	
$^{117}_{48}$Cd	
89	
161	
222	
273	
293	
314	
345	
434	
462	
565	
832	11†
880	
947	8
964	3
1009	0.7
1052	
1142	
1274	1
1303	54
1381	2
1397	6†
1577	46
1708	6
1835	
$^{106}_{49}$In	
Ann. Rad.	
530	17†
630	10
1650	
1850	
$^{107}_{49}$In	
Ann. Rad.	
220	46%
320	
730	
840	
940	
1050	
1250	
1520	
1720	
1940	
2260	
$^{108}_{49}$In	
Ann. Rad.	
383	
633	
842	
$^{108}_{49}$In	
75	
120	
150	
175	
243	
264	
325	
330	
383	

Gamma energies (KeV)	Gamma intensities
391	
411	
570	
633	
731	
753	
781	
842	
872	
924	
1013	
1033	
1054	
$^{109}_{49}$In	
Ann. Rad.	
60	
89	
94	
205	
228	
262	
289	
301	
326	
329	
340	
349	
410	
422	
428	
446	
456	
519	
533	
548	
578	
615	
623	
641	
651	
678	
725	
821	
825	
862	
920	
947	
1050	
1190	
1550	
1610	
1680	
$^{110m}_{49}$In	
121	
399	
411	
463	
562	
584	
642	
658	
707	
745	
761	
816	
844	
884	
937	
995	
1118	
1500	
$^{110}_{49}$In	
658	100†
817	
1180	3
1520	2
1780	1.3
2200	2
2800	0.6
3000	
3460	
3650	
$^{111}_{49}$In	
172.5	93†
247	100
$^{112m}_{49}$In	
156	
$^{112}_{49}$In	
Ann. Rad.	
621	6%
860	0.4%
1240	0.5%
1480	0.3%
$^{113m}_{49}$In	
391.689±0.010	100%
$^{114}_{49}$In	
311	
191.6	
$^{114m}_{49}$In	
191.6	
358	
724	
$^{114}_{49}$In	
Ann. Rad.	
1299	
$^{115m}_{49}$In	
335	
$^{116m}_{2}{}_{49}$In	
164	
$^{116m}_{1}{}_{49}$In	
138.3	3%
385	1%
417.0	36%
818.7	17%
1097.2	53%
1293.3	80%
1508	11%
1753	<0.4%
1770	1.5%
2111	20%
$^{116}_{49}$In	
434	0.1?%
950	0.1%
1293	1.2%
2200	0.01%
$^{117m}_{49}$In	
158.63±0.15	16%
315.33±0.18	
$^{117}_{49}$In	
158.63±0.15	
552.49±0.15	
$^{118}_{49}$In	
1220	
$^{118}_{49}$In	
210	3†
440	7
640	10
690	42
810	7
1050	83
1230	100
1250	6
1490	1.4
1540	1.0
2040	3
$^{119m}_{49}$In	
24	
311±0.2	
762.91±0.4	
$^{119}_{49}$In	
730	5†
820	100
$^{120}_{49}$In	
1171	
$^{120}_{49}$In	
90	12†
198	9
260	1
710	12
860	34
940	12
1020	61
1170	100
1190	7
1280	14
1470	6
1870	7
2010	6
2220	3
2630	2
$^{121}_{49}$In	
940	
$^{124}_{49}$In	
990	3†
1130	10
3210	3
$^{109}_{50}$Sn	
Ann. Rad.	
335	
521	
890	
1120	
$^{110}_{50}$Sn	
283	
$^{111}_{50}$Sn	
Ann. Rad.	
372.9±0.8	16.5±2†
458.0±0.6	11.8±1
536.2±0.6	6.5±1
563.6±0.8	10.5±1.5
616±1	≤4
635±1	≤4
762.0±0.6	46±3
953.6±0.8	18±1.5
1026.1±0.8	10±1.5
1101.2±0.8	28±3
1153.0±0.6	(100)
1542.5±0.8	27±3
1610.3±0.8	40±3
1915.1±0.8	56±4
2106.6±1	12±1
2178.2±1	9±1
2212.9±1	5.8±0.7
2324.8±1	9±1
$^{113}_{50}$Sn	
391.689±0.010	98.2%
255	1.8%
$^{117m}_{50}$Sn	
158	
159	
$^{119m}_{50}$Sn	
23.83	~20%
65	100%
$^{121m}_{50}$Sn	
37.15	
$^{123}_{50}$Sn	
159.7±0.5	100†
381±1	0.04±0.004
542±1	0.02±0.004
552±1	0.007±0.002
$^{125}_{50}$Sn	
331.7	100†
386.1	0.082
589.6	0.20
643.0	0.15
662±1	0.046
776±1	0.018
840.8	0.075
1017.5	0.093
1093±1	0.050
1152±1	0.015
1294±1	0.013
1305±1	0.012
1349±1	0.016
1368.5	0.10
1404.1	0.71
1484.2	0.22
1582±1	0.0058
1615.3	0.13
1634±1	0.024
1735.5	0.033
1913.7	0.020
1947±1	0.013
2113±1	0.0016
$^{125}_{50}$Sn	
342	10†
470	14
810	30
910	30
1070	100
1160	3.3
1410	3.3
1990	14
2230	1.1
$^{126}_{50}$Sn	
60	
67	
92	
$^{127}_{50}$Sn	
270	
440	
490	
580	
820	
990	
1100	
1480	
1600	
2000	
2140	
2320	
2580	
2680	
2820	
$^{127}_{50}$Sn	
495	
$^{128}_{50}$Sn	
44	7†
72	19
500	61
570	22
$^{112}_{51}$Sb	
Ann. Rad.	
1257	
$^{113}_{51}$Sb	
79(Sn^{113m})	
320	
640	
700	
760	
810	
950	
1030	
1200	
1520	
$^{114}_{51}$Sb	
Ann. Rad.	
900	
1299	
$^{115}_{51}$Sb	
Ann. Rad.	
499	100†
980	5
980	5
1240	5
2220	1
$^{116m}_{51}$Sb	
Ann. Rad.	
99	30%
140	30%
406	36%
545	68%
960	75%
1060	27%
1290	100%
$^{116}_{51}$Sb	
Ann. Rad.	
99	
930	26†
1290	85
2230	14
$^{117}_{51}$Sb	
Ann. Rad.	
158	100%
$^{118m}_{2}$Sb${}_{51}$	
140	4†
300	10
380	10
$^{118m}_{1}$Sb${}_{51}$	
Ann. Rad.	
40.7	29%
253.5	93%
1049	100%
1090	
1229	100%

GAMMA ENERGIES AND INTENSITIES OF RADIONUCLIDES (Continued)

Gamma energies (KeV)	Gamma intensities	Gamma energies (KeV)	Gamma intensities	Gamma energies (KeV)	Gamma intensities	Gamma energies (KeV)	Gamma intensities
$^{118}_{51}$Sb		543.2		109.6		852.3	26%
Ann. Rad.		603.8		$^{127m}_{53}$Te		1125.5	15%
830		684.9		88.7 (IT)		1206.6	12%
1229		697.9		57.6	0.19†	1645.8	1.5%
1250		721.4		591	0.0006	1887.7	1.7%
$^{119}_{51}$Sn		782.6		657	0.0043	2001.0	2.7%
23.83		$^{128}_{51}$Sb		$^{127}_{52}$Te		2240	0.7%
$^{120}_{51}$Sb		320		57.6		2330	0.3%
Ann. Rad.		743		145	0.0016†	$^{131}_{53}$Te	
1171		750		203	0.017	149.8	100†
$^{122m}_{51}$Sb		1070		214	0.012	279	1.0
26		$^{129}_{51}$Sb		360	0.046	343	1.1
61		73		417	0.31	384	1.1
75		340		$^{129m}_{52}$Te		452.4	24.0
$^{122}_{51}$Sb		460		105.7 (IT)	2.3%	493	7.0
Ann. Rad.		540			(equilib.)	544	0.4
563.9	70%	660		158.91	0.15%	603	5.9
686	3.5%	810		208.9	0.3%	654	1.8
1137	0.7%	910		250.57±0.2	0.46%	695	0.5
1258	0.7%	1024		278.4	0.7%	727	0.4
$^{124m_2}_{51}$Sb		1240		281.15	0.22%	842	0.2
25		1740		342.90	0.05%	898	0.3
$^{124m_1}_{51}$Sb		$^{130}_{51}$Sb		459.4±0.3	0.90%	933	1.2
505	100†	190		487.4	1.7%	948	3.0
602.70±0.18	100	330		556.7	0.22%	952	0.6
645.76±0.18	100	820		624.4	0.1%	997	5.1
$^{124}_{51}$Sb		940		671.9	0.05%	$^{132}_{52}$Te	
443.86±0.3		$^{131}_{51}$Sb		696.0	5%	49.3±0.3	
602.70±0.18	100†	640	37%	729.7	1.36%	111.81±0.2	
645.76±0.18	7.5	950	−48%	741.1	0.09%	116.48±0.2	
709.3±0.2		$^{116}_{52}$Te		802.2	0.24%	228.3±0.2	
713.8±0.2	4	Ann. Rad.		817.2	0.17%	$^{133m}_{52}$Te	
722.72±0.18		720	34†	833.5	0.05%	74.1±0.2	0.8†
735.2±0.5		960	6	844.9	0.07%	81.5±0.2	0.8
790.7±0.25		1080	24	982.2	0.03%	88.0±0.2	3.0
968.2±0.2	2.5	1280	32	1022.6	0.04%	94.9±0.2	6.0
1045.2±0.2	2.4	1380	32	1084.0	0.71%	164.34±0.09	1.6
1325.6±0.2	2	1580	6	1111.8	0.27%	168.87±0.09	12
1355.4±0.5		2200	2	$^{129}_{52}$Te		177.1±0.2	1.5
1368.5±0.2		$^{116}_{52}$Te		27		178.2±0.2	1.0
1489.08±0.3	2.1		93.9	208.95	2.9†	184.45±0.10	0.4
1526.4±0.3	1.1	$^{117}_{52}$Te		250.6	5	193.22±0.10	1.2
1580.1±0.2	0.6		720	278.4	9	198.2±0.2	0.6
1691.0±0.07	51	$^{118}_{52}$Te		281.1	1	213.36±0.08	4.2
2091.1±0.2	7	Sb X-rays		342.9	0.3	220.94±0.13	0.5
2183.3±0.5		$^{119m}_{52}$Te		345	0.1	224.03±0.13	0.4
2293.6±0.4	0.6	153.0	94†	459.6	100	244.28±0.10	0.7
$^{125}_{51}$Sb		164	6	487.4	19	251.49±0.10	0.6
35.6 }		207.6	38	532	2.3	257.64±0.09	1.0
109.5 } (Te^{125m})	1†	400		556.7	2.1	261.55±0.07	14
117.1		760		573	0.6	(IT)334.14±0.07	15
172.8	~1.5	917	6	624.4	1.0	344.5±0.2	1.5
176.43	20	946	10	671.9	0.5	347.22±0.09	1.3
204.2	<1	984	12	696.0	73	355.57±0.14	0.6
208.2	<1	1008		729.7	16	362.81±0.15	1.1
227.9	2.6	1050	8	741.1	0.8	376.83±0.14	0.6
321.1	5	1099	7	802.2	2.8	396.96±0.09	1.7
380.4	4	1140	10	817.2	1.8	429.02±0.11	3.6
407.8	100	1221	100	833.5	0.7	444.90±0.09	4.4
427.9		1370	2.2	844.9	0.8	462.11±0.16	3.4
443.6		2090	7	982.2	0.3	471.85±0.09	2.3
463.4	33	$^{119}_{52}$Te		1022.6	0.5	478.59±0.10	1.8
600.6	62	645	100†	1050	0.5	519.6±0.2	0.5
606.7	17	700	13	1084.0	10	534.85±0.11	2.0
635.9	36	1760	42	1112	3.5	574.04±0.10	3.6
671.4	6	$^{121m}_{52}$Te		1233	0.3	622.03±0.16	1.6
$^{126m}_{51}$Sb		81.78 } IT		1262	0.7	647.40±0.08	34
410		212.21 }		1400	0.3	702.75±0.12	4.3
670		37.15		$^{131m}_{52}$Te		731.69±0.15	1.7
$^{126}_{51}$Sb		103	0.02†	181.7 (IT)	68%	733.89±0.10	3.3
290		910	0.2		(equilibrium)	779.75±0.10	3.9
410		1000	0.2	102.2	7.5%	795.7±0.4	1.5
580		1100	3.4	149.8	24%	800.51±0.12	2.2
690		$^{121}_{52}$Te		182.4	2.1%	863.91±0.13	29
850		37.15		200.7	7.6%	882.83±0.12	4.8
990		65.58	0.4	240.9	8.4%	897.7±0.4	0.5
$^{127}_{51}$Sb		470.4	1.4	334.3	11%	912.58±0.10	100
62		507.5	18	452.4	6.5%	914.72±0.13	19
252.7		573.1	80†	665.1	5%	934.4±0.3	1.5
290.6		$^{123m}_{52}$Te		773.7	46%	978.19±0.09	9.3
392.1		88.46	100%	782.5	7.6%	980.4±0.2	2.7
412.0		159.0	100%	793.8	16%	982.9±0.2	1.3
445.2		$^{125m}_{52}$Te		822.8	6.7%	1029.8±0.2	1.8
473.2		35.3					

GAMMA ENERGIES AND INTENSITIES OF RADIONUCLIDES (Continued)

Gamma energies (KeV)	Gamma intensities
1061.83±0.11	3.1
1348.9±0.2	2.9
1459.1±0.2	2.5
1516.1±0.3	1.1
1531.6±0.4	1.0
1587.4±0.2	2.2
1683.3±0.2	6.7
1704.4±0.3	1.1
1855.7±0.3	1.3
2004.9±0.3	5.4
2027.7±0.4	1.4
2049.2±0.4	1.7
$^{133}_{52}$Te	
311.99±0.08	70%
407.63±0.07	31%
474.72±0.13	1.1%
719.65±0.10	8.3%
786.77±0.10	7.2%
844.39±0.07	4.6%
930.67±0.10	5.4%
1000.77±0.11	4.5%
1021.07±0.15	3.4%
1252.2±0.2	1.4%
1333.23±0.12	11%
1405.7±0.2	0.8%
1717.65±0.15	3.2%
1881.5±0.4	1.3%
$^{134}_{52}$Te	
80	13%
170	16%
204	21%
262	19%
$^{117}_{53}$I	
Ann. Rad.	
160	
340	
$^{118}_{53}$I	
Ann. Rad.	
550	11
600	4
1150	
$^{119}_{53}$I	
Ann. Rad.	
260	
780	
$^{120}_{53}$I	
Ann. Rad.	
560	5†
620	1
1520	
$^{121}_{53}$I	
Ann. Rad.	
213	100†
270	3
320	7
600	
740	
$^{122}_{53}$I	
Ann. Rad.	
560	100†
690	7
780	5
1250	1
1420	1
1750	2
2180	1
2540	0.2
2740	
2920	
3170	
3450	
$^{123}_{53}$I	
159.1±0.1	100
183.7±1.0	0.03±0.02
192.7±1.0	0.03±0.02
248.3±0.5	0.08±0.01
281.0±0.5	0.08±0.01
346.6±0.5	0.12±0.02
440.4±0.5	0.42±0.02
505.6±0.6	0.31±0.05
529.0±0.4	1.27±0.11
538.5±0.5	0.32±0.02
624.9±0.5	0.08±0.01
687.7±0.6	0.03±0.01
736.1±0.06	0.04±0.01
784.4±0.6	0.05±0.01
$^{124}_{53}$I	
Ann. Rad.	
602.7±0.18	100†
645.76±0.18	18
722.72±0.18	21
890	0.4
968.2±0.2	0.8
1325.6±0.2	1.5
1368.5±0.2	5
1450	1.1
1500	6
1691.0±0.07	21
1900	0.8
2091.1±0.2	3
2293.6±0.4	2.3
2740	0.9
$^{125}_{53}$I	
35.48	
$^{126}_{53}$I	
Ann. Rad.	
388.7±0.2	
491.3±0.2	
666.2±0.2	
753.8±0.2	
879.9±0.2	
1420.0±0.2	
$^{128}_{53}$I	
442.9±0.2	17%
526.6±0.2	1.8%
743.3±0.15	0.3%
696.5±0.2	0.3%
$^{129}_{53}$I	
39.58	
$^{130}_{53}$I	
419	36†
536	100
669	100
743	87
1150	12
$^{131}_{53}$I	
80.164±0.009	2.6%
177.19±0.08	0.3%
272.45±0.2	0.07%
284.31±0.08	5.9%
318.14±0.15	0.09%
325.80±0.15	0.24%
364.49±0.08	79%
503.13±0.15	0.35%
637.01±0.15	6.7%
642.80±0.17	0.2%
722.92±0.15	1.8%
$^{132}_{53}$I	
147.10±0.3	5.0†
254.9	2.0
262.75±0.3	8.42
505.80±0.3	7.87
522.48±0.3	24.55
546.95±0.4	1.70
620.95±0.3	2.49
630.16±0.3	15.44
650.51±0.3	2.80
727.04±0.3	4.96
729.08±0.25	1.44
772.49±0.25	59.77
780.24±0.25	1.0
784.92±0.4	0.35
809.77±0.2	2.04
812.30±0.3	4.19
863.26±0.4	0.40
876.96±0.4	0.63
910.44±0.3	0.57
927.82±0.4	0.37
954.66±0.25	9.72
984.38±0.4	0.40
1035.29±0.4	0.29
1136.11±0.3	1.29
1143.64±0.4	0.62
1173.22±0.4	0.48
1290.74±0.4	0.39
1295.34±0.5	0.54
1298.0±0.4	0.25
1372.0±0.3	0.70
1398.71±0.4	2.21
1442.46±0.3	0.43
1920.84±0.4	0.24
2002.08±0.4	0.21
2088.5±0.5	0.07
2175.0±1	0.02
2226.0±1	0.03
2395.0±1	0.01
$^{133}_{53}$I	
151.1±0.3	0.029
234.3±0.4	†
245.9±0.3	0.081
262.7±0.3	0.38
267.3±0.4	0.10
386.4±0.3	0.088
417.1±0.3	0.18
422.9±0.3	0.43
427.7±0.4	0.099
438.9±0.4	0.083
510.5±0.3	2.08
529.90±0.15	89% 100
600.4±0.3	0.046
608.1±0.3	0.086
618.0±0.25	0.64
680.3±0.2	0.79
706.65±0.2	1.77
768.46±0.25	0.51
789.5±0.3	0.058
820.63±0.3	0.187
835.3±0.35	
856.37±0.25	1.30
875.40±0.25	5.26
910.1±0.3	0.26
1052.3±0.3	0.64
1060.3±0.4	0.20
1236.56±0.2	1.78
1293.9±0.3	0.30
1298.32±0.2	2.72
1350.6±0.3	0.18
1460.9±0.3	0.18
1592.5±0.3	0.08
$^{134}_{53}$I	
135.4±0.3	3.5†
152.2±0.4	0.12
162.6±0.3	0.26
188.5±0.3	
216.9±0.4	0.29
235.43±0.3	1.85
319.68±0.3	0.44
356.88±0.3	
405.35±0.15	6.84
433.24±0.17	4.13
458.88±0.25	1.04
488.90±0.25	1.63
514.48±0.2	2.30
540.81±0.15	7.66
565.0±0.4	
595.29±0.15	11.24
621.77±0.17	9.93
628.12±0.25	2.60
667.58±0.25	0.74
677.26±0.17	7.70
730.77±0.2	1.79
766.67±0.18	3.98
772.2±0.4	0.60
816.7±0.3	
847.04±0.08	96% 100
857.29±0.15	7.51
875.3±0.3	0.55
884.18±0.1	67.81
947.69±0.15	4.19
967.4±0.25	0.6
974.72±0.17	4.90
1040.1±0.2	1.52
	0.5
1072.52±0.17	15.91
1100.3±0.3	0.75
1103.5±0.3	0.66
1136.15±0.18	10.68
1353.5±0.3	0.33
1455.2±0.3	2.48
1469.5±0.5	0.81
1613.54±0.3	4.61
1741.4±0.35	2.41
1806.7±0.35	5.62
2311.9±1.0	0.4
2467.1±1.0	0.21
$^{135}_{53}$I	
135.4±0.5	
197.0±0.4	
220.45±0.08	5.37†
229.64±0.18	0.64
249.56±0.08	6.24
264.16±0.08	
288.26±0.08	9.03
289.6±0.4	1.24
305.3±0.2	0.43
361.6±0.2	
403.0±0.3	
415.0±0.2	1.44
417.54±0.1	10.79
429.9±0.2	0.97
433.69±0.15	1.93
451.7±0.2	0.83
510.60±0.17	0.94
526.43±0.10	21.16
(Xe¹³⁵ᵐ)	
546.54±0.10	22.75
559.0±0.5	
648.8±0.3	
650.2±0.3	2.56
667.8±0.3	1.25
707.9±0.6	2.92
836.84±0.12	20.10
884.1±0.3	4.07
973.12±0.18	7.41
1038.74±0.15	29.86
1101.5±0.4	5.79
1123.8±0.2	13.18
1131.56±0.18	79.73
1168.9±0.4	3.02
1260.45±0.15	100 34.9%
1281.8±0.4	
1298.3±0.4	
1367.5±0.3	
1398.0±0.6	
1457.65±0.15	31.36
1470.0±0.6	
1503.0±0.5	3.92
1566.6±0.4	3.70
1678.20±0.17	30.10
1706.6±0.2	13.03
1791.36±0.2	25.16
1830.7±0.4	1.89
1927.1±0.5	
2046.1±0.5	2.76
2256.6±0.8	
2268.6±1	
2408.6±1	2.78
2465.9±1.5	
$^{136}_{53}$I	
200	12%
270	18%
390	19%
460	~2%
530	~2%
710	3%
1000	6%
1320	70%
1325	25%
1550	4%
1720	2%
1910	5%
2200	7%
2400	12%
2630	10%
2800	8%
3200	5%
$^{118}_{54}$Xe	
Ann. Rad.	

GAMMA ENERGIES AND INTENSITIES OF RADIONUCLIDES (Continued)

Gamma energies (KeV)	Gamma intensities
¹¹⁹₅₄Xe	
Ann. Rad.	
100	
¹²⁰₅₄Xe	
Ann. Rad.	
55	
73	
176	
760	
¹²¹₅₄Xe	
Ann. Rad.	
80	
95.5	
132.5	
437	
¹²²₅₄Xe	
90.3	
148.2	
239	
¹²³₅₄Xe	
Ann. Rad.	
148.6	
177.6	
328.6	
680	
900	
1100	
¹²⁵ᵐ₅₄Xe	
75	
111	
¹²⁵₅₄Xe	
54.7	
74.6	
113.1	
187.6	
242.2	
460	
¹²⁷ᵐ₅₄Xe	
125	
175	
¹²⁷₅₄Xe	
57.60	
145.22	3.9%
172.10	24%
202.84	67%
374.96	20%
¹²⁹ᵐ₅₄Xe	
39.58	
196.56	
¹³¹ᵐ₅₄Xe	
163.98±0.10	
¹³³ᵐ₅₄Xe	
232.8	
¹³³₅₄Xe	
79.55	0.25%
80.97	35%
161	0.03%
220	
302	0.003%
382	0.001%
¹³⁵ᵐ₅₄Xe	
526.43±0.10	
¹³⁵₅₄Xe	
249.56±0.1	9.2%
360	0.009%
610	0.3%
¹³⁷₅₄Xe	
455	33%
¹³⁸₅₄Xe	
160	33†
260	100
420	63
1730	44
1990	30
¹³⁹₅₄Xe	
180	4†
220	100
300	57
1150	23
¹⁴⁰₅₄Xe	
130	
¹²³₅₅Cs	
Ann. Rad.	

Gamma energies (KeV)	Gamma intensities
¹²⁵₅₅Cs	
Ann. Rad.	
112	
¹²⁶₅₅Cs	
Ann. Rad.	
385	38%
¹²⁷₅₅Cs	
Ann. Rad.	
125	13†
286	
406	100
¹²⁸₅₅Cs	
Ann. Rad.	
441	50†
528	
576	
940	1.5
1200	2.1
1660	0.4
2180	0.4
2420	0.2
¹²⁹₅₅Cs	
94	1.3†
174	1.0
270	0.8
278	6
321	8
375	100
416	52
550	9
591	1.7
900	0.5
950	0.18
¹³⁰₅₅Cs	
Ann. Rad.	
¹³²₅₅Cs	
Ann. Rad.	
464.58±0.4	
505.55±0.4	
510.58±0.4	
566.91±0.3	
630.12±0.3	
667.60±0.2	
772.53±0.3	
796.09±0.3	
1031.87±0.3	
1136.0±0.3	
1298.11±0.4	
1317.90±0.4	
1461.09±0.5	
1985.68±0.4	
¹³⁴ᵐ₅₅Cs	
~10	
127.6	
137.4	
¹³⁴₅₅Cs	
475.34±0.1	1.8%
563.20	8%
569.33	14%
604.7	98%
795.8	88%
801.9	9%
1038.6	1.1%
1167.9	1.9%
1365.1	3.4%
1400.5	0.08%
¹³⁶₅₅Cs	
66.9±0.2	13.7%
86.4	5.6%
153.38±0.2	8.2%
164.0±0.2	4.5%
176.7±0.2	13%
273.8±0.2	12%
340.6±0.15	44%
818.5±0.2	100%
1048.1±0.2	80%
1235.4±0.2	20%
¹³⁷₅₅Cs	
661.630±0.03	84.8±0.5%
¹³⁸₅₅Cs	
66.7±0.3	
75.41±0.4	0.08†
107.6±0.3	0.06

Gamma energies (KeV)	Gamma intensities
112.27±0.25	0.16
137.89±0.15	1.58
165.75±0.2	3.17
191.83±0.3	0.5
212.04±0.3	
227.60±0.18	1.56
324.53±0.2	0.43
364.1±0.3	
408.67±0.2	5.3
421.39±0.3	0.43
462.51±0.18	35.49
516.53±0.2	0.65
546.71±0.15	12.69
871.61±0.18	6.07
1009.65±0.18	37.86
1146.88±0.2	1.49
1283.4±0.8	
1343.33±0.3	1.64
1435.74±0.18	73%100
1444.59±0.25	1.32
2217.75±0.2	18.52
2316.5±1.5	
2345.0±0.5	0.81
2582.6±0.4	
2639.4±0.3	9.97
3340	0.5
¹³⁹₅₅Cs	
500	
630	
800	
1280	
1900	
2080	
¹⁴⁰₅₅Cs	
590	
880	
1140	
1620	
1850	
2060	
2320	
2720	
3150	
¹²³₅₆Ba	
Ann. Rad.	
¹²⁵₅₆Ba	
Ann. Rad.	
¹²⁶₅₆Ba	
230	100†
700	33
¹²⁸₅₆Ba	
134	
278	
¹²⁹ᵐ₅₆Ba	
Ann. Rad.	
53.2	
73.2	
129	
135.7	
176.9	
182.3	
202.3	
214.2	
220.6	
237.4	
263.6	
286.0	
293.5	
328.2	
475	
481	
501	
534	
543	
547	
554	
557	
561	
597	
679	
690	
700	

Gamma energies (KeV)	Gamma intensities
713	
738	
748	
759	
768	
779	
802	
820	
829	
833	
872	
892	
933	
957	
963	
1000	
1035	
1045	
1122	
1163	
1179	
1208	
1221	
1252	
1458	
1624	
¹³¹ᵐ₅₆Ba	
75	2†
110	100
¹³¹₅₆Ba	
54.84	
78.69	
82.24	
92.25	
123.73	
133.54	
137.3	
157.0	
216.0	
239.6	
246.8	
249.4	
294.5	
351.2	
373.2	
404.3	
427.7	
451.7	
461.3	
462.9	
480.6	
486.6	
496.3	
573	
585	
620.2	
674	
696	
832	
914	
924.4	
969.1	
1048.2	
¹³³ᵐ₅₆Ba	
275.7	
12.29	
¹³²₅₆Ba	
53.17±0.04	3.68±0.09
79.60±0.05	4.9±0.59
80.997±0.006	55.3±4.2
160.66±0.06	1.01±0.05
223.43±0.26	0.75±0.04
276.43±0.26	11.6±0.17
303.09±0.21	29.9±0.29
356.26±0.15	≡100
384.09±0.20	14.9±0.26
¹³⁵ᵐ₅₆Ba	
268	100%
¹³⁶ᵐ₅₆Ba	
163.7	
818	
1050	

GAMMA ENERGIES AND INTENSITIES OF RADIONUCLIDES (Continued)

Gamma energies (KeV)	Gamma intensities
$^{137m}_{56}$Ba	
661.630±0.030	
$^{139}_{56}$Ba	
165.85±0.010	27%
1270	
1430	
$^{140}_{56}$Ba	
13.85	
29.97	30%
132.7	1.7%
162.7	8%
304.9	7%
423.8	3%
437.6	5%
537.3	29%
$^{141}_{56}$Ba	
113.08±0.05	2†
180.70±0.09	0.7
190.33±0.05	100
276.94±0.04	52
304.18±0.04	57
343.66±0.04	31
389.66±0.07	3.1
457.52±0.05	12
462.13±0.06	10
467.26±0.06	13
625.21±0.04	7
647.78±0.04	13
739.02±0.1	10
831.70±0.04	4.1
876.01±0.14	8
929.28±0.08	1.5
981.40±0.15	1.7
1160.55±0.1	3
1197.3±0.2	13
1273.4±0.2	1.3
1310.7±0.2	2.1
1436.64±0.1	2.0
1501.28±0.18	0.7
1551.0±0.3	0.9
1740.8±0.2	7
$^{142}_{56}$Ba	
77±1	41†
231.57±0.09	59
255.33±0.07	100
286.13±0.16	7
309.5±0.3	11
364.03±0.07	28
379.42±0.16	3
425.05±0.07	30
432.42±0.16	5.6
589.95±0.2	3.5
599.61±0.1	13
769.71±0.17	4.2
840.07±0.12	19
894.86±0.1	82
944.02±0.07	47
1000.90±0.1	53
1078.52±0.07	56
1093.95±0.09	17
1126.1±0.1	10
1203.81±0.1	97
1380.3±0.2	14
$^{126}_{57}$La	
Ann. Rad.	
$^{127}_{57}$La	
Ann. Rad.	
$^{128}_{57}$La	
Ann. Rad.	
$^{129}_{57}$La	
Ann. Rad.	
$^{130}_{57}$La	
Ann. Rad.	
356	
450	
550	
720	
810	
910	
1010	
1190	
1450	
1550	

Gamma energies (KeV)	Gamma intensities
$^{131}_{57}$La	
Ann. Rad.	
115	23†
169	5
214	8
254	8
285	17
364	20
417	20
450	8
600	7
878	1
$^{132}_{57}$La	
Ann. Rad.	
470	
560	
660	
900	
1030	
1220	
1450	
1580	
1920	
$^{133}_{57}$La	
Ann. Rad.	
260	
$^{134}_{57}$La	
Ann. Rad.	
560	
604.6	
760	
820	
900	
980	
1170	
1300	
1470	
1550	
1780	
1980	
2150	
2240	
2590	
$^{135}_{57}$La	
52.4	
87.4	
117.0	
118	
132	
205	
265	
300	
379	
515	
544	
549	
569	
571	
602	
661	
679	
777	
821	
865	
901	
$^{136}_{57}$La	
Ann. Rad.	
830	
$^{140}_{57}$La	
109.6±0.2	0.22%
131.15±0.2	0.45%
241.9	0.55%
266.5	0.5%
328.7	21%
432.5	3%
487.0	46%
510.9	0.35%
751.7	0.45%
815.8	24%
867.9	5.7%
919.6	2.5%
925.2	6.7%
951.0	0.6%

Gamma energies (KeV)	Gamma intensities
1085.3	1%
1596.2	97%
2010.4	0.4%
2348.2	0.8%
2521.8	3.2%
2547.7	0.09%
$^{141}_{57}$La	
1370	2%
$^{142}_{57}$La	
106.47±0.1	0.25
178.32±0.1	0.23
418.2±0.3	0.1
430.3±0.3	0.1
510.11±0.1	0.4
578.25±0.08	2.8
641.27±0.05	100 (50%)
861.68±0.1	3.5
894.94±0.05	18.4
1011.42±0.07	7.8
1043.89±0.07	4.6
1160.40±0.1	2.4
1233.07±0.1	3.6
1363.55±0.2	3.4
1545.93±0.2	6.5
1723.23±0.3	2.4
1756.60±0.1	5.4
1901.56±0.1	11.3
2004.87±0.3	2.0
2055.6±0.2	3.5
2187.49±0.1	8.8
2397.95±0.1	24.4
2543.1±0.3	15.7
2666.5±0.3	2.0
2971.8±0.3	6.5
3314.1±0.3	0.9
3460.1±0.3	0.6
3611±1	1.2
3602.6±0.4	1.0
3719.7±0.5	0.3
3850.1±0.5	0.3
$^{143}_{57}$La	
200	8†
440	13
620	100
800	44
902	8
1070	26
1170	57
1580	28
1700	19
1980	35
2220	6
2460	13
2560	27
2850	15
$^{132}_{58}$Ce	
180	
$^{133}_{58}$Ce	
1800	
$^{135}_{58}$Ce	
52.4	
87.4	
117.0	
118.0	
132	
205	
265	
300	
379	
515	
544	
549	
569	
571	
602	
661	
679	
777	
821	
865	
901	
$^{137}_{58}$Ce	
254.5 (IT)	99.4%

Gamma energies (KeV)	Gamma intensities
168	45†
762	16
825	38
836	8.5
994	0.15
1005	2.1
1160	0.2
$^{137}_{58}$Ce	
Ann. Rad.	
10	†
433	5
436	31
446	204
479	1.7
481	~4.0
492	1.0
698	3.5
771	0.9
781	0.4
916	6.7
926	3.4
$^{139m}_{58}$Ce	
746	
$^{139}_{58}$Ce	
165.854±0.010	100%
$^{141}_{58}$Ce	
145.443±0.006	49%
$^{143}_{58}$Ce	
57.37	11.5%
142	0.4%
231.60	2%
293.22	43%
350.6±0.1	3%
374	0.4%
490.42	2%
587.38	0.5%
664.59±0.1	5.5%
721.93±0.1	6%
880.4±0.2	1%
940	0.2%
1040	0.2%
1110	0.4%
$^{144}_{58}$Ce	
33.57	
40.93	
53.41	
59.03	
80.12	
99.95	
133.53	11%
$^{146}_{58}$Ce	
50	
110	20†
142	42
220	50
270	12
320	100
$^{134}_{59}$Pr	
Ann. Rad.	
220	
300	
410	
640	
960	
$^{135}_{59}$Pr	
Ann. Rad.	
80	
220	
300	
$^{136}_{59}$Pr	
Ann. Rad.	
170	
~800	
~1100	
$^{137}_{59}$Pr	
Ann. Rad.	
$^{138}_{59}$Pr	
Ann. Rad.	
298	77†
400	9
550	~5
790	100
1040	98

GAMMA ENERGIES AND INTENSITIES OF RADIONUCLIDES (Continued)

Gamma energies (KeV)	Gamma intensities
1200	1
1830	0.4
¹³⁹₅₉Pr	
Ann. Rad.	
270	12†
1350	27
1610	16
¹⁴⁰₅₉Pr	
Ann. Rad.	
306	
1596	
1902	
¹⁴²₅₉Pr	
1575.43±0.3	4%
¹⁴³₅₉Pr	
no γ	
¹⁴⁴₅₉Pr	
696.48±0.09	1.5%
1489.14±0.07	0.35%
2185.32±0.05	0.9%
¹⁴⁵₅₉Pr	
72	
680	
750	
920	
980	
1050	
1160	
¹⁴⁶₅₉Pr	
455	77%
597	10%
750	16%
780	15%
920	6%
1020	5%
1370	6%
1510	27%
1720	4%
1970	2%
2230	4%
2390	3%
2730	1.7%
¹⁴⁷₅₉Pr	
78	17%
105	
127	
310	
340	25%
560	39%
610	10%
650	24%
1260	11%
¹⁴⁸₅₉Pr	
300	
¹³⁸₆₀Nd	
Ann. Rad.	
¹³⁹ᵐ₆₀Nd	
232	
¹³⁹₆₀Nd	
Ann. Rad.	
36.5	
114	
132.7	
200	
210	
217	
708	100
737	42
773	68†
821	
822	
828	
900	25
983	70
1030	30
1100	30
1240	20
1340	20
1480	10
1580	8
1720	4
1850	3
2050	10

Gamma energies (KeV)	Gamma intensities
¹⁴¹₆₀Nd	
Ann. Rad.	
145.443±0.0006	
1126.8±0.3	
1147.5±0.3	
1292.6±0.3	
1298.6±0.3	
¹⁴¹ᵐNd	
0.755	100%
¹⁴⁷₆₀Nd	
91.0±0.2	27%
120.5	0.5%
196.6	0.4%
275.4	0.8%
319.4	2.0%
398.1	1.6%
410.4	0.7%
439.8	1.2%
489.3	0.2%
531.0	13%
594.7	0.25%
685.8	0.8%
¹⁴⁹₆₀Nd	
30.00±0.10	
58.90±0.03	1.3%
65.37±0.05	0.041%
74.3±0.2	
74.7±0.2	
75.55±0.2	
91.12±0.05	0.063%
97.00±0.03	1.3%
107.76±0.10	
114.31±0.03	16.1%
117.0±0.2	
122.41±0.05	0.20%
126.62±0.05	0.10%
173.02±0.08	0.040%
139.21±0.05	0.41%
155.87±0.03	5.2%
177.85±0.05	0.14%
185.50±0.08	0.079%
188.64±0.03	1.7%
192.03±0.05	0.51%
198.93±0.03	1.25%
208.15±0.05	2.5%
211.32±0.03	23.4%
214.02±0.07	
226.85±0.05	0.14%
229.57±0.05	0.43%
240.23±0.03	3.4%
245.71±0.05	0.89%
250.84±0.10	0.032%
254.20±0.07	0.075%
258.07±0.05	0.33%
267.70±0.03	5.2%
270.18±0.03	9.2%
273.28±0.08	0.20%
275.46±0.05	0.51%
276.96±0.10	0.28%
282.47±0.05	0.52%
288.21±0.05	0.59%
294.82±0.05	0.50%
301.14±0.05	0.33%
310.99±0.05	0.44%
326.57±0.03	4.0%
342.85±0.10	0.029%
347.8±0.2	
349.23±0.03	1.29%
352.75±0.10	0.056%
357.05±0.10	0.047%
360.05±0.05	0.14%
366.64±0.05	0.57%
380.94±0.10	0.058%
384.69±0.05	0.29%
396.77±0.10	0.085%
423.54±0.03	8.2%
425.7±0.3	
443.53±0.03	1.29%
480.21±0.15	0.051%
483.53±0.15	0.073%
493.86±0.15	0.044%
497.81±0.15	0.047%
510.29±0.15	0.051%

Gamma energies (KeV)	Gamma intensities
515.45±0.20	
533.17±0.15	0.074%
538.00±0.20	
540.49±0.03	6.6%
556.40±0.10	1.00%
579.25±0.07	0.077%
583.07±0.07	0.072%
594.48±0.10	0.028%
598.06±0.15	0.033%
630.24±0.07	0.19%
635.47±0.07	0.096%
654.82±0.03	8.4%
661.8±0.3	0.016%
671.6±1.0	0.012%
673.6±0.5	0.014%
675.8±0.2	0.028%
681.29±0.2	0.011%
686.96±0.07	0.092%
696.29±0.05	0.17%
712.62±0.07	0.072%
718.40±0.10	0.026%
727.90±0.15	0.019%
740.61±0.10	0.017%
749.68±0.15	0.012%
754.33±0.10	0.030%
758.6±0.3	0.014%
761.45±0.15	0.020%
768.17±0.10	0.059%
781.42±0.20	0.006%
786.74±0.10	0.011%
793.43±0.10	0.022%
795.84±0.20	0.006%
808.80±0.10	0.146%
809.8±0.5	0.025%
813.2±0.5	0.008%
829.4±0.4	0.006%
832.20±0.10	0.019%
837.41±0.10	0.030%
842.86±0.10	0.048%
849.93±0.10	0.022%
859.43±0.20	0.013%
861.54±0.10	0.019%
865.03±0.15	0.018%
871.40±0.10	0.032%
873.8±0.2	0.0052%
886.59±0.3	0.0073%
907.70±0.2	0.0063%
915.3±0.3	0.0045%
923.88±0.08	0.099%
929.44±0.15	0.012%
935.94±0.2	0.0056%
938.77±0.2	0.0060%
945.82±0.10	0.021%
951.96±0.15	0.0086%
964.00±0.15	0.018%
967.44±0.15	0.0094%
971.7±0.3	0.0033%
979.03±0.08	0.096%
986.7±0.3	0.0033%
992.9±0.2	0.013%
1022.79±0.08	0.105%
1027.19±0.10	0.0098%
1031.8±0.2	0.0041%
1041.97±0.10	0.023%
1051.9±0.3	0.0037%
1076.05±0.15	0.021%
1078.78±0.10	0.058%
1100.78±0.10	0.052%
1123.6±0.4	0.014%
1125.41±0.15	0.025%
1128.5±0.3	0.0034%
1135.9±0.2	0.0021%
1141.8±0.2	0.0034%
1150.1±0.2	0.0030%
1172.1±0.3	0.0044%
1175.7±0.2	0.0044%
1190.4±0.2	0.0029%
1197.8±0.2	0.0066%
1202.3±0.2	0.0019%
1225.7±0.3	0.0014%
1234.11±0.10	0.0250%
1239.0±0.6	0.0014%
1259.6±0.2	0.0033%

Gamma energies (KeV)	Gamma intensities
1264.0±0.2	0.0060%
1280.3±0.3	0.0010%
1284.3±0.3	0.0007%
1290.10±0.15	0.0039%
1298.4±0.3	0.0015%
1312.13±0.15	0.0068%
1357.2±0.40	0.0021%
1381.5±0.3	0.0021%
1448.1±0.5	0.004%
1454.3±0.4	0.0011%
1473.9±0.5	0.0003%
1495.8±0.3	0.008%
1568.4±0.4	0.0005%
¹⁵¹₆₀Nd	
58.62±0.10	0.35%
69.19±0.05	1.27%
81.85±0.25	0.49%
85.18±0.06	1.98%
90.04±0.05	1.60%
102.89±0.13	0.48%
109.07±0.55	<0.001%
116.79±0.04	32.56%
138.88±0.04	6.31%
149.49±0.07	0.33%
158.74±0.20	0.50%
170.75±0.04	3.39%
175.07±0.05	6.12%
183.14±0.08	0.36%
197.23±0.13	0.27%
199.73±0.12	0.30%
208.10±0.14	0.31%
239.29±0.30	0.83%
249.27±0.30	0.16%
255.69±0.04	13.3%
263.56±0.07	0.67%
300.58±0.08	1.58%
312.49±0.35	0.25%
320.13±0.11	0.63%
324.56±0.35	0.48%
332.76±0.11	0.63%
347.11±0.11	0.39%
356.91±0.11	0.38%
402.33±0.06	1.41%
407.53±0.10	0.45%
414.49±0.35	0.24%
423.54±0.03	5.31%
426.60±0.30	0.50%
439.37±0.20	0.35%
460.58±0.08	0.74%
524.35±0.11	0.42%
531.99±0.15	0.16%
542.19±0.35	0.45%
549.99±0.11	0.70%
562.72±0.17	0.21%
577.41±0.11	0.26%
585.20±0.16	1.09%
589.61±0.10	0.30%
596.93±0.09	0.60%
619.02±0.12	0.35%
629.78±0.31	0.14%
658.58±0.20	0.66%
670.50±0.18	0.22%
677.95±0.06	2.22%
724.17±0.18	0.28%
736.21±0.06	5.92%
739.17±0.12	1.71%
755.56±0.11	1.08%
765.46±0.70	0.16%
768.09±0.70	0.22%
773.60±0.18	0.28%
797.48±0.06	4.65%
829.05±0.18	0.22%
841.05±0.11	0.96%
852.98±0.14	0.42%
866.44±0.26	0.32%
870.65±0.24	0.32%
876.33±0.14	0.48%
897.72±0.24	0.41%
904.69±0.16	0.35%
914.15±0.11	1.21%
925.53±0.22	0.16%
943.11±0.22	0.37%
958.09±0.11	0.46%

Gamma energies (KeV)	Gamma intensities
1016.30±0.10	2.80%
1020.21±0.50	0.19%
1032.47±0.18	0.11%
1035.98±0.14	0.19%
1041.76±0.11	0.35%
1048.05±0.09	0.67%
1066.60±0.14	0.17%
1073.00±0.22	0.085%
1077.49±0.37	0.15%
1080.09±0.21	0.30%
1100.11±0.20	0.093%
1107.20±0.11	0.44%
1122.59±0.06	4.46%
1126.56±0.25	0.28%
1132.53±0.18	0.18%
1136.54±0.15	0.23%
1145.84±0.17	0.25%
1157.02±0.13	0.21%
1180.90±0.06	14.65%
1189.33±0.10	0.31%
1201.15±0.13	0.20%
1213.34±0.14	0.093%
1217.44±0.25	0.046%
1224.39±0.27	0.051%
1232.79±0.14	0.11%
1238.30±0.18	0.065%
1250.84±1.12	0.051%
1254.53±1.12	0.051%
1270.64±0.21	0.19%
1286.17±0.12	0.26%
1293.94±0.33	0.11%
1297.75±0.21	0.19%
1314.61±0.12	0.31%
1328.36±0.12	0.31%
1333.33±0.11	0.12%
1341.61±0.23	0.11%
1346.45±0.2	0.033%
1360.20±?	0.15%
1362.07±0.20	(illegible)
1379.25±0.17	0.16%
1363.48±0.21	0.078%
1395.41±1.09	0.13%
1425.30±0.27	0.068%
1444.41±0.60	0.055%
1446.97±0.90	0.035%
1465.62±0.24	0.077%
1476.06±0.19	0.11%
1485.66±0.14	0.28%
1498.74±1.14	0.029%
1501.58±1.14	0.029%
1507.58±0.41	0.032%
1533.61±0.28	0.049%
1550.03±0.14	0.36%
1554.05±0.25	0.072%
1566.73±0.24	0.085%
1572.02±0.25	0.10%
1578.63±0.18	0.21%
1592.75±0.25	0.044%
1598.20±0.26	0.12%
1618.17±0.16	0.41%
1636.75±0.24	0.12%
1640.44±0.37	0.032%
1647.45±0.30	0.051%
1686.86±0.26	0.015%
1717.27±0.20	0.091%
1732.31±0.27	0.048%
1753.48±0.25	0.045%
1757.10±0.31	0.053%
1762.06±0.29	0.025%
1775.61±0.19	0.25%
1782.81±0.31	0.051%
1786.86±0.26	0.078%
1794.20±0.26	0.056%
1807.24±0.35	0.068%
1811.34±0.37	0.074%
1819.23±0.24	0.062%
1830.21±0.50	0.012%
1836.12±0.43	0.014%
1849.02±0.31	0.023%
1855.24±0.63	0.011%
1863.59±0.23	0.043%
1873.68±1.19	0.0074%
1877.58±0.36	0.011%

Gamma energies (KeV)	Gamma intensities
1892.45±0.19	0.018%
1903.89±0.44	0.021%
1909.40±0.35	0.037%
1912.38±1.21	0.0040%
1926.15±0.28	0.042%
1932.83±0.30	0.024%
2019.32±0.29	0.037%
2105.83±0.57	0.0058%
2254.82±0.42	0.0065%
$^{141}_{61}$Pm	
Ann. Rad.	
$^{142}_{61}$Pm	
1570	0.2%
$^{143}_{61}$Pm	
742	47%
$^{144}_{61}$Pm	
476	45%
616	100%
696.48±0.09	100%
$^{145}_{61}$Pm	
67	12%
725	8%
$^{146}_{61}$Pm	
453	65%
742	30%
749	35%
^{148m}Pm	
61.5±0.2(IT)	
75.6±0.2(IT)	
98.5±0.2	3.3%
189.5±0.3	
288.0±0.2	13%
311.7±0.3	2.8%
414.1±0.3	17%
432.7±0.3	6%
443.1±0.4	
501.1±0.4	7%
550.1±0.2	95%
599.5±0.3	8%
611.1±0.3	6%
629.9±0.2	87%
725.6±0.2	36%
896.2±0.3	
914.9±0.2	21%
1013.7±0.3	20%
1465.1±0.2	27%
^{148}Pm	
550.1±0.2	50%
914.9±0.2	16%
1465.1±0.2	24%
$^{149}_{61}$Pm	
K X-ray 22.5	2±0.7
208.1±1	0.008±0.006
242.4±1	0.009±0.006
254.4±0.5	0.045±0.02
263.3±0.4	0.091±0.04
276.9±0.4	0.27±0.10
285.90±0.05	31.0±2
304.9±0.4	0.022±0.01
327.4±0.4	0.034±0.01
350.6±0.4(doublet)	0.023±0.01
359.4±0.4	0.022±0.01
531±1	0.012±0.01
535.8±0.4	0.13±0.03
545±1	0.025±0.01
548±1	0.019±0.01
558.4±0.4	0.14±0.03
568.5±0.4	0.18±0.03
590.9±0.3	0.68±0.08
613.9±0.4	0.13±0.02
636.6±0.4	0.062±0.02
807.8±0.5	0.15±0.02
830.5±0.5	0.30±0.04
833.2±0.5	0.30±0.04
859.4±0.5	1.02±0.15
881.9±0.5	0.23±0.03
$^{150}_{61}$Pm	
334	100†
410	10
590	5
720	11

Gamma energies (KeV)	Gamma intensities
840	25
880	17
1170	32
1230	6
1330	31
1750	14
1960	3.5
2060	1.7
2530	1.3
2750	0.4
2910	0.5
3080	0.16
$^{151}_{61}$Pm	
100.00	2.7%
101.80	1.6%
104.80	3.8%
139.30	0.5%
156.20	0.2%
163.37	2.5%
167.84	9.9%
177.05	4.8%
186.60	0.2%
201.94	0.1%
208.97	2.0%
227.21	0.3%
232.36	1.2%
236.73	1.0%
240.00	3.9%
254.07	0.2%
270.41	0.1%
275.10	7.4%
280.11	0.2%
290.63	0.8%
306.60	0.2%
323.79	1.5%
329.60	0.25%
339.93	24%
344.80	2.2%
349.60	0.1%
353.10	0.1%
379.67	1.0%
406.80	0.2%
440.66	1.6%
445.51	4.3%
451.24	0.3%
490.10	0.1%
516.00	0.2%
564.90	0.4%
636.20	1.5%
654.30	0.3%
669.10	0.8%
671.40	1.0%
704.40	0.3%
717.85	4.4%
736.30	4.8%
753.00	1.4%
773.00	0.8%
808.10	0.6%
817.90	0.3%
848.90	0.3%
948.90	0.4%
953.90	0.1%
$^{152}_{61}$Pm	
120	
240	
1000	
$^{142}_{62}$Sm	
Ann. Rad.	
$^{143m}_{62}$Sm	
748	
$^{143}_{62}$Sm	
Ann. Rad.	
1050	
$^{145}_{62}$Sm	
61.3	92%(78%)
485	0.003%
$^{151}_{62}$Sm	
21.6	1.7%
$^{153}_{62}$Sm	
54.19	0.006†
68.23	0.004
69.68	17
75.43	0.61

Gamma energies (KeV)	Gamma intensities
83.4	0.75
89.5	0.58
103.23	100
151.6	0.03
172.9	0.31
411	0.01
424	0.013
438	0.01
450	0.01
463	0.05
521	0.025
531	0.23
533	0.12
539	0.06
578	0.01
603	0.014
609	0.04
636	0.018
$^{155}_{62}$Sm	
75	0.3†
104.35	73
140	1.7
246	4
450	0.18
510	0.3
650	0.2
920	0.03
990	0.04
1180	0.06
1270	0.13
$^{156}_{62}$Sm	
22.6	
25.3	
39.1	
65.0	
87.6	40%
103	
166	
204	
219	
244	
269	
291	
$^{157}_{62}$Sm	
570	
$^{144}_{63}$Eu	
Ann. Rad.	
$^{145}_{63}$Eu	
Ann. Rad.	
110.94±0.15	2.4†
191.25±0.15	0.9
542.51±0.15	7.5
653.46±0.15	24
713.7±0.3	0.5
764.65±0.15	3.1
893.57±0.20	75
974.90±0.15	3.8
1658.38±0.15	1.9
1803.90±0.15	1.9
1876.84±0.15	2.0
1996.61±0.15	12.3
$^{146}_{63}$Eu	
Ann. Rad.†	
27.18	
146.3±0.2	
267.6±0.2	
271.78±0.20	
397.32±0.20	
410.80±0.20	
430.29±0.20	
621.68±0.20	
628.7±0.20	
633.6±0.20	
665.31±0.20	
667.84±0.20	12†
743.1±0.2	
747.1±0.2	100
791.14±0.20	
865.0±0.20	
888.60±0.20	
893.61±0.20	
900.23±0.20	
909.33±0.20	

Pane 1

Gamma energies (KeV)	Gamma intensities
913.93±0.20	
702.50±0.20	
704.77±0.20	
1058.28±0.20	7
1133.02±0.20	
1150.53±0.20	
1176.26±0.20	
1296.88±0.20	6
1378.26±0.20	
1407.55±0.20	
1414.89±0.20	5
1516.96±0.20	
1522.67±0.20	
1533.55±0.20	8
1685.95±0.20	2.4
1690.94±0.20	
1756.02±0.20	3
1880	1.1
1930	2
2080	3
2160	1.8
2300	1.3
2410	1.4
2510	0.7
$^{147}_{63}$Eu	
Ann. Rad.	
76.4	1.1†
121.23±0.20	10.9
160.87±0.20	0.04
166.3±0.20	0.04
197.23±0.20	21.4
327.92±0.20	0.1
601.49±0.20	5.6
677.54±0.20	9.3
798.78±0.20	4.8
857.02±0.20	2.8
879.5±0.20	0.14
933.2±0.20	3.6
943.2±0.20	0.4
955.97±0.20	3.6
1064.0±0.20	0.7
1077.11±0.20	6.0
1119.8±0.20	0.46
1179.95±0.20	0.1
1197.0±0.20	0.1
1255.90±0.20	0.82
1318.0±0.20	0.18
1331.50±0.20	0.34
1349.93±0.20	0.16
1448.74±0.20	0.18
1459.92±0.20	0.22
$^{148}_{63}$Eu	
Ann. Rad.	% β decays
22.41±0.10	
67.63±0.10	
98.40±0.10	0.09±0.03
116.3±0.2	0.10±0.03
121.18±0.12	0.006±0.003
166.2±0.3	0.05±0.02
182.8±0.2	0.11±0.3
189.6±0.2	0.10±0.3
215.9±0.2	
241.5±0.1	1.04±0.10
243.6±0.2	0.23±0.05
252.3±0.3	0.05±0.02
279.1±0.5	0.06±0.02
287.9±0.2	0.22±0.05
296.0±0.3	
310.0±0.3	0.22±0.05
311.4±0.1	1.14±0.10
319.1±0.3	0.05±0.02
377.5±0.5	
413.9 }±0.1	} 18.6±0.5
413.9 }	
432.6±0.1	2.79±0.20
437.2±0.3	
468.4±0.3	0.49±0.08
481.2±0.5	0.16±0.05
495.2±0.3	0.25±0.06
501.3±0.1	0.75±0.13
505.2±0.5	0.28±0.07
516.7±0.3	0.40±0.10
528.5±0.3	0.19±0.05

Pane 2

Gamma energies (KeV)	Gamma intensities
550.20±0.5	99.00±0
553.2±0.1	17.1±0.5
571.90±0.05	9.1±0.3
587.6±0.5	0.08±0.03
590.1±0.2	0.32±0.08
595.2±0.3	0.20±0.06
599.5±0.2	0.51±0.10
602.6±0.3	0.23±0.16
611.26±0.05	19.3±0.5
620.0±0.1	1.0±0.1
629.90±0.05	70.9±3.0
654.3±0.1	1.31±0.15
657.0±0.5	0.17±0.06
667.7±0.5	0.13±0.05
669.9±0.2	0.34±0.08
675.4±0.5	0.08±0.03
683.2±0.1	1.06±0.10
690.7±0.5	0.08±0.03
714.8±0.1	1.54±0.12
719.6±0.3	0.11±0.03
725.7±0.1	12.2±1.0
734.8±1.0	0.06±0.03
756.6±0.5	0.19±0.05
770.4±0.2	0.38±0.07
780.2±0.5	0.08±0.03
799.2±0.2	0.34±0.07
870.0±0.1	4.9±0.5
895.8±0.1	0.55±0.10
903.9±0.2	0.34±0.07
906.9±0.3	0.15±0.04
915.3±0.1	2.4±0.2
924.9±0.2	0.27±0.06
930.5±0.1	2.4±0.2
938.2±1.0	0.11±0.04
949.7±0.2	0.21±0.06
964.2±0.4	0.20±0.07
967.3±0.1	2.9±0.3
980.0±0.5	0.13±0.04
989.7±0.2	0.39±0.08
1013.9±0.2	0.41±0.08
1034.1±0.1	7.9±0.7
1043.9±1.0	0.04±0.02
1047.5±0.3	0.20±0.07
1066.8±0.2	0.29±0.08
1069.2±0.4	0.22±0.08
1082.2±0.5	0.17±0.07
1089.3±0.5	0.20±0.07
1097.5±0.7	0.10±0.04
1104.3±0.2	0.45±0.09
1107.9±0.3	0.10±0.03
1113.8±0.3	0.13±0.04
1126.9±0.5	0.10±0.03
1146.9±0.1	1.9±0.2
1151.7±0.5	0.08±0.03
1156.4±0.5	0.04±0.02
1165.3±0.5	0.08±0.03
1180.5±0.5	0.30±0.08
1183.3±0.1	1.7±0.2
1194.1±0.5	0.13±0.04
1207.4±0.2	0.59±0.12
1219.7±0.5	0.08±0.03
1222.0±0.5	0.09±0.03
1230.0±0.5	0.03±0.02
1236.0±0.2	0.39±0.09
1267.1±0.5	0.09±0.03
1276.5±0.5	0.07±0.03
1309.8±0.2	0.46±0.09
1328.5±0.2	1.22±0.15
1344.6±0.1	3.3±0.3
1353.6±0.2	0.47±0.09
1362.6±0.2	0.55±0.10
1409.0±0.6	0.10±0.03
1454.3±0.3	0.25±0.07
1460.5±0.2	1.12±0.12
1493.2±0.6	0.07±0.03
1503.0±0.3	0.13±0.04
1512.0±0.5	0.04±0.02
1521.9±0.3	0.11±0.03
1536.1±0.3	0.08±0.03
1543.1±0.2	0.72±0.13
1547.3±0.4	0.08±0.03
1551.7±0.5	0.02±0.01
1560.7±0.2	0.87±0.10

Pane 3

Gamma energies (KeV)	Gamma intensities
1565.0±0.6	0.03±0.02
1621.5±0.1	4.6±0.5
1635.3±0.3	0.18±0.4
1650.4±0.1	3.7±0.5
1664.5±0.5	0.08±0.03
1677.8±0.1	0.42±0.08
1748.1±0.5	0.02±0.01
1776.9±0.3	0.04±0.02
1940.0±0.5	0.05±0.03
1974.2±0.5	0.03±0.02
2173.2±0.3	0.22±0.06
$^{149}_{63}$Eu	
22.5	760±40†
72.9	0.08±0.04
178.4±0.4	0.17±0.08
207.9±0.4	0.06±0.03
251.0±1	0.14±0.07
254.5±0.3	5.9±0.6
277.0±0.5	33±2
281.1±0.5	0.19±0.07
327.5±0.3	39±2
350.0±0.3	3.4±0.3
505.9±0.3	5.5±0.5
528.5±0.3	5.3±0.5
535.9±0.3	0.46±0.1
558.4±0.3	0.62±0.1
$^{150m}_{63}$Eu	
209.27±0.35	0.55±0.08†
285.23±0.35	
334.04±0.20	100±8.0
372.96±0.35	
406.47±0.20	70.9±5.7
424.96±0.35	0.2±0.04
438.67±0.30	
446.22±0.35	
620.48±0.35	0.80±0.12
632.57±0.25	
712.13±0.25	3.34±0.30
747.12±0.30	
831.77±0.3	5.0±0.4
860.2±0.6	0.23±0.06
893.6±0.4	
921.22±0.30	5.6±0.4
1046.2±0.8	0.21±0.05
1057.3±0.8	
1165.63±0.30	6.45±0.58
1193.89±0.7	0.46±0.1
1223.0±0.3	5.0±0.4
1268.9±0.5	
1452.0±0.5	0.37±0.13
1462.9±0.4	
1533.3±1.0	
1592.5±1.0	
1629.8±0.7	1.45±0.19
1963.90±0.30	2.91±0.30
1998.3±0.7	
$^{150}_{63}$Eu	
61.278±0.02	
78.830±0.02	
205.101±0.09	
284.760±0.13	
298.021±0.03	
305.438±0.15	
310.826±0.23	
315.132±0.11	
333.965±0.02	
340.369±0.04	
345.968±0.13	
382.063±0.05	
402.252±0.07	
439.402±0.01	
448.744±0.03	
461.788±0.02	
464.497±0.08	
474.544±0.14	
496.298±0.23	
505.480±0.02	
510.078±0.10	
515.776±0.03	
520.042±0.02	
542.989±0.10	
571.244±0.02	
584.253±0.07	

Pane 4

Gamma energies (KeV)	Gamma intensities
591.026±0.19	
596.334±0.09	
607.237±0.12	
612.446±0.10	
614.908±0.09	
667.031±0.06	
675.804±0.03	
712.139±0.02	
731.212±0.03	
737.405±0.02	
741.475±0.02	
748.062±0.02	
750.890±0.03	
756.473±0.11	
759.336±0.13	
828.486±0.04	
830.860±0.05	
836.574±0.09	
859.885±0.04	
869.250±0.03	
899.079±0.02	
910.989±0.16	
923.371±0.06	
1045.975±0.06	
1049.065±0.02	
1071.251±0.10	
1083.081±0.10	
1122.762±0.05	
1165.546±0.17	
1170.605±0.20	
1193.801±0.02	
1197.125±0.02	
1212.956±0.12	
1246.984±0.01	
1251.077±0.11	
1261.831±0.04	
1274.1±0.4	
1308.755±0.03	
1322.037±0.07	
1334.064±0.04	
1343.877±0.02	
1350.386±0.08	
1379.064±0.20	
1383.431±0.30	
1390.0±0.4	
1401.509±0.31	
1413.507±0.39	
1460.913±0.15	
1465.365±0.81	
1476.686±0.59	
1485.557±0.03	
1592.893±0.20	
1636.623±0.07	
1677.832±0.35	
1690.696±0.10	
1783.269±0.10	
$^{151m}_{63}$Eu	
196	
$^{152m2}_{63}$Eu	
18.3	
39.7	
89.8	
$^{152m1}_{63}$Eu	
Ann. Rad.	
73.2±0.3	†
76.8±0.3	
85.15±0.10	0.13
90.14±0.10	0.23
121.92±0.10	5.0
244.3±0.3	
270.96±0.15	0.07
344.08±0.15	1.7
562.66±0.15	0.3
688.9±0.15	
841.39±0.15	11.1
869.72±0.15	0.09
963.10±0.15	9.3
970.03±0.15	0.5
995.03±0.15	0.07
1314.33±0.15	0.8
1388.79±0.15	0.6
1411.26±0.15	0.04

GAMMA ENERGIES AND INTENSITIES OF RADIONUCLIDES (Continued)

Gamma energies (KeV)	Gamma intensities	Gamma energies (KeV)	Gamma intensities	Gamma energies (KeV)	Gamma intensities	Gamma energies (KeV)	Gamma intensities
$^{152}_{63}$Eu		60.07±05	2.8	900	26	780.8±0.7	
Ann. Rad.		86.60±0.05	72.4	950	49	788.8±0.5	16
121.83±0.10	0.3†	105.36±0.05	48.2	980	20	812.6±0.6	
147.79±0.10	0.2	**$^{156}_{63}$Eu**		1110	11	876.0±0.6	1.3
161.41±0.10		88.97±0.05	45%	1190	16	933.0±0.6	2.1
166.89±0.15	0.1	190.37±0.10	0.5%	1260	5	938.5±0.5	6
167.9±0.3	0.2	199.27±0.07	0.6%	1330	4	947.8±0.5	
244.54±0.15	7.6	434.33±0.08	0.23%	1960	3	**$^{151}_{64}$Gd**	
251.40±0.15	0.14	472.57±0.08	0.17%	2180	2	21.6	4.4†
271.01±0.15	0.09	490.35±0.08	0.17%	**$^{159}_{63}$Eu**		153.7	6
275.07±0.15	0.04	599.43±0.07	2.1%	54		175	2.7
284.42±0.15	0.08	616.23±0.07	6.3%	75		244	4.5
295.84±0.15	0.44	709.85±0.07	0.9%	90	90†	308	0.8
315.37±0.15	0.05	723.44±0.07	5.4%	130		350	0.3
324.81±0.15	0.07	768.27±0.20	0.09%	145	68	**$^{153}_{64}$Gd**	
329.23±0.15	0.13	797.67±0.10	0.09%	165	16	68.23	0.5†
344.19±0.15	28.5	811.73±0.07	9.7%	220	26	69.68	11
366.25±0.15	0.22	820.32±0.07	0.16%	670	100	75.43	0.4
367.71±0.15	1.0	836.37±0.07	0.11%	740	50	83.37	1.0
385.91±0.15	0.31	841.10±0.31	0.22%	880	64	89.48	0.3
411.06±0.15	2.8	858.19±0.13	0.27%	970	10	97.430±0.005	132
415.94±0.15	0.13	866.98±0.07	1.5%	1050	25	103.178±0.003	100
443.96	3.6	944.06±0.07	2.4%	1100	25	172.85	
488.71±0.15	0.44	960.41±0.07	1.6%	1270	8	**$^{159}_{64}$Gd**	
493.51±0.15	0.05	969.81±0.07	0.38%	1350	8	58	1.9†
503.41±0.15	0.13	1111.81±0.10	0.55%	1370	25	80	0.43
563.96±0.15	0.47	1027.53±0.07	0.38%	1440	25	225	1.9
566.25±0.15	0.15	1040.36±0.10	0.54%	1500		290	0.35
586.17±0.15	0.41	1065.08±0.07	5.3%	1600		306	0.8
674.78±0.15	0.14	1075.92±0.07	2.4%	**$^{145}_{64}$Gd**		334	0.42
678.62±0.15	0.61	1079.19±0.07	4.7%	80	9†	348	2.4
688.88±	0.90	1129.31±0.18	0.17%	1030	10	363	100
719.11±0.15	0.40	1140.51±0.07	0.31%	1750	100	536	
779.08±0.15	14.5	1153.72±0.07	12.6%	**^{146}Gd**		550	0.21
810.54±0.15	0.32	1168.94±0.15	0.31%	22.9		579	0.6
841.44±0.15	0.20	1230.66±0.07	8.7%	77.7	30†	617	0.17
867.53±0.15	4.4	1242.47±0.07	7.1%	114.7		**$^{161}_{64}$Gd**	
919.40±0.15	0.57	1238.30±0.6	0.71%	117.7		56.6	†
926.98±0.15	0.49	1277.50±0.07	3.2%	154.8	48	77.6	0.3
964.21±0.15	16.2	1366.44±0.07	1.8%	198.8		102.4	17
1005.15±0.15	1.0	1455.02±0.20	0.14%	329		105.5	1.0
1086.00±0.15	11.8	1587.06±0.20	0.20%	**$^{147}_{64}$Gd**		133.7	
1089.91±0.15	1.9	1675.92±0.50	0.17%	106.6		165.3	7
1112.20±0.15	15.4	1681.13±0.10	0.42%	111		180.6	3.5
1213.02±0.15	1.7	1857.24±0.10	0.32%	166.3		258.8	3
1234.0±0.3	0.25	1876.91±0.10	1.5%	194.0		273	2
1249.95±0.15	0.3	1937.66±0.10	2.0%	214.8		283.8	12
1292.62±0.15	0.2	1945.92±0.10	0.12%	229.4	150†	315.3	37
1299.18±0.15	1.9	1965.91±0.10	4.4%	240.5		361.0	100
1408.11±0.15	22.9	2026.68±0.10	3.8%	260.8		482	3.2
1447.93±0.15	0.14	2031.57±0.10	0.44%	297.5		531	2.6
1457.76±0.15	0.6	2097.77±0.10	5.3%	310.0		**$^{147}_{65}$Tb**	
1528.94±0.15	0.7	2110.64±0.30	0.16%	328.6		Ann. Rad.	
$^{154}_{63}$Eu		2116.46±0.15	0.17%	370.0		305	
59.75±0.10	4.0%	2181.05±0.10	4.1%	396.1		**$^{148}_{65}$Tb**	
122.91±0.10	40.0%	2186.82±0.10	6.1%	485		Ann. Rad.	
145.60±0.10	0.2%	2205.57±0.10	1.7%	559		780	
188.04±0.15	0.3%	2211.85±0.10	0.20%	609		1120	
238.34±0.15	0.1%	2269.85±0.10	2.4%	617		**$^{149m}_{65}$Tb**	
247.73±0.15	6.7%	**$^{157}_{63}$Eu**		625		Ann. Rad.	
444.22±0.15	0.7%	40.6		632	75†	**$^{149}_{65}$Tb**	
478.23±0.15	0.3%	51.8		780	60	165.5	
581.91±0.15	1.0%	54.5		940	80	187.3	
591.70±0.15	5.4%	64.0	85†	1350	6	352	
692.38±0.15	1.9%	77.0		**$^{149}_{64}$Gd**		388	
723.29±0.15	21%	131.4		149.4±0.5	100†	464	
756.78±0.15	4.6%	276.6		214.2±0.7		652	
765.74±0.15	1.0%	320.5		252.2±0.7		818	
815.71±0.15	0.6%	330.7	10	260.6±0.6		856	
845.57±0.15	0.5%	361.8		272.2±0.6		1206	
873.24±0.15	12.3%	372.8		299.3±0.5	69	**$^{150}_{65}$Tb**	
904.21±0.15	1.0%	413.1	100	346.6±0.5	42	937	100†
996.36±0.15	11.3%	477.3	20	404.2±0.7		930	35
1004.84±0.15	20%	527.0		459.8±0.7	<4	**$^{151}_{63}$Tb**	
1274.53±0.15	37%	623.3	20	478.7±0.7		108.2	
1396.00±0.15	1.9%	687.5	10	496.3±0.6	4	180.1	50†
1460.97±0.15	1.2%	729.4	5	516.5±0.6	10	192.0	
1494.14±0.15	0.6%	**$^{158}_{63}$Eu**		534.3±0.6	9	251.8	100
1593.07±0.15	0.6%	80	100†	645.2±0.6	4	287.2	100
1596.61±0.15	1.9%	180		663.3±0.8		318.5	
$^{155}_{63}$Eu		520	25	666.2±0.6		380.1	
43.37±0.05	†	610	8	748.4±0.5	19	384.6	
58.12±0.10	0.06	750	8	762.9±0.7		395.2	

GAMMA ENERGIES AND INTENSITIES OF RADIONUCLIDES (Continued)

Gamma energies (KeV)	Gamma intensities	Gamma energies (KeV)	Gamma intensities	Gamma energies (KeV)	Gamma intensities	Gamma energies (KeV)	Gamma intensities
416.0		2800		381.9±0.4		402.05±0.10	
426.4		2920		415.9±0.3		559.09±0.10	
443.8		**¹⁵³₆₅Tb**		444.6±0.4		180.06±0.10	
479.2	80†	19.4		517.9±0.3		181.59±0.10	
499.8		41.6		540.10±0.25		206.60±0.10	
512		51.8		557.2±0.2		208.22±0.10	
588		68.2	≈20	598.0±0.4		200.48±0.10	
604	100†	82.8	190	642.4±0.6		**¹⁵⁶ᵐ₆₅Tb**	
704		87.6	50	649.5±0.3		88.4(IT)	
732		88.4	45	676.4±0.4		**¹⁵⁶₆₅Tb**	
794		102.3	190	692.3±0.3		88.97±0.05	
799		109.7	180	704.6±0.5		111.9	
805		129.2	≈13	714.5±0.5		115.6	
906		139.8	<10	722.1±0.5		155.2	
939		141.9	35	756.9±0.3		170.8	
979		170.5	220	815.5±1.0		199.27±0.07	
1010		174.4	<50	873.2±0.2		262.5	
1025		183.5	25	892.7±0.3		266.9	
1053		185.9	<5	925.3±1.0		296.3	
1090		193.6	≈4	996.4±0.2		356.4	
1097		195.2	28	1004.8±0.2		374.0	
1171		208.1	<50	1060.0±1.0		381.4	
1250		212.1	1000	1148.9±1.0		422.4	
1308		249.6	72	1152.4±1.0		535.2	
1312		262	15	1258.4±1.5		687.5	
¹⁵²ᵐ₆₅Tb		266	<2	1274.9±1.0		782	
Ann. Rad.		271	<2	1289.0±1.5		844	
140	25†	275	7	1292.0±1.5		867	
230	30	278	<2	1410.0±2.0		927	
¹⁵²₆₅Tb		299	<2	1491.0±3.0		944.06±0.07	
271.2	13†	303	28	1967.0±2.0		951	
344.4	100	315	30	2186.0±1.5		963	
351.5		320	≈7	**¹⁵⁴₆₅Tb**		1012	
411.2		327	≈5	123.1		1040.36±0.1	
432.5		333	<2	141.4		1043	
496.7		340	5	225.7		1068	
527.1		349	<1	247.7		1070	
586.7		355	5	265.9		1157	
615.6		362	4	346.6		1162	
623.1		393	≈3	426.9		1225	
676		400	≈4	444.6		1338	
679		407	≈3	557.2		1425	
704	2.5	419	≈3	602.6		1649	
753		436	25	680.8		1849	
766		442	6	692.3		1951	
779		448	4	815.5		2014	
929		455	13	873.2		2090	
931		467	15	996.4		2105	
975		496	10	1124		2141	
991		507	16	1274.9		2268	
1048		526	6	1292.0		2281	
1110		531	7	1460		2310	
1138		549	6	1994		**¹⁵⁸ᵐ₆₅Tb**	
1187		566	<2	2062		0.111	100%
1206		571	<2	2116		**¹⁵⁸₆₅Tb**	
1263		599	<2	2186		79.5	12%
1300		629	10	**¹⁵⁵₆₅Tb**		99	4%
1303		636	<2	45.35±0.10		182	10%
1319		653	5	58.03±0.10		218	1%
1328		665	10	60.03±0.10	7†	782	10%
1348		689	8	72.83±0.10		928	0.2%
1352		740	7	74.99±0.10		946	47%
1361		762	<2	86.53±0.07	100	964	22%
1412		816	7	99.02±0.10		1110	2.2%
1519		836	29	101.14±0.10		1190	1.8%
1588		847	6	105.29±0.07	69	**¹⁶⁰₆₅Tb**	
1598		853	<5	120.53±0.10		86.788±0.002	
1670		858	<5	148.63±0.08	8	93.919±0.006	
1783		904	35	161.26±0.08 }	23	176.490±0.030	
1791		918	<5	163.28±0.07 }		197.035±0.008	5.1
1910		946	32	175.20±0.10		215.646±0.008	3.97
1940		974	10	203.05±0.10		230.628±0.013	0.009
2040		993	37	216.04±0.10		237.638±0.086	
2100		1102	13	220.69±0.10		246.489±0.016	0.03
2190		1203	2.4	226.94±0.10		298.582±0.010	26.3
2260		1276	1.9	239.53±0.10		309.557±0.018	0.87
2370		1364	0.9	262.30±0.10		337.324±0.030	0.34
2400		**¹⁵⁴₆₅Tb**		268.56±0.10		349.935±0.110	
2530		123.07±0.2		281.03±0.10		379.449±0.090	
2580		225.7±0.3		286.93±0.10		392.514±0.026	5.1
2600		231.7±0.3		321.74±0.10		486.075±0.080	3.97
2670		247.7±0.2		367.32±0.10	10	682.349±0.110	0.09
2710		346.6±0.2		370.70±0.10		765.194±0.110	

Gamma energies (KeV)	Gamma intensities
871.9±1.5	0.45
879.333±0.070	31.8
962.085±0.220	12.6
966.099±0.120	27.6
1003.26±0.30	1.14
1068.9±1.5	0.13
1103.22±0.25	0.7
1115.29±0.19	1.70
1178.12±0.19	16.6
1200.21±0.37	2.67
1251.3±0.5	0.11
1271.87±0.25	8.28
1311.90±0.30	3.16
1358.6±0.2	0.02
1460.9±0.2	0.05
1592.7±0.2	0.02
$^{161}_{65}$Tb	
25.55±0.2	
43.8	
57.38±0.18	16† 21%
74.68±0.15	100
77.41±0.15	0.5
86.81±0.15	2.0
103.06±0.10	0.12
105.92±0.10	0.9
131.61±0.10	<1
298.55±0.10	
340	
410	
470	
540	
$^{162}_{65}$Tb	
40	17†
80.82±0.3	8
185.18±0.25	26
260.01±0.15	100
795.80±0.5	0.8
807.53±0.2	44
882.27±025	18
888.16±0.2	36
$^{163m}_{65}$Tb	
180	
$^{163}_{65}$Tb	
25	80†
120	
160	5
235	30
330	100
360	1
440	1
510	35
$^{150}_{66}$Dy	
390	
Ann. Rad.	
$^{151}_{66}$Dy	
145	
$^{152}_{66}$Dy	
Ann. Rad.	
275.2	
$^{153}_{66}$Dy	
80.8	
82.5	
93	
99.7	
147.8	
149.0	
174.1	
218.9	
240.7	
244.5	
254.5	
274.4	
303.8	
389.8	
415.7	
512.4	
544.7	
$^{155}_{66}$Dy	
Ann. Rad.	
65.43	
90.38	2.3†
115.4	
155.8	

Gamma energies (KeV)	Gamma intensities
161.4	
181	
184.6	
205.7	
227	100
271.4	
389	
433	
451	
484	
498	
508	
550	
587	
642	
654	
723	
745	
761	
813	
837	
844	
892	
905	
929	
1000	9†
1091	8
1155	
1164	
1250	6
1390	5
1450	6
1660	4
$^{157}_{66}$Dy	
60.8	
83.01	0.5†
143.8	0.0016
182.2	2
265.3	
326.2	100
554	
577	
597	
636	
743	
769	
775	
985	
992	
$^{159}_{66}$Dy	
58.0	4%
80.0	
138	
211	4 × 10⁻⁵% (4×10^{-5}%)
290	1.2×10^{-4}%
348	9×10^{-4}%
$^{165m}_{66}$Dy	
108.16	100†
152	9
361.7	17
514	55
650	1
750	0.5
$^{165}_{66}$Dy	
94.70	370†
279.6	57
361.5	100
479	3
501	0.6
514	2.9
545	17
566	14
575	8
621	10
633	64
695	2
716	64†
995	6
1056	3
1080	10
$^{166}_{66}$Dy	
28.23	†
54.24	

Gamma energies (KeV)	Gamma intensities
82.45	100
288	0.7
344	2
371.6	8
425.8	10
$^{151}_{67}$Ho	
Ann. Rad.	
$^{151}_{67}$Ho	
Ann. Rad.	
$^{152}_{67}$Ho	
Ann. Rad.	
$^{152}_{67}$Ho	
Ann. Rad.	
$^{153}_{67}$Ho	
Ann. Rad.	
355	
$^{155}_{67}$Ho	
Ann. Rad.	
92	
117	
138	
209	
243	
326	
$^{156}_{67}$Ho	
138	100†
266.4	99
366.7	23
685.2	
890	
1200	
1410	
$^{157}_{67}$Ho	
Ann. Rad.	
87	
152	
190	
277	
710	
860	
900	
1200	
$^{158m}_{67}$Ho	
Ann. Rad.	
67.3 (IT)	
98.9	
193.7	
218.3	
309	
320	
356	
463	
471	
647	
848	
851	
946	
948	
1450	
1600	
1800	
2060	
2200	
2620	
$^{158}_{67}$Ho	
Ann. Rad.	
see Ho¹⁵⁸ᵐ	
$^{159m}_{67}$Ho	
206	
$^{159}_{67}$Ho	
69.51±0.30	
79.79±0.30	
100.55±0.30	
120.94±0.30	
131.86±0.30	
152.23±0.40	
155.88±0.40	
159.43±0.40	
173.02±0.40	
177.51±0.40	
186.09±0.40	
252.96±0.30	
258.74±0.40	

Gamma energies (KeV)	Gamma intensities
265.91±0.40	
296.59±0.40	
309.51±0.30	
538.20±0.40	
706.54±0.40	
807.05±0.40	
838.36±0.40	
$^{160m}_{67}$Ho	
Ann. Rad.	
60.1 (IT)	Equilibrium with Ho¹⁶⁰
86.91±0.30	
107.7±0.7	
196.76±0.20	
297.79±0.20	
$^{160}_{67}$Ho	
405.5±1.0	
645.20±0.20	
728.00±0.20	
753.0±0.7	
765.2±0.7	
826.1±1.0	
871.88±0.40	
879.28±0.20	
962.32±0.25	
966.24±0.25	
1004.4±0.7	
1068.8±0.7	
1131.3±1.0	
1155.4±1.0	
1198.9±1.0	
1271.7±1.0	
1285.3±1.0	
1522.6±1.0	
1594.7±1.0	
1613.2±1.0	
1654.5±1.0	
1719	
1784	
1807	
1872	
1927	
1959	
2008	
2069	
2190	
2550	
2619	
2638	
2675	
2728	
2763	
$^{161m}_{67}$Ho	
211.1	
$^{161}_{67}$Ho	
25.7	100†
60.03±0.30	4
75.03±0.40	
77.79±0.20	10
98.1±0.7	1.8
103.00±0.20	1.2
137.9±1.0	
156.90±0.30	1.7
175.08±0.30	1.6
210.8	0.6
234.5	
339.7	
416.0	
760	0.3
$^{162m}_{67}$Ho	
38.63	(IT)
58.88±0.40	
81.15±0.20	
90.28±0.40	
95.46±0.40	
184.70±0.20	
275.2±0.7	
282.48±0.20	
302.8±0.7	
333.8±0.7	
392.6±1.0	
424.3±1.0	
936.72±0.20	
1220.0±0.40	

Gamma energies (KeV)	Gamma intensities	Gamma energies (KeV)	Gamma intensities	Gamma energies (KeV)	Gamma intensities	Gamma energies (KeV)	Gamma intensities
1319.9±0.8		304.3		116.69	3%	164.4	
1374.4±0.8		320.7		124		165.4	
$^{162}_{67}$Ho		322.8		166.4	0.6%	190.1	
Ann. Rad.	~50%	331.0		210.4	0.6%	239.7	
81.15±0.2		347.7		236	0.5%	241.5	
$^{163m}_{67}$Ho		386.6		277.0	0.65%	249.2	
305	~80%	403.4		295.6	28%	275.3	
$^{164}_{67}$Ho		459.3		308.2	63%	299.9	
37.3	3.6†	531.58±0.1		371.2	0.4%	335.4	
73.4±0.2	3.3	$^{168}_{67}$Ho		404	0.03%	345.6	
91.4±0.2	3.5	79.80±0.2		418.9	0.09%	393.5	
$^{166}_{67}$Ho		184.2±0.2	2†	544	0.04%	404.2	
80.573±0.015	%	556.9±0.5	2	557	0.08%	471.2	
184.407±0.015	0.002±0.0005	741.3±0.2	100	573	0.07%	505.1	
674.991±0.040	0.020±0.002	821.1±0.3	92	606	0.1%	531.0	
705.342±0.040	0.015±0.002	$^{169}_{67}$Ho		619	0.13%	549.9	
785.949±0.040	0.013±0.002	80		670	0.3%	579.9	
1379.432±0.065	0.93±0.05	150	12†	675	0.4%	613.0	
1581.89±0.08	0.181±0.009	680	10	732	0.2%	655.5	
1662.44±0.08	0.116±0.006	760	30	783	0.31%	671.1	
1749.94±0.10	0.025±0.002	840	15	796	0.80%	$^{164}_{69}$Tm	
1830.57±0.15	0.008±0.001	920	5	842	0.04%	Ann. Rad.	
$^{166m}_{67}$Ho		$^{170}_{67}$Ho		869	0.06%	91	
80.573±0.015	12.5±0.6%	430		882	0.07%	211	
94.653±0.030	0.14±0.01%	$^{158}_{68}$Er		906	0.9%	356	
119.038±0.030	0.18±0.02%	Ann. Rad.		911	0.2%	361	
121.161±0.030	0.26±0.03%	67.2		962	0.11%	391	
135.238±0.035	0.10±0.01%	71.9		$^{172}_{68}$Er		773	
140.618±0.040	0.04±0.01%	315		29.2		862	
160.064±0.045	0.10±0.01%	250		59.6		907	
161.748±0.045	0.11±0.01%	387		67.9		930	
184.407±0.015	73.2±3.7%	875		128		$^{165}_{69}$Tm	
190.711±0.025	0.22±0.02%	906		164	0.9%	15.45	
214.763±0.045	0.55±0.07%	978		203	0.9%	27.75	
215.875±0.030	2.6±0.2%	$^{159}_{68}$Er		346	0.6%	30.05	
231.282±0.040	0.24±0.02%	205.9 (IT)		384	2.4%	35.2	
233.28±0.09	0.16±0.02%	370		407	40%	47.15	
239.99±0.08	0.05±0.01%	627		446	2.5%	53.2	
259.716±0.020	1.10±0.06%	840		476	1.0%	54.45	
280.456±0.020	29.8±1.5%	1200		535	0.3%	59.15	
300.744±0.020	3.75±0.19%	1400		610	40%	60.4	
329.12±0.08	0.16±0.02%	1800		$^{161}_{69}$Tm		70.55	
339.78±0.08	0.17±0.02%	2600		45.6		77.2	
365.739±0.025	2.52±0.13%	$^{161}_{68}$Er		59.5		82.25	
410.941±0.025	11.6±0.6%	Ann. Rad.		84.5		86.9	
451.524±0.025	3.06±0.15%	45.6		94.3		88.2	
464.825±0.040	1.52±0.08%	59.5		106.1		113.6	
529.813±0.030	10.2±0.5%	84.5		112.6		141.4	
570.992±0.030	5.75±0.29%	94.3		122.6		150.8	
594.481±0.040	0.70±0.04%	106.1		138.7		156.0	
611.522±0.065	1.39±0.08%	112.6		144		165.5	
639.770±0.060		122.6		146.7		175.7	
644.45±0.10	0.18±0.02%	138.7		156.7		181.5	
670.509±0.040	5.77±0.29%	144.0		172.0		195.6	
691.211±0.050	1.53±0.08%	146.7		172.6		197.2	
711.693±0.040	58.7±2.9%	156.5		190.6		205.2	
736.67±0.08	0.40±0.03%	172.0		207.0		209.9	
752.265±0.040	13.1±0.7%	172.6		215.9		218.2±0.2	100±5
778.817±0.040	3.30±0.17%	190.6		218.2		221.5	
810.309±0.040	62.7±3.1%	207.0		244.5		223.9	
830.560±0.040	10.6±0.6%	215.9		250.2		229	
875.64±0.05	0.79±0.06%	218.2		252.5		233.2	
1010.25±0.10	0.10±0.01%	244.5		265.3		234.6	
1120.31±0.07	0.23±0.02%	250.2		283.3		242.4±0.2	1040±30
1146.82±0.07	0.22±0.02%	252.5		354.0		248.9	
1241.44±0.06	1.00±0.05%	265.3		327.7		249.7	
1282.12±0.07	0.23±0.02%	283.3		$^{162}_{69}$Tm		264.4	17±2
1400.72±0.08	0.55±0.03%	354.0		Ann. Rad.		275.6	
1427.05±0.08	0.59±0.03%	372.7		102	20†	279.9±0.3	20±2
$^{167}_{67}$Ho		$^{163}_{68}$Er		236	10	286.1	
57.1		Ann. Rad.		$^{162}_{69}$Tm		292.3	
73.7		300	0.01%	Ann. Rad.		295.9	
79.3		430	0.06%	102		296.9±0.3	
83.4		1100	0.04%	$^{163}_{69}$Tm		306.9	4.8±1.5
133.4		$^{167m}_{68}$Er		Ann. Rad.		312.2	14±2
136.9		207.8	50%	22.2		321	
150.4		$^{169}_{68}$Er		60.2		330.5	3.6±1.8
207.78±0.08		8.42	0.1%	69.2		336	
237.7		$^{171}_{68}$Er		80.5		345.7±0.3	
241.1		5.1		83.9		346.6	89±4
260.1		12.4		85.1		356.3±0.3	81±3
266.5		86		104.3		363	
271.7		111.63	25%	145.3		365.3	16±3

GAMMA ENERGIES AND INTENSITIES OF RADIONUCLIDES (Continued)

Gamma energies (KeV)	Gamma intensities	Gamma energies (KeV)	Gamma intensities	Gamma energies (KeV)	Gamma intensities	Gamma energies (KeV)	Gamma intensities
377	w	785.91±0.08	12.7	181.55±0.08	44.0±4	93.6	6.6
384.1		810.33±0.08	1.34	203.6±0.3	~0.1	109.77	51
389.2±0.3	80±4	875.63±0.08	4.9	238.5±0.2ᵃ	~0.2	118.17	4.8
400.2	5.8±1.7	1030.37±0.03	7.9	351.8±0.2ᵃ	~0.2	130.51	32
414		1078.86±0.15	0.6	374.1±0.3	~0.1	177.18	58
420.7	10±3	1084.78±0.20	0.3	399.63±0.08	2.5±0.5	197.97	100
426.9	2	1119.15±0.13	0.2	423.0±0.2	~0.5	261.05	4.6
429.9	w	1152.32±0.05	2.4	429.0±1.0	~0.1	307.68	31.0
456	w	1161.91±0.07	1.1	436.07±0.08	4.2±0.4	336.0	0.02
442.4	24±3	1176.68±0.04	12.4	490.31±0.08	6.7±0.6	$^{175m}_{70}$Yb	
448.1	57±4	1216.08±0.09	0.9	528.73±0.10	2.0±0.3	513	
456.0	47±16	1233.41±0.04	2.4	542.0±0.4ʰ	~0.1	$^{175}_{70}$Yb	
459.9±0.3	110±9	1248.80±0.12	0.2	565.5±0.15	0.7±0.2	113.81	31†
471.8	9±2	1263.44±0.07	1.3	858.7±0.3	2.0±0.6	137.65	2.2
477.7	12±2	1273.53±0.04	19.2	912.10±0.08	25.0±3	144.85	5.9
487.0	21±3	1300.74±0.06	1.8	964.20±0.10	5.3±0.6	251.3	3.8
502	w	1313.31±0.28	0.3	1003.0±0.3	~1.0	282.6	62
507	w	1347.12±0.05	1.3	1040.1±0.3	2.6±0.6	396.1	100
513		1374.21±0.04	7.4	1076.3±0.1ᵃ	13.0±2	$^{177m}_{70}$Yb	
513.5	10±3	1432.18±0.12	1.3	1093.68±0.08	≡104.0	104	
526.8	32±5	1447.84±0.09	0.8	1116.0±0.3	~1	228	
530.9	c	1505.04±0.08	1.1	1119.2±0.3	3.8±1.0	$^{177}_{70}$Yb	
542.2	46±4	1622.48±0.20	0.7	1154.9±0.2ᵃ	3.3±1.0	121.6	0.23†
558.3	19±5	1652.80±0.07	1.4	1184±0.5	~1	138.4	~1
563.7	67±5	1868.0±0.1	5.3	1205.7±0.2	3.4±0.8	147.4	
570	w	1907.63±0.16	0.7	1289±0.15	8.4±1.0	150.8	
573.8	14±3	2008.41±0.26	0.4	1348.4±0.15	2.6±0.6	162.6	
577.3	7±4	2052.2±0.2	23.2	1387.27±0.08	95.0±6	790	0.2
589.4	57±5	2079.5±0.2	8.3	1398.17±0.08	11.0±2	940	1.1
605.6		2092.11±0.06	2.1	1402.6±0.15	2.5±0.6	1079	5
608.1		2192.29±0.13	0.2	1440.8±0.4	~0.5	1120	0.7
622.8	9±3	$^{167}_{69}$Tm		1461.0±0.1ᵃ	4.0±0.5	1239	5†
630	w	57.1		1466.07±0.08	70.0±6	$^{181}_{71}$Lu	
663	w	207.78±0.08	100†	1470.47±0.08	27.0±3	Ann. Rad.	
664.6	12±3	266.5		1476.8±0.10	4.5±0.6	29.6	
677.3	≈6	322.8		1491.0±1.0	~0.5	56.2	
680.5	≈4	531.58±0.1	4.18±0.3	1529.07±0.08	71.0±4	70.6	
698.6	41±4	$^{168}_{69}$Tm		1545.0±1.00	~0.5	100.1	
734	w	79.80±0.2	9.19	1584.2±0.10	10.0±1.5	102.5	
747.4	8±3	99.0±0.2	0.6	1592.7±0.10ᵃ	1.5±0.3	123.1	
791.0	18±4	99.3±0.2	3.3	1608.62±0.08	69.0±5	129.1	
806.4±0.1	343±12	173.5±0.3	0.054	1622.1±0.10	1.0±0.2	151.8	
821.2	w	184.2±0.2	17.2	$^{173}_{69}$Tm		178.7	
826.9		198.2±0.2	52.1	66	1.1†	181.9	
837.7	13±2	272.9±0.4	0.103	399	89	188.5	
855	3.9±0.5	348.5±0.3	0.33	465	7.6	198.9	
892.8		422.2±0.4	0.32	^{174m} Tm		213.0	
907.8		447.4±0.2	24.5	76	9†	222.5	
932.4		546.7±0.3	2.84	176	75	229.5	
952.8		556.9±0.5	0.206	273	95	235.6	
991		631.6±0.5	8.98	366	104	239.0	
1042		645.7±0.5	1.52	500	17	258.2	
1045.7		673.4±0.3	0.223	630	5	261.5	
1067		720.3±0.2	12.2	990	100	278.1	
1096		730.6±0.2	5.14	1270	8	307.8	
1131.0	48±2	741.3±0.2	12.6	$^{175}_{69}$Tm		317.2	
1184.5	74±4	748.2±0.3	0.451	513		361.4	
1232	0.9±0.3	815.9±0.2	50.0	$^{165}_{70}$Yb		371.8	
1246.4		821.1±0.3	11.7	Ann. Rad.		401.2	
1262	0.3±0.1	829.9±0.2	6.95	$^{166}_{70}$Yb		601	
1289	3.2±0.5	853.5±0.7	0.05	820		$^{168}_{71}$Lu	
1333	0.2±0.1	914.9±0.3	3.0	$^{167}_{70}$Yb		Ann. Rad.	
1339	0.4±0.1	928.9±0.5	0.05	Ann. Rad		87	7†
1344	0.3±0.1	1014.1±0.5	0.06	25.6		900	10
1365	1.8±0.4	1167.5±0.5	0.085	37.1		990	13
1379.5	8.5±1.0	1277.3±0.3	1.70	63.0		1440	
1426.9	19.5±1.6	1322.6±0.7	0.03	106	4†	1810	
$^{166}_{69}$Tm		1351.2±0.7	0.10	113.3	10	2100	
Ann. Rad.		1358.0±1.0	0.01	116.5		$^{169m}_{71}$Lu	
80.573±0.015		1461.3±0.4	0.42	131.8		29.0	
184.39±0.05	12.6	1592.5±0.4		150.3		$^{169}_{71}$Lu	
214.76±0.04	4.4	$^{170}_{69}$Tm		168.9		Ann. Rad.	
594.48±0.04	4.5	84.262±0.0005	3.4%	176.1	2.3	24.2	
598.8±0.3	2.9	$^{171}_{69}$Tm		920		62.8	
654.50±0.10	0.4	66.7	~2%	1040		70.9	
672.28±0.08	7.6	$^{172}_{69}$Tm		1250		75.0	
674.91±0.08	2.7	78.70±0.08	58.0±4	1440		87.4	
691.28±0.08	8.4	90.6±0.1	0.32±0.06	1520		92.0	
705.83±0.15	12.8	112.8±0.2ᵇ	~0.1	$^{169}_{70}$Yb		104.4	
712.83±0.15	0.6	131.8±0.2ᵃ	~0.1	8.42		110.9	
727.98±0.15	0.5	133.6±0.2ᵃ	~0.1	20.74	†	133.5	
757.84±0.10	2.9	142.50±0.10	1.8±0.2	63.12	164	144.6	
778.83±0.08	23.4	145.0±0.2	~0.2			157	

Gamma energies (KeV)	Gamma intensities	Gamma energies (KeV)	Gamma intensities	Gamma energies (KeV)	Gamma intensities	Gamma energies (KeV)	Gamma intensities
165.2		1398		323.8±0.1	1.30	1920.2±1.5	0.03
191.4		1408		330.3±0.7	0.52	1931.9±1.5	0.03
198.4		1431		337.4±1.5	~0.03	1955.3±1.5	<0.03
258.7		1453		352.4±1.5	<0.03	1994.4±0.4	0.16
291.6		1458		358.2±1.5	~0.10	2025.2±0.9	0.059
369.6		1483		366.5±0.6	0.30	2083.5±0.3	0.29
379.2		1515		372.4±0.3	2.50	2096.1±0.5	0.11
404.6		1517		373.3±1.0	<0.4	$^{173}_{71}$Lu	
456.8		1553		377.4±0.1	3.37	78.70	12.5%
470.8		1569		399.6±0.2	0.55	100.63	6.1%
484.0		1577		410.2±0.1	2.08	171.36	3.2%
489.7		1688		415.9±1.5	0.11	179.33	1.9%
549.1		1816		422.4±1.5	0.10	272.05	26%
577.1		1861		417.0±1.5	<0.1	285.40	0.8%
647.5		1939		432.5±0.3	1.65	351.0	1.0%
881		1956		437.1±1.5	~0.25	457	0.3%
891		2039		442.6±1.5	~0.18	558	0.8%
962		2124		482.3±0.5	0.49	637	0.05%
1062		2359		486.1±0.5	0.60	$^{174m}_{71}$Lu	
1076		2488		490.4±0.1	1.83	44.6 (IT)	
1173		2512		512.524.0±1.5	0.21	59.1 (IT)	
1187		2655		528.2±0.1	3.80	67.1	
1207		2684		536.2±1.2	0.56	76.50	
1273		2700		540.2±0.2	1.44	176.66	0.5%
1380		2740		551.2±1.5	0.38	272.87	0.7%
1394		2775		566.3±1.5	0.10	449.6	0.07%
$^{170m}_{71}$Lu		2836		577.0±1.5	0.27	992.1	0.08%
44.6		2872		584.8±0.8	0.32	$^{174}_{71}$Lu	
48.4		2930		594.5±0.7	0.43	76.5	8.1%
$^{170}_{71}$Lu		2955		607.2±0.7	0.50	1241.76±0.2	9%
Ann. Rad.		3023		623.1±1.5	0.22	1318.24±0.3	0.12%
84.262±0.005 (IT)		$^{171m}_{71}$Lu		631.4±1.5	0.39	$^{176m}_{71}$Lu	
152.8		71.2		681.8±0.6	0.67	88.36	0.0004%
193.5		$^{171}_{71}$Lu		697.3±0.2	6.20	1050	0.0009%
221.2		Ann. Rad.		709.1±0.6	0.79	1140	
222.7		19.3		723.2±1.1	0.40	$^{176}_{71}$Lu	
223.8		27.0		810.1±0.1	16.5	88.36	16†
228.5		46.5		816.2±1.4	1.02	203	89
236.1		55.7		900.8±0.1	29.4	306	100
241.7		66.7		910±1.0	~0.68	$^{177m}_{71}$Lu	
251.3		72.3		912.1±0.1	15.0	71.64	6.8†
283.4		75.9		927.0±1.0		105.36	100
286.8		85.5		929.1±0.1	3.10	112.97	179
301.9		91.3		967.5±1.5	0.18	*115.83	5.0
396.2		109.2		999.5±1.0	<0.2	117.2	1.8
410.6		122.2		1002.8±0.13	5.59	*121.6	52
419.8		132.2		1013.7±1.0		128.5	127
443.6		141.3		1020.2±1.0	0.16	136.7	12
455.7		154.6		1022.4±0.5	1.4	145.8	7
479.0		163.8		1039.7±1.0	~0.1	*147.1	29
492.3		170.6		1041.2±0.2	0.40	153.3	133
497.1		195.0		1080.8±0.2	0.93	159.7	5.4
540.4		518		1093.7±0.1	64.75	*171.8	37
544.6		627		1113.2±0.2	1.7	174.4	96
572.2		668		1115.7±1.0	0.4	177.0	26
579.6		690		1126.6±1.0	<0.1	182.0	0.8
689.1		714		1184.4±0.2	0.3	*195.5	7
841		741		1212.0±1.0	<0.1	204.1	114
857		768		1283.8±0.7	<0.1	208.36	485
940		782		1288.3±0.7	0.22	214.4	48
955		827		1322.8±1.0	<0.1	*218.1	27
987		842		1370.3±0.2	0.70	228.4	287
989		854		1387.2±0.1	0.88	233.8	45
1001		$^{172}_{71}$Lu		1397.1±0.4	0.23	249.7	47
1005		78.73±0.12	9.6	1399.1±1.5	0.13	*268.8	25
1030		90.60±0.15	4.5	1402.7±0.2	0.58	281.8	108
1056		112.7±0.2	1.20	1440.4±0.2	0.64	283.4	5
1063		119.1±1.17	0.050	1466.2±0.2	0.60	291.4	8
1103		134.4±0.3	0.058	1470.3±1.0	0.70	292.5	8
1136		145.8±1.3	0.10	1477.0±1.0	~0.5	296.5	38
1141		151.7±1.0	0.034	1490.0±1.0	1.10	299.0	12
1148		163.4±1.6	0.093	1532.5±1.0	0.09	305.5	14
1225		174.7±1.0	0.12	1543.0±0.2	0.90	313.7	9
1228		181.4±0.1	19.3	1579.3±1.0	0.14	*319.0	78
1231		196.2±0.4	0.090	1584.2±0.1	2.47	321.4	9
1260		203.2±0.1	4.60	1602.8±0.2	0.26	327.7	136
1266		210.2±0.5	0.10	1608.8±0.8	0.095	341.6	13
1283		228.9±0.6	0.35	1622.0±0.1	2.06	*367.4	23
1298		247.0±0.4	0.62	1666.7±0.3	0.24	378.5	222
1310		264.6±0.2	0.65	1670.6±0.2	0.51	385.0	24
1325		269.9±0.1	1.63	1724.5±0.2	0.55	*413.6	131
1344		279.5±0.1	0.98	1813.0±0.3	0.20	418.5	161
1368		318.8±0.8	0.12	1914.7±0.4	0.63	426.3	3.4

Gamma energies (KeV)	Gamma intensities	Gamma energies (KeV)	Gamma intensities	Gamma energies (KeV)	Gamma intensities	Gamma energies (KeV)	Gamma intensities
466.0	19	874		$^{177}_{73}$Ta		902	
$^{177}_{71}$Lu		878		Ann. Rad.		964	
71.64	2.4†	898		92		1000	
112.97	100	1032		115		1013	
136.7	0.74	1037		155		1038	
208.36	171	1069		208		1121	
249.7	3.3	1204		270		1146	
321.4	3.4	1209		1100		1252	
$^{178}_{71}$Lu		1213		1300		1376	
93.179		1351		$^{172}_{73}$Ta		1383	
215		1485		Ann. Rad.		1470	
325		1551		37.4		1639	
425		1781		58.0		$^{176}_{73}$Ta	
$^{178}_{71}$Lu		$^{175}_{72}$Hf		69.8		88.35	
93.179		89.36	40†	81.5		91.3	
215		113.81	3.6	90.2		99.6	
325		161.3	0.3	115		103.4	
425		229.6	7.3	160.4		125.6	
$^{179}_{71}$Lu		318.9		172.1		131.1	
90		343.4	1000	180.6		146.7	
213		433.0	16	$^{174}_{73}$Ta		156.8	
$^{168}_{72}$Hf		$^{178m2}_{72}$Hf		Ann. Rad.		158.2	
Ann. Rad.		88.878±0.015	67	90.9		175.6	
129		93.181±0.015	18	125		190.4	
170		213.444±0.015	86	160		201.9	
$^{169}_{72}$Hf		216.673±0.015	71	205		240.0	
Ann. Rad.		237.40±0.03	10	280		366	
115		238.66±0.1[b]	2.5	350		415	
$^{170}_{72}$Hf		257.62±0.02	18	$^{175}_{73}$Ta		639	
Ann. Rad.		277.35±0.05	1.6	35.8		646	
44.6		296.80±0.04	10	50.5		678	
48.4		325.562±0.015	100	70.5		686	
120.2		426.371±0.015	105	77.3		711	
164.7		454.05±0.04	19	81.5		723	
990		495.01±0.03	82	100.8		741	
1280		535.02±0.04	11	104.3		938	
1650		574.21±0.03	101	121.9		1025	
1010		$^{178m1}_{72}$Hf		126.6		1140	
2360		88.878±0.015	67%	132.0		1162	
2520		93.181±0.015	18%	142.0		1192	
2940		213.44±0.015	86%	162.0		1203	
$^{171}_{72}$Hf		325.562±0.015	96%	162.5		1225	
70.1		426.371±0.015	100%	179.1		1255	
122		$^{179m}_{72}$Hf		185.8		1272	
188		161	5%	192.7		1296	
290		217	99.5%	196.4		1343	
340		325	0.5%	207.4		1359	
470		$^{180m}_{72}$Hf		213.4		1489	
660		57.54	29†	230.8		1504	
860		93.33	10	266.9		1557	
1070		215.3	50	280.5		1586	
$^{172}_{72}$Hf		332.5	57	288.9		1618	
23.99		444	49	294.0		1633	
41.86		501	8	308.9		1672	
44.08		$^{181}_{72}$Hf		348.5		1681	
81.8		3.9		361.4		1699	
114		6.3		365.7		1707	
125.8		133.02	40†	386.0		1723	
$^{173}_{72}$Hf		136.25	6	393.2		1778	
4.65		136.86	1.7	404		1825	
17.8		345.9	13	433		1864	
123.6		476.0	~1.7	436		1908	
134.9		482.0	81	443		1958	
139.6		615.5	0.2	450		2045	
161.8		$^{182}_{72}$Hf		462		2137	
171.3		97.89±0.1	0.08%	475		2219	
297		114.33±0.02	2.6%	525		2247	
306		156.09±0.02	7.0%	533		2312	
311	18†	172.54±0.1	0.2%	540		2460	
357		270.405±0.01	80%	602		2467	
422		$^{183}_{72}$Hf		622		2519	
539		140	2†	694		2602	
549		170	0.9	732		2622	
555		210	2	751		2760	
567		280	3	762		2831	
577		320	4	810		2925	
617		400	5	815		$^{177}_{73}$Ta	
624		460	58	852		51.1	
717		610	2†	854		71.6	0.003%
759		720	6	860		96.2	
764		820	100	867		112.97	6%
853		900	6	878		136.7	
857							

GAMMA ENERGIES AND INTENSITIES OF RADIONUCLIDES (Continued)

Gamma energies (KeV)	Gamma intensities
208.38	1.0%
249.7	0.02%
257.3	
298.1	
321.3	0.024%
357.3	0.002%
396.0	
420.9	0.0018%
424.7	0.13%
452.9	
492.5	0.0015%
509	0.0024%
526.9	
549.4	
598.5	0.009%
633.1	0.039%
735.2	0.04%
737.0	0.016%
746.0	0.22%
848.2	
945.4	0.055%
1058.4	0.30%
$^{178}_{73}$Ta	
Ann. Rad.	
93	
203	
214	
970	
1100	
1170	
1180	
1200	
1260	
1330	
1340	
1390	
1430	
1440	
1460	
1480	
$^{178}_{73}$Ta	
88.8	
93.1	
213.6	10†
325.7	11
331.7	3
426.8	10
$^{179}_{73}$Ta	
No γ	
$^{180m}_{73}$Ta	
93.4±0.2	27%
103.6±0.2	3%
200	
$^{180}_{73}$Ta	
400	
$^{181m}_{73}$Ta	
6.3	
$^{182m}_{73}$Ta	
146.785±0.01	75†
171.586±0.01	≡100
184.951±0.01	51
318.401±0.04	14
356.468±0.07	0.6
$^{182}_{73}$Ta	
31.74	
42.71	
65.72	
67.75	8%
84.68	3.34%
100.10	41%
113.67	1.8%
116.41	0.4%
152.441±0.003	57%
156.386±0.003	72%
179.394±0.004	33%
222.114±0.004	97%
229.335±0.017	3.6%
264.079±0.009	24%
927.903±0.07	0.6%
1001.654±0.041	2.8%
1113.050±0.08	0.6%
1121.218±0.016	35.6%
1157.357±0.060	1.04%

Gamma energies (KeV)	Gamma intensities
1188.978±0.019	16.5%
1221.354±0.019	27.9%
1230.982±0.021	11.4%
1257.401±0.036	1.56%
1273.657±0.050	0.67%
1289.126±0.032	1.44%
1342.743±0.010	0.27%
1373.638±0.015	0.24%
1387.2±0.2	0.08%
1410.0±0.2	0.05%
1453.0±0.2	0.03%
$^{183}_{73}$Ta	
40.98	1.8†
46.48	22
52.59	22
82.92	1.4
84.71	5.1
99.08	27
101.93	1.2
102.48	0.51
103.15	0.35
107.93	44
109.73	2.4
120.73	0.23
142.27	1.4
144.12	9.5
160.53	11
161.34	35†
162.32	19
192.64	1.35
203.28	1.5
205.08	3.4
208.81	2.3
209.86	17
244.26	32
245.24	1.2
246.06	100
291.72	14.1
313.00 } 313.28	27
353.99	43
365.64	2.0
406.61	1.9
$^{184}_{73}$Ta	
111	
160	
250	
300	
410	
530	38†
660	7
790	34
900	100
950	30
1060	9
1100	1.3
1160	25
1270	10†
1320	1.8
1440	5
$^{185}_{73}$Ta	
75	8†
100	16
110	1
135	5
175	100
245	8
$^{186}_{73}$Ta	
122	25†
200	100
300	25
410	20
510	45
608	45
730	65
940	15
$^{175}_{74}$W	
260	
800	
1300	
1600	
$^{176}_{74}$W	
Ann. Rad.	

Gamma energies (KeV)	Gamma intensities
33.58	
50.55	
61.29	
84.14	
94.9	
100.2	
$^{177}_{74}$W	
30.5	
70.45	
101.8	
115.7	
142.6	
143.9	
156.0	
186.4	
203.0	
223.3	
259.2	
271.1	
290.0	
308.1	
317.5	
367.6	
377.1	
381.5	
388.9	
417.1	
427.2	
450.8	
502.0	
528.0	
562.0	
568	
615	
630	
639	
673	
712	
722	
760	
787	
797	
827	
858	
877	
1016	
1036	
1067	
1174	
1269	
1295	
$^{179m}_{74}$W	
221.8	
$^{179}_{74}$W	
30.7	
$^{180}_{74}$W	
0.102	
0.5	
0.22	
$^{181m}_{74}$W	
0.37	
$^{181}_{74}$W	
63	35%
136.3	0.11%
152.5	0.2%
$^{183m}_{74}$W	
46	†
52	
105	19
160	3.5
$^{185m}_{74}$W	
60	
131	
175	
$^{187}_{74}$W	
36.3	
49.2	
57.2	†
71.95±0.03	37.3
106.57±0.08	0.09
113.72±0.06	0.33
134.24±0.03	29.4

Gamma energies (KeV)	Gamma intensities
206.20±0.04	0.53
239.10±0.10	0.28
246.35±0.07	0.38
479.51±0.03	80.47
511.70±0.06	2.53
551.51±0.10	18.98
588.90±0.12	0.65
618.24±0.04	23.25
625.42±0.07	3.93
685.72±0.04	100
745.33±0.10	1.00
772.85±0.07	14.75
864.53±0.10	1.25
879.0±0.5	0.47
$^{188}_{74}$W	
63.58	0.9%
227	0.3%
290.5	0.6%
$^{189}_{74}$W	
94	2†
130	12
178	13
220	3
258	100
360	10
417	96
550	28
860	20
960	17
$^{177}_{75}$Re	
Ann. Rad.	
$^{178}_{75}$Re	
Ann. Rad.	
$^{179}_{75}$Re	
221.8	
280	
$^{180m}_{75}$Re	
Ann. Rad.	
110	
880	
$^{180}_{75}$Re	
Ann. Rad.	
$^{181}_{75}$Re	
19.7	
31.1	
38.1	
43.5	
65.0	
71.7	
72.7	
93.7	
102.7	
103.1	
109.9	
110.3	
113.3	
137.2	
144.3	
154.4	
163.9	
164.6	
165.8	
175.2	
177.4	
186.2	
195.0	
197.0	
252.0	
262.6	
276.2	
295.9	
318.6	
331.9	
353.6	
356.1	
360.7	
365.5	
382.3	
389.0	
408.7	
441.8	
475.6	
487.2	

GAMMA ENERGIES AND INTENSITIES OF RADIONUCLIDES (Continued)

Gamma energies (KeV)	Gamma intensities
489.0	
515.7	
522.2	
557.7	
587.6	
639.0	
651.2	
661.8	
668.2	
693.9	
738.1	
769.6	
781.4	
804	
805	
823	
840	
854	
880	
883	
908	
954	
989	
994	
1000	
1009	
1076	
1087	
1103	
1384	
1441	
1469	
1538	
$^{182}_{75}$Re	
31.7	
42.0	
42.7	
65.8	
67.8	
84.7	
100.1	
113.7	
116.4	
152.5	
156.4	
179.4	
198.4	
222.0	
229.3	
264.0	
470.0	
514.3	
536.1	
598.5	
649.5	
734.5	
786.6	
810	
836	
895	
901	
928	
960	
1002	
1045	
1122	
1158	
1189	
1222	
1231	
1257	
1274	
1289	
1374	
1387	
1438	
1955	
2014	
2023	
2054	
$^{182}_{73}$Re	
47.05±0.15	
53.04±0.15	

Gamma energies (KeV)	Gamma intensities
87.31±0.15	
100.07±0.15	
113.62±0.15	
116.45±0.15	
130.75±0.15	
133.72±0.15	
145.39±0.15	
149.00±0.15	
152.37±0.15	
156.34±0.15	
172.83±0.15	
187.47±0.15	
189.62±0.15	
191.33±0.15	
198.29±0.15	
214.24±0.15	
215.83±0.15	
217.48±0.15	
226.16±0.15	
247.44±0.15	
256.44±0.15	
264.07±0.15	
276.30±0.15	
281.45±0.15	
286.53±0.15	
295.70±0.15	
300.15±0.15	
323.36±0.15	
339.06±0.15	
341.93±0.15	
346.13±0.15	
351.07±0.15	
357.11±0.15	
1157.40±0.20	
1181.09±0.20	
1188.78±0.20	
1221.10±0.11	
1230.76±0.15	
1257.18±0.30	
1288.67±0.30	
1293.66±0.30	
1342.33±0.30	
1426.88±0.15	
1438	
$^{183}_{75}$Re	
46.73±0.15	78%
52.77±0.15	25%
67.29±0.15	
69.22±0.15	
82.99±0.15	3.1%
84.79±0.15	12%
99.11±0.15	25%
107.94±0.15	11%
109.74±0.15	11%
115.90±0.15	
117.19±0.15	
118.58±0.15	
126.49±0.15	
125.18±0.15	
144.06±0.15	0.3%
169.01±0.15	
192.54±0.15	0.5%
215.18±0.15	
220.15±0.15	
221.59±0.15	
229.29±0.15	
244.36±0.15	0.8%
245.94±0.15	1.8%
313.11±0.15	0.6%
318.09±0.15	
354.21±0.15	0.7%
$^{184m}_{75}$Re	
83.4	70%
104.7	70%
111.43±0.1	61%
161.27±0.15	10%
216.53±0.15	15%
252.82±0.15	10%
384.21±0.15	~2%
641.9±0.15	1.5%
792.83±0.10	28%
894.68±0.10	23%
903.20±0.10	36%

Gamma energies (KeV)	Gamma intensities
$^{184}_{75}$Re	
71.11±0.15	
87.50±0.2	
91.42±0.2	
104.80±0.1	
111.43±0.10	59%
117.43±0.10	
118.67±0.20	
124.47±0.20	
126.52±0.20	
161.27±0.15	
169.21±0.20	
170.57±0.20	
178.27±0.20	
180.0±0.3	
216.53±0.15	0.5%
226.72±0.15	
252.82±0.15	4.7%
317.97±0.15	1%
363.97±0.25	
384.21±0.15	
536.61±0.15	
539.24±0.20	0.3%
641.90±0.15	2.3%
769.75±0.15	5%
792.83±0.10	36%
850.78±0.15	
859.19±0.15	
894.68±0.10	19%
903.20±0.10	41%
920.88±0.15	~4%
953.64±0.20	
962.02±0.15	
970.77±0.20	
1005.88±0.20	
1010.11±0.20	
1033.49±0.15	
1109.86±0.15	
1173.72±0.15	
1274.96±0.25	
1384	
$^{186}_{75}$Re	
176	2.1
191	1.2
230	0.6
$^{188m}_{75}$Re	
2.4	
15.9	
63.58	
92.45	
105.96	
$^{188}_{75}$Re	
155.03	1550† 3%
297	1.7
321	1.6
452	7
478	100 0.27%
485	8
633	141
641	1.2
672	10
824	4
829	40
846	0.6
881	1.6
932	58
963	1.3
967	1.8†
977	0.7
1019	1.6
1132	9
1150	3.6
1171	1.1
1176	2.1
1191	1.2
1230	0.6
1307	7
1322	1.3
1368	1.1
1457	1.8
1610	11
1732	3.5
1786	2.6

Gamma energies (KeV)	Gamma intensities
1803	4.1
1866	0.6
1957	1.6
$^{189}_{75}$Re	
30.8	
36.2	
69.5	
95.2	
147.2	
150.1	
185.9	
188.6	
206.4	
211.0	
216.8	
219.4	
245.0	
$^{190}_{75}$Re	
187	10†
392	10
570	10
830	3
$^{192}_{75}$Re	
200	
290	
370	
480	
570	
$^{181}_{76}$Os	
93	
101	
$^{182}_{76}$Os	
27.6	†
55.5	
180.2	7
261.7	1.4
509.9	10
$^{183m}_{76}$Os	
67.3	6%
147	
170.7	46%
1.036	6%
1.103	26%
1.1095	22%
$^{183}_{76}$Os	
114.4	
145.4	
151.0	
167.9	
236.2	
259.8	
355.5	
381.8	
477.2	
496.1	
737	
808	
851	
887	
889	
1163	
1440	
$^{185}_{76}$Os	
71.6	1.5%
125.3	2.4%
162.6	1.1%
233.4	1.4%
592.0	1.2%
645.8	81%
718	4.2%
872	6.5%
879	6.5%
$^{189m}_{76}$Os	
30.81	
$^{190m}_{76}$Os	
38.9	<1%
186.7	100%
361.2	100%
502.5	100%
616.5	100%
$^{191m}_{76}$Os	
74.2	

Gamma energies (KeV)	Gamma intensities
$^{191}_{76}$Os	
49.5	
82.33	
96.45	
129.4	
178.9	
186.7	
$^{193}_{76}$Os	
73	1.45†
98	0.04
99	0.02
107	0.37
139	2.93
153	0.02
180	0.09
181	0.18
219	0.16
234	0.033
252	0.134
280	1.00
289	0.19
298	0.16
321	0.76
361	0.21
378	0.08
387	0.75
419	0.133
441	0.065
460	2.32
485	0.113
517	0.02
532	0.05
558	1.10
573	0.01
660	0.002
712	0.011
848	0.003
875	0.0135
$^{194}_{76}$Os	
43.0	10%
78	0.03%
$^{182}_{77}$Ir	
Ann. Rad.	
133	
278	
↓	
4000	
$^{183}_{77}$Ir	
240	
$^{184}_{77}$Ir	
Ann. Rad.	
125	100†
267	200
392	90
830	
960	
1090	
↓	
4300	
$^{185}_{77}$Ir	
24.2	
30.4	
33.8	
37.4	
60.0	
90.4	
94.5	
97.4	
100.7	
119.6	
124.9	
126.9	
127.9	
129.4	
153.6	
158.3	
160.7	
185.0	
220.4	
222.3	
223.8	
254.4	
266.5	

Gamma energies (KeV)	Gamma intensities
300.4	
307.1	
314.4	
321.5	
339.2	
352.4	
377.7	
406.8	
419.1	
431.4	
507.0	
513.7	
539.4	
745.7	
$^{186}_{77}$Ir	
Ann. Rad.	
70.9	
87.19	
102.1	
119.4	
137.15	
143.0	
160.0	
167.1	
220.0	
224.1	
234.5	
252.5	
269.0	
272.8	
276.5	
284.3	
288.8	
293.0	
296.7	
299.4	
302.9	
305.6	
309.6	
311.8	
321.2	
322.6	
326.6	
330.2	
334.0	
342.5	
351.7	
364.9	
387.9	
403.3	
406.6	
412	
420.7	
542.2	
551.4	
558.0	
565.4	
570.3	
584.4	
592	
600	
622.1	
630.3	
636.2	
650.0	
662	
679.5	
684.8	
693.6	
706	
712.6	
729.5	
745	
760.0	
767.3	
773.1	
780.8	
794	
805.5	
841.3	
885	
933.2	
943.6	

Gamma energies (KeV)	Gamma intensities
960	
1011.0	
1026.5	
1057.1	
1107	
1172	
1187.9	
1314	
1324	
1440	
1467	
1508	
1597	
1647	
1701	
1738	
1743	
1752	
1789	
1801	
1805	
1815	
1869	
1906	
1939	
1953	
1975	
2000	
2260	
2312	
2340	
2360	
2383	
2389	
2397	
2413	
2429	
2498	
2513	
2564	
2578	
2617	
2678	
2733	
2748	
2826	
2854	
2994	
3043	
$^{186}_{77}$Ir	
Ann. Rad.	
137.15	
295	
630	
770	
990	
$^{187}_{77}$Ir	
65.1	
73.7	
85.0	
125.5	
137.9	
162.6	
177.5	
187.1	
313.8	
399.2	
400.9	
426.9	
491.1	
501.5	
576.5	
610.6	
799.4	
912.6	
976.7	
987.0	
1103	
1112	
$^{188}_{77}$Ir	
Ann. Rad.	
85.0	
95.6	

Gamma energies (KeV)	Gamma intensities
115.7	
123.1	
150.5	
155.0	209
157.0	
158.0	
162.2	
175.4	
189.3	
222.3	
268.3	
279.6	
312.0	
322.9	12
332.4	
346.7	
350.0	1.6
351.7	
352.7	
371.4	
379.6	
385.3	1.4
399.0	
424.5	
447.7	
452.9	
478.0	
478.6	100
514.8	3.2
522.7	
566.7	3
586.2	
594.0	
599.0	
620.6	6
633.1 } 634.8	180
641.5	3.4
652.4	
672.3	9
736.8	
747.6	
757.2	2.0
777.5	†
781.9	
810.5	
821.2	
824.8	
828.6	
829.5	
886.0	
905.9	
940.0	
943.6	
947.4	
983.8	
987.7	13
1013	
1018	6
1053	
1097	8
1143	2
1150	4
1175	6
1193	
1206	†
1210	42
1296	1.5
1304	4
1307	
1323	10
1329	4
1333	4
1337	
1430	
1436	9
1446	
1453	4
1458	7
1463 } 1466	7
1560	4
1575	12

GAMMA ENERGIES AND INTENSITIES OF RADIONUCLIDES (Continued)

Gamma energies (KeV)	Gamma intensities
1620	4
1653	10
1689	2
1705	7
1717	25
1721	
1728	
1732	
1737	
1760	
1775	
1784	
1789	
1805	
1812	1
1904	
1931	
1945	18
1968	
2013	3
2052	22
2061	30
2098	45
2100	
2196	10
2217	80†
2294	1.5
2344	
2350	
2494	
2525	
2623	
2700	
¹⁸⁹₇₇Ir	
25.7	
30.8	
33.3	
36.2	
56.5	
59.1	
59.3	
69.6	
95.2	
97.8	
138.3	
147.1	
149.9	
164.0	
180.5	
185.9	
188.6	
197.4	
206.3	
216.7	
219.3	
233.5	
245.0	
275.8	
¹⁹⁰ᵐ²₇₇Ir	
38.9	
116.7	
148.7 (IT)	
186.7	75†
206.6	
361.2	100
502.5	115
616.5	103
¹⁹⁰ᵐ¹₇₇Ir	
26.3	
¹⁹⁰₇₇Ir	
97.8	
100.0	
137.8	
182.1	
186.7	
190.5	
196.8	
198.0	
199.3	
207.8	
223.8	
235.3	
251.4	

Gamma energies (KeV)	Gamma intensities
282.9	
288.2	
294.6	
361.2	
371.1	
380.1	
394.8	
397.3	
402.4	
407.2	
447.5	
462.5	
485.3	
490.8	
502.5	
518.4	
557.8	
569.3	
605.3	
655.9	
690.0	
726.4	
750.7	
768.5	
801	
829.0	
833	
836	
839	
915	
933	
953	
1076	
1079	
1110	
1116	
1131	
1134	
1184	
1200	
1234	
1314	
1324	
1329	
1339	
1355	
1386	
1391	
1397	
1403	
1414	
1437	
1464	
1473	
1482	
1517	
1525	
1534	
1543	
1584	
1606	
1677	
1685	
1700	
1711	
¹⁹¹ᵐ₇₇Ir	
41.8	
47	
82.3	
129.4	
¹⁹²ᵐ²₇₇Ir	
161	
¹⁹²ᵐ¹₇₇Ir	
58 (IT)	
317	
612	
¹⁹²₇₇Ir	
136.34	2.7†
201.38	5.6
205.79	38
283.4	4
295.9	360
308.43	370

Gamma energies (KeV)	Gamma intensities
316.5	1000—85%
374.6	6
416.6	≤3
468.05	600
484.6	40
489.1	4
588.6	49
604.38	105
612.43	70
884.6	5
1062	0.5
¹⁹³ᵐ₇₇Ir	
80.2	
¹⁹⁴₇₇Ir	
29.362	6.5†
300.7	0.9
328.54	33
622.1	0.8
645.3	2.7
890.1	0.10
925.3	0.09
938.9	1.5
1000.2	0.08
1048.7	0.07
1104.1	0.06
1151.3	1.6
1175.5	0.17
1184.1	0.79
1219.0	0.34
1294	0.26
1342.4	0.11
1431	0.02
1469.4	0.93
1480	<0.05
1488	0.9
1512	0.13
1623	0.24
1671	0.033
1784	0.025
1798	0.06
1805	0.06
1808	0.09
1832	0.009
1926	0.012
2044	0.022
2115	0.009
¹⁹⁵₇₇Ir	
98.5	100†
129.9	50
330	60
370	40
430	30
660	20
¹⁹⁶ᵐ₇₇Ir	
103.3±0.2	17±2
340.7±0.4	1.6±0.2
355.9±0.2	98±3
393.5±0.2	101±2
420.9±0.3	2.6±0.1
447.1±0.2	98±2
521.4±0.2	100
553.0±0.3	0.66±0.04
566.4±0.4	0.3±0.1
615.9±0.4	0.44±0.05
633.5±0.3	1.15±0.05
647.3±0.2	95±3
673.9±0.2	0.18±0.04
693.9±0.2	4.4±0.3
722.0±0.4	0.67±0.07
727.3±0.2	2.7±0.1
760.6±0.3	0.78±0.05
835.6±0.2	6.6±0.2
849.4±0.3	0.53±0.05
868.1±0.3	0.48±0.04
887.0±0.5	0.11±0.02
893.0±0.5	0.20±0.02
904.6±0.5	0.10±0.02
914.6±0.3	0.30±0.05
926.0±0.5	0.07±0.02
1024.6±0.3	0.26±0.03
1068±2	0.074±0.018
1080.5±0.5	0.12±0.02
1116.7±0.8	0.073±0.022

Gamma energies (KeV)	Gamma intensities
1281.6±0.5	0.061±0.024
1341.5±0.5	0.29±0.03
1355.8±0.5	0.06±0.01
1394.0±0.5	0.074±0.015
1482.5±0.4	2.4±0.2
¹⁹⁶₇₇Ir	
KX	40±30
332.8±0.3	23±1
355.4±0.3	100
116.6±0.3	24±2
779.4±0.3	55±2
1047.0±0.7	5±1
1228.6±1.2	1.8±0.4
1468.4±0.8	4.4±0.7
1564.2±0.8	4.7±1.0
¹⁹⁷₇₇Ir	
510	~50%
¹⁹⁸₇₇Ir	
780	
¹⁷⁴₇₈Pt	
Ann. Rad.	
¹⁷⁷₇₈Pt	
Ann. Rad.	
¹⁷⁸₇₈Pt	
Ann. Rad.	
160	
¹⁸⁰₇₈Pt	
Ann. Rad.	
¹⁸¹₇₈Pt	
Ann. Rad.	
¹⁸²₇₈Pt	
Ann. Rad.	
¹⁸³₇₈Pt	
Ann. Rad.	
¹⁸⁵₇₈Pt	
11	
63	
1560	
¹⁸⁶₇₈Pt	
140	
190	
680	
1400	
¹⁸⁷₇₈Pt	
110	
184	
2010	
¹⁸⁸₇₈Pt	
Ir KX	100†
41.9	0.4
54.8	0.3
96.7	0.2
98.4	0.4
132.8	0.30
140.3	2.2
187.6	19.2
195.1	18.4
280.4	0.22
283.0	0.11
290.6	0.11
381.6	7.0
423.6	4.4
478.3	1.9
¹⁸⁹₇₈Pt	
71.6	
82.1	
94.2	
113.8	
141.1	
176.3	
181.3	
186.7	
190.8	
203.7	
223.3	
225.6	
243.4	
252.0	
258.2	
300.4	
317.6	
403.8	
542	

GAMMA ENERGIES AND INTENSITIES OF RADIONUCLIDES (Continued)

Gamma energies (KeV)	Gamma intensities
544.8	
568.8	
607	
626.9	
644	
720.9	
735	
792	
798	
$^{191}_{78}$Pt	
82.33	
96.45	
129.39	
172.2	10.2%
178.9	3.2%
186.8	
187.8	
219.7	
221.8	
227.4	
268.8	
351.2	9.9%
360.0	16.6%
409.3	28.8%
456.5	9.6%
501.4	
539.0	40.5%
569.1	
575.6	
587.9	
624.1	
632.6	
$^{193m}_{78}$Pt	
12.58	
135.5	
$^{195m}_{78}$Pt	
30.8	
98.5	12%
129.4	2.8%
129.9	
$^{197m}_{78}$Pt	
52.9 (IT)	†
130.2	21
279.3	21
346 (IT)	100
$^{197}_{78}$Pt	
77.35	20%†
191.4	5.7%
268	0.006%
279	<0.0001%
$^{199}_{78}$Pt	
32	~5†
393	100
$^{199}_{78}$Pt	
75	†
179	38
220	w
245	16
315	11
320	22
475 D	45
540	100
715	14
790	10
960	8
$^{186}_{79}$Au	
160	
220	
330	
400	
$^{188}_{79}$Au	
Ann. Rad.	
250	
330	
630	
$^{189}_{79}$Au	
39.44	
45.7	
88.6	
166.6	
215.9	
440	
446	

Gamma energies (KeV)	Gamma intensities
$^{190}_{79}$Au	
Ann. Rad.	
295.7	
301.6	
319	
598	
623	
1400	
1400	
1760	
2080	
2400	
2700	
2940	
3250	
3460	
$^{191}_{79}$Au	
48.3	
91.03	
132.90	
136.10	
158.8	
166.5	
194.1	
206.4	
210.0	
244.4	
253.9	
279.9	
280.5	
284.0	
312.7	
336.3	
368.7	
386.9	
390.4	
399.6	
408.3	
413.8	
421.5	
451.3	
478.8	
487.5	
495.9	
525.9	
586.4	
620.3	
674.2	
702.0	
732.5	
$^{192}_{79}$Au	
Ann. Rad.	
45.1	
96.8	
104.7	
137	
157	
167	
205	
283	
295	
308	
316	
415	
437	
466	
588	
612	
783	
1158	
$^{193m}_{79}$Au	
31.9	
220	
290.6 (IT)	
$^{193}_{79}$Au	
12.7	
37.4	
44.3	
73.7	
99.8	
112.4	
114.0	
117.9	

Gamma energies (KeV)	Gamma intensities
119.4	
155.7	0.35%
173.5	2.0%
186.1	9%
187.5	
221.3	
231.8	0.5%
255.6	4.6%
258.4	
268.2	4.4%
290.0	0.2%
303.6	
317.7	1.3%
334.8	
377.6	1.3%
425	0.25%
440	3.1%
477	0.5%
490	0.6%
$^{194}_{79}$Au	
107.2	
163.94	
202.9	
216.15	
226.9	
236.3	
285.34	
293.62	
300.71	
328.54	
365.0	
482.8	
528.9	
544.7	
589.4	
593.7	
607.6	
622.1	
645.3	
668.6	
672	
676.1	
703.6	
736.5	
818.7	
843.9	
847.4	
856.1	
860	
890.1	
894.4	
925.3	
938.9	
948.4	
1002.2	
1007.5	
1030.9	
1038.8	
1048.7	
1081.6	
1104.1	
1119.7	
1141.0	
1150.9	
1156.6	
1175.5	
1183.6	
1187.1	
1195.1	
1208.5	
1219.0	
1267.4	
1292	
1293.9	
1302.6	
1308.6	
1342.4	
1421.8	
1432.3	
1442.0	
1450.5	
1463.7	
1469.4	

Gamma energies (KeV)	Gamma intensities
1479.8	
1487.9	
1492.7	
1513	
1519	
1547.9	
1563.4	
1592.9	
1596.2	
1602.0	
1618	
1622.6	
1633.5	
1671	
1675.9	
1690.2	
1735.8	
1758	
1786.3	
1798.0	
1803	
1806	
1813	
1830.1	
1835.9	
1886.8	
1893.1	
1911.1	
1924.6	
1960	
1970.1	
1983.9	
2044.1	
2085.8	
2114.0	
2164.1	
2216	
2299	
2313	
2357	
2366	
2371	
2399	
2414	
$^{195m}_{79}$Au	
318	
$^{195}_{79}$Au	
30.8	
98.82±0.15	
129.12±0.15	
$^{196}_{79}$Au	
107.13±0.1	1.7†
332.9±0.25	280
355.7±0.3	1000
425.8±0.3	5
520.9±0.5	0.5
758.5±0.4	0.4
960	0.06
1091.3±0.3	2.4
$^{197m}_{79}$Au	
130.2	
279	
$^{198}_{79}$Au	
411.795±0.009	99%
675.883±0.0018	1%
087.678±0.027	0.2%
$^{199}_{79}$Au	
49.828±0.004	
158.372±0.004	76%
208.200±0.006	16.6%
$^{200}_{79}$Au	
367.97	30%
1225	8.4%
1594	15.6%
$^{201}_{79}$Au	
530	5%
$^{203}_{79}$Au	
690	
$^{186}_{80}$Hg	
125	
270	
350	

GAMMA ENERGIES AND INTENSITIES OF RADIONUCLIDES (Continued)

Gamma energies (KeV)	Gamma intensities
440	
$^{187}_{80}$Hg	
175	
255	
400	
$^{188}_{80}$Hg	
140	
$^{189}_{80}$Hg	
Ann. Rad.	
165	
240	
320	
$^{190}_{80}$Hg	
28.8	
100.8	†
129.8	
142.7	10
155.0	
165.4	
171.5	
220	1
$^{191}_{80}$Hg	
252.6	
274.1	
$^{192}_{80}$Hg	
31.5	
40.9	
47.7	
99.4	
104.5	
105.3	
114.5	
120.1	
135.9	
142.5	
146.0	
157.2	
201.8	
245.4	
262.7	
274.8	
279.2	
306.5	
436.7	
$^{193m}_{80}$Hg	
32.2	
38.2	
39.5 (IT)	
101.2 (IT)	
165.9	
218	
220	
257	
281	
291	
299	
342	
345	
361	
364	
382	
383	
408	
462	
488	
500	
510	
535	
537	
574	
601	
624	
675	
701 D	
712	
861	
871	
879	
914	
933	
997	
1113	
1175	

Gamma energies (KeV)	Gamma intensities
1233	
1243	
1328	
1343	
1489	
1646	
$^{194}_{80}$Hg	
38.2	
186.8	
574	
762	
855	
1040	
1078	
$^{195m}_{80}$Hg	
16.2 (IT)	
37.2 (IT)	
53.3 (IT)	
56.7	
61.4	
123.0 (IT)	
172.6	
200.5	
261.5	
318.5	
368	
386.0	
387.6	
452	
467	
526	
560	
575	
680	
963	
1242	
$^{195}_{80}$Hg	
61.4	
180.2	
200.5	
207.1	
261.5	
600	
780	
821	
842	
930	
1020	
1110	
1172	
$^{197m}_{80}$Hg	
130.2	†
134.0 (IT)	1
165.3 (IT)	0.01
202	0.23
279.3	16
$^{197}_{80}$Hg	
77.34±0.005	34
191.5	1.0†
269	0.06
$^{199m}_{80}$Hg	
158	
368	
$^{203}_{80}$Hg	
279.190±0.007	
$^{205}_{80}$Hg	
203	
$^{206}_{80}$Hg	
310	
$^{191}_{81}$Tl	
Ann. Rad.	
$^{192}_{81}$Tl	
Ann. Rad.	
424	
$^{193m}_{81}$Tl	
365	
$^{193}_{81}$Tl	
241	
261	
270	
291	
309	
321	

Gamma energies (KeV)	Gamma intensities
330	
$^{194m}_{81}$Tl	
97.2	
$^{194}_{81}$Tl	
427	
$^{195m}_{81}$Tl	
99.1	
383	
$^{195}_{81}$Tl	
37.2	
225.8	
242.1	
247.3	
279	
562	
883	
$^{196m}_{81}$Tl	
33.7	
84.4	
120.1	
425.8	
$^{196}_{81}$Tl	
426	
$^{197m}_{81}$Tl	
222	
385	
$^{197}_{81}$Tl	
18.18	
133.9	
152.2	
156.3	
173.8	
260.6	
277.5	
308.5	
426.0	
433.1	
444	
451.8	
578.3	
585	
791	
857	
890	
980	
1010	
$^{198m}_{81}$Tl	
23.1 (IT)	
48.7	
259.6 (IT)	
260.9 (IT)	
282.4 (IT)	
411	13
586	10
635	10
$^{198}_{81}$Tl	
194	
227	
284.5	
400.4	
411.8	100†
586.5	
637	
675.7	
1089	3†
1202	24
1440	27
2010	17
2450	6
2780	2
$^{199}_{81}$Tl	
36.83	
49.82	4†
51.9	
158.36	54
195.1	9
208.18	119
247.2	93
284.0	13
333.9	16
403.5	4
455.1	136
492	10

Gamma energies (KeV)	Gamma intensities
$^{200}_{81}$Tl	
Ann. Rad.	
116.5	
134	
181	
227	
251	
276	
289	
309	
368	1000†
388	
545	
555	
579	120†
629	8
661	12
690	
712	8
786	15
829	87
887	5
898	9
939	
1029	
1207	330
1227	58
1265	†
1275	
1341	10
1364	45
1410	18
1479	
1517	42
1569	
1593	
1600	0
1718	7
1745	
1900	
1920	
1966	
1991	
2012	
$^{201m}_{81}$Tl	
230	
331.2	
361.3	
600	
$^{201}_{81}$Tl	
30.60	
32.19	
135.34	7%
165.9	0.3%
167.4	26%
$^{202}_{81}$Tl	
439.4±0.2	100†
509.4±0.3	
520.3±0.3	4
959.7±0.4	0.07
$^{207m}_{81}$Tl	
351	
1000	
$^{207}_{81}$Tl	
897.3	0.16%
$^{208}_{81}$Tl	
233	
252	†
277.33	7
486	0.1
510.72	23
583.169±0.013	86
763.2	2
860	12
1094	0.6
2614.611±0.0060	100
$^{209}_{81}$Tl	
120	100%
450	100%
1560	100%
$^{210}_{81}$Tl	
97	4%
296	80%

GAMMA ENERGIES AND INTENSITIES OF RADIONUCLIDES (Continued)

Gamma energies (KeV)	Gamma intensities	Gamma energies (KeV)	Gamma intensities	Gamma energies (KeV)	Gamma intensities	Gamma energies (KeV)	Gamma intensities
360	4%	381.4		259		918	
380	3%	395.1		272		984	
795	100%	406.0		275		1203	
910	3%	545.9		279		1211	
1060	12%	584.6		295	†	$^{205}_{83}$Bi	
1110	7%	692.5		352.0	36	Ann. Rad.	
1210	17%	708.6		481		26.22±0.11	
1310	21%	753.2		534		45.43±0.075	
2010	7%	767.3		549		90.03±0.024	
2090	5%	803.5		777		115.409±0.181	0.2332±0.022
2280	3%	820.3		$^{200}_{83}$Bi		122.466±0.083	0.0465±0.007
2360	8%	826.0		462		129.742±0.127	0.0314±0.006
2430	9%	907.4		1027		147.315±0.068	0.0755±0.001
$^{194}_{82}$Pb		946.8		$^{201}_{83}$Bi		149.505±0.236	0.0216±0.006
0.204		1069.8		629		157.269±0.219	0.0352±0.010
$^{195}_{82}$Pb		1098.4		$^{202}_{83}$Bi		165.113±0.143	0.0466±0.013
99.1		1115		422		185.293±0.115	0.2806±0.037
383		1149		961		205.841±0.273	0.0987±0.017
393		1239		$^{203}_{83}$Bi		221.024±0.135	0.0926±0.012
$^{196}_{82}$Pb		1309		Ann. Rad.		235.997±0.046	0.1664±0.043
191.8		$^{202m}_{82}$Pb		59.99		260.511±0.052	3.1038±0.338
240.3		129.3 (IT)		126.4		262.815±0.067	1.1908±0.172
253.2		148.9		186.5		282.261±0.103	1.1461±0.030
367		240.3 (IT)		263.9		284.124±0.103	4.8029±0.545
494		241.1		381.4		310.369±0.069	0.2648±0.044
503		335.6		626		313.052±0.092	0.1771±0.019
$^{197m}_{82}$Pb		389.9		722		349.550±0.045	1.4781±0.083
84.9 (IT)		422.1 (IT)	90%	758	0.2†	354.598±0.053	0.0767±0.002
222.4		459.8	10%	820.1	240	361.601±0.105	0.1787±0.033
234.4 (IT)		490.4	10%	826.2	100	488.349±0.183	0.1895±0.055
387		547.2 (IT)		847.4		493.701±0.033	0.9765±0.103
$^{197}_{82}$Pb		657.6 (IT)		932		498.840±0.152	0.5467±0.115
385		787.2 (IT)		1034		511.429±0.062	3.2392±0.091
387		961.4 (IT)	90%	1184		549.855±0.014	8.8773±0.236
$^{198}_{82}$Pb		$^{203m}_{82}$Pb		1256		553.723±0.276	0.3873±0.104
30.8		825.2		1510		561.264±0.203	0.1970±0.057
116.9	10†	$^{203}_{82}$Pb		1523		570.581±0.030	13.2248±0.723
122.6		279.190±0.007	99+%	1537		573.867±0.037	1.7325±0.321
173.4	100	401.28±0.3	4.6%	1846 }		576.213±0.144	1.0763±0.073
259.5	24	680.67±0.3	0.8%	1896 }	180	579.781±0.020	16.7838±0.422
290.3	50	$^{204}_{82}$Pb		$^{204}_{83}$Bi		609.782±0.198	0.3556±0.080
365.4		373.4±0.6		78.5		626.720±0.024	1.7814±0.062
382.0		899.10±0.2		80.2		646.065±0.060	0.1963±0.029
389.5		911.66±0.2		90.9		688.595±0.102	0.6343±0.058
397.7		$^{205m}_{82}$Pb		92.2		703.401±0.028	100.0000±0.000
575.0		26.22±0.11	†	100.2		717.352±0.104	0.7737±0.154
649.0		284.12±0.10	4.8	140.9		720.752±0.232	0.3716±0.093
743	8	310.37±0.07	0.26	170.2		723.673±0.118	0.5270±0.093
865.3	20	703.40±0.03	100	175.9		729.676±0.225	0.2556±0.164
$^{199m}_{82}$Pb		1014.22±0.06	3.3	213.4		744.881±0.075	2.3298±0.150
424		$^{207m}_{82}$Pb		216.0		759.120±0.049	3.7326±0.203
$^{199}_{82}$Pb		569.684±0.014	100%	219.1		761.470±0.099	2.1655±0.225
352.8	19%	1063.614±0.04	100%	221.9		772.096±0.319	0.1131±0.024
367.0	90%	$^{210}_{82}$Pb		227.5		780.837±0.081	1.5576±0.270
721.0	10%	46.52	4.1%	240.5		787.764±0.130	0.3587±0.051
Ann. Rad.		$^{211}_{82}$Pb		249.0		795.692±0.075	0.4267±0.128
$^{200}_{82}$Pb		65.5	2†	252.9		800.918±0.141	0.4230±0.064
32.74		84	1	289.3		806.431±0.036	0.5268±0.153
109.54		310	0.7	292.2		813.813±0.046	1.4921±0.196
142.30		340	1	330.5		828.213±0.126	0.9586±0.107
147.61		404.84	100	368.0		852.779±0.255	0.2291±0.043
161.4		426.99	53	374.7		860.103±0.089	1.4043±0.274
235.61		612	3	405.0		871.861±0.089	1.3232±0.132
257.15		702	13	412.2		890.087±0.021	2.0553±0.129
268.42		766.3	19	421.5		894.649±0.027	2.6969±0.215
289.1		831.8	100	438.5		901.776±0.144	0.4171±0.067
289.9		860	0.1	473.4		910.843±0.037	5.2527±0.202
302.85		1020	0.5	502		922.027±0.162	0.2010±0.031
315.30		1076	0.5	522		931.249±0.686	0.1803±0.004
450.5		1104	3.8	532		950.763±0.067	1.2799±0.076
457.6		1188	0.4	542		971.488±0.147	0.9149±0.169
605.3		1265	0.2	622		978.513±0.350	0.1241±0.027
$^{201m}_{82}$Pb		$^{212}_{82}$Pb		661		987.557±0.300	53.3456±1.175
629		115.1	0.7%	663		1001.896±0.039	1.7873±0.191
$^{201}_{82}$Pb		176.7	0.2%	671		1014.218±0.059	3.2914±0.181
Ann. Rad.		238.60±0.007	47%	692		1031.720±0.276	0.1260±0.024
58.9		300.1	3.2%	709		1043.661±0.041	23.8732±1.309
129.8		415.2	0.16%	710		1051.613±0.751	0.1545±0.015
285.0		$^{214}_{82}$Pb		719		1065.952±0.105	0.3447±0.033
309.2		53.24		724		1072.212±0.084	0.9954±0.117
331.2		196		791		1107.539±0.107	0.3714±0.081
345.3		206		899		1189.943±0.027	7.3423±0.348
361.3		241.92	3.7†	912		1199.814±0.210	0.6296±0.160

GAMMA ENERGIES AND INTENSITIES OF RADIONUCLIDES (Continued)

Column group 1

Gamma energies (KeV)	Gamma intensities
1208.569±0.075	1.6523±0.121
1216.294±0.272	0.2327±0.044
1264.422±0.310	0.8413±0.102
1351.471±0.099	3.5151±0.375
1438.606±0.236	0.3400±0.060
1500.944±0.609	0.9486±0.333
1522.121±1.091	0.6432±0.147
1550.807±0.232	2.5417±0.654
1563.112±0.130	0.7286±0.145
1577.393±0.273	0.5087±0.180
1594.466±2.001	0.3268±0.152
1614.343±0.045	6.3607±0.417
1619.059±0.207	0.7599±0.067
1764.274±0.063	93.0470±3.98
1775.795±0.038	11.0945±0.691
1818.270±0.399	0.1727±0.034
1844.759±0.400	0.0663±0.004
1861.691±0.019	17.4466±0.871
1903.346±0.056	6.7117±0.332
1965.948±0.558	0.0484±0.020
2565.467±0.844	0.1624±0.058
2606.776±0.412	0.0681±0.019
$^{206}_{83}$Bi	
107.2	
123.6	
184.0	21†
202.7	
234.4	
262.8	2.9
313.7	3.0
343.5	26
386.1	
398.1	10
407.1	18
516.3	46
537.6	34
620.7	4.0
632.4	1.5
657.3	5.3
740	3.6
754	2.7
803.1	100
816	
842	
880.8	73
895.0	19
1018.9	8
1099	13
1405	
1596	8
1720	36
1845	
1880	
1903	
$^{207}_{83}$Bi	
569.684±0.014	100†
897.3	0.16
1063.610±0.04	76
1430	0.16
1769.7	8
$^{208}_{83}$Bi	
2614.611±0.06	100%
$^{210m}_{83}$Bi	
262	10†
300	30
340	
610	
$^{211}_{83}$Bi	
350.7	15.9%
$^{212}_{83}$Bi	
39.85	
144.9	
288.2	0.28%
328.0	0.11%
434	
453	
473	
493	
616	
727	7.1%
785	1.1%
893	0.42%
953	0.1%

Column group 2

Gamma energies (KeV)	Gamma intensities
1074 } 1079 }	0.6%
1513	0.31%
1620	1.8%
1800 } 1809 }	0.1%
$^{213}_{83}$Bi	
320	
437	
$^{214}_{83}$Bi	
63	
191	
450	1%
609.3	47%
666	2%
703	0.8%
721	0.7%
768.7	5.3%
787	1.2%
806	1.5%
821	
825	
874	0.4%
934.8	3.3%
960	9.5%
1050	0.5%
1120	16%
1155	1.8%
1207	0.6%
1238	6.0%
1281	1.7%
1379	4.8%
1390—1400 complex	4%
1416	
1509	2.4%
1541	0.8%
1583	0.9%
1600	0.6%
1661	12%
1681	0.2%
1728	3.2%
1764	17%
1836	0.3%
1848	2.3%
1877	0.2%
1897	0.3%
2017	0.07%
2090	
2117	1.3%
2162	
2204	6%
2435	2.0%
2700	0.04%
2770	0.04%
2890	0.04%
2990	0.04%
3070	0.04%
$^{206}_{84}$Po	
59.9	
82.9	
107?	
117.6	
140.6	
146?	
170.8	
171.5	
181?	
282.1	
286.5	35†
311.5	
338.4	41†
354.8	
369?	
381?	
453?	
463.5	
461.1	
511.4 } 522.4 }	100†
538?	
555	
580	
646	

Column group 3

Gamma energies (KeV)	Gamma intensities
669	
678	
807.6	57†
818?	
862	
981	
1008	
1033	84†
1320	
$^{207}_{84}$Po	
100	
139.7?	
149.6	
156.1	
158.0	
177.7	
205.2	
214.4?	
222.0	
222.7	
224.0	
249.6	
288.0	
297.1?	
307.5	
330.2	
345.2	
369.3	
380.1?	
390.1?	
402.4	
405.7	
468.5?	
503.3	
531.7	
629.9	
669.5	
688	
698?	
742.7	
771	
892	
911.8	
948	
955	
992.6	
1020	
1149	
1212	
1280	
1318	
1361	
1373	
1586	
1663	
1747	
1763	
1797	
1847	
2061	
2600?	
$^{208}_{84}$Po	
285	≈0.003%
600	≈0.006%
$^{209}_{84}$Po	
260.5 } 262.8 }	≈0.4%
910	≈0.5
$^{210}_{84}$Po	
803	12×10⁻³%
$^{211m}_{84}$Po	
0.90	1.7%
560	≈80
106.0	≈80
$^{211}_{84}$Po	
880	
561	
$^{212}_{84}$Po	
$^{212m}_{84}$Po	
570	≈2%
2610	2.6
$^{214}_{84}$Po	
800	0.014%

Column group 4

Gamma energies (KeV)	Gamma intensities
$^{215}_{84}$Po	
≈443	
$^{206}_{85}$At	
68	
$^{208}_{85}$At	
180	20†
250	
660	80
120	
$^{209}_{85}$At	
90.8	
195.0	24†
545	66
780	100
$^{210}_{85}$At	
46.5	
116.5	0.68†
202.0	0.17
245.3	79
298.9	0.13
316.9	0.14
402.4	0.78
499.1	0.13
506.9	0.63
518.7	0.13
527.6	1.1
544.8	0.09
584.0	0.32
602?	
615.1	0.31
623.1	0.52
630.9	0.30
639.5	0.23
643.7	0.46
701.1	0.43
726	?
817.0	1.7
853.1	1.4
882	0.25
909	?
930.6	0.96
956.7	1.8
977	0.63
1088	0.28
1180	100
1204	?
1289	0.46
1324	0.36
1436	29
1483	48
1552	0.16
1599	14
1648	0.09
1719	0.11
1955	0.38
2002	0.11
2052	0.05
2226	0.05
2239	?
2254	1.4
2273	0.33
2285	?
2306	0.029
2353	0.13
$^{211}_{85}$At	
670	Weak
$^{212m}_{85}$At	
63	
$^{212}_{85}$At	
63	
$^{215}_{85}$At	
≈400	
$^{217}_{85}$At	
268	
455	
595	
$^{211}_{86}$Rn	
γ with EC	
32	
113.9	
168.6	
232	
264	

GAMMA ENERGIES AND INTENSITIES OF RADIONUCLIDES (Continued)

Gamma energies (KeV)	Gamma intensities	Gamma energies (KeV)	Gamma intensities	Gamma energies (KeV)	Gamma intensities	Gamma energies (KeV)	Gamma intensities
296		329.8	0.10	26.3		410	2.5
333		334.3	0.05	²²³₈₉Ac		443	
445	39†	338.7	0.2	82	0.23%	463	3.6
680	100	342.5	0.09	96	0.2	473	
865	24	369.0	0.32	120	?<0.05	484	
946	29	723	0.15	≈170	?≤0.8	510	
1130	31	746.5	0.07	²²⁴₈₉Ac		555	
1370	52	756	0.04	132	100†	573	
1800	0.8	767	0.08	216	224	619	
γ with α		776.7	1.23	²²⁵₈₉Ac		752	
68.7	100†	784	0.07	36.6		769	
169	6.5	793	0.03	38.4		783	1.4
234	6.5	798	0.03	62.8	0.5%	796	4
²¹⁸₈₆Rn		804	0.17	75	1.2	836	2.5
609.4	0.2%	813	0.06	85.5	3	831	8
²¹⁹₈₆Rn		822	0.03	87.5	≤2	912.2	23
130		826	0.14	90	≤0.5	922	
271	9%	835	0.02	96	≤1	966	20
371	≈0.002	841	0.02	100	~3	1035	
380	≈0.0005	847	0.14	109	0.6	1095	
401	6.5	860	0.01	123	0.3	1464	1.1
517	≈0.04	864	0.01	134	~0.1	1503	2.5
536	≈0.0003	876.5	0.13	146	0.2	1593	4.5
562	≈0.002	892	0.01	150	1	1642	2.4
606	≈0.0026	898	0.05	158	0.8	²³¹₈₉Ac	
673	≈0.01	908	0.04	187	0.8	185	
833	≈0.001	²²⁰₈₈Ra		196	0.3	280	
888	≈0.001	465	1%	216	1	390	
1053	≈0.0003	²²¹₈₈Ra		241	0.03	710	
²²⁰₈₆Rn		89	3.5%	254	0.3	²²⁴₉₀Th	
542	≈0.07%	152	13	324	Weak	177	9%
²²²₈₆Rn		176	2	340		235	0.4
510	0.07%	219	0.1	351	0.06	297	0.3
²¹²₈₇Fr		293	0.6	362	0.1	410	0.8
39		320	0.7	373	Weak	²²⁵₉₀Th	
62		415	0.5	384	0.06	151	1†
66		²²²₈₈Ra		453	0.3	246	5
70		324.6	3.9%	482	0.1	322	30
123		325	8×10⁻³	517	0.04	362	5
²¹⁹₈₇Fr		473	7×10⁻³	529	0.06	450	1
163		526	3×10⁻³	²²⁶₈₉Ac		490	1
189		795	3×10⁻³	γ with β⁻		²²⁶₉₀Th	
352		846	4×10⁻³	72.13		111.1	4.8%
493		²²³₈₈Ra		158.5	100†	131.3	0.34
530		31.2		230.4	151	190.6	0.15
²²¹₈₇Fr		≈68?	≈0.4%	γ with EC		206.5	0.26
63	9.1%	122.2	2	67.8		197	0.4
68	0.4	131	≈3	185	100†	242.4	1.2
91?		143.1?		255	132	²²⁷₉₀Th	
98	0.05	144.2	4.1	²²⁷₈₉Ac		29.9	
99.5	0.12	148		γ with β⁻		31.6	
118.0	0.04	154.1	55	9.3		44.5	1.4%
150.0	0.08	158.6	1.7	15.2		45.1	
171.3	0.08	179.6	0.5	24.5		48.3	
218.0	12.5	269.6	10	γ with α		50.2	10.5
268?	<0.04	288	≈0.2	12.7		56.1	<0.2
282.8	~0.01	324	2.3	70.2	12†	61.5	0.8
303	~0.03	330	≈5.1	83 }	≈27	68.8	≈0.1
324.1	0.02	338	2.8	87 }		79.8	≈2.6
359.1	0.04	362	≈0.07	99.5	20	94.0	≈3
381.8	0.04	371	0.45	107	1	100.3	<0.3
409.1	0.15	441	0.2	121.5	4	113.1	≈2
²²³₈₇Fr		446	0.4	134	2	141.6	0.13
49.8	40†	580	≤0.1	147	6	154.2	0.1
50.2	100	²²⁴₈₈Ra		160	14	173.5	<0.1
61.5	<0.8	240.98	3.7%	172	4	184.8	<0.1
79.8	25	290	9×10⁻³	190?	<0.5	205.0	0.32
100.3	≈3	410	4×10⁻³	²²⁸₈₉Ac		206.2	0.5
134.4	1.6	650	6×10⁻³	57.5		210.7	1.0
173.5	0.4	²²⁵₈₈Ra		78.1		212.0	
184.8	0.9	40	33%	99		218.8	0.08
205.0	3.2	²²⁶₈₈Ra		113?		234.8	
234.8	10	185.7	100†	129.1	5†	236.0	12.0
246	0.13	260	0.29	155?		247.7	1.4
250.2	0.13	420	0.021	178.2	6	250.2	0.7
256.2	0.13	450	0.009	184	1.7	250.4	
286.1	0.05	610	0.033	209	4.3	256.26	6.4
289.7	≈0.9	²²⁷₈₈Ra		232		273.0	0.46
300.0	0.17	27.3		270	4.1	279.8	≤0.05
304.5	0.07	291	4%	282	0.3	281.5	0.30
307	0.09	498	0.6	321		286.1	1.8
312.7	0.08	²²⁸₈₈Ra		327.5	5.3	289.7	<0.3
318	≈1.7	10.3		338	15	296.6	0.40

GAMMA ENERGIES AND INTENSITIES OF RADIONUCLIDES (Continued)

Gamma energies (KeV)	Gamma intensities
300.0	2.5
304.5	1.2
312.7	0.83
318	0.2
323	≤0.1
329.8	2.8
334.3	1.2
342.5	0.46
351?	≤0.2
371	0.03
384	0.06
404?	≤0.04
671	0.002
760	0.002
776	0.0015
$^{228}_{90}$Th	
84.5	1.6%
132	0.19
167	0.12
205	0.03
216	0.29
234?	7×10⁻⁵
$^{229}_{90}$Th	
17.3	
23.7	
25.3	
32	
42.7	
56.7	
68.9	
75.1	
86.3	
107.2	
124.1	
131.9	
137.0	
143.0	
144.2	
156.5	
179.9	
193.4	
210.7	
217.0	
242.2	
269?	
$^{230}_{90}$Th	
67.8	0.59%
110	1×10⁻⁴%
142	0.07%
184	1.4×10⁻²%
206	≈5×10⁻⁶%
235	≈5×10⁻⁶%
253	1.7×10⁻²%
257	≈8×10⁻⁴%
257	≈8×10⁻⁴%
$^{231}_{90}$Th	
17.21	
18.07	
25.6	
42.8	0.7
58.54	6
63.79	
68.50	
72.7	3.4
76.04	
81.22	13
82.07	7
84.20	100
89.94	15
99.27	2.1
102.2	7
106.8	0.34
112.6	0.29
115.4?	0.03
116.8	0.38
124.9	1.0
134.0	0.42
135.66	1.3
145.90	0.5
163.10	2.4
164.7	0.13
174.1	0.29
183.4	0.5

Gamma energies (KeV)	Gamma intensities
188.7	0.05
218.0	0.6
236.1	0.13
249.8	0.017
267.8	0.02
$^{232}_{90}$Th	
59	
$^{233}_{90}$Th	
29.35	690†
57.15	23
74.7	22
86.00	1150
131.1	24
148.2	6
151.5	4
153.6	28
162.5	96
169.1	190
179.0	16
186.8	14
190.54	55
195.0	70
201.6	13
210.1	15
211.3	8
216.6	6.3
226.1	9.6
250.5	2
252.9	5
257.3	29
278.7	3.3
285.5	9
346.6	5
359.9	50
361.4	16
368.0	2
377 ?	16
399.0	6
408.8	1.6
412.5	5.7
418.4	5
431.0	9.8
433.0	6
441.0	103
447.7	62
459.2	600
467.5	7.8
473.9	1.5
490.8	74
497.1	9
499.0	91
505.5	2.1
513.6	8.6
517.0	2.9
526.5	27
531.8	1.8
552.1	10
554.9	1.5
562.7	30
573.5	18
595.2	69
599.1	20
609.9	36
642.3	12
663.4	1
669.7	290
677.8	37
681.2	8
698.4	5
703.6	4.5
707.8	5
716.8	24
724.9	37
740.9	13
744.9	2.9
751.6	1
757.6	18
764.3	49
774.0	5.9
782.7	2.6
784.2	2.1
804.8	13
806.4	5.7

Gamma energies (KeV)	Gamma intensities
811.4	3.3
815.9	12
817.0	6.7
846.8	0.6
849.3	2
871.0	0.9
873.8	3.5
880.7	3.3
890.6	61
898.4	1.4
935.9	21
942.0	3.3
948.0	3.2
960.8	2.9
962.8	0.6
969.1	4.9
978.1	3.2
984.8	6.2
995	0.4
1001	0.5
1008	1.2
1012	1.7
1146	1.2
$^{234}_{90}$Th	
29.9	
63.2}	85†
63.6}	
69.8	
93.1}	100
93.5}	
$^{227}_{91}$Pa	
50	<20†
65	62
67	12
110	20
$^{228}_{91}$Pa	
γ with α	
130	27†
150	34
170	11
200	14
220	10
240	55
280	49
310	100
345	21
γ with EC	
57.5	
77.7	
99	13†
129.1	
137.8	
156	
178.0	
184	
191.3?	55
209	320
224	122
270	320
278	71
282	310
327.5	950
338	740
341.1	250
397	
410	1350
445.4	
462	320
463	2900
469.2	
617	
622	
641	
662	
666	
670	
680	
694	
704	
713	135
732	
739.6	
745.6	

Gamma energies (KeV)	Gamma intensities
756.2	230
773.4	375
782.6	
792.2	≈70
796	≤600
817.4	
831.4	375
835.8	
836.4	730
841.0	170
853.8	
871.0	520
889.0	
905.2	310
912.2	3100
923	930
966.0	2200
970.0	1680
976.8	620
1034.1	
1095	
1123	
1168	
1253	130
1293	83
1423	
1464	155
1489	
1503	
1563	15
1593	620
1624	
1678	
1708	
1744	230
1758	
1830	
1888	415
$^{232}_{91}$Pa	
γ with EC	
42.37	
γ with α	
26?	
40	
51	
52	
64	
65	
71	
$^{230}_{91}$Pa	
α with Ec	
53	
122	
230	
280	
320	
340	
379	
380	
398	135†
401	100
444	
456	500
464	
508	420
520	
572	100
624	
634	
730	210
783	160
901}	1600
920}	
954	3500
1012	
1027	240
α with β⁻	
51.7	
$^{231}_{91}$Pa	
18.9	
25.3	
27.3	
29.9	

GAMMA ENERGIES AND INTENSITIES OF RADIONUCLIDES (Continued)

Gamma energies (KeV)	Gamma intensities
38.1	
44.0?	
52.6	
57.1	
63.5	
74.0	
77.2	
96.7	
100.7	
102.6	
244	
257	
259.8	
274	
283.1	
299.4	
302.0	
313	
329.2	
342	
356.3	
409	
437	
439	
488	
513	
517	

$^{232}_{91}$Pa

Gamma energies (KeV)	Gamma intensities
47.6	
80.2	0.1%
81.2?	0.02
105.4	2.1
109.0	3.0
132.5	0.02
139.2	0.7
150.1	12
174.9	0.025
178	0.020
183.9	1.6
290	0.03
388.0	
422.0	
454.2	
472.8	4
516	3
564	
584	
645	
676	
687	
693	
712	
757	
820	
865	
868	
895	
971	

$^{233}_{91}$Pa

Gamma energies (KeV)	Gamma intensities
17.26	
28.54	
40.35	0.015%
41.65	
57.9	
75.28	0.8
86.59	1.7
95.8	
103.86	0.7
145.4	0.44
271.6	0.29
300.2	6.3
303.1	
311.9	34
320	
340.5	3.9
375.4	0.56
398.6	1.1
415.9	1.5

$^{234m}_{91}$Pa

Gamma energies (KeV)	Gamma intensities
43.5	
236	
255	0.05%

Gamma energies (KeV)	Gamma intensities
746	Weak
765	0.36
790	Weak
811	
1001	0.59

$^{234}_{91}$Pa

Gamma energies (KeV)	Gamma intensities
43.3	
99.8	
126.3	≈24†
153.0	9
186.0	≈3
196.7	4
208	?≤16
223.9	≈2
226.7?	
227.3	12
287	≈10
294.3?	
323	?≈3
355	?≈6
370.1	5.3
565	15
694	
727	
791	≤5
804	
822	
873	
875	
878	
920	70
922	
941	
976	
1020	≈8
1130	?3
1340	?2
1410	?6
1620	?3
1850	?1

$^{235}_{91}$Pa

No γ

$^{237}_{91}$Pa

Gamma energies (KeV)	Gamma intensities
90	50†
145	45
205	55
275	20
330	40
405	30
460	100
550	30
590	25
750	50
800	45
870	100
920	100
1040	35
1320	10
1420	15

$^{228}_{92}$U

Gamma energies (KeV)	Gamma intensities
152	20†
187	30
246	40

$^{230}_{92}$U

Gamma energies (KeV)	Gamma intensities
72.13	0.54%
154.4	0.21
158.8	0.13
230.6	0.18

$^{231}_{92}$U

Gamma energies (KeV)	Gamma intensities
18.05	
25.64	12%
58.54	
68.5	
81.3	
82.1	
84.18	7
108.2	
220	1

$^{232}_{92}$U

Gamma energies (KeV)	Gamma intensities
57.6	0.21%
129.0	0.082
270.5	3.8×10⁻³

Gamma energies (KeV)	Gamma intensities
327.8	3.4×10⁻³
500	2.2×10⁻⁵

$^{233}_{92}$U

Gamma energies (KeV)	Gamma intensities
29.0	60†
42.4	310
54.6	68
66.0	3.9
71.8	12
97.1	100
117.1	13
118.9	16
120.7	11
124.0	2.7
135.3	10
144.8	9
146.4	26
164.5	27
187.9	8
208.1	9
217.0	16
245.4	14
248.4	6
260	1.2
261.5	1.4
268.1	1.0
274.5	2.0
277.9	6
288.0	4.9
291.2	23
316.8	32
320.2	11
323.1	3.2
328.2	0.1
336.4	2.3
353.9	0.23
365.5	3.2
383.0	0.4
393.3	0.07

$^{234}_{92}$U

Gamma energies (KeV)	Gamma intensities
53.3	100†
120.9	34
460	
510	
580	

$^{235}_{92}$U

Gamma energies (KeV)	Gamma intensities
110	2.5%
143	11
163	5
180	0.5
185	54
201	0.8
204	5

$^{235m}_{92}$U

Gamma energies (KeV)	Gamma intensities
0.075	

$^{236}_{92}$U

Gamma energies (KeV)	Gamma intensities
≈50	

$^{237}_{92}$U

Gamma energies (KeV)	Gamma intensities
13.81	
26.35	2.4%
33.20	
38.54	
43.42	
51.01	0.2
59.54	36
64.83	1.3
114.09	
164.61	2.0
208.00	23
221.80	0.022
234.40	0.021
267.54	0.76
292.7	0.003
332.36	1.3
335.38	0.10
337.7	0.008
368.59	0.050
370.94	0.12

$^{238}_{92}$U

Gamma energies (KeV)	Gamma intensities
48	

$^{239}_{92}$U

Gamma energies (KeV)	Gamma intensities
43.520±0.042	104.553%

Gamma energies (KeV)	Gamma intensities
74.660±0.022	1106.989%
83.822±0.307	3.094%
86.716±0.065	1.392%
110.973±0.222	0.442%
117.658±0.030	3.204%
174.009±0.028	0.398%
186.149±0.020	0.729%
187.396±0.036	0.148%
190.345±0.067	0.067%
191.973±0.060	0.062%
196.850±0.100	0.049%
201.186±0.065	0.045%
210.960±0.091	0.072%
220.531±0.017	0.575%
228.141±0.016	
231.70±0.10	0.069%
255.27±0.123	0.063%
258.438±0.046	0.058%
260.792±0.112	0.051%
296.963±0.070	0.035%
301.862±0.074	0.036%
304.123±0.117	0.033%
306.836±0.042	0.103%
312.049±0.026	0.090%
317.182±0.107	0.076%
321.901±0.132	0.027%
326.227±0.036	0.080%
334.255±0.018*	
345.060±0.038	0.045%
357.858±0.036	0.106%
361.814±0.063	0.072%
363.064±0.217	0.020%
373.524±0.016	0.426%
378.055±0.028	0.189%
395.353±0.127	0.022%
399.3±0.2	0.011%
407.645±0.051	0.057%
434.751±0.045	0.075%
445.732±0.085	0.039%
448.141±0.024	0.177%
452.054±0.089	0.039%
462.435±0.124	0.036%
474.487±0.109	0.067%
486.865±0.016	1.18%
492.594±0.045	0.086%
499.091±0.177	0.033%
504.707±0.053	0.089%
506.462±0.109	0.039%
517.851±0.084	0.065%
522.080±0.076	0.053%
532.887±0.116	0.040%
541.221±0.078	0.054%
544.608±0.059	0.089%
548.099±0.126	0.044%
560.511±0.047	0.131%
575.295±0.026	0.269%
577.640±0.222	0.028%
587.726±0.038	0.486%
589.709±0.116	0.137%
602.659±0.076	0.101%
624.016±0.045	0.168%
631.080±0.017	1.518%
662.238±0.016	3.956%
664.178±0.093	0.203%
691.091±0.044	0.127%
695.246±0.083	0.084%
697.394±0.232	0.031%
700.748±0.079	0.062%
703.239±0.077	0.065%
707.379±0.084	0.072%
710.171±0.208	0.029%
714.155±0.073	0.092%
722.865±0.025	0.623%
727.345±0.155	0.051%
730.972±0.026	0.254%
748.108±0.016	2.234%
767.643±0.100	0.097%
772.809±0.175	0.075%
774.742±0.034	0.372%
779.78±0.200	0.010%
788.187±0.056	0.105%
791.322±0.031	0.204%

GAMMA ENERGIES AND INTENSITIES OF RADIONUCLIDES (Continued)

Gamma energies (KeV)	Gamma intensities
793.576±0.084	0.058%
812.933±0.017	1.732%
819.232±0.018	3.176%
831.834±0.075	0.079%
840.806±0.250	0.080%
844.091±0.015	3.515%
846.466±0.060	0.880%
866.89±0.300	0.040%
868.06±0.300	0.040%
869.88±0.300	0.040%
874.215±0.070	0.093%
876.029±0.140	0.048%
881.353±0.169	0.027%
884.497±0.031	0.264%
889.506±0.036	0.525%
895.412±0.130	0.035%
913.569±0.076	0.057%
917.30±0.30	0.070%
918.80±0.40	0.080%
920.70±0.40	0.070%
921.80±0.40	0.050%
922.80±0.50	0.030%
928.154±0.043	0.126%
933.057±0.037	0.791%
939.241±0.284	0.014%
943.145±0.180	0.034%
948.713±0.173	0.027%
953.677±0.190	0.030%
959.322±0.305	0.167%
961.235±0.167	0.403%
964.351±0.037	2.156%
974.609±0.079	0.082%
992.169±0.066	0.076%
1000.996±0.300	0.035%
1002.392±0.200	0.035%
1009.744±0.114	0.044%
1012.659±0.117	0.040%
1032.238±0.084	0.076%
1040.567±0.199	0.021%
1065.70±0.300	0.010%
1097.070±0.102	0.052%
1122.40±0.300	0.015%
1144.80±0.300	0.015%

$^{240}_{92}$U

Gamma energies (KeV)	Gamma intensities
440	

$^{234}_{93}$Np

Gamma energies (KeV)	Gamma intensities
43.49	
99.7	
233.6	
234.6	
238.6?	
247.9	
266?	
298?	
450	
483?	
485?	
516	
526	
557	
744	
752	
768	
788	
793	
810	
812	
854	
1003	
1105	
1196	
1240	
1395	
1439	
1531	
1562	
1575	
1606	

$^{235}_{93}$Np

Gamma energies (KeV)	Gamma intensities
26	15†
84	9

$^{236}_{93}$Np

Gamma energies (KeV)	Gamma intensities
γ with EC	
45.32	
643	
688	
γ with β⁻	
44.6	

$^{237}_{93}$Np

Gamma energies (KeV)	Gamma intensities
20?	
29.6	14†
55.0?	
56.8	
84.5?	14
85.9	
105.0	
108.4	
133.5	
144.5	0.8
160?	
207	
240	

$^{238}_{93}$Np

Gamma energies (KeV)	Gamma intensities
44.1	
101.7	
119.8	
221.0	
292.6?	
871	
885	
925	
940	
943	
986	83†
989	
1027 }	
1030 }	100
1034	
1005?	

$^{239}_{93}$Np

Gamma energies (KeV)	Gamma intensities
44.65	
49.41	
57.26	
61.48	
67.86	
88.06	
106.13	22.8†
106.47	
166.4	
181.7	0.077
209.726±0.02	3.4
226.4	
228.2±0.1	11.8
254.2	0.12
272.809±0.09	0.089
277.56±0.015	14.1
285.37±0.04	0.82
315.83±0.04	1.5
334.25±0.02	2.1
392.4	0.0016
429.5	0.0037
434.7	0.012
447.6	26×10⁻⁴
454.2	8.0×10⁻⁴
461.9	0.0016
469.8	0.0011
484.3	0.0010
492.3	0.0060
497.8	0.0031
498.7	
504.2	7.7×10⁻⁴

$^{240m}_{93}$Np

Gamma energies (KeV)	Gamma intensities
42.9	
260	1.9%
304	0.9
557	21
599	13
760	1.3
792?	
816	1.6
820	0.3
844?	

Gamma energies (KeV)	Gamma intensities
860	
863	
898	1.2
940	0.3
945	1.9
1092?	
1490	1.5
1530	1.9
1620	0.7

$^{240}_{93}$Np

Gamma energies (KeV)	Gamma intensities
85	
160	
245	
440	
560	
600	
920	
1000	
1160	

$^{234}_{94}$Pu

Gamma energies (KeV)	Gamma intensities
0.047	

$^{236}_{94}$Pu

Gamma energies (KeV)	Gamma intensities
47	0.031%
110	0.012
165	6.6×10⁻⁴
452	1.7×10⁻⁴
470	1.0×10⁻⁴
640	24×10⁻⁴

$^{237m}_{94}$Pu

Gamma energies (KeV)	Gamma intensities
145	100%

$^{237}_{94}$Pu

Gamma energies (KeV)	Gamma intensities
33.2	
43.5	
55.6?	
59.6	
76.4?	
96.0?	

$^{238}_{94}$Pu

Gamma energies (KeV)	Gamma intensities
43.49	0.038%
99.8	8×10⁻³
153.1	1×10⁻³
203	4×10⁻⁶
743	9×10⁻⁶
767	3.5×10⁻⁵
786	6×10⁻⁶
810	<1×10⁻⁶

$^{232}_{94}$Pu

Gamma energies (KeV)	Gamma intensities
38.70±0.03	(1.63±0.16)^b
41.99±0.10	(2.8±0.4)^d
46.19±0.07	(2.35±0.30)^c
51.61±0.03	(1.35±0.14)^a
54.01±0.05	(1.13±0.20)^c
56.82±0.04	(7.3±0.8)^c
59.57	^{241}Am
65.73±0.08	(2.7±0.5)^d
67.70±0.05	(1.4±0.3)^c
68.72±0.06	(4.2±0.6)^c
72.79	Pb K_{o2}
74.96	Pb K_{o1}
77.60±0.05	(4.1±0.6)^d
78.48±0.05	(1.7±0.2)^c
84.30	Pb K_{β1}
84.82	Pb K_{β2}
87.34	Pb K_{β2}
89.94±0.06	(1.1±0.2)^d
94.64	U K_{o2}
98.46	U K_{o1}
99.46	Pu K_{o2}
102.93±0.08	(2.6±0.4)^c
103.64	Pu K_{o1}
110.38	U K_{β3}
111.29	U K_{β1}
114.55	U K_{β2}
115.32±0.05	(7.4±0.9)^c
116.18±0.05	(7.1±0.8)^c
117.15	Pu K_{β1}
119.89±0.08	(2.8±0.4)^d
120.59	Pu K_{β2}
122.35±0.12	(3.0±1.5)^c
123.73±0.08	(1.9±0.3)^d
124.41±0.12	(4.8±0.8)^d

Gamma energies (KeV)	Gamma intensities
125.00±0.12	(4.8±0.8)^d
129.27±0.03	(5.6±0.7)^b
141.66±0.06	(9.8±0.9)^c
144.18±0.08	(2.7±0.4)^d
146.07±0.08	(9.8±1.0)^c
159.6±0.2	(2.0±1.9)^e
160.07±0.13	(5.0±2.0)^e
161.47±0.05	(1.0±0.2)^c
167.95±0.18	(2.0±0.1)^c
171.37±0.08	(9.8±0.9)^d
173.61±0.10	(2.9±0.4)^e
179.20±0.08	(2.9±0.3)^c
184.53±0.15	(2.0±0.1)^c
188.13±0.10	(4.2±0.6)^c
189.34±0.07	(7.4±0.9)^c
193.13±0.12	(5.9±0.8)^c
195.68±0.07	(8.9±1.0)^c
203.55±0.04	(4.8±0.4)^c
209.34±0.08	(1.1±0.3)^c
225.46±0.08	(1.1±0.2)^c
237.6±0.2	
238.2±0.2	(1.4±0.2)^d
242.08±0.08	(9.3±0.9)^c
243.38±0.07	(2.0±0.5)^d
244.90±0.09	(4.9±0.8)^c
248.82±0.08	(7.5±0.9)^c
255.38±0.07	(6.6±0.8)^d
258.20±0.10	(3.8±0.6)^c
263.92±0.07	(1.9±0.3)^c
265.68±0.12	(3.0±0.9)^c
297.48±0.07	(4.1±0.6)^d
302.84±0.10	(7.2±0.9)^c
307.71±0.10	(3.9±0.8)^c
311.74±0.06	(2.0±0.2)^c
316.44±0.08	(9.9±1.2)^c
320.83±0.06	(4.6±0.5)^d
323.81±0.06	(4.4±0.5)^d
332.80±0.06	(7.1±0.8)^d
336.14±0.05	(9.4±1.1)^d
341.51±0.06	(6.0±0.8)^c
345.02±0.03	(5.0±0.5)^c
353.60±0.10	(2.7±0.6)^c
361.90±0.08	(1.1±0.2)^d
367.08±0.08	(8.2±0.9)^d
368.57±0.08	(8.1±0.9)^c
371.67±0.10	(3.9±0.6)^c
375.04±0.03	(1.5±0.1)^c
380.19±0.05	(3.0±0.3)^c
382.77±0.05	(2.5±0.3)^c
392.7±0.2	(2.7±0.5)^c
393.1±0.2	(2.7±0.5)^c
399.54±0.09	(9.0±1.5)^c
410.92±0.06	(1.4±0.3)^d
413.71±0.03	(1.5±0.1)^b
422.62±0.05	(1.2±0.2)^c
426.79±0.08	(3.3±0.5)^d
430.12±0.12	(4.8±0.8)^c
445.78±0.12	(7.5±1.1)^c
451.49±0.05	(1.8±0.2)^c
457.57±0.10	(1.2±0.4)^c
461.09±0.10	(3.0±0.6)^c
481.58±0.10	(5.1±1.0)^c
558.31±0.12	(2.4±0.5)^c
617.45±0.10	(2.1±0.4)^c
618.6±0.2	(2.1±0.4)^c
633.3±0.2	(1.2±0.3)^c
640.05±0.08	(5.7±0.9)^c
645.80±0.06	(1.3±0.3)^d
651.98±0.10	(5.1±1.0)^c
655.20±0.20	(1.5±0.3)^c
658.82±0.10	(9.6±1.2)^c
703.52±0.10	(4.8±0.9)^c
717.59±0.15	(1.8±0.4)^c
756.10±0.20	(3.9±0.7)^c
769.12±0.15	(9.6±1.2)^c

$^{240}_{94}$Pu

Gamma energies (KeV)	Gamma intensities
45.28	
103.6	
160	
642.3	1.4×10⁻⁵%
687.8	4×10⁻⁶

Gamma energies (KeV)	Gamma intensities
$^{241}_{94}$Pu	
76.8	0.7%
103.5	4.4
148.5	9.0
160.0	0.3
$^{243}_{94}$Pu	
12.2?	
29.7?	
36.7	
42.2	1%
54	
84	21
96	
134?	
381	0.7
$^{246}_{94}$Pu	
27	13%
44	≈30
80·	3
100?	
180.4	10
224.0	25
$^{238}_{95}$Am	
360	
580	30†
950	
980	80
1350	
$^{239}_{95}$Am	
γ with α	
48	
γ with EC	
44.70	Weak
49.47	
57.31	
67.91	
181.8	
209.9	
226.5	
228.3	
277.6	
$^{240}_{95}$Am	
42.87	
98.9	
900	23%
1000	77
1400	<10
$^{241}_{95}$Am	
26.36	2.5%
33.21	0.17
43.44	0.07
55.56	
59.57	36
67.5	
99.0	0.024
102.97	0.019
106?	<0.001
114.2	4.3×10^{-4}
118.2	4×10^{-4}
123.2	1.8×10^{-3}
125.3	4.5×10^{-3}
146.6	1.6×10^{-4}
150.1	3.5×10^{-3}
154.4	8×10^{-6}
156.4	1.2×10^{-5}
158.3	9×10^{-6}
161.7	9×10^{-6}
164.6	2.5×10^{-5}
166.2	5×10^{-6}
169.6	8.5×10^{-5}
173.1	5×10^{-6}
175.0	7×10^{-6}
191.9	3×10^{-5}
197.0	1.3×10^{-5}
208.00	5.6×10^{-4}
221.4	3×10^{-5}
242.4	3×10^{-5}
245.0	3×10^{-5}
248.0	5.5×10^{-5}
267.4	3.5×10^{-5}
292.8	2×10^{-5}
294.9	5.5×10^{-5}
310.3	1.5×10^{-5}

Gamma energies (KeV)	Gamma intensities
312.1	8×10^{-5}
322.5	1.9×10^{-4}
332.4	1.4×10^{-4}
335.4	5.7×10^{-4}
343.3?	1.1×10^{-4}
351.8	9×10^{-5}
368.6	2.8×10^{-4}
371.0	6×10^{-5}
376.8	1.4×10^{-4}
383.6	5.5×10^{-3}
391.0	4×10^{-5}
419.4	6×10^{-5}
426.8	3×10^{-5}
445.1	4.5×10^{-5}
454?	$<1.7\times10^{-5}$
462	1.4×10^{-5}
511.6	8.5×10^{-5}
525.7	8×10^{-5}
570	1.0×10^{-5}
591.5?	7×10^{-5}
597	1.0×10^{-5}
609.8	2.3×10^{-4}
619.6	8×10^{-5}
641	1.0×10^{-5}
653.1	9×10^{-5}
663.0	5.3×10^{-5}
676?	$<2\times10^{-6}$
680	6×10^{-6}
688.7	4×10^{-5}
695	1.0×10^{-5}
710	9×10^{-6}
723.2	4.8×10^{-4}
737	1.2×10^{-5}
757.3	1.0×10^{-4}
767.0	8×10^{-6}
770.7	1×10^{-5}
$^{242m}_{95}$Am	
γ with IT	
48.6	
γ with α	
42	
49.3	41
66.8	4.6
67.9	1.6
69	
73.3?	1.3
86.7	8
92.5	0.9
109.6	5.3
111.1	0.6
121.8	1.3
135.5	2.3
136.1	2.1
152.9	0.3
153.9	1.0
163.4	5.2
194.9	
206.4	
$^{242}_{95}$Am	
γ with β⁻	
42.12	
γ with EC	
44.50	
$^{243}_{95}$Am	
31.42	0.03%
43.53	5.3
50.6	0.0027
55.4	0.0094
74.673	61
86.7	0.37
117.8	0.75
142.0	0.13
169	0.0012
195	0.00085
500?	
570?	
620	0.0003
650	0.0010
$^{244m}_{95}$Am	
42.9	
980?	<1%
1050	<1

Gamma energies (KeV)	Gamma intensities
$^{244}_{95}$Am	
42.9	
99.4	5%
154	19
206	
540	≈0.4
746	66
900	25
$^{245}_{95}$Am	
36	
56?	
78?	
140	
153	1.2%
240	20
252.3	0.55
296	
$^{246}_{95}$Am	
171	
228	
238	0.3%
244.6	1.5
245.4	1.5
256	0.2
262	0.14
270.3	1.0
288	0.1
402	0.3
734	1.2
752	1
759	0.8
799.0	28.6
834	2.2
987	1.1
1037	14.4
1063	19.3
1079	31.5
1086	1.7
$^{239}_{96}$Cm	
188	
$^{241}_{96}$Cm	
γ with α	
145	
γ with EC	
470	
600	
$^{242}_{96}$Cm	
44.11	0.039%
101.9	0.0035
157.7	0.0023
210	$\approx2\times10^{-5}$
562	1.8×10^{-4}
605	1.4×10^{-4}
≈900	2×10^{-5}
941	3×10^{-5}
$^{243}_{96}$Cm	
106	
210	~5%
228	~6.5
227	11
$^{244}_{96}$Cm	
42.88	0.021%
100	1.5×10^{-3}
150	1.3×10^{-3}
264	1.2×10^{-4}
~570 }	
~610 }	2.5×10^{-4}
~830	7×10^{-5}
$^{245}_{96}$Cm	
≈130	5%
173	14
$^{243}_{97}$Bk	
γ with EC	
755	10†
840	30
946	8
γ with α	
42	4†
146.4	8
187.1	41
558	13
$^{244}_{97}$Bk	
γ with EC	

Gamma energies (KeV)	Gamma intensities
144.5	7†
154.0	3.6
177.0	5.0
187.6	16
217.6	100
233.8	2.9
333.5	10
490	14
745	6
870	5
892	88
922	17
988	4.0
1153	7
1178	4.0
1233	2.8
1252	2.2
1333	0.9
1505	2.0
$^{245}_{97}$Bk	
γ with EC	
252.7	31%
380.5	3.5
384.6	1.5
165.5	
207.4	
474	
$^{246}_{97}$Bk	
734	4.6†
800	100
834	8
986	0.5
1037	2.8
1063	5
1079	5
1082	10
1124	8
$^{247}_{97}$Bk	
84	40%
≈270	30
$^{248}_{97}$Bk	
No γ	
$^{249}_{97}$Bk	
307	
327.2	2.2†
	12
$^{250}_{97}$Bk	
42.2	
98.2	
990	47%
1032	39
$^{246}_{98}$Cf	
42	0.014%
96	0.012
146	0.0035
$^{247}_{98}$Cf	
295	20†
417	13
460	9
$^{249}_{98}$Cf	
43.0	0.03%
55.0	0.20
66.7	0.04
92.6	0.4
241.0	0.3
252.8	3.0
266.7	0.7
295.9	0.17
333.4	16
388.4	72
$^{250}_{98}$Cf	
42.9	17%
$^{251}_{98}$Cf	
177	
224	
$^{252}_{98}$Cf	
43.3	0.014%
100.2	0.01
≈160	0.002
$^{252}_{99}$Es	
74	0.3%
154	0.07
198	0.08

Gamma energies (KeV)	Gamma intensities	Gamma energies (KeV)	Gamma intensities	Gamma energies (KeV)	Gamma intensities	Gamma energies (KeV)	Gamma intensities
228	0.23	145.4		~990	2.0	$^{254}_{100}$Fm	
278	0.21	368.8	0.7†	$^{254}_{99}$Es		41	0.2%
~400	1.1	381.2	10.4			98	0.028
~520	0.15	386.1		34.4		151	0.001
~570	0.14	387.2	36	35.5		$^{255}_{100}$Fm	
$^{253}_{99}$Es		389.18	55	42.6		58.6	0.9%
		428.94	11	63	2.0%	60.0	0.19
8.80		433.2	5	70.4		72.9	0.034
30.83		442.2		80.8		80.5 }	1.34
41.79		448.2	1.5	85.1		81.2 }	
42.98		475.0		150	0.02	85.7	0.009
51.96		500.2		233?	0.008	98.5	0.007
55.11		≈750		249	0.025	~120	0.03
62.09		≈900		264	0.05	178	0.007
66.85				278	0.03	~340	0.003
73.43		$^{254m}_{99}$Es		285	0.01	~370	0.005
73.82		44		304	0.07	~440	0.003
78.6		104	0.6†	316	0.15	~490	0.003
93.75		544	3	342	0.009	$^{257}_{100}$Fm	
98.09		583.3	9	348	0.007	62	
114.06		648.1	100	377	0.015	78	
121.98		688.2	~40	385	0.05	103	1%
135.5		693.1	~80			180	8
						242	10

^{a} = × 10⁻²
^{b} = × 10⁻³
^{c} = × 10⁻⁴
^{d} = × 10⁻⁵
^{e} = 10⁻⁶

PERMISSIBLE QUARTERLY INTAKES OF RADIONUCLIDES

Karl Z. Morgan

The following table lists the permissible quarterly intakes (oral and inhalation) or radionuclides and critical body organs recommended by NCPR for occupational exposure.

PERMISSIBLE QUARTERLY INTAKES (ORAL AND INHALATION) OF RADIONUCLIDES AND CRITICAL BODY ORGANS RECOMMENDED BY NCRP FOR OCCUPATIONAL EXPOSURE (1)

Radionuclide	Radioactive half-life days	Solubility state	Oral intake		Intake by inhalation	
			Critical organ	Permissible quarterly intake microcuries	Critical organ	Permissible quarterly intake microcuries
3-H	4.50E 03	Sol.	Body tissue	6.4E 03(2)	Body tissue	3.1E 03(2)
7-BE	5.36E 01	Sol.	GI LLI	3.6E 03	Tot. body	3.5E 03
		Insol.	GI LLI	3.6E 03	Lungs	7.5E 02
14-C	2.00E 06	Sol.	Fat	1.6E 03	Fat	2.2E 03
18-F	7.80E-02	Sol.	GI SI	1.7E 03	GI SI	3.3E 03
		Insol.	GI ULI	1.0E 03	GI ULI	1.6E 03
22-NA	9.50E 02	Sol.	Tot. body	8.0E 01	Tot. body	1.1E 02
		Insol.	GI LLI	6.0E 01	Lungs	5.3E 00
24-NA	6.30E-01	Sol.	GI SI	3.8E 02	GI SI	7.6E 02
		Insol.	GI LLI	5.6E 01	GI LLI	8.9E 01
31-SI	1.10E-01	Sol.	GI S	1.8E 03	GI S	3.5E 03
		Insol.	GI ULI	3.8E 02	GI ULI	6.2E 02
32-P	1.43E 01	Sol.	Bone	3.8E 01	Bone	4.4E 01
		Insol.	GI LLI	4.6E 01	Lungs	4.9E 01
35-S	8.71E 01	Sol.	Testes	1.3E 02	Testes	1.7E 02
		Insol	GI LLI	5.4E 02	Lungs	1.6E 02
36-CL	1.20E 08	Sol.	Tot. body	1.6E 02	Tot. body	2.2E 02
		Insol.	GI LLI	1.2E 02	Lungs	1.4E 01
38-CL	2.60E-02	Sol.	GI S	8.0E 02	GI S	1.6E 03
		Insol.	GI S	8.0E 02	GI S	1.3E 03
42-K	5.20E-01	Sol.	GI S	6.2E 02	GI-S	1.2E 03
		Insol.	GI LLI	4.0E 01	GI LLI	6.7E 01
45-CA	1.64E 02	Sol.	Bone	1.8E 01	Bone	2.0E 01
		Insol.	GI LLI	3.6E 02	Lungs	7.5E 01
47-CA	4.90E 00	Sol.	Bone	1.0E 02	Bone	1.1E 02
		Insol.	GI LLI	6.6E 01	GI LLI	1.1E 02
					Lungs	1.2E 02
46-SC	8.50E 01	Sol.	GI LLI	7.6E 01	Liver	1.5E 02
					GI LLI	1.5E 02
		Insol.	GI LLI	7.6E 01	Lungs	1.4E 01
47-SC	3.43E 00	Sol.	GI LLI	1.8E 02	GI LLI	3.6E 02
		Insol.	GI LLI	1.8E 02	GI LLI	2.9E 02
48-SC	1.83E 00	Sol.	GI LLI	5.4E 01	GI LLI	1.1E 02
		Insol.	GI LLI	5.4E 01	GI LLI	8.7E 01
48-V	1.61E 01	Sol.	GI LLI	5.8E 01	GI LLI	1.2E 02
		Insol.	GI LLI	5.8E 01	Lungs	3.5E 01
51-CR	2.78E 01	Sol.	GI LLI	3.2E 03	GI LLI	6.4E 03
					Tot. body	6.7E 03
		Insol.	GI LLI	3.0E 03	Lungs	1.4E 03

PERMISSIBLE QUARTERLY INTAKES (ORAL AND INHALATION) OF RADIONUCLIDES AND
CRITICAL BODY ORGANS RECOMMENDED BY NCRP FOR OCCUPATIONAL EXPOSURE (1)
(Continued)

Radionuclide	Radioactive half-life days	Solubility state	Oral intake		Intake by inhalation	
			Critical organ	Permissible quarterly intake microcuries	Critical organ	Permissible quarterly intake microcuries
52-MN	5.55E 00	Sol.	GI LLI	6.6E 01	GI LLI	1.3E 02
		Insol.	GI LLI	6.0E 01	Lungs	8.7E 01
					GI LLI	9.5E 01
54-MN	3.00E 02	Sol.	GI LLI	2.6E 02	Liver	2.4E 02
		Insol.	GI LLI	2.4E 02	Lungs	2.2E 01
56-MN	1.10E-01	Sol.	GI LLI	2.4E 02	GI LLI	4.7E 02
		Insol.	GI LLI	2.0E 02	GI LLI	3.3E 02
55-FE	1.10E 03	Sol.	Spleen	1.6E 03	Spleen	5.3E 02
		Insol.	GI LLI	4.6E 03	Lungs	6.4E 02
59-FE	4.51E 01	Sol.	GI LLI	1.2E 02	Spleen	9.3E 01
		Insol.	GI LLI	1.1E 02	Lungs	3.3E 01
57-CO	2.70E 02	Sol.	GI LLI	1.1E 03	GI LLI	2.2E 03
		Insol.	GI LLI	7.6E 02	Lungs	1.0E 02
58M-CO	3.80E-01	Sol.	GI LLI	5.6E 03	GI LLI	1.1E 04
		Insol.	GI LLI	4.0E 03	Lungs	5.5E 03
58-CO	7.20E 01	Sol	GI LLI	2.6E 02	GI LLI	5.3E 02
					Tot. body	6.0E 02
		Insol	GI LLI	1.8E 02	Lungs	3.5E 01
60-CO	1.90E 03	Sol	GI LLI	9.8E 01	GI LLI	2.0E 02
					Tot. body	2.2E 02
		Insol.	GI LLI	7.0E 01	Lungs	5.5E 00
59-NI	2.90E 07	Sol.	Bone	4.0E 02	Bone	2.9E 02
		Insol.	GI LLI	4.0E 03	Lungs	4.7E 02
63-NI	2.90E 04	Sol	Bone	5.4E 01	Bone	4.0E 01
		Insol	GI LLI	1.4E 03	Lungs	1.7E 02
65-NI	1.10E-01	Sol.	GI ULI	2.8E 02	GI ULI	5.6E 02
		Insol.	GI ULI	2.0E 02	GI ULI	3.3E 02
64-CU	5.30E-01	Sol.	GI LLI	6.6E 02	GI LLI	1.3E 03
		Insol.	GI LLI	4.2E 02	GI LLI	6.6E 02
65-ZN	2.45E 02	Sol.	Tot. body	2.0E 02	Tot. body	6.6E 01
			Prostate	2.4E 02	Prostate	8.0E 01
			Liver	2.6E 02		
		Insol.	GI LLI	3.6E 02	Lungs	3.6E 01
69M-ZN	5.80E-01	Sol.	GI LLI	1.4E 02	Prostate	2.4E 02
		Insol.	GI LLI	1.2E 02	GI LLI	2.0E 02
69-ZN	3.60E-02	Sol.	GI S	3.6E 03	Prostate	4.4E 03
		Insol.	GI S	3.6E 03	GI S	5.8E 03
72-GA	5.90E-01	Sol.	GI LLI	7.4E 01	GI LLI	1.5E 02
		Insol.	GI LLI	7.4E 01	GI LLI	1.2E 02
71-GE	1.20E 01	Sol.	GI LLI	3.2E 03	GI LLI	6.4E 03
		Insol.	GI LLI	3.2E 03	Lungs	4.0E 03
73-AS	7.60E 01	Sol.	GI LLI	9.6E 02	Tot. body	1.3E 03
		Insol.	GI LLI	9.2E 02	Lungs	2.4E 02
74-AS	1.75E 01	Sol.	GI LLI	1.1E 02	GI LLI	2.2E 02
		Insol.	GI LLI	1.0E 02	Lungs	7.8E 01
76-AS	1.10E 00	Sol.	GI LLI	4.0E 01	GI LLI	8.0E 01
		Insol.	GI LLI	3.8E 01	GI LLI	6.2E 01
77-AS	1.60E 00	Sol.	GI LLI	1.6E 02	GI LLI	3.3E 02
		Insol.	GI LLI	1.6E 02	GI LLI	2.5E 02
75-SE	1.27E 02	Sol.	Kidneys	6.0E 02	Kidneys	7.8E 02
			Tot. body	6.8E 02		
		Insol.	GI LLI	5.4E 02	Lungs	7.6E 01
82-BR	1.50E 00	Sol.	Tot. body	5.2E 02	Tot. body	7.1E 02
			GI SI	5.6E 02		
		Insol.	GI LLI	7.4E 01	GI LLI	1.2E 02
86-RB	1.86E 01	Sol.	Tot. body	1.3E 02	Tot. body	1.8E 02
			Pancreas	1.3E 02	Pancreas	1.8E 02
					Liver	2.5E 02
		Insol.	GI LLI	4.8E 01	Lungs	4.2E 01
87-RB	1.80E 13	Sol.	Pancreas	2.2E 02	Pancreas	3.1E 02
					Tot. body	4.0E 02
					Liver	4.2E 02
		Insol.	GI LLI	3.4E 02	Lungs	4.0E 01
85M-SR	4.90E-02	Sol.	GI SI	1.3E 04	GI SI	2.5E 04
		Insol.	GI SI	1.3E 04	GI SI	2.2E 04
85-SR	6.50E 01	Sol.	Tot. body	1.9E 02	Tot. body	1.4E 02
		Insol.	GI LLI	3.4E 02	Lungs	6.6E 01
89-SR	5.05E 01	Sol.	Bone	2.4E 01	Bone	1.7E 01
		Insol.	GI LLI	5.6E 01	Lungs	2.2E 01
90-SR	1.00E 04	Sol.	Bone	8.0E-01	Bone	7.3E-01
		Insol.	GI LLI	7.0E 01	Lungs	3.5E 00
91-SR	4.00E-01	Sol.	GI LLI	1.4E 02	GI LLI	2.7E 02
		Insol.	GI LLI	9.8E 01	GI LLI	1.6E 02
92-SR	1.10E-01	Sol.	GI ULI	1.4E 02	GI ULI	2.7E 02
		Insol.	GI ULI	1.2E 02	GI ULI	1.8E 02
90-Y	2.68E 00	Sol.	GI LLI	4.0E 01	GI LLI	8.0E 01
		Insol.	GI LLI	4.0E 01	GI LLI	6.4E 01
91M-Y	3.50E-02	Sol.	GI SI	6.8E 03	GI SI	1.4E 04
		Insol.	GI SI	6.8E 03	GI SI	1.1E 04
91-Y	5.80E 01	Sol.	GI LLI	5.2E 01	Bone	2.2E 01
		Insol.	GI LLI	5.2E 01	Lungs	2.0E 01
92-Y	1.50E-01	Sol.	GI ULI	1.2E 02	GI ULI	2.4E 02
		Insol.	GI ULI	1.2E 02	GI ULI	1.8E 02
93-Y	4.20E-01	Sol.	GI LLI	5.4E 01	GI LLI	1.1E 02
		Insol.	GI LLI	5.4E 01	GI LLI	8.6E 01
93-ZR	4.00E 08	Sol.	GI LLI	1.6E 03	Bone	8.0E 01
		Insol.	GI LLI	1.6E 03	Lungs	2.0E 02

Radionuclide	Radioactive half-life days	Solubility state	Oral intake Critical organ	Permissible quarterly intake microcuries	Intake by inhalation Critical organ	Permissible quarterly intake microcuries
95-ZR	6.33E 01	Sol.	GI LLI	1.3E 02	Tot. body	8.0E 01
		Insol.	GI LLI	1.3E 02	Lungs	2.0E 01
97-ZR	7.10E-01	Sol.	GI LLI	3.6E 01	GI LLI	7.3E 01
		Insol.	GI LLI	3.6E 01	GI LLI	5.6E 01
93M-NB	3.70E 03	Sol.	GI LLI	8.0E 02	Bone	7.6E 01
		Insol.	GI LLI	8.0E 02	Lungs	1.0E 02
95-NB	3.50E 01	Sol.	GI LLI	1.9E 02	Tot. body	2.9E 02
					GI LLI	3.8E 02
		Insol.	GI LLI	1.9E 02	Lungs	6.2E 01
97-NB	5.10E-02	Sol.	GI ULI	1.9E 03	GI ULI	3.8E 03
		Insol.	GI ULI	1.9E 03	GI ULI	2.9E 03
99-MO	2.79E 00	Sol.	Kidneys	3.6E 02	Kidneys	4.5E 02
			GI LLI	4.8E 02		
		Insol.	GI LLI	7.8E 01	GI LLI	1.3E 02
96M-TC	3.60E-02	Sol.	GI LLI	2.4E 04	GI LLI	4.7E 04
		Insol.	GI LLI	2.0E 04	Lungs	1.8E 04
96-TC	4.30E 00	Sol.	GI LLI	2.0E 02	GI LLI	4.0E 02
		Insol.	GI LLI	9.4E 01	GI LLI	1.5E 02
97M-TC	9.20E 01	Sol.	GI LLI	7.2E 02	GI LLI	1.4E 03
		Insol.	GI LLI	3.4E 02	Lungs	9.5E 01
97-TC	3.70E 06	Sol.	GI LLI	3.4E 03	GI LLI	6.7E 03
					Kidneys	8.0E 03
		Insol.	GI LLI	1.6E 03	Lungs	1.8E 02
99M-TC	2.50E-01	Sol.	GI ULI	1.1E 04	GI ULI	2.4E 04
		Insol.	GI ULI	5.6E 03	GI ULI	8.7E 03
99-TC	7.70E 07	Sol.	GI LLI	6.6E 02	GI LLI	1.3E 03
		Insol.	GI LLI	3.2E 02	Lungs	3.8E 01
97-RU	2.80E 00	Sol.	GI LLI	7.2E 02	GI LLI	1.4E 03
		Insol.	GI LLI	7.0E 02	GI LLI	1.1E 03
103-RU	4.10E 01	Sol.	GI LLI	1.6E 02	GI LLI	3.3E 02
		Insol.	GI LLI	1.6E 02	Lungs	5.3E 01
105-RU	1.90E-01	Sol.	GI ULI	2.2E 02	GI ULI	4.4E 02
		Insol.	GI ULI	2.0E 02	GI ULI	3.3E 02
106-RU	3.65E 02	Sol.	GI LLI	2.4E 01	GI LLI	4.7E 01
		Insol.	GI LLI	2.4E 01	Lungs	2.6E 00
103M-RH	3.80E-02	Sol.	GI S	2.4E 04	GI S	4.7E 04
		Insol.	GI S	2.4E 04	GI S	3.8E 04
105-RH	1.52E 00	Sol.	GI LLI	2.6E 02	GI LLI	5.3E 02
		Insol.	GI LLI	2.0E 02	GI LLI	3.3E 02
103-PD	1.70E 01	Sol.	GI LLI	6.0E 02	Kidneys	8.4E 02
		Insol.	GI LLI	5.4E 02	Lungs	4.7E 02
109-PD	5.70E-01	Sol.	GI LLI	1.8E 02	GI LLI	3.5E 02
		Insol.	GI LLI	1.4E 02	GI LLI	2.2E 02
105-AG	4.00E 01	Sol.	GI LLI	1.9E 02	GI LLI	3.8E 02
		Insol.	GI LLI	1.9E 02	Lungs	5.1E 01
110M-AG	2.70E 02	Sol.	GI LLI	6.0E 01	GI LLI	1.2E 02
		Insol.	GI LLI	6.0E 01	Lungs	6.4E 00
111-AG	7.50E 00	Sol.	GI LLI	8.8E 01	GI LLI	1.8E 02
		Insol.	GI LLI	8.6E 01	GI LLI	1.4E 02
109-CD	4.75E 02	Sol.	GI LLI	3.4E 02	Liver	3.3E 01
					Kidneys	3.5E 01
		Insol.	GI LLI	3.4E 02	Lungs	4.5E 01
115M-CD	4.30E 01	Sol.	GI LLI	5.0E 01	Liver	2.2E 01
		Insol.	GI LLI	5.0E 01	Lungs	2.2E 01
115-CD	2.20E 00	Sol.	GI LLI	6.8E 01	GI LLI	1.4E 02
		Insol.	GI LLI	7.2E 01	GI LLI	1.1E 02
134-CS	8.40E 02	Sol.	Tot. body	1.7E 01	Tot. body	2.4E 01
		Insol.	GI LLI	8.0E 01	Lungs	8.0E 00
135-CS	1.10E 09	Sol.	Liver	2.2E 02	Liver	2.9E 02
			Spleen	2.4E 02	Spleen	3.3E 02
			Tot. body	2.6E 02	Tot. body	3.6E 02
		Insol.	GI LLI	4.6E 02	Lungs	5.6E 02
136-CS	1.30E 01	Sol.	Tot. body	1.7E 02	Tot. body	2.4E 02
		Insol.	GI LLI	1.3E 02	Lungs	1.1E 02
137-CS	1.10E 04	Sol.	Tot. body	3.0E 01	Tot. body	4.0E 01
			Liver	3.6E 01		
			Spleen	4.4E 01		
			Muscle	4.8E 01		
		Insol.	GI LLI	8.8E 01	Lungs	9.1E 00
131-BA	1.16E 01	Sol.	GI LLI	3.6E 02	GI LLI	7.3E 02
		Insol.	GI LLI	3.4E 02	Lungs	2.2E 02
140-BA	1.28E 01	Sol.	GI LLI	5.2E 01	Bone	8.0E 01
		Insol.	GI LLI	5.0E 01	Lungs	2.7E 01
140-LA	1.68E 00	Sol.	GI LLI	4.8E 01	GI LLI	9.6E 01
		Insol.	GI LLI	4.8E 01	GI LLI	7.6E 01
141-CE	3.20E 01	Sol.	GI LLI	1.8E 02	Liver	2.7E 02
					GI LLI	3.6E 02
					Bone	3.6E 02
		Insol.	GI LLI	1.8E 02	Lungs	9.8E 01
143-CE	1.33E 00	Sol.	GI LLI	8.0E 01	GI LLI	1.6E 02
		Insol.	GI LLI	8.0E 01	GI LLI	1.3E 02
144-CE	2.90E 02	Sol.	GI LLI	2.4E 01	Bone	6.0E 00
		Insol.	GI LLI	2.4E 01	Lungs	4.0E 00
142-PR	8.00E-01	Sol.	GI LLI	6.0E 01	GI LLI	1.2E 02
		Insol.	GI LLI	6.0E 01	GI LLI	9.8E 01
143-PR	1.37E 01	Sol.	GI LLI	9.8E 01	GI LLI	2.0E 02
		Insol.	GI LLI	9.8E 01	Lungs	1.1E 02

Radionuclide	Radioactive half-life days	Solubility state	Oral intake Critical organ	Oral intake Permissible quarterly intake microcuries	Intake by inhalation Critical organ	Intake by inhalation Permissible quarterly intake microcuries
144-ND	7.30E 17	Sol.	Bone	1.3E 02	Bone	5.1E-02
		Insol.	GI LLI	1.6E 02	Lungs	1.8E-01
147-ND	1.13E 01	Sol.	GI LLI	1.2E 02	Liver	2.2E 02
					GI LLI	2.4E 02
		Insol.	GI LLI	1.2E 02	Lungs	1.4E 02
149-ND	8.30E-02	Sol.	GI LLI	5.6E 02	GI LLI	1.1E 03
		Insol.	GI ULI	5.6E 02	GI ULI	9.1E 02
147-PM	9.20E 02	Sol.	GI LLI	4.4E 02	Bone	4.0E 01
		Insol.	GI LLI	4.4E 02	Lungs	6.0E 01
149-PM	2.20E 00	Sol.	GI LLI	8.8E 01	GI LLI	1.8E 02
		Insol.	GI LLI	8.8E 01	GI LLI	1.4E 02
147-SM	4.80E 13	Sol.	Bone	1.1E 02	Bone	4.4E-02
		Insol.	GI LLI	1.4E 02	Lungs	1.6E-01
151-SM	3.70E 04	Sol.	GI LLI	7.4E 02	Bone	4.0E 01
		Insol.	GI LLI	7.4E 02	Lungs	8.7E 01
153-SM	1.96E 00	Sol.	GI LLI	1.5E 02	GI LLI	3.1E 02
		Insol.	GI LLI	1.5E 02	GI LLI	2.5E 02
152M-EU	3.80E-01	Sol.	GI LLI	1.3E 02	GI LLI	2.5E 02
		Insol.	GI LLI	1.3E 02	GI LLI	2.0E 02
152-EU	4.70E 03	Sol.	GI LLI	1.5E 02	Kidneys	7.6E 00
		Insol.	GI LLI	1.5E 02	Lungs	1.1E 01
154-EU	5.80E 03	Sol.	GI LLI	4.4E 01	Kidneys	2.4E 00
					Bone	2.4E 00
		Insol.	GI LLI	4.4E 01	Lungs	4.4E 00
155-EU	6.21E 02	Sol.	GI LLI	4.0E 02	Kidneys	5.8E 01
					Bone	6.2E 01
		Insol.	GI LLI	4.0E 02	Lungs	4.5E 01
153-GD	2.36E 02	Sol.	GI LLI	4.2E 02	Bone	1.4E 02
		Insol.	GI LLI	4.2E 02	Lungs	5.6E 01
159-GD	7.50E-01	Sol.	GI LLI	1.6E 02	GI LLI	3.1E 02
		Insol.	GI LLI	1.6E 02	GI LLI	2.5E 02
113M-IN	7.30E-02	Sol.	GI ULI	2.6E 03	GI ULI	5.3E 03
		Insol.	GI ULI	2.6E 03	GI ULI	4.2E 03
114M-IN	4.90E 01	Sol.	GI LLI	3.4E 01	Kidneys	6.4E 01
					GI LLI	6.7E 01
					Spleen	7.1E 01
		Insol.	GI LLI	3.4E 01	Lungs	1.3E 01
115M-IN	1.90E-01	Sol.	GI ULI	7.4E 02	GI ULI	1.5E 03
		Insol.	GI ULI	7.4E 02	GI ULI	1.2E 03
115-IN	2.20E 17	Sol.	GI LLI	1.8E 02	Kidneys	1.5E 02
		Insol.	GI LLI	1.8E 02	Lungs	2.2E 01
113-SN	1.12E 02	Sol.	GI LLI	1.7E 02	Bone	2.2E 02
		Insol.	GI LLI	1.6E 02	Lungs	3.3E 01
125-SN	9.50E 00	Sol.	GI LLI	3.6E 01	GI LLI	7.3E 01
		Insol.	GI LLI	3.4E 01	Lungs	5.3E 01
					GI LLI	5.5E 01
122-SB	2.80E 00	Sol.	GI LLI	5.8E 01	GI LLI	1.2E 02
		Insol.	GI LLI	5.8E 01	GI LLI	9.1E 01
124-SB	6.00E 01	Sol.	GI LLI	4.6E 01	GI LLI	9.3E 01
		Insol.	GI LLI	4.6E 01	Lungs	1.2E 01
125-SB	8.77E 02	Sol.	GI LLI	2.0E 02	Lungs	3.3E 02
					Tot. body	3.6E 02
					GI LLI	4.0E 02
					Bone	4.5E 02
		Insol.	GI LLI	2.0E 02	Lungs	1.7E 01
125M-TE	5.80E 01	Sol.	Kidneys	3.2E 02	Kidneys	2.2E 02
			GI LLI	3.6E 02		
			Testes	4.4E 02		
		Insol.	GI LLI	2.4E 02	Lungs	8.0E 01
127M-TE	1.05E 02	Sol.	Kidneys	1.2E 02	Kidneys	8.2E 01
					Testes	8.7E 01
		Insol.	GI LLI	1.1E 02	Lungs	2.5E 01
127-TE	3.90E-01	Sol.	GI LLI	5.2E 02	GI LLI	1.0E 03
		Insol.	GI LLI	3.4E 02	GI LLI	5.3E 02
129M-TE	3.30E 01	Sol.	GI LLI	6.6E 01	Kidneys	4.9E 01
					Testes	5.6E 01
		Insol.	GI LLI	4.0E 01	Lungs	2.0E 01
129-TE	5.10E-02	Sol.	GI S	1.6E 03	GI S	3.3E 03
		Insol.	GI ULI	1.6E 03	GI ULI	2.5E 03
131M-TE	1.25E 00	Sol.	GI LLI	1.2E 02	GI LLI	2.4E 02
		Insol.	GI LLI	7.4E 01	GI LLI	1.2E 02
132-TE	3.20E 00	Sol.	GI LLI	6.4E 01	GI LLI	1.3E 02
		Insol.	GI LLI	4.2E 01	GI LLI	6.6E 01
126-I	1.33E 01	Sol.	Thyroid	3.4E 00	Thyroid	4.5E 00
		Insol.	GI LLI	1.8E 02	Lungs	2.0E 02
129-I	6.30E 09	Sol.	Thyroid	7.6E-01	Thyroid	1.0E 00
		Insol.	GI LLI	4.2E 02	Lungs	4.5E 01
131-I	8.00E 00	Sol.	Thyroid	4.0E 00	Thyroid	5.3E 00
		Insol.	GI LLI	1.3E 02	GI LLI	2.0E 02
					Lungs	2.0E 02
132-I	9.70E-02	Sol.	Thyroid	1.1E 02	Thyroid	1.5E 02
		Insol.	GI ULI	3.6E 02	GI ULI	5.8E 02
133-I	8.70E-01	Sol.	Thyroid	1.5E 01	Thyroid	2.0E 01
		Insol.	GI LLI	8.2E 01	GI LLI	1.3E 02
134-I	3.60E-02	Sol.	Thyroid	2.4E 02	Thyroid	3.1E 02
		Insol.	GI S	1.2E 03	GI S	2.0E 03
135-I	2.80E-01	Sol.	Thyroid	4.8E 01	Thyroid	6.4E 01
		Insol.	GI LLI	1.4E 02	GI LLI	2.2E 02

Radionuclide	Radioactive half-life days	Solubility state	Oral intake		Intake by inhalation	
			Critical organ	Permissible quarterly intake microcuries	Critical organ	Permissible quarterly intake microcuries
131-CS	1.00E 01	Sol.	Tot. body	4.8E 03	Tot. body	6.6E 03
					Liver	8.0E 03
		Insol.	GI LLI	1.9E 03	Lungs	2.0E 03
134M-CS	1.30E-01	Sol.	GI S	1.1E 04	GI S	2.2E 04
		Insol.	GI ULI	2.2E 03	GI ULI	3.6E 03
160-TB	7.30E 01	Sol.	GI LLI	8.8E 01	Bone	6.2E 01
		Insol.	GI LLI	9.0E 01	Lungs	2.0E 01
165-DY	9.70E-02	Sol.	GI ULI	8.0E 02	GI ULI	1.6E 03
		Insol.	GI ULI	8.0E 02	GI ULI	1.3E 03
166-DY	3.40E 00	Sol.	GI LLI	7.6E 01	GI LLI	1.5E 02
		Insol.	GI LLI	7.6E 01	GI LLI	1.2E 02
166-HO	1.10E 00	Sol.	GI LLI	6.2E 01	GI LLI	1.2E 02
		Insol.	GI LLI	6.2E 01	GI LLI	1.0E 02
169-ER	9.40E 00	Sol.	GI LLI	1.9E 02	GI LLI	3.8E 02
		Insol.	GI LLI	1.9E 02	Lungs	2.4E 02
171-ER	3.10E-01	Sol.	GI ULI	2.2E 02	GI ULI	4.4E 02
		Insol.	GI ULI	2.2E 02	GI ULI	3.6E 02
170-TM	1.27E 02	Sol.	GI LLI	9.2E 01	Bone	2.2E 01
		Insol.	GI LLI	9.2E 01	Lungs	2.2E 01
171-TM	6.94E 02	Sol.	GI LLI	1.0E 03	Bone	6.9E 01
		Insol.	GI LLI	1.0E 03	Lungs	1.4E 02
175-YB	4.10E 00	Sol.	GI LLI	2.2E 02	GI LLI	4.4E 02
		Insol.	GI LLI	2.2E 02	GI LLI	3.6E 02
177-LU	6.80E 00	Sol.	GI LLI	2.0E 02	GI LLI	4.0E 02
		Insol.	GI LLI	2.0E 02	GI LLI	3.3E 02
					Lungs	4.4E 02
181-HF	4.60E 01	Sol.	GI LLI	1.4E 02	Spleen	2.4E 01
		Insol.	GI LLI	1.4E 02	Lungs	4.5E 01
182-TA	1.12E 02	Sol.	GI LLI	8.0E 01	Liver	2.4E 01
		Insol.	GI LLI	8.0E 01	Lungs	1.4E 01
181-W	1.40E 02	Sol.	GI LLI	7.2E 02	GI LLI	1.4E 03
		Insol.	GI LLI	6.4E 02	Lungs	7.8E 01
185-W	7.40E 01	Sol.	GI LLI	2.4E 02	GI LLI	4.7E 02
		Insol.	GI LLI	2.2E 02	Lungs	6.9E 01
187-W	1.00E 00	Sol.	GI LLI	1.4E 02	GI LLI	2.7E 02
		Insol.	GI LLI	1.2E 02	GI LLI	2.0E 02
183-RE	7.30E 01	Sol.	GI LLI	1.1E 03	Tot. body	1.6E 03
		Insol.	GI LLI	5.6E 02	Lungs	9.8E 01
186-RE	3.79E 00	Sol.	GI LLI	1.0E 02	GI LLI	2.0E 02
		Insol.	GI LLI	9.4E 01	GI LLI	1.5E 02
187-RE	1.80E 13	Sol.	GI LLI	5.0E 03	Skin	5.6E 03
			Skin	5.6E 03		
		Insol.	GI LLI	5.0E 03	Lungs	3.1E 02
188-RE	7.10E-01	Sol.	GI LLI	1.3E 02	GI LLI	2.5E 02
		Insol.	GI LLI	6.2E 01	GI LLI	1.0E 02
185-OS	9.50E 01	Sol.	GI LLI	1.5E 02	GI LLI	2.9E 02
		Insol.	GI LLI	1.3E 02	Lungs	2.9E 01
191M-OS	5.80E-01	Sol.	GI LLI	5.0E 03	GI LLI	1.0E 04
		Insol.	GI LLI	4.8E 03	Lungs	5.8E 03
191-OS	1.60E 01	Sol.	GI LLI	3.4E 02	GI LLI	6.7E 02
		Insol.	GI LLI	3.2E 02	Lungs	2.5E 02
193-OS	1.30E 00	Sol.	GI LLI	1.2E 02	GI LLI	2.4E 02
		Insol.	GI LLI	1.1E 02	GI LLI	1.7E 02
190-IR	1.20E 01	Sol.	GI LLI	4.0E 02	GI LLI	8.0E 02
		Insol.	GI LLI	3.6E 02	Lungs	2.5E 02
192-IR	7.45E 01	Sol.	GI LLI	8.0E 01	Kidneys	7.8E 01
		Insol.	GI LLI	7.2E 01	Lungs	1.6E 01
194-IR	7.90E-01	Sol.	GI LLI	6.8E 01	GI LLI	1.4E 02
		Insol.	GI LLI	6.0E 01	GI LLI	9.6E 01
191-PT	3.00E 00	Sol.	GI LLI	2.4E 02	GI LLI	4.7E 02
		Insol.	GI LLI	2.2E 02	GI LLI	3.5E 02
193M-PT	3.40E 00	Sol.	GI LLI	2.2E 03	GI LLI	4.4E 03
		Insol.	GI LLI	2.0E 03	GI LLI	3.3E 03
					Lungs	4.0E 03
193-PT	1.80E 05	Sol.	Kidneys	1.9E 03	Kidneys	6.4E 02
		Insol.	GI LLI	3.0E 03	Lungs	2.0E 02
197M-PT	5.60E-02	Sol.	GI ULI	2.0E 03	GI ULI	4.0E 03
		Insol.	GI ULI	1.9E 03	GI ULI	2.9E 03
197-PT	7.50E-01	Sol.	GI LLI	2.4E 02	GI LLI	4.7E 02
		Insol.	GI LLI	2.2E 02	GI LLI	3.5E 02
196-AU	5.60E 00	Sol.	GI LLI	3.2E 02	GI LLI	6.4E 02
		Insol.	GI LLI	3.0E 02	Lungs	3.8E 02
198-AU	2.70E 00	Sol.	GI LLI	1.0E 02	GI LLI	2.0E 02
		Insol.	GI LLI	9.2E 01	GI LLI	1.5E 02
199-AU	3.15E 00	Sol.	GI LLI	3.4E 02	GI LLI	6.7E 02
		Insol.	GI LLI	3.2E 02	GI LLI	5.1E 02
197M-HG	1.00E 00	Sol.	Kidneys	3.8E 02	Kidneys	4.5E 02
		Insol.	GI LLI	3.4E 02	GI LLI	5.3E 02
197-HG	2.70E 00	Sol.	Kidneys	6.0E 02	Kidneys	7.3E 02
		Insol.	GI LLI	9.8E 02	GI LLI	1.5E 03
203-HG	4.58E 01	Sol.	Kidneys	3.6E 01	Kidneys	4.4E 01
		Insol.	GI LLI	2.2E 02	Lungs	7.8E 01
200-TL	1.13E 00	Sol.	GI LLI	8.8E 02	GI LLI	1.6E 03
		Insol.	GI LLI	4.4E 02	GI LLI	7.1E 02
201-TL	3.00E 00	Sol.	GI LLI	6.2E 02	GI LLI	1.2E 03
		Insol.	GI LLI	3.4E 02	GI LLI	5.5E 02
202-TL	1.20E 01	Sol.	GI LLI	2.4E 02	GI LLI	4.7E 02
		Insol.	GI LLI	1.4E 02	Lungs	1.5E 02

PERMISSIBLE QUARTERLY INTAKES (ORAL AND INHALATION) OF RADIONUCLIDES AND
CRITICAL BODY ORGANS RECOMMENDED BY NCRP FOR OCCUPATIONAL EXPOSURE (1)
(Continued)

Radionuclide	Radioactive half-life days	Solubility state	Oral intake		Intake by inhalation	
			Critical organ	Permissible quarterly intake microcuries	Critical organ	Permissible quarterly intake microcuries
204-TL	1.10E 03	Sol.	GI LLI	2.2E 02	Kidneys	3.8E 02
					GI LLI	4.4E 02
		Insol.	GI LLI	1.2E 02	Lungs	1.6E 01
203-PB	2.17E 00	Sol.	GI LLI	7.8E 02	GI LLI	1.6E 03
		Insol.	GI LLI	7.0E 02	GI LLI	1.1E 03
210-PB	7.10E 03	Sol.	Tot. body	2.4E-01	Kidneys	7.8E-02
			Kidneys	2.8E-01		
		Insol.	GI LLI	3.6E 02	Lungs	1.5E-01
212-PB	4.40E-01	Sol.	GI LLI	3.8E 01	Kidneys	1.1E 01
			Kidneys	4.0E 01		
		Insol.	GI LLI	3.4E 01	Lungs	1.2E 01
206-BI	6.40E 00	Sol.	GI LLI	7.6E 01	Kidneys	1.2E 02
		Insol.	GI LLI	7.6E 01	Lungs	9.1E 01
207-BI	2.90E 03	Sol.	GI LLI	1.3E 02	Kidneys	1.1E 02
		Insol.	GI LLI	1.3E 02	Lungs	8.6E 00
210-BI	5.00E 00	Sol.	GI LLI	8.2E 01	Kidneys	4.0E 00
		Insol.	GI LLI	8.2E 01	Lungs	3.6E 00
212-BI	4.20E-02	Sol.	GI S	7.0E 02	Kidneys	6.0E 01
		Insol.	GI S	7.0E 02	Lungs	1.3E 02
210-PO	1.38E 02	Sol.	Spleen	1.4E 00	Spleen	3.1E-01
					Kidneys	3.1E-01
		Insol.	GI LLI	5.8E 01	Lungs	1.3E-01
211-AT	3.00E-01	Sol.	Thyroid	3.4E 00	Thyroid	4.4E 00
			Ovaries	3.4E 00		
		Insol.	GI ULI	1.5E 02	Lungs	2.2E 01
220-RN[3]	6.40E-04				Lungs	1.8E 02
222-RN[3]	3.82E 00				Lungs	1.8E 01
223-RA	1.17E 01	Sol.	Bone	1.4E 00	Bone	1.1E 00
		Insol.	GI LLI	8.4E 00	Lungs	1.5E-01
224-RA	3.64E 00	Sol.	Bone	4.6E 00	Bone	3.5E 00
		Insol.	GI LLI	1.1E 01	Lungs	4.5E-01
226-RA	5.90E 05	Sol.	Bone	2.4E-02	Bone	1.8E-02
		Insol.	GI LLI	6.4E 01	Lungs	3.3E-02
228-RA	2.40E 03	Sol.	Bone	5.6E-02	Bone	4.2E-02
		Insol.	GI LLI	5.0E 01	Lungs	2.4E-02
227-AC	8.00E 03	Sol.	Bone	3.8E 00	Bone	1.5E-03
		Insol.	GI LLI	6.0E 01	Lungs	1.6E-02
228-AC	2.60E-01	Sol.	GI ULI	1.7E 02	Liver	4.7E 01
					Bone	5.3E 01
		Insol.	GI ULI	1.7E 02	Lungs	1.1E 01
227-TH	1.84E 01	Sol.	GI LLI	3.6E 01	Bone	2.2E-01
		Insol.	GI LLI	3.6E 01	Lungs	1.1E-01
228-TH	7.00E 02	Sol.	Bone	1.5E 01	Bone	5.6E-03
		Insol.	GI LLI	2.6E 01	Lungs	3.8E-03
230-TH	2.90E 07	Sol.	Bone	3.4E 00	Bone	1.4E-03
		Insol.	GI LLI	6.4E 01	Lungs	6.4E-03
231-TH	1.07E 01	Sol.	GI LLI	4.6E 02	GI LLI	9.3E 02
		Insol.	GI LLI	4.6E 02	GI LLI	7.5E 02
232-TH	5.10E 12	Sol.	Bone	3.0E 00	Bone	1.2E-03[4]
		Insol.	GI LLI	7.6E 01	Lungs	7.3E-03
234-TH	2.41E 01	Sol.	GI LLI	3.4E 01	Bone	3.6E 01
		Insol.	GI LLI	3.4E 01	Lungs	2.0E 01
NAT-TH[5]		Sol.	Bone	2.6E 00	Bone	1.0E-03[4]
		Insol.	GI LLI	1.9E 01	Lungs	2.5E-03
230-PA	1.77E 01	Sol.	GI LLI	4.8E 02	Bone	1.1E 00
		Insol.	GI LLI	5.0E 02	Lungs	4.9E-01
231-PA	1.30E 07	Sol.	Bone	1.7E 00	Bone	7.1E-04
		Insol.	GI LLI	5.4E 01	Lungs	6.7E-02
233-PA	2.74E 01	Sol.	GI LLI	2.4E 02	Kidneys	3.8E 02
		Insol.	GI LLI	2.4E 02	Lungs	1.1E 02
230-U	2.08E 01	Sol.	Kidneys	4.8E 02	Kidneys	1.8E-01
		Insol.	GI LLI	9.2E 00	Lungs	7.1E-02
232-U	2.70E 04	Sol.	Bone	1.7E 00	Bone	6.4E-02
		Insol.	GI LLI	5.8E 01	Lungs	1.7E-02
233-U	5.90E 07	Sol.	Bone	8.4E 00	Bone	3.3E-01
		Insol.	GI LLI	6.4E 01	Lungs	7.5E-02
234-U	9.10E 07	Sol.	Bone	8.6E 00	Bone	3.5E-01
		Insol.	GI LLI	6.4E 01	Lungs	7.5E-02
235-U[5]	2.60E 11	Sol.	Kidneys	7.4F 00	Kidneys	2.9E-01
					Bone	3.6E-01
		Insol.	GI LLI	5.6E 01	Lungs	8.0E-02
236-U[5]	8.70E 09	Sol.	Bone	9.0E 00	Bone	3.6E-01
		Insol.	GI LLI	6.8E 01	Lungs	7.8E-02
238-U[5]	1.60E 12	Sol.	Kidneys	1.2E 00	Kidneys	4.5E-02
		Insol.	GI LLI	7.0E 01	Lungs	8.6E-02
240-U	5.88E-01	Sol.	GI LLI	6.8E 01	GI LLI	1.4E 02
		Insol.	GI LLI	6.8E 01	GI LLI	1.1E 02
NAT-U[5]		Sol.	Kidneys	1.2E 00	Kidneys	4.5E-02
		Insol.	GI LLI	3.2E 01	Lungs	4.0E-02
237-NP	8.00E 08	Sol.	Bone	6.2E 00	Bone	2.5E-03
		Insol.	GI LLI	7.0E 01	Lungs	7.5E-02
239-NP	2.33E 00	Sol.	GI LLI	2.6E 02	GI LLI	5.3E 02
		Insol.	GI LLI	2.6E 02	GI LLI	4.2E 02
238-PU	3.30E 04	Sol.	Bone	1.0E 01	Bone	1.2E-03
		Insol.	GI LLI	5.6E 01	Lungs	2.2E-02
239-PU	8.90E 06	Sol.	Bone	9.0E 00	Bone	1.1E-03
		Insol.	GI LLI	5.8E 01	Lungs	2.4E-02

PERMISSIBLE QUARTERLY INTAKES (ORAL AND INHALATION) OF RADIONUCLIDES AND CRITICAL BODY ORGANS RECOMMENDED BY NCRP FOR OCCUPATIONAL EXPOSURE (1)
(Continued)

Radionuclide	Radioactive half-life days	Solubility state	Oral intake		Intake by inhalation	
			Critical organ	Permissible quarterly intake microcuries	Critical organ	Permissible quarterly intake microcuries
240-PU	2.40E 06	Sol.	Bone	9.0E 00	Bone	1.1E-03
		Insol.	GI LLI	5.8E 01	Lungs	2.4E-02
241-PU	4.80E 03	Sol.	Bone	4.6E 02	Bone	5.6E-02
		Insol.	GI LLI	2.6E 03	Lungs	2.4E 01
242-PU	1.40E 08	Sol.	Bone	9.4E 00	Bone	1.1E-03
		Insol.	GI LLI	6.2E 01	Lungs	2.4E-02
243-PU	2.08E-01	Sol.	GI ULI	6.8E 02	GI ULI	1.1E 03
		Insol.	GI ULI	6.8E 02	GI ULI	1.4E 03
244-PU	2.80E 10	Sol.	Bone	8.6E 00	Bone	1.0E-03
		Insol.	GI LLI	2.2E 01	Lungs	2.0E-02
241-AM	1.70E 05	Sol.	Kidneys	7.6E 00	Bone	3.6E 03
					Kidneys	3.8E-03
		Insol.	GI LLI	5.4E 01	Lungs	6.4E-02
242M-AM	5.60E 04	Sol.	Bone	8.8E 00	Bone	3.5E-03
		Insol.	GI LLI	1.8E 02	Lungs	1.6E-01
242-AM	6.77E-01	Sol.	GI LLI	2.6E 02	Liver	2.4E 01
		Insol.	GI LLI	2.6E 02	Lungs	2.9E 01
243-AM	2.90E 06	Sol.	Bone	8.8E 00	Bone	3.5E-03
					Kidneys	3.8E-03
		Insol.	GI LLI	5.6E 01	Lungs	6.7E-02
244-AM	1.81E-02	Sol.	GI SI	9.6E 03	Bone	2.5E 03
					Kidneys	2.7E 03
		Insol.	GI SI	9.6E 03	Lungs	1.5E 04
					GI(SI)	1.5E 04
242-CM	1.62E 02	Sol.	GI LLI	4.8E 01	Liver	7.5E-02
		Insol.	GI LLI	5.0E 01	Lungs	1.0E-01
243-CM	1.30E 04	Sol.	Bone	1.0E 01	Bone	4.0E-03
		Insol.	GI LLI	5.0E 01	Lungs	6.2E-02
244-CM	6.70E 03	Sol.	Bone	1.4E 01	Bone	5.6E-03
		Insol.	GI LLI	5.2E 01	Lungs	6.2E-02
245-CM	7.30E 06	Sol.	Bone	7.0E 00	Bone	2.9E-03
		Insol.	GI LLI	5.6E 01	Lungs	6.7E-02
246-CM	2.40E 06	Sol.	Bone	7.2E 00	Bone	2.9E-03
		Insol.	GI LLI	5.6E 01	Lungs	6.6E-02
247-CM	3.30E 10	Sol.	Bone	7.2E 00	Bone	2.9E-03
		Insol.	GI LLI	4.4E 01	Lungs	6.7E-02
248-CM	1.70E 08	Sol.	Bone	8.8E-01	Bone	3.6E-04
		Insol.	GI LLI	2.6E 00	Lungs	8.2E-03
249-CM	8.90E 01	Sol.	GI S	8.8E 02	Bone	7.1E 03
		Insol.	GI S	4.4E 03	GI S	7.1E 03
249-BK	2.90E 02	Sol.	GI LLI	1.2E 03	Bone	5.8E-01
		Insol.	GI LLI	1.2E 03	Lungs	7.5E 01
250-BK	1.34E-01	Sol.	GI ULI	4.4E 02	Bone	9.1E 01
		Insol.	GI ULI	4.4E 02	GI ULI	7.1E 02
249-CF	1.70E 05	Sol.	Bone	8.2E 00	Bone	9.8E-04
		Insol.	GI LLI	4.8E 01	Lungs	6.2E-02
250-CF	3.70E 03	Sol.	Bone	2.6E 01	Bone	3.1E-03
		Insol.	GI LLI	5.0E 01	Lungs	6.2E-02
251-CF	2.90E 05	Sol.	Bone	8.6E 00	Bone	1.0E-03
		Insol.	GI LLI	5.2E 01	Lungs	6.2E-02
252-CF	8.03E 02	Sol.	GI LLI	1.5E 01	Bone	4.0E-03
		Insol.	GI LLI	1.5E 01	Lungs	2.0E-02
253-CF	1.80E 01	Sol.	GI LLI	2.8E 02	Bone	5.3E-01
		Insol.	GI LLI	2.8E 02	Lungs	4.7E-01
254-CF	5.60E 01	Sol.	GI LLI	2.4E-01	Bone	3.3E-03
		Insol.	GI LLI	2.4E-01	Lungs	3.1E-03
253-ES	2.00E 01	Sol.	GI LLI	4.6E 01	Bone	4.7E-01
		Insol.	GI LLI	4.6E 01	Lungs	3.8E-01
254M-ES	1.60E 00	Sol.	GI LLI	3.8E 01	Bone	3.3E 00
		Insol.	GI LLI	3.8E 01	Lungs	3.6E 00
254-ES	4.80E 02	Sol.	GI LLI	2.8E 01	Bone	1.2E-02
		Insol.	GI LLI	2.8E 01	Lungs	6.7E-02
255-ES	3.00E 01	Sol.	GI LLI	5.6E 01	Bone	3.1E-01
		Insol.	GI LLI	5.6E 01	Lungs	2.5E-01
254-FM	1.35E-01	Sol.	GI ULI	2.4E 02	Bone	4.0E 01
		Insol.	GI ULI	2.4E 02	Lungs	4.4E 01
255-FM	8.96E-01	Sol.	GI LLI	6.6E 01	Bone	1.0E 01
		Insol.	GI LLI	6.6E 01	Lungs	6.7E 00
256-FM	1.11E-01	Sol.	GI ULI	1.8E 00	Bone	1.7E 00
		Insol.	GI ULI	1.8E 00	Lungs	1.1E 00

[1] These quarterly intakes are calculated from values of (MPC), and (MPC), recommended by the NCRP for occupational exposure (see NBS Handbook 69 and NCRP Report No. 32).* except for uranium, ⁹⁰Sr, and certain transuranic isotopes given by the ICRP (see ICRP Publications 2 and 6).** Specifically, except where indicated in the footnotes, the quarterly intakes are calculated to two-digit accuracy as $91 \times 2 \times 10^3 \times (MPC)_w$ and $91 \times 2200 \times (MPC)_a$ where 91 ~ days/quarter. 2×10^3 ~ cc of air breathed/day, 2200 ~ cc of water intake/day, and $(MPC)_a$, $(MPC)_w$ are μCi/cc of air and water recommended as limits for occupational exposure for the 168-hour week. These quarterly intakes are calculated for standard adult man and would be expected to deliver a total dose t (= dose commitment) to the critical organ of 1/4 the **maximum permissible annual dose (MPD)** recommended as limiting by the NCRP for occupational exposure. These limiting MPD are the following: Bone, skin, thyroid—MPD = 30 rem; blood-forming organs, lenses of the eyes, gonads—MPD = 5 rem/yr; other organs—MPD = 15 rem/yr. These limits are not appropriate for exposure of the population or, specifically, for persons below the age of 18 years. For more detailed information on appropriate exposure limits recommended by the NCRP, see NBS Handbook 69 and NCRP Report No. 32.*

The abbreviations GI, S, SI, ULI, and LLI refer to gastrointestinal tract, stomach, small intestine, upper large intestine, and lower large intestine, respectively.

With few exceptions, the above intakes do not take account of chemical toxicity which should be considered separately.

[2] This intake includes absorption through the skin (see ICRP Publication 2, p. 22).**

[3] The daughter isotopes of ²²⁰Rn and ²²²Rn are assumed present to the extent they occur in unfiltered air (see NBS Handbook 69, p. 12).* For all other isotopes, the daughter elements are not considered as part of the intake and if present must be considered on the basis of the rules for mixtures.

[4] These are provisional values for ²³²Th and nat-Th (see footnote, p. 83, NBS Handbook 69).*

[5] A curie of natural uranium is here considered to correspond to 2.94 g and a curie of natural thorium to 9.0 g of these materials (see NBS Handbook 69, p. 14).* For soluble nat-U, ²³⁸U, ²³⁵U, and ²³³U these limits are based on considerations of chemical toxicity (see ICRP Publication 6, p. 13).**

* NBS Handbook 69, *Maximum Permissible Body Burdens and Maximum Permissible Concentrations of Radionuclides in Air and Water for Occupational Exposure,* 1959 (available from the Superintendent of Documents, U.S. Government Printing Office, Washington, D.C. 20402). NCRP Report No. 32, *Radiation Protection in Educational Institutions,* 1966 (NCRP Publications, P.O. Box 4867, Washington, D.C. 20008).

** *Recommendations of the International Commission on Radiological Protection,* ICRP Publication 2 (Pergamon Press, London, 1959) or with the bibliography, *Healthy Phys.* 3, 1–380 (June 1960); ICRP Publication 6 (A Pergamon Press Book, 1964, distributed by The Macmillan Co., New York).

CRYOGENIC PROPERTIES OF GASES

Property and conditions		He	Ne	Ar	Kr	Xe	H₂	CH₄	NH₃	N₂	O₂	F₂
Gas												
Density	32°F, 1 atm, lb/ft³	0.01114	0.0562	0.1113	0.234	0.368	0.00561	0.0448	0.0481	0.0781	0.0892	0.106
	0°C, 1 atm, kg/m³	0.1784	0.9002	1.783	3.748	5.895	0.0899	0.718	0.770	1.251	1.429	1.698
Boiling point	°F, 1 atm	-452.08	-410.89	-302.3	-242.1	-160.8	-423.2	-263.2	-28.03	-320.4	-297.35	-306.7
	°C, 1 atm	-268.934	-246.048	-185.7	-152.90	-107.1	-252.87	-164.0	-33.35	-195.8	-182.97	-188.14
	°K, 1 atm	4.216	27.10	87.45	120.25	166.05	20.28	109.15	239.80	77.35	90.18	85.01
Melting point	°F, 1 atm	-458.0ᵃ	-415.6	-308.6	-249.9	-169.4	-434.5	-296.46	-107.9	-345.87	-361.1	-363.3
	°C, 1 atm	-272.2ᵃ	-248.67	-189.2	-156.6	-111.9	-259.14	-182.48	-77.7	-209.86	-218.4	-219.62
	°K, 1 atm	0.95ᵃ	24.48	83.95	116.55	161.25	14.01	90.67	195.45	63.29	54.75	53.53
Vapor density at boiling point	lb/ft³	0.999	0.593	0.368	0.518	0.606	0.0830	0.1124	0.0556	0.288	0.279	
	kg/m³	16.002	9.499	5.895	8.298	9.707	1.329	1.8004	0.8906	4.613	4.4692	
Liquid density at boiling point	lb/ft³	7.803	74.91	86.77	149.8	193.5	4.37	26.47	42.58	50.19	71.23	94.4
	kg/m³	125.	1200.	1390.	2400.	3100.	70.0	424.	682.1	804.	1142	1512
Vapor pressure of solid at melting point	lb/in.²		6.25	9.98	10.6	11.8	1.04	1.35	0.87	1.86	0.038	0.002
	kg/m²		323.	516	549.	612.	54.	70.	45.2	96.4	2.0	0.12
	(N/m²) × 10⁴		4.34	6.93	7.36	8.20	0.723	0.938	0.604	1.29	0.0026	0.00014
Heat of vaporization at boiling point	Btu/lb	10.3	37.4	70.0	46.4	41.4	194.4	248.4	588.6	85.7	91.588	73.7
	kcal/kg	5.72	20.8	38.9	25.8	23.0	108	138	327	47.6	50.88	40.9
	(J/kg) × 10³	23.932	87.027	162.76	107.95	96.23	451.9	577.4	1368.2	199.2	212.9	171.1
Heat of fusion at melting point	Btu/lb	1.8	7.2	12.1	7.0	5.9	25.2	26.1	152.1	11.0	5.9	5.8
	kcal/kg	1.0	4.0	6.7	3.9	3.3	14.0	14.5	84.0	6.1	3.27	3.2
	(J/kg) × 10³	4.184	16.74	28.03	16.3	13.8	58.6	60.7	351.5	25.5	13.7	13.4
C_p	59°F, 1 atm, Btu/lb-°F or 15°C, 1 atm, kcal/kg-°C	1.25ᵇ	0.25ᶜ	0.125	0.06ᶜ	0.04ᶜ	3.39	0.528	0.523	0.248	0.220	0.180
	288.15°K, 1 atm, (J/kg) × 10³	5.23ᵇ	1.05ᶜ	0.523	0.251ᶜ	0.167ᶜ	14.2	2.21	2.188	1.038	0.9205	0.753
C_p/C_v	15-20°C, 1 atm / 288-293°K, 1 atm	1.66ᵇ	1.64ᶜ	1.67	1.68ᶜ	1.66ᶜ	1.41	1.31	1.31	1.40	1.40	
Critical temperature	°F	-450.2	-397.7	-188.5	-82.7	61.9	-399.8	-116.5	270.3	-232.8	-181.3	-200.2
	°C	-267.9	-228.7	-122.5	-63.7	16.6	-239.9	-82.5	132.4	-147.1	-118.57	-129.0
	°K	5.25	44.45	150.65	209.45	289.75	33.25	190.65	405.55	126.05	154.58	144.15
Critical pressure	lb/in.² (absolute)	33.2	394.6	705.2	798	855	188.1	672	1639	492.3	731.4	808.3
	kg/cm²	2.33	27.7	49.6	46.1	60.1	13.2	47.2	115.5	34.6	51.4	56.8
	(kg/m²) × 10³	23.3	277	496	561	601	132	472	1155	346	514	568
	(N/m²) × 10⁴	23.1	274.1	489.9	554.4	594	130.7	466.9	1139	342.	508.1	561.6

Note: For conversion factors see p. B-439.

ᵃ At 26 atmospheres.
ᵇ At -292°F or -180°C.
ᶜ Approximate.

CONVERSION FACTORS FOR TABLE OF CRYOGENIC PROPERTIES OF GASES

To convert from	To	Multiply by
lbs ft^{-3}	kg m^{-3}	16.018
lbs in.$^{-2}$	N m^{-2}	6894.8
lbs ft^{-2}	N m^{-2}	47.880
BTU lb^{-1}	J kg^{-1}	2324.4
cal g^{-1}	J kg^{-1}	4184
cal g^{-1} °F	J kg^{-1} °C	4184

VISCOSITY AND THERMAL CONDUCTIVITY OF NITROGEN AT CRYOGENIC TEMPERATURES

The viscosity and thermal conductivity of nitrogen gas for the temperature range 5 K–135 K have been computed from the second Chapman-Enskog approximation. Quantum effects, which become appreciable at the lower temperatures, are included by utilizing collision integrals based on quantum theory. A Lennard-Jones (12-6) potential was assumed. The computations yield viscosities about 20% lower than those predicted for the high end of this temperature range by the method of corresponding states, but the agreement is excellent when the computed values are compared with existing experimental data.

T, °K	η, micropoise	λ, $\dfrac{\mu cal}{cm\ sec\ °K}$
4.575	4.16639	1.10990
5.49	4.85525	1.29307
6.405	5.54773	1.47715
7.320	6.24722	1.63156
8.235	6.95025	1.85020
9.15	7.65067	2.03666
13.725	10.95231	2.91729
18.3	13.85738	3.69313
22.875	16.55204	4.41068
27.45	19.21670	5.11847
32.025	21.9423	5.84208
36.60	24.7606	6.59077
41.175	27.67445	7.36545
45.75	30.67422	8.16348
54.9	36.8747	9.81383
64.05	43.2469	11.51005
82.35	56.14634	14.94298
91.5	62.54767	16.64610
137.25	92.96598	24.74580

From Pearson, W. E., NASA Technical Note D-7565, National Aeronautics and Space Administration, 1974 (available from Superintendent of Documents, U.S. Government Printing Office, Washington, D.C.).

DEFINITIVE RULES FOR NOMENCLATURE OF ORGANIC CHEMISTRY

IUPAC 1957 Rules.

Section A. Hydrocarbons

Section B. Fundamental Heterocyclic Systems

These rules are taken from 'Definitive Rules for the Nomenclature of Organic Chemistry' which were adopted unanimously by the Commission on Nomenclature and by The Council of the International Union of Pure and Applied Chemistry at Paris 1957, and subsequently published by Butterworths Scientific Publications on behalf of the Union. The extracts are printed here by permission of the Union and of Butterworths Scientific Publications. Future 'tentative' rules will be published in the Bulletin of the Union, and when made 'definitive' in its Journal 'Pure and Applied Chemistry.'

RULES

A. HYDROCARBONS

Acyclic Hydrocarbons

A-1

1.1.—The first four saturated unbranched acyclic hydrocarbons are called methane, ethane, propane and butane. Names of the higher members of this series consist of a numerical prefix and the termination "-ane." Examples of these numerical prefixes are shown in the table below. The generic name of saturated acyclic hydrocarbons (branched or unbranched) is "alkane."

Examples:

n		n		n		n	
1	Methane	12	Dodecane	22	Docosane	32	Dotriacontane
2	Ethane	13	Tridecane	23	Tricosane	33	Tritriacontane
3	Propane	14	Tetradecane	24	Tetracosane	40	Tetracontane
4	Butane	15	Pentadecane	25	Pentacosane	50	Pentacontane
5	Pentane	16	Hexadecane	26	Hexacosane	60	Hexacontane
6	Hexane	17	Heptadecane	27	Heptacosane	70	Heptacontane
7	Heptane	18	Octadecane	28	Octacosane	80	Octacontane
8	Octane	19	Nonadecane	29	Nonacosane	90	Nonacontane
9	Nonane	20	Eicosane	30	Triacontane	100	Hectane
10	Decane	21	Heneicosane	31	Hentriacontane	132	Dotriacontahectane
11	Undecane						

1.2.—Univalent radicals derived from saturated unbranched acyclic hydrocarbons by removal of hydrogen from a terminal carbon atom are named by replacing the ending "-ane" of the name of the hydrocarbon by "-yl." The carbon atom with the free valence is numbered as 1. As a class, these radicals are called normal, or unbranched-chain, alkyls.

Examples:

Pentyl $\overset{5}{C}H_3-\overset{4}{C}H_2-\overset{3}{C}H_2-\overset{2}{C}H_2-\overset{1}{C}H_2-$

Undecyl $\overset{11}{C}H_3-[\overset{10-2}{C}H_2]_9-\overset{1}{C}H_2-$

A-2

2.1.—A saturated branched acyclic hydrocarbon is named by prefixing the designations of the side chains to the name of the longest chain present in the formula.

Example:

$$\overset{5}{C}H_3-\overset{4}{C}H_2-\overset{3}{C}H-\overset{2}{C}H_2-\overset{1}{C}H_3$$
$$|$$
$$CH_3$$

3-Methylpentane

DEFINITIVE RULES FOR NOMENCLATURE OF ORGANIC CHEMISTRY

These names are retained for unsubstituted hydrocarbons only

Isobutane	$(CH_3)_2CH—CH_3$
Isopentane	$(CH_3)_2CH—CH_2—CH_3$
Neopentane	$(CH_3)_4C$
Isohexane	$(CH_3)_2CH—CH_2—CH_2—CH_3$

Chemical Abstracts index names for the above compounds are 2-methylpropane, 2-methylbutane, 2,2-dimethylpropane, and 2-methylpentane.

2.2.—The longest chain is numbered from one end to the other by arabic numerals, the direction being so chosen as to give the lowest numbers possible to the side chains. When series of locants containing the same number of terms are compared term by term, that series is "lowest" which contains the lowest number on the occasion of the first difference. This principle is applied irrespective of the nature of the substituents.

Examples:

$$\overset{5}{C}H_3—\overset{4}{C}H_2—\overset{3}{C}H—\overset{2}{C}H_2—\overset{1}{C}H_3$$
$$|$$
$$CH_3$$

3-Methylpentane

$$\overset{6}{C}H_3—\overset{5}{C}H—\overset{4}{C}H_2—\overset{3}{C}H—\overset{2}{C}H—\overset{1}{C}H_3$$
$$\quad\;|\qquad\qquad\;|\quad\;|$$
$$\quad CH_3\qquad\quad CH_3\;CH_3$$

2,3,5-Trimethylhexane (not 2,4,5-Trimethylhexane)

$$\overset{10}{C}H_3—\overset{9}{C}H_2—\overset{8}{C}H—\overset{7}{C}H—\overset{6}{C}H_2—\overset{5}{C}H_2—\overset{4}{C}H_2—\overset{3}{C}H_2—\overset{2}{C}H—\overset{1}{C}H_3$$
$$\qquad\qquad\;|\quad\;|\qquad\qquad\qquad\qquad\qquad\;|$$
$$\qquad\qquad CH_3\;CH_3\qquad\qquad\qquad\qquad\; CH_3$$

2,7,8-Trimethyldecane (not 3,4,9-Trimethyldecane)

$$\overset{9}{C}H_3—\overset{8}{C}H_2—\overset{7}{C}H_2—\overset{6}{C}H_2—\overset{5}{C}H—\overset{4}{C}H—\overset{3}{C}H_2—\overset{2}{C}H_2—\overset{1}{C}H_3$$
$$\qquad\qquad\qquad\qquad\qquad\;|\qquad\;|$$
$$\qquad\qquad\qquad\qquad\; CH_3\; CH_2—CH_2—CH_3$$

5-Methyl-4-propylnonane (not 5-Methyl-6-propylnonane since 4,5 is lower than 5,6)

2.25.—Univalent branched radicals derived from alkanes are named by prefixing the designation of the side chains to the name of the unbranched alkyl radical possessing the longest possible chain starting from the carbon atom with the free valence, the starting atom being numbered as 1.

Examples:

1-Methylpentyl	$\overset{5}{C}H_3\overset{4}{C}H_2\overset{3}{C}H_2\overset{2}{C}H_2\overset{1}{C}H(CH_3)—$
2-Methylpentyl	$CH_3CH_2CH_2CH(CH_3)CH_2—$
5-Methylhexyl	$(CH_3)_2CHCH_2CH_2CH_2CH_2—$

These names are retained for the unsubstituted radicals only

Isopropyl	$(CH_3)_2CH—$
Isobutyl	$(CH_3)_2CHCH_2—$
sec-Butyl	$CH_3CH_2CH—$
	$\qquad\qquad\;\;|$
	$\qquad\qquad CH_3$
tert-Butyl	$(CH_3)_3C—$
Isopentyl	$(CH_3)_2CHCH_2CH_2—$
Neopentyl	$(CH_3)_3CCH_2—$
	$\qquad\qquad\quad CH_3$
	$\qquad\qquad\quad\;\;|$
tert-Pentyl	$CH_3CH_2C—$
	$\qquad\qquad\quad\;\;|$
	$\qquad\qquad\quad CH_3$
Isohexyl	$(CH_3)_2CHCH_2CH_2CH_2—$

2.3.—If two or more side chains of different nature are present, they may be cited (a) in order of increasing complexity or (b) in alphabetical order.

(a) The side chains are arranged in order of increasing complexity by applying the following criteria in series until a decision is reached:

(i) The less complex is that containing the smaller total number of carbon atoms.

Examples:

$$CH_3$$
$$|$$
$$CH_3—C—\quad \text{less complex than}$$
$$|$$
$$CH_3$$

$$\overset{5}{C}H_3—\overset{4}{C}H_2—\overset{3}{C}H_2—\overset{2}{C}H_2—\overset{1}{C}H_2—$$

C-2

(ii) The less complex is that containing the longer straight chain.

Example:

$$\overset{4}{C}H_3-\overset{3}{C}H_2-\overset{2}{C}H-\overset{1}{C}H_2- \quad \text{less complex than} \quad \overset{3}{C}H_3-\overset{2}{C}-\overset{1}{C}H_2-$$

(with CH_3 substituents above on both, and a lower CH_3 on the right structure)

(iii) The less complex is that whose longest substituent has the lower locant.

Example:

$$\overset{5}{C}H_3-\overset{4}{C}H_2-\overset{3}{C}H-\overset{2}{C}H-\overset{1}{C}H_2- \quad \text{less complex than} \quad \overset{5}{C}H_3-\overset{4}{C}H_2-\overset{3}{C}H-\overset{2}{C}H-\overset{1}{C}H_2-$$

(with CH_3, CH_2 substituents; left has CH_3 CH_2 above positions, right has CH_2 CH_3 and a CH_3 above)

(iv) The less complex is that whose next longest substituent has the lower locant.

Example:

$$\overset{5}{C}H_3-\overset{4}{C}H_2-\overset{3}{C}H-\overset{2}{C}H-\overset{1}{C}H_2- \quad \text{less complex than} \quad \overset{5}{C}H_3-\overset{4}{C}H-\overset{3}{C}H-\overset{2}{C}H_2-\overset{1}{C}H_2-$$

(with CH_3, CH_2 CH_3 substituents)

(v) The less complex is that which is the more saturated.

Example:

$$\overset{3}{C}H_3-\overset{2}{C}H_2-\overset{1}{C}H_2- \quad \text{less complex than} \quad \overset{3}{C}H_3-\overset{2}{C}H=\overset{1}{C}H-$$

(vi) The less complex is that whose multiple linkage has the lower locant.

Example:

$$\overset{3}{C}H_3-\overset{2}{C}H=\overset{1}{C}H- \quad \text{less complex than} \quad \overset{3}{C}H_2=\overset{2}{C}H-\overset{1}{C}H_2-$$

(b) The alphabetical order is decided as follows:

(i) The names of simple radicals are first alphabetized and the multiplying prefixes are then inserted.

Example:

$$\overset{7}{C}H_3-\overset{6}{C}H_2-\overset{5}{C}H_2-\overset{4}{C}H-\overset{3}{C}-\overset{2}{C}H_2-\overset{1}{C}H_3$$

(with CH_3-CH_2 and CH_3 substituents above position 4 and 3, and a CH_3 below position 3)

ethyl is cited before methyl, thus 4-Ethyl-3,3-dimethylheptane

(ii) The name of a complex radical is considered to begin with the first letter of its complete name.

Example:

$$CH_3-\overset{1}{C}H-\overset{2}{C}H-\overset{3}{C}H_2-\overset{4}{C}H_2-\overset{5}{C}H_3$$

$$\overset{13}{C}H_3-[\overset{12-8}{C}H_2]_5-\overset{7}{C}H-\overset{6}{C}H_2-\overset{5}{C}H-\overset{4}{C}H_2-\overset{3}{C}H_2-\overset{2}{C}H_2-\overset{1}{C}H_3$$

(with CH_3 above position 2, and CH_2-CH_3 below position 5)

dimethylpentyl (as a complete single substituent) is alphabetized under "d," thus
7-(1,2-Dimethylpentyl)-5-ethyltridecane

(iii) In cases where complex radicals are composed of identical words, priority for citation is given to that radical which contains the lowest locant at the first cited point of difference in the radical.

Example:

$$CH_3-CH_2-\overset{}{C}H-CH_2 \qquad \overset{}{C}H-CH_2-CH_2-CH_3$$

$$\overset{13}{C}H_3-[\overset{12-9}{C}H_2]_4-\overset{8}{C}H-\overset{7}{C}H_2-\overset{6}{C}H-\overset{5}{C}H_2-\overset{4}{C}H_2-\overset{3}{C}H_2-\overset{2}{C}H_2-\overset{1}{C}H_3$$

(with CH_3 substituents above)

6-(1-Methylbutyl)-8-(2-methylbutyl)tridecane

Chemical Abstracts has long been a proponent of the alphabetical order for prefixes. The alphabetical order as given in A-2.3(b) is the order used by Chemical Abstracts.

2.4.—If two or more side chains are in equivalent positions, the one to be assigned the lower number is that cited first in the name, whether the order of citation is based on complexity or on the alphabetical order.

Examples:

(a) Order based on complexity

$$\overset{8}{C}H_3-\overset{7}{C}H_2-\overset{6}{C}H_2-\overset{5}{C}H-\overset{4}{C}H-\overset{3}{C}H_2-\overset{2}{C}H_2-\overset{1}{C}H_3$$

CH$_3$—CH$_2$ CH$_3$

4-Methyl-5 ethyloctane

(b) Alphabetical order

$$\overset{8}{C}H_3-\overset{7}{C}H_2-\overset{6}{C}H_2-\overset{5}{C}H-\overset{4}{C}H-\overset{3}{C}H_2-\overset{2}{C}H_2-\overset{1}{C}H_3$$

CH$_3$ CH$_2$—CH$_3$

4 Ethyl-5-methyloctane

$$\overset{8}{C}H_3-\overset{7}{C}H_2-\overset{6}{C}H_2-\overset{5}{C}H-\overset{4}{C}H-\overset{3}{C}H_2-\overset{2}{C}H_2-\overset{1}{C}H_3$$

CH$_3$—CH CH$_2$

CH$_3$ CH$_2$—CH$_3$

4-Propyl-5-isopropyloctane

$$\overset{8}{C}H_3-\overset{7}{C}H_2-\overset{6}{C}H_2-\overset{5}{C}H-\overset{4}{C}H-\overset{3}{C}H_2-\overset{2}{C}H_2-\overset{1}{C}H_3$$

CH$_2$ CH—CH$_3$

CH$_3$—CH$_2$ CH$_3$

4-Isopropyl-5-propyloctane

2.5.—The presence of identical unsubstituted radicals is indicated by the appropriate multiplying prefix di-, tri-, tetra-, penta-, hexa-, hepta-, octa-, nona-, deca-, undeca-, *etc*.

Example:

CH$_3$

$$\overset{5}{C}H_3-\overset{4}{C}H_2-\overset{3}{C}-\overset{2}{C}H_2-\overset{1}{C}H_3$$

CH$_3$

3,3-Dimethylpentane

The presence of identical radicals each substituted in the same way may be indicated by the appropriate multiplying prefix bis-, tris-, tetrakis-, pentakis-, *etc*. The complete expression denoting a side chain may be enclosed in parentheses or the carbon atoms in side chains may be indicated by primed numbers.

Example:

CH$_3$

$$\overset{3*}{C}H_3-\overset{2*}{C}H_2-\overset{1*}{C}-CH_3$$

$$\overset{10}{C}H_3-\overset{9}{C}H_2-\overset{8}{C}H_2-\overset{7}{C}H_2-\overset{6}{C}H_2-\overset{5}{C}-\overset{4}{C}H_2-\overset{3}{C}H_2-\overset{2}{C}H-\overset{1}{C}H_3$$

CH$_3$—CH$_2$—C—CH$_3$ CH$_3$

CH$_3$

(a) Use of primes and order of complexity, * indicates primed numbers: 2-Methyl-5,5-bis-1′,1′-dimethylpropyldecane.
(b) Use of parentheses and alphabetical order, * indicates unprimed numbers: 5,5-Bis(1,1-dimethylpropyl)-2-methyldecane.
(c) Use of primes and alphabetical order, * indicates primed numbers: 5,5-Bis-1′,1′-dimethylpropyl-2-methyldecane.

Chemical Abstracts uses parentheses and the alphabetical order without primes (A-2.5(b)). *Chemical Abstracts* would name the example in (b) 2-Methyl-5,5-di-*tert*-pentyldecane.

CH$_3$

$$\overset{4*}{C}H_3-\overset{3*}{C}H_2-\overset{2*}{C}H_2-\overset{1*}{C}-CH_3$$

$$\overset{13}{C}H_3-\overset{12-10}{[CH_2]_3}-\overset{9}{C}H_2-\overset{8}{C}H_2-\overset{7}{C}-\overset{6}{C}H_2-\overset{5}{C}H_2-\overset{4}{C}H_2-\overset{3}{C}H_2-\overset{2}{C}H_2-\overset{1}{C}H_3$$

$$\overset{5**}{C}H_3-\overset{4**}{C}H_2-\overset{3**}{C}H_2-\overset{2**}{C}H_2-\overset{1**}{C}-CH_3$$

CH$_3$

(a) Use of primes and order of complexity, * indicates primed numbers, ** indicates doubly primed numbers: 7-1′,1′-Dimethylbutyl-7-1″,1″-dimethylpentyltridecane.
(b) Use of parentheses and alphabetical order, * and ** indicate unprimed numbers: 7-(1,1-Dimethylbutyl)-7-(1,1-dimethylpentyl)-tridecane.

2.6.—If chains of equal length are competing for selection as main chain in a saturated branched acyclic hydrocarbon, then the choice goes in series to:

(a) The chain which has the greatest number of side chains.

Example:

$$\overset{7}{C}H_3-\overset{6}{C}H_2-\overset{5}{C}H-\overset{4}{C}H-\overset{3}{C}H-\overset{2}{C}H-\overset{1}{C}H_3$$

CH$_3$ CH$_2$ CH$_3$ CH$_3$

CH$_2$—CH$_3$

2,3,5-Trimethyl-4-propylheptane

(b) The chain whose side chains have the lowest-numbered locants.

Example:

$$\overset{7}{C}H_3-\overset{6}{C}H_2-\overset{5}{C}H-\overset{4}{C}H-\overset{3}{C}H_2-\overset{2}{C}H-\overset{1}{C}H_3$$

with side chains:

$CH_3 \quad CH_2 \quad CH_3$

$CH-CH_3$

CH_3

4-Isobutyl-2,5-dimethylheptane

(c) The chain having the greatest number of carbon atoms in the smaller side chains.

$$CH_3 \qquad CH_3 \qquad CH_3 \qquad CH_2-CH_3$$
$$\underset{13}{CH_3}\underset{12}{CH_2}\underset{11}{CH}-\underset{10}{CH}-\underset{9}{CH}-\underset{8}{CH_2}\quad \underset{7}{C}\quad \underset{6}{CH_2}-\underset{5}{CH}-\underset{4}{CH_2}-\underset{3}{CH}\underset{2}{CH_2}\underset{1}{CH_3}$$

$CH_3CH_2CH-CH_2-CH-CH_2 \qquad CH_2-CH-CH_2-CHCH_2CH_3$

$CH_3 \qquad CH_3 \qquad CH_3 \qquad CH_3$

7,7-Bis(2,4-dimethylhexyl)-3-ethyl-5,9,11-trimethyltridecane

(d) The chain having the least branched side chains.

(1) Here the choice lies between two possible main chains of equal length, each containing six side chains in the same positions. Listing in increasing order, the number of carbon atoms in the several side chains of the first choice as shown and of the alternate second choice results as follows:

first choice   1, 1, 1, 2, 8, 8

second choice   1, 1, 1, 1, 8, 9

The expression, "the greatest number of carbon atoms in the smaller side chains," is taken to mean the largest side chain at the first point of difference when the size of the side chains is examined step by step. Thus, the selection in this case is made at the fourth step where 2 is greater than 1.

$$CH_3-CH_2-\overset{1}{C}H-\overset{2}{C}H_2-\overset{3}{C}H_3$$
$$\overset{14}{CH_3}-\overset{13-8}{[CH_2]_6}-\overset{7}{C}H-\overset{6}{C}H-\overset{5}{C}H_2-\overset{4}{C}H_2-\overset{3}{C}H_2-\overset{2}{C}H-\overset{1}{C}H_3$$
$$CH_3-[CH_2]_3-\overset{1}{C}H_2-\overset{2}{C}H-\overset{3}{C}H_2-\overset{4}{C}H_2-\overset{5}{C}H_2-\overset{6}{C}H_2-CH_3$$

6-(1-Ethylpropyl)-7-(1-pentylhexyl)tetradecane

A-3

3.1.—Unsaturated unbranched acyclic hydrocarbons having one double bond are named by replacing the ending "-ane" of the name of the corresponding saturated hydrocarbon with the ending "-ene." If there are two or more double bonds, the ending will be "-adiene," "-atriene," *etc*. The generic names of these hydrocarbons (branched or unbranched) are "alkene," "alkadiene," "alkatriene," *etc*. The chain is so numbered as to give the lowest possible numbers to the double bonds.

Examples:

2-Hexene $\qquad \overset{6}{C}H_3-\overset{5}{C}H_2-\overset{4}{C}H_2-\overset{3}{C}H=\overset{2}{C}H-\overset{1}{C}H_3$

1,4-Hexadiene $\qquad \overset{6}{C}H_3-\overset{5}{C}H=\overset{4}{C}H-\overset{3}{C}H_2-\overset{2}{C}H=\overset{1}{C}H_2$

These non-systematic names are retained

Ethylene $\qquad CH_2=CH_2 \qquad$ Allene $\qquad CH_2=C=CH_2$

Chemical Abstracts retains allene for the unsubstituted hydrocarbon only.

3.2.—Unsaturated unbranched acyclic hydrocarbons having one triple bond are named by replacing the ending "-ane" of the name of the corresponding saturated hydrocarbon with the ending "-yne." If there are two or more triple bonds, the ending will be "-adiyne," "-atriyne," *etc*. The generic names of these hydrocarbons (branched or unbranched) are "alkyne," "alkadiyne," "alkatriyne," *etc*. The chain is so numbered as to give the lowest possible numbers to the triple bonds.

The name "acetylene" for $HC\equiv CH$ is retained.

3.3.—Unsaturated unbranched acyclic hydrocarbons having both double and triple bonds are named by replacing the ending "-ane" of the name of the corresponding saturated hydrocarbon with the ending "-enyne," "-adienyne," "-atrienyne," "-enediyne," *etc*. Numbers as low as possible are given to double

DEFINITIVE RULES FOR NOMENCLATURE OF ORGANIC CHEMISTRY

and triple bonds even though this may at times give "-yne" a lower number than "-ene." When there is a choice in numbering the double bonds are given the lowest numbers.

Examples:

1,3-Hexadien-5-yne $\overset{6}{HC}\equiv\overset{5}{C}-\overset{4}{CH}=\overset{3}{CH}-\overset{2}{CH}=\overset{1}{CH_2}$

3-Penten-1-yne $\overset{5}{CH_3}-\overset{4}{CH}=\overset{3}{CH}-\overset{2}{C}\equiv\overset{1}{CH}$

1-Penten-4-yne $\overset{5}{HC}\equiv\overset{4}{C}-\overset{3}{CH_2}-\overset{2}{CH}=\overset{1}{CH_2}$

3.4.—Unsaturated branched acyclic hydrocarbons are named as derivatives of the unbranched hydrocarbons which contain the maximum number of double and triple bonds. If there are two or more chains competing for selection as the chain with the maximum number of unsaturated bonds, then the choice goes to (1) that one with the greatest number of carbon atoms; (2) the number of carbon atoms being equal, the one containing the maximum number of double bonds. In other respects, the same principles apply as for naming saturated branched acyclic hydrocarbons. The chain is so numbered as to give the lowest possible numbers to double and triple bonds in accordance with Rule **A-3.3.**

Examples:

3,4-Dipropyl-1,3-hexadien-5-yne
$$\begin{array}{c}CH_2-CH_2-CH_3\\ |\\ \overset{6}{CH}\equiv\overset{5}{C}-\overset{4}{C}=\overset{3}{C}-\overset{2}{CH}=\overset{1}{CH_2}\\ |\\ CH_2-CH_2-CH_3\end{array}$$

5-Ethynyl-1,3,6-heptatriene
$$\begin{array}{c}\overset{7}{CH_2}=\overset{6}{CH}-\overset{5}{CH}-\overset{4}{CH}=\overset{3}{CH}-\overset{2}{CH}=\overset{1}{CH_2}\\ |\\ C\equiv CH\end{array}$$

5,5-Dimethyl-1-hexene
$$\begin{array}{c}CH_3\\ |\\ \overset{6}{CH_3}-\overset{5}{C}-\overset{4}{CH_2}-\overset{3}{CH_2}-\overset{2}{CH}=\overset{1}{CH_2}\\ |\\ CH_3\end{array}$$

4-Vinyl-1-hepten-5-yne
$$\begin{array}{c}\overset{7}{CH_3}-\overset{6}{C}\equiv\overset{5}{C}-\overset{4}{CH}-\overset{3}{CH_2}-\overset{2}{CH}=\overset{1}{CH_2}\\ |\\ CH=CH_2\end{array}$$

The name "isoprene" is retained for the unsubstituted compound only

$$\begin{array}{c}CH_3\\ |\\ CH_2=CH-C=CH_2\end{array}$$

3.5.—The names of univalent radicals derived from unsaturated acyclic hydrocarbons have the endings "-enyl," "-ynyl," "-dienyl," *etc.*, the positions of the double and triple bonds being indicated where necessary. The carbon atom with the free valence is numbered as 1.

Examples:

Ethynyl	$CH\equiv C-$
2-Propynyl	$CH\equiv C-CH_2-$
1-Propenyl	$CH_3-CH=CH-$
2-Butenyl	$CH_2-CH=CH-CH_2-$
1,3-Butadienyl	$CH_2=CH-CH=CH-$
2-Pentenyl	$CH_3-CH_2-CH=CH-CH_2-$
2-Penten-4-ynyl	$CH\equiv C-CH=CH-CH_2-$

Exceptions: These names are retained

Vinyl (for ethenyl)	$CH_2=CH-$	
Allyl (for 2-propenyl)	$CH_2=CH-CH_2-$	
Isopropenyl	$CH_2=C-$	(for unsubstituted radical only)
(for 1-methylvinyl)	$\quad\ \ |$	
	$\quad CH_3$	

3.6.—When there is a choice for the fundamental chain of a radical, that chain is selected which contains (1) the maximum number of double and triple bonds; (2) the largest number of carbon atoms, and (3) the largest number of double bonds.

Examples:

$$\begin{array}{c}\overset{10}{CH_3}-\overset{9}{CH}=\overset{8}{CH}-\overset{7}{CH}=\overset{6}{CH}-\overset{5}{CH}-\overset{4}{CH}=\overset{3}{CH}-\overset{2}{C}\equiv\overset{1}{C}-\\ |\\ CH_2-CH_2-CH=CH-CH_3\end{array}$$

5-(3-Pentenyl)-3,6,8-decatrien-1-ynyl

$$\overset{12}{C}H_3-\overset{11}{C}H_2-\overset{10}{C}\equiv\overset{9}{C}-\overset{8}{C}H=\overset{7}{C}H-\overset{6}{C}H-\overset{5}{C}H=\overset{4}{C}H-\overset{3}{C}H=\overset{2}{C}H-\overset{1}{C}H_2-$$
$$CH=CH-CH=CH-CH_3$$

6-(1,3-Pentadienyl)-2,4 7-dodecatrien-9-ynyl

$$\overset{11}{C}H_3-\overset{10}{C}H=\overset{9}{C}H-\overset{8}{C}H=\overset{7}{C}H-\overset{6}{C}H-\overset{5}{C}H=\overset{4}{C}H-\overset{3}{C}H=\overset{2}{C}H-\overset{1}{C}H_2-$$
$$CH=CH-C\equiv C-CH_3$$

6-(1-Penten-3-ynyl)-2,4,7,9-undecatetraenyl

$$\overset{4}{C}H_3-\overset{3}{C}H=\overset{2}{C}-\overset{1}{C}H_2-$$
$$CH_2 \quad CH_2-CH_2-CH_2-CH_2-CH_2-CH_2-CH_2-CH_3$$

2-Nonyl-2-butenyl

A-4

4.1.—Bivalent and trivalent radicals derived from univalent acyclic hydrocarbon radicals whose authorized names end in "-yl" by removal of one or two hydrogen atoms from the carbon atom with the free valences are named by adding "-idene" or "-idyne," respectively, to the name of the corresponding univalent radical. The carbon atom with the free valence is numbered as 1.

The name "methylene" is retained for the radical $CH_2=$.

Examples:

Methylidyne[1]	$CH\equiv$	
Ethylidene	$CH_3-CH=$	
Ethylidyne	$CH_3-C\equiv$	
Vinylidene	$CH_2=C=$	
Isopropylidene[2]	$(CH_3)_2C=$	

4.2.—The names of bivalent radicals derived from normal alkanes by removal of a hydrogen atom from each of the two terminal carbon atoms of the chain are ethylene, trimethylene, tetramethylene, *etc.*

Examples:

Pentamethylene	$-CH_2-CH_2-CH_2-CH_2-CH_2-$
Hexamethylene	$-CH_2-CH_2-CH_2-CH_2-CH_2 \ CH_2-$

Names of the substituted bivalent radicals are derived in accordance with Rules **A-2.2** and **A-2.25.**

Example:

Ethylethylene $-\overset{2}{C}H_2-\overset{1}{C}H-$
$$CH_2-CH_3$$

The name "propylene" is retained

$$CH_3-CH-CH_2-$$

4.3.—Bivalent radicals similarly derived from unbranched alkenes, alkadienes, alkynes, *etc.*, by removing a hydrogen atom from each of the terminals carbon atoms are named by replacing the endings "-ene," "-diene," "-yne," *etc.*, of the hydrocarbon name by "-enylene," "-dienylene," "-ynylene," *etc.*, the positions of the double and triple bonds being indicated where necessary.

Example:

Propylene $-\overset{3}{C}H_2-\overset{2}{C}H=\overset{1}{C}H-$

The name "vinylene" is retained (for ethenylene)

$$-CH=CH-$$

Names of the substituted bivalent radicals are derived in accordance with Rule **A-3.4.**

Examples:

4 Propyl-2-pentenylene $-\overset{5}{C}H_2-\overset{4}{C}H-\overset{3}{C}H=\overset{2}{C}H-\overset{1}{C}H_2-$
$$CH_2-CH_2-CH_3$$

4.4.—Trivalent, quadrivalent and higher valent acyclic hydrocarbon radicals of two or more carbon atoms with the free valences at each end of a chain are named by adding to the hydrocarbon name the

---

(1) The group $=CH-$ may be referred to as the "methine" group.
(2) For unsubstituted radical only.

terminations "-yl" for a single free valence, "-ylidene" for a double and "-ylidyne" for a triple free valence on the same atom (the final "e" in the name of the hydrocarbon is elided when followed by a suffix beginning with "-yl"). If different types are present in the same radical, they are cited and numbered in the order of "-yl" "-ylidene," "-ylidyne."

Examples:

Butanediylidene	$=\overset{4}{C}H-\overset{3}{C}H_2-\overset{2}{C}H_2-\overset{1}{C}H=$
Butanediylidyne	$\equiv\overset{4}{C}-\overset{3}{C}H_2-\overset{2}{C}H_2-\overset{1}{C}\equiv$
1-Propanyl-3-ylidene	$=\overset{3}{C}H-\overset{2}{C}H_2-\overset{1}{C}H_2-$
Propadienediylidene	$=\overset{3}{C}=\overset{2}{C}=\overset{1}{C}=$
2-Pentenediylidyne	$\equiv\overset{5}{C}-\overset{4}{C}H_2-\overset{3}{C}H=\overset{2}{C}H-\overset{1}{C}\equiv$
1-Butanyliden-4-ylidyne	$\equiv\overset{4}{C}-\overset{3}{C}H_2-\overset{2}{C}H_2-\overset{1}{C}H=$

4.5.—Multivalent radicals containing three or more carbon atoms with free valences at each end of a chain and additional free valences at intermediate carbon atoms are named by adding the ending "-triyl," "-tetrayl," "-diylidene," "diylylidene," *etc.*, to the hydrocarbon name.

Examples:

$$-\overset{3}{C}H_2-\overset{2}{C}H-\overset{1}{C}H_2- \qquad\qquad -\overset{3}{C}H_2-\overset{2}{C}-\overset{1}{C}H_2-$$

1,2,3-Propanetriyl   1,3-Propanediyl-2-ylidene

Monocyclic Hydrocarbons

A-11

11.1.—The names of saturated monocyclic hydrocarbons (with no side chains) are formed by attaching the prefix "cyclo" to the name of the acyclic saturated unbranched hydrocarbon with the same number of carbon atoms. The generic name of saturated monocyclic hydrocarbons (with or without side chains) is "cycloalkane."

Examples:

Cyclopropane   Cyclohexane

11.2.—Univalent radicals derived from cycloalkanes (with no side chains) are named by replacing the ending "-ane" of the hydrocarbon name by "-yl," the carbon atom with the free valence being numbered as 1. The generic name of these radicals is "cycloalkyl."

Examples:

Cyclopropyl   Cyclohexyl

11.3.—The names of unsaturated monocyclic hydrocarbons (with no side chains) are formed by substituting "-ene," "-adiene," "-atriene," "-yne," "-adiyne," *etc.*, for "-ane" in the name of the corresponding cycloalkane. The double and triple bonds are given numbers as low as possible as in Rule **A-3.3.**

Examples:

Cyclohexene   1,3-Cyclohexadiene   1-Cyclodecen-4-yne

The names "fulvene" (for methylenecyclopentadiene) and "benzene" are retained.

11.4.—The names of univalent radicals derived from unsaturated monocyclic hydrocarbons have the endings "-enyl," "-ynyl," "-dienyl," *etc.*, the positions of the double and triple bonds being indicated according to the principles of Rule **A-3.3**. The carbon atom with the free valence is numbered as 1, except as stated in the rules for terpenes (see Rules **A-72** to **A-75**).

Examples:

2-Cyclopenten-1-yl · 2,4-Cyclopentadien-1-yl

The radical name "phenyl" is retained.

The point of attachment is numbered 1 and the double and triple bonds are given numbers as low as possible. The number 1 for the point of attachment is given in the radical name to emphasize the fact that the point of attachment takes precedence over the double and triple bonds (except with terpenes which have fixed numberings).

11.5.—Names of bivalent radicals derived from saturated or unsaturated monocyclic hydrocarbons by removal of two atoms of hydrogen from the same carbon atom of the ring are obtained by replacing the endings "-ane," "-ene," "-yne," by "-ylidene," "-enylidene" and "-ynylidene," respectively. The carbon atom with the free valences is numbered as 1, except as stated in the rules for terpenes.

Examples:

Cyclopentylidene · 2,4-Cyclohexadien-1-ylidene

11.6.—Bivalent radicals derived from saturated or unsaturated monocyclic hydrocarbons by removing a hydrogen atom from each of two different carbon atoms of the ring are named by replacing the endings "-ane," "-ene," "-diene," "-yne," *etc.*, of the hydrocarbon name by "-ylene," "-enylene," "-dienylene," "-ynylene," *etc.*, the positions of the double and triple bonds and of the points of attachment being indicated. Preference in lowest numbers is given to the carbon atoms having the free valences.

Examples:

1,3-Cyclopentylene · 3-Cyclohexen-1,2-ylene · 2,5-Cyclohexadien-1,4-ylene

The name "phenylene" is retained

Phenylene (*p*-shown)

A-12

12.1.—These names for substituted monocyclic aromatic hydrocarbons are retained

Cumene · Cymene (*p*-shown) · Mesitylene

Styrene · Toluene · Xylene (*o*-shown)

12.2.—Other substituted monocyclic aromatic hydrocarbons are named as derivatives of benzene or of one of the compounds listed in Part **1** of this rule. However, if the substituent introduced into such a compound, is identical with one already present in that compound, then the substituted compound is named as a derivative of benzene (see Rule **61.4**).

Chemical Abstracts makes certain exceptions to the last part of this rule in order to favor a larger parent, although still treating like groups alike, *e.g.*, an *ar*-methyl derivative of cymene is indexed at Cumene, dimethyl-, not at Benzene, isopropyldimethyl-.

12.3.—The position of substituents is indicated by numbers except that *o*- (*ortho*-), *m*- (*meta*-) and *p*- (*para*-) may be used in place of 1,2-, 1,3-, and 1,4-, respectively, when only two substituents are present. The lowest numbers possible are given to substituents, choice between alternatives being governed by Rule **A-2** so far as applicable, except that when names are based on those of compounds listed in Part **1** of this rule the first priority for lowest numbers is given to the substituent(s) already present in those compounds.

Examples:

1-Ethyl-4-pentylbenzene or *p*-Ethylpentylbenzene

1,4-Diethyl-benzene or *p*-Diethyl-benzene

4-Ethylsty-rene or *p*-Ethylstyrene

Chemical Abstracts limits the use of *o*-, *m*- and *p*- to like substituents when both are expressed as prefixes, *e.g.*, *p*-diethylbenzene, 1-ethyl-4-pentylbenzene. However, when one substituent is expressed in the name, *m*-, *o*- or *p*- is used to indicate the position of the second and different substituent, *e.g.*, *o*-ethylcumene.

1,4-Divinylben-zene or *p*-Divinyl-benzene, not *p*-Vinylstyrene

1,2,3-Trimethyl-benzene, not Methylxylene nor Dimethyl-toluene

1,2-Dimethyl-3-propylbenzene or 3-Propyl-*o*-xylene

1-Ethyl-2-propyl-3-butylbenzene (Order of complexity)
or 1-Butyl-3-ethyl-2-propylbenzene (Alphabetical order)

12.4.—The generic name of monocyclic and polycyclic aromatic hydrocarbons is "arene."

A-13

13.1.—Univalent radicals derived from monocyclic aromatic hydrocarbons and having the free valence at a ring atom are given the names listed below. Such radicals not listed below are named as substituted phenyl radicals. The carbon atom having the free valence is numbered as 1.

Phenyl C_6H_5-

Cumenyl (*m*-shown)

Mesityl

Tolyl (*o*-shown)

Xylyl (2,3-shown)

13.2.—Since the name phenylene (*o-*, *m-*, or *p-*) is retained for the radical —C_6H_4— (exception to Rule **A-11.6**) bivalent radicals formed from substituted benzene derivatives and having the free valences at ring atoms are named as substituted phenylene radicals. The carbon atoms having the free valences are numbered 1,2-, 1,3-, or 1,4- as appropriate.

13.3.—These trivial names for radicals having a single free valence in the side chain are retained:

Benzyl $\qquad$ C_6H_5—$\overset{\alpha}{C}H_2$—

Cinnamyl $\qquad$ C_6H_5—$\overset{\gamma}{C}H$=$\overset{\beta}{C}H$—$\overset{\alpha}{C}H_2$—

Phenethyl $\qquad$ C_6H_5—$\overset{\beta}{C}H_2$—$\overset{\alpha}{C}H_2$—

Styryl $\qquad$ C_6H_5—$\overset{\beta}{C}H$—$\overset{\alpha}{C}H$—

Trityl $\qquad$ $(C_6H_5)_3C$—

Chemical Abstracts limits trityl to the unsubstituted radical.

13.4.—Multivalent radicals of aromatic hydrocarbons with free valences in the side chain are named in accordance with Rule **A-4.**

Examples: $\qquad$ Benzylidyne $\qquad$ C_6H_5—C≡

$\qquad\qquad\quad$ Cinnamylidene $\qquad$ C_6H_5—$\overset{\gamma}{C}H$=$\overset{\beta}{C}H$—$\overset{\alpha}{C}H$=

A-14

14.1.—The generic names of univalent and bivalent aromatic hydrocarbon radicals are "aryl" and "arylene," respectively.

Fused Polycyclic Hydrocarbons
A-21

21.1.—The names of polycyclic hydrocarbons with maximum number of non-cumulative[1] double bonds end in "-ene." The names listed on pp. 11 and 12 are retained.

21.2.—The names of hydrocarbons containing five or more fused benzene rings in a straight linear arrangement are formed from a numerical prefix as specified in Rule A-1.1 followed by "-acene."

Examples:

Pentacene $\qquad\qquad\qquad\qquad\qquad$ Hexacene

The following list contains the names of polycyclic hydrocarbons which are retained (see Rule **A-21.1**).

(1) Pentalene $\qquad$ (2) Indene $\qquad$ (7) *as*-Indacene $\qquad$ (8) *s*-Indacene

(3) Naphthalene $\qquad$ (4) Azulene $\qquad$ (9) Acenaphthylene $\qquad$ (10) Fluorene

(5) Heptalene $\qquad$ (6) Biphenylene

(1) Cumulative double bonds are those present in a chain in which at least three contiguous carbon atoms are joined by double bonds; non-cumulative double bonds comprise every other arrangement of two or more double bonds in a single structure. The generic name "cumulene" is given to compounds containing three or more cumulative double bonds.

$$CH_2=C=C=C=CH_2$$
Cumulative

$$CH_3—CH=CH—CH=CH—CH=CH_2$$
or

Non-cumulative

(11) Phenalene

(12) Phenanthrene[1]

(27) Hexaphene

(28) Hexacene[2]

(13) Anthracene[1]

(14) Fluoranthene

(29) Rubicene

(30) Coronene

(15) Acephenanthrylene

(16) Aceanthrylene

(17) Triphenylene

(18) Pyrene

(31) Trinaphthylene[3]

(32) Heptaphene

(19) Chrysene

(20) Naphthacene

(33) Heptacene[2]

(21) Pleiadene

(22) Picene

(23) Perylene

(24) Pentaphene

(34) Pyranthrene

(25) Pentacene[2]

(26) Tetraphenylene[3]

(35) Ovalene

Beginning with Volume 51 Subject Index *Chemical Abstracts* uses biphenylene in place of cyclobutadibenzene and phenalene in place of benzonaphthene. The polycyclic hydrocarbons which have the trivial names listed above may be used as the base components in naming orthofused or ortho- and peri-fused polycyclic systems for which there are no acceptable trivial names.

(1) Denotes exception to systematic numbering. See Rule **A-22.5**.

(2) See Rule **A-21.2**.

(3) For isomer shown only.

DEFINITIVE RULES FOR NOMENCLATURE OF ORGANIC CHEMISTRY

21.3.—"Ortho-fused"[4] or "ortho- and peri-fused"[5] polycyclic hydrocarbons with maximum number of noncumulative double bonds which contain at least two rings of five or more members and which have no accepted trivial name such as those of **21.1** of this Rule, are named by prefixing to the name of a component ring or ring system (the base component) designations of the other components. The base component should contain as many rings as possible (provided it has a trivial name), and should occur as far as possible from the beginning of the list of Rule **A-21.1**. The attached components should be as simple as possible.

Example:

(not Naphthophenanthrene; benzo is "simpler" than naphtho, even though there are two benzo rings and only one naphtho)

Dibenzophenanthrene

(4) Polycyclic compounds in which two rings have two, and only two, atoms in common are said to be "ortho-fused." Such compounds have n common faces and $2n$ common atoms (Example I).

(5) Polycyclic compounds in which one ring contains two, and only two, atoms in common with each of two or more rings of a contiguous series of rings are said to be "ortho- and peri-fused." Such compounds have n common faces and fewer than $2n$ common atoms (Examples II and III).

Examples:

3 common faces   7 common faces   5 common faces
6 common atoms   8 common atoms   6 common atoms
"Ortho-fused" system   "Ortho- and peri-fused" systems

21.4.—The prefixes designating attached components are formed by changing the ending "-ene" of the name of the component hydrocarbon into "-eno," *e.g.*, "pyreno" (from pyrene). When more than one prefix is present, they are arranged in alphabetical order. These common abbreviated prefixes are recognized (see list in **21.1** of this Rule)

Acenaphtho	from	Acenaphthylene
Anthra	from	Anthracene
Benzo	from	Benzene
Naphtho	from	Naphthalene
Perylo	from	Perylene
Phenanthro	from	Phenanthrene

For monocyclic prefixes other than "benzo-," the following names are recognized, each to represent the form with the maximum number of non-cumulative double bonds: cyclopenta, cyclohepta, cycloocta, cyclonona, *etc.* When the base component is a monocyclic system, the ending "-ene" signifies the maximum number of noncumulative double bonds, and does not denote one double bond only.

Examples:

1*H*-Cyclopentacycloöctene   Benzocycloöctene

In forming names of polycyclic systems from the above carbocyclic prefixes the final "o" or "a" of the prefix is elided before another vowel, except in the case of anthra, phenanthro and prefixes ending in -eno from which the "a" and "o" are never elided. Examples: benz[a]anthracene, cyclopent[a]acenaphthylene, anthra[1,2 − a]anthracene, fluoreno[4,3,2-de]anthracene.

The ending "-ene" to signify the maximum number of noncumulative double bonds in all fused carbocyclic systems means that the endings -diene, triene, tetraene, *etc.*, are unnecessary when the base component is a monocyclic ring of seven or more carbon atoms. Higher stages of hydrogenation are indicated by the use of the prefixes dihydro-, tetrahydro-, *etc.* (*cf.* Rule **A-23.1**). *Chemical Abstracts* follows Rules **A-21.4** and **A-23.1** instead of Alternate Rule **A-23.5**.

21.5.—Isomers are distinguished by lettering the peripheral sides of the base component *a, b, c, etc.*, beginning with "*a*" for the side "1,2," "*b*" for "2,3" (or in certain cases "2,2a") and lettering every side around the periphery. To the letter as early in the alphabet as possible, denoting the side where fusion occurs, are prefixed, if necessary, the numbers of the positions of attachment of the other component. These numbers are chosen to be as low as is consistent with the numbering of the component, and their order conforms to the direction of lettering of the base component (see Examples II and IV). When two or more prefixes refer to equivalent positions so that there is a choice of letters, the prefixes are cited in alphabetical order according to Rule **A-21.4** and the location of the first cited prefix is indicated by a letter as early as possible in the alphabet (see Example V). The numbers and letters are enclosed in brackets and placed immediately after the designation of the attached component. This expression merely defines the manner of fusion of the components.

Examples:

Benz[*a*]anthracene

6*H*-Naphtho[2,1,8,7-*defg*]-
naphthacene

Dibenz[*a,j*]anthracene
(not Naphtho[2,1-*b*]phenanthrene)

Indeno[1,2-*a*]indene

1*H*-Benzo[*a*]cyclopent[*j*]anthracene

The completed system consisting of the base component and the other components is then renumbered according to Rule **A-22,** the numbering of the component parts being ignored.

Naphthalene + **Perylene** → **Naphtho[8,1,2-*bcd*]-
perylene**

21.6.—When a name applies equally to two or more isomeric condensed parent ring systems with the maximum number of non-cumulative double bonds and when the name can be made specific by indicating the position of one or more hydrogen atoms in the structure, this is accomplished by modifying the name with a locant, followed by italic capital *H* for each of these hydrogen atoms. Such symbols ordinarily precede the name. The said atom or atoms are called "indicated hydrogen." The same principle is applied to radicals and compounds derived from these systems.

Examples:

3*H*-Fluorene   2*H*-Indene

A-22

22.1.—For the purposes of numbering, the individual rings of a polycyclic "ortho-fused" or "ortho- and peri-fused" hydrocarbon system usually are drawn as follows:

◁ or ▷   □   ⬠ or ⬠

⬡   ⬡ or ⬡   ⬡

and the polycyclic system is oriented so that (a) the greatest number of rings are in a horizontal row and (b) a maximum number of rings are above and to the right of the horizontal row (upper right quadrant).

If two or more orientations meet these requirements, the one is chosen which has as few rings as possible in the lower left quadrant.

| Correct orientation | Incorrect orientation | Incorrect orientation |

The system thus oriented is numbered in a clockwise direction commencing with the carbon atom not engaged in ring-fusion in the most counterclockwise position of the uppermost ring, or if there is a choice, of the uppermost ring farthest to the right, and omitting atoms common to two or more rings.

Correct   Incorrect

22.2.—Atoms common to two or more rings are designated by adding roman letters "a," "b," "c," *etc.*, to the number of the position immediately preceding. Interior atoms follow the highest number, taking a clockwise sequence wherever there is a choice.

Examples: *[cf.* Note below]

Correct   Incorrect

22.3.—When there is a choice, carbon atoms common to two or more rings follow the lowest possible numbers.

Examples:

Note: I. 4, 4, 8, 9 is lower than 4, 5, 9, 9.
 II. 2, 5, 8 is lower than 3, 5, 8.
 III. 2, 3, 6, 8 is lower than 3, 4, 6, 8 or 2, 4, 7, 8.

22.4.—When there is a choice, the carbon atoms which carry an indicated hydrogen atom are numbered as low as possible.

Examples:

Correct   Incorrect

22.5.—The following are recommended exceptions to the above rules on numbering

Anthracene

Phenanthrene

Cyclopenta[a]phenanthrene
(15H- shown)
See also rules on steroids

A-23

23.1.—The names of "ortho-fused" or "ortho- and peri-fused" polycyclic hydrocarbons with less than maximum number of non-cumulative double bonds are formed from a prefix "dihydro-," "tetrahydro-," *etc.*, followed by the name of the corresponding unreduced hydrocarbon. The prefix "perhydro-" signifies full hydrogenation. When there is a choice for *H* used for indicated hydrogen it is assigned the lowest available number.

Examples: Exceptions: These names are retained

1,4-Dihydro-
naphthalene

Perhydroanthracene

Indan   Acenaphthene   Cholanthrene

6,7-Dihydro-5H-benzo-
cycloheptene

4,5,6,7,8,9-Hexahydro-
1H-cyclopentacyclooctene

Aceanthrene   Acephenanthrene

16,17-Dihydro-15H-cyclopenta[a]
phenanthrene

Violanthrene   Isoviolanthrene

Chemical Abstracts follows this Rule instead of the alternate Rule 23.5.

23.2.—When there is a choice, the carbon atoms to which hydrogen atoms are added are numbered as low as possible.

Example:

Correct   Incorrect

23.3.—Substituted polycyclic hydrocarbons are named according to the same principles as substituted monocyclic hydrocarbons (see Rules **A-12** and **A-61**).

23.5 (Alternate to part of Rule **A-23.1**).—The names of "ortho-fused" polycyclic hydrocarbons which have (a) less than the maximum number of non-cumulative double bonds, (b) at least one terminal unit which is most conveniently named as an unsaturated cycloalkane derivative, and (c) a double bond at the positions where rings are fused together, may be derived by joining the name of the terminal unit to that of the other component by means of a letter "o" with elision of a terminal "e". The abbreviations for fused aromatic systems laid down in Rule **A-21.4** are used, and the exceptions of Rule **A-23.1** apply.

Examples:

1,2-Benzo-
1,3-cycloheptadiene

1,2-Cyclopenta-
1′,3′-dienocyclooctene

1,2-Cyclopentenophenanthrene

Chemical Abstracts follows Rules **A-21.4** and **A-23.1** instead of this alternate rule.

A-24

24.1.—For radicals derived from polycyclic hydrocarbons, the numbering of the hydrocarbon is retained. The point or points of attachment are given numbers as low as is consistent with the fixed numbering of the hydrocarbon.

24.2.—Univalent radicals derived from "ortho-fused" or "ortho- and peri-fused" polycyclic hydrocarbons with names ending in "-ene" by removal of a hydrogen atom from an aromatic or alicyclic ring are named in principle by changing the ending "-ene" of the names of the hydrocarbons to "-enyl."

Examples:

2-Indenyl   1-Pyrenyl   1-Acenaphthenyl

Exceptions:

Naphthyl
(2-shown)

Anthryl
(2-shown)

Phenanthryl
(2-shown)

5,6,7,8-Tetrahydro-
2-naphthyl

The above exceptions are limited to the simple rings as shown. Radicals from names of fused derivatives of these rings are formed according to the Rule, *e.g.*, benz[a]anthracenyl.

24.3.—Bivalent radicals derived from univalent polycyclic hydrocarbon radicals whose names end in "-yl", by removal of one hydrogen atom from the carbon atom with the free valence are named by adding "-idene" to the name of the corresponding univalent radical.

Example:

1-Acenaphthenylidene

24.4.—Bivalent radicals derived from "ortho-fused" or "ortho- and peri-fused" polycyclic hydrocarbons by removing of a hydrogen atom from each of two different carbon atoms of the ring are named by changing the ending "-yl" of the univalent radical name to "-ylene."

Examples:

2,7-Phenanthrylene

3,8-Acenaphthenylene

Chemical Abstracts names multivalent radicals with three or more free valences derived from polycyclic hydrocarbons by an extension of Rule **B-5.13** to carbocyclic systems. Example: 1,4,5-anthracenetriyl.

A-28

28.1.—Radicals formed from hydrocarbons consisting of polycyclic systems and side chains are named according to the principles of the preceding rules.

Bridged Hydrocarbons

A-31. von Baeyer System

31.1.—Saturated alicyclic hydrocarbon systems consisting of two rings only, having two or more atoms in common, take the name of an open chain hydrocarbon containing the same total number of carbon atoms preceded by the prefix "bicyclo-." The number of carbon atoms in each of the three bridges[1] connecting the two tertiary carbon atoms is indicated in brackets in descending order.

Examples:

Bicyclo [1.1.0]-butane

Bicyclo [3.2.1]-octane

Bicyclo [5.2.0]nonane

31.2.—The system is numbered commencing with one of the bridge-heads, numbering proceeding by the longest possible path to the second bridge-head; numbering is then continued from this atom by the longer unnumbered path back to the first bridge-head and is completed by the shortest path.

Examples:

Bicyclo [3.2.1]octane

Bicyclo [4.3.2]undecane

Note: Longest path 1, 2, 3, 4, 5
Next longest path 5, 6, 7, 1
Shortest path 1, 8, 5

(1) A bridge is a valence bond or an atom or an unbranched chain of atoms connecting two different parts of a molecule. The two tertiary carbon atoms connected through the bridge are termed "bridge-heads."

31.3.—Unsaturated hydrocarbons are named in accordance with the principles set forth in Rule **A-11.3**. When there is a choice in numbering unsaturation is given the lowest numbers.

$$\begin{array}{ccc}
\text{CH}_2 & \text{—CH—} & \text{CH} \\
^6 & |^1 & ^2 \\
 & ^7\text{CH}_2 & \\
^5 & ^4 & ^3 \\
\text{CH}_2 & \text{—CH—} & \text{CH}
\end{array}$$

Bicyclo [2.2.1]hept-2-ene

Chemical Abstracts names the above ring system 2-norbornene by Rule **A-74.2**, but follows the above Rule for hydrocarbons which do not have terpene names.

31.4.—Radicals derived from these hydrocarbons are named in accordance with the principles set forth in Rule **A-11.4**. The numbering of the hydrocarbon is retained and the point or points of attachment are given numbers as low as is consistent with the fixed numbering of the hydrocarbon.

Example:

$$\begin{array}{ccc}
\text{CH} & \text{—CH—} & \text{CH—} \\
^6 & |^1 & ^2 \\
 & ^7\text{CH}_2 & \\
^5 & ^4 & ^3 \\
\text{CH} & \text{—CH—} & \text{CH}_2
\end{array}$$

Bicyclo[2.2.1]hept-5-en-2-yl

Chemical Abstracts names the above radical 5-norbornen-2-yl by Rule **A-75.2**, but follows the above Rule for radicals derived from hydrocarbons which do not have terpene names. Radicals derived from the saturated hydrocarbons (except terpenes) are named by replacing the ending "-ane" of the hydrocarbon name by ,"-yl." The point of attachment is given a number as low as is consistent with the fixed numbering of the hydrocarbon.

A-32

32.11.—Cyclic hydrocarbon systems consisting of three or more rings may be named in accordance with the principles stated in Rule **A-31**. The appropriate prefix "tricyclo-," "tetracyclo-," *etc.*, is substituted for "bicyclo-" before the name of the open-chain hydrocarbon containing the same total number of carbon atoms.

32.12.—A polycyclic system is regarded as containing a number of rings equal to the number of scissions required to convert the system into an open-chain compound.

32.13.—The word "cyclo" is followed by brackets containing, in decreasing order, numbers indicating the number of carbon atoms in: the two branches of the main ring, the main bridge, the secondary bridges.

Examples:

Tricyclo[2.2.1.0¹]heptane   Tricyclo[5.3.1.1¹]dodecane

(1) For location and numbering of the secondary bridge see Rules **A-32.22. A32.23, A-32.31**.

32.21.—The main ring and the main bridge form a bicyclic system whose numbering is made in compliance with Rule **A-31**.

32.22.—The location of the other or so-called secondary bridges is shown by superscripts following the number indicating the number of carbon atoms in the said bridges.

32.23.—For the purpose of numbering, the secondary bridges are considered in decreasing order. The numbering of any bridge follows from the part already numbered, proceeding from the highest-numbered bridge-head.

Numbering of the secondary bridges by proceeding from the highest-numbered bridge-head of each bridge is an improvement over the former practice of "following the shortest path from the highest previous number."

32.31.—When there is a choice, the following criteria are considered in turn until a decision is made:

(a) The main ring shall contain as many carbon atoms as possible, two of which must serve as bridge-heads for the main bridge.

Tricyclo[5.4.0.0^{2,9}]undecane
Correct numbering

Tricyclo[4.2.1.2^{7,9}]undecane
Incorrect numbering

Tricyclo[5.3.2.0^{4,9}]dodecane
Correct numbering

Tricyclo[5.2.3.0^{4,11}]dodecane
Incorrect numbering

(b) The main bridge shall be as large as possible.

Tetracyclo[5.2.2.0^{3,8}.0^{4,11}]-
undecane
Correct numbering

Tetracyclo[5.2.1.1^{4,10}.0^{2,6}]-
undecane
Incorrect numbering

(c) The main ring shall be divided as symmetrically as possible by the main bridge.

Tricyclo[4.4.1.1^{1,6}]dodecane:
Correct numbering

Tricyclo[5.3.1.1^{1,6}]dodecane:
Incorrect numbering

(d) The superscripts locating the other bridges shall be as small as possible (in the sense indicated in Rule **A-2.2**)

Tricyclo[5.5.1 0^{3,11}]tridecane
Correct numbering

Tricyclo[5.5.1.0^{5,9}]tridecane
Incorrect numbering

A-34 (Alternate to Rule A-35)

34.1.—Polycyclic hydrocarbon systems which can be regarded as "ortho-fused" or "ortho- and peri-fused" systems according to Rule **A-21** and which, at the same time, have other bridges,[1] are first named as "ortho-fused" or "ortho- and peri-fused" systems. The other bridges are then indicated by prefixes derived from the corresponding hydrocarbon by replacing the final "-ane," "-ene," *etc.*, by "-ano," "-eno," *etc.*, and their positions are indicated by the points of attachment in the parent compound. If bridges of different types are present, they are cited in alphabetical order.

Examples of bridge names

Butano	$-CH_2-CH_2-CH_2-CH_2-$	Etheno	$-CH=CH-$
Benzeno (*o-, m-, p-*)	$-C_6H_4-$	Methano	$-CH_2-$
Ethano	$-CH_2-CH_2-$	Propano	$-CH_2-CH_2-CH_2-$

1,4-Dihydro-1,4-
methanopentalene

9,10-Dihydro-9,10-(2-buteno)-
anthracene

7,14-Dihydro-7,14-ethano-
dibenz[*a,h*]anthracene

(1) The term "bridge," when used in connection with an "ortho-fused" or "ortho- and peri-fused" polycyclic system as defined in note to Rule **A-31.1**, also includes "bivalent cyclic systems."

34.2.—The parent "ortho-fused" "or ortho- and peri-fused" system is numbered as prescribed in Rule **A-22.** Where there is a choice, the position numbers of the bridge-heads should be as low as possible. The remaining bridges are then numbered in turn starting each time with the bridge atom next to the bridge-head possessing the highest number.

Example:

$$H_2C^7 \quad CH \quad CH \quad {}^{12}CH_2 \quad {}^{2}CH_2$$

Perhydro-1,4-ethanoanthracene

not

or

34.3.—When there is a choice of position numbers for the points of attachment for several individual bridges, the lowest numbers are assigned to the bridge-heads in the order of citation of the bridges and the bridge atoms are numbered according to the preceding rule.

Example:

Perhydro-1,4-ethano-5,8-methanoanthracene

34.4.—When the bridge is formed from a bivalent cyclic hydrocarbon radical, low numbers are given to the carbon atoms constituting the shorter bridge and numbering proceeds around the ring.

10,11-Dihydro-5,10-*o*-benzeno-5*H*-benzo[*b*]fluorene

Chemical Abstracts follows this rule instead of the Alternate Rule **A-35.**

A-35 (Alternate to Rule A-34)

35.1.—Polycyclic hydrocarbons which can be regarded as "ortho-fused" or "ortho- and peri-fused" according to Rule **A-21** and which have also other bridges may be named as "ortho-fused" or "ortho- and peri-fused" systems into which bivalent radicals are substituted.

35.2.—Compounds which are named in accordance with **35.1** of this rule are numbered as prescribed in Rule **A-34.** The bridge is numbered independently. Numbering of the bridge is started at the lower-numbered bridge head. Numbers assigned to the bridge atoms are primed.

Examples:

$$
\begin{array}{c}
\text{H} \\
\text{C} \\
9 \\
1'\text{CH}_2 \\
2'\text{CH} \\
3'\text{CH} \\
4'\text{CH}_2 \\
10 \\
\text{C} \\
\text{H}
\end{array}
$$

9,10-Dihydro-9,10-(2'-
butylene)anthracene

7,14-Dihydro-7,14-ethylene-
dibenz[a,h]anthracene

Chemical Abstracts follows Rule **A-34** instead of this Alternate Rule

Spiro Hydrocarbons

A "spiro union" is one formed by a single atom which is the only common member of two rings. A "free spiro union" is one constituting the only union direct or indirect between two rings.[1] The common atom is designated as the "spiro atom." According to the number of spiro atoms present, the compounds are distinguished as monospiro, dispiro, trispiro compounds, *etc.* The following rules apply to the naming of compounds containing free spiro unions.

(1) An example of a compound where the spiro union is *not* free is:

$$
\begin{array}{ccc}
\text{H}_2\text{C}{-}\text{CH}_2 & \text{H}_2\text{C}{-}\text{CH}_2 \\
\text{H}_2\text{C} & \text{C} & \text{CH}_2 \\
\text{H}_2\text{C}{-}\text{CH} & \text{HC}{-}\text{CH}_2 \\
\text{H}_2\text{C}{-}\text{CH}_2
\end{array}
$$

A-41 (Alternate to Rule A-42)

41.1.—Monospiro compounds consisting of only two alicyclic rings as components are named by placing "spiro" before the name of the normal acyclic hydrocarbon of the same total number of carbon atoms. The number of carbon atoms linked to the spiro atom in each ring is indicated in ascending order in brackets placed between the spiro prefix and the hydrocarbon name.

Examples:

Spiro[3.4]octane   Spiro[3.3]heptane

41.2.—The carbon atoms in monospiro hydrocarbons are numbered consecutively starting with a ring atom next to the spiro atom, first through the smaller ring (if such be present) and then through the spiro atom and around the second ring.

Example:

Spiro[4.5]decane

41.3.—When unsaturation is present, the same numbering pattern is maintained, but in such a direction around the rings that the double and triple bonds receive numbers as low as possible in accordance with Rule **A-11**.

Example:

$$H_2C—CH_2 \quad HC=CH$$

Spiro[4.5]deca-1,6-diene

41.4.—If one or both components of the monospiro compound are fused polycyclic systems, "spiro" is placed before the names of the components arranged in alphabetical order and enclosed in brackets. Established numbering of the individual components is retained. The lowest possible number is given to the spiro atom, and the numbers of the second component are marked with primes. The position of the spiro atom is indicated by placing the appropriate numbers between the names of the two components.

Example:

Spiro[cyclopentane-1,1'-indene]

41.5.—Monospiro compounds containing two similar polycyclic components are named by placing the prefix "spirobi" before the name of the component ring system. Established numbering of the polycyclic system is maintained and the numbers of one component are distinguished by primes. The position of the spiro atom is indicated in the name of the spiro compound by placing the appropriate locants before the name.

Example:

1,1'-Spirobiindene

41.6.—Polyspiro compounds consisting of three or more alicyclic systems are named by placing "dispiro-," "trispiro-," "tetraspiro-," etc., before the name of the normal acyclic hydrocarbon of the same total number of carbon atoms. The numbers of carbon atoms linked to the spiro atoms in each ring are indicated in brackets in the same order as the numbering proceeds about the ring. Numbering starts with a ring atom next to a terminal spiro atom and proceeds in such a way as to give the spiro atoms as low numbers as possible.

Example:

Dispiro[5.1.7.2]heptadecane

41.7.—Polycyclic compounds containing more than one spiro atom and at least one fused polycyclic component are named in accordance with **41.4** of this rule by replacing "spiro" with "dispiro," "trispiro," etc., and choosing the end components by alphabetical order.

Example:

Dispiro[fluorene-9,1'-cyclohexane-4',1''-indene]

This represents the method of naming spiro compounds which is followed by *Chemical Abstracts* instead of the Alternate Rule A-42.

DEFINITIVE RULES FOR NOMENCLATURE OF ORGANIC CHEMISTRY

A-42

42.1 (Alternate to **A-41.1** and **A-41.2**).—When two dissimilar cyclic components are united by a spiro union the name of the larger component is followed by the affix "spiro" which, in turn, is followed by the name of the smaller component. Between the affix "spiro" and the name of each component system is inserted the number denoting the spiro-position in the appropriate ring system, these numbers being as low as permitted by any fixed numbering of the component. The components retain their respective numberings but numbers for the component mentioned second are primed. Numbers 1 may be omitted when a free choice is available for a component.

Examples:

Cyclopentanespiro-cyclobutane

Cyclohexanespirocyclo-pentane

2*H*-Indene-2-spiro-1'-cyclopentane

42.2 (Alternate to **A-41.3**).—Rule **A-41.3** applies also with appropriate different numbering where nomenclature is according to Rule **A-42.1**, but the spiro-junction has priority for lowest numbers over unsaturation.

Example:

2-Cyclohexenespiro-(2'-cyclopentene)

42.3 (Alternate to **A-41.5**).—The nomenclature of Rule **A-41.5** is applied also to monocyclic components with identical saturation, the spiro-union being numbered 1.

Examples:

Spirobicyclohexane   but   2-Cyclohexenespiro-(3'-cyclohexene)

42.4 (Alternate to **A-41.6** and **A-41.7**).—Polycyclic compounds containing more than one spiro atom are named in accordance with Rule **A-42.1** starting from the senior[1] end-component irrespective of whether the components are simple or fused rings.

Examples:

Cycloöctanespirocyclopentane-3'-spirocyclohexane   Fluorene-9-spiro-1'-cyclohexane-4'-spiro-1''-indene

(1) "Seniority" in respect to spiro compounds is based on the principles: (i) an aggregate is senior to a monocycle; (ii) of aggregates, the senior is that containing the largest number of individual rings; (iii) of aggregates containing the same number of individual rings, the senior is that containing the largest ring; (iv) if aggregates consist of equal number of equal rings the senior is the first occurring in the alphabetical list of names.

Chemical Abstracts follows Rule **A-41** instead of this alternate Rule.

DEFINITIVE RULES FOR NOMENCLATURE OF ORGANIC CHEMISTRY

Hydrocarbon Ring Assemblies

A-51

51.1.—Two or more cyclic systems (single rings or fused systems) which are directly joined to each other by double or single bonds are named "ring assemblies" when the number of such direct ring junctions is one less than the number of cyclic systems involved.

Examples:

Ring assemblies

Fused polycyclic system

A-52

52.1.—Assemblies of two identical cyclic hydrocarbon systems are named in either of two ways: (a) by placing the prefix "bi-" before the name of the corresponding radical, or (b) for systems joined by a single bond by placing the prefix "bi-" before the name of the corresponding hydrocarbon. In each case, the numbering of the assembly is that of the corresponding radical or hydrocarbon, one system being assigned unprimed numbers and the other primed numbers. The points of attachment are indicated by placing the appropriate locants before the name.

Examples:

1,1'-Bicyclopropyl
or 1,1'-Bicyclopropane

1,1'-Bicyclopentadienylidene

52.2.—If there is a choice in numbering, unprimed numbers are assigned to the system which has the lower-numbered point of attachment.

Example:

1,2'-Binaphthyl
or 1,2'-Binaphthalene

52.3.—If two identical hydrocarbon systems have the same point of attachment and contain substituents at different positions, the locants of these substituents are assigned according to Rule **A-2.2**; for this purpose an unprimed number is considered lower than the same number when primed. Assemblies of primed and unprimed numbers are arranged in ascending numerical order.

Examples:

2,3,3',4',5'-
Pentamethylbiphenyl
(not 2',3,3',4,5-
Pentamethylbiphenyl)

2-Ethyl-2'-
propylbiphenyl

C-26

52.4.—The name "biphenyl" is used for the assembly consisting of two benzene rings:

Biphenyl

A-53

53.1.—Other hydrocarbon ring assemblies are named by selecting one ring system as the base component and considering the other system as substituents of the base component. Such substituents are arranged in alphabetical order. The base component is assigned unprimed numbers and the substituents are assigned numbers with primes.

53.2.—The base component is chosen by considering these characteristics in turn until a decision is reached:

(a) The system containing the larger number of rings.

Examples:

2-Phenylnaphthalene

4-Cyclooctyl-1'-cyclopentylbiphenyl

(b) The system containing the larger ring.

Examples:

2-(2'-Naphthyl)azulene

1,4-Dicyclopropylbenzene
or *p*-Dicyclopropylbenzene

(c) The system in the lowest state of hydrogenation (see also **53.3** of this rule).

Example:

Cyclohexylbenzene

(d) The order of ring systems as set forth in the list of Rule **A-21.1.**

53.3.—Compounds covered by **53.2**(c) of this rule may also be named as hydrogenation products according to Rule **A-23.**

Example:

1,2,3,3',4,4'-Hexahydro-1,1'-binaphthyl

The name of the second example in (a) which contains as a base component the ring assembly "biphenyl" indicates that the word system as used here includes ring assemblies as base components.

Chemical Abstracts does not use primed numbers with the groups which may be considered as substituent groups expressed as prefixes and attached to the name of the base component, Examples: 2-(2,4-dichlorophenyl)naphthalene; 2(2-naphthyl)-azulene.

A-54

54.1.—Unbranched assemblies consisting of three or more identical hydrocarbon ring systems are named by placing an appropriate numerical prefix before the name of the hydrocarbon corresponding to the repetitive unit. These numerical prefixes are used:

3. ter-	5. quinque-	7. septi-	9. novi-
4. quater-	6. sexi-	8. octi-	10. deci-

Example:

Tercyclopropane

54.2.—Unprimed numbers are assigned to one of the terminal systems, the other systems being primed serially. Points of attachment are assigned the lowest numbers possible.

Examples:

2,1′:5′,2″:6″,2‴-Quaternaphthalene

1,1′:3′,1″-Tercyclohexane

54.3.—As exceptions, unbranched assemblies consisting of benzene rings are named by using the appropriate prefix with the radical name "phenyl."

Examples:

p-Terphenyl
or 1,1′:4′,1″-Terphenyl

m-Terphenyl
or 1,1′:3′,1″-Terphenyl

Cyclic Hydrocarbons with Side Chains

(Note: *cf.* Rules **A-12** and **A-13**)

A-61

61.1.—Hydrocarbons more complex than those envisioned in Rule **A-12**, composed of cyclic nuclei and aliphatic chains, are named according to one of the methods given below. Choice is made so as to provide the name which is the simplest permissible or the most appropriate for the chemical intent.

61.2.—When there is no generally recognized trivial name for the hydrocarbon, then (1) the radical name denoting the aliphatic chain is prefixed to the name of the cyclic hydrocarbon, or (2) the radical name for the cyclic hydrocarbon is prefixed to the name of the aliphatic compound. Choice between these methods is made according to the more appropriate of the following principles: (a) the maximum number of substitutions into a single unit of structure; (b) treatment of a smaller unit of structure as a substituent into a larger.

61.3.—In accordance with the principle (a) of **61.2** of this rule, hydrocarbons containing several chains attached to one cyclic nucleus generally are named as derivatives of the cyclic compound; and compounds containing several side chains and/or cyclic radicals attached to one chain are named as derivatives of the acyclic compound.

Examples:

2-Ethyl-1-methylnaphthalene   Diphenylmethane

1,5-Diphenylpentane

2,3-Dimethyl-1-phenyl-1-hexene

61.4.—In accordance with principle (b) of **61.2** of this rule, a hydrocarbon containing a small cyclic nucleus attached to a long chain is generally named as a derivative of the acyclic hydrocarbon; and a hydrocarbon containing a small group attached to a large cyclic nucleus is generally named as a derivative of the cyclic hydrocarbon.

Examples:

$CH_3[CH_2]_{10}CH_2CH_2CH_2CH_2CH_2-$

1-Phenylhexadecane

$CH_3[CH_2]_{10}CH_2CH_2CH_2CH=CH-$

1-Phenyl-1-hexadecene

9-(1,2-Dimethylpentyl)-anthracene   7-(3-Phenylpropyl)-benz[a]anthracene

61.5.—Recognized trivial names for composite radicals are used if they lead to simplifications in naming.

Examples:

1-Benzylnaphthalene   1,2,4-Tris(3-p-tolylropyl)benzene

Terpene Hydrocarbons

Owing to long-established custom, terpenes are given exceptional treatment in these rules.

A-71. Acyclic Terpenes[1]

71.1.—The acyclic terpene hydrocarbons are named in a manner similar to that used for other unsaturated acyclic hydrocarbons when pure compounds are involved.

Example:

7-Methyl-3-methylene-1,6-octadiene

A-72. Cyclic Terpenes

72.1.—The following structural types with their special names and special systems of numbering are used as the basis for the specialized nomenclature of monocyclic and bicyclic terpene hydrocarbons. The name "bornane" replaces camphane and bornylane; "norbornane" replaces norcamphane and norbornylane.

(1) For a more complete discussion of terpene nomenclature see "Nomenclature for Terpene Hydrocarbons," Advances in Chemistry Series No. 14 (American Chemical Society).

Fundamental terpene types:

I
Menthane (*p*-form)

II
Thujane

III
Carane

IV
Pinane

V
Bornane

Nor-structures:

VI
Norcarane

VII
Norpinane

VIII
Norbornane

A-73. Monocyclic Terpenes

73.1.—Menthane Type: Monocyclic terpene hydrocarbons of this type (ortho, meta, and para isomers) are named menthane, menthene, menthadiene, *etc.*, are given the fixed numbering of menthane (Formula I). Such compounds substituted by additional alkyl groups are named in accordance with Rule **A-11**.

Examples:

m-Menthane   1-*p*-Menthene   1,4(8)-*p*-Menthadiene

73.2.—Tetramethylcyclohexane Type: Monocyclic terpene hydrocarbons of this type are named systematically as derivatives of cyclohexane, cyclohexene, and cyclohexadiene (see Rule **A-11**).

Examples:

1,1,2,3-
Tetramethyl-
cyclohexane

1,2,3,3-
Tetramethyl-
cyclohexene

1,5,5,6
Tetramethyl-
1,3-cyclohexadiene

A-74. Bicyclic Terpenes

74.1.—Bicyclic terpene hydrocarbons having the skeleton of Formula II or this skeleton and additional side chains except methyl or isopropyl (or methylene if one methylene group is already present) are named as thujane, thujene, thujadiene, *etc.*, and are given the fixed numbering shown for thujane (Formula II). Other hydrocarbons containing the thujane ring-skeleton are named from bicyclo[3.1.0]hexane and are given systematic bicyclo numbering (*cf.* Rule **A-31**).

Example:

4(10)-Thujene

1-Isopropyl-2,4-
dimethylenebicyclo-
[3.1.0]hexane

5-Isopropyl-
bicyclo[3.1.0]hex-
2-ene

74.2.—Bicyclic terpene hydrocarbons having the skeleton of Formula III, IV, or V and additional side chains except methyl (or methylene if one methylene group is already present) are named, respectively, as carane, carene, caradiene, *etc.*; pinane, pinene, pinadiene, *etc.*; bornane, bornene, bornadiene, *etc.* They are given, respectively, the fixed numbering shown for carane (Formula III), pinane (Formula IV), and bornane (Formula V). Other hydrocarbons containing the ring-skeleton of carane, pinane, or bornane are named, respectively, from norcarane (Formula VI), norpinane (Formula VII), or norbornane (Formula VIII). These names are preferred to those from bicyclo [4.1.0]heptane, bicyclo[3.1.1]heptane, or bicyclo [2.2.1]heptane. The nor-names are given systematic bicyclo numbering (*cf.* Rule **A-31**).

Examples:

2-Carene

7,7-Dimethyl-
2,4-norcaradiene

2(10),3-Pinadiene

4-Methylenepinane

2,4,7,7-Tetra-
methylnorcarane

6,6-Dimethyl-2-
vinyl-2-norpinene

2-Bornene   2,2-Dimethylnorbornane   2,7,7-Trimethyl-2-norbornene

74.3.—The name "camphene" is retained for the unsubstituted compound 2,2-dimethyl-3-methylene-norbornane.

Camphene

A-75. Terpene Radicals

75.1.—Simple acyclic hydrocarbon terpene radicals are named and numbered according to Rule **A-3.5.** The trivial names geranyl, neryl, linalyl and phytyl are retained for the unsubstituted radicals.

75.2.—Radicals derived from menthane, pinane, thujane, carane, bornane, norcarane, norpinane, and norbornane are named in accordance with the principles set forth in Rules **A-1.2** and **A-11.4** except that the saturated radicals of pinane are named pinanyl, pinanylene, and pinanylidene. The numbering of the hydrocarbon is retained and the point or points of attachment, whether in the ring or side chain, are given numbers as low as is consistent with the fixed numbering of the hydrocarbon.

Examples:

1-*p*-Menthen-8-yl   3-Pinanyl   4(10)-Thujen-10-yl

2-Pinen-10-ylidene   5-Norbornen-2-yl

75.3.—Radicals not named in Rules **A-75.1** and **A-75.2** are named as described in Rules **A-11** and **A-31.4.**

DEFINITIVE RULES FOR NOMENCLATURE OF ORGANIC CHEMISTRY

B. FUNDAMENTAL HETEROCYCLIC SYSTEMS

B-1. Extension of Hantzsch-Widman System

1.1.—Monocyclic compounds containing one or more hetero atoms in a three- to ten-membered ring are named by combining the appropriate prefix or prefixes from Table I (eliding "a" where necessary) with a stem from Table II. The state of hydrogenation is indicated either in the stem, as shown in Table II, or by the prefixes "dihydro-," "tetrahydro-," etc., according to Rule **B-1.2**.

TABLE I

Element	Valence	Prefix
Oxygen	II	Oxa
Sulfur	II	Thia
Selenium	II	Selena
Tellurium	II	Tellura
Nitrogen	III	Aza
Phosphorus	III	Phospha[a]
Arsenic	III	Arsa[a]
Antimony	III	Stiba[a]
Bismuth	III	Bisma
Silicon	IV	Sila
Germanium	IV	Germa
Tin	IV	Stanna
Lead	IV	Plumba
Mercury	II	Mercura

[a] When immediately followed by "in-" or "-ine," "phospha-" should be replaced by "phosphor-," "arsa-" should be replaced by "arsen-" and "stiba" should be replaced by "antimon-".

"Bora" for boron is used frequently. *Chemical Abstracts* has placed boron with the valence of III between lead and mercury in Table I. This table shows the valence of tin and lead as IV instead of II (a correction of a typographical error).

TABLE II

No. of members in the ring	Rings containing nitrogen		Rings containing no nitrogen	
	Unsaturation[a]	Saturation	Unsaturation[a]	Saturation
3	-irine	-iridine	-irene	irane[e]
4	-ete	-etidine	-ete	-etane
5	-ole	-olidine	-ole	-olane
6	-ine[b]	[c]	-ine[b]	-ane[d]
7	-epine	[c]	-epine	-epane
8	-ocine	[c]	-ocin	-ocane
9	-onine	[c]	-onin	-onane
10	-ecine	[c]	-ecin	-ecane

[a] Corresponding to the maximum number of non-cumulative double bonds, the hetero elements having the normal valences shown in Table I. [b] For phosphorus, arsenic, antimony, see the special provisions of Table I. [c] Expressed by prefixing "perhydro" to the name of the corresponding unsaturated compound. [d] Not applicable to silicon, germanium, tin and lead. In this case, "perhydro-" is prefixed to the name of the corresponding unsaturated compound. [e] The syllables denoting the size of rings containing 3, 4 or 7–10 members are derived as follows: "ir" from tr*i*-, "et" from t*e*tra, "ep" from h*e*pta, "oc" from *oc*ta, "on" from n*on*a, and "ec" from d*ec*a.

Examples:

Oxirane   Aziridine   2*H*-Azepine

1.2.—Heterocyclic systems whose unsaturation is less than the one corresponding to the maximum number of noncumulative double bonds are named by using the prefixes "dihydro-," "tetrahydro-," etc.

In the case of 4- and 5-membered rings, a special termination is used for the structures containing one double bond, when there can be more than one noncumulative double bond.

No. of members of the partly saturated rings	Rings containing nitrogen	Rings containing no nitrogen
4	-etine	-etene
5	-oline	-olene

Examples:

$$NH-AsH \quad (1,2) \quad CH=CH \quad (4,3)$$

1,2-Azarset-3-ine

3-Silolene

1.3.—Multiplicity of the same hetero atom is indicated by a prefix "di," "tri-," etc., placed before the appropriate "a" term (Table I).

Example:

1,3,5-Triazine (or *s*-Triazine)

1.4.—If two or more kinds of "a" terms occur in the same name, their order of citation is by descending group number of the Periodic Table and increasing atomic number in the group as illustrated by the sequence in Table I.

Examples:

1,2-Oxathiolane

Thiazole

1.51.—The position of a single hetero atom determines the numbering in a monocyclic compound.

Example:

Azocine

1.52.—When the same hetero atom occurs more than once in a ring, the numbering is chosen to give the lowest locants to the hetero atoms.

Example:

1,2,4-Triazine (or *as*-Triazine)

1.53.—When hetero atoms of different kinds are present, the locant 1 is given to a hetero atom which is as high as possible in Table I. The numbering is then chosen to give the lowest locants to the hetero atoms.

Examples:

6*H*-1,2,5-Thiadiazine
(not 2,1,4-Thiadiazine)
(not 1,3,6-Thiadiazine)

2*H*,6*H*-1,5,2-Dithiazine
(not 1,3,4-Dithiazine)
(not 1,3,6-Dithiazine)
(not 1,5,4-Dithiazine)

The numbering must begin with the sulfur atom. This condition eliminates 2,1,4-thiadiazine. Then the nitrogen atoms receive the lowest possible locant, which eliminates 1,3,6-thiadiazine.

The numbering must begin with a sulfur atom. The choice of this atom is determined by the set of locants which can be attributed to the remaining hetero atoms of any kind. As the set 1,2,5 is lower than 1,3,4 or 1,3,6 or 1,5,4 in the usual sense, the name is 1,5,2-dithiazine.

B-2. Trivial and Semi-trivial Names

2.11.—The following trivial and semi-trivial names constitute a partial list of such names which are retained for the compound and as a basis of fusion names. They are arranged in the inverse order of the precedence prescribed in Rule **B-3**. The names of the radicals shown are formed according to Rule **B-5**.

	Parent Compound	Radical Name		Parent Compound	Radical Name
(1)	Thiophene	Thienyl (2-shown)	(12)	Pyrrole	Pyrrolyl (3- shown)
(2)	Benzo[b]thiophene (replacing thianaphthene)	Benzo[b]thienyl (2-shown)	(13)	Imidazole	Imidazolyl (2- shown)
(3)	Naphtho[2,3-b]thiophene (replacing thiophanthrene)	Naphtho[2,3-b] thienyl (2-shown)	(14)	Pyrazole	Pyrazolyl (1- shown)
(4)	Thianthrene	Thianthrenyl (2-shown)	(15)	Pyridine	Pyridyl (3- shown)
(5)	Furan	Furyl (2-shown)	(16)	Pyrazine	Pyrazinyl
(6)	Pyran (2H-shown)	Pyranyl (2H-Pyran-3-yl shown)	(17)	Pyrimidine	Pyrimidinyl (2-shown)
(7)	Isobenzofuran	Isobenzofuranyl (1-shown)	(18)	Pyridazine	Pyridazinyl (3- shown)
(8)	Chromene (2H- shown)	Chromenyl (2H-Chromen-3-yl shown)	(19)	Indolizine	Indolizinyl (2-shown)
(9)	Xanthene[1]	Xanthenyl[1] (2- shown)	(20)	Isoindole	Isoindolyl (2- shown)
(10)	Phenoxathiin	Phenoxathiinyl (2- shown)	(21)	3H-Indole	3H-Indolyl (3H-Indol-2-yl shown)
(11)	2H-Pyrrole	2H-Pyrrolyl (2H-Pyrrol-3-yl shown)			

(1) Denotes exceptions to systematic numbering.

Parent Compound	Radical Name	Parent Compound	Radical Name

(22) Indole — Indolyl (1- shown)

(23) 1H-Indazole — Indazolyl (1H-Indazol-3-yl shown)

(24) Purine[1] — Purinyl[1] (8- shown)

(25) 4H-Quinolizine — 4H-Quinolizinyl (4H-Quinolizin-2-yl shown)

(26) Isoquinoline — Isoquinolyl (3- shown)

(27) Quinoline — Quinolyl (2- shown)

(28) Phthalazine — Phthalazinyl (1- shown)

(29) Naphthyridine (1,8- shown) — Naphthyridinyl (1,8-Naphthyridin-2-yl shown)

(30) Quinoxaline — Quinoxalinyl (2- shown)

(31) Quinazoline — Quinazolinyl (2- shown)

(32) Cinnoline — Cinnolinyl (3- shown)

(33) Pteridine — Pteridinyl (2- shown)

(34) 4aH-Carbazole[1] — 4aH-Carbazolyl[1] (4aH-Carbazol-2-yl shown)

(35) Carbazole[1] — Carbazolyl[1] (2- shown)

(36) β-Carboline — β-Carbolinyl (β-Carbolin-3-yl shown)

(37) Phenanthridine — Phenanthridinyl (3- shown)

(38) Acridine[1] — Acridinyl[1] (2- shown)

(39) Perimidine — Perimidinyl (2- shown)

(40) Phenanthroline (1,7- shown) — Phenanthrolinyl (1,7-Phenanthrolin-3-yl shown)

Parent Compound	Radical Name	Parent Compound	Radical Name
(41) Phenazine	Phenazinyl (1- shown)	(44) Phenothiazine	Phenothiazinyl (2- shown)
(42) Phenarsazine	Phenarsazinyl (2- shown)	(45) Isoxazole	Isoxazolyl (3- shown)
(43) Isothiazole	Isothiazolyl (3- shown)	(46) Furazan	Furazanyl (3- shown)
		(47) Phenoxazine	Phenoxazinyl (2- shown)

Chemical Abstracts uses 2*H*-1-benzopyran instead of chromene, and 9*H*-pyrido[3,4-*b*]indole instead of *β*-carboline.

B-2.12.—The following trivial and semi-trivial names are retained but are not recommended for use in fusion names. The names of the radicals shown are formed according to Rule **B-3.**

Parent Compound	Radical Name	Parent Compound	Radical Name
(1) Isochroman	Isochromanyl (3- shown)	(5) Imidazolidine	Imidazolidinyl (2- shown)
(2) Chroman	Chromanyl (7- shown)	(6) Imidazoline (2- shown[1])	Imidazolinyl (2-Imidazolin-4-yl[1] shown)
(3) Pyrrolidine	Pyrrolidinyl (2- shown)	(7) Pyrazolidine	Pyrazolidinyl (1- shown)
(4) Pyrroline (2- shown[1])	Pyrrolinyl (2-Pyrrolin-3-yl[1] shown)	(8) Pyrazoline (3- shown[2])	Pyrazolinyl (3-Pyrazolin-2-yl[2] shown)

(1) The "2-" denotes the position of the double bond.
(2) The "3-" denotes the position of the double bond.
(3) For 1-piperidyl use piperidino.

Parent Compound	Radical Name		Parent Compound	Radical Name
(9) Piperidine	Piperidyl[3] (2- shown)	(12)	Isoindoline	Isoindolinyl (1- shown)
(10) Piperazine	Piperazinyl (1- shown)	(13)	Quinuclidine	Quinuclidinyl (2- shown)
(11) Indoline	Indolinyl (1- shown)	(14)	Morpholine	Morpholinyl[1] (3- shown)

B-3. Fused Heterocyclic Systems

3.1.—"Ortho-fused" and "ortho- and peri-fused" ring compounds containing hetero atoms are named according to the fusion principle described in Rule **A-21** for hydrocarbons. The components are named according to Rules **A-21**, **B-1** and **B-2**. The base component should be a heterocycle. If there is a choice, the base component should be, by order of preference:

(a) A nitrogen-containing component.

Example:

Benzo[*h*]isoquinoline
not Pyrido[3,4-*a*]naphthalene

(b) A component containing a hetero atom (other than nitrogen) as high as possible in Table I.

Example:

Thieno[2,3-*b*]furan
not Furo[2,3-*b*]thiophene

(c) A component containing the greatest number of rings.

Example:

7*H*-Pyrazino[2,3-*c*]carbazole
not 7*H*-Indolo[3,2-*f*]quinoxaline

(d) A component containing the largest possible individual ring.

Example:

2*H*-Furo[3,2-*b*]pyran
not 2*H*-Pyrano[3,2-*b*]furan

(1) For 4-morpholinyl use morpholino.

(e) A component containing the greatest number of hetero atoms of any kind.

Example:

5*H*-Pyrido[2,3-*d*][1,2]oxazine
not [1,2]Oxazino[4,5-*b*]pyridine

(f) A component containing the greatest variety of hetero atoms.

Examples:

1*H*-Pyrazolo[4,3-*d*]oxazole
not 1*H*-Oxazolo[5,4-*c*]-
pyrazole

4*H*-Imidazo[4,5-*d*]thiazole
not 4*H*-Thiazolo[4,5-*d*]-
imidazole

(g) A component containing the greatest number of hetero atoms first listed in Table I.

Example:

Selenazolo[5,4-*f*]benzothiazole[1]
not Thiazolo[5,4-*f*]benzoselenazole

(1) In this example the hetero atom first listed in Table I is sulfur and the greatest number of sulfur atoms in a ring is one.

(h) If there is a choice between components of the same size containing the same number and kind of hetero atoms choose as the base component that one with the lower numbers for the hetero atoms before fusion.

Example:

Pyrazino[2,3-*d*]pyridazine

3.2.—If a position of fusion is occupied by a hetero atom, the names of the component rings to be fused are so chosen as both to contain the hetero atom.

Example:

Imidazo[2,1-*b*]thiazole

3.3.—These contracted fusion prefixes may be used: furo, imidazo, isoquino, pyrido, pyrimido, quino and thieno.

Examples:

Furo[3,4-*c*]cinnoline   4*H*-Pyrido[2,3-*c*]carbazole

3.4.—In peripheral numbering of the complete fused systems, the ring system is oriented and numbered according to the principles of Rule **A-22.** When there is a choice of orientations, it is made in the following sequence in order to:

(a) Give low numbers to hetero atoms.

Examples:

Benzo[*b*]furan   Cyclopenta[*b*]pyran   4*H*-[1,3]Oxathiolo-
[5,4-*b*]pyrrole (N.B.
1,3,4 lower than 1,3,6)

(b) Give low numbers to hetero atoms in order of Table I.

Example:

Thieno[2,3-*b*]furan

(c) Allow carbon atoms common to two or more rings to follow the lowest possible numbers (see Rules **A-22.2** and **A-22.3**). [A hetero atom common to the two rings is numbered according to Rule **B-3.4**(e).]

Examples:

not   or

Imidazo[1,2-*b*][1,2,4]triazine
(or Imidazo[1,2-*b*]-*as*-triazine)

In a compound name for a fusion prefix (*i.e.*, when more than one pair of brackets is required), the points of fusion in the compound prefix are indicated by the use of unprimed and primed numbers, the unprimed numbers being assigned to the ring attached directly to the base component, thus

Pyrido[1′,2′:1,2]imidazo-
[4,5-*b*]quinoxaline

not

or

or

(d) Give hydrogen atoms lowest numbers possible.

4*H*-1,3-Dioxolo[4,5-*d*]imidazole

(e) The ring is numbered as for hydrocarbons but numbers are given to all hetero atoms even when common to two or more rings. Interior hetero atoms are numbered last following the shortest path from the highest previous number.

B-4. "a" Nomenclature

4.1.—Names of heterocyclic compounds also may be formed by prefixing "a" terms (see Table I of Rule **B-1.1**) to the name of the corresponding homocyclic compound. The letter "a" should not be elided. There are two methods of applying this principle:

4.1(a). Stelzner Method.—In this method, the "a" term name relates to that of the hydrocarbon with the same distribution of bonds in the rings. Thus, I is not so related to benzene but to 1,4-cyclohexadiene, and II is not so related to naphthalene but to 1,4-dihydronaphthalene.

I   II

4.1(b). Chemical Abstracts Method.—If the corresponding homocyclic compound is partially or completely hydrogenated and if this state of hydrogenation is denoted in its name without the use of hydro prefixes, as indan and cyclohexane, the procedure is the same as in (a). In other cases, positions in the skeleton of the corresponding homocyclic compound which are occupied by hetero atoms are denoted by the "a" terms, and the parent heterocyclic compound is considered to be that which contains the maximum number of conjugated or isolated[1] double bonds; hydrogen is added, as necessary, as hydro prefixes and/or as *H* to the "a" name thus obtained.

Examples:	Stelzner Method	*Chemical Abstracts* Method
	Sila-2,4-cyclo-pentadiene	Sila-2,4-cyclo-pentadiene
	Sila-1,3-cyclo-pentadiene	Sila-1,3-cyclo-pentadiene
	Silabenzene	Silabenzene
	7-Azabicyclo-[2.2.1]heptane	7-Azabicyclo-[2.2.1]heptane
	1,3-Dithia-1,2,3,4-tetrahydro-naphthalene	4*H*-1,3-Dithia-naphthalene
	1,4-Dithia-1,4-di-hydronaph-thalene	1,4-Dithianaph-thalene

(1) Isolated double bonds are those which are neither conjugated nor cumulative as in

or the B ring of

	Stelzner Method	*Chemical Abstracts* Method
	2,4,6-Trithia-3a,7a-diazaper-hydroindene	2,4,6-Trithia-3a,7a-diazain-dene
	2-Oxa-1,2-dihydro-pyrene	1*H*-2-Oxapyrene
	2,7,9-Triazaphen-anthrene	2,7,9-Triazaphe-nanthrene

4.2.—In fusion names, the "a" terms precede the completed name of the parent hydrocarbon. If two or more kinds of "a" terms occur in the same name, the procedure described in Rule **B-1.4** applies. Prefixes denoting ordinary substitution procede the "a" terms.

Example:

3,4-Dimethyl-5-azabenz[a]anthracene

Chemical Abstracts follows Rule **B-4.1** (b). Since "a" names for heterocyclic systems are based on the name of the carbocyclic system, the numbering of the "a" name must agree with the numbering of the carbocyclic parent. When there is a choice in direction of numbering the ring is numbered so as to give (1) lowest numbers to the hetero atoms, (2) lowest numbers to hetero atoms as high as possible in Table 1 (Rule **B-1.1**).

B-5. Radicals

5.11.—Univalent radicals derived from heterocyclic compounds by removal of hydrogen from a ring are in principle named by adding "yl" to the names of the parent compounds (with elision of final "e" if present).

Examples:

Indolyl from indole
Pyrrolinyl from pyrroline
Triazolyl from triazole
Triazinyl from triazine

(For further examples see Rule **B-2.11**).

These exceptions are retained: furyl, pyridyl, piperidyl, quinolyl, isoquinolyl and thienyl (from thiophene) (see also Rule **B-2.12**).

As exceptions, the names "piperidino" and "morpholino" are preferred to "1-piperidyl" and "4-morpholinyl."

5.12.—Bivalent radicals derived from univalent heterocyclic radicals whose names end in "-yl" by removal of one hydrogen atom from the atom with the free valence are named by adding "-idene" to to the name of the corresponding univalent radical.

Example:

2-Pyranylidene

5.13.—Multivalent radicals derived from heterocyclic compounds by removal of two or more hydrogen atoms from different atoms in the ring are named by adding "-diyl," "-triyl," *etc.*, to the name of the ring system.

Example:

2,4-Quinolinediyl

5.21.—The use of "a" terms (Rule **B-4**) does not affect the formation of radical names. Such names are strictly analogous to those of the hydrocarbon analogs except that the "a" terms establish numbering in whole or in part.

Examples:

1,3-Dioxa-4-cyclohexyl   **1,10-Diaza-4-anthryl**

B-6. Cationic Hetero Atoms

6.1.—According to the "a" nomenclature, heterocyclic compounds containing cationic hetero atoms are named in conformity with the preceding rules by replacing "oxa-," "thia-," "aza-," *etc.*, by "oxonia-," "thionia-," "azonia-," *etc.*, the anion being designated in the usual way.

Examples:

Cl⁻   1-Oxoniaanthracene chloride

Cl⁻   4a-Azoniaanthracene chloride

Cl⁻   1-Thioniabicyclo[2.2.1]heptane chloride

Cl⁻   1-Methyl-1-oxoniacyclohexane chloride

HETEROCYCLIC SPIRO COMPOUNDS

B-10 (Alternate to B-11)

10.1.—Heterocyclic spiro compounds containing single ring units only may be named by prefixing "a" terms (see Table I, Rule **B-1.1**) to the names of the spiro hydrocarbons formed according to Rules **A-41.1, A-41.2, A-41.3** and **A-41.6.** The numbering of the spiro hydrocarbon is retained and the hetero atoms in the order of Table I are given as low numbers as are consistent with the fixed numbering of the ring. When there is a choice, hetero atoms are given lower numbers than double bonds.

Examples:

1-Oxaspiro[4.5]decane   6,8-Diazoniadispiro[5.1.6.2]hexadecane dichloride

10.2.—If at least one component of a mono- or polyspiro compound is a fused polycyclic system, the spiro compound is named according to Rule **A-41.4** or **A-41.7,** giving the spiro atom as low a number as possible consistent with the fixed numberings of the component systems.

Examples:

3,3′-Spirobi[3H-indole]   Spiro[piperidine-4,9′-xanthene]

Chemical Abstracts names heterocyclic spiro compounds according to Rules **B-10.1** and **B-10.2.**

B-11 (Alternate to B-10)

11.1.—Heterocyclic spiro compounds are named according to Rule **A-42,** the following criteria being applied where necessary: (a) spiro atoms have numbers as low as consistent with the numbering of the individual component systems; (b) heterocyclic components have priority over homocyclic components of the same size; (c) priority of heterocyclic components is decided according to Rule **B-3.** Parentheses are used where necessary for clarity in complex expressions.

Examples:

$$\begin{array}{c}
\text{H}_2\ \text{H}_2 \\
\text{C—C} \qquad \text{O—CH}_2 \\
{}^{3}\quad {}^{2}\quad {}^{1'}\ {}^{5'} \\
\text{H}_2\text{C}^{4}\qquad {}^{1}\qquad {}^{2'} \\
{}^{5}\quad {}^{6}\quad {}^{3'}\ {}^{4'} \\
\text{C—C} \qquad \text{C—CH}_2 \\
\text{H}_2\ \text{H}_2\quad \text{H}_2
\end{array}$$

Cyclohexanespiro-2'-(tetrahydrofuran)

$$\begin{array}{c}
\text{H}_2\ \text{H}_2 \qquad \text{H}_2 \\
\text{C—C} \qquad \text{O—C} \\
{}^{3'}\quad {}^{2'}\qquad {}^{1}\ {}^{6} \\
\text{H}_2\text{C}^{4'}\qquad {}^{1'}\qquad {}^{2}\qquad {}^{5}\text{CH}_2 \\
{}^{5'}\quad {}^{6'}\qquad {}^{3}\ {}^{4} \\
\text{C—C} \qquad \text{C—C} \\
\text{H}_2\ \text{H}_2\quad \text{H}_2\ \text{H}_2
\end{array}$$

Tetrahydropyran-2-
spirocyclohexane

3,3'-Spirobi(3H-indole)

1,2,3,4-Tetrahydro-
quinoline-4-spiro-4'-
piperidine

Hexahydroazepinium-1-spiro-1'-imid-
azolidine-3'-spiro-1''-piperidinium
dibromide

Chemical Abstract follows Rule **B-10**.

BRIDGED HETEROCYCLIC COMPOUNDS

Bridged heterocyclic bicyclo and polycyclo compounds are named by *Chemical Abstracts* by application of the "a" nomenclature to the names of bicyclo and polycyclo hydrocarbon systems which have been named according to Rules **A-31** and **A-32**.

Bridged heterocyclic compounds which can be regarded as ortho-fused or ortho- and peri-fused systems with additional bridges are named by extending Rule **A-34** to heterocyclic systems. Bridge names such as epoxy (—O—), epithio (—S—), epoxymethano (—O—CH₂—) are used.

INDEX

DEFINITIVE RULES FOR NOMENCLATURE OF ORGANIC CHEMISTRY

ILLUSTRATIVE PREFIXES

acetamido (acetylamino)	CH₃CONH–	disilanyl	H₃SiSiH₂–
acetimido (acetylimino)	CH₃C(=NH)–	disiloxanoxy	H₃SiOSiH₂O–
acetoacetamido	CH₃COCH₂CONH–	disulfinyl	–S(O)S(O)–
acetoacetyl	CH₃COCH₂CO–	dithio	–SS–
acetonyl	CH₃COCH₂		
acetonylidene	CH₃COCH=	enanthyl	CH₃(CH₂)₅CO–
acetyl	CH₃CO–	epoxy	–O–
acrylyl	CH₂=CHCO–	ethenyl (vinyl)	CH₂=CH–
adipyl (from adipic acid)	–OC(CH₂)₄CO–	ethinyl	HC≡C–
alanyl (from alanine)	CH₃CH(NH₂)CO–	ethoxy	C₂H₅O–
β-alanyl	H N(CH₂)₂CO–	ethyl	CH₃CH₂–
allophanoyl	H₂NCONHCO–	ethylthio	C₂H₅S–
allyl (2-propenyl)	CH₂=CHCH₂–		
allylidene (2-propenylidene)	CH₂=CHCH=	formamido (formylamino)	HCONH–
amidino (aminoiminomethyl)	H₂NC(=NH)–	formyl	HCO–
amino	H₂N–	fumaroyl (from fumaric acid)	–OCCH=CHCO–
amyl (pentyl)	CH₃(CH₂)₄–	furfuryl (2-furanylmethyl)	OC₄H₃CH₂–
anilino (phenylamino)	C₆H₅NH–	furfurylidene (2-furanylmethylene)	OC₄H₃CH=
anisidino	CH₃OC₆H₄NH–	furyl (furanyl)	OC₄H₃–
anisyl (from anisic acid)	CH₃OC₆H₄CO–		
anthranoyl (2-aminobenzoyl)	2-H₂NC₆H₄CO–	glutamyl (from glutamic acid)	–OC(CH₂)₂CH(NH₂)CO–
arsino	AsH₂–	glutaryl (from glutaric acid)	–OC(CH₂)CO–
azelaoyl (from azelaic acid)	–OC(CH₂)₇CO–	glycidyl (oxiranylmethyl)	CH₂–CHCH₂–
azido	N₃–	glycinamido	H₂NCH₂CONH–
azino	=NN=	glycolyl (hydroxyacetyl)	HOCH₂CO–
azo	–N=N–	glycyl (aminoacetyl)	H₂NCH₂CO–
azoxy	–N(O)N–	glyoxylyl (oxoacetyl)	HCOCO–
		guanidino	H₂NC(=NH)NH–
benzal	C₆H₅CH=	guanyl	H₂NC(=NH)–
benzamido (benzylamino)	C₆H₅CONH–		
benzhydryl (diphenylmethyl)	(C₆H₅)₂CH–	heptadecanoyl	CH₃(CH₂)₁₅CO–
benzimido (benzylimino)	C₆H₅C(=NH)–	heptanamido	CH₃(CH₂)₁₅CONH–
benzoxy (benzoyloxy)	C₆H₅COO–	heptanedioyl	–OC(CH₂)₅CO–
benzoyl	C₆H₅CO–	heptanoyl	CH₃(CH₂)₅CO–
benzyl	C₆H₅CH₂–	hexadecanoyl	CH₃(CH₂)₄CO–
benzylidine	C₆H₅CH=	hexamethylene	–(CH₂)₆–
benzyldyne	C₆H₅C≡	hexanedioyl	–OC(CH₂)₄CO–
biphenylyl	C₆H₅C₆H₅–	hippuryl (N-benzoylglycyl)	C₆H₅CONHCH₂CO–
biphenylene	–C₆H₄C₆H₄–	hydantoyl	H₂NCONHCH₂CO–
butoxy	C₄H₉O–	hydrazino	N₂NH–
sec-butoxy	C₂H₅CH(CH₃)O–	hydrazo	–HNNH–
tert-butoxy	(CH₃)₃CO–	hydrocinnamoyl	C₆H₅(CH₂)₂CO–
butyl	CH₃(CH₂)₃–	hydroperoxy	HOO–
iso-butyl (3-methylpropyl)	(CH₃)₂(CH₂)₂–	hydroxamino	HONH–
sec-butyl (1-methylpropyl)	C₂H₅CH(CH₃)–	hydroxy	HO–
tert-butyl (1,1, dimethylethyl)	(CH₃)₃C–		
butyryl	C₃H₇CO–	imino	HN=
		iodoso	OI–
caproyl (from caproic acid)	CH₃(CH₂)₄CO–	isoamyl (isopentyl)	(CH₃)₂CH(CH₂)₂–
capryl (from capric acid)	CH₃(CH₂)₈CO–	isobutenyl (2-methyl-1-propenyl)	(CH₃)₂C=CH–
caprylyl (from caprylic acid)	CH₃(CH₂)₆CO–	isobutoxy	(CH₃)₂CHCH O–
carbamido	H₂NCONH–	isobutyl	(CH₃)₂CHCH₂–
carbamoyl (aminocarbonyl)	H₂NCO–	isobutylidene	(CH₃)₂CHCH=
carbamyl (aminocarbonyl)	H₂NCO–	isobutyryl	(CH₃)₂CHCO–
carbazoyl (hydrazinocarbonyl)	H₂NNHCO–	isocyanato	OCN–
carbethoxy	C₂H₅O₂C–	isocyano	CN–
carbobenzoxy	C₆H₅CH₂O₂C–	isohexyl	(CH₃)₂CH(CH₂)₃–
carbonyl	–C=O–	isoleucyl (from isoleucine)	C₂H₅CH(CH₃)CH(NH₂)CO–
carboxy	HOOC–	isonitroso	HON=
cetyl	CH₃(CH₂)₁₅–	isopentyl	(CH₃)₂CH(CH₂)₂–
chloroformyl (chlorocarbonyl)	ClCO–	isopentylidene	(CH₃)₂CHCH₂CH=
cinnamyl (3-phenyl-2-propenyl)	C₆H₅CH=CHCH₂–	isopropenyl	H₂C=C(CH₃)–
cinnamoyl	C₆H₅CH=CHCO–	isopropoxy	(CH₃)₂CHO–
cinnamylidene	C₆H₅CH=CHCH=	isopropyl	(CH₃)₂CH–
cresyl (hydroxymethylphenyl)	HO(CH₃)C₆H₄–	isopropylidene	(CH₃)₂C=
crotoxyl	CH₃CH=CHCO–	isothiocyanato (isothiocyano)	SCN–
crotyl (2-butenyl)	CH₃CH=CHCH₂	isovaleryl (from isovaleric acid)	(CH₃)₂CHCH₂CO–
cyanamido (cyanoamino)	NCNH–		
cyanato	NCO–	keto (oxo)	O=
cyano	NC–		
decanedioyl	–OC(CH₂)₈CO–	lactyl (from lactic acid)	CH₃CH(OH)CO–
decanoly	CH₃(CH₂)₈CO–	lauroyl (from lauric acid)	CH₃(CH₂)₁₀CO–
diazo	N₂=	leucyl (from leucine)	(CH₃)₂CHCH₂CH(NH₂)CO–
diazoamino	–NHN=N–	levulinyl (From levulinic acid)	CH₃CO(CH₂)₂CO–

C-46

malonyl (from malonic acid)	$-OCCH_2CO-$
mandelyl (from mandelic acid)	$C_6H_5CH(OH)CO-$
mercapto	$HS-$
methacrylyl (from methacrylic acid)	$CH_2=C(CH_3)CO-$
methallyl	$CH_2=C(CH_3)CH_2-$
methionyl (from methionine)	$CH_3SCH_2CH_2CH(NH_2)CO-$
methoxy	CH_3O-
methyl	H_3C-
methylene	$H_2C=$
methylenedioxy	$-OCH_2O-$
methylenedisulfonyl	$-O_2SCH_2SO_2-$
methylol	$HOCH_2-$
methylthio	CH_3S-
myristyl (from myristic acid)	$CH_3(CH_2)_{12}CO-$
naphthal	$(C_{10}H_7)CH=$
naphthobenzyl	$(C_{10}H_7)CH_2-$
naphthoxy	$(C_{10}H_7)O-$
naphthyl	$(C_{10}H_7)-$
naphthylidene	$(C_{10}H_6)=$
neopentyl	$(CH_3)_3CCH_2-$
nitramino	O_2NNH-
nitro	O_2N-
nitrosamino	$ONNH-$
nitrosimino	$ONN=$
nitroso	$ON-$
nonanoyl (from nonanoic acid)	$CH_3(CH_2)_7CO-$
oleyl (from oleic acid)	$CH_3(CH_2)_7CH=CH(CH_2)_7CO-$
oxalyl (from oxalic acid)	$-OCCO-$
oxamido	$H_2NCOCONH-$
oxo (keto)	$O=$
palmityl (from palmitic acid)	$CH_3(CH_2)_{14}CO-$
pelargonyl (from pelargonic acid)	$CH_3(CH_2)_7CO$
pentamethylene	$-(CH_2)_5-$
pentyl	$CH_3(CH_2)_4-$
phenacyl	$C_6H_5COCH_2-$
phenacylidene	$C_6H_5COCH=$
phenanthryl	$(C_{14}H_9)-$
phenethyl	$C_6H_5CH_2CH_2-$
phenoxy	C_6H_5O-
phenyl	C_6H_5-
phenylene	$-C_6H_4-$
phenylenedioxy	$-OC_6H_4O-$
phosphino	H_2P-
phosphinyl	$H_2P(O)-$
phospho	O_2P-
phosphono	$(HO)_2P(O)-$
phthalyl (from phthalic acid)	$1,2-C_6H_4(CO-)_2$
picryl (2,4,6-trinitrophenyl)	$2,4,6-(NO_2)_3C_6H_2-$
pimelyl (from pimelic acid)	$-OC(CH_2)_5CO-$
piperidino	$C_5H_{10}N-$
piperidyl (piperidinyl)	$(C_5H_{10}N)-$
piperonyl	$3,4-(CH_2O_2)C_6H_3CH_2-$
pivalyl (from pivalic acid)	$(CH_3)_3CCO-$
prenyl (3-methyl-2-butenyl)	$(CH_3)_2C=CHCH_2-$
propargyl (2-propynyl)	$HC\equiv CCH_2-$
propenyl	$CH_2=CHCH_2-$
iso-propenyl	$(CH_3)_2C=$
propionyl	CH_3CH_2CO-
propoxy	$CH_3CH_2CH_2O-$
propyl	$CH_3CH_2CH_2-$

iso-propyl	$(CH_3)_2CH-$
propylidene	$CH_3CH_2CH=$
pyridino	C_5H_5N-
pyridyl (pyridinyl)	$(C_5H_4N)-$
pyrryl (pyrrolyl)	$(C_3H_4N)-$
salicyl (2-hydroxybenzoyl)	$2-HOC_6H_4CO-$
selenyl	$HSe-$
seryl (from serine)	$HOCH_2CH(NH_2)CO-$
siloxy	H_3SiO-
silyl	H_3Si-
silylene	$H_2Si=$
sorbyl (from sorbic acid)	$CH_3CH=CHCH=CHCO-$
stearyl (from stearic acid)	$CH_3(CH_2)_{16}CO-$
styryl	$C_6H_5CH=CH-$
suberyl (from suberic acid)	$-OC(CH_2)_6CO-$
succinamyl	$H_2NCOCH_2CH_2CO-$
succinyl (from succinic acid)	$-OCCH_2CH_2CO-$
sulfamino	$HOSO_2NH-$
sulfamyl	H_2NSO-
sulfanilyl	$4-H_2NC_6H_4SO_2-$
sulfeno	$HOS-$
sulfhydryl (mercapto)	$HS-$
sulfinyl	$OS=$
sulfo	HO_3S-
sulfonyl	$-SO_2-$
terephthalyl	$1,4-C_6H_4(CO-)_2$
tetramethylene	$-(CH_2)_4-$
thenyl	$(C_4H_3S)CH-$
thienyl	$(C_4H_3S)-$
thiobenzoyl	C_6H_5CS-
thiocarbamyl	H_2NCS-
thiocarbonyl	$-CS-$
thiocarboxy	$HOSC$
thiocyanato	$NCS-$
thionyl (sulfinyl)	$-SO-$
thiophenacyl	$C_6H_5CSCH_2-$
thiuram (aminothioxomethyl)	H_2NCS-
threonyl (from threonine)	$CH_3CH(OH)CH(NH_2)CO-$
toluidino	$CH_3C_6H_4NH-$
toluyl	$CH_3C_6H_4CO-$
tolyl (methylphenyl)	$CH_3C_6H_4-$
α-tolyl	$C_6H_5CH_2-$
tolylene (methylphenylene)	$(CH_3C_6H_3)=$
α-tolylene	$C_6H_5CH=$
tosyl [(4-methylphenyl) sulfonyl)]	$4-CH_3C_6H_4SO_2-$
triazano	$H_2NNHNH-$
trimethylene	$-(CH_2)_3-$
triphenylmethyl (trityl)	$(C_6H_5)_3C-$
tyrosyl (from tyrosine)	$4-HOC_6H_4CH_2CH(NH_2)CO-$
ureido	$H_2NCONH-$
valeryl (from valeric acid)	C_4H_9CO
valyl (from valine)	$(CH_3)_2CHCH(NH_2)CO-$
vinyl	$CH_2=CH-$
vinylidene	$CH_2=C=$
xenyl (biphenylyl)	$C_6H_5C_6H_4-$
xylidino	$(CH_3)_2C_6H_3NH-$
xylyl (dimethylphenyl)	$(CH_3)_2C_6H_3-$
xylylene	$-CH_2C_6H_4CH_2-$

ORGANIC RING COMPOUNDS

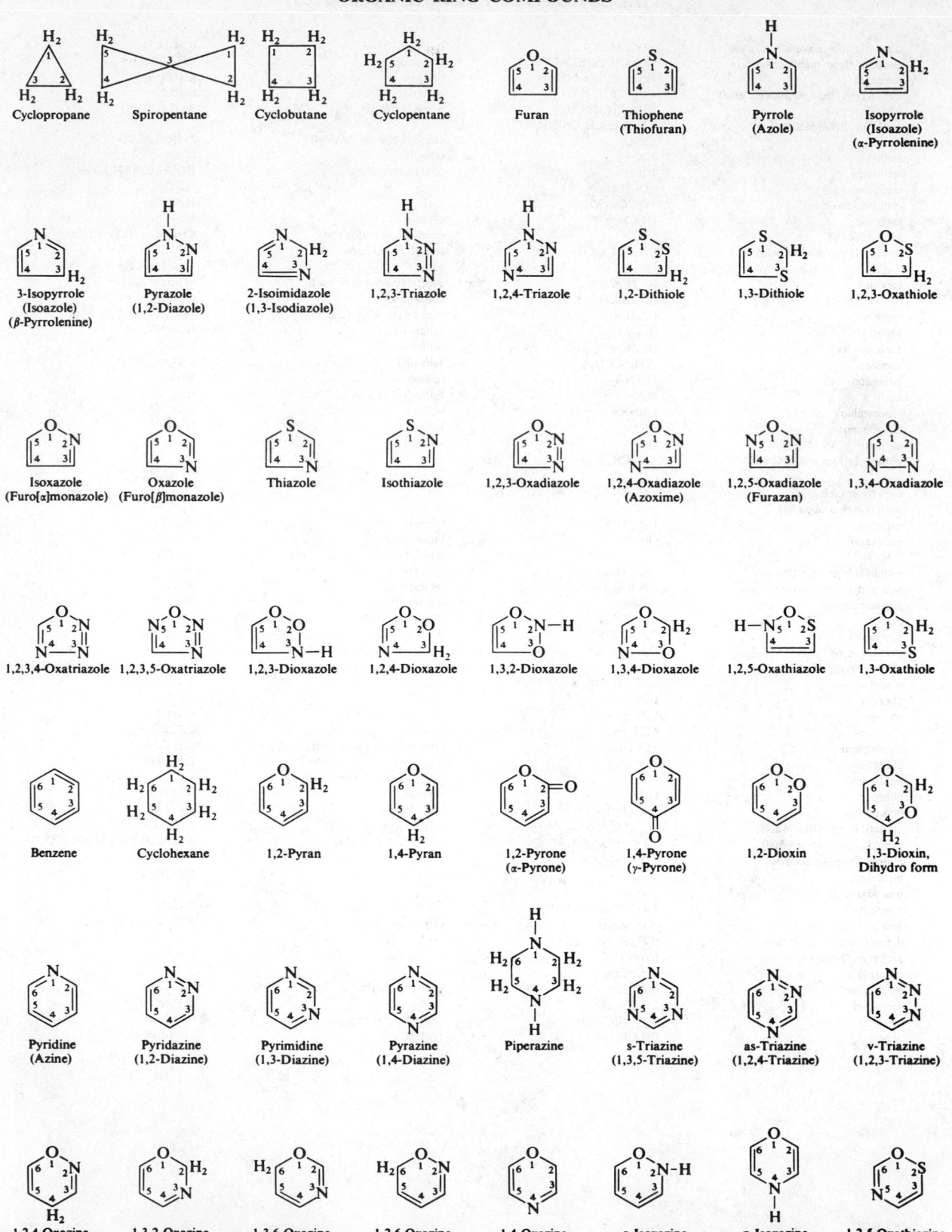

Cyclopropane Spiropentane Cyclobutane Cyclopentane Furan Thiophene (Thiofuran) Pyrrole (Azole) Isopyrrole (Isoazole) (α-Pyrrolenine)

3-Isopyrrole (Isoazole) (β-Pyrrolenine) Pyrazole (1,2-Diazole) 2-Isoimidazole (1,3-Isodiazole) 1,2,3-Triazole 1,2,4-Triazole 1,2-Dithiole 1,3-Dithiole 1,2,3-Oxathiole

Isoxazole (Furo[α]monazole) Oxazole (Furo[β]monazole) Thiazole Isothiazole 1,2,3-Oxadiazole 1,2,4-Oxadiazole (Azoxime) 1,2,5-Oxadiazole (Furazan) 1,3,4-Oxadiazole

1,2,3,4-Oxatriazole 1,2,3,5-Oxatriazole 1,2,3-Dioxazole 1,2,4-Dioxazole 1,3,2-Dioxazole 1,3,4-Dioxazole 1,2,5-Oxathiazole 1,3-Oxathiole

Benzene Cyclohexane 1,2-Pyran 1,4-Pyran 1,2-Pyrone (α-Pyrone) 1,4-Pyrone (γ-Pyrone) 1,2-Dioxin 1,3-Dioxin, Dihydro form

Pyridine (Azine) Pyridazine (1,2-Diazine) Pyrimidine (1,3-Diazine) Pyrazine (1,4-Diazine) Piperazine s-Triazine (1,3,5-Triazine) as-Triazine (1,2,4-Triazine) v-Triazine (1,2,3-Triazine)

1,2,4-Oxazine 1,3,2-Oxazine 1,3,6-Oxazine (Pentoxazole) 1,2,6-Oxazine 1,4-Oxazine o-Isoxazine p-Isoxazine 1,2,5-Oxathiazine

C-48

ORGANIC RING COMPOUNDS (Continued)

1,4-Oxazine

o-Isoxazine

p-Isoxazine

1,2,5-Oxathiazine

1,2,6-Oxathiazine

1,4,2-Oxadiazine

1,3,5,2-Oxadiazine

Morpholine
(Tetrahydro-
p-isoxazine)

Azepine

Oxepin

Thiepin

1,2,4-Diazepine

Indene

Isoindene

Benzofuran
(Coumarone)

Isobenzofuran

Thionaphthene
(Benzothiofuran)

Isothianaphthene
(Isobenzothiofuran)

Indole

Indolenine
(3-Pseudoindole)

Isoindolenine
(Pseudoisoindole)

1,5-Pyrindine
(4-Pyrindine)

Pyrano[3,1-b]
pyrrole

Isoindazole
(Benzpyrazole)

Indoxazine
(Benzisoxazine)

Benzoxazole

Anthranil

Naphthalene

Tetralin   Decalin
(Bicyclo[4,4,0]-
decane)

1,2-Benzopyran
(1,2-Chromene)

Coumarin
(1,2-Benzopyrone)

Chromone
(1,4-Benzopyrone)

Isocoumarin
(2,1-Benzopyrone)

2,3-Benzopyrone

Quinoline
(1-Benzazine)

Isoquinoline
(2-Benzazine)

Cinnoline
(1,2-Benzodiazine)

Quinazoline
(1,3-Benzodiazine)

Naphthyridine

Pyrido[3,4-b]-
pyridine

Pyrido[3,2-b]-
pyridine

Pyrido[4,3-b]-
pyridine

1,3,2-Benzoxazine

1,4,2-Benzoxazine

2,3,1-Benzoxazine

3,1,4-Benzoxazine

1,2-Benzisoxazine

1,4-Benzisoxazine

Anthracene

Phenanthrene

Benzonaphthene
(Phenalene)

Fluorene

Carbazole

Xanthene

Acridine

Norpinane

Purine

Steroid ring system

R; nearly always methyl
R'; usually methyl
R"; various groups

A more extensive listing of ring compounds and systems may be found in the following:

Chemical Abstracts, annual subject index.

The Ring Index, Patterson and Capell, Reinhold Publishing Company, 1940.

Lexikon der Kohlenstoffverbindungen, Richter, Leopold Voss, 1910.

The numbering system for the compounds listed above is that used in The Ring Index.

EXPLANATION OF TABLE PHYSICAL CONSTANTS OF ORGANIC COMPOUNDS

This table is a compilation of data on some 15,000 compounds of interest to chemists working in industrial or academic research as well as those in other areas having an occasional need for physical constant data.

An effort has been made to use names which are in most common usage with cross-reference to I.U.C. derived names. As far as possible, compounds are listed alphabetically, but derivatives are listed under the parent compound heading. Thus esters are listed under the name of the parent acid.

Frequent use is made of synonyms. As an example $C_2H_5CO_2H$ is listed under both butyric acid and butanoic acid but the derivatives, for the most part, are listed under butyric acid. As a general rule, if a derivative of a compound is not listed under the name in the first column, look for it to be listed as a derivative of the synonym.

Each compound has been given a number which is used in the formula, melting point, and boiling point indexes found at the end of the table.

References to original literature or other Compendia are given whenever possible (see Table of Abbreviations). The most important of these are, B (Berichte der Deutsche Chemische Gesselschaft), and A (Atlas for Spectral Data and Physical Constants).

Structural Formulas are given whenever possible. More complex formulas are given at the end of the Table.

Molecular Weights are obtained from recently published values of Atomic Weights.

The color of crystalline compounds is listed by abbreviations (see Table of Abbreviations) associated with the crystalline form. Solid compounds are considered as crystalline if not otherwise stated. Data on solvents of crystallization are given in brackets. (see Table of Abbreviations)

The following are specific examples:

pl	Plates
pl (al)	Plates obtained from alcohol as solvent
pl (al + 1½)	Plates obtained from alcohol with 1½ mol of alcohol of crystallization
pl (aq ace + 2w)	Solvent of crystallization is one of components of solvent, e.g. 2 mol of water
(al)	Crystalized from alcohol; crystalline form not reported.

Specific Rotation follows common usage e.g.

$$[\alpha]\frac{25}{D} = -25.8 \ (w, c = 4) \qquad w = water$$

Melting Points are rounded to the nearest 0.1°C. A second, but probably less reliable, melting point is listed in brackets for a few compounds.

Boiling Points are at atmospheric pressure (760 mm) unless otherwise indicated by a superscript.

Density (specific Gravity). In most cases specific gravities are given. Thus 0.86^{20}_4 indicates the density of the liquid at 20°C relative to the density of water at 4°C. Where only one temperature is given as a superscript, the value is in grams per milliliter at the indicated temperature.

The **Refractive Index** is reported for the D line of the sodium spectrum (η_D) at the temperature indicated by the superscript. In a few cases other spectral lines are indicated.

Common Solvents for each compound are given as a guide to solubility. No distinction is made between soluble and very soluble primarily because definitions for these terms differ widely among different investigators. Solvents in which the compounds are soluble in all proportions are underlined.

Three indexes are located at the end of the Table. These are:

1. **Empirical Formula Index** which is arranged according to increasing C, H, and remaining elements in alphabetical order. Hydrates are entered under the formula of the anhydrous compound. Salts, complexes, etc. are found under their total formula.

2. **Melting Point Index** which is arranged according to increasing melting point or freezing point. Only the lower temperature of the melting point range is given. Melting points are rounded off to the nearest unit.

3. **Boiling Point Index** which is arranged according to increasing boiling point of compounds of which the boiling point is given at a pressure near to 1 atmosphere, (i.e., 700 to 780 mm Hg). Only the lower temperature of the boiling range is given. No special entries are made for the remarks on the boiling point contained in the main table. Boiling points listed in the Index are rounded off to the nearest unit.

RULES FOR THE NAMING OF COMPOUNDS

1. Compounds are given common names when these are widely used. These compounds are usually cross-indexed to I.U.C. derived names. e.g., $HO_2CCH_2CH_2CO_2H$ is listed as succinic acid and also as butandioic acid. Physical constants are listed for both entries.

2. Derivatives of a parent compound having one principal function are listed under this compound; e.g. Ethyl butyrate is listed under butyric acid. Chloroacetic acid has a separate entry with many derivatives listed under it. However, 2-chlorobutyric acid is listed as a derivative of butyric acid because there are a limited number of chlorobutyric acid derivatives.

3. Whenever a derivative is not found under the name of the parent compound, look for it under the name of the synonym, or as a principal entry. Derivatives of derivatives are not in the table.

Aliphatic Hydrocarbons

4. I.U.C. nomenclature is used.

a) Saturated hydrocarbons are named as alkanes. Branched chain compounds are named as derivatives of the longest chain.

b) Hydrocarbons having carbon-carbon double bonds are named as alkenes, alkadienes, alkatrienes, etc., with the position of the double bonds indicated by suitable numbers. Branched chain compounds are named as derivatives of the longest chain that contains the maximum number of double bonds, e.g.

$CH_3CH=CHCH_3$ 2-Butene

$CH_2=CHCH=CH_2$ 1.3-Butadiene

$$CH_2=CHCHCH=CH_2$$
$$\underset{C_3H_7}{|}$$
3-Propyl+1.4-pentadiene

c) Hydrocarbons having carbon-carbon triple bonds are named as alkynes. Branched chain alkynes are named as derivatives of the longest chain containing the triple bonds. The position of the triple bond is indicated by suitable numbers, e.g.

$$CH_3CH_2C\equiv CH$$
1-Butyne

$$\underset{C\equiv CH}{\underset{|}{CH_3CH_2CHCH_2CH_3}}$$
3-Ethyl-1-pentyne

d) When both double and triple bonds are found in the same molecule, the double bond takes precedence in naming.

$$CH_3CH_2CH=CHC\equiv CH$$ 3 - Pentene - 1 - yne

5. Cyclic Hydrocarbons — Naphthenes — are given I.U.C. derived names.

Cyclohexane

Cyclohexene

Cyclopentadiene

Cyclobutane

6. Bridged Hydrocarbon ring systems appear under the headings bicyclo-, tricyclo-. Most such compounds however, are listed under common names.

Bicyclo [3.1.0] hexane

7. Spiro-, dispiro-, etc. hydrocarbons and derivatives appear under the spiro-headings.

Spiro-[4.5-decane]

8. Aromatic Hydrocarbons are listed under their common names, e.g. Benzene (C_6H_6), Toluene ($C_6H_5CH_3$), Xylene [$C_6H_4(CH_3)_2$,] Naphthalene ($C_{10}H_8$) etc. Aromatic hydrocarbons having aliphatic side chains are listed as derivatives, e.g.

Benzene, ethyl

Naphthalene, β-methyl

Toluene, p-chloro

When two or more benzene rings, phenyl groups, are attached to a single carbon atom or to more than one carbon atom of a hydrocarbon molecule, they are named as phenyl derivatives.

$$(C_6H_5)_2CH_2$$ Diphenyl methane

$$(C_6H_5)_3CH$$ Triphenyl methane

$$C_6H_5CH_2CH_2C_6H_5$$ 1 2,-Diphenyl ethane

9. Heterocyclic Systems. Common names are used or when these are lacking, systematic names using *oxo, azo, thio,* etc. to indicate an oxygen, nitrogen or sulfur atom in the cyclic structure.

The following common names are used; furan, pyran, pyrrole, pyrazole, imidazole, piperidine, pyridine, pyrazine, pyrimidine, pyrazidine, thiophene, etc.

10. Radicals. Radical names such as methyl, ethyl, isopropyl, phenyl, tolyl, benzyl, xylyl, allyl, styryl, etc. are commonly used. (See List of Substituent Prefixes)

Compounds Containing Functional Groups

Order of Precedence of Functions. When a compound contains one or more functional groups, the parent compound is designated according to the following order of precedence: acid, aldehyde, ketone, alcohol, amine, ether, sulfide, sulfone, and sulfoxide.

C-52

ACIDS AND DERIVATIVES

1. **Common names** have long been used in naming acids and well established names are used in this table, e.g. Acetic acid, Caprioc acid, Benzoic acid, Palmitic acid, Oleic acid, Stearic acid, Succinic acid, Tartaric acid, etc.

2. **Acids having no well-established common names** are names as alkanoic acids, alkenoic acids, alkane dioic acids, alkene dioic acids, etc., e.g.

$CH_3(CH_2)_7CO_2H$	Nonanoic acid
$CH_3CH=CHCH_2CO_2H$	3-Pentenoic acid
$(CH_3)_2CH=CHCO_2H$	2-Butenoic acid.-3-methyl
$CH_2=CHCH=CHCO_2H$	2.4-Pentadienoic acid
$HO_2CCH_2C(C_4H_9)=CHCO_2H$	2-Pentenedioic acid. -3-butyl

Use the synonym name to find derivatives not listed under the name of the acid in the first column.

The term *-iso* is used to indicate branching at the end of the hydrocarbon chain. Thus $(CH_3)_2CHCO_2H$ is named Isobutyric acid. Listing of this acid and its derivative is found following Butyric acid — alphabetically under "B".

Polycarboxylic acids use the carboxylic acid suffix added to the longest carbon chain that does not contain these groups, e.g.

$HO_2CCH_2CH(CO_2H)CH_2CO_2H$	1 2.3-Propane tricarboxylic acid

3. **Cyclic Acids.** After the name of the ring, the suffix carboxylic acid is attached.

2-Cyclohexene carboxylic acid

1,2-Cyclohexane dicarboxylic acid, -3-methyl

4. **Thioacids** are named as thio-, dithio-, thiolo-, or thiono-acids. The latter two prefixes are used only when it is certain that the hydroxyl or ketonic oxygen is replaced by the sulfur atom, e.g.,

CH_3CSOH	Thioacetic acid
$(CH_3)_2CHCS_2H$	Dithioisobutyric acid
CH_3CH_2COSH	Thiolopropionic acid
$CH_3CH_2C(S)OH$	Thionopropionic acid

5. **Imidic, Hydroxamic Acids.** Only simple compounds of these types appear in the Table. The following are examples

$CH_3CH_2CH(=NH)OH$	Butanimidic acid
$C_6H_5C(=NH)OH$	Benzimidic acid

2-Pyrrolecarboxyimidic acid

$CH_3CH_2C(=NOH)OH$	Propanehydroxamic acid
$C_6H_5C(=NOH)OH$	Benzohydroxamic acid

2-Furancarbohydroxamic acid

6. **Sulfonic and Sulfinic Acids.** Compounds containing the $-SO_3H$ group are named as sulfonic acids. When a carboxyl group is also present this group is designated by the prefix sulfo-.

Benzene sulfonic acid

$$HO_2C \underset{}{\bigcirc} SO_3H$$

Benzoic acid, 4-sulfo

Compounds containing the $-SO_2H$ group are named as sulfinic acids.

7. Keto Acids. When the carbonyl function is present in the principal chain, the compounds are designated as oxo- acids.

$CH_3COCH_2CH_2CO_2H$ Valeric acid. 4-oxo or. Pentanoic acid. 4-oxo

Common names are listed when practical, e.g., $CH_3COCH_2CO_2H$ — Acetoacetic acid.

8. Amino acids. In most cases common names are used. Otherwise they are named as aminoalkanoic acids, aminoalkenoic acids, etc.

9. Ortho acids (or esters) are listed under ortho-, e.g.,

$HC(OCH_3)_3$ Orthoformic acid. trimethyl ester

10. Acid halides and amides are listed independently from the acids.

CH_3COCl	Acetyl chloride
C_3H_7COBr	Butyryl bromide or Butanoyl bromide
C_6H_5COCl	Benzoyl chloride
$C_2H_5CONH_2$	Propionamide
$C_6H_5CONHCH_3$	Benzamide, -N-Methyl
$C_6H_5CON(C_2H_5)_2$	Benzamide, -N,N-diethyl
$C_6H_5SO_2NH_2$	Benzenesulfonamide
$C_6H_5SO_2NHCH_3$	Benzenesulfonamide, -N-Methyl
$CH_3CONHCOCH_3$	Acetamide, -N-acetyl

N-substituted amides in which the substituent is phenyl are named as amides. e.g.,

$CH_3CONHC_6H_5$	Acetanilide
$C_2H_5CONHC_6H_5$	Butyranalide

11. Esters. Esters are listed under the name of the acid, usually the common name.

$CH_3CO_2C_2H_5$	Acetic acid. ethyl ester
$C_6H_5CO_2C_4H_9$	Benzoic acid. butyl ester
$C_2H_5O_2CCH_2CH_2CO_2C_2H_5$	Succinic acid. diethyl ester
$(CH_3)_2C=CHCO_2CH$	2-Butenoic acid, 4-Methyl-methyl ester

Esters of complicated or polyhydric alcohols may appear under the name of the alcohol.

$$\begin{matrix} CH_2O_2CC_{17}H_{35} \\ | \\ CHO_2CC_{17}H_{35} \\ | \\ CH_2O_2CC_{17}H_{35} \end{matrix}$$ Glyceryl tristearate or Stearin

12. Lactams and Lactones are listed under the names of the corresponding amino- or hydroxy acids.

13. Nitriles are listed independently from the corresponding acids and are designated as nitriles, e.g.,

C_3H_7CN	Butyronitrile
$CH_3CH(NH_2)CN$	Propionitrile,-2-amino
$NCCH_2CH_2CN$	succinonitrile
$2\text{-}ClC_6H_4CN$	Benzonitrile,-o-chloro

14. **Imides** are listed independently from the corresponding acids, e.g.,

 Phthalimide

Succinimide, N-bromo

ALDEHYDES

1. Common names are listed for aldehydes. These are derived by dropping the -ic ending of the related acid and adding aldehyde, e.g.,

CH_3CHO	Acetaldehyde
$(CH_3)_2CHCHO$	Isobutyraldehyde
C_6H_5CHO	Benzaldehyde
$2-CH_3C_6H_4CHO$	o-Tolualdehyde
$CH_3CH=CHCHO$	Crotonaldehyde
$C_6H_4(CHO)_2$	Phthaldehyde

2. Where common names are not widely used, aldehydes are named as alkanals, alkenals, etc. e.g.,

$CH_3CH_2CH_2CHO$	Butanal
$(CH_3)_2C=CHCHO$	2-Butenal,-3-methyl
$CH_2=CHCH=CHCHO$	2,4-Butadienal
$OHCCH(CH_3)CHCHO$	Butanedial, 2-methyl

3. Compounds having a -CHO group attached to a cyclic structure are frequently named as carboxaldehydes, e.g.,

1-Cyclohexene carboxaldehyde

4. The -CHO group is frequently referred to as formyl when a carboxyl group is also present. The $-C=O$ group may also be designated as oxo, e.g.,

Benzoic acid,-4-formyl

$OHCCH_2CH_2COOH$ Butanoic acid,-4-oxo

5. Acetals and hemi-acetals are listed under the corresponding aldehyde or ketone, e.g.,

$CH_3CH(OC_2H_5)_2$	Acetaldehyde, diethyl acetal
$CH_3CH_2CH(CH_3)CH(OCH_3)_2$	Butanal,-2-methyl, dimethyl acetal
$CH_3CH_2CH(OH)OCH_3$	Propionaldehyde, methyl-hemi-acetal

KETONES

1. Ketones having only aliphatic groups are listed as dialkyl ketones, or as alkanones, alkenones, etc., e.g.,

CH_3COCH_3	Dimethyl ketone or acetone
$(CH_3)_2CHCOCH_3$	Methyl isopropyl ketone or 2-butanone,-3-methyl
$CH_2=CHCOCH_3$	Methyl vinyl ketone or 3-butene-2-one

When the derivative being sought is not found under the name appearing in the first column, look for it under the synonym name or as a principal entry.

2. When one of the groups attached to the carbonyl group (−C=O) is phenyl the compound is named as phenyl ketone or as a phenone, e.g.,

$C_6H_5COCH_3$ Acetophenone or methyl phenyl ketone

$C_6H_5COC_6H_5$ Benzophenone or diphenyl ketone

$C_6H_5COC_3H_7$ Butyrophenone or propyl phenyl ketone

3. When cyclic structures are attached to the carbonyl group, the ketones may be named as acetyl-, propionyl-, butyryl-, etc. derivatives of the cyclic compound.

Furan, -2-acetyl

Pyrrole, -2-propionyl

Thiophene, -3-methyl-2-butyryl

4. Thioketones are named as thiones.

$C_6H_5CSCH_3$ Acetothiophenone

$C_6H_5CSC_6H_5$ Dibenzothiophenone

5. Cyclic ketones and thiones are given the name of the parent compound followed by the ending -one or -thione.

2-Tetralone

4-Pyrone

Cyclopentanone

Cyclohexanone, 2-methyl

Cyclohexanethione

6. Quinones, the name is reserved for structures of the type

Orthoquinones or 1,2-quinones

or

Paraquinones or 1,4-quinones

Examples are:

1,4-Benzoquinone

1,2-Naphthoquinone

1,4-Benzoquinone, -2-methyl

1,4-Benzoquinone, -2-carboxylic acid

ALCOHOLS

1. Alcohols are named as alkyl alcohols, or as alkanols, alkenols, alkanediols, etc.

CH_3CH_2OH	Ethyl alcohol ethanol
$(CH_3)_2CHCH_2OH$	Isobutyl alcohol 1-propanol,-2-methyl
$CH_3CH_2CH(OH)CH_3$	Sec-butyl alcohol 2-butanol
$(CH_3)_3COH$	Tert-butyl alcohol 2-Propanol, 2-methyl

OH

Cyclohexanol

$CH_3CH(OH)CH(OH)CH_3$ 2,3-Butanediol

Phenyl substituted alcohols are named as phenyl-, diphenyl-, alkanols, etc.

$(C_6H_5)_2CHOH$ Methanol,–diphenyl

2. Iso-, sec-, and tert-alcohols and derivatives are listed sequentially after the name of the normal alcohol, e.g. isobutyl, sec-butyl and tert-butyl alcohols are listed as butyl alcohols.

3. Common names are frequently used in naming alcohols, e.g.

$CH_2=CHCH_2OH$	Allyl alcohol
$HOCH_2CH_2OH$	Ethylene glycol
$HOCH_2CH(OH)CH_2OH$	Glycerol
$C_{11}H_{23}CH_2OH$	Lauryl alcohol
$C_{17}H_{35}CH_2OH$	Stearyl alcohol

4. Thioalcohols are names as thiols or mercaptans

C_2H_5SH Ethanethiol ethyl mercaptan

HOCH₂CH₂SH   Ethanedithiol

(CH₃)₂CHCH₂SH   1-Propanethiol,-3-methyl isobutyl mercaptan

PHENOLS

1. Hydroxyaromatic compounds are named as phenols. They are much more acidic than alcohols and deserve separate listing, e.g.,

Phenol

Phenol, *o*-nitro

Phenol, *m*-amino

β-Naphthol

o-Cresol

2. Common names are given to aromatic diphenols or triphenols and some mono phenols, e.g.,

Catechol

Resorcinol

Hydroquinone

Picric acid

AMINES

1. Alkyl amines, dialkyl amines and trialkyl amines are listed alphabetically according to the alkyl groups attached. The smallest group is named first and thus determines the listing. Refer to the synonym if necessary, e.g.,

CH₃NH₂   Methyl amine

C₄H₉NH₂   Butyl amine

Cyclohexyl amine

C-58

$(CH_3)_2NH$	Dimethyl amine
$C_3H_7NHC_2H_5$	Ethyl propyl amine
$(C_2H_5)_3N$	Triethyl amine
$(C_2H_5)_2NCH_3$	Methyl diethyl amine
$H_2NCH_2CH_2CH_2NH_2$	1.3-Propane diamine

2. When more complex radical names are found the amines are names as amino alkanes, amino alkenes, etc.

$CH_3CH_2CH(NH_2)CH_3$	Butane.-2-amino
$(CH_3)_2CH_2CH(NH_2)CH_3$	Butane.-4-methyl-2-amino

3. Amino substituted heterocyclic compounds are names as amino- derivatives, e.g.,

Furan,-2-amino

Thiophene,-3-methylamino

4. Amino substituted toluene and xylene, respectively are named as toluidines and xylidines.

o-Toluidine

o-Toluidine,-4-nitro

p-Toluidine,-N-Methyl

m-Toluidine, -N-phenyl

ETHERS

1. Dialkyl ethers are listed alphabetically with the smaller group being named first, e.g.,

$C_2H_5-O-C_2H_5$	Diethyl ether
$C_2H_5-O-i-C_3H_7$	Ethyl isopropyl ether
$(CH_3)_3C-O-C_5H_{11}$	Tert-butyl pentyl ether
$CH_3-O-C_6H_5$	Methyl phenyl ether
$CH_3-O-CH=CH_2$	Methyl vinyl ether
$C_6H_5CH_2-O-CH_2C_6H_5$	Dibenzyl ether

2. Alkyl ethers of aromatic or heterocyclic compounds are named as alkoxides, e.g.,

Phenol, *o*-methoxy

Benzene, isopropoxy

Benzoic acid, *p*-ethoxy

Naphthalene, α-paenoxy

SULFIDES, DISULFIDES, SULFOXIDES AND SULFONES

1. These compounds are named in the same way as ethers. The alkyl, or aryl groups are named alphabetically, beginning with the smaller group, e.g.,

$C_2H_5-S-C_2H_5$ Diethyl sulfide

$CH_3-S-C_4H_9$ Methyl butyl sulfide

Phenyl-*o*-tolyl sulfide

$CH_3C-SO-C_2H_5$ Methyl ethyl sulfoxide

$CH_3-SS-C_2H_5$ Methyl ethyl disulfide

$C_2H_5-SO_2-C_2H_5$ Diethyl sulfone

$C_2H_5-SO_2-C_6H_5$ Ethyl phenyl sulfone

Naphthalene, -β-methylthio

There may be times when the compound can be located more readily by the name of the synonym.

SYMBOLS AND ABBREVIATIONS

Abbr.	Meaning	Abbr.	Meaning	Abbr.	Meaning	Abbr.	Meaning
α]	specific rotation	D	line in the spectrum of sodium (subscript)	sym	symmetrical	hp	heptane
>	above, more than	D, d	dextro[3]	syr	syrup	htng	heating
<	below, less than	δd	slight decomposition	ta	tablets	hx	hexane
?	unkown	dil	diluted	tcl	triclinic	hyd	hydrate
aa	acetic acid	diox	dioxane	tert	tertiary	hyg	hygroscopic
abs	absolute	distb	distillable	Tet	Tetrahedron	i	insoluble
ac	acid	dk	dark	tetr	tetragonal	i-	iso-
Ac	acetyl	Dl, dl	racemic[3]	THF	tetrahydrofuran	ign	ignites
ace	acetone	dlq	deliquescent	to	toluene	in	inactive
AFCL	Aliphatic Fluorine Compounds	DMF	dimethyl formamide	tr	transparent	inflam	inflammable
al	alcohol[1]	E	Elsevier's	trg	trigonal	infus	infusible
ALD	Aldrich Handbook of Organic Chemicals and Bio chemicals	eff	efforescent	undil	undiluted	irid	iridescent
alk	alkali	peth	petroleum ether	uns	unsymmetrical	iso	isooctaine
Am	J. Am. Chem. Soc.	pk	pink[2]	unst	unstable	J	J. Chem. Soc.
Am	amyl (pentyl)	Ph	phenyl	v	very	JOC	J. Org. Chem.
amor	amorphous	pl	plates	vac	vacuum	KHOC	Kaufman Handbook of Organometallic Compounds
anh	anhydrous	pr	prisms	var	variable	L, l	levo[3]
aqu	aqueous	Pr	propyl	vap	vapor	la	large
as	asymmetric	Prak	J. Prak. Chem.	vic	vicinal	lf	leaf
Atlas	Atlas of Spectral Data and Physical Constants for Organic Compounds	purp	purple[2]	visc	viscous	liq	ligroin
atm	atmospheres	pw	powder	volat	volatile or volatilises	liq	liquid
b	boiling	Py	pyrimidine	vt	violet[2]	lo	long
B	Beilstein	pym	pyramids	w	water	lt	light
Ber	Chem. Ber.	rac	racemic	wh	white[1]	m	melting
bipym	bipyramidal	rect	rectangular	wr	warm	m-	meta-
bk	black[2]	red	red	wx	waxy	M	molar (concentration)
bl	blue[2]	res	resinous	ye	yellow[2]	M	Merck Index, 7th Edition
BOSC	Bayant, et al., Organosilicon Compounds	rh	rhombic	xyl	xylene	mcl	monoclinic
		rhd	rhombodohedral	Et	ethyl	Me	methyl
br	brown[2]	s	soluble	eth	ether[4]	met	metallic
bt	bright	sc	scales	exp	explodes	micr	microsopic
Bu	butyl	sec	secondary	extrap	extrapolated	min	mineral
bz	Benzene	sf	softens	fl	flakes	mod	modification
CAS	Chemical Abstracts	sh	shoulder	flam	flammable	mut	mutarotatory
c	percentage concentration	silv	silvery	flr	fluorescent	n	normal chain, refractive index
ca	about (circa)	sl	slightly	fr	freezes		
chl	chloroform	so	solid	fr. p.	freezing point	N	normal (concentration)
co	columns	sol	solution	fum	fuming	N	nitrogen[5]
col	colorless	solv	solvent	gel	gelatinous	nd	needles
con	concentrated	Sol-vents	Entries in this column in medium type means soluble; entries in boldface means very soluble	gl	glacial	o-	ortho-
cor	corrected			gold	golden	oct	octahedral
cr	crystals			gr	green[3]	og	orange[2]
cy	cyclohexane	sph	sphenoidal	gran	granular	ord	ordinary
d	decomposes	st	stable	gy	gray[3]	org	organic
		sub	sublimes	h	hot	orh	orthorhombic
		suc	supercooled	H	Helv. Chim. Acta	p-	para-
		sulf	sulfuric acid	hex	hexagonal	pa	pale
				HDOC	Helibron Dictionary of Organic Compounds	par	partial
						PCHE	Egloff Physical Constants of Hydrocarbons

[1] Generally means ethyl alcohol.

[2] The abbreviation of a color ending in "sh" is to be read as ending with the suffix "-ish," e.g., grsh means greenish.

[3] D, L generally means configuration and d, l generally mean optical rotation, but there are many examples in the chemical literature for which the meaning of these symbols is ambiguous and/or interchangeable.

[4] Generally means diethyl ether.

[5] N indicates a position in the molecule.

BEILSTEIN REFERENCES

"Beilstein is the short name for "Beilsteins Handbuch der Organischen Chemie". This Handbuch contains the most complete set of data on organic compounds in the world. The 1st edition was published as two volumes during the years 1881 to 1882. That edition contained data for about 15,000 compounds described on 2200 pages. The 2nd edition of Beilstein consisted of 3 volumes and 4080 pages. The 2nd edition appeared in the years 1885 to 1889. The 3rd edition was published over the years 1892 to 1906. It consisted of 8 volumes published on approximately 11,000 pages. The 4th edition of Beilstein began to be published in 1909. This 4th edition may be described as follows:

The Series of the Beilstein Handbook (4th edition)

Series	Abbreviation	Period of literature completely covered	Colour of label on spine
Basic Series	H	up to 1909	green
Supplementary Series I	E I	1910—1919	red
Supplementary Series II	E II	1920—1929	white
Supplementary Series III	E III	1930—1949	blue
Supplementary Series III/IV	E III/IV*	1930—1959	blue/black
Supplementary Series IV	E IV	1950—1959	black

* Volumes 17 to 27 of Supplementary Series III and IV, covering the heterocyclic compounds are combined in a joint issue.

Preparations are currently in progress for the Fifth Supplementary Series (E V), which will cover the literature from 1960 to 1979.

Each of these series of the 4th edition comprises 27 volumes (or groups of volumes) in which the individual compounds are arranged according to the "Beilstein System". Which compounds are dealt with in which Beilstein volumes is shown in the following two tables:

The Main Divisions of the Beilstein Handbook

Main Division	Volume No.	System No.
A. *Acyclic* Compounds	1—4	1—449
B. *Isocyclic* Compounds	5—16	450—2358
C. *Heterocyclic* Compounds	17—27	2359—4720

Contents of the 27 Volumes of the Beilstein Handbook

Type of registry compound	Feature of the functional group	A (Acyclics)	B (Isocyclics)	C (Heterocyclics) 1O*	20*, 30*,…,	1N	2N	3N, 4N,…	1N, 1O*, 1N, 2O*, 2N, 1O*, 2N, 2O*, further heteroatoms**
1 Compounds without functional groups			5			20			
2 Hydroxy-compounds	−OH	1	6	17			23		
3 Oxo-compounds =O	=O		7			21	24		
	=O + −OH		8						
4 Carboxylic acids	$\overset{O}{\underset{OH}{\diagup}}$	2	9						
	$\overset{O}{\underset{OH}{\diagup}}$ + −OH; $\overset{O}{\underset{OH}{\diagup}}$ + =O; $\overset{O}{\underset{OH}{\diagup}}$ + =O + −OH	3	10						
5 Sulfinic acids	−SO$_2$H								
6 Sulfonic acids	−SO$_3$H								
7 Seleninic acids, Selenonic acids, Tellurinic acids	−SeO$_2$H and −SeO$_3$H −TeO$_2$H	4	11	18	19	22	25	26	27
8 Amines −NH$_2$	−NH$_2$		12						
	[−NH$_2$]$_n$; −NH$_2$ + −OH		13						
	NH$_2$ + =O, NH$_2$ + $\overset{O}{\underset{OH}{\diagup}}$		14						
9 Hydroxylamines and Dihydroxyamines	−NH−OH −N$\overset{OH}{\underset{OH}{\diagup}}$								
10 Hydrazines	−NH−NH$_2$		15						
11 Azo-compounds	−N=NH								
12 Diazonium compounds	−N≡N]$^{\oplus}$								
13 Compounds with groups of 3 or more N-atoms	−NH−NH−NH$_2$, −N(NH$_2$)$_2$, −N=N−NH$_2$, etc.								
14 Compounds containing carbon directly bonded to P, As, Sb, and Bi	e.g. −PH$_2$, PH−OH, −P(OH)$_2$, −PH$_4$, …, −PO(OH)$_2$		16						
15 Compounds containing carbon directly bonded to Si, Ge, and Sn	e.g. −SiH$_3$, −SiH$_2$(OH), …								
16 Compounds containing carbon directly bonded to elements of the 3rd−1st A-groups of the periodic table	e.g. −BH$_2$, −BH(OH), …, −Mg$^{\oplus}$								
17 Compounds containing carbon directly bonded to elements of the 1st−8th B-groups of the periodic table	e.g. −HgH, −Hg$^{\oplus}$, …								

* Instead of O also S, Se, Te.
** e.g. B, Si, P, but not S, Se, Te.

Additional informational material on Beilstein is available upon request free of charge from

Springer-Verlag KG	Springer Verlag New York Inc.
Heidelberger Platz 3	175 Fifth Avenue
D-1000 Berlin 33	New York, New York 10010

Included in such information are:

Beilstein Reference Chart ("Contents of Beilstein Handbook of Organic Chemistry")
Brochure, "What is Beilstein"
How to Use Beilstein
Beilstein Wordbook (German/English)
Beilstein Outline (List of all available volumes; revised annually)

The Beilstein references listed in this edition of the *CRC Handbook of Chemistry and Physics* were revised and updated early in 1980 by personnel of Beilstein.

PHYSICAL CONSTANTS OF ORGANIC COMPOUNDS

No.	Name, Synonyms, and Formula	Mol. wt.	Color, crystalline form, specific rotation and λ_{max} (log ε)	b.p. °C	m.p. °C	Density	n_D	Solubility	Ref.
1	Abietic acid or sylvic acid $C_{20}H_{30}O_2$	302.46	mcl pl (al-w)	250	173-4	....	....	al, et, ace, bz	B9[3], 2904
2	Abietic acid, methyl ester $C_{21}H_{32}O_2$	316.49	....	225 6[16]	....	1.0491[20/4]	1.5344	al	D9[1], 2907
3	Acenaphthanthracene or Naphtho-2',3',4,5-acenaphthene $C_{20}H_{14}$	254.36	pa ye lf (lig)	192.5-3.5	....	....	....	bz, lig	B5[3], 2468
4	Acenaphthene $C_{12}H_{10}$	154.21	....	279	96.2	1.0242[90/4]	1.6048[95]	al, bz	B5[4], 1834
5	Acenaphthene, 1-amino $C_{12}H_{11}N$	169.23	cr (peth)	135	sub	....	....	al, bz, CS_2	B12[2], 764
6	Acenaphthene, 3-amino $C_{12}H_{11}N$	169.23	pl (al), nd (peth)	....	81.5	....	....	al, chl	B12[3], 3210
7	Acenaphthene, 4-amino $C_{12}H_{11}N$	169.23	nd (al,w)	....	87	....	....	bz, lig, w	B12[3], 3212
8	Acenaphthene, 5-amino $C_{12}H_{11}N$	169.23	nd (lig) red in air	....	108	....	....	al	B12[3], 3212
9	Acenaphthene, 5-bromo $C_{12}H_9Br$	233.11	pl (al)	335	52	1.4392[52/4]	1.6565[54]	al	B5[4], 1839
10	Acenaphthene, 5-chloro $C_{12}H_9Cl$	188.66	pl or nd (al)	319.2[770],163[13]	70.5	1.1954[20/4]	1.6288[7]	al	B5[4], 1837
11	Acenaphthene, 5-iodo $C_{12}H_9I$	280.11	nd (al)	....	65	1.6738[62/4]	1.6909[65]	al, bz	B5[1], 276
12	Acenaphthene, 5,6-dinitro $C_{12}H_8N_2O_4$	244.21	....	....	220-4	....	....	....	B5[3], 1784
13	Acenaphthene, 3-nitro $C_{12}H_9NO_2$	199.21	gr-ye nd (aa)	....	151.5	....	....	aa	B5[1], 1783
14	Acenaphthene, 5-nitro $C_{12}H_9NO_2$	199.21	....	....	103.4	....	....	tr, al, xln, lig	B5[4], 1040
15	Acenaphthene, 2a,3,4,5-tetrahydro $C_{12}H_{14}$	158.24	....	246	12	....	....	chl	B5[3], 1385
16	Acenaphthene, 5-carboxylic acid or 5-Acenaphthoic acid $C_{13}H_{10}O_2$	198.22	nd (bz or lig)	....	220-1	....	....	bz, lig	B9[3], 3289
17	Acenaphthenequinone $C_{12}H_{10}O_2$	182.18	ye nd (aa)	....	261	....	....	al, bz	B7[3], 3796
18	3-Acenaphthene sulfonic acid $C_{12}H_9SO_3H$	234.28	hyg nd (bz)	....	87-9	....	....	....	B11[3], 435
19	1-Acenaphthenone or 1-oxoacenaphthene $C_{12}H_8O$	168.21	nd (al)	....	121	....	....	w, al, bz, chl	B7[3], 2046
20	Acenaphthylene $C_{12}H_8$	152.21	pr (eth), pl (al)	265-75	92-3	0.8988[16/2]	....	al, eth, bz	B5[4], 2138
21	Acetaldehyde CH_3CHO	44.05	....	20.8	-121	0.7834[18/4]	1.3316[20]	**w, al, eth, ace,** bz	B1[4], 3094
22	Acetaldehyde, 2,4-dinitrophenylhydrazone $CH_3CH=NNHC_6H_3(NO_2)_2$-2,4	224.19	ye sc (al)	....	168.5	....	....	eth, ace, bz, ch	B15[3], 426
23	Acetaldehyde, amino, diethyl acetal $H_2NCH_2CH(OC_2H_5)_2$	133.19	....	163	....	0.9159[25]	1.4170[20]	w, al, eth, chl	B4[4], 1918
24	Acetaldehyde, ammonia or 1-aminoethanol $CH_3CH(OH)NH_2$	61.09	rh (eth-al)	110d	97	....	....	w	C5[2]
25	Acetaldehyde, bromo, diethyl acetal $BrCH_2CH(OC_2H_5)_2$	197.07	....	180, 66[18]	....	1.280[20/4]	1.4376[20]	al, eth	B1[4], 3151
26	Acetaldehyde, bromo, dimethyl acetal $BrCH_2CH(OCH_3)_2$	169.02	....	148.50-50[15]	....	1.5049[184]	1.4450[20]	eth, ace, chl	B1[3], 2672
27	Acetaldehyde, chloro $ClCH_2CHO$	169.02	....	85-5.5[748]	....	....	....	eth	B1[4], 3134
28	Acetaldehyde, chloro, diethyl acetal $ClCH_2CH(OC_2H_5)_2$	152.62	....	157.4, 71-2[35]	....	1.068[20/4]	1.4150[20]	**al, eth,** bz	B1[4], 3134
29	Acetaldehyde, chloro, dimethyl acetal $ClCH_2CH(OCH_3)_2$	124.57	....	127-8	....	1.068[20/4]	1.4150[20]	al, bz	B1[4], 3134
30	Acetaldehyde, bis(2-chloroethyl) acetal $CH_3CH(OCH_2CH_2Cl)_2$	187.07	....	194-6, 106-7[14]	....	1.1737[20/4]	1.45266[20]	al, eth	B1[4], 3104
31	Acetaldehyde, diacetate or Ethylidene diacetate $CH_3CH(O_2CCH_3)_2$	146.14	....	169, 65-7[10]	18.9	1.3985[25]	1.070[25]	al, eth	B2[4], 282
32	Acetaldehyde, dichloro Cl_2CHCHO	112.94	....	90-1	....	....	....	al	B1[4], 3140
33	Acetaldehyde, dichloro, diethyl acetal $Cl_2CHCH(OC_2H_5)_2$	187.07	....	183-4, 67-71[12]	....	1.1381[14]	....	al, eth	B1[4], 3140

No.	Name, Synonyms, and Formula	Mol. wt.	Color, crystalline form, specific rotation and λ_{max} (log ε)	b.p. °C	m.p. °C	Density	n_D	Solubility	Ref.
34	Acetaldehyde, dichloro, hydrate .... $Cl_2CHCH(OH)_2$	130.96	cr (bz), ta	97	56-7	....	....	w, al, eth	B1, 614
35	Acetaldehyde, diethyl acetal .... $CH_3CH(OC_2H_5)_2$	118.18	....	103.2, 21^{22}	......	0.8314$^{20/4}$	1.3834^{20}	w, al, eth, ace	B1^4, 3103
36	Acetaldehyde, diethyl mercaptal .... $CH_3CH(SC_2H_5)_2$	150.30	....	183-5, 73^{19}	......	0.9706^{20}	1.5025^{20}	eth, ace, bz	B1^4, 3162
37	Acetaldehyde, dimethyl acetal .... $CH_3CH(OCH_3)_2$	90.12	....	64.5	−113.2	0.85015$^{20/4}$	1.3668^{20}	w, al, eth, ace, bz	B1^4, 3103
38	Acetaldehyde, dimethylamino, diethyl acetal .... $(CH_3)_2NCH_2CH(OC_2H_5)_2$	161.25	cr (bz), ye	170-1	......	0.885^7	1.4129^{20}	w, al, eth, ace	B1^3, 870
39	Acetaldehyde, 2,4-dinitrophenylhydrazone .... $CH_3CH=NNHC_6H_3(NO_2)_2$-2,4	224.19	ye sc (al)	....	168.5	....	....	eth, ace, bz	B15^3, 426
40	Acetaldehyde, diphenyl .... $(C_6H_5)_2CHCHO$	196.25	....	157.5^5	......	1.1061$^{21/4}$	1.5920^{21}	al, eth, ace	B7^3, 2117
41	Acetaldehyde, ethoxy .... $C_2H_5OCH_2CHO$	88.12	....	71-3	......	0.942$^{20/4}$	1.3956^{20}	w, al, ace	B1^3, 3181
42	Acetaldehyde, hydroxy or Glycolaldehyde........... $HOCH_2CHO$	60.05	pl	....	97	1.366^{100}	1.4772^{19}	w, al	B1^4, 3955
43	Acetaldehyde, hydroxy, diethyl acetal .... $HOCH_2CH(OC_2H_5)_2$	134.18	....	167, 57-8^8	......	0.888$^{24/4}$	1.4073^{20}	al, eth	B1^4, 3958
44	Acetaldehyde, methoxy .... CH_3OCH_2CHO	74.08	....	92.3^{770}	......	1.005$^{25/4}$	1.3950^{20}	w, al, eth, ace	B1^4, 3955
45	Acetaldehyde, methoxy, diethyl acetal .... $CH_3OCH_2CH(OC_2H_5)_2$	148.20	....	144-8	......	....	1.3990$^{20/D}$	al, eth, ace	B1^4, 3958
46	Acetaldehyde, oxime or Acetaldoxime .... $CH_3CH=NOH$	59.07	nd	115	47	0.9656$^{20/4}$	1.42567^{20}	al, eth	B1^4, 3121
47	Acetaldehyde, phenyl or α-Tolualdehyde........... $C_6H_5CH_2CHO$	120.16	....	195, 88^{18}	33-4	1.0272$^{20/4}$	1.5255^{20}	al, eth, bz	B7^3, 1003
48	Acetaldehyde, semicarbazone .... $CH_3CH=NNHCONH_2$	101.11	....	....	163	1.300$^{0/4}$	....	al	B3^4, 178
49	Acetaldehyde, phenyl, dimethyl acetal .... $C_6H_5CH_2CH(OCH_3)_2$	166.22	....	193-4	......	....	....		B7^3, 1006
50	Acetaldehyde, phenylhydrazone .... $CH_3CH=NNHC_6H_5$	134.18	....	133-6^{21}	98-101	....	....	al	B15^3, 79
51	Acetaldehyde, tribromo or bromal .... Br_3CCHO	280.74	....	174, 61^9	......	2.6650$^{25/4}$	1.5939^{20}	al, eth, ace	B1^4, 3155
52	Acetaldehyde, tribromo, hydrate or Bromal hydrate ..... $Br_3CCH(OH)_2$	298.76	mcl pr (w + 1)	d	53.5	2.5662$^{40/4}$	....	al, eth	B1^4, 3155
53	Acetaldehyde, trichloro or chloral.................... Cl_3CCHO	147.39	....	97.8	−57.5	1.5121$^{20/4}$	1.45572^{20}	w, al, eth	B1^4, 3142
54	Acetaldehyde, trichloro, diethyl acetal .... $Cl_3CCH(OC_2H_5)_2$	221.51	....	205, 84-5^{10}	......	1.266$^{25/4}$	1.4586^{25}	**al, eth**	B1^4, 3144
55	Acetaldehyde, trichloro, ethylhemiacetal .... $Cl_3CCH(OH)OC_2H_5$	193.46	....	115-6^{760}	56-7	1.143^{40}	....	w, al, eth	B1^4, 3144
56	Acetaldehyde, trichloro, hydrate or chloral hydrate ...... $Cl_3CCH(OH)_2$	165.40	mcl pl (w)	96.3$^{764/d}$	57	1.9081$^{20/4}$	....	w, al, eth, ace, bz	B1^4, 3143
57	Acetaldehyde, trimethyl or Pivaldehyde................ $(CH_3)_3CCHO$	86.14	....	77-8	6	0.7923^{17}	1.3791^{20}	al, eth	B1^4, 3295
58	Acetamide .... CH_3CONH_2	59.07	trg mcl (al-eth)	221.2,120^{20}	82.3	0.9986$^{85/4}$	1.4278^{78}	w, al	B2^4, 399
59	Acetamide, N-acetyl or diacetamide .... $CH_3CONHCOCH_3$	101.11	nd (eth)	223.5, 113^{12}	79	....	....	w, al, eth	B2^4, 416
60	Acetamide, N-acetyl-N-ethyl or Diacetylethylamine..... $CH_3CON(C_2H_5)COCH_3$	129.10	....	195-9	......	1.0092^{20}	1.4513^{20}	al	B4^4, 352
61	Acetamide, N-acetyl-N-methyl or Diacetylmethylamine .. $CH_3CON(CH_3)COCH_3$	115.33	....	194.5, 114.5^{61}	−25	1.0663$^{25/4}$	1.4502^{25}	w	B4^4, 183
62	Acetamide, N-acetyl-N-phenyl or Diacetanilide .... $(CH_3CO)_2NC_6H_5$	177.21	ta (lig)	200^{100} 142^{41}	37-8	....	....	al, bz	B12^3, 472
63	Acetamide, N-(2-aminoethyl) .... $CH_3CONHCH_2CH_2NH_2$	102.14	....	128^3	51	....	....	w, al, bz	B4^4, 1193
64	Acetamide, N-benzyl .... $CH_3CONHCH_2C_6H_5$	149.19	....	157^2	61	....	....	al, eth	B12^5, 2255
65	Acetamide, N-bromo, hydrate .... $CH_3CONHBr \cdot H_2O$	155.99	bt ye pl (w + 1)	....	108-9	....	....		B2^1, 82
66	Acetamide, N-butyl .... $CH_3CONHC_4H_9$	115.18	....	229	......	....	1.4388^{25}	....	B4^4, 565

No.	Name, Synonyms, and Formula	Mol. wt.	Color, crystalline form, specific rotation and λ_{max} (log ε)	b.p. °C	m.p. °C	Density	n_D	Solubility	Ref.
67	Acetamide, N-butyl-N-phenyl .... CH$_3$CON(C$_4$H$_9$)C$_6$H$_5$	191.13	....	281, 141[10]	24.5	0.9912[20/4]	1.5146[20]	chl	B12[3], 467
68	Acetamide, chloro or chloroacetamide .... ClCH$_2$CONH$_2$	93.51	....	224-5[717]	121	....	....	w, al	B2[4], 490
69	Acetamide, N-cyanomethyl .... CH$_3$CONHCH$_2$CN	98.10	....	....	79-81	....	....	w	B4[2], 790
70	Acetamide, N,N-diacetyl or Triacetamide .... CH$_3$CON(COCH$_3$)$_2$	143.14	nd (eth)	....	79	....	....	eth	B2[4], 416
71	Acetamide, dibenzyl .... (C$_6$H$_5$CH$_2$)$_2$CHCONH$_2$	239.32	nd (al, w)	259[18]	129	....	....	al, eth	B9[3], 3356
72	Acetamide, dichloro .... Cl$_2$CHCONH$_2$	127.96	....	233-4[745]	99.4	....	....	w, al, eth	B2[4], 505
73	Acetamide, N,N-diethyl .... CH$_3$CON(C$_2$H$_5$)$_2$	115.18	....	185-6	....	0.9130[17.4/D]	1.4374[17.4/D]	w, al, eth, ace, bz	B4[4], 349
74	Acetamide, N,N-dimethyl .... CH$_3$CON(CH$_3$)$_2$	87.12	....	165[758], 84[22]	-20	0.9366[25/4]	1.4380[20]	w, al, ace, bz, eth	B4[4], 180
75	Acetamide, N,N-dipropyl .... CH$_3$CON(C$_3$H$_7$)$_2$	143.23	....	209-10,101[16]	....	0.8992[17.4/4]	1.4419[17.4]	al	B4[4], 476
76	Acetamide, N-ethyl .... CH$_3$CONHC$_2$H$_5$	87.12	....	205,104-5[18]	....	0.942[4.5/4]	....	w, al, chl	B4[4], 347
77	Acetamide, hydroxy .... HOCH$_2$CONH$_2$	75.07	....	....	120	....	....	w	B3[4], 597
78	Acetamide, N-(2-hydroxyethyl) or N-acetylethanolamine .... CH$_3$CONHCH$_2$CH$_2$OH	103.12	....	166.7[1]	63.5	1.1073[?]	1.4674[?]	w, ace	B4[4], 1535
79	Acetamide, N-methyl .... CH$_3$CONHCH$_3$	73.10	nd	204-6,95[14]	28	0.9571[25/4]	1.4301[20]	w, al, eth, ace, bz	B4[4], 176
80	Acetamide, N-methyl-N-naphthyl .... CH$_3$CON(CH$_3$)C$_{10}$H$_7$	199.26	nd (lig)	....	93-7	....	....	al, eth	B12[3], 2867
81	Acetamide, N-methyl-N-(4-nitrophenyl) .... CH$_3$CON(CH$_3$)(C$_6$H$_4$NO$_2$-4)	194.19	pl (w)	....	95-7	....	....	al, eth	B12[3], 1595
82	Acetamide, phenoxy .... C$_6$H$_5$OCH$_2$CONH$_2$	151.16	....	....	101.5	....	....	al	B6[4], 638
83	Acetamide, phenyl .... C$_6$H$_5$CH$_2$CONH$_2$	135.17	....	....	157	....	....	al, eth	B9[3], 2193
84	Acetamide, N-phenyl or Acetanilide .... CH$_3$CONHC$_6$H$_5$	135.17	....	304	114.3	1.2190[15]	....	al, eth, ace, bz, chl	B12[3], 459
85	Acetamide, thiono .... CH$_3$CSNH$_2$	75.13	....	....	115-6	....	....	w, al	B2[4], 565
86	Acetamide, trichloro .... Cl$_3$CCONH$_2$	162.40	....	238-9[740]	142	....	....	al, eth	B2[4], 520
87	Acetamidine .... CH$_3$C(NH$_2$)=NH	58.08	....	80[10]	63-65	....	....	al	B2[4], 428
88	Acetamidine, N,N'-diphenyl .... CH$_3$C(NC$_6$H$_5$)NHC$_6$H$_5$	210.28	nd (al)	....	131-2	....	....	eth	B12[3], 471
89	Acetamidine, hydrochloride .... CH$_3$C(NH$_2$)NH·HCl	94.55	nd or pr (al)	....	177-8	....	....	w	B2[4], 429
90	Acetanilide or N-phenylacetamide .... CH$_3$CONHC$_6$H$_5$	135.17	rh or pl (w)	304	114.3	1.2190[15]	....	al, eth, ace, bz	B12[3], 459
91	Acetoacetanilide .... CH$_3$COCH$_2$CONHC$_6$H$_5$	177.21	....	....	86	....	....	al, eth	B12[3], 993
92	Acetanilide, o-amino .... 2-H$_2$NC$_6$H$_4$NHCOCH$_3$	150.18	....	....	132	....	....	w, al, eth	B13[3], 41
93	Acetanilide, m-amino .... 3-H$_2$NC$_6$H$_4$NHCOCH$_3$	150.18	nd or pl (bz)	....	87-9	....	....	w, al, bz, eth, ace	B13[3], 75
94	Acetanilide, p-amino .... 4-H$_2$NC$_6$H$_4$NHCOCH$_3$	150.18	nd (w)	267	165-8	....	....	al, eth, w	B13[3], 166
95	Acetanilide, N-bromo .... C$_6$H$_5$N(Br)COCH$_3$	214.08	ye pl (peth)	....	88	....	....	chl, lig	B12[3], 1078
96	Acetanilide, o-bromo .... 2-Br-C$_6$H$_4$NHCOCH$_3$	214.08	nd (al)	....	99	....	....	al, eth	B12[3], 1417
97	Acetanilide, m-bromo .... 3-BrC$_6$H$_4$-NHCOCH$_3$	214.08	nd (aq,al)	....	87.5	....	....	al, eth	B12[3], 1424
98	Acetanilide, p-bromo .... 4-BrC$_6$H$_4$NHCOCH$_3$	214.08	nd (60% al)	....	168	1.717	....	al	B12[3], 1437
99	Acetanilide, N-butyl .... C$_6$H$_5$N(C$_4$H$_9$)COCH$_3$	191.28	....	281, 140[10]	24.5	0.9912[20/4]	1.5146[20]	....	B12[3], 467
100	Acetanilide, p-butyl .... 4-C$_4$H$_9$C$_6$H$_4$NHCOCH$_3$	191.28	wh, pl (al)	....	105	....	....	al	B12[3], 2715

No.	Name, Synonyms, and Formula	Mol. wt.	Color, crystalline form, specific rotation and λ_{max} (log ε)	b.p. °C	m.p. °C	Density	n_D	Solubility	Ref.
101	Acetanilide, N-chloro .... $C_6H_5N(Cl)COCH_3$	169.61	nd (dil aa) pl (peth-chl)	....	91		....		B12³, 1077
102	Acetanilide, o-chloro .... $2\text{-}ClC_6H_4NHCOCH_3$	169.61	nd (dil aa)	....	87-8 (sub)		....	al, eth, bz	B12³, 1288
103	Acetanilide, 2-chloro-3-methyl or 3-acetamino-2-chloro-toluene $2\text{-}Cl\text{-}3\text{-}CH_3\text{-}C_6H_3NHCOCH_3$	183.64	nd (al)	....	133-4		....	al, bz	B12³, 1991
104	Acetanilide, 2-chloro-4-nitro .... $2\text{-}Cl\text{-}4\text{-}NO_2C_6H_3NHCOCH_3$	214.62	pr (al)	....	139-40		....		B12³, 1663
105	Acetanilide, 2-chloro-5-nitro $2\text{-}Cl\text{-}5\text{-}NO_2C_6H_3NHCOCH_3$	214.62	nd (al)	....	156		....		B12², 398
106	Acetanilide, m-chloro .... $3\text{-}Cl\text{-}6\text{-}H_4NHCOCH_3$	169.61	nd	....	79		....	al, eth, bz	B12³, 1309
107	Acetanilide, 3-chloro-2-methyl or 2-acetamino-6-chloro-toluene $3\text{-}Cl\text{-}2\text{-}CH_3\text{-}C_6H_3NHCOCH_3$	183.64	nd (dil al)	....	157-9		....	al, bz	B12², 1919
108	Acetanilide, 3-chloro-4-methyl or 4-acetamino-2-chloro-toluene $3Cl\text{-}4\text{-}CH_3\text{-}C_6H_3\text{-}NHCOCH_3$	183.64	tcl cr	....	105		....	al	B12², 530
109	Acetanilide, 3-chloro-5-methyl $3\text{-}Cl\text{-}5\text{-}CH_3C_6H_3NHCOCH_3$	183.64	nd (al)	....	151		....	al	B12, 871
110	Acetanilide, 3-chloro-6-methyl or 2-acetamino-4-chloro-toluene .... $3\text{-}Cl\text{-}6\text{-}CH_3C_6H_3NHCOCH_3$	183.64	nd (w)	....	139-40		....	al, eth, w	B12³, 1912
111	Acetanilide, 3-chloro-4-nitro $3\text{-}Cl\text{-}4\text{-}NO_2C_6H_3NHCOCH_3$	214.62	pa ye nd (al)	....	145		....	....	B12³, 1666
112	Acetanilide, p-chloro .... $4\text{-}ClC_6H_4NHCOCH_3$	169.61	nd (aq aa) ta (al ace) cr (w)	....	179	1.385^{224}	....	al, eth	B12³, 1336
113	Acetanilide, 4-chloro-2-methyl or 2-acetamino-5-chloro-toluene .... $4\text{-}Cl\text{-}2\text{-}CH_3C_6H_3NHCOCH_3$	183.64	lf (al)	....	140		....	al, bz	B12³, 1914
114	Acetanilide, 4-chloro-3-methyl or 5-acetamino-2-chloro-toluene .... $4\text{-}Cl\text{-}3\text{-}CH_3C_6H_3NHCOCH_3$	183.64	lf (al)	....	91.2-.7		....	al, bz	B12², 473
115	Acetanilide, 4-chloro-2-nitro $4\text{-}Cl\text{-}2\text{-}NO_2C_6H_3NHCOCH_3$	214.62	ye nd (al)	....	104		....		B12³, 1652
116	Acetanilide, 4-chloro-3-nitro $4\text{-}Cl\text{-}3\text{-}NO_2C_6H_3NHCOCH_3$	214.62	ye nd (al)	....	150		....		B12², 357
117	Acetanilide, α-cyano .... $C_6H_5NHCOCH_2CN$	160.19	nd (al)	....	199-200		....		B12³, 558
118	Acetanilide, 4-cyclohexyl $4\text{-}C_6H_{11}\text{-}C_6H_4NHCOCH_3$	217.31	nd (peth)	....	129		....	al, eth, chl	B12³, 2819
119	Acetanilide, 2,4-dimethyl or 2,4-Acetoxylide.......... $2,4\text{-}(CH_3)_2C_6H_3NHCOCH_3$	163.22	nd (al)	170^{10}	128-9		....	al	B12³, 2474
120	Acetanilide, 4,6-dimethyl-2-nitro .... $2\text{-}NO_2\text{-}4,6\text{-}(CH_3)_2C_6H_2NHCOCH_3$	208.22	yesh nd (w)	....	172-3		....	al, bz, w	B12³, 2491
121	Acetanilide, 2,3-dinitro $2,3\text{-}(NO_2)_2\text{-}C_6H_3\text{-}NHCOCH_3$	225.16	nd (al)	....	187		....	al, eth, bz	B12³, 1680
122	Acetanilide, 2,4-dinitro .... $2,4\text{-}(NO_2)_2C_6H_3NHCOCH_3$	225.16	ye nd (al or bz)	....	125-6		....	al, eth, bz	B12³, 1686
123	Acetanilide, 2,5-dinitro $2,5\text{-}(NO_2)_2C_6H_3NHCOCH_3$	225.16	nd (al)	....	121		....	al	B12³, 1704
124	Acetanilide, 2,6-dinitro $2,6\text{-}(NO_2)_2C_6H_3NHCOCH_3$	225.16	nd (aa)	....	197		....	al, aa	B12³, 1705
125	Acetanilide, 3,4-dinitro $3,4\text{-}(NO_2)_2C_6H_3NHCOCH_3$	225.16	ye cr (al) nd (w)	....	144-5		....	al, w	B12³, 1705
126	Acetanilide, 3,5-dinitro .... $3,5\text{-}(NO_2)_2C_6H_3NHCOCH_3$	225.16	ye nd (dil aa, w)	....	191		....	al, w, aa	B12³, 1706
127	Acetamide, N,N-dipropyl $CH_3CON(C_3H_7)_2$	143.23	....	$209\text{-}10,101^{18}$	....	$0.8992^{17.4/4}$	$1.4419^{17.4}$	al	B4⁴, 476
128	Acetanilide, 2-ethoxy or o-Acetophenetidide........... $2\text{-}(C_2H_5O)\text{-}C_6H_4NHCOCH_3$	179.22	lf (dil al)	>250	79		....	al, eth	B13³, 779
129	Acetanilide, 2-ethoxy-4-nitro .... $2\text{-}(C_2H_5O)4\text{-}(NO_2)C_6H_3NHCOCH_3$	224.22	nd (al)	....	202		....	al	B13², 194
130	Acetanilide, 2-ethoxy-5-nitro .... $2(C_2H_5O)\text{-}5(NO_2)C_6H_3NHCOCH_3$	224.22	ye nd (al)	....	199		....	bz, al	B13³, 881

No.	Name, Synonyms, and Formula	Mol. wt.	Color, crystalline form, specific rotation and λ_{max} (log ε)	b.p. °C	m.p. °C	Density	n_D	Solubility	Ref.
131	Acetanilide, 3-ethoxy or m-accetophenetidide . . . . 3-(C$_2$H$_5$O)-C$_6$H$_4$NHCOCH$_3$	179.22	gy pl (w)		97-9			al, eth, w	B13^3, 950
132	Acetanilide, 4-ethoxy or Phenacitin . . . . 4-(C$_2$H$_5$O)C$_6$H$_4$NHCOCH$_3$	179.22	mcl pr	dic	137-8		1.571	al, ace	B13^3, 1057
133	Acetanilide, 4-ethoxy-2-nitro . . . . 4-C$_2$H$_5$O-2NO$_2$C$_6$H$_4$NHCOCH$_3$	224.22	ye nd (w)		104			al, eth, ace, chl	B13^3, 1208
134	Acetanilide, 4-ethoxy-3-nitro . . . . 4-(C$_2$H$_5$O)-3-NO$_2$C$_6$H$_3$NHCOCH$_3$	224.22	nd (dil al)		123			al, ace, bz	B13^3, 1202
135	Acetanilide, 5-ethoxy-2-nitro . . . . 5-(C$_2$H$_5$O)-2-NO$_2$C$_6$H$_3$NHCOCH$_3$	224.22	nd (al)		95			al, eth, ace, bz, chl	B13^3, 976
136	Acetanilide, N-ethyl . . . . C$_6$H$_5$N(C$_2$H$_5$)COCH$_3$	163.22	(w) rh (eth)	258^{731}	55			w, eth	B12^3, 466
137	Acetanilide, N-ethyl-3-nitro . . . . 3-NO$_2$-C$_6$H$_4$N(C$_2$H$_5$)COCH$_3$	208.22	pa ye nd (dil al)		88-9			al, bz	B12, 704
138	Acetanilide, N-ethyl-4-nitro . . . . 4-NO$_2$-C$_6$H$_4$-N(C$_2$H$_5$)COCH$_3$	208.22	lf or pr (dil al)		118-9			al, bz, eth	B12^3, 1596
139	Acetanilide, o-hydroxy or o-Acetamidophenol . . . . 2-HOC$_6$H$_4$NHCOCH$_3$	151.17	pl (dil al)		208			al, eth, bz	B13^3, 778
140	Acetanilide, m-hydroxy or m-Acetamidophenal . . . . . . 3-HO-C$_6$H$_4$-NHCOCH$_3$	151.17	nd (w)		148-9			w, al, eth, bz	B13^3, 950
141	Acetanilide, p-hydroxy or p-acetamidophenol . . . . 4-HOC$_6$H$_4$NHCOCH$_3$	151.17	nd (bz-peth)		151			al, eth, bz	B13^3, 1056
142	Acetanilide, p-iodo . . . . 4-I-C$_6$H$_4$NHCOCH$_3$	261.07	la (w) pl (w, al)		184.3			al, eth, bz	B12^3, 1496
143	Acetanilide, o-methyl or o-acetotoluidide . . . . 2-CH$_3$-C$_6$H$_4$NHCOCH$_3$	149.19	nd (al)	296	110	1.168^{15}		al, eth, ace, bz	B12^3, 1853
144	Acetanilide, m-methyl or m-acetotoluidide . . . . 3-CH$_3$-C$_6$H$_4$NHCOCH$_3$	149.19	nd (w)	303,182-3^{31}	65.5	1.141		al, eth	B12^3, 1962
145	Acetanilide, p-methyl or p-acetotoluidide . . . . 4-CH$_3$-C$_6$H$_4$NHCOCH$_3$	149.19	mcl cr or nd (dil al)	307 sub	148.5	1.212^{15}		al, eth, bz	B12^3, 2051
146	Acetanilide, N-methyl-o-hydroxy . . . . 2-HOC$_6$H$_4$N(CH$_3$)COCH$_3$	165.19	nd (bz-peth)		151			al, eth, bz	B13^3, 783
147	Acetanilide, N-methyl-p-hydroxy . . . . 4-HOC$_6$H$_4$N(CH$_3$)COCH$_3$	165.19	sc (w)		245			al, eth	B13^1, 162
148	Acetanilide, o-methoxy or o-Acetanisidine . . . . 2-CH$_3$OC$_6$H$_4$-NHCOCH$_3$	163.19	nd (w)	303-5	87-8			al, eth, ace, w	B13^3, 778
149	Acetanilide, 2-methoxy-3-nitro . . . . 2-CH$_3$O-3-NO$_2$C$_6$H$_3$NHCOCH$_3$	210.19	pa ye pr (dil aa MeOH)		103.4			MeOH, al	B13^2, 195
150	Acetanilide, 2-methoxy-4-nitro . . . . 2-CH$_3$O-4-NO$_2$C$_6$H$_3$NHCOCH$_3$	210.19	pa ye cr (AcOEt)		153-4			al, bz	B13^3, 890
151	Acetanilide, 2-methoxy-5-nitro . . . . 2-CH$_3$O-5-NO$_2$C$_6$H$_3$NHCOCH$_3$	210.19	nd (w)		178				B13^3, 881
152	Acetanilide, 2-methoxy-6-nitro . . . . 2-CH$_3$O-6-NO$_2$C$_6$H$_3$NHCOCH$_3$	210.19	pa ye nd (dil al)		158-9			bz, al	B13^3, 876
153	Acetanilide, m-methoxy or m-acetanisidine . . . . 3-CH$_3$OC$_6$H$_4$NHCOCH$_3$	165.19	nd or pl (w)		81			w, al, eth, ace	B13^3, 950
154	Acetanilide, 3-methoxy-2-nitro . . . . 3-CH$_3$O-2NO$_2$C$_6$H$_3$NHCOCH$_3$	210.19	br (al) amor		265 sub			al, bz	B13^1, 136
155	Acetanilide, 3-methoxy-4-nitro . . . . 3-CH$_3$O-4NO$_2$C$_6$H$_3$NHCOCH$_3$	210.19	ye nd (w)		165			al	B13^3, 978
156	Acetanilide, 3-methoxy-5-nitro . . . . 3-CH$_3$O-5-NO$_2$C$_6$H$_3$NHCOCH$_3$	210.19	nd (aa)		201			al, eth, bz, aa	B13^2, 216
157	Acetanilide, p-methoxy or p-acetanisidine . . . . 4-CH$_3$OC$_6$H$_4$-NHCOCH$_3$	165.19	pl (w)		130-2			al, ace, eth, w	B13^3, 1056
158	Acetanilide, 4-methoxy-2-nitro . . . . 4-CH$_3$O-2-NO$_2$C$_6$H$_3$NHCOCH$_3$	210.19	ye nd (al)		116.5-7			al, eth, bz	B13^3, 1208
159	Acetanilide, 4-methoxy-3-nitro . . . . 4-CH$_3$O-3-NO$_2$C$_6$H$_3$NHCOCH$_3$	210.19	og-ye nd (w or dil al)		153			al, lig	B13^1, 186
160	Acetanilide, 5-methoxy-2-nitro . . . . 5-CH$_3$O-2-NO$_2$C$_6$H$_3$NHCOCH$_3$	210.19	wh nd (al)		125			al, lig, w	B13^3, 976
161	Acetanilide, N-methyl or Exalgin . . . . C$_6$H$_5$N(CH$_3$)COCH$_3$	149.19	nd (eth) pr (al) lf (lig)	253^{712}	102-4	1.0036$^{105/4}$	1.576	w, al, eth, lig	B12^3, 465
162	Acetanilide, N-methyl-4-hydroxy-3-nitro . . . . 3-NO$_2$-4-HOC$_6$H$_3$N(CH$_3$)NOCH$_3$	210.19	(al)		161-2			al	B13^1, 186
163	Acetanilide, N-methyl-4-nitro . . . . 4-NO$_2$-C$_6$H$_4$N(CH$_3$)COCH$_3$	194.19	pl (w)		153			al, eth	B12^1, 352

No.	Name, Synonyms, and Formula	Mol. wt.	Color, crystalline form, specific rotation and λ_{max} (log ε)	b.p. °C	m.p. °C	Density	n_D	Solubility	Ref.
164	Acetanilide, N-methyl-2-methyl or N-methyl-o-acetotoluidide. 2-CH$_3$C$_6$H$_4$N(CH$_3$)COCH$_3$	163.22	....	260	55-6		....	al	B12[3], 1854
165	Acetanilide, N-methyl-3-methyl or N-methyl-m-acetotoluidide. 3-CH$_3$-C$_6$H$_4$N(CH$_3$)COCH$_3$	163.22	cr	....	64			al	B12[3], 1963
166	Acetanilide, N-methyl-4-methyl or N-methyl-p-acetotoluidide. 4-CH$_3$-C$_6$H$_4$N(CH$_3$)COCH$_3$	163.22	lf (eth-al)	263	83			al, eth, lig	B12[3], 2052
167	Acetanilide, 4-methyl-2-nitro 4-CH$_3$-2-NO$_2$C$_6$H$_3$NHCOCH$_3$	194.19	dk ye nd (peth)		96				B12[3], 2178
168	Acetanilide, o-nitro 2-NO$_2$C$_6$H$_4$NHCOCH$_3$	180.18	ye pr (lig) lf (dil al)	100[0.1]	94	1.419[15]		w, al, eth, lig	B12[3], 1523
169	Acetanilide, m-nitro 3-NO$_2$C$_6$H$_4$NHCOCH$_3$	180.18	wh lf (al)	100[0.008]	154-6			w, al, chl	B12[3], 1551
170	Acetanilide, p-nitro 4-NO$_2$C$_6$H$_4$NHCOCH$_3$	180.18	ye pr (w)	100[0.008]	216 (217)			al, eth, aa, lig	B12[3], 1594
171	Acetanilide, p-octyl 4-(C$_8$H$_{17}$)C$_6$H$_4$NHCOCH$_3$	247.39	lf or pr (al)		94			al, eth	B12, 1185
172	Acetanilide, N-phenyl or N-acetyldiphenylamine (C$_6$H$_5$)$_2$NCOCH$_3$	211.27	rh or nd (w or lig)	sub	103			al	B12[3], 468
173	Acetanilide, N-propyl C$_6$H$_5$N(C$_3$H$_7$)COCH$_3$	177.25	mcl lf (eth, lig)	266[712]	49 (56)			al, eth	B12[3], 466
174	Acetanilide, N-isopropyl C$_6$H$_5$N(i-C$_3$H$_7$)COCH$_3$	177.22	lf (lig)	262-3[712]	39			lig	B12[3], 466
175	Acetic acid or Ethanoic acid CH$_3$CO$_2$H	60.05	rh (hyg)	117.9, 17[10]	16.6	1.0492[20/4]	1.3716[20]	w, al, ace, bz	B2[4], 94
176	Acetic acid, allyl ester or Allyl acetate CH$_3$CO$_2$CH$_2$CH=CH$_2$	100.13	....	103.5	......	0.9276[20/4]	1.4049[20]	al, eth, ace, w	B2[4], 180
177	Acetic acid, anhydride or Acetic anhydride (CH$_3$CO)$_2$O	102.09		139.55, 44[15]	-73.1	1.0820[20/4]	1.39006[20]	al, eth, bz, w	B2[4], 386
178	Acetic acid, anhydride, trifluoro or Trifluoro acetic anhydride (F$_3$CO)$_2$O	210.03	....	39.5-40.1	-65	1.490[25/4]	1.269[25]	eth, aa	B2[4], 469
179	Acetic acid, benzyl ester or Benzyl acetate CH$_3$CO$_2$CH$_2$C$_6$H$_5$	150.18	....	215.5, 93-4[10]	-51.3	1.0550[20/4]	1.5232[20]	al, eth, ace	B6[4], 2262
180	Acetic acid, dibenzyl or Dibenzylacetic acid (C$_6$H$_5$CH$_2$)$_2$CHCO$_2$H	240.3	pl (peth), dil (aa), nd (w)	235[18]	89			al, eth, bz	B9[3], 3356
181	Acetic acid, dibenzyl, methyl ester or Methyl dibenzylacetate (C$_6$H$_5$CH$_2$)$_2$CHCO$_2$CH$_3$	254.33	nd (al)	....	42-3		....	al, peth	B9[2], 475
182	Acetic acid, 2-bromo ethyl ester or (2-Bromoethyl) acetate CH$_3$CO$_2$CH$_2$CH$_2$Br	167.01	....	162-3	-13.8	1.514[20/4]	1.457[25]	w, al, eth, chl	B2[4], 136
183	Acetic acid, bromomethyl ester CH$_3$CO$_2$CH$_2$Br	152.98		130-3[750]		1.6350[20/4]	1.4520[20]	al, eth, chl	B2[4], 280
184	Acetic acid, 2-butenyl ester or Crotyl acetate CH$_3$CO$_2$CH$_2$CH=CHCH$_3$	114.15	....	132	......	0.9192[20/4]	1.4181[20]	al, eth, bz	B2[4], 183
185	Acetic acid, butyl ester or Butyl acetate CH$_3$CO$_2$C$_4$H$_9$	116.16	....	126.5	-77.9	0.8825[20/4]	1.3941[20]	al, eth, bz	B2[4], 143
186	Acetic acid, iso-butyl ester or iso-Butyl acetate CH$_3$CO$_2$-i-C$_4$H$_9$	116.16	....	117.2	-98.58	0.8712[20/4]	1.3902[20]	al, eth, ace	B2[4], 149
187	Acetic acid, sec-butyl ester d or sec-butyl acetate CH$_3$CO$_2$CH(CH$_3$)CH$_2$CH$_3$	116.16	[α]$^{20/D}$ + 25.43	112		0.8758[16/4]	1.3877[20]	al, eth, ace	B2[3], 241
188	Acetic acid, sec-butyl ester dl or sec-butyl acetate dl CH$_3$CO$_2$CH(CH$_3$)CH$_2$CH$_3$	116.16	....	112.2		0.8716[20/4]	1.3888[20]	al, eth, ace	B2[4], 148
189	Acetic acid, sec-butyl ester l or sec-butyl acetate l CH$_3$CO$_2$CH(CH$_3$)CH$_2$CH$_3$	116.16	[α]$^{19/546}$ − 20.2	116-7		0.87301[19/4]	1.3899[18]	al, eth, ace	B2[3], 241
190	Acetic acid, tert-butyl ester or tert-butyl acetate CH$_3$CO$_2$C(CH$_3$)$_3$	116.16	....	97-8		0.8665[20/4]	1.3855[20]	al, eth, aa	B2[4], 151
191	Acetic acid, (1-chloroethyl) ester CH$_3$CO$_2$CHCl-CH$_3$	122.55	cr	121.3[746]	......	1.110[20/4]	1.409[20]	eth	B2[4], 282
192	Acetic acid, (2 chloroethyl) ester CH$_3$CO$_2$CH$_2$CH$_2$Cl	122.55	....	145, 50[18]		1.178[20/4]	1.4234[20]	al, eth	B2[4], 136
193	Acetic acid, (2-chloro-iso-propyl) ester CH$_3$CO$_2$CCl(CH$_3$)$_2$	136.58		149-50		1.0788[20]	1.4223[20]	al, eth	B2[4], 286
194	Acetic acid, chloromethyl ester CH$_3$CO$_2$CH$_2$Cl	108.53	....	115-6[751]	......	1.194[20/4]	1.409[20]	al, eth	B2[4], 280

No.	Name, Synonyms, and Formula	Mol. wt.	Color, crystalline form, specific rotation and λ_{max} (log ε)	b.p. °C	m.p. °C	Density	n_D	Solubility	Ref.
195	Acetic acid, (3-chloropropyl) ester .... $CH_3CO_2CH_2CH_2CH_2Cl$	136.58		163-5[747], 62-3[10]		1.250[19]	1.431[20]	al, eth, ace	B2[4], 140
196	Acetic acid, cyclohexyl ester or Cyclohexyl acetate ...... $CH_3CO_2C_6H_{11}$	142.20		173, 62-3[12]		0.9698[20/4]	1.4401[20]	al, eth	B6[4], 36
197	Acetic acid, cyclopentyl ester or Cyclopentyl acetate .... $CH_3CO_2C_5H_9$	128.17		51.5-2.5[12]		0.9522[16]			B6[4], 7
198	Acetic acid, decyl ester or Decyl acetate .... $CH_3CO_2C_{10}H_{21}$	220.33		244, 125.8[155]	-15.05	0.8671[20/4]	1.4273[20]	al, eth, bz, aa	B2[4], 168
199	Acetic acid, 1,3-dichloro-iso-propyl ester .... $CH_3CO_2CH(CH_2Cl)_2$	171.03		202-8, 81[15]		1.281[20]	1.4542[20]	al, eth, chl	B2[3], 233
200	Acetic acid, 1,1-dimethyl butyl ester .... $CH_3CO_2C(CH_3)_2C_3H_7$	144.22		152-3[755], 34.5[10]		0.8798[18/4]	1.4068[10]	al, eth	B2[3], 257
201	Acetic acid, (2,4-dinitrophenyl) ester .... $CH_3CO_2[C_6H_3(NO_2)_2-2,4]$	226.15	cr (MeOH)		72-3				B6[4], 1380
202	Acetic acid, 3,5,-dinitro phenyl ester .... $CH_3CO_2[C_6H_3(NO_2)_2-3,5]$	226.15	cr (bz-peth)		126-7			bz, aa	B6, 258
203	Acetic acid, dodecyl ester .... $CH_3CO_2C_{12}H_{25}$	228.37		260-70			1.4439[20]		B2[4], 170
204	Acetic acid, ethyl ester or Ethyl acetate............. $CH_3CO_2C_2H_5$	88.12		77.06	-83.6	0.9003[20/4]	1.3723[20]	w, al, eth, ace, bz	B2[4], 127
205	Acetic acid, 2-ethylbutyl ester .... $CH_3CO_2CH_2CH(C_2H_5)_2$	144.22		162-3,63[20]		0.8790[20/4]	1.4109[20]	al, eth	B2[4], 161
206	Acetic acid, 2-ethyl hexyl ester .... $CH_3CO_2CH_2CH(C_2H_5)C_4H_9$	172.27		199,95[25]	-93	0.8734[20/20]	1.4204[20]	al, eth	B2[4], 166
207	Acetic acid, 2-ethoxyethyl ester or Cellosolve acetate ..... $CH_3CO_2CH_2CH_2OH$	132.16		156.4,49[12]	-61.7	0.9740[20/4]	1.4054[20]	w, al, eth, ace	B2[4], 214
208	Acetic acid, 3-ethyl-3-pentyl ester .... $CH_3CO_2C(C_2H_5)_3$	158.24		160-3				al	B2, 134
209	Acetic acid, 9-fluorenyl .... $C_{13}H_9CH_2CO_2H$	224.26	mcl nd (al)	218-20[11]	138-9				B9[3], 3448
210	Acetic acid, furfuryl ester or Furfuryl acetate .... $CH_3CO_2CH_2(C_4H_3O)$	140.15		177		1.1175[20/4]	1.4327[20]	al, eth	B17[4], 1246
211	Acetic acid, 2-furyl ester .... $CH_3CO_2C(C_4H_3O)$	126.11		102-4[0.4]	68-9			w, bz, peth, MeOH	B17[4], 1220
212	Acetic acid, 1-heptyl ester .... $CH_3CO_2C_7H_{15}$	158.24		192.4,96[28]	-50.2	0.8750[15/4]	1.4150[20]	al, eth	B2[4], 162
213	Acetic acid, 4-heptyl ester .... $CH_3CO_2CH(C_3H_7)_2$	158.24		170-2,60-70[17]		0.8742[0/0]	1.4105[19]	eth	B2[4], 163
214	Acetic acid, hexadecyl ester or Cetyl acetate .... $CH_3CO_2C_{16}H_{33}$	284.49		220-5[205]	α,-18.5, β, 24.2	0.8574[25/4]	1.4438[20]		B2[4], 171
215	Acetic acid, hexyl ester or Hexyl acetate .... $CH_3CO_2C_6H_{13}$	144.22		171.5,61.5[12]	-80.9	0.8779[15/4]	1.4092[20]	al, eth	B2[4], 159
216	Acetic acid, 2-hexyl ester d .... $CH_3CO_2CH(CH_3)C_4H_9$	144.22		158,57[20]		0.8658[18/4]	1.4014[25]	al, eth	B2[2], 145
217	Acetic acid, 2-hexyl ester dl .... $CH_3CO_2CH(CH_3)C_4H_9$	144.22		157-8		0.8651[15/15]	1.4014[25]	al, eth	B2[3], 256
218	Acetic acid, 3-hexyl ester d .... $CH_3CO_2CH(C_2H_5)C_3H_7$	144.22	$[\alpha]_D^{20/}$ +0.55	149-51		0.8672[20/4]	1.4037[20]	al, eth	B2[4], 160
219	Acetic acid, hydrazide .... $CH_3CONHNH_2$	74.08		129[18]	67			w, al, eth	B2[4], 435
220	Acetic acid, (2-hydroxyethyl) ester or Glycol mono acetate	104.12		187-9		1.108[15]		w, al, eth	B2[4], 214
	$CH_3CO_2CH_2CH_2OH$								
221	Acetic acid, menthyl ester l or Menthyl acetate l $CH_3CO_2C_{10}H_{19}$	198.3	$[\alpha]^{20/D}$- 79.42	109[10]		0.9185[20/4]	1.4469[20]		B6[4], 153
222	Acetic acid, (4-methoxybenzyl) ester .... $CH_3CO_2CH_2C_6H_4OCH_3-4$	180.20		110-7		1.104-7[25/25]			B6[3], 4549
223	Acetic acid, (2-methoxyethyl) ester .... $CH_3CO_2CH_2CH_2OCH_3$	118.13		144-5,40-1[12]		1.0090[19/19]	1.4002[20]	w, al, eth	B2[4], 214
224	Acetic acid, (2-methoxy phenyl) ester .... $CH_3CO_2C_6H_4OCH_3-2$	166.18		134-3[18]	31-2			al, eth	B6[3], 4227
225	Acetic acid, methyl ester or Methyl acetate............. $CH_3CO_2CH_3$	74.08		57	-98.1	0.9330[20/4]	1.3595[20]	al, eth, w, ace, bz, chl	B2[4], 122
226	Acetic acid, 2-methyl-2-butyl ester or tert-amyl acetate ... $CH_3CO_2C(CH_3)_2CH_2CH_3$	130.19		124-4.5		0.8740[20]	1.4010[20]	al, eth, ace	B2[4], 156
227	Acetic acid, 3-methyl butyl ester or iso Pentyl acetate..... $CH_3CO_2CH_2CH_2CH(CH_3)_2$	130.19		142	-78.5	0.8670[20/4]	1.4003[20]	al, eth, ace	B2[4], 157

No.	Name, Synonyms, and Formula	Mol. wt.	Color, crystalline form, specific rotation and λ_{max} (log ε)	b.p. °C	m.p. °C	Density	n_D	Solubility	Ref.
228	Acetic acid, methylene diester or Methanediol diacetate... $CH_2(O_2CCH_3)_2$	132.13	....	164-5	-23	$1.136^{20/4}$	1.4025^{24}	al, eth	B2⁴, 280
229	Acetic acid, (2-methyl-3-heptyl) ester .... $CH_3CO_2\text{-}CH(C_4H_9)CH(CH_3)_2$	172.27		172	......	6.875^{20}	1.4166^{20}	al	B2, 134
230	Acetic acid, 3-methyl-2-heptyl ester .... $CH_3CO_2CH(CH_3)CH(CH_3)C_4H_9$	172.27		185		$0.8545^{21/4}$	1.418^{21}	al	B2², 146
231	Acetic acid, 6-methyl-2-heptyl ester .... $CH_3CO_2\text{-}CH(CH_3)(CH_2)_3CH(CH_3)_2$	172.27		$187\text{-}8^{768}$		0.8474^{20}	1.4137^{20}	al	B2, 135
232	Acetic acid, 6-methyl-3-heptyl ester .... $CH_3CO_2CH(C_2H_5)CH_2CH_2CH(CH_3)_2$	172.27		184-5		0.8554^{20}	1.41602^{20}	al	B2, 135
233	Acetic acid, (2-methyl-1-naphthyl) ester .... $1\text{-}CH_3CO_2\text{-}C_{10}H_6\text{-}CH_3\text{-}2$	200.24	nd (eth-peth)	....	81-2	....	....	eth	B6³, 3027
234	Acetic acid, (4-methyl-1-naphthyl) ester .... $1\text{-}CH_3CO_2C_{10}H_6CH_3\text{-}4$	200.24	nd (eth-peth)		86.8				C64, 19519
235	Acetic acid, (6-methyl-1-naphthyl) ester or 5-acetoxy-2-methyl naphthalene........... $5\text{-}CH_3CO_2C_{10}H_6CH_3\text{-}(2)$	200.24		124^2					B6³, 3028
236	Acetic acid, (7-methyl-1-naphthyl) ester or 1-Acetoxy-7-methyl naphthalene .... $1\text{-}CH_3CO_2C_{10}H_6\text{-}CH_3\text{-}7$	200.24		188^{15}	39.41				B6³, 3030
237	Acetic acid, (2-methyl-3-pentyl) ester .... $CH_3CO_2CH(C_2H_5)CH(CH_3)_2$	144.22		148.5^{747}	......	0.8688^{20}		al, eth	B2⁴, 160
238	Acetic acid, (3-methyl-3-pentyl) ester .... $CH_3CO_2C(C_2H_5)_2CH_3$	144.22		148		$0.8834^{20/20}$	1.4109^{18}	al, eth	B2⁴, 161
239	Acetic acid, (4-methyl-2-pentyl) ester .... $CH_3CO_2CH(CH_3)CH_2CH(CH_3)_2$	144.22		147-8		0.8805^0	1.3980^{20}	al, eth	B2⁴, 161
240	Acetic acid, 2-methylpentyl ester .... $CH_3CO_2CH_2CH(CH_3)C_3H_7$	144.22		162.2^{746}		$0.8717^{25/25}$		al, eth	B2⁴, 160
241	Acetic acid, α-naphthyl ester or α-naphthyl acetate....... $CH_3CO_2\text{-}\alpha\text{-}C_{10}H_7$	186.21	nd or pl (al)	....	49			al, eth	B6³, 2928
242	Acetic acid, β-naphthyl ester or β-naphthyl acetate....... $CH_3CO_2\text{-}\beta\text{-}C_{10}H_7$	186.21	nd (al)		70			al, eth	B6³, 2982
243	Acetic acid, (2-nitrophenyl) ester .... $CH_3CO_2C_6H_4NO_2\text{-}2$	181.15	nd or pr (lig)	253d	40-1			al, eth, ace, bz, w	B6⁴, 1256
244	Acetic acid, 3-nitrophenyl ester .... $CH_3CO_2\text{-}C_6H_4NO_2\text{-}3$	181.15	nd (peth)		55-6			al, w, lig	B6⁴, 1273
245	Acetic acid, (4-nitro phenyl) ester .... $CH_3CO_2C_6H_4NO_2\text{-}4$	181.15	lf (dil al)	....	81-2			al, bz, w, lig	B6⁴, 1298
246	Acetic acid, octadecyl ester or Stearyl acetate .... $CH_3CO_2C_{18}H_{36}$	312.54	....	$222\text{-}3^{15}$	34.5			al	B2⁴, 171
247	Acetic acid, octyl ester or Octyl acetate............... $CH_3CO_2C_8H_{17}$	172.277	....	210, $112\text{-}3^{30}$	-38.5	$0.8705^{20/4}$	1.4150^{20}	al, eth	B2⁴, 165
248	Acetic acid, 2-octylester d .... $CH_3CO_2CH(CH_3)C_6H_{13}$	172.27	$[\alpha]^{20/}_D$ + 7.00	196	....	$0.8606^{15/4}$	1.4141^{20}	al, eth, bz	B2³, 260
249	Acetic acid, 2-octylester dl .... $CH_3CO_2CH(CH_3)C_6H_{13}$	172.27	....	194.5^{749}		$0.8626^{14/4}$	1.4146^{20}	al, eth	B2⁴, 165
250	Acetic acid, 2-octyl ester l .... $CH_3CO_2CH(CH_3)C_6H_{13}$	172.27	$[\alpha]^{14.5/D}$ - 6.0	196, $89\text{-}90^{17}$		$0.8570^{20/4}$	1.4140^{20}	al, eth	B2⁴, 165
251	Acetic acid, 3-octyl ester l .... $CH_3CO_2CH(C_2H_5)C_5H_{11}$	172.27	$[\alpha]^{20/D}$ - 4.3	....	191-1.5, 56.5^2	$0.8641^{20/4}$	1.4152^{20}	al, eth, CS_2	B2⁴, 166
252	Acetic acid, pentyl ester or Pentyl acetate............... $CH_3CO_2C_5H_{11}$	130.19	....	149.25	-70.8	$0.8756^{20/4}$	1.4023^{20}	al, eth	B2⁴, 152
253	Acetic acid, 2-pentyl ester d or See-pentyl acetate d...... $CH_3CO_2CH(CH_3)C_3H_7$	130.19	$[\alpha]^{20/D}$ + 17.16 (un-dil)	130-1		$0.8692^{18/4}$	1.3960^{20}	al, eth, ace	B2¹, 60
254	Acetic acid, 2-pentyl ester dl .... $CH_3CO_2CH(CH_3)C_3H_7$	130.18		134		$0.8692^{18/4}$	1.3960^{20}	al, eth	B2⁴, 155
255	Acetic acid, 2-pentyl ester l .... $CH_3CO_2CH(CH_3)C_3H_7$	130.19	$[\alpha]^{20/D}$ + 3.30	142		0.8803^{15}	1.4012^{20}	al, eth	B2, 132
256	Acetic acid, 3-pentyl ester .... $CH_3CO_2CH(C_2H_5)_2$	130.19		132^{741}		$0.8712^{20/4}$	1.4005^{20}	al, eth	B2⁴, 155
257	Acetic acid, phenacyl ester or ω-aceto aceto phenone..... $CH_3CO_2CH_2COC_6H_5$	178.19	rh pl	270	49-9.5	$1.1169^{65/4}$	1.5036^{65}	al, eth	B8³, 301
258	Acetic acid, (phenoxyethyl) ester .... $CH_3CO_2CH_2CH_2OC_6H_5$	180.20		$109\text{-}12^3$	......	1.1082^{20}	1.5080^{20}	....	B6⁴, 575
259	Acetic acid, phenyl ester .... $CH_3CO_2C_6H_5$	136.16		195.7, $75\text{-}6^8$		$1.0780^{20/4}$	1.5035^{20}	al, eth	B6⁴, 611

PHYSICAL CONSTANTS OF ORGANIC COMPOUNDS (Continued)

No.	Name, Synonyms, and Formula	Mol. wt.	Color, crystalline form, specific rotation and λ_{max} (log ε)	b.p. °C	m.p. °C	Density	n_D	Solubility	Ref.
260	Acetic acid, 2-phenyl ethyl ester or β-phenyl ethyl acetate CH₃CO₂CH₂CH₂C₆H₅	164.21	....	232.6, 153[76]	-31.1	1.0883[20/4]	1.5171[20]	al, eth	B6[3], 1709
261	Acetic acid, iso-propenyl ester .... CH₃CO₂-C(CH₃)=CH₂	100.12	....	92-4[732]	-92.9	0.9090[20]	1.4033[20]	al, eth, ace	B2[4], 179
262	Acetic acid, propyl ester or Propyl acetate .... CH₃CO₂C₃H₇	102.13	....	101.6	-95	0.8878[20/4]	1.3842[20]	al, eth	B2[4], 138
263	Acetic acid, iso-propyl-ester or iso-propyl acetate........ CH₃CO₂CH(CH₃)₂	102.13	....	90	-73.4	0.8718[20/4]	1.3773[20]	al, eth, ace, w	B2[4], 141
264	Acetic acid, tetrahydro furfuryl ester .... CH₃CO₂CH₂(C₄H₇O)	144.17	....	204-7		1.0624[20/4]	1.4350[25]	w, al, eth, chl	B17[4], 1103
265	Acetic acid, 2-thionyl or 2-Thiophene acetic acid .... (2-C₄H₃S)CH₂CO₂H	142.18	cr (w)	....	76			w, al, eth	B18[4], 4062
266	Acetic acid, 2-tolyl ester or o-Cresyl acetate CH₃CO₂-C₆H₄CH₃-2	150.18	....	208, 89[10]		1.0533[15]	1.5002[20]	al, eth	B6[4], 1960
267	Acetic acid, 3-tolyl ester or m-Cresyl acetate CH₃CO₂C₆H₄CH₃-3	150.18	....	212		1.043[20/4]	1.4978[20]	al, eth, bz	B6[4], 2047
268	Acetic acid, 4-tolyl ester or p-Cresyl acetate............ CH₃CO₂C₆H₄CH₃-4	150.18	....	212.5, 107-9[5]		1.0512[17/4]	1.5163[22]	al, eth, chl	B6[4], 2112
269	Acetic acid, 1,2,2-trimethyl propyl ester d .... CH₃CO₂CH(CH₃)C(CH₃)₃	144.22	[α]²⁵/D + 16.2	141[756]		0.856[20/4]	1.4001[25]	eth	B2[3], 258
270	Acetic acid, vinyl ester or Vinyl acetate............... CH₃CO₂CH=CH₂	86.09	....	72.3	-93.2	0.9317[20/4]	1.3959[20]	al, eth, ace, bz, chl	B2[4], 176
271	Aceto acetamide, n-phenyl or Acetoacetanilide......... CH₃COCH₂CONHC₆H₅	177.21	pl or nd (bz or lig)	....	86			al, eth, bz, chl, liq	B12[3], 993
272	Aceto acetamide, N-2-tolyl or Acetoacet-2-toluide .... (CH₃COCH₂CONHC₆H₄CH₃-2)	191.23	pr (AcOEt)	....	107-8			al, bz	B12[3], 1895
273	Acetoacetamide, N-3-tolyl or Acetoacet-3-toluide .... CH₃COCH₂CONH(C₆H₄CH₃-3)	191.23	pl (bz-peth)	....	57-8			al, bz	B12[3], 1983
274	Aceto acetamide, N-4-tolyl .... CH₃COCH₂CONH(C₆H₄CH₃-4)	191.23	pr (AcOEt)	....	95			al, eth	B12[3], 2129
275	Aceto acetanilide, α-bromo .... CH₃COCHBrCONHC₆H₅	256.11	lf (al)	....	138d			al, eth, chl	B12[4], 276
276	Aceto acetanilide, α-chloro .... CH₃COCHClCONHC₆H₅	211.65	nd (al)	....	137.5			al, bz, chl, MeOH	B12[3], 995
277	Acetoacetic acid or 3-Oxobutanoic acid .... CH₃COCH₂COOH	102.09	syr	<100d	....			w, al, eth	B3[4], 1527
278	Acetoacetic acid-α-acetyl, ethyl ester or Ethyl α-acetyl aceto acetate............ (CH₃CO)₂CHCO₂C₂H₅	172.19	....	209-11,104[16]		1.1045[204]	1.4690[20]	al, eth, bz	B3[4], 1781
279	Acetoacetic acid, allyl ester or Allyl acetoacetate .... CH₃COCH₂CO₂CH=CH₂	142.16	....	194-5[737], 66.5[14]	-85	1.0385[20/20]	1.4398[20]	al, bz, w, lig	B3[4], 1538
280	Aceto acetic acid, α-benzylidene, ethyl ester or Ethyl α-benzylidine acetoacetate .... CH₃COC(=CHC₆H₅)CO₂C₂H₅	218.26	rh pl or pyr (dil al)	295-7	60-1			chl, al, eth, bz	B10[3], 3158
281	Aceto-acetic acid, α-bromo, ethyl ester or Ethyl α-bromoaceto acetate............ CH₃COCHBrCO₂C₂H₅	209.05	....	210-5d, 104-10[15]	....	1.4294[16/4]	1.463[14]	al, eth	B3[4], 1551
282	Aceto acetic acid, γ-bromo, ethyl ester .... BrCH₂COCH₂CO₂C₂H₅	209.05	....	114-7[14], 56[0.005]		1.4840[20/4]	1.5281[20]	al, eth	B3[4], 1551
283	Acetoacetic acid, butyl ester or Butyl aceto acetate .... CH₃COCH₂CO₂C₄H₉	158.20	....	127[50], 85[8]	-35.6	0.9671[25/4]	1.4137[20]	al, bz, lig	B3[4], 1536
284	Acetoacetic acid, iso-butyl ester or iso-Butyl acetoacetate CH₃COCH₂CO₂CH₂CH(CH₃)₂	158.20	....	198-202		0.932[23]			B3[4], 1536
285	Acetoacetic acid, α-chloro, ethyl ester .... CH₃COCHClCO₂C₂H₅	164.59	....	197[748], 86-9[12]		1.191[14/17]	1.4414[20]	al, eth	B3[4], 1549
286	Acetoacetic acid, 2-chloro, ethyl ester .... ClCH₂COCH₂CO₂C₂H₅	164.59	....	220d, 115[14]	-8	1.2157[20/4]	1.4546[17]	al, eth, ace, bz, chl	B3[4], 1550
287	Acetoacetic acid, α,α-dibromo, ethylester or Ethyl α, α-dibromoaceto acetate .... CH₃COCBr₂CO₂C₂H₅	287.96	....	120-4[13]				al, eth	B3[2], 427
288	Acetoacetic acid, α,α-dichloro, ethyl ester .... CH₃COCCl₂CO₂C₂H₅	199.04	....	205-7[756], 91[11]		1.293[16/17]	1.4492[17]	al	B3[2], 427
289	Aceto acetic acid, α,α-dimethyl, ethyl ester .... CH₃COC(CH₃)₂CO₂C₂H₅	158.20	....	184[760]		0.9777[20/20]	1.4180[20]	al, eth	B3[4], 1594
290	Acetoacetic acid, ethyl ester or Ethylacetoacetate .... CH₃COCH₂COOC₂H₅	130.15	....	180.4, 74[14]	<-80	1.0282[20/4]	1.4194[20]	w, al, eth, bz, chl	B3[4], 1528

C-73

No.	Name, Synonyms, and Formula	Mol. wt.	Color, crystalline form, specific rotation and λ_{max} (log ε)	b.p. °C	m.p. °C	Density	n_D	Solubility	Ref.
291	Aceto acetic acid, ethylester (enol form) or Ethyl aceteoacetate . . . . CH₃C(OH)=CHCOOC₂H₅	130.15				1.0119[10]	1.4432[20]	. . . .	B3[4], 1528
292	Acetoacetic acid, ethyl ester (keto form) or Ethylaceto acetate . . . . CH₃COCH₂COOC₂H₅	130.15			-39	1.0368[10/4]	1.4171[20]	. . . .	B3[4], 1528
293	Acetoacetic acid, α-ethyl, ethyl ester or Ethyl ethyl aceto acetate . . . . CH₃COCH(C₂H₅)CO₂C₂H₅	158.20	. . . .	190, 58[3]		0.9847[16/4]	1.4214[25]	al, eth	B3[4], 1592
294	Aceto acetic acid, α-ethyl, methyl ester . . . . CH₃COCH(C₂H₅)CO₂CH₃	144.17		182, 79-80[14]		0.995[14]		al, eth, ace	B3[4], 1592
295	Aceto acetic acid, methyl ester or Methyl aceto acetate. . . . CH₃COCH₂CO₂CH₃	116.12		171.7, 60[8]	27-8	1.0762[20/4]	1.4184[20]	w, al, eth	B3[4], 1527
296	Aceto acetic acid, α-methyl, ethyl ester or Methyl ethyl aceto acetate . . . . CH₃COCH(CH₃)CO₂C₂H₅	144.17		187, 44[2]		0.9941[20/4]	1.4185[20]	al, eth, ace	B3[4], 1573
297	Aceto acetic acid, α-methyl, methyl ester . . . . CH₃COCH(CH₃)CO₂CH₃	130.15		177.4, 80[20]		1.0247[25/25]	1.416[24]	al, eth	B3[3], 1225
298	Aceto acetic acid, α-phenyl, ethyl ester . . . . CH₃COCH(C₆H₅)CO₂C₂H₅	206.24		156[22]		1.0855[20/4]	1.5176[20]	al, eth	B10[3], 3047
299	Acetoacetic acid, iso-propyl ester or iso-Propyl aceto acetate . . . . CH₃COCH₂CO₂CH(CH₃)₂	144.17		185-7, 75-6[15]	-27.3	0.9835[20/4]	1.4173[20]	al, eth, lig, w	B3[4], 1535
300	Acetoacetic acid, α-iso-propyl, ethyl ester . . . . CH₃COCH(iC₃H₇)CO₂C₂H₅	172.23		201[758], 97-8[20]		0.9648[18/4]	1.4256[18/4]	al, eth	B3[4], 1608
301	Acetoacetic acid, 4,4,4-trifluoro, ethyl ester or ethyl-trifluoro aceto acetate. . . . . F₃CCOCH₂CO₂C₂H₅	184.11		131.5[757]		1.2586[15.5]	1.37830	al, eth	B3[4], 1548
302	Acetoacetonitrile . . . . CH₃COCH₂CN	83.09		120-5				al, ace	B3[4], 1545
303	Acetoacetonitrile, α-phenyl . . . . CH₃COCH(C₆H₅)CN	159.19	pr (bz), cr (dil al or ace-peth)		90-1	. . . .		al, eth, bz, chl	B10[3], 3048
304	Aceto acetyl chloride . . . . CH₃COCH₂COCl	120.54			-50		. . . .	eth	B3[4], 1545
305	Acetoin or 3-hydroxy-2-butanone . . . . CH₃CHOHCOCH₃	88.12		143,37.1	-72	1.0062[20/20]	1.4171[20]	w, ace	B1[4], 3991
306	Acetone or 2-Propanone . . . . CH₃COCH₃	58.08		56.2	-95.35	0.7899[20/4]	1.3588[20]	w, al, eth, bz, chl	B1[4], 3180
307	Acetone azine . . . . (CH₃)₂C=NN=C(CH₃)₂	112.18	. . . .	133	-12.5	0.83899[20]	1.4535[20]	w, al, eth, ace	B1[4], 3207
308	Acetone, amino, hydrochloride or Acetonylamine hydrochloride . . . . CH₃COCH₂NH₂:HCl	109.56			75			w, al, eth	B4[3], 877
309	Acetone, bromo or Bromoacetone . . . . CH₃COCH₂Br	136.98		136.5[725], 31.5[8]	-36.5	1.634[23]	1.4697[15]	al, eth, ace	B1[4], 3223
310	Acetone, chloro or Chloroacetone. . . . . CH₃COCH₂Cl	92.53		119[763]	-44.5	1.15[20]		w, al, eth, chl	B1[4], 3215
311	Acetone, 1-chloro-3-phenyl . . . . C₆H₅CH₂COCH₂Cl	l68.63	nd (chl)	159-61[17]	72-3				B7[3], 1041
312	Acetone, 1,3-diamino,dihydrochloride . . . . (H₂NCH₂)₂CO · 2HCl	161.03	(dil al, dil aa), pr (w + l)		180d			w	B4[2], 763
313	Acetone, 1,1-dichloro . . . . CH₃COCHCl₂	126.97		120, 47[76]		1.3051[18/15]	1.4714[20]	al, eth	B1[4], 3218
314	Acetone, 1,3-dichloro . . . . (ClCH₂)₂CO	126.97	pr or nd	173.4, 86-8[12]	45	1.3826[46/4]	1.4716[46]	w, al, eth	B1[4], 3219
315	Acetone, diethyl acetal or 2,2-Diothoxypropane . . . . (CH₃)₂C(OC₂H₅)₂	132.21		114, 48[60]		0.8200[21/4]	1.3891[20]	al, eth, ace, bz	B1[4], 3200
316	Acetone, diethyl amino . . . . CH₃COCH₂N(C₂H₅)₂	129.20		155-6d, 64[16]		0.8620[20/4]	1.4249[20]	w, al, eth	B4[3], 877
317	Acetone, 1,3-dihydroxy . . . . (HOCH₂)₂CO	90.08			89-91			w, al, ace	B1[4], 4119
318	Acetone, 2,4-dinitrophenyl hydrazone . . . . (CH₃)₂C=NNH[C₆H₃(NO₂)₂-2,4]	238.21	ye nd or pl (al)		128			bz, al, eth	B15[3], 427
319	Acetone, 1,1-diphenyl . . . . CH₃COCH(C₆H₅)₂	210.27	. . . .	306-7[750], 174-6[10]	46		1.5361[16]	. . . .	B7[3], 2171

No.	Name, Synonyms, and Formula	Mol. wt.	Color, crystalline form, specific rotation and λ_{max} (log ϵ)	b.p. °C	m.p. °C	Density	n_D	Solubility	Ref.
320	Acetone, 1,3-diphenyl or Dibenzyl ketone .... $(C_6H_5CH_2)_2CO$	210.28	cr (al, peth)	331, 112-25[01]	35	1.195[0/4]	....	al, eth, peth	B7[3], 2160
321	Acetone, dipropylamino .... $CH_3COCH_2N(C_3H_7)_2$	157.76	....	70-2[6]	....	....	1.4267[25/D]		B4[3], 877
322	Acetone, fluoro .... CH_3COCH_2F	76.07	....	75	....	....	1.3700[20/D]		B1[4], 3213
323	Acetone, hexachloro or Perchloroacetone .... $(Cl_3C)_2CO$	264.75	....	202-4, 110[40]	-2	1.444[12/12]	1.5112[20]	bz	B1[4], 3223
324	Acetone, hexafluoro .... $(CF_3)_2CO$	166.02	....	-28	-129	....	....		B1[4], 3215
325	Acetone, hydroxy or Acetol .... CH_3COCH_2OH	74.08	....	145-6d, 5 4[18]	-7	1.0824[20/20]	1.4295[20]	w, al, eth	B1[4], 3977
326	Acetone, iodo or Iodoacetone .... CH_3COCH_2I	183.98	yesh (liq)	62[12]	....	2.17[15]	....	al	B1[4], 3226
327	Acetone, iodo, oxime or Iodoacetoxime .... $CH_3(C=NOH)CH_2I$	188.99	pr (peth)	....	64.5	....	....		B1, 660
328	Acetone, 3-methoxyphenyl .... $(3-CH_3OC_6H_4)CH_2COCH_3$	164.21	....	258-60, 95-7[0 7]	....	1.0812[0]	1.5230[25]		B8[3], 397
329	Acetone, 4-methoxyphenyl .... $(4-CH_3OC_6H_4)CH_2COCH_3$	164.21	....	267-9, 142[14]	<-15	1.0670[18/4]	1.5233[20]	al, eth	B8[3], 398
330	Acetone, 4-nitrophenylhydrazone .... $(CH_3)_2C=NNHC_6H_4NO_2 4$	193.21	....	....	152 (149)	....	....	al, eth	B15[3], 427
331	Acetone, oxime or Acetoxime .... $(CH_3)_2C=NOH$	73.10	pr (al)	134.8[728], 61[20]	61	0.9113[62/4]	1.4156[20]	al, ace, w, liq	B1[4], 3202
332	Acetone, pentabromo .... $CHBr_2COCBr_3$	372.40	....	....	79-80	....	....		B1[4], 3226
333	Acetone, pentachloro .... $Cl_3CCOCHCl_2$	230.3	cr (w + 4)	192[253], 98[40]	2.1 (an-kyd)	1.69[15/15]	....		B1[4], 3222
334	Acetone, phenacyl or Phenacylacetone .... $C_6H_5COCH_2CH_2COCH_3$	176.22	ye oil	162[12]	....	....	1.5250[10]	ace	B7[3], 3509
335	Acetone, phenoxy .... $C_6H_5OCH_2COCH_3$	150.18	....	229-30, 117-9[20]	....	1.0903[20/4]	1.5228[2]	eth, ace	B6[4], 604
336	Acetone, phenyl or methyl benzyl ketone .... $CH_3COCH_2C_6H_5$	134.18	....	216.5, 101[14]	-15	1.0157[20/4]	1.5168[20]	al, eth, bz	B7[3], 1036
337	Acetone, phenyl hydrazone .... $(CH_3)_2C=NNHC_6H_5$	148.21	rh	163[50]	42	....	....	al, eth	B15[3], 88
338	Acetone, semi carbazone .... $(CH_3)_2C=NNHCONH_2$	115.14	nd (w, ace)	....	190-1d	....	....	al, eth	B3[4], 179
339	Acetone, 1,1,1,3-tetrachloro .... $CH_2ClCOCCl_3$	195.86	liq (an hyd) pr (w + 4)	183, 71-2[13]	46(+ 4 w) 65	1.624[15/4]	1.497[18]	eth, ace	B1[4], 3222
340	Acetone, 1,1,3,3-tetrachloro .... $(Cl_2CH)_2CO$	195.86	....	180-2[718]	....	....	....	al, eth, ace, bz	B1[4], 3222
341	Acetone, 1,1,1-trichloro .... CH_3COCCl_3	161.42	....	149[764], 28[10]	....	1.435[20/4]	1.4635[17]	al, eth	B1[4], 3222
342	Acetonitrile .... CH_3CN	41.05	....	81.6	-45.7	0.7857[20]	1.34423[20]	w, al, eth, ace, bz	B2[4], 419
343	Acetonitrile, cyclohexylidone .... $C_6H_{10}=CHCN$	121.18	....	107-8[22]	....	0.9483[15/4]	1.4832[25]	al, eth	B9[3], 163
344	Acetonitrile, dibenzyl .... $(C_6H_5CH_2)_2CHCN$	221.31	lf or pr (al)	....	89-91	....	....	al, eth	B9[3], 3359
345	Acetonitrile, dichloro .... Cl_2CHCN	109.94	....	112-3	....	1.369[20]	1.4391[25]	al	B2[4], 506
346	Acetonitrile, diphenyl .... $(C_6H_5)_2CHCN$	193.25	pr (eth), lf (di al)	181-4[12]	72-3	....	....	al, eth	B9[3], 3304
347	Acetonitrile, methoxy .... CH_3OCH_2CN	71.08	....	118.1	....	0.9492[20/4]	1.3831[20]	al, eth, ace	B3[3], 399
348	Acetonitrolic acid .... $CH_3C(=NOH)NO_2$	104.07	ye rh (w, al, eth)	....	87-8d	....	....	w, al, eth, ace	B2[4], 434
349	Acetonyl acetate or 2-oxopropyl acetate .... $CH_3CO_2CH_2COCH_3$	116.13	....	170-1[755], 63[11]	....	1.0757[20/4]	1.4141[20]	w, al eth	B2[4], 297
350	Acetonyl acetone or 2,5-Hexanedione .... $CH_3COCH_2CH_2COCH_3$	114.15	....	194[754], 89[25]	-5.5	0.9737[20/4]	1.4421[20]	w, al, eth, ace, bz	B1[4], 3688
351	Acetophenone or Methyl phenyl ketone .... $C_6H_5COCH_3$	120.16	mcl pr or pl	202.6, 79[10]	20.5	1.0281[20/4]	1.53718[20]	al, eth, ace, bz, chl	B7[3], 936
352	Acetophenone, 2-amino or o-acetyl aniline .... $2-H_2NC_6H_4COCH_3$	135.17	ye cr	250-2d, 135[17]	20	....	1.6160[20]	eth	B14[3], 80

No.	Name, Synonyms, and Formula	Mol. wt.	Color, crystalline form, specific rotation and λ_{max} (log ε)	b.p. °C	m.p. °C	Density	n_D	Solubility	Ref.
353	Acetophenone, 3-amino or *m*-acetyl aniline............ 3-H$_2$NC$_6$H$_4$COCH$_3$	135.17	pa ye pl (al), lf (eth)	289-90	98-9	....	....	....	B14[1], 88
354	Acetophenone, 4-amino or *pp*-acetyl aniline .... 4-H$_2$NC$_6$H$_4$COCH$_3$	135.17	ye mcl pr (al)	293-5, 195-200[15]	106	....	....	al, eth	B14[1], 93
355	Acetophenone, *p*-amino, α-chloro or *p*-Amino phenacyl chloride .... 4-H$_2$NC$_6$H$_4$COCH$_2$Cl	169.62	ye pl	....	148	....	....	....	B14[1], 99
356	Acetophenone, 4-amino-3-chloro- .... 4-H$_2$N-3-Cl-C$_6$H$_3$COCH$_3$	169.62	pr (chl-peth)	....	92	....	....	....	B14, 49
357	Acetophenone, 3-amino-4-methoxy .... 3(H$_2$N)-4(CH$_3$O)C$_6$H$_3$COCH$_3$	165.19	pr (al)	....	102	....	....	al, eth, bz	B14[1], 548
358	Acetophenone, α-bromo or Phenacyl bromide .... C$_6$H$_5$COCH$_2$Br	199.05	nd (al), rh pr (dil al), pl (peth)	135[18]	50-1	1.647[20/4]	....	al, eth, bz, chl	B7[3], 979
359	Acetophenone, 2-bromo or 2-Bromophenyl methyl ketone 2-Br-C$_6$H$_4$COCH$_3$	199.05	ye	131-5[20]	....	....	1.5678[20]	....	B7[3], 976
360	Acetophenone, 3-bromo or -3-bromophenyl methyl ketone. 3-BrC$_6$H$_4$OCH$_3$	199.05	....	131[16]	7-8	....	1.5755[20]	ace, bz	B7[3], 977
361	Acetophenone, 4-bromo or 4-Bromo phenyl methyl ketone. 4-BrC$_6$H$_4$COCH$_3$	199.05	lf (al)	255.5[736], 130[11]	50-1	1.647	....	al, eth, bz, aa	B7[3], 977
362	Acetophenone, α-bromo-3-chloro or 3-Chlorophenacyl bromide 3-ClC$_6$H$_4$COCH$_2$Br	233.50	nd	....	395-40	....	....	al	B7[3], 982
363	Acetophenone, α-bromo-4-chloro or 4-Chlorophenacyl bromide 4-ClC$_6$H$_4$COCH$_2$Br	233.50	nd	....	96-7	....	....	....	B7[3], 983
364	Acetophenone, 4-bromo-α-chloro or 4-Bromophenacyl chloride 4-BrC$_6$H$_4$COCH$_2$Cl	233.50	nd (al)	....	116-7	....	....	al	B7[3], 982
365	Acetophenone, α bromo-4-methyl or 4-Methyl phenacyl bromide .... 4-CH$_3$C$_6$H$_4$COCH$_2$Br	213.08	nd or lf (al)	155-9[14]	51	....	....	al, eth	B7[3], 1060
366	Acetophenone, 4-*tert*-butyl 4-(CH$_3$)$_3$CC$_6$H$_4$COCH$_3$	176.26	....	136-8[20]	......	0.9705[0]	1.518[15/D]	....	B7[3], 1171
367	Acetophenone, 4-*tert*butyl-2,6 dimethyl, 3,5-dinitro or Musk ketone .... C$_{14}$H$_{18}$N$_2$O$_5$	294.31	ye	....	134.5-6.5	....	....	....	B7[3], 1228
368	Acetophenone, α-chloro or Phenacyl chloride......... C$_6$H$_5$COCH$_2$Cl	154.60	pl (dil al), rh, lf (peth)	247, 139-41[14]	56.5	1.324[15/4]	....	al, eth, ace, bz	B7[3], 967
369	Acetophenone, 2-chloro .... 2-ClC$_6$H$_4$COCH$_3$	154.60	....	227-8[738], 113[18]	......	1.2016[17/4]	1.685[25]	eth	B7[3], 1008
370	Acetophenone, 3-chloro .... 3-ClC$_6$H$_4$COCH$_3$	154.60	....	241-5[744], 127-31[30]	......	1.2130[04]	1.5494[20]	al, eth, ace	B7[2], 218
371	Acetophenone, 4-chloro .... 4-ClC$_6$H$_4$COCH$_3$	154.60	....	273, 106[10]	20	1.1922[20/4]	1.5550[20]	al, eth	B7[3], 1008
372	Acetophenone, α-chloro-4-methyl or 4-Methyl phenacyl chloride .... 4-CH$_3$C$_6$H$_4$COCH$_2$Cl	168.63	nd (al)	260-3, 113[4]	57-8	....	....	al, eth	B7[3], 1065
373	Acetophenone, α-chloro-2,4-dimethyl .... 2,4(CH$_3$)$_2$C$_6$H$_3$COCH$_2$Cl	182.65	nd	....	62	....	....	al, eth, ace	B7[3], 1106
374	Acetophenone, $\alpha\alpha$-dibromo or Phenacylidene bromide ... C$_6$H$_5$COCHBr$_2$	277.94	....	159-60[13]	36-7	....	....	al, eth, chl	B7[3], 985
375	Acetophenone, α,4-dibromo or 4-Bromophenacyl bromide .... 4-BrC$_6$H$_4$COCH$_2$Br	277.96	nd (al)	....	110-2	....	....	eth, al	B7[3], 985
376	Acetophenone, $\alpha\alpha$-dichloro or Phenacylidine chloride.... C$_6$H$_5$COCHCl$_2$	189.04	amor	249, 143[25]	20-1	1.340[16]	1.5686[20]	al, bz	B7[3], 972
377	Acetophenone, α,4-dichloro or *p*-Chorophenacyl chloride 4-ClC$_6$H$_4$COCH$_2$Cl	189.04	nd (al)	270	101-2	....	....	al, bz	B7[3], 972
378	Acetophenone, 2,4-dichloro .... 2,4-Cl$_2$-C$_6$H$_3$COCH$_3$	189.04	....	245-7, 140-50[15]	33-4	....	1.5640[20]	....	B7[3], 969
379	Acetophenone, 3,4-dichloro .... 3,4-Cl$_2$-C$_6$H$_3$COCH$_3$	189.04	nd (peth)	135[12]	76	....	....	....	B7[3], 971

No.	Name, Synonyms, and Formula	Mol. wt.	Color, crystalline form, specific rotation and λ_{max} (log ε)	b.p. °C	m.p. °C	Density	n_D	Solubility	Ref.
380	Acetophenone, 2,3-dihydroxy or 3-Acetylcatechol .... 2,3(HO)$_2$C$_6$H$_3$COCH$_3$	152.16	ye pr (bz-lig)		97-8			al	B8[3], 2080
381	Acetophenone, 2,4-dihydroxy or Resacetophenone ...... 2,4-(HO)$_2$C$_6$H$_3$COCH$_3$	152.16	nd or lf		147		1.1800[141]	al, bz	B8[3], 2082
382	Acetophenone, 2,5-dihydroxy or 2- Acetyl hydroquinone 2,5(HO)$_2$C$_6$H$_3$COCH$_3$	152.16	ye-gr nd (dil al or w)		204-5			al	B8[3], 2100
383	Acetophenone, 3,4-dihydroxy or 4-Acetyl catechol........ 3,4(HO)$_2$C$_6$H$_3$COCH$_3$	152.16	nd (w or chl)		115-6				B8[3], 2108
384	Acetophenone, 3,5-dihydroxy or 5-Acetyl resorcinol ..... 3,5(HO)$_2$C$_6$H$_3$COCH$_3$	152.16	cr (w)		147-8			al, eth, ace, w	B8[2], 301
385	Acetophenone, 3,4-dimethoxy or Aceto veratrone .... 3,4(CH$_3$O)$_2$C$_6$H$_3$COCH$_3$	180.21	pr (dil al)	286-8, 160-2[15]	51			w, al, chl, bz	B8[3], 2110
386	Acetophenone, 3,5-dimethoxy-4-hydroxy or Aceto syringone .... 3,5(CH$_3$O)$_2$-4-(HO)C$_6$H$_2$COCH$_3$	196.21	nd (w) pr (peth)		122-3			al, eth, ace	B8[3], 2118
387	Acetophenone, 2,4-dimethyl .... 2,4(CH$_3$)$_2$C$_6$H$_3$COCH$_3$	148.21		228, 110[13]		1.0121[15]	1.5340[20]	al	B7[3], 1105
388	Acetophenone, 2,5-dimethyl 2,5,(CH$_3$)$_2$C$_6$H$_3$COCH$_3$	148.21		232-3, 107[11]		0.9963[19/4]	1.5291[20]	al, eth, bz	B7[3], 1104
389	Acetophenone, 3,4-dimethyl 3,4-(CH$_3$)$_2$C$_6$H$_3$COCH$_3$	148.21		246-7, 213[110]		1.0090[14/4]	1.5413[15]	al, eth, bz	B7[3], 1104
390	Acetophenone, 3-dimethylamino 3-(CH$_3$)$_2$NC$_6$H$_4$COCH$_3$	163.22		148[11]	43			ace	B14, 45
391	Acetophenone, 4-dimethylamino 4-(CH$_3$)$_2$NC$_6$H$_4$COCH$_3$	163.22	nd (w, Peth)	172-5[11]	105.5			eth, w, lig	B14[3], 94
392	Acetophenone, 2-ethoxy or 2-Acetylphenetole 2-C$_2$H$_5$OC$_6$H$_4$COCH$_3$	164.21	pr (dil al), pl (lig)	243-4	43	1.0036[10]		al, eth, lig	B8[3]
393	Acetophenone, 4-ethoxy or 4-Acetylphenetole 4-C$_2$H$_5$OC$_6$H$_4$COCH$_3$	164.21	pl (eth)		39			al, eth	B8[3], 280
394	Acetophenone, 4 fluoro 4-F-C$_6$H$_4$COCH$_3$	138.14		196[760], 79[10]	-45	1.1382[25/4]	1.5081[25]		B7[3], 961
395	Acetophenone, furfurylidene C$_6$H$_5$COCH=CH(C$_4$H$_3$O)	198.22		317, 181[29]		1.1140[20]		al, eth	B17[2], 377
396	Acetophenone, α-hydroxy or Phenacylalcohol C$_6$H$_5$COCH$_2$OH	136.16	hex pl (al or eth), pl (w or dil al)	124-6[12] (sub), 56[1]	90	1.0963[99/4]		al, eth, lig	B8[3], 298
397	Acetophenone, α-hydroxy, acetate or Phenacyl acetate ... C$_6$H$_5$COCH$_2$O$_2$CCH$_3$	178.19	rh pl (eth, lig or peth)	270, 150-2[19]	49.6	1.1169[65/4]		al, eth, bz, chl	B8[3], 301
398	Acetophenone, 2-hydroxy or 2-Acetylphenol........... 2-HOC$_6$H$_4$COCH$_3$	136.16		218, 106[17]	4-6	1.1307[20/4]	1.5584[20]	al, eth, aa	B8[3], 261
399	Acetophenone, 3-hydroxy or 3-Acetylphenol........... 3-HOC$_6$H$_4$COCH$_3$	136.16	nd or lf	296[756], 153[5]	96	1.0992[109]	1.5348[109]	al, eth, bz	B8[3], 272
400	Acetophenone, 4-hydroxy or 4 Acetylphenol........... 4-HOC$_6$H$_4$COCH$_3$	136.16	nd (eth, dil al)	147-8[3]	109-10	1.1090[109]	1.5577[109]	al, eth	B8[3], 276
401	Acetophenone, α hydroxy-4-methoxy or 4-methoxy phenacyl alcohol .... 4-CH$_3$OC$_6$H$_4$COCH$_2$OH	166.18	pl (dil al)		104			al, bz	B8[3], 2120
402	Acetophenone, 2-hydroxy-3-methoxy or o Aceto vanillon 2(HO)-3(CH$_3$O)C$_6$H$_3$COCH$_3$	166.18	pa ye nd (peth or eth-peth)		53-4			eth, bz	B8[3]2, 081
403	Acetophenone, 2-hydroxy-4-methoxy or Peonol .... 2-HO-4-CH$_3$OC$_6$H$_3$COCH$_3$	166.18	nd (al)	158[20]	52-3	1.3102[81]	1.5452[81]	al, eth, bz, chl	B8[3], 2084
404	Acetophenone, 2-hydroxy- 5-methoxy 2-HO-5-CH$_3$OC$_6$H$_3$COCH$_3$	166.18	pa ye pr (dil al)		52			bz, al	B14, 477
405	Acetophenone, 3-hydroxy-4-methoxy or Iso aceto-vanillone .... 3-HO-4-CH$_3$OC$_6$H$_3$COCH$_3$	166.18	cr (eth lig) or cr (w + 1)	67-8 (+1w) 91 anhyd			eth, w	B8[3], 2109
406	Acetophenone, 4-hydroxy-α-methoxy 4-HO-C$_6$H$_4$COCH$_2$OCH$_3$	166.18	nd (bz), pr (w + 1)		130-1			al, ace, eth, bz	B8[2], 302
407	Acetophenone, 4-hydroxy-3-methoxy or Resacetophenone-2-methylether iso peonol .... 4-HO-3-CH$_3$O-C$_6$H$_3$COCH$_3$	166.18	nd (w)		138			bz, w	B8[1], 617
408	Acetophenone, 4-hydroxy-3-methoxy or Acetovanillon... 4-HO-3-CH$_3$OC$_6$H$_3$COCH$_3$	166.18	pr (w)	295-300, 233-5[15,20]	115			al, eth, ace, bz, chl	B8[3], 2108
409	Acetophenone, α-iodo or Phenacyl iodide .... C$_6$H$_5$COCH$_2$I	246.05		158-60[15]	34.4			ace	B7[3], 989

No.	Name, Synonyms, and Formula	Mol. wt.	Color, crystalline form, specific rotation and λ_{max} (log ε)	b.p. °C	m.p. °C	Density	n_D	Solubility	Ref.
410	Acetophenone, 2-iodo 2I-$C_6H_4COCH_3$	246.05	....	139-140[12]	......	1.746[20/4]	1.6180[20]	bz	B7[3], 988
411	Acetophenone, 3-iodo 3-I-$C_6H_4COCH_3$	246.05	....	129[8]	......		1.622[20]	bz	B7[3], 988
412	Acetophenone, 4-iodo 4-I-$C_6H_4COCH_3$	246.05	....	153[18]	85			al, bz, eth, lig	B7[3], 989
413	Acetophenone, 2-Methoxy or 2,Acetylanisole 2-$CH_3OC_6H_4COCH_3$	150.12	ye	245	......	1.0897[20/4]	1.5393[20]	al, ace	B8[3], 263
414	Acetophenone, 3-methoxy 3-$CH_3OC_6H_4COCH_3$	150.18	....	240, 125-6[12]	95-6	1.0343[19]	1.5410[20]	al, ace, w, eth	B8[3], 273
415	Acetophenone, 4-methoxy 4-$CH_3OC_6H_4COCH_3$	150.18	....	258, 138-9[15]	38-9	1.0818[41/4]	1.547[41]	al, eth, ace	B8[3], 277
416	Acetophenone, 2-methyl or 2-acetyl toluene 2$CH_3C_6H_4COCH_3$	134.17	....	214, 89-92[10]	......	1.026[20/4]	1.5276[20]	....	B7[3], 1052
417	Acetophenone, 3-methyl or 3 Acetyltoluene 3-$CH_3C_6H_4COCH_3$	134.17	....	220[766], 109[12]	......	1.0070[20/4]	1.5270[20]	al, eth, ace	B7[3], 1057
418	Acetophenone, 4-methyl or 4-acetyl toluene 4-$CH_3C_6H_4COCH_3$	134.17	nd	226, 113[11]	28	1.0051[20/4]	1.5335[20]	al, eth, bz	B7[3], 1060
419	Acetophenone, 2-nitro 2-$NO_2C_6H_4COCH_3$	165.15	....	158[16]	28-9		1.5468[20]	al, eth	B7[3], 990
420	Acetophenone, 3-nitro 3-$NO_2C_6H_4COCH_3$	165.15	nd (al)	202, 167[18]	81			eth, w	B7[3], 991
421	Acetophenone, 4-nitro 4-$NO_2C_6H_4COCH_3$	165.15	yesh pr (al)	....	80-2			al, eth	B7[3], 993
422	Acetophenone oxime $C_6H_5C(:NOH)CH_3$	135.17	nd (w)	245, 119[20]	60			al, eth, ace, bz, lig	B7[3], 954
423	Acetophenone, 5-*iso*-propyl-2-methyl or Carvacryl methyl ketone. 5-i-C_3H_7-2-$CH_3C_6H_3COCH_3$	176.25	....	249.50	<-20	0.956[20/4]	1.5181[20]		B7[3], 1176
424	Acetophenone, $\alpha\alpha\alpha$-trichloro $C_6H_5COCCl_3$	223.50	....	256-7, 145[25]	......	1.425[16]	....	al, eth	B7[3], 974
425	Acetophenone, $\alpha\alpha\alpha$ trifluoro $C_6H_5COCF_3$	174.12	....	152[730]	-40	1.279[20]	1.4583[20]		B7[3], 962
426	Acetophenone, 2,4,5-trimethyl 2,4,5-$(CH_3)_3C_6H_2COCH_3$	162.23	....	246-7, 137-8[20]	10-1	1.0039[15/4]	1.541[15]	al, eth, bz, aa	B7[3], 1145
427	Acetophenone, 2,4,6-trimethyl 2,4,6-$(CH_3)_3C_6H_2COCH_3$	162.23	....	240.5[735], 120[12]	......	0.9754[20/4]	1.5175[20]	al, eth, ace, bz	B7[3], 1137
428	Acetoxyacetic acid $CH_3CO_2CH_2CO_2H$	118.09	nd (bz)	144-5[12]	67-8			w, al, eth, ace, chl	B3[4], 576
429	Acetoxy acetic acid, ethylester or Ethyl acetoxyacetate $CH_3CO_2CH_2CO_2C_2H_5$	146.14	....	179	......	1.0880[20/4]	1.4112[20]	al, eth, aa	B3[4], 580
430	(2-Acetoxypropyl) trimethyl ammonium chloride *D* or *o*-acetyl-B-methyl choline chloride $CH_3CH(O_2CCH_3)CH_2N+(CH_3)_3Cl-$	195.69	hyg. $[\alpha]_D +41.9$	d	172-3	....		w, al	B4[4], 1671
431	(2-Acetoxypropyl) trimethyl ammonium chloride *L* or *o*-Acetyl-β-methyl choline chloride $CH_3CH(O_2CCH_3)CH_2N+(CH_3)_3Cl-$	195.69	$[\alpha]_D -41.9$	d	172-3				B4[4], 1671
432	Acetyl acetone or 2-4-Pentanedione $CH_3COCH_2COCH_3$	100.13	....	139[746]	-23	0.9721[25/4]	1.4494[20]	w, al, eth, ace, chl	B1[4], 3662
433	Acetyl benzoyl peroxide $CH_3CO-OO-COC_6H_5$	180.17	wb nd (lig)	85-100 (exp)	40-1	....		eth	B9[3], 1051
434	Acetyl bromide CH_3COBr	122.96	ye in air	76	-98	1.6625[16/4]	1.45376[16]	eth, ace, bz, chl	B2[4], 398
435	Acetyl chloride CH_3COCl	78.50	....	50.9	-112	1.1051[20/4]	1.38976[20]	eth, ace, bz, chl	B2[4], 395
436	Acetyl chloride, dichloro $Cl_2CHCOCl$	147.39	....	108-10	......	1.5315[16/4]	1.4591[20]	eth	B2[4], 504
437	Acetyl chloride, ethoxy $C_2H_5OCH_2COCl$	122.55	....	123-4, 49-50[37]	......	1.1170	1.4204[20]	eth, ace	B3[3], 396
438	Acetyl chloride, methoxy CH_3OCH_2COCl	108.53	....	99, 46-9[62]	......	1.1871[20/4]	1.4196[20]	eth, ace, chl	B3[3], 396
439	Acetyl fluoride CH_3COF	62.04	....	20.8	......	1.002[15/4]	....	al, eth, ace, bz	B2[4], 393
440	Acetyl iodide CH_3COI	169.95	....	108, 36[50]	......	2.0674[20/4]	1.5491[20]	eth	B2[4], 399
441	Acetyl isothiocyanate CH_3CONCS	101.12	....	132-3, 30[9]	......	1.1523[13/4]	1.5231[18]	eth	B3[3], 278

No.	Name, Synonyms, and Formula	Mol. wt.	Color, crystalline form, specific rotation and λ_{max} (log ε)	b.p. °C	m.p. °C	Density	n_D	Solubility	Ref.
442	Acetyl peroxide CH_3CO-OO-$COCH_3$	118.09	nd (eth), lf	63^{21}	30 (26)	....	....	al, eth	B2[4], 392
443	Acetylene $HC\equiv CH$	26.04	....	-84.0	-80.8	$0.6208^{-82/4}$	1.00051^0	ace, bz, chl	B1[4], 939
444	Acetylene-bromo $BrC\equiv CH$	104.94	....	4.7	......	....	....	eth	B1[3], 919
445	Acetylene-chloro $ClC\equiv CH$	60.48	....	-30	-126	....	....	....	B1[4], 957
446	Acetylene-dibromo $BrC\equiv CBr$	183.84	nd	76 (exp)	-25	....	....	al, eth, ace, bz	B1[3], 919
447	Acetylene-dichloro $ClC\equiv CCl$	94.93	....	exp	-66	....	....	al, eth, ace	B1[4], 957
448	Acetylene-diiodo $IC\equiv CI$	277.83	rh nd (lig)	32-3 (exp)	81-2	....	....	al, eth, ace, bz	B1[4], 958
449	Acetylene-diphenyl $C_6H_5C\equiv CC_6H_5$	178.24	mcl pr or pl (al)	300, 170^{19}	62.5	$0.9657^{100/4}$	....	eth	B5[3], 2119
450	Acetylene, bis(1-hydroxycyclohexyl) $(C_6H_{10}OHC)_2$	222.33	nd (ccl$_4$)	182^{13}	112-3	....	....	al, eth, ace	B6[3], 4741
451	Acetylene, 1-phenyl-2-methyl $C_6H_5C\equiv CCH_3$	116.16	....	181	......	0.942^{15}	$1.563^{15}/_D$	....	B5[2], 408
452	Aconic acid $C_5H_4O_4$	128.08	lf (eth), rh (w, al)	d	164	....	....	w, al	B18[4], 5333
453	Aconine $C_{25}H_{41}NO_9$	499.61	amor, $[\alpha]^{20/D}+23$	....	132	....	....	w, al	B21[4], 2899
454	Aconitomide $H_2NOOCH_2C(CONH_2)=CHCONH_2$	171.16	ye nd (w)	....	sinters 161	....	....	w	B2, 853
455	Aconitic acid, cis or 1, 2, 3-Propene tricarboxylic acid $HO_2CCH_2C(CO_2H)=CHCO_2H$	174.11	nd (w)	....	130	....	....	w	B2[4], 2405
456	Aconitic acid, trans or 1, 2, 3-Propene tricarboxylic acid $HO_2CCH_2C(CO_2H)=CHCO_2H$	174.11	lf (w), nd (w, eth)	....	198-9	....	....	w, al	B2[4], 2405
457	Aconitic acid, triethyl ester trans $C_2H_5O_2CH_2C(CO_2C_2H_5)=CHCO_2C_2H_5$	258.27	....	275d, 159^9	......	$1.1064^{20/4}$	1.4556^{20}	al, eth	B2[4], 2405
458	Aconitic acid, trimethylester $CH_3O_2CH_2C(CO_2CH_3)=CHCO_2CH_3$	216.19	....	270-1, 161^{14}	......	....	....	al, eth	B2[4], 2405
459	Aconitic acid, tripropylester $C_3H_7O_2CH_2C(CO_2C_3H_7)=CHCO_2C_3H_7$	300.36	....	195^{13}	......	$1.050^{25/4}$	1.4521^{20}	al, eth	B2[4], 2405
460	Aconitine $C_{34}H_{47}NO_{11}$	645.76	rh lf $[\alpha]^{20/D}+19$ (chl)	....	204	....	....	al, bz, chl	B21[4], 2901
461	Aconitine, hydrobromide $C_{34}H_{47}NO_{11}\cdot HBr\cdot 1\frac{1}{2}H_2O$	753.71	yesh pr (w), $[\alpha]_D$-31 (w,c=2.3)	....	209.10	....	....	w, al, eth	B21[4], 2903
462	Aconitine, hydrochloride $C_{34}H_{47}NO_{11}\cdot HCl\cdot 3\frac{1}{2}H_2O$	745.27	(w + 3 1/2), $[\alpha]_D$-30.5 (w)	....	170-2	....	....	w, al, eth	B21[4], 2902
463	Aconitine, nitrate $C_{34}H_{47}NO_{11}\cdot HNO_3$	708.77	$[\alpha]^{20/D}$-35 (2% aq soln)	....	ca. 200d	....	....	w	B21[4], 2903
464	Acridine or 2,3,5,6-Dibenzopyridine $C_{13}H_9N$	179.22	rh nd or pr (al)	345-6	111	$1.005^{20/4}$	....	al, eth, bz	B20[4], 3987
465	Acridine, 2-amino $C_{13}H_{10}N_2, 2$-$H_2NC_{13}H_8N$	194.24	ye nd (w or al)	....	213-4	....	....	al, eth	B22[4], 4984
466	Acridine, 2-amino-5(4-aminophenyl) or Chrysaniline $C_{19}H_{15}N_3.2H_2O$	321.38	ye nd (95% al)	....	260-7	....	....	....	B22[4], 5513
467	Acridine, 3-amino $C_{13}H_{10}N_2, 3$-$NH_2C_{13}H_8N$	194.24	ye nd (w or al + 1)	....	213-4	....	....	al, eth	B22[4], 4987
468	Acridine, 4-amino $C_{13}H_{10}N_2, 4$-$NH_2C_{13}H_8N$	194.24	red-br pr (peth), og nd (MeOH)	183-4	108	....	....	al, eth, ace, bz	B22[4], 4989
469	Acridine, 4 amino,-hydrochloride $C_{13}H_{10}N_2.HCl$	230.70	ye nd (dil al)	....	234d	....	....	w, al	B22[2], 376
470	Acridine, 9-amino $C_{13}H_{10}N_2$	194.24	ye nd (ace or al)	....	241 (cor)	....	....	....	B21[4], 4174
471	Acridine, 9-chloro $C_{13}H_8ClN$	216.68	nd (al)	sub	122	....	....	al, w	B20[4], 3995
472	Acridine, 3,6-diamino $C_{13}H_{11}N_3$	209.25	ye nd (al or w)	....	284-6	....	....	al, eth, w	B22[4], 5487

No.	Name, Synonyms, and Formula	Mol. wt.	Color, crystalline form, specific rotation and λ_{max} (log ϵ)	b.p. °C	m.p. °C	Density	n_D	Solubility	Ref.
473	Acridine, 6,9-diamino-2-ethoxy or Rivanol .... $C_{15}H_{15}N_3O$	255.31	ye nd	....	123-44d				B22[4], 6679
474	Acridine, 6,9-dichloro-2-methoxy .... $C_{14}H_8NOCl_2$	278.14			163-5				B21[4], 1553
475	Acridine, 9,10-dihydro or Acridan, carbazine .... $C_{13}H_{11}N$	278.14	pl or pr (al)	sub 300d	169-71			al, eth, ace	B20[4], 3885
476	Acridine, 9,10-dihydro-9-oxo or Acridone............. $C_{13}H_9NO$	195.22	ye lf (al)	>354	154			ace	B21[4], 4171
477	Acridine, 4-hydroxy .... $C_{13}H_9NO$	195.22			117			al, eth	B21[4], 1562
478	Acridine, 2-methyl .... $C_{14}H_{11}N$	193.25	ye nd (dil al)		134			al, eth, bz	B20[4], 4037
479	Acridine, 9 phenyl .... $C_{19}H_{13}N$	255.32	lf ye nd (al)	401-4 sub	184				B20[4], 4355
480	Acrolein or Propenal.................. $CH_2=CHCHO$	56.07		52.5-3.5	-86.9	$0.8410^{20/4}$	1.4017^{20}	w, al, eth, ace	B1[4], 3435
481	Acrolein, 2-chloro .... $CH_2=C(Cl)CHO$	90.51		40^{10}		1.199^{20}	1.463^{20}	al, eth	B1[4], 3440
482	Acrolein, diethylacetal or 3,3-diethoxy-1-propene .... $CH_2=CHCH(OC_2H_5)_2$	130.19		123.5		0.8543^{15}	1.4000^{20}	eth, CCl_4	B1[4], 3437
483	Acrolein 3(2-furyl) or Furacrolein................... $(2-C_4H_3O)-CH=CHCHO$	122.13	ye or wh nd	> 200d	54			al, eth	B17[4], 4695
484	Acrolein, 2-methyl or methacrolein.................. $CH_2=C(CH_3)CHO$	70.09		68.4		$0.837^{20/4}$	1.4144^{20}	w, al, eth	B1[4], 3455
485	Acrylamide .... $CH_2=CHCONH_2$	71.08	lf (bz)		84-5			w, al, eth, chl	B2[4], 1471
486	Acrylic acid or Propenoic acid .... $CH_2=CHCO_2H$	72.06		141.6, 48.5^{18}	13	$1.0511^{20/4}$	1.4224^{20}	w, al, eth, ace, bz	B2[4], 1455
487	Acrylic acid, allyl ester or Allyl acrylate .... $CH_2=CHCO_2CH_2CH=CH_2$	112.14		172-4		$0.9441^{20/4}$	1.4320^{20}	al, eth	B2[4], 1468
488	Acrylic acid, β-benzoyl .... $C_6H_5COCH=CHCO_2H$	176.18	nd or pr (to)		99			al, eth, to	B10[3], 3144
489	Acrylic acid, benzyl ester or Benzyl acrylate............. $CH_2=CHCO_2CH_2C_6H_5$	162.19		228		$1.0573^{20/4}$	1.5143^{20}	al, eth, ace	B6[3], 1481
490	Acrylic acid, 2-bromo, ethyl ester or Ethyl 2-bromo acrylate ..... $CH_2=C(Br)CO_2C_2H_5$	179.01		155-8		1.4581^{25}	1.4660^{25}	....	B2[4], 1487
491	Acrylic acid, butyl ester or Butylacrylate .... $CH_2=CHCO_2C_4H_9$	128.17		146-8,39^{10}	-64.6	$0.8898^{20/4}$	1.4185^{20}	al, eth, ace	B2[4], 1463
492	Acrylic acid, iso butylester or iso-Butylacrylate .... $CH_2=CHCO_2CH_2CH(CH_3)_2$	128.17		132		$0.8896^{20/4}$	1.4150^{20}	al, eth	B2[4], 1465
493	Acrylic acid, 2 chloro, ethyl ester or Ethyl-2-chloroacrylate ..... $CH_2=C(Cl)CO_2C_2H_5$	134.56		$51-3^{18}$		$1.1404^{20/4}$	1.4384^{20}	al, eth	B2[4], 1482
494	Acrylic acid, 2-chloro, methyl ester or Methyl-2-chloroacrylate..... $CH_2=C(Cl)CO_2CH_3$	120.54		52^{51}		$1.189^{20/4}$	1.4420^{20}	eth	B2[4], 1482
495	Acrylic acid, 3-chloro-cis or cis-3-chloropropenoic acid ... $ClCH=CHCO_2H$	106.51	lf or nd (HCl)	$107^{17.5}$	63-4			al, eth	B2[4], 1481
496	Acrylic acid, 3-chloro-trans or trans-3-chloropropenoic acid......... $ClCH=CHCO_2H$	106.51	lf	94^{18}	86			al, eth	B2[4], 1481
497	Acrylic acid, cyclohexyl ester or Cyclohexyl acrylate ...... $CH_2=CHCOOC_6H_{11}$	154.21		$182-4^{750}, 88^{20}$		$1.0275^{20/4}$	1.4673^{20}	al, eth, chl	B6[4], 38
498	Acrylic acid, 2,3-dichloro .... $ClCH=C(Cl)CO_2H$	140.95	mcl pr (chl)		87-8			w, al, eth, ace, chl	B2[4], 1484
499	Acrylic acid, 3,3-dichloro .... $Cl_2C=CHCO_2H$	140.95	nd (peth) pr (chl)	sub	76-7			eth, chl	B2[4], 1484
500	Acrylic acid, 2,3-diphenyl or α-phenyl cinnamic acid ..... $C_6H_5CH=C(C_6H_5)CO_2H$	224.26	nd (lig, dil al)	sub	172.5-3			al, eth	B9[3], 3414
501	Acrylic acid, ethyl ester or Ethyl acrylate .... $CH_2CO_2C_2H_5$	100.13		99.8	-71.2	$0.9234^{20/4}$	1.4068^{20}	al, eth, chl	B2[4], 1460
502	Acrylic acid, 3-(2-furyl)-cis or 2- Furanacrylic acid cis ... $(2-C_4H_3O)CH=CHCO_2H$	138.12	wh pr or pl		103-4			eth	B18[4], 4143
503	Acrylic acid, 3-(2-furyl) trans or 2-Furanacrylic acid trans $(2-C_4H_3O)CH=CHCO_2H$	138.12	nd (w)	286 (sub)	141		1.5286^{20}	al, eth, bz	B18[4], 4173

No.	Name, Synonyms, and Formula	Mol. wt.	Color, crystalline form, specific rotation and λ_{max} (log ε)	b.p. °C	m.p. °C	Density	n_D	Solubility	Ref.
504	Acrylic acid, 3-(2-furyl) cis, benzylester or Benzyl-2-furanacrylate. (2-C₄H₃O)CH=CHCH₂C₆H₅	228.25	pa ye	201-3¹²	42-5	....	1.5872²⁵	eth, ace, bz, al	B18⁴, 4147
505	Acrylic acid, 3-(2-furul) cis, butyl ester or Butyl-2-furanacrylate. (2-C₄H₃O)CH=CHCO₂C₄H₉	194.23	....	147-50¹⁵, 117-8³	......	1.045²⁰	1.5129²⁰	al, ace	B18⁴, 4146
506	Acrylic acid, 3-(2-furyl) cis, pentylester or Pentyl-2-furanacrylate. (2C₄H₃O)CH=CHCO₂C₅H₁₁	208.24	....	116-8²	......	1.0322²⁰⁴	1.5289²⁴	al, eth	B18⁴, 4146
507	Acrylic acid, 3-(2-furyl) cis, propyl ester or Propyl-2-furan-acrylate. (2-C₄H₃O)CH=CHCO₂C₃H₇	180.21	....	91-4³	......	1.07 44²⁰⁴	1.5392²⁴	al, eth, bz	B18⁴, 4146
508	Acrylic acid, 3-(2-furyl) cis, methyl ester or Methyl-2-furanacrylate. (2-C₄H₃O)CH=CHCO₂CH₃	152.16	....	227.5⁷⁷⁴, 112¹⁵	35-7	....	1.4447²⁰	al, eth, bz	B18⁴, 4144
509	Acrylic acid, methyl ester or Methyl acrylate. CH₂=CHCO₂CH₃	86.09	....	80.5	< - 75	0.9535²⁰⁴	1.4040²⁰	al, eth, ace, bz	B2⁴, 1457
510	Acrylic acid, 2 methylbutyl ester or (2-methylbutyl) acrylate. CH₂=CHCO₂CH₂CH(CH₃)CH₂CH₃	142.20	....	160, 52¹¹	......	0.8936²⁰⁴	1.4240²⁰	al, eth	B2⁴, 1525
511	Acrylic acid, 3-(α-naphthyl) cis. α-C₁₀H₇CH=CHCO₂H	198.22	cal pl (al)	....	156			al	B9³, 3284
512	Acrylic acid, 3-(α-naphthyl) trans. α-C₁₀H₇CH=CHCO₂H	198.22	nd (al,w,aa)	sub	211-2			eth, chl	B9³, 3284
513	Acrylic acid, 3 (β-naphthyl) trans. β-C₁₀H₇CH=CHCO₂H	198.22	nd (al, w)		210			al	B9³, 3288
514	Acrylic acid, 2-phenyl or 2-Phenylpropenoic acid. Atropic acid. CH₂=C(C₆H₅)CO₂H	148.17	lf (al), nd (w)	267d	106 7			al, eth, bz, chl	B9³, 2751
515	Acrylic acid, trichloro. CCl₂=CCICO₂H	193.85	....	158			1.5271¹⁸·⁵	bz	B2⁴, 1486
516	Acrylonitrile. CH₂=CHCN	53.06	....	77.5-9		0.8060²⁰⁴	1.3911²⁰	al, eth, ace, bz	B2⁴, 1473
517	Acrylonitrile, 3,3 diphenyl. (C₆H₅)₂C=CHCN	205.26	....	125-8⁰·¹⁵			1.6352²⁰	....	B9³, 3439
518	Acrylonitrile, triphenyl. (C₆H₅)₂C=C(C₆H₅)CN	281.36	nd (al), pr		166-7			eth, al	B9³, 3630
519	Acrylylchloride. CH₂=CHCOCl	90.51	....	75-6		1.1136²⁰⁴	1.4343²⁰	chl	B2⁴, 1471
520	Acrylyl chloride, trichloro. CCl₂=CCICOCl	193.85	....	158			1.5271¹⁸·⁵	bz	B2⁴, 1486
521	Actidione l or Cycloheximide. C₁₅H₁₁NO₄	281.36	pl (al) [α]²⁵ᐟᴰ-33 (chl C=l)		144-5			al	B20⁴, 1406
522	Adamantane or Tricyclo 3,3,1,1³·⁷decane. C₁₀H₁₆	136.24	nd (sub)	sub	268 (sealed)	1.07	1.568	bz	B5⁴, 469
523	Adenosine or Adenine, 9-β-D-ribofuranosyladenine. C₁₀H₁₃N₅O₄	267.25	nd (w + 1 1/2), [α]²⁰ᐟᴰ -60.0 (w,c=1)		235-6 (anh)			w	B31, 27
524	5-Adenylic acid or Adenosine-s-phosphate. C₁₀H₁₄N₅O₇P	347.23	pw, nd, (w, dil al) cr [α]⁵⁰ᐟᴰ -41.78		195-208d (sealed tube)			w, al	B31, 27
525	Adipaldehyde or 1,6-Hexanedial. OHC(CH₂)₄CHO	114.15	....	92-4⁹	-8	1.003 ¹⁹⁴	1.4350 ²⁰	al, eth, bz, aa	B1⁴, 3686
526	Adipamic acid or Adipic acid monoamide. H₂NCO(CH₂)₄CO₂H	145.16	nd (w)		161-2				B2⁴, 1972
527	Adipamide. H₂NCO(CH₂)₄CONH₂	144.18	pl		220			al	B2⁴, 1972
528	Adipic acid or Hexanedioc acid. HO₂C(CH₂)₄CO₂H	146.14	mcl pr (w, ace, lig)	265¹⁰⁰	153	1.360²⁵⁴		al, eth	B2⁴, 1956
529	Adipic acid, 2-amino -dl. HO₂C(CH₂)₃CH(NH₂)CO₂H	161.16	pl (w)		206 (anh)				B4⁴, 1555
530	Adipic acid, dibutylester or di-butyl adipate. C₄H₉O₂C(CH₂)₄CO₂C₄H₉	258.36	....	165¹⁰	-32.4	0.9615²⁰⁴	1.4369²⁰	al, eth	B2⁴, 1961
531	Adipic acid, diisobutylester or Di-isobutyl adipate. i-C₄H₉O₂C(CH₂)₄CO₂-i-C₄H₉	258.36	....	186-8¹⁵·⁷	......	0.9543¹⁹	1.4301²⁰	....	B2⁴, 1962

No.	Name, Synonyms, and Formula	Mol. wt.	Color, crystalline form, specific rotation and λ_{max} (log ε)	b.p. °C	m.p. °C	Density	n_D	Solubility	Ref.
532	Adipic acid, diethyl ester or diethyl adipate .... $C_2H_5O_2C(CH_2)_4CO_2C_2H_5$	202.25	....	245	-19.8	$1.0076^{20/4}$	1.4272^{20}	al, eth	B2[4], 1960
533	Adipic acid, di-2 ethyl butyl ester or di-2-ethyl butyl adipate .... $(C_2H_5)_2CHCH_2OC(CH_2)_4CO_2CH_2CH(C_2H_5)_2$	314.47	....	200^{10}	-15	$0934^{25/4}$	1.4434^{20}	al, ace, aa	B2[4], 1964
534	Adipic acid, di-(2 ethylhexyl) ester or di-(2-ethylhexyl) adipate .... $C_8H_{17}O_2C(CH_2)_4CO_2C_8H_{17}$	370.58	....	214^5	-67.8	$0.922^{25/4}$	1.4474^{20}	al, eth, ac, aa	B2[4], 1964
535	Adipic acid, dimethyl ester or dimethyl adipate .... $CH_3O_2C(CH_2)_4CO_2CH_3$	174.20	cr	115^{13}	10.3	$1.0600^{20/4}$	1.4283^{20}	al, eth, aa	B2[4], 1959
536	Adipic acid, dipropyl ester or dipropyl adipate .... $C_3H_7O_2C(CH_2)_4CO_2C_3H_7$	230.31	....	151^{11}	-15.7	$0.9790^{20/4}$	1.4314^{20}	al, eth, chl	B2[4], 1961
537	Adipic acid, di iso-propyl ester or di iso-propyladipate .... $i\text{-}C_3H_7O_2C(CH_2)_4CO_2\text{-}i\text{-}C_3H_7$	230.3	....	$120^{6.5}$	-1.1	0.9569^{20}	1.4247^{20}	al, eth, ace, aa	B2[4], 1961
538	Adipic acid, monoethyl ester or Ethyladipate .... $HO_2C(CH_2)_4CO_2C_2H_5$	174.20	hyg (eth-peth)	285, 170^{17}	29	$0.9796^{20/4}$	1.4311^{20}	al, eth	B2[4], 1960
539	Adipic acid, mono methyl ester or methyl adipate .... $HO_2C(CH_2)_4CO_2CH_3$	160.17	lf(Me₂N-MeOH)	158^{10}	9	1.0623^{20}	1.4283^{20}	al	B2[3], 1717
540	Adipic acid, 2-methyl .... $HO_2C(CH_2)_3CH(CH_3)CO_2H$	160.17	cr (peth-bz)	209^{13}	93	....	....	w, al, eth, chl	B2[4], 2010
541	Adipic acid, 3-methyl d .... $HO_2C(CH_2CH(CH_3)CH_2CO_2H$	160.17	cr (chl-bz)	230^{30}	94-4.5	....	....	w, al, eth, ace, bz	B2[4], 2014
542	Adipic acid, 3-methyl dl .... $HO_2CCH_2CH_2CH(CH_3)CH_2CO_2H$	160.17	nd (bz), cr (ace-bz)	$190\text{-}200^{12}$	97	....	....	w, al, eth, ace, bz	B2[4], 2014
543	Adipic acid, 2-oxo .... $HO_2C(CH_2)_3COCO_2H$	160.14	cr (al or eth)	....	127	....	....	w, al, ace	B3[4], 1821
544	Adipic acid, 3-oxo .... $HO_2CCH_2CH_2COCH_2CO_2H$	160.14	pl (ace-chl)	....	123	....	....	ace	B3[4], 1822
545	Adiponitrile .... $NC(CH_2)_4CN$	108.15	nd (eth)	295, 180^{20}	1	0.9676^{20}	1.4380^{20}	al, chl	B2[4], 1975
546	Adipoyl chloride or Adipilacid, dichloride .... $ClCO(CH_2)_4COCl$	183.05	....	126^{12}	....	....	....	....	B2[4], 1972
547	Adonitol or Adonite, Ribitol .... $HOCH_2(CHOH)_3CH_2OH$	152.15	pr (w), nd (al)	....	104	....	....	w, al	B1[4], 2832
548	Adrenaline, d or d-epinephrine .... $C_9H_{13}NO_3$	183.21	$[\alpha]^{20/D} +50.5$ (HCl)	....	215d	....	....	aa	B13[3], 2384
549	Adrenaline, l or l-epinephrine .... $C_9H_{13}NO_3$	183.21	br (in air) pw $[\alpha]^{20/D} -53$ (aq. Hcl)	....	211-2	....	....	aa	B13[3], 2384
550	Adrenalone .... $C_9H_{11}NO_3$ $3,4(HO)_2C_6H_3COCH_2NHCH_3$	181.19	nd	....	235-6d	....	....	....	B14[3], 614
551	Adrenochrome dl .... $C_9H_9NO_3$	179.18	red-br rods (MeOH-HCO₂H)	....	125 (anh)	....	....	w, al	B21[4], 6436
552	Ajmalicene or Py-tetrahydro serpentine .... $C_{21}H_{24}N_2O_3$	352.44	pr (MeOH) $[\alpha]^{24/D} -60$ (chl, C=0.5)	....	258-d 261-3(vac)	....	....	MeOH	J76, 1332
553	Ajmaline .... $C_{20}H_{26}N_2O_2$	326.44	pl (+3.5 w, aq AcOEt) $[\alpha]^{20/D} +144$(Chl)	....	158-60 (hyd) 250-7 (anh)	....	....	al, chl, eth	J1954, 1242
554	α-Alanine-D or l-α-aminopropionic acid .... $CH_3CH(NH_2)CO_2H$	89.10	nd (w, al) $[\alpha]^{25/D} -13.6$ (6N HCl,C=1)	sub	314d	....	....	w, al	B4[3], 1219
555	α Alanine-DL or dl-α-aminopropionic acid .... $CH_3CH(NH_2)CO_2H$	89.10	orh pr or nd (w)	sub 258	295-6d	1.424	....	w, al	B4[3], 1222
556	α-Alanine-L or d-α amino propioniuc acid .... $CH_3CH(NH_2)CO_2H$	89.10	rh (w), $[\alpha]^{25/D} +2.8$ (w, C=6)	sub 160-5	314d	1.432^{22}	....	w, al	B4[3], 1208
557	α-Alanine, N-alanyl-l .... $H_2NCH(CH_3)CONHC(CH_3)CO_2H$	160.17	lf $[\alpha]^{20/D} -21.6$ (w)	....	298	....	....	w	B4[3], 1218

No.	Name, Synonyms, and Formula	Mol. wt.	Color, crystalline form, specific rotation and λ_{max} (log ε)	b.p. °C	m.p. °C	Density	n_D	Solubility	Ref.
558	α-Alanine, N-benzoyl-d CH₃CH(NHCOC₆H₅)CO₂H	193.21	pl (w), [α]²⁰/D -2.4 (w,c=l)	152-4	....	....	....	....	B9³, 1141
559	α-Alanine, N-benzoyl-dl CH₃CH(NHCOC₆H₅)CO₂H	193.21	pl or pr, lf (eth)	d	165-6	....	....	w, al	B9³, 1142
560	α-Alanine, N-benzoyl-l CH₃CH(NHCOC₆H₅)CO₂H	193.21	[α]²⁰/D + 24 (w, c=1)	....	151	....	....	....	B9³, 1141
561	α-Alanine, N-(Carboxymethyl)-dl HO₂CCH₂NHCH(CH₃)CO₂H	147.8	cr (dil al or w)	....	222-3	....	....	w	B4³, 1250
562	α-Alanine, N-(4-Chlorophenyl), nitrile CH₃CH [NH(C₆H₄Cl-4)] CN	180.64	lf (eth-peth)	....	114.5	....	....	al, eth, bz, chl	B12, 617
563	**α-Alanine, N,N-diethyl, nitrile** **CH₃CH[N(C₂H₅)₂]CN**	126.20	....	81²⁷	....	0.857¹⁰/⁴	....	w, al, eth	B4³, 1238
564	α-Alanine, 3(3,4 dihydroxyphenyl)-L or l-Dopa 3,4-(HO)₂C₆H₃CH₂CH(NH₂)COOH	197.19	pl (dil al), pr or nd (w + SO₂) [α]¹⁵/D −39.5 (w, p = 1.3)	....	285.5d	....	....	w	B14³, 1629
565	α-Alanine, ethyl ester, hydrochloride CH₃CH(NH₂)CO₂C₂H₅.HCl	153.61	pr or hyg nd (al)	d	87-8	....	....	w, al, eth	B4³, 1231
566	α-Alanine, N-fumaryl-DL or Fumaro alanide HO₂CCH=CHCONHCH(CH₃)CO₂H	187.15	nd	....	229d	....	....	al	B4³, 1248
567	α-Alanine, N- methyl-dl CH₃CH(NHCH₃)CO₂H	103.12	rh pr (aks al)	sub 292	280 (d)	....	....	....	B4³, 1235
568	β Alanine H₂NCH₂CH₂CO₂H	89.09	nd, rh pr (al)	....	207d	1.437¹⁸	....	w	B4³, 1259
569	β Alanine, ethyl ester, hydrochloride H₂NCH₂CH₂CO₂H₅HCl	153.61	....	....	65.3	....	....	w	B4¹, 499
570	β Alanine, N-methyl CH₃NHCH₂CH₂CO₂H	103.12	pl (al, + 1w)	....	146 (anh)	....	....	w, al	B4³, 1264
571	β Alanine, N-methyl, ethyl ester CH₃NHCH₂CH₂CO₂C₂H₅	131.18	....	80²¹	....	1.0082²⁰/²⁰	1.4443²⁰	al, eth	B4³, 1264
572	β Alanine, N-methyl, nitrile or β Methylamino propionitrile CH₃NHCH₂CH₂CN	84.13	....	101-4⁴⁹	....	0.8992²⁰/⁴	1.4320²⁰	w, ace, bz, chl	B4³, 1264
573	β Alanine, N-phenyl C₆H₅NHCH₂CH₂CO₂H	146.19	nd (bz or dil al), lf (dil, al, bz, eth)	....	92	....	....	al, eth, bz	B12³, 933
574	**Aldrin of Octalene** **C₁₂H₈Cl₆**	364.93	....	....	104	....	....	al, eth, ace, bz	B5³, 1385
575	Alizarin or 1,2-dihydroxy anthraquinone 1,2-(HO)₂C₁₄H₆O₂	240.23	og or red tcl nd or pr (al, sub)	430 sub	289-90 (cor)	....	....	al, eth, ace, bz	B8³, 3767
576	Alkanin-l C₁₆H₁₆O₅	288.30	red,br,pr, (bz,sub), nd(eth-al) [α]²⁰/cd -157 (bz)	sub 140⁰·⁰⁰¹	149	....	....	ace, al, eth	B8³, 4089
577	Allantoic acid or Dicarbamoacetic acid (H₂NCONH)₂CHCO₂H	176.14	nd, lf (MeOH)	....	173d	....	....	al	B3⁴, 1492
578	Allantoin or Glyoxyldiureide C₄H₆N₄O₃	158.12	mcl pl or pr (w)	....	238.40	....	....	al, w	B25², 379
579	Allanturic acid or Glyoxalurea C₃H₅N₂O₂	116.08	amor, hyg pw	....	180 turns br	....	....	....	B25², 388
580	Allene or Propadiene CH₂=C=CH₂	40.07	....	-34.5	-136	....	1.4168	bz, peth	B1⁴, 966
581	Allene, tetra fluoro F₂C=C=CF₂	112.03	....	-38	....	....	....	....	B1⁴, 968
582	Allicin or S-oxodiallyl disulfide CH₂=CHCH₂SS(O)CH₂CH=CH₂	162.27	....	d	....	1.112²⁰	1.561²⁰	w	B4⁴, 7
583	D-Allitol or D-Allodulcitol HOCH₂(CHOH)₄CH₂OH	182.18	....	....	150-1	....	....	w	B1⁴, 2839

No.	Name, Synonyms, and Formula	Mol. wt.	Color, crystalline form, specific rotation and λ_{max} (log ε)	b.p. °C	m.p. °C	Density	n_D	Solubility	Ref.
584	Alloluecine-*l* .... CH₃CH₂CH(CH₃)CH(NH₂)COOH	131.18	lf (w) [α] 20/D-14.2 (w,c=2)	....	280-1d	....	....	w	B4³, 1462
585	Allomucic acid or Tetrahydroxy adipic acid............ HOOC(CHOH)₄COOH	210.14	nd or pr (w)	....	198-200(d)	....	....	w	B3³, 1116
586	α-Allonic acid-γ-lactone .... C₆O₆H₁₀	178.14	pr (al) [α] 20/D -6.8 (w,c=11)	....	....	....	....	w, al	B18⁴, 3024
587	Allocimene-A or 2,6-Dimethyl-2,3,6-Octateiene (4-trans, 6-trans)................................. CH₃CH = C(CH₃)CH = CHCH = C(CH₃)₂	136.24	....	188⁷⁵⁰,91²⁰	-35.4	0.8118²⁰/⁴	1.5446²⁰		B1⁴, 1106
588	Alloocimene-B or 2,6-Dimethyl-2,4,6-Octatriene (4-trans-6-cis)................................. CH₃CH=C(CH₃)CH=CHCH=C(CH₃)₂	136.24	....	89²⁰	-20.6	0.8060²⁰	1.5446²⁰		B1⁴, 1106
589	Allophanic acid, ethyl ester .... H₂NCONHCO₂C₂H₅	132.13	nd (w), (bz) d	....	195	....	....		B3⁴, 127
590	Allophanonitrile, 3-phenyl or 1-Cyano-3-phenyl urea .... C₆H₅NHCONHCN	161.17	nd	....	125	....	....	al	B12³, 781
591	D-Allose (β-anomer) .... C₆H₁₂O₆,HOCH₂(CHOH)₄CHO	180.16	cr (w) [α] D 0.58→14.4 (w,c=5) (mut)	....	128	....	....	w	B1⁴, 4299
592	L-Allose................ C₆H₁₂O₆	180.16	pr (dil al) [α] 20/D -0.58 (al)	....	128-9	....	....	w	B1⁴, 4300
593	Alloxan or Mesoxalurea .... C₄H₂N₂O₄	142.07	....	sub (vac)	256 d (anh)	....	....	w, al, ace, bz, aa	B24², 301
594	Alloxanic acid .... C₄H₄N₂O₅	160.09	tcl pr (eth)	....	162-3d	....	....	w, al	B25², 266
595	Alloxantin .... C₈H₄N₄O₈	286.16	rh pr (w + 2)	....	253-5d	....	....		B26², 335
596	Allyl alcohol or 1-Propene-3-ol .... CH₂=CHCH₂OH	58.08	....	97.1	-129	0.8540²⁰/⁴	1.4135²⁰/D	**w, al, eth, chl**	B1⁴, 2079
597	Allyl alcohol, 2-bromo .... CH₂=CBrCH₂OH	136.98	....	153-4⁷⁵⁵, 62¹¹	....	1.621¹⁸	1.500¹⁸	eth, chl	B1⁴, 2094
598	Allyl alcohol, 2-chloro .... CH₂=C(Cl)CH₂OH	92.52	....	136-40, 47¹⁰	....	1.1618²⁰/⁴	1.4588²⁰		B1⁴, 2091
599	Allyl alcohol, 3-chloro .... ClCH=CHCH₂OH	92.53	....	(i) 146⁷⁴⁶, (ii) 153⁷⁵⁶	....	(i) 1.1769²⁰/⁴, (ii) 1729²⁰/⁴	(i) 1.4738²⁰, (ii) 1.4664²⁰		B1⁴, 2090
600	Allyl alcohol, 3-(3,5-dimethoxy-4-hydroxyphenyl) or Sinaphyl alcohol. Syringenin................ [3,5(CH₃O)₂-4-HOC₆H₂] CH=CHCH₂OH	210.23	nd (eth-peth)	....	66-7	....	....	eth	B6³, 6690
601	Allyl alcohol, 3-(4-hydroxy phenyl) or P-coumaryl alcohol 4-HOC₆H₄CCH=CHCH₂OH	150.18	pr (dil al)	....	124	....	....	al, eth, ace, bz	C45, 3359
602	Allyl alcohol, 1-phenyl .... CH₂=CH-CH(C₆H₅)OH	134.18	....	215-6, 111¹⁸	....	1.0251²¹/⁰	1.5406²⁰	al, eth, bz, chl	B6³, 2417
603	Allyl alcohol, 3-phenyl *trans* or Cinnamyl alcohol .... C₆H₅CH=CHCH₂OH	134.18	wh nd (eth-peth)	257.5, 127-8¹⁰	34	1.0440²⁰/⁴	1.5819²⁰	al, eth	B6³, 2401
604	Allylamine or 3-Aminopropene .... H₂NCH₂CH=CH₂	57.09	....	58	....	0.7621²⁰/⁴	1.4205²⁰	**w, al, eth, chl**	B4⁴, 1057
605	Allylamine, N-methyl or Allylmethyl amine.......... CH₂CH₂NHCH₃	71.12	....	65	....	....	1.4065²⁰	w, al, eth, ace	B4⁴, <u>1058</u>
606	Allylamine, N-phenyl or N-allylaniline.............. CH₂CH₂NHC₆H₅	133.19	....	217-9⁷³⁶, l05-8¹²	....	0.982²⁵/⁴	1.563²⁰	al, eth	B12³, 277
607	Allylamine, N-iso-Propyl or Allyl-iso-propyl amine ..... CH₂CH₂NHCH(CH₃)₂	99.18	....	96-7	....	....	1.4140²⁵/	....	B4⁴, 1059
608	Allyl benzene, α-chloro .... C₆H₅CHClCH=CH₂	152.63	....	212-4, 97¹⁸	....	1.073¹⁴/⁴	1.545¹⁴	eth, ace, bz	B5⁴, 1363
609	Allylbromide or 3-Bromopropene .... BrCH₂CH=CH₂	120.98	....	70⁷⁵²	119.4	1.398²⁰/⁴	1.4697²⁰	**al, eth**	B1⁴, 754
610	Allyl chloride or 3-chloropropene .... ClCH₂CH=CH₂	76.53	....	45	134.5	0.9376²⁰/⁴	1.4157²⁰	**al, eth, ace, bz,** lig	B1⁴, 738
611	Allyl ether or Diallyl ether .... (CH₂=CHCH₂)₂O	98.15	....	94	....	0.8260²⁰/⁴	1.4163²⁰	al, eth, ace	B1⁴, 2086

No.	Name, Synonyms, and Formula	Mol. wt.	Color, crystalline form, specific rotation and λ_{max} (log ε)	b.p. °C	m.p. °C	Density	n_D	Solubility	Ref.
612	Allyl 4-chlorophenyl ether 4-ClC$_6$H$_4$OCH$_2$CH=CH$_2$	168.63	....	106-7[12]		1.131[15]	1.5348[25]	al, eth, bz	B6[4], 825
613	Allyl ethyl ether CH$_2$=CHCH$_2$OC$_2$H$_5$	86.14	....	66		0.7651[20/4]	1.3881[20]	al, eth, ace	B1[4], 2083
614	Allyl isocyanide CH$_2$=CHCH$_2$NC	67.09		98		0.794[17/4]	....	al, eth	B4, 208
615	Allyl mercaptan CH$_2$=CHCH$_2$SH	74.15		67-8		0.925[23/4]	1.4832[20]	al, eth, Chl	B1[4], 2095
616	Allyl methyl ether CH$_2$=CHCH$_2$OCH$_3$	72.12		46		0.77[11/11]	1.3778[20]	al, eth, ace	B1[4], 2083
617	Allyl isopentyl ether or Allyl-isoamyl-ether CH$_2$=CHCH$_2$OCH$_2$CH$_2$CH(CH$_3$)$_2$	128.22		120-2				al, eth	B1[2], 477
618	Allyl phenyl ether CH$_2$=CHCH$_2$OC$_6$H$_5$	134.18		191.7, 74[5]		0.9811[20/4]	1.5223[20]	al, eth	B6[4], 562
619	Allyl propyl ether CH$_2$=CHCH$_2$OC$_3$H$_7$	100.16		90-2		0.7764[20]	1.3919[20]	al, eth, ace	B1[4], 2083
620	Allyl isopropyl ether CH$_2$=CHCH$_2$OCH(CH$_3$)$_2$	100.16		83-4		0.7764[20]	1.3946[20]	al, eth, ace	B1[4], 2083
621	Allyl 2-tolyl ether (2-CH$_3$C$_6$H$_4$)OCH$_2$CH=CH$_2$	148.21		205-8, 85[12]		0.9698[15/4]	1.5188[15]	....	B6[4], 1946
622	Allyl 3-tolyl ether (3-CH$_3$C$_6$H$_4$)OCH$_2$CH=CH$_2$	148.21		211-4, 93.5[13]		0.9564[20]	1.5179[20]	bz	B6[4], 2041
623	Allyl 4-tolyl ether (4-CH$_3$C$_6$H$_4$)OCH$_2$CH=CH$_2$	148.21		214.5, 97.8[16]		0.9720[15/15]	1.5137	bz	B6[4], 2101
624	Allyl vinyl ether CH$_2$=CHCH$_2$-O-CH=CH$_2$	84.12		65[740]		0.8050[20/4]	1.4062[20]	eth, ace	B1[4], 2085
625	Allyl isothiocyanate or Allyl mustard oil CH$_2$=CHCH$_2$NCS	99.16		151, 44[12]	80	1.0126[30/11]	1.5300[11]	al, eth, bz	B4[4], 1081
626	Allyl sulfide (CH$_2$=CHCH$_2$)$_2$S	114.21		139[768], 35[6]	-85	0.8877[27/4]	1.4870[25]	al, eth	B1[4], 2097
627	Allyl sulfoxide (CH$_2$=CHCH$_2$)$_2$SO	130.20		112-5[12]	23.5	1.0261[20/4]	1.5115[20]	al, eth, ace	B1[4], 2097
628	Allyl thio cyanate CH$_2$=CHCH$_2$SCN	99.16		161		1.056[15]	....	al, eth	B3[4], 332
629	Allyl (2,4,6-tribromophenyl) ether (2,4,6-Br$_3$C$_6$H$_2$)OCH$_2$CH=CH$_2$	370.88			33-4				B6[2], 194
630	Allyl trisulfide (CH$_2$=CHCH$_2$)$_2$S$_3$	178.34		112-22[16]		1.0845[15]		eth	B1, 441
631	Aloetic Acid C$_{15}$H$_4$N$_4$O$_{11}$	450.24	og-ye nd (aa), ye cr (w + l)		285d				B8, 525
632	Aloin or Barbaloin. C$_{21}$H$_{22}$O$_9$	386.36	ye nd (al), [α]$_D$ -8.3 (dil al)		148.5			w, al, ace	B18[4], 3630
633	Alstonine C$_{21}$H$_{20}$N$_2$O$_3$	348.41	ye nd (ace)		205-10d				B27[2], 824
634	D-Altrose (β-anomer). C$_6$H$_{12}$O$_6$	180.16	pr (MeOH-al), [α]$^{20}_D$ +11.7 →33.1 (w, c = 7) (mut)		103.5			w	B1[4], 4300
635	L-Altrose (β-anomer). C$_6$H$_{12}$O$_6$	180.16	pr (al, aa) -28.55→ -32.30 (w) (mut)		107-9.5				B1[4], 4301
636	Amalic acid or Tetramethyl alloxantin C$_{12}$H$_{14}$N$_4$O$_8$	324.25	cr (w)		245d				B26[2], 336
637	Amarine C$_{21}$H$_{18}$N$_2$	298.39	pr (eth, bz-lig)	198d	136			al, eth, bz	B23[2], 274
638	Amarine hydrate C$_{21}$H$_{18}$N$_2$ · ½H$_2$O	307.40	pr (aq al + ½ w)		106			w	B23[2], 274
639	Amaron or Benzoin-imide C$_{28}$H$_{20}$N$_2$	384.49	tcl nd or pr (ace, al), nd (aa)	sub	2-46			ace, bz, chl	B23[2], 304
640	Aminoacetamide or Glycinamide H$_2$NCH$_2$CONH$_2$	74.08	hyg nd (chl)		67-8			w, al, ace	B4[2], 1118

No.	Name, Synonyms, and Formula	Mol. wt.	Color, crystalline form, specific rotation and λ_{max} (log ε)	b.p. °C	m.p. °C	Density	n_D	Solubility	Ref.
641	Aminoacetamide, N'-(4-ethoxyphenyl) or Glycine-ǫ-phenetidide. . . . . H₂NCH₂CONH-C₆H₄-OC₂H₅-4)	194.26	nd (+ lw)	. . . .	100.5 (anh)			al, eth	B13¹, 179
642	Aminoacetamide Hydrochloride . . . . H₂NCH₂CONH₂·HCl	110.54	nd (al)	. . . .	203-4			w, al	B4³, 1119
643	Aminoacetamide, N-phenyl . . . . C₆H₅NHCH₂CONH₂	151.17		. . . .	127-8			w, al	B12³, 915
644	Aminoacetic acid or Glycine. . . . . H₂NCH₂CO₂H	75.07	mcl or trg pr (dil al)		262d	1.607		w	B4³, 1097
645	Aminoacetic acid, N-acetyl or Aceturic acid . . . . CH₃CONHCH₂CO₂H	117.11	lo nd (w, MeOH)		206			al, ace, w	B4³, 1150
646	Aminoacetic acid, N-acetyl-N-phenyl or N-phenyl aceturic acid . . . . CH₃CON(C₆H₅)CH₂CO₂H	193.25	lf (w)		194			w	B12³, 919
647	Aminoacetic acid, N-(4 aminophenyl) hydrate . . . . (4-H₂N-C₆H₄)NHCH₂CO₂H·H₂0	184.20	pl (dil aa)		222-3d			w	B13¹, 34
648	Aminoacetic acid, N-benzoyl or Hippuric acid . . . . . . C₆H₅CONHCH₂COOH	179.18	pr (w or al)		190-3	1.371²⁰ᐟ⁴		w, al	B9³, 1123
649	Aminoacetic acid, N-benzyl or N-benzyl glycine . . . . C₆H₅CH₂NHCH₂CO₂H	165.19	nd (w)		198-9				B12³, 2283
650	Aminoacetic acid, N-benzyl, ethyl ester or Ethyl-N-benzyl glycinate. . . . . C₆H₅CH₂NHCH₂CO₂C₂H₅	193.25		175-9⁵⁰			1.5041²⁰		B12³, 2283
651	Aminoacetic acid, N-bromoacetyl-N-phenyl BrCH₂CON(C₆H₅)CH₂COOH	272.11	pl (w)	. . . .	153d			al, bz	B12, 477
652	Aminoacetic acid, N-(2-carboxyphenyl) or Anthranilidoacetic acid . . . . (2-HO₂CC₆H₄)NHCH₂CO₂H	195.18	nd (MeOH)	. . . .	218-20			al, eth	B14³, 938
653	Aminoacetic acid, N-chloroacetyl-N-phenyl . . . . ClCH₂CON(C₆H₅)CH₂CO₂H	227.65	pl or pr (bz)	. . . .	132-3			al, bz	B12, 476
654	Aminoacetic acid, N-chloroacetyl-N-phenyl, methyl ester ClCH₂CON(C₆H₅)CH₂CO₂CH₃	241.68	pr (lig)		59-60			al, eth, bz	B12, 477
655	Aminoacetic acid, N-ethyl of Ethyl aminoacetic acid . . . . . C₂H₅NHCH₂CO₂H	103.12	pl (al)	. . . .	181-2d			w, al	B4³, 1134
656	Aminoacetic acid, ethyl ester or Ethyl glycinate. . . . . H₂NCH₂CO₂C₂H₅	103.12		148-9⁷⁵⁰, 57-8¹⁸	. . . .	1.0275¹⁰ᐟ⁴	1.4242¹⁰	w, al, eth, ace bz, lig	B4³, 1116
657	Aminoacetic acid, ethyl ester, hydrochloride or Ethyl glycinate hydrochloride . . . . H₂NCH₂CO₂C₂H₅·HCl	139.58	nd (al)	sub	144			w, al	B4³, 1117
658	Aminoacetic acid, N,N-(dicarbethoxy),ethyl ester . . . . (C₂H₅O₂C)₂NCH₂CO₂C₂H₅	247.25	pr (peth)	152-3¹⁰	36.5			al, eth, bz	B4, 365
659	Aminoacetic acid, N,N-dimethyl . . . . (CH₃)₂NCH₂CO₂H	103.12	hyg nd (PrOH)		185-6			w, al, eth	B4³, 1124
660	Aminoacetic acid hydrazide . . . . H₂NCH₂CONHNH₂	89.10	. . . .	dec 150	80.5			chl	B4³, 1121
661	Aminoacetic acetic acid, hydrochloride or Glycine hydrochloride . . . . H₂NCH₂CO₂H·HCl	111.55	hyg rh nd (w)	. . . .	200-1			w	B4³, 1111
662	Aminoacetic acid, N,N-bis(2-hydroxyethyl) . . . . (HOCH₂CH₂)₂NCH₂CO₂H	163.18	nd (al)		193-5d			w	B4³, 1145
663	Aminoacetic acid, N-(4-hydroxyphenyl) . . . . (4-HOC₆H₄)NHCH₂CO₂H	167.17	pl (w)		245-7d				B13³, 1132
665	Aminoacetic acid, N-methyl,hydrochloride . . . . CH₃NHCH₂CO₂H · HCl	125.55			166-8d			w	B4, 345
666	Aminoacetic acid, methyl ester or Methyl glycinate, Methylamine acetate . . . . NH₂CH₂CO₂CH₃	89.10		130d, 45²⁰				al	B4³, 1115
667	Aminoacetic acid, N,N-methylene di or N,N-methylene diamino acetic acid . . . . CH₂(NHCH₂CO₂H)₂	162.15	pl	. . . .	199d			al	B4³, 1148
668	Aminoacetic acid, N-(2-naphthyl) 2-C₁₀H₇-NHCH₂COOH	201.23	(w)		134-5			al, ace, w, aa eth	B12, 1298
669	Aminoacetic acid, N-(2-nitro phenyl) . . . . (2-O₂N-C₆H₄)-NH-CH₂CO₂H	196.17	dk red pr (al)		192-3d			al	B12, 695
670	Aminoacetic acid, N-phenyl or Anilino-acetic acid . . . . C₆H₅NHCH₂CO₂H	151.17			127-8			w, al	B12³, 914

No.	Name, Synonyms, and Formula	Mol. wt.	Color, crystalline form, specific rotation and λ_max (log ε)	b.p. °C	m.p. °C	Density	n_D	Solubility	Ref.
671	Aminoacetic acid, *N*-phenyl,ethyl ester or Ethyl-*N*-phenyl amino acetate............ C$_6$H$_5$NHCH$_2$CO$_2$C$_2$H$_5$	179.22	lf (dil al)	273-4, 163[18]	58			al, eth	B12[3], 915
672	Aminoacetic acid, *N*-phenyl,methyl ester or Methyl-*N*-phenyl amino acetate............ C$_6$H$_5$NHCH$_2$CO$_2$CH$_3$	165.19	nd (al)		48			al, eth	B12[1], 263
673	Amino acetic acid, *N*-phthaloyl or Phthalimide-*N*-acetic acid............ C$_{10}$H$_7$NO$_4$	205.17	nd or pr (w or al)		193			al, eth	B21[4], 5176
674	Aminoacetic acid, *N*-phthaloyl,ethyl ester or Ethyl-*N*-phthaloyl amino acetate............ o-C$_6$H$_4$(CO)$_2$N-CH$_2$CO$_2$C$_2$H$_5$	233.23	nd (w, al or eth)	300	112-3			al, eth, bz, chl	B21[4], 5177
675	Aminoacetic acid, *N*-succinyl,ethyl ester or Ethyl succinimide-*N*-acetate............ [C$_2$H$_4$(CO)$_2$N]-CH$_2$CO$_2$C$_2$H$_5$	185.18	nd (eth)	198[32]	67			w, al	B21[2], 305
676	Amino acetonitrile............ H$_2$NCH$_2$CN	56.07		58[15]				al	B4[3], 1120
677	Amino acetonitrile-*N*,*N*-diethyl............ (C$_2$H$_5$)$_2$NCH$_2$CN	112.18		170, 70[24]			1.4260[20]	w, al	B4[3], 1136
678	Amino acetonitrile-*N*,*N*-dimethyl............ (CH$_3$)$_2$NCH$_2$CN	84.13		137-8, 42[22]		0.8650[20/0]	1.4095[20]	w, al	B4[3], 1126
679	Amino acetonitrile-*N*-ethyl............ C$_2$H$_5$NHCH$_2$CN	84.13		166-7, 81-3[29]				al, eth	B4[2], 787
680	Aminoacetonitrile-*N*-phenyl............ C$_6$H$_5$NHCH$_2$CN	132.18	pl (lig-eth)		48			al, bz	B12[3], 916
681	Aminoacetyl chloride-*N*-phthaloyl or *O*-phthalinido-*N*-acetyl chloride............ [o-C$_6$H$_4$(CO)$_2$N]CH$_2$COCl	223.62	nd (lig)		84-5			al, bz	B21[4], 5180
682	2-Aminobenzoic acid or Anthanilic acid............ 2-H$_2$NC$_6$H$_4$CO$_2$H	137.14	lf (al)	sub	146-7	1.412[20]		w, al, eth	B14[3], 879
683	3-Aminobenzamide............ 3-H$_2$NC$_6$H$_4$CONH$_2$	136.17	ye mcl nd (+lw), nd (bz)		79—80 (hyd), 113—4 (anh)			w, al, eth	B14[3], 998
684	3-Aminobenzoic acid-*N*-acetyl............ 3-CH$_3$CONHC$_6$H$_4$CO$_2$H	179.18	nd (al)		248-50			al	B14[3], 1006
685	3-Amino benzoic acid-*N*-acetyl-6-ethoxy............ 3-(CH$_3$CONH)-6-(C$_2$H$_5$O)C$_6$H$_3$CO$_2$H	223.23	nd (w)		190			al	B14, 583
686	3-Aminobenzoic acid-6-chloro............ 3-H$_2$N-6-Cl-C$_6$H$_3$CO$_2$H	171.58			188	1.519[15]		al	B14[3], 1017
687	3-Aminobenzoic acid-*N*,*N*-dimethyl............ 3-(CH$_3$)$_2$NC$_6$H$_4$CO$_2$H	165.19	nd (w)		151			w, al, eth	B14, 392
688	3-Aminobenzoic acid-*N*-ethyl............ 3-(C$_2$H$_5$NH)C$_6$H$_4$CO$_2$H	165.19	nd or pr (dil al)		154			al, eth, ace	B14, 393
689	3-Amino benzoic acid, ethyl ester or Ethyl 3-Amino benzoate............ 3-H$_2$NC$_6$H$_4$CO$_2$C$_2$H$_5$	165.19		294, 160-l[5]		1.1248[22/4]	1.5600[22]	al, eth	B14[3], 993
690	3-Amino benzoic acid, 2-hydroxy or 3-Amino salicylic acid............ 3-H$_2$N-2-HOC$_6$H$_3$CO$_2$H	153.14			235d				B14[3], 1434
691	3-Aminobenzoic acid, 4-hydroxy............ 3-H$_2$N-4-HOC$_6$H$_3$CO$_2$H	153.14	pr (w+l)		210 (anh)			w	B14[2], 360
692	3-Aminobenzoic acid, 4-hydroxy,methyl ester............ 3-H$_2$N-4-HOC$_6$H$_3$CO$_2$CH$_3$	167.17	(i) nd (bz or aa) (st) (ii) nd (chl) (unst)		(i)143, (ii)111			al, eth	B14[3], 1477
693	3-Aminobenzoic acid, methyl ester or Methyl 3-amino benzoate............ 3-H$_2$NC$_6$H$_4$CO$_2$CH$_3$	151.16		152-3[11]	39	1.232[20]		eth, al, bz, lig	B14[3], 993
694	3-Amino benzoic acid, *N*-methyl............ 3-CH$_3$NHC$_6$H$_4$CO$_2$H	151.16	pl (peth)		127			al, ace, bz, chl	B14[1], 559
695	3-Amino benzoic acid, *N*-methyl,methyl ester............ 3-CH$_3$NHC$_6$H$_4$CO$_2$CH$_3$	165.19	cr (al)		72			al, eth	B14, 392
696	3-Amino benzoic acid, 4-methyl............ 3-H$_2$N-4-CH$_3$C$_6$H$_3$CO$_2$H	151.17	nd (al)		164-6			w, al	B14[2], 291
697	3-Aminobenzoic acid, 6-methyl............ 3-H$_2$N-6-CH$_3$C$_6$H$_3$CO$_2$H	151.17	pr (w)		196			w, al	B14[2], 290

No.	Name, Synonyms, and Formula	Mol. wt.	Color, crystalline form, specific rotation and λ_{max} (log ϵ)	b.p. °C	m.p. °C	Density	n_D	Solubility	Ref.
698	3-Aminobenzoic acid, 2-nitro 3-H$_2$N-2-NO$_2$C$_6$H$_3$CO$_2$H	182.14	ye nd (w, dil al)		156-7			al, eth, ace	B14, 414
699	3-Aminobenzoic acid, 4-nitro 3-H$_2$N-4-NO$_2$C$_6$H$_3$CO$_2$H	182.14	red pl or nd (al)		298d			al, eth, ace	B14, 415
700	3-Amino benzoic acid, 5-nitro 3-H$_2$N-5-NO$_2$C$_6$H$_3$CO$_2$H	182.14	ye pr (w)		209-10			al	B14³, 565
701	3-Aminobenzoic acid, 6-nitro 3-H$_2$N-6-NO$_2$C$_6$H$_3$CO$_2$H	182.14	ye nd or pr (w)		235d			al, ace	B14³, 1021
702	3-Aminobenzoic acid, 2,4,6-tribromo 3-H$_2$N-2,4,6-Br$_3$C$_6$HCO$_2$H	373.84	nd (w)		171.5-3			al	B14³, 1019
703	3-Aminobenzomitrile or m-Cyanoaniline 3-H$_2$NC$_6$H$_4$CN	118.14	nd (dil al or CCl$_4$)	286-96	53-4			al, eth, ace	B14³, 1001
704	3-Aminobenzonitrile, 2-methyl 3-NH$_2$-2-CH$_3$C$_6$H$_3$CN	132.18	rh cr (al)		90			w	B1⁴, 477
705	3-Amino benzonitrile, 4-methyl 3NH$_2$-4-CH$_3$C$_6$H$_3$N	132.18	pr (al)		81-2			al, eth, ace, bz	B14, 487
706	3-Aminobenzo nitrile, 5-methyl 3-H$_2$N-5-CH$_3$C$_6$H$_3$CN	132.18	nd (lig)		75			al	B14¹, 600
707	3-Aminobenzonitrile, 6-methyl 3-H$_2$N-6-CH$_3$C$_6$H$_3$CN	132.17	nd (peth)	100-10²²	88			al	B14², 290
708	4-Aminobenzamide 4-H$_2$NC$_6$H$_4$CONH$_2$	136.17	ye cr (+ ¹/$_4$ w)		183			al, eth	B14³, 1061
709	4-Aminobenzamide, N(2-diethyl aminoehtyl) hydrochloride or Procainamide, hydrochloride 4-H$_2$NC$_6$H$_4$CONH[CH$_2$CH$_2$N(C$_2$H$_5$)]$_2$ · HCl	271.79	cr		165-9			w, al	C54, 428
710	4-Amino benzamide, N-phenyl 4-H$_2$NC$_6$H$_4$CONHC$_6$H$_5$	212.25			135-6			al	B14³, 1062
711	4-Aminobenzoic acid 4-H$_2$NC$_6$H$_4$CO$_2$H	137.14	mcl pr (w)		188-9	1.374²⁰/⁴		al, eth, w	B14³, 1023
712	4-Aminobenzoic acid, N-acetyl 4-(CH$_3$CONH)C$_6$H$_4$CO$_2$H	179.18	nd (aa)		256.5			al	B14³, 1112
713	4-Aminobenzoic acid, butyl ester or Butesin 4-H$_2$NC$_6$H$_4$CO$_2$C$_4$H$_9$	193.25	(al or bz)	173-4⁸	58			al, eth, bz, chl	B14³, 1027
714	4-Aminobenzoic acid, iso-butyl ester 4-H$_2$NC$_6$H$_4$CO$_2$-i-C$_4$H$_9$	193.24			64.5				B14³, 1027
715	4-Aminobenzoic acid, 2-chloro or 4-Amino-2-chlorobenzoic acid 4-H$_2$N-2-ClC$_6$H$_3$CO$_2$H	171.58			dec 213			al	B14³, 1153
716	4-Aminobenzoic acid, 3,5-dichloro 4-H$_2$N-3,5-Cl$_2$C$_6$H$_2$CO$_2$H	206.03	(al)		291			al, eth, bz	B14³, 1156
717	4-Aminobenzoic acid, (2-diethylaminoethyl) ester or Novocaine 4-H$_2$NC$_6$H$_4$CO$_2$[CH$_2$CH$_2$N(C$_2$H$_5$)$_2$]	236.31	nd (w + 2), pl (lig or eth)		51, 61 (anh)			al, eth, bz, chl	B14³, 1037
718	4-Aminobenzoic acid, 2-(diethylaminoethyl) ester, hydrochloride or Novocaine hydrochloride 4-H$_2$N-C$_6$H$_4$CO$_2$[CH$_2$CH$_2$N(C$_2$H$_5$)$_2$] · HCl	272.27	nd (al), mcl or tcl pl (w)		156	0.707¹⁷		w, al	B14³, 1037
719	4-Aminobenzoic acid, 3,5-diiodo 4-H$_2$N-3,5-I$_2$C$_6$H$_2$CO$_2$H	388.93	nd (aa-NH$_3$)		> 350				B14³, 1161
720	4-Aminobenzoic acid, 3,5-diido, ethyl ester 4-H$_2$N-3,5-I$_2$C$_6$H$_2$CO$_2$C$_2$H$_5$	406.99	nd (al)		148				B14, 439
721	4-Aminobenzoic acid, N,N-dimethyl or 4-(Dimethyl amino) benzoic acid 4-(CH$_3$)$_2$NC$_6$H$_4$CO$_2$H	165.19	nd (al)		242.5—3.5			al	B14³, 1082
722	4-Aminobenzoic acid, ethyl ester or Benzocaine 4-H$_2$NC$_6$H$_4$CO$_2$C$_2$H$_5$	165.19	nd (w), rh (eth)	310	92			al, eth, chl	B14³, 1025
723	4-Aminobenzoic acid, N-ethyl 4-(C$_2$H$_5$NH)C$_6$H$_4$CO$_2$H	165.19	cr (bz)		177-8			al, eth, ace, bz	B14³, 1086
724	4-Aminobenzoic acid, 2-hydroxy or 4-Aminosalicylic acid 4-H$_2$N-2-HOC$_6$H$_3$CO$_2$H	153.14	nd pl (al-eth)		dec 150-1			w, al, eth, ace	B14³, 1436
725	4-Aminobenzoic acid, 2-hydroxy, methyl ester 2-HO-4-H$_2$NC$_6$H$_3$CO$_2$CH$_3$	167.16			120-1				B14³, 1437
726	4-Aminobenzoic acid, methyl ester or Methyl-4-amino-benzoate 4-H$_2$NC$_6$H$_4$CO$_2$CH$_3$	151.16	lf or nd (aq MeOH)		114			Chl	B14³, 1025
727	4-Aminobenzoic acid, N-methyl 4-CH$_3$NHC$_6$H$_4$CO$_2$H	151.16	nd (bz w or dil al)		168 (159)			al, eth, w, bz	B14³, 1081
728	4-Aminobenzoic acid, N-methyl, methyl ester 4-CH$_3$NHC$_6$H$_4$CO$_2$CH$_3$	165.19	pl (dil al or lig)		95.5			al, eth	B14³, 1081

No.	Name, Synonyms, and Formula	Mol. wt.	Color, crystalline form, specific rotation and λ_{max} (log ε)	b.p. °C	m.p. °C	Density	n_D	Solubility	Ref.
729	4-Aminobenzoic acid, 2-methyl 4-H_2N-2-$CH_3C_6H_3CO_2$H	151.16	nd (al)	....	dec 165	....	....	al	B14³,1200
730	**4-Aminobenzoic acid, 3-methyl** .... **4-H_2N-3-$CH_3C_6H_3CO_2$H**	151.16	nd (w)	....	170			w	B14², 290
731	4-Aminobenzoic acid, 2-nitro 4-H_2N-2-$NO_2C_6H_3CO_2$H	182.14	rd nd (w), pr (dil aa)		239d			al, aa	B14³, 1161
732	4-Aminobenzoic acid, 3-nitro 4-H_2N-3-$NO_2C_6H_3CO_2$H	182.14	rd-ye nd (al)		284d			ace, aa	B14³, 1161
733	4-Aminobenzoic acid, phenyl ester 4-$H_2NC_6H_4CO_2C_6H_5$	213.24	nd (al)		173			al, eth	B14¹, 568
734	4-Aminobenzoic acid, propyl ester or Propaesin .... 4-$H_2NC_6H_4CO_2C_3H_7$	179.22	pr	....	75			al, eth, bz, chl	B14³, 1026
735	4-Aminobenzonitrile or p-Cyanoaniline .... 4-$H_2NC_6H_4$CN	118.14	pr or pl (w)		86			al, eth, ace, bz, aa	B14³, 1079
736	4-Aminobenzonitrile, 2-methyl 2-CH_3-4-$H_2NC_6H_3$CN	132.18	rh cr (al)		90			al	B14¹, 598
737	4-Aminobenzonitrile, 3-methyl .... 4-H_2N-3-$CH_3C_6H_3$CN	132.18	nd (w)		95				B14¹, 598
738	Ammelide or Cyanuromonoamide .... $C_3H_4N_4O_2$	128.09	mcl pr (w)		d				B26², 132
739	Ammeline or Cyanurodiamide............... $C_3H_5N_5O$	127.11	nd (aq Na_2 CO_3)		d				B26², 132
740	Amphetamine (d) or Dexedrine, β-phenyl-iso-propyl amine .... $C_6H_5CH_2CH(NH_2)CH_3$	135.21	$[\alpha]^{12}_D$ + 37.6	203-4, 80¹⁰	......	0.949¹⁵/⁴	1.4704²⁰	al, eth	B12³, 2664
741	Amphetamine (dl) or Benzedrine............... $C_6H_5CH_2CH(NH_2)CH_3$	135.21	.....	203, 97²⁰		0.9306⁷⁵/⁴	1.510¹⁸	al, lll	B12³, 2664
742	**Amphetamine, sulfate (d) or Dexedrine sulfate** .... [$C_6H_5CH_2CH(NH_2)CH_3]_2 \cdot H_2SO_4$	368.50	$[\alpha]^{20}_D$ + 22 (w, c=8)		> 300	1.15²⁵/⁴		w	B12³, 2665
743	**Amphetamine, sulfate (dl) or Benzedrine sulfate** .... [$C_6H_5CH_2CH(NH_2)CH_3]_2 \cdot H_2SO_4$	368.50			280—1	1.15²⁵/⁴		w	B12³, 2665
744	Amygdalin or Mandelonitrile-β-gentiobioside .... $C_{20}H_{27}NO_{11}$	457.44			223—6			w	B17⁴, 3614
745	Amygdalin trihydrate................. $C_{20}H_{27}NO_{11} \cdot 3H_2O$	511.48	rh (w + 3) $[\alpha]^{20}_D$ −40 (w, c=1)		200 (re-melts 125—30)			w, al	B31, 400
746	α-Amyrin or α-Amyrenol .... $C_{30}H_{50}O$	426.74	nd (al), $[\alpha]^{17}_D$ + 91.6 (bz, c = 1.3) + 83.5 (chl)	243⁰·⁵	186			al, eth, bz, aa	B6³, 2889
747	β-Amyrin or β-Amyrenol .... $C_{30}H_{50}O$	426.74	nd (lig or al), $[\alpha]^{19}_D$ + 99.3 (bz, c = 1.3) + 88.4 (chl)	260⁰·⁵	197				B6³, 2894
748	Anabasine (l) or l-2-(3-Pyridyl)piperidine............... $C_{10}H_{14}N_2$	162.24	$[\alpha]^{20}_D$ −83.1	276, 105²	9	1.0455²⁰/²⁰	1.5430²⁰	w, al, eth, bz	B23², 113
749	Anagyrine or Monolupine .... $C_{15}H_{20}N_2O$	244.34	pa ye glass, $[\alpha]^{25}_D$ −168 (al, c = 4.8)	260—70¹²				al, w, eth, bz	B24², 84
750	**Analgen or 5-benxamide-8-ethoxyquinoline** .... $C_{18}O_2N_2H_{16}$	292.34	ye nd (al)	....	206				B22, 503
751	Anatabine l or 2-(3-Pyridyl)-1,2,3,6-tetrahydropyridine .. $C_{10}H_{12}N_2$	160.22	$[\alpha]^{17}_D$− 177.8	145-6¹⁰	......	1.091¹⁹/⁴	1.5676²⁰	w, al, eth, bz	C31, 3055
752	Androstane or Etiocholane............... $C_{19}H_{32}$	260.47	lf (ace-MeOH), or (hl-ace)	75-80⁰·⁰¹(sub)	50-2			al, eth, ace, chl, lig	B5⁴, 1211
753	Androsterone or 3-α-hydroxy-17-keto androstane .... $C_{19}H_{30}O_2$	290.45	lf or nd (al, ace), $[\alpha]^{20}_D$ + 94.6 (abs al, c=0.71)		185 (cor)			al,eth, ace, bz	B7³, 586

No.	Name, Synonyms, and Formula	Mol. wt.	Color, crystalline form, specific rotation and λ_{max} (log ε)	b.p. °C	m.p. °C	Density	n_D	Solubility	Ref.
754	Anemonin or Anemone camphor .... $C_{10}H_8O_4$	192.18	rh pl (chl), nd (al or bz)	....	158	....	....	chl	B19[4], 1986
755	Anethole or 43Propenyl anisole..................... 4-$CH_3OC_6H_4CH=CHCH_3$	148.21	lf (al)	234.5[763], 115[12]	21.35	0.9882[20/4]	1.5615[20]	al, eth, ace, bz	B6[3], 2395
756	Angelic acid or cis-2-Methyl-2-butenoic acid .... $CH_3CH=C(CH_3)CO_2H$	100.13	mcl pr or nd	185, 88-9[10]	45-6	0.9834[49/4]	1.4434[47]	eth, al	B2[4], 1551
757	Anhalamine .... $C_{11}H_{15}NO_3$	209.25	nd (al)	....	187-8	....	....	al, ace	B21[4], 2521
758	Anhalonidine .... $C_{12}H_{17}NO_3$	223.28	oct (bz, eth)	....	160-1	....	....	w, al	B21[4], 2524
759	Anhalonine d .... $C_{12}H_{15}NO_3$	221.26	[α]$^{25}/_D$ + 56.7 (chl)	140[0.02]	84.5-5	....	....	w, al, eth, peth	B27[2], 542
760	Anhalonine dl .... $C_{12}H_{15}NO_3$	221.26	rh nd (peth)	....	85-5.5	....	....	al, w, eth, peth	B27[2], 542
761	Anhalonine l .... $C_{12}H_{15}NO_3$	221.26	nd (peth), [α]$^{25}_D$ − 56.3 (chl, c=4)	140[0.02]	....	....	....	w, al, eth, peth	B27[2], 542
762	Anhalonine-hydrochloride l .... $C_{12}H_{15}NO_3HCl$	257.22	rh pr, [α]$^{25}_D$ − 40.5 (al)	....	254-5d	....	....	al, w	B27[2], 542
763	Anhydroecgonine dl or Ecgonidine .... $C_9H_{13}NO_2$	167.20	cr (MeOH, MeOH-eth)	....	226-30d	....	....	w	B22[4], 284
764	Anhydroecgonine hydrochloride or Ecgonidine hydrochloride $C_9H_{13}NO_2 \cdot HCl$	203.66	rh nd (al)	....	240-1	....	....	al, eth, w	B22[4], 284
765	Aniline or Phenylamine .... $C_6H_5NH_2$	93.13	....	184, 68.3[10]	−6.3	1.02173[20/4]	1.5863[20]	al, eth, ace, bz lig	B12[3], 217
766	Aniline, benzene sulfonate .... $C_6H_5NH_2 \cdot C_6H_5SO_3H$	251.31	nd	....	240d	....	....	w, al	B12[2], 75
767	Aniline, m-benzyl or (3-aminophenyl)phenyl methane .... 3-$H_2NC_6H_4CH_2C_6H_5$	183.26	cr (lig)	....	46	....	....	lig	B12[3], 3215
768	Aniline, p-benzyl or (4-aminophenyl)phenyl methane .... 4-$(C_6H_5CH_2)C_6H_4NH_2$	183.26	mcl (lig)	300	34-5	1.038[25]	....	al, eth, lig	B12[3], 3215
769	Aniline, N-benzylidene or Benzalaniline .... $C_6H_5N=CHC_6H_5$	181.24	pa ye nd (CS₂), pl (dil al)	310	54	1.038[55/4]	1.600[100]	al, eth	B12[3], 319
770	Aniline, 2-bromo .... 2-$BrC_6H_4NH_2$	172.03	....	229, 110.5[19]	32	1.578[20/4]	1.6113[20]	al, eth	B12[3], 1415
771	Aniline, 2-bromo-4,6-dichloro .... 2-Br-4,6-$Cl_2C_6H_2NH_2$	240.92	nd (al)	273	83.5	....	....	al, bz, chl	B12[2], 335
772	Aniline, 2-bromo-3,5-dinitro .... 2-Br-3,5-$(NO_2)_2C_6H_2NH_2$	262.03	ye lf (al)	....	181	....	....	al	B12[3], 1728
773	Aniline, 2-bromo-4,5-dinitro .... 2-Br-4,5-$(NO_2)_2C_6H_2NH_2$	262.03	pa ye (al)	....	186	....	....	al	B12, 762
774	Amiline, 2-bromo-4,6-dinitro .... 2-Br-4,6-$(NO_2)_2C_6H_2NH_2$	262.03	ye nd (aa or al)	sub	153-4	....	....	w, ace	B12[3], 1724
775	Aniline, 2-bromo-4-nitro .... 2-Br-4-$(NO_2)C_6H_3NH_2$	217.03	ye nd (al)	....	104.5	....	....	al, aa	B12[2], 403
776	Aniline, 2-bromo-5-nitro .... 2-Br-5-$(NO_2)C_6H_3NH_2$	217.03	pa ye nd (al)	....	141	....	....	w, al, eth	B12[3], 1674
777	Aniline, 3-bromo .... 3-$BrC_6H_4NH_2$	172.03	....	251,130[12]	18.5	1.5793[20.4/4]	1.6260[20.4]	al, eth	B12[3], 1421
778	Aniline, 3-bromo-4-nitro .... 3-Br-4-$(NO_2)C_6H_3NH_2$	217.03	ye nd (al)	....	175-6	....	....	al	B12[3], 1676
779	Aniline, 4-bromo .... 4-$BrC_6H_4NH_2$	172.03	rh bip yrm nd (60 % al)	d	66.4	1.4970[100/4]	....	al, eth	B12[3], 1429
780	Aniline, 4-bromo-2,3-dichloro .... 4-Br-2,3-$Cl_2C_6H_2NH_2$	240.92	nd	....	77.5	....	....	al, eth, bz	B12, 653
781	Aniline, 4-bromo-2,5-dichloro .... 4-Br-2,5-$Cl_2C_6H_2NH_2$	240.92	nd	....	91	....	....	al, eth	B12[3], 1471
782	Aniline, 4-bromo-3,5-dichloro .... 4-Br-3,5-$Cl_2C_6H_2NH_2$	240.92	nd (w + al)	....	129	....	....	al, bz, chl	B12, 654

No.	Name, Synonyms, and Formula	Mol. wt.	Color, crystalline form, specific rotation and λ_{max} (log ε)	b.p. °C	m.p. °C	Density	n_D	Solubility	Ref.
783	Aniline, 4-bromo-2,5-dinitro 4-Br-2,5-$(NO_2)_2C_6H_2NH_2$	262.03	ye cr (al)	....	186	....	....	al	B12, 761
784	Aniline, 4-bromo-2,6-dinitro 4-Br-2,6-$(NO_2)_2C_6H_2NH_2$	262.03	og-red pl (abs.al)	....	163 (cor)	....	....	al	B12³, 1726
785	Aniline, 4-bromo-2-nitro 4-Br-2-$(NO_2)C_6H_3NH_2$	217.03	og ye nd (w)	sub	111.5	....	....	al	B12³, 1670
786	Aniline, 4-bromo-3-nitro 4-Br-3-$(NO_2)C_6H_3NH_2$	217.03	nd (al)	....	132	....	....	al, eth, chl, aa	B12³, 1674
787	Aniline, 5-bromo-2,4-dinitro 5-Br-2,4-$(NO_2)_2C_6H_2NH_2$	262.03	pa ye nd (dil al)	....	178.4	....	....	al, eth, bz	B12³, 1723
788	Aniline, 5-bromo-2-nitro 5-Br-2-$(NO_2)C_6H_3NH_2$	217.03	rd-ye nd (dil al)	....	151-2	....	....	al	B12³, 1671
789	Aniline, 6-bromo-2,3-dinitro 6-Br-2,3-$(NO_2)_2C_6H_2NH_2$	262.03	dk rd cr (al)	....	158	....	....		B12, 760
790	Aniline, 6-bromo-2-nitro 6-Br-2-$(NO_2)C_6H_3NH_2$	217.03	og or ye nd (dil al)	....	74.5	1.988	....	al	B12², 402
791	Aniline, N-butyl or Butyl phenyl amine $C_6H_5NHC_4H_9$	149.24	....	241.6, 118-20[15]	-14.4	0.93226[20/4]	1.53412[20]	al, eth	B12³, 267
792	Aniline, N-isobutyl or Isobutyl phenyl amine $C_6H_5NH-i-C_4H_9$	149.24	....	231-3, 109[13]		0.940[15/4]	1.5328[20]	bz, eth	B12³, 269
793	Aniline, N-sec-butyl or Sec-butyl phenyl amine $C_6H_5NHCH(CH_3)C_2H_5$	149.24	....	225,112-4[22]			1.5333[20]		B12³, 269
794	Aniline, N-tert-butyl or tert-Butyl phenyl amine $C_6H_5NHC(CH_3)_3$	149.24	....	214-6[753], 93-8[14]		1.5270[20]		al, ace, bz	B12³, 270
795	Aniline, 2-butyl or o-Butyl aniline 2-$C_4H_9C_6H_4NH_2$	149.24	ye oil	122-5[12]		0.953[20/4]		al, ace, bz	B12³, 2714
796	Aniline, 2-sec-butyl 2-$CH_3CH_2CH(CH_3)C_6H_4NH_2$	149.24	....	120-2[16]		0.9574[20]		al, ace, bz	B12³, 2721
797	Aniline, 2-tert-butyl 2-$(CH_3)_3CC_6H_4NH_2$	149.24	....	233-5		0.977[15]	1.5453[20]	al, eth, bz	B12³, 2726
798	Aniline, 4-butyl or p-Butyl aniline 4-$C_4H_9C_6H_4NH_2$	149.24	pa ye	261, 133[14]		0.945[20/4]		al, aa, bz	B12³, 2715
799	Aniline, 4-iso-butyl 4-tert-$C_4H_9C_6H_4NH_2$	149.24	pa ye	238[762],112[11]		0.949[15/4]		al, ace bz	B12³, 2715
800	Aniline, 4-sec-butyl 4-Sec-$C_4H_9C_6H_4NH_2$	149.24		238[762],118[15]		0.949[15/4]	1.5360[20]	eth, bz	B12², 635
801	Aniline, 4-tert-butyl 4-$(CH_3)_3CC_6H_4NH_2$	149.24	ye rd (peth)	240[740]	17	0.9525[15/4]	1.5380[20]	al, eth, bz	B12³, 2728
802	Aniline, 3-tert-butyl 3-$(CH_3)CC_6H_4NH_2$	149.24	....	229[708]	....		al, eth, bz		B12², 637
803	Aniline, 2-chloro-(α) 2-$ClC_6H_4NH_2$	127.57	....	208.8	-14	1.21253[20/4]	1.58951[²ls0]	al, eth, ace	B12³, 1281
804	Aniline, 2-chloro (β) 2-$ClC_6H_4NH_2$	127.57		208.8, 84.6[10]	-1.9	1.21266[20/4]	1.5889[20]	al eth, ace, bz	B12³, 1281
805	Aniline, 2-chloro,hydrochloride 2-$ClC_6H_4NH_2$·HCl	164.04	pl(w,aq.al)	....	235	1.505[18]	....	w	B12², 315
806	Aniline, 2-chloro-4-nitro 2-Cl-4-$(NO_2)C_6H_3NH_2$	172.57	ye nd (lig-CS₂)		108			al, eth	B12³, 1662
807	Aniline, 3-chloro 3-$ClC_6H_4NH_2$	127.57		229.9, 101.2[10]	-10.3	1.21606[20/4]	1.59414[20]	al, eth, ace, bz	B12³, 1303
808	Aniline, 3-chloro-2,6-dinitro 3-Cl-2,6-$(NO_2)_2C_6H_2NH_2$	217.57	og-ye nd (al)		112.			al	B12³, 1719
809	Aniline, 3-chloro, hydrochloride 3-$ClC_6H_4NH_2$·HCl	164.04	pl		222			w, al	B12², 320
810	Aniline, 3-chloro-4-nitro 3-Cl-4-$(NO_2)C_6H_3NH_2$	172.57	ye lf (bz)	....	156-7			al, eth	B12³, 1665
811	Aniline, 4-chloro 4-$ClC_6H_4NH_2$	127.57	rh pr	232	72.5	1.429[19/4]	1.5546[87]	al, eth, w	B12³, 1325
812	Aniline, 4-cloro-2,6-dinitro 4-Cl-2,6-$(NO_2)_2C_6H_2NH_2$	217.57	og-ye nd (al)		147 (cor)			al	B12³, 1720
813	Aniline, 4-chloro-2-methoxy. or 4-Chloro-o-anisidine 4-Cl-2-$(CH_3O)C_6H_3NH_2$	157.61	nd or pr (dil.al)	260	52			al, eth, bz	B13², 184
814	Aniline, 4-chloro-2-nitro 4-Cl-2-$(NO_2)C_6H_3NH_2$	172.57	dk og-ye pr (dil al)	....	116-7			al, eth, aa	B12³, 1649
815	Aniline, 4-chloro-3-nitro 4-Cl-3-$NO_2C_6H_3NH_2$	172.57	ye nd or pr (w) nd (peth)		103			al, eth, ace	B12³, 1660

No.	Name, Synonyms, and Formula	Mol. wt.	Color, crystalline form, specific rotation and λ_{max} (log ε)	b.p. °C	m.p. °C	Density	n_D	Solubility	Ref.
816	Aniline, 5-chloro-2-ethoxy or 5-Chloro-o-phenetidine .... 5-Cl-2-(C₂H₅O)C₆H₃NH₂	171.63	nd (dil.al)	....	42	....	....	al, bz, aa	B13, 383
817	Aniline, 5-chloro-2-methoxy or 5-Chloro-o-anisidine..... 5-Cl-2-(CH₃O)C₆H₃NH₂	157.61	nd (dil al)	....	84	....	....	al	B13², 183
818	Aniline, 5-chloro-2-nitro .... 5-Cl-2-(NO₂)C₆H₃NH₂	172.57	gold ye nd (CS₂) ye lf (al, bz)	sub	126.5	....	....	al, eth	B12³, 1655
819	Aniline, 5-chloro-3-nitro .... 5-Cl-3-(NO₂)C₆H₃NH₂	172.57	og.ye nd (al)	....	133-4	....	....	al, eth	B12, 732
820	Aniline, 6-chloro-2-nitro .... 6-Cl-2-(NO₂)C₆H₃NH₂	172.57	ye nd (dil al)	....	76	....	....	al	B12³, 1658
821	Aniline, 6-chloro-3-nitro .... 6-Cl-3-(NO₂)C₆H₃NH₂	172.57	ye nd (lig)	....	121	....	....	al, eth, ace	B12³, 1661
822	Aniline, N-cyclohexyl or Phenylcyclohexylamine N-cyclohexyl aniline .... C₆H₅NHC₆H₁₁	175.28	mcl pr	279⁷⁶⁴	16	1.0155²⁰ᐟ⁴	1.5610²⁰	al, eth, bz	B12², 98
823	Aniline, 4-cyclohexyl 4-C₆H₁₁C₆H₄NH₂	175.28	pl (lig)	....	55	....	....	al,bz	B12³, 2819
824	Aniline, 2,3-dibromo 2,3-Br₂C₆H₃NH₂	250.93	pl (dil al)	....	43	....	....	al, eth, aa	B12, 655
825	Aniline, 2,4-dibromo 2,4-Br₂C₆H₃NH₂	250.93	rh bi pym (chl)nd or lf (dil al)	156²⁴	79.5-80.5	2.260²⁰	....	al, eth	B12³, 1471
826	**Aniline, 2,4-dibromo-6-nitro** .... **Br₂NO₂C₆H₂NH₂**	295.93	ye cr	....	128	....	....	....	B12², 403
827	Aniline, 2,5-dibromo 2,5-Br₂C₆H₃NH₂	250.93	pr (al)	....	53-5	....	....	al, eth	B12³, 1474
828	Aniline, 2,6-dibromo 2,6-Br₂C₆H₃NH₂	250.93	nd (al)	262-4	87-8	....	....	al, eth, bz, chl	B12³, 1474
829	Aniline, 2,6-dibromo-4-nitro .... 2,6-Br₂-4-(NO₂)C₆H₂NH₂	295.93	ye nd (al or aa)	207	......	....	....	aa	B12³, 1678
830	Aniline, 3,4-dibromo 3,4-Br₂C₆H₃NH₂	250.95	lf (dil al)	100 sub	81	....	....	al	B12¹, 329
831	Aniline, 3,5-dibromo 3,5-Br₂C₆H₃NH₂	250.93	nd (dil al)	....	57	....	....	al, eth, bz	B12³, 1475
832	Aniline, 3,5-dibromo-4-methoxy 3,5-Br₂-4-CH₃OC₆H₂NH₂	280.96	pl (lig)	....	66	....	....	al, eth, ace, bz	B13³, 1195
833	Aniline, 4,6-dibromo-2-ethoxy .... 4,6-Br₂-2(C₂H₅O)C₆H₂NH₂	294.99	pr (dil al)	....	52	....	....	al, lig	B13³, 867
834	Aniline, N,N-dibutyl or Dibutylphenylamine .... C₆H₅N(C₄H₉)₂	205.34	....	274.8, 138.8¹⁰	−32.2	0.9037²⁰ᐟ⁴	1.5186²⁰	al, eth, ace, bz	B12³, 269
835	Aniline, 2,3-dichloro 2,3-Cl₂C₆H₃NH₂	162.02	nd (lig)	252	24	....	....	al, eth, ace	B12, 621
836	Aniline, 2,4-dichloro 2,4-Cl₂C₆H₃NH₂	162.02	pr (ace) nd (dil al) (lig)	245	63-4	....	....	al, eth	B12³, 1389
837	Aniline, 2,5-dichloro 2,5-Cl₂C₆H₃NH₂	162.02	nd (lig)	251	50	....	....	al, eth, bz	B12³, 1397
838	Aniline, 2,6-dichloro 2,6-Cl₂C₆H₃NH₂	162.02	....	....	39	....	....	al, eth	B12², 337
839	Aniline, 2,6-dichloro-4-ethoxy or 4-Amino-3,5-dichloro-phenetole .... 2,6-Cl₂-4-(C₂H₅O)C₆H₂NH₂	206.07	nd (dil al)	275	46	....	....	al, eth, bz	B13², 276
840	Aniline, 2,6-dichloro-4-nitro or Dichloran............. 2,6-Cl₂-4-(NO₂)C₆H₂NH₂	207.02	ye nd (al, aa)	....	191	....	....	al	B12³, 1669
841	Aniline, 3,4-dichloro 3,4-Cl₂C₆H₃NH₂	162.02	nd (lig)	272, 145¹⁵	72	....	....	al, eth	B12³, 1402
842	Aniline, 3,5-dichloro 3,5-Cl₂C₆H₃NH₂	162.02	nd (lig, dil al)	260⁷⁴¹	51-3	....	....	al, eth, bz, chl	B12², 337
843	Aniline, 3,5-dichloro-4-ethoxy or 4-Amino-2,6-dichloro-phenetole .... 3,5-Cl₂-4(C₂H₅O)C₆H₂NH₂	206.07	nd (peth)	....	105-7	....	....	al, eth, bz	B13², 275
844	Aniline, N,N-diethyl or Diethylphenylamine .... C₆H₅N(C₂H₅)₂	149.24	ye oil	216.27, 92¹⁰	−38.8	0.93507²⁰ᐟ⁴	1.5409²⁰	al, eth, ace	B12³, 260
845	Aniline, 2-diethylamino .... 2-[(C₂H₅)₂N]C₆H₄NH₂	164.25	oil	312²⁴⁴, 127²⁵	......	....	....	al, ace, bz	B13³, 34

No.	Name, Synonyms, and Formula	Mol. wt.	Color, crystalline form, specific rotation and λ_{max} (log ε)	b.p. °C	m.p. °C	Density	n_D	Solubility	Ref.
846	Aniline, 3-diethylamino 3-[(C₂H₅)₂N]C₆H₄NH₂	164.25		276-8, 117[4]				al	B13[2], 26
847	Aniline, 4-diethylamino 4-[(C₂H₅)₂N]C₆H₄NH₂	164.25	ye oil	260-2, 139-40[10]				bz	B13[3], 113
848	Aniline, N,N-diethyl-3-bromo 3-BrC₆H₄N(C₂H₅)₂	228.14		142[10]				al, ace	B12[1], 315
849	Aniline, N,N-diethyl-4-bromo 4-BrC₆H₄N(C₂H₅)₂	228.14	nd or pr	270	33			al, eth	B12[2], 347
850	Aniline, N,N-diethyl-4-chloro 4-ClC₆H₄N(C₂H₅)₂	183.68	nd (al)	251-3, 95-6[1 5]	45.5-6.5			al	B12[3], 1329
851	Aniline, N,N-diethyl-2-ethoxy or N,N-diethyl-o-phentidine 2-(C₂H₅O)C₆H₄N(C₂H₅)₂	193.29		231-3				al, eth, bz	B13, 365
852	Aniline, N,N-diethyl-3-ethoxy or N,N-diethyl-m-phenetidine 3-(C₂H₅O)C₆H₄N(C₂H₅)₂	193.29		268-70, 145[14]			1.5325[25]	al, bz, aa	B13[3], 942
853	Aniline, N,N-diethyl-3-nitro 3-NO₂C₆H₄N(C₂H₅)₂	194.24	ye	288-90					B12[3], 1545
854	Aniline, N,N-diethyl-4-nitro 4-NO₂C₆H₄N(C₂H₅)₂	194.24	ye nd (lig) pl (al)		77-8	1.225		al	B12[3], 1585
855	Aniline, N,N-diethyl-4-nitroso 4-ONC₆H₄N(C₂H₅)₂	178.24	gr mcl pr (eth) gr lf (ace)		87-8	1.24[15/4]		al, eth, ace	B12[3], 1585
856	Aniline, 3,4-dihydroxy 3,4-(HO)₂C₆H₃NH₂	125.13	br-vt nd (al-bz)		124-5d			w, al, eth	B13[2], 464
857	Aniline, 3,5-dihydroxy or Phloramin 3,5-(HO)₂C₆H₃NH₂	125.13	nd		146-52			al	B13, 707
858	Aniline, 2,4-diiodo 2,4-I₂C₆H₃NH₂	344.92	br nd or rh cr (al)		95-6	2.748		al, eth, ace, bz, chl	B12[3], 1508
859	Aniline, 2,4-diiodo-5-nitro 2,4-I₂-5(NO₂)C₆H₂NH₂	389.92	pa ye mcl pl		125				B12, 747
860	Aniline, 2,5-diiodo 2,5-I₂C₆H₃NH₂	344.92	nd (al)		88-9			al, eth, ace	B12, 675
861	Aniline, 2,6-diiodo 2,6-I₂C₆H₃NH₂	344.92	nd (al)		122			al, eth, ace, bz	B12, 675
862	Aniline, 2,6-diiodo-3-nitro 2,6-I₂-3-(NO₂)C₆H₂NH₂	389.92	ye nd (dil al)		145.5			al	B12, 747
863	Aniline, 2,6-diiodo-4-nitro 2,6-I₂-4-(NO₂)C₆H₂NH₂	389.92	pa ye lf or nd (bz)		245			bz	B12[3], 1680
864	Aniline, 3,4-diiodo 3,4-I₂C₆H₃NH₂	344.92	pa ye lf or pr (bz-peth)		74.5			al, eth, bz	B12, 675
865	Aniline, 3,5-diiodo 3,5-I₂C₆H₃NH₂	344.92	nd (al)		110			al, eth, chl	B12[1], 337
866	Aniline, 4,6-diiodo-2-nitro 4,6-I₂-2-(NO₂)C₆H₂NH₂	389.92	ye nd (ace)		154			eth, ace, bz, chl	B12[1], 361
867	Aniline, 4,6-diiodo-3-nitro 4,6-I₂-3-(NO₂)C₆H₂NH₂	389.92	pa ye nd (eth-al)		149			eth	B12, 747
868	Aniline, 2,3-dimethoxy or 3-Amino veratrole 2,3-(CH₃O)₂C₆H₃NH₂	153.18		137[15]				w	B13[3], 2104
869	Aniline, 2,4-dimethoxy 2,4-(CH₃O)₂C₆H₃NH₂	153.18	pl (lig)		33.5			al, eth, lig	B13[3], 2131
870	Aniline, 2,6-dimethoxy 2,6-(CH₃O)₂C₆H₃NH₂	153.18	pl (al) lf (peth)	146[23]	75			al, eth, bz, lig	B13[3], 2128
871	Aniline, 3,4-dimethoxy or 4-Amino veratroile 3,4-(CH₃O)₂C₆H₃NH₂	153.18	lf (eth)		87-8			eth	B13[3], 2105
872	Aniline, N,N-dimethyl C₆H₅N(CH₃)₂	121.18	pa ye	194, 77[13]	2.45	0.9557[20/4]	1.5582[20]	al, eth, ace, bz, chl	B12[3], 245
873	Aniline, 2-dimethylamino 2-[(CH₃)₂N]C₆H₄NH₂	136.20	oil	218[751], 117[25]		0.995[22]		al, eth, ace, bz	B13[3], 33
874	Aniline, 3-dimethylamino 3-[(CH₃)₂N]C₆H₄NH₂	136.20		268-70[740], 138[10]	<-20	0.995[25]		al, eth	B13[3], 68
875	Aniline, 4-dimethylamino 4-[(CH₃)₂N]C₆H₄NH₂	136.20	nd (bz)	263, 158[11]	53	1.036[20/4]		al, eth, bz, w, chl	B13[3], 109
876	Aniline, N,N-dimethyl-2-bromo 2-BrC₆H₄N(CH₃)₂	200.99		107—8[14]		1.3880[25/25]	1.5768[25]	al	B12[3], 1416

No.	Name, Synonyms, and Formula	Mol. wt.	Color, crystalline form, specific rotation and λ_{max} (log ε)	b.p. °C	m.p. °C	Density	n_D	Solubility	Ref.
877	Aniline, N,N-dimethyl-3-bromo 3BrC$_6$H$_4$N(CH$_3$)$_2$	200.09	....	239, 126[14]	11	....	....	al, aa	B12³, 1422
878	Aniline, N,N-dimethyl-4-bromo 4-BrC$_6$H$_4$N(CH$_3$)$_2$	200.09	lf (al)	264	55			al, eth	B12³, 1432
879	Aniline, N,N-dimethyl-2-chloro 2-ClC$_6$H$_4$N(CH$_3$)$_2$	155.63	....	203, 98-9[18]	......	1.1067[20/4]	1.5578[20]	al, bz	B12³, 1283
880	Aniline, N,N-dimethyl-4-chloro 4-ClC$_6$H$_4$N(CH$_3$)$_2$	155.63	nd (al)	231	35.5			al	B12³, 1329
881	Aniline, N,N-dimethjyl-4-chloro-3-nitro 4-Cl-3(NO$_2$)C$_6$H$_3$N(CH$_3$)$_2$	200.63	ye nd (dil al)	....	81.5-2.5			al, lig	B12³, 1660
882	Aniline, N,N-dimethyl, hydrochloride C$_6$H$_5$N(CH$_3$)$_2$·HCl	.157.65	hyg pl (w) (bz)	....	85-95	1.1156[19/4]	....	w, al, chl	B12³, 251
883	Aniline, N,N-dimethyl-2-nitro 2-(NO$_2$)C$_6$H$_4$N(CH$_3$)$_2$	166.18	ye-og	146[20]	fp-20	1.1794[20/4]	1.6102[20]	w, al, eth, chl	B12³, 1516
884	Aniline, N,N-dimethyl-3-nitro 3-NO$_2$C$_6$H$_4$N(CH$_3$)$_2$	166.18	og-ye or red nd pr (eth or eth-al)	280-5	60-1	....		al, eth	B12³, 1544
885	Aniline, N,N-dimethyl-4-nitro 4-NO$_2$C$_6$H$_4$N(CH$_3$)$_2$	166.18	ye nd (al)	....	164.5			al, eth, aa	B12³, 1583
886	Aniline, N,N-dimethyl-4-nitroso 4-ON-C$_6$H$_4$N(CH$_3$)$_2$	150.18	gr-pl (eth)		92.5-3.5	1.145[20]		al, eth	B12³, 1509
887	Aniline, 2,3-dimethyl or 2,3-xylidine 2,3-(CH$_3$)$_2$C$_6$H$_3$NH$_2$	121.18	....	221-2, 106[15]	<−15	0.9931[20]	1.5684[20]	al, eth	B12³, 2438
888	Aniline, 2,3-dimethyl-4-nitro or 4-Nitro-2,3-xylidine 2,3-(CH$_3$)$_2$-4-(NO$_2$)C$_6$H$_2$NH$_2$	166.18	ye pr (al)	....	114			al	B12³, 2442
889	Aniline, 2,3-dimethyl-5-nitro or 5-Nitro-2,3-xylidine 2,3-(CH$_3$)$_2$-5NO$_2$C$_6$H$_2$NH$_2$	166.18	pa ye nd (al)	....	111-2			al	B12¹, 479
890	Aniline, 2,3-dimethyl-6-nitro or 6-Nitro-2,3-xylidine 2,3-(CH$_3$)$_2$-6-NO$_2$C$_6$H$_2$NH$_2$	168.18	red pl (al)	....	118-9			al	B12³, 2442
891	Aniline, 2,4-dimethyl or 2,4-xylidine. 2,4-(CH$_3$)$_2$C$_6$H$_3$NH$_2$	121.18	....	214, 91[10]	−14.3	0.9723[20/4]	1.5569[20]	al, eth, bz	B12³, 2469
892	Aniline, 2,4-dimethyl-3-nitro or 3-Nitro-2,4-xylidine 2,4-(CH$_3$)$_2$-3-NO$_2$C$_6$H$_2$NH$_2$	166.18	ye nd	....	81-2			al, lig	B12², 612
893	Aniline, 2,4-dimethyl-5-nitro or 5-Nitro-2,4-xylidine 2,4-(CH$_3$)$_2$-5-NO$_2$C$_6$H$_2$NH$_2$	166.18	og ye nd (al)	....	123			al	B12², 612
894	Aniline, 2,4-dimethyl-6-nitro or 6-Nitro-2,4-xylidine 2,4-(CH$_3$)$_2$-6-NO$_2$C$_6$H$_2$NH$_2$	166.18	og red nd or pl (lig)	....	76			al	B12², 612
895	Aniline, 2,5-dimethyl or 2,5-Xylidine 2,5(CH$_3$)$_2$C$_6$H$_3$NH$_2$	121.18	ye lf (lig)	214, 97-100[10]	15.5	0.9790[21/4]	1.5591[21]	eth	B12³, 2503
896	Aniline, 2,6-dimethyl or 2,6-Xylidine 2,6(CH$_3$)$_2$C$_6$H$_3$NH$_2$	121.18	....	214[7,19]	11.2	0.9842[20]	1.5610[20]	al, eth	B12³, 2462
897	Aniline, 2,6-dimethyl-3-nitro or 3-Nitro-2,6-xylidine 2,6-(CH$_3$)$_2$-3-NO$_2$C$_6$H$_2$NH$_2$	166.18	ye nd (dil al)	....	81-2			al	B12², 605
898	Aniline, 2,6-dimethyl-4-nitro or 4-Nitro-2,6-xylidine 2,6-(CH$_3$)$_2$-4-NO$_2$C$_6$H$_2$NH$_2$	166.18	og-red nd or pl (lig)	....	76			al	B12³, 2469
899	Aniline, 3,4-dimethyl or 3,4-Xylidine 3,4-(CH$_3$)$_2$C$_6$H$_3$NH$_2$	121.18	pl or pr (lig)	228	51	1.076[18]	....	eth, lig	B12³, 2443
900	Aniline, 3,4-dimethyl-2-nitro or 2-Nitro-3,4-xylidine 3,5-(CH$_3$)$_2$-2-NO$_2$C$_6$H$_2$NH$_2$	166.18	red pr (al)	....	65-6			al	B12³, 2454
901	Aniline, 3,4-dimethyl-5-nitro or 5-Nitro-3,4-xylidine 3,4-(CH$_3$)$_2$-5-NO$_2$C$_6$H$_2$NH$_2$	166.18	og lf (al)	....	74-5			al	B12, 1106
902	Aniline, 3,5-dimethyl or 3,5-Xylidine 3,5-(CH$_3$)$_2$C$_6$H$_3$NH$_2$	121.18	....	220-1, 99-100[20]	9.8	0.9706[20/4]	1.5581[20]	eth	B12³, 2495
903	Aniline, 3,5-dimethyl-2-nitro or 2-Nitro-3,5-xylidine 3,5-(CH$_3$)$_2$-2-NO$_2$C$_6$H$_2$NH$_2$	166.18	ye nd (lig)	....	56			....	B12², 613
904	Aniline, 3,5-dimethyl-4-nitro or 4-Nitro-3,5-xylidine 3,5-(CH$_3$)$_2$-4-NO$_2$C$_6$H$_2$NH$_2$	166.18	og pr (bz), lf (lig)	....	133			al, bz	B12³, 2500
905	Aniline, 4,5-dimethyl-2-nitro or 2-Nitro-4,5-xylidine 4,5-(CH$_3$)$_2$-2-NO$_2$C$_6$H$_2$NH$_2$	166.18	red-br pr (al)	....	140			ace, bz	B12³, 2455
906	Aniline, 2,3-dinitro 2,3-(NO$_2$)$_2$C$_6$H$_3$NH$_2$	183.12		....	178	1.646[50]		al, eth	B12³, 1680
907	Aniline, 2,4-dinitro 2,4-(NO$_2$)$_2$C$_6$H$_3$NH$_2$	183.12	ye nd (dil ace), gr-ye ta (al)		180 (188)	1.615[14]		....	B12³, 1681
908	Aniline, 2,6-dinitro 2,6-(NO$_2$)$_2$C$_6$H$_3$NH$_2$	183.12	gold lf (50% aa), ye nd (al)	....	141-2		....	eth, bz	B12³, 1704

No.	Name, Synonyms, and Formula	Mol. wt.	Color, crystalline form, specific rotation and λ_{max} (log ε)	b.p. °C	m.p. °C	Density	n_D	Solubility	Ref.
909	Aniline, 3,5-dinitro .... 3,5-$(NO_2)_2C_6H_3NH_2$	183.12	....	....	16.3	1.601[50]	....	al, eth	B12[3], 1705
910	Aniline, N,N-dipropyl .... $C_6H_5N(C_3H_7)_2$	172.29	lf ye	243, 127[10]		0.9104[20]	1.5271[20]	al, eth, ace, bz	B12[3], 266
911	Aniline, 2-ethoxy or o-Phenetidine .... 2-$(C_2H_5O)C_6H_4NH_2$	137.18	....	232.5, 127-8[14]	<-21	....	1.5560[20]	al, eth	B13[3], 756
912	Aniline, 3-ethoxy or m-Phenetidine........ 3-$(C_2H_5O)C_6H_4NH_2$	137.18	....	248, 127-8[11]		....	1.5528[20]	al, eth	B13[3], 932
913	Aniline, 4-ethoxy or p-Phenetidine .... 4-$(C_2H_5O)C_6H_4NH_2$	137.18	....	254, 125[11]	2.4	1.06521[16/4]	1.5528[20]	al, eth	B13[3], 996
914	Aniline, N-Ethyl or Ethylphenylamine .... $C_6H_5NHC_2H_5$	121.18	....	204.7, 97.5[18]	-63.5	0.9625[20/4]	1.5559[20]	al, eth, ace, bz	B12[3], 255
915	Aniline, N-ethyl-2-chloro .... 2-$ClC_6H_4NHC_2H_5$	155.63	....	219[726]		1.104[20/4]	....		B12[3], 1284
916	Aniline, N-ethyl, hydrochloride .... $C_6H_5NHC_2H_5 \cdot HCl$	157.65	nd	....	178.5	1.0085[182]	....	w, al, chl	B12[3], 257
917	Aniline, N-ethyl-N-methyl .... $C_6H_5N(CH_3)(C_2H_5)$	135.21	....	203-5, 93-5[13]		0.9193[55/4]	....	al, eth	B12[3], 258
918	Aniline, N-ethyl-N-nitroso or N-ethylphenylnitrosamine. $C_6H_5N(NO)C_2H_5$	150.18	yesh	119-20[15]		1.0874[20]	....	aa	B12[3], 1119
919	Aniline, o-ethyl .... 2-$C_2H_4C_6H_4NH_2$	121.18	....	209-10	-43	0.983[22/4]	1.5584[22]	al, eth	B12[3], 2374
920	Aniline, m-ethyl .... 3-$C_2H_5C_6H_4NH_2$	121.18	....	211.5[762], 93-5[11]	-04	0.9896	....	al, eth	B12[3], 2379
921	Aniline, p-ethyl .... 4-$C_2H_5C_6H_4NH_2$	121.18	....	217-8, 92.3[10]	-4.87	0.9679[20/4]	1.5554[20]	al, eth	B12[3], 2380
922	Aniline, o-fluoro .... 2-$FC_6H_4NH_2$	111.12	pa ye	174.6[757]	-28.3	1.1513[11]	1.5421[16]	al, eth	B12[3], 1273
923	Aniline, m-fluoro .... 3-$FC_6H_4NH_2$	111.12	....	187-9, 82-3[18]		1.1561[19]	1.5436[20]	al, eth	B12[3], 1274
924	Aniline, p-fluoro .... 4-$FC_6H_4NH_2$	111.12	pa ye	180.5-2.5[757], 85[19]	-0.8	1.1725[20/4]	1.5195[20/4]	al, eth	B12[3], 1276
925	Anilinehydrochloride .... $C_6H_5NH_2 \cdot HCl$	129.60	lf or nd	245	198	1.2215[4]	....	w, al	B12[3], 232
926	Aniline, o-iodo .... 2-$IC_6H_4NH_2$	219.03	nd (dil al)	....	60-1	....	....	al, eth, ace	B12[3], 1487
927	Aniline, m-iodo .... 3-$IC_6H_4NH_2$	219.03	lf or nd	280	33	....	1.6811[20]	al, chl	B12[3], 1489
928	Aniline, p-iodo .... 4-$IC_6H_4NH_2$	219.03	nd (w)	....	67-8	....	....	al, eth	B12[3], 1493
929	Aniline, o-methoxy or o-Anisidine .... 2-$(CH_3O)C_6H_4NH_2$	123.16	....	224, 90[4]	6.2	1.0923[20/4]	1.5715[10]	al, eth, ace, bz	B13[3], 754
930	Aniline, 2-methoxy-3-nitro .... 2-(CH_3O)-3-NO_2-$C_6H_3NH_2$	168.16	pa ye nd (lig)	....	67	....	....	al, lig	B13[2], 195
931	Aniline, m-methoxy or m-Anisidine .... 3-$(CH_3O)C_6H_4NH_2$	123.16	....	251 (cor)	-1	1.096[20/4]	1.5794[20]	al, eth, ace, bz	B13[3], 932
932	Aniline, p-methoxy or p-Anisidine .... 4-$(CH_3O)C_6H_4NH_2$	123.16	ta (w), rh pl	243, 115[13]	57.2	1.071[57/4]	1.5559[60]	w, al, eth, ace, bz	B13[3], 994
933	Aniline, N-methyl or Methylphenyl amine............ $C_6H_5NHCH_3$	107.16	....	196.25, 86[15]	-57	0.9891[20/4]	1.5684[20]	al, eth	B12[3], 240
934	Aniline, N-methyl, hydrochloride or Methylphenyl amine hydrochloride .... $C_6H_5NHCH_3 \cdot HCl$	143.62	nd (chl-eth)	122-3		1.0660[131/4]	....	w, al	B12[3], 244
935	Aniline, N-(3-methylbutyl) or N-isopentylaniline .... C_6H_5NH-i-C_5H_{11}	163.26	....	254-5, 126-7[14]		0.8912[55/4]	1.5305[20]	al, eth	B12[3], 272
936	Aniline, N-methyl-2-chloro .... 2-$ClC_6H_4NHCH_3$	141.60	....	218, 106[20]		1.1735[12]	1.5780[25]	al, ace, bz	B12[3], 1283
937	Aniline, N-methyl-4-chloro .... 4-$ClC_6H_4NHCH_3$	141.60	....	239[764], 120[20]		1.614[20]	1.5835[20]	al, bz	B12[3], 1329
938	Aniline, N-methyl,-2-nitro or Methyl-2-nitrophenylamine 2-$O_2NC_6H_4NHCH_3$	152.16	red or og nd (peth)	d	38	....	....	al, eth, ace, bz, lig	B12[3], 1516
939	Aniline, N-methyl,3-nitro .... 3-$NO_2C_6H_4NHCH_3$	152.16	red-ye nd or pr (al), cr (lig)	....	68	....	....	al, eth, bz	B12[3], 1544
940	Aniline, N-methyl-4-nitro .... 4-$NO_2C_6H_4NHCH_3$	152.16	br-ye pr (al), cr (eth)	d	152	1.201[155/4]	....	al, bz	B12[3], 1584

No.	Name, Synonyms, and Formula	Mol. wt.	Color, crystalline form, specific rotation and λ_{max} (log ε)	b.p. °C	m.p. °C	Density	n_D	Solubility	Ref.
941	Aniline, N-methyl- N-nitroso or N-Nitroso methylphenyl-amine .... $C_6H_5N(NO)CH_3$	136.17	ye	225d, 121[13]	14.7	1.1240[20/4]	1.57688[20]	al, eth	B12[3], 1119
942	Aniline, N-methyl-4-nitroso .... 4-$ONC_6H_4NHCH_3$	136.17	bl pl (bz)	....	118			al, eth, chl	B7[3], 3370
943	Aniline, N-methyl-4-iso-Propyl or Cuminyl amine ...... 4-$(CH_3)_2CHC_6H_4NHCH_3$	149.24	oil	225-7[734], 111-2[11]				al, eth	B12[3], 2741
944	Aniline, N-methyl-2,4,5-trimethyl .... 2,4,5-$(CH_3)_3C_6H_2NHCH_3$	149.24	nd (dil al)		85			al, chl	B12, 1152
945	Aniline, N-methyl-3,4,5-trimethyl .... 3,4,5-$(CH_3)_3C_6H_2NHCH_3$	149.24	lf (w)		123			....	B12, 1176
946	Aniline, N-methyl, N,2,4,6-tetranitro or Tetryl .... 2,4,6$(NO_2)_3C_6H_2N(NO_2)CH_3$	287.15	ye pr (al)	exp 187	131-2	1.57[10]		ace, bz	B12[3], 1738
947	Aniline, 2-methyl-5-iso-propyl or Carvacryl amine ...... 2-CH_3-5-i-$C_3H_7C_6H_3NH_2$	149.24	....	241, 118[12]	-16	0.9942[20/4]	1.5387[20]	al, eth	B12[3], 2733
948	Aniline, 4-methylamino .... 4-$(CH_3NH)C_6H_4NH_2$	122.17	lf (eth-peth)	257-9, 162[20]	36			w, al, eth, ace, bz	B13[3], 108
949	Aniline, 4-methylthio .... 4-$(CH_3S)C_6H_4NH_2$	139.22	....	272-3, 140[15]		1.1379[20/4]	1.6395[20]	al, eth, ace, bz	B13[3], 1221
950	Aniline nitrate .... $C_6H_5NH_2 \cdot HNO_3$	156.15	rh	....	d190	1.356[4]	....	w, al, eth	B12[3], 232
951	Aniline, N-nitro .... $C_6H_5NHNO_2$	138.13	lf (peth)	exp	43			w, al, eth, ace, bz	B16[2], 343
952	Aniline, 2-nitro .... 2-$(NO_2)C_6H_4NH_2$	138.13	gold-ye pl or nd	284, 165—6[28]	71.5	1.442[15]		al, eth, ace, bz	B12[3], 1513
953	Aniline, 2-nitro-5-methoxy .... 2-NO_2-5-$(CH_3O)C_6H_3NH_2$	168.16	br nd	sub	131			al	B13[3], 975
954	Aniline, 3-nitro .... 3-$(NO_2)C_6H_4NH_2$	138.13	ye nd, rh bipym (w)	305-7d, 100[0.16]	114	1.1747[160/4]		al, eth, ace	B12[3], 1541
955	Aniline, 4-nitro .... 4-$(NO_2)C_6H_4NH_2$	138.13	pa ye mcl nd (w)	331.7, 106[0.03]	148-9	1.424[20/4]		al, eth, ace, chl	B12[3], 1580
956	Aniline, 5-nitro-2-propoxy .... 5-(NO_2)-2-$(C_3H_7O)C_6H_3NH_2$	196.21	og (PrOH-peth)	....	49			al	B13[3], 879
957	Aniline, 4-nitroso .... 4-$(ON)C_6H_4NH_2$	122.12	bl nd (bz)	dec	173-4			w, al	B7[3], 3370
958	Aniline, 4-octyl .... 4-$C_8H_{17}C_6H_4NH_2$	205.35	....	310-1, 170-2[17]	20			eth	B12[3], 2770
959	Aniline oxalate(mono) .... $C_6H_5NH_2 \cdot H_2C_2O_4$	183.17	ta (aq al)		150-1			w, al, bz	B12[3], 236
960	Aniline oxalate(di) .... $(C_6H_5NH_2)_2 \cdot H_2C_2O_4$	276.30	tcl pr (w)		174-5d			w	B12[3], 236
961	Aniline, pentabromo .... $C_6Br_5NH_2$	487.66	nd (al-to)		265-6			al	B12[3], 1487
962	Aniline, pentachloro .... $C_6Cl_5NH_2$	265.35	nd (al)	....	232			al, eth, lig	B12[3], 1415
963	Aniline, pentafluoro .... $C_6F_5NH_2$	183.08	....	153-4	34				CAS53, 3112
964	Aniline, pentamethyl .... $(CH_3)_5C_6NH_2$	163.26	mcl pr (al)	277-8	152-3			al, eth	B12[3], 2758
965	Aniline, N-pentyl or Pentylphenyl amine .... $C_6H_5NHC_5H_{11}$	163.26	....	260-2					B12[3], 271
966	Aniline, 2-phenoxy .... 2-$(C_6H_5O)C_6H_4NH_2$	185.23	cr (lig)	307-8[728], 172-3[14]	44-5			al, eth, ace, bz	B13[3], 758
967	Aniline, 3-phenoxy .... 3-$(C_6H_5O)C_6H_4NH_2$	185.23	pr (lig)	315, 190-1[14]	37			al, eth, ace, bz	B13[3], 933
968	Aniline, 4-phenoxy or 4-Aminodiphenylether .... 4-$(C_6H_5O)C_6H_4NH_2$	185.23	nd (w), cr (dil al)	187-9[14]	85-6			al, eth, w	B13[3], 999
969	Aniline phosphate .... $C_6H_5NH_2 \cdot H_3PO_4$	191.13	nd (al), lf (w-al)	....	180			w, al	B12, 117
970	Aniline-picrate .... $C_{12}H_{10}N_4O_7$	322.24	ye or red mcl pr (w), dk gr		181	1.558	....	al	B12[3], 236
971	Aniline, N-propyl .... $C_6H_5NHC_3H_7$	135.21	....	222 (cor), 100[11]		0.9443[20/4]	1.5428[20]	al, eth	B12[3], 264
972	Aniline, o-propyl .... 2-$C_3H_7C_6H_4NH_2$	135.21	....	226[758], 116[15]		0.9602[20/4]	1.5427[20]	al, eth	B12[3], 2657
973	Aniline, p-propyl .... 4-$C_3H_7C_6H_4NH_2$	135.21	....	224-6					B12[3], 2658

No.	Name, Synonyms, and Formula	Mol. wt.	Color, crystalline form, specific rotation and λ_{max} (log ε)	b.p. °C	m.p. °C	Density	n_D	Solubility	Ref.
974	Aniline, N-isopropyl or Phenylisopropyl amine C₆H₅NH-i-C₃H₇	135.21	....	203, 72-5[14]			1.5380[20]	al, eth	B12³, 266
975	Aniline, 2-isopropyl 2-i-C₃H₇C₆H₄NH₂	135.21	....	270-1[745], 95[13]	......	0.9760[12/4]	....	eth, bz	B12³, 2683
976	Aniline, 2-isopropyl-5-methyl or Thymylamine 2-i-C₃H₇-5-CH₃C₆H₃NH₂	149.24	oil	238-42				**al, eth**	B12³, 2741
977	Aniline, 4-isopropyl or Cumidine 4-(CH₃)₂CHC₆H₄NH₂	135.21	....	225	-63	0.953[20/4]	1.3415[20]	al, eth, bz	B12³, 2684
978	Aniline sulfate(mono) C₆H₅NH₂·H₂SO₄	200.21	lf (w + 1/2)		162				B12³, 232
979	Aniline sulfate(di) (C₆H₅NH₂)₂·H₂SO₄	284.34	lf (al)		d	1.377⁴		w	B12³, 232
980	Aniline, 2,3,4,5-tetrachloro 2,3,4,5-Cl₄-C₆HNH₂	230.91	nd (al)		118-20			al, eth, bz	B12², 340
981	Aniline, 2,3,5,6-tetrachloro 2,3,5,6-Cl₄C₆HNH₂	230.91	nd (lig, al)		108			al, eth	B12³, 1414
982	Aniline, 2,3,4,5-tetramethyl or Prehnidine 2,3,4,5-(CH₃)₄C₆HNH₂	194.24	lf (w)	259-60	70			al, eth, lig	B12³, 2743
983	Aniline, 2,3,4,6-tetramethyl or isoduridine 2,3,4,6-(CH₃)₄C₆HNH₂	149.24	....	255	23-4	0.978²⁴		al	B12³, 2744
984	Aniline, 2,4,5-tribromo 2,4,5-Br₃C₆H₂NH₂	329.83	nd (al)	....	80-1 (85)			al, eth, bz	B12, 662
985	Aniline, 2,4,6-tribromo 2,4,6-Br₃-C₆H₂NH₂	329.83	nd (al, bz)	300	122	2.35[20/20]	....	ace, chl	B12³, 1477
986	Aniline, 2,4,6-tribromo,hydrobromide 2,4,6-Br₃C₆H₂NH₂·HBr	410.75	nd	sub	195-6				B12¹, 330
987	Aniline, 3,4,5-tribromo 3,4,5-Br₃C₆H₂NH₂	329.83	nd (al)	....	123			al, eth	B12³, 1487
988	Aniline, 2,3,4-trichloro 2,3,4-Cl₃C₆H₂NH₂	196.46	nd (lig)	292⁷⁷⁴	73			al	B12, 626
989	Aniline, 2,4,5-trichloro 2,4,5-Cl₃C₆H₂NH₂	196.46	nd(lig or 50% al)	ca 270	96.5			al, eth	B12³, 1409
990	Aniline, 2,4,6-trichloro or sym-Trichloroaniline 2,4,6-Cl₃C₆H₂NH₂	196.46	cr (al), nd (lig or peth)	262⁷⁴⁶	78.5			al, eth	B12³, 1409
991	Aniline, 2,3,5-triiodo 2,3,5-I₃C₆H₂NH₂	470.82	nd		116			al, bz	B12, 676
992	Aniline, 2,3,6-triiodo 2,3,6-I₃C₆H₂NH₂	470.82	nd (al or al-eth)		116.8			al	B12, 676
993	Aniline, 2,4,6-triiodo 2,4,6-I₃C₆H₂NH₂	470.82	ye nd or pl (al), pr (aa)		185.5			aa	B12³, 1508
994	Aniline, 3,4,5-triiodo 3,4,5-I₃C₆H₂NH₂	470.82	nd (al-ace)		174.5d			al, eth, ace,bz	B12, 676
995	Aniline, 2,4,5-trimethyl or Pseudocumidine 2,4,5,-(CH₃)₃C₆H₂NH₂	135.21	nd (w)	234-5	68	0.957		al, w	B12³, 2700
996	Aniline, 2,4,6-trimethyl or Mesidine 2,4,6-(CH₃)₃C₆H₂NH₂	135.21	....	232-3	-5	1.5495[20]			B12³, 2708
997	Aniline, 2,4,6-trinitro or Picramide 2,4,6-(NO₂)₃C₆H₂NH₂	228.12	dk ye pr (aa)	exp	192-5	1.762¹²		ace, bz	B12³, 1737
998	o-Anisaldehyde or 2-Methoxybenzaldehyde 2-(CH₃O)C₆H₄CHO	136.16	pr	243-4, 124-5[18]	37-8	1.1326[20/4]	1.5600[20]	al, eth, ace, bz, chl	B8³, 143
999	m-Anisaldehyde or 3-Methoxybenzaldehyde 3-(CH₃O)C₆H₄CHO	136.16		230, 62¹		1.1187[20/4]	1.5530[20]	al, eth, ace, bz, chl	B8³, 199
1000	p-Anisaldehyde or 4-Methoxtbenzaldehyde 4-(CH₃O)C₆H₄CHO	136.16	....	249.5, 83²	0	1.1191[15/4]	1.5730[20]	w, eth, ace, bz chl	B8³, 218
1001	o-Anisamide 2-(CH₃O)C₆H₄CONH₂	151.17	nd (bz), pr (eth), pl (w)		129			eth, bz	B10³, 155
1002	p-Anisamide 4-(CH₃O)C₆H₄CONH₂	151.17	nd or ta (w)		295, 183⁴⁰	166-7	16.5	w, al	B10³, 341
1003	o-Anisic acid or 2-Methoxybenzoic acid 2-(CH₃O)C₆H₄CO₂H	152.16	pl (w)	200	101			al, eth, bz, chl	B10³ 97
1004	o-Anisic acid, ethyl ester 2-(CH₃O)C₆H₄CO₂C₂H₅	180.21	....	261, 134-6¹²	....	1.1124[20/4]	1.5224[20]	al, eth	B10³, 116

No.	Name, Synonyms, and Formula	Mol. wt.	Color, crystalline form, specific rotation and λ_{max} (log ε)	b.p. °C	m.p. °C	Density	n_D	Solubility	Ref.
1005	o-Anisic acid, methyl ester .... 2-$(CH_3O)C_6H_4CO_2CH_3$	166.18	....	245, 127[11]	......	1.1511[19/4]	1.534[19.5]	al	B10[3], 109
1006	m-Anisic acid or 3-Methoxybenzoic acid .... 3-$(CH_3O)C_6H_4CO_2H$	152.16							B10[3], 244
1007	m-Anisic acid, ethyl ester .... 3-$(CH_3O)C_6H_4CO_2C_2H_5$	180.21	....	260-1, 110[5]	......	1.0993[20/4]	1.5161[20]	al, eth	B10[3], 250
1008	m-Anisic acid, methyl ester .... 3-$(CH_3O)C_6H_4CO_2CH_3$	166.18	....	252, 121-4[10]	......	1.13101[20/4]	1.5224[20]	al	B10[3], 250
1009	p-Anisic acid or 4-Methoxybenzoic acid .... 4-$(CH_3O)C_6H_4CO_2H$	152.16	nd (w)	170-2[10]	110	....		al, eth, bz	B10[3], 280
1010	p-Anisic acid, butyl ester .... 4-$(CH_3O)C_6H_4CO_2C_4H_9$	208.24	....	183[40]	......	1.054[16.5]	1.5141[16.5]		B10[3], 307
1011	p-Anisic acid, ethyl ester .... 4-$(CH_3O)C_6H_4CO_2C_2H_5$	180.21	....	269-70, 136-7[13]	7-8	1.1038[20/4]	1.5254[20]	al, eth	B10[3], 300
1012	p-Anisic acid, methyl ester .... 4-$(CH_3O)C_6H_4CO_2CH_3$	166.18	fl (al or eth)	256, 160[20]	49	....		al, eth	B10[3], 297
1013	o-Anisidine or 2-Methoxy aniline .... 2-$(CH_3O)C_6H_4NH_2$	123.16	....	224, 90[4]	6.2	1.0923[20/4]	1.5713[20]	al, eth, ace, bz	B13[3], 754
1014	o-Anisidine, 4-nitro or 2-Methoxy-4-nitro aniline .... 2-(CH_3O)-4-$(NO_2)C_6H_3NH_2$	168.16	pa ye nd (dil al)	....	139-40	1.2112[150]		al, ace	B13[3], 887
1015	o-Anisidine, 5-nitro or 2-Methoxy-5-nitro aniline .... 2-(CH_3O)-5-NO_2-$C_6H_3NH_2$	168.16	og-red nd (al, eth, w)		118	1.2068[156]		al, eth, ace, bz, aa	B13[3], 878
1016	o-Anisidine, 6-nitro or 2-Methoxy-6-nitro aniline .... 2-(CH_3O)-6-NO_2-$C_6H_3NH_2$	168.16	ye pa red nd (al)		76			al, bz	B13[3], 875
1017	m-Anisidine or 3-Methoxy aniline .... 3-$(CH_3O)C_6H_4NH_2$	123.16	....	251		1.096[20/4]	1.5794[20]	al, ace, eth, bz	B13[3], 932
1018	m-Anisidine, 2-nitro or 3-Methoxy-2-nitro aniline .... 3-(CH_3O)-2-$(NO_2)C_6H_3NH_2$	168.16	ye nd (bz)		124 (143)			al, ace	B13[3], 974
1019	m=Anisidine, 4-nitro or 3-Methoxy-4-nitro aniline .... 3(CH_3O)-4-$(NO_2)C_6H_3NH_2$	168.16	ye nd (al)	sub	169			al, ace	B13[3], 975
1020	m-Anisidine, 5-nitro or 3-Methoxy-5-nitro aniline .... 3-(CH_3O)-5-$(NO_2)C_6H_3NH_2$	168.16	og cr (w)		120	1.2034[150]		al, ace, bz	B13[3], 977
1021	p-Anisidine or 4-Methoxy aniline .... 4-$(CH_3O)C_6H_4NH_2$	123.16	ta (w), rh pl	243 (cor), 115[13]	57.2	1.071[57/4]	1.5559[67]	al, eth, w, ace, bz	B13[3], 994
1022	p-Anisidine hydrochloride or 4-Methoxy aniline, hydrochloride .... 4-$(CH_3O)C_6H_4NH_2$·HCl	159.62	lf, nd	....	236			w, al	B13[2], 223
1023	p-Anisidine, 2-nitro or 4-Methoxy-2-nitro aniline .... 4(CH_3O)-2-$(NO_2)C_6H_3NH_2$	168.16	dk red pr (w or al)		129			w, al, eth, ace	B13[3], 1203
1024	p-Anisidine, 3-nitro or 4-Methoxy-3-nitro aniline .... 4(CH_3O)-3-$(NO_2)C_6H_3NH_2$	168.16	red (eth), og pr or pl (eth-lig)		57			al, eth, ace, bz	B13[2], 284
1025	Anisole or Methoxybenzene, Methyl phenyl ether .... $C_6H_5OCH_3$	108.15	....	155	-37.5	0.9961[20/4]	1.5179[20]	al, eth, ace, bz	B6[4], 548
1026	Anisole, 2acetyl or 2-Methoxy acetophenone............ 2-$(CH_3CO)C_6H_4OCH_3$	150.17	ye	245		1.0897[20/4]	1.5393[20]	al, ace	B8[3], 263
1027	Anisole, 3-acetyl or 3-Methoxy acetophenone .... 3-$(CH_3CO)C_6H_4OCH_3$	150.17	....	240, 125-6[12]	95-6	1.0343[19]	1.5410[20]	w, al, ace	B8[3], 273
1028	Anisole, 4-acetyl or 4-Methoxy acetophenone .... 4-$(CH_3CO)C_6H_4OCH_3$	150.17	pl (eth)	258,138-9[15]	38-9	1.0818[41/4]	1.547[41]	al, eth, ace	B8[3] 277
1029	Anisole, 4-acetyl-2-nitro or 4-Methoxy-3-nitro acetophenone .... 4-(CH_3CO)-2-$(NO_2)C_6H_3OCH_3$	195.17	nd (al)	....	99.5			al, eth, ace, bz	B8[3], 295
1030	Anisole, 2-bromo .... 2-$BrC_6H_4OCH_3$	187.04	....	216 (223), 94[10]	2.5	1.5018[20/4]	1.5727[20]	al, eth	B6[3], 736
1031	Anisole, 3-bromo .... 3-Br-$C_6H_4OCH_3$	187.04	....	210-1[752], 105[16]		1.5635[20]		al, eth, bz	B6[4], 1043
1032	Anisole, 4-bromo .... 4-$BrC_6H_4OCH_3$	187.04	....	215, 100[16]	13-4	1.4564[20/4]	1.5642[20]	al, eth, chl	B6[4], 1044
1033	Anisole, 2-chloro .... 2-$ClC_6H_4OCH_3$	142.39	....	198.5, 90-1[16]	-26.8	1.1911[20/4]	1.5480[20]	....	B6[4], 785
1034	Anisole, 3-chloro-2-nitro .... 3-Cl-2-$(NO_2)C_6H_3OCH_3$	187.58	....		56			eth, bz, chl	B6[4], 1347
1035	Anisole, 3-chloro .... 3-$ClC_6H_4OCH_3$	142.59	....	193-4, 70[9]	......	1.1759[12/4]	1.5365[20]	al, eth	B6[4], 811
1036	Anisole, 4-chloro .... 4-$ClC_6H_4OCH_3$	142.59	....	197.5, 75[10]	−18	1.201[20/4]	1.5390[20]	al, eth, chl	B6[4], 822

No.	Name, Synonyms, and Formula	Mol. wt.	Color, crystalline form, specific rotation and λ_{max} (log ε)	b.p. °C	m.p. °C	Density	n_D	Solubility	Ref.
1037	Anisole, 4-chloro-2-nitro 4-Cl-2-(NO$_2$)C$_6$H$_3$OCH$_3$	187.58	ye nd or pr (al)		98			al, MeOH	B6[4], 1348
1038	Anisole, 2-cyclohexyl 2-C$_6$H$_{11}$C$_6$H$_4$OCH$_3$	190.28		267-8.5		1.007[18]	1.5365[18]		B6[3], 2493
1039	Anisole, 4-cyclohexyl 4-C$_6$H$_{11}$C$_6$H$_4$OCH$_3$	190.28		275-6[748]	57-8				B6[3], 2502
1040	Anisole, 2,4-dichloro 2,4-Cl$_2$C$_6$H$_3$OCH$_3$	177.03	pr	235, 125[10]	28-9			al, eth	B6[4], 855
1041	Anisole, 2,3-dimethyl or 3-Methoxy-o-xylene 2,3-(CH$_3$)$_2$C$_6$H$_3$OCH$_3$	136.20		199, 85[18]	29	0.9596[40]	1.5120[40]	al, eth, ace, bz	B6[3], 1723
1042	Anisole, 2,4-dimethyl 2,4-(CH$_3$)$_2$C$_6$H$_3$OCH$_3$	136.20		192, 83-4[15]		0.9740[16/4]	1.5190[16]	al, eth, bz	B6[3], 1744
1043	Anisole, 2,5-dimethyl 2,5-(CH$_3$)$_2$C$_6$H$_3$OCH$_3$	136.20		194[773]		0.9693[13/4]	1.5182[15]	al, eth, bz, peth	B6[3], 1772
1044	Anisole, 2,6-dimethyl 2,6-(CH$_3$)$_2$C$_6$H$_3$OCH$_3$	136.20		182—3		0.9619[14/4]	1.5053[14]	al, eth, bz	B6[3], 1737
1045	Anisole, 3,4-dimethyl 3,4-(CH$_3$)$_2$C$_6$H$_3$OCH$_3$	136.20		204-5, 96-7[17]		0.9744[14/4]	1.5198[14]	al, eth, bz	B6[3], 1727
1046	Anisole, 3,5-dimethyl 3,5-(CH$_3$)$_2$C$_6$H$_3$OCH$_3$	136.20		194.5, 89[15]		0.9627[15/4]	1.5110[20]	al, eth, bz, aa	B6[3], 1756
1047	Anisole, 2,3-dinitro 2,3-(NO$_2$)$_2$C$_6$H$_3$OCH$_3$	198.14	nd (al), pl (to)		119	1.2290[137]		al, lig	B6[4], 1369
1048	Anisole, 3,4-dinitro 3,4-(NO$_2$)$_2$C$_6$H$_3$OCH$_3$	198.14	ye nd (dil al)		71	1.3332[110]		al, MeOH	B6[4], 1384
1049	Anisole, 2,6-dinitro 2,6-(NO$_2$)$_2$C$_6$H$_3$OCH$_3$	198.14	nd (al)		118	1.3000[128]		al	B6[3], 868
1050	Anisole, 2,4-dinitro 2,4-(NO$_2$)$_2$C$_6$H$_3$OCH$_3$	198.14	nd (al or w)	206-7[12], sub	94.5-5.5	1.3364[131]		al, eth, ace, bz, to	B6[4], 1372
1051	Anisole, 2,5-dinitro 2,5-(NO$_2$)$_2$C$_6$H$_3$OCH$_3$	198.14	nd (bz-lig)	136-8[2]	97	1.476[18]		ace, bz, al	B6[4], 1383
1052	Anisole, 3,5-dinitro 3,5-(NO$_2$)$_2$C$_6$H$_3$OCH$_3$	198.14	nd (al)		105.5	1.558[12]		ace, bz, MeOH	B6[4], 1385
1053	Anisole, 2-ethyl 2-C$_2$H$_5$C$_6$H$_4$OCH$_3$	136.20		186-8[755], 80[14]		0.9636[19/4]	1.5142[20]	eth, bz	B6[3] 1656
1054	Anisole, 3-ethyl 3-C$_2$H$_5$C$_6$H$_4$OCH$_3$	136.20		196-7[758], 74[10]		0.9575[18/4]	1.5102	eth, bz	B6[3], 1662
1055	Anisole, 4-ethyl 4-C$_2$H$_5$C$_6$H$_4$OCH$_3$	136.20		195-6, 83-4[16]		0.9624[15/4]	1.5120[20]	eth, bz	B6[3], 1665
1056	Anisole, 2-fluoro 2-FC$_6$H$_4$OCH$_3$	126.13		154-5, 59[13]	-39	1.5489[17]	1.4969[17.5]	eth	B6[4], 771
1057	Anisole, 4-fluoro 4-FC$_6$H$_4$OCH$_3$	126.13		154	-45	1.1781[18/4]	1.4886[18]	eth	B6[4], 773
1058	**Anisole, o-(1-hydroxyethyl) 2-[CH$_3$CH(OH)]C$_6$H$_4$OCH$_3$**	**152.20**		125[768]	-85.1	0.9647[20/4]	1.4024[20]	al, **eth, ace, bz**	B6[3], 4563
1059	Anisole, m-(1-hydroxyethyl) 3-[CH$_3$CH(OH)]C$_6$H$_4$OCH$_3$	152.20		133[15]		1.0781[19/4]	1.5325[20]	al, eth	B6[3], 4564
1060	Anisole, p-(1-hydroxyethyl) 4-[CH$_3$CH(OH)]C$_6$H$_4$OCH$_3$	152.20		310d, 140-1[17]		1.0794[20/4]	1.5310[25]	al, eth	B6[3], 4565
1061	Anisole, 2-iodo 2-IC$_6$H$_4$OCH$_3$	234.04		239-40[730], 91-2[2]		1.8[20]		al, eth, ace, bz, lig	B6[4], 1070
1062	Anisole, 3-iodo 3-IC$_6$H$_4$OCH$_3$	234.04		244-5, 123[14]				al, eth	B6[4], 1073
1063	Anisole, 4-iodo 4-IC$_6$H$_4$OCH$_3$	234.04	lf (al), nd (MeOH)	237[726], 139[25]	53			eth, al	B6[4], 1075
1064	Anisole, 4-(3-methylbutyl) or p-Isopentyl anisole 4-i-C$_5$H$_{11}$C$_6$H$_4$OCH$_3$	178.28		121[14]				bz	B6[3], 1960
1065	Anisole, 2-nitro 2-NO$_2$C$_6$H$_4$OCH$_3$	153.14		276.8[752], 144[4]	10.5	1.2540[20/4]	1.5161[20]	**al, eth**	B6[4], 1249
1066	Anisole, 3-nitro 3-NO$_2$C$_6$H$_4$OCH$_3$	153.14	nd (al), pl (bz-lig)	258	38-9	1.373[18]		al, eth	B6[4], 1270
1067	Anisole, 4-nitro 4-NO$_2$C$_6$H$_4$OCH$_3$	153.14	pr (al), nd (dil al)	274	54	1.2192[60/4]	1.5070[60]	al, eth	B6[4], 1282
1068	Anisole, 4-propenyl or Anethole 4-(CH$_3$CH=CH)$_2$-C$_6$H$_4$OCH$_3$	148.21	lf (al)	234.5[763], 115[12]	21.3	0.9882[20/4]	1.5615[20]	al, eth, ace, bz	B6[3], 2395
1069	Anisole, 4-propyl 4-C$_3$H$_7$C$_6$H$_4$OCH$_3$	150.22		210, 98[17]		0.94718[20/4]	1.5045[20]	al, eth, ace, bz, chl	B6[3], 1789

No.	Name, Synonyms, and Formula	Mol. wt.	Color, crystalline form, specific rotation and λ_{max} (log ϵ)	b.p. °C	m.p. °C	Density	n_D	Solubility	Ref.
1070	Anisole, 2,3,4,6-tetrabromo 2,3,4,6-$Br_4C_6HOCH_3$	423.75	nd (dil al or aa)	340	113-4			al	B6[3], 766
1071	Anisole, 2,3,4,5-tetrachloro 2,3,4,5-$Cl_4C_6HOCH_3$	245.92	cr (MeOH)		83			al, bz	B6[2], 182
1072	Anisole, 2,3,5,6-tetrachloro 2,3,5,6-$Cl_4C_6HOCH_3$	245.92	nd (al)		89-90			al, eth, bz	B6[2], 182
1073	Anisole, 2,3,4,6-tetrachloro 2,3,4,6-$Cl_4C_6HOCH_3$	245.92	nd (MeOH), pr (al)		64-5			al, eth, bz	B6[2], 182
1074	Anisole, 2,3,4-tribromo 2,3,4-$Br_3C_6H_2OCH_3$	344.84	nd (al)		106			al, ace, bz	B6[4], 1066
1075	Anisole, 2,3,5-tribromo 2,3,5-$Br_3C_6H_2OCH_3$	344.84	pr (dil al)	305-12	82			al, ace, bz	B6[2], 192
1076	Anisole, 1,3,5-tribromo 1,3,5-$Br_3C_6H_2OCH_3$	344.84	nd (al)	297-9	88	2.491		ace, bz	B6[4], 1067
1077	Anisole, 2,4,5-tribromo 2,4,5-$Br_3C_6H_2OCH_3$	344.84	nd (al)	306-9[775]	105			al, ace, bz	B6[4], 1067
1078	Anisole, 3,4,5-tribromo 3,4,5-$Br_3C_6H_2OCH_3$	334.84	cr (al)	300-10	91-4			al, ace, bz	B6[2] 195
1079	Anisole, 2,3,5-trichloro 2,3,5-$Cl_3C_6H_2OCH_3$	211.48	nd (al)		84			ace, al	B6[2], 180
1080	Anisole, 2,4,5-trichloro 2,4,5-$Cl_3C_6H_2OCH_3$	211.48	nd (dil al)	252-5[742]	77.5			al, ace	B6[2], 180
1081	Anisole, 2,3,6-trichloro 2,3,6-$Cl_3C_6H_2OCH_3$	211.48	pr (al)	227-9[754]	45			al, ace	B6[2], 180
1082	Anisole, 2,4,6-trichloro 2,4,6-$Cl_3C_6H_2OCH_3$	211.48	mcl nd (al)	240[738]	61-2 (65)	1.640		al, ace	B6[1], 723
1083	Anisole, 2,4,6-triiodo 2,4,6-$I_3C_6H_2OCH_3$	485.83	lf (bz), nd (eth or al)		98-9			al, ace	B6[2], 204
1084	Anisole, 2,3,4-trinitro 2,3,4-$(NO_2)_3C_6H_2OCH_3$	243.13	pa ye lf (al)	exp	155			ace, aa, al	B6[1], 129
1085	Anisole, 2,3,5-trinitro 2,3,5-$(NO_2)_3C_6H_2OCH_3$	243.13	ye nd (al)		106.8	1.618[15]		al, bz, ace	B6[3], 873
1086	Anisole, 2,4,5-trinitro 2,4,5-$(NO_2)_3C_6H_2OCH_3$	243.13	ye (al)		106-7			eth, bz, w	B6[2], 253
1087	Anisole, 2,4,6-trinitro 2,4,6-$(NO_2)_3C_6H_2OCH_3$	243.13	nd (dil MeOH)		(i) 69, (ii) 56-7	1.4947[80]		al, eth, bz	B6[4], 1456
1088	Anisole, 2-vinyl or 2-Methoxystyrene 2-$(CH_2=CH)C_6H_4OCH_3$	134.18	nd	195-200, 83-4[12]	29	1.0049[17/4]	1.5388[20]	al, eth, ace, bz	B6[3], 2383
1089	Anisole, 3-vinyl or 3-Methoxystyrene 3-$(H_2C=CH)C_6H_4OCH_3$	134.18		114-6[16-7], 90-3[15]		0.999[16/4]	1.5586[21]	al, eth, bz	B6[3], 2385
1090	Anisole, 4-vinyl or 4-Methoxystyrene 4-$(CH_2=CH)C_6H_4OCH_3$	134.18		204-5[756], 91[13]		1.0001[13/4]	1.5642[13]	al, eth, bz	B6[3], 2386
1091	o-Anisonitrile or o-Anisic acid nitrile 2-$NCC_6H_4OCH_3$	133.15		255-6, 146[20]	24.5	1.1063[20/4]		al, eth	B10[3], 159
1092	p-Anisonitrile or p-Anisic acid nitrile 4-$CH_3OC_6H_4CN$	133.15	nd (w), lf (al)	256-7	61-2			al, eth, bz	B10[3], 344
1093	o-Anisyl chloride 2-$CH_3OC_6H_4COCl$	170.60		254, 128[11]					B10[3], 151
1094	m-Anisyl chloride 3-$CH_3OC_6H_4COCl$	170.60		243-4, 123-5[15]				eth, ace, bz	B10[3], 253
1095	p-Anisyl chloride 4-$CH_3OC_6H_4COCl$	170.60	nd	262-3, 91[1]	24-5	1.261[20/4]	1.580[20]	eth, ace, bz	B10[3], 337
1095a	o-Anisyl ether or bis(2-methoxy phenyl) ether $(2-CH_3OC_6H_4)_2O$	230.27	pl (lig)	330-1	79-80			al, eth	
1096	Anthracene $C_{14}H_{10}$	178.24	ta or mcl pr (al)	340 (cor), 226.5[53] sub	216-4	1.283[25/4]		ace, bz	B5[3], 2123.
1097	Anthracene, 1-acetyl 1-$CH_3COC_{14}H_9$	220.27	pa ye (al)		107.5-9			al	B7[3], 2538
1098	Anthracene, 2-acetyl 2-$CH_3COC_{14}H_9$	220.27	ye (al), cr (AcOEt-peth)		190-2			al	B7[3], 2538
1099	Anthracene, 9-acetyl 9-$CH_3COC_{14}H_9$	220.27	pa ye (al)		76			al	B7[3], 2539
1100	Anthracene, 1-amino or 1-Anthrylamine 1-$H_2NC_{14}H_9$	193.25	gold-ye nd (al)		130			al	B12[3], 3335

No.	Name, Synonyms, and Formula	Mol. wt.	Color, crystalline form, specific rotation and λ_{max} (log ε)	b.p. °C	m.p. °C	Density	n_D	Solubility	Ref.
1101	Anthracene, 2-amino or 2-Anthrylamine .... 2-$H_2NC_{14}H_9$	193.25	ye lf (al)	sub	238-41	....		al	B12[3], 3355
1102	Anthracene, 9-amino or 9-Anthrylamine .... 9-$H_2NC_{14}H_9$	193.25	ye lf (dil al), br (bz)		145-50			al, eth, bz, chl	B7[3], 2361
1103	Anthracene, 9-benzoyl or 9-Anthraphenone .... 9-$C_6H_5COC_{14}H_9$	282.35	ye nd (bz, aa)		148			ace, bz, aa	B7[2], 502
1104	Anthracene, 1-chloro .... 1-$ClC_{14}H_9$	212.68	lf (aa)		83.5	1.1707[100/4]	1.6959[100]	al, eth, bz	B5[3], 2132
1105	Anthracene, 2-chloro 2-$ClC_{14}H_9$	212.68	nd or lf		215 (223)			CCl_4	B5[3], 2133
1106	Anthracene, 9-chloro 9-$ClC_{14}H_9$	212.68	gold-ye nd (al)		106			al, eth, bz	B5[3], 2133
1107	Anthracene, 9,10-diamino 9,10-$(NH_2)_2C_{14}H_8$	208.27	red cr		196				B14[3], 291
1108	Anthracene, 9,10-dibenzyl 9,10($C_6H_5CH_2)_2C_{14}H_8$	274.49			245			lig, bz	B5[3], 2640
1109	Anthracene, 9,10-dibromo 9,10-$Br_2C_{14}H_8$	336.05	ye nd (to or xyl)	sub	226			chl, bz	B5[3], 2135
1110	Anthracene, 9,10-dichloro 9,10-$Cl_2C_{14}H_8$	247.14	ye nd (MeCOEt or CCl_4)		212			bz	B5[3], 2134
1111	Anthracene, 9,10-dihydro or 9,10-Dihydro anthracene ... $C_{14}H_{12}$	180.23	ta or pr	305, 165-70[12], sub	111	0.8976[11/4]	....	al, eth, bz	B5[3], 1987
1112	Anthracene, 9,10-dihydro-10-chloro-9-NO_2 .... 10-Cl-9-NO_2-$C_{14}H_{10}$	259.70	nd (bz)		163			bz	B5[3], 1989
1113	Anthracene, 9,10 dihydro 9 ethyl 9-$C_2H_5C_{14}H_{11}$	208.31		320-3 (cor)		1.049[18/18]		al, eth, bz, aa	B5[2], 560
1114	Anthracene, 9,10-dihydro-1-hydroxy .... 1-HO$C_{14}H_{11}$	196.25	sl grsh flr lf or nd (bz, peth)		94			al, eth, aa	B6[2], 660
1115	**Anthracene, 9,10-dihydro-2-hydroxy** .... 2-HO$C_{14}H_{11}$	196.25	(bz-peth), lf (dil al)		129				B6[2], 660
1116	Anthracene, 9,10-dihydro-9-hydroxy or 9-Hydroanthrol.. 9-HO-$C_{14}H_{11}$	196.25	nd (peth)		76			al, eth, bz	B6[3], 3504
1117	Anthracene, 9,10-dihydro-9-oxo or 9-Anthrone .... $C_{14}H_{10}O$	194.24	nd (bz-lig, aa)		155			ace, bz	B7[3], 2359
1118	Anthracene, 9,10-dihydro-9-oxo-10-nitro $C_{14}H_9NO_3$	239.23	cr (bz-lig), nd (CS_2)		140 (148d)			al, bz	B7[3], 2371
1119	Anthracene, 1,2-dihydroxy or 1,2-Anthradiol .... 1,2-$(HO)_2C_{14}H_8$	210.24	pa gr lf		160-2			al, eth, aa	B6[2], 998
1120	Anthracene, 1,5-dihydroxy or 1,5-Anthradiol, Ruful .... 1,5-$(HO)_2C_{14}H_8$	210.24	ye nd		265d			al, eth, bz	B6[3], 5683
1121	Anthracene, 1,8-dihydroxy or 1,8-Anthradiol, Chrysazol 1,8-$(HO)_2C_{14}H_8$	210.24	ye nd (dil al), lf (al-aa)		225d			al, eth, bz, AcOEt	B6, 1033
1122	Anthracene, 2,6-dihydroxy or 2,6-Anthradiol, Flavol .... 2,6-$(HO)_2C_{14}H_8$	210.24	pa ye lf (al)		295-300d			al, eth, aa	B6[3], 5684
1123	Anthracene, 9,10-dihydroxy or 9,10-Anthradiol, Anthra-quinol .... 9,10-$(HO)_2C_{14}H_8$	210.24	br or ye nd		180			al, eth	B6[3], 5685
1124	Anthracene, 1,3-dimethyl .... 1,3-$(CH_3)_2C_{14}H_8$	206.29	pa bl fir lf (eth), cr (al)	140-5[2]	83			al, eth	B5[3], 2165
1125	Anthracene, 2,3-dimethyl 2,3-$(CH_3)_2C_{14}H_8$	206.29	bl gr fir lf (bz)		252			al, bz	B5[3], 2166
1126	Anthracene, 9-ethyl 9-$C_2H_5C_{14}H_9$	206.29	bl flr lf (al or MeOH)		59	1.0413[99/4]	1.6762[99]	al, eth	B5[3], 2165
1127	Anthracene, 1,2,3,4,5,6-hexahydro .... $C_{14}H_{16}$	184.28	lf (MeOH), cr (al)	160[15]	67 (70)			bz	B5[2], 472
1128	Anthracene, 1-hydroxy or 1-Anthrol................ 1-$(HO)C_{14}H_9$	194.24	cr (bz), br nd or lf (al or aa)	234[15]	158			al, eth	B6[3], 3551
1129	Anthracene, 2-hydroxy or 2-Anthrol................ 2-$(HO)C_{14}H_9$	194.24	ye (bz), br lf or nd (dil al)		253			al, eth, ace, bz	B6[3], 3552

No.	Name, Synonyms, and Formula	Mol. wt.	Color, crystalline form, specific rotation and λ_{max} (log ϵ)	b.p. °C	m.p. °C	Density	n_D	Solubility	Ref.
1130	Anthracene, 9-hydroxy or 9-Anthrol. 9-HOC$_{14}$H$_9$	194.24	ye red lf (dil al), pa ye nd (aa)		160-4			al, bz	B6³, 3554
1131	Anthracene, 1-methyl 1-CH$_3$C$_{14}$H$_9$	192.26	bl nd (MeOH), lf (al)	199—200	85-6	1.0471$^{99/4}$	1.6802$^{99/4}$	al, eth, bz, aa	B5³, 2149
1132	Anthracene, 2-methyl 2-CH$_3$C$_{14}$H$_9$	192.26	gr-bl flr lf (sub)	sub	209	1.81$^{0/4}$		bz, chl	B5³, 2149
1133	Anthracene, 9-methyl 9-CH$_3$C$_{14}$H$_9$	192.26	yesh nd (dil al), pr (bz, al)	196-7^{12}	81.5	1.065$^{99/4}$	1.6959^{99}	al, eth, ace, bz	B5³, 2150
1134	Anthracene, 9-nitro 9-O$_2$NC$_{14}$H$_9$	223.23	ye nd (al), pr (aa or xyl)	ca 275^{17}	146			bz, al	B5³, 2136
1135	Anthracene, octahydro or Octhracene C$_{14}$H$_{18}$	186.30	pl (al)	293-5, 167^{12}	78	0.9703$^{80/4}$	1.5372^{80}	al, bz, aa	B5⁴, 1584
1136	Anthracene, 2-phenyl 2-C$_6$H$_5$C$_{14}$H$_9$	254.34			207				B5³, 2462
1137	Anthracene, 9-phenyl 9-C$_6$H$_5$C$_{14}$H$_9$	254.34	bl flr in sol, lf (al) (aa)		207			al, ace, bz, chl	B5³, 2462
1138	Anthracene, tetradecahydro C$_{14}$H$_{24}$	192.34		128^{11}	93				B5⁴, 492
1139	Anthracene, 1,2,3,4-tetrahydro C$_{14}$H$_{14}$	182.36			106-7				B5⁴, 1909
1140	Anthracene, 1,2,9-trihydroxy 1,2,9-(HO)$_3$C$_{14}$H$_7$	226.24	og ye lf		149-51				B6³, 2728
1141	Anthracene, 1,2,10-trihydroxy 1,2,10-(HO)$_3$C$_{14}$H$_7$	226.24	ye lf, nd (al-w)		208			al, eth, ace, bz, aa	B8², 372
1142	Anthracene, 1,4,9-trihydroxy 1,4,9-(HO)$_3$C$_{14}$H$_7$	226.24	og-red nd (al)		156			al	B6³, 2800
1143	Anthracene, 1,5,9-trihydroxy 1,5,9-(HO)$_3$C$_{14}$H$_7$	226.24	gold lf (al)		200d without melt			al	B6³, 2801
1144	Anthracene, 1,8,9-trihydroxy or Anthralin 1,8,9-(HO)$_3$C$_{14}$H$_7$	226.24	ye pl or nd (lig)		178-80			al, eth, ace, bz	B6³, 2802
1145	Anthracene, 1,9,10-trihydroxy (enol form) 1,9,10-(HO)$_3$C$_{14}$H$_7$	226.24	gr nd (eth)		204-6			al, eth	B8², 372
1146	Anthracene, 1,9,10-trihydroxy (keto form) 1,9,10-(HO)$_3$C$_{14}$H$_7$	226.24	ye nd (lig)		135-7			al	B8², 372
1147	Anthracene, 2,3,9-trihydroxy 2,3,9-(HO)$_3$C$_{14}$H$_7$	226.24	ye br nd (al)		288-9			al, eth, ace, aa	B6³, 2803
1148	Anthracene-2-vinyl 2-CH$_2$=CHC$_{14}$H$_9$	204.27			125				
1149	Anthracene-9-vinyl 9-CH$_2$=CHC$_{14}$H$_9$	204.27			64-5				
1150	9-Anthraldehyde or 9-Anthracene carboxaldehyde 9-C$_{14}$H$_9$CHO	206.25	og nd (dil al)		104-5			bz, aa	B7³, 2527
1151	Anthranil or 3,4-Benzo isoxazol C$_7$H$_5$NO	119.12		215d, 99^{13}	−18	1.8127$^{20/4}$	1.5845^{20}	al, ace	B27², 17
1152	Anthranil amide 2-H$_2$NC$_6$H$_4$CONH$_2$	136.17	lf (chl, w)	300	110-1			al, AcOEt	B14³, 889
1153	Anthranilic acid or 2-Aminobenzoic acid 2-H$_2$NC$_6$H$_4$CO$_2$H	137.14	lf (al)	sub	146-7	1.412^{20}		w, al, eth, chl	B14³, 879
1154	Anthranilic acid, N-acetyl 2-CH$_3$CONHC$_6$H$_4$CO$_2$H	179.18	nd (aa)		185			al, eth, ace, bz, aa	B14³, 922
1155	Anthranilic acid, N-acetyl-4-ethoxy 2-CH$_3$CONH-4-C$_2$H$_5$O-C$_6$H$_3$CO$_2$H	223.23	nd (al or MeOH)		199d			al	B14¹, 657
1156	Anthranilic acid, butyl ester or Butyl anthranilate 2-H$_2$NC$_6$H$_4$CO$_2$C$_4$H$_9$	193.25		182					B14², 209
1157	Amthranilic acid, iso-butyl ester or iso-Butyl anthranilate 2-H$_2$NC$_6$H$_4$CO$_2$-i-C$_4$H$_9$	193.25		156-7$^{13.5}$					B14², 209
1158	Anthranilic acid, 3,4-dichloro 3,5-Cl$_2$-2-H$_2$NC$_6$H$_2$CO$_2$H	206.03	nd (aa)		237-8			al, eth, chl	B14¹, 549
1159	Anthranilic acid, 3,5-dichloro 3,5-Cl$_2$-2-H$_2$NC$_6$H$_2$CO$_2$H	206.03	nd or lf (al)		231-2d			al, ace, eth, bz	B14³, 966
1160	Anthranilic acid, 3,6-dichloro 3,6-Cl$_2$-2-H$_2$NC$_6$H$_2$CO$_2$H	206.03	nd (w or aa)	sub	155			al, eth, ace, w, bz, aa	B14, 367

PHYSICAL CONSTANTS OF ORGANIC COMPOUNDS (Continued)

No.	Name, Synonyms, and Formula	Mol. wt.	Color, crystalline form, specific rotation and λ_{max} (log ε)	b.p. °C	m.p. °C	Density	n_D	Solubility	Ref.
1161	Anthranilic acid, 4,5-dichloro 4,5-Cl$_2$-2-H$_2$NC$_6$H$_2$CO$_2$H	206.03	nd (aa)		213-4			al, eth, aa	B14[1], 549
1162	Anthranilic acid, 5,6-dichloro 5,6-Cl$_2$-2-H$_2$NC$_6$H$_2$CO$_2$H	206.03	nd (MeOH)		176-7d			al, eth, aa	B14, 368
1163	Anthranilic acid, 3,5-diido 3,5-I$_2$-2-H$_2$NC$_6$H$_2$CO$_2$H	388.93	pr (al)		232-3			eth	B14[3], 973
1164	Anthranilic acid, 3,5-diido,ethyl ester 3,5-I$_2$-2-H$_2$NC$_6$H$_2$CO$_2$C$_2$H$_5$	416.96	pr (al)		101				B14[1], 555
1165	Anthranilic acid, 4,5-diido 4,5-I$_2$-2-H$_2$NC$_6$H$_2$CO$_2$H	388.93	cr (dil NH$_3$)		210-2d			al, eth, bz	B14[1], 555
1166	Anthranilic acid, 4,5-diido,ethyl ester 4,5-I$_2$-2-H$_2$NC$_6$H$_2$CO$_2$C$_2$H$_5$	416.99	pr (al)		137			al, eth, bz	B14[1], 555
1167	Anthranilic acid, N,N-dimethyl 2-(CH$_3$)$_2$NC$_6$H$_4$CO$_2$H	165.19	pr nd (eth)	sub d	72			w, al, eth	B14[3], 896
1168	Anthranilic acid, N-ethyl 2-(C$_2$H$_5$NH)C$_6$H$_4$CO$_2$H	165.19	pr or nd (dil al)		154			al, eth, ace	B14[3], 897
1169	Anthranilic acid, ethyl ester or Ethyl anthranilate 2-H$_2$NC$_6$H$_4$CO$_2$C$_2$H$_5$	165.19		268, 145-7[15]	13	1.1374[20/4]	1.5646[20]	al, eth	B14[3], 885
1170	Anthranilic acid hydrochloride or 2-Aminobenzoic acid hydrochloride 2-H$_2$NC$_6$H$_4$CO$_2$H·HCl	173.60						w	B14[2], 207
1171	Anthranilic acid, 3-hydroxy or 2-Amino-3-hydroxy benzoic acid 3-HO-2-H$_2$NC$_6$H$_3$CO$_2$H	153.14	lf(w)		164			al, eth, chl	B14[2], 1463
1172	Anthranilic acid, 3 hydroxy or 2-Amino-3-hydroxy benzoic acid 3-HO-2-H$_2$NC$_6$H$_3$CO$_2$H	153.14	lf (w)		164			al, eth, chl	B14[2], 1463
1173	Anthranilic acid, 4-hydroxy or 2-Amino-4-hydroxy benzoic acid 4-HO-2-H$_2$NC$_6$H$_3$CO$_2$H	153.14			148d			w, al, eth, ace	B14[3], 1475
1174	Anthranilic acid, 5-hydroxy 5-HO-2-H$_2$NC$_6$H$_3$CO$_2$H	153.14	vt pr (w)		252d			al, eth, ace, bz	B14[3], 1468
1175	Anthranilic acid, N-methyl 2-CH$_3$NHC$_6$H$_4$CO$_2$H	153.17	pl (al or lig)	80[0.01]	179			al, eth, bz, chl, lig	B14[3], 895
1176	Anthranilic acid, N-methyl,ethyl ester or Ethyl-N-methyl anthranilate 2-CH$_3$NHC$_6$H$_4$CO$_2$C$_2$H$_5$	179.22		266	39			eth	B14[2], 213
1177	Anthranilic acid, N-methyl,methyl ester 2-CH$_3$NHC$_6$H$_4$CO$_2$CH$_3$	165.09	cr (peth)	255, 130-1[13]	19	1.120[15]	1.5839[15]	al, eth	B14[3], 895
1178	Anthranilic acid, methyl ester or Methyl anthranilate 2H$_2$NC$_6$H$_4$CO$_2$CH$_3$	151.17		256, 135.5[15]	24—5	1.1682[10/4]	1.5810	al, eth	B14[3], 894
1179	Anthranilic acid, 3-methyl or 2-Amino-3-methyl benzoic acid 3-CH$_3$-2-H$_2$NC$_6$H$_3$CO$_2$H	151.17	nd (al), pr (w)		172			al, eth	B14[2], 290
1180	Anthranilic acid, 5-methyl 5-CH$_3$-2-H$_2$NC$_6$H$_3$CO$_2$H	151.17	lf (al), nd (w)		175			al, eth	B14[2], 291
1181	Anthranilic acid, 6-methyl 6-CH$_3$-2-H$_2$NC$_6$H$_3$CO$_2$H	151.17	nd (MeOH)		125-6d			MeOH	B14[3], 1502
1182	Anthranilic acid, 3-nitro 3-NO$_2$-2-H$_2$NC$_6$H$_3$CO$_2$H	182.14	ye nd (w)		208-9	1.558[15]		al, eth	B14[3], 973
1183	Anthranilic acid, 4-nitro 4-NO$_2$-2-H$_2$NC$_6$H$_3$CO$_2$H	182.14	og pr (dil al)		269			al, eth, ace, xyl	B14[3], 975
1184	Anthranilic acid, 5-nitro 5-NO$_2$-2-H$_2$NC$_6$H$_3$CO$_2$H	182.14	lf (al), ye nd (w, dil al)		268-70, (280)			al, eth	B14[3], 980
1185	Anthranilic acid, 6-nitro 6-NO$_2$-2-H$_2$NC$_6$H$_3$CO$_2$H	182.14	ye nd or lf (w)		184			w, al, eth, ace, aa	B14[3], 988
1186	Anthranilic acid, N-phenyl 2-C$_6$H$_5$NHC$_6$H$_4$CO$_2$H	213.24	lf, nd or pr (al)		184d			al	B14[3], 898
1187	Anthranilic acid, phenyl ester or Phenyl anthranilate 2-H$_2$NC$_6$H$_4$CO$_2$C$_6$H$_5$	213.24	nd (al)		70			al, eth	B14[3], 885
1188	Anthranilic acid, propyl ester or Propyl anthranilate 2-H$_2$NC$_6$H$_4$CO$_2$C$_3$H$_7$	179.22		270				al, eth	B14[2], 209
1189	Anthranilonitrile or o-Cyanoaniline 2-H$_2$NC$_6$H$_4$CN	118.14	ye pr (CS$_2$), nd (peth)	263[751]	51			al, eth, ace, bz	B14[2], 210
1190	Anthranilonitrile, 4-methyl 4-CH$_3$-2-H$_2$NC$_6$H$_3$CN	132.18	lf (dil al)		94			al, ace, bz, chl	B14[3], 1208

No.	Name, Synonyms, and Formula	Mol. wt.	Color, crystalline form, specific rotation and λ_{max} (log ε)	b.p. °C	m.p. °C	Density	n_D	Solubility	Ref.
1191	Anthanilonitrile, 5-methyl 5-CH$_3$-2-H$_2$NC$_6$H$_3$CN	132.18	cr (dil al)		63			al, eth, ace, bz	B14, 482
1192	Anthanilonitrile, 6-methyl 6-CH$_3$-2-H$_2$NC$_6$H$_3$CN	132.18	ye pr (bz) (w)		128				B14^2, 290
1193	Anthranylamide 2-H$_2$NC$_6$H$_4$CONH$_2$	136.17	lf (chl or w)	300	110-1			al, AcOEt, w	B14^3, 889
1194	β-Anthraquinoline or Naphtho(2,3:5,6)quinoline C$_{17}$H$_{11}$N	229.28	lf or ta (al)	446	170			al, eth, bz	B20^4, 4290
1195	Anthraquinone or 9,10-Dioxoanthracene C$_{14}$H$_8$O$_2$	208.23	ye rh nd (al, bz)	379.8	286 (sub)	1.438^4			B7^3, 4059
1196	Anthraquinone, 1-amino 1-H$_2$NC$_{14}$H$_7$O$_2$	223.23	red nd (al), gl (aa)	sub	253-4			eth, ace, bz, aa, chl	B14^3, 407
1197	Anthraquinone, 1-amino-2-benzoyl 1-NH$_2$-2-(C$_6$H$_5$CO)C$_{14}$H$_6$O$_2$	327.34	red nd (aa)		190			aa	B14, 482
1198	Anthraquinone, 1-amino-2-bromo 1-NH$_2$-2-BrC$_{14}$H$_6$O$_2$	302.15	ye, red nd (aa), nd (xyl)		182			to, aa	B14^3, 423
1199	Anthraquinone, 1-amino-3-bromo 3-Br-1-H$_2$NC$_{14}$H$_6$O$_2$	302.15	red nd (to)		243				B14^3, 423
1200	Anthraquinone, 1-amino-4-bromo 1-H$_2$N-4-Br-C$_{14}$H$_6$O$_2$	302.13			170				B14^3, 423
1201	Anthraquinone, 1-amino-4-bromo-2-methyl C$_{15}$H$_{10}$NO$_2$Br	316.15			232			al, eth	B14^3, 494
1202	Anthraquinone, 1-amino-4-hydroxy C$_{14}$H$_9$NO$_3$	239.23			215			al, ace	B14^3, 652
1203	Anthraquinone, 1-amino-2-methyl C$_{15}$H$_{11}$NO$_2$	237.26			205-6			al, eth, bz, chl, aa	B14^3, 492
1204	Anthraquinone, 1-amino-4-methyl C$_{15}$H$_{11}$NO$_2$	237.26			181				B14^2, 123
1205	Anthraquinone, 1-amino-4-nitro C$_{14}$H$_8$N$_2$O$_4$	268.23			298-300			al	B14^3, 427
1206	Anthraquinone, 2-amino 2-NH$_2$C$_{14}$H$_8$O$_2$	223.23	red nd (al aa)	sub	303-6			ace, bz, chl	B14^3, 429
1207	Anthraquinone, 2-amino-3-benzoyl 2-NH$_2$-3(C$_6$H$_5$CO)C$_{14}$H$_6$O$_2$	327.34	ye pl (Py)		331				B14^1, 482
1208	Anthraquinone, 2-amino-3-chloro C$_{14}$H$_8$NO$_2$Cl	257.68			280-3				B14^3, 435
1209	Anthraquinone, 3-amino-1,2-dihydroxy or 3-Aminoalizarin 3-NH$_2$-1,2-(HO)$_2$C$_{14}$H$_5$O$_2$	255.23	dk red pr (aa)	sub d	>300			aq NH$_3$	B14^2, 185
1210	Anthraquinone, 2-amino-1-hydroxy 2-NH$_2$-1-HOC$_{14}$H$_6$O$_2$	239.23	red br nd (al)	sub	226-7			al, eth, bz	B14^3, 651
1211	Anthraquinone, 4-amino-1,2-dihydroxy or 4-Aminoalizarin 4-NH$_2$-1,2-(HO)$_2$C$_{14}$H$_5$O$_2$	255.23	gr bl nd (al)		d			al	B14^2, 185
1212	Anthraquinone, 1-bromo 1-BrC$_{14}$H$_7$O$_2$	287.12	ye nd (bz)	sub	188			al, bz	B7^3, 4076
1213	Anthraquinone, 1-bromo-4-methylamino 1-Br-4(CH$_3$NH)C$_{14}$H$_6$O	316.16	br-red nd (Py)		194				B14^3, 424
1214	Anthraquinone, 2-bromo 2-BrC$_{14}$H$_7$O$_2$	287.12		sub	204-5			bz	B7^3, 4076
1215	Anthraquinone, 2-bromo-1-methylamino 2-Br-1-(CH$_3$NH)C$_{14}$H$_6$O$_2$	316.16	br nd (aa)		170-2				B14^1, 446
1216	Anthraquinone, 3-bromo-1,2-dihydroxy or 3-Bromoalizarin 3-Br-1,2-(HO)$_2$C$_{14}$H$_5$O$_2$	319.12	br-red nd (to)	sub	260-1			w	B8^3, 3772
1217	Anthraquinone, 1-chloro 1-ClC$_{14}$H$_7$O$_2$	242.66	ye nd (to or al)	sub	162			eth, bz, aa	B7^3, 4064
1218	Anthraquinone, 1-chloro-2-methyl 1-Cl-2-CH$_3$C$_{14}$H$_6$O$_2$	256.69			170-1				B7^3, 4105
1219	Anthraquinone, 2-chloro 2-ClC$_{14}$H$_7$O$_2$	242.66	pa ye nd (aa or al)	sub	211			bz, al	B7^3, 4066
1220	Anthraquinone, 1,2-diamino 1,3-(NH$_2$)$_2$C$_{14}$H$_6$O$_2$	238.25	vt nd		303-4				B14^3, 438
1221	Anthraquinone, 1,3-diamino 1,3-(NH$_2$)$_2$C$_{14}$H$_6$O$_2$	238.25	red		290			Py	B14^3, 439

No.	Name, Synonyms, and Formula	Mol. wt.	Color, crystalline form, specific rotation and λ_{max} (log ε)	b.p. °C	m.p. °C	Density	n_D	Solubility	Ref.
1222	Anthraquinone, 1,4-diamino . . . . 1,4-$(NH_2)_2C_{14}H_6O_2$	238.25	dk vt nd (Py), vt cr		268			al, bz, Py	B14³, 439
1223	Anthraquinone, 1,5-diamino . . . . 1,5-$(NH_2)_2C_{14}H_6O_2$	238.25	dk red nd (al, aa)	sub	319 (cor)				B14³, 466
1224	Anthraquinone, 1,6-diamino . . . . 1,6-$(NH_2)_2C_{14}H_6O_2$	238.25	red nd (aa)		297				B14², 119
1225	Anthraquinone, 1,7-diamino . . . . 1,7-$(NH_2)_2C_{14}H_6O_2$	238.25	red, nd		290				B14¹, 470
1226	Anthraquinone, 1,8-diamino . . . . 1,8-$(NH_2)_2C_{14}H_6O_2$	238.25	red (al or aa)	265				al, aa	B14³, 478
1227	Anthraquinone, 2,3-diamino . . . . 2,3-$(NH_2)_2C_{14}H_6O_2$	238.25	red		353			Py	B14³, 480
1228	Anthraquinone, 2,6-diamino . . . . 2,6-$(NH_2)_2C_{14}H_6O_2$	238.25	red-br pr (aq Py)		320d			Py, xy, al	B14³, 480
1229	Anthraquinone, 2,7-diamino . . . . 2,7-$(NH_2)_2C_{14}H_6O_2$	238.25	og-ye nd (al), dk red nd (sub)	sub	>320				B14³, 483
1230	Anthraquinone, 2,3-dibromo . . . . 2,3-$Br_2C_{14}H_6O_2$	366.02	ye nd (to)	sub	283			bz, chl	B7³, 4077
1231	Anthraquinone, 2,7-dibromo . . . . 2,7-$Br_2C_{14}H_6O_2$	366.02	lt ye lf	sub	248			bz, aa	B7², 718
1232	Anthraquinone, 1,3-dichloro . . . . 1,3-$Cl_2C_{14}H_6O_2$	277.11	ye nd (aa)		209-10			aa	B7³, 4068
1233	Anthraquinone, 1,4-dichloro . . . . 1,4-$Cl_2C_{14}H_6O_2$	277.11	og-ye nd (aa)		187-8			Py, aa	B7³, 4068
1234	Anthraquinone, 1,5-dichloro . . . . 1,5-$Cl_2C_{14}H_6O_2$	277.11	yesh (to), ye nd		252			aa	B7³, 4068
1235	Anthraquinone, 1,6-dichloro . . . . 1,6-$Cl_2C_{14}H_6O_2$	277.11	pa ye nd (aa)		203-4			ace, to	B7¹, 4070
1236	Anthraquinone, 1,7-dichloro . . . . 1,7-$Cl_2C_{14}H_6O_2$	277.11	ye nd (aa)		213-4			bz, to	B7³, 4070
1237	Anthraquinone, 1,8-dichloro . . . . 1,8-$Cl_2C_{14}H_6O_2$	277.11	ye nd (aa)		202-3			bz	B7³, 4070
1238	Anthraquinone, 2,3-dichloro . . . . 2,3-$Cl_2C_{14}H_6O_2$	277.11	ye nd (aa)		271			bz, aa	B7³, 4071
1239	Anthraquinone, 2,6-dichloro . . . . 2,6-$Cl_2C_{14}H_6O_2$	277.11	ye nd (aa or al)		291			bz, aa, al	B7³, 4071
1240	Anthraquinone, 2,7-dichloro . . . . 2,7-$Cl_2C_{14}H_6O_2$	277.11	yesh nd		212 (231)			eth	B7³, 4072
1241	Anthraquinone, 1,2-dihydroxy or Alizarin . . . . 1,2-$(HO)_2C_{14}H_6O_2$	240.23	og or red tcl nd or pr (al, sub)	430 (sub)	289-90 (cor)			al, eth, ace, bz, Py	B8³, 3767
1242	Anthraquinone, 1,2-dihydroxy-3-iodo or β-iodo alizarin . . 1,2-$(HO)_2$-3-$IC_{14}H_5O_2$	366.11	og-red nd (xyl)		229			w	B8³, 3773
1243	Anthraquinone, 1,2-dihydroxy-3-methyl or β-methylalizarin . . . . 1,2$(HO)_2$-3-CH_3-$C_{14}H_5O_2$	254.25	og nd	sub	245			al, eth, ace	B8³, 3808
1244	Anthraquinone, 1,2-dihydroxy-3-nitro or β-nitroalizarin, Alizarin orange . . . . 1,2-$(HO)_2$-3-$(NO_2)C_{14}H_5O_2$	285.22	og-ye nd (bz), ye, pl (gl aa, al)	sub, d	244d			al, bz, aa	B8³, 3774
1245	Anthraquinone, 1,2-dihydroxy-4-nitro or 4-Nitroalizarin 1,2-$(HO)_2$-3-$(NO_2)C_{14}H_5O_2$	285.22	gold-ye nd (aa or al)	sub, d	289d			al, bz, aa, chl	B8², 491
1246	Anthraquinone, 1,3-dihydroxy or Purpuro xanthin . . . . . . 1,3-$(HO)_2C_{14}H_6O_2$	240.23	ye-red nd (sub)		268-70			al, ace, bz, aa	B8³, 3774
1247	Anthraquinone, 1,4-dihydroxy or Quinizarin . . . . 1,4-$(HO)_2C_{14}H_6O_2$	240.23	ye red lf (eth), dk red nd		200			w, al, bz, eth	B8³, 3775
1248	Anthraquinone, 1,4-dihydroxy-5,6,7,8-tetrachloro or 5,6,7,8-Tetrachloroquinizarin . . . . 1,4-$(HO)_2$-5,6,7,8-$Cl_4C_{14}H_2O_2$	378.00	red pl (aa)		270			bz, aa	B8³, 3783
1249	Anthraquinone, 1,5-dihydroxy or Anthrarufin . . . . 1,5-$(HO)_2C_{14}H_6O_2$	240.23	pa ye pl (gl aa)	sub	280			bz	B8³, 3787
1250	Anthraquinone, 1,6-dihydroxy . . . . 1,6-$(HO)_2C_{14}H_6O_2$	240.23	og-ye nd (gl aa)		276				B8³, 3791
1251	Anthraquinone, 1,7-dihydroxy . . . . 1,7-$(HO)_2C_{14}H_6O_2$	240.23	ye nd (sub)	sub	292-3			al, eth, bz, chl	B8³, 3791

No.	Name, Synonyms, and Formula	Mol. wt.	Color, crystalline form, specific rotation and λ_{max} (log ε)	b.p. °C	m.p. °C	Density	n_D	Solubility	Ref.
1252	Anthraquinone, 1,8-dihydroxy or Chrysazin 1,8-(HO)$_2$C$_{14}$H$_6$O$_2$	240.23	red or redsh-ye nd or lf (al)	sub	193	....	....	al, eth, ace, chl	B8^3, 3792
1253	Anthraquinone, 1,8-dihydroxy-3-hydroxymethyl or Aloe emedin 1,8-(HO)$_2$-3(HOCH$_2$)C$_{14}$H$_5$O$_2$	270.25	og-ye nd (to, al)	sub	223-4			al, eth, bz	B8^3, 4160
1254	Anthraquinone, 1,8-dihydroxy-3-methyl or Chryso-phenol, Chrysophanic acid 1,8-(HO)$_2$-3-CH$_3$C$_{14}$H$_5$O$_2$	254.25	ye hex or mcl nd (sub)	sub	196	0.92		ace, bz, aa	B8^2, 510
1255	Anthraquinone, 1,8-dihydroxy-2,4,5,7-tetrabromo or 2,4,5,7-Tetrabromoquinizarin 1,8-(HO)$_2$-2,4,5,7-Br$_4$C$_{14}$H$_2$O$_2$	555.82	og-ye nd (bz)		312				B8^1, 722
1256	Anthraquinone, 1,8-dihydroxy-2,4,5,7-tetranitro or Chry-sammic acid, Chrysamminic acid 1,8-(HO)$_2$-2,4,5,7-(NO$_2$)$_4$C$_{14}$H$_2$O$_2$	420.21	ye pl or lf	d	exp			al, eth	B8^1, 723
1257	Anthraquinone, 2,3-dihydroxy or Hystazin 2,3-(HO)$_2$C$_{14}$H$_6$O$_2$	240.23	ye-br nd (aa), ye nd (sub)	sub	>330				B8^3, 3794
1258	Anthraquinone, 2,6 dihydroxy or Anthraflavin 2,6-(HO)$_2$C$_{14}$H$_6$O$_2$	240.23	ye nd (al)		360d				B8^3, 3796
1259	Anthraquinone, 2,7-dihydroxy or Isoanthraflavin 2,7-(HO)$_2$C$_{14}$H$_6$O$_2$	240.23	ye nd (+ 1 w, dil al), nd (sub),	sub	350-5			al, aa	B8^3, 3798
1260	Anthraquinone, 1,8-dimethoxy 1,8(CH$_3$O)$_2$C$_{14}$H$_6$O$_2$	268.27			221			al, bz, lig	B8^3, 3793
1261	Anthraquinone, 2,6-dimethoxy 2,6-(CH$_3$O)$_2$C$_{14}$H$_6$O$_2$	268.27			256				B8^3, 3797
1262	Anthraquinone, 1,2-dimethyl 1,2-(CH$_3$)$_2$C$_{14}$H$_6$O$_2$	236.27	nd (ace or aa)		156			al, eth, ace, bz, aa	B7^3, 4130
1263	Anthraquinone, 1,3-dimethyl 1,3-(CH$_3$)$_2$C$_{14}$H$_6$O$_2$	236.27	nd (aa)		162			aa	B7^3, 4130
1264	Anthraquinone, 1,4-dimethyl 1,4-(CH$_3$)$_2$C$_{14}$H$_6$O$_2$	236.27	ye nd (al, sub)	sub	140-1			bz, xyl, aa	B7^3, 4131
1265	Anthraquinone, 2,3-dimethyl 2,3-(CH$_3$)$_2$C$_{14}$H$_6$O$_2$	236.27	ye nd (al, to or xyl), cr (aa)	sub	210			al, bz, xyl	B7^3, 4131
1266	Anthraquinone, 2,6-dimethyl 2,6-(CH$_3$)$_2$C$_{14}$H$_6$O$_2$	236.27	ye nd (aa or al)	sub	242			to	B7^3, 4132
1267	Anthraquinone, 2,7-dimethyl 2,7-(CH$_3$)$_2$C$_{14}$H$_6$O$_2$	236.27	yesh nd (al)		170			al	B7^3, 4133
1268	Anthraquinone, 1,3-dinitro 1,3-(NO$_2$)$_2$C$_{14}$H$_6$O$_2$	298.21	ye nd (HNO$_3$)		246-50				E13, 436
1269	Anthraquinone, 1,5-dinitro 1,5-(NO$_2$)$_2$C$_{14}$H$_6$O$_2$	298.21	pa ye nd (xyl), ye-cr (sub)	sub	422 (385)				B7^3, 4081
1270	Anthraquinone, 1,6-dinitro 1,6-(NO$_2$)$_2$C$_{14}$H$_6$O$_2$	298.21			257-9				B7^3, 4082
1271	Anthraquinone, 1,7-dinitro 1,7(NO$_2$)$_2$C$_{14}$H$_6$O$_2$	298.21			295				B7^3, 4082
1272	Anthraquinone, 1,8-dinitro 1,8-(NO$_2$)$_2$C$_{14}$H$_6$O$_2$	298.21			312				B7^3, 4082
1273	Anthraquinone, 1,5-disulfonic acid 1,5-(HO$_3$S)$_2$C$_{14}$H$_6$O$_2$	368.35	ye nd (HCl + 4w), pl (dil aa + 4w)		310-1d			w, al, aa	B11^3, 634
1274	Anthraquinone, 1,6-disulfonic acid 1,6-(HO$_3$S)$_2$C$_{14}$H$_6$O$_2$	368.35	ye nd (HCl + 5w), gold pr (dil aa + 5w)		215-7d			w, al, aa	B11^3, 634
1275	Anthraquinone, 1,7-disulfonic acid 1,7-(HO$_3$S)$_2$C$_{14}$H$_6$O$_2$	368.35	ye hyg pw (dil aa + 4w)		d at 120			w, al, aa	B11^3, 634
1276	Anthraquinone, 1,8-disulfonic acid 1,8-(HO$_3$S)$_2$C$_{14}$H$_6$O$_2$	368.35	ye nd (+ 5w)		293-4d			w, al	B11^3, 635
1277	Anthraquinone, 2-ethyl-1-nitro 1-(NO$_2$)-2(C$_2$H$_5$)C$_{14}$H$_6$O$_2$	281.27	yesh br (aa)		226				B7^2, 743

No.	Name, Synonyms, and Formula	Mol. wt.	Color, crystalline form, specific rotation and λ_{max} (log ϵ)	b.p. °C	m.p. °C	Density	n_D	Solubility	Ref.
1278	Anthraquinone, 1,2,3,5,6,7-hexahydroxy or Rufigalloi 1,2,3,5,6,7-(HO)$_6$C$_{14}$H$_2$O$_2$	304.23	red rh, red, ye nd (sub)	....	sub, d			ace	B8³, 4401
1279	Anthraquinone, 1-hydroxy or Erythroxyanthraquinone 1-HOC$_{14}$H$_7$O$_2$	224.25	red-og nd (al)	sub	194-5			al, eth, bz	B8³, 2906
1280	Anthraquinone, 2-hydroxy 2-HOC$_{14}$H$_7$O$_2$	224.23	ye pl or nd (al or aa)	sub	306			al, eth	B8³, 2921
1281	Anthraquinone, 2-methoxy 2-CH$_3$OC$_{14}$H$_7$O$_2$	238.24			196			bz, al	B8³, 2922
1282	Anthraquinone, 2-methyl 2-CH$_3$C$_{14}$H$_7$O$_2$	222.25	yesh nd (al aa)	sub	182-3			al, bz, aa	B7³, 4104
1283	Anthraquinone, 1-methylamino 1-(CH$_3$NH)C$_{14}$H$_7$O$_2$	237.26	ye-red nd		170			al, to, aa	B14³, 408
1284	Anthraquinone, 2-methylamino 2-(CH$_3$NH)C$_{14}$H$_7$O$_2$	237.26	red nd (aa)		226-7			al, eth, aa, to	B14³, 430
1285	Anthraquinone, 2-methyl-1-nitro 2-CH$_3$-1-NO$_2$C$_{14}$H$_6$O$_2$	267.24	pa ye nd (aa)		270-1				B7³, 4110
1286	Anthraquinone, 6-methyl-1,2,5-trihydroxy or Morindone 6-CH$_3$-1,2,5-(HO)$_3$C$_{14}$H$_4$O$_2$	270.25	og-red nd (to)		282			al, eth, bz, Py	B8³, 4151
1287	Anthraquinone, 6-methyl-1,3,8-trihydroxy or Emodin 6-CH$_3$-1,3,8-(HO)$_3$C$_{14}$H$_4$O$_2$	270.25	og-red mcl nd (aa), cr (dil aa + 1w)	sub	256-7			al, eth	B8³, 4154
1288	Anthraquinone, 1-nitro 1-NO$_2$C$_{14}$H$_7$O$_2$	253.22	yesh pr (ace), nd (aa)	270-1	232-3			acc, bz, aa	B7³, 4080
1289	Anthraquinone, 2-nitro 2-NO$_2$C$_{14}$H$_7$O$_2$	253.22	ye nd (aa or al)	sub	184-5			bz, chl	B7³, 4080
1290	Anthraquinone, 1,2,4,5,8-pentahydroxy or Alizarin cyanine R 1,2,4,5,8-(HO)$_5$C$_{14}$H$_3$O$_2$	288.21							B8³, 4379
1291	Anthraquinone, 2-sulfonamide 2-(H$_2$NO$_2$S)C$_{14}$H$_7$O$_2$	287.30	ye nd (aa)		261				B11, 339
1292	Anthraquinone, 1-sulfonic acid 1-HO$_3$SC$_{14}$H$_7$O$_2$	288.29	lf (aa), ye lf (con HCl + 3w)		214 (cor) anh			w, al	B11³, 626
1293	Anthraquinone, 1-sulfonic acid-5-chloro 1-HO$_3$S-5-Cl-C$_{14}$H$_6$O$_2$	322.73	ye rh pr (HCl or aa + 5w)		236-7			w	B11³, 627
1294	Anthraquinone, 2-sulfonic acid 2-HO$_3$SC$_{14}$H$_7$O$_2$	288.29	ye lf (+ 3w)					w, al	B11³, 628
1295	Anthraquinone-2-sulfonic acid-5-nitro 2-HO$_3$S-5-NO$_2$C$_{14}$H$_6$O$_2$	333.28	yesh pl (dil HNO$_3$)		255d			w	B11³, 629
1296	Anthraquinone, 1,2-4,6-tetrahydroxy or Hydroxyflavo-purpurin 1,2,4,6-(HO)$_4$C$_{14}$H$_2$O$_2$	272.23	dk red nd (sub)					al, Py	B8³, 4288
1297	Anthraquinone, 1,2,4,7-tetrahydroxy or 4-Hydroxyan-thrapurpurin 1,2,4,7-(HO)$_4$C$_{14}$H$_2$O$_2$	272.23	red-ye (al, Py or aa)					al	B8³, 4288
1298	Anthraquinone, 1,2,5,6-tetrahydroxy or Rufiopin 1,2,5,6-(HO)$_4$C$_{14}$H$_2$O$_2$	272.23	og-red nd (Py)	sub	340			w, al, aa	B8³, 4288
1299	Anthraquinone, 1,2,5,8-tetrahydroxy or Quinalizarin 1,2,5,8-(HO)$_4$C$_{14}$H$_2$O$_2$	272.23	og nd		>275				B8³, 4289
1300	Anthraquinone, 1,2,6,7-tetrahydroxy 1,2,6,7-(HO)$_4$C$_{14}$H$_2$O$_2$	272.23			>330				B8², 584
1301	Anthraquinone, 1,2,7,8-tetrahydroxy 1,2,7,8-(HO)$_4$C$_{14}$H$_2$O$_2$	272.23	red nd or pr (aa)		318d				B8², 585
1302	Anthraquinone, 1,3,5,7-tetrahydroxy or Anthrachrysone 1,3,5,7-(HO)$_4$C$_{14}$H$_2$O$_2$	272.23	yesh nd (al + 2w)	sub	150—60d (+ 2w), <360 (anh)			al, aa, bz, ace	B8³, 4289
1303	Anthraquinone, 1,4,5,8-tetrahydroxy 1,4,5,8-(HO)$_4$C$_{14}$H$_2$O$_2$	272.23	gr nd (aa), br nd (bz-lig)	sub	>300			al	B8³, 4290
1304	Anthraquinone, 1,2,3-trihydroxy or Anthragallol 1,2,3-(HO)$_3$C$_{14}$H$_5$O$_2$	256.23	ye nd (dil al), br (aa), og nd	sub (290)	313			al, eth, aa	B8³, 4140

No.	Name, Synonyms, and Formula	Mol. wt.	Color, crystalline form, specific rotation and λ_{max} (log ε)	b.p. $^\circ$C	m.p. $^\circ$C	Density	n_D	Solubility	Ref.
1305	Anthraquinone, 1,2,4-trihydroxy or Purpurin 1,2,4-(HO)$_3$C$_{14}$H$_5$O$_2$	256.23	og red, dk red or og-ye nd (al)	sub	259			al, eth, bz	B8[3], 4141
1306	Anthraquinone, 1,2,5-trihydroxy or 2-Hydroxyanthrarufin 1,2,5-(HO)$_3$C$_{14}$H$_5$O$_2$	256.23	red nd (gl aa)		278			eth	B8[2], 554
1307	Anthraquinone, 1,2,6-trihydroxy or Flavopurpurin 1,2,6-(HO)$_3$C$_{14}$H$_5$O$_2$.	256.23	ye nd (al)	459d	330 (sub)			al, bz, w	B8[3], 4144
1308	Anthraquinone, 1,2,7-trihydroxy or Anthrapurpurin 1,2,7-(HO)$_3$C$_{14}$H$_5$O$_2$	256.23	og nd (al)	462	374			al, bz, aa	B8[3], 4145
1309	Anthraquinone, 1,2,8-trihydroxy or 2-Hydroxychrysazin 1,2,8-(HO)$_3$C$_{14}$H$_5$O$_2$	256.23	red nd (aa-sub)	sub	239-40				B8[3], 4145
1310	Anthraquinone, 1,3,8-trihydroxy 1,3,8-(HO)$_3$C$_{14}$H$_5$O$_3$	256.23	bt red-br nd (bz), gold-ye pl (AcOEt)		287-8				B8[3], 4145
1311	Anthraquinone, 1,4,5-trihydroxy or 5-Hydroxyquinizarin 1,4,5-(HO)$_3$C$_{14}$H$_5$O$_2$	256.23	red-br nd or lf dk red nd (Py)		271				B8[3], 4146
1312	Anthraquinone, 1,4,6-trihydroxy or 6-Hydroxyquinizarin 1,4,6-(HO)$_3$C$_{14}$H$_5$O$_2$	256.23	bt red nd (al), br pw		> 300			al, Py	B8[2], 558
1313	Anthraquinone-2-carboxylic acid 2-HO$_2$CC$_{14}$H$_7$O$_2$	252.23	ye nd (aa)	sub	290-2			ace	B10[3], 3640
1314	Anthraquinone-2-carboxylic acid, 1,3-dihydroxy or Munjiston 1,3-(HO)$_2$C$_{14}$H$_5$O$_2$(CO$_2$H)-2	284.24	ye nd (al-w + 1w), lf anh					eth	B10[3], 4787
1315	Anthraquinone-2-carboxylic acid, 1,4-dihydroxy 1,4-(HO)$_2$C$_{14}$H$_5$O$_2$(CO$_2$H)-2	284.24	ye br or red nd		249-50			al, ace	B10[3], 4787
1316	Anthraquinone-2-carboxylic acid, 4,5-dihydroxy or Cassic acid 4,5-(HO)$_2$C$_{14}$H$_5$O$_2$(CO$_2$H)-2	284.24	ye or og nd (MeOH, Py)	sub	321			Py	B10[3], 4789
1317	Anthraquinone-2-carboxylic acid, 4,5-dihydroxy-7-methoxy or Parietinic acid 4,5-(HO)$_2$-7-(CH$_3$O)C$_{14}$H$_4$O$_2$(CO$_2$H)-2	314.26	red-br or ye nd (sub)		300				B10[3], 4829
1318	Anthraquinone-2-carboxylic acid, 1-nitro 1-(NO$_2$)C$_{14}$H$_5$O$_2$(CO$_2$H)-2	297.23	nd (al or aa)		288d			al, ace	B10[3], 3650
1319	Anthraquinone-2-carboxylic acid, 5-nitro 5-(NO$_2$)-C$_{14}$H$_6$O$_2$(CO$_2$H)-2	297.23	yesh nd (aa)					aa	B10[3], 3653
1321	1-Anthroic acid or 1-Anthracene carboxylic acid 1-HO$_2$CC$_{14}$H$_9$	222.25	ye nd (aa), ye pr (al, AcOEt)	sub	251-2			eth	B9[3], 3492
1322	2-Anthroic acid or 2-Anthracene carboxylic acid 2-HO$_2$CC$_{14}$H$_9$	222.25	ye lf (al), nd, lf (sub)	sub	281			aa	B9[3], 3493
1323	9-Anthroic acid or 9-Anthracene carboxylic acid 9-HO$_2$CC$_{14}$H$_9$	222.25	pa ye nd (bz, al)	sub	217d			al	B9[3], 3494
1324	Anthrone or 9-Oxodihydroanthracene C$_{14}$H$_{10}$O	194.24	nd (bz-lig, aa)		155			ace, bz	B7[3], 2359
1325	Anthrone-10-nitro 10-NO$_2$-C$_{14}$H$_9$O	239.23	cr (bz-lig), nd (CS$_2$)		140(148 d)			al, bz	B7[3], 2371
1326	Antimalarine or Plasmocid C$_{17}$H$_{25}$N$_3$O	287.41		182[1.0]		1.0569[24/4]	1.5855[24]	dil HCl	B22[4], 5787
1327	Antipyrine (α - form) or Analgesine, Phenazone C$_{11}$H$_{12}$N$_2$O	188.23	mcl lf or sc (w, bz, eth)	319[741], 211-2[10]	114	1.0747[20/4]	1.5697	w, al, bz, chl	B24[2], 11
1328	Antipyrene (β) C$_{11}$H$_{12}$N$_2$O	188.23	cr (unst)		109($\beta \rightarrow \alpha$ at 18)			bz	B24[2], 11
1329	Antipyrine, 4-acetamido 4-(CH$_3$CONH)C$_{11}$H$_{11}$N$_2$O	245.28			201			w, al	B24[2], 152
1330	Antipyrene, o-amino C$_{11}$H$_{13}$N$_3$O	203.23	nd (AcOEt-eth)		165			w, al	B24[1], 210
1331	Antipyrene, m-amino C$_{11}$H$_{13}$N$_3$O	203.25	redsh in air (bz)		148			w, al, chl	B24[1], 210
1332	Antipyrine, p-amino C$_{11}$H$_{13}$N$_3$O	203.24			109			w, al, eth, bz	B24[2], 151

No.	Name, Synonyms, and Formula	Mol. wt.	Color, crystalline form, specific rotation and λ_{max} (log ε)	b.p. °C	m.p. °C	Density	n_D	Solubility	Ref.
1333	Antipyrene, p-bromo $C_{11}H_{11}BrN_2O$	267.13	nd (w)	300[9]	122	......	......	w, al, eth, to	B24, 33
1334	Antipyrene, p-dimethylamino $C_{13}H_{17}N_3O$	231.30	pr or pl (lig or AcOEt)		134—5	......	......	w, al, bz	B24, 46
1335	Antipyrene, 2-hydroxybenzoate or Salazolon $C_{11}H_{12}N_2O\text{-}C_7H_6O_2$	326.35	pw	......	92			al, chl	B24[2], 16
1336	Aphanin $C_{40}H_{54}O$	550.88	bl blk lf (bz-MeOH)		178			chl	B7[3], 2858
1337	Aphyiline $C_{15}H_{24}N_2O$	248.37	$[\alpha]^{20}_D$ + 10.3 (MeOH, c=2)	200d	52-7			al, eth, ace, bz	C26, 2742
1338	Apiol or 2,5-Dimethoxy saffrole parsley camphor $C_{12}H_{14}O_4$	222.24	nd	294, 179[35]	29.55	1.015[20/4]	1.5360[20]	al, eth, ace, bz, lig	B19[4], 1033
1339	Apoatropine or Atropamine $C_{17}H_{21}NO_2$	271.36	pr (chl)		62			al, eth, ace, bz, chl	B21[4], 175
1340	Apoatropine hydrochloride $C_{17}H_{21}NO_2\text{-}HCl$	307.82	lf (w)		239			w	B21[4], 175
1341	Apocinchonidine $C_{19}H_{22}N_2O$	294.40	lf (al), $[\alpha]^{20}_D$ − 139.3 (chl-al, c = 2)		252			al, chl	B23[1], 131
1342	Apocinchonine or Allocinchonine $C_{19}H_{22}N_2O$	294.40	pr (al), $[\alpha]^{20}_D$ + 167.4 (abs al, c−5)		219			al, bz	B23[2], 367
1343	Apocinchonine hydrochloride $C_{19}H_{22}N_2O\text{-}HCl\text{-}2H_2O$	366.89	nd (+ 2w), $[\alpha]^{16}_n$ + 139 (w, c=0.006)					w	B23[2], 367
1344	Apocodeine or Apomorphine-3-methyl ether $C_{18}H_{19}NO_2$	281.36	pr (MeOH), $[\alpha]^{23}_D$ − 90 (abs al, c=0.449)		123.4(a nh)			al, eth, aa, lig	B21[4], 2420
1345	Apocodeine, ethanol solvate $C_{18}H_{19}NO_2\text{-}C_2H_5OH$	327.43	lf (al + 1)		104.5-6.5			al	B21[4], 2420
1346	Apocodeine, methanol solvate $C_{18}H_{19}NO_2\text{-}CH_3OH$	313.40	nd (MeOH + 1)		85			MeOH	B21[4], 2420
1347	Apocyclene C_9H_{14}	122.21	cr (al)	138-9[764]	42.5-4.3	0.8710[40/4]	1.4514[40]	al, eth, bz	B5[4], 430
1348	Apocynin or 4-Hydroxy-3-methoxy acetophenone 4-HO-3-$(CH_3O)C_6H_3COCH_3$	166.18	pr (w)	295-300, 233-5[15-20]	115			al, eth, ace, bz, chl	B8[3], 2108
1349	Apofenchocamphoric acid or 4,4-Dimethyl-1,3-cyclopentane dicarboxylic acid 4,4-$(CH_3)_2$-1,3-$(HO_2C)_2C_5H_6$	186.21	mcl		144-5			al, eth	B9[3], 3849
1350	Apomorphine $C_{17}H_{17}NO_2$	267.33	hex pl (chl-peth), rods (eth + l)	......	l95d			al, eth, ace, bz	B21[4], 2419
1351	Apomorphine-hydrochloride $C_{17}H_{17}NO_2\text{-}HCl\text{-}1/2H_2O$	312.80	gr in air, mcl pr, $[\alpha]^{25}_D$ − 48 (w, c = 1.2)		200-10				B2[4], 2419
1352	Aponal or tert-Pentylcarbamate $H_2NCOOCH_2C(CH_3)_3$	131.18	nd (dil al)		86—7			ace, bz	B3[1], 14
1353	Apoquinine (α) or Apocupreine $C_{17}H_{22}N_2O_2$	310.40	pr (eth), $[\alpha]^{20}_D$ − 214.8 (al)		190d			al	B23[2], 412
1354	Apoquinine(β) $C_{19}H_{22}N_2O_2\text{-}2H_2O$	346.43	$[\alpha]^{20}_D$ − 194 (al)		190d			al	B23[2], 412
1355	Aposafranone or Benzeneindone,l0-phenyl-2-phenazinone $C_{18}H_{12}N_2O$	272.31	br, gr (red in sl), nd (al)	......					B23[2], 364

No.	Name, Synonyms, and Formula	Mol. wt.	Color, crystalline form, specific rotation and λ_{max} (log ε)	b.p. °C	m.p. °C	Density	n_D	Solubility	Ref.
1356	D-Arabinose (α-anomer)............... $C_5H_{10}O_5$	150.13	(cr, MeOH)	....	155.5—6.5 (cor)	1.585	....	w	B1[4], 4215
1357	D-Arabinose (β-anomer)............... $C_5H_{10}O_5$	150.13	cr (MeOH), $[\alpha]^{20}_D$ −175→−108 (mut)	....	155.5—6.5 (cor)	1.625		w	B1[4], 4215
1358	L-Arabinose or pectinose............. $C_5H_{10}O_5$	150.13	pr nd (al)	....	164—5 (cor)	1.585[20/4]		w	B1[4], 4223
1359	L-Arabinose (α-anomer)............... $C_5H_{10}O_5$	150.13	orth cr (pr), cr (dil al)	....	159-60	1.585[20/4]			B1[4], 4218
1360	L-Arabinose (β-anomer)............... $C_5H_{10}O_5$	150.13	orth, $[\alpha]^{20}_D$ +190.5→ +104.5 (c=3) (mut)	....	159-60	1.625[20/4]		w	B1[4], 4218
1361	D-Arabinose-diphenylhydrazone............. $HOCH_2(CHOH)_3CH = NN(C_6H_5)_2$	316.36	orth pr (dil al)		207				B31, 33
1362	DL-Arabinose-diphenylhydrazone............. $HOCH_2(CHOH)_3CH = NN(C_6H_5)_2$	316.36	nd (aq Py)		206			Py	B31, 47
1363	L-Arabinose-diphenylhydrazone............. $HOCH_2(CHOH)_3CH = NN(C_6H_5)_2$	316.36	orth pr, nd (dil al), $[\alpha]^{20}_D$ +18.5 Py		204.5			al	B31, 44
1364	D-Arabitol or Arabite, D-Lyxitol............. 1,2,3,4,5-pentanepentol $HOCH_2(CHOH)_3CH_2OH$	152.15	pr (dil al), $[\alpha]^{20}_D$ +11.8 (borax sol, c=9.5)					w	B1[4], 2832
1365	DL-Arabitol............. $HOCH_2(CHOH)_3CH_2OH$	152.15	pr (90% al), cr (al-ace)		106			w	B1[4], 2832
1366	L-Arabitol............. $HOCH_2(CHOH)_3CH_2OH$	152.15	$[\alpha]_D$ −5.4 (borax sol)		102-3			w	B1[4], 2832
1367	D-Arabonic acid............. $HOCH_2(CHOH)_3CO_2H$	166.13	(dil aa), $[\alpha]^{25}_D$ +10.5 (c=6)		114-16			w, al	B3[4], 1205
1368	DL-Arabonic acid............. $HOCH_2(CHOH)_3CO_2H$	166.13						w	B3[2], 474
1369	L-Arabonic acid............. $HOCH_2(CHOH)_3CO_2H$	166.13	cr (al), $[\alpha]^{20}_D$ −9.6 (w, c=2.5)						B3[3], 979
1370	Arachidic acid or Eicosanoic acid $CH_3(CH_2)_{18}CO_2H$	312.54	pl (al)	328d, 203—5[1]	77	0.8240[100/4]	1.425[100]	eth, bz, chl	B2[4], 1275
1371	Arachidic acid-ethyl ester or Ethyl arachidate .... $CH_3(CH_2)_{18}CO_2C_2H_5$	340.60	....	295-7[100], 186-7[2]	50			al, eth, bz, chl	B2[3], 1067
1372	Arachidic acid, methyl ester or methyl arachidate....... $CH_3(CH_2)_{18}CO_2CH_3$	326.57	lf (MeOH)	215-6[10], 188[2]	54.5		1.4317[60]	al, eth, bz, chl	B2[4], 1276
1373	Arachidic alcohol or l-Eicosanol................. $CH_3(CH_2)_{18}CH_2OH$	298.56	wx (al), cr (chl)	309, 220-5[3]	72-3	0.8405[20/4]	1.4550[20]	ace, bz	B1[4], 1900
1374	Arachidonic acid $CH_3(CH_2)_3(CH_2CH=CH)_4(CH_2)_3CO_2H$	304.48	d		−49.5		1.4824[20]	al, eth, ace	B2[4], 1802
1375	Aramite or Niagaramite................. $C_{15}H_{23}ClO_4S$	334.87		195[2]	−37.3	1.145[20/20]	1.5100[20]	al, eth, ace, bz	M, 97
1376	Arbutin or Hydroquinone-β-d-glucopyranoside........ $C_{10}H_{16}O_7$	272.24	nd (w + l), $[\alpha]^{25}_D$ −64 (w, c=3)		199.5-200(st, anh)			w, al	B17[4], 2983
1377	Arbutin hydrate............. $C_{12}H_{16}O_7 \cdot H_2O$	290.27	$[\alpha]^{17}_D$ −60.3 (w, p=5)	....	142			al	B31, 210
1378	Arecaidine or l-Methyl-1,2,5,6-tetrahydroricotinic acid ... $C_7H_{11}NO_2$	141.16	pl (dil al), ta (dil al + lw)		232 (anh)			w	B22[3], 184
1379	Arecoline or Arecaidine methyl ester............. $C_8H_{13}NO_2$	155.20	....	209 (220), 94[1]	....	1.0504[20/20]	1.4860[20]	w, al, eth chl	B22[3], 185
1380	Arecoline hydrobromide............. $C_8H_{13}NO_2 \cdot HBr$	236.12	mcl pr (al)		172 (177)			w, al	B22[3], 185

No.	Name, Synonyms, and Formula	Mol. wt.	Color, crystalline form, specific rotation and λ_{max} (log ε)	b.p. °C	m.p. °C	Density	n_D	Solubility	Ref.
1381	Arecoline hydrochloride .... $C_8H_{13}NO_2$·HCl	191.66	nd (al)	....	157-8	....	....	w, al	B22[3], 185
1382	Arginine (DL) or DL-2-Amino-5-guanido pentanoic acid $H_2NC(=NH)NH(CH_2)_3CH(NH_2)CO_2H$	174.20		....	217-8			w	B4[3], 1359
1383	Arginine (L) .... $C_6H_{14}N_4O_2$	174.20		....	244d			w	B4[3], 1348
1384	Arginine, benzylidene (L) .... $C_{13}H_{18}N_4O_2$	262.31	lf (w)	....	204-5			MeOH	B7[3], 837
1385	Arginine diflavinate (L) .... $C_6H_{14}N_4O_2(C_{10}H_6N_2O_8S)_2$	802.67	ye nd	....	202d				B11[3], 547
1386	Arginine dipicrate (DL) .... $C_6H_{14}N_4O_2(C_6H_3N_3O_7)_2$	632.42		....	196d				B6[4], 1450
1387	Arginine dipicrate (L) .... $C_6H_{14}N_4O_2(C_6H_3N_3O_7)_2$	632.42		....	200d				B6[4], 1450
1388	Arginine picrate (D) $C_6H_{14}N_4O_2$·$C_6H_3N_3O_7$·$2H_2O$	488.44	ye-og lf	....	258-60d				B11[3], 547
1389	Arginine picrate (D) .... $C_6H_{14}N_4O_2$-$C_6H_3N_3O_7$·$2H_2O$	439.44	nd (w)	....	217-8d				B6[4], 1450
1390	Arginine picrate (DL) .... $C_6H_{14}N_4O_2$-$C_6H_3N_3O_7$·$2H_2O$	439.34	pr (w)	....	200 (223d)				B6[4], 1450
1391	Arginine picrate (L) .... $C_6H_{14}N_4O_2$-$C_6H_3N_3O_7$·$2H_2O$	439.34	ye nd (w)	....	217-8d				B6[4], 1450
1392	Arsenic acid, diphenyl $(C_6H_5)_2AsO_2H$	262.14	nd pr	....	178		....	al, chl	B16[2], 443
1393	Arsenic acid, triethyl ester or Triethyl arsenate .... $(C_2H_5O)_3AsO$	226.11		235-8, 118-20[15]		1.3023[20/0]	1.4343[20]		B1[4], 1358
1394	Arsenobenzene .... $C_{36}H_{30}As_6$	912.16			212			bz, al	B16[3], 1157
1395	Arsenous acid, triethyl ester or Triethyl arsenite .... $(C_2H_5O)_3As$	210.11		165-6, 66-7[12]		1.2239[20/4]	1.4369[13]		B1[4], 1358
1396	Arsine, bis(trifluoromethyl)iodo ... $(CF_3)_2AsI$	339.84	ye oil	92			1.425[23]	eth	B3[4], 267
1397	Arsine, chloro,diphenyl .... $(C_6H_5)_2AsCl$	264.59	rh pl (peth)	337, 193[20]	44	1.48204[16/4]	1.6332[56]	al, eth, ace, bz	B16[3], 944
1398	Arsine, dibromo,trifluoromethyl .... CF_3AsBr_2	303.75		118[745]		1.528[20]			B3[4], 267
1399	Arsine, dichloro,trifluoromethyl .... CF_3AsCl_2	303.75		133, 37[25]	-42.5	1.8358[20]	1.5677[15]	al, eth	B3[4], 267
1400	Arsine, dichloro,phenyl .... $C_6H_5AsCl_2$	222.93		254-7, 131[14]		1.6516[20/4]	1.6386[15]	al, eth, ace, bz	B16[3], 958
1401	Arsine, dichloro (trifluoromethyl) ... CF_3AsCl_2	214.83		71		1.431[20]	....	al, ace	B3[4], 268
1402	Arsine, diethyl .... $(C_2H_5)_2AsH$	134.05		105 (97)	1.1338[24/4]	1.4709		al, eth, ace, bz	B4[3], 1797
1403	Arsine, difluoro,ethyl .... $C_2H_5AsF_2$	141.98	fume in air	94.3, 74[100]	-38.7	1.708[17]			B4[3], 1799
1404	Arsine, difluoro,methyl .... CH_3AsF_2	127.95	fume in air	76.5	-29.7	1.924[18]			B4[3], 1796
1405	Arsine, difluoro,phenyl .... $C_6H_5AsF_2$	190.03	wax	110[48]	42				B16[3], 959
1406	Arsine, diiodo,trifluoromethyl .... F_3CAsI_2	397.74		183d, 100[48]					B3[4], 268
1407	Arsine, dimethyl or Cacodylhydride . $(CH_3)_2AsH$	106.00	ign in air	36		1.213[29/29]		**al, eth, ace, bz**	B4[2], 978
1408	Arsine, diphenyl .... $(C_6H_5)_2AsH$	230.14	oil	174[25]		1.30[25/25]	....	al, eth	B16[3], 918
1409	Arsine, ethyl .... $C_2H_5AsH_2$	106.00	....	36		1.217[22/22]		al, eth	B4[3], 1797
1410	Arsine, 4-methoxyphenyl,oxide .... $(4-CH_3OC_6H_4)AsO$	198.05	cr (chl-eth)- or bz-peth	....	114-6 (anh)			bz, chl	B16[2], 448
1411	Arsine, methyl .... CH_3AsH_2	91.97	....	2	-143			**al, eth, ace**	B4[3], 1795
1412	Arsine, methyl, oxide CH_3AsO	105.96	pr (CS_2) (al)	275d	95			al, bz, chl	B4[3], 1817
1413	Arsine, pheny,oxide C_6H_5AsO	168.03	cr (bz-eth or chl-eth)	....	144-6			bz, chl	B16[3], 956

No.	Name, Synonyms, and Formula	Mol. wt.	Color, crystalline form, specific rotation and λ_{max} (log ε)	b.p. °C	m.p. °C	Density	n_D	Solubility	Ref.
1414	Arsine, triethyl .... $(C_2H_5)_3As$	162.11	....	138-9	......	$1.150^{20/4}$	1.467^{20}	al, eth, ace	B4³, 1798
1415	Arsine, trifluoromethyl .... CF_3AsH_2	145.94		-11.6^{781}					B3⁴, 266
1416	Arsine, trimethyl .... $(CH_3)_3As$	120.03		52	-87.3	1.144^{15}		al, eth, bz	B4³, 1795
1417	Arsine, triphenyl .... $(C_6H_5)_3As$	306.24			61	$1.2634^{18/4}$	1.6888^{21}	eth, bz, chl	B16³, 921
1418	Arsine, triphenyl, oxide .... $(C_6H_5)_3AsO$	322.24			192				B16³, 1021
1419	Arsine, triphenyl, sulfide .... $(C_6H_5)_3AsS$	338.20			162				B16³, 1026
1420	Arsine, *tris*-(pentafluoroethyl) .... $(C_2F_5)_3As$	431.96		96					C49, 14635
1421	Arsine, *tris*-(trifluoromethyl) .... $(CF_3)_3As$	281.94		33.3				eth	B3⁴, 266
1422	Artemsic acid .... $C_{15}H_{16}O_3$	244.29	nd (dil aa)		135-6			al, eth, aa, lig	B10², 220
1423	Ascaridole .... $C_{10}H_{16}O_2$	168.24	unst, $[\alpha]_D$ − 4.14	exp^{760}, 115^{15}, $39-40^{0.2}$	3.3	$1.0103^{20/4}$	1.4769^{20}	al, ace, bz	B19⁴, 164
1424	D-Ascorbic acid .... $C_6H_8O_6$	176.14	pl or mcl nd, $[\alpha]^{18}_D$ − 48 (MeOH, c=1), $[\alpha]^{20}_D$ − 23.8 (w, c=3)		192d			w, al	B18⁴, 3046
1425	DL-Ascorbic acid .... $C_6H_8O_6$	176.14			168-9			w, al	B18⁴, 3047
1426	L-Ascorbic acid or Vitamin C .... $C_6H_8O_6$	176.14	pl or mcl nd, $[\alpha]^{20}_D$ + 24 (w, c = 1)		192d	1.65			B18⁴, 3038
1427	L-Ascorbic acid, 6-desoxy .... $C_6H_8O_5$	160.14	pr (AcOEt), $[\alpha]^{22}_D$ + 36.7 (O.1N HCl, c = 1)	sub,$160^{0.001}$	168			w, al, ace	B18⁴, 2301
1428	β-Asparagine (D) .... $H_2NCOCH_2CH(NH_2)CO_2H \cdot H_2O$	150.14	$[\alpha]^{20}_D$ + 5.4 (w, c = 1.3)		234.5 (anh)	$1.543^{15/4}$		w	B4³, 1522
1429	β-Asparagine (DL) .... $H_2NCOCH_2CH(NH_2)CO_2H \cdot H_2O$	150.14	tcl cr (w + l), pr (w al)	213-5d	182-3	$1.4540^{15/4}$		w	B4³, 1524
1430	β-Asparagine (L) .... $H_2NCOCH_2CH(NH_2)CO_2H$	132.13	rh (w + l), $[\alpha]^{20}_D$ −5.42 (w, c = 1.3)		236 (anh)	$1.543^{15/4}$		w	B4³, 1513
1431	Aspartic acid (D) or Aminosuccinic acid (D) .... $HO_2CCH_2CH(NH_2)CO_2H$	133.11	$[\alpha]^{20}_D$ −25.5 (HCl)		269-71	$1.6613^{13/13}$		w, dil, HCl	B4³, 1522
1432	Aspartic acid (DL) or Aminosuccinic acid .... $HO_2CCH_2CH(NH_2)CO_2H$	133.11	mcl pr (w)		338-9 (275 sealed tube)	$1.6632^{13/13}$		w	B4³, 1523
1433	Aspartic acid (L) or Aminosuccinic acid .... $C_4H_7NO_4$	133.11	rh lf (w)		324d, (270 sealed tube)	$1.6613^{13/13}$		w, dil, HCl	B4³, 1506
1434	Aspartic acid, N-benzoyl .... $HO_2CCH_2CH(NHCOC_6H_5)CO_2H$	237.21	nd or lf, $[\alpha]^{20}_D$ + 37 (0.76N NaOH)		171-3				B9³, 1189
1435	Aspidospermine or Vallesine .... $C_{22}H_{30}N_2O_2$	354.50	nd or pr (al), nd (peth)					al, bz, chl	J1940, 1051

No.	Name, Synonyms, and Formula	Mol. wt.	Color, crystalline form, specific rotation and λ_{max} (log ε)	b.p. °C	m.p. °C	Density	n_D	Solubility	Ref.
1436	Atabrin or Quinacrine dihydrochloride (*DL*)............ $C_{23}H_{30}ClN_3O \cdot 2HCl \cdot 2H_2O$	508.90	yesh nd (w), ye cr pw	....	248-50d			w, al, MeOH	B21[4], 6248
1437	Atisine or Anthorine................ $C_{22}H_{33}NO_2$	343.51	rh bipym		57-60			al, eth, chl	Am78, 4139
1438	Atisine-hydrochloride............ $C_{22}H_{33}NO_2 \cdot HCl$	377.96	nd (dil al), $[\alpha]_D + 28$, (w, c = 1.1)		340			w, al	C36, 4826
1439	Atrolactic acid or 2-hydroxy-2-phenyl propanoic acid .... $CH_3C(OH)(C_6H_5)CO_2H$	166.18	pr (w), $[\alpha]^{16.5}_D$ + 37.7 (al, c = 3.5)	....	116.5-7			w, ace, bz, al	B10[3], 560
1440	Atropic acid or 2-phenylpropenoic acid .... $CH_2=C(C_6H_5)CO_2H$	148.17	lf (al), nd (w)	267d	106-7			al, eth, bz, chl	B9[3], 2751
1441	Atropine or (*DL*)-hyoscyamine .... $C_{17}H_{23}NO_3$	289.36	rh nd (dil al), orth pr (ace)	sub(vac) 93-100	118-9			al, eth, bz, chl	B21[4], 183
1442	Atropine hydrochloride .... $C_{17}H_{23}NO_3 \cdot HCl$	325.84	nd (al)		165			w, al	B21[4], 184
1443	Atropine pentanoate .... $C_{17}H_{23}NO_3 \cdot C_5H_{10}O_2 \cdot \frac{1}{2}H_2O$	400.52	cr		42			w, al, eth	M, 111
1444	Atropine sulfate, hydrate .... $(C_{17}H_{23}NO_3)_2 \cdot H_2SO_4 \cdot H_2O$	694.82	nd (al, eth or al ace)	sub	194 (anh)			w, al	B21[4], 182
1445	Auramine or *bis*(p-dimethylaminophenyl)methylene imine	267.33	ye or col pl (al)		136			al	B14[3], 227
	$C_{17}H_{21}N_3$								
1446	Auramine hydrochloride or Auramine I .... $C_{17}H_{21}N_3 \cdot HCl$	321.83	ye nd (al)		267			al, chl	B14[3], 90
1447	Auramine-*N* .... $C_{18}H_{23}N_3$	281.41	ye cr (al)		133			al, ace, aa	B14, 93
1448	Aurin or Rosolic acid .... $C_{19}H_{14}O_3$	290.32	dk red lf or rh		308-10d			al, eth	B8[3], 3024
1449	Auroemycin or Chlorotetracycline .... $C_{22}H_{23}N_2O_8Cl$	478.89	gold ye, ye flr, $[\alpha]^{25}_D$ −275 (MeOH)		168-9				B14[3], 1710
1450	Auroemycin hydrochloride .... $C_{22}H_{23}N_2O_8Cl \cdot HCl$	515.35	ye or th, $[\alpha]^{20}_D$ −106.5 (dil al)	....	216d				B14[3], 1710
1451	Auxin A or Auxenetriolic acid .... $C_{18}H_{32}O_5$	328.45	hex (al-lig), $[\alpha]^{20}_D$ −3.19 (al)	....	196	1.292[19/4]	....	al	B10[3], 2045
1452	Auxin B or Auxenolonic acid .... $C_{18}H_{30}O_4$	310.44	cr (al-lig), $[\alpha]^{20}_D$ −2.8 (al)		183	1.269[20/4]		al, eth	B10[3], 4194
1453	Azelaic acid or Nonanedioic acid .... $HO_2C(CH_2)_7CO_2H$	185.22	lf or nd	287[100], 225[10]	106.5	1.225[25/4]	1.4303[111]	al, eth	B2[4], 2055
1454	Azelaic acid, diethylester or Diethyl azelate .... $C_2H_5O_2C(CH_2)_7CO_2C_2H_5$	244.34	....	291-2, 174-5[20]	-18.5	0.97294[20/4]	1.43509[20]	al, eth	B2[3], 1787
1455	Azelaic acid, di(2-ethylbutyl) ester .... $H_2C[(CH_2)_3CO_2CH_2CH(C_2H_5)_2]_2$	356.55	....	230[5]	-45	0.928[25/4]	1.443[25]	al, ace, bz	B2[3], 1787
1456	Azelaic acid, di-(2-ethylhexyl) ester .... $CH_2[(CH_2)_3CO_2CH_2CH(C_2H_5)C_4H_9]_2$	412.66	....	237[5]	-78		1.446[25]	al, ace, bz	B2[3], 1787
1457	Azelaic acid, dimethylester or Dimethyl azelate......... $CH_3O_2C(CH_2)_7CO_2CH_3$	216.28	....	156[20]		1.0082[20/4]	1.4367[20]	al, ace, bz	B2[3], 1786
1458	Azelaic acid, diphenylester or Diphenyl azelate......... $C_6H_5O_2C(CH_2)_7CO_2C_6H_5$	340.42	nd (al)	....	59-60			eth, bz	B6[3], 606
1459	Azelanitrile or Azelaic acid, dinitrile .... $NC(CH_2)_7CN$	150.23	....	198-9[25]		0.9200[19]	1.4518[19]	al, eth, bz	B2[4], 2059
1460	Azelayl chloride .... $ClOC(CH_2)_7COCl$	225.12		166[18]			1.4680[20]	eth, bz	B2[3], 1789
1461	l-Azacyclooctane, 2-methyl or α-Methyl heptamethylenimine .... $C_8H_{17}N$	127.23	nd (ace)	162-3[746]	156-7	0.853[20]	1.4620[21]		B20[1], 30
1462	Azetidine or Trimethyleneimine................ C_3H_7N	57.09	....	63[748]	....	0.8436[20]	1.4287[25]	w, al, eth, ace, bz	B20[4], 53

No.	Name, Synonyms, and Formula	Mol. wt.	Color, crystalline form, specific rotation and λ_{max} (log ε)	b.p. °C	m.p. °C	Density	n_D	Solubility	Ref.
1463	Aziridine or Ethyleneimine................ C_2H_5N	43.07	....	56[756]	....	0.8321[20/4]	....	w, al, eth, ace, bz	B20[4], 3
1464	Azobenzene (cis) .... $C_6H_5N=NC_6H_5$	182.23	og-red pl (peth)		71			al, eth, bz, aa	B16[3], 4
1465	Azobenzene (trans) .... $C_6H_5N=NC_6H_5$	182.23	og red mcl lf (al)	29.3	68.5	1.203[20/4]	1.6266[78]	al, eth, bz, aa	B16[3], 5
1466	Azobenzene, 2-acetamide-4′,5′-dimethyl .... $2(CH_3CONH)-5-CH_3-C_6H_3N=N(C_6H_4CH_3-4)$	267.33	ye nd (al-aa)		157			al, eth, chl	B16[2], 182
1467	Azobenzene, acetamido $4-(CH_3CONH)C_6H_4N=NC_6H_5$	239.28	gold-ye nd (al)		144-6				B16[3], 346
1468	Azobenzene, acetamido-2′,3-dimethyl $4(CH_3CONH)-3-CH_3-C_6H_3N=N(C_6H_4CH_3-2)$	267.33	red nd (al)		186-7			eth, chl	B16[3], 387
1469	Azobenzene, 4-acetoxy-2′-methyl $4(CH_3CO_2)C_6H_4N=N(C_6H_4-CH_3-2)$	254.29	red-ye lf (al)		68			al	B16, 105
1470	Azobenzene, 4-acetoxy-3-methyl $4-(CH_3CO_2)-3-(CH_3)C_6H_3N=NC_6H_5$	254.29	ye pl (dil al)		81-2			al, eth, bz, chl	B16, 130
1471	Azobenzene, 4-acetoxy-4′-methyl $4-(CH_3CO_2)C_6H_4N=N(C_6H_4CH_3-4)$	254.29	og nd (al, bz)		98			al	B16[2], 42
1472	Azobenzene, 2-amino or 2-Benzeneazoaniline........... $2-H_2NC_6H_4N=N-C_6H_5$	197.24	red nd or pr (al)		59			eth, ace, bz	B16[3], 334
1473	Azobenzene, 3-amino .... $3-H_2NC_6H_4N=NC_6H_5$	197.24	(i) og-ye nd (peth), (ii) br-red cr		(i) 69-70, (ii) 90-dl			al, eth, ace, bz, chl	B16[3], 335
1474	Azobenzene, amino $4-H_2NC_6H_4N=NC_6H_5$	197.24	og mcl nd (al)	>360	127			al, eth, bz, chl	B16[2], 149
1475	Azobenzene, 4-amino-2,2′-dimethyl .... $4(H_2N)-2-(CH_3)C_6H_3N=N(C_6H_4CH_3-2)$	225.30	ye nd (liq)		116-7			al, lig	B16[2], 180
1476	Azobenzene, 4-amino-2,3′-dimethyl $4-(H_2N)-2-(CH_3)-C_6H_3N=N(C_6H_4CH_3-3)$	225.30	ye-gold nd (al), ye-br nd (lig)		80			al	B16[3], 392
1477	Azobenzene, 4-amino-2′,3-dimethyl $4-(H_2N)-3-CH_3-C_6H_3N=N(C_6H_4CH_3-2)$	225.30	ye lf (al)		101.5-3			al, eth	B16[3], 386
1478	Azobenzene, 4-amino-2,4′-dimethyl $4(H_2N)-2CH_3C_6H_3N=N(C_6H_4CH_3-4)$	225.30	ye pl (al)-gold-ye (lig)		127			al, lig	B16, 348
1479	Azobenzene, 4-amino-3,3′-dimethyl $4(H_2N)-3-(CH_3)C_6H_3N=N(C_6H_4CH_3-3)$	225.30	ye br lf or nd (lig)		124			al, lig	B16, 345
1480	Azobenzene, 4-amino-3,4′-dimethyl $4(H_2N)-3-(CH_3)C_6H_3N=N(C_6H_4CH_3-4)$	225.30	og-ye nd (lig), ye pl (al)		128				B16[3], 389
1481	Azobenzene, 4-amino, hydrobromide .... $4-H_2NC_6H_4N=NC_6H_5·HBr$	278.04	bk-vt nd (dil al)		206-7				B16, 307
1482	Azobenzene, 4-amino, hydrochloride .... $4-H_2NC_6H_4N=NC_6H_5·HCl$	233.70	bl-vt or pa red, nd or pw		240			w, al	B16[2], 149
1483	Azobenzene, 4-benzoxy-4′-methyl............ $4-(C_6H_5CH_2O)C_6H_4N=N(C_6H_5CH_3-4)$	302.38	pa ye lf (lig)		128			al, eth, bz	B16, 107
1484	Azobenzene, 3,3′-bis-(dimethylamino) .... $3-(CH_3)NC_6H_4N=N(C_6H_4(OH)_2-3)$	268.27	red nd (al)		118			al, bz	B16, 305
1485	Azobenzene, 4,4′-bis-(dimethylamino) .... $4-(CH_3)_2NC_6H_4N=N[C_6H_4(CH_3)_2-4]$	268.37	og nd (bz)	sub	273			eth, bz, chl	B16[3], 375
1486	Azobenzene, 4-bromo .... $4-BrC_6H_4N=NC_6H_5$	261.13	og nd (lig)		90-1			al, eth, ace, lig	B16[3], 29
1487	Azobenzene, 2,2′-diamino $2-H_2NC_6H_4N=N(C_6H_4NH_2-2)$	212.26	red pl (al or bz)		134			eth	B16[2], 148
1488	Azobenzene, 2,4-diamino or Chrysoidine........... $2,4-(H_2N)_2C_6H_3N=NC_6H_5$	212.26	pa ye nd (w)		117.5			al, eth, bz	B16[3], 436
1489	Azobenzene, 4,4′-diamino or o-Azodianiline........... $4-H_2NC_6H_4N=N(C_6H_4NH_2-4)$	212.26	gold-ye nd (al), og-ye pr (al)		250-1			al, bz, chl	B16[3], 375
1490	Azobenzene, 2,2′-diethoxy or o-Azophenetole .... $2-C_2H_5OC_6H_4N=N(C_6H_4OC_2H_5-2)$	270.34	red pr (al)	240d	131			al, eth	B16[3], 83
1491	Azobenzene, 4,4′-diethoxy or p-Azophenetole .... $4-C_2H_5OC_6H_4N=N(C_6H_4OC_2H_5-4)$	270.34	ye-lf (al)	d	162			eth, bz, chl, aa	B16[3], 93

No.	Name, Synonyms, and Formula	Mol. wt.	Color, crystalline form, specific rotation and λ_{max} (log ε)	b.p. °C	m.p. °C	Density	n_D	Solubility	Ref.
1492	Azobenzene, 2,2'-dihydroxy or o-Azophenol $2\text{-}HOC_6H_4N=N(C_6H_4OH\text{-}2)$	214.23	gold-ye lf (bz), nd (al)	sub	172			eth, al, bz	B16², 33
1493	Azobenzene, 2,4-dihydroxy $2,4\text{-}(HO)_2C_6H_3N=NC_6H_5$	214.23	dk red nd (dil al)		170 (anh)			al, eth, bz, aa	B16³, 166
1494	Azobenzene, 2,4-dihydroxy-4'-nitro $2,4\text{-}(HO)_2C_6H_3N=N(C_6H_4NO_2\text{-}4)$	259.22	red pw (al or MeOH)		200				B16³, 162
1495	Azobenzene, 2,5-dihydroxy $2,5(HO)_2C_6H_3N=NC_6H_5$	214.22			149			eth, ace, bz	B16³, 176
1496	Azobenzene, 3,3'-dihydroxy or m-Azophenol $3\text{-}HOC_6H_4N=N(C_6H_4OH\text{-}3)$	214.23	ye lf (dil al)		207			eth, ace	B16³, 85
1497	Azobenzene, 4-dimethylamino or Butteryellow $4\text{-}(CH_3)_2NC_6H_4N=NC_6H_5$	225.30	ye lf (al)	d	117			al, eth, aa, Py	B16³, 340
1498	Azobenzene, 4,4'-dihydroxy or p-Azophenol $4\text{-}HOC_6H_4N=N(C_6H_4OH\text{-}4)$	232.24	og-ye pl (dl al, + 1w)		216-6.5 anh			al, eth, ace	B16³, 92
1499	Azobenzene, 2,2'-dimethyl or o-Azotoluene $2\text{-}CH_3C_6H_4N=N(C_6H_4CH_3\text{-}2)$	210.28	dk red mcl pr (eth)		55-6	$1.0215^{65/4}$	1.6180^{65}	al, eth, bz	B16², 19
1500	Azobenzene, 2,2'-dimethyl-4-ethoxy $2\text{-}CH_3\text{-}4\text{-}C_2H_5OC_6H_3N=N(C_6H_4CH_3\text{-}2)$	254.34	dk red nd (al)		64			al, eth, ace, bz	B16, 134
1501	Azobenzene, 2,2'-dimethyl-4-hydroxy or 4-o-Tolueneazo-m-cresol $2\text{-}CH_3\text{-}4\text{-}HOC_6H_3N=N(C_6H_4CH_3\text{-}2)$	226.28	ag-red pl (bz), red cr (w + l)		113 (anh)			al, eth, bz	B16², 61
1502	Azobenzene, 2,3-dimethyl-4-ethoxy $2\text{-}CH_3\text{-}4\text{-}C_2H_5OC_6H_3N=N(C_6H_4CH_3\text{-}3)$	254.34	red pr (al)	73				al, eth, ace, bz	B16, 135
1503	Azobenzene, 2',3-dimethyl-4-ethoxy $3\text{-}CH_3\text{-}4\text{-}C_2H_5OC_6H_3N=N(C_6H_4CH_3\text{-}2)$	254.34	red cr (lig)		35-7			al, bz, lig	B16, 131
1504	Azobenzene, 2,3-dimethyl-4-hydroxy or 4-Benzeneazo-o-xylenol $2,3(CH_3)_2\text{-}4\text{-}HO\text{-}C_6H_2N=NC_6H_5$	226.28	red pr (al, lig)		132			al, eth, ace, bz	B16², 63
1505	Azobenzene, 2,3'-dimethyl-4-hydroxy or 4-m-Tolueneazo-m-cresol $2\text{-}CH_3\text{-}4\text{-}HOC_6H_3N=N(C_6H_4CH_3\text{-}3)$	226.28	og-ye pl (bz)		106-7			eth, bz, lig	B16³, 107
1506	Azobenzene, 2,4'-dimethyl-4-ethoxy $2\text{-}CH_3\text{-}4\text{-}C_2H_5OC_6H_3N=N(C_6H_4CH_3\text{-}4)$	254.34	og-red pl (al)		64			al, eth, bz, lig	B16, 135
1507	Azobenzene, 2,4-dimehyl-5-hydroxy or 5-benzeneazo-2,4-xylenol $2,4\text{-}(CH_3)_2\text{-}5\text{-}HOC_6H_2N=NC_6H_5$	226.28	og-ye nd (lig-peth)		114			al, eth, ace, bz, aa	B16, 146
1508	Azobenzene, 2,4'dimethyl-4-hydroxy or 4-p-Tolueneazo-m-cresol $2\text{-}CH_3\text{-}4\text{-}HOC_6H_3N=N(C_6H_4CH_3\text{-}4)$	226.28	og-ye pr (bz)		135			al, eth, bz	B16², 62
1509	Azobenzene, 3,3'-dimethyl cis or m-Azotoluene $3\text{-}CH_3C_6H_4N=N(C_6H_4CH_3\text{-}3)$	210.28	red (peth)		46				B16³, 47
1510	Azobenzene, 3,3'-dimethyl trans or m-Azotoluene $3\text{-}CH_3C_6H_4N=N(C_6H_4CH_3\text{-}3)$	210.28	og-red orh		84-4.5	1.0123^{66}	1.6152^{66}	al, eth, bz	B16³, 48
1511	Azobenzene, 3,3'-bis(dimethylamino) $3(CH_3)_2N\text{-}C_6H_4N=NC_6H_4[N(CH_3)_2]\text{-}3$	268.37	og nd (bz)	sub	273			eth, bz, chl	B16³, 339
1512	Azobenzene, 3,3'-dimethyl-4-ethoxy $3\text{-}CH_3\text{-}4\text{-}C_2H_5OC_6H_3N=N(C_6H_4CH_3\text{-}3)$	254.34	red-ye pl (al)		46-7			al, eth, ace, bz, lig	B16, 131
1513	Azobenzene, 3,3'-dimethyl-4-hydroxy or 4-m-Tolueneazo-o-cresol $3\text{-}CH_3\text{-}4\text{-}HOC_6H_3N\text{-}N(C_6H_4CH_3\text{-}3)$	226.28	gold-ye nd (bz)		115			al, eth, bz	B16², 60
1514	Azobenzene, 3,4'-dimethyl-4-ethoxy $3\text{-}CH_3\text{-}4\text{-}C_2H_5OC_6H_3N=N(C_6H_4CH_3\text{-}4)$	254.34	og-ye nd (al)	251⁴²	73-4			al, eth, ace, bz	B16, 131
1515	Azobenzene, 3,4'-dimethyl-4-hydroxy or 4-p-Tolueneazo-o-cresol $3\text{-}CH_3\text{-}4\text{-}HOC_6H_3N=N(C_6H_4CH_3\text{-}4)$	226.28	og nd (bz)		163			al, eth, bz	B16³, 106
1516	Azobenzene, 3,5-dimethyl-2-hydroxy or 6-Benzeneazo-4-xylenol $3,5\text{-}(CH_3)_2\text{-}2\text{-}HOC_6H_2N=NC_6H_5$	226.28	dk red nd (al, lig)		90			al, eth, bz, lig	B16, 145
1517	Azobenzene, 3,5-dimethyl-4-hydroxy or 4-Benzeneazo-2,6-xylenol $3,5(CH_3)_2\text{-}4\text{-}HOC_6H_2N=NC_6H_5$	226.28	og-red cr (lig)		150-1			al, eth, bz	B16, 145
1518	Azobenzene, 4,4'-dimethyl cis or p-Azotoluene $4\text{-}CH_3C_6H_4N=N(C_6H_4CH_3\text{-}4)$	210.28	dk red		104			bz, lig	B16³, 49
1519	Azobenzene, 4,4'-dimethyl trans or p-Azotoluene $4\text{-}CH_3C_6H_4N=N(C_6H_4CH_3\text{-}4)$	210.28	og ye nd (al)		144			eth, bz, lig	B16³, 49
1520	Azobenzene, 4,4'-bis(dimethyl amino) $4\text{-}(CH_3)_2N\text{-}C_6H_4\text{-}N=N C_6H_4[N(CH_3)_2\text{-}4]$	268.37			273			eth, bz	B16³, 375

No.	Name, Synonyms, and Formula	Mol. wt.	Color, crystalline form, specific rotation and λ_{max} (log ε)	b.p. °C	m.p. °C	Density	n_D	Solubility	Ref.
1521	Azobenzene, 4,4′-dimethyl -2-hydroxy or 6-p-Tolueneazo-m-cresol 4-CH₃-2-HOC₆H₄N=N(C₆H₄CH₃-4)	226.28	og red cr (lig)	....	150-1	....	....	al, eth, bz, lig	B16³, 106
1522	Azobenzene, 4′,5-dimethyl-2-ethoxy 5-CH₃-2-C₂H₅OC₆H₃N=N(C₆H₄CH₃-4)	254.34	pa red nd (abs al)	253-5⁶³	43			al	B16, 141
1523	Azobenzene, 4′,5-dimethyl-2-hydroxy or 2-p-Tolueneazo-P-cresol 5-CH₃-2-HOC₆H₃N=N(C₆H₄CH₃-4)	226.28	red cr, ye pl (to)	....	112-3			al, bz, to, chl	B16³, 109
1524	Azobenzene, 2-ethoxy or o-Benzeneazo phenetole 2-C₂H₅OC₆H₄N=NC₆H₅	226.28	red pl or mcl pr (peth)		44			al, eth, ace, bz, peth	B16³, 81
1525	Azobenzene, 3-ethoxy or m-Benzeneazo phenetole 3-C₂H₅OC₆H₄N=NC₆H₅	226.28	pl (peth)	208²²	63.5-4			al, eth, ace, bz	B16, 95
1526	Azobenzene, 4-ethoxy or p-Benzeneazo phenetole 4-C₂H₅OC₆H₄N=NC₆H₅	226.28	og nd (60-70% al)	339-40	85	1.0400¹⁰⁰/₄	1.6419¹⁰⁰/₅₈₇₅	al, eth, ace, bz	B16², 40
1527	Azobenzene, 4-ethoxy-2-methyl 2-CH₃-4-C₂H₅OC₆H₃N=NC₆H₅	240.31	or red nd (dil al)		51.5-2.0			al, eth, lig	B16, 134
1528	Azobenzene, 4-ethoxy-2′-methyl 4-C₂H₅OC₆H₄N=N(C₆H₄CH₃-2)	240.31	og pl (al)		53			al, eth, bz, chl	B16, 105
1529	Azobenzene, 4-ethoxy-3-methyl 3-CH₃-4-C₂H₅OC₆H₃N=NC₆H₅	240.31	og nd or mcl pr (al)		60			al, eth, bz	B16, 130
1530	Azobenzene, 4-ethoxy-3′-methyl 4-C₂H₅OC₆H₄N=N(C₆H₄CH₃-3)	240.31	og-red pr (al)		65			al, eth, bz	B16, 106
1531	Azobenzene, 4-ethoxy-4′-methyl 4-C₂H₅OC₆H₄N=N(C₆H₄CH₃-4)	240.31	red lf (al)		121-2			al, bz, chl	B16, 107
1532	Azobenzene, 2-hydroxy or o-Benzeneazo phenol 2-HOC₆H₄N=N-C₆H₅	198.23	og-red nd (eth)		82.5-3			al, eth, ace, bz	B16³, 80
1533	Azobenzene, 2-hydroxy, benzoate 2.C₆H₅CO₂C₆H₄N=NC₆H₅	302.34	og-red nd (lig)		93			al, bz	B16³, 81
1534	Azobenzene, 2-hydroxy-4-methyl 4-CH₃-2-HOC₆H₃N=NC₆H₅	212.25	red pl (lig)		122			al, ace, bz, lig	B16¹, 241
1535	Azobenzene, 2-hydroxy-4-methyl, benzoate 2-C₆H₅CO₂-4-CH₃C₆H₃N=NC₆H₅	316.36			98			bz	B16¹, 241
1536	Azobenzene, 2-hydroxy-5-methyl or 2-benzeneazo-p-cresol 2-HO-5-CH₃C₆H₃N=NC₆H₅	212.25	og-ye lf (bz, bz-lig, gold lf) (w-al)		108-9			al, eth, bz, chl	B16, 108
1537	Azobenzene, 3-hydroxy or m-Benzeneazo phenol 3-HOC₆H₄N=NC₆H₅	198.23	ye nd (w, lig), pr (bz)		116.5-7			al, eth, ace, bz	B16³, 85
1538	Azobenzene, 3-hydroxy, acetate 3-CH₃CO₂C₆H₄N=NC₆H₅	240.27	og pl (peth)		67.5			peth	B16, 95
1539	Azobenzene, 3-hydroxy, benzoate 3-C₆H₅CO₂-C₆H₄N=N-C₆H₅	302.34	og-red pl (peth)		92			peth	B16, 95
1540	Azobenzene, 4-hydroxy or p-Benzeneazo phenol 4-HOC₆H₄N=NC₆H₅	198.23	ye lf (bz), og pr (al)	220-30²⁰/ᵈ	155-7			al, eth, bz	B16², 38
1541	Azobenzene, 4-hydroxy acetate 4-CH₃CO₂C₆H₄N=N-C₆H₅	240.27	og lf (al), nd (al, lig)	> 360 d	89			lig, al	B16¹, 236
1542	Azobenzene, 4-hydroxy, benzoate 4-C₆H₅CO₂C₆H₄N=NC₆H₅	302.34	og-ye nd (al, lig), ye-red pr (eth, al)	....	138			to, lig, al, eth	B16³, 89
1543	Azobenzene, 4-hydroxy-2′-methoxy-2-methyl 2-CH₃-4-HOC₆H₃N=N(C₆H₄OCH₃-2)	242.28	og		161			al, eth	B16, 135
1544	Azobenzene, 4-hydroxy-2′-methoxy-3-methyl 3-CH₃-4-HO-C₆H₃N=N(C₆H₄OCH₃-2)	242.28	og red pl (dil al)		68			al, eth, bz, lig	B16, 131
1545	Azobenzene, 4-hydroxy-2-methyl or 4-Benzeneazo-m-cresol 4-HO-2-CH₃C₆H₃N=NC₆H₅	212.25	ye nd (lig)		109			al, eth, bz, chl	B16², 61
1546	Azobenzene, 4-hydroxy-2′-methyl or o-Toluenwazo-p-phenol 4-HOC₆H₄N=N(C₆H₄CH₃-2)	212.25	red pl or lf (bz-lig), og-ye nd (bz)		107-8			al, eth, bz, chl	B16³, 91
1547	Azobenzene, 4-hydroxy-3-methyl or 4-Benzeneazo-o-cresol 4-HO-3-CH₃C₆H₃N=NC₆H₅	212.25	gold-ye lf or nd (al)		128-30			al, eth, bz, lig	B16³, 104
1548	Azobenzene, 4-hydroxy-3-methyl, benzoate 4-C₆H₅CO₂-3-CH₃C₆H₃N=NC₆H₅	316.36	ye nd (al)	....	110-1			eth, ace, chl	B16, 130

No.	Name, Synonyms, and Formula	Mol. wt.	Color, crystalline form, specific rotation and λ_{max} (log ε)	b.p. °C	m.p. °C	Density	n_D	Solubility	Ref.
1549	Azobenzene, 4-hydroxy-3′-methyl or m-Tolueneazo-p-phenol. $4\text{-HO-C}_6\text{H}_4\text{N}=\text{N}(\text{C}_6\text{H}_4\text{CH}_3\text{-3})$	212.25	ye-pl (al)	. . . .	144-5	. . . .	. . . .	al, bz	B16², 41
1550	Azobenzene, 4-hydroxy-4′-methyl or p-Tolueneazo-p-phenol. $4\text{-HOC}_6\text{H}_4\text{N}=\text{N}(\text{C}_6\text{H}_4\text{CH}_3\text{-4})$	212.25	og red mcl (bz-lig)		152			al, eth, bz	B16², 42
1551	Azobenzene, 4-hydroxy-4′-methyl, benzoate . . . . $4\text{-C}_6\text{H}_5\text{CO}_2\text{C}_6\text{H}_4\text{N}=\text{N}(\text{C}_6\text{H}_4\text{CH}_3\text{-4})$	316.36	og-red pr (bz), redsh-ye nd (lig)		178			al, bz, eth	B16¹, 237
1552	Azobenzene, 2-methoxy or o-Benzeneazo anisole . . . . $2\text{-CH}_3\text{OC}_6\text{H}_4\text{N}=\text{NC}_6\text{H}_5$	212.25	og-red nd (dil al)	195-7¹⁴	41			al, bz	B16³, 81
1553	Azobenzene, 3-methoxy or m-Benzeneaz anisole. . . . . $3\text{-CH}_3\text{OC}_6\text{H}_4\text{N}=\text{NC}_6\text{H}_5$	212.25	og-red pl (MeOH)	193¹⁵	32-3			al, ace	B16³, 85
1554	Azobenzene, 4-methoxy or p-Benzeneazo anisole . . . . $4\text{-CH}_3\text{OC}_6\text{H}_4\text{N}=\text{NC}_6\text{H}_5$	212.25	og-red pl, lf (al) (peth)	340	56 (64)	1.12⁷⁵		al, eth, ace	B16², 40
1555	Azobenzene, 4-methoxy-4′-methyl . . . . $4\text{-CH}_3\text{OC}_6\text{H}_4\text{N-N}(\text{C}_6\text{H}_4\text{CH}_3\text{-4})$	226.28	og-ye pr (al)		110-1			al, eth, ace, bz	B16², 236
1556	Azobenzene, 4-nitro . . . . $4\text{-NO}_2\text{C}_6\text{H}_4\text{N}=\text{NC}_6\text{H}_5$	227.23	lf, nd, (al, lig)		135 (155)			al, ace, bz, lig	B16², 17
1557	Azobenzene, 4-phenylamino or 4-Anilineazobenzene. . . . $4\text{-C}_6\text{H}_5\text{NHC}_6\text{H}_4\text{N}=\text{NC}_6\text{H}_5$	273.34	ye pl or pr		82			al, eth, lig	B16³, 343
1558	Azobenzene, 2,4,4′-triamino or Bismark's brown $2,4(\text{H}_2\text{N})_2\text{C}_6\text{H}_4\text{N}=\text{N}(\text{C}_6\text{H}_4\text{NH}_2\text{-3})$	227.27	og lf (w), red (bz)		143.5			al, eth	B16, 386
1559	2-Azobenzene carboxylic acid or o-Benzeneazobenzoic acid . . . . $2\text{-HO}_2\text{C-C}_6\text{NCH}_4\text{N}=\text{NC}_6\text{H}_5$	226.24	og-red nd or pl (al)		97-8			al, eth, ace, bz, aa	B16³, 213
1560	2-Azobenzene carboxylic acid, 4′-dimethylamino or Methyl red . . . . $2\text{-HO}_3\text{CC}_6\text{H}_4\text{N}=\text{N-}[\text{C}_6\text{H}_4\text{N}(\text{CH}_3)_2\text{-4}]$	269.31	vt or red pr (to or bz), nd (ag aa)		183			al, ace, bz, aa	B16², 164
1561	3-Azobenzene carboxylic acid or m-Benzeneazobenzoic acid . . . . $3\text{-HO}_2\text{CC}_6\text{H}_4\text{N}=\text{NC}_6\text{H}_5$	226.24	og-red pl (w), red lf (al)		170-1			al, eth, bz, chl	B16, 229
1562	3-Azobenzene carboxylic acid, 4-hydroxy or 5-Benzeneazosalicylic acid . . . . $3\text{-HO}_2\text{C-4-HOC}_6\text{H}_3\text{NC}_6\text{H}_5$	242.24	nd (bz)		230d			al, eth, ace	B16³, 237
1563	3-Azobenzene carboxylic acid, 4-hydroxy-2′-nitro or 5-o-Nitrobenzeneazosalicylic acid . . . . $3\text{-(HO}_2\text{C)-4-HOC}_6\text{H}_3\text{N}=\text{N}(\text{C}_6\text{H}_4\text{NO}_2\text{-2})$	287.23	br-red cr (al)		215-7			al, aa	B16², 101
1564	3-Azobenzene carboxylic acid, 4-hydroxy-3′-nitro or 5-m-Nitrobenzeneazosalicylic acid . . . . $3\text{-HO}_2\text{C-4-HOC}_6\text{H}_3\text{N}=\text{N}[\text{C}_6\text{H}_4\text{NO}_2\text{-3}]$	287.23	red-br nd (al)		257			al, eth, bz, aa	B16³, 238
1565	3-Azobenzene carboxylic acid, 4-hydroxy, 4′-nitro or 5-p-nitro benzeneazosalicylic acid . . . . $3\text{-(HO}_2\text{C)-4-HOC}_6\text{H}_3\text{N}=\text{N}(\text{C}_6\text{H}_4\text{NO}_2\text{-4})$	287.23	og br nd (dil aa)		257			al, aa	B16², 101
1566	3-Azobenzene carboxylic acid, 3′-sulfo . . . . $3(\text{HO}_2\text{C})\text{C}_6\text{H}_4\text{N}=(\text{C}_6\text{H}_4\text{SO}_3\text{H-3})$	. . . . . .							B16268
1567	4-Azobenzene carboxylic acid or p-Benzeneazobenzoic acid . . . . $4\text{-HO}_2\text{CC}_6\text{H}_4\text{N}=\text{NC}_6\text{H}_5$	226.24	red pl or lf (al)		249			al, eth, ace, bz	B16³, 218
1568	2,2′-Azobenzenedicarboxylic acid or o-Azobenzoic acid $2\text{-HO}_2\text{CC}_6\text{H}_4\text{N}=\text{N}(\text{C}_6\text{H}_4\text{CO}_2\text{H-2})$	270.25	dk red nd (al)		245 (cor)			eth, al	B16³, 216
1569	3,3′-Azobenzenedicarboxylic acid or m-Azobenzoic acid. $3\text{-HO}_2\text{CC}_6\text{H}_4\text{N}=\text{NC}_6\text{H}_4\text{CO}_2\text{H-3}$	270.25	ye nd (aa)		340d			al	B16³, 217
1570	4,4′-Azobenzeenedicarboxylic acid or p-Azobenzoic acid $4\text{-HO}_2\text{CC}_6\text{H}_4\text{N}=\text{N}(\text{C}_6\text{H}_4\text{CO}_2\text{H-4})$	270.25	dk ye nd (al), red nd (aa)		330d				B16³, 228
1571	3,3′-Azobenzenedisulfonamide $3\text{-(H}_2\text{NO}_2\text{S})\text{C}_6\text{H}_4\text{N}=\text{N}(\text{C}_6\text{H}_5\text{SO}_2\text{NH}_2\text{-3})$	340.38	ye nd (al)		305			al	B16, 268
1572	3,4′-Azobenzene disulfonamide . . . . $3\text{-H}_2\text{NO}_2\text{SC}_6\text{H}_4\text{N}=\text{N}(\text{C}_6\text{H}_4\text{SO}_2\text{NH}_2\text{-4})$	340.38	ye nd (al)		288			al	B16, 279
1573	4,4′-Azobenzene disulfonamide . . . . $4\text{-H}_2\text{NO}_2\text{SC}_6\text{H}_4\text{N}=\text{N}(\text{C}_6\text{H}_4\text{SO}_2\text{NH}_2\text{-4})$	340.38	og nd (al)		>250d			al, eth	B16³, 299
1574	3,3′-Azobenzene disulfonic acid or p-Azobenzoic acid . $3\text{-HO}_3\text{SC}_6\text{H}_4\text{N}=\text{N}(\text{C}_6\text{H}_5\text{SO}_3\text{H-3})$	342.35	ye lf (w + 5)					w, al, eth	B16³, 281
1575	3,4′-Azobenzne disulfonic acid . . . . $3\text{-HO}_3\text{SC}_6\text{H}_4\text{N}=\text{N}(\text{C}_6\text{H}_4\text{SO}_3\text{H-4})$	342.25	syr						B16³, 299

No.	Name, Synonyms, and Formula	Mol. wt.	Color, crystalline form, specific rotation and λ_{max} (log ε)	b.p. °C	m.p. °C	Density	n_D	Solubility	Ref.
1576	4,4'-Azobenzene disulfonic acid 4-HO$_3$SC$_6$H$_4$N=N(C$_6$H$_4$SO$_3$H-4)	342.35	red nd (w + 5)	....	169d (anh)			w	B16[3], 299
1577	3,3'-Azobenzene disulfonyl chloride .... 3-(ClO$_2$S)C$_6$H$_4$N=N(C$_6$H$_4$SO$_2$Cl-3)	379.25	red nd (eth)		166-7			eth	B16, 268
1578	3,4'-Azobenzene disulfonyl chloride 3-ClO$_2$SC$_6$H$_4$N=N(C$_6$H$_4$SO$_2$Cl-4)	379.25	red nd (eth)		123-5			eth	B16, 279
1579	4,4'-Azobenzene disulfonyl chloride 4-ClO$_2$SC$_6$H$_4$N=N(C$_6$H$_4$SO$_2$Cl-4)	379.25	br-red nd (eth-bz)					bz, chl	B16, 280
1580	3-Azobenzene sulfonic acid-4'-hydroxy-3'-nitro .... 3-NO$_2$-4-HOC$_6$H$_3$N=N(C$_6$H$_4$SO$_3$H-3)	323.29	gold-ye pr (dil HCl + w)		235 (anh), 116 (+ w)			w, al	B16, 267
1581	4-Azobenzene sulfonic acid or 4-phenyl azobenzene sulfonic acid .... 4-HO$_3$SC$_6$H$_4$N=N(C$_6$H$_5$)	262.29	og-red lf (w + 3)		127 (hyd)			w	B16[2], 281
1582	4-Azobenzene sulfonylchloride 4-ClO$_2$SC$_6$H$_4$N=NC$_6$H$_5$	280.74	og-red nd or pl (bz)		82			al, bz	B16[3], 282
1583	p,p'-Azobiphenyl or Di-p-xenyldiimide 4-C$_6$H$_5$CH$_4$N=NC$_6$H$_4$C$_6$H$_5$-4	334-42	og-red pl (bz)		256			eth	B16[3], 65
1584	Azomethane or Dimethyl diimide .... CH$_3$N=NCH$_3$	58.08	col or pa ye gas	1.5	-78	0.744[0/15]	1.4199[19]	al, eth, ace	B4[2], 1747
1585	Azomethane, hexafluoro .... F$_3$CN=NCF$_3$	166.03	pa grsh gas	-31.6	-133				B3[4], 246
1586	α,α'-Azonaphthalene or Di-α-naphthyldiimide .... α-C$_{10}$H$_7$N=NC$_{10}$H$_7$-α	282.35	red nd (gl aa)	sub	190			ace, bz	B16[2], 26
1587	α,α'-Azonaphthalene, 4-amino .... 4-H$_2$NC$_{10}$H$_6$N=NC$_{10}$H$_7$	297.36	red-br nd		183 (cor)				B16[2]
1588	1,2'-Azonaphthalene or α,β-Naphthyldiimide .... α-C$_{10}$H$_7$N = N-β-C$_{10}$H$_7$-(2)	2892.35	red nd, br lf (aa)		136			al, bz, aa	B16, 80
1589	2,2'-Azonaphthalene or Di-β-Naphthyldiimide .... β-C$_{10}$H$_7$N=N-β-C$_{10}$H$_7$	282.35	red lf (bz), og-ye nd (chl, al)	sub	208			bz	B16[2], 26
1590	Azoxybenzene cis C$_6$H$_5$N(O)=NC$_6$H$_5$	198.23	....		87	1.166[20/4]	1.633[20]		B16[3], 580
1591	Azoxybenzene trans C$_6$H$_5$N(O)=C$_6$H$_5$	198.23	bt ye nd	d	36	1.1590[26/4]	1.652[20]	al, eth, lig	B16[3], 579
1592	Azoxybenzene, 4-bromo 4-BrC$_6$H$_4$N(O)=NC$_6$H$_5$	277.13	dk ye nd		93.5-4.5	1.4138[100]		al, bz	B16[3], 584
1593	Azoxybenzene, 2,2'-dimethoxy or o-Azoxyanisole 2-CH$_3$OC$_6$H$_4$N(O)=N(C$_6$H$_4$OCH$_3$-2)	258.28	ye-og cr (al)		81			al, eth, ace, bz, chl	B16, 635
1594	Azoxybenzene, 3,3'-dimethoxy or m-Azoxyanisole 3-CH$_3$OC$_6$H$_4$N(O)=N(C$_6$H$_4$OCH$_3$-3)	258.28	ye cr (al)		51-2			al, eth	B16[2], 325
1595	Azoxybenzene, 4,4'-dimethoxy or p-Azoxyanisole 4-CH$_3$OC$_6$H$_4$N(O)=N(C$_6$H$_4$OCH$_3$-4)	258.28	ye mcl nd (al)		119-20	1.1711[115/4]		ace, bz, al	B16[3], 597
1596	Azoxybenzene, 2,2'-dimethyl trans 2-CH$_3$C$_6$H$_4$N(O)=N(C$_6$H$_4$CH$_3$-2)	226.28	....		60	1.0215[65/4]	1.61804[65]	al, eth, bz	B16[3], 587
1597	Azoxybenzene, 3,3'-dimethyl trans 3-CH$_3$C$_6$H$_4$N(O)=N(C$_6$H$_4$CH$_3$-3)	226.28	....		39	1.0123[66]	1.6152[66]	al, eth, bz, lig	B16[3], 588
1598	2,2'-Azoxybenzene dicarboxylic acid or o-Azoxybensoic acid 2-HO$_2$CC$_6$H$_4$N(O)=N(C$_6$H$_4$CO$_2$H-2)	286.25	ye pl or pr (al)		254.5			al, ace, aa	B16[3], 609
1599	3,3'-Azoxybenzene dicarboxylic acid or m-Azoxybenzoic acid 3-HO$_2$CC$_6$H$_4$N(O)=N(C$_6$H$_4$CO$_2$H-2)	286.25	pa ye nd or lf (gl aa)		320d				B16[3], 610
1600	4,4'-Azoxybenzene dicarboxylic acid or p-Azoxybenzoic acid 4-HO$_2$CC$_6$H$_4$N(O)-N(C$_6$H$_4$CO$_2$H-4)	286.25	ye am or pw		360d			Py	B16[3], 615
1601	1,1'-Azoxynaphthalene α-C$_{10}$H$_7$N(O)=N-α-C$_{10}$H$_7$	298.35	ye or red orh pl (al)		127			al	B16[3], 591
1602	2,2'-Azoxynaphthalene β-C$_{10}$H$_7$N(O)=N-β-C$_{10}$H$_7$	298.35	ye or red nd (al, eth, gl aa)		167-8			bz, aa, chl	B16[3], 592
1603	Azulene C$_{10}$H$_8$	128.19	bl or grsh-bk lf (al)	270d, 115-35[10]	99-100			al, eth, ace	B541636
1604	Azulene, 1,4-dimethyl-7-isopropyl or Guaiazulene .... C$_{15}$H$_{18}$	198.5	bl-vt pl (al)	167-8[12]	31.5			al, eth	B5[4], 1751
1605	Azulene, 1,5-dimethyl-8-isopropyl or Chamazulene ...... C$_{15}$H$_{18}$	198.31	bl lig	161[12]		0.9883[20/4]		al, eth	B5[2], 474

No.	Name, Synonyms, and Formula	Mol. wt.	Color, crystalline form, specific rotation and λ_{max} (log ε)	b.p. °C	m.p. °C	Density	n_D	Solubility	Ref.
1606	Azulene, 4,8-dimethyl-2-isopropyl or Elamazulene, Vetivazulene .... $C_{15}H_{18}$	198.31	red-vt nd (al)	140-60²	31-2	....	....	al, eth	B5⁴, 1754
1607	Azulene, 4,6,8-trimethyl .... $C_{13}H_{14}$	170.25	....	....	79-81	....	....	chl	B5⁴, 1721

No.	Name, Synonyms, and Formula	Mol. wt.	Color, crystalline form, specific rotation and λ_{max} (log ε)	b.p. °C	m.p. °C	Density	n_D	Solubility	Ref.
1608	Barbituric acid or Pyrimidinetrione................. $C_4H_4N_2O_3$	128.09	rh pr (w + 2)	260d	248			eth, w	B24², 292
1609	Barbituric acid, 5,allyl-5-butyl .... $C_{11}H_{16}N_2O_3$	224.26	(w, dil al)	....	128			w, al	B24², 292
1610	Barbituric acid, 5-allyl-5(2-cyclopenten-l-yl) or Cycclopal $C_{12}H_{14}N_2O_3$	234.26	(w, dil al)	....	139-40			w, al	C24, 5308
1611	Barbituric acid, 5-allyl-5-iso-butyl or Sandoptal .... $C_{11}H_{16}N_2O_3$	224.26	(w, dil al)	....	138			w, al, eth, ace, chl	B24², 292
1612	Barbituric acid, 5- allyl-5-iso-propyl .... $C_{10}H_{14}N_2O_3$	210.23		....	140-1				B24², 290
1613	Barbituric acid, 5-allyl-5-iso-propyl-l-methyl or Narconumal .... $C_{11}H_{16}N_2O_3$	224.26	(w, dil al)	176-8¹²	56-7			al, eth, chl, ace, bz	C32, 1052
1614	Barbituric acid, 5- allyl-5-phenyl or Alphenal .... $C_{13}H_{12}N_2O_3$	244.25		....	156-7			al, eth, chl	C26, 4828
1615	Barbituric acid, 5- amino or Uramil .... $C_4H_5N_3O_3$	143.10	nd or pl (w)	....	> 400			w, chl	B25¹, 704
1616	Barbituric acid, 5-benzylidene .... $C_{11}H_8N_2O_3$	216.21	pr (aa)	....	256			ace	B24², 299
1617	Barbituric acid, 5-(2-bromoallyl)-5-iso-propyl or Propallylonal. .... $C_{10}H_{13}BrN_2O_3$	236.27	(dil aa, dil al)	....	181			al, eth, ace, aa	B24², 291
1618	Barbituric acid, 5-butyl-5-ethyl .... $C_{10}H_{16}N_2O_3$	212.25		....	128-9				B24², 285
1619	Barbituric acid, 5-(l-cyclohexenyl)-5-ethyl or Cyclobarbitone. .... $C_{12}H_{16}N_2O_3$	236.27	lf (w)	....	172-4			al, eth	B24², 294
1620	Barbituric acid, 5,5-diallyl or Allobarbitone .... $C_{10}H_{12}N_2O_3$	208.22	pl (w or 50% al)	....	174			al, eth, bz	B24², 293
1621	Barbituric acid, 5,5-dibromo or Dibromin .... $C_4H_2Br_2N_2O_3$	285.89	pl (MeOH-bz), lf (dil HNO₃)	d	235			al, eth	B24², 272
1622	Barbituric acid, 5,5-diethyl or barbital, veronal. .... $C_8H_{12}N_2O_3$	184.20	(i) trig (w) (ii) mcl pr (iii) mcl nd (iv) tcl	....	(i) 190 (ii) 183 (iii) 181 (iv) 176	1.220		al, eth, ace, aa	B24², 279
1623	Barbituric acid, 5,5-diethyl-l-methyl or Metharbital ...... $C_9H_{14}N_2O_3$	198.22	nd	....	150-1			w	B24², 281
1624	Barbituric acid, 5,5-dipropyl or Proponal .... $C_{10}H_{16}N_2O_3$	212.25	pl (w)	....	146 (166)			al, eth, bz, chl	B24², 286
1625	Barbituric acid, 5-ethyl-5-hexyl .... $C_{12}H_{20}N_2O_3$	240.31	nd (w)	....	112-3			al, eth, bz	B24², 288
1626	Barbituric acid, 5-ethyl-5(2-methallyl)-2-thio .... $C_{10}H_{14}N_2O_2S$	226.28		....	160-1				C35, 5599
1627	Barbituric acid, 5-ethyl-5-(1-methylbutyl) .... $C_{11}H_{18}NO_3$	226.27		....	130			al, eth	B24², 287
1628	Barbituric acid, 5-ethyl-5-(1-methylbutyl)-2-thio .... $C_{11}H_{18}N_2O_2S$	242.34		....	158-9				C29, 8237
1629	Barbituric acid, 5-ethyl-5-(3-methylbutyl) .... $C_{11}H_{18}N_2O_3$	226.27		....	157-9.5				B24², 287
1630	Barbituric acid, 5-ethyl-(3-methylbutyl) or Amytal ...... $C_{11}H_{18}N_2O_3$	226.28	lf (w, dil al)	....	156-8			al, eth, bz, chl	B24², 287
1631	Barbituric acid, 5-ethyl-l-methyl-5-phenyl or Methobarbital, Prominal. .... $C_{13}H_{14}N_2O_3$	246.27	wh cr (w)	....	176			al, chl	C28, 281
1632	Barbituric acid, 5-ethyl-5-pental .... $C_{11}H_{18}N_2O_3$	226.28	cr (dil al)	....	135-6			al, eth, chl	B24², 286
1633	Barbituric acid, 5-ethyl-5-(2-pentyl) or Nembutal, Pentobarbital. .... $C_{11}H_{18}N_2O_3$	226.28	nd (w)	....	130			al, eth	B24², 287
1634	Barbituric acid, 5-ethyl-5-phenyl or Phenobarbital, Luminal. .... $C_{12}H_2N_2O_3$	232.24	pl (w)	....	174			al, eth	B24², 297
1635	Barbituric acid, 5-ethyl-5-(1-piperidyl) or Eldoral .... $C_{11}H_{17}N_3O_3$	239.28	wh cr (dil al)	....	215			al, eth, ace	C36, 3160
1636	Barbituric acid, 5-ethyl-5-iso-propyl or Ipral, Probarbital $C_9H_{14}N_2O_3$	198.22	nd (w)	....	203			eth	B24², 284

No.	Name, Synonyms, and Formula	Mol. wt.	Color, crystalline form, specific rotation and λ_{max} (log ε)	b.p. °C	m.p. °C	Density	n_D	Solubility	Ref.
1637	Barbituric acid, 5-(2-furfurylidine) $C_9H_6N_2O_4$	206.16	....		315				B27[1], 607
1638	Barbituric acid, 5-(2-furfurylidine)-2-thio $C_9H_6N_2O_3S$	222.22	ye pl	....	> 280d				B27[1], 607
1639	Barbituric acid, 5-hydroxy or Dialuric acid $C_4H_4N_2O_4$	144.09	pr or pl nd (+ lw, w)		224 (anh)			al, ace, w, aa	B25[2], 61
1640	Barbituric acid, l-methyl $C_5H_6N_2O_3$	142.11			132-3			al, ace	B24[2], 270
1641	Barbituric acid, 5-methyl-5-phenyl or Rufinal $C_{11}H_{10}N_2O_3$	218.21	cr		220			al, eth	B24[2], 296
1642	Barbituric acid, 5-nitro $C_4H_3N_3O_5$	173.09	pr, lf (w + 3)		180-1			al, w	B24[2], 273
1643	Barbituric acid, 5-(2-phenylethyl) $C_{12}H_{12}N_2O_3$	232.24	cr		212-3			al	B24[2], 296
1644	Barbituric acid, 2-thio $C_4H_4N_2O_2S$	144.15	pl (w)		235d			al, .w	B24[2], 275
1645	Bebeerine d or Chondrodendrin $C_{36}H_{38}N_2O_6$	594.72	cr (bz, eth, chl-MeOH), $[\alpha]^{25}_D$ + 297 (al)		221			al, eth, ace, chl	B27[2], 896
1646	Berbeerine l or Curine $C_{36}H_{38}N_2O_6$	594.72	pr, nd (chl-MeOH), $[\alpha]^{20}_D$ −328 p4, $[\alpha]^{28}_D$ −308 (al)		221			ace, bz, Py	B27[2], 894
1647	Behenic acid or Docosanoic acid $CH_3(CH_2)_{20}CO_2H$	340.60	nd	306[60]	80	0.8223[90]	1.4270[100]	....	B2[4], 1290
1648	Behenic acid, ethyl ester or Ethyl behenate $CH_3(CH_2)_{20}CO_2C_2H_5$	368.85	nd (al), or (ace)	240-2[10]	50			al, eth	B2[4], 1292
1649	Behenic acid, methyl ester or Methylbehenate $CH_3(CH_2)_{20}CO_2CH_3$	354.62	nd (ace)	224-5[15]	54		1.4339[60]	al, eth	B2[4], 1291
1650	Benzal acetate, 2-acetoxy $2-CH_3CO_2C_6H_4CH(O_2CCH_3)_2$	266.25	nd or pl (al), pr (al)		107			eth, bz, chl	B8[3], 148
1651	Benzal acetate, 3-acetoxy $3-CH_3CO_2C_6H_4CH(O_2CCH_3)_2$	266.25	lf (w-al)		76			al, eth	B8, 60
1652	Benzal acetate, 4-acetoxy $4-CH_3CO_2C_6H_4CH(O_2CCH_3)_2$	266.25	pr (eth or lig)		94			eth, lig, al	B8[1], 530
1653	Benzalacetophenone or $trans$ Chalcone $C_6H_5COCH=CHC_6H_5$	208.26	pa ye lf, pr, nd (peth)	345-8d, 208[25]	(i) 59 (ii) 57 (iii) 49	1.0712[62/4]		eth, bz, chl	B7[3], 2380
1654	Benzaldehyde C_6H_5CHO	106.13	....	178, 62[10]	−26(fr-56)	1.0415[10/4]	1.5463[20]	**al, eth**, ace, bz, lig	B7[3], 805
1655	Benzaldehyde, 2-acetamido $2-CH_3CONHC_6H_4CHO$	163.18			70-1			w, eth, ace	B14[3], 51
1656	Benzaldehyde, 3-acetamido $3-CH_3CONHC_6H_4CHO$	163.18	pl (bz)	....	84			al, eth, bz	B14, 29
1657	Benzaldehyde, 4-acetamido $4-CH_3CONHC_6H_4CHO$	163.18	pr (w)		156			w, bz	B14[2], 25
1658	Benzaldehyde, 3-allyl-2-hydroxy $3-(CH_2=CHCH_2)-2-HOC_6H_3CHO$	162.19		245-6[755]		1.098[15]			B8[1], 559
1659	Benzaldehyde, 2-amino or Anthranil aldehyde $2-H_2NC_6H_4CHO$	121.14	silv lf		39-42			al, eth, bz, chl	B14[3], 47
1660	Benzaldehyde, 2-amino, oxime $2-H_2NC_6H_4CH=NOH$	136.17	nd (bz)		135-6			al, eth, bz, aa	B14[3], 50
1661	Benzaldehyde, 3-amino $3-H_2NC_6H_4CHO$	121.14	nd (AcOEt)		28-30			eth	B14[3], 53
1662	Benzaldehyde, 3-amino, oxime $3-H_2NC_6H_4CH=NOH$	136.17	nd (bz)		88			al, eth	B14, 28
1663	Benzaldehyde, 4-amino $4-H_2NC_6H_4CHO$	121.14	pl (w)		71-2			w, al, eth	B14[3], 57
1664	Benzaldehyde, 4-amino, oxime $4-H_2NC_6H_4CH=NOH$	136.17	ye cr (w)		124			al, eth	B14[3], 58
1665	Benzaldehyde azine or Dibenzal hydrazine $C_6H_5CH=NN=CHC_6H_5$	208.27	ye pr (al)		93			al, eth, ace, bz	B7[3], 844
1666	Benzaldehyde, 2-bromo $2-BrC_6H_4CHO$	185.03	....	230, 118[12]	21-2		1.5925[20]	al, bz	B7[3], 882

No.	Name, Synonyms, and Formula	Mol. wt.	Color, crystalline form, specific rotation and λ_{max} (log ε)	b.p. °C	m.p. °C	Density	n_D	Solubility	Ref.
1667	Benzaldehyde, 2-bromo, diacetate .... 2-BrC$_6$H$_4$CH(O$_2$CCH$_3$)$_2$	287.12	....	....	84-6	....	....	al, eth	B7[2], 181
1668	Benzaldehyde, 2-bromo-4-hydroxy 2-Br-4-HOC$_6$H$_3$CHO	201.03	pa ye nd (w)	....	159.5	....	....	....	B8[2], 74
1669	Benzaldehyde, 2-bromo-5-hydroxy 2-Br-5-HOC$_6$H$_3$CHO	201.03	nd (w)	....	135	....	....	al, eth, ace, bz	B8[2], 57
1670	Benzaldehyde, 3-bromo .... 3-BrC$_6$H$_4$CHO	185.03	....	233-6	....	....	1.5915[20]	al, eth	B7[3], 882
1671	Benzaldehyde, 3-bromo-2-hydroxy or 3-Bromosalicylaldehyde .... 3-Br-2-HOC$_6$H$_3$CHO	201.03	nd (dil al)	....	49	....	....	al, eth, ace, bz	B8, 54
1672	Benzaldehyde, 3-bromo-4-hydroxy 3-Br-4-HOC$_6$H$_3$CHO	201.03	lf (w)	....	124	....	....	al, eth, ace, bz	B8[3], 290
1673	Benzaldehyde, 4-bromo .... 4-BrC$_6$H$_4$CHO	185.05	lf (dil al)	66-8[2]	67	....	....	al, bz	B7[3], 883
1674	Benzaldehyde, 4-bromo, diacetate 4-BrC$_6$H$_4$CH(O$_2$CCH$_3$)$_2$	287.12	ye	....	94-5	....	....	eth, al	B7[2], 182
1675	Benzaldehyde, 4-bromo-2-hydroxy or 4-Bromosalicylaldehyde .... 4-Br-2-HOC$_6$H$_3$CHO	201.03	nd (dil al)	....	52	....	....	al, eth, ace, bz	B8, 54
1676	Benzaldehyde, 4-bromo-3-hydroxy 4-Br-3-HOC$_6$H$_3$CHO	201.03	....	....	131.5	....	....	ace, bz, chl	B8[2], 56
1677	Benzaldehyde, 5-bromo-2-hydroxy or 5-Bromosalicylaldehyde .... 5-Br-2-HOC$_6$H$_3$CHO	201.03	nd (al), lf (eth)	....	105-6	....	....	al, eth	B8[3], 269
1678	Benzaldehyde, 5-bromo-4-hydroxy-3-methoxy or 5-Bromovanillan .... 5-Br-4-HO-3-CH$_3$OC$_6$H$_2$CHO	231.05	pl (aa), nd, pl (al)	....	164-6	....	....	al	B8[3], 2050
1679	Benzaldehyde, 2-chloro .... 2-ClC$_6$H$_4$CHO	140.57	nd	211.9, 84.3[10]	12.4	1.2483[20/4]	1.5662[20]	al, eth, ace, bz	B7[3], 864
1680	Benzaldehyde, 2-chloro-3-hydroxy 2-Cl-3-HOC$_6$H$_3$CHO	156.57	cr (ag aa)	....	139.5	....	....	al, aa	B8[2], 54
1681	Benzaldehyde, 2-chloro-4-hydroxy 2-Cl-4-HOC$_6$H$_3$CHO	156.57	nd (w or aa)	....	147-8	....	....	al, eth, aa	B8, 81
1682	Benzaldehyde, 2-chloro-5-hydroxy 2-Cl-5-HOC$_6$H$_3$CHO	156.57	nd (w or aa)	....	110-1	....	....	al, aa	B8[1], 526
1683	Benzaldehyde, 2-chloro, phenyl hydrazone 2-ClC$_6$H$_4$CH=NNHC$_6$H$_5$	230.70	....	....	84	....	....	....	B15[3], 86
1684	Benzaldehyde, 3-chloro .... 3-ClC$_6$H$_4$CHO	140.57	pr	213-4, 55[1]	17-8	1.2410[20/4]	1.5650[20]	al, eth, ace, bz	B7[3], 869
1685	Benzaldehyde, 3-chloro-2-hydroxy or 3-Chlorosalicylaldehyde .... 3-Cl-2-HOC$_6$H$_3$CHO	156.57	nd (MeOH)	....	55	....	....	al, eth, ace, bz	B8[3], 179
1686	Benzaldehyde, 3-chloro-4-hydroxy 3-Cl-4-HOC$_6$H$_3$CHO	156.57	nd (w)	149-50[14]	139	....	....	al, eth	B8[3], 286
1687	Benzaldehyde, 4-chloro .... 4-ClC$_6$H$_4$CHO	140.57	pl	213-4, 72-5[3]	47.5	1.196[61/4]	1.5552[61]	w, al, eth, ace, bz	B7[3], 872
1688	Benzaldehyde, 4-chloro-2-hydroxy or 4-chlorosalicylaldehyde .... 4-Cl-2-HOC$_6$H$_3$CHO	156.57	nd (al or ag aa)	....	52.5	....	....	w, eth, ace, bz	B8[2], 44
1689	Benzaldehyde, 4-chloro-3-hydroxy 4-Cl-3-HOC$_6$H$_3$CHO	156.57	nd (al)	....	121	....	....	w, al, eth, ace, bz	B8[2], 55
1690	Benzaldehyde, 4-chloro-3-nitro .... 4-Cl-3-NO$_2$C$_6$H$_3$CHO	185.57	....	....	64.5	....	....	chl	B7[3], 919
1691	Benzaldehyde, 5-chloro-2-hydroxy or 5-Chlorosalicylaldehyde .... 5-Cl-2-HOC$_6$H$_3$CHO	156.57	pl (al)	105[12]	99-100	....	....	al, eth	B8[2], 45
1692	Benzaldehyde, 5-chloro-2-hydroxy, oxime 5-Cl-2-HOC$_6$H$_3$CH=NOH	171.58	nd (w)	....	128	....	....	al	B8, 53
1693	Benzaldehyde, 5-chloro-4-hydroxy-3-methoxy or 5-Chlorovanillan .... 5-Cl-4-HO-3-CH$_3$OC$_6$H$_2$CHO	186.60	tetr	....	165	....	....	al, aa	B8[3], 2063
1694	Benzaldehyde diacetate or Benzylidene diacetate .... C$_6$H$_5$CH(O$_2$CCH$_3$)$_2$	208.22	pl (eth)	220, 154[20]	46	1.11[20]	....	al, eth, bz	B7[3], 827
1695	Benzaldehyde, 3,5-dibromo-2-hydroxy or 3,5-Dibromosalicylaldehyde .... 3,5-Br$_2$-2-HOC$_6$H$_2$CHO	279.92	....	....	82.0-3.5	....	....	....	B8[3], 269
1696	Benzaldehyde, 2,3-dichloro .... 2,3-Cl$_2$-C$_6$H$_3$CHO	175.02	cr (dil al)	....	65-7	....	....	al, eth	B7[3], 878

PHYSICAL CONSTANTS OF ORGANIC COMPOUNDS (Continued)

No.	Name, Synonyms, and Formula	Mol. wt.	Color, crystalline form, specific rotation and λ_{max} (log ε)	b.p. °C	m.p. °C	Density	n_D	Solubility	Ref.
1697	Benzaldehyde, 2,4-dichloro .... 2,4-Cl$_2$C$_6$H$_3$CHO	175.02	pr	....	72			al, eth, bz	B7^3, 878
1698	Benzaldehyde, 2,4-dichloro-3-hydroxy .... 2,4-Cl$_2$-3-HOC$_6$H$_2$CHO	191.02	cr (aa)		141			lig	B8^2, 55
1699	Benzaldehyde, 2,5-dichloro .... 2,5-Cl$_2$C$_6$H$_3$CHO	175.02	nd (al)	231-3	58			al, eth, bz, chl	B7^3, 878
1700	Benzaldehyde, 2,6-dichloro .... 2,6-Cl$_2$-C$_6$H$_3$CHO	175.02	nd (lig)		71			al, eth, lig	B7^3, 878
1701	Benzaldehyde, 2,6-dichloro-3-hydroxy .... 2,6-Cl$_2$-3-HOC$_6$H$_2$CHO	191.02	cr (w)		142			aa	B8^3, 204
1702	Benzaldehyde, 3,4-dichloro .... 3,4-Cl$_2$C$_6$H$_3$CHO	175.02		247-8	44			al, eth	B7^3, 880
1703	Benzaldehyde, 3,5-dichloro .... 3,5-Cl$_2$C$_6$H$_3$CHO	175.02	nd or lf (dil al)	235-40^{748}	65			al, eth, ace, bz	B7^3, 880
1704	Benzaldehyde, 3,5-dichloro-2-hydroxy or 3,5-dichlorsalicylaldehyde .... 3,5-Cl$_2$-2-HOC$_6$H$_2$CHO	191.02	ye rh (aa)	....	95				B8^3, 183
1705	Benzaldehyde, 3,5-dichloro-2-hydroxy, oxime .... 3,5-Cl$_2$-2-HOC$_6$H$_2$CH=NOH	206.03	nd (dil al)		195-6			al, eth, bz	B8, 54
1706	Benzaldehyde, 3,5-dichloro-4-hydroxy .... 3,5-Cl$_2$-4-HOC$_6$H$_2$CHO	191.02	nd (chl, dil al)		158-9			al, eth	B8, 81
1707	Benzaldehyde, 4,6-dichloro-3-hydroxy .... 4,6-Cl$_2$-3-HOC$_6$H$_2$CHO	191.02	nd		130			w, al, eth, bz	B8^1, 526
1708	Benzaldehyde, 2,3-diethoxy .... 2,3-(C$_2$H$_5$O)$_2$C$_6$H$_3$CHO	194.23		169^{37}				al, eth	B8^3, 1982
1709	Benzaldehyde, 3,4-diethoxy .... 3,4-(C$_2$H$_5$O)$_2$C$_6$H$_3$CHO	194.23	....	278-30				al	B8^2, 2025
1710	Benzaldehyde, 4-diethylamino .... 4-(C$_2$H$_5$)$_2$NC$_6$H$_3$CHO	177.23	ye nd (w)	174^7	41			w, al, eth, bz	B14^3, 71
1711	Benzaldehyde, 2,4-dihydroxy or β-Resorcyl aldehyde..... 2,4-(HO)$_2$C$_6$H$_3$CHO	138.12	nd (eth-lig)	220-8^{22}	201-2			w, al, eth, chl	B8^3, 1989
1712	Benzaldehyde, 2,4-dihydroxy, oxime .... 2,4-(HO)$_2$C$_6$H$_3$CH=NOH	153.14			192			w, al, eth	B8^2, 274
1713	Benzaldehyde, 2,5-dihydroxy or Gentisic aldehyde....... 2,5-(HO)$_2$C$_6$H$_3$CHO	138.12	ye nd (bz)	....	99			w, al, eth, chl	B8^3, 2003
1714	Benzaldehyde, 2,6-dihydroxy-4-methyl or Atranol....... 2,6-(HO)$_2$-4-CH$_3$C$_6$H$_2$CHO	152.16	ye nd (w)		124			w, al, eth, chl	B8^3, 2006
1715	Benzaldehyde, 3,4-dihydroxy or Protocatechualdehyde... 3,4-(HO)$_2$C$_6$H$_3$CHO	138.12	lf (w, to)		153-4			w, al, eth	B8^3, 2009
1716	Benzaldehyde, 3,4-dihydroxy, oxime .... 3,4-(HO)$_2$C$_6$H$_3$CH=NOH	153.14	nd (w)		157d			w, al	B8^2, 285
1717	Benzaldehyde, 3,5-diiodo-4-hydroxy .... 3,5-I$_2$-4-HOC$_6$H$_2$CHO	373.91	nd (al)	....	206.5			al, ace, aa	B8^3, 251
1718	Benzaldehyde, 2,4-dimethoxy .... 2,4-(CH$_3$O)$_2$C$_6$H$_3$CHO	166.18	nd (al or lig)	165^{10}	72			al, eth, bz, lig	B8^3, 1992
1719	Benzaldehyde, 2,5-dimethoxy .... 2,5-(CH$_3$O)$_2$C$_6$H$_3$CHO	166.18	....	146^{10}	52			al, eth	B8^3, 2004
1720	Benzaldehyde, 2,4-dimethoxy-6-hydroxy .... 2,4-(CH$_3$O)$_2$-6-HOC$_6$H$_2$CHO	182.18	nd, pl (dil al)	190-5^{25}, 165^{10}	71			al, eth, bz, aa	B8^3, 3364
1721	Benzaldehyde, 2,6-dimethoxy-4-hydroxy .... 2,6-(CH$_3$O)$_2$-4-HOC$_6$H$_2$CHO	182.18	nd or pl (MeOH), pr (bz)		70-1				B8^3, 3365
1722	Benzaldehyde, 3,4-dimethoxy or Veratraldehyde .... 3,4-(CH$_3$O)$_2$C$_6$H$_3$CHO	166.18	nd (eth, lig, to)	258, 172-5^{18}	44 (58)			al, eth	B8^3, 2020
1723	Benzaldehyde, 3,4-dimethoxy-5-hydroxy .... 3,4-(CH$_3$O)$_2$-5-HOC$_6$H$_2$CHO	182.18		177-80^{12}	62-3			ual, bz, aa, lig	B8^2, 437
1724	Benzaldehyde, 3,5-dimethoxy .... 3,5-(CH$_3$O)$_2$C$_6$H$_3$CHO	166.18		151^{16}	45.5			al, bz	B8^3, 2073
1725	Benzaldehyde, 3,5-dimethoxy-4-hydroxy .... 3,5-(CH$_3$O)$_2$-4-HOC$_6$H$_2$CHO	182.18	br nd (lig)	192-3^{14}	113			al, eth, bz, aa	B8^3, 3368
1726	Benzaldehyde, 2,4-dimethyl .... 2,4-(CH$_3$)$_2$C$_6$H$_3$CHO	134.18		218, 99^{10}	-9			al, eth, ace, bz	B7^3, 1072
1727	Benzaldehyde, 2,5-dimethyl .... 2,5-(CH$_3$)$_2$C$_6$H$_3$CHO	134.18		220, 100^{10}				w, al, eth, ace, bz	B7^3, 1072
1728	Benzaldehyde, 3,4-dimethyl .... 3,4-(CH$_3$)$_2$C$_6$H$_3$CHO	134.18		223-5				al, eth, ace	B7^3, 1073
1729	Benzaldehyde, 3,5-dimethyl .... 3,5-(CH$_3$)$_2$C$_6$H$_3$CHO	134.18		220-2	9			al, eth, ace, bz	B7^3, 1073

No.	Name, Synonyms, and Formula	Mol. wt.	Color, crystalline form, specific rotation and λ_{max} (log ε)	b.p. °C	m.p. °C	Density	n_D	Solubility	Ref.
1730	Benzaldehyde, 4-dimethylamino .... 4-(CH₃)₂NC₆H₄CHO	149.19	lf (w)	176-7[17]	74	....	....	al, eth, ace, bz	B14³, 58
1731	Benzaldehyde, 2,4-dinitro .... 2,4-(NO₂)₂C₆H₃CHO	196.12	pa ye pr (al), pl (bz)	190-210[10-20]	72	....	....	al, eth, bz	B7³, 923
1732	Benzaldehyde, 2,6-dinitro 2,6-(NO₂)₂C₆H₃CHO	196.12	lf (dil al)	....	123	....	....	w, al, eth, bz, aa	B7², 206
1733	Benzaldehyde, 2-ethoxy 2-C₂H₅OC₆H₄CHO	150.18	....	247-9, 143-7[25]	20-2	....	....	**al, eth**	B8³, 144
1734	Benzaldehyde, 3-ethoxy 3-C₂H₅OC₆H₄CHO	150.18	....	245.5	......	1.0768[20/4]	1.5408[20]	al, eth, bz	B8, 60
1735	Benzaldehyde, 3-ethoxy-2-hydroxy or 3-Ethoxysalicyl-aldehyde 3-C₂H₅O-2-HOC₆H₃CHO	166.18	....	263-4[740]	64-5	....	....	....	B8³, 1981
1736	Benzaldehyde, 3-ethoxy-4-hydroxy or Ethylvanillan...... 3-C₂H₅O-4-HOC₆H₃CHO	166.18	....	....	77-8	....	....	al, eth, bz, chl	B8³, 2022
1737	Benzaldehyde, 4-ethoxy 4-C₂H₅OC₆H₄CHO	150.18	....	249, 140[20]	13-4	1.08[21/21]	....	al, eth, bz	B8³, 223
1738	Benzaldehyde, 4-ethoxy-2-hydroxy or 4-Ethoxysalicyl-dehyde 4-C₂H₅O-2-HOC₆H₃CHO	166.18	nd (dil al)	....	35	....	....	al	B8³, 1992
1739	Benzaldehyde, 4-ethoxy-3-methoxy 4-C₂H₅O-3-CH₃OC₆H₃CHO	180.21	mcl pr	sub	64-5 (73-4)	....	....	**al, eth,** bz, aa	B8³, 2024
1740	Benzaldehyde, 5-ethoxy-2-hydroxy or 5-Hydroxysalicyl-dehyde 5-C₂H₅O-2-HOC₆H₃CHO	166.18	ye pr	230	51.5	....	....	al, eth, chl	B8, 245
1741	Benzaldehyde hydrazone .... C₆H₅CH=NNH₂	120.15	lf	140[14]	16	....	....	al	B7³, 863
1742	Benzaldehyde, 2-hydroxy or Salicylaldehyde.......... 2-HOC₆H₄CHO	122.13	....	197, 93[25]	-7	1.1674[20/4]	1.5740[20]	**al, eth,** ace, bz	B8³, 135
1743	Benzaldehyde-2-hydroxy-3-methoxy or o-Vanillan...... 2-HO-3-CH₃OC₆H₃CHO	152.16	lt ye-lt gr nd (w, lig)	256-6, 128[10]	44-5	....	....	al, eth, lig	B8³, 1979
1744	Benzaldehyde, 2-hydroxy-4-methoxy or 4-Methoxysalic-laldehyde 2-HO-4-CH₃OC₆H₃CHO	152.16	nd (w), cr (al)	....	40-2	....	....	al, eth, bz, lig	B8³, 1991
1745	Benzaldehyde, 2-hydroxy-5-methoxy or 5-Methoxysalic-laldehyde 2-HO-5-CH₃OC₆H₃CHO	152.16	ye (w)	247-8	4	....	....	al, eth	B8³, 2004
1746	Benzaldehyde, 2-hydroxy-3-nitro or 3-Nitrosalicylalde-hyde .... 2-HO-3-NO₂ C₆H₃CHO	167.12	nd (aa)	....	109-10	....	....	al, bz	B8³, 190
1747	Benzaldehyde, 2-hydroxy-5-nitro or 5-Nitrosalicylalde-hyde 5-NO₂-2--HO-C₆H₃CHO	167.12	(dil aa)	....	128	....	....	al, eth	B8³, 192
1748	Benzaldehyde, 2-hydroxy-5-nitro, oxime .... 2-HO-5-NO₂C₆H₃CH=NOH	182.14	....	....	225	....	....	....	B8³, 193
1749	Benzaldehyde, 3-hydroxy 3-HOC₆H₄CHO	122.13	nd (w)	240, 161[20]	108	....	....	al, eth, ace, bz	B8³, 197
1750	Benzaldehyde, 3-hydroxy, azine 3-HOC₆H₄CH=NN=CHC₆H₄OH-(3)	240.27	ye pl (al)	....	162	....	....	....	B8³, 203
1751	Benzaldehyde, 3-hydroxy-4-methoxy or Isovanillan...... 3-HO-4-CH₃OC₆H₃CHO	152.16	pl (w)	179[15]	116-7	1.198	....	al, eth, bz, chl, aa	B8³, 2019
1752	Benzaldehyde, 3-hydroxy-5-methoxy 3-HO-5-CH₃OC₆H₃CHO	152.16	....	110-20, sub	130-1	....	....	al, bz, aa	B8³, 2073
1753	Benzaldehyde, 3-hydroxy-2-nitro 3-HO-2-NO₂C₆H₃CHO	167.12	nd or pl (bz, lig), d	....	152	....	....	al, bz	B8³, 210
1754	Benzaldehyde, 3-hydroxy-2-nitro, oxime 3-HO-2-NO₂C₆H₃CH=NOH	182.14	pa ye nd	....	172.5	....	....	....	B8², 58
1755	Benzaldehyde, 3-hydroxy-4-nitro 4-NO₂-3-HOC₆H₃CHO	167.12	ye lf	....	128	....	....	al, eth, bz	B8³, 211
1756	Benzaldehyde, 3-hydroxy-4-nitro, oxime 4-NO₂-3-HOC₆H₃CH=NOH	182.14	ye nd (chl)	....	164	....	....	al, chl	B8², 58
1757	Benzaldehyde, 3-hydroxy, oxime 3-HOC₆H₄CH=NOH	137.14	(bz)	....	90	....	....	w, al, eth	B8, 61
1758	Benzaldehyde, 4-hydroxy 4-HOC₆H₄CHO	122.13	nd (w)	....	117	1.129[110/4]	1.5705[110]	al, eth, bz	B8³, 215

No.	Name, Synonyms, and Formula	Mol. wt.	Color, crystalline form, specific rotation and λ_{max} (log ϵ)	b.p. °C	m.p. °C	Density	n_D	Solubility	Ref.
1759	Benzaldehyde, 4-hydroxy, azine .... 4-HOC₆H₄CH=NN=CHC₆H₄OH(-4)	240.27	ye (al)	....	239-40, 268 (d)	....	....	al, bz	B8[1], 531
1760	Benzaldehyde, 4-hydroxy-2-iodo-3-methoxy or 2-Iodo vanillan. 4-HO-3-CH₃O-2IC₆H₂CHO	278.05	....	....	155-6			al	B8[3], 2058
1761	Benzaldehyde, 4-hydroxy-5-iodo-3-methoxy or 5-Iodo vanillan. 5-I-4-HO-3-CH₃OC₆H₂CHO	278.05	pa ye	....	180				B8[3], 2059
1762	Benzaldehyde, 4-hydroxy-2-methoxy 4-HO-2-CH₃OC₆H₃CHO	152.16	lf (bz), nd (w)	....	153			al, eth, chl	B8[3], 1990
1763	Benzaldehyde, 4-hydroxy-3-methoxy or Vanillan 4-HO-3-CH₃O-C₆H₃CHO	152.16	(i) nd (w, lig) (ii) tetr (w, lig)	285, 170[15]	(i) 77—9 (ii) 81—2	1.056		al, eth, ace, bz, chl	B8[3], 2011
1764	Benzaldehyde, 4-hydroxy-2-nitro 4-HO-2-NO₂C₆H₃CHO	167.12	ye nd	....	67			al, eth, bz	B8[3], 253
1765	Benzaldehyde, 4-hydroxy-3-nitro 4-HO-3-NO₂C₆H₃CHO	167.12	dk ye nd (al, w)	....	144.3			al, w	B8[3], 253
1766	Benzaldehyde, 4-hydroxy-3-nitro, oxime .... 4-HO-3-NO₂C₆H₃CH=NOH	182.14	pr or nd (al-chl)	....	189			al, eth, aa	B8[3], 254
1767	Benzaldehyde imine, N-ethyl or N-Benzalethylamine ..... C₆H₅CH=NC₂H₅	133.20	....	195[745]	....	0.937[20/4]	1.5378[15]	al, eth	B7[3], 831
1768	Benzaldehyde imine, 2-hydroxy-N-phenyl 2-HOC₆H₄CH=NC₆H₅	197.24	....	....	47-9	1.087	....	al	B12[3], 369
1769	Benzaldehyde imine, N-methyl C₆H₅CH=NCH₃	119.17		185[760], 90,1[10]		0.9672[14/2]	1.5526[20]	al, eth, ace, chl	B7[3], 831
1770	Benzaldehyde imine, N-phenyl .... C₆H₅CH=NC₆H₅	181.24	....	310	54	1.038[55/4]	1.600[99]	al, eth, chl	B12[3], 319
1771	Benzaldehyde imine, N-(2-tolyl) .... C₆H₅CH=NC₆H₄CH₃-(2)	195.27	....	314, 176[15]	....	1.041[20/4]	1.6310[25]	ace	B12[3], 1848
1772	Benzaldehyde imine, N-(3-tolyl) .... C₆H₅CH=N(C₆H₄CH₃-3)	195.26	....	315[775]	30-2		1.6353[2]	ace	B12[3], 1958
1773	Benzaldehyde imine, N-(4-tolyl) or N-Benzal-p-toluidine. C₆H₅CH=N(C₆H₄CH₃-4)	195.27	ye cr	318[755], 178[11]	35			ace	B12[3], 2096
1774	Benzaldehyde, 2-methoxy or o-Anisaldehyde 2-CH₃OC₆H₄CHO	136.16	pr	243-4, 124-5[18]	37-8	1.1326[20/4]	1.5600[20]	al, eth, ace, bz, chl	B8[3], 143
1775	Benzaldehyde, 2-methoxy-3-nitro or 3-Nitro-o-anisaldehyde 2-CH₃O-3-NO₂C₆H₃CHO	181.15	ye pr (dil al), nd (bz)	....	89-90			al, eth	B8[3], 190
1776	Benzaldehyde, 3-methoxy or m-Anisaldehyde.......... 3-CH₃OC₆H₄CHO	136.16	....	230, 62[1]	....	1.1187[20/4]	1.5330[20]	al, eth, ace, bz	B8[3], 199
1777	Benzaldehyde, 4-methoxy or p-Anisaldehyde .... p-CH₃OC₆H₄CHO	136.16	....	249.5, 83[2]	0	1.1191[15/4]	1.5730[20]	al, eth, ace, bz	B8[3], 218
1778	Benzaldehyde, 4-methoxy-3-methyl or 4-methoxy-m-tolualdehyde 4-CH₃O-3-CH₃C₆H₃CHO	150.18	....	80-5[1]	....		1.5670[20]	....	B8[3], 367
1779	Benzaldehyde, 4-methoxy, oxime .... 4-CH₃OC₆H₄CH=NOH	151.17	nd (bz)	....	64—5 (anti), 133 (syn)				B8[3], 235
1780	Benzaldehyde, 2-methyl or o-Tolualdehyde 2-CH₃C₆H₄CHO	120.16	....	200, 94[10]	....	1.0386[19/4]	1.5481[20]	al eth, ace, bz, chl	B7[3], 1011
1781	Benzaldehyde, 3-methyl or m-Tolualdehyde 3-CH₃C₆H₄CHO	120.16	....	199, 93-4[17]	....	1.0189[21/4]	1.5413[21]	al, eth, ace, bz, chl	B7[3], 1013
1782	Benzaldehyde, 4-methyl or p-Tolualdehyde 4-CH₃C₆H₄CHO	120.16	....	204-5, 106[10]	....	1.0194[17/4]	1.5454[20]	al, eth, ace, chl	B7[3], 1016
1783	Benzaldehyde, 2-nitro 2-NO₂C₆H₄CHO	151.12	ye nd (w)	153[23]	43.5-4	1.2844[20/4]	....	al, eth, ace, bz	B7[3], 889
1784	Benzaldehyde, 2-nitro, diacetate .... 2-NO₂C₆H₄CH(O₂CCH₃)₂	253.21	pr (lig)	....	90			eth, ace, bz	B7[3], 892
1785	Benzaldehyde, 2-nitro, dimethyl acetal .... 2-NO₂C₆H₄CH(OCH₃)₂	197.19	gr-ye	274-6, 138-9[11]	....			bz	B7[3], 891
1786	Benzaldehyde, 3-nitro 3-NO₂C₆H₄CHO	151.12	lt ye nd (w)	164[23]	58	1.2792[20/4]	....	al, eth, ace, bz	B7[3], 897
1787	Benzaldehyde, 3-nitro, diacetate .... 3-NO₂C₆H₄CH(O₂CCH₃)₂	253.21	pr or nd (al)	....	72	1.393	....	al, eth, ace, bz	B7[3], 900

No.	Name, Synonyms, and Formula	Mol. wt.	Color, crystalline form, specific rotation and λ_{max} (log ε)	b.p. °C	m.p. °C	Density	n_D	Solubility	Ref.
1788	Benzaldehyde, 3-nitro, dimethyl acetal 3-NO$_2$C$_6$H$_4$CH(OCH$_3$)$_2$	197.19	....	162-4[19]	......	1.209[15]	....	bz	B7[3], 899
1789	Benzaldehyde, 4-nitro 4-NO$_2$C$_6$H$_4$CHO	151.12	lf, pr (w)	sub	106	1.496		al, bz, aa	B7[3], 907
1790	Benzaldehyde, 4-nitro, diacetate 4-NO$_2$C$_6$H$_4$CH(O$_2$CCH$_3$)$_2$	253.21	pr (al)	....	127			al, eth, ace, bz	B7[3], 909
1791	Benzaldehyde, 4-nitro, dimethylacetal 4-NO$_2$C$_6$H$_4$CH(OCH$_3$)$_2$	197.19	....	294-6[774]	23-5			bz	B7[3], 909
1792	Benzaldehydeoxime *anti* or *anti*-Benzaldoxime C$_6$H$_5$CH=NOH	121.14	nd (eth)	....	130	1.145[20/4]		al, eth, w	B7[3], 840
1793	Benzaldehydeoxime *syn* or *syn*-Benzaldoxime C$_6$H$_5$CH=NOH	121.14	pr	200,118-9[10]	36-7	1.1111[20/4]	1.5908[20]	al, eth, bz	B7[3], 840
1794	Benzaldehyde, pentachloro C$_6$Cl$_5$CHO	278.35	nd (bz, al)		202.5			eth, ace, bz	B7[3], 881
1795	Benzaldehyde, phenylhydrazone C$_6$H$_5$CH = NNHC$_6$H$_5$	196.26	nd (lig), pr		156			ace, bz	B15[2], 57
1796	Benzaldehyde, 4-isopropyl or Cumaldehyde 4-(CH$_3$)$_2$CHC$_6$H$_4$CHO	148.21	....	235-6, 103-4[10]	......	0.9755[20/4]	1.5301[20]	al, eth	B7[3], 1095
1797	Benzaldehyde, semicarbazone C$_6$H$_5$CH=NNHCONH$_2$	163.18			224				B7[3], 854
1798	Benzaldehyde, 2,3,4,5-tetrachloro 2,3,4,5Cl$_4$C$_6$HCHO	243.91			106				B7[2], 181
1799	Benzaldehyde, 2,3,4-trichloro 2,3,4-Cl$_3$C$_6$H$_2$CHO	209.46	nd (al)		91			al	B7, 238
1800	Benzaldehyde, 2,3,5- trichloro 2,3,5-Cl$_3$C$_6$H$_2$CHO	209.46	nd (al)		56			al, eth, ace, bz	B7[2], 180
1801	Benzaldehyde, 2,3,6-trichloro 2,3,6-Cl$_3$-C$_6$H$_2$CHO	209.46	nd (lig)		86-7			eth, ace, bz	B7[3], 881
1802	Benzaldehyde, 2,4,5-trichloro 2,4,5-Cl$_3$C$_6$H$_2$CHO	209.46	nd (al)		112-3			al, eth, ace, bz	B7[2], 180
1803	Benzaldehyde, 2,4,6, trichloro 2,4,6-Cl$_3$-C$_6$H$_2$CHO	209.46	....		58-9			eth, lig	B7[3], 881
1804	Benzaldehyde, 3,4,5-trichloro 3,4,5-Cl$_3$-C$_6$H$_2$CHO	209.46	nd (al)		90-1			al, eth, ace, bz, peth	B7[2], 180
1805	Benzaldehyde, 2,4,5-trimethoxy 2,4,5-(CH$_3$O)$_3$C$_6$H$_2$CHO	196.20	....	140[4]-sub	114			w, eth, chl, lig	B8[3], 3361
1806	Benzaldehyde, 3,4,5-trimethoxy 3,4,5-(CH$_3$O)$_3$C$_6$H$_2$CHO	196.20	....		71-4			chl	B8[3], 3368
1807	Benzaldehyde, 2,4,6-trimethyl or Mesityl aldehyde 2,4,6-(CH$_3$)$_3$C$_6$H$_2$CHO	148.21	....	237-40, 192[50]	14			al, eth, ace, bz	B7[3], 1111
1808	Benzaldehyde, 2,4,6-trinitro 2,4,6-(NO$_2$)$_3$C$_6$H$_2$CHO	241.12	dk gr pl (bz)		119			al, eth, ace, bz,aa	B7[3], 926
1809	Benzamide or Benzoic acid amide C$_6$H$_5$CONH$_2$	121.14	mcl pr or pl (w)	290	132-3	1.0792[130/4], 1.341[4]		al, bz, w	B9[3], 1064
1810	Benzamide, 2-amino or Anthranilamide 2-H$_2$NC$_6$H$_4$CONH$_2$	136.15	lf (chl or w)	300	110-1.5			w, al	B14[3], 889
1811	Benzamide, 3-amino 3-H$_2$NC$_6$H$_4$CONH$_2$	136.15	ye mcl nd (+lw) nd		79—80 (hyd), 113—4 (anh)			w, al, eth	B14[3], 998
1812	Benzamide, 4-amino 4-H$_2$NC$_6$H$_4$CONH$_2$	136.15	ye cr (+ ¼w)		183			al, eth	B14[3], 1061
1813	Benzamide, 2-benzoyl 2-C$_6$H$_5$COC$_6$H$_4$COCNH$_2$	225.25	nd (to)		165 (cor)			al, bz, to	B10[3], 3295
1814	Benzamide, 2-bromo 2-BrC$_6$H$_4$COCNH$_2$	200.04	nd (w)	sub	160-1			al	B9[3], 1387
1815	Benzamide, 3-bromo 3-BrC$_6$H$_4$CONH$_2$	200.04	lf (w or al)	sub	155.3			al	B9[3], 1399
1816	Benzamide, 4-bromo 4-BrC$_6$H$_4$CONH$_2$	200.04	nd or pl (w)		192			al, w, aa	B9[3], 1418
1817	Benzamide, 2-chloro 2-ClC$_6$H$_4$CONH$_2$	155.59	rh nd (w)		142.4			al, eth	B9[3], 1338
1818	Benzamide, 3-chloro 3-ClC$_6$H$_4$CONH$_2$	155.59	nd		135-7			al, eth	B9[3], 1349

No.	Name, Synonyms, and Formula	Mol. wt.	Color, crystalline form, specific rotation and λ_{max} (log ϵ)	b.p. °C	m.p. °C	Density	n_D	Solubility	Ref.
1819	Benzamide, 4-chloro 4-ClC$_6$H$_4$CONH$_2$	155.59			179-80			al, eth, w	B9³, 1363
1820	Benzamide, 2,5-dimethoxy 2,5-(CH$_3$O)$_2$C$_6$H$_3$CONH$_2$	181.19	lf (hz-peth), nd (w)		141-2			al, eth, ace, bz, chl	B10², 258
1821	Benzamide, 3,4-dimethoxy or Veratramide 3,4-(CH$_3$O)$_2$C$_6$H$_3$CONH$_2$	181.19	cr (w)		164			w, eth, bz	B10², 261
1822	Benzamide, 3,5-dimethoxy or 3,5-Dimethoxy benzamide 3,5-(CH$_3$O)$_2$C$_6$H$_3$CONH$_2$	181.19	nd (bz)		148—9			w, al, eth, bz, lig	B10³, 1449
1823	Benzamide, N,N-diphenyl or N-Benzoyl diphenylamine C$_6$H$_5$CON(C$_6$H$_5$)$_2$	273.34	rh pr (al), nd		180				B12³, 511
1824	Benzamide, 2-ethoxy 2-C$_2$H$_5$OC$_6$H$_4$CONH$_2$	165.19	nd (w, al)		132-4			al, eth	B10², 58
1825	Benzamide, 3-ethoxy 3-C$_2$H$_5$OC$_6$H$_4$CONH$_2$	165.19	nd (w)		139			al, eth, ace, chl	B10¹, 64
1826	Benzamide, 4-ethoxy 4-C$_2$H$_5$OC$_6$H$_4$NH$_2$	165.19	pr (dil al)		206			al	B10³, 342
1827	Benzamide, 3-hydroxy 3-HOC$_6$H$_4$CONH$_2$	137.14	pl (w)		170.5			al, eth	B10³, 254
1828	Benzamide, 3-hydroxy 3-HOC$_6$H$_4$CONH$_2$	137.14	pl (w)		170.5			al, eth	B10², 82
1829	Benzamide, 3-hydroxy-N-phenyl 3-HOC$_6$H$_4$CONHC$_6$H$_5$	213.24	nd or pl (w, dil al)		156			al	B12¹, 269
1830	Benzamide, 4-hydroxy 4-HOC$_6$H$_4$CONH$_2$	137.14	nd (w + l)		162 (hyd)			al, eth, chl	B10³, 340
1831	Benzamide, 4-hydroxy-N-phenyl 4-HOC$_6$H$_4$CONHC$_6$H$_5$	213.24	pl or nd (w)		201-2			al	B12³, 947
1832	Benzamide, 2-methoxy or o-Anisamide 2-CH$_3$OC$_6$H$_4$CONH$_2$	151.17	nd (bz), pr (eth), pl (w)		129			eth, bz	B10², 58
1833	Benzamide, 4-methoxy 4-CH$_3$OC$_6$H$_4$CONH$_2$	151.17	nd or ta (w)	295	166-7			al	B10², 100
1834	Benzamide, N-methyl C$_6$H$_5$CONHCH$_3$	135.17		291⁷⁶⁵, 167¹²	82			al, ace	B9³, 1068
1835	Benzamide, N-α-naphthyl C$_6$H$_5$CONHC$_{10}$H$_7$-α	247.30	nd (al or aa)		161-2			bz, aa	B12³, 2870
1836	Benzamide, 2-nitro 2-NO$_2$C$_6$H$_4$CONH$_2$	166.14		317	176.6			al, eth	B9³, 1477
1837	Benzamide, 2-nitro-N-phenyl 2-NO$_2$C$_6$H$_4$CONHC$_6$H$_5$	242.24	nd (al or bz)		155			al, eth, bz, chl	B12³, 506
1838	Benzamide, 3-nitro 3-NO$_2$C$_6$H$_4$CONH$_2$	166.14		310—5	142.7			w, al, eth	B9³, 1515
1839	Benzamide, 4-nitro 4-NO$_2$C$_6$H$_4$CONH$_2$	166.14			201.4			al, eth	B9³, 1710
1840	Benzamide, N-(2-nitrophenyl) C$_6$H$_5$CONH(C$_6$H$_4$NO$_2$-2)	242.24	gold-ye nd (al)		98			al, eth	B12³, 1525
1841	Benzamide, N-(3-nitrophenyl) C$_6$H$_5$CONH(C$_6$H$_4$NO$_2$-3)	242.24	pl		157			eth, chl	B12³, 1552
1842	Benzamide, N-(4-nitrophenyl) C$_6$H$_5$CONH(C$_6$H$_4$NO$_2$-4)	242.24	ye nd (AcOEt)		199				B12³, 1597
1843	Benzamide, N-phenyl or Benzanilide C$_6$H$_5$CONHC$_6$H$_5$	197.24	lf (al)	117-9¹⁰, sub	163	1.315			B17³, 502
1844	Benzamide, thiono, N-phenyl C$_6$H$_5$CSNHC$_6$H$_5$	213.30	ye pl or pr (al)	d	10²			al, eth, bz	B12³, 507
1845	Benzamide, N-(2-tolyl) C$_6$H$_5$CONH(C$_6$H$_4$CH$_3$-2)	211.27	rh nd (AcOEt-ace)		145-6	1.205¹⁵		al, eth	B12³, 1858
1846	Benzamide, N-(3-tolyl) C$_6$H$_5$CONH(C$_6$H$_4$CH$_3$-3)	211.27	mcl pr (dil al)		125	1.170¹⁵		al	B12³, 1964
1847	Benzamide, N-(4-tolyl) C$_6$H$_5$CONH(C$_6$H$_4$CH$_3$-4)	211.27	rh nd (al)	232	158	1.202¹⁵		al, eth	B12³, 2062
1848	Benzamide, 3,4,5-trihydroxy or Gallamide 3,4,5-(HO)$_3$C$_6$H$_2$CONH$_2$	169.14	lf (w + l)		244-5d			al, w, chl	B10², 346
1849	Benzomide, 3,4,5-trihydroxy-N-phenyl or Gallanilide 3,4,5-(HO)$_3$C$_6$H$_2$CONHC$_6$H$_5$	245.24	lf (+2w, dil al)		207			al, aa, w	B12², 263
1850	Benzamidine C$_6$H$_5$C(=NH)NH$_2$	120.17	lf (al)		80			w, al	B9³, 1264

No.	Name, Synonyms, and Formula	Mol. wt.	Color, crystalline form, specific rotation and λ_{max} (log ε)	b.p. °C	m.p. °C	Density	n_D	Solubility	Ref.
1851	Benzamidine hydrochloride .... $C_6H_5C(:NH)NH_2 \cdot HCl$	156.62	rh pr (w + 2)		169			w, al	B9²199
1852	Benzamidine, N,N-diphenyl .... $C_6H_5C(=NH)N(C_6H_5)_2$	272.35	rh pl (eth)		112			al, eth, bz	B12³, 511
1853	Benzamidine, N,N-diphenyl, hydrochloride .... $C_6H_5C(=NH)N(C_6H_5)_2 \cdot HCl$	308.82	mcl pr or nd (al)		223d			w, al	B12, 270
1854	Benzamidine, N,N'-diphenyl .... $C_6H_5C(=NC_6H_5)NHC_6H_5$	272.35	nd (bz or al), pr (al)	146-7				al, eth	B12³, 514
1856	Benzamidine, N-(1-naphthyl) .... $C_6H_5C(=NH)NHC_{10}H_7\text{-}(\alpha)$	246.32	pl (al)		141			al, eth	B12, 1233
1857	Benzamidoxime .... $C_6H_5C(NH_2)=NOH$	136.17	mcl pr (w)		80			al, eth, bz, chl	B9³, 1308
1858	1,2-Benzanthracene or Naphthanthracene .... $C_{18}H_{12}$	228.30	ye-br flr pl or lf (al-aa)	435 (sub)	162			al, eth, ace, bz	B5³, 2374
1859	1,2-Benzanthracene, 9,10-dihydro .... $C_{18}H_{14}$	230.31	lf (bz, al, aa)		112				B5³, 2305
1860	1,2-Benzanthracene, 9,10-dimethyl .... $9,10(CH_3)_2C_{18}H_{10}$	256.35	pa ye pl (al, aa)		122-3			ace, bz	B5³, 2413
1861	1,2-Benzanthracene, 4'-hydroxy .... $4\text{-}HOC_{18}H_{11}$	244.30	pa ye or og nd (bz)		225			bz, to, aa	B6³, 3725
1862	1,2-Benzanthracene, 3-methyl .... $3\text{-}CH_3C_{18}H_{11}$	242.32	pl (bz, al), nd (bz-lig)	160⁰·⁰¹, (sub)	156-7			al, ace, bz, aa	B5³, 2388
1863	1,2-Benzanthracene, 4-methyl .... $4\text{-}CH_3C_{18}H_{11}$	242.32	nd (al)	1	126-7			al, eth	B5³, 2388
1864	1,2-Benzanthracene, 5-methyl .... $5\text{-}CH_3C_{18}H_{11}$	242.32	ye mcl pl (bz-al)		160			al, eth, bz, xyl	B5³, 2388
1865	1,2-Benzanthracene, 6-methyl .... $6\text{-}CH_3C_{18}H_{11}$	242.32	pl (al)		150-1			al, eth, chl	B5³, 2388
1866	1,2-Benzanthracene, 7-methyl .... $7\text{-}CH_3C_{18}H_{11}$	242.32	ye pl (al)		183-3.6			al, aa	2389
1867	1,2-Benzanthracene, 8-methyl .... $8\text{-}CH_3C_{18}H_{11}$	242.32	pl or nd (al, aa)		118			al, aa	B5³, 2391
1868	1,2-Benzanthracene, 9-methyl .... $9\text{-}CH_3C_{18}H_{11}$	242.32	pa ye nd or pl (aa, MeOH)		138			al, aa	B5³, 2391
1869	1,2-Benzanthracene, 10-methyl .... $10\text{-}CH_3C_{18}H_{11}$	242.32			141			al, eth, ace	B5³, 2392
1870	1,2-Benzanthracene, 10-carboxaldehyde ..... $10\text{-}OHCC_{18}H_{11}$	256.31	ye pr or nd		148			al, eth, ace, bz	B7³, 2807
1871	1,2-Benz-3,4-anthraquinone .... $C_{18}H_{10}O_2$	258.28	red nd (to)		262-3d				B7², 759
1872	1,2-Benz-9,10-anthraquinone .... $C_{18}H_{10}O_2$	258.28	ye pr (aa)	sub	169			ace, bz, chl	B7³, 4278
1873	Benzanthrene .... $C_{17}H_{12}$	216.29	lf (al)		84			al	B5³, 2289
1874	1,9-Benzanthr-10-one .... $C_{17}H_{10}O$	230.27	ye nd (xyl or al)		170				B7³, 2694
1875	1,9-Benzanthr-10-one, bz-1-bromo .... $C_{17}H_9BrO$	309.17	ye nd (aa)		178			al, MeOH	B7³, 2706
1876	1,9-Benzanthr-10-one, bz-1-chloro .... $C_{17}H_9ClO$	264.71	ye nd (aa)		182-3				B7³, 2700
1877	1,9-Benzanthr-10-one, bz-2-hydroxy .... $C_{17}H_{10}O_2$	246.27	ye nd		294				B8³, 1629
1878	1,9-Benzanthr-10-one, 4-hydroxy .... $C_{17}H_{10}O_2$	246.27	ye nd (aa)		178-9			eth, bz	B8³, 1638
1879	1,2-Benzazulene .... $C_{14}H_{10}$	178.23			176				B5³, 2122
1880	Benzedrine or Amphetamine dl .... $C_6H_5CH_2CH(NH_2)CH_3$	135.21		203, 97²⁰		0.9306²⁵/⁴	1.518²⁶	al, chl	B12³, 2665
1881	Benzestrol or 2,4-Bis(4-hydroxyphenyl)-3-ethyl hexane ... $C_{20}H_{26}O_2$	298.43	cr (al)		162-6			al, eth, ace, aa	B6³, 5538
1882	Benzene .... C_6H_6	78.12	rh pr	80.1	5.5	0.8765²⁰/⁴	1.5011²⁰	**al, eth, ace, aa**	B5⁴, 583

No.	Name, Synonyms, and Formula	Mol. wt.	Color, crystalline form, specific rotation and λ_{max} (log ε)	b.p. °C	m.p. °C	Density	n_D	Solubility	Ref.
1883	Benzene, allyl or Allylbenzene $CH_2 = CHCH_2C_6H_5$	118.18	....	156, 47[13]	-40	0.8920[20/4]	1.5131[20]	al, eth, bz	B5[4], 1362
1884	Benzene, 1-allyl-4-bromo $4\text{-}BrC_6H_4CH_2CH = CH_2$	197.08	....	222-2, 96[12]	......	1.324[15/4]	1.559[20]	eth, bz, chl	B5[4], 1363
1885	Benzene, 1-allyl-3,4-dimethoxy $3,4\text{-}(CH_3O)_2C_6H_3CH_2CH = CH_2$	178.23	cr (hx)	254.7, 104[2]	-4	1.0396[20/4]	1.5340[20]	al, eth	B6[3], 5024
1886	Benzene, 1-allyl-2,3,4,5-tetramethoxy $2,3,4,5\text{-}(CH_3)_4C_6HCH_2CH = CH_2$	238.28	....		25	1.087[25]	1.51462[25]	....	B6[3], 6691
1887	Benzene, (3-aminobutyl) $C_6H_5(CH_2CH_2CH(NH_2)CH_3)$	149	....	221-3[750], 101-2[14]	148	0.9289[15/4]	1.51520[20]	al	B12[3], 2717
1888	Benzene, (1-aminoethyl) d $C_6H_5CH(NH_2)CH_3$	121.18	$[\alpha]^{25/}_D$ +39.3 (MeOH)	187		0.9651[15]		w, al, eth	B12[3], 2386
1889	Benzene, (1-aminoethyl) dl $C_6H_5CH(NH_2)CH_3$	121.18	....	187, 87[24]		0.9395[15]	1.5238[25]	w, al, eth	B12[3], 2386
1890	Benzene, (1-aminoethyl) l $C_6H_5CH(NH_2)CH_3$	121.18	$[\alpha]^{18/}_D$ -38 (MeOH)	184-6, 77[16]		0.9520[20/4]	1.5290[25]	w, al, eth	B12[3], 2386
1891	Benzene, (2-aminoethyl) or β-Phenethylamine $C_6H_5CH_2CH_2NH_2$	121.18	....	197-8		0.9580[24/4]	1.5290[25]	w, al, eth	B12[3], 2408
1892	Benzene, 1-(2-aminoethyl)-3,4-dimethoxy $3,4\text{-}(CH_3O)_2C_6H_3CH_2CH_2NH_2$	181.23	....	163-5[14]			1.5464[20]		B13[3], 2207
1893	Benzene, 1-benzyl-4-ethyl or p-Ethyl diphenylmethane $4\text{-}C_2H_5C_6H_4CH_2C_6H_5$	196.30	....	297, 85[0.2]	-24	0.9777[20]	1.5616[20]	al, eth, chl	B5[4], 1926
1894	Benzene, 1,2-bis (bromomethyl) or o-Xylyene bromide $1,2\text{-}(BrCH_2)_2C_6H_4$	263.98	rh (chl)	128-30[4.5]	95	1.988[0]	....	al, eth, peth	B5[4], 929
1895	Benzene, 1,3-bis (bromomethyl) or m-Xylyene bromide $1,3\text{-}(BrCH_2)_2C_6H_4$	263.98	nd (chl), pr (ace)	125-40[29]	77	1.959[0]	....	al, eth, lig	B5[4], 946
1896	Benzene, 1,4-bis (bromomethyl) or p-Xylyene bromide $1,4\text{-}(BrCH_2)_2C_6H_4$	263.98	mcl pr (al), cr (chl, bz)	245, 155-8[14]	145-7			al, bz, chl	B5[4], 970
1897	Benzene, 1,3,-bis (bromomethyl)-5-methyl $1,3\text{-}(BrCH_2)_2\text{-}5\text{-}CH_3C_6H_3$	278.00	pr	....	66			al, eth, bz, lig	B5[4], 1027
1898	Benzene, 1,3-bis(carboethoxy) $1,3\text{-}C_6H_4(OCH_2CO_2H_5)_2$	226.19	nd (aa, w)		195			w, aa	B6[3], 4324
1899	Benzene, 1,3-bis (carbethoxymethoxy) $1,3\text{-}C_6H_4(OCH_2CO_2H_5)_2$	282.30	nd (eth)	278[12]	42				B6, 818
1900	Benzene, 1,2-bis (chloromethyl) or o-Xylyene chloride $1,2\text{-}(ClCH_2)_2C_6H_4$	175.07	mcl (lig)	239-41, 130-5[10]	55	1.393[0]		al, eth, lig	B5[4], 927
1901	Benzene, 1,3-bis (chloromethyl or m-Xylyene chloride $1,3\text{-}(ClCH_2)_2C_6H_4$	175.07	....	250-5	34.2			al, eth	B5[4], 944
1902	Benzene, 1,4-bis (chlormethyl) or p-Xylyene chloride $1,4\text{-}(ClCH_2)_2C_6H_4$	175.07	pl (al)	240-50d, 135[16]	100	1.417[0]		al, eth, ace, chl	B5[4], 967
1903	Benzene, 1,2-bis (cyanomethyl) or o-Xylyene dicyanide $1,2\text{-}(NCCH_2)_2C_6H_4$	156.20	(i) nd or pr (MeOH), (ii) pr (al)		(i) 18 (unst), (ii) 60 (st)			al, eth	B9[3], 4293
1904	Benzene, 1,3-bis (cyanomethyl) or m - Xylyene dicyanide $1,3\text{-}(NCCH_2)_2C_6H_4$	156.20	nd (w), pr (eth)	305-10[100], 170[20-30]	30			al, eth, bz, chl	B9[3], 4294
1905	Benzene, 1,4-bis (cyanomethyl) or p-Xylyene dicyanide $1,4\text{-}(NCCH_2)_2C_6H_4$	156.20	pr (al), nd (w)		98			al, eth, chl	B9[3], 4295
1906	Benzene, 1,2-bi(dibromomethyl) $1,2\text{-}(Br_2CH)_2C_6H_4$	421.77	mcl		116-7			chl	B5[4], 929
1907	Benzene, 1,3-bis (dibromomethyl) $1,3\text{-}(Br_2CH)_2C_6H_4$	421.77	nd (al), pr (chl)		107			al, bz, chl, lig	B5[3], 839
1908	Benzene, 1,4-bis (dibromomethyl) $1,4\text{-}(Br_2CH)_2C_6H_4$	421.77	mcl pr (chl)		172			bz	B5[3], 860
1909	Benzene, 1,2-bis (diethylamino) $1,2\text{-}[(C_2H_5)_2N]_2C_6H_4$	220.36	....	119[10]		0.9267[13/4]	1.5213[13]	al, eth	B13[2], 12
1910	Benzene, 1,3-bis (diethylamino) $1,3\text{-}[(C_2H_5)_2N]_2C_6H_4$	220.36	....	148[9]		0.9522[12/4]	1.5537[12]	al, eth	B13[2], 26
1911	Benzene, 1,4-bis (diethylamino) $1,4\text{-}[(C_2H_5)_2N]_2C_6H_4$	220.36	mcl pr, pl (al-w)	280	52			al, eth, bz, lig	B13[2], 40
1912	Benzene, 1,4-bis (dimethylamino) $1,4\text{-}[(CH_3)_2N]_2C_6H_4$	164.25	lf (dil al or lig)	260	51			al, eth, bz, lig	B13[3], 111
1913	Benzene, 1,2-bis (hydroxymethyl) or o -Xylylene glycol $1,2\text{-}C_6H_4(CH_2OH)_2$	138.17	pl (eth, peth)		65-6			w, al, eth	B6[3], 4587
1914	Benzene, 1,3-bis (hydroxymethyl) or Isophthalyl glycol $1,3\text{-}(HOCH_2)_2C_6H_4$	138.17	nd (bz)	154-9[13]	57			w, al, eth	B6[3], 4600

No.	Name, Synonyms, and Formula	Mol. wt.	Color, crystalline form, specific rotation and λ_{max} (log ε)	b.p. °C	m.p. °C	Density	n_D	Solubility	Ref.
1915	Benzene, 1,4-bis (hydroxymethyl) or Terephthalyl glycol. 1,4-(HOCH₂)₂C₆H₄	138.17	nd (w)	....	115-6	....	....	w, al, eth, ace	B6³, 4608
1916	Benzene, 1,2-bis (methylamino)-4-chloro .... 4-Cl-1,2(CN₃NH)₂C₆H₃	170.64	pr (lig)	....	61	....	....	w	B13, 25
1917	Benzene, 1,3-bis (3-methylbutoxy) .... [(CH₃)₂CHCH₂CH₂O]₂C₆H₄	250.39		....	47	....	....		B6, 815
1918	Benzene, 1,3-bis(4-tolylamino) .... 1,3-(4-CH₃C₆H₄NH)₂C₆H₄	288.40	nd (al)	....	139-9	....	....		B13, 42
1919	Benzene, 1,2-bis (trimethylsiloxy) 1,2-[(CH₃)₃SiO]₂C₆H₄	254.48	....	235⁷⁶⁰	....	....	1.4685²⁰/D		C², 249
1920	Benzene, 1,3-bis (trimethylsiloxy) 1,3-[(CH₃)₃SiO)]₂C₆H₄	254.48	....	237-40⁷⁴⁰	....	0.950²⁰/⁴	1.4748²⁰/D		C², 249
1921	Benzene, bromo or phenylbromide .... C₆H₅Br	157.02	....	156, 43¹⁸	-30.8	1.4950²⁰/⁴	1.5597²⁰	al, eth, bz	B5⁴, 670
1922	Benzene, (4-bromobutoxy) .... Br(CH₂)₄OC₆H₅	229.12	cr (al)	153-6¹⁸	41	....	....	al	B6⁴, 558
1923	Benzene, 1-bromo-4-tert butyl 4-[(CH₃)₃C]C₆H₄Br	213.13	....	231-2, 103¹⁰	19	1.2286²⁰/⁴	1.5436²⁰	eth, bz, chl	B5⁴, 1050
1924	Benzene, 1-bromo-2-chloro 2-ClC₆H₄Br	191.46	....	204⁷⁶⁵	-12.3	1.6387²⁵/⁴	1.5809²⁰	bz	B5⁴, 680
1925	Benzene, 1-bromo-3-chloro 3-ClC₆H₄Br	191.46	....	196	-21.5	1.6302²⁰/⁴	1.5771²⁰	al, eth	B5⁴, 680
1926	Benzene, 1-bromo-4-chloro 4ClC₆H₄Br	191.46	nd or pl (al, eth)	196⁷⁵⁶	68	1.576⁷¹/⁴	1.5531⁷⁰	eth, bz, chl	B5⁴, 681
1927	Benzene, 1-bromo-4-cyclohexyl 4-C₆H₁₁-C₆H₄Br	239.16	....	160²⁵	....	1.283²⁵/⁴	1.5584²⁰	bz, chl	B5³, 1258
1928	Benzene, 1-bromo-2,3-dichloro 2,3-Cl₂C₆H₃Br	225.91	pl or lf (al)	243⁷⁰⁵	60	....	....	eth, bz, chl	B5, 209
1929	Benzene, 1-bromo-3,5-dichloro 3,5-Cl₂C₆H₃Br	225.91	pr (al)	232⁷⁵⁷	82-4	....	....	eth, bz, chl, al	B5², 162
1930	Benzene, 1-bromo-2,6-dichloro 2,6-Cl₂C₆H₃Br	225.91	pr (al)	242⁷⁶⁵	65	....	....	eth, bz, chl	B5, 210
1931	Benzene, 1-bromo-2,5-dichloro 2,5-Cl₂C₆H₃Br	225.91	pr or nd (al)	235⁷³¹, 119²⁰	35	....	....	al, eth, bz, lig	B5³, 564
1932	Benzene, 1-bromo-3,4-dichloro 3,4-Cl₂C₆H₃Br	225.91	pr	237, 124³³	....	....	....	al, eth, bz, chl	B5³, 564
1933	Benzene, 1-bromo-2,3-dimethyl 2,3-(CH₃)₂C₆H₃Br	185.07	....	214, 83¹¹	....	1.365²⁰/⁴	....	eth, ace, bz	B5⁴, 928
1934	Benzene, 1-bromo-2,4-dimethyl 2,4-(CH₃)₂C₆H₃Br	185.07	....	205, 84¹³	0	....	1.5501²⁰	al, eth, bz	B5⁴, 945
1935	Benzene, 1-bromo-2,5-dimethyl 2,5-(CH₃)₂C₆H₃Br	185.07	lf or pl	199-200, 88-9¹³	9	1.3582¹⁸	1.5514¹⁸	al, bz	B5⁴, 969
1936	Benzene, 1-bromo-2,6-dimethyl 2,6(CH₃)₂C₆H₃Br	185.06	....	203-4, 98-9²⁰	....	....	1.5552²⁰	eth, ace, bz	B5⁴, 945
1937	Benzene, 1-bromo-3,4-dimethyl 3,4-(CH₃)₂C₆H₃Br	185.07	....	214.5	-0.2	1.3708²⁰/⁴	1.5530²⁰	al, eth	B5⁴, 928
1938	Benzene, 1-bromo-3,5-dimethyl 3,5-(CH₃)₂C₆H₃Br	135.06	....	204, 83-9¹²	....	1.342²⁰	1.5462²²	eth, ace, bz	B5⁴, 945
1939	Benzene, 1-bromo-2,3-dinitro .... 2,3-(NO₂)₂C₆H₃Br	247.01	pa ye pl (al)	320	101-2	....	....	al	B5⁴, 749
1940	Benzene, 1-bromo-2,4-dinitro 2,4-(NO₂)₂C₆H₃Br	247.01	ye nd (al)	....	75	....	....	al	B5⁴, 749
1941	Benzene, 1-bromo-2,5-dinitro 2,5-(NO₂)₂C₆H₃Br	247.01	nd (al), pr (al-eth)	....	70	....	....	eth, al	B5⁴, 750
1942	Benzene, 1-bromo-2,6-dinitro 2,6-(NO₂)₂C₆H₃Br	247.01	ye pr (al)	....	107	....	....	al	B5⁴, 749
1943	Benzene, 1-bromo-3,4-dinitro 3,4-(NO₂)₂C₆H₃Br	247.01	nd (al), pl (al-eth)	....	34-5 (unst)	....	....	eth, al	B5³, 640
1944	Benzene, 1-bromo-2-ethoxy 2-C₂H₅OC₆H₄Br	201.07	....	222-6	....	....	....	al, eth	B6⁴, 1038
1945	Benzene, 1-bromo-3-ethoxy 3-C₂H₅OC₆H₄Br	201.07	....	222	....	....	....	al, eth	B6³, 739
1946	Benzene, 1-bromo-4-ethoxy 4-C₂H₅OC₆H₄Br	201.07	....	230-2, 109¹⁷	4	1.4071²⁵/⁴	1.5517²⁰	al, eth	B6⁴, 1045
1947	Benzene, (1-bromoethyl) d C₆H₅(CHBrCH₃)	185.07	[α]¹⁴/D +1.5	203, 86-8¹⁵	....	1.3108²³	1.5612²⁰	al, eth	B5⁴, 906

No.	Name, Synonyms, and Formula	Mol. wt.	Color, crystalline form, specific rotation and λ_{max} (log ε)	b.p. °C	m.p. °C	Density	n_D	Solubility	Ref.
1948	Benzene, (1-bromoethyl) dl $C_6H_5(CHBrCH_3)$	185.07	....	202-3, 85[13]		1.3605[20/4]	1.5612[20]	al, eth, bz	B5[4], 906
1949	Benzene, (2-bromoethyl) $C_6H_5CH_2CH_2Br$	185.07	....	217-8[734], 92[11]		1.3587[20/4]	1.5572[20]	eth, bz	B5[4], 907
1950	Benzene, 1-bromo-2-ethyl $2-C_2H_5C_6H_4Br$	185.06	....	199			1.5473[20]_D	....	B5[4], 906
1951	Benzene, 1-bromo-4-ethyl $4-C_2H_5C_6H_4Br$	185.06	....	204			1.5445[20]		B5[4], 906
1952	Benzene, 1-bromo-4-fluoro $4-F-C_6H_4-Br$	175.01	....	152[764]	-8 (fr-17)	1.4946[20/4]	1.5604[20]	al, eth	B5[4], 678
1953	Benzene, (2-bromo-1-hydroxyethyl) $C_6H_5CH(OH)CH_2Br$	201.06	....	109-10[2]		1.4994[20/4]	1.5800[17]_D	....	B6[3], 1690
1954	Benzene, 1-bromo-2-iodo $2-I-C_6H_4-Br$	282.91	....	257[754], 120-1[15]	9-10	2.2571[25/4]	1.6618[25]	ace	B5[4], 698
1955	Benzene, 1-bromo-3-iodo $3-I-C_6H_4Br$	282.91	....	252[754], 120[18]	-9.3				B5[4], 698
1956	Benzene, 1-bromo-4-iodo $4-I-C_6H_4Br$	282.91	pr or pl (eth-al)	252[754]	92			eth	B5[4], 698
1957	Benzene, 1-(bromomethyl)-2-methyl or o-Xylylbromide $2-CH_3C_6H_4CH_2Br$	185.07	pr	216-7[742], 108[16]	21	1.3811[23]	1.5730[20]	al, eth, ace, bz	B5[4], 928
1958	Benzene, 1-(bromomethyl)-3-methyl or m-Xylylbromide $3-CH_3C_6H_4CH_2Br$	185.07	....	212-3, 105[13]	......	1.3711[23]	1.5660[20]	al, eth, chl	B5[2], 293
1959	Benzene, 1-(bromomethyl)-4-methyl or p-Xylylbromide $4-CH_3C_6H_4CH_2Br$	185.07	nd (al)	218-20[740]	35	1.324	....	al, eth, chl	B5[3], 858
1960	Benzene, 1-(bromomethyl)-3,5-dimethyl or Mesityl bromide $3,5-(CH_3)_2C_6H_3CH_2Br$	199.10	nd (eth)	229-31[740](d), 110[17]	40	....	....	al, eth, bz, chl	B5[4], 1027
1961	Benzene, 1-bromo-2-nitro $2-NO_2C_6H_4Br$	202.01	pa, ye (al)	258[756]	43	1.6245[80/4]	....	al, eth, ace, bz	B5[4], 728
1962	Benzene, 1-bromo-3-nitro $3-NO_2C_6H_4Br$	202.01	rh	265	17 (unst), 56 (st)	1.7036[20/4]	1.5979[20]	al, eth, bz	B5[4], 729
1963	Benzene, 1-bromo-4-nitro $4-NO_2C_6H_4Br$	202.01	rh or mcl pr (al)	256	127	1.948	....	al, eth, bz	B5[4], 729
1964	Benzene, 1-bromo-2-nitroso $2-ONC_6H_4Br$	186.01	nd		98			bz, chl	B5[4], 706
1965	Benzene, 1-bromo-4-nitroso $4-ONC_6H_4Br$	186.01	nd (al)		95			al, bz, chl	B5[4], 706
1966	Benzene, 3-bromo-1-propenyl) or Cinnamyl bromide $BrCH_2CH=CHC_6H_5$	197.08	nd (al, eth)	130[10]	34	1.3428[30]	1.610—1.613[20]	al	B5[3], 1188
1967	Benzene, (3-bromopropoxy) or 2-Phenoxypropyl bromide $C_6H_5-OCH_2CH_2CH_2Br$	215.10	....	127[18]	7-8	1.365[16/16]	....	eth	B6[3], 549
1968	Benzene, (1-bromopropyl)-(DL) $C_6H_5CHBrCH_2CH_3$	199.10	$d[\alpha]_D$ +5.9 (al, c=1), $l[\alpha]_D$ -5.7 (eth, c=10)	105[17]	....	1.3098[19/4]	1.5517[19]	eth, bz	B5[4], 981
1969	Benzene, (2-bromopropyl) $C_6H_5CH_2CHBrCH_3$	199.10	....	107-9[16]	......	1.2908[16]	1.5450[20]	chl	B5[4], 981
1970	Benzene, (3-bromopropyl) $C_6H_5CH_2CH_2CH_2Br$	199.10	....	110[12]	......	1.3106[25/4]	1.5440[25]	eth	B5[4], 982
1971	Benzene, (3-bromoisopropyl) $BrCH_2CH(CH_3)C_6H_5$	199.10	$[\alpha]^{25}_D(d)$ +15.6	188-9, 106-8[18]		1.3155[20]	1.5548[28] (d)	bz, chl	B5[4], 995
1972	Benzene, 1-bromo-4-isopropyl $4-i-C_3H_7C_6H_4Br$	199.10	....	218.7, 97-8[5]	-22.5	1.3145[20/4]	1.5569[20]	eth, bz, chl	B5[4], 994
1973	Benzene, 2-bromo-4-iso-propyl-1-methyl $4-i-C_3H_7-2-BrC_6H_3CH_3$	213.13	....	234.3, 99[9]		1.2689[18/4]	1.5360[20]	al, eth, chl	B5[4], 1063
1974	Benzene, 1-bromo-4-triazo $4-N_3C_6H_4Br$	198.03	pl	105[10]	20	....	....	eth, bz	B5[4], 761
1975	Benzene, 1-bromo-2,3,5-trimethyl $2,3,5-(CH_3)_3C_6H_2Br$	199.10	....	238, 117[17]	<-15	....	1.5516[20]	bz	B5[4], 1014
1976	Benzene, 1-bromo-2,4,5-trimethyl $2,4,5-(CH_3)_3C_6H_2Br$	199.10	nd (al)	233-5	73	....	....	al	B5[4], 1014
1977	Benzene, 1-bromo-2,4,6-trimethyl $2,4,6-(CH_3)_3C_6H_2Br$	199.10	....	225, 117[25]	-1	1.3191[10]	1.5510[20]	eth, bz	B5[4], 1027
1978	Benzene, butoxy or Butyl phenyl ether $C_6H_5OC_4H_9$	150.22	....	210, 95[17]	-19.4	0.9351[20/4]	1.4969[20]	ace, bz	B6[4], 558

No.	Name, Synonyms, and Formula	Mol. wt.	Color, crystalline form, specific rotation and λ_{max} (log ε)	b.p. °C	m.p. °C	Density	n_D	Solubility	Ref.
1979	Benzene, isobutoxy or Isobutyl phenyl ether .... i-$C_4H_9OC_6H_5$	150.22	....	196	......	$0.9240^{24/15}$	1.4932^{14}	ace, bz, peth	B6[4], 559
1980	Benzene, 1-iso-butoxy-2-nitro .... 2-$NO_2C_6H_4O$-i-C_4H_9	195.22	ye nd	275-80	......	1.1361^{20}	....	eth	B6, 218
1981	Benzene, sec-butoxy $C_6H_5OCH(CH_3)CH_2CH_3$	150.22	....	194.5, 70-2[5]	......	$0.9415^{20/4}$	1.4926^{25}	eth	B6[4], 558
1982	Benzene, butyl .... $C_6H_5C_4H_9$	134.22	....	183, 62.2[10]	-88	$0.8601^{20/4}$	1.4898^{20}	al, eth, ace, bz	B5[4], 1033
1983	Benzene, 1-butyl-2-methyl .... 2-$CH_3C_6H_4C_4H_9$	148.25	....	208, 81[10]	......	$0.8710^{20/4}$	1.4960^{20}	eth, ace, bz	B5[3], 999
1984	Benzene, 1-butyl-3-methyl .... 3-$CH_3C_6H_4C_4H_9$	148.25	....	205, 79[10]	......	$0.8590^{20/4}$	1.4910^{20}	eth, ace, bz	B5[3], 999
1985	Benzene, 1-butyl-4-methyl .... 4-$CH_3C_6H_4C_4H_9$	148.25	....	207, 80.6[10]	-85	$0.8586^{20/4}$	1.4916^{20}	eth, ace, bz	B5[4], 1094
1986	Benzene, iso-butyl $C_6H_5CH_2CH_2(CH_3)_2$	134.22	....	172.8, 53.1[10]	-51.5	$0.8532^{20/4}$	1.4866^{20}	al, eth, ace, bz, peth	B5[4], 1042
1987	Benzene, sec-butyl(d) $C_6H_5CH(CH_3)CH_2CH_3$	134.22	$[\alpha]^{20/}_D$ + 27.3 (undil)	173, 63[15]	-75	$0.8621^{20/4}$	1.4895^{20}	al, eth, ace, bz	B5[4], 1038
1988	Benzene, sec-butyl(dl) $C_6H_5CH(CH_3)CH_2CH_3$	134.22	....	173, 53.6[10]	-75.5	$0.8621^{20/4}$	1.4902^{20}	al, eth, ace, bz	B5[4], 1038
1989	Benzene, sec-butyl l) $C_6H_5CH(CH_3)CH_2CH_3$	134.22	$[\alpha]^{25/}_D$ −17.9 (al, c = 5)	61[18]	......	$0.868^{24/4}$	1.4891^{20}	al, eth, ace, bz	B5[4], 1038
1990	Benzene, 1-sec-butyl-4-methyl 4-$CH_3C_6H_4CH(CH_3)C_2H_5$	148.25	....	197, 73[10]	......	0.8640^{19}	1.493^{20}	eth, bz, chl	B5[4], 1095
1991	Benzene, tert-butyl $C_6H_5C(CH_3)_3$	134.22	....	169, 50.7[10]	-57.8	$0.8665^{20/4}$	1.4927^{20}	al, eth, ace, bz	B5[4], 1045
1992	Benzene, 1-tert-butyl-3,5-dimethyl,2,4,6-trinitro or Musk xylene 3,5-$(CH_3)_2$-2,4,6-$(NO_2)_3C_6$-tert-C_4H_9	297.27	pl, nd (al)	....	110	....	....	eth, al	B5[4], 1132
1993	Benzene, 2-tert-butyl-3,5-dinitro-1-iso-propyl-4-methyl or Moskene .... 2-tert-C_4H_9-3,5($NO_2)_2$-4-C_6H(i-C_3H_7)	280.33	Pa ye cr	....	132-3	....	....		C49, 3044
1994	Benzene, 2-tert-butyl-1,3-dinitro-4,5,6-trinitro or Musk tebetine .... 2-$(CH_3)_3C$-1,3-$(NO_2)_2$-4,5,6-$(CH_3)_3C_6$	266.30	pa ye pr (al)	....	135-6	....	....	eth	B5[3], 1055
1995	Benzene, 5-tert-butyl-1,3-dinitro-4-methoxy-2-methyl or Musk ambrette .... 5-$(CH_3)_3C$-1,3-$(NO_2)_2$-4-CH_3O-2-CH_3-C_6H	268.25	pa ye lf (al)	135[16]	84-6	....	....	eth	B6[3], 1984
1996	Benzene, 1-tert-butyl-2-methyl .... 2-CH_3-$C_6H_4C(CH_3)$	148.25	....	170[743]	......	....	$1.49423^{17/D}$		B5[4], 1096
1997	Benzene, 1-tert-butyl-3-methyl-2,4,6-trinitro or Artificial musk .... 3-CH_3-2,4,6-$(NO_2)_3C_6H$-tertC_4H_9	283.24	ye nd (al)	....	96-7	....	....	al, eth, bz, chl	B5[3], 1003
1998	Benzene, 1-tert-butyl-4-methyl .... 4-$CH_3C_6H_4C(CH_3)_3$	148.25	....	193, 70[10]	-52	$0.8612^{20/4}$	1.4918^{20}	eth, ace, bz	B5[4], 1097
1999	Benzene, 2-tert-butyl-4-methyl-1,3,5-trinitro or Musk baur .... 2-$(CH_3)_3C$-4-CH-1,3,5-$(NO_2)_3C_6H$	283.24	ye nd (al)	....	97	....	....	al, eth, bz, peth	B5[3], 1003
2000	Benzene, chloro or Phenyl chloride................... C_6H_5Cl	112.56	....	132, 22[10]	-45.6	$1.1058^{20/4}$	1.5241^{20}	al, eth, bz	B5[4], 640
2001	Benzene, (chloro-tert-butyl) .... $C_6H_5C(CH_3)_2CH_2Cl$	168.67	....	222[741], 104-5[18]	......	$1.047^{20/4}$	$1.5247^{20/D}$	al, eth, ace, bz	B5[4], 1048
2002	Benzene, 1-chloro-2,5-diacetamido 2,5-$(CH_3CONH)_2$-C_6H_3Cl	226.27	nd	....	196-7	....	....	al	B13, 118
2003	Benzene, 1-chloro-2,4-diamino .... 2,4-$(NH_2)_2C_6H_3Cl$	142.59	pl or nd	....	91	....	....	al	B13[3], 97
2004	Benzene, 1-chloro-2,5-diamino .... 2,5-$(NH_2)_2C_6H_3Cl$	142.59	nd (bz-lig)	....	64	....	....	w, bz	B13[2], 58
2005	Benzene, 1-chloro-2,6-diamino .... 2,6-$(NH_2)_2C_6H_3Cl$	142.29	....	....	85-6	....	....	eth	B13[1], 15
2006	Benzene, 1-chloro-3,4-diamino .... 3,4-$(NH_2)_2C_6H_3Cl$	142.59	pl (bz-lig), lf (w)	....	76	....	....	al, eth, bz, lig	B13[3], 20
2007	Benzene, 1-chloro-3,5-diamino .... 3,5-$(NH_2)_2C_6H_3Cl$	142.59	rh pr (al), nd (to-al)	....	105-6	....	....	al, eth, ace	B13[2], 29
2008	Benzene, 1-chloro-2,4-dimethyl or 4-Chloro-m-xylene.... 2,4-$(CH_3)_2$-C_6H_3Cl	140.61	....	187-8, 89[24]	......	$1.0598^{20/20}$	1.5230^{25}	ace, bz	B5[4], 943

No.	Name, Synonyms, and Formula	Mol. wt.	Color, crystalline form, specific rotation and λ_{max} (log ε)	b.p. °C	m.p. °C	Density	n_D	Solubility	Ref.
2009	Benzene, 1-chloro-2,5-dimethyl or 2-Chloro-p-xylene .... 2,5-(CH₃)₂C₆H₃Cl	140.61	....	187 (192)	1.6	1.0589[15/4]	....	ace, bz	B5[4], 965
2010	Benzene, 1-chloro-2,6-dimethyl .... 2,6-(CH₃)₂C₆H₃Cl	140.61	....	185-7, 62-3[12]	......	1.053[20]	1.526[20]	ace, bz, chl	B5[4], 943
2011	Benzene, 1-chloro-3,4-dimethyl .... 3,4-(CH₃)₂C₆H₃Cl	140.61	....	194[755]	-6	1.0692[15/15]	....	ace, bz	B5[3], 815
2012	Benzene, 1-chloro-3,5-dimethyl or 5-Chloro-m-xylene.... 3,5-(CH₃)₂C₆H₃Cl	140.61	....	187-8, 66[12]	......	....	....	ace, bz	B5[3], 834
2013	Benzene, 1-chloro-2,3-dinitro 2,3-(NO₂)₂C₆H₃Cl	202.56	pr (al), cr (MeOH)	....	78	....	....	eth	B5[4], 744
2014	Benzene, 1-chloro-2,4-dinitro 2,4-(NO₂)₂C₆H₃Cl	202.56	α: ye rh (eth), β: ye rh (eth), nd (al)	315, 158-60[2]	α: 53 (st), β: 43 (unst)	1.4982[75/4]	1.5857[60]	eth, bz	B5[4], 744
2015	Benzene, 1-chloro-2,6-dinitro 2,6-(NO₂)₂C₆H₃Cl	202.56	ye nd (al, aa)	315	88	1.6867[16]	....	al, eth, to	B5[4], 744
2016	Benzene, 1-chloro-3,4-dinitro 3,4-(NO₂)₂C₆H₃Cl	202.56	α-cr (al), β-pr (lig), γ-mcl or rh nd (eth or lig)	315d, 160[4]	α 36, β 37, γ 40-1	1.6867[20/4]	....	eth, bz	B5[4], 744
2017	Benzene, 1-chloro,3,5-dinitro 3,5-(NO₂)₂C₆H₃Cl	202.56	nd (al, peth)	....	59	....	....	al, eth	B5[3], 637
2018	Benzene, 1-chloro-2-ethoxy 2-C₂H₅OC₆H₄Cl	156.69	....	210, 97-8[15]	......	1.1288[15/4]	1.5284[25]	al, eth, bz	B6[4], 785
2019	Benzene, 1-chloro-3-ethoxy 3-C₂H₅OC₆H₄Cl	156.61	....	204-5[747]	......	1.1712[20/4]	....	al, eth, bz	B6[4], 811
2020	Benzene, 1-chloro-4-ethoxy 4-C₂H₅OC₆H₄Cl	156.61	....	212-4, 98[17]	21	1.1254[20/4]	1.5252[20]	al, eth, bz, aa	B6[4], 823
2021	Benzene, (2-chlorethoxy) C₆H₅OCH₂CH₂Cl	156.61	....	217-20, 100-2[12]	28	....	....	al, eth, ace, bz	B6[4], 556
2022	Benzene, (1-chloroethyl)(d) or d-α-Phenethyl chloride .... C₆H₅CHClCH₃	140.51	[α][20/D] +50.6 (undil)	85[20]	......	1.0631[20/4]	1.5250[25]	al, eth, bz	B5[4], 898
2023	Benzene, (1-chloroethyl)(dl) or α-Phenethyl chloride ..... C₆H₅CHClCH₃	140.61	....	81-2[17]	......	1.0620[20/4]	1.5276[20]	al, eth, bz	B5[4], 898
2024	Benzene, (1-chloroethyl)(l) or l-α-Phenethyl chloride ..... C₆H₅CHClCH₃	140.61	[α][20/D] -30.1 (un-dil)	85[20]	......	1.0632[20/4]	....	al, eth, bz	B5[4], 898
2025	Benzene, (2-chloroethyl) or β-Phenethyl chloride .... C₆H₅CH₂CH₂Cl	140.61	....	197-8, 92[20]	......	1.069[25/4]	1.5276[20]	eth, al, ace, bz, lig	B5[4], 899
2026	Benzene, (2-chloroethylthio) 2,C₆H₅SCH₂CHCl	172.68	....	117-8[11]	......	1.1769[25/4]	1.5828[20]	eth, chl	B6[4], 1469
2027	Benzene, 1-chloro-2-ethyl 2-C₂H₅C₆H₄Cl	140.61	....	178.4	-82.7	1.0569[20/4]	1.5218[20]	ace, bz, chl	B5[4], 897
2028	Benzene, 1-chloro-4-ethyl 4-C₂H₅C₆H₄Cl	140.61	....	184.4, 63.5[10]	-62.6	1.0455[20/4]	1.5175[20]	al, eth, ace, peth, bz	B5[4], 897
2029	Benzene, 1-chloro-2-fluoro 2-FC₆H₄Cl	130.55	....	137.6	-43	1.2233[30/4]	1.4968[30]	ace, bz	B5[4], 652
2030	Benzene, 1-chloro-3-fluoro 3-FC₆H₄Cl	130.55	....	127.6	......	1.221[25]	1.4911	....	B5[4], 652
2031	Benzene, 1-chloro-4-fluoro 4-FC₆H₄Cl	130.55	....	130[757]	-26.8	1.4990[15]	....	al, eth, bz	B5[4], 652
2032	Benzene, 1-chloro-3-hydroxylamino 3-(HONH)C₆H₄Cl	143.57	pl (bz)	....	49	....	....	w, bz, peth	B15[2], 8
2033	Benzene, 1-chloro-4-hydroxylamino 4-(HONH)C₆H₄Cl	143.57	lf (dil al)	....	87-8	....	....	al, bz, chl	B15[2], 9
2034	Benzene, 1-chloro-2-iodo 2-IC₆H₄Cl	238.46	....	234-5, 110[18]	......	1.9515[25]	1.6331[25]	ace	B5[4], 695
2035	Benzene, 1-chloro-3-iodo 3-IC₆H₄Cl	238.46	....	230	......	....	....	bz	B5[4], 695
2036	Benzene, 1-chloro-4-iodo 4-IC₆H₄Cl	238.46	lf (ace) (al)	227, 108[13]	57	1.886[27/4]	....	al	B5[4], 696
2037	Benzene, 1-(chloromethyl)-2,4-dimethyl 2,4-(CH₃)₂C₆H₃CH₂Cl	154.64	....	215-6, 86-7[12]	......	....	....	al, eth, bz	B5[4], 1012

PHYSICAL CONSTANTS OF ORGANIC COMPOUNDS (Continued)

No.	Name, Synonyms, and Formula	Mol. wt.	Color, crystalline form, specific rotation and λ_{max} (log ε)	b.p. °C	m.p. °C	Density	n_D	Solubility	Ref.
2038	Benzene, 1-(chloromethyl)-4-ethyl or 4-Ethylbenzyl chloride . . . . 4-C$_2$H$_5$C$_6$H$_4$CH$_2$Cl	154.64	. . . .	95-6[15]	. . . .	. . . .	1.5290[25]	al, bz, chl	B5[4], 1004
2039	Benzene, 1-(chloromethyl)-2-methyl or o-xylylchloride . . . 2-CH$_3$C$_6$H$_4$CH$_2$Cl	140.61	. . . .	197-9, 80[12]	. . . .	. . . .	1.5410[25]	**al, eth**	B5[4], 926
2040	Benzene, 1-chloromethyl-3-methyl or m-xylylchloride . . . . 3-CH$_3$C$_6$H$_4$CH$_2$Cl	140.61	. . . .	195-6, 101-2[30]	. . . .	1.064[20]	1.5345[20]	al, eth	B5[4], 943
2041	Benzene, 1-(chloromethyl)-4-methyl or p-xylylchloride . . . 4-CH$_3$C$_6$H$_4$CH$_2$Cl	140.61	. . . .	200-2, 81[15]	. . . .	1.0512[20/4]	1.5380	al, **eth**	B5[4], 966
2042	Benzene, 1-chloroisopropyl . . . . C$_6$H$_5$CCl(CH$_3$)$_2$	154.64	. . . .	98[1]	. . . .	1.192[25]	1.5290[25]	. . . .	B5[4], 992
2043	Benzene, 1-chloro-2-nitro . . . . 2-NO$_2$C$_6$H$_4$Cl	157.56	mcl nd	246, 119[8]	34-5	. . . .	. . . .	al, eth, ace, bz	B5[4], 721
2044	Benzene, 1-chloro-3-nitro . . . . 3-NO$_2$C$_6$H$_4$Cl	157.56	pa ye rh pr (al)	235-6	24 (unst) 46 (st)	1.343[50/4]	1.5374[80/α]	al, eth, bz	B5[4], 722
2045	Benzene, 1-chloro-4-nitro . . . . 4-NO$_2$-C$_6$H$_4$Cl	157.56	mcl pr	242, 113[8]	83.6	1.2979[90.5]	1.5376[100/α]	al, ace	B5[4], 723
2046	Benzene, 1-chloro-2-nitroso . . . . 2-ONC$_6$H$_4$Cl	141.56	nd (al)	. . . .	65-6	. . . .	. . . .	al, eth, bz, peth	B5[4], 705
2047	Benzene, 1-Chloro-3-nitroso . . . . 3-ONC$_6$H$_4$Cl	141.56	nd (bz)	. . . .	72 (77)	. . . .	. . . .	al, eth, ace, bz, chl	B5[4], 705
2048	Benzene, 1-chloro-4-nitroso . . . . 4-ONC$_6$H$_4$Cl	141.56	(aa or al)	. . . .	92-3	. . . .	. . . .	al	B5[4], 705
2049	Benzene, chloro-pentafluoro . . . . C$_6$ClF$_5$	202.51	. . . .	122-3[750]	. . . .	1.568[25]	1.4256[20]	. . . .	B5[4], 654
2050	Benzene, (3-chloropropoxy) . . . . C$_6$H$_5$OCH$_2$CH$_2$CH$_2$Cl	170.64	. . . .	245-55, 139[25]	12	1.1167[20]	1.5235[25]	eth, ace	B6[4], 557
2051	Benzene, (3-chloropropylthio) . . . . C$_6$H$_5$SCH$_2$CH$_2$CH$_2$Cl	186.71	. . . .	116-7[4]	. . . .	1.1536[20/4]	1.5752[20]	ace, Py	B6[3], 981
2052	Benzene, 1-chloro-4-iso-propyl or 4-Chlorocumene . . . . . . 4-(CH$_3$)$_2$CHC$_6$H$_4$Cl	154.64	. . . .	198.3, 74[10]	-12.3	1.0208[20/4]	1.5117[20]	**al, eth, ace,** bz	B5[4], 992
2053	Benzene, 1-chloro-4-triazo . . . . 4-N$_3$C$_6$H$_4$Cl	153.57	. . . .	96[20]	20	. . . .	. . . .	eth	B5[4], 760
2054	Benzene, 1-chloro 2,4,6-trimethyl . . . . . . 2,4,6(CH$_3$)$_3$C$_6$H$_2$Cl	154.64	. . . .	204-6, 104[25]	<-20	1.0337[30]	1.5212[30]	al, eth	B5[4], 1026
2055	Benzene, 1-chloro-2,4,5-trinitro . . . . 2,4,5-(NO$_2$)$_3$C$_6$H$_2$Cl	247.55	ye lf (al)	. . . .	116	. . . .	. . . .	al, bz, aa	B5[4], 757
2056	Benzene, 1-chloro-2,4,6-trinitro or Picryl chloride . . . . 2,4,6-(NO$_2$)$_3$C$_6$H$_2$Cl	247.55	wh nd or pl (chl, al-lig)	. . . .	83	1.797[20]	. . . .	al, ace, bz, to	B5[4], 757
2057	Benzene, cyclohexyl or Phenylcyclohexane . . . . C$_6$H$_5$C$_6$H$_{11}$	160.26	pl	235-6, 127-8[30]	7-8	0.9502[20/4]	1.5329[20]	al, eth	B5[4], 1424
2058	Benzene, cyclopentyl or Phenylcyclopentane . . . . C$_6$H$_5$C$_5$H$_9$	146.24	. . . .	219, 102[18]	. . . .	0.9462[20/4]	1.5280[20]	eth	B5[4], 1409
2059	Benzene, 1,2-diacetamido . . . . 1,2-(CH$_3$CONH)$_2$C$_6$H$_4$	192.22	nd (w)	. . . .	185-6	. . . .	. . . .	al, ace, aa, w	B13[3], 42
2060	Benzene, 1,4-diacetyl . . . . 1,4-(CH$_3$CO)$_2$C$_6$H$_4$	162.19	pr (al, eth)	sub	114	. . . .	. . . .	al	B7[3], 3504
2061	Benzene, 1,2-diamino or o -Phenylenediamine . . . . 1,2-(H$_2$N)$_2$C$_6$H$_4$	108.15	brsh ye lf (w), pl (chl)	256-8	102-3	. . . .	. . . .	al, eth, bz	B13[3], 28
2062	Benzene, 1,3-diamino or m-Phenylenediamine . . . . 1,3,-(H$_2$N)$_2$C$_6$H$_4$	108.15	rh (al)	282-4	63-4	1.0696[58/4]	1.6339[58]	w, al, eth, bz	B13[3], 65
2063	Benzene, 1,4-diamino or p-Phenylenediamine . . . . 1,4-(H$_2$N)$_2$C$_6$H$_4$	108.15	wh pl (bz, eth)	267	140	. . . .	. . . .	eth, chl, w, al	B13[3], 104
2064	Benzene, 1,2-dibenzoxy or Catechol dibenzyl ether . . . . 1,2-(C$_6$H$_5$CH$_2$O)$_2$C$_6$H$_4$	290.37	yesh nd or pr (al)	. . . .	63-4	. . . .	. . . .	al, eth, peth	B6[3], 4219
2065	Benzene, 1,4-dibenzoxy or Hydroquinone dibenzyl ether	290.37	pl (al)	130	. . . .	. . . .	. . . .	al, eth, aa	B6[3], 4402
	1,4-(C$_6$H$_5$CH$_2$O)$_2$C$_6$H$_4$								
2066	Benzene, 1,4-dibenzoyl . . . . 1,4-(C$_6$H$_5$CO)$_2$C$_6$H$_4$	286.33	. . . .	. . . .	161	. . . .	. . . .		B7[3], 4299
2067	Benzene, 1,4-dibenzyl . . . . 1,4-(C$_6$H$_5$CH$_2$)$_2$C$_6$H$_4$	258.36	. . . .	225[18]	87-8	. . . .	. . . .	al, bz	B5[3], 2334
2068	Benzene, 1,2-dibromo or o-Dibromobenzene . . . . 1,2-Br$_2$C$_6$H$_4$	235.92	. . . .	225, 92[10]	7.1	1.9843[20/4]	1.6155[20]	**al, eth, ace, bz**	B5[4], 682

No.	Name, Synonyms, and Formula	Mol. wt.	Color, crystalline form, specific rotation and λ_{max} (log ε)	b.p. °C	m.p. °C	Density	n_D	Solubility	Ref.
2069	Benzene, (1,2-dibromoethyl) .... $C_6H_5CHBrCH_2Br$	263.96	....	139-41[15]	74	....	....	al, eth, bz, lig	B5[4], 909
2070	Benzene, 1,2 dibromo 3-nitro .... $1,2-Br_2-3-NO_2C_6H_3$	280.91	mcl pr		85			eth, ace, chl	B5[4], 731
2071	Benzene, 1,2-dibromo-4-nitro .... $1,2-Br_2-4-NO_2C_6H_3$	280.91	nd (al, aa) mcl pr	296, 180[20]	58-9	2.354[8]	1.9835[111]	al, bz, aa	B5[4], 732
2072	Benzene, 1,3-dibromo or m-Dibromobenzene........... $1,3-Br_2C_6H_4$	235.92	....	218, 66[5]	-7	1.9523[20/4]	1.6083[17]	al, eth	B5[4], 682
2073	Benzene, 1,3-dibromo-2-nitro .... $1,3-Br_2-2-NO_2C_6H_3$	280.91	nd (al), mcl pr (al)	sub	84	1.9211[111], 2.211[8]		al, ace, bz	B5[3], 621
2074	Benzene, 1,3-dibromo-5-nitro .... $1,3-Br_2-5-NO_2C_6H_3$	280.91	pl or pr (eth), lf or nd (al)		106	1.9341[111], 2.363[8]		al, eth, bz	B5[3], 621
2075	Benzene, 1,4-dibromo or p-Dibromobenzene .... $1,4-Br_2C_6H_4$	235.92	pl	218-9	87.3	1.5742		al, eth, ace, bz	B5[4], 683
2076	Benzene, 1,5-dibromo-2,4-dimethyl or 4,6-Diamino-m-xylene................................. $1,5-Br_2-2,4-(CH_3)_2C_6H_2$	263.98	pl (al)	255-6	68	....		al	B5[3], 839
2077	Benzene, 1,4-dibromo-2-nitro .... $1,4-Br_2-2-NO_2C_6H_3$	280.91	yesh pl (ace)	....	85-6	1.9416[111], 2.368[8]		bz, ace	B5[4], 739
2078	Benzene, 2,3-dibromo 1,4,5-trimethyl .... $2,3-Br_2-1,4,5-(CH_3)_3C_6H$	278.00	nd (al)		63-4			al, eth, bz	B5[3], 910
2079	Benzene, 2,4-dibromo-1-nitro .... $2,4-Br_2C_6H_3NO_2$	280.91	ye pl or pr (al)	....	62	1.9581[111], 2.356[8]		ace, bz	B5[3], 620
2080	Benzene, 2,4-dibromo-1,3,5-trimethyl .... $2,4-Br_2-1,3,5-(CH_3)_3C_6H$	278.00	nd (al)	285	65.5			al, bz	B5[3], 921
2081	Benzene, 1,4-di-tert-butyl or p-Di-tert-butyl benzene ..... $1,4-[(CH_3)_3C]_2C_6H_4$	190.33	nd (MeOH)	237[743], 109[13]	80-1			al, eth	B5[4], 1163
2082	Benzene, 1,2-dichloro or o-Dichlorobenzene .... $1,2-Cl_2-C_6H_4$	147.01	....	180.5, 86[18]	-17	1.3048[20/4]	1.5515[20]	al, eth, ace, bz	B5[4], 654
2083	Benzene, (1,2-dichloroethyl) .... $C_6H_5CHClCH_2Cl$	175.07	....	233-4, 115[15]		1.240[15/4]	1.5544[15]	ace, bz	B5[4], 901
2084	Benzene, 1,2-dichloro-3-nitro .... $1,2-Cl_2-3NO_2C_6H_3$	192.00	mcl nd (peth, aa)	257-8	61-2	1.721[14]		al, eth, ace, bz	B5[4], 725
2085	Benzene, 1,2-dichloro-4-nitro .... $1,2-Cl_2-4-NO_2C_6H_3$	192.00	nd (al)	255-6, 189[100]	43	1.4558[75/4]		al, eth	B5[4], 726
2086	Benzene, 1,3-dichloro or m-Dichlorobenzene .... $1,3-Cl_2C_6H_4$	147.01	....	173, 53[10]	-24.7	1.2884[20/4]	1.5459[20]	al, eth, **ace**, bz	B5[4], 657
2087	Benzene, 1,3-dichloro-2-nitro .... $1,3-Cl_2-2-NO_2C_6H_3$	192.00	nd or pr (al, CS_2)	130[8]	72.5	1.603[17], 1.4094[80]		eth, al	B5[4], 726
2088	Benzene, 1,3-dichloro-5-nitro .... $1,3-Cl_2-5-NO_2C_6H_3$	192.00	mcl pr or lf (aa, al)	....	65.4	1.692[14], 1.4000[100]		eth, al	B5[4], 727
2089	Benzene, 1,4-dichloro or p-Dichlorobenzene .... $1,4-Cl_2-C_6H_4$	147.01	mcl pr, lf (ace)	174, 55[10]	53.1	1.2475[20/4]	1.5285[20]	**al**, eth, **ace**, bz	B5[4], 685
2090	Benzene, 1,4-dichloro-2-iodo .... $1,4-Cl_2-2-IC_6H_3$	272.90	pl (al)	255-6, 134[12]	23			al, eth, bz, chl	B5[4], 580
2091	Benzene, 1,4-dichloro-2-nitro .... $1,4-Cl_2-2-NO_2C_6H_3$	192.00	pl or pr (al), pl (AcOEt)	267	56	1.669[22], 1.439[75]		al, eth, bz, chl	B5[4], 726
2092	Benzene, 2,4-dichloro-1-ethoxy .... $2,4-Cl_2C_6H_3OC_2H_5$	191.07	....	237				al, eth, bz	B6[4], 885
2093	Benzene, 2,4-dichloro-1-nitro .... $2,4-Cl_2-C_6H_3NO_2$	192.00	nd (al)	258.5, 100-1[4]	34	1.4790[80]	1.5512[78/0]	al, eth	B5[4], 726
2094	Benzene, 2,4-dichloro-1-triazo .... $2,4-Cl_2-C_6H_3N_3$	188.02	ye nd (al), pr (ace or bz)	....	54			al, eth, bz, peth	B5[2], 208
2095	Benzene, 2,4-dichloro-1,3,5-trimethyl .... $2,4-Cl_2-1,3,5-(CH_3)_3C_6H$	189.08	....	243.4	59			al	B5[3], 919
2096	Benzene, 2,5-dichloro-1-ethyl .... $2,5-Cl_2-C_6H_4C_2H_5$	175.06	....	213.5		1.239[0]		bz	B5[4], 901
2097	Benzene, 1-dichlorophosphino-4-isopropryl .... $C_6H_4PCl_2-4-(CH_3)_2CH$	221.07	....	268-70, 129-30[16]		1.1917[25]	1.5677[25]	ace	B16, 773
2098	Benzene, 1,2-diethyl or o-Diethyl benzene .... $1,2-(C_2H_5)_2C_6H_4$	134.22	....	183.4, 63[10]	-31.2	0.8800[20]	1.5035[20]	al, eth, ace, bz, lig	B5[4], 1065
2099	Benzene, 1,3-diethyl or m-Diethyl benzene........ $1,3-(C_2H_5)_2C_6H_4$	134.22	....	181, 60[10]	-83.9	0.8602[20/4]	1.4955[20]	**al**, eth, **ace**, bz	B5[4], 1066
2100	Benzene, 1,3-diethyl-5-methyl .... $1,3-(C_2H_5)_2-5-CH_3C_6H_3$	148.25	....	205, 79[10]	-74.1	0.8748[20/4]	1.5027[20]	**al**, eth, **ace**, bz	B5[4], 1106

No.	Name, Synonyms, and Formula	Mol. wt.	Color, crystalline form, specific rotation and λ_{max} (log ε)	b.p. °C	m.p. °C	Density	n_D	Solubility	Ref.
2101	Benzene, 1,4-diethyl or p-Diethylbenzene............. 1,4-$(C_2H_5)_2C_6H_4$	134.22	....	183.8, 63[10]	-42.8	0.8620[20/4]	1.4967[20]	al, eth, ace, bz	B5[4], 1067
2102	Benzene, 1,2-difluoro or o-Difluorobenzene............ 1,2-$F_2C_6H_4$	114.09	....	91-2[751]	-34	1.1599[18/4]	1.44506[18]	ace, bz, chl	B5[4], 637
2103	Benzene, 1,2-dihydroxy or Catechol.................. 2-HOC_6H_4OH	110.11	cr	245	105	1.1493[22]	1.604	w, al, eth, ace, bz	B6[2], 4187
2104	Benzene, 1,3-dihydroxy or Resorcinal............... 1,3-$(HO)_2C_6H_4$	110.11	nd (bz), pl (w)	178[16]	111	1.2717	....	w, al, eth, bz	B6[3], 4292
2105	Benzene, 1,4-dihydroxy or Hydroquinone .... 1,4-$(HO)_2C_6H_4$	110.11	mcl pr (sub), nd (w), Pr (MeOH)	285[710]	173-4	1.328[15]	....	w, al, eth, ace	B6[3], 4374
2106	Benzene, 1,4-dicyclohexyl or p-Dicyclohexyl benzene..... 1,4-$(C_6H_{11})_2C_6H_4$	242.40	....	120-3	....	....	....	....	B5[4], 1607
2107	Benzene, 1,4-difluoro or p-Difluorobenzene......... 1,4-$F_2C_6H_4$	114.09	....	89[760]	-13	1.1688[20/4]	1.4422[20]	ace, bz	B5[4], 637
2108	Benzene, 1,3-difluoro-4-nitro .... 1,3-F_2-4-$NO_2C_6H_3$	159.09	....	207[760]	9.8	1.4571[14]	1.5149[14]	chl	B5[4], 720
2109	Benzene, 1,2-diiodo or o-Diiodobenzene............ 1,2-$I_2C_6H_4$	329.91	pl or pr (lig)	286[750], 109[3]	27	2.54[20]	1.7179[20]	eth	B5[4], 700
2110	Benzene, 1,3-diiodo or m-Diiodobenzene............ 1,3-$I_2C_6H_4$	329.91	rh pl or pr (eth-al)	285	40.4	2.47[25]	....	al, eth, chl	B5[4], 700
2111	Benzene, 1,4-diiodo or p-Diiodobenzene............ 1,4-$I_2C_6H_4$	329.91	rh lf (al)	285 sub	131-2	....	....	al, eth	B5[4], 700
$112	Benzene, 1,2-dimethoxy-4,5-dinitro .... 4,5-$(NO_2)_2$-1,2-$(CH_3O)_2C_6H_2$	228.17	ye nd (al)	....	130-2	1.3164[140/4]	....	....	B6[3], 4274
2113	Benzene, 1,2-dimethoxy-4-iodo or 4-Iodoveratrole....... 1,2-$(CH_3O)_2$-4-IC_6H_3	264.06	nd (dil MeOH)	163-4[26], 90[0.2]	35	....	....	al, bz	B6[4], 4262
2114	Benzene, 1,2-dimethoxy-3-nitro .... 1,2-$(CH_3O)_2$-3-$NO_2C_6H_3$	183.17	nd (al)	....	64-5	1.1404[132/4]	....	al, eth, bz, aa	B6[3], 4263
2115	Benzene, 1,2-dimethoxy-4-nitro .... 1,2-$(CH_3O)_2$-4-$NO_2C_6H_3$	183.17	ye nd (al-w)	230[15-20]	98	1.1888[133/4]	....	al, eth, chl	B6[3], 4264
2116	Benzene, 1,2-dimethoxy-4-propenyl(cis) or Isoengenol methyletha.......... 1,2-$(CH_3O)_2$-4-$CH_3CH=CH-C_6H_3$	178.23	....	270.5, 138-40[12]	....	1.0521[20/4]	1.5616[20]	ace, bz	B6[3], 4995
2117	Benzene, 1,3-dimethoxy-2-nitro .... 1,3-$(CH_3O)_2$-2-$NO_2C_6H_3$	183.17	ye nd (al or aa)	....	131	....	....	bz, chl, al	B6[3], 4344
2118	Benzene, 1,3-dimethoxy 5-nitro .... 1,3-$(CH_3O)_2$-5-$NO_2C_6H_3$	183.17	pa ye nd (AcOEt)	....	89	1.1693[133/4]	....	al, bz	B6[3], 4347
2119	Benzene, 1,4-dimethoxy-2-nitro .... 1,4-$(CH_3O)_2$-2-$NO_2C_6H_3$	183.17	gold-ye nd (dil al)	169[13] sub	72-3	1.1666[132/4]	....	al, bz, chl	B6[3], 4442
2120	Benzene, 1,4-dimethoxy-2-iso-propyl .... 1,4-$(CH_3O)_2$-2-$(CH_3)_2CHC_6H_3$	180.25	....	114-6[15]	....	1.0129[17/4]	1.5103[17]	eth, bz	B6, 929
2121	Benzene, 2,4-dimethoxy-1-nitro .... 2,4-$(CH_3O)_2$-1-$NO_2C_6H_3$	183.17	nd (al)	....	76-7	1.1876[132/4]	....	al	B6[2], 822
2122	Benzene, 1,2-dimethyl-4-iso-propyl .... 1,2$(CH_3)_2$-4-$(CH_3)_2CHC_6H_3$	148.25	....	199, 86-7[16]	....	0.8710[20/4]	1.4951[20/D]	....	B5[4], 1103
2123	Benzene, 1,3-dimethyl-5-iso-propyl .... 1,3-$(CH_3)_2$-5-$(CH_3)_2CHC_6H_3$	148.25	....	83-5[17]	....	....	1.4935[25/D]	....	B5[4], 1105
2124	Benzene, 1,2-dinitro or o-Dinitrobenzene.............. 1,2-$(NO_2)_2C_6H_4$	168.11	nd (bz) pl (al)	319[775], 194[30]	118.5	1.3119[120/4], 1.565[17]	....	al, bz, chl	B5[4], 738
2125	Benzene, 1,3-dinitro or m-Dinitrobenzene............. 1,3-$(NO_2)_2C_6H_4$	168.11	rh pl (al)	291[756], 167[14]	90	1.5751[18/4]	....	al, eth, ace, bz, Py	B5[4], 379
2126	Benzene, 1,4-dinitro or p-Dinitro benzene............. 1,4-$(NO_2)_2C_6H_4$	168.11	nd (al)	298[777], 183[34], sub	174	1.625[18/4]	....	ace, bz, to, aa	B5[4], 741
2127	Benzene, 1,4-dinitro-2,3,5,6-tetramethyl or Dinitrodurene 1,4-$(NO_2)_2$-2,3,5,6-$(CH_3)_4C_6$	224.22	pr (al)	sub	211-2	....	....	eth, bz	B5[4], 1080
2128	Benzene, 2,4-dinitro-1-fluoro .... 2,4-$(NO_2)_2C_6H_3F$	186.10	....	296, 178[25]	25.8	1.4718[84]	....	al	B5[4], 742
2129	Benzene, 2,4-dinitro-1,3,5-trimethyl .... 2,4-$(NO_2)_2$-1,3,5-$(CH_3)_3C_6H$	210.19	rh (al)	418 exp	86	....	....	al	B5[4], 1029
2130	Benzene, 1,4-dipropyl or p-Dipropyl benzene .... 1,4-$(C_3H_7)_2C_6H_4$	162.27	....	109[23]	....	0.8563[10.4/4]	1.4917[19.4]	....	B5[4], 1124
2131	Benzene, 1,2-diisopropyl or o-Diisopropyl benzene ...... 1,2-(i-$C_3H_7)_2C_6H_4$	162.28	....	204, 115.4[50]	-57	0.8701[20/4]	1.4960[20]	al, eth, ace, bz	B5[4], 1125

No.	Name, Synonyms, and Formula	Mol. wt.	Color, crystalline form, specific rotation and λ_{max} (log ϵ)	b.p. °C	m.p. °C	Density	n_D	Solubility	Ref.
2132	Benzene, 1,3-di*iso*propyl or *m*-Diisopropyl benzene...... 1,3-(i-C₃H₇)₂C₆H₄	162.28	....	203.2, 75⁹	-61	0.8559²⁰′⁴	1.4883²⁰	al, eth, ace, bz	B5⁴, 1125
2133	Benzene, 1,4-di*iso*-propyl or *p*-Diisopropyl benzene...... 1,4-(i-C₃H₇)₂C₆H₄	162.28	....	210.3, 120⁵⁰	-17	0.8568²⁰′⁴	1.4898²⁰	al, eth, ace, bz	B5⁴, 1126
2134	Benzene, 1,2-divinyl or *o*-Divinyl benzene............ 1,2(CH₂=CH)C₆H₄	130.19	....	76¹⁴		0.9325²²′⁴	1.5767²⁰	ace, bz	B5⁴, 1540
2135	Benzene, 1,3-divinyl or *m*-Divinyl benzene........... 1,3-(CH₂=CH)₂C₆H₄	130.19	....	121⁷⁶, 52³	-52.3	0.9294²⁰′⁴	1.5760²⁰	ace, bz	B5², 1367
2136	Benzene, 1,4-divinyl or *p*-Divinyl benzene........... 1,4-(CH₂=CH)₂C₆H₄	130.19	....	95-6¹⁸, 34⁰·²	31	0.913⁴⁰	1.5835²⁵	ace, bz	B5⁴, 1541
2137	Benzene, dodecyl .... C₆H₅C₁₂H₂₅	246.43	....	331, 185-8¹⁵	-7	0.8551²⁰′⁴	1.4824²⁰	....	B5⁴, 1200
2138	Benzene, (epoxy isopropyl) .... C₆H₅C(CH₃)CH₂	134.18	....	83-5¹⁵		1.0280²⁰	1.5232²⁰		B17⁴, 410
2139	Benzene, ethoxy or Phenetole .... C₆H₅OC₂H₅	122.17	....	170, 60⁹	-29.5	0.9666²⁰′⁴	1.5076²⁰	al, eth	B6⁴, 554
2140	**Benzene, ethyl or ethylbenzene**..................... C₆H₅C₂H₅	106.17	....	136.2, 25.8¹⁰	-95	0.8670²⁰′⁴	1.4959²⁰	**al, eth**	B5⁴, 885
2141	Benzene, l-ethyl-4-*iso*-butyl 4-i-C₄H₉C₆H₄C₂H₅	162.28	....	211		....		eth, ace	B5⁴, 1123
2142	Benzene, 1-ethyl-2-Iodo 2-IC₆H₄C₂H₅	232.07	....	226	....	1.6189¹⁰′⁴	1.5941²²	ace, bz	B5⁴, 970
2143	Benzene, 1-ethyl-4-iodo 4-IC₆H₄C₂H₅	232.07	....	209	-17	1.6095¹⁰′⁴	1.5909²²	ace, bz	B5⁴, 801
2144	Benzene, 1-ethyl-2-methyl or 2-Ethyl toluene............ 2-CH₃C₆H₄C₂H₅	120.20	....	165.2, 62.3²⁰	-80.8	0.8807²⁰	1.5046²⁰	al, eth, ace, bz	B5⁴, 999
2145	Benzene, 1-ethyl-3-methyl or 3-Ethyl toluene........... 3-CH₃C₆H₄C₂H₅	120.20	....	161.3, 45.6¹⁰	-95.5	0.8645²⁰′⁴	1.4966²⁰	al, eth, ace, bz peth	B5⁴, 1001
2146	Benzene, 1-ethyl-4-methyl or 4-Ethyl toluene........... 4-CH₃C₆H₄C₂H₅	120.20	....	162, 45.6¹⁰	-62.3	0.8614²⁰′⁴	1.4959²⁰	al, eth, ace, bz peth	B5⁴, 1003
2147	Benzene, l-ethyl-2-nitro 2-NO₂C₆H₄C₂H₅	151.17	....	228, 116²²	-23	1.1345⁰	....	al, eth, ace	B5⁴, 911
2148	Benzene, l-ethyl-3-nitro 3-NO₂C₆H₄C₂H₅	151.17	....	242-3	....	1.1345⁰		al, eth, ace	B5⁴, 911
2149	Benzene, l-ethyl-4-nitro 4-NO₂C₆H₄C₂H₅	151.17	....	245-6, 134-6²³	-12.3	1.1192²⁰′⁴	1.5455²⁰	al, eth, ace	B5⁴, 912
2150	Benzene, 1-ethyl-4-propyl 4-C₃H₇C₆H₄C₂H₅	148.25	....	205		0.8594²⁰′⁴	1.4921²⁰	al, eth, ace, bz	B5⁴, 1100
2151	Benzene, 1-ethyl-3-*iso*-propyl 3-i-C₃H₇C₆H₄C₂H₅	148.25	....	192, 69.2¹⁰	<-20	0.859²²′⁴	1.4921²⁰	al, eth, bz	B5³, 1105
2152	Benzene, 1-ethyl-4-*iso*-propyl 4-i-C₃H₇C₆H₄C₂H₅	148.25	....	136.6, 73¹⁰	<-20	0.8585²⁰′⁴	1.4923²⁰	eth, bz	B5⁴, 1100
2153	Benzene, fluoro or Phenyl fluoride .... C₆H₅F	96.11	....	85.1	-41.2	1.0225²⁰′⁴	1.4684⁴⁰	al, eth, ace, bz, lig	B5⁴, 632
2154	Benzene, l-fluoro-2-iodo 2-IC₆H₄F	222.00	....	188.6, 78²⁰	....	....	1.5910²⁰	ace, bz, chl	B5⁴, 693
2155	Benzene, l-fluoro-4-iodo 4-IC₆H₄F	222.0	....	182-4, 67-9¹¹	(i)-27 (ii)-18	1.9523¹⁵	1.5270²²	al, eth, ace	B5⁴, 694
2156	Benzene, l-fluoro-2-nitro 2-NO₂C₆H₄F	141.10	ye	214.6d, 86-7¹¹	-6	1.3285¹⁸′⁴	1.5489¹⁷	al, eth	B5⁴, 718
2157	Benzene, l-fluoro-3-nitro 3-NO₂C₆H₄F	141.10	ye	198-200	41	1.3254¹⁹′⁴	1.5262¹⁵	al, eth	B5⁴, 719
2158	Benzene, l-fluoro-4-nitro 4-NO₂C₆H₄F	141.10	ye nd	206-7, 87¹⁴	(i) 27 (st), (ii) 3 (unst)	1.3300²⁰′⁴	1.5316²⁰	al, eth	B5⁴, 719
2159	Benzene, l-fluoro-2,4,6-trimethyl .... 2,4,6-(CH₃)₃C₆H₂F	138.18	....	168.7	-36.7	0.9745²⁵′⁴	1.4809²⁵	ace, bz, chl	B5⁴, 1225
2160	Benzene, hexabromo .... C₆Br₆	551.52	mcl nd (bz)	....	327	....		bz, chl, peth	B5⁴, 687
2161	Benzene, hexachloro .... C₆Cl₆	284.79	nd (bz-al)	322 sub	230	1.5691²³·⁶	....	eth, bz, chl	B5⁴, 670
2162	Benzene, hexaethyl .... C₆(C₂H₅)₆	246.44	mcl pr (al or bz)	298	129	0.8305¹³⁰	1.4736¹³⁰	al, eth, bz	B5⁴, 1208
2163	Benzene, hexafluoro .... C₆F₆	186.06	....	80.5	5.3	1.6184²⁰	1.37774²⁰′ᴰ	....	B5⁴, 640
2164	Benzene, heptyl .... C₆H₅C₇H₁₅	176.30	....	245.5, 116¹²	-48	0.8567²⁰′⁴	1.4865²⁰	bz, chl	B5⁴, 1143

No.	Name, Synonyms, and Formula	Mol. wt.	Color, crystalline form, specific rotation and λ_{max} (log ε)	b.p. °C	m.p. °C	Density	n_D	Solubility	Ref.
2165	Benzene, hexahydroxy or Hexaphenol .... $C_6(OH)_6$	174.11	nd (w)	....	>300	....			B6[3], 6922
2166	Benzene, hexaiodo .... C_6I_6	833.49	red-br nd or mcl pr (bz)		350d				B5[3], 585
2167	Benzene, hexamethyl or Mellitene .... $C_6(CH_3)_6$	162.28	rh pr or nd (al)	265	166-7	1.0630^{25}		al, eth, ace, bz, aa	B5[4], 1137
2168	Benzene, 1-hydroxymethyl-4-iso-propyl or Cumic alcohol 4-i-$C_3H_7C_6H_4CH_2OH$	150.22	....	249	28	$0.9818^{20/4}$	1.5210^{20}	al, eth	B6[3], 1911
2169	Benzene, iodo or Phenyl iodide .... C_6H_5I	204.01	....	188.3, 75^{10}	−31.3	$1.8308^{20/4}$	1.6200^{20}	al, eth, ace, bz	B5[4], 688
2170	Benzene, (2-iodoethoxy) .... $C_6H_5OCH_2CH_2I$	248.07	cr (dil al)		31-2				B6[3], 548
2171	Benzene, 1-iodomethyl-2-methyl 2-$CH_3C_6H_4CH_2I$	232.07	nd (eth, (peth)		33-4			eth	B5[2], 286
2172	Benzene, 1-iodomethyl-4-methyl 4-$CH_3C_6H_4CH_2I$	232.07	nd (eth) (peth)		46-7			bz, eth, peth	B5[4], 970
2173	Benzene, 1-iodo-2-nitro .... 2-$NO_2C_6H_4I$	249.01	ye rh nd (al)	$288-9^{729}$, 162^{18}	54	1.9186^{75}		al, eth	B5[2], 190
2174	Benzene, 1-iodo-3-nitro .... 3-$NO_2C_6H_4I$	249.01	mcl pr	280, 153^{14}	(i) 38.5 (st), (ii)10 (unst)	$1.9477^{50/4}$	....	al, eth	B5[4], 733
2175	Benzene, 1-iodo-4-nitro .... 4-$NO_2C_6H_4I$	249.01	ye nd (al)	289^{772}	174			al, aa	B5[2], 191
2176	Benzene, 1-iodo-2-triazo .... 2-$N_3C_6H_4I$	245.03	ye	$90-100^{0.9}$	......	$1.8893^{25/4}$	1.6631^{25}	al, ace	B5, 278
2177	Benzene, iodoso .... C_6H_5OI	220.01	ye pw	....	210 exp			al, w	B5[4], 692
2178	Benzene, iodoxy .... $C_6H_5IO_2$	236.01	nd (w)	....	236-7 exp			w, aa	B5[4], 693
2179	Benzene, isocyano or Phenylisocyanate .... C_6H_5NCO	119.12	....	$162-3^{751}$, 55^{13}		$1.0956^{19.6/4}$	$1.5368^{19.6}$	eth, ace	B12[3], 903
2180	Benzene, methoxy or Anisole .... $C_6H_5OCH_3$	108.15	....	155	−37.5	$0.9961^{20/4}$	1.5179^{20}	al, eth, ace, bz	B6[4], 548
2181	Benzene, (3-methylbutoxy) or Phenyl isopentyl ether ..... C_6H_5O-i-C_5H_{11}	164.25	....	225	......	$0.9198^{22/4}$	1.4872^{20}	w	B6[3], 553
2182	Benzene, (1-methylbutyl) .... $C_6H_5CH(CH_3)C_3H_7$	148.25	....	$198-9^{757}$		$0.8594^{21/4}$	1.4875^{25}	al, eth	B5[4], 1087
2183	Benzene, (2-methylbutyl) .... $C_6H_5CH_2CH(CH_3)CH_2CH_3$	148.25	....	193.8, 102^{18}	......	$0.8584^{20/4}$	1.4873^{20}		B5[4], 1089
2184	Benzene, (2-methyl-1-propenyl) .... $C_6H_5CH=C(CH_3)_2$	132.20	....	$99^{43.5}$	......	$0.9029^{20/4}$	$1.5388^{20/D}$		B5[4], 1380
2185	Benzene, (2-methyl-2-butyl) .... $C_6H_5C(CH_3)_2CH_2CH_3$	148.25	....	192-4, $71-2^{12}$	......	$0.8587^{20/4}$	1.4934^{20}	al, eth	B5[4], 1090
2186	Benzene, (3-methylbutyl) or isopentyl benzene .... i-$C_5H_{11}C_6H_5$	148.25	....	199	......	$0.8558^{20/4}$	1.4853^{20}	al, eth, bz	B5[4], 1089
2187	Benzene, (4-methylpentyl) or Isohexyl benzene .... i-$C_6H_{13}C_6H_5$	162.28	....	$214-5^{740}$		0.8568^{16}		al, eth, bz	B5[4], 1116
2188	Benzene, 1-methyl-4-(nitromethyl) or 4-Nitromethyl toluene 4-$(O_2NCH_2)C_6H_4CH_3$	151.16	....	$150-1^{35}$	11-2	$1.1234^{20/4}$	1.5278^{20}	bz	B5[4], 972
2189	Benzene, 1-methyl-2-iso-propenyl or 2-iso-Propenyl toluene 2-[$CH_2=C(CH_3)$]$C_6H_4CH_3$	132.20	....	175, $59-62^{11}$	......	$0.9181^{15/0}$	$1.5112^{30/0}$		B5[3], 1214
2190	Benzene, 1-methyl-2-iso-propyl or l-cymene .... 2-i-$C_3H_7C_6H_4CH_3$	134.22	....	178.1, 57.3^{10}	−71.5	$0.8766^{20/4}$	1.5006^{20}	al, eth, ace, bz	B5[4], 1383
2191	Benzene, 1-methyl-3-iso-propyl or m-Cymene .... 3-i-$C_3H_7C_6H_4CH_3$	134.22	....	175.1, 55^{10}	-63.7	$0.8610^{20/4}$	1.4930^{20}	al, eth, ace, bz	B5[4], 1383
2192	Benzene, 1-methyl-4-iso-propyl or p-Cymene .... 4-i-$C_3H_7C_6H_4CH_3$	134.22	....	177.1, 56.3^{10}	-68	$0.8573^{20/4}$	1.4909^{20}	al, eth, ace, bz	B6[4], 1466
2193	Benzene, methylthio or Thioanisole .... $CH_3SC_6H_5$	124.22	....	193, 74^{10}	......	$1.0579^{20/4}$	1.5868^{20}	al, bz	B5[4], 708
2194	Benzene, nitro .... $C_6H_5NO_2$	123.11	....	210.8	5.7	$1.2037^{20/4}$	1.5562^{20}	al, eth, ace, bz	B5[4], 998
2195	Benzene, (α-nitro-iso-propyl) .... $C_6H_5C(NO_2)(CH_3)_2$	165.19	....	224d, $125-7^{15}$	......	1.1025^{20}	1.5209^{20}	ace, bz	B5[4], 728

No.	Name, Synonyms, and Formula	Mol. wt.	Color, crystalline form, specific rotation and λ_{max} (log ε)	b.p. °C	m.p. °C	Density	n_D	Solubility	Ref.
2196	Benzene, nitro,pentachloro or Brassicol.............. $C_6Cl_5NO_2$	295.34	cr (al)	....	144	$1.718^{25/4}$	....	bz, chl	B5[4], 728 6
2197	Benzene, 1-nitro-2,3,5,6-tetrachloro .... 2,3,5,6-$Cl_4C_6HNO_2$	260.89			99-100	$1.744^{25/4}$		al, bz, chl	B5[4], 728
2198	Benzene, 1-nitro-2-triazo .... 2-$N_3C_6H_4NO_2$	164.12	ye nd (bz-al), pr (al)		53-5			al, bz, aa, chl	B5[4], 761
2199	Benzene, 1-nitro-3-triazo .... 3-$N_3C_6H_4NO_2$	164.12	wh nd (dil al or peth)		56			al, eth, bz	B5[4], 761
2200	Benzene, 1-nitro-4-triazo .... 4-$N_3C_6H_4NO_2$	164.12	wh pl (dil al)		75			al, eth, bz, aa	B5[4], 762
2201	Benzene, 1-nitro-2,3,4-tribromo .... 2,3,4-$Br_3C_6H_2NO_2$	359.82	cr (al)		85.4			al, eth, bz	B5, 251
2202	Benzene, 1-nitro-2,3,5-tribromo .... 2,3,5-$Br_3C_6H_2NO_2$	359.82	nd		119			eth, bz	B5, 251
2203	Benzene, 1-nitro-2,3,6-tribromo .... 2,3,6-$Br_3C_6H_2NO_2$	359.82	pl or pr (eth-al)		185 (sub)			al, eth, bz	B5, 251
2204	Benzene, 1-nitro-2,4,5-tribromo .... 2,4,5-$Br_3C_6H_2NO_2$	359.82	nd (al)		95 (sub)			al, eth	B5[3], 621
2205	Benzene, 1-nitro-2,4,6-tribromo .... 2,4,6-$Br_3C_6H_2NO_2$	359.82	mcl pr (chl)	177[11]	125			eth, chl, aa	B5[4], 732
2206	Benzene, 1-nitro-3,4,5-tribromo .... 3,4,5-$Br_3C_6H_2NO_2$	359.82	tcl (eth-al)	sub	112	2.645		eth, bz	B5[4], 732
2207	Benzene, 1-nitro-2,3,4-trichloro .... 2,3,4-$Cl_3C_6H_2NO_2$	226.45	nd (al)	....	55.5			CS_2	B5[2], 186
2208	Benzene, 1-nitro-2,4,5-trichloro .... 2,4,5-$Cl_3C_6H_2NO_2$	226.45	pr (al), nd (al)	288	57-8	1.790^{23}		al, eth, bz	B5[4], 728
2209	Benzene, 1-nitro-2,3,6-trichloro .... 2,3,6-$Cl_3C_6H_2NO_2$	226.45	nd (al)		89			al	B5[2], 187
2210	Benzene, 1-nitro-2,4,6-trichloro .... 2,4,6-$Cl_3C_6H_2NO_2$	226.45	nd (al)		71			al, lig	B5[4], 728
2211	Benzene, 1-nitro-3,4,5-trichloro .... 3,4,5-$Cl_3C_6H_2NO_2$	226.45	pa ye tcl (al)		72.5	1.807		al	B5[2], 187
2212	Benzene, 1-nitro-2,3,5-triiodo .... 2,3,5-$I_3C_6H_2NO_2$	500.80	ye pr		124			bz	B5, 256
2213	Benzene, 1-nitro-2,3,6-triiodo .... 2,3,6-$I_3C_6H_2NO_2$	500.80	nd (aa)		137			aa	B5, 256
2214	Benzene, 1-nitro-2,4,5-triiodo .... 2,4,5-$I_3C_6H_2NO_2$	500.80	pa ye nd (CS_2)		178				B5, 256
2215	Benzene, 1-nitro-3,4,5-triiodo .... 3,4,5-$I_3C_6H_2NO_2$	500.80	ye pr (chl), nd (al)		167	3.256		eth, bz, aa, chl	B5[3], 626
2216	Benzene, 1-nitro-2,3,5-trimethyl .... 2,3,5-$(CH_3)_3$-$C_6H_2NO_2$	165.19	pr (al)	139-40[7]	20			al	B5[4], 1015
2217	Benzene, 1-nitro-2,4,5-trimethyl .... 2,4,5-$(CH_3)_3C_6H_2NO_2$	165.19	ye nd (al)	265	71			al, peth	B5[4], 1015
2218	Benzene, 1-nitro-2,4,6-trimethyl .... 2,4,6-$(CH_3)_3C_6H_2NO_2$	165.19	rh pr (al)	255	44	1.51		al	B5[4], 1028
2219	Benzene, nitroso .... C_6H_5NO	107.11	rh or mcl (al-eth)	57-9[18]	68-9			al, eth, bz, lig	B5[4], 702
2220	Benzene, nonyl .... $C_6H_5C_9H_{19}$	204.35		280.1		$0.8584^{20/4}$	1.4816^{20}		B5[3], 1075
2221	Benzene, pentamino .... $C_6H(NH_2)_5$	153.19	dk red nd		228d			w	B13[2], 155
2222	Benzene, pentabromo .... C_6HBr_5	472.62	wh nd (aa or al)	sub	160-1			bz, chl	B5[4], 687
2223	Benzene, pentachloro .... C_6HCl_5	250.14	nd (al)	277	86	$1.8342^{16.5}$			B5[4], 669
2224	Benzene, pentaethyl .... $C_6H(C_2H_5)_5$	218.39		277	<-20	$0.895^{19/19}$	1.5127^{20}		B5[4], 1197
2225	Benzene, pentaiodo .... C_6HI_5	707.60	nd (al)	sub	172			bz, chl, aa	B5[4], 702
2226	Benzene, pentamethyl .... $C_6H(CH_3)_5$	148.25	pr (al)	232, 100[10]	54.5	$0.917^{20/4}$	1.527^{20}	al, bz	B5[4], 1109
2227	Benzene, pentoxy or Phenyl pentyl ether .... $C_6H_5OC_5H_{11}$	164.25		200[756], 111[17]		$0.9270^{20/4}$	1.4947^{20}	al, ace	B6[3], 552
2228	Benzene, pentyl .... $C_6H_5C_5H_{11}$	148.25		205.4, 80.6[10]	-75	$0.8585^{20/4}$	1.4878^{20}	al, eth, ace, bz	B5[4], 1085

No.	Name, Synonyms, and Formula	Mol. wt.	Color, crystalline form, specific rotation and λ_{max} (log ϵ)	b.p. °C	m.p. °C	Density	n_D	Solubility	Ref.
2229	Benzene, (2-pentyl) $C_6H_5CH(CH_3)C_3H_7$	148.25	....	198-9[757]	......	0.8594[21/4]	1.4875[21]	al, eth	B5[4], 1087
2230	Benzene, propenyl (cis) $C_6H_5CH=CHCH_3$	118.18	....	69[28]	-60.5	0.9088[20/4]	1.5420[20]	al, eth, ace, bz	B5[4], 1359
2231	Benzene, propenyl (trans) $C_6H_5CH=CHCH_3$	118.18	....	175.6	-27[1]	0.9019[25]	1.5508[20]	al, eth, ace, bz	B5[4], 1360
2232	Benzene, 1-propenyl-2,4,5-trimethoxy (α form) or α Asaron 2,4,5-$(CH_3O)_3$-$C_6H_2CH=CHCH_3$	208.24	mcl nd (w)	296, 167-8[12]	67	1.165[20/4]	1.5683[20]	al, eth, peth	B6[1], 6441
2233	Benzene, iso-propyl $C_6H_5C(CH_3)=CH_2$	118.18	....	165.4, 48.5[10]	-23.2	0.9106[20/4]	1.5386[20]	al, eth, ace, bz	B5[4], 1364
2234	Benzene, propoxy or Phenyl propyl ether $C_6H_5OC_3H_7$	136.20	....	189.9	-27	0.9474[20/4]	1.5014[20]	al, eth	B6[4], 556
2235	Benzene, iso-propoxy or Phenyl-iso-propyl ether $C_6H_5OCH(CH_3)_2$	136.20	....	176.8	-33	0.9408	1.4975[20]	w, al, ace, bz	B6[1], 549
2236	Benzene, propyl $C_6H_5C_3H_7$	120.20	....	159.2, 43.3[10]	-99.5	0.8620[20/4]	1.4920[20]	al, eth, ace, bz	B5[4], 977
2237	Benzene, iso-propyl or Cumene $(CH_3)_2CHC_6H_5$	120.20	....	152.4, 38.2[10]	-96	0.8618[20/4]	1.4915[20]	al, eth, ace, bz	B5[4], 985
2238	Benzene, 1-iso-propyl-2,4,5-trimethyl 2,4,5-$(CH_3)_3C_6H_2$-i-C_3H_7	162.27	....	218-20	......	0.8795[21/4]	1.50648[21/D]	....	B5[4], 1038
2239	Benzene, 1,2,3,5-tetrabromo 1,2,3,5-$Br_4C_6H_2$	393.72	nd (al)	329	99-100	....	....	eth, bz, al	B5[4], 686
2240	Benzene, 1,2,4,5-tetrabromo 1,2,4,5-$Br_4C_6H_2$	393.72	mcl pr (CS_2)	......	182	3.072[20]	....	eth	B5[4], 687
2241	Benzene, 1,2,3,4-tetrachloro 1,2,3,4-$Cl_4C_6H_2$	215.90	nd (al)	254	47.5	....	....	eth, lig, aa	B5[4], 667
2242	Benzene, 1,2,3,5-tetrachloro 1,2,3,5-$Cl_4C_6H_2$	215.90	nd (al)	246	54.5	....	....	eth, bz, lig, al	B5[4], 686
2243	Benzene, 1,2,4,5-tetrachloro 1,2,4,5-$Cl_4C_6H_2$	215.90	nd mcl pr (eth, al or bz)	243-6	139-40	....	....	eth, bz, chl	B5[4], 686
2244	Benzene, 1,2,3,4-tetraethyl 1,2,3,4-$(C_2H_5)_4C_6H_2$	190.33	....	251[734], 121.7[14]	11.8	0.8875[20/4]	1.5125[20]	al, eth	B5[4], 1168
2245	Benzene, 1,2,4,5-tetraethyl 1,2,4,5-$(C_2H_5)_4C_6H_2$	190.33	....	250	10	0.8788[20/4]	1.5054[20]	al, eth	B5[4], 1168
2246	Benzene, 1,2,3,4-tetrafluoro 1,2,3,4-$F_4C_6H_2$	150.08	....	93-4	......	....	1.4054[20/D]	....	B5[4], 639
2247	Benzene, 1,2,4,5-tetrafluoro 1,2,4,5-$F_4C_6H_2$	150.08	....	89-90	4-5	1.4255[20]	1.4075[20]	....	B5[4], 639
2248	Benzene, 1,2,3,5-tetrahydroxy 1,2,3,5-$(HO)_4C_6H_2$	142.11	nd (w), lf (eth-AcOEt)	....	165-7	....	....	w, al, ace	B6[1], 6652
2249	Benzene, 1,2,4,5-tetrahydroxy 1,2,4,5-$(HO)_4C_6H_2$	142.11	lf (w or aa)	....	232.5	....	....	w, al, ace	B6[1], 6655
2250	Benzene, 1,2,3,4-tetraiodo 1,2,3,4-$I_4C_6H_2$	581.70	pr (eth-aa)	sub	136	....	....	al, eth, aa	B5, 229
2251	Benzene, 1,2,3,5-tetraiodo 1,2,3,5-$I_4C_6H_2$	581.70	pr (eth, aa)	sub	148	....	....	aa	B5, 229
2252	Benzene, 1,2,4,5-tetraiodo 1,2,4,5-$I_4C_6H_2$	581.70	pr (bz), nd (eth)	sub (vac)	254 (165)	....	....	aa	B5, 229
2253	Benzene, 1,2,3,4-tetramethyl or Prehnitene 1,2,3,4-$(CH_3)_4C_6H_2$	134.22	cr (peth)	205, 79.4[10]	-6.2	0.9052[20/4]	1.5203[20]	al, eth, ace, bz	B5[4], 1072
2254	Benzene, 1,2,3,5-tetramethyl or Isodurene 1,2,3,5-$(CH_3)_4C_6H_2$	134.22	....	198, 74.4[10]	-23.7	0.8903[20/4]	1.5130[20]	al, eth, ace, bz	B5[4], 1073
2255	Benzene, 1,2,4,5-tetramethyl or Durene 1,2,4,5-$(CH_3)_4C_6H_2$	134.22	....	196.8, 73.5[10]	79.2	0.8380[81/4]	1.4790[81]	al, eth, ace, bz	B5[4], 1076
2256	Benzene, 1,2,4,5-tetra-iso propyl 1,2,4,5-$(i-C_3H_7)_4C_6H_2$	246.43	....	260[775], 133[17]	118.4	0.758[150/4]	....	....	B5[4], 1207
2257	Benzene, 1,3,5-triacetyl 1,3,5-$(CH_3CO)_3C_6H_3$	204.23	nd (w,aa or al)	....	163	....	....	aa	B7[3], 4579
2258	Benzene, 1,2,3-triamino 1,2,3-$(H_2N)_3C_6H_3$	123.16	cr (dil HCl)	336	103	....	....	w, al, eth	B13[3], 551
2259	Benzene, 1,2,4-triamino 1,2,4-$(H_2N)_3C_6H_3$	123.16	pl or lf (chl)	340	95-8	....	....	w, al, chl	B13[3], 552
2260	Benzene, triazo or Phynylazide $C_6H_5N_3$	119.13	pa ye oil	70[11]	-27.5	1.0880[20/20]	1.5589[25]	....	B5[3], 759

No.	Name, Synonyms, and Formula	Mol. wt.	Color, crystalline form, specific rotation and λ_{max} (log ε)	b.p. °C	m.p. °C	Density	n_D	Solubility	Ref.
2261	Benzene, 1,2,3-tribromo 1,2,3-Br$_3$C$_6$H$_3$	314.82	pl (al)	....	87.8	2.658	....	eth	B5^3, 569
2262	Benzene, 1,2,4-tribromo 1,2,4-Br$_3$C$_6$H$_3$	314.82	nd (al or eth)	275	44-5	....		al, eth, ace	B5^4, 685
2263	Benzene, 1,3,5-tribromo 1,3,5-Br$_3$C$_6$H$_3$	314.82	nd or pr (al)	271^{765}	121-2			eth, bz, chl	B5^4, 685
2264	Benzene, 2,4,6-tribromo-1,3,5-trimethyl 1,3,5-(CH$_3$)$_3$C$_6$(Br$_3$)	356.90	tcl nd (al or bz)	....	227-8			bz	B5^3, 921
2265	Benzene, 1,2,3-trichloro 1,2,3-Cl$_3$C$_6$H$_3$	181.45	pl (al)	218-9	53-4			eth, bz	B5^4, 664
2266	Benzene, 1,2,4-trichloro 1,2,4-Cl$_3$C$_6$H$_3$	181.45	rh	213.5, 84.8^{10}	17	1.4542$^{20/4}$	1.5717^{20}	eth	B5^4, 664
2267	Benzene, 1,3,5-trichloro 1,3,5-Cl$_3$C$_6$H$_3$	181.45	nd	208^{763}	63-4			eth, bz, lig	B5^4, 666
2268	Benzene, 1,2,3-trichloro-4,5,6-trihydroxy 1,2,3-Cl$_3$C$_6$(OH)$_3$	229.45	nd (al or bz), nd (+ 3w, w)	....	185			w, al, eth	B6, 1084
2269	Benzene, 1,2,4-trichloro-3,5,6-trihydroxy 1,2,4-Cl$_3$C$_6$(OH)$_3$	229.45	nd (bz or aa)	....	160			al, eth	B6^2, 1072
2270	Benzene, 1,3,5-trichloro-2,4,6-trihydroxy 1,3,5-Cl$_3$C$_6$(OH)$_3$	229.45	cr (al)	sub	136			al	B6, 1104
2271	Benzene, 1,3,5-triethoxy or Phloroglucinol triethylether 1,3,5-(C$_2$H$_5$O)$_3$C$_6$H$_3$	210.28	cr (al dil al)	175^{24}	47.5	....		al, eth	B6^3, 6306
2272	Benzene, 1,2,4-triethyl 1,2,4-(C$_2$H$_5$)$_3$C$_6$H$_3$	162.28		217.5^{755}, 99^{15}	....	0.8738$^{20/4}$	1.5024^{20}	al, eth	B5^4, 1133
2273	Benzene, 1,3,5-triethyl 1,3,5-(C$_2$H$_5$)$_3$C$_6$H$_3$	162.28		116	66.5	1.4969^{10}	....	al, eth	B5^4, 1133
2274	Benzene, 1,2,3-trihydroxy or Pyrogallol 1,2,3-(HO)$_3$C$_6$H$_3$	126.11	lf or nd (bz)	309, 171^{12}	133-4	1.453$^{4/4}$	1.561^{114}	w, al, eth	B6^3, 2620
2275	Benzene, 1,2,5-trihydroxy, triacetate 1,2,5-(CH$_3$CO$_2$)$_3$C$_6$H$_3$	252.23	pr (al)	....	165			al	B6^3, 6282
2276	Benzene, 1,2,4-trihydroxy or Hydroxyquinol 1,2,4-(HO)$_3$C$_6$H$_3$	126.11	pl (eth), lf or pl (w)	....	140-1			w, al, eth	B6^3, 6272
2277	Benzene, 1,2,4-trihydroxy,triacetate 1,2,4 (CH$_3$CO$_2$)C$_6$H$_3$	252.23	nd (MeOH)	>300	97-8			al	B6^3, 6282
2278	Benzene, 1,3,5-trihydroxy or Phloroglucinol 1,3,5-(HO)$_3$C$_6$H$_3$	126.11	lf or pl (w + 2)	sub	117 (hyd), 218-9 (anh)	1.46		al, eth, bz	B6^3, 6301
2279	Benzene, 1,3,5-trihydroxy, triacetate or Phloroglucinol triacetate 1,3,5-(CH$_3$CO$_2$)$_3$C$_6$H$_3$	252.23	pr (w), hd (dil al)	....	105-6			al	B6^3, 6306
2280	Benzene, 1,2,3-triiodo 1,2,3-I$_3$C$_6$H$_3$	455.80	nd (al), pr (bz)	sub	116			al, eth, chl	B5^1, 122
2281	Benzene, 1,2,4-triiodo 1,2,4-I$_3$C$_6$H$_3$	455.80	nd (al)	sub	91.5			eth, chl	B5^3, 585
2282	Benzene, 1,3,5-triiodo 1,3,5-I$_3$C$_6$H$_3$	455.80	nd (aa)	sub	184.2			aa	B5^4, 702
2283	Benzene, 1,2,3-trimethoxy or Pyrogallol trimethyl ether 1,2,3-(CH$_3$O)$_3$C$_6$H$_3$	168.20	rh nd (al)	235, 140^{12}	48-9	1.1118$^{45/45}$		al, eth, bz	B6^3, 6265
2284	Benzene, 1,3,5-trimethoxy or Phloroglucinol trimethyl ether 1,3,5-(CH$_3$O)$_3$C$_6$H$_3$	168.20	pr (al), lf (peth)	255.5	54-5			al, eth, bz	B6^3, 6305
2285	Benzene, 1,2,3-trimethyl or Hemimellitene 1,2,3-(CH$_3$)$_3$C$_6$H$_3$	120.20	....	176.1, 56.7^{10}	-25.4	0.8944$^{20/4}$	1.5139^{20}	**al, eth, ace, bz**	B5^4, 1007
2286	Benzene, 1,2,4-trimethyl or Pseudocumene 1,2,4-(CH$_3$)$_3$C$_6$H$_3$	120.20		169.3, 51.6^{10}	-43.8	0.8758$^{20/4}$	1.5048^{20}	**al, eth, ace, bz**	B5^4, 1010
2287	Benzene, 1,3,5-trimethyl or Mesitylene 1,3,5-(CH$_3$)$_3$C$_6$H$_3$	120.20	....	164.7, 48.7^{10}	-44.7	0.8652$^{20/4}$	1.4994^{20}	**al, eth, ace, bz**	B5^4, 1016
2288	Benzene, 1,2,3-trimethyl-4,5,6-trinitro 4,5,6-(NO$_2$)$_3$C$_6$(CH$_3$)$_3$	255.19	pr (al)		209			al, ace, bz	B5, 400
2289	Benzene, 1,2,4-trimethyl-3,5,6-trinitro 3,5,6-(NO$_2$)$_3$C$_6$(CH$_3$)	255.19	rh pr (al)		185			bz, to	B5^4, 1015
2290	Benzene, 1,3,5-trimethyl-2,4,6-trinitro or Trinitromesitylene 2,4,6-(NO$_2$)$_3$C$_6$(CH$_3$)$_3$	255.19	tcl nd (al), pr (ace)	415 exp	238.2			ace, bz	B5^4, 1029
2291	Benzene, 1,2,3-trinitro 1,2,3-(NO$_2$)$_3$C$_6$H$_3$	213.11	ye nd or pr (MeOH)		127.5			al	B5^4, 754

No.	Name, Synonyms, and Formula	Mol. wt.	Color, crystalline form, specific rotation and λ_{max} (log ε)	b.p. °C	m.p. °C	Density	n_D	Solubility	Ref.
2292	Benzene, 1,2,4-trinitro 1,2,4-$(NO_2)_3C_6H_3$	213.11	lf (eth) pa ye pr (al)	....	61.2	....	....	al, eth, ace, bz, chl	B5[4], 754
2293	Benzene, 1,3,5-trinitro 1,3,5-$(NO_2)_3C_6H_3$	213.11	rh pl (bz), lf (w)	315, 175[2]	(i) 121-2, (ii) 61	....	....	ace, bz	B5[4], 755
2294	Benzene, 1,3,5-triphenyl 1,3,5-$(C_6H_5)_3C_6H_3$	306.41	rh nd (al or aa)	459[717]	176 (cor)	1.199[10/4]	....	al, eth, bz	B5[3], 2563
2295	Benzene, 1,3,5-tri iso-propyl 1,3,5-$(i-C_3H_7)_3C_6H_3$	204.35	....	238	-7.4	0.8545[20]	1.4882[20]	ace, bz, chl	B5[4], 1178
2296	Benzene, 1,3,5-tris(phenyl amino) or Sym-trianilino benzene 1,3,5-$(C_6H_5NH)_3C_6H_3$	351.46	nd (al)	....	193	....	....	eth	B13[2], 147
2297	Benzene, 1,3,5-tris-(4-tolylamino) 1,3,5-$(4-CH_3C_6H_4NH)_3C_6H_3$	393.54	nd (al)	....	186-7	....	....		B13, 299
2298	Benzene arsonic acid or Phenylarsonic acid $C_6H_5AsO(OH)_2$	202.02	cr (w)	....	158-62d	....	....	w, al	B16[3], 1057
2299	Benzene arsonic acid-2-amino 2-$H_2NC_6H_4AsO(OH)_2$	217.07	nd (al-eth)	....	153-4	....	....	w, al, aa	B16[3], 1099
2300	Benzene arsonic acid-3-amino 3-$H_2NC_6H_4AsO(OH)_2$	217.07	pr (w)	....	213	....	....	w	B16[3], 1104
2301	Benzene arsonic acid, 4-amino or Arsanilic acid 4$H_2NC_6H_4AsO(OH)_2$	217.07	mcl, nd, (w or al)	....	232	1.9571[10]	....	w, al, eth	B16[3]
2302	Benzenearsonic acid, 2-chloro 2-$ClC_6H_4AsO(OH)_2$	236.49	nd (w or dil al)	....	186-7	....	....	al, w	B16[2], 457
2303	Benzenearsonic acid, 3-chloro 3-$ClC_6H_4AsO(OH)_2$	236.49	cr (w)	....	175	....	....	w, al	B16[2], 457
2304	Benzenearsonic acid, 4-chloro 4-$ClC_6H_4AsO(OH)_2$	236.49	wh nd (al)	....	283-5d	....	....	al	B16[3], 1058
2305	Benzenearsonic acid, 2-hydroxy 2-$HOC_6H_4AsO(OH)_2$	218.04	nd (w)	....	190-1	....	....	w, al, bz, ace	B16[2], 464
2306	Benzenearsonic acid, 3-hydroxy 3-$HOC_6H_4AsO(OH)_2$	218.04	cr (w)	....	162-73d	....	....	w, al	B16[1], 454
2307	Benzenearsonic acid, 4-hydroxy 4-$HOC_6H_4AsO(OH)_2$	218.04	nd (aa)	....	170-4	....	....	w, al	B16[3], 1070
2308	Benzenearsonic acid, 2-nitro 2-$NO_2C_6H_4AsO(OH)_2$	246.03		....	233d	....	....	aa	B16[3], 1059
2309	Benzenearsonic acid, 3-nitro 3-$NO_2C_6H_4AsO(OH)_2$	246.03	lf ye lf (w)	....	200	....	....		B16[3], 1059
2310	Benzenearsonic acid, 4-nitro 4-$NO_2C_6H_4AsO(OH)_2$	246.03	lf or nd (w)	....	>310	....	....		B16[3], 1059
2311	Benzenearsonic acid, 4-ureido or Carbarsone $C_7H_9AsNO_4$	760.08	nd (w)	....	174	....	....	w	B16[2], 497
2312	Benzeneazoethane or Ethylphenyl diimide $C_6H_5N=NC_2H_5$	134.18	bt ye oil	175-85, 82-3[20]	....	0.9628[22/4]	α-1.5313 β-1.5579	al, eth, bz	B16[3], 4
2313	Benzeneazomethane or Methylphenyl diimide $C_6H_5N=NCH_3$	120.17	ye oil	150d, 60[15]	....	....	....	al, eth	B16[3], 3
2314	Benzeneazo-α-naphthylene $C_6H_5N=N-\alpha-C_{10}H_7$	231.30	dk red lf (al)	....	70	....	....	al, eth, bz, lig	B16[3], 57
2315	Benzene-azo-α-naphthalene, 2'-amino $C_6H_5N=N-(\alpha-C_{10}H_6NH_2-2)$	247.30	red pl (al)	....	102-4	....	....	al, aa	B16[3], 417
2316	Benzeneazo, α-naphthalene, 4-amino or Naphthyl red $C_6H_5N=N-(\alpha-C_{10}H_6NH_2-4)$	247.30	red lf (dil al), nd (dil al)	....	125-6	....	....	al, eth, bz	B16[3], 406
2317	Benzeneazo-α-naphthalene, 4-amino,hydrochloride $C_6H_5N=N-(\alpha-C_{10}H_6NH_2-4)\cdot HCl$	283.76	gr nd, pr (al, aa)	....	205-6	....	....	al, aa	B16[1], 324
2318	Benzeneazo-α-naphthalene, 2,3-dimethyl-2'-hydroxy 2,3-$(CH_3)_2C_6H_3N=N-(\alpha-C_{10}H_6OH-2)$	276.34	pa ye amor (al-bz)	....	125-30	....	....	al, to	B16[2], 71
2319	Benzeneazo-α-naphthalene, 2,4-dimethyl-2'-hydroxy 2,4-$(CH_3)_2C_6H_3N = N-(\alpha-C_{10}H_6OH-2)$	276.34	red nd (al)	....	166	....	....	al, eth	B16[2], 72
2320	Benzeneazo-α-naphthalene, 2,5-dimethyl-2'-hydroxy 2,5-$(CH_3)_2C_6H_3N=N-(\alpha-C_{10}H_6OH-2)$	276.34	nd (al)	....	153	....	....	al	B16[3], 137
2321	Benzeneazo-α-naphthalene, 3,4-dimethyl-2'-hydroxy 3,4-$(CH_3)_2C_6H_3N=N(\alpha-C_{10}H_6OH-2)$	276.34	red nd (al)	....	146	....	....	bz, chl	B16[1], 260
2322	Benzeneazo-α-naphthalene, 2'-hydroxy or Sudan yellow $C_6H_5N=N-(\alpha-C_{10}H_6OH-2)$	248.29	red-gold lf or nd (al)	133-4	....	....	....	al, eth, bz	B16[3], 129
2323	Benzeneazo-α-naphthylene, 4'-hydroxy $C_6H_5N=N-(\alpha-C_{10}H_6OH-4)$	248.29	vt-br lf (bz)	....	205-6d	....	....	al, bz	B16[3], 125

No.	Name, Synonyms, and Formula	Mol. wt.	Color, crystalline form, specific rotation and λ_{max} (log ε)	b.p. °C	m.p. °C	Density	n_D	Solubility	Ref.
2324	Benzeneazo-N = N-α-naphthalene, 2-nitro-2´-hydroxy 2-NO$_2$C$_6$H$_4$N=N(α-C$_{10}$H$_6$OH-2)	293.29	og-red nd (gl aa)	....	294	....		aa	B16^2, 70
2325	Benzeneazo-α-naphthalene, 4-nitro-2´-hydroxy or Para red Paranitroaniline red ... 4-NO$_2$C$_6$H$_4$N=N(α-C$_{10}$H$_6$OH-2)	293.29	br-og pl (to or bz)	....	257			al, bz	B16^2, 70
2326	Benzene azo-β-naphthalene C$_6$H$_5$-N=Nβ-C$_{10}$H$_7$	231.30	ye (al)		131			al, eth, bz, lig	B16^3, 60
2327	Benzeneazo-β-naphthalene, 2,4-dimethyl, 1´-hydroxy ... 2,4-(CH$_3$)$_2$C$_6$H$_3$N=N(β-C$_{10}$H$_6$OH-1)	276.34	gold-red lf or nd (al-chl)		186			bz, chl	B16^1, 249
2328	Benzeneazo-β-naphthalene, 4-dimethylamino, 1´-hydroxy 4-(CH$_3$)$_2$NC$_6$H$_4$N = N(β-C$_{10}$H$_6$OH-1)	275.36	ye-br (bz-lig)		174			bz	B16, 323
2329	Benzeneazo-β-naphthalene, 1´-hydroxy C$_6$H$_5$N=N(βC$_{10}$H$_6$OH-1)	248.29	red nd (al)	sub	138			al, aa	B16^3, 123
2330	Benzene diazonium chloride ... C$_6$H$_5$N$_2$Cl	140.57	nd (al)		exp			w, al, ace, aa	B16^3, 506
2331	Benzene diazonium cyanide ... C$_6$H$_5$N$_2$CN	131.14	ye pr (w)		69				B16, 432
2332	Benzene diazonium nitrate ... C$_6$H$_5$N$_2$ONO$_2$	167.13	nd (al-eth)		90 exp			w, al	B16^2, 268
2333	Benzene hexacarboxylic acid or Mellitic acid ... C$_6$(CO$_2$H)$_6$	342.18	nd (al)		286-8d			w, al	B9^3, 4914
2334	Benzene pentacarboxylic acid C$_6$H(CO$_2$H)$_5$	298.17	nd (w + 5)		228-30 (hyd), 238 (anh)			w, al	B9^3, 4908
2335	Benzene phosphinic acid ... C$_6$H$_5$P(OH)$_2$	142.10			82-3			w, al	B16^3, 874
2336	Benzene phosphinic acid, diethylester ... C$_6$H$_5$P(OC$_2$H$_5$)$_2$	198.20		235		1.032^{16}			B16^3, 847
2337	Benzene phosphonic acid ... C$_6$H$_5$P(O)(OH)$_2$	158.10	lf (w)		160			w, al, eth	B16^3, 883
2338	Benzenephosphonyl chloride C$_6$H$_5$POCl$_2$	194.99		258, 137-8^{15}		1.197^{25}	1.5581^{25}		B16^3, 885
2339	Benzenephosphonic acid, tetrachloride ... C$_6$H$_5$PCl$_4$	249.89			73				B16^3, 885
2340	Benzenephosphothionic acid, ethyl-4-nitro phenylester or EPN ... C$_6$H$_5$PS(OC$_2$H$_5$)OC$_6$H$_4$NO$_2$-4	323.31			36	1.27$^{25/4}$	1.5978^{30}	al, eth, bz	C48, 2742
2341	Benzeneseleninic acid ... C$_6$H$_5$SeO$_2$H	189.07	pl (w)		124-5	1.652$^{125/4}$			B11^3, 716
2342	Benzenesiliconic acid ... C$_6$H$_5$SiO$_2$H	138.20	glassy (eth)		92			al, eth	B16, 911
2343	Benzene stibonic acid ... C$_6$H$_5$SbO(OH)$_2$	248.87	nd (aa)		139				B16^3, 1178
2344	Benzene sulfenyl chloride ... C$_6$H$_5$SCl	144.62	red liq	73-5^9				bz	B6^4, 1564
2345	Benzene sulfenyl chloride, 2,4-dinitro ... 2,4-(NO$_2$)$_2$C$_6$H$_3$SCl	214.62	ye pr (bz-peth)		99			dz, chl, aa	B6^4, 1772
2346	Benzene sulfenyl chloride, 2-nitro ... 2-NO$_2$C$_6$H$_4$SCl	189.62	ye nd (bz)		75			eth, bz, chl	B6^4, 1676
2347	Benzene sulfenyl chloride, 4-nitro ... 4-NO$_2$C$_6$H$_4$SCl	189.62	ye lf (peth)	125$^{0.1}$	52			bz	B6^4, 1717
2348	Benzenesulfinic acid ... C$_6$H$_5$SO$_2$H	142.18	pr (w)	d at 100	84			al, eth, bz	B11^3, 3
2349	Benzenesulfinic acid, 3-acetamido ... 3-(CH$_3$CONH)C$_6$H$_4$SO$_2$H	199.23	cr (w)		145			w	B14^2, 427
2350	Benzenesulfinic acid, 4-acetamido ... 4-(CH$_3$CONH)C$_6$H$_4$SO$_2$H	199.23	cr (w)		160			w	B14^3, 1893
2351	Benzenesulfinic acid, 4-bromo ... 4-BrC$_6$H$_4$SO$_2$H	221.08	nd (w)		114			al, eth	B11^3, 5
2352	Benzenesulfinic acid, 4-chloro ... 4-ClC$_6$H$_4$SO$_2$H	176.62	lf or nd (w)		99			al, eth, w	B11^3, 5
2353	Benzenesulfinic acid, ethyl ester ... C$_6$H$_5$SO$_2$C$_2$H$_5$	170.23	liq	d				al, eth, bz	B11^3, 4
2354	Benzenesulfinic acid, 3-nitro ... 3-NO$_2$C$_6$H$_4$SO$_2$H	187.18	nd		98			w, al, eth, ace	B11^3, 5

No.	Name, Synonyms, and Formula	Mol. wt.	Color, crystalline form, specific rotation and λ_{max} (log ε)	b.p. °C	m.p. °C	Density	n_D	Solubility	Ref.
2355	Benzenesulfinic acid, 4-nitro 4-NO$_2$C$_6$H$_4$SO$_2$H	187.18	pr or nd (w)	....	159	....	....	al, eth, aa	B11^3, 6
2356	Benzenesulfinyl chloride C$_6$H$_5$SOCl	160.62	pl (peth)	71-2^{15}	38	1.3469^{25}	1.3470^{25}	eth, chl	B11^2, 4
2357	Benzenesulfonamide C$_6$H$_5$SO$_2$NH$_2$	157.19	lf, nd (w)	....	156			al, eth, w	B11^3, 52
2358	Benzenesulfonamide, N,N-dichloro or Dichloramine B C$_6$H$_5$SO$_2$NCl$_2$	226.08	ye mcl or pl		76			al	B11^3, 79
2359	Benzenesulfonamide, N-hydroxy C$_6$H$_5$SO$_2$NHOH	173.19	pl (w), rh		126d			w, al, eth, ace, aa	B11^3, 83
2360	Benzenesulfonamide, N-phenyl C$_6$H$_5$SO$_2$NHC$_6$H$_5$	233.29	tetr pr (al)		110			al, eth	B12^3, 1079
2361	Benzene sulfonamido, 3-acetamido 3-(CH$_3$CONH)C$_6$H$_4$SO$_2$NH$_2$	214.25	(aa)		216-9			ace, aa	B14^1, 718
2362	Benzenesulfonamido, 4-acetamido 4-(CH$_3$CONH)C$_6$H$_4$SO$_2$NH$_2$	214.25	nd (aa)		219-20			w, al, ace	B14, 702
2363	Benzenesulfonamide, 2-amino 2-H$_2$NC$_6$H$_4$SO$_2$NH$_2$	172.22	mcl pl, nd or pr (w)		152-3			al, ace, aa	B14^3, 1898
2364	Benzenesulfonamide, 3-amino 3-H$_2$NC$_6$H$_4$SO$_2$NH$_2$	172.22	lf or nd (w)		142			al	B14^3, 1908
2365	Benzenesulfonamide, 4-amino or Sulfanilamide 4-H$_2$NC$_6$H$_4$SO$_2$NH$_2$	172.22	lf (aq al)	....	165-6	1.08	....	w, al, eth, ace, MeOH	B14^3, 1919
2366	Benzenesulfonamide, 2-bromo or 2-bromo benzenesulfonamide 2-BrC$_6$H$_4$SO$_2$NH$_2$	236.09	nd (w), pr (al)		186				B11, 56
2367	Benzenesulfonamide, 3-bromo or 3-bromo sulfonamide 3-Br-C$_6$H$_4$SO$_2$NH$_2$	236.09	nd or lf (w), pr (al)		154			w, al	B11, 57
2368	Benzenesulfonamide, 4-bromo or 4-bromo sulfonamide 4-Br-C$_6$H$_4$SO$_2$NH$_2$	236.09	nd (w or dil al)		166			al	B11^3, 104
2369	Benzene sulfonamide, 2-carboxy or o-sulfamyl benzoic acid 2-HO$_2$CC$_6$H$_4$SO$_2$NH$_2$	201.20	pl or nd (w)	....	165-7			w, al, eth, bz	B11^2, 216
2370	Benzenesulfonamide, 3-carboxy or m-sulfamyl benzoic acid 3-HO$_2$CC$_6$H$_4$SO$_2$NH$_2$	201.20	pl (w)		246			w, al	B11^3, 663
2371	Benzenesulfonamide, 4-carboxy or p-sulfamyl benzoic acid 4-HO$_2$CC$_6$H$_4$SO$_2$NH$_2$	201.20	pr or lf (w)		290-2d			al	B11^3, 672
2372	Benzenesulfonamide, 2-chloro 2-Cl-C$_6$H$_4$SO$_2$NH$_2$	191.64	lf (al)		188	....		al	B11^3, 88
2373	Benzenesulfonamide, 3-chloro 3-ClC$_6$H$_4$SO$_2$NH$_2$	191.64	lf (dil al)		148			w, al, eth	B11^3, 88
2374	Benzenesulfonamide, 4-chloro 4-ClC$_6$H$_4$SO$_2$NH$_2$	191.64	pr or pl (eth)	141^{15}	55			eth, bz	B11^3, 90
2375	Benzenesulfonamide, 4-fluoro 4-FC$_6$H$_4$SO$_2$NH$_2$	175.18	pl or nd (w, al)		126			al, eth, ace	B11^3, 88
2376	Benzenesulfonamide, 4-hydroxy 4-HOC$_6$H$_4$SO$_2$NH$_2$	173.19	cr (al or w)		176-7			w, al	B11^3, 506
2377	Benzenesulfonamide, 4-ureido,amide 4-H$_2$NCONHC$_6$H$_4$SO$_2$NH$_2$	215.23	nd (al)		206-7			w, al	B14^3, 2142
2378	Benzenesulfonic acid C$_6$H$_5$SO$_3$H	158.18	nd (bz)		65-6 (anh)			w, al, aa	B11^3, 32
2379	Benzenesulfonic acid, hydrate C$_6$H$_5$SO$_3$H.1.5H$_2$O	176.20	pl (w)		45-6			w, al	B11^2, 18
2380	Benzenesulfonic acid, 2-amino or orthanilic acid 2-H$_2$NC$_6$H$_4$SO$_3$H	173.19	pr (+ ½ w)		>320d				B14^3, 1896
2381	Benzenesulfonic acid, 2-amino-4-chloro 2-H$_2$N-4-Cl-C$_6$H$_3$SO$_3$H	207.64	nd or pl (w)		310-30d				B14^3, 1900
2382	Benzenesulfonic acid, 2-amino-5-chloro 2-H$_2$N-5-ClC$_6$H$_3$SO$_3$H	207.64	nd (w)		280d			w	B14^3, 1901
2383	Benzenesulfonic acid, 2-amino-3,4-dimethyl 2-H$_2$N--3,4-(CH$_3$)$_2$C$_6$H$_2$SO$_3$H	201.25	pr (w + 1)		300d				B14^1, 731
2384	Benzenesulfonic acid, 2-amino-3,5,-dimethyl 2-H$_2$N-3,5-(CH$_3$)$_2$C$_6$H$_2$SO$_3$H	201.25	....	d					B14^2, 452
2385	Benzenesulfonic acid, 2-amino-3,6-dimethyl 2-H$_2$N-3,6-(CH$_3$)$_2$C$_6$H$_2$SO$_3$H	201.25	amor		260				B14^1, 732
2386	Benzenesulfonic acid, 2-amino-4,5-dimethyl 2-H$_2$N-4,5-(CH$_3$)$_2$C$_6$H$_2$SO$_3$H	201.25	pl		>300				B14^1, 731

PHYSICAL CONSTANTS OF ORGANIC COMPOUNDS (Continued)

No.	Name, Synonyms, and Formula	Mol. wt.	Color, crystalline form, specific rotation and λ_{max} (log ε)	b.p. °C	m.p. °C	Density	n_D	Solubility	Ref.
2387	Benzenesulfonic acid, 3-amino or Metanilic acid........ 3-H₂NC₆H₄SO₃H	173.19	nd, pr (w + l)	....	d				B14³, 1907
2388	Benzenesulfonic acid, 3-amino-2,5-dimethyl .... 3-H₂N-2,5-(CH₃)₂C₆H₂SO₃H	201.25	nd (w + l)						B14³, 2237
2389	Benzenesulfonic acid, 3-amino-4,5-dimethyl 3-H₂N-4,5-(CH₃)₂C₆H₂SO₃H	201.25	red nd (w + l)		315				B14¹, 731
2390	Benzenesulfonic acid, 3-amino-4-hydroxy 3-H₂N-4-HOC₆H₃SO₃H	189.19	rh (w + l)		>300				B14³, 2275
2391	Benzenesulfonic acid, 4-amino or Sulfanilic acid........ 4-H₂NC₆H₄SO₃H	173.19	rh pl or mcl (w + 2)		288	1.485²⁵/⁴		w	B14³, 1916
2392	Benzenesulfonic acid, 4-amino-2,3-dimethyl 4-H₂N-2,3-(CH₃)₂C₆H₂SO₃H	201.25	nd		305				B14², 452
2393	Benzenesulfonic acid, 4-amino-2,5-dimethyl 4-H₂N-2,5-(CH₃)₂C₆H₂SO₃H	201.25	pl or nd		>300			w	B14³, 2236
2394	Benzenesulfonic acid, 5-amino-2,3-dimethyl 5-H₂N-2,3-(CH₃)₂C₆H₂SO₃H	201.25	pl (w + 2)		294d			w	B14¹, 731
2395	Benzenesulfonic acid, 5-amino-2,4-dimethyl 5-H₂N-2,4-(CH₃)₂C₆H₂SO₃H	201.25	pr or nd		290d				B14³, 2235
2396	Benzenesulfonic acid, 2-bromo 2-BrC₆H₄SO₃H	237.08	nd					w, al	B11, 56
2397	Benzenesulfonic acid, 4-bromo 4-BrC₆H₄SO₃H	237.08	nd (al)	155²⁵	102-3			w, al	B11³, 97
2398	Benzenesulfonic acid, 2-carboxy or 2-sulfobenzoic acid... 2-HO₂CC₆H₄SO₃H	202.19	nd (w + 3)		141 (anh), 70 (hyd)			w, al	B11³, 658
2399	Benzenesulfonic acid, 2-carboxamide 2-H₂NCOC₆H₄SO₃H	201.20	pr (w + l)		193-4 (anh)			w, al, eth	B11², 215
2400	Benzenesulfonic acid, 2-carboxy, anhydride (endo) ...... C₇H₄SO₄	184.17	nd or pr (bz)	184-6¹⁸	129.5			eth, bz, chl	B19³, 1641
2401	Benzenesulfonic acid, 2-carboxy, imide or Saccharin, 2-Sulfobenzoic acid imide .... C₆H₅NSO₃	183.19	mcl (ace), pr (al), lf (w)	sub(vac)	228-9d	0.828		al, ace	B27², 217
2402	Benzenesulfonic acid, 3-carboxy or 3-Sulfobenzoic acid .. 3-HO₂CC₆H₄SO₃H	202.19	cr (w + 2)		191 (anh), 98 (hyd)			w, al, eth	B11³, 662
2403	Benzenesulfonic acid, 4-carboxy or 4-Sulfobenzoic acid .. 4-HO₂CC₆H₄SO₃H	202.19	nd (w + 3)		259-60 (anh)			w, al, eth	B11³, 671
2404	Benzenesulfonic acid, 4-chloro 4-ClC₆H₄SO₃H	192.62	nd (w + l)	147-8²⁵	67			w, al	B11³, 89
2405	Benzenesulfonic acid, 4-chloro,phenylester or Ovotran ... 4-ClC₆H₄SO₃C₆H₅	268.71			62			ace	C51, 1264
2406	Benzenesulfonic acid, 3,4-diamino .... 3,4-(H₂N)₂C₆H₃SO₃H	188.22	nd		d			w	B14, 717
2407	Benzenesulfonic acid, 3,5-diamino 3,5-(NH₂)₂C₆H₃SO₃H	188.22			d				B14², 446
2408	Benzenesulfonic acid, 2,5-dichloro, dihydrate .... 2,5-Cl₂C₆H₃SO₃H·2H₂O	263.11	nd (w + 2)		<100			w, al	B11², 30
2409	Benzenesulfonic acid, 3,4-dichloro, dihydrate .... 3,4-Cl₂C₆H₃SO₃H·2H₂O	263.11	nd		71-2			w, al, eth	B11, 55
2410	Benzenesulfonic acid, 3,5-diiodo-4-hydroxy, trihydrate or Sozoiodolic acid .... 3,5-I₂-4-HO-C₆H₂SO₃H·3H₂O	480.02	nd (eth), pr (w + 3)		120 (hyd), 190d			w, al, eth	B11³, 514
2411	Benzenesulfonic acid, 3,4-dimethyl .... 3,4-(CH₃)₂C₆H₃SO₃H	186.23	pl or pr (chl + 2w)		63-4			w	B11³, 337
2412	Benzenesulfonic acid, 2,4-dinitro 2,4-(NO₂)₂C₆H₃SO₃H	248.17	nd (w + 3)		130 (anh), 108			w, al	B11³, 159
2413	Benzenesulfonic acid, 3,5-dinitro 3,5-(NO₂)₂C₆H₃SO₃H	248.17	ye red (dil al)		235			al, w	B11², 36
2414	Benzenesulfonic acid, 3-ethylamino .... 3-(C₂H₅NH)C₆H₄SO₃H	201.25	nd (w)		294d			w	B14², 435
2415	Benzenesulfonic acid, 4-ethylamino .... 4-(C₂H₅NH)C₆H₄SO₃H	201.25	pl (w)		258d			w	B14³, 2025
2416	Benzenesulfonic acid, ethyl ester .... C₆H₅SO₂C₂H₅	186.23		156¹⁵		1.2167²⁰/⁴	1.5081²⁰	al, eth, chl	B11³, 36

No.	Name, Synonyms, and Formula	Mol. wt.	Color, crystalline form, specific rotation and λ_{max} (log ε)	b.p. °C	m.p. °C	Density	n_D	Solubility	Ref.
2417	Benzenesulfonic acid, 2-hydroxy or o-phenolsulfonic acid 2-HOC$_6$H$_4$SO$_3$H	174.18	cr (w + 1)	....	145d			w, al	B11^3, 488
2418	Benzenesulfonic acid, 4-hydroxy or p-phenolsulfonic acid 4-HOC$_6$H$_4$SO$_3$H	174.18	nd	....				w, al	B11^3, 498
2419	Benzenesulfonic acid, 4-hydroxy-3-nitro 4-HO-3-NO$_2$C$_6$H$_3$SO$_3$H	219.17	ye pl (AcOEt-bz), nd (w + 3)	....	141-2 (anh), 51.5 (hyd)			w, al	B11^3, 515
2420	Benzenesulfonic acid, 4-hydrazino 4-H$_2$NNHC$_6$H$_4$SO$_3$H	188.22	nd or lf (w)	....	286				B15^3, 865
2421	Benzenesulfonic acid, methyl ester C$_6$H$_5$SO$_3$CH$_3$	......	......	....					B11^3, 36
2422	Benzenesulfonic acid, 2-methyl-4-isopropyl 4-i-C$_3$H$_7$-2-CH$_3$C$_6$H$_3$SO$_3$H	214.29	pl or pr	....	88-90			w	B11, 139
2423	Benzenesulfonic acid, 2-methyl-5-isopropyl 5-i-C$_3$H$_7$-2-CH$_3$C$_6$H$_3$SO$_3$H	214.29	pl (+ 2w) mcl pr	....	220, 78-9 (+ 2w)			w	B11^3, 349
2424	Benzenesulfonic acid, 5-methyl-2-isopropyl 2-i-C$_3$H$_7$-5-CH$_3$C$_6$H$_3$SO$_3$H	214.29	......	....	130-1			w, al	B11^3, 350
2425	Benzenesulfonic acid, 2-nitro 2-NO$_2$C$_6$H$_4$SO$_3$H	203.17	......	....	85			w, al	B11^2, 31
2426	Benzenesulfonic acid, 3-nitro 3-NO$_2$C$_6$H$_4$SO$_3$H	203.17	pl	....	48			w, al	B11^3, 118
2427	Benzenesulfonic acid, 4-nitro 4-NO$_2$C$_6$H$_4$SO$_3$H	203.17	(+ 2w)	....	95, (109-11)			w	B11^3, 134
2428	Benzenesulfonic acid, 4-phenylamino or 4-Sulfodiphenylamine 4-C$_6$H$_5$NHC$_6$H$_4$SO$_3$H	249.29	pl (al-eth)	....	206			w, al	B14^1, 721
2429	Benzenesulfonic acid, propylester C$_6$H$_5$SO$_3$C$_3$H$_7$	200.26	......	162-3^{15}	......	1.1804$^{17/4}$	1.5035^{25}	al, eth, chl	B11^2, 20
2430	Benzenesulfonic acid, iso-propyl ester C$_6$H$_5$SO$_3$CH(CH$_3$)$_2$	200.26	......				1.5003^{20}	al, eth	B11^2, 20
2431	Benzenesulfonyl chloride C$_6$H$_5$SO$_2$Cl	176.62	......	251-2d, 120^{10}	14.5	1.3482$^{15/15}$		al, eth	B11^3, 51
2432	Benzenesulfonyl chloride, 3-acetamido 3-(CH$_3$CONH)C$_6$H$_4$SO$_2$Cl	233.69	nd (bz-peth)	....	88			al, eth, aa	B14^2, 435
2433	Benzenesulfonyl chloride, 4-acetamide 4-(CH$_3$CONH)C$_6$H$_4$SO$_2$Cl	233.69	nd (bz), pr (bz-chl)	....	l49			al, eth	B14^3, 2043
2434	Benzenesulfonyl chloride, 2-bromo 2-BrC$_6$H$_4$SO$_2$Cl	225.52	pr (eth)	....	5l				B11, 56
2435	Benzenesulfonyl chloride, 4-bromo 4-BrC$_6$H$_4$SO$_2$Cl	255.52	tcl or mcl pr (eth)	153^{15}	76			eth	B11^3, 104
2436	Benzenesulfonyl chloride, 3-carboxy 3-HO$_2$CC$_6$H$_4$SO$_2$Cl	220.63	pr (bz)	....	133-4			eth, bz	B11^3, 663
2437	Benzenesulfonyl chloride, 4-carboxy 4-HO$_2$CC$_6$H$_4$SO$_2$Cl	220.63	nd (ace)	....	237-8d			ace, to	B11^3, 672
2438	Benzenesulfonyl chloride, 4-chloro 4-ClC$_6$H$_4$SO$_2$Cl	211.07	pr or pl (eth)	141^{15}	55			eth, bz	B11^3, 90
2439	Benzenesulfonyl chloride, 2,5-dichloro or 2,5-dichlorobenzene sulfonyl chloride. 2,5-Cl$_2$C$_6$H$_3$SO$_2$Cl	245.51	mcl pr (bz)	....	38				B11^3, 93
2440	Benzenesulfonyl chloride, 3,4-dichloro or 3,4-Dichlorobenzenesulfonyl chloride. 3,4-Cl$_2$C$_6$H$_3$SO$_2$Cl	245.51	mcl pr	....	fr. p. 22.4				B11^3, 94
2441	Benzenesulfonyl chloride, 2,3-dimethyl or 2,3-Dimethylsulfonyl chloride. 2,3-(CH$_3$)$_2$C$_6$H$_3$SO$_2$Cl	204.68	pr (peth)	....	47			peth	B11, 120
2442	Benzenesulfonyl chloride, 3,4-dimethyl or 3,4-Dimethylbenzenesulfonyl chloride. 3,4-(CH$_3$)$_2$C$_6$H$_3$SO$_2$Cl	204.68	pr (peth)	....	51-2			eth	B11^3, 338
2443	Benzenesulfonyl chloride, 3,5-dimethyl or 3,5-Dimethylbenzenesulfonyl chloride. 3,5-(CH$_3$)$_2$C$_6$H$_3$SO$_2$Cl	204.68	nd (peth or bz)	....	94			eth	B11^1, 34
2444	Benzenesulfonyl chloride, 4-fluoro or 4-Fluorobenzenesulfonyl chloride. 4-FC$_6$H$_4$SO$_2$Cl	194.61	pl or nd	95-6^{22}	36			eth, bz, chl	B11^3, 88

No.	Name, Synonyms, and Formula	Mol. wt.	Color, crystalline form, specific rotation and λ_{max} (log ε)	b.p. °C	m.p. °C	Density	n_D	Solubility	Ref.
2445	Benzenesulfonyl chloride, 2-methoxy or 2-Methoxybenzenesulfonyl chloride . . . . 2-CH$_3$OC$_6$H$_4$SO$_2$Cl	206.65	nd (peth)	126-9[0.1]	56			eth, peth	B11, 235
2446	Benzenesulfonyl chloride, 4-methoxy or 4-Methoxybenzenesulfonyl chloride . . . . 4-CH$_3$OC$_6$H$_4$SO$_2$Cl	206.65	nd or pr (bz)	. . . .	42-3			al, eth, bz	B11[3], 504
2447	Benzenesulfonyl chloride, 2-nitro or 2-Nitrobenzenesulfonyl chloride . . . . 2-NO$_2$C$_6$H$_4$SO$_2$Cl	221.62	pr (lig, eth-peth)		68-9			eth	B11[3], 114
2448	Benzenesulfonyl chloride, 3-nitro or 3-Nitrobenzenesulfonyl chloride . . . . 3-NO$_2$C$_6$H$_4$SO$_2$Cl	221.62	mcl pr (eth), nd (lig)		64			al	B11[3], 126
2449	Benzenesulfonyl chloride, 4-nitro or 4-Nitrobenzenesulfonyl chloride . . . . 4-NO$_2$C$_6$H$_4$SO$_2$Cl	221.62	mcl pr (peth, lig)		79-80			peth	B11[3], 136
2450	Benzenesulfonyl fluoride C$_6$H$_5$SO$_2$F	160.17	. . . .	203-4, 90-1[4]	. . . . . .	1.3286[20/4]	1.4932[18]	al, eth	B11[3], 51
2451	1,3-Benzenedisulfonic acid, 4-amino 4-H$_2$NC$_6$H$_3$(SO$_3$H)$_2$-(1,3)	253.25	nd (w + 2)		120d			w, al	B14[2], 470
2452	1,3-Benzenedisulfonic acid, 4-hydroxy . . . . 4-HOC$_6$H$_3$(SO$_3$H)$_2$-(1,3)	254.24	nd (w)		>100d			w, al	B11[3], 522
2453	1,2,3,4-Benzenetetracarboxylic acid or Prehnitic acid . . . . . 1,2,3,4-C$_6$H$_2$(COOH)$_4$	254.16	pr (+ 6w)		241d			w, ace	B9[3], 4872
2454	1,2,3,4-Benzenetetracarboxylic acid, tetramethyl ester . . . . . 1,2,3,4-C$_6$H$_2$(CO$_2$CH$_3$)$_4$	310.26	nd (MeOH or w)		133.5			al, bz	B9[3], 4872
2455	1,2,3,5-Benzenetetracarboxylic acid, tetramethyl ester or Tetramethyl mellophanate . . . . 1,2,3,5-C$_6$H$_2$(CO$_2$CH$_3$)$_4$	310.26	nd (MeOH, al) MeOH		111			al	B9[3], 4872
2456	1,2,4,5-Benzenetetracarboxylic acid or Puromellitic acid . . 1,2,4,5-C$_6$H$_2$(CO$_2$H)$_4$	254.16	tcl pr (w + 2)		276 (anh), 242 (+ 2w)			al	B9[3], 4873
2457	1,2,4,5-Benzenetetracarboxylic acid, tetraethyl ester . . . . . . 1,2,4,5-C$_6$H$_2$(CO$_2$C$_2$H$_5$)$_4$	366.37	nd (al)	sub	54	. . . .	. . . .	al	B9[2], 731
2458	1,2,4,5-Benzenetetracarboxylic acid, tetramethyl ester . . . . . 1,2,4,5-C$_6$H$_2$(CO$_2$CH$_3$)$_4$	310.26	lf (al)		143-4				B9[3], 4873
2459	1,2,3-Benzenetricarboxylic acid or Hemimellitic acid . . . . . 1,2,3-C$_6$H$_3$(CO$_2$H)$_3$	210.15	tcl pl (+ 2w), nd (w)		197d (anh), 223-4 (hyd)	1.546[20]	. . . .	w	B9[3], 4791
2460	1,2,4-Benzenetricarboxylic acid or Trimellitic acid . . . . . . 1,2,4-C$_6$H$_3$(CO$_2$H)$_3$	210.25	nd (w), cr (aa or al)		238d			w, al, eth	B9[3], 4792
2461	1,2,4-Benzenetricarboxylic acid, 6-hydroxy . . . . 6-HOC$_6$H$_2$(CO$_2$H)$_3$-(1,2,4)	226.14	pr (w + 2)		278-81d			w, al	B10, 580
2462	1,3,5-Benzenetricarboxylic acid or Trimesic acid . . . . 1,3,5-C$_6$H$_3$(CO$_2$H)$_3$	210.15	pr or nd (w + l)		380 (anh)			al, eth	B9[3], 4793
2463	1,3,5-Benzenetricarboxylic acid, 2-chloro 2-Cl-C$_6$H$_2$(CO$_2$H)$_3$-(1,3,5)	244.59	nd or pl (w + l)	sub	285 (anh)			al, eth, w	B9[2], 713
2464	1,3,5-Benzenetricarboxylic acid, 2-hydroxy 2-HOC$_6$H$_2$(CO$_2$H)$_3$-(1,3,5)	226.14	pr (w + l), nd (w + 2)		306 (anh)			al, w	B10, 580
2465	1,3,5-Benzenetricarboxylic acid, triethyl ester . . . . 1,3,5-C$_6$H$_3$(CO$_2$C$_2$H$_5$)$_3$	294.31	pr or nd (al)		133-4			al, eth, bz, aa	B9[3], 4794
2466	1,3,5-Benzenetricarboxylic acid, trimethyl ester . . . . 1,3,5-C$_6$H$_3$(CO$_2$CH$_3$)$_3$	252.23	nd (dil al)		144			al	B9[3], 4794
2467	1,3,5-Benzenetrisulfonic acid . . . . 1,3,5-C$_6$H$_3$(SO$_3$H)$_3$	318.30	cr (w + 3)		d>100			w, al	B11[3], 483
2468	Benzhydrol or Diphenylmethanol . . . . (C$_6$H$_5$)$_2$CHOH	184.24	nd (lig)	297-8[748], 180[20]	69			al, eth	B6[3], 3364
2469	Benzhydrol, p-amino . . . . 4-H$_2$NC$_6$H$_4$CH(OH)C$_6$H$_5$	199.26	nd (w, bz)		121			al, ace	B13, 696
2470	Benzhydrol, 4,4′-dimethyl (4-CH$_3$C$_6$H$_4$)$_2$CHOH	212.30	nd (al)		69			al, eth, ace, chl	B6[3], 3422
2471	Benzhydrol, bis(2-dimethylamino)-α-phenyl [2-(CH$_3$)$_2$NC$_6$H$_4$]C(OH)C$_6$H$_5$	346.48	pr (lig)		105				B13, 741
2472	Benzhydrol, bis(3-dimethylamino)-α-phenyl [3-(CH$_3$)$_2$NC$_6$H$_4$]C(OH)C$_6$H$_5$	346.48	cr (eth)		128-9			eth	B13, 696

No.	Name, Synonyms, and Formula	Mol. wt.	Color, crystalline form, specific rotation and λ_{max} (log ε)	b.p. °C	m.p. °C	Density	n_D	Solubility	Ref.
2473	Benzhydrol, bis(4-dimethylamino)-α-phenyl............ [4-(CH$_3$)$_2$NC$_6$H$_4$]C(OH)C$_6$H$_5$	346.48	cr (eth, bz, lig, MeOH)	....	121-3	....	....	eth, bz	B13³, 1957
2474	Benzhydrol, 4-hydroxy or Benzaurin............... 4-HOC$_6$H$_4$CH(OH)C$_6$H$_5$	292.34	ye red pw		110-20				B8³, 1644
2475	Benzhydrol, 4-methoxy................. 4-CH$_3$OC$_6$H$_4$CH(OH)C$_6$H$_5$	214.27	nd (w, lig, dil, al)		66-8			al, bz, chl	B6³, 5417
2476	Benzhydrol, α-naphthyl............. (C$_6$H$_5$)$_2$C(OH)-α-C$_{10}$H$_7$	310.40	cr (lig, bz)	d	136.5			al, eth, bz	B6³, 3817
2477	Benzhydrol, β-naphthyl............. (C$_6$H$_5$)$_2$C(OH)-β-C$_{10}$H$_7$	310.40	pr (eth-lig)		118			al, eth, ace, bz	B6³, 3818
2478	Benzhydryl amine............... (C$_6$H$_5$)$_2$CHNH$_2$	183.26	hex pl	304⁷⁶³, 176²³	34	1.0635²⁰ᐟ²⁰	1.5963	bz	B12³, 3221
2479	o-Benzidine or 2,2′-Diamino biphenyl............. 2-H$_2$NC$_6$H$_4$-C$_6$H$_4$NH$_2$-2	184.24	mcl pr or nd (al)	162⁴	81			w, bz	B13³, 410
2480	m-Benzidine or 3,3′-Diamino biphenyl............. 3-H$_2$NC$_6$H$_4$C$_6$H$_4$NH$_2$-3	184.24	nd (w), pr (bz)	205	93—4			eth, bz	B13³, 422
2481	p-Benzidine or 4,4′-Diamino biphenyl............. (4-H$_2$NC$_6$H$_4$-)$_2$	189.24	nd (w)	400⁷⁴⁰	128			al	B12³, 425
2482	Benzil or Diphenylglyoxal................. C$_6$H$_5$COCOC$_6$H$_5$	210.25	ye pr (al)	346—8d, 188¹²	95-6	1.084¹⁰²ᐟ⁴		al, eth, ace, bz	B7³, 3804
2483	Benzil dioxime (anti) or α Benzil dioxime............. C$_6$H$_5$C=(NOH)C=(NOH)C$_6$H$_5$	240.27	lf (al), or (ace)		238d				B7², 680
2484	Benzil dioxime (syn) or β-Benzil dioxime............. C$_6$H$_5$C=(NOH)C=(NOH)C$_6$H$_5$	240.27	nd (al)		207d			al, eth	B7², 681
2485	Benzil dioxime (amphi) or α-Benzil dioxime............. C$_6$H$_5$-C=(NOH)C=(NOH)C$_6$H$_5$	240.27	nd (al + 1), (aa)		164-6			al, ace, bz	B7¹, 681
2486	Benzil monoxime (α)............. C$_6$H$_5$C=(NOH)COC$_6$H$_5$	225.25	lf (dil al, bz)	200d	137-8			al, eth, aa, chl	B7², 678
2487	Benzil monoxime (β)............. C$_6$H$_5$(C=NOH)COC$_6$H$_5$	225.25	....		70 (+ ½ bz), 113-4			al, eth, ace, bz, chl	B7², 679
2488	Benzil osazone (anti)............. C$_{26}$H$_{22}$N$_4$	390.49	ye nd (al)		230-2			bz, chl	B15³, 114
2489	Benzil osazone (syn)............. C$_{26}$H$_{22}$N$_4$	390.49	ye nd (bz-al)		210			al, bz	B15³, 114
2490	Benzilic acid, α-hydroxydiphenyl acetic acid............. (C$_6$H$_5$)$_2$C(OH)CO$_2$H	228.25	mcl nd (w)	d180	151			al, eth	B10³, 1168
2491	Benzilic acid, ethyl ester or Ethyl benzilate............. (C$_6$H$_5$)$_2$C(OH)COOC$_2$H$_5$	256.31	pr or nd	201²¹	34		1.5620²⁰	al, eth	B10³, 1171
2492	Benzilic acid, methyl ester or Methyl benzilate............. (C$_6$H$_5$)$_2$C(OH)COOCH$_3$	242.28	mcl or tcl cr (al)	187¹³	75				B10³, 1170
2493	Benzimidazole or 1,3-Benzodiazole............. C$_7$H$_6$N$_2$	118.14	rh bipym pl (w)	>360	170.5			al	B23², 151
2494	Benzimidazole, 2-amino............. 2-H$_2$N-(C$_7$H$_5$N$_2$)	133.15	pl (w)		224			w, al, ace	B24¹, 116
2495	Benzimidazole, 5,6-dimethyl............. 5,6-(CH$_3$)$_2$(C$_7$H$_4$N$_2$)	146.19	(eth)	140² sub	205-6			w, eth, al, chl	C50, 1087
2496	Benzimidazole, 2-hydroxy or o-Phenylene urea............. 2-HO(C$_7$H$_5$N$_2$)	134.14	lf (w or al)		318d			al, ace	B24², 62
2497	Benzimidazole, 2-mercapto............. 2-HS(C$_7$H$_5$N$_2$)	150.20	pl (dil al, aq NH$_3$)		298			al	B24², 65
2498	Benzimidazole, 1-methyl............. 1-CH$_3$(C$_7$H$_5$N$_2$)	132.18	nd (peth), pl (al)	286⁷⁵⁶	66	1.1254²⁰ᐟ⁴	1.6013⁷	al, peth	B23², 152
2499	Benzimidazole, 2-methyl............. 2-CH$_3$(C$_7$H$_5$N$_2$)	132.18	pr or nd (w)		176-7			w	B23², 160
2500	Benzimidazole, 2-methyl-5-nitro-1-phenyl............. 1-C$_6$H$_5$-2-CH$_3$-5-NO$_2$(C$_7$H$_3$N$_2$)	253.26	nd (al)		170-1			al	B23², 161
2501	Benzimidazole, 4-methyl............. 4-CH$_3$(C$_7$H$_5$N$_2$)	132.18	pl (w), nd (bz)		145			al, w	B23¹, 38
2502	Benzimidazole, 5-methyl............. 5-CH$_3$(C$_7$H$_5$N$_2$)	132.18	(w)		114			w	B23², 157
2503	Benzimidazole, 6-nitro............. 6-NO$_2$(C$_7$H$_4$N$_2$)	163.14	nd (w)		209—10			al	B23², 154
2504	Benzimidazole, 1-phenyl............. 1-C$_6$H$_5$(C$_7$H$_5$N$_2$) C$_{13}$H$_{10}$N$_2$	194.24	....	210-2¹⁶	98			al, w	B23², 153

No.	Name, Synonyms, and Formula	Mol. wt.	Color, crystalline form, specific rotation and λ_{max} (log ε)	b.p. °C	m.p. °C	Density	n_D	Solubility	Ref.
2505	Benzimidazole, 2-phenyl ..... 2-$C_6H_5(C_7H_5N_2)$	194.24	pl (aa), (al-w), nd (bz w)	....	293			al, aa	B23[2], 238
2506	4,5-Benzindane or 1,2-Cyclopentanonaphthalene .... $C_{13}H_{12}$	168.24	oil	294-5, 170[5]	....	1.066[20/4]	1.6290[20]	aa	B5[4], 1865
2507	1,2-Benzisothiazole ..... C_7H_5NS	135.18	....	116-8[18]					B27[2], 16
2508	1,2-Benzisothiazole, 5-hydroxy-3-phenyl 3-C_6H_5-5-HO(C_7H_3NS)	227.29	nd (dil al, aa)		159-60			al, ace, bz, aa	B27[2], 87
2509	Benzisoxazole, 3-methyl .... C_8H_7NO	133.15	....	108-10[16]				eth	B27[2], 19
2510	1,2-Benzocarbazole, 3,4-dihydro $C_{16}H_{13}N$	219.29	(lig or MeOH)	....	163-4			MeOH, lig	B20[4], 4187
2511	1,2-Benzo-1-cycloactene-3-one $C_{12}H_{14}O$	174.25		87-8[0.001]	....		1.5577[25]	....	AM76, 5462
2512	Benzodichlorofluoride $C_6H_3CCl_2F$	179.02		178-80	....	1.3138[11]	1.5180[11]	al	B5[4], 818
2513	5,6-Benzoflavone or β-Naphthoflavone ..... $C_{19}H_{12}O_2$	272.31	nd (al)		167-8			eth, bz	B17[4], 5555
2514	7,8-Benzoflavone or α-Naphthoflavone .... $C_{19}H_{12}O_2$	272.31	ye pl (al), lf or nd (dil al)		157-9 (167)				B17[4], 5551
2515	3,4-Benzofluoranthene $C_{20}H_{12}$	252.32	nd (bz)	....	168				B5[3], 2516
2516	10,11-Benzofluoranthene $C_{20}H_{12}$	252.32	ye pl (al), nd (aa)		166				B5[3], 2517
2517	11,12-Benzofluoranthene $C_{20}H_{12}$	252.32	pa ye nd (bz)	480	217			al, bz, aa	B5[3], 2516
2518	1,2-Benzofluorene or Chrysofluorene $C_{17}H_{12}$	216.29	pl (ace or aa)	413	189-90			eth, chl	B5[3], 2288
2519	1,2-Benzofluorene, 9-phenyl .... 9-$C_6H_5(C_{17}H_{11})$	292.39	nd (aa)		195.5			eth, bz	B5[3], 2558
2520	Benzofuran or Coumaron ..... C_8H_6O	118.14	....	174, 62-3[15]	<−18	1.0913[25]	1.5615[17]	al, eth	B17[4], 478
2521	Benzofuran, 2-acetyl ..... 2-$CH_3CO(C_8H_5O)$	160.17		136[11]	76			w	B17[4], 5080
2522	Benzofuran, 2-benzoyl ..... 2-$C_6H_5CO(C_8H_5O)$	222.24		360	91				B17[4], 5426
2523	Benzofuran, 2-(chloromethyl)-2,3-dihydro ..... 2-$ClCH_2$-2,3-H_2-(C_8H_4O)	168.63		118-9[11]	41-2	1.2196[7.5/16]	1.5620[7.5]	eth, ace	B17[4], 421
2524	Benzofuran, 2-methyl ..... 2-$CH_3(C_8H_5O)$	132.17	....	197-8, 93-4[20]	....	1.0540[20/4]	1.5495[22]	al, eth	B17[4], 497
2525	Benzofuran, 3-methyl ..... 3-$CH_3(C_8H_5O)$	132.17	....	196-7[742], 86[20]	....	1.0540[25/4]	1.5536[16]	al, eth	B17[4], 500
2526	Benzofuran, 5-methyl ..... 5-$CH_3(C_8H_5O)$	132.17		197-9, 83.5[17]	....	1.0603[19]	1.5570[19]	al, eth	B12[4], 502
2527	Benzofuran, 7-methyl ..... 7-$CH_3(C_8H_5O)$	132.17		190-1	....	1.0400[19]	1.5525[19]	al, eth	B17[4], 503
2528	Benzoguananine or 2,4-Diamino-6-phenyl-1,3,5-triazine 2,4-$(NH_2)_2$-6-$C_6H_5(C_3N_3)$	187.21	nd or pl (al)	....	226.4			al, eth	B26[i], 69
2529	Benzohydroxamic acid $C_6H_5CO(NHOH)$	137.14	rh ta, lf (eth)	exp	121—2			al, w	B9[3], 1304
2530	Benzohydroxamic acid, 2-hydroxy or Salicylhydroxamic acid ..... 2-HO($C_6H_4CO(NHOH)$)	153.14	nd (aa)		168			al, eth	B10[3], 160
2531	Benzoic acid ..... $C_6H_5CO_2H$	122.13	mcl lf or nd	249, 133[10]	122.13	1.0749[130], 1.2659[15/4]	1.504[12]	al, eth, ace, bz, chl	B9[3], 360
2532	Benzoic acid, 2-acetamido ..... 2-$(CH_3CONH)C_6H_4CO_2H$	179.18	nd (aa)		185			eth, ace, bz	B14[3], 922
2533	Benzoic acid, 3-acetimido ..... 3-$CH_3CONHC_6H_4CO_2H$	179.18	nd (al)		248-50			al	B14[2], 241
2534	Benzoic acid, 4-acetanido ..... 4-$(CH_3CONH)C_6H_4CO_2H$	179.18	nd (aa)		256.5			al	B14[3], 1112
2535	Benzoic acid, (4-acetimidophenyl) ester $C_6H_5CO_2(C_6H_4NHCOCH_3-4)$	255.28	nd (al)		171			al, bz, aa	B13[3], 1065
2536	Benzoic acid, 2-acetyl ..... 2-CH_3CO-$C_6H_4CO_2H$	164.17	nd (w), pr (bz)	110-2[2]	114-5			al, w	B10[3], 102

No.	Name, Synonyms, and Formula	Mol. wt.	Color, crystalline form, specific rotation and λ_{max} (log ε)	b.p. °C	m.p. °C	Density	n_D	Solubility	Ref.
2537	Benzoic acid (-2-acetylphenyl) ester $C_6H_5CO_2(C_6H_4COCH_3-2)$	240.26	nd (al)	....	88	....		al, eth, aa	B9[3], 729
2538	Benzoic acid, 3-acetyl $3-CH_3COC_6H_4CO_2H$	l64.16		....	210			w	B10[3], 248
2539	Benzoic acid, 4-acetyl $4-CH_3COC_6H_4CO_2H$	l64.17	nd (w)	sub	210			w	B10[3], 294
2540	Benzoic acid, 4-acetyl, methyl ester $4-CH_3COC_6H_4CO_2CH_3$	178.19	nd (w)	140-5[4] sub	95			w	B10, 695
2541	Benzoic acid, (4-acetylphenyl) ester $C_6H_5CO_2(C_6H_4COCH_3-4)$	240.26	nd (al, aq MeOH)	....	l35-6			al, eth, bz, aa	B9[3], 730
2542	Benzoic acid, allyl ester or Allylbenzoate $C_6H_5CO_2CH_2CH=CH_2$	162.19	....	242		1.0578[15/15]	1.5178[20]	al, eth, ace	B9[3], 402
2543	Benzoic acid, 5-allyl-2-hydroxy-3-methoxy or Eugenic acid $5-(CH_2=CHCH_2)-2-HO-3-CH_3OC_6H_2CO_2H$	208.22	pl (w + l)		85—8 (+ lw), 127 (anh)			al, eth	B10[3], 1855
2544	Benzoic acid, 2-amino or Anthranilic acid $2-H_2NC_6H_4CO_2H$	l37.14	lf (al)	sub	146-7	1.412[20]		w, al, eth, chl	B14[3], 879
2545	Benzoic acid, anhydride or Benzoic anhydride $(C_6H_5CO)_2O$	226.24	pr (eth)	360	42-3	1.989[15/4]	1.5767[15]	al, eth	B9[3], 852
2546	Benzoic acid, 2-benzamido $2-(C_6H_5CONH)C_6H_4CO_2H$	241.25	nd (al or dz)		181			al, eth	B14[3], 925
2547	Benzoic acid, 3-Benzamido $3-(C_6H_5CONH)C_6H_4CO_2H$	241.25	red pr (al)	....	252-3	1.510[4/4]		al	B14[1], 562
2548	Benzoic acid, 4-benzamido $4-(C_6H_5CONH)C_6H_4CO_2H$	241.25	nd (al)		278			al, eth	B14[1], 577
2549	Benzoic acid, 2-Benzoyl or 2-Benzophenone carboxylic acid $2-(C_6H_5CO)C_6H_4CO_2H$	226.24	tcl nd (w + l)		127-9 (anh)			al, eth, bz	B10[3], 3289
2550	Benzoic acid, 2-benzoyl-4-chloro $2-C_6H_5CO-4-ClC_6H_3CO_2H$	378.05	nd (MeOH)		92			al	B10[3], 3296
2551	Benzoic acid, 2-benzoyl, ethyl ester $2-(C_6H_5CO)C_6H_4CO_2C_2H_5$	254.29	rh pl (dil al)		58	1.221[64/4]	1.560[64]	al, eth	B10[3], 3297
2552	Benzoic acid, 2-benzoyl, methyl ester $2-(C_6H_5CO)C_6H_4CO_2CH_3$	240.26	pl or mcl pr (dil al)	350-2	52	1.1903[19/4]	1.591[20]	al, eth	B10[3], 3291
2553	Benzoic acid, benzoylmethyl ester $C_6H_5CO_2(CH_2COC_6H_5)$	240.26	pl (dil al)		118.5			eth, bz, chl, al	B9[3], 730
2554	Benzoic acid, 3-benzoyl or 3-Benzophenonecarboxylic acid $3-(C_6H_5CO)C_6H_4CO_2H$	226.24	nd (w), fl (dil al)	sub	161-2			al, eth	B10[3], 3304
2555	Benzoic acid, 4-Benzoyl or 4-Benzophenonecarboxylic acid $4-(C_6H_5CO)C_6H_4CO_2H$	226.24	nd (dil aa), pl (al), mcl lf (w)	sub	198—200 (226—7)			al, eth, aa	B10[3], 3305
2556	Benzoic acid, benzyl ester or Benzyl benzoate $C_6H_5CO_2CH_2C_6H_5$	212.25	nd or lf	323—4 (cor),170—1[11]	21	1.1121[25/4]	1.5680[20]	al, eth, ace, bz	B9[3], 428
2557	Benzoic acid, 2-benzyl $2-(C_6H_5CH_2)C_6H_4CO_2H$	212.25	nd (dil al)	sub	118			al, eth, bz, chl	B9[3], 3317
2558	Benzoic acid, 3-benzyl $3-(C_6H_5CH_2)C_6H_4CO_2H$	212.25	nd (w), lf (dil al)	sub	157-8			al, eth, bz, chl	B9[3], 676
2559	Benzoic acid, 4-benzyl $4-(C_6H_5CH_2)C_6H_4CO_2H$	212.25	nd (w), lf (dil al)	sub	157-8			al, eth, bz, chl	B9[3], 3319
2560	Benzoic acid, 4-(benzylsulfonamido) or Caronamide $4-(C_6H_5CH_2SO_2NH)C_6H_4CO_2H$	291.33			229-30			al	C45, 3418
2561	Benzoic acid, 2-bromo or 2-Bromobenzoic acid $2-BrC_6H_4CO_2H$	201.03	mcl pr (w), nd	sub	150	1.929[25/4]		al, eth, bz, chl	B9[3], 1383
2562	Benzoic acid, 2-bromo, anhydride $(2-BrC_6H_4CO)_2O$	384.02	nd (al)		79.6			al, bz, chl	B9[1], 142
2563	Benzoic acid, 2-bromo-3-chloro $2-Br-3-ClC_6H_4CO_2H$	235.47	cr (bz)		224-6			bz	B9, 355
2564	Benzoic acid, 2-bromo-5-chloro $2-Br-5-ClC_6H_3CO_2H$	235.47	cr (bz)		153			al, bz	B9[3], 1426
2565	Benzoic acid, 2-bromo-3,5-dimitro $3,5-(NO_2)_2-2-BrC_6H_2CO_2H$	291.02	ye nd (w)		213			al, bz, aa	B9[3], 1954
2566	Benzoic acid, 2-bromo, ethyl ester $2-BrC_6H_4CO_2C_2H_5$	229.08	....	254-5, 135[15]	......	1.4438[15/4]	1.5455[15]	al, eth, ace, bz	B9[3], 1385
2567	Benzoic acid, 2-bromo, methyl ester $2-BrC_6H_4CO_2CH_3$	215.05	....	244, 122[17]				al	B9[3], 1385

No.	Name, Synonyms, and Formula	Mol. wt.	Color, crystalline form, specific rotation and λ_{max} (log ε)	b.p. °C	m.p. °C	Density	n_D	Solubility	Ref.
2568	Benzoic acid, 2-bromo-3-nitro 3-NO$_2$-2-BrC$_6$H$_3$CO$_2$H	246.02	(dil al)	....	191	....	....	al	B9[2], 277
2569	Benzoic acid, 2-bromo-4-nitro 4-NO$_2$-2-BrC$_6$H$_3$CO$_2$H	246.02	nd (w or dil al)	sub>155	166-7	....	....	al, eth	B9[3], 1771
2570	Benzoic acid, 2-bromo-5-nitro 5-NO$_2$-2-BrC$_6$H$_3$CO$_2$H	246.02	nd (w)	sub	180-1	....	....	al, eth	B9[3], 1771
2571	Benzoic acid, 3-bromo 3-BrC$_6$H$_4$CO$_2$H	201.03	mcl nd (dil al)	>280	155	1.845[20]	k	al, eth	B9[3], 1392
2572	Benzoic acid, 3-bromo-2-chloro 3-Br-ClC$_6$H$_3$CO$_2$H	235.47	cr (al)	....	165	....	....	al	B9[3], 1426
2573	Benzoic acid, 3-bromo-4-chloro 3-Br-4-ClC$_6$H$_3$CO$_2$H	235.47	pl (dil aa), cr (al)	....	215-6	....	....	al, aa	B9[3], 1427
2574	Benzoic acid, 3-bromo-6-chloro 3-Br-6-ClC$_6$H$_3$CO$_2$H	235.47	nd (w), cr (aa)	....	155-6	....	....	al, aa	B9[3], 1427
2575	Benzoic acid, 3-bromo, ethyl ester or Ethyl-3-bromobenzoate 3-BrC$_6$H$_4$CO$_2$C$_2$H$_5$	229.08	....	261, 133[15]	....	1.4308[19/4]	1.5430[19]	al, eth, ace, bz	B9[3], 1393
2576	Benzoic acid, 3-bromo, methyl ester 3-BrC$_6$H$_4$CO$_2$CH$_3$	215.05	pl	122.5[15]	32	....	....	al, eth	B9[3], 1393
2577	Benzoic acid, 3-bromo-2-nitro 2-NO$_2$-3-BrC$_6$H$_3$CO$_2$H	246.02	(eth)	....	250	....	....	eth, bz	B9[3], 1770
2578	Benzoic acid, 3-bromo-4-nitro 4-NO$_2$-3-BrC$_6$H$_3$CO$_2$H	246.02	nd (dil al)	....	197	....	....	al, eth, chl	B9, 408
2579	Benzoic acid, 3-bromo-5-nitro 5-NO$_2$-3-BrC$_6$H$_3$CO$_2$H	246.02	nd (w, bz or eth), pl (ul)	....	159-60	....	....	al, eth, bz, chl	B9[3], 1771
2580	Benzoic acid, 4-bromo 4-BrC$_6$H$_4$CO$_2$H	201.03	nd (eth), lf (w), mcl pr	....	254.5	1.894[20]	....	al, eth	B9[3], 1403
2581	Benzoic acid, 4-bromo-3-chloro 4-Br-3-ClC$_6$H$_3$CO$_2$H	235.47	pl (dil aa), cr (al)	....	218	....	....	al, aa	B9[2], 236
2582	Benzoic acid, 4-bromo, ethyl ester or Ethyl-4-bromo benzoate 4-BrC$_6$H$_4$CO$_2$C$_2$H$_5$	229.08	....	262[7.37], 125[15]	....	1.4332[17/4]	1.5438[17]	al, eth, ace, bz	B9[3], 1405
2583	Benzoic acid, 4-bromo, methyl ester or Methyl-4-bromo-benzoate 4-BrC$_6$H$_4$CO$_2$CH$_3$	215.05	lf (dil al), nd (eth)	....	81	1.689	....	al, eth, ace, bz, chl	B9[3], 1405
2584	Benzoic acid, 4-bromo-2-nitro 2-NO$_2$-4-BrC$_6$H$_3$CO$_2$H	246.02	nd (w)	....	163	....	....	al, eth, bz, chl	B9[3], 1770
2585	Benzoic acid, 4-bromo-3-nitro 3-NO$_2$-4-BrC$_6$H$_3$CO$_2$H	246.02	nd (dil aa)	sub	203-4	....	....	al	B9[3], 1770
2586	Benzoic acid, 5-bromo-2-nitro 2-NO$_2$-5-BrC$_6$H$_3$CO$_2$H	246.02	cr (w, al, bz, to)	....	140	1.920[18]	....	al, bz	B9[1], 165
2587	Benzoic acid, butyl ester or Butylbenzoate C$_6$H$_5$CO$_2$C$_4$H$_9$	178.23	....	250.3	-22.4	1.000[20]	1.4940[25]	al, eth, ace	B9[3], 392
2588	Benzoic acid, iso-butyl ester or Isobutyl benzoate C$_6$H$_5$CO$_2$i-C$_4$H$_9$	178.23	....	242		0.9990[20/4]	....	al, eth, ace	B9[3], 394
2589	Benzoic acid, 2-tert-butyl 2-[(CH$_3$)$_3$C]C$_6$H$_4$CO$_2$H	178.23	pl (dil al)		80-1	....	....	al	B9[2], 365
2590	Benzoic acid, 3-tert-butyl 3-[(CH$_3$)$_3$C]C$_6$H$_4$CO$_2$H	178.23	nd (peth)		127-8	....	....	al, peth	B9[3], 2525
2591	Benzoic acid, 4-tert-butyl 4-[(CH$_3$)$_3$C]C$_6$H$_4$CO$_2$H	178.23	nd (dil al)		164-5	....	....	al, bz	B9[3], 2525
2592	Benzoic acid, (1-chloroethyl) ester or (1-Chloroethyl) benzoate C$_6$H$_5$CO$_2$CHClCH$_3$	184.62	....	134[30]	....	1.172[20]	....	al, eth	B9[2], 127
2593	Benzoic acid, (2-chloroethyl) ester or (2-Chloroethyl) benzoate C$_6$H$_5$CO$_2$CH$_2$CH$_2$Cl	184.62	....	254[729]	118-20[2]	....	....	al, eth	B9[3], 388
2594	Benzoic acid, 2-chloro 2-ClC$_6$H$_4$CO$_2$H	156.57	mcl pr (w)	sub	142	1.544[20]	....	al, eth, ace, bz	B9[3], 1330
2595	Benzoic acid, 2-chloro, ethyl ester 2-ClC$_6$H$_4$CO$_2$C$_2$H$_5$	184.62	....	243, 122-5[15]	....	1.1942[15/4]	1.5247[15]	al, eth	B9[3], 1333
2596	Benzoic acid, 2-chloro, methyl ester 2-ClC$_6$H$_4$CO$_2$CH$_3$	170.62	....	234-5[762]	....	....	....	al	B9[3], 1333
2597	Benzoic acid, 2-chloro-4-methyl 2-Cl-4-CH$_3$C$_6$H$_3$CO$_2$H	170.60	nd (al)	....	155-6	....	....	al, eth, bz, chl	B9, 497

No.	Name, Synonyms, and Formula	Mol. wt.	Color, crystalline form, specific rotation and λ_{max} (log ε)	b.p. °C	m.p. °C	Density	n_D	Solubility	Ref.
2598	Benzoic acid, 2-chloro-5-methyl 2-Cl-5-CH$_3$C$_6$H$_3$CO$_2$H	170.60	nd (w or al)	....	167	....	....	w, al	B9[3], 2327
2599	Benzoic acid, 2-chloro-6-methyl 2-Cl-6-CH$_3$C$_6$H$_3$CO$_2$H	170.60	nd (w)		102				B9[3], 2310
2600	Benzoic acid, 2-chloro-4-nitro 4-NO$_2$-2-ClC$_6$H$_3$CO$_2$H	201.57	nd (w)		140-2			al, eth, w	B9[3], 1766
2601	Benzoic acid, 2-chloro-5-nitro 5-NO$_2$-2-ClC$_6$H$_3$CO$_2$H	201.57	nd or pr (w)		165	1.608[18]		al, eth, bz	B9[3], 1765
2602	Benzoic acid, 3-chloro 3-ClC$_6$H$_4$CO$_2$H	156.57	pr (w)	sub	158	1.496[25/4]		al, eth	B9[3], 1345
2603	Benzoic acid, 3-chloro, anhydride (3-ClC$_6$H$_4$CO)$_2$O	295.14	nd (al or peth)		95.5			al, bz, chl	B9[2], 224
2604	Benzoic acid, 3-chloro, ethyl ester 3-ClC$_6$H$_4$CO$_2$C$_2$H$_5$	184.62	....	243, 121[20]	....	1.1859[15/4]	1.5223[20]	al, eth	B9[3], 1347
2605	Benzoic acid, 3-chloro, methyl ester 3-ClC$_6$H$_4$CO$_2$CH$_3$	170.60		231, 114[18]	21			al	B9[3], 1346
2606	Benzoic acid, 3-chloro-2-methyl 3-Cl-2-CH$_3$C$_6$H$_3$CO$_2$H	170.60	nd (al)		159			al, eth	B9[3], 2309
2607	Benzoic acid, 3-chloro-4-methyl 3-Cl-4-CH$_3$C$_6$H$_3$CO$_2$H	170.60	nd or lf (dil al)		200—2			al	B9[3], 2355
2608	Benzoic acid, 3-chloro-5-methyl 3-Cl-5-CH$_3$C$_6$H$_3$CO$_2$H	170.60	nd (dil al)		178			al	B9, 479
2609	Benzoic acid, 3-chloro-2-nitro 2-NO$_2$-3-ClC$_6$H$_3$CO$_2$H	201.57	nd or pl (w)		237-9	1.566[18]		al, eth	B9, 400
2610	Benzoic acid, 3-chloro-5-nitro 5-NO$_2$-3-ClC$_6$H$_3$CO$_2$H	201.57	nd (w)		147			al, eth, aa	B9[1], 165
2611	Benzoic acid, 4-chloro 4-ClC$_6$H$_4$CO$_2$H	156.57	tcl pr (al-eth)		243			al	B9[3], 1354
2612	Benzoic acid, 4-chloro, anhydride (4-ClC$_6$H$_4$CO)$_2$O	295.14	nd or lf (bz)		193-4				B9[3], 1361
2613	Benzoic acid, 2-(4-chlorobenzoyl) 2-(4-ClC$_6$H$_4$CO)-C$_6$H$_4$CO$_2$H	260.68	cr (bz, aa)		150			al, eth, bz	B10[2], 518
2614	Benzoic acid, 4-chloro, ethyl ester 4-ClC$_6$H$_4$CO$_2$C$_2$H$_5$	184.62	....	237-8, 122[15]				al	B9[3], 1356
2615	Benzoic acid, 4-chloro, methyl ester 4-ClC$_6$H$_4$CO$_2$CH$_3$	170.60	nd or mcl pr		44	1.382[20]		al	B9[3], 1356
2616	Benzoic acid, 4-chloro-2-methyl 4-Cl-2-CH$_3$C$_6$H$_3$CO$_2$H	170.60	nd (w, al, dil aa, bz)		173			al, aa	B9[2], 321
2617	Benzoic acid, 4-chloro-3-methyl 4-Cl-3-CH$_3$C$_6$H$_3$CO$_2$H	170.60	nd (w)		209-10				B9[3], 2327
2618	Benzoic acid, 4-chloro-2-nitro 2-NO$_2$-4-ClC$_6$H$_3$CO$_2$H	201.57	pl (bz-lig), pr (bz), nd (w)		142.3			al, eth, bz, w	B9[3], 1763
2619	Benzoic acid, 4-chloro-3-nitro 3-NO$_2$-4-ClC$_6$H$_3$CO$_2$H	201.57	nd or pl (w)		181-2	1.645[18]			B9[3], 1764
2620	Benzoic acid, 4-chloro-3-nitro, ethyl ester 3-NO$_2$-4-ClC$_6$H$_3$CO$_2$C$_2$H$_5$	229.63	ye nd (al)		59			al, bz, aa	B9, 402
2621	Benzoic acid, 4-chloro-3-nitro, methyl ester 3-NO$_2$-4-ClC$_6$H$_3$CO$_2$CH$_3$	215.59	nd (MeOH)		83	1.522[18]		al	B9, 402
2622	Benzoic acid, 5-chloro-2-methyl 5-Cl-2-CH$_3$C$_6$H$_3$CO$_2$H	170.60	nd (al)		168-9			al	B9[2], 320
2623	Benzoic acid, 5-chloro-2-nitro, methyl ester 2-NO$_2$-5-ClC$_6$H$_3$CO$_2$CH$_3$	215.99	pl (MeOH)		48.5	1.453[18]		MeOH	B9, 401
2624	Benzoic acid, cyclohexyl ester or Cyclohexylbenzoate C$_6$H$_5$CO$_2$C$_6$H$_{11}$	204.27	....	285	<-10	1.0429[20]	1.5200[20]	al, eth	B9[3], 404
2625	Benzoic acid, 3-cyano or Isophthalic acid mononitrile 3-NCC$_6$H$_4$CO$_2$H	147.13	nd (w)	sub d	217			al, eth	B9[3], 4243
2626	Benzoic acid, 2,3-diamino 2,3-(H$_2$N)$_2$C$_6$H$_3$CO$_2$H	152.16	nd (dil al)	d	190-1d			al, aa	B14[3], 1172
2627	Benzoic acid, 2,4-diamino 2,4-(H$_2$N)$_2$C$_6$H$_3$CO$_2$H	152.16	....	>200d	140			al, aa	B14, 448
2628	Benzoic acid, 2,5-diamino 2,5-(H$_2$N)$_2$C$_6$H$_3$CO$_2$H	152.16	br pr (w)		darkens				B14, 448
2629	Benzoic acid, 3,4-diamino 3,4-(H$_2$N)$_2$C$_6$H$_3$CO$_2$H	152.16	lf (w)		215-8d				B14[3], 1177

No.	Name, Synonyms, and Formula	Mol. wt.	Color, crystalline form, specific rotation and λ_{max} (log ε)	b.p. °C	m.p. °C	Density	n_D	Solubility	Ref.
2630	Benzoic acid, 3,5-diamino 3,5-(H₂N)₂C₆H₃CO₂H	152.16	nd (+ lw)		240 (rapid)			al, eth	B14¹, 1179
2631	Benzoic acid, 2,3-dibromo 2,3-Br₂C₆H₃CO₂H	279.95	nd (w)		149-50				B9¹, 146
2632	Benzoic acid, 2,4-dibromo 2,4-Br₂C₆H₃CO₂H	279.93	lf (w)	sub	174			al, eth	B9², 237
2633	Benzoic acid, 2,5-dibromo 2,5-Br₂C₆H₃CO₂H	279.93	nd (al or w)	sub	157			al, eth, chl, aa	B9³, 1428
2634	Benzoic acid, 2,6-dibromo 2,6-Br₂C₆H₃CO₂H	279.93	nd (w), (lig)	209—10¹⁶	150-1			al, eth, ace, chl	B9³, 1428
2635	Benzoic acid, 2,6-dibromo-3,4,5-trihydroxy or Gallo-bromol 2,6-Br₂C(OH)₃CO₂H	327.93	nd, pr or lf (w + 1)		139 (hyd)			w, al, eth	B10², 347
2636	Benzoic acid, 3,4-dibromo 3,4-Br₂C₆H₃CO₂H	279.93	nd (w), pl (al)		234-5			al, eth, MeOH	B9³, 1428
2637	Benzoic acid, 2-(N,N-dichlorosulfamyl) 2-(Cl₂NO₂S)C₆H₄CO₂H	270.09	yesh gr pl (chl)		146-8 exp				B11, 377
2638	Benzoic acid, 4-(N,N-dichlorosulfamyl) or Halazone 4-(Cl₂NO₂S)C₆H₄CO₂H	270.09	pr (aa)		213			aa	B11², 230
2639	Benzoic acid, 2,3-dichloro 2,3-Cl₂C₆H₃CO₂H	191.02	nd (w)		168.3			al, eth, w	B9², 228
2640	Benzoic acid, 2,4-dichloro 2,4-Cl₂C₆H₃CO₂H	191.02	nd (w or bz)	sub	164.3			al, eth, bz, chl	B9³, 1374
2641	Benzoic acid, 2,5-dichloro 2,5-Cl₂C₆H₃CO₂H	191.02	nd (w)	301	154.4			al, eth, w	B9³, 1376
2642	Benzoic acid, 2,6-dichloro 2,6-Cl₂C₆H₃CO₂H	191.02	nd (al) pr (w)	sub	144			al, eth, bz, w	B9³, 1377
2643	Benzoic acid, 3,4-dichloro 3,4-Cl₂C₆H₃CO₂H	191.02	nd (w, al, bz)		208-9			al, eth, w	B9³, 1378
2644	Benzoic acid, 3,5-dichloro 3,5-Cl₂C₆H₃CO₂H	191.02	nd (al or w)	sub	188			al, eth	B9³, 1379
2645	Benzoic acid, 2,3-dihydroxy or o-Pyrocatechuic acid 2,3(HO)₂C₆H₃CO₂H	154.12	pr or nd (w + 1)		204 (anh)	1.542²⁰ᐟ⁴		w, al, eth	B10³, 1363
2646	Benzoic acid, 2,4-dihydroxy or β-Resorcylic acid 2-4-(HO)₂C₆H₃CO₂H	154.12	cr + w (w)					al, eth, bz	B10³, 1370
2647	Benzoic acid, 2,4-dihydroxy-6-methyl or 4,6-dihydroxy-o-toluic acid 2,4(HO)₂-6-CH₃C₆H₂CO₂H	168.16	nd (dil aa + lw)		176d			al, eth	B10², 272
2648	Benzoic acid, 2,4-dihydroxy-6-methyl, ethyl ester 6-CH₃-2,4-(HO)₂C₆H₂CO₂C₂H₅	196.21	lf (aa), pr (al)	sub	132			al, eth	B10³, 1482
2649	Benzoic acid, 2,4-dihydroxy-6-pentyl 6-C₅H₁₁-2,4-(HO)₂C₆H₂CO₂H	224.24	wh nd		147			al, eth	B10³, 1577
2650	Benzoic acid, 2,5-dihydroxy or Gentisic acid 2,5-(HO)₂C₆H₃CO₂H	154.12	nd or pr (w)		205			w, al, eth	B10³, 1384
2651	Benzoic acid, 2,6-dihydroxy or γ-Resorcylic acid 2,6-(HO)₂C₆H₃CO₂H	154.12	nd (+ w)		167d			al, eth, w	B10³, 1401
2652	Benzoic acid, 3,4-dihydroxy or Protocatechuic acid 3,4-(HO)₂C₆H₃CO₂H	154.12	mcl nd (w + 1)		200-2d	1.524⁴		w, al, eth	B10³, 1403
2653	Benzoic acid, 3,5-dihydroxy or α-Resorcylic acid 3,5-(HO)₂C₆H₃CO₂H	154.12	pr or nd		238-40			al, eth, w	B10³, 1446
2654	Benzoic acid, 3,4-dimethoxy or Veratric acid 3,4-(CH₃O)₂C₆H₃CO₂H	182.18	nd (w or aa), rh (sub)	sub	181-2 (sub)			al, eth	B10³, 1404
2655	Benzoic acid, 2,3-dimethyl or Hemimellitic acid 2,3-(CH₃)₂C₆H₃CO₂H	150.18	pr (al)		144			al, eth	B9³, 2434
2656	Benzoic acid, 2(dimethylamino) or N,N-Dimethylanthranilic acid 2-(CH₃)₂NC₆H₄CO₂H	165.19	pr nd (eth)	sub d	72			w, al, eth	B14³, 896
2657	Benzoic acid, 3-(dimethylamino) 3-(CH₃)₂NC₆H₄CO₂H	165.19	nd (w)		151			w, al, eth	B14³, 1002
2658	Benzoic acid, 4-(dimethylamino) 4-(CH₃)₂NC₆H₄CO₂H	165.19	nd (al)		242-3			al	B14³, 1082
2659	Benzoic acid, 2,4-dimethyl or 2,4-Xylylic acid 2,4-(CH₃)₂C₆H₃CO₂H	150.18	mcl or tcl nd (w)	267⁷²⁷, sub	127 (anh), 90 (hyd)			al, ace, bz	B9², 350
2660	Benzoic acid, 2,5-dimethyl or Isoxylylic acid 2,5-(CH₃)₂C₆H₃CO₂H	150.18	nd (al)	268 sub	132	1.069²¹ᐟ⁴		al, eth, ace, bz	B9¹, 210

No.	Name, Synonyms, and Formula	Mol. wt.	Color, crystalline form, specific rotation and λ_{max} (log ϵ)	b.p. °C	m.p. °C	Density	n_D	Solubility	Ref.
2661	Benzoic acid, 2,6-dimethyl or 2,6-Xylylic acid.......... 2,6-(CH$_3$)$_2$C$_6$H$_3$CO$_2$H	150.18	nd (lig)	274-5	116	....	....	al, eth	B9^3, 2435
2662	Benzoic acid, 3,4-dimethyl or Paraxylylic acid.......... 3,4-(CH$_3$)$_2$C$_6$H$_3$CO$_2$H	150.18	pr (al)	....	166	....	....	al, eth, bz	B9^3, 2441
2663	Benzoic acid, 3,5-dimethyl or Mesitylenic acid .... 3,5-(CH$_3$)$_2$C$_6$H$_3$CO$_2$H	150.18	nd (w or al)	sub	170-1	....	....	al, eth	B9^3, 2444
2664	Benzoic acid, 2,4-dinitro .... 2,4-(NO$_2$)$_2$C$_6$H$_3$CO$_2$H	212.12	nd (w)	....	....	....	....	....	B9^3, 1776
2665	Benzoic acid, 2,4-dinitro-3-hydroxy 2,4-(NO$_2$)$_2$-3-HOC$_6$H$_2$CO$_2$H	228.12	(w)	....	204	....	....	al, eth	B10^3, 264
2666	Benzoic acid, 2,5-dinitro .... 2,5-(NO$_2$)$_2$C$_6$H$_3$CO$_2$H	212.12	pr (w)	....	177	....	....	al, eth	B9^3, 1778
2667	Benzoic acid, 2,6-dinitro .... 2,6-(NO$_2$)$_2$C$_6$H$_3$CO$_2$H	212.12	nd (w)	....	202.3	....	....	al, eth	B9^3, 1778
2668	Benzoic acid, 3,4-dinitro .... 3,4-(NO$_2$)$_2$C$_6$H$_3$CO$_2$H	212.12	nd (w)	....	165	....	....	al, eth	B9^3, 177
2669	Benzoic acid, 3,5-dinitro .... 3,5-(NO$_2$)$_2$C$_6$H$_3$CO$_2$H	212.12	mcl pr (al)	....	205	....	....	al, aa	B9^3, 1779
2670	Benzoic acid, 3,5-dinitro, benzyl ester 3,5-(NO$_2$)$_2$C$_6$H$_3$CO$_2$CH$_2$C$_6$H$_5$	302.25	nd (lig)	....	112	....	....	....	B9^3, 1848
2671	Benzoic acid, 3,5-dinitro, butyl ester ...... 3,5-(NO$_2$)$_2$C$_6$H$_3$CO$_2$C$_4$H$_9$	268.23	mcl nd (al)	....	62.5	....	....	al	B9^3, 1782
2672	Benzoic acid, 3,5-dinitro, isobutyl ester ...... 3,5-(NO$_2$)$_2$C$_6$H$_3$CO$_2$-i-C$_4$H$_9$	268.23	....	....	87-8	....	....	w, al	B9^2, 281
2673	Benzoic acid, 3,5-dinitro, ethyl ester ...... 3,5-(NO$_2$)$_2$C$_6$H$_3$CO$_2$C$_2$H$_5$	240.18	nd (al)	....	72.9	1.295^{111}	1.560	al	B9^3, 1781
2674	Benzoic acid, 3,5-dinitro-2-hydroxy 3,5-(NO$_2$)$_2$-2-HOC$_6$H$_2$CO$_2$H	228.12	ye nd or pl (+1w)	....	182 (anh)	....	....	w, al, eth, bz	B10^3, 207
2675	Benzoic acid, 3,5-dinitro-4-hydroxy 3,5-(NO$_2$)$_2$-4-HOC$_6$H$_2$CO$_2$H	228.12	ye lf (al)	....	248-9	....	....	al, eth	B10^3, 383
2676	Benzoic acid, 3,5-dinitro, methyl ester 3,5-(NO$_2$)$_2$C$_6$H$_3$CO$_2$CH$_3$	226.15	nd (w)	....	112	....	....	w, al	B9^3, 1781
2677	Benzoic acid, 3,5-dinitro, pentyl ester .... 3,5-(NO$_2$)$_2$C$_6$H$_3$CO$_2$C$_5$H$_{11}$	262.26	....	....	46.4	....	....	al	B9^2, 281
2678	Benzoic acid, 3,5-dinitro, phenyl ester .... 3,5-(NO$_2$)$_2$C$_6$H$_3$CO$_2$C$_6$H$_5$	288.23	rods (al)	....	145-6	....	....	al, bz	B9^3, 1846
2679	Benzoic acid, 3,5-dinitro, propyl ester .... 3,5-(NO$_2$)$_2$C$_6$H$_3$CO$_2$C$_3$H$_7$	254.20	mcl pl (al)	....	73	....	....	al	B9^3, 1781
2680	Benzoic acid, 3,5-dinitro, isopropyl ester 3,5-(NO$_2$)$_2$C$_6$H$_3$CO$_2$-i-C$_3$H$_7$	254.20	nd (al)	....	122	....	....	al	B9^3, 1782
2681	Benzoic acid, 3,5-dinitro, tetrahydrofurfuryl ester .... 3,5-(NO$_2$)$_2$C$_6$H$_3$CO$_2$CH$_2$(C$_4$H$_7$O)	296.24	nd (al)	....	83-4	....	....	al	B17^4, 1106
2682	Benzoic acid, 2-ethoxy .... 2-C$_2$H$_5$OC$_6$H$_4$CO$_2$H	166.18	....	211-2^{39}	20.7	....	....	....	B10^3, 98
2683	Benzoic acid, (2-ethoxyethyl)ester or Ethylcellosolve benzoate............................ C$_6$H$_5$CO$_2$CH$_2$CH$_2$OCH$_2$CH$_3$	194.23	....	260-1^{739}	......	1.0585$^{25/25}$	1.4969^{25}	al, eth, ace, bz	B9^3, 533
2684	Benzoic acid, 2-ethoxy, ethyl ester 2-C$_2$H$_5$OC$_6$H$_4$CO$_2$C$_2$H$_5$	194.23	....	251, 180-5^{111}	......	1.005^{20}	....	al, eth	B10^3, 117
2685	Benzoic acid, 3-ethoxy .... 3-C$_2$H$_5$OC$_6$H$_4$CO$_2$H	166.18	nd (w)	sub	137	....	....	w, al, eth, bz	B10^3, 245
2686	Benzoic acid, 3-ethoxy, ethyl ester 3-C$_2$H$_5$OC$_6$H$_4$CO$_2$C$_2$H$_5$	194.23	....	264, 172-3^{50}	......	1.0725$^{20/20}$	....	al, eth	B10^3, 251
2687	Benzoic acid, 4-ethoxy .... 4-C$_2$H$_5$OC$_6$H$_4$CO$_2$H	166.18	nd (w)	....	198.5	....	....	al, eth, bz	B10^3, 282
2688	Benzoic acid, 4-ethoxy, ethyl ester 4-C$_2$H$_5$OC$_6$H$_4$CO$_2$C$_2$H$_5$	194.23	....	275, 148-9^{14}	......	1.076^{12}	....	al, eth	B10^3, 301
2689	Benzoic acid, 4-ethoxy-2-hydroxy or 4-Ethoxysalicylic acid.................................. 4-C$_2$H$_5$O-2-HOC$_6$H$_3$CO$_2$H	182.18	nd (w or bz)	....	154	....	....	al, eth, bz	B10, 379
2690	Benzoic acid, ethyl ester or Ethyl benzoate.............. C$_6$H$_5$CO$_2$C$_2$H$_5$	150.18	....	213, 87^{10}	-34.6	1.0468$^{20/4}$	1.5007^{20}	al, eth, ace, bz, peth	B9^3, 384
2691	Benzoic acid, 2-ethyl .... 2-C$_2$H$_5$C$_6$H$_4$CO$_2$H	150.18	nd (w)	259	68	1.0431$^{109/4}$	1.5099^{109}	al, eth	B9^3, 2425
2692	Benzoic acid, 3-ethyl .... 3-C$_2$H$_5$C$_6$H$_4$CO$_2$H	150.18	nd (w or dil al)	....	47	1.042^{100}	1.5345^{100}	al, eth	B9^3, 2429

No.	Name, Synonyms, and Formula	Mol. wt.	Color, crystalline form, specific rotation and λ_{max} (log ε)	b.p. °C	m.p. °C	Density	n_D	Solubility	Ref.
2693	Benzoic acid, 4-ethyl . . . . 4-$C_2H_5C_6H_4CO_2H$	150.18	pr (al), pr or lf (w)		113.5			al, bz, chl	B9³, 2430
2694	Benzoic acid, 2-ethylamino . . . . . 2-($C_2H_5NH)C_6H_4CO_2H$	165.19	pr or nd (dil al)		154			al, eth, ace	B14³, 897
2695	Benzoic acid, 3-ethylamino 3-($C_2H_5NH)C_6H_4CO_2H$	165.19	nd or pl (dil al)	sub	112			al, eth, ace	B14, 393
2696	Benzoic acid, 4-ethylamino . . . . 4-($C_2H_5NH)C_6H_4CO_2H$	165.19	cr (bz)		177-8			al, eth, ace, bz	B14³, 1086
2697	Benzoic acid, 2-fluoro 2-$FC_6H_4CO_2H$	140.12	nd (w)		1.265	$1.460^{25/4}$		al, eth, chl	B9³, 1324
2698	Benzoic acid, 3-fluoro 3-$FC_6H_4CO_2H$	140.12	lf (w)		124	$1.474^{25/4}$		eth	B9³, 1327
2699	Benzoic acid, 4-fluoro 4-$FC_6H_4CO_2H$	140.12	pr (w)		185	$1.479^{25/4}$		al, eth	B9³, 1327
2700	Benzoic acid, 2-formamido . . . . 2-($HCONH)C_6H_4CO_2H$	165.15	nd (w + 1)		169			al, eth	B14³, 921
2701	Benzoic acid, 2-formyl or Phthalaldehydic acid . . . . 2-($HCO)C_6H_4CO_2H$	150.14	lf (+ w)		98-9	1.404		w, al, eth	B10³, 2986
2702	Benzoic acid, 3-formyl or Isophthalaldehydic acid . . . . 3-($HCO)C_6H_4CO_2H$	150.14	nd (w)		175			al, eth, w	B10³, 2988
2703	Benzoic acid, 4-formyl or Terephthalaldehydic acid . . . . . . 4-($HCO)C_6H_4CO_2H$	150.14	nd (w)	sub	256			al, eth, chl	B10³, 2989
2704	Benzoic acid, hexyl ester or Hexylbenzoate $C_6H_5CO_2C_6H_{13}$	206.29	. . . .	272^{770}, 139-140⁸	113-7			al, ace	B9³, 398
2705	Benzoic acid, hydrazide or Benzoyl hydrazine . . . . $C_6H_5CONHNH_2$	136.17	pl (w)		113-7			w, al	B9³, 1312
2706	Benzoic acid, 2-hydrazino . . . . 2-$H_2NNHC_6H_4CO_2H$	152.16	nd (w)		250-1			w, al	B15³, 831
2707	Benzoic acid, 2-hydrazino, hydrochloride . . . . 2-$H_2NNHC_6H_4CO_2H \cdot HCl$	188.62	nd (w)		194-5d			w	B15², 295
2708	Benzoic acid, 3-hydrazino 3-$H_2NNHC_6H_4CO_2H$	152.16	pa ye lf (w)		186d				B15³, 836
2709	Benzoic acid, 4-hydrazino 4-$H_2NNHC_6H_4CO_2H$	152.16	ye nd or pl (w)		220-5d				B15³, 837
2710	Benzoic acid, 2-hydroxy or Salicylic acid . . . . 2-$HOC_6H_4CO_2H$	138.12	nd (w), nd pr (al)	211^{20}, sub	158	$.443^{20/4}$	1.565	al, eth, ace	B10³, 87
2711	Benzoic acid, 3-hydroxy 3-$HOC_6H_4CO_2H$	138.12	nd (w), pl, pr (al)		202-3			eth, ace	B10³, 2421
2712	Benzoic acid, 3-hydroxy, ethyl ester 3-$HOC_6H_4CO_2C_2H_5$	166.18	pl (bz)	295, 211^{65}	73.5			al, eth	B10³, 250
2713	Benzoic acid, 3-hydroxy, methyl ester . . . . 3-$HOC_6H_4CO_2CH_3$	152.16	nd (bz-peth)	280^{709}, 178^{17}	71.5			al, bz	B10³, 249
2714	Benzoic acid, 4-hydroxy 4-$HOC_6H_4CO_2H$	138.12	pr or pl (w, al), cr (dil al, ace)		214-5			al, eth, ace	B10³, 277
2715	Benzoic acid, 4-hydroxy, butyl ester 4-$HOC_6H_4CO_2C_4H_9$	194.23	. . . .		68-9			al	B10³, 307
2716	Benzoic acid, 4-hydroxy, ethyl ester 4-$HOC_6H_4CO_2C_2H_5$	166.18	cr (dil al)	297-8	116-8			al, eth	B10³300
2717	Benzoic acid, 4-hydroxy, methyl ester 4-$HOC_6H_4CO_2H_3$	152.16	nd (dil al)	270-80	131			al, eth, ace	B10³, 297
2718	Benzoic acid, 4-hydroxy, propyl ester . . . . 4-$HOC_6H_4CO_2C_3H_7$	180.21	pr (eth)		96-8	$1.0630^{102/4}$	1.5050^{102}	al, eth	B10³, 306
2719	Benzoic acid, 4-(α-hydroxybenzyl) 4-[$C_6H_5CH(OH)]C_6H_4CO_2H$	228.25	nd (w)		164-5			al, eth, w	B10³, 1184
2720	Benzoic acid, 2-hydroxyethyl . . . . 2-($HOCH_2CH_2)C_6H_4CO_2H$	194.23	. . . .	$260-1^{739}$		$1.0585^{25}_{/25}$	1.4969^{25}	al, eth, ace, bz	B10³, 572
2721	Benzoic acid, 2-hydroxyethyl ester or Ethylene glycol monobenzoate . . . . $C_6H_5CO_2CH_2CH_2OH$	166.18	. . . .	$150-1^{10}$	45			al	B9³, 532
2722	Benzoic acid, 2-(hydroxymethyl) 2-($HOCH_2)C_6H_4CO_2H$	152.16	nd (w)		128			al, eth, w	B10³, 500
2723	Benzoic acid, 2-iodo . . . . 2-$IC_6H_4CO_2H$	248.02	nd (w)	233 exp	163			al, eth	B9³, 1432
2724	Benzoic acid, 2-iodo, methyl ester 2-$IC_6H_4CO_2CH_3$	262.05	. . . . . .	$277-8^{720}$, 146^{16}			1.6052^{20}	al	B9³, 1434
2725	Benzoic acid, 2-iodo-4-nitro . . . . 4-NO_2-2-$IC_6H_3CO_2H$	293.02	pa ye pr (w)		146-7			al, eth	B9³, 1774

No.	Name, Synonyms, and Formula	Mol. wt.	Color, crystalline form, specific rotation and λ_{max} (log ε)	b.p. °C	m.p. °C	Density	n_D	Solubility	Ref.
2726	Benzoic acid, 3-iodo .... 3-IC$_6$H$_4$C$_2$H	248.02	mcl pr (ace)	sub	187-8	....	....	al	B9[3], 1437
2727	Benzoic acid, 3-iodo, methyl ester .... 3-IC$_6$H$_4$CO$_2$CH$_3$	262.05	nd (dil al)	276-7[739], 50[18]	54-5	....	....	al, eth, ace	B9[3], 1438
2728	Benzoic acid, 4-iodo .... 4-IC$_6$H$_4$CO$_2$H	248.02	mcl pr (dil al), lf (sub)	sub	270	2.184[20]	....	....	B9[3], 1442
2729	Benzoic acid, 4-iodo, methyl ester .... 4-IC$_6$H$_4$CO$_2$CH$_3$	262.05	nd (eth-al)	sub	114	....	....	al, eth	B9[3], 1443
2730	Benzoic acid, 4-iodo-2-nitro 2-NO$_2$-4-IC$_6$H$_3$CO$_2$H	293.02	ye lf or pr (dil al)	....	192-3	....	....	al, eth, bz	B9[2], 278
2731	Benzoic acid, 4-iodo-3-nitro 3-NO$_2$-4-IC$_6$H$_3$CO$_2$H	293.02	ye pr (al)	....	213	....	....	al	B9[2], 278
2732	Benzoic acid, 5-iodo-3-nitro 3-NO$_2$-5-IC$_6$H$_3$CO$_2$H	293.02	nd (al), pr (peth)	....	167	....	....	al, w	B9[2], 278
2733	Benzoic acid, 2-iodoso 2-IOC$_6$H$_4$CO$_2$H	264.02	lf (w)	....	223-5d	....	....	....	B9[3], 1433
2734	Benzoic acid, 3-iodoso 3-IOC$_6$H$_4$CO$_2$H	264.02	ye amor	....	175-80	....	....	....	B9, 365
2735	Benzoic acid, 4-iodoso 4-IOC$_6$H$_4$CO$_2$H	264.02	amor	....	212d	....	....	....	B9, 366
2736	Benzoic acid, 2-mercapto or Thiosalicylic acid .... 2-HSC$_6$H$_4$CO$_2$H	154.19	lf or nd (al, w, aa)	sub	168-9	....	....	al, eth, w, aa	B10[3], 212
2737	Benzoic acid, (2-methoxyethyl) ester .... C$_6$H$_5$CO$_2$CH$_2$CH$_2$OCH$_3$	180.21	....	254-6	....	1.0891[25/25]	1.5040[25]	al, eth, ace, bz	B9[3], 533
2738	Benzoic acid, 3-methoxy-2-methylamino or Damascenine 3-CH$_3$O-2-CH$_3$NHC$_6$H$_3$CO$_2$H	195.22	pr (al)	270[750]d, 147-8[10]	27-9	....	....	al, eth, bz, lig	B14[1], 654
2739	Benzoic acid, (2-methoxyphenyl) ester or Guaiacyl benzoate .... C$_6$H$_5$CO$_2$C$_6$H$_4$OCH$_3$-2	228.25	....	....	58	....	....	al, eth, ace, chl	B9[3], 551
2740	Benzoic acid, methyl ester or Methyl benzoate.......... C$_6$H$_5$CO$_2$CH$_3$	136.16	....	199.6, 96-8[24]	−12.3	1.0888[20/4]	1.5164[20]	al, eth, MeOH	B9[3], 381
2741	Benzoic acid, 3-methylbutyl ester or iso Pentyl benzoate .. C$_6$H$_5$CO$_2$(CH$_2$)$_2$CH(CH$_3$)$_2$	192.26	....	262.3, 133[14]	....	1.0040[20/4]	1.4950[20]	al, eth	B9[3]397
2742	Benzoic acid, methylene diester or Methanediol dibenzoate .... (C$_6$H$_5$CO$_2$)$_2$CH$_2$	256.26	nd or pr (eth)	225d	99	1.275[22]	....	eth, ace, bz	B9[3], 715
2743	Benzoic acid, 2-methyl or o-Toluic acid .... 2-CH$_3$C$_6$H$_4$CO$_2$H	136.15	pr or nd (w)	258-9[751]	107-8	1.062[115]	1.512[115]	al, eth, chl	B9[3], 2298
2744	Benzoic acid, 2-methylamino .... 2-CH$_3$NHC$_6$H$_4$CO$_2$H	151.17	pl (al or lig)	80[0.01]	179	....	....	al, eth, bz	B14[3], 895
2745	Benzoic acid, 2-methylamino, ethyl ester .... 2-CH$_3$NHC$_6$H$_4$CO$_2$C$_2$H$_5$	179.22	....	266, 141-3[15]	39	....	....	eth	B14[2], 213
2746	Benzoic acid, 2-methylamino, methyl ester .... 2-(CH$_3$NH)C$_6$H$_4$CO$_2$CH$_3$	165.09	cr (peth)	255, 130-1[13]	18-19	1.120[15]	1.5839[12]	al, eth	B14[3], 895
2747	Benzoic acid, 3-methylamino .... 3-(CH$_3$NH)C$_6$H$_4$CO$_2$H	151.16	pl (peth)	....	127	....	....	al, ace, bz, chl	B14[1], 559
2748	Benzoic acid, 3-methylamino, methyl ester .... 3-(CH$_3$NH)C$_6$H$_4$CO$_2$CH$_3$	165.19	cr (al)	....	72	....	....	al, eth	B14, 392
2749	Benzoic acid, 4-methylamino .... 4-(CH$_3$NH)C$_6$H$_4$CO$_2$H	151.16	nd (bz, w or dil al)	....	168	....	....	al, eth, w, bz	B14[3], 1081
2750	Benzoic acid, 4-methylamino, methyl ester .... 4-(CH$_3$NH)C$_6$H$_4$CO$_2$CH$_3$	165.19	pl (dil al or lig)	....	93.5	....	....	al, eth	B14[3], 1081
2751	Benzoic acid, 3-(3-methylbutoxy) 3-[(CH$_3$)$_2$CH(CH$_2$)$_2$O]C$_6$H$_4$CO$_2$H	208.24	cr (al)	....	74-5	....	....	....	B10[1], 64
2752	Benzoic acid, 4-(3-methylbutoxy) 4-[(CH$_3$)$_2$CH(CH$_2$)$_2$O]C$_6$H$_4$CO$_2$H	208.24	nd	....	141-2	....	....	....	B10[1], 70
2753	Benzoic acid, 2-(α naphthoyl) .... 2-(α-C$_{10}$H$_7$CO)C$_6$H$_4$CO$_2$H	276.30	pl (al-w)	....	176.4	....	....	al, ace, bz, chl	B10[3], 3426
2754	Benzoic acid, 2-(β-naphthoyl) 2-(β-C$_{10}$H$_7$CO)C$_6$H$_4$CO$_2$H	276.30	nd (to)	....	168	....	....	al, eth, ace, bz	B10[3], 3429
2755	Benzoic acid, α napthyl ester .... C$_6$H$_5$CO$_2$-α-C$_{10}$H$_7$	248.29	pl or pr (al-eth)	....	56	....	....	al, eth	B9[3], 491
2756	Benzoic acid, β-naphthyl ester .... C$_6$H$_5$CO$_2$-β-C$_{10}$H$_7$	248.29	nd or pr (al)	....	107	....	....	al	B9[3], 492
2757	Benzoic acid, 2-nitro or 2-Nitrobenzoic acid .... 2-NO$_2$C$_6$H$_4$CO$_2$H	167.12	tcl nd (w)	....	147-8	1.575[20/4]	....	al, eth, ace	B9[3], 1466

No.	Name, Synonyms, and Formula	Mol. wt.	Color, crystalline form, specific rotation and λ_{max} (log ε)	b.p. °C	m.p. °C	Density	n_D	Solubility	Ref.
2758	Benzoic acid, (2-nitrobenzyl) ester .... $C_6H_5CO_2(CH_2C_6H_4NO_2-2)$	257.25	nd (dil al)	....	101-2			al, eth, bz, aa	B9, 121
2759	Benzoic acid, 2-nitro, ethyl ester .... $2-NO_2C_6H_4CO_2C_2H_5$	195.18	tcl (dil al)	275, 173[18]	30			al, eth	B9[3], 1469
2760	Benzoic acid, 3-nitro or 3-Nitrobenzoic acid .... $3-NO_2C_6H_4CO_2H$	167.12	mcl pr (w)	....	140-2	1.494[20]		al, eth, ace, chl	B9[3], 1489
2761	Benzoic acid, (3-nitrobenzyl) ester .... $C_6H_5CO_2(CH_2C_6H_4NO_2-3)$	257.25		....	71-2			al, eth	B9[3], 431
2762	Benzoic acid, 4-nitro .... $4-NO_2C_6H_4CO_2H$	167.12	mcl lf (w)	sub	242	1.610[20]		al, eth, chl	B9[3], 1537
2763	Benzoic acid, (4-nitrobenzyl) ester .... $C_6H_5CO_2(CH_2C_6H_4NO_2-4)$	257.25		....	94-5			al, eth	B9[3], 431
2764	Benzoic acid, 2-nitroso .... $2-ONC_6H_4CO_2H$	151.12	cr (al or aa)	....	210d				B9[3], 1462
2765	Benzoic acid, 3-nitroso .... $3-ONC_6H_4CO_2H$	151.12	cr		23d			al	B9, 369
2766	Benzoic acid, 4-nitroso .... $4-ONC_6H_4CO_2H$	151.12	ye pw		<350			al	B9[3], 1465
2767	Benzoic acid, 4-octyl .... $4-C_8H_{17}C_6H_4CO_2H$	234.34	lf (al)		139			al	B9[3], 2611
2768	Benzoic acid, pentachloro .... $C_6Cl_5CO_2H$	294.35	nd or pl (bz or dil aa)	sub (vac)	208			al, to	B9[3], 1383
2769	Benzoic acid, pentamethyl .... $C_6(CH_3)_5CO_2H$	192.24	nd (w), lf or nd (dil al)	sub	210.5			al	B9[3], 2564
2770	Benzoic acid, per or Perbenzoic acid .... C_6H_5COOOH	138.12	mcl pl (peth)	97-110[11] sub	41-3			al, eth, ace, bz	B9[3], 1049
2771	Benzoic acid, 2-phenoxy .... $2-C_6H_5OC_6H_4CO_2H$	214.22	lf (dil al)	355d	114.4			al, eth, chl	B10[3], 99
2772	Benzoic acid, 3-phenoxy .... $3-C_6H_5OC_6H_4CO_2H$	214.22			145			al, eth	B10[2], 247
2773	Benzoic acid, 4-phenoxy .... $4-C_6H_5OC_6H_4CO_2H$	214.22			161			al, eth	B10[3], 289
2774	Benzoic acid, phenyl ester or Phenylbenzoate .... $C_6H_5CO_2C_6H_5$	198.22	mcl pr (eth-al)	314	71	1.235[20/4]		al, eth	B9[3], 415
2775	Benzoic acid, (1-phenylethyl) ester .... $C_6H_5CO_2CH(C_6H_5)CH_3$	226.28		189[21]	....	1.1108[18]	1.5588[21]	al	B9[3], 431
2776	Benzoic acid, 2-phoshono .... $2-(HO_2P)C_6H_4CO_2H$	202.11	nd (w or al)	....	>300			w	B16, 820
2777	Benzoic acid, propyl ester or Propyl benzoate .... $C_6H_5CO_2C_3H_7$	164.21	....	211	-51.6	1.0230[20/4]		al, eth	B9[3], 389
2778	Benzoic acid, iso-propyl ester or Isopropyl benzoate .... $C_6H_5CO_2CH(CH_3)_2$	164.21		218		1.0172[15/15]	1.4890[20]	al, eth, ace	B9[3], 391
2779	Benzoic acid, 2-propyl .... $2-C_3H_7C_6H_4CO_2H$	164.21	lf (dil al)	272[739], 164-5[20]	58			al, eth, w	B9[1], 213
2780	Benzoic acid, 2-iso-propyl or o-Cuminic acid .... $2-[(CH_3)_2CH]C_6H_4CO_2H$	164.21	pr (w or peth)	160-1[25]	64			al, eth, bz, peth	B9[3], 2481
2781	Benzoic acid, 4-propyl .... $4-C_3H_7C_6H_4CO_2H$	164.21	pr or lf (w)		141			al, eth, w, bz, lig	B9[3], 2479
2782	Benzoic acid, 4-iso-propyl or Cumic acid .... $4-[(CH_3)_2CH]C_6H_4CO_2H$	164.21	tcl pl (al)	sub	117-8	1.162[4]		al, eth, peth	B9[3], 2482
2783	Benzoic acid, 2,3,4,5-tetrachloro .... $2,3,4,5-Cl_4C_6HCO_2H$	259.91	nd (al), cr (ace-w)		194-5			al, eth	B9[3], 1381
2784	Benzoic acid, tetrahydrofurfuryl ester .... $C_6H_5CO_2(C_5H_9O)$	206.24		300-2[750], 138-40[2]		1.137[20/4]		al, eth, chl	B17[4], 1105
2785	Benzoic acid, 2,3,4,5-tetrahydroxy .... $2,3,4,5-(HO)_4C_6HCO_2H$	186.12	pr	....	84-5			w	C50, 14644
2786	Benzoic acid, thiolo .... $C_6H_5C(=O)SH$	138.19	ye pl (aa)	85-7[10]	24		1.6040[20]	al, eth, ace, bz	B9[3], 1961
2787	Benzoic acid, 2-tolylester .... $C_6H_5CO_2C_6H_4CH_3-2$	212.25		307-8[728], 154-6[8/5]		1.114[19]		al, eth	B9[3], 423
2788	Benzoic acid, 3-tolylester .... $C_6H_5CO_2C_6H_4CH_3-3$	212.25		314, 168-70[8]	55-6			al, eth	B9[3], 424
2789	Benzoic acid, 4-tolylester .... $C_6H_5CO_2C_6H_4CH_3-4$	212.25	pl (eth-al)	316	71.5			al, eth	B9[3], 427
2790	Benzoic acid, 2-(2-tolyl) .... $2-(2-CH_3C_6H_4)C_6H_4CO_2H$	240.26	nd (w + l), cr (bz)	....	130-2 (anh)			al, eth, w	B9[3], 3323
2791	Benzoic acid, 2-(3-tolyl) .... $2-(3-CH_3C_6H_4)C_6H_4CO_2H$	240.26	nd (w + l)	....	162			eth, ace, w	B9[3], 3326

No.	Name, Synonyms, and Formula	Mol. wt.	Color, crystalline form, specific rotation and λ_{max} (log ε)	b.p. °C	m.p. °C	Density	n_D	Solubility	Ref.
2792	Benzoic acid, 2-(4-tolyl) $2\text{-}(4\text{-}CH_3C_6H_4)C_6H_4CO_2H$	240.26	pr (+ lw, al-to), nd (al)	....	146	....	....	al, eth, ace, bz	B9, 677
2793	Benzoic acid, 2-(4-tolyl), methyl ester $2\text{-}(4\text{-}CH_3C_6H_4)C_6H_4CO_2CH_3$	254.29	pl (MeOH)	....	66			al, bz	B10, 759
2794	Benzoic acid, 4-(2-tolyl) $4\text{-}(2\text{-}CH_3C_6H_4)C_6H_4CO_2H$	240.26	....	....	177			....	B9³, 3328
2795	Benzoic acid, 4-(4-tolyl) $4\text{-}(4\text{-}CH_3C_6H_4)C_6H_4CO_2H$	240.26	nd (MeOH or ace)	....	228			al, ace	B9, 677
2796	Benzoic acid, 2,3,5-triamino $2,3,5\text{-}(H_2N)_3C_6H_2CO_2H$	167.13	cr (w)		d			w	B14, 455
2797	Benzoic acid, 3,4,5-triamino $3,4,5\text{-}(H_2N)_3C_6H_2CO_2H$	167.13	nd (w + ½)		d			w	B14, 455
2798	Benzoic acid, 2,3,4-tribromo $2,3,4\text{-}Br_3C_6H_2CO_2H$	358.83	nd (bz)		197-8			al, eth, bz	B9¹, 147
2799	Benzoic acid, 2,3,5-tribromo $2,3,5\text{-}Br_3C_6H_2CO_2H$	358.83	nd (al)		193-4			al, eth, ace, bz	B9¹, 147
2800	Benzoic acid, 2,4,5-tribromo $2,4,5\text{-}Br_3C_6H_2CO_2H$	358.83	nd (al or bz)		195-6			al, eth	B9³, 1430
2801	Benzoic acid, 2,4,6-tribromo $2,4,6\text{-}Br_3C_6H_2CO_2H$	358.83	pr (w)		198			al, eth, bz	B9³, 1430
2802	Benzoic acid, 3,4,5-tribromo $3,4,5\text{-}Br_3C_6H_2CO_2H$	358.83	nd (bz or al)		240			al, eth	B9³, 1431
2803	Benzoic acid, 2,3,4-trichloro $2,3,4\text{-}Cl_3C_6H_2CO_2H$	225.46	nd (w)		187-8			....	B9, 345
2804	Benzoic acid, 2,3,5-trichloro $2,3,5\text{-}Cl_3HC_6H_2CO_2H$	225.46	nd (w)		163			al, eth, ace, bz	B9³, 1380
2805	Benzoic acid, 2,3,6-trichloro $2,3,6\text{-}Cl_3C_6H_2CO_2H$	225.46	....		124-5			eth	B9³, 1380
2806	Benzoic acid, 2,4,5-trichloro $2,4,5\text{-}Cl_3C_6H_2CO_2H$	225.46	nd (w or sub)	sub	168			al, eth, w	B9³, 1380
2807	Benzoic acid, 2,4,6-trichloro $2,4,6\text{-}Cl_3C_6H_2CO_2H$	225.46	nd (w)		164			al, eth, chl	B9³, 1381
2808	Benzoic acid, 3,4,5-trichloro $3,4,5\text{-}Cl_3C_6H_2CO_2H$	225.46	nd (dil al)		210			al, eth, ace, bz	B9², 230
2809	Benzoic acid, 2,3,4-trihydroxy $2,3,4\text{-}(HO)_3C_6H_2CO_2H$	170.12	nd (+ w)	sub	207-8d			al, eth, ace	B10³, 2057
2810	Benzoic acid, 2,3,4-trihydroxy, ethyl ester $2,3,4\text{-}(HO)_3C_6H_2CO_2C_2H_5$	198.18	cr (w + l)	....	102, (anh), 86 (+ w)			al, eth, w	B10, 467
2811	Benzoic acid, 2,3,4-trihydroxy, methyl ester $2,3,4\text{-}(HO)_3C_6H_2CO_2CH_3$	184.16	nd (w + 2)		151-2			al, w	B10³, 2058
2812	Benzoic acid, 2,4,5-trihydroxy $2,4,5\text{-}(HO)_3C_6H_2CO_2H$	170.12	nd (w + ½), nd		217-8d			al, w	B10³, 2065
2813	Benzoic acid, 2,4,6-trihydroxy $2,4,6\text{-}(HO)_3C_6H_2CO_2H$	170.12	(w + l)		100d			al, eth, w	B10², 334
2814	Benzoic acid, 2,4,6-trihydroxy, ethyl ester $2,4,6\text{-}(HO)_3C_6H_2CO_2C_2H_5$	198.18	pr or nd (w + l), pr (lig)		129			al, eth, w, bz	B10¹, 236
2815	Benzoic acid, 2,4,6-trihydroxy, methyl ester $2,4,6\text{-}(HO)_3C_6H_2CO_2CH_3$	184.16	cr (dil al)		174-6			al, eth	B10³, 2069
2816	Benzoic acid, 3,4,5-trihydroxy or Gallic acid $3,4,5\text{-}(HO)_3C_6H_2CO_2H$	170.12	pr (w + l)		253d	$1.694^{6/4}$	....	al, ace	B10³, 2070
2817	Benzoic acid, 3,4,5-trihydroxy, ethyl ester $3,4,5\text{-}(HO)_3C_6H_2CO_2C_2H_5$	198.18	mcl pr (w + 2½), nd (chl)		160-2			al, eth, w	B10², 343
2818	Benzoic acid, 3,4,5-trihydroxy, methyl ester or Gallicin $3,4,5\text{-}(HO)_3C_6H_2CO_2CH_3$	184.16	mcl pr (MeOH)		202			al	B10³, 2076
2819	Benzoic acid, 3,4,5-trihydroxy, propylester $3,4,5\text{-}(HO)_3C_6H_2CO_2C_3H_7$	212.21	nd (w)		130			....	B10³, 2078
2820	Benzoic acid, 3,4,5-trihydroxy, iso-propylester $3,4,5\text{-}(HO)_3C_6H_2CO_2CH(CH_3)_2$	212.21	....		123-4			w, al, eth	B10², 343
2821	Benzoic acid, 2,3,5-triiodo $2,3,5\text{-}I_3C_6H_2CO_2H$	499.81	pr (al)	....	224-6			al, eth	B9³, 1456
2822	Benzoic acid, 2,4,5-triiodo $2,4,5\text{-}I_3C_6H_2CO_2H$	499.81	nd (al)	....	248			eth, al	B9¹, 150

No.	Name, Synonyms, and Formula	Mol. wt.	Color, crystalline form, specific rotation and λ_{max} (log ε)	b.p. °C	m.p. °C	Density	n_D	Solubility	Ref.
2823	Benzoic acid, 3,4,5-triiodo . . . . 3,4,5-I$_3$C$_6$H$_2$CO$_2$H	499.81	pr (al)	. . . .	292-3	. . . .	. . . .	al	B9^3, 1457
2824	Benzoic acid, 2,3,4-trimethoxy . . . . 2,3,4-(CH$_3$O)$_3$C$_6$H$_2$CO$_2$H	212.21	cr (w or peth)	. . . .	100			w, al, eth	B10^3, 2058
2825	Benzoic acid, 2,4,5-trimethoxy or Asaronic acid . . . . 2,4,5-(CH$_3$O)$_3$C$_6$H$_2$CO$_2$H	212.21	nd (al or bz-peth)	300	144			w, al, bz, peth	B10^3, 2065
2826	Benzoic acid, 3,4,5-trimethoxy . . . . 3,4,5-(CH$_3$O)$_3$C$_6$H$_2$CO$_2$H	212.21	mcl nd (w)	225-7^{10}	171-2	. . . .		al, eth, chl	B10^3, 2073
2827	Benzoic acid, 2,3,4-trimethyl or Prehnitylic acid . . . . 2,3,4-(CH$_3$)$_3$C$_6$H$_2$CO$_2$H	164.21	pr (al)		167.5			w, al, eth	B9^3, 2489
2828	Benzoic acid, 2,3,5-trimethyl or α-Isodurylic acid . . . . 2,3,5-(CH$_3$)$_3$C$_6$H$_2$CO$_2$H	164.21	pl (lig)	. . . .	127			al	B9^3, 2489
2829	Benzoic acid, 2,3,6-trimethyl 2,3,6-(CH$_3$)$_3$C$_6$H$_2$CO$_2$H	164.21	nd (w or peth)		110-1			w, al, eth	B9^3, 2489
2830	Benzoic acid, 2,4,5-trimethyl or Durylic acid . . . . 2,4,5-(CH$_3$)$_3$C$_6$H$_2$CO$_2$H	164.21	nd (bz)		152-3			al, eth	B9^3, 2501
2831	Benzoic acid, 2,4,6-trimethyl or Mesitoic acid . . . . 2,4,6-(CH$_3$)$_3$C$_6$H$_2$CO$_2$H	164.21	pr (lig)		155			al, eth, ace, chl	B9^3, 2489
2832	Benzoic acid, 3,4,5-trimethyl or α-Isodurylic acid . . . . 3,4,5-(CH$_3$)$_3$C$_6$H$_2$CO$_2$H	164.21	nd (w)		215-6			al, eth	B9, 554
2833	Benzoic acid, 2,3,6-trinitro . . . . 2,3,6-(NO$_2$)$_3$C$_6$H$_2$CO$_2$H	257.12	wh nd (w + 2)		160d, 55 (+2w)			al	B9^1, 168
2834	Benzoic acid, 2,4,5-trinitro . . . . 2,4,5-(NO$_2$)$_3$C$_6$H$_2$CO$_2$H	257.12	ye cr or pl (w)		194.5d			al, eth, bz, w	B9^3, 1956
2835	Benzoic acid, 2,4,6-trinitro 2,4,6-(NO$_2$)$_3$C$_6$H$_2$CO$_2$H	257.12	rh (w)		228d			al, eth, ace	B9^3, 1956
2836	Benzoic acid, 3,4,5-trinitro . . . . 3,4,5-(NO$_2$)$_3$C$_6$H$_2$CO$_2$H	257.12	ye nd (eth + l)		168d			eth	B9^1, 168
2837	Benzoin (d) or α-Hydroxybensyl phenyl ketone . C$_6$H$_5$CH(OH)COC$_6$H$_5$	212.25	[α]$^{15}_D$ +92.8 (Py. c=1)		133-4			al, ace, Py	B8^3, 1272
2838	Benzoin (dl) C$_6$H$_5$CH(OH)COC$_6$H$_5$	212.25	. . . .	344^{768}, 194^{12}	137	1.310$^{20/4}$		al, chl, aa	B8^3, 1273
2839	Benzoin (l) C$_6$H$_5$CH(OH)COC$_6$H$_5$	212.25	nd (MeOH), [α]$^{12}_D$ −117.5 (ace, c=1.25)		133-4			al, bz	B8^3, 1272
2840	Benzoin, acetate (dl) C$_6$H$_5$CH(O$_2$CCH$_3$)COC$_6$H$_5$	254.29	pr or pl (eth)		83			al, eth	B8^2, 196
2841	Benzoin, 4,4'-dimethoxy or Anisoin . . . . 4-CH$_3$OC$_6$H$_4$CH(OH)COC$_6$H$_4$OCH$_3$-4	272.30	pr (dil al)		113			al, ace	B8^3, 3655
2842	Benzoin, ethyl ether C$_6$H$_5$CH(COC$_2$H$_5$)COC$_6$H$_5$	240.31	nd (lig)	194-5^{20}	62	1.1016$^{17/4}$	1.5727^{17}	al, eth, bz, lig	B8^2, 195
2843	Benzoin hydrazone (dl) . . . . C$_6$H$_5$CH(OH)C(=NNH$_2$)C$_6$H$_5$	226.28	pr (al)		75			al	B8, 176
2844	Benzoin methyl ether (dl) . . . . C$_6$H$_5$CH(OCH$_3$)COC$_6$H$_5$	226.28	nd (lig)	188-9^{15}	49-50	1.1278$^{14/4}$	. . . .	al, eth, bz	B8^2, 195
2845	Benzoin oxime (l) C$_6$H$_5$CH(OH)C(=NOH)C$_6$H$_5$	227.27	amor or pr, (bz) [α]$^{24}_D$ -3.2 (chl, c=0.85)		163-4			al, eth, ace	B8, 167
2846	Benzoin oxime (dl, anti) C$_6$H$_5$CH(OH)C(=NOH)C$_6$H$_5$	227.27	pr (bz)		151-2			al, eth, ace	B8^2, 196
2847	Benzoin oxime (dl, syn) C$_6$H$_5$CH(OH)C(=NOH)C$_6$H$_5$	227.27	pr (eth)		99			al, eth, ace	B8^2, 196
2848	Benzonitrile or Benzoic acid nitrile . . . . C$_6$H$_5$CN	103.13		190.7	−13	1.0102$^{15/15}$	1.5289^{20}	**al, eth,** ace, bz	B9^3, 1255
2849	Benzonitrile, 2-amino 2-H$_2$NC$_6$H$_4$CN	118.14		263^{751}	51			al, eth, ace, bz, chl	B14^2, 210
2850	Bensonitrile, 3-amino 3-H$_2$NC$_6$H$_4$CN	118.14		288-90	53-4			al, eth, ace, chl	B14^3, 1001
2851	Benzonitrile, 4-amino 4-H$_2$NC$_6$H$_4$CN	118.14			86			al, eth, ace, bz, aa	B14^3, 1079
2852	Benzonitrile, 2-bromo 2-BrC$_6$H$_4$CN	182.02	nd (w)	251-3^{754}	55.5			al	B9^3, 1387
2853	Benzonitrile, 3-bromo 3-BrC$_6$H$_4$CN	182.03	(al)	225	39-40	. . . .		al, eth	B9^3, 1399

No.	Name, Synonyms, and Formula	Mol. wt.	Color, crystalline form, specific rotation and λ_{max} (log ε)	b.p. °C	m.p. °C	Density	n_D	Solubility	Ref.
2854	Benzonitrile, 4-bromo 4-BrC₆H₄CN	182.03	nd (w or al)	235-7	114	....	....	al, eth	B9³, 1420
2855	Benzonitrile, 2-chloro 2-ClC₆H₄CN	137.57	nd	232	43-6			al, eth	B9³, 1339
2856	Benzonitrile, 3-chloro 3-ClC₆H₄CN	137.57	....	99-100¹⁵	40-2			al, eth	B9³, 1349
2857	Benzonitrile, 4-chloro 4-ClC₆H₄CN	137.57	nd (al)	223⁷⁵⁰, 95⁵	94-6			al, eth, bz, chl	B9³, 1366
2858	Benzonitrile, 3 chloromethyl or 3-Chloromethyl benzonitrile 3-ClCH₂C₆H₄CN	151.60	pr (al)	258-60	67				B9³, 2328
2859	Benzonitrile, 4-chlormethyl or 4-Chloromethyl benzonitrile 4-ClCH₂C₆H₄CN	151.60	pr (al)	263⁷⁵⁶	79.5				B9³, 2357
2860	Benzonitrile, 2,4-dimethoxy or 2,4-Dimethoxy benzonitrile 2,4-(CH₃O)₂C₆H₃CN	163.18	cr (al, lig)	....	96			aa	B10¹, 1366
2861	Benzonitrile, 2,5-dimethoxy 2,5-(CH₃O)₂C₆H₃CN	163.18	nd (al)		82			al, bz, chl	B10¹, 1388
2862	Benzonitrile, 2,6-dimethoxy or 2,6-Dimethoxy benzonitrile 2,6-(CH₃O)₂C₆H₃CN	163.18	nd or pl	310	118-20			al, ace, bz, chl	B10², 260
2863	Benzonitrile, 3,4-dimethoxy or Veratronitrile 3,4-(CH₃O)₂C₆H₃CN	163.18	nd (w)	....	67-8			bz, al, w	B10², 264
2864	Benzonitrile, 2-ethoxy 2-C₂H₅OC₆H₄CN	147.18	....	260.7, 153¹⁵	5			al, eth, lig	B10, 97
2865	Benzonitrile, 4-ethoxy 4-C₂H₅OC₆H₄CN	147.18	nd (lig)	258	61-2			al, eth, lig	B10¹, 344
2866	Benzonitrile, 4-fluoro or 4-Fluoro benzonitrile 4-FC₆H₄CN	121.12	nd (peth)	188.8	34.8	1.1070⁵⁵	1.4925⁵⁵	peth	B9², 221
2867	Benzonitrile, 3-formyl or 3-Cyanobenzaldehyde 3-HCOC₆H₄CN	131.14		210	79-81			al, eth, w, chl	B10, 671
2868	Benzonitrile, 4-formyl or 4-Cyanobenzaldehyde 4-HCOC₆H₄CN	131.14	nd (w), pr (eth, or dil al)	133¹²	101-2	....		al, eth, w, chl	B10¹, 2990
2869	Benzonitrile, 2-hydroxy 2-HOC₆H₄CN	119.12	....	149¹⁴	98	1.1052¹⁰⁰⁄⁴	1.5372¹⁰⁰	al, eth, bz, chl	B10², 60
2870	Benzonitrile, 3-hydroxy 3-HOC₆H₄CN	119.12	pr (al, eth), lf (w)	....	83-4			al, eth, bz, chl	B10³, 255
2871	Benzonitrile, 4-hydroxy 4-HOC₆H₄CN	119.12	lf (w)		113			al, eth, chl	B10³, 344
2872	Benzonitrile, 2-methoxy 2-CH₃OC₆H₄CN	133.15	....	255-6, 146²⁰	24.5	1.1063²⁰·⁵⁄⁴	....	al, eth	B10²,,60
2873	Benzonitrile, 4-methoxy 4-CH₃OC₆H₄CN	133.15	nd (w), lf (al)	256-7, 106-8⁶	61-2			al, eth, bz	B10², 101
2874	Benzonitrile, 2-methyl 2-CH₃C₆H₄CN	117.15	....	205, 90¹⁵	-13.5	0.9955²⁰⁄⁴	1.5279²⁰	al, eth	B9³, 2307
2875	Benzonitrile, 2-methyl-5-nitro 5-NO₂-2CH₃C₆H₃CN	162.15	nd (95% al)	174-5¹⁸	106			w, al, eth, ace, bz	B9³, 2314
2876	Benzonitrile, 3-methyl 3-CH₃C₆H₄CN	117.15	....	213, 84.5¹⁰	-23	1.0316²⁰⁄⁴	1.5252²⁰	al, eth	B9³, 2324
2877	Benzonitrile, 4-methyl 4-CH₃C₆H₄CN	117.15	....	217.6, 91¹¹	29.5	0.9805³⁰⁄³⁰		al, eth	B9³, 2348
2878	Benzonitrile, 3-(trifluoromethyl) 3-CF₃C₆H₄CN	171.15		189	14.5	1.28126²⁰	1.4508²⁰	....	B9³, 2327
2879	3,4-Benzophenanthrene or Benzo (c) phenanthrene C₁₈H₁₂	228.29	....		68				B9², 2379
2880	9,10-Benzophenanthrene or Triphenylene C₁₈H₁₂	228.30	nd (al, chl, bz)	425	199			al, bz, chl	B9³, 2380
2881	1,2-Benzophenazine C₁₆H₁₀N₂	230.26	pr (al), ye nd (bz)	> 360	142.5			al, eth, ace, bz, aa	B23², 259
2882	Benzophenone or Diphenyl ketone C₆H₅COC₆H₅	182.21	(α) rh pr (al, eth); (β) mcl pr	305.9	(α) 48.1, (β) 26	(α) 1.146²⁰, (β) 1.1076	α 1.6077¹⁹ β 1.6059²³	al, eth, ace, bz	B7³, 2048
2883	Benzophenone, 2-amino 2-H₂NC₆H₄COC₆H₄	197.24	pa ye lf or pr al	....	110-1			al, eth	B14³, 213
2884	Benzophenone, 2-amino-5-chloro 5-Cl-2-H₂NC₆H₃COC₆H₅	231.68	....		100-1			w, al, chl, peth	B14³, 214

No.	Name, Synonyms, and Formula	Mol. wt.	Color, crystalline form, specific rotation and λ_{max} (log ε)	b.p. °C	m.p. °C	Density	n_D	Solubility	Ref.
2885	Benzophenone, 2-amino-4′-methyl 2-H₂NC₆H₄CO(C₆H₃CH₃-4)	211.27	ye pr or pl (al)	....	96			al, eth, bz	B14³, 248
2886	Benzophenone, 2-amino-5-methyl 2-H₂N-5-CH₃C₆H₃COC₆H₅	211.27	ye nd or pl (al)	....	66			al, eth, ace, aa, lig	B14³, 246
2887	Benzophenone, 2-amino-5-nitro 5-NO₂-2H₂NC₆H₃COC₆H₅	242.23			161.5			al, aa	B14,79
2888	Benzophenone, 3-amino 3-H₂NC₆H₄COC₆H₅	197.24	ye nd (w)					al, eth	B14¹, 388
2889	Benzophenone, 3-amino-4-methyl 3-H₂N-4-CH₃C₆H₃COC₆H₅	211.27	pa ye nd (MeOH)		109			al, eth, ace, bz	B14³, 247
2890	Benzophenone, 3-amino-4′-methyl 3-H₂NC₆H₄CO(C₆H₃CH₃-4)	211.27	pr (al)		111			al, eth	B14, 107
2891	Benzophenone, 4-amino 4-H₂NC₆H₄COC₆H₅	197.24	lf (dil al)		124			al, eth, aa	B14³, 217
2892	Benzophenone, 4-amino-3-methyl 4-H₂N-3-CH₃C₆H₃COC₆H₅	211.27	ye pr (w)		112			al, eth	B14³, 245
2893	Benzophenone, 4-amino-4′-methyl 4-H₂NC₆H₄CO(C₆H₃CH₃-4)	211.27	nd (bz)		186-7			al, eth, chl	B14³, 248
2894	Benzophenone, 4,4′-bis(diethylamino) [4-(C₂H₅)₂NC₆H₄]₂CO	324.47	lf (al)		95-6				B14², 59
2895	Benzophenone, 4,4′-bis(dimethylamino) or Michler's ketone [4-(CH₃)₂NC₆H₄]₂CO	268.36	lf (al), nd (bz)	>360d	179			bz, Py	B14³, 226
2896	Benzophenone, 2-bromo 2-BrC₆H₄COC₆H₅	261.13	pl (al), nd (lig)	343	42			ace, al, lig	B7³, 2079
2897	Benzophenone, 3-bromo 3-BrC₆H₄COC₆H₅	261.13	nd (al)	185-7⁵	81			al	B7³, 2079
2898	Benzophenone, 4-bromo 4-BrC₆H₄COC₆H₅	261.13	lf (al)	330⁸¹	82.5				B7³, 2079
2899	Benzophenone, 5-tert-butyl-2-methoxy 5-t-C₄H₉-2-CH₃OC₆H₃COCH₅	268.35			62-3			chl	B8,187
2900	2-Benzophenone Carboxylic acid 2-HO₂CC₆H₄COC₆H₅	226.23			127-9			al, eth, bz	B10³, 3289
2901	2-Benzophenone carboxylic acid, ethyl ester 2-(C₂H₅O₂C)C₆H₄COC₆H₅	254.28			58	1.221⁶⁴′⁴	1.560⁶⁴	al, eth	B10², 517
2902	2-Benzophenone carboxylic acid, methyl ester 2-(CH₃O₂C)C₆H₄COC₆H₅	240.26		350-2	52	1.1903¹⁹′⁴	1.591²⁰	al, eth	B10², 517
2903	Benzophenone, 2-chloro 2-ClC₆H₄COC₆H₅	216.67	pl (chl-lig)	330,185-8¹³	52-6				B7³, 2071
2903a	Benzophenone, 2-chloro-3,5-dinitro 2-Cl-3,5-(NO₂)₂C₆H₂COC₆H₅	306.67	ye nd (aa)		148			chl, aa	B7, 428
2904	Benzophenone, 3-chloro 3-ClC₆H₄COC₆H₅	216.67	nd		82-3			bz	B7³, 2071
2905	Benzophenone, 4-chloro 4-ClC₆H₄COC₆H₅	216.67	nd (al)	332⁷⁷¹	77-8			w, al, eth	B7³, 2072
2906	Benzophenone, 2,2′-diamino (2H₂NC₆H₄)₂CO	212.25	lf (dil al), pr (bz)		134-5			al, eth	B14,87
2907	Benzophenone, 3,3′-diamino (3-H₂NC₆H₄)₂CO	212.25	nd (al)	285¹¹	173-4			al, eth	B14³, 225
2908	Benzophenone, 4,4′-diamino (4-H₂NC₆H₄)₂CO	212.25	nd (al)		244-5			al, eth	B14³, 226
2909	Benzophenone, 3,3′-dibromo (3-BrC₆H₄)₂CO	340.01			141			al, eth	B7², 361
2910	Benzophenone, 4,4′-dibromo (4-BrC₆H₄)₂CO	340.04	pl (al)	395	177			al, ace, bz, chl	B7³, 2081
2911	Benzophenone, 4,4′-dicarboxylic acid (4-HO₂CC₆H₄)₂CO	270.25	nd (al)	sub	>360			aa	B10³, 4010
2912	Benzophenone, 2,4′-dichloro 2-ClC₆H₄CO(C₆H₄Cl-4)	251.11		214-5²²	67	1.393¹⁴		al, chl	B7³, 2075
2913	Benzophenone, 3,3′-dichloro (3-ClC₆H₄)₂CO	251.11		160-6²	124			al, eth	B7³, 2076
2914	Benzophenone, 3,4′-dichloro 3-ClC₆H₄CO(C₆H₄Cl-4)	251.11			112-3				B7³, 2076
2915	Benzophenone, 4,4′-dichloro (4-ClC₆H₄)₂CO	251.13	pl (al)	353⁷⁵⁷	147-8			al, eth, ace, chl, aa	B7³, 2076
2916	Benzophenone, 3,5-dichloro-2-hydroxy 3,5-Cl₂-2HOC₆H₂COC₆H₅	267.11			116			al, bz	B8³, 1233

No.	Name, Synonyms, and Formula	Mol. wt.	Color, crystalline form, specific rotation and λ_{max} (log ε)	b.p. °C.	m.p. °C.	Density	n_D	Solubility	Ref.
2917	Benzophenone, 2,2'-dihydroxy (2-HOC$_6$H$_4$)$_2$CO	214.22	lf or pr (lig)	330-40	59.5			al, eth, chl	B8^3, 2644
2918	Benzophenone, 2,3'-dihydroxy 2-HOC$_6$H$_4$CO(C$_6$H$_4$OH-3)	214.22	nd (w)		126			al, eth	B8, 315
2919	Benzophenone, 2,4-dihydroxy 2,4-(HO)$_2$C$_6$H$_3$COC$_6$H$_5$	214.22	nd (w)		144			al, eth, aa	B8^3, 2640
2920	Benzophenone, 2,4-dihydroxy--6-methoxy or Isocotoin 2,4-(HO)$_2$-6-CH$_3$OC$_6$H$_2$COC$_6$H$_5$	244.25	ye nd (lig)		162			w, ace	B8^2, 467
2921	Benzophenone, 2,4'-dihydroxy 2-HOC$_6$H$_4$CO(C$_6$H$_4$OH-4)	214.22	pl (w)		150-1			al, eth, bz	B8^3, 2646
2922	Benzophenone, 2,5-dihydroxy 2,5-(HO)$_2$C$_6$H$_3$COC$_6$H$_5$	214.22	ye nd (dil al)		125-6			al, eth, bz, w	B8^3, 2643
2923	Benzophenone, 2,6-dihydroxy-4-methoxy or Cotoin 2,6-(HO)$_2$-4-CH$_3$OC$_6$H$_2$COC$_6$H$_5$	244.25	yesh pr (chl), lf or nd (w)		130-1			al, eth, ace, bz	B8^3, 3638
2924	Benzophenone, 3,3'-dihydroxy (3-HOC$_6$H$_4$)$_2$CO	214.22	nd (w)		170			al, w	B8^3, 2647
2925	Benzophenone, 3,4'-dihydroxy 3-HOC$_6$H$_4$CO(C$_6$H$_4$OH-4)	214.22	nd (w)		208			al, eth, w	B8^3, 2648
2926	Benzophenone, 4,4'-dihydroxy (4-HOC$_6$H$_4$)$_2$CO	214.22	nd (lig), cr (w)		210	1.133$^{1.31}$		al, eth, ace	B8^3, 2646
2927	Benzophenone, 4,4'-diiido (4-IC$_6$H$_4$)$_2$CO	434.03	pl (to), nd (bz)	281^{12}	238.5				B7,425
2928	Benzophenone, 2,4-dimethoxy 2,4-(CH$_3$O)$_2$C$_6$H$_3$COC$_6$H$_5$	242.28	pr (dil al)	218^{10}	87-8			al, chl	B8^3, 2641
2929	Benzophenone, 2,4'-dimethoxy 2-CH$_3$OC$_6$H$_4$CO(C$_6$H$_4$OCH$_3$-4)	242.28	nd (al)		100			al, eth, bz, aa	B8^1, 640
2930	Benzophenone, 2,5-dimethoxy 2,5(CH$_3$O)$_2$C$_6$H$_3$COC$_6$H$_5$	242.28	(lig)	225^{38}	51				B8^3, 2644
2931	Benzophenone, 3,4-dimethoxy 3,4-(CH$_3$O)$_2$C$_6$H$_3$COC$_6$H$_5$	242.28	nd or pl (al)		103-4			al	B8^2, 354
2932	Benzophenone, 3,4'-dimethoxy 3-CH$_3$OC$_6$H$_4$CO(C$_6$H$_4$OCH$_3$-4)	242.28	pr (al)		58-9			al	B8^2, 354
2933	Benzophenone, 4,4'-dimethoxy (4-CH$_3$OC$_6$H$_4$)$_2$CO	242.28	nd (al)		148			al, eth, ace, bz	B8^2, 355
2934	Benzophenone, 2,2'-dimethyl (2-CH$_3$C$_6$H$_4$)$_2$CO	210.27		310^{740}, 175-80^{17}	72			al, ace, bz	B7^3, 2178
2935	Benzophenone, 4,4'-dimethyl (4-CH$_3$C$_6$H$_4$)$_2$CO	210.28	rh (al)	333^{725}	95			al, eth, ace, bz	B7^3, 2181
2936	Benzophenone, 3-(dimethylamino) 3-(CH$_3$)$_2$NC$_6$H$_4$COC$_6$H$_5$	225.29	pa ye pl (al)	216^{15}	47			al	B14^1, 388
2937	Benzophenone, 4-(dimethylamino) 4-(CH$_3$)$_2$NC$_6$H$_4$COC$_6$H$_5$	225.29	ye lf (al), nd (peth)		92-3			al, eth, peth, chl	B14^3, 218
2938	Benzophenone, 2,2'-dinitro (2-NO$_2$C$_6$H$_4$)$_2$CO	272.21	nd (to or aa)		188-9			to, aa	B7^3, 2085
2939	Benzophenone, 3,3'-dinitro (3-NO$_2$C$_6$H$_4$)$_2$CO	272.22			157			ace	B7^3, 2085
2940	Benzophenone, 2-hydroxy 2-HOC$_6$H$_4$COC$_6$H$_5$	198.22	pl (dil al)	250^{560}	39			al, eth, bz, aa	B8^3, 1227
2941	Benzophenone, 3-hydroxy 3-HOC$_6$H$_4$COC$_6$H$_5$	198.22	lf or pl (al)		116			al, eth	B8^3, 1235
2942	Benzophenone, 4-hydroxy 4-HOC$_6$H$_4$COC$_6$H$_5$	198.22	nd (al), pr (dil al)		135 (st), 122 (unst)			al, eth, aa, w	B8^3, 1237
2943	Benzophenone, 2-hydroxy-4-methoxy 4-CH$_3$O-2-HOC$_6$H$_3$COC$_6$H$_5$	228.25			65.6				B8^3, 2640
2944	Benzophenone, 2-hydroxy-4'-methoxy 2-HOC$_6$H$_4$CO(C$_6$H$_4$OCH$_3$-4)	228.25			98				
2945	Benzophenone, 2-hydroxy-5-methoxy 5-CH$_3$O-2-HOC$_6$H$_3$COC$_6$H$_5$	228.25			84				B8^3, 2644
2946	Benzophenone, 2-hydroxy-4'-methyl 2-HOC$_6$H$_4$CO(C$_6$H$_4$CH$_3$-4)	212.25			40-1				B8^3, 1301
2947	Benzophenone, 2-hydroxy-4-methyl 4-CH$_3$-2-HOC$_6$H$_3$COC$_6$H$_5$	212.25			63				B8^3, 1300
2948	Benzophenone, 2-hydroxy-5-methyl 5-CH$_3$-2-HOC$_6$H$_3$COC$_6$H$_5$	212.25			84				B8^3, 1295

No.	Name, Synonyms, and Formula	Mol. wt.	Color, crystalline form, specific rotation and λ_{max} (log ε)	b.p. °C	m.p. °C	Density	n_D	Solubility	Ref.
2949	Benzophenone, 2-hydroxy-5-nitro 5-NO$_2$-2-HOC$_6$H$_3$COC$_6$H$_5$	243.22			124-5				B8, 157
2950	Benzophenone, 4-hydroxy-3-methyl 3-CH$_3$-4-HOC$_6$H$_3$COC$_6$H$_5$	212.25			173-4				B8^3, 1294
2951	Benzophenone, 4-hydroxy-4'-nitro 4-HOC$_6$H$_4$CO(C$_6$H$_4$NO$_2$-4)	243.22			190-2			al, eth, aa	B8, 163
2952	Benzophenone imine (C$_6$H$_5$)$_2$C=NH	181.24		282, 158^{12}		1.0847$^{19/4}$	1.6191^{19}	eth	B7^3, 2061
2953	Benzophenone, 2-methoxy 2-CH$_3$OC$_6$H$_4$COC$_6$H$_5$	212.25		194-6^{18}	41			al, bz, aa	B8^3, 1227
2954	Benzophenone, 3-methoxy 3-CH$_3$OC$_6$H$_4$COC$_6$H$_5$	212.25		342-3^{730}, 201^{17}	44			al, bz, aa	B8^3, 1236
2955	Benzophenone, 4-methoxy 4-CH$_3$OC$_6$H$_4$COC$_6$H$_5$	212.25	pr (eth)	354-5^{729}, 168^{12}	61-2			al, eth, ace, bz, aa	B8^3, 1238
2956	Benzophenone, 2-methyl 2-CH$_3$C$_6$H$_4$COC$_6$H$_5$	196.25		309.5^{762}, 128^{12}	< -18			al	B7^3, 2123
2957	Benzophenone, 3-methyl 3-CH$_3$C$_6$H$_4$COC$_6$H$_5$	196.25	oil	314-5^{725}, 170^9		1.088$^{17.5}$		al, eth, bz, chl, aa	B7^3, 2126
2958	Benzophenone, 4-methyl 4-CH$_3$C$_6$H$_4$COC$_6$H$_5$	196.25	mcl pr	327-8	59-60			eth, bz, chl	B7^3, 2127
2959	Benzophenone, 4-methyl, diphenyl acetal 4-CH$_3$C$_6$H$_4$C(OC$_6$H$_5$)$_2$C$_6$H$_5$	366.47	(dil al, eth-peth)		134				B7^2, 372
2960	Benzophenone, 4-methyl, imine 4-CH$_3$C$_6$H$_4$C(=NH)C$_6$H$_5$	195.27		147^5	37	1.0617$^{30/11}$	1.6097^{20}		B7^3, 2129
2961	Benzophenone, 4-methyl-3-nitro 2-NO$_2$-4-CH$_3$C$_6$H$_3$COC$_6$H$_5$	241.25	nd or pl (al)	sub	126-7			al, bz, chl	B7, 442
2962	Benzophenone, 4-methyl-2'-nitro 4-CH$_3$C$_6$H$_4$CO(C$_6$H$_4$NO$_2$-2)	241.25	pr (aa or al)		111			bz, chl, ace	B7^3, 2132
2963	Benzophenone, 4-methyl-3-nitro 3-NO$_2$-4-CH$_3$C$_6$H$_3$COC$_6$H$_5$	241.25	pa ye pl (al or aa)		130-2			al, eth, ace, bz, aa	B7^3, 2131
2964	Benzophenone, 4-methyl-3'-nitro 4-CH$_3$C$_6$H$_4$CO(C$_6$H$_4$NO$_2$-3)	241.25	lf (al)		111			eth, bz, chl	B7^2, 375
2965	Benzophenone, 4-methyl-4'-nitro 4-CH$_3$C$_6$H$_4$CO(C$_6$H$_4$NO$_2$-4)	241.25	nd (al)	sub	122-4			al, eth, bz, chl, aa	B7^2, 375
2966	Benzophenone, 4-methyl-4'-nitro, oxime 4-CH$_3$C$_6$H$_4$C(=NOH)(C$_6$H$_4$NO$_2$-4)	256.26	nd (eth-lig)		145			al, eth, bz	B7, 443
2967	Benzophenone, 2-nitro 2-NO$_2$C$_6$H$_4$COC$_6$H$_5$	227.22	mcl (al)		105			al	B7^3, 2082
2968	Benzophenone, 3-nitro 3-NO$_2$C$_6$H$_4$COC$_6$H$_5$	227.22	ye nd (al)	234^{18}	95			al	B7^3, 2082
2969	Benzophenone, 4-nitro 4-NO$_2$C$_6$H$_4$COC$_6$H$_5$	227.22	nd or lf (al)		138	1.406		bz	B7^3, 2082
2970	Benzophenone oxime (C$_6$H$_5$)$_2$C=NOH	197.24	nd (al)		144			al, eth, ace, chl	B7^3, 2063
2971	Benzophenone phenyl hydrazone (C$_6$H$_5$)$_2$C=NNHC$_6$H$_5$	272.35	pr or nd (al)		137			eth, bz, aa	B15^2, 63
2972	Benzophenone, 2,2',3,4-tetrahydroxy 2,3,4-(HO)$_3$C$_6$H$_2$CO(C$_6$H$_4$OH-2)	246.22	ye lf or pl (w + l)		149, 102 (+ w)			al, eth, w, aa	B8^2, 539
2973	Benzophenone, 2,2',4,4'-tetrahydroxy [2,4-(HO)$_2$C$_6$H$_3$]$_2$CO	246.22	ye nd (w + l)		196-8			al, eth, ace, w, bz, aa	B8^3, 4064
2974	Benzophenone, 2,2',4,6'-tetrahydroxy or Isoeuxanthonic acid 2,4-(HO)$_2$C$_6$H$_4$CO[C$_6$H$_3$(OH)$_2$-2,6]	246.22	(w + l)		200d			w, al, eth	B8, 496
2975	Benzophenone, 2,2',5,6'-tetrahydroxy or Euxanthoic acid 2,5-(HO)$_2$C$_6$H$_4$CO[C$_6$H$_3$(OH)$_2$-2,6]	246.22	ye nd (w)		200-2d			w, al	B8^2, 541
2976	Benzophenone, 2,3',4,4'-tetrahydroxy 2,4-(HO)$_2$C$_6$H$_4$CO[C$_6$H$_3$(OH)$_2$-3,4]	246.22	nd (w + 2)		202 (anh)			w, al, eth, ace, aa	B8^2, 541
2977	Benzophenone, 2,3',4,6-tetrahydroxy 2,4,6(HO)$_3$C$_6$H$_2$CO(C$_6$H$_4$OH-3)	246.22	pa ye lf (w)		246d			w, al	B8^2, 540
2978	Benzophenone, 2,4,4',6-tetrahydroxy 2,4,6-(HO)$_3$C$_6$H$_2$CO(C$_6$H$_4$OH-4)	246.22	pr or nd (w + 2)		210			w, al, eth	B8^2, 540
2979	Benzophenone, 3,3',4,4'-tetrahydroxy [3,4-(HO)$_2$C$_6$H$_3$]$_2$CO	246.22	(w)		227-8			w, al, bz	B8^2, 541
2980	Benzophenone, 2,2',4,4'-tetramethoxy (2,4-(CH$_3$O)$_2$C$_6$H$_3$)$_2$CO	246.22			130			al, bz	B8^3, 4064
2981	Benzophenone, 2,2',5,5'-tetramethoxy [2,5-(CH$_3$O)$_2$C$_6$H$_3$]$_2$CO	302.33	ye (aa or al)		109			al, eth, bz, chl, aa	B8^2, 541

No.	Name, Synonyms, and Formula	Mol. wt.	Color, crystalline form, specific rotation and λ_{max} (log ε)	b.p. °C	m.p. °C	Density	n_D	Solubility	Ref.
2982	Benzophenone, 2,2′,6,6′-tetramethoxy . . . . [2,6-(CH₃O)₂C₆H₃]₂CO	302.33	pl (bz)	. . . .	204	. . . .		chl	B8³, 4065
2983	Benzophenone, 2,3′,4,4′-tetramethoxy . . . . 2,4-(CH₃O)₂C₆H₃CO[C₆H₃(OCH₃)₂-3,4]	302.33	nd, lf or pr (al)	. . . .	126			al	B8³, 4065
2984	Benzophenone, 2,3′,4,5′-tetramethoxy . . . . 2,4-(CH₃O)₂C₆H₃CO[C₆H₃(OCH₃)₂-3,5]	302.33	nd (bz-peth)	. . . .	73-4			al, eth	B8¹, 735
2985	Benzophenone, 2,3′4′,5-tetramethoxy . . . . 2,5-(CH₃O)₂C₆H₃CO[C₆H₃-(OCH₃)₂-3,4]	302.33	pr (dil al)	. . . .	101-2			al, eth	B8, 497
2986	Benzophenone, 2,3,4,6-tetramethoxy . . . . 2,3,4,6-(CH₃O)₄C₆HCOC₆H₅	302.33	nd (lig)	. . . .	125-6			al, ace, bz	B8¹, 734
2987	Benzophenone, 2,4,4′,5-tetramethoxy . . . . 2,4,5-(CH₃O)₃C₆H₂CO(C₆H₄OCH₃-4)	302.33	ye pw (al)	. . . .	122-4			al, ace, bz, chl	B8¹, 734
2988	Benzophenone, 2,4,4′,6-tetramethoxy . . . . 2,4,6(CH₃O)₃C₆H₂CO(C₆H₄OCH₃-4)	302.33	pr (al)	. . . .	146			al, eth	B8, 496
2989	Benzophenone, 3,3′,4,4′-tetramethoxy or Veratrophenone [3,4-(CH₃O)₂C₆H₃]₂CO	302.33	pr (al)	. . . .	145			w, bz	B8³, 4066
2990	Benzophenone, 3,3′,4,5′-tetramethoxy . . . . 3,4-(CH₃O)₂C₆H₃CO[C₆H₃(OCH₃)₂-3,5]	302.33	nd (bz)	. . . .	114-5			al, eth, bz, chl	B8¹, 735
2991	Benzophenone, 2,2′,4,4′-tetramethyl [2,4-(CH₃)₂C₆H₃]₂CO	238.33	. . . .	190[10]		1.043[15]	1.5790[25]	. . . .	B7³, 2247
2992	Benzophenone, thio or Thiobenzophenone . . . . C₆H₅CSC₆H₅	198.28	. . . .	174[14]	53-4			bz	B7³, 2087
2993	Benzophenone, 2,2′,6-trihydroxy . . . . 2,6-(HO)₂C₆H₃CO(C₆H₄OH-2)	230.22	ye nd (dil al)	. . . .	133-4			al, eth, bz	B8², 468
2994	Benzophenone, 2,3,4-trihydroxy or Callobenzophenone . . 2,3,4-(HO)₃C₆H₂COC₆H₅	230.22	ye nd (dil al)	. . . .	140-1			w, al, eth, ace, aa	B8³, 3635
2995	Benzophenone, 2,4,4′-trihydroxy . . . . 2,4-(HO)₂C₆H₃CO(C₆H₄OH-4)	230.22	ye nd (w + 2)	. . . .	200-1			w, al	B8³, 3639
2996	Benzophenone, 2,4,6-trihydroxy . . . . 2,4,6-(HO)₃C₆H₂COC₆H₅	230.22	ye nd (w + l, dil al)	. . . .	165			w, al, eth	B8³, 3637
2997	Benzophenone, 3,4,5-trihydroxy . . . . 3,4,5-(HO)₃C₆H₂COC₆H₅	230.22	ye pl (+ lw), col (chl)	. . . .	177-8 (anh)			w, al, eth, ace	B8, 422
2998	Benzophenone, 2,3,4-trimethoxy . . . . 2,3,4-(CH₃O)₃C₆H₂COC₆H₅	272.30	pr (dil al)	. . . .	55			al, eth	B8,418
2999	Benzophenone, 2,4,4′-trimethoxy . . . . 2,4-(CH₃O)₂C₆H₃CO(C₆H₄OCH₃-4)	272.30	nd (al)	. . . .	73-4			al, eth, bz, aa	B8¹, 702
3000	Benzophenone, 2,4,5-trimethoxy . . . . 2,4,5-(CH₃O)₃C₆H₂COC₆H₅	272.30	ye nd (w)	. . . .	97			w, al, ace, bz	B8¹, 701
3001	Benzophenone, 2,4,6-trimethoxy . . . . 2,4,6-(CH₃O)₃C₆H₂COC₆H₅	272.30	mcl pr or rh pl (al)	. . . .	115			al, eth, chl	B8³, 3638
3002	Benzophenone, 3,3′,4-trimethoxy . . . . 3,4-(CH₃O)₂C₆H₃CO(C₆H₄OCH₃-3)	272.30	nd (MeOH)	. . . .	83-4			al, ace, bz	B8², 468
3003	Benzophenone, 3,4,4′-trimethoxy . . . . 3,4-(CH₃O)₂C₆H₃CO(C₆H₄OCH₃-4)	272.30	nd (al)	. . . .	98-9			al, ace, bz	B8², 469
3004	Benzophenone, 3,4′,5-trimethoxy . . . . 3,5-(CH₃O)₂C₆H₃CO(C₆H₄OCH₃-4)	272.30	nd (bz)	. . . .	97-8			al, eth, bz, chl	B8¹, 702
3005	Benzopinacolone (α-form) or α,α,α-triphenylacetophenone C₆H₅COC(C₆H₅)₃	348.45	nd	. . . .	206-7			bz, chl	C55, 22234
3006	Benzopinacolone (β form) or α,α,α-triphenylacetophenone C₆H₅COC(C₆H₅)₃	348.45	nd (al)	. . . .	182			eth, bz, chl	B7³, 2941
3007	1,2-Benzopyrene . . . . C₂₀H₁₂	252.31			177.8				B5³, 2520
3008	2,3-Benzopyrene . . . . C₂₀H₁₂	252.32			176.5-7.5			chl	E14,457
3009	3,4-Benzopyrone . . . . C₂₀H₁₂	252.32			176-7				B5³, 2517
3010	3,4-Benzopyrene, 5-amino . . . . C₂₀H₁₃N	267.33	ye pl (bz-lig)		239-41			al, ace, bz, chl	B12³, 3404
3011	3,4-Benzopyrene, 5-hydroxy . . . . C₂₀H₁₂O	268.32	nd (eth-lig)		207-9			al, bz, aa	B6³, 3810
3012	3,4-Benzopyrene, 8-hydroxy . . . . C₂₀H₁₂O	268.32	ye nd (bz-peth)		226-7d			al, eth, ace, bz	E14s, 704
3013	5,6-Benzoquinoline or β-Naphthoquinoline . . . . C₁₃H₉N	179.22	lf (peth or w)	350[721], 202—5[8]	94			al, eth, ace, bz	B20⁴, 4009

No.	Name, Synonyms, and Formula	Mol. wt.	Color, crystalline form, specific rotation and λ_{max} (log ε)	b.p. °C	m.p. °C	Density	n_D	Solubility	Ref.
3014	7,8-Benzoquinoline or α-Naphthoquinoline............ $C_{13}H_9N$	179.22	lf (eth), pl (peth)	338[719], 233[47]	52				al, eth, ace, bz	B20[4], 4003
3015	7,8-Benzoquinoline, 2-methyl .... $C_{14}H_{11}N$	193.25	....	324-6		1.1464[20]	1.6738[20]	al	B20[4], 4043
3016	1,2-Benzoquinone or o-Quinone .... $1,2\text{-}O=C_6H_4=O$	108.10	red pl or pr		60-70d			eth, ace, bz	B7[3], 3352
3017	1,2-Benzoquinone, 3-chloro .... $3\text{-}Cl\text{-}(1,2\text{-}O=C_6H_3=O)$	142.54	pa ye-red pr (hx)		68d			al, eth	B7[1], 338
3018	1,2-Benzoquinone, 4-chloro .... $4\text{-}Cl\text{-}(1,2\text{-}O=C_6H_3=O)$	142.54	pa ye-red nd (hx)		78			eth	B7[1], 3354
3019	1,2-Benzoquinone, 4,5-dichloro .... $4,5\text{-}Cl_2\text{-}(1,2\text{-}O=C_6H_2=O)$	176.99	yesh red pr or pl		94d			eth, bz	B7[1], 338
3020	1,2-Benzoquinone, 4-methoxy,1-oxime .... $4\text{-}(H_3O\text{-}(1\text{-}HON=C_6H_3=O\text{-}2)$	153.14	ye pr		158-9			al, bz, aa	B8[3], 1966
3021	1,2-Benzoquinone, 3-methoxy .... $3\text{-}CH_3O(1,2\text{-}O=C_6H_3=O)$	138.12	br-red pl, pr, nd		115-20			w, al, eth, bz, chl	B8[3], 1964
3022	1,2-Benzoquinone, 5-isopropyl-4-methyl-1-oxime ...... $C_{10}H_{13}NO_2$	179.22	nd (bz-chl)		165-7d			bz, chl	B7[2], 595
3023	1,4-Benzoquinone or p-Quinone .... $1,4\text{-}O=C_6H_4=O$	108.10	ye mcl pr (w)	sub	115-7	1.318[20]		al, eth	B7[3], 3356
3024	1,4-Benzoquinone, 2-bromo-6-methyl .... $2\text{-}Br\text{-}6\text{-}CH_3\text{-}(1,4\text{-}O=C_6H_2=O)$	201.03	ye nd (al), pr (eth or lig)	sub	95			al, eth, chl	B7[3], 3393
3025	1,4-Benzoquinone, 5-bromo-2-methyl $5\text{-}Br\text{-}2\text{-}CH_3(1,4\text{-}O=C_6H_2=O)$	201.03	ye lf (lig)		106			al	B7[3], 591
3026	1,4-Benzoquinone, 2-chloro .... $2\text{-}Cl\text{-}(1,4\text{-}O=C_6H_3=O)$	142.54	ye-red rh (hx)		57	....	....	al, eth, chl	B7[3], 3373
3027	1,4-Benzoquinone, 2-chloro,oxime .... $2\text{-}Cl\text{-}1,4(1\text{-}HON=C_6H_3=O)$	157.56	gr-ye nd (bz-aa)		184d			al, eth	B7[3], 3374
3028	1,4-Benzoquinone, 2-chloro-3-methyl .... $2\text{-}Cl\text{-}3\text{-}CH_3(1,4\text{-}O=C_6H_2=O)$	156.57	cr (lig)		55			al, chl, lig	B7[3], 3391
3029	1,4-Benzoquinone, 2-chloro-5-methyl .... $2\text{-}Cl\text{-}5\text{-}CH_3\text{-}(1,4\text{-}O=C_6H_2=O)$	156.57	ye nd (w or al)		105			al, eth, chl	B7[3], 3391
3030	1,4-Benzoquinone, 2-chloro-6-methyl .... $2\text{-}Cl\text{-}6\text{-}CH_3(1,4\text{-}O=C_6H_2=O)$	156.57	ye nd (w)		90			al, eth, chl	B7[3], 3392
3031	1,4-Benzoquinone, 2,5-diaminoanil or Bandrowski's base $C_{18}H_{18}N_4$	318.38	dk red or br lf		238			dil HCl	B14[3], 363
3032	1,4-Benzoquinone, 2,6-dibromo-1-imine, N-chloro ...... $C_6H_2Br_2ClNO$	299.36	ye pr (aa, al)		85-6				B7[2], 584
3033	1,4-Benzoquinone, 2,6-dibromo-4 imine- N-chloro ...... $2,6\text{-}Br_2\text{-}(1\text{-}O=C_6H_2=NCl\text{-}4)$	299.36	ye pr (al or aa)		83			al	B7[2], 584
3034	1,4-Benzoquinone, 3,5-dibromo,2,6-dimethyl .... $2,6\text{-}(CH_3)_2(1,4\text{-}O=C_6Br_2=O)$	293.95	ye lf (al)	sub	176			al	B7[3], 3401
3035	1,4-Benzoquinone, 3,5-dibromo-2-methyl .... $3,5\text{-}Br_2\text{-}2\text{-}CH_3(1,4\text{-}O=C_6H=O)$	279.93	ye		117			al, eth, chl	B7[3], 3393
3036	1,4-Benzoquinone, 2,5-di-tert-butyl .... $2,5\text{-}t\text{-}C_4H_9(1,4\text{-}O=C_6H_2=O)$	220.32	ye (al)		152.5			eth, bz, aa	B7[3], 3428
3037	1,4-Benzoquinone, 2,6-di-tert-butyl .... $2,6\text{-di-}t\text{-}C_4H_9(1,4\text{-}O=C_6H_2=O)$	222.23			102-4				
3038	1,4-Benzoquinone, 2,3-dichloro .... $2,3\text{-}Cl_2\text{-}(1,4\text{-}O=C_6H_2=O)$	176.99	ye lf		100-1			eth, bz	B7[3], 3375
3039	1,4-Benzoquinone, 2,5-dichloro .... $2,5\text{-}Cl_2(1,4\text{-}O=C_6H_2=O)$	176.99	pa ye mcl pr (al)		161-2			eth, chl	B7[3], 3376
3040	1,4-Benzoquinone, 2,5-dichloro-3,6-dihydroxy or Chloranilic acid................ $3,6\text{-}(HO)_2(1,4\text{-}O=C_6Cl_2=O)$	208.99	red lf (w + 2)		283-4			w	B8[3], 3350
3041	1,4-Benzoquinone, 2,6-dichloro .... $2,6\text{-}Cl_2(1,4\text{-}O=C_6H_2=O)$	176.99	ye rh (lig, bz)		120-1			chl	B7[3], 3376
3042	1,4-Benzoquinone, 2,6-dichloro-1-imine, N-chloro ...... $2,6\text{-}Cl_2\text{-}(1\text{-}ClN=C_6H_2=O\text{-}4)$	210.45	ye nd (al)		67-8			eth, chl	B7[2], 581
3043	1,4-Benzoquinone, 2,5-dihydroxy .... $2,5\text{-}(HO)_2\text{-}(1,4\text{-}O=C_6H_2=O)$	140.10	dk ye nd	215d sub	211			aa	B8[3], 3348
3044	1,4-Benzoquinone, 2,5-dihydroxy-3-methoxy-6-methyl or Spinulosin .... $C_8H_8O_5$	284.16	red-bl	120[1] sub	202-3				B8[3], 3983
3045	1,4-Benzoquinone, 2,5-dihydroxy-3-undecyl or Embelin . $3\text{-}C_{11}H_{23}\text{-}2,5\text{-}(HO)_2\text{-}(1,4\text{-}O=C_6H=O)$	308.42	og red pl (al-bz)	sub	143				B8[2], 452

No.	Name, Synonyms, and Formula	Mol. wt.	Color, crystalline form, specific rotation and λ_{max} (log ε)	b.p. °C	m.p. °C	Density	n_D	Solubility	Ref.
3046	1,4-Benzoquinone, diimine 1,4-HN=C₆H₄=NH	106.13	ye nd	....	124	....	....	bz, chl	B7², 574
3047	1,4-Benzoquinone, diimine, N,N-dichloro 1,4-ClN=C₆H₄=NCl	175.02	nd (w)	....	126d	....	....	al, eth, bz	B7², 574
3049	1,4-Benzoquinone, 2,6-dimethoxy 2,6-(CH₃O)₂-(1,4-O=C₆H₂=O)	168.16	ye mcl pr (aa)	sub	256	....	....	aa	B8³, 3354
3050	1,4-Benzoquinone, 2,3-dimethyl or o-Xyloquinone 2,3-(CH₃)₂-(1,4-O=C₆H₂=O)	136.16	ye nd	sub	55	....	....	al, eth	B7³, 3397
3051	1,4 Benzoquinone, 2,5-dimethyl 2,5-(CH₃)₂-(1,4-O=C₆H₂=O)	136.15		....	125	....	....	eth, bz	B7³, 3397
3052	1,4-Benzoquinone, 2,6-dimethyl 2,6-(CH₃)₂-(1,4-O=C₆H₂=O)	136.16	ye nd	sub	72-3	1.0479⁷⁸/⁴	....	....	B7³, 3399
3053	1,4-Benzoquinone dioxime 1,4-C₆H₄(=NOH)₂	138.13	pa ye nd (w)	....	240d	....	....	....	B7,627
3054	1,4-Benzoquinone, 2,5-diphenyl 2,5-(C₆H₅)₂-(1,4-O=C₆H₂=O)	260.30	og-ye pl (bz or aa)	....	214	....	....	bz, aa	B7³, 4251
3055	1,4-Benzoquinone, 2-hydroxy-3,5,6-trimethyl C₉H₁₀O₃	166.18		....	90-2	....	....	chl	B8³, 2183
3056	1,4-Benzoquinone, 3-hydroxy-2-methoxy-5-methyl or Fumigatin C₈H₈O₄	168.16	br nd or pl (peth)	....	116	....	....	al, eth, ace, bz, chl	B8³, 3374
3057	1,4-Benzoquinone, 2-methoxy 2-CH₃O-(1,4-O=C₆H₃=O)	138.12	ye nd (w)	sub	145	....	....	al	B8³, 1969
3058	1,4-Benzoquinone, 2-methyl or Toluquinone 2-CH₃-(1,4-O=C₆H₃=O)	122.13	ye pl or nd	sub	69	1.08⁷⁵·⁵/⁴	....	al, eth	B7³, 3387
3059	1,4-Benzoquinone, 2-methyl-6-propyl-4-oxime 2-CH₃-6-C₃H₇-(1,4-O=C₆H₂=NOH)	179.22	br nd (lig)	....	93-4	....	....	al, bz, aa	B7², 595
3060	1,4-Benzoquinone, 2-methyl-3,5,6-tribromo 2-CH₃-(1,4-O=C₆Br₃=O)	358.83	ye pl (al)	....	235-6	....	....	eth, bz	B7², 592
3061	1,4-Benzoquinone monoimine, N-chloro 1-ClN=C₆H₄=O-4	141.56	ye nd (peth)	....	85	....	....	w, al, eth	B7,619
3062	1,4-Benzoquinone, 2-phenyl 2-C₆H₅-(1,4-O=C₆H₃=O)	184.21	ye lf (peth, al)	....	114	....	....	al, bz, chl	B7³, 3764
3063	1,4-Benzoquinone, 2-isopropyl-5-methyl 2-i-C₃H₇-5-CH₃[1,4-O=C₆H₂=O]	164.20		....	45-7	....	....	chl	B7³, 3411
3064	1,4-Benzoquinone, 2-isopropyl-5-methyl-dioxime or Thymoquinone dioxime 2-i-C₃H₇-5-CH₃[1,4-HON=C₆H₂=NOH]	179.22	nd (bz-chl)	....	165-7d	....	....	bz, chl	B7³, 3414
3065	1,4-Benzoquinone, tetrachloro or Chloranil 1,4-O=C₆Cl₄=O	245.88	ye mcl pr (bz), ye lf (aa)	sub	290 (sealed tube)	....	....	eth	B7³, 3378
3066	1,4-Benzoquinone, tetrahydroxy 1,4-O=C₆(OH)₄=O	172.10	bl-bk cr	....	....	....	....	al, w	B8³, 4204
3067	1,4-Benzoquinone, tetramethyl or Duroquinone 1,4-O=C₆(CH₃)₄=O	164.21	ye nd (al or lig)	....	111-2	....	....	al, eth, ace, bz, aa	B7³, 3417
3068	1,4-Benzoquinone, trichloro 1,4-O=C₆Cl₃H=O	211.43	ye pl (al)	....	169-70	....	....	al, eth	B7³, 3377
3069	α-Benzosuberone C₁₁H₁₂O	160.22	....	124-5⁷, 108¹	....	1.0780²⁰/⁴	1.5698²⁰	al	B7³, 1442
3070	Benzotrichloride C₆H₅CCl₃	195.48	....	220.6, 150¹⁰⁰	-4.75	1.3723²⁰	1.5580²⁰	al, eth, bz	B5⁴, 820
3071	Benzotrichloride, 2-chloro 2-ClC₆H₄CCl₃	229.92	....	264.3, 129.5¹³	30	1.5187²⁰/⁴	1.5836²⁰	eth, ace	B5⁴, 823
3072	Benzotrichloride, 3-chloro 3-ClC₆H₄CCl₃	229.92	....	255	....	1.495¹⁴	1.4461²⁰	eth, ace	B5⁴, 823
3073	Benzotrichloride, 4-chloro 4-ClC₆H₄CCl₃	229.92	....	245, 108-12⁸	....	....	1.4463²⁰	eth, ace	B5⁴, 823
3074	Benzotrifluoride C₆H₅CF₃	146.11	....	102, 10¹⁰	-29.1	1.1884²⁰	1.4146²⁰	al, eth, ace, bz	B5⁴, 802
3075	Benzotrifluoride, 3-amino 3-H₂NC₆H₄CF₃	161.13	....	187.5⁷⁶⁴, 74-5¹⁰	....	....	1.4787²⁰	al, eth	B12³, 1988
3076	Benzotrifluoride, 2-nitro 2-NO₂C₆H₄CF₃	191.11	cr (al)	216.3⁷⁶⁵	32.5	....	....	al, bz, aa	B5³, 744
3077	Benzotrifluoride, 3-nitro 3-NO₂C₆H₄CF₃	191.11	....	202.8, 81.6¹⁰	-2.4	1.4357¹⁵/⁴	1.4719²⁰	al, eth	B5³, 744
3078	Benzothiazole C₇H₅NS	135.19	....	231, 131³⁴	2	1.2460²⁰/⁴	1.6379²⁰	al, eth, ace, bz	B27², 17

No.	Name, Synonyms, and Formula	Mol. wt.	Color, crystalline form, specific rotation and λ_{max} (log ε)	b.p. °C	m.p. °C	Density	n_D	Solubility	Ref.
3079	Benzothiazole, 2-amino $C_7H_6N_2S$	150.20	pl (w)	....	132	....	....	al, eth, chl	B27[2], 225
3080	Benzothiazole, 2-amino-6-chloro $C_7H_5N_2ClS$	184.64	....	....	199-201	....	....		B27[2], 230
3081	Benzothiazole, 2-amino-6-ethoxy $C_9H_{10}N_2OS$	194.26	nd (al)	....	163-4	....	....	al	B27[2], 335
3082	Benzothiazole, 2-amino-4-methyl $C_8H_8N_2S$	164.24	nd (w), pl (al)	....	145	....	....	al	B27[2], 237
3083	Benzothiazole, 2-amino-5-methyl $C_8H_8N_2S$	164.24	pl (dil al)	....	171-2	....	....	al	B27[2], 240
3084	Benzothiazole, 2-amino-6-methyl $C_8H_8N_2S$	164.24	nd (w), pr (dil al)	....	142	....	....	al, w	B27[2], 241
3085	Benzothiazole, 5-amino-2-mercapto $C_7H_6N_2S_2$	182.27	nd	....	216	....	....	aniline	B27[2], 475
3086	Benzothiazole, 6-amino $C_7H_6N_2S$	150.20	pr (w)	....	87	....	....	al	B27,366
3087	Benzothiazole, 6-amino-2-mercapto $C_7H_6N_2S_2$	182.27	....	....	263	....	....		B27[2], 475
3088	Benzothiazole, 2-chloro C_7H_5ClNS	169.63	....	248, 136-6[28]	24	1.3715[10/4]	1.6338[10]	al, eth, ace	B27[2], 18
3089	Benzothiazole, 5-chloro-2-methyl C_8H_6ClNS	183.66	pl (eth)	70-90[0.5] sub	69	....	....	al, peth	B27[2], 22
3090	Benzothiazole, 6-dimethylamino-2-mercapto $C_9H_{10}N_2S_2$	210.32	ye nd (bz)	....	230d	....	....	ace	B27[2], 475
3091	Benzothiazole, 2-(2,4-dinitrophenylthio) $C_{13}H_7N_3O_4S_2$	333.35	ye	....	162	1.24[20/4]	....		C48,1443
3092	Benzothiazole, 2-hydroxy C_7H_5NOS	151.19	pr (dil al) nd	360	138	....	....	al, eth	B27[2], 225
3093	Benzothiazole, 2-(hydroxyphenyl) $C_{13}H_9NOS$	227.29	nd or lf (al)	175-93[3]	132-3	....	....	al	B27[2], 91
3094	1,2-Benzisothiazole, 5-hydroxy-3-phenyl $C_{13}H_9NOS$	227.29	nd (dil al or aa)	....	159-60	....	....	al, ace, bz, aa	B27[2], 87
3095	Benzothiazole, 6-hydroxy-2-phenyl $C_{13}H_9NOS$	227.29	wh nd (dil al)	....	227 (cor)	....	....	al, ace, bz, aa	B27[2], 88
3096	Benzothiazole, 2-mercapto $C_7H_5NS_2$	167.25	nd (al or dil MeOH)	....	180-2	1.42[20/4]	....	al	B27, 185
3097	Benzothiazole, 2-mercapto;benzoate $C_{14}H_9NOS_2$	271.36	ye	....	132	....	....		C51,3491
3098	Benzothiazole, 2-mercapto-4-methyl $C_8H_7NS_2$	181.28	nd (dil aa)	....	186	....	....	aa	B27[2], 240
3099	Benzothiazole, 2-mercapto-5-methyl $C_8H_7NS_2$	181.28	nd (to)	....	171-3	....	....	al, ace, bz	B27[2], 241
3100	Benzothiazole, 2-mercapto-6-methyl $C_8H_7NS_2$	181.28	....	....	181	....	....	al, ace, bz	B27[2], 242
3101	Benzothiazole, 2-mercapto-7-methyl $C_8H_7NS_2$	181.28	....	....	184	....	....	ace, bz, to	B27[2], 242
3102	Benzothiazole, 2-mercapto-6-nitro $C_7H_4N_2O_2S_2$	212.25	ye nd (aa)	....	255-7	....	....	ace, aa	B27[2], 234
3103	Benzothiazole, 2-methyl C_8H_7NS	149.22	....	238, 150-1[15]	14	1.1763[19/4]	1.6092[19]	al	B27[2], 21
3104	Benzothiazole, 2-methyl-6-nitro $C_8H_6N_2O_2S$	194.21	....	....	175	....	....		B27, 47
3105	Benzothiazole, 2-(methylthio) $C_8H_7NS_2$	181.28	pr (dil al)	....	52	....	....	al	B27[2], 71
3106	Benzothiazole, 2-phenyl $C_{13}H_9NS$	211.29	nd (dil al)	>360	114	....	....	al, eth	B27[2], 37
3107	Benzothiazole, 2-phenylamino $C_{13}H_{10}N_2S$	226.30	nd (al)	....	161	....	....	AcOEt	B27[2], 226
3108	Benzothiazoline, 3-methyl-2-imino $C_8H_8N_2S$	164.24	pl (w), nd (al)	....	128	....	....	al, eth, chl	B27[2], 228
3109	2-Benzothiazolinethione, 3-methyl $C_8H_7NS_2$	181.28	nd (al), pr (aa)	335[757]	90	....	....	al, bz, chl	B27[2], 233
3110	Benzothiophene or Thionaphthene C_8H_6S	134.20	lf	221, 103-5[20]	32	1.1484[32/4]	1.6374[37]	al, eth, ace, bz	B17[3], 482
3111	Benzothiophene, 3-hydroxy C_8H_6OS	150.20	nd (w)	....	71	....	....	al, eth, ace, bz	B17[4], 1458

No.	Name, Synonyms, and Formula	Mol. wt.	Color, crystalline form, specific rotation and λ_{max} (log ε)	b.p. °C	m.p. °C	Density	n_D	Solubility	Ref.
3112	Benzothiophene, 4-hydroxy .... C_8H_6OS	150.20	nd (sub), cr (peth)	sub	78-9	....	....	al	B17[4], 1467
3113	2,3-Benzothiophene, 5-methyl .... C_9H_8S	148.22	....	111-5[12]	19-22	1.111[22/D]	1.615[22/D]	....	B17[4], 502
3114	2,3-Benzothiophene quinone or Thioisatin............. $C_8H_4O_2S$	164.18	gold-ye pr (al)	247	121	....	....	al, bz, aa	B17[4], 6129
3115	1,2,3-Benzotriazole or Azimidobenzene .... $C_6H_5N_3$	119.13	nd (chl or bz)	204[15]	100	....	....	al, bz, chl	B26[2], 17
3116	1,2,3-Benzoxadiazole, 5,7-dinitro .... $C_6H_2N_4O_5$	210.11	ye pl (al)	....	158	....	....	al	B16[3], 549
3117	2,3-Benzoxazin-1-one .... $C_7H_5NO_2$	147.14	cr (bz)	....	120d	....	....	....	B27[2], 249
3118	Benzoxazole .... C_7H_5NO	119.12	pr (dil al)	182.5, 45[4]	31	....	1.5594[20]	al	B27[2], 17
3119	Benzoxazole, 2-chloro .... C_7H_4ClNO	153.57	....	201-2	7	1.3453[18/4]	1.5678[20]	....	B27[2], 17
3120	Benzoxazole, 2,5-dimethyl .... C_9H_9NO	147.18	....	218-9	....	....	1.5412[20]	....	B27[2], 25
3121	Benzoxazole, 2-hydroxy or 2(3)-Benzoxazolone .... $C_7H_5NO_2$	135.13	ng (bz or w + 1)	230[30]	141—2 (anh), 97—8 (+w)	....	....	al, eth	B27[2], 223
3122	Benzoxazole, 2-(2-hydroxyphenyl) .... $C_{13}H_9NO_2$	211.22	pink nd (al or aa)	338	123-4	....	....	al, eth, ace, bz	B27[2], 91
3123	Benzoxazole, 4-hydroxy-2-phenyl .... $C_{13}H_9NO_2$	211.22	nd (bz)	....	138-9	....	....	al, eth, aa	B27[2], 88
3124	Benzoxazole, 5-hydroxy-2-phenyl .... $C_{13}H_9NO_2$	211.22	nd (lig or dil al)	....	175	....	....	al, bz, chl, aa	B27[2], 88
3125	Benzoxazole, 6-hydroxy-2-phenyl .... $C_{13}H_9NO_2$	211.22	nd	....	216-7	....	....	ace, bz, aa	B27[2], 88
3126	Benzoxazole, 7-hydroxy-2-phenyl .... $C_{13}H_9NO_2$	211.22	nd (bz or dil al)	....	191-2	....	....	al, eth, bz, chl, lig	B27[2], 91
3127	Benzoxazole, 2-mercapto .... C_7H_5NOS	151.19	nd (w)	....	196	....	....	eth, aa	B27[2], 224
3128	Benzoxazole, 2-methyl .... C_8H_7NO	133.16	....	200—1, 59-60[12]	8.5-10	1.1211[20/4]	1.5497[20]	al, eth	B27[2], 20
3129	Benzoyl acetamide or α-Benzoyl acetanilide............ $C_6H_5COCH_2CONHC_6H_5$	239.28	lf (bz)	....	108	....	....	al	B12[3], 1005
3130	Benzoyl acetic acid .... $C_6H_5COCH_2CO_2H$	164.17	nd (bz-peth)	....	103-4d	....	....	al, eth, bz	B10[3], 2990
3131	Benzoyl acetic acid, methyl-ester or Ethylbenzoyl acetate $C_6H_5COCH_2CO_2C_2H_5$	192.22	....	265-70, 165[14]	<0	1.1220[20/4]	1.5312[16]	al, eth	B10[3], 2991
3132	Benzoyl acetic acid, methyl-ester .... $C_6H_5COCH_2CO_2CH_3$	178.19	pa ye	265d, 151.5[12]	....	1.158[29/4]	1.537[20]	al, eth, ace	B10[3], 2991
3133	Benzoyl acetone or 1-Phenyl-1,3-butanedione........... $C_6H_5COCH_2COCH_3$	162.19	pr	261-2	56	1.0599[74/4]	1.5678[78]	eth	B7[3], 34821
3134	Benzoyl acetonitrile .... $C_6H_5COCH_2CN$	145.16	pr or lf (w)	160[10]	80-1	....	....	al, eth, bz	B10[3], 2994
3135	Benzoyl azide .... $C_6H_5CON_3$	147.14	pl (ace)	exp	32	....	....	al, eth	B9[3], 1324
3136	Benzoyl bromide .... C_6H_5COBr	185.03	....	218-9, 48-50[0.05]	−24	1.570[15]	1.5868[25]	eth	B9[3], 1064
3137	Benzoyl chloride .... C_6H_5COCl	140.57	....	197.2, 71[9]	....	1.2120[20/4]	1.5537[20]	eth	B9[3], 1058
3138	Benzoyl chloride, 2-bromo .... 2-BrC_6H_4COCl	219.47	nd	245, 118[10]	11	....	1.5963[20]	....	B9[3], 1387
3139	Benzoyl chloride, 4-bromo .... 4-Br-C_6H_4COCl	219.47	nd (peth)	245-7, 123-6[15]	42	....	....	al, eth, bz, lig	B9[3], 1418
3140	Benzoyl chloride, 2-chloro .... 2-ClC_6H_4COCl	175.02	....	238, 110[15]	−4	....	1.5726[20]	....	B9[3], 1338
3141	Benzoyl chloride, 3-chloro .... 3-ClC_6H_4COCl	175.02	....	225, 103-4[14]	....	....	1.5677[20]	....	B9[3], 1348
3142	Benzoyl chloride, 4-chloro .... 4-ClC_6H_4COCl	175.02	....	222, 111[18]	16	1.3770[20/4]	1.5756[20]	....	B9[3], 1362
3143	Benzoyl chloride, 2,4-dichloro .... 2,4-Cl_2C_6H_3COCl	209.47	lig	150[34], 111[7.5]	15-8	....	1.5895[20]	....	B9[3], 1375
3144	Benzoyl chloride, 3,4-dichloro .... 3,4-Cl_2C_6H_3COCl	209.47	lig	242, 160[42]	24-6	....	....	....	B9[3], 1379

No.	Name, Synonyms, and Formula	Mol. wt.	Color, crystalline form, specific rotation and λ_{max} (log ε)	b.p. °C	m.p. °C	Density	n_D	Solubility	Ref.
3145	Benzoyl chloride, 3,5-dinitro 3,5-$(NO_2)_2C_6H_3COCl$	230.57	ye nd (bz)	....	74	....	....	eth	B9[3], 1779
3146	Benzoyl chloride, 3-hydroxy 3-HOC_6H_4COCl	156.57	....	110-2[0.5]	<-15	....	....	chl	B10[2], 82
3147	Benzoyl chloride, 2-methoxy or o-Anisoyl chloride 2-$CH_3OC_6H_4COCl$	170.60	....	254, 128[11]	....	....	....		B10[3], 151
3148	Benzoyl chloride, 3-methoxy 3-$CH_3OC_6H_4COCl$	170.60	....	243-4, 123-5[15]	....	....	....	eth, ace, bz	B10[3], 252
3149	Benzoyl chloride, 4-methoxy 4-$CH_3OC_6H_4COCl$	170.60	nd	262-3, 91[1]	24-5	1.261[20/4]	1.580[20]	eth, ace, bz	B10[3], 337
3150	Benzoyl chloride, 2-nitro 2-$NO_2C_6H_4COCl$	185.57	....	275-8, 154[18]	....	....	....	eth	B9[3], 1477
3151	Benzoyl chloride, pentachloro C_6Cl_5COCl	312.80	pl (al)	5	87	....	....	al	B9[2], 230
3152	Benzoyl chloride, 2,4,6-trimethyl or Mesitoyl chloride 2,4,6-$(CH_3)_3C_6H_2COCl$	182.65	....	143-6[60]	....	....	....	ace	B9[3], 2469
3153	Benzoyl fluoride C_6H_5COF	124.12	....	154-5	>1	....	....	al, eth	B9[3], 1058
3154	Benzoyl glycolic acid $C_6H_5COCH(OH)CO_2H$	180.17	lo pr (lig)	....	112	....	....	al, eth, chl	B10[2], 677
3155	Benzoyl iodide C_6H_5COI	232.02	nd	128[20]	3	1.748[18/18]	1.137[20]	al, eth	B9[3], 1064
3156	Benzoyl peroxide or Dibenzoyl peroxide $C_6H_5CO-OO-COC_6H_5$	242.23	rh (eth), nr	exp	100-8	....	1.543	al, eth, ace, bz	B9[3], 1052
3157	Benzoyl peroxide, 3,3'-dinitro 3-$O_2NC_6H_4CO-OO-OO(COC_6H_4NO_2-3)$	332.24	nd (al)	....	139-40d	....	....	eth, ace, bz	B9[2], 252
3158	Benzoyl peroxide, 4,4'-dinitro 4-$O_2NC_6H_4CO-OO(COC_6H_4NO_2-4)$	332.24	ye cr (al r) nd (to)	....	136d	....	....	to	B9[2], 270
3159	β-Benzoyl propionic acid $C_6H_5COCH_2CH_2CO_2H$	178.19	lf (dil al)	....	116	....	....	al, eth, bz, chl	B10[2], 482
3160	Benzyl alcohol or α-Hydroxy toluene $C_6H_5CH_2OH$	108.15	....	205.3, 93[10]	-15.3	1.0419[24/4]	1.5396[20]	w, al, eth, ace, bz	B6[4], 2222
3161	Benzyl alcohol, 2-amino 2-$H_2NC_6H_4CH_2OH$	123.15	....	270-80, 160[5-10]	83-4	....	....	w, al, eth, bz, chl	B13[3], 1615
3162	Benzyl alcohol, 2-amino-3-methyl 3-CH_3-2-$H_2NC_6H_3CH_2OH$	137.18	nd (bz)	135-45[12]	71	....	....		B13[2], 367
3163	Benzyl alcohol, 2-amino-4-methyl 4-CH_3-2-$H_2NC_6H_3CH_2OH$	137.18	nd (bz)	140-50[13]	141	....	....		B13[2], 369
3164	Benzyl alcohol, 2-amino-5-methyl 2-NH_2-5-$CH_3C_6H_3CH_2OH$	137.18	nd (bz)	145-50[12]	123	....	....		B13[2], 367
3165	Benzyl alcohol, 3-amino 3-$H_2NC_6H_4CH_2OH$	123.15		....	97	....	....	w, al, eth, bz, chl	B13[3], 1617
3166	Benzyl alcohol, 4-amino 4-$H_2NC_6H_4CH_2OH$	123.15		....	65	....	....	al, eth, bz	B13[3], 1619
3167	Benzyl alcohol, 2-bromo 2-$BrC_6H_4CH_2OH$	187.04	nd (lig)	....	80	....	....	w, al, eth	B6[4], 2600
3168	Benzyl alcohol, 4-bromo 4-$BrC_6H_4CH_2OH$	187.04	nd (lig)	....	77	....	....	al, eth, bz	B6[4], 2602
3169	Benzyl alcohol, 5-bromo-2-hydroxy or 5-Bromosaligenin 5-Br-2-$HOC_6H_3CH_2OH$	203.04	lf (bz)	....	113	....	....	al, eth, bz, chl	B6[3], 4541
3170	Benzyl alcohol, 2-chloro 2-$ClC_6H_4CH_2OH$	142.59	lf or nd (dil al)	230, 100-5[28]	74	....	....	al, eth, lig	B6[4], 2589
3171	Benzyl alcohol, 4-chloro 4-$ClC_6H_4CH_2OH$	142.59	nd (w), pr (bz or bz-lig)	235	75	....	....	al, eth, bz	B6[4], 2593
3172	Benzyl alcohol, 2,5-dihydroxy or Gentisyl alcohol 2,5-$(HO)_2C_6H_3CH_2OH$	140.15	nd (chl)	sub vac	100	....	....	w, al, eth, chl	B6[3], 6322
3173	Benzyl alcohol, 2,3-dimethoxy 2,3-$(CH_3O)_2C_6H_3CH_2OH$	168.19	....	257-8, 155-60[17]	50	....	....		B6[2], 1082
3174	Benzyl alcohol, 3,4-dimethoxy 3,4-$(CH_3O)_2C_6H_3CH_2OH$	168.19	....	135-8[0.5]		1.179[17/17]	1.555[17/0]	al, w	B6[3], 6324
3175	Benzyl alcohol, α,α-dimethyl $C_6H_5C(OH)(CH_3)_2$	138.20	pr	202, 93[11]	35-7	0.9735[20/4]	1.5325[20]	al, eth, bz, aa	B6[3], 1813
3176	Benzyl alcohol, α-ethyl $C_6H_5CH(OH)C_2H_5$	136.20	....	213-5, 98[10]		0.9938[22/4]	1.5210[22]	al, eth	B6[3], 1792
3177	Benzyl alcohol, hexahydro $C_6H_{11}CH_2OH$	114.19	....	183, 83[14]	-43	0.9297[20/4]	1.4644[20]	al, eth	B6[4], 106

No.	Name, Synonyms, and Formula	Mol. wt.	Color, crystalline form, specific rotation and λ_{max} (log ε)	b.p. °C	m.p. °C	Density	n_D	Solubility	Ref.
3178	Benzyl alcohol, 2-hydroxy or Saligenin Salicyl alcohol.... 2-HOC₆H₄CH₂OH	124.15	lf (bz), nd or pl (w, eth)	sub	87	1.1613²⁵	....	w, al, eth, bz, chl	B6³, 4537
3179	Benzyl alcohol, 2-hydroxy, glucoside or Saligenin-β-D-glucoside .... C₁₃H₁₈O₇	286.29	rh nd or lf (w), [α]²⁰_D +62.6 (w, c=3)	240d	205—9	1.434²⁶	....	w, al, aa	B6³, 4537
3180	Benzyl alcohol, 3-hydroxy 3-HO-C₆H₄CH₂OH	124.15	nd (bz), cr (CCl₄)	300d	73	1.161²⁵	....	al, eth, w	B6³, 4545
3181	Benzyl alcohol, 4-hydroxy 4-HOC₆H₄CH₂OH	124.15	pr or nd (w)	252	124-5	....	....	w, al, eth	B6³, 4546
3182	Benzyl alcohol, 4-hydroxy-3-methoxy or Vanillyl alcohol 3-CH₃O-4-HOC₆H₃CH₂OH	154.17	pr (w), nd (bz)	d	115	....	....		B6³, 6323
3183	Benzyl alcohol, 2-methoxy or Saligenin, 2-methylether ... 2-CH₃OC₆H₄CH₂OH	138.17	....	249, 119⁸	......	1.0395²⁵/¹⁵	1.5455²⁰	al, eth	B6³, 4538
3184	Benzyl alcohol, 4-methoxy or Anisyl alcohol 4-CH₃OC₆H₄CH₂OH	138.17	nd	259.1, 134-5¹²	25	1.109²⁶/⁴	1.5420²⁵	al, eth, w	B6³, 4547
3185	Benzyl alcohol, 2-methyl 2-CH₃C₆H₄CH₂OH	122.17	nd	223⁷⁵⁰, 117-9²⁰	37-9	1.023⁴⁰	....	al, eth, chl	B6³, 1733
3186	Benzyl alcohol, 3-methyl 3-CH₃C₆H₄CH₂OH	122.17	....	215-6	<−20	0.9157¹⁷	....	al, eth	B6³, 1768
3187	Benzyl alcohol, 4-methyl 4-CH₃C₆H₄CH₂OH	122.17	nd (hp)	217, 116-8²⁰	61-2	0.978²²/⁴	....	al, eth	B6³, 1779
3188	Benzyl alcohol, α-1-naphthyl .... C₆H₅CH(OH)-1-C₁₀H₇	234.30	cr (al, lig)	ca 360	86.5	....	....	al, eth, bz	B6³, 3608
3189	Benzyl alcohol, α-2-naphthyl .... C₆H₅CH(OH)-2-C₁₀H₇	234.30	nd (al, lig)	....	87-8	....	....	al, eth, bz	B6³, 3609
3191	Benzyl alcohol, 2-nitro 2-NO₂C₆H₄CH₂OH	153.14	nd (w)	270, 168²⁰	74	....	....	al, eth	B6³, 1563
3192	Benzyl alcohol, 3-nitro 3-NO₂C₆H₄CH₂OH	153.14	rh nd (w)	175-80³	30.5	1.296¹⁹/¹⁵	....	al, eth, w	B6³, 1565
3193	Benzyl alcohol, 4-nitro 4-NO₂C₆H₄CH₂OH	153.14	nd (w)	250-60d, 185²²	96-7	....	....	al, eth	B6³, 1567
3194	Benzyl alcohol, α-isopropyl .... C₆H₅CH(OH)CH(CH₃)₂	150.22	d: [α]²⁰_D +47.7, l: [α]²⁰_D −25.2	222.4, 112-3¹⁵		0.9869¹⁴	1.5193¹⁴	al, ace	B6³, 1859
3195	Benzyl alcohol, 4-isopropyl 4-i-C₃H₇C₆H₄CH₂OH	150.22	....	246, 122.5¹³	28	0.9401²⁰/⁴	1.519²⁰	al, eth, bz	B6³, 1911
3196	Benzyl amine or α-Aminotoluene .... C₆H₅CH₂NH₂	107.16		185⁷⁷⁰, 90¹²		0.9813²⁰/⁴	1.5401²⁰	w, al, eth, ace, bz	B12³, 2194
3197	Benzyl amine, 3-bromo 3-BrC₆H₄CH₂NH₂	186.07		244-5, 84¹⁵				eth	B12³, 2349
3198	Benzyl amine, 4-bromo 4-BrC₆H₄CH₂NH₂	186.07		127¹⁵	20			eth	B12³, 2351
3199	Benzyl amine, N-tert-butyl or Benzyl tert-butyl amine..... C₆H₅NHC(CH₃)₃	163.26		73.7⁴⁴			1.4951²⁵		B12³, 2208
3200	Benzyl amine, N-(3-butynyl)-N-methyl .... C₆H₅CH₂N(CH₃)CH₂CH₂C≡CH	173.26		127¹⁶		0.9372²⁰	1.5202²⁰		C51, 7295
3201	Benzyl amine, 2-chloro 2-ClC₆H₄CH₂NH₂	141.60		72-3²			1.5594²⁵/_D		B12³, 2340
3202	Benzyl amine, 3-chloro 3-ClC₆H₄CH₂NH₂	141.60		89²			1.5570²⁵/_D		B12³, 2342
3203	Benzyl amine, 4-chloro 4-ClC₆H₄CH₂NH₂	141.60		109-10¹³			1.5566²⁵		B12³, 2343
3204	Benzyl amine, N-cyclopropyl .... C₆H₅CH₂NHC₃H₅	147.22		80-1⁵		1.5222²⁵			C59, 509
3205	Benzyl amine, 2,4-dichloro 2,4-Cl₂C₆H₃CH₂NH₂	176.04		124-6¹³			1.5762²⁵/_D	chl	....
3206	Benzyl amine, 3,4-dimethoxy .... 3,4-(CH₃O)₂C₆H₃CH₂NH₂	167.21		154-8¹²	1.43⁰			chl	B13³, 2183
3207	Benzyl amine, N,N-dimethyl or Benzyl dimethyl amine .. C₆H₅CH₂N(CH₃)₂	135.21		180-2, 73-4¹⁵		0.915⁰/⁴	1.5011²⁰	al, eth	B12³, 2203
3208	Benzyl amine, 2,4-dimethyl .... 2,4-(CH₃)₂C₆H₃CH₂NH₂	135.21		86-90¹⁰					B12³, 2709
3209	Benzyl amine, 2,5-dimethyl .... 2,5-(CH₃)₂C₆H₃CH₂NH₂	135.21		225-6			1.5377²⁰/_D		ALD 12695-0

No.	Name, Synonyms, and Formula	Mol. wt.	Color, crystalline form, specific rotation and λ_{max} (log ϵ)	b.p. °C	m.p. °C	Density	n_D	Solubility	Ref.
3210	Benzylamine, *N,N*-diphenyl or Benzyl diphenyl amine.... $C_6H_5CH_2N(C_6H_5)_2$	259.36	nd (al)		95			eth, ace, bz	B12³, 2220
3211	Benzylamine, *N*-ethyl or Benzyl ethyl amine	135.21		194, 82[15]		0.9350[17/15]	1.5117[20]	al, eth, bz	B12³, 2204
	$C_6H_5CH_2NHC_2H_5$								
3212	Benzyl amine, *N*-ethyl-*N*-phenyl or Benzyl ethyl phenyl amine.......... $C_6H_5CH_2N(C_2H_5)C_6H_5$	211.31	pa ye	285-6d	34-6	1.034[19/4]	1.5930[20]	al, eth	B12³, 2218
3213	Benzyl amine hydrochloride $C_6H_5CH_2NH_2\cdot HCl$	143.62			255-8			w, al	B12³, 2197
3214	Benzyl amine, 2-hydroxy--5-nitro $5\text{-}NO_2\text{-}2\text{-}HOC_6H_3CH_2NH_2$	168.16	ye nd or lf (w)		253d			w, al	B13, 587
3215	Benzyl amine, 4-hydroxy-3-nitro $3\text{-}NO_2\text{-}4\text{-}HOC_6H_3CH_2NH_2$	168.16	og red nd (w + l)		225d			w	B13, 610
3216	Benzylamine, *N*-(3-hydroxy propyl)-*N*-methyl $C_6H_5CH_2N(CH_3)CH_2CH_2CH_2OH$	179.26		132-5[4]			1.425[20]	CS_2	B12³, 2234
3217	Benzyl amine, 3-methoxy $3\text{-}CH_3OC_6H_4CH_2NH_2$	137.18		141[30]				al, eth	B13³, 1569
3218	Benzyl amine, 4-methoxy $4\text{-}CH_3OC_6H_4CH_2NH_2$	137.18		236-7, 133-4[33]		1.050[15]	1.5462[20]	w, al, eth	B13³, 1594
3219	Benzyl amine, 2-methyl $2\text{-}CH_3C_6H_4CH_2NH_2$	121.18		66-8[2]			1.5408[25]		B12³, 2460
3220	Benzyl amine, *N*-nitroso-*N*-phenyl $C_6H_5CH_2N(NO)C_6H_5$	212.25	ye nd (al)		58			al, eth, chl, lig	B12³, 2233
3221	Benzyl amine, *N*-phenyl or *N*-Benzyl aniline $C_6H_5CH_2NHC_6H_5$	183.25		306-7, 171.5[10]	37-8	1.0298[65/4]	1.6118[25]	al, eth	B12³, 2215
3222	Benzyl amine, *N*-2-tolyl $C_6H_5CH_2NH(C_6H_4CH_3\text{-}2)$	197.28		300-5, 176[10]	60	1.0142[65/4]	1.5861[65]	al, ace, chl	B12³, 2220
3223	Benzyl arsonic acid $C_6H_5CH_2ASO(OH)_2$	216.07	nd (al)		167-8			chl	B16³, 1057
3224	Benzyl azide $C_6H_5CH_2N_3$	133.15		108[23], 74[11]		1.0655[25/4]	1.53414[25]	al, eth	B5⁴, 759
3225	Benzyl boric acid $C_6H_5CH_2B(OH)_2$	135.96	cr (w, bz)		140, 104 (hyd)			al, bz	B16³, 1278
3226	Benzyl bromide $C_6H_5CH_2Br$	171.04	pr	201, 114[15]	-3	1.4380[25]	1.5752[20]	al, eth	B5⁴, 829
3227	Benzyl bromide, 2-bromo $2\text{-}BrC_6H_4CH_2Br$	249.94	cr (al, lig)	129[19]	31			al, eth, aa	B5⁴, 836
3228	Benzyl bromide, 3-bromo $3\text{-}BrC_6H_4CH_2Br$	249.94	nd or lf		41			al, eth, aa	B5⁴, 836
3229	Benzyl bromide, 4-bromo $4\text{-}BrC_6H_4CH_2Br$	249.94	nd (al)		63			al, eth, bz, aa	B5⁴, 836
3230	Benzyl bromide, 2-chloro $2\text{-}ClC_6H_4CH_2Br$	205.49		120[10]				bz, chl	B5⁴, 833
3231	Benzyl bromide, 2,4-dibromo $2,4\text{-}Br_2C_6H_3CH_2Br$	328.85			40-1			al	B5², 240
3232	Benzyl bromide, 3,5-dibromo $3,5\text{-}Br_2C_6H_3CH_2Br$	328.85	pl or nd (al)	173[19]	96			al	B5³, 719
3233	Benzyl bromide, 2-nitro $2\text{-}O_2NC_6H_4CH_2Br$	216.04	pl (dil al)		46-7			al, eth, bz, lig	B5³, 752
3234	Benzyl bromide, 3-nitro $3\text{-}O_2NC_6H_4CH_2Br$	216.04	nd or pl (al)	153-4[8]	58-9			al	B5⁴, 860
3235	Benzyl bromide, 4-nitro $4\text{-}O_2NC_6H_4CH_2Br$	216.04	nd (al)		99-100			al, eth, aa	B5⁴, 861
3236	Benzyl butyl ether $C_6H_5CH_2C_4H_9$	164.25		223, 92[10]		0.9227[20/4]	1.4833[20]	al, eth, ace	B6⁴, 2231
3237	Benzyl-*iso*-butyl ether $C_6H_5CH_2O\text{-}i\text{-}C_4H_9$	164.25		211-2[743]		0.9233[20/4]	1.4826[20]	eth, chl	B6³, 1457
3238	Benzyl chloride or α-chloro toluene $C_6H_5CH_2Cl$	126.59		179.3, 66[11]	-39	1.1002[20/20]	1.5391[20]	al, eth, chl	B5⁴, 809
3239	Benzyl chloride, 2-bromo $2\text{-}BrC_6H_4CH_2Cl$	205.49		124-6[20]				al, eth	B5⁴, 832
3240	Benzyl chloride, 3-bromo $3\text{-}BrC_6H_4CH_2Cl$	205.49	nd (al or peth)	119[18]	22-3			al	B5⁴, 832
3241	Benzyl chloride, 4-bromo $4\text{-}BrC_6H_4CH_3$	205.49	nd (al or peth)	236, 110-1[9]	50			al, eth, peth	B5⁴, 832
3242	Benzyl chloride, 2-chloro $2\text{-}ClC_6H_4CH_2Cl$	161.03		217, 94-5[10]	-17	1.2699[0/4]	1.5530[20]		B5⁴, 816

No.	Name, Synonyms, and Formula	Mol. wt.	Color, crystalline form, specific rotation and λ_{max} (log ε)	b.p. °C	m.p. °C	Density	n_D	Solubility	Ref.
3243	Benzyl chloride, 3-chloro 3-ClC₆H₄CH₂Cl	161.03	....	215-6[753], 110-1[25]		1.2695[15/4]	1.5554[20]	al	B5⁴, 816
3244	Benzyl chloride, 4-chloro 4-ClC₆H₄CH₂Cl	161.03	nd (dil al)	222, 117[20]	31	....	....	eth, bz, aa	B5⁴, 816
3245	Benzyl chloride, 2,6-dichloro 2,6-Cl₂C₆H₃CH₂Cl	195.48	cr (lig, eth, al-eth)	117-9[14]	39-40			al, eth, lig	B5⁴, 820
3246	Benzyl chloride, 3,4-dichloro 3,4-Cl₂C₆H₃CH₂Cl	195.48	....	241	37.5			al	B5⁴, 820
3247	Benzyl chloride, 3,5-dichloro 3,5-Cl₂C₆H₃CH₂Cl	195.48	cr(MeOH)	60[0.35]	36			al	B5³, 699
3248	Benzyl chloride, 4-hydroxy-3-nitro 3-NO₂-4-HOC₆H₃CH₂Cl	187.58	ye nd (lig or al), lf (peth)		75			al, bz	B6, 413
3249	Benzyl chloride, 4-methoxy 4-CH₃OC₆H₄CH₂Cl	156.61	....	116-20[15], 83-4[2]	....	1.159[20]	1.553		B6⁴, 2137
3250	Benzyl chloride, 2-nitro 2-NO₂C₆H₄CH₂Cl	171.58	cr (lig)		50-2		1.5557[62]	al, eth, ace, bz, aa	B5⁴, 854
3251	Benzyl chloride, 3-nitro 3-NO₂C₆H₄CH₂Cl	171.58	pa ye nd (lig)	175-83[30-5]	45-7		1.5577[62]	al, eth, ace, bz, aa	B5⁴, 855
3252	Benzyl chloride, 4-nitro 4-NO₂C₆H₄CH₂Cl	171.58	pl or nd (al)	....	71		1.5647[62]	al, eth, ace, bz	B5⁴, 856
3253	Benzyl chloride, 2,4,5-trichloro 2,4,5-Cl₃C₆H₂CH₂Cl	229.92	....	273		1.547[20]	....	al, eth, ace	B5⁴, 822
3254	Benzyl chloromethyl ether C₆H₅CH₂OCH₂Cl	156.61	....	103[13]		1.1350[20/4]	1.5192[20]	....	B6⁴, 2253
3255	Benzyl ether or Dibenzyl ether .... (C₆H₅CH₂)₂O	198.27	....	298, 160[11]	3.6	1.0428[20/4]	1.5168[20]	al, eth	B6⁴, 2240
3256	Benzyl diphenyl methanol .... (C₆H₅)₂C(OH)CH₂C₆H₅	274.37	nd (bz-lig), pr (peth)	222[11]	89-90			al	B6³, 3680
3257	Benzyl ethyl ether .... C₆H₅CH₂OC₂H₅	136.20	....	185, 70[15]		0.9490[20/4]	1.4955[20]	al, eth	B6⁴, 2229
3258	Benzyl ethyl sulfide C₆H₅CH₂SC₂H₅	152.25	....	218-20					B6⁴, 2635
3259	Benzyl fluoride C₆H₅CH₂F	110.13	nd (fr)	139.8[753], 40[14]	-35	1.0228[25/4]	1.4892[25]		B5⁴, 800
3260	N-Benzyl hydroxy, 1-amine .... C₆H₅CH₂NHOH	123.16	nd (peth or lig)		57			al, lig	B15³, 20
3261	Benzyl iodidie or α-Iodo toluene .... C₆H₅CH₂I	218.04	col or ye nd (MeOH)	93[10]	24.5	1.7335[25]	1.6334[25]	al, eth, bz	B5⁴, 842
3262	Benzyl isocyanide C₆H₅CH₂NC	117.15	....	198-200d, 93-4[55]		0.972[15]			B12³, 2241
3263	Benzyl isothiocyanate or Benzyl mustard oil ..... C₆H₅CH₂NCS	149.22	ye oil	243, 124-5[12]		1.1246[16/4]	1.6049[15]	al, eth	B12², 567
3264	Benzylmercaptan or α-Mercapto toluene........ C₆H₅CH₂SH	124.22	....	194-5		1.058[20]	1.5751[20]	al, eth	B6⁴, 2632
3265	Benzyl methyl ether .... C₆H₅CH₂OCH₃	122.17	....	170, 59-60[12]	-52.6	0.9634[20/4]	1.5008[20]	al, eth, bz	B6⁴, 2229
3266	Benzyl (2-methylbutyl) ether .... C₆H₅CH₂OCH(CH₃)C₂H₅	178.28	....	231[722]		0.911[22/4]	1.4854[22]	al, eth	B6, 341
3267	Benzyl methyl sulfide .... C₆H₅CH₂SCH₃	138.23	....	195-8			1.5620[20/D]		B6⁴, 2633
3268	Benzyl α-naphthyl ketone .... α-C₁₀H₇COCH₂C₆H₅	246.31	ta, lf (al)	194-6[0.05]	66-7			eth, chl	B7², 461
3269	Benzyl β-naphthyl ketone .... β-C₁₀H₇COCH₂C₆H₅	246.31	nd (al)		99.5			al, eth, bz, chl	B7², 461
3270	Benzyl phenyl ketone .... C₆H₅COCH₂C₆H₅	196.25	pl (al)	320, 177[20]	60	1.201[0/4]		al, eth, chl	B7², 368
3271	Benzyl isopentyl ether .... C₆H₅CH₂O-i-C₅H₁₁	178.28	....	236-7[748], 117-9[19]		0.9098[20/4]	1.4792[20]	al, eth	B6², 410
3272	Benzyl phenyl methanol d CH₅CH₂CH(OH)C₆H₅	198.27	nd (eth-peth or dil al), [α][25/D] + 53 (al)	167-70[10]	67-8	1.0358[70/4]	....	al, eth	B6³, 3390
3273	Benzyl phenyl methanol dl .... C₆H₅CH₂CH(OH)C₆H₅	198.27	nd (bz-peth)	177[15]	69			al, eth	B6³, 3390
3274	Benzyl phenyl methanol l .... C₆H₅CH₂CH(OH)C₆H₅	198.27	[α][20/D] -9.4 (w, c=10)	....	67	1.0358[70/4]	....	al, eth	B6³, 3390

No.	Name, Synonyms, and Formula	Mol. wt.	Color, crystalline form, specific rotation and λ_{max} (log ε)	b.p. °C	m.p. °C	Density	n_D	Solubility	Ref.
3275	Benzyl phenyl ketone or α-Phenylacetophenone .... $C_6H_5CH_2COC_6H_5$	196.25	pl (al)	320, 177[20]	60	1.201[0/4]	....	al, eth, chl	B7[2], 368
3276	Benzyl phenyl ketone, α-chloro or Desyl chloride ...,,, $C_6H_5COCHClC_6H_5$	230.70	nd (al)	d	68.5	....			B7[2], 369
3277	Benzyl phenyl sulfide .... $C_6H_5CH_2SC_6H_5$	200.21	lf (al)	197[27]	42-3.5	....		al, eth	B6[4], 2644
3278	Benzyl phenyl sulfone .... $C_6H_5CH_2SO_2C_6H_5$	232.30	nd (al)	....	146	1.1261[153/4]			B6[4], 2647
3279	Benzyl-4-tolyl sulfone .... $(4-CH_3C_6H_4)SO_2CH_2C_6H_5$	246.33	nd (al)	....	144-5			al, bz, aa	B6[4], 2649
3280	Benzyl sulfide or Dibenzyl sulfide .... $(C_6H_5CH_2)_2S$	214.33	pl (eth or chl)	d	49-50	1.071[50/50]		al, eth	B6[4], 2649
3281	Benzyl sulfonamide $C_6H_5CH_2SO_2NH_2$	171.22	pr or nd (w), nd (al)		105			w, al	B11[3], 331
3282	Benzyl sulfonamide, N-methyl $C_6H_5CH_2SO_2NHCH_3$	185.25	nd or lf (aa-lig)		108-9			al, eth	B11[2], 73
3283	Benzyl sulfonamide, N-2-tolyl $C_6H_5CH_2SO_2NH(C_6H_4CH_3-2)$	261.35	cr (dil al)		83			al	B12[2], 452
3284	Benzyl sulfonamide, N-3-tolyl $C_6H_5CH_2SO_2NH(C_6H_4CH_3-3)$	261.35	cr (al)		75			al	B12[2], 473
3285	Benzyl sulfonamide, N-4-tolyl $C_6H_5CH_2SO_2NH(C_6H_4CH_3-4)$	261.35	pr (al)		113		,,	al, chl	B12[2], 528
3286	Benzyl sulfonyl chloride ...,,,,, ... $C_6H_5CH_2SO_2Cl$	190.65	pr (eth), nd (hz)		93			eth, bz	B11[3], 331
3287	Benzyl sulfoxide or Dibenzyl sulfoxide $(C_6H_5CH_2)_2SO$	230.33	lf (al, w)	210d	134-5	.....	, ,	al, chl	B6[4], 2651
3288	Benzyl thiocyanate ,,,,, ...,,,,,, ... $C_6H_5CH_2SCN$	149.22	pr (al)	256	43			al, eth	B6[4], 2680
3289	Benzyl 4-tolyl sulfide .... $C_6H_5CH_2S(C_6H_4CH_3-4)$	246.33	nd (al)		144-5			al, bz, aa	B6[4], 2649
3290	Benzyltrimethylammonium bromide .... $C_6H_5CH_2N^+(CH_3)_3Br^-$	230.16	pl (al-lig), (w)		235			w, al	B12[3], 2204
3291	Benzyltrimethylammonium chloride .... $C_6H_5CH_2N^+(CH_3)_3Cl^-$	185.70	(ace)		243			w	B12[3], 2203
3292	Benzyltrimethylammonium iodide .... $C_6H_5CH_2N^+(CH_3)_3I^-$	277.15	(al)		180			al	B12[3], 2204
3293	Benzyltrimethylammonium nitrate .... $C_6H_5CH_2N^+(CH_3)_3NO_3$	212.25			151-60			w, al	B12[2], 546
3294	Benzyldimethylphenylammonium chloride .... $C_6H_5CH_2N^+(CH_3)_2C_6H_5Cl^-$	247.77			134-8			w	B12[3], 2218
3295	Benzyltriethylammonium bromide .... $C_6H_5CH_2N^+(C_2H_5)_3Br^-$	272.23			194d			w	B12[2], 547
3296	Benzylidene bromide or α,α-Dibromotoluene, Benzal bromide .... $C_6H_5CHBr_2$	249.94	....	156[23]	....	1.51[15]	1.6147[20]	al, eth	B5[4], 836
3297	Benzylidene bromide, 4-nitro $4-NO_2C_6H_4CHBr_2$	294.94	nd (al)		84			al, eth	B5[4], 862
3298	Benzylidene chloride or α,α-Dichlorotoluene........... $C_6H_5CHCl_2$	161.03		205.2	-16.4	1.2557[14]	1.5502[20]	al, eth	B5[4], 817
3299	Benzylidene chloride, 2-chloro .... $2-ClC_6H_4CHCl_2$	195.48		228.5		1.399[15]			B5[3], 699
3300	Benzylidene chloride, 2,5-dichloro .... $2,5-Cl_2C_6H_3CHCl_2$	229.92	cubic cr (chl)	....	42			al, eth, bz	B5[3], 702
3301	Benzylidene chloride, 3,4-dichloro .... $3,4-Cl_2C_6H_3CHCl_2$	229.92		257	....	1.518[22/22]	....	al, eth, bz, aa	B5[3], 702
3302	Benzylidene chloride, 3,5-dichloro $3,5-Cl_2C_6H_3CHCl_2$	229.92	cr (MeOH or dil aa)		36.5			eth, ace	B5[3], 703
3303	Benzylidene chloride, 3-nitro $3-NO_2C_6H_4CHCl_2$	206.03	mcl (al)	....	65			al, eth	B5[3], 750
3304	Benzylidene chloride, 4-nitro $4-NO_2C_6H_4CHCl_2$	206.03	pr (al)		46			al, eth	B5[4], 859
3305	Benzylidene chloride, pentachloro .... $C_6Cl_5CHCl_2$	333.26	lf (al)	334, 199[13]	119.5				B5[4], 824
3306	Benzylidene chloride, 2,3,4-trichloro .... $2,3,4-Cl_3C_6H_2CHCl_2$	264.37	cr (lig)	275-85	84			bz	B5[1], 153

No.	Name, Synonyms, and Formula	Mol. wt.	Color, crystalline form, specific rotation and λ_{max} (log ϵ)	b.p. °C	m.p. °C	Density	n_D	Solubility	Ref.
3307	Benzylidene chloride, 2,3,6-trichloro 2,3,6-Cl$_3$C$_6$H$_2$CHCl$_2$	264.37	nd (MeOH)	145-50^{12}	83	....	....	bz	B5^4, 823
3308	Benzylidene chloride, 2,4,5-trichloro 2,4,5-Cl$_3$C$_6$H$_2$CHCl$_2$	264.37	....	280-1, 153-5^{15}	<0	1.5956$^{20/4}$	1.5992^{20}	bz	B5^4, 824
3309	Benzylidene chloride, 2,4,6-trichloro 2,4,6-Cl$_3$C$_6$H$_2$CHCl$_2$	264.37	cr (MeOH)	158^{15}	27	....	....	bz	B5^3, 704
3310	Benzylidene diacetate or Benzaldehyde diacetate C$_6$H$_5$CH(O$_2$CCH$_3$)$_2$	208.22	pl (eth)	220, 154^{20}	46	1.11^{20}	....	al, eth, bz	B7^2, 161
3311	Benzylidene ethyl amine C$_6$H$_5$CH=NC$_2$H$_5$	133.20	....	195^{740}, 117^{12}		0.9370$^{20/4}$	1.5365^{20}	al, eth, ace	B7^2, 163
3312	Benzylidene methylamine C$_6$H$_5$CH=N-CH$_3$	119.17	....	185, 90-1^{30}		0.9672$^{14/2}$	1.5526^{20}	al, eth, ace	B7^2, 162
3313	Benzylidene fluoride or α,α-Difluoro toluene C$_6$H$_5$CHF$_2$	128.12	....	139.9		1.1357^{20}	1.4577^{20}	al	B5^4, 801
3314	Berbamine C$_{37}$H$_{40}$N$_2$O$_6$	608.74	lf (+2w, al), cr (peth), [α]$^{20/}_D$ +109 (chl)		197-200 (anh), 156 (hyd)			al, eth, peth, chl	B27^2, 891
3315	Berberine C$_{20}$H$_{19}$NO$_5$	353.38	red-ye nd (w+6), cr (chl+1)		145 (anh), 110 (+6w)			al, eth	B27^2, 567
3316	Berberine hydrochloride C$_{20}$H$_{22}$ClNO$_6$	407.86	ye cr (w+2), nd (w+4)					w	B27^1, 514
3317	Berberine nitrate C$_{20}$H$_{18}$N$_2$O$_7$	398.38	red-ye nd (al)		155d			w	B27, 500
3318	Berberine sulfate, trihydrate C$_{40}$H$_{44}$N$_2$O$_{15}$S	822.28	red-ye nd					w, al, chl	B27, 500
3319	Berberine, tetrahydro (dl) C$_{20}$H$_{21}$NO$_4$	339.40	mcl nd (al)		173-4			al, chl	B27^2, 557
3320	Berbine (dl) C$_{17}$H$_{17}$N	235.33	nd (eth or MeOH)		89			al, eth, ace	B20^4, 4108
3321	Betaine (CH$_3$)$_3$N$^+$CH$_2$COO$^-$	117.15	(w+1), pr or lf (al)		293d			w, al	B4^3, 1127
3322	Betonicine C$_7$H$_{13}$NO$_3$	159.19	pr (dil al +1w), [α]$^{21/}_D$-37 (w, c=4.8)		252d			al	B22^4, 2054
3323	Betulin or Lupenediol C$_{30}$H$_{50}$O$_2$	442.73	nd (al+1), [α]$^{15/}_D$ +20 (Py, c=2)	170-80$^{0.08/d}$	251-2			eth, aa	B6^3, 5234
3324	Betulinic acid or Betulic acid C$_{30}$H$_{48}$O$_3$	456.71	pr or nd (al+1), [α]$^{22/546}$ +7.9 (Py)					Py	B10^3, 1059
3325	Biacene or Biacenaphthene C$_{24}$H$_{16}$	304.39	red-ye pl or nd (bz)		277			bz	B5^3, 2595
3326	Biacetyl or 2,3-Butanedione CH$_3$COCOCH$_3$	86.09	....	88	-2.4	0.9808$^{18.5/4}$	1.3951^{20}	w, al, eth, ace, bz	B1^4, 3644
3327	Biacetyl dioxime or Dimethyl glyoxime CH$_3$C(:NOH)C(:NOH)CH$_3$	116.13	nd (to or dil al)	sub 234-5	245-6			al, eth	B1^4, 3647
3328	Biacetyl monoxime CH$_3$COC(:NOH)CH$_3$	101.11	pr (chl), lf (w)	185-6	77-8			al, eth, chl	B1^4, 3646
3329	10,10'-Bianthronyl or Bianthrone C$_{28}$H$_{18}$O$_2$	386.45	pl (ace)		256-8d			chl	B7^3, 4486
3330	Biarsine, tetraethyl or Ethylcacodyl (C$_2$H$_5$)$_2$As-As(C$_2$H$_5$)$_2$	266.09	....	185-7		1.1388$^{24/4}$	1.4709	al, eth	B4^3, 1832
3331	Biarsine, tetrakis(trifluoromethyl) (CF$_3$)$_2$AsAs(CF$_3$)$_2$	425.87	....	106-7			1.372^{19}		B3^4, 269
3332	Bibenzyl, 4,4'-diamino 4-H$_2$NC$_6$H$_4$CH$_2$CH$_2$(C$_6$H$_4$NH$_2$-4)	212.30	pl (w)	sub	135-6 sub			al	B13^3, 470
3333	Bicyclo[2,2,1]-hepta-2,5-diene C$_7$H$_8$	92.15	....	89.5	-19.1	0.9064$^{20/4}$	1.4702^{20}	al, eth, ace, bz lig	B5^4, 879

PHYSICAL CONSTANTS OF ORGANIC COMPOUNDS (Continued)

No.	Name, Synonyms, and Formula	Mol. wt.	Color, crystalline form, specific rotation and λ_{max} (log ε)	b.p. °C	m.p. °C	Density	n_D	Solubility	Ref.
3334	Bicyclo[2,2,1]heptane or Norbornane.............. C_7H_{12}	96.17	....	sub	87.5	....		al, eth, ace, bz	B5[4], 258
3335	Bicyclo[4,1,0]-heptane, 7-azo $C_6H_{11}N$	97.16	....,	48-51[33]	20-2			al, eth, bz	B20[4], 1937
3336	Bicyclo[2,2,1]-heptane-2,3-dicarboxylic acid or 2,3-Nor-camphane dicarboxylic acid .... $C_9H_{12}O_4$	184.19			192-3				B9[3], 3970
3337	Bicyclo[2,2,1]-heptane-2-carboxaldehyde .... $C_8H_{12}O$	122.17	cr (w)	70-2[22]		1.0227[19/4]	1.4760[25]	eth	B7[3], 267
3338	Bicycloheptyl .... $C_{14}H_{26}$	194.36		290-1[728]		0.9069[20/0]			B5[4], 344
3339	Bicyclo[3,1,0]hex-2-ene-4-one, 5-iso-propyl-2-methyl .... $C_{10}H_{14}O$	150.22	$[\alpha]_D$ − 36.5	219.20[749]		0.9581[15/15]	1.48325		B7[3], 582
3340	Bicyclohexyl or Dodecahydrobiphenyl (cis,cis).......... C_6H_{11}-C_6H_{11}	166.31	....	238	4	0.8914[20/4]	1.4766[20]	al, eth	B5[3], 273
3341	Bicyclohexyl (trans,trans).............. $C_{12}H_{22}$	166.31		217-8, 95-6[9]	4.2	0.8592[20/4]	1.4663[20]	al, eth	B5[4], 334
3342	Bicyclo[3,1,1]-heptane, 2,4,6-trimethyl .... $C_{10}H_{18}$	138.25		169-70[768]		0.8467[21/4]	1.4605[21]		PCHE2, 238
3343	Bicyclo[3,3,1]-nonane C_9H_{16}	124.23	cr (MeOH)	169-70 sub	145-6			al, aa	B5[4], 293
3344	Bicyclo[2,2,2]octane C_8H_{14}	110.20	....		169-71			......	B5[4], 370
3345	Bicyclo[3,2,1]octane .... C_8H_{14}	110.20	....		139-41				B5[4], 278
3346	Bicyclo[3,3,0]octane (cis) C_8H_{14}	110.20		137[765]	< − 80	0.8638[25/4]	1.4595[25]	al	B5[4], 277
3347	Bicyclo[3,3,0]octane (trans) .... C_8H_{14}	110.20		132[755]	− 30	0.8624[18/4]	1.4625[18]	al	B5[4], 277
3348	Bicyclo[3,3,0]octane, 2,6-dione .... $C_8H_{10}O_2$	138.17		86-8[0.2]	45	1.1290[60/4]	1.4877[54]	....	B7[3], 3279
3349	Bicyclo[2,2,2]octane, 2-methyl .... $C_8H_{13}CH_3$	124.23		158[740]	33-4	0.8664[40.5/4]	1.4608[40.5]		B5[4], 295
3350	Bicyclo[3,3,0]octane, 2-one (cis) $C_8H_{12}O$	124.19		72[13]		1.0097[20/4]	1.4790[20]	al, ace	B7[3], 264
3351	9,9'-Bifluorenyl .... $C_{26}H_{18}$	330.43	nd (bz-al)		247			aa, Py	B5[3], 2626
3352	9,9'-Bifluorenyl, 9,9'-diphenyl $C_{38}H_{26}$	482.63	pl (bz)		256 (under CO_2)	1.266[0/4]			B5[3], 2782
3353	Bifluorenylidene .... $C_{26}H_{16}$	328.42	red nd (bz)		194-5			eth, bz, chl	B5[3], 2652
3354	Biguanide .... $H_2NC(:NH)NHC(:NH)NH_2$	101.11	pr or nd (al)	d 142	136			w, al	B3[2], 76
3355	Biguanide, 1-phenyl $C_6H_5NHC(=NH)NHC(=NH)NH_2$·HCl	177.21			143				B12[3], 807
3356	Biguanide, 1-(2-tolyl) .... $C_9H_{13}N_5$	191.24	nd or pl (w + 1)		144			al, ace	B12[3], 1873
3357	2-2'-Biindane, 1,1',3,3'-tetraoxo or Bisdiketohydrindene. $C_{18}H_{10}O_4$	290.28	red nd (bz)	....	297			bz	B7[2], 863
3358	Bikhaconitine $C_{36}H_{51}NO_{11}$	673.81	$[\alpha]_D$ + 12 (al)		118-23			al, eth, chl	B21[4], 2869
3359	Bilifucsin $C_{16}H_{20}N_2O_4$	304.35	dk br pw		183			al, ace	C30, 1936
3360	Bilirubin or Haematoidine .... $C_{33}H_{36}N_4O_6$	584.68	red mcl pr or pl (chl)					bz, chl	C38, 1230
3361	Biliverdin or Dehydrobilirubin.............. $C_{33}H_{34}N_4O_6$	582.66	dk gr pl or pr (MeOH)		>300			al, bz	J 1961, 2284
3362	α,α'-Binaphthyl or α,α'-Dinaphthyl .... α-$C_{10}H_7$-α-$C_{10}H_7$, $C_{20}H_4$	254.34	(i) pl (aa), (ii) rh (peth)	>360, 240-2[12]	(i) 144, (ii) 160			eth, ace, bz	B5[3], 2465
3363	α,α'-Binaphthyl, 4,4'-diamino-3,3'-dimethyl .... (3-CH_3-4-H_2N-α-$C_{10}H_5$)$_2$	312.42			213			al, bz	B13[3], 542
3364	α,α'-Binaphthyl, 2,2'-dihydroxy or β-Dinaphthol .... [2-HO-α-$C_{10}H_6$]$_2$	286.33	nd (al), cr (w)		220			al, eth	B6[3], 5877
3365	α,α'-Binaphthyl, 4,4'-dihydroxy or α-Dinaphthol .... [4-HO-α-$C_{10}H_6$]$_2$	286.33	pl	sub	300			al, eth	B6[3], 5878

No.	Name, Synonyms, and Formula	Mol. wt.	Color, crystalline form, specific rotation and λ_{max} (log ε)	b.p. °C	m.p. °C	Density	n_D	Solubility	Ref.
3366	β,β'-Binaphthyl or β,β'-Dinaphthyl β-$C_{10}H_7$-β-$C_{10}H_7$	254.34	bl fluor pl (al)	452[753] sub	187-8			eth, bz	B5[3], 2467
3367	Biotin or Vitamin H-Coenzyme R $C_{10}H_{16}N_2O_3S$	244.32	nd (w)		232d			w	Am67, 2096
3368	Biotin, methyl ester $C_{11}H_{18}N_2O_3S$	258.34	pl (MeOH-eth), [α][22]_D + 57 chl	sub	166.7			al, ace, chl	C43, 1810
3369	Biphenyl or Phenylbenzene C_6H_5-C_6H_5	154.21	lf (dil al)	255.9, 145[22]	71	0.8660[20/4]	1.475[20], 1.588[75]	al, eth, bz	B5[4], 1807
3370	Biphenyl, 2-acetamide 2-(CH_3CONH)C_6H_4-C_6H_5	211.27	pr or nd (dil al or peth)	355	121			al, eth	B12[3], 3125
3371	Biphenyl, 3-acetamido 3-(CH_3CONH)C_6H_4-C_6H_5	211.27	nd (al)		149			al, bz	B12[2], 751
3372	Biphenyl, 4-acetamido-3-NO_2 3-NO_2-4-(CH_3CONH)C_6H_3-C_6H_5	211.27	cr (dil MeOH)		172			al, ace	B12[3], 3156
3373	Biphenyl, 4-acetamido,3-nitro 4-(CH_3CONH)-3-$NO_2C_6H_3C_6H_5$	256.26	ye nd (al)		132			al, eth, aa	B12[3], 3199
3374	Biphenyl, 4-acetyl or 4-Phenylacetophenone 4-(CH_3CO)C_6H_4-C_6H_5	196.25	pr (ace), cr (al)	325-7	121			al, bz	B7[2], 377
3375	Biphenyl, 2-amino 2-$H_2NC_6H_4$-C_6H_5	169.23	lf (dil al)	299, 170[15]	51-3			al, eth, bz	B12[3], 3124
3376	Biphenyl, 2-amino-4'-nitro 2-$H_2NC_6H_4$-$C_6H_4NO_2$-4	214.23	og-red nd (al)		159			al	B12[3], 3141
3377	Biphenyl, 2-amino-5-nitro 5-NO_2-2-$H_2NC_6H_3$-C_6H_5	214.23	ye nd (al)		125			al	B12[3], 3140
3378	Biphenyl, 3-amino 3-$H_2NC_6H_4$-C_6H_5	169.23	nd	254[135]	30			al, eth, ace, bz	B12[3], 3146
3379	Biphenyl, 3-amino-4-hydroxy 3-H_2N-4-HO-C_6H_3-C_6H_5	185.23	nd (chl)		208			al, eth, bz	B13[3], 1946
3380	Biphenyl, 3-amino-4-nitro 4-NO_2-3-$H_2NC_6H_3$-C_6H_5	214.23	og nd (dil al)		116			al	B12[3], 3149
3381	Biphenyl, 3-amino-4'-nitro 3-$H_2NC_6H_4$-$C_6H_4NO_2$-4	214.23	og nd (al)		137			al, aa	B12[2], 753
3382	Biphenyl, 4-amino or Xenylamine 4-$H_2NC_6H_4$-C_6H_5	169.23	lf (dil al)	302, 191[15]	53-4			al, eth, chl	B12[3], 3152
3383	Biphenyl, 4-amino-2'-hydroxy 4-$H_2NC_6H_4$-C_6H_4OH-2	185.23	nd (to)		181-2			al, aa	B13[3], 1943
3384	Biphenyl, 4-amino-4'-hydroxy 4-$H_2NC_6H_4$-C_6H_4OH-4	185.23	pl (dil al)		275				B13[2], 420
3385	Biphenyl, 4-amino-2'-nitro 4-$H_2NC_6H_4$-$C_6H_4NO_2$-2	214.23	red mcl pr (al)		99			al	B12[3], 3199
3386	Biphenyl, 4-amino-3-nitro 3-NO_2-4-$H_2NC_6H_3$-C_6H_5	214.23	red nd (al)		170-1			al, eth, chl, aa	B12[3], 3198
3387	Biphenyl, 4-amino-4'-nitro 4-$H_2NC_6H_4$-$C_6H_4NO_2$-4	214.23	red nd (al)		203-4			al, aa	B12[3], 3200
3388	Biphenyl, 5-amino-2-hydroxy 5-H_2N-2-HOC_6H_3-C_6H_5	185.23	nd (al or bz)		201			al, bz	B13[3], 1940
3389	Biphenyl, 2-benzyl or o-Biphenylphenyl methane 2-($C_6H_5CH_2$)C_6H_4-C_6H_5	244.34	mcl nd (al)	283-7[110]	54-6			al, eth, bz	B5[3], 2323
3390	Biphenyl, 4-benzyl or p-Biphenylphenyl methane 4-($C_6H_5CH_2$)C_6H_4-C_6H_5	244.34	lf	285-6[110]	85	1.171[0/4]		al, eth, bz	B5[3], 2324
3391	Biphenyl, 4,4'-bis(diethylamino) [4-(CH_3)$_2NC_6H_4$-]$_2$	296.46	nd (al)		85			al, eth	B13[3], 430
3392	Biphenyl, 2,4'-bis(dimethylamino) 2-(CH_3)$_2N$-$C_6H_4C_6H_4$[$N(CH_3)_2$-4]	240.35	pl (al)	206-7[11]	51-2			al, eth	B13[2], 88
3393	Biphenyl, 4,4'-bis(dimethylamino) 4-(CH_3)$_2N$-C_6H_4-C_6H_4[$N(CH_3)_2$-4]	240.35	nd (al or bz-lig)	>360	198			bz, chl	B13[3], 429
3394	Biphenyl, 4,4'-bis(ethylamino) 4-$C_2H_5NHC_6H_4$-$C_6H_4(NHC_2H_5$-4)	240.35	nd or pl (al)		120.5			al, eth, bz	B13, 222
3395	Biphenyl, 4,4'-bis(methylamino) 4-$CH_3NHC_6H_4$-$C_6H_4(NHCH_3$-4)	212.30	lf (al, w, or lig)		91			al, lig	B13[2], 97
3396	Biphenyl, 4,4'-bis(phenylamino) 4-$C_6H_5NHC_6H_4$-$C_6H_4(NHC_6H_5$-4)	336.24	lf (to)		244-5			aa, to	B13[3], 431
3397	Biphenyl, 2-bromo 2-BrC_6H_4-C_6H_5	233.12		296-8, 160[11]	1-2	1.2175[26]	1.6248[25]	al, eth	B5[4], 1818
3398	Biphenyl, 3-bromo 3-BrC_6H_4-C_6H_5	233.12		299-30[1], 169-73[17]			1.6411[20]		B5[4], 1818

No.	Name, Synonyms, and Formula	Mol. wt.	Color, crystalline form, specific rotation and λ_{max} (log ε)	b.p. °C	m.p. °C	Density	n_D	Solubility	Ref.
3399	Biphenyl, 3-bromo-4-hydroxy 3-Br-4-HOC$_6$H$_3$-C$_6$H$_5$	249.11	nd (chl-peth)	....	96	....	....	al, chl, aa	B6^3, 3333
3400	Biphenul, 4-bromo 4-BrC$_6$H$_4$-C$_6$H$_5$	233.12	pl (al)	310	91.2	0.9327$^{25/4}$	....	al, eth, bz, aa	B5^4, 1819
3401	Biphenyl, 4-(bromoacetyl) 4-(BrCH$_2$CO)C$_6$H$_4$-C$_6$H$_5$	275.15	nd (95% al)	....	127	....	....	....	B7^3, 2137
3402	Biphenyl, 4-bromo-4′-hydroxy 4-BrC$_6$H$_4$-C$_6$H$_4$(OH-4)	249.11	pl (al)	....	164-6	....	....	al, eth, ace, bz	B6^3, 3334
3403	Biphenyl, 2-chloro 2-ClC$_6$H$_4$-C$_6$H$_5$	188.66	nd (dil al)	274, 154^{12}	34	1.1499$^{32.5}$	....	al, eth, lig	B5^4, 1816
3404	Biphenyl, 3-chloro 3-ClC$_6$H$_4$-C$_6$H$_5$	188.66	....	284-5, 150-60^6	16	1.1579$^{25/4}$	1.6181^{25}	al, eth, ace	B5^4, 1816
3405	Biphenyl, 3-chloro-2-hydroxy 3-Cl-2-HOC$_6$H$_3$-C$_6$H$_5$	204.66	....	317-8d	6	1.24$^{25/4}$	1.6237^{30}	al, eth, ace, bz	B6^3, 3297
3406	Biphenyl, 4-chloro 4-ClC$_6$H$_4$-C$_6$H$_5$	188.66	lf (lig or al)	291, 180-95$^{20-30}$	77.7	....	....	al, eth, lig	B5^4, 1816
3407	Biphenyl, 4-(chloroacetyl) 4-(ClCH$_2$CO)C$_6$H$_4$-C$_6$H$_5$	230.70	pl (al)	....	125	....	....	al	B7, 443
3408	Biphenyl, 4-chloro-4′-hydroxy 4-Cl-C$_6$H$_4$-(C$_6$H$_4$OH-4)	204.66	cr (dil al)	....	146-7	....	....	al, eth, ace, bz	B6^3, 3332
3409	Biphenyl, 5-chloro-2-hydroxy 5-Cl-2-HOC$_6$H$_3$-C$_6$H$_5$	204.66	....	319^{745}, 128-30^2	11	....	....	....	B6^3, 3299
3410	Biphenyl, 2,2′-diacetamido [2-(CH$_3$CONH)C$_6$H$_4$-]$_2$	268.32	pr (al)	....	164-5	....	....	bz, aa	B13^3, 412
3411	Biphenyl, 2,4-diacetamido 2,4-(CH$_3$CONH)$_2$C$_6$H$_3$-C$_6$H$_5$	268.32	nd (al)	....	202	....	....	al	B13^3, 417
3412	Biphenyl, 4,4′-diacetamido [4-(CH$_3$CONH)C$_6$H$_4$-]$_2$	268.32	nd (aa)	....	328.3	....	....		B13^3, 437
3413	Biphenyl, 2,2′-diamino or o-Benzidine 2-H$_2$NC$_6$H$_4$-C$_6$H$_4$NH$_2$-2	184.24	mcl pr or nd (al)	162^4	81	....	....	w, bz	B13^3, 410
3414	Biphenyl, 2,4′-diamino or Diphenyline 2-H$_2$NC$_6$H$_4$-C$_6$H$_4$NH$_2$-4	184.24	nd (dil al)	363	54.5	....	....	al, eth	B13^3, 416
3415	Biphenyl, 3,3′-diamino or m-Benzidine 3-H$_2$NC$_6$H$_4$-C$_6$H$_4$NH$_2$-3	184.24	nd (w), pr (bz)	205	93-4	....	....	eth, bz	B13^3, 422
3416	Biphenyl, 3,4-diamino 3,4-(H$_2$N)$_2$C$_6$H$_3$-C$_6$H$_5$	184.24	lf (eth or al)	....	103	....	....	al, eth	B13^2, 89
3417	Biphenyl, 4,4′-diamino or Benzidine 4-H$_2$NC$_6$H$_4$-C$_6$H$_4$NH$_2$-4	184.24	nd (w)	400^{740}	125	....	....	al	B13^3, 425
3418	Biphenyl, 4,4′-diamino-3,3′-dimethoxy [3-CH$_3$O-4-H$_2$NC$_6$H$_3$-]$_2$	244.30	lf or nd (w)	....	137	....	....	al, eth, ace, bz, chl	B13^3, 2310
3419	Biphenyl, 4,4′-diamino-2,2′-methyl [2-CH$_3$-4-H$_2$NC$_6$H$_3$-]$_2$	212.30	pr (w)	....	108-9	....	....	al, eth	B13, 255
3420	Biphenyl, 4,4′-diamino-3,3′-dimethyl [3-CH$_3$-4-H$_2$NC$_6$H$_3$-]$_2$	212.30	lf (dil al)	....	131-2	....	....	al, eth	B13^3, 484
3421	Biphenyl, 4,4′-diamino-3-ethoxy 3-C$_2$H$_5$O-4-H$_2$NC$_6$H$_3$-C$_6$H$_4$-NH$_2$-4	228.30	nd (w)	....	134	....	....	al	B13^2, 419
3422	Biphenyl, 4,4′-dibromo 4-BrC$_6$H$_4$-C$_6$H$_4$Br-4	312.03	mcl pr (MeOH)	355-60	164	....	....	bz	B5^4, 1820
3423	Biphenyl, 3,3′-dichloro 3-ClC$_6$H$_4$-C$_6$H$_4$Cl-3	223.10	....	322-4	29	....	....	al, eth, bz	B5^3, 1739
3424	Biphenyl, 4,4′-dichloro 4-ClC$_6$H$_4$-C$_6$H$_4$Cl-4	223.11	pr or nd (al or to-peth)	315-9	148-9	....	....	bz	B5^4, 1817
3425	Biphenyl, 4,4′-dichloro-2,2′-dinitro [4-Cl-2-NO$_2$C$_6$H$_3$-]$_2$	313.10	ye cr (al)	....	140	....	....	bz, aa	B5^4, 1828
3426	Biphenyl, 2,2′-diethoxy-3,3′-dimethyl [3-CH$_3$-2-C$_2$H$_5$O-C$_6$H$_3$-]$_2$	270.38	lf (al)	....	85	....	....		B6^2, 974
3427	Biphenyl, 2,4′-diethoxy-3,3′-dimethyl 3-CH$_3$-2-C$_2$H$_5$O-C$_6$H$_3$-C$_6$H$_3$-OC$_2$H$_5$(4)-CH$_3$-3	270.38	nd (al)	....	53	....	....		B6^2, 974
3428	Biphenyl, 4,4′-diethoxy-3,3′-diethyl [3-C$_2$H$_5$-4-C$_2$H$_5$OC$_6$H$_3$-]$_2$	298.43	lf (al)	....	120	....	....		B6, 1015
3429	Biphenyl, 4,4′-diethoxy-3,3′-dimethyl [3-CH$_3$-4-C$_2$H$_5$OC$_6$H$_3$-]$_2$	270.38	pl (al)	....	156	....	....		B6, 1010
3430	Biphenyl, 3,3′-diethyl-6,6′-dihydroxy [3-C$_2$H$_5$-6-HOC$_6$H$_3$-]$_2$	242.32	nd (dil al)	....	131	....	....		B6^2, 981
3431	Biphenyl, 2,2′-difluoro 2-F-C$_6$H$_4$-C$_6$H$_4$-F-2	190.19	....	....	118.5-9.5	1.393$^{20/4}$	....	al	B5^3, 1735

No.	Name, Synonyms, and Formula	Mol. wt.	Color, crystalline form, specific rotation and λ_{max} (log ε)	b.p. °C	m.p. °C	Density	n_D	Solubility	Ref.
3432	Biphenyl, 3,3′-difluoro . . . . 3-FC$_6$H$_4$C$_6$H$_4$F-3	190.19	. . . .	130[14]	8	1.192[25/4]	1.5678[20]	. . . .	B5[3], 1736
3433	Biphenyl, 4,4′-difluoro . . . . 4-FC$_6$H$_4$-C$_6$H$_4$F-4	190.21	mcl pr (al), lf (w)	254—5, 119[14]	94-5	. . . .	. . . .	al, eth, ace, bz, chl	B5[3], 1736
3434	Biphenyl, 2,2′-dihydroxy or o,o′-Biphenol . . . . 2-HOC$_6$H$_4$-C$_6$H$_4$OH-2	186.2[1]	lf (w + 1), pr (to)	325-6	110-2 (anh)	. . . .	. . . .	al, eth, ace, bz, aa	B6[3], 5374
3435	Biphenyl, 2,2′-dihydroxy-3,3′-dimethyl . . . . [3-CH$_3$-2-HOC$_6$H$_3$-]$_2$	214.27	nd (peth)	sub	113	. . . .	. . . .	al, eth, bz	B6[3], 5445
3436	Biphenyl, 2,2′-dihydroxy-5,5′-dimethyl . . . . [5-CH$_3$-2-HOC$_6$H$_3$-]$_2$	214.27	nd (bz or w)	sub	153.5	. . . .	. . . .	al, eth, ace, bz	B6[3], 5447
3437	Biphenyl, 2,2′-dihydroxy-6,6′-dimethyl . . . . [6-CH$_3$-2-HOC$_6$H$_3$-]$_2$	214.27	pl (dil al)	. . . .	164	. . . .	. . . .	al	B6[3], 5445
3438	Biphenyl, 2,2′-dihydroxy-3,3′,5,5′-tetramethyl . . . . [3,5-(CH$_3$)$_2$-2-HOC$_6$H$_2$-]$_2$	242.32	nd or pl (eth or lig)	140-60[0.05]	137-8	. . . .	. . . .	al, eth, lig	B6[3], 5483
3439	Biphenyl, 2,4′-dihydroxy or o,p′-Biphenol . . . . 2-HOC$_6$H$_4$-C$_6$H$_4$OH-4	186.21	mcl pr or nd (dil al)	342, 206-10[11]	162-3	. . . .	. . . .	eth	B6[3], 5387
3440	Biphenyl, 2,5-dihydroxy . . . . 2,5-(HO)$_2$C$_6$H$_3$C$_6$H$_5$	186.21	nd (dil al)	. . . .	97-8	. . . .	. . . .	al	B6[3], 5371
3441	Biphenyl, 2,5′-dihydroxy-2′,5-dimethyl . . . . 5-CH$_3$-2-HOC$_6$H$_3$C$_6$H$_3$OH(5)-CH$_3$-2	214.27	. . . .	. . . .	158	. . . .	. . . .	al, eth	B6[3], 5445
3442	Biphenyl, 3,3′-dihydroxy or m,m′-Biphenol . . . . 3-HOC$_6$H$_4$-C$_6$H$_4$OH-3	186.21	nd (w)	247[18]	123-4	. . . .	. . . .	al, eth, bz, chl	B6[3], 5388
3443	Biphenyl, 3,4-dihydroxy . . . . 3-HOC$_6$H$_4$-C$_6$H$_4$OH-4	186.21	. . . .	>360	145	. . . .	. . . .	al, eth, ace, bz, chl	B6[3], 5387
3444	Biphenyl, 4,4′-dihydroxy or p,p′-Biphenol . . . . 4-HOC$_6$H$_4$-C$_6$H$_4$OH-4	186.21	nd or pl (al)	sub	274-5	. . . .	. . . .	al, eth	B6[3], 5389
3445	Biphenyl, 4,4′-dihydroxy-3,3′-diethyl . . . . [3-C$_2$H$_5$-4-HOC$_6$H$_3$-]$_2$	270.38	nd (aa)	. . . .	148	. . . .	. . . .	al, ace, bz, aa	B6[3], 5482
3446	Biphenyl, 4,4′-dihydroxy-3,3′-dimethyl . . . . [3-CH$_3$-4-HOC$_6$H$_3$-]$_2$	214.27	lf (w), nd	. . . .	161	. . . .	. . . .	al, eth	B6[3], 5445
3447	Biphenyl, 4,4′-diacetoxy-3,3′-dimethyl . . . . [3-CH$_3$-4-CH$_3$CO$_2$C$_6$H$_3$-]$_2$	298.34	wh nd (al, aa)	. . . .	135.3	. . . .	. . . .	. . . .	B6[3], 5446
3448	Biphenyl, 4,4′-dihydroxy-3,3′,5,5′-tetramethoxy or Hydrocerulignone . . . . [3,5-(CH$_3$)$_2$-4-HOC$_6$H$_2$-]$_2$	306.32	mcl pr (al)	. . . .	190	. . . .	. . . .	al	B6[1], 593
3449	Biphenyl, 4,4′-dihydroxy-3,3′,5,5′-tetramethyl . . . . [3,5-(CH$_3$)$_2$-4-HOC$_6$H$_2$-]$_2$	242.32	pa ye nd or pr (aa)	sub	222-3	. . . .	. . . .	al, aa	B6, 1015
3450	Biphenyl, 5,5′-dihydroxy-2,2′-dimethyl . . . . [2-CH$_3$-5-HOC$_6$H$_3$-]$_2$	214.27	pr (al)	. . . .	229	. . . .	. . . .	al, eth	B6[3], 5444
3451	Biphenyl, 4,4′-dihydroxy-3,3′,5,5′-tetranitro . . . . [3,5-(NO$_2$)$_2$-4-HOC$_6$H$_2$-]$_2$	366.20	ye nd	. . . .	223	. . . .	. . . .	. . . .	B6[3], 5399
3452	Biphenyl, 2,2′-dimethoxy or o,o′-Bianisole . . . . (2-CH$_3$OC$_6$H$_4$-)$_2$	214.27	rh bipyr pr (al)	307-8[766]	155	1.268	. . . .	al, bz, chl	B6[3], 5375
3453	Biphenyl, 3,3′-dimethoxy or m, m′ Bianisole . . . . (3-CH$_3$OC$_6$H$_4$-)$_2$	214.27	nd (dil al)	328, 211-20[15]	36	. . . .	. . . .	al, eth, ace, bz, chl	B6[3], 5388
3454	Biphenyl, 4,4′-dimethoxy or p, p′-Bianisole . . . . (4-CH$_3$OC$_6$H$_4$-)$_2$	214.27	lf (bz)	sub	173	. . . .	. . . .	al, bz, chl	B6[3], 5391
3455	Biphenyl, 2,2′-dimethoxy-5,5′-dimethyl . . . . [5-CH$_3$-2-CH$_3$OC$_6$H$_3$-]$_2$	242.32	nd (dil al)	188[12]	71	. . . .	. . . .	al, eth, ace, bz	B6[3], 5447
3456	Biphenyl, 2,5-dimethoxy-2′,5-dimethyl . . . . 5-CH$_3$-2-CH$_3$OC$_6$H$_3$-C$_6$H$_3$OCH$_3$(5)-CH$_3$-2	242.32	pr (al)	168[4]	86	. . . .	. . . .	bz, peth	B6[2], 973
3457	Biphenyl, 4,4′-dimethoxy-3,3′-dimethyl . . . . [3-CH$_3$-4-CH$_3$OC$_6$H$_3$-]$_2$	242.32	pr (al)	. . . .	145.5	. . . .	. . . .	. . . .	B6[3], 5445
3458	Biphenyl, 2,2′-dimethyl or o, o′-Bitolyl . . . . (2-CH$_3$C$_6$H$_4$-)$_2$	182.27	cr (al)	256	19—20	0.9906[20]	1.5752[20]	al, eth, ace, bz	B5[4], 1897
3459	Biphenyl-2,3-dimethyl . . . . 2,3-(CH$_3$)$_2$C$_6$ H$_3$C$_6$H$_5$	182.26	. . . .	141[14]	42	. . . .	1.5845[23]	eth	B5[4], 1897
3460	Biphenyl, 2,3′-dimethyl or o,m′-Bitolyl . . . . 2-CH$_3$C$_6$H$_4$-C$_6$H$_4$CH$_3$-3	182.27	. . . .	270	. . . . . .	0.9924[20]	1.5810[20]	al, eth, ace, bz	B5[4], 1902
3461	Biphenyl, 2,4-dimethyl . . . . 2,4-(CH$_3$)$_2$C$_6$H$_3$-C$_6$H$_5$	182.26	. . . .	270-6[767]	0.9947 [20/4]	1.5844[20]	. . . .	. . . .	B5[4], 1897
3462	Biphenyl, 2,4′-dimethyl or o,p′-Bitolyl . . . . 2-CH$_3$C$_6$H$_4$-C$_6$H$_4$CH$_3$-4	182.27	. . . .	273-6, 137[12.5]	. . . . . .	0.9924[20]	1.5826[20]	al, eth, ace, bz	B5[4], 1903
3463	Biphenyl, 2,5-dimethyl . . . . 2,5-(CH$_3$)$_2$C$_6$H$_3$-C$_6$H$_5$	182.26	. . . .	140[14.5]	. . . . . .	0.9931[20/4]	1.5819[20]	. . . .	B5[4], 1897
3464	Biphenyl, 2,6-dimethyl . . . . 2,6-(CH$_3$)$_2$C$_6$H$_3$-C$_6$H$_5$	182.26	. . . .	260-5, 132[16.5]	− 5	0.9907[20/4]	1.5745[20]	. . . .	B5[4], 1897

No.	Name, Synonyms, and Formula	Mol. wt.	Color, crystalline form, specific rotation and λ_{max} (log ε)	b.p. °C	m.p. °C	Density	n_D	Solubility	Ref.
3465	Biphenyl, 3,3′-dimethyl or m, m′-Bitolyl .... (3-CH₃C₆H₄-)₂	182.27	....	280, 150[18]	9	0.9995[20/4]	1.5946[20]	al, eth, ace, bz	B5[4], 1903
3466	Biphenyl, 3,3′-dimethyl-4,4′-dipropoxy .... (3-CH₃-4-C₃H₇OC₆H₃-)₂	298.43	lf	.........,	115				B6, 1010
3467	Biphenyl, 3,4-dimethyl .,.., 3,4-(CH₃)₂C₆H₃C₆H₅	182.26	....	281-3, 139-40[8]	29.2-9.7	1.0087[20/4]	1.6036[20]	bz	B5[4], 1903
3468	Biphenyl, 3,4′-dimethyl 3-CH₃C₆H₄-C₆H₄CH₃-4	182.26		288-9[752], 153[15]	14-15	0.9978[20/4]	1.5968[20]	bz	B5[4], 1905
3469	Biphenyl, 3,5-dimethyl 3,5-(CH₃)₂C₆H₃-C₆H₅	182.26		273-6	22-3	0.9990[20/4]	1.5952[20]	....	B5[4], 1903
3470	Biphenyl, 4,4′-dimethyl (4-CH₃C₆H₄-)₂	182.27	mcl pr (eth)	295	125	0.917[121/4]	....	eth, ace, bz	B5[4], 1906
3471	Biphenyl, 2,2′-dinitro (2-NO₂C₆H₄-)₂	244.22	ye mcl pr or nd (al)	....	127-8	1.45 (sol)		al, eth, bz, aa	B5[4], 1826
3472	Biphenyl, 2,3′-dinitro 2-NO₂C₆H₄-C₆H₄NO₂-3	244.21		....	118-9.5			al	B5[4], 1826
3473	Biphenyl, 2,4′-dinitro 2-NO₂C₆H₄-C₆H₄NO₂-4	244.22	mcl pr (al)	....	93-4	1.474		al, eth, bz, aa	B5[4], 1827
3474	Biphenyl, 3,3′-dinitro (3-NO₂C₆H₄-)₂	244.22	ye og nd (al or aa)	....	200			bz, aa	B5[4], 1827
3475	Biphenyl, 4,4′-dinitro (4-NO₂C₆H₄-)₂	244.22	nd (al)	....	240-3			bz, aa	B5[4], 1827
3476	Biphenyl, 2-ethoxy 2-C₂H₅OC₆H₄-C₆H₅	198.27	pr (peth)	276, 132[6]	34	....		al, eth, ace, bz, chl	B6[3], 3284
3477	Biphenyl, 3-ethoxy 3-C₂H₅OC₆H₄-C₆H₅	198.27	(peth)	305, 158[8]	35			al, eth, bz, ace	B6[3], 3312
3478	Biphenyl, 3-ethyl .,. 3-C₂H₅C₆H₄-C₆H₅	182.26		283-4[763]		1.013[0]			B5[4], 1896
3479	Biphenyl, 2-fluoro 2-FC₆H₄-C₆H₅	172.20		248	73.5	1.2452[25/4]		al, eth	B5[4], 1815
3480	Biphenyl, 4-fluoro 4-FC₆H₄-C₆H₅	172.20		253	74.5			eth	B5[4], 1815
3481	Biphenyl, 2,2′,4,4′,6,6′-hexamethyl [2,4,6-(CH₃)₃C₆H₂-]₂	238.37		296[735]	103—4	1.023[50]		eth, bz	B5[4], 1989
3482	Biphenyl, 2-hydroxy-2′-methoxy-5,5′-dimethyl C₁₅H₁₆O₂	228.29		205[12]	....			al, eth, bz, chl	B6[2], 974
3483	Biphenyl, 2-iodo 2-IC₆H₄-C₆H₅	280.11		189-92[36]		1.6038[25/25]	1.6620[20]	al, eth, bz, aa	B5[4], 1820
3484	Biphenyl, 4-iodo 4-IC₆H₄-C₆H₅	280.11	nd (al or aa)	320d, 183[11]	113-4			al, eth, bz, aa	B5[4], 1821
3485	Biphenyl, 2-methoxy or 2-Phenylanisole 2-CH₃OC₆H₄-C₆H₅	184.24	pr (peth)	274, 150[13]	29	1.0233[99/4]	1.5641[99]	al, peth	B6[3], 3284
3486	Biphenyl, 4-methoxy or 4-Phenylanisole 4-CH₃OC₆H₄-C₆H₅	184.24	pl (al)	157[10]	90	1.0278[100/4]	1.5744[100]	al, eth	B6[3], 3321
3487	Biphenyl, 2-methyl or 2-Phenyl toluene 2-CH₃C₆H₄-C₆H₅	168.24		255.5, 130-6[27]	-0.2	1.010[22/4]	1.5914[20]	al, eth	B5[4], 1855
3488	Biphenyl, 3-methyl 3-CH₃C₆H₄C₆H₅	168.24		272.7, 148.50[20]	4.5	1.0182[17/4]	1.5972[20]	al, eth	B5[4], 1858
3489	Biphenyl, 4-methyl 4-CH₃C₆H₄C₆H₅	168.24	pl (lig, MeOH)	267-8, 134-6[15]	49-50	1.015[27]	....	al, eth	B5[4], 1860
3490	Biphenyl, 2-nitro 2-NO₂C₆H₄-C₆H₅	199.21	pl (al or MeOH)	320, 201[30]	37.2	1.44	....	al, eth	B5[4], 1823
3491	Biphenyl, 3-nitro 3-NO₂C₆H₄-C₆H₅	199.21	ye pl or nd (dil al)	225-30[35]	62			al, eth, aa, lig	B5[4], 1823
3492	Biphenyl, 4-nitro 4-NO₂C₆H₄-C₆H₅	199.21	ye nd (al)	340, 224[30]	114			eth, bz, chl, aa	B5[4], 1823
3493	Biphenyl, 2,2′,4,4′-tetrahydroxy [2,4-(HO)₂C₆H₃-]₂	218.21		....	226-7			w, al, eth, ace	B6[3], 6705
3494	Biphenyl, 3,3′,5,5′-tetrahydroxy or Diresorcinol........ [3,5-(HO)₂C₆H₃-]₂	218.21	pl or nd (w + 2)	....	310 (anh)			al, eth, w	B6[2], 1129
3495	Biphenyl, 2,2′,4,4′-tetramethyl [2,4-(CH₃)₂C₆H₃-]₂	210.32		....	41			al	B5[3], 1891
3496	Biphenyl, 2,2′,5,5′-tetramethyl [2,5-(CH₃)₂C₆H₃-]₂	210.32		284[732]	50			eth, bz	B5[3], 1892
3497	Biphenyl, 2,2′,4,4′-tetranitro [2,4-(NO₂)₂C₆H₃]₂	334.20	ye pr (bz)	....	165-6			bz, aa	B5[3], 1772

No.	Name, Synonyms, and Formula	Mol. wt.	Color, crystalline form, specific rotation and λ_{max} (log ε)	b.p. °C	m.p. °C	Density	n_D	Solubility	Ref.
3498	Biphenyl, 2,4,4'-triamino 2,4-(H$_2$N)$_2$C$_6$H$_3$-C$_6$H$_4$NH$_2$-4	199.26	nd	....	134				B13[3], 560
3499	Biphenyl, 4-vinyl 4-(CH$_2$=CH)C$_6$H$_4$-C$_6$H$_5$	180.25	....	136-8[6]	119				B5[3], 1987
3500	2-Biphenyl carboxylic acid 2-(HO$_2$C)C$_6$H$_4$-C$_6$H$_5$	198.22	lf (dil al)	343-4, 199[10]	113-4			al, bz, aa	B9[3], 3268
3501	2-Biphenylcarboxylonitrile 2-NCC$_6$H$_4$-C$_6$H$_5$	179.22	nd	170-2[15]	41			al, eth	B9[3], 3269
3502	3-Biphenylcarboxylic acid 3-(HO$_2$C)C$_6$H$_4$C$_6$H$_5$	198.22	lf (al)	....	165—6			al, eth, bz, aa	B9[3], 3274
3503	4-Biphenylcarboxylic acid 4-(HO$_2$C)C$_6$H$_4$C$_6$H$_5$	198.22	nd (bz or al)	sub	228			al, eth, bz	B9[3], 3276
3504	3-Biphenylcarboxylic acid, 2-hydroxy 2-HO-3-(HO$_2$C)C$_6$H$_3$-C$_6$H$_5$	214.22	cr (bz)	....	186-7			al	B10[3], 1159
3505	2,2'-Biphenyldicarboxylic acid or Diphenic acid [2-(HO$_2$C)C$_6$H$_4$-]$_2$	242.23	mcl pr or lf (w), cr (aa)	sub	233.5			al, eth	B9[3], 4496
3506	2,2'-Biphenyldicarboxylic acid anhydride C$_{14}$H$_8$O$_3$	224.23	nd (aa or bz)	sub	217				B17[4], 6425
3507	2,2'-Biphenyldicarboxylyl chloride [2-(ClCO)C$_6$H$_4$-]$_2$	279.14	....	sub	94 (97)				B9[2], 657
3508	2,2'-Biphenyldicarboxylic acid, diethyl ester [2-(C$_2$H$_5$O$_2$C)C$_6$H$_4$-]$_2$	298.34	....		42			eth	B9[2], 656
3509	2,2'-Biphenyldicarboxylic acid, dimethyl ester [2-(CH$_3$O$_2$C)C$_6$H$_4$-]$_2$	270.29	mcl pr (MeOH)	204-6[14]	74			al, eth, bz	B9[3], 4497
3510	2,2'-Biphenyldicarboxylic acid, imide C$_{14}$H$_9$NO$_2$	223.23	nd (al)	....	219-20			chl	B21[4], 5601
3511	2,3'-Biphenyldicarboxylic acid or Isodiphenic acid 2-(HO$_2$C)-C$_6$H$_4$-C$_6$H$_4$(CO$_2$H)-3	242.22	nd (w or dil aa)	....	216			al	B9[2], 663
3512	2,4'-Biphenyldicarboxylic acid 2-(HO$_2$C)-C$_6$H$_4$-C$_6$H$_4$(CO$_2$H)-4	242.23	lf (al)	....	272-3			al, bz, aa	B9[3], 4514
3513	3,3'-Biphenyldicarboxylic acid [3-(HO$_2$C)-C$_6$H$_4$-]$_2$	242.23	lf (al)	....	356-7			chl	B9[3], 4517
3514	3,3'-Biphenyl dicarboxylic acid, dimethyl ester [3-CH$_3$O$_2$C-C$_6$H$_4$-]$_2$	270.29	lf (MeOH)		104			al, eth, bz	B9[3], 4517
3515	3,4'-Biphenyl dicarboxylic acid [3-(HO$_2$C)-C$_6$H$_4$-C$_6$H$_4$-(CO$_2$H)-4]	242.23	nd	....	334-5				B9[3], 4518
3516	3,4'-Biphenyldicarboxylic acid, dimethyl ester 3-(CH$_3$O$_2$C)-C$_6$H$_4$-C$_6$H$_4$CO$_2$CH$_3$-4	270.29	nd (lig or MeOH)		98-9			lig, MeOH	B9, 927
3517	3,5-Biphenyldicarboxylic acid 3,5-(HO$_2$C)$_2$C$_6$H$_3$C$_6$H$_5$	242.23	lf (aa)	....	>310			al, eth, ace, bz	B9, 926
3518	3,5-Biphenyldicarboxylic acid, dimethyl ester 3-5-(CH$_3$O$_2$C)$_2$C$_6$H$_3$C$_6$H$_5$	270.29	lf (MeOH)		214				B9[2], 665
3519	2,2'-Biphenyldicarboxylic acid, 3,3'-dimethyl-5-nitro C$_{16}$H$_{13}$NO$_6$	287.23	lf (w or dil al)	....	267			al, eth, bz	B9[2], 659
3520	2,2'-Biphenyldicarboxylic acid, 3,3'-dimethyl-6-nitro dl C$_{16}$H$_{13}$NO$_6$	287.23	lf (w)	....	248-50d			al, eth, ace, aa	B9[2], 659
3521	2,2'-Biphenyldicarboxylic acid, 3,3'-dimethyl-4-nitro C$_{16}$H$_{13}$NO$_6$	287.23	lf, pr or wh nd (w)	....	217 (250)			al, eth	B9[2], 659
3522	2,2'-Biphenyldicarboxylic acid, 3,3'-dimethyl [3-CH$_3$-2-(HO$_2$C)C$_6$H$_3$-]$_2$	270.29		....	230			al, eth, bz	B9[1], 407
3523	2,2'-Biphenyl disulfonyl chloride [2-ClO$_2$S-C$_6$H$_4$-]$_2$	351.24	pr (chl), cr (aa)	....	142-4			eth, bz, chl	B11[3], 469
3524	3,3'-Biphenyl disulfonamide [3-(H$_2$NO$_2$S)-C$_6$H$_4$-]$_2$	312.37	nd (ace)		285			al	B11, 219
3525	3,3'-Biphenyl disulfonyl chloride [3-(ClO$_2$S)-C$_6$H$_4$-]$_2$	351.24	nd (chl)		128			eth, bz	B11[3], 471
3526	4,4'-Biphenyl disulfonic acid [4-(HO$_3$S)-C$_6$H$_4$-]$_2$	314.34	pr	>200	72.5			w	B11[3], 472
3527	4,4'-Biphenyl disulfonamide [4-(H$_2$NO$_2$S)-C$_6$H$_4$-]$_2$	312.37	nd (w)		300			w, eth	B11, 220
3528	4,4'-Biphenyl disulfonyl chloride [4-(ClO$_2$S)C$_6$H$_4$-]$_2$	351.24	pr (aa)	....	205—7			eth, bz, aa	B11[3], 472
3529	2,2'-Biphenyl disulfonic acid, 4,4'-diamino [2-(HO$_3$S)-4-H$_2$N-C$_6$H$_3$-]$_2$	344.37	lf		175d				B14[3], 2264
3530	2,2'-Bipyridyl 2-NC$_5$H$_4$-C$_5$H$_4$N-2	156.20	pr (peth)	272-5	71-3			al, eth, bz, chl, lig	B23[3], 211

No.	Name, Synonyms, and Formula	Mol. wt.	Color, crystalline form, specific rotation and λ_{max} (log ϵ)	b.p. °C	m.p. °C	Density	n_D	Solubility	Ref.
3531	2,3'-Bipyridyl or Isonicoteine 2-NC$_5$H$_4$-C$_5$H$_4$N-(3)	156.20	....	295-6	....	1.140[20/4]	1.6223[20]	al, eth, bz, chl	B23[2], 212
3532	2,4'-Bipyridyl 2-NC$_5$H$_4$-C$_5$H$_4$N-(4)	156.20		280-2, 148-50[1]	61.5	....		al, eth, chl	B23, 200
3533	3,3'-Bipyridyl 3-NC$_5$H$_4$-C$_5$H$_4$N-(3)	156.20	....	291-2, 190-2[25]	68	1.1635[20/20]	....	w, al	B23[2], 212
3534	3,4'-Bipyridyl 3-NC$_5$H$_4$-C$_5$H$_4$N-(4)	156.20	lf (peth)	297	62			w, al, peth	B23, 212
3534a	4,4'-Bipyridyl 4-NC$_5$H$_4$-C$_5$H$_4$N-(4)	156.20	nd (w + 2)	305 sub	114, 171-2			al, eth, bz, chl, lig	B23[2], 212
3535	2,2'-Biquinolyl (C$_9$H$_6$N)$_2$	256.31	pl or lf (al)	....	196			al, eth, ace, bz	B23[2], 267
3536	2,3'-Biquinolyl (C$_9$H$_6$N)$_2$	256.31	lf (al), ye pl or nd (bz)	>400	176-7			al, eth, bz, chl	B23[2], 267
3537	2,6'-Biquinolyl (C$_9$H$_6$N)$_2$	256.31	pl (al)	....	144			al, ace, bz	B23, 294
3538	2,7'-Biquinolyl(higher melting) (C$_9$H$_6$N)$_2$	256.31	mcl pl (al)	....	160			al	B23, 294
3539	2,7'-Biquinolyl(lower melting) C$_{18}$H$_{12}$N$_2$	256.31	tcl	....	115			al, eth, bz	B23, 294
3540	3,4'-Biquinolyl (C$_9$H$_6$N)$_2$	256.31	pw (peth)	....	83-4			al, bz	B23[2], 267
3541	3,7'-Biquinolyl (C$_9$H$_6$N)$_2$	256.31	lf or nd (al or bz)	....	190			al, bz, chl	B23[2], 268
3542	4,4'-Biquinolyl (C$_9$H$_6$N)$_2$	256.31	pr (peth)	....	171			al, ace, bz	B23[2],268
3543	4,6'-Biquinolyl (C$_9$H$_6$N)$_2$	256.31	cr (bz)	....	177			al, bz, chl	B23,294
3544	6,6'-Biquinolyl (C$_9$H$_6$N)$_2$	256.31	lf (al)	....	181			al, eth, bz	B23, 295
3545	6,8'-Biquinolyl (C$_9$H$_6$N)$_2$	256.31	lf (al)	....	148			al, bz	B23, 296
3546	8,8'-Biquinolyl (C$_9$H$_6$N)$_2$	256.31	lf or pl (al or aa)	....	205-7			al, ace, bz, chl	B23[2], 268
3547	Bismuthine, triphenyl (C$_6$H$_5$)$_3$Bi	440.30		242[14]	77.6	1.715[78/4]	1.7040[75]	eth, ace, bz, lig	B16[3], 1188
3548	Biquinone, 3,3'-dihydroxy-5,5'-dimethyl or Phenicin (C$_7$H$_5$O$_3$)$_2$	274.23	yesh-br (al)	....	230-1			al, chl, aa	B8[3], 4251
3549	2,2-Bithiophene or 2,2-Bithienyl (C$_4$H$_3$S)$_2$	166.27	lf (al)	260, 103[3]	33			al, eth, aa	B19[4], 265
3550	2,2-Bithiophene, hexabromo C$_8$Br$_6$S$_2$	639.67	nd (bz)	....	257-8			bz	B19[4], 266
3551	3,3'-Bithiophene (3-C$_4$H$_3$S-3)$_2$	166.26		....	132			al, eth, bz, chl, lig	B19[4], 267
3552	Biuret or Carbamoylurea H$_2$NCONHCONH$_2$	103.08	pl (al), nd (w + 1)	....	190d			al, w	B3[4], 141
3553	Biuret, acetyl CH$_3$CONHCONHCONH$_2$	145.12	nd (w or al)	....	193-4			w, al	B3[4], 142
3554	Biuret, 1,5-diamino H$_2$NNHCONHCONHNH$_2$	133.11	pr (dil al), nd (aa)	....	199-200d			w, aa	B3[4], 178
3555	Bixin C$_{25}$H$_{30}$O$_4$	394.52	vt pr (ace)	....	198			al, ace, chl	B2[3], 2020
3556	Boric acid, tributyl ester or Tributoxyborine B(OC$_4$H$_9$)$_3$	230.16	oil	230—1, 114-5[25]	....	0.8567[20/4]	1.4106[18]	al, eth, bz	B1[4], 1544
3557	Boric acid, isobutyl i-C$_4$H$_9$B(OH)$_2$	101.94	lo pl (w)	....	112			al, eth	B4[3], 1965
3558	Boric acid, isopentyl i-C$_5$H$_{11}$B(OH)$_2$	115.97	pl (w)	....	169			al, eth, ace	B4[2], 1023
3559	Boric acid, triethyl ester or Triethoxyborine B(OC$_2$H$_5$)$_3$	146.00	....	120		0.8546[20/4]	1.3749[20]	al, eth	B1[4], 1365
3560	Boric acid, trimethyl ester or Trimethoxyborine B(OCH$_3$)$_3$	103.92	....	67-9	−29.3	0.915[20]	1.3568[20]	al, eth, bz	B1[4], 1269
3561	Boric acid, tri isopentyl ester or Triisopentyl borate B-(O-i-C$_5$H$_{11}$)$_3$	272.24		254-5, 132-3[12]	....	0.8518[20/4]	1.4156[20]	al, eth	B1[4], 1688
3562	Boric acid, tripropyl ester or Tripropoxyborine B(OC$_3$H$_7$)$_3$	188.08		179-80, 64[9]	....	0.8576[20/4]	1.3948[20]	al, eth	B1[4], 1436
3563	Boric acid, tri-iso-propyl ester or Tri-iso-propoxyborine B-(O-i-C$_3$H$_7$)$_3$	188.08		164, 52-3[32]	....	0.8251[20/4]	1.3772[20]	al, eth, bz	B1[4], 1488

No.	Name, Synonyms, and Formula	Mol. wt.	Color, crystalline form, specific rotation and λ_{max} (log ε)	b.p. °C	m.p. °C	Density	n_D	Solubility	Ref.
3564	Boric acid, tri-isopentyl ester B-(O-i-C$_5$H$_{11}$)$_3$	272.24	......	254-5, 132-3[12]	......	0.8518[20/4]	1.4156[20]	al, eth	B1[4], 1688
3565	Borine, bis(dimethylamino)-fluoro [(CH$_3$)$_2$N]$_2$BF	117.96	......	106	−44.3	......	......	eth, ace	B4[4], 303
3566	Borine, bis(methylthio) methyl (CH$_3$S)$_2$BCH$_3$	120.05	......	100[147]	−59	......	......	eth, ace	Am78,1523
3567	Borine, tri-(methylthio) B(SCH$_3$)$_3$	152.11	......	218.2	5	1.126[20]	1.5788[20]	eth, ace	C55,9346
3568	Borine, difluoro-(dimethyl amino) (CH$_3$)$_2$NBF$_2$	92.89	rh	sub 132	165-8d	......	......		B4[4]303
3569	Borine, phenyl-difluoro C$_6$H$_5$BF$_2$	125.92	......	97-8[747]	−36.2	1.087[25]	1.4441[25]	eth, bz	B16[2], 638
3570	Borine, 4-tolyl, difluoro (4-CH$_3$C$_6$H$_4$)BF$_2$	139.94	......	127-8[747]	......	1.055[25]	1.4535[25]	eth, bz	C18,992
3571	Borine, dimethyl-(dimethyl amino) (CH$_3$)$_2$BN(CH$_3$)$_2$	84.99	......	65	−92	......	......	eth, ace	B4[3], 1960
3572	Borine, dimethyl,methoxy (CH$_3$)$_2$BOCH$_3$	71.92	......	21	......	......	......	eth, ace	Am75,3872
3573	Borine, (methylthio)-dimethyl CH$_3$SB(CH$_3$)$_2$	87.98	......	71	−84	......	......	eth, ace	Am76,3307
3574	Borine, triethyl B(C$_2$H$_5$)$_3$	98.00	......	95-6	−92.9	0.6961[23]	......	al, eth	B4[3], 1957
3575	Borine, tri-iso-butyl B(i-C$_4$H$_9$)$_3$	182.16	......	188[766], 86[20]	......	0.7380[25/4]	1.4188[23]	al, eth, bz	B4[3], 1958
3576	Borine, trimethyl B(CH$_3$)$_3$	55.92	......	20	−161.5	......	......	al, eth	B4[3], 1955
3577	Borine, triisopentyl (i-C$_5$H$_{11}$)$_3$B	224.24	......	119.14	......	0.7600[25/4]	1.4321	al, eth, ace	B4[2], 1023
3578	Borine, triphenyl B(C$_6$H$_5$)$_3$	242.13	wh cr	245—50[15]	142	......	......	bz, lig	B16[3], 1271
3579	Borine, tri-isopentyl (i-C$_5$H$_{11}$)$_3$B	224.24	......	119.14	......	0.7600[23/4]	1.4321	al, eth, ace, w	B4[2], 1023
3580	Borine, tripropyl B(C$_3$H$_7$)$_3$	140.08	......	159, 43-4[17]	−56	0.7204[25/4]	1.4135[22.5]	......	B4[3], 1957
3581	3-Bornanone-(d) or Epicamphor C$_{10}$H$_{16}$O	152.24	[α][17/D] +45.4 (bz)	......	182	......	......	al, eth, peth	B7[3], 420
3582	3-Bornanone (dl) C$_{10}$H$_{16}$O	152.24	cr (peth)	......	175-7	......	......	al, eth, peth	B7[3], 421
3583	3-Bornanone oxime-(d) C$_{10}$H$_{17}$NO	167.25	nd (MeOH), [α]$_D$ −98.9	......	103	......	......	eth, ace	B7[1], 86
3584	3-Bornanone oxime-(dl) C$_{10}$H$_{17}$NO	167.23	nd (dil al)	......	98-100	......	......	eth, ace	B7[1], 87
3585	3-Bornanone oxime-(l) C$_{10}$H$_{17}$NO	167.25	nd (dil MeOH), [α]$_D$ +100.5	......	103-4	......	......	eth, ace	B7[1], 86
3586	3-Bornanone semicarbazide (d) C$_{11}$H$_{19}$N$_3$O	209.29	nd (al)	......	237-8	......	......	......	B7[3], 421
3587	3-Bornanone semicarbazide (l) C$_{11}$H$_{19}$N$_3$O	209.29	nd (al), [α][20/D] +145 (MeOH, c=0.73)	......	237-8d	......	......	......	B7[3], 421
3588	Borneol (d) C$_{10}$H$_{18}$O	154.26	lf or hex pl (peth), [α][20/D] +37.7 (al)	......	208	1.011[20/4]	......	al, eth, bz, lig	B6[4], 281
3589	Borneol (dl) C$_{10}$H$_{18}$O	154.26	lf (lig)	sub	210.5	1.011[20/4]	......	al, eth, bz	B6[4], 281
3590	Borneol (l) C$_{10}$H$_{18}$O	154.26	hex pl, [α][20/D] −37.74 (al)	210[779]	208.6	1.1011[20/4]	......	al, eth, ace, bz	B6[4], 281
3591	Borneol acetate (d) C$_{12}$H$_{20}$O$_2$	196.28	rh, [α][20/D] +44.4 (al)	223-4, 107[15]	29	0.9920[20/4]	......		B6[3], 302
3592	Borneal acetate (dl) C$_{12}$H$_{20}$O$_2$	196.29	......	223-4	<−17	1.4630[20]	......		B6[3],303

No.	Name, Synonyms, and Formula	Mol. wt.	Color, crystalline form, specific rotation and λ_{max} (log ε)	b.p. °C	m.p. °C	Density	n_D	Solubility	Ref.
3593	Borneol acetate (l) . . . . $C_{12}H_{20}O_2$	196.29	$[\alpha]^{20/D}$ −44.45 (undil)	223-4, 107[15]	29	0.9920[20/4]	1.4634[20]	al, eth	B6[1]303
3594	Borneol formate . . . . $C_{11}H_{18}O_2$	182.27	$[\alpha]_D$ +48.75 (undil)	90[10]		1.009[22]	1.4700[15]	. . . .	B6[3], 301
3595	Bornylamine (d) . . . . $C_{10}H_{17}NH_2$	153.27	$[\alpha]^{20/D}$ +47.2 (al)	200 sub	163			al, eth, ace, bz	B12[3], 193
3596	Bornyl chloride (d) . . . . $C_{10}H_{17}Cl$	172.70	nd	207-8 sub	132	. . . .		al, eth, bz, peth	B5[4], 319
3597	Bornylene (d) or 2-Bornene . . . . $C_{10}H_{16}$	136.24	cr (al), $[\alpha]_D$ +30.5 (to)	146[750] sub	109-10	. . . .			B5[4], 460
3598	Bornylene (l) . . . . $C_{10}H_{16}$	136.24	cr (al), $[\alpha]^{19/D}$ −23.9 (bz)	146[746]	113			al, eth, bz	B5[4], 460
3599	Brazilein . . . . $C_{16}H_{12}O_5$	284.27	red-br nd or lf (w + l)	. . . .	250			al, w, bz, chl, aa	B18[4], 2770
3600	Bromcresolgreen or 3,3′,5,5′-Tetrabromo-m-Cresol sulfophthalein . . . . $C_{21}H_{14}O_5BrS$	698.04	wh or red (+7w), ye (aa)	. . . .	218-9			al, eth, bz, aa	B19[4], 1133
3601	Bromo acetamide or Bromo acetamide . . . . $BrCH_2CONH_2$	137.97	nd (al or bz)	. . . .	91	. . . .	. . . .	w, bz	B2[4], 530
3602	Bromoacetic acid . . . . $BrCH_2CO_2H$	138.95	hex or rh	208, 127.5[30]	50	1.9335[50/4]	1.4804[50]	w, al, eth, ace, bz	B2[4], 526
3603	Bromoacetic acid, iso-butyl ester or iso-Butyl bromoacetate . . . . $BrCH_2CO_2-iC_4H_9$	195.06	. . . .	188[752], 74.5[10]	. . . .	1.3260[20/4]	. . . .	al, eth, ace	B2[3], 482
3604	Bromoacetic acid, tert-butyl ester or tert-Butyl bromoacetate . . . . $BrCH_2CO_2C(CH_3)_3$	195.06	. . . .	73-4[25]		1.4430[20]		al, eth	B2[3], 482
3605	Bromoacetic acid, ethyl ester or Ethyl bromoacetate . . . . . $BrCH_2CO_2C_2H_5$	167.01	. . . .	168-9, 58.9[15]	. . . .	1.5059[20/20]	1.4489[20]	al, eth, ace	B2[4], 527
3606	Bromoacetic acid, methyl ester or Methyl bromoacetate . . $BrCH_2CO_2CH_3$	152.98	. . . .	144d, 64[33]				al, eth, ace	B2[4], 527
3607	Bromoacetic acid, phenyl ester . . . . $BrCH_2CO_2C_6H_5$	215.05	pl (al)	140[20]	32	. . . .		al, eth	B6[1], 87
3608	Bromoacetic acid, propyl ester or Propyl bromoacetate . . . $BrCH_2CO_2C_3H_7$	181.04	. . . .	176[762]		1.4099[20/4]	1.4518[20]	al, eth, ace	B2[4], 528
3609	Bromoacetonitrile . . . . $BrCH_2CN$	119.95	pa ye	150-1[752], 46[13]				eth	B2[4], 531
3610	Bromo acetyl bromide . . . . $BrCH_2COBr$	201.86	. . . .	130		2.317[22/22]	1.5449[20]	ace	B2[4], 530
3611	Bromochloro acetic acid . . . . $BrClCHCO_2H$	173.40	. . . .	215d, 103-4[11]	38	1.9848[31/4]	1.5014[31]	w, al, eth, ace	B2[4], 532
3612	Bromochloro acetic acid, ethyl ester or Ethyl bromochloro acetate . . . . $BrClCHCO_2C_2H_5$	201.45	. . . .	174d		1.5890[22/4]	1.4639[24]	al, eth	B2[4], 532
3613	Bromo difluoro acetic acid . . . . BrF_2CCO_2H	174.94	lf (chl)	145-6, 87[82]	40	. . . .	. . . .	w, al, chl	B2[4], 532
3614	Bromo diphenyl acetic acid . . . . $(C_6H_5)_2CBrCO_2H$	291.15	(chl-peth)	. . . .	133-4	. . . .		chl, to	B9[2], 471
3615	Bromodiphenyl acetyl bromide . . . . $(C_6H_5)_2CBrCOBr$	354.95	nd (lig)	. . . .	65-6			al, eth, bz, chl	B9[1], 283
3616	bis-(2-Bromoethyl) ether . . . . $(BrCH_2CH_2)_2O$	231.91	. . . .	115[32]		1.8222[27/4]	1.5131[27/D]	. . . .	B1[4], 1386
3617	Bromofluoro acetamide . . . . $BrCHFCONH_2$	155.96	nd (CCl₄)	. . . .	44			w, al, eth	B2,217
3618	Bromofluoro acetic acid . . . . $BrCHFCO_2H$	156.95	. . . .	183, 102[20]	48			w, al, chl	B2[4]531
3619	Bromofluoro acetic acid, ethyl ester . . . . $BrFCHCO_2C_2H_5$	184.99	. . . .	154		1.5587[17]			B2[4], 531
3620	Bromophenol blue or 3,3′,5,5′-Tetrabromophenol sulfonphthalein . . . . $C_{19}H_{10}O_5Br_4S$	670.02	hex pr (aa-ace)	279d				al, bz, aa	B19[4], 1129
3621	bis-(4-Bromophenyl) ether . . . . $(4-BrC_6H_4)_2O$	328.03	lf (al)	338-40, 210[11]	60.5	1.8 (sol)	. . . .	al, eth, bz	B6[4], 1048

No.	Name, Synonyms, and Formula	Mol. wt.	Color, crystalline form, specific rotation and λ_{max} (log ε)	b.p. °C	m.p. °C	Density	n_D	Solubility	Ref.
3622	4-Bromophenyl isocyanate (4-BrC$_6$H$_4$)NCO	198.03	nd	226, 158[14]				eth	B12[1], 321
3623	4-Bromphenyl isothiocyanate 4-BrC$_6$H$_4$NCS	214.09	nd		60-1			al	B12[3], 1463
3624	Brucine C$_{23}$H$_{26}$N$_2$O$_4$	394.44	mcl pr (w + 4), [α]$^{20}_{546.1}$ −149.5 (chl, c=1		178 (anh) 105 (hyd)			al, chl	B27[2], 797
3625	Brucine hydrochloride C$_{23}$H$_{26}$N$_2$O$_4$·HCl	430.94	pr					w, al	B27[2], 801
3626	Brucine nitrate, dihydrate C$_{23}$H$_{26}$N$_2$O$_4$.HNO$_3$.2H$_2$O	493.52	pr	sub	230d				B27[2], 797
3627	Brucine sulfate, heptahydrate C$_{46}$H$_{68}$N$_4$O$_{19}$S·7H$_2$O	1013.1	nd [α]$_D$ −24.4 (w)					w, MeOH	B27[2], 797
3628	Bufotalin C$_{26}$H$_{36}$O$_6$	444.58	cr (+ 1 al), [α]$^{20}_D$ + 5.4 (chl, c=0.5)		223d			al, chl	B18[4], 2557
3629	Bulbocapnine (d) C$_{19}$H$_{19}$NO$_4$	325.37	pr (al), [α]$_D$ + 237.1 (chl, c=4)					al, chl	B27[2], 554
3630	Bulbocapnine (l) C$_{19}$H$_{19}$NO$_4$	325.37		209-10					B27[2], 554
3631	1,2-Butadiene or Methylallene CH$_2$=C=CHCH$_3$	54.09		10.8	−136.2	0.676$^{0.4}$	1.4205$^{1.1}$	al, eth, bz	B1[4], 975
3632	1,2-Butadiene, 4-bromo CH$_2$=C=CH-CH$_2$Br	133.00		109-11		1.4255$^{20/4}$	1.5248^{20}	ace	B1[3], 929
3633	1,2-Butadiene, 4-chloro CH$_2$=C=CH-CH$_2$Cl	88.54		88		0.9891$^{20/4}$	1.4775^{20}	eth, ace, bz	B1[4], 975
3634	1,2-Butadiene, 4-iodo CH$_2$=CCCH-CH$_2$I	179.99		130		1.7129$^{20/4}$	1.5709^{20}		B1[3], 929
3635	1,2-Butadiene, 4-methoxy CH$_2$=C=CH-CH$_2$OCH$_3$	84.13		87-9		0.8286$^{20/4}$	1.435^{20}	al	B1[4], 2221
3636	1,2-Butadiene, 3-methyl CH$_2$=C=C(CH$_3$)$_2$	68.13		40	−120	0.6804$^{20/4}$	1.4166^{20}	al, eth, ace, bz, peth	B1[4], 1006
3637	1,2-Butadiene, 4-ol CH=C=CH-CH$_2$OH	70.09		126-8, 68-9^{45}		0.9164$^{20/4}$	1.4759^{20}	w, al, ace, eth, bz	B1[4], 2221
3638	1,3-Butadiene or Bivinyl CH$_2$=CH-CH=CH$_2$	54.09		−4.4	−108.9	0.6211$^{20/4}$	1.4292$^{−25}$	al, eth, ace, bz	B1[4], 976
3639	1,3-Butadiene, 2-bromo or Bromoprene CH$_2$=CBr-CH=CH$_2$	133.00		42-3^{165}		1.397$^{20/4}$	1.4988^{20}	al, eth	B1[4], 989
3640	1,3-Butadiene, 1-chloro CH$_2$=CH-CH=CHCl	88.54		68		0.9606$^{20/4}$	1.4712^{20}	al, eth, chl	B1[3], 949
3641	1,3-Butadiene, 1-chloro-2-methyl ClCH=C(CH$_3$)CH=CH$_2$	102.57		107, 50.4^{100}		0.9710$^{20/4}$	1.4792^{20}	al, ace	B1[3], 974
3642	1,3-Butadiene, 1-chloro-3-methyl ClCH=CH-C(CH$_3$)=CH$_2$	102.57		99-100		0.9543$^{20/4}$	1.4719^{20}	al, eth, ace, chl	B1[3], 975
3643	1,3-Butadiene, 2-chloro or Chloroprene CH$_2$=C(Cl)CH=CH$_2$	88.54		59.4, 6.4^{100}		0.9583$^{20/4}$	1.4583^{20}	eth, ace, bz	B1[4], 984
3644	1,3-Butadiene, 2-chloro-3-methyl CH$_2$=CCl-C(CH$_3$)=CH$_2$	102.57		93		0.9593$^{20/4}$	1.4686^{20}	al, eth, ace, chl	B1[4], 1004
3645	1,3-Butadiene, 4-cyano or 2,4-Pentadieno nitrile CH$_2$=CH-CH=CHCN	79.10		135-8	0.8444^2	1.4880			B2[4], 1692
3646	1,3-Butadiene, 1,1-dichloro Cl$_2$C=CH-CH=CH$_2$	122.98		42-3^{90}		1.1831$^{20/4}$	1.5022^{20}	eth, bz	B1[4], 985
3647	1,3-Butadiene, 1,2-dichloro ClCH=CCl-CH=CH$_2$	122.98		60-5^{105}, 35^{40}		1.1991$^{20/44}$	1.4960^{20}	CCl$_4$	B1[4], 985
3648	1,3-Butadiene, 2,3-dichloro CH$_2$=CCl-CCl=CH$_2$	122.98		98		1.1829$^{20/4}$	1.4890^{20}	chl	B1[4], 986
3649	1,3-Butadiene, 2,3-dimethyl (cis,cis) or Biisopropenyl CH$_2$=C(CH$_3$-C)CH$_3$=CH$_2$	82.15		68-78	−76	1.4394$^{20/4}$			B1[4], 1023
3650	Butadiene dioxide CH$_2$CHCHCH$_2$	86.09		144	4	1.113^{20}	1.435^{20}	w, al	B19[4], 111
3651	1,3-Butadiene, 1,4-diphenyl (cis,cis) or cis,cis-Bistyryl C$_6$H$_5$CH=CH-CH=CHC$_6$H$_5$	206.29	lf or nd (al or MeOH)		70.5	0.9697^{100}	1.6183^{100}	al, eth, bz, chl, peth	B5[2], 2159

No.	Name, Synonyms, and Formula	Mol. wt.	Color, crystalline form, specific rotation and λ_{max} (log ε)	b.p. °C	m.p. °C	Density	n_D	Solubility	Ref.
3652	1,3, Butadiene, 1,4-diphenyl (trans,trans) $C_6H_5CH=CH-CH=CHC_6H_5$	206.29	lf (al or aa)	350[720]	152.5	....	....	al, eth, bz, chl, peth	B5[3], 2159
3653	1,3-Butadiene, 1-ethoxy $C_2H_5OCH=CH-CH=CH_2$	98.15	....	109-12	....	0.8154[20/4]	1.4529[20]	al, eth, ace, bz, chl	B1[4], 2222
3654	1,3-Butadiene, 2-ethoxy $CH_2=C(OC_2H_5)-CH=CH_2$	98.15	....	94-5	....	0.8177[20/4]	1.4400[20]	al, eth, ace, bz	B1[4], 2223
3655	1,3-Butadiene, 2-fluoro or Fluoroprene $CH_2=CF-CH=CH_2$	72.08	....	12	....	0.843[4/4]	1.4004	....	B1[4], 982
3656	1,3-Butadiene, hexachloro $CCl_2=CCl-CCl=CCl_2$	260.76	....	215, 101[20]	−21	1.5542[20]	....	al, eth	B1[1], 955
3657	1,3-Butadiene, hexafluoro $CF_2=CF-CF=CF_2$	162.04	....	6	−132	1.553[20/4]	1.378[-20]	....	B1[4], 983
3658	1,3-Butadiene, 2-iodo or Iodoprene $CH_2=CI-CH=CH_2$	179.99	....	111-3	....	1.7278[20/4]	1.5616	....	B1[3], 956
3659	1,3-Butadiene, 1-methoxy $CH_3OCH=CH-CH=CH_2$	84.12	....	91-2	....	0.8296[20/4]	1.4594[20]	w, al	B1[4], 2221
3660	1,3-Butadiene, 2-methoxy $CH_2=C(OCH_3)-CH=CH_2$	84.13	....	75	....	0.8272[20/4]	1.4442[20]	al, eth, ace, bz	B1[4], 2223
3661	1,3-Butadiene, 2-methyl or Isoprene $CH_2=C(CH_3)-CH=CH_2$	68.13	....	34	−146	0.6810[20/4]	1.4219[20]	al, eth, ace, bz	B1[4], 1001
3662	1,3-Butadiene, pentafluoro-2-trifluoro methyl $CF_2=C(CF_3)-CF=CF_2$	212.05	....	39	....	1.527[0/4]	1.3000[0]	eth, ace, bz	C49,2479
3663	1,3-Butadiene, 1-phenyl (trans) $C_6H_5CH=CH-CH=CH_2$	130.19	....	76[11]	4.5	0.9286[20/4]	1.6009[11]	al, eth, ace, bz	B5[4], 1536
3664	1,3-Butadiene, 2-phenyl $CH_2=C(C_6H_5)-CH=CH_2$	130.19	....	60-1[17]	....	0.9266[20/4]	1.5489[20]	eth, bz	B5[4], 1539
3665	1,3-Butadiene, 1,2,3,4-tetrachloro (liquid) $ClCH=CCl-CCl=CHCl$	191.87	....	188, 67[10]	−4	1.516[18/11]	1.5453[11]	al, eth, ace, bz	B1[4], 987
3666	1,3-Butadiene, 1,2,3,4-tetrachloro (solid) $ClCH=CCl-CCl=CHCl$	191.87	....	....	52	1.4961[20]	1.5438[20]	al, eth, ace, bz, chl	B1[4], 987
3667	1,3-Butadiene, 1,2,3-trichloro $ClCH=CCl-CCl=CH_2$	157.44	....	33-4[7]	....	1.4060[20/4]	1.5262[20]	eth, chl	B1[3], 954
3668	Butadiyne or Biacetylene $CH\equiv C-C\equiv CH$	50.06	....	10.3	−36.4	0.7364[0/4]	1.4189[5]	al, eth, ace, chl	B1[4], 1116
3669	Butadiyne, 1,4-dichloro $ClC\equiv C-C\equiv CCl$	118.95	nd	....	1-3	....	....	chl	B1[3], 1057
3670	Butadiyne, 1,4-bis(1-hydroxycyclohexyl) $(C_6H_9O)_2$	246.35	cr (bz)	174	....	....	....	MeOH	B6[3], 5178
3671	Butanal or Butyraldehyde $CH_3CH_2CH_2CHO$	72.12	....	75.7	−99	0.8170[20/4]	1.3843[20]	w, al, eth, ace, bz	B1[4], 3229
3672	Butane C_4H_{10}	58.12	....	−0.5	−138.4	0.6012[0/4], 0.5788[20/4]	1.3543[-13], 1.3326[0]	al, eth, chl, w	B1[4], 236
3673	Butane, 1-amino or n-Butylamine $CH_3CH_2CH_2CH_2NH_2$	73.14	....	77.8	−49.1	0.7414[20/4]	1.4031[20]	w, al, eth	B4[4], 540
3674	Butane, 1-amino-3-methyl or iso-Pentylamine $(CH_3)_2CH_2CH_2CH_2NH_2$	87.17	....	95-7[761]	....	0.7505[20/4]	1.4083[20]	w, al, eth, ace, chl	B4[4], 696
3675	Butane, 2-amino or sec-Butylamine (d) $CH_3CH_2CH(NH_2)CH_3$	73.14	$[\alpha]^{20/}_D$ +7.4 (w)	63	−104.5	0.724[20/4]	1.344[20]	w, al, eth, ace, chl	B4[4], 617
3676	Butane, 2-amino (dl) or sec-Butylamine $CH_3CH_2CH(NH_2)CH_3$	73.14	....	63.5[764]	<−72	0.7246[20/4]	1.3932[20]	w, al, eth, ace, chl	B4[4], 617
3677	Butane, 2-amino (l) or sec-Butylamine $CH_3CH_2CH(NH_2)CH_3$	73.14	$[\alpha]^{20/}_D$ -7.4 (w, c=4.7)	63	....	0.7205[20/4]	....	w, al, eth, ace, chl	B4[4], 617
3678	Butane, 2-amino-2,3-dimethyl $(CH_3)_2CH-C(NH_2)(CH_3)_2$	101.19	....	104-5	....	0.7683[0/4]	1.4096[17]	....	B4[4], 733
3679	Butane, 2-amino-2-methyl $CH_3CH_2C(CH_3)_2NH_2$	87.17	....	77	-105	0.731[25/4]	1.3954[25]	w, al, eth, ace	B4[4], 694
3680	Butane, 2-amino-3-methyl $(CH_3)_2CH-CH(CH_3)NH_2$	87.17	....	84-7	....	0.7574[19]	1.4096[18]	w, al	B4[4], 695
3681	Butane, 3-amino-2,2 dimethyl $CH_3CH(NH_2)-C(CH_3)_3$	101.19	....	102	-20	....	....	w	B4[4], 730
3682	Butane, 3-amino 2,2-dimethyl, hydrochloride $(CH_3)_3CCH(NH_2)CH_3 \cdot HCl$	137.65	nd	sub 245	300-1 (cor)	....	....	w	B4[4], 730
3683	Butane, 1-bromo $CH_3CH_2CH_2CH_2Br$	137.03	....	101.6, 18.8[30]	-112.4	1.2758[20/4]	1.4401[20]	al, eth, ace, chl	B1[4], 258
3684	Butane, 1-bromo-4-chloro $BrCH_2CH_2CH_2CH_2Cl$	171.48	....	174-5[756], 63-4[10]	....	1.488[20/4]	1.4885[20]	al, eth, chl	B1[4], 264

No.	Name, Synonyms, and Formula	Mol. wt.	Color, crystalline form, specific rotation and λ_{max} (log ε)	b.p. °C	m.p. °C	Density	n_D	Solubility	Ref.
3685	Butane, 1-bromo-3,3-dimethyl .... BrCH₂CH₂C(CH₃)₃	165.08	....	138, 54⁴⁰	....	1.556²⁰ᐟ⁴	1.4440²⁰	al, eth, chl	B1³, 409
3686	Butane, 1-bromo-2,4-diphenyl (d) C₆H₅CH₂CH₂CH(C₆H₅)CH₂Br	289.21	[α]²⁰ 16.8 (chl, c=10)	122-30⁰·⁰²			1.5812²⁵		
3687	Butane, 1-bromo-2,4-diphenyl (dl) C₆H₅CH₂CH₂CH(C₆H₅)CH₂Br	289.21	....	126-8⁰·⁰¹			1.5812²⁵		
3688	Butane, 1-bromo-4-fluoro .... Br(CH₂)₄F	155.01	lf	134-5⁷⁴⁰	....		1.4370²⁵	al, eth	B1⁴, 263
3689	Butane, 1-bromo-2-methyl (d) or act-Amylbromide ...... BrCH₂CH(CH₃)CH₂CH₃	151.05	[α]²⁰ + 3.68	121.6	....	1.2234²⁰ᐟ⁴	1.4451²⁰	al, eth, chl	B1⁴, 327
3690	Butane, 1-bromo-3-methyl (d) BrCH₂CH(CH₃)CH₂CH₃	151.05	....	120.5, 12.3¹⁰	....	1.2205²⁰ᐟ⁴	1.4452²⁰	al, eth, chl	B1⁴, 327
3691	Butane, 1-bromo-3-methyl or Isoamylbromide .... BrCH₂CH₂CH(CH₃)₂	151.05	....	120.4, 12.3¹⁰	-112	1.2071²⁰ᐟ⁴	1.4420²⁰	al, eth, chl	B1⁴, 328
3692	Butane, 2-bromo (dl) or sec-butylbromide .... CH₃CH₂CHBrCH₃	137.03	....	91.2	-111.9	1.2585²⁰ᐟ⁴	1.4366²⁰	eth, ace, chl	B1⁴, 261
3693	Butane, 2-bromo (l) or sec-butylbromide .... CH₃CH₂CHBrCH₃	137.03	[α]²²ᐟ_D - 23.13 (un-dil)	90-1	....	1.2536²⁵ᐟ⁴	1.4359¹⁹	al, eth, ace, chl	B1⁴, 261
3694	Butane, 2-bromo-1-chloro .... ClCH₂CHBrCH₂CH₃	171.48	....	146-7⁷⁵⁸	....	1.468²⁰ᐟ⁴	1.4880²⁰	al, eth, bz, chl	B1⁴, 264
3695	Butane, 2-bromo-2,3-dimethyl CH₃CHBr(CH₃)CH(CH₃)₂	165.08	....	132-3⁷⁴², 87¹⁸⁰	24-5	1.1772¹⁰	1.4517	eth, chl	B1⁴, 374
3696	Butane, 2-bromo-2-methyl (CH₃)₂CBrCH₂CH₃	151.05	....	108⁷⁶⁵	....	1.198¹⁸ fˢᐟ¹⁵	1.44207	....	B1⁴, 327
3697	Butane, 1,4-bis(dicarbethoxy amino) .... C₂H₅O₂CNH-(CH₂)₄-NHCO₂C₂H₅	232.28	nd (lig)	....	85-6			al, eth, bz, chl	B4⁴, 1292
3698	Butane, 1-chloro or n-Butylchloride .... CH₃CH₂CH₂CH₂Cl	92.57	....	78.4	-123.1	0.8862²⁰ᐟ⁴	1.4021²⁰	al, eth	B1⁴, 246
3699	Butane, 1-chloro-2,3-dimethyl ClCH₂CH(CH₃CH(CH₃)₂	120.62	....	116⁷³⁵	....		1.4200²⁰	al, eth, chl	B1⁴, 373
3700	Butane, 1-chloro-3,3-dimethyl .... ClCH₂CH₂C(CH₃)₃	120.62	....	115,41⁵⁰	....	0.8670²⁰ᐟ⁴	1.4161²⁰	al, eth, chl	B1⁴, 369
3701	Butane, 1-chloro-4-fluoro .... Cl(CH₂)₄F	110.57	....	114.7	....	1.0627²⁵ᐟ⁴	1.4020²⁵	al, eth	B1⁴, 249
3702	Butane, 1-chloro-2-methyl (d) or act-Amyl chloride ...... ClCH₂CH(CH₃)CH₂CH₃	106.60	[α]²⁰ᐟ₅₈₉₂ + 1.64	100.5, 43¹⁰⁰	....	0.8857²⁰ᐟ⁴	1.4126²⁰	al, eth	B1⁴, 324
3703	Butane, 1-chloro-2-methyl (dl) ClCH₂CH(CH₃)CH₂CH₃	106.60	....	99.9, 52.2⁵⁰	....	0.8818¹⁵ᐟ¹⁵	1.4102²⁵	al, eth	B1⁴, 324
3704	Butane, 1-chloro-3-methyl or Isopentyl chloride .... (CH₃)₂CHCH₂CH₂Cl	106.60	....	98.5	-104.4	0.8704²⁰ᐟ⁴	1.4084²⁰	al, eth, chl	B1⁴, 325
3705	Butane, 1-chloro-2,2,3,3-tetramethyl .... (CH₃)₃CC(CH₃)₂CH₂Cl	148.68	....	80-1⁴⁰	52-3	....		eth	B1³, 502
3706	Butane, 2-chloro (d) or sec-Butyl chloride .... CH₃CH₂CHClCH₃	92.57	....	68.2	-131.3	0.8732²⁰ᐟ⁴	1.3971²⁰	al, eth, bz, chl	B1⁴, 248
3707	Butane, 2-chloro (l) .... CH₃CH₂CHClCH₃	92.57	[α]²⁰ᐟ_D -8.48	68	-140.5	0.8950⁰ᐟ⁴	....	al, eth, bz, chl	B1⁴, 248
3708	Butane, 2-chloro-2,3-dimethyl CH₃CCl(CH₃)CH(CH₃)₂	120.62	....	112	-10.4	0.8780²⁰ᐟ⁴	1.4191²⁰	al, ace	B1⁴, 373
3709	Butane, 2-chloro-2-methyl or tert-Amyl chloride .... CH₃CH₂CCl(CH₃)₂	106.60	....	85.6	-73.5	0.8653²⁰ᐟ⁴	1.4055²⁰	al, eth	B1⁴, 324
3710	Butane, 2-chloro-3-methyl (dl) .... CH₃CHClCH(CH₃)₂	106.60	....	92.8, 25.7⁶⁰	....	0.8620²⁰ᐟ⁴	1.4020²⁰	al, eth	B1³, 358
3711	Butane, 2-(chloromethyl)-1,3-dichloro (dl) .... CH₃CHClCH(CH₂Cl)₂	175.49	....	79-81¹⁵	....	1.2793¹⁵ᐟ⁴		al, eth, chl	C31, 1003
3712	Butane, 2-(chloromethyl)-1,2,3-trichloro CH₃CHClCCl(CH₂Cl)₂	209.94	....	102-3¹³	....	1.3977¹⁸ᐟ⁴	1.5012¹⁸	eth, chl	B1³, 362
3713	Butane, 2-chloro-2,3,3-trimethyl .... (CH₃)₃CClC(CH₃)₃	134.65	....	sub	136	....		eth	B1⁴, 411
3714	Butane, 3-chloro-2,3-dimethyl or Pinacolyl chloride .... (CH₃)₃CCHClCH₃	120.62	....	111, 7¹⁰	0.9	0.8767²⁰ᐟ⁴	1.4182²⁰	eth	B1⁴, 373
3715	Butane, decafluoro or Perfluoro butane .... C₄F₁₀	238.03	....	3.96	-128	1.6484	....	bz, chl	B1⁴, 245
3716	Butane, 1,4-diamino or Putrescine .... H₂N(CH₂)₄NH₂	88.15	lf	158-9	27-8	0.877²⁵	1.4969²⁰	w	B4⁴, 1283
3717	Butane, 1,4-diamino, dihydrochloride.................. H₂N(CH₂)₄NH₂·2HCL	161.08	nd or lf (al or w)	sub	315d	....	....	w, al	B4⁴, 1284

No.	Name, Synonyms, and Formula	Mol. wt.	Color, crystalline form, specific rotation and λ_{max} (log ϵ)	b.p. °C	m.p. °C	Density	n_D	Solubility	Ref.
3718	Butane, 1,2-dibromo CH$_3$CH$_2$CHBrCH$_2$Br	215.94	....	166.3	-65.4	1.7915[20/4]	1.4025[20]	eth, chl	B1[4], 266
3719	Butane, 1,3-dibromo CH$_3$CHBrCH$_2$CH$_2$Br	215.94		174, 72[20]	,,, ...	1.800[10]	1.507[20]	eth, chl	B1[4], 266
3720	Butane, 1,4-dibromide or Tetramethylene dibromide Br(CH$_2$)$_4$Br	215.94		197, 79[10]	-16.5	1.7890[20/4]	1.5190[20]	chl	B1[4], 267
3721	Butane, 2,3-dibromo CH$_3$CH$_2$BrCH$_2$BrCH$_3$	215.94		161	<-80	1.7893[22/4]	1.5133[22]	eth, chl	B1[4], 268
3722	Butane, 2,3-dibromo (meso) CH$_3$CHBrCHBrCH$_3$	215.94		157.3		1.7913[15/4]	1.5132[15.4]		B1[4], 268
3723	Butane, 1,1-dichloro or Butylidene dichloride CH$_3$CH$_2$CH$_2$CHCl$_2$	127.03		113.8		1.0863[20/4]	1.4355[20]	chl	B1[4], 250
3724	Butane, 1,1-dichloro-3-methyl (dl) (CH$_3$)$_2$CHCH$_2$CHCl$_2$	141.04		130, 48-9[40]		1.0473[20]	1.4344[20]	al, eth	B1[4], 326
3725	Butane, 1,2-dichloro CH$_3$CH$_2$CHClCH$_2$Cl	127.03		124		1.1116[25/4]	1.4450[20]	eth, chl	B1[4], 250
3726	Butane, 1,2-dichloro-2-methyl CH$_3$CH$_2$ClCl(CH$_3$)CH$_2$Cl	141.04		133-5, 71.5[100]		1.0785[20/4]	1.4432[21.5]	al, eth, chl	B1[4], 325
3727	Butane, 1,3-dichloro CH$_3$CHClCH$_2$CH$_2$Cl	127.03		134		1.1158[20/4]	1.4445[20]	eth, chl	B1[4], 250
3728	Butane, 1,3-dichloro-3-methyl (d) (CH$_3$)$_2$ClCH$_2$CH$_2$Cl	141.04		145-6, 39[10]		1.0654[20/4]	1.4455[20]	eth, chl	B1[4], 326
3729	Butane, 1,4-dichloro or Tetramethylene dichloride Cl(CH$_2$)$_4$Cl	127.03		153.9, 39.7[10]	-37.3	1.1408[20/4]	1.4542[20]	chl	B1[4], 250
3730	Butane, 1,4-dichloro-2-methyl ClCH$_2$CH$_2$CH(CH$_3$)CH$_2$Cl	141.04		168-9, 50[12]		1.1003[25/4]	1.4562[21]	chl	B1[4], 360
3731	Butane, 2,2-dichloro CH$_3$CH$_2$CCl$_2$CH$_3$	127.03		104	-74	....	1.4295	chl	B1[4], 251
3732	Butane, 2,2-dichloro-3,4-dimethyl (CH$_3$)$_2$CCCl$_2$CH$_3$	155.07		....	151-2			al, eth	B1[4], 369
3733	Butane, 2,3-dichloro CH$_3$CHClCHClCH$_3$	127.03		116, 49.5[80]	-80	1.1134[20/4]	1.4420[20]	chl	B1[4], 251
3734	Butane, 2,3-dichloro-2,3-dimethyl (CH$_3$)$_2$CClCCl(CH$_3$)$_2$	155.07	pr (dil al)	....	164			al, eth	B1[4], 374
3735	Butane, 2,3-dichloro-2-methyl CH$_3$CHClCHClCH(CH$_3$)$_2$	141.04		138, 37.5[20]		1.0696[15/4]	1.4450[18]	al, eth	B1[4], 325
3736	Butane, 1,1-dicyclohexyl CH$_3$CH$_2$CH$_2$CH(C$_6$H$_{11}$)$_2$	222.41	....	280.2		0.8842[16/0]	1.485[16]	....	B5[4], 359
3737	Butane, 1,2-dicyclohexyl CH$_3$CH$_2$CH(C$_6$H$_{11}$)CH$_2$C$_6$H$_{11}$	222.41		276.8		0.9084[18/0]	1.475[21]	chl, eth	B5[4], 358
3738	Butane, 1,2,3,4-diepoxy (dl) or Butadiene dioxide CH$_2$CHCHCH$_2$	86.09		144	4	1.113[20]	1.435[20]	w, al	B19[4], 111
3739	Butane, 1,2,3,4-diepoxy (meso) CH$_2$CHCHCH$_2$	86.09		138[767]	-16	1.1157[20/4]	1.4330[20]	w, al	B19[4], 110
3740	Butane, 1,1-diethoxy CH$_3$CH$_2$CH$_2$CH(OC$_2$H$_5$)$_2$	146.23		143		0.841[25/4]	....	....	B1[4], 3232
3741	Butane, 1,4-diethoxy C$_2$H$_5$O(CH$_2$)$_4$OC$_2$H$_5$	146.23		155-7[730]					B1[4], 2517
3742	Butane, 1,4-difluoro-octachloro FCCl$_2$CCl$_2$CCl$_2$CCl$_2$F	369.67		152.5[20]	4-5	1.9272[20/4]	1.5256[20]	....	B1[4], 258
3743	Butane, 2,2-di(2-furyl) CH$_3$CH$_2$C(C$_4$H$_3$O)$_2$CH$_3$	190.25		64-6[1]		1.0330[20/4]	1.4970[20]	al, eth	B19[4], 287
3744	Butane, 1,4-diido or Tetramethylene diiodide I(CH$_2$)$_4$I	309.92		125-6[15/d]	5.8	2.349[26/4]	1.619[25]		B1[4], 276
3745	Butane, 2,2-dimethyl or neo-Hexane CH$_3$CH$_2$C(CH$_3$)$_3$	86.18		49.7	-99.9	0.6485[20/4]	1.3688[20]	al, eth, ace, bz	B1[4], 367
3746	Butane, 2,3-dimethyl (CH$_3$)$_2$CH-CH(CH$_3$)$_2$	86.18		58	-128.5	0.6616[20/4]	1.3750[20]	al, eth, ace, bz	B1[4], 371
3747	Butane, 2,3-dimethyl-2,3-epoxy (CH$_3$)$_2$C-C(CH$_3$)$_2$	100.16		90-4[745]		0.8156[16/4]	1.3984[16.4/D]	w	B17[4], 89
3748	Butane, 1,4-bis (dimethylamino) or Tetramethyl putrescine (CH$_3$)$_2$N(CH$_2$)$_4$N(CH$_3$)$_2$	144.26		168, 78-80[28]		0.7942[15]	1.4621[25]	w, al, eth	B4[4], 1284
3749	Butane, 1,4-dinitro O$_2$N(CH$_2$)$_4$NO$_2$	148.13	pl (al)	176-8[13]	33-4	....	....	eth, bz	B1[4], 280
3750	Butane, 2,3-dinitro (dl) CH$_3$CH(NO$_2$)CH(NO$_2$)CH$_3$	148.13			48-9			eth	B1[4], 280

No.	Name, Synonyms, and Formula	Mol. wt.	Color, crystalline form, specific rotation and λ_{max} (log ϵ)	b.p. °C	m.p. °C	Density	n_D	Solubility	Ref.
3751	Butane, 2,3-dinitro *(meso)* $CH_3CH(NO_2)CH(NO_2)CH_3$	148.13	pr (eth)	76-7[1]	......	......	......	eth	B1[4], 280
3752	Butane, 1,1-diphenyl $(C_6H_5)_2CH-CH_2CH_2CH_3$	210.32	....	286-8, 161-3[20]	27	0.9928[20/4]	1.5664[20]	al, eth, bz, chl	B5[4], 1944
3753	Butane, 1,2-diphenyl $C_6H_5CH_2CH(C_6H_5)CH_2CH_3$	210.32	....	289[750], 152[11]		0.9777[20]	1.5554[20]	al, eth, bz, chl	B5[4], 1939
3754	Butane, 1,3-diphenyl *(l)* $C_6H_5CH_2CH_2CH(C_6H_5)CH_3$	210.32	$[\alpha]^{20/}_D$ -15.6 (chl, c=10)	68-70[0.01]	......		1.5503[25]	chl	B5[4], 1938
3755	Butane, 1,4-diphenyl $C_6H_5(CH_2)_4C_6H_5$	210.32	....	317, 108-9[0.1]	52.5			al, eth, chl	B5[4], 1937
3756	1,4-Butane dithiol $HS(CH_2)_4SH$	122.25	....	195-6	-53.9	1.0621[0/4]	1.5290[20]	al	B1[4], 2523
3757	Butane, 1,2-epoxy or 1,2-Butylene oxide $CH_3CH_2CHCH_2$	72.12		63.3		0.837[17/4]	1.3851[20]	al, **eth**, ace	B17[4], 45
3758	Butane, 2,3-epoxy *(cis)* $CH_3CHCHCH_3$	72.12		59.7[742]	-80	0.8226[25/4]	1.3802[20]	eth, ace, bz	B17[4], 48
3759	Butane, 2,3-epoxy *(trans)* $CH_3CHCHCH_3$	72.12		56-7	-85	0.8010[25/4]	1.3736[20]	eth, ace, bz	B17[4], 49
3760	Butane, 2,2-bis(ethylsulfonyl) $CH_3CH_2C(SO_2C_2H_5)_2CH_3$	242.36	pl (w)	d	76	1.199[85/4]		al, eth, bz, peth, lig	B1[3], 2790
3761	Butane, 1-fluoro or *n*-Butyl fluoride $CH_3CH_2CH_2CH_2F$	76.12		32.5	-134	6.7789[20/4]	1.3396[20]	al	B1[4], 244
3762	Butane, 1,1,2,2,3,3,4,4,-heptachloro $Cl_2CHCHClCCl_2CHCl_2$	299.24		137.5[13.5]		1.742[20/20]	1.5407[20]	ace, CCl_4	B1[4], 257
3763	Butane, 1,1,2,3,4,4-hexachloro (liquid) $Cl_2CHCHClCHClCHCl_2$	264.79		111[10]		1.6460[20/4]	1.5258[20]	bz, CCl_4	B1[3], 288
3764	Butane, 1,1,2,3,4,4-hexachloro (solid) $Cl_2CHCHClCHClCHCl_2$	264.79	nd (al), pr (bz, aa)	......	109-10			bz, CCl_4	B1[3], 288
3765	Butane, 1,1-bis(4-hydroxyphenyl) $(4-HOC_6H_4)_2CHC_3H_7$	242.32	nd (to)	270[12]	137			al, bz, to	B6[3], 5476
3766	Butane, 2,2-bis(4-hydroxyphenyl) $(4-HOC_6H_4)_2C(CH_3)C_2H_5$	242.32	nd or pr (w)	250-3	133-4			al, eth, ace, bz	B6[3], 5477
3767	Butane, 1-iodo or *n*-Butyl iodide $CH_3CH_2CH_2CH_2I$	184.02	....	130.5, 19.2[10]	-103	1.6154[20/4]	1.5001[20]	al, eth, chl	B1[4], 271
3768	Butane, 1-iodo-2-methyl *(d)* or *act*-Amyliodide $CH_3CH_2CH(CH_3)CH_2I$	198.05	$[\alpha]^{15}_D$ +5.78 (undil)	148[20], 47.1[20]		1.5253[20/4]	1.4977[20]	al, eth	B1[3], 366
3769	Butane, 1-iodo-2-methyl *(dl)* $CH_3CH_2CH(CH_3)CH_2I$	198.05		144-7			1.497[20]	al, eth	B1[3], 366
3770	Butane, 1-iodo-3-methyl or Isopentyl iodide $(CH_3)_2CHCH_2CH_2I$	198.05		147		1.5118[20/4]	1.4939[20]	**al, eth**	B1[4], 331
3771	Butane, 2-iodo *(dl)* or *sec*-Butyl iodide $CH_3CH_2CHICH_3$	184.02	....	120, 33[45]	-104.2	1.5920[20/4]	1.4991[20]	**al, eth**, chl	B1[4], 272
3772	Butane, 2-iodo *(l)* $CH_3CH_2CHICH_3$	184.02	$[\alpha]_D$ -12.15 (al, c=20)	117-8		1.585[20/4]	1.4945[19]	**al, eth**, chl	B1[4], 272
3773	Butane, 2-iodo-2-methyl or *tert*-Amyliodide $(CH_3)_2CICH_2CH_3$	198.05	....	124.5		1.4937[20/4]	1.4981[20]	**al, eth**	B1[4], 331
3774	Butane, 2-iodo-3-methyl $(CH_3)_2CHCHICH_3$	198.05		138-9				al, eth, ace	B1[3], 367
3775	Butane, 1-isocyano or Butyl carbylamine $CH_3CH_2CH_2CH_2NC$	83.13	....	118	......	1.4061[20]		**al, eth**	B4[4], 562
3776	Butane, 1-isocyano-3-methyl or Isoamylcarbylamine $(CH_3)_2CHCH_2CH_2NC$	97.16		140		0.806[20/4]	1.406[20]	**al, eth**	B4, 184
3777	Butane, 2-methyl or Isopentane $CH_3CH_2CH(CH_3)_2$	72.15		27.8	-159.9	0.6201[20/4]	1.3537[20]	al, eth	B1[4], 320
3778	Butane, 3-methyl-2-phenyl $CH_3CH(C_6H_5)CH(CH_3)_2$	148.25		186-8		0.8672[16/4]	1.4972[16]	....	B5[4], 1118
3779	Butane, 2-methyl-1,2,3-trichloro $CH_3CHClCCl(CH_3)CH_2Cl$	175.49		183-5[762], 65.5[11]	......	1.2527[20/4]		chl	B1[3], 361
3780	Butane, 2-methyl-2,3,3-trichloro $(CH_3)_2CClCCl_2CH_3$	175.459		182-3		1.215[15/4]	1.472[21]	chl, aa	B1[3], 361
3781	Butane, 1-nitro $CH_3CH_2CH_2CH_2NO_2$	103.12		153	......	0.9710[20/4]	1.4303[20]	**al, eth**	B1[4], 277
3782	Butane, 2-nitro *(dl)* $CH_3CH_2CH(NO_2)CH_3$	103.12		140	-132	0.9854[17/4]	1.4044[20]	....	B1[4],,278
3783	Butane, 1,1,2,2,3,3,3,4,4-octachloro $Cl_2CHCCl_2CCl_2CHCl_2$	333.69	cr (al)	......	81			al, eth, ace, bz	B1[4], 257

No.	Name, Synonyms, and Formula	Mol. wt.	Color, crystalline form, specific rotation and λ_{max} (log ε)	b.p. °C	m.p. °C	Density	n_D	Solubility	Ref.
3784	Butane, 1,1,2,3,4-pentachloro (liquid) ClCH₂(CHCl)₃CHCl₂	230.35	....	95.5[11]	......	1.561[18]	1.5140[18]	al, CCl₄	B1⁴, 256
3785	Butane, 1,1,2,3,4-pentachloro (solid) ClCH₂CHClCHClCHCl₂	230.35	lf (al)	230, 102[11]	49	1.539[53]	1.5065[53]	al, CCl₄	B1⁴, 256
3786	Butane, 1,2,2,3,4-pentachloro ClCH₂CHClCCl₂CH₂Cl	230.35	....	85[10]	......	1.5543[20/4]	1.5157[20]	ace, chl	B1⁴, 256
3787	Butane, 1,1,4,4-tetrabromo Br₂CHCH₂CH₂CHBr₂	373.73	....	138-45[10]	......	2.529[20/4]	1.6077[20]	eth, bz, chl	B1³, 298
3788	Butane, 1,2,2,3-tetrabromo CH₃CHBrCBr₂CH₂Br	373.73	....	128-30[14]	-2	2.5100[20/4]	1.6070[20]	ace	B1³, 298
3789	Butane, 1,2,2,4-tetrabromo BrCH₂CH₂CBr₂CH₂Br	373.73	nd (lig)	....	72-3	....	....	lig	B1³, 298
3790	Butane, 1,2,3,4-tetrabromo (dl) BrCH₂CHBrCHBrCH₂Br	373.73	lf (peth)	....	40-1	....	....	al, eth, ace, lig	B1⁴, 271
3791	Butane, 1,2,3,4-tetrabromo (meso) BrCH₂CHBrCHBrCH₂Br	373.73	nd (al or lig)	180-1[60]	118-9	....	....	al, ace, chl	B1⁴, 271
3792	Butane, 2,2,3,3-tetrabromo CH₃CBr₂CBr₂CH₃	373.73	lf (lig), pr (eth-lig)	....	243	....	....	eth, ace, bz, chl lig	B1³, 298
3793	Butane, 1,1,1,2-tetrachloro CH₃CH₂CHClCCl₃	195.91	....	134-5[742]	......	1.3932[20/20]	1.4920[25]	ace, chl	B1⁴, 254
3794	Butane, 1,2,2,3-tetrachloro CH₃CHClCCl₂CH₂Cl	195.91	....	182, 85[10]	-48	1.4276[18/4]	1.491[20]	ace, chl	B1³, 286
3795	Butane, 1,2,3,3-tetrachloro CH₃CCl₂CHClCH₂Cl	195.91	....	90[32], 55-7[10]	......	1.4204[20/4]	1.4938[20]	eth, ace, chl	B1⁴, 254
3796	Butane, 2,2,3,3-tetramethyl CH₃C(CH₃)₂C(CH₃)₂CH₃	114.23	lf (eth)	106.5, 13.1[10]	100.7	0.8242[20], 0.6485[110]	1.4695[20]	eth	B1⁴, 117
3797	Butane, 1,1,2-tribromo CH₃CH₂CHBrCHBr₂	294.83	....	216.2, 98[14]	......	2.1936[20/11]	1.5626[11]	al, eth, chl	B1², 84
3798	Butane, 1,2,2-tribromo CH₃CH₂CBr₂CH₂Br	294.83	......	213.8, 90.1[14]	......	2.1692[20/4]	1.5624[10]	al, eth, chl	B1², 85
3799	Butane, 1,2,3-tribromo CH₃CHBrCHBrCH₂Br	294.83	......	220, 97[10]	-19	2.1908[20/4]	1.5680[20]	al, eth, chl	B1⁴, 271
3800	Butane, 1,2,4-tribromo BrCH₂CH₂CHBrCH₂Br	294.83	....	215, 93[10]	-18	2.170[20/4]	1.5608[20]	al, eth, chl	B1⁴, 271
3801	Butane, 1,3,3-tribromo CH₃CBr₂CH₂CH₂Br	294.83	....	200-5, 70[8]	......	1.444[20/4]	1.5564[20]	al, eth, chl	B1³, 298
3802	Butane, 2,2,3-tribromo CH₃CHBrCBr₂CH₃	294.83	....	200, 86[10]	1.8	2.1724[20/4]	1.5602[20]	al, eth, chl	B1³, 298
3803	Butane, 1,1,3-trichloro CH₃CHClCH₂CHCl₂	161.46	....	152[753]-	......	1.317[15]-	1.4600[15]	al, eth, chl	B1⁴, 252
3804	Butane, 1,2,3-trichloro CH₃CHClCHClCH₂Cl	161.46	....	165-8[725], 63[28]	......	1.3164[20/4]	1.4790[20]	al, eth, chl	B1³, 285
3805	Butane, 1,2,4-trichloro ClCH₂CHClCH₂CH₂Cl	161.46	....	61-2.5[10]	......	1.3175[20]	1.4820[20]	bz, chl	B1⁴, 253
3806	Butane, 2,2,3-trichloro CH₃CHClCCl₂CH₃	161.46	....	143-5	......	1.2699[20/4]	1.4645[20]	chl	B1³, 285
3807	Butane, 2,2,3-trimethyl or Triptane (CH₃)₃CHC(CH₃)₃	100.21	....	80.9	-24.2	0.6901[20/4]	1.3864[20]	al, eth, ace, bz	B1⁴, 410
3808	Isobutane or 2-Methylpropane (CH₃)₂CHCH₃	58.12	....	-11.7	-159.4	0.549[10]	....	al, eth, chl	B1⁴, 282
3809	Isobutane, 1,2-dibromo or 1,2-dibromo-2-methylpropane (CH₃)₂CBrCH₂Br	215.94	....	149-51, 61[40]	9-12	1.759[20/4]	1.509	al, eth, bz, chl	B1⁴, 298
3810	Isobutane, 1,1-dichloro or 1,1-dichloro-2-methyl propane (CH₃)₂CHCHCl₂	127.03	....	105-6	......	1.0111[12/12]	1.4330[25]	al, eth, bz, chl	B1⁴, 292
3811	Isobutane, 1,2-dichloro or 1,2-dichloro-2-methyl propane (CH₃)₂CClCH₂Cl	127.03	....	108, 38-9[70]	......	1.093[20/4]	1.4370[20]	al, eth, ace, bz	B1⁴, 292
3812	Isobutane, 1,3-dichloro or 1,3-dichloro-2-methyl propane CH₃CH(CH₂Cl)₂	127.03	....	134.6, 60[49]	......	1.1325[25/4]	1.4488[25]	al, eth, bz	B1⁴, 293
3813	Isobutane, 1,2-epoxy or Isobutylene oxide (CH₃)₂C-CH₂	72.12	....	52	......	0.8650[0]	1.3712[22]	al, eth	B17⁴, 46
3814	Isobutane, 1,2-epoxy-3-chloro ClCH₂C(CH₃)CH₂	106.55	....	12[2], 51[55]	......	1.1011[20/4]	1.4340[20]	eth, w	B17⁴, 47
3815	Isobutane, 1-nitro (CH₃)₂CHCH₂NO₂	103.12	....	140.5, 61-2[45]	......	0.9625[25/25]	1.4066[20]	al, eth	B1⁴, 301
3816	Isobutane, 2-nitro (CH₃)₃CNO₂	103.12	....	127.2	26.23	0.9501[28]	1.4015[20]	al, eth, ace, bz	B1⁴, 301

No.	Name, Synonyms, and Formula	Mol. wt.	Color, crystalline form, specific rotation and λ_{max} (log ε)	b.p. °C	m.p. °C	Density	n_D	Solubility	Ref.
3817	Isobutane, 1,1,1,2,3-pentachloro .... $ClCH_2CCl(CH_3)CCl_2$	230.35	(al)	215[757], 90-3[10]	73.5	1.5686[25/4]	1.5165[25]	chl	B1[4], 294
3818	Isobutane, 1,1,1,2-tetrabromo .... $(CH_3)_2CBrCBr_3$	373.73	lf	....	217	....	....	al, eth	B1, 128
3819	Isobutane, 1,1,2,3-tetrabromo .... $BrCH_2CBr(CH_3)CHBr_2$	373.73	....	134[11]	....	2.4545[20]	1.5990[20]	chl	B1[3], 325
3820	Isobutane, 1,1,1,2-tetrachloro .... $(CH_3)_2CClCCl_3$	195.91	cr (al)	192[117.5] sub	178-9	....	....	al, eth, chl	B1[4], 293
3821	Isobutane, 1,1,2,3-tetrachloro .... $ClCH_2CCl(CH_3)CHCl_2$	195.91	....	190-1, 69[12]	-46	1.4393[25/4]	1.4963[20]	eth, bz, chl	B1[4], 294
3822	Isobutane, 1,1,2-tribromo .... $(CH_3)_2CBrCHBr_2$	294.83	....	208-15d, 96[14]	....	2.0169[20/4]	....	eth, chl	B1[3], 325
3823	Isobutane, 1,2,3-tribromo .... $(BrCH_2)_2CBrCH_3$	294.83	....	88.5[9]	....	2.1750[20/4]	1.5652[20]	eth, chl	B1[4], 299
3824	Isobutane, 1,1,2-trichloro .... $(CH_3)_2CClCHCl_2$	161.46	....	145-6, 46-7[18]	6	1.2588[20/4]	1.4666[20]	eth, chl	B1[4], 293
3825	Isobutane, 1,2,3-trichloro .... $(ClCH_2)_2CClCH_3$	161.46	....	162-3, 81[50]	....	1.3012[25/4]	1.4765[20]	chl	B1[4], 293
3826	1,2-Butanediol-(d) .... $CH_3CH_2CHOHCH_2OH$	90.12	$[α]^{20}_D$ + 14.5 (al, c=6)	192.4, 68[0.4]	....	1.0059[17.5/0]	1.4375[20]	w, al, ace	B1[4], 2507
3827	1,2-Butanediol (dl) or α-Butylene glycol .... $CH_3CH_2CHOHCH_2OH$	90.12	....	190.5, 96.5[10]	....	1.0024[20/4]	1.4378[20]	w, al, ace	B1[4], 2507
3828	1,2-Butanediol (l) .... $CH_3CH_2CHOHCH_2OH$	90.12	$[α]^{22}_D$ -7.4 (al, c=4)	94-6[12]	....	....	....	w, al, ace	B1[4], 2507
3829	1,3-Butanediol (d) or β-Butyleneglycol .... $CH_3CH(OH)CH_2CH_2OH$	90.12	$[α]^{22}_D$ + 18.5 (al, c=4)	204, 60-5[0.8]	....	1.0053[20/4]	1.4418[20]	w, al	B1[4], 2508
3830	1,3-Butanediol (dl) .... $CH_3CH(OH)CH_2CH_2OH$	90.12	....	207.5, 103-4[8]	....	1.0053[20/4]	1.4410[20]	w, al	B1[4], 2508
3831	1,3-Butanediol (l) .... $CH_3CH(OH)CH_2CH_2OH$	90.12	$[α]^{25}_D$ -18.8 (al, c=4)	107-10[23]	....	1.005[20/4]	....	w, al	B1[4], 2508
3832	1,3-Butanediol sulfite .... $C_4H_8O_3S$	135.16	....	185, 76-7[17]	5	1.2352[20/4]	1.4661[20]	al, eth, ace, bz	B1[4], 2510
3833	1,4-Butanediol or Tetramethylene glycol .... $HOCH_2CH_2CH_2CH_2OH$	90.12	....	235, 120[10]	20.1	1.0171[20/4]	1.4460[20]	w, al	B1[4], 2515
3834	1,4-Butanediol diacetate .... $CH_3CO_2(CH_2)_4CO_2CH_3$	174.20	....	229[765]	12	1.0479[15]	1.4251[15]	....	B2[4], 224
3835	1,4-Butanediol dibenzoate .... $C_6H_5CO_2(CH_2)_4CO_2C_6H_5$	298.34	....	....	81-2	....	....	eth	B9[3], 540
3836	1,4-Butanediol, 2-hexyl .... $HOCH_2CH_2CH(C_6H_{13})CH_2OH$	174.28	....	118[0.05]	....	....	....	....	....
3837	1,4-Butanediol, 2-methyl .... $HOCH_2CH_2CH(CH_3)CH_2OH$	104.15	$[α]^{22}_D$ + 11.65	131-3	....	0.9929[22/4]	....	....	B1[4], 2546
3838	2,3-Butanediol (d) .... $CH_3CHOHCHOHCH_3$	90.12	$[α]^{25}_D$ + 12.5 (undil)	180-2	34 (anh), 16.86 (+ sw)	0.9872[25]	1.4306[25]	**w, al**, eth, ace	B1[4], 2524
3839	2,3-Butyldiol (dl) .... $CH_3CHOHCHOHCH_3$	90.12	cr (i-Pr₂O)	182.5, 86[10]	7.6	1.0033[20/4]	1.4310[25]	**w, al**, eth, ace	B1[4], 2524
3840	2,3-Butanediol (l) .... $CH_3CHOHCHOHCH_3$	90.12	$[α]^{25}_D$ -13.0 (undil)	178-81, 77.5[10]	19.7	0.9869[25/4]	1.4340[18]	**w, al**, eth, ace	B1[4], 2524
3841	2,3-Butanediol (meso) .... $CH_3CHOHCHOHCH_3$	90.12	cr (i-Pr₂O)	181.7, 83.5[10]	34.4	1.0003[20/4]	1.4367[20]	w, al, eth	B1[4], 2524
3842	2,3-Butanediol, 2,3-dimethyl or Pinacol .... $(CH_3)_2C(OH)C(OH)(CH_3)_2$	118.17	nd (al or eth)	174.4	43	....	....	al, eth	B1[4], 2575
3843	2,3-Butanediol, 2,3-dimethyl, hexahydrate or Pinacol hydrate .... $(CH_3)_2C(OH)C(OH)(CH_3)_2 \cdot 6H_2O$	226.27	pl (w + 6)	....	47	0.967[15]	....	al, eth, w	B1[4], 2575
3844	2,3-Butanediol, 2,3-diphenyl or Acetophenone Pinacol ... $C_6H_5COH(CH_3)COH(CH_3)C_6H_5$	242.32	pr (dil al)	....	121-2	....	....	al, eth	B6[3], 5474
3845	1,2-Butanediol, 3-methyl or α-Isopentylene glycol .... $(CH_3)_2CHCHOHCH_2OH$	104.15	....	206, 81-3[5]	....	0.9987[0/4]	....	al, eth	B1[4], 2549
3846	1,3-Butanediol, 3-methyl or α-Isopentylene glycol .... $(CH_3)_2C(OH)CH_2CH_2OH$	104.15	....	202-3, 108[16]	....	0.9448[20/4]	1.4452[20]	w, al	B1[4], 2549
3847	2,3-Butanediol, 2-methyl or β-Isopentyl glycol .... $CH_3CHOHC(CH_3)OHCH_3$	104.15	....	175, 68[5]	....	0.9920[25/4]	1.4375[20]	w, al, eth	B1[4], 2547

No.	Name, Synonyms, and Formula	Mol. wt.	Color, crystalline form, specific rotation and λ_{max} (log ε)	b.p. °C	m.p. °C	Density	n_D	Solubility	Ref.
3848	2,3-Butanedione or Biacetyl, Dimethyl glycol .... $CH_3COCOCH_3$	86.09	....	88	-2.4	$0.9808^{18.5/4}$	1.3951^{20}	w, al, eth, ace, bz	B1[4], 3644
3849	1,4-Butanedithiol .... $HS(CH_2)_4SH$	122.25	....	195-6, 110-12[50]	-53.9	$1.0621^{0/4}$	1.5290^{20}	al	B1[4], 2523
3850	1,2,3,4-Butane tetracarboxylic acid (dl) .... $HO_2CCH_2CH(CO_2H)CH(CO_2H)CH_2CO_2H$	234.16	lf (w), cr (ace)	....	236-7			w, al	B2[4], 2419
3851	1-Butanethiol or n-Butyl mercaptan .... $CH_3CH_2CH_2CH_2SH$	90.19	....	98.4	-115.7	$0.8337^{20/4}$	1.4440^{20}	al, eth	B1[4], 1555
3852	1-Butanethiol, 2-methyl- (d) or act-Amyl mercaptan ..... $CH_3CH_2CH(CH_3)CH_2SH$	104.22	$[\alpha]^{23/}_D$ +3.21	118.2		$0.8420^{20/4}$	1.4440^{20}	....	B1[4], 1668
3853	1-Butanethiol, 3-methyl or Isopentyl mercaptan .... $(CH_3)_2CHCH_2CH_2SH$	104.22	....	118		$0.8350^{20/4}$	1.4418^{20}	al, eth	B1[4], 1688
3854	2-Butanethiol, (d) or sec-Butyl mercaptan .... $CH_3CH_2CHSHCH_3$	90.19	$[\alpha]^{20/}_D$ +15.7	85-95, 37.4[134]		$0.8299^{20/4}$	1.43385^{25}	al, eth, bz, peth	B1[3], 1549
3855	2-Butanethiol, (dl) or sec-Butyl mercaptan .... $CH_3CH_2CHSHCH_3$	90.19	$[\alpha]^{17/}_D$ -17.35	85		$0.8295^{20/4}$	1.4366^{20}	al, eth, bz, peth	B1[4], 1584
3856	2-Butanethiol, (l) .... $CH_3CH_2CHSHCH_3$	90.19	$[\alpha]^{17/}_D$ -17.35	83-4		$0.8300^{17/4}$		al, eth, bz, peth	B1[3], 1549
3857	2-Butanethiol, 2-methyl $CH_3CH_2C(SH)(CH_3)_2$	104.22	....	99-100			1.4385^{20}		B1[4], 1674
3858	1,2,3-Butanetriol or 1-Methyl glycerol .... $CH_3CHOHCHOHCH_2OH$	106.12		170^{20}			1.4462^{20}	w, al	B1[4], 2771
3859	1,2,4-Butanetriol $HOCH_2CH_2CH(OH)CH_2OH$	106.12		$172-4^{12}$		1.018^{20}	1.4688^{20}	w, al	B1[4], 2775
3860	1-Butanol or n-Butyl alcohol .... $CH_3CH_2CH_2CH_2OH$	74.12	$[\alpha]^{20/}_D$ +9.8 (w)	117.2	-89.5	$0.8098^{20/4}$	1.3993^{20}	w, al, eth, ace, liz	D1[4], 1506
3861	1-Butanol, 2-amino- (d) $CH_3CH_2CHNH_2CH_2OH$	89.14	$[\alpha]^{20/}_D$ +9.8 (w)	80^{11}		$0.94^{14/}$	1.4518^{20}	w, al, eth	B4[4], 1705
3862	1-Butanol, 2-amino- (dl) .... $CH_3CH_2CHNH_2CH_2OH$	89.14	....	178	-2	0.9162^{20}	1.4489^{25}	w, al, eth	B4[4], 1705
3863	1-Butanol, 2-amino-1-phenyl $C_6H_5CH(CH_2)CH(OH)C_6H_5$	165.24	pl (bz-eth)	....	79-80			al, bz, chl	B13[3], 1791
3864	1-Butanol, 3-amino $CH_3CHNH_2CH_2CH_2OH$	89.14	....	$82-5^{19}$			1.4534^{25}	w, al	B4[4], 1710
3865	1-Butanol, 4-amino $H_2NCH_2CH_2CH_2CH_2OH$	89.14		206^{776}		0.967^{12}	1.4625^{20}	w, al	B4[4], 1711
3866	1-Butanol, 2-chloro $CH_3CH_2CHClCH_2OH$	108.57	....	$74-6^{25}$		$1.062^{25/4}$	1.4438^{20}	al, eth	B1[4], 1548
3867	1-Butanol, 3-chloro $CH_3CHClCH_2CH_2OH$	120.58	....	170-80, 73[20]		$1.0883^{20/4}$	1.4518^{20}	al, eth	B1[4], 1549
3868	1-Butanol, 4-chloro $ClCH_2CH_2CH_2CH_2OH$	108.57	....	$84-5^{16}$		$1.0883^{20/4}$	1.4518^{20}	al, eth	B1[4], 1550
3869	1-Butanol, 2,2-dimethyl $CH_3CH_2C(CH_3)_2CH_2OH$	102.18	....	136.7	<-15	$0.8283^{20/4}$	1.4208^{20}	al, eth	B1[4], 1726
3870	1-Butanol, 2,3-dimethyl- (d) .... $CH_3CH(CH_3)CH(CH_3)CH_2OH$	102.18	$[\alpha]^{25}_D$ +1.9	142		$0.823^{25/4}$	....	al, eth, ace	B1[4], 1729
3871	1-Butanol, 2,3-dimethyl- (dl) .... $CH_3CH(CH_3)CH(CH_3)CH_2OH$	102.18	....	144-5		$0.8297^{20.5/4}$	$1.4195^{20.5}$	al, eth, ace	B1[4], 1729
3872	1-Butanol, 3,3-dimethyl $(CH_3)_3CCH_2CH_2OH$	102.18		143	-60	1.4323^{15}	....	al, eth, ace	B1[4], 1729
3873	1-Butanol, 2,4-diphenyl .... $C_6H_5CH_2CH_2CH(C_6H_5)CH_2OH$	226.32	$[\alpha]^{20}_D$ -17.5 (chl, c=10)	$145-6^{0.1}$	51-2		1.5686^{25}	chl	B6[3], 3428
3874	1-Butanol, 2-ethyl $(C_2H_5)_2CHCH_2OH$	162.18	....	146.3	<-15	$0.8326^{20/4}$	1.4220^{20}	al, eth	B1[4], 1725
3875	1-Butanol, 4-fluoro $FCH_2CH_2CH_2CH_2OH$	92.12	....	58^{15}			1.3942^{15}	al, eth, ace	B1[4], 1546
3876	1-Butanol, 2,2,3,3,4,4,4-heptafluoro $CF_3CF_2CF_2CH_2OH$	200.06	....	95		$1.600^{20/4}$	1.294^{20}	al, ace	B1[4], 1547
3877	1-Butanol, 3-methoxy $CH_3CH(OCH_3)CH_2CH_2OH$	104.15	....	160		0.923^{23}	1.4148^{25}	al, eth, ace	B1[4], 2509
3878	1-Butanol, 2-methyl- (d) or act-Amyl alcohol- (d) ..... $CH_3CH_2CH(CH_3)CH_2OH$	88.15	lf (w)	128, 65.7[50]		$1.8191^{20/4}$	1.4102^{20}	al, eth, ace	B1[4], 1666
3879	1-Butanol, 2-methyl- (dl) .... $CH_3CH_2CH(CH_3)CH_2OH$	88.15	....	127-8, 70[60]		$0.8152^{25/4}$	1.4092^{20}	al, eth, ace	B1[4], 1666
3880	1-Butanol, 2-methyl- (l) .... $CH_3CH_2CH(CH_3)CH_2OH$	88.15	$[\alpha]^{18/}_D$ +3.75	129		$0.816^{18/4}$	1.4098^{20}	al, eth, ace	B1[4], 1666

No.	Name, Synonyms, and Formula	Mol. wt.	Color, crystalline form, specific rotation and λ_{max} (log ϵ)	b.p. °C	m.p. °C	Density	n_D	Solubility	Ref.
3881	1-Butanol, 2-methyl-4-phenyl (l) $C_6H_5CH_2CH_2CH(CH_3)CH_2OH$	164.25		135[11]		0.9719[20/4]	1.5173[16]	al eth, ace, bz	B6[3], 1962
3882	1-Butanol, 3-methyl or iso-Butyl alcohol $(CH_3)_2CHCH_2CH_2OH$	88.15		128.5[750]		0.8092[20/4]	1.4053[20]	**al**, eth, ace	B1[4], 1677
3883	1-Butanol, 3-methyl-1-phenyl or α-Hydroxy isopentyl benzene $(CH_3)_2CHCH_2CH(OH)C_6H_5$	164.25		235-6[746], 112[9]		0.9537[19/4]	1.5080[18]	al, eth, ace, bz	B6[2], 505
3884	1-Butanol, 3-methyl-2-phenyl $(CH_3)_2CHCH(C_6H_5)CH_2OH$	164.25		130[15]		0.9694[25/4]	1.5137[20]	al, eth, ace, bz	B6[2], 506
3885	1-Butanol, 2-nitro $CH_3CH_2CH(NO_2)CH_2OH$	119.12		105[10]	-47	1.1332[25/4]	1.4390[20]	w, al, eth, ace, aa	B1[4], 1555
3886	1-Butanol, 1-phenyl $CH_3CH_2CH_2CH(C_6H_5)OH$	150.22		232, 113-5[17]	16	0.9740[20/4]	1.5139[20]	al, eth	B6[3], 1845
3887	1-Butanol, 2,2,3-trichloro $CH_3CHClCCl_2CH_2OH$	177.46	pr (dil al)	199-200, 97-8[18]	62			al, eth, ace	B1[3], 1518
3888	2-Butanol-(d) or sec-Butyl alcohol $CH_3CH_2CH(OH)CH_3$	74.12	$[\alpha]^{20}_D$ +13.9	99.5		0.8080[20/4]	1.3954[20]	**al, eth**, ace, bz	B1[4], 1566
3889	2-Butanol-(dl) or sec-Butyl alcohol $CH_3CH_2CHOHCH_3$	74.12		99.5, 45.5[60]		0.8063[20/4]	1.3978[20]	**al, eth**, ace, bz	B1[4], 1567
3890	2-Butanol-(l) $CH_3CH_2CHOHCH_3$	74.12	$[\alpha]^{20}_D$ +13.9	99.5		0.8070[20/4]	1.3975[20]	**al, eth**, ace, bz	B1[4], 1566
3891	2-Butanol, 3-amino $CH_3CHNH_2CHOHCH_3$	89.14		159-60[745], 70[20]	18-20 (44)	0.9299[25/4]	1.4502[20]	w, al, ace	B1[4], 1726
3892	2-Butanol, 3-bromo-(dl) $CH_3CHBrCHOHCH_3$	153.03		154, 46-50[8]		1.4550[20/4]	1.4786[20]	al, eth	B1[4], 1580
3893	2-Butanol, 1-chloro $CH_3CH_2CHOHCH_2Cl$	108.57		141, 52[15]		1.068[25/4]	1.4400[20]	al, eth	B1[4], 1578
3894	2-Butanol, 1-chloro-2-methyl $CH_3CH_2C(CH_3)OHCH_2Cl$	122.60		150-2		1.0161[20]	1.4469[20]	al	B1[4], 1673
3895	2-Butanol, 3-chloro $CH_3CHClCHOHCH_3$	108.57		138-40, 52-4[30]		1.0669[20/4]	1.4432[20]	al, eth	B1[3], 1538
3896	2-Butanol, 3-chloro-(erythro, dl) $CH_3CHClCHOHCH_3$	108.57		135.4[748], 56.1[30]		1.0610[25/4]	1.4397[25]	al, eth, chl	B1[3], 1538
3897	2-Butanol, 3-chloro-(threo) $CH_3CH(OH)CHClCH_3$	108.57		130.8[748]		1.0586[25/4]	1.4386[25]		B1[3], 1538
3898	2-Butanol, 3-chloro-2-methyl $CH_3CHClC(CH_3)OHCH_3$	122.60		141-2, 55-6[30]		1.0295[20/4]	1.4436[20]	al, eth	B1[3], 1627
3899	2-Butanol, 4-chloro $ClCH_2CH_2CHOHCH_3$	108.57		67[20]			1.4408[20]	al, eth	B1[4], 1578
3900	2-Butanol, 4-cyclohexyl $C_6H_{11}CH_2CH_2CH(OH)CH_3$	156.27		112[14]		0.903[21]	1.464[21]		B6[3], 123
3901	2-Butanol, 1,4-dibromo $BrCH_2CH_2CH(OH)CH_2Br$	231.91		114-5[15]		2.023[0]	1.544[20]		B1[4], 1581
3902	2-Butanol, 1,3-dichloro $CH_3CHClCHOHCH_2Cl$	143.02		63-4[10]		1.2860[15/4]	1.4766[20]	al, eth	B1[3], 1541
3903	2-Butanol, 2,3-dimethyl $(CH_3)_2C(OH)CH(CH_3)_2$	102.18		118.4	-14	0.8236[20/4]	1.4176[20]	**al, eth**	B1[4], 1729
3904	2-Butanol, 3,3-dimethyl or Pinacolyl alcohol $(CH_3)_3CCHOHCH_3$	102.18		120.4	5.6	0.8122[25]	1.4148[20]	al, **eth**	B1[4], 1727
3905	2-Butanol, 2-methyl $CH_3CH_2C(CH_3)(OH)CH_3$	88.15		102, 50[60]	-8.4	0.8059[25/4]	1.4052[20]	**al, eth**, ace, bz	B1[4], 1668
3906	2-Butanol, 2-methyl-1-phenyl-(dl) $CH_3CH_2C(CH_3)(OH)CH_2C_6H_5$	164.25		215-25[747], 103-5[11]		0.9754[20/0]	1.5182[20]	al, eth, lig	B6[3], 1958
3907	2-Butanol, 2-methyl-3-phenyl $CH_3CH(C_6H_5)C(OH)(CH_3)_2$	164.25		196-8, 118[24]		0.9794[20/4]	1.5193[20]	al, eth, ace, bz	B6[3], 1971
3908	2-Butanol, 2-methyl-4-phenyl $C_6H_5CH_2CH_2CH(OH)(CH_3)_2$	164.25	nd	121[13]	24.5	0.9626[21/4]	1.5077[21]	al, eth, ace, bz	B6[3], 1962
3909	2-Butanol, 3-methyl-(d) $(CH_3)_2CHCHOHCH_3$	88.15	$[\alpha]^{20}_D$ +5.34 (al)	112[734]		0.8225[16/4]	1.4089[20]	al, eth, ace, bz, chl	B1[1], 196
3910	2-Butanol, 3-methyl-(dl) $(CH_3)_2CHCHOHCH_3$	88.15		112.9		0.8180[20/4]	1.4089[20]	al, eth, ace, bz	B1[4], 1675
3911	2-Butanol, 3-methyl-3-phenyl $(CH_3)_2C(C_6H_5)CHOHCH_3$	164.25		196-8		0.9653[13/4]	1.5161[13]	al, eth, ace, bz	B6[1], 269
3912	2-Butanol, 1-nitro $CH_3CH_2CH(OH)CH_2NO_2$	119.12		204[767], 75[2]		1.1353[20/4]	1.4435[20]	al, eth, ace, bz	B1[4], 1582
3913	2-Butanol, 3-nitro $CH_3CH(NO_2)CH(OH)CH_3$	119.12		55[0.5]			1.4414[20]		B1[4], 1583

No.	Name, Synonyms, and Formula	Mol. wt.	Color, crystalline form, specific rotation and λ_{max} (log ε)	b.p. °C	m.p. °C	Density	n_D	Solubility	Ref.
3914	2-Butanol, 2-phenyl-(d)........ CH₃CH₂C(OH)(C₆H₅)CH₃	150.22	[α]²²/D +17.5	112-4[21]	-13	0.984[25/4]	1.5185[20]	al, eth	B6³, 185
3915	2-Butanol, 2-phenyl-(dl)........ CH₃CH₂C(OH)(C₆H₅)CH₃	150.22	....	211-2, 90-1[4]	,,,,	0.984[31/11]	1.5150[20]	al, eth	B6³, 1854
3916	2-Butanol, 1,1,1-trichloro........ CH₃CH₂CHOHCCl₃	177.46	....	169-71[738], 82-4[22]	......	1.3670[25/25]	1.4800[20]	al, eth, ace, bz, chl	B1⁴, 1579
3917	2-Butanol, 2,3,3-trimethyl CH₃C(CH₃)₂C(CH₃)(OH)CH₃	116.21	cr (dil al + ½ w)	131-2, 40-1[15]	83-4	0.8380[25/4]	1.4233[22]	al, eth, ace	B1⁴, 1755
3918	2-Butanone or Methyl ethyl ketone............. CH₃CH₂COCH₃	72.12	....	79.6, 30[119]	-86.3	0.8054[20/4]	1.3788[20]	w, al, eth, ace, bz	B1⁴, 3243
3919	2-Butanone, 1-chloro .... CH₃CH₂COCH₂Cl	106.55	....	137-8, 34-5[10]	......	1.0850[20/4]	1.4372[20]	MeOH	B1⁴, 3255
3920	2-Butanone, 3-chloro CH₃CHClCOCH₃	106.55	....	115, 40[30]	......	1.0554[0]	1.4219[20]	al, eth	B1⁴, 3256
3921	2-Butanone, 3-chloro-3-methyl (CH₃)₂CHClCOCH₃	120.58	....	117.2[758]	......	1.0083[20/4]	1.4204[20]	al, eth	B1³, 2818
3922	2-Butanone, 4-chloro ClCH₂CH₂COCH₃	106.55	....	120-1d, 48[15]	......	1.0680[23]	1.4284[23]	al, eth	B1⁴, 3256
3923	2-Butanone, 3,4-dibromo-4-phenyl or Benzalacetone dibromide C₆H₅CHBrCHBrCOCH₃	306.91	nd (al)	....	124-5	......	......	al, chl	B7³, 1085
3924	2-Butanone, 1,3-dichloro CH₃CHClCOCH₂Cl	141.00	....	166-7, 55.5[10]	......	1.3116[20/4]	1.4686[20]	al, eth, ace, bz	B1³, 2786
3925	2-Butanone, 4-(diethylamino) (C₂H₅)₂NCH₂CH₂COCH₃	143.23	....	84[30]	......	0.8630[20/4]	1.4333[24]	al, eth, ace, bz	B4⁴, 1930
3926	2-Butanone, 3,3-dimethyl or Pinacolone........ (CH₃)₃CCOCH₃	100.16	....	106	-49.8	0.8012[25/4]	1.3952[20]	al, eth, ace	B1⁴, 3310
3927	2-Butanone, 3,3-diphenyl .... CH₃C(C₆H₅)₂CH₂COCH₃	224.31	pr (al)	300-1, 176[11]	41	1.069[20/4]	1.5748[20]	al, eth, chl, aa	B7³, 2209
3928	2-Butanone, 1-hydroxy .... CH₃CH₂COCH₂OH	88.12	....	160, 48[9]	......	1.0272[20/4]	1.4189[20]	w, al, eth	B1⁴, 3989
3929	2-Butanone, 3-hydroxy or Acetoin......... CH₃CHOHCOCH₃	88.12	....	143, 37[11]	-72	1.0062[20/20]	1.4171[20]	w, ace	B1⁴, 3991
3930	2-Butanone, 3-methyl or Methyl isopropyl ketone .... (CH₃)₂CHCOCH₃	86.14	....	94-5	-92	0.8051[20/4]	1.3880[20]	al, eth, ace	B1⁴, 3287
3931	2-Butanone, 3-methyl, oxime or Methyl isopropyl ketoxime (CH₃)₂CHC(:NOH)CH₃	86.14	....	....	157-8	......	......	al, eth	B1⁴, 3289
3932	2-Butanone oxime CH₃CH₂C(:NOH)CH₃	87.12	....	152-3, 59-60[15]	-29.5	0.9232[20/4]	1.4410[20]	w, al, eth	B1⁴, 3250
3933	2-Butanone, 1-phenyl or Ethyl benzyl ketone CH₃CH₂COCH₂C₆H₅	148.28	....	230[755]	......	1.002[20/4]	......	al, eth, ace	B7³, 1080
3934	2-Butanone, 4-phenyl or Benzylacetone......... C₆H₅CH₂CH₂COCH₃	148.21	....	233-4, 115[13]	......	0.9849[22/4]	1.511[22]	al, eth, ace	B7³, 1081
3935	2-Butenal or Crotonaldehyde........ CH₃CH=CHCHO	70.09	....	104-5	-74	0.8495[25/4]	1.4366[20]	al, eth, ace, bz	B1⁴, 3447
3936	2-Butenal diethylacetal or Crotonaldehyde diethylacetal .. CH₃CH=CHCH(OC₂H₅)₂	144.22	....	147-8	......	0.8473[18/4]	1.4097[20]	al, eth, ace, bz	B1⁴, 3450
3937	2-Butenal, 2-bromo, diethyl acetal .... CH₃CH=CBrCH(OC₂H₅)₂	223.12	....	86[15]	......	1.2255[21]	1.4565[21]	....	B1⁴, 3452
3938	2-Butenal, 2-chloro CH₃CH=CClCHO	104.54	....	147-8, 53-4[20]	......	1.1404[23/4]	1.4780[25]	al, eth, chl	B1³, 2981
3939	2-Butenal, 3-ethoxy, diethylacetal or 1,1,3-triethoxy-2-butene CH₃C(OC₂H₅)=CH-CH(OC₂H₅)₂	188.27	....	190-5, 79-82[10]	......	0.908[21/0]	1.430[21]	al	B1⁴, 4082
3940	2-Butenal, 2-methyl or Tiglaldehyde......... CH₃CH=C(CH₃)CHO	84.13	....	116.7[738], 63-5[110]	......	0.8710[20/4]	1.4475[20]	w, al, eth	B1⁴, 3464
3941	2-Butenal, 3-methyl or β,β-Dimethylacrolein......... (CH₃)₂C=CHCHO	84.13	....	133[730]	......	0.8722[20/4]	1.4528[20]	w, al, eth	B1⁴, 3464
3942	1-Butene CH₃CH₂CH=CH₂	56.12	....	-6.3	-185.3	0.5951[20/4]	1.3962[20]	al, eth, bz	B1⁴, 765
3943	1-Butene, 1-bromo-(cis) CH₃CH₂CH=CHBr	135.00	....	86.15	......	1.3265[15/4]	1.4536[20]	eth, ace, bz	B1³, 174
3944	1-Butene, 1-bromo-(trans)......... CH₃CH₂ = CHBr	135.0	....	94.7	-100.3	1.3209[15/4]	1.4527[20]	eth, ace, bz	B1³, 726
3945	1-Butene, 2-bromo CH₃CH₂CBr=CH₂	135.0	....	88, 25[10]	-133.4	1.3209[15/4]	1.4527[20]	eth, ace, bz	B1⁴, 775

No.	Name, Synonyms, and Formula	Mol. wt.	Color, crystalline form, specific rotation and λ_{max} (log ε)	b.p. °C	m.p. °C	Density	n_D	Solubility	Ref.
3946	1-Butene, 2-bromo-3-methyl $(CH_3)_2CHCBr=CH_2$	149.04	....	105^{757}	......	$1.2328^{20/4}$	1.4504^{20}	eth, bz, chl	B1³, 800
3947	1-Butene, 2-bromo-4-phenyl $C_6H_5CH_2CH_2CBr=CH_2$	211.11	....	$117\text{-}8^{21}$, $90\text{-}1^5$	......	$1.2907^{20/4}$	1.5450^{20}	ace	B5⁵, 1209
3948	1-Butene, 4-bromo $BrCH_2CH_2CH=CH_2$	135.01	....	98.5		$1.3230^{20/4}$	1.4622^{20}	al, eth, bz	B1⁴, 775
3949	1-Butene, 1-chloro-(cis) $CH_3CH_2CH=CHCl$	90.55	....	63.5		$0.9153^{15/4}$	1.4194^{15}	al, eth, ace, chl	B1³, 723
3950	1-Butene, 1-chloro-(trans) $CH_3CH_2\text{-}CH=CHCl$	90.55	....	68		$0.9205^{15/4}$	1.4223^{15}	al, eth, ace, chl	B1³, 723
3951	1-Butene, 1-chloro-2-methyl $CH_3CH_2C(CH_3)=CHCl$	104.58	....	96-7		$0.9170^{20/4}$	1.4141^{20}	eth, ace	B1³, 787
3952	1-Butene, 1-chloro-3-methyl $(CH_3)_2CH\text{-}CH=CHCl$	104.58	....	$86\text{-}8^{756}$		....	1.4229^{20}	eth, ace, chl	B1³, 799
3953	1-Butene, 2-chloro $CH_3CH_2CCl=CH_2$	90.55	....	58.5		$0.9107^{15/4}$	1.4165^{21}	al, eth, ace, bz	B1⁴, 769
3954	1-Butene, 2(chloromethyl)-1,3-dichloro $CH_3CHClC(CH_2Cl)=CHCl$	174.47	....	$68\text{-}70^8$		$1.2775^{19/4}$	....	CCl₄	B1³, 788
3955	1-Butene, 3-chloro $CH_3CHClCH=CH_2$	90.55	....	64-5		$0.8978^{20/4}$	1.4149^{20}	eth, ace, chl	B1⁴, 769
3956	1-Butene, 3-chloro-2-chloromethyl $CH_3CHClC(CH_2Cl)=CH_2$	139.04	....	155, $31\text{-}3^7$		$1.1233^{20/4}$	1.4724^{20}	ace, chl	B1³, 787
3957	1-Butene, 3-chloro-2-methyl $CH_3CHClC(CH_3)=CH_2$	104.58	....	94		$0.9088^{20/4}$	1.4304^{20}	eth, ace, chl	B1⁴, 819
3958	1-Butone, 4-chloro $ClCH_2CH_2CH=CH_2$	90.55	....	75^{773}		$0.9211^{20/4}$	1.4233^{20}	eth, ace, chl	B1⁴, 771
3959	1-Butene, 1,3-dichloro $CH_3CHClCH=CHCl$	125.00	....	125, $58\text{-}60^{25}$		$1.1341^{24/4}$	1.4647^{20}	al, eth, ace, chl	B1⁴, 772
3960	1-Butene, 2,3-dichloro $CH_3CHClCCl=CH_2$	125.00	....	112		$1.1340^{20/4}$	1.4580^{20}	eth, ace, chl	B1⁴, 772
3961	1-Butene, 3,3-dichloro-2-methyl $CH_3CCl_2C(CH_3)=CH_2$	139.04	....	151-3		$1.1276^{18/4}$	1.4737^{18}	eth, ace, chl	B1³, 787
3962	1-Butene, 3,4-dichloro $ClCH_2CHClCH=CH_2$	125.00	....	115-7, 45.5^{40}	......	1.1170^{20}	1.4475^{20}, 1.4641^{20}	al, eth, bz	B1⁴, 772
3963	1-Butene, 2,3-dimethyl $CH_3CH(CH_3)C(CH_3)=CH_2$	84.16	....	55.67	-157.3	$0.6803^{20/4}$	1.3995^{20}	al, eth, ace	B1⁴, 852
3964	1-Butene, 3,3-dimethyl $(CH_3)_3CCH=CH_2$	84.16	....	41.2	-115.2	$0.6529^{20/4}$	1.3763^{20}	al, eth	B1⁴, 850
3965	1-Butene-3,4-diol $HOCH_2CHOHCH=CH_2$	88.12	....	196.5, 98^{16}		$1.0470^{20/4}$	1.4628^{21}	w, al	B1⁴, 2658
3966	1-Butene, 1,3-diphenyl $CH_3CH(C_6H_5)CH=CHC_6H_5$	208.30	....	$175\text{-}6^{14}$		1.016^{15}	1.590	....	B5⁵, 2205
3967	1-Butene, 3,4-epoxy or Butadiene monoxide $CH_2CHCH=CH_2$	70.09	....	70		0.9006^0	1.4168^{20}	al, eth, bz	B17⁴, 145
3968	1-Butene, 2-ethyl $CH_3CH_2C(C_2H_5)=CH_2$	84.16	....	64.7	-131.5	$0.6894^{20/4}$	1.3969^{20}	eth, ace, bz, chl	B1⁴, 850
3969	1-Butene, 2-ethyl-3-methyl $(CH_3)_2CH\text{-}C(C_2H_5)=CH_2$	98.19	....	89		$0.7150^{20/4}$	1.410^{20}	eth, ace, bz, chl	B1⁴, 871
3970	1-Butene, 2-methyl $CH_3CH_2CCH_3=CH_2$	70.14	....	31.2	-137.5	$0.6504^{20/4}$	1.3378^{20}	al, eth, bz	B1⁴, 818
3971	1-Butene, 3-methyl $(CH_3)_2CHCH=CH_2$	70.14	....	20	-168.5	$0.6272^{20/4}$	1.3643^{20}	al, eth, bz	B1⁴, 825
3972	1-Butene, perfluoro $F_3CCF_2CF=CF_2$	200.03	....	4.8		$1.615^{-20/4}$, 1.5443^0		....	B1⁴, 769
3973	1-Butene, 1-phenyl (cis) $CH_3CH_2CH=CHC_6H_5$	132.20	....	196.2^{755}, $84\text{-}5^{23}$		$0.9106^{20/4}$	1.5381^{16}	al, eth, bz	B5⁵, 1205
3974	1-Butene, 1-phenyl-(trans) $CH_3CH_2CH=CHC_6H_5$	132.20	....	198.7, $91\text{-}2^{23}$	-43.1	$0.9019^{20/4}$	1.5420^{20}	al, eth, bz	B5⁴, 1374
3975	1-Butene, 4-phenyl $C_6H_5CH_2CH_2CH=CH_2$	132.20	....	177, 64^{10}	-70	$0.8831^{20/4}$	1.5059^{20}	eth, bz	B5⁴, 1378
3976	1-Butene, 1,3,4,4-tetrachloro $Cl_2CHCHClCH=CHCl$	193.89	....	88^{20}		$1.0711^{20/4}$	1.4773^{20}	chl	B1⁴, 774
3977	1-Butene, 2,3,3,4-tetrachloro $ClCH_2CCl_2CCl=CH_2$	193.89	....	$41\text{-}2^7$	......	$1.4602^{20/4}$	1.5135^{20}	ace, chl	B1³, 726
3978	1-Butene, 2,3,4-trichloro $ClCH_2CHClCCl=CH_2$	159.44	....	60^{20}		$1.3430^{20/4}$	1.4944^{20}	ace, chl	B1³, 725

No.	Name, Synonyms, and Formula	Mol. wt.	Color, crystalline form, specific rotation and λ_{max} (log ε)	b.p. °C	m.p. °C	Density	n_D	Solubility	Ref.
3979	1-Butene, 2,3,3-trimethyl or Triptene $(CH_3)_3CC(CH_3)=CH_2$	98.19	....	77.9	-109.9	$0.7050^{20/4}$	1.4025^{20}	eth, bz, MeOH	B1[4], 873
3980	2-Butene (cis) $CH_3CH=CHCH_3$	56.12	....	3.7	-138.9	$0.6213^{20/4}$	1.3931^{-25}	al, eth, bz	B1[4], 778
3981	2-Butene (trans) $CH_3CH=CHCH_3$	56.12	....	0.9	-105.5	$0.6042^{20/4}$	1.3848^{-25}	al, eth, bz	B1[4], 781
3982	2-Butene, 1-bromo $CH_3CH=CHCH_2Br$	135.00	....	103-6, 13[10]	......	$1.3371^{25/4}$	1.4822^{20}	al, eth, bz	B1[4], 789
3983	2-Butene, 1-bromo-3-methyl $(CH_3)_2C=CH-CH_2Br$	149.04	....	129-33d, 50-1[40]	......	$1.2819^{20/0}$	1.4930^{15}	al, eth, ace, bz	B1[4], 824
3984	2-Butene, 1-bromo-4-phenyl $C_6H_5CH_2CH=CHCH_2Br$	211.11	....	126-30[10.5]	......	$1.2660^{20/4}$	1.5678^{20}	eth	B5[4], 1377
3985	2-Butene, 2-bromo (cis) $CH_3CH=CBrCH_3$	135.00	....	93.9	-111.5	$1.3416^{15/4}$	1.4631^{19}	al, eth, bz	B1[4], 790
3986	2-Butene, 2-bromo-3-methyl $(CH_3)_2C=CBrCH_3$	149.04	....	119-20	......	$1.2773^{20/4}$	1.4738^{20}	eth, ace, chl	B1[4], 824
3987	2-Butene, 2-bromo-3-phenyl $CH_3C(C_6H_5)=CBr-CH_3$	211.11	....	120-30[11]	......	$1.3348^{20/4}$	1.5811^{20}	eth, bz, chl	B5[4], 1379
3988	2-Butene, 1-chloro (cis) $CH_3CH=CHCH_2Cl$	90.55	....	84.1	......	$0.9426^{20/4}$	1.4390^{20}	al, ace, chl	B1[4], 783
3989	2-Butene, 1-chloro (trans) $CH_3CH=CHCH_2Cl$	90.55	....	84.8^{752}	......	$0.9295^{20/4}$	1.4350^{20}	ace, chl	B1[4], 784
3990	2-Butene, 1-chloro-2,3-dimethyl $(CH_3)_2C=C(CH_3)CH_2Cl$	118.61	....	$111-2^{756}$	......	$0.9355^{20/4}$	1.4605^{20}	eth, ace, chl	B1[4], 856
3991	2-Butene, 1-chloro-2-methyl $CH_3CH=C(CH_3)CH_2Cl$	104.58	....	110, 36.4[11]	......	$0.9327^{20/4}$	1.4481^{20}	al, eth, ace, chl	B1[4], 822
3992	2-Butene, 1-chlor-3-methyl $CH_3C(CH_3)=CHCH_2Cl$	104.58	....	109, 54.3^{95}	......	$0.9273^{20/4}$	1.4485^{20}	al, eth, ace, chl	B1[4], 823
3993	2-Butene, 2-chloro (cis) $CH_3CH=CClCH_3$	90.55	....	70.6	-117.3	$0.9239^{20/4}$	1.4240^{20}	al, ace, chl	B1[4], 785
3994	2-Butene, 2-chloro (trans) $CH_3CH=CClCH_3$	90.55	....	62.8	-105.8	$0.9138^{20/4}$	1.4190^{20}	al, ace, chl	B1[4], 785
3995	2-Butene, 2-chloro-3-methyl $(CH_3)_2C=CClCH_3$	104.58	....	94	......	$0.9324^{20/4}$	1.4320^{20}	al, eth, ace, chl	B1[3], 794
3996	2-Butene, 1,4-dibromo (trans) $BrCH_2CH=CHCH_2Br$	213.91	pl (peth)	203, 85[10]	53.4	....	....	al, ace, peth	B1[4], 791
3997	2-Butene, 1,1-dichloro $CH_3CH=CHCHCl_2$	125.00	....	125-7	......	1.1310^{20}	1.466^{18}	eth, ace, chl	B1[4], 787
3998	2-Butene, 1,2-dichloro (high b.p.) $CH_3CH=CClCH_2Cl$	125.00	....	130-1	......	$1.1601^{20/4}$	1.4734^{20}	eth, ace, chl	B1[3], 741
3999	2-Butene, 1,2-dichlor (low b.p.) $CH_3CH=CClCH_2Cl$	125.00	....	$116-8^{765}$	......	$1.1544^{20/4}$	1.4642^{20}	bz, CCl_4	B1[3], 741
4000	2-Butene, 1,3-dichloro (cis) $CH_3CCl=CHCH_2Cl$	125.00	....	129.9^{745}, 34^{20}	......	$1.1605^{20/4}$	1.4735^{20}	al, eth, ace, bz	B1[4], 786
4001	2-Butene, 1,3-dichloro (trans) $CH_3CCl=CHCH_2Cl$	125.00	....	130^{745}, 53^{50}	......	$1.1585^{20/4}$	1.4719^{20}	al, eth, ace, bz	B1[4], 786
4002	2-Butene, 1,3-dichloro-2-methyl $CH_3CCl=C(CH_3)CH_2Cl$	139.04	....	151-3	......	$1.1293^{20/4}$	....	ace, chl	B1[3], 795
4003	2-Butene, 1,4-dichloro (cis) $ClCH_2.CH=CHCH_2Cl$	125.00	....	152.5, 22.5^3	-48	$1.188^{25/4}$	1.4887^{25}	al, eth, ace, bz	B1[4], 787
4004	2-Butene, 1,4-dichloro (trans) $ClCH_2CH=CHCH_2Cl$	125.00	....	155.5^{758}, 55.5^{20}	1-3	$1.183^{25/4}$	1.4871^{25}	al, eth, ace, bz	B1[4], 787
4005	2-Butene, 1,4-dichloro-2-methyl $ClCH_2CH=C(CH_3)CH_2Cl$	139.04	....	93^{50}	1	$1.1526^{20/4}$	1.4932^{20}	ace, chl	B1[3], 795
4006	2-Butene, 2,3-dichloro (cis) $CH_3CCl=CClCH_3$	125.00	....	$125-6^{758}$	......	$1.1618^{20/4}$	1.4590^{20}	al, eth, ace, bz	B1[4], 787
4007	2-Butene, 2,3-dichloro (trans) $CH_3CCl=CClCH_3$	125.00	....	$101-3^{758}$	......	$1.1416^{20/4}$	1.4582^{20}	al, eth, ace, bz	B1[4], 787
4008	2-Butene, 2,3-dimethyl $(CH_3)_2=C(CH_3)_2$	84.16	....	73.2	-74.3	$0.7080^{20/4}$	1.4122^{20}	al, eth, ace, chl	B1[4], 853
4009	2-Butene-1,4-diol (cis) $HOCH_2CH=CHCH_2OH$	88.12	....	235, 132[16]	4	$1.0698^{20/4}$	1.4782^{20}	w, al	B1[4], 2660
4010	2-Butene-1,4-diol (trans) $HOCH_2CH=CHCH_2OH$	88.12	....	131[13]	25	$1.0700^{20/4}$	1.4755^{20}	w, al	B1[4], 2660
4011	2-Butene, 1,1,2,3,4,4-hexachloro (liquid) $Cl_2CHCCl=CClCHCl_2$	262.78	....	$97-8^{10}$	-19	$1.651^{15/15}$	1.5331	chl	B1[4], 789

No.	Name, Synonyms, and Formula	Mol. wt.	Color, crystalline form, specific rotation and λ (log ε)	b.p. °C	m.p. °C	Density	n_D	Solubility	Ref.
4012	2-Butene, 1,1,2,3,4,4-hexachloro (solid) .... $Cl_2CHCCl=CClCHCl_2$	262.78	lf (al)	....	80	....	....	al, eth, bz, chl	B1[4], 789
4013	2-Butene, 2-methyl .... $CH_3CH=C(CH_3)_2$	70.14	....	38.6	-133.8	0.6623[20/4]	1.3874[20]	al, eth, bz, lig	B1[4], 820
4014	2-Butene, perfluoro .... $CF_3CF=CFCF_3$	200.03	....	0–3	-129	1.5297[0]	....	....	B1[4], 783
4015	2-Butene, 1,1,1,4,4-pentachloro .... $Cl_2CHCH=CHCCl_3$	228.33	....	78-80[11]	....	1.612[21/21]	1.5538[21]	al, chl	B1[3], 744
4016	2-Butene, 1,2,4-trichloro .... $ClCH_2CH=CClCH_2Cl$	159.44	....	67-9[10]	....	1.3843[20/4]	1.5175[20]	al, eth, bz, chl	B1[3], 744
4017	2-Butenoic acid (cis) or Isocrotonic acid .... $CH_3CH=CHCO_2H$	86.09	nd or pr (peth)	169.3, 74[15]	15.5	1.0267[20/4]	1.4483[14]	w, al	B2[4], 1497
4017a	2-Butenoic acid (trans) or Crotonic acid .... $CH_3CH=CHCO_2H$	86.09	mcl pr or nd (w or lig)	185	71.5	1.1018[15/4]	1.4249[77]	w, al, eth, ace	B2[4], 1509
4018	2-Butenoic acid, 2-chloro (cis) .... $CH_3CH=CClCO_2H$	120.54	nd (w)	....	67	....	....	w, al, lig	B2[4], 1510
4019	2-Butenoic acid, 2-chloro, ethyl ester (cis) .... $CH_3CH=CClCO_2C_2H_5$	148.59	....	75[30]	....	1.1021[18/4]	....	al, eth	B2[4], 1510
4020	2-Butenoic acid, 3-chloro (cis) .... $CH_3CCl=CHCO_2H$	120.54	....	195 sub	61	1.1995[66/4]	1.4704[66]	al, peth	B2[4], 1511
4021	2-Butenoic acid, 3-chloro, ethyl ester (cis) .... $CH_3CCl=CHCO_2C_2H_5$	148.59	....	161.4, 50[10]	....	1.0860[20/4]	1.4542[19/4]	al, eth	B2[4], 1510
4022	2-Butenoic acid, 3-chloro, methyl ester (cis) .... $CH_3CCl=CHCO_2CH_3$	134.57	....	142.4, 42-3[13]	....	1.138[20/4]	1.4573[19]	eth, MeOH	B2[4], 1500
4023	2-Butenoic acid, ethyl ester (cis) or Ethyl isocrotonate .... $CH_3CH=CHCO_2C_2H_5$	114.15	....	136	....	0.9182[20/4]	1.4242[20]	al, eth, ace	B2[4], 1551
4024	2-Butenoic acid, 2-methyl (cis) or Angelic acid .... $CH_3CH=C(CH_3)CO_2H$	100.13	mcl pr or nd	185, 88-9[10]	45-6	0.983[49/4], 0.9539[76/4]	1.4434[47]	al, eth	B2[4], 1497
4026	2-Butenoic acid, allyl ester (trans) .... $CH_3CH=CHCO_2CH_2=CH_2$	126.15	....	88-9[70]	....	0.9440[20/4]	1.4465[20]	....	B2[4], 1503
4027	2-Butenoic acid, 3-amino, ethyl ester (trans) .... $CH_3C(NH_2)=CHCO_2C_2H_5$	129.16	mcl pr	210-5d, 105[15]	34 (st), 20-1 (unst)	1.0219[19/4]	1.4988[22]	al, eth, bz, chl, lig	B3[3], 1199
4028	2-Butenoic acid, 3-bromo (trans) or 3-Bromo-crotonic acid .... $CH_3CBr=CHCO_2H$	165.00	nd (lig), lf (w)	....	97	....	....	al, eth, bz, aa	B2[3], 1275
4029	2-Butenoic acid, 4-bromo, ethyl ester (trans) or Ethyl-4-bromocrotonate .... $BrCH_2CH=CHCO_2C_2H_5$	193.05	....	97-8[15]	....	1.402[16/4]	1.4925[20]	al	B2[4], 1517
4030	2-Butenoic acid, 2-chloro (trans) or 2-Chlorocrotonic acid $CH_3CH=CClCO_2H$	120.54	nd (w or peth)	212, 111-2[14]	100.5	....	....	al, eth	B2[4], 1509
4031	2-Butenoic acid, 2-chloro, ethyl ester (trans) or Ethyl 2-chlorocrotonate .... $CH_3CH=CHCClCO_2C_2H_5$	148.59	....	176-8, 61[10]	....	1.1135[20/4]	1.4538[20]	al, eth	B2[4], 1510
4032	2-Butenoic acid, 2-chloro, methyl ester (trams) or Methyl 2-chlorocrotonate .... $CH_3CH=CClCO_2CH_3$	134.57	....	161.5[762], 59.5[16]	....	1.160[20/4]	1.4569[23]	eth	B2[4], 1510
4033	2-Butenoic acid, 3-chloro (trans) or 3-Chlorocrotonic acid $CH_3CCl=CHCO_2H$	120.54	....	206-11d	94-5	....	....	al, CS_2	B2[4], 1510
4034	2-Butenoic acid, 3-chloro, ethyl ester (trans) .... $CH_3CCl=CHCO_2C_2H_5$	148.59	....	184, 66[10]	....	1.1062[20/4]	1.4592[20]	al, eth	B2[4], 1511
4035	2-Butenoic acid, 3-chloro, methyl ester (trans) .... $CH_3CCl=CHCO_2CH_3$	134.57	....	64-7[14]	....	1.157[20/4]	1.4630[20]	al, eth	B2[4], 1510
4036	2-Butenoic acid, 4-chloro- (trans) .... $ClCH_2CH=CHCO_2H$	120.54	cr (peth-eth)	117-8[13]	83	....	....	eth	B2[4], 1511
4037	2-Butenoic acid, ethyl ester (trans) or Ethyl crotonate .... $CH_3CH=CHCO_2C_2H_5$	114.15	....	136.5, 58-9[48]	....	0.9175[20/4]	1.4243[20]	al, eth	B2[4], 1500
4038	2-Butenoic acid, 2-ethyl (trans) .... $CH_3CH=C(C_2H_5)CO_2H$	114.15	mcl pr (peth)	209, 109[13]	45-6	0.9578[50/4]	1.4475[50]	al, eth	B2[3], 1329
4039	2-Butenoic acid, methyl ester (trans) or Methyl crotonate $CH_3CH=CHCO_2CH_3$	100.13	....	121	-42	0.9444[20/4]	1.4242[20]	al, eth	B2[4], 1500
4040	2-Butenoic acid, 2-methyl (trans) or Tiglic acid .... $CH_3CH=C(CH_3)CO_2H$	100.13	ta (w)	198.5	64.5-5	0.9641[76/4]	1.4330[76]	al, eth	B2[4], 1552
4041	2-Butenoic acid, 2-methyl, ethyl ester (trans) or Ethyl tiglate .... $CH_3CH=C(CH_3)CO_2C_2H_5$	128.17	....	156, 55.5[11]	....	0.9200[20/4]	1.4340[20]	al, bz	B2[4], 1553

PHYSICAL CONSTANTS OF ORGANIC COMPOUNDS (Continued)

No.	Name, Synonyms, and Formula	Mol. wt.	Color, crystalline form, specific rotation and λ_max (log ε)	b.p. °C	m.p. °C	Density	n_D	Solubility	Ref.
4042	3-Butenoic acid or Vinyl acetic acid.......... $CH_2=CHCH_2CO_2H$	86.09	....	169, 69-70[12]	-35	1.0091[20/4]	1.4239[20]	w, **al**, eth	B2[4], 1491
4043	3-Butenoic acid-ethyl ester or Ethyl, 3-butenoate .... $CH_2=CHCH_2CO_2C_2H_5$	114.14	....	119		0.9122[20]	1.4105[18]	al	B2[4], 1491
4044	3-Butenoic acid, 2-hydroxy, ethyl ester .... $CH_2=CHCHOHCO_2C_2H_5$	130.14	....	173d, 68[15]	......	1.0470[15/4]	1.436[11]	w, al, eth	B2[3], 685
4045	3-Butenoic acid, 2-hydroxy-4-phenyl or Benzollactic acid $C_6H_5CH=CHCHOHCO_2H$	178.19	nd (w)		137	....	....		B10[3], 862
4046	3-Butenoic acid, 4-phenyl or Styrylacetic acid .... $C_6H_5CH=CHCH_2CO_2H$	162.19	nd (w), pr (CS₂)	302	87	....	....	al, eth	B9[3], 2756
4047	3-Butenonitrile or Allyl cyanide.................... $CH_2=CHCH_2CN$	67.09	....	119	-84	0.8329[20/4]	1.4060[20]	**al**, eth	B2[4], 1491
4048	2-Butene-1-ol or Crotyl alcohol .... $CH_3CH=CHCH_2OH$	72.12	....	121.2	<-30	0.8521[20/4]	1.4288[20]	**al**, eth	B1[4], 2107
4049	2-Butene-1-ol, 2-chloro .... $CH_3CH=CClCH_2OH$	106.55	....	159		1.1180[20/4]	1.4682[20]	w, al	B1[3], 1900
4050	2-Butene-1-ol, 4-chloro .... $ClCH_2CH=CHCH_2OH$	106.55	....	64-5[2]		....	1.4845[20]	al, eth	B1[4], 2110
4051	3-Butene-1-ol $CH_2=CHCH_2CH_2OH$	72.12	....	113.5		0.8424[20/4]	1.4224[20]	w, **al**, eth, ace	B1[4], 2105
4052	3-Buten-1-ol, 2-chloro .... $CH_2=CHCHClCH_2OH$	106.55	....	66-7[30]		1.1044[20/4]	1.4665[20]	al, eth	B1[4], 2106
4053	3-Buten-2-ol (d) .... $CH_2=CH-CHOHCH_3$	72.12	$[\alpha]^{20}_D$ +33.9 (undil)	96.5[745]		0.8362[15/4]	1.4120[20]	....	B1[4], 2102
4054	3-Buten-2-ol (dl) $CH_2=CH-CHOHCH_3$	72.12	....	97.3	<-100	0.8318[20/4]	1.4137[20]	....	B1[4], 1102
4055	3-Buten-2-ol, 1-chloro .... $CH_2=CH-CHOHCH_2Cl$	106.55	....	144-7, 63.3[30]		1.11[18]	1.4643[20]	chl	B1[3], 1893
4056	3-Buten-2-ol, 3-chloro .... $CH_2=CClCHOHCH_3$	106.55	....	53-7[19]		1.1138[23/4]	....	al, eth	B1[3], 1893
4057	3-Buten-2-one or Methyl vinyl ketone .... $CH_2=CHCOCH_3$	70.09	....	81.4, 33-4[130]		0.8636[20/4]	1.4081[20]	w, al, eth, ace, bz	B1[4], 3444
4058	3-Butene-2-one, 4-bromo-4-phenyl $C_6H_5CBr=CHCOCH_3$	225.09	....	150-1[10]				bz	B7[3], 1407
4059	3-Butene-2-one, 4-(2-hydroxyphenyl) .... 2-HOC₆H₄CH=CHCOCH₃	162.19	nd (al or lig), pr (bz)		140			eth, al, bz	B8[3], 810
4060	3-Butene-2-one, 4-(3-hydroxyphenyl) 3-HOC₆H₄CH=CHCOCH₃	162.19	ye pr (bz)		97-8			bz	B8[3], 812
4061	3-Butene-2-one, 4-(4-hydroxyphenyl) 4-HOC₆H₄CH=CHCOCH₃	162.19	nd (w)		114-5			al, aa	B8[3], 812
4062	3-Butene-2-one, 4-(4-methoxy phenyl) or Anisylidene acetone 4-CH₃OC₆H₄CH=CHCOCH₃	176.22	lf (al, eth, aa)		73			al, eth, bz, aa	B8[3], 812
4063	3-Butene-2-one, 3-methyl or Isopropenyl methyl ketone .. $CH_2=C(CH_3)COCH_3$	84.13	....	98	-54	0.8527[20/4]	1.4220[20]	**al**	B1[4], 3462
4064	3-Butene-2-one, 4-phenyl (trans) or Benzal acetone ...... $C_6H_5CH=CHCOCH_3$	146.19	pl	26[2], 140[16]	42	1.097[45]	1.5836[45]	al, eth, ace, bz, chl	B7[3], 1399
4065	1-Butene-3-one or Vinyl acetylene.................. $CH_2=CH-C≡CH$	52.08	....	5.1		0.7095[0/0]	1.4161[1]	bz	B1[4], 1083
4066	1-Butene-3-one, 4-chloro .... $CH_2=CH-C≡CCl$	86.52	....	55-7		1.0022[20/4]	1.4656[20]	chl	B1[4], 1085
4067	1-Butene-3-one, 4-methoxy .... $CH_3OCH=CH-C≡CH$	82.10	....	122-5d, 30-2[15]	......	0.906[20/4]	1.4818[20]	....	B1[4], 2300
4068	1-Butene-3-one, 2-methyl .... $CH_2=C(CH_3)C≡CH$	66.10	....	34		0.6801[11/4]	1.4105[20]	....	B1[4], 1089
4069	n-Butyl alcohol or 1-Butanol.................... $CH_3CH_2CH_2CH_2OH$	74.12	....	117.2	-89.5	0.8098[20/4]	1.3993[20]	w, **al**, eth, ace, bz	B1[4], 1506
4070	Isobutyl alcohol or 2-methyl-1-propanol .... $(CH_3)CHCH_2OH$	74.12	....	108.1		0.8018[20/4]	1.3955[20]	al, eth, ace	B1[4], 1588
4071	Isobutyl alcohol, 2-amino .... $(CH_3)_2C(NH_2)CH_2OH$	89.14	....	165.5, 69-70[10]	25-6	0.934[20/4]	1.449[20]	w	B4[4], 1740
4072	Isobutyl alcohol, 2-chloro or β Isobutylene chlorohydrin $(CH_3)_2CClCH_2OH$	108.57	....	132-3d, 59-61[50]	......	1.0472[20/4]	1.4388[20]	al, eth	B1[4], 1603
4073	Isobutyl alcohol, 2-nitro .... $(CH_3)_2C(NO_2)CH_2OH$	119.12	nd or pl (MeOH)	94-5[10]	89-90	....	....	al, eth	B1[4], 1604

No.	Name, Synonyms, and Formula	Mol. wt.	Color, crystalline form, specific rotation and λ_{max} (log ε)	b.p. °C	m.p. °C	Density	n_D	Solubility	Ref.
4074	Isobutyl alcohol, 3-chloro .... ClCH₂CH(CH₃)CH₂OH	108.57	....	76-8[11]	......	1.083[25/4]	1.4460[25]	al, eth	B1[3], 1564
4075	sec-Butyl alcohol or 2-Butanol .... CH₃CH₂CH(OH)CH₃	74.12	....	99.5, 45.5[60]	......	0.8063[20/4]	1.3978[20]	al, eth, ace, bz	B1[4], 1566
4076	tert-Butyl alcohol or 2-Methyl-2-propanol (CH₃)₃COH	74.12	....	82.3, 20[31]	25.5	0.7887[20/4]	1.3878[20]	w, al, eth	B1[4], 1609
4077	tert-Butyl alcohol, methoxy .... (CH₃)₂C(OH)CH₂OCH₃	104.15	....	116.6[747]	......	0.9021[15/15]	....	....	B1[4], 1615
4078	tert-Butyl alcohol, 1,1,1-tribromo or Brometone .... (CH₃)₂C(OH)CBr₃	310.83	nd (lig), cr (dil al)	sub	168-70	....	....	al, eth	B1[3], 1588
4079	tert-Butyl alcohol, 1,1,1-trichloro or Chloreton .... (CH₃)₂C(OH)CCl₃	177.46	hyg nd (w + 1)	167	98-9, 77 (hyd)	....	....	al, eth, ace, bz	B1[4], 1629
4080	n-Butyl amine .... CH₃CH₂CH₂CH₂NH₂	73.14	....	77.8	-49.1	0.7414[20/4]	1.4031[20]	w, al, eth	B4[4], 540
4081	Butyl dimethyl amino C₄H₉N(CH₃)₂	101.19	....	95	......	0.7206[20/4]	1.3970[20]	w, al, eth, ace, bz	B4[4], 546
4082	Butyl ethyl amine C₄H₉NHC₂H₅	101.19	....	108-9	......	0.7398[20/4]	1.4040[20]	al, eth, ace, bz	B4[4], 547
4083	Butyl bis(2-hydroxyethyl) amine C₄H₉N(CH₂CH₂OH)₂	161.25	....	273-5[741], 80[35]	......	0.9692[20/4]	1.4625[20]	w, al, eth, ace	B4[4], 1520
4084	Butyl phenyl amine or N-butylanilene .... C₄H₉NHC₆H₅	149.24	....	241.6, 118-20[15]	-14.4	0.9322[20/4]	1.5341[20]	al, eth	B12[3], 267
4085	sec-Butyl amine (d) or 2-Aminobutane (d) .... CH₃CH₂CH(NH₂)CH₃	73.14	[α][20/D] + 7.4 (w)	63	-104.5	0.724[20/4]	1.344[20]	w, al, eth, ace	B4[4], 617
4086	sec-Butyl amine (dl) or 2-Amino butane (dl) .... CH₃CH₂CH(NH₂)CH₃	73.14	....	63.5[764]	<-72	0.7246[20/4]	1.3932[20]	al, eth, ace	B4[4], 618
4087	sec-Butyl amine (l) or 2-Amino butane (l) .... CH₃CH₂CH(NH₂)CH₃	73.14	[α][20/D] -7.4 (w, c=4.7)	63	......	0.7205[20/4]	....	w, al, eth, ace	B4[4], 617
4088	sec-Butyl ethyl amine (d) sec-C₄H₉NC₂H₅	101.19	[α][15/D] + 18	98	......	0.7396[15/4]	1.4043[15]	al, eth, ace, bz	B4[3], 307
4089	sec-Butyl ethyl amine (dl) or Ethyl-sec-butyl lamine (dl) sec-C₄H₉NHC₂H₅	101.19	....	97-8[741]	-104.3	0.7358[20/4]	....	al, eth, ace, bz	B4[2], 636
4090	sec-Butyl phenyl amine or N-sec-Butylaniline .... CH₃CH₂CH(CH₃)NHC₆H₅	149.24	....	223, 112-4[22]	......	....	1.5333[20]	....	B12[3], 269
4091	tert-Butyl amine (CH₃)₃CNH₂	73.14	....	44.4	-67.5	0.6958[20/4]	1.3784[20]	w, al, eth	B4[4], 657
4092	tert-Butyl phenyl amine or N-tert-Butylaniline .... (CH₃)₃CNHC₆H₅	149.24	....	214-6[753], 93-8[19]	......	....	1.5270[20]	al, ace, bz, chl	B12[3], 270
4093	Butyl 4-aminophenyl ketone .... (4-H₂NC₆H₄)COC₄H₉	177.25	cr (bz-peth)	160-3[3]	74-5	....	....	al, eth	B14[2], 43
4094	Butylarsonic acid .... C₄H₉AsO(OH)₂	182.05	....	......	160	....	....	w, al	B4[3], 1824
4095	n-Butyl bromide or 1-Bromobutane .... CH₃CH₂CH₂CH₂Br	137.03	....	101.6, 18.8[30]	-112.4	1.2758[20/4]	1.4401[20]	al, eth, ace, chl	B1[4], 258
4096	Isobutylbromide or l-Bromo-2-methyl propane .... (CH₃)₂CHCH₂Br	137.02	....	91.7, 41-3[135]	-117.4	1.2532[20/4]	1.4348[20]	al, eth, ace, bz	B1[4], 294
4097	sec-Butyl bromide (dl) or 2-Bromobutane .... CH₃CH₂CHBrCH₃	137.03	....	91.2	-111.9	1.2585[20/4]	1.4366[20]	eth, ace, chl	B1[4], 261
4098	tert-Butyl bromide or 2-Bromo-2-methylpropane .... (CH₃)₃CBr	137.02	....	73.25	-16.2	1.2209[20/4]	1.4278[20]	....	B1[4], 295
4099	n-Butyl chloride or 1-Chlorobutane .... CH₃CH₂CH₂CH₂Cl	92.57	....	78.44	-123.1	0.8862[20/4]	1.4021[20]	al, eth	B1[4], 246
4100	Isobutyl chloride or 1-Chloro-2-methylpropane .... (CH₃)₂CHCH₂Cl	92.57	....	68-70	-130.3	0.8810[20/4]	1.39841[20]	eth, ace, chl	B1[4], 287
4101	sec-Butyl chloride or 2-Chlorobutane............... CH₃CH₂CHClCH₃	92.57	....	68.2	-131.3	0.8732[20/4]	1.3971[20]	al, eth, bz, chl	B1[4], 248
4102	tert-Butyl chloride or 2-Chloro-2-methyl propane .... (CH₃)₃CCl	92.57	....	-25.4	0.8420[20/4]	1.3857[20]	al, eth, bz, chl	B1[4], 288
4103	Butyl (β-chloroethyl) ether C₄H₉OCH₂CH₂Cl	136.62	....	154.5d, 49-50[11]	......	0.9335[20/4]	1.4155[20]	eth	B1[4], 1519
4103a	1,2-Butylene oxide CH₃CH₂CH-CH₂ ⌐O⌐	72.12	....	63.3	......	0.837[17/4]	1.3851[20]	al, eth, ace	B1[4], 796
4104	Isobutylene or 2-methylpropene .... (CH₃)₂C=CH₂	56.11	....	-6.9	-140.3	0.5942[20/4]	1.3926[-25]	al, eth, bz	B1[3], 762
4105	Isobutylene (trimer) or Trisobutylene .... (C₄H₈)₃	168.33	....	174-81, 56[10]	-76	0.7590[20/4]	1.4314[20]	....	B1[3], 763

No.	Name, Synonyms, and Formula	Mol. wt.	Color, crystalline form, specific rotation and λ_{max} (log ε)	b.p. °C	m.p. °C	Density	n_D	Solubility	Ref.
4106	Isobutylene (tetramer) .... $(C_4H_8)_4$	224.44	....	242-6, 109.5[15]	-98	0.7944[20/4]	1.4482[20]	....	B1[4], 803
4107	Isobutylene, 1-chloro or 1-chloro-2-methylpropene ...... $(CH_3)_2C=CHCl$	90.55	....	68[754]	....	0.9186[20/4]	1.4221[20]	al, eth, ace, chl	B1[4], 803
4108	Isobutylene, 3-chloro or 3-chloro-2-methyl propene ...... $ClCH_2C(CH_3)=CH_2$	90.55	....	71-2	....	0.9165[20/4]	1.4291[20]	al, eth, ace, chl	B1[4], 804
4109	Isobutylene, 1,1-dichloro or 1,1-dichloro-2-methylpropene $(CH_3)_2C=CCl_2$	125.00	....	108-9, 42-3[75]	....	1.1449[20/0]	1.4580[20]	eth, bz, chl	B1[4], 2533
4110	Isobutylene glycol or 2-methyl-1,2-propane diol .... $(CH_3)_2C(OH)CH_2OH$	90.12	....	176,79-80[12]	....	1.0024[20/4]	1.4350[20]	w, al, eth	B1[4], 2536
4111	Isobutylene glycol, 3-chloro or 3-chloro-2-methyl propyl- ene glycol $ClCH_2C(OH)(CH_3)CH_2OH$	124.57	....	114-7[20]	....	1.2362[20/4]	1.4788[20]	w, al, eth	B1[4], 805
4112	Isobutylene, 3,3-dichloro or 3,3-dichloro-2-methyl pro- pene $Cl_2CHC(CH_3)=CH_2$	125.00	....	108-12, 49-50[120]	....	1.3631[24/4]	1.4523[24]	eth, bz, chl	B1[3], 1904
4113	Isobutylene, 3-methoxy or 3-methoxy-2-methyl propene .. $CH_3OCH_2C(CH_3)=CH_2$	86.13	....	68[773]	....	0.7698[20/4]	1.3964[20]	....	B17[4], 45
4115	Isobutylene, 1,1,1-trichloro or 1,1,1-trichloro-2-methyl propene $ClCH_2C(CH_3)=CCl_2$	159.44	....	156, 45-6[12]	....	1.346[20]	1.4990[20]	ace, bz, chl	B1[4], 805
4116	Isobutylene, 3,3,3-trichloro or 3,3,3-trichloro-2-methyl propene $Cl_3CC(CH_3)_2=CH_2$	159.44	....	....	132-4	1.293[20]	1.4770[20]	ace, bz, chl, aa	B1[4], 805
4117	Butyl ether or Dibutylether...................... $C_4H_9OC_4H_9$	130.23	....	142	-95.3	0.7689[20/4]	1.3992[20]	al, eth, ace	B1[4], 1520
4118	sec-Butyl ether (dl) or Di-sec-Butyl ether... $(sec-C_4H_9)_2O$	130.23	....	120-1	....	0.756[25]	1.393[25]	al, eth, ace	B1[3], 1533
4119	Isobutyl ether α, β-dichloro or α, β-Diclorodiisobutyl ether $(CH_3)_2CCHCl-O-i-C_4H_9$	199.12	....	192.5, 83[15]	....	1.031[5/4]	....	eth, ace	B1, 675
4120	Butyl isobutyl ether .... $(CH_3)_2CHCH_2OC_4H_9$	130.23	....	148-52[730]	....	0.7980[22/4]	1.4077[21]	al, eth, ace	B1[4], 1594
4121	Butyl ethyl ether .... $C_4H_9OC_2H_5$	102.18	....	96	....	0.7490[20/4]	1.3818[20]	al, eth, ace	B1[4], 1518
4122	sec-Butyl ethyl ether .... $sec-C_4H_9OC_2H_5$	102.18	....	81	....	0.7503[20/4]	1.3802[20]	al, eth	B1[4], 1572
4123	tert-Butyl ethyl ether .... $(CH_3)_3COC_2H_5$	102.18	....	73.1	-94	0.7519[25]	1.3794[20]	al, eth	B1[4], 1615
4124	Butyl ethynyl ether or Butoxy acetylene .... $C_4H_9OC\equiv CH$	98.15	....	102-4,exp ca 100	....	0.8200[20/4]	1.4020[20]	al, eth	B1[4], 2213
4125	Butyl ethyl sulfide .... $C_4H_9SC_2H_5$	118.24	....	144.2, 33.3[10]	-95.1	0.8376[20/4]	1.4491[20]	al, chl	B1[4], 1558
4126	Butyl furfuryl ether .... $C_4H_9OCH_2(C_4H_3O)$	154.21	....	189.9[765]	....	0.9516[20/4]	1.4522[20]	al, eth	B17[2], 115
4127	tert-Butyl hydro peroxide .... $(CH_3)_3COOH$	90.12	....	d at89, 35-7[17]	6	0.8960[20]	1.4015[20]	w, al, eth, chl	B1[4], 1616
4128	tert-Butyl hypochlorite .... $(CH_3)_3COCl$	108.57	ye lig	77-8	....	0.9583[18/4]	1.403[20]	eth, ace, bz	B1[4], 1621
4129	tert-Butyl iodide or 2-Iodo-2-methyl propane .... $(CH_3)_3CI$	184.02	....	100, 20.8[30]	-38.2	1.5445[20/4]	1.4918[20]	al, eth	B1[4], 300
4130	Isobutyl isocyanate .... $i-C_4H_9NCO$	99.13	....	106	....	....	....	....	B4[4], 653
4131	tert-Butyl isocyanate .... $(CH_3)_3CNCO$	99.13	....	85.5	....	0.8670[9]	1.4061[20]	....	B4[4], 669
4132	tert-Butyl isocyanide or 2-Isocyano-2-methyl propane .... $(CH_3)_3CNC$	83.13	....	167-70, 91[38]	....	....	....	eth, al	B4[4], 562
4133	Butyl isothiocyanate .... C_4H_9NCS	115.20	....	168, 64-6[12]	....	0.9546[20/4]	1.501[20]	al, eth	B4[4], 596
4134	sec-Butyl isothiocyanate (d) or sec-Butyl mustard oil .... $sec-C_4H_9NCS$	115.20	$[\alpha]^{20}_D$ + 61.88	159	....	0.943[20/4]	....	al, eth	B4[1], 372
4135	sec-Butyl isothiocyanate (dl) .... $sec-C_4H_9NCS$	115.20	....	159.5	....	0.944[12]	....	al, eth	B4[4], 624
4136	sec-Butyl isothiocyanate (l) .... $sec-C_4H_9NCS$	115.20	$[\alpha]^{20}_D$ -61.8	159	....	0.942[20/4]	....	al, eth	B4, 161
4137	Isobutyl isothiocyanate or Isobutyl mustard oil .... $i-C_4H_9NCS$	115.20	....	160 (cor)	....	0.9638[14/4]	1.5005[14]	al, eth	B4[4], 653

No.	Name, Synonyms, and Formula	Mol. wt.	Color, crystalline form, specific rotation and λ_{max} (log ε)	b.p. °C	m.p. °C	Density	n_D	Solubility	Ref.
4138	tert-Butyl isothiocyanate $(CH_3)_3CNCS$	115.20	....	140[770]	10-11	0.9187[20/4]	....	eth	B4[4], 669
4139	n-Butyl mercaptan or l-Butanethiol $CH_3CH_2CH_2CH_2SH$	90.19	....	98.4	-115.7	0.8337[20/4]	1.4440[20]	al, eth	B1[4], 1555
4140	Isobutyl mercaptan or 2-Methyl-l-propanethiol $(CH_3)_2CHCH_2SH$	90.19	....	88.7	<-70	0.8339[20/4]	1.4387[20]	al, eth, ace	B1[4], 1605
4141	sec-Butyl mercaptan (d) or 2-Butanethiol $CH_3CH_2CHSHCH_3$	90.19	$[\alpha]^{20}_D$ + 15.7	85-95, 37.4[114]		0.8299[20/4]	1.4338[25]	al, eth, bz, peth	B1[3], 1549
4142	sec-Butyl mercaptan (dl) $CH_3CH_2CH(SH)CH_3$	90.19	....	85		0.8295[20/4]	1.4366[20]	al, eth, bz, peth	B1[4], 1584
4143	sec-Butyl mercaptan (l) $CH_3CH_2CH(SH)CH_3$	90.19	$[\alpha]^{17}_D$ - 17.35	83-4		0.8300[17/4]	....	al, eth, bz, peth	B1[3], 1549
4144	tert-Butyl mercaptan or 2-Methyl-2-propanethiol $(CH_3)_3CSH$	90.19	....	64.2	1.11	0.8002[20/4]	1.4232[20]	....	B1[4], 1634
4145	Butyl methyl ether $C_4H_9OCH_3$	88.15	....	71	-115.5	0.7443[20/4]	1.3736[20]	al, eth, ace	B1[4], 1518
4146	Isobutyl methyl ether i-$C_4H_9OCH_3$	88.15	....	58		0.7311[20/4]	....	al, eth	B1[4], 1593
4147	sec-Butyl methyl ether sec-$C_4H_9OCH_3$	88.15	....	60		0.7415[20/4]	1.3680[25]	al, eth, ace	B1[4], 1572
4148	tert-Butyl methyl ether $(CH_3)_3COCH_3$	88.15	....	55.2	-109	0.7405[20/4]	1.3690[20]	al, eth	B1[4], 1615
4149	Butyl methyl sulfide $C_4H_9SCH_3$	104.23	....	123.2	-97.8	0.8426[20/4]	1.4477[20]	al	B1[4], 1557
4150	Isobutyl methyl sulfide i-$C_4H_9SCH_3$	104.22	....	112.5		0.8335[20/4]	1.4433[20]	al, eth, ace	B1[4], 1606
4151	n-Butyl nitrate $C_4H_9ONO_2$	119.12	....	135.5, 70-1[86]		1.0228[30]	1.4013[23]	al, eth	B1[4], 1524
4152	sec-Butyl nitrate sec-$C_4H_9ONO_2$	119.12	....	124, 59[80]		1.0264[20/4]	1.4015[20]	al, eth	B1[4], 1573
4153	iso-butyl nitrite i-C_4H_9ONO	103.12	....	67		0.8699[22/4]	1.3715[22]	al, eth	B1[4], 1595
4154	tert-Butyl nitrite $(CH_3)_3CONO$	103.12	pa ye	63, 34[250]		0.8670[20/4]	1.368[20]	al, eth, chl	B1[4], 1622
4155	tert-Butyl peroxide or Di-tert-butyl peroxide $(CH_3)_3COOC(CH_3)_3$	146.23	....	111, 70[197]	-40	0.704[20]	1.3890[20]	ace, lig	B1[4], 1619
4156	Butyl phenyl ether $C_4H_9OC_6H_5$	150.22	....	210, 95[17]	-19.4	0.9351[20/4]	1.4969[20]	eth, ace	B6[4], 558
4157	sec-Butyl phenyl ether sec-$C_4H_9OC_6H_5$	150.22	....	194-5, 70-2[5]		0.9415[20/4]	1.4926[25]	eth	B6[4], 558
4158	tert-Butyl phenyl ether $(CH_3)_3COC_6H_5$	150.22	....	185-6		0.9214[20]	....	....	B6[4], 559
4159	Butyl phenyl ketone or Valerophenone $C_4H_9COC_6H_5$	162.23	....	248.5, 131-3[13]		0.988[20/20]	1.5158[20]	al, eth	B7[3], 1114
4160	Butylphosphonic acid $C_4H_9PO(OH)_2$	138.10	pl (bz)	d	106	....	....	w, al, eth	B4[3], 1782
4161	Isobutyl phosphonic acid i-$C_4H_9PO(OH)_2$	138.10	pl (xyl)	d	119	....	....	w, al, eth, xyl	B4[1], 573
4162	sec-Butyl phosphonic acid sec-$C_4H_9PO(OH)_2$	137.12	lf (eth-lig)		48	....	....	w, al, eth, bz	Am75, 3379
4163	tert-Butyl phosphonic acid $tert$-$C_4H_9PO(OH)_2$	138.10	wh nd (xyl, aa-lig)	d	192	....	....	w, al	Am75, 3379
4164	Butyl propyl ether $C_4H_9OC_3H_7$	116.21	....	117.1		0.7773[0/0]	....	al, eth	B1[4], 1519
4165	Isobutyl propyl ether i-$C_4H_9OC_3H_7$	116.21	....	105-6[720]		0.7549[20]	1.3852[25]	al, eth	B1[4], 1594
4166	Butyl isopropyl ether i-$C_3H_7OC_4H_9$	116.21	....	108[738]		0.7594[15/4]	1.3870[15]	al, eth, ace	B1[4], 1519
4167	Butyl sulfate $(C_4H_9O)_2SO_2$	210.30	....	109.5[4]	1.4192[20]	1.0616[20]	....	....	B1[4], 1523
4168	Butyl sulfide (α-form) or α-Dibutyl ssulfide $(C_4H_9)_2S$	146.30	....	185	-79.7	0.8386[20/4]	1.4530[20]	al, eth, chl	B1[4], 1559
4169	Butyl sulfide (β-form) or β-Dibutyl sulfide $(C_4H_9)_2S$	146.30	....	190-230d		....	....	al, eth, ace	Ber62, 2168
4170	Butyl sulfide-2,2′-dimethyl (d) $[C_2H_5CH(CH_3)CH_2]_2S$	146.30	....	165		0.8348[20/4]	1.4506[20]	al, eth	B1, 387

No.	Name, Synonyms, and Formula	Mol. wt.	Color, crystalline form, specific rotation and λ_{max} (log ε)	b.p. °C	m.p. °C	Density	n_D	Solubility	Ref.
4171	Isobutyl sulfide or Di isobutyl sulfide $(i\text{-}C_4H_9)_2S$	146.29		170.5[752]	-105.5	0.8363[10]			B1[4], 1607
4172	sec-Butyl sulfide or Di-sec-Butyl sulfide $(sec\text{-}C_4H_9)_2S$	146.29		165		0.8348[10/15]	1.4506[20]	al, eth	B1[4], 1586
4173	Butyl sulfite $(C_4H_9O)_2SO$	194.30		230, 116[19]		0.9957[20/4]	0.4310[20]	al, eth	B1[4], 1522
4174	Isobutyl sulfite $(i\text{-}C_4H_9O)_2SO$	194.30		209[741], 92-4[13]		0.9862[20/4]	1.4268[20]		B1[4], 1607
4175	n-Butyl sulfonamide $C_4H_9SO_2NH_2$	137.12	lf (eth-lig)		48			w, al, eth, bz	B4[4], 45
4176	n-Butyl sulfonyl chloride $C_4H_9SO_2Cl$	156.63		75[10]			1.4559[20]		B4[4], 45
4177	Isobutyl sulfonyl chloride $i\text{-}C_4H_9SO_2Cl$	156.63		189-91, 87[15]			1.4520[25]		B4[4], 49
4178	Butyl sulfoxide or Di butyl sulfoxide $(C_4H_9)_2SO$	162.30	nd (dil al)	d	32.6	0.8317[21/4]	1.4669[20]	al, eth	B1[4], 1561
4179	Butyl thiocyanate C_4H_9SCN	115.20		185[743]		0.9563[15]	1.4360[20]	al, eth	B3[4], 329
4180	Isobutyl thiocyanate $(CH_3)_2CHCH_2SCN$	115.20		175.4, 66[15]	-59			al, eth	B3[4], 330
4181	tert-butyl thiocyanate $(CH_3)_3CSCN$	115.20		140[770]d, 39-40[10]	10.5	0.9187[10]			B3[4], 330
4182	Butyl 2-tolyl ether $(2\text{-}CH_3C_6H_4)OC_4H_9$	164.25		223		0.943[0/0]			B6[3], 1247
4183	Butyl 3-tolyl ether or 3-Butoxy toluene $(3\text{-}CH_3C_6H_4)OC_4H_9$	164.25		229.2		0.9407[70/0]	1.4970[20]	eth	B6[4], 2040
4184	Butyl 4-tolyl ether or 4-Butoxy toluene $(4\text{-}CH_3C_6H_4)OC_4H_9$	164.25		229.5, 88[3]		0.9237[25/25]	1.4970[20]	eth	B6[3], 1354
4185	Butyl vinyl ether $C_4H_9OCH=CH_2$	100.16		93.8	-92	6.7888[20/4]	1.4026[20]	al, eth, ace, bz	B1[4], 2052
4186	Isobutyl vinyl ether $i\text{-}C_4H_9OCH=CH_2$	100.16		83	-112	0.7645[20/4]	1.3966[20]	al, eth, ace, bz	B1[4], 2054
4187	Butyl nitrite C_4H_9ONO	103.12		-77.8, 27[88]		0.8823[20/4]	1.3762[20]	al, eth	B1[4], 1523
4188	sec-Butyl nitrite $sec\text{-}C_4H_9ONO$	103.12		68-9, 28[180]		0.8726[20/4]	1.3710[20]	al, eth, chl	B1[4], 1573
4189	Isobutylene or 2-Methylpropene $(CH_3)_2CH=CH_2$	56.11	gas	-6.9	-140.3	0.5942[20/4]	1.3926[25]	al, eth, bz	B1[4], 796
4190	Isobutylene trimer or Triisobutylene $(C_4H_8)_3$	168.33		179.8, 56[16]	-76	0.7590[204]	1.4314[20]		B1[3], 762
4191	Isobutylene tetramer or Tetraiso butylene $(C_4H_8)_4$	224.44		243-6, 109.5[22]	-98	0.7444[20]	1.4482[20]		B1[3], 763
4192	Isobutylene, 1-chloro $(CH_3)_2C=CHCl$	90.55		68[754]		0.9186[20/4]	1.4221[20]	al, eth, ace, chl	B1[4], 803
4193	Isobutylene, 3-chloro or 3-Chloro-2-methylpropene $ClCH_2C(CH_3)=CH_2$	90.55		71-2		0.9165[20/4]	1.4291[20]	al, eth, ace, chl	B1[4], 803
4194	Isobutylene oxide or 1,2-Epoxy-2-methylpropane $(CH_3)_2CCH_2$	72.12		52		0.8650[0]	1.3712[22]	al, eth	B17[4], 46
4195	1-Butyne $CH_3CHC\equiv CH$	54.09		8.1	-125.7	0.6784[0/0]	1.3962[20]	al, eth	B1[4], 969
4196	1-Butyne, 3-chloro $CH_3CHClC\equiv CH$	88.54		68.5		0.9466[25]	1.4218[25]		B1[4], 970
4197	1-Butyne, 3-chloro-3-methyl $(CH_3)_2CClC\equiv CH$	102.57		77-9		0.9061[20/4]			B1[4], 1000
4198	1-Butyne, 3,3-dimethyl or tert-Butyl acetylene $(CH_3)_3CC\equiv CH$	82.15		39-40	-81.2	0.6695[20/4]	1.3738[20]		B1[4], 1022
4199	1-Butyne, 3-methyl or Isopropyl acetylene $(CH_3)_2CHC\equiv CH$	68.13		29.5	-89.7	0.6660[20/4]	1.3723[20]	al, eth	B1[4], 999
4200	1-Butyne, 4-phenyl $C_6H_5CH_2CH_2C\equiv CH$	130.19		190		0.9258[20/4]	1.5208[20]		B5[4], 1535
4201	2-Butynal $CH_3C\equiv CCHO$	68.07		106-7, 27-8[34]	-26	0.9265[17/0]	1.446[19]	eth, ace	B1[4], 3540
4202	2-Butyne or Dimethyl acetylene $CH_3C\equiv CCH_3$	54.09		27	-32.2	0.6910[20/4]	1.3921[20]	al, eth	B1[4], 971
4203	2-Butyne, 1-chloro $CH_3C\equiv CCH_2Cl$	88.54		104-6		1.0152[20]	1.4581[20]	al, eth, ace	B1[4], 973

No.	Name, Synonyms, and Formula	Mol. wt.	Color, crystalline form, specific rotation and λ_{max} (log ε)	b.p. °C	m.p. °C	Density	n_D	Solubility	Ref.
4204	2-Butyne, 1,4-dibromo BrCH₂C≡CCH₂Br	211.90	....	92[15]	......	2.014[18]	1.588[18]	eth, ace, chl	B1[4], 974
4205	2-Butyne, 1,4-dichloro ClCH₂C≡CCH₂Cl	122.98		165-6, 73[24]		1.258[20/4]	1.5058[20]	eth, ace, chl	B1[4], 973
4206	2-Butyne, 1,4-diiodo ICH₂C≡CCH₂I	305.89	nd (al)	70-2[0 1]	53	....		al, eth, ace, chl	B1[4], 975
4207	2-Butyne dinitrile NCC≡CCN	76.06		76-6.5[78.3]	20.5-1	0.9703[25/4]	1.46471[25]	....	B2[4], 2295
4208	2-Butyne, 1,4-diol HOCH₂C≡CCH₂OH	86.09	pl (bz, AcOEt)	238, 145[15]	58	....	1.4804[20]	w, al, ace	B1[4], 2687
4209	2-Butyne, 1,4-diol,diacetate CH₃CO₂CH₂-C≡C-CH₂O₂CCH₃	170.16		122-3[10]			1.4611[20]		B2[4], 244
4210	3-Butyne, 1,2-diol HC≡CCHOHCH₂OH	86.09		64-6[0 2]	40			w, al	B1[4], 2689
4211	2-Butyne, perfluoro CF₃C≡CCF₃	162.04	....	− 24.6	− 117.4			al, eth, ace, aa	B1[4], 972
4212	2-Butynedioic acid or Acetylene dicarboxylic acid HO₂CC≡CCO₂H	114.06	pl (eth)		179			w, al, eth	B2[4], 2290
4213	2-Butynedioic acid, diethyl ester H₅C₂O₂CC≡CCO₂C₂H₅	170.17		184[200]	1-2	1.0675[20/4]	1.4425[20]	al, eth	B2[4], 2294
4214	2-Butynedioic acid, dimethyl ester CH₃O₂CC≡CCO₂CH₃	142.11		195-8d, 98[20]		1.1564[20/4]	1.4434[20]	al, eth	B2[4], 2291
4215	2-Butynoic acid or Tetrolic acid CH₃C≡CCO₂H	84.08	pl (eth, peth)	203, 99-100[18]	78	0.9641[20/4]	....	w, al, eth, chl	B2[4], 1690
4216	2-Butynoic acid, ethyl ester CH₃C≡CCO₂C₂H₅	112.14		163, 105[190]	......	0.9641[20/4]	1.4372[20]	....	B2[4], 1691
4217	2-Butyn-1-ol CH₃C≡CCH₂OH	70.09		143, 52-3[14]	− 2.2	0.9370[20/4]	1.4530[20]	al, eth	B1[4], 2220
4218	3-Butyn-1-ol HC≡CCH₂CH₂OH	70.09		129	− 63.6	0.9257[20/4]	1.4409[20]	w, al	B1[4], 2219
4219	3-Butyn-2-ol HC≡C-CHOHCH₃	70.09		107		0.8858[20]	1.4265[20]	w, al, eth	B1[4], 2218
4220	3-Butyn-2-ol-2, methyl (CH₃)₂COHC≡CH	84.13		104, 56[97]	+ 3	0.8618[20/4]	1.4207[20]	w, al	B1[4], 2229
4221	2-Butynyl methyl ether CH₃C≡CCH₂OCH₃	84.13		99-100, 33[27]		0.8496[20/4]	1.4262[20]	al, eth, bz	B1[3], 1973
4222	Butyraldehyde or Butanal CH₃CH₂CH₂CHO	72.12		75.7	− 99	0.8170[20/4]	1.3843[20]	w, al, eth, ace, bz	B1[4], 3229
4223	Butyraldehyde phenyl hydrazone C₃H₇CH=NNHC₆H₅	162.23		190-5[80], 152[14]	93-5				B15[3], 80
4224	Butyraldehyde oxime or Butyraldoxime CH₃CH₂CH₂CH=NOH	87.12		152[715]	− 29.5	0.923[20/4]		al, eth, ace, bz	B1[4] 3234
4225	Butyraldehyde, 2-bromo CH₃CH₂CHBrCHO	151.00		33[17]	......	1.469[20]	1.4683[20]	eth, ace, bz	B1[4], 3241
4226	Butyraldehyde, 3-chloro, diethyl acetal CH₃CHClCH₂CH(OC₂H₅)₂	180.68		70-1[12]		0.9709[20/4]	1.4210[20]	al, eth, ace, bz	B1[4], 3239
4227	Butyraldehyde, 4-chloro ClCH₂CH₂CH₂CHO	106.55		50-1[13]		1.107[8.15/15]	1.4466[8.5]	al, eth, ace	B1[4], 3240
4228	Butyraldehyde, 2,3-dichloro CH₃CHClCHClCHO	141.00		58-60[26]		1.2666[21/4]	1.4618[21]	al, eth, ace, chl	B1[4], 3240
4229	Butyraldehyde, 2-ethyl (C₂H₅)₂CHCHO	100.16		117-9[160]		0.8110[20/4]	1.4025[20]	al, eth	B1[4], 3310
4230	Butyraldehyde, 3-hydroxy or Aldol CH₃CH(OH)CH₂CHO	88.12		83[20], d85		1.103[20/4]	1.4238[20]	w, al, eth, ace	B1[4], 3984
4231	Butyraldehyde, 2-methyl (dl) CH₃CH₂CH(CH₃)CHO	86.14		92-3, 54[200]		0.8029[20/4]	1.3869[20]	al, eth, ace	B1[4], 3286
4232	Butyraldehyde, 3-methyl or Isovaleraldehyde (CH₃)₂CH₂CH₂CHO	86.14		92.5	− 51	0.7977[20/4]	1.3902[20]	al, eth	B1[4], 3291
4233	Butyraldehyde, 3-methyl,oxime or Isovaleraldoxime (CH₃)₂CH₂CH₂CH=NOH	101.15		161.3	48.5	0.8934[20/4]	1.4367[20]	al, eth, ace	B1[4], 3293
4234	Butyraldehyde, 2,2,3-trichloro or n-Butyl chloral CH₃CHClCCl₂CHO	175.44		163-5, 49[8]		1.3956[20/4]	1.4755[20]	w, al eth	B1[4], 3241
4235	Butyraldehyde, 2,2,3-trichloro,hydrate CH₃CHClCCl₂CH(OH)₂	192.46	rh pl or lf (w)	d	78	1.694[20/4]		w, al, eth	B1[3], 2768
4236	Isobutyraldehyde (CH₃)₂CHCHO	72.11	....	64.2-4.6	......	0.7938[20/4]	1.3730[20]	w, eth, ace, chl	B1[4], 3262

No.	Name, Synonyms, and Formula	Mol. wt.	Color, crystalline form, specific rotation and λ_{max} (log ε)	b.p. °C	m.p. °C	Density	n_D	Solubility	Ref.
4237	Isobutyraldehyde, 2-chloro $(CH_3)_2CClCHO$	106.55	....	90	....	$1.053^{15/4}$	1.4160^{16}	al, eth	B1[4], 3267
4238	Isobutyraldehyde, 3-(4-isopropylphenyl) or Cyclamenaldehyde $4-(CH_3)_2CHC_6H_4CH_2CH(CH_3)CHO$	190.29	....	$133-7^{99}$, 115^9	....	0.951^{15}	1.5068^{20}	al, eth, bz	B7[3], 1200
4239	Isobutyraldehyde, 2-chloro $(CH_3)_2CClCHO$	106.55							B1[4], 3267
4240	Butyramide or Butanoic acid, amide $CH_3CH_2CH_2CONH_2$	87.12	lf (bz)	216	114.8	0.8850^{120}	1.4087^{130}	al	B2[4], 804
4241	Butyramide, α-bromo $CH_3CH_2CHBrCONH_2$	166.03	lf (bz), nd (ace)	....	112—3			w, al, eth, ace, bz	B2[4], 834
4242	Butyramide, 2-bromo-2-ethyl or Neuronal $(C_2H_5)_2CBrCONH_2$	194.08	....		67			al, eth, bz	B2[3], 755
4243	Butyramide, 2-bromo-3-methyl (dl) $(CH_3)_2CHCHBrCONH_2$	180.05	lf (bz)		133			w, lig	B2[4], 906
4244	Butyramide, 2-bromo-2-iso-propyl or Neodorme $(CH_3)_2CHCBr(C_2H_5)CONH_2$	208.11	nd (sub)	sub	50-1			al, eth, ace, bz	B2[4], 979
4244a	Butyramide, 2-bromo-N-methyl-N-phenyl $CH_3CH_2CHBrCON(CH_3)C_6H_5$	256.15	cr (lig)		44			to, lig	B12, 254
4245	Butyramide, N,N-diethyl $CH_3CH_2CH_2CON(C_2H_5)_2$	143.23	....	206, 97^{16}	....		1.4403^{25}	w, al	B4[4], 354
4246	Butyramide, N,N-dimethyl $CH_3CH_2CH_2CON(CH_3)_2$	115.18	....	185-8, $124-5^{100}$	−40	$0.9064^{25/4}$	1.4391^{25}	w, al, eth, ace, bz	B4[4], 185
4247	Butyramide, 2,3-dimethyl $(CH_3)_2CHCH(CH_3)CONH_2$	115.18	pl (ace-peth)		130.9			al, eth	B2[3], 761
4248	Butyramide, 3,3-dimethyl or tert-Butylacetamide $(CH_3)_3CCH_2CONH_2$	115.18	lf (w or ace-peth)		134			al	B2[4], 956
4249	Butyramide, 2-ethyl-N,N-diethyl $(C_2H_5)_2CHCON(C_2H_5)_2$	171.29	....	$??0-1$, 100^{11}	....			al, eth, bz	B4, 111
4250	Butyramide, 3-methyl $(CH_3)_2CHCH_2CONH_2$	101.15	mcl lf (al)	224-8	137			w, al, eth, peth	B2[4], 902
4251	Butyramide, 3-methyl-2-phenyl $(CH_3)_2CHCH(C_6H_5)CONH_2$	177.25	nd (dil al)	$180—2^{14}$	111-2			al, eth	B9[3], 2518
4252	Butyramide, 2-phenoxy $CH_3CH_2CH(OC_6H_5)CONH_2$	179.22	nd (w or al)		123			al, eth, ace, chl	B6[4], 645
4253	Butyramide, 4-phenoxy $C_6H_5OCH_2CH_2CH_2CONH_2$	179.22	lf (dil al), nd (bz)		80			al	B6[3], 617
4254	Butyramide, N-phenyl or Butyranilide $C_3H_7CONHC_6H_5$	163.22	mcl pr (al, bz, eth)	189^{15}	97	1.134		al, eth	B12[3], 474
4255	Butyramide, 2-phenyl $CH_3CH_2CH(C_6H_5)CONH_2$	163.22	cr	185^{16}	86			al	B9[3], 2465
4256	Butyramide, 3-phenyl $CH_3CH(C_6H_5)CH_2CONH_2$	163.22	nd (dil al)		106-7			al	B9[3], 2459
4257	Butyramide, 4-phenyl $C_6H_5CH_2CH_2CH_2CN$	163.22	pl (w)		84.5			al, eth	B9[3], 2453
4258	Butyramide, 2,2,3,3-tetramethyl $CH_3C(CH_3)_2C(OH)_2CONH_2$	143.23	nd (peth-al)		201-2			al	B2[4], 1018
4259	Isobutyramide $(CH_3)_2CHCONH_2$	87.12	....		127-9			chl	B2[4], 852
4260	Isobutyramide, 2-bromo or 2-Bromo-2-methylpropionamide $(CH_3)_2CBrCONH_2$	166.03	pr (chl)	145^{17}	148			al, chl	B2[4], 863
4262	Isobutyramide, N-phenyl $(CH_3)_2CHCONHC_6H_5$	163.22	mcl pr (al, eth), nd (lig)		106-7			al, eth	B12[3], 475
4263	Butyric acid or Butanoic acid $CH_3CH_2CH_2CO_2H$	88.12	....	165.5	−4.5	$0.9577^{20/4}$	1.3980^{20}	al, eth	B2[4], 779
4264	Butyric acid, 2-acetyl, ethyl ester $CH_3CH_2CH(OCCH_3)CO_2C_2H_5$	158.20	....	$85-7^{13}$	....	$0.9924^{15/15}$	1.4237^{15}	al, eth	B2[4], 1592
4265	Butyric acid, 2-acetyl-3-oxo-ethyl ester or Ethyl-α-acetyl aceto acetate $(CH_3CO)_2CHCO_2C_2H_5$	172.19	....	209^{11}, 104^{16}	....	$1.1045^{20/4}$	1.4690^{20}	al, eth, bz	B2[4], 1781
4266	Butyric acid, allyl ester or Allylbutyrate, allylbutanoate $CH_3CH_2CH_2CO_2CH_2CH=CH_2$	128.17	....	$142-3^{772}$, 44.5^{15}	....	$0.9017^{20/4}$	1.4158^{20}	al, eth	B2[4], 793
4267	Butyric acid, 2-amino- (d) or 2-Aminobutanoic acid $C_2H_5CHNH_2CO_2H$	103.12	lf (dil al), $[\alpha]^{16}_D$ +8.4	....	292d			w	B4[3], 1294

No.	Name, Synonyms, and Formula	Mol. wt.	Color, crystalline form, specific rotation and λ_{max} (log ε)	b.p. °C	m.p. °C	Density	n_D	Solubility	Ref.
4268	Butyric acid, 2-amino- (dl) $C_2H_5CHNH_2CO_2H$	103.12	lf (w)	sub	304d			w	B4[3], 1296
4269	Butyric acid, 2-amino- (l) $C_2H_5CHNH_2CO_2H$	103.12	lf (w-al), cr (al), $[\alpha]^{20}_D$ − 14.9		292d			w	B4[3], 1296
4270	Butyric acid, 3-amino- (d) $CH_3CH(NH_2)CH_2CO_2H$	103.12	pr (MeOH), $[\alpha]^{20}_D$ + 35.3		d at 220				B4[3], 1312
4271	Butyric acid, 3-amino- (dl) $CH_3CH(NH_2)CH_2CO_2H$	103.12	nd (al)		193-4			w	B4[3], 1312
4272	Butyric acid, 3-amino- (l) $CH_3CHNH_2CH_2CO_2H$	103.12	pr (MeOH), $[\alpha]^{20}_D$ − 35.2 (w, c=10)		d at 220			w	B4[3], 1312
4273	Butyric acid, 3-amino-2-hydroxy or 3-Methyl iso-serine $CH_3CHNH_2CHOHCO_2H$	119.12	pr (dil al)		200d			w	B4, 513
4274	Butyric acid, 3-amino-3-methyl $(CH_3)_2CNH_2CO_2H$	117.15	pr (w + l), cr (dil al), nd (eth-al)		217			w	B4[3], 1364
4275	Butyric acid, 4-amino or Piperidinic acid $H_2NCH_2CH_2CH_2CO_2H$	103.12	pr or nd (dil al), lf (MeOH-eth)		203d			w	B4[3], 1316
4276	Butyric acid, 4-amino-2-hydroxy $H_2NCH_2CH_2CHOHCO_2H$	119.12	pr (w or dil al)		214			w	B4[1], 548
4277	Butyric acid, 4-amino-3-hydroxy (dl) $H_2NCH_2CHOHCH_2CO_2H$	119.12	pr (w), cr (dil al)		218				B4[3], 1635
4278	Butyric acid anhydride or Butyric anhydride $(C_3H_7CO)_2O$	158.20		199-201	− 75	0.9668[20/4]	1.4070[20]	eth	B2[4], 802
4279	Butyric acid, 2-benzoyl, ethyl ester $CH_3CH_2CH(COC_6H_5)CO_2C_2H_5$	220.27		152[7]		1.0706[15/4]	1.509[15]	eth	B10[3], 3065
4280	Butyric acid, 4-benzoyl $C_6H_5CO(CH_2)_3CO_2H$	192.22	pl (w)		128-9			w	B10[3], 3060
4281	Butyric acid, benzyl ester or Benzyl butyrate $C_3H_7CO_2CH_2C_6H_5$	178.23		238-40, 105[7]		1.0111[20/4]	1.4920[20]	al, eth	B6[4], 2266
4282	Butyric acid, 2-bromo (d) or α-Bromobutyric acid $CH_3CH_2CHBrCO_2H$	167.01	$[\alpha]^{20}_D$ + 35.2 (eth, c=20)	105-7[15]		1.568[20/4]	1.4483[20]	al	B2[3], 630
4283	Butyric acid, 2-bromo (dl) $CH_3CH_2CHBrCO_2H$	167.01		217d, 108[13]	− 4	1.5669[20/20]		al, eth	B2[4], 833
4284	Butyric acid, 2-bromo, ethyl ester $CH_3CH_2CHBrCO_2C_2H_5$	195.06		177.5[765], 43-4[5.5]		1.3297[20/20]	1.4475[20]	al, eth	B2[4], 834
4285	Butyric acid, 2-bromo, methyl ester $CH_3CH_2CHBrCO_2CH_3$	181.04		170-2, 75-8[18]		1.4528[20]	1.4029[25]	al	B2[4], 834
4286	Butyric acid, 2-bromo-3-methyl- (d) $(CH_3)_2CHCHBrCO_2H$	181.04	pr (peth), $[\alpha]^{20}_D$ + 22.8 (bz, p=4)	230, 95-100[2]	44-5			al, eth, ace, bz	B2[3], 705
4287	Butyric acid, 2-bromo-3-methyl- (dl) $(CH_3)_2CHCHBrCO_2H$	181.04	pr (eth or chl)	230d, 136-40[25]	44	1.459[20]		al, eth, ace, bz	B2[4], 905
4288	Butyric acid, 2-bromo-3-methyl- (l) $(CH_3)_2CHCHBrCO_2H$	181.04	cr (peth), $[\alpha]^{20}_D$ − 21.6 (bz, p=4)	150[40], 119-20[14]	43-4			al, eth, ace, bz	B2[4], 905
4289	Butyric acid, 2-bromo-3-methyl, ethyl ester $(CH_3)_2CHCHBrCO_2C_2H_5$	209.09		186, 73-4[12]		1.2760[20/4]	1.4496[20]	al, eth	B2[4], 906
4290	Butyric acid, 2-bromo-3-methyl, methyl ester $(CH_3)_2CHCHBrCO_2CH_3$	195.06		176-8, 64-5[11]		1.353[13/13]	1.4530[20]	al, eth	B2[3], 706
4291	Butyric acid, 3-bromo-3-methyl $(CH_3)_2CBrCH_2CO_2H$	181.04	nd (lig)		73.5			al, eth, bz	B2[4], 905
4292	Butyric acid, 4-bromo $BrCH_2CH_2CH_2CO_2H$	167.01		124-7[7]	33				B2[4], 835
4293	Butyric acid, 4-bromo, methyl ester $BrCH_2CH_2CH_2CO_2CH_3$	181.04		186-7, 86[15]		1.371[25]	1.4567[25]	al	B2[4], 835
4294	Butyric acid, butyl ester or Butyl butyrate $C_3H_7CO_2C_4H_9$	144.22		166.6, 55[11]	− 91.5	0.8700[20/4]	1.4075[20]	al, eth	B2[4], 789

No.	Name, Synonyms, and Formula	Mol. wt.	Color, crystalline form, specific rotation and λ_{max} (log ε)	b.p. °C	m.p. °C	Density	n_D	Solubility	Ref.
4295	Butyric acid, sec-butyl ester-(d) or sec-Butyl butyrate.... $C_3H_7CO_2$-sec-C_4H_9	144.22	$[\alpha]^{20}_D$ + 22	151.5[747], 54[18]	− 91.5	0.8737[13/4]	1.4011[20]	al, bz, Py	B2[3], 600
4296	Butyric acid, iso-butyl ester or iso-Butyl butyrate....... $C_3H_7CO_2$-i-C_4H_9	172.27	....	157	......	0.8364[18/4]	1.4032[20]	al, eth	B2[4], 790
4297	Butyric acid, sec-butyl ester-(dl).................... $C_3H_7CO_2$-sec-C_4H_9	144.22	....	152.5, 52[16]	......	0.8609[20/4]	1.4019[20]	....	B2[4], 790
4298	Butyric acid, tert-butyl ester or tert-Butyl butyrate...... C_3H_7-$CO_2C(CH_3)_3$	144.22	....	145-7	......	....	1.4007[17.5]	al, eth, ace	B2[4], 790
4299	Butyric acid, 2-chloro .... $CH_3CH_2CHClCO_2H$	122.55	....	189[627], 101[15]	......	1.1796[20/4]	1.4411[20]	al, eth	B2[4], 821
4300	Butyric acid, 2-chloro, ethyl ester .... $CH_3CH_2CHClCO_2C_2H_5$	150.61	....	163-4, 63[70]	......	1.0560[20]	1.4248[20]	al, eth	B2[4], 822
4301	Butyric acid, 2-chloro, methyl ester or Methyl 2-chloro-butyrate .... $CH_3CH_2CHClCO_2CH_3$	136.58	....	145-6[756]	......	1.0979[14]	1.4247[20]	al, eth	B2[4], 821
4302	Butyric acid, 2-chloro-2-methyl .... $CH_3CH_2CCl(CH_3)CO_2H$	136.58	....	200-5[754/d]	......	1.1204[20/4]	1.4445[20]	al, eth	B2[3], 688
4303	Butyric acid, 2-chloro-2-methyl, ethyl ester .... $CH_3CH_2C(CH_3)ClCO_2C_2H_5$	164.63	....	175[747]	......	1.069[14]	1.4388[11]	al, eth	B2[3], 688
4304	Butyric acid, 2-chloro-2-methyl-3-oxo, ethyl ester or Ethyl α-chloro-α-acetopropionate.................... $CH_3COC(CH_3)ClCO_2C_2H_5$	164.63	....	178-9[756]	......	1.021[13]	....	al, eth	B3[2], 433
4305	Butyric acid, 2-chloro-3-methyl or 2-chloroisovaleric acid . $(CH_3)_2CHCHClCO_2H$	136.58	....	210-2[756], 126[12]	20-2	1.135[11]	1.4450[11]	al, eth	B2[4], 904
4306	Butyric acid, 3-chloro (dl) .... $CH_3CHClCH_2CO_2H$	122.55	cr (eth)	116[22]	16	1.1898[20/4]	1.4221[20]	al, eth	B2[4], 823
4307	Butyric acid, 3-chloro, ethyl ester .... $CH_3CHClCH_2CO_2C_2H_5$	150.61	....	109, 65[15]	......	1.0517[10/4]	1.4246[20]	al	B2[4], 824
4308	Butyric acid, 3-chloro, methyl ester .... $CH_3CHClCH_2CO_2CH_3$	136.58	....	155-6	......	1.0996[20/4]	1.4258[20]	eth	B2[4], 824
4309	Butyric acid, 4-chloro .... $ClCH_2CH_2CH_2CO_2H$	122.55	....	196[22], 68[0.2]	16	1.2236[20/4]	1.4642[20]	eth	B2[4], 825
4310	Butyric acid, 4-chloro, ethyl ester .... $ClCH_2CH_2CH_2CO_2C_2H_5$	150.61	....	186,77[10]	......	1.0756[20/4]	1.4311[20]	al, eth, ace	B2[4], 825
4311	Butyric acid, 4-chloro, methyl ester .... $ClCH_2CH_2CH_2CO_2CH_3$	136.58	....	175-6[764], 55[4]	......	1.1201[20/4]	1.4321[20]	al, eth, ace	B2[4], 825
4312	Butyric acid, cyclohexyl ester or Cyclohexyl butyrate ..... $C_3H_7CO_2C_6H_{11}$	170.25	....	212[750]	......	0.9572[20/4]	....	al	B6[4], 37
4313	Butyric acid, 2,3-dibromo (high m.p.) .... $CH_3CHBrCHBrCO_2H$	245.91	nd (eth)	100-10[20]	87	....	....	al, eth, bz	B2[4], 837
4314	Butyric acid, 2,3-dibromo (low m.p.) .... $CH_3CHBrCHBrCO_2H$	245.91	nd (lig)	....	59-60	....	....	al, eth	B2[4], 837
4315	Butyric acid, 2,3-dibromo, ethyl ester .... $CH_3CHBrCHBrCO_2C_2H_5$	273.96	nd	113[30]	58-9	....	....	al, eth	B2[4], 837
4316	Butyric acid, 2,4-dibromo, ethyl ester .... $BrCH_2CH_2CHBrCO_2C_2H_5$	273.96	....	149-50[52]	......	1.6990[20/0]	1.4960[20]	al, eth	B2[4], 838
4317	Butyric acid, 2,2-dichloro .... $CH_3CH_2CCl_2CO_2H$	157.00	....	107-10[14]	......	1.389[20/4]	....	al, eth	B2[2], 254
4318	Butyric acid, 2,3-dichloro (high m.p.) .... $CH_3CHClCHClCO_2H$	157.00	pr (dil al)	131.5[20]	78	....	....	al, eth	B2[4], 827
4319	Butyric acid, 2,3-dichloro (low m.p.) .... $CH_3CHClCHClCO_2H$	157.00	pr (dil al)	124-5[20]	63	....	....	al, eth, bz, chl	B2[4], 827
4320	Butyric acid, 2,2-diethyl-3-oxo, ethyl ester .... $CH_3COC(C_2H_5)_2CO_2C_2H_5$	186.25	....	215-6[744], 64[3]	......	0.9717[18/4]	1.4326[17]	**al, eth**	B3[4], 1625
4321	Butyric acid, 2,2-dimethyl .... $CH_3CH_2C(CH_3)_2CO_2H$	116.16	....	186, 80[11]	− 14	0.9276[20/4]	1.4145[20]	....	B2[4], 954
4322	Butyric acid, (2,2-dimethylpropyl) ester or neo-pentyl butyrate .... $C_3H_7CO_2CH_2C(CH_3)_3$	158.24	....	165-6	......	0.8719[0]	....	....	B2, 272
4323	Butyric acid, 2,2-dimethyl-3-oxo, ethyl ester .... $CH_3COC(CH_3)_2CO_2C_2H_5$	158.20	....	184, 40-1[3]	......	0.9773[20/20]	1.4180[20]	al, eth	B3[4], 1594
4324	Butyric acid, 2,3-dimethyl .... $(CH_3)_2CH-CH(CH_3)CO_2H$	116.16	....	191.7	− 1.5	0.9275[20/4]	1.4146[20]	al, eth	B2[4], 958
4325	Butyric acid, 3,3-dimethyl or tert-Butylacetic acid ...... $(CH_3)_3CCH_2CO_2H$	116.16	....	190, 96[26]	6-7	0.9124[20/4]	1.4096[20]	al, eth	B2[4], 955
4326	Butyric acid, 3,3-dimethyl-2-oxo .... $(CH_3)_3CCOCO_2H$	130.14	....	189[747], 80[15]	90-1	....	....	eth, bz, chl	B3[4], 1595

No.	Name, Synonyms, and Formula	Mol. wt.	Color, crystalline form, specific rotation and λ_{max} (log ε)	b.p. °C	m.p. °C	Density	n_D	Solubility	Ref.
4327	Butyric acid, 2,3-epoxy-3-phenyl, ethyl ester .... $CH_3C(C_6H_5)CH\text{-}CO_2C_2H_5$	206.24	....	272—5, 147-9[22]	....	1.0442[20]	1.5182[20]	....	B18[2], 275
4328	Butyric acid, ethyl ester or Ethylbutyrate .... $C_3H_7CO_2C_2H_5$	116.16	....	121-6, 48.8[50]	− 100.8	0.8785[20/4]	1.4000[20]	al, eth	B2[4], 787
4329	Butyric acid, 2-ethyl or Diethyl acetic acid .... $(C_2H_5)_2CHCO_2H$	116.10	....	194, 90[13]	− 31.8	0.9239[20/4]	1.4132[20]	**al, eth**	B2[4], 951
4330	Butyric acid, 2-ethyl-2-methyl .... $(C_2H_5)_2C(CH_3)CO_2H$	130.19	....	208.5, 104[13]	<−20	....	1.4250[20]	al	B2[4], 981
4331	Butyric acid, 4-fluoro .... $FCH_2CH_2CH_2CO_2H$	106.10	....	76-8[5]	....	....	1.3993[25]	al, eth	B2[4], 809
4332	Butyric acid, furfuryl ester .... $C_3H_7CO_2CH_2(C_4H_3O)$	168.20	....	212-3[764]	....	1.0530[20/4]	....	al, eth	B17[4], 1247
4333	Butyric acid, heptafluoro .... $F_3CCF_2CF_2CO_2H$	214.04	....	120[735]	− 17.5	1.651[20/4]	1.295[15]	w, eth, to	B2[4], 810
4334	Butyric acid, heptafluoro, ethyl ester .... $F_3CCF_2CF_2CO_2C_2H_5$	242.09	....	95	....	....	1.3011[20]	eth, ace	B2[4], 813
4335	Butyric acid, heptafluoro, methyl ester .... $CF_3CF_2CF_2CO_2CH_3$	....	....	80	....	1.483[20/4]	1.295[20]	eth, ace	B2[4], 812
4336	Butyric acid, heptyl ester or Heptyl butyrate .... $C_3H_7CO_2C_7H_{15}$	186.30	....	225.8, 105[10]	− 57.5	0.8637[20]	1.4231[20]	al	B2[4], 791
4337	Butyric acid, hexyl ester or Hexyl butyrate............... $C_3H_7CO_2C_6H_{13}$	172.27	....	208	− 78	0.8652[20]	1.4160[15]	al	B2[4], 791
4338	Butyric acid, 2-hydroxy (dl) .... $CH_3CH_2CHOHCO_2H$	104.12	nd (CCl$_4$)	266d, 140[14]	44-4.5	1.125[20]	....	w, al, eth	B3[4], 754
4339	Butyric acid, 2-hydroxy, ethyl ester (d) .... $CH_3CH_2CHOHCO_2C_2H_5$	132.16	[α]$^{22}_D$ + 8.4	165-70	....	0.978[15]	1.4101	al	B3[4], 756
4340	Butyric acid, 2-hydroxy, ethyl ester (dl) .... $CH_3CH_2CHOHCO_2C_2H_5$	132.16	....	167, 74.5[24]	....	1.0069[20/4]	1.4179[20]	al	B3[4], 756
4341	Butyric acid, 2-hydroxy-3-methyl (d) .... $(CH_3)_2CHCHOHCO_2H$	118.13	cr (eth-peth), [α]$^{20}_D$ − 1.81 (w, c=12)	124-5[13]	69.5	....	....	w, al, eth, ace	B3[4], 830
4342	Butyric acid, 2-hydroxy-3-methyl (dl) .... $(CH_3)_2CHOHCO_2H$	118.13	rh bipyr	....	86	....	....	w, al, eth, ace	B3[4], 830
4343	Butyric acid, 3-hydroxy-(dl) .... $CH_3CHOHCH_2CO_2H$	104.12	....	130[12-14], 94-6[0.1]	48-50	....	1.4424[20]	w, al, eth	B3[4], 760
4344	Butyric acid, 3-hydroxy-(l) .... $CH_3CHOHCH_2CO_2H$	104.12	[α]$^{25}_D$ − 24.5 (w, c=5)	....	49-50	....	....	w, al, eth	B3[4], 760
4345	Butyric acid, 3-hydroxy, ethyl ester (dl) .... $CH_3CHOHCH_2CO_2C_2H_5$	132.16	....	184-5[755], 76-7[15]	....	1.017[20/4]	1.4182[20]	al, w	B3[4], 762
4346	Butyric acid, 3-hydroxy, methyl ester (l) .... $CH_3CHOHCH_2CO_2CH_3$	118.13	[α]$^{20}_D$ − 21.09	76-7[20]	....	1.058[20/20]	....	w, al, eth, bz	B3[3], 569
4347	Butyric acid, 3-hydroxy-3-methyl .... $(CH_3)_2COHCH_2CO_2H$	118.13	....	162[12]	<−32	0.9384[20/4]	1.5081[20]	w, al, eth	B3[4], 827
4348	Butyric acid, 4-hydroxy .... $4HOCH_2CH_2CH_2CO_2H$	104.12	....	d at 178-80	<−17	....	....	....	B3[4], 774
4349	Butyric acid, 4-hydroxy, lactone or γ-Butyrolactone...... $\overline{CH_2CH_2CH_2CO}$	86.09	....	206, 89[12]	− 42	1.1286[16/0]	1.4341[20]	w, al, eth, ace, bz	B17[4], 4159
4350	Butyric acid, 4-hydroxy-2-methylene, lactone or γ-Butyrolactone $\overline{CH_2CH_2C(=CH_2)CO}$	98.10	....	85-6[10]	....	1.1206[20]	1.4650[20]	w, al, eth, ace, bz	B17[4], 4304
4351	Butyric acid, methyl ester or Methyl butyrate .... $C_3H_7CO_2CH_3$	102.13	....	102.3	− 84.8	0.8984[20/4]	1.3878[20]	**al, eth**	B2[4], 786
4352	Butyric acid, 2-methyl (d) .... $CH_3CH_2CH(CH_3)CO_2H$	102.13	[α]$^{15}_D$ + 19.2 (w)	176, 77[12]	....	0.9419[20/4]	1.4058[20]	al, eth	B2[4], 888
4353	Butyric acid, 2-methyl (dl) .... $CH_3CH_2CH(CH_3)CO_2H$	102.13	....	177	<−80	0.9410[20/4]	1.4051[20]	al, eth	B2[4], 889
4354	Butyric acid, 2-methyl (l) .... $CH_3CH_2CH(CH_3)CO_2H$	102.13	[α]$^{20}_D$ − 24 (w, c=0.9)	176-7, 71-2[12]	....	0.9340[20/4]	1.4042[25]	al, eth	B2[4], 888
4355	Butyric acid, 2-methyl butyl ester (d) .... $C_3H_7CO_2CH_2CH(CH_3)C_2H_5$	158.24	[α]$^{20}_D$ + 3.5	179[765]	....	0.8620[20/4]	1.4135[20]	....	B2[3], 304
4356	Butyric acid, 2-methyl, ethyl ester (d) .... $CH_3CH_2CH(CH_3)CO_2C_2H_5$	130.19	[α]$^{26}_{5892}$ + 5.16	131-3[730], 35[16]	....	0.8689[25/4]	1.3964[20]	al, bz	B2[4], 890
4357	Butyric acid, 2-methyl butyl ester (dl) .... $C_3H_7CO_2CH_2CH(CH_3)C_2H_5$	158.24	....	166-7	....	0.862[20/4]	1.4100[25]	....	B2, 304

No.	Name, Synonyms, and Formula	Mol. wt.	Color, crystalline form, specific rotation and λ_{max} (log ε)	b.p. °C	m.p. °C	Density	n_D	Solubility	Ref.
4358	Butyric acid, 2-(2-methyl butyl) ester or *tert*-Pentyl butyrate $C_3H_7CO_2C(CH_3)_2C_2H_5$	158.24	....	164	......	$0.8646^{15/0}$	....	....	B2[3], 602
4359	Butyric acid, 3-methyl or Isovaleric acid $(CH_3)_2CHCH_2CO_2H$	102.13	....	176.7	−29.3	$0.9286^{20/4}$	1.4033^{20}	al, eth, chl	B2[4], 895
4360	Butyric acid, 3 methyl, anhydride $[(CH_3)_2CHCH_2CO]_2O$	186.25	....	215^{762}, $102-3^{15}$	......	$0.9327^{20/4}$	1.4043^{20}	eth	B2[4], 901
4361	Butyric acid, 3-methyl, isopropyl ester $(CH_3)_2CHCH_2CO_2CH(CH_3)_2$	144.22	....	142^{756}, $69-70^{55}$	......	0.8538^{17}	$1.3960^{20.5}$	al, eth, ace	B2[3], 698
4362	Butyric acid, 3-methylbutyl ester or Isopentyl butyrate $(CH_3)_2CHCH_2CO_2CH_2CH_2CH(CH_3)_2$	158.24	....	178.5, $65-8^{12}$	......	$0.8651^{20/4}$	1.4110^{20}	al, eth	B2[4], 791
4363	Butyric acid, 3-methyl, isobutyl ester $(CH_3)_2CHCH_2CO_2$-*i*-C_4H_9	158.24	....	171.4, $60-2^{12}$	......	$0.8736^{20/4}$	1.4057^{20}	**al, eth**, ace	B2[4], 790
4364	Butyric acid, 3-methyl, ethyl ester or Ethyl isovalerate $(CH_3)_2CHCH_2CO_2C_2H_5$	130.19	....	134.7	−99.3	$0.8656^{20/4}$	1.3962^{20}	al, eth	B2[4], 898
4365	Butyric acid, 3-methyl, methyl ester or Methyl isovalerate $(CH_3)_2CHCH_2CO_2CH_3$	116.16	....	116.7	......	$0.8808^{20/4}$	1.3927^{20}	al, eth, ace	B2[4], 897
4366	Butyric acid, 3-methyl-2-oxo $(CH_3)_2CHCOCO_2H$	......	....	170.5, 73^{11}	31.0 — 1.5	$0.9968^{20/4}$	1.3850^{16}	w, al, eth	B3[4], 1577
4367	Butyric acid, 3-methyl, propyl ester or Propyl isovalerate $(CH_3)_2CHCH_2CO_2C_3H_7$	144.22	....	155.7, 40.5^{13}	......	$0.8617^{20/4}$	1.4031^{20}	al, eth	B2[4], 898
4368	Butyric acid, 3-methyl-2-phenyl $(CH_3)_2CHCH(C_6H_5)CO_2H$	178.23	pr (lig)	$159-60^{14}$	63	....	....	al, eth	B9[3], 2518
4369	Butyric acid, octyl ester or Octyl butyrate $C_3H_7CO_2C_8H_{17}$	200.33	....	244.1	−55.6	0.8629^{20}	$1.4267^{18/H9}$	al	B2[4], 791
4370	Butyric acid, 2-oxo or α-Ketobutyric acid $CH_3CH_2COCO_2H$	102.09	pl	$80-2^{16}$	31-2	$1.200^{17/4}$	1.3972^{20}	w, al	B3[4], 1524
4371	Butyric acid, 2-oxo, oxime $CH_3CH_2C(H)CO_2H$	117.11	nd (w) (diox)	....	164	....	....	al	B3[4], 1525
4372	Butyric acid, pentyl ester or Pentyl butyrate, amyl butyrate $C_3H_7CO_2C_5H_{11}$	158.24	....	186.4	−73.2	$0.8713^{15/4}$	1.4123^{20}	al, eth	B2[4], 790
4373	Butyric acid, 1-methylpentyl ester (d) or sec-Hexyl butyrate $C_3H_7CO_2CH(CH_3)C_4H_9$	172.27	$[\alpha]_D$ + 10.16	85^{20}	......	$0.8744^{21/4}$	....	....	B2[1], 120
4374	Butyric acid, 2-phenoxy $CH_3CH_2CH(OC_6H_5)CO_2H$	180.21	nd (w), pl (lig)	258	98	....	....	al, eth, ace, bz	B6[4], 644
4375	Butyric acid, 2-phenoxy, ethyl ester $CH_3CH_2CH(OC_6H_5)CO_2C_2H_5$	208.24	....	$250-1^{749}$, $87-90^2$	......	1.0388^{21}	....	**al, eth**, ace, chl	B6[4], 645
4376	Butyric acid, 4-phenoxy $C_6H_5OCH_2CH_2CH_2CO_2H$	180.21	pl (lig), cr (w)	$192-7^{15}$	64-5	....	....	al, eth, ace, bz	B6[4], 645
4377	Butyric acid, 4-phenoxy, ethyl ester $C_6H_5OCH_2CH_2CH_2CO_2C_2H_5$	208.24	....	$170-3^{25}$	......	$1.045^{35/25}$	1.491^{33}	al, eth	B6[2], 159
4378	Butyric acid, phenyl ester or Phenyl butyrate $C_3H_7CO_2C_6H_5$	164.21	....	$227-8$, 85^8	......	$1.0382^{15/4}$	$1.0267^{15/15}$	al, eth	B6[4], 615
4379	Butyric acid, 2-phenyl $CH_3CH_2CH(C_6H_5)CO_2H$	164.21	pl (eth)	$270-2$, $145-50^{14}$	47.5	....	....	eth, bz	B9[3], 2461
4380	Butyric acid, 2-phenyl, methyl ester $CH_3CH_2CH(C_6H_5)CO_2CH_3$	178.23	nd (dil al)	228	77-8	....	....	al, eth	B9[3], 2461
4381	Butyric acid, 3-phenyl (dl) $CH_3CH(C_6H_5)CH_2CO_2H$	164.20	....	$140-5^3$	46-7	1.0701^{20}	1.5155^{20}	....	B9[3], 2457
4382	Butyric acid, 4-phenyl $C_6H_5CH_2CH_2CH_2CO_2H$	164.21	lf (w)	290, 171^{15}	52	....	....	al, eth	B9[3], 2451
4383	Butyric acid, propyl ester or Propyl butyrate $C_3H_7CO_2C_3H_7$	130.19	....	143, 39.2^{14}	−97.2	$0.8730^{20/4}$	1.4001^{20}	**al, eth**	B2[4], 788
4384	Butyric acid, *iso*-propyl ester or *iso*-Propyl butyrate $C_3H_7CO_2CH(CH_3)_2$	130.19	....	130-1	......	$0.8588^{20/4}$	1.3936^{20}	al	B2[4], 789
4385	Butyric acid, 3-isopropyl-3-oxo, ethyl ester $CH_3COCH(i-C_3H_7)CO_2C_2H_5$	172.23	....	201^{758}, $97-8^{20}$	......	$0.9648^{18/4}$	$1.4256^{18.5}$	**al, eth**	B3[4], 1608
4386	Butyric acid, 3-thioxo, ethyl ester $CH_3CSCH_2CO_2C_2H_5$	146.21	dk red	75^{15}	......	$1.0554^{31/4}$	1.4712^{26}	al, eth	B3[4], 1552
4387	Butyric acid, 2,2,3-trichloro $CH_3CHClCCl_2CO_2H$	191.44	lf or nd (peth)	236-8	60	....	....	eth	B2[3], 629
4388	Butyric acid, 2,2,3-trichloro, ethyl ester $CH_3CHClCCl_2CO_2C_2H_5$	219.50	....	212, 101.9^{17}	......	$1.3138^{20/20}$	....	al, eth	B2, 281
4389	Butyric acid, 2,2,4-trichloro $ClCH_2CH_2CCl_2CO_2H$	191.44	cr (peth)	....	73-5	....	....	al, eth	B2, 281

No.	Name, Synonyms, and Formula	Mol. wt.	Color, crystalline form, specific rotation and λ_{max} (log ε)	b.p. °C	m.p. °C	Density	n_D	Solubility	Ref.
4390	Butyric acid, 2,3,3-trichloro .... $CH_3CCl_2CHClCO_2H$	191.44	pl (lig)	....	52	....	....	al, eth, ace, bz	B2, 281
4391	Butyric acid, 4,4,4-trichloro .... $Cl_3CCH_2CH_2CO_2H$	191.44	nd (w)	....	35	....	....	al, eth, chl	B2[4], 830
4392	Isobutyric acid or 2-Methyl propionic acid............ $(CH_3)_2CHCO_2H$	88.11	....	153.2, 53.7[10]	−46.1	0.9681[20/4]	1.3930[20]	al, eth, w	B2[4], 843
4393	Isobutyric acid, allyl ester or Allyl isobutyrate.......... $(CH_3)_2CHCO_2CH_2CH=CH_2$	128.17	....	133-5[755]	......	....	....	al, eth, ace	B2[3], 650
4394	Isobutyric acid anhydride or Isobutyric anhydride ...... $[(CH_3)_2CHCO]_2O$	158.20	....	181.5[734], 89-90[32]	−53.5	0.9535[20/4]	1.4061[19]	eth, chl	B2[4], 851
4395	Isobutyric acid, 2-amino .... $(CH_3)_2C(NH_2)CO_2H$	103.12	ta or pr (w)	sub 280	337 (cor)	....	....	w	B4[3], 1322
4396	Isobutyric acid, benzyl ester .... $(CH_3)_2CHCO_2CH_2C_6H_5$	178.23	....	114-5[20]	......	1.0075[15]	1.4883[20]	....	B6[4], 2267
4397	Isobutyric acid, 2-bromo or 2-Bromo-2-methyl propionic acid .... $(CH_3)_2CBrCO_2H$	167.01	cr (peth)	198-200, 115[24]	48-9	1.5225[60/60]	....	al, eth	B2[4], 862
4398	Isobutyric acid, 2-bromo, ethyl ester .... $(CH_3)_2C(Br)CO_2C_2H_5$	195.06	....	164[762], 70[20]	......	1.3182[20/4]	1.4446[20]	al, eth	B2[4], 862
4399	Isobutyric acid, isobutyl ester or Isobutyl isobutyrate..... $(CH_3)_2CHCO_2CH_2CH(CH_3)_2$	144.22	....	148.6, 36-40[11]	−80.6	0.8750[0/4]	1.3999[20]	al, eth, ace	B2[4], 847
4400	Isobutyric acid, tert-butyl ester or tert-Butyl isobutyrate $(CH_3)_2CHCO_2C(CH_3)_3$	144.22	....	126.7	......	....	1.3921[20]	al, eth, ace	B2[3], 648
4401	Isobutyric acid, 2-chloro .... $(CH_3)_2CClCO_2H$	122.55	....	118[50]	31	....	1.450[20]	al	B2[4], 858
4402	Isobutyric acid, 2-chloro, ethyl ester .... $(CH_3)_2CClCO_2C_2H_5$	150.61	....	148-9	......	1.062[0]	1.4109[16]	al, eth	B2[4], 859
4403	Isobutyric acid, 2-chloro, methyl ester .... $(CH_3)_2CClCO_2CH_3$	136.58	....	135, 42-4[17]	......	1.0893[15/15]	1.4122[21]	eth	B2[4], 859
4404	Isobutyric acid, 3-chloro .... $ClCH_2CH(CH_3)CO_2H$	122.55	....	128-33[50]	......	1.0153[20]	1.4310[20]	al, eth	B2[4], 860
4405	Isobutyric acid, cyclohexyl ester or Cyclohexyl isobutyrate $(CH_3)_2CHCO_2C_6H_{11}$	170.25	....	204[750]	......	0.9489[0/4]	....	al, eth	B6[3], 24
4406	Isobutyric acid, 3,3-dichloro-2-hydroxy .... $Cl_2CHC(CH_3)(OH)CO_2H$	173.00	pr (al-eth)	d	82-3	....	....	w, al, eth	B3[2], 224
4407	Isobutyric acid, ethyl ester or Ethyl isobutyrate........ $(CH_3)_2CHCO_2C_2H_5$	116.16	....	111.0	−88.2	0.8693[20/4]	1.3869[18]	al, eth, ace	B2[4], 846
4408	Isobutyric acid, furfuryl ester or Furfuryl isobutyrate .... $(CH_3)_2CHCO_2CH_2(C_4H_3O)$	168.20	....	85-6[15]	......	1.0313[20/4]	....	al, eth	B17[2], 115
4409	Isobutyric acid, 2-hydroxy or Acetonic acid............ $(CH_3)_2C(OH)CO_2H$	104.12	hyg pr (eth), nd (bz)	212, 108-11[8]	82-3	....	....	al, eth, w	B3[4], 782
4410	Isobutyric acid, 2-hydroxy, ethyl ester .... $(CH_3)_2C(OH)CO_2C_2H_5$	132.18	....	150 (cor), 46[14]	......	0.987[20]	1.4080[20]	w, al	B3[4], 783
4411	Isobutyric acid, 2-hydroxy, methyl ester .... $(CH_3)_2C(OH)CO_2CH_3$	118.13	....	137, 62-4[12]	......	....	1.4056[20]	w, al	B3[4], 783
4412	Isobutyric acid, methyl ester or Methyl isobutyrate....... $(CH_3)_2CHCO_2CH_3$	102.13	....	92.3	−84.7	0.8906[20/4]	1.3840[20]	al, eth, ace	B2[4], 846
4413	Isobutyric acid, 2-nitropentyl ester or 2-Nitropentyl isobutyrate .... $(CH_3)_2CHCO_2CH_2CH(NO_2)C_3H_7$	202.24	....	248-51, 122[10]	......	1.0329[20/20]	1.4315[20]	eth, ace	B2[3], 648
4414	Isobutyric acid, isopentyl ester or Isopentyl isobutyrate... $(CH_3)_2CHCO_2-i-C_5H_{11}$	158.24	....	168.9	......	0.8627[20]	....	al, eth, ace	B2[3], 649
4415	Isobutyric acid, 3-phenyl $C_6H_5CH_2CH(CH_3)CH_2CO_2H$	164.21	pl (dil al)	272, 155-6[11]	36.5	....	....	al, eth	B9[3], 2472
4416	Isobutyric acid, propyl ester or Propyl isobutylrate ...... $(CH_3)_2CHCO_2C_3H_7$	130.19	....	135-6	......	0.8843[0/4]	1.3955[20]	al, eth, ace	B2[4], 847
4417	Isobutyric acid, isopropyl ester or Isopropyl isobutylrate $(CH_3)_2CHCO_2CH(CH_3)_2$	130.19	....	120.7	......	0.8471[21/4]	....	al, eth, ace	B2[4], 847
4418	Isobutyric acid, vinyl ester or Vinyl isobutyrate .... $(CH_3)_2CHCO_2CH=CH_2$	114.14	....	104-5	......	0.8932[20]	1.4061[20]	....	B2[4], 848
4419	γ-Butyrolactone .... $CH_2CH_2CH_2CO$ (ring)	86.09	....	206, 89[12]	−42	1.1286[16/0]	1.4341[20]	w, al, eth, ace, bz	B17[4], 4159

No.	Name, Synonyms, and Formula	Mol. wt.	Color, crystalline form, specific rotation and λ_{max} (log ϵ)	b.p. °C	m.p. °C	Density	n_D	Solubility	Ref.
4420	γ-Butyrolactone, 2-methylene or Butanoic acid-4-hydroxy-2-methylene, lactone . . . . CH$_2$CH$_2$C(=CH$_2$)CO	98.10	. . . .	85-6[10]	. . . .	1.1206[20]	1.4650[20]	w, al, eth, acc, bz	B1/[a], 4304
4421	Butyronitrile . . . . CH$_3$CH$_2$CH$_2$CN	69.11	. . . .	118	−112	0.7936[20/4]	1.3842[10]	al, eth, bz	B2[4], 806
4422	Butyronitrile, 4-bromo BrCH$_2$CH$_2$CH$_2$CN	148.01	. . . .	205-7, 91[12]	. . . . . .	1.4967[20/4]	1.4818[20]	al, eth	B2[4], 836
4423	Butyronitrile, 4-chloro . . . . ClCH$_2$CH$_2$CH$_2$CN	103.55	. . . .	189-91, 75[11]	. . . . . .	1.0934[15]	1.4413[20]	al, eth	B2[4], 827
4424	Butyronitrile, 3,4-epoxy or Epicyanohydrin . . . . CH$_2$CHCH$_2$CN	83.09	pr	. . . .	162	. . . .	. . . .	al	B18[4], 3822
4425	Butyronitrile, 2-ethyl . . . . (C$_2$H$_5$)$_2$CHCN	97.16	. . . .	145-6	. . . . . .	. . . .	1.3891[24]	al, eth	B2[4], 953
4426	Butyronitrile, 2-methyl CH$_3$CH$_2$CH(CH$_3$)CN	83.13	. . . .	125	. . . . . .	0.7913[15/4]	1.3933[20]	al, eth	B2[4], 892
4427	Butyronitrile, 3-methyl or Isovaleronitrile . . . . (CH$_3$)$_2$CHCH$_2$CN	83.13	. . . .	130.5, 53[50]	−100.8 f.p.	0.7914[20/4]	1.3927[20]	al, eth, ace	B2[4], 902
4428	Butyronitrile, 3-methyl-2-phenyl (CH$_3$)$_2$CHCH(C$_6$H$_5$)CN	159.23	. . . .	245-9[765]	. . . . . .	0.967[15.5]	1.5038[25]	al, bz	B9[2], 364
4429	Butyronitrile, 2-phenoxy CH$_3$CH$_2$CH(OC$_6$H$_5$)CN	161.21	. . . .	228-30[748]$_d$	. . . . . .	. . . .	. . . .	al, eth	B6, 164
4430	Butyronitrile, 4-phenoxy C$_6$H$_5$OCH$_2$CH$_2$CH$_2$CN	161.21	nd	287-9[765], 170.5[22]	45-6	. . . .	. . . .	eth	B6[4], 646
4431	Butyronitrile, 2-phenyl CH$_3$CH$_2$CH(C$_6$H$_5$)CN	145.21	. . . .	238-40[763], 141-3[8]	. . . . . .	. . . .	. . . .	al, eth, bz	B9[3], 2468
4432	Butyronitrile, 4-phenyl C$_6$H$_5$CH$_2$CH$_2$CH$_2$CN	145.21	. . . .	142-5[16]	. . . . . .	. . . .	. . . .	. . . . . .	B9[3], 2454
4433	Isobutyronitrile . . . . (CH$_3$)$_2$CHCN	69.11	. . . .	103.8	−71.5	0.7608[20/4]	1.3720[20]	al, eth, ace, chl	B2[4], 853
4434	Isobutyronitrile, 2-bromo (CH$_3$)$_2$CBrCN	148.01	. . . .	139-40, 61-2[5]	. . . . . .	1.4796[15/4]	1.4739[15]	al, eth, bz	B2[4], 863
4435	Isobutyronitrile, 2-hydroxy or Acetone cyanohydrin . . . . . (CH$_3$)$_2$C(OH)CN	85.11	. . . .	82[23]	−19	0.932[20/4]	1.3996[20]	w, al, eth, ace, bz	B3[4], 785
4436	Isobutyronitrile, 2-hydroxy-3-chloro ClCH$_2$C(OH)(CH$_3$)CN	119.55	. . . .	110[27]	. . . . . .	1.2027[15]	1.4356[11]	w, al, ace	B3[3], 599
4437	Butyrophenone or n-Propyl phenyl ketone . . . . CH$_3$CH$_2$CH$_2$COC$_6$H$_5$	148.21	. . . .	228-9	11-3	0.988[20/4]	1.5203[20]	al, eth, ace	B7[3], 1075
4438	Butyrophenone, p-methyl or p-Totyl-propyl ketone . . . . . . (4-CH$_3$C$_6$H$_4$)COC$_3$H$_7$	162.24	. . . .	251.5[758]	12	0.9745[20/4]	1.5232[20]	al, eth	B7[3], 1127
4439	Isobutyrophenone C$_6$H$_5$COCH(CH$_3$)$_2$	148.21	. . . .	221, 86[4]	. . . . . .	0.9863[11/4]	1.5172[20]	al, eth	B7[3], 1088
4440	Butyryl bromide C$_3$H$_7$COBr	151.01	. . . .	128	. . . . . .	1.4162[17/4]	1.1596[17]	. . . .	B2[4], 804
4441	Butyrylbromide, 2-bromo CH$_3$CH$_2$CHBrCOBr	229.91	. . . .	172-4, 57-60[10]	. . . . . .	. . . .	. . . .	eth	B2[4], 834
4442	Butyrylbromide, 3-methyl or Isovaleryl bromide . . . . (CH$_3$)$_2$CHCH$_2$COBr	165.03	mcl pl (al)	143	. . . . . .	. . . .	. . . .	eth	B2[4], 902
4443	Isobutyryl bromide (CH$_3$)$_2$CHCOBr	151.01	. . . .	116-8	. . . . . .	1.4067[15/4]	1.4552[15]	. . . .	B2[4], 852
4444	Isobutyryl bromide, 2-bromo (CH$_3$)$_2$CBrCOBr	229.91	. . . .	162-4, 91-8[100]	. . . . . .	1.4067[14/4]	1.4552[14]	ace, CS$_2$	B2[3], 661
4445	Butyryl chloride CH$_3$CH$_2$CH$_2$COCl	106.55	. . . .	102	−89	1.0277[20/4]	1.4121[20]	eth	B2[4], 803
4446	Butyryl chloride, 2-bromo CH$_3$CH$_2$CHBrCOCl	185.45	. . . .	150-2, 41[12]	. . . . . .	1.5320[20]	. . . .	eth	B2[4], 834
4447	Butyryl chloride, 2-chloro CH$_3$CH$_2$CHClCOCl	141.00	. . . .	130-1, 51-2[41]	. . . . . .	1.2360[17]	1.4475[20]	eth	B2[4], 822
4448	Butyryl chloride, 2-chloro-2-methyl CH$_3$CH$_2$C(CH$_3$)ClCOCl	155.04	. . . .	144[730]	. . . . . .	1.187[14]	. . . .	eth	B2[4], 893
4449	Butyryl chloride, 2-chloro-3-methyl (CH$_3$)$_2$CHCHClCOCl	155.04	. . . .	148-9	. . . . . .	1.135[13]	. . . .	eth	B2[4], 904
4450	Butyryl chloride, 3-chloro CH$_3$CHClCH$_2$COCl	141.00	. . . .	40-1[12]	. . . . . .	1.2163[20/4]	1.4509[20]	CS$_2$	B2[4], 824
4451	Butyryl chloride, 4-chloro ClCH$_2$CH$_2$CH$_2$COCl	141.00	. . . .	173-4, 60-1[12]	. . . . . .	1.2581[20/4]	1.4616[20]	eth	B2[4], 826
4452	Butyryl chloride, 4-chloro-3-oxo . . . . ClCH$_2$COCH$_2$COCl	154.99	. . . .	117-9[17]	. . . . . .	1.4397[20/4]	1.4860[20]	bz	B3[3], 1207

No.	Name, Synonyms, and Formula	Mol. wt.	Color, crystalline form, specific rotation and λ_{max} (log ϵ)	b.p. °C	m.p. °C	Density	n_D	Solubility	Ref.
4453	Butyryl chloride, 2,2-dimethyl .... CH$_3$CH$_2$C(CH$_3$)$_2$COCl	134.61	....	132, 27[11]	....	0.9801[20/4]	1.4245[20]	eth	B2[4], 955
4454	Butyryl chloride, 2,3-dimethyl .... (CH$_3$)$_2$CHCH(CH$_3$)COCl	134.61	....	135-6[751], 38-9[18]	....	0.9795[20/4]	....	eth	B2[3], 761
4455	Butyryl chloride, 3,3-dimethyl .... (CH$_3$)$_3$CCH$_2$COCl	134.61	....	128-30[745], 68[100]	....	0.9696[20/4]	1.4210[20]	eth	B2[4], 956
4456	Butyryl chloride, 2-ethyl or Diethylacetyl chloride .... (C$_2$H$_5$)$_2$CHCOCl	134.61	....	140, 40[20]	....	0.9825[20/4]	1.4234[20]	eth	B2[4], 952
4457	Butyryl chloride, 3-methyl or Isovaleryl chloride......... (CH$_3$)$_2$CHCH$_2$COCl	120.58	....	114-5[771]	....	0.9844[20/4]	1.4149[20]	eth	B2[4], 901
4458	Butyryl chloride, 2-methyl- (dl) CH$_3$CH$_2$CH(CH$_3$)COCl	120.58	....	116	....	0.9917[20/4]	1.4170[20]	....	B2[4], 891
4459	Butyryl chloride, 2,2,3-trimethyl .... (CH$_3$)$_2$CHC(CH$_3$)$_2$COCl	148.63	....	148-50	....	....	....	eth	B2[4], 982
4460	Isobutyryl chloride .... (CH$_3$)$_2$CHCOCl	106.55	....	92	− 90	1.0174[20/4]	1.4079[20]	eth	B2[4], 852
4461	Isobutyryl chloride, 2-chloro .... (CH$_3$)$_2$CCICOCl	141.00	....	126-7	....	....	1.4369[20]	eth	B2[4], 859
4462	Isobutyryl chloride, 3-chloro .... (CH$_3$)$_2$CClCH$_2$COCl	106.55	....	92	− 90.0	1.0174[20/4]	1.4079[20]	eth	B2[4], 860

No.	Name, Synonyms, and Formula	Mol. wt.	Color, crystalline form, specific rotation and λ_{max} (log ε)	b.p. °C	m.p. °C	Density	n_D	Solubility	Ref.
4463	Cacodyl or Tetramethyl biarsine $(CH_3)_2AsAs(CH_3)_2$	209.98	pl	165	−6	1.447[15]	....	al, eth	B4[3], 1831
4464	Cacodyl chloride or Dimethyl chloraisine $(CH_3)_2AsCl$	140.45	....	109	< −45	1.5046[12/4]	1.5203[12]	al	B4[3], 1797
4465	Cacodyl oxide or bis-Dimethyl arsenous oxide $(CH_3)_2AsOAs(CH_3)_2$	225.98	....	150	−25	1.4816[15]	1.5225[9]	al, eth	B4[3], 1814
4466	β-Cadinene-(l) or 3,9-Cadinadiene $C_{15}H_{24}$	204.36	$[\alpha]^{27}_D$ −15.9 (chl, c=1)	274, 149[20]	......	0.9230[20/4]	1.5059[20]	eth, lig	B5[3], 1086
4467	Caffeine or 1,3,7-Trimethylxanthene $C_8H_{10}N_4O_2$	194.20	wh nd (w + 1), hex pr (sub)	sub 178, sub 89[15]	238 (anh)	1.23[19]	....	Py, chl	B26[2], 266
4468	Caffeine benzoate $C_8H_{10}N_4O_2 \cdot C_6H_5CO_2H$	316.32	wh so pw	....				al, w	B26[2], 268
4469	Caffeine citrate $C_8H_{10}N_4O_2 \cdot C_6H_8O_7$	386.32	mcl cr						B26[2], 269
4470	Caffeine hydrobromide $C_8H_{10}N_4O_2 \cdot HBr \cdot 2H_2O$	311.14	ye	....	d 80-100			w	B26[1], 137
4471	Caffeine hydrochloride $C_8H_{10}N_4O_2 \cdot HCl \cdot 2H_2O$	266.69	mcl pr		d 80-100				B26[2], 268
4472	Caffeine, 2-hydroxybenzoate or Caffeine salicylate $C_8H_{10}N_4O_2 \cdot C_7H_6O_3$	322.32	wh nd (w)		137			al	B26[1], 269
4473	Caffeine, 3-methylbutanoate or Caffeine isovalerate $C_{10}H_8N_4O_2 \cdot C_5H_{10}O_2$	296.33	unst nd	....	....			w	B26, 467
4474	Caffeine sulfate $C_8H_{10}N_4O_2 \cdot H_2SO_4$	292.27	wh nd					.....	B26, 466
4475	Caffeine, 0-ethoxy or 1,3,7-Trimethyl-2,6-dioxo-8-ethoxy purine $C_{10}H_{14}N_4O_3$	238.25	wh or yesh nd (w)	... .	143			al	B26[2], 322
4476	Caffeine, 8-methoxy $C_9H_{12}N_4O_3$	224.22	wh nd (al or w)	....	176	1.399[25/4]	....	al, bz	B26[2], 322
4477	Calciferol or Vitamin D2 $C_{28}H_{44}O$	396.67	pr (ace), $[\alpha]^{20}_D$ +102.5 (al)	sub	115—8			al, eth, ace	B6[3], 3089
4478	Camphane or Bornane $C_{10}H_{18}$	138.25	hex pl (al), pr (MeOH)	sub 161	158-9			al, eth	B5[4], 319
4479	Camphane-3-carboxylic acid $C_{10}H_{17}CO_2H$	182.27	pl (dil aa), $[\alpha]^{20}_D$ +56 (al, p=8)	153[13]	90-1			al, aa	B9[3], 244
4480	Camphene-(d) $C_{10}H_{16}$	136.24	nd $[\alpha]^{17}_D$ +103.9 (eth, c=4)	160-2, 52[17]	52	0.8450[50/4]	1.4570[25]	eth	B5[4], 461
4481	Camphene (dl) $C_{10}H_{16}$	136.24	nd (sub)	158-9	51-2	0.879[20/4]	1.4551[54]	al, eth	B5[4], 462
4482	Camphene-(l) $C_{10}H_{16}$	136.24	$[\alpha]^{19}_D$ −106.1 (eth, c=4)	158	52	0.8446[50/4]	1.4564[54]	eth	B5[4], 461
4483	Camphenilone $C_9H_{14}O$	138.21	$[\alpha]^{20}_D$ +70.4 (al)	193[751], 76[12]	41			eth	B7[3], 306
4484	3-Camphanol or Epiborneol. $C_{10}H_{15}OH$	154.26	nd $[\alpha]^{17}$ +11.1 (al)	213[742]	181-2			peth	B6[4], 288
4485	3-Camphanol acetate $C_{12}H_{20}O_2$	196.29	$[\alpha]^{10}_D$ +15.63	101[11]	< −15	0.9872[14/4]	1.4651[14]	....	B6[2], 92
4486	Campholic acid-(d) $C_{10}H_{18}O_2$	170.25	pr $[\alpha]^{20}_D$ +59.3 (bz)	255[768], 146[12]	106			al, eth	B9[3], 98
4487	Campholic acid-(dl) $C_{10}H_{18}O_2$	170.25	tcl pr	....	109			al, eth	B9[3], 99
4488	Campholic acid-(l) $C_{10}H_{18}O_2$	170.25	pr (dil al), $[\alpha]^{15}_D$ −49.1 (al)	250	106-7			al, eth	B9[3], 99
4489	Campholytic acid-(α,l) $C_9H_4O_2$	154.21	$[\alpha]^{13}_D$ −60.4	240-3, 140[15]	......	1.0145[18]	1.4712[17]	lig	B9[3], 187
4490	Camphor-(d) or 2-Camphanone (d). $C_{10}H_{16}$	152.24	pl $[\alpha]^{20}_D$ +44.26 (al)	sub 204	179.8	0.990[25/4]	1.5462	al, eth, ace, bz	B7[3], 400

No.	Name, Synonyms, and Formula	Mol. wt.	Color, crystalline form, specific rotation and λ_{max} (log ε)	b.p. °C	m.p. °C	Density	n_D	Solubility	Ref.
4491	Camphor-(dl) .... $C_{10}H_{16}O$	152.24	wh	sub	178.8	....	....	al, eth, ace, bz	B7³, 406
4492	Camphor-(l) .... $C_{10}H_{16}O$	152.24	$[\alpha]^{16}_D$ − 44.2 (al, c=16.5)	204 sub	178.6	0.9853¹⁸	....	al, eth, ace	B7³, 405
4493	Camphor oxime-(d) .... $C_{10}H_{16}NOH$	167.25	pr (lig-eth), $[\alpha]^{22}_D$ + 42.5 (al)	....	115			al, eth	B7³, 408
4494	Camphor oxime-(dl) .... $C_{10}H_{16}NOH$	167.25	cr (peth)	....	118			al, eth	B7³, 409
4495	Camphor oxime-(l) .... $C_{10}H_{16}NOH$	167.25	mcl nd or pr (dil al), $[\alpha]^{20}_D$ − 42.4 (al)	249-54d	118	1.01¹¹⁶/⁴	....	al, eth	B7³, 408
4496	Camphor, 3-amino-(d) or 3-Camphoryl amine .... $C_{10}H_{17}NO$	167.25	wx	244	110-15d			al, eth	B14³, 15
4497	Camphor, 3-bromo-(d,α) .... $C_{10}H_{15}BrO$	231.14	pr (al), $[\alpha]^{20}_D$ + 129.3 (MeOH, c=4.6)	274(d) sub	76			al, eth, bz, chl	B7³, 414
4498	Camphor, 3-bromo-(d,α´) .... $C_{10}H_{15}BrO$	231.13	$[\alpha]^{20}_D$ 29.4	d265	78	1.484¹⁴		al, chl	B7³, 415
4499	Camphor, 3-bromo-(dl,α) .... $C_{10}H_{15}BrO$	231.14			51			al, eth, bz, chl	B7³, 415
4500	Camphor, 3-bromo-(l,α) .... $C_{10}H_{15}BrO$	231.14	mcl nd (al), $[\alpha]^{18}_D$ − 138.8 (ace, c=6)	....	76			al, eth, bz, chl	B7³, 415
4501	Camphor, 3-bromo-(d,α´) .... $C_{10}H_{15}BrO$	231.14	nd (dil al)	265d	78	1.484¹⁴		al, chl, aa	B7², 101
4502	Camphor, 5-bromo-(exo) .... $C_{10}H_{15}BrO$	231.13	$[\alpha]^{16}_D$ bz 2%	100¹⁵	114			al, eth, ace, bz	B7³, 415
4503	Camphor, 8-bromo-(d) .... $C_{10}H_{15}BrO$	231.14	tetr pr (lig), $[\alpha]^{19'}_D$ + 122.2 (chl)	sub	93			al, eth, ace, bz	B7³, 416
4504	Camphor, 8-bromo-(dl) .... $C_{10}H_{15}BrO$	231.14	pr (eth-peth), pym (eth)	sub	92.7			al, eth, ace, bz	B7³, 417
4505	Camphor, 10-bromo-(d) .... $C_{10}H_{15}BrO$	231.14	pr (peth), $[\alpha]^{20}_D$ + 19.2 (abs al), + 15.7 (bz)	265d	78			al, eth, ace, bz	B7³, 416
4506	Camphor, 10-bromo-(dl) .... $C_{10}H_{15}BrO$	231.14	cr	....	77	....			B7³, 416
4507	Camphor, 3-chloro-(d,α) .... $C_{10}H_{15}ClO$	186.68	lf $[\alpha]^{20}_D$ + 71.1	244-7	94			al, eth, ace, bz	B7³, 411
4508	Camphor, 3-chloro-(d,α´) .... $C_{10}H_{15}ClO$	186.68	$[\alpha]^{20}_D$ + 35 (al, c=5)	231d	118			eth, chl	B7³, 411
4509	Camphor, 8-chloro-(d) .... $C_{10}H_{15}ClO$	186.68	pr (al), $[\alpha]_D$ + 99.9 (chl)		139				B7³, 414
4510	Camphor, 8-chloro-(dl) .... $C_{10}H_{15}ClO$	186.68	....	sub	138				B7, 136
4511	Camphor, 10-chloro-(d) .... $C_{10}H_{15}ClO$	186.68	pr (al), $[\alpha]^{14}_D$ + 40.7 (al)		132-5			al, eth, bz, chl	B7², 100
4512	Camphor, 3,3-dibromo-(d) .... $C_{10}H_{14}Br_2O$	310.04	wh ye rh pr (al, peth), $[\alpha]^{20}_D$ + 40 (chl) + 39.2 (al)	sub	64	1.8954²¹·⁶/⁴		al, eth, bz, chl	B7², 101
4513	Camphor, 3-nitro-(l) .... $C_{10}H_{15}NO_3$	197.24	mcl pr (bz), $[\alpha]^{13}_D$ − 26→-9 (al), (mut)	....				al, eth, bz, chl	B7³, 419

No.	Name, Synonyms, and Formula	Mol. wt.	Color, crystalline form, specific rotation and λ_{max} (log ε)	b.p. °C	m.p. °C	Density	n_D	Solubility	Ref.
4514	Camphor-3-carboxylic acid-(d) .... $C_{11}H_{16}O_3$	196.25	pr (eth, 50% al), [α]²⁰_D + 34.9 (bz)	....	128d			w, al, eth, bz	B10³, 2925
4515	Camphor-3-carboxylic acid-(dl) , ... $C_{11}H_{16}O_3$	196.25	cr (bz)	....	136-7			al, eth, bz	B10³, 2927
4516	Camphor-3-carboxylic acid .... $C_{11}H_{16}O_3$	196.25	pr (eth), cr (bz), [α]²⁰_D -64 (al)	....	127-8d			al, eth, bz	B10³, 2927
4517	β-Camphor-(l) or l-EpiCamphor, 1,3-Bornanone .... $C_{10}H_{16}O$	152.24	[α]¹⁹_D -58.21 (bz, c=13)	213	184			al, eth, peth	B7³, 421
4518	Camphoric acid-(d) or 1,2,2-Trimethylcyclopentane-1,3-dicarboxylic acid $C_{10}H_{16}O_4$	200.24	Pr, lf (w), [α]²⁰_D + 47.7 (al)		188.2	1.186²⁰ᐟ⁴		al, eth, ace	B9³, 3876
4519	Camphoric acid-(dl) $C_{10}H_{16}O_4$	200.24	pr (al, aa), mcl nd		208	1.228²⁰ᐟ⁴		al, ace	B9³, 3878
4519a	Camphoric acid-(l) .... $C_{10}H_{16}O_4$	200.24	Cr (w), [α]¹⁸ᐟ_D -48.1 (abs. al, c=8)	223	270d	1.190		al, eth, ace	B9³, 3878
4520	Camphoric anhydride-(l) .... $C_{10}H_{14}O_3$	182.22	rh (al), pr (bz)	7270d	223.5	1.194²⁰		hz	B17⁴, 5957
4521	Camphoric anhydride-(dl) $C_{10}H_{14}O_3$	182.22	rh (al)	270	221	1.194²⁰ᐟ⁴	....	bz	B17¹, 238
4522	Camphoric anhydride-(l) .... $C_{10}H_{14}O_3$	182.22	[α]_D -77 (bz)	>270	221	1.194²⁰ᐟ⁴		hz	B17⁴, 5958
4523	Camphoric acid, diethyl ester-(d) .... $C_{14}H_{24}O_4$	256.35	[α]¹⁵_D + 7.5 (al), + 9 (bz)	286⁷⁵², 164²⁰	....	1.0298²⁰ᐟ⁴	1.4535²⁰	al, eth, bz	B9², 536
4524	Camphoric acid, dimethyl ester-(d) .... $C_{12}H_{20}O_4$	228.29	[α]_D + 49.07 (al)	264⁷⁵⁸, 155¹⁵	< -16	1.0747²⁰ᐟ⁴	1.4627¹⁹	al, eth	B9³, 3880
4525	Camphoric acid, 1-monoamide or β-Camphoramic acid .. $C_{10}H_{17}NO_3$	199.25	pl (nd), [α]²⁰_D + 74 (al)	....	183			al, ace	B9³, 3886
4526	Camphoric acid, 3-monoamide-(d) or α-Camphoramic acid .... $C_{10}H_{17}NO_3$	199.25	nd or lf (w), [α]²⁰_D + 25 (al)	....	176-7				B9³, 3887
4527	Camphoric acid, 3-monoamide-(dl) or α-Camphoramic acid .... $C_{10}H_{17}NO_3$	199.25	nd (w)	....	198				B9, 761
4528	Camphoronic acid-(d) or 2,4-Dimethylpentane-1,2,3-tricarboxylic acid. $(CH_3)_2C(CO_2H)C(CH_3)(CO_2H)CH_2CO_2H$	218.21	nd (w), [α]¹⁹ + 27.05 (w)					w, al, eth, ace	B2, 837
4529	Camphoronic acid-(dl) .... $(CH_3)_2C(CO_2H)C(CH_3)(CO_2H)CH_2CO_2H$	218.21	nd or pr (w)		172d			al, eth, ace	B2³, 2045
4530	Camphoronic acid-(l) $(CH_3)_2C(CO_2H)C(CH_3)(CO_2H)CH_2CO_2H$	218.21	nd (w), [α]¹⁹ᐟ_D -26.9		164-5d			w, al, eth, ace	B2³, 2045
4531	Camphorpinacol-(l) or 2,2′-Bicamphane-2,2′-diol .... $(C_{10}H_{17}O)_2$	306.49	rh [α]_D -27.2 (bz)		158			al, eth	B6³, 4767
4532	Camphor quinone or 2,3-Camphor dione............. $C_{10}H_{14}O_2$	166.22	ye nd (dil al, w), pr (eth) [α]²⁰_D -113.2 (bz)	sub	199			al, eth, bz, chl	B7³, 3297
4533	Camphor-3-sulfonic acid, methyl ester-(d) .... $C_{11}H_{17}SO_3$	246.33	cr (MeOH), nd (peth), [α]²⁰ᐟ_D + 98.6 (chl, c=5)					al	B11², 179
4534	Camphor-10-sulfonic acid-(d) .... $C_{10}H_{16}O_4S$	232.30	pr (aa), [α]²⁰ᐟ_D + 32.8 (AcOEt, + c=3), + 24 (w)	....	195d			w	B11³, 585

No.	Name, Synonyms, and Formula	Mol. wt.	Color, crystalline form, specific rotation and λ_{max} (log ε)	b.p. °C	m.p. °C	Density	n_D	Solubility	Ref.
4535	Camphor-10-sulfonic acid-(dl) .... $C_{10}H_{16}O_4S$	232.30	cr (aa)	....	202d			w	B11[3], 587
4536	Camphor-10-sulfonic acid-(l) $C_{10}H_{16}O_4S$	232.30	cr (aa), nd (AcoEt), $[\alpha]_D^{20}$ −20.75 (w)		194-5d			w	B11[3], 585
4537	α-Camphyl amine or 4-(2-Aminoethyl)-1,5,5-trimethyl cyclopentene .... $C_{10}H_{18}N$	153.27	$[\alpha]_D$ +6	194-6, 95[12]	......	0.8688[20]	1.4728[18]	....	B12[3], 176
4538	β-Camphyl amine or 2-(2-Aminoethyl)-1,5,5-trimethyl cyclopentene .... $C_{10}H_{18}N$	153.27	....	206		0.8697[20/20]			B12, 40
4539	Canadine-(d) or d-Tetrahydroberberine............ $C_{20}H_{21}NO_4$	339.40	ye nd (dil al), $[\alpha]_D^{29/}$ +299 (chl, c=1)		132 (140)			al, eth, bz, chl	B27[2], 557
4540	Canadine-(dl) .... $C_{20}H_{21}NO_4$	339.40	mcl nd (al)		174			chl	B27[2], 557
4541	Canadine-(l) .... $C_{20}H_{21}NO_4$	339.40	ye nd (al) $[\alpha]_D^{20/}$ −299.2 (chl)		134			al, eth, bz, chl	B27[2], 557
4542	Canaline $H_2NOCH_2CH_2CH(NH_2)CO_2H$	134.14	nd (al), $[\alpha]_D^{23/}$ −8.31 (w)		214d			w, al	B4[3], 1636
4543	Canavanine-(L). $HN:C(NH_2)NHOCH_2CH_2CH(NH_2)CO_2H$	176.18	cr (al), $[\alpha]_D^{20/D}$ + 7.9 (w, C=2)					w	B4[3], 1636
4544	Cannabidiol $CH_3C_6H_4CCH_3CH_2C_6H_2(OH)_2(CH_2)_4CH_3$	314.47	rods (peth)	187-90[2]	67			al, eth, bz, chl	B6[3], 5362
4545	Cannabinol .... $C_{21}H_{26}O_2$	310.44	pl, lf (peth) $[\alpha]_D^{20/}$ −148 (al)	185[0.05]	77			al, eth, ace, bz	B17[4], 1652
4546	Cantharidin .... $C_{10}H_{12}O_4$	196.21	rh pl or sc	sub 84	218				B19[4], 1958
4547	Capraldehyde or Decanal .... $CH_3(CH_2)_8CHO$	156.27	....	208-9, 81[9]	−5	0.830[15/4]	1.4287[20]	al, eth, ace	B1[4], 3366
4548	Capradehyde oxime or Capraldoxime .... $CH_3(CH_2)_8CHH$	171.29	lf (dil MeOH)		69			al, eth	B1, 711
4549	Capramide .... $CH_3(CH_2)_8CONH_2$	171.29	lf (eth)		108	0.999[20/4]	1.4261[110]	al, eth, ace	B2[4], 1050
4550	Capric acid or Decanoil acid............ $CH_3(CH_2)_8CO_2H$	172.27	nd	270, 148-50[11]	fr 31.5	0.8858[40/4]	1.4288[40]	al, eth, ace, bz, peth	B2[4], 1041
4551	Capric acid, 2-acetyl, ethyl ester .... $C_8H_{15}CH(COCH_3)CO_2C_2H_5$	242.36	....	280-2		0.9354[18.5/11 7.5]			B3[3], 1278
4552	Capric anhydride .... $(C_9H_{19}CO)_2O$	326.53	lf		24.7	0.8865[25/4]	1.400[25]	al, eth	B2[4], 1049
4553	Capric acid, 2-bromo .... $C_8H_{17}CHBrCO_2H$	251.17	....	140-1[2]	4	1.1912[24]	1.4595[24]	eth	B2[4], 1054
4554	Capric acid, decyl ester .... $C_9H_{19}CO_2C_{10}H_{21}$	312.54	....	219[15]	9.7	0.8586[20]	1.4423[20]	eth	B2[4], 1045
4555	Capric acid, ethyl ester or Ethyl caprate .... $C_8H_{19}CO_2C_2H_5$	200.33	....	241.5, 122-4[13]	−20	0.8650[20/4]	1.4256[20]	al, eth, chl	B2[4], 1044
4556	Capric acid, 10-fluoro .... $F(CH_2)_9CO_2H$	190.26	....	135-8[10]	49			al, eth, lig	B2[4], 1051
4557	Capric acid, methyl ester or Methyl caprate............ $C_9H_{19}CO_2CH_3$	186.30	....	224, 114[15]	−18	0.8730[20/4]	1.4259[20]	al, eth, chl	B2[4], 1044
4558	Capric acid, 2-octyl or 9-Heptadecane carboxylic acid .... $(C_8H_{17})_2CHCO_2H$	284.49	nd or lf (al)	212-8[13]	38.5			al, eth	B2[4], 1254
4559	Capric acid, 4-oxo or γ-Ketocapric acid .... $C_6H_{13}COCH_2CH_2CO_2H$	186.25	(dil al)		70-1			al	B3[4], 1642
4560	Capric acid, propyl ester or Propyl caprate .... $C_9H_{19}CO_2C_3H_7$	214.35	....	128.5[10]	0.8623[2 0]	1.4280[20]		al, eth	B2[4], 1045
4561	Capric acid, isopropyl ester or Isoporpyl caprate .... $C_9H_{19}CO_2CH(CH_3)_2$	214.35	....	121[10]		0.8543[20]	1.4221[25]		B2[4], 1045
4562	Caprinitrile .... $C_9H_{19}CN$	153.27	....	243, 106[10]	fr−17.9	0.8199[20/4]	1.4296[20]	**al, eth, ace, chl**	B2[4], 1051

No.	Name, Synonyms, and Formula	Mol. wt.	Color, crystalline form, specific rotation and λ_{max} (log ε)	b.p. °C	m.p. °C	Density	n_D	Solubility	Ref.
4563	Capryl chloride .... $C_9H_{19}COCl$	190.72	....	232, 114[15]	−34.5	0.973[8/4]	....	eth	B2[4], 1050
4564	Caproaldehyde or Hexanal........... $CH_3(CH_2)_4CHO$	100.16	....	128, 28[12]	−56	0.8139[20/4]	1.4039[20]	al, eth, ace, bz	B1[4], 3296
4565	Caproaldehyde oxime or Capraldoxime, Hexanaldoxime $CH_3(CH_2)_4CH=NOH$	115.18	cr (MeOH)	....	51				B1[2], 745
4566	Caproaldehyde, 2-ethyl .... $CH_3(CH_2)_3CH(C_2H_5)CHO$	128.22	....	163, 65[15]	<−100	0.8540[20]	1.4142[20]	al, eth	B1[4], 3345
4567	Caproaldehyde, 3-methyl or 3-Methylhexanal........... $CH_3(CH_2)_2CH(CH_3)CH_2CHO$	114.19	....	142-3[755]	....	0.8203[20/4]	1.4122[20]	al, eth	B1[2], 756
4568	Caproamide or Hexananoamide .... $CH_3(CH_2)_4CONH_2$	115.18	cr (ace)	255	101	0.999[20/4]	1.4200[110]	al, eth, bz, chl	B2[4], 929
4569	Caproamide, 2-ethyl .... $C_4H_9CH(C_2H_5)CONH_2$	143.23	nd (w)	....	102-3			w	B2[4], 1007
4570	Caproamide, N-phenyl or Capranilide .... $CH_3(CH_2)_4CONHC_6H_5$	191.28	nd (peth), pr (al)	....	95	1.112		al, eth	B12[3], 478
4571	Caproic acid or Hexanoic acid.................... $CH_3(CH_2)_4CO_2H$	116.16	....	205	−2	0.9274[20/4]	1.4163[20]	al, eth	B2[4], 917
4572	Caproic acid, 2-acetyl, ethyl ester or Ethyl 2-acetyl caproate $C_4H_9CH(COCH_3)CO_2C_2H_5$	186.25	....	219-24, 104[12]	....	0.9523[20/4]	1.4301[20]	eth, ace	B3[4], 1616
4573	Caproic acid, 2-amino .... $C_4H_9CH(NH_2)CO_2H$	131.18		....	297-300			w	B4[4], 1386
4574	Caproic acid, 6-amino .... $H_2NCH_2(CH_2)_4CO_2H$	131.18	lf (eth)	202-3	....			w	B4[3], 1393
4575	Caproic acid, 6-amino-ε-lactam or ε-Caprolactam .... $C_6H_{11}NO$	113.16	lf (lig)	139[12]	69-71			w, al, lth, chl	B21[4], 3196
4576	Caproic acid, 6-amino-3-methyl, lactam-(l) .... $CH_2(CH_2)_3CH(CH_3)CH_2CO$ ─NH─	127.19	cr (bz, peth), [α][20/D] −36.1	....	105-6			w, bz	B21, 243
4577	Caproic acid, 6-amino-5-methyl, lactam-(l) $CH_2CH(CH_3)(CH_2)_3CO$ ─NH─	127.19	cr (peth, bz-lig), [α][20/D] −22.2	....	68-9			w, eth	B21, 242
4578	Caproic acid anhydride .... $(C_5H_{11}CO)_2O$	214.31	....	254-7d, 143[15]	−41	0.9240[15/4]	1.4297[20]	al, eth	B2[4], 928
4579	Caproic acid, 6-benzoylamino .... $C_6H_5CONH(CH_2)_5CO_2H$	235.29	nd (al-eth)	....	79-80			AcOEt	B9[3], 1157
4580	Caproic acid, 6-benzoylamino-2-bromo-(dl) .... $C_6H_5CONH(CH_2)_3CHBrCO_2H$	314.19	cr (dil al)	....	166			al	B9[3], 1158
4581	Caproic acid, 6-benzoylamino-2-bromo-(l) $C_6H_5CONH(CH_2)_3CHBrCO_2H$	314.19	nd (dil al), [α][18/D] −29.2 (al)	....	129			al	B9[3], 1157
4582	Caproic acid, 2-bromo-(dl) $CH_3(CH_2)_3CHBrCO_2H$	195.06	....	240, 140-2[23]	4			al, eth	B2[4], 938
4583	Caproic acid, 2-bromo-(l) $CH_3(CH_2)_3CHBrCO_2H$	195.06	[α][20/D] −27 (eth, C=5)	129[14]	....			al, eth	B2[3], 736
4584	Caproic acid, 2-bromo, ethyl ester-(dl) .... $CH_3(CH_2)_3CHBrCO_2C_2H_5$	223.12	....	205-10, 95-6[9]	....			al	B2[4], 938
4585	Caproic acid, 3-bromo $C_3H_7CHBrCH_2CO_2H$	195.06	nd (dil al)	....	35			al, bz, chl, lig	B2[4], 939
4586	Caproic acid, 6-bromo .... $Br(CH_2)_5CO_2H$	195.06	cr (peth)	165-70[20]	35			peth	B2[4], 940
4587	Caproic acid, butyl ester or Butyl caproate............. $C_5H_{11}CO_2C_4H_9$	172.27	....	208	−64.3	0.8653[20/4]	1.4152[20]	al, eth	B2[4], 922
4588	Caproic acid, 6-cyclohexyl $C_6H_{11}(CH_2)_5CO_2H$	198.31	....	180[11]	33-5	0.9626[20/4]	1.4750[20]	eth	B9[3], 112
4589	Caproic acid, ethyl ester or Ethyl caproate............. $C_5H_{11}CO_2C_2H_5$	144.22	....	168	−67	0.8710[20/4]	1.4073[20]	al, eth	B2[4], 921
4590	Caproic acid, 2-ethyl $C_4H_9CH(C_2H_5)CO_2H$	144.22	....	228[755], 120[13]	....	0.9031[25/4]	1.4241[20]	eth	B2[4], 1003
4591	Caproic acid, 6-fluoro .... $F(CH_2)_5CO_2H$	134.15	....	138[28], 67-8[0.6]	....		1.4166[25]	al, eth	B2[4], 932
4592	Caproic acid, heptyl ester .... $C_5H_{11}CO_2C_7H_{15}$	214.35	....	261	−34.4	0.8611[20]	1.4293[15]	al, eth, ace, bz	B2[4], 923

PHYSICAL CONSTANTS OF ORGANIC COMPOUNDS (Continued)

No.	Name, Synonyms, and Formula	Mol. wt.	Color, crystalline form, specific rotation and λ_{max} (log ε)	b.p. °C	m.p. °C	Density	n_D	Solubility	Ref.
4593	Caproic acid, hexyl ester or Hexyl caproate............ $C_5H_{11}CO_2C_6H_{13}$	200.33	....	246	−55	0.865[18]	1.4264[15]	al, eth, ace, bz	B2[4], 922
4594	Caproic acid, 2-hydroxy-(d)..................... $C_4H_9CH(OH)CO_2H$	132.16	[α][20/D] +0.7 (w, C=14)	....	60			w, al, eth	B3[4], 838
4595	Caproic acid, 2-hydroxy-(dl) $CH_3(CH_2)_3CH(OH)CO_2H$	132.16	pr (eth-al, peth)	....	60-1			w, al, eth, chl	B3[4], 838
4596	Caproic acid, 2-hydroxy-(l) $C_4H_9CH(OH)CO_2H$	132.16	pr (eth), [α][20/D] −3.8 (w, C=4.5)	....	60-1			w, al, eth, chl	B3[4], 838
4597	Caproic acid, 4-hydroxy, lactone or γ-Caprolactone...... $CH_3CH_2CHCH_2CH_2CO$	114.15	....	215-6, 103[14]	−18		1.4495[20]	w, al	B17[4], 4194
4598	Caproic acid, 6-hydroxy-ε-lactone .... $C_6H_{10}O_2$	114.14	....	108[10]	−1.3	1.0693[20/4]	1.4611[20]	al, eth, ace	B17[4], 4186
4599	Caproic acid, methyl ester or Methyl caproate........... $C_5H_{11}CO_2CH_3$	130.19	....	151, 52[15]	−71	0.8846[20/4]	1.4049[20]	al, eth ace, bz	B2[4], 921
4600	Caproic acid, 2-methyl-(d) or 2-methyl hexanoic acid..... $C_4H_9CH(CH_3)CO_2H$	130.19	[α][22/D] +19.6 (eth)	105[5]	....	0.909[25]	1.4189[20]	al, eth, ace, bz	B2[3], 773
4601	Caproic acid, 2-methyl-dl or 2-Methylhexanoic acid...... $C_4H_9CH(CH_3)CO_2H$	130.19	....	215-6, 100[11]	....	0.9612[20/4]	1.4195[20]	al, eth, ace, bz	B2[4], 969
4602	Caproic acid, 2-methyl-(l) or 2-Methylhexanoic acid...... $C_4H_9CH(CH_3)CO_2H$	130.19	[α][25/D] -4.3 (w, c=26)	121[20]	....	0.909[25/4]	1.4189[25]	al, eth, ace, bz	B2[3], 773
4603	Caproic acid, 4-methyl-(d) or 4-Methylhexanoic acid..... $C_2H_5CH(CH_3)CH_2CH_2CO_2H$	130.19	[α][20/D] +7.6 (MeOH)	221, 115[16]	....	0.9228[20/4]	1.4198[20]	al, ace, bz	B2[4], 973
4604	Caproic acid, 4-methyl-(dl) or 4-Methylhexanoic acid .... $C_2H_5CH(CH_3)(CH_2)_2CO_2H$	130.19	....	217-8, 85[2]	−80	0.9215[20/4]	1.4211[20]	al, eth, ace, bz	B2[4], 974
4605	Caproic acid, 5-methyl or 5-Methylhexanoic acid ...... $(CH_3)_2CH(CH_2)_3CO_2H$	130.19	....	216, 109[16]	<−25	0.9138[21/4]	1.4220[20]	al, eth, ace, bz	B2[4], 970
4606	Caproic acid, octyl ester or Octyl caproate .... $C_5H_{11}CO_2C_8H_{17}$	228.38	....	275	−28	0.8603[20]	1.4326[15]	al, eth, ace, bz	B2[3], 727
4607	Caproic acid, 4-oxo or Homolevulinic acid .... $C_2H_5COCH_2CH_2CO_2H$	130.15	hyg ta or lf (eth-peth)	183[20], 89[0.4]	41-2	....		w, al, eth	B3[4], 1581
4608	Caproic acid, pentyl ester or Pentyl caproate........... $C_5H_{11}CO_2C_5H_{11}$	186.30	....	226, 116.6[20]	−47	0.8612[25/4]	1.4202[25]	al, eth, ace	B2[4], 922
4609	Caproic acid, isopentyl ester or Isopentyl caproate....... $C_5H_{11}CO_2CH_2CH_2CH(CH_3)_2$	186.30	....	224-7	....	0.861[20/4]		al, eth	B2[3], 727
4610	Caproic acid, propyl ester or Propyl caproate .... $C_5H_{11}CO_2C_3H_7$	158.24	....	187	−68.7	0.8672[20/4]	1.4170[20]	al, eth	B2[4], 922
4611	Caprolactam or Hexanoic acid-6-amino-ε-lactam .... $CH_2(CH_2)_4CO$	113.16	lf (lig)	139[12]	69-71			w, al, bz, chl	B21[4], 3196
4612	Caprolactam, 6-amino-3-methyl .... $C_7H_{13}NO$	127.19	cr (bz-peth), [α][20/D] -36.1	....	105-6			w, bz	B21, 243
4613	Caprolactam, 6-amino-5-methyl .... $C_7H_{13}NO$	127.19	cr (peth, bz lig), [α][20/D] -22.2	....	68-9			w, eth	B21, 242
4614	2-Caprolactone .... $C_6H_{10}O_2$	114.15	....	215-6, 103[14]	−18	....	1.4495[20]	w, al	B17[4], 4194
4615	Capronitrile .... $CH_3(CH_2)_4CN$	97.16	....	163.6, 47.3[10]	−80.3	0.8051[20/4]	1.4068[20]	al, eth, chl	B2[4], 930
4616	Caprophenone or Pentyl phenyl ketone .... $C_5H_{11}COC_6H_5$	176.21	fl	265, 122-4[15]	27	0.9576[20/4]	1.5027[25]	al, eth, ace	B7[3], 1151
4617	Caproyl chloride .... $C_5H_{11}COCl$	134.61	....	153	−87	0.9754[20/4]	1.4264[20]	eth, ace	B2[4], 928
4618	Caproyl chloride, 3-methyl .... $C_4H_9CH(CH_3)CH_2COCl$	148.63	....	163[751], 82[50]	....	0.967[20/4]	1.4293[25]	bz	B2[3], 777
4619	Caproyl chloride, 4-methyl .... $C_2H_5CH(CH_3)(CH_2)_2COCl$	148.63	....	167[767]	....	0.9677[20/4]	....	eth	B2[2], 299
4620	Caproyl chloride, 5-methyl .... $(CH_3)_2CH(CH_2)_3COCl$	148.63	....	168[739], 76-82[34]	....			eth	B2[4], 971
4621	Caprylaldehyde or Octanal....................... $CH_3(CH_2)_6CHO$	128.22	....	171, 72[20]	....	0.8211[20/4]	1.4217[20]	al, eth, ace, bz	B1[4], 3337
4622	Caprylaldehyde oxime or Caprylaldoxime .... $CH_3(CH_2)_6CH=NOH$	143.25	nd (peth, dil al)	112[9]	60			al, ace	B1[2], 758

No.	Name, Synonyms, and Formula	Mol. wt.	Color, crystalline form, specific rotation and λ_{max} (log ε)	b.p. °C	m.p. °C	Density	n_D	Solubility	Ref.
4623	Caprylamide .... $C_7H_{15}CONH_2$	143.23	lf, pl	239	fr 106-10	0.8450^{110}	....	al, eth, ace	B2[4], 992
4624	Caprylic acid or Octanoic acid .... $CH_3(CH_2)_6CO_2H$	144.22	....	239.3, 140^{23}	16.5	0.9088^{20}	1.4285^{20}	al, chl	B2[4], 982
4625	Caprylic acid-2-amino-(d) , ... $C_5H_{11}CH(NH_2)CO_2H$	159.23	$[\alpha]^{26}_D$ +23.5 (6N HCl, c=1)	....	....	....	....	aa	B4[2], 886
4626	Caprylic acid, 2-amino-(dl) .... $C_6H_{13}CH(NH_2)CO_2H$	159.23	lf (w)	(sub, d)	270	....	....	aa	B4[3], 1472
4627	Caprylic acid, 2-amino-(l) .... $C_6H_{13}CH(NH_2)CO_2H$	159.23	$[\alpha]_D$ -23 (5N HCl)	....	276	....	....	aa	B4[2], 886
4628	Caprylic acid, 8-amino .... $H_2N(CH_2)_7CO_2H$	159.23	....	....	172	....	....	al	B4[1], 527
4629	Caprylic anhydride .... $(C_7H_{15}CO)_2O$	270.42	....	280-5, 186^{15}	-1	$0.9065^{18/4}$	1.4358^{18}	al, eth, ace	B2[3], 796
4630	Caprylic acid, 2-bromo .... $C_6H_{13}CHBrCO_2H$	223.12	....	140^5	......	1.2785^{24}	1.4613^{24}	....	B2[4], 1000
4631	Caprylic acid, butyl ester or Butyl caprylate............ $C_7H_{15}CO_2C_4H_9$	200.33	....	240.5, $121-2^{20}$	-42.9	0.8628^{20}	1.4232^{25}	al, eth, ace	B2[4], 987
4632	Caprylic acid, ethyl ester or Ethyl caprylate........... $C_7H_{15}CO_2C_2H_5$	172.27	....	208.5, 104^{80}	-43.1	$0.8693^{20/4}$	1.4178^{20}	al, eth	B2[4], 987
4633	Caprylic acid, 8-fluoro .... $F(CH_2)_7CO_2H$	162.21	....	$132-3^4$	35	....	....	al, eth	B2[4], 994
4634	Caprylic acid, heptyl ester or Heptyl caprylate .... $C_7H_{15}CO_2C_7H_{15}$	242.41	....	290.5, 160^{14}	10.6	0.8596^{20}	1.4340^{20}	al, eth, ace	B2[3], 794
4635	Caprylic acid, hexyl ester or Hexyl caprylate .... $C_7H_{15}CO_2C_6H_{13}$	228.38	....	277.4	30.6	0.8603^{20}	1.4323^{25}	al, eth, ace	B2[3], 794
4636	Caprylic acid, 2-hydroxy ...., ... $C_6H_{13}CH(OH)CO_2H$	160.22	pl	$160-5^{10}$	70	....	....	al, eth	B3[4], 874
4637	Caprylic acid, 4-hydroxy, lactone or 2-Caprylolactone... $C_4H_9CH-CH_2CH_2CO$	142.20	....	$132-3^{20}$	......	$0.9796^{19/4}$	1.4451^{19}	al	B17[4], 4228
4638	Caprylic acid, methyl ester or Methyl caprylate........ $C_7H_{15}CO_2CH_3$	158.24	....	192.9, 83^{15}	-40	$0.8775^{20/4}$	1.4170^{20}	al, eth	B2[4], 986
4639	Caprylic acid, 2-methyl-3-oxo, ethyl ester .... $C_5H_{11}COCH(CH_3)CO_2C_2H_5$	200.28	....	$128-9^{12}$	......	$0.963^{0/4}$	....	al	B3, 713
4640	Caprylic acid, 2-methallyl ester .... $C_7H_{15}CO_2CH_2(CH_3)=CH_2$	198.31	....	147.8^{50}	......	0.8703	1.4308	al	B2[3], 795
4641	Caprylic acid, octyl ester or Octyl caprylate .... $C_7H_{15}CO_2C_8H_{17}$	256.43	....	306.8, 192.5^{30}	-18.1	$0.8554^{20/4}$	1.4352^{20}	al, eth, ace	B2[4], 988
4642	Caprylic acid, pentyl ester or Pentyl caprylate .... $C_7H_{15}CO_2C_5H_{11}$	214.35	....	260.2, $124-6^{20}$	-34.8	0.8613^{20}	1.4262^{25}	al, eth, ace	B2[3], 794
4643	Caprylic acid, perfluoro .... $CF_3(CF_2)_6CO_2H$	414.07	....	187-9	53	....	....	....	B2[4], 994
4644	Caprylic acid, perfluoro, methyl ester .... $C_7F_{15}CO_2CH_3$	428.10	....	158	......	$1.684^{20/4}$	1.304^{27}	....	B2[4], 995
4645	Caprylic acid, propyl ester or Propyl caprylate .... $C_7H_{15}CO_2C_3H_7$	186.30	(peth)	226.4, 112^{20}	-46.2	0.8659^{20}	1.4191^{25}	al, eth, ace	B2[4], 987
4646	Caprylic acid, isopropyl ester or Isopropyl caprylate ...... $C_7H_{15}CO_2CH(CH_3)_2$	186.30	....	93.8^{10}	......	0.8555^{20}	1.4147^{25}	....	B2[4], 987
4647	Caprylonitrile .... $C_7H_{15}CN$	125.22	....	205.2, $77-8^{10}$	-45.6	$0.8136^{20/4}$	1.4203^{20}	eth	B2[4], 993
4648	Caprylyl chloride .... $C_7H_{15}COCl$	162.66	....	195.6, 89^{20}	-63	$0.9535^{15/4}$	1.4335^{20}	eth	B2[4], 992
4649	Capsaicin .... $C_{18}H_{27}NO_3$	305.42	mcl pr or sc (peth)	$210-20^{0.01}$	65	....	....	al, eth, bz, peth	B13[3], 2192
4650	Carbamic acid, benzyl ester .... $H_2NCO_2CH_2C_6H_5$	151.17	pl (to), lf (w)	220d	91	....	....	al	B6[3], 1485
4651	Carbamic acid, N-benzyl, ethyl ester .... $C_6H_5CH_2NHCO_2C_2H_5$	179.22	lf (lig)	230d	49	....	....	al, eth, bz, chl	B12[3], 2271
4652	Carbamic acid, N-nitro, ethyl ester .... $C_6H_5CH_2N(NO_2)CO_2C_2H_5$	224.22	ye	d	......	$1.213^{20/20}$	1.5203^{20}	**al, eth**	C55, 24616
4653	Carbamic acid, butyl ester .... $H_2NCO_2C_4H_9$	117.15	pr	204d	54	....	....	al	B3[4], 54
4654	Carbamic acid, N-butyl, butyl ester .... $C_4H_9NHCO_2C_4H_9$	173.26	....	88^3	......	$0.9238^{20/20}$	1.4359^{20}	**al, eth**	B4[4], 577

No.	Name, Synonyms, and Formula	Mol. wt.	Color, crystalline form, specific rotation and λ_{max} (log ε)	b.p. °C	m.p. °C	Density	n_D	Solubility	Ref.
4655	Carbamic acid, N-butyl-N-nitro, butyl ester .... $C_4H_9N(NO_2)CO_2C_4H_9$	218.26	....	98[3]	......	1.048[20/20]	1.4359[20]	**al, eth**	Am73, 5449
4656	Carbamic acid, isobutyl ester or Isobutyl carbamate...... $H_2NCO_2CH_2CH(CH_3)_2$	117.15	lf	207	67	....	1.4098[76]	al, eth	B3[4], 56
4657	Carbamic acid, N-isobutyl, ethyl ester or Isobutyl urethane.............. $(CH_3)_2CHCH_2NHCO_2C_2H_5$	145.20	....	110[30]	>-65	0.9432[20/4]	1.4288[20]	al, eth	B4[4], 647
4658	Carbamic acid, N-tert-butyl-N-nitro, ethyl ester ..... $(CH_3)_3CN(NO_2)CO_2C_2H_5$	190.20	....	56[2]d	......	1.051[20/20]	1.4331[20]	al, eth	Am83, 1191
4659	Carbamic acid, N,N-diethyl .... $(C_2H_5)_2NCO_2H$	117.15	nd (eth)	171	-15d	0.9276[20/4]	1.4206[20]	w, al, eth	B4[3], 222
4660	Carbamic acid, N,N-diphenyl, ethyl ester or Diphenyl urethane.................... $(C_6H_5)_2NCO_2C_2H_5$	241.29	pr (lig)	360	72	....	....	w, eth, bz	B12[3], 888
4661	Carbamic acid, ethyl ester or Urethane.............. $H_2NCO_2C_2H_5$	89.10	pr (bz, to)	185	48-50	0.9862[21/4]	1.4144[51]	al, w, eth, bz	B3[4], 40
4662	Carbamic acid, N-ethyl,butyl ester .... $C_2H_5NHCO_2C_4H_9$	145.20	....	66[3]	......	0.9413[20/20]	1.4301[20]	w, al, eth	C55, 2334
4663	Carbamic acid, N-ethyl, ethyl ester or Ethyl urethane..... $C_2H_5NHCO_2C_2H_5$	117.15	....	176, 75[14]	......	0.9813[20/4]	1.4215[20]	w, al, eth	B4[4], 365
4664	Carbamic acid, N-ethyl-N-nitro,butyl ester .... $C_2H_5N(NO_2)CO_2C_4H_9$	190.20	....	79[3]	......	1.091[20/20]	1.4455[20]	w, **al, eth**	Am73, 5043
4665	Carbamic acid, N-ethyl-N-nitro,ethyl ester .... $C_2H_5N(NO_2)CO_2C_2H_5$	162.15	....	107[31]	......	1.163[20/20]	1.4432[20]	al, eth	Am73, 5449
4666	Carbamic acid, N-ethyl-N-nitro,methyl ester .... $C_2H_5N(NO_2)CO_2CH_3$	148.13	....	72[11]	......	1.233[20/20]	1.4483[20]	al, eth	Am73, 5449
4667	Carbamic acid, N-ethylidine, diethyl ester or Ethylidine diurethane .... $CH_3CH(NHCO_2C_2H_5)_2$	204.23	nd (eth)	170-8[20]	126	....	....	w, al, ace, chl	B3[1], 11
4668	Carbamic acid, methyl ester or Urethylan .... $H_2NCO_2CH_3$	75.07	nd	177, 82[14]	54	1.1361[56/4]	1.4125[56]	w, al, eth	B3[4], 37
4669	Carbamic acid, 2-methyl-2-butyl ester or tert-Pentylcarbamate, Aponal.................... $H_2NCO_2C(CH_3)_2CH_2CH_3$	131.18	nd (dil al)	....	85-7	....	....	ace, bz	B3[1], 14
4670	Carbamic acid, 3-methylbutyl ester or Isopentyl carbamate .... $H_2NCO_2CH_2CH_2CH(CH_3)_2$	131.18	nd (w)	220, 114-5[16]	64	0.9438[71/4]	1.4175[71]	al, eth	B3[4], 58
4671	Carbamic acid, N-methyl, ethyl ester or Methyl urethane $CH_3NHCO_2C_2H_5$	103.12	....	170, 80[15]	......	1.0115[20/4]	1.4183[20]	w, al	B4[4], 200
4672	Carbamic acid, N-nitro,ethyl ester .... $O_2NNHCO_2C_2H_5$	134.09	pl (eth, lig)	140d	64	1.0074[20/4]	....	w, al, ace, lig	B3[4], 247
4673	Carbamic acid, N-phenyl,ethyl ester or Ethylcarbanilate $C_6H_5NHCO_2C_2H_5$	165.19	wh nd (w), pl (dil al)	237d	53	1.1064[30/4]	1.5376[30]	al, eth, bz	B12[3], 612
4674	Carbamic acid, N-phenyl,isobutyl ester .... $C_6H_5NHCO_2CH_2CH(CH_3)_2$	193.25	nd (dil al)	216	86	....	....	al, eth, bz	B12[3], 614
4675	Carbamic acid, N-phenyl,propyl ester .... $C_6H_5NHCO_2C_3H_7$	179.22	wh nd (dil al)	....	57-9	....	....	al, eth, bz	B12[3], 613
4676	Carbamic acid, N-phenyl, isopropyl ester .... $C_6H_5NHCO_2CH(CH_3)_2$	179.22	wh nd (al)	....	90	1.09[20]	1.4989[91]	al, bz	B12[3], 613
4677	Carbamic acid, propyl ester .... $H_2NCO_2C_3H_7$	103.12	pr	196, 92[12]	60	....	....	w, al, eth, ace	B3[4], 52
4678	Carbamic acid, isopropyl ester .... $H_2NCO_2CH(CH_3)_2$	103.12	nd	181[711]	92-4	0.9951[66]	....	....	B3[4], 53
4679	Carbamic acid, N-propyl,ethyl ester .... $C_3H_7NHCO_2C_2H_5$	131.18	....	192[751], 92[22]	......	0.9921[15]	....	al	B4[4], 480
4680	Carbamic acid, N-propyl-N-nitro, ethyl ester .... $C_3H_7N(NO_2)CO_2C_2H_5$	176.17	....	66[3]	......	1.123[20/20]	1.4431[20]	al, eth	Am73, 5449
4681	Carbamic acid, N-propyl-N-nitro,methyl ester .... $C_3H_7N(NO_2)CO_2CH_3$	162.15	....	....	......	1.2585[15/15]	....	al, w, eth	B4, 146
4682	Carbamic acid, N-isopropyl, ethyl ester or Isopropylurethane .... $(CH_3)_2CHNHCO_2C_2H_5$	131.18	lf	79[15]	......	0.9548[20/20]	1.4229[20]	al, eth	B4[4], 520
4683	Carbamic acid, N-isopropyl-N-nitro, ethyl ester .... $(CH_3)_2CHN(NO_2)CO_2C_2H_5$	176.17	....	72[7]	......	1.112[20/20]	1.4381[20]	al, eth	Am73, 5449
4684	Carbamic acid, thiolo, ethyl ester .... $H_2NCOSC_2H_5$	105.16	pl (w)	sub d	109	....	....	al, eth	B3[4], 294
4685	Carbamic acid, thiono, ethyl ester or Thiourethane ...... $H_2NCSOC_2H_5$	105.16	nd lf or pyr	d	41	1.069[20/4]	1.520[20]	**al, eth, chl**	B3[4], 294

No.	Name, Synonyms, and Formula	Mol. wt.	Color, crystalline form, specific rotation and λ_{max} (log ε)	b.p. °C	m.p. °C	Density	n_D	Solubility	Ref.
4686	Carbamonitrile, N-ethyl-N-phenyl $C_6H_5N(CH_2CH_3)CN$	146.19	....	271,153[19]				....	B12[3]423
4687	Carbamyl chloride H_2NCOCl	78.49	....	62d					B3[3], 65
4688	Carbamyl chloride, N,N-diethyl $(C_2H_5)_2NCOCl$	135.60	....	186					B4[4], 379
4689	Carbamyl chloride, N,N-diphenyl $(C_6H_5)_2NCOCl$	231.68	lf (al)		85				B12[3], 893
4690	Carbamyl chloride, N-methyl-N-phenyl $C_6H_5N(CH_3)COCl$	169.62	pl (al)	280	88-9			al, eth	B12[3], 874
4691	Carbazic acid, methyl ester $H_2NNHCO_2CH_3$	90.08	....	108[12]	73			w, al, bz	B3[2], 78
4692	Carbazide $H_2NNHCONHNH_2$	90.09	nd (dil al)		154	1.616[20]		w, al	B3[4], 240
4693	Carbazide, 1,5-diphenyl $(C_6H_5NHNH)_2CO$	242.28	cr (al + 1), cr (aa)	d	170			bz, aa	B15[3], 187
4694	Carbazide, 1-phenyl $C_6H_5NHNHCONHNH_2$	166.18	nd (al)		151			w	B15[3], 187
4695	Carbazide, 1,1,5,5-tetraphenyl $[(C_6H_5)_2NNH]_2CO$	394.48	(al), nd (aa)		242			al, aa	B15[2], 115
4696	Carbazide, 3-thio $(H_2NNH)_2CS$	106.15	nd, pl (w)		170d			w	B3[3], 310
4697	Carbazole or Dibenzopyrrole $C_{12}H_9N$	167.21	pl or lf	355, 200[147]	247-8			ace	B20[4], 3824
4698	Carbazole, 9-acetyl $C_{14}H_{11}NO$	209.25	(eth), nd (w)	190[6]	69	1.161[100/24]	1.640[100]	....	B20[4], 3834
4699	Carbazole, 9-benzoyl $C_{19}H_{13}NO$	271.32	nd or pr (al)	....	98.5			al, bz	B20[4], 3838
4700	Carbazole, 9-benzyl $C_{19}H_{15}N$	257.34	nd (al)	267-8[24]	118-20			bz	B20[4], 3831
4701	Carbazole, 9-butyl $C_{16}H_{17}N$	223.32	nd (al)	218-9[19]	58			eth	B20[4], 3829
4702	Carbazole, 9-ethenyl $C_{14}H_{11}N$	193.25	cr (al)	....	66			eth	B20[4], 3830
4703	Carbazole, 9-ethyl $C_{14}H_{13}N$	195.27	nd (al)	190[10]	68	1.059[80/4]	1.6394[80]	al, eth	B20[4], 3829
4704	Carbazole, 9-methyl $C_{13}H_{11}N$	181.24	nd, lf (al)	195[12]	88			eth	B20[4], 3828
4705	Carbazole, 1-nitro $C_{12}H_8N_2O_2$	212.22	ye nd (aa)		187			aa	B20[4], 3863
4706	Carbazole, 3-nitro $C_{12}H_8N_2O_2$	212.22	ye		214				B20[4], 3865
4707	Carbazole, 3-nitro-9-nitroso $C_{12}H_7N_3O_3$	241.21	ye nd (al)		169			chl	B20[4], 3868
4708	Carbazole, 1-oxo-1,2,3,4-tetrahydro $C_{12}H_{11}NO$	185.23	nd (dil al)		170			al, bz, aa	B21[4], 4075
4709	Carbazole, 9-phenyl $C_{18}H_{13}N$	243.31	nd or pl (al)		95			al, eth, bz, aa	B20[4], 3830
4710	Carbazole, 9-propionyl $C_{15}H_{13}NO$	223.28	....		90			al, eth	B20[4], 3837
4711	Carbazole, 9-propyl $C_{15}H_{15}N$	209.29	nd (al)		50			eth	B20[4], 3829
4712	Carbazole, 1,2,3,4-tetrahydro $C_{12}H_{13}N$	171.25	lf (dil al)	325-30, 190[10]	120			al, eth, bz	B20[4], 3566
4713	Carbazone, 1,5-diphenyl $C_6H_5N=NCONHNHC_6H_5$	240.27	og nd (bz), pr (al)	....	157d			al, bz, chl	B16[3], 18
4714	Carbodiimide, diphenyl $C_6H_5N=C=NC_6H_5$	194.24	....	331, 218[21]	168-70			bz	B12[3], 906
4715	Carbon dioxide CO_2	44.00	....	−78.6 sub	−56.6 (5.2 atm)	1.0310[−20]			B3, 4
4716	Carbon diselenide CSe_2	169.93	ye	126, 46[50]	−45.5	2.6824[20/4]	1.8454[20]	....	B3[4], 436
4717	Carbon disulfide CS_2	76.14	....	46.2	−111.5	1.2632[20/4]	1.6319[20]	al, eth, chl	B3[4], 395
4718	Carbonic acid, bis(2-chloroethyl) ester or bis-(2-Chloroethyl) carbonate $(ClCH_2CH_2O)_2CO$	187.03	....	241	8	1.3506[20/4]	1.461[20]	....	B3[4], 6

No.	Name, Synonyms, and Formula	Mol. wt.	Color, crystalline form, specific rotation and λ_{max} (log ε)	b.p. °C	m.p. °C	Density	n_D	Solubility	Ref.
4719	Carbonic acid, bis (3-chloropropyl) ester $(ClCH_2CH_2CH_2O)_2CO$	215.08	....	265-70[740]	......				B3[3], 8
4720	Carbonic acid, bis(2-ethoxyethyl) ester $(C_2H_5OCH_2CH_2O)_2CO$	206.24	....	245-6[758], 112-3[5]	......	1.0439[20/4]	1.4227[20]	al, eth, ace	B3[3], 17
4721	Carbonic acid, bis(2 methoxyethyl) ester $(CH_3OCH_2CH_2O)_2CO$	178.19	....	230-2, 99-100[5]	......	1.0988[20/4]	1.4204[20]	al, eth, ace	B3[3], 17
4722	Carbonic acid, bis(2-methoxyphenyl) ester or Duotal $(2-CH_3OC_6H_4O)_2CO$	274.28	cr (al)	....	89			eth, chl	B6[3], 4233
4723	Carbonic acid, bis(trichloromethyl) ester $[Cl_3CO]_2CO$	296.75	cr (eth, peth)	203d	79				B3[4], 33
4724	Carbonic acid, dibutyl ester or Dibutyl carbonate $(C_4H_9O)_2CO$	174.24	....	207, 96-7[16]	......	0.9251[20/4]	1.4117[20]	al, eth	B3[4], 8
4725	Carbonic acid, diisabutyl ester $(i-C_4H_9O)_2CO$	174.24	....	190, 85[16]	......	0.9138[20/4]	1.4072[20]	al, eth	B3[4], 9
4726	Carbonic acid, di-*tert*-butyl ester or di-*tert*-Butyl carbonate $[(CH_3)_3CO]_2CO$	174.24	cr (al)	158[767]	40 (sub)			al	B3[4], 9
4727	Carbonic acid, diethyl ester or Diethyl carbonate $(C_2H_5O)_2CO$	118.13	....	126	-43	0.9752[20/4]	1.3845[20]	al, eth	B3[4], 5
4728	Carbonic acid, dimethyl ester or Dimethyl carbonate $(CH_3O)_2CO$	90.08	....	90-1	2-4	1.0694[20/4]	1.3687[20]		B3[4], 3
4729	Carbonic acid, diphenyl ester or Diphenyl carbonate $(C_6H_5O)_2CO$	214.22	nd (al, bz)	306, 168[15]	83 (88)	1.1215[87/4]		eth	B6[4], 629
4730	Carbonic acid, di-isopentyl ester or Diisopentyl carbonate $(i-C_5H_{11}O)_2CO$	202.30	....	232-5[751], 122[16]	......	0.9067[20/4]	1.4174[20]		B3[3], 11
4731	Carbonic acid, di-propyl ester or Dipropyl carbonate $(C_3H_7O)_2CO$	146.19	....	168, 59.5[15]	......	0.9435[20/4]	1.4008[20]	al, eth	B3[4], 6
4732	Carbonic acid, Diisopropyl ester $(i-C_3H_7O)_2CO$	146.19	....	147, 43[12]	......	0.9162[20/4]	1.3932[20]	al	B3[3], 9
4733	Carbonic acid, dithiolo, diethyl ester $(C_2H_5S)_2CO$	150.26	ye	197, 85-7[19]	......	1.085[20]	1.5237[18]	al, eth	B3[3], 339
4734	Carbonic acid, di-2-tolyl ester $(2-CH_3C_6H_4O)_2CO$	242.28	nd (al)	144-5[0.5]	60			aa	B6[3], 1256
4735	Carbonic acid, di-3-tolyl ester $(3-CH_3C_6H_4O)_2CO$	242.28	cr (al)	....	50-1			bz, chl	B6[3], 1307
4736	Carbonic acid, di-4-tolyl ester $(4-CH_3C_6H_4O)_2CO$	242.28	....	....	115			chl	B6[3], 1366
4737	Carbonic acid, ethyl-2-butoxyethyl ester $C_2H_5O-CO-OCH_2CH_2OC_4H_9$	190.24	....	224[759]	......	0.9756[25/4]	1.4143[25]	al, eth, ace	B3[3], 14
4738	Carbonic acid, ethyl-methyl ester $CH_3O-CO-OC_2H_5$	104.12	....	107-8	-14	1.012[20/4]	1.3778[20]	al, eth	B3[4], 4
4739	Carbonic acid, trithio $(HS)_2CS$	110.22	red	57d	-30	1.47[17/4]	....	al, chl, to	B3[4], 428
4740	Carbon suboxide or 1,3-Dioxoallene $OC=C=CO$	68.03	gas	6.8	-107	1.114[0/4]	1.4538[0]	eth, bz	B1[4], 3764
4741	Carbonyl fluoride COF_2	66.01	....	-83	-114	1.139[-114]	....		B3[4], 21
4742	Carbonyl sulfide COS	60.08	....	-50	-138	1.028[17/4]	....	al	B3[4], 271
4743	Carbothialdine $C_5H_{10}N_2S_2$	162.28	cr (al)	....	120d			al	B27[2], 687
4744	Δ^3-Carene-*(dl)* $C_{10}H_{16}$	136.24	....	167[732], 44-5[8]	......	0.8602[20/4]	1.4759[20]	eth, ace, bz, aa	B5[2], 362
4745	Δ^3-Carene-*(l)* $C_{10}H_{16}$	136.24	[α]$^{20/}_D$-5.72	168-9[705], 123-4[200]	......	0.8586[30/30]	1.4684[30]	eth, ace, bz, aa	B5[4], 449
4746	Δ^4-Carene-*(l)* $C_{10}H_{16}$	136.24	[α]$^{20/}_D$ + 62.2	167[707], 64[20]	......	0.8441[30/4]	1.4740[30]	eth, ace, bz, aa	B5[4], 451
4747	Carminic acid $C_{22}H_{20}O_{13}$	492.40	red mcl pr (ag MeOH)	....	d 136			al	B10[3], 4874
4748	Carnaubyl alcohol $CH_3(CH_2)_{19}OH$	438.81	lf (al)	....	69			al, ace	B1[2], 472
4749	Carnosine-*(D,-)* or Ignotine. β-Alanyl-*(D,-)*-histidine $C_9H_{14}N_4O_3$	226.24	[α]$^{18/}_D$ -20.4 (w, c=1.5)	....	260			w	B25[2], 408

No.	Name, Synonyms, and Formula	Mol. wt.	Color, crystalline form, specific rotation and λ_{max} (log ε)	b.p. °C	m.p. °C	Density	n_D	Solubility	Ref.
4750	Carnosine-(L +) .... $C_9H_{14}N_4O_3$	226.24	nd (w-al), $[\alpha]^{20}_D$ +24.1 (w, c=1.5)	....	246-50d	....	....	w	B25[2], 408
4751	α-Carotene or α-Carotin................... $C_{40}H_{36}$	536.90	red pl or pr (peth, bz-MeOH)		187.5	$1.00^{20/20}$	....	eth, bz, chl	B5[3], 2457
4752	β-Carotene or Provitamin A.................. $C_{40}H_{36}$	536.90	red br hex pr (bz-MeOH)		184	$1.00^{20/20}$		eth, ace, bz, peth	B5[3], 2453
4753	γ-Carotene or γ-Carotin................... $C_{40}H_{36}$	536.90	red br (bz-MeOH) vt pr (bz-eth)		178			bz, chl	B5[3], 2451
4754	Carpaine-(d) .... $C_{28}H_{50}N_2O_4$	478.72	mcl pr (al or ace) $[\alpha]^{21}_D$ +24.7 (al, c=1.07)	sub $120^{0.01}$	121			al, eth, ace, bz, chl	B27[2], 209
4755	Carpaine hydrochloride .... $C_{28}H_{50}N_2O_4 \cdot HCl$	515.18	wh mcl nd or pl	....	225d			w, al, eth	B27[2], 210
4756	Carpiline or Carpidine, Pilosine .... $C_{26}H_{18}N_2O_3$	286.33	pl (al), pr (dil al or w), $[\alpha]^{20}_D$ +35.9 (nl)		187			al	B27[1], 612
4757	Carvacrol or 2-Methyl-5-iso-propyl phenol............ 2-CH₃-5-(CH₃)₂CHC₆H₃OH	150.22	nd	237.7, $101-2^{10}$	1	$0.9772^{20/4}$	1.5230^{20}	al, eth, ace	B6[3], 1885
4758	Carvacrol-acetate or 2-Methyl-5-iso-propyl phenyl acetate .... 2-CH₃-5-(CH₃)₂CHC₆H₃O.CCH₃	192.24	....	245-8	...	0.9806^{11}	1.4913^{28}	al, eth	B6[2], 494
4759	Carvacrol, 4-amino or 2-Methyl-5-iso-propyl-4-amino-phenol................... 2-CH₃-4-H₂N-5-CH(CH₃)₂-C₆H₂OH	165.24	cr (MeOH)	....	134			MeOH	B13[3], 1801
4760	Carvenone-(dl)................... $C_{10}H_{16}O$	152.24	....	$235-6^{762}$, 104^{10}	....	$0.9263^{20/4}$	1.4826^{20}	ace	B7[3], 332
4761	Carvenone-(l)................... $C_{10}H_{16}O$	152.24	$[\alpha]_D$ -2.08	232-4		$0.9290^{20/4}$	1.4805	ace	B7[1], 66
4762	Carveol, dihydro-(d)................... $C_{10}H_{16}O$	154.26	$[\alpha]^{18/}_D$ +34.2	225, 107^{15}		$0.9274^{20/4}$	1.4780^{20}	eth	B6[3], 256
4763	Carveol, dihydro-(l)................... $C_{10}H_{16}O$	154.26	$[\alpha]^{20/}_D$ -33.3	107^{14}		0.9368^{15}	1.4836^{20}	ace	B6[3], 256
4764	β-Carveol, dihydro................... $C_{10}H_{16}O$	154.26	$[\alpha]_D$ +7.64	130^{20}		$0.9266^{20/4}$	1.4809^{20}	....	B6, 64
4765	Carvomenthane-(d)................... $C_{10}H_{16}$	138.25	$[\alpha]_{5780}$ +118	175-7, 77^{24}	....	$0.8246^{18/4}$	1.4563^{18}	al, bz, peth	B5[4], 301
4766	Carvomenthol-(d) or Hexahydrocarvacrol............ $C_{10}H_{20}O$	156.27	$[\alpha]^{21/}_D$ 31.4	222, 102^{14}		$0.8995^{20/4}$	1.4617^{20}	al, eth	B6[4], 148
4767	Carvomenthol-(l-neo)................... $C_{10}H_{20}O$	156.27	$[\alpha]^{21/}_D$ -41.7	217-8, 102^{18}		$0.9012^{20/4}$	1.4632^{20}	al, eth	B6[4], 148
4768	Carvomenthone................... $C_{10}H_{18}O$	154.26	$[\alpha]^{21/}_D$ +17.15	$218-20^{745}$, $95-9^{15}$		0.9075^{15}	1.4544^{20}	al, ace, chl	B7[3], 146
4769	Carvone-(d) or Carvol................... $C_{10}H_{14}O$	150.22	$[\alpha]^{20/}_D$ +69.1	231, 104^{11}		$0.9608^{20/4}$	1.4999^{18}	al, eth, chl	B7[3], 561
4770	Carvone-(dl)................... $C_{10}H_{14}O$	150.22		231, 85^5		$0.9645^{15/15}$	1.5003^{20}	al, eth, chl	B7[3], 564
4771	Carvone-(l)................... $C_{10}H_{14}O$	150.22	$[\alpha]^{20/}_D$ -62.46	231, 98^9		$0.9593^{20/4}$	1.4988^{20}	al, eth, chl	B7[3], 561
4772	Carvone oxime-(a,d) or D-Carotime-(a)............ $C_{10}H_{15}NO$	165.24	lf (al) $[\alpha]^{17/}_D$ +39.71 (al, p=8.45)	....	72			al, eth, ace	B7[3], 564
4773	Carvone oxime-(l,a) or L-Carvoximine-(a)............ $C_{10}H_{15}NO$	165.24	mcl (dil al) $[\alpha]^{18/}_D$ -39.43 (al, p=4.33)	....	73.5	1.0140^{73}		al, eth, ace	B7[3], 564

No.	Name, Synonyms, and Formula	Mol. wt.	Color, crystalline form, specific rotation and λ_{max} (log ε)	b.p. °C	m.p. °C	Density	n_D	Solubility	Ref.
4776	Carvone oxime-(dl) or dl-Carvoxime .... $C_{10}H_{15}NO$	165.24			93-4			al, eth, ace	B7³, 564
4777	Carvone, dihydro-(d) .... $C_{10}H_{16}O$	152.24	[α]_D +17.5	221-2		0.928¹⁹	1.4724	eth, ace	B7³, 337
4778	Carvone, dihydro-(l) .... $C_{10}H_{16}O$	152.24	[α]²⁰/_D-19	221-2, 104¹⁸		0.9253²⁰/⁴	1.4717²⁰	eth, ace	B7³, 337
4779	α-Caryophyllene .... $C_{15}H_{24}$	204.36	[α]²⁰/⁴ +1±0.3 (chl, c=9.26)	123¹⁰		0.8905²⁰/⁴	1.5038²⁰		B5⁴, 1171
4780	β-Caryophyllene .... $C_{15}H_{24}$	204.36	[α]²⁰/_D -9.08	122¹³·⁵		0.9075²⁰/⁴	1.4988²⁰	bz	B5⁴, 1182
4781	γ-Caryophyllene or Isocaryophyllene ... $C_{15}H_{24}$	204.36	[α]¹⁹/_D -26.2	130-1²⁴		0.8953²⁰/⁴	1.4967¹⁹	bz	B5³, 1083
4782	Caryophyllenic acid-(l,cis) .... $C_9H_{14}O_4$	186.21	pr (w) [α]_546-7.4		77-8			w, al, eth, ace, bz	B9³, 3850
4783	Caryophyllenic acid-(d,trans) .... $C_9H_{14}O_4$	186.21	nd [α] ²⁰/_D +35.3 (bz)		81-2			w, al, eth, ace, bz	B9³, 3850
4784	Caryophyllin or Oleanolic acid .... $C_{30}H_{48}O_3$	456.72	nd or pr (al) [α]²⁰/_D +83.3 (chl, c=0.9)	280-308, sub (vac)	310d			aa, Py	B10³, 1049
4785	Catechin-(cis´d) or 3,5,7,3´,4´-Flavanpentol........... $C_{15}H_{14}O_8$	290.28	nd (w + 4) [α]¹⁸/_D +18.4 (w, c=0.9)	240-5	96 (hyd) 177 (anh)	1.344⁴⁴		al, ace, aa	B17⁴, 3841
4786	Catechin-(cis,dl) .... $C_{15}H_{14}O_6$	290.28	nd (w + 3)		212-4d			al, ace	B17⁴, 3842
4787	Catechin-(cis,l) .... $C_{15}H_{14}O_6$	290.28	nd (+ 4, w), [α]_5780 (w ace, p=3)		96 (hyd) 177 (anh)			al, ace, aa	B17², 255
4788	Catechol or 1,2-Dihydroxybenzene .... 2-HOC_6H_4OH	110.11	cr	245⁷⁵⁰	105	1.1493²¹	1.604	w, al, eth, ace	B6³, 4187
4789	Catechol, 3-acetyl or 2,3-Dihydroxyacetophenone .... 2, 3-(HO)_2C_6H_3COCH_3	152.16	ye pr (bz-lig)		97-8			al	B8³, 2080
4790	Catechol, 4-acetyl or 3,4-Dihydroxyacetophenone .... 3,4(HO)_2C_6H_3COCH_3	152.16	nd (w or chl)		115-6				B8³, 2108
4791	Catechol, 4-bromo or 4-Bromo-1,2-dihydroxybenzene ... 1,2-(HO)_2-4-Br-C_6H_3	189.02	pr or nd (chl)		87			w, al, eth, bz	B6³, 4253
4792	Catechol, 3-chloro or 3-Chloro-1,2-dihydroxybenzene.... 3-Cl-1,2-(HO)_2C_6H_3	144.56	cr (lig)	110-1¹¹	46-8			lig	B6³, 4249
4793	Catechol, 4-chloro .... 4-Chloro-1,2-(HO)_2C_6H_3	144.56	lf (bz-peth)	139¹⁰·⁵	90-1			w, al, eth, ace, aa	B6³, 4249
4794	Catechol, diacetate .... 2-CH_3CO_2-C_6H_4-O_2CCH_3	194.19	nd (al)	142-3⁹	64-5			al, eth, chl	B6³, 4228
4795	Catechol, dibenzoate .... 1,2-(C_6H_5CO_2)C_6H_4	318.33	lf (eth-al)		86			al, eth, bz	B9³, 552
4796	Catechol, dibenzyl ether .... 1,2-(C_6H_5CH_2O)_2C_6H_4	290.37	yesh nd or pr (al)		63-4			eth, peth	B6³, 4219
4797	Catechol, dibutyl ether or 1,2-Dibutoxybenzene .... 1,2-(C_4H_9O)_2C_6H_4	222.23	ye	241⁷⁶⁵, 135-8¹²					B6³, 4210
4798	Catechol, 3,5-dichloro .... 3,5-Cl_2C_6H_2(OH)_2-1,2	179.00	pr		83-4			al, ace	B6, 783
4799	Catechol, 4,5-dichloro .... 4,5-Cl_2C_6H_2(OH)_2-1,2	179.00	pr (chl-CS_2), nd (bz-peth)		116-7			w, al, bz	B6³, 4252
4800	Catechol, diethyl ether or 1,2-Diethoxybenzene .... 1,2-(C_2H_5O)_2C_6H_4	166.22	pr (peth, dil al)	219	43-5	1.0075²⁰/⁴	1.5083²⁵	al, eth	B6³, 4208
4801	Catechol, dimethylether or Veratrole-1-2 dimethoxy benzene.............. 1,2-(CH_3O)_2C_6H_4	138.17	cr (lig)	206⁷⁵⁰, 90¹⁰	22.5	1.0842²⁵/²⁵	1.5827²¹/₄	al, eth	B6³, 4205
4802	Catechol, 3,5-dimethyl .... 3,5-(CH_3)_2C_6H_2(OH)_2-1,2	158.17	pr (w), nd (peth-bz)		73-4			w, al, eth	B6³, 4591
4803	Catechol, 4,5-dimethyl .... 4,5-(CH_3)_2C_6H_2(OH)_2-1,2	138.17	mcl pr or nd (peth)	sub	87-8			w, al, eth	B6³, 4584

No.	Name, Synonyms, and Formula	Mol. wt.	Color, crystalline form, specific rotation and λ_{max} (log ε)	b.p. °C	m.p. °C	Density	n_D	Solubility	Ref.
4804	Catechol, dipropylether or 1,2-Dipropoxy benzene........ 1,2$(C_3H_7O)_2C_6H_4$	194.28	....	234-7, 117-20[12]		0.9554[33/4]	1.4950[27]	....	B6[3], 4209
4805	Catechol, dithio or 1,2-Dimercaptobenzene............ 1,2-$(HS)_2C_6H_4$	142.25	....	238-9, 120[17]	28-9			al, eth, bz	B6[3], 4286
4806	Catechol, 4-iodo 4-$IC_6H_3(OH)_2$-1,2	236.01	lf (CCl_4)	sub	92			al, eth, ace, bz	B6[3], 4262
4807	Catechol, 3-methoxy .... 3-$CH_3OC_6H_4(OH)_2$-1,2	140.15	nd	129[10]	43-4				B6[3], 6264
4808	Catechol, 3-methyl 3-$CH_3C_6H_3(OH)_2$-1,2	124.15	lf (bz)	241, 127[12]	68			w, al, bz, chl	B6[3], 4492
4809	Catechol, monoacetate .... 2-$(CH_3CO_2)C_6H_4OH$	152.16	pl	189-91[102], 148[25]	57-8			w, al, ace, peth	B6[3], 4227
4810	Catechol, monobenzoate 2-$(C_6H_5CO_2)C_6H_4OH$	214.22	nd (w)		130-1			al	B9[3], 551
4811	Catechol, monobenzylether .... 2-$(C_6H_5CH_2O)C_6H_4OH$	200.24	....	173-4[13]		1.154[22]	1.5906[18]	al, eth	B6[3], 4218
4812	Catechol, monobutyl ether .... 2-$(C_4H_9O)C_6H_4OH$	166.22	....	231-4, 159[69]		1.026[25]	1.5113[25]	....	B6[3], 4209
4813	Catechol, monoethyl ether or 2-Ethoxy phenol 2-$(C_2H_5O)C_6H_4OH$	138.7	....	217, 68[4]	29			al, eth	B6[3], 4207
4814	Catechol, monomethyl ether or 2-Methoxy phenol, Guaiacol 2-$(CH_3O)C_6H_4OH$	124.15	hex pr	205	32	1.1287[21/4]	1.5429[20]	al, eth, chl	B6[3], 4200
4815	Catechol, monomethyl ether, acetate .... 2-$CH_3OC_6H_4OCOCH_3$	166.18	....	123-4[11]		1.1285[25/4]	1.5101[25]	al, eth	B6[3], 4227
4816	Catechol, monopropyl ether or 2-Propoxy phenol........ 2-$(C_3H_7O)C_6H_4OH$	152.21	....	228-9, 80-3[4]		1.0523[25]	1.5176[25]	al	B6[3], 4209
4817	Catechol, 4-nitro .... 4-$NO_2C_6H_3(OH)_2$-1,2	155.11	....	174				w, al, eth, chl	B6[3], 4263
4818	Catechol, 3-iso-propyl 6-methyl .... 3-i-C_3H_7-6-CH_3-$C_6H_2(OH)_2$-1,2	166.22	....	270	48			eth, ace	B6[3], 4673
4819	Catechol, 4-propyl .... 4-$C_3H_7C_6H_3(OH)_2$-1,2	152.20	pr (w, bz)	152[13]	60	1.100[18/4]	1.4440[18]	al, eth, ace	B6[3], 4613
4820	Catechol, 4-iso-propyl .... 4-i-$C_3H_7C_6H_3(OH)_2$-1,2	152.20	lf (lig)	270-2, 168[26]	78				B6[3], 4632
4821	Catechol, tetrabromo .... $Br_4C_6(OH)_2$-1,2	425.72	nd (bz, al) lf (bz-peth)		192-3			al, bz	B6[3], 4261
4822	Catechol, tetrachloro $Cl_4C_6(OH)_2$-1,2	247.89	cr (dil al, bz) cr (+3w) (aq aa)		110 (anh), 94 (+3w)				B6[3], 4253
4823	Catechol, 3,4,5-trichloro .... 3,4,5-$Cl_3C_6H(OH)_2$-1,2	213.46	(i) pr (+1w, aa), (ii) pr (+$^1/_2$w, bz)	115, 134-5				al, eth, aa	B6[1], 389
4824	Cedrene .... $C_{15}H_{24}$	204.36	$[\alpha]^{20}_D$ −91.3	262-3, 124-6[12]		0.9342[20/4]	1.5034[20]	bz, lig	B5[3], 1095
4825	Cedrol .... $C_{14}H_{26}O$	222.37	$[\alpha]$ +10.08 (chl, c=10)		86	0.9496[90/20]	1.4824[90/0]	....	B6[3], 424
4826	β-Cellobiose or 4-o-β-D-Glucopyranosyl-β-D-glucose .... $C_{12}H_{22}O_{11}$	342.30	cr (dil al), $[\alpha]^{20}_D$ +14.2→ +34.6 (mut) (w, c=8, 15 hr)		d 225			w	B17[4], 3061
4827	Cellobiose, octa-acetate (α-anomer) .... $C_{28}H_{38}O_{19}$	678.61	nd (al), $[\alpha]^{20/D}$ +43.6 (chl, c=6)		229.5			chl, aa	B17[4], 3589
4828	Cellobiose, octa-acetate (β-anomer) .... $C_{28}H_{38}O_{19}$	678.61	nd (al), $[\alpha]^{20}_D$ −14.7 (chl, c=5)		202			chl	B17[4], 3590
4829	Cellulose or Polycellobiose .... $(C_6H_{10}O_5)_n$	(162.14)$_n$	wh amor		260-70d	1.27-1.60			

No.	Name, Synonyms, and Formula	Mol. wt.	Color, crystalline form, specific rotation and λ_{max} (log ε)	b.p. °C	m.p. °C	Density	n_D	Solubility	Ref.
4830	Cellulose, hexanitrate or Gun cotton................ $(C_{12}H_{14}N_6O_{22})_x$	(594.27)$_x$	wh amor		160-70 (ign)	1.66		ph NO_2	
4831	Cellulose, pentanitrate .... $(C_{12}H_{15}N_5O_{20})_x$	(549.28)$_x$	wh amor			1.66		eth-al	
4832	Cellulose, tetranitrate or in Collodion............... $(C_{12}H_{16}N_4O_{18})_x$	(504.26)$_x$	wh amor			1.66		eth-al	
4833	Cellulose, triacetate $(C_{12}H_{16}O_8)_x$	(288.26)$_x$	yesh fl [α]$_D$ −22.5 (chl)					aa	
4834	Cellulose, triethylether or Ethylcellulose .... $(C_{12}H_{22}O_5)_x$	(246.31)$_x$	wh nd (bz) [α]$^{20}_D$ + 26.1 (bz)		240-55			eth	
4835	Cellulose, trinitrate or in Collodion................ $(C_{12}H_{27}N_3O_{16})_x$	(459.28)$_x$	wh			1.66		ace, aa	
4836	Cepharanthine $C_{37}H_{38}N_2O_6$	606.78	ye amor pw [α]$^{20}_D$ + 277 (chl)		145-55			al, eth, ace, bz	C49, 1745
4837	Cerane or Isohexacosane $CH_3(CH_2)_{24}CH_3$	366.72	pl (eth), sc (w)	207$^{0.7}$	61			al, eth	B1², 143
4838	Cerulignone. $[(C_6H_2O(CH_3O)_2]_2$	304.30	bl gr						B8², 573
4839	Cetane or Hexadecane............... $CH_3(CH_2)_{14}CH_3$	226.45	lf (ace)	287, 149^{10}	18.2	0.7733$^{20/4}$	1.4345	eth	B1⁴, 537
4840	Cetene or 1-Hexadecene $CH_3(CH_2)_{13}CH=CH_2$	224.41	lf	284.4, 155^{15}	4.1	0.7811$^{20/4}$	1.4412^{20}	al, eth, peth	B1⁴, 927
4841	Cetyl alcohol or 1-Hexadecanol................ $CH_3(CH_2)_{14}CH_2OH$	242.45	fl (AcOEt)	344, 190^{15}	50	0.8176$^{50/4}$	1.4283^{79}	eth, ace, bz, chl	B1⁴, 1876
4842	Cetylamine or 1-Amino hexadecane .... $CH_3(CH_2)_{14}CH_2NH_2$	241.47	lf	322.5, 144²	46.8	0.8129$^{20/4}$	1.4496^{20}	al, eth, ace, bz, chl	B4⁴, 818
4843	Cetyl Phenyl Ether or Hexadecyl phenyl ether .... $C_{16}H_{33}OC_6H_5$	318.55	lf (al)	200¹	41.8	0.8434^{82}	1.4556^{82}		B6³, 555
4844	Cetyl sulfate $(C_{16}H_{33}O)_2SO_2$	546.95			66.2			w	B1⁴, 1879
4845	Cevagenine $C_{27}H_{43}NO_8$	509.65	nd (MeOH-eth), [α] $_{20/D}$ −47.5 (al)		246-8				B21⁴, 6815
4846	Chalcone dibromide-(threo) .... $C_6H_5CHBrCHBrCOC_6H_5$	386.08	nd (al)		122-3			al	B7³, 2155
4847	Chalcone dibromide-(erythro) .... $C_6H_5CHBrCHBrCOC_6H_5$	386.08	pr or nd (al)		159-60				B7³, 2154
4848	Chalcone-(trans) or Benzalacetophenone .... $C_6H_5COCH=CHC_6H_5$	208.26	pa ye lf, pr, nd (peth)	345-8d, 208²⁵	(i)59 (ii)57 (iii)49	1.0712$^{62/4}$		eth, bz, chl	B7³, 2380
4849	Chalcone, 4,4-dimethyl .... (4-$(CH_3C_6H_4)COCH=CH(C_6H_4CH_3$-4)	236.32	cr (MeOH)		127-9			al	B7², 441
4850	Chalcone, 3,3′-dinitro .... (3-$O_2NC_6H_4)COCH=CH(C_6H_4NO_2$-3)	298.26	pa ye nd (aa)		210-1			bz	B7³, 2407
4851	Chalcone, 2-methoxy or 2-Anisylidene acetophenone..... (2-$CH_3OC_6H_4)CH=CHCOC_6H_5$	238.29	yesh nd (peth or eth-lig)		64-5			al, eth, bz, chl	B8³, 1456
4852	Chalcone, 3-methoxy (3-$CH_3OC_6H_4)CH=CHCOC_6H_5$	238.29	yesh pl or pr (MeOH)	247^{12}	65			al, eth, ace, bz	B8³, 1463
4853	Chalcone, 4-methoxy (4-$CH_3OC_6H_4)CH=CHCOC_6H_5$	238.29	ye nd (al)	187-8^{18}	79			al, eth, chl, aa	B8³, 1464
4854	Chalcone, 3,4-methylene dioxy or Piperonylidine aceto-phenone [3,4-$(CH_2O_2)C_6H_3]CH=CHCOC_6H_5$	252.27	ye nd (al)		128			al, aa	B19⁴, 1866
4855	Chalcone, 2-nitro (2-$O_2NC_6H_4)CH=CHCOC_6H_5$	253.26	pa br nd (al)		125			al, eth, aa	B7³, 2399
4856	Chalcone, 2′-nitro $C_6H_5CH=CHCO(C_6H_4NO_2$-2)	253.26	nd (al)		128-9			al, eth	B7³, 2402
4857	Chalcone, 3-nitro (3-$O_2NC_6H_4)CH=CHCOC_6H_5$	253.26	ye nd (al or bz)		145-6			al, bz, chl, aa	B7³, 2400
4858	Chalcone, 4-nitro (4-$O_2NC_6H_4)CH=CHCOC_6H_5$	253.26	pa ye nd (al), pl (bz)		164			al, chl	B7³, 2401

No.	Name, Synonyms, and Formula	Mol. wt.	Color, crystalline form, specific rotation and λmax (log ε)	b.p. °C	m.p. °C	Density	n_D	Solubility	Ref.
4859	Chalcone, α-nitro .... $C_6H_5C(NO_2)=CHCOC_6H_5$	253.26	ye pl (eth or bz-lig) cr aa	....	90	....	....	eth, ace, bz	B7[3], 2403
4860	Chaulmoogric acid-(d) or d-13(2-cyclopentenyl)tridecanoic acid .... $C_{18}H_{32}O_2$	280.46	pl or lf (al, aa), $[\alpha]_D$ +62 (chl)	247-8[20]	68.5	....	....	eth, chl	B9[3], 284
4861	Chaulmoogric acid-(dl) .... $C_{18}H_{32}O_2$	280.46	cr (peth)	247-8[20]	....	....	....	eth, chl	B9[3], 285
4862	Chloroacetaldehyde, trimer or 2,4,6-tris (chloromethyl) 1,3,5-trioxane .... $C_6H_9Cl_3O_3$	235.50	nd (eth)	142-4[10]	87	....	....	eth	B19[4], 4718
4863	Chelerythrine .... $C_{19}H_{21}NO_5$	365.39	cr (chl-MeOH), cr (al + l)	207	....	....	....	chl	B27[2], 563
4864	Chelidonic acid or 4-Pyrone-2,6-dicarboxylic acid .... $C_6H_4O_6$	184.11	rose mcl nd (al-w + 1w)	....	262	....	....	....	B18[4], 6136
4865	Chelidonic acid, diethyl ester .... $C_{11}H_{12}O_6$	240.22	pr, nd	....	69	....	....	eth	B18[4], 6137
4866	Chelidonine-(d) .... $C_{20}H_{19}NO_5$	353.38	mcl pl (al + 1w), $[\alpha]_D$ +151 (al, 1%)	....	136-40d	....	....	al, eth, chl	B27[2], 615
4867	Chelidonine, hydrochloride-(d) .... $C_{20}H_{19}NO_5 \cdot HCl$	389.84	wh (w)	....	....	....	....	....	B27, 557
4868	Chloral or Trichloroacetaldehyde .... Cl_3CCHO	147.39	....	97.8	−57.5	1.51214[20/4]	1.4111[20]	w, al, chl	B1[4], 3142
4869	Chloral ammonia .... $Cl_3C CH(OH)NH_2$	164.42	nd (al)	100d	72-4	....	....	al, eth, bz	B1[4], 3147
4870	Chloral hydrate .... $Cl_3CCH(OH)_2$	165.40	mcl pl (w)	96.3[764]/d	57	1.9081[20/4]	....	w, al, eth, ace, bz, chl, Py	B1[4], 3143
4871	Chloralide or 2,5-bis(trichloromethyl)1,3-dioxolan-4-one .... $C_5H_2Cl_6O_3$	322.79	pr (al or eth)	272-3, 147-8[12]	116	....	....	eth, aa	B19[4], 1571
4872	Chloroacetamide .... $ClCH_2CONH_2$	93.51	mcl pr	224.5[743]	121	....	....	w, al	B2[4], 490
4873	Chloroacetamide, N-allyl-N-phenyl .... $ClCH_2CON(C_6H_5)CH_2CH=CH_2$	217.74	....	119[0.15]	....	....	1.5079[25]	....	Am78, 2556
4874	Chloroacetamide, N,N-bis(2-chloroallyl) .... $ClCH_2CON(CH_2C(Cl)=CH_2)_2$	242.53	....	161—3[12]	....	....	1.5220[25]	....	B4[4], 1089
4875	Chloroacetamide-N,N-bis(3-chloroallyl) .... $ClCH_2CON[CH_2C(Cl)=CH_2]_2$	242.53	....	140.5[1]	....	....	1.5220[25]	....	Am78, 2556
4876	Chloroacetamide, bis(2-chloropropyl) .... $ClCH_2CON(CH_2CHClCH_3)_2$	246.57	....	134[0.7]	....	....	1.5018[25]	....	B4[4], 499
4877	Chloroacetamide, N,N-bis(2-ethylhexyl) .... $ClCH_2CON[CH_2CH(C_2H_5)C_4H_9]_2$	317.95	....	154[0.8]	....	....	1.4622[25]	....	Am78, 2556
4878	Chloroacetamide, N,N-bis(2-methylallyl) .... $ClCH_2CON[CH_2C(CH_3)=CH_2)]_2$	20.70	....	133-5[20]	....	....	1.4882[2]	....	B4[4], 1105
4879	Chloroacetamide, N,N-bis(3-Methylbutyl) .... $ClCH_2CON[CH_2CH_2CH(CH_3)_2]_2$	233.79	....	109[0.6]	....	....	1.4625[25]	....	B4[4], 702
4880	Chloroacetamide, N-butyl .... $ClCH_2CONHC_4H_9$	149.62	....	110[7]	....	....	1.4665[25]	....	B4[4], 567
4881	Chloroacetamide, N-butyl-iso-propyl .... $ClCH_2CON(C_4H_9)CH(CH_3)_2$	168.02	....	101[1.4]	....	....	1.5078[25]	....	B4[4], 567
4882	Chloroacetamide, N-sec-butyl .... $ClCH_2CONHCH(CH_3)C_2H_5$	149.62	....	68[0.7]	45-6	....	....	peth	B4[4], 621
4883	Chloroacetamide, N-tert-butyl .... $ClCH_2CONHC(CH_3)_3$	149.62	cr (peth)	....	84	....	....	peth	B4[4], 662
4884	Chloroacetamide, N-butyl-N-ethyl .... $ClCH_2CON(C_2H_5)C_4H_9$	177.68	....	90[1.5]	....	....	1.4665[25]	....	B4[4], 567
4885	Chloroacetamide, N-2-chloroallyl .... $ClCH_2CONHCH_2CCl=CH_2$	168.02	....	101[1.4]	....	....	1.5078[25]	....	B4[4], 1089
4886	Chloroacetamide, N-2-chloroallyl-N-phenyl .... $ClCH_2CON(C_6H_5)CH_2CCl=CH_2$	243.12	....	138[0.7]	....	....	1.5602[25]	....	Am78, 2556
4887	Chloroacetamide, N-3-chloroallyl .... $ClCH_2CONHCH_2CH=CHCl$	168.02	cr (peth)	112[0.5]	52-3	....	....	peth	B4[4], 1087
4888	Chloroacetamide, N-(2-chloro-4-nitrophenyl) .... $(4-NO_2-2-Cl-C_6H_3)NCOCH_2Cl$	263.08	cr (peth)	....	118-9	....	....	....	B12[3], 1664

No.	Name, Synonyms, and Formula	Mol. wt.	Color, crystalline form, specific rotation and λ_{max} (log ε)	b.p. °C	m.p. °C	Density	n_D	Solubility	Ref.
4889	Chloroacetamide, N-(4-chlorophenyl)-N-ethyl $(4\text{-}ClC_6H_4)N(C_2H_5)COCH_2Cl$	323.11	cr (peth)		70-1			peth	Am78, 2557
4890	Chloroacetamide, N-2-chloropropyl $ClCH_2CONHCH_2CHClCH_3$	170.04		88[1.5]			1.4942[25]		B4[4], 499
4891	Chloroacetamide, N-3-chloropropyl $ClCH_2CONHCH_2CH_2CH_2Cl$	170.04			36-7				B4[4], 501
4892	Chloroacetamide, N-N-diallyl $ClCH_2CON(CH_2CH=CH_2)_2$	173.64		92[0.7]			1.4932[25]		B4[4], 1064
4893	Chloroacetamide, N,N-dibenzyl $ClCH_2CON(CH_2C_6H_5)_2$	273.77		190[1.8]			1.5837[25]		Am78, 2556
4894	Chloroacetamide, N,N-di-iso-butyl $ClCH_2CON(i\text{-}C_4H_9)_2$	205.73		99[2]			1.4642[25]		B4[4], 633
4895	Chloroacetamide, N,N-di-sec-butyl $ClCH_2CON(sec\text{-}C_4H_9)_2$	205.73		92[0.7]			1.4681[25]		Am78, 2556
4896	Chloroacetamide, N-(2,3-dichloroallyl) $ClCH_2CONHCH_2CCl=CHCl$	202.47		126-31[1.8]			1.5311[25]		Am78, 2556
4897	Chloroacetamide, N-(2,4-dichlorobenzyl) $ClCH_2CONH\text{-}(CH_2C_6H_3Cl_2\text{-}2,4)$	252.54	cr (bz)		96-7			bz	Am78,2556
4898	Chloroacetamide, N-(2,4-dichlorobenzyl) $ClCH_2CONH\text{-}(CH_2C_6H_3Cl_2\text{-}2,4)$	252.54	cr (dil al)		105-6			al	Am78, 2556
4899	Chloroacetamide, N-(2,4-dichlorophenyl) $ClCH_2CONH(C_6H_3Cl_2\text{-}2,4)$	238.50	cr (dil al)		101-2			al, peth	Am78, 2556
4900	Chloroacetamide, N-(2,5-dichlorophenyl) $ClCH_2CONH(C_6H_3Cl_2\text{-}2,5)$	238.50	cr (dil al)		116-7			al	Am78, 2556
4901	Chloroacetamide, N(2,3-dichloropropyl) $ClCH_2CONHCH_2CHClCH_2Cl$	204.50	cr (peth)		65-6			peth	B4[4], 502
4902	Chloroacetamide, N,N-diethyl $ClCH_2CON(C_2H_5)_2$	149.62		190-5[25]					B4[4], 350
4903	Chloroacetamide, N,N-dihexyl $ClCH_2CON(C_6H_{13})_2$	261.84	cr (peth)		114-5				B4[4], 713
4904	Chloroacetamide, N-(2,4-dinitrophenyl) $[2,4\text{-}(NO_2)_2C_6H_3]NHCOCH_2Cl$	259.61	nd (al)		114-5				B9[2], 315
4905	Chloroacetamide, N,N-dipentyl $ClCH_2CON(C_5H_{11})_2$	233.79		126[1]			1.4651[25]		B4[4], 679
4906	Chloroacetamide, N,N-dipropyl $ClCH_2CON(C_3H_7)_2$	177.68		120[8], 90-2[0.8]			1.4670[20]		B4[4], 476
4907	Chloroacetamide, N,N-di-iso-propyl $ClCH_2CON(i\text{-}C_3H_7)_2$	177.68	cr (peth)	86[2.7]	48-9		1.4619[25]		B4[4], 516
4908	Chloroacetamide, N-ethyl-N-hexyl $ClCH_2CON(C_2H_5)C_6H_{13}$	205.73		120[11]			1.4978[25]		Am78, 2556
4909	Chloroacetamide, N-furfuryl $ClCH_2CONH(CH_2C_4H_3O)$	173.61	cr(peth)		58			peth	B18[4], 7080
4910	Chloroacetamide, N-hexyl $ClCH_2CONHC_6H_{13}$	177.68	cr (peth)	95-105[0.2]	108-9			peth	B4[4], 713
4911	Chloroacetamide, N-hexyl-N-methyl $ClCH_2CON(CH_3)C_6H_{13}$	191.70		134[3.8]			1.5005[25]		Am78, 2556
4912	Chloroacetamide, 3-methoxypropyl $ClCH_2CONH(CH_2)_3COCH_3$	165.62		88[0.5]	30		1.4712[25]		B4[4], 1645
4913	Chloroacetamide, N-(2-methylallyl) $ClCH_2CONHCH_2C(CH_3)=CH_2$	147.61		96[1]			1.4860[25]		B4[4], 1105
4914	Chloroacetamide, N-(3-methylbutyl) $ClCH_2CONH\text{-}CH_2CH_2CH(CH_3)_2$	163.65		134-5[13]	−15				B4[2], 647
4915	Chloroacetamide, N-pentyl $ClCH_2CONHC_5H_{11}$	163.65		82[0.5]			1.4665[25]		B4[4], 678
4916	Chloroacetamide, N-propyl $ClCH_2CONHC_3H_7$	135.59		105-6[10.5]	62			al, eth, ace, chl	B4[4], 476
4917	Chloroacetamide, N-iso-propyl $ClCH_2CONHCH(CH_3)_2$	136.60	cr (peth)		62			peth	B4[4], 515
4918	Chloroacetamide, N-tetradecyl $ClCH_2CONHC_{14}H_{29}$	289.89	cr(peth)		64-5				B4[4], 814
4919	Chloroacetamide, N-tetrahydrofurfuryl $ClCH_2CONH(C_5H_7O)$	177.63	cr (peth)		62-3			peth	B18[4], 7039
4920	Chloroacetic acid (α) $ClCH_2CO_2H$	94.50	mcl pr	187.8, 104[20]	63	1.4043[40/4]	1.4351[55]	w, al, eth, bz, chl	B2[4], 474
4921	Chloroacetic acid (β) $ClCH_2CO_2H$	94.50	mcl pr	187.9, 104[20]	56.2	1.4043[40/4]	1.4351[50]	w, al, eth, bz, chl	B2[4], 474

No.	Name, Synonyms, and Formula	Mol. wt.	Color, crystalline form, specific rotation and λ_{max} (log ε)	b.p. °C	m.p. °C	Density	n_D	Solubility	Ref.
4922	Chloroacetic acid (γ) .... ClCH₂CO₂H	94.50	....	187.8, 104²⁰	52.5	1.4043⁴⁰/⁴	1.4351⁵⁵	w, al, eth, bz, chl	B2⁴, 474
4923	Chloroacetic acid, anhydride .... (ClCH₂CO)₂O	170.98	pr (bz)	203	46	1.5497²⁰	....		B2⁴, 487
4924	Chloroacetic acid, benzyl ester or Benzyl chloracetate .... ClCH₂CO₂CH₂C₆H₅	184.62	....	147.5°, 84-6⁰·⁴	....	1.2223⁴/⁴	1.5426¹⁸	al, eth	B6³, 1479
4925	Chloroacetic acid, butyl ester or Butyl chloroacetate ..... ClCH₂CO₂C₄H₉	150.61	....	183, 94³⁸	....	1.0704²⁰/⁴	1.4297²⁰	al, eth	B2⁴, 482
4926	Chloroacetic acid, iso-butyl ester or iso-Butyl chloroacetate .... ClCH₂CO₂CH₂CH(CH₃)₂	150.61	....	170	....	1.0612²⁰/⁴	1.4255²⁰	eth, ace	B2⁴, 483
4927	Chloroacetic acid, sec-butyl ester or sec-Butylchloroacetate .... ClCH₂CO₂-sec-C₄H₉	150.61	....	163-4	....	1.062²⁰/²⁰	1.4251¹⁹	al, eth	B2³, 443
4928	Chloroacetic acid, 2-chloroethyl ester or 2-chloroethyl chloroacetate .... ClH₂CO₂CH₂CH₂Cl	157.00	....	202, 89¹⁰	....	1.3600²⁵/⁴	1.4619²⁵	eth	B2⁴, 481
4929	Chloroacetic acid, ethyl ester or Ethyl chloroacetate...... ClCH₂CO₂C₂H₅	122.55	....	144⁷⁴⁰, 52²⁰	−26	1.1585²⁰/⁴	1.4215²⁰	al, eth, ace, bz	B2⁴, 481
4930	Chloroacetic acid, 2-hydroxyethyl ester or 2-Hydroxyethyl chloroacetate .... ClCH₂CO₂CH₂CH₂OH	138.55	....	240d, 86¹·⁵	....	1.3300²⁰/⁴	1.4609²⁰	w, al	B2³, 447
4931	Chloroacetic acid, hydrazide .... ClCH₂CONHNH₂	108.53	....	....	93	....	....		
4932	Chloroacetic acid, 2-methoxyethyl ester or 2-Methoxyethyl chloroacetate .... ClCH₂CO₂CH₂CH₂OCH₃	152.58	....	85-6°	....	1.2015²⁰/⁴	1.4382²⁰	eth	B2³, 447
4933	Chloroacetic acid, methyl ester or Methyl chloroacetate... ClCH₂CO₂CH₃	108.53	....	129.8, 29¹⁰	−32.1	1.2337²⁰/⁴	1.4218²⁰	al, eth, ace, bz	B2⁴, 480
4934	Chloroacetic acid, phenyl ester or Phenyl chloroacetate... ClCH₂CO₂C₆H₅	170.60	nd or pl (al)	230-5, 114⁸	44-5	1.2202⁴⁴/⁴	1.5146⁴⁴	al, eth	B6³, 598
4935	Chloroacetic acid, propylester or Propyl chloroacetate ... ClCH₂CO₂C₃H₇	136.58	....	161⁷⁶⁴	....	1.1033²⁰/⁴	1.4261²⁰	eth	B2⁴, 482
4936	Chloroacetic acid, iso-propyl ester or iso-Propyl chloroacetate. .... ClCH₂CO₂-i-C₃H₇	136.58	....	150-1	....	1.0888²⁰/⁴	1.4382²⁰	eth	B2⁴, 482
4937	Chloroacetic acid, 4-tolyl ester .... ClCH₂CO₂(C₆H₄CH₃-4)	184.62	pl	162⁴¹	32	....	....	al, eth	B6², 378
4938	Chloroacetone cyanohydrin .... ClCH₂C(CH₃)(OH)CN	119.55	....	110²⁷	....	1.2027¹⁵	1.4536¹¹	w, al, ace	B3³, 599
4939	Chloroacetonitrile .... ClCH₂CN	75.50	....	126-7, 30-2¹⁵	....	1.1930²⁰	1.4202²⁵	al, eth	B2⁴, 492
4940	Chloroacetyl chloride .... ClCH₂COCl	112.94	....	107	....	1.4202²⁰/⁴	1.4541²⁰	eth, ace	B2⁴, 488
4941	2-Chloroally isotthiocyanate .... CH=CClCH₂NCS	133.60	....	182	....	1.27¹²	....		B4, 219
4942	Chlorodifluoro acetic acid .... ClF₂CCOOH	130.48	....	....	22.9	....	....	chl	B2⁴, 497
4943	Chlorodiphenyl acetamide .... (C₆H₅)₂CClCONH₂	245.71	cr (to)	....	115	....	....	al, eth, bz, chl	B9³, 3308
4944	Chlorodiphenyl acetic acid .... (C₆H₅)₂CClCO₂H	246.70	pl (bz-lig)	118-9d	....	....	....	al, eth, ace, bz	B9³, 3307
4945	Chlorodiphenyl acetic acid, ethyl ester .... (C₆H₅)₂CClCO₂C₂H₅	274.75	pl (chl), cr (al)	185¹⁴	43-4	....	....	al, eth	B9³, 3307
4946	Chlorodiphenyl acetyl chloride .... (C₆H₅)₂CClCOCl	265.14	cr (lig)	180¹⁴	50-1	....	....		B9³, 3308
4947	bis-(2-Chloroethyl) methylamine, hydrochloride .... (ClCH₂CH₂)₂NCH₃·HCl	192.52	hyg nd	....	111-2	....	....	w, al	B4⁴, 446
4948	tris-(2-Chloroethyl) amine or Nitrogen mustard gas ...... (ClCH₂CH₂)₃N	204.53	Pa ye	143-4¹⁵	−4	....	....	al, eth, bz	B4⁴, 447
4949	(α-Chloroethyl)ether or bis-(α-chloroethyl) ether .... (CH₃CHCl)₂O	143.02	....	116-7	....	1.1060²⁵/⁴⁵	1.4186²⁵	al, eth, chl	B1⁴, 3120
4950	(α-Chloroethyl)(β-Chloroethyl) ether .... CH₃CHClOCH₂CH₂Cl	143.02	....	d⁷⁶⁰, 55-7¹⁷	....	1.1867²⁰/⁴	1.4473²⁰	al, eth, chl	B1³, 2655
4951	(β-Chloroethyl)diethyl amminium chloride .... ClCH₂CH₂NH(C₂H₅)₂Cl⁻	172.10	nd (al-eth)	....	210-1	....	....	w, al	B4⁴, 447

No.	Name, Synonyms, and Formula	Mol. wt.	Color, crystalline form, specific rotation and λ_{max} (log ϵ)	b.p. °C	m.p. °C	Density	n_D	Solubility	Ref.
4952	α-Chloroethyl methyl ether .... CH₃OCHClCH₃	94.54	....	72-3[751]	......	0.9902[20/4]	1.4004[20]	eth	B1[4], 3119
4953	α-Chloroethyl pentyl ether .... CH₃CH₂Cl-O-C₅H₁₁	150.65	....	63-6[8]	......	0.9200[20/4]	1.4218[20]	eth	B1[3], 2656
4954	α Chloroethyl propyl ether .... CH₃CHClOC₃H₇	122.60	....	112-5[731]	......	0.9322[20/4]	1.4013[20]	eth	B1[3], 2655
4955	(β-Chloroethyl)ether or bis-(β-Chloroethyl)ether........ (ClCH₂CH₂)₂O	143.04	....	178, 75[20]	−24.5	1.2199[20/4]	1.4575[20]	al, eth, ace, bz	B1[4], 1375
4956	α-Chloroethyl methyl ether .... ClCH₂CH₂OCH₃	94.54	....	92-3	......	1.0345[20/4]	1.4111[20]	w, eth	B1[4], 1375
4957	β-Chloroethyl vinyl ether .... ClCH₂CH₂OCH=CH₂	106.55	....	108	......	1.0475[20/4]	1.4378[20]	al, eth	B1[4], 2051
4958	Chloroformic acid, benzyl ester or Carbobenzoxy chloride ClCO₂CH₂C₆H₅	170.60	....	103[20]	......	1.20	1.5150[20]	eth, ace, bz	B6[3], 1485
4959	Chloroformic acid, butyl ester .... ClCO₂C₄H₉	136.58	....	138[750], 35.5[13]	......	1.0513[20/4]	1.4121[20]	eth, ace	B3[4], 25
4960	Chloroformic acid, isobutyl ester or Isobutyl chlorofor-mate .... ClCO₂CH₂CH(CH₃)₂	136.58	....	128.8	......	1.0426[18/4]	1.4071[18/Hc]	eth, bz, chl	B3[4], 26
4961	Chloroformic acid, β-chloroethyl ester .... ClCO₂CH₂CH₂Cl	142.97	....	156	......	1.3847[20/4]	1.4483[20]	al, eth, ace, bz	B3[4], 24
4962	Chloroformic acid, chloromethyl ester .... ClCO₂CH₂Cl	128.54	....	107	......	1.465[15]	1.4286[22]	eth, ace	B3[4], 30
4963	Chloroformic acid, (3-chloropropyl) ester .... ClCO₂CH₂CH₂CH₂Cl	157.00	....	177	......	1.2949[25/20]	1.4456[20]	....	B3[3], 25
4964	Chloroformic acid, cyclohexyl ester .... ClCO₂C₆H₁₁	162.62	....	87.5[27]	......	....	....	eth	B6[4], 43
4965	Chloroformic acid, dichloromethyl ester .... ClCO₂CHCl₂	163.39	....	110-1	......	....	....	ace	B3[4], 31
4966	Chloroformic acid, β-ethoxyethyl ester or Cellosolve chloroformate .... ClCO₂CH₂CH₂OC₂H₅	152.58	....	67.2[14]	......	1.1341[25]	1.4169[25]	al	B3[2], 29
4967	Chloroformic acid, ethyl ester or Ethyl chloroformate.... ClCO₂C₂H₅	......	....	95	−80.6	1.1352[20/4]	1.3974[20]	eth, bz, chl	B3[4], 23
4968	Chloroformic acid, 2-methoxyethyl ester .... ClCO₂CH₂CH₂OCH₃	138.55	....	58.7[13]	......	1.1905[25]	1.4163[20]	eth	B3[2], 29
4969	Chloroformic acid, methyl ester or Methyl chloroformate ClCO₂CH₃	94.50	....	70-1	......	1.2231[20/4]	1.3868[20]	al, eth, bz, chl	B3[4], 23
4970	Chloroformic acid, isopentyl ester or Isopentyl chlorofor-mate .... ClCO₂-i-C₅H₁₁	150.61	....	154.3, 60[15]	......	1.0288[17/4]	1.4176[20]	al, eth	B3[4], 26
4971	Chloroformic acid, pentyl ester .... ClCO₂C₅H₁₁	150.61	....	60-2[15]	......	....	1.4181[18]	eth	B3[4], 26
4972	Chloroformic acid, isopropenyl ester .... ClCO₂C(CH₃)=CH₂	120.53	....	100	......	1.103[20/20]	....	eth	B3[3], 28
4973	Chloroformic acid, propyl ester or Propyl chloroformate ClCO₂C₃H₇	122.55	....	115.2	......	1.0901[20/4]	1.4035[20]	al, eth	B3[4], 24
4974	Chloroformic acid, isopropyl ester or Isopropyl chloro-formate .... ClCO₂CH(CH₃)₂	122.55	....	105, 66.3[200]	......	....	1.4013[20]	....	B3[4], 24
4975	Chloroformic acid, trichloro methyl ester or Diphosgene ClCO₂CCl₃	197.83	....	128, 49[50]	−57	1.6525[14]	1.4566[22]	al, eth	B3[4], 33
4976	Chloromethyl ether or bis-(Chloromethyl) ether .... (ClCH₂)₂O	114.96	....	104	−41.5	1.328[15/4]	1.435[21]	al, eth	B1[4], 3051
4977	Chloromethyl propyl ether .... ClCH₂OC₃H₇	108.57	....	109	......	0.9884[20/4]	1.4125[20]	al, eth	B1[3], 2589
4978	Chloromethyl thiocyanate .... ClCH₂SCN	107.56	....	185	......	1.37[15]	....	....	B3[2], 124
4979	Chloromycetin or Chloroamphenicol.................. C₁₁H₁₁ClN₂O₅	323.14	pa ye pl or nd (w), [α][25/D] + 19 (al, c=5) -25.5 (AcOEt)	sub vac	150-1	....	....	al, ace, chl	B13[3], 2268
4980	(4-Chlorophenyl) ether or bis-(4-Chlorophenyl) ether..... (4-ClC₆H₄)₂O	239.11	nd (al)	312-4, 168.7[7]	30	1.1231[20]	1.611[20]	....	B6[4], 826

No.	Name, Synonyms, and Formula	Mol. wt.	Color, crystalline form, specific rotation and λ_{max} (log ε)	b.p. °C	m.p. °C	Density	n_D	Solubility	Ref.
4981	4-Chlorophenyl phenyl ether (4-ClC$_6$H$_4$)OC$_6$H$_5$	204.66	....	284-5	....	1.2026[15]	1.599$_D$	....	B6[4], 826
4982	2-Chlorophenyl isocyanate 2-ClC$_6$H$_4$NCO	153.37	....	115-7[43]	30-1				B12[3], 1296
4983	3-Chlorophenyl isocyanate 3-ClC$_6$H$_4$NCO	153.57	....	113-4[43]					B12[3], 1316
4984	4-Chlorophenyl isocyanate 4-ClC$_6$H$_4$NCO	153.57	....	115-7[45]					B12[3], 1376
4985	4-Chlorophenyl isothiocyanate 4-ClC$_6$H$_4$NCS	169.63	nd (al)	249-50	45				B12[3], 1376
4986	Chlorophyll-a C$_{55}$H$_{72}$MgN$_4$O$_5$	893.53	bl bk hex pl	....	150-3			al, eth, lig	M, 245
4987	Chlorophyll b C$_{55}$H$_{70}$MgN$_4$O$_6$	907.51	bl bk gr pw	....	120-30			al, eth, lig	M, 245
4988	Chloroprene or 2-Chloro-1,3-Butadiene CH$_2$=CCl-CH=CH$_2$	88.54	....	59.4, 6.4[100]		0.9583[20/4]	1.4583[20]	eth, ace, bz	B1[4], 984
4989	Chlorpromazine or Thorazine C$_{17}$H$_{19}$ClN$_2$S	318.87	....	200-5[0.8]				al, eth, bz, chl	C50, 1951
4990	Chlorpromazin, hydrochloride C$_{17}$H$_{19}$ClN$_2$S·HCl	355.33	....	194-7d				w, al, chl	C50, 1931
4991	(β-Chloropropyl) ether or bis-(β-chloropropyl) ether CH$_3$CHClCH$_2$OCH$_2$CHClCH$_3$	171.07	....	188		1.109[20/4]	1.4467[20]	al, eth	B1[4], 1442
4992	(γ-Chloropropyl) ether or bis-(α-Chloropropyl)ether (ClCH$_2$CH$_2$OH)$_2$O	171.07	....	215, 90.5[11]		1.140[20/20]	1.4158[20]	al, eth	B1[2], 370
4993	(β-Chloroisopropyl) ether or bis-(β-Chloroisopropyl) ether [ClCH$_2$CH(CH$_3$)]$_2$O	171.07	....	187		1.103[20/4]	1.4505[20]	al, eth, ace, bz	B1[3], 1470
4994	Bis-(1,2,2,2-Tetrachloroethyl) ether (Cl$_3$CCHCl)$_2$O	400.80	cr (al, MeOH)	130-1[11]	40.2			bz, peth	B1[3], 2672
4995	Chlorosulfinic acid, ethyl ester or Ethyl chlorosulfinate (C$_2$H$_5$O)SOCl	128.58	....	52.5[44], 32[16]		1.2766[25/4]	1.4550[25]	eth	B1[3], 1316
4996	Chlorosulfonic acid, ethyl ester or Ethyl chlorosulfonate C$_2$H$_5$OSO$_2$Cl	144.58	....	151-4, 52[14]		1.3502[25/4]	1.416[20]	eth, chl, lig	B1[4], 1326
4997	Chlorosulfonic acid, methyl ester or Methyl chlorosulfonate CH$_3$OSO$_2$Cl	130.51	....	133-5, 48[29]		1.4805[25/4]	1.4138[18]	eth, ace, bz	B1[4], 1252
4998	Cholanic acid or Ursocholamic acid C$_{24}$H$_{40}$O$_2$	360.59	nd (al), cr (aa), [α]$^{20}_D$ +21.7 (chl)	....	163-4			al, chl, aa	B9[3], 2656
4999	Cholanic acid, 3α, 6α-dihydroxy or Hyodeoxycholic acid C$_{24}$H$_{40}$O$_4$	392.56	cr (AcOEt), [α]$^{20}_D$ +37.2 (MeOH)	....	198-9			al, aa	B10[3], 1631
5000	Cholanic acid, 3α,7α-dihydroxy or Chenodeoxycholic acid C$_{24}$H$_{40}$O$_4$	392.59	nd (AcOEt-hp), [α]$^{20}_D$ +11.1 (al, c=2.1)	....	143			al, eth, ace, aa	B10[3], 1635
5001	Cholanic acid, ethyl ester C$_{26}$H$_{44}$O$_2$	388.64	lf, nd (dil al), [α]$^{19}_D$ +21 (chl)	273[12]	93-4			chl	B9[3], 2658
5002	Cholanic acid, methyl ester C$_{25}$H$_{42}$O$_2$	374.61	nd [α]$_D$ +23±2 (diox)	....	87-8			diox	B9[3], 2658
5003	Clolanthrene or Benz(j)aceanthrylene C$_{20}$H$_{14}$	254.34	pa ye lf (bz-al)	sub, 210[0.2]	174-5d			al, bz, aa, lig	B5[3], 2469
5004	Cholanthrene, 6,7-dihydro-20-methyl C$_{21}$H$_{18}$	270.37	lf (MeOH)	....	155			MeOH	B5[3], 2429
5005	Cholanthrene, 11,14-dihydro-20-methyl C$_{21}$H$_{18}$	270.37	nd (PrOH)	....	138-9				B5[3], 2429
5006	Cholanthrene, 20-methyl C$_{21}$H$_{16}$	268.38	yesh nd (bz)	....	180				B5[3], 2484
5007	$\Delta^{2,4}$-Cholestadiene C$_{27}$H$_{44}$	368.65	cr (eth-ace), [α]$^{23}_D$ +168.5 (eth, c=1.5)		68.5			al, eth, chl	B5[3], 1428

No.	Name, Synonyms, and Formula	Mol. wt.	Color, crystalline form, specific rotation and λ_{max} (log ε)	b.p. °C	m.p. °C	Density	n_D	Solubility	Ref.
5008	$\Delta^{3,5}$-Cholestadiene or Cholesterilene .... $C_{27}H_{44}$	368.65	wh nd (al). $[\alpha]^{20}_D$ -129.6 (chl, c=3)	260[13]	80	0.925[100/4]	....	al, eth, bz, chl	B5[4], 1620
5009	$\Delta^{3,7}$-Cholestadiene-3β-ol or 7-Dehydrocholesterol .... $C_{27}H_{44}O$	384.65	pl (+1w, eth-MeOH), $[\alpha]^{25}_D$ -115 (chl, C=2.5)	....	150-1			eth, ace	B6[3], 2819
5010	$\Delta^{4,6}$-Cholestadiene-3-one .... $C_{27}H_{42}O$	382.64	ye pr $[\alpha]^{20}_D$ +31 (chl)	....	78				B7[3], 1760
5011	Cholestane .... $C_{27}H_{48}$	372.69	sc or pl (eth-al, ace), $[\alpha]^{20}_D$ +30.2 (chl, c=2)	250[1]	80	0.9090[88/4]	1.4887[88]	eth, bz, chl	B5[4], 1227
5012	3β-Cholestane carboxylic acid .... $C_{28}H_{48}O_2$	416.70	nd $[\alpha]^{25}_D$ +28.8 (chl, c=1.7)		210-1				B9[3], 2668
5013	3,6-Cholestane dione .... $C_{27}H_{44}O_2$	400.65	nd $[\alpha]^{20}_D$ +8.9		171-2				B7[3], 3282
5014	3α-Cholestanol or Epidihydro cholesterol .... $C_{27}H_{40}O$	388.69	nd (al), $[\alpha]^{20}_D$ +34 (chl, c=1.1)		188				B6[3], 2135
5015	3β-Cholestanol or Dihydrocholesterol .... $C_{27}H_{48}O$	388.69	lf (al+1w), pr, pl (MeOH), $[\alpha]^{22}_D$ +24.2 (chl, c=1.3		141-3			al, eth, chl	B6[3], 2131
5016	3-Cholestanone or Zymostanone .... $C_{27}H_{46}O$	386.67	nd or lf (al), $[\alpha]_D$ +42 (chl, c=2.12)		128-30			al	B7[3], 1330
5017	6-Cholestanone, 3β-hydroxy or 6-Keto Cholestanol ...... $C_{27}H_{46}O_2$	402.67	nd (al), $[\alpha]_D$ -3.0 (chl)		150-1			al	B8[3], 658
5018	7-Cholestanone, 3β-hydroxy or 7-Oxocholestanol. .... $C_{27}H_{48}O_2$	402.67	pl $[\alpha]^{22}_D$ -34 (chl)		165-8			al	B8[3], 665
5019	2-Cholestene or Neocholestene .... $C_{27}H_{46}$	370.67	nd (eth-ace or al), $[\alpha]_D$ +66 (c=1.65)		75-6				B5[4], 1507
5020	3-Cholestene .... $C_{27}H_{46}$	370.67	wh pr or nd (al), $[\alpha]^{18}_D$ -56.3 (chl)		93-4				B5[4], 1508
5021	5-Cholestene, 3β-bromo or Cholestyryl bromide........ $C_{27}H_{45}Br$	449.57	mcl lf (al), $[\alpha]^{20}_D$ -19 (bz, c=0.4)		100-2			bz, chl	B5[4], 1512
5022	5-Cholestene, 3β-chloro or Cholesteryl chloride .... $C_{27}H_{45}Cl$	405.12	nd (al or ace), $[\alpha]^{20}_D$ -26.4 (bz) -33.3 (chl)	....	96			bz, chl	B5[4], 1510
5023	5-Cholestene, 3β,7β-diol or 7β-Hydroxycholesterol .... $C_{27}H_{46}O_2$	402.67	wh nd (eth), $[\alpha]^{20}_D$ +7.2 (chl, c=2)	sub, 145[0.005]	177-8				B6[3], 5130
5024	5-Cholestene, 3β-iodo or Cholestyryl iodide............ $C_{27}H_{45}I$	496.57	nd (ace or AcOEt)	....	106-7			bz, chl, lig	B5[4], 1512
5025	1-Cholesten-3-one .... $C_{27}H_{44}O$	384.65	nd (dil ace or al), $[\alpha]^{25}_D$ +88.2 (chl)	....	99-101		....	bz	B7[3], 1592

No.	Name, Synonyms, and Formula	Mol. wt.	Color, crystalline form, specific rotation and λ_{max} (log ϵ)	b.p. °C	m.p. °C	Density	n_D	Solubility	Ref.
5026	4-Cholestene-3-one $C_{27}H_{44}O$	384.65	nd or pl (al), $[\alpha]^{25}_D$ + 92 (chl, c=2.01)	....	81-2	....	....	eth, bz, peth	B7[3], 1594
5027	5-Cholesten-3-one $C_{27}H_{44}O$	384.65	lf (al), $[\alpha]^{20}_D$ -4.3 (chl)	....	127	....	....	eth, al	B7[3], 1608
5028	Cholesterol $C_{27}H_{44}OH$	386.67	rh or tcl lf (al + 1w) nd (eth), $[\alpha]^{20}_D$ -31.5 (eth), $[\alpha]^{20}_D$ -31.5 (eth, c=2) -39.5 (chl, c=2)	360d, 233[0.5]	148.5 (anh)	1.067[20/4]	....	eth, bz, chl, aa	B6[2], 2607
5029	Cholesterol, acetate or Cholestyryl acetate $C_{29}H_{48}O_2$	428.71	wh nd (ace or al), $[\alpha]^{20}_D$ -47.7 (chl, c=2)	....	115-6	....	....	eth, ace, bz, chl	B6[3], 2630
5030	Cholesterol, benzoate $C_{34}H_{50}O_2$	490.78	wh nd, $[\alpha]$ -13.7 (chl, c=0.9)	....	152-3	....	....	eth, chl	B9[3], 460
5031	Cholesterol, hexadecanoate or Cholesteryl palmitate $C_{43}H_{76}O_2$	625.09	wh nd (eth or al), $[\alpha]^{20}$ -25.4 (chl, c=2)	....	80	....	....	bz, chl	B6[3], 2640
5032	Cholestrophane or Dimethylparabanic acid Oxalyldimethylurea $C_5H_6N_2O_3$	142.12	lf or pl (w or al)	275-7, 148-50[13]	155.5	....	....	w, eth	B24[2], 265
5033	Cholic acid $C_{24}H_{40}O_5$	408.59	rh (eth), tetr rh (w or dil al), cr (al + 1), $[\alpha]^{20}_D$ + 37 (al, c=0.6)	....	198 (anh)	....	....	al, eth, ace, chl, aa	B10[3], 2162
5034	Choline or Trimethyl(2-hydroxymethyl) ammonium hydroxide $(CH_3)_3N^+CH_2CH_2OH \cdot OH^-$	121.18	syr	....	....	....	....	w, al	B4[4], 1443
5035	Choline, O-acetyl, bromide or Acetylcholine bromide $(CH_3)_3N^+CH_2CH_2O_2CCH_3 \cdot Br^-$	226.12	hex pr (dil al)	d	143	....	....	w, al	B4[4], 1446
5036	Choline, O-acetyl, chloride or Acetyl choline chloride $(CH_3)_3N^+CH_2CH_2O_2CCH_3 \cdot Cl^-$	181.66	yesh nd	....	153	....	....	w, al	B4[4], 1446
5037	Choline, O-benzyl, chloride $(CH_3)_3N^+CH_2CHOO_2CC_6H_5Cl^-$	243.74	pr (ace-al)	....	200	....	....	al, ace	B9[3], 877
5038	Choline, carbonate, chloride $(CH_3)_3N^+CH_2CH_2O_2CNH_2 \cdot Cl^-$	182.65	hyg pr or pw	....	210-2	....	....	w	B4[4], 1455
5039	Chroman or Dihydrobenzopyran $C_9H_{10}O$	134.18	....	214-5[742], 98-9[18]	....	1.0610[20]	1.5444[20]	....	B17[4], 413
5040	Chroman, 2,2-dimethyl $C_{11}H_{14}O$	162.23	....	225[769], 98[11]	....	1.009[15/15]	....	....	B17[4], 445
5041	Chromanone $C_9H_8O_2$	148.16	....	160[50], 128[13]	38-9	1.1291[100/4]	1.5460[100]	al, eth, ace, bz	B17[4], 4957
5042	3-Chromene C_9H_8O	132.16	....	91[13], 50[1]	....	....	1.5879[20]	....	B17[4], 496
5043	Chromone or α-Benzopyrone $C_9H_6O_2$	146.15	nd (peth or w)	sub	59	....	....	al, eth, bz, chl	B17[4], 5052
5044	cis-Chrysanthemucic acid-(d) or Cyclopropane carboxylic acid-2,2-dimethyl-3-(2-methyl propenyl) $C_{10}H_{16}O_2$	168.24	pr $[\alpha]^{22}_D$ + 83.3 (chl, c=1.6)	95[0.1]	40-2	....	....	al, eth	B9[3], 210
5045	trans-Chrysanthemucic acid-(d) or Cyclopropane carboxylic acid-2,2-dimethyl-3-(2 methyl propenyl) (trans) $C_{10}H_{16}O_2$	168.24	pr $[\alpha]^{19}_D$ + 25.8 (chl, c=2.5)	245d, 135[12]	18-21	....	....	al, eth, chl	B9[3], 211
5046	Chrysazin or 1,8-Dihydroxyanthraquinone $1,8-(HO)_2C_{14}H_6O_2$	240.23	red or redsh-ye nd or lf (al)	sub	193	....	....	al, eth, ace, chl	B8[3], 3792

No.	Name, Synonyms, and Formula	Mol. wt.	Color, crystalline form, specific rotation and λ_{max} (log ε)	b.p. °C	m.p. °C	Density	n_D	Solubility	Ref.
5047	Chrysene or 1,2-Benzophenanthrene.............. $C_{18}H_{12}$	228.30	red bl flr rh pl (bz-aa)	448	255-6	1.274^{20}	....	....	B5[3], 2380
5048	Chrysene, 5,6-dimethyl .... $C_{20}H_{16}$	256.35	pl or nd (bz-al)	$200^{0.5}$ (sub 140 vac)	128-9			al, aa	B5[3], 2418
5049	Chrysene, 1-methyl .... $C_{19}H_{14}$	242.32	lf (hx, bz, to)	sub 130-140 (vac)	256-7			al	B5[3], 2395
5050	Chrysene, 2-methyl .... $C_{19}H_{14}$	242.32	lf (bz-al)	....	229-30			al, aa	B5[3], 2395
5051	Chrysene, 3-methyl .... $C_{19}H_{14}$	242.32	lf (bz-peth)		172-3			al	B5[3], 2395
5052	5,6-Chrysoquinone $C_{18}H_{10}O_2$	258.28	red nd (bz or to), lf or pl (aa)	sub	239.5			al, bz	B7[3], 4285
5053	6,12-Chrysoquinone $C_{18}H_{10}O_2$	258.25	red ye nd (aa)	....	288-90 d			al, aa	B7[3], 4285
5054	Cinchonamine .... $C_{19}H_{24}N_2O$	296.42	rh nd (al), orh pr (MeOH), $[\alpha]^{20}_D$ + 123 (al, c=0.66		186 (194)			al, eth, bz, chl	B23[2], 358
5055	Cinchonicine or Cinchotoxine .... $C_{19}H_{22}N_2O$	294.40	nd or pr (eth), $[\alpha]^{15}_D$ + 48 (al, c=1)	....	58-60			al, eth, ace, bz, chl	B24[2], 100
5056	Cinchonidine or Cinchovatine .... $C_{19}H_{22}N_2O$	294.40	orh pl or pr (al), $[\alpha]^{20}_D$ - 109.2 (al, p=1)	sub	210.5			al	B23[2], 373
5057	Cinchonidine, hydrochloride .... $C_{19}H_{22}N_2O \cdot HCl \cdot H_2O$	348.88	wh pr (w), $[\alpha]^{20}_D$ - 117.6 (w, c=1.2)	....	242d (anh)			w, al, chl	B23[2], 373
5058	Cinchonidine, sulfate .... $(C_{19}H_{22}N_2O)_2 \cdot H_2SO_4 \cdot 6H_2O$	794.97	mcl nd (al +2w), $[\alpha]^{18.5}_D$ - 97.9 (w, c=1.2)	....	205 (anh)				B23[2], 373
5059	β-Cinchonidine .... $C_{19}H_{22}N_2O$	294.40	pr or lf (al or dil al)		241			al, chl	B23[1], 131
5060	Cinchonine .... $C_{19}H_{22}N_2O$	294.40	nd or mcl cr (al)	sub	255 (265)			chl	B23[2], 369
5061	Cinchonine, dihydrochloride $C_{19}H_{22}N_2O \cdot 2HCl$	367.32	pl $[\alpha]^{24}_D$ + 205.5 (w, c=3.6)					w, al	B23[1], 133
5062	Cinchonine, hydrochloride $C_{19}H_{22}N_2O \cdot HCl \cdot 2H_2O$	366.89	mcl $[\alpha]^{25}_D$ + 133.6 (chl, c=1.4)		ca 215 (anh)	1.234		w, al, chl	B23[2], 370
5063	Cinchonine, sulfate $(C_{19}H_{22}N_2O)_2 \cdot H_2SO_4$	722.91	rh pr (w), $[\alpha]^{15}_D$ + 169 (w, c=1.4)		206-7			al, w	B23[2], 371
5064	Cinchotine or Hydrocinchonine.............. $C_{19}H_{24}N_2O$	296.42	pr $[\alpha]^{21}_D$ + 203.4		268-9			w	B23[2], 356
5065	1,4-Cineole or 1,4-epoxy-p-menthane.............. $C_{10}H_{18}O$	154.26	....	173-4	1	0.8997^{20}	1.4562^{20}	al, eth, bz, lig	B17[4], 213
5066	1,8-Cineole or Eucalyptol.............. $C_{10}H_{18}O$	154.26	....	176.4, 61^{14}	1.5	0.9267^{20}	1.4586^{20}	al, eth, chl	B17[4], 273
5067	Cineolic acid-(d) .... $C_{10}H_{16}O_5$	216.24	cr (w + 1), $[\alpha]^{20}_D$ + 18.6 (w, p=8.21)		79 (+ w), 138-9 (anh)			al	B18, 322
5068	Cineolic acid-(dl) .... $C_{10}H_{16}O_5$	216.24	....	(i)197.5 (ii)208			al, eth	B18[4], 4446
5069	Cineolic acid-(l) .... $C_{10}H_{16}O_5$	234.25	rh (w + 1), $[\alpha]^{20}_D$ - 19.1(w)		79 (+ w), 138.9 (anh)			al	B18, 322

No.	Name, Synonyms, and Formula	Mol. wt.	Color, crystalline form, specific rotation and λ_{max} (log ε)	b.p. °C	m.p. °C	Density	n_D	Solubility	Ref.
5070	Cinnamaldehyde-(trans) or β-Phenyl acrolein .... $C_6H_5CH=CHCHO$	132.17	yesh	253d, 127[16]	-7.5	1.0497[20/4]	1.6195[20]	al, eth, chl	B7[3], 1364
5071	Cinnamaldehyde, β-bromo-(cis) .... $C_6H_5CBr=CHCHO$	211.06	....	144-6[12]		1.492[20/4]	1.6368[20]	eth	B7[3], 1383
5072	Cinnamaldehyde, α-ethyl .... $C_6H_5CH=C(C_2H_5)CHO$	160.22	....	157-8[5]	....	1.0201[22/4]	1.578[20]	....	B7[3], 1435
5073	Cinnamaldehyde, 4-hydroxy-3-methoxy or Coniferaldehyde [4-(HO)-3-(CH_3O)C_6H_3]CH=CHCHO	178.19	cr (bz)	157[2.5]	84			al, eth, bz	B8[3], 2331
5074	Cinnamaldehyde, α-methyl .... $C_6H_5CH=C(CH_3)CHO$	146.19	ye	150[100]	....	1.0407[17/4]	1.6057[17]		B7[3], 1412
5075	Cinnamaldehyde, 4-methyl ...... $4-CH_3C_6H_4CH=CHCHO$	146.19	ye lf (dil al)	154[25]	41.5			al	B7[3], 1414
5076	Cinnamaldehyde, 2-nitro ...... $2-O_2NC_6H_4CH=CHCO$	177.16	nd (eth or al)		127.5			al, eth, chl	B7[3], 1387
5077	Cinnamaldehyde, 3-nitro ...... $3-O_2NC_6H_4CH=CHCHO$	177.16	ye nd (w), pr (al), cr (aa)		116			bz, aa	B7[2], 282
5078	Cinnamaldehyde, 4-nitro ...... $4-O_2NC_6H_4CH=CHCHO$	177.16	nd (w or al)		141-2			al, eth, ace, bz	B7[3], 1388
5079	Cinnamaldehyde, oxime-(trans) .... $C_6H_5CH=CHCH=NOH$	147.18	nd		138.5			bz	B7[3], 1376
5080	Cinnamamide-(trans) .... $C_6H_5CH=CHCONH_2$	147.18	nd (bz)		148			al, eth	B9[3], 2711
5081	Cinnamic acid-(cis)-(1st form) or Phenyl acrylic acid..... $C_6H_5CH=CHCO_2H$	148.17	mcl pr (w)		42		....	al, eth, bz	B9[3], 2670
5082	Cinnamic acid-(cis)-(2nd form) .... $C_6H_5CH=CHCO_2H$	148.17	mcl pr (lig)	265	58			al, eth, ace, chl	B9[3], 2670
5083	Cinnamic acid-(cis)-(3rd form) or Allocinnamic acid..... $C_6H_5CH=CHCO_2H$	148.17	mcl pr		68			al, eth, lig	B9[3], 2670
5084	Cinnamic acid -(trans) .... $C_6H_5CH=CHCO_2H$	148.17	mcl pr (dil al)	300 (cor)	135-6	1.2475[4/4]	....	al, eth, ace, bz, chl	B9[3], 2671
5085	Cinnamic acid, α-acetamido .... $C_6H_5CH=C(NHOCH_3)CO_2H$	205.22	cr (+2w)		193-4			w	B10[3], 3001
5086	Cinnamic acid, allyl ester or Allylcinnamate .... $C_6H_5CH=CHCO_2CH_2CH=CH_2$	188.23	....	268d, 163[17]		1.048[23]	1.530[20]	al, eth	B9[2], 387
5087	Cinnamic acid, 2-amino-(trans) ...... $2-H_2NC_6H_4CH=CHCO_2H$	163.18	ye nd (w)		158-9d			al, eth	B14[3], 1304
5088	Cinnamic acid, 3-amino-(trans) ...... $3-H_2NC_6H_4CH=CHCO_2H$	163.18	nd (al)		191-3			aa	B14[3], 1305
5089	Cinnamic acid, 4-amino ...... $4-H_2NC_6H_4CH=CHCO_2H$	163.18	ye nd (w or al)		175-6d			al, eth	B14[3], 1305
5090	Cinnamic acid-(trans), anhydride .... $(C_6H_5CH=CHCO)_2O$	278.31	nd (bz or al) pr, (al)		138			bz	B9[3], 2703
5091	Cinnamic acid, 2-benzamido-(trans) ...... $2-(C_6H_5CONH)C_6H_4CH=CHCO_2H$	267.29	nd (al)		191-3			aa	B14[3], 1304
5092	Cinnamic acid, 3-benzamido-(trans) ...... $3-(C_6H_5CONH)C_6H_4CH=CHCO_2H$	267.29	nd (AcOEt)		229			ace, aa	B14[3], 1305
5093	Cinnamic acid, 4-benzamido-(trans) ...... $4-(C_6H_5CONH)C_6H_4CH=CHCO_2H$	267.29	lf (aa), nd (ace)		274d			ace, bz, aa	B14[3], 1312
5094	Cinnamic acid, benzyl ester-(trans) or Benzyl cinnamate . $C_6H_5CH=CHCO_2CH_2C_6H_5$	238.29	pr	350d, 244[5]	39	1.109[15]	....	al, eth	B9[3], 2691
5095	Cinnamic acid, α-bromo-(cis) .... $C_6H_5CH=CBrCO_2H$	227.06	lf (w), pr (chl)		120-1			al, bz	B9[3], 2734
5096	Cinnamic acid, α-bromo-(trans) .... $C_6H_5CH=CBrCO_2H$	227.06	nd (w)		131-2			al, eth, bz	B9[3], 2734
5097	Cinnamic acid, β-bromo-(cis) .... $C_6H_5CBr=CHCO_2H$	227.06	nd (bz), pl (al)		159-60			eth, chl	B9[3], 2732
5098	Cinnamic acid, β-bromo-(trans) .... $C_6H_5CBr=CHCO_2H$	227.06	pa ye nd or pl (w), pr (chl)		135			al, bz	B9[3], 2733
5099	Cinnamic acid, 2-carboxy-(trans) ...... $(2-HO_2CC_6H_4)CH=CHCO_2H$	192.18	pr or nd (w)		208-9			al	B9[3], 4384
5100	Cinnamic acid, 3-carboxy-(trans) ...... $(3-HO_2CC_6H_4)CH=CHCO_2H$	192.18	nd (ace-lig)		275			aa	B9[2], 642
5101	Cinnamic acid, 4-carboxy-(trans) ...... $4-HO_2CC_6H_4CH=CHCO_2H$	192.18	pw	sub> 350	358d				B9[2], 642

No.	Name, Synonyms, and Formula	Mol. wt.	Color, crystalline form, specific rotation and λ_{max} (log ε)	b.p. $^{\circ}C$	m.p. $^{\circ}C$	Density	n_D	Solubility	Ref.
5102	Cinnamic acid, α-chloro-*(cis)* $C_6H_5CH=CClCO_2H$	182.61	....	....	111	....	....	al	B9[3], 2729
5103	Cinnamic acid, α-chloro,methyl ester-*(trans)* $C_6H_5CH=CHClCO_2CH_3$	196.63	....	108-9[0.5]	33			ace	B9[2], 396
5104	Cinnamic acid, β-chloro-*(cis)* $C_6H_5CCl=CHCO_2H$	182.61	....	....	133			al, eth	B9[3], 2728
5105	Cinnamic acid, β-chloro-*(trans)* $C_6H_5CCl=CHCO_2H$	182.61	....	....	143			al, eth	B9[3], 2728
5106	Cinnamic acid, β-chloro,methyl ester $C_6H_5CCl=CHCO_2CH_3$	196.63	....	113-4[0.5]	29	1.2248[21/4]	1.5781[21]	ace	B9[2], 396
5107	Cinnamic acid, α,β-dibromo-*(cis)* $C_6H_5CBr=CBrCO_2H$	305.97	ye pr or pl (chl or lig)	124[0.5]	100			eth, chl, aa	B9[3], 2736
5108	Cinnamic acid, 2,4-dihydroxy-*(trans)* or Umbellic acid $2,4\text{-}(HO)_2C_6H_3CH=CHCO_2H$	180.17	ye nd or pl	....	260d 240 (darkens)			al	B10[3], 1830
5109	Cinnamic acid, 2,5-dihydroxy-*(trans)* $2,5\text{-}(HO)_2C_6H_3CH=CHCO_2H$	180.17	ye cr (w), cr (dil al + 1w)	....	207d			al	B10[3], 1833
5110	Cinnamic acid, 3,4-dihydroxy-*(trans)* or Caffeic acid $3,4\text{-}(HO)_2C_6H_3CH=CHCO_2H$	180.17	ye pr or pl (w)	....	225d			al	B10[3], 1834
5111	Cinnamic acid, 3,5-dihydroxy-*(trans)* $3,5\text{-}(HO)_2C_6H_3CH=CHCO_2H$	180.17	nd (w + ½)	....	245-6			al, eth	B10[2], 297
5112	Cinnamic acid, 2,3-dimethoxy-*(trans)* $2,3\text{-}(CH_3O)_2C_6H_3CH=CHCO_2H$	208.22	....	....	180-1			....	B10[3], 1829
5113	Cinnamic acid, 2,4-dimethoxy-*(cis)* $2,4\text{-}(CH_3O)_2C_6H_3CH=CHCO_2H$	208.22	nd (al)	....	138			al, eth, bz	B10[3], 1831
5114	Cinnamic acid, 2,4-dimethoxy-*(trans)* $2,4\text{-}(CH_3O)_2C_6H_3CH=CHCO_2H$	208.22	nd (w or dil al)	....	187-9			al, eth, bz, chl	B10[3], 1831
5115	Cinnamic acid, 2,5-dimethoxy-*(trans)* $2,5\text{-}(CH_3O)_2C_6H_3CH=CHCO_2H$	208.22	pa ye or ye-gr nd (w)	....	148-9			al, eth	B10[3], 1833
5116	Cinnamic acid, 3,4-dimethoxy-*(trans)* $3\text{-}4\text{-}(CH_3O)_2C_6H_3CH=CHCO_2H$	208.22	nd (w, dil al), pa ye pw (dil aa)	....	183			al, eth	B10[3], 1835
5117	Cinnamic acid, 3,4-dimethoxy, methyl ester $3,4\text{-}(CH_3O)_2C_6H_3CH=CHCO_2CH_3$	222.24	....	....	68-70			chl	B10[3], 1838
5118	Cinnamic acid, 3,5-dimethoxy-*(trans)* $3,5\text{-}(CH_3O)_2C_6H_3CH=CHCO_2H$	208.22	nd (w)	....	175-6			al, bz	B10[2], 297
5119	Cinnamic acid, 3,5-dimethoxy-4-hydroxy-*(trans)* or 5-Methoxyferulic acid $4\text{-}HO\text{-}3,5\text{-}(CH_3O)_2C_6H_2CH=CHCO_2H$	224.22	pa ye nd (al)	....	192			al	B10, 508
5120	Cinnamic acid, α-ethyl-*(cis)* or α-Benzal butyric acid $C_6H_5CH=C(C_2H_5)CO_2H$	176.22	nd (w)	....	82			al, eth, bz	B9[3], 2783
5121	Cinnamic acid, α-ethyl-*(trans)* or α-Benzal butyric acid $C_6H_5CH=C(C_2H_5)CO_2H$	176.22	nd (w)	....	106 (114)			al, eth	B9[3], 2783
5122	Cinnamic acid, ethyl ester-*(trans)* or Ethyl cinnamate $C_6H_5CH=CHCO_2C_2H_5$	176.22	....	271.5, 144[15]	12	1.0491[20/4]	1.5598[20]	al, eth, ace, bz	B9[3], 2682
5123	Cinnamic acid, α-fluoro $C_6H_5CH=CFCO_2H$	166.15	....	290	157.6			al, eth	B9[1], 237
5124	Cinnamic acid, 2-hydroxy-*(trans)* or o-Coumaric acid $2\text{-}HOC_6H_4CH=CHCO_2H$	164.17	nd (w)	....	217d			al	B10[3], 833
5125	Cinnamic acid, 3-hydroxy-*(trans)* or m-Coumaric acid $3\text{-}HOC_6H_4CH=CHCO_2H$	164.17	pr (w)	....	193			al, eth, bz	B10[3], 840
5126	Cinnamic acid-4-hydroxy-*(trans)* or p-Coumaric acid $4\text{-}HOC_6H_4CH=CHCO_2H$	164.17	nd (w + 1), cr (w)	....	215d			eth	B10[3], 844
5127	Cinnamic acid-4-hydroxy-3-methoxy or Ferulic acid $4\text{-}(HO)\text{-}3\text{-}CH_3OC_6H_3CH=CHCO_2H$	194.19	pr or nd (w)	....	171			al, chl	B10[3], 1834
5128	Cinnamic acid-2-methoxyphenyl ester-*(trans)* or o-Anisyl cinnamate $C_6H_5CH=CHCO_2(C_6H_4OCH_3\text{-}2)$	254.29	wh nd (al)	....	130			ace, bz, chl	B9[3], 2698
5129	Cinnamic acid-4-methoxy $4\text{-}CH_3OC_6H_4CH=CHCO_2H$	178.19	wh nd (al)	....	172-5			aa	B10[3], 845
5130	Cinnamic acid-methyl ester-*(trans)* $C_6H_5CH=CHCO_2CH_3$	162.19	cr (peth or dil al)	261.9, 127[10]	36.5	1.0911[20/4]	1.5766[22]	al, eth, ace, bz	B9[3], 2680
5131	Cinnamic acid-α-methyl-*(cis)* or α-Benzal propionic acid $C_6H_5CH=C(CH_3)CO_2H$	162.19	nd (bz)	288, 190[21]	74	....	....	al, eth, bz, peth	B9[3], 2764

No.	Name, Synonyms, and Formula	Mol. wt.	Color, crystalline form, specific rotation and λ_{max} (log ε)	b.p. °C	m.p. °C	Density	n_D	Solubility	Ref.
5132	Cinnamic acid-α-methyl- (trans) or α-Benzal propionic acid $C_6H_5CH=C(CH_3)CO_2H$	162.19	pr (aa, eth or dil al)	....	81-2	....	....	al, eth, bz, peth	B9³, 2764
5133	Cinnamic acid-α-methyl-2-nitro $2-O_2NC_6H_4CH=C(CH_3)CO_2H$	207.19	mcl pr (al)	....	164-5			al, eth	B9³, 2766
5134	Cinnamic acid-α-methyl-3-nitro $3-O_2NC_6H_4CH=C(CH_3)CO_3H$	207.19	nd or pw	....	203.5			eth, bz, aa	B9³, 2767
5135	Cinnamic acid-α-methyl-4-nitro $4-O_2NC_6H_4CH=C(CH_3)CO_2H$	207.19	ye rh (aa), tcl pym (aa, al-eth)		208			al, eth, bz	B9³, 2767
5136	Cinnamic acid-2-methyl-4-nitro $[2-(CH_3)-4-(O_2N)C_6H_3]CH=CHCO_2H$	207.19	nd (al)		256				B9¹, 256
5137	Cinnamic acid-4-methyl-3-nitro- (trans) $[4-(CH_3)-3-(O_2N)C_6H_3]CH=CHCO_2H$	207.19	ye pl or nd (al)		173.5			al, eth	B9³, 2770
5138	Cinnamic acid-2-nitro- (cis) $2-O_2NC_6H_4CH=CHCO_2H$	193.16	yesh (bz or chl)		146-7			al, bz, chl	B9¹, 246
5139	Cinnamic acid-2-nitro- (trans) $2-O_2NC_6H_4CH=CHCO_2H$	193.16	nd (al)	sub	242-3				B9³, 2739
5140	Cinnamic acid-2-nitro, ethyl ester- (trans) $2-O_2NC_6H_4CH=CHCO_2C_2H_5$	221.22	ye rh bipym (al)		44			al, eth, bz	B9³, 2739
5141	Cinnamic acid-2-nitro, methyl ester- (trans) $2-O_2NC_6H_4CH=CHCO_2CH_3$	207.19	wh nd (w)	187-9¹⁵	73			al	B9³, 2739
5142	Cinnamic acid-3-nitro-(cis) $3-O_2NC_6H_4CH=CHCO_2H$	193.16	nd	....	158				B9¹, 247
5143	Cinnamic acid-3-nitro-(trans) $3-NO_2C_6H_4CH=CHCO_2H$	193.16	nd (al)	....	204-1	....	al	B9, 2741
5144	Cinnamic acid-3-nitro, ethyl ester- (trans) $3-NO_2C_6H_4CH=CHCO_2C_2H_5$	221.22	nd (al), pr (aa)		78-9				B9³, 2742
5145	Cinnamic acid-3-nitro, methyl ester- (trans) $3-O_2NC_6H_4CH=CHCO_2CH_3$	207.19	pa ye pr (MeOH)	d	123-4			eth, bz, chl	B9³, 2742
5146	Cinnamic acid-4-nitro- (trans) $4-O_2NC_6H_4CH=CHCO_2H$	193.16	yesh-wh pr (al)		286				B9³, 2744
5147	Cinnamic acid-4-nitro, ethyl ester- (trans) $4-O_2NC_6H_4CH=CHCO_2C_2H_5$	221.22	pl (aa)		141-2				B9³, 2744
5148	Cinnamic acid-4-nitro, methyl ester- (trans) $4-O_2NC_6H_4CH=CHCO_2CH_3$	207.19	wh nd (al)	281-6	162			al	B9³, 2744
5149	Cinnamic acid-2-octyl ester- (d) $C_6H_5CH=CHCO_2CH(CH_3)C_6H_{13}$	260.38	$[\alpha]^{17}_D$ + 40.2	218²⁸	....	0.9645²⁰/⁴	1.5145²⁰	bz, chl	B9³, 8687
5150	Cinnamic acid-2-octyl ester- (trans, dl) $C_6H_5CH=CHCO_2CH(CH_3)C_6H_{13}$	260.38		240⁶⁰	....	0.9715¹⁷/⁴		bz, chl	B9¹, 230
5151	Cinnamic acid, 2-octyl ester (trans, l) $C_6H_5CH=CHCO_2CH(CH_3)C_6H_{13}$	260.38	$[\alpha]^{17}_D$ - 39.78	211²⁸	....	0.9692¹⁷/⁴		bz, chl	B9³, 2687
5152	Cinnamic acid, α-phenyl or α-Phenyl cinnamic acid $C_6H_5CH=C(C_6H_5)CO_2H$	224.26	nd (lig, dil al)	sub	134-5			al, eth	B9³, 3414
5153	Cinnamic acid, phenylester- (trans) $C_6H_5CH=CHCO_2H_5$	224.26	....	205-7¹⁵	72.5			eth, bz	B9³, 2689
5154	Cinnamic acid, propyl ester- (trans) $C_6H_5CH=CHCO_2C_3H_7$	190.25		285		1.0435⁰/⁰			B9³, 2684
5155	Cinnamic acid, iso-propyl ester- (trans) or Isopropyl cinnamate $C_6H_5CH=CHCO_2CH(CH_3)_2$	190.25		268-70, 153-5²⁰		1.0320²⁰	1.5455²⁰	al, eth, ace	B9³, 2685
5156	Cinnamic acid, 4-isopropyl or Cumiliden acetic acid $4-i-C_3H_7C_6H_4CH=CHCO_2H$	190.25	pr (bz)	....	165			al	B9³, 2814
5157	Cinnamic acid, 3,4,5-trimethoxy $3,4,5-(CH_3O)_3C_6H_2CH=CHCO_2H$	238.24	nd (w)		126-7			chl	B10³, 2200
5158	Cinnamonitrile, (cis) or Allocinnamonitrile $C_6H_5CH=CHCN$	129.16		249, 139³⁰	-4.4		1.5843²⁰	al, bz	B9³, 2720
5159	Cinnamonitrile, (trans) $C_6H_5CH=CHCN$	129.16		263.8, 134-6²⁸	22	1.0304²⁰/⁴	1.6013²⁰	al, ace	B9³, 2721
5160	Cinnamylalcohol, (cis) or 3-phenyl-2-propen-1-ol $C_6H_5CH=CHCH_2OH$	134.18	wh nd (eth-peth)	257.5, 127-8¹⁰	34	1.0440²⁰/⁴	1.5819²⁰	al, eth	B6³, 2401
5161	Cinnamyl alcohol, acetate- (trans) or Cinnamyl acetate $C_6H_5CH=CHCH_2O_2CCH_3$	176.22	....	145-6¹⁵, 114¹	....	1.0567²⁰	1.5425²⁰	al, eth, ace, bz	B6³, 2406
5162	Cinnamyl chloride, (trans) $C_6H_5CH=CHCOCl$	166.61	....	257.5, 131¹¹	37-8	1.1617⁴⁵/⁴	1.614⁴²·⁵	lig	B9³, 2710

No.	Name, Synonyms, and Formula	Mol. wt.	Color, crystalline form, specific rotation and λ_{max} (log ε)	b.p. °C	m.p. °C	Density	n_D	Solubility	Ref.
5163	Citraconic acid or Methyl maleic acid .... $HO_2CC(CH_3)=CHCO_2H$	130.10	tcl pr or pl (eth-bz) nd (eth-lig)	....	93-4	1.617	....	w	B2[4], 2230
5164	Citraconic anhydride or Methylmaleic anhydride .... $C_5H_4O_3$	112.09	....	213-4, 99-100[15]	7-8	1.2469[16/4]	1.4716[21]	al, eth, ace	B17[4], 5912
5165	Citraconic acid, diethyl ester or Diethyl citraconate ...... $C_2H_5O_2CC(CH_3)=CHCO_2C_2H_5$	186.21	....	228[766], 120[20]	......	1.0491[20/4]	1.4467[20]	al, eth, aa	B2[4], 2232
5166	Citraconic acid, dimethyl ester or Dimethyl citraconate ... $CH_3O_2CC(CH_3)=CHCO_2CH_3$	158.16	....	210.5[758], 92.8[10]	......	1.1153[20/4]	1.4473[20]	al, eth, ace, aa	B2[4], 2232
5167	Citral a or Geranial .... $C_{10}H_{16}O$	152.24	....	229, 118-9[20]	......	0.8888[20]	1.4898[20]	al, eth	B1[4], 3569
5168	Citral b or Neral................ $C_{10}H_{16}O$	152.24	....	120[20]	......	0.8869[20]	1.4869[20]	al, eth	B1[4], 3569
5169	β-Citraurin .... $C_{30}H_{40}O_2$	432.65	pl (bz-peth), cr (al)	....	147	......	......	al, eth, ace, bz	B8[3], 1599
5170	Citric acid or 2-Hydroxy-1,2,3-propane tricarboxylic acid $HOC(CH_2CO_2H)_2CO_2H$	192.14	rh (w + 1)	d	153 anh	1.665[20/4]	......	w, al, eth	B3[4], 1272
5171	Citric acid, anhydro-methylene or 1,3-Dioxolan-4-one-5,5-diacetic acid................ $C_8H_{11}O_7$	204.15	(w)	....	298	......	......	bz, chl	B19[2], 324
5172	Citric acid, tribenzyl ester or Benzyl citrate .... $HOC(CH_2CO_2CH_2C_6H_5)_2CO_2CH_2C_6H_5$	462.51	nd (al)	....	51	......	......	al	B6[3], 1537
5173	Citric acid, triethyl ester or Ethyl citrate .... $HOC(CH_2CO_2C_2H_5)_2CO_2C_2H_5$	276.29	....	294, 185[17]	......	1.1369[20/4]	1.4455[20]	al, eth	B3[4], 1276
5174	Citric acid, trimethyl ester or Methyl citrate .... $HOC(CH_2CO_2CH_3)_2CO_2CH_3$	234.21	tcl	287d, 176[16]	78-9	......	......	al, eth	B3[4], 1276
5175	Citric acid, triphenyl ester or Phenyl citrate .... $HOC(CH_2CO_2C_6H_5)_2CO_2C_6H_5$	420.42	nd (al)	....	124.5	......	......	eth	B6, 170
5176	Citric acid, tripropyl ester or Propyl citrate .... $HOC(CH_2CO_2C_3H_7)_2CO_2C_3H_7$	318.37	....	198[18]	......	......	......	al, eth	B3, 568
5177	Citric acid, triethyl ester, acetate .... $CH_3CO_2C(CH_2CO_2C_2H_5)CO_2C_2H_5$	318.32	....	131-2	....	1.135	1.4380	....	......
5178	Citric triamide .... $HOC(CH_2CONH_2)_2CONH_2$	189.17	(w)	....	210-5d	......	......	w	B3[1], 197
5179	Citrinin .... $C_{13}H_{14}O_5$	250.25	ye nd (MeOH), $[\alpha]^{111}_D$ -37	....	178-9d	......	......	ace, bz, chl	B18[4], 6329
5180	Citronellal-(d) or d-Rhodinal .... $(CH_3)_2C=CH(CH_2)_2CH(CH_3)CH_2CHO$	154.26	$[\alpha]^{18}_D$ + 13.09	207.8, 92[14]	......	0.8573[20/4]	1.4456[20]	al, eth	B1[4], 3515
5181	Citronellal-(dl) or dl-Rhodinal .... $C_{10}H_{18}O$	154.26	....	207-8, 79-81[10]	......	0.8535[17/4]	1.4473[20]	al, eth	B1[4], 3515
5182	Citronellal-(l) or l-Rhodinal................ $C_{10}H_{18}O$	154.26	$[\alpha]^{20}_D$ -2.5	205-6, 87[10]	......	0.8567[17/4]	1.4479[20]	al, eth	B1[4], 3515
5183	Citronellol-(d) or d-Rhodinol .... $(CH_3)_2C=CH(CH_2)_2CH(CH_3)CH_2CH_2OH$	156.27	$[\alpha]^{17}_D$ + 6.8	244.4, 118[17]	......	0.8590[20]	1.4565[20]	al, eth	B1[4], 2188
5184	Citronellol-(dl) or Dihydrogeranoil .... $C_{10}H_{20}O$	156.27	....	99[10]	......	0.8560[20/4]	1.4543[20]	al, eth	B1[4], 2188
5185	Citronellol-(l) or l-Rhodinol .... $C_{10}H_{20}O$	156.27	$[\alpha]^{18}_D$ -5.3	108-9[10]	......	0.859[18/4]	1.4576[18]	al, eth	B1[4], 2188
5186	Citrulline-(L) or α-Amino-δ-ureido valeric acid .... $H_2NCONH(CH_2)_3CH(NH_2)CO_2H$	175.19	pr (aq MeOH), $[\alpha]^{20}_D$ + 3.7 (w, c=2)	......	234-7	......	......	w	B4[3], 1347
5187	Clovene .... $C_{15}H_{24}$	204.35	$[\alpha]_D$ + 2.84	259-60	......	0.9241[18]	1.4999[18]	....	B5[2], 356
5188	Cocaine-(d) or Benzoyl methyl ecgonine............... $C_{17}H_{21}NO_4$	303.36	mcl pr (eth), $[\alpha]^{20}_D$ + 15.8 (chl, p=10)	......	98	......	......	al, eth, ace, bz	B22[2], 150
5189	Cocaine-dl .... $C_{17}H_{21}NO_4$	303.36	rh bipym pr (peth)	......	79-80	......	......	al, eth, ace, bz, lig	B22[4], 2103

No.	Name, Synonyms, and Formula	Mol. wt.	Color, crystalline form, specific rotation and λ_{max} (log ε)	b.p. °C	m.p. °C	Density	n_D	Solubility	Ref.
5190	Cocaine-(l) .... $C_{17}H_{21}NO_4$	303.36	mcl pr (al), $[\alpha]^{20}_D$ -16.3 (chl, c=4)	187-8[0.1], (sub, vac)	98	....	1.5022[98]	al, eth, ace, bz, chl	B22[4], 2101
5191	Cocaine, hydrochloride-(dl) .... $C_{17}H_{21}NO_4 \cdot HCl$	339.82	pl (al)	....	187 (cor)	....		w, al, ace	B22[2], 156
5192	Cocaine, hydrochloride-(l) $C_{17}H_{21}NO_4 \cdot HCl$	339.82	mcl pr (al), cr (w + 2), $[\alpha]^{20}_D$ -71.95		197			w, al, ace	B22[4], 2102
5193	Coclaurine-(l) .... $C_{17}H_{19}NO_3$	285.35	pl (al), $[\alpha]^{20}_D$ 17.01	....	220-1				B21[4], 2605
5194	Codamine $C_{20}H_{25}NO_4$	343.43	pr (bz or eth)		127			al, eth, chl	B21[4], 2704
5195	Codeine or Morphine-3-methylether .... $C_{18}H_{21}NO_3$	299.37	rh oct (+1w, w or dil al), cr (eth)	250[22], sub 140[1.5]	157-8	1.32		al, eth, bz, chl	B27[2], 137
5196	Codeine, hydrate $C_{18}H_{21}NO_3 \cdot H_2O$	317.39	rh oct (w, aq al), $[\alpha]^{25}_D$ -136 (al, c=2.8)			1.31		al, eth, ace, bz, chl	B27[2], 137
5197	Codeine, hydrochloride $C_{18}H_{21}NO_3 \cdot HCl \cdot 2H_2O$	371.87	nd, $[\alpha]^{15}_D$ 108.2		287d			w, al	B27[2], 143
5198	Codeine, phosphate .... $C_{18}H_{21}NO_3 \cdot H_3PO_4 \cdot 1\frac{1}{2} H_2O$	424.39	lf or pr (dil al)		220-35d	....		eth, chl	B27[2], 144
5199	Codeine, sulfate $(C_{18}H_{21}NO_3)_2 \cdot H_2SO_4 \cdot 5H_2O$	786.90	pr, $[\alpha]^{15}_D$ -100.9 (w, p=3)		278 (anh)			w	B27[2], 144
5200	β-Codeine or Neopine $C_{18}H_{21}NO_3$	299.37	nd (peth), $[\alpha]^{23}_D$ -28 (chl, c=7.5)		127.5			al, eth, bz, chl	B27[2], 176
5201	Codeine, dihydro-(d) or Dihydroneopine .... $C_{18}H_{23}NO_3$	301.39	cr (+1w, dil MeOH)	248[15]	112-3		....		B27[2], 103
5202	Colchiceine or N-Acetyltrimethyl colchicinic acid .... $C_{21}H_{23}NO_6$	385.42	pa ye nd (diox)	....	178-9	1.24		al, chl	B14[3], 692
5203	Colchicine .... $C_{22}H_{25}NO_6$	399.45	ye pl (w + 1½), pa ye nd (AcOEt), ye cr (bz), $[\alpha]^{17}_D$ -121 (chl, c=0.9), -429 (w, c=1.72)		155-7			w, al	B14[3], 693
5204	Colchicinic acid, trimethyl .... $C_{19}H_{21}NO_5$	343.39	pa ye nd (al), $[\alpha]^{25}_D$ -184.5 (chl, c=1)	....	155-7			w, al	Am75, 5292
5205	Conessine or Neriine .... $C_{24}H_{40}N_2$	356.60	lf or pl (ace), $[\alpha]^{20}_D$ +25.3 (al, c=0.7)	165-7[0.1]	125-6	....		aa	B22[4], 4382
5206	Congo red $C_{32}H_{22}N_6Na_2O_6S_2$	696.68	pw					al	B16[3], 474
5207	Conhydrine-(d) or 2(1-Hydroxypropyl)piperidine .... $C_8H_{17}NO$	143.23	lf (eth), $[\alpha]_D$ +10 (w)	226	121			w, al, eth, chl	B21[4], 122
5208	Conhydrine-(dl)(lower m.p.) .... $C_8H_{17}NO$	143.23	nd (peth)	sub	69-70			w, al, eth, bz	B21[4], 122
5209	Conhydrine-(dl)(higher m.p.) or 2-(1-Hydroxypropyl) piperidine .... $C_8H_{17}NO$	143.23	nd (eth)	sub	98-9			w, al, eth, bz	B21[4], 123

No.	Name, Synonyms, and Formula	Mol. wt.	Color, crystalline form, specific rotation and λ_{max} (log ε)	b.p. °C	m.p. °C	Density	n_D	Solubility	Ref.
5210	α-Coniceine-(d) or 2-Methylconidine .... $C_8H_{15}N$	125.22	$[\alpha]^{15}_D$ + 18.4 (al, c=2)	158	-16	$0.891^{15.5/4}$	....	al	B20, 152
5211	Corticosterone, 11-dehydro .... $C_{21}H_{28}O_4$	344.46	pr (ace-w, al or ace-eth), $[\alpha]^{25}_D$ + 258 (al)	....	183-4			al, ace, bz	B8³, 3624
5212	α-Coniceine-(dl) or 2-Methyl conidine .... $C_8H_{15}N$	125.22	....	156-9	......	$0.890^{15.5/4}$		al	B20, 153
5213	β-Coniceine-(dl) or 2-Propenyl piperidine .... $C_8H_{15}N$	125.22	nd $[\alpha]^{45}_D$ +49.9	168-9	38-9			al, eth	B20, 146
5214	β-Coniceine-(dl) or 2-Propenylpiperidine .... $C_8H_{15}N$	125.22	nd	168-70⁷⁵³	8	$0.8716^{15/4}$		al, eth	B20, 146
5215	β-Coniceine-(l) or 2-Propenylpiperidine .... $C_8H_{15}N$	125.22	nd $[\alpha]^{45}_D$ - 50.5	168-9	41	$0.8520^{50/4}$		al, eth	B20, 146
5216	γ-Coniceine $C_8H_{15}N$	125.22		173-4, 64-5¹⁴	......	$0.8720^{20/4}$	1.4607^{18}	al	B20⁴, 1970
5217	ε-Coniceine-(d) or d-2-Methyl coniceine .... $C_8H_{15}N$	125.22	$[\alpha]^{15}_D$ +67.4	152-4		$0.8856^{15/4}$		al, eth	B20, 151
5218	ε-Coniceine-(dl) $C_8H_{15}N$	125.22		150-1		$0.8836^{15/4}$		al, eth	B20, 152
5219	ε-Coniceine-l $C_8H_{15}N$	125.22	$[\alpha]^{15}_D$ - 87.34	151-3.5		$0.8642^{15/4}$			B20, 151
5220	α-Conidendrin or Tsugalactone...................... $C_{20}H_{20}O_6$	356.38	cr (al), $[\alpha]^{20}_D$ - 54.5 (ace, c=2.1)	....	255-6			al, eth, bz	B18⁴, 3346
5221	α-Conidine, 3-methyl-(l) .... $C_8H_{15}N$	125.22	$[\alpha]^{8}_D$ + 16	158		$0.8856^{15/4}$		al, eth	B20, 153
5222	α-Conidine, 3-methyl-(dl) .... $C_8H_{15}N$	125.22	....	158		$0.8946^{15/4}$		al, eth	B20, 153
5223	Conidine, 3-methyl-(l) $C_8H_{15}N$	125.22	$[\alpha]^{17}_D$ -17.1	158		$0.8856^{15/4}$		al, eth	B20, 153
5224	Coniferin or 4-(3-hydroxypropenyl)-3-Methoxyphenyl-D-glucoside............ $C_{16}H_{22}O_8$	342.35	nd (w + 2), $[\alpha]^{20}_D$ -68 (w, c=0.5)		186 (anh)				B17⁴, 2999
5225	Coniferyl alcohol .... $C_{10}H_{12}O_3$	180.21	pr (eth-lig)	163-5³	74			al, eth	B6³, 6442
5226	Coniine-(d) or 2-Propyl piperidine.................. $C_8H_{17}N$	127.23	$[\alpha]^{20}_D$ +15.6	166-7, 64¹⁸	-2	0.8440^{20}	1.4512^{22}	**al**, eth, bz	B20⁴, 1611
5227	Coniine-(dl) or 2-Propyl piperidine .... $C_8H_{17}N$	127.23	....	166-7⁷⁴⁵, 59-63¹⁷	......	0.8447^{20}	1.4513^{23}	**al**, eth, bz	B20⁴, 1611
5228	Coniine-(l) $C_8H_{17}N$	127.23	$[\alpha]^{15}_D$ -15.6	166, 64¹⁸		$0.845^{15/4}$	1.4512^{22}	**al**, eth, bz	B20², 62
5229	Coniine, hydrobromide-(d) $C_8H_{17}N\cdot HBr$	208.15	pr	....	211			w, al, eth, chl	B20, 112
5230	Coniine, hydrochloride-(d) $C_8H_{17}N\cdot HCl$	163.69	orth (w), $[\alpha]^{20}_D$ + 10.1 (lig, NH_3)	....	221			w, al	B20⁴, 1611
5231	Coniine, hydrochloride-(dl) $C_8H_{17}N\cdot HCl$	163.69	nd (al-eth)	....	216-7			w, al	B20¹, 31
5232	Coniine, hydrochloride-(l) $C_8H_{17}N\cdot HCl$	163.69	nd	....	220-1			w, al	B20, 118
5233	Coniine, N-methyl-(d) .... $C_8H_{19}N$	141.26	$[\alpha]^{24}_D$ +82.4	173-4⁷⁵⁷	......	$0.8326^{23/4}$	1.4538^{13}	al, ace	B20⁴, 1612
5234	Coniine, picrate-(d) $C_8H_{17}N\cdot C_6H_3N_3O_7$	356.34	ye pr (w)		75			al, eth	B20, 112
5235	Coniine, picrate-(l) $C_8H_{17}N\cdot C_6H_3N_3O_7$	356.34	ye pr (w)		74			al, eth	B20², 62
5236	Conquinamine .... $C_{19}H_{24}N_2O_2$	312.41	ye tetr, $[\alpha]^{15}_D$ + 200 (al, c=0.5)	....	123			al, eth, chl	B27², 667
5237	Copaene .... $C_{15}H_{24}$	204.36	$[\alpha]_D$ -25.8	246-51, 119-20¹⁰	......	$0.8996^{20/4}$	1.4894^{20}	eth, ace, aa, lig	B5⁴, 1189

No.	Name, Synonyms, and Formula	Mol. wt.	Color, crystalline form, specific rotation and λ_{max} (log ε)	b.p. °C	m.p. °C	Density	n_D	Solubility	Ref.
5238	Coproergostane or Pseudoergostane............ $C_{28}H_{50}$	386.71	nd (ace), $[\alpha]^{19}_D$ + 25.3 (chl, c=2)	....	64	....	....	eth, chl	B5³, 1143
5239	Coprostane or Pseudocholestane.............. $C_{27}H_{48}$	372.69	orth nd (al, ace or eth-al), $[\alpha]^{20}_D$ + 25.1 (chl, c=2)	....	72	$0.9119^{87.7/4}$	$1.4884^{88/4}$	eth, chl	B5⁴, 1226
5240	3β-Coprostanol or Coprosterol $C_{27}H_{48}O$	388.69	nd (MeOH), $[\alpha]^{18}_D$ + 28 (chl, c=1.8)	....	102 (105)	....	....	al, eth, bz, chl	B6³, 2128
5241	Coprostenol or Allocholesterol.............. $C_{27}H_{46}O$	386.67	nd (eth-MeOH), $[\alpha]_D$ + 43.7 (bz, c=1)	....	....	....	....	al, eth, ace, bz, chl	B6³, 2604
5242	Coprostenone or Δ⁴-Cholesten-3-one.............. $C_{27}H_{44}O$	384.65	pl (MeOH), $[\alpha]'_D$ + 88.6 (chl)	....	81-2	....	....	eth, bz, lig	B7³, 1592
5243	Coronene or Hexabenzobenzene............ $C_{24}H_{12}$	300.36	ye nd (bz)	525	438-40 (cor)	1.371	....	....	B5³, 2651
5244	Corticosterone or 11,21-Dihydroprogesterone.......... $C_{21}H_{30}O_4$	346.47	nd (al), pl (ace), $[\alpha]^{11}_D$ + 223 (al, c=1.1)	sub 190⁰·⁰¹	180-2	....	....	al, eth, ace	B8³, 3574
5245	Corticosterone, 17-hydroxy or Cortisol Hydrocortisone .. $C_{21}H_{30}O_5$	362.47	pr(al or i-prOH), $[\alpha]^{22}_D$ + 167 (al)	....	220	....	....	aa, al	B8³, 4036
5246	Cortisone-(d)............ $C_{21}H_{28}O_5$	360.45	$[\alpha]^{25}_D$ + 209 E + OH, c=1.2)	....	220-4	....	....	al, ace	B8³, 4057
5247	Cortisone, 21-acetate............. $C_{23}H_{30}O_6$	402.49	nd (ace), rods (chl), $[\alpha]^{24}_D$ + 164 (ace, c=0.5)	....	239-40	....	....	ace, chl	B8³, 4058
5248	Corybulbine-(d) or Corydalis-6 $C_{21}H_{25}NO_4$	355.44	nd (al), $[\alpha]^{20}_D$ 303 (chl, c=1.4)	....	237-8	....	....	ace, chl	B21⁴, 2779
5249	Corybulbine-(dl)............. $C_{21}H_{25}NO_4$	355.44	cr (chl-al)	....	220-2	....	....	al, ace, chl, bz	B21, 217
5250	Corycavamine $C_{21}H_{21}NO_3$	367.41	pr (eth or al), $[\alpha]^{20}_D$ + 166.6 (chl, c=2.2)	....	149	....	....	al, chl	B27², 621
5251	Corycavine .... $C_{21}H_{21}NO_3$	367.41	orh pl (al)	....	221-2	....	....	chl	B27², 621
5252	Corydaldine .... $C_{11}H_{13}NO_3$	207.23	mcl pr (w or al)	....	175	....	....	w, al, eth, bz, chl	B21⁴, 6443
5253	Corydaline-(d) or Corydolis-A............ $C_{22}H_{27}NO_4$	369.47	pr (al), $[\alpha]^{20}_D$ + 311 (al, c=0.8)	....	136	....	....	al, eth	B21⁴, 2779
5254	Corydaline-(dl)............ $C_{22}H_{27}NO_4$	369.47	cr (al)	....	135-6	....	....	eth, al, w	B21⁴, 2780
5255	Corydaline-(meso,d)............. $C_{22}H_{27}NO_4$	369.47	pr (eth), $[\alpha]_D$ + 180 (chl, c=3)	....	155-6	....	....	eth, al	B21⁴, 2779
5256	Corydaline-(meso,dl)............ $C_{22}H_{27}NO_4$	369.47	cr (al)	....	163-4	....	....	eth, al, w	B21⁴, 2779

No.	Name, Synonyms, and Formula	Mol. wt.	Color, crystalline form, specific rotation and λ_{max} (log ε)	b.p. °C	m.p. °C	Density	n_D	Solubility	Ref.
5257	Corydaline-(meso,l) $C_{22}H_{27}NO_4$	369.47	pr (eth), $[\alpha]^{20}_D$ -181 (chl, c=3)	....	155-6	....	....	eth, al, w	B21[1], 257
5258	Corynantheine $C_{22}H_{28}N_2O_3$	384.48	$[\alpha]^{18}_D$ +28.8 (MeOH, c=28.8)		165-6			al	B25[2], 212
5259	Cotarnine $C_{12}H_{15}NO_4$	237.26	nd (bz), cr (eth)	....	132-3d			al, eth, bz, chl	B27[2], 543
5260	Cotarnine, chloride or Stypticin $C_{12}H_{14}ClNO_3 \cdot 2H_2O$	291.74	ye pw or nd (al-AcOEt)		197			w, al	B27[1], 456
5261	Cotarnine, O-phthalate or Styptol $(C_{12}H_{14}NO_3)_2C_6H_4(CO_2)_2$	604.62	og cr or pw		103-5			w	B27, 476
5262	Coumalic acid or α-Pyrone-5-carboxylic acid $C_6H_4O_4$	140.10	pr (MeOH)	218[120]	205-10d			al, aa	B18[4], 5382
5263	Coumaran or 2,3-Dihydrobenzofuran C_8H_8O	120.16	....	188-9, 76[14]	-21.5	1.0576[24/4]	1.5426[20]	al, eth, chl	B17[4], 404
5264	3-Coumaranone $C_8H_6O_2$	134.14	red nd (al)	152-4[10]	102-3			bz	B17[4], 1456
5265	Coumarin or 1,2-Benzopyrone $C_9H_6O_2$	146.15	rh pym (eth)	301.7	71	0.935[20/4]	....	al, eth, chl	B17[4], 5055
5266	Coumarin, 6-amino $C_9H_7NO_2$	161.16			168-70			w, al	B18[4], 7920
5267	Coumarin, 4-chloro $C_9H_5ClO_2$	180.59	nd (al)		165			al, eth, bz	B17[4], 5058
5268	Coumarin, 7-diethylamino-4-methyl $C_{14}H_{17}NO_2$	231.30	cr (al, bz-lig)	....	89 (al), 135 (bz-lig)			al, eth, ace	B18, 612
5269	Coumarin, 3,4-dihydro or Hydrocoumarin $C_9H_8O_2$	148.17	lf	272, 145[18]	25	1.169[18]	1.5563[20]	chl	B17[4], 4956
5270	Coumarin, 5,7-dihydroxy-4-methyl $C_{10}H_8O_4$	192.18	nd (al), lf (aa)	....	282-4			al	B18[4], 1367
5271	Coumarin, 6,7-dihydroxy or Aesculetin $C_9H_6O_4$	178.15	nd (w + 1), pr (aa), lf (sub)	sub	276			al, ace, chl	B18[4], 1322
5272	Coumarin, 7,8-dihydroxy or Daphnetin $C_9H_6O_4$	178.15	yesh (dil al)	sub	261-3			al	B18[4], 1330
5273	Coumarin, 6,7-dihydroxy-4-methyl or 4-Methyl aesculetin $C_{10}H_8O_4$	192.18	ye nd (dil al)	....	274-6			al	B18[4], 1371
5274	Coumarin, 7,8-dihydroxy-6-methoxy or Fraxetin $C_{10}H_8O_5$	208.18	pl (dil al)	....	230-2			al	B18[4], 2371
5275	Coiumarin, 5,7-dimethoxy or Citropten $C_{11}H_{10}O_4$	206.20	pr or nd (al)	200d	148-50			al, ace, chl, aa	B18[4], 1322
5276	Coumarin, 7-ethoxy-4-methyl or Maraniol $C_{12}H_{12}O_3$	204.23	wh	....	114			al	B18[4], 334
5277	Coumarin, 3-hydroxy $C_9H_6O_3$	162.14			154			w, al, eth, ace	B17[4], 6152
5278	Coumarin, 4-hydroxy or Benzotetronic acid $C_9H_6O_3$	162.15	nd (w)		213-4 (232)			al, eth	B17[4], 6153
5279	Coumarin, 4-hydroxy-3-(1-phenyl-3-oxobutyl) or Coumadin, Warfarin $C_{19}H_{16}O_4$	308.34	cr (al)		161			al, bz, diox	B17[4], 6794
5280	Coumarin, 5-hydroxy $C_9H_6O_3$	162.14			224-7			al, eth, bz	B18[3], 291
5281	Coumarin, 6-hydroxy $C_9H_6O_3$	162.15	nd (dil HCl)	....	250	1.25		al	B18[4], 293
5282	Coumarin, 7-hydroxy or Umbelliferone $C_9H_6O_3$	162.15	nd (w)	sub	230-1			al, chl, aa	B18[4], 294
5283	Coumarin, 7-hydroxy-6-methoxy or Chrysatropic acid $C_{10}H_8O_4$	132.18	nd or pr (al)		204			chl, aa	B18[4], 1323
5284	Coumarin, 7-hydroxy-4-methyl or 4-Methyl umbelliferone $C_{10}H_8O_3$	176.18	nd (al)	....	185-7			al, aa	B18[4], 332
5285	Coumarin, 8-hydroxy $C_9H_6O_3$	162.15	nd (dil al)		160			al, aa	B18[4], 304

PHYSICAL CONSTANTS OF ORGANIC COMPOUNDS (Continued)

No.	Name, Synonyms, and Formula	Mol. wt.	Color, crystalline form, specific rotation and λ_{max} (log ε)	b.p. °C	m.p. °C	Density	n_D	Solubility	Ref.
5286	Coumarin, 7-methoxy or Herniarin.............. $C_{10}H_8O_3$	176.18	lf (w or MeOH)	....	117-8	....	....	al, eth	B18[4], 295
5287	Coumarin, 8-methoxy $C_{10}H_8O_3$	176.17	....		89	....	....	al, eth, bz	B18[4], 304
5288	Coumarin, 3-methyl .... $C_{10}H_8O_2$	160.18	rh bipym (al)	292.5	91	....	....	al	B17[4], 5073
5289	Coumarin, 4-methyl $C_{10}H_8O_2$	160.18	nd (w), pr (bz)	....	83-4	....	....	al, bz	B17[4], 5074
5290	Coumarin, 5-methyl $C_{10}H_8O_2$	160.17	....	173-4[12]	65.8	....	....	al, eth, bz	B17[4], 5075
5291	Coumarin, 6-methyl $C_{10}H_8O_2$	160.17	....	303[725], 174[14]	75-6	....	....	al, eth, bz	B17[4], 5076
5292	Coumarin, 7-methyl $C_{10}H_8O_2$	160.17	....	171.5[11]	128	....	....	al, eth	B17[4], 5077
5293	Coumarin, 8-methyl $C_{10}H_8O_2$	160.17	....	178[20]	109-10	....	....	al, eth, bz	B17[4], 5079
5294	Coumarin, 6,7,8-trimethoxy or Fraxetin dimethyl ether... $C_{12}H_{12}O_5$	236.23	rh bipym pl (dil al)	90-100[0.2]	103-4	....	....	al, eth	B18[4], 2372
5295	3-Coumarin carboxylic acid .... $C_{10}H_6O_4$	190.16	nd (w or bz)	....	190d	....	....	al	B18[4], 5569
5296	2-Coumarone carboxylic acid or Coumarilic acid .... $C_9H_6O_3$	162.15	nd (w)	310-5d	192-3	....	....	al	B18[4], 4247
5297	2-Coumarone carboxylic acid, ethyl ester .... $C_{11}H_{10}O_3$	190.20	....	274[740], 161[13]	30-1	1.1656[28.8/4]	1.564[27.8]		B18[1], 442
5298	2-Coumarone carboxylic acid, 3-methyl	176.18	nd (dil al)	....	188-9	....	....	al	B17[4], 5074
5299	P-Coumaryl alcohol or 2(4-Hydroxyphenyl)-2-propeno 1-ol $4-HOC_6H_4CH=CH-CH_2OH$	150.18	pr (dil al)	....	124	....	....	al, eth, ace, bz	C41, 1119
5300	Coumestrol .... $C_{15}H_8O_5$	268.24	gy micr rods	....	385d	....	....		B19[4], 2870
5301	Coumestrol, diacetate .... $C_{19}H_{12}O_7$	352.30	pl	....	234	....	....		B19[4], 2872
5302	Creatine or (α-Methylguanido) acetic acid .... $H_2NC(=NH)N(CH_3)CH_2CO_2H$	131.14	mcl pr (w + 1)	....	303	1.33	....	w	B4[3], 1170
5303	Creatinine or l-Methylglylocy amidine .... $C_4H_7N_3O$	113.12	rh pr (w + 2), lf (w)	....	ca 300d	....	....		B24[2], 128
5304	o-Cresol or 2-Hydroxytoluene............ $2-CH_3C_6H_4OH$	108.15	....	191, 74.9[10]	30.9	1.0273[20/4]	1.5361[20]	al, eth, ace, bz	B6[4], 1940
5305	o-Cresol, 3-amino $3-H_2N-2-CH_3C_6H_3OH$	123.16	nd (w)	....	129	....	....	al	B13, 579
5306	o-Cresol, 4-amino $4-H_2N-2-CH_3C_6H_3OH$	123.16	nd or lf (bz)	sub	175	....	....	al, eth	B13[3], 1531
5307	o-Cresol, 4-amino-6-nitro $6-NO_2-4-NH_2-2-CH_3C_6H_2OH$	168.16	br-red nd (al)	....	118	....	....	al	B13, 578
5308	o-Cresol, 6-amino $6-H_2N-2-CH_3C_6H_3OH$	123.16	pl (w)	....	89	....	....	al, eth, ace, bz	B13[3], 1527
5309	o-Cresol, 6-amino-4-nitro $4-NO_2-6-H_2N-2-CH_3C_6H_2OH$	168.16	red-br nd (bz)	....	176	....	....	al, bz	B13[3], 1528
5310	o-Cresol, 3-bromo $3-Br-2-CH_3C_6H_3OH$	187.04	nd (peth)	55-7[4]	95	....	....	al, eth, ace, bz	B6, 360
5311	o-Cresol, 4-bromo $4-Br-2-CH_3C_6H_3OH$	187.04	nd (al or peth)	235 sub, 137-43[18]	64	....	....	al, eth, ace	B6[4], 2006
5312	o-Cresol, 5-bromo $5-Br-2-CH_3C_6H_3OH$	187.04	nd (lig or peth)	....	80	....	....	al, eth, ace	B6[2], 333
5313	o-Cresol, 3-chloro $3-Cl-2-CH_3C_6H_3OH$	142.59	lo nd (w)	225	86	....	....	al, eth, bz	B6[4], 2000
5314	o-Cresol, 4-chloro $4-Cl-2-CH_3C_6H_3OH$	142.59	nd (peth)	223	51	....	....		B6[4], 1987
5315	o-Cresol, 5-chloro $5-Cl-2-CH_3C_6H_3OH$	142.59	nd (peth)	....	73-4	....	....	al, bz	B6[4], 1986
5316	o-Cresol, 6-chloro $6-Cl-2-CH_3H_6H_3OH$	142.59	....	188-9[740], 80-1[20]	....	....	1.5449[20]	eth	B6[4], 1984
5317	o-Cresol, 3,5-dibromo $3,5-Br_2-2-CH_3C_6H_2OH$	265.94	nd (peth)	283-7[758]	98-101	....	....	peth	B6[2], 334
5318	o-Cresol, 3,6-dibromo $3,6-Br_2-2-CH_3C_6H_2OH$	265.94	cr	255-60	38	....	....		B6[1], 176

No.	Name, Synonyms, and Formula	Mol. wt.	Color, crystalline form, specific rotation and λ_{max} (log ε)	b.p. °C	m.p. °C	Density	n_D	Solubility	Ref.
5319	o-Cresol, 4,6-dibromo 4,6-Br₂-2-CH₃C₆H₂OH	265.94	nd (peth)	263-6[745]/d	58			al, eth, bz	B6[3], 1271
5320	o-Cresol, 4,5-dichloro 4,5-Cl₂-2-CH₃C₆H₂OH	177.03	nd (peth)		101			al, bz, aa	B6[2], 333
5321	o-Cresol, 4,5dichloro 4,6-Cl₂-2-CH₃C₆H₂OH	177.03	nd (w or peth)	266.5, 73-8[4]	55			al, eth, chl	B6[4], 2001
5322	o-Cresol, 3,5-dinitro 3,5-(NO₂)₂-2-CH₃C₆H₂OH	198.14	ye pr (al)		85.8			al, eth, ace	C51, 10414
5323	o-Cresol, 4,6-dinitro 4,6-(NO₂)₂-2-CH₃C₆H₂OH	198.14	ye pr or nd (al)		86.5			al, eth, ace	B6[4], 2014
5324	o-Cresol, 3-nitro 3-NO₂-2-CH₃C₆H₃OH	153.14	pa ye nd (w)		147			al, eth	B6[1], 178
5325	o-Cresol, 4-nitro 4-NO₂-2-CH₃C₆H₃OH	153.14	ye or col nd (w or aq al)	186-90[9]	96 (anh)			al, eth, bz, aa	B6[4], 2011
5326	o-Cresol, 5-nitro 5-NO₂-2-CH₃C₆H₃OH	153.14	ye nd (lig)		118			al, eth, bz	B6[4], 2010
5327	o-Cresol, 6-nitro 6-NO₂-2-CH₃C₆H₃OH	153.14	ye pr (dil al or peth)	250-60d, 185[12]	70			al, eth	B6[4], 2009
5328	o-Cresol, 4-nitroso 4-ON-2-CH₃C₆H₃OH	137.14	nd (w)		134-5d			al, eth, bz, chl	B7[3], 3388
5329	o-Cresol-4-isopropyl 4-i-C₃H₇-2-CH₃C₆H₃OH	150.22	....	230[766], 83[3]	8.6	0.9793[25]	1.5253[20]	al, bz, chl	B6[3], 1884
5330	o-Cresol, 5-isopropyl 5-i-C₃H₇-2-CH₃-C₆H₃OH	150.22	....	117-20	0-3	0.976[20/4]	1.523[20]	....	B6[3], 1885
5331	o-Cresol, 6-isopropyl 6-i-C₃H₇-2-CH₃C₆H₃OH	150.22	....	225, 104[14]	-14.5	0.9789[25]	1.5239[20]	al, bz, chl	B6[3], 1882
5332	o-Cresol-3,4,5,6-tetrabromo 2-CH₃C₆Br₄OH	423.75	ye nd (chl or aa)	d	208			al, eth, bz, chl, aa	B6[3], 1272
5333	o-Cresol-3,4,5,6-tetrachloro 2-CH₃C₆Cl₄OH	245.92	nd (lig)		190			al, eth, bz, aa	B6[3], 333
5334	o-Cresol-4,5,6-trinitro 4,5,6-(NO₂)₃-2-CH₃C₆HOH	243.13	og-ye pr (ace)		102			al, eth, ace, chl	B6, 369
5335	o-Cresylphosphate or Tri-o-cresylphosphate (2-CH₃C₆H₄O)₃PO	368.37	col or pa ye	410, 283-5[20]	11	1.1955[20/4]	1.5575[20]	al, eth, aa	B6[4], 1979
5336	m-Cresol or 3-Hydroxy toluene. 3-CH₃C₆H₄OH	108.15	....	202.2, 86[10]	11.5	1.0336[20/4]	1.5438[20]	**al, eth, ace, bz**	B6[4], 2035
5337	m-Cresol-2-amino 2-NH₂-3-CH₃C₆H₃OH	123.16	pl (w)	sub	150			eth	B13[2], 324
5338	m-Cresol-4-amino 4-H₂N-3-CH₃C₆H₃OH	123.16	pr (dil al), cr (bz)		179			al, eth	B13[3], 1559
5339	m-Cresol-5-amino 5-H₂N-3-CH₃C₆H₃OH	123.16	cr (dil MeOH)	245	139			al	C47, 9300
5340	m-Cresol-5-amino-2-nitro 2-NO₂-H₂-5-H₂N-3CH₃C₆H₂OH	168.16	red-br nd (al)		201			al	B13, 595
5341	m-Cresol-6-amino 6-H₂N-3-CH₃C₆H₃OH	123.16	nd (bz, dil al)		162d			al, eth, ace	B13[3], 1552
5342	m-Cresol-4-bromo 4-Br-3-CH₃C₆H₃OH	187.04	nd (peth or w)	137-43[16]	63.5			eth, Py	B6[4], 2072
5343	m-Cresol-5-bromo 5-Br-3-CH₃C₆H₃OH	187.04	nd (w)	161-2[28]	56-7			al, eth	B6[2], 357
5344	m-Cresol-6-bromo 6-Br-3-CH₃C₆H₃OH	187.04	cr (peth)	206-8[731], 81-2[4]	38			al, eth, ace	B6[3], 1320
5345	m-Cresol-2-chloro 2-Cl-3-CH₃C₆H₃OH	142.59	lf or nd (dil al)	230, 100-5[28]	74			al, eth	B6[4], 2064
5346	m-Cresol-4-chloro 4-Cl-3-CH₃C₆H₃OH	142.59	nd (peth)	235	66-8			al, eth	B6[4], 2064
5347	m-Cresol-6-chloro 6-Cl-3-CH₃C₆H₃OH	142.59	pr (peth)	196	45-6	1.215[15]		al, w	B6[3], 1315
5348	m-Cresol-2,4-dichloro 2,4-Cl₂-3-CH₃C₆H₃OH	177.03	pr (peth)	241-2	27			eth, chl	B6[3], 1319
5349	m-Cresol-2,6-dichloro 2,6-Cl₂-3-CH₃C₆H₃OH	177.03	....	235-6[745], 75-80[4]	58-9			eth, chl	B6[3], 1319
5350	m-Cresol-4,6-dichloro 4,6-Cl₂-3-CH₃C₆H₂OH	177.03	pr (peth)	235-6, 110[18]	72-4		1.572[20]	chl, peth	B6[4], 2069
5351	m-Cresol, 4-nitro 4-NO₂-3-CH₃C₆H₃OH	153.14	nd or pr (w)		129			al, eth, bz, chl	B6[4], 2075

No.	Name, Synonyms, and Formula	Mol. wt.	Color, crystalline form, specific rotation and λ_{max} (log ϵ)	b.p. °C	m.p. °C	Density	n_D	Solubility	Ref.
5352	m-Cresol, 5-nirto 5-NO₂-3-CH₃C₆H₃OH	153.14	pa ye cr (bz)	....	90-1			eth, bz	B6², 361
5353	m-Cresol, 6-nitro 6-NO₂-3-CH₃C₆H₃OH	153.14	ye mcl nd (eth or bz)		56			al, eth, bz	B6³, 1326
5354	m-Cresol, 4-nitroso 4-ON-3-CH₃C₆H₃OH	137.14	nd (w or bz), pr (aa)		165d			al, eth, bz, aa	B7³, 3389
5355	m-Cresol, 2,4,5,6-tetrabromo 3-CH₃C₆Br₄OH	423.75	nd (chl aa)		194			eth	B6³, 1324
5356	m-Cresol, 2,4,5,6-tetrachloro 3-CH₃C₆Cl₄OH	245.92	nd (peth)		189-90			al, eth, ace, bz	B6⁴, 2071
5357	m-Cresol, 2,4,5,6-tribromo 2,4,6-Br₃-3-CH₃C₆H₂OH	344.83			84				B6³, 1324
5358	m-Cresol, 2,4,6-trinitro or Methyl picric acid 2,4,6-(NO₂)₃-3-CH₃C₆HOH	243.13	pa ye nd (w or al)	150 exp	109-10			al, eth, ace, bz, chl	B6⁴, 2079
5359	m-Cresyl phosphate or Tri-m-Cresyl phosphate (3-CH₃C₆H₄O)₃PO	368.37	wax	260¹⁵	25-6	1.150²⁵	1.5575²⁰	eth, aa	B6⁴, 2057
5360	p-Cresol or 4-Hydroxytoluene 4-CH₃C₆H₄OH	108.15	pr	201.9, 85.7¹⁰	34.8	1.0178²⁰/⁴	1.5312²⁰	al, eth, ace, bz	B6⁴, 2093
5361	p-Cresol, 2-amino 2-NH₂-4-CH₃C₆H₃OH	123.16	cr (w), rh (bz), lf or nd (sub)	sub	137			al, eth, chl	B13³, 1576
5362	p-Cresol, 2-amino-5-nitro 5-NO₂-2-NH₂-4-CH₃C₆H₂OH	168.16	ye-og cr (al)		199-200d			al, dil HCl	B13², 346
5363	p-Cresol, 2-amino-6-nitro 6-NO₂-2-NH₂-4-CH₃C₆H₂OH	168.16	red-br (al)		119			al	B13², 345
5364	p-Cresol, 3-amino 3-H₂N-4-CH₃C₆H₃OH	123.16	cr (w or eth), lf (sub)	sub	156-7			eth	B13², 337
5365	p-Cresol, 2-bromo 2-Br-4-CH₃C₆H₃OH	187.04	nd (peth)	213-4	56-7	1.5468²⁵/²⁵	1.5772²⁰	al, bz	B6⁴, 2143
5366	p-Cresol, 3-bromo 3-Br-4-CH₃-C₆H₃OH	187.04	nd (peth)	245-7	56			al, eth, ace, bz	B6⁴, 2143
5367	p-Cresol, 2-chloro 2-Cl-4-CH₃C₆H₃OH	142.59		195-6		1.1785²⁵/⁴	1.5200²⁷	al, eth, bz, aa	B6⁴, 2135
5368	p-Cresol, 3-chloro 3-Cl-4-CH₃C₆H₃OH	142.59	nd (al)	228	55-6			al, eth, bz, aa	B6³, 1374
5369	p-Cresol, 2,6-dichloro 2,6-Cl₂-4-CH₃C₆H₂OH	177.03	nd (lig)	138-9²⁸	39 (42)			al, eth, aa	B6⁴, 2141
5370	p-Cresol, 2,6-dinitro 2,6-(NO₂)₂-4-CH₃C₆H₂OH	198.14	ye nd (eth or peth)		85			al, eth, bz	B6⁴, 2152
5371	p-Cresol, 2-methoxy or Cresolol 2-CH₃O-4-CH₃C₆H₃OH	138.17	pr	221, 113.5²²	5.5	1.098²⁰/⁴	1.5353²⁵	al, eth	B6², 865
5372	p-Cresol, 3-methylamino 3(CH₃NH)-4-CH₃C₆H₃OH	137.18	cr (bz-lig)		108			al, eth, bz	B13, 599
5373	p-Cresol, 2-nitro 2-NO₂-4-CH₃C₆H₃OH	153.14	ye nd (al or w)	125²²	36.5	1.2399²⁰/⁴	1.574⁴⁰	al, eth, ace, bz	B6⁴, 2149
5374	p-Cresol, 3-nitro 3-NO₂-4-CH₃C₆H₃OH	153.14	ye pr (eth)		79			al, eth	B6³, 1384
5375	p-Cresol, 2-isopropyl 2-p-C₃H₇-4-CH₃C₆H₃OH	150.22		228-9, 82³	36-7	0.9910²⁰/⁴	1.5275²⁰	al, bz, chl	B6³, 1882
5376	p-Cresol, 2,3,5,6-tetrabromo 4-CH₃C₆Br₄OH	423.75	nd (al or chl)		198-9			al, eth, chl	B6³, 1383
5377	p-Cresol, 2,3,5,6-tetrachloro 4-CH₃C₆Cl₄OH	245.92	nd (dil al, aa bz-lig)		190			al, bz, chl, aa	B6⁴, 2142
5378	p-Cresyl phosphate or Tri-p-Cresyl phosphate (4-CH₃C₆H₄O)₃PO	368.37	nd (al), ta (eth)	224³·⁵	77-8	1.247²⁵		al, eth, bz, chl	B6⁴, 2130
5379	Crocetin-(trans) or Gardenin C₂₀H₂₄O₄	328.41	brick red rh		285-7 (cor)			Py	B2³, 2018
5380	Croconic acid or Crocic acid C₅H₂O₅	142.07	pa ye nd (+ 3w al-diox)		sub d> 150			w, al	B8³, 3977
5381	Crotonaldehyde or 2-Butenal CH₃CH=CHCHO	70.09	....	104-5	-74	0.8495²⁵/⁴	1.4355²⁰	al, eth, ace, bz	B1⁴, 3447
5382	Crotonaldehyde, diethylacetal or 2-Butenaldiethylacetal CH₃CH=CHCH(OC₂H₅)₂	144.22		147-8, 49¹⁷		0.8473¹⁸/⁴	1.4097²⁰	al, eth, ace, bz	B1⁴, 3450
5383	Crotonamide CH₃CH=CHCONH₂	85.11	nd (ace)	sub at 140¹³	161.5 (cor)		1.4420¹⁶⁵	al, bz	B2⁴, 1506

No.	Name, Synonyms, and Formula	Mol. wt.	Color, crystalline form, specific rotation and λ_{max} (log ε)	b.p. °C	m.p. °C	Density	n_D	Solubility	Ref.
5384	Crotonic acid or *(trans)*-2-Butenoic acid .... $CH_3CH=CHCO_2H$	86.09	mcl pr or nd (w or lig)	185	71.5	$1.018^{15/4}$	1.4249^{77}	w, al, eth, ace	B2[4], 1498
5385	Crotonic acid, allyl ester .... $CH_3CH=CHCO_2CH_2CH=CH_2$	126.15		$88-9^{70}$	....	$0.9440^{20/4}$	1.4465^{20}	....	B2[4], 1503
5386	Crotonic acid, anhydride or Crotonic anhydride........ $(CH_3CH=CHCO)_2O$	154.17		$246-8, 129^{19}$		1.0397^{20}	1.4745^{20}	eth	B2[4], 1505
5387	Crotonic acid, ethyl ester .... $CH_3CH=CHCO_2C_2H_5$	114.15		$136.5, 58-9^{48}$	....	$0.9175^{20/4}$	1.4243^{20}	al, eth	B2[4], 1500
5388	Crotonic acid, methyl ester or Methyl crotonate .... $CH_3CH=CHCO_2CH_3$	100.13		121	-42	$0.9444^{20/4}$	1.4242^{20}	al, eth	B2[4], 1500
5389	Crotononitrile or 2-Butenonitrile................. $CH_3CH=CHCN$	67.09		$12-1^{762}$	-51.5	$0.8239^{20/4}$	1.4225^{20}	eth, ace	B2[4], 1507
5390	Crotonyl acetate .... $CH_3CO_2CH_2CH=CHCH_3$	114.15		132	....	$0.9192^{20/4}$	1.4181^{20}	al, eth, ace	B2[4], 183
5391	Crotonyl chloride .... $CH_3CH=CHCOCl$	104.54		$124-5, 35^{18}$		1.0905^{20}	1.460^{18}	ace	B2[4], 1506
5392	Cryptopine or Cryptocavine...................... $C_{21}H_{23}NO_5$	369.42	pr or pl (bz), nd (chl-MeOH)	....	223 (cor)	$1.315^{20/4}$	....	chl, aa	B27[2], 578
5393	Cryptoxanthin or β-Caroten-3-ol.................... $C_{40}H_{56}O$	552.90	garnet red pr (bz-MeOH)	....	169			bz, chl	B6[3], 3772
5394	Cumene or *iso*-Propylbenzene.................... $(CH_3)_2CHC_6H_5$	120.20	....	$152.4, 38.2^{10}$	-96	$0.8618^{20/4}$	1.4915^{20}	al, eth, ace, bz	B5[4], 985
5395	Cumene, 2-nitro or 1-*iso*-Propyl-2-nitro benzene .... $2-NO_2C_6H_4CH(CH_3)_2$	165.19	pa ye	103^9		1.101^{12}	1.5259^{20}	ace, bz, aa	B5[4], 997
5396	Cumene, 4-nitro or 4-Nitro-1-*iso*-propyl benzene .... $4-NO_2C_6H_4CH(CH_3)_2$	165.19	pa ye oil	122^9		$1.0830^{20/4}$	1.5367^{20}	ace, bz, lig	B5[4], 997
5397	Cubebin .... $C_{20}H_{20}O_6$	356.38	nd (al or bz), $[α]_D^{25}$ -45.6 (chl, c=5)	....	131-2			al, eth, chl	B19[4], 5967
5398	Cumic alcohol or 1-Hydroxymethyl-4-isopropylbenzene .. $4-i-C_3H_7C_6H_4CH_2OH$	150.22	....	289	28	$0.9818^{18/4}$	1.5210^{20}	al, eth	B6[3], 1911
5399	Cupreine or Hydroxycinchonine .... $C_{19}H_{22}N_2O_2$	310.40	pr (eth), $[α]_D^{17}$ -175.5 (al)	....	198 (anh)			al	B23[4], 416
5400	C-Curarine-III-hydroxide or C-flurorcuraninehydroxide.. $C_{20}H_{24}N_2O_2$	328.46	cr (MeOH-eth)	....	212			w, al	H36, 102
5401	ar-Curcumene .... $C_{15}H_{22}$	202.34	$[α]_D^{18}$ +35.8	140^{19}		$0.8821^{20/20}$	1.4989^{20}	bz	B5[4], 1465
5402	Curcumin .... $C_{21}H_{20}O_6$	368.39	or ye pr, rh pr (MeOH)	....	183			al, aa	B8[3], 4312
5403	Cuscohygrine or α,α′ Bis(*N*-methyl-α-pyrrolidyl) acetone	224.35	....	185^{22}		$0.9782^{16/4}$		w, al, eth, bz	B24[2], 36
	$C_{13}H_{24}N_2O$								
5404	Cuscohygrine, hydrate .... $C_{13}H_{24}N_2O.3½H_2O$	287.40	nd (a)	....	40-1			eth, bz	B24, 78
5405	Cusparine .... $C_{19}H_{17}NO_2$	307.35	(i) wh, nd (peth), (ii) ye, nd, (iii) pr	....	(i) 92, (ii) 92, (iii) 110-22			al, eth, ace, bz, chl	B27[2], 545
5406	Cyamelide or *sym*-Trioxane triimine .... $C_3H_3N_3O_3$	129.08	am or pw	d	....	$1.127^{15/4}$	....		B3[3], 30
5407	Cyanamide or Carbamonitrile...................... H_2NCN	42.04	nd	140^{19}	42 (46)	$1.282^{20/4}$	1.4418^{48}	w, al, eth, ace, bz, chl	B3[4], 145
5408	Cyanamide, benzyl .... $C_6H_5CHNHCN$	132.18	pl (eth)		43			al, eth	B12, 1051
5409	Cyanamide, diallyl .... $(CH_2=CHCH_2)_2NCN$	122.17	....	$140-5^{90}, 95^9$				al, eth, ace, bz	B4[4], 1078
5410	Cyanamide, dibutyl .... $(C_4H_9)_2NCN$	154.26	....	$187-91^{190}, 146-51^{35}$				al, eth, ace, bz	B4[4], 592
5411	Cyanamide, diethyl .... $(C_2H_5)_2NCN$	98.15	....	$188, 62^{10}$		$0.854^{20/4}$	1.4126^{25}	al, eth	B4[4]381
5412	Cyanamide, dimethyl .... $(CH_3)_2NCN$	70.10	....	$163.5, 56^{15}$	....		1.4089^{19}	al, eth, ace	B4[4], 226

No.	Name, Synonyms, and Formula	Mol. wt.	Color, crystalline form, specific rotation and λ_{max} (log ε)	b.p. °C	m.p. °C	Density	n_D	Solubility	Ref.
5413	Cyanamide, diphenyl .... $(C_6H_5)_2NCN$	194.24	pr (al)	235-40[60]	73-4	....	....	al, lig	B12[3], 895
5414	Cyanamide, methyl-α-naphthyl .... α-$C_{10}H_7N(CH_3)CN$	182.23	yesh	185-7[2]	....	....	....	al, eth	B12[2], 697
5415	Cyanamide, phenyl or Carbanilonitrile......... C_6H_5NHCN	127.15	cr (w, eth), lf (aa)	....	47 (hyd)	....	....	al, eth	B12[3], 805
5416	Cyanic acid .... HOCN	43.03	gas	23.5	-81	1.140[20/4]	....	w, eth, bz, chl aa	B3[4], 80
5417	Cyanic acid, ethyl ester or Ethyl cyanate.............. C_2H_5OCN	71.08	....	162d, 30[12]	....	0.89[20/4]	1.3788[25]	al, eth	Tet[1], 964, 2829
5418	Cyano acetamide $NCCH_2CONH_2$	84.08	pl (w)	....	121-2	....	....	w	B2[4], 1891
5419	Cyano acetamide, N-phenyl or α-Cyano acetanilide..... $NCCH_2CONHC_6H_5$	160.19	nd (al)	....	199-200	....	....	....	B12[2], 167
5420	Cyanoacetic acid .... $NCCH_2CO_2H$	85.06	....	108[0 15]/d	70-1	....	....	w, al, eth	B2[4], 1888
5421	Cyano acetic acid, benzal or α-Cyanocinnamic acid ...... $C_6H_5CH=C(CN)CO_2H$	173.17	cr (al)	....	183	....	....	....	B9[3], 4379
5422	Cyanoacetic acid, benzal, ethyl ester .... $C_6H_5CH=C(CN)CO_2C_2H_5$	201.23	(i) nd, (al), (ii) oil	(ii) 188[15]	(i) 51	(ii) 1.1076	(ii) 1.5033	eth, bz, chl	B9[1], 4380
5423	Cyano acetic acid, 1-cyclo hexenyl .... $C_6H_9CH(CN)CO_2H$	165.19	nd (bz)	....	109-10	....	....	al, ace, bz	B9[2], 560
5424	Cyanoacetic acid, diethyl, ethyl ester .... $(C_2H_5)_2C(CN)CO_2C_2H_5$	169.23	....	214-5, 100-1[11]	....	....	1.4200[11]	al, eth	B2[1], 1761
5425	Cyanoacetic acid, ethyl ester or Ethyl cyanoacetate ...... $NCCH_2CO_2C_2H_5$	113.12	....	205, 99[15]	-22.5	1.0654[20/4]	1.4175[20]	al, eth	B2[4], 1889
5426	Cyanoacetic acid, methyl ester or Methyl cyanoacetate .. $NCCH_2CO_2CH_3$	99.09	....	200-1, 113[17]	-22.3	1.1128[17]	1.4170[17]	al, eth	B2[1], 1889
5427	Cyanoacetic acid, phenyl, ethyl ester or Ethyl phenyl cyenate $C_6H_5CH(CN)CO_2C_2H_5$	189.22	oil	275d, 165[20]	....	1.091[20/4]	1.5012[25]	al, eth, ace, bz	B9[3], 4262
5428	α-Cyanocaproic acid, ethyl ester $C_4H_9CH(CN)CO_2C_2H_5$	216.28	....	245-50[762], 105[9]	....	0.988[15]	1.4248[20]	al, eth	B2[3], 1746
5429	Cyanogen or Oxalodinitrile ...... NCCN	52.04	gas	-21.2	-27.9	0.9537[-21]	....	w, al, eth	B2[4], 1863
5430	Cyanogen bromide .... BrCN	105.93	nd	61.4	52	2.015[20/4]	....	w, al, eth	B3[4], 92
5431	Cyanogen chloride .... ClCN	61.47	gas	12.7	-6	1.186[20/4]	....	w, al, eth	B3[4], 90
5432	Cyanogen iodide .... ICN	152.92	nd (al or eth)	sub>45	146-7	2.84[18]	....	al, eth	B3[4], 93
5433	Cyanogen sulfide .... $S(CN)_2$	84.10	rh pl	sub 30-40	65	....	....	w, al, eth	B3[4], 339
5434	Cyanuric acid, dihydrate or 2,4,6,-Triazinetriol ...... $C_3H_3N_3H_3 \cdot 2H_2O$	165.11	mcl (w + 2)	d	>360d	2.500[20/4]	....	w	B26[2], 131
5435	Cyanuric acid, tribenzyl ester or Benzyl cyanurate .... $C_{24}H_{21}N_3O_3$	399.46	nd (al)	>320	159	....	....	al	B26[1], 76
5436	Cyanuric chloride .... $C_3N_3Cl_3$	184.41	cr (eth or bz)	190[720]	154	....	....	al	B26[2], 16
5437	Cyclamen aldehyde .... 4-i-$C_3H_7C_6H_4CH_2CH(CH_3)CHO$	190.29	....	133-7[99], 115[5]	....	0.951[15]	1.5068[20]	al, eth, bz	B7[3], 1200
5438	Cyclobutane or Tetramethylene...................... C_4H_8	56.12	....	12	-50	0.720[5/4]	1.4260[10]	al, eth, ace, bz	B5[4], 6
5439	Cyclobutane, benzoyl or Cyclobutyl phenyl ketone ...... $C_6H_5COC_4H_7$	160.22	....	260, 122[10]	....	1.0457[25/25]	1.5472[20]	....	B7, 374
5440	Cyclobutane, ethyl .... $C_2H_5C_4H_7$	84.16	....	70.7	-142.9	0.7284[20/4]	1.4020[20]	al, eth, ace, bz	B5[4], 87
5441	Cyclobutane, methyl .... $CH_3C_4H_7$	70.14	....	36.3	....	0.6884[20/4]	1.3866[20]	al, eth, ace, bz, peth	B5[4], 21
5442	Cyclobutane, octafluoro or Perfluro cyclobutane ...... C_4F_8	200.03	....	-4[764]	-38.7	....	....	eth	B5[4], 8
5443	Cyclobutane carboxylic acid .... $C_4H_7CO_2H$	100.13	....	190[754], 74-5[2]	-2	1.0599[20/4]	1.4400[20]	al, eth	B9[3], 6
5444	1,1-Cyclobutane dicarboxylic acid .... $1,1-C_4H_6(CO_2H)_2$	144.14	pr (eth or w)	....	156.6 (cor)	....	....	w, al, eth, bz	B9[3], 3797
5445	1,1-Cyclobutane dicarboxylic acid, diethyl ester .... $1,1-C_4H_6(CO_2C_2H_5)_2$	200.24	....	229[735], 104[12]	....	1.0456[20/4]	1.4344[20]	al	B9[3], 3798

No.	Name, Synonyms, and Formula	Mol. wt.	Color, crystalline form, specific rotation and λ_{max} (log ε)	b.p. °C	m.p. °C	Density	n_D	Solubility	Ref.
5446	1,2-Cyclobutane dicarboxylic acid *(cis, dl)* 1,2-$C_4H_6(CO_2H)_2$	144.14	pl (w), pr (bz)	....	138	....	....	w, al, eth	B9[3], 3798
5447	1,2-Cyclobutane dicarboxylic acid *(trans, d)* 1,2-$C_4H_6(CO_2H)_2$	144.14	$[\alpha]^{30/}_D$ + 123.3 (w, c=2)	....	105	....	....	w, al	B9[3], 3799
5448	1,2-Cyclobutane dicarboxylic acid *(trans-dl)* 1,2-$C_4H_6(CO_2H)_2$	144.14	rh nd (bz)	....	131	....	....	w, al	B9[3], 3799
5449	1,2-Cyclobutane dicarboxylic acid *(trans-l)* 1,2-$C_4H_6(CO_2H)_2$	144.14	nd (HCl), $[\alpha]^{30/}_D$ + 123.3 (w, c=0.8)	....	105	....	....	w, al	B9[3], 3798
5450	1,3-Cyclobutane dicarboxylic acid *(cis)* 1,3-$C_4H_6(CO_2H)_2$	144.14	pr (w)	252	143-4	....	....	w, al	B9[3], 3801
5451	1,3-Cyclobutane dicarboxylic acid *(trans)* 1,3-$C_4H_6(CO_2H)_2$	144.14	pr (w), nd (sub)	sub	171	....	....	w, al	B9[3], 3802
5452	Cyclobutane, acetyl $CH_3COC_4H_7$	98.14		137-9	....	0.9020^{20}	1.4322^{19}	....	B7[3], 45
5453	Cyclobutane, 1,2,bis (amino methyl) 1,2-$(CH_3NH)_2C_4H_6$	114.19		....	193-4	....	$1.4778^{27.5}$	....	
5454	Cyclobutanone C_4H_6O	70.09		96-7	....	$0.9548^{0/0}$	1.4215^{20}	al, eth, bz, chl	B7[3], 4
5455	Cyclobutene C_4H_6	54.09		2	....	$0.733^{0/4}$	....	ace, bz, peth	B5[4], 207
5456	Cyclobutene, perfluoro C_4F_6	162.04		3	-60	$1.602^{-20/4}$	1.298^{-20}	....	B5[4], 208
5457	Cyclobutyl phenyl ketone $C_4H_7COC_6H_5$	160.22		260, 121-2[10]	....	$1.0457^{25/25}$	1.5472^{20}	....	B7, 374
5458	Cyclocamphene or Epicyclene $C_{10}H_{16}$	136.24		150-1	117-8	0.7948^{121}	....	al, aa	B5[3], 393
5459	1,6-Cyclodecanediol *(trans)* $C_{10}H_{20}O_2$	172.27	cr (chl, AcOEt)	....	151-3	....	....	eth	B6[3], 4106
5460	1,6-Cyclodecadione $C_{10}H_{16}O_2$	168.24	cr (eth)	....	100	....	....	ace, aa	B7[3]3246
5461	Cyclodecanol $C_{10}H_{19}OH$	156.27		125[12]	40-1	$0.9606^{20/4}$	1.4926^{20}	al	B6[4], 138
5462	Cyclodecanone $C_{10}H_{18}O$	154.26	amor pw	106-7[13]	28	$0.9654^{20/4}$	1.4806^{20}	eth, bz, chl	B7[3], 134
5463	Cyclofenchene $C_{10}H_{16}$	136.24		144-6	....	$0.859^{20/4}$	1.4503^{22}	....	B5[4], 468
5464	9-Cycloheptadecen-1-one or Civetone $C_{17}H_{30}O$	250.43		342[742], 159[2]	32.5	$0.9170^{33/4}$	1.4830^{33}	al, bz	B7[3], 524
5465	Cycloheptane or Suberane C_7H_{14}	98.19		118.5	-12	$0.8098^{20/4}$	1.4436^{20}	al, eth, bz, lig, chl	B5[4], 92
5466	Cycloheptane, 1-aza $C_6H_{13}N$	99.18		138[749]	....	$0.8643^{22/4}$	1.4631^{20}	al, eth	B20[4], 1406
5467	Cycloheptane, bromo or Suberyl bromide $C_7H_{13}Br$	177.09		101.5[40], 75[12]	....	$1.2887^{22/4}$	1.4996^{20}	eth, chl	B5[4], 93
5468	Cycloheptane, methyl $C_7H_{13}CH_3$	112.22		134	....	$0.8001^{20/4}$	1.4401^{20}	al, eth, bz, peth	B5[4], 114
5469	1,3-Cycloheptanedione $C_7H_{10}O_2$	126.16	ye	107-9[17]	-40	$1.0607^{22/22}$	1.4689^{22}	al	C50, 6327
5470	Cycloheptanol or Suberol $C_7H_{13}OH$	114.19		185, 95[24]	2	0.9554^{20}	1.4705^{20}	al, eth	B6[4], 94
5471	Cycloheptanone or Suberone $C_7H_{12}O$	112.17		178-9, 71[9]	....	$0.9508^{20/4}$	1.4608^{20}	al, eth	B7[3], 46
5472	Cycloheptane carboxylic acid $C_7H_{13}CO_2H$	142.20		254-8[711], 130-1[8]	....	$1.0423^{20/4}$	1.4753^{20}	al	B9[3], 47
5473	Cycloheptasiloxane, tetradicamethyl $C_{14}H_{42}O_7Si_7$	519.09		154[20]	-26	$0.9703^{20/4}$	1.4040^{20}	....	B4[3], 1886
5474	1,3,5-Cycloheptatriene or Tropilidene C_7H_8	92.15	cubic (at-80)	117, 60.5[122]	-79.5	$0.8875^{19/4}$	1.5343^{20}	al, eth, bz, chl	B5[4], 765
5475	2,4,6-Cycloheptatriene-1-one or Tropone C_7H_6O	106.13		113[15]	-7	$1.095^{22/4}$	1.6172^{22}	....	Am73, 876
5476	2,4,6-Cycloheptatriene-1-one, 2-amino 2-$H_2N(C_7H_5O)$	121.14	ye pl (bz)	....	106-7	....	....	al, bz, chl	C46, 7559
5477	2,4,6-Cycloheptatriene-1-one, 3-bromo-2-hydroxy 3-Br-2-HO(C_7H_4O)	201.03	ye pl or nd	....	107-8	....	....	al, eth	C49, 2405

No.	Name, Synonyms, and Formula	Mol. wt.	Color, crystalline form, specific rotation and λ_{max} (log ε)	b.p. °C	m.p. °C	Density	n_D	Solubility	Ref.
5478	2,4,6-Cycloheptatrien-1-one, 2-hydroxy or Tropolone .... 2-HO(C₇H₅O)	122.13	nd	sub 40[4]	51-2	....		w, eth, ace	Am74, 4456
5479	2,4,6-Cycloheptatriene-1-one, 2-hydroxy-4-isopropyl C₁₀H₁₂O₂	164.21	pa ye (peth)	....	50-1				B8³, 440
5480	2,4,6-Cycloheptatriene-1-one, 2-hydroxy-4-methyl .... 2-HO-4-CH₃(C₇H₅O)	136.15	nd (peth)		75-6			eth, chl	B8³, 260
5481	2,4,6-Cycloheptatriene-1-one, 2-methoxy 2-CH₃O-(C₇H₅O)	136.15	pa, ye nd (+ ½w)	128[5]	41(+ ½ w)			al, bz	C46, 4521
5482	Cycloheptene or Suberene .... C₇H₁₂	96.17	....	115	-56	0.8228²⁰ᐟ⁴	1.4552²⁰	al, eth, bz, peth	B5⁴, 244
5483	1,3-Cyclohexadiene or 1,2-Dihydrobenzene............ C₆Hₓ	80.14	....	80.5	-89	0.8405²⁰ᐟ⁴	1.4755²⁰	al, eth, bz, chl	B5⁴382
5484	1,3-Cyclohexadiene, 5-methyl (dl) 5-CH₃C₆H₇	94.16	....	101.5⁷⁶²		0.8354²⁰ᐟ⁴	1.4763²⁰	al, eth, bz, lig	B5³, 318
5485	1,3-Cyclohexadiene, perfluoro C₆Fₓ	224.05			62-4	1.601²⁰	1.3149²⁰		B5⁴384
5486	1,4-Cyclohexadiene or 1,4-Dihydrobenzene C₆Hₓ	80.14	....	85.6	-49.2	0.8471²⁰ᐟ⁴	1.4725²⁰	al, eth, bz, peth, chl	B5⁴, 385
5487	1,4-Cyclohexadiene, octafluoro C₆Fₓ	224.05			57-8		1.318¹⁸	....	B5⁴, 386
5488	1,4-Cyclohexadiene-1,2-dicarboxylic acid or 3,6-Dihydrophthalic acid .... 1,4-(HO₂C)₂C₆H₆	168.16	mcl pr (w)		153			al	B9³, 4047
5489	2,4-Cyclohexadiene-1,2-dicarboxylic acid or 2,3-Dihydrophthalic acid 1,2-(HO₂C)₂C₆H₆	168.16	pr (w or al)		179.80			al	B9³, 4047
5490	2,6-Cyclohexadiene-1,2-dicarboxylic acid or 4,5-Dihydrophthalic acid 1,2-(HO₂C)₂C₆H₆	168.16	tcl (w)		215			al, ace	B9³, 575
5491	Cyclohexane C₆H₁₂	84.16	....	80.7	6.5	0.7785²⁰ᐟ⁴	1.4266²⁰	al, eth, ace, bz, lig	B5⁴, 27
5492	Cyclohexane, acetyl or Cyclohexyl methyl ketone....... CH₃COC₆H₁₁	126.20	....	180-1, 69¹²		0.9176²⁰ᐟ⁴	1.4565¹⁶	eth	B7³, 84
5493	Cyclohexane, allyl (CH₂=CHCH₂)C₆H₁₁	124.23	....	131.5⁷⁵⁷		0.8135²⁰	1.4500²⁰	al, eth, ace, bz, chl	B5⁴, 283
5494	Cyclohexane, amino or Cyclohexyl amine .... C₆H₁₁NH₂	99.18	....	134.5, 30.5¹⁵	-17.7	0.8191²⁰ᐟ⁴	1.4372²⁰	w, al, eth, ace, bz	B12³, 10
5495	Cyclohexane, bromo or Cyclohexyl bromide.......... C₆H₁₁Br	163.06	....	166.2, 45.5¹⁰	-56.5	1.3359²⁰ᐟ⁴	1.4957²⁰	al, eth, ace, bz	B5⁴, 67
5496	Cyclohexane, 1-bromo-1-methyl 1-Br-1-CH₃(C₆H₁₀)	177.09		156-60, 65-9¹⁰		1.2510²⁰	1.4866²⁰	al, chl	B5⁴, 100
5497	Cyclohexane, 1-bromo-2-methyl 1-Br-2-CH₃(C₆H₁₀)	177.08		90-2¹⁰					B5², 12
5498	Cyclohexane, 1-bromo-3-methyl (dl)................ 1-Br-3-CH₃(C₆H₁₀)	177.09		181, 60¹¹		1.275²⁵ᐟ⁴	1.4979²⁰	eth, bz	B5³, 76
5499	Cyclohexane, 1-bromo-4-methyl 1-Br-4-CH₃(C₆H₁₀)	177.09		130²⁰⁰, 55¹⁵				eth, bz	B5⁴, 100
5500	Cyclohexane, (bromomethyl) BrCH₂C₆H₁₁	177.09		76-7²⁶		1.2763²⁵ᐟ⁴	1.4907²⁰	eth, bz, chl	B5⁴, 100
5501	Cyclohexane, butyl or Cyclohexyl butane. C₄H₉C₆H₁₁	140.27		181, 59¹⁰	−74.7	0.7992²⁰ᐟ⁴	1.4408²⁰	....	B5⁴, 146
5502	Cyclohexane, isobutyl i-C₄H₉C₆H₁₁	140.27		171.3	−95	0.7952²⁰ᐟ⁴	1.4386²⁰	al, eth, ace, bz, chl	B5⁴, 147
5503	Cyclohexane, sec-butyl CH₃CH₂CH(CH₃)C₆H₁₁	140.27		179.3		0.8131²⁰ᐟ⁴	1.4467²⁰	ace	B5⁴, 146
5504	Cyclohexane, tert-butyl (CH₃)₃CC₆H₁₁	140.27		171.5	−41.2	0.8127²⁰ᐟ⁴	1.4469²⁰	....	B5⁴, 147
5505	Cyclohexane, butylamino C₄H₉NHC₆H₁₁	155.29		207				al, eth	B12³, 15
5506	Cyclohexane, chloro or Cyclohexyl chloride .... C₆H₁₁Cl	118.61		143	−43.9	1.000²⁰ᐟ⁴	1.4626²⁰	al, eth, ace, bz, chl	B5⁴, 48
5507	Cyclohexane, cyclopentyl C₅H₉C₆H₁₁	152.28	....	215.1		0.8758²⁰ᐟ⁴	1.4725²⁰		B5⁴, 328
5508	Cyclohexane, 1,2-dibromo (cis) 1,2-Br₂C₆H₁₀	241.97		115¹⁴	9.7	1.803²⁵ᐟ²⁵	1.5514²⁵	eth, ace, bz, chl, lig	B5⁴, 70
5509	Cyclohexane, 1,2-dibromo (trans, dl) 1,2-Br₂C₆H₁₀	241.97		145—6¹⁰⁰, 105²⁰	−4	1.7759²⁰ᐟ⁴	1.5445¹⁹	al, eth, ace, bz	B5⁴, 71

No.	Name, Synonyms, and Formula	Mol. wt.	Color, crystalline form, specific rotation and λ_{max} (log ε)	b.p. °C	m.p. °C	Density	n_D	Solubility	Ref.
5510	Cyclohexane, 1,3-dibromo (cis) .... 1,3-$Br_2C_6H_{10}$	241.97	rods (al)	....	112	....	....	bz, al	B5[4], 71
5511	Cyclohexane, 1,3-dibromo (trans) .... 1,3-$Br_2C_6H_{10}$	241.97	....	116[16]	1	....	1.5480[20]	al, bz	B5[4], 72
5512	Cyclohexane, 1,4-dibromo (cis) .... 1,4-$Br_2C_6H_{10}$	241.97	....	137-8[25]	......	1.7834[20/4]	1.5531[20]	eth	B5[4], 72
5513	Cyclohexane, 1,4-dichloro (cis) .... 1,4-$Cl_2C_6H_{10}$	153.05	....	80.3[25]	18	1.1900[20/4]	1.4942[20]	eth	B5[4], 51
5514	Cyclohexane, 1,4-dibromo (trans) .... 1,4-$Br_2C_6H_{10}$	241.97	cr (eth)	....	......	....	....	eth	B5[4], 72
5515	Cyclohexane, 1,.2-dichloro (cis) .... 1,2-$Cl_2C_6H_{10}$	153.05	....	206-9[762], 91[20]	-1.5	1.2021[20/4]	1.4967[20]	bz	B5[4], 50
5516	Cyclohexane, 1,2-dichloro-(trans,dl) .... 1,2-$Cl_2C_6H_{10}$	153.05	....	189, 78[20]	-6.3	1.1839[20/4]	1.4902[20]	....	B5[4], 51
5517	Cyclohexane, (diethyl amino) .... $(C_2H_5)_2NC_6H_{11}$	155.29	....	192-3[740], 85-6[20]	......	0.872[0/0]	....	al	B12[3], 14
5518	Cyclohexane, (difluoramino) decafluoro .... $C_6F_{11}NF_2$	333.05	....	75-6	......	1.787[25/4]	1.286[25]	....	J1950, 1966
5519	Cyclohexane, 1,1-dimethyl .... 1,1-$(CH_3)_2C_6H_{10}$	112.22	....	119.5, 10[10]	-33.5	0.7809[20/4]	1.4290[20]	al, eth, ace, bz, lig	B5[4], 117
5520	Cyclohexane, 1,2-dimethyl (cis) .... 1,2-$(CH_3)_2C_6H_{10}$	112.22	....	129.7, 18.3[10]	-50.1	0.7963[20/4]	1.4360[20]	al, eth, ace, bz, lig	B5[4], 118
5521	Cyclohexane, 1,2-dimethyl (trans) .... 1,2-$(CH_3)_2C_6H_{10}$	112.22	....	123.4, 12.9[10]	-89.2	0.7760[20/4]	1.4270[20]	al, eth, ace, bz, lig	B5[4], 118
5522	Cyclohexane, 1,2-dimethyl,perfluoro .... 1,2-$(CF_3)_2C_6F_{10}$	400.06	....	101.5	-56	1.829[25/4]	1.283[25]	....	B5[3], 98
5523	Cyclohexane, 1,3-dimethyl (cis) .... 1,3-$(CH_3)_2C_6H_{10}$	112.22	....	120.1, 11.1[10]	-75.6	0.7660[20/4]	1.4229[20]	al, eth, ace, bz	B5[4], 121
5524	Cyclohexane, 1,3-dimethyl (trans, d) .... 1,3-$(CH_3)_2C_6H_{10}$	112.22	$[\alpha]_{549}$ + 1.33	124.4, 15[10]	-90	0.7847[20/4]	1.4309[20]	al, eth, ace, bz, lig	B5[4], 121
5525	Cyclohexane, 1,4-dimethyl (cis) .... 1,4-$(CH_3)_2C_6H_{10}$	112.22	....	124.3, 14.4[10]	-87.4	0.7829[20/4]	1.4230[20]	al, eth, ace, bz, lig	B5[4], 122
5526	Cyclohexane, 1,4-dimethyl (trans, dl) .... 1,4-$(CH_3)_2C_6H_{10}$	112.22	....	119.3, 10[10]	-37.0	0.7626[20/4]	1.4209[20]	al, eth, ace, bz, lig	B5[4], 123
5527	Cyclohexane, 1,2-dimethylene .... (1,2-$CH_2)C_6H_8$	108.19	....	124[740], 60—1[90]	......	0.8229[25/4]	1.4718[25]	al, eth, ace, bz, chl	B5[4], 409
5528	Cyclohexane, 1,2-epoxy or Cyclohexene oxide .... $C_6H_{10}O$	98.15	....	131.5, 54-5[10]	<-10	0.9663[20]	1.4519[20]	al, eth, ace, bz	B17[4], 164
5529	Cyclohexane, 1,2-epoxy-4 (epoxy ethyl) or 4-Vinylcyclohexene dioxide .... $C_8H_{12}O_2$	140.18	....	227, 92[5]	<-55	1.0986[20/20]	1.4787[20]	w	B19[4], 161
5530	Cyclohexane, 1,2-epoxy-4-vinyl .... $C_8H_{12}O$	124.19	....	169, 20[2]	<-100	0.9598[20/20]	1.4700[20]	....	B17[4], 314
5531	Cyclohexane, ethyl .... $C_2H_5C_6H_{11}$	112.22	....	131.8, 20.5[10]	-111.3	0.7880[20/4]	1.4330[20]	al, eth, ace, bz, lig	B5[4], 115
5532	Cyclohexane, ethylamino .... $C_2H_5NHC_6H_{11}$	127.23	....	164, 62-5[15]	......	0.868[0/0]	....	al, eth	B12[3], 14
5533	Cyclohexane, fluoro or Cyclohexyl fluoride .... $C_6H_{11}F$	102.15	....	100.2, 48[100]	13	0.9279[20/4]	1.4146[20]	Py	B5[4], 44
5534	Cyclohexane, 1,1,2,3,4,5,6-heptachloro .... $C_6H_4Cl_7$	325.28	rods	....	55-6	....	....	....	B5[4], 63
5535	Cyclohexane, 1,2,3,4,5,6-hexabromo or Benzene-β-hexabromide (β or cis) .... $C_6H_6Br_6$	557.57	pr	....	253d	....	....	....	B5[4], 78
5536	Cyclohexane, 1,2,3,4,5,6-hexabromo (α or trans) .... $C_6H_6Br_6$	557.57	mcl pr (xyl)	....	212	....	....	....	B5[4], 78
5537	Cyclohexane, 1,2,3,4,5,6-hexachloro (α dl) or Benzene-trans-hexachloride .... $C_6H_6Cl_6$	290.83	mcl pr (al or aa)	288	159-60	....	....	al, bz, chl	B5[4], 60
5538	Cyclohexane, 1,2,3,4,5,6-hexachloro-(β) or Benzene-cis-hexachloride .... $C_6H_6Cl_6$	290.83	cr (bz, al or xyl)	60[0.50]	314-5 sub	1.89[19]	....	....	B5[4], 61
5539	Cyclohexane, 1,2,3,4,5,6-hexachloro -γ- or Benzene-γ-hexachloride, Lindane .... $C_6H_6Cl_6$	290.83	nd (al)	323.4, 176.2[10]	112-3	....	....	ace, bz	B5[4], 58
5540	Cyclohexane, 1,2,3,4,5,6-hexachloro-(δ) .... $C_6H_6Cl_6$	290.83	pl	60[0.36]	141-2	....	....	....	B5[4], 57

PHYSICAL CONSTANTS OF ORGANIC COMPOUNDS (Continued)

No.	Name, Synonyms, and Formula	Mol. wt.	Color, crystalline form, specific rotation and λ_{max} (log ε)	b.p. °C	m.p. °C	Density	n_D	Solubility	Ref.
5541	Cyclohexane, α-hydroxyethyl $C_6H_{11}[CH(OH)CH_3]$	128.22	....	189, 81-2[15]	....	0.9250[20/4]	1.4677[20]	al, eth	B6[4], 117
5542	Cyclohexane, β-hydroxy ethyl $C_6H_{11}CH_2CH_2OH$	128.22	....	207-9[757], 97-9[15]	....	0.9229[20/4]	1.4641[20]	al, eth, bz	B6[4], 119
5543	Cyclohexane, hydroxymethyl $C_6H_{11}CH_2OH$	114.19	....	183, 83[14]	−43	0.9297[20/4]	1.4644[20]	al, eth	B6[4], 106
5544	Cyclohexane, iodo or Cyclohexy iodide $C_6H_{11}I$	210.06	....	180d, 81.5[20]	....	1.6244[20/4]	1.5477[20]	al, eth, ace, bz, lig, chl	B5[4], 78
5545	Cyclohexane, methyl or Hexahydrotoluene $CH_3C_6H_{11}$	98.19	....	100.9, 16.3[10]	−126.6	0.7694[20/4]	1.4231[20]	al, eth, ace, bz	B5[4], 94
5546	Cyclohexane, methylamino $CH_3NHC_6H_{11}$	113.20	....	145-7, 76-7[18]	....	0.8660[23]	1.4530[23]	al, eth	B12[3], 13
5547	Cyclohexane, 1-methyl-4-ethyl 4-$C_2H_4(C_6H_4)CH_3$	126.24	....	150-1	....	0.791[20/20]	1.435[20]	....	B5[4], 136
5548	Cyclohexane, 1-methyl-4-methylene 4-$CH_2=(C_6H_9)CH_3$	110.20	....	122	....	0.7923[19/19]	1.4465[18]	....	B5[4], 271
5549	Cyclohexane, 1-methyl-2-pentyl 2-$C_5H_{11}(C_6H_{10})CH_3$	168.32	....	216-9	....	0.816[20/20]	1.4487[20]	....	B5[4], 171
5550	Cyclohexane, methyl, perfluror $CF_3C_6F_{11}$	350.06	....	76.1	−44.7	1.7878[25/4]	1.285[17]	ace, bz	B5[4], 97
5551	Cyclohexane, methylene $CH_2:(C_6H_{10})$	96.17	....	102—3[764]	−106.7	0.8074[20/4]	1.4523[20]	eth, bz, lig	B5[4], 250
5552	Cyclohexane, nitro $O_2NC_6H_{11}$	129.16	....	205.5[768], 95[22]	fr-34	1.0610[20/4]	1.4612[19]	al, lig	B5[4], 81
5553	Cyclohexane, 1,2,3,4,5-pentahydroxy (d) or d-Quercitol $1,2,3,4,5$-$(HO)_5C_6H_7$	164.16	pr (w or dil al) $[\alpha]^{18}_D$ + 25 (w,c=10)	....	235-7	1.5845[13]	....	w	B6[3], 6873
5554	Cyclohexane, 1,2,3,4,5-pentahydroxy (l) or Viboquercetol $1,2,3,4,5$-$(HO)_5C_6H_7$	164.16	pr (w), nd (al), nd (w + 1), $[\alpha]^{20}_D$ −50 (w,c=4)	....	180-1 (anh)	....	....	w	B6[3], 6873
5555	Cyclohexane, pentyl $C_5H_{11}C_6H_{11}$	154.30	....	202.8, 75.3[10]	−57.5	0.8037[20/4]	1.4437[20]	al, eth, ace, bz	B5[4], 164
5556	Cyclohexane, iso-pentyl i-$C_5H_{11}C_6H_{11}$	154.30	....	196.5	....	0.8023[20/4]	1.4420[20]	bz, lig	B5[3], 143
5557	Cyclohexane, phenyl or Cyclohexyl benzene $C_6H_5C_6H_{11}$	160.26	....	235-6, 127-8[30]	7-8	0.9502[20/4]	1.5329[20]	al, eth	B5[4], 1424
5558	Cyclohexane, propyl $C_3H_7C_6H_{11}$	126.24	....	156.7, 40.1[10]	−94.9	0.7936[20/4]	1.4370[20]	al, eth, ace, bz, peth	B5[4], 134
5559	Cyclohexane, iso-propyl or Hexahydrocumene $(CH_3)_2CHC_6H_{11}$	126.24	....	154.5, 38.3[10]	90	0.8023[20/4]	1.4410[20]	al, eth, ace, bz	B5[4], 134
5560	Cyclohexane, 1,1,3-trimethyl $1,1,3$-$(CH_3)_3C_6H_9$	126.24	....	138-94	....	0.7664[20/0]	1.4237[15]	....	B5[4], 137
5561	Cyclohexane, 1,3,5-trimethyl (cis) or Hexahydromesitylene $1,3,5$-$(CH_3)_3C_6H_9$	126.24	....	138.5	−49.7	0.7708[20/4]	1.4269[20]	eth, bz, lig	B5[4], 138
5562	Cyclohexane, 1,3,5-trimethyl (trans) or Hexahydromesitylene $1,3,5$-$(CH_3)_3C_6H_9$	126.24	....	140.5	−107.4	0.7794[20/4]	1.4307[20]	eth, bz, lig	B5[4], 138
5563	Cyclohexanecarboxaldehyde or Hexahydrobenzaldehyde $C_6H_{11}CHO$	112.17	....	159.3, 36[10]	....	0.9035[20/4]	1.4496[20]	eth	B7[3], 66
5564	Cyclohexane carboxylic acid or Hexahydrobenzoic acid $C_6H_{11}CO_2H$	128.17	mcl pr	232-3, 120-1[13]	31-2	1.0334[22/4]	1.4599[22]	al, bz, chl	B9[3], 15
5565	Cyclohexane carboxylic acid, ethyl ester $C_6H_{11}CO_2C_2H_5$	156.23	....	196, 63[12]	....	0.9362[20/4]	1.4501[15]	al, eth, ace, chl	B9[3], 17
5566	Cyclohexane carboxylic acid, 2-hydroxy or Hexahydrosalicylic acid 2-$HOC_6H_{10}CO_2H$	144.17	nd (AeOEt)	....	111	....	....	w, al, eth	B10[3], 14
5567	Cyclohexane carboxylic acid, methyl ester $C_6H_{11}CO_2CH_3$	142.20	....	183, 73[15]	....	0.9954[15/4]	1.4433[20]	al, eth, ace, chl	B9[3], 16
5568	Cyclohexane carboxylic acid, propyl ester $C_6H_{11}CO_2C_3H_7$	170.25	....	215.5	....	0.9530[15/4]	1.4486[15]	al, eth, ace, chl	B9[3], 17

No.	Name, Synonyms, and Formula	Mol. wt.	Color, crystalline form, specific rotation and λ_{max} (log ε)	b.p. °C	m.p. °C	Density	n_D	Solubility	Ref.
5569	Cyclohexane carboxylic acid, 1,3,4,5-tetrahydroxy (d) or d-Quinic acid 1,3,4,5-(HO)₄C₆H₇CO₂H	192.17	mcl pr (w), $[\alpha]^{20/D} +44$ (w,c=10)	d	164	1.637	....	w	B10, 538
5570	Cyclohexane carboxylic acid, 1,3,4,5-tetra hydroxy (dl) or dl-Quinic acod .... 1,3,4,5-(HO)₄C₆H₇CO₂H	192.17	pr (w)	....	142		....	w	B10³, 2408
5571	Cyclohexane carboxylic acid, 1,3,4,5-tetrahydroxy (l) or l-Quinic acid 1,3,4,5-(HO)₄(C₆H₇)CO₂H	192.17	pr (w), $[\alpha]^{18/D} -44.1$ (w, c=12)	d	172	1.64	....	w	B10³, 2407
5572	Cyclohexane carboxylonitrile, epoxy .... C₇H₉NO	123.16	....	244.5, 110¹⁰	−33	1.0929²⁰/²⁰	1.4763²⁰	w, eth	B18⁴, 3891
5573	Cyclohexanecarboxylyl chloride .... C₆H₁₁COCl	146.62	....	180, 75-7¹⁵		1.0962¹⁵/⁴	1.4711²⁰		B9³, 27
5574	1,2-Cyclohexanedicarboxylic acid (cis) or Hexahydro-phthalic acid 1,2-C₆H₁₀(CO₂H)₂	172.19	tcl nd (al)	d	192			al, eth, ace, bz	B9³, 3812
5575	1,2-Cyclohexane dicarboxylic acid (trans, d) .... 1,2-C₆H₁₀(CO₂H)₂	172.19	pw (w), $[\alpha]_D +18.2$		179-83			w	B9³, 3812
5576	Cyclohexanehexone, octahydrate C₆O₆·8H₂O	312.19	mic nd (dil HNO₃)	....	100-1				B7³, 4857
5577	Cyclohexanol C₆H₁₁OH	100.16	hyg nd	161.1	25.1	0.9624²⁰/⁴	1.4641²⁰	w, al, ace, eth, bz	B6⁴, 20
5578	Cyclohexanol, 1-acetyl 1-CH₃COC₆H₁₀OH	142.20	....	125-6, 91¹¹		1.0248²⁵/⁴	1.4670²⁵	al, eth	B8³, 15
5579	Cyclohexanol, 2-allyl (trans) .... 2-(CH₂=CHCH₂)C₆H₁₀OH	140.23	....	94-6¹⁵		0.947²⁰/⁴	1.4778²⁰	aa	B6⁴, 235
5580	Cyclohexanol, 2-amino (trans, dl) 2-H₂NC₆H₁₀OH	115.18	hyg	105¹⁰	68			bz, chl, aa	B13³, 704
5581	Cyclohexanol, 2-butyl (trans) 2-C₄H₉C₆H₁₀OH	156.27	....	111-2¹⁶		0.9020²⁰/⁴	1.4641²⁰	eth, ace, bz	B6³, 121
5582	Cyclohexanol, 2-chloro (cis, dl) 2-ClC₆H₁₀OH	134.61	hyg (peth)	93-4²⁶	36-7	1.1261²⁵	1.4894²⁵	al, bz, chl	B6⁴, 64
5583	Cyclohexanol, 2-chloro (cis, l) 2-ClC₆H₁₀OH	134.61	hyg $[\alpha]_{549}$ 19.5	87¹⁵		1.137¹⁵	1.4894²⁵	w, al, bz, chl	B6³, 39
5584	Cyclohexanol, 2-chloro (trans)................ 2-ClC₆H₁₀OH	134.61	pr (bz-lig)	93²⁶	29	1.146¹⁶/⁴	1.4899²⁰	al, eth, bz, chl	B6⁴, 64
5585	Cyclohexanol, 4-chloro (trans) 4-ClC₆H₁₀OH	134.16	pl (cy)	106¹⁴	82-3	1.1435¹⁷/⁴	1.4930¹⁷	al, eth, bz, chl	B6⁴, 68
5586	Cyclohexanol, 3-dimethylamino .... 3-(CH₃)₂NC₆H₁₀OH	143.23	....	231, 126-7²²	73	0.9766²⁵/²⁵	1.4852²⁰	al	B13³, 719
5587	Cyclohexanol, 1-ethyl 1-C₂H₅C₆H₁₀OH	128.22	pr	166, 67¹⁰	34-5	0.9227²⁵	1.4633²⁰	bz, peth	B6⁴, 115
5588	Cyclohexanol, 2-ethyl (cis, dl) 2-C₂H₅C₆H₁₀OH	128.22	....	180-2, 74¹²		0.9274²⁰/⁴	1.4655²¹	eth, ace, bz, peth	B6³, 85
5589	Cyclohexanol, 2-ethyl (trans, dl) 2-C₂H₅C₆H₁₀OH	128.22	....	79¹²		0.9193²¹/⁴	1.4640²¹	eth, ace, bz, peth	B6⁴, 117
5590	Cyclohexanol, 1-ethynyl 1-(HC≡C)C₆H₁₀OH	124.19	cr (peth)	174, 73¹²	31-2	0.9873²⁰/⁴	1.4822²⁰	al, bz, peth	B6⁴, 348
5591	Cyclohexanol, 2-(l-hydroxyethyl) 2[CH₃CH(OH)]C₆H₁₀OH	144.22	....	140¹²		0.976²⁰/⁷	1.4900²⁰		C50, 3299
5592	Cyclohexanol, 1-methyl 1-CH₃C₆H₁₀OH	114.19	....	155, 70²⁵	25	0.9194²⁰/⁴	1.4595²⁰	al, bz, chl	B6⁴, 95
5593	Cyclohexanol, 2-methyl (cis, dl) 2-CH₃C₆H₁₀OH	114.19	....	165, 60¹²	7	0.9360²⁰/⁴	1.4640²⁰	al, eth	B6⁴, 100
5594	Cyclohexanol, 2-methyl (trans, d) 2-CH₃C₆H₁₀OH	114.19	$[\alpha]^{20/}_D +17.19$ (un-dil)	166, 78²⁰		0.9454²⁰	1.4610²⁰	al, eth	B6³, 62
5595	Cyclohexanol, 2-methyl (trans, dl) 2-CH₃C₆H₁₀OH	114.19	....	167-8, 78²⁰	−4	0.9247²⁰/⁴	1.4616²⁰	al, eth	B6⁴, 100
5596	Cyclohexanol, 2-methyl- (trans, l) 2-CH₃C₆H₁₀OH	114.19	$[\alpha]^{20}_D -35.5$ (un-dil)	166, 78²⁰		0.9454²⁰	1.4610²⁰	al, eth	B6³, 62
5597	Cyclohexanol, 3-methyl (cis, l) 3-CH₃C₆H₁₀OH	114.19	$[\alpha]^{22/}_D -4.75$ (un-dil)	174-5, 94¹²	−4.7	0.9155²⁰/⁴	1.4574²⁰	al, eth	B6³, 67

No.	Name, Synonyms, and Formula	Mol. wt.	Color, crystalline form, specific rotation and λ_{max} (log ε)	b.p. °C	m.p. °C	Density	n_D	Solubility	Ref.
5598	Cyclohexanol, 3-methyl (trans, l) 3-CH₃C₆H₁₀OH	114.19	[α]²⁰_D − 7.3 (undil)	174-5, 84[13]	−1	0.9214²⁰ᐟ⁴	1.4590²⁰	al, eth	B6³, 68
5599	Cyclohexanol, 4-methyl (cis) 4-CH₃C₆H₁₀OH	114.19	. . . .	173-4, 78-9²⁰	−9.2	0.9170²⁰ᐟ⁴	1.4614²⁰	al, eth	B6⁴, 105
5600	Cyclohexanol, 4-methyl (trans) C₇H₁₄O	114.19	. . . .	173-4, 54³	. . . .	0.9118²⁰ᐟ⁴	1.4561²⁰	al, eth	B6⁴, 105
5601	Cyclohexanol, 1-phenyl 1-C₆H₅C₆H₁₀OH	172.26	. . . .	157.5²⁸, 112-3⁵	63-3.5	1.035¹⁶	1.5415¹⁶	. . . .	B6³, 2510
5602	Cyclohexanol, 2-phenyl (cis, dl) 2-C₆H₅C₆H₁₀OH	176.24	. . . .	140-1¹⁶	41-2 (56)	1.035¹⁶	1.5415¹⁶	. . . .	B6³, 2510
5603	Cyclohexanol, 2-phenyl (trans, dl) 2-C₆H₅C₆H₁₀OH	176.24	cr (peth)	152-5¹⁶	56-7	. . . .	. . . .	al, chl	B6³, 2511
5604	Cyclohexanol, 2-isopropyl (cis) 2-(CH₃)₂CHC₆H₁₀OH	142.24	. . . .	77¹³	52-3	0.9223²⁵	1.4665²⁵	. . . .	B6⁴, 131
5605	Cyclohexanol, 2,2,6,6-tetrakis (hydroxymethyl) . . . . 2,2,6,6,(HOCH₂)₄C₆H₇OH	220.27	pl (al)	. . . .	131	. . . .	. . . .	w, al	B6³, 6877
5606	Cyclohexanol, 1,2,2-trimethyl (dl) 1,2,2-(CH₃)₃C₆H₈OH	142.24	cr (+ ½ w)	81-2²⁰	41 (hyd)	0.9230²⁰ᐟ⁴	1.4682²⁰	al, eth, bz	B6³, 114
5607	Cyclohexanol, 1,2,6-trimethyl . . . . 1,2,6-(CH₃)₃C₆H₈OH	142.24	. . . .	78²²	. . . .	0.9126¹⁵ᐟ⁴	1.4598¹⁵	al, eth, ace, bz	B6¹, 17
5608	Cyclohexanol, 1,3,3-trimethyl 1,3,3-(CH₃)₃C₆H₈OH	142.24	pr (dil al)	. . . .	74	. . . .	. . . .	al, eth, ace, bz	B6¹, 16
5609	Cyclohexanol, 1,3,5-trimethyl 1,3,5-(CH₃)₃C₆H₈OH	142.24	. . . .	181, 82-3¹⁴	. . . .	0.8876¹⁷ᐟ⁴	1.454¹⁶·³	al, eth, chl	B6¹, 117
5610	Cyclohexanol, 1,4,4-trimethyl 1,4,4-(CH₃)₃C₆H₈OH	142.24	hyg nd (dil al)	79-80¹⁵	58	. . . .	. . . .	al, eth, chl	B6¹, 16
5611	Cyclohexanol, 2,2,3-trimethyl 2,2,3-(CH₃)₃C₆H₈OH	142.24	. . . .	85-7¹⁵	. . . .	. . . .	. . . .	al, eth	B6¹, 16
5612	Cyclohexanol, 2,2,5-trimethyl or Pulenol. . . . . 2,2,5-(CH₃)₃C₆H₈OH	142.24	. . . .	187-9, 90-2²³	. . . .	0.8955²²ᐟ⁴	1.4569²⁰	al	B6, 22
5613	Cyclohexanol, 2,2,6-trimethyl (liquid) . . . . 2,2,6-(CH₃)₃C₆H₈OH	142.24	. . . .	186-7⁷⁵²	. . . .	0.9128²⁰ᐟ⁴	1.4600²⁰	al, eth, chl	B6⁴, 135
5614	Cyclohexanol, 2,2,6-trimethyl (solid) . . . . 2,2,6-(CH₃)₃C₆H₈OH	142.24	cr (peth or al)	87²⁸	51	. . . .	. . . .	al, eth, chl	B6⁴, 135
5615	Cyclohexanol, 2,3,3-trimethyl 2,3,3-(CH₃)₃C₆H₈OH	142.24	nd	197, 97¹⁹	28	. . . .	. . . .	al, ace	B6¹, 16
5616	Cyclohexanol, 2,3,6-trimethyl 2,3,6-(CH₃)₃C₆H₈OH	142.24	. . . .	193-5⁷⁴⁷	. . . .	0.9117¹⁷ᐟ⁴	. . . .	al, chl	B6, 22
5617	Cyclohexanol, 2,4,5-trimethyl (cis) . . . . 2,4,5-(CH₃)₃C₆H₈OH	142.24	hyg	191-3, 84¹⁷	. . . .	0.9120²⁰ᐟ⁴	1.463²⁰	al, eth, chl	B6², 36
5618	Cyclohexanol, 2,4,5-trimethyl (trans) 2,4,5-(CH₃)₃C₆H₈OH	142.24	hyg	196, 112³⁵	. . . .	0.906²⁰ᐟ⁴	1.461²⁰	al, eth, chl	B6², 36
5619	Cyclohexanol, 3,3,5-trimethyl (cis) . . . . 3,3,5-(CH₃)₃C₆H₈OH	142.24	. . . .	201-3⁷⁵⁰, 92¹²	37.3	0.9006¹⁶ᐟ⁴	1.4550¹⁶	al, eth, chl	B6⁴, 135
5620	Cyclohexanol, 3,3,5-trimethyl (trans) 3,3,5-(CH₃)₃C₆H₈OH	142.24	cr (eth)	189.2	55.8	0.8647⁶⁰ᐟ²⁰	. . . .	al, eth, chl	B6⁴, 135
5621	Cyclohexanone C₆H₁₀O	98.15	. . . .	155.6, 47¹⁵	−16.4	0.9478²⁰ᐟ⁴	1.4507²⁰	al, eth, ace, bz, chl	B7³, 14
5622	Cyclohexanone, 2-acetyl 2-CH₃CO(C₆H₉O)	140.19	. . . .	111-2¹⁸	. . . .	1.078²⁰	1.5138²⁰	. . . .	B7³, 3223
5623	Cyclohexanone, 2,6-dibromo (cis) . . . . 2,6-Br₂(C₆H₈O)	255.96	cr (eth, aa)	. . . .	106-7	. . . .	. . . .	. . . .	B7³, 38
5624	Cyclohexanone, 2-butyl 2-C₄H₉(C₆H₉O)	154.26	. . . .	70²	. . . .	0.905²⁰ᐟ⁴	1.4545²⁰	. . . .	B7³, 140
5625	Cyclohexanone, 2-butylidene . . . . C₁₀H₁₆O	152.24	. . . .	98-100¹⁰	. . . .	0.935²⁰ᐟ⁴	1.4800²⁰	al, eth, ace, bz	C49, 1598
5626	Cyclohexanone, 2-chloro 2-Cl(C₆H₉O)	132.59	. . . .	82¹⁵	23	1.161²⁰ᐟ¹⁵	1.4825²⁰	eth, bz	B7³, 36
5627	Cyclohexanone, 3-chloro 3-Cl(C₆H₉O)	132.59	. . . .	91-2¹⁴	. . . .	. . . .	. . . .	eth	B7, 10
5628	Cyclohexanone, 4-chloro 4-Cl(C₆H₉O)	132.59	. . . .	95¹⁷	. . . .	. . . .	1.4867²⁰	eth	B7², 11
5629	Cyclohexanone, 2-β-cyanoethyl 2-NCCH₂CH₂-(C₆H₉O)	151.21	. . . .	138-42¹⁰	. . . .	1.0181²⁰ᐟ⁴	1.4755²⁰	. . . .	B10³, 2835
5630	Cyclohexanone, cyanohydrin . . . . C₆H₁₀(OH)CN	125.17	. . . .	109-13⁹	34-6	. . . .	1.4643²⁰	eth, w	B10³, 11

No.	Name, Synonyms, and Formula	Mol. wt.	Color, crystalline form, specific rotation and λ_{max} (log ε)	b.p. °C	m.p. °C	Density	n_D	Solubility	Ref.
5631	Cyclohexanone, 2-cyclohexyl 2-$C_6H_{11}(C_6H_9O)$	180.29	....	264	−32	0.9752[25/25]	1.4877[25]	....	B7[3], 474
5632	Cyclohexanone, 2,6-dibenzyl 2,6-$(C_6H_5CH_2)(C_6H_8O)$	274.37	ye nd (al)	185-95[20]	117-8	....	....	bz, aa	B7[3], 2661
5633	Cyclohexanone, 2,4-dimethyl (trans, d) 2,4-$(CH_3)_2(C_6H_8O)$	126.20	[α][24]/D 64.8 al 6%	178.7[766], 69[17]	....	0.9004[16/4]	1.4488[22]	eth, ace, bz	B7[3], 93
5634	Cyclohexanone, 2,5-dimethyl (d) 2,5-$(CH_3)_2(C_6H_8O)$	126.20	[α][20] + 11.5 (un-dil)	172-4[750], 51[10]	....	0.8985[20/4]	1.4445[20]	al, eth	B7[3], 97
5635	Cyclohexanone, 2,5-dimethyl (trans, dl) 2,5-$(CH_3)_2(C_6H_8O)$	126.20	....	171-3, 76-7[27]	....	0.9025[20]	1.4446[20]	al, eth	B7[3], 97
5635a	Cyclohexanone, 2-(dimethylaminomethyl) 2-$(CH_3)_2NCH_2(C_6H_9O)$	155.24	....	92[10.5]	....	0.9504[20/4]	1.4672[20]	al, eth	B14[3], 5
5636	Cyclohexanone, 2-ethylidene (2-$CH_3CH=)(C_6H_8O)$	124.19	....	92[20]	....	0.962[20/4]	1.4882[20]	....	B7[2], 58
5637	Cyclohexanone, 2-hydroxy or Adipoin 2-$HO(C_6H_9O)$	114.15	nd (al or MeOH)	....	113	....	1.4785[21]	w, al	B8[3], 4
5638	Cyclohexanone, 2-methyl (d) 2-$CH_3(C_6H_9O)$	112.17	[α][25]/D + 14.21 (chl)	167-8[735]	....	0.9262[18/4]	1.4440[25]	....	B7[3], 49
5639	Cyclohexanone, 2-methyl (dl) 2-$CH_3(C_6H_9O)$	112.17	....	165[757], 90[20]	−13.9	0.9250[20/4]	1.4483[25]	al, eth	B7[3], 49
5640	Cyclohexanone, 2-methyl (l) 2-$CH_3(C_6H_9O)$	112.17	[α][25]/D − 15.22 (un-dil)	59-60[20]	....	0.9230[25/4]	1.4440[25]	al, eth	B7[3], 49
5641	Cyclohexanone, 3-methyl (d) 3-$CH_3(C_6H_9O)$	112.17	[α][20]/D + 12.7 (un-dil)	169	....	0.9155[20/4]	1.4493[20]	al, eth	B7[3], 55
5642	Cyclohexanone, 3-methyl (dl) 3-$CH_3(C_6H_9O)$	112.17	....	168-9[738], 65[15]	−73.5	0.9136[20/4]	1.4456[20]	al, eth	B7[3], 57
5643	Cyclohexanone, 4-methyl 4-$CH_3(C_6H_9O)$	112.17	....	170	−40.6	0.9138[20/4]	1.4451[20]	al, eth	B7[3], 63
5644	Cyclohexanone, oxime $C_6H_{10}=NOH$	113.16	hex pr (lig)	206-10	90	....	....	w, al, eth	B7[3], 32
5645	Cyclohexanone, 2-propyl 2-$C_3H_7(C_6H_9O)$	140.23	....	195, 70[6]	....	0.927[20/4]	1.4538[20]	al, eth, ace, bz	B7[3], 115
5646	Cyclohexanone, 2-isopropyl 2-i-$C_3H_7(C_6H_9O)$	140.23	....	....	72-3	0.922[16/4]	1.4564[15]	al, eth, ace, bz	B7[3], 117
5647	Cyclohexanethiol or Cyclohexyl mercaptan $C_6H_{11}SH$	116.23	....	158, 41[12]	....	0.9782[20/4]	1.4921[20]	al, eth, ace, bz, chl	B6[4], 72
5648	Cyclohexanethione $C_6H_{10}S$	114.20	....	74[11]	....	....	1.5375[20]	al, eth, ace	B7[3], 39
5649	1,2-Cyclohexane dicarboxylic acid (trans, dl) 1,2-$C_6H_{10}(CO_2H)_2$	172.19	lf or pr (w)	....	222	....	....	w	B9[3], 3813
5650	1,2-Cyclohexane dicarboxylic acid (trans, l) 1,2-$C_6H_{10}(CO_2H)_2$	172.19	pw (w)	....	179-82	....	....	....	B9, 732
5651	1,2-Cyclohexane dicarboxylic acid, diethyl ester (cis) 1,2-$C_6H_{10}(CO_2HC_{25})_2$	228.29	....	133[10]	....	1.0540[22/4]	1.45512[14]	eth	B9[3], 3813
5652	1,2-Cyclohexane dicarboxylic acid, diethyl ester (trans, dl) 1,2-$C_6H_{10}(CO_2C_2H_5)_2$	228.29	....	135[11]	....	1.040[20/4]	1.4522[13]	eth	B9[3], 3813
5653	1,3-Cyclohexane dicarboxylic acid (cis) or cis-Hexahydro isophthalic acid 1,3-$C_6H_{10}(CO_2H)_2$	172.19	nd (con HCl), cr (w)	....	167-8	....	....	w, al, eth, bz	B9[3], 3817
5654	1,3-Cyclohexane dicarboxylic acid (trans, d) 1,3-$C_6H_{10}(CO_2H)_2$	172.19	cr (w), [α][22]/D + 23.8 (w, c=4)	....	134	....	....	w, al, eth	B9[2], 523
5655	1,3-Cyclohexane dicarboxylic acid (trans, dl) 1,3-$C_6H_{10}(CO_2H)_2$	172.19	nd (w)	....	150.5	....	....	w, al, eth	B9[3], 3817
5656	1,3-Cyclohexane dicarboxylic acid (trans, l) 1,3-$C_6H_{10}(CO_2H)_2$	172.19	w, [α][22]/D − 23.2 (w, c=2)	....	134	....	....	w, al, eth	B9[3], 523
5657	1,3-Cyclohexane dicarboxylic acid, diethyl ester (cis) 1,3-$C_6H_{10}(CO_2C_2H_5)_2$	228.29	....	288, 142[11]	....	1.0450[20/4]	1.4521[20]	eth, al, ace	B9[3], 3817
5658	1,3-Cyclohexane dicarboxylic acid, diethyl ester (trans, dl) 1,3-$C_6H_{10}(CO_2C_2H_5)_2$	228.29	....	286[756], 142[12]	....	1.0485[21/4]	1.4530[20]	eth, al, ace	B9[3], 3817
5659	1,4-Cyclohexane dicarboxylic acid (cis) or cis-Hexahydro terephthalic acid 1,4-$C_6H_{10}(CO_2H)_2$	178.19	lf (w)	....	170-1	....	....	al, eth, chl	B9[3], 3818

No.	Name, Synonyms, and Formula	Mol. wt.	Color, crystalline form, specific rotation and λ_{max} (log ε)	b.p. °C	m.p. °C	Density	n_D	Solubility	Ref.
5660	1,4-Cyclohexane dicarboxylic acid (trans) .... 1,4-C$_6$H$_{10}$(CO$_2$H)$_2$	172.19	pr (w), pl (ace)	300 sub	312-3	....	....	al, ace	B9[3], 3818
5661	1,4-Cyclohexane dicarboxylic acid, diethyl ester (cis) ..... 1,4-C$_6$H$_{10}$(CO$_2$C$_2$H$_5$)$_2$	228.29	....	151[13]		1.0516[21/4]	1.4522[21]	eth	B9[2], 524
5662	1,4-Cyclohexane dicarboxylic acid, diethyl ester (trans) ... 1,4-C$_6$H$_{10}$(CO$_2$C$_2$H$_5$)$_2$	228.29	nd	....	43-4	1.0110[20/4]	1.4337[64]	eth	B9[2], 524
5663	1,2-Cyclohexanediol (cis) .... 1,2-(HO)$_2$C$_6$H$_{10}$	116.16	cr (eth), pl (bz)	120[15]	99-101	1.0297[101/4]	....	al, ace, bz	B6[3], 4058
5664	1,2-Cyclohexanediol (trans) .... 1,2-(HO)$_2$C$_6$H$_{10}$	116.16	cr (ace)	117[13]	105	1.147[24/4]	....	w, al	B6[3], 4060
5665	1,4-Cyclohexanediol (cis) or cis-Quinitol .... 1,4-(HO)$_2$C$_6$H$_{10}$	116.16	pr (ace)		113-4		....	w, al	B6[3], 4080
5666	1,4-Cyclohexanediol (trans) .... 1,4-(HO)$_2$C$_6$H$_{10}$	116.16	mcl pr (ace)		143	1.18[20/4]		w, al	B6[3], 4081
5667	1,2-Cyclohexanedione .... C$_6$H$_8$O$_2$	112.14	cr (peth)	193-5, 96-7[25]	38-40	....	1.4995[20]	w, al, eth, bz	B7[3], 3209
5668	1,2-Cyclohexanedione, 3,5-dimethyl .... C$_8$H$_{12}$O$_2$	140.19	cr (dil MeOH)		71-2	....		w	B7[1], 314
5669	1,2-Cyclohexanedione, 5,5-dimethyl or Dimedone .... C$_8$H$_{12}$O$_2$	140.19	yesh nd (w, aq ace), mcl pr (al-eth)		150			ace, chl, aa	B7[3], 3225
5670	1,2-Cyclohexanedione, dioxime or Nioxime .... C$_6$H$_{10}$N$_2$O$_2$	142.16	nd (w or ace)		191-3			ace	B7[3], 3210
5671	1,3-Cyclohexanedione .... C$_6$H$_8$O$_2$	112.14	pr (bz)	....	105-6	1.0861[91]	1.4576[102]	w, al, ace, chl	B7[3], 3210
5672	1,3-Cyclohexanedione, 2-bromo .... C$_6$H$_7$BrO$_2$	191.03	micr nd		169-70	....	....	al	B7, 556
5673	1,3-Cyclohexanedione, dioxime .... C$_6$H$_2$N$_2$O$_2$	142.16	cr (w)		156-7			w, al, aa	B7[3], 3211
5674	1,4-Cyclohexanedione .... C$_6$H$_8$O$_2$	112.14	mcl pl (w), nd (peth)	sub 100	78			w, al, eth, ace, bz	B7[3], 3211
5675	1,4-Cyclohexanedione, dioxime .... C$_6$H$_{10}$N$_2$O$_2$	142.16	cr (w)		188			w	B7[3], 3212
5676	Cyclohexasiloxane, dodecamethyl .... C$_{12}$H$_{36}$O$_6$Si$_6$	444.93	....	245, 128[20]	−3	0.9672	1.4015[20]	....	B4[3], 1886
5677	Cyclohexasiloxane, 2,4,6,8,10,12-hexamethyl .... C$_6$H$_{24}$O$_6$Si$_6$	360.77			−79	1.006[20/4]	1.3944[20]		B4[3], 1874
5678	Cyclohexene .... C$_6$H$_{10}$	82.15	....	83	−103.5	0.8102[20/4]	1.4465[20]	al, eth, ace, bz	B5[4], 218
5679	Cyclohexene, 1-acetyl .... 1-(CH$_3$CO)C$_6$H$_9$	124.19		201-2, 63-4[6]	....	0.9655[20/4]	1.4881[20]	al, eth	B7[4], 244
5680	1-Cyclohexene, 1-bromo .... 1-BrC$_6$H$_9$	161.04		164-6, 69[35]		1.3901[20/4]	1.5134[20]	eth, ace, bz	B5[4], 236
5681	Cyclohexene, 3-bromo .... 3-BrC$_6$H$_9$	161.05		80-2[40]		1.3890[20/4]	1.5230[20]	eth, bz, chl	B5[4], 237
5682	Cyclohexene, 1-chloro .... 1-ClC$_6$H$_9$	116.59		142-3, 35[13]		1.0361[19/4]	1.4797[20]	eth, ace, chl	B5[4], 230
5683	Cyclohexene, 1,2-dimethyl .... 1,2-(CH$_3$)$_2$C$_6$H$_8$	110.20		136		0.823[20/4]	1.4580[21]	....	B5[4], 268
5684	Cyclohexene, 1-ethyl .... 1-C$_2$H$_5$C$_6$H$_9$	110.20		135-6[755]		0.8238[19]	1.4567[19]	....	B5[4], 266
5685	Cyclohexene, 1-methyl .... 1-CH$_3$C$_6$H$_9$	96.17		110, 24.6[30]	−121	0.8102[20/4]	1.4503[20]	eth, bz	B5[4], 245
5686	Cyclohexene, 3-methyl (d) .... 3-CH$_3$C$_6$H$_9$	96.17	$[\alpha]^{20}_D$ + 110	104	....	0.8010[20/4]	1.4414[20]	eth, bz, peth, chl	B5[3], 200
5687	Cyclohexene, 3-methyl (dl) .... 3-CH$_3$C$_6$H$_9$	96.17		104	−115.5	0.7990[20/4]	1.4414[20]	eth, bz, peth, chl	B5[4], 247
5688	Cyclohexene, 4-methyl .... 4-CH$_3$C$_6$H$_9$	96.17		102.7, 19[30]	−115.5	0.7991[20/4]	1.4414[20]	al, eth	B5[4], 248
5689	Cyclohexene, 1,3,4,5,6-pentachloro (γ) .... C$_6$H$_5$Cl$_5$	254.37		115-6[4]			1.5630[20]		B5[4], 234
5690	Cyclohexene, 1,3,4,5,6-pentachloro (δ) .... C$_6$H$_5$Cl$_5$	254.37		68-9	1.80		....	al	B5[4], 234
5691	Cyclohexene-per fluoro .... C$_6$F$_{10}$	262.05		52-3[750]			1.293[20]	....	B5[4], 229

No.	Name, Synonyms, and Formula	Mol. wt.	Color, crystalline form, specific rotation and λ_{max} (log ε)	b.p. °C	m.p. °C	Density	n_D	Solubility	Ref.
5692	Cyclohexene, 1-phenyl 1-$C_6H_5C_6H_9$	158.25		251-3, 125-6[14]	−11	0.9939[20/4]	1.5718[20]	MeOH	B5[4], 1557
5693	Cyclohexene, 1-isopropyl-4-methylene i-C_3H_7-4(CH_2=)C_6H_7	136.24		173-4	......	0.838[22]	1.4754[22]	....	B5[4], 437
5694	Cyclohexene, 4-isopropyl-1-methyl 4-i-$C_3H_7C_6H_8CH_3$	136.24		174.5	......	0.8465[15.5/15.5]	1.4735[20]	....	B5[4], 300
5695	Cyclohexene, 1-vinyl 1-(CH_2=CH)C_6H_9	108.19		145, 50-2[22]	......	0.8623[15/4]	1.4915[20]	eth, bz, MeOH	B5[4], 405
5696	Cyclohexene, 4-vinyl 4-(CH_2=CH)C_6H_9	108.18		128.9, 66-7[100]	......	0.8299[20/4]	1.4639[20]	eth, bz, peth	B5[4], 406
5697	Cyclohexene, 1-carboxaldehyde C_6H_9CHO	110.16		72[15]	......	0.9694[20/4]	1.5005[20]	al, eth	B7[3], 234
5698	3-Cyclohexene, 1-carboxaldehyde C_6H_9CHO	110.16		164, 52[13]	fr−96.1	0.9709[20/4]	1.4725[19]	ace, MeOH	B7[3], 237
5699	Cyclohexene, 1-carboxylic acid 2-$C_6H_9CO_2H$	126.16		240-2, 138[14]	38	1.109[20/4]	1.4902[20]	al, ace	B9[3], 144
5700	3-Cyclohexene, 1-carboxylic acid 4-$C_6H_9CO_2H$	126.16		237[748], 132-3[20]	17	1.0815[20/4]	1.4812[20]	w, al, ace	B9[3], 148
5701	1-Cyclohexene-1,2-dicarboxylic acid or Δ′-Tetrahydrophthalic acid $C_8H_{10}O_4$	170.17	nd pr (w)	....	126	....	....	w	B9[3], 3939
5702	1-Cyclohexene-1,2-dicarboxylic anhydride $C_8H_8O_3$	152.16	pl (eth)		74	....	....	al, eth, ace, chl	B17[4], 5995
5703	2-Cyclohexene-1,2-dicarboxylic anhydride $C_8H_8O_3$	152.16	pr (eth)		78-9	....	....	al, eth, chl	B17[4], 5994
5704	4-Cyclohexene-1,2-dicarboxylic anhydride (cis) $C_8H_8O_3$	152.16	pl (eth or lig)		103-4	....	....	al, ace, chl	B17[4], 5996
5705	4-Cyclohexene-1,2-dicarboxylic anhydride (trans, d) $C_8H_8O_3$	152.16	lf [α][25/D] + 6.6 (al)		128	....	....	al, bz	B17, 462
5706	4-Cyclohexene-1,2-dicarboxylic anhydride (trans, dl) $C_8H_8O_3$	152.16	cr (bz-lig)		141	....	....	al, bz, chl	B17[4], 5996
5707	2-Cyclohexene-1-ol C_6H_9OH	98.15	....	164-6, 63-5[12]	......	0.9923[15/4]	1.4790[22]	al, ace	B6[4], 196
5708	2-Cyclohexene-1-ol, 5-methyl (cis, d) $CH_3C_6H_8$OH	112.17	[α][30/D] + 6.95	83[25]	......	0.9391[25/4]	1.4727[25]	eth, lig	B6[4], 206
5709	2-Cyclohexene-1-ol, 5-methyl-(cis, l) $CH_3C_6H_8$OH	112.17	[α][25/D] − 7	82[25]	......	0.9391[25/4]	1.4727[25]	eth, lig	B6[4], 206
5710	2-Cyclohexene-1-ol, 5-methyl (trans, d) $CH_3C_6H_8$OH	112.17	[α][27/D] + 127 (ace, c=19.4)	68-9[24]	......	0.9430[20/4]	1.4737[25]	eth, lig	B6[4], 207
5711	2-Cyclohexene-1-ol, 5-methyl-(trans, l) $CH_3C_6H_8$OH	112.17	[α][27/D] − 163.9	82-3[24]	......	0.9430[20/4]	1.4737[25]	eth, lig	B6[4], 207
5712	3-Cyclohexene-1-ol C_6H_9OH	98.14	....	164, 68-9[16]	......	0.9845[20/4]	1.4851[20]	eth, ace	B6[4], 200
5713	2-Cyclohexene-1-one C_6H_8O	96.14	....	169-71, 61-2[10]	......	0.9620[25]	1.4883[20]	al, ace	B7[3], 224
5714	2-Cyclohexene-1-one, 2,3-dimethyl $(CH_3)_2(C_6H_7O)$	124.18		93-6[20]	......	0.9695[20]	1.4995[20]	al, eth	B7[3], 250
5715	2-Cyclohexene-1-one, 2,5-dimethyl $(CH_3)_2(C_6H_6O)$	124.19		189-90	......	0.938[22]	1.4753[22]	al, eth	B7[1], 51
5716	2-Cyclohexene-1-one, 3,5-dimethyl $(CH_3)_2(C_6H_6O)$	124.19		208-9, 94[17]	......	0.9400[20/4]	1.4812[20]	al, eth	B7[3], 255
5717	2-Cyclohexene-1-one, 3,5-dimethyl-4-carboxyethyl $C_{11}H_{16}O_3$	196.25		157-8[18]	......	1.0493[20/4]	1.4773[20]	ace	B10[3], 2902
5718	2-Cyclohexene-1-one, 3,6-dimethyl $(CH_3)_2(C_6H_6O)$	124.19		75[19]	......	1.008[18/18]	1.4805[18]	al, eth	B7[3], 257
5719	2-Cyclohexene-1-one, 2-methyl $CH_3(C_6H_7O)$	110.16		178-9, 56[9]	......	0.9667[20/4]	1.4833[20]	bz	B7[3], 233
5720	2-Cyclohexene-1-one, 3-methyl $CH_3(C_6H_7O)$	110.16		200-2, 78-9[12]	−21	0.9693[20/4]	1.4947[20]	bz	B7[3], 230
5721	3-Cyclohexene-1-one, 4-methyl $CH_3(C_6H_7O)$	110.16		169-72[755], 74[17]	......	0.9551[20/4]	1.4652[20]	al, ace, bz	B7[3], 232
5722	2-Cyclohexene-1-one, 5-methyl (dl) $CH_3(C_6H_7O)$	110.16		179-83, 60[8]	......	0.947[20/4]	1.4739[25]	al, ace, bz	B7[3], 236
5723	2-Cyclohexene-1-one, 5-isopropyl-3-methyl or Hexeton $C_{10}H_{16}O$	152.24	pa ye	244, 124[15]	......	0.9340[21]	1.4865[21]	al, ace	B7[3], 323

No.	Name, Synonyms, and Formula	Mol. wt.	Color, crystalline form, specific rotation and λ_{max} (log ε)	b.p. °C	m.p. °C	Density	n_D	Solubility	Ref.
5724	3-Cyclohexene-1-one, 4,6-dimethyl $(CH_3)_2(C_6H_6O)$	124.19	....	194	......	0.9539^0	....	eth	B7[3], 256
5725	Cycloheximide $C_{15}H_{23}NO_4$	281.36	wh pl (w), $[\alpha]^{20}_D -3.4$ (al)	....	119-21			al, eth, ace	B21[4], 6632
5726	Cyclohexyl acetic acid $C_6H_{11}CH_2CO_2H$	142.20	nd (HCO_2H)	224-6, 135^{13}	33	$1.0423^{18/4}$	1.4775^{20}	al, eth	B9[3], 47
5727	Cyclohexylamine $C_6H_{11}NH_2$	99.18	....	134.5, 30.5^{15}	−17.7	$0.8191^{20/4}$	1.4372^{20}	al, eth, ace, bz	B12[3], 10
5728	Cyclohexylamine, hydrochloride $C_6H_{11}NH_2.HCl$	135.64	nd (w or al-eth)	....	206-7			w, al	B12[3], 12
5729	Cyclohexyl bromide $C_6H_{11}Br$	163.06	....	166.2, 45.5^{10}	−56.5	$1.3359^{20/4}$	1.4957^{20}	al, eth, ace, bz, lig	B5[4], 67
5730	Cyclohexyl chloride $C_6H_{11}Cl$	118.61	....	143	$1.000^{20/4}$	1.4626^{20}	....	al, eth, ace, bz, chl	B5[4], 48
5731	Cyclohexyl ether or Dicyclohexyl ether $(C_6H_{11})_2O$	182.31	....	242-3	......	$0.9227^{20/4}$	1.4741^{20}	....	B6[3], 19
5732	Cyclohexyl 2-furyl ether $C_6H_{11}O(C_4H_3O)$	166.22	....	118.9^{28}	......	$1.0200^{28/4}$	1.4861^{28}	al, eth, ace	B17[4], 1219
5733	Cyclohexyl hydro peroxide $C_6H_{11}OOH$	116.16	....	$42^{0.1}$	−20	$1.019^{20/4}$	1.4645^{25}	al, eth, aa	B6[4], 53
5734	Cyclohexyl methyl ether or Hexahydro anisole $C_6H_{11}OCH_3$	114.19	....	133	−74.4	$0.8756^{20/4}$	1.4355^{20}	al, eth	B6[4], 26
5735	Cyclohexyl methyl ether, 2-bromo (dl, trans) $2\text{-}BrC_6H_{10}OCH_3$	193.09	....	$78\text{-}9^{12}$	......	$1.3314^{20/4}$	1.4871^{20}	al, eth	B6[3], 43
5736	Cyclohexyl phenyl ether $C_6H_{11}OC_6H_5$	176.26	....	128^{17}	......	$1.0077^{20/4}$	1.520^{22}	ace, bz	B6[4], 565
5737	Cyclohexyl isothiocyanate C_5H_9NCS	141.24	....	219^{746}	......		1.5375^{20}	al, eth	B12[3], 53
5738	Cyclononanone $C_9H_{16}O$	140.23	....	148.5^{24}	34	$0.9560^{20/4}$	1.4729^{20}	al	B7[3], 111
5739	Cyclononasiloxane, octadecamethyl $C_{18}H_{54}O_9Si_9$	667.40	....	188^{20}	......		1.4070^{20}	bz, lig	B4[3], 1886
5740	Cyclononene (cis) C_9H_{16}	124.23	....	167-9, $73\text{-}4^{30}$	......	$0.8671^{20/4}$	1.4805^{20}	bz	B5[4], 280
5741	Cyclononene-(trans) C_9H_{16}	124.23	....	$94\text{-}6^{30}$	......	$0.8615^{20/4}$	1.4799^{20}	bz	B5[4], 281
5742	1,5-Cyclooctadiene (cis, cis) C_8H_{12}	108.19	....	150.8^{757}, $51\text{-}2^{28}$	−70	$0.8818^{25/4}$	1.4905^{25}	bz	B5[4], 403
5743	1,3-Cyclooctadiene (cis, cis) C_8H_{12}	108.18	....	$54\text{-}5^{35}$	−57	$0.8699^{25/4}$	1.4940^{25}	eth, bz, chl	B5[4], 401
5744	Cyclooctane C_8H_6	112.22	....	$148\text{-}9^{749}$, 63^{45}	14.3	$0.8349^{20/4}$	1.4586^{20}	bz, lig	B5[4], 111
5745	Cyclooctanol $C_8H_{15}OH$	128.22	....	99^{16}	25.1	$0.9740^{20/4}$	1.4871^{20}	al	B6[4], 113
5746	Cyclooctanone or Azelaone $C_8H_{14}O$	126.20	....	194-8, 74^{12}	28-30	$0.9581^{20/4}$	1.4694^{20}	al, ace, bz	B7[3], 77
5747	Cyclooctanone, semicarbazone $C_8H_{14}NHCONHNH_2$	183.26	lf (dil MeOH)	....	170-1				B7[3], 78
5748	Cyclooctasiloxane, hexadecamethyl $C_{16}H_{48}O_8Si_8$	593.25	....	290, 175^{20}	31.5	1.177	1.4060^{20}	bz, lig	B4[3], 1886
5749	Cyclooctatetraene C_8H_8	104.16	ye or wh	140.5, 29.1^{10}	−4.7	$0.9206^{20/4}$	1.5381^{20}	al, eth, ace, bz	B5[4], 1331
5750	Cyclooctatetraene, chloro C_8H_7Cl	138.60	....	$50\text{-}1^{5.5}$	......	$1.1199^{25/4}$	1.5542^{25}	ace, bz	B5[4], 1334
5751	Cyclooctatetraene, methyl $CH_3C_8H_7$	118.18	....	84.5^{67}	......	$0.8978^{25/4}$	1.5249^{25}	al, eth, bz, chl	B5[4], 1358
5752	Cyclooctene (cis) C_8H_{14}	110.20	....	138, 42^{18}	−12	$0.8472^{20/4}$	1.4698^{20}	al, eth	B5[4], 262
5753	Cyclooctene (trans) C_8H_{14}	110.20	....	143, 75^{78}	−59	$0.8483^{20/4}$	1.4741^{25}	al, chl	B5[4], 263
5754	Cyclopentadecanone or Exaltone $C_{15}H_{28}O$	224.39	....	$120^{0.3}$	63	0.8895	1.4637^{66}	al, ace	B7[3], 203
5755	Cyclopentadecanone, 3-methyl or Muscone, muskone $3\text{-}CH_3C_{15}H_{27}O$	238.42	ye $[\alpha]^{17}_D -13$ (undil)	327-30[752], $130^{0.5}$	......	$0.9221^{17/4}$	1.4802^{17}	al, eth, ace	B7[3], 208

No.	Name, Synonyms, and Formula	Mol. wt.	Color, crystalline form, specific rotation and λ_{max} (log ε)	b.p. °C	m.p. °C	Density	n_D	Solubility	Ref.
5756	Cyclopentadiene C_5H_6	66.10	....	40.0	−97.2	0.8021[20/4]	1.4440[20]	**al, eth, ace, bz**	B5[4], 377
5757	Cyclopentadiene, perchloro C_5Cl_6	272.77	ye gr liq	239[753], 48-9[0.3]	−9	1.7019[25/4]	1.5658[20]	....	B5[4], 381
5758	1,3-Cyclopentadiene, 5-isopropylidene $[CH_2=C(CH_3)]C_5H_5$	106.17	....	49-50[11]		0.881[20/4]	1.5474[20]	....	B5[4], 974
5759	Cyclopentadiene benzoquinone $C_{11}H_{10}O_2$	174.20	gr-ye lf (MeOH)	....	77-8	....		**al, eth, ace, bz**	B6[3], 5311
5760	Cyclopentadienone, tetraphenyl or Tetracyclone $(C_6H_5)_4C_5O$	384.48	bk-vt lf, cr (aa or xyl)		220-1	....		al, bz	B7[3], 2997
5761	Cyclopentasiloxane, 2,4,6,8,10-pentamethyl $C_5H_{20}O_5Si_5$	300.64	....	169	−108	0.9985[20/4]	1.3912[20]		B4[3], 1874
5762	Cyclopentane or Pentamethylene C_5H_{10}	70.14	....	49.2	−93.9	0.7457[20/4]	1.4065[20]	**al, eth, ace, bz**	B5[4], 14
5763	Cyclopentane, acetyl or Cyclopentyl methyl ketone $CH_3COC_5H_9$	112.17	....	158-9		0.918[20/20]	1.4409[20]	eth	B7[3], 71
5764	Cyclopentane, allyl $(CH_2=CHCH_2)C_5H_9$	110.20	....	124-6		0.793[25/4]	1.4412[20]		B5[4], 272
5765	Cyclopentane, amino or Cyclopentyl amine $H_2NC_5H_9$	85.15	....	108	−85.7	0.8689[20/4]	1.4778[25]	ace, bz	B12[3], 5
5766	Cyclopentane, bromo or Cyclopentyl bromide C_5H_9Br	149.04	....	137-8, 56[49]	......	1.3873[20/4]	1.4886[20]	....	B5[4], 19
5767	Cyclopentane, butyl $C_4H_9C_5H_9$	126.23	....	156.7, 41.6[10]	−108	0.7846[20/4]	1.4316[20]	**al, eth, ace, bz, peth**	B5[4], 139
5768	Cyclopentane, chloro or Cyclopentyl chloride C_5H_9Cl	104.58	....	113-4[752]	......	1.0051[20/4]	1.4510[20]	eth, ace, bz	B5[4], 18
5769	Cyclopentane, 1,2-diethyl (trans) $1,2-(C_2H_5)_2C_5H_8$	124.24	....	153.6	−95.6	0.7832[20/4]	1.4295[20]	eth, bz, peth	B5[4], 141
5770	Cyclopentane, 1,1-dimethyl $1,1-(CH_3)_2C_5H_8$	98.19	....	87.5		0.7552[20/0]	1.4139[20]	....	B5[4], 105
5771	Cyclopentane, 1,2-dimethyl (cis) $1,2-(CH_3)_2C_5H_8$	98.19	....	99.25	−62	0.7718[20/4]		....	B5[4], 106
5772	Cyclopentane, 1,2-dimethyl (trans) $1,2-(CH_3)_2C_5H_8$	98.19	....	91.8	−120	0.7495[20/4]		....	B5[4], 107
5773	Cyclopentance, ethyl $C_2H_5C_5H_9$	98.19	....	103.5, 19.4[30]	−138.4	0.7665[20/4]	1.4198[20]	**al, eth, ace, bz, peth**	B5[4], 104
5774	Cyclopentane, β-hydroxyethyl $C_5H_9CH_2CH_2OH$	114.19	....	183-4[770], 96-7[24]		0.9180[20/4]	1.4577[20]	eth	B6[4], 110
5775	Cyclopentane, iodo or Cyclopentyl iodide C_5H_9I	196.03	....	166-7, 52[12]	......	1.7096[20/4]	1.5447[20]	eth, bz	B5[4], 20
5776	Cyclopentane, methyl $CH_3C_5H_9$	84.16	....	71.8	−142.4	0.7486[20/4]	1.4097[20]	**al, eth, ace, bz**	B5[4], 84
5777	Cyclopentane, 1-methly-2-propyl (trans) $1-CH_3-2-C_3H_7(C_5H_8)$	126.24	....	152.6	−104.9	0.7921[20/4]	1.4321[20]	....	B5[4], 140
5778	Cyclopentane, nitro $C_5H_9NO_2$	115.13	....	90-1[40]	......	1.0776[23/4]	1.4538[20]	bz	B5[4], 20
5779	Cyclopentane, phenyl $C_6H_5C_5H_9$	146.23	....	219, 102[18]		0.9462[20/4]	1.5309[20]	....	B5[4], 1409
5780	Cyclopentane, propyl $C_3H_7C_5H_9$	112.22	....	131, 21.2[10]	−117.3	0.7763[20/4]	1.4266[20]	**al, eth, ace, bz**	B5[4], 125
5781	Cyclopentane, isopropyl $(CH_3)_2CHC_5H_9$	112.22	....	126.4, 16.3[10]	−111.4	0.7765[20/4]	1.4258[20]	**al, eth, ace, bz**	B5[4], 125
5782	Cyclopentane, 1,1,2-trimethyl $1,1,2-(CH_3)_3C_5H_7$	112.21	....	113-4[749]		0.7661[20/D]	1.4199[20]	....	B5[4], 127
5783	Cyclopentane, 1,1,3-trimethyl $1,1,3-(CH_3)_3C_5H_7$	112.21	....	115-6		0.7703[20/4]	1.4223[20]	....	B5[4], 128
5784	Cyclopentane carboxaldehyde C_5H_9CHO	98.15	....	133-4		0.9371[20/4]	1.1432[20]	w, al, eth	B7[3], 43
5785	Cyclopentane carboxylic acid $C_5H_9CO_2H$	114.15	....	212-3[752], 104[11]	−7	1.0527[20/4]	1.4532[20]		B9[3], 11
5786	Cyclopentanecarboxylic acid, 3-formyl-2,2,3-trimethyl,methyl ester (d) $C_{11}H_{18}O_3$	198.27	$[\alpha]_D$ + 51.4 (al)	130-2[8]		1.048[20/4]	1.4160[22]	al	B10[3], 2868
5787	Cyclopentanecarboxylic acid, 2,oxo,ethyl ester $C_8H_{12}O_3$	156.19	....	218[704], 110[16]		1.0781[21/4]	1.4519[20]	eth, bz	B10[3], 2808
5788	Cyclopentanecarboxylic acid, 3-oxo $C_6H_8O_3$	....	197[30]	64-5		....			B10[3], 2812

No.	Name, Synonyms, and Formula	Mol. wt.	Color, crystalline form, specific rotation and λ_{max} (log ε)	b.p. °C	m.p. °C	Density	n_D	Solubility	Ref.
5789	1,2-Cyclopentanedicarboxylic acid (cis) .... 1,2-C₅H₈(CO₂H)₂	158.16	nd (w)	....	140	....	....	w	B9³, 3807
5790	1,2-Cyclopentanedicarboxylic acid (trans d) .... C₇H₁₀O₄	158.16	cr (w), [α]_D + 87.6 (w, c=0.9)		181			w, al	B9³, 3807
5791	1,2-Cyclopentanedicarboxylic acid (trans, dl) .... C₇H₁₀O₄	158.16	cr (w)		162-3			w, al	B9³, 3807
5792	1,2-Cyclopentanedicarboxylic acid (trans,l) .... C₇H₁₀O₄	158.16	cr (w)[α]_D −85.9 (w, c=1.2)		180-1			w, al	B9³, 3807
5793	1,3-Cyclopentanedicarboxylic acid (cis) or Norcamphoric acid .... 1,3-C₅H₈(CO₂H)₂	158.16	pr (w)	>300d	121			al, eth, ace, chl	B9³, 3808
5794	1,3-Cyclopentanedicarboxylic acid (trans, d) .... 1,3-C₅H₈(CO₂H)₂	158.16	cr (CCl₄), [α]_D + 5.9 (w, c=5)		93.5			w	B9², 519
5795	1,3-Cyclopentanedicarboxylic acid (trans, dl) .... C₇H₁₀O₄	158.16	pr (CCl₄)		88			w	B9³, 3808
5796	1,3-Cyclopentanedicarboxylic acid (trans, l) .... C₇H₁₀O₄	158.16	cr (CCl₄), [α]_D − 5.3 (w, c=5)		93			w	B9², 519
5797	1,3-Cyclopentanedicarboxylic acid, 4,4-dimethyl or Apo-fenchocamphoric acid .... 4,4 (CH₃)₂C₅H₆(CO₂H)₂-1,3	186.21	mcl	....	144-5			w, al, eth	B9³, 3849
5798	Cyclopentanol .... C₅H₉OH	86.14	....	140.8, 53¹⁰	−19	0.9478²⁰/⁴	1.4530²⁰	al, eth, ace	B6⁴, 5
5799	Cyclopentanol, 2-acetyl-1,3,3,4,4-pentamethyl (s) or Deoxymesityl oxide .... C₁₂H₂₂O₂	190.31	cr (petli-eth)		45			bz	B8³, 32
5800	Cyclopentanol, 1-methyl 1-CH₃C₅H₈OH	100.16	....	53-4³⁰	35-7	0.9044²³·⁵/⁴	1.4429²³·⁵	....	B6⁴, 86
5801	Cyclopentanone .... C₅H₈O	84.12	....	130.6	−51.3	0.9487²⁰/⁴	1.4366²⁰	al, eth, ace	B7³, 5
5802	Cyclopentanone, 2-methyl .... 2-CH₃(C₅H₇O)	98.15	....	139.5, 44¹⁸	−75	0.9139²⁰	1.4364²⁰	al, eth, ace	B7³, 40
5803	Cyclopentanone, 3-methyl (d) .... 3-CH₃(C₅H₇O)	98.15	[α]²⁵/_D + 143.7 (un-dil)	143.5⁷⁴², 43-4¹²	−58.4	0.9140¹⁹/⁴	1.4340¹⁹	w, al, eth, ace, aa	B7³, 42
5804	Cyclopentanone, 3-methyl (dl) .... 3-CH₃(C₅H₇O)	98.15	....	144, 38¹¹	....	0.913²²	1.4329²⁰	w, al, eth, ace	B7³, 43
5805	Cyclopentasiloxane, decamethyl .... C₁₀H₃₀O₅Si₅	370.78	....	210, 101²⁰	−38	0.9593²⁰/⁴	1.3982²⁰		B4³, 1885
5806	Cyclopentene .... C₅H₈	68.13	....	44.2	−135	0.7720²⁰/⁴	1.4225²⁰	al, eth, bz, peth	B5⁴, 209
5807	Cyclopentene, 3-chloro 3-ClC₅H₇	102.57	....	25-31³⁰	....	1.0577¹⁵	1.4708²⁶	al, eth, chl	B5⁴, 212
5808	Cyclopentene, perchloro .... C₅Cl₈	343.68	nd (al)	283, 140¹⁰	41	1.8200⁵⁰/⁴	1.5660⁵⁰	al	B5⁴, 213
5809	Cyclopentene, 1,2-diemthyl .... 1,2(CH₃)₂C₅H₆	96.17		103⁷⁵⁷		0.7992¹³·⁵	1.4447¹³·⁵	....	B5⁴, 254
5810	Cyclopentene, 2,3-dimethyl-4-isopropyl .... 2,3-(CH₃)₂-4-(CH₃)₂CHC₅H₅	138.25		164-6		0.8085²²/⁴	1.4503²²		PCHE 2, 319
5811	Cyclopentene, 2,4-dimethyl .... 2,4-(CH₃)₂C₅H₆	96.17		93.2			1.4283²⁰		B5⁴, 255
5812	Cyclopentene, 1-ethyl 1-C₂H₅C₅H₇	96.17		108	−123.3	0.8000²⁰/⁴	1.4429²¹		B5⁴, 252
5813	Cyclopentene, 3-ethyl 3-C₂H₅C₅H₇	96.17		99-103⁷⁵⁸		0.7874²⁰/⁴	1.4303²⁰		B5⁴, 252
5814	Cyclopentene, 1,2-dichloro,perfluoro .... C₅Cl₂F₆	244.95		90.7	−105.8	1.6546²⁰/⁴	1.3676²⁰		B5⁴, 213
5815	Cyclopentene, 1-methyl 1-CH₃C₅H₇	82.14		75.5	−127.2	0.7851¹⁵/⁴	1.4347¹⁵		B5⁴, 239
5816	Cyclopentene, 3-methyl 3-CH₃C₅H₇	82.14		69-71		0.9705²⁰/⁴	1.42476²⁰		B5⁴, 240
5817	Cyclopentene, 1,2,3-trimethyl .... 1,2,3(CH₃)₃C₅H₅	110.20		121.6		0.8039¹⁵/⁴	1.4464¹⁶·⁵		B5³, 219
5818	2-Cyclopentene-1-one, 3-phenyl .... C₁₁H₁₀O	158.20	....	234.2	−23	0.9711²⁰	1.5440²⁰	al, ace, chl	B5³, 1654

No.	Name, Synonyms, and Formula	Mol. wt.	Color, crystalline form, specific rotation and λ_{max} (log ϵ)	b.p. °C	m.p. °C	Density	n_D	Solubility	Ref.
5819	3-Cyclopentene-1-one, 3,4-bis(4-methoxyphenyl) $(CH_3OC_6H_4)_2C_5H_4O$	294.35	ye br	....	129	....	....	al, eth, ace	B8, 355
5820	1,2-Cyclopentenophenanthrene $C_{17}H_{14}$	218.30	nd (al), cr (peth)	....	135-6			al	B5³, 2241
5821	1,2-Cyclopentenophenanthrene, 3-methyl $C_{18}H_{16}$	232.33	cr (aa)	....	126-7				B5³, 2252
5822	2,3-Cyclopentenophenanthrene $C_{17}H_{14}$	218.30	pl or pr (al), nd (MeOH)	....	84-5			al	B5³, 2240
5823	9,10-Cyclopentenophenanthrene $C_{17}H_{14}$	218.30	pl (xyl), nd (i-ProH)	....	155-6			al, bz	B5³, 2242
5824	Cyclopropane C_3H_6	42.08	....	−32.7	−127.6	$0.720^{-79/4}$	$1.3799^{-42.5}$	al, eth, bz, peth	B5⁴, 3
5825	Cyclopropane, acetyl or Cyclopropyl methyl ketone $CH_3COC_3H_5$	84.13	....	114⁷⁷²	fp −68.4	$0.8984^{20/4}$	1.4251^{20}	w, al, eth	B7³, 13
5826	Cyclopropane, amino or Cyclopropyl amine $H_2NC_3H_5$	57.09	....	50-1		$0.8240^{20/4}$	1.4210^{20}	w, al, eth	B12³, 3
5827	Cyclopropane, 1,1-dimthyl $1,1\text{-}(CH_3)_2C_3H_4$	70.14	....	20.6	−109	$0.6589^{20/4}$	1.3668^{20}	al, eth	B5⁴, 24
5828	Cyclopropane, 1,2-dimethyl (cis) $1,2\text{-}(CH_3)_2C_3H_4$	70.13	....	37		0.6928^{20}	1.3822^{20}	al, eth	B5⁴, 25
5829	Cyclopropane, 1,2-dimethyl (trans, dl) $1,2\text{-}(CH_3)_2C_3H_4$	70.13	....	29		0.6769^{20}	1.3713^{20}	al, eth	B5⁴, 26
5830	Cyclopropane, 1,2-dimethyl (trans, l) $1,2\text{-}(CH_3)_2C_3H_4$	70.13	$[\alpha]^{22}_D$ − 2.39 diglyme 13.9%	28-9	......	1.3699^{16}	....	al, eth	TETRA 20, 1965
5831	Cyclopropane, ethyl $C_2H_5C_3H_5$	70.13	....	34.5		$0.677^{20/4}$	1.379^{20}	....	B5⁴, 23
5832	Cyclopropane, methoxy $CH_3OC_3H_5$	72.12	....	44.7	−119	$0.8100^{20/4}$	1.3802^{20}	w, al, eth, bz	B6³, 3
5833	Cyclopropane, methyl $CH_3C_3H_5$	56.12	....	4-5	−117.2	$0.6912^{-20/4}$	....	al, eth	B5⁴, 13
5834	Cyclopropane, 1-methyl-1-phenyl $1\text{-}CH_3\text{-}1\text{-}C_6H_5C_3H_4$	132.18	....	91⁵⁰	......		1.5160^{20}	ace, bz, chl	B5⁴, 1388
5835	Cyclopropane, phenyl $C_6H_5C_3H_5$	118.18	....	173.6⁷⁵⁸, 80³⁷	−31	$0.9317^{20/4}$	1.5285^{20}	eth, ace, chl	B5³, 1200
5836	Cyclopropane, isopropenyl $[CH_2=CH(CH_3)]C_3H_5$	82.14	....	69.5-70⁷⁵¹	......	$0.7500^{20/4}$	1.4252^{20}	....	B5⁴, 243
5837	Cyclopropane, 1,1,2-trimethyl (dl) $1,1,2\text{-}(CH_3)_3C_3H_3$	84.10	....	52.6	−138.3	$0.6974^{20/4}$	1.3864^{20}	eth, bz	B5⁴, 91
5838	Cyclopropane carbonitrile C_3H_5CN	67.09	....	135, 69-70⁸⁸	......	$0.8946^{20/4}$	1.4229^{20}	eth	B9³, 6
5839	Cyclopropane carboxylic acid $C_3H_5CO_2H$	86.09	....	182-4	18-9	$1.0885^{20/4}$	1.4390^{20}	al, eth	B9³, 3
5840	Cyclopropane carboxylic acid, 2,2-dimethyl-3-(2-methyl-propenyl) or d-cis-Chrysanthemamic acid $C_{10}H_{16}O_2$	168.24	pr $[\alpha]^{22}_D$ + 83.3 (chl, c=1.6)	95⁰·¹	40-2			al, eth, chl	B9³, 210
5841	Cyclopropane carboxylic acid, 2,2-dimethyl-3-(2-methyl-propenyl) cis,dl $C_{10}H_{16}O_2$	168.24	pr (AcOEt), cr (peth)	....	115-6			al, eth	B9³, 211
5842	Cyclopropane carboxylic acid, 2,2-dimethyl-3-(2-methyl-propenyl) (cis,l) $C_{10}H_{16}O_2$	168.24	pr $[\alpha]^{19}_D$ − 83.3 (chl, c=1.6)	95⁰·¹	41-3			al, eth	B9³, 211
5843	Cyclopropane carboxylic acid, 2,2-dimethyl-3-(2-methyl-propenyl) trans,d or trans-Chrysanthemucic acid (d) $C_{10}H_{16}O_2$	168.24	pr $[\alpha]^{20}_D$ + 25.8 (chl, c=2.5)	245d, 135¹²	18-21			al, eth, chl	B9³, 211
5844	Cyclopropane carboxylic acid, 2,2-dimethyl-3-(2-methyl-propenyl) trans,dl $C_{10}H_{16}O_2$	168.24	pr (AcOEt)	145-6¹³	54			al, eth, chl	B9³, 212
5845	Cyclopropane carboxylic acid, 2,2-dimethyl-3-(2-methyl-propenyl) trans,l $C_{10}H_{16}O_2$	168.24	$[\alpha]^{20}_D$ − 25.8 (chl 2.9%)	99-100⁰·²	17-21			al	B9³, 211
5846	Cyclopropane carboxylic acid, methyl ester $C_3H_5CO_2CH_3$	100.12	....	119	......	$0.9848^{20/4}$	1.4144^{19}	....	B9³, 3
5847	1,1-Cyclopropane dicarboxylic acid $1,1\text{-}C_3H_4(CO_2H)_2$	130.10	pr or nd (chl), pr (w + 1)	....	140-1			w, eth	B9³, 3795

No.	Name, Synonyms, and Formula	Mol. wt.	Color, crystalline form, specific rotation and λ_{max} (log ε)	b.p. °C	m.p. °C	Density	n_D	Solubility	Ref.
5848	1,1-Cyclopropane dicarboxylic acid, diethyl ester $1,1\text{-}C_3H_4(CO_2C_2H_5)_2$	186.21	....	$214\text{-}6^{748}$, $99\text{-}100^{12}$	......	$1.0566^{25/25}$	1.4345^{18}	al, eth	$B9^3$, 3795
5849	1,2-Cyclopropane dicarboxylic acid (cis) $1,2\text{-}C_3H_4(CO_2H)_2$	130.10	pr (eth or w)		139			w, al, eth	$B9^3$, 3796
5850	1,2-Cyclopropane dicarboxylic acid (trans, d) $1,2\text{-}C_3H_4(CO_2H)_2$	130.10	$[\alpha]^{27}_D +$ 84.87 (w)		175			w, al, eth	$B9^3$, 3796
5851	1,2-Cyclopropane dicarboxylic acid (trans, dl) $1,2\text{-}C_3H_4(CO_2H)_2$	130.10	nd (eth), pl (ace-bz)	210^{30}	175			w, al, eth	$B9^3$, 3797
5852	1,2-Cyclopropane dicarboxylic acid (trans, l) $1,2\text{-}C_3H_4(CO_2H)_2$	130.10	$[\alpha]^{27}_D -$ 84.40 (w)		175			w, al, eth	$B9^3$, 3797
5853	1,2-Cyclopropane dicarboxylic acid, 1-bromo $1\text{-}BrC_3H_3(CO_2H)_2\text{-}1,2$	209.00	pr (eth-chl), cr (ace-bz)		175			eth, ace	$B9^2$, 514
5854	1,2-Cyclopropane dicarboxylic acid, diethyl ester-(cis) $1,2\text{-}C_3H_4(CO_2C_2H_5)_2$	186.21	....	106-7		$1.062^{12/4}$	1.4450^{20}	al, eth	$B9^2$, 513
5855	1,2-Cyclopropane dicarboxylic acid, dimethyl ester $C_7H_{10}O_4$	158.16	....	219-20, 110^3		$1.1584^{16/4}$	1.4472^{14}	al, eth	$B9^2$, 513
5856	1,2,3-Cyclopropane tricarboxylic acid $C_3H_3(CO_2H)_3$	174.11	nd (w), cr (HCl)		220			w, al	$B9^3$, 4746
5857	Cyclopropyl methyl ketone $CH_3COC_3H_5$	84.13	....	114^{772}	fp − 68.4	$0.8984^{20/4}$	1.4251^{20}	w, al, eth	$B7^3$, 13
5858	Cyclotetrasiloxane, octamethyl $C_8H_{24}O_4Si_4$	296.62	....	175.8, 74^{20}	17.5	0.9561^{20}	1.3968^{20}	....	$B4^3$, 1885
5859	Cyclotetrasiloxane, octaphenyl $(C_6H_5)_8O_4Si_4$	793.20	nd (bz-al or aa)	$330\text{-}4^1$	200-1			bz, aa	$B16^2$, 1213
5860	Cyclotetrasiloxane, 1,3,5,7-tetramethyl-1,3,5,7-tetra-phenyl $C_{28}H_{32}O_4Si_4$	544.91	cr (aa)	$237^{1.5}$	99	$1.1183^{20/4}$	1.5161^{20}	aa	$B16^2$, 1211
5861	Cyclotetrasiloxane, 2,4-6,8-tetramethyl $C_4H_{16}O_4Si_4$	240.51	....	134-5	−65	$0.9912^{20/4}$	1.3870^{20}	....	$B4^3$, 1874
5862	Cyclotrisiloxane, hexaphenyl $(C_6H_5)_6O_3Si_3$	594.90	pl (bz-al or aa)	$290\text{-}300^1$	190	$1.23^{25/4}$		bz, aa	$B16^3$, 1213
5863	Cyclotrisiloxane, 1,3,5-triethyl,1,3,5-triphenyl $(C_2H_5)_3(C_6H_5)_3O_3Si_3$	450.77	....	$166^{0.025}$	177.5	$1.0952^{25/4}$	1.5402^{25}	bz	$B16^3$, 1211
5864	Cymarose or 4,5-Dihydroxy-3-methoxyhexanal or 2,6-didesoxy-3-O-methyl-D-allose $CH_3(CHOH)_2CH(OCH_3)CH_2CHO$	162.19	pr (eth-peth), nd (ace), $[\alpha]^{21}_D +$ 53.4 (w, c=2.2)		100-2			w, al, ace	$B1^4$, 4193
5865	o-Cymene or 1-Methyl-2-iso-propyl benzene $2\text{-}(CH_3)_2CHC_6H_4CH_3$	134.22	....	178.1, 57.3^{10}	−71.5	$0.8766^{20/4}$	1.5006^{20}	**al, eth, ace, bz**	$B5^4$, 1057
5866	m-Cymene or 1-Methyl-3-iso-propyl benzene $3\text{-}(CH_3)_2CHC_6H_4CH_3$	134.22	....	175.1, 55^{10}	−63.7	$0.8610^{20/4}$	1.4930^{20}	**al, eth, ace, bz**	$B5^4$, 1058
5867	p-Cymene or 1-Methyl-4-iso-propyl benzene $4\text{-}(CH_3)_2C HC_6H_4CH_3$	134.22	....	177.1, 56.3^{10}	−67.9	$0.8573^{20/4}$	1.4909^{20}	**al, eth, ace, bz**	$B5^4$, 1060
5868	p-Cymene, 2-chloro or 2-Chloro-4-isopropyltoluene $2\text{-}Cl\text{-}4\text{-}i\text{-}C_3H_7C_6H_3CH_3$	168.67	....	216-7, 104^{20}		$1.0104^{25/4}$	1.50782^{20}	ace, bz	$B5^4$, 1062
5869	p-Cymene, 2-nitro $4\text{-}(CH_3)_2CH\text{-}2\text{-}NO_2\text{-}C_6H_3CH_3$	179.22	....	126^{10}		$1.0744^{20/4}$	1.5301^{20}	al, eth	$B5^4$, 1064
5870	Cysteic acid (d) or 1,2-amino-3-sulfopropanoic acid $HO_2CCH(NH_2)CH_2SO_3H$	169.16	oct cr or nd (dil al), pr or nd (w + 1), $[\alpha]^{20}_D +$ 8.66 D (w)		260d			w	$B5^3$, 1713
5871	Cysteic acid (dl) $HO_2CCH(NH_2)CH_2SO_3H$	169.16	pr (w)		272-4d			w	$B4^3$, 1714
5872	Cysteine (L) or L-β-Mercaptoalanine $HSCH_2CH(NH_2)CO_2H$	121.16	cr (w), $[\alpha]^{30}_D +$ 9.8 (w, c=1.3)		240d			w, al, aa	$B4^3$, 1580
5873	Cystine (D) or D-Dicystine $[HO_2CCH(NH_2)CH_2S]_2$	240.30	cr (dil NH3), $[\alpha]^{20}_D +$ 224 (1N HCl, c=1)		247-9				$B4^3$, 1618
5874	Cystine (DL) $[HO_2CCH(NH_2)CH_2S\text{-}]_2$	240.30	....		260				$B4^3$, 1621

No.	Name, Synonyms, and Formula	Mol. wt.	Color, crystalline form, specific rotation and λ_{max} (log ε)	b.p. °C	m.p. °C	Density	n_D	Solubility	Ref.
5875	Cystine (L)................ [HO₂CCH(NH₂)CH₂S-]₂	240.30	hex pl or pr (w), $[\alpha]^{20}_D$ − 223.4 (1N HCl, c=1)	....	260-1d	1.677	....		B4³, 1593
5876	Cystine (meso) .... [HO₂CCH(NH₂)CH₂S-]₂	240.30	....	....	200-21d	....	....		B4³, 1593
5877	Cytidine or 1-β-D-ribofuranosylcytosine................ C₉H₁₃N₃O₅	243.22	nd (dil al), $[\alpha]^{20.5}_D$ + 35.3 (w, c=1)		230-1d			w	B31, 24
5878	Cytidylic acid or 3-Cytosylic acid................ C₉H₁₄N₃O₈P	323.20	orh nd $[\alpha]^{20}_D$ + 49.4 (w, c=1)	....	233-4d	....	....	w, al	B31, 25
5879	Cytisine (l) or Sophorine, Baptitoxine................ C₁₁H₁₄N₂O	190.25	orh pr (aa or ace)	218²	154.5			w, al, ace, bz, chl	B24², 70
5880	Cytisine, N-methyl or Caulophylline .... C₁₃H₁₆N₂O	204.28	cr (w + 2), nd (al, bz or lig), pr (al)		137			w, al, ace, bz	B24², 70
5881	Cytosine or 4-amino-1,2-dihydro-1,3-diazin-2-one .... C₄H₅N₃O	111.10	mcl or tcl pl (w + 1)	....	320-5d			w	B24, 314
5882	Cytosine, 5-methyl .... C₅H₇N₃O	125.13	pr (w + ½)	....	270d	....		w	B24, 355

No.	Name, Synonyms, and Formula	Mol. wt.	Color, crystalline form, specific rotation and λ_{max} (log ε)	b.p. °C	m.p. °C	Density	n_D	Solubility	Ref.
5883	1,3-Decadiene .... $CH_2(CH_2)_5CH=CHCH=CH_2$	138.25	....	168-70	....	0.752^{20}	....	bz	B1[4], 1056
5884	3,7-Decadiene-5-yne, 4,7-dipropyl .... $C_2H_5CH=C(C_3H_7)C\equiv CCH=C(C_3H_7)C_2H_5$	218.39		125-7[18]	....	$0.8131^{19/4}$	1.4890^{20}	bz	B1[4], 1134
5885	2,4-Decadienoic acid, ethyl ester .... $C_5H_{11}CH=CHCH=CHCO_2C_2H_5$	196.29		130[0.6]	....		1.5020^{26}	....	B2[4], 17314
5886	4,6-Decadiyne or Dipentyne.................... $C_3H_7C\equiv C-C\equiv CC_3H_7$	134.22		88[12]	....	$0.8695^{19/4}$			B1[3], 1064
5887	Decalin (cis) or Decahydronaphthalene .... $C_{10}H_{18}$	138.25		195.6, 69.4[10]	−43	$0.8965^{20/4}$	1.4810^{20}	al, eth, ace, bz, chl	B5[4], 310
5888	Decalin (trans) or Decahydronaphthalene .... $C_{10}H_{18}$	138.25		187.2, 63[10]	−30.4	$0.8699^{20/4}$	1.4695^{20}	al, eth, ace, bz, chl	B5[4], 311
5889	Decalin, 1-amino (cis) .... $1-H_2NC_{10}H_{17}$	153.27		100[12]	(i) 8, (ii) − 2				B12[3], 178
5890	Decalin, 1-amino (trans) .... $1-H_2NC_{10}H_{17}$	153.27		106[16]	(i) − 18, (ii) − 1				B12[2], 35
5891	Decalin, 1-chloro .... $C_{10}H_{17}Cl$	172.70		d[760], 114-6[20]					B5[4], 313
5892	Decalin, 2-methylene (trans) .... $2-CH_2:C_{10}H_{16}$	150.27		200-1[756], 82[10]		$0.8897^{20/4}$	1.4841^{22}		B5[3], 397
5893	Decalin-1,3-dione (cis) .... $C_{10}H_{14}O_2$	166.22	nd (bz, ace, dil al)	....	124-5			al, ace, bz	B7[3], 3288
5894	Decalin-1,3-dione (trans) .... $C_{10}H_{14}O_2$	166.22	nd (bz, dil al)	....	152-3		....	al	B7[3], 3289
5895	Decalin-2,3-dione (cis) .... $C_{10}H_{14}O_2$	166.22	rh (al, ace, lig)	....	88-9	....	....	al	B7[2], 552
5896	Decalin-2,3-dione (trans) .... $C_{10}H_{14}O_2$	166.22	nd (w), lf (dil al)		100-1			al, lig	B7[3], 3291
5897	2-Decalincarboxylic acid (cis) or Decahydro-2-naphthoic acid .... $2-C_{10}H_{17}CO_2H$	182.27	cr (hx)	150[15]	81			al, eth, bz, chl	B9[3], 233
5898	Decanal or Capraldehyde .... $CH_3(CH_2)_8CHO$	156.27		208-9, 81[7]	ca −5	$0.830^{15/4}$	1.4287^{20}	al, eth, ace	B1[4], 3366
5899	Decane .... $C_{10}H_{22}$	142.29		174.1, 57.6[10]	−29.7	$0.7300^{20/4}$	1.4102^{20}	al, eth	B1[4], 464
5900	Decane, 1-amino or Decylamine .... $CH_3(CH_2)_8CH_2NH_2$	157.30		220.5, 95.8[10]	17	$0.7936^{20/4}$	1.4369^{20}	al, eth, ace, bz, chl	B4[4], 783
5901	Decane, 1-bromo or Decyl bromide.................... $CH_3(CH_2)_8CH_2Br$	221.19		240.6, 110[10]	−29.2	$1.0702^{20/4}$	1.4557^{20}	eth, chl	B1[4], 470
5902	Decane, 1-bromo-10-fluoro .... $F(CH_2)_{10}Br$	239.18		131-2[11]	......	$1.152^{20/4}$	1.4512^{25}	....	B1[4], 471
5903	Decane, 2-bromo (dl) .... $CH_3(CH_2)_7CHBrCH_3$	221.19		111[11]	......	1.0512^{20}	1.4526^{25}	eth, chl	B1[3], 523
5904	Decane, 1-chloro or Decyl chloride .... $CH_3(CH_2)_8CH_2Cl$	176.73		223.4, 97[10]	−31.3	$0.8705^{20/4}$	1.4379^{20}	eth, chl	B1[4], 469
5905	Decane, 1-chloro-10-fluoro .... $F(CH_2)_{10}Cl$	194.72		115[9]		$0.957^{20/4}$	1.4333^{25}	al, eth	B1[4], 469
5906	Decane, 1,10-dibromide or Decamethylene dibromide .... $Br(CH_2)_{10}Br$	300.09	pl (al)	160[15]d, 127-30[4]	28	1.335^{30}	1.4905^{20}	eth	B4[4], 1368
5907	Decane, 1,10-dibromide or Decamethyiene dibromide .... $Br(CH_2)_{10}Br$	300.09	pr (al)	160[15]d, 127-30[4]	28	1.335^{30}	1.4905^{20}	eth	B1[4], 471
5908	Decane, 2,5-dimethyl .... $(CH_3)_2CH(CH_2)_2CHC_5H_{11}$	170.34	$[\alpha]^{25}/_D$ −0.05	122[100]		$0.739^{25/4}$		....	B1[4], 506
5909	Decane, 1-fluoro or Decyl fluoride .... $CH_3(CH_2)_8CH_2F$	160.28	....	186.2, 69[10]	−35	$0.8194^{20/4}$	1.4085	eth	B1[4], 468
5910	Decane, 1-iodo or Decyl iodide .... $CH_3(CH_2)_8CH_2I$	268.18		132[15]	−16.3	$1.2546^{20/4}$	1.4858^{20}	al, eth	B1[4], 472
5911	Decane, 2-methyl .... $(CH_3)_2CH(CH_2)_7CH_3$	156.31	....	189.2	−48.86	0.7368^{20}	1.4154^{20}	....	B1[4], 491
5912	Decane, 3-methyl .... $C_7H_{15}CH(CH_3)C_2H_5$	156.31	....	188.1, 71-2[12]	−92.9	$0.7422^{20/4}$	1.4177^{20}	....	B1[4], 491
5913	Decane, 1-nitro .... $CH_3(CH_2)_8CH_2NO_2$	187.28	....	86[1]			1.4387^{20}		B1[4], 4738536
5914	Decandioic acid or Sebacic acid .... $HO_2C(CH_2)_8CO_2H$	202.25	lf	295[100], 232[10]	134.5	$1.2705^{20/4}$	1.422^{133}	al, eth	B2[4], 2078
5915	1-10-Decanediol or Decamethylene glycol .... $HO(CH_2)_{10}OH$	174.29	nd (w)	175-6[14]	72-5	....	....	al	B1[4], 2613

No.	Name, Synonyms, and Formula	Mol. wt.	Color, crystalline form, specific rotation and λ_{max} (log ϵ)	b.p. °C	m.p. °C	Density	n_D	Solubility	Ref.
5916	1-Decanethiol or Decylmercaptan .... CH$_3$(CH$_2$)$_8$CH$_2$SH	174.35	....	240.6, 125-7[19]	−26	0.8443[20/4]	1.4569[20]	al, eth	B1[4], 1821
5917	Decanoic acid or Capric acid .... CH$_3$(CH$_2$)$_8$CO$_2$H	172.27	nd	270, 148-50[11]	fr 31.5	0.8858[40/4]	1.4288[40]	al, eth, ace, bz, chl	B2[4], 1041
5918	Decanoic acid, 10-fluoro or 10-Fluoro capric acid....... F(CH$_2$)$_9$CO$_2$H	190.26	....	135-8[10]	49			al, eth, lig	B2[4], 1052
5919	Decanoic acid, 2-octyl or 9-Heptadecane carboxylic acid . (C$_8$H$_{17}$)$_2$CHCO$_2$H	284.49	nd or lf (al)	212-8[13]	38.5			al, eth	B2[4], 1254
5920	Decanoic acid, 4-oxo or γ-Ketocapric acid .... C$_6$H$_{13}$COCH$_2$CH$_2$CO$_2$H	186.25	(dil al)	....	70-1			al	B3[4], 1642
5921	1-Decanol or Decyl alcohol................ CH$_3$(CH$_2$)$_8$CH$_2$OH	158.29	....	229, 107-8[7]	fr 7	0.8297[20/4]	1.4372[20]	**al, eth, ace, bz, chl**	B1[4],1815
5922	1-Decanol, 10-chloro .... Cl-(CH$_2$)$_{10}$OH	192.73	....	185-9[15]	12-3	0.9630[25]	1.4578[20]	al, eth	B1[4], 1821
5923	1-Decanol, 10-fluoro .... F(CH$_2$)$_{10}$OH	176.28	....	136-7[15]	ca 22	0.919[20/4]	1.4322[25]	al, eth	B1[4], 1821
5924	2-Decanol (dl).................... CH$_3$(CH$_2$)$_7$CHOHCH$_3$	158.29	....	211, 110-11[10]	−2.4	0.8250[20/4]	1.4326[25]	al, eth, ace, bz	B1[4], 1823
5925	4-Decanol .... C$_3$H$_7$CHOHC$_6$H$_{13}$	158.29	....	210-1, 96[13]	−11	0.8262[20/0]	1.4320[20]	al	**B1[4], 1824**
5926	2-Decanone or Methyl-m-octyl ketone................ CH$_3$(CH$_2$)$_7$COCH$_3$	156.27	nd	210-1[767], 95-7[12]	14	0.8248[20]	1.4255[20]	al, eth	B1[4], 3367
5927	3-Decanone or Ethyl-M-heptyl ketone................ CH$_3$(CH$_2$)$_6$COCH$_2$CH$_3$	156.27	....	203[754]	1-4	0.8251[20/4]	1.4252[20]	al, eth	B1[4], 3368
5928	4-Decanone or Propyl hexyl ketone.................. C$_6$H$_{13}$COC$_3$H$_7$	156.27	nd	206-7, 87-9[11]	-9	0.824[20/D]	1.4240[21]	**al, eth**	B1[4], 3368
5929	Decasiloxane, dicosamethyl .... CH$_3$[Si(CH$_3$)$_2$O]$_9$Si(CH$_3$)$_3$	755.63	....	183[4]	....	0.925[20/4]	1.3988[20]	bz, lig	C47, 4679
5930	1,4,9-Decatriene (trans) .... CH$_2$=CHCH$_2$CH=CH(CH$_2$)$_3$CH=CH$_2$	136.24	....	164-6	....		1.4496[20]	....	
5931	1-Decene .... CH$_3$(CH$_2$)$_6$CH=CH$_2$	140.27	....	170.5, 54.3[10]	fr-66.3	0.7408[20/4]	1.4215[20]	al, eth	B1[3], 858
5932	1-Decene, 2-bromo .... CH$_3$(CH$_2$)$_7$CBr=CH$_2$	219.17	....	115-6[22]	....	1.0844[20/4]	1.4629[20]	....	B1[3], 859
5933	1-Decene-3-yne .... CH$_3$(CH$_2$)$_5$C≡C-CH=CH$_2$	136.24	....	76[20]	....	0.7873[20]	1.4620[20]	....	B1[4], 1105
5934	1-Decene-4-yne .... CH$_3$(CH$_2$)$_4$C≡CCH$_2$CH=CH$_2$	136.24	....	73-4[22]	....	0.7880[20]	1.445[20]	....	B1[3], 1049
5935	2-Decene, 1-bromo .... CH$_3$(CH$_2$)$_6$CH=CHCH$_2$Br	219.17	....	121[17]	....	1.074[18/4]	1.4716[18]	lig	B1[4], 902
5936	2-Decen-4-yne .... CH$_3$(CH$_2$)$_4$C≡C-CH=CHCH$_3$	136.24	....	55[5]	....	0.7850[25/4]	1.4609[25]	....	B1[3], 1049
5937	5-Decene (cis) .... CH$_3$(CH$_2$)$_3$CH=CH(CH$_2$)$_3$CH$_3$	140.27	....	170[739], 73[20]	−112	0.7445[20/4]	1.4258[20]	al, eth	B1[4], 902
5938	5-Decene (trans) .... CH$_3$(CH$_2$)$_3$CH=CH(CH$_2$)$_3$CH$_3$	140.27	....	170.2[739]	−73	0.7401[20/4]	1.4243[20]	al, eth	B1[4], 902
5939	n-Decyl nitrate .... C$_{10}$H$_{21}$ONO$_2$	203.28	....	127-8[11], 88-9[1]	....	0.951[0/4]		al, eth	B1[4], 1819
5940	Decyl nitrite .... C$_{10}$H$_{21}$ONO	187.29	yesh	105-8[12]	....		1.4247[20]	al, eth	B1[4], 1819
5941	Decyl sulfate .... (C$_{10}$H$_{21}$O)$_2$SO$_2$	378.62	....	....	37-8			....	B1[4], 1819
5942	1-Decyne or n-Octyl acetylene .... CH$_3$(CH$_2$)$_7$C≡CH	138.25	....	174, 57[10]	−36	0.7655[20/4]	1.4265[20]	al, eth	B1[4], 1054
5943	3-Decyne .... CH$_3$CH$_2$C≡C(CH$_2$)$_5$CH$_3$	138.25	....	175-6	....	0.765[21/4]	1.433[21]	....	B1[4], 1055
5944	4-Decyne .... C$_4$H$_9$C≡CC$_3$H$_7$	138.25	....	74.5[19]	....	0.772[17/4]	1.436[17]	....	B1[3], 1017
5945	4-Decyne, 3,3-dimethyl .... CH$_3$CH$_2$C(CH$_3$)$_2$C≡C(CH$_2$)$_4$CH$_3$	166.31	....	86[20]	....	0.7731[20/4]	1.4399[20]	....	B1[3], 1026
5946	5-Decyne or Dibutyl acetylene .... C$_4$H$_9$C≡CC$_4$H$_9$	138.25	....	177[751], 788[25]	−73	0.7690[20/4]	1.4331[20]	al, eth	B1[4], 1055
5947	Dehydroacetic acid .... C$_8$H$_8$O$_4$	168.16	nd (w), rh nd or pr (al)	270, 132-3[5]	109			eth, w	B17[2], 524

No.	Name, Synonyms, and Formula	Mol. wt.	Color, crystalline form, specific rotation and λ_{max} (log ε)	b.p. °C	m.p. °C	Density	n_D	Solubility	Ref.
5948	Dehydrochloric acid or 3,7,12-Trioxocholanic acid ...... $C_{24}H_{34}O_5$	402.54	(ace), $[\alpha]^{20}_D$ + 26 (al, c=1.4)	....	237			ace, chl	B10³, 3986
5949	Dehydro ergosterol $C_{28}H_{42}O$	394.65	lf (al + 1w), pl (al), nd (eth), $[\alpha]^{15}_D$ + 149.2 (chl,c=1.9)	230⁰·⁵	146			al, eth, ace, bz, chl	B6³, 3479
5950	Delphinidine chloride or 3,3′,4′,5′,7′-hexahydroxy flavinium chloride. .... $C_{15}H_{11}ClO_7$	338.70	br pr, nd or pl (HCl)	....	>350			w, al	B18², 247
5951	Delphinine $C_{33}H_{45}NO_9$	599.73	orh (al), $[\alpha]^{25}_D$ + 25 (al)		198-200d			al, eth, ace, chl	B2⁴, 2867
5952	Demissine $C_{50}H_{83}NO_{20}$	1050.22	nd (al), $[\alpha]$ −20 Py		276-9			al	B21⁴, 844
5953	Derritol $C_{21}H_{22}O_6$	370.41	ye nd (MeOH)	220-5⁰·⁰⁶	164				B18⁴, 3352
5954	Deserpidine or Canescine .... $C_{32}H_{38}N_2O_8$	578.67	nd or pr $[\alpha]^{24.5}_D$ 137 (chl)	....	229-32			al, chl	Am77, 4335
5955	Desoxycholic acid or 3,12-Dihydroxy cholamic acid ...... $C_{24}H_{40}O_4$	392.59	(al) $[\alpha]^{24}_D$ + 57 (al)		176			al	B10³, 1641
5956	Desoxycorticosterone or Δ⁴-Pregnene-3,20-dion-21-ol.... $C_{21}H_{30}O_2$	330.47	pl (eth), $[\alpha]^{20}_D$ + 178 (al,c=1)		141-2			al, eth, ace	B8³, 2506
5957	Desthiobiotin .... $C_{10}H_{18}N_2O_3$	214.27	lo nd (w), $[\alpha]^{21}_D$ + 10.7 (w,c=2)		156-8			w	J1948, 1552
5958	Desthiobiotin, methyl ester .... $C_{11}H_{20}N_2O_3$	228.29	cr (MeOH), $[\alpha]^{28}_D$ + 2.6 (chl,c=2)	194-7⁰·⁰³	69-70			al	C44, 4934
5959	Dextrin (starch) or Amylin .... $(C_6H_{10}O_5)_x$	(162.14)ₓ	amor $[\alpha]'_D$ > + 200		chars	1.0384²⁰/⁴		w	J1925, 636
5960	Dextropimaric acid, methyl ester .... $C_{21}H_{32}O_2$	316.49	$[\alpha]_D$ + 60.5 (MeOH)	140⁰·⁰³	69	1.030¹⁹/⁴	1.5208¹⁹	al, eth	B9³, 2912
5961	Diacetone alcohol or 4-Methyl-2-pentanon-4-ol .... $(CH_3)_2C(OH)CH_2COCH_3$	116.16	....	164, 67-9¹⁹	−44	0.9387²⁰/⁴	1.4213²⁰	w, al, eth	B1⁴, 4023
5962	Diacetyl disulfide $CH_3COSSCOCH_3$	150.22		105-6¹⁸	20			al, eth	B2⁴, 564
5963	Diallyl amine ...... $(CH_2=CHCH_2)_2NH$	97.16		111			1.4387²⁰	al, eth	B4⁴, 1060
5964	Diallyl trisulfide $(CH_2=CHCH_2)_2S_3$	178.34		112-22¹⁶		1.0845¹⁵		eth	B1, 441
5965	Di(4-aminophenyl)amine .... $(4-H_2NC_6H_4)_2NH$	199.26	lf (w)	d	158			al, eth	B13³, 256
5966	Diaziridine, 3-ethyl-3-methyl $C_4H_{10}N_2$	86.14		32¹⁷			1.4390²⁰		
5967	Diazoacetic acid, ethyl ester or Ethyl diazo acetate .... $N_2CH_2CO_2C_2H_5$	114.10	ye rh	140-1⁷⁵⁰/_d	−22	1.0852¹⁸/⁴	1.4605²⁰	**al, eth, bz, lig**	B3⁴, 1495
5968	Diazoamino benzene or 1,3-Diphenyltriazine .... $C_6H_5N=N-NHC_6H_5$	197.24	ye lf or pr (al)	....	98			al, eth, bz, Py	B16³, 643
5969	Diazoamino benzene, 2,2′-dimethyl $(2-CH_3C_6H_4)N=N-NH(C_6H_4CH_3-2)$	225.30	og (al)		51			al, eth, lig	B16³, 653
5970	Diazoamino benzene, 2′,3-dimethyl $(3-C_6H_3C_6H_4)N=N-NH(C_6H_4CH_3-2)$	225.30	ye cr (lig)		74			ace, bz, lig	B16³, 654
5971	Diazoamino benzene, 2,4′-dimethyl $(2-CH_3C_6H_4)N=N-NH(C_6H_4-CH_3-4)$	225.30	ye nd (lig)		120			lig	B16³, 655
5972	Diazoamino benzene, 3,3′-dimethyl $(3-CH_3C_6H_4)N=N-NH(C_6H_4CH_3-3)$	225.30	ye nd (peth)		52			eth, ace, bz	B16³, 655

No.	Name, Synonyms, and Formula	Mol. wt.	Color, crystalline form, specific rotation and λ_{max} (log ε)	b.p. °C	m.p. °C	Density	n_D	Solubility	Ref.
5973	Diazoamino benzene, 3,4′-dimethyl (3-CH₃C₆H₄)N=NNH(C₆H₄CH₃-4)	225.30	ye nd (lig)	....	97				B16[3], 656
5974	Diazoamino benzene, 4,4′-dimethyl (4-CH₃C₆H₄)N=N-NH(C₆H₄CH₃-4)	225.30	red-ye nd (lig) pr (al)	....	118			al, lig	B16[3], 656
5975	Diazoamino benzene, 4,4′-dinitro (4-O₂NC₆H₄)N=NNH(C₆H₄NO₂-4)	287.24	ye nd (al), lf (bz)	....	240d			eth	B16[3], 650
5976	Diazoamino benzene, 3-methyl (3-CH₃C₆H₄)N=N-NHC₆H₅	211.27	ye nd (lig)	....	86				B16[3], 654
5977	Diazoamino benzene, 4-methyl (4-CH₃C₆H₄)N=NNHC₆H₅	211.27	ye pl (lig)	....	86-7				B16[3], 655
5978	α,α′-Diazoamino naphthalene α-C₁₀H₇N=N-NH-α-C₁₀H₇	297.36	ye lf (al)	....	exp > 100				B16, 716
5979	β,β′-Diazoamino naphthalene β-C₁₀H₇N=N-NH-β-C₁₀H₇	297.36	red nd (xyl)	....	156				Ber19, 1282
5980	Diazoethane, 1,1,1-trifluoro CF₃CHN₂	110.04	ye	13[752]				w, eth	B1[3], 2659
5981	1,4-Diazopine, 1-methyl-perhydro C₆H₁₄N₂	114.19	....	152-5[742]		0.9111[20]	1.4769[20]	....	CAS 59, 7370
5982	1,2:3,4-Dibenzanthracene C₂₂H₁₄	278.36	nd (aa or al)		205			bz	B5[3], 2555
5983	1,2:5,6-Dibenzanthracene C₂₂H₁₄	278.36	pl (dil ace)	....	269-70			ace, bz, aa	B5[3], 2553
5984	1,2:5,6-Dibenzanthracene, 4′, 4″ dihydroxy C₂₂H₁₄O₂	310.36	og (bz)	sub	415-8			bz	B6[3], 5914
5985	1,2:6,7-Dibenzanthracene C₂₂H₁₄	278.36	ye lf or nd (xyl)	sub 275[2-4]	263-4				B5[3], 2552
5986	1,2:7,8-Dibenzanthracene C₂₂H₁₄	278.36	og lf or nd (bz)	....	197-8			peth	B5[3], 2553
5987	Dibenzanthrone or Violanthrone C₃₄H₁₆O₂	456.51	vt-bl or bk nd (PhNO₂)		490-5d				B7[3], 4539
5988	2,3:6,7-Dibenzocycloheptadiene-5-one C₁₅H₁₂O	208.26	....	203-4[7]	30	1.1635[20]	1.6324[20]	....	....
5989	1,2:5,6-Dibenzofluorene C₂₁H₁₄	266.35	pl (bz-al)	195-200[0.3]	174-5			bz	B5[3], 2527
5990	1,2:7,8-Dibenzo fluorene C₂₁H₁₄	266.35	lf (bz), pl (aa)		234			bz	B5[3], 2527
5991	1,2:6,7-Dibenzo-9-fluorenone C₂₂H₁₂O	280.33	og-ye pl or pr (aa or xyl)	sub 190-200[0.04]	214				B7[3], 2897
5992	Dibenzofuran or Diphenylene oxide C₁₂H₈O	168.21	lf or nd (al)	287	86-7	1.0886[99/4]	1.6079[99]	al, eth, ace, aa	B17[4], 585
5993	Dibenzofuran, 1-amino 1-H₂N(C₁₂H₇O)	183.21	br nd (dil MeOH)	....	85				B18[4]7183
5994	Dibenzofuran, 2-amino 2-H₂N(C₁₂H₇O)	183.21	pl (dil al)		128			al, eth	B18[4]7184
5995	Dibenzofuran, 3-amino 3-H₂N(C₁₂H₇O)	183.21	(dil al)	....	94 (99)				B18[4], 7191
5996	Dibenzofuran, 4-amino 4-H₂N(C₁₂H₇O)	183.21	(al)	....	85				B18[4], 7211
5997	Debenzofuran, 2-bromo 2-Br(C₁₂H₇O)	247.0	nd (al), lf (aa)	220[40]	110			al	B17[4], 588
5998	Dibenzofuran, 3-bromo 3-Br(C₁₂H₇O)	247.10	lf (al)	220[40]	120			al, eth	B17[4], 588
5999	Dibenzofuran, 4-bromo 4-Br(C₁₂H₇O)	247.10	....		67				B17[4], 589
6000	Dibenzofuran, 2,8-dibromo 2,8-Br₂(C₁₂H₆O)	326.00	lf (al)	....	199-200			al, eth, bz, aa	B17[4], 589
6001	Dibenzofuran, 3,6-dinitro 3,6-(NO₂)₂(C₁₂H₆O)	258.19		....	245				B17[4], 595
6002	Dibenzofuran, 3,8-dinitro 3,8-(NO₂)₂(C₁₂H₆O)	258.19			225-6			ace, bz	B17[4], 594
6003	Dibenzofuran, 1-nitro 1-O₂N(C₁₂H₇O)	213.20	lt ye nd (al)		120-1				B17[4], 591
6004	Dibenzofuran, 3-nitro 3-O₂N(C₁₂H₇O)	213.20	ye nd (aa)	180-5[3]	181-2			aa	B17[4], 591

No.	Name, Synonyms, and Formula	Mol. wt.	Color, crystalline form, specific rotation and λ$_{max}$ (log ε)	b.p. °C	m.p. °C	Density	n_D	Solubility	Ref.
6005	Dibenzofuran, 4-nitro 4-O$_2$N(C$_{12}$H$_7$O)	213.20	ye nd	190-205[15]	138-9				B17[4], 591
6006	1-Dibenzofuran carboxylic acid 1-(C$_{12}$H$_7$O)CO$_2$H	212.22	nd (50% al)		232-3				B18[4], 4340
6007	2-Dibenzo furan carboxylic acid 2-(C$_{12}$H$_7$O)CO$_2$H	212.22	nd (dil al aa)		246-7 (252)			al, eth	B18[4], 4341
6008	4-Dibenzofuran carboxylic acid 4-(C$_{12}$H$_7$O)CO$_2$H	212.22	nd (al)		209-10				B18[4], 4346
6009	1,2:6,7-Dibenzophenanthrene C$_{22}$H$_{14}$	278.36	pa gr-ye lf (xyl)		294			bz, diox	B5[3], 2552
6010	2,3:6,7-Dibenzophenanthrene C$_{22}$H$_{14}$	278.36	ye-gr nd or lf (xyl)		257			al, bz	B5[3], 2552
6011	1,2:4,5-Dibenzopyrene C$_{24}$H$_{14}$	302.38	pa ye nd (xyl)		233-4				B5[3], 2621
6012	Dibenzothiophene or Diphenylene sulfide C$_{12}$H$_8$S	184.27	nd (dil al or lig)	332-3, 152-4[3]	99-100			al, bz	B17[4], 601
6013	Dibenzothiophene, 2-amino 2-H$_2$N(C$_{12}$H$_7$S)	199.28	lf (dil al)		122-3				B18[4], 7186
6014	Dibenzothiophene, 3-amino 3-H$_2$N(C$_{12}$H$_7$S)	199.28	(dil al)		129-31				B18[4], 7204
6015	Dibenzothiophene, 2-bromo 2-Br(C$_{12}$H$_7$S)	263.16	nd (al)		125-6				B17[4], 605
6016	Dibenzothiophene, 2-bromo-monoxide 2-Br(C$_{12}$H$_7$SO)	279.16			171-2				B17[2], 71
6017	Dibenzothiophene, 3-bromo, dioxide 3-Br (C$_{12}$H$_7$SO$_2$)	295.16			224-5				B17[4], 605
6018	Dibenzothiophene, 4-bromo 4-Br(C$_{10}$H$_7$S)	263.16	(al)		84				B17[4], 605
6019	Dibenzothiophene, 2,8-diamino 2,8-(NH$_2$)$_2$(C$_{12}$H$_6$S)	214.29	nd (al)		194-6			al	B18[4], 7286
6020	Dibenzothiophene, 3,7-diamino 3,7-(NH$_2$)$_2$(C$_{12}$H$_6$S)	214.29	pa ye cr		169-70				B18[4], 7289
6021	Dibenzothiophene, 3,7-diamino, dioxide 3,7-(NH$_2$)$_2$(C$_{12}$H$_6$SO$_2$)	246.29	ye nd (al)		327-8				B18[4], 7289
6022	Dibenzothiophene, 2,8-dibromo 2,8-Br$_2$(C$_{12}$H$_6$S)	342.06	(aa)		229				B17[4], 605
6023	Dibenzothiophene, 2,8-dibromo, dioxide 2,8-Br$_2$(C$_{12}$H$_6$SO$_3$)	374.06	(aa)		361-2				B17[4], 606
6024	Dibenzothiophene, 3,7-dinitro, dioxide 3,7-(NO$_2$)$_2$(C$_{12}$H$_6$SO$_2$)	306.26	(ace)		273-5				B17[4], 608
6025	Dibenzothiophene, 2-nitro 2-O$_2$N(C$_{12}$H$_7$S)	229.26	pa ye nd		186				B17[4], 606
6026	Dibenzothiophene, 2-nitro, dioxide 2-O$_2$N(C$_{12}$H$_7$SO$_2$)	261.26	(ace)		257-8				B17[4], 607
6027	Dibenzothiophene, 3-nitro 3-O$_2$N(C$_{12}$H$_7$S)	229.26	pa ye (dil al)		153-4				B17[4],607
6028	Dibenzothiophene, 3-nitro, monoxide 3-O$_2$N(C$_{12}$H$_7$SO)	245.26	(al)		210				B17[4], 607
6029	2-Dibenzothiophene carboxylic acid 2-(C$_{12}$H$_7$S)CO$_2$H	228.28	(al)		255				B18[4], 4344
6030	4-Dibenzothiophene carboxylic acid 4-(C$_{12}$H$_7$S)CO$_2$H	228.28	(dil MeOH)		261-2				B18[4], 4349
6031	Dibenzoyl disulfide C$_6$H$_5$COSSCOC$_6$H$_5$	224.36	pr (al) sc (chl-peth)	d	136				B9[3], 1977
6032	Dibenzoylmethane (C$_6$H$_5$CO)$_2$CH$_2$	224.26			70-1			eth, chl	B7[3], 3838
6033	Dibenzoyl methane, (enol form) C$_6$H$_5$COCH=C(OH)C$_6$H$_5$	224.26	rh bipym (eth)	219-21[18]	78-9			al, eth, chl	B7[3], 3838
6034	Dibenzoylmethane, keto form (C$_6$H$_5$CO)$_2$CH$_2$	224.26	nd or pl (eth)		81			al, eth, chl	B7[3], 3838
6035	Dibenzoylmethane, α,α-dibromo (C$_6$H$_5$CO)$_2$CBr	382.06	pr (eth)		95				B7[3], 3846
6036	Dibenzoylmethane, oxo or Diphenyl triketone (C$_6$H$_5$CO)$_2$CO	238.25	ye nd (lig)	289[175], 248[60]	68-70			eth	B7[3], 4620
6037	Dibenzyl acetamide (C$_6$H$_5$CH$_2$)$_2$CHCONH$_2$	239.32	nd (al, w)	259[18]	129			al, eth	B9[2], 476

No.	Name, Synonyms, and Formula	Mol. wt.	Color, crystalline form, specific rotation and λ_{max} (log ε)	b.p. °C	m.p. °C	Density	n_D	Solubility	Ref.
6038	Dibenzylacetic acid .... $(C_6H_5CH_2)_2CHCO_2H$	240.31	pl (peth, dil aa) nd (w)	235[18]	89			al, eth, bz, chl, aa	B9[3], 3356
6039	Dibenzyl acetic acid, methyl ester or Methyldibenzyl acetate .... $(C_6H_5CH_2)_2CHCO_2CH_3$	254.33	nd (al)	....	42-3				B9[2], 475
6040	Dibenzyl acetonitrile .... $(C_6H_5CH_2)_2CHCN$	221.31	lf or pl (al)		89-91				B9[3], 3359
6041	Dibenzyl amine .... $(C_6H_5CH_2)_2NH$	197.28		300d, 270[250]	-26	1.0256[22/4]	1.5731[20]	al, eth	B12[2], 2221
6042	Dibenzyl disulfide .... $C_6H_5CH_2SSCH_2C_6H_5$	246.40	lf (MeOH or al) nd (aa)		71-2			al, eth, bz, MeOH	B6[4], 2760
6043	Dibenzylphenyl amine .... $(C_6H_5CH_2)_2NC_6H_5$	273.38	nd or pr (al)	> 300d, 226[10]	71-2	1.0444[80/4]	1.6065[80]	eth, bz	B12[3], 2225
6044	Dibenzyl sulfone or Benzylsulfene .... $(C_6H_5CH_2)_2SO_2$	246.33	nd (al-bz)	290d	155			ace, bz, aa	B6[4], 2651
6045	Dibenzyl sulfoxide .... $(C_6H_5CH)_2SO$	230.33	lf (al or w)	210d	134-5			al, eth	B6[4], 2651
6046	Diborane, methylthio .... $CH_3SB_2H_5$	73.76		53	-101.5			ace, bz	B1[4], 1287
6047	Dibromo acetamide, N,N-dimethyl .... $Br_2CHCON(CH_3)_2$	244.93	pr (w or eth)	128[16]	79-80				B4, 59
6048	Dibromoacetic acid .... Br_2CHCO_2H	217.86	dlq cr	195[250]	48			w, al, eth	B2[4], 533
6049	Dibromoacetic acid, ethyl ester or Ethyl dibromo acetate .... $Br_2CHCO_2C_2H_5$	245.91		194, 121[74]		1.9025[20/20]	1.5017[13]	al, eth	B2[4], 533
6050	Dibromoacetic acid, methyl ester or Methyl dibromo acetate .... $Br_2CHCO_2CH_3$	231.88		182-3				al, eth	B2, 219
6051	Dibutyl amine .... $(C_4H_9)_2NH$	129.25		159, 48[13]	-60	0.7670[20/4]	1.4177[20]	w, al, eth, ace, bz	B4[4], 550
6052	Diisobutyl amine .... $(i-C_4H_9)_2NH$	129.25		139-40	-73.5		1.4090[20]	al, eth, ace, bz	B4[4], 630
6053	Di-sec-butyl amine .... $(Sec-C_4H_9)_2NH$	129.25		135[765]		0.7534[20/4]	1.4162[20]	w, al	B4[4], 620
6054	Dibutyl disulfide .... $C_4H_9SSC_4H_9$	178.36		226, 85[3]		0.9383[20/4]	1.4926[20]	al, eth	B1[4], 1562
6055	Dibutyl sulfone or Butyl sulfone .... $(C_4H_9)_2SO_2$	178.30	pl (w or al)		46			al, eth	B1[4], 1561
6056	Dibutyl sulfoxide .... $(C_4H_9)_2SO$	162.30	nd (dil al)	d	32.6	0.8317[23/4]	1.4669[20]	al, eth	B1[4], 1561
6057	Dichloroacetaldehyde .... Cl_2CHCHO	112.94		90-1				al	B1[4], 3140
6058	Dichloroacetaldehyde, diethyl acetal .... $Cl_2CHCH(OC_2H_5)_2$	187.07		183-4, 67-71[12]		1.1383[14]		al, eth	B1[4], 3140
6059	Dichloroacetamide .... $Cl_2CHCONH_2$	127.96	mcl pr (w)	233-4[745] sub	99.4			al, eth	B2[4], 505
6060	Dichloroacetic acid .... Cl_2CHCO_2H	128.94		194	13.5	1.5634[20/4]	1.4658[20]	w, al, eth, ace	B2[4], 498
6061	Dichloroacetic acid, anhydride or Dichloroacetic anhydride .... $(Cl_2CHCO)_2O$	239.87		214-6d		1.574[24]			B2[4], 503
6062	Dichloroacetic acid, butyl ester or Butyl dichloroacetate . $Cl_2CHCO_2C_4H_9$	185.05		193-4, 102[37]		1.1820[20/4]	1.4420[20]	al, eth	B2[4], 502
6063	Dichloroacetic acid, ethyl ester or Ethyl dichloroacetate . $Cl_2CHCO_2C_2H_5$	157.00		155.5[764], 56[10]		1.2827[20/4]	1.4386[20]	al, eth, ace	B2[4], 501
6064	Dichloroacetic acid, 2-hydroxy ethyl ester or (2-Hydroxyethyl) dichloroacetate .... $Cl_2CHCO_2CH_2CH_2OH$	173.00		81-2[0.5]		1.438[20/4]	1.4735[20]	al	B2[3], 460
6065	Dichloroacetic acid, methyl ester or Methyl dichloroacetate .... $Cl_2CHCO_2CH_3$	142.97		142.8, 38[10]	-51.9	1.3774[20/4]	1.4429[20]	al	B2[4], 501
6066	Dichloroacetic acid, propyl ester or Propyl dichloro acetate .... $Cl_2CHCO_2C_3H_7$	171.03		176		1.2240[20]	1.4398[20]	al, eth	B2[4], 502
6067	Dichloroacetic acid, iso-propyl ester or iso-Propyl dichloro acetate .... $Cl_2CHCO_2CH(CH_3)_2$	171.03		163-4		1.2053[20/4]	1.4328[20]	al, eth	B2[4], 502

No.	Name, Synonyms, and Formula	Mol. wt.	Color, crystalline form, specific rotation and λmax (log ε)	b.p. °C.	m.p. °C	Density	n_D	Solubility	Ref.
6068	Dichloro acetonitrile Cl₂CHCN	109.94	....	112-3	......	1.369[20]	1.4391[25]	al	B2[4], 506
6069	Dichloroacetyl chloride Cl₂CHCOCl	147.39	....	108-10		1.5315[16/4]	1.4591[20]	eth	B2[4], 504
6070	Di-(β-chloroethyl) methyl amine, hydrochloride (ClCH₂CH₂)NCH₃.HCl	192.52	hyg nd	....	111-2			w, al	B4[4], 446
6071	Dichloro fluoro acetic acid Cl₂FCCO₂H	146.93	....		162.5				B2[4], 507
6072	Dicoumarin-(cis) C₁₈H₁₀O₄	290.27	lf (aa)		262				B19[4], 2106
6073	Dicoumarin-(trans) C₁₈H₁₀O₄	290.27	nd or pl (aa)		>275				B19[4], 2106
6074	Dictamnine Cl₂H₉NO₂	199.21	pr (al)		133-4			al, eth, chl	B27[2], 79
6075	Dicumarol or 4,4′-Dihydroxy-3,3′-methylene bis coumarin C₁₉H₁₂O₆	336.31	nd		288-92				B19[4], 2261
6076	Dicyclohexadiene C₁₂H₁₆	160.26	....	229-30, 104[16]		0.9950[20/4]	1.5267[20.5]	eth, ace, bz, aa	B5[3], 1267
6077	Dicyclohexylamine (C₆H₁₁)₂NH	181.33		255.8d, 113-5[9]	-0.1	0.9123[20/4]	1.4842[20]	al, eth, bz	B12[3], 19
6078	Dicyclohexyl ketone (C₆H₁₁)₂CO	194.32		159[20]	....	0.986[20]	1.1060[20]	eth, ace	B7[4], 494
6079	Dicyclohexyl methane, 4,4′-diamino (cis,cis) (4-H₂NC₆H₁₀)₂CH₂	210.37	....	141[2]	60-2		1.5014[27]		Am73, 641
6080	Dicyclohexyl methane, 4,4′-diamino (cis,trans) (4-H₂NC₆H₁₀)₂CH₂	210.37	rr	141.9[4]	36.7	0.9600[35/4]	1.5016[35]		Am73, 741
6081	Dicyclohexyl methane, 4,4′-diamino (trans,trans) C₁₃H₂₆N₂	210.37	cr (peth)	130-1[0.3]	64-5	....	1.5032[25]		Am73, 741
6082	a-Dicyclopentadiene (endo form) C₁₀H₁₂	132.31	....	170d, 64-5[14]	32	0.9302[35/4]	1.5050[35]	al, eth, aa	B5[4], 1399
6083	a-Dicyclopentadiene, 3,4,5,6,7,8,8a-heptachloro or Heptachlor C₁₀H₅Cl₇	373.32	wh	....	95-6	1.57[9]	....	al, eth, bz, lig	B5[3], 1236
6084	a-Dicyclopentadiene, tetrahydro or Tricyclodecane C₁₀H₁₆	136.24	(al or aa)	193[769], 86-7[12]	77	0.9128[79]	1.4726[79]	al, aa	B5[4], 467
6085	Didodecylamine or Dilaurylamine (C₁₂H₂₅)₂NH	353.68			55-6			al, eth, bz, chl	B4[4], 801
6086	Dieldrin or Octalox C₁₂H₈OCl₆	380.95			175-6	1.75		ace, bz	B17[4], 526
6087	Diethanol amine (HOCH₂CH₂)₂NH	105.14	....	271, 154-5[10]	28	1.0966[20/4]	1.4776[20]	w, al	B4[4], 1514
6088	Diethanol amine, N-phenyl (HOCH₂CH₂)₂NC₆H₅	181.24	pl (al)	228[15]	58			al, eth, ace, bz	B12[3], 299
6089	Diethoxyacetic acid, ethyl ester or Ethyldiethoxy acetate (C₂H₅O)₂CHCO₂C₂H₅	176.22		199, 83-5[13]		0.994[18]	1.4089[25]	al, eth	B3[4], 1494
6090	Diethoxy disulfide C₂H₅OSSOC₂H₅	154.25	....	67-8[16]		1.0913[20/4]	1.4766[20]	....	B1[4], 1324
6091	Diethoxy sulfide (C₂H₅O)₂S	122.19	....	117[733]		0.9940[20/4]	1.4234[20]	w, al, eth, ace, bz	B1[3], 1314
6092	Diethyl amine (C₂H₅)₂NH	73.14	....	56.3	-48	0.7056[20/4]	1.3864[20]	w, al, eth	B4[4], 313
6093	Diethyl amine, hydrochloride (C₂H₅)₂NH.HCl	109.60	lf (al-eth)	320-30	227-30	1.0477[22/4]	....	w, al	B4[4], 318
6094	Diethyl amine, N-nitro (C₂H₅)₂N.NO₂	118.14	....	206.5[757], 93[16]		1.057[15]		al, eth	B4[3], 233
6095	Diethyl amine, N-nitroso (C₂H₅)₂N.NO	102.14	ye	176.9		0.9422[20/4]	1.4386[20]	w, al, eth	B4[3], 233
6096	Diethyl (B-bromo ethyl) amine, hydrobromide (C₂H₅)₂NC H₂CH₂Br.HBr	261.01	nd (al-eth)	....	209			w	B4[3], 249
6097	Diethyl (2,2-diethoxy ethyl) amine or Diethylamino acetal (C₂H₅)₂NCH₂CH(OC₂H₅)₂	189.30	....	194-5		0.863	1.4189[20]	w, al, eth	B4[4], 1919
6098	Diethyl disulfide C₂H₅SSC₂H₅	122.25	....	154	-101.5	0.9931[20/4]	1.5073[20]	al, eth	B1[4], 1397
6099	Diethyl (methoxymethyl) amine (C₂H₅)₂NCH₂OCH₃	117.19	....	117[763]				w, al, eth	B4[3], 203
6100	Diethyl methyl amine (C₂H₅)₂NCH₃	87.17	....	66	......	0.703[25/4]	1.3879[25]	w, al, eth	B4[4], 321

No.	Name, Synonyms, and Formula	Mol. wt.	Color, crystalline form, specific rotation and λ_{max} (log ε)	b.p. °C	m.p. °C	Density	n_D	Solubility	Ref.
6101	Diethyl selenide $(C_2H_5)_2Se$	137.06	pa ye	108	......	$1.2300^{20/4}$	1.4768^{20}	al, eth, bz	B1[4], 1411
6102	Diethyl sulfide $(C_2H_5)_2S$	90.19	....	92.1	-103.8	$0.8362^{20/4}$	1.4430^{20}	al, eth	B1[4], 1394
6103	Diethyl sulfide, 2,2-diamino or bis(2-Aminoethyl) sulfide $(H_2NCH_2CH_2)_2S$	120.22	ye	$231-3^{755}$, $118-20^{17}$	......			......	B4[4], 1577
6104	Diethyl sulfide, 2,2'-dichloro or bis-(2-Chloroethyl) sulfide mustard gas $(ClCH_2CH_2)_2S$	159.09	ye pr	217, 95^{10}	13-4	$1.2741^{20/4}$	1.5312^{20}	al, eth, ace, bz	B1[4], 1407
6105	Diethyl sulfide, 2,2'-dihydroxy or bis-(2-Hydroxyethyl) sulfide $(HOCH_2CH_2)_2S$	122.19	....	$164-6^{20}$	-10	$1.1819^{20/4}$	1.5203^{20}	w, al, eth, chl	B1[4], 2437
6106	Diethyl sulfide, 2,2'-diphenoxy or bis-(2-Phenoxyethyl) sulfide $(C_6H_5OCH_2CH_2)_2S$	274.39	nd (al)	....	42			al, eth	B6[4], 581
6107	Diethyl sulfone or Ethyl sulfone $(C_2H_5)_2SO_2$	122.19	rh pl	248	73-4	$1.357^{20/4}$	....	w, bz	B1[4], 1396
6108	Diethyl sulfoxide $(C_2H_5)_2SO$	106.19	syr	104^{25}	14			w, al, eth	B1[4], 1395
6109	bis (Diethylthiocarbamyl) disulfide or Antabuse $[(C_2H_5)_2NCS]_2S_2$	296.54	(al)	117^{17}	71-2			al, chl	B4[4], 398
6110	Diethyl trisulfide $C_2H_5S_3C_2H_5$	154.32	ye	$96-7^{26}$	......	1.114^{20}	1.5689^{13}	....	B1[4], 1398
6111	Diethylene glycol $HOCH_2CH_2OCH_2CH_2OH$	106.12	....	245, 133^{14}	-10.5	$1.1197^{15/4}$	1.4472^{20}	w, al, eth	B1[4], 2390
6112	Diethylene glycol, monobutyl ether or Butylcarbitol $C_4H_9OCH_2CH_2OCH_2CH_2OH$	162.29	....	231, 118^{12}	-68.1	$0.9553^{20/4}$	1.4321^{20}	w, al, eth, ace, bz	B1[4], 2394
6113	Diethylene glycol, monobutyl ether, acetate $C_4H_9OCH_2CH_2OCH_2CH_2O_2CCH_3$	204.27	....	245	-32	$0.985^{20/4}$	1.4262^{20}	w, al, eth, ace	B2[3], 308
6114	Diethylene glycol, diacetate $CH_3CO_2CH_2CH_2OCH_2CH_2O_2CCH_3$	190.20	....	245-51, $110-35^{16}$	......	$1.1078^{15/15}$	1.4348^{20}	w, al, eth	B2[4], 216
6115	Diethylene glycol, dibenzyl ether $(C_6H_5CH_2OCH_2CH_2)_2O$	314.34	....	$279-81^{24}$, 250^1	33.5	$1.1701^{15/15}$		w, al	B9[3], 535
6116	Diethylene glycol, diisobutyrate $(i-C_3H_7CO_2CH_2CH_2)_2O$	246.30	....	$178-80^{40}$	......		1.4282^{20}	....	B2[3], 652
6117	Diethylene glycol, diethyl ether or Diethyl carbitol $(C_2H_5OCH_2CH_2)_2O$	162.23	....	189	......	$0.9063^{20/4}$	1.4115^{20}	w, al, eth	B1[4], 2394
6118	Diethylene glycol, dimethyl ether $(CH_3OCH_2CH_2)_2O$	134.17	....	162	-68	$0.9451^{20/20}$		w, al, eth	B1[4], 2393
6119	Diethylene glycol, di-octadecanoate or Diethylene glycol disfearate $(C_{17}H_{35}CO_2CH_2CH_2)_2O$	611.02	wax	....	54-5	$0.9333^{20/4}$			B2[4], 1223
6120	Diethylene glycol, dioleate $(C_{17}H_{33}CO_2CH_2CH_2)_2O$	635.04	pa ye oil	....	......	$0.9310^{20/4}$		al, eth	C47, 4618
6121	Diethylene glycol, monododecanoate or Diethylene glycol monolaurate $C_{11}H_{23}CO_2CH_2CH_2OCH_2CH_2OH$	288.43	lt ye	> 270	17-8	$0.96^{25/25}$		al, eth, ace, bz	B2[3], 887
6122	Diethylene glycol monoethyl ether or Ethylcellosolve $C_2H_5OCH_2CH_2OCH_2CH_2OH$	134.19	hyg lig	195	......	$0.9881^{20/4}$	1.4300^{20}	w, al, eth, ace, bz	B1[4], 2377
6123	Diethylene glycol, monoethyl ether, acetate or Carbitol acetate $C_2H_5OCH_2CH_2OCH_2CH_2O_2CCH_3$	176.22	....	218	-25	$1.0096^{20/4}$	1.4230^{25}	w, al, eth, ace	B2[3], 308
6124	Diethylene glycol, (2-hydroxypropyl) ether $[(CH_3CH(OH)CH_2]OCH_2CH_2OCH_2CH_2OH$	164.20	....	277-9	......	$1.0789^{20/4}$	1.4498^{20}	w, al, bz	C52, 17693
6125	Diethylene glycol, methyl ether or Methyl carbitol $CH_3OCH_2CH_2OCH_2CH_2OH$	120.15	....	193	......	$1.0270^{20/4}$	1.4264^{20}	w, al, eth, ace	B1[4], 2392
6126	Diethylene triamine or bis-(2-Aminoethyl) amine $H_2NCH_2CH_2NHCH_2CH_2NH_2$	103.17	ye hyg liq	207	-39	$0.9586^{20/20}$	1.4810^{25}	w, al, lig	B4[3], 1238
6127	Difluoroacetic acid F_2CHCO_2H	96.03	....	134.2, $67-70^{20}$	-0.3	1.5255^{20}	1.3420^{20}	w, al, eth, ace, bz	B2[4], 455
6128	Difluoroacetic acid, ethyl ester or Ethyl difluoro acetate $F_2CHCO_2C_2H_5$	124.09	....	99.2^{750}	......	$1.1893^{9.8}$			B2[4], 455
6129	Difurfuryl amine $(OC_4H_3CH_2)_2NH$	177.21	....	$135-42^{15}$	......	$1.1045^{20/4}$	1.5168^{20}	eth	B18[4], 7092
6130	Di(2-furfuryl) disulfide $[(2-OC_4H_3)CH_2]_2S_2$	226.32	....	$112-3^{0.5}$	10	....		al	B17[4], 1258
6131	m-Digallic acid or Gallicacid-3-monogallate $C_{14}H_{10}O_9$	332.23	nd (dil al + 1w)	....	268-70d	....		al, ace	B10[3], 2086

No.	Name, Synonyms, and Formula	Mol. wt.	Color, crystalline form, specific rotation and λ_{max} (log ε)	b.p. °C	m.p. °C	Density	n_D	Solubility	Ref.
6132	Digitalose or 3-o-Methyl-D-fucose .... $C_7H_{14}O_5$	178.19	nd (AcOEt) $[\alpha]_D^{22}$ +109→ +126 (mut)		106→ 119			w	B1[4], 4270
6133	Digitogenin or 5α,22α-Spirostan-2,3,15-triol .... $C_{27}H_{44}O_5$	448.65	nd (al) $[\alpha]_D$ −18 (chl, c=1.4)		280-3			chl	B19[4], 1242
6134	Digitoxigenin .... $C_{23}H_{34}O_4$	374.53	(dil MeOH) $[\alpha]_D^{20}$ +119.1 (MeOH, c=1.36)		253 (256)			al	B18[4], 1468
6135	Digitoxin .... $C_{41}H_{64}O_{13}$	764.96	wh (chl-eth) pr (dil al) $[\alpha]_D^{20}$ +4.8 (diox, c=1.2)		255-6			al, eth, chl	B18[4], 1478
6136	Digitoxose or 2-Deoxy-D-$altro$-Methylose .... $C_6H_{12}O_4$	148.16	cr (MeOH-eth) $[\alpha]_D^{15}$ +27.9→ 43.3 (Pyr=1, mut)		112			w, ace, py	B1[4], 4191
6137	Diglycolic acid or Oxydiethanoic acid.................... $O(CH_2CO_2H)_2$	132.11	mcl pr (w + 1)	u	148			w, al, eth	B3[4], 577
6138	Diglycolyl chloride .... $O(CH_2COCl)_2$	170.99	....	116[15]				chl	B3, 240
6139	Diglycolic acid, hydrate .... $O(CH_2CO_2H)_2 \cdot H_2O$	152.11	mcl pr (w + 1)	d	148			w, al, eth	B3, 234
6140	Digoxigenin .... $C_{23}H_{34}O_5$	390.53	pr (AcoEt)		222			al, MeOH	B18[4], 2450
6141	Diheptyl amine .... $(C_7H_{15})_2NH$	213.41	nd	271[750], 134-6[9]	30			al, eth	B4[4], 736
6142	Diheptyl sulfide .... $(C_7H_{15})_2S$	230.46	....	298, 164[20]	....	0.8416[20/4]	1.4606[20]	eth	B1[4], 1739
6143	Dihexyl amine .... $(C_6H_{13})_2NH$	185.36	....	192-5, 112-4[17]			1.4339[20]	al, eth	B4[4], 711
6144	Dihexyl sulfide .... $(C_6H_{13})_2S$	202.41	....	230, 113.5[4]	....	0.8411[20/4]	1.4586[20]	....	B1[4], 1706
6145	Dihydrosamidin .... $C_{21}H_{24}O_7$	388.42	$[\alpha]_D$ +19(al)	....	117-9			al, eth	B19[4], 2788
6146	3,4-Dihydroxybenzoic acid, 5-bromo .... 5-Br-3,4-(HO)$_2$C$_6$H$_2$CO$_2$H	233.03	nd (w, aa, dil al)		230			aa	B10[1], 192
6147	2,3-Dihydroxybenzoic acid, 5-bromo .... 5-Br-2,3-(HO)$_2$C$_6$H$_2$CO$_2$H	233.03	pr (w + 1) nd (w)		187 (pr) 215 (nd)			al, eth, aa	B10[3], 1367
6148	2,4-Dihydroxybenzoic acid, 3-bromo .... 3-Br-2,4-(HO)$_2$C$_6$H$_2$CO$_2$H	233.03	br or ye nd (w)	....	202			aa	B10[2], 254
6149	2,4-Dihydroxybenzoic acid, 5-bromo .... 5-Br-2,4-(HO)$_2$C$_6$H$_2$CO$_2$H	233.03	micr pr (w + 1)		212			al, eth	B10[3], 1378
6150	Di-(β-Hydroxyethyl) methyl amine .... $(HOCH_2CH_2)_2NCH_3$	119.17	....	246-8[747], 123-5[4]	....	1.0377[20]	1.4642[20]	w, al	B4[4], 1571
6151	Diiodoacetic acid .... I_2CHCO_2H	311.85	lf ye cr or wh nd (bz)		110			w, al, eth, bz	B2[4], 537
6152	Diisoeugenol .... $C_{24}H_{32}O_4$	328.41	nd(bz liq or al)		180-1			al, eth, chl	B6[3], 6765
6153	Dilactic acid .... $[HO_2CCH(CH_3)]_2O$	162.14	rh		112-3			w, eth	B3[3], 468
6154	Dilauryl amine .... $(C_{12}H_{25})_2NH$	353.68	....		55-6			al, eth, bz, chl	B4[4], 801
6155	Dimethisoquin hydrochloride or Quotane .... $C_{17}H_{24}N_2O$·HCl	308.36	....		146			w, al	B21[4], 1335
6156	Dimethyl amine .... $(CH_3)_2NH$	45.09	....	7.4	−93	0.6804[0/4]	1.350[17]	w, al, eth	B4[4], 128

No.	Name, Synonyms, and Formula	Mol. wt.	Color, crystalline form, specific rotation and λ_{max} (log ε)	b.p. °C	m.p. °C	Density	n_D	Solubility	Ref.
6157	Dimethyl amine hydrochloride or Dimethylammonium chloride $(CH_3)_2NH.HCl$	81.56	rh, nd (al)	....	171	....		w, al, chl	B4[4], 132
6158	Dimethyl amine, hexafluoro .... $(CF_3)_2NH$	153.03	....	−6.7	−130				B3[4], 78
6159	Dimethylamine, perfluoro $(CF_3)_2NF$	171.02	....	−37					B3[4], 170
6160	Dimethylamine, N-nitro .... $(CH_3)_2N-NO_2$	90.08	nd(eth)	187	58	$1.1090^{72/4}$	1.4462^{72}	w, al, eth, ace, bz	B4[3], 167
6161	Dimethyl amine, N-nitroso or Dimethylnitrosamine ..... $(CH_3)_2$ -N-NO	74.08	ye	154		$1.0059^{20/4}$	1.4358^{20}	w, al, eth	B4[3], 166
6162	3,4-Dimethylbenzoic acid, ethyl ester $3,4-(CH_3)_2C_6H_3CO_2C_2H_5$	178.23		$127-8^{10}$			1.5144^{20}		B9[2], 353
6163	Dimethyl isobutyl amine .... $i-C_4H_9N(CH_3)_2$	101.19		80-1		$0.7097^{20/4}$	1.3907^{20}	w	B4[4], 627
6164	Dimethyl disulfide CH_3SSCH_3	94.20		109.7, 6.4^{10}	−84.7	$1.0625^{20/4}$	1.5289^{20}	**al, eth**	B1[4], 1281
6165	Dimethyl disulfide, hexafluoro .... CF_3SSCF_3	202.14		34.6	>1			al, peth	B3[4], 278
6166	Dimethyl ethyl amine $(CH_3)_2NC_2H_5$	73.14		36-7		1.3705^{25}			B4[4], 312
6167	Dimethyl glyoxime $CH_3C(:NOH)C(:NOH)CH_3$	116.13	nd (to or dil al)	sub 234	245-6			al eth	B1[4], 3647
6168	Dimethyl hexesterol or 3,4-bis(4-hydroxy-3-methylphenyl) hexane $(C_{10}H_{13}O)_2$	298.43	cr (dil al)		145				B6[3], 5541
6169	Dimethyloxonium bromide .... $[(CH_3)_2OH]^+Br^-$	126.99	....		−13			eth	B1[4], 1247
6170	Dimethyloxonium chloride .... $[(CH_3)_2OH]^+Cl^-$	82.93	gas	−2	−97			eth, lig, HCl	B1[4], 1247
6171	Dimethyl pentyl amine .... $(CH_3)_2NC_5H_{11}$	115.22		123		$0.743^{20/4}$	1.4083^{20}	**eth**	B4[4], 675
6172	Dimethyl phosphinic acid, ethyl ester $(CH_3)_2PO_2C_2H_5$	122.11		89^{15}		$1.0278^{25/4}$	1.4281^{25}	al, eth	Am73, 5466
6173	bis-(Dimethyl phosphino) amine or Amino-bis(dimethylphosphine) .... $[(CH_3)_2P]_2NH$	137.10	....	$33.5^{5.4}$ sub	39.5				Am75, 3869
6174	Dimethyl sulfone .... $(CH_3)_2SO_2$	218.28	pr	238	110	$1.1702^{110/0}$	1.4226	w, al, bz	B1[4], 1279
6175	Dimethyl sulfoxide $(CH_3)_2SO$	78.13	....	$189, 85-7^{20}$	18.4	$1.1014^{20/4}$	1.4770^{20}	w, al, eth, ace	B1[4], 1277
6176	bis-(Dimethylthiocarbamyl) disulfide or Arasan .... $[(CH_3)_2NCS]_2S_2$	240.43	wh or ye mcl (chl-al)	129^{20}	155.6			chl	B4[4], 242
6177	Di-α-Naphthyl amine .... $(\alpha-C_{10}H_7)_2NH$	269.35	....	$310-5^{15}$	115			al, eth, ace, bz, chl	B12[3], 2859
6178	Diβ-Naphthyl amine .... $(\beta-C_{10}H_7)_2NH$	269.35	lf (bz)	471	172.2			eth	B12[3], 3003
6179	α,α′-Dinaphthyl disulfide .... $(\alpha-C_{10}H_7)_2S_2$	318.46	pl (al) nd (lig)		91	1.144^{20}		eth	B6[3], 2947
6180	β,β′-Dinaphthyl disulfide .... $(\beta-C_{10}H_7)_2S_2$	318.46	nd		139-40	$0.8409^{20/4}$	1.4555^{20}	al, eth	B6[3], 3013
6181	α,α′-Dinaphthyl ketone .... $\alpha-C_{10}H_7CO-\alpha-C_{10}H_7$	282.35	wh nd yesh pr (aa)		104			al, eth, bz	B7[2], 503
6182	α,β′-Dinaphthyl ketone .... $\alpha-C_{10}H_7CO-\beta-C_{10}H_7$	282.35	nd (al bz-lig)	$235^{0.06}$	136-7			chl	B6[3], 2869
6183	β,β′-Dinaphthyl ketone .... $\beta-C_{10}H_7CO-\beta-C_{10}H_7$	282.35	(i) nd (eth), (ii) lf (chl-eth)		(i) 125.5, (ii) 164.5			chl	B7[2], 504
6184	Di(α-naphthyl)methane .... $(\alpha-C_{10}H_7)_2CH_2$	268.38	pr or nd (al)	$>360, 270^{14}$	109			eth, bz, chl	B5[3], 2480
6185	Di(β-Naphthyl)methane .... $(\beta-C_{10}H_7)_2CH_2$	268.38	nd (al eth)		93			bz	B5[1], 360
6186	2,4-Dinitrobenzoic acid .... $2,4-(NO_2)_2C_6H_3CO_2H$	212.12	nd(w)		183			bz	B9[3], 1776
6187	2,5-Dinitrobenzoic acid .... $2,5-(NO_2)_2C_6H_3CO_2H$	212.12	pr (w)		177			al, eth	B9[3], 1778

PHYSICAL CONSTANTS OF ORGANIC COMPOUNDS (Continued)

No.	Name, Synonyms, and Formula	Mol. wt.	Color, crystalline form, specific rotation and λ_{max} (log ε)	b.p. °C	m.p. °C	Density	n_D	Solubility	Ref.
6188	2,6-Dinitrobenzoic acid 2,6-$(NO_2)_2C_6H_3CO_2H$	212.12	nd(w)		202-3			al, eth	B9³, 1778
6189	3,4-Dinitrobenzoic acid 3,4-$(NO_2)_2C_6H_3CO_2H$	212.12	nd (w)		165			al, eth	B9³, 1778
6190	3,5-Dinitrobenzoic acid 3,5-$(NO_2)_2C_6H_3CO_2H$	212.12	mcl pr (al)		205			al, aa	B9³, 1779
6191	3,5-Dinitrobenzoic acid, benzyl ester 3,5-$(NO_2)_2C_6H_3CO_2CH_2C_6H_5$	302.25	nd (lig)		112				B9³, 1848
6192	3,5-Dinitrobenzoic acid, 2-bromo 3,5-$(NO_2)_2$-2-$BrC_6H_2CO_2H$	291.02	ye nd (w)		213			al, bz, aa, lig	B9³, 1954
6193	3,5-Dinitrobenzoic acid, butyl ester 3,5-$(NO_2)_2C_6H_3CO_2C_4H_9$	268.23	mcl nd (al)		62.5		1.488		B9³, 1782
6194	3,5-Dinitrobenzoic acid, isobutyl ester 3,5-$(NO_2)_2C_6H_3CO_2$-i-C_4H_9	268.23	mcl pl, nd (al)		87-8			al	B9³, 1783
6195	3,5-Dinitrobenzoic acid, ethyl ester 3,5-$(NO_2)_2C_6H_3CO_2C_2H_5$	240.18	nd (al)		92.9	1.295[111]	1.560	al	B9³, 1781
6196	3,5-Dintrobenzoic acid, furfuryl ester 3,5-$(NO_2)_2C_6H_3CO_2CH_2(C_4H_3O)$	292.22	(bz-py)		78-81				B17⁴, 1249
6197	3,5-Dinitrobenzoic acid, methyl ester 3,5-$(NO_2)_2C_6H_3CO_2CH_3$	226.15	nd(w)		112			al, w	B9³, 1781
6198	3,5-Dinitrobenzoic acid, tetrahydrofurfuryl ester 3,5-$(NO_2)_2C_6H_3CO_2CH_2(C_4H_7O)$	296.24	nd (al)		83-4			al	B17⁴, 1106
6199	3,5-Dinitrobenzoic acid, pentyl ester 3,5-$(NO_2)_2C_6H_3CO_2C_5H_{11}$	282.26			46.4			al	B9³, 281
6200	3,5-Dinitrobenzoic acid, phenyl ester 3,5-$(NO_2)_2C_6H_3CO_2C_6H_5$	288.23	rods (al)		145.6			al, bz	B9³, 1846
6201	3,5-Dinitrobenzoic acid, propyl ester 3,5-$(NO_2)_2C_6H_3CO_2C_3H_7$	254.20	mcl pl (al)		73			al	B9³, 1781
6202	3,5-Dintrobenzoic acid, isopropyl ester 3,5-$(NO_2)_2C_6H_3CO_2$-i-C_3H_7	254.20	nd (al)		122			al	B9³, 1782
6203	3,5-Dinitrobenzoyl chloride 3,5-$(NO_2)_2C_6H_3COCl$	230.51	ye nd bz	196¹²	74			eth	B9³, 1936
6204	Dioctadecyl amine or Distearyl amine $(C_{18}H_{37})_2NH$	522.01			73-4			chl	B4⁴, 829
6205	Dioctyl amine $(C_8H_7)_2NH$	241.47	nd	297-8, 175¹⁴	35.6	0.7968²⁶/⁴	1.4415²⁶	al, eth	B4⁴, 753
6206	Di-2-Octyl amine $[C_5H_{11}CH(CH_3)]_2NH$	241.47		281.5⁷³⁹		0.7948²⁰/⁴		al, eth	B4⁴, 765
6207	1,3-Dioxane or m-Dioxane $C_4H_8O_2$	88.12		105⁷⁵⁵	−42	1.0342²⁰/⁴	1.4165²⁰	w, al, eth, ace, bz	B19⁴, 8
6208	1,3-Dioxane, 2,4-dimethyl $C_6H_{12}O_2$	116.16		115-8		0.9392²⁰/⁴	1.4136²⁰		B19⁴, 61
6209	1,3-Dioxane, 5-ethyl-4-propyl $C_9H_{18}O_2$	158.24		196		0.9305²⁰/⁴	1.4370²⁰		C52, 2795
6210	1,3-Dioxane, 5-hydroxy-2-methyl $C_5H_{10}O_3$	118.13		176		1.0705¹⁷/⁴	1.4375¹⁷	w	B19⁴, 624
6211	1,3-Dioxane, 4-methyl $C_5H_{10}O_2$	102.13		114		0.9758²⁰	1.4159²⁰		B19⁴, 49
6212	1,3-Dioxane, 4-methyl-4-phenyl $C_{11}H_{14}O_2$	178.23		256, 102⁴	35-40	1.0864²⁰	1.5240²⁰		B19⁴, 233
6213	1,3-Dioxane, 2-phenyl $C_{10}H_{12}O_2$	164.21	nd (peth)	252-4, 98-9⁶	41			al, eth	B19⁴ 215
6214	1,3-Dioxane, 4-phenyl $C_{10}H_{12}O_2$	164.21		245, 128-30¹⁰		1.1038²⁰/⁴	1.5306¹⁸		B19⁴, 218
6215	1,4-Dioxane or Diethylene dioxide $C_4H_8O_2$	88.12		101⁷⁵⁰	11.8	1.0337²⁰/⁴	1.4224²⁰	w, al, eth, ace, bz	B19⁴, 9
6216	1,4-Dioxane, -2,3-dichloro $C_4H_6Cl_2O_2$	157.00		80-2¹⁰	30	1.468²⁰/⁴	1.4928²⁰	eth, ace, bz	B19⁴, 30
6217	1,4-Dioxane, heptachloro $C_4HCl_7O_2$	329.22		123-8⁸	54-6			eth, ace, bz, lig	B19⁴, 32
6218	1,4-Dioxene $C_4H_6O_2$	86.09		94.1		1.0836²⁰/⁴	1.4372²⁰	eth, ace, bz	B19⁴, 108
6219	1,4-Dioxine $C_4H_4O_2$	84.07		74.6⁷⁴⁸		1.115²⁰/⁴	1.4350²⁰	eth, ace, bz	B19⁴, 154
6220	1,3-Dioxolane or Glycol methylene ether $C_3H_6O_2$	74.08		78⁷⁶⁵	−95	1.0600²⁰/⁴	1.3974²⁰	w, al, eth, ace	B19⁴, 5

No.	Name, Synonyms, and Formula	Mol. wt.	Color, crystalline form, specific rotation and λ_{max} (log ε)	b.p. °C	m.p. °C	Density	n_D	Solubility	Ref.
6221	1,3-Dioxolane, 4-(hydroxymethyl)-2-methyl or Glycerolethylidine ether.................................... $C_5H_{10}O_3$	118.13	....	187, 68-70[1]	......	1.1243[17/4]	1.4413[17]	al	B19[4], 631
6222	1,3-Dioxolane, 2-methyl or Glycol ethylidine ether...... $C_4H_8O_2$	88.12	81-2			0.9811[20/4]	1.4035[17]	w, al, eth	B19[4], 42
6223	1,3-Dioxlane-4-carboxaldehyde, 2,2-dimethyl .... $C_6H_{10}O_3$	130.15	....	74[50]			1.4189[25]	w	B19[4], 1579
6224	1,3-Dioxolan-2-one or 1,2-Ethanediol carbonate .... $C_3H_4O_3$	88.06	mcl pl (al)	248[760]	39-40	1.3214[39/4]	1.4158[50]	w, al, eth, bz, chl	B19[4], 1556
6225	1,3-Dioxolane-2-one, 4-methyl or Propylene carbonate ... $C_4H_6O_3$	102.09	....	24.2	-48.8	1.2069[20/20]	1.4189[20]	w, al, eth, ace, bz	B19[4], 1564
6226	Dipentyl amine $(C_5H_{11})_2NH$	157.30	....	202-3, 91-3[14]		0.7771[20/4]	1.4272[20]	al, %eth,] ace	B4[4], 676
6227	Dipentylamine, 2,2'-dimethyl-2,2'-dihydroxy .... $[C_3H_7C(OH)(CH_3)CH_2]_2NH$	217.36	....	165-70[15]		0.9264[20/4]	1.4585[20]	al, %eth,] ace	C53, 16829
6228	Diisopentyl amine $(i-C_5H_{11})_2NH$	157.30	....	188	-44	0.7672[21/4]	1.4235[20]	al, %eth]	B4[4], 699
6229	Dipentyl disulfide or [m]-Amyldi sulfide.............. $C_5H_{11}SSC_5H_{11}$	206.42	....	119[7]		0.9221[20/4]	1.4889[20]	....	B1[4], 1654
6230	Di-[iso]-Pentyl disulfide or Isoamylsulfide .... $([i]-C_5H_{11})_2S_2$	206.42	....	250		0.9192[20/4]	1.4864[20]	....	B1[4], 1689
6231	Diphenadione or 2-Diphenyl acetyl-1,3-indanedione ..... $C_{23}H_{16}O_3$	340.38	pa ye mcl (al)	....	146-7	....	1.670	ace, aa	C49, 3264
6232	Diphenyl acetamide .... $(C_6H_5)_2CHCONH_2$	211.25	pl (al)	....	167-8			al	B9[3], 3301
6232a	Diphenyl acetamide, α-chloro .... $(C_6H_5)_2CClCONH_2$	245.7	cr (to)		115			al, eth, bz, chl	....
6233	Diphenyl acetic acid .... $(C_6H_5)_2CHCO_2H$	212.25	nd (w) lf (al)	194[25] sub	148	1.258[15/?15]		al, eth, chl	B9[3], 3290
6234	Diphenyl acetic acid-anhydride .... $[(C_6H_5)_2CHCO]_2O$	406.49	nd (eth,)	220-5[15]	98			bz, chl	B9[1], 281
6235	Diphenyl acetic acid, α-bromo .... $(C_6H_5)_2C(Br)CO_2H$	291.15	(chl-peth)	....	133-4			chl, to	B9[2], 471
6236	Diphenyl acetic acid, α-chloro .... $(C_6H_5)_2C(Cl)CO_2H$	246.70	pl (bz-liq)	....	118-9d			al, eth, ace, bz	B9[3], 3307
6237	Diphenyl acetic acid, α-chloro, ethyl ester or Ethyl-diphenyl chloroacetate.............................. $(C_6H_5)_2C(Cl)CO_2C_2H_5$	274.75	pl (chl) cr (al)	185[14]	43-4			al, eth	B9[3], 3307
6238	Diphenl acetic acid, ethyl ester or Ethyldiphenyl acetate .. $(C_6H_5)_2CHCO_2C_2H_5$	240.31	nd (al) rh (AcOEt)	195[25]	59	....	....	al, eth	B9[3], 3291
6239	Diphenyl acetic acid, α-hydroxy or Benzilic acid .... $(C_6H_5)_2C(OH)CO_2H$	228.25	mcl nd (w)	d180	151			al, eth	B10[3], 1168
6240	Diphenyl acetic acid, α-hydroxy, ethyl ester .... $(C_6H_5)_2C(OH)CO_2C_2H_5$	256.31	pr or nd	201[21]	34		1.5620[20]	al, eth	B10[3], 1171
6241	Diphenyl acetic acid, α-hydroxy, methyl ester or Methyl benzilate.................................... $(C_6H_5)_2C(OH)CO_2CH_3$	242.28	mcl or tcl cr (al)	187[13]	75	....	....	al, eth, aa	B10[3], 1170
6242	Diphenyl acetic acid, methyl ester or Methyl diphenyl acetate.................................... $(C_6H_5)_2CHCO_2CH_3$	226.28	mcl pl (AcOEt) lf (dil al)	....	60			al, eth	B9[3], 3291
6243	Diphenyl acetic acid, 2,2',4,4'-tetranitro,ethyl ester ...... $[2,4-(NO_2)C_6H_3]_2CHCO_2C_2H_5$	420.30	lf (al) nd (bz, aa)	....	154				B9[3], 3315
6244	Diphenyl acetic acid, 2,2',4,4'-tetranitro,methyl ester ...... $[2,4-(NO_2)C_6H_3]_2CHCO_2CH_3$	406.27	lf (chl-MeOH)		159			chl	B9, 675
6245	Diphenyl acetonitrile .... $(C_6H_5)_2CHCN$	193.25	pr (eth) lf (dil al)	181-4[12]	72-3			al, eth	B9[3], 3304
6246	Diphenyl acetyl bromide, α-bromo .... $(C_6H_5)_2CBrCOBr$	354.05	nd (lig)		65-6			al, eth, bz, chl	B9[1], 283
6247	Diphenyl acetyl chloride .... $(C_6H_5)_2CHCOCl$	230.70	pl (lig)	178[15]	56-7				B9[3], 3300
6248	Diphenyl acetyl chloride, α-chloro .... $(C_6H_5)_2CClCOCl$	265.14	cr (lig)	180[14]	50-1				B9[3], 3308
6249	Diphenyl amine .... $(C_6H_5)_2NH$	169.23	mcl lf (dil al)	302, 179[22]	54-5	1.160[22/?20]		al, eth, ace, bz	B12[3], 284
6250	Diphenyl amine, 4,4'-dimethylamino .... $[4-(CH_3)_2NC_6H_4]_2NH$	255.37	tetr pl (CS$_2$)	....	119	....		eth	B13[2], 56

No.	Name, Synonyms, and Formula	Mol. wt.	Color, crystalline form, specific rotation and λ_{max} (log ε)	b.p. °C	m.p. °C	Density	n_D	Solubility	Ref.
6251	Diphenyl amine, 2,2'-dinitro $(2\text{-}NO_2C_6H_4)_2NH$	259.22	lf (al ace) ye cr (al aa)	....	169	....	....	al, ace, aa	B12³, 1518
6252	Diphenyl amine, 2,4-dinitro $2,4\text{-}(NO_2)_2C_6H_3NHC_6H_4$	259.22	ye red nd (al)		157			al, ace, chl, py	B12³, 1683
6253	Diphenyl amine, 2,4'-dinitro $2\text{-}NO_2C_6H_4NH(C_6H_4NO_2\text{-}4)$	259.22	red nd (aa)		222-3			chl, to	B12³, 1587
6254	Diphenyl amine, 2,6-dinitro $2,6\text{-}(NO_2)_2C_6H_3NHC_6H_5$	259.22	og lf (al, aa)		107-8			al, aa	B12³, 1705
6255	Diphenyl amine, 3,4'-dinitro $3\text{-}NO_2C_6H_4NH(C_6H_4NO_2\text{-}4)$	259.22	pa ye (chl aq py)		217			ace, chl	B12³, 1587
6256	Diphenyl amine, 4,4'-dinitro $(4\text{-}NO_2C_6H_4)_2NH$	259.22	ye nd (al)		216			ace, aa	B12³, 1587
6257	Diphenyl amine, 2,4-dinitro-4'-hydroxy $2,4\text{-}(NO_2)_2C_6H_3NH(C_6H_4OH\text{-}4)$	275.22	red lf		195-6				B13³, 1019
6258	Diphenyl amine, 2,4-dinitro-5-hydroxy $2,4\text{-}(NO_2)_2\text{-}5\text{-}HOC_6H_2NHC_6H_5$	275.22	dk ye nd (al)		139			al	B13¹, 138
6259	Diphenyl amine, 2,6-dinitro-2'-hydroxy $2,6\text{-}(NO_2)_2C_6H_3NH(C_6H_4OH\text{-}2)$	275.22	red vt nd (al)		191			al, eth, bz	B13, 365
6260	Diphenyl amine, 2,6-dinitro-3-hydroxy $2,6\text{-}(NO_2)_2\text{-}3\text{-}HOC_6H_2NHC_6H_5$	275.22	ye br nd (MeOH)		124-5			al	B13², 216
6261	Diphenyl amine, 2,2',4,4',6,6'-hexanitro or Dipicryl amine $[2,4,6\text{-}(NO_2)_3C_6H_2]_2NH$	439.21	pa ye pr (aa)		244d			py	B12³, 1734
6262	Diphenyl amine, hydrobromide $(C_6H_5)_2NH.HBr$	250.15	pl (dil al)		230d			w	B12, 180
6263	Diphenyl amine, N-nitroso $(C_6H_5)_2N\text{-}NO$	198.23	ye pl (lig)		66.5			al, bz	B12³, 1120
6264	Diphenyl amine, 2,2',4,4'-tetrabromo $(2,4\text{-}Br_2C_6H_3)_2NH$	484.83	nd (chl or bz)		187.5			bz, chl	B12³, 1472
6265	Diphenyl amine, 2,2',4,4'-tetranitro $[2,4\text{-}(NO_2)_2C_6H_3]_2NH$	349.22	ye nd or cr (aa)		201			py	B12³, 1684
6266	Diphenyl diselenide $(C_6H_5)_2Se_2$	312.13	ye nd		63-4	$1.557^{80/4}$		al, eth, xyl	B6³, 1110
6267	Diphenyl disulfide $(C_6H_5)_2S_2$	218.34	nd (al) or rh	310, 192¹⁵	61-2	$1.353^{20/4}$		al, eth, bz	B6³, 1027
6268	1,2-Dipehnylethane or Bibenzyl $C_6H_5CH_2CH_2C_6H_5$	182.27	nd (al)	285, 95-6¹	52.2	$0.9583^{60/4}$	1.5478^{60}	al, eth	B5⁴, 1868
6269	1,2-Diphenylethane, 4,4'-diamino $(4\text{-}H_2NC_6H_4)CH_2CH_2(C_6H_4NH_2\text{-}4)$	212.30	pl (w)	sub	135-6 sub			al	B13³, 470
6270	1,2-Diphenylethane, 2,2'-dibromo $(2\text{-}BrC_6H_4)CH_2CH_2(C_6H_4Br\text{-}2)$	340.07	pl (al)	138-40⁰·⁰¹²	84.5			al	B5⁴, 1875
6271	1,2-Diphenylethane, 4,4'-dibromo $(4\text{-}BrC_6H_4)CH_2CH_2(C_6H_4Br\text{-}4)$	340.07	pr (al)	ca 198¹⁰	115				B5⁴, 1875
6272	1,2-Diphenyl ethane, α,α'-dinitro(dl) $C_6H_5CH(NO_2)CH(NO_2)C_6H_5$	272.26	(al)pr(aa)		154-5			al, eth, ace, bz, chl	B5⁴, 1878
6273	1,2-Diphenyl ethane, α,α'-dinitro-(meso) $C_6H_5CH(NO_2)CH(NO_2)C_6H_5$	272.26	nd (aa)		235-6			ace	B5⁴, 1878
6274	1,2-Diphenyl ethane, 2,2'-dinitro $(2\text{-}O_2NC_6H_4)CH_2CH_2(C_6H_4NO_2\text{-}2)$	272.26	pr (aa)		127			eth, bz, aa	B5⁴, 1878
6275	1,2-Diphenyl ethane, 4,4'-dinitro $(4\text{-}O_2NC_6H_4)CH_2CH_2(C_6H_4NO_2\text{-}4)$	272.26	yesh nd (al or bz)		180.5				B5⁴, 1878
6276	1,2-Diphenyl ethane, α-hydroxy-(l) $C_6H_5CH(OH)CH_2C_6H_5$	198.27	nd (eth-peth), $[\alpha]^{20/}_D$ -9.4 (w, c=10)		67	$1.0358^{20/4}$		al, eth	B6², 637
6277	1,2-Diphenyl ethane, α-hydroxy-(dl) $C_6H_5CH(OH)CH_2C_6H_5$	198.27	nd (bz-peth)	177¹⁵	69			al, eth	B6³, 3390
6278	1,2-Diphenyl ethane, α-hydroxy-(d) $C_6H_5CH(OH)CH_2C_6H_5$	198.27	nd (eth-peth or dil al), $[\alpha]^{25/}_D$ + 53 (al)	167-70¹⁰	67-8	$1.0358^{20/4}$		al, eth	B6³, 3390
6279	Diphenyl ether, 2,3'-dimethoxy $(2\text{-}CH_3OC_6H_4)O(C_6H_4OCH_3\text{-}3)$	230.27	pr(bz-peth)	326-9, 152²	54			al, eth, bz	B6³, 4318
6280	Diphenyl glyoxal or Benzil $C_6H_5COCOC_6H_5$	210.23	ye pr (al)	346-8d, 188¹²	95-6	$1.084^{102/4}$		al, eth, ace, bz	B7³, 3804
6281	Diphenyl glyoxime (anti) or Benzildioxime (anti) $C_6H_5C(=NOH)C(=NOH)C_6H_5$	240.27	lf (al or ace)	....	238d				B7³, 3816

No.	Name, Synonyms, and Formula	Mol. wt.	Color, crystalline form, specific rotation and λ_{max} (log ε)	b.p. °C	m.p. °C	Density	n_D	Solubility	Ref.
6282	Diphenyl methane or Ditan .... $(C_6H_5)_2CH_2$	168.24	pr nd	264.3, 125.5[10]	25.3	1.0060[20/4]	1.5753[20]	al, eth, chl	B5[4], 1841
6283	Diphenyl methane, α-amino .... $(C_6H_5)_2CHNH_2$	183.26	hex pl	304[763], 176[23]	34	1.0675[20/0]	1.5963	bz	B12[3], 3221
6284	Diphenyl methane, 3-amino .... $(3-H_2NC_6H_4)CH_2C_6H_5$	183.26	cr(lig)		46	....	....	lig	B12[3], 3215
6285	Diphenyl methane, 4-amino .... $(4-H_2NC_6H_4)CH_2C_6H_5$	183.26	mcl (lig)	300	34-5	1.038[55]	....	al, eth, lig	B12[3], 3215
6286	Diphenyl methane, α-bromo .... $(C_6H_5)_2CHBr$	247.14	tcl (peth)	193[26], 111[0.3]	45	....	....	al, bz	B5[4], 1850
6287	Diphenyl methane, α-chloro or Benzydryl chloride...... $(C_6H_5)_2CHCl$	202.69	nd	173[19]	20.5	1.1398[20/4]	1.5959[20]	....	B5[4], 1847
6288	Diphenyl methane, 4-chloro .... $(4-ClC_6H_4)CH_2C_6H_5$	202.69	....	298[742], 147-8[8]	7.5	1.1247[20/4]	....	ace	B5[4], 1847
6289	Diphenyl methane, 4,4′-diamino or 4,4′-Diamino ditan ... $(4-H_2NC_6H_4)_2CH_2$	198.27	pl or nd (w) pl (bz)	398-9[768], 257[18]	92-3	....	....	al, eth, bz	B13[3], 454
6290	Diphenyl methane, diazo .... $(C_6H_5)_2CN_2$	194.24	bl-red nd (peth)	exp	30-2	....	....	al, eth, ace	B7[3], 2068
6291	Diphenyl methane, α,α-dichloro or Benzophenone dichloride .... $(C_6H_5)_2CCl_2$	237.13	....	305d, 190[21]		1.235[18]	....	eth, bz	B5[4], 1848
6292	Diphenyl methane, 4,4′-dichloro .... $(4-ClC_6H_4)_2CH_2$	237.13	....	186-90[18]	55-6	1.365[17]	....	al	B5[4], 1848
6293	Diphenyl methane, 5,5′-dichloro-2,2′-dihydroxy or Dichlorophene .... $(5-Cl-2-HOC_6H_3)_2CH_2$	269.13	cr (bz peth)	....	177-8	....	....	al, ace	B6[3], 5406
6294	Diphenyl methane, 2,4′-dihydroxy .... $(2-HOC_6H_4)CH_2(C_6H_4OH-4)$	200.24	nd (dil al, bz or w)	....	119-20	....	....	al, eth	B6[3], 5409
6295	Diphenyl methane, 3,3′-dihydroxy or m,m′-Methylene diphenol .... $[3-HOC_6H_4]_2CH_2$	200.24	nd (dil aa)	230-40[3]	102-3	....	....	al, eth, aa	B6[3], 5411
6296	Diphenyl methane, 4,4′-dihydroxy or p,p′-Methylene diphenol .... $(4-HOC_6H_4)_2CH_2$	200.24	lf or nd (w)	sub	162-3	....	....	al, eth, chl	B6[3], 5412
6297	Diphenyl methane, 4,4′-dimethylamino .... $[4-(CH_3)_2NC_6H_4]_2CH_2$	254.38	pl or ta (al lig)	390d, 182-5[3]	91-2	....	....	eth, bz	B13[3], 454
6298	Diphenyl methane, α,α-dinitro .... $(C_6H_5)_2C(NO_2)_2$	258.24	pl (dil al)	....	79-80	....	....	al, eth, bz, chl	B5[3], 1797
6299	Diphenyl methane, 2,2′-dinitro .... $(2-O_2NC_6H_4)_2CH_2$	258.24	cr	....	83.5	....	....	al, eth	B5[3], 1796
6300	Diphenyl methane, 2,2′-dinitro,4,4′-diamino .... $[2-NO_2-4-H_2NC_6H_3]_2CH_2$	288.27	og pl (al)	....	205	....	....	al, aa	B13[2], 113
6301	Diphenyl methane, 2,4′-dinitro .... $(2-O_2NC_6H_4)CH_2(C_6H_4NO_2-4)$	258.24	ye mcl pr(bz)	....	118	....	....	....	B5[3], 1797
6302	Diphenyl methane, 3,3′-dinitro .... $(3-O_2NC_6H_4)_2CH_2$	258.24	lf(aa)	....	175.5	....	....	al, bz, aa	B5[3], 1797
6303	Diphenyl methane, 3,3′-dinitro-4,4′-diamino .... $[3-NO_2-4-NH_2C_6H_3]_2CH_2$	288.27	red nd	....	232-3	....	....	....	B13[3], 113
6304	Diphenyl methane, 3,4′-dinitro .... $(3-O_2NC_6H_4)CH_2(C_6H_4NO_2-4)$	258.24	nd(al)	....	103-4	....	....	....	B5[3], 1797
6305	Diphenyl methane, 4,4′-dinitro .... $(4-O_2NC_6H_4)_2CH_2$	258.24	nd (bz, peth, aa)	....	188	....	....	bz, aa	B5[3], 1797
6306	Diphenyl methane, 4-ethyl .... $(4-C_2H_5-C_6H_4)CH_2C_6H_5$	196.30	....	297 85[0.2]	−24	0.9777[20]	1.5618[20]	al, eth, chl	B5[3], 1870
6307	Diphenyl methane, 4-hydroxy or 4-Benzylphenol .... $(4-HOC_6H_4)CH_2C_6H_5$	184.24	nd or pl(al)	325-30, 198-200[10]	84	....	....	al, eth, bz, chl, aa	B6[3], 3357
6308	Diphenyl methane, 2-methyl or Phenyl-2-tolyl methane .. $(2-CH_3C_6H_4)CH_2C_6H_5$	182.26	....	280.5	6.6	....	1.5763[20/D]	....	B5[1], 1855
6309	Diphenyl methane, 3-methyl or Phenyl-3-tolyl methane... $(3-CH_3C_6H_4)CH_2C_6H_5$	182.27	....	279.2, 120[0.2]	−28	0.9913[20/4]	1.5712[20]	al, eth, bz, aa, chl	B5[4], 1858
6310	Diphenyl methane, 4-methyl .... $(4-CH_3C_6H_4)CH_2C_6H_5$	182.27	....	286, 114-5[3]	−30	0.9976[20/4]	1.5712[20]	al, eth, bz, chl, aa	B5[4], 1860
6311	Diphenyl methane, 4-propyl .... $(4-C_3H_7C_6H_4)CH_2C_6H_5$	210.32	....	152-5[10]		0.9739[18/4]	1.5552[20]	....	B5[4], 1950
6312	Diphenyl methane, 2,2′,4,4′-tetramethyl .... $[2,4-(CH_3)_2C_6H_3]_2CH_2$	222.34	....	140-2[3]		....	1.5635[15]	....	B5[4], 1971
6313	Diphenyl methane, 2,2′,4,4′-tetranitro .... $[2,4-(NO_2)_2C_6H_3]_2CH_2$	348.24	ye pr (aa)	....	181	....	....	....	B5[3], 1798

No.	Name, Synonyms, and Formula	Mol. wt.	Color, crystalline form, specific rotation and λ_{max} (log ε)	b.p. °C	m.p. °C	Density	n_D	Solubility	Ref.
6314	Diphenyl methanol or Benzhydrol $(C_6H_5)_2CHOH$	184.24	nd(lig)	297-8[748], 180[20]	69	....		al, eth, chl	B6[3], 3364
6315	Diphenyl methanol, 4,4′-dimethyl or di-(4-tolyl)methanol $(4\text{-}CH_3C_6H_4)_2CHOH$	212.30	nd(al)	....	69			al, eth, ace, chl, aa	B6[3], 3422
6316	Di-(α-Phenylethyl) amine $[CH_3CH(C_6H_5)]_2NH$	225.34	ye	295-8, 190[10]		1.018[13]	1.573		B12[2], 589
6317	Di-(β-Phenylethyl)amine $(C_6H_5CH_2CH_2)_2NH$	225.34	....	335-7[603], 190[15]	28-30		1.5550[25]	al, eth	B12[2], 593
6318	Diphenyl ethyl amine $(C_6H_5)_2NC_2H_5$	197.28		295-6, 148[11]		1.0396[20/20]	1.6095[20]	al, eth	B12[3], 291
6319	Diphenyl methyl amine $(C_6H_5)_2NCH_3$	183.26		293-4, 145[10]	-7.5	1.0476[20/4]	1.6193[20]	....	B12[3], 290
6320	Diphenyl propanetrione or Diphenyl triketone $C_6H_5COCOCOC_6H_5$	238.25	ye nd(lig)	289[175], 248[60]	68-70			eth	B7[3], 4620
6321	Diphenyl selenide $(C_6H_5)_2Se$	233.17	ye nd(bz)	301-2, 126-7[5]	2.5	1.351[20/4]	1.5500[20]	al, eth, bz, xyl	B6[4], 1779
6322	Diphenyl selenonium dichloride $(C_6H_5)_2SeCl_2$	304.08	pa ye pr(xyl, ace)nd(al)	....	183			w, al, ace	B6[3], 1107
6323	Diphenyl sulfide $(C_6H_5)_2SO_2$	218.28	mcl pr (bz) pl (al) nd (w)	379, 232[18]	128-9	1.252[20/4]		eth, bz	B6[4], 1490
6324	Diphenyl sulfoxide $(C_6H_5)_2SO$	202.28	pr(lig)	340d, 210[15]	70.5			al, eth, bz, aa	B6[4], 1489
6325	Diphosphine, tetrakis(trifluoromethyl) $(F_3C)_2PP(CF_3)_2$	337.97		84		>1.0			B3[4], 265
6326	Dipicryl amine $[2,4,6\text{-}(NO_2)_3C_6H_2]_2NH$	439.21	pa ye pr (aa)	....	244d			py	B12[2], 422
6327	Diploicin $C_{18}H_{10}Cl_4O_5$	424.07	(bz)	....	232			...	B19[, 7347
6328	Dipropanol amine, N-methyl $(HOCH_2CH_2CH_2)_2NCH_3$	147.22	hyg	164-5[12]				w, al, bz	B4[3], 1644
6329	Di-iso-Propanol amine $[CH_3CH(OH)CH_2]_2NH$	133.19	cr	249-50[745], 151[23]	44-5			w, al	B4[3], 761
6330	Di-iso-Propanol amine, N-2-hydroxy ethyl $[CH_3CH(OH)CH_2]_2NCH_2CH_2OH$	177.25		155-6[1]		1.0458[20/4]	1.4708[20]	w, al, ace	B4[3], 764
6331	Dipropyl amine $(C_3H_7)_2NH$	101.19		109-10	-39.6	0.7400[20/4]	1.4050[20]	w, al, eth, ace, bz	B4[4], 469
6332	Dipropyl amine, N-nitroso $(C_3H_7)_2N\text{-}NO$	130.19	gold	206, 89[13]		0.9163[20/4]	1.4437[20]	al, eth	B4[3], 264
6333	Di-m-Propyl-iso-butyl amine, perfluoro $i\text{-}C_4F_9N(C_3F_7)_2$	571.08		146-8		1.84[25/4]	1.283[25]	al, eth	B2[4], 858
6334	Di-iso-Propyl amine $(i\text{-}C_3H_7)_2NH$	101.19		84	-61	0.7169[20/4]	1.3924[20]	al, eth, ace, bz	B4[4], 510
6335	Di-iso-Propyl amine, N-nitroso $[(CH_3)_2CH]_2N\text{-}NO$	130.19	cr(eth w)	194.5, 76-81[14]	48	0.9422[20/4]	....	al, eth, bz	B4[3], 281
6336	Dipropyl disulfide $(C_3H_7)_2S_2$	150.31		193.5		0.9599[20/4]	1.4981[20]	....	B1[4], 1454
6337	Diisopropyl disulfide $(i\text{-}C_3H_7)_2S_2$	150.31		177.2, 56.8[10]		0.9435[20/4]	1.4916[20]	....	B1[4], 1503
6338	Dipropylene glycol $(CH_3CH(OH)CH_2)_2O$	134.18		229-32		1.0224[20/20]		w, al	B1[4], 2473
6339	Dipropyl sulfide $(C_3H_7)_2SO_2$	150.24	sc		29-30	1.0278[50/4]	1.4456[30]	al, eth	B1[4], 1453
6340	Di-isopropyl sulfone $(i\text{-}C_3H_7)_2SO_2$	150.24	eth	....	36			w, eth	B1[4], 1502
6341	Diisopropyl ketone, cyanohydrin $(i\text{-}C_3H_7)_2C(OH)CN$	141.22	rh(eth or peth)	111[18]	59			al, eth, ace, bz	B3[2], 239
6342	Di-(α-Pyrryl)methane $[2\text{-}C_4H_3NH]_2CH_2$	146.19	lf or nd(al)	163-7[12]	73			al, eth, bz	B23, 167
6343	Disilane, 1,2-dichloro-1,1,2,2-tetramethyl $(CH_3)_2SiClSiCl(CH_3)_2$	187.22		49-50[18]	....	1.010[20]	1.4548[20]	....	KHOC, 366
6344	Disilane, 1,2-difluoro-1,1,2,2-tetramethyl $(CH_3)_2SiFSiF(CH_3)_2$	154.31		92-9		0.9120[20/4]	1.3837[20]		BOSC2[2], 183
6345	Disilane, 1,2-diphenyl-1,1,2,2-tetramethyl $C_6H_5Si(CH_3)_2Si(CH_3)_2C_6H_5$	270.52		111[1], 73[0.1]	34.5	0.9892[20/4]	1.5161[20]		BOSC2[2], 285
6346	Disilane, hexamethyl $(CH_3)_3SiSi(CH_3)_3$	518.80		112.5-4.3	12.8-14	0.7247[22.5/4]	1.4229[20]		KHOC, 597

No.	Name, Synonyms, and Formula	Mol. wt.	Color, crystalline form, specific rotation and λ_{max} (log ε)	b.p. °C	m.p. °C	Density	n_D	Solubility	Ref.
6347	Disiloxane, 1,3-dichlo-1,1,3,3,-tetramethyl $(CH_3)_2SiClOSiCl(CH_3)_2$	203.22	......	138	−37.5	$1.038^{20/4}$	......		KHOC, 593
6348	Disiloxane, dichloromethyl-pentamethyl $Cl_2CHSi(CH_3)_2OSi(CH_3)_3$	231.27	......	200-5	......	1.046^{20}	1.4382^{20}		KHOC, 596
6349	Disiloxane, 1,3-diethenyl-1,1,3,3-tetramethyl $CH_2=CHSi(CH_3)_2-O-Si(CH_3)_2CH=CH_2$	186.40	......	39	−99.7	0.811^{20}	1.4123^{20}		KHOC, 602
6350	Disiloxane, 1,3-dimethoxy-1,1,3,3-tetramethyl $CH_3OSi(CH_3)_2O-Si(CH_3)_2OCH_3$	194.38	......	139	......	$0.9048^{20/4}$	1.3835^{20}		BOSC2[2], 197
6351	Disiloxane, 1,3-diphenyl-1,1,3,3-tetramethyl $C_6H_5Si(CH_3)_2-O-Si(CH_3)_2C_6H_5$	286.53	......	110^2	......	$0.9763^{20/4}$	1.5176^230		Am79, 1437
6352	Disiloxane, hexaethyl $[(C_2H_5)_3Si]_2O$	246.55	......	$233^{756}, 129^{30}$	......	$0.8590^{0/0}$	1.4340^{20}		B4, 627
6353	Disiloxane, hexakis (2-ethylbutoxy) $([(C_2H_5)_2CHCH_2]_3Si)_2O$	679.20	......	220^1	>-54	$0.9219^{20/20}$	1.4330^{20}	eth, bz	C48, 3761
6354	Disiloxane, hexakis(2-ethylhexoxy) $([CH_3(CH_2)_3CH(C_2H_5)CH_2O]_3Si)_2O$	847.52	......	$253^{0.9}$	......	$0.9044^{20/20}$	1.4402^{20}	eth, bz	C48, 3761
6355	Disiloxane, hexamethyl $(CH_3)_3SiOSi(CH_3)_3$	162.38	......	99.5-100	-66	0.7638^{20}	$1.3774^{20}/_D$		KHOC, 598
6356	Disiloxane, 1,1,3,3-tetramethyl $(CH_3)_2SiH-O-SiH(CH_3)_2$	134.33	......	$70.5-71^{731}$	......	$0.7572^{20/4}$	$1.3700^{20}/_D$		BOSC2[2], 184
6357	1,2-Dithiane $C_4H_8S_2$	120.23	......	89^{14}	32-3	......	1.5981^{25}	eth, bz, chl	B19[4], 8
6358	1,4-Dithiane or Diethylene disulfide $C_4H_8S_2$	120.25	mcl pr	199-200	111-2			al, eth, aa	B19[4], 35
6359	1,3,5-Dithiazine, 4,5-dihydro-5-methyl or Methylthioformaldine. $C_4H_9NS_2$	135.25	nd(eth)	185d	65			al, eth, aa	B27[2], 524
6360	1,4-Dithiine, 2,5-diphenyl $C_{16}H_{12}S_2$	268.40	ye pr (al)	......	118-9			al, bz	B19[4], 404
6361	1,3-Dithiolane or Trimethylene-1,3-disulfide. $C_3H_6S_2$	106.21	......	175	−50	1.259^{17}	1.5975^{15}	al, eth, xyl	B19[4], 6
6362	di(2-Thienyl)ketone or Thienone. $C_3H_3SCOC_3H_3S$	194.28	nd(al)	326	90			eth, ace	B19[4], 1745
6363	Dithizone or Diphenyl thiocarbazone. $C_6H_5N=NCSNHNHC_6H_5$	256.33	bl-bk(chl-al)	......	165-9d			chl	B16[3], 19
6364	Dithioacetic acid or Thiolo-thionoacetic acid. CH_3CS_2H	92.18	ye-red oil	$66^{85}, 37^{15}$	......	1.24^{20}		w, al, eth, ace, bz	B2[4], 572
6365	Di-2-tolylamine $(2-CH_3C_6H_4)_2NH$	197.28	bl flr, wh cr	$312^{727}, 192^{23}$	52-3				B12[2], 437
6366	Di-3-tolylamine $(3-CH_3C_6H_4)_2NH$	197.28	pa ye (peth)	319-20	53			al, eth, peth	B12[2], 467
6367	Di-4-tolylamine $(4-CH_3C_6H_4)_2NH$	197.28	nd(peth)	330.5	79			eth, peth	B16[3], 2033
6368	3,4'-Ditolylamine hydrochloride $(3-CH_3C_6H_4)NH(C_6H_4CH_3-4).HCl$	233.74	cr	......	202-3			al, ace, bz, chl	B12[2], 414
6369	Di-2-tolyl disulfide $(2-CH_3C_6H_4)_2S_2$	246.40	lf(al)	......	38-9			al, eth, ace	B6[4], 2027
6370	Di-4-tolyl disulfide $(4-CH_3C_6H_4)_2S_2$	246.40	nd or lf(al)	$210-15^{20}$	47-8	1.114^{51}		al, eth, ace	B6[4], 2206
6371	2,2'-Ditolylsulfide or o-Totylsulfide. $(2-CH_3C_6H_4)_2S$	214.33	pl(al)	$285, 174^{15}$	64			al, eth, chl	B6[4], 2019
6372	3,4'-Ditolylsulfide $(3-CH_3C_6H_4)S(C_6H_4CH_3-4)$	214.33	nd(al)	179^{11}	28			al, eth	B6[4], 2173
6373	4,4'-Ditolylsulfide or p-Totylsulfide. $(4-CH_3C_6H_4)_2S$	214.33	nd(al)	$>300, 179^{11}$	57.3			al, eth, ace, bz	B6[4], 2173
6374	2,4'-Ditolylsulfide $(2-CH_3C_6H_4)S(C_6H_4CH_3-4)$	214.33	......	173^{11}	......	$1.0774^{15/4}$		al, eth	B6[4], 2172
6375	2,2'-Ditolylsulfone or o-Tolylsulfone. $(2-CH_3C_6H_4)_2SO_2$	246.33	nd(al)	......	134-5			al, eth, bz, chl	B6[4], 2020
6376	3,4'-Ditolylsulfide $(3-CH_3C_6H_4)SO_2(C_6H_4CH_3-4)$	246.32		......	116				B6[3], 1404
6377	4,4'-Ditolylsulfide $(4-CH_3C_6H_4)_2SO_2$	246.33	pr(bz)nd(w, al)pl(al)	405^{714}	159			bz, chl	B6[4], 2174
6378	4,4'-Ditolylsulfoxide or bis-(4-tolyl)sulfoxide $(4-CH_3C_6H_4)_2SO$	230.33	cr(lig)	......	94			al, eth, bz, chl, aa	B6[4], 2173
6379	Diurea or Dicarbamide $H_2NCONHNHCONH_2$	116.08	pr(w)	......	270				B3[4], 236

No.	Name, Synonyms, and Formula	Mol. wt.	Color, crystalline form, specific rotation and λ_{max} (log ϵ)	b.p. °C	m.p. °C	Density	n_D	Solubility	Ref.
6380	Djenkoic acid or β,β'-Methylene dithio dialanine [HO$_2$CCH(NH$_2$)CHS]$_2$CH$_2$	254.33	nd(w)	....	300-50d	....	....	w	B4³, 1591
6381	Docosane CH$_3$(CH$_2$)$_{20}$CH$_3$	310.61	pl(to) cr(eth)	368.6, 213^{10}	44.4	0.7944$^{20/4}$	1.4455^{20}	al, eth, chl	B1⁴, 572
6382	Docosanoic acid or Behenic acid CH$_3$(CH$_2$)$_{20}$CO$_2$H	340.60	nd	306^{60}	80	0.8223^{90}	1.4270^{100}	....	B2⁴, 1290
6383	Docosanoic acid, ethyl ester or Ethyl behenate C$_{21}$H$_{43}$CO$_2$C$_2$H$_5$	368.65	nd (al) cr (ace)	240-2^{10}	50	0.8820^{51}	....	al, eth	B2⁴, 1292
6384	Docosanoic acid, methyl ester or Methyl behenate C$_{21}$H$_{43}$CO$_2$CH$_3$	354.62	nd(ace)	224-5^{12}	54	....	1.4339^{60}	al, eth	B2⁴, 1291
6385	1-Docosanol or Docosyl alcohol CH$_3$(CH$_2$)$_{20}$CH$_2$OH	326.61	(ace chl)	180$^{0.22}$	71(87)	....	....	al, chl	B1⁴, 1906
6386	4,7,11-Docosatriene-18-ynoic acid or Clupanodonic acid C$_{22}$H$_{34}$O$_2$	330.52	pa ye	236^5	<−78	0.9290^{20}	1.4868^{20}	eth	B1³, 1528
6387	13-Docosenoic acid-(cis) or Erucic acid CH$_3$(CH$_2$)$_7$CH=CH(CH$_2$)$_{11}$CO$_2$H	338.58	nd(al)	265^{15}	33-4	0.860$^{55/4}$	1.4758^{20}	al, eth	B2⁴, 1676
6388	13-Docosenoic acid-(trans) or Brassidic acid CH$_3$(CH$_2$)$_7$CH=CH(CH$_2$)$_{11}$CO$_2$H	338.58	pl(al)	282^{30}	61.5	0.8585$^{57/4}$	1.4472^{64}	....	B2⁴, 1677
6389	13-Docosenoic acid, anhydride-(trans) or Brassidic anhydride (C$_{21}$H$_{41}$CO)$_2$O	659.15	nd (al) pl (eth)	....	64	0.835$^{70/4}$	1.4366^{100}	eth, ace, peth	B2², 448
6390	13-Docosynoic acid or Behenolic acid CH$_3$(CH$_2$)$_7$C≡C(CH$_2$)$_{11}$CO$_2$H	336.57	mcl pr or nd(al)	....	59.5	....	....	al, eth, chl	B2⁴, 1764
6391	1,11-Dodecadiene CH$_2$=CH(CH$_2$)$_8$CH=CH$_2$	166.31	....	208-9^{758}		0.7702$^{20/4}$	1.4400^{20}		B1⁴, 1067
6392	Dodecanal or Lauraldehyde CH$_3$(CH$_2$)$_{10}$CHO	184.32	lf	185^{100}, 100$^{3.5}$	44.5	0.8262$^{15/4}$	1.4367^{22}	al, eth	B1⁴, 3380
6393	Dodecane C$_{12}$H$_{26}$	170.34	....	216.3, 91.5^{10}	−9.6	0.7487$^{20/4}$	1.4216^{20}	al, eth, ace, chl	B1⁴, 498
6394	Dodecanone, 1-amino or Lauryl amine CH$_3$(CH$_2$)$_{10}$CH$_2$NH$_2$	185.36	....	280, 126.5^{10}	27.5	0.8015$^{20/4}$	1.4421^{20}	al, eth, bz, chl	B1⁴, 1791
6395	Dodecane, 1-bromo or Lauryl bromide CH$_3$(CH$_2$)$_{10}$CH$_2$Br	249.24	....	276, 139^{10}	−9.5	1.0399$^{20/4}$	1.4583^{20}	al, eth, ace	B1⁴, 502
6396	Dodecane, 1-bromo-12-fluoro F(CH$_2$)$_{12}$Br	267.29	....	85.8$^{0.15}$		....	1.4524^{25}	al, eth, ace	B1⁴, 502
6397	Dodecane, 1-chloro or Lauryl chloride CH(CH$_2$)$_{10}$CH$_2$Br	204.79	....	260, 126.4^{10}	fr−9.3	0.8687$^{20/4}$	1.4433^{20}	al, ace, bz, lig	B1⁴, 501
6398	Dodecane, 1,12-dibromo Br(CH$_2$)$_{12}$Br	328.14	nd(aa al)	215^{15}	41	....	....	al, eth, chl, aa	B1⁴, 503
6399	Dodecane, 1-iodo or Lauryl iodide CH$_3$(CH$_2$)$_{10}$CH$_2$I	296.24	....	298.2, 153^{10}	0.3	1.1999$^{20/4}$	1.4840^{20}	al, eth, ace, chl	B1⁴, 503
6400	6-Dodecanol C$_5$H$_{11}$CH(OH)C$_6$H$_{13}$	186.34	peth	119^9	30	....	....	al, eth	B1³, 1794
6401	2-Dodecanone or n-Decyl methyl Ketone CH$_3$COC$_{10}$H$_{21}$	184.33	....	246-7, 144^{11}	21	0.8198$^{20/4}$	1.4330^{20}	al, eth, ace	B1⁴, 3382
6402	6-Dodecanone C$_6$H$_3$COC$_5$H$_{11}$	184.32	....	112^9	9	....	1.4302$^{20/}{}_D$		B1⁴, 3383
6403	1-Dodecene CH$_3$(CH$_2$)$_9$CH=CH$_2$	168.33	....	213.4, 88.7^{10}	−35.2	0.7584$^{20/4}$	1.4300^{20}	al, eth, ace, bz	B1⁴, 914
6404	Dodecanedioc acid HO$_2$C(CH$_2$)$_{10}$CO$_2$H	230.29	....		128	....	....		B2⁴, 2126
6405	Dodecanedioic acid, dimethyl ester CH$_3$O$_2$C(CH$_2$)$_{10}$CO$_2$CH$_3$	258.36	pr	167-9^9	31.3	....	....		B2⁴, 2126
6406	1-Dodecanethiol or Lauryl mercaptan CH$_3$(CH$_2$)$_{10}$CH$_2$SH	202.41	....	142-5^{15}		0.8450$^{20/20}$	1.4589^{20}	al, eth	B1⁴, 1851
6407	Dodecanoic acid or Lauric acid CH$_3$(CH$_2$)$_{10}$CO$_2$H	200.33	nd(al)	131^1	44	0.8679$^{50/4}$	1.4304^{50}	al, eth, ace, bz, peth	B2⁴, 1082
6408	Dodecanoic acid, 2-bromo CH$_3$(CH$_2$)$_9$CHBrCO$_2$H	279.23	pl	157-9^2	32	1.1474^{74}	1.4585^{24}	al, eth, bz, chl, lig	B2⁴, 1106
6409	1-Dodecanol or Lauryl alcohol CH$_3$(CH$_2$)$_{10}$CH$_2$OH	186.32	lf(dil al)	255-9, 150^{20}	26	0.8309$^{24/4}$	....	al, eth	B1⁴, 1844
6410	2-Dodecanol CH$_3$(CH$_2$)$_9$CH(OH)CH$_3$	186.34	....	252	19	0.8286^{20}	1.4400$^{20/}{}_D$		B1³, 1793
6411	2-Dodecenedioic acid-(cis) or Traumatic acid HO$_2$CCH=CH(CH$_2$)$_8$CO$_2$H	228.29	(al, ace)	....	67-8	....	....	al, eth, bz, chl	B2³, 1979
6412	1-Dodecene-3-yne CH$_3$(CH$_2$)$_7$C≡C-CH=CH$_2$	164.29	....	78^4		0.7858$^{25/4}$	1.4510^{25}	....	B1⁴, 1112

No.	Name, Synonyms, and Formula	Mol. wt.	Color, crystalline form, specific rotation and λ_{max} (log ε)	b.p. °C	m.p. °C	Density	n_D	Solubility	Ref.
6413	2-Dodenedioic acid-*(trans)* .... $HO_2CCH=CH(CH_2)_8CO_2H$	228.29	(al, ace)	....	165-6	....	....	al, eth, chl	B2[4], 2279
6414	Dodecyl sulfate .... $(C_{12}H_{25}O)_2SO_2$	434.73		....	48.5				B1[4], 1849
6415	Dodecyltrimethyl ammonium chloride .... $C_{12}H_{25}N^+(CH_3)_3Cl^-$	263.90		....	246d			w, al, ace, chl	B4[4], 798
6416	1-Dodecyne .... $CH_3(CH_2)_9C{\equiv}CH$	166.31	....	215, 89[10]	−19	0.7788[20/4]	1.4340[20]	....	B1[4], 1066
6417	2-Dodecyne .... $CH_3(CH_2))_8C{\equiv}CCH_3$	166.31		105[15]	−9	0.7917[15/4]	1.4828[20]		B1, 261
6418	3-Dodecyne .... $CH_3(CH_2)_7C{\equiv}CCH_2CH_3$	166.31		95[12]	....	0.7871[20/4]	1.4442[20]	eth, ace	B1[3], 1025
6419	6-Dodecyne .... $CH_3(CH_2)_4C{\equiv}C(CH_2)_4CH_3$	166.31		209[745], 100[14]	....	0.7871[20/4]	1.4442[20]	al, eth, ace	B1[4], 1067
6420	Dotriacontane or Dicetyl .... $CH_3(CH_2)_{30}CH_3$	450.89	pl (bz, chl, aa, eth)	467, 292.7[10]	69.7	0.8124[20/4]	1.4550[20]	eth, bz	B1[4], 595
6421	1-Dotriacontanol .... $CH_3(CH_2)_{30}CH_2OH$	466.89	pl(bz)	sub 200-50[1]	89.4				B1[4], 1919
6422	Durene or 1,2,4,5-tetramethyl benzene .... $1,2,4,5-(CH_3)_4C_6H_2$	134.22	....	196.8, 73.5[10]	79.2	0.8380[81/4] 0.8875[20/4]	1.4790[81], 1.5116[20]	**al, eth, ace, bz**	B5[4], 1076

No.	Name, Synonyms, and Formula	Mol. wt.	Color, crystalline form, specific rotation and λ_{max} (log ε)	b.p. °C	m.p. °C	Density	n_D	Solubility	Ref.
6423	Ecgonidine-(l) or Anhydroecgonine $C_9H_{13}NO_2$	167.21	cr(MeOH-eth), $[\alpha]^{14}_D$ -84.6 (w, p=1.7)	....	225d			al	B22⁴, 284
6424	Ecgonine-(dl) $C_9H_{15}NO_3$	185.23	pl(w + 3)	....	203			w, al	B22², 156
6425	Ecgonine, Ecqonine $C_9H_{15}NO_3$	125.23	mcl pr, $[\alpha]^{15}_D$ -45.5 (w, c=5)		205			w, al	B22⁴, 2097
6426	Ecgonine hydrate-(l) $C_9H_{15}NO_3 \cdot H_2O$	206.23	mcl pr (al) eff 120-30		198				B22, 196
6427	Ecgonine benzoate-(l) or O-Benzoyl ecqonine $C_{16}H_{19}NO_4$	289.34	nd(w), $[\alpha]^{14}_D$ -63.5 (w, p=1.7)		195			al, bz	B22⁴, 2098
6428	Ecgonine benzoate, ethyl ester (l) or Homococaine $C_{18}H_{23}NO_4$	317.39	pr(eth)		109			al, eth	B22, 202
6429	Ecognine benzoate, tetrahydrate (l) $C_{16}H_{19}NO_4 \cdot 4H_2O$	361.40	pr(w)		92			al, bz	B22, 197
6430	Ecgonine hydrochloride (l) $C_9H_{15}NO_3 \cdot HCl$	221.69	rh(al), $[\alpha]^{25}_D$ -59 (w, c=10)		246			w, al	B22⁴, 2098
6431	Echinochrome A $C_{12}H_{10}O_7$	266.21	dk red nd (to)	sub 120·⁰⁰⁰¹	220d			al, eth, ace	B8³, 4365
6432	Echinochrome A, 3,6,7-trimethyl ether $C_{15}H_{16}O_7$	308.29	dk red lf (aq diox)	....	133 vac			al	B8³, 4361
6433	Echinopsine or N-Methyl-α-quinoline $C_{10}H_9NO$	159.19	α-nd(bz) β-nd(al)	....	α152 β155			w, al, bz, chl	B21⁴, 3722
6434	Echitamidine $C_{20}H_{26}N_2O_3$	342.44	pl(eth), $[\alpha]^{16}_D$ -515 (al)	....	244d			w, al	J1932, 2628
6435	Echitamine or Ditaine $C_{22}H_{28}N_2O_4 \cdot 4H_2O$	456.54	pr (al + 4w) eff −105 (-3w), $[\alpha]^{20}_D$ -29 (al)		206 (+ 1w)			w, al, eth	J1925, 1640
6436	Echitamine, hydrochloride $C_{22}H_{28}N_2O_4 \cdot HCl$	420.94	nd(w), $[\alpha]^{15}_D$ -58 (w, c=1)		295-300d			al	J1925, 1640
6437	Echitin $C_{23}H_{32}O_3$	468.77	lf, $[\alpha]_D$ + 73 (eth)		170			al, eth, ace, bz	M, 397
6438	Egonol $C_{19}H_{18}O_5$	326.35	pl(BuOH)	228-30⁰·¹⁵	118			chl	B19⁴, 4882
6439	Eicosane or Didecyl $C_{20}H_{42}$	282.56	lf(al)	343, 195.7¹⁰	36.8	0.7886²⁰/⁴	1.4425²⁰	eth, ace, bz, peth	B1⁴, 563
6440	Eicosane, 1-cyclohexyl $C_{26}H_{52}$	364.70	....	422	48.5	0.8318²⁰	1.4622²⁰/D	....	B5⁴, 198
6441	Eicosane, 9-octyl $C_{11}H_{23}CH(C_8H_{15})_2$	394.77	....	257¹⁰, 199⁰·⁵	0.5	0.8075²⁰/⁴	1.4515²⁰/D	....	B1⁴, 590
6442	Eicosane, 1-phenyl $CH_3(CH_2)_{19}C_6H_5$	358.65	....	212¹	42.3	0.8235⁶⁰/⁴	1.4725⁴⁰/D	....	B5⁴, 1222
6443	Eicosane, 2-phenyl $CH_3(CH_2)_{17}CH(CH_3)C_6H_5$	358.65	....	204.5¹	29.0	0.8547²⁰/⁴	1.4795²⁰	....	B5⁴, 1222
6444	Eicosone, 3-phenyl $CH_3(CH_2)_{16}CH(C_2H_5)C_6H_5$	358.65	....	202.0¹	29.3	0.8546²⁰/⁴	1.4796²⁰/D	....	B5⁴, 1222
6445	Eicosane, 4-phenyl $CH_3(CH_2)_{15}CH(C_3H_7)C_6H_5$	358.65	....	199.0¹	31.4	0.8546²⁰/⁴	1.4794²⁰/D	....	B5⁴, 1223
6446	Eicosane, 5-phenyl $CH_3(CH_2)_{14}CH(C_4H_9)C_6H_5$	358.65	....	197¹	30.2	0.8549²⁰/⁴	1.4796²⁰/D	....	B5⁴, 1223
6447	Eicosane, 9-phenyl $CH_3(CH_2)_7CH(C_6H_5)(CH_2)_{10}CH_3$	358.65	....	196¹⁰	17.9	0.8534²⁰/⁴	1.4790²⁰	....	B5⁴, 1223
6448	Eicosanedioic acid or Octadecane dicarboxylic acid $HO_2C(CH_2)_{18}CO_2H$	342.53	cr(bz or al)	233-4²	125-6	....	....	eth	B2⁴, 2185
6449	Eicosanedioic acid, diethyl ester or Diethyl eicosanedioate $C_2H_5O_2C(CH_2)_{18}CO_2C_2H_5$	370.58	....	240¹²	54-5			al, eth	B2³, 1881
6450	Eicosanoic acid or Arachidic acid $CH_3(CH_2)_{18}CO_2H$	312.54	pl(al)	328d, 203-5¹	77	0.8240¹⁰⁰/⁴	1.425¹⁰⁰	eth, chl	B2⁴, 1275

No.	Name, Synonyms, and Formula	Mol. wt.	Color, crystalline form, specific rotation and λ_{max} (log ϵ)	b.p. °C	m.p. °C	Density	n_D	Solubility	Ref.
6451	Eicosanoic acid, ethyl ester or Ethyleicosonoate .... $C_{19}H_{39}CO_2C_2H_5$	340.60	....	295-7[100], 186-7[2]	50	....	....	al, eth, bz, chl	B2[3], 1067
6452	Eicosanoic acid, methyl ester or Methyl eicosanoate..... $C_{19}H_{39}CO_2CH_3$	326.57	lf(MeOH)	215-6[10]	54.5	....	1.4317[60]	al, eth, bz, chl	B2[4], 1276
6453	1-Eicosanol or Arachidic alcohol.................... $CH_3(CH_2)_{18}CH_2OH$	298.56	wx(al) cr(chl)	309, 220-5[3]	72-3	0.8405[20/4]	1.4550[20]	ace, bz	B1[4], 1900
6454	2-Eicosonal .... $CH_3(CH_2)_{17}CH(OH)CH_3$	298.56	cr(MeOH)	357	63-4	0.8378[20/4]	1.4312[80]	ace, bz	B1[4], 1901
6455	2-Eicosanone or Methyl octadecyl ketone................ $C_{18}H_{35}COCH_3$	296.54	lf(MeOH)	....	58			al, ace, bz	B1[4], 3402
6456	3-Eicosanone or Ethyl heptadecyl ketone .... $C_{17}H_{35}COC_2H_5$	296.54	lf(al)	....	60-1			eth, ace, bz, chl, aa	B1[4], 3403
6457	3-Eicosanone oxime .... $C_{17}H_{35}C(H)C_2H_5$	311.56	nd(al)		α55-6, β64-5			al, eth	B1[3], 2932
6458	7-Eicosanone or Hexyl tridecyl ketone .... $C_{13}H_{27}COC_6H_{13}$	296.54	cr	210-1[11]	52-3		1.4258[20]	eth, ace	B1[4], 3403
6459	5,8,11,14-Eicosatetraenoic acid or Arachidonic acid .... $CH_3(CH_2)_3(CH_2CH=CH)_4(CH_2)_3CO_2H$	304.48	....	d	−49.5		1.4824[20]	al, eth, ace, chl	B2[4], 1802
6460	1-Eicosene .... $CH_3(CH_2)_{17}CH=CH_2$	280.54	....	341, 151[1.5]	28.5	0.7882[20/4]	1.440[30]	bz, peth	B1[4], 934
6461	1-Eicosyne .... $CH_3(CH_2)_{17}C\equiv CH$	278.53	....	340, 191.8[10]	36	0.8073[20/4]	1.4501[20]	bz, peth	B1[4], 1077
6462	Elaidamide .... $CH_3(CH_2)_7CH=CH(CH_2)_7CONH_2$	281.49			93-4			al	B2[4], 1668
6463	Elaidic acid or 9-Octadecenoic acid (trans)............. $CH_3(CH_2)_7CH=CH(CH_2)_7CO_2H$	282.47	pl(al)	288[100], 234[15]	45	0.8734[45]	1.4499[45]	al, eth, bz, chl	B2[4], 1647
6464	Elaidic acid, dibromide or 9,10-Dibromo octadecanedioic acid $CH_3(CH_2)_7CHBrCHBr(CH_2)_7CO_2H$	442.29	col or ye	....	29-30	1.2458[30/4]	1.4893[42]	eth	B2[3], 1048
6465	Elaidic acid, ethyl ester or Ethylelaidate .... $C_{17}H_{33}CO_2C_2H_5$	310.53		217-9[15]	5.8	0.8664[25]	1.4480[25]	al, eth	B2[4], 1652
6466	Elaidic acid, methyl ester .... $C_{17}H_{33}CO_2CH_3$	296.50		213-5[15]	......	0.8730[20]	1.4513[20]	al, eth	B2[4], 1651
6467	Elaidyl alcohol or 9-Octadecen-1-ol (trans).............. $C_8H_{17}CH=CH(CH_2)_7CH_2OH$	268.48		333, 198[10]	36-7	0.8338[40/4]	1.4552[40]	al, eth, ace	B1[4], 2204
6468	α-Elaterin .... $C_{32}H_{44}O_8$	556.67	cr(chl-MeOH or al), [α][20/D] -64.3 (chl, c=1.6)	....	234			eth, bz, chl	B8[3], 4377
6469	β-Elaterin .... $C_{20}H_{28}O_5$	348.45	nd(al), [α][25/D] + 13.9	....	195.5			chl	B8[3], 4377
6470	Elemane or Dihydroelemene .... $C_{15}H_{30}$	210.41		115-9[10]	......	0.8509[20/4]	1.4640[20]	eth, bz, peth	B5[2], 117
6471	α-Elemene (d) .... $C_{15}H_{24}$	204.35	[α]_D 116 (chl, 14.85%)	120-30[7]	......	0.8782[20/4]	1.5130[26]	ace, bz	B5[3], 1083
6472	Elemenonic acid or Dihydro-β-elemonic acid.......... $C_{30}H_{48}O_3$	436.72	nd(al or AcOEt)	....	249-50				B10[3], 3227
6473	α-Elemol .... $C_{15}H_{26}O$	222.38	cr, [α]_D + 43.7 (chl, c=1.9)	142-3[12]	52-3	0.9345[18/4]	1.4980[18]	....	B6[3], 410
6474	α-Eleostearic acid or 9,11,13-Octadecatrienoic acid (cis) . $C_4H_9[CH=CH]_3(CH_2)_7CO_2H$	278-44	[α]_D -5.8 (chl, c=3.4) nd (al)	235[12]d, 170[1]	49	0.9028[50/4]	1.5112[50]	al, eth	B2[4], 1787
6475	β-Eleostearic acid or 9,11,13-Octadecatrienoic acid (trans) $C_4H_9[CH=CH]_3(CH_2)_7CO_2H$	278.44	lf (al, MeOH)	188[1]	71-2	0.8839[80/4]	1.5000[80]	MeOH, al	B2[4], 1787
6476	Eluetherin .... $C_{16}H_8O_4$	272.30	[α][15/D] 346 (chl)	....	175				B18[4], 1642
6477	Ellagene or Indino-2′,3′:2,3-fluorene .... $C_{20}H_{14}$	254.34	pl(bz)	....	216				E14s, 498
6478	Ellagic acid, dihydrate .... $C_{14}H_6O_8.2H_2O$	338.23	pa ye nd(py)	....	450-80d				B19[4], 3164
6479	Elliptic acid .... $C_{20}H_{18}O_8$	386.36	nd(aq al)	....	190				B18[4], 3342

No.	Name, Synonyms, and Formula	Mol. wt.	Color, crystalline form, specific rotation and λ_{max} (log ε)	b.p. °C	m.p. °C	Density	n_D	Solubility	Ref.
6480	Emeraldine $C_{48}H_{40}N_8$	726.90	indigo-bl pw	....	....			aa(80)	B13³, 257
6481	Emetine (l) or Cephaline-O-methyl ether $C_{29}H_{40}N_2O_4$	480.66	amor pw $[\alpha]^{20}_D$ -50 (chl, c=2)		74 (cor)			al, eth, ace	B23², 449
6482	Emetine, hydrochloride (l) $C_{29}H_{40}N_2O_4 \cdot 2HCl \cdot 7H_2O$	179.69	nd(w), $[\alpha]_D$ + 11 (w, c=1)		269-70d			w, al	B23², 451
6483	Emicymarin $C_{30}H_{46}O_9$	550.70	nd or pr (+ MeOH), $[\alpha]^{20}_D$ + 12.5 (al, c=2.5)		Ca 207			....	B18⁴, 2440
6484	Enneaphyllin $C_{90}H_{154}$	1236.24	rods (bz)		295-6			al	C32, 2686
6485	Eosin or 2,4,5,7-tetrabromfluorescein $C_{20}H_8O_5Br_4$	647.93	ye-red		295-6			al	B19⁴, 2917
6486	Ephedrine (d) $C_6H_5CH(OH)CH(CH_3)NHCH_3$	165.24	pl(w)	225	40			w, al, eth, bz, chl	B13³, 1723
6487	Ephedrine, hydrochloride-(d) $C_{10}H_{15}NO \cdot HCl$	201.70	pl(abs al), $[\alpha]^{20}_D$ + 35.8 (w, c=11.5)		218			w, al	B13³, 1723
6488	Ephedrine (dl) or Racephedrine $C_6H_5CH(OH)CH(CH_3)NHCH_3$	183.24	nd(eth or peth)	133-7¹	76-7			w, al, eth, bz, chl	B13³, 1723
6489	Ephedrine, p-amino dihydrochloride $CH_3CH(NH_2)CH(OH)C_6H_4NH_2$-4.2HCl	239.15	lf(al-eth)		192-3d			w, al	C27, 2762
6490	Ephedrine, hydrochloride (dl) or Ephotonin $C_{10}H_{15}NO.HCl$	201.70	pl(al)		189-90			w, al	B13³, 1724
6491	Ephedrine (l) or Natural ephedrine $C_6H_5CH(OH)CH(CH_3)NHCH_3$	183.25	pl(w + l)	225	40			w, al, eth, bz, chl	B13³, 1720
6492	Ephedrine, hydrochloride (l) $C_{10}H_{15}NO.HCl$	201.70	orb nd, $[\alpha]^{20}_D$ - 36.6		218-20			w, al	B13³, 1721
6493	Ephedrine, N-methyl $C_6H_5CH(OH)CH(CH_3)N(CH_3)_2$	179.26	nd or pl (al or eth), $[\alpha]_D$ -29.5 (MeOH, c=4.5)		87-8			al, eth, MeOH	B13³, 1726
6494	Ephedrine, N-(4-nitrobenzoyl) (dl) $C_6H_5CH(OH)CH(CH_3)NH(CH_3)COC_6H_4NO_2$-(4)	314.35	pa ye pl(al)		162			....	B13², 384
6495	Ephedrine sulfate (l) $(C_{10}H_{15}NO)_2.H_2SO_4$	428.54	hex pl or orh nd(w)		245-8d			w	C21, 2169
6496	Epi-β-amyrin acetate $C_{30}H_{50}O$	426.72	cr(MeOH)		225 (cor)			....	B6³, 2896
6497	Epi-α-amyrin acetate $C_{32}H_{52}O_2$	468.76	nd(chl-MeOH), $[\alpha]_D$ + 39 (chl)		135 (cor)			....	B6³, 2890
6498	Epiandrosterone $C_{19}H_{30}O_2$	290.45	cr(bz-peth ace), $[\alpha]^{20}_D$ + 108 (MeOH)		177-9			....	B8³, 584
6499	Epiborneol (l) or 3-Camphanol $C_{10}H_{18}OH$	154.26	nd(peth)	213⁷⁴²	181-2			....	B6⁴, 288
6500	Epiborneol, acetate or 3-Camphanol acetate $C_{12}H_{20}O_2$	196.29	$[\alpha]^{19}_D$ + 15.63	101¹¹	<-15	0.9872¹⁴/⁰	1.4651¹⁴	....	B6³, 318
6501	Epibreinonol or Breinonol A $C_{28}H_{44}O_2$	440.72	pl (MeOH) nd (chl), $[\alpha]^{17}_D$ + 37 (chl)		204			chl	B8³, 1093
6502	Epicamphor (d) or 3-Bornanone $C_{10}H_{16}O$	152.24	$[\alpha]^{17}_D$ + 45.4 (bz)		182			al, eth, peth	B7³, 420
6503	Epicamphor, oxime (d) or d-3-Bornanone oxime $C_{10}H_{16}NOH$	167.25	nd(MeOH)		103			eth, ace	B7¹, 86
6504	Epicamphor, semicarbazone (d) $C_{10}H_{16}NNHCONH_2$	209.29	nd(al)		237-8			al	B7³, 421

No.	Name, Synonyms, and Formula	Mol. wt.	Color, crystalline form, specific rotation and λ_{max} (log ε)	b.p. °C	m.p. °C	Density	n_D	Solubility	Ref.
6505	Epicamphor (dl) or 3-Bornanone............ $C_{10}H_{16}O$	152.24	cr(peth)	....	175-7	....		al, eth, peth	B7³, 421
6506	Epicamphor, oxime (dl) .... $C_{10}H_{16}NOH$	167.25	nd(dil al)		98-100			eth, ace	B7¹, 87
6507	Epicamphor-(l) or β-Camphor............ $C_{10}H_{16}O$	152.24	....	213	184 (187)			al, eth, bz, peth	B7³, 421
6508	Epicamphor, bromo (l) .... $BrC_{10}H_{15}O$	231.14	nd or pw (peth), $[\alpha]_D$ -86.6 (AcOGt ,c=3.6)		133-4			bz, chl, lig	B7³, 422
6509	Epicamphor, oxime (l) or β-camphor oxime............ $C_{10}H_{16}NOH$	167.25	nd(dil MeOH), $[\alpha]_D$ + 100.5 (bz, c=6.3)		103-4			al, eth, ace	B7¹, 86
6510	Epicamphor, semicarbazone (1)............ $C_{10}H_{16}NNHCONH_2$	209.29	nd(al)	....	237-8			al	B7³, 421
6511	α-Epicamphyl amine .... $C_{10}H_{15}NH_2$	153.27	$[\alpha]_D$ + 17.6 (bz, c=6.5)	127-8¹⁰⁰	....			w	B12³, 176
6512	Epicatechin (dl) or Epicatechol............ $C_{15}H_{14}O_6$	290.28	nd(w + 1)pr (w + 4)		224-6d			al, ace	B17², 258
6513	Epicatechin (l) or Epicatechol............ $C_{15}H_{14}O_6$	290.28	cr(w + 4), $[\alpha]^{25}_D$ -69 (al)		245d			al, ace	B17⁴, 3841
6514	Epicholestan-3-ol or 3-α-hydroxycholestane............ $C_{27}H_{48}O$	388.69	nd (al, MeOH), $[\alpha]^{20}_D$ + 34 (chl)		185-6			eth, chl	B6³, 2135
6515	Epicholestan-4-ol or 4α-Hydroxycholestane............ $C_{27}H_{48}O$	388.69	lf(al MeOH-ace), $[\alpha]^{21}_D$ + 29.0 (chl)		187-8			eth, chl	B6³, 2155
6516	Epicholesterol or 5-Cholesten-3α-ol .... $C_{27}H_{46}O$	386.67	cr(al chl MeOH), $[\alpha]^{30}_D$ - 37.5 (al)		141.5				B6³, 2622
6517	Epicoprostanol or 3-α-Coprostanol............ $C_{27}H_{48}O$	388.69	cr(al ace), $[\alpha]^{20}_D$ + 31.6 (chl)		117-8			al, eth, bz, chl	B6³, 2130
6518	Epicoprostenol or 4-Cholesten-3α-ol............ $C_{27}H_{46}O$	386.67	nd(ace), $[\alpha]^{24}_D$ + 120.8 (bz)		84			al, eth, bz, chl	B6³, 2605
6519	Epidicentrin (dl) or dl-Domesticine methyl ether............ $C_{20}H_{21}NO_4$	339.40	pr(MeOH)		142			al, eth, chl	B27², 553
6520	Epidicentrin (l) or l-Domesticin methyl ether............ $C_{20}H_{21}NO_4$	339.40	pr(MeOH), $[\alpha]^{18}_D$ - 101.3 (chl, c=0.5)		138-9			al, eth, chl	B27², 533
6521	Epiergosterol (d) or Δ-7,9,(11),22-Ergostatrien-3α-ol..... $C_{28}H_{44}O$	396.67	nd(eth-MeOH), $[\alpha]^{19}_D$ + 36.2 (chl)	....	203-4				B6³, 3096
6522	D-epi-Fucitol or 6-deoxy-D-glucitol............ $C_6H_{14}O_5$	166.18	cr(eth w), $[\alpha]^{21}_D$ + 2.2 (w, c=1)		105-7			w, eth	Am74, 4373
6523	L-epi-Fucitol............ $C_6H_{14}O_5$	166.18	cr(w), $[\alpha]^{20}_D$ -2.3 (w, c=1)		105-7			w, eth	C24, 2431
6524	L-epi-Fucose or L-Quinovose............ $CH_3(CHOH)_4CHO$	164.16	cr(AcOEt), $[\alpha]_D$ -36.9 (w, c=6)		135-45			w	B1⁴, 4267

No.	Name, Synonyms, and Formula	Mol. wt.	Color, crystalline form, specific rotation and λ_{max} (log ε)	b.p. °C	m.p. °C	Density	n_D	Solubility	Ref.
6525	Epiisofenchol or 4,6,6-Trimethyl-2-norbornanol .... $C_{10}H_{17}OH$	154.26	nd(sub), $[\alpha]^{20}_D$ -7.35	....	71-2				B6[3], 292
6526	Epiisofenchone or 4,6,6-Trimethyl norbornanone (d) ... $C_{10}H_{16}O$	152.24	$[\alpha]_D$ +19.6	195	......	$0.934^{20/4}$	1.459^{20}	eth, ace	B7[3], 396
6527	Epiisofenchone (dl) or 4,6,6,-Trimethyl norbornanone .. $C_{10}H_{16}O$	152.24	....	195-8, 90-3[21]			1.4625^{25}	eth, ace	B7[3], 397
6528	Epilupinine or d-isolupinine .... $C_{11}H_{19}O$	169.27	nd(peth), $[\alpha]^{17}_D$ +32 (al, c=1.5)		76-8			eth, ace, bz	B21[4], 290
6529	Epinine .... $C_9H_{13}NO_2$	167.21	nd(al)	....	188-9				B13[3], 2209
6530	Epiquinidine .... $C_{20}H_{24}N_2O_2$	324.43	cr (AcOEt) lf (eth), $[\alpha]^{20}_D$ +103.7 (al, c=1.86)		113			al, eth	B23, 505
6531	Epirhodanhydrin or 2,3-Epoxy propyl rhodanine .... CH₂CHCH₂SCN	115.16	dk red liq (garlic odor)	d				al, chl	B17, 106
6532	Episarsapogenin .... $C_{27}H_{44}O_3$	416.65	nd(ace), $[\alpha]_D$ -71		204-6				B19[4], 825
6533	Epitruxillic acid or 2,4-cis-Diphenyl cyclobutane-1,3-trans-dicarboxylic acid .... $C_{18}H_{16}O_4$	296.35	cr(dil al or bz-aa)		285-7				B9[3], 4625
6534	2,3-Epoxypropyl ethyl ether .... CH₂CHCH₂OC₂H₅	102.13	....	128	......	0.9700^{20}	1.4320^{20}	w, al, eth	B17[4], 987
6535	2,3-Epoxypropyl phenyl ether .... CH₂CHCH₂OC₆H₅	150.10	....	242.5^{755}		$1.1109^{21.7/4}$	$1.5307^{21/D}$		B17[4], 990
6536	Equilenin (d) .... $C_{18}H_{18}O_2$	266.34	$[\alpha]^{16}_D$ +87 (diox)	sub 170-80[0.1]	258-9				B8[3], 1523
6537	Equilenin (dl) .... $C_{18}H_{18}O_2$	266.34	cr(bz)		276-8				B8[3], 1525
6538	Equilenin (l) .... $C_{18}H_{18}O_2$	266.34	$[\alpha]^{20}_D$ -85 (diox)	sub 170-80[1]	258-9			al	B8[3], 1522
6539	Equilin .... $C_{18}H_{20}O_2$	268.36	orh sph pl(AcOEt)	170-200 (sub vac)	238-40			al, ace	B8[3], 1415
6540	Equilin, α-dihydro .... $C_{18}H_{22}O_2$	270.38	cr(ace), $[\alpha]_D$ +220 (diox, c=1)	....	174-6			al, ace, diox	B6[3], 5530
6541	Equisetrin .... $C_{27}H_{30}O_{16}$	610.53	ye nd (+2w)	....	195-6			al	B18, 3296
6542	Equol or 4,7-Isoflavandiol .... $C_{15}H_{14}O_3$	242.28	cr(aq al)		189-90				B17[4], 2186
6543	Eremophilol .... $C_{15}H_{24}O$	220.36	visc oil	164.5^{13}			1.5202^{20}		B6[3], 2080
6544	Eremophilone .... $C_{15}H_{22}O$	218.34	nd(MeOH)	171^{15}	42-3	$0.9994^{25/25}$	1.5182^{25}		B7[3], 1266
6545	Eremophilone, 8,9-epoxy .... $C_{15}H_{22}O_2$	234.34	nd(peth)	....	63-4			al, eth, ace, bz	B17[4], 4767
6546	Ergine or Ergonovine .... $C_{16}H_{17}N_3O$	283.33	cr (MeOH) Pr (aq ace)		135-40				H32, 506
6547	Ergocornine .... $C_{31}H_{39}N_5O_5$	561.70	cr(MeOH), $[\alpha]^{20}_D$ -188 (chl, c=1)	....	182-4d			al, ace, bz, chl	H52, 1549
6548	Ergocorninine .... $C_{31}H_{39}N_5O_5$	561.70	lo pr (al), $[\alpha]^{20}_D$ +409 (chl, c=1)		228d			al, ace, bz, chl	H52, 1549
6549	Ergocristine .... $C_{35}H_{39}N_5O_5$	609.63	rh (bz+2), $[\alpha]^{20}_D$ -183 (chl, c=1) -93 Py		175d			al, ace, chl	H34, 1944

No.	Name, Synonyms, and Formula	Mol. wt.	Color, crystalline form, specific rotation and λ_{max} (log ε)	b.p. °C	m.p. °C	Density	n_D	Solubility	Ref.
6550	Ergocristinine $C_{35}H_{39}N_5O_5$	609.63	pr(al), $[\alpha]^{20}_D$ +366 (chl, c=0.68)	....	237-8d	....		....	B27², 860
6551	Ergocryptine $C_{32}H_{41}N_5O_5$	575.72	pr(al), $[\alpha]^{20}_D$ -187 (chl, c=1)	....	212-4d	....		al, chl	B27², 860
6552	Ergocryptinine $C_{32}H_{41}N_5O_5$	575.72	lo pr (al) $[\alpha]^{20}_D$ +408 (chl, c=1) +479 Py		245d,			ace, chl	B27², 860
6553	Ergometrine (l) or Ergobasine $C_{19}H_{23}N_3O_2$	325.42	nd(bz), $[\alpha]^{20}_D$ -89 (w)	....	159-62d			al, ace	Am60, 1701
6554	Ergometrinine (d) or Ergobasinine $C_{19}H_{23}N_3O_2$	325.42	pr(ace), $[\alpha]^{20}_D$ +416 (chl, c=0.26)		195-7d			chl, py	Am60, 1701
6555	Ergometrinine (l) $C_{19}H_{23}N_3O_2$	325.42	pr(ace), $[\alpha]^{20}_D$ -41.5 (chl, c=0.26)		196			chl, py	Am60, 1701
6556	Ergopinacol II or Bisergostadienol $C_{56}H_{36}O_2$	791.31	nd(bz-al), $[\alpha]_D$ -155 (Py, c=0.8)		205				B6³, 5897
6557	Ergosine $C_{30}H_{37}N_5O_5$	547.66	pr(MeOH, AcOEt),		228d			ace, chl	H34, 1544
6558	Ergosinine $C_{30}H_{37}N_5O_5$	547.66	Pr (al), (aq ace, bz) nd MeOH, $[\alpha]^{20}_D$ +420 (chl, c=1)	....	220d			ace, chl	J1937, 396
6559	$\Delta^{5:6,7}:^8$-Ergostadien-3β-ol or Provitamin D₄ $C_{28}H_{46}O$	398.88	nd(MeOH-AcOEt), $[\alpha]^{19}_D$ -109 (chl)	....	152-3				B6³, 2836
6560	$\Delta^{14,22}$-Ergostadien-3β-ol $C_{28}H_{46}O$	398.68	cr(MeOH), $[\alpha]^{20}_D$ -9 (chl)		116				E14s, 1762
6561	α-Ergostadienone $C_{28}H_{44}O$	396.67	lf(al), $[\alpha]^{19}_D$ +2 (chl)		182-3				B7³, 1767
6562	Ergostane or 24-Methyl-5α-cholestane $C_{28}H_{50}$	386.71	lf or pl(eth-MeOH or ace), $[\alpha]^{25}_D$ +21 (chl, c=2)	....	85			eth, ace, chl	B5⁴, 1234
6563	Ergostanol or Ergostan-3β-ol $C_{28}H_{50}O$	402.71	nd(MeOH-eth), $[\alpha]_D$ +15.4 (chl, c=1.8)		144-5			eth, chl	B6³, 2161
6564	$\Delta^{3,5,7,22}$-Ergostatetraene $C_{28}H_{42}$	378.65	pl(al)$[\alpha]^{20}_D$ -40.5 (chl)	....	104				B5³, 1927
6565	$\Delta^{4,6,22}$-Ergostatrienone or Isoergosterone $C_{28}H_{42}O$	394.65	nd(ace-eth)		110				B7³, 2031
6566	α-Ergostenol or α-Tetrahydroergosterol $C_{28}H_{46}O$	400.70	lf or nd (MeOH) nd (gl aa), $[\alpha]^{16}_D$ +11 (MeOH, c=0.9)		131 (135)			eth, bz, chl	B6³, 2685

No.	Name, Synonyms, and Formula	Mol. wt.	Color, crystalline form, specific rotation and λ_{max} (log ε)	b.p. °C	m.p. °C	Density	n_D	Solubility	Ref.
6567	β-Ergostenol or β-Tetrahydroergosterol $C_{28}H_{46}O$	400.70	pl or ta(al), $[\alpha]^{20}_D$ +21.2 (chl, c=0.9)	....	141-2			....	B6³, 2689
6568	α-Ergostenol or α-Tetrahydroergosterol $C_{28}H_{46}O$	400.70	nd (MeOH) cr (ProH) lf (w)		148 (152)				B6³, 2683
6569	δ-Δ-⁸:⁹-Ergostenol $C_{28}H_{46}O$	400.70	$[\alpha]_D$ +39		155			chl	B6³, 2685
6570	Ergosterol or Δ⁵,⁷,²²-Ergostatrien-2β-ol $C_{28}H_{44}O$	396.67	pl (+ w, al) nd (eth), $[\alpha]^{20}_D$ -135 (chl, c=1.2)	250⁰·⁰¹	168 (+ w)			bz, chl	B6³, 3099
6571	Ergosterol D $C_{28}H_{44}O$	396.77	nd(al), $[\alpha]^{17}_D$ +24.6		167			bz, chl	B6³, 3120
6572	Ergosterol-5,6-dihydro or α-Dihydroergosterol $C_{28}H_{46}O$	398.68	lf (ace) pl (chl-MeOH), $[\alpha]^{20}_D$ -19 (chl)	....	176-7				B6³, 2841
6573	Ergosterone $C_{?}H_{?}O$	394.65	nd(ace-MeOH), $[\alpha]^{20}_D$ -4.52 (chl)		132				B7³, 2032
6574	Ergotamine $C_{33}H_{35}N_5O_5$	581.68	nd (al) pr (bz) pl (aq ace), $[\alpha]^{20}_D$ -160 (chl, c=1)	....	213-4d			eth, bz, chl	B27², 860
6575	Ergotaminine $C_{33}H_{35}N_5O_5$	581.68	rh pl (MeOH) pl (al), $[\alpha]^{20}_D$ +369 (chl, c=0.5)	....	252d			chl, py	B27², 860
6576	Ergothioneine or Thiasine $C_9H_{15}N_3O_2S$	229.30	pl(w + 2)nd or lf (dil al), $[\alpha]_D$ +115 (w, c=1)		290d				B25², 413
6577	Erucic acid or cis-13-Docosenoic acid $CH_3(CH_2)_7CH=CH(CH_2)_{11}CO_2H$	338.58	nd(al)	265¹⁵	33-4	0.860⁵⁵/⁴	1.4758²⁰	al, eth	B2⁴, 1676
6578	Erysocine $C_{18}H_{21}NO_3$	299.37	nd(eth), $[\alpha]_D$ +238.1	....	162			al, eth, chl	B21⁴, 2624
6579	Erysodine $C_{18}H_{21}NO_3$	299.37	nd(al), $[\alpha]^{27}_D$ +248 (al)	....	204-5			al, eth	B21⁴, 2623
6580	Erysonine $C_{17}H_{19}NO_3$	285.35	cr(al)$[\alpha]^{25}_D$ +285 (aq.HCl)		236-7d				B21⁴, 2623
6581	Erysopine $C_{17}H_{19}NO_3$	285.35	cr(al), $[\alpha]^{25}_D$ +265.2 (al-glycerol)	....	241-2				B21⁴, 2622
6582	Erysothiopine $C_{19}H_{21}NO_3S$	407.45	cr(al-w), $[\alpha]^{25}_D$ +194 (al)	....	168-9				B21⁴, 2624
6583	Erysovine $C_{18}H_{21}NO_3$	299.37	pr(eth), $[\alpha]_D$ +252 (al)	....	178-9			al, eth, chl	B21⁴, 2623
6584	Erythraline $C_{18}H_{19}NO_3$	297.34	cr(al), $[\alpha]^{27}_D$ +211.8 (al)	....	106-7			al, chl	Am73, 589

No.	Name, Synonyms, and Formula	Mol. wt.	Color, crystalline form, specific rotation and λ_{max} (log ε)	b.p. °C	m.p. °C	Density	n_D	Solubility	Ref.
6585	Erythramine or Dihydroerythraline................. $C_{18}H_{21}NO_3$	299.37	cr(eth-peth), $[\alpha]^{20}_D$ +228 (al, c=0.19)	125[4×10-4]	103-4	....	....	al, eth, ace, bz	Am73, 589
6586	Erythraline .... $C_{18}H_{19}NO_4$	315.37	cr(eth-peth), $[\alpha]^{28}_D$ +145.5 (al)	....	170	....	....	....	Am73, 589
6587	meso-Erythritol .... $HOCH_2(CHOH)_2CH_2OH$	122.12	bipym tetr pr,	329-31	121.5	1.451[20/4]	....	w, py	B1[4], 2807
6588	Erythritol anhydride.....................	86.09	....	138	......	1.113[18/4]	....	....	....
6589	Erythritol, tetranitrate or Cardilate	302.12	leaflets	....	61	....	....	al, w	....
6590	β-Erythroidine .... $C_{16}H_{19}NO_3$	273.34	cr(abs al), $[\alpha]^{25}_D$ +88.8 (w)	....	99-100	....	....	w, al, eth, bz, chl	Am80, 3905
6591	D-Erythronic acid-γ-lactone.................... $C_4H_6O_4$	118.09	pr, $[\alpha]^{20}_D$ -73.2 (w, c=4)	....	104-5	....	....	w, al	B18[4], 1099
6592	L-Erythronic acid-γ-lactone.................... $C_4H_6O_4$	118.09	nd (AcOEt), $[\alpha]^{20}_D$ +73 (w, c=4)	....	105	....	....	al	B18[4], 1099
6593	D-Erythrose .... $C_4H_8O_4$	120.12	syr $[\alpha]^{20}_D$ +1→-14.3 (w, c=11)	....	....	....	....	w, al	B1[4], 4172
6594	L-Erythrose .... $C_4H_8O_4$	120.12	syr $[\alpha]^{24}_D$ +11.5→ +30.5 (w, c=3)	....	....	....	....	....	B1[4], 4172
6595	Erythrosin or 2,4,5,7-Tetraiodo fluorescein............ $C_{20}H_8O_5I_4$	835.91	og-ye (eth)	....	....	....	....	al, eth	B19[4], 2923
6596	L-Erythrulose.................... $HOCH_2CH(OH)COCH_2OH$	120.12	syr $[\alpha]D$ + 11.2	d	....	....	....	w, al	B1[4], 4176
6597	Escholerine .... $C_{41}H_{61}NO_{13}$	775.95	pl (ace-w), $[\alpha]^{25}_D$ -30 (Py, c=1)	....	235d	....	....	ace	B21[4], 6844
6598	Esculin or Aesculin (αβ-glucoside) .... $C_{15}H_{16}O_9$	340.29	pr (w + 2) $[\alpha]^{18}_D$-78.4 (50% aq diox)	230d	205d	....	....	aa, py	B18[4], 1326
6599	α-Estradiol .... $C_{18}H_{24}O_2$	272.39	nd(+ ½w) (80% al), $[\alpha]^{20}_D$ +56 (diox, c=0.9)	....	220-3	....	....	al, ace, peth	B6[3], 5332
6600	β-Estradiol .... $C_{18}H_{24}O_2$	272.39	pr (80% al), $[\alpha]^{25}_D$ +76 (diox)	....	178-9	....	....	al, ace, diox	B6[3], 5337
6601	Estriol .... $C_{18}H_{24}O_3$	288.39	lf(al) mcl (dil al), $[\alpha]_D$ +61 (al) +30 (Py)	....	288d	1.27	....	....	B6[3], 6520
6602	Estrone .... $C_{18}H_{22}O_2$	270.38	α mcl(al) βγ orth(al)	....	260.2	β-1.236, γ-1.228	....	diox, py	B8[3], 1171
6603	Ethane .... CH_3CH_3	30.07	gas, hex cr	-88.6	-183.3	0.572[-100/4]	1.0377[0/546 mm]	bz	B1[4], 108
6604	Ethane, 1,2-bis (ethylthio) .... $C_2H_5SCH_2CH_2SC_2H_5$	150.31	....	217, 95.5[12]	......	0.9815[20/4]	1.5118[20]	al, eth	B1[4], 2452

No.	Name, Synonyms, and Formula	Mol. wt.	Color, crystalline form, specific rotation and λ_{max} (log ε)	b.p. °C	m.p. °C	Density	n_D	Solubility	Ref.
6605	Ethane, 2,2-bis (4-methoxyphenyl)-1,1,1-trichloro or Methoxychlor .... (4-CH₃OC₆H₄)₂CHCCl₃	345.66	cr (dil al)	....	94	....	....	al, eth, bz	B6³, 5436
6606	Ethane, 1,2 bis(methyl thio) .... CH₃SCH₂CH₂SCH₃	122.25	....	182.5⁷⁵⁰, 78-80¹¹	....	1.0371²⁰/⁴	1.5292²⁰	w, al, eth, ace, chl	B1⁴, 2457
6607	Ethane, 1,2-bis (phenyl sulfonyl) C₆H₅SO₂CH₂CH₂O₂SC₆H₅	310.39	nd or lf (al)	....	180	....	....	al, bz, aa	B6⁴, 1495
6608	Ethane, 1,2-bis (phenylthio) .... C₆H₅SCH₂CH₂SC₆H₅	246.40	ta (al)	....	70	....	....	ace	B6⁴, 1493
6609	Ethane, 2,2-bis (4 chlorophenyl)-1,11-trichloro or DDT, Dichloro diphenyl trichloro ethane.............. C₁₄H₉Cl₅	354.49	nd (al)	260	108-9	....	....	eth, ace, bz, chl, peth	B5³, 1833
6610	Ethane, bromo or Ethyl bromide............... C₂H₅Br	108.97	....	38.4	−118.6	1.4604²⁰/⁴	1.4239²⁰	al, eth, chl	B1⁴, 150
6611	Ethane, 1-bromo-2-chloro or Ethylene chloro bromide ... ClCH₂CH₂Br	143.42	....	107	−16.7	1.7392²⁰/⁴	1.4908²⁰	al, eth	B1⁴, 155
6612	Ethane, 1-bromo-2-fluoro .... FCH₂CH₂Br	126.96	....	71-2	....	1.7044²⁵/⁴	1.4236²⁰	al, eth	B1⁴, 154
6613	Ethane, 1-bromo-2-methoxy .... BrCH₂CH₂OCH₃	139.99	....	110.3	....	1.4623²⁰/⁴	1.44753²⁰	....	B1⁴, 1386
6614	Ethane, chloro or Ethyl chloride .... C₂H₅Cl	64.52	....	12.3	−136.4	0.8978²⁰/⁴	1.3676²⁰	al, eth	B1⁴, 124
6615	Ethane, 1-chloro-2-fluoro .... FCH₂CH₂Cl	82.51	....	59⁷⁵⁰	....	1.1747²⁰/⁴	1.3775²⁰	al, eth	B1⁴, 127
6616	Ethane, 1-chloro-2-iodo .... ICH₂CH₂Cl	190.41	....	140	−15.6	2.1643⁰⁄	....	....	B1⁴, 167
6617	Ethane, 1-chloro-1-nitro .... CH₃CHCl(NO₂)	109.51	....	124-5	....	1.2860²⁰/²⁰	....	....	B1⁴, 172
6618	Ethane, chloro pentafluoro .... ClCF₂CF₃	154.47	gas	−38	−106	....	....	al, eth	B1⁴, 129
6619	Ethane, 2-chloro-1,1,1-trifluoro .... ClCH₂CF₃	118.49	....	6.93	−105.5	1.389⁰/⁴	1.3090⁰/₀	....	B1⁴, 128
6620	Ethane, 1-(2 chloroethoxy)-2-phenoxy .... ClCH₂CH₂OCH₂CH₂OC₆H₅	200.67	....	149¹⁰	....	1.149¹⁵/¹⁵	....	....	B6³, 568
6621	Ethane, 1,1-dibromo or Ethylidene bromide .... CH₃CHBr₂	187.87	....	108, 9.0¹⁰	−63	2.0555²⁰/⁴	1.5128²⁰	al, eth, ace, bz	B1⁴, 157
6622	Ethane, 1,2-dibromo or Ethylene dibromide .... CH₂BrCH₂Br	187.87	....	131.3, 29.1¹⁰	9.8	2.1792²⁰/⁴	1.5387²⁰	al, eth, ace, bz	B1⁴, 158
6623	Ethane, 1,2-dibromo-1,1-dichloro .... BrCH₂CBrCl₂	256.76	....	178.3, 58.8¹⁰	−66.8	2.2623²⁰/⁴	1.5567²⁰	al, eth, ace, bz	B1⁴, 161
6624	Ethane, 1,2-dibromo-1,2-dichloro .... BrCHCl-CHBrCl	256.76	....	195, 84⁴⁵	−26	2.135²⁰/⁴	1.5662²⁰	al, eth, ace, bz	B1⁴, 161
6625	Ethane, 1,2-dibromo-tetrafluoro .... BrCF₂CF₂Br	259.82	....	46.4	−112	2.149²⁵/⁴	....	....	B1⁴, 160
6626	Ethane, 1,1-dichloro or Ethylidene chloride........... CH₃CHCl₂	98.96	....	57.3	−97	1.1757²⁰/⁴	1.4164²⁰	al, eth, ace, bz	B1⁴, 130
6627	Ethane, 1,1-dichloro-1-fluoro .... CH₃CCl₂F	116.95	....	32	−103.5	1.250¹⁰/⁴	1.3600¹⁰/_D	....	B1⁴, 134
6628	Ethane, 1,1-dichloro-1,2,2,2 tetrafluoro .... F₃CCCl₂F	170.92	....	3.6	−94	1.455²⁵/⁴	1.3092⁰	al, eth, bz, chl	B1⁴, 136
6629	Ethane, 1,2-dichloro or Ethylene dichloride........... ClCH₂CH₂Cl	98.96	....	83.5	−35.3	1.2351²⁰	1.4448²⁰	al, eth, ace, bz	B1⁴, 131
6630	Ethane, 1,2-dichloro-1,1-difluoro .... CH₂ClClCClF₂	134.97	....	46.8	−1012	1.4163²⁰/⁴	1.36193²⁰/_D	....	B1⁴, 135
6631	Ethane, 1,2-dichloro-1-fluoro .... ClCH₂CHClF	116.95	....	73.7	....	1.3814²⁰/⁴	1.41132²⁰/_D	....	B1⁴, 134
6632	Ethane, 1,1-difluoro or Ethylidene fluoride .... CH₃CHF₂	66.05	gas	−24.7	−117	0.95²⁰ (sat pr)	1.3011⁻⁷²	....	B1⁴, 120
6633	Ethane, 1,1-difluoro-1,2,2,2-tetrachloro .... Cl₃CCClF₂	203.83	....	91.5	40.6	....	....	al, eth, chl	B1⁴, 146
6634	Ethane, 1,2-difluoro or Ethylene difluoride .... CH₂FCH₂F	66.05	....	30.7	....	....	....	eth, bz, chl	B1⁴, 121
6635	Ethane, 1,2-difluoro-1,1,2,2-tetrachloro .... FCCl₂-CFCl₂	203.83	....	93	25	1.6447²⁵/⁴	1.4130²⁵	al, eth, chl	B1⁴, 146
6636	Ethane, 1,1-diiodo or Ethylidine iodide .... CH₃CHI₂	281.86	....	179-80, 60-1¹²	....	2.84⁰	1.673²⁰	al, eth, chl, ace	B1⁴, 169

No.	Name, Synonyms, and Formula	Mol. wt.	Color, crystalline form, specific rotation and λ_{max} (log ε)	b.p. °C	m.p. °C	Density	n_D	Solubility	Ref.
6637	Ethane, 1,2-diiodo or Ethylene diiodo .... CH_2ICH_2I	281.86	ye mcl pr or rh (eth) d in lt	200, 74[10]	83	3.325[20/4]	1.871[20]	al, eth, ace, chl	B1[4], 169
6638	Ethane, 1,2-di-N-morpholyl .... $C_{10}H_{20}N_2O_2$	200.28	wh-yesh (eth or lig)	160-3[25]	75	....	....	w, al, ace, bz	B27, 7
6639	Ethane, 1,1-dinitro .... $CH_3CH(NO_2)_2$	120.07	ye mcl (bz or MeOH)	185-6, 72[12]	......	1.3503[24/24]	....	al, eth	B1[4], 174
6640	Ethane, 1,2-dinitro-1,1,2,2-tetrafluoro .... $O_2NCF_2CF_2(NO_2)$	192.03	....	58-9	−41.5	1.6024[25/4]	1.3265[25]	ace	B1[4], 175
6641	Ethane, 1,1-diphenyl or α-Methylditon............. $CH_3CH(C_6H_5)_2$	182.27	....	286, 148[15]	−21.5	0.9997[20/4]	1.5756[20]	al, eth, bz	B5[4], 1880
6642	Ethane, 1,2-diphenyl or Bibenzyl................. $C_6H_5CH_2CH_2C_6H_5$	182.27	nd (al)	285, 95-6[1]	52.2	0.9583[60/4]	1.5478[60]	al, eth	B5[4], 1868
6643	Ethane, 1,2-di-(4-tolyl) .... $(4-CH_3C_6H_4)CH_2CH_2(C_6H_4CH_3-4)$	210.32	lf(MeOH or dil al) pl (lig)	296-8, 178[18]	82-3	....	....	bz, peth	B5[4], 1943
6644	Ethane, 2,2-di-4-tolyl-1,1,1-trichloro .... $(4-CH_3C_6H_4)_2CHCCl_3$	313.56	mcl pr (al eth-al)	....	92	....	....	al, eth, ace	B5[4], 1949
6645	Ethane, 1-ethoxy-2-methylamino or Ethyl-β-methyl amino ethyl ester.............................. $C_2H_5OCH_2CH_2NHCH_3$	103.17	....	114-5[744]	......	0.8363[20/4]	1.4147[20]	w, al, eth, ace, bz	B4[3], 647
6646	Ethane, fluoro or Ethyl fluoride .... CH_3CH_2F	48.06	gas	−37.7	−143.2	0.7182[20/4] (liq)	1.2656[20]	al, eth	B1[4], 120
6647	Ethane, 1-fluoro-1,2,2-trichloro .... $Cl_2CHCHFCl$	151.40	....	101-3	......	1.54968[17]	1.4390[20/D]		B1[4], 141
6648	Ethane, fluoro penta chloro .... FCl_2CCCl_3	220.29	....	134-6	101.3	....	....	al, eth	B1[4], 148
6649	Ethane, hexabromo or Perbromo ethane .... C_2Br_6	503.48	rh pr (bz)	d200-10	d	2.823[20/4]	1.863	....	B1[3], 193
6650	Ethane, hexachloro or Perchloro ethane........... C_2Br_6	236.74	rh (al-eth)	186[777]	186-7 (sealed tube)	2.091[20/4]	....	al, eth, bz	B1[4], 148
6651	Ethane, hexafluoro or Perfluoro ethane .... C_2F_6	138.01	gas	−79	−94	1.590[-78]	....		B1[4], 123
6652	Ethane, hexaphenyl .... $(C_6H_5)_3CC(C_6H_5)_3$	486.67	cr (ace)	d	145-7d	....	....	eth, ace, chl, MeOH	B5[3], 2746
6653	Ethane, iodo or Ethyl iodide .... C_2H_5I	155.97	....	72.3	−108	1.9358[20/4]	1.5133[20]	al, eth	B1[4], 163
6654	Ethane, isocyano or Ethyl carbylamine............ CH_3CH_2NC	55.08	....	79[775]	<−66	0.7402[20/4]	1.3622[20]	al, eth, ace	B4[4], 342
6655	Ethane-1-(4-methoxyphenyl)-1-phenyl or 1-p-Anisyl-1-phenyl ethane. $CH_3CH(C_6H_5)(C_6H_4OCH_3-4)$	212.30	....	180-2[19]	......	1.0473[20/4]	1.5725[20]	eth, ace, bz	B6[2], 639
6656	Ethane, nitro .... $C_2H_5NO_2$	75.07	....	115	−50	1.0448[25/4]	1.3917[20]	al, eth, ace	B1[4], 170
6657	Ethane, nitro-pentafluoro .... $CF_3CF_2NO_2$	165.02	....	0	......	....	....	eth	B1[4], 172
6658	Ethane, 1,nitro-2,2,2-trifluoro .... $F_3CCH_2NO_2$	129.04	....	96	......	1.3914[20/4]	1.3394[20]	eth	B1[4], 172
6659	Ethane, nitroso-pentafluoro .... CF_3CF_2NO	149.02	....	−42	......	....	....		B1[4], 169
6660	Ethane, pentabromo .... Br_2CHBr_3	424.58	mcl pr (dil al)	210[300]	56-7	3.312[20/4]	....	al, eth	B1[3], 193
6661	Ethane, pentachloro .... $CHCl_2CCl_3$	202.30	....	162	−29	1.6796[20/4]	1.5025[20]	al, eth	B1[4], 147
6662	Ethane, pentaiodo .... CHI_2CI_3	659.55	mcl pr (aa)	....	182-4	....	....	al, eth, bz, aa	B1[1], 31
6663	Ethane, perfluoro .... CF_3CF_3	138.01	....	−79	−100.6	1.590[-98]	....		B1[4], 123
6664	Ethane, 1,1,1,2-tetrabromo .... CH_2BrCBr_3	345.67	....	112[18]d	0.0	2.8748[20/4]	1.6277[20]	al, eth, ace, bz, chl	B1[4], 162
6665	Ethane, 1,1,2,2,-tetrabromo .... $CHBr_2CHBr_2$	345.67	yesh	243.5, 114.8[10]	0	2.9656[20/4]	1.6353[20]	al, eth, ace, bz, aa	B1[4], 162
6666	Ethane, 1,1,1,2--tetrachloro .... CH_2ClCCl_3	167.85	yesh red	130.5, 22.1[10]	−70.2	1.5406[20/4]	1.4821[20]	al, eth, ace, bz, chl	B1[4], 143
6667	Ethane, 1,1,2,2-tetrachloro .... $CHCl_2CHCl_2$	167.85	....	146.2, 33.9[10]	−36	1.5953[20/4]	1.4940[20]	al, eth, ace, bz	B1[4], 144

No.	Name, Synonyms, and Formula	Mol. wt.	Color, crystalline form, specific rotation and λ_{max} (log ε)	b.p. °C	m.p. °C	Density	n_D	Solubility	Ref.
6668	Ethane, 1,1,1,2-tetrafluoro CH_2FCF_3	102.03	....	-26.5^{736}	....			eth	B1[4], 123
6669	Ethane, 1,1,1,2-tetraphenyl $C_6H_5CH_2CCC_6H_5)_3$	334.47	mcl (eth-peth)	277-80^{21}	143-4			bz	B5[3], 2575
6670	Ethane, 1,1,2,2-tetraphenyl $(C_6H_5)_2CHCH(C_6H_5)_2$	334.47	cr (bz + l) rh nd (chl)	358-62, 260^{16}	214-5			aa	B5[3], 2574
6671	Ethane, 1,1,2-tribromo $CH_2BrCHBr_2$	266.77	....	188.9, 73.1^{10}	-29.3	$2.6211^{20/4}$	1.5933^{20}	al, eth, bz, chl	B1[4], 161
6672	Ethane, 1,1,1-trichloro or Methyl chloroform CH_3CCl_3	133.41	....	74.1	-30.4	$1.3390^{20/4}$	1.4379^{20}	al, eth, chl	B1[4], 138
6673	Ethane, 1,1,1-trichloro-2,2,2-trifluoro CF_3CCl_3	187.38		45.8	14.2	$1.5790^{20/4}$	1.3610^{25}	al, eth, chl	B1[4], 142
6674	Ethane, 1,1,2-trichloro $CH_2ClCHCl_2$	133.41		113.8, 9.5^{10}	-36.5	$1.4397^{20/4}$	1.4714^{20}	al, eth, chl	B1[4], 139
6675	Ethane, 1,1,2-trichloro-1,2,2-trifluoro CF_2ClCCl_2F	187.38		47.7	-36.4	$1.5635^{25/4}$	1.3557^{25}	al, eth, bz	B1[4], 142
6676	Ethane, 1,1,1-trifluoro CH_3CF_3	84.04	gas	-47.3	-111.3			eth, chl	B1[4], 122
6677	Ethane, 1,1,1-triiodo CH_3CI_3	407.76	ye oct (al)		95			eth, bz	B1[3], 199
6678	Ethane, 1,1,1-triphenyl or α-Methyl tritan $CH_3C(C_6H_5)_3$	258.37	nd (al, eth)	205-10^{18}	95			eth, bz	B5[3], 2331
6679	Ethane, 1,1,2-triphenyl $C_6H_5CH_2CH(C_6H_5)_2$	258.37	mcl lf (dil al) nd (al)	348-9^{751}	57			al, eth, bz	B5[3], 2329
6680	Ethanediol or Ethylene glycol $HOCH_2CH_2OH$	62.07	....	198, 93^{11}	-11.5	$1.1088^{20/4}$	1.4318^{20}	w, al, eth, ace	B1[3], 2369
6681	1,2-Ethanediol, 1,2-dicylohexyl (dl) or Cyclohexanone pinacol $C_6H_{11}CH(OH)\text{-}CH(OH)C_6H_{11}$	226.36	nd		129-30			bz, peth	B6[3], 4156
6682	1,2-Ethanediol, 1,2-diphenyl (d) $C_6H_5CH(OH)CH(OH)C_6H_5$	214.27	nd (w) lf or pr (abs al) [α] + 9₂ (abs al c=1.2) + 128 Cbz, c=0.3		148-9d			al	B6[3], 5431
6683	1,2-Ethanediol, 1,2-diphenyl (dl) $C_6H_5CH(OH)CH(OH)C_6H_5$	214.27	nd (w or al) ta (eth)	>300, $133^{0.023}$	122-3			al, eth, chl	B6[3], 5431
6684	1,2-Ethanediol, 1,2-diphenyl (l) $C_6H_5CH(OH)CH(OH)C_6H_5$	214.27	lf (eth abs al or bz) pr(bz or abs al)		148-9			al, eth, ace, bz	B6[3], 5431
6685	1,2-Ethanediol, 1,2-diphenyl- (meso) $C_6H_5CH(OH)CH(OH)C_6H_5$	214.27	nd or lf (w or bz-peth) mcl lf (al, w)	>300, $139^{0.023}$	139-40			al, chl	B6[3], 5429
6686	1,2-Ethanediol, phenyl-(dl) or Phenyl ethylene glycol $C_6H_5CH(OH)CH_2OH$	138.17	nd (lig)	272-4^{755}	69-70			w, al, eth, bz	B6[3], 4572
6687	1,2-Ethanol, tetraphenyl or Bensopinacol $(C_6H_5)_2C(OH)C(OH)(C_6H_5)_2$	366.47	pr (bz + l) cr (ace)		182			eth, ace, bz, chl	B6[3], 5923
6688	Ethanedione, di-(2-furyl) $(OC_4H_3)COCO(C_4H_3O)$	190.16	ye nd (al) cr (bz)		165-6			al, eth, bz, chl	B19[4], 2007
6689	1,2-Ethane disulfonic acid $HO_3SCH_2CH_2SO_3H$	190.19	hyg nd (gl aa)		111-2 (+2w) 174 (anh)			w, al, diox	B4[4], 78
6690	1,2-Ethanedithiol or Dithioglycol $HSCH_2CH_2SH$	94.20	....	146, 46-7^6	-41.2	$1.1243^{20/4}$	1.5590^{20}	al, eth, ace, bz	B1[4], 2450
6691	Ethane dithiolic acid, diethyl ester or Diethyl dithio oxalate $C_2H_5SCO\text{-}COSC_2H_5$	178.28	ye nd (eth)	235, 80-2^{32}	27	$1.0565^{21/4}$	....	eth	B1[1], 244
6692	Ethane phosphonic acid $C_2H_5PO(OH)_2$	110.05	pl or nd (w)		61-2			w, al, eth	B4[3], 1779
6693	Ethane phosphonic acid, diethyl ester $C_2H_5PO(OC_2H_5)_2$	166.16		198, 83^{13}		$1.0259^{20/4}$	1.4163^{20}	al, eth	B4[3], 1779
6694	Ethane phosphonic acid, dimethyl ester $C_2H_5PO(OCH_3)_2$	138.10	....	82^{18}	....	$1.1029^{30/4}$	1.4128^{30}	w, al, bz	J1954, 3222
6695	Ethane sulfonic acid or Ethyl sulfonic acid $CH_3CH_2SO_3H$	110.13	hyg	123^1	-17	1.3341^{25}	1.4335^{20}	w, al	B4[4], 33

No.	Name, Synonyms, and Formula	Mol. wt.	Color, crystalline form, specific rotation and λ_{max} (log ε)	b.p. °C	m.p. °C	Density	n_D	Solubility	Ref.
6696	Ethane sulfonyl chloride $C_2H_5SO_2Cl$	128.56	pa ye	171, 65[13]	....	1.357[22.5]	1.4531[20]	eth	B4[4], 34
6697	Ethane sulfonyl chloride, 2-bromo $BrCH_2CH_2SO_2Cl$	207.48	pa ye	102[13]		1.921[20]	1.5242[20]	....	B4[4], 37
6698	Ethane sulfonyl chloride, 1-chloro $CH_3CHClSO_2Cl$	163.02		80-1[22]			1.4782[20]		B1[4], 3121
6699	Ethane sulfonyl chloride, 2-chloro $ClCH_2CH_2SO_2Cl$	163.02		200-3, 93-7[17]		1.555[20/4]	1.4920[20]		B4[4], 36
6700	Ethane sulfonic acid, 2-hydroxy, dihydrate or Isethionic acid $HOCH_2CH_2SO_3H.H_2O$	152.17	hyg cr (aa-AC$_2$O)		111-2			w, al	B4[4], 84
6701	1,1,2,2-Ethane tetracarboxylic acid, 1,2-diethyl ester $C_2H_5O_2CCH(CO_2H)CH(CO_2H)CO_2C_2H_5$	262.22	hyg lf ($+\,^1/_2$ w)		132-3d			w, al, eth	B2, 858
6702	1,1,2,2-Ethane tetracarboxylic acid, tetraethyl ester $(C_2H_5O_2C)_2CHCH(CO_2C_2H_5)_2$	318.32		d 305	77	1.064[80]	1.4105[80]	al	B2[4], 2415
6703	1,1,2,2-Ethane tetracarboxylic acid, tetramethyl ester $(CH_3O_2C)_2CHCH(CO_2CH_3)_2$	262.22	cr (eth al bz)		138			al	B2[3], 2076
6704	Ethanethiol or Ethyl mercaptan C_2H_5SH	62.13	....	35	-144.4	0.8391[20/4]	1.4310[20]	al, eth, ace	B1[4], 1390
6705	Ethanethiol, 2-amino or Cysteamine $H_2NCH_2CH_2SH$	77.15	cr (sub)	d[760] Sub (vac)	99-100			w, al	B4[4], 1570
6706	Ethanethiol, 2-chloro $ClCH_2CH_2SH$	96.58		113		1.1826[20/4]	1.4929[20]	al, eth, diox	B1[4], 1406
6707	Ethanethiol, 1-phenyl (l) $CH_3CH(C_6H_5)SH$	138.23	$[\alpha]^{20}_D$ −89 al, c=6	199-200, 83[10]		1.022[20/4]	1.5593[20]	al, eth, bz	B6[3], 1697
6708	Ethanol or Ethyl alcohol CH_3CH_2OH	46.07		78.5	-117.3	0.7893[20/4]	1.3611[20]	**w, eth, ace, bz**	B1[4], 1289
6709	Ethanol, β-(4-amino phenyl) $(4-NH_2C_6H_4)CH_2CH_2OH$	137.18	nd (al)		108				B13[3], 1679
6710	Ethanol, 2-bromo or Ethylene bromohydrin $BrCH_2CH_2OH$	124.97		149-50[750], 51[4]		1.7629[20/4]	1.4915[20]	**w, al, eth**	B1[4], 1385
6711	Ethanol-2-chloro or Ethylene chlorohydrin $ClCH_2CH_2OH$	80.52		128, 44[20]	-67.5	1.2003[20/4]	1.4419[20]	**w, al**	B1[4], 1372
6712	Ethanol, 2-chloro-1-phenyl or Styrene Chlorohydrin $ClCH_2CH(OH)C_6H_5$	156.61		128[17]		1.1926[20/4]	1.5523[20]	al, eth	B6[3], 1683
6713	Ethanol, 2,2-dichloro Cl_2CHCH_2OH	114.96		146, 37-8[6]		1.4040[25/4]	1.4626[25]	al, eth	B1[4], 1383
6714	Ethanol, 2,2-diphenyl $(C_6H_5)_2CHCH_2OH$	198.26		195[20], 144-5[1]	64.5	....		eth, al, ace	B6[3], 3397
6715	Ethanol, 2-ethylthio $C_2H_5SCH_2CH_2OH$	106.19		184	Ca-100	1.0166[20/4]	1.4867[20]	al, ace	B1[4], 2430
6716	Ethanol, 2-fluoro or Ethylene fluorohydrin FCH_2CH_2OH	64.06		103.5	-26.4	1.1040[20/4]	1.3647[18]	**w, al, eth, ace**	B1[4], 1366
6717	Ethanol, 2-iodo or Ethylene iodohydrin ICH_2CH_2OH	171.97		176-7d, 85-8[25]		2.1968[20/4]	1.5713[20]	w, al, eth	B1[4], 1387
6718	Ethanol, 2-mercapto or Monothio ethylene glycol $HSCH_2CH_2OH$	78.13		157-8[742], 55[13]		1.1143[20/4]	1.4996[20]	w, al, eth, bz	B1[4], 2428
6719	Ethanol, 2-methylthio $CH_3SCH_2CH_2OH$	92.17		68-70[20]		1.6640[20/20]	1.4867[30]	w, al, eth	B1[4], 2429
6720	Ethanol, 2-nitro or β-Nitro ethanol $O_2NCH_2CH_2OH$	91.07		194[765], 102[10]	-80	1.270[15/4]	1.4438[19]	**w, al, eth**	B1[4], 1388
6721	Ethanol, 2-triazo $N_3CH_2CH_2OH$	87.08		75[40]	....	1.149[24/24]	....	**w**	B1[4], 1389
6722	Ethanol, 2,2,2-tribromo or Avertin, bromethol CBr_3CH_2OH	282.77	nd or pr (peth)	92-3[10]	81	....		al, eth, bz	B1[3], 1362
6723	Ethanol, 2,2,2-trichloro CCl_3CH_2OH	149.41	hyg rh ta or pl	151[737], 52[11]	19		1.4861[20]	**al, eth**	B1[4], 1383
6724	Ethanol, 2,2,2-trifluoro CF_3CH_2OH	100.04	....	74	-43.5	1.4680[20], 1.3739[22/4]	1.2907[22]	al, eth, ace, bz	B1[4], 1370
6725	Ethanol, 1,1,2-triphenyl $C_6H_5CH_2C(OH)(C_6H_5)_2$	274.37	nd (bz-lig) pr (peth)	222[11]	89-90			al	B6[3], 3680
6726	Ethanol, 2,2,2-triphenyl $(C_6H_5)_3CCH_2OH$	274.37	cr (al eth lig)	....	110.5d			al, eth, bz, lig	B6[3], 3683
6727	Ethanolamine or 2-Amino ethanol $H_2NCH_2CH_2OH$	61.09	....	170, 58[5]	10.3	1.0180[20/4]	1.4541[20]	**w, al, chl**	B4[4], 1406

No.	Name, Synonyms, and Formula	Mol. wt.	Color, crystalline form, specific rotation and λ_{max} (log ε)	b.p. °C	m.p. °C	Density	n_D	Solubility	Ref.
6728	Ethanolamine, N-acetyl $CH_3CONHCH_2CH_2OH$	103.12	nd (ace)	166-7[8]	63-5	1.1079[25/4]	1.4674[20]	w	B4[4], 1535
6729	Ethanolamine, N-butyl $C_4H_9NHCH_2CH_2OH$	117.19	....	199-200, 91-2[11]	......	0.8907[20/4]	1.4437[20]	w, al, eth	B4[3], 682
6730	Ethanolamine, N-isobutyl $i-C_4H_9NHCH_2CH_2OH$	117.19	....	199-200, 90[16]	......	0.8818[20/4]	1.4402[20]	w, al, eth	B4[3], 683
6731	Ethanolamine, N,N-diethyl $(C_2H_5)_2NCH_2CH_2OH$	117.19	hyg	163, 56-7[15]	......	0.8921[20/4]	1.4412[20]	w, al, eth, ace, bz	B4[4], 1471
6732	Ethanolamine, N,N-dimethyl $(CH_3)_2NCH_2CH_2OH$	89.14	....	134	......	0.8866[20/4]	1.4300[20]	w, al, eth	B4[4], 1424
6733	Ethanolamine, N-ethyl $C_2H_5NHCH_2CH_2OH$	89.14	....	169-70, 78-80[27]	......	0.914[20/4]	1.444[20]	w, al, eth	B4[4], 1465
6734	Ethanolamine, N-methyl $CH_3NHCH_2CH_2OH$	75.11	....	158, 52[6]	......	0.937[20]	1.4385[20]	w, al, eth	B4[4], 1422
6735	Ethanolamine, N-methyl-N-phenyl $C_6H_5N(CH_3)CH_2CH_2OH$	151.21	yesh	150[14]	......	0.9995[15/0]		al, eth, ace, bz	B12[3], 296
6736	Ethanolamine, N-β-naphthyl $β-C_{10}H_7NHCH_2CH_2OH$	187.24	lf (eth or al)	197-8[3]	52			eth	B12[2], 717
6737	Ethanolamine, N-phenyl $C_6H_5NHCH_2CH_2OH$	137.18	....	286, 167[17]	......	1.0945[20]	1.5760[20]	al, eth, chl	B12[3], 293
6738	Ethanolamine, 1-phenyl $H_2NCH_2CH(OH)C_6H_5$	137.18	nd (al-eth-peth)	160[17]	56-7			w, al	B13[4], 1657
6739	Ethanolamine, N-isopropyl $i-C_3H_7NHCH_2CH_2OH$	103.17	....	172-4, 76-7[15]	......	0.8970[20/4]	1.4395[20]	w, al, eth	B4[3], 681
6740	Ethoxyacetic acid $C_2H_5OCH_2CO_2H$	104.12	....	206-7, 111[25]	......	1.1021[20]	1.4194[20]	w, al, eth	B2[3], 374
6741	Ethoxyacetic acid, ethyl ester or Ethyl ethoxy acetate $C_2H_5OCH_2CO_2C_2H_5$	132.16	....	158, 52[12]	......	0.9702[20/4]	1.4029[20]	al, eth, ace	B3[4], 581
6742	Ethoxyacetic acid, l menthyl ester or l-Menthyl ethoxy acetate $C_2H_5OCH_2CO_2C_{10}H_{19}$	242.36	$[α]_B^{18}$ −66.35	155[11]	......	0.9343		al, eth, chl	B6[2], 47
6743	Ethoxyacetic acid, methyl ester or Methylethoxy acetate $C_2H_5OCH_2CO_2CH_3$	118.13	....	147-8[734]	......	1.0112[15]		al, eth, ace	B3[4], 578
6744	2-Ethoxybenzoic acid $2-C_2H_5OC_6H_4CO_2H$	166.18	....	211-2[35]	20.7				B10[3], 98
6745	2-Ethoxybenzoic acid, ethyl ester $2-C_2H_5OC_6H_4CO_2C_2H_5$	194.23	....	251, 180-5[113]	......	1.005[20]		al, eth	B10[3], 117
6746	3-Ethoxybenzoic acid $3-C_2H_5OC_6H_4CO_2H$	166.18	nd (w or sub)	sub	137			al, eth, bz	B10[3], 245
6747	3-Ethoxybenzoic acid, ethyl ester or Ethyl-3-ethoxy benzoate $3-C_2H_5OC_6H_4CO_2C_2H_5$	194.23	....	264, 172-3[50]	......	1.0725[20/20]		al, eth	B10[3], 251
6748	4-Ethoxybenzoic acid $4-C_2H_5OC_6H_4CO_2H$	166.18	nd (w)		198.5			al, eth, bz	B10[3], 282
6749	4-Ethoxybenzoic acid, ethyl ester or Ethyl-4-ethoxy benzoate $4-C_2H_5OC_6H_4CO_2C_2H_5$	194.23	....	275 148-9[14]	: lf	1.076[14]		al, eth	B10[3], 301
6750	Ethoxyl amine or α-Ethyl hydroxyl amine $C_2H_5ONH_2$	61.05			68	0.8872[8/8]		w, al, eth	B1[4], 1326
6751	Ethylaceto acetate $CH_3COCH_2CO_2C_2H_5$	130.15	....	180.4, 74[14]	<-80	1.0282[20/4]	1.4194[25]	al, eth, bz, chl	B3[4], 1528
6752	Ethylacetoacetate (enol form) $CH_3C(OH)=CHCO_2C_2H_5$	130.15				1.0119[10]	1.4432[20]		B3[4], 1528
6753	Ethylacetoacetate (keto form) $CH_3COCH_2CO_2C_2H_5$	130.15			-39	1.0368[10/4]	1.4171[20]		B3[4], 1528
6754	Ethyl amine $C_2H_5NH_2$	45.09		16.6	-81	0.6829[20/4]	1.3663[20]	w, al, eth	B4[4], 307
6755	Ethyl amine, hydrobromide $C_2H_5NH_2\cdot HBr$	126.01	mcl nd or pl (al)		159.3			w, al	B4[4], 310
6756	Ethyl amine, hydrochloride $C_2H_5NH_2\cdot HCl$	81.56	mcl pl (al)	d 315	109-10			w, al	B4[4], 310
6757	Ethyl amine, hydroiodide $C_2H_5NH_2HI$	173.01	mcl nd (w)		188.5	2.100		w	B4[4], 310
6758	Ethyl amine, 2-bromo, hydrochloride $BrCH_2CH_2NH_2\cdot HCl$	160.45	lf (al-AcOEt)		174-5d			w, al, bz	B4[3], 248
6759	Ethyl amine, 2-chloro, hydrochloride $ClCH_2CH_2NH_2\cdot HCl$	115.9	hyg cr (al-eth)		144 (148)			w, al, ace	B4[3], 236

No.	Name, Synonyms, and Formula	Mol. wt.	Color, crystalline form, specific rotation and λ_{max} (log ε)	b.p. °C	m.p. °C	Density	n_D	Solubility	Ref.
6760	Ethyl amine, 2-ethoxy $C_2H_5OCH_2CH_2NH_2$	89.14	....	108[758]	......	0.8512[20/4]	1.4101[20]	w, al, eth, ace, bz	B4[4], 1411
6761	Ethyl amine, 2-methoxy $CH_3OCH_2CH_2NH_2$	75.11	....	95[756]				w, al	B4[4], 1411
6762	Ethyl amime, perfluoro $CF_3CF_2NF_2$	171.02	....	−35					B2[4], 473
6763	Ethyl β-amino, ethyl ether $C_2H_5OCH_2CH_2NH_2$	89.14	....	108[758]		0.8512[20/4]	1.4101[20]	w, al, eth, ace, bz	B4[4], 1411
6764	Ethyl benzene $C_6H_5CH_2CH_3$	106.17	....	136.2, 25.8[10]	-95	0.8670[20/4]	1.4959[20]	al, eth	B5[3], 776
6765	2-Ethyl benzoic acid 2-$C_2H_5C_6H_4CO_2H$	150.18	nd (w)	259	68	1.0413[100/4]	1.5099[100]	al, eth	B9[2], 2425
6766	3-Ethyl benzoic acid 3-$C_2H_5C_6H_4CO_2H$	150.18	nd (w or dil al)	....	47	1.042[100]	1.5345[100]	al, eth	B9[1], 208
6767	4-Ethyl benzoic acid 4-$C_2H_5C_6H_4CO_2H$	150.18	pr (al) pl or lf (w)	....	113.5			al, bz, chl	B9[2], 349
6768	Ethyl benzyl ketone or 1-Phenyl-2-butanone $CH_3CH_2COCH_2C_6H_5$	148.28	....	230[755]		1.002[20/4]	....	al, eth, ace	B7[3], 1080
6769	Ethylboric acid $C_2H_5B(OH)_2$	73.89	pl (eth)	d	40 (sub)			w, al, eth	B4[3], 1964
6770	Ethyl bromide CH_3CH_2Br	108.97	....	38.4	-118.6	1.4604[20/4]	1.4239[20]	al, eth, chl	B1[4], 150
6771	Ethyl 2-bromoethyl ether $C_2H_5OCH_2CH_2Br$	153.03	ye nd (al)	127-8[755], 40[24]		1.3572[20/4]	1.4447[20]	al, eth	B1[4], 1386
6772	Ethyl bromomethyl ether $C_2H_5OCH_2Br$	139.00	....	109[746]		1.4402[20/4]	1.4515[20]	eth	B1[3], 2594
6773	Ethyl isobutyl ether C_2H_5-O-i-C_4H_9	102.18	....	81		0.751[20/4]	1.3739[25]	al, eth, ace, chl	B1[4], 1593
6774	Ethyl isobutyl sulfide i-$C_4H_9SC_2H_5$	118.24	....	134.2, 24.8[10]		0.8306[20/4]	1.4450[20]	al, eth	B1[4], 1606
6775	Ethyl chloride CH_3CH_2Cl	64.52	....	12.3	-136.4	0.8978[20/4]	1.3676[20]	al, eth	B1[4], 124
6776	Ethyl (l-chloroethyl) ether $C_2H_5OCHClCH_3$	108.57	....	92-5		0.950[20/4]	1.4053[20]	....	B1[4], 3119
6777	Ethyl β-chloroethyl ether $C_2H_5OCH_2CH_2Cl$	108.57	....	107-8		0.9894[20/4]	1.4113[20]	eth, chl	B1[4], 1375
6778	Ethyl β-chloroethyl sulfide $C_2H_5SCH_2CH_2Cl$	124.63	....	156, 63-5[47]		1.0663[20/4]	1.4878[20]	chl	B1[4], 1707
6779	Ethyl chloromethyl ether $C_2H_5OCH_2Cl$	94.54	....	83[763]		1.0372[0/4]	1.4040[20]	al, eth	B1[4], 3047
6780	Ethyl chloro sulfinate C_2H_5OSOCl	128.58	....	52.5[44]		1.2766[25/4]	1.4550[25]	eth	B1[3], 1316
6781	Ethyl chloro sulfonate $C_2H_5OSO_2Cl$	144.58	....	151-4, 52[14]		1.3502[25/4]	1.416[20]	eth, chl, lig	B1[4], 1326
6782	Ethyl cyanoacetate $NCCH_2CO_2C_2H_5$	113.12	....	205, 99[15]	-22.5	1.0654[20/4]	1.4175[20]	al, eth	B2[4], 1889
6783	Ethyl cyanoformate $NCCO_2C_2H_5$	99.09	....	116-8[765]		1.0034[20/4]	1.3821[20]	al, eth	B2[4], 1862
6784	Ethyl di-benzylamine $C_2H_5N(CH_2C_6H_5)_2$	225.34	....	306, 131[11]			....	al, eth	B12[3], 2222
6785	Ethyl (1,2-dibromoethyl) ether $C_2H_5OCH_2BrCH_2Br$	231.94	....	80[20]		1.7320[20/4]	1.5044[20]	al, chl	B1[4], 3152
6786	Ethyl (1,2-dichloroethyl) ether $C_2H_5OCHClCH_2Cl$	143.02	....	145, 66-8[45]		1.1370[20/4]	1.4435[20]	al, eth	B1[4], 3136
6787	Ethyl (1,2-dichlorovinyl) ether $C_2H_5OCCl=CHCl$	141.00	....	128.2		1.1972[25/4]	1.4558[17]		B1[4], 3425
6788	Ethyl (diethylaminomethyl) ether $(C_2H_5)_2NCH_2OC_2H_5$	131.22	....	136, 76[11]			....	al, eth, ace	B4[4], 336
6789	Ethyl-di-isopropylamine $C_2H_5N(i-C_3H_7)_2$	129.24	....	126.5			1.4138[20]	....	B4[4], 511
6790	Ethyl 2,3-epoxypropyl ether $C_2H_5OCH_2CHCH_2$	102.13	....	128		0.9700[20]	1.4320[20]	w, al, eth	B17, 105
6791	Ethyl ether or Diethyl ether $C_2H_5OC_2H_5$	74.12	....	34.5	fr−116.2	0.7138[20/4]	1.3526[20]	al, ace, bz, chl	B1[4], 1314
6792	Ethyl ether borofluoride $(C_2H_5)_2O.BF_3$	141.93	....	125-6 60[20]	-60.4	1.3572[20/4]	1.4447[20]	al, eth	B1[4], 1321

No.	Name, Synonyms, and Formula	Mol. wt.	Color, crystalline form, specific rotation and λ_{max} (log ε)	b.p. °C	m.p. °C	Density	n_D	Solubility	Ref.
6793	Ethyl ethynyl ether or Ethoxy acetylene .... $C_2H_5OC\equiv CH$	70.00	....	50 exp 100	.....	$0.8000^{20/4}$	1.3796^{20}	....	B1[4], 2211
6794	Ethyl furfuryl ether .... $C_2H_5OCH_2(C_4H_3O)$	126.16	....	149-50[770]		$0.9844^{20/4}$	1.4523^{20}	al, eth	B17[2], 114
6795	Ethyl-2-furyl ether .... $(2-C_4H_3O)OC_2H_5$	112.14	....	125-6		$0.9849^{23/4}$	1.4500^{23}	....	B17[4], 1219
6796	Ethyl heptyl ether .... $C_2H_5OC_7H_{15}$	144.26	....	166.6		$0.790^{16/4}$	1.4111^{20}	al, eth	B1[4], 1733
6797	Ethyl hexyl ether .... $C_2H_5OC_6H_{13}$	130.23	....	142-3[773], 42[14]		$0.7722^{20/4}$	1.4008^{20}	al, eth	B1[4], 1697
6798	Ethyl hydrogensulfate .... $C_2H_5OSO_3H$	126.13	....	280d		$1.3657^{20/4}$	1.4105^{20}	w	B1[4], 1324
6799	Ethyl hydroperoxide .... C_2H_5OOH	62.07	....	93-7 ext >100	-100	$0.9332^{20/4}$	1.3800^{20}	w, al, eth, bz	B1[4], 1323
6800	Ethyl-bis (2-hydroxyethyl) amine .... $C_2H_5N(CH_2CH_2OH)_2$	133.19	ye	246-8, 118[1]		$1.0135^{20/4}$	1.4663^{20}	al	B4[3], 693
6801	Ethyl (2-hydroxy--5-methylphenyl) ketone ($5-CH_3-2-HOC_6H_3$)COC_2H_5	164.20	....	129-30[16·5]	2	$1.0841^{14/4}$	$1.549^{13·8}$	chl	B8[2], 120
6802	Ethyl (2-hydroxy--3-methylphenyl) ketone ($3-CH_3-2-HOC_6H_3$)COC_2H_5	164.20	....	127-9[15]	22-3	....	....	chl	B8[2], 119
6803	Ethyl (2-hydroxy--4-methylphenyl) ketone ($4-CH_3-2-HOC_6H_3$)COC_2H_5	164.20	....	115-20[10]	41.5-2.5	....	....	chl	B8[3], 471
6804	Ethyl hypochlorite .... C_2H_5OCl	80.52	ye lig	36[732]		$1.013^{-6/4}$	....	al, eth, bz, chl	B1[4], 1324
6805	Ethyl iodide .... CH_3CH_2I	155.97	....	72.3	-108	$1.9358^{20/4}$	1.5133^{20}	al, eth	B1[4], 163
6806	Ethyl isocyanate .... C_2H_5NCO	71.08	....	60		$0.9031^{20/4}$	1.3808^{20}	al, eth	B4[4], 402
6807	Ethyl methyl amine or Methylethylamine .... $CH_3NHC_2H_5$	59.11	....	36.7		....	....	w, al, eth, ace	B4[4], 312
6808	Ethyl methyl amine, hydrochloride .... $C_2H_5NHCH_3.HCl$	95.57	pl (al-eth)	....	126-30	$1.0874^{20/4}$	....	w, al, eth, ace	B4[2], 589
6809	Ethyl methyl ether .... $C_2H_5OCH_3$	60.10	....	10.8		$0.7252^{0/0}$	1.3420^4	w, al, eth, ace, chl	B1[4], 1314
6810	Ethyl methyl sulfide .... $C_2H_5SCH_3$	76.16	....	66.6	-105.9	$0.8422^{20/4}$	1.4404^{20}	al, eth	B1[4], 1392
6811	Ethyl α-naphthyl ether or 1-Ethoxynaphthalene .... $\alpha-C_{10}H_9OC_2H_5$	172.23	nd	280.5, 136-8[14]	5.5	$1.060^{20/4}$	1.5953^{25}	al, eth	B6[3], 2924
6812	Ethyl β-naphthyl ether .... $\beta-C_{10}H_7OC_2H_5$	172.23	pl (al)	282, 148[10]	37-8	$1.0640^{20/20}$	1.5975^{36}	al, eth, lig, to	B6[3], 2972
6813	Ethyl nitrate .... $C_2H_5ONO_2$	91.07	flam	87.2	-94.6	$1.1084^{20/4}$	1.3852^{20}	w, al, eth	B1[4], 1327
6814	Ethyl nitrite .... C_2H_5ONO	75.07	yesh	16-7[725]		$0.90^{15/15}$	1.3418^{10}	al, eth	B1[4], 1327
6815	Ethyl octyl ether .... $C_2H_5OC_8H_{17}$	158.29	....	186.3, 74[9]	12.5	$0.7847^{20/4}$	1.4127^{20}	al	B1[4], 1759
6816	Ethyl pentyl ether .... $C_2H_5OC_5H_{11}$	116.21	....	119-20		$0.7622^{20/4}$	1.3927^{20}	al, eth	B1[3], 1602
6817	Ethyl isopentyl ether .... $C_2H_5O-i-C_5H_{11}$	116.21	....	112-3		$0.7695^{21/15}$	....	al, eth	B1[4], 1681
6818	Ethyl tert-pentyl ether .... $C_2H_5O-tert-C_5H_{11}$	116.21	....	101		$0.7657^{20/4}$	1.3912^{20}	al, eth	B1[3], 1626
6819	Ethylarsonic acid .... $C_2H_5AsO(OH)_2$	154.00	nd (al) rh nd (w)	209-11[12]	99.5	....	....	w, al	B4[3], 1823
6820	Ethyl perchlorate .... $C_2H_5OClO_3$	128.51	oil	....	89	....	....	al, eth	B1[3], 1314
6821	Ethyl Diethylperoxide .... $C_2H_5OOC_2H_5$	90.12	....	65	-70	0.8240^{19}	1.3715^{17}	al, eth	B1[4], 1323
6822	Ethyl phenyl sulfide or (Ethylthio)benzene .... $C_6H_5SC_2H_5$	138.23	....	205, 84[10]	.....	$1.0211^{20/4}$	1.5670^{20}	al	B6[4], 1468
6823	Ethyl propyl amine .... $C_2H_5NHC_3H_7$	73.14	....	61-2		0.7204^{17}	1.3858^{25}	w, al, ace	B4[4], 468
6824	Ethyl isopropyl amine .... $i-C_3H_7NHC_2H_5$	87.16	....	70-1		....	1.3872^{25}		B4[4], 508
6825	Ethyl propyl ether .... $C_2H_5OC_3H_7$	88.15	....	63.6	<-79	$0.7386^{20/4}$	1.3695^{20}	al, eth, aa	B1[4], 1421

No.	Name, Synonyms, and Formula	Mol. wt.	Color, crystalline form, specific rotation and λ_{max} (log ε)	b.p. °C	m.p. °C	Density	n_D	Solubility	Ref.
6826	Ethyl iso propyl ether $C_2H_5OCH(CH_3)_2$	88.15	....	63-4	......	$0.720^{25/4}$	1.3698^{25}	al, eth, ace, chl	B1[4], 1471
6827	Ethyl propyl sulfide $C_2H_5SC_3H_7$	104.22	....	118.5, 13.5^{10}	-117	$0.8370^{20/4}$	1.4462^{20}	al	B1[4], 1451
6828	Ethyl 1-propynyl ether $C_2H_5OC\equiv CCH_3$	84.13	....	84	......	$0.8276^{20/4}$	1.4039^{20}	al, eth	B1[3], 1969
6829	Ethyl 2-propynyl ether or Ethyl propargyl ether $C_2H_5OCH_2C\equiv CH$	84.13	....	82	......	$0.8326^{20/4}$	1.4039^{20}	al, eth	B1[4], 2215
6830	N-Ethyl quinolinium iodide $C_9H_7N^+(C_2H_5)I^-$	285.13	ye pr (al)	....	158	....	....	w, al, chl	B20[4], 3358
6831	Ethyl pyridinium bromide $(C_5H_5N^+)C_2H_5Br^-$	188.08	cr (al)	....	111-2	....	....	w, al	B20[4], 2309
6832	Ethyl sulfate $(C_2H_5O)_2SO_2$	154.19	....	208d, 96^{15}	-24.5	$1.1774^{20/4}$	1.4004^{20}	al, eth	B1[4], 1326
6833	Ethyl sulfite $(C_2H_5O)_2SO$	138.19	....	230, 116^{10}	......	$0.9957^{20/4}$	1.4310^{20}	al, eth	B1[4], 1324
6834	Ethyl sulfoxide or Diethyl sulfoxide $(C_2H_5)_2SO$	106.19	syr	104^{25}	14	....	....	w, al, eth	B1[4], 1395
6835	Ethyl telluride or Diethyl telluride $(C_2H_5)_2Te$	185.73	red-ye	137-8	......	$1.599^{15/4}$	1.5182^{15}	al	B1[4], 1412
6836	Ethyl thioacetic acid $C_2H_5SCH_2CO_2H$	120.18	....	164^{83}, 109^5	-8.5	$1.1497^{20/4}$	....	w, al, eth	B3[4], 603
6837	Ethyl thiocyanate C_2H_5SCN	87.14	....	145^{758}	-85.5	$1.0071^{22/4}$	1.4684^{15}	al, eth	B3[4], 328
6838	Ethyl isothiocyanate or Ethyl mustard oil C_2H_5NCS	87.14	....	131-2	-5.9	$0.9990^{20/4}$	1.5130^{20}	al, eth	B4[4], 403
6839	Ethyl vinyl ether $C_2H_5OCH=CH_2$	72.12	....	35-6	-115.8	$0.7589^{20/4}$	1.3767^{20}	al, eth	B1[4], 2049
6840	Ethylene $CH_2=CH_2$	28.05	gas, mcl pr	-103.7	-169	....	1.363^{100}	eth	B1[4], 677
6841	Ethylene, amino or Vinylamine $CH_2=CHNH_2$	43.07	....	$55-6^{750}$	......	0.8321^{24}	....	w, al, eth	B4, 203
6842	Ethylene, 1,2-bis(trimethylsilyl)- (trans) $(CH_3)_3SiCH=CHSi(CH_3)_3$	172.42	....	145.5	......	0.7589^{20}	$1.4310^{20/}{}_D$	....	KHOC, 603
6843	Ethylene, l-bromo-2-chloro $BrCH=CHCl$	141.39	....	84.6	-86.7	1.7972^{15}	1.4982	....	B1[3], 671
6844	Ethylene, 1-bromo-1,2,2-triphenyl $C_6H_5CBr=C(C_6H_5)_2$	335.25	nd (aa)	....	116-7	....	....	aa	B5[3], 2400
6845	Ethylene, 1-chloro-2-dichloroarsino (trans) or Lewisite $ClCH=CHAsCl_2$	207.32	....	196d, 93^{20}	0.1	$1.888^{20/4}$	....	al, eth	B4[3], 1810
6846	Ethylene chlorohydrin $ClCH_2CH_2OH$	80.52	....	128, 44^{20}	-67.5	$1.2003^{20/4}$	1.4419^{20}	w, al	B1[4], 1372
6847	Ethylene, 1-chloro-1,2,2-trifluoro $FCCl=CF_2$	116.47	....	-26.2	-157.5	$1.54^{-60/4}$	1.38^0	bz	B1[4], 704
6848	Ethylene, 1-chloro-1,2,2-triphenyl $(C_6H_5)_2C=CClC_6H_5$	290.80	....	....	117	....	....	al, eth, ace, bz, chl	B5[3], 2400
6849	Ethylenediamine $H_2NCH_2CH_2NH_2$	60.11	....	116.5	8.5	$0.8995^{20/20}$	1.4568^{20}	w, al	B4[4], 1166
6850	Ethylene diamine, N,N'-dibenzoyl $C_6H_5CONHCH_2CH_2NHCOC_6H_5$	268.32	pr or nd (al)	....	247	....	....	aa	B9[3], 1210
6851	Ethylenediamine, N,N'-diethyl $C_2H_5NHCH_2CH_2NHC_2H_5$	116.21	....	144, $38-40^{15}$	......	$0.8280^{20/4}$	1.4340^{20}	w, al, eth, to	B4[4], 1174
6852	Ethylenediamine, N,N'-dimethyl $CH_3NHCH_2CH_2NHCH_3$	88.15	....	120	......	$0.828^{15/4}$	....	al, eth, dil HCl	B4[4], 1171
6853	Ethylenediamine, N,N'-diphenyl $C_6H_5NHCH_2CH_2NHC_6H_5$	212.30	lf (dil al)	$178-82^2$	74	....	....	al, eth	B12[3], 1042
6854	Ethylenediamine, hydrate $H_2NCH_2CH_2NH_2.H_2O$	78.12	....	118	10	$0.964^{20.5/4}$	$1.4500^{20.5}$	w	B4[4], 1168
6855	Ethylenediamine, hydrochloride $H_2NCH_2CH_2NH_2.2HCl$	133.02	mcl pr (w)	sub	300-30 sub	....	1.633	w	B4[4], 1168
6856	Ethylenediamine, N-β-hydroxyethyl $HOCH_2CH_2NHCH_2CH_2NH_2$	104.15	....	238-40, 123^{10}	......	$1.0254^{25/D}$	1.4861^{20}	w, al, ace	B4[4], 1558
6857	Ethylene, dibenzoyl (cis) $C_6H_5COCH=CHCOC_6H_5$	236.27	nd (al)	....	134	....	....	eth, ace, bz, chl	B3[7], 4115
6858	Ethylene, dibenzoyl (trans) $C_6H_5COCH=CHCOC_6H_5$	236.27	ye nd (al or bz)	....	111	....	....	bz, chl, aa	B3[7], 4116
6859	Ethylene, 1,1-dibromo or Vinylidene bromide $CH_2=CBr_2$	185.85	....	92	......	$2.1780^{21/4}$	....	al, eth, ace, bz	B1[4], 720

No.	Name, Synonyms, and Formula	Mol. wt.	Color, crystalline form, specific rotation and λ_{max} (log ε)	b.p. °C	m.p. °C	Density	n_D	Solubility	Ref.
6860	Ethylene, 1,1-dibromo--2-ethoxy .... $C_2H_5OCH=CBr_2$	229.91	....	$170\text{-}2^{747}$, $73\text{-}5^{15}$	......	$1.7697^{18/4}$	....	eth	B1[2], 473
6861	Ethylene, 1,2-dibromo (cis) .... $CHBr=CHBr$	185.86	....	112.5^{760}	-53	2.2464^{20}	1.5428^{20}	al, eth, ace, bz, chl	B1[4], 720
6862	Ethylene, 2,2-dibromo (trans) .... $CHBr=CHBr$	185.86	....	108	-6.5	2.2308^{20}	1.5505^{18}	al, eth, ace, bz, chl	B1[4], 721
6863	Ethylene, 1,1-dichloro or Vinylidene chloride .... $CH_2=CCl_2$	96.94	....	37	-122.1	1.218^{20}	1.4249^{20}	al, eth, ace, bz, chl	B1[4], 706
6864	Ethylene, 1,1-dichloro-2-fluoro .... $CCl_2=CHF$	114.93	....	37.5	......	$1.37324^{16.4}$	$1.4031^{16.4}{}_D$	....	B1[4], 711
6865	Ethylene, 1,2-dichloro (cis)...................... $CHCl=CHCl$	96.94	....	60.3	-80.5	$1.2837^{20/4}$	....	**al, eth, ace**, bz, chl	B1[4], 707
6866	Ethylene, 1,2-dichloro (trans) .... $CHCl=CHCl$	96.94	....	47.5	-50	$1.2565^{20/4}$	1.4454^{20}	**al, eth, ace**, bz, chl	B1[4], 709
6867	Ethylene, 1,2-dichloro-1,2-difluoro .... $CFCl=CFCl$	132.92	....	21.1	-130.5	$1.4950^{0/4}$	$1.3777^{0/}{}_D$	....	B1[4], 712
6868	Ethylene, 1,1-difluoro or Vinylidene fluoride ...... $CH_2=CF_2$	64.04	gas	....	<-84	....	....	al, eth	B1[4], 696
6869	Ethylene, 1,2-diiodo .... $CHI=CHI$	279.85	....	72.5^{16}	-14	3.0625^{20}	....	eth, chl	B1[4], 724
6870	Ethylene, 1,1-diphenyl .... $CH_2=C(C_6H_5)_2$	180.25	....	$277, 94\text{-}5^{11}$	8.2	$1.0281^{16/4}$	1.6100^{20}	eth, chl	B5[3], 1975
6871	Ethylene, fluoro-trichloro .... $CFCl=CCl_2$	149.38	....	71	-108.9	$1.5460^{20/4}$	$1.4379^{20/}{}_D$	chl	B1[4], 715
6872	Ethylene, nitro .... $CH_2=CHNO_2$	73.05	....	$98.5, 38\text{-}9^{80}$	-55.5	$1.2212^{14/4}$	1.4282^{20}	al, eth, ace, bz, chl	B1[4], 725
6873	Ethylene, tetrabromo .... $Br_2C=CBr_2$	343.66	pl (dil al) nd (al)	$225\text{-}7, 100^{15}$	56.5	....	....	al, eth, ace, bz	B1[4], 722
6874	Ethylene, tetrachloro .... $Cl_2C=CCl_2$	165.83	....	$121, 14^{10}$	-19	$1.6227^{20/4}$	1.5053^{20}	**al, eth, bz**	B1[4], 715
6875	Ethylene, tetracyano .... $(NC)_2C=C(CN)_2$	128.09	....	223	198-200	1.348^{28}	1.560^{25}	ace	B2[4], 2450
6876	Ethylene, tetrafluoro .... $F_2C=CF_2$	100.02	gas	-76.3	-142.5	$1.519^{-76.3}$	....	....	B1[4], 698
6877	Ethylene, tetraiodo .... $I_2C=CI_2$	531.64	ye lf pr (eth)	sub	192	2.983^{20}	....	bz, chl	B1[4], 724
6878	Ethylene, tetraphenyl .... $(C_6H_5)_2C=C(C_6H_5)_2$	332.45	mcl or rh (bz-eth or chl-al)	415-25	225	$1.155^{0/4}$	....	bz	B5[3], 2598
6879	Ethylene tetracarboxylic acid, tetraethyl ester .... $(C_2H_5O_2C)_2C=C(CO_2C_2H_5)_2$	316.31	tcl pr (eth)	$325\text{-}8d, 210^{22}$	58	....	....	al, eth	B2[4], 2450
6880	Ethylene, tribromo .... $BrCH=CBr_2$	264.76	....	$163\text{-}4, 75^{15}$	....	$2.708^{20.5/4}$	1.6045^{16}	al, eth, ace, chl	B1[4], 722
6881	Ethylene, trichloro .... $ClCH=CCl_2$	131.29	....	87	−73	$1.4642^{20/4}$	1.4773^{20}	al, eth, ace, chl	B1[4], 712
6882	Ethylene, triphenyl .... $C_6H_5CH=C(C_6H_5)_2$	756.35	lf (al or MeOH)	$220\text{-}1^{14}$	72-3	$1.0373^{78/4}$	1.6292^{78}	al, eth	B5[3], 2398
6883	Ethylene glycol or Ethanediol .... $HOCH_2CH_2OH$	62.07	....	$198, 93^{13}$	-11.5	$1.1088^{20/4}$	1.4318^{20}	**w, al, eth, ace**	B1[4], 2369
6884	Ethylene glycol, bis(chloroacetate) .... $ClCH_2CO_2CH_2CH_2O_2CCH_2Cl$	215.03	pr (eth-peth)	$142\text{-}4^2$	45-6	....	....	eth	B2[3], 448
6885	Ethylene glycol, bis(2-chloroethyl)ether .... $ClCH_2CH_2OCH_2CH_2OCH_2CH_2Cl$	187.07	....	$230, 118^{10}$	......	$1.197^{20/20}$	1.4592^{25}	....	B1[4], 2379
6886	Ethylene glycol, diacetate or Ethylene glycol diacetate .... $CH_3CO_2CH_2CH_2O_2CCH_3$	146.14	....	190	-31	$1.1063^{20/20}$	1.4159^{20}	**w, al, eth, ace**, bz	B2[4], 217
6887	Ethylene glycol, dibenzoate .... $C_6H_5CO_2CH_2CH_2O_2CC_6H_5$	270.29	rh pr (eth)	>360d	73-4	....	....	eth	B9[3], 536
6888	Ethylene glycol, dibutyrate .... $C_3H_7CO_2CH_2CH_2O_2CC_3H_7$	202.25	....	$240, 118\text{-}21^{11}$	......	$1.0005^{20/4}$	1.4262^{20}	al, eth	B2[4], 796
6889	Ethane-1,1-di-4-tolyl .... $(4CH_3C_6H_4)_2CHCH_3$	210.32	....	$298\text{-}9, 153\text{-}6^{11}$	<-20	$0.974^{20/4}$	....	bz	B5[4], 1948
6890	Ethylene glycol, diethylether or 1,2-diethoxyethane ...... $C_2H_5OCH_2CH_2OC_2H_5$	118.18	....	123.5	......	0.8484^{20}	1.3860^{20}	al, eth, ace, bz	B1[4], 2379
6891	Ethylene glycol, diformate .... $HCO_2CH_2CH_2O_2CH$	118.09	....	174	......	$1.193^{0/4}$	1.3580	al, eth	B2[4], 37
6892	Ethylene glycol, dilaurate or Ethyleneglycoldidodecanoate .... $C_{11}H_{23}CO_2CH_2CH_2O_2CC_{11}H_{23}$	426.69	pl (al)	188^{20}	56.6	....	....	al, eth	B2[4], 1094

No.	Name, Synonyms, and Formula	Mol. wt.	Color, crystalline form, specific rotation and λ_{max} (log ε)	b.p. °C	m.p. °C	Density	n_D	Solubility	Ref.
6893	Ethylene glycol, dimethylether or 1,2-dimethoxyethane... $CH_3OCH_2CH_2OCH_3$	90.12	....	83-4	-58	0.8628[20/4]	1.3796[20]	w, al, eth, ace, bz	B1⁴, 2376
6894	Ethylene glycol, dimyristate or Ethylene glycol ditetradecylate $C_{13}H_{27}CO_2CH_2CH_2O_2CC_{13}H_{27}$	482.80	cr (eth or ace)	....	65			eth, ace, bz	B2⁴, 1133
6895	Ethylene glycol, dinitrate .... $O_2NOCH_2CH_2ONO_2$	152.06	ye	197-200	-22.3	1.4918[20/4]	....	al, eth	B1⁴, 2413
6896	Ethylene glycol, dinitrite .... $ONOCH_2CH_2ONO$	120.07		98	<-15	1.2156[0/4]	....	al, eth	B1⁴, 2411
6897	Ethylene glycol, dipalmitate or Ethanediol dihexadecylate $C_{15}H_{31}CO_2CH_2CH_2O_2CC_{15}H_{31}$	538.91	lf or nd (al-chl)	226 (vac)	72	0.8594[78]	....	eth, ace	B2⁴, 1169
6898	Ethylene glycol, di-phenyl ether .... $C_6H_5OCH_2CH_2OC_6H_5$	214.27	lf (al)	180-5[12]	98			eth, chl	B6⁴, 573
6899	Ethylene glycol, dipropionate $C_2H_5CO_2CH_2CH_2O_2CC_2H_5$	174.20		211		1.020[15]		al, eth, ace	B2⁴, 715
6900	Ethylene glycol, distearate or Ethylene glycol di-octadecylate $C_{17}H_{35}CO_2CH_2CH_2O_2CC_{17}H_{35}$	595.02	lf	241[20]	79			eth, ace	B2⁴, 1223
6901	Ethylene glycol, dithiocyanate .... $NCSCH_2CH_2SCN$	144.22	rh pl or nd (w) ta (al or eth)	d	90			al, eth, ace	B3⁴, 333
6902	Ethylene glycol, methyl, ethyl ether .... $CH_3OCH_2C_2OC_2H_5$	104.13	....	102	....	0.8529[20/4]	1.3868[20]	w, al, eth ace, bz	B1³, 2078
6903	Ethylene glycol, monoacetate or 2-Hydroxyethyl acetate. $CH_3CO_2CH_2CH_2OH$	104.22	....	187-9	....	1.108[15]		w, al, eth	B2⁴, 154
6904	Ethylene glycol, monoallyl ether .... $HOCH_2CH_2OCH_2CH=CH_2$	102.13	....	159[755], 64[15]		0.9580[20/4]	1.4358[20]	w, al, bz	B1⁴, 2388
6905	Ethylene glycol, monobenzoate .... $C_6H_5CO_2CH_2CH_2OH$	166.18	....	150-1[10]	45			al	B9³, 532
6906	Ethylene glycol, monobenzyl ether or Benzyl cellosolve... $HOCH_2CH_2OCH_2C_6H_5$	152.20	....	256,138[15]	<-75	1.0640[20/4]	1.5233[20]	w, al, eth	B6⁴, 2241
6907	Ethylene glycol, monobutyl ether or Butyl cellosolve..... $HOCH_2CH_2OC_4H_9$	118.18	....	171, 50⁴	....	0.9015[20/4]	1.4198[20]	w, al, eth	B1⁴, 2380
6908	Ethylene glycol, mono-iso-butyl ether or Isobutyl cellosolve .... $HOCH_2CH_2OCH_2CH(CH_3)_2$	118.18	....	159[745]	....	0.8900[20/4]	1.4143[20]		B1⁴, 2382
6909	Ethylene glycol, mono-β-chloroethyl ether or β-Chloroethyl cellosolve $HOCH_2CH_2OCH_2CH_2Cl$	124.57	....	180-85, 91-2[13]			1.4805[19]	w, al, eth	B1³, 2078
6910	Ethylene glycol, monoethyl ether or Ethyl cellosolve ..... $HOCH_2CH_2OC_2H_5$	90.12	....	135, 35[10]	....	0.9297[20/4]	1.4080[20]	w, al, eth, ace	B1⁴, 2377
6911	Ethylene glycol, monohexyl ether or Hexyl cellosolve..... $HOCH_2CH_2OC_6H_{13}$	146.23	....	208,96[13]	-45.1	0.8894[20/20]	1.4291[20]	al, eth	B1⁴, 2383
6912	Ethylene glycol, monomethyl ether or Methyl cellosolve .. $HOCH_2CH_2OCH_3$	76.11	....	125[768]	-85.1	0.9647[20/4]	1.4024[20]	w, al, eth, ace, bz	B1⁴, 2375
6913	Ethylene glycol, monophenyl ether or Phenyl cellosolve .. $HOCH_2CH_2OC_6H_5$	138.17	....	237, 134-5[18]		1.1020[20]	1.5340[20]	al, eth	B6⁴, 571
6914	Ethylene glycol, monopropyl ether or Propyl cellosolve... $HOCH_2CH_2OC_3H_7$	104.15	....	150[743]		0.9112[20/4]	1.4133[20]	w, al, eth	B1⁴, 2379
6915	Ethylene glycol, mono isopropyl ether or Isopropyl cellosolve .... $i-C_3H_7OCH_2CH_2OH$	104.15	....	144[743]		0.9030[20/4]	1.4095[20]	w, al, eth, ace	B1⁴, 2380
6916	Ethylene glycol, monostearate $C_{17}H_{35}CO_2CH_2CH_2OH$	328.54	(peth)	189-91³	60-1	0.8780[60/4]	1.4310[60]	eth	B4², 1222
6917	Ethylene glycol sulfate .... $C_2H_4O_4SO_2$	124.12	nd or pr (bz -lig)	sub	99	....	....	al, eth, ace, bz	B1³, 2110
6918	Ethylene glycol sulfite .... $C_2H_4O_3SO$	108.12		173, 70-1[20]	-11	1.4402[20/4]	1.4463[20]	w, al, eth, ace, bz	B1⁴, 2409
6919	Ethylene imine or Aziridine .... C_2H_5N	43.07		50[756]		0.8321[20/4]		w, al, eth, ace, bz	B20⁴, 3
6920	Ethylene oxide or Epoxyethane .... C_2H_4O	44.05		13.2[746]	-111	0.8824[10/10]	1.3597⁷	w, al, eth, ace, bz	B17⁴, 3
6921	Ethylidene diacetate .... $CH_3CH(O_2CCH_3)_2$	146.14		169, 65-7[10]	18.9	1.070[25]	1.3985[25]	al, eth	B2⁴, 282
6922	Ethynyl methyl ether .... $CH_3OC\equiv CH$	56.07		50		0.8001[20/4]	1.3812[20]	al, eth	B1⁴, 2211
6923	Ethynyl phenyl ether .... $C_6H_5C\equiv CH$	118.14	....	61-2[25]	-36	1.0614[20/4]	1.5125[20]	al, eth	B6⁴, 565

No.	Name, Synonyms, and Formula	Mol. wt.	Color, crystalline form, specific rotation and λ_{max} (log ϵ)	b.p. °C	m.p. °C	Density	n_D	Solubility	Ref.
6924	Ethynyl propyl ether .... $C_3H_7OC\equiv CH$	84.13	....	75	......	$0.8080^{20/4}$	1.3935^{20}	al, 6924	B1[4], 2213
6925	α-Eucaine .... $C_{19}H_{27}NO_4$	333.43	pr (eth or al)	....	104-5	....	....	w, al, eth, bz, chl	B22, 194
6926	α-Eucaine, hydrochloride .... $C_{19}H_{27}NO_4 \cdot HCl$	369.89	pl (w + 1) pr	....	ca 200d	....	....	w, al	B22, 194
6927	β-Eucaine-(d) .... $C_{15}H_{21}NO_2$	247.34	pr (peth)	....	57-8	....	....	al, eth, bz, chl, peth	B21[2], 14
6928	β-Eucaine (dl) or Betacaine .... $C_{15}H_{21}NO_2$	247.34	pr (peth)	....	57-8	....	....	al, eth, bz, chl, peth	B21[2], 13
6929	β-Eucaine (l) .... $C_{15}H_{21}NO_2$	247.34	pr (peth)	....	57-8	....	....	al, eth, bz, chl, peth	B21[2], 14
6930	β-Eucaine, hydrochloride-(dl) .... $C_{15}H_{21}NO_2.HCl$	283.80	pl (w)	....	277-9	....	....	w, al, eth, chl	B21[2], 13
6931	Eugenol or 5-Allylguaiacol .... $4-(CH_2=CHCH_2)-2-CH_3OC_6H_3OH$	164.29	cr (hx)	253.2, 130.5[10]	-7.5	$1.0652^{20/4}$	1.5405^{20}	al 6931	B6[3], 5021
6932	Eugenol-acetate .... $4-(CH_2=CHCH_2)-2-CH_3OC_6H_3O_2CCH_3$	206.25	pr (al)	127-8[6]	30-1	$1.0806^{20/4}$	1.5205^{20}	al	B6[3], 5029
6933	Eupitone or Eupittonic acid .... $C_{25}H_{26}O_9$	470.48	nd (al-eth)	....	200	....	....	a: 6933	B8[3], 4427
6934	Euxanthic acid .... $C_{19}H_{16}O$	404.43	ye nd (w + 1) α−108 (+ lw)	....	130d (+ w), 162d (anh)	....	....	al	B31, 277
6935	Eucarvone .... $C_{10}H_{16}O$	150.22	....	99-100[15], 88[16]	......	$0.9490^{20/4}$	1.50872^{20}	eth, ace	JOCEA 26, 1609
6936	Evernic acid .... $C_{17}H_{16}O_7$	332.31	nd (w or ace) pr (al)	....	170	....	....		B10[3], 1488
6937	Evodiamine (d) or Rhetsine .... $C_{19}H_{17}N_3O$	303.37	yesh lf (al) $[\alpha]^{15}_D + 352$ (ace,c=0.5)	....	278	....	....		B26[2], 103
6938	Evodiamine, hydrate (d) .... $C_{19}H_{17}N_3O.H_2O$	321.38	pl (al)	....	146-7	....	....		B24[2], 72

No.	Name, Synonyms, and Formula	Mol. wt.	Color, crystalline form, specific rotation and λ_{max} (log ε)	b.p. °C	m.p. °C	Density	n_D	Solubility	Ref.
6939	Fagaramide $C_{14}H_{17}NO_3$	247.30	nd (bz, dil al or peth)	....	119.5	....	....	al, bz	B19², 299
6940	β-Fagarine or Skimmianine $C_{14}H_{13}NO_4$	259.27	pym oct (al)	....	177	....	....	al, chl	B27, 134
6941	α-Fagarine or Haplophine $C_{13}H_{11}NO_3$	229.24	pr (al)	....	142	....	....	al, eth, bz, chl	C51, 4402
6942	α-Farnesine or 3,7,11-trimethyl-1,3,6,10-dodecatetracene $C_{15}H_{24}$	204.36	....	129-32¹²	....	0.8410²⁰/⁴	1.4836²⁰	eth, ace, peth, lig	B1³, 1067
6943	β-Farnesine $C_{15}H_{24}$	204.36	....	121-2⁹	....	0.8363²⁰/⁴	1.4899²⁰	eth, ace, chl, aa	B1⁴, 1133
6944	Farnesol (trans, trans) $(CH_3)_2C=CH(CH_2)_2C(CH_3)=CH(CH_2)_2C(CH_3)=CHCH_2OH$	222.37	....	160¹⁰	....	0.8846²⁰/⁴	1.4877²⁰	al, eth, ace	B1⁴, 2335
6945	Farnesol (cis, trans) $C_{15}H_{26}O$	222.37	....	120⁰·³	....	0.8846²⁰/⁴	1.4877²⁰	al, eth, ace	B1⁴, 2335
6946	Fenchane (d) $C_{10}H_{18}$	138.25	....	151-2⁷⁶⁵	....	0.8345²⁰/⁴	1.44714²⁰	al, eth	B5³, 256
6947	Fenchane (dl) $C_{10}H_{18}$	138.25	....	151-2²⁶⁵	....	0.8345²⁰/⁴	1.4471²⁰	al, eth	B5³, 256
6948	α-Fenchene (d) $C_{10}H_{16}$	136.24	$[\alpha]^{14/}{}_D + 29$	155-6	....	0.8660²⁰/⁴	1.4713²⁰	al, eth, ace	B5¹, 86
6949	α-Fenchene (dl) or Isopinene $C_{10}H_{16}$	136.24	....	154-6	....	0.8660²⁰/⁴	1.4705²⁰	al, eth, ace	B5³, 389
6950	α-Fenchene (l) $C_{10}H_{16}$	136.24	$[\alpha]^{20/}{}_D$-43.8	158-9	....	0.8670²⁰/⁴	1.4713²⁰	al, eth, ace	B5⁴, 466
6951	β-Fenchene $C_{10}H_{16}$	136.24	$[\alpha]'_D$ 62.91 (al)	151-3	....	0.8591²⁰/⁴	1.4645²⁵/$_D$	....	B5⁴, 465
6952	Fenchone (d) or Trimethyl norcamphor $C_{10}H_{16}O$	152.24	$[\alpha]^{20/}{}_D$ + 66.9 (al)	193.5, 80²⁰	6	0.9465²⁰/⁴	1.4623²⁰	al, eth, ace	B7³, 392
6953	Fenchone-(dl) $C_{10}H_{16}O$	152.24	....	193-4, 72-3¹²	-18	0.9501¹⁵/¹⁵	1.4702²⁰	al, eth, ace	B7³, 393
6954	Fenchone (l) $C_{10}H_{16}O$	152.24	$[\alpha]^{23}{}_D$ -66.94 (al)	192-4	5(8.5)	0.948²⁰	1.4636²⁰	al, eth, ace	B7³, 392
6955	Fenchyl alcohol (dl) $C_{10}H_{18}O$	154.26	....	α202-301, β201	α38-9, β6	....	....	al, eth	B6;3, 288
6956	Filixic acid BBB or Filicin $C_{36}H_{44}O_{12}$	668.74	cr (AcOEt-ace)	....	172-4	....	....	bz, chl	B8³, 4436
6957	Ferrocene or Dicyclopentadienyl iron $C_{10}H_{10}Fe$	186.04	....	249	172.5-3.0	....	....	....	KHOC, 152
6958	Flavaniline or 2-(p-Aminophenyl) lepidine $C_{16}H_{14}N_2$	234.30	pr (bz)	133-41¹⁵	97	....	....	al, bz	B22, 469
6959	Flavanone or 2,3-dihydro-2-phenyl-1,4-benzopyrone $C_{15}H_{12}O_2$	224.26	nd (lig)	....	76	....	....	ace, bz	B17⁴, 5338
6960	Flavanone, 4'-methoxy-3',5,7-trihydroxy or Hesperetin $C_{16}H_{14}O_6$	302.29	pl (dil al + ½ w)	sub 205⁰·⁰⁰⁴	227-8	....	....	al, eth	B18⁴, 3215
6961	Flavanone, 3',4',5,7-tetra hydroxy or Eriodictyol $C_{15}H_{12}O_6$	288.26	Pa br nd (dil al + 1.5 w)	....	267 d	....	....	al, aa	B18⁴, 3214
6962	Flavone or 2-Phenyl-α-benzopyrone $C_{15}H_{10}O_2$	222.25	nd (lig)cr (30% al)	....	100	....	....	al, eth, ace, bz, chl	B17⁴, 5413
6963	Flavone, 6-bromo $C_{15}H_9BrO_2$	301.15	nd (al)	....	191-2	....	....	al	B17⁴, 5417
6964	Flavone, 5,7-dihydroxy or Chrysin $C_{15}H_{10}O_4$	254.25	Pa ye pl or pr (MeOH) nd (sub)	sub	275	....	....	al, ace	B18⁴, 1766
6965	Flavone, 5,7-dihydroxy 4'-methoxy or Acacetin $C_{16}H_{12}O_5$	284.27	Pa ye nd (al)	....	261	....	....	ace	B18⁴, 2683
6966	Flavone, 5,7-dihydroxy-6-methoxy $C_{16}H_{12}O_5$	284.27	ye nd (al)	....	231-2	....	....	al, eth, ace, aa	B18⁴, 2671
6967	Flavone, 3-hydroxy or Flavanol $C_{15}H_{10}O_3$	238.25	Pa ye nd (al)	....	169-70	....	....	al	B17⁴, 6428
6968	Flavone, 2',3,3',5,7-pentahydroxy $C_{15}H_{10}O_7$	302.24	ye nd (aa + 1.5 w)	....	300	....	....	al	B18⁴, 3468
6969	Flavone, 2',3,4',5,7-pentahydroxy or Morin $C_{15}H_{10}O_4$	302.24	Pa ye nd (+ 1w, dil al)	....	303-4	....	....	al, bz	B18⁴, 3468

No.	Name, Synonyms, and Formula	Mol. wt.	Color, crystalline form, specific rotation and λ_{max} (log ε)	b.p. °C	m.p. °C	Density	n_D	Solubility	Ref.
6970	Flavone, 2′,3,5,5′,7-pentahydroxy .... $C_{15}H_{10}O_7$	302.24	red ye cr (dil al + 1 w)	....	306-8	....	....	....	B18⁴, 3470
6971	Flavone, 3,3′,4′,5,7-pentahydroxy or Quercitin......... $C_{15}H_{10}O_7$	302.24	ye rd (dil al + 2w)	sub	316-7	....	....	al, ace, aa	B18⁴, 3470
6972	Flavone, 3,3′,4′,7,8-pentahydroxy .... $C_{15}H_{10}O_7$	302.24	ye nd (dil al + 1w)	....	308d	....	....	al, ace	B18⁴, 3506
6973	Flavone, 3,3′,5,5′,7-pentahydroxy .... $C_{15}H_{10}O_7$	302.24	ye nd	....	>300	....	....	....	B18², 239
6974	Flavone, 3′,4′,5,5′,7-pentahydroxy or Tricetin .... $C_{15}H_{10}O_7$	302.24	ye nd(dil al + w)	....	>330d	....	....	....	B18⁴, 3454
6975	Flavone, 2′,3,5,7-tetrahydroxy or Datiscetin .... $C_{15}H_{10}O_6$	286.24	Pa ye nd (al, aq aa)	....	277-8	....	....	al, eth, ace	B18⁴, 3281
6976	Flavone, 3,3′,4′,7-tetrahydroxy or Fisetin .... $C_{15}H_{10}O_6$	286.24	lf ye nd (dil al + 1w)	sub ⁷⁶⁰	330	....	....	al, ace	B18⁴, 3304
6977	Flavone, 3′,4′,5,7-tetrahydroxy or Luteolin............. $C_{15}H_{10}O_6$	286.24	ye nd (dil al + 1w)	sub ⁷⁶⁰	329-30 d	....	....	al, eth	B18⁴, 3261
6978	Flavone, 3,4′,5,7-tetrahydroxy or Kaempferol .... $C_{15}H_{10}O_6$	286.24	ye nd (al + 1w) (aa)	....	276-8	....	....	al, ace	B18⁴, 3283
6979	Flavone, 4′,5,7-trihydroxy or Apigenin .... $C_{15}H_{10}O_5$	270.25	ye nd (py-w) lf (al))	sub ⁷⁶⁰	347-8	....	....	al, py	B18⁴, 2682
6980	Flavone, 5,6,7-trihydroxy or Baicalein .... $C_{15}H_{10}O_5$	270.25	ye pr (al)	....	264-5d	....	....	al, eth, ace	B18⁴, 2671
6981	Floridoside or 2-o-α-D-Galactosyl glycerol................. $C_9H_{18}O_8$	254.24	pr (al) [α]′ᴅ + 13.1 (w)	....	86-7	....	....	w	B17¹, 2006
6982	Fluoran or 9-Hydroxy-9-xanthene-o-benzoic acid acetone $C_{20}H_{12}O_4$	300.32	nd (al + 2 al)	....	182-3	....	....	al	B19⁴, 2903
6983	Fluoran, 1,6-dihydroxy .... $C_{20}H_{12}O_5$	332.32	ye nd	....	>260	....	....	....	B19⁴, 2903
6984	Fluoran, 2,6-dihydroxy .... $C_{20}H_{12}O_5$	332.32	ye nd (al)	....	177	....	....	....	B19⁴, 2902
6985	Fluoran, 3,5-dihydroxy .... $C_{20}H_{12}O_5$	332.32	ye gr nd or pl (aa)	....	179	....	....	....	B19⁴, 2903
6986	Fluoranthene or 1,2-Benzacenaphthene .... $C_{16}H_{10}$	202.26	pa ye nd or pl (al)	375, 217⁵⁰	....	1.252⁰ᐟ⁴	....	al, eth, bz, aa	B5³, 2276
6987	Fluorene or 2,3-Benzindene .... $C_{13}H_{10}$	166.23	lf (al)	293-5	116-7	1.203⁰ᐟ⁴	....	eth, ace, bz	B5³, 1936
6988	Fluorene, 2-acetamido or N(2-Fluorenyl) acetamide ..... $C_{15}H_{13}NO$	223.28	nd (50% al or 50% aa)	....	194	....	....	al, eth, aa	B12³, 3287
6989	Fluorene, 9-acetamido .... $C_{15}H_{13}NO$	223.28	nd (aa)	....	262	....	....	....	B12³, 3299
6990	Fluorene, 1-amino-9-hydroxy .... $C_{13}H_{11}NO$	197.24	dk red rd (w)	....	142	....	....	al, eth, bz, aa	B13², 435
6991	Fluorene, 2-amino .... $C_{13}H_{11}N$	181.24	lo pl or nd (dil al)	....	131-2	....	....	al, eth	B12³, 3285
6992	Fluorene, 2-amino-9-hydroxy .... $C_{13}H_{11}NO$	197.24	irid nd (al)	....	200-1	....	....	al	B13³, 2023
6993	Fluorene, 4-amino-9-hydroxy .... $C_{13}H_{11}NO$	197.24	ye (60% al)	....	183-4	....	....	al	B13³, 2023
6994	Fluorene, 9-amino .... $C_{13}H_{11}N$	181.24	nd (lig)	....	64-5	....	....	al, eth, ace, bz, chl	B12³, 3297
6995	Fluorene, 9-benzhydrylidene or ω, ω -Diphenyl dibenzo fulvene................. $C_{26}H_{18}$	330.43	ye (bz)	....	229.5	....	....	chl	B5³, 2625
6996	Fluorene, 9-benzylidene or ω -Phenyl dibenzo fulvene.... $C_{20}H_{14}$	254.34	lf (al)	....	76	....	....	al, bz	B5³, 2464
6997	Fluorene, 2-bromo .... $C_{13}H_9Br$	245.13	nd or pl (al)	185¹³⁵	113-4	....	....	al, chl, aa	B5³, 1943
6998	Fluorene, 9-bromo .... $C_{13}H_9Br$	245.13	(lig or al)	....	104-5	....	....	al, ace	B5³, 1944
6999	Fluorene, 9 (3-bromobenzylidene) .... $C_{20}H_{13}Br$	333.24	ye nd (aa)	....	92-3	....	....	al, MeOH	B5¹, 358
7000	Fluorene, 9 (4-bromobenzylidene) .... $C_{20}H_{13}Br$	333.24	ye nd (aa, AcOEt-chl)	....	147-8	....	....	aa	B5², 640

No.	Name, Synonyms, and Formula	Mol. wt.	Color, crystalline form, specific rotation and λ_{max} (log ε)	b.p. °C	m.p. °C	Density	n_D	Solubility	Ref.
7001	Fluorene, 9 (2-chlorobenzylidene) $C_{20}H_{13}Cl$	288.78	ye nd (aa, MeOH)	180[0.7]	69-70			al, aa	B5[3], 2464
7002	Fluorene, 9-(3-chlorobenzylidene) $C_{20}H_{13}Cl$	288.78	Pa ye pr or pym (MeOH)		90.5			al, aa	B5[1], 358
7003	Fluorene, 9-(4 chlorobenzylidene) $C_{20}H_{13}Cl$	288.78	ye nd (aa,al)		151			al, aa	B5[1], 358
7004	Fluorene, 9-Cinnamylidene (trans) $C_{22}H_{16}$	280.37	pa ye nd (aa)		155			al, chl	B5[3], 2533
7005	Fluorene, 2,7-diamino $C_{13}H_{13}N_2$	196.26	nd (w) pr (bz) pl (eth)		165-7			al, chl	B13[3], 507
7006	Fluorene, 2,7-dichloro $C_{13}H_8Cl_2$	235.13	pl or nd (bz)	sub	128			bz, chl	B5[3], 1942
7007	Fluorene, 9,9-dichloro $C_{13}H_8Cl_2$	235.13	rh pr (bz-eth) nd (peth)		103			al, eth, ace, bz	B5[3], 1942
7008	Fluorene, 1,8-dimethyl-9-(2-tolyl) $C_{22}H_{20}$	299.40			168-9			al, eth, ace	C50, 11293
7009	Fluorene, 2-hydroxy $C_{13}H_{10}O$	182.23	lf (w) nd (chl)		171-4			al, eth, ace, aa	B6[3], 3487
7010	Fluorene, 9-hydroxy or 9-Fluorenol $C_{13}H_{10}O$	182.23	hex nd (w or peth)		154			eth, ace, bz	B6[3], 3489
7011	Fluorene, 9-hydroxy-9-phenyl $C_{19}H_{14}O$	258.32	ye or col pr (lig)		108-9			bz, aa	B6[3], 3732
7012	Fluorene, 9-methyl $C_{14}H_{12}$	180.25	pr	154-6[15]	46-7	1.0263[66/4]	1.610[66]	al, eth, ace, bz, chl	B5[3], 1992
7013	Fluorene, 9-(2-methylbenzylidene) $C_{21}H_{16}$	268.38	nd or pr (aa)		109.5			al, bz	B5[1], 359
7014	Fluorene, 9-(4-methylbenzylidene) $C_{21}H_{16}$	268.38	nd (aa) pr (al)		97.5			al, bz	B5[1], 359
7015	Fluorene, 9-methylene or Dibenzofulvene $C_{14}H_{10}$	178.24			53			al, eth, ace, bz	B5[3], 2147
7016	Fluorene, 2-nitro $C_{13}H_9NO_2$	211.22	nd (50% aa or ace)		158			ace, bz	B5[3], 1948
7017	Fluorene, 3-nitro $C_{13}H_9NO_2$	211.22	pa ye nd (al, chl-peth)		106			al, ace	B5[3], 1949
7018	Fluorene, 9-nitro $C_{13}H_9NO_2$	211.22	gr ye nd (al) lf (bz)		181-2d			eth, ace,bz, chl, aa	B5[3], 1949
7019	Fluorene, 9-phenyl $C_{19}H_{14}$	242.32	nd or lf (al or bz)		148			al, bz, aa, chl	B5[3], 2385
7020	2-Fluorene sulfonic acid $C_{14}H_{10}SO_3$	246.29	nd (aa)		155 (+1w)			w, al, ace, chl	B11[3], 440
7021	9-Fluorenone or 9-Oxofluorene $C_{13}H_8O$	180.22	ye rh bipym (al, bz-peth)	341.5	84	1.1300[99/4]	1.6369[99]	al, eth, ace, bz	B7[3], 2330
7022	9-Fluorenone, 1-amino $C_{13}H_9NO$	195.22	ye nd (dil al)		118-20			al, eth, ace	B14[3], 285
7023	9-Fluorenone, 2-amino $C_{13}H_9NO$	195.22	red vt pr (al)		163			al, eth, bz, aa	B14[3], 286
7024	9-Fluorenone, 3-amino $C_{13}H_9NO$	195.22	ye nd (w or dil al)		158-9			al	B14[3], 289
7025	9-Fluorenone, 4-amino $C_{13}H_9NO$	195.22	red nd (al)		145			al, eth, ace, chl, aa	B14[3], 289
7026	9-Fluorenone, 2-bromo $C_{13}H_7BrO$	259.11	ye nd (al or aa)		149			ace, bz, chl, aa	B7[3], 2340
7027	9-Fluorenone, 2-chloro $C_{13}H_7ClO$	214.65	og ye nd (dil al)	sub	125-6			al	B7[3], 2338
7028	9-Fluorenone, 1,8-dimethyl $C_{15}H_{12}O$	208.24	ye		197-8			chl, aa	C62, 7611
7029	9-Fluorenone, 2-nitro $C_{13}H_7NO_3$	225.26	ye nd or lf (aa)	sub	222-3			ace	B7[3], 2344
7030	9-Fluorenone, oxime $C_{13}H_9NO$	195.22	nd (chl-peth or bz)		195-6			al, chl	B7[3], 2335
7031	Fluorenone, 2,3,7-trinitro $C_{13}H_5N_3O_7$	315.20	Pa ye nd (aa)		180-1			bz, chl	B7[3], 2348

No.	Name, Synonyms, and Formula	Mol. wt.	Color, crystalline form, specific rotation and λ_{max} (log ϵ)	b.p. °C	m.p. °C	Density	n_D	Solubility	Ref.
7032	9-Fluorenone, 2,4,7-trinitro .... $C_{13}H_5N_3O_7$	315.20	pa ye nd (aa or bz)	....	176	....	....	ace, bz, chl	B7³, 2348
7033	9-Fluorenone-1-carboxamide .... $(C_{11}H_7O)CONH_2$	223.23	ye nd (al)		229-30			ace	B10³, 3369
7034	9-Fluorenone-1-carboxylic acid .... $(C_{13}H_7O)CO_2H$	224.23	og, red nd (dil al)		192-4			al, eth	B10³, 3368
7035	9-Fluorenone-1-carboxylic acid, ethyl ester .... $(C_{13}H_7O)CO_2C_2H_5$	252.27	ye nd (dil al)		84-5			al, eth	B10³, 3369
7036	9-Fluorenone-1-carboxylyl chloride .... $(C_{13}H_7O)COCl$	242.66	pa ye nd (bz)		140			al, eth, bz	B10, 774
7037	9-Fluorenone-2-carboxylic acid .... $(C_3H_7O)_2CO_2H$	224.23	ye nd (al or aa)	sub 340	338			al, aa	B10³, 3370
7038	9-Fluorenone-2-carboxylic acid, methyl ester $C_{15}H_{10}O_3$ 2-$(C_3H_7O)CO_2CH_3$	238.25	ye nd (MeOH)		181			al, eth, ace	B10³, 3370
7039	9-Fluorenone-3-carboxylic acid .... $C_{14}H_8O_3$	224.23	ye (aa, MeOH)		299			al	B10³, 3372
7040	9-Fluorenone 4-carboxylic acid .... $C_{14}H_7O_3$	224.23	ye nd (al)		227			al, eth	B10³, 3372
7041	Fluorescein or 3',4'-Dehydroxy fluoran .... $C_{20}H_{12}O_5$	332.32	red rh pr		314-6d (sealed tube)			ace, py, MeOH	B19⁴, 2904
7042	Fluorescin or 2 (3,6-Dihydroxyxanthyl) benzoic acid ..... $C_{20}H_{14}O_5$	334.33	col or ye nd (aa or eth) pl (bz)		125-7			al, eth, ace	B19⁴, 2904
7043	Fluoroacetic acid .... FCH_2CO_2H	78.04	nd	165	35.2	1.3693³⁶	....	al, w	B2⁴, 446
7044	Fluorophosphoric acid, diisopropyl ester .... $(i-C_3H_7O)_2POF$	184.15	....	62⁹		1.055	1.3830²⁵	eth	B1⁴, 1480
7045	Folic acid or Pteroylglutamic acid, Vitamin Bc .... $C_{19}H_{19}N_7O_6$	441.41	ye og nd (w) α^{25}_D + 23 (0.1 N NaOH, c=0.5)		250d (dark-ens)			al, aa, py	Am69, 1476
7046	Folinic acid or 5-Formyl-5,6,7,8-tetrahydropteroyl-L-glutamic acid .... $C_{20}H_{23}N_7O_7$	473.45	cr (w + 3) $[\alpha]^{25}_D$ + 16.76 (5%Na₂CO₃, c=3.5)		248-50d				Am73, 1979
7047	Formaldehyde or Methanal .... HCHO	30.05	gas	−21	−92	0.815²⁰/⁴	....	w, al, eth, ace, bz	B1⁴, 3017
7048	Formaldehyde, bis-4-chlorophenyl acetal .... $H_2C(OC_6H_4Cl-4)$	269.13		189-94⁶	70			eth, ace, bz	B6⁴, 833
7049	Formaldehyde, dibutyl acetal .... $H_2C(OC_4H_9)$	160.26		179.2	−58.1	0.834²⁰/ᴰ	1.4072¹⁷·²/ᴰ	....	B1⁴, 3029
7050	Formaldehyde, β,β'-(dichloro isopropyl) ethyl acetal .... $[(ClCH_2)_2CHO]CH_2[OC_2H_5]$	187.07		96-8¹⁶	....	1.182¹⁷/¹⁷	1.4491¹⁷	eth	B1³, 2574
7051	Formaldehyde, diethyl acetal or Ethylal .... $CH_2(OC_2H_5)_2$	104.13		89	−66.5	0.8319²⁰/⁴	1.3748¹⁸	w, al, eth, ace bz	B1⁴, 3027
7052	Formaldehyde, dimethyl acetal or Methylal .... $CH_2(OCH_3)_2$	76.11		45.5	−104.8	0.8593²⁰/⁴	1.3513²⁰	w, al, eth, ace, bz	B1⁴, 3026
7053	Formaldehyde, 2,4-dinitrophenyl hydrazone .... $[2,4-(NO_2)_2C_6H_3]NHN=CH_2$	210.15	ye cr (al) pr (lig)		167				B15³, 426
7054	Formaldehyde, dipropyl acetal .... $CH_2(OC_3H_7)_2$	132.21		140.5	−97.3	0.8345²⁰	1.3939¹⁹	w, al, eth, ace, bz	B1⁴, 3029
7055	Formaldehyde, fluoro or Formyl fluoride .... FCHO	48.00	gas	−24					B2⁴, 42
7056	Formaldehyde, oxime or Formaldoxime .... $H_2C=NOH$	45.04		109¹⁵	2.5	1.133	....	w, al, eth	B1⁴, 3055
7057	Formaldomedone .... $C_{17}H_{22}O_4$	292.38	nd (al or bz)		189-90				B7³, 4736
7058	Formamide .... $HCONH_2$	45.04	....	111²⁰	2.5	1.1334²⁰/⁴	1.4472²⁰	w, al, ace	B2⁴, 45
7059	Formamide, N,N-diethyl .... $HCON(C_2H_5)_2$	101.15		177-8, 68¹⁵	....	0.9080¹⁹	1.4321²⁵	w, al, eth, ace, bz	B4⁴, 346
7060	Formamide, N,N-dimethyl .... $HCON(CH_3)_2$	73.09	....	149-56, 39.9¹⁰	−60.5	0.9487²⁰/⁴	1.4305²⁰	w, al, eth, ace, bz, chl	B4⁴, 171

No.	Name, Synonyms, and Formula	Mol. wt.	Color, crystalline form, specific rotation and λ_{max} (log ε)	b.p. °C	m.p. °C	Density	n_D	Solubility	Ref.
7061	Formamide, N,N-diphenyl .... HCON(C₆H₅)₂	197.24	rh (dil al)	337.5, 190¹³	73-4	....	....	al, eth, bz	B12³, 455
7062	Formamide, N-ethyl .... HCONHC₂H₅	73.09	....	197-9, 109.6³⁰	....	0.9552²⁰/⁴	1.4320²⁰	w, al, eth	B4⁴, 346
7063	Formamide, N-(1-hydroxy-2,2,2-trichloro ethyl) or Chloral formamide .... HCONH[CH(OH)CCl₃]	192.43	cr	118		....	....	w, al, eth, ace	B2², 37
7064	Formamide, N-methyl .... HCONHCH₃	59.07	....	180-5, 102-3²⁰		1.011¹⁹	1.4319²⁰	w, al, ace	B4⁴, 170
7065	Formamide, N-phenyl or Formanilide .... HCONHC₆H₅	121.14	mcl pr (lig-xyl)	271, 166¹⁴	50	1.1322⁵⁰/⁵⁰	....	al, eth, bz	B12³, 453
7066	Formamide, N-2-tolyl .... HCONH(C₆H₄CH₃-2)	135.17	lf (al)	288	62	1.086⁵⁵/⁴	....	al	B12³, 1852
7067	Formamide, N-3-tolyl .... HCONH(C₆H₄CH₃-3)	135.17	....	278⁷²⁴/d, 176-8¹⁷	<-18			w	B12³, 1962
7068	Formamide, N-4-tolyl .... HCONH(C₆H₄CH₃-4)	135.17	nd		53			al, ace	B12³, 2050
7069	Formamidine .... HN=CHNH₂	44.06	pr	d	81	....	....	w, al	B2⁴, 82
7070	Formamidine, N,N'-diphenyl .... C₆H₅N=CHNHC₆H₅	196.26	nd (al)	> 250	142			al, eth, ace, bz, chl	B12³, 456
7071	Formamidoxime or Isoretin .... HON=CHNH₂	60.06	rh nd (al)	d	114-5	....	....	al	B2⁴, 84
7072	Formic acid or Methanoic acid .... HCO₂H	46.03	....	100.7, 50¹²⁰	8.4	1.220²⁰/⁴	1.3714²⁰	w, al, eth, ace bz	B2⁴, 3
7073	Formic acid, allyl ester or Allyl formate .... HCO₂CH₂CH=CH₂	86.08	....	83.6		0.9460²⁰/⁴		al, eth,	B2¹, 46
7074	Formic acid, benzyl ester or Benzyl formate .... HCO₂CH₂C₆H₅	136.16	....	202-3⁷⁴⁷, 84-5¹⁰		1.081²⁰/⁴	1.5154²⁰	al, eth, ace	B6⁴, 2262
7075	Formic acid, butyl ester or Butyl formate .... HCO₂C₄H₉	102.13	....	106.8	−91.3	0.8885²⁰/⁴	1.3912²⁰	al, eth, ace	B2⁴, 28
7076	Formic acid, isobutyl ester or Isobutyl formate .... HCO₂-i-C₄H₉	102.13	....	98.4	−95.8	0.8854²⁰/⁴	1.3857²⁰	al, eth, ace	B2⁴, 29
7077	Formic acid, sec-butyl ester-(dl) or sec-Butyl formate.... HCO₂-sec-C₄H₉	102.13	....	97		0.8846²⁰/⁴	1.3865²⁰	al, eth, ace	B2⁴, 29
7078	Formic acid, cyclohexyl ester or Cyclohexyl formate .... HCO₂C₆H₁₁	128.17	....	162.5⁷⁵⁰		1.0057⁷/⁴	1.4430²⁰	al, eth, aa	B6⁴, 35
7079	Formic acid, ethyl ester or Ethyl formate .... HCO₂C₂H₅	74.08	....	54.5	−80.5	0.9168²⁰	1.3598¹⁰	w, al, eth, ace	B2⁴, 23
7080	Formic acid, heptyl ester or Heptyl formate .... HCO₂C₇H₁₅	144.22	....	178.1, 83³⁰		0.8784²⁰	1.4140²⁰	al, eth	B2⁴, 31
7081	Formic acid, hexyl ester or Hexyl formate .... HCO₂C₆H₁₃	130.19	....	155.5	−62.6	0.8813²⁰/⁴	1.4071²⁰	al, eth	B2⁴, 31
7082	Formic acid, hydrazide .... HCONHNH₂	60.06	ye lf or nd (al)	....	54	....	....	al, eth, bz, chl	B2⁴, 85
7083	Formic acid, methyl ester or Methyl formate .... HCO₂CH₃	60.05	....	31.5	−99	0.9742²⁰/⁴	1.3433²⁰	w, al, eth	B2⁴, 20
7084	Formic acid, octyl ester or Octyl formate .... HCO₂C₈H₁₇	158.24	....	198.8	−39.1	0.8744²⁰	1.4208¹⁵	al, eth	B2⁴, 31
7085	Formic acid, pentyl ester .... HCO₂C₅H₁₁	116.16	....	132.1	−73.5	0.8853²⁰/⁴	1.3992²⁰	al, eth	B2⁴, 30
7086	Formic acid, isopentyl ester .... HCO₂-i-C₅H₁₁	116.16	....	124.2	−93.5	0.8857²⁰/⁴	1.3976²⁰	al, eth	B2⁴, 30
7087	Formic acid, propyl ester or Propyl formate .... HCO₂C₃H₇	88.12	....	81.3	−92.9	0.9058²⁰/⁴	1.3779²⁰	al, eth	B2⁴, 26
7088	Formic acid, isopropyl ester or Isopropyl formate ...... HCO₂CH(CH₃)₂	88.12	....	68.2		0.8728²⁰/⁴	1.3678²⁰	al, eth, ace	B2⁴, 27
7089	Formimidic acid, N-phenyl, ethyl ether .... C₆H₅N=CHOC₂H₅	149.19	....	213-5		1.0051²⁰/⁴	1.5279²⁰	eth, bz	B12³, 455
7090	Formonitrolic acid or Methyl nitrolic acid .... HON=CHNO₂	90.04	nd (eth or eth-peth)	....	68d	....	....	w, al, eth	B2⁴, 85
7091	Frangulin A (αβ-L-Rhamnoside) .... C₂₁H₂₀O₉	416.39	ye or red (al or AcOEt)	....	228			al, bz, aa	B17⁴, 2535
7092	Fraxin or αβ-Glucoside of fraxetin .... C₁₆H₁₈O₁₀	370.32	ye nd (al) (w + 3)		205				B18⁴, 2373

No.	Name, Synonyms, and Formula	Mol. wt.	Color, crystalline form, specific rotation and λ_{max} (log ε)	b.p. °C	m.p. °C	Density	n_D	Solubility	Ref.
7093	β-D-Fructose or Levulose .... $C_6H_{12}O_6$	180.16	pr or nd (w) orh, pr (al) [α]$^{20/}$$_D$(mut) - 133 →-92 (w, c=2)	....	103-5d	1.60$^{20/4}$	....	w, al, ace	B1⁴, 4401
7094	α-L-Fucose or 6-Deoxy-L-galactose .... $C_6H_{12}O_5$	164.16	nd (al) [α]$^{20/}$$_D$(mut) - 124.1→- 75.6 (w, c=9)						B1⁴, 4265
7095	Fucoxanthin .... $C_{40}H_{60}O_6$	632.89	red br pl (eth-peth) hex pl c + 2w, dil al) [α]$^{18/}$$_D$ + 72.5 (chl)		168			al, eth	B18⁴, 2820
7096	Fulvene ...... C_6H_6	78.11	....	7-8⁵⁶	....		1.4920^{20}	bz	B5⁴, 764
7097	Fulvene-6-vinyl (trans) .... 6-(CH_2=CH)C_6H_5	104.15		45^{12}	−35	0898^{20}	....	bz, chl	HCACA 47, 1022
7098	Fumagacin .... $C_{33}H_{44}O_8$	554.69	nd (dil aa) [α]$^{18/}$$_D$ - 125 (chl, c=1)		212 d			eth, ace, bz, aa	Am78, 5275
7099	Fumaric acid or trans-Butenedioic acid .... HO_2CCH=$CHCO_2H$	116.07	nd mcl pr or lf (w)	165$^{1.7}$ sub	300-2 (sealed tube)	1.635$^{20/4}$	....	al	B2⁴, 2202
7100	Fumaric acid, bromo or Bromofumaric acid .... HO_2CCBr=$CHCO_2H$	194.98	pr (AcOEt)	d 200	185-6			w, al	B2⁴, 2224
7101	Fumaric acid, dibutyl ester or Dibutyl fumarate .... $C_4H_9O_2CCH$=$CHCO_2C_4H_9$	228.29	....	150⁴	......	0.9869$^{20/4}$	1.4469^{20}	ace, chl	B2⁴, 2210
7102	Fumaric acid, diisobutyl ester .... i-$C_4H_9O_2CCH$=$CHCO_2$-c-C_4H_9	228.29		170^{160}, 122⁵		0.9760$^{20/4}$	1.4432^{20}	al, eth, ace	B2⁴, 2211
7103	Fumaric acid, chloro .... HO_2CCCl=$CHCO_2H$	150.52	pl (aa)	sub	192-3			al, eth	B2⁴, 2221
7104	Fumaric acid, chloro, diethyl ester .... $C_2H_5O_2CCCl$=$CHCO_2C_2H_5$	206.63	......	250d, 127^{10}	......	1.1880$^{20/4}$	1.4571^{20}	al, eth	B2³, 1909
7105	Fumaric acid, chloro, dimethyl ester .... CH_3O_2CCCl=$CHCO_2CH_3$	178.57		224, 108^{15}		1.2899$^{25/4}$	1.4720^{18}	al, eth	B2³, 1909
7106	Fumaric acid, diethyl ester .... $C_2H_5O_2CCH$=$CHCO_2C_2H_5$	172.18		214, 98^{10}	1-2	1.0452$^{20/4}$	1.4412^{20}	ace, chl	B2⁴, 2207
7107	Fumaric acid, dimethyl .... $HO_2CC(CH_3)$=$C(CH_3)CO_2H$	144.14	nd (w)	....	241				B2⁴, 2243
7108	Fumaric acid, dimethyl ester .... CH_3O_2CCH=$CHCO_2CH_3$	144.13		193	103-4	1.37$^{20/4}$	1.40625^{111}	ace, chl	B2⁴, 2205
7109	Fumaric acid, diphenyl ester .... $C_6H_5O_2CCH$=$CHCO_2C_6H_5$	268.27	nd (al)	219^{14}	161-2				B6⁴, 628
7110	Fumaric acid, dipropyl ester .... $C_3H_7O_2CCH$=$CHCO_2C_3H_7$	200.23		110⁵		1.0129$^{20/4}$	1.4435^{20}	al, eth	B2⁴, 2209
7111	Fumaric acid, diisopropyl ester .... $(CH_3)_2CHO_2CCH$=$CHCO_2CH(CH_3)_2$	200.24		225-6				al, eth, ace	B2⁴, 2209
7112	Fumaric acid, methyl or Mesaconic acid................ $HO_2CC(CH_3)$=$CHCO_2H$	130.10	rh nd or mcl pr (eth, AcOEt)	sub	204-5	1.466$^{20/4}$		al, eth	B2⁴, 2231
7113	Fumaronitrile .... NCCH=CHCN	78.07	nd (bz-peth)	186	96.8	0.9416^{111}	1.4349^{111}	w, al, eth, ace, bz	B2⁴, 2219
7114	Fumaryl chloride .... ClOCCH=CHCOCl	152.97	pa ye liq	158-60, 63^{13}		1.408^{20}	1.5004^{18}	....	B2⁴, 2217
7115	Fumaryl chloride, chloro .... ClOCCCl=CHCOCl	187.41	pa gr	184-7d, 73-5^{20}	....	1.564$^{20/4}$	1.5206^{20}	eth, aa	B2³, 1909
7116	Furan .... C_4H_4O	68.08	....	31.4	−85.6	0.9514$^{20/4}$	1.4214^{20}	al, eth, ace, bz	B17⁴, 225
7117	Furan, 2-acetyl or 2-Furyl methyl ketone .... 2-$CH_3CO(C_4H_3O)$	110.11	cr (lig)	175, 67^{10}	33	1.098^{20}	1.5017^{20}	al, eth	B17⁴, 4500
7118	Furan, 2-benzoyl or 2-Furyl phenyl ketone............. $C_6H_5CO(2$-$C_4H_3O)$	172.20	....	285, 164^{19}	<-15	1.1732^{20}	1.6055^{20}	al, eth	B17⁴, 5184

No.	Name, Synonyms, and Formula	Mol. wt.	Color, crystalline form, specific rotation and λ_{max} (log ε)	b.p. °C	m.p. °C	Density	n_D	Solubility	Ref.
7119	Furan, 2-bromo or α-Furyl bromide .... 2-Br(C_4H_3O)	146.98	....	102[744]	....	1.6500[20]	1.4980[20]	al, eth, ace, bz	B17[4], 231
7120	Furan, 3-bromo or β Furyl bromide 3-Br(C_4H_3O)	146.98	....	103	....	1.6606[20/4]	1.4958[20]	al, eth, ace, bz	B17[4], 232
7121	Furan, 2-tert butyl 2-(CH_3)$_3$C(C_4H_3O)	124.19	....	119-20	....	0.869[20/4]	1.4373[20]	al, eth, ace	B17[4], 307
7122	Furan, 2-chloro or α-Furyl chloride .... 2-Cl(C_4H_3O)	102.52	....	77.5[744]	....	1.1923[70/4]	1.4569[20]	al, eth, ace	B17[4], 230
7123	Furan, 3-chloro or β-Furyl chloride 3-Cl(C_4H_3O)	102.52	....	79[742]	....	1.2094[20/4]	1.4601[20]	eth, ace	B17[4], 230
7124	Furan, 2,5-dibromo .... 2,5-Br$_2$(C_4H_2O)	225.88	pl	164-5[764], 62[13]	....	2.27[20/20]	1.5455[20]	....	B17[4], 232
7125	Furan, 2,5-di-tert-butyl 2,5-[tert-(C_4H_9)$_2$](C_4H_2O)	180.29	....	210, 61-2[17]	....	0.837[20/4]	1.4369[20]	al, eth,	B17[4], 336
7126	Furan, 2,5-dichloro 2,5-Cl$_2$(C_4H_2O)	136.97	....	115	....	1.371[25]	....	....	B17[4], 230
7127	Furan, 2,5-dimethoxy 2,5(CH_3O)$_2$(C_4H_2O)	132.16	....	145-7	....	....	1.4168[20/]$_D$	....	B17[4], 1993
7128	Furan, 2,4-dimethyl .... 2,4(CH_3)$_2$(C_4H_2O)	96.13	....	94	....	0.8993[20/4]	1.4371[20]	....	B17[4], 287
7129	Furan, 2,5-dimethyl 2,5-(CH_3)$_2$(C_4H_2O)	96.14	....	93-4	−62.8	0.8883[20/4]	1.4363[20]	al, eth, ace, bz	B17[4], 289
7130	Furan, 2,5-dinitro 2,5-(NO_2)$_2$(C_4H_2O)	158.07	nd (w) pr (al)	....	102	....	....	eth	B17[4], 234
7131	Furan, 2,5-diphenyl 2,5-(C_6H_5)$_2$(C_4H_2O)	220.27	nd or lf (dil al)	343-5	91	....	....	al, eth, ace, bz	B17[4], 682
7132	Furan, 2-ethyl 2-C_2H_5(C_4H_3O)	96.14	....	92-3[768]	....	0.912[15/15]	1.4466[23]	al, eth, ace, bz	B17[4], 284
7133	Furan, 2-furfuryl (C_4H_3O)CH_2(C_4H_3O)	148.17	....	94[22]	....	1.102[20/4]	1.5049[20]	al, eth, ace	B19[4], 269
7134	Furan, 2-iodo or α-Furyl iodide .... 2-I-(C_4H_3O)	193.97	....	43-5[15]	....	2.024[20/4]	1.5661[20]	eth	B17[4], 232
7135	Furan, 3-iodo or β Furyl iodide .... 3-I(C_4H_3O)	193.97	....	132.2[732], 37-8[22]	....	2.045[20/4]	1.5610[20]	eth	B17[4], 232
7136	Furan, 2-methyl or Sylvan .... 2-CH_3(C_4H_3O)	82.10	....	63[737]	....	0.9132[20/4]	1.4342[20]	al, eth	B17[4], 265
7137	Furan, 3-methyl 3-CH_3(C_4H_3O)	82.10	....	65.5[749]	....	0.923[18/4]	1.4330[19]	al, eth	B17[4], 276
7138	Furan, 2-nitro 2-O_2N(C_4H_3O)	113.07	yesh mcl cr (peth)	133-5[123]	29	....	....	al, eth	B17[4], 233
7139	Furan, 2-phenyl .... 2-C_6H_4(C_4H_3O)	144.19	....	107-8[18]	....	1.083[20/4]	1.5920[20]	ace, bz	B17[4], 542
7140	Furan, 2-propionyl or Ethyl Furyl Ketone.............. 2-CH_3CH_2CO(C_4H_3O)	124.15	cr	88[14]	28	1.0626[28]	1.4922[25]	eth	B17[4], 4537
7141	Furan, 2-propyl 2-C_3H_7(C_4H_3O)	110.16	....	114-5[750]	....	0.8876[20/4]	1.4549[20]	al, eth, ace	B17[4], 296
7142	Furan, tetrahydro C_4H_8O	72.12	....	67	fr-108	0.8892[20/4]	1.4050[20]	al, eth, ace, bz	B17[4], 24
7143	Furan, tetrahydro-2,2-diethyl 2,2-(C_2H_5)$_2$(C_4H_6O)	128.22	....	146	....	0.8703[20/4]	1.4317[20]	al, eth, ace,,bz	B17[4], 107
7144	Furan, tetrahydro,2-ethyl 2-C_2H_5(C_4H_7O)	100.16	....	109	....	0.8570[19/4]	1.4147[19]	al, eth, ace, bz	B17[4], 78
7145	Furan, tetrahydro-2-methyl .... 2-CH_3(C_4H_7O)	86.14	....	80	....	0.8552[20/4]	1.4059[21]	al, eth, ace, bz, chl	B17[4], 60
7146	Furan, tetrahydro-3-methyl 3-CH_3(C_4H_7O)	86.14	....	86-7	....	0.8642[20/4]	1.4122[20]	al, eth, ace, bz	B17[2], 21
7147	Furan, tetrahydro-2-propyl 2-C_3H_7(C_4H_7O)	114.19	....	139.4	....	0.8547[20/4]	1.4242[20]	....	B17[4], 93
7148	Furan, tetraiodo C_4I_4O	571.66	nd	....	165	....	....	al	B17[4], 233
7149	Furan, tetraphenyl .... (C_6H_5)$_4$(C_4O)	372.47	....	220	175	....	....	al, eth, ace,bz, aa	B17[4], 810
7150	Furan, 2,3,5-trichloro 2,3,5-Cl$_3$(C_4HO)	171.41	....	147	....	1.50[25]	....	....	B17[4], 231
7151	2-Furancarboxylic acid or α-Furoic acid .... 2-(C_4H_3O)CO_2H	112.09	mcl nd or lf (w)	230-2, 141-4[20]	133-4	....	....	w, al, eth	B18[4], 3914

No.	Name, Synonyms, and Formula	Mol. wt.	Color, crystalline form, specific rotation and λ_{max} (log ε)	b.p. °C	m.p. °C	Density	n_D	Solubility	Ref.
7152	2-Furancarboxylic acid, 5-ethoxy 5-C₂H₅O(C₄H₂O)CO₂H-2	156.15	....		140-1	....		al	B18[4], 4827
7153	2-Furancarboxylic acid, 5-iodo 5-I(C₄H₂O)CO₂H-2	237.98			197d				B18[4], 3992
7154	2-Furancarboxylic acid, 5-methoxy 5-CH₃O(C₄H₂O)CO₂H-2	142.11	....		136-8d	....		al	B18[4], 4827
7155	3-Furancarboxylic acid or β-Furoic acid 3-(C₄H₃O)CO₂H	112.09	nd (w)	105-10[12] sub	122-3			al, eth	B18[4], 4052
7156	3-Furancarboxylic acid, 4-methyl 4-CH₃(C₄H₂O)CO₂H-3	126.11	nd (bz-peth)		138-9			w	B18[4], 4089
7157	3-Furancarboxylic acid, 5-methyl 5-CH₃(C₄H₂O)CO₂H-3	126.11	(w)	sub	119			w, eth	B18[4], 4076
7158	2,3-Furandicarboxylic acid 2,3-(C₄H₂O)(CO₂H)₂	156.10	pr (aa or sub)	sub	226			w, al	B18[4], 4477
7159	2,3-Furandicarboxylic acid 2,3-(C₄H₂O)(CO₂CH₃)₂	184.16	(MeOH)		39			al, eth	B18[4], 4477
7160	2,4-Furandicarboxylic acid 2,4-(C₄H₂O)(CO₂H)₂	156.10	lf (w + 1)	sub	266			al, ace	B18[4], 4479
7161	2,4-Furandicarboxylic acid, dimethyl ester 2,4-(C₄H₂O)(CO₂CH₃)₂	184.16	pr (MeOH)		109-10				B18[4], 4480
7162	2,5-Furandicarboxylic acid or Dehydromucic acid 2,5-(C₄H₂O)(CO₂H)₂	156.10	nd (w) lf (al)	sub	> 320				B18[4], 4481
7163	2,5-Furandicarboxylic acid, dimethyl ester 2,5-(C₄H₂O)(CO₂CH₃)₂	184.16	nd (w) cr (MeOH)	154-6[15]	112			al, eth, chl	B18[4], 4482
7164	3,4-Furandicarboxylic acid 3,4-(C₄H₂O)(CO₂H)₂	156.10	....		217-8				B18[4], 4497
7165	2-Furanone, tetrahydro C₄H₆O₂	86.09	....	206, 89[12]	-42	1.1286[16/0]	1.4341[20]	w, al, eth, ace, bz	B17[2], 4159
7166	2-Furanone, 5-methyl (d), tetrahydro C₅H₈O₂	100.12	[α][20]_D + 13.5 undil	86-90[14]					B17[2], 288
7167	2-Furanone, 5-methyl (dl), tetrahydro C₅H₈O₂	100.12	....	206, 83-4[13]	-31	1.0465[23]	1.4328[20]		B17[4], 4176
7168	2-Furanone, 5-methyl (l), tetrahydro C₅H₈O₂	100.12	[α][20]_D -4.6 eth 10%	78-80[8]			1.4322[20]		B17[4], 4176
7169	Furazan, 3,4-dimethyl or 3,4-dimethyl-1,2,5-oxadiazole C₄H₆N₂O	98.11	....	156[764]	-7	1.0528[14/4]	1.4237[20]	al, eth	B27[2], 628
7170	Furfural or α-Furaldehyde (OC₄H₃)CHO	96.09	....	161.7, 90[65]	-38.7	1.1594[20/4]	1.5261[20]	al, eth, ace, bz, chl	B17[4], 4403
7171	Furfural acetone (C₄H₃O)CH=CHCOCH₃	138.16	nd	229d, 112-3[10]	39-40	1.0496[57/4]	1.5788[25]	al, eth, chl, peth	B17[4], 4714
7172	Furfural, 5-bromo 5-Br(OC₄H₂)CHO-2	174.99	cr (50% al)	112[16]	82			al, eth	B17[4], 4456
7173	Furfural, 5-chloro 5-Cl(C₄H₂O)CHO-2	130.53	....	70[1]	31-3				B17[4], 4454
7174	Furfural, diacetate or Furfurylidene diacetate (C₄H₃O)CH(O₂CCH₃)₂	198.17	nd or pl (eth-peth)	220, 143-4[20]	52-3			al, eth, bz	B17[4], 4413
7175	Furfural, diethyl acetal (C₄H₃O)CH(OC₂H₅)₂	170.21	....	191-2, 62-4[5]		0.9994[20/20]	1.4451[20]	al	B17[4], 4412
7176	Furfural, 5-hydroxymethyl 5-HOCH₂(C₄H₂O)CHO-2	126.11	nd (eth-peth)	114-6[0.5]	35	1.2062[25/4]	1.5627[18]	w, al, eth, bz, chl	B18[4], 100
7177	Furfural, 5-methyl 5-CH₃(C₄H₂O)CHO-2	110.11	....	187, 79-81[12]		1.1072[18/4]	1.5262[20]	w, al, eth	B17[4], 4523
7178	Furfural, 5-nitro 5-NO₂(C₄H₂O)CHO-2	141.08	pa ye (peth)	128-32[10]	35-6			peth	B17[4], 4459
7179	Furfural, 5-nitro, semicarbazone C₆H₆N₄O₄	198.14	pa ye nd (w) darkens in light		237d				B17[4], 4467
7180	Furfural, oxime, (anti) or α-Furfuraldoxime 2-(C₄H₃O)CH=NOH	111.10	nd (lig)		75-6			al, eth, bz, aa	B17[4], 4430
7181	Furfural, oxime (syn) or β-Furfuraldoxime 2-(C₄H₃O)CH=NOH	111.10	nd (lig)	201-8d, 98[9]	91-2			al, eth, bz, chl	B17[4], 4428
7182	Furfural, phenylhydrazone 2-(C₄H₃O)CH=NNHC₆H₅	186.22	ye lf (al)		97-8			al, eth	B17[4], 4432
7183	Furfural, tetrahydro 2-(C₄H₇O)CHO	100.13	....	142-3[779], 45-7[29]		1.0727[20/4]	1.4366[20]	w, eth,	B17[4], 4179
7184	Furfuryl alcohol or 2-Hydroxymethyl furan 2-(C₄H₃O)CH₂OH	98.10	col ye	171[750], 68-9[20]		1.1296[20/4]	1.4868[20]	w, al, eth	B17[4], 1242

No.	Name, Synonyms, and Formula	Mol. wt.	Color, crystalline form, specific rotation and λ_{max} (log ε)	b.p. °C	m.p. °C	Density	n_D	Solubility	Ref.
7185	Furfuryl alcohol, 5- methyl 5-CH$_3$(C$_4$H$_2$O)CH$_2$OH-2	112.14	....	194-6[744]/[d] 81[23]	......	1.0769[20/4]	1.4853[20]	al, eth	B17[4], 1279
7186	Furfuryl alcohol, tetrahydro 2-(C$_4$H$_7$O)CH$_2$OH	102.13	....	177-8[750], 80-2[20]	......	1.0544[20/4]	1.4517[20]	eth, ace	B17[4], 1095
7187	Furfuryl amine 2-(C$_4$H$_3$O)CH$_2$NH$_2$	97.12	....	145-6, 80[84]	......	1.0995[20/4]	1.4908[20]	w, al, eth	B18[4], 7068
7188	Furfuryl amine, tetrahydro 2-(C$_4$H$_7$O)CH$_2$NH$_2$	101.15	....	151-2[735]	......	0.9770[20/20]	1.4551[20]	w, al, eth	B18[4], 7034
7189	Furfuryl bromide 2-(C$_4$H$_3$O)CH$_2$Br	161.00	pa ye	33-4[2]	......	1.560[20/20]	1.5380[20]	eth	B17[4], 268
7190	2-Furfuryl bromide, tetrahydro 2-(C$_4$H$_7$O)CH$_2$Br	165.03	....	168-70[744], 69-70[22]	......	1.3653[20/4]	1.4850[20]	al, eth	B17[4], 62
7191	Furfuryl chloride 2-(C$_4$H$_3$O)CH$_2$Cl	116.55	....	49[26]	......	1.1783[20/4]	1.4941[20]	al, eth, ace, bz	B17[4], 268
7192	Furfuryl mercaptan 2-(C$_4$H$_3$O)CH$_2$SH	114.17	....	155, 47[12]	......	1.1319[20/4]	1.5329[20]	....	B17[4], 1255
7193	Furfurin C$_{15}$H$_{12}$N$_2$O$_3$	268.28	lt br nd or rh pr (w or eth)	....	116-7	....	....	al, eth	B27[2], 918
7194	Furfuryl ether or Difurfuryl ether [(C$_4$H$_3$)CH$_2$]$_2$O	178.19	....	101[2]	......	1.1405[20/4]	1.5088[20]	....	B17[2], 116
7196	Furfuryl methyl ether (C$_4$H$_3$O)CH$_2$OCH$_3$	112.14	....	131-3	......	1.0163[20/4]	1.4570[20]	al, eth	B17[4], 1243
7197	α -Furoic acid or 2-Furancarboxylic acid 2-(C$_4$H$_3$O)CO$_2$H	112.09	mcl nd or lf (w)	230-2, 141-4[20]	133-4	....	....	w, al, eth	B18[4], 3914
7198	α -Furoic acid, allyl ester 2-(C$_4$H$_3$O)CO$_2$(CH$_2$CH=CH$_2$)	152.16	....	206-9	......	1.118[25/25]	1.4945[20]	eth, ace	B18[4], 3919
7199	α -Furoic acid, benzyl ester 2-(C$_4$H$_3$O)CO$_2$CH$_2$C$_6$H$_5$	202.21	ye	179-81[18]	......	1.1623[22/4]	1.5550[20]	eth, ace	B18[4], 3921
7200	α -Furoic acid, butyl ester or Butyl furoate 2-(C$_4$H$_3$O)CO$_2$C$_4$H$_9$	168.20	....	233, 83-4[1]	......	1.0555[20/4]	1.4740	al, eth, bz, peth	B18[4], 3918
7201	α-Furoic acid, isobutyl ester or Isobutyl α-furoate 2-(C$_4$H$_3$O)CO$_2$-i-C$_4$H$_9$	168.20	....	221-3, 97[13.5]	......	1.0388[20/4]	1.4676[20]	al, eth, ace, bz	B18[2], 266
7202	α-Furoic acid, secbutyl ester 2-(C$_4$H$_3$O)CO$_2$CH(CH$_3$)C$_2$H$_5$	168.20	....	67-9[1]	......	1.0465[20/4]	....	al, eth	B18[2], 266
7203	α-Furoic acid, 5-ethoxy 5-C$_2$H$_5$O(C$_4$H$_2$O)CO$_2$H-2	156.15	....	....	140-1	....	....	al	B18[4], 4827
7204	α-Furoic acid, ethyl ester or Ethyl-α-furoate 2-(C$_4$H$_3$O)CO$_2$C$_2$H$_5$	140.15	lf or pr	196.8, 128[95]	34-5	1.1174[21/4]	1.4797[21]	al, eth, ace, bz, peth	B18[4], 3917
7205	α-Furoic acid, furfuryl ester 2-(C$_4$H$_3$O)CO$_2$CH$_2$(C$_4$H$_3$O)	192.18	dimorphic	122[2]	27.5	1.2384[25/25]	1.5280[20]	al, eth, ace, bz, peth	B18[4], 3936
7206	α-Furoic acid, heptyl ester or Heptyl-α-furoate 2-(C$_4$H$_3$O)CO$_2$C$_7$H$_{15}$	210.28	....	116-7[1]	......	1.0005[20/4]	....	al, eth	B18[2], 267
7207	α-Furoic acid, hexylester or Hexyl-α-Furoate 2-(C$_4$H$_3$O)CO$_2$C$_6$H$_{13}$	196.25	....	105-7[1]	......	1.0170[20/4]	....	al, eth, bz	B18[2], 267
7208	α-Furoic acid, 5-iodo 5-I(C$_4$H$_2$O)CO$_2$H-2	237.98	....	....	197d	....	....	....	B18[4], 3992
7209	α-Furoic acid, 5-methoxy 5-CH$_3$O(C$_4$H$_2$O)CO$_2$H-2	142.11	....	....	136-8d	....	....	al	B18[4], 4827
7210	α-Furoic acid, methyl ester or Methyl-α-Furoate 2-(C$_4$H$_3$O)CO$_2$CH$_3$	126.11	....	181.3	......	1.1786[21/4]	1.4860[20]	al, eth, bz	B18[4], 3916
7211	α-Furoic acid, 3-methyl 3-CH$_3$(C$_4$H$_2$O)CO$_2$H-2	126.11	nd (w)	sub	134	....	....	....	B18[4], 4067
7212	α-Furoic acid, 3-methyl, ethyl ester 3-CH$_3$(C$_4$H$_2$O)(CO$_2$C$_2$H$_5$)-2	154.17	pl	205	47-8	....	....	....	B18[4], 4067
7213	α-Furoic acid, 3-methyl, methyl ester 3-CH$_3$(C$_4$H$_2$O)(CO$_2$CH$_3$)-2	140.15	pl (al)	72-6[2]	36-8	....	....	....	B18[4], 4067
7214	α-Furoic acid, 4-methyl 4-CH$_3$(C$_4$H$_2$O)CO$_2$H-2	126.11	nd (bz-peth)	....	131-2	....	....	....	B18[4], 4073
7215	α-Furoic acid, 5-methyl 5-CH$_3$(C$_4$H$_2$O)CO$_2$H-2	176.11	pl or nd (w)	105[1]	109-10	....	....	al, eth, chl	B18[4], 4076
7216	α-Furoic acid, 5-methyl, methyl ester 5-CH$_3$(C$_4$H$_2$O)CO$_2$CH$_3$-2	140.14	....	205, 98[15]	......	....	....	....	B18[4], 4076
7217	α-Furoic acid, 5-nitro 5-O$_2$N(C$_4$H$_2$O)CO$_2$H-2	157.08	pa ye pl (w)	sub	184	....	....	al, eth	B18[4], 3992
7218	α-Furoic acid, octyl ester or Octyl-α-furoate 2-(C$_4$H$_3$O)CO$_2$C$_8$H$_{17}$	224.30	....	126-7[1]	......	0.9885[20/4]	....	al	B18[2], 267

No.	Name, Synonyms, and Formula	Mol. wt.	Color, crystalline form, specific rotation and λ_{max} (log ε)	b.p. °C	m.p. °C	Density	n_D	Solubility	Ref.
7219	α-Furoic acid, pentyl ester or Pentyl-α-Furoate . . . . 2-(C₄H₃O)CO₂C₅H₁₁	182.22	. . . .	95-7[1]		1.0335[20/4]		al	B18[2], 266
7220	α-Furoic acid, isopentyl ester or Isopentyl-α-Furoate . . . . 2-(C₄H₃O)CO₂-i-C₅H₁₁	182.22		282		1.030[20/4]	1.4274[20]	al, bz, peth	B18[2], 226
7221	α-Furoic acid, 5-phenoxy . . . . 5-C₆H₅O(C₄H₂O)CO₂H-2	204.19			122-3			al	B18[4], 4827
7222	α-Furoic acid, propylester or Propyl-α-Furoate . . . . 2-(C₄H₃O)CO₂C₃H₇	154.17		210.9		1.0745[20/4]	1.4737[20]	al, eth, ace, bz, peth	B18[4], 3918
7223	α-Furoic acid, isopropyl ester or isopropyl-α-furoate . . . . . 2-(C₄H₃O)CO₂CH(CH₃)₂	154.17		198-9		1.0655[24/4]	1.4682[24]	al, eth, ace, bz	B18, 275
7224	α-Furoic acid, tetrahydro . . . . 2-(C₄H₇O)CO₂H	116.13		145[25]	21	1.1933[20/20]	1.4612[20]	w	B18[4], 3824
7225	α-Furonitrile . . . . 2-(C₄H₃O)CN	93.09		146[738]		1.0822[20/4]	1.4798[20]	al, eth	B18[4], 3964
7226	α-Furoyl chloride or 2-Furan carboxylyl chloride . . . . 2-(C₄H₃O)COCl	130.53		173, 66[10]	-2			eth, chl	B18[4], 3938
7227	β-Furoic acid or 3-Furan carboxylic acid . . . . 3-(C₄H₃O)CO₂H	112.09	nd (w)	105-10[12], sub	122-3			al, eth	B18[4], 4052
7228	β-Furoic acid, 4-bromo . . . . 4-Br(C₄H₂O)CO₂H-3	190.99	nd (w)		129			al, eth, bz, chl	B18[4], 3980
7229	β-Furoic acid, 5-bromo . . . . 5-Br(C₄H₂O)CO₂H-3	190.99	lf (w)		190-1			al, eth, chl	B18[4], 3980
7230	β-Furoic acid, 5-bromo, ethyl ester . . . . 5-Br(C₄H₂O)(CO₂C₂H₅)-3	219.04	pr	235[767], 134-6[34]	17	1.528[20]		al, eth	B18[4], 3981
7231	β-Furoic acid, 4-chloro . . . . 4-Cl(C₄H₂O)CO₂H-3	146.53	pl or pr (w)		149			al, eth, bz	B18[4], 3977
7232	β-Furoic acid, 5-chloro . . . . 5-Cl(C₄H₂O)CO₂H-3	146.53	lf (w)		179-80			al, eth, bz	B18[4], 3977
7233	β-Furoic acid, 5-chloro, methyl ester . . . . 5-Cl(C₄H₂O)CO₂CH₃-3	160.56			40-1			eth	B18[4], 3978
7234	β-Furoic acid, 2,5-dimethyl or Pyrotritaric acid . . . . 2,5-(CH₃)₂(C₄HO)CO₂H-3	140.15	nd (w)	sub	135			al, eth	B18[4], 4099
7235	β-Furoic acid, methyl ester . . . . 3-(C₄H₃O)CO₂CH₃	126.11		160, 79[42]		1.1744[15/15]	1.4676[20]	al, ace	B18[4], 4052
7236	β-Furoic acid, 2-methyl . . . . 2-CH₃(C₄H₂O)CO₂H-3	126.11	cr (w)		102-3			al, eth	B18[4], 4072
7237	β-Furoic acid, 2-methyl, ethyl ester . . . . 2-CH₃(C₄H₂O)CO₂C₂H₅-3	154.17		85-7[20]		1.0102[25/4]	1.4620[25]	eth	B18[4], 4072
7238	β-Furoic acid, 4-methyl . . . . 4-CH₃(C₄H₂O)CO₂H-3	126.11	nd (bz-peth)		138-9				B18[4], 4089
7239	β-Furoic acid, 5-methyl . . . . 5-CH₃(C₄H₂O)CO₂H-3	126.11	(w)	sub	119			w, eth	B18[4], 4076
7240	Furoin . . . . C₁₀H₈O₄	192.18	nd (al)		138-9			eth, MeOH	B19[4], 2543
7241	2-Furylacetic acid or 2-Furanacetic acid . . . . (2-C₄H₃O)CH₂CO₂H	126.11	lf (bz peth)	102-4[0.4]	68-9			w, MeOH	B18[4], 4061
7242	2-Furylacetonitrile or Furfuryl cyanide . . . . (2-C₄H₃O)CH₂CN	107.11		75-80[20]		1.0854[25/4]	1.4693[20]	al, eth	B18[4], 4062
7243	2-Furyl octyl ether . . . . (2-C₄H₃O)OC₈H₁₇	196.29		129-30[18]		0.9214[28/4]	1.4520[28]	eth	B17[4], 1219
7244	2-Furyl phenyl ether . . . . (2-C₄H₃O)OC₆H₅	160.18		105-6[18]		1.1010[23/4]	1.5418[23]	eth	B17[4], 1219
7245	2-Furyl phenyl ketone or 2-Benzoyl furan . . . . (2-C₄H₃O)COC₆H₅	172.20		285, 164[19]	<-15	1.1732[20]	1.6055[20]	al, eth, bz	B17[4], 5184
7246	2-Furyl phenyl sulfide . . . . (2-C₄H₃O)SC₆H₅	176.23		119-20[8]		1.1341[26]	1.5976[20]	al, eth	B17[4], 1220
7247	2-Furyl isopropyl ether or 2-isopropoxy furan . . . . (2-C₄H₃O)O-i-C₃H₇	126.16		135-6		0.9689[20/4]	1.4419[20]		B17[4], 1219
7248	2-Furyl methyl ether . . . . (2-C₄H₃O)OCH₃	98.10		110-1		1.0646[25/4]	1.4468[25]		B17[4], 1219

No.	Name, Synonyms, and Formula	Mol. wt.	Color, crystalline form, specific rotation and λ_{max} (log ε)	b.p. °C	m.p. °C	Density	n_D	Solubility	Ref.
7249	Galacitol or D-dulcitol. . . . . $CH_2OH(CHOH)_4CH_2OH$	182.18	mcl pr (w)	275-8¹ sub	189	1.466	. . . .	w	B1⁴, 2844
7250	D-Galactonic acid-γ-lactone . . . . $C_6H_{10}O_4$	178.14	nd (w + 1) nd (al or AcOEt) $[\alpha]^{20}_D$ −65.5 (w)	. . . .	112 (anh), 66 (hyd)	. . . .	. . . .	w	B18⁴, 3026
7251	D-Galactose . . . . $C_6H_{12}O_6$	180.16	pl or pr (al) pr or nd (w + 1) [α] + 83.3 (w)	. . . .	. . . .	. . . .	. . . .	w, py	B1⁴, 4336
7252	D-Galactose, 3,6-anhydro- $C_6H_{10}O_5$	162.14	lf (PrOH) $[\alpha]_D$ + 24 (w)	. . . .	123-5	. . . .	. . . .	w	B18⁴, 2278
7253	α-D-Galactose, 2,3,4,6-tetra-O-methyl . . . . $C_9H_{20}O_6$	236.27	$[\alpha]^{20}_D$ + 150 → + 114 (w)	172¹²	71-3	. . . .	. . . .	w, al, eth	B1⁴, 4371
7254	D-Galacturonic acid . . . . $C_6H_{10}O_7$	194.14	nd (w + 1) β-$[\alpha]^{20}_D$ + 27 (w) α-$[\alpha]_D$ + 98 → + 53 (w, c=10)	. . . .	α 156, β 150	. . . .	. . . .	w, al	B3⁴, 2000
7255	Galegine or 4-Guanidino-2-methyl-2-butene . . . . $(CH_3)_2C=CHCH_2N=C(NH_2)_2$	127.19	hyg	d	60-5	. . . .	. . . .	w, al	B4³, 465
7256	Galipine or Galipoline methyl ether. . . . . $C_{20}H_{21}NO_2$	323.40	pr (al eth) nd (peth)	. . . .	115.5	. . . .	. . . .	al, eth, ace, bz	B21⁴, 2656
7257	Gallein or 4,5-Dihydro fluorescein . . . . $C_{20}H_{12}O_5$	164.32	br red pw (+ 1.5w) red (anh)	. . . .	>300	. . . .	. . . .	al, ace	B19⁴, 3147
7258	Gallin or 4,5-Dihydroxyfluorescin. . . . . $C_{20}H_{14}O_7$	365.33	nd (eth) turns red in air	. . . .	. . . .	. . . .	. . . .	al, ace, aa	B18, 368
7259	Gelsemine (d) or Gelseminine. . . . . $C_{20}H_{22}N_2O_2$	322.41	cr (ace)	. . . .	178	. . . .	. . . .	al, eth, ace, bz,chl	B27², 720
7260	Gelsemine, hydrochloride . . . . $C_{20}H_{22}N_2O_2.HCl$	358.57	pr (w)	. . . .	326	. . . .	. . . .	w	B27², 720
7261	Geranial or Citral a . . . . $C_{10}H_{16}O$	152.24	$[\alpha]^{20}_D$ + 2.5 (w)	229, 118-9²⁰	. . . . . .	0.8888²⁰	1.4898²⁰	**al, eth**	B1⁴, 3569
7262	Geraniol or 2,7-Dimethyl-2,6-octadiene-1-ol . . . . $(CH_3)_2C=CHCH_2CH_2C(CH_3)=CH_2CH_2OH$	154.26	. . . .	230, 121¹⁸	>-15	0.8894²⁰	1.4766²⁰	al, eth, ace	B1⁴, 2277
7263	Geraniol formate . . . . $C_{11}H_{18}O_2$	182.27	. . . .	229d, 113-4²⁵	. . . . . .	0.9086²⁵/⁴	1.4659²⁰	al, eth, ace	B2⁴, 35
7264	Geraniol, tetrahydro (d) or 3,7-Dimethyl-1-octanol . . . . . . $(CH_3)_2CH(CH_2)_3CH(CH_3)CH_2CH_2OH$	158.29	$[\alpha]^{20}_D$ + 4.09	212-3, 105-6¹⁰	. . . . . .	0.8285²⁰/⁴	1.4355²⁰	eth	B1⁴, 1830
7265	Geraniol, tetrahydro or 3,7-Dimethyl-1-octanol . . . . $(CH_3)_2CH(CH_2)_3CH(CH_3)CH_2CH_2OH$	158.29	. . . .	202-3, 106¹²	. . . . . .	0.8308¹⁰/⁴	1.4367²⁰	al, eth, ace, bz	B1⁴, 1830
7266	Geraniol, tetrahydro (l) or 3,7-Dimethyl-1-octanol . . . . . . $(CH_3)_2CH(CH_2)_3CH(CH_3)CH_2CH_2OH$	158.29	$[\alpha]^{27}_{548}$ − 3.67	212-3, 109¹⁵	. . . . . .	0.830¹⁸	1.4370¹⁵	eth	B1³, 1768
7267	Geranyl bromide . . . . $(CH_3)_2C=CHCH_2CH_2C(CH_3)CH_2CH_2Br$	217.16	. . . .	101-2¹², 47-8⁰·⁰⁰⁵	. . . . . .	1.0940²²/⁴	1.5027²⁰	al, eth	B1⁴, 1059
7268	Germanidine . . . . $C_{37}H_{57}NO_{10}$	675.87	nd $[\alpha]^{24}_D$ − 30 (al)	. . . .	221-2 (vac)	. . . .	. . . .	. . . .	B21⁴, 6806
7269	Germanitrine . . . . $C_{39}H_{59}NO_{11}$	717.91	nd (aq aw) $[\alpha]^{24}_D$ − 61 (Py, c=1)	. . . .	228-9 (vac)	. . . .	. . . .	. . . .	B21⁴,6807
7270	Germerine . . . . $C_{37}H_{59}NO_{11}$	693.88	cr (ace) lf (bz) $[\alpha]^{25}_D$ + 15.7 (chl, c=1.02)	. . . .	193-5d	. . . .	. . . .	al, ace, bz, chl	B21⁴, 6808
7271	Germidine . . . . $C_{34}H_{53}NO_{10}$	635.81	pl (aq MeOH) $[\alpha]^{25}_D$ + 13 (chl)	. . . .	242-4, 203-3	. . . .	. . . .	al, bz	B21⁴, 6802

No.	Name, Synonyms, and Formula	Mol. wt.	Color, crystalline form, specific rotation and λ_{max} (log ε)	b.p. °C	m.p. °C	Density	n_D	Solubility	Ref.
7272	Germine $C_{27}H_{43}NO_8$	509.65	pr or cr (MeOH) $[\alpha]^{25}_D$ + 5 (95% al)		220			bz	B21⁴, 6796
7273	Germinitrine $C_{39}H_{57}NO_{11}$	715.89	pr (dil ace) $[\alpha]^{24}_D$ −36 (Py, c=1.12)		175 (vac)				Am75, 4925
7274	Germitrine $C_{43}H_{67}NO_{13}$	735.92	cr (dil al) $[\alpha]^{25}_D$ + 11 (chl)		197-9			bz, chl	B21⁴, 6809
7275	Gitogenin or Digine $C_{27}H_{44}O_4$	432.65	lf (bz) nd (eth) $[\alpha]^{20}_D$ −75 (chl, c=1)		271-2			al, chl	B19⁴, 1050
7276	Gitoxigenin or Hydroxy digitoxin $C_{23}H_{34}O_5$	390.53	pr (AcOEt) pr (+ w, dil al) $[\alpha]^{20}_{548}$ + 38.5 (MeOH, c=0.7)		234 (anh)			chl	B18⁴, 2456
7277	Gitoxin or Anhydrogitalin $C_{41}H_{64}O_{14}$	780.95	pr (chl MeOH) $[\alpha]^{24}_D$ + 5 (py, c=1)		285d				B18⁴, 2462
7278	D-Glucoascorbic acid or 3-keto-D-keto-heptonofuranolactone $C_7H_{10}O_7$	206.15	leus (+ w) (ace MeOH Peth) 140 (hyd) $[\alpha]^{20}_B$ −?? (MeOH, c=1)		131 (anh)			w, al	B10⁴, 3100
7279	D-gluco-Heptose $C_7H_{14}O_7$	210.19	rh pl (w) $[\alpha]^{20}_D$ −19.7 (w) (mut)		193 (210)			w	B1⁴, 4436
7280	D-gluco-Methylose or D-Isorhamnose $C_6H_{12}O_5$	164.16	cr (AcOEt) $[\alpha]^{20}_D$ + 73→ + 29.7 (w, mut)		139.40			w, al	B1⁴, 4260
7281	D-Gluconic acid or Dextronic acid $C_6H_{12}O_7$	196.16	nd (al eth) $[\alpha]^{25}_D$ −3.49→ + 12.95 (w, mut)		131			w	B3⁴, 1255
7282	D-Gluconic acid-γ-lactone $C_6H_{10}O_6$	178.14	nd (al) $[\alpha]_D$ + 67.5→ + 6.2 (w)		134-6			al	B18⁴, 3024
7283	D-Gluconic acid-δ-lactone $C_6H_{10}O_6$	178.14	nd (al) $[\alpha]^{25}_D$ + 63.5→ + 6.2 (w)						B18⁴, 3018
7284	D-Gluconitrile $C_6H_{11}NO_5$	177.16	(i) cr (al, aa) (ii) pl (al) (i) $[\alpha]^{24}_D$ + 10.0 (w, c=1.8) (ii) $[\alpha]^{21}_D$ + 8.8 (w)		(i)146.8 (ii)120 .5			w, py	B18⁴, 3024
7285	D-Gluconic acid, phenylhydrazide $C_{12}H_{18}N_2O_6$	286.29	pr (w) $[\alpha]$ + 12(w)		204-5			w	B15², 122
7286	D-Gluconic acid, 5-oxo or 5-keto-D-gluconic acid $C_6H_{10}O_7$	194.14	cr or syr $[\alpha]_D$ −14.5 (w)		125-6			w, al	B3⁴, 1993
7287	Gluconol (d) $C_6H_{10}O_4$	146.14	hyg nd $[\alpha]^{22}_D$ −7 (w)		60				B17⁴, 2332

No.	Name, Synonyms, and Formula	Mol. wt.	Color, crystalline form, specific rotation and λ_{max} (log ε)	b.p. °C	m.p. °C	Density	n_D	Solubility	Ref.
7288	D-Glucose (equilib mixt) or Dextrose.......... $C_6H_{12}O_6$	180.16	$[\alpha]^{20/}_D$ + 52.7(w)	....	146 (150)			w	B1⁴, 4302
7289	α-D-Glucose .... $C_6H_{12}O_6$	180.16	reds cubes orh nd (al) $[\alpha]^{20/}_D$ + 112.2→ + 52.7 (w, c=4)	....	146d	1.5620¹⁸ᐟ⁴		w	B1⁴, 4304
7290	α-D-Glucose, monohydrate.............. $C_6H_{12}O_6.H_2O$	198.18	lf pl or orh (w) $[\alpha]^{20/}_D$ + 102→ + 47.9	....	86	1.54²⁵ᐟ⁴		w	B1⁴, 4306
7291	β-D-Glucose .... $C_6H_{12}O_6$	180.16	nd (al) (w + 1) $[\alpha]^{20/}_D$ + 17.5→ + 52.7 (w, c=4)	....	150	1.5620¹⁸ᐟ⁴		w	B1⁴, 4306
7292	α-D-Glucose, pentaacetate .... $C_{16}H_{22}O_{11}$	390.35	pl or nd (al) $[\alpha]^{20/}_D$ + 100.9 (al, c=0.5)	sub	112-3	....		eth	B17⁴, 3276
7293	β-D-Glucose, pentaacetate .... $C_{16}H_{22}O_{11}$	390.35	nd (al) $[\alpha]'_D$ + 3.9 (chl, c=6)	sub (vac)	134			bz, chl	B17⁴, 3278
7294	D-Glucose, pentamethyl ether .... $C_{11}H_{22}O_6$	250.30	$[\alpha]^{20/}_D$ + 147.4 (w, p=10)	180⁰·⁴		1.0944²⁰ᐟ⁴	1.4466²⁰	w, al, eth, ace	B17⁴, 2928
7295	D-Glucose, phenylhydrazone (α type) .... $C_{12}H_{18}N_2O_5$	270.29	pl (al) $[\alpha]^{25/}_D$ - 8.7→- 52.2 (w, c=2) (mut)	....	160			w	B31, 173
7296	D-Glucose, phenylhydrazone (β type) .... $C_{12}H_{18}N_2O_5$	270.29	pr nd (al) $[\alpha]^{19/}_D$ -4.5→- 53.7 (w py) (mut)	....	140-1			w	B31, 173
7297	D-Glucose, phenylosazone or D-Glucosazone, D-Fructosazone, D-Mannosazone.............. $C_{18}H_{22}N_4O_4$	358.40	ye nd (dil al) $[\alpha]'_D$-41 (MeOH)	d 213	210			....	B31, 350
7298	DL-Glucose, phenylosazone .... $C_{18}H_{22}N_4O_4$	358.40	ye nd (al)					....	B31, 355
7299	β-D-Glucose, 2-amino or D-Glucosamine .... $C_6H_{13}NO_5$	179.17	nd (al MeOH) $[\alpha]_D$ + 28→ + 47.5 (w, c=0.4)		110d			w	B4⁴, 2019
7300	D-Glucose, 2-amino, hydrochloride.............. $C_6H_{13}NO_5 \cdot HCl$	215.64	mcl (w or dil al) $[\alpha]_{20D}$ + 25→ + 72.6 (w)					w	B4⁴, 2019
7301	D-Glucose, 2-(methylamino) or N-methyl-D-glucosamine. $C_7H_{14}NO_5$	193.20	gummy, $[\alpha]^{25/}_4$ -65 (MeOH, c=1)	....	130-2d			w	B18⁴, 7526
7302	D-Glucothiose .... $C_6H_{12}SO_5$	196.22	hyg pw(w + 1), $[\alpha]^{30/}_D$ + 48.7 (w, c=1.4)		70		....	w	B1⁴, 4391
7303	D-Glucoside, α-methyl or Methyl-α-D-glucopyranoside $C_7H_{14}O_6$	194.18	rh nd, (al), $[\alpha]^{20/}_D$ + 158.9 (w)	200⁰·²	168	1.46³⁰ᐟ⁴	....	w	B17⁴, 2909

No.	Name, Synonyms, and Formula	Mol. wt.	Color, crystalline form, specific rotation and λ_{max} (log ε)	b.p. °C	m.p. °C	Density	n_D	Solubility	Ref.
7304	D-Glucoside, β-methyl or Methyl-β-D-glucopyranoside $C_7H_{14}O_6$	194.19	tetr pr (al), $[\alpha]^{20}_D$ -34.2 (w, p=10)		115-6			w	B17[4], 2911
7305	β-D-Glucuronic acid $C_6H_{10}O_7$	194.14	nd(al AcOEt), $[\alpha]^{20}_D$ +11.7 +36.3 (w, c=6) (mut)		165			w, al	D3[4], 1996
7306	D-Glucuronic acid, γ-lactone or D-glucurone $C_6H_{10}O_7$	176.13	mcl pl (w) cr (al), $[\alpha]^{25}_D$ +19.8 (w, c=5.2)		177-8	1.76[20/4]		w	B18[4], 3055
7307	Glutaconic acid, diethyl ester or Ethyl glutaconate $C_2H_5O_2CCH_2CH=CHCO_2C_2H_5$	186.21		236-8, 125[12]		1.0496[20/4]	1.4411[20]	al, eth	B2[4], 2227
7308	Glutamic acid (D) or 2-Aminopentanedioic acid $HO_2CCH_2CH_2CH(NH_2)CO_2H$	147.13	lf (w), $[\alpha]^{25}_D$ -31.7 (1.7 NHCl)		213d				B4[3], 1549
7309	Glutamic acid (DL) $HO_2CCH_2CH_2CH(NH_2)CO_2H$	147.13	rh(al w)		199d, (225-7)	1.4601[20/4]			B4[3], 1550
7310	Glutamic acid (L+) $HO_2CCH_2CH_2CH(NH_2)CO_2H$	147.13	orh(dil al) $[\alpha]^{22}_D$ +31.4 (dil HCl, c=1)	sub 175[10]	224-5d	1.538[20/4]			B4[3], 1530
7311	Glutamic acid, hydrochloride (L) or Acidulin $HO_2CCH_2CH_2CH(NH_2)CO_2H.HCl$	183.59	rh pl (w), $[\alpha]^{19}_D$ +31.1 (dil HCl)		214d			w, al	B4[3], 1537
7312	Glutamic acid, N-acetyl (L) $HO_2CCH_2CH_2CH(NHCOCH_3)CO_2H$	189.17	pr (w), $[\alpha]_D$ -15.3 (w, c=2)		199				B4[3], 1544
7313	Glutamic acid, 3-hydroxy (D) $HO_2CCH_2CH(OH)CH(NH_2)CO_2H$	163.13	hyg pr(w)	d	135			w, aa	B4[3], 1676
7314	Glutamic acid, 3-hydroxy (DL) $HO_2CCH_2CH(OH)CH(NH_2)CO_2H$	163.13	rh pr or nd(w)	d	198d			w	B4[3], 1676
7315	Glutamine (L+) or α-Amino glutaramic acid $HO_2CCH(NH_2)CH_2CH_2CONH_2$	146.15	nd (w or dil al), $[\alpha]^{25}_D$ +6.5 (w, c=2)		185-6d			w	B4[3], 1540
7316	Glutyraldehyde or 1,5-Pentanedial $OHC(CH_2)_3CHO$	100.13		187-9d, 71-2[10]				**w, al**, bz	B1[4], 3111
7317	Glutyraldehyde, dioxime or Glutyraldoxime $HON=CH(CH_2)_3CH=NOH$	130.15	nd(w or Py)	sub	178				B1[4], 3660
7318	Glutaric acid or Pentanedioic acid $HO_2C(CH_2)_3CO_2H$	132.13	nd(bz)	302-4d, 200[20]	99	1.424[25/4]	1.4188[106]	w, al, eth, chl	B2[4], 1934
7319	Glutaric acid, 2-acetyl, diethyl ester $C_2H_5O_2CCH_2CH(COCH_3)CO_2C_2H_5$	230.26		271-2d, 119-2[0.1]		1.0712[20/4]	1.4420[15]	al, eth	B3[4], 1835
7320	Glutaric anhydride, 2-phenyl $C_{11}H_{10}O_3$	190.20	nd(eth)	218-30[13]	95			al	B17[4], 6185
7321	Glutaric anhydride, 3-phenyl $C_{11}H_{10}O_3$	190.20	cr(bz)	217-9[15]	105			eth, bz, chl	B17[4], 6185
7322	Glutaric acid, diethyl ester or Diethylglutarate $C_2H_5O_2C(CH_2)_3CO_2C_2H_5$	188.23	syr	236-7, 103-4[7]	-24.1	1.0220[20/4]	1.4241[20]	eth	B2[4], 1937
7323	Glutaric acid, dimethyl ester or Dimethyl glutarate $CH_3O_2C(CH_2)_3CO_2CH_3$	160.17		214[751], 109[21]		1.0876[20/4]	1.4242[20]	al, eth	B2[4], 1937
7324	Glutaric acid, 2,3-dimethyl $HO_2CCH_2CH_2C(CH_3)_2CO_2H$	160.17	nd(bz-lig)		85			al, chl, aa	B2[4], 2018
7325	Glutaric acid, 3,3-dimethyl $(CH_3)_2C(CH_2CO_2H)_2$	160.17	mcl pl nd(bz)	126-7[4.5]	103-4	1.4278[20/4]		w, al, eth	B2[4], 2023
7326	Glutaric acid, diphenyl ester $C_6H_5O_2C(CH_2)_3CO_2C_6H_5$	284.32	nd(lig)	236.5[15]	54			al, ace, lig	B6[4], 626
7327	Glutaric acid, 2-ethyl-3-methyl $HO_2CCH_2CH(CH_3)CH(C_2H_5)CO_2H$	175.20	(i) pr (w), (ii) pr (chl-lig)		(i)100 (ii)88			w, eth	B2[4], 2047

No.	Name, Synonyms, and Formula	Mol. wt.	Color, crystalline form, specific rotation and λ_{max} (log ε)	b.p. °C	m.p. °C	Density	n_D	Solubility	Ref.
7328	Glutaric acid, 2-ethyl-4-methyl (dl) or Paramethyl ethyl glutaric acid........... $HO_2CCH(CH_3)CH_2CH(C_2H_5)CO_2H$	175.20	nd(w)		107	....	....	eth, lig	B2[4], 2042
7329	Glutaric acid, 2-ethyl-4-methyl (meso) .... $HO_2CCH(CH_3)CH_2CH(C_2H_5)CO_2H$	175.20	nd(al)		83-4			eth	B2[4], 2042
7330	Glutaric acid, 3-ethyl-3-methyl .... $(HO_2CCH_2)_2C(CH_3)C_2H_5$	175.20	pl(bz-peth)	260[740]	87			al, eth, bz	B2[3], 1781
7331	Glutaric acid, 2-hydroxy (d) $HO_2CCH_2CH_2CH(OH)CO_2H$	148.13	$[\alpha]^{19}{}_D$ + 1.76 (w, c=1)		72				B3[4], 1146
7332	Glutaric acid, 2-hydroxy (dl) $HO_2CCH_2CH_2CH(OH)CO_2H$	148.13	pr(AcOEt)		72			w, al	B3[4], 1146
7333	Glutaric acid, 2-hydroxy (l) $HO_2CCH_2CH_2CH(OH)CO_2H$	148.13	$[\alpha]_D$ -1.98 (w)		72-3			w, al	B3[4],1146
7334	Glutaric acid, 3-methyl .... $CH_3CH(CH_2CO_2H)_2$	146.14	....	165-7[0.5]	87			w, al, eth	B2[4], 1992
7335	Glutaric acid, 2-oxo $HO_2CCH_2CH_2COCO_2H$	146.10	cr(ace-bz)		115-6			w, al, eth, ace	B3[4], 1813
7336	Glutaric acid, 3-oxo or Acetone dicarboxylic acid........ $(HO_2CCH_2)_2C=O$	146.10	nd(al AcOEt) rh(w)	d	135d			w, al	B3[4], 1816
7337	Glutaric acid, 3-oxo, diethyl ester .... $(C_2H_5O_2CCH_2)_2C=O$	202.21	....	250, 140[12]		1.113[20/4]		al	B3[4], 1817
7338	Glutaric acid, 2-phenyl $HO_2CCH_2CH_2CH(C_6H_5)CO_2H$	208.22	cr(bz or eth-peth)		82-3				B9[3], 4298
7340	Glutaric acid, 2,3,4-trihydroxy (d) $HO_2C(CHOH)_3CO_2H$	180.13	cr(ace) pl(w), $[\alpha]^{20}{}_D$ + 22.2		128			w, al	B3[4], 1265
7341	Glutaric acid, 2,3,4-trihydroxy (dl) $HO_2C(CHOH)_3CO_2H$	180.13	cr(ace)		154d			w, al, ace	B3, 553
7342	Glutaronitrile .... $NC(CH_2)_3CN$	94.12	....	286, 160.4[22]	−29	0.9911[15/4]	1.4295[20]	al, chl	B2[4], 1941
7343	Glutaryl chloride $ClCO(CH_2)_3COCl$	169.01	....	216-8		1.324[30/4]	1.4728[20]	eth	B2[4], 1939
7344	Glutathione or α-Glutamyl cysteinyl glycine .... $C_{10}H_{17}N_3O_6S$	307.33	rh (w) cr (50% al), $[\alpha]^{27}{}_D$ -21.3 (w, c=2)		195			w, DMF	B4[3], 1612
7345	D-Glyceraldehyde or D-2,3-Dihydroxy propanal .... $HOCH_2CH(OH)CHO$	90.08	syr, $[\alpha]_D$ + 14 (w)					w	B1[4], 4114
7346	DL-Glyceraldehyde $HOCH_2CH(OH)CHO$	90.08	nd or pr(40% MeOH)	140-50[0.8]	145	1.455[18/18]		w	B1[4], 4114
7347	L-Glyceraldehyde............ $HOCH_2CH(OH)CHO$	90.08	$[\alpha]$ -13.8 (w, c=1)					w	B1[4], 4114
7348	D-Glyceraldehyde, diethyl acetal or 3,3-Diethoxy-1,2- .... $HOCH_2CH(OH)CH(OC_2H_5)_2$	164.20	$[\alpha]^{15}{}_D$ + 21.2 (w, c=18)	127-9[17]				w, al, eth, ace	B1[2], 888
7349	DL-Glyceric acid or 2,3-Dihydroxy propionic acid............ $HOCH_2CH(OH)CO_2H$	106.08	syr	d				w, al, eth	B3[4], 1050
7350	D-Glyceric acid, ethyl ester or D-Ethyl glycerate .... $HOCH_2CH(OH)CO_2C_2H_5$	134.13	$[\alpha]^{11}{}_D$ -22.73					w, al, eth	B3[4], 1052
7351	DL-Glyceric acid, ethyl ester .... $HOCH_2CH(OH)CO_2C_2H_5$	134.13	....	230-40, 120-1[16]		1.1908[15/15]		w, al, eth	B3[4], 1052
7352	DL-Glyceric acid, methyl ester or DL-Methyl glycerate .... $HOCH_2CH(OH)CO_2CH_3$	120.12	....	239-44, 119-20[14]		1.2814[15/15]	1.4502[20]	w, al	B3[4], 1052
7353	L-Glyceric acid, methyl ester or L-Methyl glycerate............ $HOCH_2CH(OH)CO_2CH_3$	120.12	$[\alpha]^{15}{}_D$ -6.44	119-20[14], 74-5[0.2]		1.2798[15/15]		w, al	B3[4], 1052
7354	Glycerol or Glycerin, 1,2,3-trihydroxy propane......... $HOCH_2CH(OH)CH_2OH$	92.11	syr rh pl	290d, 182[20]	20	1.2613[20/4]	1.4746[20]	w, al	B1[4], 2751
7355	Glycerol, 1-monoacetate or α-Monoacetin............ $HOCH_2(CHOH)CH_2O_2CCH_3$	134.13	....	158[165], 129-31[3]		1.2060[20/4]	1.4157[20]	w, al	B2[4], 251
7356	Glycerol borate .... $(C_3H_5BO_3)n$	(99.88)n	glass	....	Ca 150				B1, 519
7357	Glycerol, 1-monobutyrate (dl) or α-Monobutyrin .... $HOCH_2CH(OH)CH_2O_2CC_3H_7$	162.19	....	289-71, 163[16]	....	1.129[18]	1.4531[20]	w, al	B2[4], 798

No.	Name, Synonyms, and Formula	Mol. wt.	Color, crystalline form, specific rotation and λ_{max} (log ε)	b.p. °C	m.p. °C	Density	n_D	Solubility	Ref.
7358	Glycerol, 1(2-chlorophenyl) ether .... (2-ClC₆H₄)OCH₂CH(OH)CH₂OH	202.64	nd (bz)	250[19]	71-2	....	....	eth	B6[4], 790
7359	Glycerol, 1-(4-chlorophenyl) ether .... (4-ClC₆H₄)OCH₂CH(OH)CH₂OH	202.64	nd (eth peth) lf (bz)	214-5[19]	76	....	....	al, eth, ace	B6[4], 831
7360	Glycerol, 1,3-diacetate or Diacetin .... HOCH(CH₂O₂CCH₃)₂	176.17	....	280, 155-6[15]	40	1.1779[15/4]	1.4395[20]	w, al	B2[4], 252
7361	Glycerol, 1,2-dibutryrate (d) or α,β-Dibutyrin . HOCH₂CH(O₂CC₃H₇)CH₂(O₂CC₃H₇)	232.38	[a]_D +1.7 (py,c=7)	273.5, 167[20]	....	....	1.4422[20]	al	B2[4], 799
7362	Glycerol, 1,3-dilaurate or α,γ-Dilarurin .... HOCH(CH₂O₂CC₁₁H₂₃)₂	456.72	pl (al) nd (eth-al)	....	[α] −49.5 (unst) β −56.5 (st)	....	....	al, eth, bz, chl lig	B2[4], 1098
7363	Glycerol, 1,2-dimethyl ether (dl) or 2,3-Dimethoxy-1-propanol .... HOCH₂CH(OCH₃)CH₂OCH₃	120.15	....	180, 100[40]	....	1.016[25/4]	1.4200[20]	w, al, eth	B1[3], 2317
7364	Glycerol, 1,3-dimethyl ether or 1,3-Dimethoxy-2-propanol .... HOCH(CH₂OCH₃)₂	120.15	....	169, 88[40]	....	1.0085[20/4]	1.4192[20]	w, al, eth	B1[3], 2318
7365	Glycerol, 1,3-dinitrate .... HOCH₂(CH₂ONO₂)₂	191.10	pr (w) cr (eth + 1w)	148[15], 116[0.6]	26 (hyd)	1.523[20/4]	1.4715[20]	w, al, eth	B1[2], 591
7366	Glycerol, 1,3-dipalmitate .... HOCH(CH₂O₂CC₁₅H₃₁)₂	568.93	cr (al, chl)	....	72-4	....	....	eth	B2[4], 1174
7367	Glycerol, 1,3-diphenylether .... HOCH(CH₂OC₆H₅)₂	244.29	lf (al)	224.5[17.5], 175[1]	81-2	1.179[24/4]	....	al, eth, bz, chl	B6[4], 590
7368	Glycerol, 1,3-diphenylether-2-acetate .... CH₃CO₂CH(CH₂OC₆H₅)₂	286.33	(dil al)	190[160]	70-1	....	....	al, eth, bz, chl	B6[3], 583
7369	Glycerol, 1,3-dipropionate .... HOCH(CH₂O₂CC₂H₅)₂	204.22	....	170-3[10]	....	....	....	al	B2[4], 717
7370	Glycerol, 1,3-distearate or α,γ-Distearin .... HOCH(CH₂O₂CC₁₇H₃₅)₂	625.04	nd or pl (eth chl lig)	....	79.1	....	....	eth	B2[4], 1231
7371	Glycerol, 1-mono (2-hydroxybenzoate) or α-Glyceryl salicylate .... (2-HOC₆H₄)CO₂CH₂CH(OH)CH₂OH	212.21	nd (eth)	....	76	....	....	al, bz	B10[3], 142
7372	Glycerol, 1-monolaurate (dl) or α-Monolaurin (dl) .... HOCH₂CH(OH)CH₂O₂CC₁₁H₂₃	274.41	lf (peth)	186[2]	63	0.9248[97]	1.4350[86]	eth, ace, bz, chl	B2[4], 1096
7373	Glycerol, 1-monolaurate (l) or α-Monolaurin (l) .... HOCH₂CH(OH)CH₂O₂CC₁₁H₂₃	274.41	cr (eth or peth)	....	54-5	....	....	eth, ace, bz, chl	B2[4], 1096
7374	Glycerol, 1-linoleate (dl) or Glycerol-1-(9,12-actadecadienoate) .... C₁₇H₃₁CO₂CH₂CH(OH)CH₂OH	354.54	cr (bz)	....	14-5	....	1.4758[20]	eth, bz, chl	B2[4], 1758
7375	Glycerol, 1-(2-methoxyphenyl) ether or Guaiacol-α-glyceryl ether .... (2-CH₃OC₆H₄)OCH₂CH(OH)CH₂OH	198.22	rh pr (eth, eth-peth)	215[19], 126[0.2]	78-9	....	....	al, bz, chl, w	B6[3], 4224
7376	Glycerol, 1-methyl ether or 3-Methoxy-1,2-propanediol .. HOCH₂CH(OH)CH₂OCH₃	106.12	hyg liq	220, 110-2[13]	....	1.830[20/4]	1.442[25]	w, al, eth, ace	B1[4], 2755
7377	Glycerol, 2-methyl ether or 2-Methoxy-1,3-propandiol ... CH₃OCH(CH₂OH)₂	106.12	hyg liq	232, 119-20[9]	....	1.124[25/4]	1.4505[12]	w, al, eth, ace	B1[3], 2317
7378	Glycerol, 1-mononitrate .... HOCH₂CH(OH)CH₂ONO₂	137.09	pr (w, al, eth)	155-60, 102[1]	61	1.4164[20/4]	1.4698[20]	w, al	B1[4], 2761
7379	Glycerol, 2-mononitrate .... ONO₂CH(CH₂OH)₂	137.09	lf (w)	155-60	54	1.40[22/4]	....	w, al, eth	B1[2], 591
7380	Glycerol, 1-octadecyl ether (d) or Batyl alcohol ... HOCH₂CH(OH)CH₂OC₁₈H₃₇	344.59	pl (bz, aa)	215-20[2]	70-1	....	....	eth	B1[4], 2758
7381	Glycerol, 1-oleate or α-Monoolein Glycerol-1-(9-octadeceneoate) .... C₁₇H₃₃CO₂CH₂CH(OH)CH₂OH	365.55	pl (al)	238-40[3]	35	0.9420[20/4]	1.4626[20]	al, eth, chl	B2[4], 1657
7382	Glycerol, monopalmitrate or α-Monopalmitin Glycerol-mono hexadicaneate .... C₁₅H₃₁CO₂CH₂CH(OH)CH₂OH	330.51	pl or lf(eth lig)	....	77	....	....	al	B2[4], 1170
7383	Glycerol, monopalmitate (l) or α-Monopalmitate (l) .... HOCH₂CH(OH)CH₂O₂CC₁₅H₃₁	330.51	[a]_D −4.37 (py)	....	71-2	....	....		B2[4], 1170
7384	Glycerol, 1-palmityl ether or Glycerol-1-hexadecylether... HOCH₂CH(OH)CH₂OC₁₆H₃₃	316.53	lf (hex) [a][20]_D +3 (chl)	120[0.005]	64	....	....	ace, chl, peth	B1[3], 2322
7385	Glycerol, 1-phenyl ether or Antodyne .... HOCH₂CH(OH)CH₂OC₆H₅	168.20	nd (eth peth)	200[22]	67-8	1.225[20/4]	....	w, al, eth, bz	B6[4], 589

No.	Name, Synonyms, and Formula	Mol. wt.	Color, crystalline form, specific rotation and λ_{max} (log ε)	b.p. °C	m.p. °C	Density	n_D	Solubility	Ref.
7386	Glycerol, 1-ricinoleate or 1-Mono (12-hydroxy-9-octade-canoate) . . . . . . HOCH₂CH(OH)CH₂[O₂C(CH₂)₇CH=CHCH₂(OH)CHC₆H₁₃]	372.55	ye	. . . .	. . . . . .	1.028²⁰⁄⁴	. . . . . .	al, eth, ace, bz	B3⁴, 1030
7387	Glycerol, 1-stearate or α-Monostearin . . . . C₁₇H₃₅CO₂CH₂CH(OH)CH₂OH	358.37	pl (MeOH)	. . . .	81	0.9841²⁰⁄⁴	1.4400⁸⁶	lig	B2⁴, 1225
7388	Glycerol, 1-stearate (l) . . . . C₁₇H₃₅CO₂CH(OH)CH₂OH	358.57	cr (eth or peth) [α]_D −3.58 (py)	. . . .	76-7	. . . .	. . . . . .	. . . .	B2⁴, 1225
7389	Glycerol, 1-(2-tolyl) ether . . . . (2-CH₃C₆H₄)OCH₂CH(OH)CH₂OH	182.22	nd (bz-peth)	. . . .	70-1d	. . . .	. . . . . .	al	B6³, 1952
7390	Glycerol, triacetate or Triacetin . . . . (CH₃CO₂CH₂)₂CHO₂CCH₃	218.21	cr (al)	258-60, 130.5⁷	4.1	1.1596²⁰⁄⁴	1.4301²⁰	al, eth, ace, bz, chl	B2⁴, 253
7391	Glycerol, tribenzoate or Tribenzoin . . . . C₆H₅CO₂CH(CH₂O₂CC₆H₅)₂	404.43	nd (MeOH)	. . . .	76	1.228¹²⁄⁴	. . . . . .	eth, ace, bz, chl	B9³, 666
7392	Glycerol, tributyrate or Tributyrin . . . . C₃H₇CO₂CH(CH₂O₂CC₃H₇)₂	302.37	. . . .	305-10, 190¹⁵	−75	1.0350²⁰⁄⁴	1.4359²⁰	al, eth, ace, bz	B2⁴, 799
7393	Glycerol, tricaproate or Tricaproin Glycerol trihexanoate C₅H₁₁CO₂CH(CH₂O₂CC₅H₁₁)₂	386.54	. . . .	>200	−60	0.9867²⁰⁄⁴	1.4427²⁰	al, eth, ace, bz, peth	B2⁴, 926
7394	Glycerol, tricaprylate or Tricuprylin Glycerol trioctoate . C₇H₁₅CO₂CH(CH₂O₂CC₇H₁₅)₂	470.70	. . . .	233.1	10 (st) −22 (unst)	0.9540²⁰⁄⁴	1.4482²⁰	al, eth, bz, chl, lig	B2⁴, 991
7395	Glycerol, trielaidate or Trielaidin Glycerol tri (trans-9-oc-tacteceneate) . . . . C₁₇H₃₃CO₂CH(CH₂O₂CC₁₇H₃₃)₂	885.47	. . . .	. . . .	α-16.6β-42.8	. . . .	. . . . . .	eth, bz, chl	B2⁴, 1664
7396	Glycerol, trilaurate or Trilaurin Glycerol tridodecanoate . C₁₁H₂₃CO₂CH(CH₂O₂CC₁₁H₂₃)₂	639.03	nd(al)	. . . .	46	0.8986⁵⁵	1.4404⁶⁰	al, eth, ace, bz, chl	B2⁴, 1098
7397	Glycerol, trimethylether or Trimethylol Propane . . . . CH₃OCH(CH₂OCH₃)₂	134.18	. . . .	. . . .	148	0.9460¹⁵⁄⁴	1.4055¹⁵	w, eth, ace, bz	B1⁴, 2755
7398	Glycerol, trimyristate or Glycerol tritetradecanoate . . . . . . C₁₃H₂₇CO₂CH(CH₂O₂CC₁₃H₂₇)₂	723.19	poly-morphic (al-eth)	311	56.5 (st) 32 (unst)	.08848⁶⁰⁄⁴	1.4428⁶⁰	eth, ace, bz, chl	B2⁴, 1134
7399	Glycerol, trinitrate or Nitroglycerin . . . . O₂NOCH(CH₂ONO₂)₂	227.09	pa ye fcl or rh	256 exp, 125²	13	1.5931²⁰⁄⁴	1.4786¹²	al, eth, ace, bz, chl	B1⁴, 2762
7400	Glycerol, trioleate or Trioleen, Glycerol-(cis-9-octadece-neate) . . . . C₁₇H₃₃CO₂CH(CH₂O₂CC₁₇H₃₃)₂	885.47	poly-morphic	235-40¹⁸	. . . . . .	0.8988⁴⁰	1.4621⁴⁰	eth, chl, peth	B2⁴, 1664
7401	Glycerol, tripalmatate or Tripalmitin Glycerol trihexade-canoate . . . . C₁₅H₃₁CO₂CH(CH₂O₂CC₁₅H₃₁)₂	807.35	nd (eth)	310-20	66 (st) 44.7 (unst)	0.8752⁷⁰⁄⁴	1.4381⁸⁰	eth, bz, chl	B2⁴, 1176
7402	Glycerol, tripropionate . . . . C₂H₅CO₂CH(CH₂O₂CC₂H₅)₂	260.29	. . . .	175-6²⁰	. . . . . .	1.100²⁰⁄¹⁸	1.4318¹⁹	al, eth, chl	B2⁴, 717
7403	Glycerol, tristearate or Tristearin Glycerol triactadeca-noate . . . . C₁₇H₃₅CO₂CH(CH₂O₂CC₁₇H₃₅)₂	891.51	. . . .	. . . .	α-55, β-73	0.8559⁹⁰⁄⁴	1.4395⁸⁰	ace	B2⁴, 1233
7404	Glycerol, triisovalerate or Glycerol tri (3-methylbutyrate) i-C₄H₉CO₂CH(CH₂O₂C-i-C₄H₉)	344.45	. . . .	330-5, 194¹⁵	. . . . . .	0.9984²⁰⁄⁴	1.4354²⁰	al, eth	B2⁴, 900
7405	Glycidic acid or 2,3-Epoxypropionic acid . . . . CH₂CHCO₂H	88.06	. . . .	. . . .	. . . . . .	. . . .	. . . . . .	w, al, eth	B18¹, 435
7406	Glycidol (d) or 3-Hydroxypropylene oxide . . . . HOCH₂CHCH₂	74.08	α + 15 (undil)	56.5¹¹	. . . . . .	1.117²⁰⁄⁴	1.4293¹⁶	w, al, eth, ace, bz	B17⁴, 985
7407	Glycidol dl or 2,3-Epoxy-1-propanol (dl) . . . . CH₂CHCH₂OH	74.08	. . . .	65-6²·⁵	. . . . . .	1.1143²⁵	1.4287²⁰	w, al, eth, ace, bz	B17⁴, 985
7408	Glycidol (l) or 3-Hydroxy propylene oxide . . . . HOCH₂CHCH₂	74.08	[α]²⁵⁄_D −8.6 (indil)	56¹¹	. . . . . .	1.1050¹⁸	1.4293¹⁶	w, al, eth, ace, bz	B17⁴, 985
7409	Glycidol, phenyl . . . . C₆H₅CHCHCH₂OH	150.18	. . . .	138³	26.5	1.512²⁷	1.5432²⁷	al, eth	B6³, 1800
7410	Glycine or Aminoacetic acid . . . . H₂NCH₂CO₂H	75.07	mcl or trg pr (dil al)	. . . .	262d	1.607	. . . . . .	w	B4³, 1097
7411	Glycine, N-Leucyl (DL) or DL-Leucyl Glycine . . . . (CH₃)₂CHCH₂-CH(NH₂)CONHCH₂CO₂H	188.23	. . . .	. . . .	243d	. . . .	. . . . . .	w	B4³, 1434
7412	Glycine, -Leucyl (L) or L-Leucyl glycine . . . . (CH₃)₂CHCH₂CH(NH₂)CONHCH₂CO₂H	188.23	lf or nd (w al)	. . . .	248d	. . . .	. . . . . .	w	B4³, 1414
7413	Glycyl glycine . . . . H₂NCH₂CONHCH₂CO₂H	132.12	[α]²⁰⁄_D + 85.8 (w,c=2)	. . . .	dec 215	. . . .	. . . . . .	. . . .	B4³, 1191

No.	Name, Synonyms, and Formula	Mol. wt.	Color, crystalline form, specific rotation and λ_{max} (log ε)	b.p. °C	m.p. °C	Density	n_D	Solubility	Ref.
7414	Glycocholic acid or Cholyglycine............ $C_{26}H_{43}NO_6$	465.64	nd (w) $[\alpha]^{23}_D$ +32.3 (al,c=1)	....	165-8 (anh) 132-4 (+w)				B10[3], 2176
7415	Glycocyamine or N-Guanyl glycine $NH=C(NH)_2NHCH_2CO_2H$	117.11	pl or nd (w)	....	>300				B4[3], 1165
7416	Glycogen or Animal starch............... $(C_6H_{10}O_5)n$	(162.14) n	wh pw $[\alpha]^{25}_D$ +196.5 (w)					w	C25, 4940
7417	Glycolic acid or Hydroxyacetic acid .... $HOCH_2COOH$	76.05	rh nd (w) lf (eth)	d	80			w, al, eth	B3[4], 571
7418	Glycolic acid, benzoyl $C_6H_5COCH(OH)CO_2H$	180.17	lo pr (lig)		112			al, eth, chl	B9, 954
7419	Glycolide or 2,5-p-dioxane dione .... $C_4H_4O_4$	116.07	lf (al-chl or al)		86-7			ace	B19[4], 1922
7420	Glycoluril or Glyoxaldiurene .... $C_4H_6N_4O_2$	142.12	nd or pr (w)		300d			eth	B26[2], 260
7421	18α-Glycyrrhetinic acid or Glycyrhetic acid .... $C_{30}H_{46}O_4$	470.70	α-pl (dil al) β-nd (al peth)	α283 β296				al	B10[3], 4392
7422	Glycyrrhizic acid or Glycyrhizin .... $C_{42}H_{62}O_{16}$	822.96	pl or pr (aa)		220d			w	B18[4], 5156
7423	Glyoxal or Ethanedial .... OHCCHO	58.04	ye pr	50.4	15	1.14[20]	1.3826[20]	w, al, eth	B1[4], 3625
7424	Glyoxal, dioxine HON=CHCH=NOH	88.07	rh pl (w)	sub	170d	....	....	w, al, eth	B1[4], 3629
7425	Glyoxal, methyl, phenyl or Methylphenyl glyoxal........ $CH_3COCOC_6H_5$	148.17	ye oil	228, 101[12]	....	1.0065[20/4]	1.537[10]	w, al, eth	B7[3], 3463
7426	Glyoxal, methylphenyl, dioxime or Methylphenyl glyoxime .... $CH_3C(=NOH)C(=NOH)C_6H_5$	178.19	nd (dil al)	....	140-1	....	....	al	B7[3], 3461
7427	Glyoxal, phenyl or Benzoyl formaldehyde .... C_6H_5COCHO	134.11	nd (+w)	142[125]	91	(hyd)		w, al, eth, ace, bz	B7[3], 3443
7428	Glyoxal, phenyl, hydrate .... $C_6H_5COCH(OH)_2$	152.15	nd (w, chl, al, lig)	93.4				al, eth, chl	B7[3], 3443
7429	Glyoxal, phenyl-1-oxime .... $C_6H_5COCH=NOH$	149.15	mcl pr or lf (chl, w)		129			chl	B7[3], 3447
7430	Glyoxylic acid or Oxoacetic acid .... $HCOCO_2H$	74.04	rh pr (w + 1/2)		98			w	B3[4], 1489
7431	Glyoxylic acid, phenyl or Phenylglyoxylic acid, Benzoyl formic acid............. $C_6H_5COCO_2H$	150.14	pr (CCl$_4$)	147-51[12]	66			w, al, eth	B10[3], 2972
7432	Glyoxylic acid, phenyl, methyl ester or Methyl phenyl glyoxylate............. $C_6H_5COCO_2CH_3$	164.16	ye	246-8, 137[14]	....		1.5268[20]	....	B10[3], 2973
7433	Glyoxylonitrile, phenyl or Benzoyl cyanide .... C_6H_5CON	131.14	ta	206-8, 99[19]	32-3			al, eth	B10[3], 2976
7434	Gramine or 3-(Dimethylaminomethyl indole) .... $C_{11}H_{14}N_2$	174.25	lf (eth) nd (ace)		138-9			al, eth, chl	B22[4], 4302
7435	Griseofulvin or Fulvicin .... $C_{17}H_{17}ClO_6$	352.78	oct or rh (bz) $[\alpha]^{17}_D$ +376 (chl,Sat sul)		220				B18[4], 3160
7435a	Guaiol or champicol .... $C_{15}H_{26}O$	228.38	trq Pr (al), $[\alpha]^{-30}$ $c=4$	165[17]	91	0.9074[100/4]	1.4716[100/6]	al, eth	
7436	Guanamine or 2,4-Diamino-1,3,5-triazine .... $2,4(H_2N)_2(C_3HN_3)$	111.11	nd (w)		329d			w	B26[1], 65
7437	Guanidine or Aminomethanamidine, Carbaniedine ..... $HN=C(NH_2)_2$	59.07	cr		Ca 50			al, w	B3[4], 148
7438	Guanidine acetate .... $HN=C(NH_2)_2 \cdot CH_3CO_2H$	119.13	nd (al-eth)		229-30			w, al	B3[4], 152
7439	Guanidine, amine or Guanyl hydrazine .... $HN=C(NH_2)NHNH_2$	74.09	cr		d			w, al	B3[4], 236
7440	Guanidine, carbonate .... $[HN=C(NH_2)_2]_2 \cdot H_2CO_3$	180.17	oct tetr pr (w)		198	1.24[4]		w	B3[4], 152
7441	Guanidine, 1-cyano .... $HN=C(NH_2)NHCN$	84.08	rh lf or pl (al)	d	211-2	1.404[14]		w, al, ace	B3[4], 160

No.	Name, Synonyms, and Formula	Mol. wt.	Color, crystalline form, specific rotation and λ_{max} (log ε)	b.p. °C	m.p. °C	Density	n_D	Solubility	Ref.
7442	Guanidine, 1,3-diphenyl or Melaniline .... $HN=C(NHC_6H_5)_2$	211.27	mcl nd (al or to)	d 170	150	$1.13^{20/4}$	....	al, eth	B12³, 805
7443	Guanidine, 1,3-di(2-totyl) .... $HN=C[NH(C_6H_4CH_3-2)]$	239.32	cr (dil al)	....	179	$1.10^{20/4}$		eth, chl	B12³, 1871
7444	Guanidine, hydrochloride .... $HN=C(NH_2)_2).HCl$	95.53	rh bipym (al)		178-85	$1.354^{20/4}$		w, al	B3⁴, 150
7445	Guanidine, nitrate .... $HN=C(NH_2)_2.HNO_3$	122.08	lf (w)	d	217			w, al	B3⁴, 151
7446	Guanidine, 1-nitro .... $HN=C(NH_2)NHNO_2$	104.07	nd or pr (w)		239d				B3⁴, 249
7447	Guanidine, picrate .... $HN=C(NH_2)_2.C_6H_3N_3O_7$	288.19	og ye pl or nd (w)		333d				B6³, 960
7448	Guanidine, thiocyanate .... $HN=C(NH_2)_2HCNS$	118.16	lf		118			w	B3², 121
7449	Guanidine, tetraphenyl .... $HN=C[N(C_6H_5)_2]_2$	363.47	rh (lig)		130-1			al, eth, bz	B12, 430
7450	Guanidine, 1,1,3-triphenyl .... $HN=C(NHC_6H_5)N(C_6H_5)_2$	287.37	ta (dil al)		134			al, eth	B12³, 895
7451	Guanidine, 1,2,3-triphenyl .... $C_6H_5N=C(NHC_6H_5)_2$	287.37	nd or pr(al)	d	146-7	$1.163^{20/4}$		al	B12³, 907
7452	Guanidine, 1-ureido or Dicyandiamidine .... $HN=C(NH_2)NHCONH_2$	162.20	pr(al)	d160	105			py	B3⁴, 155
7453	Guanine or 2-Aminohypoxanthine .... $C_5H_5N_5O$	151.13	nd or pl(aq NH₃)	sub	360d				B26², 262
7454	Guanosine or 9-D-Ribosidoguanine .... $C_{10}H_{13}N_5O_5$	283.25	nd (w), $[\alpha]^{20/}_D$ -60.5 (0.1 NNaOH, p=3)		239d			aa	B31, 28
7455	Guanylic acid or Guanosine phosphoric acid .... $C_{10}H_{14}N_5O_8P$	363.23	nd or pr (w + 2), $[\alpha]^{20/}_D$ -7.5 (w, p=1) $^{25/}_D$ -65 (5% NaOH, c=2)		208d			w	B31, 29
7456	Guiacol or Catechol monomethyl ether .... $2-CH_3OC_6H_4OH$	124.15	hex pr	205, 106.5²⁴	32	$1.1287^{21/4}$	1.5429^{20}	al, eth, chl	B6³, 4200
7457	Guiacol, 3-nitro or 2-Methoxy-6-nitro phenol .... $2-CH_3O-6-NO_2C_6H_3OH$	169.14	og ye nd(sub)	sub	62			w, al	B6³, 4263
7458	Guiacol, 4-nitro or 2-Methoxy-5-nitro phenol .... $2-CH_3O-5-NO_2C_6H_3OH$	169.14	pa ye nd(w)	....	105			al, eth	B6³, 4264
7459	Guiacol, 5-nitro or 2-Methoxy-4-nitro phenol .... $2-CH_3O-4-NO_2C_6H_3OH$	169.14	ye nd(w)		103-4			al, eth	B6³, 4264
7460	Guaiacol, 6-nitro or 2-Methoxy-3-nitro phenol .... $2-OCH_3-3-NO_2C_6H_3OH$	169.14	yesh rh pr(peth)		102-3			al	B6³, 4263
7461	Guaiacol, 5-vinyl or 2-Methoxy-4-vinyl phenol .... $2-CH_3O-4-(CH_2=CH)C_6H_3OH$	150.18	cr		57			al, eth	B6³, 4981
7462	D-Gulonic acid-γ-lactone .... $C_6H_{10}O_5$	178.14	pr ta (w), $[\alpha]^{20}_D$ -57.1 (w)		180-1			w	B18⁴, 3025
7463	L-Gulonic acid-γ-lactone .... $C_6H_{10}O_5$	178.14	rh pr (w), $[\alpha]^{20}_D$ + 55.1 (w)		185			w	B18⁴, 3026
7464	D-Gulonic acid, phenylhydrazide...................... $C_{12}H_{18}N_2O_6$	286.29	(w), $[\alpha]^{20/}_D$ + 13.45	d 195	147-9			w, al	B15¹, 82
7465	D-Gulose .... $C_6H_{12}O_6$	180.16	syr $[\alpha]^{20}_D$ -20.4 (w)	d	....			w	B1⁴, 4333
7466	L-Gulose .... $C_6H_{12}O_6$	180.16	sy $[\alpha]^{20}_D$ + 61.6 (w)	d	....			w	B1⁴, 4334
7467	Guvacine or 1,2,5,6-Tetrahydronicotinic acid .... $C_6H_9NO_2$	127.14	pr(w) rods(+ 1w dil al)	....	295d			w	B22¹, 489
7468	β-Gurjunene .... $C_{15}H_{24}$	204.35	$[\alpha]_D$ + 74.5	120-3¹³	......	0.9348	1.5028	....	B5³, 1093

No.	Name, Synonyms, and Formula	Mol. wt.	Color, crystalline form, specific rotation and λ_{max} (log ε)	b.p. °C	m.p. °C	Density	n_D	Solubility	Ref.
7469	Halostachine (l) $C_6H_5CH(OH)CH_2NHCH_3$	151.21	[α] -47	....	43-5	....		w, al, eth	B13³, 1658
7470	Harmaline or 3,4-Dihydroharmine $C_{13}H_{14}N_2O$	214.27	ta(MeOH) th pr(al)	....	250d				B23², 345
7471	Harmine or Banisterine, Telepathine $C_{13}H_{12}N_2O$	212.55	rh(al) pr(MeOH)	sub	272-4			py	B23³, 348
7472	Hecogenin $C_{27}H_{42}O_4$	430.64	pl (eth), [α]²²/$_D$ + 7		265-8			al, eth, ace	B19⁴, 2581
7473	Hecogenin, acetate $C_{29}H_{44}O_5$	472.67	cr(MeOH)		243 (252)			al	B19⁴, 2584
7474	Hederagenin $C_{30}H_{48}O_4$	472.72	pr (al), [α]²⁰/$_D$ + 70.1 (chl-MeOH)		332-4			al	B10³, 1923
7475	Helenine or Alantolactone $C_{15}H_{20}O_2$	232.33	nd	275, 197¹⁰	76	....		al, eth, bz	B17⁴, 5030
7476	Helicin or Salicylaldehyde-β-D- glucoside $C_{13}H_{16}O_7$	284.27	nd (a), [α]²⁰/$_D$ -60.4 (w)		175			w, al	B17⁴, 3010
7476a	Helvolic acid or Fumagacin $C_{33}H_{44}O_8$	554.69	nd (dil al), [α]¹⁸/$_D$ -125 (chl, c=1)		212d			eth, ace, bz, aa	Am 78, 5275
7477	Hematein or Haematein $C_{16}H_{12}O_6$	300.27	red br cr	....	250d				B18⁴, 3343
7478	Hematin or Ferriporphyrin hydroxide $C_{34}H_{35}N_4O_5$FeOH	633.51	br pw(py)	....	>200				M, 508
7479	Hematommic acid $C_9H_{10}O_5$	244.17	nd(aa)		172 d	....		aa	H10, 787
7480	Hematoporphyrin or Photodyn $C_{34}H_{38}N_4O_6$	598.71	red	....	172-3			aa	M, 509
7481	Hematoxylin $C_{16}H_{14}O_6 \cdot 3H_2O$	356.33	yesh cr, [α] + 11 (w, c=3.7)		140			al	B17², 273
7482	Hemimellitic acid or 1,2,3-Benzene tricarboxylic acid 1,2,3-$C_6H_3(CO_2H)_3$	210.15	tcl pl(+2w) nd(w)	....	197d	1.546²⁰		w	B9³, 4791
7483	Heneicosane or Uneicosane $CH_3(CH_2)_{19}CH_3$	296.59	cr (w)	356.5, 203¹⁰	40.5	0.7919²⁰/⁴	1.4441²⁰	Peth	
7484	Heptachlor or 3,4,5,6,7,8,8a Heptachloro-α-Dicyclopentadiene $C_{10}H_5Cl_7$	373.32	wh	....	95-6	1.57⁹		al, eth, bz, lig	B5³, 1236
7485	Heptacosane $CH_3(CH_2)_{25}CH_3$	380.75	cr(al bz) lf(AcOEt)	442, 270¹⁵	59.5	0.7796⁶⁰/⁴	1.4345⁶⁵	....	B1⁴, 586
7486	7,10-Heptadecadiyne $CH_3(CH_2)_5C\equiv CCH_2C\equiv C(CH_2)_5CH_3$	232.41	....	150⁶		0.84¹⁹/⁴	1.4700¹⁹		B1³, 1068
7487	Heptadecanal or Margaraldehyde $CH_3(CH_2)_{15}CHO$	254.46	nd (peth) cr (al +1)	204²⁶	36			eth, bz, aa	B1⁴, 3395
7488	Heptadecane $CH_3(CH_2)_{15}CH_3$	240.48	hex lf	301.8, 161.7¹⁰	22	0.7780²⁰/⁴	1.4369²⁰	eth	B1⁴, 548
7489	Heptadecane, 1-amino or Heptadecyl amine $CH_3(CH_2)_{15}CH_2NH_2$	255.49		336, 189¹⁰	49	0.8510²⁰/⁴	1.4510²⁰	al, eth	B4⁴, 824
7490	Heptadecane, 1-bromo or Heptadecyl bromide $CH_3(CH_2)_{15}CH_2Br$	319.38		349, 199¹⁰	32	0.9916²⁰/⁴	1.4625²⁰	chl	B1⁴, 549
7491	Heptadecane, 1,17-dibromo $Br(CH_2)_{17}Br$	398.28	lf(al)	208-10³	38			chl	B1³, 564
7492	Heptadecane, 9-hexyl $CH_3(CH_2)_7CH[(CH_2)_5CH_3]_2$	324.63	....	213.0¹⁰, 151⁰·⁵	-19.4	0.7976²⁰/⁴	1.4465²⁰/$_D$	....	B1⁴, 578
7493	Heptadecane-9-octyl $CH[(CH_2)_8CH_3]_3$	352.69		231.5¹⁰	-13.8	0.8020²⁰/⁴	1.4487²⁰/$_D$		B1⁴, 583
7494	Heptadecane-9-Phenethyl $C_6H_5(CH_2)_2CH(C_8H_{17})_2$	344.62		189¹⁰	-26.7	0.8560²⁰/⁴	1.4806²⁰/$_D$	....	PCHG 3, 180
7495	Heptadecanoic acid or Margaric acid $CH_3(CH_2)_{15}CO_2H$	270.46	pl(peth)	227¹⁰⁰	62-3	0.8532⁶⁰	1.4342⁶⁰	eth, ace, bz, chl	B2⁴, 1193
7496	Heptadecanoic acid, ethyl ester or Ethyl heptadecanoate $CH_3(CH_2)_{15}CO_2C_2H_5$	298.52	pl(dil al)	185⁵	28			al, eth, ace, bz	B2⁴, 1194

No.	Name, Synonyms, and Formula	Mol. wt.	Color, crystalline form, specific rotation and λ_{max} (log ε)	b.p. °C	m.p. °C	Density	n_D	Solubility	Ref.
7497	Heptadecanoic acid, methyl ester .... $CH_3(CH_2)_{15}CO_2CH_3$	284.49	pl(al)	184-7[9]	30	....	....	al, eth, ace, bz	B2[4], 1194
7498	Heptadecanonitrile or Margaronitrile............ $CH_3(CH_2)_{15}CN$	251.46	cr(al)	349, 183[10]	34	0.8315[20/4]	1.4467[20]	eth	B2[4], 1195
7499	1-Heptadecanol .... $CH_3(CH_2)_{15}CH_2OH$	256.48	lf(al) cr(ace)	308	54	0.8475[20/4]	....	al, eth	B1[4], 1884
7500	2-Heptadecanol .... $CH_3(CH_2)_{14}CH(OH)CH_3$	256.48	pl(dil al)	140[0.5]	54	....	1.4407[37]	al, eth	B1[4], 1885
7501	9-Heptadecanol .... $(C_8H_{17})_2CHOH$	256.48	pl(dil al)	174[9]	61	....	1.4262[80]	al, eth, ace, bz	B1[4], 1884
7502	2-Heptadecanone or Methyl penta decylketone .... $C_{15}H_{31}COCH_3$	254.46	pl(dil al)	320, 246[110]	48	0.8140[48/48]	....	eth, ace, bz	B1[4], 3395
7503	9-Heptadecanone or Pelargone .... $(C_8H_{17})_2CO$	254.46	pl(MeOH)	250-3, 142[1]	53	....	....	....	B1[4], 3396
7504	1-Heptadecene .... $CH_3(CH_2)_{14}CH=CH_2$	238.46	....	300, 160[10]	11.2	0.7852[20/4]	1.4432[20]	eth, bz, lig	B1[4], 927
7505	8-Heptadecene, 9-octyl .... $C_7H_{15}C=C(C_8H_{17})_2$	350.67	....	227[10]		0.8086[20/4]	1.4554[20/D]	....	B1[4], 936
7506	Heptadecyl amine .... $CH_3(CH_2)_{15}CH_2NH_2$	255.49	....	336, 189[10]	49	0.8510[20/4]	1.4510[20]	al, eth	B4[4], 827
7507	1,4-Heptadiene .... $CH_2CH_2CH=CHCH_2CH=CH_2$	96.17	....	93[772]	....	0.7270[20/4]	1.4370[20]	eth, bz, peth	B1[3], 999
7508	1,5-Heptadiene .... $CH_3CH=CHCH_2CH_2CH=CH_2$	96.17	....	94	....	0.7186[20/4]	1.4200[20]	al, eth, ace, bz	B1[3], 999
7509	1,5-Heptadiene-4-ol .... $CH_3CH=CHCH(OH)CH_2CH=CH_2$	112.17	....	155-6[742], 68[24]		0.8598[20/4]	1.4510[25]	ace	B1[4], 2248
7510	1,6-Heptadiene-3-yne .... $CH_2=CHCH_2C≡CCH=CH_2$	92.14	....	110[950]		0.787[25/4]	1.4694[25]	bz, peth	B1[3], 1061
7511	2,4-Heptadiene .... $CH_3CH_2CH=CHCH=CHCH_3$	96.17	....	108		0.7384[20/4]	1.4578[20]	al, eth, ace, bz	B1[4], 1029
7512	2,5-Heptdien-4-one, 2,6-dimethyl or Phorone.......... $CH_3C(CH_3)=CHCOCH=C(CH_3)_2$	138.21	ye gr pr	197.8	28	0.8850[20/4]	1.4998[20]	al, eth, ace	B1[4], 3051
7513	3,5-Heptadiene-2-one or Crotonylidene acetone .... $CH_3CH=CHCH=CHCOCH_3$	110.16	....	88[28]		0.8946[19/4]	1.5177[19]	eth	B1[4], 3549
7514	1,5-Heptadiyne .... $CH_3C≡CCH_2CH_2C≡CH$	92.14	....	26[30]		0.8100[21/4]	1.4521[21]	bz, peth	B1[2], 247
7515	1,6-Heptadiyne .... $HC≡C(CH_2)_3C≡CH$	92.14	....	112, 36[20]	−85	0.8164[17/4]	1.451[17]	bz, aa	B1[4], 1121
7516	Heptamethyleneimine, α-methyl or 1-Azacyclooctane-2-methyl. .... $C_8H_{17}N$	127.21	nd(ace)	162-3[746]	156-7	0.853[30]	1.4620[21]	w, al, eth, ace, bz	B20[1], 30
7517	Heptanal or Enanthaldehyde. Heptaldehyde .... $CH_3(CH_2)_5CHO$	114.19	....	152.8, 59.6[30]	−43.3	0.8495[20/4]	1.4113[20]	al, eth	B1[4], 3314
7518	Heptanal, 2-benzylidene or α-Pentylcinnamaldehyde..... $C_5H_{11}C(=CHC_6H_5)CHO$	202.30	ye oil	174-5[20]	80	0.9711[20]	1.5381[20]	ace	B7[3], 1517
7519	Heptanal, oxime or Heptaldoxime .... $CH_3(CH_2)_5CH=NOH$	129.20	pl(al)	195, 100.5[14]	57-8	0.8583[55]	1.4210[20]	al, eth	B1[4], 3316
7520	1,6-Heptadiene-4-ol, 4-methyl .... $CH_3C(OH)(CH_2CH=CH)_2$	126.20	....	158.4, 54-6[11]	....	0.86258[20/20]	1.4500[23]	al, ace	B1[4], 2260
7521	1,6-Heptadiene-3-yne .... $CH_2=CHC≡CCH_2CH=CH_2$	92.14	....	110[750]		0.787[25/4]	1.4694[25]	bz, peth	B1[3], 1061
7522	Heptane .... C_7H_{16}	100.21	....	98.4	−90.6	0.6837[20/4]	1.3878[20]	al, eth, ace, chl, peth	B1[4], 376
7523	Heptane, 1-amino or Heptyl amine.......... $CH_3(CH_2)_5CH_2NH_2$	115.22	....	156.9, 45.6[20]	−18	0.7754[20/4]	1.4251[20]	al, eth	B4[4], 734
7524	Heptane, 2-amino .... $CH_3(CH_2)_4CHNH_2CH_3$	115.22	....	142	....	0.7665[19]	1.4199[19]	al, eth, peth	B4[4], 743
7525	Heptane, 1-bromo or Heptyl bromide.......... $CH_3(CH_2)_5CH_2Br$	179.11	....	178-9, 59.7[10]	−56.1	1.1400[20/4]	1.4502[20]	al, eth, chl	B1[4], 391
7526	Heptane, 1-bromo-7-fluoro .... $Br(CH_2)_7F$	197.10	....	85[11]		....	1.4463[20]	al, eth	B1[4], 392
7527	Heptane, 2-bromo .... $C_5H_{11}CHBrCH_3$	179.10	....	165-7, 63.8[20]	....	1.1277[20/4]	1.4503[20]	bz	B1[3], 431
7528	Heptane, 3-bromo .... $CH_3CH_2CHBr(CH_2)_3CH_3$	179.10	....	62[18]	....	1.1362[20/4]	1.4503[20]	bz, chl	B1[4], 392

No.	Name, Synonyms, and Formula	Mol. wt.	Color, crystalline form, specific rotation and λ_{max} (log ε)	b.p. °C	m.p. °C	Density	n_D	Solubility	Ref.
7529	Heptane, 4-bromo .... $CH_3(CH_2)_2CHBrCH_2CH_3$	179.10	....	84.6[72]	......	1.1351[20/4]	1.4495[20]	bz, chl	B1[4], 392
7530	Heptane, 1-chloro or Heptyl chloride .... $CH_3(CH_2)_6CH_2Cl$	134.65	....	159, 45[10]	−69.5	0.8758[20/4]	1.4256[20]	al, eth	B1[4], 389
7531	Heptane, 1-chloro-7-fluoro .... $F(CH_2)_7Cl$	152.64	....	70[10]	......	0.993[20]	1.4222[25]	al, eth, bz, chl	B1[4], 390
7532	Heptane, 2-chloro .... $CH_3(CH_2)_4CHClCH_3$	134.65	....	46[19]	......	0.8672[20]	1.4221[20]	eth, bz, chl, aa	B1[4], 390
7533	Heptane, 2-chloro-2-methyl .... $C_5H_{11}CCl(CH_3)_2$	148.68	....	50[15]	......	0.8568[25/4]	1.4240[25]	al, eth, bz, chl	B1[4], 428
7534	Heptane, 2-chloro-6-methyl .... $(CH_3)_2CH(CH_2)_3CHClCH_3$	148.68	....	74[35]	......	....	1.4260[15]	al, eth, bz, chl	B1[3], 472
7535	Heptane, 3-chloro .... $CH_3(CH_2)_3CHClCH_2CH_3$	134.65	....	144[751], 48.3[20]	......	0.8960[20/4]	1.4228[20]	eth, bz	B1[4], 390
7536	Heptane, 3-chloro-2,3-dimethyl .... $C_4H_9CCl(CH_3)CH(CH_3)_2$	162.71	....	54[8]	......	0.8395[20]	1.4391[20]	eth, chl	B1[3], 511
7537	Heptane, 3-chloro-3-ethyl .... $C_4H_9CCl(C_2H_5)_2$	162.71	....	46[3]	......	0.8856[20/4]	1.4400[20]	eth, chl	B1[3], 510
7538	Heptane, 3-chloromethyl .... $CH_3CH_2CH(CH_2Cl)(CH_2)_2CH_3$	148.68	....	174	......	0.8769[20/4]	1.4319[20]	al, eth, ace, bz	B1[4], 430
7539	Heptane, 3-chloro-3-methyl .... $C_4H_9CCl(CH_3)C_2H_5$	148.68	....	64[27]	......	0.8764[20/4]	1.4317[20]	al, eth, bz, chl	B1[4], 430
7540	Heptane, 4-chloro .... $(C_3H_7)_2CHCl$	134.65	....	144[758], 48.9[21]	......	0.8710[20/4]	1.4237[20]	eth, bz	B1[4], 390
7541	Heptane, 4-chloro-4-ethyl .... $(C_3H_7)_2C(Cl)C_2H_5$	162.71	....	6[11]	......	0.8821[20/4]	1.4438	eth, bz, chl	B1[4], 511
7542	Heptane, 4-chloro-4-methyl .... $(C_3H_7)_2C(Cl)CH_3$	148.68	....	50[12]	......	0.8690[20/4]	1.4310[15]	al, eth, bz, chl	B1[3], 477
7543	Heptane, 5-chloro-2,5-dimethyl .... $CH_3CH_2CCl(CH_3)CH_2CH_2CH(CH_3)_2$	162.71	....	63[15]	......	0.8692[10/4]	1.4346[10]	bz, chl	B1[4], 64
7544	Heptane, 1,7-diamino or Heptamethylenediamine ...... $H_2N(CH_2)_7NH_2$	130.23	....	223.5, 104-5[12]	28.9	....	....	al, eth, ace, bz	B4[4], 1354
7545	Heptane, 1,7-dibromo .... $Br(CH_2)_7Br$	258.00	....	263, 132[11]	−41.7	1.5306[20/4]	1.5034[20]	eth, ace, bz	B1[4], 393
7546	Heptane, 1,1-dichloro .... $CH_3(CH_2)_5CHCl_2$	169.10	....	187, 82[20]	......	1.0008[20/4]	1.4440[20]	eth, bz, chl	B1[4], 390
7547	Heptane, 1,2-dichloro .... $CH_3(CH_2)_4CHClCH_2Cl$	169.10	....	68-72[7]	......	1.064[20/4]	1.4490[20]	eth, bz, chl	B1[3], 430
7548	Heptane, 2,2-dichloro .... $CH_3(CH_2)_4CCl_2CH_3$	169.10	....	77[25]	......	1.012[20/4]	1.4440[20]	eth, bz, chl	B1[4], 430
7549	Heptane, 2,6-dichloro-2,6-dimethyl .... $(CH_3)_2CCl(CH_2)_3CCl(CH_3)_2$	197.15	....	93[16]	43	....	....	bz, chl	B1[3], 513
7550	Heptane, 4,4-dichloro .... $(C_3H_7)_2CCl_2$	169.10	....	86[27]	......	1.0008[17]	1.448[17]	eth, chl	B1[2], 117
7551	Heptane, 3,3-di(hydroxymethyl) .... $C_2H_5C(CH_2OH)_2C_4H_9$	160.26	wh	262, 123[15]	43.8	0.929[50/20]	1.4587[25]	al	B1[3], 2228
7552	Heptane, 2,2-dimethyl .... $(CH_3)_3C(CH_2)_3CH_3$	128.26	....	132.7	−113	0.7105[20/4]	1.4016[20]	eth, ace, bz	B1[4], 457
7553	Heptane, 2,3-dimethyl .... $CH_3(CH_2)_3CH(CH_3)CH(CH_3)_2$	128.26	....	140.5, 29.4[10]	−116	0.7260[20/4]	1.4088[20]	al, eth, ace, bz	B1[4], 457
7554	Heptane, 2,4-dimethyl .... $C_3H_7CH(CH_3)CH_2CH(CH_3)_2$	128.26	....	133.5, 23.8[10]	......	0.7143[20/4]	1.4031[20]	al, eth, ace, bz	B1[4], 457
7555	Heptane, 2,5-dimethyl (d) .... $C_2H_5CH(CH_3)(CH_2)_2CH(CH_3)_2$	128.26	....	136	......	0.7198[20/4]	1.4033[20]	al, eth, ace, bz	B1[4], 457
7556	Heptane, 2,6-dimethyl .... $(CH_3)_2CH(CH_2)_3CH(CH_3)_2$	128.26	....	135.2, 25.5[10]	−102.9	0.7089[20/4]	1.4011[20]	....	B1[4], 458
7557	Heptane, 3,3-dimethyl .... $C_4H_9C(CH_3)_2CH_2CH_3$	128.26	....	137.3, 26[10]	......	0.7254[20/4]	1.4087[20]	al, eth, ace, bz	B1[4], 458
7558	Heptane, 3,4-dimethyl .... $CH_3CH_2CH(CH_3)CH(CH_3)C_3H_7$	128.26	....	140.1	......	0.7314[20/4]	1.4108[20]	eth, ace, bz	B1[3], 514
7559	Heptane, 3,5-dimethyl .... $CH_3CH_2CH(CH_3)CH_2CH(CH_3)CH_2CH_3$	128.26	....	136	......	0.7225[20/4]	1.4083[20]	eth, ace, bz	B1[4], 458
7560	Heptane, 4,4-dimethyl .... $(C_3H_7)_2C(CH_3)_2$	128.26	....	135.2	......	0.7221[20/4]	1.4076[20]	eth, ace, bz	B1[4], 458
7561	Heptane, 4-ethyl .... $(C_3H_7)_2CHC_2H_5$	128.26	....	141.2, 31[10]	......	0.7270[20/4]	1.4096[20]	al, eth, ace, bz, chl	B1[4], 457

No.	Name, Synonyms, and Formula	Mol. wt.	Color, crystalline form, specific rotation and λ_{max} (log ε)	b.p. °C	m.p. °C	Density	n_D	Solubility	Ref.
7562	Heptane, 1-fluoro or Heptyl fluoride $CH_3(CH_2)_5CH_2F$	118.20	....	117.9	−73	$0.8062^{20/4}$	1.3854^{20}	eth, ace, bz, peth	B1[4], 387
7563	Heptane, perfluoro C_7F_{16}	388.05		82.4	−78	1.7333^{20}	1.2618^{20}	al, eth, ace, chl	B1[4], 388
7564	Heptane, 1-iodo or Heptyl iodide $CH_3(CH_2)_5CH_2I$	226.10		204, 76.1^{10}	−48.2	$1.3791^{20/4}$	1.4904^{20}	al, eth, ace, chl	B1[4], 393
7565	Heptane, 2-iodo $CH_3CHI(CH_2)_4CH_3$	226.10		98^{50}	......	1.304^{20}	1.4826	ace, bz	B1[4], 393
7566	Heptane, 2-methyl $C_5H_{11}CH(CH_3)_2$	114.23		117.6, 12.3^{10}	−109	$0.6980^{20/4}$	1.3949^{20}	al eth, ace, bz, chl	B1[4], 428
7567	Heptane, 2-methhylamino $C_5H_{11}CH(CH_3)NHCH_3$	129.25		155	....	....	....	....	B4[4], 743
7568	Heptane, 3-methyl (d) $C_4H_9CH(CH_3)C_2H_5$	114.23	$[\alpha]^{26/}_D$ +9.34	115-8	......	$0.7075^{16/4}$	1.4002^{18}	al, eth, ace, bz, chl	B1[4], 429
7569	Heptane, 3-methyl (dl) $C_4H_9CH(CH_3)C_2H_5$	114.23		119, 13.3^{10}	−120.5	$0.7058^{20/4}$	1.3985^{20}	al, eth, ace, bz, chl	B1[4], 429
7570	Heptane, 3-methyl (l) $C_4H_9CH(CH_3)C_2H_5$	114.23		$117\text{-}8^{745}$			1.3990^{20}	al, eth, ace, bz, chl	B1[3], 476
7571	Heptane, 4-methyl $(C_3H_7)_2CHCH_3$	114.23		117.7, 12.4^{10}	−121	$0.7046^{20/4}$	1.3979^{20}	al, eth ace, bz, chl	B1[4], 431
7572	Heptane, 2,2,4,4,6-pentamethyl $(CH_3)_2CHCH_2CH(CH_3)_2CH_2C(CH_3)_3$	388.05		177.8	−67	$0.7463^{20/4}$	$1.4440^{20/}_D$	....	B1[4], 510
7573	Heptane, 2,2,4-trimethyl $(CH_3)_3CCH_2CH(CH_3)(C_3H_7)$	142.28		147.7, 32.9^{10}		$0.7275^{20/4}$	1.4092^{20}	bz, chl	B1[4], 481
7574	Heptane, 3,3,5-trimethyl $CH_3CH_2C(CH_3)_2CH_2CH(CH_3)CH_2CH_3$	142.28		155.7, 38.9^{10}		$0.7248^{20/4}$	1.4170^{20}	bz, chl	B1[4], 483
7575	Heptanedioic acid or Pimelic acid $HO_2C(CH_2)_5CO_2H$	160.17	pr(w)	272^{100}sub, 212^{10}	106	1.329^{15}	....	w, al, eth	B2[4], 2003
7576	1,7-Heptanediol or Heptamethylene glycol $HO(CH_2)_7OH$	132.21		262, 151^{14}	22	$0.9569^{25/4}$	1.4520^{25}	w, al	B1[4], 2580
7577	2,4-Heptanediol, 3-methyl $C_3H_7CH(OH)CH(CH_3)CH(OH)CH_3$	146.23		115^{3}		$0.928^{20/4}$	1.4459^{20}	al	B1, 491
7578	2,4-Heptanedione $C_3H_7COCH_2COCH_3$	128.17		174, 70^{20}		$0.9411^{25/4}$			B1[4], 3698
7579	1-Heptanethiol $CH_3(CH_2)_5CH_2SH$	132.27		177	−43	$0.8427^{20/4}$	1.4521^{20}	al, eth	B1[4], 1738
7580	1,4,7-Heptanetriol $(HOCH_2CH_2CH_2)_2CHOH$	148.20		$230\text{-}2^{25}$, 146^{1}	−35	1.075^{18}	1.4725^{20}	w, al, ace	B1[4], 2787
7581	2,4,6-Heptanetrione or Diacetyl acetone $(CH_3COCH_2)_2CO$	142.16	lf	121^{10}	49	$1.0681^{40/40}$	1.4930^{20}	w, al, eth	B1[4], 3783
7582	Heptano amide $CH_3(CH_2)_5CONH_2$	129.20	nd(al)lf(w)	250-8	96	$0.852^{110/4}$	1.4217^{110}	w, al, eth	B2[4], 963
7583	Heptanoic acid or Enanthic acid $CH_3(CH_2)_5CO_2H$	130.19		223, 116^{11}	−7.5	$0.9200^{20/4}$	1.4170^{20}	al, eth, ace	B2[4], 958
7584	Heptanoic acid, 7-amino $H_2N(CH_2)_6CO_2H$	145.20	cr (w, MeOH-peth)	....	195		....	w, al	B4[3], 1467
7585	Heptanoic anhydride $(C_6H_{13}CO)_2O$	242.36		268-71, $164^{12.5}$	−12.4	$0.9321^{20/4}$	1.4335^{15}	al, eth	B2[4], 962
7586	Heptanoic acid, 2-bromo $CH_3(CH_2)_4CHBrCO_2H$	209.09		250d, 147^{12}	....	1.319^{15}	1.471^{18}	eth, ace	B2[4], 967
7587	Heptanoic acid, 7-bromo $Br(CH_2)_6CO_2H$	209.09	wh cr(dil al)	280	31			al, eth, ace, bz	B2[4], 968
7588	Heptanoic acid, butyl ester $CH_3(CH_2)_5CO_2C_4H_9$	186.30		226.2	−67.5	0.8638^{20}	1.4204^{20}	al, eth, ace, bz	B2[3], 768
7589	Heptanoic acid, iso-butyl ester $C_6H_{13}CO_2\text{-}i\text{-}C_4H_9$	186.30		208		0.8593^{20}		al, eth, ace, bz	B2[1], 145
7590	Heptanoic acid, ethyl ester or Ethyl heptanoate $CH_3(CH_2)_5CO_2C_2H_5$	158.24		187, 78^{14}	−66.1	$0.8817^{20/4}$	1.4100^{20}	al, eth	B2[4], 960
7591	Heptanoic acid, 7-fluoro $F(CH_2)_6CO_2H$	148.18		133^{10}		1.039^{20}	1.4207^{25}	....	B2[4], 964
7592	Heptanoic acid, heptyl ester $C_6H_{13}CO_2C_7H_{15}$	228.38		276-8	−33	$0.8649^{20/4}$	1.4320^{20}	al, eth	B2[4], 961
7593	Heptanoic acid, hexyl ester $C_6H_{13}CO_2C_6H_{13}$	214.35		261	−48	0.8611^{20}	1.429^{15}	al, eth, ace, bz	B2[3], 768
7594	Heptanoic acid, 7-iodo $I(CH_2)_6CO_2H$	256.09	lf(dil al)	....	49-51		....	al, eth, ace, bz	B2[4], 969

No.	Name, Synonyms, and Formula	Mol. wt.	Color, crystalline form, specific rotation and λ_{max} (log ϵ)	b.p. °C	m.p. °C	Density	n_D	Solubility	Ref.
7595	Heptanoic acid, methyl ester .... $C_6H_{13}CO_2CH_3$	144.22	....	172	−56	0.8815[20/4]	1.4152[20]	al, eth, ace	B2[4], 960
7596	Heptanoic acid, octyl ester or Octyl heptanoate......... $C_6H_{13}CO_2C_8H_{17}$	242.41	....	290	−22.5	0.8596[20]	1.4349[15]	al, eth, ace, bz	B2[3], 768
7597	Heptanoic acid, 6-oxo .... $CH_3CO(CH_2)_4CO_2H$	144.17	....	250-3[280], 135[1]	40.2	....	1.4306[25]	w, al, eth, ace	B3[4], 1598
7598	Heptanoic acid, pentyl ester .... $C_6H_{13}CO_2C_5H_{11}$	200.33	....	245.4	−50	0.8623[20]	1.4263[15]	al, eth, ace, bz	B2[4], 960
7599	Heptanoic acid, propyl ester or Propyl heptanoate .... $C_6H_{13}CO_2C_3H_7$	172.27	cr(peth)	207.9	−63.5	0.8641[15/4]	1.4183[15]	al, eth, ace, bz	B2[3], 767
7600	Heptanonitrile .... $CH_3(CH_2)_5CN$	111.19	....	183[765], 70-2[10]		0.8107[20/0]	1.4104[30]	eth, ace, bz, aa	B2[4], 963
7601	Heptanoyl chloride .... $CH_3(CH_2)_5COCl$	148.63	....	125.2	−83.8	0.9590[20]	1.4345[18]	eth, lig	B2[4], 963
7602	1-Heptanol .... $CH_3(CH_2)_5CH_2OH$	116.21	....	176	−34.1	0.8219[20/4]	1.4249[20]	al, eth	B1[4], 1731
7603	1-Heptanol, 7-chloro .... $Cl(CH_2)_7OH$	150.65	cr(peth or bz)	150[20]	11	0.9998[15/4]	1.4537[25]	al, peth	B1[4], 1738
7604	1-Heptanol, 7-fluoro .... $F(CH_2)_7OH$	134.20	....	98-9[12]		0.956[20/4]	1.4197[25]	al, eth	B1[4], 17369
7605	1-Heptanol, 4-methyl .... $C_3H_7CH(CH_3)CH_2CH_2OH$	130.23	....	183, 71.6[20]		0.8065[25/4]	1.4258[20]	al	B1[4], 1789
7606	1-Heptanol, 6-methyl .... $(CH_3)_2CH(CH_2)_4OH$	130.23	....	188[764], 95.8[20]	−106	0.8176[25/4]	1.4251[25]	al, eth	B1[4], 1782
7607	1-Heptanol, 1-phenyl .... $CH_3(CH_2)_5CH(C_6H_5)OH$	192.31	....	275, 153-5[18]		0.948	1.5024[21]	....	B0[1], 513
7608	2-Heptanol (d) .... $C_5H_{11}CHOH)CH_3$	116.21	$[\alpha]^{20}_D$ +11.4 (al)	160-2, 73[20]		0.8190[20/4]	1.4209[20]	al, eth	B1[4], 1740
7609	2-Heptanol (dl) .... $C_5H_{11}CH(OH)CH_3$	116.21	....	160, 66[20]		0.8167[20/4]	1.4210[20]	al, eth	B1[4], 1740
7610	2-Heptanol (l) .... $C_5H_{11}CH(OH)CH_3$	116.21	$[\alpha]^{12}_D$ -10.5	74[23]		0.8184[20/4]	1.4201[20]	al, eth	B1[3], 1687
7611	2-Heptanol, 6-amino-2-methyl, hydrochloride .... $CH_3CH(NH_2)(CH_2)_3C(OH)(CH_3)_2 \cdot HCl$	181.71	cr	....	154-5	....	....	w, al	B4[4], 1809
7612	2-Heptanol, 1-chloro .... $C_5H_{11}CH(OH)CH_2Cl$	150.65	....	93[13]		0.9885[20/4]	1.4499[20]	al, eth, ace	B1[4], 1741
7613	2-Heptanol, 2-methyl .... $C_5H_{11}C(OH)(CH_3)_2$	130.23	....	156, 66-8[15]		0.8142[20/4]	1.4250[20]	al, eth	B1[4], 1780
7614	2-Heptanol, 3-methyl .... $C_4H_9CH(CH_3)CH(OH)CH_3$	130.23	....	166.1, 68.1[20]		0.8177[25/4]	1.4199[25]	al, eth	B1[3], 1729
7615	3-Heptanol (d) .... $C_4H_9CH(OH)C_2H_5$	116.21	....	157[750], 66[18]	−70	0.8227[20/4]	1.4201[20]	al, eth	B1[4], 1741
7616	3-Heptanol, 2,6-dimethyl .... $(CH_3)_2CHCH(OH)CH_2CH_2CH(CH_3)_2$	144.26	....	175		0.8212[20/4]	1.4246[20]	....	B1[3], 1753
7617	3-Heptanol, 2-methyl (d) .... $C_4H_9CH(OH)CH(CH_3)_2$	130.23	$[\alpha]^{20}_D$ +27.7 (al)	72[12]		0.8235[20/4]	1.4265[20]	al, eth	B1[1], 209
7618	3-Heptanol, 2-methyl (dl) .... $C_4H_9CH(OH)CH(CH_3)_2$	130.23	....	167.2, 73[19]		0.8235[20/4]	1.4265[20]	al, eth	B1[4], 1781
7619	3-Heptanol, 2-methyl (l) .... $C_4H_9CH(OH)CH(CH_3)_2$	130.23	$[\alpha]_D$ -21.08	87[36]		0.8235[20/4]	1.4265[20]	al, eth	B1[1], 209
7620	3-Heptanol, 3-methyl .... $C_4H_9C(OH)(CH_3)CH_2CH_3$	130.23	....	163, 64-5[10]	−83	0.8282[20/4]	1.4279[20]	al, eth	B1[4], 1783
7621	4-Heptanol .... $(C_3H_7)_2CHOH$	116.21	....	161, 63.8[16]	−41.2	0.8183[20/4]	1.4205[20]	al, eth	B1[4], 1743
7622	4-Heptanol, 2,6-dimethyl .... $[(CH_3)_2CHCH_2]_2CHOH$	144.26	....	176-7		0.809[21/4]	1.4242[20]	al, eth	B1[4], 1810
7623	4-Heptanol, 4-ethyl .... $(C_3H_7)_2C(OH)C_2H_5$	144.26	....	182		0.8350[20]	1.4332[20]	al, eth	B1[4], 1809
7624	4-Heptanol, 3-methyl (dl) .... $C_2H_5CH(CH_3)CH(OH)C_3H_7$	130.23	....	164.7, 67.3[20]		0.8335[25/4]	1.4211[25]	al, eth	B1[3], 1730
7625	4-Heptanol, 4-methyl .... $(C_3H_7)_2C(CH_3)OH$	130.23	....	161, 61-3[12]	−82	0.8248[20/4]	1.4258[20]	al, eth	B1[4], 1789
7626	4-Heptanol, 4-propyl .... $(C_3H_7)_3COH$	158.29	....	190-2, 89-90[15]		0.8338[21/0]	1.4355[21]	al, eth, bz	B1[4], 1831
7627	2-Heptanone or Methyl hexyl ketone .... $CH_3(CH_2)_4COCH_3$	114.19	....	151.4, 111[21]	−35.5	0.8111[20/4]	1.4088[20]	al, eth	B1[4], 3318

No.	Name, Synonyms, and Formula	Mol. wt.	Color, crystalline form, specific rotation and λ_{max} (log ϵ)	b.p. °C	m.p. °C	Density	n_D	Solubility	Ref.
7628	2-Heptanone, 1-chloro . . . . C₅H₁₁COCH₂Cl	184.63	. . . .	83[16]	. . . . . .	0.802[20]	1.4371[20]	**al, eth**	B1[4], 3320
7629	2-Heptanone, 3-methyl . . . . C₄H₉CH(CH₃)COCH₃	128.22	. . . .	167	. . . . . .	0.8218[20/4]	1.4172[20]	al, eth, ace, bz	B1[3], 2878
7630	2-Heptanone, 6-methyl . . . . (CH₃)₂CH(CH₂)₃COCH₃	128.21	. . . .	167, 51-3[12]	. . . . . .	0.8151[20/4]	1.4162[20]	al, eth, ace, bz	B1[4], 3344
7631	2-Heptanone, 4-isopropyl . . . . i-C₃H₇CH(C₃H₇)CH₂COCH₃	156.27	. . . .	82-4[14]	. . . . . .	. . . . . .	. . . . . .	al, eth, ace, bz	B1[3], 2902
7632	3-Heptanone or Butyl ethyl ketone . . . . C₄H₉COC₂H₅	114.19	. . . .	147[765]	−39	0.8183[20/4]	1.4057[20]	**al, eth**	B1[4], 3321
7633	3-Heptanone, 6-dimethylamino-4,4-diphenyl (l) or l-Methadone . . . . (CH₃)₂NCH(CH₃)CH₂C(C₆H₅)₂COC₂H₅	309.46	[α]²⁰_D -32 (al)	. . . .	99-100	. . . . . .	. . . . . .	al	B14[3], 278
7634	3-Heptanone, 6-dimethylamino-4,4-diphenyl,hydrochloride (dl) or Physopeptone . . . . (CH₃)₂NCH(CH₃)CH₂C(C₆H₅)₂COC₂H₅.HCl	345.92	pl(al-eth)	. . . .	236	. . . . . .	. . . . . .	w, al, eth, chl	B14[3], 279
7635	3-Heptanone, 6-dimethylamino-4,4-diphenyl hydrochloride (l) . . . . (CH₃)₂NCH(CH₃)CH₂C(C₆H₅)₂COC₂H₅.HCl	345.92	[α]²⁰_D -169 (al, c=2.1)	. . . .	245-6	. . . . . .	. . . . . .	w, al, chl	B14[3], 278
7636	3-Heptanone, 2-methyl . . . . C₄H₉COCH(CH₃)₂	128.21	. . . .	158, 63-5[25]	. . . . . .	0.8163[20/4]	1.4115[20]	al, eth, ace	B1[4], 3343
7637	3-Heptanone, 6-methyl . . . . (CH₃)₂CHCH₂CH₂COC₂H₅	128.22	. . . .	163[734]	. . . . . .	0.8304[20]	1.4209[20]	al, eth, bz	B1[4], 3344
7638	4-Heptanone or Depropyl ketone . . . . (C₃H₇)₂CO	114.19	. . . .	144	−33	0.8174[20/4]	1.4069[20]	**al, eth**	B1[4], 3323
7639	4-Heptanone, 2,6-dimethyl or Isovalerone . . . . (i-C₄H₉)₂CO	142.24	. . . .	168, 60-1[18]	. . . . . .	0.8053[20/4]	1.412[20]	**al, eth**	B1[4], 3360
7640	4-Heptanone, 2-methyl . . . . C₃H₇COCH₂CH(CH₃)₂	128.22	. . . .	155[750]	. . . . . .	0.813[22/0]	. . . .	al, eth	B1[4], 3343
7641	Heptano phenone . . . . C₆H₁₃COC₆H₅	190.29	lf	283.3, 155[15]	16.4	0.9516[20/4]	1.5060[20]	al, eth, ace	B7[3], 1188
7642	Heptasiloxane, hexadecamethyl . . . . CH₃[Si(CH₃)₂O-]₆Si(CH₃)₃	533.06	. . . .	270, 165[20]	−78	0.9012[20/4]	1.3965[20]	bz, lig	B4[3], 1880
7643	1-Heptene . . . . CH₃(CH₂)₄CH=CH₂	98.19	. . . .	93.6	−119	0.6970[20/4]	1.3998[20]	al, eth	B1[4], 857
7644	1-Heptene, 1-chloro . . . . C₅H₁₁CH=CHCl	132.64	. . . .	155, 78-82[75]	. . . . . .	0.8948[20]	1.4380[20]	eth, ace, bz, chl	B1[3], 823
7645	1-Heptene, 2-chloro . . . . C₅H₁₁CCl=CH₂	132.64	. . . .	138[748]	. . . . . .	0.8895[20/4]	1.4349[20]	al, eth, chl	B1[3], 823
7646	1-Heptene, 2-methyl . . . . C₅H₁₁C(CH₃)=CH₂	112.21	. . . .	118.2	−90.1	0.2206[20/4]	1.4120[20/D]	. . . .	B1[4], 881
7647	2-Heptene (cis) . . . . CH₃(CH₂)₃CH=CHCH₃	98.19	. . . .	98.5	. . . . . .	0.708[20/4]	1.406[20]	al, eth, ace, bz, chl	B1[3], 824
7648	2-Heptene (trans) . . . . CH₃(CH₂)₃CH=CHCH₃	98.19	. . . .	98	−109.5	0.7012[20/4]	1.4045[20]	al, eth, ace, bz, chl	B1[4], 860
7649	2-Heptene, 4-chloro . . . . C₃H₇CHClCH=CHCH₃	132.64	. . . .	140-5, 49[21]	. . . . . .	0.879[18/4]	1.4430[23]	al, eth, chl	B1[3], 825
7650	2-Heptene, 6-chloro-2-methyl . . . . CH₃CHCl(CH₂)₂CH=C(CH₃)₂	147.67	. . . .	60-1[15]	. . . . . .	0.8931[18/4]	1.4458[18]	al, eth, ace, bz	B1[2], 200
7651	2-Heptene, 2-methyl . . . . C₄H₉CH=C(CH₃)₂	112.21	. . . .	122.6	. . . . . .	0.7241[20/0]	1.4170[20]	eth, bz	B1[4], 882
7652	2-Heptene, 2-methyl-6-methylamino . . . . CH₃NHCH(CH₃)CH₂CH₂CH=C(CH₃)₂	141.26	. . . .	176-8, 58-9[17]	. . . . . .	. . . . . .	. . . . . .	al, eth	B4[3], 467
7653	3-Heptene (cis) . . . . C₃H₇CH=CHC₂H₅	98.19	. . . .	95.8	. . . . . .	0.7030[20/4]	1.4059[20]	al, eth, ace, bz, chl	B1[4], 861
7654	3-Heptene (trans) . . . . C₃H₇CH=CHC₂H₅	98.19	. . . .	95.7	−136.6	0.6981[20/4]	1.4043[20]	al, eth, ace, bz, peth	B1[4], 861
7655	3-Heptene, 4-chloro . . . . C₃H₇CCl=CHC₂H₅	132.64	. . . .	139	. . . . . .	0.883[14]	1.437[14]	al, eth, chl	B1[3], 827
7656	3-Heptene, 4-propyl . . . . C₂H₅CH=C(C₃H₇)₂	140.27	. . . .	160.5	. . . . . .	0.7518[17/8]	1.4302[17/8]	. . . .	B1[4], 906
7657	1-Heptene-2-carboxaldehyde, 1-phenyl or Jasminaldehyde C₅H₁₁C(CHO)=CHC₆H₅	203.30	. . . .	174-5[20], 140[5]	. . . . . .	0.9718[20]	1.5381[20]	al, eth	B7[3], 1517
7658	1-Heptene-4-ol, 4-methyl . . . . C₃H₇COH(CH₃)CH₂CH=CH₂	128.22	. . . .	159-60	. . . . . .	0.8345[20/0]	1.4479[18]	al, eth	B1[3], 1945
7659	2-Heptene-1-ol (trans) . . . . C₄H₉CH=CHCH₂OH	114.19	. . . .	177-9, 75[10]	. . . . . .	0.8516[20]	1.4460[20]	al, ace	B1[3], 1936

No.	Name, Synonyms, and Formula	Mol. wt.	Color, crystalline form, specific rotation and λ_{max} (log ε)	b.p. °C	m.p. °C	Density	n_D	Solubility	Ref.
7660	2,Heptene-4-ol (dl) $C_3H_7CH(OH)CH=CHCH_3$	114.19	....	152-4, 64[14]	......	0.8445[20/4]	1.4373[20]	al, eth	B1[4], 2155
7661	3-Heptene-2-one (trans) $C_3H_{11}CH=CHCOCH_3$	112.17		62[15]	......	0.8496[20/4]	1.4436[20]	al, eth	B1[4], 3481
7662	5-Heptene-2-one, 6-methyl $(CH_3)_2C=CHCH_2CH_2COCH_3$	126.20		173, 58.6[10]	−67	0.8546[16/4]	1.4445[20]	al, eth	B1[4], 3493
7663	1-Heptene-3-one $C_3H_7C\equiv C-CH=CH_2$	94.16		110, 44[75]	......	0.7603[20/4]	1.4520[25]	al, ace, bz, peth	B1[4], 1097
7664	1-Heptene-4-one, 6,6-dimethyl $(CH_3)_3CC\equiv CCH_2CH=CH_2$	122.21		125, 68[100]	......	0.758[20]	1.4312[20]	al, ace, bz, peth	C55, 23329
7665	6-Heptene-4-one-3-ol, 3-ethyl $CH_2=CCHC\equiv CC(OH)(C_2H_5)_2$	138.21		62[4]	......	0.8875[20/4]	1.4800[20]	al, eth	B1[3], 2034
7666	Heptyl amine $CH_3(CH_2)_6CH_2NH_2$	115.22		156.9, 45.6[10]	−18	0.7754[20/4]	1.4251[20]	al, eth	B4[4], 734
7667	Heptyl ether or Diheptyl ether $(C_7H_{15})_2O$	214.40		258.5	......	0.8008[20/4]	1.4275[20]	al, eth	B1[4], 1733
7668	Heptyl methyl ether $C_7H_{15}OCH_3$	130.23		151	......	0.7869[15/15]	1.4073[20]	al, eth, ace	B1[3], 1682
7669	Heptyl nitrite $C_7H_{15}ONO$	145.20		155-8, 44[18]	......	0.8939[0/4]	1.4032[20]	eth	B1[4], 1735
7670	Heptyl phenyl ether $C_7H_{15}OC_6H_5$	192.31		267, 128-30[12]	......	0.9178[15/15]	1.4912[20]	al, eth, ace	B6[4], 560
7671	Heptyl sulfate $(C_7H_{15}O)_2SO_2$	294.46	cr(peth)	146.6[1.5]	13	0.9819[25/25]	1.4362[25]	....	B1[3], 1683
7672	Heptyl thiocyanate $C_7H_{15}SCN$	157.28	....	234[40], 130	.....	0.91[20]	....	al, eth	B2[4], 221
7673	1-Heptyne $CH_3(CH_2)_4C\equiv CH$	96.17	....	99.7, 6[10]	−81	0.7328[20/4]	1.4087[20]	al, eth, bz, chl, peth	B1[4], 1025
7674	1-Heptyne, 1-bromo $CH_3(CH_2)_4C\equiv CBr$	175.08		164[144], 69[15]	......	1.2120[11/4]	1.4670[11]	al, eth, ace, chl	B1[4], 997
7675	1-Heptyne, 1-chloro $CH_3(CH_2)_4C\equiv CCl$	130.62		141	......	0.9250[24/4]	1.4411[24]	al, eth	B1[3], 997
7676	1-Heptyne, 1-iodo $CH_3(CH_2)_4C\equiv CI$	222.07		90-2[17]	......	1.4701[19/4]	1.5123[19/D]	....	B1[3], 998
7677	2-Heptyne $CH_3(CH_2)_3C\equiv CCH_3$	96.17		112	......	0.7480[20/4]	1.4230[20]	al, eth, bz, chl, peth	B1[4], 1026
7678	2-Heptyne, 1-bromo $CH_3(CH_2)_3C\equiv CCH_2Br$	175.08		104[55], 84[20]	......		1.4878[25]	al, eth, ace	B1[4], 1026
7679	2-Heptyne, 1-chloro $CH_3(CH_2)_3C\equiv CCH_2Cl$	130.62		167, 73[24]	......		1.4570[25]		B1[4], 1026
7680	2-Heptyne, 7-chloro $Cl(CH_2)_4C\equiv CCH_3$	130.62		166	......		1.4507[25]	al, eth	B1[3], 998
7681	3-Heptyne $C_3H_7C\equiv CC_2H_5$	96.17		105-6	......	0.7527[20/4]	1.4220[20]	al, eth, bz, chl, peth	B1[4], 1027
7682	3-Heptyne, 1-chloro $C_3H_7C\equiv CCH_2CH_2Cl$	130.62		162, 90-93[20]	......		1.4520[25]		B1[3], 999
7683	3-Heptyne, 7-chloro $Cl(CH_2)_3C\equiv CC_2H_5$	130.62		164, 74-5[31]	......		1.4517[20]	al, eth	B1[3], 999
7684	3-Heptyne, 2,6-dimethyl $(CH_3)_2CHCH_2C\equiv CCH(CH_3)_2$	124.23		130-6	......	0.785[20/4]		eth, ace	B1[3], 1015
7685	3-Heptyne, 5,5-dimethyl $CH_3CH_2C(CH_3)_2C\equiv CCH_3$	124.23		69[100]	......	0.7610[20/4]	1.4360[20]	eth, ace	B1[3], 1015
7686	3-Heptyne, 5-ethyl-5-methyl $(C_2H_5)_2C(CH_3)C\equiv CC_2H_5$	138.25		88[100]	......	0.7714[20/4]	1.4386[20]	eth, ace	B1[3], 1021
7687	2-Heptyne-1-ol or Butyl propargyl alcohol $C_4H_9C\equiv CCH_2OH$	112.17		94[22]	......		1.4523[25]		B1[4], 2247
7688	Heroin or o,o-Diacetyl morphine $C_{21}H_{23}NO_5$	369.42	rh, $[\alpha]^{15/D}$ -166 (MeOH)	272-4[12]	173	1.56-1.61	....	bz, chl	B27[2], 151
7689	Hesperidin or Hesperitin-7-(6-o-L-rhamnopyransoyl)-β-D-glucoside $C_{28}H_{34}O_{15}$	610.58	wh nd(dil MeOH or aa), $[\alpha]^{20/D}$ -77.5 (py, c=4.2	....	261-3			al, py, aa	B18[4], 3219
7690	Hexacene or Anthraceno-2′:3′,2:3-anthracene $C_{26}H_{16}$	328.42	dk bl-gr cr(sub)	sub	Ca 380				B5[3], 2654

No.	Name, Synonyms, and Formula	Mol. wt.	Color, crystalline form, specific rotation and λ_{max} (log ϵ)	b.p. °C	m.p. °C	Density	n_D	Solubility	Ref.
7691	Hexachlorophene or Bis(2-hydroxy-3,5,6-trichlorophenyl) methane . . . . (2-HO-3,5,6-Cl$_3$C$_6$H)$_2$CH$_2$	406.91	nd(bz)	. . . .	166-7	. . . .			al, eth, ace, chl	B6[1], 5407
7692	Hexacosane or Cerane . . . . CH$_3$(CH$_2$)$_{24}$CH$_3$	366.72	mcl tcl or rh (bz) cr (eth)	412.2, 248.2[22]	56.4	0.7783[60], 0.8032[20/4]	1.4357[60]	bz, lig, chl	B1[4], 583
7693	Hexacosane, 13-dodecyl . . . . C$_{13}$H$_{27}$CH(C$_{12}$H$_{25}$)$_2$	353.03	. . . .	272[1]	13.7	0.8188[20/4]	1.4577[20]		B1[4], 600
7694	1-Hexacosanol or Ceryl alcohol Cerotin . . . CH$_3$(CH$_2$)$_{24}$CH$_2$OH	382.72	rh pl (dil al)	305[20] d	80	. . . .		al, eth	B1[4], 1912
7695	1,15-Hexadecadiyne . . . . HC≡C(CH$_2$)$_{12}$C≡CH	218.39	fl(al)	152-5[12]	44-5				B1[2], 249
7696	6,9-Hexadecadiyne . . . . CH$_3$(CH$_2$)$_5$C≡CCH$_2$C≡C(CH$_2$)$_4$CH$_3$	218.39	. . . .	169[15]		0.845[18/4]	1.4694[18]		B1[3], 1067
7697	6,10-Hexadecadiyne . . . . CH$_3$(CH$_2$)$_4$C≡CCH$_2$CH$_2$C≡C(CH$_2$)$_4$CH$_3$	218.39	. . . .	157[10]		0.7907[20/4]	1.4523[20]		B1[3], 1067
7698	Hexadecanal or Palmitaldehyde . . . . CH$_3$(CH$_2$)$_{14}$CHO	240.43	pl (eth) nd (peth)	200-2[29]	34			al, eth, ace, bz	B1[4], 3393
7699	Hexadecanal, dimethyl acetal . . . . CH$_3$(CH$_2$)$_{14}$CH(OCH$_3$)$_2$	286.50	. . . .	144[2]	10	0.8542[20]	1.4382[25]	al, eth, ace	B1[4], 3393
7700	Hexadecanal, oxime . . . . CH$_3$(CH$_2$)$_{14}$CH=NOH	255.45	nd(dil al)		88			al, chl	B1[3], 2923
7701	Hexadecane or Cetane . . . . CH$_3$(CH$_2$)$_{14}$CH$_3$	226.45	lf(ace)	287, 149[10]	18.2	0.7733[20/4]	1.4345	eth	B1[4], 537
7702	Hexadecane, 1-amino or Cetyl amine . . . . CH$_3$(CH$_2$)$_{14}$CH$_2$NH$_2$	241.47	lf	322.5, 144[2]	46.8	0.8129[20/4]	1.4496[20]	al, eth, ace, bz, chl	B4[4], 818
7703	Hexadecane, 1-bromo or Cetyl bromide . . . . CH$_3$(CH$_2$)$_{14}$CH$_2$Br	305.35	. . . .	336, 188[10]	17-9	0.9991[20/4]	1.4618[25]	. . . .	B1[4], 542
7704	Hexadecane, 1-chloro or Cetyl chloride . . . . CH$_3$(CH$_2$)$_{14}$CH$_2$Cl	260.90	. . . .	322, 177[10]	17.9	0.8652[20/4]	1.4505[20]		B1[4], 542
7705	Hexadecane, 1,16-dibromo . . . . Br(CH$_2$)$_{16}$Br	384.25	lf(al)	204[4]	56			chl	B1[2], 138
7706	Hexadecane, 6,11-dipentyl . . . . (C$_5$H$_{11}$)$_2$CH(CH$_2$)$_4$CH(C$_5$H$_{11}$)$_2$	366.71	. . . .	231[10]	−16.2	0.8072[20/4]	1.4502[20/D]		B1[4], 586
7707	Hexadecane, 1-fluoro or Cetyl fluoride . . . . CH$_3$(CH$_2$)$_{14}$CH$_2$F	244.44	. . . .	289, 152.6[10]	18	0.8321[20/4]	1.4317[20]	eth, lig	B1[4], 542
7708	Hexadecane, 1-iodo or Cetyl iodide . . . . CH$_3$(CH$_2$)$_{14}$CH$_2$I	352.35	lf(al)	357, 202[10]	24.7	1.1257[20/4]	1.4818[20]	eth, ace, bz, chl	B1[4], 543
7709	Hexadecane, 1-phenyl or Cetylbenzene . . . . CH$_3$(CH$_2$)$_{14}$CH$_2$C$_6$H$_5$	302.55	. . . .	237[16]	27	0.8560[20/4]	1.4814[20]	eth, bz, lig	B5[4], 1216
7710	Hexadecanedioic acid or Thapsic acid . . . . HO$_2$C(CH$_2$)$_{14}$CO$_2$H	286.42	pl(al, AcOEt)		126			al, ace	B2[4], 2162
7711	Hexadecanoic acid or Palmitic acid . . . . CH$_3$(CH$_2$)$_{14}$CO$_2$H	256.43	nd(al)	350, 267[100]	63	0.8527[62/4]	1.4335[60]	al, eth, ace, bz, chl	B2[4], 1157
7712	1-Hexadecanol or Cetyl alcohol . . . . CH$_3$(CH$_2$)$_{14}$CH$_2$OH	242.45	fl(AcOEt)	344, 190[15]	50	0.8176[50/4]	1.4283[20]	al, eth, ace, bz	B1[4], 1876
7713	2-Hexadecanol . . . . CH$_3$(CH$_2$)$_{13}$CHOHCH$_3$	242.44	. . . .	314	44	0.8338[20]	1.4479[20/D]	. . . .	B1[4], 1882
7714	1-Hexadecanethiol or Cetyl mercaptan . . . . CH$_3$(CH$_2$)$_{14}$CH$_2$SH	258.52	(lig)	123-8[0.5]	18-20			eth	B1[4], 1881
7715	1-Hexadecene or Cetene . . . . CH$_3$(CH$_2$)$_{13}$CH=CH$_2$	224.44	lf	284.4, 155[15]	4.1	0.7811[20/4]	1.4412[20]	al, eth, peth	B1[4], 927
7716	2-Hexadecenoic acid (form I) or Gaidic acid . . . . CH$_3$(CH$_2$)$_{12}$CH=CHCO$_2$H	254.42	lf(al)	. . . .	39			al	B2, 461
7717	2-Hexadecenoic acid (form II) or Δ-α,β-Hypogeic acid . . CH$_3$(CH$_2$)$_{12}$CH=CHCO$_2$H	254.42	fl(al)		49			al, eth, chl, peth	B2[4], 1629
7718	7-Hexadecenoic acid . . . . CH$_3$(CH$_2$)$_7$CH=CH(CH$_2$)$_5$CO$_2$H	254.42	cr	230[10]	33			al, eth	B2, 460
7719	Hexadecyl ether or Dicetyl ether . . . . (C$_{16}$H$_{33}$)$_2$O	466.89	lf(al)	270d	55	0.978[19]	. . . .	al, eth	B1[4], 1878
7720	Hexadecylpyridinium chloride . . . . (C$_5$H$_5$N$^+$)C$_{16}$H$_{33}$Cl$^-$	340.00	wh pw	. . . .	77-83			w, chl	B20[4], 2316
7721	1-Hexadecyne . . . . CH$_3$(CH$_2$)$_{13}$C≡CH	222.42	. . . .	284, 147.8[10]	15	0.7965[20/4]	1.4440[20]	bz	B1[4], 1073
7722	2-Hexadecyne . . . . CH$_3$(CH$_2$)$_{12}$C≡CCH$_3$	222.42	fl	160[15]	20	0.8039[20/4]	. . . .		B1[3], 1028
7723	7-Hexadecynoic acid . . . . C$_8$H$_{17}$C≡C(CH$_2$)$_5$CO$_2$H	252.40	nd(w) cr(al)	214[15]	47			al, eth	B2[3], 1474

No.	Name, Synonyms, and Formula	Mol. wt.	Color, crystalline form, specific rotation and λ_{max} (log ε)	b.p. °C	m.p. °C	Density	n_D	Solubility	Ref.
7724	2,4-Hexadienal or Sorbaldehyde .... CH₃CH=CHCH=CHCHO	96.14	....	173-4[756], 76[30]	......	0.898[20]	1.5384[20]	....	B1[4], 3545
7725	1,2-Hexadiene or Propylallene............... CH₃CH₂CH₂CH=C=CH₂	82.15	....	76		0.7149[20]	1.4282[20]	eth, chl	B1[4], 1011
7726	1,3-Hexadiene CH₃CH₂CH=CH-CH=CH₂	82.15	....	73		0.7050[20/4]	1.4380[20]	eth	B1[4], 1011
7727	1,3-Hexadiene, 3-chloro .... CH₃CH₂CH=CClCH=CH₂	116.59	....	68[117]		0.9390[20/4]	1.4770[20]	eth, chl	B1[4], 1012
7728	1,4-Hexadiene CH₃CH=CHCH₂CH=CH₂	82.15	....	65		0.7000[20/4]	1.4150[20]	eth	B1[4], 1013
7729	1,4-Hexadiene, 5-chloro-2-isopropyl .... CH₃CCl=CHCH₂C(i-C₃H₇)=CH₂	158.67	....	95[18]		0.9310[25/4]	1.4370[25]	ace, chl	B1[3], 1014
7730	1,4-Hexadiene, 3,3,6-trichloro .... ClCH₂CH=CHCCl₂CH=CH₂	185.48	....	100-3[4]		1.3036[20/4]	1.5585[20]	eth, chl	B1[3], 982
7731	1,5-Hexadiene or Biallyl .... H₂C=CHCH₂CH₂CH=CH₂	82.15	....	59.5	−141	0.6880[20/4]	1.4042[20]	al, eth, bz, chl	B1[4], 1013
7732	1,5-Hexadiene, 2,5-dimethyl .... CH₂=C(CH₃)CH₂CH₂C(CH₃)=CH₂	110.20	....	134	−75.6	0.7512[20]	1.4399[21]	ace, chl	B1[4], 1042
7733	1,5-Hexadiene, 2-methyl .... CH₂=C(CH₃)CH₂CH₂CH=CH₂	96.17	....	90-3	−128.8	0.7198[20/4]	1.4183[20/D]	....	B1[4], 1030
7734	1,5-Hexadiene, perchloro .... Cl₂C=CClCClCClCCl=CCl₂	426.60	cr(ace)	121[0.03]	49	1.905[52/4]	1.6012[51]	eth, ace, chl	B1[4], 1016
7735	1,5-Hexadeiene, 3,4-diol (d) or Divinyl glycol .... CH₂=CHCH(OH)CH(OH)CH=CH₂	114.15	[α][17/D] +94.8 (al)	198, 97[13]	−60	1.006[20/4]	1.4700[20]	al, eth, chl	B1[4], 2693
7736	1,5-Hexadiene-3,4-diol (dl) CH₂=CHCH(OH)CH(OH)CH=CH₂	114.15	hyg	90-1[8]	21.7	1.017[19/4]	1.4790[19]	al, eth, chl	B1[3], 2272
7737	1,5-Hexadiene-3,4-diol (meso) .... CH₂=CHCH(OH)CH(OH)CH=CH₂	114.15	hyg	100[14]	18	1.023[19/4]	1.4810[19]	w, al, eth, chl	B1[4], 2693
7738	2,4-Hexadiene CH₃CH=CHCH=CHCH₃	82.15	....	80	−79	0.7196[20/4]	1.4500[20]	al, eth, chl	B1[4], 1016
7739	2,4-Hexadiene, 6-chloro-2-methyl .... ClCH₂CH=CHCH=C(CH₃)₂	130.62	....	57[11]		0.9416[20/4]	1.5120[20]	ace, chl	B1[3], 1000
7740	2,4-Hexadiene, 1,3-dichloro .... CH₃CH=CHCCl=CHCH₂Cl	151.05	....	80-2[17]		1.1456[20/4]	1.5271[20]	bz, chl	B1[4], 1017
7741	2,4-Hexadiene, 2,5-dimethyl .... (CH₃)₂C=CHCH=C(CH₃)₂	110.20	....	134, 75[100]	14-5	0.7625[20/4]	1.4785[20]	al, eth, bz, chl	B1[4], 1043
7742	2,4-Hexadiene, 1,3,4,6-tetrachloro .... ClCH₂CH=CClCCl=CHCH₂Cl	219.93	....	84-9[2]		1.4013[20/4]	1.5465[20]	chl, MeOH	B1[3], 987
7743	2,4-Hexadienedioic acid or cis-Muconic acid .... HO₂CCH=CHCH=CHCO₂H	142.11	....	194-5		....		aa	B2[4], 2297
7744	2,4-Hexadienedioic acid (trans) or trans-Muconic acid ... HO₂CCH=CHCH=CHCO₂H	142.11	nd(al)	320	305d	....		aa, AcOEt	B2[4], 2298
7745	2,4-Hexadienoic acid or Sorbic acid .... CH₃CH=CHCH=CHCO₂H	112.14	nd(dil al)	228d, 153[50]	134.5	1.204[19/4]		al, eth	B2[4], 1701
7746	2,4-Hexadien-1-ol or Sorbyl alcohol .... CH₃CH=CHCH=CHCH₂OH	98.15	nd	76[12]	30-1	0.8967[23/4]	1.4981[20]	al, eth	B1[4], 2239
7747	3,5-Hexadien-2-ol CH₂=CHCH=CHCH(OH)CH₃	98.15	....	77-8[26]		0.8678[20]	1.4816[20]	al	B1[4], 2237
7748	3,5-Hexadiene, 2-one-6-phenyl or Cinnamylidene acetone C₆H₅CH=CHCH=CHCOCH₃	172.23	wh lf (eth)	170-2[15]	68	....		al, eth, ace, bz, chl	B7[3], 1656
7749	1,5-Hexadiene-3-one or Divinyl acetylene .... CH₂=CHC≡CCH=CH₂	78.12	....	85	−88	0.7851[20/4]	1.5035[20]	bz	B1[4], 1120
7750	1,5-Hexadiene, 3-one-2,5-dimethyl .... CH₂=C(CH₃)C≡CC(CH₃)=CH₂	106.17	ye	123		0.7863[25/4]	1.4845[20]	bz, chl	B1[4], 1124
7751	3,5-Hexadiene, 1-yne .... CH₂=CHCH=CH₂C≡CH	78.12	....	83-4, 32[100]		0.7806[20/4]	1.5095[20]	bz	B1[4], 1120
7752	1,4-Hexadiyne CH₃C≡CCH₂C≡CH	78.12	....	78-83	<-80	0.825[0/4]		bz, chl	B1[3], 1057
7753	1,5-Hexadiyne or Dipropargyl .... HC≡CCH₂CH₂C≡CH	78.12	....	86, 20[46]	−6	0.8049[20/4]	1.4380[23]	al, eth, ace, bz	B1[4], 1118
7754	1,5-Hexadiyne, 1,6-diamino H₂NC≡CCH₂CH₂C≡CNH₂	108.13	....		104-5	....		bz	B4[4], 1399
7755	2,4-Hexadiyne CH₃C≡C-C≡C-CH₃	78.12	pr(sub)	129-30	68.5	....		al, eth	B1[4], 1119
7756	Hexaethyl tetraphosphate .... [(C₂H₅O)₂P(O)O]₂PO	506.26	hyg	>150d	ca-40	1.2917[27/4]	1.4273[27]	al, ace, bz	B1[3], 1331

No.	Name, Synonyms, and Formula	Mol. wt.	Color, crystalline form, specific rotation and λ_{max} (log ϵ)	b.p. °C	m.p. °C	Density	n_D	Solubility	Ref.
7757	Hexamethylene diamine or 1,6-Diamino hexane $H_2N(CH_2)_6NH_2$	116.21	rh bi pym pl	204-5, 100[20]	41-2	....	....	w, al, bz	B4[4], 1320
7758	Hexamethylene tetramine or Hexamin. Urotropine $C_6H_{12}N_4$	140.19	rh (al)	sub	285-95 sub	1.331[-5]	....	w, al, ace, chl	B26[2], 200
7759	Hexanal or Caproaldehyde $CH_3(CH_2)_4CHO$	100.16	....	128, 28[12]	-56	0.8139[20/4]	1.4039[20]	al, eth, ace, bz	B1[4], 3296
7760	Hexane C_6H_{14}	86.18	....	69	-95	0.6603[20/4]	1.3751[20]	al, eth, chl	B1[4], 338
7761	Hexane, 1-amino or n-Hexyl amine $CH_3(CH_2)_4CH_2NH_2$	101.19	....	130	-19	0.7660[20]	1.4180[20]	**al, eth**	B4[4], 709
7762	Hexane, 2-amino (d) $CH_3(CH_2)_3CHNH_2CH_3$	101.19	....	114-5, 64[90]	....	0.755[27/4]	....	al, eth	B4[3], 361
7763	Hexane, 2-amino (dl) $CH_3(CH_2)_3CHNH_2CH_3$	101.19	....	117-8	-19	0.7534[20/0]	1.4080[25]	al, eth	B4[4], 721
7764	Hexane, 2-amino-4-methyl $CH_3CH_2CH(CH_3)CH_2CH(NH_2)CH_3$	115.22	....	130-5	....	0.7655[20]	1.4150[25]	al, eth, chl	B4[4], 747
7765	Hexane, 1-bromo or n-Hexyl bromide $CH_3(CH_2)_4CH_2Br$	165.08	....	155.3, 41[10]	-84.7	1.1744[20/4]	1.4478[20]	**al, eth**, ace, chl	B1[4], 352
7766	Hexane, 1-bromo-6-fluoro $F(CH_2)_6Br$	183.07	....	67-8[11]	....	1.293[20/4]	1.4435[25]	al, eth, ace, chl	B1[4], 353
7767	Hexane, 2-bromo or sec-Hexyl bromide $CH_3(CH_2)_3CHBrCH_3$	165.08	....	144[749], 78[90]	....	1.1658[20/4]	1.4832[25]	al, eth, ace, chl	B1[4], 353
7768	Hexane, 3-bromo $C_3H_7CHBrC_2H_5$	165.08	....	141-3	....	1.1799[20/4]	1.4472[20]	al, eth, ace, chl	B1[4], 353
7769	Hexane, 1-chloro or n-Hexyl chloride $CH_3(CH_2)_4CH_2Cl$	120.62	....	134.5	-94	0.8785[20/4]	1.4199[20]	al, eth, ace, bz, chl	B1[4], 349
7770	Hexane, 1-chloro-3-ethyl (d) $C_3H_7CH(C_2H_5)CH_2CH_2Cl$	148.68	$[\alpha]^{27}_D$ + 1.15	85[40]	....	0.879[21/4]	1.4335[25]	eth	B1[4], 432
7771	Hexane, 1-chloro-6-fluoro $F(CH_2)_6Cl$	138.61	....	167[740], 62[15]	....	1.015[20/4]	1.4168[25]	al, eth, chl	B1[4], 350
7772	Hexane, 1-chloro-3-methyl $C_3H_7CH(CH_3)CH_2CH_2Cl$	134.65	....	150-2[758]	....	0.8766[20/4]	1.4274[20]	al, eth, chl	B1[2], 119
7773	Hexane, 2-chloro or sec-Hexyl chloride $CH_3(CH_2)_3CHClCH_3$	120.62	....	122-3, 61[100]	....	0.8694[21/4]	1.4142[22]	al, eth, ace, bz, chl	B1[4], 349
7774	Hexane, 2-chloro-2,5-dimethyl $(CH_3)_2CHCH_2CH_2CCl(CH_3)_2$	148.68	....	86[100]	....	0.8476[18/4]	1.4232[20]	al, eth, ace, bz	B1[4], 434
7775	Hexane, 2-chloro-2-methyl $C_4H_9CCl(CH_3)_2$	134.65	....	135d, 59.5[52]	....	0.8635[20/4]	1.4200[20]	al, eth, chl	B1[4], 398
7776	Hexane-2-chloro-5-methyl $(CH_3)_2CHCH_2CH_2CHClCH_3$	134.65	....	138[735]d	....	0.863[20/4]	....	al, eth, chl	B1[3], 436
7777	Hexane-3-chloro $C_3H_7CHClC_2H_5$	120.62	....	123, 60[95]	....	0.8700[20/20]	1.4163[20]	al, eth, ace, bz, chl	B1[4], 349
7778	Hexane-3-chloro-2,3-dimethyl $C_3H_7CCl(CH_3)CH(CH_3)_2$	148.68	....	41-3[12]	....	0.8869[20/4]	1.4333[25]	al	B1[3], 481
7779	Hexane-3-chloro-3-ethyl $C_3H_7CCl(C_2H_5)_2$	148.68	....	155d, 62-3[24]	....	0.9018	1.4358[20]	eth, chl	B1[3], 479
7780	Hexane-3-chloro-3-methyl $C_3H_7CCl(CH_3)C_2H_5$	134.65	....	135	....	0.8787[20/4]	1.4250[20]	al, eth, chl	B1[3], 438
7781	Hexane-3-chloro-2,2,3-trimethyl $C_3H_7CCl(CH_3)C(CH_3)_3$	162.71	....	64-5[13]	....	0.9010[20/4]	1.4465[20]	al, eth, chl	B1[3], 515
7782	Hexane-1,6-diamino or Hexamethylene diamine $H_2N(CH_2)_6NH_2$	116.21	rh bipym pl	204-5, 100[20]	41-2	....	....	w, al, bz	B4[4], 1320
7783	Hexane-1,6-diamino, dihydrochloride $H_2N(CH_2)_6NH_2.2HCl$	189.13	nd(al-eth)	....	248-50	....	....	w	B4[4], 1320
7784	Hexane-2,5-diamino $CH_3CH(NH_2)CH_2CH_2CH(NH_2)CH_3$	116.21	....	175	....	....	....	w, al, eth	B4, 269
7785	Hexane-1,2-dibromo $C_4H_9CHBrCH_2Br$	243.98	....	103-5[36]	....	1.5774[20/4]	1.5024[20]	al, **eth**, bz, chl	B1[3], 392
7786	Hexane-1,6-dibromo $Br(CH_2)_6Br$	243.97	....	245-6, 110[12]	-2.3	1.5948[15]	1.5037[20]	eth, ace	B1[4], 354
7787	Hexane-2,5-dibromo (dl) $CH_3CHBrCH_2CH_2CHBrCH_3$	243.97	....	108-9[30]	-44.64	1.5788[20]	1.5007[20]	eth, ace, chl	B1[4], 354
7788	Hexane-1,2-dichloro $C_4H_9CHClCH_2Cl$	155.07	....	172-4, 73-4[0.03]	....	1.085[15]	....	eth, chl	B1[4], 350
7789	Hexane-1,6-dichloro $Cl(CH_2)_6Cl$	155.07	....	203-5, 94[22]	....	1.0677[20/4]	1.4572[20]	eth, chl	B1[4], 350

No.	Name, Synonyms, and Formula	Mol. wt.	Color, crystalline form, specific rotation and λ_{max} (log ε)	b.p. °C	m.p. °C	Density	n_D	Solubility	Ref.
7790	Hexane-2,2-dichloro C₄H₉CCl₂CH₃	155.07	....	68[49]	......	1.0150[25/4]	1.4353[25]	eth, chl	B1[3], 390
7791	Hexane-2,3-dichloro C₃H₇CHClCHClCH₃	155.07	....	162-5	......	1.0527[11]	....	eth, chl	B1[4], 350
7792	Hexane-2,5-dichloro (dl) CH₃CHClCH₂CH₂CHClCH₃	155.07	....	177[751], 106[71]	fp-38.4	1.0474[20]	1.4491[20]	chl	B1[4], 350
7793	Hexane-2,5-dichloro (meso) CH₃CHClCH₂CH₂CHClCH₃	155.07	....	178[752], 109[99]	19.9	1.0474[20/4]	1.4484[20]	chl	B1[4], 350
7794	Hexane-2,5-dichloro-2,5-dimethyl (CH₃)₂CClCH₂CH₂CCl(CH₃)₂	183.12	lf, nd		67-8	0.9543[70]	....	al, eth, bz, chl	B1[4], 435
7795	Hexane-3,4-dichloro (dl) CH₃CH₂CHClCHClCH₂CH₃	155.07	....	167.7, 62[20]	......	1.0617[20]	1.4541[20]	ace, chl	B1[4], 351
7796	Hexane-3,4-dichloro-3,4-dimethyl C₂H₅CCl(CH₃)CCl(CH₃)C₂H₅	183.12	....	165, 114-5[18]	......			chl	B1[4], 436
7797	Hexane-1,6-diiodo I(CH₂)₆I	337.97	nd	141-2[10]	10	2.03[22/4]	1.585[20]	al, eth	B1[4], 356
7798	Hexane-2,2-dimethyl C₄H₉C(CH₃)₃	114.23	....	106.8	−121	0.6953[20/4]	1.3935[20]	al, eth, ace, bz, chl	B1[4], 432
7799	3-Hexane-2,3-dimethyl-(dl) C₃H₇CH(CH₃)CH(CH₃)₂	114.23	....	115.6, 9.9[0]	......	0.7121[20/4]	1.4011[20]	al, eth, ace, bz, chl, lig	B1[4], 432
7800	Hexane-2,3-dimethyl (l) C₃H₇CH(CH₃)CH(CH₃)₂	114.23	[α]²⁵/D -0.92	113	......			al, eth, ace, bz, lig, chl	B1[3], 482
7801	Hexane, 2,4-dimethyl (d) C₂H₅CH(CH₃)CH₂CH(CH₃)₂	114.23	[α]³⁰/D +2.99	111	......	0.696[20/4]	1.3810[20]	al, eth, ace, bz, chl, lig	B1[3], 483
7802	Hexane, 2,4-dimethyl (dl) C₂H₅CH(CH₃)CH₂CH(CH₃)₂	114.23	....	109.4, 5.2[10]	......	0.7004[20/4]	1.3953[20]	al, eth, ace, bz, lig, chl	B1[4], 433
7803	Hexane, 2,4-dimethyl (l) C₂H₅CH(CH₃)CH₂CH(CH₃)₂	114.23	[α]²¹/D -10.85	110	......	0.703[21/4]	....	al, eth, ace, bz, chl, lig	B1[3], 483
7804	Hexane, 2,5-dimethyl (CH₃)₂CHCH₂CH₂CH(CH₃)₂	114.23	....	109, 5.3[10]	−91.2	0.6935[20/4]	1.3925[20]	al, eth, ace, bz, lig, chl	B1[4], 434
7805	Hexane, 3,3-dimethyl C₃H₇C(CH₃)₂C₂H₅	114.23	....	112, 6.1[10]	−126.1	0.7100[20/4]	1.4001[20]	al, eth, ace, bz	B1[4], 435
7806	Hexane, 3,4-dimethyl C₂H₅CH(CH₃)CH(CH₃)C₂H₅	114.23	....	117.7, 11.3[10]	......	0.7200[20/4]	1.4046[20]	al, eth, ace, bz, chl, lig	B1[4], 436
7807	Hexane, 1,6-dinitro O₂N(CH₂)₆NO₂	176.17	cr(MeOH)	100-3[0.3]	37.5			aa	B1[4], 357
7808	Hexane, 3-ethyl C₃H₇CH(C₂H₅)₂	114.23	....	118.5, 12.8[10]	......	0.7136[20/4]	1.4018[20]	al, eth, ace, bz, lig	B1[4], 431
7809	Hexane, 3-ethyl-2-methyl (CH₃)₂CHCH(C₂H₅)CH₂CH₂CH₃	128.26	....	138	......		1.4106[20/D]		B1[4], 459
7810	Hexane, 3-ethyl-3-methyl (CH₃CH₂)₂C(CH₃)C₃H₇	128.26	....	140.6	......		1.4140[20/D]		B1[4], 459
7811	Hexane, 3-ethyl-4-methyl (CH₃CH₂)₂CHCH(CH₃)C₂H₅	128.26	....	140.4	......		1.4134[20/D]		B1[4], 459
7812	Hexane, 4-ethyl-2-methyl (C₂H₅)₂CHCH₂CH(CH₃)₂	128.26	....	133.8	......		1.4063[20/D]		B1[4], 459
7813	Hexane, 1-fluoro or n-Hexyl fluoride CH₃(CH₂)₄CH₂F	104.17	....	91.5	−103	0.7995[20]	1.3738[20]	eth, bz	B1[4], 348
7814	Hexane, 3,4-bis(4-hydroxy-3-methylphenyl) or Dimethyl-lexesterol [4-HO-3-CH₃C₆H₃CH(C₂H₅)₂-]₂	298.43	cr(dil aa)	....	145				B6[3], 5541
7815	Hexane, 2,4-bis(4-hydroxyphenyl)-3-ethyl or Benzestrol 4-HOC₆H₄CH(C₂H₃)CH(C₂H₅)CH(CH₃)(C₆H₄OH-4)	298.43	cr(al)	....	162-6			al, eth, ace, aa	B6[3], 5538
7816	Hexane, 1-iodo or n-Hexyl iodide CH₃(CH₂)₄CH₂I	212.08	....	181.3, 58.2[10]	−75	1.4397[20/4]	1.4929[20]	....	B1[4], 355
7817	Hexane, 2-iodo (l) CH₃CHIC₄H₉	212.07	[α]¹⁷/D -38.35	90-1[70]. 45[9]	......	1.4354[17/4]	1.4878[25]	ace, chl	B1[3], 395
7818	Hexane, 2-methyl or Isoheptane CH₃(CH₂)₃CH(CH₃)₂	100.21	....	90	−118.3	0.6787[20/4]	1.3848[20]	al, eth, ace, bz, lig, chl	B1[4], 397
7819	Hexane, 3-methyl (d) C₃H₇CH(CH₃)C₂H₅	100.23	[α]²⁰/D +9.5	92	−119	0.6860[20/4]	1.3887[20]	al, eth, ace, bz, chl, lig	B1[4], 400
7820	Hexane, 3-methyl (dl) C₃H₇CH(CH₃)C₂H₅	100.21	....	92	ca−173	0.6872[20]	1.3885[20]	al, eth, ace, bz, lig, chl	B1[4], 400
7821	Hexane, 3-methyl (l) C₃H₇CH(CH₃)C₂H₅	100.21	[α]²¹/D -7.75	92	......	0.687[21/4]	1.3854[25]	al, eth, ace, bz, chl, lig	B1[3], 440
7822	Hexane, 1-nitro CH₃(CH₂)₄CH₂NO₂	131.17	....	193-4[765], 84[21]	......	0.9396[20/4]	1.4270[20]	al, eth, ace, bz	B1[4], 356
7823	Hexane, 1,1,1,2,2,pentachloro CH₃(CH₂)₃CCl₂CCl₃	258.40	....	129-31[10]	......	1.370[25]	1.4872[25]	eth, chl	B1[3], 390

PHYSICAL CONSTANTS OF ORGANIC COMPOUNDS (Continued)

No.	Name, Synonyms, and Formula	Mol. wt.	Color, crystalline form, specific rotation and λ_{max} (log ε)	b.p. °C	m.p. °C	Density	n_D	Solubility	Ref.
7824	Hexane, perfluoro .... C_6F_{14}	338.04	....	57.11	−87.1	1.6995[20/4]	1.2515[20]	eth, bz, chl	B1[4], 348
7825	Hexane, 1-phenyl or Hexylbenzene .... $CH_3(CH_2)_4CH_2C_6H_5$	162.28	....	227	−62	0.8613[20]	1.4900[20]	eth, bz, peth	B5[4], 1115
7826	Hexane, 2-phenyl .... $CH_3(CH_2)_3CH(C_6H_5)CH_3$	162.27	....	208	....	0.869[15/4]	1.492[15]	....	B5[4], 1116
7827	Hexane, 3-phenyl .... $CH_3CH_2CH(C_6H_5)C_3H_7$	162.27	....	209-12	....	0.8254[25/20]	1.4859[20/D]	....	B5[4], 1117
7828	Hexane, 1,1,2,2,-tetrachloro .... $CH_3(CH_2)CCl_2CHCl_2$	223.96	....	99-101[14]	....	1.3096[25/4]	1.488[25]	ace, bz, chl	B1[3], 390
7829	Hexane, 2,2,3-trimethyl .... $C_3H_7CH(CH_3)C(CH_3)_3$	128.26	....	131.7	....	....	1.4100[20]	....	B1[4], 459
7830	Hexane, 2,2,4-trimethyl .... $C_2H_5CH(CH_3)CH_2C(CH_3)_3$	128.26	....	126.5	120	0.711[20/4]	1.40328[20/D]	....	B1[4], 459
7831	Hexane, 2,2-5-trimethyl .... $(CH_3)_2CHCH_2CH_2C(CH_3)_3$	128.26	....	124, 16.2[10]	−105.8	0.7072[20/4]	1.3997[20]	al, eth, ace, bz, lig	B1[4], 460
7832	Hexane, 2,3,3-trimethyl .... $C_3H_7C(CH_3)_2CH(CH_3)_2$	128.26	....	137.7	−116.8	....	1.4141[20/D]	....	B1[4], 461
7833	Hexane, 2,3,4-trimethyl .... $C_2H_5CH(CH_3)CH(CH_3)CH(CH_3)_2$	128.26	....	139	....	....	1.4144[20/D]	....	B1[4], 461
7834	Hexane, 2,3,5-trimethyl .... $CH_3CH(CH_3)CH(CH_3)CH_2CH(CH_3)_2$	128.26	....	131.3	−s127.9	0.7818[20/4]	1.4051[20/D]	....	B1[4], 461
7835	Hexane, 2,4,4-trimethyl .... $C_2H_5C(CH_3)_2CH_2CH(CH_3)_2$	128.26	....	126.5	−123.4	0.711[20/4]	1.40328[20/D]	....	B1[4], 461
7836	Hexane, 3,3,-4-trimethyl .... $CH_3CH_2C(CH_3)_2CH(CH_3)CH_2CH_3$	128.26	....	140.5	−101.2	....	1.4178[20/D]	....	B1[4], 462
7837	1,6-Hexanedial or Adipaldehyde .... $OHC(CH_2)_4CHO$	114.25	....	92-4[9]	ms;8	1.003[19/4]	1.4350[20]	al, eth, bz, aa	B1[4], 3686
7838	Hexanedioic acid or Adipic acid .... $HOOC(CH_2)_4COOH$	146.14	mcl pr(w, ace-lig)	265[100], 205[10]	153	1.360[25/4]	....	al, eth	B2[4], 1956
7839	1,3-Hexanediol, 2-ethyl .... $C_3H_7CH(OH)CH(C_2H_5)CH_2OH$	146.23	....	244	−40	0.9325[22/4]	1.4497[20]	al, eth	B1[4], 2597
7840	1,6-Hexanediol or Hexamethylene glycol .... $HO(CH_2)_6OH$	118.18	nd(w)	250, 132[9]	43(59)	....	....	w, al, ace	B1[4], 2556
7841	2,3-Hexanediol .... $C_3H_7CH(OH)CH(OH)CH_3$	118.18	cr	204-6, 102[0.8]	....	0.9900[15]	1.4510[15]	w, al, eth	B1[4], 2561
7842	2,5-Hexanediol .... $CH_3CH(OH)CH_2CH_2CH(OH)CH_3$	118.18	cr(eth)	216-8[750], 85-7[1]	43	0 9610[20/4]	1.4475[20]	w, al, eth	B1[4], 2562
7843	3,4-Hexanediol, 3,4-diethyl .... $(C_2H_5)_2C(OH)C(OH)(C_2H_5)_2$	174.29	cr(eth)	230, 112[10]	28	0.9630[13/25]	1.467[13]	al, eth	B1[4], 2621
7844	2,5-Hexanediol, 2,5-dimethyl .... $(CH_3)_2C(OH)CH_2CH_2C(OH)(CH_3)_2$	146.23	pr (AcOEt) fl (peth)	214, 118[15]	92	0.898[20]	....	w, al, bz, chl	B1[4], 2600
7845	2,3-Hexanedione, 3-oxime .... $C_3H_7C(=NOH)COCH_3$	129.16	cr(al)	....	60	....	....	al	B1[3], 3127
7846	2,5-Hexanedione or Acetonyl acetone .... $CH_3COCH_2CH_2COCH_3$	114.15	....	194[754], 89[25]	−5.5	0.9737[20/4]	1.4421[20]	w, al, eth, ace, bz	B1[4], 3688
7847	2,5-Hexanedione, dioxime .... $CH_3C(=NOH)CH_2CH_2C(=NOH)CH_3$	144.18	pl(bz)	....	137	....	....	al, eth	B1[3], 3130
7848	2,5-Hexanedione, 3-hydroxy .... $CH_3COCH_2CH(OH)COCH_3$	130.15	....	62-7[0.5]	....	....	1.4497[25]	eth, bz	B1[3], 3317
7849	3,4-Hexanedione, 2,2,5,5-tetramethyl or Di-*tert*-butyl glyoxal .... $(CH_3)_3CCOCOC(CH_3)_3$	170.25	....	168[745]	−2	0.8776[20]	1.4157[20]	eth	B1[4], 3726
7850	1,2,3,4,5,6-Hexanehexol or Dulcitol .... $HOCH_2(CHOH)_4CH_2OH$	182.18	mcl pr	257-80[1]	189	1.466[15]	....	w	B1[4], 2844
7851	1,3,4,5-Hexanetetrol or Digitoxit .... $CH_3(CHOH)_4CH_2OH$	150.18	pr, [α][15/D] −86.2	....	88	....	....	al	B1[2], 603
7852	1-Hexanethiol or Hexyl mercapton .... $CH_3(CH_2)_4CH_2SH$	118.24	....	151	−81	0.8424[20/4]	1.4496[20]	al, eth	B1[4], 1705
7853	2-Hexanethiol or *sec*-Hexyl mercaptan .... $CH_3(CH_2)_3CH(SH)CH_3$	118.24	....	142, 60-6[50]	−147	0.8345[20/4]	1.4451[20]	al, eth, bz	B1[4], 1711
7854	3-Hexanethiol .... $C_3H_7CH(SH)C_2H_5$	118.24	....	57[25]	....	0.9206[20]	1.4496[20]	eth	B1[4], 1713
7855	1,2,3-Hexanetriol (threo) .... $C_3H_7CH(OH)CH(OH)CH_2OH$	134.18	....	130[0.05]	64-5	1.089[26]	1.472[26]	w, al	B1[4], 2784
7856	1,2,4-Hexanetriol .... $C_2H_5CH(OH)CH_2CH(OH)CH_2OH$	134.18	....	190-2[30]	....	....	....	w, al	B1, 521

No.	Name, Synonyms, and Formula	Mol. wt.	Color, crystalline form, specific rotation and λ_{max} (log ϵ)	b.p. °C	m.p. °C	Density	n_D	Solubility	Ref.
7857	1,2,5-Hexanetriol .... $CH_2CH(OH)CH_2CH_2CH(OH)CH_2OH$	134.18	....	181[10]	......	1.1012[20/4]	....	w, al	B1[4],2784
7858	2,3,4-Hexanetriol .... $CH_3CH_2(CHOH)_3CH_3$	134.18	....	256-7, 155-6[20]	......	....	....	w, al	B1[3], 2349
7859	Hexanoic acid or Caproic acid............... $CH_3(CH_2)_4CO_2H$	116.16	....	205	−2	0.9274[20/4]	1.4163[20]	al, eth	B2[4], 917
7860	Hexanoic acid, 2-acetyl, ethyl ester .... $C_4H_9CH(COCH_3)CO_2C_2H_5$	186.25	....	219-24, 104[12]	......	0.9523[20/4]	1.4301[20]	eth, ace	B3[4], 1616
7861	Hexanoic acid, 2,4-dioxo or Propionyl pyravic acid ...... $CH_3CH_2COCH_2COCO_2H$	144.14	cr(al w + l)	....	83	....	....	w, al, eth	B3[3], 1334
7862	Hexanoic acid, 2,4-dioxo, ethyl ester .... $CH_3CH_2COCH_2COCO_2C_2H_5$	172.19	....	163-5, 108-11[11]	......	....	....	al, eth, ace	B3[4], 1779
7863	Hexanoic acid, 2-ethyl .... $C_4H_9CH(C_2H_5)CO_2H$	144.22	....	228[755], 120[13]	......	0.9031[25/4]	1.4241[20]	eth	B2[4], 1003
7864	Hexanoic acid, 2-methyl (dl) or 2-Methyl caproic acid .... $C_4H_9CH(CH_3)CO_2H$	130.19	....	215-6, 100[12]	......	0.9612[20/4]	1.4193[20]	al, eth, ace, bz, chl	B24, 969
7865	Hexanoic acid, 4-methyl (dl) or 4-Methyl caproic acid .... $C_2H_5CH(CH_3)(CH_2)_2CO_2H$	130.19	....	217-8, 85[2]	−80	0.9215[20/4]	1.4211[20]	al, eth, ace, bz	B2[4], 974
7866	Hexanoic acid, 5-methyl or 5-Methyl caproic acid....... $(CH_3)_2CH(CH_2)_3CO_2H$	130.19	....	216	<−25	0.9138[21/4]	1.4220[20]	al, eth, ace, bz	B2[4], 970
7867	Hexanoic acid, 5-oxo .... $CH_3CO(CH_2)_3CO_2H$	130.14	....	155[12]	13-14	....	1.4445[20]	w, al, eth	B3[4], 1583
7868	1-Hexanol or n-Hexyl alcohol .... $CH_3(CH_2)_5CH_2OH$	102.18	....	158	−46.7	0.8136[20/4]	1.4178[20]	al, eth, ace, bz, chl	B1[4], 1694
7869	1-Hexanol, 6-chloro .... $Cl(CH_2)_6OH$	136.22	....	107[11]	......	1.0241[20/4]	1.4550[20]	al, eth	D1[4], 1704
7870	1-Hexanol, 3,5-dimethoxy .... $CH_3CH(OCH_3)CH_2CH(OCH_3)CH_2CH_2OH$	162.23	....	114[13]	......	0.9631[25]	1.4329[25]	....	B1[4], 2785
7871	3-Hexanol, 2-ethyl .... $CH_3(CH_2)_3CH(C_2H_5)CH_2OH$	130.23	....	185, 84-6[15]	<−76	0.8328[20/4]	1.4328[20]	al, eth, ace, bz	B1[4], 1783
7872	1-Hexanol, 6-fluoro .... $F(CH_2)_6OH$	120.17	....	85-6[14]	......	0.975[20/4]	1.4141[27]	al, eth	B1[4], 1703
7873	1-Hexanol, 2-methyl (d) .... $C_4H_9CH(CH_3)CH_2OH$	116.21	$[\alpha]^{25}/_D$ + 2.45(al)	164-5, 70-2[15]	......	0.8313[13/4]	1.4245[17]	al, eth	B1[2], 444
7874	1-Hexanol, 2-methyl (dl) .... $C_4H_9CH(CH_3)CH_2OH$	116.21	....	164[750]	......	0.8270[20/4]	1.4226[20]	al, eth	B1[4], 1745
7875	1-Hexanol, 3-methyl (dl) .... $C_3H_7CH(CH_3)CH_2CH_2OH$	116.20	....	168-9[754], 91[13]	......	0.8258[20]	1.4245[20]	al, eth, ace	B1[4], 1748
7876	1-Hexanol, 3-methyl (l) .... $C_3H_7CH(CH_3)CH_2CH_2OH$	116.211	$[\alpha]^{27}/_D$ −1.67 (chl)	161-2[740], 80[25]	......	0.8208[25/4]	1.4204[30]	al, eth, ace	B1[4], 1748
7877	1-Hexanol, 4-methyl (d) .... $C_2H_5CH(CH_3)(CH_2)_2OH$	116.21	$[\alpha]^{28}/_D$ + 2.2	77[20]	......	0.809[23/4]	1.4233[25]	al, eth, ace, bz	B1[4], 1749
7878	1-Hexanol, 4-methyl (dl) .... $C_2H_5CH(CH_3)(CH_2)_2OH$	116.21	....	173, 83[24]	......	0.8239[20/4]	1.4219[20]	al, eth, ace, bz	B1[4], 1749
7879	1-Hexanol, 5-methyl .... $(CH_3)_2CH(CH_2)_3CH_2OH$	116.21	....	170[755], 53-5[15]	......	0.8119[20/4]	1.4175[20]	al, eth	B1[4], 1748
7880	1-Hexanol, 1-phenyl .... $C_5H_{11}CH(OH)C_6H_5$	178.28	....	170[50]	......	0.9477[25/4]	1.5105[20]	al, eth	B6[3], 1994
7881	1-Hexanol, 2-isopropyl-5-methyl .... $(CH_3)_2CHCH_2CH_2CH(i-C_3H_7)CH_2OH$	158.29	....	211	......	0.8345[20/4]	1.4369[20]	al, eth	B1[4], 1833
7882	2-Hexanol (d) .... $CH_3(CH_2)_3CH(OH)CH_3$	102.18	$[\alpha]^{25}/_D$ + 14.1 (eth, c=1)	138	......	0.8104[25/4]	1.4126[25]	al, eth	B1[3], 1663
7883	2-Hexanol (dl) .... $CH_3(CH_2)_3CH(OH)CH_3$	102.18	....	140	......	0.8159[20/4]	1.4144[20]	al, eth	B1[4], 1708
7884	2-Hexanol (l) .... $C_4H_9CH(OH)CH_3$	102.18	$[\alpha]^{18}/_{5780}$ −12.04	136-8[754]	......	0.8178[18/4]	....	al, eth	B1[3], 1663
7885	2-Hexanol, 1-chloro .... $C_4H_9CH(OH)CH_2Cl$	136.62	....	73-5[12]	......	0.0139[20/4]	1.4478[20]	al, eth, ace	B1[4], 1710
7886	2-Hexanol, 2-methyl .... $C_4H_9C(OH)(CH_3)_2$	116.21	....	143, 53-5[15]	......	0.8119[20/4]	1.4175[20]	al, eth	B1[4], 1745
7887	2-Hexanol, 3-methyl .... $C_3H_7CH(CH_3)CH(OH)CH_3$	116.20	....	79-81[52]	......	0.8820[25/4]	1.4198[18]	al, eth, ace	B1[3], 1693
7888	2-Hexanol, 5-methyl .... $(CH_3)_2CHCH_2CH_2CH(OH)CH_3$	116.21	....	150[744], 78[28]	......	0.814[20/4]	1.4180[20]	al, eth	B1[4], 1747
7889	2-Hexanol, 2,3,4-trimethyl .... $C_2H_5CH(CH_3)CH(CH_3)C(OH)(CH_3)_2$	144.26	....	57[5]	......	0.853[15/4]	1.4415[15]	al, eth, ace, peth	B1[3], 1756

No.	Name, Synonyms, and Formula	Mol. wt.	Color, crystalline form, specific rotation and λ_{max} (log ϵ)	b.p. °C	m.p. °C	Density	n_D	Solubility	Ref.
7890	2-Hexanol, 2,3,5-trimethyl .... $(CH_3)_2CHCH_2CH(CH_3)C(OH)(CH_3)_2$	144.26	....	171[755], 72[21]	......	0.8271[20/20]	1.4321[20]	al, eth, ace	B1[1], 212
7891	3-Hexanol (d) .... $C_3H_7CH(OH)C_2H_5$	102.18	[α][20/D] +6.8 (chl)	131-3	......	0.8213[20/4]	1.4150[20]	al, eth, ace	B1[3], 1665
7892	3-Hexanol (dl) $C_3H_7CH(OH)C_2H_5$	102.18		135		0.8182[20/4]	1.4167[20]	al, eth, ace	B1[4], 1711
7893	3-Hexanol (l) .... $C_3H_7CH(OH)C_2H_5$	102.18	[α][20/D]-7.17	135	......	0.8213[20/4]	1.4140[20]	al, eth, ace	B1[3], 1665
7894	3-Hexanol, 1-chloro $C_3H_7CH(OH)CH_2CH_2Cl$	136.62	....	120[35], 78[6]		1.003[25/4]	1.446[25]	al, eth, ace, bz	B1[4], 1712
7895	3-Hexanol, 2-chloro $C_3H_7CH(OH)CHClCH_3$	136.62	....	171	......	1.0143[11]	....	al, eth, ace, bz	B1[2], 438
7896	3-Hexanol, 5-chloro $CH_3CHClCH_2CH(OH)C_2H_5$	136.62	....	78-9[13]		1.0012[15/4]	1.4433[19]	al, eth, ace, bz	B1[4], 1712
7897	3-Hexanol, 2,4-dimethyl $C_2H_5CH(CH_3)CH(OH)CH(CH_3)_2$	130.23	....	61-2[18]		0.8371[20/4]	1.4309[20/D]	....	B1[4], 1791
7898	3-Hexanol, 3-ethyl $C_3H_7C(OH)(C_2H_5)_2$	130.23	....	160		0.8373[20/4]	1.4300[20]	al, eth, ace, bz	B1[4], 1790
7899	3-Hexanol, 3-ethyl-5-methyl $(CH_3)_2CHCH_2C(OH)(C_2H_5)_2$	144.26	....	272		0.8396[22/4]	1.4346[13]	al, eth	B1[3], 1755
7900	3-Hexanol, 3-methyl .... $C_3H_7C(OH)(CH_3)C_2H_5$	116.21	....	143, 56[18]		0.8234[20/0]	1.4231[20]	al, eth	B1[4], 1749
7901	3-Hexanol, 5-methyl (d) $(CH_3)_2CHCH_2CH(OH)C_2H_5$	116.21	[α][35/D] +21.2	81[60]		....	1.4171[25]	....	B1[3], 1692
7902	3-Hexanol, 5-methyl (dl) $(CH_3)_2CHCH_2CH(OH)C_2H_5$	116.21	....	147-8[756]		0.827[0]	1.4128[20]	al, eth	B1[4], 1747
7903	3-Hexanol, 5-methyl (l) $(CH_3)_2CHCH_2CH(OH)C_2H_5$	116.21	[α][23/D] -3.88	93-6[105], 63[19]		....	1.4171[25]	....	B1[3], 1692
7904	3-Hexanol, 2,2,5,5-tetramethyl .... $(CH_3)_3CCH_2CH(OH)C(CH_3)_3$	158.29	cr(peth)	166-70	52-3	....	....	al, eth, ace, peth	B1[3], 1772
7905	3-Hexanol, 2,2,3-trimethyl .... $C_3H_7C(OH)(CH_3)C(CH_3)_3$	144.26	....	170		0.8474[20/4]	1.4402[20]	al, eth, ace	B1[3], 1755
7906	3-Hexanol, 2,3,5-trimethyl $(CH_3)_2CHCH_2C(OH)(CH_3)CH(CH_3)_2$	144.26	....	72[21]		0.8271[20/20]	1.4321[20]	al, eth, ace	B1[4], 1813
7907	3-Hexanol, 2,4,4-trimethyl .... $C_2H_5C(CH_3)_2CH(OH)CH(CH_3)_2$	144.26	....	170		0.8489[20]	1.4395[20]	al, eth, ace	B1[3], 1756
7908	3-Hexanol, 2,5,5-trimethyl $(CH_3)_3CCH_2CH(OH)CH(CH_3)_2$	144.26	....	77[32]		0.8250[20]	1.4286[20]	al, eth, ace	B1[4], 1813
7909	3-Hexanol, 3,4,4-trimethyl $C_2H_5C(CH_3)_2C(OH)(CH_3)C_2H_5$	144.26	....	165-6		0.8323[21/0]	1.4341[21]	al, eth, ace	B1, 425
7910	3-Hexanol, 3,5,5-trimethyl $(CH_3)_3CCH_2C(OH)(CH_3)C_2H_5$	144.26	....	62[14]		0.8350[20]	1.4352[20]	al, eth, ace	B1[3], 1755
7911	2-Hexanone or Methyl butyl ketone $C_4H_9COCH_3$	100.16	....	128	−57	0.8113[20/4]	1.4007[20]	al, eth, ace	B1[4], 3298
7912	2-Hexanone, 3,3-dimethyl $C_3H_7C(CH_3)_2COCH_3$	128.22	....	149[765]		0.838[0/4]	1.4098[20]	al, eth, ace	B1[4], 3350
7913	2-Hexanone, 3,4-dimethyl $C_2H_5CH(CH_3)CH(CH_3)COCH_3$	128.22	....	158		0.8295[22/4]	1.4193[20]	al, eth, ace	B1[3], 2882
7914	2-Hexanone, 3-methyl .... $C_3H_7CH(CH_3)COCH_3$	114.19	....	142-5		0.828[25]	1.4035[20]	al, eth, ace, bz	B1[3], 2863
7915	2-Hexanone, 4-methyl $C_2H_5CH(CH_3)CH_2COCH_3$	114.19	....	142, 35-7[11]		....	1.4081[24]	al, eth, ace, bz	B1[4], 3329
7916	2-Hexanone, 5-methyl .... $(CH_3)_2CH(CH_2)_2COCH_3$	114.19	....	144		0.888[20/4]	1.4062[20]	al, eth, ace, bz	B1[4], 3329
7917	2-Hexanone, 5-methyl, oxime $(CH_3)_2CH(CH_2)_2C(=NOH)CH_3$	129.20	....	195-6		0.8881[20/4]	1.4448[20]	....	B1, 701
7918	3-Hexanone or Ethyl propyl ketone.... $C_3H_7COC_2H_5$	100.16	....	125		0.8118[20/4]	1.4004[20]	al, eth, ace	B1[4], 3301
7919	3-Hexanone, 2,2-dimethyl or tert-Butyl propyl ketone ... $C_3H_7COC(CH_3)_3$	128.22	....	145-8[745]		0.8105[25/4]	1.4119[20]	al, eth, ace	B1[4], 3347
7920	3-Hexanone, 2,2-dimethyl, oxime ...... $C_3H_7C(=NOH)C(CH_3)_3$	143.23	nd(al)	....	78	....	....	....	B1[3], 2881
7921	3-Hexanone, 2,5-dimethyl .... $(CH_3)_2CHCH_2COCH(CH_3)_2$	128.22	....	147-8		0.8270[0/0]	1.4049[20]	al, eth, ace	B1[4], 3349
7922	3-Hexanone, 4,4-dimethyl $C_2H_5C(CH_3)_2COC_2H_5$	128.22	....	151		0.8285[20]	1.4203[25]	al, bz, chl	B1[4], 3350

No.	Name, Synonyms, and Formula	Mol. wt.	Color, crystalline form, specific rotation and λ_{max} (log ϵ)	b.p. °C	m.p. °C	Density	n_D	Solubility	Ref.
7923	3-Hexanone, 6-dimethylamino-4,4-diphenyl-5-methyl (l) or l-Isomethadone.............................. $(CH_3)_2NCH_2CH(CH_3)C(C_6H_5)_2COC_2H_5$	309.46	....	162-5[0.5]	....	....	....	al	B14[3], 287
7924	3-Hexanone, 4-hydroxy or Propioin.............. $CH_3CH_2CH(OH)COC_2H_5$	116.16	....	132-5[117], 73[30]	....	0.956[81/4]	1.4340[81]	al, ace	B1[1], 4021
7925	3-Hexanone, 4-hydroxy-2,2,5,5-tetramethyl .... $(CH_3)_3CCH(OH)COC(CH_3)_3$	172.27	....	80[10]	81(sub)	....	....	eth	B1[4], 4060
7926	3-Hexanone, 2-methyl or Propyl isopropyl ketone .... $C_3H_7COCH(CH_3)_2$	114.19	....	134-6	......	0.8091[20]	1.4042[20]	al, eth, ace, chl	B1[4], 3328
7927	3-Hexanone, 4-methyl .... $C_2H_5CH(CH_3)COC_2H_5$	114.19	....	134-5	......	0.8162[20]	1.4069[20]	al, eth, ace, bz	B1[4], 3329
7928	3-Hexanone, 5-methyl .... $(CH_3)_2CHCH_2COC_2H_5$	114.19	....	134[735]	......	0.8090[20]	1.4047[20]	**al, eth**	B1[4], 3328
7929	Hexaphenyl ethane .... $(C_6H_5)_3CC(C_6H_5)_3$	486.67	cr(ace)	d	145-7d	....	....	eth, ace, chl, MeOH	B5[3], 2746
7930	Hexasiloxane, tetradicamethyl .... $(CH_3)_3SiO[-Si(CH_3)_2O-]_4Si(CH_3)_3$	443.97	....	245.5, 142[20]	−59	0.8910[20/4]	1.3948[20]	bz	B4[3], 1880
7931	1,2,3,5 Hexatetracene, 4-chloro .... $CH_2=CHCCl=C=C=CH_2$	112.56	....	127d, 55[54]	......	0.9997[20/4]	1.5280[20]	eth, ace, chl	B1[3], 1061
7932	1,2,4,5-Hexatetracene, 3,4-dichloro .... $CH_2=C=CClCCl=C=CH_2$	147.01	....	38-40[8]	......	1.1819[20/4]	1.5456[20]	eth, ace, chl	B1[3], 1061
7933	1,2,4-Hexatriene, 3,4,6-trichloro .... $ClCH_2CH=CClCCl=C=CH_2$	183.47	....	50[1]	......	1.3132[20/4]	1.5517[20]	ace, bz, chl	B1[3], 1041
7934	1,3,4-Hexatriene, 3,6-dichloro .... $ClCH_2CH=C=CClCH=CH_2$	149.02	....	45-6[1]	......	1.1807[20/4]	1.5195[20]	al	B1[3], 1041
7935	1,3,5-Hexatriene (cis) or Divinyl ethylene.............. $CH_2=CHCH=CHCH=CH_2$	80.14	....	70	12	0.7175[10/4]	1.4577[20]	al, ace, chl, peth	B1[3], 1041
7936	1,3,5-Hexatriene (trans) .... $CH_2=CHCH=CHCH=CH_2$	80.14	....	78.5	−17	0.7389[20/4]	1.5133[22]	al, ace, chl, peth	B1[3], 1094
7937	1,3,5-Hexatriene, 2,5-dimethyl .... $CH_2=C(CH_3)CH=CHC(CH_3)=CH_2$	108.19	....	145[747]	−9	0.7822[20/4]	1.5122[20]	ace, lig, MeOH	B1[4], 1102
7938	1,3,5-Hexatriene, 1,6-diphenyl .... $C_6H_5CH=CHCH=CHCH=CHC_6H_5$	232.33	lf(ace)	....	200-3	....	....	....	B5[3], 2243
7939	2-Hexenal (trans) .... $C_3H_7CH=CHCHO$	98.15	....	146-7, 43[12]	......	0.8491[20/4]	1.4480[20]	....	B1[4], 3468
7940	3-Hexenal .... $CH_3CH_2CH=CHCH_2CHO$	98.14	....	42-3[28]	......	0.8455[22/4]	1.4275[21.5]	eth, ace	B1[4], 3469
7941	1-Hexene .... $C_4H_9CH=CH_2$	84.16	....	63.3	−139.8	0.6731[20/4]	1.3837[20]	al, eth, bz, chl, peth	B1[4], 828
7942	1-Hexene, 5-amino-4-methyl .... $CH_3CH(NH_2)CH(CH_3)CH_2CH=CH_2$	113.20	....	133-6	......	0.793[15/0]	....	w	B4, 226
7943	1-Hexene, 1-chloro .... $C_4H_9CH=CHCl$	118.61	....	121	......	0.8872[22]	1.4300[22]	eth, ace, bz, chl	B1[4], 831
7944	1-Hexene, 2-chloro .... $C_4H_9CCl=CH_2$	118.61	....	113, 63[118]	......	0.8886[25/4]	1.4278[25]	ace, bz, chl	B1[3], 803
7945	1-Hexene, 5-chloro .... $CH_3CHCl(CH_2)_2CH=CH_2$	118.61	....	120.7, 28-30[13]	......	0.8891[25/4]	1.4305[20]	eth, ace, bz, chl	B1[4], 832
7946	1-Hexene, 1,2-dichloro (cis) .... $C_4H_9CCl=CHCl$	153.05	....	88[30]	......	1.0812[25/4]	1.4631[25]	bz, chl	B1[3], 803
7947	1-Hexene, 1,2-dichloro (trans) .... $C_4H_9CCl=CHCl$	153.05	....	63-5[22]	......	1.1167[25/4]	1.4576[25]	bz, chl	B1[3], 803
7948	1-Hexene, 2-ethyl .... $C_4H_9C(C_2H_5)=CH_2$	112.22	....	120	......	0.7270[20/4]	1.4157[20]	eth, bz, peth	B1[4], 884
7949	1-Hexene, 2-methyl .... $C_4H_9C(CH_3)=CH_2$	98.19	....	91.1	......	0.7000[20/4]	1.4040[20/D]	....	B1[4], 863
7950	1-Hexene, 3-methyl .... $C_3H_7CH(CH_3)CH=CH_2$	98.19	....	84	......	0.6945[20/4]	1.3970[20]	....	B1[4], 865
7951	1-Hexene, 4-methyl .... $C_2H_5CH(CH_3)CH_2CH=CH_2$	98.19	....	87.5	......	1.6969[20/4]	1.3985[20/D]	....	B1[4], 867
7952	1-Hexene, perfluoro .... $C_4F_9CF=CF_2$	300.05	....	57	......	....	....	chl	B1[4], 831
7953	1-Hexene, 1,1,2-trichloro .... $C_4H_9CCl=CCl_2$	187.50	....	90-93[10]	......	1.125[25]	1.4760[25]	eth	B1[3], 803
7954	2-Hexene (cis) .... $C_3H_7CH=CHCH_3$	84.16	....	68.8	−141.3	0.6869[20/4]	1.3977[20]	al, eth, bz, lig, chl	B1[4], 833
7955	2-Hexene (trans) .... $C_3H_7CH=CHCH_3$	84.16	....	68[750]	−133	0.6784[20/4]	1.3935[20]	al, eth, bz, lig, chl	B1[4], 834

No.	Name, Synonyms, and Formula	Mol. wt.	Color, crystalline form, specific rotation and λ_{max} (log ϵ)	b.p. °C	m.p. °C	Density	n_D	Solubility	Ref.
7956	2-Hexene, 4-chloro .... $C_2H_5CHClCH=CHCH_3$	118.61	....	123, 30[10]	......	0.8934[20/4]	1.4400[20]	eth, ace, bz, chl	B1[4], 835
7957	2-Hexene, 2,3-dimethyl .... $C_3H_7C(CH_3)=C(CH_3)CH_3$	112.21		122.1		0.7405[20/4]	1.4269[20/D]	....	B1[4], 887
7958	2-Hexene, 2,5-dimethyl .... $(CH_3)_2CHCH_2CH=C(CH_3)_2$	112.21		112.6		0.7182[20/4]	1.4135[20/D]		B1[4], 888
7959	2-Hexene, 2-methyl .... $C_3H_7CH=C(CH_3)_2$	98.19		95-8			1.4040[20·5/D]	....	B1[4], 863
7960	3-Hexene (cis) .... $C_2H_5CH=CHC_2H_5$	84.16		66.4	−137.8	0.6796[20/4]	1.3947[20]	al, eth, bz, lig, chl	B1[4], 837
7961	3-Hexene (trans) .... $C_2H_5CH=CHC_2H_5$	84.16		67.1	−113.4	0.6772[20/4]	1.3943[20]	al, eth, bz, lig, chl	B1[4], 237
7962	3-Hexene, 1-chloro .... $C_2H_5CH=CH(CH_2)_2Cl$	118.61		61[60]	......	0.900[24/4]	1.435[24]	eth, ace, bz, chl	B1[4], 838
7963	3-Hexene, 1-chloro-4-ethyl .... $(CH_3)_2C=CHCH_2CH_2Cl$	146.66		173		0.9102[20/4]	1.4524[20]	bz, chl	B1[2], 201
7964	3-Hexene, 2-chloro-2,5-dimethyl .... $(CH_3)_2CHCH=CHCCl(CH_3)_2$	146.66		45-60[15]			1.450[20]	bz, chl	Am63, 3474
7965	3-Hexene, 3-chloro (cis) .... $C_2H_5CH=CClC_2H_5$	118.61		119.6		0.9009[20/4]	1.4360[20]	eth, ace, bz, chl	B1[4], 838
7966	3-Hexene, 1,2,3,4,5,6-hexachloro .... $ClCH_2CHClCCl=CClCHClCH_2Cl$	290.83	cr (peth)	110-12[2]	58-9			chl, MeOH	B1[4], 839
7967	3-Hexene, 3-(4-hydroxyphenyl)-4-(4-methoxyphenyl) .... $CH_3CH_2C(4-CH_3OC_6H_4)=C(4-HOC_6H_4)CH_2CH_3$	282.39	nd (bz-lig) lf (70% al)	185-95[0·3]	117-8			al, eth, ace	B6[3], 5623
7968	3-Hexene, perfluoro .... $C_2F_5CF=CFC_2F_5$	300.05	....	49	......			chl	B1[4], 838
7969	2-Hexenoic acid, 2-methyl (trans) .... $C_3H_7CH=C(CH_3)CO_2H$	128.17		204-6, 118[11]	......	0.9627[20/4]	1.4601[20]	eth, ace	B2[4], 1581
7970	2-Hexenoic acid, 3-phenyl .... $C_3H_7C(C_6H_5)=CHCO_2H$	190.25	cr (peth)	183-4[14]	94	....		bz	B9[3], 2811
7971	3-Hexenoic acid or Hydrosorbic acid .... $C_2H_5CH=CHCH_2CO_2H$	114.15		208, 81-2[2]	12	0.9640[23/4]	1.4935[20]	....	B2[4], 1566
7972	5-Hexenoic acid .... $CH_2=CH(CH_2)_3CO_2H$	114.15		203, 107[17]	−37	0.9610[20/4]	1.4343[20]	al, eth	B2[4], 1562
7973	1-Hexene-3-ol .... $C_3H_7CH(OH)CH=CH_2$	100.16		134, 50[20]	......	0.834[22/4]	1.4297[18]	al, eth, ace	B1[4], 2136
7974	2-Hexene-1-ol (cis) .... $C_3H_7CH=CHCH_2OH$	100.16		156-8, 58-60[15]	......	0.8472[20/4]	1.4397[20]	al, eth, ace	B1[4], 2138
7975	3-Hexene-1-ol (cis) .... $C_2H_5CH=CHCH_2CH_2OH$	100.16		156-7, 58[12]	......	0.8478[22/4]	1.4380[20]	al, eth	B1[4], 2139
7976	3-Hexene-2-one .... $C_2H_5CH=CHCOCH_3$	98.14		140, 36-8[18]	......	0.86554[20/4]	1.4418[20]	al, eth, ace	B1[4], 3468
7977	5-Hexene-2-one, 5-methyl .... $CH_2=C(CH_3)CH_2CH_2COCH_3$	112.17		150	......	0.8475[20/20]	1.4348[20]	al, eth, ace	B1[4], 3482
7978	1-Hexene-3-one, 1-phenyl-5-methyl or Benzal pinacolone $(CH_3)_2CHCH_2COCH=CHC_6H_5$	188.27	cr	154[25]	43	0.9509[46]	1.5523[25]	al, bz, chl	B7[3], 1487
7979	4-Hexene-3-one, 2-methyl .... $CH_3CH=CHCO(CH_3)_2$	112.17		147-8.5[739]	......	0.843[20/4]	1.4345[20]	al, eth, ace	B1[4], 3483
7980	2-Hexene-4-one .... $CH_3CH_2COCH=CHCH_3$	98.14		138-9	......	0.8559[20/4]	1.4388[20]	al, eth, ace	B1[4], 3468
7981	1-Hexene-3-yne .... $C_2H_5C≡CCH=CH_2$	80.14		85[758]		0.7492[20/4]	1.4522[20]	eth, bz, peth, chl	B1[4], 1091
7982	1-Hexene-3-yne, 5-chloro-5-methyl .... $(CH_3)_2CClC≡C-CH=CH_2$	128.60		48[28]		0.9375[15]	1.4778[20]	al, eth, ace, bz, peth	B1[4], 1098
7983	1-Hexene-5-yne .... $HC≡CCH_2CH_2CH=CH_2$	80.14		70		0.7650[20/4]	1.4318[20]	eth, bz, peth, chl	B1[4], 1092
7984	3-Hexene-1-yne, 3-propyl .... $C_2H_5CH=C(C_3H_7)C≡CH$	122.21		136		0.7799[25/4]	1.4432[25]	bz, peth, chl	B1[3], 1049
7985	4-Hexene-1-yn-3-ol (d) .... $CH_3CH=CHCH(OH)C≡CH$	96.14	[α][20/D] +16.06	157-9		0.9090[20/4]	1.4645[17]	al, ace, bz	B1[4], 2307
7986	4-Hexene-1-yn-3-ol (dl) .... $CH_3CH=CHCH(OH)C≡CH$	96.14		154-6, 60[18]		0.9148[25/4]	1.4651[23]	al, ace, bz	B1[4], 2307
7987	Hexylamine or 1-aminohexane .... $CH_3(CH_2)_4CH_2NH_2$	101.19	....	130	−19	0.7660[20]	1.4180[20]	**al, eth**	B4[4], 709
7988	Hexyl nitrite .... $C_6H_{13}ONO$	131.18	ye	129-30[774], 52[44]	......	0.8778[20/4]	1.3987[20]	al, eth	B1[4], 1699

No.	Name, Synonyms, and Formula	Mol. wt.	Color, crystalline form, specific rotation and λ_{max} (log ε)	b.p. °C	m.p. °C	Density	n_D	Solubility	Ref.
7989	Hexyl phenyl ether .... $C_6H_{13}OC_6H_5$	178.28	....	240, 130[22]	-19	0.9174[20/4]	1.4921[20]	eth	B6[4], 560
7990	Hexyl sulfate .... $(C_6H_{13}O)_2SO_2$	266.40	....	125.3[2]	....	1.0036[21/0]	1.433[21]	....	B1[4], 1699
7991	1-Hexyne or n-Butyl acetylene. $C_4H_9C{\equiv}CH$	82.15	....	71.3	-131.9	0.7155[20/4]	1.3989[20]	al, eth, bz, peth, chl	B1[4], 1006
7992	1-Hexyne, 5-methyl or isopentyl acetylene .... $(CH_3)_2CHCH_2CH_2C{\equiv}CH$	96.17	....	92	-125	0.7274[20/4]	1.4059[20]	al, eth, bz, peth, chl	B1[3], 1000
7993	2-Hexyne $C_3H_7C{\equiv}CCH_3$	82.15	....	84	-89.6	0.7315[20/4]	1.4138[20]	al, eth, bz, peth, chl	B1[4], 1009
7994	2-Hexyne, 5-methyl .... $(CH_3)_2CHCH_2C{\equiv}CCH_3$	96.17	....	102.5	-92.9	0.7378[20/4]	1.4176[20]	eth, ace, bz, chl	B1[4], 1030
7995	3-Hexyne or Diethyl acetylene .... $C_2H_5C{\equiv}CC_2H_5$	82.15	....	81.5[764]	-103	0.7231[20/4]	1.4115[20]	al, eth, bz, peth, chl	B1[4], 1009
7996	3-Hexyne, 2,5-dimethyl-2,5-dichloro .... $(CH_3)_2CClC{\equiv}C-CCl(CH_3)_2$	181.11	....	175-8[745]	29	....	....	chl	B1[4], 1042
7997	3-Hexyne, 1:2,5:6-diepoxy .... $CH_2CHC{\equiv}C-CH-CH_2$	110.11	....	98[20]	-16	1.1189[23]	1.4871[23]	chl	B19[2], 19
7998	3-Hexyne, 2,5-dimethyl-2,5-diol or Acetylene pinacol .... $(CH_3)_2C(OH)C{\equiv}CC(OH)(CH_3)_2$	142.20	nd(w)	205	95	0.949[20/20]	....	al, eth, ace, bz, chl	B1[4], 2699
7999	3-Hexyne, 2-methyl .... $CH_3CH_2C{\equiv}CCH(CH_3)_2$	96.17	....	95.2	-116.7	0.7263[20/4]	1.4114[20]	eth, bz, peth, chl	B1[4], 1029
8000	1-Hexyn-3-ol, 3-methyl $C_3H_7C(OH)(CH_3)C{\equiv}CH$	112.17	....	137	....	0.8620[20/4]	1.4338[20]	al, eth	B1[4], 2252
8001	3-Hexyne-2,5-diol $CH_3CH(OH)C{\equiv}CCH(OH)CH_3$	114.14	....	120[11]	....	1.0180[20]	1.4691[20]		B1[3], 2271
8002	3-Hexyn-2-ol, 2-methyl $C_2H_5C{\equiv}CC(OH)(CH_3)_2$	112.17	....	145-7, 46[7]	....	0.962[0]	1.4392[25]	al, eth	B1[4], 2250
8003	Hippuric acid or N-benzoylaminoacetic acid .... $C_6H_5CONHCH_2CO_2H$	179.18	pr (w or al)	....	190-3	1.371[20/4]	....	al, w	B9[3], 1123
8004	Hippuric acid-p-amino .... $(4-H_2NC_6H_4)CONHCH_2CO_2H$	194.19	pr or nd (w)	....	198-9	....	....	al	B14[3], 1069
8005	Hippuric acid, o-bromo .... $(2-BrC_6H_4)CONHCH_2CO_2H$	258.09	nd (w)	....	192-3	....	....	AcOEt	B9[3], 1387
8006	Hippuric acid, m-bromo .... $(3-BrC_6H_4)CONHCH_2CO_2H$	258.09	nd (w)	....	146-7	....	....	al, MeOH	B9[2], 233
8007	Hippuric acid-p-bromo .... $(4-BrC_6H_4)CONHCH_2CO_2H$	258.09	nd (w)	....	162	....	....	al	B9[2], 236
8008	Histamine or 4-Imidoazol ethylamine................ $C_5H_9N_3$	111.15	wh nd (chl)	209[18]	86	....	....	w, al	B25[2], 302
8009	Histamine, dihydrochloride .... $C_5H_9N_3.2HCl$	184.07	pl (eth-ace) pr (w)	....	249-52	....	....	w, MeOH	B25[2], 303
8010	Histidine (d) $C_6H_9N_3O_2$	155.16	ta (w) $[\alpha]_D^{20}$ + 40.2 (w)	....	287d	....	....	w	B25[2], 404
8011	Histidine (dl) $C_6H_9N_3O_2$	155.16	ta, tetr pr (w)	....	285d	....	....	w	B25[2], 409
8012	Histidine (l) $C_6H_9N_3O_2$	155.16	nd or pl (dil al) $[\alpha]_D^{20}$ −39.7 (w, c=1.13)	....	287d	....	....	w	B25[2], 404
8013	Histidine, bis(3,4-dichlorobenzene sulfonate) (d) .... $C_6H_9N_3O_2-2Cl_2C_6H_4O_3S$	609.29	rh nd(w)	....	280d	....	....		C51, 5184
8014	Histidine, diflavianate .... $C_6H_9N_3O_2.2C_{10}H_6N_3O_8S$	783.62	nd(w)	....	251-4d	....	....		B25[2], 407
8015	Histidine, dihydrochloride (dl) .... $C_6H_9N_3O_2.2HCl$	228.08	....	....	237d	....	....	w, al	B25[1], 718
8016	Histidine, dihydrochloride (l) .... $C_6H_9N_3O_2.2HCl$	228.08	rh pl	....	252d	....	....	w, al	B25, 513
8017	Histidine, monohydrochloride, hydrate .... $C_6H_9N_3O_2.HCl.H_2O$	209.63	pl (w) $[\alpha]_D^{26}$ + 8.0 (3NHCl, c=2)	....	259d	....	....		B25[2], 407
8018	Holocaine or N,N-bis(p-ethoxyphenyl) acetamidine...... $C_{18}H_{22}N_2O_2$	298.39	nd (al)	....	117-8	....	....	al, eth, ace, bz	B13[3], 1069

No.	Name, Synonyms, and Formula	Mol. wt.	Color, crystalline form, specific rotation and λ_{max} (log ε)	b.p. °C	m.p. °C	Density	n_D	Solubility	Ref.
8019	Holocaine, hydrochloride or Phenacaine $C_{18}H_{22}N_2O_2 \cdot HCl$	334.85	cr(w + 1)		190-2			w, al, chl	B13[3], 1069
8020	Homatropine or Mandelyl tropine $C_{16}H_{21}NO_3$	275.35	pr (al or eth)		99-100			al, eth, ace, chl	B21[4], 179
8021	Homatropine, hydrobromide $C_{16}H_{21}NO_3 \cdot HBr$	356.27	rh pym or pl(w)		217-8d			w, al	B21[4], 179
8022	Homatropine, hydrochloride $C_{16}H_{21}NO_3 \cdot HCl$	311.81	wh pr(w)		220-7d			w, al	B21[4], 179
8023	Homocysteine (dl) $HSCH_2CH_2CH(NH_2)CO_2H$	135.18			270-5 d			w	B4[3], 1647
8024	Homocystine (d) $HO_2CCH(NH_2)CH_2CH_2SSCH_2CH_2CH(NH_2)CO_2H$	268.35	$D,^{26}_D$ −79 in HCl		281-4 d			w	B4[3], 1646
8025	Homogentisic acid or 2,5-dihydroxyphenyl acetic acid $(2,5-(HO)_2C_6H_3)CH_2CO_2H$	168.16	pr (w + 1) lf (al-chl)		152-4			w, al, eth	B10[3], 1456
8026	Homoveratric acid or 3,4-dimethoxyphenyl acetic acid $[3,4-(CH_3O)_2C_6H_3]CH_2CO_2H$	196.21	nd (w + 1) cr (bz-peth)		80-2 (hyd) 98-9 (anh)			w, al, eth	B10[3], 1459
8027	Hordenine or 1-(dimethylamino)-2-(4-hydroxyphenyl)ethane 4-HOC$_6$H$_4$CH$_2$CH$_2$N(CH$_3$)$_2$	165.24	rh pr (al or bz-peth) nd (w)	173-4[11] sub	117-8			al, eth, bz, lig, chl	B13[3], 1640
8028	Hordenine, sulfate $2(C_{10}H_{15}NO) \cdot H_2SO_4$	464.58	fl		210-11			w	B13[3], 1641
8029	Hordenine, sulfate, dihydrate $(C_{10}H_{15}NO)_2 \cdot H_2SO_4 \cdot 2H_2O$	464.58	pr or pl		197			w	B13[3], 1641
8030	Humulon or α-Lupulic acid $C_{21}H_{30}O_5$	362.47	yesh cr (eth) $[\alpha]^{20}_D$ −232 (bz)		66.5			al, eth, ace, bz	B8[3], 4034
8031	Hydantoic acid or N-Carbomoyl glycine $H_2NCONHCH_2CO_2H$	118.09	mcl pr		180d			w, al	B4[3], 1163
8032	Hydantoic acid, ethyl ester or Ethyl hydantoate $H_2NCONHCH_2CO_2C_2H_5$	146.15	nd (w)		135			w, al	B4[3], 1167
8033	Hydantoic acid, phenylthio or N-phenylpseudothio hydantoic acid $C_6H_5N=C(NH_2)SCH_2CO_2H$	210.26	wh (al)	175	159-61				B12[3], 868
8034	Hydantoin or Glycol urea $C_3H_4N_2O_2$	110.18	nd (MeOH) lf (w)		220			w, al	B24[2], 127
8035	Hydantoin, 1-acetyl-2-thio $C_5H_6N_2O_2S$	158.18	pl (al)		175-6			al	B24[1], 293
8036	Hydantoin, 1-benzoyl-2-thio $C_{10}H_8N_2O_2S$	220.26	pr (al)		165d			w	B24[1], 294
8037	Hydantoin, 5-benzylidene-2-thio $C_{10}H_8N_2OS$	204.26	ye nd (al)		258d			al	B24[1], 355
8038	Hydantoin, 5,5-dimethyl $C_5H_8N_2O_2$	128.14	pr (w-al)	sub	178			al, w, eth, ace, bz	B24[2], 157
8039	Hydantoin, 5,5-diphenyl $C_{15}H_{12}N_2O_2$	252.28	nd (al)		286			al, ace, aa	B24[2], 227
8040	Hydantoin, 3,5-diphenyl-2-thio $C_{15}H_{12}N_2OS$	268.34			233			al, eth, bz	B24, 385
8041	Hydantoin, 5-ethyl-5-phenyl (d) or d-Nirvanol $C_{11}H_{12}N_2O_2$	204.23	pl (10 al) $[\alpha]_D$ + 123 (al)		237			al	B24[2], 206
8042	Hydantoin, 5-ethyl-5-phenyl (dl) or dl-Nirvanol $C_{11}H_{12}N_2O_2$	204.23	pr (dil al)		199-200			al, aa	B24[2], 206
8043	Hydantoin, 5(2-hydroxybenzylidene)-2-thio $C_{10}H_8N_2O_2S$	220.26	nd (aa)		248			aa	B25[1], 502
8044	Hydantoin, 1-methyl $C_4H_6N_2O_2$	114.11	cr (w) pl (al)		157-9			w, al, chl	B24[2], 128
8045	Hydantoin, 5-methyl (dl) or α-Lactylurea $C_4H_6N_2O_2$	114.11	pr (w)		145-6 (anh)			w, al	B24[2], 155
8046	Hydantoin, 5-methyl (l) $C_4H_6N_2O_2$	114.11	(w) $[\alpha]^{20}_D$ −50.6		175			w, al	B24[1], 304
8047	Hydantoin, 5-methyl, hydrate (dl) $C_4H_6N_2O_2 \cdot H_2O$	132.12	rh (w)		155-6			w, al, ace	B24[2], 155
8048	Hydantoin, 5-phenyl $C_9H_8N_2O_2$	176.17			184-5				B24[2], 201

No.	Name, Synonyms, and Formula	Mol. wt.	Color, crystalline form, specific rotation and λ_{max} (log ε)	b.p. °C	m.p. °C	Density	n_D	Solubility	Ref.
8049	Hydantoin, 2-thio or Glycol thiourea .... $C_3H_4N_2OS$	116.14	wh nd (w)	....	229-31d	....		w, al, eth	B24[2], 138
8050	β(-)Hydrastine .... $C_{21}H_{21}NO_6$	383.41	yesh pr (al) $[\alpha]^{17}_D$ −67.8 (chl,c=2.5)		132 (135)			eth, ba, chl	B27[2], 603
8051	β(-)Hydrastine, hydrochloride .... $C_{21}H_{21}NO_6.HCl$	419.87	micr pw $[\alpha]'_D$ + 158.0 (w, c=2)		116			w	B27[2], 604
8052	Hydrastinine .... $C_{11}H_{13}NO_3$	207.23	nd (lig) cr (eth)		116-7			al, eth, chl	B27[2], 530
8053	Hydrastinine, bisulfate .... $C_{11}H_{13}NO_3.H_2SO_4$	287.29	gr-flr ye cr (al)		216d			w, al	B27[2], 530
8054	Hydrastinine, hydrochloride .... $C_{11}H_{13}NO_3.HCl$	225.68	pa ye nd		212d			w, al	B27[2], 530
8055	Hydratropamide or 2-Phenyl propionamide .... $C_6H_5CH(CH_3)CONH_2$	149.19	lo nd (w dil al) $[\alpha]^{28}_D$ + 57.9 (chl, c=1.6)		100.5			al, chl	B9[3], 2420
8056	Hydratropic acid (d) or 2-Phenyl propionic acid .... $CH_3CH(C_6H_5)CO_2H$	150.18	$[\alpha]^{20}_D$ + 81.1 (al, c=3)	152[16]	....				B9[3], 2417
8057	Hydratropic acid (dl) or 2-Phenyl propionic acid .... $C_6H_5CH(CH_3)CO_2H$	150.18		260-2, 160[25]	< -20	1.1[0/4]	1.5237[20]	....	B9[3], 2418
8058	Hydratropic acid (l) or 2-Phenyl propionic acid .... $C_6H_5CH(CH_3)CO_2H$	150.18	$[\alpha]^{20}_D$ 50 (al)	152[10]					B9[3], 2417
8059	Hydratroponitrile or 2-Phenyl propionitrile .... $C_6H_5CH(CH_3)CN$	131.18		230-2, 116-7[20]		0.9854[20/4]	1.5095[25]	al, eth	B9[3], 2421
8060	Hydrazine, 1-acetyl-2-phenyl .... $CH_3CONHNHC_6H_5$	150.18	hex pr (eth)		130-2			w, al, bz, chl	B15[2], 92
8061	Hydrazine-allyl .... $CH_2=CHCH_2NHNH_2$	72.12		122-4[757]				w, eth, chl	B4[1], 562
8062	Hydrazine, 1,2-bis-(3-aminophenyl) or m,m′-hydrazino dianiline .... $3-H_2NC_6H_4NHNHC_6H_4NH_2-(3)$	214.27	pym (al)		151				B15[3], 876
8063	Hydrazine, 1,2-bis-(4-aminophenyl) or p,p′-Hydrazino di-aniline .... $4-H_2NC_6H_4NHNHC_6H_4NH_2-(4)$	214.27	ye cr		145			al, eth	B15[3], 879
8064	Hydrazine, benzyl .... $4-CH_3C_6H_4NHNHCH_2C_6H_5$	212.29	fl or pr (al)	103[41]	26			w, al, eth	B15[3], 699
8065	Hydrazine, 1-benzyl-2-(4-tolyl) .... $4-H_2NC_6H_4NHNHCH_2C_6H_5$	212.29		212[17]					B15, 533
8066	Hydrazine, 2-bromophenyl .... $2-BrC_6H_4NHNH_2$	187.05	nd		48				B15[1], 117
8067	Hydrazine, 4-bromophenyl .... $4-BrC_6H_4NHNH_2$	187.05	nd (w) lf (lig) cr (al)		108			al, eth, lig	B15[3], 289
8068	Hydrazine, 1-butyl-1-phenyl .... $C_6H_5N(C_4H_9)NH_2$	164.25		250[763]					B15[1], 28
8069	Hydrazine, 1,2-diallyl .... $(CH_2=CHCH_2)NHNH(CH_2CH=CH_2)$	112.18		145[752]					B4[3], 1737
8070	Hydrazine, 1,2-dibenzoyl .... $C_6H_5CONHNHCOC_6H_5$	240.27	nd (al)		241				B9[3], 1318
8071	Hydrazine, 1,2-dibenzoyl-1,2-dimethyl .... $C_6H_5CON(CH_3)N(CH_3)COC_6H_5$	268.32	pr (al)		85-6			al	B9[2], 217
8072	Hydrazine, 1,1-dibenzyl .... $(C_6H_5CH_2)_2NNH_2$	212.30	cr (peth)		65			al, eth	B15[2], 245
8073	Hydrazine, 1,2-dibenzyl .... $C_6H_5CH_2NHNHCH_2C_6H_5$	212.30	lf (dil al)		47				B15[3], 701
8074	Hydrazine, 1,2-diisobutyl .... $i-C_4H_9NHNH-i-C_4H_9$	144.26		170[735], 63[10]		0.8002[20/4]	1.4276	al, eth, ace, bz	B4[2], 962
8075	Hydrazine, (2,4-dichlorophenyl) .... $2,4-Cl_2C_6H_3NHNH_2$	177.03	nd (eth or peth)		94			al, eth, aa	B15[2], 152
8076	Hydrazine, 1,1-diethyl .... $(C_2H_5)_2NNH_2$	88.15		98-9[750]		0.8804[20/4]	1.4214[20]	w, al, eth, bz, chl	B4[2], 959

No.	Name, Synonyms, and Formula	Mol. wt.	Color, crystalline form, specific rotation and λ_{max} (log ε)	b.p. °C	m.p. °C	Density	n_D	Solubility	Ref.
8077	Hydrazine, 1,2-diethyl $C_2H_5NHNHC_2H_5$	88.15	....	85-6	......	0.797^{26}	1.4204^{20}	al, eth, bz	B4³, 1730
8078	Hydrazine, 1,1-dimethyl $(CH_3)_2NNH_2$	60.11	....	63^{752}		0.7914^{22}	1.4075^{22}	w, al, eth	B4³, 1726
8079	Hydrazine, 1,2-dimethyl $CH_3NHNHCH_3$	60.11	....	81^{753}		$0.8274^{20/4}$	1.4209^{20}	w, al, eth	B4³, 1727
8080	Hydrazine, 1,2-dimethyl, dihydrochloride $CH_3NHNHCH_3.2HCl$	133.02	pr (w)	....	170d			w, al	B4³, 1727
8081	Hydrazine, (2,3-dimethylphenyl) $[2,3-(CH_3)_2C_6H_3]NHNH_2$	136.20	nd (al)		111-2			al, eth	B15³, 718
8082	Hydrazine, (2,4-dimethylphenyl) $[2,4-(CH_3)_2C_6H_3]NHNH_2$	136.20	nd (eth)		85			al, eth	B15², 249
8083	Hydrazine, (2,5-dimethylphenyl) $[2,5-(CH_3)_2C_6H_3]NHNH_2$	136.20	nd (dil aa)		78			al, eth, ace, bz	B15³, 719
8084	Hydrazine, (2,6-dimethylphenyl) $[2,6-(CH_3)_2C_6H_3]NHNH_2$	136.20	nd (lig)		46			lig	B15², 552
8085	Hydrazine, (3,4-dimethylphenyl) $[3,4-(CH_3)_2C_6H_3]NHNH_2$	136.20	yesh nd (eth)		57				B15², 249
8086	Hydrazine, 1,2-di-α-naphthyl or 1,1′-Hydrazonaphthalene $β-C_{10}H_7NHNH-α-C_{10}H_7$	284.37	lf (bz) pl (peth)		153			eth, bz	B15³, 729
8087	Hydrazine, 1,2-di-β-naphthyl or 2,2′-Hydrazonaphthalene $β-C_{10}H_7NHNH-β-C_{10}H_7$	284.37	red pl (bz)		140-1				B15³, 734
8088	Hydrazine, (2,4-dinitrophenyl) $[2,4-(O_2N)_2C_6H_3]NHNH_2$	198.14	blsh-red (al)		194 (198d)			al	B15³, 425
8089	Hydrazine, (2,6-dinitrophenyl) $[2,6-(O_2N)_2C_6H_3]NHNH_2$	198.14	red nd (dil al)		145				B15², 219
8090	Hydrazine, 1,1-diphenyl $(C_6H_5)_2NNH_2$	184.24	ta (lig)	220^{40-50}	49-52	$1.190^{16/4}$		al, eth, bz, chl	B15³, 74
8091	Hydrazine, 1,2-diphenyl or Hydrazobenzene $C_6H_5NHNHC_6H_5$	184.24	ta (al-eth)		131	$1.158^{16/4}$		al	B15³, 76
8092	Hydrazine, 1,2-diisopropyl $(CH_3)_2CHNHNHCH(CH_3)_2$	116.21	....	125, 63^{84}	......	$0.7894^{20/4}$	1.4173^{20}	al, eth, ace, bz	B4³, 1732
8093	Hydrazine, 1,1-di-4-tolyl $(4-CH_3C_6H_4)_2NNH_2$	212.30	lf (al)		93			al	B15¹, 154
8094	Hydrazine, 1,2-di-(2-tolyl) or o-Hydrazotoluene $(2-CH_3C_6H_4)NHNH(C_6H_4CH_3-2)$	212.30	lf (al)		165			al, eth, bz	B15³, 655
8095	Hydrazine, 1,2-di-(3-tolyl) or m-Hydrazotoluene $(3-CH_3C_6H_4)NHNH(C_6H_4CH_3-3)$	212.30	cr (peth)	224	38			al, eth, bz	B15³, 670
8096	Hydrazine, 1,2-di-(4-tolyl) or p=Hydrazotoluene $(4-CH_3C_6H_4)NHNH(C_6H_4CH_3-4)$	212.30	lf (lig) cr (bz-al) pl (al-eth)		135	$0.957^{20/4}$		al, eth, bz	B15³, 677
8097	Hydrazine, 1-isobutyl-1-phenyl $i-C_4H_9N(C_6H_5)NH_2$	164.25	....		240-5	$0.9633^{15/4}$			B15³, 75
8098	Hydrazine, ethyl $C_2H_5NHNH_2$	60.11	....	101	......			w, al, eth, ace, bz	B4³, 1730
8099	Hydrazine, 1-ethyl-1-phenyl $C_2H_5N(C_6H_5)NH_2$	136.20	....	237, $115-9^{19}$	......	$1.0181^{21/4}$	1.5711^{21}	al, eth, ace, bz	B15³, 74
8100	Hydrazine, 1-ethyl-2-phenyl $C_2H_5NHNHC_6H_5$	136.20	....	240^{750}, 110^{14}	......	$1.0150^{20/4}$	1.5676^{20}	al, eth, bz, chl	B15³, 74
8101	Hydrazine, methyl CH_3NHNH_2	46.07	....	87.5^{760}	−52.4	0.874^{25}	$1.4325^{20/D}$	w, al, eth	B4³, 1726
8102	Hydrazine-1-methyl, 1-phenyl $C_6H_5N(CH_3)NH_2$	122.17	....	227^{745}, 131^{35}	......	$1.0404^{20/4}$	1.5691^{20}	**al, eth, bz, chl**	B15³, 73
8103	Hydrazine-1-methyl, 2-phenyl $CH_3NHNHC_6H_5$	122.17	....	230^{728}, 112^{14}	......	$1.0320^{20/4}$	1.5733^{20}	al, eth, bz, chl	B15³, 73
8104	Hydrazine-1-methyl, 2-(3-tolyl) $(3-CH_3C_6H_4)NHNHCH_3$	136.20	ye		59-61	$1.0265^{100/4}$		al, bz	B15², 229
8105	Hydrazine-1-methyl, 2-(4-tolyl) $(4-CH_3C_6H_4)NHNHCH_3$	136.20	ye nd (eth) pl (lig)		91			al, eth, bz	B15², 154
8106	Hydrazine, α-naphthyl $α-C_{10}H_7NHNH_2$	158.21	....	203^{20}	117			al, eth, bz, chl	B15³, 728
8107	Hydrazine, β-naphthyl $β-C_{10}H_7NHNH_2$	158.21	lf (w)		124-5			al, bz	B15³, 734
8108	Hydrazine, (2-nitrophenyl) $(2-O_2NC_6H_4)NHNH_2$	153.14	red nd (bz)		90-2			w	B15³, 316

No.	Name, Synonyms, and Formula	Mol. wt.	Color, crystalline form, specific rotation and λ_{max} (log ε)	b.p. °C	m.p. °C	Density	n_D	Solubility	Ref.
8109	Hydrazine, (3-nitrophenyl) .... (3-O₂NC₆H₄)NHNH₂	153.14	red nd or pr (ace) ye nd (al)	....	93	....	....	chl, aa	B15³, 326
8110	Hydrazine, (4-nitrophenyl) .... (4-O₂NC₆H₄)NHNH₂	153.14	og-red lf or nd (al)	....	158d	....	....	eth, chl	B15³, 331
8111	Hydrazine, 1-pentyl-2-phenyl *(d)* C₅H₁₁NHNHC₆H₅	178.28	$[\alpha]'_D$ +4.45	173-5⁵⁰	......	0.986²⁰	1.5523²⁰	eth	B15, 121
8112	Hydrazine, 1-isopentyl-1-phenyl .... *i*-C₅H₁₁N(C₆H₅)NH₂	178.28	....	....	236	0.9588¹⁵	....	....	B15, 121
8113	Hydrazine, phenyl .... C₆H₅NHNH₂	108.15	mcl pr or pl	243, 115¹⁰	19.8	1.0986²⁰/⁴	1.6084¹⁰	al, eth, ace, bz, chl	B15³, 67
8114	Hydrazine, phenyl, hemihydrate C₆H₅NHNH₂·½H₂O	117.15	fl (w)	120¹²	24	1.0970²⁵/²⁵	1.6081²⁰	....	B15, 68
8115	Hydrazine, phenyl, hydrochloride .... C₆H₅NHNH₂·HCl	144.61	lf (al)	sub	243-6d	....	....	w, al	B15³, 71
8116	Hydrazine, 1-phenyl-2-(2-tolyl) (2-CH₃C₆H₄)NHNHC₆H₅	198.27	pl (al)	....	101-2	....	....	eth, bz	B15³, 655
8117	Hydrazine, 1-phenyl-2-(3-tolyl) (3-CH₃C₆H₄)NHNHC₆H₅	198.27	ye cr (peth)	....	61	1.0265¹⁰⁰/⁴	....	al, bz, lig	B15², 229
8118	Hydrazine, 1-phenyl-2-(4-tolyl) (4-CH₃C₆H₄)NHNHC₆H₅	198.27	pl (lig), cr (al)	....	91	....	....	al, bz	B15³, 677
8119	Hydrazine, propyl C₃H₇NHNH₂	74.13	....	....	119	....	....	....	B4³, 1731
8120	Hydrazine, *iso*-propyl (CH₃)₂CHNHNH₂	74.13	....	107⁷⁵⁰	....	....	....	w, al, bz	B4³, 1731
8121	Hydrazine, 1-isopropyl-2-methyl (CH₃)₂CHNHNHCH₃	88.15	cr (eth)	100	....	....	....	....	B4², 960
8122	Hydrazine, tetraphenyl (C₆H₅)₂NN(C₆H₅)₂	336.44	pr (chl-al)	....	149d	....	....	eth, ace, bz, chl	B15³, 77
8123	Hydrazine, (2-tolyl) (2-CH₃C₆H₄)NHNH₂	122.17	nd (dil al)	....	59	....	....	al, eth, chl	B15³, 654
8124	Hydrazine, (3-tolyl) (3-CH₃C₆H₄)NHNH₂	122.17	....	244d	....	1.057²⁰/⁴	....	al, eth, bz	B15³, 669
8125	Hydrazine, (4-tolyl) .... (4-CH₃C'6H₄)NHNH₂	122.17	lf (w or eth)	244d	66	....	....	al, eth, bz	B15³, 676
8126	Hydrazine, (2,4,6-tribromophenyl) (2,4,6-Br₃C₆H₂)NHNH₂	344.85	nd (peth) cr (lig)	....	146	....	....	bz, chl, lig	B15¹, 126
8127	Hydrazine, (2,4,6-trichlorophenyl) (2,4,6-Cl₃C₆H₂)NHNH₂	211.48	cr (bz)	....	143	....	....	....	B15³, 281
8128	Hydrazine, (2,4,6-trinitrophenyl) [2,4,6(O₂N)₃C₆H₂]NHNH₂	243.14	red pl (al)	....	186	....	....	al, aa	B15³, 652
8129	Hydrazine, triphenyl (C₆H₅)₂NNHC₆H₅	260.34	nd (bz-peth) cr (al)	....	142d	0.869⁷⁰/⁴	....	al, bz	B15², 54
8130	Hydrazine carboxylic acid, ethyl ester or *N*-Amino urethane. H₂NNHCO₂C₂H₅	104.12	cr	198d, 93⁹	46	....	....	al, eth	B3⁴, 174
8131	1,1-Hydrazine dicarboxylic acid, diethyl ester H₂NN(CO₂C₂H₅)₂	176.17	pr (w)	138¹²	29	....	....	al	B3³, 79
8132	1,2-Hydrazine dicarbon amide H₂NCONHNHCONH₂	118.10	pl (w)	....	257-9	....	....	....	B3⁴, 236
8133	1,2-Hydrazine dicarboxylic acid, diethyl ester C₂H₅O₂CNHNH CO₂C₂H₅	176.17	nd (chl) pr (w)	250d	135	1.324⁸	....	al, eth	B3⁴, 175
8134	Hydrazobenzene or 1,2-Diphenyl hydrazine C₆H₅NHNHC₆H₅	184.24	ta (al-eth)	....	131	1.158¹⁶/⁴	....	al	B15³, 76
8135	Hydrazo diformic acid OCHNHNHCHO	88.07	pr (al)	....	160	....	....	w	B2⁴, 86
8136	α,α'-Hydrazo naphthalene α-C₁₀H₇NHNH-α-C₁₀H₇	284.37	lf (bz) pl (peth)	....	153	....	....	eth, bz	B15³, 729
8137	β,β'-Hydrazo naphthalene β-C₁₀H₇NHNH-β-C₁₀H₇	284.37	red pl (bz)	....	140-1	....	....	....	B15³, 734
8138	Hydrindane *(trans, l)* or Hexahydro indane C₉H₁₆	124.23	....	161, 71.7⁴⁰	....	0.8627²⁰/⁴	1.4636²⁰	eth, bz, peth	B5⁴, 292
8139	2-Hydrindanone *(cis)* C₉H₁₄O	138.21	....	225⁷⁵⁴, 108²³	10	....	1.4830²⁰	al, bz, lig	B7³, 294
8140	2-Hydrindanone *(trans)* C₉H₁₄O	138.21	....	218⁷⁵⁴	-12	0.9807¹⁷/⁴	1.4769¹⁷	al, bz, lig	B7³, 294

No.	Name, Synonyms, and Formula	Mol. wt.	Color, crystalline form, specific rotation and λ_{max} (log ε)	b.p. °C	m.p. °C	Density	n_D	Solubility	Ref.
8141	Hydrobenzamide or Tribenzal diamine . . . . $C_6H_5CH(N=CHC_6H_5)_2$	298.39	nd (bz) cr (al w)	130	110	. . . .	. . . .	al, eth	B7[3], 838
8142	Hydroberberine (d) or d-Canadine . . . . $C_{20}H_{21}NO_4$	339.40	$[\alpha]^{20}_D$ +297.4 (chl, c=1)	. . . .	132	. . . .	. . . .	al, eth, bz, chl	B27[2], 556
8143	Hydroberberine (l) . . . . $C_{20}H_{21}NO_4$	339.40	nd (al) $[\alpha]^{20}_D$ −298.2 (chl,c=1)	. . . .	154	. . . .	. . . .	al, eth, bz, chl	B27[2], 557
8144	Hydrocinchonidine or Cinchamidine . . . . $C_{19}H_{24}N_2O$	296.42	lf (al) $[\alpha]^{20}_D$ −98.4 (al)	(al)	229	. . . .	. . . .	al	B23[2], 357
8145	Hydrocinnamaldehyde or 3-Phenylpropionaldehyde . . . . . $C_6H_5CH_2CH_2CHO$	134.18	mcl	223[745], 104-5[13]	47	. . . .	. . . .	al, eth	B7[3], 1046
8146	Hydrocinnamide or 3-Phenylpropionamide . . . . $C_6H_5CH_2CH_2CONH_2$	149.19	nd (w)	. . . .	106-8	. . . .	. . . .	al, eth	B9[3], 2393
8147	Hydrocinnamic acid or 3-Phenylpropionic acid . . . . $C_6H_5CH_2CH_2CO_2H$	150.18	pr (peth)	279.8, 169-70[28]	48.6	1.0712[49/4]	. . . .	al, eth, bz	B9[3], 2382
8148	Hydrocinnamic acid, 3-amino (d) or 3-Amino-3-phenyl-propionic acid . . . . $C_6H_5CH(NH_2)CH_2CO_2H$	165.19	pl (w) $[\alpha]^{20}_D$ +7.0 (w,p=1)	. . . .	234-5d	. . . .	. . . .	. . . .	B14[3], 1218
8149	Hydrocinnamic acid, 3-amino (dl) . . . . $C_6H_5CH(NH_2)CH_2CO_2H$	165.19	cr (w)	. . . .	231ld	. . . .	. . . .	. . . .	B14[3], 1218
8150	Hydrocinnamic acid, 3-amino (l) . . . . $C_6H_5CH(NH_2)CH_2CO_2H$	165.19	cr (w) $[\alpha]^{25}_D$ −7.5 (w,c=1)	. . . .	234-5d	. . . .	. . . .	. . . .	B14[3], 1218
8151	Hydrocinnamic acid, benzyl ester or Benzyl hydrocinnamate . . . . $C_6H_5CH_2CH_2CO_2CH_2C_6H_5$	240.13	. . . .	310-40, 198-9[20]	. . . .	1.090[15]	. . . .	eth	B9[2], 339
8152	Hydrocinnamic acid, ethyl ester or Ethyl hydrocinnamate $C_6H_5CH_2CH_2CO_2C_2H_5$	178.23	. . . .	247.2, 123[16]	. . . .	1.0147[20]	1.4954[20]	al, eth	B9[3], 2385
8153	Hydrocinnamic acid, methyl ester or Methyl hydrocinnamate . . . . $C_6H_5CH_2CH_2CO_2CH_3$	164.21	. . . .	238-9[757]	. . . .	1.0455[0]	. . . .	al, eth, bz	B9[3], 2384
8154	Hydrocinnamic acid, propyl ester or Propyl hydrocinnamate . . . . $C_6H_5CH_2CH_2CO_2C_3H_7$	192.26	. . . .	262.1, 135[16]	. . . .	1.008[12]	. . . .	. . . .	B9[3], 2386
8155	Hydrocinnamic acid, isopropyl ester or Isopropyl hydrocinnamate. . . . . . . $C_6H_5CH_2CH_2CO_2CH(CH_3)_2$	192.24	. . . .	126[11]	. . . .	0.9860[25/4]	. . . .	al, eth	B9[2], 339
8156	Hydrocinnamonitrile or 3-Phenyl propionitrile . . . . $C_6H_5CH_2CH_2CN$	131.17	. . . .	261, 125-6[15]	. . . .	1.0016[20]	1.5266[28]	al, eth	B9[3], 2395
8157	Hydrocinnamyl chloride or 3-Phenyl propionyl chloride . . $C_6H_5CH_2CH_2COCl$	168.63	. . . .	225d, 105[10]	. . . .	1.135[21]	. . . .	eth	B9[3], 2393
8158	Hydroconiferyl alcohol or 3-(4-hydroxy-3-methoxy-phenyl)-1-propanol . . . . $(3-CH_3O-4-HOC_6H_3)CH_2CH_2CH_2OH$	182.22	. . . .	197[5]	65	. . . .	1.5545[25]	al, eth	B6[3], 6347
8159	Hydrocotamine, hemihydrate . . . . $C_{12}H_{15}NO_3 \cdot \frac{1}{2}H_2O$	230.27	pr (eth)	. . . .	56	. . . .	. . . .	al, eth, ace, chl, aa	B27[2], 541
8160	Hydrocupreine . . . . $C_{19}H_{24}N_2O_2$	312.42	pl (dil al) $[\alpha]^{23}_D$ −159.2 (abs, al)	. . . .	230	. . . .	. . . .	eth, chl	B23[2], 399
8161	Hydrocyanic acid or Hydrogen cyanide . . . . HCN	27.04	. . . .	25.7	−13.2	0.6876[20/4]	1.2614[20]	w, al, eth	B2[4], 50
8162	Hydrofuramide or Furfur amide . . . . $C_{15}H_{12}N_2O_3$	268.28	nd (al)	. . . .	117	. . . .	. . . .	al, eth	B17[4], 4428
8163	Hydrohydrastinine . . . . $C_{11}H_{13}NO_2$	191.23	nd (lig) cr (peth)	303[752]	66	. . . .	. . . .	al, eth, ace, bz, aa	B27[2], 528
8164	Hydrolaphacol . . . . $C_{15}H_{16}O_3$	244.29	ye nd (al) cr (peth)	. . . .	94	. . . .	. . . .	al	B8[3], 2595
8165	Hydroquinidine . . . . $C_{20}H_{26}N_2O_2$	326.44	nd (al) $[\alpha]^{18}_D$ +229.26	. . . .	168-9	. . . .	. . . .	al, eth, ace, chl	B23,411
8166	Hydroquinine (d) or Quinotine . . . . $C_{20}H_{26}N_2O_2$	326.44	$[\alpha]^{18}_D$ +143.5	. . . .	171	. . . .	. . . .	al, eth, ace, chl	B23[2], 400

No.	Name, Synonyms, and Formula	Mol. wt.	Color, crystalline form, specific rotation and λ_{max} (log ε)	b.p. °C	m.p. °C	Density	n_D	Solubility	Ref.
8167	Hydroquinine (dl) $C_{20}H_{26}N_2O_2$	326.44	nd (al, chl)		175-7			al, eth, ace, chl	B23[2], 400
8168	Hydroquinine (l) $C_{20}H_{26}N_2O_2$	326.44	nd (eth, chl) $[\alpha]^{20/}_D$ −142.2		172.3			al, eth, ace, chl	B23[2], 400
8169	Hydroquinone or 1,4-Dehydroxy benzene 1,4-$(HO)_2C_6H_4$	110.11	mcl pr (sub) nd (w)	285[750]	173-4	1.328[15]		w, al, eth, ace	B6[3], 4374
8170	Hydroquinone, 2-acetyl or 2,5-Dihydroxy acetophenone 2,5-$(HO)_2C_6H_3COCH_3$	152.16	ye-gr nd (dil al or w)		204-5			al	B8[3], 2100
8171	Hydroquinone, monobenzoate 4-$(C_6H_5CO_2)C_6H_4OH$	214.22	nd (al)		163-4			al, eth, bz, chl	B9[3], 558
8172	Hydroquinone, 2-bromo or Adurol-1,4-dehydroxy-2-bromobenzene 1,4-$(HO)_2$-2-Br-C_6H_3	189.02	lf (lig) cr (chl)	sub	110-1			w, al, eth, bz	B6[3], 4436
8173	Hydroquinone, 2-chloro 2-Cl-1,4-$(HO)_2C_6H_3$	144.56	red lf (chl) nd (bz)	263	108			w, al, eth, bz	B6[3], 4432
8174	Hydroquinone, diacetate 1,4-$(CH_3CO_2)_2C_6H_4$	194.19	pl (w, al)		123-4	0.8731[25/4]		al, eth, lig, chl	B6[3], 4414
8175	Hydroquinone, dibenzoate 1,4-$(C_6H_5CO_2)_2C_6H_4$	318.33	mcl nd (al or to)		204			w	B9[3], 559
8176	Hydroquinone, monobenzyl ether 4-$C_6H_5CH_2OC_6H_4OH$	200.24	pl (w)		122			al, eth, bz	B6[3], 4402
8177	Hydroquinone, dibenzyl ether 1,4-$(C_6H_5CH_2O)_2C_6H_4$	290.37	pl (al)		130			eth, aa	B6[3], 4402
8178	Hydroquinone, dibenzoyl 1,4-$(C_6H_5CO)_2C_6H_4$	286.33			161				B7[3], 4299
8179	Hydroquinone, 2,5-di-tert-butyl 2,5-$(t-C_4H_9)_2C_6H_2(OH)_2$-1,4	222.33	cr (aq aa)		213.4				B6[3], 4741
8180	Hydroquinone, 2,3-dichloro 2,3-$Cl_2C_6H_2(OH)_2$-1,4	179.00	cr (sub) nd (w + 2)		146 8 (anh)			al	B6[3], 4434
8181	Hydroquinone, 2,5-dichloro 2,5-Cl_2-C_6H_2-$(OH)_2$-1,4	179.00	nd or pr (w, ace, bz)		172.5			al, eth, ace	B6[3], 4434
8182	Hydroquinone, 2,6-dichloro 2,6-$Cl_2C_6H_2(OH)_2$-1,4	179.00	nd or lf (w, bz)		164			al, ace	B6[3], 4435
8183	Hydroquinone, diethyl ether or 1,4-diethoxybenzene 1,4-$(C_2H_5O)_2C_6H_4$	166.22	pl (dil al)	246	72			al, eth, bz, chl	B6[3], 4387
8184	Hydroquinone, 2,3-dimethyl 2,3-$(CH_3)_2C_6H_2(OH)_2$-1,4	138.17	cr (w)		224-5d			w, al, eth	B6[3], 4582
8185	Hydroquinone, 2,5-dimethyl 2,5-$(CH_3)_2C_6H_2(OH)_2$-1,4	138.17	lf (w, al, bz)		217 sub			al, eth, chl	B6[3], 4601
8186	Hydroquinone, 2,6-dimethyl 2,6-$(CH_3)_2$-$C_6H_2(OH)_2$-1,4	138.17	nd (xyl) cr (w)		153-4			w, al, eth	B6[3], 4588
8187	Hydroquinone, dimethyl ether or 1,4-dimethoxybenzene 1,4-$(CH_3O)_2C_6H_4$	138.17	lf (w)	212.6, 109[20]	58-60	1.0526[55/55]		al, eth, bz	B6[3], 4385
8188	Hydroquinone, dithio or 1,4-dimercapto benzene 1,4-$(HS)_2C_6H_4$	142.25	lf (al)		98			al, bz, aa	B6[3], 4472
8189	Hydroquinone, monoethyl ether or 4-ethoxy phenol 4-$C_2H_5OC_6H_4OH$	138.17	pr or lf (w)	246-7	66-7			al, eth	B6[3], 4387
8190	Hydroquinone, monoheptyl ether 4-$C_7H_{15}OC_6H_4OH$	208.30	cr (lig)		60				B6[3], 4391
8191	Hydroquinone, monohexyl ether 4-$C_6H_{13}OC_6H_4OH$	194.28	cr (lig)		48				B6[3], 4390
8192	Hydroquinone, 2-iodo 2,5-$(HO)_2C_6H_3I$	236.01			115-6				B6[3], 4440
8193	Hydroquinone, 2-methyl 2,5-$(HO)_2C_6H_3CH_3$	124.34		sub 163[11]	128			w, al, eth, ace	B6[3], 4498
8194	Hydroquinone, monomethyl ether or 4-methoxyphenol 4-$CH_3OC_6H_4OH$	124.15	pl	243	57			w, al, eth, bz	B6[3], 4383
8195	Hydroquinone, 2-nitro 2,5-$(HO)_2C_6H_3NO_2$	155.11	og-red rh (w)		133-4			al, eth, bz	B6[3], 4442
8196	Hydroquinon, monooctyl ether 4-$C_8H_{17}OC_6H_4OH$	222.33	cr (lig)		60-1			al	B6[3], 4391
8197	Hydroquinone, phenyl or 2,5-dihydroxybiphenyl 2,5-$(HO)_2C_6H_3.C_6H_5$	186.21	nd (dil al)		97-8			al	B6[3], 5371

No.	Name, Synonyms, and Formula	Mol. wt.	Color, crystalline form, specific rotation and λ_{max} (log ϵ)	b.p. °C	m.p. °C	Density	n_D	Solubility	Ref.
8198	Hydroquinone, monopropyl ether or 4-propoxyphenol ... 4-C$_3$H$_7$OC$_6$H$_4$OH	152.20	cr (w, lig, al)	....	56-7				B6[3], 4388
8199	Hydroquinone, 2-iso-propyl .... 2,5-(HO)$_2$C$_6$H$_3$-i-C$_3$H$_7$	152.20	nd (w)	....	130-1				B6[3], 4632
8200	Hydroquinone, 2-iso-propyl-5-methyl or Thynohydro quinone [2,5-(HO)$_2$-4-CH$_3$]C$_6$H$_2$-i-C$_3$H$_7$	166.22	pr (dil al)	290 sub	148			al, eth	B6[3], 4673
8201	Hydroquinone, tetrabromo 1,4-(HO)$_2$C$_6$Br$_4$	425.72	mcl pr (al-eth)		244	3.023[21]		al, eth, aa	B6[3], 4440
8202	Hydroquinone, tetrachloro .... 1,4-(HO)$_2$C$_6$Cl$_4$	247.89	nd (aa)	sub	232			al, eth	B6[3], 4436
8203	Hydroquinone, tetraiodo .... l,4-(HO)$_2$C$_6$I$_4$	613.70	(aa)		258			eth, chl	B6[1], 417
8204	Hydroquinone, tetra methyl or Durohydroquinone ...... 1,4-(HO)$_2$C$_6$(CH$_3$)$_4$	166.22	nd (al)		233				B6[3], 4682
8205	Hydroquinone, 2,3,5-tribromo 2,3,5-Br$_3$C$_6$H(OH)$_2$-1,4	346.82	nd (chl)		136-7			al, eth, bz, aa, chl	B6[3], 4439
8206	Hydroquinone, 2,3,5-trimethyl 2,3,5-(CH$_3$)$_3$C$_6$H(OH)$_2$-1,4	152.20	nd (w)		168-70d			al, eth, bz	B6[2], 897
8207	Hydroquinonephthalein or 2,7-Dihydroxy fluoran .... C$_{20}$H$_{12}$O$_5$	332.32	nd (eth)		228-9			al, eth, ace, aa	B19[2], 247
8208	Hydroxyacetamide .... HOCH$_2$CONH$_2$	75.07	lf (al) rh (AcOEt)		120	1.415[13]		w	B3[4], 597
8209	Hydroxyacetic acid or Glycolic acid .... HOCH$_2$COOH	76.05	rh, nd (w) lf (eth)	d	80			w, al, eth	B3[4], 571
8210	Hydroxyacetic acid, acetate or Acetoxyacetic acid .... CH$_3$CO$_2$CH$_2$COOH	118.19	nd (bz)	144-5[12]	67-8			w, al, eth, ace, chl	B3[4], 576
8211	Hydroxyacetic acid, anhydride or Glycolic anhydride .... (HOCH$_2$CO)$_2$O	134.09	pw	d	128-30				B3[1], 92
8212	Hydroxyacetic acid, 4-bromophenyl ester (dl) .... (4-BrC$_6$H$_4$)O$_2$CCH$_2$OH	231.05	nd (bz)		118-20			al, eth, bz	B16[2], 125
8213	Hydroxyacetic acid, ethyl ester or Ethyl glycolate .... HOCH$_2$CO$_2$C$_2$H$_5$	104.11	....	160, 69[25]	....	1.0826[23/4]	1.4180[20]	al, eth	B3[4], 580
8214	Hydroxyacetic acid, methyl ester or Methyl glycolate ..... HOCH$_2$CO$_2$CH$_3$	90.08	....	151.1		1.1677[18/4]	....	w, al, eth	B3[4], 578
8215	Hydroxyacetic acid, propyl ester or Propyl glycolate ..... HOCH$_2$CO$_2$C$_3$H$_7$	118.13	....	170-1		1.0631[18/4]	1.4231[18]	....	B3[4], 588
8216	Hydroxyacetonitrile or Glycolonitrile.................. HOCH$_2$CN	57.05	....	183d, 119[24]	<-72		1.4117[19]	w, al, eth	B3[4], 598
8217	Hydroxy amphetamine or Paredrine .... C$_9$H$_{13}$NO	151.21	....		125-6			al, chl	B13[3], 1709
8218	Hydroxy amphetamine, hydrobromide .... C$_9$H$_{13}$NO.HBr	232.13	....		189			w, al, ace	B13[3], 1709
8219	Hydroxycitronellal .... (CH$_3$)$_2$COH(CH$_2$)$_3$CH(CH$_3$)CH$_2$CHO	172.27	....	103[3]		0.9220[20]	1.4494[20]	al, ace	B1[4], 4058
8220	bis [3-Hydroxy-sec-butyl] amine .... [CH$_3$CH(OH)CH(CH$_3$)]$_2$NH	161.25	yesh	112-5[3]		0.9775[20/4]	1.4162[20]	w, al, ace	B4[4], 1726
8221	β-Hydroxyethyl 2-tolyl amine .... (2-CH$_3$C$_6$H$_4$)NHCH$_2$CH$_2$OH	151.21	....	285-6, 149[4]		1.0794[20/4]	1.5675[20]	al, eth	B12[3], 1846
8222	bis (2-Hydroxyethyl) methyl amine .... (HOCH$_2$CH$_2$)$_2$NCH$_3$	119.17	....	246-8[747], 123-5[4]		1.0377[20]	1.4642[20]	w, al	B4[4], 1517
8223	bis-(2-Hydroxyethyl) ethyl amine .... (HOCH$_2$CH$_2$)$_2$NC$_2$H$_5$	133.19	ye	246-8, 118[3]		1.0135[20/4]	1.4663[20]	w, al	B4[3], 693
8224	β-Hydroxyethyl-4-tolyl amine .... (4-CH$_3$C$_6$H$_4$)NHCH$_2$CH$_2$OH	151.21	pl (eth-lig)	286-8, 153-5[4]	42-3	....	....	al, eth, bz, chl	B12[3], 2034
8225	2-(Hydroxyethyl)-2-(hydroxybutyl) amine .... (HOCH$_2$CH$_2$)NH(CH$_2$CH(OH)CH$_2$CH$_3$)	133.19	yesh	137[9]		1.0310[204]	1.4690[30]	w, al, ace	B4[4], 1725
8226	(2-Hydroxyethyl)-3-(hydroxybutyl) amine .... (HOCH$_2$CH$_2$)NH(CH$_2$CH$_2$CH(OH)CH$_3$)	133.19	yesh	107-9[i]		1.0331[20/4]	1.4718[20]	w, ace	C59, 6397
8227	(2-Hydroxypropyl)trimethylammonium chloride or β-Methyl choline chloride CH$_3$CH(OH)CH$_2$N$^+$(CH$_3$)$_3$Cl$^-$	153.65	pr (BuOH)	d	165			w, al	B4[3], 754
8228	tris-(2-Hydroxyethyl) amine or Triethanolamine........ (HOCH$_2$CH$_2$)$_3$N	149.19	hyg cr	277[150]	21.2	1.1242[20/4]	1.4852[20]	w, al, chl	B4[4], 1524
8229	2-Hydroxybenzoic acid or Salicylic acid .... 2-HOC$_6$H$_4$CO$_2$H	138.12	nd (w), mcl pr (al)	211[20] sub	159	1.443[20/4]	1.565	al, eth, ace, bz	B10[3], 87

No.	Name, Synonyms, and Formula	Mol. wt.	Color, crystalline form, specific rotation and λ_{max} (log ε)	b.p. °C	m.p. °C	Density	n_D	Solubility	Ref.
8230	3-Hydroxybenzomide 3-HOC$_6$H$_4$CONH$_2$	137.14	pl (w)	....	170.5	....		al, eth	B10³, 255
8231	3-Hydroxybenzamide-N-phenyl 3-HOC$_6$H$_4$CONHC$_6$H$_5$	213.24	nd or pl (w, dil al)		156			al	B12¹, 269
8232	3-Hydroxybenzoic acid 3-HOC$_6$H$_4$CO$_2$H	138.12	nd (w) pl or pr (al)		201-3			eth, ace, MeOH	B10³, 242
8233	3-Hydroxybenzoic acid, 2-bromo 2-Br-3-HOC$_6$H$_3$CO$_2$H	217.03	nd (w)		160-1			eth	B10², 83
8234	3-Hydroxybenzoic acid, 4-bromo 4-Br-3-HOC$_6$H$_3$CO$_2$H	217.03	pl, nd (w)		214			al	B10³, 258
8235	3-Hydroxybenzoic acid, 6-bromo 6-Br-3-HOC$_6$H$_3$CO$_2$H	217.03	(w)		185d			eth	B10³, 258
8236	3-Hydroxybenzoic acid, 2-chloro 2-Cl-3-HO-C$_6$H$_3$CO$_2$H	172.52	lf (w or bz)		157-8			bz	B10³, 257
8237	3-Hydroxybenzoic acid, 4-chloro 4-Cl-3-HOC$_6$H$_3$CO$_2$H	172.57	nd (w)		219-20				B10², 83
8238	3-Hydroxybenzoic acid, 6-chloro 6-Cl-3-HOC$_6$H$_3$CO$_2$H	172.57	(w)		178-9			al, ace	B10³, 258
8239	3-Hydroxybenzoic acid, 4,5-dimethoxy 4,5-(CH$_3$O)$_2$-3-HOC$_6$H$_2$CO$_2$H	198.18	nd (aa or w)		197-8			al, aa	B10³, 2073
8240	3-Hydroxybenzoic acid, 5,6-dimethoxy 5,6-(CH$_3$O)$_2$-3-HOC$_6$H$_2$CO$_2$H	198.18	pl or lf (w)		186-8			al	B10³, 2060
8241	3-Hydroxybenzoic acid, 2,4-dinitro 2,4-(NO$_2$)$_2$-3-HOC$_6$H$_2$CO$_2$H	228.12	(w)		204			al, eth	B10³, 264
8242	3-Hydroxybenzoic acid, ethyl ester or Ethyl 3-hydroxy-benzoate 3-HOC$_6$H$_4$CO$_2$C$_2$H$_5$	166.18	pl (bz)	295, 211⁶⁵	73.8			al, eth	B10³, 250
8243	3-Hydroxybenzoic acid, 4-formyl 4-HCO-3-HOC$_6$H$_3$CO$_2$H	166.14	nd (w)	sub	2 ?			al, eth	B10, 951
8244	3-Hydroxybenzoic acid, 2-iodo 2-I-3-HOC$_6$H$_3$CO$_2$H	264.02	nd (chl)	158-9				al, eth	B10², 84
8245	3-Hydroxybenzoic acid, 4-iodo 4-I-3-HOC$_6$H$_3$CO$_2$H	264.02	nd (w)		226-8			al, eth	B10³, 261
8246	3-Hydroxybenzoic acid, 6-iodo 6-I-3-HOC$_6$H$_3$CO$_2$H	264.02	nd (w)	sub 160	198			al, eth	B10³, 261
8247	3-Hydroxybenzoic acid, 4-methoxy or Isovanillic acid 4-CH$_3$O-3-HOC$_6$H$_3$CO$_2$H	168.16	nd, pr, pl (w)	sub	255-7			al, eth	B10³, 1404
8248	3-Hydroxybenzoic acid, methyl ester 3-HOC$_6$H$_4$CO$_2$CH$_3$	152.16	nd (bz-peth)	280⁷⁰⁹, 178¹⁷	71.5			al	B10³, 249
8249	3-Hydroxybenzoic acid, 2-methyl or 2,3-Cresotic acid 2-CH$_3$-3-HOC$_6$H$_3$CO$_2$H	152.16	nd (w, dil al)		145-6			al, eth	B10³, 494
8250	3-Hydroxybenzoic acid, 4-methyl or 3,4-Cresotic acid 4-CH$_3$-3-HOC$_6$H$_3$CO$_2$H	152.16	nd or pr (w)	sub	208.5			al, eth	B10³, 527
8251	3-Hydroxybenzoic acid, 5-methyl or 3,5-Cresotic acid 5-CH$_3$-3-HOC$_6$H$_3$CO$_2$H	152.16	nd (w)	sub	210			al, eth	B10, 227
8252	3-Hydroxybenzoic acid, 6-methyl or 3,6-Cresotic acid 6-CH$_3$-3-HOC$_6$H$_3$CO$_2$H	152.16	nd or pr (w)		185			al, eth	B10, 215
8253	3-Hydroxybenzoic acid, 2-nitro 2-NO$_2$-3-HOC$_6$H$_3$CO$_2$H	183.12	pl or pr (w + l)		180-1			al, eth	B10², 84
8254	3-Hydroxybenzoic acid, 4-nitro 4-NO$_2$-3-HOC$_6$H$_3$CO$_2$H	183.12	ye lf (w)		235			al, eth	B10³, 263
8255	3-Hydroxybenzoic acid, 5-nitro 5-NO$_2$-3-HOC$_6$H$_3$CO$_2$H	183.12	ye lf or pl (25% HCl)		167			al, eth	B10², 85
8256	3-Hydroxybenzoic acid, 6-nitro 6-NO$_2$-3-HOC$_6$H$_3$CO$_2$H	183.12	ye nd or pr (w + l)		172			al, eth	B10³, 263
8257	3-Hydroxybenzoic acid, 4-sulfo 4-HO$_3$S-3-HOC$_6$H$_3$CO$_2$H	218.19	ye-gr nd (w + 2)		208 (213)			w, al	B11³, 707
8258	3-Hydroxybenzoic acid, 5-sulfo 5-HO$_3$S-3-HOC$_6$H$_3$CO$_2$H	218.19	nd (w + 1)		120d			w, al, eth	B11³, 708
8259	3-Hydroxybenzonitrile 3-HOC$_6$H$_4$CN	119.12	pr (al or eth) lf (w)		83-4			al, eth, bz, chl	B10³, 255
8260	3-Hydroxybenzoyl chloride 3-HOC$_6$H$_4$COCl	156.57	....	110-3⁰·⁵	<-15			chl	B10², 82
8261	4-Hydroxybenzamide 4-HOC$_6$H$_4$CONH$_2$	137.14	nd (w + 1)		162 (hyd)			al, eth	B10³, 340

No.	Name, Synonyms, and Formula	Mol. wt.	Color, crystalline form, specific rotation and λ_{max} (log ε)	b.p. °C	m.p. °C	Density	n_D	Solubility	Ref.
8262	4-Hydroxybenzamide, N-phenyl .... 4-HOC$_6$H$_4$CONHC$_6$H$_5$	213.24	pl or nd (w)	....	201-2	....	....	al	B12^3, 947
8263	4-Hydroxybenzoic acid .... 4-HOC$_6$H$_4$CO$_2$H	138.12	pr or pl (w, al, xyl-al) cr (dil al or ace)	....	214-5	....	....	eth, ace	B10^3, 277
8264	4-Hydroxybenzoic acid, 2-bromo .... 2-Br-4-HOC$_6$H$_3$CO$_2$H	217.03	nd (w)	....	151	....	....	eth	B10^2, 103
8265	4-Hydroxybenzoic acid, 3-bromo .... 3-Br-4-HOC$_6$H$_3$CO$_2$H	217.03	nd or pr (+ w) (w)	....	177	....	....	al, eth, aa	B10^3, 363
8266	4-Hydroxybenzoic acid, butyl ester or Butyl 4-hydroxy- benzoate.................. 4-HOC$_6$H$_4$CO$_2$C$_4$H$_9$	194.23	....	....	68-9	....	....	al	B10^3, 307
8267	4-Hydroxybenzoic acid, 2-chloro .... 2-Cl-4-HOC$_6$H$_3$CO$_2$H	172.57	nd (w)	....	159	....	....	ace	B10^3, 360
8268	4-Hydroxybenzoic acid, 3-chloro .... 3-Cl-4-HOC$_6$H$_3$CO$_2$H	172.57	nd (w)	sub	170-2	....	....	al, eth, ace	B10^2, 102
8269	4-Hydroxybenzoic acid, 3,5-dichloro .... 3,5-Cl$_2$-4-HOC$_6$H$_2$CO$_2$H	207.01	nd (dil al or dil aa)	sub d	269	....	....	al, eth	B10^3, 362
8270	4-Hydroxybenzoic acid, 3,5-diiodo .... 3,5-I$_2$-4-HOC$_6$H$_2$CO$_2$H	389.92	nd (dil al)	d 260	237	....	....	al, eth	B10^3, 370
8271	4-Hydroxybenzoic acid, 3,5-diiodo, ethyl ester .... 3,5-I$_2$-4-HOC$_6$H$_2$CO$_2$C$_2$H$_5$	417.98	nd (dil al)	....	123	....	....	al	B10^3, 372
8272	4-Hydroxybenzoic acid, 2,3-dimethoxy .... 2,3-(CH$_3$O)$_2$-4-HOC$_6$H$_2$CO$_2$H	198.18	pl or lf (w or al)	....	154-5	....	....	al, ace, chl	B10^2, 332
8273	4-Hydroxybenzoic acid, 2,6-dimethoxy .... 2,6-(CH$_3$O)$_2$-4-HOC$_6$H$_2$CO$_2$H	198.18	pl (w)	....	175	....	....	bz, py	B10^1, 235
8274	4-Hydroxy benzoic acid, 3,5-dimethoxy .... 3,5-(CH$_3$O)$_2$-4-HOC$_6$H$_2$CO$_2$H	198.18	nd (w)	....	204-5	....	....	al, eth, ace, chl	B10^3, 2073
8275	4-Hydroxybenzoic acid, 3,5-dinitro .... 3,5-(NO$_2$)$_2$-4-HOC$_6$H$_2$CO$_2$H	228.12	ye lf (al)	....	248-9	....	....	al, eth	B10^3, 383
8276	4-Hydroxybenzoic acid, ethyl ester .... 4-HOC$_6$H$_4$CO$_2$C$_2$H$_5$	166.18	cr (dil al)	297-8	116-8	....	....	al, eth	B10^3, 300
8277	4-Hydroxybenzoic acid, 3-formyl .... 3-HCO-4-HOC$_6$H$_3$CO$_2$H	166.14	pr (w)	sub	244	....	....	al, eth	B10^2, 675
8278	4-Hydroxybenzoic acid, 2-iodo .... 2-I-4-HOC$_6$H$_3$CO$_2$H	264.02	nd (w)	....	215d	....	....	al, eth	B10^2, 104
8279	4-Hydroxybenzoic acid, 3-iodo .... 3-I-4-HOC$_6$H$_3$CO$_2$H	264.02	nd (w + ½)	sub	173-4	....	....	al, eth, aa	B10^3, 367
8280	4-Hydroxybenzoic acid, 3-methoxy or Vanillic acid ...... 3-(CH$_3$O)-4-HOC$_6$H$_3$CO$_2$H	168.16	nd (w)	sub	213-5	....	....	eth	B10^3, 1403
8281	4-Hydroxybenzoic acid, 3-methoxy, ethyl ester or Ethyl 3- methoxy-4-hydroxybenzoate................ 3-CH$_3$O-4-HOC$_6$H$_3$CO$_2$C$_2$H$_5$	196.21	nd (dil al)	291-3	44	....	....	al, eth	B10, 397
8282	4-Hydroxybenzoic acid, 3-methoxy methyl ester .... 3-CH$_3$O-4-HOC$_6$H$_3$CO$_2$CH$_3$	182.18	nd (dil al)	285-7, 118^2	64	....	....	al, chl	B10^3, 1410
8283	4-Hydroxybenzoic acid, methyl ester or Methyl 4-hy- droxy- benzoate.................. 4-HOC$_6$H$_4$CO$_2$CH$_3$	152.16	nd (dil al)	270-80d	131	....	....	al, eth, ace	B10^3, 296
8284	4-Hydroxybenzoic acid, 2-methyl or 4,2-Cresotic acid .... 2-CH$_3$-4-HOC$_6$H$_3$CO$_2$H	152.16	nd (w + ½)	236-7 sub	177-8	....	....	al, eth	B10^3, 494
8285	4-Hydroxybenzoic acid, 3-methyl or 4,3-Cresotic acid .... 3-CH$_3$-4-HOC$_6$H$_3$CO$_2$H	152.16	nd (w + ½)	....	l74-5 sub	....	....	al, eth	B10^3, 512
8286	4-Hydroxybenzoic acid, 3-nitro .... 3-NO$_2$-4-HOC$_6$H$_3$CO$_2$H	183.12	nd or lf (w)	....	186-7	....	....	al, eth	B10^3, 376
8287	4-Hydroxybenzoic acid, propyl ester .... 4-HOC$_6$H$_4$CO$_2$C$_3$H$_7$	180.21	pr (eth)	....	96-8	1.0630$^{102/4}$	1.5050^{102}	eth	B10^3, 306
8288	4-Hydroxybenzoic acid, 5-iso-propyl-2-methyl or p-Thy- motinic acid.................. 5-(CH$_3$)$_2$CH-2-CH$_3$-4-HOC$_6$H$_2$CO$_2$H	194.23	pl (dil al)	....	157	....	....	al, eth, bz, chl	B10^3, 631
8289	4-Hydroxybenzoic acid, 3-sulfo .... 3-HO$_3$S-4-HOC$_6$H$_3$CO$_2$H	218.19	nd or lf (w)	....	d	....	....	w, al	B11^3, 709
8290	4-Hydroxybenzonitrile .... 4-HOC$_6$H$_4$CN	119.12	lf (w)	....	113	....	....	al, eth, chl	B10^3, 344
8291	Hydroxylamine, N-ethyl or β-Ethyl hydroxyl amine ..... C$_2$H$_5$NHOH	61.09	nd (lig)	....	59-60d	0.9079$^{20/4}$	1.4152^{66}	w, al	B4^3, 1717

No.	Name, Synonyms, and Formula	Mol. wt.	Color, crystalline form, specific rotation and λ_{max} (log ε)	b.p. °C	m.p. °C	Density	n_D	Solubility	Ref.
8292	Hydroxylamine, *N*-methyl .... CH$_3$NHOH	47.06	hyg nd	62.5^{15}	87-8	1.0003$^{20/4}$	1.4164^{20}	w, al	B4^3, 1715
8293	Hydroxyquinol or 1,2,4-Trihydroxybenzene .... 1,2,4-(HO)$_3$C$_6$H$_3$	126.11	pl (eth), lf or pl (w)	....	140-1			w, al, eth	B6^3, 6276
8294	Hydroxyquinol, triacetate or 1,2,4-Triacetoxy benzene ... 1,2,4-(CH$_3$CO)$_3$C$_6$H$_3$	252.23	nd (MeOH)	>300	97-8			al	B6^3, 6282
8295	Hydroxyquinol, 3,5,6-trichloro or 3,5,6-trichloro-1,2,4-trihydroxybenzene .... C$_6$H$_3$Cl$_3$O$_3$	229.45	nd (bz, aa)	....	160			al, eth	B6^2, 1072
8296	Hyenic acid .... C$_{25}$H$_{50}$O$_2$	382.68	cr (eth) nd (bz)		77-8			eth	B2^2, 380
8297	Hygrine-*(l)* or 2-acetonyl-1-methylpyrrolidine .... C$_8$H$_{15}$N	141.22	[α]$_D$ −1.3	193-5, 92-4^{20}		0.935$^{17/4}$	....	al, chl	B21^4, 3257
8298	Hyoscine *(dl)* or Scopalamine...................... C$_{17}$H$_{21}$NO$_4$	303.36	syr					al, eth, ace, bz	B27^1, 248
8299	Hyoscine, hydrobromide *(d)* .... C$_{17}$H$_{21}$NO$_4$.HBr	384.28	[α]$_D$ + 26.3 (w)		195			w, al	B27^1, 247
8300	Hyoscine, hydrobromide *(dl)* .... C$_{17}$H$_{21}$NO$_4$.HBr	384.28	eff (ace)		185			w, al	B27^1, 248
8301	Hyoscine, hydrobromide *(l)* .... C$_{17}$H$_{21}$NO$_4$.HBr	384.28	lf (al) [α]$'_D$ −26		209			w	B27^2, 64
8302	Hyoscine, hydrobromide, trihydrate *(d)* .... C$_{17}$H$_{21}$NO$_4$.HBr.3H$_2$O	438.33	ta (w) [α]$'_D$ + 26.3 (w,c=3)		55				B27^1, 247
8303	Hyoscine, hydrobromide, trihydrate *(dl)* .... C$_{17}$H$_{21}$NO$_4$.HBr.3H$_2$O	438.33			55-8				B27^1, 248
8304	Hyoscine, hydrobromide, trihydrate-*(l)* .... C$_{17}$H$_{21}$NO$_4$.HBr-3H$_2$O	438.33	ta (w + 3) [α]$_D$-22.8 (w,c=2)					w	B27^2, 64
8305	Hyoscine, hydrochloride, dihydrate *(l)* .... C$_{17}$H$_{21}$NO$_4$.HCl.2H$_2$O	375.85	pr (w)	....	80			w, al	B27, 101
8306	Hyoscine, monohydrate *(dl)* .... C$_{17}$H$_{21}$NO$_4$.H$_2$O	321.38	cr (w + 1)		56-7			al, eth, chl	B27, 102
8307	Hyoscine, monohydrate-*(l)* .... C$_{17}$H$_{21}$NO$_4$.H$_2$O	321.38	cr (w + 1) [α]$^{20/}_D$-28 (w,c=2.7)		59			al, eth, ace, bz, chl	B27^2, 63
8308	Hyoscine, hydrochloride *(l)* .... C$_{17}$H$_{21}$NO$_4$.HCl	339.82	cr (al)		200			w, al	B27^2, 64
8309	Hyoscyamine *(d)* .... C$_{17}$H$_{23}$NO$_3$	289.38	nd (dil al) [α]$'_D$ + 31.3 (al,c=4)		106			al, eth, bz, chl	B21^4, 181
8310	Hyoscyamine *(l)* or Daturine .... C$_{17}$H$_{23}$NO$_3$	289.38	tetr nd (dil al)[α]$^{20/}_D$-1 (al,c=1)		108.5			al, chl	B21^4, 181
8311	Hyoscyamine, hydrobromide-*(l)* .... C$_{17}$H$_{23}$NO$_3$.HBr	370.30	pr	....	152			w, al, chl	B21^4, 182
8312	Hyoscyamine, sulfate *(l)* .... (C$_{17}$H$_{23}$NO$_3$)$_2$.H$_2$SO$_4$	712.86	dlq nd (al or w)		206 anh			w, al	B21^4, 182
8313	Hyoscyamine, hydrochloride-*(l)* .... C$_{17}$H$_{23}$NO$_3$.HCl	325.84	[α]$'_D$ −23.2 (w,c=0.5)		149-51			w, al	B21^4, 182
8314	Hyoscyamine, sulfate *(l)* .... (C$_{17}$H$_{23}$NO$_3$)$_2$.H$_2$SO$_4$.2H$_2$O	712.86	dlq nd (al, w) [α]$'_D$ −28.3 (w)		206			w, al, bz	B21^4, 182
8315	Hypaphorine *(d)* or *N,N*-Dimethyl-*L*-tryptophane betaine .... C$_{13}$H$_{17}$N$_2$O$_2$	246.31	cr (dil al) [α]$^{25/}_D$ + 113.4 (w,c=1.6)	....	255d			w, al	B22^2, 469
8316	Hypnal or Antipyrine chloral hydrate................. C$_{11}$H$_{12}$N$_2$O·Cl$_3$CCH(OH)$_2$	353.64	wh		68			w, al	B24^1, 196
8317	Hypoxanthene or 6-hydroxy purine................... C$_5$H$_4$N$_4$O	136.11	oct nd (w)	....	150 d				B26^2, 252

No.	Name, Synonyms, and Formula	Mol. wt.	Color, crystalline form, specific rotation and λ_{max} (log ϵ)	b.p. °C	m.p. °C	Density	n_D	Solubility	Ref.
8318	D-Iditol or 1,2,3,4,5,6-Hexane hexol HOCH$_2$(CHOH)$_4$CH$_2$OH	182.18	mcl pr (al) [α]$^{20}_D$ + 3.5 (w)	....	73.5			w	B1[4], 2843
8319	D-Idonic acid or 2,3,4,5,6-pentahydroxy hexanoic acid .... HOCH$_2$(CHOH)$_4$CO$_2$H	196.16	nd (w, al) α^{20}_D + 5.2→- 13,7 (mut)	....	205d			w	B3[4], 1257
8320	D-Idonic acid-γ-lactone........................... C$_6$H$_{10}$O$_6$	178.14	pl α^{20}_D 52.6 (w)	....	174			w	B18[4], 3026
8321	D-Idonic acid, phenylhydrazide .... HOCH$_2$(CHOH)$_4$CONHNHC$_6$H$_5$	286.29	[α]$^{20}_D$ −15.3 (w,c=1)	....	115-7			w, al	B15[3], 208
8322	D-Idose .... C$_6$H$_{12}$O$_6$	180.16	syr [α]$^{3/}_D$ + 16 (w)	....				w	B1[4], 4335
8323	L-Idose C$_6$H$_{12}$O$_6$	180.16	syr [α]$^{20/}_D$ −17.4 (w,c = 6.2)	....				w	B1[4], 4336
8324	Imesatin or Isatin-3-imide....................... C$_8$H$_6$N$_2$O	146.15	dk ye pr (dil al)	....	175-6			al	B21[4], 4984
8325	Imidazole or 1,3-Diazole. Glyoxaline .... C$_3$H$_4$N$_2$	68.08	ncl pr (bz)	257, 138.2[12]	90-1	1.0303[101/4]	1.4801[101]	w, al, eth, ace, chl	B23[2], 34
8326	Imidazole, 2-mercapto-1-methyl or Methimazole .... C$_4$H$_6$N$_2$S	114.17	lf (al)	280d	142	....	....	w, al, chl	B24, 17
8327	Imidazole, l-ethyl .... C$_5$H$_8$N$_2$	96.13	....	209-10	....	0.999	....	w	B23, 46
8328	Imidazole, 1-methyl or Oxalmethylene................ C$_4$H$_6$N$_2$	82.11	....	195-6, .94-5	−6	1.0325[20]	1.4970[20]	w, al, eth, ace	B23[2], 35
8329	Imidazole, 4-methyl C$_4$H$_6$N$_2$	82.10	....	263, 120[0.02]	56	1.0416[14.3/]	1.5037[14.3]	w, al	B23[2], 60
8330	Imidazole, 1-phenyl C$_9$H$_8$N$_2$	144.18	....	276,153-4[23]	13	....	1.6025[25]	eth, ace	B23[2], 36
8331	4,5-Imidazoledicarboxylic acid or 1,3-Diazole-4,5-dicarboxylic acid .... C$_5$H$_4$N$_2$O$_4$	156.10	pr	288d		1.749	....		B25[2], 159
8332	2-Imidazolidine thione or N,N-Ethylene-thiourea ...... C$_3$H$_6$N$_2$S	102.16	nd (al), pr (al)		200-3			w, al	B24[2], 4
8333	2-Imidazolidone or Ethylene urea .... C$_3$H$_6$N$_2$O	86.09	nd (chl)		131-3			w, al	B24[2], 3
8334	2-Imidazoline, 2-methyl or 2-methyl-2-glyoxalidine ..... C$_4$H$_8$N$_2$	84.13	hyg	195-8	107			w, al, chl	B23[2], 26
8335	Imino-di-acetic acid or Diglycolamidic acid............ HN(CH$_2$CO$_2$H)$_2$	133.11	rh pr		247.5				B4[3], 1176
8336	Imperatorin or Ammidin .... C$_{16}$H$_{14}$O$_4$	270.29	cr (al)		102			al, eth, ace, bz	B19[4], 2635
8337	Indaconitine or Acetylbenzoyl pseudoaconine.......... C$_{34}$H$_{47}$NO$_{10}$	629.76	cr	....	202-3d			al, eth, chl	B21[4], 2889
8338	Indan or 2,3-dihydroindene .... C$_9$H$_{10}$	118.18	....	178, 73[13]	−51.4	0.9639[20/4]	1.5978[20]	al, eth	B5[4], 1371
8339	Indan, 1-amino or dl-1-hydrindamine .... C$_9$H$_9$NH$_2$	133.20	oil	220.5[747], 96-7[8]	....	1.038[15/4]	1.5613[20]	eth, ace, bz	B12[3], 2798
8340	Indan-5-amino or 5-Hydrindamine.................. 5-H$_2$NC$_9$H$_9$	133.20	nd (peth)	247-9[745], 131[15]	37-8			eth, ace, bz	B12[3], 2800
8341	Indan, 2,3-dibromo or Indene dibromide.............. 2,3-Br$_2$C$_9$H$_8$	275.99	....	144[10]	31-2	1.747[25/4]	1.6290[25]	eth	B5[3], 1203
8342	Indan, 2,3-dichloro or Indene dichloride .... 2,3-Cl$_2$C$_9$H$_8$	187.07	....	87-90[2]		1.254[25/4]	1.5715[23]	....	B5[3], 1202
8343	Indan, 1,1-dimethyl .... C$_{11}$H$_{14}$	146.23	....	191		0.919[20]	1.5135[25/D]	....	B5[4], 1415
8344	Indan, 1,2-dimethyl .... 1,2-(CH$_3$)$_2$C$_9$H$_8$	146.23	....	79-80[10]		0.927[20/4]	1.5186[20/D]	....	B5[4], 1415
8345	Indan, 4,6-dimethyl .... 4,6-(CH$_3$)$_2$C$_9$H$_8$	146.23	....	52.9[1]			1.5325[20/D]	....	B5[2], 395
8346	Indan, 4,7-dimethyl .... 4,7-(CH$_3$)$_2$C$_9$H$_8$	146.23	....	94-7[10]		0.949[20/4]	1.5342[20/D]	....	B5[4], 1416

PHYSICAL CONSTANTS OF ORGANIC COMPOUNDS (Continued)

No.	Name, Synonyms, and Formula	Mol. wt.	Color, crystalline form, specific rotation and λmax (log ε)	b.p. °C	m.p. °C	Density	n_D	Solubility	Ref.
8347	Indan, 5,6-dimethyl 5-6-(CH₃)₂C₉H₈	146.23	....	94[10]	......	0.9449[20]	1.5360[20]/D	....	B5[4], 1416
8348	Indan, 1-ethyl 1-C₂H₅C₉H₉	146.23	....	222	......	0.9348[25]	1.5121[25]/D		B5[4], 1414
8349	Indan, 5-hexyl 5-C₆H₁₃C₉H₉	202.34	....	292.1		0.9114[20]	1.5122[20]D		B5[4], 1474
8350	Indan, 1-methyl 1-CH₃C₉H₉	132.21		188-90, 60[10]		0.9402[20/4]	1.5260[20]		B5[4], 1397
8351	Indan, 2-methyl 2-CH₃C₉H₉	132.20		187, 70[10]		0.9034[20/4]	1.5070[20]		B5[4], 1397
8352	Indan, 4-methyl 4-CH₃C₉H₉	132.20		205.5		0.9577[20]	1.5356[20]/D		B5[4], 1397
8353	Indan, 5-methyl 5-CH₃C₉H₉	132.20		101-2[32], 82[17]		0.9494[20/4]	1.5316[20]		B5[4], 1398
8354	Indan, 4-nitro 4-O₂NC₉H₉	163.18	wh cr (al)	139[10]	44				B5[3], 1204
8355	Indan, per hydro C₉H₁₆	124.23		165-6		0.8334[20]	1.4629[20]/D		B5[4], 292
8356	Indan, 1-phenyl-1,3,3-trimethyl 1-C₆H₅-1,3,3-(CH₃)₃C₆H₆	236.36	tcl pr (al)	307-10, 161-5[12]	52-3	1.0009[20/4]	1.5681[20]	bz, MeOH	B5[4], 2246
8357	Indan, 1,1,4,7-tetramethyl 1,1,4,7(CH₃)₄C₉H₆	174.29		114.2[15]		0.934[25]	1.5216[25]/D		B5[3], 1282
8358	Indan, 1,1,3-trimethyl 1,1,3-(CH₃)₃C₉H₇	160.26		204[748]			1.5082[20]		B5[4], 1433
8359	Indan, 1,1,4-trimethyl 1,1,4-(CH₃)₃C₉H₇	160.26		52.5[1]			1.5157[20]/D		B5[4], 1433
8360	Indan, 1,1,5-trimethyl 1,1,5-(CH₃)₃C₉H₇	160.26		86.[10]		0.9119[20]	1.5126[20]		B5[4], 1433
8361	Indan, 1,1,6-trimethyl 1,1,6-(CH₃)₃C₉H₇	160.26		55[1]			1.5134[20]		API 23, 2(35.5202)
8362	Indan, 1,4,7-trimethyl (CH₃)₃C₉H₇	160.26		95[10]		0.938[20]	1.5252[20]		B5[4], 1434
8363	Indan, 1,5,7-trimethyl 1,5,7-(CH₃)₃C₉H₇	160.26		106.1[15]			1.5231[25]		B5[3], 1267
8364	Indan, 4,5,7-trimethyl 4,5,7-(CH₃)₃C₉H₇	160.26		59.9[1]			1.5322[20]		API 23, 2(35.5202)
8365	1,2-Indandione or α,β-Dioxohydrindene C₉H₆O₂	146.15	gold ye pl or lf (bz, eth)		114-6			al, chl	B7[3], 3593
8366	1,2-Indandione, β-oxime C₉H₆O(=NOH)	161.16	nd (al), (bz)		215-20d			al, bz	B7[3], 3593
8367	1,3-Indandione or 1,3-Dioxohydrindene C₉H₆O₂	146.15	nd (eth, lig)		131-2d	1.37[21]		al, eth, bz	B7[3], 3594
8368	1,3-Indandione, 2,2-dimethylpropoxy or Pivalyl indandione C₁₄H₁₄O₃	230.27	(dil al)		108-10			al, eth, ace	B7[3], 4599
8369	1,3-Indandione, dioxime 1,3-C₉H₆(=NOH)₂	176.19	nd (w)		ca 225d				B7, 695
8370	1,3-Indandione, 2(3-methylbutoxy) or Valone C₁₄H₁₄O₃	230.27	ye (dil al)		67-8			al, eth, ace	B7[3], 4598
8371	1,3-Indandione, 2-phenyl or Danilone C₁₅H₁₀O₂	222.25	lf (al, bz)		149-51			al, eth, ace, bz, chl	B7[3], 4100
8372	1-Indanol or -1-hydroxyindan 1-HOC₉H₉	134.18	pl (peth)	255, 128[12]	54			al, eth, bz, chl	B6[3], 2423
8373	4-Indanol or 4-Hydroxy Indan 4-HOC₉H₉	134.18	(i) tcl pr (peth) (ii) nd (peth)	120[12]	(i) 50 (ii) 40				B6[3], 2427
8374	5-Indanol or 5-Hydroxy indan 5-HOC₉H₉	134.18	nd (peth)	225, 110[8]	56			al, eth	B6[3], 2428
8375	1-Indanone or α-Hydrindone C₉H₈O	132.17	ta, nd (w + 3)	241-2[739], 129[12]	42	1.1028[40/40]	1.561[25]	al, eth, ace, chl, lig	B7[3], 1392
8376	1-Indanone, 3,3-dimethyl C₁₁H₁₂O	160.22		130-1[18]		1.0320[14.5/]	1.5453[14.5]		B7[3], 1450
8377	1-Indanone, 2-nitro C₉H₇NO₃	177.16	ye nd (bz-lig)		117d			w, al, eth, ace, bz	B7[1], 192
8378	1-Indanone, 6-nitro C₉H₇NO₃	177.16	ye lf or nd (peth, al)		74			al, eth, ace, bz, chl	B7[2], 285

No.	Name, Synonyms, and Formula	Mol. wt.	Color, crystalline form, specific rotation and λ_{max} (log ε)	b.p. °C	m.p. °C	Density	n_D	Solubility	Ref.
8379	2-Indanone or β-Hydrindone................. C_9H_8O	132.17	nd (al or eth)	218d	59(61)	$1.0712^{69/4}$	1.538^{67}	al, eth, ace, chl	B7³, 1397
8380	2-Indanone, 5-nitro .... $C_9H_7NO_3$	177.16	br nd (al)	....	141			al, eth, aa	B7, 364
8381	Indanthrene or Dihydroanthraquinonazine-indanthrone.. $C_{28}H_{14}N_2O_4$	442.43	bl nd	....	470-500d				B24², 317
8382	Indazole or 1,2-Benzodiazole................. $C_7H_5N_2$	118.14	nd (al or w)	$267-70^{743}$	147-9			al, eth	B23², 117
8383	Indazole, 1-benzhydrylidene or ω,ω-Diphenyl benzofulvene................. $C_{22}H_{16}$	280.37	og-ye (al)	....	114.5			eth, ace, bz	B5³, 2532
8384	Indazole, 3-chloro .... $C_7H_5ClN_2$	152.50	nd (w or lig)	sub	148			al, eth, bz	B23², 139
8385	Indazole, 4-chloro .... $C_7H_5ClN_2$	152.58	nd (to)	....	156			w, al, eth, ace	B23², 139
8386	Indazole, 2-methyl .... $C_8H_8N_2$	132.16		$261, 135^{16}$	56			al, eth, ace	B23², 118
8387	Indazole, 1-methyl .... $C_8H_8N_2$	132.16		$231, 109^{17}$	60-1			al, ace, eth	B23², 118
8388	Indazole, 3-methyl .... $C_8H_8N_2$	132.16		$280-1, 169-74^9$	113			al, eth, ace	B23², 155
8389	Indazole, 5-methyl .... $C_8H_8N_2$	132.16		$293-4^{747}$	117				B23², 157
8390	Indazole, 4-nitro .... $C_7H_5N_3O_2$	163.14	nd (w)	....	205-7			al, eth, ace, bz, aa	B23², 144
8391	Indazole, 5-nitro .... $C_7H_5N_3O_2$	163.14	yesh nd or col nd (al)	....	208			al, eth, ace, bz, aa	B23², 145
8392	Indazole, 6-nitro .... $C_7H_5N_3O_2$	163.14	nd (w, al, aa)	....	181d			al, eth, ace, bz	B23², 146
8393	Indazole, 7-nitro .... $C_7H_5N_3O_2$	163.14	cr (al)	....	188-90 sub			eth, ace	B23², 150
8394	3-Indazolinone or Benzo pyrazolone................. $C_7H_6N_2O$	134.14	nd or lf (w or MeOH) pl or nd (al)	....	250-2				B24², 59
8395	Indene or Indonaphthene................. C_9H_8	116.16	(aa)	182.6	−1.8	$0.9960^{25/4}$	1.5768^{20}	al, eth, ace, bz, py	B5⁴, 1532
8396	Indene, 1,2-Diphenyl .... $1,2-(C_6H_5)_2C_9H_6$	268.38	lo nd (aa)	....	177-8			eth	B5³, 2473
8397	Indene, 1,3-diphenyl .... $1,3-(C_6H_5)_2C_9H_6$	268.38	(i) nd (aa), (ii) pym (aa)	230^{15}	(i) 68-9 (ii) 85			eth, ace	B5³, 2474
8398	Indene, 2,3-diphenyl .... $2,3-(C_6H_5)_2C_9H_6$	268.38	pr (aa)	$235-40^{12}$	108-9			eth, ace, bz	B5³, 2473
8399	Indene, 2-methyl .... $2-CH_3C_9H_7$	130.19	oil	$187, 62-5^{20}$	....	$0.9034^{20/4}$	1.5070^{20}	eth, ace, bz	B5⁴, 1545
8400	Indene, 3-methyl .... $3-CH_3C_9H_7$	130.19		$198.5d, 70^{10}$	....	$0.9640^{20/4}$	1.5591^{27}	eth, ace, bz	B5⁴, 1545
8401	1-Indenecarboxylic acid .... $1-C_9H_7CO_2H$	160.18	pa ye nd(bz)	$193-5^{12}$	161			eth	B9³, 3068
8402	2-Indenecarboxylic acid .... $2-C_9H_7CO_2H$	160.18	nd or lf(bz)	....	234 sub			al, eth	B9³, 3069
8403	1-Indenone, 2,3-diphenyl .... $C_{21}H_{14}O$	282.35	og-red (lig or al)	....	153-5			al, ace, bz	B7³, 2861
8404	Indican or Indoxyl-β-glucoside .... $C_{14}H_{17}NO_6.3H_2O$	349.34	orh nd (w+3), $[\alpha]^{19}_{546}$ −65.6 (w, c=1)	....	57-8 (hyd) 178-80d (anh)			w, al, ace	B21⁴, 748
8405	Indigo white or 2,2′-Diindoxyl leucoindigo .... $C_6H_{12}N_2O_2$	264.29	ye cr (dil al)	....	....			al, eth	B23², 429
8406	Indigotin .... $C_{16}H_{10}N_2O_2$	262.27		....	390-2d	1.35		....	B24², 233
8407	4,4′-Indigotin dicarboxylic acid .... $C_{18}H_{10}N_2O_2$	350.29	bl nd	....	....				B25, 273
8408	5,5′-Indigotin disulfonic acid, sodium Salt or Indigo carmine .... $C_{16}H_8N_2O_8Na_2S_2$	466.4	dk bl amor or br-red cr	....	....			w, al	B25², 298

No.	Name, Synonyms, and Formula	Mol. wt.	Color, crystalline form, specific rotation and λ_{max} (log ε)	b.p. °C	m.p. °C	Density	n_D	Solubility	Ref.
8409	Indigotin sulfonic acid $C_{16}H_{10}N_4S_2O_7$	342.33	amor	....	200d	....	....	w, al	B25[2], 246
8410	Indirubin or Indigo red $C_{16}H_{10}N_2O_2$	262.27	red or br rh nd (sub)	....	sub	....	....	eth	B24[2], 246
8411	Indole or 1-Benzo [b] pyrrole C_8H_7N	117.15	lf (w, peth) cr (eth)	254, 123-4[5]	52.5	1.22	....	al, eth, bz, lig	B20[4], 3176
8412	Indole, 1-acetyl 1-CH₃CO(C_8H_6N) $C_{10}H_9NO$	159.19		152-3[14] 100[0.001]				eth, ace	B20[4], 3182
8413	Indole, 3-(2-aminoethyl) or Tryptamine 3-(H₂NCHCH₂) $C_8H_7NC_{10}H_{12}N_2$	160.22	nd (al-bz or liq)	137[0.15]	120 (146)			al, ace	B22[4], 4319
8414	Indole, 1,3-dimethyl or N-Methyl skatole 1,3-(CH₃)₂C_8H_5N	145.21	nd	257-60, 119[7]	141-3			eth	B20[4], 3208
8415	Indole, 2,3-dimethyl $C_{10}H_{11}N$ 2,3-(CH₃)₂C_8H_5N	145.26	....	285	105-7				B20[4], 3226
8416	Indole, 2-hydroxy-3-nitroso or β-Isatoxime $C_8H_6N_2O_2$	162.15	gold-ye nd	....	214			al	B21[4], 4988
8417	Indole, 3-hydroxy or Indoxyl 3-HOC₈H₆N	133.15	bt ye pr	....	85			w, al, eth, ace, bz	B21[3], 746
8418	Indole, 1-methyl 1-CH₃C₈H₆N	131.18	....	240-1, 70-5[2]		1.0707[0]		al, eth, bz	B20[4], 3180
8419	Indole, 2-methyl 2-CH₃C₈H₆N	131.18	pl(dil al)nd or lf(w)	272	61	1.07[20/4]		al, eth, ace	B20[4], 3202
8420	Indole, 3-methyl or Skatole 3-CH₃C₈H₆N	131.18	lf(liq)	265-6[755]	97-8			w, al, eth, ace, bz	B20[4], 3206
8422	Indole, 3-methyl-2-phenyl 3-CH₃-2-C₆H₅C₈H₅N	207.27	....	280-90[120]	91-2			al, bz	B20, 474
8423	Indole, 2-phenyl 2-C₆H₅C₈H₅N	193.25	....	290[11]	189			eth, bz	B20[2], 302
8424	Indole, 1,2,3-trimethyl 1,2,3-(CH₃)₃C₈H₄N	159.23	....	283-4[750]	18				B20[4], 3227
8425	3-Indolylacetic acid, 2-methyl $C_{11}H_{11}NO_2$	189.22	ace	195-200				al, eth, ace	B22[4], 1117
8426	2-Indolecarboxylic acid $C_9H_7NO_2$	161.16	ye pl(bz peth)	....	205-8			al, eth	B22[4], 1059
8427	2-Indolecarboxylic acid, 3-hydroxy or Indoxylic acid $C_9H_7NO_3$	177.16	cr(sub)	122-3 sub					B22[2], 168
8428	Indoline or 2,3-Dihydroindole C_8H_9N	119.17	....	228-30, 70-5[2]		1.069[20/4]	1.5923[20]	eth, ace, bz	B20[4], 2896
8429	3-Indolylacetic acid or Heterouxin (C_8H_5N)CH₂CO₂H	175.19	lf(bz)pl(chl)	....	165-6			al, eth, ace	B20[4], 1088
8430	Indone-2,3-dibromo $C_9H_6Br_2O^-$	287.95	og-ye nd(al aa)	....	123			al, eth, chl	B7[3], 1647
8431	Indophenin $C_{24}H_{14}N_2O_2S_2$	426.20	bl nd or pw	....	d				B21[2], 330
8432	Indopenol $C_{12}H_9NO_2$	199.21	red br pl (ace-peth)	....	160			w, al, eth, bz, chl	B13[3], 1047
8433	Indoxazene or 4,5-Benzoisoxazol C_7H_5NO	119.12	oil	100[26]	....	1.1727[21/4]	1.5570[20]	eth	B27[2], 15
8434	Inosine or Hyoxanthosine-β- -ribose $C_{10}H_{12}N_4O_5$	268.23	pl (w + 2) nd (80% al)	....	90 (+ 2w) 218d (anh)			al	B31, 25
8435	D-Inositol or 1,2,3,4,5,6-cyclohexanehexol $C_6H_{12}O_6$	180.16	pr (w + 2) (al), $[\alpha]_D$ + 65.0 (w, 12%)	....	249-50			w, aa	B6[3], 6925
8436	DL-Inositol or Phaseo mannitol $C_6H_{12}O_6$	180.16	mcl pr (w)cr(gl aa)	319(vac)	253	1.752[15]		w, aa	B6[3], 6925
8437	L-Inositol $C_6H_{12}O_6$	180.16	nd(w + 2)	250 vac	247	1.598[20]		w, aa	B6[3], 6925
8438	Inulin or Plant starch ($C_6H_{10}O_5$)n	7000	wh amor or pw, $[\alpha]^{21}\Sigma_D$ - 38.3	....	178d	1.35[20/4]		....	J1952, 2384

No.	Name, Synonyms, and Formula	Mol. wt.	Color, crystalline form, specific rotation and λ_{max} (log ε)	b.p. °C	m.p. °C	Density	n_D	Solubility	Ref.
8439	Iodoacetamide .... ICH$_2$CONH$_2$	184.97	cr(w)	....	95	....	....	w	B2[4], 536
8440	Iodoacetic acid .... ICH$_2$COOH	185.95	pl(w,peth)	d	83	....	....	w, al	B2[4], 534
8441	Iodoacetic acid, ethyl ester or Ethyliodoacetate......... ICH$_2$COOC$_2$H$_5$	214.00	oil	178-80, 73[16]		1.8173[13/4]	1.5079[13]	al, eth	B2[4], 535
8442	Iodogorgoic acid (d) or 3,5-Diidotyrosine(d) .... C$_9$H$_9$I$_2$NO$_3$	432.99	yesh nd(w or 70% al), [α]^{20}$_D$ +2.9 (HCl, c=5)		213				B14[3], 1563
8443	Iodogorgoic acid (dl) .... C$_9$H$_9$I$_2$NO$_3$	432.99	nd (50%) al, pl (w)		200d				B14[3], 1563
8444	Iodogorgoic acid (l) .... C$_9$H$_9$I$_2$NO$_3$	432.99	nd(w or 70%), [α]^{20}$_D$ -2.98 (4% HCl, c=5)	213d				B14[2], 366	B14[3], 1563
8445	Ionene .... C$_{13}$H$_{18}$	174.29	....	238-9[730], 90-1[4]		0.9356[20/4]	1.5257[20]	al, eth, bz, chl	B5[4], 1445
8446	α-Ionol .... C$_{13}$H$_{22}$O	194.32	oil	127[15]		0.9474[20/4]	1.4735[20]	al, eth, ace	B6[3], 402
8447	β-Ionol .... C$_{13}$H$_{22}$O	194.32		131[15], 89[0.7]		0.9243[20/4]	1.4969[20]	al, eth, ace	B6[3], 401
8448	α-Ionone (d,trans) or 4-(2,6,6-trimethyl-2-cyclohenyl)-3-buturic-2-one............ C$_{13}$H$_{20}$O	192.31	[α]^{25}$_D$ +347	....			1.5061[20]		B7[3], 640
8449	α-Ionone (dl) .... C$_{13}$H$_{20}$O	192.31	....	146-7[28]		0.9298[21]	1.5041[20]	al, eth, ace	B7[3], 641
8450	α-Ionone (l) .... C$_{13}$H$_{20}$O	192.31	[α]^{27}$_D$ -406	73-7[0.15]			1.5000[25]	al, eth, ace	B7[3], 640
8451	α-Ionone, semicarbazone (dl) .... C$_{14}$H$_{23}$N$_3$O	249.36	(60% al)	....	(i) 107-8, (ii) 143				B7[3], 644
8452	β-Ionone .... C$_{13}$H$_{20}$O	192.31		140[18], 72-4[0.1]		0.9462[20/4]	1.5198[20]	al, eth	B7[3], 634
8453	β-Ionone, semicarbazone .... C$_{14}$H$_{23}$N$_3$O	249.36	nd(al)		149			al, eth, bz, chl	B7[3], 639
8454	Irene or anhydro irone (?)............. C$_{14}$H$_{20}$	188.31	....	120-5[10]		0.9332[20/4]	1.5217[20]	ace, bz	B5[4], 1460
8455	β-Irone or 4-(2,5,6,6-tetramethyl-1-cyclohexenyl)-3-butene-2-one .... C$_{14}$H$_{22}$O	206.32		85-90[0.1]		0.9434[21/4]	1.5017[20]	al, eth, bz, chl, lig	B7[3], 666
8456	Isatic acid or 2-Aminobenzoyl formic acid .... H$_2$NC$_6$H$_4$COCO$_2$H	165.15	pw	....	d			w	B14[3], 1650
8457	Isatin or 2,3-Indolinedione.................... C$_8$H$_5$NO$_2$	147.14	yesh red pr(sub)	sub	203-5	....		al, ace, bz	B21[4], 4981
8458	Isatin, 1-acetyl .... C$_{10}$H$_7$NO$_3$	189.17	ye pr or nd		144-5			al, ace	B21[4], 4998
8459	Isatin, chloride or 2-chloro-3-indolone .... C$_8$H$_4$ClO	165.58	br nd		180d			al, eth, aa	B21[4], 3720
8460	Isatin, 1-methyl .... C$_9$H$_7$NO$_2$	161.16	red-ye orh nd (w)	....	134			al, eth, ace, bz	B21[4], 4991
8461	Isatin, 5-methyl .... C$_9$H$_7$NO$_2$	161.16	red pl (w)nd (w or al)	....	187			al	B21[4], 5451
8462	Isatin, 7-methyl .... C$_9$H$_7$NO$_2$	161.16		....	267				B21[4], 5454
8463	Isatin, 5-nitro .... C$_8$H$_4$N$_2$O$_4$	192.13	ye nd al		254-5d				B21[4], 5015
8464	Isatin, 2-oxime .... C$_8$H$_6$N$_2$O$_2$	162.15	ye-og nd al or w		198-200d			eth, ace, aa	B21[4], 4987
8465	Isatin, 3-oxime .... C$_8$H$_6$N$_2$O$_2$	162.15	gold ye nd		225d			al	B21[4], 4988
8466	Isatoic acid, anhydride or N-carboxyanthranilic anhydride C$_8$H$_5$NO$_3$	163.13	pr(al or gl aa)cr (al)	....	243d				B27[2], 299

No.	Name, Synonyms, and Formula	Mol. wt.	Color, crystalline form, specific rotation and λ_{max} (log ε)	b.p. °C	m.p. °C	Density	n_D	Solubility	Ref.
8467	d-α-Isatropic acid or d,α-1-phenyl-1,4-tetralindicarboxylic acid . . . . C₁₈H₁₆O₄	296.33	pr, [α]25/D +9.44 (al, c=12.6)	. . . .	239d	. . . .	. . . .		B9[1], 417
8468	dl-α-Isatropic acid or dl-α-1-phenyl-1,4-tetralindicarboxylic acid . . . . C₁₈H₁₆O₄	296.3	cr(chl-peth)	d	238-9	. . . .	. . . .	aa	B9[3], 4635
8469	dl-β-Isatropic acid . . . . C₁₈H₁₆O₄	296.33	pl w	. . . .	208-9	. . . .		al, aa	B9[3], 4635
8470	l,β-Isatropic acid C₁₈H₁₆O₄	296.33	[α]D -8.8 (al, c=5)	. . . .	197	. . . .		al, aa	B9[3], 4634
8471	Isoapiol . . . . C₂H₁₄O₄	222.24	mcl pr lf or nd (al)	303-4, 189[33]	56	. . . .		al, eth, ace, bz	B19[4], 1033
8472	Isobergaptene . . . . C₁₂H₈O₄	216.21	cr(al)	. . . .	222-3	. . . .		al, diox	B19[4], 2638
8473	Isoborneol (d) or 2-Hydroxybornane C₁₀H₁₈O	154.26	cr(peth)	. . . .	212 (sealed tube)	. . . .		al, eth, bz, chl	B6[3], 299
8474	Isoborneol (dl) or α,β-Camphol . . . . C₁₀H₁₈O	154.26	ta (peth)	sub	212(sealed tube)	. . . .		al, eth, chl	B6[4], 282
8475	Isoborneol (l) or β-Camphol . . . . C₁₀H₁₈	154.26	(peth), [α]D +33.9 (al)		214 (218)	. . . .		al, eth, chl	B6[4], 282
8476	Isoborneol, acetate (d) CH₃CO₂C₁₀H₁₇	196.29	[α]20/D -50.2 (al)	112[17]	. . . . . .	0.9905[20/4]	1.4633[20]	al, ace	B6[2], 90
8477	Isoborneol, acetate (dl) CH₃CO₂C₁₀H₁₇	196.29	. . . .	115-7[21]	. . . . . .	0.9841	1.4640[20]	al, ace	B6[4], 283
8478	Isoborneol, acetate (l) CH₃CO₂C₁₀H₁₇	196.29	. . . .	225, 123-7[..]	<30	1.003[..]		al, ace	B6, 89
8479	Isoborneol, formate (d) HCO₂C₁₀H₁₇	182.27	[α]20/D +129.5 (al, c=5)	94[15]	. . . . . .	1.0136[20/4]	1.4678[22]	al, ace	B6[3], 301
8480	Isobornylamine C₁₀H₁₇NH₂	153.27	pw, [α]D -47.7 (4% al)	. . . .	184	. . . .		eth, ace	B12[3], 195
8481	Isocalycanthine C₂₂H₂₈N₄.H₂O	183.26	rh	. . . .	235-6	. . . .		al, eth, ace, chl	Am32, 1305
8482	Isocamphane (d) or Dihydrocamphene C₁₀H₁₈	138.25	cr (MeOH), [α]20/D +8.68 (bz, p=20)	166[750]	62-3	. . . .		al, ace	B5[3], 263
8483	Isocamphane (dl) . . . . C₁₀H₁₈	138.25	cr (MeOH)	165-6[730]	65-7	0.8276[67/4]	1.4419[67]	al, ace, bz	B5[3], 263
8484	Isocamphane (l) . . . . C₁₀H₁₈	138.25	cr (al), [α]20/D -8.5 (al)	164-5[757], 62-3[17]	64	. . . .		al, ace	B5[3], 263
8485	Isocamphoric acid (d) or α-trans-1,2,2-Trimethyl-1,3-cyclopentane-dicarboxylic acid . . . . C₁₀H₁₆O₄	200.24	lf (w), [α]20/D +48.6	. . . .	171-2	. . . .		al, aa	B9, 762
8486	Isocamphoric acid (dl) C₁₀H₁₆O₄	200.24	pl(al or gl aa)cr(w)	. . . .	197	1.249	. . . .	eth	B9[3], 3879
8487	Isocamphoric acid (l) C₁₀H₁₆O₄	200.24	tetr, [α]17/D -48.4 (MeOH, p=9.9)	. . . .	173	. . . .		al, aa	B9[3], 3878
8488	Isocarotene or Dehydro-β-carotene . . . . C₄₀H₅₄	534.88	vt pr(bz-MeOH) vt nd lf(bz)	. . . .	192-3	. . . .			B5[3], 2515
8489	Isocarvomenthol (d) . . . . C₁₀H₂₀O	156.27	[α]D + 20.2	110[20]	. . . . . .	0.904[20/4]	1.4669[18]	al, eth, ace	B6[3], 131
8490	Isocarvomenthol (l) . . . . C₁₀H₂₀O	156.27	[α]16/D -17.7	106[17]	. . . . . .	0.9109[20/4]	1.4662[20]	al, eth, ace	B6[4], 149
8491	Isocodeine . . . . C₁₈H₂₁NO₃	299.37	pl (bz) pr (AcoEt or aa), [α]15/D -152 (chl, c=2)	d	171-2	1.87[4]	1.675	. . . .	B27[2], 175
8492	Isocorybulbine . . . . C₂₁H₂₅NO₄	355.44	lf (al), [α]15/D +301 (chl, c=1)	. . . .	187-8	1.045[20/4]	. . . .	al, chl	B21[4], 4779

No.	Name, Synonyms, and Formula	Mol. wt.	Color, crystalline form, specific rotation and λ_{max} (log ε)	b.p. °C	m.p. °C	Density	n_D	Solubility	Ref.
8493	Isocorydine or Corytuberine methyl ether . . . . $C_{20}H_{23}NO_4$	341.41	pl, $[\alpha]^{20}_D$ +195.3 (chl)	. . . .	185	. . . .	. . . .	chl	B21[4], 2755
8494	Isocoumarin or o-(β-Hydroxyvinyl) benzoic acid lactone $C_9H_6O_2$	146.15	pl(bz)	285-6[719]	47	. . . .	. . . .	al, eth, bz	B17[4], 5062
8495	Isocyanuric acid, trimethyl ester . . . . $C_6H_9N_3O_3$	171.16	mcl pr (w or al)	274	176-7	. . . .	. . . .	al	B26[2], 134
8496	Isoderritol . . . . $C_{21}H_{21}O_6$	370.41	ye lf (MeOH)	. . . .	150	. . . .	. . . .	. . . . . .	B18[4], 3351
8497	Isodurene or -1,2,3,5-tetramethylbenzene . . . . 1,2,3,5-$(CH_3)_4C_6H_2$	134.22	. . . .	198, 24.4[10]	-23.7	0.8903[20/4]	1.5130[20]	al, eth, ace, bz	B5[4], 1073
8498	8-Isoestradiol . . . . $C_{18}H_{24}O_2$	272.39	cr (dil MeOH chl), $[\alpha]^{20}_D$ +18 (diox)	. . . .	181	. . . .	. . . .	al, diox	B6[3], 5332
8499	8-Isoestrone or 8-Epiestrone. . . . . $C_{18}H_{22}O_2$	270.38	cr (MeOH), $[\alpha]^{20}_D$ +94 (diox)	. . . .	247	. . . .	. . . .	eth, diox	B8[3], 1170
8500	Iso-β-eucaine (dl) or 4-Benzoyloxy-2,2,6-trimethyl piperi- dine . . . . $C_{15}H_{21}NO_2$	247.34	. . . .	188[19]	<-5	1.0467[22/22]	. . . .	al	B21[4], 131
8501	Iso-β-eucaine, hydrochloride (l) $C_{15}H_{21}NO_2 \cdot HCl$	283.80	nd (w) $[\alpha]_{5461}$ +17 (w,c=1)	. . . .	271-3	. . . .	. . . .	w	B21[2], 16
8502	Iso-β-eucaine, hydrochloride (dl) . . . . $C_{15}H_{21}NO_2.HCl$	283.80	ta (w) pl (aq al or al-eth)	. . . .	269-71	. . . .	. . . .	w	B21[2], 15
8503	Iso-β-eucaine, hydrochloride (l) . . . . $C_{15}H_{21}NO_2.HCl$	283.80	nd α_{5461} -16.3 (w,c=1)	. . . .	271-3	. . . .	. . . .	w	B21[2], 15
8504	Isoeugenol (cis) or 2-Methoxy-4-propenylphenol 2-CH_3O-4-$(CH_3CH=CH)$-C_6H_3OH	164.21	. . . .	134-5[13], 80-1[0.5]	. . . .	1.0837[20/4]	1.5726[20]	. . . .	B6[3], 4992
8505	Isoeugenol (trans) 2-CH_3O-4-$(CH_3CH=CH)C_6H_3OH$	164.21	. . . .	141-2[13]	33-4	1.0852[20/4]	1.5784[20]	al, eth	B6[3], 4992
8506	Isoeugenol (cis), acetate 2-CH_3O-4-$(CH_3CH=CH)C_6H_3O_2CCH_3$	206.24	. . . .	160-2[13]	. . . .	1.0947[19/4]	1.5418[20]	eth	B6[3], 5007
8508	Isoeugenol, acetate (trans) . . . . 2-CH_3O-4-$(CH_3CH=CH)C_6H_3O_2CCH_3$	206.24	nd (al bz-lig)	282-3	80-1	1.0251[100/4]	1.5052[100]	. . . .	B6[3], 5008
8509	Isoflavone, 4',7-dihydroxy or Daidzein . . . . $C_{15}H_{10}O_4$	254.25	pa ye pr (50% al)	sub	323d	. . . .	. . . .	al, eth	B18[4], 1805
8510	Isoflavone, 4',5,7-trihydroxy or Genistein . . . . $C_{15}H_{10}O_5$	270.25	nd (eth) pr (dil al)	. . . .	301-2d	. . . .	. . . .	al, eth	B18[4], 2724
8511	Isofurfurine . . . . $C_{15}H_{12}N_2O_3$	268.28	nd (w)	. . . .	143	. . . .	. . . .	. . . . . .	B27, 674
8512	Isogeraniolene or 2,6-Dimethyl-1,3-heptadiene . . . . $CH_2=C(CH_3)CH=CHCH_2CH(CH_3)_2$	124.23	. . . .	143-4[755], 31[7]	. . . .	0.7561[20/4]	1.4520[20]	eth, bz	B1, 1015
8513	Isoleucine (d) $C_2H_5CH(CH_3)CH(NH_2)CO_2H$	131.17	$[\alpha]^{20}_D$ -12.2H$_2$O -(3,2%)	. . . .	283-4 d	. . . .	. . . .	w	B4[3], 1458
8514	Isoleucine (l) $C_2H_5CH(CH_3)CH(NH_2)CO_2H$	131.17	$[\alpha]^{25}_D$ +12.2 (3.2% in H$_2$O) 36.7 (lNHCl)	. . . .	285-6 d	. . . .	. . . .	w	B4[3], 1454
8515	Isoleucine, allo (l) $C_2H_5CH(CH_3)CH(NH_2)CO_2H$	131.17	$[\alpha]^{20}_D$ -14.2H$_2$O 2	. . . .	dec 280-1	. . . .	. . . .	w	B4[3], 1462
8516	Isolysergic acid (d) . . . . $C_{16}H_{16}N_2O_2$	268.32	cr (w + 2) $[\alpha]^{20}_D$ +281 (py,c=1)	. . . .	218d	. . . .	. . . .	py	B27[2], 860

No.	Name, Synonyms, and Formula	Mol. wt.	Color, crystalline form, specific rotation and λ_{max} (log ε)	b.p. °C	m.p. °C	Density	n_D	Solubility	Ref.
8517	D-Isomannide or 1,4,3,6-Dianhydro-D-mannitol .... $C_6H_{10}O_4$	146.14	mcl cr $[\alpha]^{26}_D$ +62.2 (chl) +91 (w)	274d	87-9	....		w	B19[4], 990
8518	Isomenthol (d) or 2-isopropyl-5-methyl cyclohexanol..... $C_{10}H_{20}O$	156.27	nd (dil al) $[\alpha]^{20}_D$ +26.5 (al,c=4)	218.6, 96.5[10]	82 (85)	....		al, eth, aa	B6[4], 151
8519	Isomenthol (dl) .... $C_{10}H_{20}O$	156.27	nd	218.5, 97.4[16.5]	53-4	0.9040[30]	1.4510[60]	al, eth, aa	B6[4], 152
8520	Isomenthol (l) .... $C_{10}H_{20}O$	156.27	$[\alpha]^{25}_D$ −24.1		82.5			al, eth, aa	B6[4], 152
8521	Isomenthone $C_{10}H_{18}O$	154.25	....	89-90[15]		0.8995[20/4]	1.4527[20/]$_D$	....	B7[3], 151
8522	α-Isomorphine $C_{17}H_{19}NO_3$	285.35	nd (MeOH-AcOEt) $[\alpha]^{15}_D$ −167 (MeOH,c=3)		248			al, MeOH	B27[2], 174
8523	Isonicotine or 4-(4-pyridyl) piperidine............ $C_{10}H_{14}N_2$	162.24	hyg wh nd	292	80			al, eth, bz, lig	B23, 119
8524	Isonicotinaldehyde or 4-Pyridine carboxaldehyde........ 4-C_5H_4NCHO	107.11	....	77-8[12]	......		1.5423[20]	w, al, aa, chl	B21[4], 3529
8525	Isonicotinic acid or 4-Pyridine carboxylic acid.......... 4-$C_5H_4NCO_2H$	123.11	nd (w)	sub 260[15]	319			......	B22[], 318
8526	Isonicotinic acid-2,6-dihydro or Citrazinic acid........ 4-$C_5H_4NCO_2H$	155.12	yesh-gr pw (w)	....	>330d			......	B22[4], 2459
8527	Isonicotinic acid, ethyl betaine .... $C_8H_9NO_2$	151.17	nd		241d			w, al	B22, 47
8528	Isonicotinic acid, ethyl ester or Ethylisonicotinate ...... 4-$C_5H_4NCO_2C_2H_5$	151.17	nd	220, 110[15]	23	1.1052[20]	1.5177[20]	al, eth, bz, chl	B22[4], 521
8530	Isonicotinic acid, hydrazide or Isoniazid .... 4-$C_5H_4NCONHNH_2$	137.14	nd (al)	....	171				B22, 47
8531	Isonicotinic acid, methyl ester or Methyl isonicotinate .... 4-$C_5H_4NCO_2CH_3$	137.14	nd	209d, 104[21]	8.5	1.1599[20/4]	1.5315[20]	al, eth, bz	B22[4], 545
8532	Isonicotinonitrile .... 4-C_5H_4NCN	104.11	nd (liq-eth)		83			w, al, eth, bz	B22[4], 520
8533	Isopapaverine, N-benzyl $C_{27}H_{27}NO_4$	429.52	ye lf (al)	....	139-40			eth	B22[4], 542
8534	Isopapaverine, N-ethyl .... $C_{22}H_{25}NO_4$	367.45	pr(al)		ca 101			eth, bz	B21, 229
8535	Isopapaverine, N-methyl $C_{21}H_{23}NO_4$	353.24	ye hg mcl pr (al)		129-31			w	B21[4], 2799
8536	Isopelletierine or 2-Acetonyl piperidine $C_8H_{15}NO$	141.22	oil	....	91-2[14]	0.9624[20/4]	1.4683[20]	al, chl	B21[4], 3264
8537	Isopelletriene, N-methyl .... $C_9H_{17}NO$	155.24	....	96-8[13]	......	0.9478[20/4]	1.4674[20]	w, lig	B21[4], 3265
8538	Isophenolphthalein or 3-(o-hydroxy phenyl)-3-(p-hydroxy phenyl) phthalide $C_{20}H_{14}O_4$	318.33	(dil aa)		189-90			al, ace	B10[3], 2013
8539	3-Isophenothiazin, 3-one or Azthione thiazone $C_{12}H_7NO_5$	213.26	red (dil al)		164			al, bz, chl	B27[1], 251
8540	3-Isophenothiazin-3-one, 7-hydroxy or Thionol .... $C_{12}H_7NO_2S$	229.26	red br pw or nd (aa)		>360			al	B27[2], 109
8541	Isophorone or 3,5,5-trimethyl-2-cyclohexene-1-one ..... $C_9H_{14}O$	138.21	....	214[754], 99[18]		0.9229[20]	1.4759[20]	al, eth, ace	B7[3], 283
8542	Isophthaldehyde or 1,3-Benzene dicarboxaldehyde....... 1,3-$(OHC)_2C_6H_4$	134.14	nd(dil al)	245-8[77]	89-90			al, ace, bz	B7[3], 3459
8543	Isophthalamide 1,3-$(H_2NOC)_2C_6H_4$	164.18	pl (w)	....	280				B9[1], 372
8544	Isophthalamide, N,N,N′,N′-tetraethyl .... 1,3-$C_6H_4[CON(C_2H_5)_2]_2$	276.38	....	242[12]	85			al, eth, ace, bz	B9[3], 4242
8545	Isophthalic acid or 1,3-Benzene dicarboxylic acid .... 1,3-$C_6H_4(CO_2H)_2$	166.14	nd (w or al)	sub	348			al, aa	B9[3], 4240
8546	Isophthalic acid, 2-amino 2-$H_2NC_6H_3(CO_2H)_2$-1,3	181.15	pl (al), nd (aa)	sub 267	>260			al, eth	B14[2], 337

No.	Name, Synonyms, and Formula	Mol. wt.	Color, crystalline form, specific rotation and λ_{max} (log ε)	b.p. °C	m.p. °C	Density	n_D	Solubility	Ref.
8547	Isophthalic acid, 2-amino, dimethyl ester . . . . . 2-$H_2NC_6H_3(CO_2CH_3)_2$-1,3	209.20	nd (al)	. . . .	103-4				B14[2], 337
8548	Isophthalic acid, 4-amino . . . . 4-$H_2NC_6H_3(CO_2H)$-1,3	181.15	nd (w)	. . . .	336-7			al, eth, ace, aa	B14[1], 633
8549	Isophthalic acid, 4-amino, dimethyl ester . . . . . 4-$H_2NC_6H_3(CO_2CH_3)_2$-1,3	209.20	nd(al)	. . . .	131.5				B14[2], 337
8550	Isophthalic acid, 4-amino . . . . 5-$H_2NC_6H_3(CO_2H)_2$-1,3	181.15	pr (al) pl (w)	sub	>360				B14[1], 636
8551	Isophthalic acid, 5-amino, dimethyl ester . . . . . 5-$H_2NC_6H_3(CO_2CH_3)_2$-1,3	209.20	lf or pl (MeOH)		176			eth	B14, 556
8552	Isophthalic acid, 4-bromo . . . . 4-$BrC_6H_3(CO_2H)_2$-1,3	245.04	nd (al)		287			al	B9[3], 4247
8553	Isophthalic acid, 4-chloro . . . . 4-$ClC_6H_3(CO_2H)_2$-1,3	200.58	nd (w)		295			al	B9[3], 4245
8554	Isophthalic acid, 5-chloro . . . . 5-$ClC_6H_3(CO_2H)_2$-1,3	200.58	nd (w + ½)		278 (anh)			al	B9[3], 4246
8555	Isophthalic acid, 4,6-dichloro . . . . 4,6-$Cl_2C_6H_2(CO_2H)_2$-1,3	235.03	nd (w, dil al)		280			al, eth, chl	B9[3], 4246
8556	Isophthalic acid, 4,5-dimethoxy or Isohemipinic acid . . . . . 4,5-$(CH_3O)_2C_6H_2(CO_2H)_2$-1,3	226.19	nd (w)		245-6			al, eth	B10[3], 2435
8557	Isophthalic acid, diethyl ester or Diethyl isophthalate . . . . . 1,3-$C_6H_4(CO_2C_2H_5)_2$	222.24	. . . .	302, 170[2·4]	11.5	1.1239[17/4]	1.508[18]	. . . .	B9[3], 4241
8558	Isophthalic acid, dimethyl ester or Dimethylisophthalate . . 1,3-$C_6H_4(CO_2CH_3)_2$	194.19	nd (dil al)	282, 124[12]	67-8	1.194[20/4]	1.5168[20]		B9[3], 4241
8559	Isophthalic acid, 4,6-dimethyl or α-Cumidic acid . . . . 4,6-$(CH_3)_2C_6H_2(CO_2H)_2$-1,3	194.19	nd (w) pr (al-bz) lf (sub)	sub	266			al	B9[3], 4298
8560	Isophthalic acid, 2-hydroxy . . . . 2-$HOC_6H_3(CO_2H)_2$-1,3	182.13	nd (w + 1)		244-5			al, eth, chl	B10[3], 2192
8561	Isophthalic acid, 4-hydroxy . . . . 4-$HOC_6H_3(CO_2H)_2$-1,3	182.13	nd (w) lf (dil al)		310			al, eth	B10[3], 2193
8562	Isophthalic acid, 5-hydroxy . . . . 5-$HOC_6H_3(CO_2H)_2$-1,3	182.13	nd (w + 2) cr (aq al)	sub	293			al, eth, bz	B10[3], 2195
8563	Isophthalic acid, 5-methyl or Uvitic acid . . . . 5-$CH_3C_6H_3(CO_2H)_2$-1,3	180.17	nd (w)		298			al, eth, ace	B9[3], 4274
8564	Isophthalic acid, 2-nitro, dimethyl ester . . . . . 2-NO_2-$C_6H_3(CO_2CH_3)_2$-1,3	239.19	nd (w or al)		135			al, eth	B9[1], 373
8565	Isophthalic acid, 5-nitro . . . . 5-$NO_2C_6H_3(CO_2H)_2$-1,3	211.13	gr lf (+ ³/₂w)		260-1 (anh)			al, eth	B9[1], 373
8566	Isophthalic acid, 5-nitro, diethyl ester . . . . . . 5-$NO_2C_6H_3(CO_2C_2H_5)_2$-1,3	267.24	nd (al)		83.5			al, eth	B9, 840
8567	Isophthalic acid, 5-nitro, dimethyl ester . . . . . 5-$NO_2C_6H_3(CO_2CH_3)_2$-1,3	239.19	nd (dil al)		123			al, eth	B9[2], 611
8568	Isophthalic acid, tetrabromo . . . . 1,3-$(HO_2C)_2C_6Br_4$	481.74	nd (w)		288-92				B9, 839
8569	Isophthalonitrile . . . . 1,3-$C_6H_4(CN)_2$	128.14	nd (al)		162			al, eth, bz, chl	B9[3], 4243
8570	Isophthalylalcohol or m-Xylylene glycol . . . . . . 1,3-$(HOCH_2)_2C_6H_4$	138.17	nd (bz)	154-9[13]	57	1.1359[53]		w, al, eth	B6[3], 4600
8571	Isophthalyl chloride . . . . 1,3-$C_6H_4(COCl)_2$	203.03	pr (eth)	276	43-4	1.3880[17/4]	1.570[47]	eth	B9[3], 4242
8572	Isopilocarpine or N-Methyl isopilocarpidine . . . . $C_{11}H_{16}N_2O_2$	208.26	pr	261[10]				w, al, eth, bz, chl	B27[2], 697
8573	Isopimpinellin . . . . $C_{13}H_{10}O_5$	246.22	ye nd (MeOH)	. . . .	151			MeOH	B19[4], 2811
8574	Isopomiferin . . . . $C_{25}H_{24}O_6$	270.25	nd (dil al)	. . . .	265d			al	B19[4], 5231
8575	Isoprene or 2-Methyl-1,3-butadiene . . . . CH_2=$C(CH_3)$-CH=CH_2	68.13	. . . .	34	−146	0.6810[20/4]	1.4219[20]	al, eth, ace, bz	B1[4], 1001
8576	Isopropenyl methyl ketone or 3-Methyl-3-buten-2-OH. . . CH_2=$C(CH_3)COCH_3$	84.13	. . . .	98	-54	0.8527[20/4]	1.4220[20]	al	B1[4], 3462
8577	Isopulegol (d) or Δ[8(9)]-p-Menthenol-3 . . . . $C_{10}H_{18}O$	154.26	$[α]_{5461} + 29.3$	212, 93-4[14]	. . . . . .	0.9110[20/4]	1.4723[20]	al, eth	B6[3], 257
8578	Isopulegol (l) . . . . $C_{10}H_{18}O$	154.26	$[α]^{20}_{5461} -25.9$	212, 94[14]		0.9110[20/4]	1.4723[20]	al, eth	B6[3], 257

No.	Name, Synonyms, and Formula	Mol. wt.	Color, crystalline form, specific rotation and λ_{max} (log ε)	b.p. °C	m.p. °C	Density	n_D	Solubility	Ref.
8579	α-Isoquinine .... $C_{20}H_{24}N_2O_2$	324.43	(bz-peth) $[\alpha]^{18}_D$ −245 (al,c=1)	....	196.5			al, eth	B23², 414, 423
8580	β-Isoquinine .... $C_{20}H_{24}N_2O_2$	324.43	pr (dil al or amor) $[\alpha]^{17}_D$ −187 (97% al,c=1)	....	190-1			al, bz, chl	B23², 413
8581	Isoquinoline .... C_9H_7N	129.16	hyg pl	242.2⁷⁴³, 142⁴⁰	26.5	1.0986²⁰	1.6148²⁰	al, eth, ace, bz	B20⁴, 3410
8582	Isoquinoline, hydrochloride .... $C_9H_7N \cdot HCl$	165.63	pr or pl (al)	....	209			w	B20⁴, 3412
8583	Isoquinoline, hydrogen sulfate .... $C_9H_7N \cdot H_2SO_4$	227.15	pr or pl (al)	....	209			w	B20⁴, 3413
8584	Isoquinoline, 1-amino 1-$H_2N(C_9H_6N)$	144.19	pl (w)		123			al	B22⁴, 4736
8585	Isoquinoline, 3-amino 3-$H_2N(C_9H_6N)$	144.19			178-9				B22⁴, 4744
8586	Isoquinoline, 5-amino 5-$H_2N(C_9H_6N)$	144.19	pa ye nd (peth)	sub	128				B22⁴, 4747
8587	Isoquinoline, 4-bromo 4-$Br(C_9H_6N)$	208.09	(peth)	280-5	40-3			eth	B20⁴, 3448
8588	Isoquinoline, 1-chloro .... 1-$Cl(C_9H_6N)$	163.61	....	274-5, 135-40¹⁰	37 0			bz	B20⁴, 3444
8589	Isoquinoline, 6,7-dimethoxy-1,2-dimethyl-1,2,3,4-tetrahydro or Carnegine .... $C_{13}H_{19}NO_2$	221.30	pa br syr $[\alpha]^{25}_D$ +20 (w)	170¹	....			w, al, eth, chl	B21⁴, 2125
8590	Isoquinoline, 1-hydroxy or Isocarbostyril .... 1-$HO(C_9H_6N)$	145.16	mcl (bz) nd (bz al w)	240 sub	209-10			al	B21⁴, 1208
8591	Isoquinoline, 7-hydroxy 7-$HO(C_9H_6N)$	145.16			230			al	B21⁴, 1214
8592	Isoquinoline, 7-methoxy 7-$CH_3O(C_9H_6N)$	159.19	....	182-6³⁴	49			al, lig	B21⁴, 1214
8593	Isoquinoline, 1-methyl 1-$CH_3(C_9H_6N)$	143.19		248, 124-5¹⁰	10	1.0777²⁰/⁴	1.6095²⁰	eth, ace, bz	B20⁴, 3505
8594	Isoquinoline, 3-methyl 3-$CH_3(C_9H_6N)$	143.19	cr(eth)	246	68			eth, ace	B20⁴, 3507
8595	Isoquinoline, 4-methyl 4-$CH_3(C_9H_6N)$	143.19		256				eth, ace, bz	B20⁴, 3510
8596	Isoquinoline, 6-methyl 6-$CH_3(C_9H_6N)$	143.19	cr	265.5	85-6			al, eth, ace, bz	B20², 247
8597	Isoquinoline, 7-methyl 7-$CH_3(C_9H_6N)$	143.19		245	67-8			eth, ace	B20⁴, 3511
8598	Isoquinoline, 8-methyl 8-$CH_3(C_9H_6N)$	143.19		258				eth, ace, bz	B20, 404
8599	Isoquinoline, 5-nitro 5-$O_2N(C_9H_6N)$	174.16	nd (w + 1)	sub	110			al, eth, chl, aa	B20⁴, 3450
8600	Isoquinoline, 1,2,3,4-tetrahydro or 2-Azatetralin .... $C_9H_{11}N$	133.20		232-3	<−15	1.0642²⁴/⁴	1.5668²⁰	al	B20⁴, 2949
8601	1-Isoquinoline carbolonitrile or 1-Cyanoisoquinoline.. 1-$NC(C_9H_6N)$	154.17	nd (peth MeOH)	....	78(93)			al, eth, bz	B22⁴, 1205
8602	5-Isoquinoline carbolonitrile or 5-Cyanoisoquinoline.. 5-$NC(C_9H_6N)$	154.17	nd (w or dil al)	sub 100-120	135			al, eth	B22⁴, 1208
8603	Isoraunescine .... $C_{31}H_{36}N_2O_8$	564.64	wh nd		241-2			chl, aa	C50, 1267
8604	Isoreserpilline .... $C_{23}H_{28}N_2O_5$	412.49	wh pr $[\alpha]^{20}_D$ −84 (py)	....	210-2				C53, 14134
8605	Isorubijervine or Δ⁵-β-18-Dihydroxy solanidene .... $C_{27}H_{43}NO_2$	413.65	dr (al) $[\alpha]_D$ +9.2 (al)	....	241-4			bz, chl	B21⁴, 2312
8606	Isorubijervosine .... $C_{33}H_{53}NO_7$	575.80	wh nd $[\alpha]^{24}_D$ −20 (py,c=1.45)		279-80				B21⁴, 2313
8607	Isosaccharic acid or 3,4-Dihydroxytetrahydro-2,5-furan-dicarboxylic acid .... $C_6H_8O_7$	192.14	rh $[\alpha]^{20}_D$ +46.7 (w,p=4.2)	d	185			w, al	B18², 309

No.	Name, Synonyms, and Formula	Mol. wt.	Color, crystalline form, specific rotation and λ_{max} (log ε)	b.p. °C	m.p. °C	Density	n_D	Solubility	Ref.
8608	Isosafrole (trans) .... $C_{10}H_{10}O_2$	162.19	....	253, 111-2[6]	fp 6.8	1.1224[20/4]	1.5782[20]	al, eth, ace, bz, chl	B19[4], 273
8609	Isoserine (l) $H_2NCH_2CH(OH)CO_2H$	105.10	cr (dil al w) $[\alpha]^{20}_D$ −32.6	....	199-201d	....	....	w	B4[3], 1566
8610	Isothebaine (d) .... $C_{19}H_{21}NO_3$	311.39	rh cr (al) $[\alpha]^{18}_D$ + 285 (al,c=2)	....	203-4	....	....	al, chl	B21[4], 2646
8611	Isothebaine, sulfate .... $(C_{19}H_{21}NO_3)_2,H_2SO_4$	720.85	nd	....	120-1d	....	....	w	B21[1], 250
8612	Isovaleric acid or 3-Methylbutyric acid .... $(CH_3)_2CHCH_2CO_2H$	102.13	....	176.7	−29.3	0.9286[20/4]	1.4033[20]	al, eth, chl	B2[4], 895
8613	Isovalerophenone or Isobutyl phenyl ketone $(CH_3)_2CHCH_2COC_6H_5$	162.23	....	236.5, 137-8[38]	....	0.9701[16.4/4]	1.5139[15.3]	al, eth, ace	B7[3], 1121
8614	Isovaline (d) or 2-Amino-2-methyl butyric acid .... $CH_3CH_2C(CH_3)(NH_2)CO_2H$	117.15	nd (aq al) $[\alpha]^{?}_D$ + 13 (w,c=2)	sub	ca 300	....	....	w	B7[3], 1361
8615	Isovaline (dl) or 2-Amino-2-methyl butyric acid .... $CH_3CH_2CH(CH_3)(NH_2)CO_2H$	117.15	rh nd (al-eth) mcl pr	sub 300	315 (sealed tube)	....	....	w, al	B4[3], 1361
8616	Isovaline (l) or 2-Amino-2-methyl butyric acid .... $CH_3CH_2CH(CH_3)(NH_2)CO_2H$	117.15	lo nd (w, ace) $[\alpha]^{20/}_D$ −9.1 (w,c=2)	....	....	....	....	w, al	B4[2], 851
8617	Isoxanthen-3-one, 9-phenyl-2,6,7-trihydroxy or 9-Phenyl-2,6,7-trihydroxy-fluorone .... $C_{19}H_{12}O_5$	320.31	og red (al-HCl)	....	>300	....	....	....	B18[4], 2824
8618	Isoxazole .... C_3H_3NO	69.06	....	95-6	....	1.078[20/4]	1.4298[17]	....	B27[2], 9
8619	Isoxazole, 5-methyl .... 5-$CH_3(C_3H_2NO)$	84.08	....	122	....	1.4386[20/0]	....	....	B27[2], 9
8620	Itaconic acid or Methylene succinic acid .... $CH_2=C(CO_2H)CH_2CO_2H$	130.10	rh (bz)	d	175	1.632	....	w, al, ace, chl	B2[4], 2228
8621	Itaconic anhydride or Methylene succinic anhydride...... $C_5H_4O_3$	112.09	rh bipym pr (eth chl)	139-40[30], 114-5[18]	68-70	....	....	chl	B17[4], 5913
8622	Itaconic acid, diethyl ester or Diethylitaconate .... $CH_2=C(CO_2C_2H_5)CH_2CO_2C_2H_5$	186.21	....	228, 111[13]	58-9	1.0467[20/4]	1.4377[20]	al, eth, ace, bz	B2[4], 2230
8623	Itaconic acid, dimethyl ester or Dimethylitaconate .... $CH_2=C(CO_2CH_3)CH_2CO_2CH_3$	158.16	hyg mcl (MeOH)	208, 108[11]	38	1.1241[18/4]	1.4457[20]	al, eth, ace	B2[4], 2229
8624	Iticonyl chloride or Methylene succinyl chloride .... $CH_2=C(COCl)CH_2COCl$	166.99	....	89[17]	....	....	1.4919[20]	ace	B2[3], 1934

No.	Name, Synonyms, and Formula	Mol. wt.	Color, crystalline form, specific rotation and λ_{max} (log ε)	b.p. °C	m.p. °C	Density	n_D	Solubility	Ref.
8625	Jacareubin .... $C_{18}H_{14}O_6$	326.32	ye pr (MeOH)	....	256-7d	....	....	al, ace	B19[4], 3041
8626	Jaconecic acid .... $C_{10}H_{16}O_6$	232.24	nd (eth) $[\alpha]_D^{25}$ + 28.1 (95% al)	....	183-4	....	....	....	B18[4], 5079
8627	Japaconitine-A or Acetyl benzoyl aconitine.......... $C_{34}H_{47}NO_{11}$	645.76	rh $[\alpha]_D^{18}$ + 20.7 (chl)		202-3d			ace, chl	B21[4], 2901
8628	Japaconitine-Al .... $C_{34}H_{47}NO_{11}$	645.76	rh (MeOH) $[\alpha]_D^{23}$ + 26.4 (chl)		208-9			al, eth, chl	Ber57, 1462
8629	Japaconitine-B .... $C_{34}H_{47}NO_{11}$	645.76	rh (MeOH) $[\alpha]_D^{21}$ + 26.9		208-9d			al, eth, chl	Ber57, 1462
8630	Jasmin aldehyde .... $C_5H_{11}C(CHO)=CHC_6H_5$	203.30		174-5[20], 140[5]	....	0.9718[20]	1.5381[20]	al, eth	B7[3], 1517
8631	Jasmone or 3-Methyl-2-(2-pentenyl)-2-cyclopenten-1-one $C_{11}H_{16}O$	164.25	ye oil	257-8[755], 134-5[12]		0.9437[22/4]	1.4979[22]	al, eth, lig	B7[3], 601
8632	Javanicin .... $C_{15}H_{14}O_6$	290.28	red (al)		208d				B8[3], 4231
8633	Jervine .... $C_{27}H_{39}NO_3.2H_2O$	461.65	nd (w + 2) (MeOH w) $[\alpha]_D^{23}$ −158.5 (al, c 0.33)		243-4d			al, ace, chl	Am73, 2970
8634	Julolidine .... $C_{12}H_{15}N$	173.25	....	280d, 155-6[17]	40	1.003[20]	1.568[25]	....	B20[4], 3281
8635	Julolidine 1,6 dione or 1,6-Diketo julolidine.......... $C_{12}H_{11}NO_2$	201.23	ye (al)	190-210[0.3]	145-6			al, MeOH	B21[4], 5525
8636	Junipal or 5-(α-propynyl)-2-formyl thiophene C_8H_6OS	150.20	nd (peth or dil al)		80				B17[4], 4946
8637	Juniperol or Macrocarpol......... $C_{15}H_{26}O$	220.38	tcl (al) $[\alpha]_D^{20}$ + 25.4 (al)	286-8d	112	1.0460[20/20]	1.519	....	B6[2], 426
8638	Junipic acid or 5-(α-propynyl)-2-thiophencarboxylic acid $C_8H_6O_2S$	166.20	ye nd (peth) cr (aq al)	....	180			al	B18[4], 4198
8639	Junipic acid, methyl ester or Methyl janipate.......... $C_9H_8O_2S$	180.22	(aq MeOH)	sub 50-5	62			ace	B18[4], 4199

No.	Name, Synonyms, and Formula	Mol. wt.	Color, crystalline form, specific rotation and λ_{max} (log ε)	b.p. °C	m.p. °C	Density	n_D	Solubility	Ref.
8640	Ketene . . . . CH$_2$=CO	42.04	. . . .	−56	−151	. . . .	. . . .	. . . .	B1[4], 3418
8641	Ketene, diethyl acetal or 1,1-diethoxyethylene. . . . . CH$_2$=C(OC$_2$H$_5$)$_2$	116.16	. . . .	68[100]	. . . . . .	0.7932[20/4]	1.3643[21]	. . . .	B1[4], 3420
8642	Ketene, dimethyl . . . . (CH$_3$)$_2$C=C=O	70.09	ye	34	−97.5	. . . .	. . . .	. . . .	B1[4], 3453
8643	Ketene, diphenyl . . . . (C$_6$H$_5$)$_2$C=C=O	193.24	ye-red	265-70d, 146[12]	. . . . . .	1.1107[14/4]	1.615[14]	eth, bz	B7[3], 2356
8644	Ketene, methyl . . . . CH$_3$CH=C=O	56.07	. . . .		−80	. . . .	. . . .	eth	B1[4], 3433
8645	Khellin . . . . C$_{14}$H$_{12}$O$_5$	260.25	(MeOH or eth)	180-200° 05	154-5 (vac)	. . . .	. . . .	ace, MeOH	B19[4], 2816
8646	Kynurenine or 3-Anthranyloyl alanine . . . . C$_{10}$H$_{12}$N$_2$O$_3$	208.22	lf (+ ½w) [α]$_D^{20/}$ −29 (w,c=4)	. . . .	191d	. . . .	. . . .		B14[4], 1656

No.	Name, Synonyms, and Formula	Mol. wt.	Color, crystalline form, specific rotation and λ_{max} (log ϵ)	b.p. °C	m.p. °C	Density	n_D	Solubility	Ref.
8647	Lactamide (d) .... $CH_3CH(OH)CONH_2$	89.09	cr (AcOEt) $[\alpha]^{8/}_{578}$		49-51	....	....	w, al	B3[3], 450
8648	Lactamide (dl) .... $CH_3CH(OH)CONH_2$	89.09	pl (AcOEt)		75.5	$1.1381^{80/4}$	....	w, al	B3[4], 674
8649	Lactamide, N-(4-ethoxyphenyl) or N-Lactyl-β-phenetidide $CH_3CH(OH)CONH(C_6H_5OC_2H_5\text{-}4)$	209.25	nd (w)	....	118			al	B13[3], 1135
8650	D-Lactic acid or D-2-Hydroxypropionic acid .... $CH_3CH(OH)CO_2H$	90.08	pl (chl aa) $[\alpha]^j_D$ −2.26 (w, c=1.24)	103^2	53	....	....	w, al	B3[4], 633
8651	DL-Lactic acid .... $CH_3CH(OH)CO_2H$	90.08	ye	122^{15}	18	$1.2060^{21/4}$	1.4392^{20}	w, al, eth	B3[4], 633
8652	L-Lactic acid............ $CH_3CH(OH)CO_2H$	90.08	hyg pr (eth), $[\alpha]^{15/}_D$ +3.8 (w, c=10.5)	....	....			w, al	B3[4], 633
8653	DL-Lactic acid acetate .... $CH_3CO_2CH(CH_3)CO_2H$	132.13	dlq	$167\text{-}70^{78}$, 127^{11}	57-60	$1.1758^{20/4}$	1.4240^{20}	al, bz	B3[4], 638
8654	DL-Lactic acid, allyl ester or Allyl lactate .... $CH_3CH(OH)CO_2CH_2CH=CH_2$	130.15	....	$56\text{-}60^8$		$1.0452^{20/4}$	1.4369^{20}	py	B3[3], 486
8655	DL-Lactic acid anhydride $[CH_3CH(OH)CO]_2O$	162.14	pa ye amor or syr	250d		....	....	al, eth	B3[3], 494
8656	D-Lactic acid, butyl ester or D-Butyl lactate .... $CH_3CH(OH)CO_2C_4H_9$	146.19	$[\alpha]^{27/}_D$ +13.6	77^{10}		$0.9744^{27/4}$	....	al, eth	B3[2], 188
8657	DL-Lactic acid, butyl ester or DL-Butyl lactate............ $CH_3CH(OH)CO_2C_4H_9$	146.19	....	83^{12}	−49	$0.9807^{22/4}$	1.4217^{20}	al, eth	B3[4], 649
8658	D-Lactic acid, ethyl ester or D-Ethyl lactate .... $CH_3CH(OH)CO_2C_2H_5$	118.13	$[\alpha]^{18/}_D$ +14.5	58^{20}		$1.0324^{20.4/4}$	1.4125^{20}	w, al, eth	B3[3], 449
8659	DL-Lactic acid, ethyl ester............ $CH_3CH(OH)CO_2C_2H_5$	118.13	....	154.5, 58^{19}		$1.0302^{20/4}$	1.4124^{20}	w, al, eth	B3[4], 643
8660	L-Lactic acid, ethyl ester............ $CH_3CH(OH)CO_2C_2H_5$	118.13	$[\alpha]^{19/}_D$ −11.3	$69\text{-}70^{36}$		$1.0314^{20/4}$	1.4156^{20}	w, al, eth	B3[3], 446
8661	D-Lactic acid, methyl ester or D-Methyl lactate .... $CH_3CH(OH)CO_2CH_3$	104.12	$[\alpha]^{20/}_D$ +7.5	40^{11}		$1.0857^{25/4}$	....	w, al, eth	B3[3], 449
8662	DL-Lactic acid, Methyl ester............ $CH_3CH(OH)CO_2CH_3$	104.12	....	144.8		$1.0928^{20/4}$	1.4141^{20}	w, al, eth	B3[4], 640
8663	L-Lactic acid, Methyl ester............ $CH_3CH(OH)CO_2CH_3$	104.12	$[\alpha]^{20/}_D$ −8.3	58^{19}		$1.0895^{20/4}$	1.4139^{20}	w, al, eth	B3[3], 445
8664	Lactic acid, isopentyl ester or Isopentyl lactate .... $CH_3CH(OH)CO_2\text{-}i\text{-}C_5H_{11}$	160.22	....	202.4, 82^7		$0.9617^{25/25}$	1.4240^{25}	al, eth	B3[3], 483
8665	D-Lactic acid or Atrolactic acid .... $CH_3C(OH)(C_6H_5)CO_2H$	166.18	pr (w), $[\alpha]^{16.5/}_D$ +37.7 (al, c=3.5)	....	116-7	....	....	al, ace, bz	B10[3], 560
8666	DL-Lactic acid, 2-phenyl............ $CH_3C(OH)(C_6H_5)CO_2H$	166.18	nd pl(liq)	....	93-5			w, al, ace, bz	B10[3], 560
8667	L-Lactic acid, 2-phenyl .... $CH_3C(OH)(C_6H_5)CO_2H$	166.18	nd (bz, w), $[\alpha]^{13.8/}_D$ −37.7 (al, c=3.4)		116-7			w, al, ace, bz	B10[3], 560
8668	Lactic acid, 3-phenyl (d) or α-Hydroxy hydrocinnamic acid .... $C_6H_5CH_2CH(OH)CO_2H$	166.18	nd (w), $[\alpha]^{20/}_D$ +22.2 (w, c=2.2)		124-6			w, al, ace	B10[3], 554
8669	Lactic acid, 3-phenyl (dl) .... $C_6H_5CH_2CH(OH)CO_2H$	166.18	cr(chl bz) pr (w)	$148\text{-}50^{15}$	98			al, eth, ace	B10[3], 554
8670	Lactic acid, 3-phenyl (l) .... $C_6H_5CH_2CH(OH)CO_2H$	166.18	nd (w), $[\alpha]^{20/}_D$ −19.9 (w, c=3.2)		124-5			al, eth, ace	B10[3], 554
8671	Lactic acid, isopropyl ester or Isopropyl lactate.......... $CH_3CH(OH)CO_2CH(CH_3)_2$	132.16	....	166-8, $75\text{-}80^{12}$		$0.9980^{20/4}$	1.4082^{25}	w, al, eth, bz	B3[3], 479
8672	Lactic acid, 3,3,3-trichloro .... $CCl_3CH(OH)CO_2H$	193.42	pr (eth)	$140\text{-}70^{45}$	125	....	....	w, al, eth, chl	B3[4], 680

PHYSICAL CONSTANTS OF ORGANIC COMPOUNDS (Continued)

No.	Name, Synonyms, and Formula	Mol. wt.	Color, crystalline form, specific rotation and λ_{max} (log ε)	b.p. °C	m.p. °C	Density	n_D	Solubility	Ref.
8673	Lactide (d) or 2,5-Dimethyl, 3,6-dioxo-1,4-dioxane ...... C$_6$H$_8$O$_4$	144.14	hyg rh (eth), [α]$^{18}_D$ -298 (bz, c=1.17)	150^{25}	95	....	....	....	B19^4, 1927
8674	Lactide (dl) ...... C$_6$H$_8$O$_4$	144.14	pa ye tcl pr or nd (al)	255^{757}, 138-42^8	124.5	0.862$^{10/4}$	....	ace, bz	B19^4, 1927
8675	Lactide (l) ...... C$_6$H$_8$O$_4$	144.14	rh (eth), [α]$^{26}_D$ + 281.6 (bz, c=0.82)	150^{25}	95	....	....	....	B19^4, 1927
8676	D-Lactobionic acid ...... C$_{12}$H$_{22}$O$_{12}$	358.30	syr	....	....	....	....	w	B17^4, 3392
8677	D-Lactonitrile or Acetaldehyde cyanohydrin ...... CH$_3$CH(OH)CN	71.08	ye liq	182-4d, 102^{30}	−40	0.9877$^{20/4}$	1.4058^{18}	**w, al**, eth	B3^4, 675
8678	D-Lactonitrile, acetate ...... CH$_3$CO$_2$CH(CH$_3$)CN	113.12	....	172-3, 76-7^{25}	....	1.0278$^{20/4}$	1.4027^{20}	w	B3^4, 676
8679	D-Lactonitrile-3,3,3-trichloro ...... Cl$_3$CCH(OH)CN	174.42	pl(w)	215-20d	61	....	....	w, al, eth	B3^4, 680
8680	Lactose or Milk sugar (α-anomer) ...... C$_{12}$H$_{22}$O$_{11}$	342.30	pw, [α]$^{20}_D$ +92.6→ +52.3 (w, c=4.5)	....	222.8	....	....	w	B17^4, 3066
8681	Lactose (β-anomer) ...... C$_{12}$H$_{22}$O$_{11}$	342.30	[α]$^{20}_D$ +34.2→ +52.3 (mut, w)	....	253	1.59^{20}	....	w	B17^4, 3068
8682	Lactose, monohydrate (α-anomer) ...... C$_{12}$H$_{22}$O$_{11}$.H$_2$O	360.31	mcl (w), [α]$^{20}_D$ +83.5 (w, 10min)	d	201-2	1.525^{20}	....	w	B17^4, 3067
8683	Lactyl chloride, acetate (dl) ...... CH$_3$CO$_2$CH(CH$_3$)COCl	150.56	....	150d, 56^{11}	....	1.1920$^{17/4}$	1.4241^{17}	....	B3^4, 674
8684	Lactyl chloride, acetate (d+) ...... CH$_3$CO$_2$CH(CH$_3$)COCl	150.56	[α]$^{18}_{578}$ + 32.4	51-3^{11}	....	1.177^{20}	....	....	B3^2, 189
8685	Lanosterol or Isocholesterol ...... C$_{30}$H$_{50}$O	426.72	nd (eth) cr (MeOH-ace), [α]$^{20}_D$ + 62 (chl, c=1)	....	140-1	....	....	al, eth, chl	B6^3, 2880
8686	Lanosterol, benzoate or Isocholesterol benzoate ...... C$_{34}$H$_{50}$O$_2$	490.77	pw or nd (eth), [α]$^{17}_D$ + 72.2	....	191.5	....	....	al, eth	B9^3, 485
8687	Lanthionine (D) or d-β,β-Thio di-alanine ...... [HO$_2$CCH(NH$_2$)CH$_2$]$_2$S	208.24	hex pl, [α]$^{21}_D$ -8.0 (2.4N NaOH, c=5)	....	293-5d darkens 245	....	....	....	B4^3, 1618
8688	Lanthionine (DL) ...... [HO$_2$CCH(NH$_2$)CH$_2$]$_2$S	208.24	hex pl	....	282-95d chars 240	....	....	....	B4^3, 1620
8689	Lanthionine (L) ...... [HO$_2$CCH(NH$_2$)CH$_2$]$_2$S	208.24	hex pl, [α]$^{22}_D$ + 8.6 (2.4N NaOH, c=5)	....	293-5d darkens 245	....	....	....	B4^3, 1593
8690	Lanthionine (meso) ...... [HO$_2$CCH(NH$_2$)]$_2$S	208.24	hex pl (aq NH$_3$)	....	304d softens 270	....	....	....	B4^3, 1620
8691	Lanthopine ...... C$_{23}$H$_{25}$NO$_4$	379.46	pw	....	200	....	....	chl	C20, 2715
8692	Lapachol or Taiguic acid ...... C$_{15}$H$_{14}$O$_3$	242.28	ye pr (eth,bz) pl (aa,al)	....	139-40	....	....	al, eth, bz, aa	B8^2, 365
8693	Lapachol-δ-hydroxy ...... C$_{15}$H$_{14}$O$_4$	258.28	ye nd (bz,w)	....	127	....	....	al, eth, bz	B8^3, 3661

No.	Name, Synonyms, and Formula	Mol. wt.	Color, crystalline form, specific rotation and λ_{max} (log ϵ)	b.p. °C	m.p. °C	Density	n_D	Solubility	Ref.
8694	Lappaconitine .... $C_{32}H_{44}N_2O_9$	600.72	hex pl (al), $[\alpha]^{18}_D$ + 27 (chl)	....	223	....	....	bz, chl	B21⁴, 2850
8695	Laudanidine (d) .... $C_{20}H_{25}NO_4$	343.43	(MeOH), $[\alpha]^{17}_D$ + 93.5 (chl, c=1)	....	184-5	....	....	....	B21⁴, 2703
8696	Laudanidine (l) or Tritopine......................... $C_{20}H_{25}NO_4$	343.43	hex pr (al), $[\alpha]^{17}_D$ -94.8 (chl, c=2)	....	184-5	....	....	w, bz	B21⁴, 2703
8697	Laudanine (dl) .... $C_{20}H_{25}NO_4$	343.43	ye wh pr (dil al or al-chl)	....	167	1.26²⁰/⁴	....	al, bz, chl	B21⁴, 2703
8698	Laudanosine (d) or N-Methyl tetrahydro papaverine ..... $C_{21}H_{27}NO_4$	357.46	nd (peth) pr (al), $[\alpha]^{16}_D$ + 106 (al, c=1.6)	....	89	....	....	al, eth, ace, chl	B21⁴, 2704
8699	Laudanosine (dl) .... $C_{21}H_{27}NO_4$	357.46	nd(al)	....	115-6	....	....	al, eth, ace, bz, chl	B21⁴, 2704
8700	Laudanosine (l) .... $C_{21}H_{27}NO_4$	357.46	cr (al), $[\alpha]^{15}_D$ -105.4 (al, c=31)	....	89	....	....	....	B21⁴, 2704
8701	Laureline .... $C_{19}H_{19}NO_3$	309.35	ta (al) cubes (peth), $[\alpha]^{18}_D$ -99 (abs al, c=0.7)	....	114	....	....	al, eth	B27⁴, 161
8702	Lauraldehyde or Dodecanal.......................... $CH_3(CH_2)_{10}CHO$	184.33	lf	185¹⁰⁰, 100³·⁵	44.5	0.8352¹⁵/⁴	1.435²²	al, eth	B1⁴, 3380
8703	Lauraldehyde, dimethylacetal .... $CH_3(CH_2)_{10}CH(OCH_3)_2$	230.40	....	132-4⁵	....	....	1.4310²⁵	al, eth	B1⁴, 3381
8704	Lauramide .... $CH_3(CH_2)_{10}CONH_2$	199.34	nd	199¹²	110	....	1.4287¹¹⁰	al, ace	B2⁴, 1103
8705	Lauramide, N-Phenyl .... $CH_3(CH_2)_{10}CONHC_6H_5$	275.44	nd (dil al)	....	78	....	....	al, eth, ace, bz, chl	B12³, 485
8706	Lauric acid or Dodecanoic acid .... $CH_3(CH_2)_{10}CO_2H$	200.33	nd (al)	131¹	44	0.8679⁵⁰/⁴	1.4304⁵⁰	al, eth, ace, bz, peth	B2⁴, 1082
8707	Lauric anhydride .... $(C_{11}H_{23}CO)_2O$	382.64	lf (al, eth)	....	41.8	0.8533⁷⁰/⁴	1.4292⁷⁰	al	B2⁴, 1100
8708	Lauric acid, benzyl ester or Benzyl laurate .... $C_{11}H_{23}CO_2CH_2C_6H_5$	290.45	....	209-11¹²	8.5	0.9457²⁵/²⁵	1.4812²⁴	al, eth, bz, chl, peth	B6⁴, 2267
8709	Lauric acid, 2-bromo .... $CH_3(CH_2)_9CHBrCO_2H$	279.23	pl	157-9²	32	1.1474²⁴	1.4585²⁴	al, eth, bz, chl, lig	B2⁴, 1106
8710	Lauric acid, ethyl ester or Ethyl laurate .... $C_{11}H_{23}CO_2C_2H_5$	228.38	....	273⁷⁶⁴, 154¹⁵	fr-1.8	0.8618²⁰/⁴	1.4311²⁰	al, eth	B2⁴, 1092
8711	Lauric acid, 12-fluoro .... $F(CH_2)_{11}CO_2H$	218.32	....	....	60-1	....	....	al, eth	B2⁴, 1105
8712	Lauric acid, methyl ester or Methyl laurate .... $C_{11}H_{23}CO_2CH_3$	214.35	....	262⁷⁶⁶, 141¹⁵	fr 5.2	0.8702²⁰/⁴	1.4319²⁰	al, eth, ace, bz	B2⁴, 1090
8713	Lauric acid, phenyl ester or Phenyl laurate............ $C_{11}H_{23}CO_2C_6H_5$	276.42	lf (al)	210¹⁵	24.5	....	....	al, eth, ace	B6⁴, 618
8714	Lauric acid, propyl ester or Propyl laurate............ $C_{11}H_{23}CO_2C_3H_7$	242.31	....	205⁶⁰, 124²	....	0.8600²⁰	1.4335²⁰	....	B2⁴, 1092
8715	Lauric acid, isopropyl ester or iso-Propyl laurate....... $C_{11}H_{23}CO_2CH(CH_3)_2$	242.41	....	196⁶⁰, 117²	....	0.8536²⁰	1.4280²⁵	al, eth	B2⁴, 1092
8716	Laurone or 12-Tricosanone Diundecyl ketone.......... $(C_{11}H_{23})_2CO$	338.63	lf (al)	....	69.3	0.8086⁶⁹/⁴	1.4283⁸⁰	eth, bz, chl	B1⁴, 3408
8717	Lauronitrile .... $C_{11}H_{23}CN$	181.33	....	277, 131¹⁰	fr 4	0.8240²⁰/⁴	1.4361²⁰	al, eth, ace, bz, chl	B2⁴, 1104
8718	Laurophenone or n-undecyl phenyl ketone .... $C_{11}H_{23}COC_6H_5$	260.43	og cr	222-3²¹	46-7	0.8969⁵²/⁴	1.4850⁵²	ace	B7³, 1291
8719	Lauroyl peroxide .... $C_{11}H_{23}CO-OO-COC_{11}H_{23}$	398.63	wh pl	....	49	....	....	....	B2⁴, 1102
8720	Lauryl alcohol or 1-Dodecanol .... $CH_3(CH_2)_{10}CH_2OH$	186.32	lf (dil al)	255-9, 150²⁰	26	0.8309²⁴/⁴	....	al, eth	B1⁴, 1844

No.	Name, Synonyms, and Formula	Mol. wt.	Color, crystalline form, specific rotation and λ_{max} (log ε)	b.p. °C	m.p. °C	Density	n_D	Solubility	Ref.
8721	Lauryl amine or 1-Amino dodecane .... $CH_3(CH_2)_{10}CH_2NH_2$	185.36	....	259, 126.5[10]	28.3	0.8015[20/4]	1.4421[20]	al, eth, bz, chl	B4[4], 794
8722	Lauryl amine, acetate .... $C_{12}H_{25}NH_2 \cdot CH_3CO_2H$	245.41	....	....	fr 69.5	....	....	w, al	B4[4], 797
8723	Lauryl amine, hydrochloride .... $C_{12}H_{25}NH_2 \cdot HCl$	221.82	....	....	98	....	....	w, al	B4[4], 795
8724	Lauroyl chloride .... $C_{11}H_{23}COCl$	218.77	....	145[18]	−17	....	1.4458[20]	eth	B2[4], 1103
8725	Lauryl mercaptan .... $CH_3(CH_2)_{11}SH$	202.41	....	142-5[15]	......	0.8450[20/20]	1.4589[20]	al, eth	B1[4], 1851
8726	Lauryl sulfate .... $(C_{12}H_{25}O)_2SO_2$	434.73	....	....	48.5	....	....	....	B1[4], 1849
8727	Lead, tetraphenyl .... $Pb(C_6H_5)_4$	515.62	....	126[13]	227-8	....	....	....	B16[3], 1252
8728	Lecithin $RCO_2CH_2(CHOCOR)CH_2OPOCH_2CH_2N(CH_3)_3$	......	$[\alpha]^{24'}_D$ +7.0	....	236-7	....	....	eth, chl, peth	B4[4], 1462
8729	Ledol .... $C_{15}H_{26}O$	222.38	nd (al), $[\alpha]^{20'}_D$ +28 (chl, c=10)	292 sub	105-6 sub	0.9094[100/20]	1.4667[110]	al, eth, ace	B6[4], 426
8730	p-Leucaniline, N,N,N',N'-tetramethyl or bis(4-dimethylaminophenyl)4-aminophenylmethane	345.49	(al)	....	151-2	....	....	....	B13[3], 566
8731	Leucic acid (D) or 2-Hydroxy-4-methyl valeric acid ...... $(CH_3)_2CHCH_2CH(OH)CO_2H$	132.16	nd (bz) pr (eth-peth), $[\alpha]^{125'}_D$ +10.7 (w, c=5)	....	80-1	....	....	w, al, eth	B3[4], 850
8732	Leucic acid-(DL) .... $(CH_3)_2CHCH_2CH(OH)CO_2H$	132.16	pl (eth-peth)	....	77	....	....	w, al, eth	B3[4], 851
8733	Leucic aicd (l) .... $(CH_3)_2CHCH_2CH(OH)CO_2H$	132.16	rh (eth), $[\alpha]^{20'}_D$ −11.3 (w, c=1)	....	81-2	....	....	w, al, eth	B3[4], 851
8734	Leucin amide (dl) or α-Amino isocaproamide .... $(CH_3)_2CHCH_2CH(NH_2)CONH_2$	130.19	pr (bz)	....	106-7	....	....	w, al, ace	B4[3], 1434
8735	Leucine (D) or D-α-Amino isocaproic acid.............. $(CH_3)_2CHCH_2CH(NH_2)CO_2H$	131.18	pl (al), $[\alpha]^{20'}_D$ +10.34	sub	293 (sealed tube)	....	....	....	B4[3], 1424
8736	Leucine (DL) .... $(CH_3)_2CHCH_2CH(NH_2)CO_2H$	131.18	lf (w)	sub	293-5 (sealed tube)	1.293[18/4]	....	w	B4[3], 1430
8737	Leucine (L) .... $(CH_3)_2CHCH_2CH(NH_2)CO_2H$	131.18	hex pl (dil al), $[\alpha]^{25'}_D$ −10.4 (w, p=22)	sub	293-5 (sealed tube)	1.293[18/4]	....	....	B4[3], 1408
8738	Leucine, N-acetyl-(dl) .... $(CH_3)_2CHCH_2CH(NHCOCH_3)CO_2H$	173.21	nd (dil al)	....	161	....	....	al	B4[3], 1439
8739	Leucine, N-benzoyl .... $(CH_3)_2CHCH_2CH(NHCOC_6H_5)CO_2H$	235.29	nd (dil al)	....	137-41	....	....	w, al, eth, chl	B9[3], 1159
8740	Leucine, N-glycyl (dl) .... $(CH_3)_2CHCH_2CH(NHOCH_2NH_2)CO_2H$	188.21	tetr (dil al)	....	242d	1.181	....	w	B4[3], 1443
8741	Leucine, N-glycyl (l) .... $(CH_3)_2CHCH_2CH(NHCONH_2)CO_2H$	188.21	pl (dil al), $[\alpha]^{20'}_D$ −35.2 (w)	....	256d	....	....	w	B4[3], 1420
8742	Leucomethylene blue .... $C_{16}H_{19}N_3S$	285.42	ye nd (eth,al)	....	185	....	....	al	B27[2], 448
8743	Levopimaric acid, methyl ester .... $C_{21}H_{32}O_2$	316.49	(MeOH or eth), $[\alpha]_D$ −190.4 (al) −268 (eth)	166-9[0.5]	63-4	1.0312[22/4]	1.5232[22]	al	B9[3], 2904
8744	Levulin or Fructosin .... $(C_6H_{10}O_5)n$	(162.14)n	amor, $[\alpha]$ −52.1	....	140-5d	....	....	w, al	B1, 925
8745	Levulinaldehyde or 4-Oxovaleraldehyde............. $CH_3COCH_2CH_2CHO$	100.13	....	186-8d, 70[12]	<−21	1.0184[21/4]	1.4257[22]	w, al, eth, ace, bz	B1[4], 3659
8746	Levulinic acid or 4-Oxovalericacid .... $CH_3COCH_2CH_2CO_2H$	116.13	lf or pl	245-6d, 139-40[8]	37.2	1.1335[20/4]	1.4396[20]	w, al, eth	B3[4], 1560

No.	Name, Synonyms, and Formula	Mol. wt.	Color, crystalline form, specific rotation and λ_{max} (log ε)	b.p. °C	m.p. °C	Density	n_D	Solubility	Ref.
8747	Levulinic acid, benzyl ester or Benzyl levulinate .... $CH_3COCH_2CH_2CO_2CH_2C_6H_5$	206.24	....	132-4[2]	......	1.0935[20/4]	1.5090[20]	to	B6[4], 2481
8748	Levulinic acid, butyl ester or Butyl levulinate.......... $CH_3COCH_2CH_2CO_2C_4H_9$	172.23		237.8	......	0.9735[20/4]	1.4290[20]	al, eth, ace, bz	B3[4], 1563
8749	Levulinic acid, iso butyl ester or Butyl levulinate........ $CH_3COCH_2CH_2CO_2\text{-}i\text{-}C_4H_9$	172.23		231	......	0.9705[20/4]	1.4268[20]	al, eth, ace, bz	B3[4], 1563
8750	Levulinic acid, Sec-butyl ester or Sec-Butyl levulinate.... $CH_3COCH_2CH_2CO_2CH(CH_3)C_2H_5$	172.23		225.8	......	0.9669[20/4]	1.4249[20]	al, eth, ace	B3[3], 1221
8751	Levulinic acid, methyl ester or Methyl levulinate .... $CH_3COCH_2CH_2CO_2CH_3$	130.15		196, 85-6[14]	......	1.0511[20/4]	1.4233[20]	al, eth, ace, bz	B3[4], 1562
8752	Levulinic acid-propyl ester or Propyl levulinate......... $CH_3COCH_2CH_2CO_2C_3H_7$	158.20		221.2		0.9896[20/4]	1.4258[20]	al, eth, ace, bz	B3[4], 1563
8753	Levulinic acid-isopropyl ester or Isopropyl levulinate.... $CH_3COCH_2CH_2CO_2CH(CH_3)_2$	158.20		209.3		0.9842[20/4]	1.4420[20]	al, eth, ace, bz	B3[4], 1563
8754	Limonene (d) or 4-isopropenyl-1-methyl cyclohexene..... $C_{10}H_{16}$	136.24	$[\alpha]^{20}_D$ +125.6 (undil)	178, 61[12]	-74.3	0.8411[20/4]	1.4730[20]	al, eth	B5[4], 438
8755	Limonene (dl) or Dipentene .... $C_{10}H_{16}$	136.24	....	178, 64.4[15]	-95.5	0.8402[21/4]	1.4727[20]	....	B5[4], 440
8756	Limonene (l) .... $C_{10}H_{16}$	136.24	$[\alpha]^{20}_D$ -122.1 (undil)	177-8[755], 64.4[15]	......	0.8422[20/4]	1.4746[20]	al, eth	B5[4], 440
8757	Linalool (d) or 3,7-Dimethyl-1,6-octadien-3-ol .... $CH_2=CHC(OH)(CH_3)CH_2CH_2CH=C(CH_3)_2$	154.26	$[\alpha]^{20}_D$ +19 18	198-200, 87-8[12]	......	0.8700[20/4]	1.4636[20]	al, eth	B1[3], 2012
8758	Linalool, acetate (l) or Linalyl acetate, Bergamol .... $C_{12}H_{20}O_2$	196.29	$[\alpha]^{20}_D$ -9.45	220[762]		0.8951[20/4]	1.4544[21]	al, eth	B2[4], 204
8759	Linalool formate or Linalyl formate .... $CH_2=CHC(CH_3)(O_2CH)CH_2CH_2CH=C(CH_3)_2$	182.27	......	100-3[10]	......	0.915[25/4]	1.456[20]	al	B2[4], 35
8760	Linalool, tetrahydro-(dl) or 3,7-dimethyl-3-octanol-(dl) . $(CH_3)CH(CH_3)_2C(OH)(CH_3)CH_2CH_3$	158.29		196-7, 87-8[10]	31-2	0.8280[20]	1.4335[20]	al	B1[4], 1829
8761	Linaloolen or 2,6-Dimethyl-2,7-octadiene .... $(CH_3)_2C=CHCH_2CH_2CH(CH_3)CH=CH_2$	138.25		168, 58[12]		0.7882[20]	1.4561[20]	al	B1, 261
8762	β-Linaloolene or Dihydromyrcene $CH_2=CHCH(CH_3)CH_2CH_2CH=C(CH_3)_2$	138.25		165-8		0.7601[20/4]	1.4362[20]	....	B1[4], 1060
8763	Linamarin or Acetone cyanohydrin-β-d-glycopyranoside $C_{10}H_7NO_6$	247.25	nd (w al), $[\alpha]^{18}_D$ -29.1 (w, p=5)	....	145			ace	B17[4], 3340
8764	Linoleic acid or 9,12-Octadecadienoic acid .... $CH_3(CH_2)_4CH=CHCH_2CH=CH(CH_2)_7CO_2H$	280.46		229-30[16]	-5	0.9022[204]	1.4699[20]	al, eth, ace, bz, chl	B2[4], 1754
8765	Linoleic acid, ethyl ester or Ethyl linoleate .... $CH_3(CH_2)_4CH=CH\text{-}CH_2\text{-}CH=CH\text{-}(CH_2)_7CO_2C_2H_5$	308.49	ye or col	270-5[180], 212[12]		0.8865[20/4]		al, eth	B2[4], 1757
8766	Linoleic acid, methyl ester or Methyl linoleate .... $CH_3(CH_2)_4CH=CHCH_2CH=CH(CH_2)_7CO_2CH_3$	294.48	....	215[20]	-35	0.8886[18/4]	1.4638[20]	al, eth	B2[4], 1756
8767	10,12-Linoleic acid or 10,12-Octadecadienoic acid .... $CH_3(CH_2)_4CH=CHCH=CH(CH_2)_7CO_2H$	280.48			56-7	0.8686[70/4]	1.4689[60]		B2[4], 1752
8768	α-Linolenic acid or 9,12,15-Octadecatrienoic acid (cis,cis,cis) $CH_3[CH_2CH=CH]_3(CH_2)_7CO_2H$	278.44	....	230-2[17], 129[0.05]	-11.3	0.9164[20/4]	1.4800[20]	al, eth	B2[4], 1781
8769	α-Linolenic acid, ethyl ester or Ethyl linolenate .... $CH_3[CH_2CH=CH]_3(CH_2)_7CO_2C_2H_5$	306.49		218[15]		0.8919[20/4]	1.4694[20]	....	B2[4], 1782
8770	α-Lipoic acid (d) or 6,8-Epidithiooctanoic acid .... $C_8H_{14}O_2S_2$	206.33	pa ye pl, $[\alpha]^{25}_D$ +96.7	....	47.5			bz, MeOH	B19[4], 3459
8771	Lithocholic acid or 3α-Hydroxycholanic acid $C_{24}H_{40}O_3$	376.59	hex lf (al) pr (dil al or aa), $[\alpha]^{20}_D$ +32.14 (al)		186			al, chl, aa	B10[3], 687
8772	Lobelanidine $C_{22}H_{29}NO_2$	339.48	sc(al eth)	distb vac	150	....		al, ace, bz, chl, py	B21[4], 2380
8773	Lobelanine $C_{22}H_{25}NO_2$	335.45	nd(peth or eth)	....	99			al, ace, bz, aa, chl	B21[4], 5630
8774	Lobeline (dl) $C_{23}H_{27}NO_2$	337.47	ye pl (eth) pr (al)	....	110			al, eth, bz, chl	B21[4], 6346

No.	Name, Synonyms, and Formula	Mol. wt.	Color, crystalline form, specific rotation and λ_{max} (log ε)	b.p. °C	m.p. °C	Density	n_D	Solubility	Ref.
8775	Lobeline (l) $C_{22}H_{27}NO_2$	337.47	nd (al eth bz), $[\alpha]^{15}_D$ -42.8 (al, c=1)	....	130.1	....	....	al, eth, ace, bz, chl	B21⁴, 6345
8776	Longifolene (d) $C_{15}H_{24}$	204.36	$[\alpha]^{18}_D$ +42.7	254-6⁷⁰⁶, 126-7¹⁵	....	0.9319¹⁰/⁴	1.5040²⁰	bz	B5⁴, 1192
8777	Lophine or 2,4,5-Triphenyl imidazole $C_{21}H_{16}N_2$	296.38	nd (al)	sub	275	....	....	al, eth	B23², 280
8778	Luciculine $C_{22}H_{35}NO_3$	361.51	cr (+1w ace), $[\alpha]^{11.5}_D$ -11.4 (al)	165⁰·⁰²	148-50	....	....	al	B21⁴, 2584
8779	Luciculine, hydrochloride $C_{22}H_{35}NO_3 \cdot HCl$	397.97	cr (w + ³/₂), $[\alpha]_D$ -9.4 (w)	....	198-203	....	....	al	B21⁴, 2584
8780	Luminol or 3-Aminophthalhydrazide $C_8H_7N_3O_2$	177.16	ye nd (al)	....	329-32	....	....	....	B25², 389
8781	Lumisterol $C_{28}H_{44}O$	396.67	nd (ace MeOH), $[\alpha]^{18}_D$ +197 (chl)	....	118	....	....	al, eth, ace, bz, aa	B6³, 3108
8782	Lupanine (d) $C_{15}H_{24}N_2O$	248.37	hyg nd, $[\alpha]^{20}_D$ +82.4 (w, c=3)	190-3³	40-4	....	1.544²⁶	w, al, eth, chl	B24², 53
8783	Lupanine (dl) $C_{15}H_{24}N_2O$	248.37	nd(peth) rh pr(ace)	233-4¹⁸	98-9	....	....	w, al, eth, chl, peth	B24², 55
8784	Lupeol $C_{30}H_{50}O$	426.74	nd (al or ace), $[\alpha]^{20}_D$ +27.2 (chl)	....	215-7	0.9457²¹⁸/⁴	1.4910²¹⁸	al, eth, ace, bz, chl	B6³, 2901
8785	Lupinine (l) $C_{10}H_{19}NO$	169.27	rh (peth), $[\alpha]^{17}_D$ -20.3 (al)	269-70⁷⁵⁴	70	....	....	w, al, eth, bz, chl	B21⁴, 291
8786	Lupinine, hydrochloride $C_{10}H_{19}NO \cdot HCl$	205.73	pr (dil al), $[\alpha]_D$ -14 (w)	....	212-3	....	....	w, al	B21⁴, 292
8787	Lupulone or β-Lupulinic acid $C_{26}H_{38}O_4$	414.59	pr (MeOH)	....	93	....	....	al, peth	B7³, 4753
8788	2,3-Lutidine or 2,3-dimethyl pyridine 2,3-$(CH_3)_2(C_5H_3N)$	107.15	....	163-4	....	0.9319²⁵/⁴	1.5057²⁰	w, al, eth	B20⁴, 2765
8789	2,4-Lutidine or 2,4-dimethyl pyridine 2,4-$(CH_3)_2(C_5H_3N)$	107.16	....	159	....	0.9309²⁰/⁴	1.5010²⁰	w, al, eth, ace	B20⁴, 2768
8790	2,5-Lutidine or 2,5-dimethyl pyridine 2,5-$(CH_3)_2(C_5H_3N)$	107.16	....	157-9	−16	0.9297²⁰/⁴	1.5006²⁰	al, eth, ace	B20⁴, 2774
8791	2,6-Lutidine or 2,6-dimethyl pyridine 2,6-$(CH_3)_2(C_5H_3N)$	107.16	....	145.7	−6.1	0.9226²⁰/⁴	1.4953²⁰	w, eth, ace	B20⁴, 2776
8792	3,4-Lutidine or 3,4-dimethyl pyridine 3,4-$(CH_3)_2(C_5H_3N)$	107.16	....	163-4	....	0.9281²⁰/⁴	1.5096²⁰	al, eth, ace, chl	B20⁴, 2787
8793	3,5-Lutidine or 3,5-dimethyl pyridine 3,5-$(CH_3)_2(C_5H_3N)$	107.16	....	171.6	....	0.9419²⁰/⁴	1.5061²⁰	w, al, eth, ace	B20⁴, 2788
8794	Lycaconitine $C_{27}H_{34}N_2O_6$	566.61	amor, $[\alpha]_D$ +31.5	....	111-4	....	....	al, bz, chl, peth	B21⁴, 4564
8795	Lycomarasmine $HO_2CCH_2CH(CO_2H)NHCH_2CH(CO_2H)NHCH_2CONH_2$	277.24	$[\alpha]^{20}_D$ -48 (w)	....	227-9d	....	....	....	B4³, 1521
8796	Lycopene $C_{40}H_{56}$	536.90	red pr or nd (peth)	....	175	....	....	eth, bz, chl	B1⁴, 1166
8797	Lycorine (l) or Narcissine $C_{16}H_{17}NO_4$	287.32	pr (al py), $[\alpha]^{16}_D$ -129 (al, c=0.16)	sub	280	....	....	....	B27², 547
8798	Lycoxanthin $C_{40}H_{56}O$	552.93	red pl (bz-MeOH)	....	168	....	....	bz	B1⁴, 2368

No.	Name, Synonyms, and Formula	Mol. wt.	Color, crystalline form, specific rotation and λ_{max} (log ϵ)	b.p. °C	m.p. °C	Density	n_D	Solubility	Ref.
8799	Lysergic acid . . . . $C_{16}H_{16}N_2O_2$	268.32	lf or hex sc(w), $[\alpha]^{20}_D$ + 40 (py, c=0.5)	. . . .	240d	. . . .	. . . .	al, py	J1955, 1626
8800	Lysine(L) or L-α-ω-Diaminocaproic acid . . . . $H_2N(CH_2)_4CH(NH_2)CO_2H$	146.19	nd (w, dil al), $[\alpha]^{20}_D$ + 14.6 (w, c=6)	. . . .	224-5d dark-ens 210	. . . .	. . . .	w	B4³, 1400
8801	Lysine, dihydro chloride (L) . . . . $H_2N(CH_2)_4CH(NH_2)CO_2H.2HCl$	219.11	(al-eth or aq HCl), $[\alpha]^{20}_D$ + 15.3 (w)	. . . .	201-2	. . . .	. . . .	w, al	B4³, 1402
8802	Lysine, α-N-benzoyl (dl) . . . . $H_2N(CH_2)_4CH(NHCOC_6H_5)CO_2H$	250.30	nd (w)	. . . .	235	. . . .	. . . .	w	B9³, 1235
8803	Lysine, ω-N-benzoyl (dl) . . . . $C_6H_5CONH(CH_2)_4CH(NH_2)CO_2H$	250.30	cr (w)	. . . .	268	. . . .	. . . .	w	B9³, 1238
8804	Lysine, ω-N-benyoyl-(l) . . . . $C_6H_5CONH(CH_2)_4CH(NH_2)CO_2H$	250.30	lf (w), $[\alpha]^{19}_D$ + 20.1 (aq HCl)	. . . .	240d	. . . .	. . . .	w	B9³, 1237

No.	Name, Synonyms, and Formula	Mol. wt.	Color, crystalline form, specific rotation and λ_{max} (log ε)	b.p. °C	m.p. °C	Density	n_D	Solubility	Ref.
8805	Malathion or S(1,2-Dicarboxymethyl)-0,0-dimethyldithio-phosphate.................. $C_{10}H_{19}PS_2O_6$	330.36	ye-br	156-7[0.7]d	2.8	1.2076[20]	1.4960[20]	al, eth, bz	B3[4], 1136
8806	Maleicacid-cis-butenedioic acid $HO_2CCH=CHCO_2H$	116.07	mcl pr (w)	....	139-40	1.590[20]	....	w, al, eth, ace, aa	B2[4], 2199
8807	Maleic acid, bromo or Bromo maleicacid............. $HO_2CCBr=CHCO_2H$	194.98	nd or pr	d	136-40	....	....	w, al, eth	B2[4], 2224
8808	Maleic acid, chloro $HO_2CCl=CHCO_2H$	150.52	pr(eth-chl)	....	108 (114)	....	....	al, eth, ace, aa	B2[4], 2221
8809	Maleic acid, chloro,diethyl ester or Diethyl chloromaleate $C_2H_5O_2CCCl=CHCO_2C_2H_5$	206.63	....	235d, 125[19]	......	1.1741[20/4]	....	al, eth, ace, aa	B2[3], 1928
8810	Maleic acid, chloro, dimethyl ester .... $CH_3O_2CCCl=CHCO_2CH_3$	178.57	....	106.5[18]	......	1.2775[25/4]	....	al, eth	B2[3], 1928
8811	Maleic acid, dichloro $HO_2CCCl=CClCO_2H$	184.96	nd (liq eth)	....	119-20	....	....	w, al, eth	B2[4], 2223
8812	Maleic acid, diethyl ester or Diethyl maleate .... $C_2H_5O_2CCH=CHCO_2C_2H_5$	172.19	....	223, 105-6[14]	−88	1.0662[20/4]	1.4416[20]	al, eth	B2[4], 2207
8813	Maleic acid, dihydroxy $HO_2CC(OH)=C(OH)CO_2H$	148.07	pl (w + 2)	....	155 (anh)	....	....	....	B3[4], 1975
8814	Maleic acid, dimethyl ester or Dimethyl maleate .... $CH_3O_2CCH=CHCO_2CH_3$	144.14	....	202, 102[17]	−19	1.1606[20/4]	1.4416[20]	eth	B2[4], 2204
8815	Maleic acid, diphenyl ester $C_6H_5O_2CCH=CHCO_2C_6H_5$	286.27	pl(lig)	226.15	73	....	....	al, eth, ace, bz, chl	B6[4], 628
8816	Maleic acid, dipropyl ester or Dipropylmaleate $C_3H_7O_2CCH=CHCO_2C_3H_7$	200.23	....	126[12]	......	1.0245[20/4]	1.4434[20]	al, eth, ace, bz	B2[4], 2209
8817	Maleic acid, hydrazide $C_4H_4N_2O_2$	112.09	cr (w)	....	>300d	....	....	....	B2[4], 2220
8818	Maleic acid, methyl or Citraconic acid $HO_2CC(CH_3)=CHCO_2H$	130.10	tcl pr or pl (eth-bz) nd (eth-lig)	....	93-4	1.617	....	w	B2[4], 2230
8819	Maleic acid, monoamide or Maleamic acid .... $HO_2CCH=CHCONH_2$	115.09	lf(w)	....	178	....	....	w, al	B2[4], 2218
8820	Maleic anhydride .... $C_4H_2O_3$	98.06	nd(chl eth)	197-9, 82[14]	60	1.314[60]	....	eth, ace, chl	B17[4], 5897
8821	Maleic anhydride, dimethyl or Pyrocinchonic anhydride . $OC(CH_3)=C(CH_3)CO$	126.11	pl or lf(dil al)	223, 105[12]	96	1.107[100/4]	....	al, eth, bz, chl	B17[4], 5919
8822	Maleic anhydride, methyl or Citraconic anhydride .... $C_5H_4O_3$	112.08	....	213-4, 99-100[15]	7-8	1.2469[16/4]	1.4710[21]	al, eth, ace	B17[4], 5912
8823	Maleic anhydride, phenyl $C_{10}H_6O_3$	174.16	ye nd (CS₂)	....	122	....	....	al, eth	B17[4], 6279
8824	Maleimide or Maleic acid, imide $OCC=CH-CO$	97.07	pl (bz)	sub	93-5	....	....	w, al, eth	B21[4], 4627
8825	Maleimide, N-ethyl .... $C_{10}H_7NO_2$	125.13	cr (bz)	....	45.5	....	....	al, eth	B21[4], 4629
8826	Maleimide, N-phenyl or Maleanil .... $C_{10}H_7NO_2$	173.17	ye nd (bz-lig)	162[12]	90-1	....	....	al, eth, bz	B21[4], 4631
8827	Malonamide .... $H_2C(CONH_2)_2$	102.09	mcl pr (w)	....	171-2	....	....	w	B2[4], 1887
8828	Malonamide, benzyl .... $C_6H_5CH_2CH(CONH_2)_2$	192.22	nd (al)	....	225	....	....		B9[3], 4285
8829	Malonamide, N,N′-diphenyl or Malonanilide .... $H_2C(CONHC_6H_5)_2$	254.29	nd (al)	....	226	....	....	al, bz, aa	B12[3], 558
8830	Malonic acid or Propanedioic acid .... $HO_2CCH_2CO_2H$	104.06	tcl(al)	d140	135.6	1.619[16]	....	al, eth, py	B2[4], 1874
8831	Malonic acid, acetamide,diethyl ester .... $CH_3CONHCH(CO_2C_2H_5)_2$	217.22	cr(al bz-peth)	185[20]	95-6	....	....	al	B4[3], 1503
8832	Malonic acid, acetyl, diethyl ester .... $(CH_3CO)CH(CO_2C_2H_5)_2$	202.21	....	232, 120[17]	......	1.0834[26/4]	1.4435[25]	ace	B3[4], 1819
8833	Malonic acid, allyl or 3-Butene-1,1-dicarboxylic acid ..... $(CH_2=CHCH_2)CH(CO_2H)_2$	144.14	tcl (eth)	>180d	105	....	....	w, al, eth	B2[3], 1945
8834	Malonic acid, allyl,diethyl ester .... $(CH_2=CHCH_2)CH(CO_2C_2H_5)_2$	200.24	....	222-3, 93[6]	......	1.0098[20/4]	1.4305[20]	al, eth	B2[4], 2240
8835	Malonic acid, amino,monohydrate .... $H_2NCH(CO_2H)_2 \cdot H_2O$	137.09	pr (w + 1)	....	112d	....	....		B4[3], 1501
8836	Malonic acid, amino,diethyl ester .... $H_2NCH(CO_2C_2H_5)_2$	175.19	....	122-3[16]	......	1.100[16/4]	1.4353[16]	w, al, eth, ace, bz	B4[3], 1501

No.	Name, Synonyms, and Formula	Mol. wt.	Color, crystalline form, specific rotation and λ_{max} (log ε)	b.p. °C	m.p. °C	Density	n_D	Solubility	Ref.
8837	Malonic acid, benzoyl amino,diethyl ester .... (C_6H_5CONH)$CH(CO_2C_2H_5)_2$	279.30	nd(peth)	....	61	....	....	al, eth	B9³, 1186
8838	Malonic acid, benzyl .... $C_6H_5CH_2CH(CO_2H)_2$	194.19	pr(bz eth chl peth)	....	121-2	....	....	w, al, eth, bz	B9³, 4283
8839	Malonic acid, benzyl,diethyl ester .... $C_6H_5CH_2CH(CO_2C_2H_5)_2$	250.30	....	300, 169⁴¹²	....	1.0750²⁰/⁴	1.4872²⁰	....	B9³, 4284
8840	Malonic acid, benzyl,hydroxy .... $C_6H_5CH_2C(OH)(CO_2H)_2$	210.19	pr	....	147d	....	....	w, al, eth	B10³, 2216
8841	Malonic acid, benzylidene .... $C_6H_5CH=C(CO_2H)_2$	192.18	pr(w)	....	195-6d	....	....	al, ace	B9³, 4378
8842	Malonic acid, benzylidene,diethyl ester .... $C_6H_5CH=C(CO_2C_2H_5)_2$	248.28	....	308-12d, 180¹⁰	32	1.1045²⁰/⁴	1.5389²⁰	al, eth, ace, bz	B9³, 4379
8843	Malonic acid, bromo .... $BrCH(CO_2H)_2$	182.96	nd(eth) pl(ace-bz)	....	113d	....	....	al, eth	B2⁴, 1904
8844	Malonic acid, bromo,diethyl ester .... $BrCH(CO_2C_2H_5)_2$	239.07	....	253-5d, 123²⁰	−54	1.4022²⁵/⁴	1.4521²⁰	al, eth, ace	B2⁴, 1904
8845	Malonic acid, butyl or 1,1-Pentane dicarboxylic acid ..... $C_4H_9CH(CO_2H)_2$	160.17	pr(w)	....	104-5	....	....	w, al, eth	B2⁴, 2011
8846	Malonic acid, butyl,diethyl ester .... $C_4H_9CH(CO_2C_2H_5)_2$	216.28	....	235-40, 122¹²	....	....	1.4250²⁰	al, eth	B2⁴, 2011
8847	Malonic acid, isobutyl .... i-$C_4H_9CH(CO_2H)_2$	160.17	cr(bz)	d	115d	....	....	w, al, eth	B2⁴, 2023
8848	Malonic acid, isobutyl,diethyl ester .... i-$C_4H_9CH(CO_2C_2H_5)_2$	216.28	....	255, 119-20¹⁶	....	0.9804²⁰/⁴	1.4236²⁰	al, eth	B2³, 1756
8849	Malonic acid, sec-Butyl,diethyl ester .... $C_2H_5CH(CH_3)CH(CO_2C_2H_5)_2$	216.28	....	245-50, 105⁹	....	0.988¹⁸	1.4248²⁰	al, eth	B2⁴, 2019
8850	Malonic acid, acetyl or Hexadecyl malonicacid .... $C_{16}H_{33}CH(CO_2H)_2$	328.50	nd(lig) lf(aa)	....	121-2	....	....	eth, bz, aa	B2⁴, 2183
8851	Malonic acid, acetyl, diethyl ester .... $C_{16}H_{33}CH(CO_2C_2H_5)_2$	384.61	amor	238-40¹⁴	(i)25.1 (ii)12.7	....	1.4433²⁰	al	B2⁴, 2183
8852	Malonic acid, chloro .... $ClCH(CO_2H)_2$	138.51	pr (w)	d	133	....	....	w, sl, eth	B2³, 1637
8853	Malonic acid, chloro, diethyl ester or Ethyl chloromalonate .... $ClCH(CO_2C_2H_5)_2$	194.62	....	222, 118¹⁶	....	1.2040²⁰/⁴	1.4327²⁰	al, eth, chl	B2⁴, 1903
8854	Malonic acid, (3-chloropropyl), diethyl ester .... $Cl(CH_2)_3CH(CO_2C_2H_5)_2$	236.71	....	147-9¹⁰	....	....	1.4429²⁰	al, eth, chl	B2⁴, 1992
8855	Malonic acid, cinnamylidene-(trans) or Cinnamal malonic acid .... $C_6H_5CH=CHCH=C(CO_2H)_2$	218.21	dk ye nd	....	212d	....	....	al, chl	B9³, 4432
8856	Malonic acid, cyclohexyl, diethyl ester .... $C_6H_{11}CH(CO_2C_2H_5)_2$	242.32	....	163-5²⁰	....	1.0281¹⁹/⁴	1.4478²⁵	al, eth, ace, bz	B9³, 3834
8857	Malonic acid, 2-cylopentenyl, diethyl ester .... (2-C_5H_7)$CH(CO_2C_2H_5)_2$	226.28	....	141¹⁰	....	1.0507²⁰/⁴	1.4536²⁰	eth, ace	B9³, 3951
8858	Malonic acid, cyclopentylidene, diethyl ester .... (C_5H_8)=$C(CO_2C_2H_5)_2$	226.28	....	140¹⁰	....	1.0616²⁰/⁴	1.4724²⁰	eth, ace	B9³, 3952
8859	Malonic acid, diallyl, diethyl ester .... ($CH_2=CHCH_2$)$_2C(CO_2C_2H_5)_2$	240.3	....	243-4, 128¹⁶	....	0.9943²⁰/⁴	1.4445²²	al, eth, ace	B2³, 2003
8860	Malonic acid, dibenzyl, diethyl ester .... ($C_6H_5CH_2$)$_2C(CO_2C_2H_5)_2$	340.42	....	234-5²³	14	1.093²⁰/⁴	....	al, eth	B9³, 4553
8861	Malonic acid, dibromo, diethyl ester .... $Br_2C(CO_2C_2H_5)_2$	317.97	....	250-6d, 154²⁸	....	....	....	al, eth, ace	B2⁴, 1905
8862	Malonic acid, dibutyl ester .... $CH_2(CO_2C_4H_9)_2$	216.28	....	251-2, 137¹⁴	−83	0.9824²⁰/⁴	1.4262²⁰	al, eth, ace, bz, aa	B2⁴, 1884
8863	Malonic acid, dibutyl,diethyl ester .... (C_4H_9)$_2C(CO_2C_2H_5)_2$	272.39	....	153-4¹⁴	....	0.9457²⁰/⁴	1.4341²⁰	al, eth	B2⁴, 2119
8864	Malonic acid, diethyl or 3,3-Pentanedicarboxylic acid .... (C_2H_5)$_2C(CO_2H)_2$	160.17	pr (w bz)	d	127	....	....	w, al, eth	B2⁴, 2026
8865	Malonic acid, diethyl, diethyl ester .... (C_2H_5)$_2C(CO_2C_2H_5)_2$	216.28	....	230, 100¹²	....	0.9643³⁰	1.4240²⁰	al, eth	B2⁴, 2026
8866	Malonic acid, diethyl ester or Diethyl malonate ...... $CH_2(CO_2C_2H_5)_2$	160.17	....	199.3, 96²²	−48.9	1.0551²⁰/⁴	1.4139²⁰	al, eth, ace, bz, chl	B2⁴, 1881
8867	Malonic acid, dimethyl ester or Dimethyl malonate ...... $CH_2(CO_2CH_3)_2$	132.13	....	181.4, 78.4¹⁵	−61.9	1.528²⁰	1.4135²⁰	al, eth, ace, bz, chl	B2⁴, 1880
8868	Malonic acid, dimethyl, diethyl ester .... (CH_3)$_2C(CO_2C_2H_5)_2$	188.23	....	197, 97-8²²	−30.4	0.9964²⁰/⁴	1.4129²⁰	al, eth	B2⁴, 1955

No.	Name, Synonyms, and Formula	Mol. wt.	Color, crystalline form, specific rotation and λ_{max} (log ε)	b.p. °C	m.p. °C	Density	n_D	Solubility	Ref.
8869	Malonic acid, dipropyl ester .... $CH_2(CO_2C_3H_7)_2$	188.23	glass	229, 113[13]	−77.1	1,0097[20/4]	1.4206[20]	al, eth, ace, bz	B2[3], 1619
8870	Malonic acid, ethoxymethylene),diethyl ester .... $(C_2H_5OCH=)C(CO_2C_2H_5)_2$	216.24	....	279-81d, 165[19]	......	......	1.4600[20]	al, eth	B3[4], 1192
8871	Malonic acid, ethyl or 1,1-Propanedicarboxylic acid .... $C_2H_5CH(CO_2H)_2$	132.13	pr (w + 1)	160d	114	......	......	w, al, eth, bz, chl	B2[4], 1952
8872	Malonic acid, ethyl, diethyl ester .... $C_2H_5CH(CO_2C_2H_5)_2$	188.23	....	207-9[755], 98-9[12]	......	1.0047[20/4]	1.4166[20]	al, eth, ace, chl	B2[4], 1953
8873	Malonic acid, ethyl (methyl) or 2,2-Butane dicarboxylic acid .... $C_2H_5C(CH_3)(CO_2H)_2$	146.14	pr or nd (eth)	......	122	......	......	w, al, eth, ace	B2[4], 1977
8874	Malonic acid, ethyl (phenyl), diethyl ester .... $C_6H_5C(C_2H_5)(CO_2C_2H_5)_2$	264.33	....	170[19]	......	1.071[20/4]	1.4896[25]	al, eth	B9[3], 4304
8875	Malonic acid, ethyl (isopropyl), diethyl ester .... $(CH_3)_2CHC(C_2H_5)(CO_2C_2H_5)_2$	230.31	....	232-4[742], 108-10[11]	......	......	1.4280[25]	al, eth, ace	B2[4], 2052
8876	Malonic acid, ethylidene,diethyl ester .... $(CH_3CH=)C(CO_2C_2H_5)_2$	186.21	....	115-8[17]	......	1.0194[17/4]	1.4308[17]	al, eth	B2[4], 2234
8877	Malonic acid, formamido,diethyl ester .... $HCONHCH(CO_2C_2H_5)_2$	203.20	pl (MeOH)	173-4[11]	48-9	......	......	al	B4[3], 1503
8878	Malonic acid, heptyl or 1,1-Octanedicarboxylic acid .... $C_7H_{15}CH(CO_2H)_2$	202.25	pr (bz-peth)	......	96-8	......	......	al, eth, ace	B2[4], 2094
8879	Malonic acid, hexyl,diethyl ester .... $C_6H_{13}CH(CO_2C_2H_5)_2$	244.34	....	268-70, 143[15]	......	0.9577[21/4]	1.4278[21]	al, eth, ace, bz	B2[3], 1791
8880	Malonic acid, hydroxy or Tartronic acid .... $HOCH(CO_2H)_2$	120.06	pr (w + 1)	sub	156-8	......	......	w, al	B3[4], 1120
8881	Malonic acid, hydroxy,diethyl ester .... $HOCH(CO_2C_2H_5)_2$	176.17	....	222-5, 121[15]	−2.5	......	......	al, eth, ace, bz	B3[3], 905
8882	Malonic acid, hydroxy,dimethyl ester .... $HOCH(CO_2CH_3)_2$	148.13	cr (eth-peth)	122[19]	45	......	......	w, al, eth, ace, bz	B3[3], 905
8883	Malonic acid, (2-hydroxycyclohexyl) lactone,ethyl ester $C_{11}H_{15}O_4$	212.25	pa ye	199[30]	......	1.0735[19]	......	......	B18[4], 5351
8884	Malonic acid, hydroxy-(methyl) or α-Isomalic acid...... $CH_3C(OH)(CO_2H)_2$	134.09	....	170d	142d	......	......	w, al, eth	B3[3], 927
8885	Malonic acid, methyl or 1,1-Ethane dicarbozylic acid .... $CH_3CH(CO_2H)_2$	118.09	nd (AcOEt-bz) pr(eth-bz)	......	135d	1.455[20/4]	......	w, al, eth, aa	B2[4], 1932
8886	Malonic acid, methyl,diethyl ester or Diethyl methylmalonate $CH_3CH(CO_2C_2H_5)_2$	174.20	....	201, 94[16]	......	1.0225[20/4]	1.4126[20]	al, eth, ace, chl	B2[4], 1932
8887	Malonic acid, methyl,dimethyl ester .... $CH_3CH(CO_2CH_3)_2$	146.14	....	176.5	......	1.0977[20/4]	1.4128[20]	**al, eth**, ace, chl	B2[3], 1681
8888	Malonic acid, monoamide-N-Phenyl or Malonanilic acid $HO_2CCH_2CONHC_6H_5$	179.18	cr(w eth al)	......	132	......	......	al, eth	B12[3], 557
8889	Malonic acid, monochloride,ethyl ester or Carbethoxy acetyl chloride .... $ClCOCH_2CO_2C_2H_5$	150.56	....	170-80d, 75-7[15]	......	......	......	eth	B2[4], 1887
8890	Malonic acid, monochloride,methyl ester .... $ClCOCH_2CO_2CH_3$	136.54	....	71[15]	......	......	......	eth	B2[4], 1886
8891	Malonicacid, octyl or 1,1-Nonane dicarboxylic acid ...... $C_8H_{17}CH(CO_2H)_2$	216.28	pr(bz-peth)	......	116	1.173[17]	......	al, ace	B2[4], 2114
8892	Malonic acid, pentyl,diethyl ester .... $C_5H_{11}CH(CO_2C_2H_5)_2$	230.31	....	134-6[14]	......	0.9652[20/4]	1.4253[20]	al, eth	B2[4], 2034
8893	Malonic acid, isopentyl,diethyl ester .... $i-C_5H_{11}CH(CO_2C_2H_5)_2$	230.31	....	240-2, 137.5[19]	......	0.9580[25/4]	1.4255[25]	al, eth, ace	B2[3], 1777
8894	Malonic acid, (3-pentyl), diethyl ester .... $[(C_2H_5)_2CH]CH(CO_2C_2H_5)_2$	230.31	....	130[16]	......	......	1.4291[20]	al, eth	B2[3], 1781
8895	Malonic acid, phenyl,diethyl ester .... $C_6H_5CH(CO_2C_2H_5)_2$	236.27	....	205d, 168[12]	16-7	1.0950[20/4]	1.4977[20]	al, ace	B9[3], 4260
8896	Malonic acid, phenyl,dimethyl ester .... $C_6H_5CH(CO_2CH_3)_2$	208.22	cr(lig)	145-7[13]	51	......	......	al, eth	B9[3], 4260
8897	Malonic acid, phenyl amino,diethyl ester .... $C_6H_5NHCH(CO_2C_2H_5)_2$	251.29	cr(al or lig)	......	45	......	......	al, eth, bz, chl	B12[3], 971
8898	Malonic acid, phthalimido,diethyl ester .... $C_{15}H_{15}NO_6$	305.29	pr(al)	......	74	......	......	al, eth, ace, bz, chl	B21[4], 5264
8899	Malonic acid, propyl .... $C_3H_7CH(CO_2H)_2$	146.14	pl(bz)	d	96-7	......	......	w, al, eth, chl	B2[4], 1991
8900	Malonic acid, propyl,diethyl ester .... $C_3H_7CH(CO_2C_2H_5)_2$	202.25	....	221[767], 114[22]	......	0.9873[20]	1.4197[20]	al, eth	B2[4], 1991

No.	Name, Synonyms, and Formula	Mol. wt.	Color, crystalline form, specific rotation and λ_{max} (log ε)	b.p. °C	m.p. °C	Density	n_D	Solubility	Ref.
8901	Malonic acid, isopropyl,diethyl ester .... $(CH_3)_2CHCH(CO_2C_2H_5)_2$	202.25	....	235, 107-9[18]	......	0.9970[20/15]	1.4188[21]	al, eth, chl	B2[4], 2011
8902	Malonic acid, isopropylidene .... $(CH_3)_2C=C(CO_2H)_2$	144.14	cr (ace-chl)	d	170-1	....	....	al	B2[3], 1948
8903	Malonic acid, isopropylidene,diethyl ester .... $(CH_3)_2C=C(CO_2C_2H_5)_2$	200.24	....	175-8, 140-1[20]	......	1.0282[18/4]	1.4486[17]	al, ace	B2[4], 2244
8904	Malononitrile .... $CH_2(CN)_2$	66.06	....	218-9, 109[20]	32	1.1910[20/4]	1.4146[34]	w, al, eth, ace, bz	B2[4], 1892
8905	Malononitrile, benzyl .... $C_6H_5CH_2CH(CN)_2$	156.20	pl(al) nd(w lig)	174[23]	91	....	....	al, eth, bz	B9, 870
8906	Malononitrile, phenyl .... $C_6H_5CH(CN)_2$	142.16	cr (dil al)	152-3[21]	70-1	....	....	al	B9[3], 4263
8907	Malonyl chloride .... $CH_2(COCl)_2$	140.95	....	58[26]	......	1.4509[20/4]	1.4639[20]	eth	B2[4], 1887
8908	Maltose (pyranose form, α-anomer, anhydrous) or 4-O-α-D-glucopyranosyl-D-glucose .... $C_{12}H_{22}O_{11}$	342.30	nd (abs al), $[\alpha]_D$ +140.7 (w, c=10)	....	160-5	....	....	w	B17[4], 3057
8909	Maltose, monohydrate (β-anomer) .... $C_{12}H_{22}O_{11} \cdot H_2O$	360.32	nd (w), $[\alpha]^{20/}_D$ +111.7→ +130.4 (c=4, mut)	....	102-3	1.54	....	w	B17[4], 3057
8910	Malvidine chloride or Syringidine chloride............. $C_{17}H_{15}ClO_7$	366.76	rh ta or pr(al-aq HCl + 1w)	....	>300	....	....	al	B18[4], 3561
8911	Mandelic acid-(D) or α-Hydroxyphenyl acetic acid ...... $C_6H_5CH(OH)CO_2H$	152.16	ta, $[\alpha]^{20/}_D$ -158 (w, c=2.5)	....	133-5	1.341	....	w, al, eth, aa, chl	B10[3], 445
8912	Mandelic Acid (DL) or α-Hyroxyphenyl acetic acid ..... $C_6H_5CH(OH)COOH$	152.16	pl (wh) rh (bz)	d	121.3	1.300[20/4]	....	w, al, eth	B10[3], 448
8913	Mandelic acid (L+) or L+-α-Hydroxyphenylacetic acid. $C_6H_5CH(OH)CO_2H$	152.16	pl(w), $[\alpha]^{20/}_D$ +156.6 (w, c=2.9)	....	....	....	....	w, al, eth, chl	B10[3], 447
8914	Mandelic acid, 2,4-dimethyl or (2,4-Dimethylphenyl)hydroxyacetic acid................ $[2,4-(CH_3)_2C_6H_3]CH(OH)CO_2H$	180.21	rh(w) nd(bz) lf(to peth al)	....	119	....	....	al, eth, chl	B10[3], 613
8915	Mandelic acid, 2,5-dimethyl or (2,5-Dimethylphenyl)-α-Hydroxyacetic acid.......... $[2,5-(CH_3)_2C_6H_3]CH(OH)CO_2H$	180.21	nd or pr(bz)	....	116-7	....	....	al, eth, chl	B10[3], 612
8916	Mandelic acid, 3,4-dimethyl or (3,4-Dimethylphenyl)-α-hydroxyacetic acid $[3,4-(CH_3)_2C_6H_3]CH(OH)CO_2H$	180.21	lf(bz)	....	135	....	....	w, bz	B10[3], 611
8917	Mandelic acid, p-iodo-(dl) .... $4=IC_6H_4CH(OH)CO_2H$	278.05	....	....	135	....	....	w, al, eth	B10[3], 481
8918	Mandelic acid, p-isopropyl-(dl) .... $[4-i-C_3H_7C_6H_4]CH(OH)CO_2H$	194.23	nd(w)	....	159-60	....	....	al, eth	B10[3], 629
8919	Mandelic acid, p-isopropyl-(dl) .... $[4-i-C_3H_7C_6H_4]CH(OH)CO_2H$	194.23	lf (w), $[\alpha]^{17/}_D$ +134.9 (abs, al, c=4)	....	153-4	....	....	al, eth	B10, 279
8920	Mandelic acid, p-isopropyl-(l) $[4-i-C_3H_7C_6H_4]CH(OH)CO_2H$	194.23	ta (20% al), $[\alpha]^{17/}_D$ -135 (abs, al, c=4.09)	....	153-4	....	....	al, eth	B10, 279
8921	D-Mannitane or 3,6-(1,4)Anhydro-D-mannitol .... $C_6H_{12}O_5$	164.16	amor pl	....	146-7	....	....	w, al	B17[4], 2641
8922	D-Mannitol or D-Mannite .... $C_6H_{14}O_6$	182.18	rh nd or pr (w), $[\alpha]^{25/}_D$ -0.49 (w)	295[3.5]	168	1.489[20/4]	1.3330	w	B1[4], 2841
8923	DL-Mannitol or α-Acritol .... $C_6H_{14}O_6$	182.18	cr(al)	....	168	....	....	w	B1[3], 2405
8924	L-Mannitol or L-Mannite .... $C_6H_{14}O_6$	182.18	nd(al)	....	163-4	....	....	w	B1[4], 2843
8925	Mannitol, hexanitrate or Nitro mannitol .... $O_2NOCH_2(CHONO_2)_4CH_2ONO_2$	452.17	nd (al), $[\alpha]^{25/}_{546}$ +46.8	120 exp	112.3	1.8[20/4]	....	al, eth, bz, aa	B1[4], 2849

No.	Name, Synonyms, and Formula	Mol. wt.	Color, crystalline form, specific rotation and λ_{max} (log ϵ)	b.p. °C	m.p. °C	Density	n_D	Solubility	Ref.
8926	α-D-Mannoheptose $C_7H_{14}O_7$	210.19	nd (al) $[\alpha]^{20}_D$ +85→ +68.6 (w,c=1)	....	134-5			w	B1⁴, 4438
8927	β-D-Mannoheptose, monohydrate $C_7H_{14}O_7 \cdot H_2O$	244.20	cr (w + 1) $[\alpha]^{20}_D$ +45.7		83			w	B1⁴, 4441
8928	D-Mannoic acid-γ-lactone $C_6H_{10}O_6$	178.14	pr (abs al) $[\alpha]^{20}_D$ +54 (w,c=2)		151			w	B18⁴, 3018
8929	D-Mannoheptose (β-anomer), hydrate $C_7H_{14}O_7$	178.14	rh nd (w) $[\alpha]_D$ -51.8		151			w	B18², 199
8930	D-Mannosaccharic acid, γ,γ'-dilactone $C_6H_6O_6$	174.11	nd (w + 2) $[\alpha]^{23}_D$ +202 (w)		190d			al	B19⁴, 2962
8931	L-Mannosaccharic acid, γ,γ'-dilactone $C_6H_6O_6$	174.11	nd (w al) $[\alpha]^{20}_D$ -202.5 (w)		183-5d, 68 (hyd)			w	B19², 266
8932	D-Mannose (β-anomer) $C_6H_{12}O_6$	180.16	nd or orh pr (al or aa) $[\alpha]^{20}_D$ -17→ +14.6 (w,p=3)		132d	1.539²⁰ᐟ⁴			B1⁴, 4328
8933	DL-Mannose $C_6H_{12}O_6$	180.16	cr (al)		132-3			w	B31, 294
8934	L-Mannose (β-anomer) $C_6H_{12}O_6$	180.16	cr (al) $[\alpha]_D$ +14→ -14		132			w	B1⁴, 4333
8935	D-Mannose, phenylhydrazone $C_{12}H_{18}N_2O_5$	270.29	ye pr (w) nd (al) $[\alpha]_D$ +26.3→ +33.8 (py)		199-200				B31, 290
8936	L-Mannose, phenylhydrazone $C_{12}H_{18}N_2O_5$	270.29	ye pr (w)		195				B31, 294
8937	D-Mannuronic acid (α-anomer), monohydrate $C_6H_{10}O_7$	194.14	hyg nd (al eth) $[\alpha]^{25}_D$ +16→ -6 (w,mut)		120-30d			w	B3⁴, 1998
8938	D-Mannuronic acid (β-anomer) $C_6H_{10}O_7$	194.14	cr (w ace eth) $[\alpha]^{25}_D$ -47.9→ -23.9 (w)		165-7			w	B3⁴, 1999
8939	Margaric acid or Heptadecanoic acid $CH_3(CH_2)_{15}CO_2H$	270.46	pl (peth)	227¹⁰⁰	62-3	0.8532⁶⁰	1.4342⁶⁰	eth, ace, bz, chl	B2⁴, 1193
8940	Matrine or Sophocarpidine $C_{15}H_{24}N_2O$	248.37	α, nd or pl, β, orh pr γ, lig; δ, pr or lf (peth), $[\alpha]^{15}_D$ α40.9 (w), β -28.7 (w)	γ-223⁶	α,76-7 β, 87 δ, 84	γ-1.088²⁰ᐟ⁴	γ-1.5286⁸⁵	al, eth, ace, bz	B24², 58
8941	Meconic acid or 3-hydroxy-γ-pyrone-2,6-dicarboxylic acid $C_7H_4O_7$	200.11	rh pl (w dilHCl) (+3w)	d at 120	-w at 100			al, bz	B18⁴, 6203
8942	Meconidine $C_{21}H_{23}NO_4$	3:3.42	ye amor		58			al, eth, ace, bz, chl	M, 641
8943	Meconin or 6,7-Dimethoxyphthalide $C_{10}H_{10}O_4$	194.19	wh nd (w)	155sub	102-3			al, eth, ace, bz, chl	B18⁴, 1226
8944	Medicagenic acid $C_{30}H_{46}O_6$	502.70	pr or nd		352-3				Am76, 2271
8945	Medicagenic acid, diacetate $C_{34}H_{50}O_8$	586.78	mcl $[\alpha]^{22}_D$ +92 chl		210-2	1.190²⁵ᐟ²⁵			Am79, 5292
8946	Melam $C_6H_8N_{11}$	235.21	pw						B3⁴, 319
8947	Melamine or 2,4,6-triamino-1,3,5-triazine $C_3H_6N_6$	126.12	mcl pr (w)	sub	345d	1.573¹⁶	1.872²⁰		B26², 132

No.	Name, Synonyms, and Formula	Mol. wt.	Color, crystalline form, specific rotation and λ_{max} (log ε)	b.p. °C	m.p. °C	Density	n_D	Solubility	Ref.
8948	Melene $C_{30}H_{60}$	420.82	nd (ace) cr (peth)	380	62-3	$0.9037^{25/25}$	1.4228^{90}	....	B1³, 885
8949	Melissic acid $CH_3(CH_2)_{28}CO_2H$	466.84	sc or nd (al or ace)		93			bz	B2³, 1097
8950	Melezitose $C_{18}H_{32}O_{16}$	504.45	cr (w + 2) $[\alpha]^{20/}_D$ + 88 (w,c=4)		153-4 (anh)	1.5565^0		w	B17⁴, 3015
8951	Melibiose (α-anomer) or 6-O-α-D-galactopyranosyl-α-D-glucose $C_{12}H_{22}O_{11}$	342.30	amor (anh) $[\alpha]^{15}$ + 145.8 → 141.6 (w,c=2,mut)					w	B17⁴, 3075
8952	Melibiose (β-anomer), dihydrate $C_{12}H_{22}O_{11}.2H_2O$	378.33	mcl cr (w + 2) $[\alpha]^{20/}_D$ + 111.7 → + 129.5 (w,c=4)		84-5			w, MeOH	B17⁴, 3075
8953	Mellitic acid or Benzene hexacarboxylic acid $C_6(CO_2H)_6$	342.18	nd (al)		286-8			w, al	B9³, 4914
8954	P-Menthane (cis) or 1-isopropyl-4-methyl cyclohexane $C_{10}H_{20}$	140.27	....	170.9^{725}	-89.9	$0.8039^{20/4}$	1.4431^{20}	al, eth, bz, peth	B5⁴, 151
8955	P-Menthane (trans) $C_{10}H_{20}$	140.27	....	170.6, 58.6¹²		$0.7928^{20/4}$	1.4366^{20}	al, eth, bz, lig	B5⁴, 151
8956	p-Menthane, 1,8-epoxy $C_{10}H_{18}O$	154.25	....	176-7	1.5	0.9267^{20}	$1.4584^{15/}_D$	al, eth, chl	B17⁴, 213
8957	Δ¹-p-Menthene(d) (one form) or 1-isopropyl-4-methyl cyclohexene-(3) $C_{10}H_{18}$	138.25	$[\alpha]^{20/}_D$ + 115.6 (undil)	168		$0.8118^{20/4}$	1.4524^{20}	al, eth, bz, aa	B5⁴, 301
8958	Δ¹-p-methene(d) (one form) $C_{10}H_{18}$	138.25	$[\alpha]_D$ + 29.6 → + 54.4	167-8		$0.8078^{20/4}$		al, eth, bz, peth	B5⁴, 299
8959	Δ³-p-methene(dl) $C_{10}H_{18}$	138.25		168^{754}, 60.5^{12}		$0.8069^{20/4}$	1.4503^{15}	al, eth, bz, peth	B5⁴, 301
8960	Δ³-p-Menthene $C_{10}H_{18}$	138.25	$[\alpha]_D$ -13.5 (al)	167-8	1	$0.8112^{19/19}$	1.4511^{20}	al, eth, bz, peth	B5⁴, 301
8961	Menthol-(d) or 3-p-Menthol, Hexahydro thymol $C_{10}H_{20}O$	156.27	$[\alpha]_D$ + 49.2 (al,c=5)	$103-4^9$	42-3			al, eth, ace, bz	B6⁴, 150
8962	Menthol-(dl) $C_{10}H_{20}O$	156.27	nd (peth)	216, $103-5^{16}$	(i)28 (ii)38	$0.904^{15/15}$	1.4615^{20}	al, eth, ace, bz	B6⁴, 151
8963	Menthol-(l) $C_{10}H_{20}O$	156.27	nd (MeOH) $[\alpha]^{20/}_D$ -48 (al,c=2.5)	216.4, 111^{20}	44	$0.904^{15/15}$	1.460^{22}	al, eth, ace, chl	B6⁴, 151
8964	Menthol, 3-isovalerate $C_{15}H_{28}O_2$	240.39	$[\alpha]^{20/}_D$ -64 (bz,c=10)	129^9	....	0.9089^{15}	1.4486^{20}	al, ace	B6³, 145
8965	Menthol, acetate(l) $C_{12}H_{22}O_2$	198.30	$[\alpha]^{20/}_D$ -79.42	109^{10}		$0.9185^{20/4}$	1.4469^{20}		B6³, 142
8966	Menthone(d) or 2-Isopropyl-5-methyl cyclohexanone $C_{10}H_{18}O$	154.26	$[\alpha]^{18/}_D$ + 24.8	204^{750}, 85^{14}		$0.8963^{20/20}$	1.4503^{20}	al, eth, ace, bz, aa	B7³, 154
8967	Menthone (dl) $C_{10}H_{18}O$	154.26		210.5		0.911^0		al, eth, ace, bz, aa	B7³, 154
8968	Menthone (l) $C_{10}H_{18}O$	154.26	$[\alpha]^{20/}_D$ -29.6	209.6, 96^{20}		$0.8954^{20/4}$	1.4505^{20}	al, eth, ace, bz	B7³, 152
8969	Menthoxy acetic acid (l) $C_{12}H_{23}O_3$	214.31	cr (eth) $[\alpha]^{20/}_D$ -92.9 (MeOH)	171^{11}	53-5			al, eth, ace	B6³, 155
8970	Menthoxyacetyl chloride (l) $C_{12}H_{22}ClO_2$	232.75	$[\alpha]^{13/}_D$ -84.8 (chl)	$128-31^{11}$				eth	B6³, 156
8971	Mercaptoacetamide, N-Phenyl $HSCH_2CONHC_6H_5$	167.23	nd (al)		110-1			al, eth	B12³, 925
8972	Mercaptoacetamide, N-β-naphthyl or Thionalide $HSCH_2CONH-β-C_{10}H_7$	217.29	nd		111-2			al	B12³, 3052

No.	Name, Synonyms, and Formula	Mol. wt.	Color, crystalline form, specific rotation and λ_{max} (log ε)	b.p. °C	m.p. °C	Density	n_D	Solubility	Ref.
8973	Mercaptoacetic acid or Thioglycolic acid .... HSCH$_2$COOH	92.12	....	120[20]	−16.5	1.3253[20]	1.5030[20]	**w, al, eth**	B3[4], 600
8974	Mercaptoacetic acid, acetate or Acetyl thioglycolic acid... (CH$_3$COS)CH$_2$CO$_2$H	134.16	yo	158-9[17], 115-8[2.5]				w	B3[4], 610
8975	Mercaptoacetic acid, ethyl ester or Ethylmercapto acetate HSCH$_2$CO$_2$C$_2$H$_5$	120.18		156-8, 55[17]		1.0964[15]	1.4582[20]	al, eth	B3[4], 617
8976	Mercaptoacetic acid, butyl ester or Butyl mercapto acetate HSCH$_2$CO$_2$C$_4$H$_9$	148.22		85-8[16]		1.03[20/4]			B3[4], 620
8977	Mercaptoacetic acid, dodecyl ester or Lauryl mercapto-acetate .... HSCH$_2$CO$_2$C$_{12}$H$_{25}$	260.43		....	3	0.43[20/4]			B3[3], 432
8978	Mercaptoacetic acid, 2-ethylhexyl ester or 2-Ethylhexyl mercaptoacetate .... HSCH$_2$CO$_2$[CH$_2$CH(C$_2$H$_5$)(CH$_2$)$_3$CH$_3$]	204.33		133.5		0.97[20/4]			
8979	Mercaptoacetic acid, hexadecyl ester .... HSCH$_2$CO$_2$C$_{16}$H$_{33}$	316.54			20-5	....		chl	B3[4], 622
8980	Mercaptoacetic acid, isopropyl ester or isopropyl mercap-toacetate.................. HSCH$_2$CO$_2$CH(CH$_3$)$_2$	134.19		80-8[45]		1.05[20/4]			B3[3], 430
8981	Mercaptoacetic acid, methyl ester or Methyl-mercaptoace-tate .... HSCH$_2$CO$_2$CH$_3$	106.14		42-3[10]			1.4657[20]	al, eth	B3[4], 614
8982	Mercaptoacetic acid, octyl ester or Octyl mercaptoacetate HSCH$_2$CO$_2$C$_8$H$_{17}$	204.33	....	125[17]			1.4606[21]	....	B3[3], 431
8983	Mesaconic acid or Methylfumaric acid .... HO$_2$CC(CH$_3$)=CHCO$_2$H	130.10	rh nd or mcl pr (eth, AcOEt)	sub	204-5	1.466[20/4]		al, eth	B2[4], 2231
8984	Mesaconic acid, diethyl ester or Diethyl mesaconate...... C$_2$H$_5$O$_2$CC(CH$_3$)=CCO$_2$C$_2$H$_5$	186.21		229, 93-5[10]		1.0453[20/20]	1.4488[20]	al, eth, ace, bz	B2[4], 2232
8985	Mesaconic acid, dimethyl ester or Dimethyl mesaconate .. CH$_3$O$_2$CC(CH$_3$)=CHCO$_2$CH$_3$	158.16		203.5, 100[10]		1.0914[20/4]	1.4512[20]	al, eth, ace	B2[4], 2232
8986	Mescaline or 3,4,5-trimethoxy-β-phenethylamine .... C$_{11}$H$_{17}$NO$_3$	211.26	cr	180[12]	35-6			w, al, bz, chl	B13[3], 2375
8987	Mesitylene or 1,3,5-trimethylbenzene .... 1,3,5-(CH$_3$)$_3$C$_6$H$_3$	120.20		164.7, 48.7[10]	−44.7	0.8652[20/4]	1.4994[20]	**al, eth, ace, bz**	B5[4], 1016
8988	Mesitylene, acetyl or 2,4,6-trimethyl acetophenone 2,4,6-(CH$_3$)$_3$C$_6$H$_2$COCH$_3$	162.22		240.5[735], 120[12]		0.9754[20/4]	1.5175[20]	al, eth, ace, bz	B7[3], 1137
8989	Mesitylene, γ-bromo isobutylryl .... 2,4,6-(CH$_3$)$_3$C$_6$H$_2$COCBr(CH$_3$)$_2$	269.19	gold-ye oil	160-70[24]	27			eth	B7[3], 1209
8990	Mesitylene, vinyl 2,4,6(CH$_3$)$_3$C$_6$H$_2$CH=CH$_2$	146.23		208-10, 83[12]		0.9057[20/4]	1.5296[20]	....	B5[4], 1408
8991	Mesityl oxide or 4-Methyl-3-penten-2-one (CH$_3$)$_2$C=CHCOCH$_3$	98.15		129.7, 41[25]	−52.9	0.8653[20/4]	1.4440[20]	w, **al, eth**, ace	B1[4], 3471
8992	Mesoxalic acid, hydrate or Dihydroxymalonic acid ...... (HO)$_2$C(CO$_2$H)$_2$	136.06	dlq nd (w)		120-1			w, al, eth	B3[3], 1355
8993	Mesoxalic acid, diethyl ester .... O=C(CO$_2$C$_2$H$_5$)$_2$	174.16	pa ye gr oil	208[220], 105[19]	ca −30	1.1419[16/4]	1.4310[22]	w, al, eth, chl	B3[3], 1356
8994	Mesoxalic acid, diethyl ester, hydrate or Ethyl dihydroxy malonate .... (HO)$_2$C(CO$_2$C$_2$H$_5$)$_2$	192.17	pl (bz)	ca 200	57			w, al, eth, ace, bz	B3[3], 1356
8995	Mesoxalic acid, diethyl ester, oxime .... HON=C(CO$_2$C$_2$H$_5$)$_2$	189.17		172[12]		1.1821[18/4]	1.4544[18]	al, eth, ace, bz	B3[4], 1805
8996	Mesoxalonitrile or Oxomalonic acid, dinitrile .... O:C(CN)$_2$	80.05		65.5	−36	1.124[20/4]	1.3919[20]	eth, ace	B2[4], 1806
8997	Mestilbol or 3-(4-hydroxyphenyl)-4(4-methoxyphenyl)-3-Hexene .... CH$_3$CH$_2$C(4-CH$_3$OC$_6$H$_4$)=C(4-HOC$_6$H$_5$)CH$_2$CH$_3$	282.39	nd (bz-lig) lf (70% al)	185-95[0.3]	117-8			al, eth, ace	B6[3], 5623
8998	Metacrolein or 2,4,6-triethynyl-1,3,5-trioxane.......... C$_9$H$_{12}$O$_3$	168.20	pl (al)	170	50			al, eth	B1[4], 3435
8999	Metaldehyde or Metacetaldehyde (C$_2$H$_4$O)$_{4-6}$	(44.05)$_{4-6}$	tetr nd or pr (al)	sub 115	246.2 (sealed tube)				B19[4], 5643
9000	Metaldehyde II(tetramer) .... (C$_2$H$_4$O)$_4$	176.22		110, 65[15]	47			al, eth, ace, bz	B19[4], 5643
9001	Metameconin or 5,6-Dimethoxy phthalide............. C$_{10}$H$_{10}$O$_4$	194.19	cr (dil al)		155-7				B18[4], 1223

No.	Name, Synonyms, and Formula	Mol. wt.	Color, crystalline form, specific rotation and λ_{max} (log ε)	b.p. °C	m.p. °C	Density	n_D	Solubility	Ref.
9002	Methacrolein .... CH$_2$=C(CH$_3$)CHO	70.09	....	68.4	....	0.837[20/4]	1.4144[20]	w, al, eth,	B1[4], 3455
9003	Methacryl amide .... CH$_2$=C(CH$_3$)CONH$_2$	85.11	cr (bz)		110-1			al	B2[4], 1538
9004	Methacrylic acid or 2-Methylpropenoic acid .... CH$_2$=C(CH$_3$)COOH	86.09	pr	162-3[757], 60[12]	16	0.0153[20/4]	1.4314[20]	w, al, eth	B2[4], 1518
9005	Methacrylic acid, anhydride .... [CH$_2$=C(CH$_3$)CO]$_2$O	154.17	....	89[5]			1.4540[20]	al, eth	B2[4], 1537
9006	Methacrylic acid, butyl ester or Butyl methacrylate ...... CH$_2$=C(CH$_3$)CO$_2$C$_4$H$_9$	142.20		160, 52[11]		0.9836[20/4]	1.4240[20]	al, eth	B2[4], 1525
9007	Methacrylic acid, isobutyl ester or isobutyl methacrylate . CH$_2$=C(CH$_3$)CO$_2$-i-C$_4$H$_9$	142.20		155, 45[11]		0.8858[20/4]	1.4199[20]	al, eth	B2[4], 1526
9008	Methacrylic acid, ethyl ester or Ethyl methacrylate ...... CH$_2$=C(CH$_3$)CO$_2$C$_2$H$_5$	114.15		117, 30[18]		0.9135[20/4]	1.4147[20]	al, eth	B2[4], 1523
9009	Methacrylic acid, methyl ester or Methyl methacrylate .... CH$_2$=C(CH$_3$)CO$_2$CH$_3$	100.13		100-1, 24[32]	−48	0.9440[20/4]	1.4142[20]	al, eth, ace	B2[4], 1519
9010	Methacrylic acid, propyl ester or Propyl methacrylate .... CH$_2$=C(CH$_3$)CO$_2$C$_3$H$_7$	128.17		141		0.9022[20/4]	1.4190[20]	al, eth	B2[4], 1524
9011	Methacrylic acid, isopropyl ester or Isopropyl methacrylate CH$_2$=C(CH$_3$)CO$_2$-i-C$_3$H$_7$	128.17		125		0.8847[20/4]	1.4122[20]	al, eth, ace, bz	B2[4], 1525
9012	Methacrylonitrile .... CH$_2$=C(CH$_3$)CN	67.09		90.3	−35.8	0.7998[20/4]	1.4190[20]	al, eth	B2[4], 1539
9013	Methacrylyl chloride .... CH$_2$=C(CH$_3$)COCl	104.54		96, 50[135]		1.0871[20/4]	1.4435[20]	eth, ace, chl	B2[4], 1537
9014	Methallyl alcohol or 2-Methyl-2-propen-1-ol .... CH$_2$=C(CH$_3$)CH$_2$OH	72.11		114.5		0.8515[20]	1.4255[20]	w, al, eth	B2[4], 2114
9015	Methane .... CH$_4$	16.04	gas	−164	−182	0.466[−164], 0.55[179]	....	al, eth, bz	B1[4], 3
9016	Methane, bis (dimethyl amino) .... [(CH$_3$)$_2$N]$_2$CH$_2$	102.18		82.5		0.7491[18.7]		w	B4[4], 153
9017	Methane, bis-(2-hydroxy-3,5,6-trichlorophenyl) or Hexachlorophene (2-HO-3,5,6-Cl$_3$C$_6$H)$_2$CH$_2$	406.91	nd (bz)	....	166-7			al, eth, ace, chl	B6[3], 5407
9018	Methane, bis (trichlorosilyl) .... (Cl$_3$Si)$_2$CH$_2$	282.92		182.7[745], 64[10]	....	1.5567[20/4]	1.4740[20]	al, eth	B1[4], 3077
9019	Methane, bis (4-chlorophenoxy) or Oxythane .... (4-ClC$_6$H$_4$O)$_2$CH$_2$	269.13		189-94[6]	69-70			eth, ace, bz	B6[4], 833
9020	Methane, bromo or Methyl bromide .... CH$_3$Br	94.94	....	3.6	−93.6	1.6755[20/4]	1.4218[20]	al, eth	B1[4], 68
9021	Methane, bromochloro .... CH$_2$BrCl	129.39		68.1	−86.5	1.9344[20/4]	1.4838[20]	al, eth, ace, bz	B1[4], 74
9022	Methane, bromochloro-dinitro .... BrCCl(NO$_2$)$_2$	219.38		75-6[15]	9.3	2.0394[20/4]	1.4793	al	B1[3], 115
9023	Methane, bromo-chloro-fluoro .... BrHClF	147.38		36.1[756]	−115	1.9771[0/4]	1.4144[25]	eth, ace, chl	B1[4], 75
9024	Methane, bromo-dichloro .... BrHCl$_2$	163.83		90	−57.1	1.980[20/4]	1.4964[20]	al, eth, ace, bz, chl	B1[4], 76
9025	Methane, bromo-difluoro .... BrCHF$_2$CHF$_2$	130.93			−14.5	1.55[16]	....	w, al	B1[4], 72
9026	Methane, bromo difluoro nitroso .... BrF$_2$CNO	159.92	bl gas	−12					B1[4], 100
9027	methane, bromo-diiodo .... BrCHI$_2$	346.74	ye (peth)	110[25]	60				B1, 72
9028	Methane, bromo-diphenyl or Benzydryl bromide .... (C$_6$H$_5$)$_2$CHBr	247.14	tcl (peth)	193[26]d, 111[0.3]	45			al, bz	B5[4], 1850
9029	Methane, bromo fluoro .... BrCH$_2$F	112.94		18-20				al, chl	B1[4], 72
9030	Methane, bromo-iodo .... BrCH$_2$I	220.84		138-41		2.926[17]	1.6410[20]	chl	B1[4], 95
9031	Methane, bromo-nitro .... BrCH$_2$NO$_2$	139.94		148-9[742]			1.4880[20]	al	B1[4], 106
9032	Methane, bromo-trichloro .... BrCCl$_3$	198.28		104.7, 0.6[10]	−5.6	2.0122[20/4]	1.5063[20]	al, eth	B1[4], 77
9033	Methane, bromo-trifluoro .... BrCF$_3$	148.92	gas	−59[740]				chl	B1[4], 73
9034	Methane, bromo-trinitro .... BrC(NO$_2$)$_3$	229.94		56[10]	17-8	2.0313[20/4]	1.4808[20]	al, chl	B1[3], 116

No.	Name, Synonyms, and Formula	Mol. wt.	Color, crystalline form, specific rotation and λ_{max} (log ε)	b.p. °C	m.p. °C	Density	n_D	Solubility	Ref.
9035	Methane, chloro or Methyl chloride . . . . CH₃Cl	50.49	gas	−24.2	−97.1	0.9159²⁰/⁴	1.3389²⁰	al,eth, ace, bz, chl	B1⁴, 28
9036	Methane, chloro-dibromo . . . . ClCHBr₂	208.29	. . . .	119-20⁷⁴⁸	. . . .	2.451²⁰/⁴	1.5482²⁰	al, eth, ace, bz	B1⁴, 81
9037	Methane, chloro-difluoro or Freon 22 . . . . ClCHF₂	86.47	gas	−40.8	−146	. . . .	. . . .	eth, ace, chl	B1⁴, 32
9038	Methane, chloro difluoro nitro ClF₂CNO₂	131.47	. . . .	25	. . . .	. . . .	. . . .	chl	B1⁴, 106
9039	Methane, chloro difluoro nitroso ClF₂CNO	115.47	bl gas	ca -35	. . . .	. . . .	. . . .		B1⁴, 99
9040	Methane, chloro diiodo . . . . ClCHI₂	302.28	. . . .	200d, 88³⁰	−4	. . . .	. . . .	eth, ace, chl	B1⁴, 97
9041	Methane, chloro dinitro ClCH(NO₂)₂	140.48	. . . .	34-6¹³	. . . .	1.6125²⁰	1.4575²⁰		B1³, 115
9042	Methane, chloro diphenyl or Benzhydryl chloride . . . . (C₆H₅)₂CHCl	202.69	nd	173¹⁹	20.5	1.1398²⁰/⁴	1.5959²⁰		B5⁴, 1847
9043	Methane, chloro-fluoro ClCH₂F	68.48	gas	. . . .	−9.1	. . . .	. . . .	chl	B1⁴, 32
9044	Methane, chloro-iodo . . . . ClCH₂I	176.39	. . . .	109	. . . .	2.422²⁰/⁴	1.5822²⁰	al, eth, ace, bz, chl	B1⁴, 94
9045	Methane, chloro-nitro . . . . ClCH₂NO₂	95.49	. . . .	122-3	. . . .	1.466¹⁵		w	B1⁴, 106
9046	Methane, chloro tribromo ClCBr₃	287.19	lf (eth)	158-9	55	2.71¹⁵	. . . .	eth	B1⁴, 85
9047	Methane, chloro trifluoro or Freon 13 . . . . ClCF₃	104.46	gas	−81.1	−181	. . . .	. . . .		B1⁴, 34
9048	Methane, chloro trinitro ClC(NO₂)₃	185.48	. . . .	133-5d, 56⁴⁰	4.5	1.6769²⁰/⁴	1.4500²⁰	al, eth, chl	B1³, 116
9049	Methane, chloro triphenyl or Trityl chloride . . . . (C₆H₅)₃CCl	278.79	nd or pr (bz-peth)	310, 230-5²⁰	113-4	. . . .	. . . .	eth, ace, bz, chl	B5³, 2315
9050	Methane, deutero-trichloro CDCl₃	120.39	. . . .	61-2	-64.1	1.5004²⁰/⁴	1.4450²⁰		B1⁴, 54
9051	Methane, diazo or Acomethylene . . . . CH₂N₂	42.04	ye gas	ca -0	−145	. . . .	. . . .	eth	B1⁴, 3056
9052	Methane, diazo diphenyl . . . . (C₆H₅)₂CN₂	194.24	bl-red nd (peth)	exp	30-2	. . . .	. . . .	al, eth, ace, bz	B7³, 2068
9053	Methane, dibromo or Methylene bromide . . . . CH₂Br₂	173.85	. . . .	97	−52.5	2.4970²⁰/⁴	1.5420²⁰	al, eth, ace	B1⁴, 78
9054	Methane, dibromo chloro fluoro Br₂CClF	226.27	. . . .	80.3	. . . .	2.3173²²	1.4570²⁰/_D	. . . .	B1⁴, 82
9055	Methane, dibromo dichloro Br₂CCl₂	242.74	. . . .	150.2	38	. . . .	. . . .	al, eth, ace, bz	B1⁴, 82
9056	Methane, dibromo difluoro Br₂CF₂	209.83	. . . .	24.5	. . . .	. . . .	. . . .	al, eth, ace, bz	B1⁴, 80
9057	Methane, dibromo dinitro . . . . Br₂C(NO₂)₂	263.84	nd	158d, 77²¹	5.5	2.4440²⁰/⁴	1.5280²⁵	al	B1³, 115
9058	Methane, dibromo fluoro . . . . FCHBr₂	191.84	. . . .	64.9⁷⁵⁷	. . . .	2.421²⁰/⁴	1.4685²⁰	al, eth, ace, bz, chl	B1⁴, 80
9059	Methane, dibromo iodo . . . . ICHBr₂	299.74	pl (peth)	91⁴²	22.5	. . . .	. . . .		B1³, 99
9060	Methane, dichloro or Methylene chloride . . . . CH₂Cl₂	84.93	. . . .	40	−95.1	1.3266²⁰/⁴	1.4242²⁰	al, eth	B1⁴, 35
9061	Methane, dichloro difluoro or Freon 12 . . . . Cl₂CF₂	120.91	. . . .	−29.8	−158	1.75⁻¹¹⁵, 1.1834⁵⁷	. . . .	al, eth, aa	B1⁴, 40
9062	Methane, dichloro-dinitro Cl₂C(NO₂)₂	174.93	. . . .	121-2, 46²⁰	. . . .	1.6124²⁰/⁴	1.4575²⁰	al, eth, bz, chl	B1³, 115
9063	Methane, dichloro diphenyl . . . . (C₆H₅)₂CCl₂	237.13	. . . .	305d, 190²¹	. . . .	1.235¹⁸	. . . .		B5⁴, 1848
9064	Methane, dichloro-fluoro or Freon 21 . . . . Cl₂CHF	102.92	. . . .	9	−135	1.405⁹	1.3724⁹	al, eth, chl, aa	B1⁴, 39
9065	Methane, dichloro-iodo . . . . Cl₂CHI	210.83	. . . .	132, 40³⁰	. . . .	2.392²⁰/⁴	1.5840²⁰	al, eth, ace, bz, chl	B1⁴, 95
9066	Methane, dichloro-nitro . . . . Cl₂CHNO₂	129.93	. . . .	107	. . . .	. . . .	. . . .		B1³, 113
9067	Methane, difluoro or Methylene fluoride . . . . CH₂F₂	52.02	. . . .	−51.6	. . . .	0.909²⁰	1.190²⁰	al	B1⁴, 24

No.	Name, Synonyms, and Formula	Mol. wt.	Color, crystalline form, specific rotation and λ_{max} (log ϵ)	b.p. °C	m.p. °C	Density	n_D	Solubility	Ref.
9068	Methane, difluoro-iodo .... F_2CHI	177.92	....	21.6	−122	3.238[−19]	....		B1[4], 92
9069	Methane, di-2-furyl .... $(2-C_4H_3O)_2CH_2$	148.17	....	94[22.5]		1.102[20/4]	1.5049[20]	al, eth, ace	B19[4], 269
9070	Methane, diiodo or Methylene iodide .... CH_2I_2	267.84	ye nd or lf	182; 60[10]	6.1	3.3254[20/4]	1.7425[20]	al, eth, bz, chl	B1[4], 96
9071	Methane, diiodo-fluoro .... $FCHI_2$	285.83	pa ye	100-1, 50[50]	−34.5	3.1969[22]	....	al, eth	B1[4], 97
9072	Methane, dinitro .... $CH_2(NO_2)_2$	106.04	ye nd	100 exp	<−15			al, eth	B1[4], 107
9073	Methane, dinitro diphenyl .... $(C_6H_5)_2C(NO_2)_2$	258.24	pl (dil al)	....	79-80			al, eth, bz, chl	B5[3], 1797
9074	Methane, diphenyl or Diphenyl methane .... $(C_6H_5)_2CH_2$	188.24	pr nd	264.3, 125.5[10]	25.3	1.0060[20/4]	1.5753[20]	al, eth, chl	B5[4], 1841
9075	Methane, diphenyl 3-tolyl .... $(3-CH_3C_6H_4)CH(C_6H_5)_2$	258.37	pr (al, MeOH)	354	62	1.07[16]	....	eth, bz, chl, aa	B5[3], 2332
9076	Methane, di(α-pyrryl) or 2, 2′-Methylene dipyrrole ...... $(2-C_4H_4N)_2CH_2$	146.19	lf or nd (al)	163-7[12]	73			al, eth, bz	B23[2], 167
9077	Methane, disilano .... $(SiH_3)_2CH_2$	76.25	....	14.7[754]	....	0.6979[4/4]	1.4115[4]		B1[4], 3072
9078	Methane, di-(4-tolyl) phenyl .... $(4-CH_3C_6H_4)_2CHC_6H_5$	272.40	nd (MeOH)	218-20[12]	56			al, eth, bz, chl	B5[3], 2342
9079	Methane, fluoro or Methyl fluoride .... CH_3F	34.03	....	−78.4	−141.8	0.8428[−60]	1.1727[20]	al, eth, bz, chl	B1[4], 22
9080	Methane, fluoro-iodo .... FCH_2I	159.93	....	53.4	....	2.366[20/4]	1.5256[20]	eth, ace, bz, chl	B1[4], 92
9081	Methane, fluoro-tribromo .... $FCBr_3$	270.74	hyg nd	62.5[15]	42	1.0003[20/4]	1.4161[20]	al	B1[4], 85
9082	Methane, iodo or Methyl iodide..................... CH_3I	141.94	....	42.4	−66.4	2.279[20/4]	1.5380[20]	al, eth ace, bz	B1[4], 87
9083	Methane, iodo-trichloro or .... $ICCl_3$	245.27	....	142	....	2.355[20/4]	1.5854[20]	eth, ace, bz, chl	B1[4], 95
9084	Methane, iodo-trifluoro .... CF_3I	195.91	....	−22.5		2.3608[−32/4]	1.3790[−42/D]	....	B1[4], 92
9085	Methane, nitro .... CH_3NO_2	61.04	....	100.8	fr -17	1.1371[20/4]	1.3817[20]	al, eth, ace	B1[4], 100
9086	Methane, nitro-tribromo or Bromopicrin .... Br_3CNO_2	297.74	....	exp[760], 89-90[20]	10.2	2.7930[20/4]	1.5790[20]	al, eth, ace, bz, aa	B1[4], 106
9087	Methane, nitro-trichloro or Chloropicrin............ Cl_3CNO_2	164.38	....	111.8	−64.5	1.6566[20/4]	1.4622[20]	al, ace, bz, aa	B1[4], 106
9088	Methane, nitro-trifluoro or Fluoropicrin .... F_3CNO_2	115.01	....	−31.1		....	....		B1[4], 105
9089	Methane, nitroso-trifluoro .... F_3CNO	99.01	....	−84	−197	....	....		B1[4], 99
9090	Methane, (pentafluorothio)trifluoro .... $F_3C(SF_5)$	196.06	....	−20		....	....		B1[4], 35
9091	Methane, tetrabromo or Carbon tetrabromide .... CBr_4	331.65	mcl ta (dil al)	189-90, 102[50]	90-4	2.9609[100/4]	1.5942[100]	al, eth	B1[4], 85
9092	Methane, tetrachloro or Carbon tetrachloride.......... CCl_4	153.82	....	76.5	−23	1.5940[20/4]	1.4601[20]	al, eth, ace, bz, chl	B1[4], 56
9093	Methane, tetrafluoro or Carbon tetrafluoride......... CF_4	88.01	....	−129[754]	−150	3.034[0]		bz, chl	B1[4], 26
9094	Methane, tetraiodo or Carbon tetraiodide .... CI_4	519.63	red lf (bz,chl)	130-40[1-2]	171d	4.23[20]	....	chl, py	B1[4], 98
9095	Methane, tetranitro .... $C(NO_2)_4$	196.03	....	126, 21-3[22]	14.2	1.6380[20/4]	1.4384[20]	al, eth	B1[4], 107
9096	Methane, tetraphenyl .... $C(C_6H_5)_4$	320.44	rh nd (bz,sub)	431 sub		....	....		B5[3], 2568
9097	Methane, tribromo or Bromoform .... $CHBr_3$	252.75	hex sc	149.5, 46[15]	8.3	2.8899	1.5976[20]	al, eth, bz, chl, lig	B1[4], 82
9098	Methane, trichloro or Chloroform .... $CHCl_3$	119.38	....	61.7	−63.5	1.4832[20/4]	1.4459[20]	al, eth, ace, bz, lig	B1[4], 42
9099	Methane, tricyclohexyl .... $(C_6H_{11})_3CH$	262.48	....	322-9	48	0.9274[50/0]	1.4986[40/D]	eth, bz	B5[4], 504
9100	Methane, trifluoro or Fluroform .... CHF_3	70.01	....	−82.2	−160	1.52[−100]		al, ace, bz	B1[4], 24

No.	Name, Synonyms, and Formula	Mol. wt.	Color, crystalline form, specific rotation and λ_{max} (log ε)	b.p. °C	m.p. °C	Density	n_D	Solubility	Ref.
9101	Methane, triiodo or Iodoform .... CHI_3	393.73	ye hex pr or nd(ace)	ca 218	123	4.008[20/4]	....	eth, ace, chl, aa	B1[4], 97
9102	Methane, trinitro or Nitroform .... $CH(NO_2)_3$	151.04		exp[760], 45-7[22]	19	1.479[20/4]	1.4451[24]	al, ace	B1[4], 107
9103	Methane, triphenyl or Triphenyl methane .... $(C_6H_5)_3CH$	244.34	rh (al)	358-9[754], 190-215[10]	94	1.014[99/4]	1.5839[99]	eth, ba, chl	B5[3], 2307
9104	Methane, tris (4-aminophenyl) or p-Leucaniline........ $(4-NH_2C_6H_4)_3CH$	289.38	lf (w,al,bz)	....	208			al, eth	B13[3], 566
9105	Methane, tris (4-dimethylaminophenyl) or Leucocrystal violet................ $[4-(CH_3)_2NC_6H_4]_3CH$	373.55	lf (al) nd (bz, lig)	....	175			eth, bz, chl, aa	B13[3], 566
9106	Methane, tris (2-tolyl) .... $(2-CH_3C_6H_4)_3CH$	286.42	nd (al)	....	130-1			al, eth	B5[3], 2347
9107	Methane arsonic acid .... $CH_3AsO(OH)_2$	139.99	lf (al)	....	160-1			w, al	B4[3], 1822
9108	Methanedisulfonic acid, dihydrate or Methionic acid.... $CH_2(SO_3H)_2.2H_2O$	212.18	hyg nd (w + 2)	220-70[15-20] d				w, al	B1[4], 3054
9109	Methane phosphonic acid or Methyl phosphonic acid ... $CH_3PO(OH)_2$	96.02	hyg pl	d	108-9			w, al, eth	B4[3], 1778
9110	Methanephosphonic acid, diethyl ester or Diethyl methyl-phosphonate .... $CH_3PO(OC_2H_5)_2$	152.13		194, 85[15]		1.0406[30/4]	1.4101[30]	w, al, eth	B4[4], 1778
9111	Methanephosphonic acid, dimethyl ester or Dimethyl methylphosphonate .... $CH_3PO(OCH_3)_2$	124.08	....	181[754], 79.5[20]				w	B4[3], 1778
9112	Methanesulfenyl chloride, trichloro or Perchloromethyl-mercaptan .... Cl_3CSCl	185.89	ye oil	147-8, 51[25]		1.6947[20/4]	1.5484[20]	eth	B3[4], 290
9113	Methanesulfenylchloride, trifluoro or Perfluoromethyl-mercaptan .... F_3CSCl	136.52		-0.7					B3[4], 289
9114	Methane sulfinic acid, amino, imino $HN=C(NH_2)SO_2H$	108.12	nd (al)	....	144d			w	B3[4], 145
9115	Methane sulfonamide, trifluoro .... $F_3CSO_2NH_2$	149.09			119			w, chl	B3[4], 35
9116	Methanesulfonic acid .... CH_3SO_3H	96.11		167[10]	20	1.4812[18/4]	1.4317[18]	w, al, eth	B4[4], 10
9117	Methanesulfonic acid, trifluoro .... F_3CSO_3H	150.08	hyg liq	162, 81[37.5]				eth	B3[4], 34
9118	Methane sulfonamide, trifluoro-N,N-diethyl .... $F_3CSO_2N(C_2H_5)_2$	205.20		55[7]				eth	B4[4], 415
9119	Methane sulfonyl chloride, trifluoro .... F_3CSO_2Cl	168.52	....	31.6				eth	B3[4], 35
9120	Methane sulfonic acid, trifluoro, ethyl ester .... $F_3CSO_3C_2H_5$	178.13		115, 42[40]				eth	B3[4], 34
9121	Methane sulfonyl chloride .... CH_3SO_2Cl	114.55	....	161[730], 55[11]		1.4805[18/4]	1.4573[20]	al, eth	B4[4], 27
9122	Methane sulfonyl chloride, trichloro .... Cl_3CSO_2Cl	217.89	cr (al-w)	170 (sub)	140-1			al, eth	B3[4], 36
9123	Methane sulfonyl fluoride, trifluoro .... F_3CSO_2F	152.07		-21.7					B3[4], 34
9124	Methanethiol or Methyl mercaptan.............. CH_3SH	48.11		6.2	-123	0.8665[20/4]	....	al, eth	B1[4], 1273
9125	Methanetricarboxylic acid, trimethyl ester .... $HC(CO_2CH_3)_3$	190.15	pr (MeOH)	242.7, 128[15]	46-7			al, eth, bz, chl	B2[3], 2023
9126	Methantheline bromide or Banthine bromide .... $C_{21}H_{26}NO_3Br$	420.36	cr (i-prOH)	....	172-7			w, al, eth	B18[4], 4352
9127	Methanol or Methyl alcohol.............. CH_3OH	32.04		65, 15[73]	-93.9	0.7914[20/4]	1.3288[20]	w, al, eth, ace, bz, chl	B1[4], 1227
9128	Methanol-d or o-Deutero methanol .... CH_3OD	33.05		65.5	-100	0.8127[20/4]	....	w, al, eth, ace, bz	B1[4], 1244
9129	Methapyrilene (base) or Histadyl base .... $C_{14}H_{19}N_3S$	261.39		173-5[3]			1.5915[20]		B22[4], 3950
9130	Methionine (DL) or dl-2-amino-4-(methylthio) butyric acid .... $CH_3SCH_2CH_2CH(NH_2)CO_2H$	149.21	pl	....	281d	1.340		w	B4[3], 1647

No.	Name, Synonyms, and Formula	Mol. wt.	Color, crystalline form, specific rotation and λ_{max} (log ϵ)	b.p. °C	m.p. °C	Density	n_D	Solubility	Ref.
9131	Methionine (L) CH$_3$SCH$_2$CH$_2$CH(NH$_2$)CO$_2$H	149.21	hex pl (dil al $[\alpha]^{25}_D$ -8.2 (w,c=1) $+22.5$ (INHCl)	sub 186	283d	....	....	w	B4^3, 1639
9132	Methoxyacetic acid or Methyl glycolic acid .... CH$_3$OCH$_2$CO$_2$H	90.08	hyg	203-4, 96^{13}	......	1.1768$^{20/4}$	1.4168^{20}	w, al, eth	B3^4, 574
9133	Methoxyacetic acid, ethyl ester or Ethyl methoxy acetate CH$_3$OCH$_2$CO$_2$C$_2$H$_5$	118.13	....	142, 44-5^9	......	1.0118^{15}	1.4050^{20}	al, eth	B3^3, 381
9134	Methoxyacetic acid, methyl ester or Methyl methoxy acetate CH$_3$OCH$_2$CO$_2$CH$_3$	104.12	....	131^{763}, 57^{50}	......	1.0511$^{20/4}$	1.3962^{20}	al, eth, ace	B3^4, 578
9135	2-Methoxybenzoic acid or o-Anisic acid .... 2-CH$_3$OC$_6$H$_4$CO$_2$H	152.16	pl (w), fl (al)	200	101	....	....	al, eth, bz, chl	B10^3, 97
9136	2-Methoxybenzoic acid, 5-chloro .... 5-Cl-2-CH$_3$OC$_6$H$_3$CO$_2$H	186.60	nd (w)	......	81-2	....	....	al, ace	B10^3, 165
9137	2-Methoxybenzoic acid, ethyl ester .... 2-CH$_3$OC$_6$H$_4$CO$_2$C$_2$H$_5$	180.21	....	261, 135-6^{12}	......	1.1124$^{20/4}$	1.5224^{20}	al, eth	B10^3, 116
9138	2-Methoxybenzoic acid, methyl ester .... 2-CH$_3$OC$_6$H$_4$CO$_2$CH$_3$	166.18	....	245, 127^{11}	......	1.1571$^{19/4}$	1.534$^{19.5}$	al	B10^3, 109
9139	3-Methoxybenzoic acid or m-Anisic acid .... 3-CH$_3$OC$_6$H$_4$CO$_2$H	152.16	nd (w)	170-2^{10}	110	....	....	al, eth, bz, chl	B10^3, 244
9140	3-Methoxybenzoic acid, ethyl ester .... 3-CH$_3$OC$_6$H$_4$CO$_2$C$_2$H$_5$	180.21	....	260-1, 110^5	......	1.0993$^{20/4}$	1.5161^{20}	al, eth	B10^3, 260
9141	3-Methoxybenzoic acid, methyl ester .... 3-CH$_3$OC$_6$H$_4$CO$_2$CH$_3$	166.18	....	232, 121-4^{10}	......	1.1310$^{20/4}$	1.5224^{20}	al	B10^3, 250
9142	4-Methoxybenzoic acid or p-Anisic acid .... 4-CH$_3$OC$_6$H$_4$CO$_2$H	152.16	pr or nd (w)	275-80	185	....	....	al, eth, chl	B10^3, 280
9143	4-Methoxybenzoic acid, butyl ester .... 4-CH$_3$OC$_6$H$_4$CO$_2$C$_4$H$_9$	208.24	....	183^{40}	......	1.054$^{16.5}$	1.5141$^{18.5}$	....	B10^3, 307
9144	4-Methoxybenzoic acid, ethyl ester .... 4-CH$_3$OC$_6$H$_4$CO$_2$C$_2$H$_5$	180.21	....	269-70	7-8	1.1038$^{20/4}$	1.5254^{20}	al, eth	B10^3, 300
9145	4-Methoxybenzoic acid, methyl ester .... 4-CH$_3$OC$_6$H$_4$CO$_2$CH$_3$	166.18	fl (al,eth)	256, 160^{20}	49	....	....	al, eth	B10^3, 297
9146	Methoxylamine, hydrochloride or o-Methyl hydroxylamine hydrochloride........... CH$_3$ONH$_2$.HCl	83.52	pr	149	......	....	....	w, al	B1^4, 1252
9147	2-Methoxyphenyl isothiocyanate 2-CH$_3$OC$_6$H$_4$NCS	165.21	....	131-2^{11}	......	1.1878^{20}	1.6458$^{20/}_D$	....	B13^3, 823
9148	Methylamine or Amino methane........... CH$_3$NH$_2$	31.06	gas	-6.3	-93.5	0.699^{-4}, 0.6628^{20}	....	w, al, eth, ace, bz	B4^4, 118
9149	Methylamine, hydrochloride .... CH$_3$NH$_2$.HCl	67.52	diq tetr ta (al)	sub 30^{15}	227-8	....	....	w, al	B4^4, 122
9150	Methyl isobutyl ketone or 4-methyl-2-pentanone .... i-C$_4$H$_9$COCH$_3$	100.16	....	116.8, 35-40^{16}	-84.7	0.7978^{20}	1.3962^{20}	al, eth, ace, bz, chl	B1^4, 3305
9151	β-Methyl chalcone C$_6$H$_5$C(CH$_3$)=CHCOC$_6$H$_5$	222.29	....	340-5d, 225^{22}	......	1.108$^{20/0}$	1.6312^{20}	eth	B7^3, 2422
9152	Methyl β-chloroethyl sulfide .... CH$_3$SCH$_2$CH$_2$Cl	110.61	....	140, 44^{20}	......	1.1097$^{25/4}$	1.4902^{20}	al, eth, ace	B1^4, 1406
9153	Methyl chlorosulfonate .... CH$_3$OSO$_2$Cl	130.55	....	133-5, 48^{29}	......	1.4805$^{25/4}$	1.4138^{18}	eth, ace, bz	B1^4, 1252
9154	β-Methyl choline chloride or (2-hydroxypropyl) trimethyl ammonium chloride .... CH$_3$CH(OH)CH$_2$N$^+$(CH$_3$)$_3$ Cl$^-$	153.65	pr (BROH)	d	165	....	....	w, al	B4^3, 754
9155	Methyl ether or Dimethyl ether .... CH$_3$OCH$_3$	46.07	gas	-25	-138.5	....	....	w, al, eth ace, chl	B1^4, 1245
9156	Methyl ether, boro fluoride .... (CH$_3$)$_2$O.BF$_3$	113.88	....	127d	-14	1.2410$^{20/4}$	1.302^{20}	....	B1^4, 1248
9157	Methyl chloromethyl ether .... CH$_3$OCH$_2$Cl	80.52	....	59.1	-103.5	1.0605$^{20/4}$	1.3974^{20}	al,, eth, ace, chl	B1^4, 3046
9158	Methyl (2,3-dibromopropyl) ether .... CH$_3$OCH$_2$CHBrCH$_2$Br	231.91	....	185, 84^{15}	......	1.8320$^{12/4}$	1.5123^{20}	eth	B1^4, 1447
9159	Methyl (2,3-epoxypropyl) ether or Epimethylin........... CH$_3$OCH$_2$CHCH$_2$	89.11	....	115-8	......	0.9890^{20}	1.4320^{20}	w, al, eth, ace	B17^4, 986
9160	Methyl ethyl amine or Ethyl methyl amine........... CH$_3$NHC$_2$H$_5$	59.11	....	36.7	......	....	....	w, al, eth, ace	B4^4, 312

No.	Name, Synonyms, and Formula	Mol. wt.	Color, crystalline form, specific rotation and λ_{max} (log ε)	b.p. °C	m.p. °C	Density	n_D	Solubility	Ref.
9161	Methyl ethyl ketone or 2-Butanone .... $CH_3CH_2COCH_3$	72.12	....	79.6, 30[119]	−86.3	0.8054[20/4]	1.3788[20]	w, **al**, **eth**, **ace**, bz	B1[4], 3243
9162	Methyl ethynyl ether or Methoxy acetylene .... $CH_3OC≡CH$	56.07	....	50		0.8001[20/4]	1.3812[20]	al, eth	B1[4], 2211
9163	Methyl glyoxal or 2-oxopropionaldehyde.............. CH_3COCHO	72.06	ye hyg liq	72	....	1.0455[24]	1.4002[18]	al, eth, bz	B1[4], 3631
9164	Methylglyoxal, dioxime or Methyl glyoxime .... $CH_3C(NOH)CH=NOH$	102.09	nd (w, sub) Pr (al)	sub	157			al, eth	B1[4], 3633
9165	Methylglyoxal, 1-oxime .... $CH_3COCH=NOH$	87.08	nd (CCl_4) lf (eth-peth)	sub	69	1.0744[67]	....	w, eth	B1[4], 3632
9166	Methyl green or Heptamethyl pararosaniline chloride .... $C_{26}H_{33}N_3Cl_2$	458.48	gr pw (al)	....				w	B13[3], 2074
9167	Methyl hydrogen sulfate .... CH_3OSO_3H	112.11	....	130-40d	<−30			w, al, **eth**	B1[4], 1250
9168	Methyl hydroperoxide .... CH_3OOH	48.04	....	38-40[65]		1.9967[15/4]	1.3641[15]	**w**, **al**, **eth**, bz	B1[4], 1249
9169	Methyl isocyanate .... CH_3NCO	57.05	....	59.6	−45	0.9230[27/4]	1.3419[18]	w	B4[4], 247
9170	Methyl isothiocyanate or Methyl mustard oil........... CH_3NCS	73.12	....	119[758]	36	1.0691[37/4]	1.5258	**al**, eth	B4[4], 248
9171	Methyl α-naphthyl ether .... CH_3O-α -$C_{10}H_7$	158.20	....	269, 135[10]	<−10	1.0964[14/2]	1.6227[22]	al, eth, bz, chl	B6[3], 2922
9172	Methyl β-naphthyl ether .... CH_3O-β -$C_{10}H_7$	158.20	lf (eth) pl (peth) sub	274, 138[10]	73-4			eth, bz, chl	B6[3], 2969
9173	Methyl nitrate .... CH_3ONO_2	77.04	exp.vapor	64.6 exp	−82.3	1.2075[20/4]	1.3748[20]	al, eth	B1[4], 1254
9174	Methyl nitrite .... CH_3ONO	61.04	gas	−12	−16	0.991[15] (liq)	....	al, eth	B1[4], 1253
9175	Methyl nitro amine .... CH_3NHNO_2	76.05	....	80-5[10]	38	1.2433[49]	1.4616[49]	w, al, eth, bz	B4[3], 1753
9176	Methyl (2-mitrophenyl) sulfide .... (2-$O_2NC_6H_4SCH_3$)	169.20	....		64.5	1.2628[78/4]	1.6246[78]	al, bz, chl	B6[4], 1661
9177	Methyl (4-nitrophenyl) sulfide .... (4-$O_2NC_6H_4SCH_3$)	169.20	....	135-40[2]	72	1.2391[80/4]	1.64008[20]	....	B6[4], 1687
9178	Methyl (2-octyn-1-yl) ether .... $CH_3O[CH_2C≡C(C_5H_{11})]$	140.23	....	77[19]		0.8370[25/4]	1.4380[20]	eth	B1[3], 1996
9179	Methyl orange or Sodium 4´-dimethylamino azobenzene- 4-sulfonate .... $C_{14}H_{14}N_3O_3NaS$	327.34	og,ye pl or sc (w)	....	d	....	....		B16[3], 371
9180	Methyl pentyl ether .... $CH_3OC_5H_{11}$	102.18	....	99-100		0.767[19]	1.3855[19]	al, eth, ace	B1[4], 1643
9181	Methyl iso pentyl ether .... CH_3O-i-C_5H_{11}	102.18	....	91[765]		0.7517[20]	1.3830[20]	al, eth	B1[4], 1681
9182	Methyl-tert-pentyl ether .... $CH_3OC(CH_3)_2CH_2CH_3$	102.18	....	86.3		0.7703[20/4]	1.3885[20]	al, **eth**	B1[4], 1671
9183	Methyl pentyl sulfide .... $CH_3SC_5H_{11}$	118.24	....	145	−94	0.8431[20/4]	1.4506[20]	al, eth, ace, bz, chl	B1[3], 1608
9184	Methyl perchlorate .... CH_3OClO_3	114.49	oil	ca 52		....	....	**al**, **eth**	B1[4], 1249
9185	Methyl phenyl glyoxal .... $C_6H_5COCOCH_3$	148.17	ye oil	222, 101[12]		1.0065[20/4]	1.537[10]	w, al, eth	B7[3], 3463
9186	Methylphenylglyoxal, dioxime or Methylphenylglyoxime $C_6H_5C(=NOH)C(=NOH)CH_3$	178.19	nd (dil al)	....	238-40			al	B7[3], 3465
9187	Methylphenylglyoxal, 1-oxime .... $C_6H_5C(=NOH)COCH_3$	163.18	pa ye nd or lf (aa, al)	....	166-7			....	B7[3], 3464
9188	Methylphenylglyoxal, 2-oxime .... $C_6H_5COC(=NOH)CH_3$	163.18	wh nd (w)	....	115				B7[3], 3464
9189	Methyl phenyl sulfide .... $CH_3SC_6H_5$	124.22	....	193, 74[10]		1.0579[20/4]	1.5868[20]	al, bz	B6[4], 1466
9190	Methyl propyl ether .... $CH_3OC_3H_7$	74.12	....	38-9		0.738[20/4]	1.3579[25]	w,al, eth, ace	B1[4], 1421
9191	Methyl isopropyl ether .... $(CH_3)_2CHOCH_3$	74.12	....	32.5[777]		0.7237[15/4]	1.3576[20]	al, eth	B1[4], 1471
9192	Methyl isopropyl ketone or 3-methyl-2-butanone .... i-$C_3H_7COCH_3$	86.14	....	94-5	−92	0.8051[20/4]	1.3880[20]	**al**, **eth**, ace	B1[4], 3287
9193	Methyl propyl sulfide .... $CH_3SC_3H_7$	90.19	....	95.5, −4[10]	−113	0.8424[20]	1.4442[20]	w, al, eth, ace	B1[4], 1450

No.	Name, Synonyms, and Formula	Mol. wt.	Color, crystalline form, specific rotation and λ_{max} (log ε)	b.p. °C	m.p. °C	Density	n_D	Solubility	Ref.
9194	Methyl isopropyl sulfide i-C$_3$H$_7$SCH$_3$	90.19	....	84.7	−101.5	0.8291$^{20/4}$	1.4932^{20}	al, eth, ace	B1^4, 1500
9195	N-Methyl quinolinium chloride C$_{10}$I$_0$NCl	179.63	cr (+ w, al)	....	126			w, chl	B20^4, 3357
9196	Methyl red or 4′ dimethylamino azobenzene-2-carboxylic acid C$_{15}$H$_{15}$N$_3$O$_2$	269.31	vt or red pr (to, bz) nd (aq, aa) lf (dil al)	....	183			al, ace, bz, chl, aa	B16^3, 367
9197	Methyl selenide or Dimethyl selenide (CH$_3$)$_2$Se	109.03	....	54-5^{753}	....		1.4077$^{15/4}$	al, eth, chl	B1^4, 1288
9198	Methyl sulfate or Dimethyl sulfate (CH$_3$O)$_2$SO$_2$	126.13		188.5d, 76^{15}	−31.7	1.3283^{20}	1.3874^{20}	w, al, eth, bz	B1^4, 1251
9199	Methyl sulfide or Dimethyl sulfide (CH$_3$)$_2$S	62.13		37.3	−98.3	0.8483$^{20/4}$	1.4438^{20}	al, eth	B1^4, 1275
9200	Methyl sulfite or Dimethyl sulfite (CH$_3$O)$_2$SO	110.13		126, 52^{45}	....	1.2129$^{20/4}$	1.4093^{20}	al, eth	B1^4, 1250
9201	Methyl sulfone or Dimethyl sulfone (CH$_3$)$_2$SO$_2$	94.13	pr	238	110	1.1707$^{110/0}$	1.4226	w, al, bz	B1^4, 1279
9202	Methyl sulfoxide or Dimethyl sulfoxide (CH$_3$)$_2$SO	78.13		189, 85-7^{20}	18.4	1.1014$^{20/4}$	1.4770^{20}	w, al, eth, ace	B1^4, 1277
9203	Methyl telluride or Dimethyl telluride (CH$_3$)$_2$Te	157.67	pa ye	93.5^{749}	>1		....	al	B1^4, 1288
9204	Methyl thiocyanate CH$_3$SCN	73.12		132.9^{757}	−51	1.0678$^{25/4}$	1.4669^{25}	al, eth	B3^4, 327
9205	Methyl vinyl ether CH$_3$OCH=CH$_2$	58.08	....	12	−122	0.7723	1.3730	al, eth, ace, bz	B1^4, 2049
9206	Methyl vinyl ketone or 3-buten-2-one CH$_3$=CH$_2$COCH$_3$	70.09		81.4, 33-4^{130}		0.8636$^{20/4}$	1.4081^{20}	w, al, eth, ace, bz	B1^4, 3444
9207	Methyl vinyl sulfide CH$_3$SCH=CH$_2$	74.14		69-70		0.9028	1.4837		B1^4, 2003
9208	Methyl vinyl sulfone CH$_3$SO$_2$CH=CH$_2$	106.14		122-4^{24}		1.2117$^{20/4}$	1.4636^{20}	eth, ace	B1^4, 2065
9209	Methylene blue or 3,9-bis Dimethylamino phenazothionium chloride C$_{16}$H$_{18}$N$_3$ClS	319.86	dk gr cr or pw (chl-eth)	....				w, al, chl	B27^2, 448
9210	5,5′-Methylene disalicylic acid or Bix-(3-carboxy-4-hydroxy phenyl) methane [4-HO-3HO$_2$CC$_6$H$_3$]$_2$CH$_2$	288.26	nd (bz)	....	243-4			al, eth, ace	B10^3, 2507
9211	Methysticin or Kavatin C$_{15}$H$_{14}$O$_5$	274.28	nd (MeOH) pr (ace) $[\alpha]^{20/}_D$ + 94.34 (aa,p=5)	....	137			al, ace, bz, chl	B19^4, 5161
9212	Metrazol or Cardiazole Leptazolo C$_6$H$_{10}$N$_4$	138.17	cr (bz-liq)	194^{12}	59-60			w, al, eth, ace, bz	B26^2, 213
9213	Metycaine or Piperocaine hydrochloride C$_{16}$H$_{23}$NO$_2$Cl	292.78			172-5			w, al, chl	B20^4, 1460
9214	Michler's hydrol (4-(CH$_3$)$_2$NC$_6$H$_4$)$_2$CHOH	270.38	lt gr lf or pr (bz)	....	98			al, eth, bz, aa	B13^3, 1958
9215	Mimosine (l) or l-Leucenol C$_8$H$_{10}$N$_2$O$_4$	198.18	ta (w) $[\alpha]^{22/}_D$ −21 (w,c=0.5)	228-9d					B21^4, 4648
9216	Morphine C$_{17}$H$_{19}$NO$_3$	285.35	pr	....	254-6			py, MeOH	B27^2, 118
9217	Morphine, acetate,trihydrate (l) C$_{17}$H$_{19}$NO$_3$.CH$_3$CO$_2$H.3H$_2$O	398.44	cr (dil al) $[\alpha]^{25/}_D$ −77 (w)	....	200d			w	B27^2, 134
9218	Morphine, hydrate C$_{17}$H$_{19}$NO$_3$.H$_2$O	303.36	orh pr (dil al) $[\alpha]^{25/}_D$ −132 (MeOH, c=1)	d	254-6 anh	1.32$^{20/4}$	1.56-1.64		B27^2, 122
9219	Morphine, hydrochloride,trihydrate C$_{17}$H$_{19}$NO$_3$.HCl.3H$_2$O	375.85	nd or fl (dil HCl) $[\alpha]^{25/}_D$ −113.5 (w,c=2.2)		200d			w, al	B27^2, 132

No.	Name, Synonyms, and Formula	Mol. wt.	Color, crystalline form, specific rotation and λ_{max} (log ε)	b.p. °C	m.p. °C	Density	n_D	Solubility	Ref.
9220	Morphine, N-oxide or Genomorphine .... $C_{17}H_{19}NO_4$	301.35	pr (50% al)	....	274-5	....	....		B27[2], 159
9221	Morphine, sulfate,pentahydrate .... $2(C_{17}H_{19}NO_3).H_2SO_4.5H_2O$	758.85	pw or cubes [α]$^{25/}_D$ −107.8 (w,c=4)	....	ca 250d	....	....	w	B27[2], 133
9222	Morphine, o,o-diacetyl or Heroin.................... $C_{21}H_{23}NO_5$	369.42	rh,o [α]$^{15/}_D$ −166 (MeOH)	272-4[12]	173	1.56-1.61	....	bz, chl	B27[2], 151
9223	Morphine-o,o-diacetyl, hydrochloride, monohydrate .... $C_{21}H_{23}NO_5.HCl.H_2O$	423.90	[α]$^{20/}_D$ −153 (w,c=1.17)	....	231-2	....	....	w, al, chl	B27[2], 153
9224	Morphine-3-ethyl ether hydrochloride, dihydrate or Dionin .... $C_{19}H_{23}NO_3.HCl.2H_2O$	385.89	cr	....	123-5d, 170 (anh)	....	....	w, al	B27[2], 148
9225	Morpholine or Tetrahydro-1,4-isoxazine .... C_4H_9NO	87.12	hyg	128.3, 24.8[10]	−4.7	1.0005[20/4]	1.4548[20]	w, al, eth, ace, bz	B27[2], 3
9226	Morpholine, 4-acetyl .... 4-$CH_3CO(C_4H_8NO)$	129.16	....	152[50], 118[12]	14.5	1.1165[20/20]	1.4827[20]	w, al, ace	C50, 7112
9227	Morpholine, 4-(2-aminoethyl) .... 4-$(H_2NCH_2CH_2)(C_4H_8NO)$	130.19	....	116[50]	25.6	0.9915[20/20]	1.4715[20]	**w, al,** ace, **bz,** lig	C42, 6747
9228	Morpholine, 4-(3 aminopropyl) .... 4-$[H_2N(CH_2)_3](C_4H_8NO)$	144.22	....	219[733], 134[50]	−15	0.9872[20/20]	1.4762[20]	**w, al,** ace, **bz,** lig	Am66, 725
9229	Morpholine, 4-benzyl .... 4-$C_6H_5CH_2(C_4H_8NO)$ $C_{11}H_{15}NO$	177.25	....	260-1, 128-9[13]	....	1.0387[20/4]	1.5302[20]	ace, bz	B27[1], 203
9230	Morpholine, 4-butyl .... 4-$C_4H_9(C_4H_8NO)$ $C_8H_{17}NO$	143.23	....	213-4, 67-8[10]	−57.1	0.9068[20/4]	1.4451[20]	w, al, ace, bz	Am61, 171
9231	Morpholine, 2,6,-dimethyl .... 2,6-$(CH_3)_2(C_4N_7NO)$	115.18	....	146.6, 58[30]	fr −85	0.9346[20/20]	1.4460[20]	**w, al,** ace bz, lig	Am80, 1257
9232	Morpholine, 4-(2-ethoxy ethyl) .... 4-$(C_2H_5OCH_2CH_2)(C_4H_8NO)$ $C_8H_{17}NO_2$	159.23	....	206, 93-7[14]	−100	0.963[20]	....	w, eth, ace, bz	Am63, 298
9233	Morpholine, 4-ethyl .... 4-$C_2H_5(C_4H_8NO)$ $C_6H_{13}NO$	115.18	....	138-9[763]	....	0.9886[20/4]	1.4400[20]	**w, al,** eth, ace, bz	B27[1], 203
9234	Morpholine, 4-(2-hydroxy ethyl) or 4-(β-morpholinol) ethanol .... $C_6H_{13}NO_2$	131.18	....	227[757]	....	1.0710[20/4]	1.4763[20]	w, al	B27, 7
9235	Morpholine, 4-(2-hydroxy propyl) .... $C_7H_{15}NO_2$	145.20	....	92-4[13]	....	1.0174[20/4]	1.4638[20]	w, al, eth, ace, bz	Am64, 970
9236	Morpholine, 4-methyl .... $C_5H_{11}NO$ 4-$CH_3(C_4H_8NO)$	101.15	....	115-6[750]	....	0.9051[20/4]	1.4332[20]	w, al, eth	B27[1], 203
9237	Morpholine, 4-phenyl .... 4-$C_6H_5(C_4H_8NO)C_{10}H_{13}NO$	163.22	cr (al-eth)	259-60[745], 165-70[45]	57-8	....	....	eth	B27[2], 3
9238	Morpholine, 4(4-tolyl) .... 4-(4-$CH_3C_6H_4)(C_4H_8NO)$ $C_{11}H_{15}NO$	177.25	cr (dil al)	167[30]	51	....	....	al, eth	B27[2], 4
9239	Mucic acid or 2,3,4,5-Tetrahydroxy hexanedioic acid or Galactaric acid .... $HO_2C(CHOH)_4CO_2H$	210.14	pr (w)	....	255 (rapid htng)	....	....	aa	B3[4], 1292
9240	cis-Muconic acid or 2,4-Hexadiendioic acid .... $HO_2CCH=CHCH=CHCO_2H$	142.11	....	....	194-5	....	....	aa	B2[4], 2297
9241	trans-Muconic acid .... $HO_2CCH=CHCH=CHCO_2H$	142.11	nd (al)	320	305d	....	....		B2[4], 2298
9242	Murexide or Ammonium purpurate .... $C_8H_{10}N_6O_7$	302.21	red-gr pr (aq.NH₄Cl)	....	....	....	....		B25[1], 709
9243	Mycophenolic acid .... $C_{17}H_{20}O_6$	320.35	nd (w)	....	141	....	....	al, eth, chl	B18[4], 6513
9244	Myrcene, dihydro or β-Linaloolene................ $CH_2=CHCH(CH_3)CH_2CH_2CH=C(CH_3)_2$	138.25	....	165-8	....	0.7601[20/4]	1.4362[20]	....	B1[3], 1020
9245	Myrcene or -7-methyl-3-methylene-1,6-octadiene .... $(CH_3)_2C=CH(CH_2)_2-C(=CH_2)CH=CH_2$	136.24	....	167, 65[20]	....	0.8013[15/4]	1.4722[20]	al, eth, bz, chl, aa	B1[4], 1108
9246	Myricyl alcohol or 1-Triacontanol................ $CH_3(CH_2)_{28}CH_2OH$	438.83	nd (eth) pl (bz)	....	88	0.777[95]	....	al, eth, bz	B1[3], 1850
9247	Myristaldehyde or Tetradecanal .... $C_{13}H_{27}CHO$	212.38	lf	166[24]	30	....	....	al, eth, ace	B1[4], 3389
9248	Myristaldehyde, dimethyl acetal .... $C_{13}H_{27}CH(OCH_3)_2$	258.43	....	134-6[4]	....	....	1.4342[25]	al, eth	B1[4], 3389
9249	Myristaldehyde, oxime .... $C_{13}H_{27}CH=NOH$	227.39	lf or nd (al)	....	82-3	....	....	al	B1[2], 770
9250	Myristamide .... $C_{13}H_{27}CONH_2$	227.39	lf (ace)	217[12]	105-7	....	....	al	B2[4], 1138

No.	Name, Synonyms, and Formula	Mol. wt.	Color, crystalline form, specific rotation and λ_{max} (log ε)	b.p. °C	m.p. °C	Density	n_D	Solubility	Ref.
9251	Myristic acid or Tetradecanoic acid................... $C_{13}H_{27}CO_2H$	228.38	lf (eth, 80% aa)	250.5^{100}, 149.3^1	58	$0.8439^{80/4}$	1.4305^{60}	al, ace, bz, chl	B2[4], 1126
9252	Myistic acid, anhydride .... $(C_{13}H_{27}CO)_2O$	438.74	lf (peth)	vac distb	53.4	$0.8502^{70/4}$	1.4335^{70}	al, eth	B2[4], 1138
9253	Myristic acid, benzyl ester or Benzyl myristate.......... $C_{13}H_{27}CO_2CH_2C_6H_5$	318.51	....	229.3^{11}	20.5	$0.9321^{25/25}$	....	al, eth, bz, chl	B6[7], 417
9254	Myristic acid, ethyl ester or Ethyl myristate .... $C_{13}H_{27}CO_2H_5$	256.43	....	295, 162.5^9	12.3	$0.8573^{25/4}$	1.4362^{20}	al	B2[4], 1131
9255	Myristic acid, methyl ester or Methyl myristate .... $C_{13}H_{27}CO_2CH_3$	242.41	....	295^{751}, $155-7^7$	19	....	1.425^{45}	**al, eth, ace, bz, chl**	B2[4], 1131
9256	Myristic acid, propyl ester or Propyl myristate .... $C_{13}H_{27}CO_2C_3H_7$	270.46	....	147^2	....	0.8592^{20}	1.4356^{25}	al, eth, ace, bz	B2[4], 1132
9257	Myristic acid, isopropyl ester or Isopropyl myristate ..... $C_{13}H_{27}CO_2CH(CH_3)_2$	270.46	....	192.6^{20}, 140.2^2	....	0.8532^{20}	1.4325^{25}	al, eth, ace, bz,	B2[4], 1132
9258	Myristicin or 1,3-Benzodioxide,4-methoxy-6-(2-propenyl) $C_{11}H_{12}O_3$	192.22	....	$276-7$, 157^{21}	<−20	$1.1437^{20/20}$	1.5403^{20}	eth, bz	B19[4], 801
9259	Myristonitrile .... $C_{13}H_{27}CN$	209.38	....	226.5^{100}, 119^1	19.2	$0.8281^{19/4}$	1.4392^{23}	**al, eth, ace, bz, chl**	B2[4], 1139
9260	Myristoyl chloride .... $C_{13}H_{27}COCl$	246.83	....	174^{16}	−1	....	....	eth	B2[4], 1138
9261	Myristyl alcohol or 1-Tetradecanol................... $C_{13}H_{27}CH_2OH$	214.40	lf	263.2, 167^{15}	39-40	$0.8236^{38/4}$	....	al, eth, ace, bz, chl	B1[4], 1864

No.	Name, Synonyms, and Formula	Mol. wt.	Color, crystalline form, specific rotation and λ_{max} (log ϵ)	b.p. °C	m.p. °C	Density	n_D	Solubility	Ref.
9262	Naphthacene or 2,3-Benzanthracene.............. $C_{18}H_{12}$	128.30	og-ye lf (bz, xyl)	sub	357				B5[2], 2372
9263	Naphthacene, 9,10,-dihydro .... $C_{18}H_{14}$	230.31	nd (xyl) lf (bz)	ca 400	212			bz, aa	B5[3], 2304
9264	Naphthacene, 9,10,-diphenyl .... $C_{30}H_{20}$	380.50	og (eth)		207-8			eth	B5[3], 2703
9265	Naphthacene, 9,11-diphenyl .... $C_{30}H_{20}$	380.50	ye		301-2			eth, bz	B5[3], 2702
9266	Naphthacene, 9,10,11-triphenyl .... $C_{36}H_{24}$	456.60	og (eth) (bz)		236-7, 177 (+ 1 bz)			eth, bz	B5[3], 2766
9267	9,10-Naphthacene quinone or 2,3-Benzanthraquinone.... $C_{18}H_{10}O_2$	258.28	ye nd (aa)	sub	294			al	B7[3], 4273
9268	9,11-Naphthacene quinone .... $C_{18}H_{10}O_2$	258.28	dk red (aa,xyl)		322				B7[3], 4276
9269	1-Naphthaldehyde .... α-$C_{10}H_7$CHO	156.20	pa ye	292, 160[15]	33-4	1.1503[20/4]	1.6507[20]	al, eth, ace, bz	B7[3], 1953
9270	1-Naphthaldehyde, 2-ethoxy .... 2-C_2H_5O-α-$C_{10}H_4$CHO	200.24	yesh nd (al,aa)	185-7[25]	115			al, aa	B8[3], 1110
9271	1-Naphthaldehyde, 4-ethoxy .... 4-C_2H_5O-α-$C_{10}H_4$CHO	200.24	yesh cr (aa)		75			al, eth	B8[2], 174
9272	1-Naphthaldehyde, 2-hydroxy .... 2-HO-α-$C_{10}H_4$CHO	172.20	pr (al) nd (AcOEt)	192[27]	82			al, eth, peth	B8[3], 1108
9273	2-Naphthaldehyde .... β-$C_{10}H_7$CHO	156.20	lf (w)	160[19]	61-3	1.0775[99/4]	1.6211[99]	al, eth, ace	B7[3], 1957
9274	2-Naphthaldehyde, 1-hydroxy .... 1-HO-β-$C_{10}H_6$CHO	172.20	grsh-ye nd (dil al,dil aa,lig)		60			al, eth, aa	B8[3], 1118
9275	Naphthalene .... $C_{10}H_8$	128.19	mcl pl (al)	218, 87.5[10]	80.5	0.9625[100/4] 1.0253[20]	1.5898[85], 1.4003[24]	al, eth, ace, bz	B5[3], 1549
9276	Naphthalene, 1-acetoxy or α-Naphthyl acetate .... α-$CH_3CO_2C_{10}H_7$	186.21	nd or pl (al)		49			al, eth	B6[3], 2928
9277	Naphthalene, 1-acetoxy-2-methyl .... 1-CH_3CO_2-2-$CH_3C_{10}H_6$	200.24	nd(eth-peth)		81-2			eth	B6[3], 3027
9278	Naphthalene, 1-acetoxy-4-methyl .... 1-CH_3CO_2-4-$CH_3C_{10}H_6$	200.24	nd(eth-peth)		86-8				C64, 19519
9279	Naphthalene, 1-acetoxy-7-methyl .... 1-CH_3CO_2-7-$CH_3C_{10}H_6$	200.24		188[15]	38-41				B6[3], 3030
9280	Naphthalene, 2-acetoxy or β-Naphthyl acetate .... β-$CH_3CO_2C_{10}H_7$	186.21	nd (al)		70			al, eth, chl	B6[3], 2982
9281	Naphthalene, 5-acetoxy-2-methyl .... 5-CH_3CO_2-2-$CH_3C_{10}H_6$	200.24		124[2]					B6[3], 3028
9282	Naphthalene, 1-acetyl or Methyl-α-naphthyl ketone..... α-$CH_3COC_{10}H_7$	170.21		296-8, 170[20]	34	1.1171[21.5/4]	1.6280[22]	al, eth, ace	B7[3], 1960
9283	Naphthalene, 2-acetyl or Methyl-β-naphthyl ketone..... β-$CH_3COC_{10}H_7$	170.21	nd (liq, dil al)	301-3, 171-3[11]	56				B7[3], 1967
9284	Naphthalene, 1-allyl .... α-$(CH_2=CHCH)C_{10}H_7$	168.24		265-7, 129-30[10]		1.0228[20/4]	1.6140[20]	al, bz, chl	B5[4], 1740
9285	Naphthalene, 1(2-aminoethyl) .... α-$(H_2NCH_2CH_2)C_{10}H_7$	171.25		182-3[18]				al, xyl	B12[3], 3112
9286	Naphthalene, β-(2-aminoethyl) .... β-$(H_2NCH_2CH_2)C_{10}H_7$	171.25		174[25]				al, aa	B12[3], 3114
9287	Naphthalene, 1-aminomethyl .... α-$(H_2NCH_2)C_{10}H_7$	157.22	yesh turns red in air	294-5, 162-3[12]				al, eth	B12[3], 3097
9288	Naphthalene, 2-aminomethyl .... β-$(H_2NCH_2)C_{10}H_7$	157.22	pr (eth)	180[24]	59-60			al, eth	B12[3], 3109
9289	Naphthalene, α-benzyl .... α-$(C_6H_5CH_2)C_{10}H_7$	218.30	mcl lf or ta (al)	350, 217-20[19]	59-60	1.166[17]		al, eth, bz, chl	B5[3], 2236
9290	Naphthalene, β-benzyl or β-Benzyl naphthalene .... β-$(C_6H_5CH_2)C_{10}H_7$	218.30	mcl pr (al, MeOH)	350	58	1.176[0]		eth, bz, chl,	B5[3], 2237
9291	Naphthalene, α-bromo .... α-$BrC_{10}H_7$	207.08	pr (β form)	281, 139[16]	α-6.2, β2-3	1.4826[20/4]	1.658[20]	**al, eth**, ace, **bz**, chl	B5[4], 1665
9292	Naphthalene, 1-bromo-2-(bromomethyl) .... 2-$BrCH_2$-1-$BrC_{10}H_6$	300.02	nd (al), cr (peth)		107-8			al, bz, liq	B5[3], 1633
9293	Naphthalene, α-(bromo methyl) .... α-$(BrCH_2)C_{10}H_7$	221.10	cr (peth, al)	183[18], 145-50[2]	56			al, eth, ace, bz	B5[4], 1693

No.	Name, Synonyms, and Formula	Mol. wt.	Color, crystalline form, specific rotation and λ_{max} (log ε)	b.p. °C	m.p. °C	Density	n_D	Solubility	Ref.
9294	Naphthalene, β-bromo .... β-BrC₁₀H₇	207.08	pl or rh lf (al)	287-2, 147[18]	59	1.605⁰	1.6382⁶⁰	al, eth, bz, chl	B5⁴, 1667
9295	Naphthalene, β-bromomethyl β-(BrCH₂)C₁₀H₇	221.10	lf(al)	213¹⁰⁰ 165-9¹⁴	56			al, eth, chl	B5⁴, 1698
9296	Naphthalene, α-butyl .... α-C₄H₉C₁₀H₇	184.28	....	289.34, 151-2¹⁴	−19.76	0.9738²⁰/⁴	1.5819²⁰	al, eth, ace, bz,	B5⁴, 1737
9297	Naphthalene, β-butyl .... β-C₄H₉C₁₀H₇	184.28	....	292, 146¹²	−5	0.9673²⁰/⁴	1.57774²⁰	al, ace, bz	B5⁴, 1738
9298	Naphthalene, β-tert-butyl .... β-[(CH₃)₃C]C₁₀H₇	184.28	....	274-7⁷⁵⁶	−4	0.9674²⁰/⁴	1.5685²⁰/D	al, ace, bz	B5³, 1805
9299	Naphthalene, α-chloro .... α-ClC₁₀H₇	162.62	cr (al, ace)	258.8⁷⁵³, 106.5⁵	−2.3	1.1938²⁰/⁴	1.6326²⁰	al, eth, bz	B5⁴, 1658
9300	Naphthalene, 1-chloro-2-nitro 1-Cl-2-(O₂N)C₁₀H₆	207.62	pa ye nd (al, liq)	....	81	....	....	al	B5⁴, 1677
9301	Naphthalene, 1-chloro-3-nitro 1-Cl-3-(NO₂)C₁₀H₆	207.62	ye rd (al)	....	129.5	....	....	al	B5⁴, 1677
9302	Naphthalene, 1-chloro-4-nitro 1-Cl-4(NO₂)C₁₀H₆	207.62	br ye nd (peth,al)	....	87			al, eth	B5⁴, 1676
9303	Naphthalene, 1-chloro-5-nitro 1-Cl-5-(NO₂)C₁₀H₆	207.62	nd (dil al,aa)	>360, 181²	111			al	B5³, 1597
9304	Naphthalene, 1-chloro-6-nitro 1-Cl-6-(NO₂)C₁₀H₆	207.62	ye nd (dil al)	....	131			al, eth, ace, bz, chl	B5³, 1598
9305	Naphthalene, 1-chloro-8-nitro 1-Cl-8-(NO₂)C₁₀H₆	207.62	lt ye nd (gl aa bz liq)	175²	94-5	....	....	al, aa	B5⁴, 1676
9306	Naphthalene, β-chloro β-ClC₁₀H₇	162.62	pl (dil al)lf	256, 121-2¹⁵	61	1.1377⁶⁰/⁶⁰	1.6079⁶⁰	al, eth, bz	B5³, 1660
9307	Naphthalene, 2-chloro-1-nitro 1-NO₂-2-ClC₁₀H₆	207.62	nd (peth) ye nd (al)	>360	99-100	....	....	al, eth, ace, bz	B5⁴, 1676
9308	Naphthalene, 2-chloro-3-nitro 2-Cl-3(NO₂)C₁₀H₆	207.62	br cr or nd (al)	....	94.5				B5⁴, 1677
9309	Naphthalene, 2-chloro-6-nitro 2-Cl-6-(NO₂)C₁₀H₆	207.62	ye nd	180-90¹⁵	170				B5³, 1598
9310	Naphthalene, 2-chloro-7-nitro 2-Cl-7(NO₂)C₁₀H₆	207.62	pa ye nd	....	136				B5⁴, 1677
9311	Naphthalene, 2-chloro-8-nitro 8-(NO₂)-2-ClC₁₀H₆	207.62	ye nd (al)	....	116			al, eth	B5⁴, 1676
9312	Naphthalene, 3-chloro-1-nitro 3-Cl-1-(NO₂)C₁₀H₆	207.62	grsh br nd	....	105				B5⁴, 1676
9313	Naphthalene, 3-chloro-8-nitro 8-(NO₂)-3-ClC₁₀H₆	207.62	nd (aq ace)	....	100.5			ace	B5⁴, 1676
9314	Naphthalene, α-(chloromethyl) .... α-(ClCH₂)C₁₀H₇	176.65	pr	291-2, 135-6⁶	32			al, bz	B5⁴, 1692
9315	Naphthalene, β-(chloromethyl) .... β-(ClCH₂)C₁₀H₇	176.65	lf (al)	170²⁰	48-9			al, bz	B5⁴, 1697
9316	Naphthalene, decahydro (cis) or Decalin .... C₁₀H₁₈	138.25	....	155.6, 69.4¹⁰	−43	0.8965²²/⁴	1.4810²⁰	al, eth, ace, bz, chl	B5⁴, 310
9317	Naphthalene, 1,2-diamino or 1,2-Naphthalenediamine ... 1,2-(H₂N)₂C₁₀H₆	158.21	lf (w) (red in air)	214¹³	98.5			al, eth, chl	B13³, 377
9318	Naphthalene, 1,3-diamino-2-phenyl 2-C₆H₅-1,3-(CH₂N)₂C₁₀H₅	234.30	pl, (MeOH, bz) (red in air)	....	116			al, bz	B13², 131
9319	Naphthalene, 1,4-diamino or 1,4-Naphthylene diamine .... 1,4-(H₂N)₂C₁₀H₆	158.21	ye nd (w)	....	120		1.6441¹⁸	....	B13³, 383
9320	Naphthalene, 1,4-diamino-2-methyl .... 2-CH₃-1,4-(H₂N)₂C₁₀H₅	172.23	ye cr (peth)	....	113-4				B13³, 406
9321	Naphthalene, 1,4-diamino-2-methyl,dihydrochloride or Vitamin K₅ 2-CH₃-1,4-(H₂N)₂C₁₀H₅.2HCl	245.15	cr (dil HCl)	....	300d			w	B13³, 406
9322	Naphthalene, 1,5-diamino or 1,5-Naphthylenediamine ... 1,5-(H₂N)₂C₁₀H₆	158.21	pr (eth,al,w)	sub	190	1.4		al, eth, chl	B13³, 390
9323	Naphthalene, 1,5-diamino-2,methyl .... 2-CH₃-1,5-(H₂N)₂C₁₀H₅	172.23	red-ye lf (dil al)	....	136			al, eth, bz	B13³, 407
9324	Naphthalene, 1,6-diamino or 1,6-Naphthylene diamine ... 1,6-(H₂N)₂C₁₀H₆	158.21	nd (w,eth)	....	85-6	1.1477⁹⁹/⁴	1.7083⁹⁹	al, bz	B13², 85
9325	Naphthalene, 1,7-diamino or 1,7-Naphthylenediamine ... 1,7-(H₂N)₂C₁₀H₆	158.21	lf (bz)nd (w)	....	117.5			al, bz	B13³, 398

No.	Name, Synonyms, and Formula	Mol. wt.	Color, crystalline form, specific rotation and λ_{max} (log ε)	b.p. °C	m.p. °C	Density	n_D	Solubility	Ref.
9326	Naphthalene, 1,8-diamino or 1,8-Naphthylenediamine 1,8-$(CH_2N)_2C_{10}H_6$	158.21	nd(dil al)	205[12],sub	66.5	1.1265[99/4]	1.6828[99]	al, eth	B13[3], 398
9327	Naphthalene, 2,3-diamino or 2,3-Naphthylenediamine 2,3-$(H_2N)_2C_{10}H_6$	158.21	lf (eth or w)	. . . .	199	1.0968[26/4]	1.6342[26]	al, eth	B13[3], 402
9328	Naphthalene, 2,6-diamino or 2,6-naphthylenediamine 2,6-$(H_2N)_2C_{10}H_6$	158.21	nd or lf (w) lf (al)	. . . .	222d	. . . .	. . . .	. . . .	B13[3], 402
9329	Naphthalene, 1,2-dichloro 1,2-$Cl_2C_{10}H_6$	197.07	pl (al)	295-8, 151-3[19]	35-7	1.3147[49/4]	1.5338[49]	al, eth	B5[4], 1661
9330	Naphthalene, 1,3-dichloro 1,3-$Cl_2C_{10}H_6$	197.07	nd or pr (al)	291[775]	61.5	. . . .	. . . .	al	B5[4], 1661
9331	Naphthalene, 1,4-dichloro 1,4-$Cl_2C_{10}H_6$	197.07	nd or pr (al,aa,ace)	286-7[740], 147[12]	68	1.2997[76/4]	1.6228[76]	eth, ace, bz, aa	B5[4], 1661
9332	Naphthalene, 1,5-dichloro 1,5-$Cl_2C_{10}H_6$	197.07	nd or lf (al,aa) pr (sub)	sub	107	. . . .	. . . .	al	B5[4], 1662
9333	Naphthalene, 1,6-dichloro 1,6-$Cl_2C_{10}H_6$	197.07	nd or pr (al,peth,sub)	sub	49	. . . .	. . . .	. . . .	B5[4], 1662
9334	Naphthalene, 1,7-dichloro 1,7-$Cl_2C_{10}H_6$	197.07	nd or pr (al, aa)	285-6	63-4	1.2611[100/4]	1.6092[100]	al, eth, bz, aa	B5[4], 1662
9335	Naphthalene, 1,8-dichloro 1,8-$Cl_2C_{10}H_6$	197.07	rh pl (hx) nd (al,sub)	d sub	89	1.2924[100/4]	1.6236[100]	al, peth	B5[4], 1662
9336	Naphthalene, 2,3-dichloro 2,3-$Cl_2C_{10}H_6$	197.07	rh lf (al)	. . . .	120	. . . .	. . . .	eth	B5[4], 1662
9337	Naphthalene, 2,6-dichloro 2,6-$Cl_2C_{10}H_6$	197.07	pr (aa), nd or lf (al) pl (eth,bz)	285	140-1	. . . .	. . . .	eth, bz, chl, aa	B5[4], 1662
9338	Naphthalene, 2,7-dichloro 2,7-$Cl_2C_{10}H_6$	197.07	pl or lf (al)	. . . .	114	. . . .	. . . .	al	B5[4], 1662
9339	Naphthalene, 2,3-dihydrazino 2,3-$(H_2NNH)_2C_{10}H_6$	188.23	red-br (al,w) col nd (bz)	. . . .	167-8d	. . . .	. . . .	al	B15, 583
9340	Naphthalene, 1,2-dihydro $C_{10}H_{10}$	130.19	lf, pl	206-7, 78[9]	-8	0.9974[20/4]	1.5814[20]	. . . .	B5[4], 1543
9341	Naphthalene, 1,2-dihydro-3-methyl $C_{11}H_{12}$	144.22	. . . .	105[13]	. . . . . .	0.9837[20/20]	1.5751[20]	eth, bz	B5[4], 1552
9342	Naphthalene, 1,2-dihydro-4-methyl $C_{11}H_{12}$	144.22	. . . .	112[18]	. . . . . .	0.9895[20/4]	1.5758[20]	eth, bz	B5[4], 1551
9343	Naphthalene, 1,4-dihydro $C_{10}H_{10}$	130.19	pl	211-2, 94[17]	25(30)	0.9928[33/4]	1.5577[20]	aa	B5[4], 1544
9344	Naphthalene, 2,3-dihydro-1,4-dimethyl $C_{12}H_{16}$	160.26	. . . .	234	. . . . . .	0.940[20]	1.528[20/D]	. . . .	35.5214) 35.5214)
9345	Naphthalene, 1,2-dihydroxy or 1,2-Naphthalenediol 1,2-$(HO)_2C_{10}H_6$	160.18	lf or nd (CS_2) lf (w + 1) nd (liq)	. . . .	103-4 (anh), 58-60 (+ 1w)	. . . .	. . . .	eth	B6[3], 5240
9346	Naphthalene, 1,4-dihydroxy or 1,4-Naphthalenediol 1,4-$(HO)_2C_{10}H_6$	160.18	mcl nd (bz,w)	. . . .	192	. . . .	. . . .	al, eth, aa	B6[3], 5260
9347	Naphthalene, 1,4-diacetoxy-2-methyl 2-CH_3-1,4-$(CH_3CO_2)_2C_{10}H_5$	258.28	pr (al)	. . . .	113	. . . .	. . . .	al	B6[3], 5302
9348	Naphthalene, 1,5-dihydroxy or 1,5-Napthalene diol 1,5$(HO)_2C_{10}H_6$	160.18	pr (w) nd (sub)	sub	265d	. . . .	. . . .	eth, ace, aa	B6[3], 5265
9349	Naphthalene, 1,5-diacetoxy 1,5-$(CH_3CO_2)_2C_{10}H_6$	244.25	nd (bz)	. . . .	161	. . . .	. . . .	bz,aa	B6[3], 5267
9350	Naphthalene, 1,6-dihydroxy or 1,6-Naphthylene diol 1,6-$(HO)_2C_{10}H_6$	160.18	pr (bz)	sub	138	. . . .	. . . .	eth, ace, bz, MeOH	B6[3], 5279
9351	Naphthalene, 1,7-dihydroxy 1,7$(HO)_2C_{10}H_6$	160.18	nd (bz or sub)	sub	178-81	. . . .	. . . .	al, eth, bz, aa	B6[3], 5281
9352	Naphthalene, 1,8-dihydroxy or 1,8-Napthalenediol 1,8-$(HO)_2C_{10}H_6$	160.18	lf or nd (w)	. . . .	144	. . . .	. . . .	al, eth, bz	B6[3], 5283
9353	Naphthalene, 2,3-dihydroxy or 2,3 -Napthalenediol 2,3-$(HO)_2C_{10}H_6$	160.18	lf(w)	. . . .	163-4	. . . .	. . . .	al, eth, bz, aa, lig	B6[3], 5287
9354	Naphthalene, 2,6-dihydroxy or 2,6-Naphthalenediol 2,6-$(HO)_2C_{10}H_6$	160.18	rh pl (w)	sub	222	. . . .	. . . .	al, eth, ace, aa	B6[3], 5287
9355	Naphthalene, 2,7-dihydroxy or 2,7-naphthalenediol 2,7-$(HO)_2C_{10}H_6$	160.18	nd (w,dil al) pl (dil al)	sub	190-4	. . . .	. . . .	al, eth, bz, chl	B6[3], 5291

PHYSICAL CONSTANTS OF ORGANIC COMPOUNDS (Continued)

No.	Name, Synonyms, and Formula	Mol. wt.	Color, crystalline form, specific rotation and λ_{max} (log ε)	b.p. °C	m.p. °C	Density	n_D	Solubility	Ref.
9356	Naphthalene, 1,5-dimercapto or 1,5-naphthylenedithiol 1,5-(HS)₂C₁₀H₆	192.31	ye lf (al,eth,bz)	....	119	....		al, eth, bz	B6³, 5276
9357	Naphthalene, 1,2-dimethyl 1,2 (CH₃)₂C₁₀H₆	156.23	....	266-7, 132¹²	−1.6	1.0179²⁰/⁴	1.61656²⁰	eth, bz	B5⁴, 1708
9358	Naphthalene, 1,3-dimethyl 1,3-(CH₃)₂C₁₀H₆	156.23		263, 138-40¹²	−6	1.0144²⁰/⁴	1.6140²⁰	eth, bz	B5⁴, 1708
9359	Naphthalene, 1,4-dimethyl 1,4-(CH₃)₂C₁₀H₆	156.23		268, 129¹⁰	7.6	1.0166²⁰/⁴	1.6127²⁰	al,eth, ace, bz, lig	B5⁴, 1709
9360	Naphthalene, 1,5-dimethyl 1,5-(CH₃)₂C₁₀H₆	156.23		265	82	....		eth, bz	B5⁴, 1710
9361	Naphthalene, 1,6-dimethyl 1,6-(CH₃)₂C₁₀H₆	156.23		264, 126¹³	−16.9	1.0021²⁰	1.61656²⁰	eth, bz	B5⁴, 1711
9362	Naphthalene, 1,6-dimethyl-4-isopropyl or Cadalene 1,6-(CH₃)₂C₁₀H₅-i-C₃H₇-4	198.31		291-2⁷²⁰, 165²⁰		0.9792¹⁹/⁴	1.5851¹⁹	....	B5⁴, 1758
9363	Naphthalene, 1,7-dimethyl 1,7-(CH₃)₂C₁₀H₆	156.27		263, 148¹⁵	−13.9	1.0115²⁰/⁴	1.60831²⁰	eth, bz	B5⁴, 1711
9364	Naphthalene, 1,8-dimethyl 1,8-(CH₃)₂C₁₀H₆	156.23		270, 140¹⁸	65	....		eth, bz	B5⁴, 1712
9365	Naphthalene, 2,3-dimethyl or Guaiene 2,3-(CH₃)₂C₁₀H₆	156.23	lf (al)	268, 128.7¹⁰	105	1.003²⁰/⁴	1.5060²⁰	eth, bz	B5⁴, 1713
9366	Naphthalene, 1,2-dinitro 1,2-(NO₂)₂C₁₀H₆	218.17			162-3				B5³, 1605
9367	Naphthalene, 1,3-dinitro 1,3-(NO₂)₂C₁₀H₆	218.17	ye nd (bz,py-w)	sub	147-9			al, ace	B5⁴, 1680
9368	Naphthalene, 1,5-dinitro 1,5-(NO₂)₂C₁₀H₆	218.17	hex nd (aa,ace)	sub	219			eth, bz	B5⁴, 1680
9369	Naphthalene, 1,7-dinitro 1,7-(NO₂)₂C₁₀H₆	218.17			156			ace, bz	B5⁴, 1680
9370	Naphthalene, 1,8-dinitro 1,8-(NO₂)₂C₁₀H₆	218.17	ye rh pl (chl)	445d	173			ace, py	B5⁴, 1681
9371	Naphthalene, 2,3-dinitro 2,3-(NO₂)₂C₁₀H₆	218.17			172-4			al, bz	B5⁴, 1681
9372	Naphthalene, 2,4-dinitro-1-triazo 1-N₃-2,4-(NO₂)₂C₁₀H₅	259.18	ye rh nd (al)	....	105d			eth, bz, chl	B5², 460
9373	Naphthalene, 2,6-dinitro 2,6-(NO₂)₂C₁₀H₆	218.17			278			al, bz	B5³, 1609
9374	Naphthalene, 2,7-dinitro 2,7(NO₂)₂C₁₀H₆	218.17			234			al, eth, bz	B5⁴, 1681
9375	Naphthalene, 1-ethoxy or Ethyl α-naphthyl ether α-C₁₀H₇OC₂H₅	172.23	nd	280.5, 136-8¹⁴	5.5	1.060²⁰/⁴	1.5953²⁵	al, eth	B6³, 2924
9376	Naphthalene, 2-ethoxy or Ethyl β-naphthyl ether β-C₁₀H₇OC₂H₅	172.23	pl (al)	282, 148¹⁰	37-8	1.0640²⁰/²⁰	1.5975³⁶	al, eth, lig, to	B6³, 2972
9377	Naphthalene, α-ethyl α-C₂H₅C₁₀H₇	156.23		258.6, 120¹⁰	−13.9	1.0082²⁰/⁴	1.6062²⁰	al, eth	B5⁴, 1705
9378	Naphthalene, β-ethyl β-C₂H₅C₁₀H₇	156.23		258, 119¹⁰	−7.4	0,9922²⁰/⁴	1.5999²⁰	al, eth	B5⁴, 1707
9379	Naphthalene, α-fluoro α-FC₁₀H₇	146.17		215⁷⁵⁶, 80¹¹	−9	1.1322²⁰	1.5939²⁰	al, eth, bz, chl, aa	B5⁴, 1657
9380	Naphthalene, β-fluoro β-FC₁₀H₇	146.17	nd (al)	211.5⁷³⁷, 90¹⁶	61			al, eth, bz, chl, aa	B5⁴, 1658
9381	Naphthalene, 1,2,3,4,9,10-hexahydro C₁₀H₁₄	134.22		200, 82²		0.934²³	1.5260¹⁶	eth, bz	B5⁴, 1082
9382	Naphthalene, 1-(α-hydroxyethyl)- (dl) α-C₁₀H₇CH(OH)CH₃	172.23	nd (peth)	178¹⁵	66	1.1190¹⁴/⁴	1.6188²⁵	al, ace, bz, chl	B6³, 3034
9383	Naphthalene, 1-(α-hydroxy ethyl (l) α-C₁₀H₇-CH(OH)CH₃	172.23	[α]²⁰/D -78.9 (al, c=5)	166¹¹	47	1.1190¹⁴/⁴	1.6180²⁵	al, ace, bz, chl	B6³, 3034
9384	Naphthalene, α-hydroxymethyl α-(HOCH₂)C₁₀H₇	158.20	nd (w,al) cr (bz-liq)	301⁷¹⁵,163¹²	64	1.1039⁸⁰/⁴		al, eth	B6³, 3024
9385	Naphthalene, α-iodo α-IC₁₀H₇	254.07	....	302	4.2	1.7399²⁰/⁴	1.7026²⁰	al, eth, bz	B5⁴, 1670
9386	Naphthalene, β-iodo β-IC₁₀H₇	254.07	lf (dil al)	308, 172²¹	54.5	1.6319⁹⁹/⁴	1.6662⁹⁹	al, eth, aa	B5⁴, 1671
9387	Naphthalene, α-mercapto or Naphthalene thiol α-C₁₀H₇SH	160.24	....	285d, 161²⁰	......	1.1607²⁰/⁴	1.6802²⁰	al, eth	B6³, 2943
9388	Naphthalene, β-mercapto or β-Naphthalene thiol β-C₁₀H₇SH	160.24	pl(al)	288, 162.7²⁰	81	1.550	....	al, eth, lig	B6³, 3007

No.	Name, Synonyms, and Formula	Mol. wt.	Color, crystalline form, specific rotation and λ_{max} (log ε)	b.p. °C	m.p. °C	Density	n_D	Solubility	Ref.
9389	Naphthalene, α-methoxy or Methyl α-naphthyl ether ..... α-C₁₀H₇OCH₃	158.20	....	269, 135[10]	<−10	1.0964[14/2]	1.6940[25]	al, eth, bz, chl	B6[3], 2922
9390	Naphthalene, β-methoxy or Methyl β-naphthyl ether ..... β-C₁₀H₇OCH₃	158.20	lf (eth) pl (peth)	274, 138[10]	73-4			eth, bz, chl	B6[3], 2969
9391	Naphthalene, α-methyl .... α-CH₃C₁₀H₇	142.20	....	244.6, 107.4[10]	−22	1.0202[20/4]	1.6170[20]	al, eth, bz	B5[4], 1687
9392	Naphthalene, 1-methyl-2-nitro 1-CH₃-2-NO₂C₁₀H₆	187.20	lt ye nd (al)	....	58-9	....	....	al	B5[4], 1693
9393	Naphthalene, 1-methyl-3-nitro 1-CH₃-3NO₂C₁₀H₆	187.20	ye nd (al)	....	81-2	....	....	al	B5[3], 1624
9394	Naphthalene, 1-methyl-4-nitro 1-CH₃-4-NO₂C₁₀H₆	187.20	pa ye nd (al)	182-3[18]	71-2	....	....	al, eth, ace	B5[3], 1624
9395	Naphthalene, 1-methyl-5-nitro 1-CH₃-5-NO₂C₁₀H₆	187.20	brsh nd (al)	....	82-3	....	....	al, ace	B5[3], 1624
9396	Naphthalene, 1-methyl-6-nitro 1-CH₃-6NO₂C₁₀H₆	187.20	ye nd (dil al)	....	76-7	....	....	al	B5[3], 1625
9397	Naphthalene, 1-methyl-7-nitro 1-CH₃-7-NO₂C₁₀H₆	187.20	ye nd (al)	....	98-9	....	....	al	B5[3], 1625
9398	Naphthalene, 1-methyl-8-nitro 1-CH₃-8-NO₂C₁₀H₆	187.20	br lf (al)	....	65	....	....	al	B5[3], 1625
9399	Naphthalene, 1-methyl-7-isopropyl 1-CH₃-7-i-C₃H₇C₁₀H₆	184.28	....	152[18]	....	0.9740[20/4]	1.5833[20]	eth, ace, bz	B5[4], 1741
9400	Naphthalene, β-methyl .... β-CH₃C₁₀H₇	142.20	mcl (al)	241, 104.7[10]	34.6	1.0058[20/4]	1.6015[40]	al, eth, bz	B5[4], 1693
9401	Naphthalene, 2-methyl-1-nitro 2-CH₃-1-NO₂C₁₀H₆	187.20	yesh pr or nd (al)	188[20]	81-2	....	....	al, ace	B5[4], 1698
9402	Naphthalene, 2-methyl-3-nitro 2-CH₃-3-NO₂C₁₀H₆	187.20	yesh pl (al)	....	117-8	....	....	al	B5[4], 1698
9403	Naphthalene, 2-methyl-4-nitro 2-CH₃-4-NO₂C₁₀H₆	187.20	pa ye nd (al)	....	49-50	....	....	al	B5[3], 1634
9404	Naphthalene, 2-methyl-5-nitro 2-CH₃-5-NO₂C₁₀H₆	187.20	yd nd (al)	....	61-2	....	....	al	B5[3], 1634
9405	Naphthalene, 2-methyl-6-nitro 2-CH₃-6-NO₂C₁₀H₆	187.20	ye nd (al)	....	119	....	....	al	B5[3], 1634
9406	Naphthalene, 2-methyl-7-nitro 2-CH₃-7-NO₂C₁₀H₆	187.20	yesh pl (al)	....	105	....	....	al	B5[3], 1634
9407	Naphthalene, 2-methyl-8-nitro 2-CH₃-8-NO₂C₁₀H₆	187.20	ye nd (al)	....	36-8	....	....	al	B5[3], 1635
9408	Naphthalene, α-nitramino .... α-(O₂NNH)C₁₀H₇	188.20	lt ye nd (w)	....	123-4	....	....	bz	B16[2], 346
9409	Naphthalene, β-nitramino .... β-(O₂NNH)C₁₀H₇	188.20	lf or nd	....	131-6	....	....		B16, 675
9410	Naphthalene, α-nitro .... α-NO₂C₁₀H₇	173.17	ye nd (al)	304 sub, 30-40[0.01]	61.5	1.332[20/4]	....	al, eth, bz, py	B5[4], 1673
9411	Naphthalene, 1-nitro-5-triazo 1-NO₂-5-N₃C₁₀H₆	214.19	gold-ye nd (al)	....	121	....	....	al, ace	B5[2], 459
9412	Naphthalene, β-nitro .... β-NO₂C₁₀H₇	173.17	ye rh nd or pl (al)	312.5[734], 165[15]	79	....	....	al, eth	B5[4], 1675
9413	Naphthalene, 2-nitro-1-triazo 2-NO₂-1-N₃C₁₀H₆	214.19	ye nd (dil ace)	....	103-4d	....	....	al, ace, bz, ace	B5, 565
9414	Naphthalene α-(nitrosohydroxyl amino) .... α-C₁₀H₇(N(NO)OH)	188.20	nd (peth)	....	54-5	....	....	chl	B16[3], 639
9415	Naphthalene, β-(nitrosohydroxylamino) .... β-C₁₀H₇(N(NO)OH)	188.20	nd (AcOEt-peth)	....	88-92	....	....	eth, aa	B16[1], 396
9416	Naphthalene, octachloro or Perchloro naphthalene ...... C₁₀Cl₈	403.74	nd (bz-CCl₄)	440-2[7.4]	197-8	....	....	bz, chl, lig	B5[4], 1665
9417	Naphthalene, α-isopentoxy or Isopentyl-α-naphthyl ether α-C₁₀H₇O-i-C₅H₁₁	214.31	....	317-9[742], 148-53[3]	....	1.0069[14/4]	1.5705[16]	....	B6[3], 2925
9418	Naphthalene, β-Isopentoxy or Isopentyl-β-naphthyl ether i-C₅H₁₁O-β-C₁₀H₇	214.31	lf	323-6d, 155-60[6]	26.5	1.0155[12/4]	1.5768[12]	al, eth	B6[3], 2974
9419	Naphthalene, α-phenyl .... α-C₆H₅C₁₀H₇	204.28	cr	334, 190[12]	ca 45	1.096[20/4]	1.6664[20]	al, eth, bz, aa	B5[3], 2230
9420	Naphthalene, β-phenyl .... β-C₆H₅C₁₀H₇	204.28	lf (al)	345-6, 185-90[5]	103-4	....	....	al, eth, bz, aa, chl	B5[3], 2231

No.	Name, Synonyms, and Formula	Mol. wt.	Color, crystalline form, specific rotation and λ_{max} (log ε)	b.p. °C	m.p. °C	Density	n_D	Solubility	Ref.
9421	Naphthalene, picrate $C_{10}H_8 \cdot C_6H_3N_3O_7$	357.28	ye pr or pl (aa) cr (eth)	....	152	1.53	....	eth, bz	B5[3], 1568
9422	Naphthalene, 1-propoxy or Propyl α-naphthyl ether α-$C_{10}H_7OC_3H_7$	186.26	....	293.5, 167[18]		1.0447[18/4]	1.5928[18]		B6[3], 2924
9423	Naphthalene, 2-propoxy or Propyl β-naphthyl ether β-$C_{10}H_7OC_3H_7$	186.26	nd (al)	305, 144[10]	41			al	B6[3], 2973
9424	Naphthalene, α-propyl α-$C_3H_7C_{10}H_7$	170.25	....	274-5	−8.6				B5[4], 1721
9425	Naphthalene, β-isopropyl β-i-$C_3H_7C_{10}H_7$	170.25	....	268[2], 129-30[14]	......	0.9753[20/4]	1.58482[20]	al, eth, bz	B5[4], 1722
9426	Naphthalene, β-isopropyl β-i-$C_3H_7C_{10}H_7$	192.17	....		225			w, al, ace, bz	B6[3], 6699
9427	Naphthalene, 1,2,3,4-tetrahydroxy 1,2,3,4-$(HO)_4C_{10}H_4$	184.28	....		106-7			al, ace, bz	B5[4], 1745
9428	Naphthalene, 1,2,5,6-tetramethyl 1,2,5,6-$(CH_3)_4C_{10}H_4$	184.28	....	150-5[12]	118			al, bz	B5[4], 1746
9429	Naphthalene, 1,3,6,8-tetramethyl 1,3,6,8-$(CH_3)_4C_{10}H_4$	184.28	....	115-6[2]	84-5			al	B5[4], 1746
9430	Naphthalene, 1,3,5,8-tetranitro or γ-Tetranitro naphthalene 1,3,5,8-$(NO_2)_4C_{10}H_4$	308.17	lt ye tetr (ace)	....	194-5			ace	B5[4], 1683
9431	Naphthalene, 1,3,6,8-tetranitro or β-tetranitro naphthalene 1,3,6,8-$(NO_2)_4C_{10}H_4$	308.17	ye nd (al, bz)	exp	207			bz, aa	B5[4], 1683
9432	Naphthalene, α-triazo or α-Naphthyl azide α-$C_{10}H_7N_3$	169.19	pa ye pr	d	12	1.1713[25]	1.6550[25]	al, eth, ace	B5[4], 1684
9433	Naphthalene, β-triazo or β-Naphthyl azide β-$C_{10}H_7N_3$	169.19	pl (al nd (peth) ye in air	....	33	....	....	al, eth, ace, bz	B5[4], 1684
9434	Naphthalene, 1,2,3-trimethyl 1,2,3-$(CH_3)_3C_{10}H_5$	170.25	....	125-30[12]	27-8				B5[4], 1725
9435	Naphthalene, 1,2,4-trimethyl 1,2,4-$(CH_3)_3C_{10}H_5$	170.25	....	146[12]	55-6			al, eth, bz, chl	B5[4], 1725
9436	Naphthalene, 1,2,5-trimethyl 1,2,5-$(CH_3)_3C_{10}H_5$	170.26	nd (al)	140[12]	33.5	1.0103[22/4]	16093[22]	eth, bz	B5[4], 1726
9437	Naphthalene, 1,2,6-trimethyl 1,2,6-$(CH_3)_3C_{10}H_5$	170.26	lf	146[10]	14		1.6010[20]	eth, bz	B5[4], 1726
9438	Naphthalene, 1,2,7-trimethyl 1,2,7-$(CH_3)_3C_{10}H_5$	170.26	....	147-8[16]	......	1.0087[20/4]	1.6097[20]	eth, bz	B5[4], 1727
9439	Naphthalene, 2,3,6-trimethyl 2,3,6-$(CH_3)_3C_{10}H_5$	170.26	....	263-4, 146-8[14]	100-2			eth, bz	B5[4], 1730
9440	Naphthalene, 1,2,5-trinitro 1,2,5-$(NO_2)_3C_{10}H_5$	263.17	lt ye nd (al)	....	112-3			al	B5[3], 1614
9441	Naphthalene, 1,3,5-trinitro 1,3,5-$(NO_2)_3C_{10}H_5$	263.17	ye rh (chl)	364exp	122			al, ace, aa, chl	B5[4], 1682
9442	Naphthalene, 1,3,8-trinitro 1,3,8-$(NO_2)_3C_{10}H_5$	263.17	yesh mcl pr (al,ace,aa)	....	218			ace, py	B5[4], 1682
9443	Naphthalene, 1,4,5-trinitro 1,4,5-$(NO_2)_3C_{10}H_5$	263.17	ye lf or rh pl (al,aa,bz)	....	154			ace	B5[2], 458
9444	Naphthalene, β-vinyl β-$(CH_2{=}CH)C_{10}H_7$	154.21	....	136-8[17]	66			al, ace, bz	B5[4], 1833
9445	1,2-Naphthalene dicarboxylic acid 1,2-$C_{10}H_6(CO_2H)_2$	216.21	nd (al) cr (w)	....	175d			al, eth, aa	B9[3], 4462
9446	1,2-Naphthalene dicarboxylic acid, 3,4-dihydro, anhydride 1,2-$C_{10}H_8(CO)_2O$	200.21	pa ye nd (lig,al)	227-30[23]	126-7			bz, aa, MeOH	B9[4], 650
9447	1,4-Naphthalene dicarboxylic acid 1,4-$C_{10}H_6(CO_2H)_2$	216.21	rods (aa)	....	309 (320)			al, aa	B9[3], 4463
9448	1,5-Naphthalene dicarboxylic acid 1,5-$C_{10}H_6(CO_2H)_2$	216.21	nd	....	320-2d				B9[3], 4465
9449	1,6-Naphthalene dicarboxylic acid 1,6-$C_{10}H_6(CO_2H)_2$	216.21	nd (aa)	....	310 sinters			al,aa	B9[2], 651
9450	1,7-Naphthalene dicarboxylic acid 1,7-$C_{10}H_6(CO_2H)_2$	216.21	ye pw (dil al,aa)	....	308d			al, eth, ace, aa	B9[3], 4466
9451	2,3-Naphthalene dicarboxylic acid 2,3-$C_{10}H_6(CO_2H)_2$	216.21	pr (aa,w) pr (sub)	....	246			....	B9[3], 4470

No.	Name, Synonyms, and Formula	Mol. wt.	Color, crystalline form, specific rotation and λ_{max} (log ε)	b.p. °C	m.p. °C	Density	n_D	Solubility	Ref.
9452	2,6-Naphthalene dicarboxylic acid .... 2,6-$C_{10}H_6(CO_2H)_2$	216.21	nd (al or sub)	....	>300d	....	....	al	B9³, 4471
9453	2,7-Naphthalene dicarboxylic acid 2,7-$C_{10}H_6(CO_2H)_2$	216.21	nd (w,al,dil HCl)	....	>300d			al	B9², 653
9454	1,3-Naphthalene disulfonic acid, 7-amino or Amino-G-acid 7-NH_2-1,3-$C_{10}H_5(SO_3H)_2$	303.32	mcl pr or nd (w + 2)	....	273-5			w, al	B14³, 2263
9455	1,3-Naphthalene disulfonic acid, 7-hydroxy or G-acid ... 7-HO-1,3-$C_{10}H_5(SO_3H)_2$	304.31						w	B11³, 560
9456	1,5-Naphthalene disulfonic acid 1,5-$C_{10}H_6(SO_3H)_2$	288.31	pl (+ 4w,dil aa)		240-5d	1.493		w, al	B11³, 17
9457	1,6-Naphthalene disulfonic acid 1,6-$C_{10}H_6(SO_3H)_2$	288.31	og pr (+ 4w,aa or w)		125d (anh)			w, al	B11³, 467
9458	2,7-Naphthalene disulfonic acid 2,7-$C_{10}H_6(SO_3H)_2$	288.31	hyg nd (con HCl)		199			w	B11³, 468
9459	2,7-Naphthalene disulfonic acid, 4-amino-5-hydroxy or H acid .... 4-NH_2-5-HO-2,7-$C_{10}H_4(SO_3H)_2$	319.31							B14³, 2292
9460	2,7-Naphthalene disulfonic acid, 4,5-dihydroxy or Chromotropic acid 4,5-$(HO)_2$-2,7-$C_{10}H_4(SO_3H)_2$	356.33	nd or lf (w + 2)					w	B11³, 576
9461	2,7-Naphthalene disulfonic acid, 3-hydroxy or R-acid .... 3-HO-2,7-$C_{10}H_5(SO_3H)_2$	304.31	dlq nd	....	d			w, al	B11³, 559
9462	α-Naphthalene phosphonic acid α-$C_{10}H_7PO(OH)_2$	208.16	cr (w)		189			al	B16³, 896
9463	α-Naphthalene phosphonyl chloride .... α-$C_{10}H_7PO(Cl)_2$	245.05			ca 60				B16², 392
9464	α-Naphthalene sulfinic acid .... α-$C_{10}H_7SO_2H$	192.25	nd (w)		104			w, al	B11³, 12
9465	β-Naphthalene sulfinic acid .... β-$C_{10}H_7SO_2H$	192.25	nd (w)		98(105)			w, al, eth	B11³, 14
9466	α-Naphthalene sulfonic acid .... α-$C_{10}H_7SO_3H$	208.25	pr (+ 2w,dil HCl)		90 (hyd) 139-40 (anh)			w, al	B11³, 382
9467	α-Naphthalene sulfonic acid, 4-amino or Naphthionic acid 4-H_2N-α-$C_{10}H_6SO_3H$	232.26	wh nd (w + ½) red-br cr		d	1.6703²⁵ᐟ⁴		py, MeOH	B14³, 2241
9468	α-Naphthalene sulfonic acid, 4-amino-5-hydroxy or S-acid 4-H_2N-5-HO-α-$C_{10}H_5SO_3H$	239.25	nd	....	....				B14³, 2290
9469	α-Naphthalene sulfonic acid, 7-amino or Bayer's acid, Cassela's acid............ 7-H_2N-α-$C_{10}H_6SO_3H$	241.27	nd (w + 1) pl (aq ace)					aa	B14³, 2245
9470	α-Naphthalene sulfonic acid, 4-hydroxy or Nenile-winther acid .... 4-HO-α-$C_{10}H_6SO_3H$	224.25	ta or pl (w)		170d (rapid htg)				B11³, 540
9471	α-Naphthalene sulfonic acid, 5-hydroxy or α-Naphtholsulfonic acid L .... 5-HO-α-$C_{10}H_6SO_3H$	224.25	dlq		120			w, aa	B11³, 541
9472	α-Naphthalene sulfonic acid, 7-hydroxy or Croceic acid . 7-HO-α-$C_{10}H_6SO_3H$	224.25						w	B11³, 556
9473	α-Naphthalene sulfonic acid, 8-hydroxy or α-Naphthol sulfonic acid S 8-HO-α-$C_{10}H_6SO_3H$	224.25	cr (w + 1)		106-7			w	B11², 157
9474	α-Naphthalene sulfonic acid, 8-hydroxy,lactone or 1-Naphthol-8-sulfonic acid sultone .... $C_{10}H_6O_3S$	206.22	pr (bz)	>360	154			bz, chl	B19⁴, 323
9475	α-Naphthalene sulfonyl chloride .... α-$C_{10}H_7SO_2Cl$	226.68	lf (eth)	195¹¹, 147.5⁰ᐧ⁹	68			al, eth, bz	B11³, 383
9476	β-Naphthalene sulfonic acid .... β-$C_{10}H_7SO_3H$	208.25	dlq pl (+ 1w) cr (+ 3w,dil HCl)	d	124-5 (+ 1w) 83(+ 3 w)	1.441²⁵ᐟ⁴		w, al, eth	B11³, 397
9477	β-Naphthalene sulfonic acid, 4-amino or Cleve's acid..... 4-H_2N-β-$C_{10}H_6SO_3H$	241.27	md (w + 1)	....	....				B14³, 2248
9478	β-Naphthalene sulfonic acid, 5,7-dinitro-8-hydroxy or Flavianic acid............ 5,7-$(NO_2)_2$-8-HO-β-$C_{10}H_4SO_3H$	314.23	pa ye nd (con HCl, + 3w) cr (w)		100 (+ 3w) 151 (anh)			w, al	B11³, 542

No.	Name, Synonyms, and Formula	Mol. wt.	Color, crystalline form, specific rotation and λ_{max} (log ε)	b.p. °C	m.p. °C	Density	n_D	Solubility	Ref.
9479	β-Naphthalene sulfonic acid, 1-hydroxy or α-Naphthol sulfonic acid . . . . 1-HO-β-$C_{10}H_6SO_3H$	224.25	pl (w)	. . . .	>250	. . . .	. . . .	w, al	B11³, 540
9480	β-Naphthalene sulfonic acid, 6-hydroxy or Schaeffer acid 6-HO-β-$C_{10}H_6SO_3H$	224.25	lf cr (w + 1)	. . . .	167 (anh) 129 (+1w) 118 (+2w)	. . . .	. . . .	w, al, aa	B11³, 553
9481	β-Naphthalene sulfonic acid, 7-hydroxy or F acid . . . . 7-HO-β-$C_{10}H_6SO_3H$	224.25	nd (HCl) cr (w + 1,2 or 4)	. . . .	115-6 (anh) 108-9 (+1w)	. . . .	. . . .	w, al	B11³, 555
9482	β-Naphthalene sulfonyl chloride . . . . β-$C_{10}H_7SO_2Cl$	226.68	pw or lf (bz-peth)	201¹³, 148⁰·⁵	79	. . . .	. . . .	al, eth, bz, chl	B11³, 399
9483	1,4,5,8-Naphthalene tetracarboxylic acid . . . . 1,4,5,8-$C_{10}H_4(CO_2H)_4$	304.23	lf or nd (w, dil HCl)	. . . .	320	. . . .	. . . .	ace, aa	B9³, 4889
9484	1,4,5,8-Naphthalene tetracarboxylic acid, 1,8: 4,5-dianhydride . . . . $C_{14}H_4O_6$	268.19	nd (al)	sub 320³	>300	. . . .	. . . .		B19⁴, 2258
9485	1,2,5-Naphthalene tricarboxylic acid . . . . 1,2,5-$C_{10}H_5(CO_2)_3$	260.22	nd (MeOH or sub)	. . . .	270-2	. . . .	. . . .	w, MeOH	B9⁴, 4835
9486	1,4,5-Naphthalene tricarboxylic acid . . . . 1,4,5-$C_{10}H_5(CO_3)_3$	260.22	cr (eth or con HCl)	. . . .	266-8	. . . .	. . . .	al, eth	B9³, 4835
9487	Naphthalic acid or 1,8-Naphthalene dicarboxylic acid . . . . 1,8-$C_{10}H_6(CO_2H)_2$	216.21	nd (al)	. . . .	260	. . . .	. . . .		B9³, 4466
9488	Naphthalic anhydride . . . . 1,8-$C_{10}H_6(CO)_2O$	198.18	nd (al) pr (aa) lf (sub)	sub	274	. . . .	. . . .	aa	B17⁴, 6392
9489	Naphthalic acid, diethyl ester or Diethyl naphthalate . . . 1,8-$C_{10}H_6(CO_2C_2H_5)_2$	272.30	yesh mcl or lf (dil al)	238-9¹⁹	59-60	1.1399⁷⁰·⁴	1.5586⁷⁰	al, eth	B9³, 4467
9490	Naphthalic acid, dimethyl ester or Dimethyl naphthalate . . . 1,8-$C_{10}H_6(CO_2CH_3)_2$	244.25	nd (al) pr MeOH)	. . . .	104	. . . .	. . . .	al, aa, MeOH	B9³, 4467
9491	Naphthalic acid, 3,6-dinitro or 3,6-Dinitronaphthalic acid 3,6-$(NO_2)_2$-1,8-$C_{10}H_4(CO_2H)_2$	306.19	silvery lf (w)	. . . .	212	. . . .	. . . .	al, aa	B9³, 4470
9492	Naphthalic acid, imide . . . . 1,8-$C_{10}H_6(CO)_2NH$	197.20	nd (chl-al)	. . . .	300	. . . .	. . . .		B21⁴, 5557
9493	Naphthalic anhydride, 3-nitro . . . . 3-NO_2-1,8-$C_{10}H_5(CO)_2O$	243.18	yesh nd (aa)	. . . .	252-3	. . . .	. . . .	aa	B17⁴, 6396
9494	Naphthalic acid, 4-nitro . . . . 4-NO_2-1,8-$C_{10}H_5(CO_2H)_2$	261.19	ye nd	. . . .	d140-50	. . . .	. . . .	aa	B9², 653
9495	Naphthaloyl chloride . . . . 1,8-$C_{10}H_6(COCl)_2$	253.09	pr (CS_2)	195-200⁰·²	84-6	. . . .	. . . .	bz, chl	B9³, 4468
9496	1-Naphthamidine . . . . $C_{11}H_{10}N_2$	170.22	lf (dil al)	. . . .	154	. . . .	. . . .	al, ace, chl	B9³, 3147
9497	2-Naphthamidine . . . . $C_{11}H_{10}N_2$	170.22	cr (bz)	. . . .	133-6	. . . .	. . . .	al, bz, aa	B9³, 3187
9498	α-Naphthamide . . . . α-$C_{10}H_7CONH_2$	171.20	nd or pl (al,gl aa)	sub	204-5	. . . .	. . . .	aa	B9³, 3145
9499	Naphthocaine, hydrochloride . . . . $C_{17}H_{23}N_2O_2Cl$	322.84	pa ye	. . . .	212-6	. . . .	. . . .		B14³, 1333
9500	α-Naphthoic acid or α-Naphthalene carboxylic acid . . . . . . α-$C_{10}H_7CO_2H$	172.20	nd (aa-w,w,al)	>300, 231⁵⁰	161	1.398	. . . .	al, eth, chl	B9³, 3136
9501	α-Naphthoic acid-2-amino . . . . 2-H_2N-α -$C_{10}H_6CO_2H$	187.20	nd (dil al)	. . . .	126d	. . . .	. . . .	al, eth, aa	B14³, 1330
9502	α-Naphthoic acid-3-amino . . . . 3-H_2N-α-$C_{10}H_6CO_2H$	187.20	ye or pksh nd (eth)	. . . .	181-2	. . . .	. . . .	al, MeOH	B14³, 1330
9503	α-Naphthoic acid-4,amino . . . . 4-H_2N-α -$C_{10}H_6CO_2H$	187.20	brsh nd (w,al)	. . . .	177	. . . .	. . . .	al, eth, ace, aa	B14³, 1332
9504	α-Naphthoic acid-5-amino . . . . 5-H_2N-α -$C_{10}H_6CO_2H$	187.20	og nd (w, dil al) (sub) nd	211-2, 196 (sub)	. . . .	. . . .	. . . .	al, aa	B14, 553
9505	α-Naphthoic acid-6-amino . . . . 6-H_2N-α-$C_{10}H_6CO_2H$	187.20	pa ye nd (al, w)	. . . .	205-6	. . . .	. . . .	eth, ace, aa	B14³, 1340
9506	α-Naphthoic aicd-7-amino . . . . 7-H_2N-α $C_{10}H_6CO_2H$	187.20	pa br pr (al)	. . . .	223-4	. . . .	. . . .	al, eth, ace, aa	B14³, 1340

No.	Name, Synonyms, and Formula	Mol. wt.	Color, crystalline form, specific rotation and λ_{max} (log ε)	b.p. °C	m.p. °C	Density	n_D	Solubility	Ref.
9507	α-Naphthoic anhydride $(\alpha\text{-}C_{10}H_7CO)_2O$	326.35	pr (bz)	....	145-6	....	....	eth, bz	B9[2], 450
9508	α-Naphthoic acid-4-bromo 4-Br-α-$C_{10}H_6CO_2H$	251.09	nd (aa, dil al, xyl)		220			al, eth, ace, bz	B9[3], 3153
9509	α-Naphthoic acid-5-bromo 5-Br-α-$C_{10}H_6CO_2H$	251.09	nd (aa, al)	sub	261			bz	B9[3], 3153
9510	α-Naphthoic acid-8-Bromo 8-Br-α -$C_{10}H_6CO_2H$	251.09	pr (w, bz)		178			al, eth, bz, aa	B9[3], 3154
9511	α-Naphthoic acid-2-chloro 2-Cl-α-$C_{10}H_6CO_2H$	206.63	cr (w, bz)		153			al, eth	B9[3], 3148
9512	α-Naphthoic acid-4-chloro 4-Cl-α-$C_{10}H_6CO_2H$	206.63	nd (al)		210 (223)			al, aa	B9[3], 3148
9513	α-Naphthoic acid-5-chloro 5-Cl-α-$C_{10}H_6CO_2H$	206.63	nd (dil al)	sub	245			al	B9[3], 3149
9514	α-Naphthoic acid-8-chloro 8-Cl-α -$C_{10}H_6CO_2H$	206.63	pl (al, w, bz)	sub	171-2			ace, aa	B9[3], 3151
9515	α-Naphthoic acid-1,2-dihydro α-$C_{10}H_9CO_2H$	174.20	cr (50% al)		138			w	B9[3], 3076
9516	α-Naphthoic acid-1,4-dihydro α-$C_{10}H_9CO_2H$	174.20	nd or pl (lig)		91			al, eth, ace	B9[3], 3075
9517	α-Naphthoic acid-3,4-dihydro α-$C_{10}H_9CO_2H$	174.20	nd (w, AcOEt) cr (peth)	305-6[748]	125			al, chl	B9[3], 3076
9518	α-Naphthoic acid-4,5-dinitro 4,5-$(NO_2)_2$-α -$C_{10}H_5CO_2H$	262.18	yesh nd or pl (al)	....	267 sub			al, eth	B9[3], 3168
9519	α-Naphthoic acid-ethyl ester or Ethyl α-naphthoate α-$C_{10}H_7CO_2C_2H_5$	200.24		310, 183-6[70]		1.1274[15/15]	1.5966[15]	al	B9[3], 3139
9520	α-Naphthoic acid-5-hydroxy 5-HO -α-$C_{10}H_6CO_2H$	188.20	pr or nd (w) cr (bz)		236			al, eth, chl, aa	B10[3], 1070
9521	α-Naphthoic acid-6-hydroxy 6-HO-α -$C_{10}H_6CO_2H$	188.20	nd or pr (w)		212-3			al, eth, ace	B10[3], 1070
9522	α-Naphthoic acid-7-hydroxy 7-HO -α -$C_{10}H_6CO_2H$	188.20	nd (w, al aa)		256			al, aa	B10[3], 1072
9523	α-Naphthoic acid-8-iodo 8-I-α -$C_{10}H_6CO_2H$	298.08	br pr (w)		164-5			al, eth, bz, aa	B9[3], 3157
9524	α-Naphthoic acid-methyl ester or Methyl α-naphthoate α-$C_{10}H_7CO_2CH_3$	186.21		167-9[20], 100-2[0.04]	59.5	1.1290[20/4]	1.6086[20]	al, bz	B9[3], 3138
9525	α-Naphthoic acid-3-nitro 3-NO_2-α -$C_{10}H_6CO_2H$	217.18	cr (al)	....	271-5			al	B9[3], 3158
9526	α-Naphthoic acid, 4-nitro 4-NO_2-α -$C_{10}H_6CO_2H$	217.18	yesh nd (al,aa)		225-6			al, chl, aa	B9[3], 3160
9527	α-Naphthoic acid, 5-nitro 5-NO_2-α -$C_{10}H_6CO_2H$	217.18	yesh nd (al)	sub	241-2			aa	B9[3], 3162
9528	α-Naphthoic acid, 8-nitro 8-NO_2-α -$C_{10}H_6CO_2H$	217.18	nd (w) pr (al)		217			al	B9[3], 3164
9529	α-Naphthoic acid, 1,2,3,4-tetrahydro α-$C_{10}H_{11}CO_2H$	176.22	tcl pr (AcOEt)		85			al, eth, ace, bz	B9[3], 2801
9530	α-Naphthoic acid, 5,6,7,8-tetrahydro α-$C_{10}H_{11}CO_2H$	176.22	pr (w) wh nd (dil aa)		150			al, bz, chl	B9[3], 2794
9531	α-Naphthonitrile α-$C_{10}H_7CN$	153.19	nd (lig)	299, 148[12]	37.5	1.1113[25/25]	1.6298[18]	al, eth, lig	B9[3], 3146
9532	α-Naphthoyl chloride α-$C_{10}H_7COCl$	190.63	....	297.5, 172[15]	20				B9[3], 3145
9533	β-Naphthamide β-$C_{10}H_7CONH_2$	171.20	lf (al)		195			al, eth, ace, chl, lig	B9[2], 454
9534	β-Naphthoic acid or 2-Naphthalene carboxylic acid β-$C_{10}H_7CO_2H$	172.20	nd (lig, chl, sub) pl (ace)	>300	185.5	1.077[100/4]		al, eth, chl	B9[3], 3174
9535	β-Naphthoic acid, 4-acetyl-3-hydroxy 4-CH_3CO-3-HO-β-$C_{10}H_5CO_2H$	230.22	ye pr (aa)		194			al	B10[3], 4408
9536	β-Naphthoic acid, 1-amino 1-H_2N-β -$C_{10}H_6CO_2H$	187.20	nd (dil al,aa)	....	205 (rapid htg)			al, eth, bz	B14[3], 1340
9537	β-Naphthoic acid, 3-amino 3-H_2N-β -$C_{10}H_6CO_2H$	187.20	ye lf (dil al)	....	216-7			al, eth	B14[3], 1341
9538	β-Naphthoic acid, 4-amino 4-H_2N-β -$C_{10}H_6CO_2H$	187.20	nd (dil al)		215-6			al, eth, ace, bz	B14[3], 1345

No.	Name, Synonyms, and Formula	Mol. wt.	Color, crystalline form, specific rotation and λ_{max} (log ε)	b.p. °C	m.p. °C	Density	n_D	Solubility	Ref.
9539	β-Naphthoic acid, 5-amino 5-H$_2$N-β-C$_{10}$H$_6$CO$_2$H	187.20	ye lf (al)	....	291-2			al, eth, ace, bz	B14³, 1345
9540	β-Naphthoic acid, 6-amino 6-H$_2$N-β-C$_{10}$H$_6$CO$_2$H	187.20	pa ye nd (w, dil al)		225			al, eth, ace, bz, aa	B14², 324
9541	β-Naphthoic acid, 7-amino 7-H$_2$N-β-C$_{10}$H$_6$CO$_2$H	187.20	pa ye nd or lf (al)		243			al, eth, ace, aa	B14², 324
9542	β-Naphthoic acid, 8-amino 8-H$_2$N-β-C$_{10}$H$_6$CO$_2$H	187.20	grsh-ye nd (aa)		220			al, eth, ace	B14², 324
9543	β-Naphthoic anhydride (β-C$_{10}$H$_7$CO)$_2$O	326.36	nd (eth)	....	135			bz, aa	B9³, 3184
9544	β-Naphthoic acid, 1-bromo 1-Br-β-C$_{10}$H$_6$CO$_2$H	251.09	nd (aa,bz)		191			al, bz, aa	B9³, 3195
9545	β-Naphthoic acid, 5-bromo 5-Br-β-C$_{10}$H$_6$CO$_2$H	751.09	nd (al,sub)	sub	270			al, eth, bz, aa	B9³, 3196
9546	β-Naphthoic acid, 1-chloro 1-Cl-β-C$_{10}$H$_6$CO$_2$H	206.63	nd (bz)		196			al, ace	B9³, 3191
9547	β-Naphthoic acid, 3-chloro 3-Cl-β-C$_{10}$H$_6$CO$_2$H	206.63	cr (dil MeOH)		216.5			al, eth, ace, bz, chl	B9³, 3192
9548	β-Naphthoic acid, 5-chloro 5-Cl-β-C$_{10}$H$_6$CO$_2$H	206.63	nd (al,aa)		270			bz, aa	B9³, 3192
9549	β-Naphthoic acid, 1,2-dihydro β-C$_{10}$H$_9$CO$_2$H	174.20	nd or pr (dil al,w,peth)		105-6			al, chl, aa	B9³, 3076
9550	β-Naphthoic acid, 1,4-dihydro β-C$_{10}$H$_9$CO$_2$H	174.20	pl (bz,w,dil al)		162-3			al, eth	B9³, 3076
9551	β-Naphthoic acid, 3,4-dihydro β-C$_{10}$H$_9$CO$_2$H	174.20	nd (dil aa,dil al)		120			bz, chl, aa	B9³, 3076
9552	β-Naphthoic acid, 1,3-dihydroxy,ethyl ester 1,3 (HO)$_2$ β-C$_{10}$H$_6$CO$_2$C$_2$H$_5$	232.24	nd (dil al, dil aa)		83-4			al, eth, lig	B10³, 1932
9553	β-Naphthoic acid, ethyl ester or Ethyl β-naphthoate β-C$_{10}$H$_7$CO$_2$C$_2$H$_5$	200.24		308-9, 224⁷⁴	32	1.1143²³/⁴	1.5951²³	al, eth, chl, aa	B9³, 3177
9554	β-Naphthoic acid, 1-hydroxy 1-HO-β-C$_{10}$H$_6$CO$_2$H	188.20	cr (dil al,w,aa) nd (al,eth,bz)	....	195			al, th, bz	B10³, 1075
9555	β-Naphthoic acid, 1-hydroxy-4-chloro 1-HO-4-Cl-β-C$_{10}$H$_5$CO$_2$H	222.63	nd (al,aa)	....	234			al, ace	B10³, 1079
9556	β-Naphthoic acid, 1-hydroxy,phenyl ester 1-HO-β-C$_{10}$H$_6$CO$_2$C$_6$H$_5$	264.28			96			al, bz	B10, 332
9557	β-Naphthoic acid, 3-hydroxy 3-HO-β-C$_{10}$H$_6$CO$_2$H	188.20	ye lf (dil al) nd (dil al)		222.3			al, eth, bz, chl	B10³, 1084
9558	β-Naphthoic acid, 3-hydroxy-4-chloro 4-Cl-3-HO-β-C$_{10}$H$_5$CO$_2$H	222.63	ye nd		231d				B10³, 1091
9559	β-Naphthoic acid, 3-hydroxy,ethyl ester 3-HO-β-C$_{10}$H$_6$CO$_2$C$_2$H$_5$	216.24	nd or mcl pr (aa)	291	85			ace, chl	B10³, 1086
9560	β-Naphthoic acid, 3-hydroxy,hexadecyl ester 3-HO-β-C$_{10}$H$_6$CO$_2$C$_{16}$H$_{33}$	412.62	grsh-wh pr		72-3			bz, lig	B10³, 1087
9561	β-Naphthoic acid, 3-hydroxy,methyl ester 3-HO-β-C$_{10}$H$_6$CO$_2$CH$_3$	202.21	pa ye rh nd (dil MeOH)	205-7	75-6			al	B10³, 1086
9562	β-Naphthoic acid, 5-hydroxy 5-HO-β-C$_{10}$H$_6$CO$_2$H	188.20	wh nd (w,dil al)	....	213			al, eth, ace, aa	B10³, 1098
9563	β-Naphthoic acid, 7-hydroxy 7-HO-β-C$_{10}$H$_6$CO$_2$H	188.20	lf or pl (dil al) pa ye nd (al)		269-70			al, eth, ace, aa	B10³, 1101
9564	β-Naphthoic acid, methyl ester or Methyl β-naphthoate β-C$_{10}$H$_7$CO$_2$CH$_3$	186.21	lf (MeOH)	290, 141-3⁴	77			al, eth, bz, chl	B9³, 3176
9565	β-Naphthoic acid, 5-nitro 5-NO$_2$-β-C$_{10}$H$_6$CO$_2$H	217.18	yesh nd (al)	....	295			ace	B9³, 3203
9566	β-Naphthoic acid, 8-nitro 8-NO$_2$-β-C$_{10}$H$_6$CO$_2$H	217.18	yesh nd (al)	sub	295				B9², 455
9567	β-Naphthoic acid, 1,2,3,4-tetra hydro β-C$_{10}$H$_{11}$CO$_2$H	178.22	nd (dil al)	168-70¹⁵	97			al, eth, bz, chl	B9³, 2805
9568	β-Naphthoic acid, 5,6,7,8-tetra hydro β-C$_{10}$H$_{11}$CO$_2$H	176.22	nd (al) cr (aa,bz)	216¹⁴	154			al, bz	B9³, 2802
9569	β-Naphthonitrile β-C$_{10}$H$_7$CN	153.19	lf (lig)	306.5, 156-8¹²	66	1.0939⁶⁰/⁶⁰		al, eth, lig	B9³, 3186
9570	β-Naphthoyl chloride β-C$_{10}$H$_7$COCl	190.63	cr (peth)	304-6, 142-3⁵	51			eth, bz, chl aa	B9³, 3185

No.	Name, Synonyms, and Formula	Mol. wt.	Color, crystalline form, specific rotation and λ_{max} (log ε)	b.p. °C	m.p. °C	Density	n_D	Solubility	Ref.
9571	α-Naphthol or 1-Hydroxy naphthalene................ α -C$_{10}$H$_7$OH	144.19	ye mcl nd (w)	288 sub	96	1.0989[99/4]	1.6224[99]	al, eth, ace, bz, chl	B6[3], 2193
9572	α-Naphthol, 2-acetyl .... 2-CH$_3$CO-α -C$_{10}$H$_6$OH	186.21	(i)pr(bz,lig) (ii)gr-ye nd(al)	325d	(i)98 (ii)103			bz, aa	B8[3], 1130
9573	α-naphthol, 2-acetyl-4-bromo .... 2-CH$_3$CO-4-Br-α -C$_{10}$H$_5$OH	265.11	ye nd (al)	....	126-7			al, eth, bz, chl	B8[3], 1134
9574	α-Naphthol, 2-acetyl-4-nitro .... 2-CH$_3$CO-4-NO$_2$-α -C$_{10}$H$_5$OH	231.21	ye nd (al)		159			eth, bz	B8[3], 1135
9575	α-Naphthol, 3-acetyl .... 3-CH$_3$CO-α -C$_{10}$H$_6$OH	186.21	nd (bz)	....	173-4			al, aa	B8, 150
9576	α-naphthol, 4-acetyl .... 4-CH$_3$CO-C$_{10}$H$_6$OH	186.21	yesh pr (aa,al,to)		198			al, bz, aa	B8[3], 1124
9577	α-Naphthol, 2-benzyl .... 2-C$_6$H$_5$CH$_2$-α -C$_{10}$H$_6$OH	234.30	nd (lig) pr (bz)	237-40[12]	73-4				B6[3], 3608
9578	α-Naphthol, 4-benzyl .... 4-C$_6$H$_5$CH$_2$-α -C$_{10}$H$_6$OH	234.30	pl or nd (aa-lig) pr (bz)	237[10]	125-6				B6[2], 680
9579	α-Naphthol, 4-bromo .... 4-Br-α -C$_{10}$H$_6$OH	223.08	nd (dil al,hex)		128			chl, aa	B6[3], 2935
9580	α-Naphthol, 5-bromo .... 5-Br-α -C$_{10}$H$_6$OH	223.08	nd (w)		137			eth	B6[3], 2936
9581	α-Naphthol, 6-bromo .... 6-Br-α -C$_{10}$H$_6$OH	223.08	nd (w)		129-30			w	B6[3], 2936
9582	α-Naphthol, 7-bromo .... 7-Br-α -C$_{10}$H$_6$OH	223.08	cr (w)		105-6			w	B6[2], 583
9583	α-Naphthol, 8-bromo .... 8-Br-α -C$_{10}$H$_6$OH	223.08	pl (peth)		61			al	B6, 614
9584	α-Naphthol, 4-methyl .... 4-CH$_3$-α -C$_{10}$H$_6$OH	158.20	....	165-7[13]	86-7			al, eth, ace, bz	B6[3], 3022
9585	α-Naphthol, 8-nitro or 8-nitro-1-naphthol .... 8-NO$_2$-α -C$_{10}$H$_6$OH	189.17	grsh ye nd (al,chl,bz-hx)		130-3			al, eth, ace, bz	B6[3], 2939
9586	α -Napthol, 2-propionyl .... 2-CH$_3$CH$_2$CO- α -C$_{10}$H$_6$OH	200.26	grsh ye lf or pl (al)		81			al, eth	B8[3], 1144
9587	α-Naphthol, 2,3,4-trichloro .... 2,3,4-Cl$_3$-α-C$_{10}$H$_4$OH	247.51	nd (aa or lig)		168			eth, al	B6[2], 582
9588	β-Naphthol or 2-Hydroxy naphthalene................ β -C$_{10}$H$_7$OH	144.19	mcl lf (w)	295	123-4	1.28[20]		al, eth, bz, chl	B6[3], 2955
9589	β-Naphthol, 1-acetoamido .... 1-CH$_3$CONH-β-C$_{10}$H$_6$OH	201.23	lf (w,dil al)	sub	235d			al, eth, ace, bz	B13[2], 414
9590	β-Naphthol, acetate or 2-Acetoxy naphthalene .... β -C$_{10}$H$_7$(O$_2$CCH$_3$)	186.21	nd (al)	132-4[2]	71-2			al, eth, chl	B6[3], 2982
9591	β-Naphthol, 1-acetyl .... 1-CH$_3$CO-β -C$_{10}$H$_6$OH	186.21	pa ye lf (peth) rh (lig)		64-5			al, eth, bz	B8[3], 1122
9592	β -Napthol, 3-acetyl .... 3-CH$_3$CO-β -C$_{10}$H$_6$OH	186.21	ye lf or nd (al,peth)		112			ace, bz	B8[3], 1135
9593	β -Napthol, 1-amino .... 1-H$_2$N-β -C$_{10}$H$_6$OH	159.19	silvery lf (bz, eth)		150d			al	B13[3], 1891
9594	β-Naphthol, 3-amino .... 3-H$_2$N-β -C$_{10}$H$_6$OH	159.19	silvery lf (bz) nd (al,w)		235			al	B13[3], 1897
9595	β-Naphthol, 5-amino .... 5-H$_3$N-β -C$_{10}$H$_6$OH	159.19	nd or og pr (w)		190.6			al, eth, ace, w	B13[3], 1901
9596	β-Naphthol, 6-amino .... 6-H$_2$N-β -C$_{10}$H$_6$OH	159.19	pr (w)		192-4, 212d			al, w	B13[3], 1902
9597	β-Naphthol, 7-amino .... 7-H$_2$N-β -C$_{10}$H$_6$OH	159.19	nd or lf (al)		201 (208)			al, eth	B13[2], 416
9598	β-Naphthol, 8-amino .... 8-H$_3$N-β -C$_{10}$H$_6$OH	159.19	nd (w,al)	sub	205-7			eth	B13[3], 1907
9599	β-Naphthol, benzoate or 2-Benzoyloxy naphthalene...... β-C$_{10}$H$_7$(O$_2$CC$_6$H$_5$)	248.29	nd or pr (al) cr (lig)	....	108			al, eth	B9[3], 492
9600	β-Naphthol, 1-benzyl .... 1-C$_6$H$_5$CH$_2$-β -C$_{10}$H$_6$OH	234.30	nd (bz)	247-50[13]	115			al, eth, ace, bz, chl	B6[3], 3607
9601	β-Naphthol, 1-bromo .... 1-Br-β-C$_{10}$H$_6$OH	223.08	rh pr(bz-lig) nd(aa lig)	d130	84			al, eth, bz, aa, lig	B6[3], 2994

No.	Name, Synonyms, and Formula	Mol. wt.	Color, crystalline form, specific rotation and λ_{max} (log ε)	b.p. °C	m.p. °C	Density	n_D	Solubility	Ref.
9602	β-Naphthol, 3-bromo 3-Br-β-C₁₀H₆OH	223.08	nd(lig)	....	84-5		....	al, bz	B6³, 2995
9603	β-Naphthol, 5-Bromo 5-Br-β-C₁₀H₆OH	223.08	nd(w)		105			al	B6³, 2996
9604	β-Naphthol, 6-bromo 6-Br-β-C₁₀H₆OH	237.10	nd(bz)		127			al, bz	B6³, 3020
9605	β-Naphthol, 6-bromo-1-methyl 6-Br-1-CH₃-β-C₁₀H₅OH	237.10	nd(bz)		129			al, eth, ace, bz, chl	B6³, 3020
9606	β-Naphthol, 7-bromo 7-Br-β-C₁₀H₆OH	223.08	cr(peth)		132-3			eth, ace	B6², 605
9607	β-Naphthol, 1-chloro 1-Cl-β-C₁₀H₆OH	178.62	nd(lig) pr(chl) pl(w)		71			al, bz, chl	B6³, 2990
9608	β-Naphthol, 1,6-dibromo 1,6-Br₂-β-C₁₀H₅OH	301.98	nd(peth bz) lf(al)		106			al, eth, ace	B6³, 2998
9609	β-Naphthol, 1,6-dinitro 1,6-(NO₂)₂-β-C₁₀H₅OH	234.17	pa ye nd(chl)		195d			al, eth, chl, py	B6³, 3005
9610	β-Naphthol, 1-methyl 1-CH₃-β-C₁₀H₆OH	158.20	nd(w bz-lig dil aa)	180¹² sub	112			al, eth, ace, bz, aa	B6³, 3019
9611	β-Naphthol, 1-nitro or 1-nitro-2-naphthol 1-NO₂-β-C₁₀H₆OH	189.17	ye nd lf or pr(al)	115⁰·⁰⁵	104			al, eth	B6³, 3002
9612	β-Naphthol, 5-nitro 5-NO₂-β-C₁₀H₆OH	189.17	lt ye nd(w)		147-9			eth, ace	B6³, 3004
9613	β-Naphthol, 1,3,6-tribromo 1,3,6-Br₃-β-C₁₀H₄OH	380.88	nd(aa al)		133			al, bz	B6³, 3000
9614	β-Naphthol, 1,4,6-tribromo 1,4,6-Br₃-β-C₁₀H₄OH	380.88	nd(bz)		157-8			al, bz, chl, aa	B6³, 3000
9615	β-Naphthol, 3,4,6-tribromo 3,4,6-Br₃-β-C₁₀H₄OH	380.88	nd(bz)		137-8			al, bz	B6³, 607
9616	β-Naphthol, 1,3,4-trichloro 1,3,4-Cl₃-β-C₁₀H₄OH	247.51	nd		162			al, aa	B6², 604
9618	1,2-Naphthoquinone C₁₀H₆O₂	158.16	ye-red nd (eth) og lf (bz)		146	1.450		w, a;, eth	B7³, 3686
9619	1,2-Naphthoquinone, 3-bromo C₁₀H₅BrO₂	237.06	red nd or pl(aa al)	sub	178			al, bz	B7³, 3692
9620	1,2-Naphthoquinone, 4-bromo C₁₀H₅BrO₂	237.06	red nd(bz-lig)		154			al, bz, aa	B7³, 3691
9621	1,2-Naphthoquinone, 6-bromo C₁₀H₅BrO₂	237.06	og-red or ye pr or pl (bz) nd (w)		168d			al, ace	B7³, 3691
9622	1,2-Naphthoquinone, 3-chloro C₁₀H₅ClO₂	192.60	red nd(al aa bz chl)		172d			bz	B7³, 3690
9623	1,2-Naphthoquinone, 4-chloro C₁₀H₅O₂Cl	192.60	....		134-6			al	B7³, 3690
9624	1,2-Naphthoquinone, 3,4-dibromo C₁₀H₄Br₂O₂	315.96	red lf or pl(aa bz)		172-4			bz	B7³, 3693
9625	1,2-Naphthoquinone, 3,6-dibromo C₁₀H₄Br₂O₂	315.96	red pr(AcOEt)		176			al	B7³, 3692
9626	1,2-Naphthoquinone, 4,6-dibromo C₁₀H₄Br₂O₂	315.96	og-red pr (bz peth) nd (AcOEt)		153			bz, aa	B7³, 3692
9627	1,2-Naphthoquinone, 3,4-dichloro C₁₀H₄Cl₂O₂	227.05	red lf or pl(aa bz) nd(bz chl)	sub	184			bz, chl	B7³, 3691
9628	1,2-Naphthoquinone, dioxime C₁₀H₈N₂O₂	188.20	ye nd(bz-lig dil al)		169			al, bz, diox	B7³, 3690
9629	1,2-Naphthoquinone, 6-hydroxy C₁₀H₆O₃	174.16	red lf(ace)		165d			al, eth, ace, aa	B8³, 2542
9631	1,2-Naphthoquinone, 7-hydroxy C₁₀H₆O₃	174.16	br nd		194			al	B8³, 2542
9632	1,2-Naphthoquinone, 3-methyl C₁₁H₈O₂	172.20	red or og nd(abs al) lf(bz peth)		116 (122)			al	B8³, 3709
9633	1,2-Naphthoquinone, 4-methyl C₁₁H₈O₂	172.20	og nd (MeOH) nd (aa)		248-50d			al	B7³, 3708

PHYSICAL CONSTANTS OF ORGANIC COMPOUNDS (Continued)

No.	Name, Synonyms, and Formula	Mol. wt.	Color, crystalline form, specific rotation and λ_{max} (log ε)	b.p. °C	m.p. °C	Density	n_D	Solubility	Ref.
9634	1,2-Naphthoquinone, 3-nitro .... $C_{10}H_5NO_4$	203.16	red pl(aa)	....	158	....	....	bz, aa	B7², 651
9635	1,2-Naphthoquinone, 1-oxime .... $C_{10}H_7NO_2$	173.17	ye nd(bz) og pr or pl(al)	....	112	....	....	al, eth, ace, bz	B7³, 3688
9636	1,4-Naphthoquinone .... $C_{10}H_6O_2$	158.16	bt ye nd(al peth) ye(sub)	sub	128.5	....	....	al, eth, bz, aa	B7³, 3696
9637	1,4-Naphthoquinone, 2-amino .... $C_{10}H_7NO_2$	173.17	....	....	207	....	....	al, eth	B14³, 388
9638	1,4-Naphthoquinone, 4-anil,2-anilino .... $C_{22}H_{16}N_2O$	324.39	ye red nd(bz al)	....	182-3	....	....	aa, bz	B14³, 390
9639	1,4-Naphthoquinone, 2-bromo .... $C_{10}H_5BrO_2$	237.06	ye pl or nd (al, dil aa)	....	132	....	....	aa, bz, chl, aa	B7³, 3705
9640	1,4-Naphthoquinone, 2-bromo-3-methyl .... $C_{11}H_7BrO_2$	251.09	ye-br nd(al)	sub 100	151	....	....	al, eth, ace, bz, chl	B7³, 3714
9641	1,4-Naphthoquinone, 5-bromo-2,3-dichloro .... $C_{10}H_3BrCl_2O_2$	305.95	ye pr(al)	....	180	....	....	ace, bz	B7², 655
9642	1,4-Naphthoquinone, 6-bromo .... $C_{10}H_5BrO_2$	237.06	og-red or ye pr(bz AcOEt) nd(w)	....	168d	....	....	al, ace	B7, 722
9643	1,4-Naphthoquinone, 2-chloro .... $C_{10}H_5ClO_2$	192.60	ye nd(w al aa)	....	117-8	....	....	al, ace, bz	B7³, 3702
9644	1,4-Naphthoquinone, 5-chloro .... $C_{10}H_5ClO_2$	192.60	ye nd(lig)	sub	163	....	....	al, aa, lig	B7³, 3701
9645	1,4-Naphthoquinone, 6-chloro .... $C_{10}H_5ClO_2$	192.60	red-br(al eth) ye cr(dil MeOH)	....	109-10	....	....	bz	B7³, 3702
9646	1,4-Naphthoquinone, 2,3-dibromo .... $C_{10}H_4Br_2O_2$	315.96	ye nd(aa)	....	218	....	....	aa	B7³, 3705
9647	1,4-Naphthoquinone, 5,8-dibromo .... $C_{10}H_4Br_2O_2$	315.96	ye nd(al)	....	171-3	....	....	....	B7, 732
9648	1,4-Naphthoquinone, 2,3-dichloro .... $C_{10}H_4Cl_2O_2$	227.05	ye nd(al)	....	195	....	....	chl	B7³, 3703
9649	1,4-Naphthoquinone, 2,6-dichloro .... $C_{10}H_4Cl_2O_2$	227.05	dk ye nd(al)	....	148-9	....	....	al	B7, 730
9650	1,4-Naphthoquinone, 5,6-dichloro .... $C_{10}H_4Cl_2O_2$	227.05	ye nd(al)	sub	181	....	....	eth	B7, 730
9651	1,4-Naphthoquinone, 5,8-dichloro .... $C_{10}H_4Cl_2O_2$	227.05	ye nd(al)	sub	173-4	....	....	eth	B7³, 3702
9651a	1,4-Naphthoquinone, 2,3-dihydro .... $C_{10}H_8O_2$	160.18	lf (hx) nd (peth)	....	98-9	....	....	al	E12B, 2806
9652	1,4-Naphthoquinone, 2,3-dihydroxy or Isonaphthazarin. $C_{10}H_6O_4$	190.16	red og nd or lf (sub)	sub	282	....	....	ace	B8³, 3596
9653	1,4-Naphthoquinone, 3,5-dihydroxy-2-methyl or Prose-rone. $C_{11}H_8O_4$	204.19	og-ye nd(al aa)	sub 100³	181	....	....	al, eth, peth	B8³, 3604
9654	1,4-Naphthoquinone, 5,8-dihydroxy or Naphthazarin.... $C_{10}H_6O_4$	190.16	dk red mcl pr (bz) red-br nd (al)	sub	276-80	....	....	aa	B8³, 3600
9655	1,4-Naphthoquinone, 5,8-dihydroxy-2-methyl .... $C_{11}H_8O_4$	204.19	gr pl	....	173	....	....	al	B8³, 3605
9656	1,4-Naphthoquinone, 2,3-dimethyl .... $C_{12}H_{10}O_2$	186.21	ye pr(al)	....	127	....	....	bz, aa	B7³, 3717
9657	1,4-Naphthoquinone, 2,5-dimethyl .... $C_{12}H_{10}O_2$	186.21	ye nd(peth eth)	....	95	....	....	diox	B7³, 3716
9658	1,4-Naphthoquinone, 2,6-dimethyl .... $C_{12}H_{10}O_2$	186.21	ye pr or nd (AcOEt)	....	136-7	....	....	al, eth, bz	B7³, 3720
9659	1,4-Naphthoquinone, 2,8-dimethyl .... $C_{12}H_{10}O_2$	186.21	pr (peth) nd (MeOH)	....	135-6	....	....	al	B7³, 3716
9660	1,4-Naphthoquinone, dioxime .... $C_{10}H_8N_2O_2$	188.20	nd(dil al)	....	207d	....	....	w, al	B7², 653
9661	1,4-Naphthoquinone, 2-ethyl .... $C_{12}H_{10}O_2$	186.21	pr(al) nd aa peth MeOH	....	88-9	....	....	al, aa	B7³, 3715

No.	Name, Synonyms, and Formula	Mol. wt.	Color, crystalline form, specific rotation and λ_{max} (log ε)	b.p. °C	m.p. °C	Density	n_D	Solubility	Ref.
9662	1,4-Naphthoquinone, 2-ethyl-3-hydroxy $C_{12}H_{10}O_3$	202.21	ye nd(dil MeOH)	....	141			eth, ace	B8[3], 2586
9663	1,4-Naphthoquinone, 2-ethyl-3,5,6,7,8-pentahydroxy or Echinochrome A $C_{12}H_{10}O_7$	266.21	red nd(diox-w)	sub 120[10]	220d			w, al, eth, ace, bz	B8[3], 4360
9664	1,4-Naphthoquinone, 2-hydroxy or Lawsone $C_{10}H_6O_3$	174.16	redsh-br(aa)	....	192d			al, aa	B8[3], 2543
9665	1,4-Naphthoquinone, 2-hydroxy,acetate $C_{12}H_8O_4$	216.21	ye lf(al)	....	131			al, eth, chl, aa	B8[3], 2547
9666	1,4-Naphthoquinone, 2-hydroxy-3-methyl or Phthiocol $C_{11}H_8O_3$	188.20	ye pr(eth-peth)	sub	173-4			eth, ace	B8[3], 2569
9667	1,4-Naphthoquinone, 2-hydroxy-3-phenyl $C_{16}H_{10}O_3$	250.26	gold-ye or og pr or nd(al bz MeOH)	....	147			al, eth, bz, chl, lig	B8[3], 2981
9668	1,4-Naphthoquinone, 3-hydroxy-2-bromo $C_{10}H_5BrO_3$	253.06	ye mcl pr (al) nd (al, w)	sub	202			ace	B8[3], 2552
9669	1,4-Naphthoquinone, 3-hydroxy-2-chloro $C_{10}H_5ClO_3$	208.60	ye nd(al aa)	sub	213			al, eth, bz	B8[2], 347
9670	1,4-Naphthoquinone, 5-hydroxy or Juglon, Nucin $C_{10}H_6O_3$	174.16	redsh-ye nd or pr(chl bz)	sub	154 (161)			al, eth, bz, chl, aa	B8[3], 2558
9671	1,4-Naphthoquinone, 5-hydroxy-2-methyl or Plumbagin $C_{11}H_8O_3$	188.20	gold pr or og-ye nd(dil al)	sub	78-9			al, eth, ace, bz, chl	B8[3], 2576
9672	1,4-Naphthoquinone, 6-hydroxy $C_{10}H_6O_3$	174.16	gold-ye or red ye nd(al bz al)	....	170d			al, eth, ace, MeOH	B8[2], 348
9673	1,4-Naphthoquinone, 2-methyl or Menadione, Vitamin K$_3$ $C_{11}H_8O_2$	172.20	ye nd(al peth)	....	107			ace, bz	B7[3], 3709
9674	1,4-Naphthoquinone, 2-phenyl $C_{16}H_{10}O_2$	234.26	gold-ye nd(al)	....	111			al, eth, bz, chl	B7[3], 4209
9675	1,4-Naphthoquinone, 2-phenylamino or Lawsone anilide $C_{16}H_{11}NO_2$	249.27	red nd(dil al)	sub	193			eth, bz	B14[3], 389
9676	1,4-Naphthoquinone, 5,6,7,8-tetrahydro $C_{10}H_{10}O_2$	162.19	gold-ye nd(peth)	....	55-6			al, eth	B7[3], 3506
9677	1,4-Naphthoquinone, 2,5,8-trihydroxy $C_{10}H_6O_5$	206.16	red nd(bz MeOH)	....	195			al, aa	B8[3], 4043
9678	1,4-Naphthoquinone, 2-oxime $C_{10}H_7NO_2$	173.17	pa ye nd(bz) nd(dil al)	....	198			al, eth, ace	B7[3], 3700
9679	2,6-Naphthoquinone $C_{10}H_6O_2$	158.16	ye-red pr(bz bz-peth)	....	135d			al, MeOH	B7[3], 3706
9680	2,6-Naphthoquinone, 1,5-dichloro $C_{10}H_4Cl_2O_2$	227.05	og pr(chl) gold-ye nd(al)	....	206d			ace, bz, chl, aa	B7[3], 3706
9681	α-Naphthoxy acetic acid α-$C_{10}H_7OCH_2CO_2H$	202.21	pr	....	190			al, eth	B6[3], 2930
9682	β-Naphthoxy acetic acid β-$C_{10}H_7OCH_2CO_2H$	202.21	pr(w)	....	156			al, eth, aa	B6[3], 2985
9683	α-Naphthy acetamide α-$C_{10}H_7CH_2CONH_2$	185.23	nd(w al)	....	180-1 sub			eth, bz, aa	B9[3], 3208
9684	α-Naphthyl acetic acid α-$C_{10}H_7CH_2CO_2H$	186.21	wh nd(w)	d	133			eth, ace, bz, chl, aa	B9[3], 3206
9685	α-Naphthyl acetonitrile α-$C_{10}H_7CH_2CN$	167.21	wx	162-4[12]	32-3		1.6192[20]	....	B9[3], 3209
9686	β-Naphthyl acetamide β-$C_{10}H_7CH_2CONH_2$	185.23	lf(w)	....	202-4d			al, eth	B9[3], 3212
9687	β-Naphthyl acetonitrile β-$C_{10}H_7CH_2CN$	167.21	nd or lf(dil al)	145.50[2]	85-6			eth, bz, chl	B9[3], 3212
9688	α-Naphthyl amine α-$H_2NC_{10}H_7$	143.19	nd(dil al eth)	300.8, 160[12] sub	50	1.1229[25/25]	1.6703[51]	al, eth	B12[3], 2846
9689	α-Naphthyl amine, 2-acetyl,hydrochloride 2-$CH_3CO-α-C_{10}H_6NH_2.HCl$	186.21	cr(w)	....	220d			w, aa	B14[3], 208
9690	α-Naphthyl amine, N-acetyl α-$C_{10}H_7NHCOCH_3$	185.23	cr(al)	....	160			....	B12[3], 2866

No.	Name, Synonyms, and Formula	Mol. wt.	Color, crystalline form, specific rotation and λ_{max} (log ε)	b.p. °C	m.p. °C	Density	n_D	Solubility	Ref.
9691	α-Naphthyl amine, N-acetyl-2-nitro 2-O₂N-α-C₁₀H₆NHCOCH₃	230.23	lt ye nd(aa al)		200			al, aa	B12³, 2969
9692	α-Naphthyl amine, N-acetyl-4-nitro 4-O₂N-α-C₁₀H₆NHCOCH₃	230.23	pa ye nd(ace)		192-3			al, ace	B12³, 2972
9693	α-Naphthyl amine, N-acetyl-5-nitro 5-O₂N-α-C₁₀H₆NHCOCH₃	230.23	br pr(aa) ye cr(al)		220			al	B12³, 2973
9694	α-Naphthyl amine, N-acetyl-8-nitro 8-NO₂-α-C₁₀H₆NHCOCH₃	230.23	nd(w)		191				B12³, 2976
9695	α-Naphthyl amine, N-2-aminoethyl α-C₁₀H₇(NHCH₂CH₂NH₂)	186.26	ye	320d, 204⁹		1.114²⁵/⁴	1.6648²⁵	al, ace	B12³, 2955
9696	α-Naphthyl amine, N-benzylidene α-C₁₀H₇N=CHC₆H₅	231.30	ye lf(al)		73.5			al, eth, bz, MeOH	B12³, 2861
9697	α-Naphthyl amine, 4-bromo 4Br-α-C₁₀H₆NH₂	222.10	nd(al bz peth)		102			al, ace, bz, lig	B12³, 2964
9698	α-Naphthyl amine, 4-bromo-2-nitro 4-Br-2-NO₂-α-C₁₀H₅NH₂	267.09	og cr(al aa)		200			al, bz, chl	B12³, 2979
9699	α-Naphthyl amine, 5-bromo 5-Br-α-C₁₀H₆NH₂	222.10	lf or pl(w lig)	sub	69			al, eth, ace, bz, lig	B12³, 2966
9700	α-Naphthyl amine, 2-chloro 2-Cl-α-C₁₀H₆NH₂	177.65	nd(peth dil al)		60			al, ace	B12³, 2960
9701	α-Naphthyl amine, 4-chloro 4-Cl-α-C₁₀H₆NH₂	177.65	nd(al bz lig)		99-100			al, eth, bz	B12³, 2961
9702	α-Naphthyl amine, 2,4-dibromo 2,4-Br₂-α-C₁₀H₅NH₂	300.99	nd or pl(dil al)		118-9			al, eth, bz, chl, lig	B12³, 2966
9703	α-Naphthyl amine, 2,4-dichloro 2,4-Cl₂-α-C₁₀H₅NH₂	212.08	nd(al)		83-4			al	B12³, 2963
9704	α-Naphthyl amine, N,N-diethyl or 1-Diethylamino naphthalene α-(C₂H₅)₂NC₁₀H₇	199.30		285, 155-65³⁰		1.015²⁰/²⁰	1.5961²⁰	al, eth, aa	B12³, 2855
9705	α-Naphthyl amine, 5,8-dihydro α-C₁₀H₉NH₂	145.21	pl or nd(to) (pink in air)	247⁴⁰⁸	37.5			al, chl	B12³, 2837
9706	α-Naphthyl amine, N,N-dimethyl α-C₁₀H₇N(CH₃)₂	171.25	vt flr	274.5⁷¹¹, 139-40¹³		1.0423²⁰/⁴	1.624¹⁵	al, eth	B12³, 2854
9707	α-Naphthyl amine, 2,4-dinitro 2,4-(NO₂)₂-α-C₁₀H₅NH₂	233.19	ye nd(al) pr(aa) cr(ace)		242			ace	B12³, 2983
9708	α-Naphthyl amine, N-ethyl α-C₁₀H₇NHC₂H₅	171.25		303⁷²³, 191¹⁶		1.060²⁰	1.6477¹⁵		B12³, 2854
9709	α-Naphthyl amine, 4-fluoro 4-F-α-C₁₀H₆NH₂	161.19	lt ye	162¹⁶	48				B12³, 2960
9710	α-Naphthyl amine, N-formyl α-C₁₀H₇NHCHO	171.20	nd(w)		137.5			al, eth, ace	B12³, 2866
9711	α-Naphthyl amine, hydrochloride α-C₁₀H₇NH₂.HCl	179.65	nd	sub				al, w	B12³, 2849
9712	α-Naphthyl amine, N-methyl α-C₁₀H₇NHCH₃	157.22	oil	293, 165-7¹⁵			1.6722²⁰	al, eth	B12³, 2854
9713	α-Naphthyl amine, 2-methyl 2-CH₃-α-C₁₀H₆NH₂	157.22	nd (peth) (turns red in air)		32			eth, lig	B12³, 3702
9714	α-Naphthyl amine, 3-methyl 3-CH₃-α-C₁₀H₆NH₂	157.22	cr(peth)		51-2			al, eth, lig	B12³, 3106
9715	α-Naphthyl amine, 4-methyl 4-CH₃-α-C₁₀H₆NH₂	157.22	nd(peth)	176¹²	51-2			eth	B12³, 3093
9716	α-Naphthyl amine, 2-nitro 2-O₂N-α-C₁₀H₆NH₂	188.20	ye-red mcl pr(al)		144			al	B12³, 2968
9717	α-Naphthyl amine, 3-nitro 3-O₂N-α-C₁₀H₆NH₂	188.20	og-ye nd(50% al)		137			al, bz, chl	B12³, 2969
9718	α-Naphthyl amine, 4-nitro 4-O₂N-α-C₁₀H₆NH₂	188.20	og-ye nd(al)		195			al, aa	B12³, 2971
9719	α-Naphthyl amine, 5-nitro 5-O₂N-α-C₁₀H₆NH₂	188.20	red nd(w)		118-9			eth, aa	B12³, 2973
9720	α-Naphthyl amine, 6-nitro 6-O₂N-α-C₁₀H₆NH₂	188.20	og-red nd(chl)		172-3			al	B12³, 2974
9721	α-Naphthyl amine, 8-nitro 8-O₂N-α-C₁₀H₆NH₂	188.20	red lf(peth)		96-7			eth	B12³, 2975

No.	Name, Synonyms, and Formula	Mol. wt.	Color, crystalline form, specific rotation and λ_{max} (log ε)	b.p. °C	m.p. °C	Density	n_D	Solubility	Ref.
9722	β-Naphthyl amine β-C₁₀H₇NH₂	143.19	lf(w)	306.1	113	1.0614[98/4]	1.6493[98]	al, eth	B12³, 2989
9723	β-Naphthyl amine, N-acetyl β-C₁₀H₇(NHCOCH₃)	185.23	lf(al w)	....	134				B12³, 3014
9724	β-Naphthyl amine, N-acetyl-1-nitro 1-NO₂-β-C₁₀H₆NHCOCH₃	230.23	ye nd(al)		126			al, eth, bz, aa	B12², 731
9725	β-Naphthyl amine, N-acetyl-5-nitro 5-NO₂-β-C₁₀H₆NHCOCH₃	230.23	ye-br rh(al) ye nd(bz)		186			al, aa	B12³, 732
9726	β-Naphthyl amine, N-acetyl-6-nitro 6-NO₂-β-C₁₀H₆NHCOCH₃	220.23	lf ye nd(dil al)		224			al, bz, aa	B12², 733
9727	β-Naphthyl amine, N-acetyl-8-nitro 8-NO₂-β-C₁₀H₆NHCOCH₃	230.23	ye nd(al)		195.5			aa	B12², 733
9728	β-Naphthyl amine, N-benzylidene β-C₁₀H₇N=CHC₆H₅	231.30	yesh nd(al)		103			al, bz, chl, aa	B12³, 3006
9729	β-Naphthyl amine, 1-bromo 1-Br-β-C₁₀H₆NH₂	222.10	rh nd(dil al lig)		63-4			al, eth, bz, chl	B12³, 3070
9730	β-Naphthyl amine, 3-bromo 3-Br-β-C₁₀H₆NH₂	222.10	pl(al)		169			al, aa	B12³, 3072
9731	β-Naphthyl amine, 4-bromo 4-Br-β-C₁₀H₆NH₂	222.10	nd(bz-peth or 90% aa)		72			al, eth, bz	B12³, 3073
9732	β-Naphthyl amine, 5-bromo 5-Br-β-C₁₀H₆NH₂	222.10	cr	207-10[16]	38			al, ace	B12³, 3074
9733	β-Naphthyl amine, 6-bromo 6-Br-β-C₁₀H₆NH₂	222.10	lf(al m peth)		129			al, ace, bz	B12³, 3074
9734	β-Naphthyl amine, 1-chloro 1-Cl-β-C₁₀H₆NH₂	177.65	nd(al peth)	....	60			al, ace	B12³, 3065
9735	β-Naphthyl amine, 1,4-dibromo 1,4-Br₂-β-C₁₀H₅NH₂	300.99	nd(al ba)		106 7			al, eth, bz	B12, 1311
9736	β-Naphthyl amine, 1,6-dibromo 1,6-Br₂-β-C₁₀H₅NH₂	300.99	nd(al peth)		121			al, bz, aa	B12¹, 544
9737	β-Naphthyl amine, N,N-dimethyl β-C₁₀H₇N(CH₃)₂	171.25	dk red nd	305, 160-2[12]	52-3	1.0455[60/60]	1.6443[53]	al, eth	B12³, 2995
9738	β-Naphthyl amine, 1,4-dimethyl 1,4-(CH₃)₂-β-C₁₀H₅NH₂	171.24	....	333	75			al, eth	B12, 1317
9739	β-Naphthyl amine, 1,6-dinitro 1,6-(NO₂)₂-β-C₁₀H₅NH₂	233.19	gold-ye nd(al aa) ye pw(py)		248				B12³, 3087
9740	β-Naphthyl amine, N-ethyl β-C₁₀H₇NHC₂H₅	171.25	....	316-7, 191[25]	<15	1.0545[21]	1.6544[21]	....	B12³, 2996
9741	β-Naphthyl amine, N-formyl β-C₁₀H₇NHCHO	171.20	lf(bz-peth)	....	129			al, bz	B12³, 3013
9742	β-Naphthyl amine, hydrochloride β-C₁₀H₇NH₂.HCl	179.65	lf		254			w, al	B12³, 2992
9743	β-Naphthyl amine, 3-iodo 3-I-β-C₁₀H₆NH₂	269.10	cr(al)		137			al, bz, chl, aa	B12³, 3078
9744	β-Naphthyl amine, N-methyl β-C₁₀H₇NHCH₃	157.22	dk in air	317, 165-70[12]			1.6722[20]	....	B12³, 2995
9745	β-Naphthyl amine, 1-methyl 1-CH₃-β-C₁₀H₆NH₂	157.22	nd(lig) pr(peth)		51			al, eth, ace, bz	B12³, 3091
9746	β-Naphthyl amine, 6-methyl 6-CH₃-β-C₁₀H₆NH₂	157.22	lf(peth w)(turns red in air)		129-30				B12², 743
9747	β-Naphthyl amine, 1-nitro 1-NO₂-β-C₁₀H₆NH₂	188.20	og-ye nd(al)		126-7			al, ace, aa	B12³, 3079
9748	β-Naphthyl amine, 5-nitro 5-NO₂-β-C₁₀H₆NH₂	188.20	red nd(al)		143.5			al, bz, aa	B12¹, 3082
9749	β-Naphthyl amine, 6-nitro 6-NO₂-β-C₁₀H₆NH₂	188.20	lt og pl(al) ye pl(aa)		207.5			bz, aa, diox	B12³, 3082
9750	β-Naphthyl amine, 8-nitro 8-NO₂-β-C₁₀H₆NH₂	188.20	red nd(aa)		104.5			al, eth, bz	B12³, 3084
9751	β-Naphthyl amine, 1-nitroso 1-ON-β-C₁₀H₆NH₂	172.20	gr nd(dil al bz)		150.2			al	B7³, 3690
9752	β-Naphthyl amine, 1,3,6-tribromo 1,3,6-Br₃-β-C₁₀H₄NH₂	379.89	pa red cr (chl al-eth)		143			eth, chl	B12³, 3078
9753	α-Naphthyl ether or α-Naphthyl ether (α-C₁₀H₇)₂O	270.34	lf(al al-eth)	280-5[22]	110			eth, bz	B6³, 2926

No.	Name, Synonyms, and Formula	Mol. wt.	Color, crystalline form, specific rotation and λ_{max} (log ε)	b.p. °C	m.p. °C	Density	n_D	Solubility	Ref.
9754	β-Naphthyl ether or bis-(2-Naphthyl) ether (β-C$_{10}$H$_7$)$_2$O	270.34	nd or lf(al)	250d	105	....	....	eth, bz	B6³, 2976
9755	α-β-Naphthyl ether α-C$_{10}$H$_7$O-β-C$_{10}$H$_7$	270.34	lf(al or al-eth)	264^{15}	81	....	....	eth, bz	B6², 600
9756	α-Naphthyl isocyanate α-C$_{10}$H$_7$NCO	169.19	....	269-70	......	1.1774$^{20/4}$	....	eth, bz	B12³, 2948
9757	β-Naphthyl isocyanate β-C$_{10}$H$_7$NCO	169.19	lf	....	55-6	....	....	eth, bz	B12³, 3051
9758	α-Naphthyl isothiocyanate or α-Naphthyl mustard oil α-C$_{10}$H$_7$NCS	185.25	wh nd(al)	....	58	....	....	al, eth, ace, bz, chl	B12³, 2948
9759	β-Naphthyl isothiocyanate β-C$_{10}$H$_7$NCS	185.25	yesh nd(al)	....	62-3	....	....	al, eth, bz, chl	B12³, 3052
9760	α-Naphthyl pentyl ether α-C$_{10}$H$_7$OC$_5$H$_{11}$	214.31	nd(al)	322	30	....	....	al, eth, bz, chl	B6³, 2925
9761	β-Naphthyl pentyl ether β-C$_{10}$H$_7$OC$_5$H$_{11}$	214.31	lf(al)	335	24.5	....	1.5587^{30}	al, eth, bz, chl	B6³, 2973
9762	α-Naphthyl phenyl amine α-C$_{10}$H$_7$NHC$_6$H$_5$	219.29	lf(lig) pr or nd(al)	335^{528}, 226^8	62	....	....	al, eth, bz, chl, aa	B12³, 2856
9763	α-Naphthyl propyl amine α-C$_{10}$H$_7$NHC$_3$H$_7$	185.27	ye	316-8^{771}	......	....	....	....	B12, 1224
9764	β-Naphthyl phenyl amine β-C$_{10}$H$_7$NHC$_6$H$_5$	219.20	nd(MeOH)	395-9, 237-13	108	....	....	al, eth, bz, aa	B12³, 2999
9765	α-Naphthyl sulfide (α-C$_{10}$H$_7$)$_2$S	286.40	nd or pr(al)	289-90^{15}	110	....	....	bz, aa	B6³, 2945
9766	β-Naphthyl sulfide (β-C$_{10}$H$_7$)$_2$S	286.40	pl(al) lf(bz)	295-6^{15}	151	....	....	bz	B6³, 3009
9767	α-Naphthyl thiocyanate α-C$_{10}$H$_7$SCN	185.25	cr(peth)	....	55	....	....	peth	B6², 588
9768	β-Naphthyl thiocyanate 2-C$_{10}$H$_7$SCN	185.25	....	....	35	....	....	....	B6², 611
9769	α-Naphthyl (2-tolyl)amine (2-CH$_3$C$_6$H$_4$)NH-α-C$_{10}$H$_7$	233.32	nd(lig)	198-202^9	94-5	....	....	al, eth, bz	B12³, 2857
9770	α-Naphthyl (4-tolyl) amine (4-CH$_3$C$_6$H$_4$)NH-α-C$_{10}$H$_7$	233.32	pr(al)	360^{528}, 236^{15}	79	....	....	eth, bz	B12³, 2857
9771	β-Naphthyl 2-tolyl amine or Yellow OB (2-CH$_3$C$_6$H$_4$)NH-β-C$_{10}$H$_7$	233.32	lf(lig)	400-5, 235-7^{14}	95-6 (105)	....	....	al, eth, ace, bz, lig	B12³, 3001
9772	β-Naphthyl 4-tolyl amine (4-CH$_3$C$_6$H$_4$)NH-β-C$_{10}$H$_7$	233.32	red lf(al)	....	103	....	....	eth, bz	B12³, 3001
9773	α-Naphthyl 2-tolyl ketone α-C$_{10}$H$_7$CO(C$_6$H$_4$CH$_3$-2)	246.31	(al)	365	64	....	....	al, eth	B7³, 2639
9774	α-Naphthyl 3-tolyl ketone α-C$_{10}$H$_7$CO(C$_6$H$_4$CH$_3$-3)	246.31	....	....	74-5	....	....	al	B7³, 2639
9775	Narceine or Pseudoarcene C$_{23}$H$_{27}$NO$_8$.3H$_2$O	499.52	nd or pr (w + 3)	....	145.2 (+ 3w) 176-7 (anh)	....	....	al	B19⁴, 4382
9776	Narceine, bisulfate, decahydrate C$_{23}$H$_{27}$NO$_8$.H$_2$SO$_4$.10H$_2$O	723.71	nd(sulf)	....	d	....	....	w, al, eth	B19, 372
9777	Narceine, hydrochloride, trihydrate C$_{23}$H$_{27}$NO$_8$.HCl.3H$_2$O	535.98	pr(HCl)	....	192 (anh)	....	....	al	B19⁴, 4383
9778	α-Narcotine (dl) C$_{22}$H$_{23}$NO$_7$	413.43	nd(al chl MeOH)	....	232-3	....	....	....	B27², 607
9779	α-Narcotine (l) C$_{22}$H$_{23}$NO$_7$	413.43	pr or nd (al), [α] -200 (chl)	....	176	....	....	al, ace, bz, chl	B27², 605
9780	α-Narcotine, hydrochloride C$_{22}$H$_{23}$NO$_7$·HCl	449.89	(w + 3) [α] + 100	....	193 anh	....	....	w	B27², 606
9781	β-Narcotine-(dl) or β-Gnoscopine C$_{22}$H$_{23}$NO$_7$	413.43	nd pr(MeOH al)	....	180	....	....	al	B27¹, 559
9782	β-Narcotine, hydrochloride C$_{22}$H$_{23}$NO$_7$.HCl	449.89	pr	....	86-8, 224 (on standing)	....	....	....	B27¹, 559
9783	Naringin C$_{27}$H$_{32}$O$_{14}$·2H$_2$O	616.58	nd (w + 8), [α]$^{111/}_D$ -82.1 (al)	....	82 (+ 8w), 17 (+ 2w)	....	....	....	B18⁴, 2637

No.	Name, Synonyms, and Formula	Mol. wt.	Color, crystalline form, specific rotation and λ_{max} (log ε)	b.p. °C	m.p. °C	Density	n_D	Solubility	Ref.
9784	Neoabietic acid, methyl ester or Methyl neoabietate ...... $C_{21}H_{32}O_2$	316.49	cr(MeOH)	....	61-2	....	....	MeOH	B9[2], 433
9785	Neoamygdalin or L-Mandelonitrile-β-gentobioside............ $C_{20}H_{27}NO_{11}$	457.44	cr (al), $[\alpha]^{88}_D$ -61.4 (w, c=8.5)		212			w, al	B31, 404
9786	Neobornyl amine or Isobornyl amine .... $C_{10}H_{19}N$	153.27	pw, $[\alpha]_D$ -47.7 (4%al)		184			eth, ace	B12[3], 195
9787	Neocarvomenthol (dl) .... $C_{10}H_{20}O$	156.27					1.4637[20]	al	B6[3], 130
9788	Neocarvomenthol (l) .... $C_{10}H_{20}O$	156.27	$[\alpha]^{21}_D$ -41.7	102[18]		0.9012[20/4]	1.4632[20]	al	B6[4], 148
9789	Neoergesterol .... $C_{27}H_{40}O$	380.62	pr or nd (al), $[\alpha]^{17}_D$ -12 (chl, c=2)	....	155-7			eth, ace	B6[3], 3474
9790	Neogermitrine................. $C_{36}H_{55}NO_{11}$	677.84	wh nd, $[\alpha]^{25}_D$ -79.2 (py)		237-9			chl	B21[4], 6803
9791	Neoisocarvomenthol (l) .... $C_{10}H_{20}O$	156.27	$[\alpha]^{17}_D$ -34.7	87-8[4]	<-25	0.9102[20/4]	1.4676[20]	al, ace	B6[4], 148
9792	Neoisamenthol (d) or p-menthol-3................... $C_{10}H_{20}O$	156.27	$[\alpha]^{15}_D$ +3.3 (al, c=2)	214.6, 91.6[11]	-8	0.9131[18/4]	1.4670[20]	al, ace	B6[4], 149
9793	Neoisomenthol (dl) .... $C_{10}H_{20}O$	156.27	....	214.5, 81[6]	14	0.8854[55/4]	1.4649[20]	al, ace	B6[4], 149
9794	Neomenthol (d) or d-p-Dolcomenthol .... $C_{10}H_{20}O$	156.27	$[\alpha]^{40}_D$ +19.6 (al)	211.7, 95[12]	15	0.897[11]	1.4600[11]	al, ace	B6[3], 149
9795	Neomenthol (dl) .... $C_{10}H_{20}O$	156.27	pl or pr(peth)	211.7, 103-5[16]	52	0.903[15/15]	1.4600[20]	al, ace	B6[4], 150
9796	Neomenthol-(l) .... $C_{10}H_{20}O$	156.27	$[\alpha]^{18}_D$ -19.6 (al)	211.7, 97.6[10]			1.4603[20]	al, ace	B6[3], 140
9797	Neopentane or 2,2-Dimethylpropane .... $(CH_3)_4C$	72.15	gas	9.5	-16.5	0.6135[20]	1.3476[6]	al, eth	B1[4], 333
9798	Neopentyl alcohol or 2,2-Dimethyl-1-propanol........... $(CH_3)_3CCH_2OH$	88.15	....	113-4	52-3	0.812	....	al, eth	B1[4], 1690
9799	Neopentyl alcohol, 1-phenyl .... $(CH_3)_3CCH(OH)C_6H_5$	164.25	nd	114-6[16]	45			al, eth	B6[3], 1972
9800	Neopentyl alcohol, 3-phenyl .... $C_6H_5CH_2C(CH_3)_2CH_2OH$	164.25	nd	125-6[14.5]	34-5			al, eth	B6[2], 507
9801	Neopentyl amine or 1-amino-2,2-dimethyl propane ...... $(CH_3)_3CCH_2NH_2$	87.17	....	81-2[741]	....	0.7455[20/4]	1.4023[20]	eth	B4[4], 707
9802	Neopentyl bromide or 1-bromo-2,2-dimethyl propane .... $(CH_3)_3CCH_2Br$	151.05	....	106, 34.6[100]	....	1.1997[20/4]	1.4370[20]	al, eth, ace, bz, chl	B1[4], 337
9803	Neopentyl chloride or 1-Chloro-2,2-dimethyl propane .... $(CH_3)_3CCH_2Cl$	106.60	....	84.3	-20	0.86604[20/4]	1.4044[20]	al, eth, bz, chl	B1[4], 336
9804	Nepentylene glycol or 2,2-Dimethyl-1,3-propanediol ..... $(HOCH_2)_2C(CH_3)_2$	104.15	nd (bz)	206[747], 120-30[15]	130	....	....	w, al, eth	B1[4], 2551
9805	Neopentyl iodide or 1-Iodo-2,2-dimethyl propane .... $(CH_3)_3CCH_2I$	198.05	....	127-9d, 42-4[20]	....	1.4940[20]	1.4890[20]	al, eth	B1[4], 338
9806	Neral or Citral b............... $C_{10}H_{16}O$	152.24	....	120[20]		0.8869[20]	1.4869[20]	**al, eth**	B1[4], 3569
9807	Nerol or 3,7-Dimethyl-2,6-octadiene-1-ol .... $(CH_3)_2C=CHCH_2CH_2C(CH_3)=CHCH_2OH$	154.26	....	224-5[745], 125[25]	<-15	0.8756[20/4]	1.4746[20]	al	B1[4], 2276
9808	Nerolidol (d) or α-3,7,11-trimethyl-1,6,10-dodecatriene-3-ol................... $(CH_3)_2C=CHCH_2CH_2C(CH_3)=CH(CH_2)_2C(CH_3)(OH)CH=CH_2$	222.38	$[\alpha]^{20}_D$ +15.5 (undil)	276, 128-9[6]		0.8778[20/4]	1.4898[20]	al, eth, ace, aa	B1[4], 2336
9809	Nerolidol (dl) .... $C_{15}H_{26}O$	222.38	....	145-6[12], 75-6[0.1]		0.8756[19/4]	1.4801[16]	al, eth, ace	B1[4], 2336
9810	Nerolidol (l) .... $C_{15}H_{26}O$	222.38	$[\alpha]_D$ -6.5 (undil)	124-6[3]		0.8881[15/15]	1.4799[20]	al, eth, ace, aa	B1[3], 2042
9811	Neurine or Trimethyl ethenyl ammonium hydroxide ..... $CH_2=CHN^+(CH_3)_3OH^-$	103.17	syr					w, al, eth	B4[4], 1053
9812	Nicotine (d) or α-N-Methyl-d-β-pyridylpyrrolidine ...... $C_{10}H_{14}N_2$	162.24	hyg $[\alpha]^{20}_D$ +163.2	245-6[729]		1.0094[20/4]	1.5280[20]	**w**, al, eth, chl, lig	B23, 117

No.	Name, Synonyms, and Formula	Mol. wt.	Color, crystalline form, specific rotation and λ_{max} (log ε)	b.p. °C	m.p. °C	Density	n_D	Solubility	Ref.
9813	Nicotine (dl) or Tetrahydronicotyrine............... $C_{10}H_{14}N_2$	162.24	....	242-3	.....	$1.0082^{20/4}$	1.5289^{20}	w, al, eth, chl, lig	B23[2], 111
9814	Nicotine (l) $C_{10}H_{14}N_2$	162.24	hyg (br in air) $[\alpha]^{20}_D$ −169	246.7^{745}, $124-5^{18}$	−79	$1.0097^{20/4}$	1.5282^{20}	w, al, eth, chl, lig	B23[2], 107
9815	Nicotine-hydrochloride (d) .... $C_{10}H_{14}N_2.HCl$	198.70	diq $[\alpha]^{20}_D$ + 104 (w, p=10)	....	.....	1.0337		w	B23, 114
9816	Nicotine amide or Niacin amide............ $3\text{-}C_5H_4NCONH_2$	122.13	wh pw nd (bz)	$150\text{-}60^{0.0005}$	129-31	1.400	1.466	w, al	B22[4], 389
9817	Nicotinamide-N,N-diethyl or Coramine........... $3\text{-}C_5H_4NCON(C_2H_5)_2$	178.24	yesh	280d, 175^{25}	24-6	$1.060^{25/4}$	1.525^{20}	w, al, eth, ace, chl	B22[4], 393
9818	Nicotinic acid or Niacin 3-Pyridine carboxylic acid...... $3\text{-}C_5H_4NCO_2H$	123.11	nd (w al)	sub	236-7	1.473	....	....	B22[4], 348
9819	Nicotinic acid-6-amino .... $6\text{-}N_2N\text{-}(C_5H_3N)CO_2H\text{-}3$	138.13	cr (dil aa + 2w) aa	....	312				B22[4], 6726
9820	Nicotinic acid-ethyl betaine .... $C_8H_9NO_2$	151.17	hyg pl	....	84-6			w	B22, 43
9821	Nicotinic acid-ethyl ester or Ethyl nicotinate .... $3\text{-}(C_5H_4N)CO_2C_2H_5$	151.17	....	224, $103\text{-}5^5$	8-9	1.1070^{20}	1.5024^{20}	w, al, eth, bz	B22[4], 357
9823	Nictinic acid-hydrochloride .... $3\text{-}(C_5H_4N)CO_2H.HCl$	159.57	pr or pl rh bipym(w)	....	274			w, al	B22[4], 354
9824	Nicotinic acid-2-hydroxy .... $2\text{-}HO(C_5H_3N)CO_2H\text{-}3$	139.11	nd (w)		$\alpha 259\text{-}61d$, $\beta 301\text{-}2d$				B22[4], 2139
9825	Nicotinic acid-4-hydroxy .... $4\text{-}HO(C_5H_3N)CO_2H\text{-}3$	139.11	nd (w + 2) cr (al)		254-5				B22[4], 2145
9826	Nicotinic acid, 6-hydroxy .... $6\text{-}HO\text{-}3(C_5H_3N)CO_2H\text{-}3$	139.11	nd (w)	sub	304d				B22[4], 2147
9827	Nicotinic acid, N-methyl or Trigoneiline .... $C_7H_7NO_2$	137.14	pr (aq al + 1w)		218d anh			w	B22[4], 462
9828	Nicotinic acid, methyl ester or Methyl nicotinate........ $3\text{-}(C_5H_4N)CO_2CH_3$	137.14	cr	204, 118.5^{23}	42-3			w, al, bz	B22[4], 356
9829	Nicotinonitrile .... $3\text{-}C_5H_4N)CN$	104.11	nd (lig peth-eth)	240-5	50-2			w, al, eth, bz	B22[4], 434
9830	2,2′-Nicotyrine (solid) or N-Methyl-2-(2-pyridyl)pyrrole . $C_{10}H_{10}N_2$	158.21	cr, br in air	....	43-4			al, eth, bz, dil HCl	B23[2], 192
9831	2,2′-Nicotyrine (liquid) .... $C_{10}H_{10}N_2$	158.21	....	273^{764}, $149\text{-}50^{22}$	−28			ace, bz	B23[2], 191
9832	3,2′-Nicotyrine or N-Methyl-2-(3-pyridyl)pyrrole .... $C_{10}H_{10}N_2$	158.21	br in air	$280\text{-}1^{744}$, 150^5	....	$1.2111^{20/4}$	1.6057^{20}	al, eth, ace	B23[2], 192
9833	Ninhydrin or 1,2,3-triketo hydrindene monohydrate ..... $C_9H_6O_4$	178.15	pr (w)	....	241-3d			w, al	B7[3], 4592
9834	Nitranilic acid or 2,5-Dihydroxy-3,6-dinitro-p-benzoqui-none .... $C_6H_2N_2O_8$	230.09	gold-ye pl (+ w dil HNO_3)	exp	100d			w, al	B8[3], 3351
9835	Nitro acetic acid .... $O_2NCH_2CO_2H$	105.05	nd (chl)	....	92-3d			al, eth, bz, chl	B2[4], 537
9836	Nitro acetic acid, ethyl ester or Ethyl nitro acetate ...... $O_2NCH_2CO_2C_2H_5$	133.11	....	$105\text{-}7^{25}$		$1.1953^{20/4}$	1.4250^{20}	al, eth	B2[4], 537
9837	2-Nitrobenzamide .... $2\text{-}NO_2C_6H_4CONH_2$	166.14	nd (dil al)	317	176-6			al, eth	B9[3], 1477
9838	2-Nitrobenzamide, N-phenyl or N-Phenyl-2-nitrobenza-mide .... $2\text{-}NO_2C_6H_4CONHC_6H_5$	242.24	nd (al bz)	....	155			al, bz, chl	B12[3], 506
9839	2-Nitrobenzoic acid .... $2\text{-}NO_2C_6H_4CO_2H$	167.12	tcl nd (w)	....	147-8	$1.575^{20/4}$		al, eth, ace	B9[3], 1466
9840	2-Nitrobenzoic acid, azide .... $2\text{-}NO_2C_6H_4CON_3$	192.14	ye pr (eth)	....	37.5			eth, bz, chl	B9[3], 1489
9841	2-Nitrobenzoic acid, 3-bromo .... $3\text{-}Br\text{-}2\text{-}NO_2C_6H_3CO_2H$	246.02	(eth)	....	250			eth, bz	B9[3], 1770
9842	2-Nitrobenzoic acid , 4-bromo .... $4\text{-}Br\text{-}2\text{-}NO_2C_6H_3CO_2H$	246.02	nd (w)	....	163			al, eth, bz, chl	B9[3], 1770
9843	2-Nitrobenzoic acid, 5-bromo .... $5\text{-}Br\text{-}2\text{-}NO_2C_6H_3CO_2H$	246.02	cr (w al bz)	....	140	1.920^{18}	....	al, bz	B9[1], 165
9844	2-Nitrobenzoic acid, 3-chloro .... $3\text{-}Cl\text{-}2\text{-}NO_2C_6H_3CO_2H$	201.57	nd or pl (w)	....	237-9	1.566^{18}	....	al, eth	B9, 400

No.	Name, Synonyms, and Formula	Mol. wt.	Color, crystalline form, specific rotation and λ_{max} (log ε)	b.p. °C	m.p. °C	Density	n_D	Solubility	Ref.
9845	2-Nitrobenzoic acid, 4-chloro 4-Cl-2-NO$_2$C$_6$H$_3$CO$_2$H	201.57	pl (bz-lig) pr (bz) nd (w)	....	142.3	....	....	al, eth	B9[3], 1763
9846	2-Nitrobenzoic acid, 5-chloro, methyl ester 5-Cl-2-NO$_2$C$_6$H$_3$CO$_2$CH$_3$	215.59	pl (MeOH)	....	48.5	1.453[18]	....	MeOH	B9, 401
9847	2-Nitrobenzoic acid, ethyl ester or Ethyl 2-nitrobenzoate 2-NO$_2$C$_6$H$_4$CO$_2$C$_2$H$_5$	195.18	tcl (dil al)	275, 173[18]	30	....	....	al, eth	B9[3], 1469
9848	2-Nitrobenzoic acid, hydrazide 2-NO$_2$C$_6$H$_4$CONHNH$_2$	181.15	ye-br pr (w)	....	123	....	....	al	B9[3], 1481
9849	2-Nitrobenzoic acid, 4-iodo 4-I-2-NO$_2$C$_6$H$_3$CO$_2$H	293.02	ye lf or pr (dil al)	....	192-3	....	....	al, eth	B9[2], 278
9850	2-Nitrobenzoic acid, methyl ester 2-NO$_2$C$_6$H$_4$CO$_2$CH$_3$	181.15	....	275, 176[21]	−13	1.2855[20]	....	al, eth, bz, chl	B9[3], 1469
9851	2-Nitrobenzonitrile 2-NO$_2$C$_6$H$_4$CN	148.12	nd (w, aa)	sub	111	....	....	al, eth, ace, bz, aa	B9[3], 1479
9852	2-Nitrobenzonitrile, 4-chloro 4-Cl-2-NO$_2$C$_6$H$_3$CN	182.57	nd (w)	....	100-1	....	....	al, eth	B9[3], 1763
9853	3-Nitrobenzamide 3-NO$_2$C$_6$H$_4$CONH$_2$	166.14	ye mcl nd (w)	310-5	142.7	....	....	al, eth	B9[3], 1515
9854	3-N-trobenzamide, N-phenyl or N-Phenyl-3-nitrobenza-mide 3-NO$_2$C$_6$H$_4$CONHC$_6$H$_5$	242.24	lf (w, al)	sub	153-4	....	....	al, eth, bz	B12[3], 506
9855	3-Nitrobenzoic acid 3-NO$_2$C$_6$H$_4$CO$_2$H	167.12	mcl pr (w)	....	140-2	1.494[20/4]	....	al, eth, ace	B9[3], 1489
9856	3-Nitrobenzoic acid, azide 3-NO$_2$C$_6$H$_4$CON$_3$	192.14	pl (dil al)	....	68	....	....	al, eth, bz, aa	B9[3], 1536
9857	3-Nitrobenzoic acid, 2-bromo 2-Br-3-NO$_2$C$_6$H$_3$CO$_2$H	246.02	(dil al)	....	191	....	....	al	B9[2], 277
9858	3-Nitrobenzoic acid, 4-bromo 4-Br-3-NO$_2$C$_6$H$_3$CO$_2$H	246.02	nd (dil aa)	sub	203-4	....	....	al	B9[3], 1770
9859	3-Nitrobenzoic acid, 5-bromo 5-Br-3-NO$_2$C$_6$H$_3$CO$_2$H	246.02	nd (w, bz eth), pl (al)	....	159-60	....	....	al, eth, bz, aa, chl	B9[3], 1771
9860	3-Nitrobenzoic acid, 6-bromo 6-Br-3-NO$_2$C$_6$H$_3$CO$_2$H	246.02	nd (w)	sub	180-1	....	....	al, eth, chl	B9[3], 1771
9861	3-Nitrobenzoic acid, 2-chloro 2-Cl-3-NO$_2$C$_6$H$_3$CO$_2$H	201.57	....	....	185	1.662	....	al	B9[2], 275
9862	3-Nitrobenzoic acid, 4-chloro 4-Cl-3-NO$_2$C$_6$H$_3$CO$_2$H	201.57	nd or pl (w)	....	181-2	1.645[18]	....	....	B9[3], 1764
9863	3-Nitrobenzoic acid, 4-chloro, ethyl ester 4-Cl-3-NO$_2$C$_6$H$_3$CO$_2$C$_2$H$_5$	229.63	ye nd (al)	....	59	....	....	al, bz, aa	B9, 402
9864	3-Nitrobenzoic acid, 4-chloro, methyl ester 4-Cl-3-NO$_2$C$_6$H$_3$CO$_2$CH$_3$	215.59	nd (MeOH)	....	83	1.522[18]	....	al	B9, 402
9865	3-Nitrobenzoic acid, 5-chloro 5-Cl-3-NO$_2$C$_6$H$_3$CO$_2$H	201.57	nd (w)	....	147	....	....	al, eth, aa	B9[1], 165
9866	3-Nitrobenzoic acid, 6-chloro 6-Cl-3-NO$_2$C$_6$H$_3$CO$_2$H	201.57	nd or pr (w)	....	165	1.608[18]	....	al, eth, bz	B9[3], 1765
9867	3-Nitrobenzoic acid, ethyl ester or Ethyl-3-nitrobenzoate 3-NO$_2$C$_6$H$_4$CO$_2$C$_2$H$_5$	195.18	mcl pr	296-8, 156[10]	47	....	....	al, eth	B9[3], 1493
9868	3-Nitrobenzoic acid hydrazide 3-NO$_2$C$_6$H$_4$CONHNH$_2$	181.15	nd (w)	....	153-4	....	....	....	B9[3], 1524
9869	3-Nitrobenzoic acid, 2-iodo 2-I-3-NO$_2$C$_6$H$_3$CO$_2$H	293.02	pr (w)	....	206	....	....	al, eth	B9[2], 278
9870	3-Nitrobenzoic acid, 4-iodo 4-I-3-NO$_2$C$_6$H$_3$CO$_2$H	293.02	ye pr (al)	....	213	....	....	al	B9[2], 278
9871	3-Nitrobenzoic acid, 5-iodo 5-I-3-NO$_2$C$_6$H$_3$CO$_2$H	293.02	nd (al), pr (peth)	....	167	....	....	al	B9[2], 278
9872	3-Nitrobenzoic acid, methyl ester 3-NO$_2$C$_6$H$_4$CO$_2$CH$_3$	181.15	nd	....	78	....	....	....	B9[3], 1493
9873	3-Nitrobenzonitrile 3-NO$_2$C$_6$H$_4$CN	148.12	nd (w)	sub	118	....	....	al, eth, ace, aa	B9[3], 1521
9874	3-Nitrobenzoyl chloride 3-NO$_2$C$_6$H$_4$COCl	185.57	ye cr	275-8, 154-5[18]	35	....	....	eth	B9[3], 1514
9875	4-Nitrobenzamide 4-NO$_2$C$_6$H$_4$CONH$_2$	166.14	nd (w)	....	201-4	....	....	al, eth	B9[3], 1710

No.	Name, Synonyms, and Formula	Mol. wt.	Color, crystalline form, specific rotation and λ_{max} (log ε)	b.p. °C	m.p. °C	Density	n_D	Solubility	Ref.
9876	4-Nitrobenzamide, N-phenyl or N-Phenyl-4-nitrobenzamide .... 4-NO$_2$C$_6$H$_4$CONHC$_6$H$_5$	242.24	lf (eth)	....	211	....	....	al, eth	B12^3, 506
9877	4-Nitrobenzoic acid .... 4-NO$_2$C$_6$H$_4$CO$_2$H	167.12	mcl lf (w)	sub	242	1.610^{20}	....	al, eth, chl	B9^3, 1537
9878	4-Nitrobenzoic acid, 2-bromo .... 2-Br-4-NO$_2$C$_6$H$_3$CO$_2$H	246.02	nd (w, dil al)	sub >155	166-7			al, eth	B9^3, 1771
9879	4-Nitrobenzoic acid, 3-bromo .... 3-Br-4-NO$_2$C$_6$H$_3$CO$_2$H	246.02	nd (dil al)	....	197			al, eth, chl	B9, 408
9880	4-Nitrobenzoic acid, butyl ester or Butyl-4-nitrobenzoate 4-NO$_2$C$_6$H$_4$CO$_2$C$_4$H$_9$	223.23	nd	160^8	35.3			eth, bz	B9^3, 1544
9881	4-Nitrobenzoic acid, 2-chloro .... 2-Cl-4-NO$_2$C$_6$H$_3$CO$_2$H	201.57	nd (w)	....	140-2			al, eth	B9^3, 1766
9882	4-Nitrobenzoic acid, ethyl ester ...... 4-NO$_2$C$_6$H$_4$CO$_2$C$_2$H$_5$	195.18	tcl lf (al)	186.3	57			al, eth	B9^3, 1541
9883	4-Nitrobenzoic acid, hydrazide ...... 4-NO$_2$C$_6$H$_4$CONHNH$_2$	181.15	yesh nd (w)		214				B9^3, 1751
9884	4-Nitrobenzoic acid, 2-iodo .... 2-I-4-NO$_2$C$_6$H$_3$CO$_2$H	293.02	pa ye pr (w)		146-7			al, eth	B9^2, 278
9885	4-Nitrobenzoic acid, methyl ester ...... 4-NO$_2$C$_6$H$_4$CO$_2$CH$_3$	181.15	ye mcl lf		96			al, eth, chl	B9^3, 1541
9886	4-Nitrobenzoic acid, 2,2,2-trichloroethyl ester ...... 4-NO$_2$C$_6$H$_4$CO$_2$CH$_2$C(Cl)$_3$	253.51	pr (al)	106-7^1	71	1.5343^{26}			B9^3, 1542
9887	4-Nitrobenzonitrile .... 4-NO$_2$C$_6$H$_4$CN	148.12	lf (al), nd (bz)	sub	149			chl	B9^3, 1748
9888	4-Nitrobenzoyl chloride .... 4-NO$_2$C$_6$H$_4$COCl	185.57	ye nd (lig)	202-5^{105}, 150-2^{15}	75			eth	B9^3, 1709
9889	2-Nitrophenyl isocyanate .... 2-O$_2$NC$_6$H$_4$NCO	164.12	wh nd (peth)	135^7	41			eth, bz, chl	B12^3, 1535
9890	3-Nitrophenyl isocyanate .... 3-O$_2$NC$_6$H$_4$NCO	164.12	wh lf (lig)	130-1^{11}	51			eth, bz, chl	B12^3, 1573
9891	4-Nitrophenyl isocyanate .... 4-O$_2$NC$_6$H$_4$NCO	164.12	pa ye nd	137-8^{11}	57			eth, bz, chl	B12^3, 1630
9892	bis-(2-nitrophenyl)trisulfide .... (2-NO$_2$C$_6$H$_4$)$_2$S$_3$	340.31	ye nd (al)	....	175-6				B6^3, 1062
9893	2-Nonanol (d) .... CH$_3$CH(OH)C$_7$H$_{15}$	144.26	[α]^{20}/$_D$	105^{19}		0.8230$^{20/4}$	1.4299^{20}	al, eth	B1^1, 211
9894	2-Nonanol (dl) .... CH$_3$CH(OH)C$_7$H$_{15}$	144.26	....	193-4, 91^{12}	....	0.84708$^{20/4}$	1.43533^{20}	al, eth	B1^4, 1803
9895	3-Nonanol (s) .... C$_2$H$_5$CH(OH)C$_6$H$_{13}$	144.26	[α]^{25}/$_D$ 7.08	97^{17}		0.8281$^{17/4}$	1.4308^{20}	al, eth	B1^3, 1746
9896	3-Nonanol (dl) .... C$_2$H$_5$CH(OH)C$_6$H$_{13}$	144.26	....	195^{750}, 93^{18}	-22	0.8250$^{20/4}$	1.4289^{20}	al, eth	B1^4, 1803
9897	4-Nonanol (s) .... C$_3$H$_7$CH(OH)C$_5$H$_{11}$	144.26	[α]^{25}/$_D$ 0.57 eth 0.9%	192-3, 94-5^{18}		0.8282$^{20/4}$	1.41971^{20}	al, eth	B1^3, 1747
9898	5-Nonanol .... (C$_4$H$_9$)$_2$CHOH	144.26	....	193^{750}, 97^{20}		0.8356$^{20/4}$	1.4289^{20}	al	B1^4, 1803
9899	Nitron or 4,5-Dihydro-1,4-diphenyl-3,5-phenylamino-1,2,4-triazole .... C$_{19}$H$_{16}$N$_4$	312.38	ye lf (al), nd (+ chl)	189d				al, ace, bz, chl	B26^2, 76,199
9900	Nonacosane .... CH$_3$(CH$_2$)$_{27}$CH$_3$	408.80	rh cr (peth)	440.8, 271.4^{10}	63.7	0.8083$^{20/4}$, 0.7630$^{100/4}$	1.4529^{20}	al, eth, ace, bz	B1^4, 591
9901	Nonacosane, 2-methyl .... (CH$_3$)$_2$CH(CH$_2$)$_{26}$CH$_3$	422.83	pl (lig)	222$^{0.3}$	73-4			eth, ace, bz	B1^1, 72
9902	1-Nonacosanol .... CH$_3$(CH$_2$)$_{27}$CH$_2$OH	424.80	cr (al)	sub 200-50^1	84-5				B1^4, 1916
9904	Nonadecane .... CH$_3$(CH$_2$)$_{17}$CH$_3$	268.53	wax	329.7, 193^{15}	32.1	0.7855$^{20/4}$	1.4409^{20}	eth, ace	B1^4, 560
9905	1,2,3-Nonadecane tricarboxylic acid, 2-hydroxy or Agaric acid, Laricic acid .... CH$_3$(CH$_2$)$_{15}$CH(CO$_2$H)C(OH)(CO$_2$H)CH$_2$CO$_2$H	416.56	lf (+ ^{3}/$_2$ w, dil al), [α]^{19}/$_D$ -8.8 (NaOH)	....	142d				B3^4, 1284
9906	1,2,3-Nonadecane tricarboxylic acid, 2-hydroxy,triethyl ester .... C$_{28}$H$_{52}$O$_7$	500.73	nd	36-7				al, bz	B3^2, 373
9907	1,2,3-Nonadecane tricarboxylic acid, 2-hydroxy,trimethyl ester .... C$_{25}$H$_{46}$O$_7$	458.64	nd (al)	....	63-4			bz	B3^2, 373

No.	Name, Synonyms, and Formula	Mol. wt.	Color, crystalline form, specific rotation and λ_{max} (log ε)	b.p. °C	m.p. °C	Density	n_D	Solubility	Ref.
9908	Nonadecanoic acid or n-Nonadecylic acid............ $C_{18}H_{37}CO_2H$	298.52	lf (al)	297-8[100], 227-30[10]	69.4	....	....	al, eth, bz, chl, lig	B2[4], 1256
9909	1-Nonadecanol .... $C_{18}H_{37}CH_2OH$	284.53	cr (ace)	166-7[0.32]	62-3	....	1.4328[75]	eth, ace	B1[4], 1898
9910	2-Nonadecanone or Methyl-n-heptadecyl ketone........ $C_{17}H_{35}COCH_3$	282.52	pr (al)	266.5[110], 165[2]	57	0.8108[56]	....	eth, ace, chl	B1[4], 3400
9911	2-Nonadecanone oxime .... $C_{17}H_{35}C(=NOH)CH_3$	297.55	cr (al)	....	76-7	....	....	al	B1, 718
9912	4-Nonadecanone or Propyl pentadecyl ketone.......... $C_{15}H_{31}COC_3H_7$	282.52	lf (al)	211[11]d	50.5	....	....	eth, ace	B1[4], 3401
9913	10-Nonadecanone or Caprinone-dinonyl ketone........ $(C_9H_{19})_2CO$	282.52	lf (al)	>350, 155.6[11]	65.5	....	....	eth, ace, bz, chl, lig	B1[4], 3402
9914	1-Nonadecyne .... $C_{17}H_{35}C\equiv CH$	264.50	....	327, 181.6[10]	37-8	0.8054[20/4]	1.4488[20]	eth, ace, bz	B1[3], 1030
9915	1,8-Nonadiyne .... $HC\equiv C(CH_2)_5C\equiv CH$	120.20	....	162, 55[13]	−27.3	0.8158[20/4]	1.4490[20]	eth, ace	B1[4], 1125
9916	Nonanal................................... $C_8H_{17}CHO$	142.24	....	190-2, 93.5[23]	......	0.8264[22/4]	1.4273[20]	eth	B1[4], 3352
9917	Nonanal oxime .... $C_8H_{17}CH=NOH$	157.26	lf (dil al)	....	64	....	....	al, eth, ace	B1[2], 761
9918	Nonane .... $CH_3(CH_2)_7CH_3$	128.26	....	150.8, 39[10]	−51	0.7176[20/4]	1.4054[20]	al, eth, ace, bz, chl	B1[4], 447
9919	Nonane, 1-amino or Nonyl amine .... $C_8H_{17}CH_2NH_2$	143.28	....	202.2, 80.8[10]	−1	0.7886[20/4]	1.4336[20]	al, eth	B4[4], 777
9920	Nonane, 1-bromo .... $C_9H_{19}Br$	207.15	....	88[4]	......	1.0183[20/20]	1.4533[20]	....	B1[4], 451
9921	Nonane, 5-butyl .... $(C_4H_9)_2CH$	184.36	....	217-8	......	0.7635[18.5/4]	1.4273[18.5]	....	B1[4], 517
9922	Nonane, 1-chloro or Nonyl chloride .... $CH_3(CH_2)_7CH_2Cl$	162.71	....	203.4, 80.5[10]	−39.4	0.8720[20/4]	1.4345[20]	eth, chl	B1[4], 450
9923	Nonane, 1-chloro-9-fluoro .... $F(CH_2)_9Cl$	180.70	....	102[11]	......	0.966[20/4]	1.4301[25]	al, eth	B1[4], 450
9924	Nonane, 2-chloro .... $CH_3(CH_2)_6CHClCH_3$	162.71	....	190[764]	......	0.8790[20]	1.4420[20]	chl	B1, 166
9925	Nonane, 5-chloro .... $C_4H_9CHClC_4H_9$	162.71	....	85-7[14]	......	0.8639[15/4]	1.4314[15]	eth	B1[2], 128
9926	Nonane, 2-methyl .... $CH_3(CH_2)_6CH(CH_3)_2$	142.28	....	166.8	−74.3	0.7281[20/4]	1.4099[20]	eth, bz, chl	B1[4], 473
9927	Nonane, 3-methyl (dl) .... $C_2H_5CH(CH_3)(CH_2)_5CH_3$	142.28	....	167.8	−84.6	0.7354[20/4]	1.4125[20]	eth, bz, chl	B1[4], 474
9928	Nonane, 4-methyl .... $C_3H_7CH(CH_3)(CH_2)_4CH_3$	142.28	....	165.7	−101.6	0.7323[20/4]	1.4123[20]	eth, bz, chl	B1[4], 475
9929	Nonane, 5-methyl .... $CH_3CH[(CH_2)_3CH_3]_2$	142.28	....	165.1	−86.5	0.7326[20/4]	1.4116[20]	eth, bz, chl	B1[4], 475
9930	Nonanedioic acid or Azelaic acid................. $HO_2C(CH_2)_7CO_2H$	188.23	lf or nd	>360d, 287[100], 225[10]	106.5	1.225[25/4]	1.4303[111]	al	B2[4], 2055
9931	1,9-Nonanediol or Nonamethylene glycol .... $HOCH_2(CH_2)_7CH_2OH$	160.26	cr (bz)	173-5[20]	45.8	....	....	al, eth, bz	B1[4], 2607
9932	Nonanoic acid or Pelargonic acid .... $CH_3(CH_2)_7CO_2H$	158.24	....	255, 150[20]	15	0.9057[20/4]	1.4343[19]	al, eth, chl	B2[4], 1018
9933	1-Nonanol or Nonyl alcohol.................... $CH_3(CH_2)_7CH_2OH$	144.26	....	213.5, 118[15]	−5.5	0.8273[20/4]	1.4333[20]	al, eth	B1[4], 1798
9934	5-Nonanol, 5-butyl or Tri-n-butyl carbinol .... $(C_4H_9)_3COH$	200.37	....	230-5d, 118-20[17]	20	0.8408[20/4]	1.4445[20]	al	B1[4], 1863
9935	1-Nonanol, 9-chloro .... $Cl(CH_2)_8CH_2OH$	178.71	....	146-8[14]	28	....	1.4575[20]	al, eth	B1[4], 1802
9936	1-Nonanol, 9-fluoro .... $F(CH_2)_8CH_2OH$	162.25	....	125-6[15]	......	0.928[20/4]	1.4279[25]	al, eth	B1[4], 1801
9937	2-Nonanone or Heptyl methyl ketone .... $C_7H_{15}COCH_3$	142.24	....	195.3, 73.8[10]	−7.5	0.8208[20/4]	1.4210[20]	al, eth, ace, bz, chl	B1[4], 3353
9938	4-Nonanone or Propyl pentyl ketone .... $C_5H_{11}COC_3H_7$	142.24	....	187-8, 75-6[20]	......	0.8190[25/4]	1.4189[20]	al, eth, ace, chl	B1[4], 3354
9939	5-Nonanone or Dibutyl ketone.................... $(C_4H_9)_2CO$	142.24	....	188.4, 88[22]	−4.8	0.8217[20/4]	1.4195[20]	al, eth, chl	B1[4], 3355
9940	Nonasiloxane, eicosa methyl .... $CH_3[Si(CH_3)_2O]_8Si(CH_3)_3$	681.47	....	307.5, 198.8[16]	......	0.9173[20]	1.3980[20]	bz	B4[3], 1881

No.	Name, Synonyms, and Formula	Mol. wt.	Color, crystalline form, specific rotation and λ_{max} (log ϵ)	b.p. °C	m.p. °C	Density	n_D	Solubility	Ref.
9941	1,3,6,8-Nonatetraen-5-one, 1,9-diphenyl or Dicinnamylidene acetone .... $(C_6H_5CH=CHCH=CH)_2CO$	286.38	ye nd (abs al)	....	144			....	B7[3], 2756
9942	1-Nonene $C_7H_{15}CH=CH_2$	126.24	....	146		0.730[21]	1.414[21]	....	B1[4], 894
9943	3-Nonene (trans) $C_2H_5CH=CHC_5H_{11}$	126.24		147-8		0.732[21/4]	1.4181[21]	eth, bz, chl	B1[4], 895
9944	3-Nonene, 2-methyl $CH_3CH(CH_3)CH=CHC_5H_{11}$	140.27		161.0		0.7340[20/4]	1.4202[20]	....	B1[4], 903
9945	4-Nonene (trans) $C_3H_7CH=CHC_4H_9$	126.24		....		0.7318[20/4]	1.4205[20]	eth, bz, chl	B1[4], 896
9946	4-Nonene, 5-butyl $C_3H_7CH=C(C_4H_9)_2$	182.35		215-6		0.7745[20/4]	1.4375[20]	....	B1[4], 922
9947	1-Nonene-3-yne $C_5H_{11}C\equiv CCH=CH_2$	122.21		27-8[4]		0.7602[25/4]	1.4487[25]	eth, ace	B1[4], 1103
9948	1-Nonene-4-yne $C_4H_9C\equiv C-CH_2CH=CH_2$	122.21		58[22]		0.777[25/4]	1.4413[25]	eth, ace	B1[4], 1103
9949	2-Nonene-4-yne $C_4H_9C\equiv C-CH=CHCH_3$	122.21		70[20]		0.7832[25/4]	1.4590[25]	eth, ace	B1[3], 1048
9950	1-Nonyne $C_7H_{15}C\equiv CH$	124.23		150.8, 33.3[10]	−50	0.7568[20/4]	1.4217[20]	eth, bz	B1[4], 1047
9951	1-Nonyne, 1-chloro $C_7H_{15}C\equiv CCl$	158.67		75-7[15]		0.906[20]	1.450[20]	eth	B1[3], 1012
9952	2-Nonyne $C_6H_{13}C\equiv C-CH_3$	124.23		158-9		0.7690[20/4]	1.4337[20]	eth, lig	B1[3], 1013
9953	3-Nonyne $C_5H_{11}C\equiv CC_2H_5$	124.23		153-5[740], 92[97]		0.7616[20/4]	1.4299[20]	eth, lig	B1[3], 1013
9954	4-Nonyne $C_4H_9C\equiv CC_3H_7$	124.23		150-4[750]		0.757[25/4]	1.4296[25]	eth, ace	B1[4], 1047
9955	4-Nonyne, 3,3-dimethyl $C_4H_9C\equiv C-C(CH_3)_2C_2H_5$	152.28		82[40]		0.7667[20/4]	1.4317[20]	eth, bz	B1[3], 1024
9956	4-Nonyne, 8-methyl $(CH_3)_2CHCH_2CH_2C\equiv CC_3H_7$	138.25		104.5[97]		0.7681[20/4]	1.4311[20]	eth, bz	B1[3], 1018
9957	n-Nonyl amine or 1-Amino nonane .... $C_8H_{17}CH_2NH_2$	143.18		202.2, 80.8[10]	−1	0.7886[20/4]	1.4336[20]	al, eth	B4[4], 777
9958	Nopinone (d) .... $C_9H_{14}O$	138.21	$[\alpha]^{20/}_D$ + 34 (chl)	209, 87-8[14]	0	0.9807[20/4]	1.4787[20]	w, al, eth	B7[3], 303
9959	Noradrenaline-(l) or 1-Norepinephrine $C_8H_{11}NO_3$	169.18	$[\alpha]^{25/}_D$ −37.5 (dil HCl)	....	216-8d	....		dil HCl	B13[3], 2382
9960	Norbornane or Norcamphane, bicyclo [2,2,1] heptane.... C_7H_{12}	96.17	....	sub	87-8	....		al, eth, ace, bz	B5[4], 258
9961	Norbornyl amine or Isobornyl amine $C_{10}H_{17}NH_2$	153.27	pw $[\alpha]'_D$ −47.7 (4% al)	....	184			eth, ace	B12[3], 195
9962	Norcamphane-2-carboxaldehyde .... $C_8H_{12}O$	122.17	cr (w)	70-2[22]	......	1.0227[19/4]	1.4760[25]	eth	B7[3], 545
9963	2,3-Norcomphane dicarboxylic acid or Bicyclo [2,2,1]-heptane dicarboxylic acid.......... $C_9H_{12}O_4$	184.19	....		192-5				B9[3], 3971
9964	2,3-Norcamphane dicarboxylic anhydride .... $C_9H_{10}O_3$	166.18	....		165-7			bz	B17[4], 6005
9965	Nordihydroguaiaretic acid or NDGA .... $C_{18}H_{22}O_4$	302.38	nd (w, al, aa)		185-6			al, eth, ace	B6[3], 6731
9966	Norephedrine hydrochloride-(dl) .... $C_6H_5CH(OH)CH(CH_3)NH_2 \cdot HCl$	187.67	pl (abs al), cr (dil HCl, al)	....	194			w, al	B13[2], 371
9967	Norephedrine, N,N-diethyl, hydrochloride.......... $C_6H_5CH(OH)CH(CH_3)N(C_2H_5)_2 \cdot HCl$	243.78	cr (al-ace)		205.6			al	B13[2], 380
9968	Norephedrine, N-ethyl $C_6H_5CH(OH)CH(CH_3)NHC_2H_5$	179.26	cr (lig)	143[18]	51.5			bz	B9[3], 1728
9969	Normorphine or Desmethylmorphine.......... $C_{16}H_{17}NO_3$	271.32	(w + 3/2)	....	273 (+ 3/2 w), 263-4 (anh)			....	B27[2], 117

No.	Name, Synonyms, and Formula	Mol. wt.	Color, crystalline form, specific rotation and λ_{max} (log ε)	b.p. °C	m.p. °C	Density	n_D	Solubility	Ref.
9970	Nornicotine-(l) or *l*−3-(2-Pyrrolidyl) pyridine.......... $C_9H_{12}N_2$	148.21	hyg $[\alpha]^{22}_D$ −88.8 (undil)	270, 130-1[11]	....	1.0737[19.5]	1.5378[18.5]	w, al, eth, ace, chl	B23[2], 107
9971	Ocimene or 3,7-Dimethyl-1,3,7-octatriene............. $CH_2=C(CH_3)CH_2CH_2CH=C(CH_3)CH=CH_2$	136.24	....	176-8d, 73-4[21]	....	0.8000[20]	1.4862[20]	al, eth, bz, chl, aa	B1[4], 1108
9972	Ocimene, dihydro or 2,6-Dimethyl-2,6-octadiene........ $CH_3CH=C(CH_3)CH_2CH_2CH=C(CH_3)_2$	138.25	....	168, 75[30]	....	0.775[20/4]	1.4498[20]	al, eth, ace, aa	B1[4], 1058
9973	Octacosane .... $CH_3(CH_2)_{26}CH_3$	394.78	mcl or rh (bz-al)	431.6, 264[10]	64.5	0.8067[20/4], 0.7750[70]	1.4520[20], 1.4330[70]	ace, bz, chl	B1[4], 588
9974	Octacosanoic acid .... $CH_3(CH_2)_{26}CO_2H$	424.76	(ace or aa)	....	90.4	0.8191[100]	1.4313[100]	....	B2[4], 1318
9975	1-Octacosanol .... $CH_3(CH_2)_{26}CH_2OH$	410.78	(ace or peth)	sub, 200-50[1]	83.3	....	....	....	B1[4], 1915
9976	9,12-Octadecadienoic acid or -Linoleic acid *(cis,cis)*...... $CH_3(CH_2)_4CH=CHCH_2CH=CH(CH_2)_7CO_2H$	280.46	....	229.30[16]	−5	0.9022[20/4]	1.4699[20]	al, eth, ace, bz, chl	B2[4], 1754
9977	10,12-Octadecadienoic acid *(trans,trans)* or 10,12-Linoleic acid .... $CH_3(CH_2)_4CH=CH-CH=CH(CH_2)_8CO_2H$	280.45	(bz or al)	....	56-7	0.8686[70/4]	1.4689[60]	....	B2[4], 1752
9978	7,11-Octadecadiyne .... $CH_3(CH_2)_5C≡CCH_2CH_2C≡C(CH_2)_5CH_3$	246.44	....	167-8[7]	....	0.841[19/4]	1.4698[19]	....	B1[3], 1068
9979	Octadecanal or Stearaldehyde .... $CH_3(CH_2)_{16}CHO$	268.49	nd (peth)	261, 212-13[12]	55	....	....	....	B1[4], 3397
9980	Octadecanal, dimethyl acetal or 1,1-Dimethyoxy octadecane .... $CH_3(CH_2)_{17}CH(OCH_3)_2$	314.56	....	167-70[3]	....	....	1.4410[25]	al, eth	B1[4], 3397
9981	Octadecane .... $CH_3(CH_2)_{16}CH_3$	254.51	nd (al, eth-MeOH)	316.1, 173.5[20]	28.2	0.7768[20/4]	1.4390[20]	eth, ace, lig	B1[4], 553
9982	Octadecane, 1-amino or Stearyl amine .... $C_{18}H_{37}NH_2$	269.52	(w)	348.8, 199.5[10]	tr 52.9	0.8618[20/4]	1.4522[30]	al, eth, bz, chl	B4[4], 825
9983	Octadecane, 1-amino, hydro chloride .... $C_{18}H_{37}NH_2 \cdot HCl$	305.98	orh pl (al)	....	162-3	....	....	....	B4[4], 826
9984	Octadecane, 1-bromo .... $CH_3(CH_2)_{16}CH_2Br$	333.41	cr (al)	210[10]	28.2	0.9848[20/4]	1.4631[20]	al, eth, peth	B1[4], 555
9985	Octadecane, 1-chloro .... $CH_3(CH_2)_{16}CH_2Cl$	288.95	....	348, 199[10]	28.6	0.8641[20]	1.4531[20]	....	B1[4], 554
9986	Octadecane, 1,18-dibromo .... $Br(CH_2)_{18}Br$	412.31	nd or lf (al)	205-7[15]	64	....	....	chl	B1[4], 556
9987	Octadecane, 1-iodo .... $CH_3(CH_2)_{16}CH_2I$	380.40	lf (lig), nd (ace, al-ace)	383, 223[10]	34	1.0994[20/4]	1.4810[20]	....	B1[4], 556
9988	Octadecane, 9-(4-tolyl) .... $C_8H_{17}CH(4-C_6H_4CH_3)C_9H_{19}$	344.62	....	185[10]	....	0.8549[20/4]	1.4811[20]	....	B5[4], 1221
9989	Octadecandioic acid, diethyl ester or Diethyl eicosanedioate .... $C_2H_5O_2C(CH_2)_{16}CO_2C_2H_5$	370.58	....	240[12]	54-5	....	....	al, eth	B2[4], 2176
9990	1,18-Octadecanediol .... $HO(CH_2)_{18}OH$	286.50	lf (al, bz), nd (bz, diox)	210-1[2]	97-9	....	....	....	B1[4], 2639
9991	1-Octadecanethiol or *n*-Octadecyl mercaptan .... $CH_3(CH_2)_{16}CH_2SH$	286.57	....	188[1-2]	24-8	0.8475[20]	1.4645[20]	eth	B1[4], 1894
9992	Octadecanoic acid or Stearic acid .... $CH_3(CH_2)_{16}CO_2H$	284.50	mcl lf (al)	360d, 232[15]	71-2	0.9408[20/4]	1.4299[80]	eth, ace, chl	B1[4], 1206
9993	1-Octadecanol or Stearyl alcohol.......... $CH_3(CH_2)_{16}CH_2OH$	270.50	lf (al)	210.5[15]	59-60	0.8124[59/4]	....	al, eth, chl	B1[4], 1888
9994	3-Octadecanone or Ethyl pentadecyl ketone.......... $C_{15}H_{31}COC_2H_5$	268.48	....	....	48-50	....	....	....	B1[4], 3398
9995	9,11,13-Octadecatrienoic acid *(cis)* or α-Eleostearic acid . $C_4H_9[CH=CH]_3(CH_2)_7CO_2H$	278.44	nd (al)	235[12_d], 170[1]	49	0.9028[50/4]	1.5112[50]	al, eth	B2[4], 1787
9996	9,12-15-Octadecatrienoic acid *(cis,cis,cis)* or α-Linolenic acid .... $CH_3[CH_2CH=CH]_3(CH_2)_7CO_2H$	278.44	....	230-2[17], 125[0.05]	−11.3	0.9164[20/4]	1.4800[20]	al, eth	B2[4], 1781
9997	9-Octadecenal or Olealdehyde .... $CH_3(CH_2)_7CH=CH(CH_2)_7CHO$	266.47	ye nd	168-9[3]	....	0.8509[20/4]	1.4558[20]	....	B1[4], 3533
9998	1-Octadecene .... $C_{16}H_{33}CH=CH_2$	252.49	....	179[15], 145[8]	17.5	0.7891[20/4]	1.4448[20]	bz	B1[4], 930

PHYSICAL CONSTANTS OF ORGANIC COMPOUNDS (Continued)

No.	Name, Synonyms, and Formula	Mol. wt.	Color, crystalline form, specific rotation and λ_{max} (log ε)	b.p. °C	m.p. °C	Density	n_D	Solubility	Ref.
9999	9-Octadecene $C_8H_{17}CH=CHC_8H_{17}$	252.49	....	162[9]	−30.5	0.7916[20/4]	1.4470[20]	....	B1[4], 932
10000	9-Octadecenoic acid (trans) or Elaidic acid............. $CH_3(CH_2)_7CH=CH(CH_2)_7CO_2H$	282.47	pl (al)	288[100], 234[15]	45	0.8734[41]	1.4499[45]	al, eth, bz, chl	B2[4], 1647
10001	6-Octadecenoic acid, 6,7-diiodo $CH_3(CH_2)_{10}CI=CI(CH_2)_4CO_2H$	534.77	nd (al)	d	48.5	....	....	eth, bz, chl	B2[4], 1638
10002	9-Octadecenoic acid (cis), 12-hydroxy or Ricinoleic acid.. $C_6H_{13}CH(OH)CH=CH(CH_2)_8CO_2H$	298.47	$[\alpha]^{22}_D$ + 5.05	226-8[10]	α:7.7 β:16 γ:5.5	0.9450[21/4]	1.4716[21]	al, eth	B3[4], 1026
10003	11-Octadecenoic acid (trans) or Vaccenic acid .... $C_6H_{13}CH=CH(CH_2)_9CO_2H$	282.47	....	....	44	....	1.4439[60]	ace	B2[4], 1640
10004	9-Octadecene, 1-ol (cis) or Oleylalcohol .... $CH_3(CH_2)_7CH=CH(CH_2)_7CH_2OH$	268.49	....	205-10[15]	6-7	0.8489[20/4]	1.4606[20]	al, eth	B1[4], 2204
10005	9-Octadecenl-1-ol (trans) or Elaidyl alcohol........... $CH_3(CH_2)_7CH=CH(CH_2)_7CH_2OH$	268.49	(al or ace)	198[10]	36-7	0.8338[40/4]	1.4552[40]	al, eth	B1[4], 2204
10006	Octadecyl sulfate $(C_{18}H_{37}O)_2SO_2$	603.06	....	....	70.5	....	....	....	B1[4], 1892
10007	1-Octadecyne $C_{16}H_{33}C\equiv CH$	250.47	(al)	313, 180[15]	22.5	0.8025[20/4]	1.4774[20]	....	B1[4], 1075
10008	2-Octadecyne $C_{15}H_{31}C\equiv CCH_3$	250.47	(al)	184[15]	30	0.8016[30/4]	....	....	B1[4], 1075
10009	9-Octadecyne $C_8H_{17}C\equiv CC_8H_{17}$	250.47	....	163-4[7]	3	0.8012[20/4]	1.4488[25]	....	B1[4], 1076
10010	9-Octadecynoic acid or Stearolic acid .... $CH_3(CH_2)_7C\equiv C(CH_2)_7CO_2H$	280.46	pr (al, peth), nd (dil al)	189-90[1.8]	48	....	1.4510[54]	eth	B2[4], 1751
10011	1,3-Octadiene, 3-chloro $C_4H_9CH=CClCH=CH_2$	144.65	....	64-5[18]	....	0.9366[20/4]	1.4794[20]	eth	B1[4], 1038
10012	1,6-Octadiene, 7-methyl-3-methylene $(CH_3)_2C=CHCH_2CH_2C(=CH_2)CH=CH_2$	136.24	....	56-7[12]	....	0.7982[20/4]	1.47065[20]	....	B1[4], 1108
10013	1,7-Octadiene $CH_2=CH(CH_2)_4CH=CH_2$	110.20	....	113-8	....	0.735[20/20]	1.424[20]	....	B1[4], 1038
10014	2,4-Octadiene, 7-methyl $(CH_3)_2CHCH_2CH=CHCH=CHCH_3$	124.23	....	149	....	0.7521[18/4]	1.4543[18]	....	B1[2], 239
10015	2-6-Octadiene $CH_3CH=CHCH_2CH_2CH=CHCH_3$	110.20	....	118-20	....	0.748[17]	1.4292[17]	....	B1[4], 1039
10016	2,6-Octadiene, 2,6-dimethyl (cis,cis) .... $CH_3CH=C(CH_3)CH_2CH_2CH=C(CH_3)_2$	138.25	....	168, 75[30]	....	0.775[21/4]	1.4498[20]	al, eth, ace	B1[4], 1058
10017	2,6-Octadiene, 2,6-dimethyl or Dihydroocimene .... $CH_3CH=C(CH_3)CH_2CH_2CH=C(CH_3)_2$	138.25	....	168, 75[30]	....	0.775[21/4]	1.4498[20]	al, eth, ace, aa	B1[4], 1058
10018	2,7-Octadiene, 2,6-dimethyl or Linaloolen............. $(CH_3)_2CHCH_2CH_2CH(CH_3)CH=CH_2$	138.25	....	168, 58[12]	....	0.7882[20]	1.4561[20]	al	B1, 261
10019	1,5-Octadien, 3-yn-5-propyl $C_2H_5CH=C(C_3H_7)C\equiv CCH=CH_2$	148.25	....	57-8[6]	....	0.8047[20/4]	1.4949[20]	eth, ace	B1[4], 1129
10020	2,6-Octadien-4-yne, 3,6-diethyl .... $CH_3CH=C(C_2H_5)C\equiv CC(C_2H_5)=CHCH_3$	162.28	....	169-71, 99[12]	....	0.8196[20/4]	1.4965[20]	eth, ace	B1[3], 1066
10021	2,6-Octadien-4-yne, 3,6-dimethyl .... $CH_3CH=C(CH_3)C\equiv CC(CH_3)=CHCH_3$	134.22	....	170	−45	0.8071[22/4]	1.4998[20]	eth, ace	B1[4], 1129
10022	1,7-Octadiyne .... $HC\equiv C(CH_2)_4C\equiv CH$	106.17	....	135-6, 93-5[16]	....	0.8169[21/4]	1.4521[18]	eth	B1[4], 1122
10023	2,6-Octadiyne $CH_3C\equiv CCH_2CH_2C\equiv CCH_3$	106.17	....	62[19]	27	0.828[80/4]	1.4658[30]	eth	B1[4], 1123
10024	3,5-Octadiyne $CH_3CH_2C\equiv CC\equiv CCH_2CH_3$	106.17	....	163-4, 78[34]	....	0.826[0/4]	1.4968[0]	eth	B1[4], 1123
10025	3,5-Octadiyne, 2,7-dimethyl .... $(CH_3)_2CHC\equiv CC\equiv CCH(CH_3)_2$	134.22	....	74[12]	....	0.8090[20/4]	....	eth	B1[1], 128
10026	Octanal or Caprylaldehyde....................... $CH_3(CH_2)_6CHO$	128.22	....	171, 72[20]	....	0.8211[20/4]	1.4217[20]	al, **eth, ace**, bz	B1[4], 3337
10027	Octane $CH_3(CH_2)_6CH_3$	114.23	....	125.7, 19.2[10]	−56.8	0.7025[20/4]	1.3974[20]	**al, eth, ace, bz, chl, peth**	B1[4], 412
10028	Octane, 1-amino or Octyl amine .... $CH_3(CH_2)_6CH_2NH_2$	129.25	....	179.6, 63.2[10]	0	0.7826[20/4]	1.4924[20]	al, eth	B4[4], 751
10029	Octane, 2-amino (d) $C_6H_{13}CH(NH_2)CH_3$	129.25	$[\alpha]^{17}_D$ + 8.6 (un-dil)	70[25]	....	0.771[25/4]	1.4220[25]	al, eth	B4[4], 762
10030	Octane, 2-amino (dl) .... $C_6H_{13}CH(NH_2)CH_3$	129.25	....	163-5, 58-9[13]	....	0.7745[20/0]	1.4232[25]	al, eth	B4[4], 763

No.	Name, Synonyms, and Formula	Mol. wt.	Color, crystalline form, specific rotation and λ_{max} (log ε)	b.p. °C	m.p. °C	Density	n_D	Solubility	Ref.
10031	Octane, 1-bromo or Octyl bromide . . . . CH$_3$(CH$_2$)$_6$CH$_2$Br	193.13	. . . .	200.8, 77.3[10]	−55	1.1122[20/4]	1.4524[20]	al, eth	B1[4], 422
10032	Octane, 1-Bromo-8-fluoro . . . . Br(CH$_2$)$_8$F	211.13	. . . .	118-20[22.5]			1.4500[20]	al, eth	B1[4], 424
10033	Octane, 2-bromo (d) C$_6$H$_{13}$CHBrCH$_3$	193.13	$[\alpha]^{25}/_D$ + 34.2	71[14]		1.0982[25/4]	1.4500[20]	al, eth	B1[4], 423
10034	Octane, 2-bromo (dl) C$_6$H$_{13}$CHBrCH$_3$	193.13	. . . .	188-9, 72[14]	. . . .	1.0878[25/4]	1.4442[25]	al, eth	B1[4], 423
10035	Octane, 2-bromo (l) C$_6$H$_{13}$CH(Br)CH$_3$	193.13	$[\alpha]^{25}/_D$ −37.5	72-3[18], 46[l]		1.0920[20/4]	1.4475[25]	al, eth	B1[4], 423
10036	Octane, 1-chloro or Octyl chloride . . . . CH$_3$(CH$_2$)$_6$CH$_2$Cl	148.68	. . . .	182, 78[15]	−57.8	0.8738[20/4]	1.4305[20]	. . . .	B1[4], 419
10037	Octane, 1-chloro-8-fluoro F(CH$_2$)$_8$Cl	166.67	. . . .	87[10]		0.978[20/4]	1.4266[25]	al, eth, ace	B1[4], 420
10038	Octane, 2-chloro (d) C$_6$H$_{13}$CHClCH$_3$	148.68	$[\alpha]^{20}/_D$ + 33.7	171-3, 75[28]		0.8658[17/4]	1.4273[21]	al, eth	B1[4], 419
10039	Octane, 3-chlor-3-methyl C$_5$H$_{11}$CCl(CH$_3$)CH$_2$CH$_3$	162.71	. . . .	73-4[15]		0.8680[25/4]	1.4351[20]	eth, ace, bz, chl	B1[3], 508
10040	Octane, 4-chloro (d) C$_4$H$_9$CHClC$_3$H$_7$	148.68	$[\alpha]^{25}/_D$ + 0.28 (undil)	92[50]		. . . .	. . . .	al, eth, chl	B1[3], 466
10041	Octane, 4-chloro-4-methyl C$_4$H$_9$CCl(CH$_3$)C$_3$H$_7$	162.71	. . . .	71[14.5]		0.8723[20/4]	1.4360[20]	eth, chl	B1[3], 509
10042	Octane, 1-cyclohexyl C$_8$H$_{17}$C$_6$H$_{11}$	196.38	. . . .	263.6	−19.7	0.8138[20]	1.4504[20]		B5[4], 179
10043	Octane, 1,2-dibromo C$_6$H$_{13}$CHBrCH$_2$Br	272.04	. . . .	240-2, 118.5[15]		1.4580[20/4]	1.4970[20]		B1[3], 468
10044	Octane, 1,8-dibromo . . . . Br(CH$_2$)$_8$Br	272.04	. . . .	270-7, 92-3[0.45]	11-1	1.4190[17]	1.4971[20]	eth, chl	B1[4], 424
10045	Octane, 2,3-dimethyl C$_6$H$_{11}$CH(CH$_3$)CH(CH$_3$)$_2$	142.28	. . . .	164.7		0.7377[20/4]	1.4146[20]	. . . .	B1[4], 476
10046	Octane, 2,6-dimethyl CH$_3$CH$_2$CH(CH$_3$)(CH$_2$)$_3$CH(CH$_3$)$_2$	142.28	. . . .	160-1, 90-3[26]		0.7313[20/4]	1.4097[20]	. . . .	B1[4], 477
10047	Octane, 2,7-dimethyl (CH$_3$)$_2$CH(CH$_2$)$_4$CH(CH$_3$)$_2$	142.29	. . . .	159.6	−54.6	0.7240[20/4]	1.4092[20]	eth	B1[4], 478
10048	Octane, 1-fluoro or Octyl fluoride. . . . . CH$_3$(CH$_2$)$_6$CH$_2$F	132.22	. . . .	142-3		0.8103[20/4]	1.3935[20]	. . . .	B1[4], 418
10049	Octane, 1-iodo or Octyl iodide. . . . . CH$_3$(CH$_2$)$_7$CH$_2$I	240.13	. . . .	225.5, 86.5[5]	−45.7	1.3297[20/4]	1.4889[20]	al, eth	B1[4], 425
10050	Octane, 2-iodo (D-) C$_6$H$_{13}$CHICH$_3$	240.13	$[\alpha]^{28}/_D$ −45.5	92[12]		1.3219[20/4]	1.4863[25]	al, eth, lig	B1[4], 425
10051	Octane, 2-iodo (dl) C$_6$H$_{13}$CHICH$_3$	240.13	. . . .	210, 95-6[16]		1.3251[20/4]	1.4896[20]	al, eth, lig	B1[4], 425
10052	Octane, 2-iodo (l, +) C$_6$H$_{13}$CHICH$_3$	240.13	$[\alpha]^{26}/_D$ + 46.3	101[22]		1.3314[17/4]	1.4877[22]	al, eth, lig	B1[4], 425
10053	Octane, 2-methyl C$_6$H$_{13}$CH(CH$_3$)$_2$	128.26	. . . .	142.8	−80.1	0.7107[20/4]	1.4029[20]	al, eth, lig	B1[4], 454
10054	Octane, 3-methyl (d) C$_5$H$_{11}$CH(CH$_3$)C$_2$H$_5$	128.26	$[\alpha]^{17}/_D$ + 9.4	143-4	−107.6	0.7206[17]	1.4068[20]	ace, bz	B1[4], 455
10055	Octane, 3-methyl (l) C$_5$H$_{11}$CH(CH$_3$)C$_2$H$_5$	128.26	$[\alpha]^{27}/_D$ −8.5	143		0.714[27/4]	1.4052[25]	ace, bz	B1[3], 509
10056	Octane, 4-methyl (dl) C$_4$H$_9$CH(CH$_3$)C$_3$H$_7$	128.26	. . . .	142.4, 32[10]	−113.2	0.7199[20/4]	1.4061[20]	al, eth, ace, bz	B1[4], 456
10057	Octane, 4-methyl (l) C$_4$H$_9$CH(CH$_3$)C$_3$H$_7$	128.26	$[\alpha]^{19}/_D$ −1.06	141		0.717[19/4]	. . . .	al, eth, ace, bz	B1[3], 510
10058	Octane, 1-phenyl or Octylbenzene. . . . . C$_7$H$_{15}$C$_6$H$_5$	190.33	. . . .	264-5, 131-4[12]	−7	0.8582[20/4]	1.4851[20]	bz, eth	B5[4], 1157
10059	Octane, 2-phenyl C$_6$H$_{13}$CH(C$_6$H$_5$)CH$_3$	190.33	. . . .	123-5[20]		0.8611[20/4]	1.4837[20]	. . . .	B5[4], 1157
10060	Octanedial or Suberaldehyde. . . . . OHC(CH$_2$)$_6$CHO	142.20	. . . .	230-40d, 96-8[3]			1.4439[20]	w, al	B1[4], 3706
10061	Octanedioic acid or Suberic acid . . . . HO$_2$C(CH$_2$)$_6$CO$_2$H	174.20	lo nd or pl (w)	300 sub, 219.5[10]	144	. . . .	. . . .	al	B2[4], 2028
10062	1,7-Octanediol, 3,7-dimethyl (CH$_3$)$_2$C(OH)(CH$_2$)$_3$CH(CH$_3$)CH$_2$CH$_2$OH	174.28	. . . .	265		0.937[20]	1.4599[20]	. . . .	B1[3], 2233
10063	1,8-Octanediol HO(CH$_2$)$_8$OH	146.23	nd (bz-lig), pr	172[20]	63	. . . .	. . . .	al, eth, ace	B1[4], 2592

No.	Name, Synonyms, and Formula	Mol. wt.	Color, crystalline form, specific rotation and λ_{max} (log ε)	b.p. °C	m.p. °C	Density	n_D	Solubility	Ref.
10064	4,5-Octanediol (dl) or 1,2-Dipropyl ethylene glycol....... $C_3H_7CHOHCHOHC_3H_7$	146.23	....	110^8	28		1.4419^{25}	....	B1[4], 2593
10065	4,5-Octanediol (meso) .... $C_3H_7CHOHCHOHC_3H_7$	146.23			123-4					B1[4], 2593
10066	2,3-Octanedione............... $C_5H_{11}COCOCH_3$	142.20		$172\text{-}3^{733}$				al	B1[2], 3138
10067	2,3-Octanedione dioxime or Methyl pentyl glyoxime ..... $C_5H_{11}C(=NOH)C(=NOH)CH_3$	172.23	nd (dil al)		173			al, ace	B1[4], 3706
10068	2,3-Octanedione, 3-oxime .... $C_5H_{11}C(=NOH)COCH_3$	157.22	(liq)	133^{15}	59			eth	B1, 795
10069	2,7-Octanedione .... $CH_3CO(CH_2)_4COCH_3$	142.20	pl (bz)	114^{10}	44			al	B1[4], 3707
10070	2,7-Octanedione dioxime .... $CH_3C(=NOH)(CH_2)_4C(=NOH)CH_3$	172.23	(al)		158			al	B1, 795
10071	3,6-Octanedione .... $CH_3CH_2COCH_2CH_2COCH_2CH_3$	142.20	pl (al)	98^{14} sub	35-6			al	B1[4], 3708
10072	3,6-Octanedione, 2,2,7,7-tetramethyl .... $(CH_3)_3CCOCH_2CH_2COC(CH_3)_3$	198.31	....	$115\text{-}7^{17}$, $55\text{-}6^{0.5}$	2.5	$0.900^{27/4}$	1.4400^{20}	al	B1[4], 3734
10073	4,5-Octanedione .... $C_3H_7COCOC_3H_7$	142.20	ye oil	168, 60^{12}		$0.934^{0/4}$		al, eth, ace	B1[4], 3708
10074	4,5-Octanedione dioxime .... $C_3H_7C(=NOH)C(=NOH)C_3H_7$	172.23	....	sub	186-7			al, eth	B1[2], 846
10075	1-Octanethiol or Octyl mercaptan............... $CH_3(CH_2)_6CH_2SH$	146.30		199.1, 86^{15}	-49.2	$0.8433^{20/4}$	1.4540^{20}	al	B1[4], 1767
10076	2-Octanethiol (dl) .... $C_6H_{13}CHSHCH_3$	146.30		186.4, 88.9^{30}	-79	$0.8366^{20/4}$	1.4504^{20}	al, eth, bz	B1[4], 1777
10077	2-Octanethiol (l) .... $C_6H_{13}CHSHCH_3$	146.30	$[\alpha]^{25}_{546}$ -36.4	$78\text{-}80^{22}$		$0.830^{25/4}$		al, eth, bz	B1[4], 1777
10078	Octanoic acid or Caprylic acid .... $CH_3(CH_2)_6CO_2H$	144.22		239.3, 140^{23}	16.5	0.9088^{20}	1.4285^{20}	al, chl	B1[4], 982
10079	1-Octanol or Octyl alcohol .... $CH_3(CH_2)_6CH_2OH$	130.23		194.4, 98^{19}	-16.7	$0.8270^{20/4}$	1.4295^{20}	al, eth	B1[4], 1756
10080	1-Octanol, 8-chloro .... $Cl(CH_2)_8OH$	164.68		139^{19}			1.4563^{25}	al, eth	B1[4], 1766
10081	1-Octanol, 3,7-dimethyl (d) or Tetrahydrogeraniol...... $(CH_3)_2CH(CH_2)_3CH(CH_3)CH_2CH_2OH$	158.29	$[\alpha]^{20}_D$ +4.1	212-3, $105\text{-}6^{10}$		$0.8285^{20/4}$	1.4355^{20}	eth	B1[4], 1830
10082	1-Octanol, 3,7-dimethyl (l) .... $(CH_3)_2CH(CH_2)_3CH(CH_3)CH_2CH_2OH$	158.29	$[\alpha]^{11}_{546}$ -3.7	212-3, 109^{15}		0.830^{18}	1.4370^{15}	eth	B1[3], 1768
10083	1-Octanol, 8-fluoro .... $F(CH_2)_8OH$	148.22		$106\text{-}7^{10}$		$0.945^{20/4}$	1.4248^{25}	al, eth	B1[4], 1766
10084	2-Octanol (d) .... $C_6H_{13}CH(OH)CH_3$	130.23	$[\alpha]^{17}_D$ +9.9	86^{20}		$0.8216^{20/4}$	1.4264^{20}	al, eth, ace	B1[4], 1770
10085	2-Octanol (dl) .... $C_6H_{13}CH(OH)CH_3$	130.23		180, 87^{20}	-31.6	$0.8193^{20/4}$	1.4203^{20}	al, eth, ace	B1[4], 1770
10086	2-Octanol (l) .... $C_6H_{13}CH(OH)CH_3$	130.23	$[\alpha]^{17}_D$ -9.9	86^{20}		$0.8201^{20/4}$	1.4264^{20}	al, eth, ace	B1[3], 1721
10087	2-Octanol, 2-methyl .... $C_6H_{13}C(OH)(CH_3)_2$	144.26		178, $81\text{-}3^{16}$		$0.8210^{20/4}$	1.4280^{20}	al, eth	B1[4], 1805
10088	3-Octanol .... $C_2H_5CH(OH)C_5H_{11}$	130.23		166-72					B1[4], 1779
10089	3-Octanol, 3,6-dimethyl .... $C_2H_5CH(CH_3)CH_2CH_2C(OH)(CH_3)C_2H_5$	158.28		202.2	-67.5		1.4370^{20}		B1[1], 214
10090	3-Octanol, 3,7-dimethyl (dl) or Tetrahydrolinalool ...... $(CH_3)_2CH(CH_2)_3C(OH)(CH_3)CH_2CH_3$	158.29		196-7, $87\text{-}8^{10}$	31-2	0.8280^{20}	1.4335^{20}	al	B1[4], 1829
10091	3-Octanol, 3-ethyl .... $C_5H_{11}C(OH)(C_2H_5)_2$	158.29		199, 84^{12}		$0.8361^{25/4}$	1.4390^{20}	al	B7[3], 1766
10092	4-Octanol (d) .... $C_3H_7CH(OH)C_4H_9$	130.23	$[\alpha]^{22}_D$ 0.74	79^{16}		$0.8159^{25/4}$	1.4275^{25}	al	B1[2], 452
10093	4-Octanol (dl) .... $C_3H_7CH(OH)C_4H_9$	130.23		176.3, 81.3^{20}	-40.7	$0.8186^{20/4}$	1.4248^{20}	al	B1[4], 1779
10094	2-Octanone or Methyl hexyl ketone............... $C_6H_{13}COCH_3$	128.22		173, $59\text{-}60^{11}$	-16	$0.8202^{20/4}$	1.4151^{20}	al, eth	B1[4], 3339
10095	3-Octanone or Ethyl pentyl ketone .... $C_5H_{11}COC_2H_5$	128.22		167^{749}		$0.8221^{20/4}$	1.4153^{20}	al, eth	B1[4], 3341
10096	4-Octanone or Butyl propyl ketone .... $C_4H_9COC_3H_7$	128.22		163, 70^{26}		$0.8146^{25/4}$	1.4173^{14}	al, eth	B1[4], 3342

No.	Name, Synonyms, and Formula	Mol. wt.	Color, crystalline form, specific rotation and λ_{max} (log ε)	b.p. °C	m.p. °C	Density	n_D	Solubility	Ref.
10097	4-Octanone, 5-hydroxy or Butyroin .... $C_3H_7CH(OH)COC_3H_7$	144.22	....	180-90, 95[20]	−10	0.9231[20]	1.4290[20]	al, eth, ace	B1[4], 4042
10098	4-Octanone, 7-methyl or Propyl-*iso*-pentyl ketone ...... $(CH_3)_2CHCH_2CH_2COC_3H_7$	142.24	....	177-9	......	0.8239[20/4]	1.4210[20]	al, eth	B1[4], 3356
10099	Octasiloxane, octadecamethyl .... $CH_3[-Si(CH_3)_2O.]_7Si(CH_3)_3$	607.32	....	153[5.1]		0.913	1.3970[20]	bz, lig, peth	B4[3], 1881
10100	1,3,5,7-Octatetraene .... $CH_2=CHCH=CHCH=CHCH=CH_2$	106.17	(bz)	sub	ca 50	....	....	aa	B1[4], 1124
10101	2,4,6-Octatriene *(trans,trans,trans)* .... $CH_3CH=CHCH=CHCH=CHCH_3$	108.19	lf	147-8, 43[10]	52	0.7961[23/4]	1.5131[27]	al, chl, lig	B1[4], 1101
10102	1,3,7-Octatriene, 3,7-dimethyl or Ocimene .... $CH_2=C(CH_3)CH_2CH_2CH=C(CH_3)CH=CH_2$	136.24	....	176-8d, 73-4[21]	......	0.8000[20]	1.4862[20]	al, eth, bz, chl, aa	B1[4], 1108
10103	2,4,6-Octatriene, 2,6-dimethyl (4-*trans*,6-*trans*) or Alloocimene A .... $CH_3CH=C(CH_3)CH=CH-CH=C(CH_3)_2$	136.24	....	188[750], 91[20]	−35.4	0.8118[20/4]	1.5446[20]	....	B1[4], 1106
10104	2,4,6-Octatriene, 2,6-dimethyl (4-*trans*,6-*cis*) or Alloocimene B .... $CH_3CH=C(CH_3)CH=CHCH=C(CH_3)_2$	136.24	....	89[20]	−20.6	0.8060[20]	1.5446[20]	....	B1[4], 1106
10105	1-Octene .... $C_6H_{13}CH=CH_2$	112.22	....	121.3, 15.4[10]	−101.7	0.7149[20/4]	1.4087[20]	al, eth, ace, bz, chl	B1[4], 874
10106	1-Octene, 2-chloro .... $C_6H_{13}CCl=CH_2$	146.66	....	168-70	......	0.9274[0/0]	....	eth, ace, bz	B1[1], 840
10107	1-Octene, 3,7-dimethyl .... $(CH_3)_2CH(CH_2)_3CH(CH_3)CH=CH_2$	140.27	....	154	......	0.7396[20/4]	1.4212[20]	....	B1[4], 905
10108	1-Octene, 2-methyl .... $C_6H_{13}C(CH_3)=CH_2$	126.24	....	114.8	−77.8	0.7343[20/4]	1.4184[20]	....	B1[4], 896
10109	2-Octene *(cis)* .... $C_5H_{11}CH=CHCH_3$	112.22	....	125.6, 16.5[10]	−100.2	0.7243[20/4]	1.4150[20]	al, eth, ace, bz, chl	B1[4], 878
10110	2-Octene *(trans)* .... $C_5H_{11}CH=CHCH_3$	112.22	....	125, 16[10]	−87.7	0.7199[20/4]	1.4132[20]	al, eth, ace, bz, chl	B1[4], 879
10111	2-Octene, 2-chloro .... $C_5H_{11}CH=CClCH_3$	146.66	....	167-8	......	0.8923[16/16]	1.4424[16]	al, eth, ace, bz	B1[4], 879
10112	2-Octene, 4-chloro .... $C_4H_9CHClCH=CHCH_3$	146.66	....	153, 65-6[15]	......	0.8924[20/4]	1.4452[20]	eth, ace, bz, chl	B1[1], 840
10113	2-Octene, 2,6-dimethyl .... $C_2H_5CH(CH_3)CH_2CH_2CH_2C(CH_3)=CH_2$	140.27	....	163	......	0.746[22/4]	1.425[22]	....	B1[4], 904
10114	3-Octene *(cis)* .... $C_4H_9CH=CHC_2H_5$	112.22	....	122.9, 14.3[10]	−126	0.7189[20/4]	1.4135[20]	al, eth, ace, bz, lig	B1[4], 880
10115	3-Octene *(trans)* .... $C_4H_9CH=CHC_2H_5$	112.22	....	123.3, 14.6[10]	−110	0.7152[20/4]	1.4126[20]	al, eth, ace, bz, lig	B1[4], 880
10116	4-Octene *(cis)* .... $C_3H_7CH=CHC_3H_7$	112.22	....	122.5, 14[10]	−118.7	0.7212[20/4]	1.4148[20]	al, eth, ace, bz, lig	B1[4], 880
10117	4-Octene *(trans)* .... $C_3H_7CH=CHC_3H_7$	112.22	....	122.3, 13.7[10]	−93.8	0.7141[20/4]	1.4114[20]	al, eth, ace, bz, lig	B1[4], 880
10118	4-Octene, 4-chloro *(cis)* .... $C_3H_7CH=CClC_3H_7$	146.66	....	165.3	......	0.8912[20/4]	1.4447[20]	eth	B1[4], 881
10119	1-Octen-3-yne .... $C_4H_9C\equiv C-CH=CH_2$	108.19	....	62[60]	......	0.7830[20/4]	1.4592[20]	eth	B1[4], 1100
10120	1-Octen-3yne-5-ol, 5-methyl .... $C_3H_7C(OH)(CH_3)C\equiv C-CH=CH_2$	138.21	....	80[13]	......	0.8851[15/4]	1.4735[20]	al	B1[3], 2034
10121	Octyl ether or Dioctyl ether .... $(C_8H_{17})_2O$	242.45	....	286-7	......	0.8063[20/4]	1.4327[20]	al, eth	B1[4], 1760
10122	Octyl nitrate .... $C_8H_{17}ONO_2$	175.23	....	110-2[20]	......	0.8419[17/17]	....	al, eth	B1[4], 1762
10123	Octyl nitrite .... $C_8H_{17}ONO$	159.23	ye	174-5, 60[10]	......	0.862[17]	1.4127[20]	al, eth	B1[4], 762
10124	Octyl phenyl ether .... $C_8H_{17}OC_6H_5$	206.33	....	285, 164-7[20]	8	0.9319[15/15]	1.4875[20]	al, eth	B6, 144
10125	Octyl thiocyanate .... $C_8H_{17}SCN$	171.31	....	141-2[19]	105	0.9149[25/4]	1.4649[20]	al, eth	B3[3], 282
10126	1-Octyne or Hexyl acetylene.......... $C_6H_{13}C\equiv CH$	110.20	....	125.2, 19.7[10]	−79.3	0.7461[20]	1.4159[20]	al, eth	B1[4], 1034
10127	1-Octyne, 1-chloro .... $C_6H_{13}C\equiv CCl$	144.65	....	61-2[17]	......	0.912[20]	1.445[20]	al, eth	B1[3], 1005
10128	1-Octyne-3-ol, 3-methyl .... $C_5H_{11}C(OH)(CH_3)C\equiv CH$	140.22	....	75[10]	......	0.863[10/10]	1.443[10]	....	B1[4], 2268

No.	Name, Synonyms, and Formula	Mol. wt.	Color, crystalline form, specific rotation and λ_{max} (log ε)	b.p. °C	m.p. °C	Density	n_D	Solubility	Ref.
10129	2-Octyne .... $C_5H_{11}C \equiv CCH_3$	110.20	....	138	−61.6	0.7596[20/4]	1.4278[20]	al, eth	B1[4], 1035
10130	2-Octyne-1-ol .... $C_5H_{11}C \equiv CCH_2OH$	126.20		98-9[15]	−18	0.8805[20/4]	1.4556[20]	eth	B1[4], 2256
10131	3-Octyne .... $C_4H_9C \equiv CC_2H_5$	110.20		133, 85[169]	−103.9	0.7529[20/4]	1.4250[20]	al, eth	B1[4], 1036
10132	3-Octyne, 2-chloro-2-methyl .... $C_4H_9C \equiv CCCl(CH_3)_2$	158.67		68[15]	......	0.8929[20/4]	1.4480[20]	eth	B1[4], 1048
10133	3-Octyne, 2,2-dimethyl .... $C_4H_9C \equiv CC(CH_3)_3$	138.25	....	79[60]	......	0.7491[20/4]	1.4270[20]	eth	B1[3], 1018
10134	3-Octyne, 7-methyl .... $(CH_3)_2CHCH_2CH_2C \equiv CC_2H_5$	124.23		87[99]	......	0.7599[20/4]	1.4280[20]	eth	B1[3], 1014
10135	4-Octyne .... $C_3H_7C \equiv CC_3H_7$	110.20		131.5	−102.5	0.7509[20/4]	1.4248[20]	al, eth	B1[4], 1037
10136	Olealdehyde or 9-Octadecenal .... $CH_3(CH_2)_7CH=CH(CH_2)_7CHO$	266.47	ye nd	168-9[3]	......	0.8509[20/4]	1.4558[20]	....	B1[4],3533
10137	Oleamide .... $CH_3(CH_2)_7CH=CH(CH_2)_7CONH_2$	281.49			76	....	....	eth	B2[4], 1668
10138	Oleamide, N-phenyl or Oleanilide .... $CH_3(CH_2)_7CH=CH(CH_2)_7CONHC_6H_5$	357.59	nd	143.5[10]	41	....	....	eth, bz, aa	B12[2], 150
10139	Oleic acid or 9,10-Octadecenoic acid (cis)............. $CH_3(CH_2)_7CH=CH(CH_2)_7CO_2H$	282.47	....	286[100], 228-9[15]	16.3	0.8935[20/4]	1.4582[20]	al, eth, ace, bz, chl	B2[4], 1641
10140	Oleic acid, benzyl ester or Benzyl oleate .... $C_{17}H_{33}CO_2CH_2C_6H_5$	372.60		237[1]	......	0.9330[25/25]	1.4875[25]	al, eth	B6[4], 2269
10141	Oleic acid, butyl ester or Butyl oleate .... $C_{17}H_{33}CO_2C_4H_9$	338.58	ye	227-8[15]	−26.4	0.8704[15]	1.4480[25]	al	B2[4], 1653
10142	Oleic acid, ethyl ester or Ethyl oleate............. $C_{17}H_{33}CO_2C_2H_5$	310.53		207[13]	0.8720[2]	1.4515[20]		al, eth	B2[4], 1651
10143	Oleic acid, methyl ester or Methyl oleate .... $C_{17}H_{33}CO_2CH_3$	296.50		218.5[20]	−19.9	0.8739[20]	1.4522[20]	....	B2[4], 1649
10144	Oleic acid, isopentyl ester or Isopentyl oleate .... $C_{17}H_{33}CO_2\text{-}i\text{-}C_5H_{11}$	352.61		223-4[10]	......	0.897[15]	....	al, eth	B2[3], 1414
10145	Oleonitrile .... $C_{17}H_{33}CN$	263.47	....	330-5d, 204[12]	−1	0.848[17/17]	1.4566[20]		B2[4], 1668
10146	Oleyl alcohol or 9-Octadecen-1-ol .... $CH_3(CH_2)_7CH=CH(CH_2)_7CH_2OH$	268.49		205-10[15]	6-7	0.8489[20/4]	1.4606[20]	al, eth	B1[4], 2204
10147	Orcein or Orcin .... $C_{28}H_{24}N_2O_7$	500.51	br-red pw	....	....	....	....	al, ace	B6[3], 4531
10148	Ornithine (L) or L-2,5-Diamino valeric acid............. $H_2N(CH_2)_3CH(NH_2)CO_2H$	132.16	cr (al-eth), $[\alpha]^{25/}_D$ + 11.5 (w, c=6.5)		140			w, al	B4[3], 1346
10149	Ornithine monohydrochloride (L) $H_2N(CH_2)_3CH(NH_2)CO_2H.HCl$	168.63	nd $[\alpha]^{25/}_D$ + 11.0 (w, c=5.5)		215 (230)			w	B4[3], 1347
10150	Ornithine-sulfate (L) $H_2N(CH_2)_3CH(NH_2)CO_2H.H_2SO_4$	230.24	$[\alpha]^{25/}_D$ + 8.4 (w)		234d			w	B4[3], 1347
10151	Orthoacetic acid, triethyl ester or 1,1,1-triethoxyethane, triethyl ortho acetate............. $CH_3C(OC_2H_5)_3$	162.23	....	144-6, 66.5[41]	......	0.8847[25/4]	1.3980[20]	al, eth, chl	B2[4], 137
10152	Orthoacetic acid, trimethyl ester or 1,1,1-trimethoxy ethane,trimethyl ortho acetate............. $CH_3C(OCH_3)_3$	120.15		107-9		0.9438[25/4]	1.3859[25]	al, eth	B2[4], 127
10153	Orthocarbonic acid, tetraethyl ester or Ethyl ortho carbonate,Tetraethoxy methane............. $C(OC_2H_5)_4$	192.26		160-1, 62[28]		0.9186[20/4]	1.3928[20]	al, eth	B3[4], 6
10154	Orthocarbonic acid, tetrapropyl ester or Propyl ortho carbonate............. $C(OC_3H_7)_4$	248.37		224.2		0.897[20/4]	1.4100[20]	al, eth	B3[4], 7
10155	Orthoformic acid, triisobutyl ester or Isobutyl ortho formate............. $HC(O\text{-}i\text{-}C_4H_9)_3$	232.37		224-6		0.8582[20/4]	1.4120[20]	al, eth	B2[4], 29
10156	Orthoformic acid, triethyl ester or Ethyl ortho formate,Triethoxy methane............. $HC(OC_2H_5)_3$	148.20		143[765], 60[20]		0.8909[20/4]	1.3922[20]	al, eth	B2[4], 25
10157	Orthoformic acid, trimethyl ester or Methyl ortho formate $HC(OCH_3)_3$	106.12	....	103-5		0.9676[20/4]	1.3793[20]	al, eth	B2[4], 22

No.	Name, Synonyms, and Formula	Mol. wt.	Color, crystalline form, specific rotation and λ_{max} (log ε)	b.p. °C	m.p. °C	Density	n_D	Solubility	Ref.
10158	Orthoformic acid, triisopentyl ester or isopentyl ortho formate HC(O-i-C₅H₁₁)₃	274.45	....	267-9d, 166[25]	......	0.8628[20/4]	1.4233[20]	....	B2[4], 30
10159	Orthoformic acid, triphenyl ester or Phenyl ortho formate HC(OC₆H₅)₃	292.34	....	269-70[10] d	76-7			al, eth	B6[4], 611
10160	Orthoformic acid, tripropyl ester or Propyl ortho formate HC(OC₃H₇)₃	190.29	....	190-1[745], 93[30]	......	0.8805[20/4]	1.4072[20]	al, eth	B2[4], 27
10161	Orthoformic acid, triisopropyl ester or isopropyl ortho formate. HC[OCH(CH₃)₂]₃	190.29	....	166-8	......	0.8621[20/4]	1.4000[20]	al, eth	B2[4], 28
10162	Orthoformic acid, trithio, triethyl ester or Ethyl orthothio formate. HC(SC₂H₅)₃	196.40	....	235d, 127-8[12]	......	1.053[20/4]	1.5410[15]	al, eth	B2[4], 93
10163	Orthopropionic acid, triethyl ester or Ethyl ortho propionate CH₃CH₂C(OC₂H₅)₃	176.26	....	171, 44[9]			1.4000[25]	al, eth	B2[4], 707
10164	Orthosilicic acid, tetraethyl ester or Ethyl ortho silicate Si(OC₂H₅)₄	432.77	liq	168.8	−82.5	0.9320[20/4]	1.3928[20]	al, eth	B1[4], 1360
10165	Orthosilicic acid, tetrakis(2-ethylbutyl)ester Si[OCH₂CH(C₂H₅)₂]₄	432.77	liq	358		0.8920[20/4]	1.4307[20]	eth, bz	B1[4], 1725
10166	Orthosilicic acid, tetrakis(2-ethylhexyl)ester Si[OCH₂CH(C₂H₅)C₄H₉]₄	544.99	....	419, 227[5]	−90	0.8803[20/4]	1.4388[20]	eth, bz	B1[4], 1787
10167	Orthosilicic acid, tetramethyl ester or Methyl ortho silicate Si(OCH₃)₄	152.22	nd	121, 25-7[12]	−2	1.0232[20]	1.3683[20]	al	B1[4], 1266
10168	1,3,4-Oxadiazole, 2,5-dimethyl C₄H₆N₂O	98.11		178-9				w, al, eth	B27, 565
10169	Oxalic acid or Ethanedioic acid. HO₂CCO₂H	90.04	mcl ta or pr (+ 2w w), orh (anh)	157 sub	a:189.5, β:182 (anh), 101.5 (hyd)	α:1.900[17/4] β:1.895	....	w, al	B2[4], 1819
10170	Oxalic acid, diallyl ester or Diallyl oxalate C₃H₅O₂CCO₂C₃H₅	170.17	....	217, 86[2]	......	1.1582[20]	1.4481[20]	al, ace, bz	B2[4], 1851
10171	Oxalic acid, dibutyl ester or Dibutyl oxalate. C₄H₉O₂CCO₂C₄H₉	202.25	....	242[773], 96[2]	−30.5	0.9873[20/4]	1.4234[20]	al, eth	B2[4], 1850
10172	Oxalic acid, diisobutyl ester (i-C₄H₉O₂C-)₂	202.25	....	229, 143[20]	......	0.9737[20/4]	1.4180[20]	al, eth, ace	B2[4], 1850
10173	Oxalic acid, di(2-chloroethyl)ester [ClCH₂CH₂O₂C-]₂	215.04	lf (dil al)	132[3]	45			al, bz	B2[4], 1849
10174	Oxalic acid, dicyclohexyl ester or Dicyclohexyl oxalate (C₆H₁₁O₂C-)₂	254.33	(MeOH)	190-1[73]	42 (47)			al, eth	B6[3], 26
10175	Oxalic acid, diethyl ester or Diethyl oxalate (C₂H₅O₂C-)₂	146.14	....	185.7, 97[20]	−38.5	1.0785[20/4]	1.4101[20]	al, eth, ace	B2[4], 1848
10176	Oxalic acid dihydrazide H₂NNHCOCONHNH₂	118.10	nd (w)		243d	1.458[22.5]			B2[4], 1868
10177	Oxalic acid, dimethyl ester or Dimethyl oxalate CH₃O₂CCO₂CH₃	118.09	mcl ta	164.5	54	1.148[15], 1.1716[60]	1.379[82]	al, eth, ace	B2[4], 1847
10178	Oxalic acid, di-isopentyl ester or Diisopentyl oxalate (i-C₅H₁₁O₂C-)₂	230.30	....	267-8, 144[14]		0.968[11/11]		al, eth	B2[4], 1851
10179	Oxalic acid, dipropyl ester or Dipropyl oxalate C₃H₇O₂CCO₂C₃H₇	174.20	....	211, 78-80[3]	−44.3	1.0188[20/4]	1.4158[20]	al, eth	B2[4], 1849
10180	Oxalic acid, diisopropyl ester (i-C₃H₇O₂C-)₂	174.20	....	191[765]		1.0010[20/4]	1.4100[20]	al, eth	B2[4], 1850
10181	Oxalic acid, di(2-tolyl)ester [(2-CH₃C₆H₄)O₂C-]₂	270.29	nd (al)	distb	91			al, eth, ace, bz, chl	B6[2], 330
10182	Oxalic acid, di-(3-tolyl)ester [(3-CH₃C₆H₄)O₂C-]₂	270.29	nd (al)	distb	106			al, eth, ace, bz, chl	B6[2], 353
10183	Oxalic acid, di-(4-tolyl)ester [(4-CH₃C₆H₄)O₂C-]₂	270.29	lf or pl, (al-eth)	....	148			al, eth, ace, bz, chl	B6[4], 2114
10184	Oxalic acid, monochloride, monoethyl ester ClCOCO₂C₂H₅	136.54	hyg	137, 30[10]		1.2226[20/4]		eth, bz	B2[4], 1853
10185	Oxalic acid, monoethyl, monomethyl ester CH₃O₂CCO₂C₂H₅	132.13	....	173.7		1.5505[0/0]		al, eth	B2[1], 232
10186	Oxaluric acid or Oxalic acid ureide H₂NCONHCOCO₂H	132.08	cr	....	208-10d			w	B3[4], 121
10187	Oxalyl chloride ClCOCOCl	126.93	nd (eth, peth)	63-4	−16	1.4785[20/4]	1.4316[20]	eth	B2[4], 1853

No.	Name, Synonyms, and Formula	Mol. wt.	Color, crystalline form, specific rotation and λ_{max} (log ε)	b.p. °C	m.p. °C	Density	n_D	Solubility	Ref.
10188	Oxamic acid or Oxalic acid monoamide .... HO_2CCONH_2	89.05	cr (w)	....	210d	....	....	....	B2⁴, 1857
10189	Oxamic acid, N-acetyl, ethyl ester .... $CH_3CONHCOCO_2C_2H_5$	159.14	pl (eth)	....	54-5	....	....	al, eth	B2², 509
10190	Oxamic acid, N-sec-butyl .... Sec-$C_4H_9NHCOCO_2H$	145.16	cr (eth)	....	88-9	....	....	eth	B4, 162
10191	Oxamide, dithio or Dithiooxamide .... $H_2NCSCSNH_2$	120.20	og-red	sub	....	....	....	....	B2⁴, 1871
10192	Oxamic acid, ethyl ester or Ethyl oxamate .... $H_2NCOCO_2C_2H_5$	117.11	....	....	114-5	....	....	w, eth	B2⁴, 1857
10193	Oxamic acid, N-phenyl or Oxanilic acid .... $HO_2CCONHC_6H_5$	165.15	nd (bz)	....	150	....	....	al, eth, chl	B12³, 550
10194	Oxamic acid, N-phenyl, ethyl ester .... $C_6H_5NHCOCO_2C_2H_5$	193.21	pl or pr (al), nd (w)	260-300	65-6	....	....	al, eth, ace, bz	B12³, 550
10195	Oxamide .... $H_2NOCCONH_2$	88.07	nd (w)	....	419d	....	....	....	B2⁴, 1860
10196	Oxamide, N,N'-diethyl .... $C_2H_5NHCOCONHC_2H_5$	144.18	nd (al)	....	175 (180)	1.169⁴	....	al	B4⁴, 359
10197	Oxamide, N,N'-dimethyl .... $CH_3NHCOCONHCH_3$	116.13	pl or nd (al)	sub	217	1.3⁴/⁴	....	w, chl	B4⁴, 246
10198	Oxamide, N,N'-dephenyl or Oxanilide .... $C_6H_5NHCONHC_6H_5$	240.27	lf (bz)	>360	254	....	....	bz	B12³, 551
10199	Oxamide, N,N'-diisopropyl .... [i-C_3H_7NHCO-]$_2$	172.23	nd (al)	....	212	....	....	al	B4⁴, 518
10200	1,4-Oxathiane .... C_4H_8OS	104.07	....	147⁷⁵⁵	....	1.1174²⁰/⁴	....	....	B19⁴, 33
10201	Oxazole .... C_3H_3NO	69.06	....	69-70	....	....	1.4285¹⁷·⁵	....	B27², 9
10202	Oxazole, 2,4-dimethyl .... 2,4$(CH_3)_2(C_3HNO)$	97.12	....	108	....	0.9352¹⁵/⁴	1.4166¹⁵	w, al, eth	B27², 10
10203	Oxazole, 2,5-dimethyl .... 2,5-$(CH_3)_2(C_3HNO)$	97.12	....	117-8	....	0.9958²¹/⁴	1.4385²¹	w	B27², 10
10204	Oxazole, 2,4-diphenyl .... 2,4-$(C_6H_5)_2(C_3HNO)$	221.26	lf (al)	338-40	103	....	....	al, eth, bz	B27, 78
10205	Oxazole, 2,5-diphenyl .... 2,5-$(C_6H_5)_2(C_3HNO)$	221.26	nd (lig)	360	74	1.0940¹⁰⁰/⁴	1.6231¹⁰⁰	al, eth	B27², 43
10206	Oxazole, 4,5-diphenyl .... 4,5-$(C_6H_5)_2(C_3HNO)$	221.26	pl or pr (lig)	192-5¹⁵	44	....	1.6283¹⁰⁰	....	B27, 79
10207	Oxazole, 2,4,5-triphenyl or Azobenzil .... 2,4,5-$(C_6H_5)_3(C_3NO)$	297.36	pr	....	116	....	....	bz	B27², 56
10208	2,4-Oxazolidenedione, 5,5-dipropyl .... $C_9H_{15}NO_3$	185.23	....	148-50³	42-3	....	....	....	Am67, 522
10209	Oxetane or Trimethylene oxide .... $CH_2CH_2CH_2$	58.08	....	47.8	....	0.8930²⁵/⁴	1.3961²⁰	**w, al**, eth, ace	B17⁴, 13
10210	Oxetane, 3-chloro or 2-Chloro-1,3-epoxy propane .... C_3H_5ClO	92.53	....	....	132-4	....	....	al, eth, chl	B17⁴, 14
10211	Oximide .... O=C–C=O NH	71.04	pr (al)	....	....	....	....	....	B21, 368
10212	Oxindole .... C_8H_7NO	133.15	nd (w)	227²³	127	....	....	al, eth	B21⁴, 3611
10213	Oxindole, 1-ethyl .... 1-$C_2H_5(C_8H_6NO)$	161.21	nd (ace, w)	....	97-8	....	....	ace	B21⁴, 3615
10214	Oxindole, 3-hydroxy or Diozindole .... 3-HO(C_8H_6NO)	149.15	cr (w, al)	....	180	....	....	....	B21⁴, 6076
10215	Oxindole, 1-methyl-3-ethyl .... 1-CH_3-3-$C_2H_5(C_8H_5NO)$	175.23	....	280-5⁷⁴⁵, 103-7⁰·⁵	....	....	1.557²⁵	al, eth	B21², 258
10216	Oxacanthine or Vinetine .... $C_{37}H_{40}N_2O_6$	608.71	nd (al, eth), $[\alpha]^{20}_D$ +131.5 (chl, c=1)	....	216-7	....	....	al, eth, bz, chl	B27², 892
10217	Oxyacanthine hydrochloride .... $C_{37}H_{40}N_2O_6 \cdot HCl$	645.17	nd, $[\alpha]^{15}_D$ +163.8 (w, c=3)	....	270-1	....	....	w	B27², 893
10218	Oxyacanthine nitrate, dihydrate .... $C_{37}H_{40}N_2O_6 \cdot HNO_3 \cdot 2H_2O$	705.74	nd	....	195-200	....	....	....	B27², 893

No.	Name, Synonyms, and Formula	Mol. wt.	Color, crystalline form, specific rotation and λ_{max} (log ε)	b.p. °C	m.p. °C	Density	n_D	Solubility	Ref.
10219	Oxynarcotine or a-Narcotine-N-oxide .... $C_{22}H_{23}NO_8$	429.43	hyg nd, $[a]_D$ +135 (chl)	....			....	w, al, chl	B27², 607
10220	Oxysparteine or Isolupanine. .... $C_{15}H_{24}N_2O$	248.36	ye to col hyg nd (peth), $[a]^{18/}_D$ −10.0 (al, c=18)	209¹²	111	....	....	w, al, eth, chl	B24², 56
10221	Oxysparteine monohyrochloride, tetrahydrate .... $C_{15}H_{24}N_2.HCl.4H_2O$	356.89	wh cr (w)	....	48-50	....	....	w, al	B24², 57

No.	Name, Synonyms, and Formula	Mol. wt.	Color, crystalline form, specific rotation and λ_{max} (log ε)	b.p. °C	m.p. °C	Density	n_D	Solubility	Ref.
10222	Palmitaldehyde or Hexadecanal $CH_3(CH_2)_{14}CHO$	240.43	pl (eth), nd (peth)	200-2[29]	34			al, eth, ace, bz	B1[4], 3393
10223	Palmitamide $CH_3(CH_2)_{14}CONH_2$	255.45	lf	236[12]	107				B2[4], 1182
10224	Palmitamide, N-phenyl or Palmitanilide $CH_3(CH_2)_{14}CONHC_6H_5$	331.55	nd (al)	282-4[17]	90.5			al, ace, bz, chl	B12[3], 486
10225	Palmitic acid or Hexadecanoic acid $CH_3(CH_2)_{14}CO_2H$	256.43	nd (al)	350, 267[100]	63	0.8527[62/4]	1.4335[60]	al, eth, ace, bz, chl	B2[4], 1157
10226	Palmitic acid anhydride $(C_{15}H_{31}CO)_2O$	494.85	lf (peth)		64	0.8383[83/4]	1.4364[68]	eth	B2[4], 1181
10227	Palmitic acid, benzyl ester or Benzyl palmitate $C_{15}H_{31}CO_2CH_2C_6H_5$	346.56	cr (al)		36	0.9136[35/25]	1.4689[50]	al, eth, bz, chl	B6[3], 1481
10228	Palmitic acid, butyl ester or Butyl palmitate $C_{15}H_{31}CO_2C_4H_9$	312.54	cr (dil al)		16.9		1.4312[50]	al, eth	B2[4], 1167
10229	Palmitic acid, ethyl ester or Ethyl palmitate $C_{15}H_{31}CO_2C_2H_5$	284.49	nd	191[10]	α: 24, β: 19.3	0.8577[25/4]	1.4347[34]	al, eth, ace, bz, chl	B2[4], 1165
10230	Palmitic acid, hexadecyl ester or Cetyl palmitate $C_{15}H_{31}CO_2C_{16}H_{33}$	480.87	pl (eth)	360	53-4	0.8324[50/4]	1.4425[50]	eth, bz, chl	B2[4], 1168
10231	Palmitic acid, 16-hydroxy $HO(CH_2)_{15}CO_2H$	272.43	cr (bz-eth)		95			al, ace	B3[3], 664
10232	Palmitic acid, (2-hydroxy ethyl) ester or β-hydroxy ethyl palmitate $C_{15}H_{31}CO_2CH_2CH_2OH$	300.49		173-4[3]	51	0.8768[60/4]		al	B2[4], 1168
10233	Palmitic acid, methyl ester or Methyl palmitate $C_{15}H_{31}CO_2CH_3$	270.46		415-8[747], 148[2]	30			al, eth, ace, bz, chl	B2[4], 1165
10234	Palmitic acid, myricyl ester or Myricyl palmitate $C_{15}H_{31}CO_2C_{31}H_{63}$	691.28	cr (eth)		72			eth	B2, 373
10235	Palmitic acid, propyl ester $C_{15}H_{31}CO_2C_3H_7$	298.52	nd	190[12]	20.4	0.8455[88]	1.4392[25]		B2[4], 1167
10236	Palmitic acid, isopropyl ester $C_{15}H_{31}CO_2CH(CH_3)_2$	298.52		160[2]	13-4	0.8404[38]	1.4364[25]	al, eth, ace, bz, chl	B2[4], 1167
10237	Palmitic acid, 9,10,16-trihydroxy or Aleuritic acid $HO(CH_2)_6CH(OH)CH(OH)(CH_2)_7CO_2H$	304.43	lf (dil al), nd (w)		102				B2[4], 1118
10238	Palmitonitrile $C_{15}H_{31}CN$	237.43	hex	333, 251[100]	31	0.8303[20/4]	1.4450[20]	al, eth, ace, bz, chl	B2[4], 1183
10239	Palmityl chloride $C_{15}H_{31}COCl$	274.88		199[20]	12		1.4514[20]	eth	B2[4], 1182
10240	Palmitone $C_{15}H_{31}COC_{15}H_{31}$	450.34	lf (al)		83	0.7947[91/4]	1.4297[94]	eth	B1[4], 3413
10241	Paludrine or Proganil $C_{11}H_{16}ClN_5$	253.74	pl (dil al)		130-1			al	C40, 2931
10242	Paludrine hydrochloride $C_{11}H_{16}ClN_5 \cdot HCl$	290.20	nd (w)		245			al	J1946, 729
10243	Pamelonitrile $NC(CH_2)_8CN$	122.17		175[14]	−31.4	0.949[18]	1.4472[20]	al, eth, chl	B2[4], 2006
10244	Panthesin $C_{18}H_{32}N_2O_5S$	388.53	pa ye pw (al)		157-9			w, al	B3[14], 1058
10245	Pantothenic acid (d) $HOCH_2C(CH_3)_2CH(OH)CONH(CH_2)_2CO_2H$	219.24	ye visc oil, $[\alpha]^{25}_D$ + 37.5 (w)					w, eth, bz	B4[3], 1283
10246	Pantothenic acid, calcium salt (d) or Calcium pantothenate $(C_9H_{16}NO_5)_2Ca$	476.55	wh (MeOH), $[\alpha]^{26}_D$ + 28.2 (w)		195-6			w	B4[3], 1286
10247	Pantothenic acid, calcium salt-(l) or Calcium pantothenate $(C_9H_{16}NO_5)_2Ca$	476.55	cr (MeOH), $[\alpha]^{26}_D$ -27.8 (w)		187-9			w	B4[3], 1288
10248	Pantothenyl alcohol or Panthenol $HOCH_2C(CH_3)_2CH(OH)CONH(CH_2)_3OH$	205.26	hyg oil	d, 118-20[0.02]		1.2[20/20]	1.497[20]	w, al	B4[4], 1652
10249	Papaveraldine or Xanthaline $C_{20}H_{19}NO_5$	353.38	nd (al) cr, (bz, peth)	210-1				bz, chl, aa	B21[4], 6738
10250	Papaverine or Papaveroline tetramethyl ether $C_{20}H_{21}NO_4$	339.40	wh pr (al-eth) nd, (chl-peth)	d, sub, 135-40[11]	147-8	1.337[20/4]	1.625	al, ace, chl, Py	B21[4], 2788
10251	Papaverine hydrochloride $C_{20}H_{21}NO_4 \cdot HCl$	375.86	wh mcl pr (w)		224-5			w, al	B21[4], 2790

No.	Name, Synonyms, and Formula	Mol. wt.	Color, crystalline form, specific rotation and λ_{max} (log ε)	b.p. °C	m.p. °C	Density	n_D	Solubility	Ref.
10252	Parabanic acid or oxalurea . . . . $C_3H_2N_2O_3$	114.06	mcl nd (w)	sub 100	243-5d	. . . .	. . . .	w, al	B24[2], 263
10253	Parabutyraldehyde or 2,4,6-Tripropyl-1,3,5-trioxane . . . $C_{12}H_{24}O_3$	216.32	. . . .	98-100[35]		0.918			B19[4], 4725
10254	Para isobutyraldehyde or 2,4,6-Triisopropyl-1,3,5-triox-ane . . . . $C_{12}H_{24}O_3$	216.32	nd (al)	195 sub	59-60			al, eth	B19[4], 4726
10255	Paraconic acid or Hydroxymethyl succinic acid-γ-lactone $C_5H_6O_4$	130.10	dlq	. . . .	57-8			w	B18[4], 5264
10256	Paracyanogen . . . . $(CN)_x$		br pw	. . . .	sub				B2[4], 1864
10257	Paraldehyde or 2,4,6-Trimethyl-1,3,5-trioxane . . . . $C_6H_{12}O_3$	132.16	. . . .	128	12.6	0.9943[20/4]	1.4049[20]	al, eth, chl	B19[4], 4715
10258	Paraldol . . . . $C_8H_{16}O_4$	176.22	wh tcl pr	90[15]	89-91	1.116[20]	1.4610[20]	w, al, eth	B1[4], 3987
10259	Parasorbic acid . . . . $C_6H_8O_2$	112.14	oily liq, $[\alpha]^{19}_D$ + 210 (al, c=3)	100[15]	. . . . . .	1.079[18/4]	1.4730[20]	w, al, eth	B17[4], 4305
10260	Parathion or Diethyl-p-nitro phenyl mono thiophosphate $C_{10}H_{14}NO_5PS$	291.27	ye liq	375, 157-62[0.5]	6.1	1.2704[20/20]	1.5370[25]	al, eth, ace, chl	B6[4], 1337
10261	Patulin or Clavacin . . . . $C_7H_6O_4$	154.12	pl or pr (eth, chl)	. . . .	111			w, al, eth, ace, bz	B18[4], 1184
10262	Patchouli alcohol . . . . $C_{15}H_{26}O$	222.37	. . . .	. . . .	56	0.9924[65/20]	1.5029[65]	al, eth	B6[3], 426
10263	Paucin, hydrate . . . . $C_{27}H_{39}N_5O_5.6\frac{1}{2}H_2O$	630.75	ye lf	. . . .	126d				B10[2], 1841
10264	Pelargonaldehyde . . . . $C_8H_{17}CHO$	142.24	. . . .	190-2, 93.5[23]	. . . . . .	0.8264[22/4]	1.4275[20]	eth	B1[4], 3352
10265	Pelargonaldehyde osime or Pelargonaldoxime . . . . $C_8H_{17}CH=NOH$	157.26	lf (dil al)	. . . .	64			al, eth, ace	B1[2], 761
10266	Pelargonamide . . . . $CH_3(CH_2)_7CONH_2$	157.26	. . . .	sub	99-100	0.8394[110]	1.4248[110]		B2[4], 1023
10267	Pelargonic acid or Nonanoic acid . . . . $CH_3(CH_2)_7CO_2H$	158.24	. . . .	255, 150[20]	fp 12.2	0.9057[20/4]	1.4343[19]	al, eth	B2[4], 1018
10268	Pelargonic acid, 9-amino . . . . $H_2N(CH_2)_8CO_2H$	173.26	. . . .	185-7				w, al	B4[3], 1479
10269	Pelargonic acid, ethyl ester or Ethyl pelargonate . . . . . . $CH_3(CH_2)_7CO_2C_2H_5$	186.30	. . . .	227, 96-8[10]	−36.7	0.8657[20/4]	1.4220[20]	al, eth, ace	B2[4], 1019
10270	Pelargonic acid, 9-fluoro . . . . $F(CH_2)_8CO_2H$	176.23	. . . .	88-90[0.2]	ca 18		1.4289[25]	al, eth	B2[4], 1024
10271	Pelargonic acid, methyl ester or Methyl pelargonate . . . . . $CH_3(CH_2)_7CO_2CH_3$	172.27	. . . .	213-4[757], 104-6[23]	. . . . . .	0.8799[15]	1.4214[20]	al, eth	B2[4], 1019
10272	Pelargononitrile . . . . $CH_3(CH_2)_7CN$	139.24	. . . .	224.4, 91.9[10]	−34.2	0.8178[20/4]	1.4255[20]	al, eth	B2[4], 1024
10273	Pelargonyl chloride . . . . $CH_3(CH_2)_7COCl$	176.69	. . . .	215.3, 98[15]	−60.5	0.9463[15/4]	. . . .	eth, ace	B2[4], 1023
10274	Pelargonidin chloride . . . . $C_{15}H_{11}ClO_5$	306.70	red br hyg (anh) pr, or pl (dil HCl)	. . . .	>350 (anh)			w, al	B18[4], 3198
10275	Pellotine or N-Methyl anhalonidine . . . . $C_{13}H_{19}NO_3$	237.30	pl (al, peth)	. . . .	111.5			al, eth, ace, chl, peth	B21[4], 2525
10276	Penicilic acid . . . . $C_8H_{10}O_4$	170.17	rh or hex pl (+1w), nd (peth)	. . . .	64-5 (+1w) 8-7 (anh)			w, al, eth, ace, bz	B3[2], 519
10277	Pentacene or 2,3,6,7-Dibenzanthracene . . . . $C_{22}H_{14}$	278.36	deep vt-bl nd or lf (ph NO₂), cr (bz)	290-300 sub (vac)	270-1				B5[3], 2551
10278	Pentacene, 6,13-diphenyl . . . . $C_{34}H_{22}$	430.56	vt-bl nd	318-20	. . . . . .			bz, aa	B5[3], 2738
10279	Pentacosane . . . . $CH_3(CH_2)_{23}CH_3$	352.69	. . . .	401.9, 239.9[10]		0.8012[20/4]	1.4491[20]	bz, chl	B1[4], 582
10280	Pentacosane, 13-phenyl . . . . $C_6H_5CH(C_{12}H_{25})_2$	428.78	. . . .	235[10]	31.7	0.8537[20/4]	1.4787[20/4]		B5[4], 1241
10281	Pentacosane, 13-undecyl . . . . $(C_{12}H_{25})_2CHC_{11}H_{23}$	506.98	. . . .	307[10]	9.7	0.8168[20/4]	1.4567[20]	. . . .	B1[4], 600

No.	Name, Synonyms, and Formula	Mol. wt.	Color, crystalline form, specific rotation and λ_{max} (log ε)	b.p. °C	m.p. °C	Density	n_D	Solubility	Ref.
10282	6,9-Pentadecadiyne $H_2C(C{\equiv}C{-}C_5H_{11})_2$	204.34	....	135-6[4]	......	0.840[21/4]	1.4693[21]	....	B1[4], 1132
10283	Pentadecanal $CH_3(CH_2)_{13}CHO$	226.41	nd	185[25]	24-5	....	......	al, eth, ace	B1[4], 3391
10284	Pentadecanal oxime $CH_3(CH_2)_{13}CH{=}NOH$	241.42	nd (dil al)		86			eth	B1[2], 770
10285	Pentadecane $CH_3(CH_2)_{13}CH_3$	212.42	....	270.6, 136.10	10	0.7685[20/4]	1.4315[20]	al, eth	B1[4], 529
10286	Pentadecane, 1-amino or Pentadecylamine $CH_3(CH_2)_{13}CH_2NH_2$	227.44	fl	307.6, 165.8[10]	37.3	0.8104[20/4]	1.4480[20]	al, eth	B4[4], 817
10287	Pentadecane, 1-bromo or Pentadecyl bromide $CH_3(CH_2)_{13}CH_2Br$	291.33	....	322, 177[10]	19	1.0675[20/4]	1.4611[20]	ace, chl	B1[4], 531
10288	Pentadecane, 1,15-dibromo $Br(CH_2)_{15}Br$	370.23	lf (al)	215-25[15], 192[2]	27	....		chl	B1[4], 531
10289	Pentadecane, 1(2,3-dihydroxyphenyl) or Tetra hydra uru-shiol $(2,3{-}(HO)_2C_6H_3)CH_2(CH_2)_{13}CH_3$	320.52	nd (to, eth, peth)	....	59			al, eth, bz, aa, chl	B6[3], 4771
10290	Pentadeconoic acid or Pentadecylic acid $CH_3(CH_2)_{13}CO_2H$	242.41	pl (aq, al, aa), cr (peth)	257[100], 158[1]	53-4	0.8423[80]	1.4254[80]	al, eth, ace, bz, chl	B2[4], 1147
10291	Pentadecanoic acid, methyl ester or Methyl pentadecon-oate $CH_3(CH_2)_{13}CO_2CH_3$	256.43	nd (dil al)	153.5	18.5	0.8618[25/4]	1.4390[25]	al, eth	B2[3], 936
10292	2-Pentadecanol $C_{13}H_{27}CH(OH)CH_3$	228.42	....	299	35	0.8328[20]	1.4463[20]	....	B1[4], 1871
10293	2-Pentadecanone $C_{13}H_{27}COCH_3$	226.40	....	294	39.5	0.8182[39]	......	....	B1[4], 3391
10294	8-Pentadecanone or Diheptyl ketone $(C_7H_{15})_2CO$	226.43	cr (al)	291, 178[20]	43	....		al, eth, bz, chl	B1[4], 3392
10295	1-Pentadecene $CH_3(CH_2)_{12}CH{=}CH_2$	210.41	....	268.2, 133.7[10]	2-8	0.7764[20/4]	1.4389[20]	ace	B1[4], 926
10296	1-Pentadecyne $CH_3(CH_2)_{12}C{\equiv}CH$	208.39	....	268, 129.8[10]	10	0.7928[20/4]	1.4419[20]	ace	B1[4], 1072
10297	2,4-Pentadienal, 5-phenyl or Cinnamylidene acetaldehyde $C_6H_5CH{=}CHCH{=}CHCHO$	158.20	....	155-65[3], 92-5[0.05]	42-3			**al, eth, bz**	B7[3], 1653
10298	1,2-Pentadiene or Ethyl allene $CH_3CH_2CH{=}C{=}CH_2$	68.13	....	44.9	−137.3	0.6926[20/4]	1.4209[20]	**al, eth, ace, bz**	B1[4], 993
10299	1,2-Pentadiene, 1-chloro-3-ethyl $(C_2H_5)_2C{=}C{=}CHCl$	130.62	....	85-8[100]		0.9297[19/4]	......	eth	B1[3], 1002
10300	1,2-Pentadiene, 1-chloro-3-methyl $CH_3CH_2C(CH_3){=}C{=}CHCl$	116.59	....	68-70[100]		0.9562[20/4]	......	eth	B1[3], 990
10301	1,3-Pentadiene or Piperylene $CH_3CH{=}CHCH{=}CH_2$	68.13	....	42	−87.5	0.6760[20/4]	1.4301[20]	**al, eth, ace, bz**	B1[4], 994
10302	1,3-Pentadiene, 1-chloro-3-methyl $CH_3CH{=}C(CH_3)CH{=}CHCl$	116.59	....	62-3[100]		0.9574[20/4]	......	eth, chl	B1[3], 990
10303	1,3-Pentadiene, 2-chloro-3-methyl $CH_3CH{=}C(CH_3)CCl{=}CH_2$	116.59	....	57-60[95]		0.9437[20/4]	1.4671[20]	eth, chl	B1[3], 991
10304	1,3-Pentadiene, 3-chloro or Methyl chloroprene $CH_3CH{=}C(Cl)CH{=}CH_2$	102.57	....	99-101		0.9576[20/4]	1.4785[20]	eth, ace, bz, chl	B1[3], 962
10305	1,3-Pentadiene, 2,4-dimethyl $(CH_3)_2C{=}CHC(CH_3){=}CH_2$	96.17	....	93[758]		0.7343[23/4]	1.43904[23]	....	B1[4], 1034
10306	1,3-Pentadiene, 4-methyl $(CH_3)_2C{=}CHCH{=}CH_2$	82.14	....	76.5		0.7181[20/4]	1.4532[20]	....	B1[4], 1020
10307	1,4-Pentadiene $CH_2{=}CHCH_2CH{=}CH_2$	68.13	....	26	−148.3	0.6608[20/4]	1.3888[20]	al, eth, ace, bz	B1[4], 998
10308	2,3-Pentadiene $CH_3CH{=}C{=}CHCH_3$	68.13	....	48.2	−125.6	0.6950[20/4]	1.4284[20]	**al, eth, ace, bz**	B1[4], 999
10309	2,4-Pentadienoic acid or β-Vinyl acrylic acid $CH_2{=}CHCH{=}CHCO_2H$	98.10	hyg pr (eth)	d110-5	80			w, al, eth, bz, peth	B2[4], 1694
10310	2,4-Pentadienoic acid, 4-hydroxy-γ-lactone $CH_2{=}C{-}CH{=}CHCO$	96.09	pa ye oil	73[11]				chl	B17[4], 4495
10311	2,4-Pentadienoic acid, 5-(3,4-methylene dioxy)phenyl or Piperic acid $(3,4{-}CH_2O_2)C_6H_3CH{=}CHCH{=}CHCO_2H$	218.21	ye in light nd (al), ye nd (sub)	sub	215			al	B19[4], 3565
10312	2,4-Pentadienoic acid, 5-phenyl or Cinnamylidenacetic acid $C_6H_5CH{=}CHCH{=}CHCO_2H$	174.20	pl (al), pr (bz)	....	166-7			eth, bz	B9[3], 3070

No.	Name, Synonyms, and Formula	Mol. wt.	Color, crystalline form, specific rotation and λ_{max} (log ε)	b.p. °C	m.p. °C	Density	n_D	Solubility	Ref.
10313	2,4-pentadienoic acid, 5-phenyl,ethyl ester . . . . $C_6H_5CH=CHCH=CHCO_2C_2H_5$	202.26	ye oil	149-50[4]	25-7	1.0469[20/4]	1.5768[80]	al, eth	B9[2], 441
10314	2,4-Pentadienoic acid, 5-phenyl,methyl ester . . . . $C_6H_5CH=CHCH=CHCO_2CH_3$	188.23	lf or pl	185[20]	71	. . . .	. . . .	al	B9[3], 3071
10315	2,4-Pentadienonitrile (cis) . . . . $CH_2=CHCH=CHCN$	79.10	. . . .	49.5[32], 32.5[13]	−64	0.8541[26/4]	1.4855[20]	eth, ace	B2[4], 1695
10316	2,4-Pentadienonitrile (trans) . . . . $CH_2=CHCH=CHCN$	79.10	. . . .	41[13]	−43	0.8576[20/4]	1.4986[20]	eth, ace	B2[4], 1696
10317	1,4-Pentadien-3-one, 1,5-bis-(2-ethoxyphenyl) $[(2-C_2H_5OC_6H_4)CH=CH]_2CO$	322.41	ye lf (dil al)	. . . .	89			al	B8[1], 666
10318	1,4-Pentadiene-3-one, 1,5-bis(2-hydroxyphenyl) $[(2-HOC_6H_4)CH=CH]_2CO$	266.30	ye nd (dil al)		168d			al, eth , ace, bz, Py	B8[3], 2954
10319	1,4-Pentadien-3-one, 1,5-bis-(4-hydroxyphenyl) or Bis-(4-hydroxystyryl)ketone . . . . $[(4-HOC_6H_4)CH=CH]_2CO$	266.30	ye-og nd or lf (dil al)		237-8			ace, al	B8[3], 2958
10320	1,4-Pentadien-3-one, 1,5-bis (2-methoxyphenyl) or Bis (2-methoxystyryl)ketone . . . . $[(2-CH_3OC_6H_4)CH=CH]_2CO$	294.35	ye nd or lf (al)		127				B8[2], 405
10321	1,4-Pentadien-3-one, 1,5-bis-(3-methoxyphenyl) or Bis (3-methoxystyryl)ketone . . . . $[(3-CH_3OC_6H_4)CH=CH]_2CO$	294.35	nd (chl-MeOH)		55-6			ace, chl	B81, 666
10322	1,4-Pentadien-3-one, 1,5-bis-(4-methoxyphenyl) or Dianisal acetone . . . . $[(4-CH_3OC_6H_4)CH=CH]_2CO$	294.35	ye lf (aa)		129-30			bz, aa, chl	B8[3], 2958
10323	1,4-Pentadien-3-one, 1,5-bis(3,4-methylene dioxyphenyl) or Dipiperonylidene acetone . . . . $C_{19}H_{14}O_5$	322.32	ye nd (bz)		196	. . . .	. . . .	ace, chl	B19[1], 3031
10324	1,4-Pentadien-3-one, 1,5-bis-(4-nitrophenyl) or Bis (2-nitrostyryl)ketone . . . . $[(2-O_2NC_6H_4)CH=CH]_2CO$	324.30	ye nd (aa)		170-1			chl	B7[2], 455
10325	1,4-Pentadien-3-one, 1,5-bis-(3-nitrophenyl) $[(3-O_2NC_6H_4)CH=CH]_2CO$	324.30	ye br (Ac_2O)		238			ace	B7[2], 455
10326	1,4-Pentadien-3-one, 1,5-bis-(4-nitrophenyl) or Bis(4-nitrosytyryl) ketone . . . . $[(4-O_2NC_6H_4)CH=CH]_2CO$	324.30	ye (Ac_2O)		254			bz	B7[2], 455
10327	1,4-Pentadiene-3-one, 1(2-chlorphenyl)-5-(3-chlorophenyl) $(2-ClC_6H_4)CH=CH-CO-CH=CH(C_6H_4Cl-3)$	303.19	ye nd (dil al)		67-8				B7[2], 454
10328	1,4-Pentadiene-3-one, 1(2-chlorophenyl)-5-(4-chlorophenyl) . . . . $(2-ClC_6H_4)CH=CH-CO-CH=CH(C_6H_4Cl-4)$	303.19	ye nd (al)		109				B7[2], 454
10329	1,4-Pentadien-3-one, 1,5-di(2-furyl) . . . . $[2-(C_4H_3O)CH=CH]_2CO$	214.22	dlq pr (peth), ye pr (lig)	d	60-1			al, eth, chl	B19[4], 1822
10330	1,4-Pentadien-3-one, 1,5-diphenyl or Dibenzal acetone . . . $[C_6H_5CH=CH]_2CO$	234.30	pl or lf (ace, AcOEt)	d	113d			ace, chl	B7[3], 2559
10331	1,3-Pentadiyne $CH_3C\equiv CC\equiv CH$	64.09		55-6	. . . .	0.7375[20/4]	1.4431[21]	eth, bz, chl	B1[4], 1117
10332	Pentaerythritol or Tetramethylol methane . . . . $C(CH_2OH)_4$	136.15	cr (dil HCl)	sub	269	. . . .	1.548	w	B1[4], 2812
10333	Pentaerythritol tetra acetate . . . . $(CH_3CO_2CH_2)_4C$	304.30	tetr nd (w, bz)	. . . .	83-4	1.273[18/4]	. . . .	w, al, eth	B1[4], 264
10334	Pentaerythrityl tetrabromide or 2,2-bis(bromo ethyl)-1,3-dibromo propane . . . . $C(CH_2Br)_4$	387.76	cr (ace), nd (lig)	305-6	163	2.596[15]	. . . .	al, bz	B1[4], 337
10335	Pentaerythrityl tetrachloride or 2,2-bis(chloromethyl)-1,3-dichloropropane . . . . $C(CH_2Cl)_4$	209.94	. . . .	110[12]	97			eth, chl	B1[4], 336
10336	Pentaerythrityl tetraiodide or 2,2-bis(iodomethyl)-1,3-diiodo methane . . . . $C(CH_2I)_4$	575.75	nd (to)	. . . .	225				B1[4], 338
10337	Pentaerythritol tetranitrate or PETN . . . . $C(CH_2ONO_2)_4$	316.15	tetr (ace), pr (ace-al)	. . . .	140-1	1.773[20/4]	. . . .	ace, bz	B1[4], 2816
10338	Pentamethylene sulfide or Tetrahydro thiopyran . . . . $C_5H_{10}S$	102.20	. . . .	141.7, 93[82]	19	0.9861[20/4]	1.5067[20]	al, eth, ace, bz	B17[4], 55
10339	Pentanal or Valeraldehyde . . . . $CH_3(CH_2)_3CHO$	86.14	. . . .	103	−91.5	0.8095[20/4]	1.3944[20]	al, eth	B1[4], 3268

No.	Name, Synonyms, and Formula	Mol. wt.	Color, crystalline form, specific rotation and λ_{max} (log ε)	b.p. $^\circ C$	m.p. $^\circ C$	Density	n_D	Solubility	Ref.
10340	Pentane . . . . $CH_3(CH_2)_3CH_3$	72.15	. . . .	36.1	−130	$0.6262^{20/4}$	1.3575^{20}	al, eth, ace, bz, chl	B1[4], 303
10341	Pentane, 1-amino or Pentylamine . . . . $CH_3(CH_2)_3CH_2NH_2$	87.17	. . . .	104.4, 5.9[10]	−55	$0.7547^{20/4}$	1.4118^{20}	al, eth, ace, bz	B4[4], 675
10342	Pentane, 2-amino . . . . $C_3H_7CH(NH_2)CH_3$	87.17	. . . .	91.5^{755}	. . . .	0.7384^{20}	1.4027^{20}	w, al, eth, ace, bz	B4[4], 689
10343	Pentane, 3-amino . . . . $H_2NCH(C_2H_5)_2$	87.17	. . . .	91	. . . .	$0.7487^{20/4}$	1.4063^{20}	al	B4[4], 692
10344	Pentane, 3,3-bis(ethyl sulfonyl) . . . . $(C_2H_5)_2C(SO_2C_2H_5)_2$	256.39	lf (dil al)	. . . .	85	. . . .	. . . .	al, eth, w	B4, 681
10345	Pentane, 1-bromo or Pentyl bromide . . . . $CH_3(CH_2)_3CH_2Br$	151.05	. . . .	129.6, 21^{10}	−87.9	$1.2182^{20/4}$	1.4447^{20}	al, eth, bz, chl	B1[4], 312
10346	Pentane, 1-bromo-5-fluoro . . . . $F(CH_2)_5Br$	169.04	. . . .	162	. . . .	1.3604^{25}	1.4406^{25}	al, eth	B1[4], 313
10347	Pentane, 1-bromo-2-methyl . . . . $C_3H_7CH(CH_3)CH_2Br$	165.08	. . . .	142-5, $51-3^{25}$	. . . .	$1.1624^{20/4}$	1.4495^{20}	eth, chl	B1[4], 361
10348	Pentane, 1-bromo-3-methyl . . . . $C_2H_5CH(CH_3)CH_2CH_2Br$	165.08	. . . .	$148-9^{766}$	. . . .	$1.1829^{20/4}$	1.4496^{20}	eth, chl	B1[4], 365
10349	Pentane, 1-bromo-4-methyl or Isohexyl bromide . . . . $(CH_3)_2CH(CH_2)_3Br$	165.08	. . . .	147-8	. . . .	1.1683^{20}	1.4490	eth, chl	B1[4], 361
10350	Pentane, 2-bromo . . . . $C_3H_7CHBrCH_3$	151.05	. . . .	117.4, 58.4^{100}	−95.5	$1.2075^{20/4}$	1.4413^{20}	al, eth, bz, chl	B1[4], 312
10351	Pentane, 2-bromo-2-methyl . . . . $C_3H_7CBr(CH_3)_2$	165.08	. . . .	142-3, 70^{100}	. . . .	. . . .	1.442^{23}	eth, chl	B1[4], 361
10352	Pentane, 3-bromo . . . . $BrCH(C_2H_5)_2$	151.05	. . . .	118.6, 10.8^{10}	−126.2	$1.2124^{20/4}$	$1.4441^{20/}$	al, eth, bz, chl	B1[4], 313
10353	Pentane, 3-bromo-3-methyl . . . . $(C_2H_5)_2C(Br)CH_3$	165.08	. . . .	129-31, $82-3^{145}$	. . . .	1.1835^{20}	1.4525^{20}	eth, chl	B1[4], 365
10354	Pentane, 1-chloro or Pentyl chloride . . . . $CH_3(CH_2)_3CH_2Cl$	106.60	. . . .	107.8, 5^{10}	−99	$0.8818^{20/4}$	1.4127^{20}	al, eth, bz, chl	B1[4], 309
10355	Pentane, 1-chloro-5-fluoro . . . . $F(CH_2)_5Cl$	124.59	. . . .	143.2	. . . .	1.0325^{25}	1.4120^{23}	al, eth	B1[4], 309
10356	Pentane, 2-chloro . . . . $C_3H_7CHClCH_3$	106.60	$[\alpha]_D$ +34.1	96.9	−137	$0.8698^{20/4}$	1.4069^{20}	al, eth, bz, chl	B1[4], 309
10357	Pentane, 2-chloro-2,3-dimethyl . . . . $C_2H_5CH(CH_3)CCl(CH_3)_2$	134.65	. . . .	$38-9^{20}$	. . . .	. . . .	1.4264^{20}	eth, bz	B1[4], 406
10358	Pentane, 2-chloro-2,4-dimethyl . . . . $(CH_3)_2CHCH_2CCl(CH_3)_2$	134.65	. . . .	$127-8^{733}, 33-4^{20}$	. . . .	$0.861^{20/4}$	1.4180^{20}	eth	B1[4], 408
10359	Pentane, 2-chloro-3-ethyl . . . . $(C_2H_5)_2CHCH(Cl)CH_3$	134.65	. . . .	83.5^{100}	. . . .	$0.8951^{25/25}$	1.4318^{20}	eth, chl	B1[2], 120
10360	Pentane, 2-chloro-2-methyl . . . . $C_3H_7CCl(CH_3)_2$	120.62	. . . .	$110-1^{734/}d$, $36-7^{15}$	. . . .	$0.863^{20/4}$	1.4126^{20}	eth	B1[4], 360
10361	Pentane, 2-chloro-4-methyl . . . . $(CH_3)_2CHCH_2CHClCH_3$	120.62	. . . .	$111-2^{733}$	. . . .	$0.8610^{20/4}$	1.4113^{20}	eth	B1[3], 399
10362	Pentane, 2-chloro-2,4,4-trimethyl . . . . $(CH_3)_2CClCH_2C(CH_3)_3$	148.68	. . . .	145-50d, 44^{16}	−26	0.8746^{20}	1.4308^{20}	al	B1[4], 444
10363	Pentane, 3-chloro . . . . $ClCH(C_2H_5)_2$	106.60	. . . .	97.8, $38-9^{20}$	−105	$0.8731^{20/4}$	1.4082^{20}	al, eth, bz, chl	B1[4], 309
10364	Pentane, 3-chloro-2,2-dimethyl-3-ethyl . . . . $(C_2H_5)_2CClC(CH_3)_3$	162.71	. . . .	d $53-4^6$	. . . .	. . . .	1.4528^{25}	eth, chl	B1[4], 462
10365	Pentane, 3-chloro-2,3-dimethyl . . . . $C_2H_5C(CH_3)ClCH(CH_3)_2$	134.65	. . . .	$135-8^{757}d$, $41-2^{20}$	. . . .	$0.884^{22/22}$	1.4318^{20}	eth, chl	B1[4], 406
10366	Pentane, 3-chloro-3-ethyl . . . . $(C_2H_5)_3CCl$	134.65	. . . .	143-4, $43-4^{20}$	. . . .	$0.8856^{20/4}$	1.4400^{20}	eth	B1[4], 403
10367	Pentane, 3-chloro-3-ethyl-2-methyl . . . . $(C_2H_5)_2CClCH(CH_3)_2$	148.68	. . . .	150-5d, $98-100^{98}$	. . . .	1.0325^{25}	1.4120^{23}	al, eth	B1[4], 437
10368	Pentane, 3-chloro-2-methyl . . . . $C_2H_5CHClCH(CH_3)_2$	120.62	. . . .	$115-7^{752}d$	. . . .	. . . .	. . . .	eth, chl	B1[2], 111
10369	Pentane, 3-chloro-3-methyl . . . . $(C_2H_5)_2C(Cl)CH_3$	120.62	. . . .	116, 35^{25}	. . . .	$0.8900^{20/4}$	1.4210^{20}	eth, bz, chl	B1[4], 365
10370	Pentane, 3-(chloromethyl) . . . . $(C_2H_5)_2CHCH_2Cl$	120.62	. . . .	125-7	. . . .	$0.8914^{20/4}$	1.4222^{20}	eth, bz, chl	B1[3], 403
10371	Pentane, 4-chloro-2,2-dimethyl . . . . $CH_3CHClCH_2C(CH_3)_3$	134.65	. . . .	93^{250}	. . . .	$0.855^{20/4}$	1.4180^{20}	eth, chl	B1[4], 404
10372	Pentane, 1,5-diamino or Pentamethylendiamine Cadaverine . . . . $H_2N(CH_2)_5NH_2$	102.18	. . . .	178-80	9	0.867^{25}	1.4561^{25}	w, al	B4[4], 1310

No.	Name, Synonyms, and Formula	Mol. wt.	Color, crystalline form, specific rotation and λ_{max} (log ε)	b.p. °C	m.p. °C	Density	n_D	Solubility	Ref.
10373	Pentane, 1,5-dibromo .... Br(CH$_2$)$_5$Br	229.95	....	222.3, 98.6[20]	−39.5	1.7018[20/4]	1.5126[20]	bz, chl	B1[4], 314
10374	Pentane, 1,2-dichloro .... C$_3$H$_7$CHClCH$_2$Cl	141.04	....	148-9, 58-9[20]	......	1.0872[20/4]	1.4485[20]	al, chl	B1[3], 341
10375	Pentane, 1,2-dichloro-4,4-dimethyl (CH$_3$)$_3$CCH$_2$CHClCH$_2$Cl	169.10	....	173-5[745], 58-9[12]	......	1.0259[20]	1.4489[20]	bz, chl	B1[3], 445
10376	Pentane, 1,3-dichloro .... C$_2$H$_5$CHClCH$_2$CH$_2$Cl	141.04	....	80.4[60]	......	1.0834[20/4]	1.4485[20]	ace, chl	B1[3], 342
10377	Pentane, 1,4-dichloro .... CH$_3$CHClCH$_2$CH$_2$CH$_2$Cl	141.04	....	161-3, 58-60[15]	......	1.0840[20/4]	1.4503[20]	ace, chl	B1[4], 309
10378	Pentane, 1,5-dichoro .... Cl(CH$_2$)$_5$Cl	141.04	....	180, 59[10]	−72.8	1.1006[22/4]	1.4564[20]	al, eth, bz, chl	B1[4], 310
10379	Pentane, 1,5-dichloro-3,3-dimethyl (ClCH$_2$CH$_2$)$_2$C(CH$_3$)$_2$	169.10	....	135[80], 58-9[8]	......	1.0563[20/4]	1.4652[20]	chl	B1[4], 409
10380	Pentane, 2,2-dichloro .... C$_3$H$_7$CCl$_2$CH$_3$	141.04	....	128-9, 36-7[20]	......	1.040[20]	1.434[20]	eth, bz, chl	B1[3], 342
10381	Pentane, 2,4-dichloro .... CH$_3$(CHClCH$_3$)$_2$	141.04	....	147-50, 62[12]	......	1.0634[15/4]	1.447[18]	eth, bz, chl	B1[3], 343
10382	Pentane, 2,4-dichloro-2,4-dimethyl CH$_2$[CCl(CH$_3$)$_2$]$_2$	169.10	....	51-7[8]	23-4	1.0292[20/4]	1.4537[20]	....	B1[4], 408
10383	Pentane, 3,3-dichloro .... (C$_2$H$_5$)$_2$CCl$_2$	141.04	....	131-2[750], 32[14]	......	1.053[20]	1.442[20]	bz, chl	B1[3], 343
10384	Pentane, 3,3-dichloro-2,4-dimethyl [(CH$_3$)$_2$CH]$_2$CCl$_2$	169.10	....	118-20d	......	0.9513[9]	....	eth, chl	B1, 158
10385	Pentane, 3,3-diethyl or Tetraethyl methane C(C$_2$H$_5$)$_4$	128.26	....	146.2, 20.7[10]	−33.1	0.7536[20/4]	1.4206[20]	eth, bz	B1[4], 462
10386	Pentane, 1,5-diiido .... I(CH$_2$)$_5$I	323.94	....	149[20], 101-2[3]	9	2.1903[15]	1.6046[15]	eth, chl	B1[4], 317
10387	Pentane, 2,2-dimethyl .... C$_3$H$_7$C(CH$_3$)$_3$	100.21	....	79.2	−123.8	0.6739[20/4]	1.3822[20]	al, eth, ace, bz, chl	B1[4], 403
10388	Pentane, 2,2-dimethyl-3-ethyl (CH$_3$)$_3$CCH(C$_2$H$_5$)$_2$	128.26	....	133.83	99.3	0.74378[20/4]	1.41227[20]	....	B1[4], 462
10389	Pentane, 2,3-dimethyl .... C$_2$H$_5$CH(CH$_3$)CH(CH$_3$)$_2$	100.21	....	89.8	......	0.6951[20/4]	1.3919[20]	al, eth, ace, bz, chl	B1[4], 405
10390	Pentane, 2,3-dimethyl-3-ethyl (CH$_3$)$_2$CHC(C$_2$H$_5$)$_2$CH$_3$	128.26	....	141.6	......	....	1.4186[20]	....	B1[4], 462
10391	Pentane, 2,4-dimethyl .... CH$_2$[CH(CH$_3$)$_2$]$_2$	100.21	....	80.5	−119.2	0.6727[20/4]	1.3815[20]	al, eth, ace, bz, chl	B1[4], 406
10392	Pentane, 2,4-dimethyl-3-ethyl CH$_3$CH(CH$_3$)CH(C$_2$H$_5$)CH(CH$_3$)$_2$	128.26	....	136.73	−122.4	0.7365[20/4]	1.4131[20]	....	B1[4], 462
10393	Pentane, 2,4-dimethyl-3-phenyl (CH$_3$)$_2$CHCH(C$_6$H$_5$)CH(CH$_3$)$_2$	176.30	....	220-5	......	0.8822[10/0]	1.512	....	B5[1], 214
10394	Pentane, 3,3-dimethyl .... (C$_2$H$_5$)$_2$C(CH$_3$)$_2$	100.21	....	86.1	−134.4	0.6936[20/4]	1.3909[20]	al, eth, ace, bz, chl	B1[4], 409
10395	Pentane, 1,5-dinitro .... O$_2$N(CH$_2$)$_5$NO$_2$	162.14	....	134[1.2]	......	....	1.461[20]	bz	B1[4], 319
10396	Pentane, 1,1-diphenyl .... C$_4$H$_9$CH(C$_6$H$_5$)$_2$	224.35	....	307.9	−12	0.9659[20]	1.5511[20]	....	B5[4], 1966
10397	Pentane, 1,5-diphenyl .... C$_6$H$_5$(CH$_2$)$_5$C$_6$H$_5$	224.35	....	330.6, 187-9[10]	......	0.9814[19/0]	1.559[19]	....	B5[4], 1963
10398	Pentane, 1,2-epoxy-2,4,4-trimethyl (CH$_3$)$_3$CCH$_2$C(CH$_3$)CH$_2$	128.22	....	140.9, 20[5.4]	−64	0.8287[20/20]	1.4097[20]	eth, bz	B17[4], 112
10399	Pentane, 3-ethyl or Triethyl methane (C$_2$H$_5$)$_3$CH	100.21	....	93.5	−118.6	0.6982[20/4]	1.3934[20]	al, eth, ace, bz, chl	B1[4], 402
10400	Pentane, 3-ethyl-2-methyl .... (C$_2$H$_5$)$_2$CHCH(CH$_3$)$_2$	114.23	....	115.6, 9.5[10]	−115	0.7193[20/4]	1.4040[20]	al, eth, ace, bz, chl	B1[4], 437
10401	Pentane, 3-ethyl-3-methyl .... (C$_2$H$_5$)$_3$CCH$_3$	114.25	....	118.2, 9.9[10]	−90.9	0.7274[20/4]	1.4078[20]	**al**, eth, **ace, bz,** chl	B1[4], 438
10402	Pentane, 1-fluoro or Pentyl fluoride........ C$_5$H$_{11}$F	90.14	....	62.8	−120	0.7907[20/4]	1.3591[20]	al, eth	B1[4], 308
10403	Pentane, 1-iodo or Pentyl iodide.......... C$_5$H$_{11}$I	198.05	....	157, 39.3[10]	−85.6	1.5161[20/4]	1.4959[20]	al, eth	B1[4], 315
10404	Pentane, 2-iodo .... C$_3$H$_7$CHICH$_3$	198.05	....	144-5	......	1.5096[20/4]	1.4961[20]	eth, ace, bz	B1[4], 316
10405	Pentane, 3-iodo .... (C$_2$H$_5$)$_2$CHI	198.05	....	145-6, 68[50]	......	1.5176[20/4]	1.4974[20]	eth, ace, bz	B1[3], 349

No.	Name, Synonyms, and Formula	Mol. wt.	Color, crystalline form, specific rotation and λ_{max} (log ε)	b.p. °C	m.p. °C	Density	n_D	Solubility	Ref.
10406	Pentane, 2-methyl $C_3H_7CH(CH_3)_2$	86.18	....	60.3	−153.7	$0.6532^{20/4}$	1.3715^{20}	al, eth, **ace**, bz, chl	B1[4], 358
10407	Pentane, 2-methyl-2-phenyl $C_6H_5C(CH_3)_2CH_2CH_2CH_3$	162.27	....	205-6	......	$0.8796^{10/4}$	$1.4955^{16.5}$	....	B5[3], 1019
10408	Pentane, 3-methyl $(C_2H_5)_2CHCH_3$	86.18	....	63.3	......	$0.6645^{20/4}$	1.3765^{20}	al, eth, **ace**, bz	B1[4], 363
10409	Pentane, 3-methyl-1-phenyl $C_6H_5CH_2CH_2CH(CH_3)C_2H_5$	162.27	....	220^{757}	......	$0.8644^{14.5/4}$	1.4896	....	B5[4], 1116
10410	Pentane, 3-nitro $(C_2H_5)_2CHNO_2$	117.15		153-5	......	$0.957^{0/4}$	....	al, eth, ace	B1[4], 318
10411	Pentane, perfluoro C_5F_{12}	288.04	....	57.73, −30.5[10]	......	$1.7326^{20/4}$	1.2564^{22}	bz	B1[4], 308
10412	Pentane, 3-phenyl $C_6H_5CH(C_2H_5)_2$	148.25	....	187.5, 83-5[22]	......	$0.8649^{20/4}$	1.4880^{20}	....	B5[4], 1090
10413	Pentane, 1,1,1,5-tetrachloro $CCl_3(CH_2)_3CH_2Cl$	209.93	....	112^{24}	......	1.3416^{25}	1.4859^{25}	....	B1[4], 311
10414	Pentane, 2,2,3,3,-tetramethyl $(CH_3)_3CC(CH_3)_2C_2H_5$	128.26	....	140.27^{760}	......	....	1.4236^{20}	....	B1[4], 463
10415	Pentane, 2,2,3,4-tetramethyl $(CH_3)_3CCH(CH_3)CH(CH_3)_2$	128.26	....	133, 70.6[104]	......	$0.7389^{20/4}$	1.4147^{20}	....	B1[4], 463
10416	Pentane, 2,2,4,4-tetramethyl $CH_2[C(CH_3)_3]_2$	128.26	....	122.7, 12.5[10]	−66.5	$0.7195^{20/4}$	1.4069^{20}	al, bz	B1[4], 464
10417	Pentane, 2,3,3,4-tetramethyl $(CH_3)_2C[CH(CH_3)_2]_2$	128.26	....	141.5, 78[104]	-102.14	$0.7547^{20/4}$	1.4222^{20}	....	B1[4], 464
10418	Pentane, 2,2,3-trimethyl $C_2H_5CH(CH_3)C(CH_3)_3$	114.23	....	110, 3.9[10]	−112.3	$0.7161^{20/4}$	$1.4030^{20/4}$	al, eth, **ace**, bz, chl	B1[4], 438
10419	Pentane, 2,2,4-trimethyl or Isooctane $(CH_3)_3CCH_2C(CH_3)_3$	114.23	....	99.2, 4.3[10]	−107.4	$0.6919^{20/4}$	1.3915^{20}	al, eth, **ace**, bz, chl	B1[4], 439
10420	Pentane, 2,3,3-trimethyl $C_2H_5C(CH_3)_2CH(CH_3)_2$	114.23	....	114.7, 6.9[10]	−100.7	$0.7262^{20/4}$	1.4075^{20}	al, eth, **ace**, bz, chl	B1[4], 445
10421	Pentane, 2,3,4-trimethyl $CH_3CH[CH(CH_3)_2]_2$	114.23	....	113.4, 7.1[10]	−109.2	$0.7191^{20/4}$	1.4042^{20}	al, eth, **ace**, bz, chl	B1[4], 446
10422	Isopentane or 2-Methyl butane $CH_3CH_2CH(CH_3)_2$	72.15	....	27.8	−159.9	$0.6201^{20/4}$	1.3537^{20}	**al**, eth	B1[4], 320
10423	1,5-Pentanedial or Glutyraldehyde $OHC(CH_2)_3CHO$	100.13		187-9d, 71-2[10]	......			**w, al**, bz	B1[4], 3659
10424	Pentanedioic acid or Glutaric acid $HO_2C(CH_2)_3CO_2H$	132.13	nd (bz)	302-4d, 200[20]	99	$1.424^{25/4}$	1.4188^{106}	w, al, eth, chl	B2[4], 1934
10425	1,2-Pentanediol (d) or 1,2-Pentylene glycol $C_3H_7CHOHCH_2OH$	104.15	[α] +0.95	$210-12^{751}$, 99-102[13]	......	$0.9802^{20/20}$	1.4412^{19}	....	B1[4], 2538
10426	1,2-Pentanediol, 2,4,4-trimethyl $(CH_3)_3CCH_2C(OH)(CH_3)CH_2OH$	146.23	pr or pl (bz)	....	62-3			w, al, eth	B1[4], 2604
10427	1,3-Pentanediol, 2,2-dimethyl $C_2H_5CH(OH)C(CH_3)_2CH_2OH$	132.21	(eth)	212-4, 119[21]	60-3			al	B1, 490
10428	1,3-Pentanediol, 2,2,4-trimethyl $(CH_3)_2CHCHOHC(CH_3)_2CH_2OH$	146.23	pl (bz)	234^{737}, 81-2[1]	51-2	$0.937^{15/15}$	1.4513^{15}	al, eth	B1[4], 2604
10429	1,4-Pentanediol or γ-Pentylene glycol $CH_3CHOHCH_2CH_2CH_2OH$	104.15	....	220^{715}, 124-6[10]	......	$0.9883^{20/4}$	1.4452^{23}	**w, al**, chl	B1[4], 2539
10430	1,4-Pentanediol, 2,2,4-trimethyl $(CH_3)_2C(OH)CH_2C(CH_3)_2CH_2OH$	146.23	cr (eth)	209.11, 114-5[13]	86	....	....	al, eth	B1, 493
10431	1,5-Pentanediol or Pentamethylene glycol $HO(CH_2)_5OH$	104.15	....	260, 137-8[12]	−18	$0.9939^{20/20}$	1.4494^{20}	**w, al**	B1[4], 2540
10432	1,5-Pentanediol, diacetate or 1,5-Diacetoxy pentane $CH_3CO_2(CH_2)_5O_2CCH_3$	188.22	....	122-3[3]	......	1.0296^{20}	1.4261^{19}	....	B2[4], 226
10433	1,5-Pentanediol, 2,2-dimethyl $HOCH_2CH_2CH_2C(CH_3)_2CH_2OH$	132.20	....	130[12]	......			al, eth	B1[1], 254
10434	2,3-Pentanediol $C_2H_5CHOHCHOHCH_3$	104.15	....	187.5, 97[17]	......	$0.9800^{19/0}$	1.4412^{25}	**w, al**	B1[4], 2543
10435	2,3-Pentanediol, 2,4,4-trimethyl $(CH_3)_3CCH(OH)C(CH_3)(OH)CH_3$	146.23	mcl pr (lig)	....	65-6			al, eth	B1[4], 2604
10436	2,4-Pentanediol, 2-methyl $CH_3CH(OH)CH_2C(OH)(CH_3)_2$	188.18	....	197	......	$0.9254^{17/4}$	1.4250^{20}	w, al, eth	B1[4], 2565
10437	2,4-Pentanediol, 3-methyl $CH_3CH[CH(OH)CH_3]_2$	118.18	....	211-2, 91[3]	......	0.9640^{20}	1.4433^{20}	w, al	B1[4], 2572
10438	2,3-Pentanedione $C_2H_5COCOCH_3$	100.13	dk ye liq	108	......	$0.9565^{19/4}$	1.4014^{19}	w, **al, eth**, ace	B1[4], 3660

No.	Name, Synonyms, and Formula	Mol. wt.	Color, crystalline form, specific rotation and λ_{max} (log ϵ)	b.p. °C	m.p. °C	Density	n_D	Solubility	Ref.
10439	2,3-Pentanedione, dioxime $C_2H_5C(NOH)C(NOH)CH_3$	130.15	ye nd (al), pl (to,al)	sub	172-3	....	....	al	B1[4], 3661
10440	2,3-Pentanedione, 2-oxime $C_2H_5COC)NOH)CH_3$	115.13	lf (dil al)	....	69-72			al, eth	B1[4], 3661
10441	2,3-Pentanedione, 3-oxime $C_2H_5C(NOH)COCH_3$	115.13	pl (lig)	183-7d	58-9			al, eth, chl	B1[4], 3661
10442	2,4-Pentanedione or Acetylacetone $CH_3COCH_2COCH_3$	100.13	....	139[746]	−23	0.9721[25/4]	1.4494[20]	w, al, eth, ace, chl	B1[4], 3662
10443	2,4-Pentanedione, 3,3-dimethyl $(CH_3CO)_2C(CH_3)_2$	128.17		173, 58[10]	19	0.9575[20/4]	1.4306[20]	eth	B1[4], 3705
10444	2,4-Pentanedione, dioxime $CH_2[C(=NOH)CH_3]_2$	130.15	pr (eth)	....	149-50			al	B1[3], 3123
10445	2,4-Pentanedione, 3-ethyl $(CH_3CO)_2CHC_2H_5$	128.17		177-80, 69-70[13]		0.9531[19/4]	1.4408[19]	al, eth, chl	B1[4], 3703
10446	2,4-Pentanedione, monoimide $CH_3COCH_2C(NH)CH_3$	99.13		209	43			w, eth	B1[4], 3678
10447	1-Pentanethiol or Pentylmercaptan $CH_3(CH_2)_3CH_2SH$	104.22		126.6[460], 99.5[10]	−75.7	0.8421[20/4]	1.4469[20]	al, eth	B1[4], 1653
10448	2-Pentanethiol $C_3H_7CH(SH)CH_3$	104.22		112.9, 63.9[150]	−169	0.8327[20/4]	1.4412[20]	al, lig	B1[4], 1662
10449	3-Pentanethiol $(C_2H_5)_2CHSH$	104.22		105	−110.8	0.8410[20/4]	1.4447[20]	al	B1[4], 1665
10450	3-Pentanecarboxylic acid-1,5-dinitrile, 3-acetyl, ethyl ester $CH_3COC(CH_2CN)_2CO_2C_2H_5$	236.27	cr	190-200[2]	83				B3[2], 512
10451	1,2,3-Pentanetriol $C_2H_5(CHOH)_2CH_2OH$	120.15	syr	192[63]		1.0851[34/0]	....	w, al, eth	B1[4], 2778
10452	1,2,5-Pentanetriol $HOCH_2CHOH(CH_2)_2OH$	120.15		190-1[13]		1.136[20/15]	1.4730[20]	w, al	B1[4], 2779
10453	1,3,5-Pentanetriol $(HOCH_2CH_2)_2CHOH$	120.15		188-9[11]		1.1291[20/4]	1.4785[20]	w, al, eth, ace	B1[4], 2779
10454	Pentanoic acid or Valeric acid $CH_3(CH_2)_3CO_2H$	102.13		186, 82.7[10]	−33.8	0.9391[20/4]	1.4085[20]	w, al, eth	B2[4], 868
10455	1-Pentanol or n-Pentyl alcohol $CH_3(CH_2)_3CH_2OH$	88.15		137.3[748], 50[13]	−79	0.8144[20/4]	1.4101[20]	al, eth, ace	B1[4], 1640
10456	1-Pentanol, 5-amino $HO(CH_2)_5NH_2$	103.16		221-2, 126-8[22]	38-9	0.9488[17/4]	1.4618[17]	w, al, ace	B4[4], 1750
10457	1-Pentanol, 5-chloro $Cl(CH_2)_5OH$	122.60		112[12]			1.4518[20]	al, eth	B1[4], 1650
10458	1-Pentanol, 2,4-dimethyl((dl) $(CH_3)_2CHCH_2CH(CH_3)CH_2OH$	116.21		160-2, 65-7[1b]		0.793[20/4]	1.427[20]	al, eth	B1[4], 1753
10459	1-Pentanol, 2,4-dimethyl (l) $(CH_3)_2CHCH_2CH(CH_3)CH_2OH$	116.21	$[\alpha]^{22}_D$ −1.1 (undil)	157		0.816[25/4]	....	al, eth	B1[4], 1753
10460	1-Pentanol, 5-fluoro $F(CH_2)_5OH$	106.14		70-1[11]			1.4057[25]	al, eth	B1[4], 1647
10461	1-Pentanol, 2-methyl $C_3H_7CH(CH_3)CH_2OH$	102.18		148		0.8263[20/4]	1.4182[20]	al, eth, ace	B1[4], 1713
10462	1-Pentanol, 3-methyl(-(dl) $C_2H_5CH(CH_3)CH_2CH_2OH$	102.18		152.4, 51-3[8]		0.8242[20/4]	1.4112[23]	al, eth	B1[4], 1722
10463	1-Pentanol, 4-methyl or Isohexyl alcohol $(CH_3)_2CH(CH_2)_3OH$	102.18		151.6		0.8131[20/4]	1.4134[25]	al, eth	B1[4], 1721
10464	1-Pentanol, 1-phenyl $C_4H_9CH(OH)C_6H_5$	164.25	$[\alpha]^{20}_D$ +40.8	140-2[25]		0.9672[20/20]	1.4086[25]	al, eth, ace	B6[4], 1952
10465	1-Pentanol, 5-phenyl $C_6H_5(CH_2)_5OH$	164.25		155[20]		0.9725[20]	1.5156[20]	al, eth	B6[4], 1954
10466	2-Pentanol $C_3H_7CH(OH)CH_3$	88.15	....	118.9, 62[60]		0.8103[20/4]	1.4053[20]	w, al, eth	B1[4], 1655
10467	2-Pentanol, 1-chloro $C_3H_7CH(OH)CH_2Cl$	122.60		157-60[735], 59-62[14]		1.037[20/20]	1.4404[25]	al, eth	B1[3], 1613
10468	2-Pentanol, 2,4-dimethyl $(CH_3)_2CHCH_2C(OH)(CH_3)_2$	116.21		133.1, 53-4[25]	>−20	0.8103[20/4]	1.4172[20]	al, eth	B1[4], 1753
10469	2-Pentanol, 2-methyl $C_3H_7C(OH)(CH_3)_2$	102.18	col	120-2, 49.5[27.5]	−103	0.8350[16/4]	1.4100[20]	al, eth	B1[4], 1714
10470	2-Pentanol, 3-methyl $C_2H_5CH(CH_3)CH(OH)CH_3$	102.18		134.3, 75.6[50]		0.8307[20/4]	1.4182[20]	al, eth	B1[3], 1672
10471	2-Pentanol, 4-methyl $(CH_3)_2CHCH_2CH(OH)CH_3$	102.18		133, 50-5[25]		0.8075[20/4]	1.4100[20]	al, eth	B1[4], 1717

No.	Name, Synonyms, and Formula	Mol. wt.	Color, crystalline form, specific rotation and λ_{max} (log ϵ)	b.p. °C	m.p. °C	Density	n_D	Solubility	Ref.
10472	2-Pentanol, 4-methyl,acetate or 4-Methyl-2-pentyl acetate $(CH_3)_2CHCH_2CH(CH_3)O_2CCH_3$	144.22	....	147-8, 76.5[47]	......	0.8805[0/0]	1.4066[20]	ace, bz	B2,133
10473	2-Pentanol, 2-phenyl .... $C_3H_7C(OH)(CH_3)C_6H_5$	164.25		216, 112[14]	......	0.9723[22/4]	....	al	B6[2], 505
10474	3-Pentanol .... $(C_2H_5)_2CHOH$	88.15	....	116.1, 30[12]	......	0.8212[20/4]	1.4104[20]	al, eth, ace	B1[4], 1662
10475	2-Pentanol, 2,4,4-trimethyl .... $(CH_3)_3CCH_2C(OH)(CH_3)_2$	130.23		147.5, 42-4[7]	−20	0.8225[20/4]	1.4284[20]	eth	B1[4], 1796
10476	3-Pentanol, 4-amino-2-methyl .... $CH_3CH(NH_2)CH(OH)CH(CH_3)_2$	117.19		174[745]	35-6	....	....	al, eth, ace, bz	B4[4], 796
10477	3-Pentanol, 1-chloro $C_2H_5CH(OH)CH_2CH_2Cl$	122.60		173, 77[20]	......	1.0327[25/4]	1.448[25]	al, eth	B1[4], 1665
10478	3-Pentanol, 2,2-dimethyl $C_2H_5CH(OH)C(CH_3)_3$	116.20		135, 44-5[15]	−5	0.8253[20/4]	1.4223[20]	al, eth	B1[4], 1751
10479	3-Pentanol, 2,3-dimethyl $C_2H_5C(OH)(CH_3)CH(CH_3)_2$	116.21		139.7, 44-5[14]	<−30	0.833[20/4]	1.4287[20]	al, eth	B1[4], 1752
10480	3-Pentanol, 2,4-dimethyl $(i-C_3H_7)_2CHOH$	116.21		138.7, 87.5[125]	<−70	0.8288[20/4]	1.4250[20]	al, eth	B1[4], 1754
10481	3-Pentanol, 2,4-dimethyl-3-phenyl $(i-C_3H_7)_2C(OH)C_6H_5$	192.31	ye	229, 157[60]	......	0.9755[20/4]	1.5239[20]	eth	B6[1], 273
10482	3-Pentanol, 3-ethyl .... $(C_2H_5)_3COH$	116.21		143.1, 73[52]	......	0.8407[22/4]	1.4294[20]	al, eth	B1[4], 1750
10483	3-Pentanol, 3-ethyl-2-methyl $(C_2H_5)_2C(OH)CH(CH_3)_2$	130.23		159-61[750], 55-7[48]	......	0.8295[20/20]	1.4372[10]	al, eth	B1[4], 1794
10484	3-Pentanol, 2-methyl $C_2H_5CH(OH)CH(CH_3)_2$	102.18		126.7	......	0.8243[20/4]	1.4175[20]	al, eth	B1[4],1716
10485	3-Pentanol, 3-methyl .... $(C_2H_5)_2C(OH)CH_3$	102.18	$[\alpha]^{20}_D$ +40.8	122.4	−23.6	0.8286[20/4]	4186[20]	**al, eth**	B1[4], 1723
10486	3-Pentanol, 1-phenyl (d) $C_2H_5CH(OH)CH_2CH_2C_6H_5$	164.25	$[\alpha]^{20}_D$ +26.8 (al)	143[19]	38	0.9687[20/4]	....	al	B6[4], 1953
10487	3-Pentanol, 3-phenyl $(C_2H_5)_2C(OH)C_6H_5$	164.25	....	223-4[762], 110[12]	<−17	0.9831[20/4]	1.5165[20]	vs	B6[4], 1963
10488	3-Pentanol, 2,3,4-trimethyl .... $(CH_3)_2CHC(OH)(CH_3)CH(CH_3)_2$	130.23		156.5	......	0.8507[20/20]	1.4353[20]	....	B1[4], 1798
10489	1-Pentanol-2-one $C_3H_7COCH_2OH$	162.13		152, 62-4[18]	......	0.9860[20/4]	1.4234[12]	w, al, eth	B1[4], 4004
10490	1-Pentanol-4-one .... $HO(CH_2)_3COCH_3$	102.13		208[730], 116-8[33]	......	1.0071[20/4]	1.4390[20]	al, eth, w	B2[4], 704
10491	2-Pentanol-4-one, 2-methyl or Diacetone alcohol .... $(CH_3)_2C(OH)CH_2COCH_3$	116.16		164, 67-9[19]	−44	0.9387[20/4]	1.4213[20]	**w, al, eth**	B1[4], 4023
10492	3-Pentanol-2-one $C_2H_5CH(OH)COCH_3$	102.13		147-8, 59[27]	......	0.9500[20/4]	1.4350[10]	al, eth, ace, bz	B1[3], 3220
10493	2-Pentanol-3-one $C_2H_5COCH(OH)CH_3$	102.13		152.5, 63[20]	......	0.9742[20/4]	1.4128[20]	al, eth	B1[4], 4008
10494	4-Pentanol-2-one $CH_3CH(OH)CH_2COCH_3$	102.13		177, 62-4[12]	......	1.0071[20/4]	1.4265[20]	al, eth	B1[4], 4005
10495	2-Pentanone or Methyl propyl ketone.................... $C_3H_7COCH_3$	86.14	....	102	−77.8	0.8089[20/4]	1.3895[20]	**al, eth**	B1[4], 3271
10496	2-Pentanone, 4-amino-4-methyl .... $(CH_3)_2C(NH_2)CH_2COCH_3$	115.18		25[0.14]	......	<1	....	al, eth	B4[3], 894
10497	2-Pentanone, 3-benzylidene .... $C_6H_5CH=C(C_2H_5)COCH_3$	174.24		136-8[12]	......	1.0005[22/4]	1.5650[22]	....	B7[1], 198
10498	2-Pentanone, 1-chloro .... $C_3H_7COCH_2Cl$	120.58		154-6d, 58-9[17]	......	....	....	MeOH	B1[4], 3276
10499	2-Pentanone, 5-chloro $Cl(CH_2)_3COCH_3$	120.58		76[34]	......	1.0523[20/4]	1.4375[20]	eth, bz	B1[4], 3277
10500	2-Pentanone, 3-ethyl-4-methyl $(CH_3)_2CHCH(C_2H_5)COCH_3$	128.22		154-5	......	0.812[20/4]	1.4105[20]	al, eth, bz, chl, aa	B1[3], 2883
10501	2-Pentanone, 3-methyl (dl) or sec butyl methyl ketone .... $C_2H_5CH(CH_3)COCH_3$	100.16		118[758]	......	0.8130[30/4]	1.4002[20]	al, eth, chl	B1[4], 3309
10502	2-Pentanone, 4-methyl or Methyl isobutyl Ketone ...... $(CH_3)_2CHCH_2COCH_3$	100.16		116.8, 35-40[16]	−84.7	0.7978[20]	1.3962[20]	al, eth, ace, bz, chl	B1[4], 3305
10503	2-Pentanone oxine or Methyl propyl ketoxime .... $C_3H_7C(=NOH)CH_3$	101.15		167[748]	......	0.9095[20/4]	1.4450[20]	w, al, eth	B1[4], 3274
10504	3-Pentanone or Diethyl ketone.................... $C_2H_5COC_2H_5$	86.14		101.7	−39.8	0.8138[20/4]	1.3924[20]	w, al, ace	B1[4], 3282

No.	Name, Synonyms, and Formula	Mol. wt.	Color, crystalline form, specific rotation and λ$_{max}$ (log ε)	b.p. °C	m.p. °C	Density	n_D	Solubility	Ref.
10505	3-Pentanone, 1-chloro C$_2$H$_5$COCH$_2$CH$_2$Cl	120.58	....	68^{20}	......		1.4361^{20}	al, eth	B1^4, 3284
10506	3-Pentanone, 2-chloro C$_2$H$_5$COCHClCH$_3$	120.58	....	135				al, eth	B1^4, 3284
10507	3-Pentanone, 2,2-dimethyl or Ethyl *tert*-butyl ketone .... C$_2$H$_5$COC(CH$_3$)$_3$	114.19		124.5^{730}	−45	0.8125$^{20/4}$	1.4065^{20}	al, eth, ace, chl	B1^4, 3331
10508	3-Pentanone, 2,4 dimethyl or Diisopropyl ketone .... (i-C$_3$H$_7$)$_2$CO	114.19		124-5	−69	0.8108$^{20/4}$	1.3999^{20}	**al, eth**, bz	B1^4, 3334
10509	3-Pentanone, 2-methyl or Ethyl isopropyl ketone .... (CH$_3$)$_2$CHCOC$_2$H$_5$	100.16		114-5^{745}	......	0.830$^{0/0}$	1.3975^{20}	al, **eth, ace**, bz, chl	B1^4, 3304
10510	3-Pentanone, 2,2,4,4-tetramethyl or Pivalone........... [(CH$_3$)$_3$C]$_2$CO	142.24		152, 70^{43}		0.8240^{18}	1.4194^{20}	al, eth, ace, chl, aa	B1^4, 3334
10511	Pentaguine C$_{18}$H$_{27}$N$_3$O	301.44		165-70$^{0.02}$			1.5785^{25}	dil HCl	Am68, 1524
10512	Pentasiloxane, dodecamethyl CH$_3$[Si(CH$_3$)$_2$O-]$_4$Si(CH$_3$)$_3$	384.85		229^{710}, 103-7^{12}	−80	0.8755^{20}	1.3925^{20}	bz, lig	Am68, 2284
10513	Pentatriacontane CH$_3$(CH$_2$)$_{33}$CH$_3$	492.97	cr (al)	490, 311^{10}	75	0.8157$^{20/4}$	1.4568^{20}	ace	B1^4, 598
10514	18-Pentatriacontanone or Stearone C$_{17}$H$_{35}$COC$_{17}$H$_{35}$	506.95	lf (lig)	....	88.4	0.793$^{95/4}$	....		B1^4, 3413
10515	2-Pentenal, 2-methyl C$_2$H$_5$CH=C(CH$_3$)CHO	98.15		136-7, 38-9^{18}		0.8581$^{20/4}$	1.4488^{20}	al, eth, bz	B1^4, 3471
10516	4-Pentenal CH$_2$=CHCH$_2$CH$_2$CHO	84.12		96	......	0.852$^{20/4}$	1.4191^{20}	eth, ace	B1^4, 3459
10517	1-Pentene CH$_3$CH$_2$CH$_2$CH=CH$_2$	70.14		30	−138	0.6405$^{20/4}$	1.3715^{20}	al, eth, bz	B1^4, 808
10518	1-Pentene, 1-bromo C$_3$H$_7$CH=CHBr	149.04		121-2, 43.5^{30}	......	1.2606$^{20/4}$	1.4572^{20}	eth, bz, chl	B1^3, 775
10519	1-Pentene, 2-bromo C$_3$H$_7$CBr=CH$_2$	149.04		107-8		1.228^{20}	1.4535^{20}	eth, bz, chl, lig	B1^3, 775
10520	1-Pentene, 3-bromo CH$_3$CH$_2$CHBrCH=CH$_2$	149.04		30.5^{30}		1.2417$^{25/4}$	1.4626^{25}	ace, bz, chl	B1^3, 775
10521	1-Pentene, 2-chloro CH$_3$CH$_2$CH$_2$CCl=CH$_2$	104.58		95-7		0.872^5	....	al, eth	B1^3, 774
10522	1-Pentene, 2-chloro-3-ethyl-3-methyl .... (C$_2$H$_5$)$_2$C(CH$_3$)CCl=CH$_2$	146.66		147^{742}, 53^{20}		0.9147$^{20/4}$	1.4450^{25}	bz, chl	B1^3, 847
10523	1-Pentene, 3-chloro CH$_3$CH$_2$CHClCH=CH$_2$	104.58		93-4^{764}		0.8978$^{20/4}$	1.4254^{20}	al, eth, ace	B1^3, 774
10524	1-Pentene, 3-chloro-2-methyl C$_2$H$_5$CHClC(CH$_3$)=CH$_2$	118.61		121-4		....	1.4422^{20}	eth, bz, chl	B1^3, 809
10525	1-Pentene, 4-chloro CH$_3$CHClCH$_2$CH=CH$_2$	104.58		97-100		0.934^{15}	1.417^{15}	eth, chl	B1^3, 774
10526	1-Pentene, 5-Chloro ClCH$_2$CH$_2$CH$_2$CH=CH$_2$	104.58		103-4^{773}		0.9125$^{20/4}$	1.4297^{20}	eth, ace	B1^4, 810
10527	1-Pentene, 2,3-dimethyl C$_2$H$_5$CH(CH$_3$)C(CH$_3$)=CH$_2$	98.19		84.3	−134.8	0.7051$^{20/4}$	1.4033^{20}	**al, eth**	B1^4, 870
10528	1-Pentene, 2,4-dimethyl (CH$_3$)$_2$CHCH$_2$C(CH$_3$)=CH$_2$	98.19		81.6	−123.8	0.6943$^{20/4}$	1.3986^{20}	**al, eth**, bz, chl	B1^4, 870
10529	1-Pentene, 3,3-dimethyl C$_2$H$_5$C(CH$_3$)$_2$CH=CH$_2$	98.19		77.5	−134.3	0.6974$^{20/4}$	1.3984^{20}	**al, eth**, bz, chl	B1^4, 873
10530	1-Pentene, 4,4-dimethyl (CH$_3$)$_3$CCH$_2$CH=CH$_2$	98.19		72.5	−136.6	0.6827$^{20/4}$	1.3918^{20}	**al, eth**, bz	B1^4, 869
10531	1-Pentene, 2-ethyl C$_3$H$_7$C(C$_2$H$_5$)=CH$_2$	98.19		94		0.7079$^{20/4}$	1.405^{20}	al, eth, bz	B1^4, 867
10532	1-Pentene, 3-ethyl (C$_2$H$_5$)$_2$CHCH=CH$_2$	98.19		85		0.6948^{22}	1.3966^{23}	....	B1^4, 867
10533	1-Pentene, 2-methyl C$_3$H$_7$C(CH$_3$)=CH$_2$	84.16		60.7	−135.7	0.6799$^{20/4}$	1.3920^{20}	al, bz, chl, peth	B1^4, 841
10534	1-Pentene, 3-methyl C$_2$H$_5$CH(CH$_3$)CH=CH$_2$	84.16		51.1	−153	0.6675$^{20/4}$	1.3841^{20}	al, bz, chl, peth	B1^4, 847
10535	1-Pentene, 4-methyl (CH$_3$)$_2$CHCH$_2$CH=CH$_2$	84.16		53.9	−153.6	0.6642$^{20/4}$	1.3828^{20}	al, bz, chl, peth	B1^4, 846
10536	1-Pentene, 2-methyl, parfluoro F$_3$C(CF$_2$)$_2$C(CF$_3$)=CF$_2$	300.05		60		....	....	bz	B1^4, 842
10537	1-Pentene, perfluoro CF$_3$CF$_2$CF$_2$CF=CF$_2$	250.04		29-30^{740}		....	1.2571^{25}	chl	B1^4, 810

PHYSICAL CONSTANTS OF ORGANIC COMPOUNDS (Continued)

No.	Name, Synonyms, and Formula	Mol. wt.	Color, crystalline form, specific rotation and λ_{max} (log ε)	b.p. °C	m.p. °C	Density	n_D	Solubility	Ref.
10538	1-Pentene, 2,4,4-trimethyl or Diisobutylene. $(CH_3)_3CCH_2C(CH_3)=CH_2$	112.22	....	101.4	−93.5	$0.7150^{20/4}$	1.4086^{20}	eth, bz, lig, chl	B1[4], 892
10539	2-Pentene (cis) $CH_3CH_2CH=CHCH_3$	70.14		36.9	−151.4	$0.6556^{20/4}$	1.3830^{20}	al, eth, bz	B1[4], 814
10540	2-Pentene-(trans) $CH_3CH_2CHCHCH_3$	70.14		36.3	−136	$0.6482^{20/4}$	1.3793^{20}	al, eth, bz	B1[4], 814
10541	2-Pentene, 1-bromo $CH_3CH_2CH=CHCH_2Br$	149.04		123-4, 35^{25}	......	1.2545^{20}	1.4731^{20}	ace, bz, chl	B1[4], 816
10542	2-Pentene, 2-bromo $CH_3CH_2CH=CBrCH_3$	149.04		110.5^{750}	......	$1.277^{20/20}$	1.4580^{20}	ace, bz, chl	B1[1], 784
10543	2-Pentene, 3-bromo $CH_3CH_2CBr=CHCH_3$	149.04		115.2^{750}	......	$1.273^{20/20}$	1.4628^{20}	ace, bz, chl	B1[1], 784
10544	2-Pentene, 4-bromo $CH_3CHBrCH=CHCH_3$	149.04		117d, 22^9	......	1.2312^{21}	1.4752^{21}	ace, bz, chl	B1[4], 816
10545	2-Pentene, 5-bromo $BrCH_2CH_2CH=CHCH_3$	149.04		121.7^{621}	......	$1.2715^{20/4}$	1.4695^{20}	ace, bz, chl	B1[4], 816
10546	2-Pentene, 1-chloro $CH_3CH_2CH=CHCH_2Cl$	104.58		109.5, 62^{148}	......	$0.908^{22/4}$	1.4352^{22}	al, eth, ace, chl	B1[3], 781
10547	2-Pentene, 2-chloro $CH_3CH_2CH=CClCH_3$	104.58		95-7, 45^{130}	......	$0.9067^{20/4}$	1.4261^{20}	eth, ace	B1[3], 782
10548	2-Pentene, 3-chloro $CH_3CH_2CCl=CHCH_3$	104.58		90-2		1.423^{24}	0.9125^{20}	eth, ace	B1[3], 782
10549	2-Pentene, 3-chloro-2,4-dimethyl $(CH_3)_2CHCCl=C(CH_3)_2$	132.64		118-20, $44-5^{30}$	......	$0.9513^{9/9}$	....	eth, bz, chl	B1, 221
10550	2-Pentene, 4-chloro $CH_3CHClCH=CHCH_3$	104.58		103, $18-20^{12}$	......	$0.9004^{20/20}$	1.4322^{20}	eth, ace, chl	B1[4], 815
10551	2-Pentene, 5-chloro $ClCH_2CH_2CH=CHCH_3$	104.58		$107-8^{755}$	......	$0.9043^{20/4}$	1.4310^{20}	ace, bz, chl	B1[4], 816
10552	2-Pentene, 5-chloro-2-methyl $ClCH_2CH_2CH=C(CH_3)_2$	118.61		$132-3^{756}$	......	0.9135^{20}	....	ace, chl	B1[4], 843
10553	2-Pentene, 2,5-dichloro $ClCH_2CH_2CH=CClCH_3$	139.04		$40-1^8$	......	$1.1182^{15/4}$	....	chl	B1[4], 816
10554	2-Pentene, 2,3-dimethyl $C_2H_5C(CH_3)=C(CH_3)$	98.19		97.5	−118.3	0.7277^{20}	1.4208^{20}	al, eth, bz, chl	B1[4], 870
10555	2-Pentene, 2,4-dimethyl $(CH_3)_2CHCH=C(CH_3)_2$	98.19		83.4	−127.7	$0.6954^{20/4}$	1.4040^{20}	al, eth, bz, chl	B1[4], 872
10556	2-Pentene, 3,4-dimethyl $(CH_3)_2CH(CH_3)=CHCH_3$	98.19		87	......	$0.7126^{20/4}$	1.4070^{20}	....	B1[4], 871
10557	2-Pentene, 4,4-dimethyl (trans) $(CH_3)_3CCH=CHCH_3$	98.19		76.7	−115.2	$0.6889^{20/4}$	1.3982^{20}	al, eth, bz, chl	B1[4], 868
10558	2-Pentene, 3-ethyl $(C_2H_5)_2C=CHCH_3$	98.19		94	......	$0.7079^{20/4}$	1.405^{20}	al, eth, bz, chl	B1[4], 868
10559	2-Pentene, 2-methyl $C_2H_5CH=C(CH_3)_2$	84.16		67.3	−135	$0.6863^{20/4}$	1.4004^{20}	al, bz, chl, peth	B1[4], 842
10560	2-Pentene, 3-methyl (cis) $C_2H_5C(CH_3)=CHCH_3$	84.16		67.6	−138.4	$0.6986^{20/4}$	1.4045^{20}	al, bz, chl, peth	B1[4], 848
10561	2-Pentene, 3-methyl (trans) $C_2H_5C(CH_3)=CHCH_3$	84.16		70.4	−134.8	$0.6942^{20/4}$	1.4016^{20}	al, bz, chl, peth	B1[4], 848
10562	2-Pentene, 4-methyl (cis) $(CH_3)_2CHCH=CHCH_3$	84.16		56.3	−134.4	$0.6690^{20/4}$	1.3800^{20}	al, bz, chl, peth	B1[4], 844
10563	2-Pentene, 4-methyl (trans) $(CH_3)_2CHCH=CHCH_3$	84.16		58.5	−140.8	$0.6686^{20/4}$	1.3889^{20}	al, bz, chl, peth	B1[4], 844
10564	2-Pentene, 2,3-dimethyl. $(CH_3)_2CHC(CH_3)=C(CH_3)_2$	112.21		116.5	-113.38	0.74342^{204}	1.4274^{20}	....	B1[4], 894
10565	2-Pentene, 2,4,4-trimethyl $(CH_3)_3CCH=C(CH_3)_2$	112.22		104.9	−106.3	$0.7218^{20/4}$	1.4160^{20}	eth, bz, lig, chl	B1[4], 891
10566	2-Pentenedioc acid, diethyl ester (trans) or Ethyl glutaconate. $C_2H_5O_2CCH_2CH=CHCO_2C_2H_5$	186.21	....	236-8, 125^{12}	......	$1.0496^{20/4}$	1.4411^{20}	al, eth	B2[4], 2227
10567	2-Pentenoic acid, 4-hydroxy, lactone or β-Angelica lactone. $CH_3CHCH=CHCO$	98.10	....	$208-9^{751}$, 98^{15}	<−17	$1.0810^{20/4}$	1.4454^{20}	w, al, eth	B17[4], 4302
10568	2-Pentenoic acid, 2-methyl (trans) $C_2H_5CH=C(CH_3)CO_2H$	114.14		213^{750}, 112^{12}	24.4	0.9751^{20}	1.4513^{20}	eth, chl	B2[4], 1568
10569	2-Pentenoic acid, 4-methyl $(CH_3)_2CHCH=CHCO_2H$	114.15		217, $115-6^{20}$	35	$0.9529^{21/4}$	1.4489^{21}	al, eth, ace	B2[4], 1569

No.	Name, Synonyms, and Formula	Mol. wt.	Color, crystalline form, specific rotation and λ_{max} (log ε)	b.p. °C	m.p. °C	Density	n_D	Solubility	Ref.
10570	3-Pentenoic acid, 4-hydroxy, γ-lactone or γ-Angelica lactone. . . . . CH₃-C=CHCH₂CO	98.10	. . . .	167, 53[12]	18	1.084[20/4]	1.4476[20]	w, al, eth	B17[4], 4300
10571	3-Pentenonitrile, 2-hydroxy (cis) . . . . CH₃CH=CHCH(OH)CN	97.12	. . . .	139[70]		0.9675[15/4]	1.4460[21]	al, eth, bz, chl	B3[4], 1001
10572	4-Pentenamide, 2,2-diethyl or Novonal . . . . CH₂=CHCH₂C(C₂H₅)₂CONH₂	155.24	wh pw, cr (eth-peth)	155[10]	75-6	. . . .	. . . .	al, eth	B2[3], 1351
10573	4-Pentenoic acid or Allyl acetic acid . . . . CH₂=CHCH₂CH₂CO₂H	100.13	. . . .	188-9, 93[20]	−22.5	0.9809[20/4]	1.4281[20]	al, eth	B2[4], 1542
10574	4-Pentenoic acid, 2-acetyl ethyl ester . . . . CH₂=CHCH₂CH(COCH₃)CO₂C₂H₅	170.21	. . . .	211-2, 102[12]		0.9898[20/4]	1.4388[18]	al, eth, bz	B3, 738
10575	4-Pentenonitrile . . . . CH₂=CHCH₂CH₂CN	81.12	. . . .	140, 60-1[40]		0.8239[24]	1.4213[14]	al, eth	B2[4], 1593
10576	1-Penten-3-ol . . . . C₂H₅CH(OH)CH=CH₂	86.14	. . . .	114-6, 37[20]		0.8935[22/4]	1.4239[20]	al, eth	B1[4], 2117
10577	2-Penten-1-ol (cis) . . . . C₂H₅CH=CHCH₂OH	86.13	. . . .	138, 41.2[7]		0.8529[20/4]	1.4354[20]	al, eth, ace	B1[4], 2121
10578	2-Penten-1-ol (trans) . . . . C₂H₅CH=CHCH₂OH	86.13	. . . .	139.5, 42[7]		0.8471[20/4]	1.4341[20]	al, eth, ace	B1[4], 2121
10579	3-Pentene-2-ol (d) . . . . CH₃CH=CHCH(OH)CH₃	86.13	[α]D + 0.413	118-21, 33[10]		0.8354[20/4]	1.4250[25/20]	al, eth, ace	B1[4], 2122
10580	3-Pentene-2-ol (dl) . . . . CH₃CH=CHCH(OH)CH₃	86.13	. . . .	121.6, 60.6[55]		0.8328[25]	1.4280[20]	al, eth, ace	B1[4], 2122
10581	3-Penten-2-ol (l) . . . . CH₃CH=CHCH(OH)CH₃	86.13	[α]D −3.3 (H₂O 1.5%)	119-21	. . . . . .	0.8331[20/4]	1.4290[20]	al, eth, ace	B1[4], 2124
10582	4-Penten-2-ol, 2-methyl . . . . CH₃CH=CHC(OH)(CH₃)₂	100.16	. . . .	121-2[757]		0.8347[20/4]	1.4302[20]	al, eth	B1[4], 2145
10583	4-Pentene-1-ol . . . . CH₂=CHCH₂CH₂CH₂OH	86.14	. . . .	140-2		0.8457[20/4]	1.4309[20]	eth	B1[4], 2119
10584	4-Penten-2-ol . . . . CH₂=CHCH₂CH(OH)CH₃	86.14	. . . .	115-6[750]		0.8367[20/4]	1.4225[20]	al, eth	B1[4], 2118
10585	4-Penten-2-ol, 2-methyl . . . . CH₂=CHCH₂C(OH)(CH₃)₂	100.16	. . . .	119.5		0.8300[20/4]	1.4263[20]	al, eth	B1[4], 2146
10586	1-Penten-3-one or Ethyl vinyl ketone . . . . C₂H₅COCH=CH₂	84.12	. . . .	102[740], 44[90]		0.8468[20/4]	1.4192[20]	al, eth, ace, bz	B1[4], 3457
10587	1-Penten-3-one, 5,5-dimethyl-1-phenyl or Benzalpinacolone. . . . . (CH₃)₃CHCH₂COCH=CHC₆H₅	188.27	cr	154[25]	43	0.9508[46]	1.5523[25]	al, bz, chl	B7[3], 1487
10588	1-Penten-3-one, 1-(4 methoxy phenyl) or Ethyl-4-methoxy styryl ketone . . . . C₂H₅COCH=CH(C₆H₄OCH₃-4)	190.24	col-lt ye pl (eth-peth)	. . . .	60	. . . .	. . . .	al, eth	B8[3], 832
10589	1-Penten-3-one, 2-methyl or Ethyl isopropenyl ketone . . . . C₂H₅COC(CH₃)=CH₂	98.15	. . . .	118.5	−69.5	1.8530[20/4]	1.4289[20]	al, ace	B1[4], 3470
10590	1-Penten-3-one, 1-Phenyl . . . . C₂H₅COCH=CHC₆H₅	160.22	lf (lig)	142[12]	38-9	0.8697[20/4]	1.5684[20]	al, eth, bz	B7[2], 298
10591	3-Penten-2-one (trans) or Methyl propenyl ketone . . . . CH₃CH=CHCOCH₃	84.13	. . . .	122	. . . . . .	0.8624[20/4]	1.4350[20]	eth, ace	B1[4], 3460
10592	3-Penten-2-one, 4-methyl or Mesityl oxide . . . . (CH₃)₂C=CHCOCH₃	98.15	. . . .	129.7, 41[23]	−52.8	0.8653[20/4]	1.4440[20]	w, al, eth, ace	B1[4], 3471
10593	1-Penten-3-yne or Puryline . . . . CH₃C≡C-CH=CH₂	66.10	. . . .	59-60	. . . . . .	0.7401[20/4]	1.4496[20]	eth, bz	B1[4], 1087
10594	1-Penten-3-yne, 2-methyl . . . . CH₃C≡CC(CH₃)=CH₂	80.14	. . . .	81-2[100]	. . . . . .	. . . .	1.4002[20]	eth, bz	B1[4], 1095
10595	1-Penten-4-yne or Allyl acetylene . . . . CH₂=CHCH₂C≡CH	66.10	. . . .	42-3	. . . . . .	0.777[22/22]	1.3653[22]	eth, bz	B1[4], 998
10596	3-Penten-1-yne, 3-ethyl . . . . CH₃CH=C(C₂H₅)C≡CH	94.16	. . . .	96.5, 41-3[100]	. . . . . .	0.7886[25]	1.4338[25]	eth, bz	B1[4], 1099
10597	3-Penten-1-yne, 3-methyl . . . . CH₃CH=C(CH₃)C≡CH	80.14	. . . .	66-7	. . . . . .	0.789[20]	1.4332[20]	eth, bz	B1[4], 1096
10598	Peucedanin . . . . C₁₅H₁₄O₄	258.28	pr or pl (bz-peth), yesh cr (eth)	276-81[17]	109	. . . .	. . . .	eth, chl, aa	B19[4], 2647
10599	Pentyl amine or 1-Amino pentane . . . . CH₃(CH₂)₃CH₂NH₂	87.17	. . . .	104.4, 5.9[10]	−55	0.7547[20/4]	1.4118[20]	al, eth, ace, bz	B4[4], 674
10600	iso-Pentyl amine . . . . (CH₃)₂CHCH₂CH₂NH₂	87.17	. . . .	95-7	. . . . . .	0.7505[20/4]	1.4083[20]	w, al, eth, ace, chl	B4[4], 696

No.	Name, Synonyms, and Formula	Mol. wt.	Color, crystalline form, specific rotation and λ_{max} (log ε)	b.p. °C	m.p. °C	Density	n_D	Solubility	Ref.
10601	Isopentyl boric acid i-C$_5$H$_{11}$B(OH)$_2$	115.97	pl (w)	....	169	....	....	w, al, eth, ace	B4[2], 1023
10602	Pentyl ether or Dipentyl ether C$_5$H$_{11}$OC$_5$H$_{11}$	158.29	....	190, 70[12]	−69	0.7833[20/4]	1.4119[20]	al, eth	B1[4], 1643
10603	Isopentyl ether or Diisopentyl ether (i-C$_5$H$_{11}$)$_2$O	158.28	....	172-3, 60[10]	......	0.7777[20/4]	1.4085[20]	al, eth, chl	B1[4], 1682
10604	Pentyl isocyanide C$_5$H$_{11}$NC	97.16	....	155.5, 50[45]	−51.1	0.806[20/4]	....	al	B4[3], 331
10605	Pentyl isothiocyanate C$_5$H$_{11}$NCS	129.23	br ye	....	193.4			al, eth	B4[4], 685
10606	Isopentyl isothiocyanate (CH$_3$)$_2$CHCH$_2$CH$_2$NCS	129.33	....	182-4	......	0.9419[17]		al, eth	B4[4], 707
10607	Isopentyl α-naphtyl ether i-C$_5$H$_{11}$O-α-C$_{10}$H$_7$	214.31	....	317[742], 148-53[3]	......	1.0069[14/4]	1.5705[14]	....	B6[4], 2925
10608	Isopentyl β-naphthyl ether i-C$_5$H$_{11}$O-β-C$_{10}$H$_7$	214.31	lf	323-6d, 155-60[6]	26.5	1.0155[12/4]	1.5768[12]	al, eth	B6[4], 2974
10609	Isopentyl nitrate (CH$_3$)$_2$CHCH$_2$CH$_2$ONO$_2$	133.15	....	147-8	......	0.9961[22/4]	1.4122[22]	al, eth	B1[4], 1683
10610	Pentyl nitrite C$_5$H$_{11}$ONO	117.15	ye	104-5, 29[60]	......	0.8817[20/4]	1.3851[20]	al, eth	B1[4], 1644
10611	Isopentyl nitrite (CH$_3$)$_2$CHCH$_2$CH$_2$ONO	117.15	....	99.2, 30[60]	......	0.8828[20/4]	1.3918[20]	al, eth	B1[4], 1683
10612	Pentyl sulfide or Dipentyl sulfide (C$_5$H$_{11}$)$_2$S	174.35	....	230, 84.5[4]	−51.3	0.8409[20/4]	1.4556[20]	eth	B1[4], 1654
10613	Isopentyl sulfide or Disopentyl sulfide (i-C$_5$H$_{11}$)$_2$S	174.35	col pa-ye	216, 85.5[6]	......	0.8323[20/4]	1.4520[20]	al, eth	B1[4], 1689
10614	Pentyl sulfate (C$_5$H$_{11}$O)$_2$SO$_2$	238.35	....	117[3.5]	14	1.029[20/0]	1.4290[20]	....	B1[2], 418
10615	Isopentyl sulfate (i-C$_5$H$_{11}$O)$_2$SO$_2$	238.35	....	139-41[12]	−20				B1[2], 434
10616	Isopentyl thiocyanate (CH$_3$)$_2$CHCH$_2$CH$_2$SCN	129.23	....	197	......			al, eth	B3[3], 282
10617	Iso-Pentyl vinyl ether i-C$_5$H$_{11}$OCH=CH$_2$	114.19	....	112-3	......	0.7826[20/4]	1.4072[20]	al, eth	B1[4], 2055
10618	1-Pentyne C$_3$H$_7$C≡CH	68.13	....	40.2	−90	0.6901[20/4]	1.3852[20]	al, eth, bz, chl	B1[4], 990
10619	1-Pentyne, 1-bomo C$_3$H$_7$C≡CBr	147.01	....	44-6[52]	......	1.281[13/4]	1.4579[13]	....	B1[3], 958
10620	1-Pentyne, 3-chloro-3-ethyl (C$_2$H$_5$)$_2$C(Cl)C≡CH	130.62	....	73-6[100]	......	0.9230[19/4]	1.4437[19]	eth, bz, chl	B1[4], 1032
10621	1-Pentyne, 3-chloro-3-methyl C$_2$H$_5$CCl(CH$_3$)C≡CH	116.59	....	102.3, 55[130]	......	0.9163[20/4]	1.4330[20]	eth, bz, chl	B1[4], 1021
10622	1-Pentyne, 4,4-dimethyl (CH$_3$)$_3$CCH$_2$C≡CH	96.17	....	76.1	−75.7	0.7142[20/4]	1.3983[20]	eth, bz, chl	B1[4], 1033
10623	1-Pentyne, 3-ethyl (C$_2$H$_5$)$_2$CHC≡CH	96.17	....	87-9	......	0.7246[25/4]	1.4043[25]	eth, bz, chl	B1[3], 1002
10624	1-Pentyne, 3-ethyl-3-methyl (C$_2$H$_5$)$_2$C(CH$_3$)C≡CH	110.20	....	101-2	......	0.7422[20/4]	1.4110[20]	eth, bz, chl	B1[4], 1046
10626	1-Pentyne, 1-iodo C$_3$H$_7$C≡CI	194.01	....	54[23]	......	1.6127[19/4]	1.5148[19]	....	B1[4], 992
10627	1-Pentyne, 4-methyl (CH$_3$)$_2$CHCH$_2$C≡CH	82.15	....	61-2	−105.1	0.7092[15/4]	1.3936[15]	bz, chl	B1[4], 1019
10628	2-Pentyne C$_2$H$_5$C≡CCH$_3$	68.13	....	56	−101	0.7107[20/4]	1.4039[20]	al, eth, bz, chl	B1[4], 992
10629	2-Pentyne, 4-chloro-4-methyl (CH$_3$)$_2$C(Cl)C≡CCH$_3$	116.59	....	55[70]	......		1.4143[20]	ace, bz, chl	B1[4], 1018
10630	2-Pentyne, 4,4-dimethyl (CH$_3$)$_3$CC≡CCH$_3$	96.17	....	83	−82.4	0.7176[20/4]	1.4071[20]	eth, bz, chl	B1[4], 1032
10631	2-Pentyne, 4-methyl (CH$_3$)$_2$CHC≡CCH$_3$	82.15	....	72.5	−110.4	0.716[19/4]	1.4078[19]	bz, chl	B1[4], 1018
10632	2-Pentynoic acid CH$_3$CH$_2$C≡CCO$_2$H	98.10	cr (peth)	122[10]	50	0.978[20]	1.4619[20]	w	B2[4], 1693
10633	4-Pentynoic acid HC≡CCH$_2$CH$_2$CO$_2$H	98.10	....	102[17]	57.7	....	....	w, al, eth	B2[4], 1693

No.	Name, Synonyms, and Formula	Mol. wt.	Color, crystalline form, specific rotation and λ_{max} (log ε)	b.p. °C	m.p. °C	Density	n_D	Solubility	Ref.
10634	1-Pentyn-3-ol, 3,4-dimethyl .... (CH$_3$)$_2$CHC(OH)(CH$_3$)C≡CH	112.17	....	133[735]	......	0.8691[20/4]	1.4372[20]	w, al, eth	B1[4], 2255
10635	1-Pentyn-3-ol, 3-methyl .... C$_2$H$_5$C(OH)(CH$_3$)C≡CH	98.15	....	120-1, 61[70]	30-1	0.8688[20/4]	1.4310[20]	....	B1[4], 2242
10636	Perbenzoic acid .... C$_6$H$_5$COOOH	138.12	mcl pl (peth)	97-110[13-5] sub	41-3	....	....	al, eth, ace, bz, chl	B9[3], 1049
10637	Pereirine .... C$_{19}$H$_{26}$N$_2$O	298.43	pa ye amor pw, [α] + 137.5 (al)	....	135d	....	....	al, eth, chl	C28, 5459
10638	Perimidine or Peri-naphthimidazole .... C$_{11}$H$_8$N$_2$	168.21	gr cr (dil al)	....	222	....	....	al, eth, ace, bz	B23[2], 209
10639	Perseitol or α-Mannoheptitol .... C$_7$H$_{16}$O$_7$	212.20	nd, [α][20] + 4.53 (w)	....	188	....	....	w	B1[4], 2854
10640	Perylene or peri-Dinaphthalene .... C$_{20}$H$_{12}$	252.32	gold-br ye pl (bz, aa)	sub 350-400	277-9	1.35	....	ace, bz, chl	B5[3], 2521
10641	Perylene-3-carboxylic acid .... 3-C$_{20}$H$_{11}$CO$_2$H	296.33	og-br nd (PhNO$_2$)	....	330	....	....	....	B9[3], 3664
10642	α-Phellandrene (d) or 5-isopropyl-2-methyl-1,3-cyclohexadiene .... C$_{10}$H$_{16}$	136.24	[α]$_D$ + 49.1 (undil)	175-6, 61[11]	....	0.8463[25/4]	1.4777[22]	eth	B5[4], 436
10643	β-Phellandrene or 3-isopropyl-6-methylene cyclohexene .. C$_{10}$H$_{16}$	136.24	[α][20/D] + 65.2 (undil)	171-2, 57[11]	....	0.8520[20/4]	1.4788[20]	eth	B5[4], 436
10644	Phenacyl acetate or α-Hydroxy acetophene acetate....... C$_6$H$_5$COCH$_2$O$_2$CCH$_3$	178.19	rh pl (eth, lig)	270, 150-2[10]	49	1.1169[65/4]	....	al, eth, bz, chl	B8[3], 301
10645	Phenacyl alcohol or α-Hydroxy acetophenone .... C$_6$H$_5$COCH$_2$OH	136.16	hex pl (al, eth) pl (w, dil al), pr (lig)	124-6[12] (sub 56[1])	90.5 (anh)	1.0963[99/4]	....	al, eth, chl	B8[3], 298
10646	Phenacyl bromide or α-Bromo acetophenone .... C$_6$H$_5$COCH$_2$Br	199.05	nd (al), rh pr (dil al), pl (peth)	135[18]	50-1	1.647[20/4]	....	al, eth, bz, chl	B7[3], 979
10647	Phenacyl chloride or α-Chloro acetophenone .... C$_6$H$_5$COCH$_2$Cl	154.60	pl (dil al), rh, lf (peth)	247, 139-41[14]	56-5	1.324[15/4]	....	al, eth, ace, bz	B7[3], 967
10648	Phenacyl chloride-2,4-dimethyl .... 2,4(CH$_3$)$_2$C$_6$H$_3$COCH$_2$Cl	182.65	nd	....	62	....	....	al, eth, bz, chl	B7[2], 172
10649	Phenanthrene .... C$_{14}$H$_{10}$	178.24	mcl pl (al), lf (sub)	340, 210-15[12]	101	0.9800[4]	1.5943	al, eth, ace, bz, aa	B5[4], 2297
10650	Phenanthrene, 2-acetyl .... 2-CH$_3$COC$_{14}$H$_9$	220.27	nd (MeOH)	....	144-5	....	....	al, bz	B7[3], 2543
10651	Phenanthrene, 3-acetyl .... 3-CH$_3$COC$_{14}$H$_9$	220.27	nd (MeOH)	....	72	....	....	al, bz, chl, aa	B7[3], 2544
10652	Phenanthrene, 9-acetyl .... 9-CH$_3$COC$_{14}$H$_9$	220.27	nd (MeOH)	....	74.5	....	....	al, eth, bz	B7[3], 2549
10653	Phenanthrene, 2-amino or 2-Phenanthryl amine........ 2-NH$_2$C$_{14}$H$_9$	193.25	lt ye (lig)	....	85	....	....	....	B12[3], 3339
10654	Phenanthrene, 3-amino or 3-Phenanthryl amine........ 3-NH$_2$C$_{14}$H$_9$	193.25	α:lf (lig) β:cr (lig)	....	α:143, β:87.5	....	....	al	B12[3], 3339
10655	Phenanthrene, 4-amino .... 4-H$_2$NC$_{14}$H$_9$	193.25	....	....	104-5	....	....	al, eth, bz, chl	B12[3], 3341
10656	Phenanthrene, 9-amino (a form) or 9-Phenanthryl amine	193.25	lt ye cr (al)	sub	137-8	....	....	al, eth, bz, chl	B12[3], 3341
10657	Phenanthrene, 9-amino (b form) .... 9-NH$_2$C$_{14}$H$_9$	193.25	lt ye nd (al)	sub	104	....	....	al, eth, bz, chl	B12[1], 555
10658	Phenanthrene, 9-bromo .... 9-BrC$_{14}$H$_9$	257.14	pr (al)	>360, 190[12] sub	64-5	1.4093[10/4]	....	al, eth	B5[3], 2145
10659	Phenanthrene, 9,10-diamino .... 9,10-(NH$_2$)$_2$C$_{14}$H$_8$	208.27	pa ye lf	....	166	....	....	....	B13[3], 524
10660	Phenanthrene, 9,10-dihydro .... C$_{14}$H$_{12}$	180.25	nd (MeOH)	168-9[15]	34-5	1.0757[40/4]	1.6415[20/4]	al, eth	B5[3], 1989
10661	Phenanthrene, 3,4-dihydroxy or 3,4-Phenanthrenediol ... 3,4-(HO)$_2$C$_{14}$H$_8$	210.24	col nd (peth), dk in air	sub 130 (vac)	143	....	....	al, eth	B6[3], 5689
10662	Phenanthrene, 3,4-dimethoxy .... 3,4-(CH$_3$O)$_2$C$_{14}$H$_8$	238.29	bt ye lf (MeOH)	298-303[112]	45	....	....	al, eth	B6[3], 5689

No.	Name, Synonyms, and Formula	Mol. wt.	Color, crystalline form, specific rotation and λ_{max} (log ε)	b.p. °C	m.p. °C	Density	n_D	Solubility	Ref.
10663	Phenanthrene, 4,5-dimethyl 4,5-(CH₃)₂C₁₄H₈	206.28	pr (MeOH)	....	76-7	....	....	chl	B5[3], 2175
10664	Phenanthrene, 9,10-dimethyl 9,10-(CH₃)₂C₁₄H₈	206.29	lt red pr (aa), nd (MeOH)	sub	144	....	....	bz, chl, aa	B5[3], 2175
10665	Phenanthrene, 9,10-diphenyl 9,10-(C₆H₅)₂C₁₄H₈	330.44	nd (eth, bz)	sub 270	240	....	....	eth, bz	B5[3], 2624
10666	Phenanthrene, 1,2,3,4,5,6,7,8,9,10,11,12-dodecahydro C₁₄H₂₂	190.33	....	81-2[1.5]	....	0.9674[20/4]	1.5102[20]	ace, bz	B5[3], 1074
10667	Phenanthrene, 9-ethyl 9-C₂H₅C₁₄H₉	206.29	....	199-200[1]	62-3 (66)	1.0603[78/4]	1.6582[78]	al, bz	B5[3], 2170
10668	Phenanthrene, 1,2,3,4,11,12-hexadydro C₁₄H₁₆	184.28	....	307	−3	1.045[20]	1.5810[15]	eth, bz, aa, chl, peth	B5[3], 1676
10669	Phenanthrene, 1-hydroxy or 1-Phenanthrol 1-HOC₁₄H₉	194.24	nd (peth, bz-lig, eth)	....	157	....	....	....	B6[3], 3557
10670	Phenanthrene, 2-hydroxy or 2-Phenanthrol 2-HOC₁₄H₉	194.24	pl, lf (al, eth, lig)	....	168	....	....	al, eth, bz	B6[3], 3557
10671	Phenanthrene, 3-hydroxy or 3-Phenanthrol 3-HOC₁₄H₉	194.24	nd (al, lig)	....	122-3	....	....	al, eth	B6[3], 3558
10672	Phenanthrene, 9-hydroxy or 9-Phenanthrol 9-HOC₁₄H₉	194.24	nd (lig, bz)	....	158	....	....	al, eth, bz, chl, lig	B6[3], 3560
10673	Phenanthrene, 1-methyl 1-CH₃C₁₄H₉	192.26	lf, pl (dil al)	....	123	....	....	al	B5[3], 2151
10674	Phenanthrene, 3-methyl 3-CH₃C₁₄H₉	192.26	pr or nd (al)	140-50[6]	65	....	....	ace	B5[3], 2154
10675	Phenanthrene, 4-methyl 4-CH₃C₁₄H₉	192.26	....	175-80[10]	52.5	....	....	al	B5[3], 2154
10676	Phenanthrene, 9-methyl 9-CH₃C₁₄H₉	192.26	....	....	90-1	....	....	....	B5[3], 2154
10677	Phenanthrene, 2-nitro 2-O₂NC₁₄H₉	223.23	....	....	119-20	....	....	al, eth, ace	B5[3], 2146
10678	Phenanthrene, 3-nitro 3-O₂NC₁₄H₉	223.23	....	....	172-4	....	....	ace, bz, chl	B5[3], 2146
10679	Phenanthrene, 9-nitro 9-NO₂C₁₄H₉	223.23	ye nd (al)	....	116-7	....	....	al, eth, bz	B5[3], 2147
10680	Phenanthrene, 1,2,3,4,5,6,7,8-octahydro or Octathrene C₁₄H₁₈	186.30	....	295, 169[15]	16.7	1.026[20/4]	1.5569[17]	ace, bz, aa	B5[4], 1585
10681	Phenanthrene, 1,2,8-trimethyl 1,2,8-(CH₃)₃C₁₄H₇	220.31	....	210-20[15]	144-5	....	....	....	B5[3], 2187
10682	Phenanthrene, 1,4,7-trimethyl 1,4,7-(CH₃)₃C₁₄H₇	220.31	....	....	72.3	....	....	....	B5[3], 2188
10683	Phenanthrene, 1,2,3,4,9,10,11,12-octahydro (cis) C₁₄H₁₈	186.30	....	129[6], 88-90[0.1]	....	1.0072[25/4]	1.5549[21]	bz	B5[4], 1585
10684	Phenanthrene, 1,2,3,4,9,10,11,12-octahydro (trans) C₁₄H₁₈	186.30	nd	94-5[15]	23-4	1.0060[20/4]	1.5528[21]	bz	B5[3], 1402
10685	Phenanthrene, perhydro C₁₄H₂₄	192.35	....	86-9[2]	....	0.9447[20/4]	1.5011[20]	eth, ace, bz	B5[4], 492
10686	Phenanthrene, 7-isopropyl-1-methyl 1-CH₃-7-i-C₃H₇C₁₄H₈	234.34	pl (al)	390, 290[10]	100-1	1.035	....	bz, lig	B5[3], 2199
10687	Phenanthrene, 1,2,3,4-tetrahydro or Tetranthrene C₁₄H₁₄	182.27	lf (MeOH)	173[11]	33-4	1.0601[40/4]	....	al, eth, ace, bz, chl, lig	B5[4], 1909
10688	Phenanthrene, 3,4,5-trihydroxy or 3,4,5-Phenanthrene-triol 3,4,5-(HO)₃C₁₄H₇	226.24	lf or pl (w)	....	148	....	....	al, eth, chl	B6[1], 1411
10689	Phenanthrene, 1,6,7-trimethyl 1,6,7-(CH₃)₃C₁₄H₇	220.31	....	....	123-4	....	....	....	B5[3], 2188
10690	1-Phenanthrene carboxylic acid or 1-Phenanthroic acid 1-C₁₄H₉CO₂H	222.25	nd (al)	....	232-3	....	....	al, bz	B9[3], 3496
10691	2-Phenanthrene carboxylic acid or 2-Phenanthroic acid 2-C₁₄H₉CO₂H	222.25	nd (aa)	....	258-60	....	....	al, bz, aa	B9[3], 3497
10692	2-Phenanthronitrile 2-C₁₄H₉CN	203.25	cr (bz-lig, al)	....	108-10	....	....	al, eth, ace	B9[3], 3498
10693	3-Phenanthrene carboxylic acid or 3-Phenanthroic acid 3-C₁₄H₉CO₂H	222.25	nd (aa)	sub	270	....	....	al, eth, aa	B9[3], 3498
10694	3-Phenanthronitrile 3-C₁₄H₉CN	203.27	nd (abs al)	....	102	....	....	al, eth	B9[3], 3499

No.	Name, Synonyms, and Formula	Mol. wt.	Color, crystalline form, specific rotation and λ_{max} (log ε)	b.p. °C	m.p. °C	Density	n_D	Solubility	Ref.
10695	9-Phenanthrene carboxylic acid or 9-Phenanthroic acid.. 9-$C_{14}H_9CO_2H$	222.25	nd (aa), lf (sub)	sub	256-7	....	....	al, eth, bz, aa	B9³, 3501
10696	9-Phenanthronitrile 9-$C_{14}H_9CN$	203.25	nd (al)	....	103			eth, ace	B9³, 3502
10697	2-Phenanthrene sulfonic acid .... 2-$C_{14}H_9SO_3H$	258.30	cr (bz), cr (w + 1)	....	ca 150			w, al, bz	B11³, 445
10698	3-Phenanthrene sulfonic acid .... 3-$C_{14}H_9SO_3H$	258.30	lf (bz), cr (w)	....	88-9 (+ 2w), 120-1 (+ 1w), N175-6 (anh)			w	B11³, 445
10699	9-Phenanthrene sulfonic acid 9-$C_{14}H_9SO_3H$	258.30	lf or nd (bz, w + 2)	....	134 hyd, 174 sub			w, al, aa	B11², 111
10700	Phenanthridine or 3-4-Benzoquinoline .... $C_{13}H_9N$	179.22	nd (dil al)	349	106-7	....	....	al, eth, ace, bz, chl	B20⁴, 4016
10701	Phenanthridine, 6-hydroxy .... $C_{13}H_9NO$	195.22	nd (al, sub)	sub	293-4			....	B21⁴, 1572
10702	1,7-Phenanthrolene or 1,7-Diazaphenanthrene .... $C_{12}H_8N_2$	180.22	pl (anh), nd (w + 2)	>360	78 (anh) 65.5 (+ 2w)			al	B23¹, 61
10703	1,10-Phenanthroline or 4,5-Diazaphenanthrene .... $C_{12}H_8N_2$	180.22	wh nd (bz), cr (w + 1)	>300	117 (anh)	....		al, ace, bz	B23, 227
10704	4,7-Phenanthroline or 1,8-Dizaphenanthrene .... $C_{12}H_8N_2$	180.22	nd (w)	sub 100	177	....		al, chl	B23¹, 61
10705	4,7-Phenanthroline hydrate .... $C_{12}H_8N_2.H_2O$	198.23	wh nd (w)	....	100-3			....	B23², 235
10706	1,7-Phenanthroline, 9-nitro .... $C_{12}H_7N_3O_2$	225.21	nd (dil al)		168				B23², 236
10707	9,10-Phenanthroquinone......................... $C_{14}H_8O_2$	208.23	og nd (to), og-red pl (sub)	>360 sub	208-10	1.405²²/⁴	....	eth	B7³, 4084
10708	9,10-Phenanthroquinone-2-bromo .... $C_{14}H_7BrO_2$	287.12	red-ye cr (aa)	....	233-4				B7³, 4093
10709	9,10-Phenanthroquinone, 3-bromo .... $C_{14}H_7BrO_2$	287.12	ye nd (aa)		268-9			bz	E13, 925
10710	9,10-Phenanthraquinone, 1-chloro $C_{14}H_7O_2Cl$	242.66			229			al, bz	B7³, 4092
10711	9,10-Phenanthraquinone, 2-Chloro $C_{14}H_7ClO_2$	242.66	ye-red nd (aa)		252-3			al	B7³, 4093
10712	9,10-Phenanthraquinone, 3-chloro $C_{14}H_7ClO_2$	242.66	og-ye nd (aa, bz-al)		264-5			al, bz, aa	B7³, 4093
10713	9,10-Phenanthraquinone, 1,2-dihydroxy .... $C_{14}H_8O_4$	240.23	dk red nd (ace)	....	d			al, ace, aa	B8², 506
10714	9,10-Phenanthraquinone, 2,5-dihydroxy $C_{14}H_8O_4$	240.23	dk red nd (w)	....	400d			....	B8², 507
10715	9,10-Phenanthraquinone, 2,7-dihydrocy $C_{14}H_8O_4$	240.23	dk red or br nd		>400d			al, eth, ace, bz, aa	B8², 507
10716	9,19-Phenanthraquinone, 4,5-dihydroxy $C_{14}H_8O_4$	240.23	dk red (al), nd (w)	d>400	....			al	B8², 508
10717	9,10-Phenanthroquinone, 2,5-dinitro .... $C_{14}H_6N_2O_6$	298.21	red ye pr (aa)	....	228			aa	B7³, 4096
10718	9,10-Phenanthroquinone, 2,7-dinitro .... $C_{14}H_6N_2O_6$	298.21	gold ye nd (aa)		301-3			aa	B7², 730
10719	9,10-Phenanthroquinone, 2-hydroxy .... $C_{14}H_8O_3$	224.23	br-red or bk-vt nd (aa)	sub	283			w	B8³, 2929
10720	9,10-Phenanthroquinone, 3-hydroxy .... $C_{14}H_8O_3$	224.23	ye-red or red nd (aa MeOH)	sub	330d			w, al	B8³, 2930
10721	9,10-Phenanthroquinone, 2-nitro .... $C_{14}H_7NO_4$	253.22	ye lf or nd (aa)		260			....	B7³, 4095

No.	Name, Synonyms, and Formula	Mol. wt.	Color, crystalline form, specific rotation and λ_{max} (log ε)	b.p. °C	m.p. °C	Density	n_D	Solubility	Ref.
10722	9,10-Phenanthroquinone, 7-isopropyl-1-methyl or Reten-oquinone . . . . $C_{18}H_{16}O_2$	264.33	og nd (chl-al)	sub	197-8	. . . .	. . . .	al, eth	B7[3], 4165
10723	9,10-Phenanthroquinone, 1,2,4-trihydroxy . . . . $C_{14}H_8O_5$	302.29	red (al + 1)	. . . .	d	. . . .	. . . .	al	B8[2], 558
10724	9,10-Phenanthroquinone, 2,3,4-trihydroxy . . . . $C_{14}H_8O_5$	256.29	red br pw	. . . .	185d	. . . .	. . . .	w	B8[2], 559
10725	Phenazine or Azophenylene . . . . $C_{12}H_8N_2$	180.22	ye-red nd (aa)	>360 sub	176-7	. . . .	. . . .	bz	B23[2], 233
10726	Phenazine, 1,4-dihydroxy-di-(N-oxide) . . . . $C_{12}H_8N_2O_4$	244.22	purp (chl)	. . . .	236d	. . . .	. . . .	chl	C50, 358
10727	Phenazine, 9,10-dihydro or Hydrazophenylene . . . . $C_{12}H_{10}N_2$	182.23	rh lf		317 (sealed tube)	. . . .	. . . .	bz	B23[2], 225
10728	Phenazine, 1-hydroxy or 1-Phenazinol . . . . $C_{12}H_8N_2O$	196.22	ye nd (bz, dil MeOH) lf (dil MeOH)	sub	158	. . . .	. . . .	Py	B23[2], 360
10729	Phenazine, 2-methyl . . . . $C_{13}H_{10}N_2$	194.24	lt ye nd or pr	350d	117	. . . .	. . . .	al, eth, chl	B23[2], 239
10730	β-Phenethyl amine $C_6H_5CH_2CH_2NH_2$	121.18		197-8		0.9580[24/4]	1.5290[25]	al, eth	B12[3], 2408
10731	β-Phenethyl amine, hydrochloride $C_6H_5CH_2CH_2NH_2.HCl$	157.65	pl or lf (al)		218-9	. . . .	. . . .	w, al	B12[3], 2409
10732	o-Phenetidine or 2-Ethoxy aniline $2-C_2H_5OC_6H_4NH_2$	137.18	. . . .	232.5, 127-8[14]	<−20		1.5560[20]	al, eth	B13[3], 756
10733	o-Phenetidine, 4-nitro or 2-Ethoxy-4-nitro aniline $4-NO_2-2-C_2H_5OC_6H_3NH_2$	182.18	ye nd (dil al)	. . . .	91	. . . .	. . . .	al, eth, ace	B13[3], 888
10734	o-Phenetidine, 5-nitro or 2-Ethoxy-5-nitro aniline $5-NO_2-2-C_2H_5OC_6H_3NH_2$	182.18	ye nd (dil al)	205-6[14]	96-7	. . . .	. . . .	al, eth	B13[3], 878
10735	o-Phenetidine, 6-nitro or 2-Ethoxy-6-nitro aniline $6-NO_2-2-C_2H_5OC_6H_3NH_2$	182.18	ye or og cr (w)	. . . .	60	. . . .	. . . .	w	B13[3], 876
10736	m-Phenetidine or 3-Ethoxy aniline $3-C_2H_5OC_6H_4NH_2$	137.18	. . . .	248, 127-8[11]		. . . .	. . . .	al, eth	B13[3], 932
10737	m-Phenetidine, 4-nitro or 3-Ethoxy-4-nitro aniline $4-NO_2-3-C_2H_5OC_6H_3NH_2$	182.18	nd (dil al)	. . . .	122-3	. . . .	. . . .	al, eth, ace, bz	B13[3]978
10738	m-Phenetidine, 5-nitro or 5-Ethoxy-3-nitro aniline $3-NO_2-5-C_2H_5OC_6H_3NH_2$	182.18	ye og red nd (al)	. . . .	115	. . . .	. . . .	al, ace, bz	B13[3], 977
10739	m-Phenetidine, 6-nitro or 5-Ethoxy-2-nitro aniline $5-(C_2H_5O)-2-(NO_2)C_6H_3NH_2$, $2-NO_2-5-C_2H_5OC_6H_3NH_2$	182.18	ye nd (dil al)	. . . .	105-6	. . . .	. . . .	ace, bz	B13[3]975
10740	p-Phenetidine or 4-Ethoxy aniline $4-C_2H_5OC_6H_4NH_2$	137.18	. . . .	254, 125[12]	2.4	1.0652[16/4]	1.5528[20]	al, eth	B13[3], 996
10741	p-Phenetidine, 2-nitro or 4-Ethoxy-2-nitro aniline $2-NO_2-4-C_2H_5OC_6H_3NH_2$	182.18	red pr (al)	. . . .	113	. . . .	. . . .	eth, chl	B13[3], 1203
10742	p-Phenetidine, 3-nitro or 4-Ethoxy-3-nitro aniline $3-NO_2-4-C_2H_5OC_6H_3NH_2$	182.18	og-ye nd (bz, dil al)	. . . .	41	. . . .	. . . .	ace, bz	B13[2], 284
10743	Phenetole or Ethoxy benzene $C_6H_5OC_2H_5$	122.17	. . . .	170, 60[9]	−29.5	0.9666[20/4]	1.5076[20]	al, eth	B6[4], 554
10744	Phenetole, 2-bromo or 1-Bromo-2-ethoxybenzene $2-BrC_6H_4OC_2H_5$	201.07	. . . .	222-6		. . . .	. . . .	al, eth	B6[4], 1038
10745	Phenetole, 3-bromo $3-BrC_6H_4OC_2H_5$	201.07	. . . .	222		. . . .	. . . .	al, eth	B6[3], 739
10746	Phenetole, 4-bromo $4-BrC_6H_4OC_2H_5$	201.07	. . . .	230-2, 109[17]	4	1.4071[25/4]	1.5517[20]	al, eth	B6[4], 1045
10747	Phenetole, 2-chloro or 1-Chloro-2-ethoxy benzene $2-Cl-C_6H_4OC_2H_5$	156.61	. . . .	210, 97-8[15]		1.1288[25/4]	1.5284[25]	al, eth, bz	B6[4], 785
10748	Phenetole, 3-chloro or 1-Chloro-3-ethoxy benzene $3-ClC_6H_4OC_2H_5$	156.61	. . . .	204-5[7]		1.1712[20/4]	. . . .	al, eth, bz, aa	B6[4], 811
10749	Phenetole, 4-chloro or 4-Chloro-1-ethoxy benzene $4-ClC_6H_4OC_2H_5$	156.61	. . . .	212-4, 98[17]	21	1.1254[20/4]	1.5252[20]	al, eth, bz, aa	B6[4], 823
10750	Phenetole, 2,4-dichloro $2,4-Cl_2-C_6H_3OC_2H_5$	191.07	. . . .	237		. . . .	. . . .	al, eth, bz	B6[4], 885
10751	Phenetole, 2,4-dinitro $2,4-(NO_2)_2C_6H_3OC_2H_5$	212.17	nd or lf (al)	. . . .	86-7	. . . .	. . . .	ace	B6[4], 1373
10752	Phenetole, 2,5-dinitro $2,5-(NO_2)_2C_6H_3OC_2H_5$	212.17	lf (al)	. . . .	96-8	. . . .	. . . .	. . . .	B6[2], 245

No.	Name, Synonyms, and Formula	Mol. wt.	Color, crystalline form, specific rotation and λ_{max} (log ε)	b.p. °C	m.p. °C	Density	n_D	Solubility	Ref.
10753	Phenetole, 2,6-dinitro . . . . . 2,6-(NO$_2$)$_2$C$_6$H$_3$OC$_2$H$_5$	212.17	nd (eth)	137-9[3]	60-1	. . . . . .	. . . . . .	eth	B6[3], 868
10754	Phenetole, 3,5 dinitro . . . . . 3,5-(NO$_2$)$_2$C$_6$H$_3$OC$_2$H$_5$	212.17	nd (al)	. . . . . .	97.5	. . . . . .	. . . . . .	. . . .	B6[3], 869
10755	Phenetole, 4-ethyl . . . . . 4-C$_2$H$_5$C$_6$H$_4$OC$_2$H$_5$	150.22	. . . . . .	211, 92-3[12]	. . . . . .	0.9385[17/4]	. . . . . .	al, ace, bz	B6[3], 1665
10756	Phenetole, 2-fluoro . . . . . 2-FC$_6$H$_4$OC$_2$H$_5$	140.16	. . . . . .	171.4, 64[11]	−16.7	1.0874[17]	1.4932[17]	ace, bz	B6[4], 771
10757	Phenetole, 3-fluoro . . . . . 3-FC$_6$H$_4$OC$_2$H$_5$	140.16	. . . . . .	171.4[755], 65.2[15]	−27.5	1.0716[16/4]	1.4847[17]	bz	B6[3], 669
10758	Phenetole, 4-fluoro . . . . . 4-FC$_6$H$_4$OC$_2$H$_5$	140.16	. . . . . .	173[766], 54[7]	−8.5	1.0715[18]	1.4826[18]	bz, chl	B6[4], 774
10759	Phenetole, 2-iodo . . . . . 2-IC$_6$H$_4$OC$_2$H$_5$	248.07	. . . . . .	245[736], 121-31[18]	. . . . . .	. . . . . .	. . . . . .	al, eth, ace, bz	B6[3], 769
10760	Phenetole, 4-iodo . . . . . 4-IC$_6$H$_4$OC$_2$H$_5$	248.07	cr (dil MeOH)	249-50[729]	29	. . . . . .	. . . . . .	al, eth, bz, chl	B6[4], 1077
10761	Phenetole, 3-mercapto . . . . . 3-HSC$_6$H$_4$OC$_2$H$_5$	154.23	. . . . . .	238-9	. . . . . .	. . . . . .	. . . . . .	al, eth, ace	B6, 833
10762	Phenetole, 4-merapto . . . . . 4-HSC$_6$H$_4$OC$_2$H$_5$	154.23	. . . . . .	238	1.6	. . . . . .	. . . . . .	al, eth, ace, bz	B6[2], 852
10763	Phenetole, 2-nitro . . . . . 2-NO$_2$C$_6$H$_4$OC$_2$H$_5$	167.17	br ye	267, 149[15]	2.1	1.1903[15]	1.5425[20]	al, eth	B6[4], 1250
10764	Phenetole, 3-nitro . . . . . 3-NO$_2$C$_6$H$_4$OC$_2$H$_5$	167.17	br ye	284d, 169[70] d	36	. . . . . .	. . . . . .	al, eth	B6[4], 1271
10765	Phenetole, 4 nitro . . . . . 4-NO$_2$C$_6$H$_4$OC$_2$H$_5$	167.17	pr (dil al, eth)	283, 168[15]	60	1.1176[100/4]	. . . . . .	eth, ace, bz	B6[4], 1283
10766	Phenicin or 3,3′-Dihydroxy-5,5′-dimethyl biguinone . . . . . C$_{16}$H$_{10}$O$_6$	274.25	yesh-br (al)	. . . . . .	230-1	. . . . . .	. . . . . .	al, chl, aa	B8[3], 4251
10767	Phenol . . . . . C$_6$H$_5$OH	94.11	. . . . . .	181.7, 70.9[10]	43	1.0576[20/4]	1.5408[41]	w, al, eth, ace, bz, chl	B6[4], 531
10768	Phenol, 2-acetyl or 2-Hydroxy acetophenone . . . . . 2-CH$_3$COC$_6$H$_4$OH	136.16	. . . . . .	218, 106[17]	4-6	1.1307[20/4]	1.5584[20]	al, eth, aa	B8[3], 261
10769	Phenol, 3-acetyl or 3-Hydroxy acetophenone . . . . . 3-CH$_3$COC$_6$H$_4$OH	136.16	nd or lf	296[756], 153[5]	96	1.0992[109]	1.5348[109]	al, eth, bz, chl	B8[3], 272
10770	Phenol, 4-acetyl or 4-Hydroxy acetophenone . . . . . 4-CH$_3$COC$_6$H$_4$OH	136.16	nd (eth, dil al)	147-8[3]	109-10	1.1090[109]	1.5577[109]	. . . .	B8[3], 276
10771	Phenol, 2-allyl . . . . . 2-(CH$_2$=CHCH$_2$)C$_6$H$_4$OH	134.18	. . . . . .	220, 93-4[8]	−6	1.0255[15/15]	1.5181[20]	eth	B6[2], 528
10772	Phenol, 2-allyl-4-chloro . . . . . 2-(CH$_2$=CHCH$_2$)-4-ClC$_6$H$_3$OH	168.63	pr (lig)	130-2[15]	48	1.171[15]	. . . . . .	al, bz	B6[1], 282
10773	Phenol, 2-allyl-6-methoxy . . . . . 2-C$_3$H$_5$-6-(CH$_3$O)C$_6$H$_3$OH	164.20	. . . . . .	250-1, 115[9]	. . . . . .	1.2090[20/4]	1.5545[20]	ace, bz	B6[3], 5013
10774	Phenol, 4-allyl or Chavicol . . . . . 4-(CH$_2$=CHCH$_2$)C$_6$H$_4$OH	134.18	. . . . . .	235-6, 120[12]	16	1.033[18/4]	1.5448[20]	al, eth, chl, peth	B6[3], 2415
10775	Phenol, 5-allyl-2-methoxy or Chavibetol . . . . . 5-(CH$_2$=CHCH$_2$)-2-(CH$_3$O)C$_6$H$_3$OH	164.21	. . . . . .	253-4, 111[8]	8.5	1.0613[25/4]	1.5413[20]	al, eth	B6[3], 5024
10776	Phenol, o-amino . . . . . 2-H$_2$NC$_6$H$_4$OH	109.13	wh rh bipym nd (bz)	sub 153[11]	174	1.328	. . . . . .	al, eth, w	B13[3], 752
10777	Phenol, 2-amino-3-chloro . . . . . 3-Cl-2-H$_2$NC$_6$H$_3$OH	143.57	nd	. . . . . .	122	. . . . . .	. . . . . .	w	B13[2], 182
10778	Phenol, 2-amino-4-chloro-5-nitro . . . . . 5-NO$_2$-4-Cl-2-H$_2$NC$_6$H$_2$OH	188.57	ye nd	. . . . . .	225d (dk at 200)	. . . . . .	. . . . . .	al	B13[2], 196
10779	Phenol, 2-amino-4-chloro-6-nitro . . . . . 6-NO$_2$-4-Cl-2H$_2$NC$_6$H$_2$OH	188.57	. . . . . .	. . . . . .	152	. . . . . .	. . . . . .	al	B13[2], 196
10780	Phenol, 2-amino-6-chloro-4-niro . . . . . 4-NO$_2$-6-Cl-2-H$_2$NC$_6$H$_2$OH	188.57	ye nd (w + 1)	. . . . . .	160	. . . . . .	. . . . . .	. . . .	B13[3], 895
10781	Phenol, 2-amino-5-chloro . . . . . 5-Cl-2-H$_2$NC$_6$H$_3$OH	143.57	nd (al), pr (dil al)	. . . . . .	154-5	. . . . . .	. . . . . .	al, eth	B13[3], 850
10782	Phenol, 2-amino-3,5-dibromo . . . . . 3,5-Br$_2$-2-H$_2$NC$_6$H$_2$OH	266.93	nd (lig)	. . . . . .	145	. . . . . .	. . . . . .	w	B13[2], 187
10783	Phenol, 2-amino-4,6-dibromo . . . . . 4,6-Br$_2$-2-H$_2$NC$_6$H$_2$OH	266.93	ye nd (dil al)	. . . . . .	99	. . . . . .	. . . . . .	al, eth, bz, chl	B13[2], 188
10784	Phenol, 2-amino-3,5-dichloro . . . . . 3,5-Cl$_2$-2-H$_2$NC$_6$H$_2$OH	178.02	nd (bz, w)	. . . . . .	132-3	. . . . . .	. . . . . .	al, ace, bz	B13[2], 185

No.	Name, Synonyms, and Formula	Mol. wt.	Color, crystalline form, specific rotation and λ_{max} (log ε)	b.p. °C	m.p. °C	Density	n_D	Solubility	Ref.
10785	Phenol, 2-amino-4,6-dinitro or Picramic acid .... 4,6-(NO₂)₂-2-H₂NC₆H₂OH	199.12	dk red nd (al), pr (chl)	....	169		....	al, bz, aa	B13³, 899
10786	Phenol, 2-(2-aminoethyl) 2-(H₂NCH₂CH₂)C₆H₄OH	137.18	rh (al-eth)	....	152-3				B13³, 1624
10787	Phenol, 2-amino-3-nitro 3-NO₂-2-H₂NC₆H₃OH	154.13	red nd (w)	sub	216-7			w	B13³, 875
10788	Phenol, 2-amino-4-nitro 4-NO₂-2-H₂NC₆H₃OH	154.13	og pr (+w)		80-90 (+w), 145-7 (anh)			al, eth, aa	B13³, 877
10789	Phenol, 2-amino-5-nitro 5-NO₂-2-NH₂C₆H₃OH	154.12	....		207-8			w, al, bz	B13³, 887
10790	Phenol, 2-amino-6-nitro 6-NO₂-2-H₂NC₆H₃OH	154.13	red nd (dil al)		111-2			al, eth, bz, chl, aa	B13², 195
10791	Phenol, 3-amino 3-H₂NC₆H₄OH	109.13	pr (to)	164¹¹	123			al, eth	B13³, 931
10792	Phenol, m-amino,hydrobromide 3-H₂NC₆H₄OH.HBr	190.06	pr (w)		224			w	B13, 403
10793	Phenol, m-amino,hydrochloride 3-H₂NC₆H₄OH.HCl	145.60	pr (w)		229			w	B13³, 932
10794	Phenol, m-amino,hydroiodide 3-H₂NC₆H₄OH.HI	237.05	pr (w)		209			w	B13, 403
10795	Phenol, 3-(2-aminoethyl), hydrochloride 3-(H₂NCH₂CH₂)C₆H₄OH.HCl	173.65	cr (al-eth)		145		....		B13³, 1630
10796	Phenol, 3-amino-2-chloro 2-Cl-3-H₂NC₆H₃OH	143.57			85-7			al, eth	B13, 420
10797	Phenol, 3-amino-4,6-dichloro 4,6-Cl₂-3-H₂NC₆H₂OH	178.02	ye br pr (w)		135-6			al, eth, ace, bz, chl	B13³, 970
10798	Phenol, 3-amino-4-nitro 4-NO₂-3-H₂NC₆H₃OH	154.13	og nd (w)		185-6			al, eth, bz, chl	B13¹, 136
10799	Phenol, p-amino 4-H₂NC₆H₄OH	109.13	wh pl (w)	110⁰·³	186-7			al	B13³, 991
10800	Phenol, 4-(2 aminoethyl) or Tyramine .... 4-(H₂NCH₂CH₂)C₆H₄OH	137.18	pl or nd (bz), cr (al), nd (w)	205-7²⁵	164-5		....	al, xyl	B13³, 1637
10801	Phenol, 4-(2 aminopropyl) (1) or L-Paredrine .... 4-[CH₃CH(NH₂)CH₂]C₆H₄OH	151.21	[α]¹⁷/D −52 (al)	....	111			al, eth, chl	H34, 2202
10802	Phenol, 4-amino-2-chloro 2-Cl-4-H₂NC₆H₃OH	143.57	nd (al, eth, w)	....	153			al, eth	B13³, 1180
10803	Phenol, 4-amino-2-chloro-6-nitro 6-NO₂-2-Cl-4-H₂NC₆H₂OH	188.57			130			al	B13, 524
10804	Phenol, 4-amino-3-chloro 3-Cl-4-H₂NC₆H₃OH	143.57	nd	....	160			al, eth	B13³, 1184
10805	Phenol, 4-amino-2,6-dibromo 2,6-Br₂-4-H₂NC₆H₂OH	266.93	nd (al, bz)		192-3			al, bz	B13³, 1195
10806	Phenol, 4-amino 2,5-dichloro 2,5-Cl₂-4-H₂NC₆H₂OH	178.02	cr (bz)		178-9			al, eth, aa	B13², 274
10807	Phenol, 4-amino -2,6-dichloro 2,6-Cl₂-4-H₂NC₆H₂OH	178.02	nd or lf (w, bz)	sub	167			al, eth	B13², 274
10808	Phenol, 4-amino-3,5-dichloro 3,5-Cl₂-4-H₂NC₆H₂OH	178.02	nd (w, bz)		154			al, eth, chl, aa	B13³, 1189
10809	Phenol, 4-amino-2-nitro 2-NO₂-4-H₂NC₆H₃OH	154.13	dk red pl or nd (w, al)	....	131			al, eth	B13², 284
10810	Phenol, 4-amino-3-nitro 3-NO₂-4-H₂NC₆H₃OH	154.13	dk red pr (eth)		154			w, al, eth, chl	B13³, 1203
10811	Phenol, 4-(2-aminopropyl) .... 4-CH₃CH(NH₂)CH₂C₆H₄OH	151.21	cr (bz)	....	125-6			w, al, chl	B13³, 1709
10812	Phenol, 2-benzyl .... 2-(C₆H₅CH₂)C₆H₄OH	184.24	cr (peth)	312, 159-62¹²	51-3				B6³, 3349
10813	Phenol, 4-benzyl .... 4-(C₆H₅CH₂)C₆H₄OH	184.24	nd (al)	320-2, 198-200¹⁰	84			al, eth, bz, chl, aa	B6³, 3357
10814	Phenol, 2-bromo 2-BrC₆H₄OH	173.02	....	194-5, 87.3¹³	5.6	1.4924²⁰/⁴	1.589²⁰	al, eth	B6⁴, 1037
10815	Phenol, 2-bromo-3-nitro 2-Br-3-NO₂C₆H₃OH	218.01	pa ye nd (HCl)	sub	147-8	....	....	al, eth	B6³, 844

No.	Name, Synonyms, and Formula	Mol. wt.	Color, crystalline form, specific rotation and λ_{max} (log ε)	b.p. °C	m.p. °C	Density	n_D	Solubility	Ref.
10816	Phenol, 2-bromo-4-nitro 2-Br-4-NO$_2$C$_6$H$_3$OH	218.01	cr (to, w), nd (chl, eth, dil al)		114			al, eth, chl	B6[3], 845
10817	Phenol, 2-bromo-5-nitro 2-Br-5-NO$_2$C$_6$H$_3$OH	218.01	pa ye nd (w), cr (peth)		129-30			al, eth, ace, bz	B6[4], 1365
10818	Phenol, 2-bromo-6-nitro 2-Br-6-NO$_2$C$_6$H$_3$OH	218.01	pa ye nd (chl, al)		68			al, aa	B6[3], 844
10819	Phenol, 3-bromo 3-BrC$_6$H$_4$OH	173.02		236.5, 135-40[12]	33			al, eth, chl	B6[4], 1042
10820	Phenol, 3-bromo-5-chloro 3-Br-5-ClC$_6$H$_3$OH	207.46	nd (peth)	256-60[756]	70			al, aa	B6[2], 187
10821	Phenol, 3-bromo-2,4-dinitro 2,4-(NO$_2$)$_2$-3-BrC$_6$H$_2$OH	263.02	pa ye nd (w)		175			eth	B6[2], 249
10822	Phenol, 3-bromo-2,6-dinitro 2,6-(NO$_2$)$_2$-3-BrC$_6$H$_2$OH	263.02	nd (peth)		131			al, eth	B6[2], 250
10823	Phenol, 3-bromo-2-nitro 3-Br-2-NO$_2$C$_6$H$_3$OH	218.01	ye nd (peth), col nd (+ w)		65-7 (anh)			al	B6[3], 842
10824	Phenol, 3-bromo-4-nitro 3-Br-4-NO$_2$C$_6$H$_3$OH	218.01	pa ye nd (bz, peth)		129-30			al, eth, bz	B6[4], 1365
10825	Phenol, 3-bromo-5-nitro 3-Br-5-NO$_2$C$_6$H$_3$OH	218.01	cr (w)		145			al, eth	B6[2], 233
10826	Phenol, 4-bromo 4-BrC$_6$H$_4$OH	173.02		238, 118.2[11]	66.4	1.840[15]		w, al, eth, chl	B6[4], 1043
10827	Phenol, 4-bromo-2-chloro 4-Br-2-ClC$_6$H$_3$OH	207.46	nd (lig, bz, aa)	233-4, 127-30[12]	50-1	1.6170[20/4]	1.5859[20]	al, eth, ace, bz	B6[3], 750
10828	Phenol, 4-bromo-2,6-dinitro 2,6-(NO$_2$)$_2$-4-BrC$_6$H$_2$OH	263.02	pa ye nd (w), nd (al), pr (aa)	sub	78			al, eth, bz, chl	B6[3], 872
10829	Phenol, 4-bromo-2-nitro 4-Br-2-NO$_2$C$_6$H$_3$OH	218.01	ye nd or lf (al), pr (eth)	sub	92			al, eth, bz, chl	B6[4], 1363
10830	Phenol, 4-bromo-3-nitro 4-Br-3-NO$_2$C$_6$H$_3$OH	218.01	ye nd (w)		147			eth, bz	B6[4], 1364
10831	Phenol, 5-bromo-2,4-dinitro 2,4(NO$_2$)$_2$-5-BrC$_6$H$_2$OH	263.02	pr (al, eth)		92			al, eth	B6[3], 871
10832	Phenol, 5-bromo-2-nitro 5-Br-2-NO$_2$C$_6$H$_3$OH	218.01	ye pr or nd (lig)		44			al, eth, lig	B6[3], 844
10833	Phenol, 6-bromo-2,4-dinitro 2,4-(NO$_2$)$_2$-6-BrC$_6$H$_3$OH	263.02	pa ye cr (al), pr (eth)	sub	118-9			eth, bz, lig	B6[3], 871
10834	Phenol, o-butyl 2-C$_4$H$_9$C$_6$H$_4$OH	150.22		235, 106.5[10]		0.975[20/4]	1.5180[25.5]	al, eth	B6[3], 1843
10835	Phenol, o-sec-butyl 2-CH$_3$CH$_2$CH(CH$_3$)C$_6$H$_4$OH	150.22		227-8[751], 116[21]	16	0.9804[25]	1.5200[25]		B6[3], 1852
10836	Phenol, o-tert-butyl or 2-tert-butyl phenol 2-(CH$_3$)$_3$CC$_6$H$_4$OH	150.22		221, 99[10]		0.9783[20/4]	1.5160[20]	al, eth	B6[3], 1861
10837	Phenol, m-butyl or 3-Butyl phenol 3-C$_4$H$_9$C$_6$H$_4$OH	150.22		248, 123[10]		0.974[20/4]		al, eth	B6[2], 485
10838	Phenol, m-tert-butyl 3-(CH$_3$)$_3$CC$_6$H$_4$OH	150.22	nd (peth)	240, 132.5[20]	41-2			al, eth	B6[3], 1862
10839	Phenol, p-butyl or 4-Butyl phenol 4-C$_4$H$_9$C$_6$H$_4$OH	150.22		248, 138-9[18]	22	0.978[20/4]	1.5165[25.5]	al, eth	B6[3], 1844
10840	Phenol, p-isobutyl 4-i-C$_4$H$_9$C$_6$H$_4$OH	150.22		235-9	51-2	0.9796[20/20]	1.5319[25]	al, eth, ace	B6[3], 1859
10841	Phenol, p-sec-butyl-(d) 4-sec C$_4$H$_9$C$_6$H$_4$OH	150.22	nd [α]$^{20}_d$ + 13.3 (xyl)	240-2	61-2	0.9883[20]	1.5182[21]	al, eth	B6[2], 487
10842	Phenol, p-tert-butyl or 4-tert-butyl phenol 4-(CH$_3$)$_3$CC$_6$H$_4$OH	150.22	nd (lig)	239.5, 114[10]	101	0.908[80/4]	1.4787[114]	al, eth	B6[3], 1862
10843	Phenol, o-chloro 2-ClC$_6$H$_4$OH	128.56		174.9, 56.4[10]	9.0	1.2634[20/4]	1.5524[20]	al, eth, bz	B6[4], 782
10844	Phenol, 2-chloro-4-cyclohexyl 2-Cl-4-C$_6$H$_{11}$C$_6$H$_3$OH	210.70		122-3[3]	39				B6[3], 2508

No.	Name, Synonyms, and Formula	Mol. wt.	Color, crystalline form, specific rotation and λ_{max} (log ε)	b.p. °C	m.p. °C	Density	n_D	Solubility	Ref.
10845	Phenol, 2-chloro-3,4-dimethyl .... 3,4-(CH₃)₂-2-Cl-C₆H₂OH	156.61	cr (peth)	187-9	27	1.5538[20]	....	peth	B6[2], 456
10846	Phenol, 2-chloro-4,6-dinitro 2-Cl-4,6-(NO₂)₂C₆H₂OH	218.55	....	....	114-6			al, eth, ace, bz	B6[4], 1385
10847	Phenol, 2-chloro-4-nitro 2-Cl-4-NO₂C₆H₃OH	173.55	wh nd (50% al)	....	111			al, eth, chl	B6[4], 1353
10848	Phenol, 2-chloro-5-nitro 2-Cl-5-NO₂C₆H₃OH	173.55	ye nd or pr (w)	....	121-2			....	B6[4], 1353
10849	Phenol, 2-chloro-6-nitro 2-Cl-6-NO₂C₆H₃OH	173.56			70			al, chl	B6[3], 837
10850	Phenol, m-chloro 3-ClC₆H₄OH	128.56	....	214	33	1.268[25]	1.5565[40]	al, eth, bz	B6[4], 810
10851	Phenol, 3-chloro-4-nitro 3-Cl-4-NO₂C₆H₃OH	173.56			121-2			al, bz	B6[4], 1357
10852	Phenol, 4-Chloro or p-Chloro phenol... 4-ClC₆H₄OH	128.56	....	219.7, 125[18]	43-4	1.2651[40/4]	1.5579[40]	al, eth, bz	B6[4], 820
10853	Phenol, 4-chloro-2-allyl .... 2(CH₂=CHCH₂)-4-ClC₆H₃OH	168.63	pr (lig)	130-2[15]	48	1.171[15]		al, bz	B6[1], 282
10854	Phenol, 4-chloro-2,6-dimethyl 2,6-(CH₃)₂-4-ClC₆H₂OH	156.61	nd (w)		83			al, bz, aa	B6[3], 1738
10855	Phenol, 4-chloro-2,3-dinitro .... 4-Cl-2,3(NO₂)₂C₆H₂OH	218.55	pr	....	127			al, eth, bz, chl	B6, 259
10856	Phenol, 4-chloro-2-nitro 4-Cl-2-NO₂C₆H₃OH	173.55	ye mcl pr (al)	....	88-9			al, eth, chl	B6[4], 1349
10857	Phenol, 5-chloro-2,4-dinitro 5-Cl-2,4-(NO₂)₂C₆H₂OH	218.55	nd (al, peth)		92	1.74[22]		al, eth, chl, peth	B6[4], 1385
10858	Phenol, 5-chloro-2-nitro 5-Cl-2-NO₂C₆H₃OH	173.55	ye pr or nd (w)	sub	41			al, eth, aa	B6[1], 836
10859	Phenol, o-cyclohexyl 2-C₆H₁₁C₆H₄OH	176.24	nd (lig)	283, 147[17]	56-7			al, aa	B6[3], 2492
10860	Phenol, p-cyclohexyl 4-C₆H₁₁C₆H₄OH	176.24	nd (bz)	293-5[752], 132-5[4]	133			al, eth, bz	B6[3], 2501
10861	Phenol, 2,4-diamino 2,4-(H₂N)₂C₆H₃OH	124.14	lf	....	78-80d			al, ace, w	B13[3], 1338
10862	Phenol, 2,4-diamino, dihydrochloride or Amidol .... 2,4-(H₂N)₂C₆H₃OH·2HCl	197.07	nd		230-40d			w	B13[3], 1338
10863	Phenol, 2,5-diamino 2,5-(H₂N)₂C₆H₃OH	124.15	nd		68			w	B13[2], 312
10864	Phenol, 3,4-diamino 3,4-(H₂N)₂C₆H₃OH	124.15	nd		170-2			w	B13[1], 210
10865	Phenol, 3,5-diamino 3,5-(H₂N)₂C₆H₃OH	124.15	nd or pr (chl)	....	168-70 (180)			w	B13[3], 1370
10866	Phenol, 2,4-dibromo 2,4-Br₂C₆H₃OH	251.92	nd(peth)	238-9, 177[17]	40			al, eth, bz	B6[4], 1061
10867	Phenol, 2,4-dibromo-6-nitro 2,4-Br₂-6-NO₂C₆H₂OH	296.90	....		118			eth, bz, chl	B6[3], 848
10868	Phenol, 2,6-dibromo 2,6-Br₂C₆H₃OH	251.92	nd (w)	162[21] (sub)	56-7			al, eth	B6[4], 1064
10869	Phenol, 2,6-dibromo-4-nitro 2,6-Br₂-4-NO₂C₆H₂OH	296.92	pa ye pr or lf (al)	d>144	145-6			al, eth	B6[4], 1366
10870	Phenol, 3,5-dibromo 3,5-Br₂C₆H₃OH	251.90		120-2[3]	81			al, eth	B6[4], 1065
10871	Phenol, 2,4-di-tert-butyl 2,4-[(CH₃)₃C]₂-C₆H₃OH	206.33	....	263.5, 146[20]	56.5		1.5080[20]	....	B6[3], 2062
10872	Phenol, 2,4-di-tert-butyl-5-methyl 2,4(t-C₄H₉)₂-5-CH₃C₆H₂OH	220.36	....	282, 167[20]	62.1	0.912[80/4]	....	al, eth, ace, bz	B6[3], 2072
10873	Phenol, 2,4-di-tert-butyl-6-methyl 2,4-(t-C₄H₉)₂-6-CH₃C₆H₂OH	220.36		269, 138.5[10]	51	0.891[80/4]			B6[3], 2073
10874	Phenol, 2,6-di-sec-butyl.............. 2,6-di-(sec-C₄H₉)₂C₆H₃OH	206.33	....	255-60	-42		1.5080[20]		ALD 11971, 7
10875	Phenol, 2,6-di-tert-butyl 2,6-[(CH₃)₃C]₂C₆H₃OH	206.33	pr (al)	133[20]	39	....	1.5001[20]	al	B6[3], 2061
10876	Phenol, 2,6-di-tert-butyl-4-ethyl 2,6-(t-C₄H₉)₂-4-C₂H₅C₆H₂OH	234.39	....	272, 140[10]	44				B6[3], 2087
10877	Phenol, 3,6-di-tert-butyl-4-(2-methyl butyl) .... 4-[CH₃CH₂CH(CH₃)CH₂]-2,6-(t-C₄H₉)₂C₆H₂OH	276.47	....	135-8[6]	47				B6[3], 2097

No.	Name, Synonyms, and Formula	Mol. wt.	Color, crystalline form, specific rotation and λ_{max} (log ε)	b.p. °C	m.p. °C	Density	n_D	Solubility	Ref.
10878	Phenol, 2,6-di-*tert*-butyl-4-methyl or Ionol 2,6-(t-C$_4$H$_9$)$_2$-4-CH$_3$C$_6$H$_2$OH	220.36	....	265, 136[10]	71	0.8937[75/4]	1.4859[75]	al, ace, bz, chl	B6[3], 2073
10879	Phenol, 2,3 dichloro 2,3-Cl$_2$C$_6$H$_3$OH	163.00	cr (lig, bz)	....	57-9			al, eth	B6[4], 883
10880	Phenol, 2,3-dichloro-4,5-dimethyl 2,3-Cl$_2$-4,5-(CH$_3$)$_2$C$_6$HOH	191.07	nd (peth)	....	102.5			eth, ace, bz	B6[2], 456
10881	Phenol, 2,3-dichloro-5,6-dimethyl 2,3-Cl$_2$-5,6-(CH$_3$)$_2$C$_6$HOH	191.07	cr (peth)		90			eth, ace, bz, chl	B6[2], 454
10882	Phenol, 2,4-dichloro 2,4-Cl$_2$C$_6$H$_3$OH	163.00	hex nd (bz)	210, 145-7[110]	45			al, eth, bz, chl	B6[4], 885
10883	Phenol, 2,4-dichloro-3,5-dimethyl 2,4-Cl$_2$-3,5-(CH$_3$)$_2$C$_6$HOH	191.07	....		83			eth	B6[3], 1760
10884	Phenol, 2,5-dichloro 2,5-Cl$_2$C$_6$H$_3$OH	163.00	pr (bz, peth)	211[764]	59			al, eth, bz	B6[4], 942
10885	Phenol, 2,5-dichloro-3,4-dimethyl 2,5-Cl$_2$-3,4-(CH$_3$)$_2$C$_6$HOH	191.07	nd (peth)		84			eth	B6[3], 1730
10886	Phenol, 2,6-dichloro 2,6-Cl$_2$C$_6$H$_3$OH	163.00	nd (peth)	219-20[740]	68-9			al, eth, bz	B6[4], 949
10887	Phenol, 2,6-dichloro-3,4-dimethyl 2,6-Cl$_2$-3,4-(CH$_3$)$_2$C$_6$HOH	191.07	cr (peth)		52			eth	B6[2], 456
10888	Phenol, 2,6-dichloro-3,5-dimethyl 2,6-Cl$_2$-3,5-(CH$_3$)$_2$C$_6$HOH	191.07	cr (peth)	105-10[1]	87-8			eth, chl	B6[4], 3157
10889	Phenol, 3,4-dichloro 3,4-Cl$_2$C$_6$H$_3$OH	163.00	nd (bz-peth)	253.5[767]	68			al, eth, bz	B6[4], 952
10890	Phenol, 3,5-dichloro 3,5-Cl$_2$C$_6$H$_3$OH	163.00	pr (peth)	233[757], 122-4[8]	68			al, eth	B6[4], 957
10891	Phenol, 2,6-dichloro-4-nitro 2,6-Cl$_2$-4-NO$_2$C$_6$H$_2$OH	208.00	br nd (w)		127d (anh p <100)	1.822		eth, chl	B6[4], 1361
10892	Phenol, 3-(diethyl amino) 3-(C$_2$H$_5$)$_2$NC$_6$H$_4$OH	165.24	rh bipym (CS$_2$-lig)	276-80, 170[15]	78			al, eth, w	B13[3], 942
10893	Phenol, 2,4-diiido 2,4-I$_2$C$_6$H$_3$OH	345.91	nd (w)	sub 100	72-3			al, eth	B6[4], 1082
10894	Phenol, 2,3-dimethoxy or Pyrogallol-1,2-dimethyl ether 2,3(CH$_3$O)$_2$C$_6$H$_3$OH	154.17	....	232-4, 124-5[17]	....		1.5392[20]	....	B6[3], 6264
10895	Phenol, 3-(dimethyl amino) 3-(CH$_3$)$_2$NC$_6$H$_4$OH	137.18	nd (lig)	265-8, 152-3[15]	87		1.5895[26]	al, eth, ace, bz	B13[3], 934
10896	Phenol, 4-(dimethyl amino) 4-(CH$_3$)$_2$NC$_6$H$_4$OH	137.18	....	165[30]	76			al, eth	B13[3], 1009
10897	Phenol, 2,3-dimethyl or *o*-3-Xylenol 2,3-(CH$_3$)$_2$C$_6$H$_3$OH	122.17	nd (w, dil al)	218, 95.4[10]	75		1.5420[20]	al, eth	B6[3], 1722
10898	Phenol, 2,3-dimethyl-4-*tert*-butyl 2,3-(CH$_3$)$_2$-4-(CH$_3$)$_3$C-C$_6$H$_2$OH	178.28	....	259, 145[20]					B6[3], 2019
10899	Phenol, 2,3-dimethyl-4-chloro 2,3-(CH$_3$)$_2$-4-Cl-C$_6$H$_2$OH	156.61	nd (peth)	....	85			al, eth, ace	B6[3], 1724
10900	Phenol, 2,3-dimethyl-6-chloro 2,3-(CH$_3$)$_2$-6-Cl-C$_6$H$_2$OH	156.61	....	221-3, 100[17]				al, ace, bz	B6[3], 1747

No.	Name, Synonyms, and Formula	Mol. wt.	Color, crystalline form, specific rotation and λ_{max} (log ϵ)	b.p. °C	m.p. °C	Density	n_D	Solubility	Ref.
10901	Phenol, 2,3 dimethyl-5-nitro 2,3-$(CH_3)_2$-5-$NO_2C_6H_2OH$	167.17	og-ye nd (bz, w)	....	109 (120)	....	....	al, ace, aa	B6[2], 455
10902	Phenol, 2,3-dimethyl-4,5,6-trichloro 2,3-$(CH_3)_2$-4,5,6-Cl_3C_6OH	225.50	nd (dil al, peth)	....	180-1	....	....	al	B6[2], 454
10903	Phenol, 2,4-dimethyl 2,4-$(CH_3)_2C_6H_3OH$	122.17	nd (w)	210, 89.3[10]	27-8	0.9650[20/4]	1.5420[14]	al, eth	B6[3], 1741
10904	Phenol, 2,4-dimethyl, acetate or 2,4-Dimethyl phenyl acetate 2,4-$(CH_3)_2C_6H_3O_2CCH_3$	164.20	....	226, 108[13]	....	1.0298[15.5/4]	1.4990[15]	al, eth	B6[2], 459
10905	Phenol, 2,4-dimethyl-6-tert-butyl 2,4-$(CH_3)_2$-6-$(CH_3)_3C$-C_6H_2OH	178.28	....	249, 115[10]	22.3	0.917[80/4]	1.5183[20]	....	B6[3], 2020
10906	Phenol, 2,4-dimethyl-5-chloro 2,4-$(CH_3)_2$-5-Cl-C_6H_2OH	156.61	nd (lig, w)	....	90-1	....	....	al, eth, ace	B6[3], 1747
10907	Phenol, 2,4-dimethyl-6-(hydroxy methyl) 2,4-$(CH_3)_2$-6-$(HOCH_2)C_6H_2OH$	152.20	nd (bz-peth)	57-8		....	....	al, eth	B6[3], 4653
10908	Phenol, 2,4-dimethyl-6-nitro 2,4$(CH_3)_2$-6-$NO_2C_6H_2OH$	167.17	ye nd (al)	....	73	....	....	al, eth	B6[3], 1750
10909	Phenol, 2,4-dimethyl-3,5,6-trichloro 2,4-$(CH_3)_2$-3,5,6-Cl_3C_6OH	225.50	pa ye nd	....	174	....	....	eth	B6[2], 460
10910	Phenol, 2,5-dimethyl 2,5-$(CH_3)_2C_6H_3OH$	122.17	nd (w), pr (al-eth)	211.5	75	....	....	al, eth	B6[3], 1769
10911	Phenol, 2,5-dimethyl, acetate or 2,5-Dimethyl phenyl acetate 2,5$(CH_3)_2C_6H_3O_2CCH_3$	164.21	....	237[768]	<−20	1.0624[15]	....	al, eth	B6[2], 467
10912	Phenol, 2,5-dimethyl-4-tert-butyl 2,5$(CH_3)_2$-4-$(CH_3)_3C$-C_6H_2OH	178.28	....	264, 136[10]	71-2	0.939[80/4], 1.001[27/4]	1.5311[20]	....	B6[3], 2019
10913	Phenol, 2,5-dimethyl-4-chloro 2,5$(CH_3)_2$-4-Cl-C_6H_2OH	156.61	silv-gr nd (lig)	....	74-5	....	....	al, bz, aa, peth	B6[3], 1773
10914	Phenol, 2,5-dimethyl-3-nitro 2,5-$(CH_3)_2$-3-$NO_2C_6H_2OH$	167.17	ye lf (peth)	....	91	....	....	al, eth	B6, 497
10915	Phenol, 2,5-dimethyl-6-nitro 2,5-$(CH_3)_2$-6-$NO_2C_6H_2OH$	167.17	nd (peth)	236d, 150[15]	34.5	....	....	al, eth, ace, bz	B6[2], 246
10916	Phenol, 2,5-dimethyl-4-nitro 2,5-$(CH_3)_2$-4-$NO_2C_6H_2OH$	167.17	pa ye nd (dil al)	....	122-3	....	....	al, eth, ace	B6[3], 1775
10917	Phenol, 2,5-dimethyl-2,4,6-trichloro 2,5-$(CH_3)_2$-2,4,6-Cl_3C_6HOH	225.50	pa gr nd (al, aq MeOH)	....	175-6	....	....	al, eth, bz, chl	B6[2], 467
10918	Phenol, 2,6-dimethyl 2,6-$(CH_3)_2C_6H_3OH$	122.17	lf or nd (al)	212, 91.2[10]	49	....	....	al, eth	B6[3], 1735
10919	Phenol, 2,6-dimethyl-4-tert-butyl 2,6-$(CH_3)_2$-4-$(CH_3)_3C$-C_6H_2OH	178.28	pr (peth)	248, 119[10]	82.4	0.916[80/4]	....	....	B6[3], 2019
10920	Phenol, 2,6-dimethyl-3-nitro 2,6-$(CH_3)_2$-3-$NO_2C_6H_2OH$	167.17	lf pr (bz), nd (lig)	....	99-100	....	....	al, chl	B6, 485
10921	Phenol, 2,6-dimethyl-4-nitro 2,6-$(CH_3)_2$-4-NO_2-C_6H_2OH	167.17	pr (MeOH)	....	171	....	....	al, ace, chl	B6, 486
10922	Phenol, 2,6-bis(1,1-dimethylpropyl)-4-methyl 4-CH_3-2,6-$[CH_3CH_2C(CH_3)_2]_2C_6H_2OH$	248.41	....	283, 165[20]	....	0.931[25/4]	1.4950[20]	....	B6[3], 2071
10923	Phenol, 3,4-dimethyl 3,4-$(CH_3)_2C_6H_3OH$	122.17	nd (w)	225, 106-8[10]	66-8	0.9830[20/4]	....	al, eth	B6[3], 1725
10924	Phenol, 3,4-dimethyl, acetate or 3,4-Dimethyl phenyl acetate 3,4-$(CH_3)_2C_6H_3O_2CCH_3$	164.21	....	235, 140[80]	22	....	....	al, eth, bz	B6[3], 1728
10925	Phenol, 3,4-dimethyl-6-tert-butyl 3,4-$(CH_3)_2$-6-$(CH_3)_3C$-C_6H_2OH	178.28	....	145[20]	46.0	0.920[80/4], 0.973[27/4]	1.5222[20]	....	B6[3], 2019

No.	Name, Synonyms, and Formula	Mol. wt.	Color, crystalline form, specific rotation and λ_{max} (log ε)	b.p. °C	m.p. °C	Density	n_D	Solubility	Ref.
10926	Phenol, 3,4 dimethyl-5-chloro .... 3,4-(CH$_3$)$_2$-5-Cl-C$_6$H$_2$OH	156.61	nd (peth)	....	98	....	....	al, ace, chl	B6[2], 456
10927	Phenol, 3,4-dimethyl-6-chloro .... 3,4-(CH$_3$)$_2$-6-Cl-C$_6$H$_2$OH	156.61	nd (peth)	....	72	....	....	al, bz, aa	B6[3], 1729
10928	Phenol, 3,4-dimethyl-6-nitro .... 3,4-(CH$_3$)$_2$-6-NO$_2$C$_6$H$_2$OH	167.17	ye rh (al)	....	87-9	....	....	al, eth, bz, chl	B6[3], 1731
10929	Phenol, 3,4-dimethyl-2,5,6-tribromo .... 3,4-(CH$_3$)$_2$-2,5,6-Br$_3$-C$_6$-OH	358.87	nd (al)	....	173-4	....	....	al	B6[3], 1730
10930	Phenol, 3,4-dimethyl-2,5,6-trichloro .... 3,4-(CH$_3$)$_2$-2,5,6-Cl$_3$C$_6$OH	225.50	nd (peth)	....	182.5	....	....	al	B6[2], 456
10931	Phenol, 3,5-dimethyl .... 3,5-(CH$_3$)$_2$C$_6$H$_3$OH	122.17	nd (w, peth)	219.5 sub, 102-3[10]	68	0.9680[20/4]	....	al	B6[3], 1753
10932	Phenol, 3,5-dimethyl-4-chloro .... 3,5-(CH$_3$)$_2$-4-Cl-C$_6$H$_2$OH	156.61	....	246	115-8	....	....	al, eth, bz	B6[3], 1759
10933	Phenol, 3,5-dimethyl-6-chloro .... 3,5-(CH$_3$)$_2$-6-Cl-C$_6$H$_2$OH	156.61	nd	....	49-50	....	....	eth	B6[2], 464
10934	Phenol, 3,5-dimethyl-2-nitro .... 3,5-(CH$_3$)$_2$-2-NO$_2$C$_6$H$_2$OH	167.17	ye nd (lig, dil MeOH)	....	66	....	....	al, eth, ace, bz	B6[3], 1765
10935	Phenol, 3,5-dimethyl-2,4,6-tribromo .... 3,5-(CH$_3$)$_2$-2,4,6-Br$_3$C$_6$OH	358.87	nd (al)	....	166	....	....	al	B6[3], 1763
10936	Phenol, 3,5-dimethyl-2,4,6-trichloro .... 3,5-(CH$_3$)$_2$-2,4,6-Cl$_3$C$_6$OH	225.50	ye nd (peth)	....	117-8	....	....	peth	B6[3], 1761
10937	Phenol, 2,3-dinitro 2,3-(NO$_2$)$_2$C$_6$H$_3$OH	184.11	ye nd (w)		144-5	1.681[20]		al, eth, bz	B6[4], 1369
10938	Phenol, 2,4-dinitro .... 2,4-(NO$_2$)$_2$C$_6$H$_3$OH	184.11	pa ye pl or lf (w)	sub	115-6	1.683[24]	....	al, eth, ace, bz, chl	B6[4], 1369
10939	Phenol, 2,5-dinitro 2,5(NO$_2$)$_2$C$_6$H$_3$OH	184.11	ye nd pl or nd (dil al, w, lig)		108			eth, bz	B6[4], 1383
10940	Phenol, 2,6-dinitro 2,6-(NO$_2$)$_2$C$_6$H$_3$OH	184.11	pa ye rh nd or lf (dil al)		63-4	....	....	al, eth, ace, bz, Py	B6[4], 1383
10941	Phenol, 3,4-dinitro .... 3,4-(NO$_2$)$_2$C$_6$H$_3$OH	184.11	tcl nd (w)	....	134	1.672	....	al, eth, bz	B6[4], 1384
10942	Phenol, 3,5-dinitro 3,5-(NO$_2$)$_2$C$_6$H$_3$OH	184.11	lf (w)	....	126.1	1.702	....	al, eth, bz, chl	B6[4], 1385
10943	Phenol, 2,6-dipropyl 2,6(C$_3$H$_7$)$_2$C$_6$H$_3$OH	178.28	....	256, 114-6[5]	28		....	eth	B6[3], 2013
10944	Phenol, 2-ethyl .... 2-C$_2$H$_5$C$_6$H$_4$OH	122.17	....	207, 84.1[10]	<−18	1.0371[0]	1.5367[20]	al, eth, ace, bz	B6[3], 1655
10945	Phenol, 2-ethylamino 2-C$_2$H$_5$NHC$_6$H$_4$OH	137.18	pl (bz)	....	113-4	....	....	al	B13[3], 763
10946	Phenol, 2-ethyl-4-*tert*-butyl .... 2-C$_2$H$_5$-4-(CH$_3$)$_3$C-C$_6$H$_3$OH	178.28	....	257, 141[20]			....	....	B6[3], 2012

No.	Name, Synonyms, and Formula	Mol. wt.	Color, crystalline form, specific rotation and λ_{max} (log ϵ)	b.p. °C	m.p. °C	Density	n_D	Solubility	Ref.
10947	Phenol, 3-ethyl .... 3-$C_2H_5C_6H_4OH$	122.17	....	214, 99.3[10]	−4	1.0283[20/4]	....	al, eth	B6[3], 1660
10948	Phenol, 3-(ethylamino) .... 3-$C_2H_5NHC_6H_4OH$	137.18	cr (bz-peth)	176[12]	62			al, eth, bz, chl	B13, 408
10949	Phenol, 4-(ethylamino) .... 4-$C_2H_5NHC_6H_4OH$	137.18	nd (w)		110-2			al, eth	B13[3], 1012
10950	Phenol, 4-ethyl .... 4-$C_2H_5C_6H_4OH$	122.17	nd	219, 99.5[10]	47-8		0.5239[25]	al, eth, ace, bz	B6[3], 1663
10951	Phenol, 4-ethyl-2-tert-butyl .... 4-C_2H_5-2-$(CH_3)_3C$-C_6H_3OH	178.28	....	250, 123[10]	23			al	B6[2], 2012
10952	Phenol, o-fluoro .... 2-F-C_6H_4OH	112.10	....	151-2	16.1			w	B6[4], 770
10953	Phenol, p-fluoro .... 4-F-C_6H_4OH	112.10	....	185.5, 87[23]	48			ace, peth	B6[4], 773
10954	Phenol, 4-fluoro-2-nitro .... 4-F-2-$NO_2C_6H_3OH$	157.10	....		73.7			al	B6[1], 121
10955	Phenol, 4-[bis-(2-hydroxyethyl) amino .... 4[(HOCH_2CH_2)_2N]C_6H_4OH	197.24	cr (w)		140			w	B13[3], 1033
10956	Phenol, 2-hydroxymethyl-4-methyl or Homosaligenine ... 2-$HOCH_2$-4-$CH_3C_6H_3OH$	138.17	lf (w, chl)		106-7			w, al, eth	B6[3], 4598
10957	Phenol, o-iodo .... 2-IC_6H_4OH	220.01	nd	186-7[160], 91-2[2]	43	1.8757[80]		al, eth	B6[4], 1070
10958	Phenol, m-iodo .... 3-IC_6H_4OH	220.01	nd (lig)	d	118			al, eth	B6[4], 1073
10959	Phenol, p-iodo .... 4-IC_6H_4OH	220.01	nd (w or sub)	138-40[5]d	93-4	1.8573[112]		al, eth	B6[4], 1074
10960	Phenol, 2-mercapto .... 2-HSC_6H_4OH	126.18	oil	216-7[751], 88-90[8]	5-6	1.2373[0/0]		eth	B6[3], 4276
10961	Phenol, 3-mercapto .... 3-HSC_6H_4OH	126.18	cr	168[35]	16-7			al	B6[2], 827
10962	Phenol, 4-mercapto .... 4-HSC_6H_4OH	126.18	cr	166-8[45], 133-7[11]	29-30	1.1285[25/4]	1.5101[25]	w, al	B6[3], 4445
10963	Phenol, 2-methoxy-3-nitro or 6-Nitroguaiacol .... 2-CH_3O-3-$NO_2C_6H_3OH$	169.14	yesh rh pr (peth)		102-3			al	B6[3], 4263
10964	Phenol, 2-methoxy-4-nitro or 5-Nitroguaiacol .... 2-CH_3O-4-$NO_2C_6H_3OH$	169.14	ye nd (w)		103-4			al, eth	B6[3], 4264
10965	Phenol, 2-methoxy-5-nitro or 4-Nitroguaiacol .... 2-CH_3O-5-$NO_2C_6H_3OH$	169.14	pa ye nd (w)		105			al, eth	B6[3], 4264
10966	Phenol, 2-methoxy-6-nitro or 3-Nitroguaiacol .... 2-CH_3O-6-$NO_2C_6H_3OH$	169.14	og-ye nd (sub)	sub	62			w, al	B6[2], 789
10967	Phenol, 2-methoxy-4-propenyl (cis) or Isoeugenol (cis) ... 2-CH_3O-4-$(CH_3CH=CH)C_6H_3OH$	164.21	....	134-5[13], 80-1[0.5]	....	1.0837[20/4]	1.5726[20]	al, eth	B6[3], 4992
10968	Phenol, 3-methoxy-4-nitro .... 3-CH_3O-4-$NO_2C_6H_3OH$	169.14	ye nd (al)		144			al	B6[2], 822
10969	Phenol, 3-methoxy-5-nitro .... 3-CH_3O-5-$NO_2C_6H_3OH$	169.14	ye cr (al)		144			al	B6[3], 4347
10970	Phenol, 4-methoxy-2-nitro .... 4-CH_3O-2-$NO_2C_6H_3OH$	169.14	og ye nd or mcl cr (al, lig)		80 (83)			al	B6[3], 4442
10971	Phenol, 4-methoxy-3-nitro .... 4-CH_3O-3-$NO_2C_6H_3OH$	169.14	pa ye nd (w), cr (bz)		98-100			al	B6[2], 848
10972	Phenol, 5-methoxy-2-nitro .... 5-CH_3O-2-$NO_2C_6H_3OH$	169.14	yesh nd (al)		95			al	B6[3], 4345
10973	Phenol, 2-(methylamino) .... 2-$CH_3NHC_6H_4OH$	123.16	pl (bz-peth)		96-7			al, bz	B13[3], 761
10974	Phenol, 4-methyl amino,sulfate or Metol, Pictol .... (4-$CH_3NHC_6H_4OH)_2H_2SO_4$	344.39	wh nd(w)		250-60d			w, al	B13[3], 1007
10975	Phenol, 4-(methyl amino)-2-nitro .... 2-NO_2-4-$(CH_3NH)C_6H_3OH$	168.15	dk red-br nd (al)		113-4			al	B13[1], 186
10976	Phenol, 4-(2-methylaminopropyl) .... 4-[$CH_3CH(NHCH_3)CH_2]C_6H_4OH$	165.24	cr (MeOH)		161			al, eth	B13[3], 1710
10977	Phenol, 4-(2-methyl-2-butyl) .... 4-[H_5C_2-$C(CH_3)_2$]-C_6H_4OH	164.25	nd	262.5, 138[15]	94-6			....	B6[3], 1965
10978	Phenol, 2-methyl-4-tert-butyl .... 2-CH_3-4-$(CH_3)_3C$-C_6H_3OH	164.25	yesh	235-7[740], 132[20]	27-8	0.965[20/4]	1.5230[20]	eth, ace, bz	B6[3], 1979

No.	Name, Synonyms, and Formula	Mol. wt.	Color, crystalline form, specific rotation and λ_{max} (log ε)	b.p. °C	m.p. °C	Density	n_D	Solubility	Ref.
10979	Phenol, 2-methyl-4(3-methyl-3-pentyl) 2-CH₃-4-[(C₂H₅)₂C(CH₃)]-C₆H₃OH	192.31	....	145-6[11]	35	....	1.5200[25]	ace	C48, 1328
10980	Phenol, 2-methyl-5-isopropyl or Carvacrol 2-CH₃-5-i-C₃H₇C₆H₃OH	150.22	....	237.7, 101-3[10]	1	0.9772[20/4]	1.5230[20]	al, eth, ace	B6³, 1885
10981	Phenol, 2-methyl-5-isopropyl-3-nitro 2-CH₃-5-i-C₃H₇-3-ONC₆H₂OH	179.22	yesh pr (bz), nd (dil al)	....	153			al, eth, bz, chl	B7³,3412
10982	Phenol, 2-methyl-3,4,5-trichloro 2-CH₃-3,4,5-Cl₃C₆HOH	211.48	nd (lig)	....	77			ace	B6³, 1268
10983	Phenol, 3-methyl-6-tert-butyl 3-CH₃-6-(CH₃)₃C-C₆H₃OH	164.25	....	224, 127[11]	46-7	0.922[80/4]	1.5250[20]	al, eth, ace	B6³, 1982
10984	Phenol, 4-methyl-2-tert-butyl 4-CH₃-2-(CH₃)₃C-C₆H₃OH	164.25	nd (peth)	237, 111[10]	55	0.9247[75/4]	1.4969[75]	eth, ace, bz	B6³, 1978
10985	Phenol, 4-(3-methyl-1-butyl) 4-(CH₃)₂CHCH₂CH₂C₆H₄OH	164.25	nd (w)	255, 126[14]	93	0.9579[23/20]	1.5050[27]	al, eth	B6³, 1960
10986	Phenol, 3-methyl-2,4,6-trichloro 3-CH₃-2,4,6-Cl₃-C₆HOH	211.48	nd (w), pl (peth)	265, 162-3[28]	47			al, eth, chl	B6⁴, 2070
10987	Phenol, 4-methyl-2,3,5-trichloro 4-CH₃-2,3,5-Cl₃C₆HOH	211.48	nd (aa, lig)	....	66-7			ace	B6¹, 204
10988	Phenol, o-nitro 2-O₂NC₆H₄OH	139.11	ye nd or pr (eth, al)	216, 96-7[10]	45-6	1.2942[40], 1.485[14]	1.5723[50]	al, eth, ace, bz, chl	B6⁴, 1246
10989	Phenol, m-nitro 3-O₂NC₆H₄OH	139.11	ye mcl (eth, aq HCl)	194[70]	97			al, eth, ace, bz	B6⁴, 1269
10990	Phenol, p-nitro 4-O₂NC₆H₄OH	139.11	ye mcl pr (to)	279d sub	114.6	1.479[20]		al, eth, ace, Py	B6⁴, 1279
10991	Phenol, p-nitroso 4-ONC₆H₄OH	123.11	pa ye rh nd (ace, bz)	....	d 144			al, eth, ace, bz	B7³, 3367
10992	Phenol, o-octyl 4-C₈H₁₇C₆H₄OH	206.33	....	160[19]	41.7				B6³, 2046
10993	Phenol, penta bromo Br₅C₆OH	488.62	mcl pr (aa), nd (al)	sub	229.5			al, bz	B6⁴, 1069
10994	Phenol, penta chloro Cl₅C₆OH	266.34	mcl pr (al + 1w), nd (bz)	309-10[754]d	174 (+1w), 191 (anh)	1.978[22/4]		al, eth, bz	B6⁴, 1025
10995	Phenol, penta fluoro C₆F₅OH	184.07	....	72-3[48]			1.4263[26]	....	B6⁴, 782
10996	Phenol, penta methyl (CH₃)₅C₆OH	164.25	nd (al, peth, ace)	267	128			al	B6¹, 1991
10997	Phenol, p-pentyl 4-C₅H₁₁C₆H₄OH	164.25	....	250.5, 119-20³	23	0.960[20/4]	1.5272[25]	al, eth	B6¹, 1950
10998	Phenol, o-phenyl or 2-Hydroxy biphenyl 2-HOC₆H₄C₆H₅	170.21	nd (peth)	286, 145[14]	58-60	1.213[25/4]		al, eth, ace, bz, lig	B6³, 3281
10999	Phenol, m-phenyl or 3-hydroxy biphenyl 3-HOC₆H₄C₆H₅	170.21	nd (w or peth)	>300	78			al, eth, bz, peth	B6³, 3311
11000	Phenol, p-phenyl or 4-Hydroxy biphenyl 4-HOC₆H₄C₆H₅	170.21	nd or pl (dil al)	305-8 sub	165-7			al, eth, chl	B6³, 3319
11001	Phenol, 4-propenyl 4-(CH₃CH=CH)C₆H₄OH	134.18	pl (w), lf	250d, 140-5[15]	93-4			al, eth	B6³, 2394
11002	Phenol, o-propionyl 2(CH₃CH₂CO)C₆H₄OH	150.18	....	150[80], 115[15]	....		1.5501[20]	al, eth	B8³, 373
11003	Phenol, p-propionyl 4-(CH₃CH₂CO)C₆H₄OH	150.18	wh nd or pr (w)	....	149			al, eth	B8³, 379
11004	Phenol, 2-propyl 2-C₃H₇C₆H₄OH	136.20	....	220, 106.7[10]	....	1.015[20/4]		al, eth	B6³, 1784
11005	Phenol, 3-propyl 3-C₃H₇C₆H₄OH	136.20	....	228, 111.2[10]	26	0.987[20]	1.5223[20]	al, eth	B6³, 1787
11006	Phenol, 4-propyl 4-C₃H₇C₆H₄OH	136.20	....	232.6, 111.7[10]	22	1.009[20/4]	1.5379[25]	al	B6³, 1783
11007	Phenol, 2-isopropyl or o-Cumenol 2-i-C₃H₇C₆H₄OH	136.20	....	213-4	15-6	1.012[20/4]	1.5315[20]	al, eth, bz	B6³, 1807
11008	Phenol, 2-isopropyl-5-methyl-4-nitroso or 4-Nitroso carvacrol 2-i-C₃H₇-5-CH₃-4-ON-C₆H₂OH	179.22	yesh red	....	175			al, eth, chl	B7³, 3412
11009	Phenol, 3-isopropyl or m-Cumenol 3-i-C₃H₇C₆H₄OH	136.20	....	228	26	....	1.5261[20]	eth	B6³, 1810

No.	Name, Synonyms, and Formula	Mol. wt.	Color, crystalline form, specific rotation and λ_{max} (log ε)	b.p. °C	m.p. °C	Density	n_D	Solubility	Ref.
11010	Phenol, 4-isopropyl or p-Cumenol 4-i-C₃H₇C₆H₄OH	136.20	nd (peth)	228-30[745], 109-10[10]	62-3	0.990[20]	1.5228[20]	al	B6[3], 1810
11011	Phenol, seleno or Selenyl benzene C₆H₅SeH	157.08	....	183.6	......	1.4865[15]	....	al, eth	B6[4], 1777
11012	Phenol, 2,3,4,6-tetrabromo 2,3,4,6-Br₄C₆HOH	409.72	nd (al, aa)	sub	113-4 (120)	....	....	al, bz	B6[3], 766
11013	Phenol, 2,3,4,5-tetrachloro 2,3,4,5-Cl₄C₆HOH	231.89	nd (peth, sub)	sub	116-7	....	....	al	B6[4], 1020
11014	Phenol, 2,3,4,6-tetrachloro 2,3,4,6-Cl₄C₆HOH	231.89	nd (lig, aa)	150[15]	70	....	....	al, bz, chl, lig	B6[4], 1021
11015	Phenol, 2,3,5,6-tetrachloro 2,3,5,6-Cl₄C₆HOH	231.89	lf (lig)	....	115	....	....	bz	B6[4], 1025
11016	Phenol, 2,3,4,5-tetramethyl 2,3,4,5-(CH₃)₄C₆HOH	150.22	....	266	86-7	....	....	al, eth	B6[3], 1918
11017	Phenol, 2,3,4,6-tetramethyl or Isodurenol 2,3,4,6-(CH₃)₄C₆HOH	150.22	cr (peth)	230-50	80-1	....	....	al	B6[3], 1919
11018	Phenol, 2,3,5,6-tetramethyl or Durenol 2,3,5,6-(CH₃)₄C₆HOH	150.22	nd (lig), pr (al)	247	118-9	....	....	aa, peth	B6[3], 1919
11019	Phenol, 2,3,4,6-tetranitro 2,3,4,6-(NO₂)₄C₆HOH	274.10	lt ye nd (chl)	exp	140d	....	....	w	B6[3], 973
11020	Phenol, 2,4,6-triamino 2,4,6-(H₂N)₃C₆H₂OH	139.16	....	257	......	....	....	w, al, eth	B13[3], 1373
11021	Phenol, 2,3,4-tribromo 2,3,4-Br₃C₆H₂OH	330.82	nd or pl (w, lig)	....	94-5	....	....	al, eth, ace, lig	B6[2], 192
11022	Phenol, 2,4,5-tribromo 2,4,5-Br₃C₆H₂OH	330.82	nd (lig)	....	87	....	....	al, lig	B6[3], 760
11023	Phenol, 2,4,6-tribromo or Bromol 2,4,6-Br₃C₆H₂OH	330.82	nd (al), pr (bz), cr (aa + 1)	282-90[764] (sub)	95-6	2.55[20/20]	....	al, eth	B6[4], 1067
11024	Phenol, 3,4,5-tribromo 3,4,5-Br₃C₆H₂OH	330.82	tcl (bz-lig)	....	129	....	....	al, bz, aa	B6[2], 195
11025	Phenol, 2,4,6-tri-tert-butyl 2,4,6-(t-C₄H₉)₃C₆H₂OH	262.44	cr (al, peth)	278, 130[15]	131	0.864[27/4]	....	ace	B6[3], 2094
11026	Phenol, 2,3,4-trichloro 2,3,4-Cl₃C₆H₂OH	197.45	nd (bz, lig, sub)	sub	83.5	....	....	al, eth, bz, aa	B6[3], 716
11027	Phenol, 2,3,5-trichloro 2,3,5-Cl₃C₆H₂OH	197.45	nd (al)	248-9[250]	62	....	....	al, eth	B6[3], 716
11028	Phenol, 2,3,6-trichloro 2,3,6-Cl₃C₆H₂OH	197.45	nd (dil al, lig)	....	58	....	....	al, eth, bz, aa, lig	B6[4], 962
11029	Phenol, 2,4,5-trichloro 2,4,5-Cl₃C₆H₂OH	197.45	nd (al, peth)	sub	68-70	....	....	al, lig	B6[4],962
11030	Phenol, 2,4,6-trichloro 2,4,6-Cl₃C₆H₂OH	197.45	rh nd (aa)	246	69.5	1.4901[75/4]	....	al, eth	B6[4], 1005
11031	Phenol, 3,4,5-trichloro 3,4,5-Cl₃C₆H₂OH	197.45	nd(lig)	271-7[746]	101	....	....	eth	B6[3], 729
11032	Phenol, 2,3,5-triiodo 2,3,5-I₃C₆H₂OH	471.80	nd (peth, bz-lig)	....	114	....	....	al, eth, ace, bz	B6[4], 1085
11033	Phenol, 2,4,6-triiodo 2,4,6-I₃C₆H₂OH	471.80	nd (dil al)	sub, d	158-9	....	....	eth, ace	B6[4], 1085
11034	Phenol, 2,4,5-trimethyl or Psuedocumenol 2,4,5-(CH₃)₃C₆H₂OH	136.20	nd (lig)	232	72	....	....	al, eth	B6[3], 1831
11035	Phenol, 2,4,5-trimethyl, acetate 2,4,5-(CH₃)₃C₆H₂(O₂CCH₃)	178.23	nd (peth)	245-6	34	....	....	al, eth	B6[2], 482
11036	Phenol, 2,4,6-trimethyl or Mesitol 2,4,6-(CH₃)₃C₆H₂OH	136.20	nd (peth, MeOH)	221 sub	72	....	....	al, eth	B6[3], 1835
11037	Phenol, 2,3,6-trinitro 2,3,6-(O₂N)₃C₆H₂OH	229.11	ye nd (w)	....	119	....	....	al, eth, bz, aa	B6[2], 253
11038	Phenol, 2,4,5-trinitro 2,4,5-(O₂N)₃C₆H₂OH	229.11	wh nd (w, dil al)	....	96	....	....	al, eth, bz, aa	B6[4], 1388
11039	Phenol, 2,4,6-trinitro or Picric acid 2,4,6-(O₂N)₃C₆H₂OH	229.11	ye lf (w), pr (eth), pl (al)	sub exp>300	122-3	....	1.763	al, eth, ace, bz, aa	B6[4], 1388
11040	Phenol, o-vinyl 2-(CH₂=CH)C₆H₄OH	120.16	nd	101[14]	29-30	1.0609[18/4]	1.5851[20]	al, eth	B6[3], 2383
11041	Phenol, m-vinyl 3-(CH₂=CH)C₆H₄OH	120.16	....	114-6[16]	......	1.0353[21/4]	1.5804[21]	....	B6[3], 2385

No.	Name, Synonyms, and Formula	Mol. wt.	Color, crystalline form, specific rotation and λ_{max} (log ε)	b.p. °C	m.p. °C	Density	n_D	Solubility	Ref.
11042	Phenolphthalein or 2,2-bis(4-hydroxyphenyl)phthalide .. $C_{20}H_{14}O_4$	318.33	wh rh nd	....	262-3	$1.277^{32/4}$	$1.277^{32/4}$	al, eth, ace, chl, Py	B18⁴, 1945
11043	Phenolphthalein-3′,3″,5′,5″-tetrabromo .... $C_{20}H_{10}Br_4O_4$	633.94	nd (al, eth, aa)	....	295-7			eth	B18⁴, 1948
11044	Phenolphthalein-3′,3″,5′,5″-tetrachloro .... $C_{20}H_{10}Cl_4O_4$	456.11	cr (bz, aa)	....	225			al, ace, bz	B18², 123
11045	Phenolphthalein-3′,3″,5′,5″-tetraiodo or Iodophen ..... $C_{20}H_{10}I_4O_4$	821.92	amor	....	227-9d	$2.0246^{22/22}$		....	B18⁴, 1949
11046	Phenolphthalin .... $C_{20}H_{16}O_4$	320.35	nd (w)	....	229-32			al	B10³, 2013
11047	Phenolsulfonphthalein or Phenol red .... $C_{19}H_{14}O_5S$	354.37	dk red nd or pl	....	>300			....	B19⁴, 1128
11048	Phenothiazine .... $C_{12}H_9NS$	199.28	ye pr (al), ye lf or pl (to)	371, 290⁴⁰ (sub 130¹)	186-9			al, eth, ace, bz	B27², 32
11049	Phenothiazine, 2,8-dinitro-5-oxide $C_{12}H_7N_3O_5S$	305.27	ye-red lf (aa)	....	d			....	B27¹, 229
11050	Phenothiazine, 10-(2-diethylaminoethyl) $C_{18}H_{22}N_2S$	298.46	ye oil	195-208⁴⁻⁵	....			dil HCl	Am66, 888
11051	Phenothiazine, 10-(2-dimethylaminopropyl) or Promethazine .... $C_{17}H_{20}N_2S$	284.43	....	190-3⁰·⁵	60			....	C42, 575
11052	Phenothiazine, 10-(2-dimethylaminopropyl), hydrochloride .... $C_{17}H_{20}N_2S.HCl$	320.89	....	....	230-2			w, al, chl	C42, 575
11053	Phenothiazine, 10-phenyl .... $C_{18}H_{13}NS$	275.38	ye pr (al)	....	89-90			al	B27¹, 227
11054	Phenothiazine, 10-octadecyl .... $C_{30}H_{45}NS$	451.71		290-300⁰·¹	51.0⁷·⁵			....	CAS39, 2095
11056	Phenothiazine, N-propyl .... $C_{15}H_{15}NS$	241.35	....	155-65⁰·⁸	48-9			....	CAS53, 18045
11057	Phenoxanthin .... $C_{12}H_8OS$	200.27	nd, cr (MeOH)	311⁷⁴⁵, 183-4¹²	59-60			al, eth, ace	B19⁴, 341
11058	Phenoxazine or Dibenzoxazine .... $C_{12}H_9NO$	183.21	lf (dil al, bz)	d	156			al, eth, bz, aa	B27¹, 223
11059	Phenoxy acetamide .... $C_8H_9OCH_2CONH_2$	151.17	nd (w al)	....	101.5			al	B6⁴, 638
11060	Phenoxyacetic acid .... $C_8H_9OCH_2CO_2H$	152.16	nd or pl (w)	285d	98-9			w, al, eth, bz, aa	B6⁴, 634
11061	Phenoxyacetic acid, anhydride .... $(C_8H_9OCH_2CO)_2O$	268.29	lf (eth)	....	67-9			bz	B6, 162
11062	Phenoxyacetic acid, 2-bromo .... 2-$BrC_6H_4OCH_2CO_2H$	231.05	nd (dil al), cr (w)	....	142.5			al, eth	B6⁴, 1040
11063	Phenoxyacetic acid, 4-bromo .... 4-$Br-C_6H_4OCH_2COOH$	231.05	pr (al), cr (w)	....	161-2			al, eth	B6⁴, 1052
11064	Phenoxyacetic acid, 2-carboxy or Salicyl acetic acid .. (2-$HO_2CC_6H_4O)CH_2CO_2H$	196.17	nd (w)	....	190-2			al, eth, ace, aa, chl	B10³, 106
11065	Phenoxyacetic acid, 3-carboxy .... (3-$HO_2CC_6H_4O)CH_2CO_2H$	196.17		....	206-7			....	B10¹, 65
11066	Phenoxyacetic acid, 4-carboxy .... (4-$HO_2CC_6H_4O)CH_2CO_2H$	196.17	nd (ace, w)	....	280-2			al, eth, ace, bz, chl	B10², 94
11067	Phenoxyacetic acid, 2-chloro .... 2-$ClC_6H_4OCH_2CO_2H$	186.60	nd (w, al)	....	148-9			w, al	B6⁴, 796
11068	Phenoxyacetic acid, 3-chloro .... 3-$ClC_6H_4OCH_2CO_2H$	186.60	cr (w)	....	110			....	B6⁴, 816
11069	Phenoxyacetic acid, 4-chloro .... 4-$ClC_6H_4OCH_2COOH$	186.60	pr or nd (w)	....	156-7			....	B6⁴, 845
11070	Phenoxyacetic acid, 4-chloro-2-methyl .... [4-Cl-2($CH_3)C_6H_3O]CH_2CO_2H$	200.62		....	120			al, eth, bz	B6⁴, 1991
11071	Phenoxyacetic acid, 2,4-dichloro .... (2,4-$Cl_2-C_6H_3O)CH_2CO_2H$	221.04	cr (bz)	160⁰·⁴	140-1			al	B6⁴, 908
11072	Phenoxyacetic acid, 2,4-dinitro .... (2,4-$(NO_2)_2C_6H_3O)CH_2CO_2H$	242.15	pa ye pr (w)	....	147-8			al	B6², 244
11073	Phenoxyacetic acid, 3,5-dinitro .... (3,5-$(NO_2)_2C_6H_3O)CH_2CO_2H$	242.15	pa br cr pw	....	207			al	B6, 259

No.	Name, Synonyms, and Formula	Mol. wt.	Color, crystalline form, specific rotation and λ_{max} (log ϵ)	b.p. °C	m.p. °C	Density	n_D	Solubility	Ref.
11074	Phenoxyacetic acid, ethyl ester or Ethyl phenoxy acetate . . $C_6H_5OCH_2CO_2C_2H_5$	180.21	. . . .	250-1, 136[19]	. . . . . .	1.104[18]		al, eth	B6[4], 635
11075	Phenoxyacetic acid, 3-hydroxy . . . . (3-HOC$_6$H$_4$O)CH$_2$CO$_2$H	168.16	nd or pr (w, to)		158-9			al	B6, 817
11076	Phenoxy acetic acid, 3-hydroxy, ethyl ester or Ethyl-3-hydroxy phenoxy acetate . . . . (3-HOC$_6$H$_4$O)CH$_2$CO$_2$C$_2$H$_5$	196.21	pr (w, bz)	274d, 170-3[11]	55				B6, 817
11077	Phenoxy acetic acid, 4-hydroxy . . . . (4-HOC$_6$H$_4$O)CH$_2$CO$_2$H	168.16	nd (to), pr (+ ½w)		154 (hyd)				B6, 847
11078	Phenoxy acetic acid, 2-methoxy . . . . (2-CH$_3$OC$_6$H$_4$O)CH$_2$CO$_2$H	182.18	nd (w)		123-5 (129)			w, al, eth, bz, aa	B6[3], 4234
11079	Phenoxy acetic acid, 3-methoxy . . . . (3-CH$_3$OC$_6$H$_4$O)CH$_2$CO$_2$H	182.18	nd (w)		118				B6[3], 4323
11080	Phenoxy acetic acid, methyl ester or Methyl phenoxy acetate . . . . $C_6H_5OCH_2CO_2CH_3$	166.18	. . . .	245, 130[14]	. . . . . .	1.1493[20/4]	1.5155[20]	al, eth	B6[4], 635
11081	Phenoxy acetic acid, 2-methyl or o-Cresoxyacetic acid . . . . (2-CH$_3$C$_6$H$_4$O)CH$_2$CO$_2$H	166.18	lf (w)		157			al	B6[2], 331
11082	Phenoxy acetic acid, 3-methyl or m-Cresoxyacetic acid . . . (3-CH$_3$C$_6$H$_4$O)CH$_2$CO$_2$H	166.18	nd (w)		103-4			al, bz	B6[2], 353
11083	Phenoxy acetic acid, 4-methyl or p-Cresoxyacetic acid . . . . (4-CH$_3$C$_6$H$_4$O)CH$_2$CO$_2$H	166.18	nd (w)		136			al, bz	B6[2], 380
11084	Phenoxy acetic acid, 2-nitro . . . . (2-O$_2$NC$_6$H$_4$O)CH$_2$CO$_2$H	197.15	pr (w)		158			al, eth	B6[3], 804
11085	Phenoxy acetic acid, 2-iso-propyl-5-methyl or Thymoxy acetic acid. . . . . (2-i-C$_3$H$_7$-5-CH$_3$C$_6$H$_4$O)CH$_2$CO$_2$H	208.24	nd (dil al), cr (bz)		149-50			al, eth	B6[3], 1902
11086	Phenoxy acetic acid, 2,4,6-tribromo . . . . [2,4,6-(Br)$_3$C$_6$H$_2$O]CH$_2$CO$_2$H	388.85	nd (dil al)		200			al, eth	B6[4], 1068
11087	Phenoxy acetic acid, 2,4,5-trichloro . . . . [2,4,5-(Cl)$_3$-C$_6$H$_2$O]CH$_2$CO$_2$H	255.49	cr (bz)		157-8			al	B6[4], 973
11088	Phenoxy acetic acid, 2,4,6-trichloro . . . . [2,4,6-(Cl)$_3$-C$_6$H$_2$O]CH$_2$CO$_2$H	255.49	cr (al)		177			al	B6[4], 1011
11089	Phenoxy acetonitrile . . . . $C_6H_5OCH_2CN$	133.15	. . . .	239-40, 128[17]	. . . . . .	1.0991[20/4]	1.5246[20]	al, eth	B6[4], 640
11090	Phenoxy acetyl chloride . . . . $C_6H_5OCH_2COCl$	170.60	. . . .	225-6, 111[13]				eth	B6[3], 613
11091	β-Phenoxy ethyl ether . . . . (C$_6$H$_5$OCH$_2$CH$_2$)$_2$O	258.32	nd (dil al)		66-7			al	B6[4], 573
11092	Phenyl acetaldehyde or α-Tolualdehyde . . . . $C_6H_5CH_2CHO$	120.16	(w)	195, 88[18]	33-4	1.0272[20/4]	1.5255[20]	al, eth, ace	B7[3], 1003
11093	Phenyl acetamide . . . . $C_6H_5CH_2CONH_2$	135.17	pl or lf (w)		157				B9[3], 2193
11094	Phenyl acetamide, 3,4-dimethoxy . . . . 3,4-(CH$_3$O)$_2$C$_6$H$_3$CH$_2$CONH$_2$	195.22	cr (w)		145-7			al, eth	B10[3], 1463
11095	Phenyl acetamide, α-hydroxy (dl) or Mandelamide . . . . . . $C_6H_5CH(OH)CONH_2$	151.17	pl (bz, al)		134-5			al, bz	B10[3], 469
11096	Phenyl acetamide, α-hydroxy-N-ethyl (d) or l-N-Ethoxy mandelamide . . . . $C_6H_5CH(OH)CONHC_2H_5$	179.22	pl (chl-peth), [α][18/]$_D$ −103.6 (ace)		65-6			al, eth, ace, bz, chl	B10[2], 117
11097	Phenyl acetamide, α-hydroxy-N-ethyl (dl) or dl-N-Ethyl mandelamide . . . . $C_6H_5CH(OH)CONHC_2H_5$	179.22	pl (bz-peth)		53-4			w, al, eth, ace, bz	B10[1], 89
11098	Phenyl acetamide, 2-hydroxy . . . . 2-HOC$_6$H$_4$CH$_2$CONH$_2$	151.17	lf (al-chl)		118				B10, 188
11099	Phenyl acetamide, N-methyl . . . . $C_6H_5CH_2CONHCH_3$	149.10	cr (bz)		58			al, eth, chl	B9[3], 2195
11100	Phenyl acetamide, N-phenyl or α-Phenyl acetamide . . . . . . $C_6H_5CH_2CONHC_6H_5$	211.27	pr (al)		117-8			al, eth	B12[3], 521
11101	Phenyl acetamide, 4-isopropyl . . . . 4-i-C$_3$H$_7$C$_6$H$_4$CH$_2$CONH$_2$	177.25	pl (bz)		170			al	B9[3], 2529
11102	Phenyl acetic acid . . . . $C_6H_5CH_2CO_2H$	136.16	lf, pl (peth)	265.5, 144-5[12]	77	1.091[77/4], 1.228[6]		al, eth, ace	B9[3], 2169
11103	Phenyl acetic acid, anhydride . . . . (C$_6$H$_5$CH$_2$CO)$_2$O	254.29	pr or nd (eth)	195-8[12]	71-2			eth, chl	B9[3], 2190

No.	Name, Synonyms, and Formula	Mol. wt.	Color, crystalline form, specific rotation and λ_{max} (log ε)	b.p. °C	m.p. °C	Density	n_D	Solubility	Ref.
11104	Phenyl acetic acid, 4-amino or Aminophenyl acetic acid p-$H_2NC_6H_4CH_2CO_2H$	151.17	pl (w)		199-200d				B14[3], 1182
11105	Phenyl acetic acid, α-bromo or α-Bromophenyl acetic acid $C_6H_5CH(Br)CO_2H$	215.05		82-3					B9[3], 2276
11106	Phenyl acetic acid, α-bromo, ethyl ester or Ethyl-α-bromophenyl acetate $C_6H_5CH(Br)CO_2C_2H_5$	243.10		145-55[18]					B9[3], 2276
11107	Phenyl acetic acid, 2-bromo or 2-Bromotoluic acid 2-$BrC_6H_4CH_2CO_2H$	215.05	cr (aa)		105-6			al, eth, aa	B9[3], 2273
11108	Phenyl acetic acid, 3-bromo or 3-Bromotoluic acid 3-$BrC_6H_4CH_2CO_2H$	215.05	nd (w)		100-1				B9[3], 2274
11109	Phenyl acetic acid, 4-bromo 4-$BrC_6H_4CH_2CO_2H$	215.05	nd (w)	sub	116			al, eth	B9[3], 2275
11110	Phenyl acetic acid, iso-butyl ester or iso-Butyl phenyl acetate, Eglantine $C_6H_5CH_2CO_2$-i-C_4H_9	192.24		247, 123-5[14]		0.999[18]		al, eth	B9[3], 2180
11111	Phenyl acetic acid, 2-carboxy 2-$(HO_2C)C_6H_4CH_2CO_2H$	180.17	cr (w, eth)		185-7			al, w	B9[3], 4266
11112	Phenyl acetic acid, 3-carboxy 3-$(HO_2C)C_6H_4CH_2CO_2H$	180.17	nd or pl (w)		184-5			al, eth	B9[3], 4269
11113	Phenyl acetic acid, 4-carboxy 4-$(HO_2C)C_6H_4CH_2CO_2H$	180.17	cr (dil al)		239-41			al, eth, bz	B9[3], 4269
11114	Phenyl acetic acid, α-chloro (d) d-$C_6H_5CHClCOOH$	170.60	cr (peth), $[\alpha]_D^{20}$ +192 (bz)		60-1			al, eth, bz, chl	B9[3], 2265
11115	Phenyl acetic acid, α-chloro (dl) dl-$C_6H_5CHClCOOH$	170.60	lf (peth)		78			al, eth	B9[3], 2265
11116	Phenyl acetic acid, α-chloro (l) l-$C_6H_5CHClCOOH$	170.60	nd (peth), $[\alpha]_D^{18}$ -191.3 (bz)		61			al, eth, bz, chl	B9[2], 307
11117	Phenyl acetic acid, 2-chloro 2-$ClC_6H_4CH_2CO_2H$	170.60	nd (w)		96			al	B9[3], 2262
11118	Phenyl acetic acid, 3-chloro or 3-Chloro phenyl acetic acid 3-$ClC_6H_4CH_2CO_2H$	170.60	pl (dil al), nd (hep)		77-8			eth	B9[3], 2263
11119	Phenyl acetic acid, 4-chloro 4-$ClC_6H_4CH_2CO_2H$	170.60	nd (w)		105-6			al, eth, bz	B9[3], 2264
11120	Phenyl acetic acid, 2,5-dihydroxy or Homogentisic acid 2,5-$(HO)_2C_6H_3CH_2CO_2H$	168.16	pr (w + 1), lf (al-chl)		152-4			al, eth, w	B10[3], 1456
11121	Phenyl acetic acid, 3,4-dihydroxy 3,4-$(HO)_2C_6H_3CH_2CO_2H$	168.16			131-2			w, al, eth	B10[3], 1458
11122	Phenyl acetic acid, 3,4-dimethoxy or Homoveratric acid 3,4-$(CH_3O)_2C_6H_3CH_2CO_2H$	196.23	nd (w + 1), cr (bz-peth)		98-9 (anh), 80-2 (+ w)			w, al, eth	B10[3], 1459
11123	Phenyl acetic acid, 2,4-dimethyl-α-hydroxy or 2,4-Dimethyl mandelic acid 2,4-$(CH_3)_2C_6H_3$-$CH(OH)CO_2H$	180.23	rh (w), lf (peth-chl)		119			al, eth, chl, w	B10[3], 613
11124	Phenyl acetic acid, 2,5-dimethyl-α-hydroxy or 2,5-Dimethyl mandelic acid 2,5-$(CH_3)_2C_6H_3CH(OH)CO_2H$	180.21	nd or pr (bz)		116-7			al, eth, chl	B10[3], 612
11125	Phenyl acetic acid, 3,4-dimethyl-α-hydroxy or 3,4-Dimethyl mandelic acid 3,4-$(CH_3)_2C_6H_3CH(OH)CO_2H$	180.21	lf (bz)		135			al, bz	B10[3], 611
11126	Phenyl acetic acid, 2,4-dinitro or 2,4-Dinitrophenyl acetic acid 2,4-$(NO_2)_2C_6H_3CH_2CO_2H$	226.15			179-80d			al, eth	B9[3], 2292
11127	Phenyl acetic acid, 2,4-dinitro, ethyl ester 2,4-$(NO_2)_2C_6H_3CH_2CO_2C_2H_5$	254.20	nd (w)		37			al, eth	B9[3], 2292
11128	Phenyl acetic acid, 2,6-dinitro 2,6-$(NO_2)_2C_6H_3CH_2CO_2H$	226.15	ye lf (aa)		201-2d			al	B9[1], 185
11129	Phenyl acetic acid, 2-ethoxy 2-$(C_2H_5O)C_6H_4CH_2CO_2H$	180.21	nd (lig), cr (w)		103-4				B10[1], 82
11130	Phenyl acetic acid, ethyl ester or Ethyl phenyl acetate $C_6H_5CH_2CO_2C_2H_5$	164.21		227, 120-1[20]		1.0333[20/4]	1.4980[20]	al, eth	B9[3], 2176
11131	Phenyl acetic acid, α-hydroxy-(D) or l-Mandelic acid $C_6H_5CH(OH)CO_2H$	152.46	ta $[\alpha]_D^{20}$ -158 (w, c=2.5)		133-5	1.341		w, al, eth, aa, chl	B10[3], 447

No.	Name, Synonyms, and Formula	Mol. wt.	Color, crystalline form, specific rotation and λ_{max} (log ϵ)	b.p. °C	m.p. °C	Density	n_D	Solubility	Ref.
11132	Phenyl acetic acid, α-hydroxy (DL) or dl-Mandelic acid.. C₆H₅CH(OH)CO₂H	152.16	pl (w), rh (bz)	d	121-3	1.300²⁰/⁴	....	al, eth	B10³, 448
11133	Phenyl acetic acid, α-hydroxy (L +) or d-Mandelic acid . C₆H₅CH(OH)CO₂H	152.16	pl (w), [α]²⁰/$_D$ + 156.6 (w, c=2.9)	....	134-5			w, al, eth, chl	B10³, 445
11134	Phenyl acetic acid, α-hydroxy,acetate(D) or l-Acetyl mandelic acid . C₆H₅CH(O₂CCH₃)CO₂H	194.19	nd (w + 1) [α]²⁰/$_D$ −156−4 (ace)	....	96-8 (anh)			al, eth, ace, bz, chl	B10³, 453
11135	Phenyl acetic acid, α-hydroxy,acetate (DL) or dl-acetyl mandelic acid . C₆H₅CH(CO₂CH₃)COOH	194.19	cr (bz), amor (chl-peth), cr (w + 1)	....	38-9 (+ 1w), 79-80 (anh)			al, eth, bz, chl	B10³, 453
11136	Phenyl acetic acid, α-hydroxy-4-bromo or p-Bromo mandelic acid . 4-Br-C₆H₄CH(OH)CO₂H	231.05	nd (bz)	....	118-20			al, eth, bz, chl	B10³, 480
11137	Phenyl acetic acid, α-hydroxy-4-chloro or 4-Chloro mandelic acid . 4-Cl-C₆H₄CH(OH)CO₂H	186.60	nd (bz)	....	119-20			w, al	B10³, 479
11138	Phenyl acetic acid, α-hydroxy, ethyl ester (D-) or l-Ethyl mandelate. C₆H₅CH(OH)CO₂C₂H₅	180.21	(peth), [α]²⁰/$_D$ −128.4 (chl, c=6.7)	150²⁰	35	1.1270²⁰/⁴		al, eth	B10³, 456
11139	Phenyl acetic acid, α-hydroxy, ethyl ester (DL) or dl-Ethyl mandelate. C₆H₅CH(OH)CO₂C₂H₅	180.21	nd (peth)	253-5, 141¹⁵	37	....		al, eth, lig	B10³, 457
11140	Phenyl acetic acid, α-hydroxy, ethyl ester (L +) or d-Ethyl mandelate. C₆H₅CH(OH)CO₂C₂H₅	180.21	(peth), [α]²⁰/$_D$ + 205 (CS₂, c=0.7)	....	33	1.1270²⁰/⁴		al, eth	B10³, 456
11141	Phenyl acetic acid, α-hydroxy, methyl ester (D-) or l-Methyl mandelate. C₆H₅CH(OH)CO₂CH₃	166.18	cr (peth), [α]²⁰/$_D$ −131.5 (w)	160²²	55	1.1756²⁰	....	w, al, ace, bz, chl	B10³, 454
11142	Phenyl acetic acid, α-hydroxy, methyl ester (DL) or dl-Methyl mandelate. C₆H₅CH(OH)CO₂CH₃	166.18	pl (bz-lig)	250d, 144²⁰	58	1.1756²⁰		al, chl	B10³, 454
11143	Phenyl acetic acid, α-hydroxy, methyl ester (L +) or d-Methyl mandelate. C₆H₅CH(OH)CO₂CH₃	166.11	cr (peth), [α]²⁰/$_D$ −133.6 (w)	160³²	55.5	1.1756²⁰		w, al, ace, bz, chl	B10³, 454
11144	Phenyl acetic acid, α-hydroxy-4-iso-propyl (d) or d-4-iso-Propyl mandelic acid. 4-i-C₃H₇C₆H₄CH(OH)CO₂H	194.23	lf (w), [α]¹⁷/$_D$ + 135 (abs al, c = 4)	....	153-4			al, eth	B10, 279
11145	Phenyl acetic acid, α-hydroxy-4-iso-propyl (dl) or dl-4-iso-Propyl mandelic acid. 4-i-C₃H₇C₆H₄CH(OH)CO₂H	194.23	nd (w)	....	159-60		....	al, eth	B10³, 629
11146	Phenyl acetic acid, α-hydroxy-4-iso propyl (l) or l-4-iso-propyl mandelic acid. 4-i-C₃H₇C₆H₄CH(OH)CO₂H	194.23	ta (20% al), [α]¹⁷/$_D$ −135 (abs al, c=4)		153-4		....	al, eth	B10, 279
11147	Phenyl acetic acid, 2-hydroxy . 2-HOC₆H₄CH₂CO₂H	168.15	....	240-3d	147-9		....	eth	B10³, 422
11148	Phenyl acetic acid, 2-hydroxy-5-nitro . 2-(HO)-5(NO₂)C₆H₃CH₂CO₂H	197.15	nd	....	160-2			w, al, eth	B10, 189
11149	Phenyl acetic acid, 2-hydroxy-5-nitro, ethyl ester or Ethyl-2-hydroxy-2-nitro phenyl acetate . 5-NO₂-2-HOC₆H₃CH₂CO₂C₂H₅	225.20	pr or pl (al)		154-5			bz, chl	B10, 189
11150	Phenyl acetic acid, 2-hydroxy, hydrazide . 2-HO-C₆H₄-CH₂CONHNH₂	166.18	lf (chl, bz), nd (al)		154			bz, al	B10², 112
11151	Phenyl acetic acid, 3-hydroxy . 3-HOC₆H₄CH₂CO₂H	152.16	nd (bz-lig)	190¹¹	131-4			w, al, eth, bz, chl	B10³, 428
11152	Phenyl acetic acid, 4-hydroxy . 4-HOC₆H₄CH₂CO₂H	152.16	nd (w)	sub	149-51		....	al, eth	B10³, 430
11153	Phenyl acetic acid, 3-hydroxy-4-methoxy or Homoisovanillic acid . 3-(HO)-4-(CH₃O)C₆H₃CH₂CO₂H	182.17	....	....	130-1		....	w, al, eth	B10³, 1458

No.	Name, Synonyms, and Formula	Mol. wt.	Color, crystalline form, specific rotation and λ_{max} (log ε)	b.p. °C	m.p. °C	Density	n_D	Solubility	Ref.
11155	Phenyl acetic acid, 2-mercapto 2-HSC$_6$H$_4$CH$_2$CO$_2$H	168.23	pl (w, bz-lig)	....	96-7			al, eth, bz	B10[1], 1782
11156	Phenyl acetic acid, α-methoxy (D-) C$_6$H$_5$CH(OCH$_3$)CO$_2$H	166.18	nd (peth), [α][13][D] -150 (al)	....	63-4			w, al, ace	B10[3], 451
11157	Phenyl acetic acid, α-methoxy (DL) C$_6$H$_5$CH(OCH$_3$)CO$_2$H	166.18	pl (lig)	....	71-2			al, eth	B10[3], 452
11158	Phenyl acetic acid, 2-methoxy 2-CH$_3$OC$_6$H$_4$CH$_2$CO$_2$H	166.18	nd (w)	100-1[2]	123			al, eth, ace, bz, chl	B10[3], 422
11159	Phenyl acetic acid, 4-methoxy or Homoanisic acid 4-CH$_3$OC$_6$H$_4$CH$_2$CO$_2$H	166.18	pl (w)	138-40[2-3]	86			al, eth	B10[3], 431
11160	Phenyl acetic acid, 4-methoxy-2-nitro 4-(CH$_3$O)-2-(NO$_2$)C$_6$H$_3$CH$_2$CO$_2$H	211.28	ye nd (50% al)	....	157-8d			al, aa	B10[2], 113
11161	Phenyl acetic acid, methyl ester or Methyl phenyl acetate C$_6$H$_5$CH$_2$CO$_2$CH$_3$	150.18	....	218, 131-2[50]		1.0633[16/16]	1.5075[20]	al, eth, ace	B9[3], 2175
11162	Phenyl acetic acid, 2-methyl or 2-Tolylacetic acid (2-CH$_3$C$_6$H$_4$)CH$_2$CO$_2$H	150.18	nd (w)		88-90				B9[3], 2426
11163	Phenyl acetic acid, 3-methyl or 3-Methyl-α-toluic acid 3-CH$_3$C$_6$H$_4$CH$_2$CO$_2$H	150.18	nd (w)	120-3[26]	62				B9[3], 2429
11164	Phenyl acetic acid, 4-methyl or 4-Methyl phenyl acetic acid 4-CH$_3$C$_6$H$_4$CH$_2$CO$_2$H	150.18	nd or pl (al, w)	265-7 sub	91-3			al, eth, bz, chl	B9[3], 2432
11165	Phenyl acetic acid, 2-nitro 2-NO$_2$C$_6$H$_4$CH$_2$CO$_2$H	181.15	nd (w), pl (dil al)	....	141-2			al	B9[3], 2282
11166	Phenyl acetic acid, 3-nitro 3-NO$_2$C$_6$H$_4$CH$_2$CO$_2$H	181.15	nd (w)	....	122			al	B9[3], 2283
11167	Phenyl acetic acid, 4-nitro 4-NO$_2$C$_6$H$_4$CH$_2$CO$_2$H	181.15	pa ye nd (w)	....	153-5			al, eth, bz	B9[3], 2284
11168	Phenyl acetic acid, phenethyl ester C$_6$H$_5$CH$_2$CO$_2$CH$_2$CH$_2$C$_6$H$_5$	240.30	....	177-8[4.5]	26.5	1.080[25/25]	....	al	B9[3], 2183
11169	Phenyl acetonitrile C$_6$H$_5$CH$_2$CN	117.15	....	234, 107[17]	-23.8	1.0157[20/4]	1.5230[20]	**al, eth**, ace	B9[3], 2252
11170	Phenyl acetonitrile, α-amino (dl) C$_6$H$_5$CH(NH$_2$)CN	132.18	hyg lf (lig)	....	55				B14[3], 1189
11171	Phenyl acetonitrile, 2-amino 2-H$_2$NC$_6$H$_4$CH$_2$CN	132.18	lf (dil al)		72			al, bz	B14[3], 1181
11172	Phenyl acetonitrile, 4-amino or 4-Amino phenyl acetonitrile 4-H$_2$NC$_6$H$_4$CH$_2$CN	132.18	lf (w)	312, 177[11]	46			al	B14[3], 1182
11173	Phenyl acetonitrile, α-bromo C$_6$H$_5$CH(Br)CN	196.05	yesh cr (dil al)	242d, 132-4[12]	29	1.539[29/4]		al, eth, ace, bz	B9[3], 2278
11174	Phenyl acetonitrile, 2-bromo or 2-Bromo phenyl acetonitrile 2-Br-C$_6$H$_4$CH$_2$CN	196.05	....	145-7[14]	1			al	B9[3], 2274
11175	Phenyl acetonitrile, 4-bromo or 4-Bromo phenyl acetonitrile 4-BrC$_6$H$_4$CH$_2$CN	196.05	pa ye cr (al)	....	47			al, bz	B9[3], 2275
11176	Phenyl acetonitrile, 2-carboxy 2-(HO$_2$C)C$_6$H$_4$CH$_2$CN	161.16	cr (aa), nd (w, bz)		116d			al, eth, bz	B9[3], 4267
11177	Phenyl acetonitrile, 2-chloro 2-ClC$_6$H$_4$CH$_2$CN	151.50	grsh-ye nd	251, 170[120]	24	1.1737[18/4]	1.534[18]	....	B9[3], 2263
11178	Phenyl acetonitrile, α-hydroxy (D+) or d-Mandelonitrile C$_6$H$_5$CH(OH)CN	133.15	nd, [α][25/546.1] +46.9 (bz)	....	28-9			al, eth	B10[3], 473
11179	Phenyl acetonitrile, α-hydroxy (DL) or dl-Mandelonitrile C$_6$H$_5$CH(OH)CN	133.15	ye pr	170d	22	1.1165[20/4]	1.5201[20]	al, eth	B10[3], 474
11180	Phenyl acetonitrile, α-hydroxy (L-) or l-Mandelonitrile C$_6$H$_5$CH(OH)CN	133.15	....					al, eth	B10[1], 86
11181	Phenyl acetonitrile, 2-hydroxy 2-HOC$_6$H$_4$CH$_2$CN	133.15	nd (bz-lig)	....	117-9			w, eth, ace, bz	B10[3], 425
11182	Phenyl acetonitrile, 3-hydroxy 3-HOC$_6$H$_4$CH$_2$CN	133.15	pl (w)		52-3			w, al, eth	B10, 189
11183	Phenyl acetonitrile, 4-hydroxy 4-HOC$_6$H$_4$CH$_2$CN	133.15	pl (w), mcl pr	330[756], 210[10]	69-70			al, eth	B10[3], 438
11184	Phenyl acetonitrile, 2-methoxy or 2-Methoxy phenyl acetonitrile 2-CH$_3$OC$_6$H$_4$CH$_2$CN	147.18	pr (bz-lig)	141-3[11]	68.5			bz	B10[3], 425

No.	Name, Synonyms, and Formula	Mol. wt.	Color, crystalline form, specific rotation and λ_{max} (log ϵ)	b.p. °C	m.p. °C	Density	n_D	Solubility	Ref.
11185	Phenyl acetonitrile, 4-methoxy .... 4-CH$_3$OC$_6$H$_4$CH$_2$CN	147.18	....	286-7, 152[16]	......	1.0845[20/4]	1.5309[20]	al, eth	B10[3], 439
11186	Phenyl acetonitrile, 2-methyl or 2-Totyl acetonitrile..... 2-CH$_3$C$_6$H$_4$CH$_2$CN	131.18	....	244	......	1.0156[22]	1.5252[20]	al, eth, bz	B9[3], 2427
11187	Phenyl acetonitrile, 3-methyl or 3-Totyl acetonitrile..... 3-CH$_3$C$_6$H$_4$CH$_2$CN	131.18	....	245-7[745]d, 133[15]	......	1.0022[22]	1.5233[20]	al, eth, bz	B9[2], 349
11188	Phenyl acetonitrile, 4-methyl or 4-Totyl acetonitrile..... 4-CH$_3$C$_6$H$_4$CH$_2$CN	131.18	....	242-3, 122[13]	18	0.9922[22]	1.5167[20]	al, eth, bz	B9[3], 2433
11189	Phenyl acetonitrile, α-nitro .... C$_6$H$_5$CH(NO$_2$)CN	162.15	....		39-40	....	....	al	B9[3], 2291
11190	Phenyl acetonitrile, 2-nitro .... 2-NO$_2$C$_6$H$_4$CH$_2$CN	162.15	nd (al-w), pr (aa, al)	178[12]	84	....	....	al, eth, ace, bz, chl	B9[3], 2283
11191	Phenyl acetonitrile, 3-nitro .... 3-NO$_2$C$_6$H$_4$CH$_2$CN	162.15	(eth-lig)	180[1.5]	63	....	....	al, eth, bz, chl	B9[2], 312
11192	Phenyl acetonitrile, 4-nitro .... 4-NO$_2$C$_6$H$_4$CH$_2$CN	162.15	pl	195-7[12]	116-7	....	....	al, eth, bz, chl	B9[3], 2291
11193	Phenyl acetyl chloride .... C$_6$H$_5$CH$_2$COCl	154.60	....	170[250], 104-5[24]	......	1.1682[20/4]	1.5325[20]	eth	B9[3], 2192
11194	Phenyl acetyl chloride, α-acetyl (DL) or dl-Acetyl mandelyl chloride.......... C$_6$H$_5$CH(CO$_2$CH$_3$)COCl	212.64	....	150-5[33]	......	....	....	eth, bz, chl	B10[1], 89
11195	Phenyl acetyl chloride, 2,4-dinitro .... 2,4-(NO$_2$)$_2$C$_6$H$_3$CH$_2$COCl	244.59	ye lf (CS$_2$)		77	....	....	eth, bz, chl	B9[1], 185
11196	Phenyl acetylene.............. C$_6$H$_5$C≡CH	102.14	....	142-4, 44[18]	−44.8	0.9281[20/4]	1.5485[20]	**al, eth**, ace	B5[4], 1525
11197	α-Phenyl acrylic acid or Atropic acid .... CH$_2$=C(C$_6$H$_5$)CO$_2$H	148.17	lf (al), nd (w)	267d	106-7	....	....	al, eth, bz, chl	B9[3], 2751
11198	α-Phenyl acrylic acid, 3-(2-nitro phenyl)(trans) (3-NO$_2$C$_6$H$_4$)CH=C(C$_6$H$_5$)CO$_2$H	269.26	ye cr (al)	......	196-7	....	....	eth, bz	B9[3], 3423
11199	α-Phenyl acrylic acid, 3-(3-nitro phenyl)-(cis) .... (3-NO$_2$C$_6$H$_4$)CH=C(C$_6$H$_5$)CO$_2$H	269.26	nd (al, aa)		195-6	....	....		B9[3], 3423
11200	α-Phenyl acrylic acid, 3-(3-nitrophenyl)-(trans) .... (3-NO$_2$C$_6$H$_4$)CH=C(C$_6$H$_5$)CO$_2$H	269.26	ye pr (eth), nd (dil al)		182	....	....	al, ace, bz, chl	B9[3], 3424
11201	α-Phenyl acrylic acid, 3-(4-nitrophenyl)-cis .... (4-NO$_2$C$_6$H$_4$)CH=C(C$_6$H$_5$)CO$_2$H	269.26	ye cr pr (dil al + 1w), lf (bz + ½)		144	....	....	al, eth	B9[3], 3424
11202	α-Phenyl acrylic acid, 3-(4-nitrophenyl)-trans (4-NO$_2$C$_6$H$_4$)CH=C(C$_6$H$_5$)CO$_2$H	269.26	ye pr or nd (al)	......	213-4	....	....	al	B9[3], 3425
11203	Phenylalanine .... C$_6$H$_5$CH$_2$CH(NH$_2$)CO$_2$H	165.19	pr (w), [α]$^{18}_D$ + 70 (w)	sub at 295	283-4d	....		w	B14[3], 1228
11204	Phenylalanine (D) .... C$_6$H$_5$CH$_2$CH(NH$_2$)CO$_2$H	165.19	nd or pr (w), [α]^{20} -69.5 (w)	sub at 295	283d	....	1.600	w	B14[3], 1228
11205	Phenylalanine (DL) .... C$_6$H$_5$CH$_2$CH(NH$_2$)CO$_2$H	165.19	red-br lf (dil al), pr or nd, (w)	sub d	284-8d	....		w	B14[3], 1229
11206	Phenylalanine, N-acetyl (d,t) C$_6$H$_5$CH$_2$CH(NHCOCH$_3$)CO$_2$H	207.23	cr (w), [α]$^{20}_D$ -50.9		172	....		al	B14[2], 297
11207	Phenylalanine, N-acetyl (dl) .... C$_6$H$_5$CH$_2$CH(NHCOCH$_3$)CO$_2$H	207.23	nd or pl (w), hex tab (ace)		152-3	....		w, al	B14[3], 1238
11208	Phenylalanine, N-acetyl (l) .... C$_6$H$_5$CH$_2$CH(NHCOCH$_3$)CO$_2$H	207.23	[α]$^{26}_D$ + 35.1		172	....		al	B14[2], 298
11209	Phenylalanine, ethyl ester .... C$_6$H$_5$CH$_2$CH(NH$_2$)CO$_2$C$_2$H$_5$	193.24	....	148[15]	......	1.065[15]	....		B14[3], 1232
11210	Phenylboric acid or Benzeneboronic acid.............. C$_6$H$_5$B(OH)$_2$	121.93	nd (w)	....	218-20	....		al, eth, bz	B16[2], 638
11211	o-Phenylenediamine or 1,2-Diamino benzene.......... 1,2-(H$_2$N)$_2$C$_6$H$_4$	108.15	brsh-ye lf (w), pl (chl)	256-8	102-3	....		al, eth, bz, chl	B13[3], 28
11212	o-Phenylenediamine, 3,5-dichloro .... 1,2-(H$_2$N)$_2$-3,5-Cl$_2$C$_6$H$_2$	177.03	nd (al)	....	60.5	....			B13, 27
11213	o-Phenylenediamine, 3,6-dichloro .... 1,2-(H$_2$N)$_2$-3,6-Cl$_2$C$_6$H$_2$	177.03	nd (50% al)		100	....		w, al, eth, ace	B13[2], 20

No.	Name, Synonyms, and Formula	Mol. wt.	Color, crystalline form, specific rotation and λ_{max} (log ε)	b.p. °C	m.p. °C	Density	n_D	Solubility	Ref.
11214	o-Phenylenediamine, 4,5-dimethoxy 1,2-(H₂N)₂-4,5-(CH₃O)₂C₆H₂	168.20	bl pr	....	131-2			w, al	B13³, 2127
11215	o-Phenylenediamine, 3-methoxy-6-methyl 1,2-(H₂N)₂-3-CH₃O-6-CH₃-C₆H₂	152.20	pr (eth-bz)		75-6			al, eth	B13², 349
11216	o-Phenylenediamine, 4-methoxy 1,2-(H₂N)₂C₆H₃OCH₃-4	138.17	gr pl	167-70¹¹	50-2			eth	B13³, 1362
11217	o-Phenylenediamine, 4-methyl 1,2-(H₂N)₂-4-CH₃C₆H₃	122.17	pl (lig)	265, 92¹	89-90			w	B13³, 292
11218	o-Phenylene diamine, 3-nitro 1,2-(H₂N)₂-3-NO₂-C₆H₃	153.14	dk red nd (dil al)		158-9				B13³, 61
11219	o-Phenylenediamine, 4-nitro 1,2-(H₂N)₂-4-NO₂-C₆H₃	153.14	dk red nd		199-200				B13³, 62
11220	o-Phenylenediamine, N-phenyl 2-H₂NC₆H₄NHC₆H₅	184.24	nd (w)	312.5⁷⁴⁴	80-1			ace, bz, chl	B13³, 34
11221	m-Phenylenediamine or 1,3-Diamino benzene 1,3-(H₂N)₂C₆H₄	108.15	rh (al)	282-4	63-4	1.0696⁵⁸/⁴	1.6339⁵⁸	w, al, eth, bz	B13³, 65
11222	Phenylenediamine, N-acetyl 3-(CH₃CONH)C₆H₄NH₂	150.18	lf (ace-bz), nd or pl (bz)	d at 100	87-9			w, al, eth, ace	B13³, 75
11223	m-Phenylenediamine, 2,5-dichloro 2,5-Cl₂-1,3-(H₂N)₂C₆H₂	177.03	nd (w)		100			w, al, bz	B13², 29
11224	m-Phenylenediamine, 4,6-dichloro 4,6-Cl₂-1,3-(H₂N)₂C₆H₂	177.03	nd (dil al)		136-7			al	B13³, 98
11225	m-Phenylenediamine, 6-methoxy 6-CH₃O-1,3-(NH₂)₂C₆H₃	138.17	nd (eth)	....	87-8	....		al, eth	B13², 308
11226	p-Phenylenediamine or 1,4-Diamino benzene 1,4-(H₂N)₂C₆H₄	108.15	wh pl (bz, eth)	267	140			eth, chl	B13³, 104
11227	p-Phenylenediamine, N-acetyl, sulfate 4-(CH₃CONH)C₆H₄NH₂.H₂SO₄	248.26	nd (eth-al)		283 d			w, al	B13³, 65
11228	p-Phenylenediamine, 2,5-dichloro 2,5-Cl₂-1,4-(H₂N)₂C₆H₂	177.03	pr (w)		170				B13, 118
11229	p-Phenylenediamine, 2,6-dichloro 2,6-Cl₂-1,4-(H₂N)₂C₆H₂	177.03	nd, pr (dil al)		124-6			al, eth, ace, bz	B13³, 269
11230	p-Phenylenediamine, 2-methoxy-5-methyl 2-CH₃O-5-CH₃-1,4-(H₂N)₂C₆H₂	152.20			166			al, eth	B13², 349
11231	p-Phenylenediamine, N-phenyl 4-H₂NC₆H₄NHC₆H₅	184.24	nd (al), cr (lig)	354, 155⁰·⁰²⁶	66 (al), 75 (lig)			al, eth, lig	B13³, 115
11232	p-Phenylenediamine, 2,3,5,6-tetramethyl or Diamino durene 1,4-(H₂N)₂-2,3,5,6-(CH₃)₄C₆	164.25	nd (w)	....	148			al, eth, chl	B13³, 360
11233	bis-(1-phenylethyl) amine [C₆H₅CH(CH₃)]₂NH	225.34	ye	295-8, 190¹⁰	......	1.018¹⁵	1.573		B12², 589
11234	Phenyl-2-aminophenyl sulfide 2-H₂NC₆H₄SC₆H₅	201.29	pl (al)	257.5¹⁰⁰	35-6			al	B13³, 904
11235	Phenyl-2-aminophenyl sulfone (2-H₂NC₆H₄)SO₂C₆H₅	233.29	lf (dil al)		122-4			al, bz, aa	B3³, 905
11236	Phenyl-4-aminophenyl sulfide (4-H₂NC₆H₄)SC₆H₅	201.39	nd (dil al), cr (lig)	242.5²⁹	95.8			al, eth	B13³, 1224
11237	Phenyl-4-aminophenyl sulfone (4-H₂NC₆H₄)SO₂C₆H₅	233.29	nd (al)		176			al, bz, aa	B13³, 1226
11238	Phenyl-4-aminophenyl sulfoxide (4-H₂NC₆H₄)SOC₆H₅	217.29	nd (w)		152			al, eth	B13³, 1226
11239	Phenyl-4-bromophenyl ether 4-BrC₆H₄OC₆H₅	249.11		310.1, 163¹⁰	18.7	1.4208²⁰/⁴	1.6084²⁰	eth	B6⁴, 1047
11240	Phenyl tert-butyl ketone or Pivalophenone C₆H₅COC(CH₃)₃	162.23	....	219-21, 97-8¹⁶		0.963²⁶	1.5086¹⁹	ace	B7³, 1125
11241	Phenyl-α-chlorobenzyl ketone or Desylchloride C₆H₅CO(CHClC₆H₅)	230.70	nd (al)	d	68.5			al	B7³, 2106
11242	Phenyl-4-chlorophenyl sulfone or Sulphenone (4-ClC₆H₄)SO₂C₆H₅	252.72	cr (al)		98			eth, ace, bz	B6⁴, 1587
11243	Phenyl-2,4-diaminophenyl sulfone [2,4-(NH₂)₂C₆H₃]SO₂C₆H₅	248.31	nd (al)		188			al	B13, 553
11244	Phenyl-2,5-dihydroxyphenyl sulfone [2,5-(HO)₂C₆H₃]SO₂C₆H₅	250.28	pr (w), nd (dil al)		196			al	B6², 1072
11245	Phenyl-2,4-dinitrophenyl ether [2,4-(NO₂)₂C₆H₃]OC₆H₅	260.22	pl (al), nd (al-ace)	230-50²⁷	71			al, eth	B6⁴, 1375

No.	Name, Synonyms, and Formula	Mol. wt.	Color, crystalline form, specific rotation and λ_{max} (log ε)	b.p. °C	m.p. °C	Density	n_D	Solubility	Ref.
11246	Phenyl-2,4-dinitrophenyl sulfide .... [2,4-$(O_2N)_2C_6H_3$]SC_6H_5	276.28	pa ye nd, (ace, bz-al)	....	121	....	....	ace, bz, aa	B6[4], 1746
11247	Phenyl-2,6-dinitrophenyl ether .... [2,6-$(NO_2)_2C_6H_3$]OC_6H_5	260.22	lf (al)	....	99-100				B6[2], 245
11248	Phenyl-3,4-dinitrophenyl ether .... [3,4-$(NO_2)_2C_6H_3$]OC_6H_5	260.22	lf (al)	....	99-100				B6[1], 127
11249	Phenyl ether or Diphenyl ether................ $(C_6H_5)_2O$	170.21	....	257.9, 121[10]	......	1.0748[20]	1.5787[25]	al, eth, bz, aa	B6[4], 568
11250	α-Phenyl ethyl alcohol (d) .... $C_6H_5CH(OH)CH_3$	122.17	[α]$^{19}_D$ +42.9 (undil)	203, 100[18]	......	1.0129[20/4]	1.5272[20]	al, chl	B6[3], 1671
11252	α-Phenyl ethyl alcohol (dl) .... $C_6H_5CH(OH)CH_3$	122.17	glassy	203.4, 87.2[10]	20	1.0135[20/4]	1.5275[20]	al, eth	B6[3], 1673
11253	α-Phenyl ethyl alcohol (l) .... $C_6H_5CH(OH)CH_3$	122.17	[α]$^{20}_D$ -45.5 (MeOH, c=5)	202-4, 93[14]	......	1.0129[20/4]	1.5272[20]	al, eth	B6[3], 1672
11254	α-Phenyl ethyl alcohol, 3-methyl .... [3-$CH_3C_6H_4$]$CH(OH)CH_3$	136.20	....	112[12]	......	0.9974[15/4]	1.5240[20]	al, eth	B6[3], 1823
11255	α-Phenyl ethyl alcohol, 4-methyl .... (4-$CH_3C_6H_4$)$CH(OH)CH_3$	136.20	....	219[756], 120[19]	......	0.9944[20/4]	1.5246[20]	al, eth	B6[3], 1826
11256	β-Phenyl ethyl alcohol .... $C_6H_5CH_2CH_2OH$	122.17	glass	218.2, 97.4[10]	fr-27	1.0202[20/4]	1.5325[20]	al, eth	B6[3], 1703
11257	β-Phenyl ethyl alcohol, 2-amino .... (2-$H_2NC_6H_4$)CH_2CH_2OH	137.18	ye in air	152-3[6]	......		1.5849[19]	w	B13[3], 1679
11258	β-Phenyl ethyl alcohol, 4-amino .... (4-$H_2NC_6H_4$)CH_2CH_2OH	137.18	nd (al)	....	108				B13[3], 1679
11259	β-Phenyl ethyl alcohol, 2-methoxy or 2-Anisyl methyl carbinol .... (2-$CH_3OC_6H_4$)CH_2CH_2OH	152.20	....	128[17]	......	1.0862[15/4]	1.5312[25]	al, eth	B6[3], 4569
11260	β-Phenyl ethyl alcohol, 3-methoxy or 3-Anisyl methyl carbinol .... (3-$CH_3OC_6H_4$)CH_2CH_2OH	152.20	....	133[15]	......	1.0781[19/4]	1.5325[20]	al, eth	B6[3], 4570
11261	β-Phenyl ethyl alcohol, 4-methoxy or 4-Anisyl methyl carbinol .... (4-$CH_3OC_6H_4$)CH_2CH_2OH	152.20	....	140-1[17]	......	1.0794[20/4]	1.5310[25]	al, eth	B6[3], 4571
11262	bis-(2-Phenyl ethyl)amine .... [$C_6H_5CH_2CH_2$]$_2NH$	225.34	....	335-7[603], 190[15]	28-30		1.5550[25]	al, eth	b12[3], 2415
11263	bis-(α-Phenyl ethyl) ether (dl) .... [$C_6H_5CH(CH_3)$]$_2O$	226.32	....	280.2, 167-8[23]	......	1.0058[15/4]	1.5454[21]	eth, chl	B6[3], 1677
11264	bis-(β-Phenyl ethyl) ether .... $(C_6H_5CH_2CH_2)_2O$	226.32	vt-bl flr	317-20, 194.5[20]	......	1.0141[18/4]	1.5488[18]	eth, chl	B6[3], 1707
11265	Phenyl ethynyl ether or Phenoxy acetylene................ $C_6H_5OC{\equiv}CH$	118.14	....	61-2[25]	-36	1.0614[20/4]	1.5125[20]	al, eth	B6[4], 565
11266	Phenyl glycidol .... $C_6H_5CHCHCH_2OH$	150.18	....	138[3]	26.5	1.512[27]	1.5432[27]	al, eth	B17[4], 1349
11267	Phenyl glyoxal .... C_6H_5COCHO	134.14	nd (+w)	142[125], 95-7[25]	91 (+w)	....	....	al, eth, ace, bz, chl	B7[3], 3443
11268	Phenyl glyoxime................ $C_6H_5C(NOH)C(NOH)H$	164.18	nd (chl)	....	180			w, al, eth	B7[3], 3448
11269	Phenyl glyoxylic acid, 2-nitro .... (2-$NO_2C_6H_4$)$COCO_2H$	195.13	pr (w + 1)	....	123			w, al, ace, aa	B10[1], 315
11270	Phenyl glyxylonitrile .... C_6H_5COCN	131.14	ta	206-8, 99[19]	32-3			al, eth	B10[3], 2976
11271	Phenylhydrazine, 1-Benzoyl .... $C_6H_5NHNHCOC_6H_5$	212.25	pr (al), nd (w), lf (dil al)	314	168			bz, chl	B15[3], 163
11272	Phenylhydrazine, o-bromo .... 2-$BrC_6H_4NHNH_2$	187.05	nd	....	48				B15[1], 117
11273	Phenyl hydrazine, p-bromo .... 4-$BrC_6H_4NHNH_2$	187.05	nd (w), lf (lig), cr (al)	....	108			al, eth	B15[3], 289
11274	Phenyl hydrazine, 1-butyl .... $H_2NN(C_4H_9)C_6H_5$	164.25	....	250[76.1]	....				B15[1], 28
11275	Phenylhydrazine, 2,4,6-tribromo .... (2,4,6-$Br_3C_6H_2$)$NHNH_2$	344.85	nd (peth), cr (lig)	....	146			bz, chl, lig	B15[1], 126

PHYSICAL CONSTANTS OF ORGANIC COMPOUNDS (Continued)

No.	Name, Synonyms, and Formula	Mol. wt.	Color, crystalline form, specific rotation and λmax (log ε)	b.p. °C	m.p. °C	Density	n_D	Solubility	Ref.
11276	Phenylhydrazine, 2,4,6-trichloro (2,4,6-Cl₃C₆H₂)NHNH₂	211.48	cr (bz)		143				B15³, 281
11277	Phenylhydroxylamine C₆H₅NHOH	109.13	nd (w, bz, peth)		83-4			al, eth, bz, chl	B15³, 5
11278	Phenylhydroxylamine, N-nitroso C₆H₅N(NO)OH	138.13	nd (lig)		59			al, eth	B16³, 638
11279	Phenyl, 2-hydroxyphenyl amine (2HOC₆H₄)NHC₆H₅	185.23	pr (w)	180-9²⁰	69-70			al, eth, aa	B13³, 764
11280	Phenyl, 3-hydroxyphenyl amine (3-HOC₆H₄)NHC₆H₅	185.23	lf (w)	340	81-2			al, eth, ace, bz	B13³, 931
11281	Phenyl, 4-hydroxyphenyl amine (4-HOC₆H₄)NHC₆H₅	185.23	lf (w)	330, 215-6¹²	73			al, eth, bz, chl	B13³, 991
11282	Phenyl isocyanate C₆H₅NCO	119.12		162-3⁷⁵⁰, 55¹³		1.0956²⁰ᐟ⁴	1.5368²⁰	eth	12³, 903
11283	Phenyl isothiocyanate or Phenyl mustard oil C₆H₅NCS	135.19		221, 95¹²	−21	1.1303²⁰ᐟ⁴	1.6492²³	al, eth	12³, 908
11284	Phenyl, 2-methoxyphenyl ether (2-CH₃OC₆H₄)OC₆H₅	200.24	cr (MeOH), nd (lig)	288⁷⁴⁵, 91-2⁷	79			al, eth, bz	B6³, 4215
11285	Phenyl nitromethane or α-Nitrotoluene C₆H₅CH₂NO₂	137.14	ye liq	225-7, 118-9¹⁶		1.1598²⁰ᐟ⁰	1.5323²⁰	eth, ace	B5⁴, 850
11286	Phenyl 2-nitrophenyl amine (2-NO₂C₆H₄)NHC₆H₅	214.23	og pl (dil al), rh bi-pym		75.5			al	B12³, 1517
11287	Phenyl 3-Nitrophenyl amine (3-NO₂C₆H₄)NHC₆H₅	214.23	red liq or pl (dil al)		114			al, eth, bz	B9³, 3585
11288	Phenyl 4-nitrophenyl amine (4-NO₂C₆H₄)NHC₆H₅	214.23	ye nd, tab (CCl₄)	211²⁰	133-4			al, aa	B12³, 1586
11289	Phenyl 4-nitrosophenyl amine (4-ONC₆H₄)NHC₆H₅	198.23	ye pw pl (bz)		143			al, eth, bz, chl	B11³, 147
11290	Phenyl 2-nitrophenyl ether (2-O₂NC₆H₄)OC₆H₅	215.21	ye liq	235⁶⁰, 183-5⁸	<−20	1.2539²²	1.575²⁰	al, eth, bz, chl, aa	B6⁴, 1252
11291	Phenyl 4-nitrophenyl ether (4-O₂NC₆H₄)OC₆H₅	215.21	pl (peth, MeOH)	188-90⁸	61			eth, bz	B6⁴, 1287
11292	Phenyl 2-nitrophenyl sulfide (2-NO₂C₆H₄)SC₆H₅	231.28	ye-og nd (lig, al-eth)	210¹⁵	82			al, eth	B6⁴, 1663
11293	Phenyl 4-nitrophenyl sulfide (4-O₂NC₆H₄)SC₆H₅	231.28	pa ye mcl pr (lig)	240²⁵	55			al, eth	B6⁴, 1694
11294	Phenyl isopentyl ketone or Isocaprophenone i-C₅H₁₁COC₆H₅	176.24		255-6, 145-7³⁰	−2	0.9623¹⁵ᐟ⁴	1.533²⁰	al, eth, ace, bz, chl	B7³, 1158
11295	Phenyl phosphinic acid C₆H₅P(OH)₂	142.60			82-3			w, al	B16³, 874
11296	2-Phenyl propionic acid, 2-hydroxy (D) or 2-Hydroxy hydratropic acid C₆H₅C(OH)(CH₃)CO₂H	166.18	[α]¹⁰·⁵ᐟD +37.7 (al, c=3.5)		116-7			ace, bz	B10³, 560
11297	2-Phenyl propionic acid, 2-hydroxy (DL) or 2-Hydroxy hydratropic acid C₆H₅C(OH)(CH₃)CO₂H	166.18	nd pl (lig)		93-5			ace, bz	B10³, 560
11298	2-Phenyl propionic acid, 3-hydroxy (d) or 3-Hydroxy hydratropic acid HOCH₂CH(C₆H₅)CO₂H	166.18	nd (w, bz), pr (eth, w)		130			al, eth	B10², 158
11299	2-Phenyl propionic acid, 3-hydroxy (dl) or 3-Hydroxy hadratropic acid HOCH₂CH(C₆H₅)CO₂H	166.18	nd, pl (al, bz, w)	d	118			w, al, eth	B10³, 564
11300	2-Phenyl propionic acid, 3-hydroxy-(l) or 3-Hydroxy hydratropic acid HOCH₂CH(C₆H₅)CO₂H	166.18	pl (AcOEt), nd (w), [α]¹⁵ᐟD −81.2 (w, c=1.5)					al, eth, AcOEt	B10³, 564
11301	2-Phenyl propionic acid, 2-hydroxy (L) or 2-Hydroxy hydratropic acid C₆H₅C(OH)(CH₃)CO₂H	166.18	nd (bz, w), [α]¹³·⁸ᐟD −37.7 (al, c=3.4)		116-7			w, al, ace, bz	B10³, 560
11302	3-Phenyl propionic acid, 2-hydroxy (d) or 2-Hydroxy hydrocinnamic acid C₆H₅CH₂CH(OH)CO₂H	166.18	nd (w), [α]²⁰ᐟD +22.2 (w, c=2.2)		124-6			w, al, ace	B10³, 554

No.	Name, Synonyms, and Formula	Mol. wt.	Color, crystalline form, specific rotation and λ_{max} (log ε)	b.p. °C	m.p. °C	Density	n_D	Solubility	Ref.
11303	3-Phenyl propionic acid, 2-hydroxy-(dl) or 2-Hydroxy hydrocinnamic acid. . . . . C₆H₅CH₂CH(OH)CO₂H	166.18	cr (chl, bz) pr (w)	148-50[15]	98	. . . .	. . . . . .	al, eth, ace	B10³, 554
11304	3-Phenyl propionic acid, 2-hydroxy (l) or 2-Hydroxy hydrocinnamic acid. . . . . C₆H₅CH₂CH(OH)CO₂H	166.18	nd (w), [α]²⁰_D -19.9 (w, c=3.2)	. . . .	124.5	. . . .	. . . . . .	al, eth, ace	B10³, 554
11305	3-Phenyl propionic acid, 3-hydroxy-(d) or 3-Hydroxy hydrocinnamic acid. . . . . C₆H₅CH(OH)CH₂CO₂H	166.18	cr (bz), [α]¹⁸_D + 20.6 (MeOH, c=5)	. . . .	116	. . . .	. . . . . .	al, eth	B10³, 545
11306	3-Phenyl propionic acid, 3-hydroxy-(dl) or 3-Hydroxy hydrocinnamic acid. . . . . C₆H₅CH(OH)CH₂CO₂H	166.18	pr (w)	. . . .	96	. . . .	. . . . . .	al, ace, chl	B10³, 546
11307	3-Phenyl propionic acid, 3-hydroxy (l) or 3-Hydroxy hydrocinnamic acid. . . . . C₆H₅CH(OH)CH₂CO₂H	166.18	nd (bz), [α]¹⁸_D -19.8 (al, c=4.7)	. . . .	115-6	. . . .	. . . . . .	. . . .	B10³, 546
11308	Phenyl sulfide or Diphenyl sulfide (C₆H₅)₂S	186.28	. . . .	296, 145[8]	−25.9	1.1136²⁰/⁴	1.6334²⁰	eth, bz	B6⁴, 1488
11309	Phenyl sulfide, 2,4'-diamino (2-H₂NC₆H₄)S(C₆H₄NH₂-4)	216.31	nd (w, al), pr (dil al)	. . . .	62.5	. . . .	. . . . . .	al, eth, bz	B13³, 1245
11310	Phenyl sulfide, 4,4'-diamino or bis-(4-Amino phenyl)sulfide (4-H₂NC₆H₄)₂S	216.31	nd (w)	. . . .	108-9	. . . .	. . . . . .	al, eth, bz	B13³, 1246
11311	Phenyl sulfide, 4,4'-dibromo (4-BrC₆H₄)₂S	344.06	. . . .	268.5⁴⁰	115	1.84	. . . . . .	eth, chl	B6⁴, 1651
11312	Phenyl sulfide, 4,4'-dichloro-2,2'-dinitro (4-Cl₁-2-NO₂C₆H₃)₂S	345.16	br-ye nd (90% aa)	. . . .	149-50	. . . .	. . . . . .	bz	B6², 312
11313	Phenyl sulfide, 4,4'-dihydroxy or bis-(4-Hydroxy phenyl)sulfide (4-HOC₆H₄)₂S	218.28	mcl pr or lf (al)	. . . .	151	. . . .	. . . . . .	al, eth	B6³, 4455
11314	Phenyl sulfide, 2,2'-dimethoxy or o-Anisyl disulfide (2-CH₃OC₆H₄)₂S	246.33	lf (al)	252-3¹⁰	73	. . . .	. . . . . .	al, eth, bz	B6, 794
11315	Phenyl sulfide, 4,4'-dinitro or bis-(4-Nitro phenyl)sulfide (4-O₂NC₆H₄)₂S	276.28	og pl (aa)	. . . .	160-1	. . . .	. . . . . .	. . . .	B6⁴, 1696
11316	Phenyl sulfide, 2,2'-dinitro or bis-(2-Nitro phenyl)sulfide (2-O₂NC₆H₄)₂S	276.28	gold-ye pl (aa)	sub d	122-3	. . . .	. . . . . .	. . . .	B6⁴, 1665
11317	Phenyl sulfide, 2,2',4,4'-tetranitro or bis-(2,4-Dinitro phenyl)sulfide . . . . [2,4-(NO₂)₂C₆H₃]₂S	366.27	ye nd or pl (aa)	. . . .	197	. . . .	. . . . . .	. . . .	B6⁴, 1748
11318	Phenyl sulfone . . . . (C₆H₅)₂SO₂	218.28	mcl pr (bz), pl (al), nd (w)	379, 232¹⁸	128-9	1.252²⁰/⁴	. . . . . .	eth, bz	B6⁴, 1490
11319	Diphenyl sulfone, 4,4'-diacetamide . . . . (4-CH₃CONHC₆H₄)₂SO₂	332.38	pa ye nd (eth, dil aa), lf (dil al)	. . . .	282-5	. . . .	. . . . . .	al	B13³, 1286
11320	Phenyl sulfone, 3,3'-diamino or bis-(3-Amino phenyl)sulfone. . . . . (3-H₂NC₆H₄)₂SO₂	248.31	pr	. . . .	168	. . . .	. . . . . .	w, al	B13³, 984
11321	Phenyl sulfone, 4,4'-diamino or bis-(4-Amino phenyl)sulfone. . . . . (4-H₂NC₆H₄)₂SO₂	248.31	lf (dil al)	. . . .	178	. . . .	. . . . . .	al	B13³, 1246
11322	Phenyl sulfone, 4,4'-dichloro or bis-(4-Chloro phenyl)sulfone. . . . . (4-ClC₆H₄)₂SO₂	287.18	mcl	sub	148-9	. . . .	. . . . . .	. . . .	B6⁴, 1587
11323	Phenyl sulfone, 4,4'-diethoxy or bis-(4-Ethoxy phenyl)sulfone. . . . . (4-C₂H₅OC₆H₄)₂SO₂	306.38	pl (al, aa)	. . . .	163	. . . .	. . . . . .	al, eth	B6³, 4459
11324	Phenyl sulfone, 4,4'-diethyl . . . . (4-C₂H₅C₆H₄)₂SO₂	274.38	. . . .	. . . .	102	. . . .	. . . . . .	eth, bz	B6, 475
11325	Phenyl sulfone, 2,2'-dihydroxy or bis-(2-Hydroxy phenyl)sulfone. . . . . (2-HOC₆H₄)₂SO₂	250.28	nd (bz)	. . . .	164-5 (179)	. . . .	. . . . . .	w, al, eth, aa	B6³, 4278
11326	Phenyl sulfone, 3,3'-dihydroxy or bis-(3-Hydroxy phenyl)sulfone. . . . . (3-HOC₆H₄)₂SO₂	290.28	190-1	. . . .	. . . .	. . . .	. . . . . .	al, eth	B6³, 4365

No.	Name, Synonyms, and Formula	Mol. wt.	Color, crystalline form, specific rotation and λ_{max} (log ε)	b.p. °C	m.p. °C	Density	n_D	Solubility	Ref.
11327	Phenyl sulfone, 4,4′-dihydroxy or bis-(4-Hydroxy phenyl)sulfone................... (4-HOC₆H₄)₂SO₂	250.28	nd (w), rh bipym	....	240-1	1.3663[15]		al, eth	B6³, 4456
11328	Phenyl sulfone, 2,2′-dimethoxy or bis-(2-Methoxy phenyl)sulfone................... (2-CH₃OC₆H₄)₂SO₂	278.33	nd (bz)	....	157-8			al, aa	B6³, 4278
11329	Phenyl sulfone, 4,4′-dimethoxy or bis-(4-Methoxy phenyl)sulfone................... (4-CH₃OC₆H₄)₂SO₂	278.33	lf or pr (al), nd (al-eth)	sub	130			al	B6³, 4458
11330	Phenyl-o-tolyl sulfide (2-CH₃C₆H₄)SC₆H₅	200.30	....	304.5²²⁴, 164¹²		1.0893²⁰/⁴		ace, bz	B6⁴, 2018
11331	Phenyl-m-tolyl sulfide................ (3-CH₃C₆H₄)SC₆H₅	200.30	....	309.5, 164.5¹¹	-6.5	1.0937¹⁵/⁴		ace, bz	B6⁴, 2080
11332	Phenyl-p-tolyl sulfide (4-CH₃C₆H₄)SC₆H₅	200.30	....	317, 167.5¹¹	15.7	1.0986²⁵/⁴	1.6225²⁵	ace, bz	B6⁴, 2169
11333	Phenyl-2-tolyl sulfone (2-CH₃C₆H₄)SO₂C₆H₅	232.30	pl (al)	....	81			al, eth, bz	B6⁴, 2018
11334	Phenyl-4-tolyl sulfone (4-CH₃C₆H₄)SO₂C₆H₅	232.30	pl (al)		127-8				B6⁴, 2171
11335	Phenyl sulfoxide or Diphenyl sulfoxide (C₆H₅)₂SO	202.28	pr (lig)	340d, 210¹⁵	70.5			al, eth, bz, aa	B6⁴, 1489
11336	Phenyl sulfoxide, 4-amino-4′-nitro (4-H₂NC₆H₄)SO(C₆H₄NO₂-4)		ye (al)		132			al	B13³, 1226
11337	Phenyl sulfoxide, 4,4′-diamino or bis(4-Amino phenyl)sulfoxide (4-H₂NC₆H₄)₂SO	232.31	pr (w, al)		175d				B13³, 1246
11338	Phenyl sulfoxide, 4,4′-dibromo (4-BrC₆H₄)₂SO	360.06	....		153-4			al, bz, chl	B6⁴, 1651
11339	Phenyl sulfoxide, 4,4′-dichloro or bis-(4-Chloro phenyl)sulfoxide................ (4-ClC₆H₄)₂SO	271.16	....		143-4			chl	B6⁴, 1497
11340	Phenyl sulfoxide, 4,4′-dihydroxy or bis-(4-Hydroxy phenyl)sulfone................ (4-HOC₆H₄)₂SO	234.28	nd (ace)	....	195			al, ace	B6³, 4456
11341	Phenyl thiocyanate C₆H₅SCN	135.19	....	232-3, 71-3¹·⁵	....	1.155¹⁸/¹⁸		al, eth	B6⁴, 1536
11342	Phenyl thiocyanate, 4-amino (4-H₂NC₆H₄)SCN	150.20	nd (w), cr (dil al)		57-8			al, eth, bz	B13³, 1239
11343	Phenyl thiocyanate, 4-chloro (4-ClC₆H₄)SCN	169.83	nd (al)		35-6			al	B6⁴, 1601
11344	Phenyl thiocyanate, 4-(dimethyl amino) [4-(CH₃)₂NC₆H₄]SCN	178.26	nd (lig, w, al)		73-4			eth	B13³, 1251
11345	Phenyl trimethyl ammonium bromide .... C₆H₅N⁺(CH₃)₃Br⁻	216.13	hygm pr (al, al-eth)		213-4			w	B12³, 254
11346	Phenyl trimethyl ammonium iodide C₆H₅N⁺(CH₃)₃I⁻	263.12	lf (al)		224			w, al, aa	B12³, 254
11347	Phenyl vinyl ether C₆H₅OCH=CH₂	120.16		155-6	......	0.9770²⁰/⁴	1.5224²⁰	eth	B6⁴, 561
11348	Phloretin or Dihydronaringenin C₁₅H₁₄O₅	274.28	nd (dil al), cr (dil ace)		262-4d			**al, bz**	B8³, 4076
11349	Phlorhizin, dihydrate or Asebotin (a β-glucoside)........... C₂₁H₂₄O₁₀·2H₂O	472.45	nd (w), [α]²⁵/D -52 (96% al)		108 (+2w), 170d (anh)	1.4298		al, ace, Py	B17⁴, 3042
11350	Phloroglucinol or 1,3,5-Trihydroxybenzene .... 1,3,5-(HO)₃C₆H₃	126.11	lf or pl (w + 2)	sub	218-9 (anh)	1.46	....	al, eth, bz, Py	B6³, 6301
11351	Phloroglucinol, 2-acetyl or 2,4,6-Trihydroxy acetophen-one .... 2,4,6(HO)₃C₆H₂COCH₃	168.15	nd (w + 1)	....	222-4 (anh)			al, eth, ace, aa	B8³, 3386
11352	Phloroglucinol diacetate .... C₁₀H₁₀O₅	210.19	pl (w)	....	104			al, eth, ace	B6¹, 547
11353	Phloroglucinol dimethyl ether or 3,5-Dimethoxy phenol . 3,5-(CH₃O)₂C₆H₃OH	154.17	cr (bz-lig)	172-5¹⁷	36-8			eth, bz	B6³, 6305
11354	Phloroglucinol triacetate 1,3,5(CH₃CO)₃C₆H₃	252.23	pr (w), nd (dil al)		105-6			al	B6³, 6306
11355	Phloroglucinol, 2,4,6-trichloro or 2,4,6-Trichloro-1,3,5-trihydroxy benzene .... C₆H₃Cl₃O₃	229.45	cr (al)	sub	136			al	B6, 1104

PHYSICAL CONSTANTS OF ORGANIC COMPOUNDS (Continued)

No.	Name, Synonyms, and Formula	Mol. wt.	Color, crystalline form, specific rotation and λmax (log ε)	b.p. °C	m.p. °C	Density	n_D	Solubility	Ref.
11356	Phloroglucinol, triethyl ether or 1,3,5-Triethoxy benzene 1,3,5-$(C_2H_5O)_3C_6H_3$	210.28	cr (al, dil al)	175[24]	43.5	....	....	al, eth	B6[3], 6306
11357	Phloroglucinol-trimethyl ether or 1,3,5-Trimethoxy benzene. 1,3,5-$(CH_3O)_3C_6H_3$	168.20	pr (al), lf (peth)	255.5	54-5	....	....	al, eth, bz	B6[3],6305
11358	Phorone or 2,6-Dimethyl, 2,5-heptadiene-4-one $CH_3C(CH_3)=CHCOCH=C(CH_3)_2$	138.21	ye gr pr	197.8	28	0.8850[20/4]	1.4998[20]	al, eth, ace	B[4], 3564
11359	Phosgene or Carbonyl chloride $COCl_2$	98.92	....	7.6	−118	1.381[20/4]	....	bz, chl, aa	B3[4], 31
11360	Phosphine, bis trifluoromethyl $(F_3C)_2PH$	169.99	spont flam	1	−137	....	....		B3[4], 255
11361	Phosphine, bis(trifluoromethyl) chloro $(F_3C)_2PCl$	204.44	spont flamm	21	....	....	....		B3[4], 257
11362	Phosphine, bis(trifluoromethyl) cyano $(F_3C)_2PCN$	195.00	spont. flamm	48	....	....	1.3248[20]		B3[4], 256
11363	Phosphine, bis(trifluoromethyl) iodo $(F_3C)_2PI$	295.89	....	73	....	....	1.403[15]		B3[4], 257
11364	Phosphine, dichloro-2,4-(dimethyl phenyl). $[2,4-(CH_3)_2C_6H_3]PCl_2$	207.04	....	256-8	....	....	....		B16, 773
11365	Phosphine, dichloro-(2,5-dimethyl phenyl) $[2,5-(CH_3)_2C_6H_3]PCl_2$	207.04	....	253-4	−30	1.25[18/18]	....		B16[3], 848
11366	Phosphine, dichloro(4-ethyl phenyl) $[4-C_2H_5C_6H_4]PCl_2$	207.04	....	250-2, 85[0.4]	....	1.237[20/4]	1.584[20]		B16[3], 848
11367	Phosphine, dichloro-phenyl $C_6H_5PCl_2$	178.99	....	224-6, 99-101[11]	....	1.356[20/4]	1.6030[20]	bz	B16[3], 847
11368	Phosphine, dichloro-trifluoromethyl F_3CPCl_2	170.89	....	37	....	....	....		B3[4], 258
11369	Phosphine, diethyl $(C_2H_5)_2PH$	90.11	....	85	....	....	....		B4[3], 1761
11370	Phosphine, diiodo-trifluoromethyl F_3CPI_2	353.79	ye fum	d 760, 73[37]	....	....	1.630[20]		B3[4], 258
11371	Phosphine, dimethyl $(CH_3)_2PH$	62.05	....	25	<1	....	....	al, eth	B4[3], 1759
11372	Phosphine, diphenyl-ethyl $(C_6H_5)_2PC_2H_5$	214.25	....	293	....	....	....		B16[3], 833
11373	Phosphine, ethyl $C_2H_5PH_2$	62.05	....	25	<1	....	....		B4[3], 1761
11374	Phosphine, methyl CH_3PH_2	48.02	gas	−14	....	....	....	eth	B4[3], 1759
11375	Phosphine, phenyl $C_6H_5PH_2$	110.10	....	160-1, 40[10]	....	1.001[15]	1.5796[20]		B16[3], 831
11376	Phosphine, triethyl $(C_2H_5)_3P$	118.16	....	129[762]	−88	0.8006[19/4]	1.458[15]	al, eth	B4[3], 1761
11377	Phosphine, triethyl, oxide $(C_2H_5)_3PO$	134.16	wh hyg nd	243	50	....	....	al, eth, w	B4[3], 1775
11378	Phosphine, triethyl, sulfide $(C_2H_5)_3PS$	150.23	cr (al)	....	94	....	....	w	B4[3], 1775
11379	Phosphine, trifluoromethyl. F_3CPH_2	102.00	spont. flamm	−26.5	....	....	....		B3[4], 255
11380	Phosphine, trimethyl $(CH_3)_3P$	76.08	....	37.8	−85	<1	....	eth	B4[3], 1759
11381	Phosphine, triphenyl $(C_6H_5)_3P$	262.29	....	188[1]	80	1.0749[80/4]	1.6358[80]	al, eth, bz, chl	B16[3], 833
11382	Phosphine, triphenyl, oxide $(C_6H_5)_3PO$	278.29	pr	>360	156-7	1.2124[23/4]	....	al, bz	B16[3], 864
11383	Phosphine, tris (chloromethyl) $(ClCH_2)_3P$	179.41	....	100[7]	....	1.414[20]	....	al, eth, ace, bz	B1[3], 2608
11384	Phosphine, tris (trichloromethyl) $(Cl_3C)_3P$	386.08	....	....	53	....	....		B3[4], 259
11385	Phosphine, tris(trifluoromethyl) $(F_3C)_3P$	237.9	spont, flamm.	17.3	−112	....	....		B3[4], 255
11386	Phosphine, tris(trifluoromethyl) oxide $(F_3C)_3PO$	253.99	....	23.6	....	....	....		J1955, 574
11387	Phosphinic acid, diethyl $(C_2H_5)_2PO(OH)$	122.11	....	320, 134[0.7]	18.5	....	....	w, al, eth	B4[3], 1776
11388	Phosphinic acid, diethyl, anhydride $[(C_2H_5)_2PO]_2O$	226.20	....	188[14]	....	1.1053[20/4]	1.4647[20]	....	Am73, 5466

No.	Name, Synonyms, and Formula	Mol. wt.	Color, crystalline form, specific rotation and λ_{max} (log ε)	b.p. °C	m.p. °C	Density	n_D	Solubility	Ref.
11389	Phosphinic acid, diethyl, ethyl ester .... $(C_2H_5)_2PO(OC_2H_5)$	150.16	....	95[14]	....	0.9908[20/4]	1.4337[20]	al, eth	AM73, 5466
11390	Phosphinic acid, dimethyl $(CH_3)_2PO(OH)$	94.05	cr (bz)	377	92	....	....	w, al, eth	B4[3], 1776
11391	Phosphinic acid, dimethyl, anhydride $[(CH_3)_2PO]_2O$	170.09	nd (bz)	192[15]	119-21	....	....	....	Am73, 5466
11392	Phosphinic acid, bis(trifluoromethyl) .... $(F_3C)_2PO(OH)$	201.99	visc liq	182, 137-8[238]	....	<1.9	....	w	J1955, 563
11393	Phosphinyl chloride dimethyl............... $(CH_3)_2POCl$	112.50	hyg nd (bz, peth)	204	67-8	....	....	....	Am73, 5466
11394	Phosphinyl chloride, diethyl .... $(C_2H_5)_2POCl$	140.55	....	104[15]	....	1.1394[20/4]	1.4647[20]	....	B4[3], 1776
11395	Phosphonic acid, acetyl, diethyl ester .... $CH_3COPO(OC_2H_5)_2$	180.14	....	114-5[20]	....	....	1.4200[26]	....	B2[4], 440
11396	Phosphonic acid, benzyl, diethyl ester .... $C_6H_5CH_2PO(OC_2H_5)_2$	228.23	....	110[2]	....	....	1.4930[20]	....	B16[3], 889
11397	Phosphonic acid, carboxy methyl .... $HO_2CCH_2PO(OH)_2$	140.03	....	....	142-3	....	....	....	B4[2], 975
11398	Phosphonic acid, diethyl ester .... $(C_2H_5O)_2P(O)H$	138.10	....	53.0-55[6]	....	....	....	....	B1[4], 1329
11399	Phosphonic acid, diisobutyl ester .... $(i-C_4H_9O)_2P(O)H$	194.21	....	235-6	....	0.9759[20/4]	....	....	B1[4], 1596
11400	Phosphonic acid, diisopropyl ester .... $(i-C_3H_7O)_2P(O)H$	166.16	....	76-7[10]	....	0.9972[18/0]	....	....	B1[4], 1475
11401	Phosphonic acid, diphenyl ester .... $(C_6H_5O)_2P(O)H$	234.19	....	218-9[16]	12	....	1.5564[25]	....	B6[4], 703
11402	Phosphonic acid, methyl, diisopropyl ester .... $CH_3PO(O-i-C_3H_7)_2$	180.18	....	66[3]	....	....	1.4120[16.5]	....	B4[3], 1778
11403	Phosphonic acid, methyl, diphenyl ester .... $CH_3P(O)(OC_6H_5)_2$	248.22	....	205[13]	35	1.2051[20/0]	....	....	B6[2], 164
11404	Phosphoric acid, methyl, ethyl ester .... $CH_3P(O)(OC_2H_5)(OH)$	124.08	....	106-7[0.1]	....	1.1800[20]	1.4258[20]	....	CAS58, 5720
11405	Phosphoric acid, diethyl ester or Diethyl phosphate..... $(C_2H_5O)_2PO(OH)$	154.10	syr	203d	....	1.186[25/4]	1.4170[20]	eth	B1[4], 1339
11406	Phosphoric acid, dimethyl ester or Dimethyl phosphate... $(CH_3O)_2PO(OH)$	126.05	....	172-6d	....	1.335[25]	1.408[25]	w, al, ace	B1[4], 1259
11407	Phosphoric acid, monoethyl ester .... $C_2H_5OPO(OH)_2$	126.05	hyg cr	d	....	1.430[25/4]	1.427	w, al, eth, ace	B1[4], 1338
11408	Phosphoric acid, phenyl ester or Phenyl phosphate ..... $C_6H_5OPO(OH)_2$	174.09	pl (chl), nd (w)	....	99.5	....	....	w, al, eth, bz, chl	B6[4], 708
11409	Phosphoric acid triamide, hexamethyl............... $[(CH_3)_2N]_3PO$	179.20	....	98-100[6]	....	1.024[25/25]	1.4579[20]	al, eth	B4[4], 284
11410	Phosphoric acid, tributyl ester or Tributyl phosphate... $(C_4H_9O)_3PO$	266.32	....	289, 160-2[15]	....	0.9727[25/4]	1.4224[25]	w, al, eth, bz	B1[4], 1531
11411	Phosphoric acid, tris (2,4-dimethylphenyl) ester .... $[2,4-(CH_3)_2C_6H_3O]_3PO$	410.46	glassy	232-5	....	1.142[38/4]	1.5550[20]	bz, hx	B6, 488
11412	Phosphoric acid, tris (-2,5-dimethylphenyl) ester........ $[2,5-(CH_3)_2C_6H_3O]_3PO$	410.46	cr (dil al)	260-5[8]	78-81	1.197[25]	....	eth, bz	B6[3], 1773
11413	Phosphoric acid, tris(2,6-dimethylphenyl) ester .... $[2,6-(CH_3)_2C_6H_3O]_3PO$	410.46	wax	262-4[6]	136-8	....	....	bz	J1956, 3043
11414	Phosphoric acid, tris (3,4-dimethylphenyl) ester .... $[3,4-(CH_3)_2C_6H_3O]_3PO$	410.46	wax	260-3[7]	71-2	....	....	bz	B6, 482
11415	Phosphonic acid, tris (3,5-dimethylphenyl) ester .... $[3,5-(CH_3)_2C_6H_3O]_3PO$	410.46	wax	290[10]	46	....	....	aa	C31, 187
11416	Phosphoric acid, triisobutyl ester .... $(i-C_4H_9O)_3PO$	266.32	....	264, 138[10]	....	0.9681[20/4]	1.4193[20]	w, al, eth, bz	B1[4], 1598
11417	Phosphoric acid, triethyl ester or Triethyl phosphate ..... $(C_2H_5O)_3PO$	182.16	....	215-6, 103[25]	−56.4	1.0695[20/4]	1.4053[20]	al, eth, bz	B1[4], 1339
11418	Phosphoric acid, trimethyl ester or Trimethyl phosphate. $(CH_3O)_3PO$	140.08	....	197.2, 85[26]	α −46 (st), β-62	1.2144[20/4]	1.3967[20]	w, eth	B1[4], 1259
11419	Phosphoric acid, tripentyl ester or Tripentyl phosphate... $(C_5H_{11}O)_3PO$	308.40	....	225[50], 167[5]	....	0.9608[20/4]	1.4319[20]	al, eth, to	B1[4], 1645
11420	Phosphoric acid, triphenyl ester or Triphenyl phosphate . $(C_6H_5O)_3PO$	326.29	cr (abs al-lig), pr, (al), nd (eth-lig)	245[11]	50-1	1.2055[50/4]	....	al, eth, bz, chl	B6[4], 720

No.	Name, Synonyms, and Formula	Mol. wt.	Color, crystalline form, specific rotation and λ_{max} (log ε)	b.p. °C	m.p. °C	Density	n_D	Solubility	Ref.
11421	Phosphoric acid, tripropyl ester or Tripropyl phosphate (C₃H₇O)₃PO	224.24	....	252, 107.5[5]	......	1.0121[20/4]	1.4165[20]	al, eth, to	B1[4], 1428
11422	Phosphoric acid, triisopropyl ester or Triisopropyl phosphate (i-C₃H₇O)₃PO	224.24	....	218-20, 95-6[8]	......	0.9867[20/4]	1.4057[20]	al	B1[4], 1478
11423	Phosphoric acid, tri(2-tolyl) ester or Tri-o-cresyl phosphate (2-CH₃C₆H₄O)₃PO	369.37	col or pa ye	410, 283-5[20]	11	1.955[20/4]	1.5575[20]	al, eth, aa, to	B6[4], 1979
11424	Phosphoric acid, tri(3-tolyl) ester or Tri-m-cresyl phosphate (3-CH₃C₆H₄O)₃PO	368.37	wax	260[15]	25-6	1.150[25]	1.5575[20]	eth	B6[4], 2057
11425	Phosphoric acid, tri (4-tolyl) ester or Tri-p-cresyl phosphate T.P.C. (4-CH₃C₆H₄O)₃PO	368.37	nd (al), ta (eth)	244[3.5]	77-8	1.247[25]	....	al, eth, bz, chl, aa	B6[4], 2130
11426	Phosphoric acid, thiono, tri(2-tolyl) ester (2-CH₃C₆H₄O)₃PS	384.44	nd (al)	260-5[1]	45-6			al, aa	B6[4], 1980
11427	Phosphonic acid, thiono, tri-3-tolyl ester (3-CH₃C₆H₄O)₃PS	384.44	nd (al)	270-2[1]	40-1			aa	B6[4], 2058
11428	Phosphonic acid, thiono, tri-4-tolyl ester (4-CH₃C₆H₄O)₃PS	384.44	nd (al)	....	93-4				B6[4], 2132
11430	Phosphorous acid, diethyl ester or Diethyl phosphite (C₂H₅O)₂P(OH)	138.10	....	87[20]	......	1.0720[20/4]	1.4101[20]	al, eth	B1[4], 1329
11431	Phosphorous acid, dimethyl ester or Dimethyl phosphite (CH₃O)₂P(OH)	110.05	....	170-1, 70[25]	......	1.2004[20/0]	1.4036[20]	al, Py	B1[4], 1255
11432	Phosphorous acid, tributyl ester or Tributyl phosphite (C₄H₉O)₃P	250.32	....	122[12]	......	0.9259[20/4]	1.4321[19]	al, eth	B1[4], 1527
11433	Phosphorous acid, triethyl ester or Triethyl phosphite (C₂H₅O)₃P	166.16	....	157.9, 49[12]	......	0.9629[20/4]	1.4127[20]	....	B1[4], 1333
11434	Phosphorous acid, trimethyl ester or Trimethyl phosphite (CH₃O)₃P	124.08	....	111-2, 22[23]	......	1.0520[20/0]	1.4095[20]	al, eth	B1[4], 1256
11435	Phosphorous acid, triphenyl ester or Triphenyl phosphite (C₆H₅O)₃P	310.29	....	360, 200-1[5]	ca 25	1.1844[20/0]	1.5900[20]	al	B6[4], 695
11436	Phosphorous acid, tripropyl ester or Tripropyl phosphite (C₃H₇O)₃P	208.24	....	206-7, 92[14]	......	0.9417[20]	1.4282[20]	al, eth	B1[4], 1426
11437	Phosphorous acid, trisopropyl ester or Trisopropyl phosphite (i-C₃H₇O)₃P	208.24	....	60-1[10]	......	0.9687[18.5/0]	1.4085[25]	al, eth	B1[4], 1476
11438	Phosphorous acid, tris (2,2,2-trichloro ethyl) ester (Cl₃CCH₂O)₃P	476.16	....	263, 127-31[0.1]	......	....	1.5174[20]	....	B1[4], 1384
11439	Phosphorous acid, tri(2-tolyl) ester or Tri-o-cresyl phosphite (2-CH₃C₆H₄O)₃P	352.38	ye	238[11]	......	1.1423[20/4]	1.5740[28]	eth	B6[4], 1977
11440	Phosphorous acid, tri(4-tolyl) ester or Tri-p-cresyl phosphite (4-CH₃C₆H₄O)₃P	352.38	pa ye	250-5[10]	......	1.1313[25/25]	1.5703[28]	eth	B6[4], 2128
11441	Phthaladehyde or o-Phthalic aldehyde 1,2-C₆H₄(CHO)₂	134.14	ye cr or nd (lig)	....	53-6			al, eth	B7[3], 3457
11442	Phthalamic acid or Phthalic acid monoamide 2-HO₂CC₆H₄CONH₂	165.15	pr	....	148-9			al	B9[3], 4191
11443	Phthalamic acid, N-(1-naphthyl) or Alanap-1 2-HO₂CC₆H₄CONH-α-C₁₀H₇	291.31			185				B12[3], 2876
11444	Phthalamide 1,2 C₆H₄(CONH₂)₂	164.18	cr	d	222				B9[3], 4197
11445	Phthalmide, N,N,N′N′-tetraethyl 1,2-C₆H₄[CON(C₂H₅)₂]₂	276.38	....	204[16]	36				B9[3], 4198
11446	Phthalazine C₈H₆N₂	130.15	....	175[17]	90-1			w, al, bz	B23, 174
11447	Phthalazine, 1,2-dihydro-1-oxo or Phthalazone C₈H₆N₂O	146.15	nd (w), pr (sub)	337[755], sub	184-5			al, bz	B24[2], 70
11448	Phthalhydrazide or Phthalic acid hydrazide C₈H₆N₂O₂	162.15	mcl nd (w, dil al, aa)	....	342-4			aa	B24[2], 194
11449	Phthalic acid or 1,2 Benzene dicarboxylic acid 1,2 C₆H₄(CO₂H)₂	166.14	pl (w)	d	210-11d, 191 (sealed tube)	1.593	....	al	B9[3], 4094
11450	Phthalic acid , 3-amino 3-H₂NC₆H₃(CO₂H)₂-1,2	181.15	nd	....	231-2				B14[3], 1393

No.	Name, Synonyms, and Formula	Mol. wt.	Color, crystalline form, specific rotation and λ_{max} (log ε)	b.p. °C	m.p. °C	Density	n_D	Solubility	Ref.
11451	Phthalic acid, 4-amino, dimethyl ester.................. $4\text{-}H_2NC_6H_3(CO_2CH_3)_2\text{-}1,2$	209.20	pl (al, bz), pr (w)		54			al, chl, Py	B14[3], 1397
11452	Phthalic acid, 3-benzoyl or 2,3-Benzophenone dicarboxylic acid.................. $3\text{-}C_6H_5COC_6H_3(CO_2H)_2\text{-}1,2$	270.25	pl, nd (w + 1)	d	140-1			al, bz	B10[3], 4007
11453	Phthalic acid, 4-benzoyl or 3,4-Benzophenone dicarboxylic acid.................. $4\text{-}C_6H_5COC_6H_3(CO_2H)_2\text{-}1,2$	270.25	lf (xyl)		177			al	B10[3], 4010
11454	Phthalic acid, 3-bromo .... $3\text{-}BrC_6H_3(CO_2H)_2\text{-}1,2$	245.04	nd		188d			al, eth	B9[3], 4212
11455	Phthalic acid, 4-bromo .... $4\text{-}BrC_6H_3(CO_2H)_2\text{-}1,2$	245.04			173-5			al, eth	B9[3], 4213
11456	Phthalic acid, butyl ester (mono) or mono-Butyl phthalate.................. $2\text{-}HO_2CC_6H_4CO_2C_4H_9$	222.24	pl (ace, al)		73-4			al, chl	B9[3], 4101
11457	Phthalic acid, sec-butyl ester (mono) (d) or mono-sec-Butyl phthalate.................. $2\text{-}sec\text{-}C_4H_9O_2CC_6H_4CO_2H$	222.24	$[\alpha]^{20}/_D$ + 38.4 (al)		48			al, chl	B9[3], 4104
11458	Phthalic acid, sec-butyl ester (mono) (dl) or mono-sec-butyl phthalate.................. $2\text{-}HO_2CC_6H_4CO_2\text{-}sec\text{-}C_4H_9$	222.24	cr (peth)		63			al, chl	B9[3], 4104
11459	Phthalic acid, 3-chloro.................. $3\text{-}ClC_6H_3(CO_2H)_2\text{-}1,2$	200.58	nd (w)		186-7			al, eth	B9[3], 4202
11460	Phthalic acid, 4-chloro.................. $4\text{-}ClC_6H_3(CO_2H)_2\text{-}1,2$	200.58	nd (dil al)		157			al, eth	B9[3], 4202
11461	Phthalic acid, dibenzyl ester or Dibenzyl phthalate.................. $1,2\text{-}C_6H_4(CO_2CH_2C_6H_5)_2$	346.39	pr (al)	277[15]	42-3			al, eth	B9[3], 4158
11462	Phthalic acid, dibutyl ester or Dibutyl phthalate.................. $1,2\text{-}C_6H_4(CO_2C_4H_9)_2$	278.35		340, 206[20]		$1.047^{20/20}$	1.4911^{20}	al, eth, bz	B9[3], 4102
11463	Phthalic acid, diisobutyl ester or Diisobutyl phthalate.................. $1,2\text{-}C_6H_4(CO_2\text{-}i\text{-}C_4H_9)_2$	278.35		295-8, 182-4[10]		1.0490^{15}			B9[3], 4105
11464	Phthalic acid, 3,4-dichloro.................. $3,4\text{-}Cl_2C_6H_2(CO_2H)_2\text{-}1,2$	235.03	pl (w)		195			al, eth	B9, 817
11465	Phthalic acid, 3,5-dichloro.................. $3,5\text{-}Cl_2C_6H_2(CO_2H)_2\text{-}1,2$	235.03	nd, ta (aq HCl)	sub	164d			al, eth, ace	B9[3], 4204
11466	Phthalic acid, 3,6-dichloro.................. $3,6\text{-}Cl_2C_6H_2(CO_2H)_2\text{-}1,2$	235.03	pl (w)		d 100			al, eth	B9[3], 4204
11467	Phthalic acid, 4,5-dichloro.................. $4,5\text{-}Cl_2C_6H_2(CO_2H)_2\text{-}1,2$	235.03	nd (w)		ca 200d			eth	B9[3], 4205
11468	Phthalic acid, dicyclohexyl ester or Dicyclohexyl phthalate $1,2\text{-}C_6H_4(CO_2C_6H_{11})_2$	330.43	pr (al)		66	$1.383^{20/4}$	1.451^{20}	al, eth	B9[3], 4123
11469	Phthalic acid, di-(2-ethoxyethyl) ester.................. $1,2\text{-}C_6H_4(CO_2CH_2CH_2OC_2H_5)_2$	310.35		345, 233-5[23]	34	1.1229^{21}			B9[2], 597
11470	Phthalic acid, diethyl ester or Diethylphthalate.......... $1,2\text{-}C_6H_4(CO_2C_2H_5)_2$	222.24		298, 172[12]		$1.1175^{20/4}$	1.5000^{21}	**al, eth**, ace, bz	B9[3], 4099
11471	Phthalic acid, 4,5-dimethoxy or m-Hemipic acid.......... $4,5\text{-}(CH_3O)_2C_6H_2(CO_2H)_2\text{-}1,2$	226.19	nd (w), pr (w + 2)		174-5				B10[3], 2431
11472	Phthalic acid, di-(2-methoxyethyl) ester .... $1,2\text{-}C_6H_4(CO_2CH_2CH_2OCH_3)_2$	282.30		230[10]		1.1708^{15}			B9[3], 4173
11473	Phthalic acid, dimethyl ester or Dimethyl phthalate ..... $1,2\text{-}C_6H_4(CO_2CH_3)_2$	194.19	pa ye	283.8	0-2	$1.1905^{20/4}$	1.5138^{20}	**al, eth**, bz	B9[3], 4098
11474	Phthalic acid, diisopentyl ester or Diisopentylphthalate....... $1,2\text{-}C_6H_4(CO_2\text{-}i\text{-}C_5H_{11})_2$	306.41		330-8d, 225[40]		$1.0220^{16/16}$	1.4871^{20}	al	B9[3], 4107
11475	Phthalic acid, diphenyl ester or Diphenyl phthalate...... $1,2\text{-}C_6H_4(CO_2C_6H_5)_2$	318.33	pr (al, lig)	250-7[14], sub	73				B9[3], 4157
11476	Phthalic acid, dipropyl ester or Dipropyl phthalate...... $1,2\text{-}C_6H_4(CO_2C_3H_7)_2$	250.30		304-5				al, eth	B9[3], 4101
11477	Phthalic acid, ethyl ester or Mono-ethylphthalate....... $2\text{-}HO_2CC_6H_4CO_2C_2H_5$	194.19		d	2	$1.1877^{22/4}$	1.509^{22}	al, eth	B9[3], 4099
11478	Phthalic acid, 3-hydroxy.................. $3\text{-}HOC_6H_3(CO_2H)_2\text{-}1,2$	182.13	nd, pr (eth-peth, w)	sub	150d			al, eth	B10[3], 2189
11479	Phthalic acid, 4-hydroxy.................. $4\text{-}HOC_6H_3(CO_2H)_2\text{-}1,2$	182.13	rosettes (w)		204-5d			al, eth	B10[3], 2190
11480	Phthalic acid mono amide or Phthalamic acid.......... $2\text{-}HO_2CC_6H_4\text{-}CONH_2$	165.15	pr		148-9			al	B9[3], 4191
11481	Phthalic acid mononitrile .... $2\text{-}NCC_6H_4CO_2H$	147.14	nd (al)		187d			ace, al	B9[3], 4199

No.	Name, Synonyms, and Formula	Mol. wt.	Color, crystalline form, specific rotation and λ_{max} (log ε)	b.p. °C	m.p. °C	Density	n_D	Solubility	Ref.
11482	Phthalic acid mononitrile, monoamide or Phthalamonitrile 2-NCC$_6$H$_4$CONH$_2$	146.15	nd (MeOH), cr (aa)	....	175	....	....	al, ace	B9³, 4199
11483	Phthalic acid, 3-nitro .... 3-NO$_2$C$_6$H$_3$(CO$_2$H)$_2$-1,2	211.13	pa ye pr (w)	....	218		....	al	B9³, 4215
11484	Phthalic acid, 3-nitro, diethyl ester 3-NO$_2$C$_6$H$_3$(CO$_2$C$_2$H$_5$)$_2$-1,2	267.24	pr (al), nd (peth)	....	46		....	al, eth	B9³, 4216
11485	Phthalic acid, 4-nitro .... 4-NO$_2$1,2-C$_6$H$_4$(CO$_2$H)$_2$	211.13	pa ye nd (w, eth)	....	165-6		....	al, w	B9³, 4234
11486	Phthalic acid, 4-nitro, dimethyl ester 4-NO$_2$C$_6$H$_3$(CO$_2$CH$_3$)$_2$-1,2	239.19	cr (dil al)	....	69-71		....	al, MeOH	B9², 607
11487	Phthalic acid, 2-octyl ester (D) or 2-Octyl phthalate (mono) .... 2-HO$_2$CC$_6$H$_4$CO$_2$[CH(CH$_3$)C$_6$H$_{13}$]	278.35	pr (peth), $[\alpha]^{20}_D$ + 48.7 (al)	....	75	1.027$^{72/4}$	....	al, bz, chl	B9³, 4112
11488	Phthalic acid, 2-octyl ester-mono (dl) 2-HO$_2$CC$_6$H$_4$CO$_2$[CH(CH$_3$)C$_6$H$_{13}$]	278.33	pr (peth)	....	55		....	al, ace, bz, chl	B9¹, 358
11489	Phthalic acid, 2-octyl ester-mono (l) 2-HO$_2$CC$_6$H$_4$CO$_2$[CH(CH$_3$)C$_6$H$_{13}$]	278.35	pr (peth)	....	75		....	al, bz, chl	B9³, 4112
11490	Phthalic acid, tetrabromo .... C$_6$Br$_4$(CO$_2$H)$_2$-1,2	481.74	nd (w)	....	266d (anh)		....		B9¹, 367
11491	Phthalic acid, tetrachloro .... C$_6$Cl$_4$(CO$_2$H)$_2$-1,2	303.92	pl (w)	2	250d (anh)		....	ace	B9³, 4205
11492	Phthalic acid, tetrachloro, monoethyl ester .... HO$_2$CC$_6$Cl$_4$(CO$_2$C$_2$H$_5$)	331.97	pr (dil al)	d 250	94-5		....	al, eth	B9³, 4206
11493	Phthalic anhydride .... 1,2-C$_6$H$_4$(CO)$_2$O	148.12	wh nd (al, bz)	295 sub	131.6		....	al	B17⁴, 6135
11494	Phthalic anhydride, 3-chloro .... 3-Cl-1,2-C$_6$H$_3$(CO)$_2$O	182.56	nd (sub)	sub	124-5		....		B17⁴, 6142
11495	Phthalic anhydride, 4-chloro .... 4-Cl-1,2-C$_6$H$_3$(CO)$_2$O	182.56	pr	294-5^{720}	98.5		....	al, eth	B17⁴, 6142
11496	Phthalic anhydride, 3,4-dichloro .... 3,4-Cl$_2$-1,2-C$_6$H$_2$(CO)$_2$O	217.01	pl	329	121		....	al, chl	B17⁴, 6142
11497	Phthalic anhydride, 3,5-dichloro .... 3,5-Cl$_2$-1,2-C$_6$H$_2$(CO)$_2$O	217.00	nd	....	89		....	bz, chl	B17⁴, 6143
11498	Phthalic anhydride, 3,6-dichloro .... 3,6-Cl$_2$-1,2-C$_6$H$_2$(CO)$_2$O	217.01	nd	339	194.5		....		B17⁴, 6143
11499	Phthalic anhydride, 4,5-dichloro .... 4,5-Cl$_2$-1,2-C$_6$H$_2$(CO)$_2$O	217.01	ta or pr (to)	313	187-9		....	al, eth, to	B17⁴, 6143
11500	Phthalic anhydride, 3-nitro .... 3-(NO$_2$)-1-,2-C$_6$H$_3$(CO)$_2$O	193.12	nd (aa, ace, al)	....	164		....	ace, al	B17⁴, 6149
11501	Phthalic anhydride, tetra bromo .... 1,2-C$_6$Br$_4$(CO)$_2$O	463.72	nd (aa-xyl)	....	279-81		....		B17⁴, 6147
11502	Phthalic anhydride, tetrachloro .... 1,2-C$_6$Cl$_4$(CO)$_2$O	285.90	pr, nd (sub)	sub	255-7		....		B17⁴, 6144
11503	Phthalic anhydride, tetra iodo .... 1,2-C$_6$I$_4$(CO)$_2$O	651.71	ye pr, nd (aa), nd (sub)	sub	327-8		....	al	B17⁴, 6149
11504	Phthalinide .... 1,2-C$_6$H$_4$(CO)$_2$NH	147.14	nd (w), pr (aa), lf (sub)	....	238		....	bz	B21⁴, 5017
11505	Phthalinide, N-acetyl .... 1,2-C$_6$H$_4$(CO)$_2$NCOCH$_3$	189.17	nd (bz), cr (al, aa)	....	135-6		....	eth, chl	B21⁴, 5171
11506	Phthalimide, N-benzyl .... 1,2-C$_6$H$_4$(CO)$_2$NCH$_2$C$_6$H$_5$	237.26	ye nd (al), cr (aa)	....	116	1.343^{18}	....	al	B21⁴, 5053
11507	Phthalimide, N(2-bromo-isobutyl) .... 1,2-C$_6$H$_4$(CO)$_2$N[CH$_2$CBr(CH$_3$)$_2$]	282.14	nd (al), lf (chl)	....	97		....	al, bz, chl	B21², 349
11508	Phthalimide, N(2-bromo ethyl) .... 1,2-C$_6$H$_4$(CO)$_2$NCH$_2$CH$_2$Br	254.10	nd (w)	....	82-4		....	eth	B21⁴, 5033
11509	Phthalimide, N-bromo methyl .... 1,2-C$_6$H$_4$(CO)$_2$NCH$_2$Br	240.06	pr (chl, bz, aa)	....	151.5		....	ace	B21⁴, 5110
11510	Phthalimide, N-(2-bromo propyl) .... 1,2-C$_6$H$_4$(CO)$_2$N(CH$_2$CHBrCH$_3$)	268.12	nd (al, MeOH)	....	110-1		....	al, eth	B21², 349
11511	Phthalimide, N-(3-bromo propyl) .... 1,2-C$_6$H$_4$(CO)$_2$N[(CH$_2$)$_3$Br]	268.12	nd (lig)	....	72-3		....	al, eth	B21⁴, 5033
11512	Phthalimide, N-isobutyl .... 1,2-C$_6$H$_4$(CO)$_2$N-i-C$_4$H$_9$	203.24	....	293-5	93		....		B21⁴, 5036
11513	Phthalimide, 3,6-dihydroxy .... 3,6-(HO)$_2$-1,2-C$_6$H$_2$(CO)$_2$NH	179.13	gr-ye nd (w + 3)	....	273-4		....	w	B21¹, 478

No.	Name, Synonyms, and Formula	Mol. wt.	Color, crystalline form, specific rotation and λ_{max} (log ε)	b.p. °C	m.p. °C	Density	n_D	Solubility	Ref.
11514	Phthalimide, N-ethyl .... 1,2-$C_6H_4(CO)_2NC_2H_5$	175.19	nd (al)	285.6	79			eth	B21[4], 5032
11515	Phthalimide, N-(2 hydroxy ethyl) .... 1,2-$C_6H_4(CO)_2NCH_2CH_2OH$	191.19	nd (al), lf (w)	....	129.5				B21[4], 5063
11516	Phthalimide, N-(hydroxy methyl) .... 1,2-$C_6H_4(CO)_2NCH_2OH$	177.16	lf, pr (to)		141-2			to	B21[4], 5108
11517	Phthalimide, N-methyl .... 1,2-$C_6H_4(CO)_2NCH_3$	161.16	nd (al), lf (sub)	285-7	134				B21[4], 5030
11518	Phthalimide, N-α-naphthyl .... 1,2-$C_6H_4(CO)_2N$-α-$C_{10}H_7$	273.30	pr, pl (al-aa)		180-1				B21[4], 5059
11519	Phthalimide, N-β-naphthyl .... 1,2-$C_6H_4(CO)_2N$-β-$C_{10}H_7$	273.30	nd (aa)		216			al, aa	B21[4], 5060
11520	Phthalimide, 4-nitro .... 4-NO_2-1,2-$C_6H_3(CO)_2NH$	192.13	col nd (w), ye lf (al-ace)		202			ace, aa	B21[2], 373
11521	Phthalimide, N-(4-nitrophenyl) .... 1,2-$C_6H_4(CO)_2N$-($C_6H_4NO_2$-4)	268.24	cr (aa)		271-2				B21[4], 5049
11522	Phthalimide, N-phenyl .... 1,2-$C_6H_4(CO)_2NC_6H_5$	223.23	wh nd (al)	sub	210			chl	B21[4], 5047
11523	Phthalonitrile .... 1,2-$C_6H_4(CN)_2$	128.14	nd (w, lig)		141			al, eth, ace, bz	B9[3], 4199
11524	Phthalonitrile, 3,6-dihydroxy .... 3,6-$(HO)_2C_6H_2(CN)_2$-1,2	160.13	yesh lf (w + 2)		230			al, eth	B10[3], 2430
11525	Phthalyl alcohol or o-Xylylene glycol......... 1,2-$C_6H_4(CH_2OH)_2$	138.17	pl (alh, peth)	....	66			w, al, eth	B6[3], 4587
11526	Phthalyl chloride .... 1,2-$C_6H_4(COCl)_2$	203.03		281.1, 131-3[9]	15-6	$1.4089^{20/4}$	1.5684^{20}		B9[3], 4190
11527	Phthalyl fluoride .... 1,2-$C_6H_4(COF)_2$	170.12		227-8, 84[13]	42-3			peth	B9[3], 4190
11528	Phthalide or α-Hydroxy-o-toluic acid lactone..... $C_8H_6O_2$	134.14	nd or pl (w)	290	75	$1.1636^{99/4}$	1.536^{99}	al, eth	B17[4], 4948
11529	Phthalide, 3-benzylidene (trans) .... $C_{15}H_{10}O_2$	222.25	mcl pr		108			al	B17[4], 5433
11530	Phthalide, 3,3-diphenyl or Phthalophenone........... $C_{20}H_{14}O_2$	286.33	lf (al)	235[15]	120				B17[4], 5561
11531	Phthalide, 6-nitro .... $C_8H_5NO_4$	179.13	ye nd (al, aa)		145			al, eth, bz, aa	B17[4], 4953
11532	Phthalocyanine or Tetrabenzoporphyrazine .... $C_{32}H_{18}N_8$	514.55	grsh-bl mcl (guino-line)	sub 55d (vac)					J1938, 1151
11533	Physostigmine or Eserine .... $C_{15}H_{21}N_3O_2$	275.35	orth pr (eth, bz), $[\alpha]^{17}_D$ -82 (chl, c=1.3)		(i) 105-6 (st), (ii) 86-7 (unst)			al, eth, bz, chl	B23[2], 330
11534	Physostigmine, 2-hydroxybenzoate or Eserin salicylate ... $C_{22}H_{27}N_3O_5$	413.48	pr (al)		185-7			al, chl	B23[2], 332
11535	Physostigmine sulfate .... $(C_{15}H_{21}N_3O_2)_2.H_2SO_4$	648.79	dlq sc (ace-eth), $[\alpha]_D$ -130 (w)		140-2 (anh)			w, al, ace	B23[2], 332
11536	Phytadiene-(d) or 2,6,10,14-Tetramethyl-13,15-hexacadiene........... $C_{20}H_{38}$	278.53	$[\alpha]$ +0.89 (undil)	186-8[14]		$0.826^{0/4}$		peth, aa, MeOH	B1[4], 1078
11537	Phytol (dl) or 3,7,11,15-Tetramethyl-2-hexadecene-1-ol .. $C_{20}H_{40}O$	296.54		202-4[10], 140-1[0.03]		$0.8497^{25/4}$	1.4595^{25}		B1[4], 2208
11538	Picein or p-Hydroxyacetophenone-D-glucoside........ $C_{14}H_{18}O_7$	298.30	nd (w + 1), nd (MeOH), $[\alpha]_D$ -86.5		195-6			al, eth, al	B17[4], 3013
11539	Picene or 3,4-Benzochrysene .... $C_{22}H_{14}$	278.36	lf, pl (xyl, Py, sub)	518-20, sub 300[2]	367-9				B5[3], 2555
11540	α-Picoline or 2-Methyl pyridine .... 2-$CH_3(C_5H_4N)$	93.13		128.8	-66.8	$0.9443^{20/4}$	1.4957^{20}	w, al, eth, ace	B20[4], 2679
11541	α-Picoline, 6-amino or 6-Amino-2-methylpyridine .... 6-NH_2-2-$CH_3(C_5H_3N)$	108.15	hyg (lig)	208-9	41			w, al, eth, ace, bz	B22[4], 4133
11542	α-Picoline, 6-dimethyl amino or 6-Dimethyl amino-2-methyl pyridine .... 6-$(CH_3)_2N$-2-$CH_3(C_5H_3N)$	136.20		198-200, 88[15]				al, eth	B22[4], 4134

No.	Name, Synonyms, and Formula	Mol. wt.	Color, crystalline form, specific rotation and λ_{max} (log ε)	b.p. °C	m.p. °C	Density	n_D	Solubility	Ref.
11543	α-Picoline, 4-ethyl 4-C_2H_5-2-$CH_3(C_5H_3N)$	121.18	....	179	....	$0.9130^{25/4}$	....	w, al, eth, ace, bz	B20[4], 2798
11544	α-Picoline, 5-ethyl or 5-ethyl-2-methyl pyridine 5-C_2H_5-2-$CH_3(C_5H_3N)$	121.18	....	178.3, 65-6[17]	....	$0.9219^{20/20}$	1.4971^{20}	al, eth, ace, bz	B20[4], 2798
11545	α-Picoline, 6-ethyl 6-C_2H_5-2-$CH_3(C_5H_3N)$	121.18	....	160-1, 73-6[12]	....	$0.9207^{25/4}$	1.4920^{25}	al, eth, ace	B20[4], 2803
11546	β-Picoline or 3-Methylpyridine 3-$CH_3(C_5H_4N)$	93.13	....	144.1	−18.3	$0.9566^{20/4}$	1.5040^{20}	**w, al, eth**, ace	B20[4], 2710
11547	β-Picolene, 2-amino or 2, Amino-3-methyl pyridine 2-NH_2-3-$CH_3(C_5H_3N)$	108.15	hyg	221.5[748], 95[8]	33.5	....	....	w, al, eth, ace, bz	B22[4], 4154
11548	γ-Picoline or 4-Methylpyridine 4-$CH_3(C_5H_4N)$	93.13	....	144.9	3.6	$0.9548^{20/4}$	1.5037^{20}	**w, al, eth**, ace	B20[4], 2732
11549	γ-Picoline, 2-amino or 2-Amino-4-methyl pyridine 2-NH_2-4-$CH_3(C_5H_3N)$	108.15	lf or pl (lig)	115-7[11]	100	....	....	w, al, eth, ace, bz	B22[4], 4172
11550	γ-Picoline, 3-Amino or 3-Amino-4-methyl pyridine 3-NH_2-4-$CH_3(C_5H_3N)$	108.15	pr (bz-peth)	254[735]	106	....	....	w, al, eth, ace, bz	B22[4], 4181
11551	γ-Picoline-2-ethyl or 2-Ethyl-4-methylpyridine 2-C_2H_5-4-$CH_3(C_5H_3N)$	121.18	....	173-5[748]	....	$0.9239^{20/0}$	....	al, eth, ace	B20[4], 2798
11552	γ-Picoline-3-ethyl or β-Collidine 3-C_2H_5-4-$CH_3(C_5H_3N)$	121.18	....	198, 76[12]	....	$0.9286^{17/4}$	....	al, eth, ace, chl	B20[4], 2804
11553	Picolinamide or 2-Pyridine carboxamide 2-$(C_5H_4N)CONH_2$	122.13	mcl pr (w)	....	107-8	....	....	al, bz	B22[4], 311
11554	Picolinic acid or 2-Pyridine carboxylic acid 2-$(C_5H_4N)CO_2H$	123.11	nd (w, al, bz)	sub	136-7	....	....	al, aa	B22[4], 303
11555	Picolinic acid, ethyl ester or Ethyl picolinate 2-$(C_5H_4N)CO_2C_2H_5$	151.17	ye in air	243, 122[13]	0-2	1.1194^{20}	1.5104^{20}	**w, al, eth**	B22[4], 308
11556	Picolinic acid, N-Methyl or Trigonelline $C_7H_7NO_2$	137.14	pr (aq al + w)	....	218d (anh), 130 (+1w)	....	....	w	B22[4], 462
11557	Picolinic acid, 2,3,6-tetrahydro or Baikiain 2-$(C_5H_8N)CO_2H$	127.14	pr (MeOH), $[\alpha]^{20/}_D$ −201.6	....	274d	....	....	w	B22[4], 182
11558	Picolino nitrile 2-$(C_5H_4N)CN$	104.11	nd or pr (eth)	222-7	29	$1.0810^{25/4}$	1.5242^{25}	w, al, eth, bz chl	B22[4], 320
11559	Picric acid or 2,4,6-Trinitro phenol 2,4,6-$(NO_2)_3C_6H_2OH$	229.11	ye lf (w), pr (eth), pl (al)	sub exp >300	122-3	....	1.763	al, eth, ace, bz, aa, Py	B6[4], 1388
11560	Picrolonic acid $C_{10}H_8N_4O_5$	264.21	ye nd (al)	....	116-7, (d 125)	....	....	....	B24[2], 25
11561	Picropodophyllin $C_{22}H_{22}O_8$	414.42	col nd (al, bz)	....	228	....	....	al, eth, ace, bz, chl	B19[4], 5298
11562	Picrotoxin or Cocculin $C_{30}H_{34}O_{13}$	602.60	rh lf, $[\alpha]^{16/}_D$ −29.3 (abs al, c=4)	....	203-4	....	....	al, Py	B19[4], 5245
11563	Pilocarpidine (d) $C_{10}H_{14}N_2O_2$	194.24	syr, $[\alpha]^{20/}_D$ +81.3 (w, al)	....	....	....	....	w, al	B27[2], 694
11564	Pilocarpidine nitrate $C_{10}H_{14}N_2O_2.HNO_3$	257.25	pr (w), $[\alpha]^{20/}_D$ +73.2 (w)	....	137	....	....	w, al	B27[2], 694
11565	Pilocarpine $C_{11}H_{16}N_2O_2$	208.26	nd $[\alpha]^{20/}_D$ +100.5 (w)	260[5]	34	....	....	w, al, chl	B27[2], 694
11566	Pilocarpine hydrochloride $C_{11}H_{16}N_2O_2.HCl$	244.72	hyg cr	....	204-5	....	....	w, al	B27[2], 695
11567	Pilocarpine, 2-hydroxybenzoate or Pilocarpine salicylate $C_{18}H_{22}N_2O_5$	346.49	nd or lf (al), $[\alpha]'_D$ +63	....	120	....	....	w, al, eth	B27[2], 696
11568	Pilocarpine nitrate $C_{11}H_{16}N_2O_2.HNO_3$	271.28	wh pw or cr (al), $[\alpha]_D$ +80 (w, c=4)	....	178	....	....	w	B27[2], 695

PHYSICAL CONSTANTS OF ORGANIC COMPOUNDS (Continued)

No.	Name, Synonyms, and Formula	Mol. wt.	Color, crystalline form, specific rotation and λ_{max} (log ε)	b.p. °C	m.p. °C	Density	n_D	Solubility	Ref.
11569	Pilocarpine, sulfate .... 2($C_{11}H_{16}N_2O_2$)$_2$,H_2SO_4	514.60	hyg cr (al-eth), $[\alpha]^{20}_D$ +85 (w, c=7)	....	132	....		w, al	B27, 635
11570	Pimaric acid (d) or Dextropimaric acid .... $C_{20}H_{30}O_2$	302.46	orh (ace), pr (al) $[\alpha]^{20}_D$ +87.3 (chl)	282[18]	218-9	....		al, eth, Py	B9[3], 2911
11571	Pimelic acid or Heptanedioic acid .... $HO_2C(CH_2)_5CO_2H$	160.17	pr (w)	272[100] (sub), 212[10]	106	1.329[15]	....	w, al, eth	B2[4], 2003
11572	Pimelic acid, diethyl ester or Diethyl pimelate .... $C_2H_5O_2C(CH_2)_5CO_2C_2H_5$	216.28	....	252-5[745], 139-41[15]	−24	0.9945[20]	1.4305[20]	al, eth	B2[4], 2004
11573	Pimelic acid, dimethyl ester or Pimethyl pamelate .... $CH_3O_2C(CH_2)_5CO_2CH_3$	188.23	....	120[10], 80[1]	−21	1.0625[20/4]	1.4309[20]	al, eth	B2[4], 2004
11574	Pimelic acid, ethyl ester, mono .... $HO_2C(CH_2)_5CO_2C_2H_5$	188.23	cr (eth)	182[18], 160[4]	10	....	1.4415[20]	....	B2[3], 1742
11575	Pimelic acid, 4-oxo .... $OC(CH_2CH_2CO_2H)_2$	174.16	rh pl (w)	....	143	....		al, w	B3[3], 1380
11576	Pimelonitrile or 1,5-Dicyanopentane .... $NC(CH_2)_5CN$	122.17	....	175[14]	−31.4	0.949[18]	1.4472[20]	al, eth, chl	B2[4], 2006
11577	Pimelyl chloride or Heptanedioyl chloride .... $ClOC(CH_2)_5COCl$	197.06	....	137[15]					B2[4], 2005
11578	Pinacolone or 3,3 Dimethyl-2-butanone .... $(CH_3)_3CCOCH_3$	100.16	....	106	−49.8	0.8012[20/4]	1.3952[20]	al, eth, ace	B1[1], 634
11580	Pinacolyl alcohol or 3,3-Dimethyl-2-butanol .... $(CH_3)_3CCH(OH)CH_3$	102.18	....	120.14	5.6	0.8122[25]	1.4148[20]	al, eth	B1[4], 1727
11581	Pinane (d-cis) .... $C_{10}H_{18}$	138.26	$[\alpha]^{20}_D$ +23.3 (undil)	169, 60.1[10]	−53	0.8560[20/4]	1.4629[20]	eth, bz	B3[1], 310
11582	Pinane (dl) or Pincocamphane .... $C_{10}H_{18}$	138.26	....	164-5	......	0.8551[20/4]	1.4609[20]	eth, bz	B5[1], 48
11583	Pinane (l, cis) .... $C_{10}H_{18}$	138.26	$[\alpha]^{20}_D$ −47 (un-dil)	167-8[757]	......	0.8556[21]	1.4645[20]	eth, bz	B5[4], 318
11584	α-Pinene (dl) .... $C_{10}H_{16}$	136.24	oil	156.2, 51.4[20]	−55	0.8582[20/4]	1.4658[20]	al, eth, chl	B5[4], 456
11585	β-Pinene (d) or Nopinene .... $C_{10}H_{16}$	136.24	$[\alpha]_D$ +28.6	164-6, 59.7[20]		0.8654[20/4]	1.4789[20]	al, eth, bz, chl	B5[4], 456
11586	β-Pinene (l) .... $C_{10}H_{16}$	136.24	$[\alpha]^{25}_D$ −21.5	164, 59.7[20]		0.8694[20/4]	1.4762[20]	al, eth, bz, chl	B5[4], 457
11587	Pinic acid (dl)(α-form) or 3-Carboxy-2,2-dimethyl cyclobutyl acetic acid .... $C_9H_{14}O_4$	186.21	wh pr (w)	214-6[9]	101-2	1.0925[109/4]	1.4458[109]	....	B9[3], 3852
11588	Pinic acid-(dl)(β form) .... $C_9H_{14}O_4$	186.21	....	214-6[9]	68				B9[3], 3852
11589	Pinic acid (dl)(c form) .... $C_9H_{14}O_4$	186.21	pl	214-6[9]	58				B9[3], 3852
11590	Pinic acid (l) .... $C_9H_{14}O_4$	186.21	nd (w)	....	135-6				B9[3], 3853
11591	Pinol (dl) or (dl)-Sobrerone-6,8-epoxy-1-p-menthene .... $C_{10}H_{16}O$	154.24	$[\alpha]_D$ −7.1 (ace)	183-4, 76-7[14]		0.9515[20/4]	1.4695[20]	al, eth	B17[4], 327
11592	Pinol hydrate (trans, dl) .... $C_{10}H_{16}O.H_2O$	170.25	pl or nd (w)	270-1, 157-8[12]	132			w, al, eth	B6[3], 4136
11593	Piperitenone or 3-Terpinolenone .... $C_{10}H_{14}O$	150.22	$[\alpha]_{546}$ −0.1	120-2[14]		0.9774[20]	1.5294[20]	al, eth	B7[3], 559
11594	Piperazine or Hexahydropyrazine .... $C_4H_{10}N_2$	86.14	hyg pl or lf (al)	146	106	1.446[113]		w, al	B23[2], 3
11595	Piperazine, N-benzyl .... $C_{11}H_{16}N_2$	176.26	....	145-7[12]			1.5430[28]	w, al, eth	Am72, 753
11596	Piperazine, N,N′-bis-(4 methoxybenzoyl) or N,N′-Dianisoylpiperazine .... $C_{20}H_{22}N_2O_4$	354.41	wh	....	192-4				Am56, 150
11597	Piperazine, N,N′-bis-(phenylacetyl) or N,N′-Di-(α-toluyl) piperazine .... $N,N′-(C_6H_5CH_2CO)_2(C_4H_8N_2)$	322.41	wh	....	150-1				Am56, 150

No.	Name, Synonyms, and Formula	Mol. wt.	Color, crystalline form, specific rotation and λ_{max} (log ε)	b.p. °C	m.p. °C	Density	n_D	Solubility	Ref.
11598	Piperazine, N,N'-bis-(3-phenyl propionyl) or N,N'-Bis-(hydrocinnanmyl)-piperazine N,N'-(C₆H₅CH₂CH₂CO)₂(C₄H₈N₂)	350.47	wh	. . . .	122-3	. . . .	. . . .	. . . .	Am56, 150
11599	Piperazine, dihydrobromide C₄H₁₀N₂.2HBr	247.97	wh nd	. . . .	d	. . . .	. . . .	w	J1957, 1881
11600	Piperazine, dihydrochloride, monohydrate C₄H₁₀N₂·2HCl·H₂O	177.07	nd (dil al)	. . . .	82-3	. . . .	. . . .	w	B23, 5
11601	Piperazine, N,N'-dimethyl or N,N'-Dimethylpiperazine C₆H₁₄N₂	114.19	. . . .	131-2[764]	. . . . . .	0.8600[20/4]	1.4474[20]	w, al, eth	B23², 5
11602	Piperazine, 2,5-dimethyl (cis) 2,5-(CH₃)₂C₄H₈N₂)	114.19	rh bipym nd or pr (chl)	162 sub	114	. . . .	1.4720[20]	w, al, chl	B23², 21
11603	Piperazine, 2,5-dimethyl (trans) 2,5-(CH₃)₂(C₄H₈N₂)	114.19	mcl pl or pr (bz, chl)	162 sub	118-9	. . . .	. . . .	w, al, chl	B23², 19
11604	Piperazine, 2,6-dimethyl (cis) 2,6-(CH₃)₂(C₄H₈N₂)	114.19	lf or pl (bz)	162	111-3	. . . .	. . . .	w, al, peth, chl	B23¹, 8
11605	Piperazine, N,N'-dinitroso or N,N'-Dinitrosopiperazine C₄H₈N₄O₂	144.14	pa ye pl (w)	. . . .	158	. . . .	. . . .	al	B23¹, 7
11606	Piperazine, N,N'-diphenyl C₁₆H₁₈N₂	238.33	. . . .	230-5[112], dec 300	164	. . . .	. . . .	. . . .	B23², 5
11607	Piperazine, N-ethyl C₆H₁₄N₂	114.19	. . . .	154[753]	. . . . . .	. . . .	. . . .	w, al, eth	B23², 5
11608	Piperazine hexahydrate C₄H₁₀N₂·6H₂O	194.23	. . . .	125-3	44-5	. . . .	. . . .	w, al	CAS47, 10014
11609	Piperazine, N-methyl N-CH₃(C₄H₉N₂)	100.17	. . . .	138	. . . . . .	. . . .	1.4378[20]	w, al, eth	C51, 10538
11610	Piperazine, 2-methyl 2-CH₃(C₄H₉N₂)	100.17	hyg lf (al)	155[763]	62	. . . .	. . . .	w, al, eth, bz, chl	B23², 16
11611	Piperazine, N-phenyl N-C₆H₅(C₄H₉N₂)	162.24	pa ye oil	286.5, 156-7[10]	. . . . . .	1.0621[20/4]	1.5875[20]	al, eth	C49, 11662
11612	Piperazine, 1,2,4-trimethyl 1,2,4-(CH₃)₃C₄H₇N₂	128.22	. . . .	149-50	. . . . . .	. . . .	1.4433[20]	. . . .	. . . .
11613	N-Piperazine carboxaldehyde N-OHC(C₄H₉N₂)	114.15	. . . .	94-7[0.5]	. . . . . .	. . . .	1.5094[20/D]	. . . .	. . . .
11614	N-Piperazine carboxylic acid, ethyl ester C₇H₁₄N₂O₂	158.20	. . . .	237, 116-7[12]	. . . . . .	. . . .	1.4760[25]	w, al, eth	B23², 9
11615	N,N'-Piperazine dicarboxylic acid-diethyl ester N,N'-(C₂H₅O₂C)₂C₄H₈N₂	230.27	nd (hex)	315, 131-3[3]	49	. . . .	. . . .	al, eth	B23², 9
11616	2,5-Piperazinedione C₄H₆N₂O₂	114.11	ta or pl (w)	. . . .	318-20d, sub 260	. . . .	. . . .	al	B24², 141
11617	Piperidine or hexahydropyridine C₅H₁₀NH	85.15	. . . .	106, 17.7[20]	-9	0.8606[20/4]	1.4530[20]	w, al, eth, ace, bz, chl	B20⁴, 287
11618	Piperidine, N-acetyl N-CH₃CO(NC₅H₁₀)	127.19	. . . .	226-7, 109[18]	. . . . . .	1.011[9]	1.4790[25]	w, al	B20⁴, 965
11619	Piperidine, 2-allyl 2-CH₂=CHCH₂(C₅H₉NH)	125.22	. . . .	. . . .	170-1	0.8823[15/4]	. . . .	. . . .	B20, 147
11620	Piperidine, N-benzoyl C₆H₅CO(NC₅H₁₀)	189.26	tcl	320-1, 180[15]	49	. . . .	. . . .	al, eth	B20⁴, 972
11621	Piperidine, 4-benzoyl-N-methyl C₁₃H₁₇NO	203.29	nd	160-3[13], 130-7[2]	35-7	. . . .	1.5430[23]	al, eth, ace, bz	B21⁴, 3689
11622	Piperidine, 4-benzyl 4-C₆H₅CH₂(C₅H₉NH)	175.27	. . . .	279, 150-2[17]	6-7	0.9972[20/0]	1.5337[25]	al, eth	B20⁴, 3055
11623	Piperidine, N-butyl C₉H₁₉(NC₅H₁₀)	141.26	. . . .	175-7, 47-8[20]	. . . . . .	0.8245[20/4]	1.4467[20]	. . . .	B20⁴, 311
11624	Piperidine, 2-butyl-(d) 2-C₄H₉(C₅H₉NH)	141.26	[α]'_D +15.7	. . . .	. . . . . .	0.8512	. . . .	. . . .	B20, 127
11625	Piperidine, 2-butyl (dl) 2-C₄H₉(C₅H₉NH)	141.26	. . . .	191-3, 75[14]	. . . . . .	0.8529[15/4]	. . . .	. . . .	Am20⁴, 1633
11626	Piperidine, 2-butyl (l) 2-C₄H₉(C₅H₉NH)	141.26	[α]¹⁶/_D -18.7	. . . .	. . . . . .	0.8533	. . . .	. . . .	B20, 127
11627	Piperidine, 3-butyl 3-C₄H₉(C₅H₉NH)	141.26	. . . .	196-7	. . . . . .	. . . .	. . . .	al, eth, chl	B20⁴, 1635
11628	Piperidine, N-isobutyl i-C₄H₉(NC₅H₁₀)	141.26	. . . .	160-1, 87[43]	. . . . . .	0.8161[25]	1.4382[25]	w	B20⁴, 314
11629	Piperidine, 2-isobutyl 2-i-C₄H₉(C₅H₉NH)	141.26	. . . .	181-2	. . . . . .	0.8510[22/4]	1.4553[22]	al, eth	B20⁴, 1636

No.	Name, Synonyms, and Formula	Mol. wt.	Color, crystalline form, specific rotation and λ_{max} (log ε)	b.p. °C	m.p. °C	Density	n_D	Solubility	Ref.
11630	Piperidine, N-sec-butyl (dl) $CH_3CH_2CH(CH_3)(NC_5H_{10})$	141.26		175-6		$0.8378^{20/4}$	1.4506^{20}	al, eth	B20[4], 313
11631	Piperidine, N-sec-butyl (l) $CH_3CH_2CH(CH_3)(NC_5H_{10})$	141.26	$[\alpha]^{25}/_D$ −54.6	175		$0.835^{25/4}$	1.4486^{21}	al, eth	B20[4], 313
11632	Piperidine, N-tert-butyl $(CH_3)_3C(NC_5H_{10})$	141.26		166		$0.8465^{20/4}$	1.4532^{20}	al, eth	B20[4], 315
11633	Piperidine, 2,4-diethyl $2,4-(C_2H_5)_2(C_5H_8NH)$	141.26		174-9		0.8722^{0}			B20, 128
11634	Piperidine, 2,5-diethyl $2,5(C_2H_5)_2(C_5H_8NH)$	141.26		190, 100-5²²		0.8722^{0}		eth, chl	B20, 128
11635	Piperidine, 3,4-diethyl (cis) $3,4-(C_2H_5)_2(C_5H_8NH)$	141.26	$[\alpha]^{22}/_D$ + 26.0 (90% al, c=4.35)	70¹²					B20[4], 1637
11636	Piperidine, 3,4-diethyl (trans) $3,4-(C_2H_5)_2(C_5H_8NH)$	141.26		193⁷²⁰					B20[4], 1637
11637	Piperidine, N, 2-dimethyl (d) or N-Methyl-α -pipecoline $N-2-(CH_3)_2(NC_5H_9)$	113.20	$[\alpha]^{15}/_D$ + 68.8 (undil)	127		0.825^{16}	1.4395^{20}		B20[4], 1444
11638	Piperidine, N,2-dimethyl-(dl) $N,2-(CH_3)_2(NC_5H_9)$	113.20		127.5		$0.824^{15/4}$	1.4395^{20}	w, al, eth	B20[4], 1444
11639	Piperidine, N, 3-dimethyl- (dl) or N-Methyl)-β -pipecoline $N, 3-(CH_3)_2(NC_5H_9)$	113.20		124-6		0.818^{15}			B20[4], 1499
11640	Piperidine, 2,3-dimethyl or α, β -Lupetidine $2,3-(CH_3)_2(C_5H_8NH)$	113.20		138-40⁷²⁰				w, al	B20[4], 1573
11641	Piperidine, 2,4-dimethyl (d) or α, γ -Lupetidine $2,4-(CH_3)_2(C_5H_8NH)$	113.20	$[\alpha]_D$ + 23.2	140-2		0.843			B20[4], 1574
11642	Piperidine, 2,4-dimethyl (dl) $2,4-(CH_3)_2(C_5H_8NH)$	113.20		140-2		0.8615^{0}	1.4366^{25}	al, eth	B20[4], 1574
11643	Piperidine, 2,5-dimethyl or α, β′ -Lupetidine $2,5-(CH_3)_2(C_5H_8NH)$	113.20		138-40			1.4452^{25}	al	B20[4], 1574
11644	Piperidine, 2,6-dimethyl or α, α′ -Lupetidine $2,6-(CH_3)_2(C_5H_8NH)$	113.20		127-8⁷⁶⁸		$0.8158^{25/4}$	1.4377^{20}	w, al, eth	B20[4], 1579
11645	Piperidine, 3,3-dimethyl or β, β′ -Lupetidine $3,3-(CH_3)_2(C_5H_8NH)$	113.20		137, 45-6²⁰			1.4452^{25}	al	B20[4], 1595
11646	Piperidine, 4,4-dimethyl or γ, γ -Lupetidine $4,4-(CH_3)_2(C_5H_8NH)$	113.20		145-6, 30-2¹²			1.4489^{25}	w	B20[4], 1596
11647	Piperidine, N-(2,2-dimethyl propyl) or N-Neopentyl piperdine $CH_3C(CH_3)_2CH_2(NC_5H_{10})$	155.29		188		$0.8608^{20/4}$	1.4593^{20}	al, eth	B20[2], 13
11648	Piperidine, N-dodecyl $C_{12}H_{25}(NC_5H_{10})$	253.48	pa ye	161⁵, 114-6⁰·⁶		$0.8378^{20/4}$	1.4588^{20}		B20[4], 321
11649	Piperidine, N-ethyl $C_2H_5(NC_5H_{10})$	113.20		130.8		$0.8237^{20/4}$	1.4480^{20}		B20[4], 307
11650	Piperidine, 2-ethyl (d) $2-C_2H_5(C_5H_9NH)$	113.20	$[\alpha]_D$ + 17.1		142-4	0.8680^{4}			B20[4], 1564
11651	Piperidine, 2-ethyl (dl) $2-C_2H_5(C_5H_9NH)$	113.20		142-3,73-5⁵²		$0.8650^{0/0}$	1.4494^{21}		B20[1], 28
11652	Piperidine, 2-ethyl (l) $2-C_2H_5(C_5H_9NH)$	113.20	$[\alpha]_D$ −14.9	143⁷²⁰		0.8680^{4}	1.4544^{20}		B20[4], 1564
11653	Piperidine, 3-ethyl (dl) $3-C_2H_5(C_5H_9NH)$	113.20	fum in air	152.6		$0.8565^{23/4}$	1.4531^{20}	ace	B20[4], 1567
11654	Piperidine, 3-ethyl (l) $3-C_2H_5(C_5H_9NH)$	113.20	$[\alpha]^{15}/_D$ −4.5	155				ace	B20[4], 1567
11655	Piperidine, 4-ethyl $4-C_2H_5(C_5H_9NH)$	113.20		156-8		0.8759^{0}	1.4503^{25}		B20[4], 1569
11656	Piperidine, N-heptyl $C_7H_{15}(NC_5H_{10})$	183.34		259.5, 100-3⁹		0.8316^{20}	1.4531^{20}		B20[4], 317
11657	Piperidine, N-hexyl $C_6H_{13}(NC_5H_{10})$	169.31		219.2, 103-4²⁰		$0.8292^{20/4}$	1.4522^{20}	al, eth	B20[4], 317
11658	Piperidine, 2-(1-hydroxyethyl) $2-[CH_3CH(OH)](C_5H_9NH)$	129.20		106-10¹⁸					B21[4], 85
11659	Piperidine, N-(2-hydroxyethyl) $HOCH_2CH_2(NC_5H_{10})$	129.20		200-2⁷⁴², 90¹²		$0.9732^{25/25}$	1.4749^{20}	w, al	B20[4], 387
11660	Piperidine, 2-(2-hydroxyethyl) $2-(HOCH_2CH_2)(C_5H_9NH)$	129.20		234.5, 145-6³⁶	39-40	1.01^{27}		w, al, eth	B21[4], 85
11661	Piperidine, 3-(2-hydroxyethyl) $3-(HOCH_2CH_2)(C_5H_9NH)$	129.20		121-3⁶		$1.0106^{25/4}$	1.4888^{25}	w, al, eth	B21[4], 92

No.	Name, Synonyms, and Formula	Mol. wt.	Color, crystalline form, specific rotation and λ_{max} (log ε)	b.p. °C	m.p. °C	Density	n_D	Solubility	Ref.
11662	Piperidine, 4-(2-hydroxyethyl) 4-(HOCH$_2$CH$_2$)(C$_5$H$_9$NH)	129.20	syr	227-8, 120-5[15]	132-3	1.0059[15/4]	1.4907[20]	w, al, eth	B21[4], 93
11663	Piperidine, N-methyl CH$_3$(NC$_5$H$_{10}$)	99.18	....	107	......	0.8159[20/4]	1.4355[20]	w, **al**, eth	B20[4], 305
11664	Piperidine, 2-methyl (d) or d-α-Pipecoline 2-CH$_3$(C$_5$H$_9$NH)	99.18	[α][22/]$_D$ + 5.6 (al)	117[745]	......	....	1.4459[20]	....	B20[4], 1444
11665	Piperidine, 2-methyl (dl) or α-Pipecoline 2-CH$_3$(C$_5$H$_9$NH)	99.18	....	117-8[747]	-4.9	0.8436[24/4]	1.4459[20]	w, al, eth	B20[4], 1444
11666	Piperidine, 3-methyl (dl) or β-Pipecoline 3-CH$_3$(C$_5$H$_9$NH)	99.18	....	125-6[763]	0-5	0.8446[26/4]	1.4470[20]	w	B20[4],1499
11667	Piperidine, 3-methyl (l) 3-CH$_3$(C$_5$H$_9$NH)	99.18	[α][25/]$_D$ -4	124	....	....	....	w	B20[4], 1498
11668	Piperidine, 4-methyl or γ-Pipecoline 4-CH$_3$(C$_5$H$_9$NH)	99.18	....	132-4	......	0.8674[0]	1.4458[20]	w	B20[4], 1511
11669	Piperidine, N-methyl-4-benzoyl C$_{13}$H$_{17}$NO	203.29	....	160-3[13]	35-7	....	1.5430[25]	al, eth, ace, bz	B21[4], 3689
11670	Piperidine, N-(2-methyl-2-pentyl) C$_{11}$H$_{23}$N	169.31	....	205-7	......	0.8517[20/4]	1.4592[20]	ace	B20[2], 14
11671	Piperidine, N-(3-methyl-3-pentyl) C$_{11}$H$_{23}$N	169.31	....	214[752]	......	0.8614[20/4]	1.4637[20]	....	B20[2], 14
11672	Piperidine, N-nitroso N-ON(C$_5$H$_{10}$N)	114.15	pa ye	217[721], 109[20]	......	1.0631[18.5/4]	1.4933[18.5]	w	B20[4], 1371
11673	Piperidine, N-nonyl C$_9$H$_{19}$(NC$_5$H$_{10}$)	211.40	....	135-7[11]	......	0.8313[25]	1.4538[25]	....	B20[4], 319
11674	Piperidine, N-octyl C$_8$H$_{17}$(NC$_5$H$_{10}$)	197.37	....	136-8[13], 89[1]	......	0.8324[20/4]	1.4544[20]	....	B20[4], 318
11675	Piperidine, N-pentyl C$_5$H$_{11}$(NC$_5$H$_{10}$)	155.29	....	198.2, 80[8]	......	0.8282[20]	1.4498[20]	....	B20[4], 315
11676	Piperidine, N-isopentyl i-C$_5$H$_{11}$(NC$_5$H$_{10}$)	155.29	....	188-9, 76-9[20]	......	....	....	ace	B20[4], 316
11677	Piperidine, N-phenyl C$_6$H$_5$(NC$_5$H$_{10}$)	161.25	....	257-8[752], 126.5[15]	......	....	....	al, eth, bz, chl	B20[4], 339
11678	Piperidine, N-phenyl-4-dimethyl amino or Irenal 4-(CH$_3$)$_2$N(C$_5$H$_9$NC$_6$H$_5$)	204.32	lf (bz)	123-6[0.5]	47-8	....	....	bz	B22[4], 3751
11679	Piperidine, 2-propenyl (d) or β-Coniceine 2-(CH$_2$=CHCH$_2$)C$_5$H$_9$NH	125.22	nd, [α][45/]$_D$ + 49.9	168-9	39	....	....	al, eth	B20, 146
11680	Piperidine, 2-propenyl-(dl) 2-(CH$_2$=CHCH$_2$)C$_5$H$_9$NH	125.22	nd	168-70[753]	8	0.8716[25/4]	....	al, eth	B20, 146
11681	Piperidine, 2-propenyl (l) 2-(CH$_2$=CHCH$_2$)C$_5$H$_9$NH	125.22	nd, [α][15/]$_D$ -50.5	168-9	39-40	0.8672[15/4]	....	....	B20, 146
11682	Piperidine, N-propyl C$_3$H$_7$(NC$_5$H$_{10}$)	127.23	....	151.2	......	0.8231[20]	1.4446[20]	w, al, eth	B20[4], 309
11683	Piperidine, 3-propyl (d) 3-C$_3$H$_7$(C$_5$H$_9$NH)	127.23	[α][16/]$_D$ + 5.9	174[752]	......	0.8517[19/4]	....	w, al	B20, 120
11684	Piperidine, 3-propyl (dl) 3-C$_3$H$_7$(C$_5$H$_9$NH)	127.23	....	174[758]	......	0.8475[26/4]	....	al, w	B20, 119
11685	Piperidine, 3-propyl (l) 3-C$_3$H$_7$(C$_5$H$_9$NH)	127.23	oil, [α][16/]$_D$ -6.6	174[752]	......	0.8517[19/4]	....	w, al	B20, 120
11686	Piperidine, 4-propyl 4-C$_3$H$_7$(C$_5$H$_9$NH)	127.23	....	172[748]	......	0.864[22]	1.4465[23]	al	B20[4], 1616
11687	Piperidine, N-isopropyl (CH$_3$)$_2$CH(NC$_5$H$_{10}$)	127.23	....	149-50[757]	......	0.8389[20/4]	1.4491[20]	w	B20[4], 311
11688	Piperidine, 2-isopropyl (dl) 2-(CH$_3$)$_2$CH(C$_5$H$_9$NH)	127.23	....	162	......	0.8668[0]	....	....	B20[4], 1620
11689	Piperidine, 2-isopropyl (l) 2-(CH$_3$)$_2$CH(C$_5$H$_9$NH)	127.23	[α]$_D$ -13.1	161.5	......	0.8503[19]	....	....	B20, 121
11690	Piperidine, 4-isopropyl 4-(CH$_3$)$_2$CH(C$_5$H$_9$NH)	127.23	....	168-71, 66-70[15]	......	....	....	....	B20[4], 1620
11691	Piperidine, 2,2,6,6-tetramethyl 2,2,6,6-(CH$_3$)$_4$(C$_5$H$_6$NH)	141.26	....	155-7	......	0.8367[16/4]	1.4455[20]	eth	B20[4], 1639
11692	Piperidine, 2,2,4-trimethyl 2,2,4-(CH$_3$)$_3$(C$_5$H$_7$NH)	127.23	....	148	......	0.832[15]	1.4458[20]	al, eth	B20[4], 1624
11693	Piperidine, 2,3,6-trimethyl 2,3,6-(CH$_3$)$_3$(C$_5$H$_7$NH)	127.23	....	36[5]	......	0.8302[20/4]	1.4434[20]	....	B20[4], 1625
11694	Piperidine, 2,4,6-trimethyl 2,4,6-(CH$_3$)$_3$(C$_5$H$_7$NH)	127.23	....	165-6	......	0.8315[19/4]	1.4412[20]	**al, eth**	B20[4], 1625
11695	N-Piperidine carboxaldehyde or N-Formylpiperidine (C$_5$H$_{10}$N)CHO	113.16	pa ye	222, 106-10[17]	......	1.0205[25/4]	1.4700[20]	w, al, eth, bz, chl, lig	B20[4], 964

No.	Name, Synonyms, and Formula	Mol. wt.	Color, crystalline form, specific rotation and λ_{max} (log ε)	b.p. °C	m.p. °C	Density	n_D	Solubility	Ref.
11696	4-Piperidine carboxylic acid or Hexahydroisonicotinic acid 4-(HNC₅H₉)CO₂H	129.16	nd (w)	. . . .	ca 326	. . . .	. . . .	w	B22⁴, 128
11697	4-Piperidine carboxylic acid, ethyl ester 4-(HNC₅H₉)CO₂C₂H₅	157.22	col oil	100-1¹⁰			1.4591²⁰	w, al, eth, bz	B22², 128
11698	4-Piperidine carboxylic acid, methyl ester 4-(HNC₅H₉)CO₂CH₃	143.19	col oil	107-10²²			1.4635²⁵	w, al, eth, bz	B22⁴, 128
11699	4-Piperidine carboxylic acid, N-methyl, methyl ester N-CH₃(NC₅H₉)CO₂CH₃-4	157.22		96-100²⁰			1.4539²⁴	w, al, eth, bz	B22⁴, 131
11700	4-Piperidine carboxylic acid, N, methyl-4-phenyl,ethyl ester hydrochlo or Demerol hydrochloride N-CH₃(NC₅H₈)(C₆H₅)(CO₂C₂H₅)-4,4	283.80	cr (al)	. . . .	186-9			w, ace	B22⁴, 1004
11701	2,4-Piperidine dione, 3,3-diethyl or Piperidone C₉H₁₅NO₂	169.23	nd (w)		103-5			w, al	B21⁴, 4614
11702	2-Piperidone or 5-Amino pentanoic acid lactone C₅H₉NO	99.13	hyg	256, 137¹⁴	39-40			w, al, eth	B21⁴, 3170
11703	2-Piperidone, 3-hydroxy (dl) C₅H₉NO₂	115.13	nd or pr (AcOEt)		141-2			w, eth	B21⁴, 6022
11704	2-Piperidone, 5-hydroxy C₅H₉NO₂	115.13	(al, al-AcOEt)		145-6			w, al	B21⁴, 6023
11705	4-Piperidone hydrochloride C₅H₉NO.HCl	135.60	cr (+ ³/₂ al, al-eth)	. . . .	147-9			w	B21⁴, 3183
11706	4-Piperidone, 2,3,6,6-tetramethyl or Triacetone amine C₉H₁₇NO	155	rh pl (moist eth), nd (eth)	205	58			w, al, eth	B21⁴, 3278
11707	Piperine or 1-Piperylpiperidine C₁₇H₁₉NO₃	285.35	pr (AcOEt), pl or mel pr (al), cr (bz-lig)	. . . .	130-3			al, bz, chl, aa, Py	B20⁴, 1341
11708	Piperitone (d) or 1-Methyl-4-isopropyl-3-cyclohexenone C₁₀H₁₆O	152.24	yesh in air, [α]²⁰ᴅ +49.1 (undil)	222-30, 116-8²⁰	. . . . . .	0.9344²⁰/⁴	1.4843⁷⁰	eth	B7³, 326
11709	Piperitone- (dl) C₁₀H₁₆O	152.24		232-3, 113¹⁸		0.9331²⁰/⁴	1.4845²⁰	. . . .	B7³, 326
11710	Piperitone (l) C₁₀H₁₆O	152.24	[α]²⁰ᴅ -51.5 (un-dil)	235, 109-10¹⁵		0.9324²⁰/⁴	1.4848²⁰		B7³, 324
11711	Piperoin C₁₆H₁₂O₆	300.27	. . . .		120	. . . .		al, chl	B19⁴, 5958
11712	Piperolidine (dl) or δ -Coniceine C₈H₁₅N	125.22	. . . .	161.5⁷⁵⁰, 65-7¹⁰	. . . . . .	0.9012¹⁵/⁴	1.4748	al, eth	B20⁴, 1989
11713	Piperolidine (l) C₈H₁₅N	125.22	[α]²⁷ᴅ -7.9	158⁷²⁹, 65-7¹⁸		0.8976²⁰/⁴	1.4748²⁰	al, eth	B20⁴, 1989
11714	Piperonal or 3,4-Methylenedioxy benzaldehyde (3,4-CH₂O₂)C₆H₃CHO	150.14	wh-ye (w)	263, 140¹⁵	37	. . . .		al, eth, ace	B19⁴, 1649
11715	Piperonal oxime (3,4-CH₂O₂)C₆H₃CH=NOH	165.15			146			eth	B19⁴, 1666
11716	Piperonyl alcohol or 3,4-Methylenedioxy benzyl alcohol (3,4-CH₂O₂)C₆H₃CH₂OH	152.16	nd (peth)	157¹⁶	58	. . . .		al, eth, bz, chl	B19⁴, 734
11717	Piperonylic acid or 3,4-Methylene dioxy benzoic acid (3,4-CH₂O₂)C₆H₃CO₂H	166.14	nd (al), pr (sub)	sub	229-31			al, eth	B19⁴, 3493
11718	Piperonylic acid, ethyl ester or Ethyl piperonylate (3,4-CH₂O₂)C₆H₃CO₂C₂H₅	194.19	pr	285-6, 164-5¹¹	18.5			al, eth, peth	B19², 293
11719	Piperonylic acid, methyl ester or Methyl piperonylate (3,4-CH₂O₂)C₆H₃CO₂CH₃	180.17	nd or lf (peth)	273-4d	53			al, eth	B19⁴, 3493
11720	Piperonyl chloride (3,4-CH₂O₂)C₆H₃COCl	184.58	cr	155²⁵	80				B19⁴, 3497
11721	Piperylene or 1,3,-Pentadiene CH₃CH=CHCH=CH₂	68.13	. . . .	42	−87.5	0.6760²⁰/⁴	1.4301²⁰	al, eth, ace, bz	B1⁴, 994
11722	Pivaldehyde or 2,2-Dimethyl propionaldehyde (CH₃)₃CCHO	86.14	. . . .	77-8	6	0.7923¹⁷	1.3791²⁰	al, eth	B1⁴, 3295
11723	Pivaldehyde oxime or Pivaldoxime (CH₃)₃CCH=NOH	101.15		65²⁰	48			al, eth	B1³, 2824
11724	Pivalamide, N,N-diethyl (CH₃)₃CCON(C₂H₅)₂	157.26		203	. . . . . .	0.891¹⁵		al, eth	B4³, 211
11725	Pivalic acid or 2,2-Dimethyl propionic acid (CH₃)₃CCO₂H	102.13	nd	164, 70¹⁴	35	0.905⁵⁰	1.3931³⁰·⁵	al, eth	B2⁴, 908

No.	Name, Synonyms, and Formula	Mol. wt.	Color, crystalline form, specific rotation and λ_{max} (log ε)	b.p. °C	m.p. °C	Density	n_D	Solubility	Ref.
11726	Pivalic acid, chloro or 3-Chloro-2,2-dimethyl propionic acid . . . . ClCH$_2$C(CH$_3$)$_2$CO$_2$H	136.58	. . . .	108-12[10]	41-2	. . . .	. . . .	. . . .	B2[4], 914
11728	Pivalic acid, ethyl ester or Ethyl pivalate . . . . (CH$_3$)$_3$CCO$_2$C$_2$H$_5$	130.19		118	−89.5	0.856[20/4]	1.3906[20]	al, eth	B2[4], 910
11729	Pivalic acid, methyl ester or Methyl pivalate . . . . (CH$_3$)$_3$CCO$_2$CH$_3$	116.16		101	. . . . . .	0.891[0/4]	1.3880[20]	al, eth	B2[4], 909
11730	Pivalonitrile or tert-Butyl cyanide. . . . . (CH$_3$)$_3$CCN	83.13		105-6	15-6	0.7586[25/4]	1.3774[20]		B2[4], 913
11731	Pivalophenone or tert-Butyl phenyl ketone . . . . C$_6$H$_5$COC(CH$_3$)$_3$	162.23		219-21, 97-8[16]		0.963[26]	1.5086[19]	ace	B7[3], 1125
11732	Pivalyl chloride or 2,2-Dimethyl propionyl chloride . . . . . . (CH$_3$)$_3$CCOCl	120.58		107, 48[100]		1.003[20]	1.4139[20]	eth	B2[4], 912
11733	Pivalyl chloride-chloro . . . . ClCH$_2$C(CH$_3$)$_2$COCl	155.04		85-6[60]			1.4539[20]		B2[4], 914
11734	Podophyllotoxin C$_{22}$H$_{22}$O$_8$	414.41	[α]$^{20}_D$ -132.7 (CHCl$_3$, 2%)		114-8			al, ace, bz, chl	B19[4], 5299
11735	Polyglycolid [O-CH$_2$CO]$_n$	. . . . . .			223				B19[1], 679
11736	Polyporic acid C$_{18}$H$_{12}$O$_4$	292.29			310-2				B8[3], 3854
11737	Popalin or Salicinbenzoate (an α-glucoside) . . . . C$_{20}$H$_{22}$O$_8$	390.40	nd (w + 2), pr (al), [α]$_D$ -2.0 (Py, c=5)		180			eth, aa	B17[4], 3298
11738	Porphin or Tetramethene tetrapyrrole . . . . C$_{20}$H$_{14}$N$_4$	310.36	red or og lf (chl-MeOH)	sub 300[12]	darkens 360	1.336	. . . .	bz, diox	B26[2], 228
11739	5-α-Pregnane C$_{21}$H$_{36}$	288.52	[α]$^{19}_D$ (CHCl$_3$, c=1.69)		84-5				B5[4], 1215
11740	5α-Pregnane-20β -ol-3 one . . . . C$_{21}$H$_{34}$O$_2$	318.50	[α]$_D$ (20 CHCl$_3$, c=1.2)		185				B8[3], 614
11741	5β -Pregnane or 17β -Ethyletiocholane . . . . C$_{21}$H$_{36}$	288.52	mcl sc or pl (MeOH), [α]$^{19}_D$ + 21.2 (chl, c=2)		83.5	1.032[15/4]		chl	B5[4], 1215
11742	5β -Pregnane-3α, 20 α -diol . . . . C$_{21}$H$_{36}$O$_2$	320.52	pl (ace), [α]$^{20}_D$ + 27.4 (al, c=0.7)		243-4	1.15			B6[3], 4779
11743	5 β -Pregnane 3 α -20 β-diol . . . . C$_{21}$H$_{36}$O$_2$	320.52	cr (al), [α]$^{20}_D$ + 10		244-6			al	B6[3], 4779
11744	5 β -Pregnane-3 β -20 α -diol . . . . C$_{21}$H$_{36}$O$_2$	320.52	cr (al, ace)		182			al	B6[3], 4778
11745	5 β -Pregnane-3 β -20 β -diol . . . . C$_{21}$H$_{36}$O$_2$	320.52	cr (AcOEt-peth, dil al)		174-6			AcOEt	B6[3], 4778
11746	5 β -Pregnane-3,20-dione . . . . C$_{21}$H$_{32}$O$_2$	316.49	nd (dil al), cr (dil ace)		123			al, eth, ace	B7[3], 3568
11747	5 β -Pregnan-3 α -ol-20-one . . . . C$_{21}$H$_{34}$O$_2$	318.51	nd (bz), cr (dil al)		149.5			al	B8[3], 618
11748	5 β -Pregnan-3 β -ol-20-one . . . . C$_{21}$H$_{34}$O$_2$	318.51	cr (dil al)		149			al	B8[3], 617
11749	5 β -Pregnan-20 α -ol-3-one . . . . C$_{21}$H$_{34}$O$_2$	318.51	pr (ace)		152				B8[3], 614
11750	5 β-Pregnan-20 β -ol-3-one . . . . C$_{21}$H$_{34}$O$_2$	318.51	cr (dil MeOH)		172			MeOH	B8[3], 614
11751	5-Pregnen-3 β -ol-20-one . . . . C$_{21}$H$_{32}$O$_2$	316.49	nd (dil al), [α]$^{20}_D$ + 28 (al)		192				B8[3], 949

No.	Name, Synonyms, and Formula	Mol. wt.	Color, crystalline form, specific rotation and λ_{max} (log ε)	b.p. $^\circ C$	m.p. $^\circ C$	Density	n_D	Solubility	Ref.
11752	4-Pregnen-11 β, 17 α, 20 β, 21-tetrol-3-one or Reichstein's substance E .... $C_{21}H_{32}O_5$	364.49	cr (aq, ace)	....	ca 125d			al, ace	B8³, 4027
11753	4-Pregnen-17 α, 20 β, 21-triol-3-one $C_{21}H_{32}O_4$	348.49	cr (MeOH), $[\alpha]_D$ +63 (diox, c=1)		190			diox, chl, MeOH	B8³, 3533
11754	Prehnitene or 1,2,3,4-Tetramethyl benzene .... 1,2,3,4-$(CH_3)_4C_6H_2$	134.22	cr (peth)	205, 79.4¹⁰	-6.2	0.9052²⁰ᐟ⁴	1.5203²⁰	al, eth, ace, bz	B5⁴, 1072
11755	Primeverose or 6-O-β-D-xylopyranosyl-α-D-glucose......... $C_{11}H_{20}O_{10}$	312.28	cr (MeOH), $[\alpha]^{20}_D$ +23→-3.2 (w, c=5) (mut)		210			w, MeOH	B17⁴, 2447
11756	α -Progesterone or 17 α-Progesterone .... $C_{21}H_{30}O_2$	314.47	orh pr (dil al), $[\alpha]_D$ +192		129-31	1.166²³			B7³, 3648
11757	β -Progesterone or 17 β-Progesterone .... $C_{21}H_{30}O_2$	314.47	nd (peth), $[\alpha]^{20}_D$ +172 (diox, c=2)		121-2	1.171²⁰		al, ace, diox	B7³, 3648
11758	Progesterone dioxime .... $C_{21}H_{32}N_2O_2$	344.50	pl (dil al)	....	243			al	B7³, 3654
11759	Proline (D) or d-2-Pyrrolidene carboxylic acid .... $C_5H_9NO_2$	115.13	hyg pr (al-eth), $[\alpha]^{20}_D$ +81.9 (w)		215-20d	....		w, al	B22⁴, 8
11760	Proline -(DL) .... $C_5H_9NO_2$	115.13	hyg nd (al-eth), cr (+ w)		205d (anh)			w, al	B22⁴, 12
11761	Proline (L) .... $C_5H_9NO_2$	115.13	nd (al-eth), pr (w), $[\alpha]^{20}_D$ -80.9 (w, c=1)		220-2d			w	B22⁴, 8
11762	Proline, 4-hydroxy (D-cis) or Allo-4-hydroxyproline ..... $C_5H_9NO_3$	131.13	nd (w + 1), $[\alpha]^{18}_D$ +58.6 (w, p=5)		237-41			w	B22⁴, 2045
11763	Proline-4-hydroxy (DL, cis) or Allo -4-hydroxyproline ... $C_5H_9NO_3$	131.1	cr (w, dil al)	....	250			w	B22⁴, 2046
11764	Proline-4-hydroxy (L, cis) or Allo-4-hydroxyproline ..... $C_5H_9NO_3$	131.13	nd (w + 1), $[\alpha]^{18}_D$ -58.1 (w, p=5.2)		238-41			w, al	B22⁴, 2046
11765	Proline-4-hydroxy (D-trans) or α-4-Hydroxyproline...... $C_5H_9NO_3$	131.13	lf (dil al), $[\alpha]^{21}_D$ +75.2 (w)		274			w	B22⁴, 2047
11766	Proline-4-hydroxy (DL,trans) or α -4-Hydroxyproline .... $C_5H_9NO_3$	131.13	pl (MeOH)	....	270			w	B22⁴, 2049
11767	Proline-4-hydroxy-(L,trans) or α-4-Hydroxyproline...... $C_5H_9NO_3$	131.13	lf (dil al), pr (w), $[\alpha]_D$ -76.5 (w, c=2.5)		274			w	B22⁴, 2047
11768	Proline, 4-hydroxy, betaine (d) .... $C_7H_{13}NO_3$	159.19	pr (w + 1), $[\alpha]^{21}_D$ +36		249d				B22⁴, 2053
11769	Prontosil .... $C_{12}H_{14}N_5O_2ClS$	327.80	og-red pw		248-51			al, ace, oils	B16³, 439
11770	Propadiene or Allene............ $CH_2=C=CH_2$	40.07	gas	-34.5	-136	0.787	1.4168	bz, peth	B1⁴, 966
11771	Propanal or Propionaldehyde .... CH_3CH_2CHO	58.08	....	48.8	-81	0.8058²⁰ᐟ⁴	1.3636²⁰	w, al, eth	B1⁴, 3165
11772	Propane $CH_3CH_2CH_3$	44.11	....	-42.1	-189.7	0.5853⁻⁴⁵ᐟ⁴	1.2898²⁰	al, eth, bz, chl	B1⁴, 176
11773	Propane, 1-amino or n-Propyl amine............ $CH_3CH_2CH_2NH_2$	59.11	....	47.8	-83	0.7173²⁰ᐟ⁴	1.3870²⁰	w, al, eth, ace, bz, chl	B4⁴, 464

No.	Name, Synonyms, and Formula	Mol. wt.	Color, crystalline form, specific rotation and λ_{max} (log ε)	b.p. °C	m.p. °C	Density	n_D	Solubility	Ref.
11774	Propane, 1-amino-2,2-dimethyl or Neopentyl amine .... $(CH_3)_3CCH_2NH_2$	87.17	....	81-2[741]		0.7455[20/4]	1.4023[20]	eth	B4[4], 707
11775	Propane, 1-amino-3-dodecyloxy .... $C_{12}H_{25}O(CH_2)_3NH_2$	243.44	....	140[3]		0.8439[20/4]	1.4487[20]	ace, bz, chl, MeOH	C55, 22136
11776	Propane, 1-amino-3-methoxy .... $CH_3O(CH_2)_3NH_2$	89.14	....	116-9		0.8727[20/4]	1.4391[20]	w, al, bz, MeOH, chl	B4[4], 1623
11777	Propane, 2-amino or Isopropyl amine .... $(CH_3)_2CHNH_2$	59.11	....	32.4	−95.2	0.6891[20]	1.3742[20]	w, **al, eth**, ace, bz, chl	B4[4], 504
11778	Propane, 2,2-bis(4-aminophenyl) $(4-H_2NC_6H_4)_2C(CH_3)_2$	226.32	....	200-31[5.5]	128.2-9.6	....	....		B13[3], 495
11779	Propane, 1,3-bis (dimethylamino) $CH_2[CH_2N(CH_3)_2]_2$	130.23	....	144		0.7837[18/4]	....	w, al, eth	B4[4], 1259
11780	Propane, 1,1-bis(4-hydroxyphenyl) or 4,4′-Propylidenediphenol $CH_3CH_2CH(C_6H_4OH-4)_2$	228.29	nd (w)	275[20]	130	....	....	al, eth, aa	B6[3], 5457
11781	Propane, 2,2-bis(4-hydroxyphenyl) or Bis-phenol A -4,4′-isopropylidene diphenol .... $(4-HOC_6H_4)_2C(CH_3)_2$	228.29	pr (dil aa), nd (w)	250-2[13]	152-3	....	....	al, eth, bz, aa	B6[3], 5459
11782	Propane, 1-bromo or n-Propylbromide............. $CH_3CH_2CH_2Br$	123.00	....	71	−110	1.3537[20/4]	1.4343[20]	al, eth, ace, bz	B1[4], 205
11783	Propane, 1-bromo-2-chloro .... $CH_3CHClCH_2Br$	157.44	....	118		1.531[20/4]	1.4745[20]	al, eth, ace, bz	B1[4], 212
11784	Propane, 1-bromo-3-chloro .... $Cl(CH_2)_3Br$	157.44	....	143.3, 32.4[10]	−58.9	1.5969[20/4]	1.4864[20]	al, eth, chl	B1[4], 212
11785	Propane, 1-bromo 2,2-dimethyl or Neopentyl bromide.... $(CH_3)_3CCH_2Br$	151.05	....	106, 34.6[100]		1.1997[20/4]	1.4370[20]	al, eth, ace, bz, chl	B1[4], 337
11786	Propane, 1-bromo-3-fluoro .... $F(CH_2)_3Br$	140.99	....	101.4		1.542[25/4]	1.4290[25]	al, eth, bz, chl	B1[4], 210
11787	Propane, 1-bromo-1-nitro $CH_3CH_2CH(Br)NO_2$	168.00	....	160-5, 82.5[50]		....	....	al, eth	B1[3], 260
11788	Propane, 2-bromo or Isopropyl bromide .... $(CH_3)_2CHBr$	123.00	....	59.4	−89	1.3140[20/4]	1.4251[20]	**al, eth**, ace, bz, chl	B1[4], 208
11789	Propane, 2-bromo-1-chloro .... $CH_3CHBrCH_2Cl$	157.44	....	118[756]		1.537[20/4]	1.4795[20]	al, eth, ace, bz, chl	B1[4], 212
11790	Propane, 2-bromo-2-chloro .... $(CH_3)_2CClBr$	157.44	....	93-5[745]		1.474[22]	1.4575[20]	al, eth, ace, bz, chl	B1[4], 213
11791	Propane, 3-bromo-1,2-epoxy (d) or Epibromohydrin..... $BrCH_2CHCH_2$	136.98	$[\alpha]^{16}_D$ +45.4 (undil)	134-6[50]		....	....	eth, bz, chl	B17[4], 22
11792	Propane, 3-bromo-1,2-epoxy (dl) or Epibromohydrin .... $BrCH_2CHCH_2$	136.98	....	138-40, 61-2[50]		1.615[14]	1.4841[20]	eth, bz, chl	B17[4] 22
11793	Propane, 2-bromo-2-nitro $(CH_3)_2CBrNO_2$	168.00	....	152[745], 73-5[50]		1.6562[0]	....	al, eth	B1[4], 233
11794	Propane, 2-(bromomethyl)-1,2,3-tribromo .... $(BrCH_2)_3CBr$	373.73	....	150-1[14]	25	2.5595[20/4]	1.6246[20]	eth, ace	B1[2], 91
11795	Propane, 1-chloro or n-Propyl chloride............. $CH_3CH_2CH_2Cl$	78.54	....	46.6	−122.8	0.8909[20/4]	1.3879[20]	**al, eth**, bz, chl	B1[4], 189
11796	Propane, 1-chloro-2,2, difluoro $CH_3CF_2CH_2Cl$	114.52	....	55	−56.2	1.2001[20/4]	1.3520[20]	eth, bz, chl	B1[4], 193
11797	Propane, 1-chloro-2,2-dimethyl or Neopentyl chloride ... $(CH_3)_3CCH_2Cl$	106.60	....	84.3	−20	0.8660[20/4]	1.4044[20]	al, eth, bz, chl	B1[4], 336
11798	Propane, 1-chloro-3-fluoro .... $F(CH_2)_3Cl$	96.53	....	79.5[740]		....	1.3871[25]	al, eth, bz, chl	B1[4], 193
11799	Propane, 1-chloro-3-iodo .... $ICH_2CH_2CH_2Cl$	204.44	....	170-2, 57.10		1.904[20]	1.5472[20]	eth, bz, chl	B1[4], 226
11800	Propane, 1-chloro-1-nitro $CH_3CH_2CH(Cl)NO_2$	123.54	....	141-3, 67[56]		1.209[20/20]	1.4251[20]	al, eth	B1[4], 232
11801	Propane, 1-chloro-2-nitro $CH_3CH(NO_2)CH_2Cl$	123.54	....	172-3, 94[46]		1.245[22]	1.4432[25]	al, eth, chl	B1[4], 233
11802	Propane, 1-chloro-3-nitro $ClCH_2CH_2CH_2NO_2$	123.54	....	197d, 115-6[40]		1.267[20]	....	al, eth	B1[3], 259
11803	Propane, 1-chloro-2-phenyl (dl) .... $CH_3CH(C_6H_5)CH_2Cl$	154.64	....	85[13]		1.0484[70]	1.5245[20]	....	B5[4], 992
11804	Propane, 1-chloro-3-phenyl $C_6H_5(CH_2)_3Cl$	154.64	....	219-20, 110[21]		1.056[21/4]	1.5160[25]	....	B5[4], 980
11805	Propane, 2-chloro or Isopropyl chloride............. $(CH_3)_2CHCl$	78.54	....	35.7	−117.2	0.8617[20/4]	1.3777[20]	**al, eth**, bz, chl	B1[4], 191

No.	Name, Synonyms, and Formula	Mol. wt.	Color, crystalline form, specific rotation and λ_{max} (log ϵ)	b.p. °C	m.p. °C	Density	n_D	Solubility	Ref.
11806	Propane, 2-chloro-1-nitro . . . . CH$_3$CHClCH$_2$NO$_2$	123.54	. . . .	172, 75[15]	. . . . . .	1.2361[15]	1.4447[20]	al, eth	B1[4], 232
11807	Propane, 2-chloro-2-nitro . . . . (CH$_3$)$_2$CClNO$_2$	123.54	. . . .	134d, 57[50]	. . . . . .	1.230[19]	1.4378[19]	al, eth	B1[4], 233
11808	Propane, 2-chloro-1-phenyl (d) . . . . CH$_3$CHClCH$_2$C$_6$H$_5$	154.64	[α]$_{4359}$, + 21.2	94[17]	. . . . . .	1.038[19/4]	1.5198[20]	ace, bz, chl	B5[4], 980
11809	Propane, 2-chloro-1-phenyl (dl) . . . . CH$_3$CHClCH$_2$C$_6$H$_5$	154.64	. . . .	79[10]	. . . . . .	1.0367[17/4]	1.5134[20]	ace, bz, chl	B5[4], 980
11810	Propane, 2-chloro-1-Phenyl (l) . . . . CH$_3$CHClCH$_2$C$_6$H$_5$	154.64	[α]$_{5461}$ -24.9	94[17]	. . . . . .	1.038[19/4]	1.5198[22]	ace, bz, chl	B5[4], 980
11811	Propane, 2-(chloromethyl)-1,1,2,3-tetrachloro (ClCH$_2$)$_2$CClCHCl$_2$	230.35	. . . .	226[737], 95[9]	. . . . . .	1.5686[25/4]	1.5165[25]	al, bz, chl	B1[3], 321
11812	Propane, 2-(chloro methyl)-1,2,3-trichloro . . . . . . (ClCH$_2$)$_3$CCl	195.91	. . . .	209-10[737] 87[9]	. . . . . .	1.5036[25/4]	1.508[20]	chl	B1[3], 321
11813	Propane, 3-chloro-1,2-epoxy (dl) or α-Epichlorohydrin . . ClCH$_2$CHCH$_2$	92.53	. . . .	116.5, 60-1[100]	−48	1.1801[20/4]	1.4361[20]	al, eth, bz	B17[4], 22
11814	Propane, 3-chloro-1,2,-epoxy (l) or α-Epichlorohydrin . . . ClCH$_2$CHCH$_2$	92.53	[α]^{18}/$_D$ -25.6	92-3[360]	. . . . . .	1.2007	. . . .	al, eth, bz	B17[2], 12
11815	Propane, 3-chloro-2-methyl-1,2-epoxy . . . . ClCH$_2$C(CH$_3$)CH$_2$	106.55	. . . .	122, 51[55]	. . . . . .	1.1011[20/4]	1.4340[20]	w, eth	B17[4], 47
11816	Propane, 1,2-diamino (d) or 1,2-Propanediamine . . . . . . CH$_3$CH(NH$_2$)CH$_2$NH$_2$	74.13	[α]^{25}/$_D$ + 29.8	120.5	. . . . . .	0.8584[25/4]	. . . .	w, chl	B4[4], 1255
11817	Propane, 1,2-diamino-N,N -diacetyl or 1,2-Diacetamido propane . . . . CH$_3$CH(NHCOCH$_3$)CH$_2$NHCOCH$_3$	158.20	nd (bz)	190[18]	138-9			w, al, chl	B4[3], 552
11818	Propane, 1,3-diamino or Trimethylene diamine . . . . . . H$_2$N(CH$_2$)$_3$NH$_2$	74.13	. . . .	135.5	. . . . . .	0.884[25/4]	1.4600[20]	w, al, eth	B4[4], 1258
11819	Propane, 1,1-dibromo or Propylidene bromide . . . . . . CH$_3$CH$_2$CHBr$_2$	201.90	. . . .	133.5, 20.4[11]	. . . . . .	1.982[20/4]	1.5100[20]	al, eth, chl	B1[4], 215
11820	Propane, 1,1-dibromo-2,2-dimethyl or Neopentylidene bromide . . . . (CH$_3$)$_3$CCHBr$_2$	229.95	. . . .	180, 66[20]	14	1.6695[20/4]	1.5047[20]	eth, bz, chl	B1[3], 371
11821	Propane , 1,2-dibromo . . . . CH$_3$CHBrCH$_2$Br	201.90	. . . .	140, 35.7[10]	−55.2	1.9324[20/4]	1.5201[20]	al, eth, chl	B1[4], 215
11822	Propane, 1,2-dibromo-2-methyl . . . . (CH$_3$)$_2$CBrCH$_2$Br	215.94	. . . .	149-50, 61[40]	9-12	1.759[20/4]	1.509	al, eth, bz, chl	B1[4], 298
11823	Propane, 1,3,-dibromo . . . . Br(CH$_2$)$_3$Br	201.90	. . . .	167.3, 56.6[20]	fr−34.2	1.9822[20/4]	1.5232[20]	al, eth	B1[4], 216
11824	Propane, 1,3-dibromo-2,2-dimethyl . . . . (CH$_3$)$_2$C(CH$_2$Br)$_2$	229.95	. . . .	185-90d, 72[14]	. . . . . .	1.6934[20/4]	1.5050[20]	al, eth, bz, chl	B1[4], 337
11825	Propane, 2,2-dibromo . . . . (CH$_3$)$_2$CBr$_2$	201.90	. . . .	114-5[740]	. . . . . .	1.7825[20/4]	. . . .	al, eth, chl	B1[4], 217
11826	Propane, 1,1-dichloro or Propylidene dichloride . . . . . . CH$_3$CH$_2$CHCl$_2$	112.99	. . . .	88.1	. . . . . .	1.1321[20/4]	1.4289[20]	al, eth, bz, chl	B1[4], 195
11827	Propane, 1,1-dichloro-2-methyl or Isobutylidene chloride	127.03	. . . .	105-6	. . . . . .	1.0111[12/12]	1.4330[15]	al, eth, bz, chl	B1[4], 292
	(CH$_3$)$_2$CHCHCl$_2$								
11828	Propane, 1,2-dichloro . . . . CH$_3$CHClCH$_2$Cl	112.99	. . . .	96.4, −3.7[10]	−100.4	1.1560[20/4]	1.4394[20]	al, eth, bz, chl	B1[4], 195
11829	Propane, 1,2-dichloro-2-fluoro . . . . CH$_3$CClFCH$_2$Cl	130.98	. . . .	88.6	−91.7	1.2624[20/4]	1.4099[20]	ace, bz	B1[4], 197
11830	Propane, 1,2-dichloro-2-methyl or Isobutylene chloride . . (CH$_3$)$_2$CClCH$_2$Cl	127.03	. . . .	108, 38-9[70]	. . . . . .	1.093[20/4]	1.4370[20]	al, eth, ace, bz	B1[4], 292
11831	Propane, 1,3-dichloro . . . . Cl(CH$_2$)$_3$Cl	112.99	. . . .	120.4, 14[10]	−99.5	1.1876[20/4]	1.4487[20]	al, eth, bz, chl	B1[4], 196
11832	Propane, 1,3-dichloro-2-methyl . . . . CH$_3$CH(CH$_2$Cl)$_2$	127.03	. . . .	134.6, 60[49]	. . . . . .	1.1325[25/4]	1.4488[25]	al, eth, bz	B1[4], 293
11833	Propane, 2,2-dichloro . . . . (CH$_3$)$_2$CCl$_2$	112.99	. . . .	69.3	−33.8	1.1136[20/4]	1.4148[20]	al, eth, bz, chl	B1[4], 196
11834	Propane, 1,1-dicyclohexyl . . . . CH$_3$CH$_2$CH(C$_6$H$_{11}$)$_2$	208.39	. . . .	270-1	. . . . . .	0.8887[23/0]	1.485[23]	. . . .	B5[4], 350
11835	Propane, 1,2-dicyclohexyl . . . . CH$_3$CH(C$_6$H$_{11}$)CH$_2$C$_6$H$_{11}$	208.39	. . . .	272-3	. . . . . .	0.8725[21/0]	1.479[21]	. . . .	B5[4], 350
11836	Propane, 1,3-dicyclohexyl . . . . CH$_2$(CH$_2$C$_6$H$_{11}$)$_2$	208.39	. . . .	291-2	−17	0.8752[24/24]	1.4736[24]	. . . .	B5[4], 349
11837	Propane, 1,3-dicyclohexyl-2-ethyl . . . . C$_2$H$_5$CH(CH$_2$C$_6$H$_{11}$)$_2$	236.44	. . . .	296	. . . . . .	0.8846[21/0]	1.483[21]	. . . .	B5[4], 363
11838	Propane, 1,3-dicylohexyl-2-methyl . . . . CH$_3$CH(CH$_2$C$_6$H$_{11}$)$_2$	222.41	. . . .	295.2	0.6	0.8715[20]	1.4756[20]	. . . .	B5[4], 358

No.	Name, Synonyms, and Formula	Mol. wt.	Color, crystalline form, specific rotation and λ_{max} (log ε)	b.p. °C	m.p. °C	Density	n_D	Solubility	Ref.
11839	Propane, 2,2-dicylohexyl $(C_6H_{11})_2C(CH_3)_2$	208.39	....	273-4	......	$0.9002^{23/0}$	1.490^{23}	....	B5[4], 350
11840	Propane, 2,2-di(ethyl sulfonyl) or Sulfonal $(CH_3)_2C(SO_2C_2H_5)_2$	228.33	mcl (w), pr (al)	300d	125.8	....	....	al, bz, chl	B1[3], 2754
11841	Propane, 1,1-difluoro-1,2,2,3,3-pentachloro $CClF_2CCl_2CCl_2H$	252.20		168.4	.....	$1.73162^{20/4}$	1.46241^{20}	....	B1[4], 204
11842	Propane, 1,3-difluoro $F(CH_2)_3F$	80.08	....	41.6	.....	$1.0057^{25/4}$	1.3190^{26}	bz	B1[3], 218
11843	Propane, 2,2-difluoro $(CH_3)_2CF_2$	80.08		−0.4	−104.8	$0.9205^{20/4}$	1.2904^{20}	....	B1[4], 187
11844	Propane, 1,2-diiido CH_3CHICH_2I	295.89		......	......	....	$2.490^{18.5}$	al, eth	B1, 115
11845	Propane, 1,3-diiido $I(CH_2)_3I$	295.89		227d, 110[19]	fr-20	$2.5755^{20/4}$	1.6423^{20}	eth, chl	B1[4], 228
11846	Propane, 2,2-diiido $(CH_3)_2CI_2$	295.89		173, 53[10]	......	$2.5755^{20/4}$	1.651^{20}	eth, chl	B1[3], 255
11847	Propane, 2,2-dimethyl or Neopentane $(CH_3)_4C$	72.15	gas	9.5	−16.5	0.6135^{20}	1.3476^0	al, eth	B1[4], 333
11848	Propane, 1,1-dinitro $CH_3CH_2CH(NO_2)_2$	134.09		184	−42	1.2610^{25}	1.4339^{20}	....	B1[4], 234
11849	Propane, 1,3-dinitro $O_2N(CH_2)_3NO_2$	134.09		103[1]	−21.4	$1.353^{26/4}$	1.4654^{20}	eth	B1[4], 234
11850	Propane, 2,2-dinitro $(CH_3)_2C(NO_2)_2$	134.09		185.5, 48-50[2]	53	1.30^{25}	....	....	B1[4], 234
11851	Propane, 1,2-diphenoxy $CH_3CH(OC_6H_5)CH_2OC_6H_5$	228.29	rh (MeOH)	175-8[12]	32	$1.0748^{33.3/4}$	$1.5542^{33.3}$	al, eth, ace, bz, chl	B6[2], 151
11852	Propane, 1,3-diphenoxy $C_6H_5O(CH_2)_3OC_6H_5$	228.29	lf (al)	338-40, 160[25]	61	....	....	al, eth	B6[4], 577
11853	Propane, 1,1-diphenyl $CH_3CH_2CH(C_6H_5)_2$	196.13	....	280	.....	$0.9951^{14/4}$	1.5681^{14}	....	B5[4], 1920
11854	Propane, 1,2-diphenyl $CH_3CH(C_6H_5)CH_2C_6H_5$	196.29	....	280-1[758], 109[2]	52	$0.9807^{20/4}$	1.5700^{20}	....	B5[4], 1914
11855	Propane, 1,3-diphenyl $C_6H_5(CH_2)_3C_6H_5$	196.29	....	300.3, 124[2]	6	$1.007^{20/4}$	1.5760^{20}	....	B5[4], 1913
11856	Propane, 2,2-diphenyl $(C_6H_5)_2C(CH_3)_2$	196.29	....	282-3	29	....	....	....	B5[4], 1925
11857	Propane, 1,2-epoxy or Propylene oxide CH_3CHCH_2	58.08		34.3	......	$0.859^{0/4}$	1.3670^{20}	w, al, eth	B17[4], 16
11858	Propane, 1,2-epoxy-3-fluoro FCH_2CHCH_2	76.07		85.0-6.5	.....	$1.090^{20/20}$	1.3730^{20}	....	B17[4], 20
11859	Propane, 1,2-epoxy-3 iodo or Epiiodohydrin ICH_2CHCH_2	183.98		160-2	....	1.982^{24}	....	al, eth	B17[4], 23
11860	Propane, 1,2-epoxy,3,3,3-trichloro or 3,3,3-Trichloro propylene oxide Cl_3CCHCH_2	161.42		149, 44-5[13]	....	$1.495^{20/4}$	1.4737^{25}	eth	B17[4], 22
11861	Propane, 1-ethoxy-3-phenoxy $C_6H_5O(CH_2)_3OC_2H_5$	180.25	oil	328-30	.....	....	....	al, eth	B6, 147
11862	Propane, 1-fluoro or Propyl fluoride $CH_3CH_2CH_2F$	62.09		2.5	−159	$0.7956^{20/4}$	1.3115^{20}	al, eth	B1[4], 187
11863	Propane, 1,1,1,2,2,3,3-heptachloro $Cl_2CHCCl_2CCl_3$	285.21	amor	247-8, 132[30]	29.4	$1.8048^{24/4}$	....	chl	B1[4], 205
11864	Propane, 1,1,1,2,3,3,3,-heptachloro $(Cl_3C)_2CHCl$	258.21	....	249, 93[2]	11	$1.7921^{34/4}$	1.5427^{21}	chl	B1[4], 205
11865	Propane, heptafluoro-1-nitro $CF_3CF_2CF_2NO_2$	215.03		25	.....	....	....	....	B1[4], 232
11866	Propane, heptafluoro-1-nitroso $CF_3CF_2CF_2NO$	199.03	deep bl	−12	−150	....	....	....	B1[4], 229
11867	Propane, 1,1,1,2,2,3-hexafluoro $CH_2FCF_2CF_3$	152.04		1.2	.....	....	....	....	B1[4], 188
11868	Propane, 1-iodo or n-Propyl iodide $CH_3CH_2CH_2I$	169.99		102.4	−101	$1.7489^{20/4}$	1.5058^{20}	al, eth, bz, chl	B1[4], 222
11869	Propane, 1-iodo-2,2-dimethyl or Neopentyl iodide $(CH_3)_3CCH_2I$	198.05	....	127-9d, 42-4[20]	.....	1.4940^{20}	1.4890^{20}	al, eth	B1[4], 338
11870	Propane, 2-iodo or Isopropyl iodide $(CH_3)_2CHI$	169.99		89.4	−90.1	$1.7033^{20/4}$	1.5028^{20}	al, eth, bz, chl	B1[4], 223
11871	Propane, 1-methoxy-3-phenoxy $C_6H_5O(CH_2)_3OCH_3$	166.22	oil	230-1	......	....	....	....	B6, 147

No.	Name, Synonyms, and Formula	Mol. wt.	Color, crystalline form, specific rotation and λ_{max} (log ε)	b.p. °C	m.p. °C	Density	n_D	Solubility	Ref.
11872	Propane, 2-methyl or Isobutane . . . . (CH$_3$)$_2$CHCH$_3$	58.12	. . . .	−11.633	−159.4	0.549[20]	. . . .	al, eth, chl	B1[4], 282
11873	Propane, 1-nitro . . . . CH$_3$CH$_2$CH$_2$NO$_2$	89.09	. . . .	130-1	−108	1.0081[24/4]	1.4016[20]	al, eth chl	B1[4], 229
11874	Propane, 2-nitro . . . . (CH$_3$)$_2$CHNO$_2$	89.09	. . . .	120	−93	0.9876[20/4]	1.3944[20]	chl	B1[4], 230
11875	Propane, 3-nitro-1,1,1-trifluoro . . . . F$_3$CCH$_2$CH$_2$NO$_2$	143.07	. . . .	135	. . . . . .	1.4203[20/20]	1.3549[20]	eth	B1[4], 232
11876	Propane, 1,1,1,2,3-pentachloro . . . . CH$_2$ClCHClCCl$_3$	216.32	. . . .	179-80	. . . . . .	. . . .	1.5130[20]	al, eth	B1[4], 203
11877	Propane, 1,1,2,3,3-pentachloro . . . . Cl$_2$CHCHClCHCl$_2$	216.32	. . . .	198-200, 78-100[20]	. . . . . .	1.6086[14/4]	1.5131[17]	al, eth, chl	B1[4], 204
11878	Propane, 1,1,2,3,3-pentafluoro-1,2,3-trichloro . . . . . . CF$_2$ClCClFC.ClF$_2$	237.38	. . . .	73.7	−72	1.6631[20/4]	1.3512[20]	. . . .	B1[4], 200
11879	Propane, perchloro . . . . C$_3$Cl$_8$	319.66	. . . .	268-9[7,34]	160	. . . .	. . . .	al, eth	B1[4], 205
11880	Propane, perfluoro . . . . C$_3$F$_8$	188.02	. . . .	−36	−183	. . . .	. . . .	. . . . . .	B1[4], 189
11881	Propane, 1,1,1,2-tetrachloro . . . . CH$_3$CHClCCl$_3$	181.88	. . . .	150, 37[10]	. . . . . .	1.473[20/4]	1.4867[20]	al, eth, chl	B1[4], 201
11882	Propane, 1,1,1,3-tetrachloro . . . . CH$_2$ClCH$_2$CCl$_3$	181.88	. . . .	159, 59[24]	. . . . . .	1.4510[20/6]	1.4825[20]	al, eth, bz, chl	B1[4], 201
11883	Propane, 1,1,2,2-tetrachloro . . . . CH$_3$CCl$_2$CHCl$_2$	181.88	. . . .	153	. . . . . .	1.47[12]	1.4850[25]	al, eth, chl	B1[4], 201
11884	Propane, 1,1,2,3-tetrachloro . . . . CH$_3$ClCHClCHCl$_2$	181.88	. . . .	179-80	. . . . . .	1.513[11]	1.5037[20]	al, eth	B1[4], 202
11885	Propane, 1,2,2,3-tetrachloro . . . . CH$_2$ClCCl$_2$CH$_2$Cl	181.88	. . . .	165, 51[12]	. . . . . .	1.500[18]	1.4940[18]	al, eth, chl	B1[4], 202
11886	Propane, 1,2,2,3-tetrachloro,-1,1,3,3,-tetrafluoro . . . . CF$_2$ClCCl$_2$CF$_2$Cl	253.84	. . . .	112	−42.9	1.7199[20/4]	1.39584[20]	. . . .	B1[4], 203
11887	Propane, 1,1,2-tribromo . . . . CH$_3$CHBrCHBr$_2$	280.30	. . . .	200-1, 83[6]	. . . . . .	2.3548[20/4]	1.5790[20]	al, eth, chl, aa	B1[3], 251
11888	Propane, 1,2,2-tribromo . . . . CH$_3$CBr$_2$CH$_2$Br	280.80	. . . .	190-1, 81[20]	. . . . . .	2.2985[20/4]	1.5670[20]	al, eth, chl, aa	B1[2], 77
11889	Propane, 1,2,3-tribromo . . . . CH$_2$BrCHBrCH$_2$Br	280.80	. . . .	222.1, 98.5[10]	16.9	2.4209[20/4]	1.5862[20]	al, eth	B1[4], 221
11890	Propane, 1,1,1-trichloro . . . . CH$_3$CH$_2$CCl$_3$	147.43	. . . .	107-9	. . . . . .	1.287[23/4]	. . . .	al, eth, chl	B1[4], 198
11891	Propane, 1,1,2-trichloro . . . . CH$_3$CHClCHCl$_2$	147.43	. . . .	140	. . . . . .	1.372[15]	. . . .	al, eth, chl	B1[4], 199
11892	Propane, 1,1,3-trichloro . . . . CH$_2$ClCH$_2$CHCl$_2$	147.43	. . . .	145.5, 33.4[10]	−59	1.3557[20/4]	1.4718[20]	al, eth, chl, aa	B1[4], 199
11893	Propane, 1,2,2-trichloro . . . . CH$_3$CCl$_2$CH$_2$Cl	147.43	. . . .	123-5	. . . . . .	1.318[25]	1.4609[20]	al, eth, chl	B1[4], 199
11894	Propane, 1,2,3-trichloro . . . . CH$_2$ClCHClCH$_2$Cl	147.43	. . . .	156.8, 41.9[10]	−14.7	1.3889[20/4]	1.4852[20]	al, eth, chl	B1[4], 199
11895	Propane, 1,1,1-triphenyl . . . . CH$_3$CH$_2$C(C$_6$H$_5$)$_3$	272.40	pr	. . . .	51	. . . .	. . . .	. . . . . .	B5[3], 2340
11896	1-Propanearsonic acid . . . . CH$_3$CH$_2$CH$_2$AsO$_3$H$_2$	168.03	nd (al), pl (w)	. . . .	134-5	. . . .	. . . .	w, al	B4[3], 1824
11897	1-Propane boronic acid or n-Propylboric acid . . . . CH$_3$CH$_2$CH$_2$B(OH)$_2$	87.92	wh nd	d	107	. . . .	. . . .	w, al, eth	B4[3], 1964
11898	1,2-Propanediol or Propylene glycol . . . . CH$_3$CH(OH)CH$_2$OH	76.11	. . . .	189, 96-8[21]	. . . . . .	1.0361[20/4]	1.4324[20]	w, al, eth, bz	B1[4], 2468
11899	1,2-Propanediol, 3-amino or 3-Amino-1,2-propylene glycol . . . . H$_2$NCH$_2$CH(OH)CH$_2$OH	91.11	. . . .	265d, 145[9]	. . . . . .	1.1752[20/4]	1.4910[25]	w, al	B4[4], 1865
11900	1,2-Propanediol carbonate or 1,2-Propylene glycol carbonate . . . . C$_4$H$_6$O$_3$	102.09	. . . .	240, 110[10]	−48.8	1.2041[20/4]	1.4189[20]	w, al, ace, bz	B19[4], 1564
11901	1,2-Propanediol, 3-chloro . . . . ClCH$_2$CH(OH)CH$_2$OH	110.54	yesh liq	213d, 116[11]	. . . . . .	1.326[18/15]	1.4809[20]	w, al, eth	B1[4], 2484
11902	1,2-Propanediol, 3-chloro, diacetate . . . . ClCH$_2$CH(OCOCH$_3$)CH$_2$OCOCH$_3$	194.62	. . . .	245, 116[22]	. . . . . .	1.199[25/4]	1.4407[20]	al, eth	B2[3], 313
11903	1,2, Propanediol, 3-chloro-2-methyl . . . . ClCH$_2$C(OH)(CH$_3$)CH$_2$OH	124.57	. . . .	114-7[20]	. . . . . .	1.2362[20/4]	1.4748[20]	w, al, eth	B1[4], 2536
11904	1,2-Propanediol diacetate . . . . CH$_3$CH(O$_2$CCH$_3$)CH$_2$O$_2$CCH$_3$	160.17	. . . .	190-1[762]	. . . . . .	1.059[20/4]	1.4173[20]	w, al, eth	B2[4], 220

No.	Name, Synonyms, and Formula	Mol. wt.	Color, crystalline form, specific rotation and λ_{max} (log ε)	b.p. °C	m.p. °C	Density	n_D	Solubility	Ref.
11905	1,2-Propane diol, 3-(diethyl amino) $(C_2H_5)_2NCH_2CH(OH)CH_2OH$	147.22	syr	233-5	......	......	......	w, al, eth, chl	B4[3], 840
11906	1,2-Propanediol, 3-(dimethylamino) $(CH_3)_2NCH_2CH(OH)CH_2OH$	119.17	......	220[749]	......	......	......	w, al, eth, chl	B4[3], 840
11907	1,2-Propanediol, 3-mercapto or 1-Thioglycerol $HSCH_2CH(OH)CH_2OH$	108.17	visc	100-1[1]	......	1.2455[20]	1.5268[20]	w, al, ace	B1[3], 2339
11908	1,2-Propanediol, 2-methyl or Isobutylene glycol $(CH_3)_2C(OH)CH_2OH$	90.12	......	176, 79-80[12]	......	1.0024[20/4]	1.5268[20]	w, al, eth	B1[4], 2533
11909	1,2-Propanediol-sulfite $C_3H_6O_3S$	122.14	......	175, 85[25]	<-60	1.2960[20/4]	1.4370[20]	w, al, eth, ace, bz	B1[4], 2476
11910	1,3-Propanediol or Trimethylene glycol $HOCH_2CH_2CH_2OH$	76.11	......	213.5, 110[12]	......	1.0597[20/4]	1.4398[20]	**w, al,** eth	B1[4], 2493
11911	1,3-Propanedioll, 2-amino-2-ethyl $C_2H_5C(CH_2OH)_2NH_2$	119.17	ye	143-5[10]	37-8	1.099[20/4]	1.490[20]	w	B4[4], 1883
11912	1,3-Propanediol, 2-amino-2-(hydroxymethyl) $(HOCH_2)_3CNH_2$	121.14	nd or fl (MeOH)	219-20[18]	170-1	......	......	w	B4[4], 1903
11913	1,3,-Propanediol, 2-amino-2-methyl $(HOCH_2)_2C(CH_3)NH_2$	105.14	......	151.2[10]	109-11	......	......	w, al	B4[4], 1881
11914	1,3-Propanediol, 2-butyl-2-ethyl or 3,3-Di(hydroxymethyl) heptane $(HOCH_2)_2C(C_2H_5)(C_4H_9)$	160.26	wh	262, 123[15]	43.8	0.929[20/20]	1.4587[25]	al	B1[4], 2611
11915	1,3-Propanediol, 2-chloro $(HOCH_2)_2CHCl$	110.54	......	146[18]	......	1.3219[20/4]	1.4831[20]	w, al, ace	B1[4], 2499
11916	1,3 Propanediol diacetate $CH_3CO_2CH_2CH_2CH_2O_2CCH_3$	160.17	......	209-10, 84.5[10]	......	1.070[14]	1.4192	w, al	B2[4], 221
11917	1,3-Propanediol, 2,2-diethyl $(HOCH_2)_2C(C_2H_5)_2$	132.21	wh	240-1, 131[13]	61-2	1.052[20/20]	......	w, al, eth	B1[4], 2589
11918	1,3-Propanediol, 2,2-dimethyl or Neopentylene glycol $(HOCH_2)_2C(CH_3)_2$	104.15	nd (bz)	206[747], 120-30[15]	130	......	......	w, al, eth	B1[4], 2551
11919	1,3-Propanediol, 2,2-dinitro $(HOCH_2)_2C(NO_2)_2$	166.09	wh pl (bz)	......	142	......	......	w, al, eth	B1[4], 2501
11920	1,3-Propanediol, 2-ethyl-2-hydroxymethyl or TMP Trimethylol propane $CH_3CH_2C(CH_2OH)_3$	134.18	wh pw or pl	160[5]	58	......	......	**w, al**	B1[4], 2786
11921	1,3-Propanediol, 2-ethyl-2-nitro $(HOCH_2)_2C(NO_2)C_2H_5$	149.15	nd (w)	d	57-8	......	......	w, al, eth	B1[4], 2550
11922	1,3-Propanediol, 2-hydroxymethyl-2-nitro $(HOCH_2)_3CNO_2$	151.12	nd or pr	d	165	......	......	w, al, eth	B1[4], 2777
11923	1,3,Propanediol, 2-methyl-2-(hydroxy methyl) or Trimethylol ethane $CH_3C(CH_2OH)_3$	120.15	wh pw or nd (al)	135-7[15]	204	......	......	**w, al**	B1[4], 2780
11924	1,3-Propanediol, 2-methyl-2-nitro $(HOCH_2)_2C(CH_3)NO_2$	135.12	mcl	d	149-50	......	......	w, al	B1[4], 2537
11925	1,3-Propane diol, 2-methyl-2-propyl or 2,2-Bis-(hydroxymethyl) pentane $(HOCH_2)_2C(CH_3)C_3H_7$	132.31	cr (hx)	234	62-3	......	......	w	B1[4], 2585
11926	1,2 Propanedithiol $CH_3CH(SH)CH_2SH$	108.22	......	41-3[11]	......	1.08[20/4]	1.532[20]	chl	B1[4], 2492
11927	1,3-Propanedithiol $HSCH_2CH_2CH_2SH$	108.24	......	172.9, 63[15]	-79	1.0783[20/4]	1.5392[20]	**al, eth, bz, chl**	B1[4], 2503
11928	1-Propane phosphonic acid or n-Propylphosphonic acid $CH_3CH_2CH_2PO(OH)_2$	124.08	pl (bz)	d	23	......	......	w, al, eth	B4[3], 1781
11929	2-Propane phosphonic acid or Isopropyl phosphonic acid $(CH_3)_2CHPO_3H_2$	124.08	pl (bz)	d	74-5	......	......	w, al, eth	B4[3], 1781
11930	1-Propane sulfonamide $C_3H_7SO_2NH_2$	123.18	pr (eth), cr (bz)	......	53.5	......	......	w, al	B4[4], 39
11931	1-Propanesulfonyl chloride $C_3H_7SO_2Cl$	142.61	......	180d, 77[12]	......	1.2826[15/4]	1.452[20]	......	B4[4], 39
11932	2-Proponesulfonamide or Isopropyl sulfonamide $(CH_3)_2CHSO_2NH_2$	123.18	cr (eth-peth)	......	67.5	......	......	w, al, eth	B4[4], 42
11933	2-Propanesulfonic acid or Isopropyl sulfonic acid $(CH_3)_2CHSO_3H$	124.17	......	159[1.4]	-37	1.187[25]	1.4332[20]	w	B4[4], 42
11934	1,1,2,3-Propane tetra carboxylic acid, tetra ethyl ester $C_2H_5O_2CCH_2CH(CO_2C_2H_5)CH(CO_2C_2H_5)_2$	332.36	......	203-4[18]	......	1.1184[20/4]	1.4395[20]	al	B2[3], 2077
11935	1,1,3,3-Propane tetra carboxylic acid, tetra ethyl ester or Ethyl methane dimalonate $[(C_2H_5O_2C)_2CH]_2CH_2$	332.36	......	300-10d, 195[8]	-30	1.116[20]	1.4398[20]	al	B2[4], 2417

No.	Name, Synonyms, and Formula	Mol. wt.	Color, crystalline form, specific rotation and λ_{max} (log ε)	b.p. °C	m.p. °C	Density	n_D	Solubility	Ref.
11936	1-Propanethiol or n-Propyl mercaptan .... CH₃CH₂CH₂SH	76.17	....	67-8	−113.3	0.8411[20]	1.4380[20]	al, eth, ace, bz	B1[4], 1449
11937	2- Propane thiol or Isopropyl mercaptan .... (CH₃)₂CHSH	76.17	....	52.5	−130.5	0.8143[20/4]	1.4255[20]	al, eth, ace	B1[4], 1498
11938	1,2,3-Propane tricarboxylic acid or Tricarballic acid ..... (HO₂CCH₂)₂CHCO₂H	176.14	orh (w, eth)	....	166	....	....	w, al	B2[4], 2366
11939	1,2,3-Propanetricarboxylic acid-1,2-dihydroxy (l) .... HO₂CCH₂C(OH)(CO₂H)CH(OH)(CO₂H)	208.14	nd, [α]ₐ 17.7 (acid)	....	159-60	1.39[25]	....	w, eth	B3[4], 1298
11940	1,2,3-Propanetricarboxylic acid, 1-hydroxy or Isocitric acid HO₂CCH₂CH(CO₂H)CH(OH)CO₂H	192.14	yesh syr	....	105	....	....	....	B3[4], 1270
11941	Propanoic acid or Propionic acid .... CH₃CH₂CO₂H	74.08	....	141, 41.6[10]	−20.8	0.9930[20]	1.3869[20]	w, al, eth	B2[4], 695
11942	1-Propanol or n-Propyl alcohol .... CH₃CH₂CH₂OH	60.11	....	97.4	−126.5	0.8035[20/4]	1.3850[20]	w, al, eth, ace, bz	B1[4], 1413
11943	1-Propanol, 2-amino (dl) .... CH₃CH(NH₂)CH₂OH	75.11	....	173-6, 80[18]	....	....	1.4502[20]	w, al, eth	B4[4], 1615
11944	1-Propanol, 3-amino or Propanolamine .... H₂NCH₂CH₂CH₂OH	75.11	....	187-8	....	0.9824[26/4]	1.4617[20]	w, al, eth	B4[4], 1623
11945	1-Propanol, 3-benzyloxy .... C₆H₅CH₂OCH₂CH₂CH₂OH	166.22	....	172[43]	....	1.0474[20/4]	....	al, eth	B6[4], 2243
11946	1-Propanol, 3-bromo .... BrCH₂CH₂CH₂OH	139.00	....	98-112[185], 62[5]	....	1.5374[20/4]	1.4834[25]	w, al, eth	B1[4], 1446
11947	1-Propanol, 2-chloro or Propylene chlorohydrin .... CH₃CHClCH₂OH	94.54	....	133-4	....	1.103[20]	1.4390[20]	w, al, eth	B1[4], 1440
11948	1-Propanol, 2-chloro 2 methyl or 2-Chloro isobutyl alcohol .... (CH₃)₂CClCH₂OH	108.57	visc	132-3d, 59-61[50]	....	1.0477[20/4]	1.4388[20]	....	B1[4], 1003
11949	1-Propanol, 3-chloro or Trimethylene chlorohydrin .... ClCH₂CH₂CH₂OH	94.54	....	165-53[6]	....	1.1309[20/4]	1.4459[20]	w, al, eth	B1[4], 1441
11950	1-Propanol, 2,3-dibromo (d) .... CH₂BrCHBrCH₂OH	217.90	[α]ₐ +7.3	219d	....	2.11	....	al, eth, ace, bz	B1[2], 371
11951	1-Propanol, 2,3-dibromo (dl) .... CH₂BrCHBrCH₂OH	217.90	....	219d, 118[17]	....	2.0739[20/4]	1.5466[20]	al, eth, ace, bz	B1[4], 1446
11952	1-Propanol, 2,3-dichloro .... CH₂ClCHClCH₂OH	128.99	visc	183-5, 70-80[17]	....	1.3607[20/4]	1.4819[20]	al, eth, ace, bz	B1[4], 1442
11953	1-Propanol, 2-diethylamino .... (C₂H₅)₂NCH(CH₃)CH₂OH	131.22	....	166-9[749], 78[12]	....	0.8665[27/20]	1.4332[20]	al, eth, ace, bz	B4[4], 1618
11954	1-Propanol, 3-diethylamino .... (C₂H₅)₂NCH₂CH₂CH₂OH	131.22	....	189.5, 87[16]	....	0.8600[20/4]	1.4439[20]	al, eth, ace, bz	B4[4], 1633
11955	1-Propanol, 2,3-dimercapto or 1,2 Dithioglycerol....... HSCH₂CH(SH)CH₂OH	124.24	visc liq	120[15]	....	1.2463[20/4]	1.5733[20]	al, eth, oils	B1[4], 2770
11956	1-Propanol, 3-(3,5-dimethoxy-4-hydroxyphenyl) or 2-Syringyl ethanol. Hydrosinapyl alcohol.................. (3,5-(CH₃O)₂-4-HOC₆H₂)CH₂CH₂CH₂OH	212.25	wh nd (ethpeth)	....	75-6	....	....	w	B6[3], 6669
11957	1-Propanol, 2,2-dimethyl or Neopentyl alcohol......... (CH₃)₃CCH₂OH	88.15	....	113-4	52-3	0.812	....	al, eth	B1[4], 1690
11958	1-Propanol, 2,3-epoxy or Glycidol .... CH₂CHCH₂OH	74.08	....	166-7d, 65-6[2.5]	....	1.1143[25]	1.4287[20]	w, al, eth, ace, bz, chl	B17[4], 985
11959	1-Propanol, 2-ethoxy .... CH₃CH(OC₂H₅)CH₂OH	104.15	....	140-1	....	0.9044[20/4]	1.4122[20]	w, al, eth	B1[3], 2147
11960	1-Propanol, 3-fluoro .... FCH₂CH₂CH₂OH	78.09	....	127.8	....	1.0390[25/4]	1.3771[25]	w, al, eth	B1[4], 1437
11961	1-Propanol, 3-(4-hydroxy-3-methoxyphenyl) or Hydroconiferyl alcohol .... (3-CH₃O-4-HOC₆H₃)CH₂CH₂CH₂OH	182.22	....	197[15]	65	....	1.5545[25]	al, eth	B6[3], 6347
11962	1-Propanol, 2-methoxy .... CH₃CH(OCH₃)CH₂OH	90.12	....	130[758]	....	0.938[20/4]	1.4070[20]	....	B1[4], 2471
11963	1-Propanol, 2-methyl or Isobutyl alcohol......... (CH₃)₂CHCH₂OH	74.12	....	108	....	0.8018[20/4]	1.3955[20]	w, al, eth, ace	B1[4], 1588
11964	1-Propanol, 2-methyl-2-nitro .... (CH₃)₂C(NO₂)CH₂OH	119.12	nd or pl (MeOH)	94-5[10]	89-90	....	....	al, eth	B1[4], 1604
11965	1-Propanol, 2-nitro .... CH₃CH(NO₂)CH₂OH	105.10	....	100[12]	....	1.1841[25/4]	1.4379[20]	al, eth	B1[4], 1448
11966	1-Propanol, 2-phenoxy .... CH₃CH(OC₆H₅)CH₂OH	152.30	....	244, 124-6[20]	....	0.9830[25/25]	1.4760[25]	al, eth	B6[4], 582
11967	1-Propanol, 3-phenoxy .... C₆H₅OCH₂CH₂CH₂OH	152.20	oil	249-50[764], 170[60]	....	....	1.491[20]	....	B6[4], 584

No.	Name, Synonyms, and Formula	Mol. wt.	Color, crystalline form, specific rotation and λ_{max} (log ε)	b.p. °C	m.p. °C	Density	n_D	Solubility	Ref.
11968	1-Propanol, 1-phenyl- *(dl)* .... $CH_3CH_2CH(OH)C_6H_5$	136.20	....	213-5[740], 98[10]	......	0.9938[23/4]	1.5210[23]	al, eth	B6[3], 1793
11969	1-Propanol, 3-phenyl $C_6H_5CH_2CH_2CH_2OH$	136.30	....	236-7[750], 132[21]	<−18	1.008[20/4]	1.5278[20]	w, al, eth	B6[3], 1880
11970	1-Propanol, 2,2,3,3-tetrafluoro .... $CHF_2CF_2CH_2OH$	132.06	....	109-1	−15	1.4853[20/4]	1.3197[20]	al, ace, chl	B1[4], 1438
11971	2-Propanol or Isopropyl alcohol .... $(CH_3)_2CHOH$	60.11	....	82.4	−89.5	0.7855[20/4]	1.3776[20]	w, al, eth, ace, bz	B1[4], 1461
11972	2-Propanol, 1-amino *(dl)* or 150 Propanol amine $CH_3CH(OH)CH_2NH_2$	75.11	....	159.4, 59.5[10]	1.7	0.9611[20/4]	1.4479[20]	w, al, eth, ace, bz	B4[4], 1665
11973	2-Propanol, 1-amino *(l)* .... $CH_3CH(OH)CH_2NH_2$	75.11	....	156-8[758]	......	0.973[18]	....	w, al, eth, ace, bz	B4[4], 1664
11974	2-Propanol, 1-amino-3-(diethylamino) .... $(C_2H_5)_2NCH_2CH(OH)CH_2NH_2$	146.24	....	223, 116-8[25]	......	1.937[20/4]	1.465[20]	w	B4[3], 767
11975	2-Propanol, 2-benzyl .... $C_6H_5CH_2C(OH)(CH_3)_2$	150.22	....	214-6, 104-5[10]	24	0.9790[20/25]	1.5174[20]	al	B6[3], 1860
11976	2-Propanol, 1,3-bis (dimethylamino) .... $[(CH_3)_2NCH_2]_2CHOH$	146.23	....	178-85, 79-81[18]	......	0.8788[20/4]	1.4418[20]	w	B4[4], 1695
11977	2-Propanol, 1-bromo $CH_3CH(OH)CH_2Br$	139.00	....	145-8, 49.6[12]	......	1.5585[10]	1.4801[20]	w, al, eth	B1[4], 1495
11978	2-Propanol, 1-butoxy or 1,2-Propylene glycol-1-monobutyl ether .... $CH_3CH(OH)CH_2OC_4H_9$	132.21	....	168.75	......	1.0035[20/4]	1.4168[20]	al, eth, bz	B1[4], 2471
11979	2-Propanol, 1-chloro $CH_3CH(OH)CH_2Cl$	94.54	....	126-7[750]	......	1.115[20/20]	1.4392[20]	w, al, eth	B1[4], 1490
11980	2-Propanol, 1-chloro-2-methyl $(CH_3)_2C(OH)CH_2Cl$	108.57	....	128-9, 71[100]	−20	1.0628[20/4]	1.4380[24]	w, al	B1[4], 1628
11981	2-Propanol, 1-chloro-3-propoxy $C_3H_7OCH_2CH(OH)CH_2Cl$	152.63	....	92-5[15]	......	1.0526[25/4]	1.4378[25]	eth, ace	B1[3], 2153
11982	2-Propanol, 1-chloro-3-isopropoxy $(CH_3)_2CHOCH_2CH(OH)CH_2Cl$	152.62	....	87[20]	......	1.0530[25]	1.4370[25]	al, eth	B1[4], 2485
11983	2-Propanol, 1,3-diamino .... $(H_2NCH_2)_2CHOH$	90.13	cr	235, 93-5[2]	42-5	....	....	w	B4[4], 1694
11984	2-Propanol, 1-3 diamino, dihydrochloride .... $(H_2NCH_2)_2CHOH.2HCl$	163.05	hyg pr or nd (dil al)	......	184.5	....	....	w	B4[2], 739
11985	2-Propanol, 1,3-dibromo $(BrCH_2)_2CHOH$	217.90	yesh liq	219d, 105[16]	......	2.1202[25/4]	1.5495[25]	al, eth, ace	B1[4], 1496
11986	2-Propanol, 1-1,dichloro $CH_3CH(OH)CHCl_2$	128.99	....	146-8[765]	......	1.3334[22/4]	....	al, eth, ace	B1[2], 383
11987	2-Propanol, 1,1-dichloro-2-methyl $(CH_3)_2C(OH)CHCl_2$	143.02	....	150-5, 52[10]	8	1.2363[19/4]	1.4598[19]	al, bz	B1[4], 1629
11988	2-Propanol, 1,3-dichloro $(ClCH_2)_2CHOH$	128.99	....	176, 69[12]	......	1.3506[17/4]	1.4837[20]	w, al, eth, ace	B1[4], 1491
11989	2-Propanol, 1,3-dichloro-2-methyl $(ClCH_2)_2C(OH)CH_3$	143.02	....	174-5, 55-6[10]	......	1.2745[20/4]	1.4744[21]	al, ace	B1[4], 1629
11990	2-Propanol, 1-(diethylamino) $CH_3CH(OH)CH_2N(C_2H_5)_2$	131.22	....	158-9[756], 63[22]	......	0.8511[20/0]	1.4255[20]	w, al, ace	B4[3], 759
11991	2-Propanol, 1-ethoxy .... $CH_3CH(OH)CH_2OC_2H_5$	104.15	....	131	......	0.9028[20/4]	1.4075[20]	w, al, eth	B1[4], 2471
11992	2-Propanol, 1-methoxy .... $CH_3CH(OH)CH_2OCH_3$	90.12	....	118	......	0.9620[20/4]	1.4034[20]	....	B1[4], 2471
11993	2-Propanol, 1-methoxy-2-methyl $CH_3OCH_2C(OH)(CH_3)_2$	104.15	....	116.6[747]	......	0.9021[15/15]	....	....	B1[4], 2533
11994	2-Propanol, 2-methyl or *tert*-Butyl alchol .... $(CH_3)_3COH$	74.12	....	82.3, 20[11]	25.5	0.7887[20/4]	1.3878[20]	w, al, eth	B1[4], 1609
11995	2-Propanol, 1-nitro $CH_3CH(OH)CH_2NO_2$	105.09	....	112[13], 68[1]	......	1.1906[20]	1.4383[20]	w, al, eth, bz	B1[4], 1496
11996	2-Propanol, 1-nitro-3,3,3-trichloro .... $O_2NCH_2CH(OH)CCl_3$	208.43	pr or pl (chl)	105-6[3.5]	45-6	....	....	al, eth	B1[3], 1497
11997	2-Propanol, 1-phenoxy .... $CH_3CH(OH)CH_2OC_6H_5$	152.20	....	134.5[20]	......	1.0622[20/4]	1.5210[22]	al, eth	B6[4], 582
11998	2-Propanol, 2-phenyl $(CH_3)_2C(OH)C_6H_5$	138.20	pr	202, 93[13]	35-7	0.9735[20/4]	1.5325[20]	al, eth, bz, aa	B6[3], 1813
11999	2-Propanol, 1-propoxy $CH_3CH(OH)CH_2OC_3H_7$	118.18	....	148-9[730]	......	0.8886[20/4]	1.4130[20]	....	B1[3], 2147
12000	2-Propanol, 1-isopropoxy .... $CH_3CH(OH)CH_2OCH(CH_3)_2$	118.18	....	137-8	......	0.879[20/4]	1.4070[20]	....	B1[4], 2471

No.	Name, Synonyms, and Formula	Mol. wt.	Color, crystalline form, specific rotation and λ_{max} (log ϵ)	b.p. °C	m.p. °C	Density	n_D	Solubility	Ref.
12001	2-Propanol, 1,1,1,3-tetrachloro .... CH₂ClCH(OH)CCl₃	197.88	....	95-6[17]	....	1.610[20/4]	1.5145[20]	eth	B1[3], 1474
12002	2-Propanol, 1,1,3,3 tetrachloro .... (Cl₂CH)₂CHOH	197.88	....	80-90[14]	....	1.612[20/4]	1.5133[20]	eth	B1[3], 1474
12003	2-Propanol, 1,1,1-trichloro .... CH₃CH(OH)CCl₃	163.43	....	161.8[773], 53-5[12]	50-1	....	....	al, eth, ace, bz	B1[3], 1474
12004	2-Propanol, 1,1,1-trifluoro (dl) .... F₃CCH(OH)CH₃	114.07	....	77.7[754]	−52	1.2799[15/4]	1.3172[15]	al, eth, ace, bz	B1[4], 1489
12005	2-Propanone or Acetone, Dimethyl Ketone .... CH₃COCH₃	58.08	....	56.2	−95.3	0.7899[20/4]	1.3588[20]	w, al, eth, bz, chl	B1[4], 3180
12006	Propargyl aldehyde or Propynal .... HC≡CCHO	54.05	....	59-61	....	....	1.4033[25]	w, al, eth, ace, bz	B1[4], 3537
12007	Propargyl aldehyde, diethyl acetyl .... HC≡CCH(OC₂H₅)₂	128.17	....	130.4, 37[11]	....	0.8942[22/4]	1.4140[20]	al, eth, ace, chl	B1[4], 3538
12008	Propargyl aldehyde, phenyl .... C₆H₅C≡CCHO	130.15	....	127-8[28] 65[0.1]	....	1.0639[16/4]	1.6079[18]	....	B7[3], 1644
12009	Propargyl bromide or 3-Bromo propyne .... BrCH₂C≡CH	118.97	....	88-90, 33[130]	....	1.579[19]	1.4922[20]	al, eth, bz, chl	B1[4], 964
12010	Propargyl alcohol or 2-Propyn-1-ol .... HC≡CCH₂OH	56.07	....	113.6, 30[21]	−48	0.9485[20]	1.4322[20]	w, al, eth	B1[4], 2214
12011	Propargyl alcohol, acetate or Propargyl acetate .... CH₃CO₂CH₂C≡CH	98.10	....	124-5	....	1.0082[20/4]	1.4205[20]	al, eth	B2[4], 197
12012	Propargyl alcohol, 3-phenyl .... C₆H₅C≡CCH₂OH	132.16	....	137-8[15]	....	1.07[18/18]	1.5873[18]	eth, ace, bz	B6[3], 2736
12013	Propargylic acid or Propiolic acid .... HC≡CCO₂H	70.09	cr (CS₂)	144d, 83-4[60]	18 (alm)	1.1380[20/4]	1.4306[20]	w, al, eth, ace, alal	B2[4], 1687
12014	Propargylic acid, ethyl ester or Ethyl propargylate .... HC≡CCO₂C₂H₅	98.10	....	119[745]	....	0.9583[25/25]	1.4105[20]	al, eth, chl	B2[4], 1688
12015	Propargylic acid, (2-chlorophenyl) .... (2-ClC₆H₄)C≡CCO₂H	180.59	cr (bz, 50% aa)	....	133-4	....	....	aa	B9[3], 3066
12016	Propargylic acid, 3-chlorophenyl .... (3-ClC₆H₄)C≡CCO₂H	180.59	cr (aa, bz-peth)	....	144-5	....	....	aa	B9[3], 3066
12017	Propargylic acid, (4-chlorophenyl) .... (4-ClC₆H₄)C≡CCO₂H	180.59	cr (aa), pl (bz)	....	192-3	....	....	aa	B9[3], 3066
12018	Propargylic acid, (2-nitrophenyl) .... (2-O₂NC₆H₄)C≡CCO₂H	191.15	....	exp	157d	....	....	al, eth	B9[3], 3067
12019	Propargylic acid, (4-nitrophenyl) .... (4-O₂NC₆H₄)C≡CCO₂H	191.15	nd (eth, al)	....	181	....	....	eth, aa, chl	B9[3], 3067
12020	**Propargylic acid, phenyl** .... C₆H₅C≡CCO₂H	146.15	....	sub	137	....	....	al, eth	B9[3], 3061
12021	Propargylic acid, phenyl, ethyl ester .... C₆H₅C≡C-CO₂C₂H₅	174.20	....	260-70d, 144[11]	....	1.0550[25/4]	1.5535[20]	eth	B9[3], 3063
12022	Propargylonitrile or Propynonitrile .... HC≡CCN	51.05	....	42.5	5	0.8167[17/4]	1.3868[25]	al	B2[4], 1689
12023	**Propenal or Acrolein** .... **CH₂=CHCHO**	56.07	....	52-3	−87	0.8410[20/4]	1.4017[20]	w, al, eth, ace	B1[4], 3435
12024	2-Propen-1-arsonic acid or Allylarsonic acid .... CH₂=CHCH₂AsO₃H₂	166.01	....	....	130-1	....	....	w, al	B4[3], 1826
12025	**Propene or Propylene** .... CH₃CH=CH₂	42.08	gas	−47.4	−185.2	0.5193[20/4] liq	1.3567[−70]	al, aa	B1[4], 725
12026	1-Propene-3,3, diol, diacetate .... (CH₃CO)₂CHCH=CH₂	158.16	....	180, 76[13]	−37.6	1.0760[20/4]	1.4193[20]	**al, eth, ace, bz, lig**	B2[4], 291
12027	1,2,3-Propene tricarboxylic acid (*cis*) or *cis*-Aconitic acid .... HO₂CCH₂C(CO₂H)=CHCO₂H	174.11	nd (w)	....	130	....	....	w	B2[4], 2405
12028	1,2,3-Propene tricarboxylic acid (*trans*) or *trans*-Aconitic acid .... HO₂CCH₂C(CO₂H)=CHCO₂H	174.11	lf (w), nd (w, eth)	....	198-9	....	....	w, al	B2[4], 2405
12029	Propenyl benzene (*cis*) .... (CH₃CH=CH)C₆H₅	118.18	....	69[28]	−60.5	0.9088[20/4]	1.5420[20]	al, eth, ace, bz	B5[4], 1359
12030	Propenyl benzene-(*trans*) .... (CH₃CH=CH)C₆H₅	118.18	....	175-6	−27.1	0.9019[25]	1.5508[20]	al, eth, ace, bz	B5[4], 1359
12031	Propenyl benzene, α-chloro .... CH₃CH=CClC₆H₅	152.63	....	90.5[0.9]	....	1.085[20/4]	1.5635[15]	ace, bz	B5[2], 372
12032	Propenyl benzene, β-chloro .... CH₃CH=CClCH₃	152.63	....	118-23[20], 61-2[1]	....	1.0738[19/4]	1.5565[19]	ace, bz, chl	B5[3], 1186
12033	Propenyl benzene, γ-chloro-(*trans*) or *trans*-Cinnamyl chloride .... C₆H₅CH=CHCH₂Cl	152.63	....	106-7[13]	8-9	1.0926[20/4]	1.5851[20]	al, eth, ace, bz	B5[4], 1360

No.	Name, Synonyms, and Formula	Mol. wt.	Color, crystalline form, specific rotation and λ_{max} (log ε)	b.p. °C	m.p. °C	Density	n_D	Solubility	Ref.
12034	Propenyl benzene, γ-ethoxy $C_2H_5OCH_2CH=CHC_6H_5$	162.23	....	127-8[22]	.....	0.970[15/4]	1.547[15]	....	B6[3], 2404
12035	Propenyl benzene, β-nitro $C_6H_5CH=C(NO_2)CH_3$	163.18	ye nd (peth)	....	65-6			al	B5[4], 1360
12036	Isopropenyl methyl ketone $CH_2=C(CH_3)COCH_3$	84.13	....	98	−54	0.8527[20/4]	1.4220[20]	al	B1[4], 3462
12037	Propenoic acid or Aerylic acid $CH_2=CHCO_2H$	72.06	....	141.6, 48.5[15]	13	1.0511[20/4]	1.4224[20]	w, al, eth, ace, bz	B2[4], 1455
12038	Propenioc acid, 2-methyl or Methacrylic acid $CH_2=C(CH_3)CO_2H$	86.09	pr	162-3[757], 60[12]	16	1.0153[20/4]	1.4314[20]	w, al, eth	B2[4], 1518
12039	2-Propene-1-ol, 2-methyl or Methallyl alcohol $CH_2=C(CH_3)CH_2OH$	72.11	....	114.5	......	0.8515[20]	1.4255[20]	w, al, eth	B1[4], 2114
12040	2-Propene-1-ol, 1-phenyl or α-Vinyl benzyl alcohol $CH_2=CH-CH(OH)C_6H_5$	134.18	....	215-6, 111[18]	......	1.0251[21/0]	1.5406[20]	al, eth, bz, chl	B6[3], 2417
12041	β-Propiolactone $\overline{CH_2CH_2CO}$	72.06	....	162d, 51[10]	−33.4	1.1460[20/5]	1.4105[20]	eth, chl	B17[4], 4157
12042	Propionaldehyde or Propanal CH_3CH_2CHO	58.08	....	48.8	−81	0.8058[20/4]	1.3636[20]	w, al, eth	B1[4], 3165
12043	Propionaldehyde, 2-Bromo $CH_3CHBrCHO$	136.98	....	109-10, 52-4[80]	......	1.592[20]	1.4813[20]	eth	B1[4], 3177
12044	Propionaldehyde, 2-chloro $CH_3CHClCHO$	92.53	....	86	......	1.182[15/4]	1.431[17]	eth, bz	B1[3], 2691
12045	Propionaldehyde, 3-chloro $ClCH_2CH_2CHO$	92.53	....	130-1, 40[19]	......	1.268[15]	1.475[25]	al, eth	B1[4], 3174
12046	Propionaldehyde, 3-chloro, diethyl acetal $ClCH_2CH_2CH(OC_2H_5)_2$	166.65	....	84[25]	......	0.9951[19/4]	1.4268[20]	ace, bz	B1[3], 2692
12047	Propionaldehyde, 2,3-dibromo $BrCH_2CHBrCHO$	215.88	pa ye fum liq	73-5[10]	......	2.198[15]	1.5082[20]	eth	B1[4], 3178
12048	Propionaldehyde, 2,2-dichloro CH_3CCl_2CHO	126.97	(peth)	80	38-9	....			B1[4], 3174
12049	Propionaldehyde, 2,3-dichloro $ClCH_2CHClCHO$	126.97	hyd (w)	73[50]	......	1.400[20]	1.4762[20]		B1[4], 3174
12050	Propionaldehyde diethyl acetal or 1,1-Diethoxypropane $CH_3CH_2CH(OC_2H_5)_2$	132.21	....	122.8[744]	......	0.8232[20/4]	1.3924[19]	w, al, eth, ace, bz	B1[4], 3168
12051	Propionaldehyde, 2,3-epoxy $\overline{CH_2CHCHO}$	72.06	....	112-3	−62	....	1.4265[20]		B17[4], 4159
12052	Propionaldehyde, 3-ethoxy, diethyl acetal or 1,1,3-Triethoxy propane $C_2H_5OCH_2CH_2CH(OC_2H_5)_2$	176.26	....	184-6d, 78[14]	......	0.898[19/4]	1.4067[20]	al	B1[4], 3971
12053	Propionaldehyde, 3-hydroxy-2-oxo (enol form) or Reductone $HOCH=C(OH)CHO$	88.06	ye nd (w)	....	200-20d			w, al	B1[4], 4145
12054	Propionaldehyde oxime or Propionaldoxime $CH_3CH_2CH=NOH$	73.10	....	131.5	40	0.9258[20/4]	1.4287[20]		B1[4], 3170
12055	Propionaldehyde, 2-phenoxy $CH_3CH(OC_6H_5)CHO$	150.18	....	229-30, 99-101[19]	......			al, eth, bz	B6, 151
12056	Propionaldehyde, 2-phenoxy, oxime $CH_3CH(OC_6H_5)CH=NOH$	165.19	nd (dil al)	....	110			al, eth	B6, 151
12057	Propionaldehyde, 2-phenyl or Hydratropic aldehyde $CH_3CH(C_6H_5)CHO$	134.18	....	202-5, 92.5[10]	......	1.0089[20/4]	1.5176[20]	al	B7[3], 1050
12058	Propionaldehyde, 3-phenyl or Hydrocinnamaldehyde $C_6H_5CH_2CH_2CHO$	134.18	mcl	223[745], 104-5[13]	47			al, eth	B7[3], 1046
12059	Propionaldehyde, 3-(4-tolyl) $(4-CH_3C_6H_4)CH_2CH_2CHO$	148.20	....	122[15]	......	0.999[14/4]	1.525[14]		B7[2], 247
12060	Propionaldehyde, 2,2,3-trichloro $ClCH_2CCl_2CHO$	161.42	....	63-5[45]	......	1.470[25]	1.473[25]	eth	B1[3], 2693
12061	Propionamide $CH_3CH_2CONH_2$	73.10	rh, pl (bz)	213	81.3	0.9262[110]	1.4180[110]	w, al, ace, chl	B2[4], 725
12062	Propionamide, 3-bromo $CH_2BrCH_2CONH_2$	152.00	cr (w)	....	111			al, eth, ace	B2[4], 766
12063	Propionamide, 3-chloro-N-(2-tolyl) $ClCH_2CH_2CONH(C_6H_4CH_3-2)$	197.67	(w, dil al)	....	78			al	B12[2], 441
12064	Propionamide, N,N-diethyl $CH_3CH_2CON(C_2H_5)_2$	129.20	....	191, 81-5[20]	......		1.4425[20]	al	B4[4], 353
12065	Propionamide, 2-hydroxy $CH_3CH(OH)CONH_2$	89.09	....	....	75.5	1.1381[80/4]		w, al	B3[4], 674
12066	Propionamide, 2-phenoxy $CH_3CH(OC_6H_5)CONH_2$	165.19	nd or pl (to, w)	....	132-3			al, eth, aa	B6[4], 643

PHYSICAL CONSTANTS OF ORGANIC COMPOUNDS (Continued)

No.	Name, Synonyms, and Formula	Mol. wt.	Color, crystalline form, specific rotation and λ_{max} (log ε)	b.p. °C	m.p. °C	Density	n_D	Solubility	Ref.
12067	Propionamide, 3-phenoxy .... $C_6H_5OCH_2CH_2CONH_2$	165.19	nd (w)	....	119	....	....	al, eth	B6[3], 616
12068	Propionamide, N-phenyl or Propionanilide........... $CH_3CH_2CONHC_6H_5$	149.19	pl (eth, al, bz)	222.2	105-6	1.175	....	al, eth	B12[3], 472
12069	Propionamide, N-propionyl or Dipropionamide........ $(CH_3CH_2CO)_2NH$	129.16	nd (w, eth)	210-20 sub	154	....	....	....	B2[4], 727
12070	Propionamide, N-(2-tolyl) .... $CH_3CH_2CONH(C_6H_4CH_3-2)$	163.22	nd (bz)	298-9	89.5	....	....	al, eth, chl, aa	B12[2], 440
12071	Propionamide, N-(3-tolyl) .... $CH_3CH_2CONH(C_6H_4CH_3-3)$	163.22	nd (eth)	....	81	....	....	al, eth, lig	B12, 861
12072	Propionic acid or Propanoic acid .... $CH_3CH_2CO_2H$	74.08	....	141, 41.6[10]	-20.8	0.9930[20]	1.3809[20]	**w, al, eth**	B2[4], 695
12073	Propionic acid, allyl ester or Allyl propionate .... $CH_3CH_2CO_2CH_2CH=CH_2$	114.15	....	124-5[774]	....	0.9140[20]	1.4105[20]	al, eth, ace	B2[4], 711
12074	Propionic acid anhydride or Propionic anhydride..... $(CH_3CH_2CO)_2O$	130.15	....	168[71.2], 67.5[18]	-45	1.0110[20/4]	1.4038[20]	**eth**	B2[4], 722
12075	Propionic acid, β-benzoyl or β-Benzoyl propionic acid.... $C_6H_5COCH_2CH_2CO_2H$	178.19	lf (dil al)	....	116	....	....	al, eth, bz, chl	B10[3], 3035
12076	Propionic acid, 2-bromo-(dl) .... $CH_3CHBrCO_2H$	152.98	pr	203.5, 96[10]	25.7	1.7000[20/4]	1.4753[20]	w, al, eth	B2[4], 761
12077	Propionamide, 2-bromo-N-(2-tolyl) .... $CH_3CHBrCONH(C_6H_4CH_3-2)$	242.12	nd	....	131	....	....	al, eth, bz, chl	B12, 794
12078	Propionic acid, 2-bromo ethyl ester .... $CH_3CHBrCO_2C_2H_5$	181.04	....	159-61d, 71[11]	....	1.4135[20/4]	1.4490[20]	**al, eth**, chl	B2[4], 762
12079	Propionic acid, 2-bromo,methyl ester(d) .... $CH_3CHBrCO_2CH_3$	167.01	$[\alpha]^{17}_D$ +42.65	144, 61-2[26]	....	1.482[17]	....	al, eth, chl	B2, 253
12080	Propionic acid, 2-bromo methyl ester (l) .... $CH_3CHBrCO_2CH_3$	167.01	$[\alpha]^{20}_D$ -35.5	61-3[32]	....	1.484[20]	....	al, eth	B2[2], 229
12080a	Propionic acid, 2-bromo,methyl ester-(dl) .... $CH_3CHBrCO_2CH_3$	167.01	....	143-5, 51.5[19]	....	1.4966[25/4]	1.4451[22]	al, eth	B9[3], 2423
12081	Propionic acid, 2-bromo-2-phenyl (dl) .... $CH_3CBr(C_6H_5)CO_2H$	229.08	pl (CS_2)	....	93-4	....	....	bz	B2[4], 764
12082	Propionic acid, 3-bromo .... $BrCH_2CH_2CO_2H$	152.98	pl (CCl_4)	140-2[45]	62.5	1.48	....	al, eth, bz, chl	B2[4], 765
12083	Propionic acid, 3-bromo, butyl ester .... $CH_2BrCH_2CO_2C_4H_9$	209.09	....	130[26]	....	1.4549[20]	1.3051[20]	al, eth, ace	B2[4], 765
12084	Propionic acid, 3-bromo, ethyl ester .... $CH_2BrCH_2CO_2C_2H_5$	181.04	....	179, 70[12]	....	....	1.4516[20]	al, eth, ace	B2[4], 765
12085	Propionic acid, 3-bromo,methyl ester .... $CH_2BrCH_2CO_2CH_3$	167.01	....	105.5[60]	....	1.4897[15]	1.4542[20]	al, eth, ace	B2[4], 762
12087	Propionic acid, 3-bromo,isopentyl ester .... $CH_2BrCH_2CO_2-i-C_5H_{11}$	232.12	....	110-1[11]	....	1.2217[15/4]	1.4556[9]	al, eth	B2[2], 231
12088	Propionic acid, 3-bromo-2-phenyl .... $CH_2BrCH(C_6H_5)CO_2H$	229.08	pr (CS_2)	....	93-4	....	....	al, eth, bz	B9, 526
12089	Propionic acid, 3-bromo-3-phenyl .... $C_6H_5CHBrCH_2CO_2H$	229.08	mcl pr (al)	....	137	....	....	al	B9[3], 2401
12090	Propionic acid, 3-bromo-3-phenyl,methyl ester .... $C_6H_5CHBrCH_2CO_2CH_3$	243.11	pr	....	37-8	....	....	al	B9[1], 201
12091	Propionic acid, butyl ester or Butyl propionate .... $CH_3CH_2CO_2C_4H_9$	130.19	....	145.5	-89.5	0.8754[20/4]	1.4014[20]	**al, eth**	B2[4], 708
12092	Propionic acid, isobutyl ester .... $CH_3CH_2CO_2-i-C_4H_9$	130.19	....	136.8, 66.5[60]	-71.4	0.8687[20/4]	1.3973[20]	**al, eth**	B2[4], 709
12093	Propionic acid, sec-butyl ester or sec-Butyl propionate... $C_2H_5CO_2CH(CH_3)C_2H_5$	130.19	....	132-1	....	0.8657[20]	1.3952[20]	al, eth	B2[4], 709
12094	Propionic acid, 2-chloro .... $CH_3CHClCO_2H$	108.53	....	186, 84[12]	....	1.2585[20/4]	1.4380[20]	w, **al, eth**, ace	B2[4], 745
12095	Propionyl chloride, 2-chloro-(d) .... $CH_3CHClCOCl$	126.97	$[\alpha]^{18}_D$ +0.2	110[744]	....	1.2394[7.5]	....	....	B2[4], 747
12096	Propionic acid, 2-chloro, butyl ester .... $CH_3CHClCO_2C_4H_9$	164.63	....	183-5, 72-3[10]	....	1.0253[20/4]	1.4263[20]	eth	B2[4], 746
12097	Propionic acid, 2-chloro, ethyl ester .... $CH_3CHClCO_2C_2H_5$	136.58	....	147-8, 52-4[18]	....	1.0793[20/4]	1.4178[20]	**al, eth**	B2[4], 746
12098	Propionic acid, 2-chloro, isobutyl ester (d) .... $CH_3CHClCO_2-i-C_4H_9$	164.63	$[\alpha]_D$ +5.2	175-7	....	1.0312[20/4]	1.4247[20]	al, eth, ace	B2, 248
12099	Propionic acid, 2-chloro, isopropyl ester .... $CH_3CHClCO_2CH(CH_3)_2$	150.61	....	151-2, 46-7[12]	....	1.0315[20/4]	1.4149[20]	al, eth	B2[4], 746

No.	Name, Synonyms, and Formula	Mol. wt.	Color, crystalline form, specific rotation and λ_{max} (log ε)	b.p. °C	m.p. °C	Density	n_D	Solubility	Ref.
12100	Propionic acid, 2-chloro, methyl ester (d) .... CH$_3$CHClCO$_2$CH$_3$	122.55	$[\alpha]_D$ +19.9 (undil)	133-4, 50^{35}	....	1.1815$^{20/4}$	....	al	B2, 248
12101	Propionic acid, 2-chloro, methyl ester-(dl) .... CH$_3$CHClCO$_2$CH$_3$	122.55	....	132.5	......	1.209$^{20/4}$	1.4182^{20}	al	B2^4, 746
12102	Propionic acid, 2-chloro, methyl ester-(l) .... CH$_3$CHClCO$_2$CH$_3$	122.55	$[\alpha]_D$ −26.9 (un-dil)	79-80^{120}	......	1.158$^{5/4}$	....	al	B2^3, 553
12103	Propionamide, 2-chloro-N-(2-tolyl) .... CH$_3$CHClCONH(C$_6$H$_4$CH$_3$-2)	197.67	nd (abs al)	....	111			al	B12, 794
12104	Propionic acid, 3-chloro .... CH$_2$ClCH$_2$CO$_2$H	108.53	lf (w), hyg cr (lig)	204d	41 (61)			w, al, eth	B2^4, 748
12105	Propionic acid, 3-chloro, butyl ester .... CH$_2$ClCH$_2$CO$_2$C$_4$H$_9$	164.63	....	104^{22}	......	1.0370$^{20/4}$	1.4321^{20}	w, eth	B2^4, 749
12106	Propionic acid, 3-chloro, isobutyl ester .... CH$_2$ClCH$_2$CO$_2$-i-C$_4$H$_9$	164.63	....	191.3	......	1.0323$^{20/4}$	1.4295^{20}	al, eth	B2^4, 749
12107	Propionic acid, 3-chloro, ethyl ester .... CH$_2$ClCH$_2$CO$_2$C$_2$H$_5$	136.58	....	162, 56^{11}	......	1.1086$^{20/4}$	1.4254^{20}	al, eth	B2^4, 749
12108	Propionic acid, 3-chloro, methyl ester .... CH$_2$ClCH$_2$CO$_2$CH$_3$	122.55	....	155-7, 40-2^{10}	......	1.1861$^{15/4}$	1.4263^{20}	al	B2^4, 748
12109	Propionic acid, 3-chloro, isopentyl ester .... CH$_2$ClCH$_2$CO$_2$-i-C$_5$H$_{11}$	178.66	....	207-8^{740}, 87^{12}	......	1.0171$^{20/4}$	1.4343^{20}	al, eth	B2^4, 749
12110	Propionic acid, 3-chloro, propyl ester .... CH$_2$ClCH$_2$CO$_2$C$_3$H$_7$	150.61	....	180, 77-8^{12}	......	1.0656$^{20/4}$	1.4290^{20}	al, eth	B2^4, 749
12111	Propionic acid, 3-cyclohexyl .... C$_6$H$_{11}$CH$_2$CH$_2$CO$_2$H	156.23		275-8, 143^{11}	16	0.9966$^{20/4}$	1.4634^{20}	eth	B9^3, 64
12112	Proprionic acid, cyclohexyl ester or Cyclohexyl propionate .... CH$_3$CH$_2$CO$_2$C$_6$H$_{11}$	156.23	....	193^{750}, 93^{35}	......	0.9359$^{20/4}$	1.4403^{20}	al, eth, ace	B6^4, 37
12113	Propionic acid, 2,3-diamino .... H$_2$NCH$_2$CH(NH$_2$)CO$_2$H	104.12	hyg rosettes	....	ca 110-20				B4^3, 1292
12114	Propionic acid, 2,3-dibromo .... CH$_2$BrCHBrCO$_2$H	231.88	....	220-40d, 160^{20}	66-7			al, eth, bz	B2^4, 767
12115	Propionic acid, 2,3-dibromo, ethyl ester .... CH$_2$BrCHBrCO$_2$C$_2$H$_5$	259.95	....	214-5, 112^{23}	......	1.7966$^{20/4}$	1.5007^{20}	al, eth	B2^4, 767
12116	Propionic acid, 2,3-dibromo, methyl ester .... CH$_2$BrCHBrCO$_2$CH$_3$	245.91	....	206, 115^{25}	......	1.9333$^{20/4}$	1.5127^{20}	al	B2^4, 767
12117	Propionic acid, 2,3-dibromo-3-phenyl (d) or d-Cinnamic acid dibromide........ C$_6$H$_5$CHBrCHBrCO$_2$H	307.99	pr (chl), $[\alpha]^{15}_D$ +45.8 (abs al)	182				al, eth	B9^3, 2404
12118	Propionic acid, 2,3-dibromo-3-phenyl (dl) .... C$_6$H$_5$CHBrCHBrCO$_2$H	307.99	mcl pr (chl)	sub	240d			al, eth	B9^3, 2405
12119	Propionic acid, 2,3-dibromo-3-phenyl (meso) .... C$_6$H$_5$CHBrCHBrCO$_2$H	307.99	nd	....	91.3			bz	B9^3, 2406
12120	Propionic acid, 2,3-dibromo-3-phenyl, ethyl ester-(d) .... C$_6$H$_5$CHBrCHBrCO$_2$C$_2$H$_5$	336.04	cr (CS$_2$)		71			al	B9, 518
12121	Propionic acid, 2,3-dibromo-3-phenyl, ethyl ester-(dl) .... C$_6$H$_5$CHBrCHBrCO$_2$C$_2$H$_5$	336.04	mcl pr or pl		15-6				B9^3, 2406
12122	Propionic acid, 2,2-dichloro .... CH$_3$CCl$_2$CO$_2$H	142.97	....	185-90, 90-2^{14}	......	1.389$^{12/4}$	....	al, eth	B2^4, 753
12123	Propionic acid, 2,2-dichloro-3,3,3-trifluoro, methyl ester CF$_3$CCl$_2$CO$_2$CH$_3$	210.97	....	116-7^{625}	......	1.5092^{20}	1.3806^{20}	al, eth	B2^4, 758
12124	Propionic acid, 2,2-dichloro-3,3,3-trifluoro, propyl ester CF$_3$CCl$_2$CO$_2$C$_3$H$_7$	239.02	....	144.5^{625}	......	1.3531^{20}	1.3888^{20}	al, eth	B2^4, 758
12125	Propionic acid, 2,3-dichloro .... CH$_2$ClCHClCO$_2$H	142.97	hyg nd (peth)	210, 113^{12}	50		1.4650^{20}	al, eth, w	B2^4, 756
12126	Propionic acid, 2,3-dichloro, ethyl ester .... CH$_2$ClCHClCO$_2$C$_2$H$_5$	171.03	....	183-4, 76-7^{15}	......	1.2461$^{20/4}$	1.4482^{20}	al, eth	B2^4, 756
12127	Propionic acid, 2,3-dichloro, methyl ester-(d) .... CH$_2$ClCHClCO$_2$CH$_3$	157.00	$[\alpha]^{20}_D$ +1.7	92^{50}	......	1.3282$^{20/4}$	....	al, eth, ace	B2^1, 111
12128	Propionic acid, 2,3-dichloro, isopropyl ester .... CH$_2$ClCHClCO$_2$CH(CH$_3$)$_2$	185.05	....	61-2^5	......	1.2010	1.4470	....	B2^4, 756
12129	Propionic acid, 3,3-dichloro .... CHCl$_2$CH$_2$CO$_2$H	142.97	pr	....	56	....	....	al, ace, bz, chl	B2^4, 758
12130	Propionic acid, 2,3-dihydroxy or Glyceric acid .... HOCH$_2$CH(OH)CO$_2$H	106.08	syr	d	......	....	....	**w, al, ace**	B3^4, 1050

No.	Name, Synonyms, and Formula	Mol. wt.	Color, crystalline form, specific rotation and λ_{max} (log ε)	b.p. °C	m.p. °C	Density	n_D	Solubility	Ref.
12131	Propionic acid, 2,3-dihydroxy, propyl ester .... HOCH$_2$CH(OH)CO$_2$C$_3$H$_7$	148.16	....	132-4[3]	....	1.1537[20/4]	1.4503[20]	....	B3[3], 854
12132	Propionic acid, 2,2-dimethyl or Pivalic acid .... (CH$_3$)$_3$CCO$_2$H	102.13	nd	164, 70[14]	35	0.905[50]	1.3931[36.5]	al, eth	B2[4], 908
12133	Propionic acid, 2,2-diphenyl .... CH$_3$C(C$_6$H$_5$)$_2$CO$_2$H	226.28	pl (bz-peth), nd (w), lf (dil al)	sub >300	175-7			al, eth, bz, chl	B9[3], 3342
12134	Propionic acid, 2,3-diphenyl (d) .... C$_6$H$_5$CH$_2$CH(C$_6$H$_5$)CO$_2$H	226.28	cr (dil al), [α]$^{20}_D$ + 94 (bz)	....	83-9			al, eth, bz	B9[3], 3333
12135	Propionic acid, 2,3-diphenyl (dl) .... C$_6$H$_5$CH$_2$CH(C$_6$H$_5$)CO$_2$H	226.28	(i) pr (chl) (ii) pl (chl) (iii) cr (MeOH)	330-40	(i)88-9 (ii)95-6 (iii)82			al, eth, bz	B9[3], 3333
12136	Propionic acid, 2,3-diphenyl (l) .... C$_6$H$_5$CH$_2$CH(C$_6$H$_5$)CO$_2$H	226.28	nd (dil al), [α]$^{20}_D$ -85.1 (bz)	....	83-9			al, eth, bz	B9[3], 3333
12137	Propionic acid, 2,3-diphenyl-2-hydroxy or α-Benzylmandelic acid .... C$_6$H$_5$CH$_2$C(C$_6$H$_5$)(OH)CO$_2$H	242.28	nd (bz), cr (dil al)		165.6			al, eth, aa	B10[3] 1193
12138	Propionic acid, 3,3-diphenyl or Benzhydryl acetic acid ... (C$_6$H$_5$)$_2$CHCH$_2$CO$_2$H	226.28	nd (dil al)		155			al, eth	B9[3], 3338
12139	Propionic acid, 3,3-diphenyl-2-hydroxy or β,β-Diphenyl lactic acid .... (C$_6$H$_5$)$_2$CHCH(OH)CO$_2$H	242.28	nd (w)		159d			al, eth	B10[2], 228
12140	Propionic acid, 3,3-diphenyl-3-hydroxy .... (C$_6$H$_5$)$_2$C(OH)CH$_2$CO$_2$H	242.28	nd (dil al)		212			al, ace, aa	B10[3], 1196
12141	Propionic acid, 3,3-diphenyl-3-hydroxy, ethyl ester ...... (C$_6$H$_5$)$_2$C(OH)CH$_2$CO$_2$C$_2$H$_5$	270.33	pl (dil al)		67			al	B10[4], 477
12142	Propionic acid, 2,3-epoxy or Glycidic acid............. CH$_2$CHCO$_2$H (with O bridge)	88.06						w, al, eth	B18[1], 435
12143	Propionic acid, ethyl ester or Ethyl propionate .... CH$_3$CH$_2$CO$_2$C$_2$H$_5$	102.13	....	99.1	-73.9	0.8917[20]	1.3839[20]	al, eth, ace	B2[4], 705
12144	Propionic acid, 3-fluoro .... FCH$_2$CH$_2$CO$_2$H	92.07		83-4[14]			1.3889[25]	w, al, eth	B2[4], 734
12145	Propionic acid, 3-(2-furyl) or Furfuryl acetic acid .... (2-C$_4$H$_3$O)CH$_2$CH$_2$CO$_2$H	140.15	cr (chl-lig, w, peth)	229, 108-10[10]	58			w, eth, chl	B18[4], 4090
12146	Propionic acid, 3-(2-furyl), ethyl ester .... (2-C$_4$H$_3$O)CH$_2$CH$_2$CO$_2$C$_2$H$_5$	168.20		212, 108-10[10]			1.4812[25]	....	B18[4], 4090
12147	Propionic acid, furfuryl ester or Furfuryl propionate..... CH$_3$CH$_2$CO$_2$CH$_2$(C$_4$H$_3$O)	154.17		195-6[762]	....	1.1085[20/4]		al, eth, ace	B17[2], 115
12148	Propionic acid, heptyl ester or Heptyl propionate .... CH$_3$CH$_2$CO$_2$C$_7$H$_{15}$	172.27		210, 124-5[16]	-50.9	0.8679[20/4]	1.4201[15]	al, eth, ace	B2[4], 710
12149	Propionic acid, hexyl ester or Hexyl propionate .... CH$_3$CH$_2$CO$_2$C$_6$H$_{13}$	158.24		190, 73-4[10]	-57.5	0.8698[20]	1.4162[15]	al, eth, ace	B2[4], 709
12150	Propionic acid, 2-hydroxy (D) or D-Lactic acid......... CH$_3$CH(OH)CO$_2$H	90.08	pl (chl aa), [α]$_D$ -2.3 (w, c=1.24)	103[2]	53			w, al	B3[4], 633
12151	Propionic acid, 3-hydroxy or Hydracrylic acid .... HOCH$_2$CH$_2$CO$_2$H	90.08	syr	d	....		1.4489[20]	w, al, eth	B3[4], 689
12152	Propionic acid, 3-hydroxy, lactone or β-Propiolactone ... CH$_2$CH$_2$CO (lactone)	72.06		162d, 51[10]	-33.4	1.1460[20/5]	1.4105[20]	eth, chl	B17[4], 4157
12153	Propionic acid, 3-Hydroxy, methyl ester................ HOCH$_2$CH$_2$CO$_2$CH$_3$	104.11		121[94]	....	1.118[25]	1.4306[23]	w, al, eth	B3[3], 525
12154	Propionic acid, 3-(2-hydroxyphenyl) or o-Hydrocoumaric acid.............. (2-HOC$_6$H$_4$)CH$_2$CH$_2$CO$_2$H	166.18	pr (w)	....	82-4			w, al, eth	B10[3], 534
12155	Propionic acid, 3-(4-hydroxy phenyl) .... (4-HOC$_6$H$_4$)CH$_2$CH$_2$CO$_2$H	166.18		208-10[14]	129-30			w, al, eth, bz	B10[3], 539
12156	Propionic acid, 3-(4-hydroxy-3-methoxy phenyl) or Hydroferulic acid.............. (3-CH$_3$O-4-HOC$_6$H$_3$)CH$_2$CH$_2$CO$_2$H	196.25	pl (w)	....	89-90			al, eth	B10[3], 1517
12157	Propionic acid, 2-hydroxy-3,3,3-trichloro .... Cl$_3$CCH(OH)CO$_2$H	193.42	pr (eth)	140-7[49]	125			al, eth, chl	B3[4], 680

No.	Name, Synonyms, and Formula	Mol. wt.	Color, crystalline form, specific rotation and λ_{max} (log ε)	b.p. °C	m.p. °C	Density	n_D	Solubility	Ref.
12158	Propionic acid, 2,2´-imino-di-(dl) .... HN[CH(CH₃)CO₂H]₂	161.16	(i) nd (w), pr (eth) (ii) cr	....	(i) 234-5d (ii) 254-5			w	B4³, 1251/1252
12159	Propionic acid, 3,3-imino-di, N-methyl, diethyl ester ..... CH₃N[CH₂CH₂CO₂C₂H₅]₂	231.30	....	136-8⁴		1.0190²⁰′²⁰	1.4421²⁰	al, eth	B4², 829
12160	Propionic acid, β-(3-indolyl) .... C₁₁H₁₁NO₂	189.22	pl (w)		134			al, eth, ace, bz, chl	B22⁴, 1112
12161	Propionic acid, β-(3-indolyl), methyl ester .... C₁₂H₁₃NO₂	203.24	pr (MeOH)		79-80			al	B22², 53
12162	Propionic acid, 2-iodo (dl) CH₃CHICO₂H	199.98	nd (bz, w, al)	93-6⁰·²	45-7	2.073¹⁸′⁴		al, eth	B2³, 573
12163	Propionic acid, 3-iodo .... ICH₂CH₂CO₂H	199.98	lf (w)	....	85			al, eth, ace	B2⁴, 770
12164	Propionic acid, 3-iodo, methyl ester .... ICH₂CH₂CO₂CH₃	214.00	....	188⁷⁵⁶	......	1.8408⁷	....	al	B2³, 574
12165	Propionic acid, 2-mercapto CH₃CH(SH)CO₂H	106.14	....	106-7¹⁵	10-4	1.1938²⁰′⁴	1.4810²⁰	w, al, eth	B3⁴, 682
12166	Propionic acid, 3-mercapto HSCH₂CH₂CO₂H	106.14	amor	110-12¹⁵	17-9	1.218²¹	1.4911²⁰	w, al, eth	B3⁴, 726
12167	Propionitrile, 2-methoxy CH₃-CH(OCH₃)CN	85.11	....	118⁷⁴⁰		0.8928²⁰′⁴	1.3818²⁰	al	B3⁴, 675
12168	Propionitrile, 3-methoxy CH₃OCH₂CH₂CN	85.11	....	163.5⁷⁶·¹, 85.5⁴⁹		0.9379²⁰′⁴	1.4043²⁰	al, eth	B3⁴, 708
12169	Propionic acid, methyl ester or Methyl propionate .... CH₃CH₂CO₂CH₃	88.12	....	79.9	−87.5	0.9150²⁰′⁴	1.3775²⁰	....	B2⁴, 704
12170	Propionic acid, 2-methyl or Isobuturic acid .... (CH₃)₂CHCO₂H	88.11	....	153.2, 53.7¹⁰	−46.1	0.9681²⁰′⁴	1.3930²⁰	al, eth, w	B2⁴, 843
12171	Propionic acid, 3-(α-naphthyl) (α-C₁₀H₇)CH₂CH₂CO₂H	200.24	cr (bz), nd (al)	179¹¹	156-7			al	B9³, 3219
12172	Propionic acid, 3-(β-naphthyl) (β-C₁₀H₇)CH₂CH₂CO₂H	200.24	lf or nd (w, al)	....	135			al	B9³, 3221
12173	Propionic acid, octyl ester or Octyl propionate .... CH₃CH₂CO₂C₈H₁₇	186.30	....	228	−42.6	0.8663²⁰	1.4221¹⁵′₀	al, eth, bz	B2⁴, 710
12174	Propionic acid, 2-oxo or Pyruvic acid.......... CH₃COCO₂H	88.06	....	165d, 54¹⁰	13.8	1.2272²⁰′⁴	1.4280²⁰	w, al, eth, ace	B3⁴, 1505
12175	Propionic acid, 3-oxo-2-phenyl, ethyl ester .... C₆H₅CH(CHO)CO₂C₂H₅	192.22	pl (chl)	136¹⁶	70-1	1.1204²⁰′²⁰	1.532²¹	al, eth	B10³, 3023
12176	Propionic acid, pentachloro .... CCl₃CCl₂CO₂H	246.31	cr (CCl₄)	....	200-15d				B2², 228
12177	Propionic acid, pentyl ester or Pentyl propionate .... CH₃CH₂CO₂C₅H₁₁	144.22	....	168.6	−73.1	0.8761²⁵′⁴	1.4096¹⁵	al, eth, bz	B2⁴, 709
12178	Propionic acid, isopentyl ester or Isopentyl propionate ... CH₃CH₂CO₂-i-C₅H₁₁	144.22	....	160.7	......	0.8697²⁰′⁴	1.4069²⁰		B2⁴, 709
12179	Propionic acid, 2-phenoxy (D) .... CH₃CH(OC₆H₅)CO₂H	166.18	nd (w), [α]³¹′_D + 39.3 (al,c=1.2)	265-6⁷⁵⁸	87			al, eth	B6⁴, 642
12180	Propionic acid, 2-phenoxy (DL) CH₃CH(OC₆H₅)CO₂H	166.18	nd (w)	265-6, 105-6⁵	115-6	1.1865²⁰′⁴	1.5184²⁰	al, eth	B6⁴, 642
12181	Propionic acid, 2-phenoxy, ethyl ester .... CH₃CH(OC₆H₅)CO₂C₂H₅	194.23	....	243-4, 120-5⁶		1.360¹⁷′⁴			B6⁴, 643
12182	Propionic acid, 3-phenoxy .... C₆H₅OCH₂CH₂CO₂H	166.18	nd (w), lf (lig)	234-45⁷⁷¹, 188-9²⁴	97-8			al, eth	B6⁴, 643
12183	Propionic acid, 3-phenoxy, ethyl ester.......... C₆H₅OCH₂CH₂CO₂C₂H₅	194.23	nd (peth)	170⁴⁰	24	1.0821²⁵′²⁵	1.5007¹⁸	al, eth	B6⁴, 644
12184	Propionic acid, phenyl ester or Phenyl propionate .... CH₃CH₂CO₂C₆H₅	150.18	pr	211, 100¹⁶	20	1.0467²⁵′²⁵	1.4980²⁰	al, eth, bz	B6⁴, 615
12185	Propionic acid, 2-phenyl (d) or Hydratropic acid .... CH₃CH(C₆H₅)CO₂H	150.18	[α]²⁰′_D + 81.1 (al,c=3)	152¹⁶					B9³, 2417
12186	Propionic acid, 3-phenyl or Hydrocinnamic acid .... C₆H₅CH₂CH₂CO₂H	149.19	nd (w)	....	106-8			al, eth	B9³, 2382
12187	Propionic acid, 3-(3-pyrenyl) .... (3-C₁₆H₉)CH₂CH₂CO₂H	274.32	pl (aa)	....	180			ace, bz	E14, 441
12188	Propionic acid, propyl ester or Propyl propionate .... CH₃CH₂CO₂C₃H₇	116.16	....	122.3	−75.9	0.8809²⁰′⁴	1.3935²⁰	al, eth, ace, chl	B2⁴, 707

No.	Name, Synonyms, and Formula	Mol. wt.	Color, crystalline form, specific rotation and λ_{max} (log ϵ)	b.p. °C	m.p. °C	Density	n_D	Solubility	Ref.
12189	Propionic acid, isopropyl ester or Isopropyl propionate... $CH_3CH_2CO_2CH(CH_3)_2$	116.16	....	109-10	....	$0.8660^{20/4}$	1.3872^{20}	al, eth	B2[4], 708
12190	Propionic acid, 2,2,3,3-tetrachloro .... $Cl_2CHCCl_2CO_2H$	211.86	cr (CS_2-chl)	....	76				w	B2[4], 760
12191	Propionic acid, 2,3,3,3 tetrafluoro, ethyl ester ..... $CF_3CHFCO_2C_2H_5$	174.09		108-9	....	$1.289^{20/4}$	1.3260^{20}	....	B2[4], 737
12192	Propionic acid, tetrahydrofurfuryl ester .... $CH_3CH_2CO_2CH_2(C_4H_7O)$	158.20		$204-7^{756}, 85-7^3$	....	$1.044^{20/0}$		al, eth, chl	B17[4], 1104
12193	Propionic acid, 3-(2-tetrahydrofuryl) ($2-C_4H_7O)CH_2CH_2CO_2H$	144.17		$263, 118-20^2$	....	$1.1155^{20/20}$	1.4578^{25}	al, ace	B18[4], 3846
12194	Propionic acid, 3-(2-tetrahydrofuryl), ethyl ester ($2-C_4H_7O)CH_2CH_2CO_2C_2H_5$	172.23		$221-2^{750}, 73^2$	....	$1.024^{7/15}$	1.440^{20}	al, ace	B18[4], 3846
12195	Propionic acid, 2,2,3-trichloro .... $CH_2ClCCl_2CO_2H$	177.42	hyg pr (CS_2)	140^{40}	65-6			al, bz	B2[4], 759
12196	Propionic acid, 2,3,3-triphenyl ($C_6H_5)_2CHCH(C_6H_5)CO_2H$	302.38	nd (dil al, peth)	....	222-3			al, eth	B9[3], 3598
12197	Propionic acid, 3,3,3-triphenyl ($C_6H_5)_3CCH_2CO_2H$	302.38	pr (al)	....	179-80			al, eth	B9[3], 3603
12198	Propionitrile CH_3CH_2CN	55.08		97.3	−92.9	$0.7818^{20/4}$	1.3655^{20}	w, al, eth, ace, bz	B2[4], 728
12199	Propionitrile, 2-bromo $CH_3CHBrCN$	133.98		59^{24}	....	$1.5505^{20/4}$	1.4585^{20}	eth, ace	B2[4], 764
12200	Propionitrile, 3-bromo $BrCH_2CH_2CN$	133.98		$92^{25}, 69^7$	....	$1.6152^{20/4}$	1.4800^{20}	al, eth	B2[4], 766
12201	Propionitrile, 2-chloro .... $CH_3CHClCN$	89.53		$123-4, 73^{144}$	....	1.0792^{10}			B2[4], 748
12202	Propionitrile, 3-chloro .... $ClCH_2CH_2CN$	89.53		$175-6, 65.7^{10}$	....	1.1573^{20}	1.4360^{20}		B2[4], 751
12203	Propionitrile, 2,3-dichloro $CH_2ClCHClCN$	123.97		$58-9^8$	....	1.3500^{20}	1.4640^{20}		B2[4], 757
12204	Propionitrile, 2,2-dimethyl .... $(CH_3)_3CCN$	83.13		105-6	15-6	$0.7586^{25/4}$	1.3774^{20}	....	B2[3], 913
12205	Propionitrile, 2-(dimethylamino) $[(CH_3)_2N]CH(CH_3)CN$	98.15		144				al, eth	B4[3], 1236
12206	Propionitrile, 2-ethoxy .... $CH_3CH(OC_2H_5)CN$	99.13		136^{765}		$0.8743^{20/4}$	1.3890^{22}	al, eth, aa	B3[4], 675
12207	Propionitrile, 3-ethoxy $C_2H_5OCH_2CH_2CN$	99.13		$171, 65^{15}$		$0.9285^{15/4}$	1.4068^{20}	al, eth	B3[4], 708
12208	Propionitrile, 2-hydroxy or Acetaldehyde cyanohydrin ... $CH_3CH(OH)CN$			$182-4, 102^{30}$	−40	$0.9877^{20/4}$	1.4058^{18}	w, al, eth, chl	B3[4], 675
12209	Propionitrile, 2-hydroxy-3,3,3-trichloro or Chloral cyanohydrin. $Cl_3CCH(OH)CN$	174.42	pl (w)	215-20d	61			w, al, eth	B3[4], 680
12210	Propionitrile, 3-hydroxy or Hydracrylonitrile .... $HOCH_2CH_2CN$	71.08	....	$230, 110^{15}$		1.0588^0	1.4240^{20}	w, al	B3[4], 708
12211	Propionitrile, 2,2'-imino-di- or 2,2'-imino dipropionitrile $HN[CH(CH_3)CN]_2$	123.16	nd (eth)	....	68			al, eth	B4[3], 1251
12212	Propionitrile, 3-(2-oxocyclohexyl) .... $C_9H_{13}NO$	151.21		$138-42^{10}$		$1.0181^{20/4}$	1.4755^{20}	....	B10[3], 2835
12213	Propionitrile, 2-propoxy $CH_3CH(OC_3H_7)CN$	113.16		150^{727}		$0.866^{20/4}$	1.398^{20}		B3, 285
12214	Propionitrile, 3-phenyl $C_6H_5CH_2CH_2CN$	131.18		$261, 125-6^{15}$		1.0016^{20}	1.5266^{20}	al, eth	B9[3], 2395
12215	Propionyl bromide CH_3CH_2COBr	136.98		$103-1^{770}$		$1.5210^{16/4}$	1.4578^{16}	eth	B2[4], 724
12216	Propionyl bromide, 2-bromo $CH_3CHBrCOBr$	215.88		152-4		$2.0612^{16/4}$			B2[4], 764
12217	Propionyl chloride CH_3CH_2COCl	92.53		80	−94	1.0646^{20}	1.4032^{20}	eth	B2[4], 724
12218	Propionyl chloride, 2-bromo $CH_3CHBrCOCl$	171.43		131-3	....	1.697^{11}		eth, chl	B2[4], 764
12219	Propionyl chloride, 3-chloro CH_2ClCH_2COCl	126.97	yesh	$143-5^{763}, 82^{102}$		1.3307^{13}	1.4549^{20}	al, eth, chl	B2[4], 750
12220	Propionyl chloride, 2,2-dichloro CH_3CCl_2COCl	161.42		$117-8^{750}, 68-72$		$1.4062^{20/4}$	1.4524^{20}		B2[4], 755
12221	Propionyl chloride, 2,3-dichloro $CH_2ClCHClCOCl$	161.42		$52-4^{16}$		$1.4757^{20/4}$	1.4764^{20}	eth	B2[4], 757

No.	Name, Synonyms, and Formula	Mol. wt.	Color, crystalline form, specific rotation and λ_{max} (log ϵ)	b.p. °C	m.p. °C	Density	n_D	Solubility	Ref.
12222	Propionyl chloride, 3,3-dichloro .... CHCl$_2$CH$_2$COCl	161.42	....	43-4[10]	......	1.4557[20/4]	1.4738[20]	eth, diox	B2[3], 562
12223	Propionyl chloride, pentachloro .... CCl$_3$CCl$_2$COCl	264.75	nd		42			bz	B2[4], 761
12224	Propionyl chloride, 2-phenoxy .... CH$_3$CH(OC$_6$H$_5$)COCl	184.62	....	146-7[55], 115-7[10]	......	1.1865[20/4]	1.5178[20]	eth	B6[4], 643
12225	Propionyl fluoride .... CH$_3$CH$_2$COF	76.07	....	44		0.972[15/4]	1.329[13]	....	B2[4], 724
12226	Propionyl iodide .... CH$_3$CH$_2$COI	183.98	....	127-8				....	B2[3], 542
12227	Propioryl urea .... CH$_3$CH$_2$CONHCONH$_2$	116.13	cr (w)		210-11			w, al	B3[3], 121
12228	Propiophenone or Ethyl phenyl ketone..... C$_6$H$_5$COC$_2$H$_5$	134.18	....	217.5, 91.6[10]	18.6	1.0096[20]	1.5269[20]	al, eth	B7[3], 1022
12229	Propiophenone, α-amino, hydrochloride .... C$_6$H$_5$COCH(NH$_2$)CH$_3$.HCl	185.66	nd (al-eth)		187			w	B14[3], 147
12230	Propiophenone, o-amino (2-H$_2$NC$_6$H$_4$)COCH$_2$CH$_3$	149.19	pa ye lf (peth) pl (dil al)	93[0.8]	46-7			w, al, eth, ace	B14[3], 143
12231	Propiophenone, [p] -amino (4-H$_2$NC$_6$H$_4$)COCH$_2$CH$_3$	149.19	pl (al, w), nd (w)		140			w, al, chl	B14[3], 146
12232	Propiophenone, α-bromo C$_6$H$_5$COCHBrCH$_3$	213.07	....	245-50, 134-5[18]		1.4298[20/4]	1.5720[20]	al, eth, ace, bz, chl	B7[3], 1033
12233	Propiophenone, [p] -bromo (4-BrC$_6$H$_4$)COC$_2$H$_5$	213.08	nd	169[15]	48			al, eth, ace	B7[3], 1032
12234	Propiophenone, β-chloro C$_6$H$_5$COCH$_2$CH$_2$Cl	168.63	lf (eth), cr (peth al)		49-50				B7[3], 1030
12235	Propiophenone, [p]- chloro (4-ClC$_6$H$_4$)COC$_2$H$_5$	168.63	....	134-7[31]	36-7			al	B7[3], 1029
12236	Propiophenone, [p] -chloro, oxime 4-ClC$_6$H$_4$C(=NOH)C$_2$H$_5$	183.64	pl (al)		62-3			al	B7, 301
12237	Propiophenone, α,β-dibromo-β-phenyl or [threo] -Chalone dibromide C$_6$H$_5$COCHBrCHBrC$_6$H$_5$	368.07	nd (al)		122-3			al	B7[3], 2155
12238	Propiophenone, [m] -methoxy (3-CH$_3$OC$_6$H$_4$)COCH$_2$CH$_3$	164.21	....	258-60, 95-7[0.7]		1.0812[0]	1.5230[25]	al, eth	B8, 106
12239	Propiophenone, β,β-diphenyl .... C$_6$H$_5$COCH$_2$CH(C$_6$H$_5$)$_2$	286.38	nd (al)		96			ace, bz, chl	B7[3], 2756
12240	Propiophenone, [p] -methoxy (4-CH$_3$OC$_6$H$_4$)COCH$_2$CH$_3$	164.21	....	267-9, 142[14]	<−15	1.0670[18/4]	1.5253[20]	al, eth	B8[3], 381
12241	Propiophenone, [m] -methyl or Ethyl-3-tolyl ketone ..... (3-CH$_3$C$_6$H$_4$)COC$_2$H$_5$	148.21	....	234[745], 130-5[33]	......	1.0059[0/4]		al, eth, ace, bz	B7[3], 1092
12242	Propiophenone, [p] -methyl or Ethyl-4-tolyl ketone ...... (4-CH$_3$C$_6$H$_4$)COC$_2$H$_5$	148.21	....	238-9, 120[18]	......	0.9926[20/4]	1.5278[20]	al, eth, ace, bz	B7[3], 1093
12243	Propiophenone oxime or Propiophenoxime C$_6$H$_5$C(=NOH)C$_2$H$_5$	149.19	pl (peth)	245-6d, 165[38]	53-5			al, eth	B7[3], 1025
12244	n-Propyl alcohol or 1-Propcenol........... CH$_3$CH$_2$CH$_2$OH	60.11	....	97.4	−126.5	0.8035[20/4]	1.3850[20]	w, al, eth, ace, bz	B1[4], 1413
12245	Isopropyl alcohol or 2-Propanol .... (CH$_3$)$_2$CHOH	60.11	....	82.4	−89.5	0.7855[20/4]	1.3776[20]	w, al, eth, ace, bz	B1[4], 1461
12246	n-Propyl amine or l-Amino propane........... CH$_3$CH$_2$CH$_2$NH$_2$	59.11	....	47.8	−83	0.7173[20/4]	1.3870[20]	w, al, eth, ace, bz, chl	B4[4], 464
12247	Propyl amine, 3-(diethyl amino)............ (C$_2$H$_5$)$_2$NCH$_2$CH$_2$CH$_2$NH$_2$	130.23	....	165-72	0.825[20/ 20]	1.443[20]$_D$			B4[4], 1260
12248	n-Propyl amine, N,N-dimethyl or -l-(dimethyl amino)propane CH$_3$CH$_2$CH$_2$N(CH$_3$)$_2$	87.17	....	65.5[752]		0.7152[20/4]	1.3860[20]	al, eth, bz, chl	B4[4], 467
12249	Propyl amine, N-nitro .... CH$_3$CH$_2$CH$_2$NHNO$_2$	104.12	....	128-9[40]	−21	1.1046[15]	1.4610[20]	al, eth	B4[1], 569
12250	Isopropyl amine or 2-Amino propane............ (CH$_3$)$_2$CHNH$_2$	59.11	nd	32.4	−95.2	0.6891[20]	1.3742[20]	w, al, eth, ace, bz, chl	B4[4], 504
12251	n-Propyl bromide or l-Bromo propane .... CH$_3$CH$_2$CH$_2$Br	123.00	....	71	−110	1.3537[20/4]	1.4343[20]	al, eth, ace, bz	B1[4], 205
12252	Isopropyl bromide or 2-Bromo propane............ (CH$_3$)$_2$CHBr	123.00	....	59.4	−89	1.3140[20/4]	1.4251[20]	al, eth, ace, bz, chl	B1[4], 208
12253	n-Propyl chloride or l-Chloro propane .... CH$_3$CH$_2$CH$_2$Cl	78.54	....	46.6	−122.8	0.8909[20/4]	1.3879[20]	al, eth, bz, chl	B1[4], 189

No.	Name, Synonyms, and Formula	Mol. wt.	Color, crystalline form, specific rotation and λ_{max} (log ε)	b.p. °C	m.p. °C	Density	n_D	Solubility	Ref.
12254	Isopropyl chloride or 2-Chloro propane . . . . (CH$_3$)$_2$CHCl	78.54	. . . .	35.7	−117.2	0.8617[20/4]	1.3777[20]	**al, eth**, bz, chl	B1[4], 191
12255	Propyl, 1,2-dichloro propyl ether . . . . C$_3$H$_7$O(CHClCHClCH$_3$)	171.07		176		1.129[15/4]	1.447[16]	eth	B1[3], 2691
12256	Propyl, 1,3-dichloro propyl ether . . . . C$_3$H$_7$O(CHClCH$_2$CH$_2$Cl)	171.07		65[12]		1.112[20]	1.4476[20]	al, eth	B1[3], 690
12257	Propylene or Propene . . . . CH$_3$CH=CH$_2$	42.08	gas	−47.4	−185.2	0.5193[20/4] liq	1.3567[-70]	al, aa	B1[4], 725
12258	Propylene, 3-amino or Allyl amine . . . . H$_2$NCH$_2$CH=CH$_2$	57.09		58		0.7621[20/4]	1.4205[20]	**w, al, eth**, chl	B4[4], 1057
12259	Propylene, 1-bromo (cis) or 1-Bromopropene, Propenyl bromide . . . . CH$_3$CH=CHBr	120.98		57.8	−113	1.4291[20/4]	1.4560[20]	eth, ace, chl	B1[4], 754
12260	Propylene, 2-bromo or Isopropenyl bromide . . . . CH$_3$CBr=CH$_2$	120.98		48.4[748]	−124.8	1.362[20/4]	1.4440[20]	eth, ace, chl	B1[4], 754
12261	Propylene, 2-bromo-3-cyclohexyl . . . . C$_6$H$_{11}$CH$_2$CBr=CH$_2$	203.13		90[15]		1.215[17]	1.495[17]	eth	B5[3], 224
12262	Propylene, 3-bromo or Allyl bromide . . . . BrCH$_2$CH=CH$_2$	120.98		70[752]	−119.4	1.398[20/4]	1.4697[20]	**al, eth**	B1[4], 754
12263	Propylene, 3-bromo-3,3-difluoro . . . . BrCF$_2$CH=CH$_2$	156.96		42		1.543[25/4]	1.3773[25]	. . . .	B1[4], 756
12264	Propylene, 1-chloro or [cis]-Propenyl chloride . . . . CH$_3$CH=CHCl	76.53		32.8	−134.8	0.9347[20/4]	1.4055[20]	eth, ace, bz, chl	B1[4], 737
12265	Propylene, 1-chloro ([trans]) or [trans]-Propenyl chloride	76.53	. . . .	37.4	−99	0.9350[20/7]	1.4054[20]	eth, ace, bz, chl	B1[4], 737
12266	Propylene, 2-chloro or Isopropenyl chloride . . . . CH$_3$CCl=CH$_2$	76.53		22.6	−137.4	0.9014[20/4]	1.3973[20]	eth, ace, bz, chl	B1[4], 737
12267	Propylene, 3-chloro or Allyl chloride . . . . ClCH$_2$CH=CH$_2$	76.53	. . . .	45	−134.5	0.9376[20/4]	1.4157[20]	**al, eth**, ace, bz, liq	B1[4], 738
12268	Propylene, 3-chloro-2-(chloro methyl) . . . . (ClCH$_2$)$_2$C=CH$_2$	125.00		138, 30-1[9]	−15	1.1782[20/3]	1.4754[20]	al, chl	B1[4], 805
12269	Propylene, 2-cyclopropyl . . . . CH$_2$=C(C$_3$H$_5$)CH$_3$	82.14		69.5-70[751]		0.7500[20/4]	1.4252[20]	. . . .	B5[4], 243
12270	Propylene, 1,1-dibromo . . . . CH$_3$CH=CBr$_2$	199.87		127.4, 41.5[30]		1.9803[20/20]	1.5260[20]	chl, bz	B1[4], 759
12271	Propylene, 2,3-debromo or 2,3-Dibromo propene . . . . . . CH$_2$BrCBr=CH$_2$	199.88		141, 37-7[11]		2.0346[25/4]	1.5416[25]	eth, ace, chl	B1[4], 760
12272	Propylene, 1,1-dichloro or 1,1-Dichloro propene . . . . CH$_3$CH=CCl$_2$	110.97		76-7		1.1864[25/4]	1.4430[25]	eth, ace, chl	B1[4], 742
12273	Propylene, 1,1-dichloro-2-fluoro . . . . CH$_3$CF=CCl$_2$	128.96		77.7[745]		1.3026[25/4]	1.4196[25]	. . . .	B1[4], 745
12274	Propylene, 1,1-dichloro-3-phenyl or Cinnamylidene chloride . . . . C$_6$H$_5$CH$_2$CH=CCl$_2$	187.08	cr (eth, chl), pl (peth)	142-3[30]	59	. . . .	. . . .	eth, bz, chl	B5[4], 1363
12275	Propylene, 1,2-dichloro ([cis]) . . . . CH$_3$CCl=CHCl	110.97	. . . .	92.5[742]			1.4549[20]	ace, bz, chl	B1[4], 742
12276	Propylene, 1,2-dichloro ([trans]) or 1,2-Dichloro propene	110.97	. . . .	77[757]		1.1818[20/4]	1.4471[20]	al	B1[4], 742
12277	Propylene, 1,2-dichloro-1,3,3,3-tetrafluoro . . . . CF$_3$CCl=CFCl	182.93		47.3	−137	1.5468	1.3511[20]	. . . .	B1[4], 747
12278	Propylene, 1,2-dichloro-3,3,3-trifluoro . . . . CF$_3$CCl=CHCl	164.94		53.7	−109.2	1.4653[20/4]	1.3670[20]	. . . .	B1[4], 746
12279	Propylene, 1,3-dichloro ([cis]) or 1,3-Dichloro propene . . ClCH$_2$CH=CHCl	110.97		104.3		1.217[20/4]	1.4730[20]	eth, bz, chl	B1[4], 743
12280	Propylene, 1,3-dichloro-([trans]) . . . . ClCH$_2$CH=CHCl	110.97		112		1.224[20/4]	1.4682[20]	eth, bz, chl	B1[4], 744
12281	Propylene, 2,3-dichloro . . . . CH$_2$ClCCl=CH$_2$	110.97		94		1.211[20/4]	1.4603[20]	**al**, eth, bz, chl	B1[4], 744
12282	Propylene, 3,3-dichloro or 3,3-Dichloro propene . . . . CHCl$_2$CH=CH$_2$	110.97		84.4		1.175[20/4]	1.4510[20]	al, eth, bz, chl	B1[4], 745
12283	Propylene, 1,1-diphenyl or 1,1-Diphenyl propene . . . . . . CH$_3$CH=C(C$_6$H$_5$)$_2$	194.28	lf (al)	280-1, 149[11]	52	1.0250[20/4]	1.5880[20]	al, bz	B5[3], 1998
12284	Propylene, 2,3-diphenyl . . . . C$_6$H$_5$CH$_2$C(C$_6$H$_5$)=CH$_2$	194.28		289[757]	48	1.10143[20/4]	1.5903[20]	. . . .	B5[2], 553
12285	Propylene, 1,2-epoxy or Methyl oxirene . . . . CH$_3$C=CH	56.07		63				**al, eth**	B17, 20
12286	Propylene, 3-fluoro or Allyl fluoride . . . . FCH$_2$CH=CH$_2$	60.07	gas	−3		. . . .	. . . .	al, eth, chl	B1[4], 733

No.	Name, Synonyms, and Formula	Mol. wt.	Color, crystalline form, specific rotation and λ_{max} (log ε)	b.p. °C	m.p. °C	Density	n_D	Solubility	Ref.
12287	Propylene glycol or 1,2-Propanediol CH₃CH(OH)CH₂OH	76.11		189, $96\text{-}8^{21}$		$1.0361^{20/4}$	1.4324^{20}	w, al, eth, bz	B1⁴, 2468
12288	Propylene, 3-iodo or Allyl iodide CH₂=CHCH₂I	167.98		102	−99.3	$1.8494^{20/4}$	1.5530^{20}	al, eth, chl	B1⁴, 761
12289	Propylene, 1-nitro CH₃CH=CHNO₂	87.08		$59\text{-}60^{34}$		$1.0661^{20/4}$	1.4527^{20}	eth, ace, chl	B1⁴, 763
12290	Propylene, 2-nitro CH₂=C(NO₂)CH₃	87.08		57^{100}, 32^{30}		$1.0643^{20/4}$	1.4358^{20}	eth, ace, chl	B1⁴, 764
12291	Propylene, 1,1,2,3,3-pentachloro CHCl₂CCl=CCl₂	214.31		185, 116^{9}		$1.6317^{34/4}$	1.5313^{20}	eth	B1⁴, 753
12292	Propylene, perchloro CCl₃CCl=CCl₂	248.75		209-10, 140^{100}		$1.7652^{20/4}$	1.5455^{20}	chl	B1⁴, 753
12293	Propylene, perfluoro CF₃CF=CF₂	150.02	gas	−29.4	−156.2	$1.583^{-40/4}$			B1⁴, 735
12294	Propylene, 2-phenyl or α-Methyl styrene C₆H₅C(CH₃)=CH₂	118.18		163-4, 60^{17}	24.3	$0.9082^{20/4}$	1.5303^{20}	eth, bz, chl	B5⁴, 364
12295	Propylene, 1,2,3,3-tetrachloro CHCl₂CCl=CHCl	179.86		165, 50.6^{12}		$1.537^{19/4}$	1.5121^{19}	bz, chl	B1⁴, 752
12296	Propylene, 1,1,2-trichloro CH₃CCl=CCl₂	145.42		118, 41^{52}		$1.382^{20/4}$	1.4827^{20}	al, eth, bz, chl	B1⁴, 747
12297	Propylene, 1,1,2-trichloro-3,3,3-trifluoro CF₃CCl=CCl₂	199.39		88.1	−114.6	$1.617^{20/4}$			B1⁴, 751
12298	Propylene, 1,2,3-trichloro CH₂ClCCl=CHCl	145.42		142, $32\text{-}3^{14}$		$1.414^{20/20}$	1.5030^{20}	al, eth, bz, chl	B1⁴, 748
12299	Propylene, 3,3,3-trichloro CCl₃CH=CH₂	145.42		114-5, 57^{102}	−30	$1.369^{20/20}$	1.4827^{20}	al, eth, bz, chl	B1⁴, 749
12300	Propylene oxide or 1,2-Epoxy propane CH₃CHCH₂	58.08		34.3		$0.859^{0/4}$	1.3670^{20}	w, al, eth	B17⁴, 16
12301	Propyl ether or Dipropyl ether C₃H₇OC₃H₇	102.18		91	fr−122	$0.7360^{20/4}$	1.3809^{20}	al, eth	B1⁴, 1422
12302	Isopropyl ether or Diisopropyl ether (CH₃)₂CHOCH(CH₃)₂	102.18		68	−85.9	$0.7241^{20/4}$	1.3679^{20}	al, eth, ace	B1⁴, 1471
12303	Propyl ethynyl ether C₃H₇OC≡CH	84.13		75		$0.8080^{20/4}$	1.3935^{20}	al, eth	B1⁴, 2213
12304	n-Propyl fluoride or 1-Fluoro propane CH₃CH₂CH₂F	62.09		250	−159	$0.7956^{20/4}$	1.3115^{20}	al, eth	B1⁴, 187
12305	Propyl hexedrine or 1-Cyclohexyl-2-methyl amino propane C₁₀H₂₁N	155.29		205, $92\text{-}3^{20}$		$0.8501^{20/4}$	1.4600^{20}	al	B12³, 108
12306	n-Propyl iodide or 1-Iodo propane CH₃CH₂CH₂I	169.99		102.4	−10.3	$1.7489^{20/4}$	1.5058^{20}	al, eth, bz, chl	B1⁴, 222
12307	Isopropyl iodide or 2-Iodo propane (CH₃)₂CHI	169.99		891.4	−90.1	$1.7033^{20/4}$	1.5028^{20}	al, eth, bz, chl	B1⁴, 223
12308	Propyl isocyanide CH₃CH₂CH₂NC	69.11		99.5				al, eth	B4⁴, 474
12309	Isopropyl isocyanide i-C₃H₇NC	69.11		87		0.7596^{0}		al, eth	B4, 154
12310	Propyl isothiocyanate C₃H₇NCS	101.17		153		$0.9781^{16/4}$	1.5085^{16}	al, eth	B4⁴, 491
12311	Isopropyl isothiocyanate or Isopropyl mustard oil (CH₃)₂CHNCS	101.17		138, $29\text{-}30^{10}$				al, eth	B4⁴, 532
12312	Isopropyl methyl ketone or 3-Methyl-2-butanone i-C₃H₇COCH₃	86.14		94-5	−92	$0.8051^{20/4}$	1.3880^{20}	al, eth, ace	B1⁴, 3287
12313	Propyl-α-naphthyl ether C₃H₇O-α-C₁₀H₇	186.26		293.5, 167^{18}		$1.0447^{18/4}$	1.5928^{18}		B6³, 2924
12314	Propyl-β-naphthyl ether C₃H₇O-β-C₁₀H₇	186.26	nd (al)	305, 144^{10}	41			al	B6³, 2973
12315	Propyl nitrate C₃H₇ONO₂	105.10		110^{762}		$1.0538^{20/4}$	1.3973^{20}	al, eth	B1⁴, 1424
12316	Isopropyl nitrate (CH₃)₂CHONO₂	105.10		100-1		$1.036^{19/19}$	1.3912^{16}	al, eth	B1⁴, 1475
12317	Propyl nitrite C₃H₇ONO	89.10		79		0.935^{21}	1.3604^{20}	al, eth	B1⁴, 1424
12318	Isopropyl nitrite (CH₃)₂CHONO	89.10		45		$0.8684^{15/4}$		al, eth	B1⁴, 1474
12319	Propyl isopropyl ether i-C₃H₇OC₃H₇	102.18		83		$0.7370^{20/4}$	1.376^{21}	al, eth, ace	B1⁴, 1471

No.	Name, Synonyms, and Formula	Mol. wt.	Color, crystalline form, specific rotation and λ_{max} (log ε)	b.p. °C	m.p. °C	Density	n_D	Solubility	Ref.
12320	Propyl red or 4′ Dipropyl aminoazobenzene-2-carboxylic acid........... $C_{16}H_{17}N_3O_2$	325.42	vt-bl or purp-red cr (al)	....	....	....	....	w, al	B16³, 368
12321	Propyl sulfide or Dipropyl sulfide $(C_3H_7)_2S$	118.24	....	142.4, 32.3¹⁰	−102.5	0.8377²⁰/⁴	1.4487²⁰	al, eth	B1³, 1432
12322	Propyl sulfide, 2,2′-dichloro or bis(2-chloropropyl)sulfide $(CH_3CHClCH_2)_2S$	187.13	....	122²³	−40	1.1569²⁵/⁴	1.5020²⁰	al	B1³, 1437
12323	Propyl sulfide, 3,3′-dichloro or bis(3-chloropropyl)sulfide $(ClCH_2CH_2CH_2)_2S$	187.13	....	162⁴³	....	1.1774²⁵/⁴	1.5075²⁰	al, eth, to	B1³, 1438
12324	Isopropyl sulfide or Diisopropyl sulfide $[(CH_3)_2CH]_2S$	118.24	....	120	−78.1	0.8142²⁰/⁴	1.4438²⁰	al, eth	B1⁴, 1502
12325	Propyl sulfone or Dipropyl sulfone $(C_3H_7)_2SO_2$	150.24	sc	....	29-30	1.0278⁵⁰/⁴	1.4456³⁰	al, eth	B1⁴, 1453
12326	Isopropyl sulfone or Diisopropyl sulfone $(i\text{-}C_3H_7)_2SO_2$	150.24	(eth)	....	36	....	....	w	B1⁴, 1502
12327	Propyl sulfate or Dipropyl sulfate $(C_3H_7O)_2SO_2$	182.24	....	121²⁰	....	1.1064²⁰/⁴	1.4135²⁰	peth	B1⁴, 1423
12328	Propyl sulfoxide or Dipropyl sulfoxide $(C_3H_7)_2SO$	134.24	nd	80²	22-3	0.9654²⁰/⁴	1.4663²⁰	al, eth	B1⁴, 1453
12329	Propyl thiocyanate C_3H_7SCN	101.17	....	163	....	....	....	al, eth	B3⁴, 329
12330	Isopropyl thiocyanate $(CH_3)_2CHSCN$	101.17	....	152-3⁷⁵⁴	....	0.9784²⁰	....	al, eth	B3⁴, 329
12331	Isopropyl vinyl ether $i\text{-}C_3H_7OCH{=}CH_2$	86.14	....	55-6	....	0.7534²⁰/⁴	1.3840²⁰	al, eth, ace, bz	B1⁴, 2052
12332	Propynal or Propargyl aldehyde $HC{\equiv}CCHO$	54.05	....	59-61	....	....	1.4033²⁵	w, al, eth, ace, liq	B1⁴, 3537
12333	Propyne or Methyl acetylene $CH_3C{\equiv}CH$	40.07	gas	−23.2	−101.5	0.7062⁻⁵⁰	1.3863⁻⁴⁰	al, bz, chl	B1⁴, 958
12334	Propyne, 3-bromo or Propargyl bromide $BrCH_2C{\equiv}CH$	118.97	....	88-90, 33¹³⁰	....	1.579¹⁸	1.4922²⁰	al, eth, bz, chl	B1⁴ 964
12335	Propyne, 3-chloro or Propargyl chloride $ClCH_2C{\equiv}CH$	74.51	....	65	....	1.0297²⁰/⁴	1.4320²⁰	al, eth, bz, chl	B1⁴, 963
12336	Propyne, 3-cyclohexyl $C_6H_{11}CH_2C{\equiv}CH$	122.21	....	157-8, 48¹¹	....	0.8449²⁰/⁴	1.4605²⁰	eth	B5⁴, 419
12337	Propyne, 1,3-dibromo $BrCH_2C{\equiv}CBr$	197.88	....	73-4³⁰	....	2.1894²⁰	1.5690²⁰	eth, chl	B1⁴, 965
12338	Propyne, 1-iodo $CH_3C{\equiv}CI$	165.96	....	110	....	2.08²²	....	eth, al	B1⁴, 965
12339	Propyne, 3-iodo or Propargyl iodide $ICH_2C{\equiv}CH$	165.96	....	115	....	2.0177⁰	....	eth	B1⁴, 965
12340	Propyne, 3-methoxy or Methyl propargyl ether $CH_3OCH_2C{\equiv}CH$	70.09	....	63	....	0.83¹²	1.5035²⁰	al, eth	B1⁴, 2215
12341	Propyne, 1-phenyl $CH_3C{\equiv}CC_6H_5$	116.17	....	183, 77¹⁷	....	0.9388²⁰/⁴	1.5650²⁰	eth	B5⁴, 1530
12342	Propyne, 3,3,3-trifluoro $F_3CC{\equiv}CH$	94.04	gas	−48.3	....	....	....	....	B1⁴, 962
12343	Propynoic acid or Propargylic acid $HC{\equiv}CCO_2H$	70.09	cr (CS₂)	144d, 83-4⁵⁰	18 (anh)	1.1380²⁰/⁴	1.4306²⁰	w, al, eth, ace, chl	B2⁴, 1687
12344	2-Propyn-1-ol or Propargyl alcohol $HC{\equiv}CCH_2OH$	56.07	....	113.6, 30²¹	−48	0.9485²⁰	1.4322²⁰	w, al, eth	B1⁴, 2214
12345	Prostigmine bromide or Neostigmine bromide $C_{12}H_{19}O_2N_2Br$	303.21	(al-eth)	....	167d	....	....	w, al	B13³, 939
12346	Protopine or Fumarine $C_{20}H_{19}NO_5$	353.38	mcl pr (al-chl)	....	208	....	....	chl	B27², 625
12347	Protoveratridine $C_{32}H_{51}NO_9$	593.76	cr (al-chl), $[\alpha]^{20}/_D$ −14 (Py, c=1)	....	272-3	....	....	al	B21⁴, 6801
12348	Protoveratrine $C_{39}H_{63}NO_{13}$	751.93	pl (ace), $[\alpha]^{24}/_D$ −8.3 (chl)	....	225d	....	....	....	B21⁴, 6845
12349	Protoveratrine A $C_{41}H_{63}NO_{14}$	793.96	lf (al), $[\alpha]^{20}/_D$ −44.1 (Py, c=1.12)	....	305d	....	....	....	B21⁴, 6845

No.	Name, Synonyms, and Formula	Mol. wt.	Color, crystalline form, specific rotation and λ_{max} (log ε)	b.p. °C	m.p. °C	Density	n_D	Solubility	Ref.
12350	Protoveratrine B or Neoprotoveratrine............. $C_{41}H_{63}NO_{15}$	809.96	lf (chl-al), $[\alpha]^{20}_D$ −39.8 (Py, c=1.24)	....	285-90d	....			B21⁴, 6847
12351	Protoverine .... $C_{27}H_{43}NO_9$	525.65	nd (MeOH)	....	220-2			al, bz	B21⁴, 6841
12352	Prulaurasin or dl-Mandelonitrile glucoside .... $C_{14}H_{17}NO_6$	295.30	wh nd or pl (al), $[\alpha]_D$ −54	....	122			w, al	B31, 240
12353	Prunasin or d-Mandelonitrile-β-d-glucoside............. $C_{14}H_{17}NO_6$	295.30	$[\alpha]^{22}_D$ −27.0	....	149-50			w, al, ace	B17⁴, 3356
12354	Pseudoaconitine .... $C_{36}H_{51}NO_{12}$	687.79	tcl (MeOH), $[\alpha]^{15}_D$ + 18.4 (al)	....	214			al, eth	B21⁴, 2890
12355	Pseudocodeine or Neoisocodeine............. $C_{18}H_{21}NO_3$	299.37	wh nd, $[\alpha]_D$ −96.6 (al)	....	181-2	1.290⁸⁰	1.574	al	B27², 112
12356	Pseudoconhydrine or 5-Hydroxyconine............. $C_8H_{17}NO$	143.23	hyg nd (eth), $[\alpha]^{20}_D$ + 11.0 (al, c=10)	236	106 (anh), 60 (+1 w)			w, al, eth	B21⁴, 121
12357	Psuedoconiceine (L) or γ-Coniceine .... $C_8H_{17}N$	125.22	hyg, $[\alpha]^{15}_D$ + 122.6	171-2		0.8776¹⁵/⁴	1.4607¹⁰	al, eth, chl	B20, 146
12358	Pseudocumene or 1,2,4-Trimethyl benzene .... $1,2,4\text{-}(CH_3)_3C_6H_3$	120.20	....	169.3, 51.6¹⁰	−43.8	0.8758²⁰/⁴	1.5048²⁰	al, eth, ace, bz, peth	B5⁴, 1010
12359	Pseudocumene-5-acetyl or 2,4,5-Trimethyl acetophenone $2,4,5\text{-}(CH_3)_3C_6H_2COCH_3$	162.22	....	246-7, 137-8²⁰	10-1	1.0039¹⁵/⁴	1.541¹⁵	al, eth, bz, aa	B7³, 1145
12360	Pseudoephedrine(d) or Isoephedrine............. $C_{10}H_{15}NO$	165.24	pr or lf (eth), $[\alpha]^{20}_D$ + 51.9 (abs al, c=0.6)		118.7			al, eth, bz	B13¹, 1719
12361	Pseudoephedrine (dl) .... $C_{10}H_{15}NO$	165.24	nd (eth)	130¹⁶	118.2			al, eth, bz	B13¹, 1720
12362	Pseudoephedrine (l) .... $C_{10}H_{15}NO$	165.24	lf or pr (eth), $[\alpha]^{20}_D$ −51.9 (abs al, c=0.6)	....	118.7			al, eth, bz	B13³, 1719
12363	Pseudoephedrine hydrochloride (d) .... $C_{10}H_{15}NO \cdot HCl$	201.70	rh pl or nd (al), $[\alpha]^{20}_D$ + 62 (w, c=0.8)		181-2				B13¹, 1719
12364	Pseudoephedrine hydrochloride (dl) .... $C_{10}H_{15}NO \cdot HCl$	201.70	nd (abs al)	....	164			w, al, eth	B13¹, 1720
12365	Pseudoephedrine hydrochloride (l) .... $C_{10}H_{15}NO.HCl$	201.70	nd (abs al or AcOEt), $[\alpha]^{20}_D$ −62.1 (w, c=1.8)	....	182			w	B13², 377
12366	Pseudohyoscyamine or Norhyoscyanine............. $C_{16}H_{21}NO_3$	275.34	nd, $[\alpha]_D$ −22 (al)	....	140.5			al, chl	B21⁴, 167
12367	Pseudoionone or Citrylidenacetone .... $(CH_3)_2CCH_2CH_2C(CH_3)COCH_3$	140.22	pa ye oil	143-5¹²	....	0.8984²⁰	1.5335²⁰	al, eth, chl	B1⁴, 3598
12368	Pseudojervine .... $C_{33}H_{49}NO_8$	587.73	wh nd or hex pl, $[\alpha]_D$ −133 (al-chl, ⅓)		304-5d				C39, 1413
12369	Pseudomorphine or 2,2′-Bimorphine .... $C_{34}H_{36}N_2O_6$	568.68	cr (aq NH₃ + 3w), $[\alpha]^{24}_D$ + 44.8 (in HCl, c=0.86)	....	282-3			Py	B27², 886

No.	Name, Synonyms, and Formula	Mol. wt.	Color, crystalline form, specific rotation and λ_{max} (log ε)	b.p. °C	m.p. °C	Density	n_D	Solubility	Ref.
12370	Pseudomorphine hydrochloride .... $C_{34}H_{36}N_2O_6 \cdot 2HCl \cdot 2H_2O$	677.64	pw, $[\alpha]_D$ -114.76						B27[2], 886
12371	Pseudopelletierine or Pseudoplenicine .... $C_9H_{15}NO$	153.23	orh pr (peth)	246	54	1.001[100]	1.4760[100]	w, al, eth, chl	B21[4], 3315
12372	Pseudoreserpine .... $C_{33}H_{38}N_2O_9$	594.67	$[\alpha]^{24}_D$ -N-65 (chl)		257-8			ace	C52, 2876
12373	Pseudothiohydantiol or 2-Imino-4-thiazolidine.......... $C_3H_4N_2OS$	116.14	pr or nd (w)		255-8d				B27[2], 284
12374	Pseudotropine or Pseudotropanol............. $C_8H_{15}NO$	141.22	rh ta or pr (eth), rh bipym (peth-bz)		109			w, al, bz, chl	B21[4], 169
12375	D-Psicose or D-Allulose................. $C_6H_{12}O_6$	180.16	$[\alpha]^{25}_D$ + 4.7 (w, c=4.3)		58			w, al	B1[4], 4400
12376	Pteridine or Pyrimido[(4,5)]pyrazine .... $C_6H_4N_4$	132.13	ye pl (bz, sub)	sub 125-30[20]	139-40			w, al	J1951, 474
12377	Pteridine, 2-amino-4,6-dihydroxy or Uropterin......... $C_6H_5N_5O_2$	179.14	og-ye (w + 1)		>410d dark- ens 360				B26[2], 313
12378	Pteridine, 2-amino-4-hydroxy or 2-Amino-4-pteridol..... $C_6H_5N_5O$	163.14	ye		>360				C47, 5945
12379	Rukotoino (l), or 4-Hydroxy-5,6-methylenedioxyaporphin $C_{18}H_{17}NO_3$	295.34	cr (al, eth) $[\alpha]^{15}_D$ 200 (ul, c=6)	210-5[2]	200				B27[2], 461
12380	Pulegenone or 4-Methyl-1-isopropyl cyclopentinen-5-one $C_9H_{14}O$	138.21	....	188.0		0.9144[20/0]	1.4660[20]	al, eth, ace	B7[3], 291
12381	Pulegone or 4(8)p-Menthen-3-one.................... $C_{10}H_{16}O$	152.24	$[\alpha]^{20}_D$ + 23.4 (undil)	224, 103[17]		0.9346[45]	1.4894[20]	al, eth, chl	B7[3], 334
12382	Purine or 7-Imidazo[4,5]pyrimidine .... $C_5H_4N_4$	120.11	nd (to al)	sub	216-7			w, al, ace	B26, 354
12383	Purine, 6-amino or Adenine......................... $C_5H_5N_5$	135.13	nd or lf (w + 3)	sub at 220	360-5d (anh)				B26[2], 252
12384	Purine, 6-mercapto or 6-Purinethiol................. $C_5H_4N_4S$	152.18	ye pr (w + 1)		313-4d				Am74, 411
12385	Purine, 2,6,8-trichloro .... $C_5HCl_3N_4$	223.45	nd (al)		159-61			al, eth, ace, bz	B26, 356
12386	Pyocyanine .... $C_{13}H_{10}N_2O$	210.24	dk bl nd (w + 1), (chl-peth)	sub	133d			ace, chl, Py	B23[2], 361
12387	Pyraconitine .... $C_{32}H_{43}O_9N$	585.70	nd		171			al	B21[4], 6794
12388	γ-Pyran .... C_5H_6O	82.10	col oil	80			1.4559[20]	al, eth, bz	B17, 36
12389	γ-Pyran, 2,3-dihydro .... C_5H_8O	84.13	....	86-7		0.922[19/15]	1.4402[19]	w, al	B17[4], 148
12390	γ-Pyran, tetrahydro or Pentamethylene oxide .... $C_5H_{10}O$	86.14	....	88		0.8810[20/4]	1.4200[20]	al, eth, bz	B17[4], 51
12391	Pyranthrone or 8,16-Pyranthenedione .... $C_{30}H_{14}O_2$	406.45	red-ye or red-br nd (PhNO₂)	sub, vac	d				B7[3], 4514
12392	Pyrazine or 1,4-Diazine..................... $C_4H_4N_2$	80.09	pr (w)	115-6[768]	54	1.0311[61/4]	1.4953[61]	w, al, eth, ace	B23[2], 80
12393	Pyrazine, 2,3-dimethyl .... $2,3(CH_3)_2C_4H_2N_2$	108.14	....	156		1.0281[0/4]		w, al, eth	B23[2], 80
12394	Pyrazine, 2,5-dimethyl-3-ethyl .... $2,5-(CH_3)_2-3-C_2H_5(C_4HN_2)$	136.20	....	180-1		0.9657[24/4]	1.5014[24]		B23, 99
12395	Pyrazine, 2,5-dimethyl or Ketine .... $2,5-(CH_3)_2(C_4H_2N_2)$	108.15	....	155	15	0.9887[20/4]	1.4980[20]	w, al, eth, ace	B23[2], 80
12396	Pyrazine, 2,6-dimethyl .... $2,6-(CH_3)_2(C_4H_2N_2)$	108.14	....	155.6	47-8	0.9647[50/4]		w, al, eth	B23, 97
12397	Pyrazine, 2-methyl .... $2-CH_3(C_4H_3N_2)$	94.12	....	136-7		1.0290[20/4]	1.5067[19]	w, al, eth, ace	B23, 94
12398	Pyrazime carboxamide .... $(C_4H_3N_2)CONH_2$	123.12	wh nd (w, al)	sub	191-3				C48, 2074

No.	Name, Synonyms, and Formula	Mol. wt.	Color, crystalline form, specific rotation and λ_{max} (log ϵ)	b.p. °C	m.p. °C	Density	n_D	Solubility	Ref.
12399	2,3-Pyrazine dicarboxylic acid .... 2,3-$(C_4H_2N_2)(CO_2H)_2$	168.11	pr (w + 2)	....	193d	....	....	w, ace, MeOH	B25[2], 164
12400	2,3-Pyrazine dicarboxylic acid, 5-methyl .... 5-CH_3-2,3-$(C_4HN_2)(CO_2H)_2$	182.14	(w, ace, dil al-eth)	....	174-5	....	....	w, al	B25[2], 165
12401	2,5-Pyrazine dicarboxylic acid .... 2,5-$(C_4H_2N_2)(CO_2H)_2$	168.11	nd (w + 2)	....	255-6d (sealed tube)	....	....	....	B25[2], 164
12402	2,6-Pyrazine dicarboxylic acid .... 2,6-$(C_4H_2N_2)(CO_2H)_2$	168.11	mcl nd (w + 2)	....	217-8d	....	....	al	B25, 168
12403	Pyrazole or 1,2-Diazole................. $C_3H_4N_2$	68.08	nd or pr (lig)	186-8	69-70	....	1.4203	w, al, eth, bz	B23[2], 33
12404	Pyrazole, 4-bromo-1,3-dimethyl .... 4-Br-1,3-$(CH_3)_2(C_3HN_2)$	175.03	....	76-7[10]	....	1.4975[15/4]	1.5214[15]	al, eth, ace	B23[2], 54
12405	Pyrazole, 4-bromo-1,5-dimethyl .... 4-Br-1,5-$(CH_3)_2(C_3HN_2)$	175.03	cr	85[10]	38-9	....	....	....	B23[2], 54
12406	Pyrazole, 4-bromo-3,5-dimethyl .... 4-Br-3,5-$(CH_3)_2(C_3HN_2)$	175.03	nd (dil al)	....	123	....	....	eth, ace	B23[2], 68
12407	Pyrazole, 4-bromo-3-methyl .... 4-Br-3-$CH_3(C_3H_2N_2)$	161.00	cr (dil al)	....	76-7	1.5638[100/4]	1.5182[100]	al, eth, ace	B23[2], 54
12408	Pyrazole, 3-chloro-1,5-dimethyl .... 3-Cl-1,5-$(CH_3)_2(C_3HN_2)$	130.58	pl (w)	210-2, 138[72]	47-8	1.0823[100/4]	1.4648[100]	al, eth, ace, bz	B23[2], 49
12409	Pyrazole, 4-chloro-3,5-dimethyl .... 4-Cl-3,5-$(CH_3)_2(C_3HN_2)$	130.58	pr (w), cr (al)	220-2	117-8	....	....	al, eth, ace, bz	B23[2], 67
12410	Pyrazole, 5-chloro-1,3-dimethyl .... 5-Cl-1,3-$(CH_3)_2(C_3HN_2)$	130.58	....	157-8	....	1.1367[18/4]	1.4877[18]	w	B23[2], 49
12411	Pyrazole, 5-chloro-3-methyl .... 5-Cl-3-$CH_3(C_3H_2N_2)$	116.55	cr (eth, lig)	258, 138[15]	118-9	....	....	w	B23[2], 49
12412	Pyrazole, 1,3-dimethyl .... 1,3-$(CH_3)_2(C_3H_2N_2)$	96.14	....	136-8, 31[1]	....	0.9561[17/4]	1.4734[15]	w	B23[2], 44
12413	Pyrazole, 1,5-dimethyl .... 1,5-$(CH_3)_2(C_3H_2N_2)$	96.14	....	153	....	0.9813[17/4]	1.4782[16]	w, al, eth	B23[2], 44
12414	Pyrazole, 3,4-dimethyl .... 3,4-$(CH_3)_2(C_3H_2N_2)$	96.14	(peth)	111[10]	58	....	....	al, eth, ace, bz	B23[2], 64
12415	Pyrazole, 3,5-dimethyl .... 3,5-$(CH_3)_2(C_3H_2N_2)$	96.14	cr (peth, al)	218[758]	107-8	0.8839[16/4]	....	al, eth, ace, bz, chl	B23[2], 65
12416	Pyrazole, 1-ethyl .... 1-$C_2H_5(C_3H_3N_2)$	96.14	....	139	....	0.9537[20/4]	1.4700[20]	al, eth, ace, bz, Py	C55, 22291
12417	Pyrazole, 3-methyl .... 3-$CH_3(C_3H_3N_2)$	82.11	....	204[752], 108[25]	36-7	1.0203[16/4]	1.4915[20]	**w, al, eth**	B23[2], 44
12418	Pyrazole, 1-isopropyl .... 1-i-$C_3H_7(C_3H_3N_2)$	110.16	....	143	....	....	....	w, al	C53, 21043
12419	2-Pyrazoline or 4,5-Dihydropyrazole .... $C_3H_6N_2$	70.10	....	144	....	1.0200[17/4]	1.4796[17]	**w, al, eth**	B23[2], 24
12420	2-Pyrazoline, 1-phenyl .... 1-$C_6H_5(C_3H_5N_2)$	148.19	pl (lig)	273[754], 151[17]	52	1.0689[58/4]	1.6015[58]	al, eth, bz	B23[2], 25
12421	5-Pyrazolinone, 3-methyl .... $C_4H_6N_2O$	98.10	....	....	215	....	....	w	B24[2], 8
12422	5-Pyrazolinone, 1-methyl-3-phenyl .... $C_{10}H_{10}N_2O$	174.20	....	330-40, 235[68]	213-7	....	....	al, eth	B24, 148
12423	5-Pyrazolinone, 3-methyl-1-phenyl .... $C_{10}H_{10}N_2O$	174.20	[α] + 1.637	287[105], 191[17]	127	....	....	w, al	B24[2], 9
12424	5-Pyrazolone or 2-Pyrazolin-5-on .... $C_3H_4N_2O$	84.08	nd (to, w)	sub	165	....	....	w, al	B24[2], 6
12425	5-Pyrazolone, 3-methyl .... $C_4H_6N_2O$	98.11	pr (w), nd (al), lf (sub)	sub	215 (219)	....	....	w	B24[2], 8
12426	5-Pyrazolone, 3-methyl-1(nitrophenyl) .... $C_{10}H_9N_3O_3$	219.20	ye amor (aa)	....	185	....	....	....	B24[1], 191
12427	5-Pyrazolone, 3-methyl-1-phenyl .... $C_{10}H_{10}N_2O$	174.20	mcl pr (w)	287[105], 191[17]	127	....	1.637	....	B24[2], 9
12428	5-Pyrazolone, 3-methyl-1-(4-sulfophenyl) .... $C_{10}H_{10}N_2O_4S$	254.27	nd (w + 1)	....	290-320d	....	....	....	B24[2], 20
12429	5-Pyrazolone-3-carboxylic acid, 1-phenyl .... $C_{10}H_8N_2O_3$	204.20	nd (w, al)	....	261	....	....	al	B25[2], 219
12430	Pyrene or Benzo[d,e,f]phenanthrene .... $C_{16}H_{10}$	202.26	pa ye pl (to, sub)	393, 260[60]	156	1.271[23/4]	....	al, eth, bz, lig	B5[4], 2467

No.	Name, Synonyms, and Formula	Mol. wt.	Color, crystalline form, specific rotation and λ_{max} (log ε)	b.p. °C	m.p. °C	Density	n_D	Solubility	Ref.
12431	Pyrene, 1-acetyl 1-(CH₃CO)C₁₆H₉	244.30	gr-ye lf (al, MeOH)		89-90			eth, bz	B7[3], 2726
12432	Pyrene, 1-amino 1-NH₂C₁₆H₉	217.27	ye nd (hx), lf (dil al)		117-8			al, ace	B12[3], 3368
12433	Pyrene, 1,2-dihydro C₁₆H₁₂	204.27			132				BCHE, 4239
12434	Pyrene, 2,7-dimethyl C₁₈H₁₄	230.31			228-32				B5[3], 2307
12435	Pyrene perhydro (3a,8a-cis) C₁₆H₂₆	218.38		318.5	90				B5[4], 1199
12436	3-Pyrene carboxylic acid 3-(C₁₆H₉)CO₂H	246.27	ye nd (eth-al, sub)	sub	274			eth, chl	B9[3], 3575
12437	4-Pyrene carboxylic acid 4-(C₁₆H₉)CO₂H	246.27	gr lf or nd (Ph NO₂)		327-8			chl	C5[2], 11081
12438	Pyrethrin I or Chrysanthemum monocarboxylic acid C₂₁H₂₈O₃	328.46	visc liq, [α]²⁵ᴅ −32.3 (eth, c=5.66)	170⁰·¹/d			1.5192[18]	al, eth, peth	B9[3], 215
12439	Pyrethrin II or Chrysanthemum dicarboxylic acid, methyl ester C₂₂H₂₈O₅	372.47	visc liq, [α]²⁰ᴅ −6 (eth, c=5)	200⁰·¹/d			1.5258[20]	al, eth, peth	B9[3], 3988
12440	Pyridazine or 1,2-Diazine C₄H₄N₂	80.09		208, 47-8[1]	−8	1.1035[20/4]	1.5218[20]	w, al, eth, ace, DL	B23[1], 28
12441	Pyridine C₅H₅N	79.10		115.5	−42	0.9819[20]	1.5095[20]	w, al, eth, ace, bz	B20[4], 2205
12442	Pyridine, 3-acetamido 3-CH₃CONH(C₅H₄N)	136.15		126.7	133			w, al	B22[4], 4073
12443	Pyridine, 2-acetyl or Methyl 2-pyridyl ketone 2-CH₃CO(C₅H₄N)	121.14	ye in air	192, 78[12]			1.5203[20]	al, eth	B21[4], 3544
12444	Pyridine, 3-acetyl or Methyl-3-pyridyl ketone 3-CH₃CO(C₅H₄N)	121.14	ye in air	220, 106[12]	13-4		1.5341[20]	w, al, eth	B21[4], 3548
12445	Pyridine, 3-acetyl,oxime 3-CH₃C(=NOH)(C₅H₄N)	136.17	cr (al, bz)		113 (130.5)				B21[4], 3549
12446	Pyridine, 2-allyl 2-(CH₂=CHCH₃)(C₅H₄N)	119.17		190, 58[10]		0.959[20]		al, eth	B20[4], 2890
12447	Pyridine, 2-amino or α-Pyridyl amine 2-H₂N(C₅H₄N)	94.12	lf (liq)	204 (sub), 104-6[20]	57-8			al, eth, ace, bz	B22[4], 3840
12448	Pyridine, 2-amino-5-bromo 5-Br-2-H₂N(C₅H₃N)	173.02	cr (bz)		137			al, bz	B22[4], 4031
12449	Pyridine, 2-amino-5-chloro 5-Cl-2-H₂N(C₅H₃N)	128.56	pl	127-8[11]	136-8			w, al	B22[4], 4019
12450	Pyridine, 2-amino-3,5-dibromo 3,5-Br₂-2-H₂N(C₅H₂N)	251.92	nd (dil al, peth)		104			al, eth	B22[4], 4041
12451	Pyridine, 2-amino-3,5-dichloro 3,5-Cl₂-2-H₂N(C₅H₂N)	163.01	nd or pr (dil al)		84-5			al, ace, lig	B22[4], 4028
12452	Pyridine, 2-amino-5-iodo 5-I-2-H₂N(C₅H₃N)	220.01	nd or lf (dil al)		129			al, eth	B22[4], 4045
12453	Pyridine, 2-amino-3-nitro 3-NO₂-2-H₂N(C₅H₃N)	139.12	ye nd (dil al)		165-7			al	B22[4], 4053
12454	Pyridine, 2-amino-5-nitro 5-NO₂-2-H₂N(C₅H₃N)	139.12	ye lf (dil al)		188			al	B22[4], 4054
12455	Pyridine, 3-amino or β-Pyridyl amine 3-H₂N(C₅H₄N)	94.12	lf (bz-lig)	252, 131-2[12]	64-5			w, al, eth	B22[4], 4067
12456	Pyridine, 4-amino or γ-Pyridyl amine 4-H₂N(C₅H₄N)	94.12	nd (bz)	180[13]	158-9			w, al, eth, bz	B22[4], 4098
12457	Pyridine, 4-amino-2,6-dichloro 2-6-Cl₂-4-H₂N(C₅H₂N)	163.01	nd (dil al)		176			al, eth, ace, bz	B22[4], 4119
12458	Pyridine, 4-amino-3,5-dinitro 3,5-(NO₂)₂-4-H₂N(C₅H₂N)	184.11	ye pl (dil al)		170-1				B22[4], 4128
12459	Pyridine, 4-amino-3-nitro 3-NO₂-4-H₂N(C₅H₃N)	139.12	ye nd (w)		200			al	B22[4], 4122
12460	Pyridine, 5-amino-2-butoxy 5-NH₂-2-C₄H₉O(C₅H₃N)	166.23		148-50[12]		1.037[25]	1.5373[20]	al, eth	B22[4], 5574
12461	Pyridine, 2-benzoyl or Phenyl-2-pyridyl ketone 2-C₆H₅CO(C₅H₄N)	183.21		114-5[0.01]	41-3			chl	B21[4], 4119

No.	Name, Synonyms, and Formula	Mol. wt.	Color, crystalline form, specific rotation and λ_{max} (log ε)	b.p. °C	m.p. °C	Density	n_D	Solubility	Ref.
12462	Pyridine, 4-benzoyl or Phenyl-4-pyridyl ketone. 4-$C_6H_5CO(C_5H_4N)$	183.21	nd (peth), pl (w)	314[742], 170-2[10]	71-3	....	....	al, eth, bz	B21[4], 4125
12463	Pyridine, 2-benzyl 2-$C_6H_5CH_2(C_5H_4N)$	169.23	nd	276[742], 149[16]	11-4	1.067[0/0]	1.5785[20]	al, eth	B20[4], 3649
12464	Pyridine, 3-benzyl 3-$C_6H_5CH_2(C_5H_4N)$	169.23	nd	286[740]	34	1.061	....	al, eth	B20[4], 3651
12465	Pyridine, 4-benzyl 4-$C_6H_5CH_2(C_5H_4N)$	169.23		287[742], 180-1[31]		1.0614[20/0]	1.5818[20]	al, eth	B20[4], 3652
12466	Pyridine, 2-bromo 2-Br(C_5H_4N)	158.00		193-4[764], 74-5[13]		1.657[15]	1.5734[20]	al, eth	B20[4], 2503
12467	Pyridine, 3-bromo 3-Br(C_5H_4N)	158.00	nd (al)	172-3[751], 68-70[18]	142-3	1.645[0/4]	1.5694[20]	w, al, eth	B20[4], 2505
12468	Pyridine, 3-bromo-6-hydroxy 3-Br-6-HO(C_5H_3N)	174.00	pr (w, al)	....	177-8			al	B21[4], 362
12469	Pyridine, 4-bromo 4-Br(C_5H_4N)	158.00		172-3[752], 68-70[18]	0-1	1.6450[0/4]	1.5694[20]	w, al, eth	B20[4], 2512
12470	Pyridine, 5-bromo-2-hydroxy 2-HO-5-Br(C_5H_3N)	174.00	pr (w, al)		177-8			al	B21[4], 2512
12471	Pyridine, 2-chloro 2-Cl(C_5H_4N)	113.55	oil	170, 54-8[10]		1.205[15]	1.5320[20]	al, eth	B20[4], 2493
12472	Pyridine, 2-chloro-5-nitro 2-Cl-5-$NO_2(C_5H_3N)$	158.54			108-10				B20[4], 2533
12473	Pyridine, 3-chloro 3-Cl(C_5H_4N)	113.55		148[764], 85-7[100]			1.5304[20]		B20[4], 2497
12474	Pyridine, 4-chloro 4-Cl(C_5H_4N)	113.55		147-8, 85-7[100]	−42.5			w, al	B20[4], 2499
12475	Pyridine, 5-chloro-2-trichloro methyl 5-Cl-2-$Cl_3C(C_5H_3N)$	230.91		139-42[25]	45.5-6.5				B20[4], 2704
12476	Pyridine, 2,3-diamino-(one form) 2,3-$(NH_2)_2(C_5H_3N)$	109.13	lf or pl (dil al)	148-50[5]	122	....		al	B22[2], 395
12477	Pyridine, 2,3-diamino (one form) 2,3-$(NH_2)_2(C_5H_3N)$	109.13	nd (bz)	sub	116			w, al, bz	B22[2], 394
12478	Pyridine, 2,4-diamino 2,4-$(NH_2)_2(C_5H_3N)$	109.13	hyg lf or nd		107			w	B22[2], 394
12479	Pyridine, 2,5-diamino 2,5-$(NH_2)_2(C_5H_3N)$	109.13	nd	180-5[12]	109-10			w, al	B22[2], 394
12480	Pyridine, 3,4-diamino 3,4-$(NH_2)_2(C_5H_3N)$	109.13	nd or lf		218-9				B22[2], 395
12481	Pyridine, 3,5-diamino 3,5-$(NH_2)_2(C_5H_3N)$	109.13	hyg nd or lf		119-20			w	B22[1], 648
12482	Pyridine, 2,6-dibromo 2,6-$Br_2(C_5H_3N)$	236.90		255	118				B20[4], 2515
12483	Pyridine, 3,5-dibromo 3,5-$Br_2(C_5H_3N)$	236.90	nd (al)	222, sub 100	112			eth	B20[4], 2516
12484	Pyridine, 1,2-dihydro-1-methyl-2-oxo C_6H_7NO	109.13	nd	250, 121[10]	7			w	B21[4], 3348
12485	Pyridine, 2,4-dihydroxy or 2,4-Pyridinediol 2,4-$(HO)_2(C_5H_3N)$	111.10	rh bipym (al, w)		260-5d			w, al	B21[4], 2058
12486	Pyridine, 2,5-dihydroxy 2,5-$(HO)_2(C_5H_3N)$	111.10			248			w, al	B21[4], 2062
12487	Pyridine, 3,5-diiodo-2-hydroxy 3,5-I_2-2-HO(C_5H_2N)	346.89	br nd (al)		261-2				B21[4], 365
12488	Pyridine, 2,3-dimethyl or 2,3-Lutidine 2,3-$(CH_3)_2(C_5H_3N)$	107.15		163-4		0.9319[25/4]	1.5057[20]	w, al, eth	B20[4], 2765
12489	Byridine, 2,6-dimethyl-3-ethyl 2,6-$(CH_3)_2$-3-$C_2H_5(C_5H_2N)$	135.21		83-5[23]		0.9120[18]			B20[4], 2828
12490	Pyridine, 2,6-dimethyl-4-ethyl 2,6-$(CH_3)_2$-4-$C_2H_5(C_5H_2N)$	135.21		187.5-8[758]		0.9089[25/4]	1.4964[25]		B20[4], 2828
12491	Pyridine, 3,5-dimethyl-2-ethyl or α-Parvoline 3,5-$(CH_3)_2$-2-$C_2H_5(C_5H_2N)$	135.21		198-9[764], 85-7[15]		0.9338[0]	....	al, eth, ace	B20[4], 2828
12492	Pyridine, 2-(dimethylamino) 2-$(CH_3)_2N(C_5H_4N)$	122.17		196, 88[15]		1.0157[14/14]	1.5663[20]	al, eth, bz	B22[4], 3847
12493	Pyridine, 4-(dimethylamino) 4-$(CH_3)_2N(C_5H_4N)$	122.17	pl (eth)	....	114			w, al, bz, chl	B22[4], 4101

No.	Name, Synonyms, and Formula	Mol. wt.	Color, crystalline form, specific rotation and λ_{max} (log ε)	b.p. °C	m.p. °C	Density	n_D	Solubility	Ref.
12495	Pyridine, 2-ethyl 2-C$_2$H$_5$(C$_5$H$_4$N)	107.16	....	148.6	−63.1	0.9502[0]	1.4964[20]	**al**, eth, ace, w	B20[4], 2755
12496	Pyridine, 2-ethyl-4-methyl or 2 Ethyl-γ-picoline 2-C$_2$H$_5$-4-CH$_3$(C$_5$H$_4$N)	121.18	....	173-5[748]		0.9239[20/0]	....	al, eth, ace	B20[4], 2798
12497	Pyridine, 3-ethyl 3-C$_2$H$_5$(C$_5$H$_4$N)	107.16	....	165	−76.9	0.9539[0]	1.5021[20]	w, al, eth, ace	B20[4], 2758
12498	Pyridine, 4-ethyl 4-C$_2$H$_5$(C$_5$H$_4$N)	107.16	....	167.7	−90.5	0.9417[20]	1.5009[20]	w, al, eth, ace	B20[4], 2761
12499	Pyridine, 2-fluoro 2-F(C$_5$H$_4$N)	97.09	....	125[758]			1.4574[20]	....	B20[4], 2491
12500	Pyridine hydrochloride C$_5$H$_5$N.HCl	115.56	hyg pl or sc (al)	281-9	82			w, al, chl	B20[4], 2230
12501	Pyridine, 2-hydroxy or 2-Pyridol 2-HO(C$_5$H$_4$N)	95.10	nd (bz)	280-1	106-7			w, al, bz, chl	B21[4], 344
12502	Pyridine, 2-(β-hydroxyethyl) 2-(HOCH$_2$CH$_2$)(C$_5$H$_4$N)	123.16	hyg	118-21[15]		1.1111[4/0]	1.5368[20]	w, al, chl	B21[4], 512
12503	Pyridine, 2-hydroxy-4-methyl 4-CH$_3$-2-HO(C$_5$H$_4$N)	109.13	....	307-9	130			w, al, chl	B21[4], 505
12504	Pyridine, 2-hydroxymethyl 2-HOCH$_2$(C$_5$H$_4$N)	109.13	....	112-3[16], 102.5[8]		1.1317[20/4]	1.5444[20]	w, al, eth, ace, bz	B21[4], 487
12505	Pyridine, 2-(1-hydroxypropyl) 2-[CH$_3$CH$_2$CH(OH)](C$_5$H$_4$N)	137.18	pa ye	213-6, 112-3[12]		1.0501[20/4]	1.5197[20]	w, al	B21[4], 544
12506	Pyridine, 2-(β-hydroxypropyl) 2-[CH$_3$CH(OH)CH$_2$](C$_5$H$_4$N)	137.18	pr	123.5[20]	32			w, al, chl	B21[4], 545
12507	Pyridine, 2-(β-hydroxyisopropyl) 2-[HOCH$_2$CH(CH$_3$)](C$_5$H$_4$N)	137.18	....	128-31[17]				w, al	B21, 57
12508	Pyridine, 3-hydroxy or 3-Pyridol 3-HOC$_5$H$_4$N	95.10	nd (bz)	....	129			w, al	B21[4], 402
12509	Pyridine, 3-(α-hydroxy isopropyl) 3-[(CH$_3$)$_2$C(OH)](C$_5$H$_4$N)	137.18	cr	140-1[12]	58			w, al, chl	B21[4], 552
12510	Pyridine, 4-hydroxy or 4-Pyridol 4-HOC$_5$H$_4$N	95.10	pr or nd (w + 1)	>350, 257-60[10]	148.5 (anh), 65 (hyd)			w, al	B21[4], 446
12511	Pyridine, 4-(hydroxymethyl) 4-HOCH$_2$(C$_5$H$_4$N)	109.13	....	140-2[12]	51-5			chl	B21[4], 507
12512	Pyridine, 2-iodo 2-I-C$_5$H$_4$N	205.00	....	93[13]	......	1.9735[20]	1.6366[20]	al, eth, ace, bz	B20[4], 2522
12513	Pyridine, 2-mercapto 2-HS(C$_5$H$_4$N)	111.16	....		128			w, al, bz, chl	B21[4], 373
12514	Pyridine, 2-methoxy 2-CH$_3$O(C$_5$H$_4$N)	109.13	....	142-3			1.5042.[20]	....	B21[4], 345
12515	Pyridine, 4-methoxy 4-CH$_3$O(C$_5$H$_4$N)	109.13	....	191[738], 95[45]				**w**	B21[4], 447
12516	Pyridine, 2-methyl or α-Picoline 2-CH$_3$C$_5$H$_4$N	93.13	....	128.8	−66.8	0.9443[20/4]	1.4957[20]	w, **al, eth**, ace	B20[4], 2679
12517	Pyridine, 2-methylamino 2-CH$_3$NH(C$_5$H$_4$N)	108.15	....	200-1, 90[9]	15	1.052[29/29]	....	w, al, eth, bz, aa	B22[4], 3847
12518	Pyridine, 4-methylamino 4-CH$_3$NH(C$_5$H$_4$N)	108.15	pl (eth)		117-8			w, al, eth, ace, bz	B22[4], 4100
12519	Pyridine-N-oxide C$_5$H$_5$NO	95.10	....	146-7[13]	65-6				B20[4], 2305
12520	Pyridine, pentachloro C$_5$Cl$_5$N	251.33	....	sub 280	125-6			al, bz, lig	B20[4], 2503
12521	Pyridine, 2-phenyl or α-Pyridyl benzene 2-C$_6$H$_5$(C$_5$H$_4$N)	155.20	....	270-2, 146[15]	......	1.0833[25]	1.6210[20]	**al, eth**	B20[4], 3639
12522	Pyridine, 3-phenyl or β-Pyridyl benzene 3-C$_6$H$_5$(C$_5$H$_4$N)	155.20	pa ye oil	273-4			1.6123[25]	al, eth	B20[4], 3643
12523	Pyridine, 4-phenyl 4-C$_6$H$_5$(C$_5$H$_4$N)	155.20	pl (w)	280-2	77-8			al, eth	B20[4], 3645
12524	Pyridine picrate C$_5$H$_5$N.C$_6$H$_3$N$_3$O$_7$	308.22	ye nd (al)		167-8				B20[4], 2287
12525	Pyridine, 2-propyl or Conyrine 2-C$_3$H$_7$(C$_5$H$_4$N)	121.18	....	166-8, 60[11]	2	0.9119[20/4]	1.4925[20]	**al, eth**, ace	B20[4], 2790
12526	Pyridine, 4-propyl 4-C$_3$H$_7$(C$_5$H$_4$N)	121.18	....	184-6, 80[2]	......	0.9381[15]	1.4966[20]	al, eth	B20[4], 2792

No.	Name, Synonyms, and Formula	Mol. wt.	Color, crystalline form, specific rotation and λmax (log ε)	b.p. °C	m.p. °C	Density	n_D	Solubility	Ref.
12527	Pyridine, 2-isopropyl 2-i-C₃H₇(C₅H₄N)	121.18	....	159.8	......	0.9342[0]	1.4915[20]	**al**, eth, ace	B20⁴, 2794
12528	Pyridine, 4-isopropyl 4-i-C₃H₇(C₅H₄N)	121.18	....	178	−54.9	0.9382[25/4]	1.4962[20]	**al**, eth, ace	B20⁴, 2795
12529	Pyridine, 2,3,4,6-tetramtehyl or β-Parvoline 2,3,4,6-(CH₃)₄(C₅HN)	135.21		203[750]		0.9322[25/4]	1.5087[25]	al, eth	B20⁴, 2830
12530	Pyridine, 2-(trichloromethyl) 2-Cl₃C(C₅H₄N)	196.46	....	125-6[25]	−10	1.4526[25]	1.5596[25]	....	B20⁴, 2703
12531	Pyridine, 2,4,6-trihydroxy 2,4,6(HO)₃(C₅H₂N)	127.10	ye nd or pw	....	230d			al, eth, ace	B21⁴, 2504
12532	Pyridine, 2,3,4-trimethyl 2,3,4-(CH₃)₃(C₅H₂N)	121.18		192-3, 79-82[14]	......	0.9127[15/4]	1.5150[20]	w, al, eth	B20⁴, 2806
12533	Pyridine, 2,3,5-trimethyl 2,3,5-(CH₃)₃(C₅H₂N)	121.18		186.7		0.9352[19/4]	1.5057[25]	al, eth, ace, bz	B20⁴, 2807
12534	Pyridine, 2,3,6-trimethyl 2,3,6-(CH₃)₃(C₅H₂N)	121.18		176-8		0.9220[25/4]	1.5053[20]	al, eth, ace, bz	B20⁴, 2808
12535	Pyridine, 2,4,5-trimethyl 2,4,5-(CH₃)₃(C₅H₂N)	121.18		188		0.9330[25/4]	1.5054[25]	al, eth, ace, bz	B20⁴, 2809
12536	Pyridine, 2,4,6-trimethyl or γ-Collidine 2,4,6-(CH₃)₃(C₅H₂N)	121.18		170.5[762]	−44.5	0.9166[22/4]	1.4959[25]	w, al, eth, ace	B20⁴, 2810
12537	Pyridine, 2-vinyl 2-CH₂=CH(C₅H₄N)	105.14		159-60, 50-5[4]		0.9985[20/0]	1.5495[20]	al, eth, ace, chl	B20⁴, 2884
12538	Pyridine, 3-vinyl 3-CH₂=CH(C₅H₄N)	105.14		160, 67-8[18]			1.5530[20]	al, chl	B20⁴, 2886
12539	Pyridine, 4-vinyl 4-CH₂=CH(C₅H₄N)	105.14	red to dk-br	65[15]		0.9800[20/4]	1.5449[20]	....	B20⁴, 2887
12540	2-Pyridine carboxaldehyde 2-(C₅H₄N)CHO	107.11	....	180[7.50], 81.5[25]		1.1201[20/4]	1.53653[20]	w, al, eth	B21⁴, 3495
12541	2-Pyridine carboxylic acid or Picolinic acid 2-(C₅H₄N)CO₂H	123.11	nd (w, al, bz)	sub	136-7			al, aa	B22⁴, 303
12542	3-Pyridine carboxaldehyde 3-(C₅H₄N)CHO	107.11		89.5[14]		1.415[20/4]		w, al, ace, chl	B21⁴, 3517
12543	3-Pyridine carboxylic acid or Nicotinic acid, Niacin 3-(C₅H₄N)CO₂H	123.11	nd (w, al)	sub	236-7	1.473		....	B22⁴, 348
12544	4-Pyridine carboxaldehyde or Isonicotinaldehyde 4-OHC(C₅H₄N)	107.11		77-8[12]			1.5423[20]	w, eth	B21⁴, 3529
12545	4-Pyridine carboxylic acid or Isonicotinic acid 4-(C₅H₄N)CO₂H	123.11	nd (w)	sub at 260[15]	319			....	B22⁴, 518
12546	4-Pyridine carboxylic acid anhydride C₁₂H₈N₂O₃	228.22	nd		103-4			....	B22⁴, 526
12547	4-Pyridine carboxylic acid, ethyl betaine C₈H₉O₂N	151.17	nd	....	241d			w, al	B22, 47
12548	2,3-Pyridine dicarboxylic acid or Quinolinic acid 2,3-(C₅H₃N)(CO₂H)₂	167.12	mcl pr (w)		228-9			....	B22⁴, 1618
12549	2,4-Pyridine dicarboxylic acid or Lutidinic acid 2,4-(C₅H₃N)(CO₂H)₂	167.12	lf (w + 1)	....	248-50 (anh)	0.942		....	B22⁴, 1630
12550	2,4-Pyridine dicarboxylic acid, 6-methyl or Uvitonic acid 6-CH₃-2,4-(C₅H₂N)(CO₂H)₂	181.15	cr (w)		282d			....	B22², 107
12551	2,5-Pyridine dicarboxylic acid or Isocinchomeronic acid 2,5-(C₅H₃N)(CO₂H)₂	167.12	lf (w + 1)	sub	256-8d			....	B22⁴, 1632
12552	2,6-Pyridine dicarboxylic acid or Dipicolinic acid 2,6-(C₅H₃N)(CO₂H)₂	167.12	nd (w + 3'/₂)		252 (anh)			....	B22⁴, 1635
12553	3,4-Pyridine dicarboxylic acid or Cinchomeronic acid 3,4-(C₅H₃N)(CO₂H)₂	167.12	pr, nd or lf (w)	sub	262d			....	B22⁴, 1641
12554	3,4-Pyridine dicarboxylic acid, 4,5-dihydro-2,6-dimethyl,diethyl ester C₁₃H₁₉NO₄	253.30	bl flr pl (eth)	....	85			al, ace	B22², 98
12555	3,4-Pyridine dicarboxylic acid, 2,6-dimethyl,diethyl ester 2,6-(CH₃)₂(C₅HN)(CO₂C₂H₅)₂-3,4	251.29	....	270d, 163[13]	16			....	B22², 109
12556	3,4-Pyridine dicarboxylic acid, 5-methoxy-6-methyl 6-CH₃-5-CH₃O-(C₅HN)(CO₂H)₂-3,4	211.28	cr (w-ace)	....	213-5			....	B22⁴, 2591
12557	3,5-Pyridine dicarboxylic acid or Dinicotinic acid 3,5-(C₅H₃N)(CO₂H)₂	167.12	cr (w)	sub	325d			....	B22⁴, 1643
12558	3,5-Pyridine dicarboxylic acid, 1,4-dihydro-2,6-dimethyl,diethyl ester C₁₃H₁₉NO₄	253.30	ye nd or lf (al)	....	184-5			chl	B22⁴, 1582

No.	Name, Synonyms, and Formula	Mol. wt.	Color, crystalline form, specific rotation and λ_{max} (log ε)	b.p. °C	m.p. °C	Density	n_D	Solubility	Ref.
12559	3,5-Pyridine dicarboxylic acid, 1,4-dihydro-2,4,6-trimethyl,diethyl est . . . . $C_{14}H_{21}NO_4$	267.33	lt bl flr pl (al)	. . . .	131	. . . .	. . . .	chl	B22[4], 1594
12560	3,5-Pyridine dicarboxylic acid, 2,6-dimethyl,diethyl ester 2,6-$(CH_3)_2(C_5HN)(CO_2C_2H_5)_2$-3,5	251.29	nd (al), pr (eth)	301-2, 180[16]	75-6	. . . .	. . . .	al, eth, bz, chl, lig	B22[4], 1656
12561	2,6-Pyridine diol hydrate . . . . $C_5H_5NO_2 \cdot H_2O$	129.12	ye pr	. . . .	202-3	. . . .	. . . .	w, al	B21[4], 2063
12562	Pyridine sulfate . . . . $C_5H_5NH_2SO_4$	177.18	cr (al)	. . . .	. . . . .	. . . .	. . . .	**w, al**	B20[4], 2233
12563	4-Pyridine sulfonic acid . . . . 4-$C_5H_4NSO_3H$	159.16		. . . .	333-4	. . . .	. . . .	w, al	B22[4], 3465
12564	Pyridine pentacarboxylic acid-dihydrate . . . . $C_5N(CO_2H)_5 \cdot 2H_2O$	335.18	cr (eth, w)	. . . .	220d	. . . .	. . . .	w	B22[2], 142
12565	3-Pyridine sulfonic acid . . . . 3-$C_5H_4NSO_3H$	159.16	orh	. . . .	375d	$1.718^{25/25}$	. . . .	w	B22[4], 3458
12566	2,3,4,5-Pyridine tetracarboxylic acid 2,3,4,5-$(C_5HN)(CO_2H)_4$	225.14	cr	. . . .	160d	. . . .	. . . .	w	B22, 188
12567	2,3,4,6-Pyridine tetracarboxylic acid 2,3,4,6-$(C_5HN)(CO_2H)_4$	255.14	nd (w + 3)	. . . .	236d	. . . .	. . . .	w, aa	B22[2], 142
12568	2,3,5,6-Pyridine tetracarboxylic acid 2,3,5,6-$(C_5HN)(CO_2H)_4$	255.14	cr (w + 2)	. . . .	200d	. . . .	. . . .	w	B22[1], 544
12569	2,3,4-Pyridine tricarboxylic acid or γ-Carbocinchomeronic acid . . . . 2,3,4-$(C_5H_2N)(CO_2H)_3$	211.13	lf (w + $^3/_2$)	. . . .	250 (anh)	. . . .	. . . .		B22[4], 1777
12570	2,3,5-Pyridine tricarboxylic acid or Carbodinicotinic acid 2,3,5-$(C_5H_2N)(CO_2H)_3$	211.13	pl, lf or nd (w, dil al)	. . . .	323 (anh)	. . . .	. . . .	w	B22[2], 136
12571	2,4,5-Pyridine tricarboxylic acid or Berberonic acid 2,4,5-$(C_5H_2N)(CO_2H)_3$	211.13	tcl pr (dil HCl + aq)	. . . .	243 (anh)	. . . .	. . . .		B22[4], 1778
12572	2,4,6-Pyridine tricarboxylic acid or Trimesic acid . . . . 2,4,6 $(C_5H_2N)(CO_2H)_3$	211.13	nd (w + 2)	sub	227d	. . . .	. . . .		B22[4], 1778
12573	3,4,5-Pyridine tricarboxylic acid or β-Carbocinchomeronic acid 3,4,5-$(C_5H_2N)(CO_2H)_3$	211.13	lf or pl (w + 3), ta (w)	. . . .	261d —w115	. . . .	. . . .	al	B22[4], 1779
12574	2,4,6-Pyridine triol . . . . 2,4,6-$(HO)_3(C_5H_2N)$	127.10	ye nd or pw	. . . .	230d	. . . .	. . . .	al, eth, ace	B21[4], 2504
12575	Pyridoxal hydrochloride . . . . $C_8H_9NO_3 \cdot HCl$	203.63	rh	. . . .	165d	. . . .	. . . .	w	B21[4], 6419
12576	Pyridoxal oxime . . . . $C_8H_{10}N_2O_3$	182.18	cr (al)	. . . .	225-6d	. . . .	. . . .	al	B21[4], 6427
12577	Pyridoxamine dihydrochloride . . . . $C_8H_{12}N_2O_2 \cdot 2HCl$	241.12	pr (al)	. . . .	226-7d	. . . .	. . . .	w	C51, 14833
12578	Pyrimidine or 1,3-Diazin . . . . $C_4H_4N_2$	80.09	. . . .	123-4	22	. . . .	1.4998^{20}	w, al	B23, 89
12579	Pyrimidine, 2-amino . . . . 2-$H_2N(C_4H_3N_2)$	95.11	nd (AcOEt)	sub	127-8	. . . .	. . . .	w	B24, 80
12580	Pyrimidine, 2-amino-4,6-dihydroxy . . . . 4,6-$(HO)_2$-2-$H_2N(C_4HN_2)$	127.10	pr (w + 1)	. . . .	>330	. . . .	. . . .		B24, 468
12581	Pyrimidine, 2-amino-4,5-dimethyl 4,5-$(CH_3)_2$-2-$H_2N(C_4HN_2)$	123.16	nd (w)	sub	214-5	. . . .	. . . .	al, ace, bz, chl	B24, 91
12582	Pyrimidine, 2-amino-4,6-dimethyl 4,6-$(CH_3)_2$-2-$H_2N(C_4HN_2)$	123.16		. . . .	153-4	. . . .	. . . .	w, al, ace, bz, chl	B24[1], 234
12583	Pyrimidine, 2-amino-4-methyl . . . . 4-CH_3-2-$H_2N(C_4H_2N_2)$	109.13	pl (w), nd (sub)	sub	159-60	. . . .	. . . .	al	B24, 84
12584	Pyrimidine, 2-amino-5-methyl . . . . 5-CH_3-2-$H_2N(C_4H_2N_2)$	109.13	pl (sub), pr (w)	sub	193.5	. . . .	. . . .	al	B24, 87
12585	Pyrimidine, 2-amino-5-nitro . . . . 5-NO_2-2-$H_2N(C_4H_2N_2)$	140.10	nd (al)	. . . .	236-7	. . . .	. . . .	al, ace	B24[1], 231
12586	Pyrimidine, 4-amino . . . . 4-$H_2N(C_4H_3N_2)$	95.11	pl (AcOEt)	. . . .	151-2	. . . .	. . . .	w, al	B24, 81
12587	Pyrimidine, 4-amino-2,6-dimethyl or Kyanmethin . . . . . . 2,6-$(CH_3)_2$-4-$H_2N(C_4HN_2)$	123.16	nd (al), pl (bz)	sub	183 (192)	. . . .	. . . .		B24[2], 46
12588	Pyrimidine, 4-amino-2-methyl . . . . 2-CH_3-4-$H_2N(C_4H_2N_2)$	109.13	rh (ace)	. . . .	205	. . . .	. . . .	w, al, ace, bz	B24, 84
12589	Pyrimidine, 4-amino-5-methyl . . . . 5-CH_3-4-$H_2N(C_4H_2N_2)$	109.13	pl (al, AcOEt)	. . . .	176	. . . .	. . . .	al, w	B24, 87
12590	Pyrimidine, 4-amino-6-methyl . . . . 6-CH_3-4-$H_2N(C_4H_2N_2)$	109.13	pr (w), nd, lf (sub)	sub	197	. . . .	. . . .	w, al	B24, 85

No.	Name, Synonyms, and Formula	Mol. wt.	Color, crystalline form, specific rotation and λ_{max} (log ε)	b.p. °C	m.p. °C	Density	n_D	Solubility	Ref.
12591	Pyrimidine, 6-amino-2,4-dihydroxy or 4-Amino uracil ... 2,4-(HO)$_2$-6-H$_2$N(C$_4$HN$_2$)	127.10	cr (w)	....	d	....	....	w	B24, 469
12592	Pyrimidine, 6-amino-4,5-dimethyl .... 4,5-(CH$_3$)$_2$-6-NH$_2$(C$_4$HN$_2$)	123.16	nd (w)	....	230	....	....	bz, ace, chl	B24, 92
12593	Pyrimidine, 2-chloro-4-dimethylamino-6-methyl .... 4-[(CH$_3$)$_2$N]-6-CH$_3$-2-Cl(C$_4$HN$_2$)	171.63	br wax	140-7^4	87	....	....	al	C36, 911
12594	Pyrimidine, 2,4-dichloro-5-methyl .... 5-CH$_3$-2,4-Cl$_2$(C$_4$HN$_2$)	163.01	pl (al)	235	25-7	....	....	al, eth, bz, chl	B23, 93
12595	Pyrimidine, 2,4-dichloro-6-methyl .... 6-CH$_3$-2,4-Cl$_2$(C$_4$HN$_2$)	163.01	nd (lig)	219	46-7	....	....	al, eth, bz, chl	B23, 92
12596	Pyrimidine, 2,4-dichloro-5-nitro .... 2,4-Cl$_2$-5-NO$_2$(C$_4$HN$_2$)	193.98	....	153-5^{58}	29.3	....	....	....	B23, 90
12597	Pyrimidine, 2-hydroxy,hydrochloride .... 2-HO(C$_4$H$_3$N$_2$).HCl	132.55	rods (al)	....	205	....	....	w	B24^1, 231
12598	Pyrimidine, 2-methyl .... 2-CH$_3$(C$_4$H$_3$N$_2$)	94.12	....	138^{758}	-4	....	....	w	B23, 92
12599	Pyrimidine, 4-methyl .... 4-CH$_3$(C$_4$H$_3$N$_2$)	94.12	....	142	32	1.031$^{16/16}$	1.5000^{20}	w	B23, 92
12600	Pyrimidine, 5-methyl .... 5-CH$_3$(C$_4$H$_3$N$_2$)	94.12	....	152-4	30.5	....	....	w	B23, 93
12601	Pyrimidine, 4-methylamino .... 4-CH$_3$NH(C$_4$H$_3$N$_2$)	109.13	lf (al), nd (peth)	142-4^{16}	74-5	....	....	w	B24^2, 38
12602	Pyrocalciferol or 9α-Lumisterol..................... C$_{28}$H$_{44}$O	396.67	nd (MeOH), $[\alpha]^{20}_D$ +512 (al)	....	93-5	....	....	al	B6^3, 3098
12603	Pyrocoll .... C$_{10}$H$_6$N$_2$O$_2$	186.17	ye mcl pl (aa)	sub	268	....	....	al, eth, bz, chl	B24^1, 360
12604	Pyrogallol or 1,2,3-Trihydroxy benzene .... 1,2,3-(HO)$_3$C$_6$H$_3$	126.11	lf or nd (bz)	309, 171^{12}	133-4	1.453$^{4/4}$	1.561^{134}	w, al, eth	B6^3, 6260
12605	Pyrogallol, 4-acetyl or 2,3,4-Trihydroxy acetophenone ... 2,3,4-(HO)$_3$C$_6$H$_2$(COCH$_3$-1)	168.15	pa ye nd or lf (w)	....	173	....	....	al, eth, ace, aa	B8^3, 3376
12606	Pyrogallol, 1,2-dimethyl ether or 2,3-Dimethoxy phenol . 2,3(CH$_3$O)$_2$C$_6$H$_3$OH	154.17	....	232-4, 124-5^{17}	....	1.5392^{20}	....	....	B6^3, 6264
12607	Pyrogallol, 1,3-dimethyl ether or 2,6-Dimethoxy phenol . 2,6-(CH$_3$O)$_2$C$_6$H$_3$OH	154.17	mcl pr (w)	262-7	56-7	....	....	al, eth	B6^3, 6264
12608	Pyrogallol triacetate .... 1,2,3-(CH$_3$CO$_2$)$_3$C$_6$H$_3$	252.23	pr (al)	....	165	....	....	al	B6^3, 6269
12609	Pyrogallol, 4,5,6-trichloro or 4,5,6-Trichloro-1,2,3-hydroxy benzene .... C$_6$H$_3$Cl$_3$O$_3$	229.45	nd (al, bz)	....	185	....	....	w, al, eth	B6, 1084
12610	Pyrogallol trimethyl ether or 1,2,3-Trimethoxy benzene .. 1,2,3-(CH$_3$O)$_3$C$_6$H$_3$	168.20	rh nd (al)	235, 140^{12}	48-9	1.1118$^{45/45}$	....	al, eth, bz	B6^3, 6265
12611	2-Pyrone, 4,6-dimethyl .... 4,6-(CH$_3$)$_2$(C$_5$H$_2$O$_2$)	124.15	lf (eth)	245, 126^{11}	51.5	....	....	w, al, eth	B17^4, 4531
12612	2-Pyrone, 3-hydroxy .... 3-HO-(C$_5$H$_3$O$_2$)	112.09	nd (w + 2)	112^{20}	95 (anh), 80-5 (hyd)	....	....	w, al, eth, chl	B17^4, 5908
12613	4-Pyrone or 4H-Pyran-4-one .... C$_5$H$_4$O$_2$								
12614	4-Pyrone, 2,6-dimethyl .... 2,6-(CH$_3$)$_2$(C$_5$H$_2$O$_2$)	96.09	hyg cr	215-7^{742}, 105^{23}	32.5	1.190	1.5238	w, al, eth, bz, chl	B17^4, 4399
12615	4-Pyrone, 3-Hydroxy-2-methyl or Larixinic acid, Maltol . 2-CH$_3$-3-HO(C$_5$H$_2$O$_2$)	124.15	pl, nd (sub)	248-9^{713}, 139-40^{25} sub	132	....	....	w, al, eth, ace	B17^4, 4532
12616	4-Pyrone, 5-hydroxy-2-(hydroxymethyl) or Kojic acid ... 2-HOCH$_2$-5-HO(C$_5$H$_2$O$_2$)	126.11	red mcl pr (chl)	sub at 93	162.4	....	....	chl	B17^4, 5916
12617	4-Pyrone-2-carboxylic acid-5,6-dihydro-6,6-dimethyl, butyl ester .... 6,6-(CH$_3$)$_2$-2-C$_5$H$_3$O$_2$CO$_2$C$_4$H$_9$	226.28	ye or pa red-br	256-70, 113-4^{14}	....	1.052^{25}	1.4745^{29}	al, eth, bz, chl	B18^4, 5345
12618	4-Pyrone-2-carboxylic acid-5-hydroxy or Comenic acid ... C$_6$H$_4$O$_5$	156.10	ye cr	....	>270d	....	....	....	B18^4, 5985
12619	Pyrophosphoramide, octamethyl .... [(CH$_3$)$_2$N]$_4$P$_2$O$_3$	286.25	....	118-22$^{0.3}$	....	1.1343^{25}	1.462^{25}	w, al, chl	B4^4, 288
12620	Pyrophosphoric acid, tetraethyl ester or Tetraethyl pyrophosphate.................................... (C$_2$H$_5$O)$_4$P$_2$O$_3$	290.19	....	155^3	....	1.1847$^{20/4}$	1.4180^{20}	al, eth, ace, chl	B1^4, 1340

No.	Name, Synonyms, and Formula	Mol. wt.	Color, crystalline form, specific rotation and λ_{max} (log ε)	b.p. °C	m.p. °C	Density	n_D	Solubility	Ref.
12621	Pyrophosphoric acid, dithiono, tetraethyl ester or Sulfo-tep.............. $(C_2H_5O)_4P_2OS_2$	322.32	col oil	136-9[2]	......	1.196[25/4]	1.4753[25]	al	B1[4], 1351
12622	Pyrotartaric acid (dl) or Methyl succinic acid........... $HO_2CCH(CH_3)CH_2CO_2H$	132.13	pr	d	115		1.4303	w, al, eth	B2[4], 1948
12623	Pyrrole C_4H_5NH	67.09	......	130-1	......	0.9691[20/4]	1.5085[20]	al, eth, ace, bz	B20[4], 2072
12624	Pyrrole, N-acetyl $C_4H_5N(COCH_3)$	109.13		181-2					B20[4], 2089
12625	Pyrrole, 2-acetyl $2\text{-}CH_3CO(C_4H_3NH)$	109.13	mcl nd (w)	220	90			w, al, eth	B21[4], 3437
12626	Pyrrole, 2-acetyl-N-methyl $2\text{-}CH_3CO(C_4H_3N)CH_3$	123.15		200-2[252], 92-4[22]		1.0445[15/4]	1.5403[15]	al, bz, chl	B21[4], 3439
12627	Pyrrole, N-benzyl $(C_4H_4N)CH_2C_6H_5$	157.22		247, 138[27]	15		1.5655[24]	al, eth	B20[4], 2086
12628	Pyrrole, N-butyl $(C_4H_4N)C_4H_9$	123.20	oil	170-1, 53-4[11]			1.4727[20]		B20[4], 2082
12629	Pyrrole, 2,3-dimethyl-4-ethyl or Hemopyrrole $2,3\text{-}(CH_3)_2\text{-}4\text{-}C_2H_5(C_4HNH)$	123.20		198[725], 113[16]	16-7	0.915[20/4]		w	B20[4], 2152
12630	Pyrrole, 2,4-dimethyl $2,4\text{-}(CH_3)_2(C_4H_2NH)$	95.15	pa bl flr	171, 62-3[10]		0.9236[20/4]	1.5048[20]	al, eth, bz	B20[4], 2107
12631	Pyrrole, 2,4-dimethyl-3-ethyl or Krypto pyrrole $2,4\text{-}(CH_3)_2\text{-}3\text{-}C_2H_5(C_4HNH)$	123.20	pr	197[710], 96[16]	0	0.913[20/4]	1.4961[20]	al, eth, bz, chl	B20[4], 2153
12632	Pyrrole, 2,5-dimethyl $2,5\text{-}(CH_3)_2(C_4H_2NH)$	95.15		170.2[765], 50-3[8]		0.9353[20/4]	1.5036[20]	al, eth	B20[4], 2152
12633	Pyrrole, 1,3-diphenyl $3\text{-}C_6H_5(C_4H_3N)C_6H_5$	219.29	pl (al)		122-3			eth, bz, chl	B20[1], 148
12634	Pyrrole, 2,5-diphenyl $2,5\text{-}(C_6H_5)_2(C_4H_2NH)$	219.29	lf (aa, dil al)		143.5		....	al, eth, bz, aa	B20[4], 4162
12635	Pyrrole, N-ethyl $(C_4H_4N)C_2H_5$	95.15		129-30[762]		0.9009[20/4]	1.4841[20]	al	B20[4], 2082
12636	Pyrrole, 2-ethyl $2\text{-}C_2H_5(C_4H_3NH)$	95.15		163-5, 59-60[15]		0.9042[20/4]	1.4942[20]	al	B20[4], 2106
12637	Pyrrole, 3-ethyl-4-methyl or Opsopyrrole $3\text{-}C_2H_5\text{-}4\text{-}CH_3(C_4H_3NH)$	109.17	ye oil	70[11]	3	0.9059[20/4]	1.4913[20]	al, eth	B20[4], 2145
12638	Pyrrole, 3-ethyl-2,4,5-trimethyl or Phyllopyrrole $3\text{-}C_2H_5\text{-}2,4,5\text{-}(CH_3)_3(C_4NH)$	137.23	pl (sub), lf (eth), wh lf (peth)	213[725], 92-3[12]	67-8			al, eth	B20[4], 2174
12639	Pyrrole, N-methyl $(C_4H_4N)CH_3$	81.12		114-5[747]	......	0.9145[15/4]	1.4875[20]	al, eth	B20[4], 2080
12640	Pyrrole, 2-methyl $2\text{-}CH_3(C_4H_3NH)$	81.12		147-8[750]		0.9446[15/4]	1.5035[16]	al, eth	B20[4], 2103
12641	Pyrrole, 3-methyl $3\text{-}CH_3(C_4H_3NH)$	81.12		142-3[743], 45[11]			1.4970[20]	al, eth	B20[4], 2105
12642	Pyrrole, N-Phenyl $(C_4H_4N)C_6H_5$	143.19	pl (sub), red in air	234, 140[38] sub	62			al, eth, ace, bz, peth	B20[4], 2084
12643	Pyrrole, 2-phenyl $2\text{-}C_6H_5(C_4H_3NH)$	143.19	pl (al, sub)	271-2[726]	129			al, eth, bz, chl	B20[4], 3452
12644	Pyrrole, N-Propyl $(C_4H_4N)C_3H_7$	109.17		145-6		0.8833[20/4]		al, eth	B20[4], 3082
12645	Pyrrole, 3-Propyl $3\text{-}C_3H_7(C_4H_3NH)$	109.17		90[30], 49[2]			1.4900[25]	eth, ace	B20[4], 2143
12646	Pyrrole, 2-isopropyl $2\text{-}i\text{-}C_3H_7(C_4H_3NH)$	109.17		171-2[741]		0.908[25/4]	1.491[25]	al, eth	B20[4], 2143
12647	Pyrrole, tetraiodo or Iodol C_4I_4NH	570.68	ye nd (al)		150d			eth, ace, chl, aa	B20[4], 2098
12648	Pyrrole, tetramethyl $C_4(CH_3)_4NH$	123.20	lf (dil al, peth)	130[7]	111	......		al, eth, ace, bz	B20[4], 2156
12649	2-Pyrrole carboxaldehyde $(HNC_4H_3)CHO\text{-}(2)$	95.10	rh pr (peth)	217-9, 114[15]	46-7		1.5939[16]		B21[4], 3419
12650	2-Pyrrole carboxylic acid $(HNC_4H_3)CO_2H\text{-}2$	111.10	lf (w)		208d			w, al, eth	B22[4], 225
12651	2-Pyrrole carboxylic acid, 3,5-dimethyl, ethyl ester $3,5\text{-}(CH_3)_2(C_4HNH)CO_2C_2H_5\text{-}2$	167.21	cr (al)		125			al, ace	B22[4], 260
12652	3-Pyrrole carboxylic acid $(HNC_4H_3)CO_2H\text{-}3$	111.10	nd (lig)		161-2				B22[4], 235

No.	Name, Synonyms, and Formula	Mol. wt.	Color, crystalline form, specific rotation and λ_{max} (log ε)	b.p. °C	m.p. °C	Density	n_D	Solubility	Ref.
12653	3-Pyrrole carboxylic acid, 2,4-dimethyl, ethyl ester 2,4-$(CH_3)_2$-3-$(C_4HNH)CO_2C_2H_5$	167.21	cr (eth-lig, peth)	291, 181-2[35]	78-9	....	....	al, eth	B22[4], 255
12654	3-Pyrrole carboxylic acid, 2,5-dimethyl, ethyl ester 2,5-$(CH_3)_2$-3-$(C_4HNH)CO_2C_2H_5$	167.21	rh (al)	290[731], 130[15]	117-8	....	....	al	B22[4], 267
12655	3-Pyrrole carboxylic acid, 4,5-dimethyl, ethyl ester 4,5-$(CH_3)_2$-3-$(C_4HNH)CO_2C_2H_5$	167.21	cr (dil al)	....	110-1	....	....	al, eth, chl	B22[4], 260
12656	2,4-Pyrrole dicarboxylic acid 2,4-$(C_4H_2NH)(CO_2H)_2$	155.11	cr (w)	....	295d	....	....	w	B22[4], 1542
12657	2,4-Pyrrole dicarboxylic acid, 3,5-dimethyl, diethyl ester 3,5-$(CH_3)_2$-2,4-$(C_4NH)(CO_2C_2H_5)_2$	239.27	nd (dil al)	....	136-7	....	....	ace, bz, chl	B22[4], 1568
12658	2,4-Pyrrole dicarboxylic acid, 3,5-dimethyl-N-ethyl, diethyl ester $C_{14}H_{21}NO_4$	267.33	cr (dil al)	....	40-1	....	....	al	B22[4], 1574
12659	2,4-Pyrrole dicarboxylic acid, 5-formyl-3-methyl, diethyl ester 3-CH_3-5-HCO-2,4-$(C_4NH)(CO_2C_2H_5)_2$	253.26	nd	....	124-5	....	....	al, to	B22[4], 3265
12660	2-Pyrone $C_5H_4O_2$	96.09	....	206-9, 120[30]	8-9	1.2000[20/4]	1.5270[25]	w, ace	B17[4], 4399
12661	2-Pyrone, tetrahydro $C_5H_8O_2$	100.12	....	218-20, 113-4[14]	−12.5	1.0794[20]	1.4503[20]	w, al, eth	B17[4], 4169
12662	Pyrrolidine or Tetrahydropyrrol C_4H_8NH	71.12	....	88-9	....	0.8520[22/4]	1.4431[20]	al, eth, chl	B20[4], 61
12663	Pyrrolidine, N-butyl $(C_4H_8N)C_4H_9$	127.23	....	154-5	....	0.816[25]	1.4373[25]	al	B20[4], 67
12664	Pyrrolidine, 2-butyl 2-$C_4H_9(C_4H_7NH)$	127.23	....	173-4[741], 67[18]	....	0.8277[20/4]	1.4490[20]	w, al	B20[4], 1627
12665	Pyrrolidine, N-(chloroacetyl) $(C_4H_8N)COCH_2Cl$	147.61	....	112[0.5]	44-6	....	....	....	B20[4], 186
12666	Pyrrolidine, N,2-dimethyl 2-$CH_3(C_4H_7N)CH_3$	99.18	....	96	....	0.7994[20/4]	1.4252[20]	w, al, eth	B20[4], 1384
12667	Pyrrolidine, 2,4-dimethyl 2,4-$(CH_3)_2(C_4H_6NH)$	99.18	....	115-7	....	0.8297[20/4]	1.4325[20]	w, al, eth	B20[4], 1537
12668	Pyrrolidine, 2,5-dimethyl (cis) 2,5-$(CH_3)_2(C_4H_6NH)$	99.18	....	....	106-7	0.8205[20/4]	1.4299[20]	**w, al, eth**	B20[4], 1538
12669	Pyrrolidine, N-methyl $(C_4H_8N)CH_3$	85.15	....	81.3	....	0.8188[20/4]	1.4247[20]	w, eth	B20[4], 64
12670	Pyrrolidine, N-phenyl $(C_4H_8N)C_6H_5$	147.22	....	119-20[12]	....	1.0260[25]	1.5813[20]	eth	B20[4], 78
12671	2-Pyrrolidone or γ-Butyrolactam C_4H_7NO	85.11	cr (peth)	250.5[742], 133[12]	24.6	1.120[20/4]	1.4806[30]	w, al, eth, bz	B21[4], 3142
12672	2-Pyrrolidone, N,5-dimethyl $C_6H_{11}NO$	113.16	....	215-7[743], 87.5[10]	....	....	1.4650[20]	**w, eth**	B21[4], 3192
12673	2-Pyrrolidone, 3,3-dimethyl $C_6H_{11}NO$	113.16	lf (bz)	237	65-7	....	....	w, al, eth, ace	B21, 242
12674	2-Pyrrolidone, N-methyl C_5H_9NO	99.13	....	202, 84-5[14]	−23	1.0260[25/25]	1.4684[20]	w, eth, ace	B21[4], 3145
12675	3-Pyrroline or Dihydropyrrole C_4H_6NH	69.11	....	90-1	....	0.9097[20/4]	1.4664[20]	al, w, eth, ace	B20[4], 1906
12676	Pyruvic acid or 2-oxo-Propionic acid CH_3COCO_2H	88.06	....	165d, 54[10]	13.8	1.2272[20/4]	1.4280[20]	w, al, eth, ace	B3[4], 1505
12677	Pyruvic acid-acetyl $CH_3COCH_2COCO_2H$	130.10	nd (bz-chl), pr (bz)	130[37] sub	101 (anh) 55-63 (hyd)	....	....	w, al, eth, ace, bz, chl	B3[3], 1331
12678	Pyruvic acid, acetyl, ethyl ester $CH_3COCH_2COCO_2C_2H_5$	158.16	....	213-5, 111-2[16]	18	1.1251[20]	1.4757[17]	al, eth	B3[4], 1777
12679	Pyruvic acid, 3-(2,5-dimethoxyphenyl) $[2,5-(CH_3O)_2C_6H_3]CH_2COCO_2H$	224.22	yesh cr (aa)	....	166-70d	....	....	ace, MeOH	B10[2], 723
12680	Pyruvic acid, 3-(3,4-dimethoxy phenyl) $[3,4-(CH_3O)_2C_6H_3]CH_2COCO_2H$	224.22	lf (aa)	....	ca 187d	....	....	al, ace	B10[3], 4521
12681	Pyruvic acid, ethyl ester or Ethyl pyruvate $CH_3COCO_2C_2H_5$	116.13	....	155, 69-71[42]	−50	1.0596[15 b/4]	1.4052[20]	al, eth, ace	B3[4], 1513
12682	Pyruvic acid, methyl ester or Methyl pyruvate $CH_3COCO_2CH_3$	102.09	....	134-7, 53[15]	....	1.154[0/4]	1.4046[25]	al, eth, ace	B3[3], 1160
12683	Pyruvic acid, 3-phenyl $C_6H_5CH_2COCO_2H$	164.17	lf (chl, bz)	157-8	....	....	....	al, eth	B10[3], 3000
12684	Pyruvonitrile CH_3COCN	69.06	rh	92-3	....	0.9745[20/4]	1.3764[20]	eth, ace	B3[4], 1515
12685	Pyruvic acid, 3-(2-nitro phenyl) $(2-NO_2C_6H_4)CH_2COCO_2H$	209.16	ye nd or lf (w, al, bz)	....	130	....	....	al, eth, aa	B10[3], 3017

No.	Name, Synonyms, and Formula	Mol. wt.	Color, crystalline form, specific rotation and λ_{max} (log ε)	b.p. °C	m.p. °C	Density	n_D	Solubility	Ref.
12686	o,o'-Quaterphenyl or o,o'-Diphenyl biphenyl $C_{24}H_{18}$	306.41	pr (al)	420	118-9			al, eth, ace, bz, chl	B5³, 2561
12687	p,p'-Quaterphenyl or Tetraphenyl $C_{24}H_{18}$	306.41	lf (bz)	428¹⁸	320				B5³, 2562
12688	Quercitrin or 3-D-α-L-rhamnopyranosylquercitrin $C_{21}H_{26}O_{11}$	448.39	pa ye nd or pl (+2w dil al), $[\alpha]_{578}$ −73.5 (al, c=4)		182-5 (hyd), 250-2 (anh)			al	B18⁴, 3491
12689	Quillaic acid or Quillaja sapogenin $C_{30}H_{46}O_5$	486.70	nd (dil al), $[\alpha]^{20}_D$ +56.1 (py, c=2.91)		294			al, eth, ace, aa, Py	B10³, 4652
12690	Quinacrine dihydrochloride (dl) or Atebrine $C_{25}H_{30}ClN_3O.2HCl.2H_2O$	532.94	yesh nd (w), ye cr pw		248-50d			w, MeOH	B21⁴, 6247
12691	Quinamine $C_{19}H_{24}N_2O_2$	312.42	pr (bz), nd (80%al), $[\alpha]^{15}_D$ +93 (chl, c=2)		185-6			eth, ace, bz, lig	B27², 667
12692	Quinazoline or 1,3-Benzodiazine $C_8H_6N_2$	130.15	ye pl (peth)	241.5, 117-20¹⁵	48			w, al, eth, ace, bz	B23², 177
12693	Quinazoline, 3,4-dihydro-4-oxo or Quinazolinone $C_8H_6N_2O$	146.15	nd (dil aa)	360	216.8				B24², 71
12694	Quinazoline, 3,4-dihydro-2-phenyl $C_{14}H_{12}N_2$	208.27	lf (dil al)		142			al, eth, chl, aa	B23, 239
12695	Quinazoline, 3,4-dihydro-3-phenyl or Phenzoline $C_{14}H_{12}N_2$	208.27	pl (eth lig)	d	96.7	1.290⁴		al, eth, bz, chl	B23², 155
12696	Quinazoline, 3,4-dihydro-4-phenyl $C_{14}H_{12}N_2$	208.27	pl (al, AcOEt)		166-7			al, eth	B23, 239
12697	Quinazoline, 2,4-dihydroxy-6,8-dinitro $C_8H_4N_4O_6$	252.15	ye gr pr (aa)		274-5d				B24¹, 344
12698	2,4-Quinazolinedione or Benzoylene urea $C_8H_6N_2O_2$	162.15	nd (w, al), lf (aa)		356				B24², 197
12699	Quinhydrone or Benzoquinhydrone $C_{12}H_{10}O_4$	218.20	red br nd	sub	171	1.401²⁰/⁴		al, eth	B7³, 3363
12700	Quinicine or Quinotoxin $C_{20}H_{24}N_2O_2$	324.43	red ye amor, $[\alpha]^{15}_D$ +44.1 (chl)		ca 60			al, eth, chl	B25², 20
12701	Quinicine oxalate (d) $(C_{20}H_{24}N_2O_2)_2H_2C_2O_4.9H_2O$	901.03	pr (chl), nd (al), $[\alpha]^{15}_D$ +19.5 (al-chl), (+9.5w)		149			w, al, chl	B25², 20
12702	Quinidine $C_{20}H_{24}N_2O_2$	324.43	cr (+2.5w, dil al)		174-5 (anh)			al, bz, chl	B23², 414
12703	Quinidine hydrate or Conquinine $C_{20}H_{24}N_2O_2.2½H_2O$	369.47	cr (dil al)		171d			al, eth, bz, chl	B23², 414
12704	Quinidine hydrochloride $C_{20}H_{24}N_2O_2.HCl.H_2O$	378.91	pr (w), $[\alpha]^{20}_D$ +200 (w, c=1)		258-9d (anh)			chl	B23², 415
12705	Quinidine sulfate (d) $(C_{20}H_{24}N_2O_2)_2.H_2SO_4.2H_2O$	782.95	pr, nd (w), $[\alpha]^{25}_D$ +212 (al)					w, al, chl	B23², 415
12706	Quinine $C_{20}H_{24}N_2O_2.3H_2O$	378.47	cr (+3w, eth), nd (+3w, al), rh nd (abs al), $[\alpha]^{15}_D$ −145.2 (al)		177 (anh), 57 (hyd)		1.625	al, eth, chl, Py	B23², 416

No.	Name, Synonyms, and Formula	Mol. wt.	Color, crystalline form, specific rotation and λ_{max} (log ε)	b.p. °C	m.p. °C	Density	n_D	Solubility	Ref.
12707	Quinine bisulfate $C_{20}H_{24}N_2O_2.H_2SO_4.7H_2O$	548.62	pr (+ w, w), pr (+ 5w, al), $[\alpha]^{15}_D$ + 168.4 (al)	....	160d	....	....	chl	B23², 416
12708	Quinine, o-ethyl carbonate $C_{23}H_{28}N_2O_4$	396.49	nd (w), cr (dil al)		95			al, eth, chl	B23², 424
12709	Quinine formate or Quinoform $C_{20}H_{24}N_2O_2.HCO_2H$	370.45	nd $[\alpha]^{20}_D$ −144.2 (w, c=1)		149-50 (anh), 126 (+ 1w)			w, al, chl	B23¹, 169
12710	Quinine hydrobromide $C_{20}H_{24}N_2O_2.HBr.H_2O$	423.36	silky efflor nd (w)		ca 208 soft-ens 152			w, al, chl	B23¹, 168
12711	Quinine hydrochloride $C_{20}H_{24}N_2O_2.HCl$	360.89	silky efflor nd (w), $[\alpha]^{15}_D$ −145 (al)		158-60 (anh)			w, al, chl	B23², 420
12712	Quinine hydrochloride, hydrate $C_{20}H_{24}N_2O_2.HCl.2H_2O$	396.92	silky efflor nd (w), $[\alpha]^{20}_D$ −149.8 (w, c=1.3)		156-90			w, al, chl	B23², 420
12713	Quinine salicylate or Quinine-2-hydroxy benzoate $C_{20}H_{24}N_2O_2.C_7H_6O_3.H_2O$	480.55	wh pr (al), cr (+ 2w, w)		195			al, chl	B23², 422
12714	Quinine sulfate $2(C_{20}H_{24}N_2O_2).H_2SO_4$	746.93	silky nd (w)		235.2			al	B23², 420
12715	Quinine sulfate, dihydrate $2(C_{20}H_{24}N_2O_2)_2.H_2SO_4.2H_2O$	782.97	silky nd (w), $[\alpha]^{15}_D$ −220 (0.5N HCl)		205			al	B23², 420
12716	Quinine valerate $C_{20}H_{24}N_2O_2.C_4H_9CO_2H.H_2O$	444.58	wh	....	ca 96			al, eth	M, 892
12717	Quininone $C_{20}H_{22}N_2O_2$	322.41	nd lf (eth), $[\alpha]^{20}_D$ + 75.5 (al, c=2)		108 (rapid htng)			al, eth, bz, chl	B25², 23
12718	Quinizarin or 1,4-Dihydroxy anthraquinone $1,4-(HO)_2C_{14}H_6O_2$	240.23	ye red lf (eth), dk red nd (al), red cr (to, aa)		200-2 (aa), 194 (to)				B8¹, 3775
12719	Quinoline or Benzo[b]-pyridine C_9H_7N	129.16	....	238, 114¹⁷	fr−15.6	1.0929²⁰/⁴	1.6268²⁰	al, eth, ace, bz	B20⁴, 3334
12720	Quinoline, 3-acetamido $3-(CH_3CONH)(C_9H_6N)$	186.22	cr (w)		166-7			al	B22⁴, 4607
12721	Quinoline, 4-acetamide $4-(CH_3CONH)(C_9H_6N)$	186.22	nd (w + 1)	sub	176			al	B22⁴, 4614
12722	Quinoline, 5-acetamido $5-(CH_3CONH)(C_9H_6N)$	186.22	pl (w), pr (dil al)		178			al	B22⁴, 4671
12723	Quinoline, 6-acetamido $6-(CH_3CONH)(C_9H_6N)$	186.22	nd (w)		138			w, al	B22², 355
12724	Quinoline, 8-acetamido $8-(CH_3CONH)(C_9H_6N)$	186.22	cr (w, al)		167.5			al	B22², 356
12725	Quinoline, 3-acetamido $8-(CH_3CONH)(C_9H_6N)$	186.22	nd (al)		103			al	B22¹, 640
12726	Quinoline, 2-amino or α-Quinolyl amine $2-H_2N(C_9H_6N)$	144.19	lf (w)	sub	131.5			al, eth, ace, chl, aa	B22⁴, 4587
12727	Quinoline, 2-amino-4-hydroxy $4-HO-2-H_2N(C_9H_5N)$	160.19	nd (w + 1), rh (al)		303-4				B22⁴, 5718
12728	Quinoline, 2-amino-4-methyl or 2-Aminolepidine $4-CH_3-2-H_2N(C_9H_5N)$	158.25	cr pw (bz)	320	133			al, eth, chl, aa	B22⁴, 4801
12729	Quinoline, 3-amino or β-Quinolyl amine $3-H_2N(C_9H_6N)$	144.19	rh (w, dil al)		94			al, eth, chl	B22⁴, 4605

No.	Name, Synonyms, and Formula	Mol. wt.	Color, crystalline form, specific rotation and λ_{max} (log ϵ)	b.p. °C	m.p. °C	Density	n_D	Solubility	Ref.
12730	Quinoline, 3-amino-2-methyl or 3-Amino quinaldine ..... 2-CH₃-3-H₂N(C₉H₅N)	158.21	pa ye nd (eth, peth)	278d, 198[16]	160			al, eth, bz, chl	B22[4], 4756
12731	Quinoline, 4-amino or γ-Quinolyl amine 4-H₂N(C₉H₅N)	144.19	nd (w + 1), nd (bz, dil al)	180[12]	154 (anh), 70 (+ w)			w, al, eth, chl	B22[4], 4611
12732	Quinoline, 4-amino-2-methyl or 4-Amino quinaldine ..... 2-CH₃-4-H₂N(C₉H₅N)	158.21	nd (bz-lig), pr (eth-bz)	333	168			al, eth, ace, bz	B22[4], 4559
12733	Quinoline, 5-amino or 5-Quinalyl amine .... 5-H₂N(C₉H₅N)	144.19	ye nd (al), lf (eth)	310 sub, 184[10]	110			al, eth, bz	B22[4], 4669
12734	Quinoline, 5-amino-6-hydroxy .... 6-HO-5-H₂N(C₉H₅N)	160.19	gr nd (w + 2)		185			al	B22, 501
12735	Quinoline, 5-amino-8-hydroxy 8-HO-5-H₂N(C₉H₅N)	160.19	nd (bz)		143				B22, 5865
12736	Quinoline, 5-amino-2-methyl or 5-Amino quinaldine..... 2-CH₃-5-H₂N(C₉H₅N)	158.21	grsh pl or nd (w + 1)		117-8			w, al, bz, lig	B22[2], 360
12737	Quinoline, 5-amino-8-methyl 8-CH₃-5-H₂N(C₉H₅N)	158.21	yesh nd (w, dil al)		143			al	B22[4], 4828
12738	Quinoline, 6-amino or 6-Quinolyl amine .... 6-H₂N(C₉H₅N)	144.19	cr (w + 2), pr (eth)	187[12]	114 (anh)			al	B22[4], 4681
12739	Quinoline, 6-amino-2-methyl or 6-Amino quinaldine..... 2-CH₃-6-H₂N(C₉H₅N)	158.21	pa br (w, dil al)		187.5			w, al, chl	B22[4], 4780
12740	Quinoline, 6-amino-4-methyl .... 4-CH₃-6-H₂N(C₉H₅N)	158.21	nd (w)		169-70			al, eth, chl	B22[4], 4812
12741	Quinoline, 7-amino or 7-Quinolyl amine .... 7-H₂N(C₉H₅N)	144.19	ye nd (+ 1w)		93-4 (anh), 74-5 (+ 1w)			al	B22[4], 4704
12742	Quinoline, 7-amino-8-hydroxy .... 8-HO-7-H₂N(C₉H₅N)	160.19	br pr (eth, dil al)		124			al, eth, bz, chl	B22[4], 5873
12743	Quinoline, 7-amino-2-methyl 2-CH₃-7-H₂N(C₉H₅N)	158.21	pr (dil al)	304	129			al, eth, ace, bz	B22[4], 4784
12744	Quinoline, 7-amino-8-methyl 8-CH₃-7-H₂N(C₉H₅N)	158.21	pr (dil al)	304	125			al, eth, ace, bz	B22, 456
12745	Quinoline, 8-amino or 8-Quinolyl amine .... 8-H₂N(C₉H₅N)	144.19	pa ye nd (sub), cr (al, lig)	157-8[19]	70			w, al	B22[4], 4708
12746	Quinoline, 8-amino-6-methoxy 6-CH₃O-8-H₂N(C₉H₅N)	174.20	cr	137-8[1]	51				B22[4], 5747
12747	Quinoline, 8-amino-2-methyl or 8-Amino quinaldine..... 2-CH₃-8-H₂N(C₉H₅N)	158.21	pr (lig)		57-8			al, eth, ace, bz	B22[4], 4785
12748	Quiniline, 8-amino-6-methyl...................... 6-CH₃-8-H₂N(C₉H₅N)	158.21	nd	sub	73			w, al, eth, ace, bz	B22[4], 4822
12749	Quinoline, 2-bromo .... 2-Br(C₉H₆N)	208.06	nd (al)		49			al, eth, bz, chl	B20[4], 3387
12750	Quinoline, 3-bromo .... 3-Br(C₉H₆N)	208.06	ye oil	274-6, 98[0.5]	13-5		1.6641[20]	aa	B20[4], 3388
12751	Quinoline, 3-bromo-2-hydroxy or β-Bromo carbostyril ... 3-Br-2-HO(C₉H₅N)	224.06	pr (al)	sub	253				B21[4], 1063
12752	Quiniline, 4-bromo .... 4-Br(C₉H₆N)	208.06	cr	270d	29-30				B20[4], 3389
12753	Quinoline, 4-bromo-2-hydroxy or γ-Bromo carbostyril ... 4-Br-2-HO(C₉H₅N)	224.06	nd (al)	sub	266-7			al	B21[4], 1063
12754	Quinoline, 5-bromo .... 5-Br(C₉H₆N)	208.06	nd	280[756], 105-7[1,2]	52				B20[4], 3389
12755	Quinoline, 5-bromo-2-hydroxy or 5-Bromo carbostyril ... 5-Br-2-HO(C₉H₅N)	224.06	nd (al)		300				B21[4], 1063
12756	Quinoline, 5-bromo-6-hydroxy .... 5-Br-6-HO(C₉H₅N)	224.06	nd (dil al)		186				B21[4], 1119
12757	Quinoline, 5-bromo-8-hydroxy .... 5-Br-8-HO(C₉H₅N)	224.06	nd (al), nd or lf (sub)	sub	124			w, al, bz, chl	B21[4], 1184
12758	Quinoline, 6-bromo .... 6-Br(C₉H₆N)	208.06		278, 155-6[15]	24			al, eth	B20[4], 3390
12759	Quinoline, 6-bromo-2-hydroxy .... 6-Br-2-HO(C₉H₅N)	224.06	ye nd (al)		269			w, al, eth, chl	B21[4], 1063
12760	Quinoline, 7-bromo .... 7-Br(C₉H₆N)	208.06	nd	290	34 (52)			al, eth	B20[4], 3390

No.	Name, Synonyms, and Formula	Mol. wt.	Color, crystalline form, specific rotation and λ_{max} (log ϵ)	b.p. °C	m.p. °C	Density	n_D	Solubility	Ref.
12761	Quinoline, 7-bromo-2-hydroxy or 7-Bromo carbostyril ... 7-Br-2-HO(C$_9$H$_5$N)	224.06	nd (aa), pl (al)	sub	228	....	....	al, eth, chl	B21[4], 1064
12762	Quinoline, 8-Bromo .... 8-Br(C$_9$H$_6$N)	208.06	....	302-4, 165-6[20]	<−10 (80)	....	....	al	B20[4], 3391
12763	Quinoline, 8-bromo-5-hydroxy .... 8-Br-5-HO(C$_9$H$_5$N)	224.06	nd (al)		190d	....	....	al, chl	B21, 85
12764	Quinoline, 2-chloro .... 2-Cl(C$_9$H$_6$N)	163.61	nd (aq al)	265-6[750], 153-4[22]	38	1.2464[25/4]	1.6342[25]	al, eth, bz, lig	B20[4], 3376
12765	Quinoline, 2-chloro-4-methyl or 2-Chloro lepidine ...... 4-CH$_3$-2-Cl(C$_9$H$_5$N)	177.65	nd (dil al)	296	59	....	....	al, eth, chl	B20[4], 3482
12766	Quinoline, 2-chloro-6-methyl .... 6-CH$_3$-2-Cl(C$_9$H$_5$N)	177.65	nd (dil al)		112 (116)	....	....	al, eth, bz, chl	B20[4], 3492
12767	Quinoline, 2-chloro-8-methyl .... 8-CH$_3$-2-Cl(C$_9$H$_5$N)	177.65	nd (eth)	286[734]	61	....	....	al, eth, bz, chl	B20[1], 152
12768	Quinoline, 3-chloro .... 3-Cl(C$_9$H$_6$N)	163.61	hyg cr	255, 141[15]	....	....	....		B20[4], 3377
12769	Quinoline, 3-chloro-2-methyl or 3-Chloro quinaldine ..... 2-CH$_3$-3-Cl(C$_9$H$_5$N)	177.65	nd (dil al)		71-2	....	....	al, eth	B20[4], 3461
12770	Quinoline, 3-chloro-4-methyl or 3-Chloro lepidine .... 4-CH$_3$-3-Cl(C$_9$H$_5$N)	177.65	nd (dil al)	....	55	....	....	al	B20[1], 150
12771	Quinoline, 3-chloro-6-methyl .... 6-CH$_3$-3-Cl(C$_9$H$_5$N)	177.65	nd (dil MeOH)		85.5	....	....	al, MeOH	B20[2], 246
12772	Quinoline, 4-chloro .... 4-Cl(C$_9$H$_6$N)	163.61	cr	261[764], 130[15]	34-5	1.251	....	al, eth	B20[4], 3377
12773	Quinoline, 4-chloro-2-methyl or 4-Chloro quinaldine ...... 2-CH$_3$-4-Cl(C$_9$H$_5$N)	177.65	nd (+1w)	269-70	42-3	....	....	al, eth, bz, chl	B20[4], 3461
12774	Quinoline, 5-chloro .... 5-Cl(C$_9$H$_6$N)	163.61	cr (al)	256-7[756]	45	....	....	al	B20[4], 3379
12775	Quinoline, 5-chloro-7-iodo-8-hydroxy or Vioform .... 5-Cl-7-I-8-HO(C$_9$H$_4$N)	305.50	ye br nd (al, aa)		178-9	....	....		B21[4], 1190
12776	Quinoline, 6-chloro .... 6-Cl(C$_9$H$_6$N)	163.61	pr (eth), nd (al)	262-4	44-5	....	1.6110[56]		B20[4], 3379
12777	Quinoline, 6-chloro-2-methyl or 6-Chloro quinaldine ...... 2-CH$_3$-6-Cl(C$_9$H$_5$N)	177.65	lf or nd (dil al)	....	91	....	....	eth	B20[4], 3462
12778	Quinoline, 6-chloro-4-methyl or 6-Chloro lepidine .... 4-CH$_3$-6-Cl(C$_9$H$_5$N)	177.65	nd (al)		70-2	....	....	al, eth, ace, bz	B20[4], 3482
12779	Quinoline, 7-chloro .... 7-Cl(C$_9$H$_6$N)	163.61	nd or pr	267-8	31-2	1.2158[58/4]	1.6108[58/4]	al, eth, ace, bz, chl	B20[4], 3381
12780	Quinoline, 7-chloro-4-hydroxy .... 7-Cl-4-HO(C$_9$H$_6$N)	179.60	nd (al-w)		276-80	....	....		B21[4], 1084
12781	Quinoline, 7-chloro-2-methyl or 7-Chloro quinaldine ...... 2-CH$_3$-7-Cl(C$_9$H$_5$N)	177.65	nd (eth), cr (lig)	87[0.5]	75-6	....	....		B20[4], 3462
12782	Quinoline, 8-chloro .... 8-Cl(C$_9$H$_6$N)	163.61	....	288-9	fr−20	1.2834[14/4]	1.6408[14.3]	al, eth, ace, bz, chl	B20[4], 3377
12783	Quinoline, 8-chloro-5-methyl .... 5-CH$_3$-8-Cl(C$_9$H$_5$N)	177.65	nd (w)		49	....	....	al, eth, bz	B20[2], 246
12784	Quinoline, 8-(chloromethyl) .... 8-ClCH$_2$(C$_9$H$_5$N)	177.65	nd (w)		49	....	....	al, eth, bz	B20[4], 3502
12785	Quinoline, decahydro (cis) .... C$_9$H$_{17}$N	139.24	....	205[735], 90[20]	−40	0.9426[20/4]	1.4926[20]	al, eth	B20[4], 2017
12786	Quinoline, decahydro (trans, d) .... C$_9$H$_{17}$N	139.24	$[\alpha]^{25}_D$ +4.8 (al, c=3)	200-2	75	....	....	al, eth, ace, bz	B20[4], 2017
12787	Quinoline, decahydro (trans, dl) .... C$_9$H$_{17}$N	139.24	pr (lig, sub)	203[735] sub	48	0.9610[22]	1.4692[56]	al, eth, ace, bz	B20[4], 2017
12788	Quinoline, decahydro (trans, l) .... C$_9$H$_{17}$N	139.24	$[\alpha]^{25}_D$ −4.5 (al, c=3)	200-1	74-5	....	....		B20[4], 2017
12789	Quinoline, 5,7-dibromo-8-hydroxy .... 5,7-Br$_2$-8-HO(C$_9$H$_4$N)	302.96	nd (al)	sub	196	....	....	al, ace, bz, aa, chl	B21[4], 1180
12790	Quinoline, 6,8-dibromo-2-hydroxy or 6,8-Dibromo carbostyril .... 6,8-Br$_2$-2-HO(C$_9$H$_4$N)	302.96	nd (dil al)		230	....	....	al	B21[4], 1064
12791	Quinoline, 2,3-dichloro .... 2,3-Cl$_2$(C$_9$H$_5$N)	198.05	(dil al)	....	104-5	....	....	al, eth, bz	B20[4], 3382
12792	Quinoline, 2,4-dichloro .... 2,4-Cl$_2$(C$_9$H$_5$N)	198.05	nd (dil al)	280-2	67-8	....	....	al, eth, bz, chl	B20[4], 3382

No.	Name, Synonyms, and Formula	Mol. wt.	Color, crystalline form, specific rotation and λ_{max} (log ε)	b.p. °C	m.p. °C	Density	n_D	Solubility	Ref.
12793	Quinoline, 2,6-dichloro .... 2,6-Cl$_2$(C$_9$H$_5$N)	198.05	nd (eth)	....	156 (161)	....	....	al, eth	B20[4], 3383
12794	Quinoline, 2,7-dichloro .... 2,7 Cl$_2$(C$_9$H$_5$N)	198.05	nd (al)	sub 100[2]	120	....	....	al, eth	B20[4], 3383
12795	Quinoline, 3,4-dichloro .... 3,4-Cl$_2$(C$_9$H$_5$N)	198.05	peth	....	69-70	....	....	al, eth	B20[4], 3383
12796	Quinoline, 4,5-dichloro .... 4,5-Cl$_2$(C$_9$H$_5$N)	198.05	....	134[5.5]	118	....	....	al, eth	B20[4], 3383
12797	Quinoline, 4,6-dichloro .... 4,6-Cl$_2$(C$_9$H$_5$N)	198.05	(peth)	....	104	....	....	al, eth	B20[4], 3383
12798	Quinoline, 4,7-dichloro .... 4,7-Cl$_2$(C$_9$H$_5$N)	198.05	cr (MeOH), nd (80% al)	148[10]	93	....	....	....	B20[4], 3384
12799	Quinoline, 4,8-dichloro .... 4,8-Cl$_2$(C$_9$H$_5$N)	198.05	....	....	155-6	....	....	al, eth	B20[4], 3385
12800	Quinoline, 5,6-dichloro .... 5,6-Cl$_2$(C$_9$H$_5$N)	198.05	nd (al)	....	85	....	....	al, eth, peth	B20, 361
12801	Quinoline, 5,7-dichloro .... 5,7-Cl$_2$(C$_9$H$_5$N)	198.05	nd (al)	....	117	....	....	al, eth	B20[4], 3385
12802	Quinoline, 5,7-dichloro-8-hydroxy .... 5,7-Cl$_2$-8-HO(C$_9$H$_4$N)	214.05	nd (al)	....	183	....	....	bz, peth	B21[4], 1180
12803	Quinoline, 5,8-dichloro .... 5,8-Cl$_2$(C$_9$H$_5$N)	198.05	nd (al), pl (eth)	sub	97-8	....	....	al, eth	B20[4], 3385
12804	Quinoline, 6,8-dichloro .... 6,8-Cl$_2$(C$_9$H$_5$N)	198.05	nd (al)	....	104-5	....	....	eth	B20[4], 3385
12805	Quinoline, 7,8-dichloro .... 7,8-Cl$_2$(C$_9$H$_5$N)	198.05	nd	....	85.5	....	....	al, eth	Prak 48, Prak 279
12806	Quinoline, 1,2-dihydro-1-methyl-2-oxo or N-Methyl carbostyril C$_{10}$H$_9$NO	159.19	nd (lig)	324[728]	71	....	....	al, eth, ace, bz, chl	B21[4], 3737
12807	Quinoline, 2,4-dihydroxy .... 2,4-(HO)$_2$(C$_9$H$_5$N)	161.16		....	355	....	....		B21[4], 2219
12808	Quinoline, 5,7-diiodo-8-hydroxy or Diiodoquin 5,7-I$_2$-8-HO(C$_9$H$_4$N)	396.96	yesh nd (aa, xyl)	....	210d	....	....	al	B21[4], 1191
12809	Quinoline, 5,8-diiodo-6-hydroxy .... 5,8-I$_2$-6-HO(C$_9$H$_4$N)	396.96	yesh	....	191	....	....	al, eth	B21[2], 54
12810	Quinoline, 2,3-dimethyl or 3-Methyl quinaldine 2,3-(CH$_3$)$_2$(C$_9$H$_5$N)	157.22	ye rh (eth)	261[729]	68-9	1.1013	....	al, eth, lig	B20[4], 3521
12811	Quinoline, 2,4-dimethyl or 4-Methyl quinaldine 2,4-(CH$_3$)$_2$(C$_9$H$_5$N)	157.22	rh pr (eth)	264-6, 143[15]	......	1.0611[15]	1.6075[20]	al, eth	B20[4], 3522
12812	Quinoline, 2,4-dimethyl-6-hydroxy .... 2,4-(CH$_3$)$_2$-6-HO(C$_9$H$_4$N)	173.22	pr or pl (al)	360d	214	....	....	al, ace	B21[2], 68
12813	Quinoline, 2,4-dimethyl-7-hydroxy .... 2,4-(CH$_3$)$_2$-7-HO(C$_9$H$_4$N)	173.22	nd (al)	....	218	....	....	al	B2[1], 116
12814	Quinoline, 2,4-dimethyl-8-hydroxy .... 2,4-(CH$_3$)$_2$-8-HO(C$_9$H$_4$N)	173.22	pr (eth)	281 sub	65	....	....	al, eth, ace, bz, chl	B21[4], 1292
12815	Quinoline, 2,3-dimethyl-4-hydroxy .... 2,3-(CH$_3$)$_2$-4-HO(C$_9$H$_4$N)	173.22	pr (w + 1)	sub	319-20	....	....	....	B21[4], 1293
12816	Quinoline, 2,6-dimethyl or 6-Methyl quinaldine 2,6-(CH$_3$)$_2$(C$_9$H$_5$N)	157.22	rh pr (eth)	266-7, 152-5[13]	60	....	....	bz	B20[4], 3525
12817	Quinoline, 2,6-dimethyl-4-hydroxy .... 2,6-(CH$_3$)$_2$-4-HO(C$_9$H$_4$N)	173.22	nd (w + 1)	....	279	....	....	w	B21[4], 1294
12818	Quinoline, 2,7-dimethyl .... 2,7-(CH$_3$)$_2$(C$_9$H$_5$N)	157.21	....	264-5, 115-6[7]	61	....	....	al, eth, chl	B20[4], 3528
12819	Quinoline, 2,8-dimethyl or o-Toluquinaldine 2,8-(CH$_3$)$_2$(C$_9$H$_5$N)	157.22	....	255.3, 103-4[5]	27	1.0394[20/4]	1.6022[20]	al, eth	B20[4], 3529
12820	Quinoline, 2,8-dimethyl-4-hydroxy .... 2,8-(CH$_3$)$_2$-4-HO(C$_9$H$_4$N)	173.22	lf or pl (w + 1)	sub	260-1	....	....	al	B21[4], 1296
12821	Quinoline, 3,4-dimethyl or 3-Methyl lepidine 3,4-(CH$_3$)$_2$(C$_9$H$_5$N)	157.22	cr (eth)	290[737]	73-4	....	....	al, eth	B20[4], 3530
12822	Quinoline, 4,6-dimethyl-2-hydroxy .... 4,6-(CH$_3$)$_2$-2-HO(C$_9$H$_4$N)	173.22	pr (al)	....	249-50	....	....	....	B21[4], 1301
12823	Quinoline, 4,7-dimethyl-2-hydroxy .... 4,7-(CH$_3$)$_2$-2-HO(C$_9$H$_4$N)	173.22	cr (aa)	....	220	....	....	al, aa	B21[4], 1303
12824	Quinoline, 4,8-dimethyl-2-hydroxy .... 4,8-(CH$_3$)$_2$-2-HO(C$_9$H$_4$N)	173.22	pl (aq aa)	....	217-8	....	....	al	B21[4], 1304
12825	Quinoline, 5,8-dimethyl .... 5,8-(CH$_3$)$_2$(C$_9$H$_5$N)	157.22	nd	265[736]	4-5	1.070[21]	....	al, eth	B20[4], 3538

No.	Name, Synonyms, and Formula	Mol. wt.	Color, crystalline form, specific rotation and λ_{max} (log ε)	b.p. °C	m.p. °C	Density	n_D	Solubility	Ref.
12826	Quinoline, 6,8-dimethyl or β-Cytisolidine . . . . 6,8-(CH$_3$)$_2$(C$_9$H$_5$N)	157.22	. . . .	268-9, 133-4[14]	. . . . . .	1.0665[4]	. . . .	al, eth	B20[4], 3539
12827	Quinoline, 6,8-dimethyl-2-hydroxy . . . . 6,8-(CH$_3$)$_2$-2-HO(C$_9$H$_4$N)	173.22	nd (al)	. . . .	201-2			al	B21[1], 225
12828	Quinoline, 6,8-dimethyl-5-hydroxy or Cytisoline . . . . 6,8-(CH$_3$)$_2$-5-HO(C$_9$H$_4$N)	173.22	pl (chl), cr (al)	sub	197-8			al, bz, chl	B21, 117
12829	Quinoline, 6-(dimethylamino)-2-methyl or 6-Dimethyl amino quinaldine . . . . 2-CH$_3$-6-[(CH$_3$)$_2$N](C$_9$H$_5$N)	186.26	ye pr (aa, AcOEt)	319	101			al, eth, bz	B22[2], 361
12830	Quinoline, 3-ethyl-2-hydroxy or 3-Ethyl carbostyril . . . . . 3-C$_2$H$_5$-2-HO(C$_9$H$_4$N)	173.22	(dil HCl)	. . . .	168			bz, chl	B21, 115
12831	Quinoline, 2-hydrazino . . . . 2-H$_2$NNH(C$_9$H$_6$N)	159.19	(bz)	. . . .	142-3			al	B22[1], 690
12832	Quinoline, 5-hydrazino 5-H$_2$NNH(C$_9$H$_6$N)	159.19	ye nd (w)		150-1			al	B22, 565
12833	Quinoline hydrochloride or Quinilirium chloride . . . . . . C$_9$H$_7$N.HCl	165.63	pr (w)		134.5			w, al, chl	B20[2], 226
12834	Quinoline hydrogen sulfate . . . . C$_9$H$_7$N.H$_2$SO$_4$	227.24	cr (al, aa)		164			w, aa	B20[2], 226
12835	Quinoline, 2-hydroxy or 2-Quinolinol . . . . 2-HO(C$_9$H$_6$N)	145.16	pr (al, dil al + 1w), nd (sub)	sub	199-200 (anh)			al, eth	B21[4], 1057
12836	Quinoline, 2-hydroxy-3-methyl or 3-Methyl carbostyril . . . 3-CH$_3$-2-HO(C$_9$H$_5$N)	159.19	yesh nd (ace, dil al)	sub	234-5			al	B21[4], 1246
12837	Quinoline, 2-hydroxy-4-methyl or 2-Hydroxy lepidine . . . . 4-CH$_3$-2-HO(C$_9$H$_5$N)	159.19	nd (w)	>360, 270[17]	245			al	B21[4], 1252
12838	Quinoline, 2-hydroxy-6-methyl or 6-Methyl carbostyril . . . 6-CH$_3$-2-HO(C$_9$H$_5$N)	159.19	nd (al)	240-1[12]	237			al, eth, ace, bz	B21[4], 1271
12839	Quinoline, 3-hydroxy or 3-Quinolinol . . . . 3-HO(C$_9$H$_6$N)	145.16	cr (bz, dil al)		200-1			al, bz	B21[4], 1075
12840	Quinoline, 3-hydroxy-2-methyl or 3-Hydroxy quinaldine 2-CH$_3$-3-HO(C$_9$H$_5$N)	159.19	nd (al)		203-5			al, eth, chl	B21[4], 1216
12841	Quinoline, 4-hydroxy or 4-Quinolinol . . . . 4-HO(C$_9$H$_6$N)	145.16	nd (w + 3)		210 (anh), 100 (hyd)			al	B21[4], 1079
12842	Quinoline, 4-hydroxy-2-methyl or 4-Hydroxy quinaldine 2-CH$_3$-4-HO(C$_9$H$_5$N)	159.19	pr (w + 2)	300d	232			al	B21[4], 1218
12843	Quinoline, 4-hydroxy-2-phenyl . . . . 2-C$_6$H$_5$-4-HO(C$_9$H$_5$N)	221.26	pl or pr (al), nd (aa)		256-7			al	B21[4], 1625
12844	Quinoline, 5-hydroxy or 5-Quinolinol . . . . 5-HO(C$_9$H$_6$N)	145.16	nd (al), pl	sub	224d			. . . .	B21[4], 1103
12845	Quinoline, 5-hydroxy-2-methyl or 5-Hydroxy quinaldine 2-CH$_3$-5-HO(C$_9$H$_5$N)	159.19	pl (al)		246-7			eth	B21[2], 63
12846	Quinoline, 5-hydroxy-6-methyl 6-CH$_3$-5-HO(C$_9$H$_5$N)	159.19	nd (al, sub)	sub	230			al, eth, ace	B21, 111
12847	Quinoline, 5-hydroxy-8-methyl 8-CH$_3$-5-HO(C$_9$H$_5$N)	159.19	nd (dil al, sub)	sub	762-3			. . . .	B21, 112
12848	Quinoline, 6-hydroxy or 6-Quinolinol . . . . 6-HO(C$_9$H$_6$N)	145.16	pr (al, eth)	>360	193			al	B21[2], 53
12849	Quinoline, 6-hydroxy-2-methyl or 6-Hydroxy quinaldine 2-CH$_3$-6-HO(C$_9$H$_5$N)	159.19	(w)	304-5, 186[35]	213	1.1665[0]		al, eth	B21, 106
12850	Quinoline, 6-hydroxy-4-methyl or 6-Hydroxy lepidine . . . . 4-CH$_3$-6-HO(C$_9$H$_5$N)	159.19	nd (w, dil al)		222-4			al, ace, chl	B21[4], 1265
12851	Quinoline, 6-hydroxy-8-methyl 8-CH$_3$-6-HO(C$_9$H$_5$N)	159.19	nd (dil al)	. . . .	200			al	B21, 113
12852	Quinoline, 7-hydroxy or 7-Quinolinol . . . . 7-HO(C$_9$H$_6$N)	145.16	pr (al), nd (dil al-eth)	sub	238-40			al	B21[4], 1130
12853	Quinoline, 7-hydroxy-6-methyl . . . . 6-CH$_3$-7-HO(C$_9$H$_5$N)	159.19	nd (al)	240[22] sub	244			al, eth, bz	B21, 111
12854	Quinoline, 8-hydroxy or 8-Quinolinol . . . . 8-HO(C$_9$H$_6$N)	145.16	nd (dil al)	266.6[7.5.2] sub	75-6	1.034[209]		al, ace, bz, chl	B21[4], 1135
12855	Quinoline, 8-hydroxy-2-methyl or 8-Hydroxy quinaldine 2-CH$_3$-8-HO(C$_9$H$_5$N)	159.19	pr (dil al)	267, 145-60[22]	74-5			eth, bz	B21[4], 1232
12856	Quinoline, 8-hydroxy-4-methyl or 8-Hydroxy lepidine . . . . 4-CH$_3$-8-HO(C$_9$H$_5$N)	159.19	nd (lig)	. . . .	141			ace, bz, chl, aa	B21[4], 1266

No.	Name, Synonyms, and Formula	Mol. wt.	Color, crystalline form, specific rotation and λ_{max} (log ε)	b.p. °C	m.p. °C	Density	n_D	Solubility	Ref.
12857	Quinoline, 8-hydroxy-5-methyl 5-CH₃-8-HO(C₉H₅N)	159.19	nd (dil al)	sub 100	122-4				B21⁴, 1269
12858	Quinoline, 8-hydroxy-6-methyl 6-CH₃-8-HO(C₉H₅N)	159.19	nd (chl, bz)	sub	95-6			al	B21⁴, 1274
12859	Quinoline, 8-hydroxy-7-methyl 7-CH₃-8-HO(C₉H₅N)	159.19	nd (dil al)	sub 100	72-4			al	B21⁴, 1279
12860	Quinoline, 8-hydroxy-5-nitroso 5-NO-8-HO(C₉H₅N)	174.16	nd (al)		245d				B21¹, 405
12861	Quinoline, 8-hydroxy, sulfate or Chinosol (8-HOC₉H₆N)₂·H₂SO₄	388.40			177.5			w, al	B21, 92
12862	Quinoline, 2-iodo 2-I(C₉H₆N)	255.06	nd (dil al)		52-3			al, eth, ace	B20⁴, 3393
12863	Quinoline, 4-iodo 4-I(C₉H₆N)	255.06	nd or pr	sub	97 (100)			al, eth	B20, 370
12864	Quinoline, 5-iodo 5-I(C₉H₆N)	255.06	nd (al, eth)	sub	101-2				B20¹, 141
12865	Quinoline, 6-iodo 6-I(C₉H₆N)	255.06	lf (w), nd (sub)	sub	91			al, eth	B20, 370
12866	Quinoline, 8-iodo 8-I(C₉H₆N)	255.06	lo nd (al)		36			al, eth, ace, bz, lig	B20¹, 141
12867	Quinoline, 2-methoxy-6-nitro 6-NO₂-2-CH₃O(C₉H₅N)	204.20	nd (dil aa, bz, sub)		189-90			al, eth, bz, chl, lig	B21, 81
12868	Quinoline, 2-methoxy-8-nitro 8-NO₂-2-CH₃O(C₉H₅N)	204.20	nd (dil al)		124-5			al, bz	B21, 82
12869	Quinoline, 4-methoxy 4-CH₃O(C₉H₆N)	159.19		243, 107	41			al, eth, bz	B21⁴, 1080
12870	Quinoline, 5-methoxy 5-CH₃O(C₉H₆N)	159.19		282⁷⁵⁰				al, eth	B21⁴, 1103
12871	Quinoline, 6-methoxy or p-Quinanisole 6-CH₃O(C₉H₆N)	159.19	hyg lf	303¹⁵, 153¹²	26.5	1.154¹⁰/¹⁰, 1.000²⁰⁹		al, eth	B21⁴, 1108
12872	Quinoline, 6-methoxy-5-nitro 5-NO₂-6-CH₃O(C₉H₅N)	204.20	cr (al)		104-5			ace	B21⁴, 1121
12873	Quinoline, 6-methoxy-8-nitro 8-NO₂-6-CH₃O(C₉H₅N)	204.20	yesh nd (al)		159-60			chl	B21⁴, 1121
12874	Quinoline, 6-methoxy-1,2,3,4-tetrahydro or Thalline 6-CH₃O(C₉H₁₀N)	163.22	pr (peth, al), rh pym (w)	283⁷³⁵, 127-30¹	42-3		1.5718²⁰	al, eth, bz	B22, 61
12875	Quinoline, 8-methoxy or o-Quinanisole 8-CH₃O(C₉H₆N)	159.19	nd (peth)	282⁷⁵⁰, 164¹⁴	49-50	1.034²⁹		al, eth, bz, peth	B21⁴, 1154
12876	Quinoline, 8-methoxy-5-nitro 5-NO₂-8-CH₃O(C₉H₅N)	204.20	cr (al)		151.5			al	B21⁴, 1193
12877	Quinoline, 1-methyl-1,2,3,4-tetrahydro or Kairoline (C₉H₁₀N)CH₃	147.22		247-50, 123-6¹		1.022²⁰/⁴	1.5802²³	al	B20², 174
12878	Quinoline, 2-methyl or Quinaldine 2-CH₃(C₉H₆N)	143.19		247.6, 118¹⁰	-2	1.0585²⁰/⁴	1.8116²⁰	al, eth, ace, chl	B20⁴, 3454
12879	Quinoline, 2-methyl-5-nitro or 5-Nitro quinaldine 5-NO₂-2-CH₃(C₉H₅N)	188.20	nd (dil al)		82			al, eth	B20⁴, 3468
12880	Quinoline, 2-methyl-6-nitro or 6-Nitro quinaldine 6-NO₂-2-CH₃(C₉H₅N)	188.20	cr (80% MeOH)		165 (173)			al	B20⁴, 3468
12881	Quinoline, 2-methyl-8-nitro or 8-Nitro quinaldine 8-NO₂-2-CH₃(C₉H₅N)	188.20	pa ye nd (dil al)		137			al, eth, bz	B20⁴, 3468
12882	Quinoline, 3-methyl or β-Methyl quinoline 3-CH₃(C₉H₆N)	143.19	pr	259.8, 140-2²⁵	16-7	1.0673²⁰/⁴	1.6171²⁰	al, eth, ace	B20⁴, 3472
12883	Quinoline, 4-methyl or Lepidine 4-CH₃(C₉H₆N)	143.19	red br	264.2, 133¹⁵	9-10	1.0862²⁰/⁴	1.6206²⁰	**al, eth**, ace, bz, lig	B20⁴, 3477
12884	Quinoline, 4-methyl-3-nitro or 3-Nitro lepidine 3-NO₂-4-CH₃(C₉H₅N)	188.20	pr (w)		118			al	B20⁴, 3485
12885	Quinoline, 4-methyl-8-nitro 8-NO₂-4-CH₃(C₉H₅N)	188.20	lf (abs al)		126-7			eth	B20⁴, 3485
12886	Quinoline, 5-methyl 5-CH₃(C₉H₆N)	143.19		262.7	19	1.0832²⁰/⁴	1.6219²⁰	**al, eth**, ace	B20⁴, 3488
12887	Quinoline, 6-methyl or p-Toluquinoline 6-CH₃(C₉H₆N)	143.19		258.6, 130¹⁵	ca -22	1.0654²⁰/⁴	1.6157²⁰	al, eth, ace	B20⁴, 3489
12888	Quinoline, 6-methyl-5-nitro 5-NO₂-6-CH₃(C₉H₅N)	188.20	pa ye nd (al)		116-7			al, eth, ace, bz	B20⁴, 3495
12889	Quinoline, 6-methyl-8-nitro 8-NO₂-6-CH₃(C₉H₅N)	188.20	pa ye nd (w)		122			al, eth, ace, bz	B20, 400

No.	Name, Synonyms, and Formula	Mol. wt.	Color, crystalline form, specific rotation and λ_{max} (log ε)	b.p. °C	m.p. °C	Density	n_D	Solubility	Ref.
12890	Quinoline, 7-methyl or m-Toluquinoline 7-CH$_3$(C$_9$H$_6$N)	143.19	ye	257.6, 144[18]	39	1.0609[20/4]	1.6150[20]	al, eth, ace	B20[4], 3497
12891	Quinoline, 8-methyl or o-Toluquinoline 8-CH$_3$(C$_9$H$_6$N)	143.19	....	247[8], 143[14]	......	1.0719[20/4]	1.6164[20]	**al, eth**, ace	B20[4], 3500
12892	Quinoline, 8-methyl-5-nitro 5-NO$_2$-8-CH$_3$(C$_9$H$_5$N)	188.20	pa ye nd (al)	....	93			al, eth, ace, bz	B20[4], 3504
12893	Quinoline, 8-methyl-6-nitro 6-NO$_2$-8-CH$_3$(C$_9$H$_5$N)	188.20	cr (al)	....	129			al	B20[4], 3504
12894	Quinoline, 3-nitro 3-NO$_2$(C$_9$H$_6$N)	174.16	nd (dil al)	....	128			al, ace	B20[4], 3395
12895	Quinoline, 4-nitro, oxide 4-NO$_2$(C$_9$H$_6$NO)	190.16	ye nd pl (ace)	....	154				B20[4], 3396
12896	Quinoline, 5-nitro 5-NO$_2$(C$_9$H$_6$N)	174.16	pl (w or al), nd (+ w, w)	sub	73-5 (sub)			bz	B20[4], 3397
12897	Quinoline, 6-nitro 6-NO$_2$(C$_9$H$_6$N)	174.16	ye pl (HCl-aa)	sub	153-4			bz	B20[4], 3397
12898	Quinoline, 7-nitro 7-NO$_2$(C$_9$H$_6$N)	174.16	nd or lf, w al pl (sub)	sub	132-3			eth, chl	B20[4], 3398
12899	Quinoline, 8-nitro 8-NO$_2$(C$_9$H$_6$N)	174.16	mcl pr (al)	....	91-2			al, eth	B20[4], 3399
12900	Quinoline, 2-oxo-1,2,3,4-tetrahydro C$_9$H$_9$NO	147.18	pr (al, eth)	201[45]	163-4			al, eth	B21[4], 3638
12901	Quinoline, 2-phenyl 2-C$_6$H$_5$(C$_9$H$_6$N)	205.26	nd (dil al)	363, 310[187]	86			al, eth, ace, bz	B20[4], 4137
12902	Quinoline, 3-phenyl 3-C$_6$H$_5$(C$_9$H$_6$N)	205.26	pl (eth)	205-7[12]	52			al, eth, ace, bz, chl	B20[4], 4145
12903	Quinoline, 4-phenyl 4-C$_6$H$_5$(C$_9$H$_6$N)	205.26	....	260[77]	61-2			al, ace, bz, chl	B20[4], 4149
12904	Quinoline, 5-phenyl 5-C$_6$H$_5$(C$_9$H$_6$N)	205.26	nd (dil al)	....	83			al, eth, ace	B20[4], 4150
12905	Quinoline, 6-phenyl 6-C$_6$H$_5$(C$_9$H$_6$N)	205.26	pl (al, bz)	260[77]	110	1.945[20]	....	al, ace, bz, chl	B20[4], 4151
12906	Quinoline, 8-phenyl 8-C$_6$H$_5$(C$_9$H$_6$N)	205.26	ye gr oil	283[187]	......			al, ace, eth, bz, chl	B20[4], 4153
12907	Quinoline, 1,2,3,4-tetrahydro C$_9$H$_{11}$N	133.20	nd	251	20	1.0588[20/4]	1.6062[19]	**al, eth**	B20[4], 2923
12908	Quinoline, 5,6,7,8-tetrahydro C$_9$H$_{11}$N	133.20	....	222, 92-5[12]	......	1.0304[13]	1.5435[20]	al, eth, ace, bz	B20[4], 2922
12909	Quinoline, 2,3,4-trimethyl or 3,4-Dimethyl quinaldine 2,3,4-(CH$_3$)$_3$(C$_9$H$_4$N)	171.25	....	285, 156-8[12]	92				B20[4], 3557
12910	Quinoline, 2,3,8-trimethyl 2,3,8-(CH$_3$)$_3$(C$_9$H$_4$N)	171.24	....	285	86-7			al, eth, chl	B20[4], 3557
12911	Quinoline, 2,4,6-trimethyl or 4,6-Dimethyl quinaldine 2,4,6-(CH$_3$)$_3$(C$_9$H$_4$N)	171.25	nd (w, dil al + 1w)	281-2, 146-8[13]	65.5 (anh), 40 (+ w)			al, eth, ace, bz, chl	B20[4], 3558
12912	Quinoline, 2,4,7-trimethyl or 4,7-Dimethyl quinaldine 2,4,7-(CH$_3$)$_3$(C$_9$H$_4$N)	71.25	nd (w)	200-1	63-4	1.0337[20/4]	1.5973[24]	al, eth, ace, bz	B20[4], 3558
12913	Quinoline, 2,4,8-trimethyl or 4,8-Dimethyl quinaldine 2,4,8-(CH$_3$)$_3$(C$_9$H$_4$N)	171.25	....	287	50-1		1.5855[50]	....	B20[4], 3559
12914	Quinoline, 2,5,6-trimethyl 2,5,6(CH$_3$)$_3$(C$_9$H$_4$N)	171.25	nd	....	69-70			al, eth, bz	B20, 415
12915	Quinoline, 2,5,7-trimethyl or Tetracoline 2,5,7-(CH$_3$)$_3$(C$_9$H$_4$N)	171.25	pr	286.6[746], 107-8[719]	......			al, eth, bz, peth	B20[4], 3559
12916	Quinoline, 2,6,8-trimethyl 2,6,8-(CH$_3$)$_3$(C$_9$H$_4$N)	171.25	pr (peth), lf (dil al)	260[719]	46			al, eth, bz, peth	B20[4], 3560
12917	Quinoline, 4,5,8-trimethyl 4,5,8-(CH$_3$)$_3$(C$_9$H$_4$N)	171.25	....	155[13]	73-4			al, eth, bz	B20[4], 3562
12918	2-Quinoline carbonitrite 2-NC(C$_9$H$_6$N)	154.17	nd (lig, chl)	100-70[20-23]	94			al, eth, bz, chl, lig	B22[4], 1157
12919	2-Quinoline carboximide 2-H$_2$NCO(C$_9$H$_6$N)	172.20	nd (dil al, bz-lig)	....	133			al, bz, chl	B22[4], 1154
12920	2-Quinoline carboxylic acid or Quinaldinic acid 2-HO$_2$C(C$_9$H$_6$N)	173.17	nd (+ 3w), (bz)	....	157 (anh)			w, bz	B22[4], 1149
12921	2-Quinoline carboxylic acid, 4,8-dihydroxy or Xanthurenic acid 2-HO$_2$C-4,8-(HO)$_2$(C$_9$H$_4$N)	705.17	ye micr cr (w)	....	289 (297)			al	B22[4], 2513

No.	Name, Synonyms, and Formula	Mol. wt.	Color, crystalline form, specific rotation and λ_{max} (log ε)	b.p. °C	m.p. °C	Density	n_D	Solubility	Ref.
12922	2-Quinoline carboxylic acid, 4-hydroxy or Kynurenic acid 4-HO-2-HO$_2$C(C$_9$H$_5$N)	189.17	ye nd (+ w dil aa)	....	282-3				B22⁴, 2245
12923	2-Quinoline carboxylic acid, methyl ester 2-CH$_3$O$_2$C(C$_9$H$_6$N)	187.20	nd (lig)	....	86			al, lig	B22⁴, 1152
12924	2-Quinoline carboxylyl chloride 2-ClCO(C$_9$H$_6$N)	191.62	nd (eth, lig)		(i)97-8, (ii)175-6			eth, bz	B22⁴, 509
12925	3-Quinoline carbonitrile or 3-Cyanoquinoline 3-NC(C$_9$H$_6$N)	154.17	(al or sub)	....	108			eth, ace, bz	B22⁴, 1171
12926	3-Quinoline carboxylic acid 3-HO$_2$C(C$_9$H$_6$N)	173.17	pl (al, dil al)	....	275d			al	B22⁴, 1167
12927	3-Quinoline carboxylic acid, 2-phenyl 2-C$_6$H$_5$-3-HO$_2$C(C$_9$H$_5$N)	249.27	nd (al)	....	230d			al, ace, aa	B22⁴, 1355
12928	4-Quinoline carbonitrile or Cinchoninonitrile 4-NC(C$_9$H$_6$N)	154.17	cr (chl, lig, eth) nd(sub)	240-5	103-4			al, eth, ace, bz	B22⁴, 1181
12929	4-Quinoline carboxylic acid or Cinchoninic acid 4-HO$_2$C(C$_9$H$_6$N)	173.17	mcl pr (w), nd (+ 1w), mcl or tcl (+ 2w)		257-8 (anh)				B22⁴, 1177
12930	4-Quinoline carboxylic acid, 2-(3-carboxy-4-hydroxy phenyl) or Hexophan C$_{17}$H$_{11}$NO$_5$	309.28	vesh pw		283-4d			chl	B22², 206
12931	4-Quinoline carboxylic acid, 6-methoxy or Quininic acid 6-CH$_3$O-4-HO$_2$C(C$_9$H$_5$N)	203.20	pa ye pr (dil al)	sub	285d				B22⁴, 2297
12932	4-Quinoline carboxylic acid, 8-methoxy-2-phenyl or Isatophan 8-CH$_3$O-2-C$_6$H$_5$-4-HO$_2$C(C$_9$H$_4$N)	279.20	ye nd (al)	....	216			al, chl	B22⁴, 559
12933	4-Quinoline carboxylic acid, 6-methyl-2-phenyl,ethyl ester or Novatophan 6-CH$_3$-2-C$_6$H$_5$-4-(C$_3$H$_5$O$_2$C)(C$_9$H$_4$N)	291.35	ye cr (al)		75-6			eth, ace, bz, chl	B22⁴, 1393
12934	4-Quinoline carboxylic acid, 2-phenyl or Cinchophene 2-C$_6$H$_5$-4-HO$_2$C(C$_9$H$_5$N)	249.27	nd (MeOH, dil al), ye in air		218			eth	B22⁴, 1358
12935	4-Quinoline carboxylic acid, 2-phenyl,allyl ester or Ato-quinol 2-C$_6$H$_5$-4-(C$_3$H$_5$O$_2$C)(C$_9$H$_5$N)	289.34	nd (dil al)	265¹⁵	30			al, eth, ace, bz	B22⁴, 1361
12936	5-Quinoline carbonitrile 5-NC(C$_9$H$_6$N)	154.17	nd (lig), nd(+ 3/2 w, dil al)		89, 70 (+ w)			al, bz	B22⁴, 1195
12937	5-Quinoline carboxylic acid 5-HO$_2$C(C$_9$H$_6$N)	173.17	cr (aa, sub)	sub<338	342				B22⁴, 1195
12938	6-Quinoline carboxylic acid 6-HO$_2$C(C$_9$H$_6$N)	173.17	nd, pr or pl (sub)	sub<290	291-2				B22⁴, 1196
12939	7-Quinoline carboxylic acid 7-HO$_2$C(C$_9$H$_6$N)	173.17	nd (w al)	sub	249-50			al	B22⁴, 1199
12940	8-Quinoline carbonitrile or 8-Cyano quinoline 8-NC(C$_9$H$_6$N)	154.17	nd (dil al)	....	84			al	B22⁴, 1203
12941	8-Quinoline carboxylic acid 8-HO$_2$C(C$_9$H$_6$N)	173.17	nd (w)	sub	187			al	B22⁴, 1200
12942	5-Quinoline sulfonic acid, 8-hydroxy-7-iodo C$_9$H$_6$NO$_4$IS	351.12			260d				B22⁴, 3497
12943	5-Quinoline sulfonic acid, 6-hydroxy 6-HO-5-(HO$_3$S)(C$_9$H$_5$N)	225.23	ye nd (+ ½ w, w, al)		270d			w	B22, 407
12944	5-Quinoline sulfonic acid, 8-hydroxy 8-HO-5-HO$_3$S(C$_9$H$_5$N)	225.23	ye lf, nd (+ 1w, dil HCl)		322-3				B22⁴, 3493
12945	5-Quinoline sulfonic acid, 8-hydroxy-7-iodo or Loretin 8-HO-7-I-5-HO$_3$S(C$_9$H$_4$N)	351.12	ye pr or lf (al)		260d				B22⁴, 3497
12946	7-Quinoline sulfonic acid, 8-hydroxy 8-HO-7-HO$_3$S(C$_9$H$_5$N)	225.23	ye nd (al)		314-5			w, al	B22, 408
12947	5-Quinoline sulfonic acid, 5-hydroxy-6-iodo or Lorenite 5-HO-6-I-8-HO$_3$S(C$_9$H$_4$N)	351.13	ye nd or lf		210-30d				B22, 406
12948	2-Quinolinone, 1-methyl C$_{10}$H$_9$NO	159.19		324⁷²⁸	74			al, eth, ace, bz, chl	B20⁴, 3505
12949	Quinovic acid or Quinovaic acid C$_{30}$H$_{46}$O$_5$	486.67	pl or nd [α]¹⁶⁰$_D$ + 87 (aq KOH)	....	298d				B10³, 2307

No.	Name, Synonyms, and Formula	Mol. wt.	Color, crystalline form, specific rotation and λ_{max} (log ε)	b.p. °C	m.p. °C	Density	n_D	Solubility	Ref.
12950	Quinoxaline or 1,4-Benxodiazine................ $C_8H_6N_2$	130.15	cr (peth)	229.5, 108-11[12]	28	1.1334[48/4]	1.6231[48]	w, al, eth, ace, bz	B23[2], 177
12951	Quinoxaline, 6-amino .... 6-$H_2N(C_8H_5N_2)$	145.17	ye nd (eth)	sub	159	....	....	w, al, eth, chl	B25, 326
12952	Quinoxaline, 6-chloro .... 6-$Cl(C_8H_5N_2)$	164.60	nd (w)	117-9[10] sub	64	....	....	....	B23[2], 177
12953	Quinoxaline, 2,3-dichloro .... 2,3-$Cl_2(C_8H_4N_2)$	199.04	cr (al, bz)	....	151-3	....	....	al, bz, chl, aa	B23[2], 177
12954	Quinoxaline, 2,3-dihydroxy .... 2,3-$(HO)_2(C_8H_4N_2)$	162.15	nd (w)	....	410	....	....	w	B24[2], 200
12955	Quinoxaline, 2,3-dimethyl .... 2,3-$(CH_3)_2(C_8H_4N_2)$	158.21	nd (w + 3) (ace)	....	106 (anh), 85 (+ w)	....	....	al, eth, ace, bz	B23[2], 197
12956	Quinoxaline, 2,6-dimethyl .... 2,6-$(CH_3)_2(C_8H_4N_2)$	158.21	....	267-9	54	....	....	w, al, ace, bz	B23, 192
12957	Quinoxaline, 2-hydroxy or 2-Quinoxalinol............ 2-$HO(C_8H_5N_2)$	146.15	lf (al)	sub 200[0.5]	271	....	....	....	B24[2], 72
12958	Quinoxaline, 6-methoxy .... 6-$CH_3O(C_8H_5N_2)$	160.17	nd (w)	128[7] sub	57.5	....	....	eth, ace	B23, 387
12959	Quinoxaline, 2-methyl .... 2-$CH_3(C_8H_5N_2)$	144.19	ye	245-7, 118[16]	180-1	....	....	w, al, eth, ace, bz	B23[2], 190
12960	Quinoxaline, 6-methyl .... 6-$CH_3(C_8H_5N_2)$	144.19	....	248[748], 141.5[29]	218-9	1.1164[20/4]	1.6211[18.4]	w, al, eth, ace, bz	B23, 184
12961	Quinoxaline, 1,2,3,4-tetrahydro .... $C_8H_{10}N_2$	134.18	lf (w, eth, peth)	289, 153-4[14]	99	....	....	al, eth, bz, chl	B23[2], 106
12962	Quinoxaline, 5-hydroxy or 5-Quinoxalol............ 5-$HO(C_8H_5N_2)$	146.15	(eth)	184[7], sub 90[2.5]	101-2	....	....	....	C48, 8232
12963	Quinuclidine or 1,4-Ethylene piperidine............ $C_7H_{13}N$	111.19	(eth)	....	158 (sealed tube)	....	....	w, al, eth, ace, bz	B20[4], 1966
12964	Quinuclidine, 3-hydroxy .... $C_7H_{13}NO$	127.19	cr (bz)	....	221-1	....	....	ace	B21[4], 237
12965	2-Quinuclidine carboxylic acid .... $C_8H_{13}NO_2$	155.20	col hyg cr (al-ace)	....	280d	....	....	....	B22[4], 200

No.	Name, Synonyms, and Formula	Mol. wt.	Color, crystalline form, specific rotation and λ_{max} (log ε)	b.p. °C	m.p. °C	Density	n_D	Solubility	Ref.
12966	Raffinose or Melitriose $C_{18}H_{32}O_{16}$	594.50	wh pw, pr or nd (+5w, dil al) $[\alpha]^{20}_D$ +101, +123	d 130	118-9 (anh), 80 (hyd)	1.465°		w, MeOH, Py	B17⁴, 3801
12967	Raunescine hydrate $C_{31}H_{36}N_2O_8$	582.61	wh hex pr (90% MeOH), $[\alpha]^{25}_D$ −74 (chl)		160-70			chl, aa	Am79, 250
12968	Reductone or 3-Hydroxy-2-oxo-propionaldehyde (enol-form) HOCH=C(OH)CHO	88.06	ye nd (w)	200-20d				w, al	B1⁴, 4145
12969	Resazurin $C_{12}H_7NO_4$	229.18	dk red to grsh pr or pl (aa, AcOEt)	sub vac	d				B27, 128
12970	Rescinnamine or Moderil $C_{35}H_{42}N_2O_9$	634.74	nd (bz) $[\alpha]^{24}_D$ −98 (chl, c=0.1)		238-9			ace, chl	Am77, 2241
12971	Reserpic acid or Reserpinolic acid $C_{22}H_{26}N_2O_5$	400.48	(MeOH), $[\alpha]^{23}_D$ −91 (of hydrochloride)		241-3				Am78, 2023
12972	Reserpine or Rivasin-Serparsin $C_{33}H_{40}N_2O_9$	608.70	lo pr (dil al) $[\alpha]^{27}_D$ −117.7 (chl, c=1)		264-5, 277 (sealed tube)			bz, chl	H37, 67
12973	Reserpinine or Raubasinine $C_{22}H_{26}N_2O_4$	382.47	pa ye pl (aq ace), $[\alpha]^{20}_D$ −131 (chl, c=1.18)		243-4d			al, chl	C49, 11672
12974	Resorcinol or 1,3-Dihydroxy benzene 1,3-(HO)₂C₆H₄	110.11	nd (bz), pl (w)	178¹⁶	111	1.2717		w, al, eth, aa	B6³, 4292
12975	Resorcenol monoacetate 3-(CH₃CO₂)C₆H₄OH	152.18	ye	283d				al	B6³, 4319
12976	Resorcinol, 4-acetyl or 2,4-Dihydroxy acetophenone. Resacetophenone 2,4-(HO)₂C₆H₃COCH₃	152.16	nd or lf		147	1.1800¹⁴¹		aa, Py	B8³, 2082
12977	Resorcinol, 5-acetyl or 3,5-Dihydroxy acetophenone 3,5-(HO)₂C₆H₃COCH₃	152.16	cr (w)		147-8			w, al, eth, ace	B8², 301
12978	Resorcinol benzoate (mono) 3-C₆H₅CO₂C₆H₄OH	214.22	pr (dil al)		135-6			chl, aa	B9³, 555
12979	Resorcinol, 4-(2-aminoethyl), hydrochloride 4(H₂NCH₂CH₂)-1,3-(HO)₂C₆H₃.HCl	189.64	nd (w)		237d			w, al	B13², 486
12980	Resorcinol acetate (mono) or 3-Hydroxy phenyl acetate 3-HO-C₆H₄O₂CCH₃	152.16	ye	283d				al	B6³, 4319
12981	Resorcinol, monobenzyl ether 3-C₆H₅CH₂OC₆H₄OH	200.24	cr (CCl₄)	200⁵	69.2			w	B6³, 4314
12982	Resorcinol, 2-bromo or 1,3-Dihydroxy-2-bromo benzene 1,3-(HO)₂-2-BrC₆H₃	189.02	nd (chl), cr (peth)		102-3			w, al, eth, chl	B6³, 4336
12983	Resorcinol, 4-bromo or 1,3-Dihydroxy-4-bromo benzene 1,3-(HO)₂-4-BrC₆H₃	189.02		138-41¹²	103			w, eth	B6³, 4336
12984	Resorcinol, 5-bromo or 1,3-Dihydroxy-5-bromo benzene 1,3-(HO)₂-5-BrC₆H₃	189.02	nd (bz), pr(+1)		87, 79 (+w)			w, eth, bz	B6², 820
12985	Resorcinol, 4-butyl 4-C₄H₉-1,3-(HO)₂C₆H₃	166.22		164-6⁶⁻⁷	47-8				B6³, 4657
12986	Resorcinol, 4-isobutyl 4-i-C₄H₉-1,3-(HO)₂C₆H₃	166.22		166-8⁶⁻⁷	62-3			al, eth	B6², 899
12987	Resorcinol, 2-chloro or 2-Chloro-1,3-dihydroxy benzene 2-Cl-1,3-(HO)₂C₆H₃	144.56			97-8				B6³, 4333
12988	Resorcinol, 4-chloro or 4-Chloro-1,3-dihydroxy benzene 4-Cl-1,3-(HO)₂C₆H₃	144.56		259, 147¹⁸	(i)89 (ii)105			w, al, eth, ace, bz	B6³, 4333
12989	Resorcinol, 5-chloro or 5-Chloro-1,3-dihydroxy benzene 5-Cl-1,3-(HO)₂C₆H₃	144.56	hyg nd (sub)	sub	117			w, eth, ace	B6², 819
12990	Resorcinol, 4-cyclohexyl or 4-Cyclohexyl-1,3-dihydroxy benzene 4-C₆H₁₁C₆H₃(OH)₂(1,3)	192.24	nd (bz chl), cr (bz-peth)		127-8				B6³, 5057

No.	Name, Synonyms, and Formula	Mol. wt.	Color, crystalline form, specific rotation and λ_{max} (log ε)	b.p. °C	m.p. °C	Density	n_D	Solubility	Ref.
12991	Resorcinol diacetate 1,3-$(CH_3CO_2)_2C_6H_4$	194.19	....	278, 153-4[12]	....			al, eth	B6[3], 4320
12992	Resorcinol, 4,6-diacetyl or Resodiacetophenone 4,6-$(CH_3CO)_2C_6H_2(OH)_2$-(1,3)	194.19	nd (al)	....	185			eth, ace, bz	B8[3], 3511
12993	Resorcinol dibenzoate 3-$(C_6H_5CO_2)C_6H_4(O_2CC_6H_5)$	318.33	pl (dil al)	....	117			al, eth	B9[3], 555
12994	Resorcinol, 4,6-dichloro 4,6-Cl_2-1,3-$(HO)_2C_6H_2$	179.00	....	254 sub	113			w, al, eth, ace	B6[3], 4335
12995	Resorcinol diethyl ether or 1,3-Diethoxy benzene 1,3-$(C_2H_5O)_2C_6H_4$	166.22	pr	235[756]	12.4			al, eth	B6[3], 4307
12996	Resorcinol, 2,5-dimethyl 2,5-$(CH_3)_2$-1,3-$(HO)_2C_6H_2$	138.17	nd (bz), pr (w)	277-80	163			al, eth	B6[3], 4606
12997	Resorcinol, 4,5-dimethyl 4,5-$(CH_3)_2$-1,3-$(HO)_2C_6H_2$	138.17	nd (bz), pr (w + 1)	sub	136-7, 115-7 (+ w)			w, al, eth, aa	B6[3], 4581
12998	Resorcinol, 4,6-dimethyl 4,6-$(CH_3)_2$-1,3-$(HO)_2C_6H_2$	138.17	mcl pr (w + 1), nd (sub)	276-9 sub	125			w, al, eth	B6[3], 4595
12999	Resorcinol dimethyl ether or 1,3-Dimethoxy benzene 1,3-$(CH_3O)_2C_6H_4$	138.17	....	217-8	−52	1.0552[25/25]	1.5231[20]	al, eth, bz	B6[3], 4305
13000	Resorcinol, 2,4-dinitro 2,4-$(NO_2)_2$-1,3-$(HO)_2C_6H_2$	200.11	ye lf (al)		147-8				B6[3], 4351
13001	Resorcinol, 2,4-dinitroso 2,4-$(ON)_2$-1,3-$(HO)_2C_6H_2$	168.11	ye rh pl (50% al), lf (aq MeOH)		168			w, al	B7[3], 4732
13002	Resorcinol dipropyl ether or 1,3-Dipropoxy benzene 1,3-$(C_3H_7O)_2C_6H_4$	194.28	....	251, 127-8[12]	......	1.035[20/21]	1.5138[83]	....	B6[3], 4308
13003	Resorcinol, dithio or 1,3-Dimercapto benzene 1,3-$(HS)_2C_6H_4$	142.25	lf (dil al)	245, 123[17]	27			al, eth, ace, bz	B6[3], 4366
13004	Resorcinol, 4-ethyl 4-C_2H_5-1,3-$(HO)_2C_6H_3$	138.17	pr (chl, bz)	131[15]	98-9				B6[3], 4554
13005	Resorcinol ethyl ether(mono) or 3-Ethoxy phenol 3-$C_2H_5OC_6H_4OH$	138.17	ye	246-7, 117[5.5]	......	1.0705[4/4]	....	al, eth, bz	B6[3], 4307
13006	Resorcinol, 4-hexyl 4-C_6H_{13}-1,3-$(HO)_2C_6H_3$	194.28	pa ye nd (lig)	333, 178[8]	68-9			al, eth, ace, bz, chl	B6[3], 4712
13007	Resorcinol, 4-iodo 4-I-1,3-$(HO)_2C_6H_3$	236.01	pr, nd (+ 1w, w)	....	67			al, eth, ace, chl	B6[3], 4341
13008	Resorcinol, 5-iodo 5-I-1,3-$(HO)_2C_6H_3$	236.01	nd (bz, sub)	sub	92.3			al	B6[3], 821
13009	Resorcinol, 2-methoxy 2-CH_3O-1,3-$(HO)_2C_6H_3$	140.15	cr (bz)	154-5[24]	85-7				B6[3], 6264
13010	Resorcinol, 5-methoxy 5-CH_3O-1,3-$(HO)_2C_6H_3$	140.15	pl (bz)	213[16]	78-81			al, eth	B6[3], 6304
13011	Resorcinol, 2-methyl 2-CH_3-1,3-$(HO)_2C_6H_3$	124.15	pr (bz)	264, 168[16]	119-21			w, al, eth, bz	B6[3], 4512
13012	Resorcinol methyl ether(mono) or 3-Methoxy phenol 3-$CH_3OC_6H_4OH$	124.15	....	244.3, 144[25]	<−17	....	1.5520[20]	**al, eth**	B6[3], 4303
13013	Resorcinol, 4-(4-methyl pentyl) 4-$[(CH_3)_2CH(CH_2)_3]$-1,3-$(HO)_2C_6H_3$	194.28	lf (peth)	190[12]	71		1.5292[15]	al, eth	B6[3], 4717
13014	Resorcinol, 2-nitro 2-NO_2-1,3-$(HO)_2C_6H_3$	155.11	og-red pr (al)	....	87-8			al	B6[3], 4343
13015	Resorcinol, 4-nitro 4-NO_2-1,3-$(HO)_2C_6H_3$	155.11			122			al, eth, bz	B6[3], 4344
13016	Resorcinol, 4-nitroso or 2-Hydroxy-4-benzoquinone oxime 4-ON-1,3-$(HO)_2C_6H_3$	139.11	ye nd (+ w), br nd (chl)		150d			w, al, eth, ace, chl	B8[3], 1965
13017	Resorcinol, 4-pentyl 4-C_5H_{11}-1,3-$(HO)_2C_6H_3$	180.25	....	168-70[6-7]	72-3			al, eth, bz	B6[3], 4693
13018	Resorcinol, 4-isopentyl 4-i-C_5H_{11}-1,3-$(HO)_2C_6H_3$	180.25	....	177-8[8]	68-70			al, eth, bz	B6[3], 4700
13019	Resorcinol, 5-pentyl or Olivetol 5-C_5H_{11}-1,3-$(HO)_2C_6H_3$	180.25	nd (+ w), pr (bz-lig)	164[5]	49			w, al, eth, ace, bz	B6[3], 4695
13020	Resorcinol, 4-propionyl 4-$(CH_3CH_2CO)C_6H_3(OH)_2$-1,3	166.18	ye nd (al)	176-8[6]	97 (anh)			al, eth, bz, aa	B8[3], 2144
13021	Resorcinol monopropyl ether or 3-Propoxy phenol 3-$C_3H_7OC_6H_4OH$	152.20	cr (w, al)	256-7, 120[5]	55				B6[3], 4308

No.	Name, Synonyms, and Formula	Mol. wt.	Color, crystalline form, specific rotation and λ_{max} (log ε)	b.p. °C	m.p. °C	Density	n_D	Solubility	Ref.
13022	Resorcinol, 4-propyl 4-C$_3$H$_7$-1,3-(HO)$_2$C$_6$H$_3$	152.20	pr (bz), nd (peth)	172[14]	82-3			w, al, eth	B6[3], 4611
13023	Resorcinol, 4-iso-propyl 4-(CH$_3$)$_2$CH-1,3-(HO)$_2$C$_6$H$_3$	152.20	cr (aq aa)	265-81, 114[0.2]	105			al, eth	B6[2], 896
13024	Resorcinol, 5-propyl or Divarinol 5-C$_3$H$_7$-1,3-(HO)$_2$C$_6$H$_3$	152.20	lf (+1w), lf (bz)	148-9[3]	83-4, 51 (+w)			al, eth, ace, bz, aa	B6[3], 4622
13025	Resorcinol, tetrachloro 1,3-(HO)$_2$C$_6$Cl$_4$	247.89	nd (w)		141			al, eth, bz, aa	B6[2], 819
13026	Resorcinol, 2,4,6-tribromo 2,4,6-Br$_3$C$_6$H(OH)$_2$-1,3	346.82	nd (w)		112			al, eth	B6[3], 4340
13027	Resorcinol, 2,4,6-trichloro 2,4,6-Cl$_3$C$_6$H(OH)$_2$-1,3	213.46	nd (w)		83			al, eth	B6[3], 4336
13028	Resorcinol, 2,4,5-trimethyl 2,4,5-(CH$_3$)$_3$C$_6$H(OH)$_2$-1,3	152.20	cr		156				B6[3], 4643
13029	Resorcinol, 2,4,6-trimethyl or Mesorcinol 2,4,6-(CH$_3$)$_3$-C$_6$H(OH)$_2$-1,3	152.20	pl (al), lf	275	149-50			al, eth, bz	B6[3], 4653
13030	Resorcinol, 4,5,6-trimethyl 4,5,6-(CH$_3$)$_3$C$_6$H(OH)$_2$-1,3	152.20	nd or lf		163-4			al, eth, ace, bz, chl	B6[3], 4643
13031	Resorcinol, 2,4,6-trinitro or Styphnic acid 2,4,6-(NO$_2$)$_3$C$_6$H(OH)$_2$-1,3	245.11	hex (al), ye cr (aa)	sub	179-80			al, eth	B6[3], 4354
13032	Resorufin or 7-Hydroxy-2-phenoxazone C$_{12}$H$_7$NO$_3$	213.20	br nd (PhNH$_2$), pr (HCl)						B27[2], 108
13033	Retene or 7-isopropyl-1-methyl phenanthrene	234.34	pl (al)	390, 200[11]	100-1	1.033		bz, lig	B5[], 2199
13034	Retromecine or Senecifolinene C$_7$H$_{11}$NO$_3$	155.20	pr (ace) [α]$'_D$ +27.4 (w), ^{26}[$'_D$] +50.2 (al)	80[0.01]	121-2 (130)			w, al, ace	B21[4], 2048
13035	Rhamnetin C$_{16}$H$_{12}$O$_7$	316.27	ye nd (al)		294-6			ace	B18[4], 3474
13036	D-Rhamnitol or Rhamnite C$_6$H$_{14}$O$_5$	166.18	pr (ace), [α]$^{20}_D$ −12.4 (w,c=0.5)		123			w, al, Py	B1[4], 2837
13037	D-Rhamnose, hydrate (α-anomer) C$_6$H$_{12}$O$_5$.H$_2$O	182.18	(wh), [α]$^{18.5}_D$ −8.25 (w)		90-1			w	B1[4], 4260
13038	DL-Rhamnose C$_6$H$_{12}$O$_5$	164.16	(w)		151-3 (anh)			w, al	B1[4], 4261
13039	L-Rhamnose, hydrate C$_6$H$_{12}$O$_5$.H$_2$O	182.18	mcl pl (al w + l) [α]$^{20}_D$ −77→ +8.9 (mut)	105[2] sub	92	1.4708[20]		w, al	B1[4], 4261
13040	L-Rhamnose (β-anomer) C$_6$H$_{12}$O$_5$	164.16	nd (ace), [α]$^{20}_D$ +38.4 +8.9 (w,mut)		122-6			w, al	B1[4], 4262
13041	Rheadin or Rhoedine C$_{21}$H$_{21}$NO$_6$	383.41	nd (chl, eth, al), [α]$^{17.5}_D$ +232 (al)	sub	256-8				C38, 6060
13042	Rhizopterin or 12-Formylpteroic acid C$_{15}$H$_{12}$N$_6$O$_4$	340.31	lt ye pl (w)		>300 darkens 285				Am69, 2751
13043	Rhodamine β or Tetraethyl rhodamine C$_{28}$H$_{30}$N$_2$O$_3$	442.57	gr lf (w + 4), col pr (al, xyl)		165 (anh)			w, al, eth, bz	B19[4], 4301
13044	Rhodamine, β-hydrochloride C$_{28}$H$_{30}$N$_2$O$_3$.HCl	479.01	gr or red vt pw (w), lf (dil HCl)					w, al, bz	B19[4], 4301
13045	Rhodamine or 4-Thioxo-4-thiazolidone C$_3$H$_3$NOS$_2$	133.19	lt ye pr (al, w, aa)		170	0.868		al, eth	B27[2], 288

No.	Name, Synonyms, and Formula	Mol. wt.	Color, crystalline form, specific rotation and λ_{max} (log ε)	b.p. °C	m.p. °C	Density	n_D	Solubility	Ref.
13046	Rhodanine, 3-amino $C_3H_4N_2OS_2$	148.20			100				B27², 289
13047	Rhodanine, 3-amino-5-benzylidene $C_{10}H_8N_2OS_2$	236.31			194-7				C61, 7190
13048	Rhodanine, 5-(dimethylamino) amino-benzylidene $C_{12}H_{12}N_2OS_2$	264.37	red or og nd (al, Py-w)		296			eth, ace	B27², 484
13049	Rhodanine, 3-ethyl $C_5H_7NOS_2$	161.25			37-9				B27¹, 309
13050	Rhodanind, 5-ethyl $C_5H_7NOS_2$	161.25	ye amor (dil al)		105			al, eth, ace, aa	B27¹, 313
13051	Rhodanine, 3-methyl $C_4H_5NOS_2$	147.21			69-71				B27², 288
13052	Rhodanine, 3-phenyl $C_9H_7NOS_2$	209.29	ye pl (aa), nd or pr (al)		194-5				B27², 288
13053	Rhodanine, 5-isopropylidene $C_6H_7NOS_2$	173.25			197-8				C44, 2979
13054	Rhodizonic acid or 5,6-Dihydroxy-5-cyclohexene-1,2,3,4-tetrone $C_6H_2O_6$	170.08	dk og nd (sub)		155-60			al	B8³, 4214
13055	Riboflavin or Lactoflavin Vitamin B2 $C_{17}H_{20}N_4O_6$	376.37	ye or og-ye nd (w, dil aa), $[\alpha]^{25}_D$ −112→ −122 (0.02N NaOH, c = 0.5), (mut)		280d				C28, 2036
13056	D-Ribose $C_5H_{10}O_5$	150.13	pl (abs al), $[\alpha]_D$ −21.5 (w)		95			w	B1⁴, 4211
13057	Ricinidine or 1-Methyl, 2-pyridone-3-carbonitrile $C_7H_6N_2O$	134.14	nd (sub), (chl, al)	243²⁸	140			al	B22², 222
13058	Ricinine or Ricidine $C_8H_8N_2O_2$	164.18	pr or lf (w al)	sub 170-80²⁰	201.5			Py	B22⁴, 3354
13059	Ricinoleic acid or 12-Hydroxy-9-octadecenoic acid (cis) $C_6H_{13}CH(OH)CH_2CH=CH(CH_2)_7CO_2H$	298.47	$[\alpha]^{12}_D$ + 5.05	226.8¹⁰	α:7.7 β:16 γ:5.5	0.9450²¹/⁴	1.4716²¹	al, eth	B3⁴, 1026
13060	Ricinoleic acid, butyl ester (cis) or Butyl ricinoleate $C_6H_{13}CH(OH)CH_2CH=CH(CH_2)_7CO_2C_4H_9$	354.58	$[\alpha]^{25}_D$ + 3.7	278¹²		0.9058²²	1.4566²²	eth	B3³, 711
13061	Ricinoleic acid, isobutyl ester or Isobutyl ricinoleate $C_6H_{13}CHOHCH_2CH=CH(CH_2)_7CO_2-i-C_4H_9$	354.58	$[\alpha]$ + 4.01	282⁹		0.9078²²	1.4538²²	eth	B3¹, 138
13062	Ricinoleic acid, ethyl ester (cis) or Ethyl ricinoleate $C_6H_{13}CH(OH)CH_2CH=CH(CH_2)_7CO_2C_2H_5$	326.53	$[\alpha]^{22}_D$ + 5.3	258¹³		0.9045²²	1.4618²²		B3⁴, 1029
13063	Rosaniline $C_{20}H_{21}N_3O$	319.41	nd (w)		186d			al	B13³, 2078
13064	Rosinduline, anhydro base $C_{22}H_{15}N_3$	321.39	red-br lf (eth-bz)		198-9			al, eth, bz	B25², 322
13065	Rotenone or Tubotoxine $C_{23}H_{22}O_6$	394.43	nd or lf (al, aq ace), $[\alpha]^{29.5}_D$ −225.2 (bz)	210-20⁰·⁵	(i)163 (ii) 176			al, ace, bz, chl, aa	B19⁴, 5227
13066	Rubicene $C_{26}H_{14}$	326.40	red nd (xyl)		306			CS_2	B5³, 2673
13067	Rubignol $C_5H_4O_4$	128.08			203.5			w, al, eth, ace	B17⁴, 6679

No.	Name, Synonyms, and Formula	Mol. wt.	Color, crystalline form, specific rotation and λ_{max} (log ε)	b.p. °C	m.p. °C	Density	n_D	Solubility	Ref.
13068	Rubijervine $C_{27}H_{43}NO_2$	413.65	nd (+1w dil al), $[\alpha]^{25}_D$ +19 (al, c=1)	....	242	....	....	al, bz, chl	B21⁴, 2310
13069	Rubixanthin or 3-Hydroxy-γ-carotene $C_{40}H_{56}O$	552.90	dk red nd (bz-MeOH), og-red (bz-peth)	....	160	....	....	bz, chl	B6³, 3772
13070	Rubrene or 9,10,11,12-Tetraphenyl naphthacene $C_{42}H_{28}$	532.69	og red (bz-lig)	....	331			bz	B5³, 2803
13071	Rufol or 1,5-Dihydroxy anthracene 1,5-(HO)₂C₁₄H₈	210.34	ye nd	....	265d			al, eth, bz	B6, 1032
13072	Rutaecarpine $C_{18}H_{13}N_3O$	287.33	yesh nd (al, AcOEt)	....	259-60				B26², 104
13073	Rutinose or 6-D-β-L-rhamnopyransoyl-D-glucose $C_{12}H_{22}O_{10}$	376.30	hyg pw (al, eth), $[\alpha]^{20}_D$ −10 (al)	....	189-92d			w, al	B17⁴, 2536
13074	Sabadine $C_{29}H_{48}NO_8$	541.73	nd (eth), $[\alpha]_D$ −11 (al)		256-60			al, ace	B21⁴, 2883
13075	Sabinane or d-Thujane $C_{10}H_{18}$	138.26	$[\alpha]_D$ +79.1	157⁷⁵⁸	....	0.8139²⁰ᐟ⁴	1.4376²⁰	al, eth, ace, bz	B5⁴, 317

No.	Name, Synonyms, and Formula	Mol. wt.	Color, crystalline form, specific rotation and λ_{max} (log ε)	b.p. °C	m.p. °C	Density	n_D	Solubility	Ref.
13076	Sabinene (d) or 4(10)-Thujene .... $C_{10}H_{16}$	136.24	$[\alpha]^{20}_D$ +101.4	163-5[758], 49[13]	....	0.8437[20/4]	1.4676[20]	al, eth, bz, chl	B5[3], 365
13077	Sabinene (l) .... $C_{10}H_{16}$	136.24	$[\alpha]^{15}_D$ -42.5	162-6		0.8468[20]	1.4674[17]	al, eth, bz	B5[3], 365
13078	Sabinol (d) or 4(10)Thujene-3-01 .... $C_{10}H_{16}O$	152.24	$[\alpha]^{18}_D$ +3.94	208, 90[11]		0.9488[19/4]	1.4871[25]	eth	B6[4], 382
13079	Sabinol acetate(d) .... $C_{12}H_{18}O_2$	194.27	$[\alpha]_D$ +79	81-2[3]		0.972[15]			B6[3], 384
13080	D-Saccaharic acid .... $C_6H_{10}O_8$	210.14	nd (95%al), $[\alpha]_D$ +6.9→ + 20.6 (w, c=2.5, mut)	....	125-6			w, al	B3[4], 1291
13081	Saccharin or 2-Sulfobenzoic acid imide .... $C_7H_5NO_3S$	183.19	mcl (ace), pr (al), lf (w)	sub vac	229d	0.828		al, ace	B27[2], 214
13082	Safrole or 1-Allyl-3,4-(methylene dioxy) benzene .... $C_{10}H_{10}O_2$	162.19	mcl	234.5, 104-5[6]	11.2	1.000[20/4]	1.5381[20]	al, eth, chl	B19[4], 275
13083	Salicylaldehyde or 2-Hydroxybenzaldehyde............ 2-HOC$_6$H$_4$CHO	122.18	....	197, 93[25]	-7	1.1674[20/4]	1.5740[20]	al, eth, ace, bz	B8[3], 135
13084	Salicylaldehyde azine or Salazine..................... $C_{14}H_{12}N_2O_2$	240.27	ye nd or lf (al)	....	214			al, bz, chl	B8[2], 43
13085	Salicylaldehyde oxime or Salicylaldoxime .... 2-HOC$_6$H$_4$CH=NOH	137.14	pr (bz-peth)	....	63			al, eth, bz	B8[1], 168
13086	Salicylamide or 2-Hydroxy benzamide .... 2-HOC$_6$H$_4$CONH$_2$	137.14	ye lf (dil al)	181.5[14]	142	1.175[140/4]		al	B10[3], 152
13087	Salicylamide, N-phenyl or N-Phenyl salicylamide........ 2-HOC$_6$H$_4$CONHC$_6$H$_5$	213.24	pr (w, al)	....	136-7			al	B12[3], 944
13088	Salicylamide, N-(2-tolyl) or N-(2-Tolyl) salicylamide ..... 2-HOC$_6$H$_4$CONH(C$_6$H$_4$CH$_3$-2)	227.27	nd (al)	....	144			al	B12[3], 1888
13089	Salicylic acid or 2-Hydroxybenzoic acid .... 2-HOC$_6$H$_4$CO$_2$H	138.12	nd (w), mcl pr (al)	211[20] sub	159	1.443[20/4]	1.565	al, eth, ace	B10[3], 87
13090	Salicylic acid, allyl ester or Allyl salicylate .... 2-HOC$_6$H$_4$CO$_2$CH$_2$CH=CH$_2$	178.19		247-50, 105-6[5]	....	1.1000[15]			B10[3], 125
13091	Salicylic acid, 4-acetamido, phenyl ester or Salophene .... 4-CH$_3$CONH-2-HOC$_6$H$_3$CO$_2$C$_6$H$_5$	271.28	pl (w), lf (al)	....	187-8			al, eth, bz	B13[3], 1066
13092	Salicylic acid acetate or Aspirin .... 2-CH$_3$CO$_2$C$_6$H$_4$CO$_2$H	180.17	nd (w), mcl ta (w)		135			al, eth, chl	B10[3], 102
13093	Salicylic acid, 3-amino or 3-Amino-2-hydroxybenzoic acid 3-NH$_2$-2-HOC$_6$H$_3$CO$_2$H	153.14	....		235d				B14[3], 1434
13094	Salicylic acid, 4-amino or 4-Amino-2-hydroxybenzoic acid 4-H$_2$N-2-HOC$_6$H$_3$CO$_2$H	153.14	....		150-1d			w, al, eth, ace	B14[3], 1436
13095	Salicylic acid, 5-amino or 5-Amino-2-hydroxybenzoic acid 5-H$_3$N-2-HOC$_6$H$_3$CO$_2$H	153.14			283				B14[3], 1456
13096	Salicylic acid anhydride .... (2-HOC$_6$H$_4$CO)$_2$O	258.23	amor		255d			al, eth	B19[4], 2059
13097	Salicylic acid, benzyl ester or Benzyl salicylate............ 2-HOC$_6$H$_4$CO$_2$CH$_2$C$_6$H$_5$	228.25	....	320, 170[5]	....	1.1799[20/4]	1.5805[20]	al, eth	B10[3], 132
13098	Salicylic acid, benzyl ester, acetate 2-(CH$_3$CO$_2$)C$_6$H$_4$CO$_2$CH$_2$C$_6$H$_5$	270.29		197-200[7]	26			al, eth, ace, bz	B10[2], 52
13099	Salicylic acid, 3-bromo or Hydroxy-3-bromobenzoic acid 3-Br-2-HOC$_6$H$_3$CO$_2$H	217.03	nd (dil al)		184-5			al, eth, ace	B10[3], 174
13100	Salicylic acid, 4-bromo 4-Br-2-HO-C$_6$H$_3$CO$_2$H	217.03	pl, nd (w)		214			al	B10[2], 63
13101	Salicylic acid, 5-bromo 5-Br-2-HOC$_6$H$_3$CO$_2$H	217.03	nd (w, dil al)	sub>100	168-9			w, al, eth	B10[3], 176
13102	Salicylic acid, 5-bromo,acetate .... 5-Br-2(CH$_3$CO$_2$)C$_6$H$_3$CO$_2$H	259.06	nd (al)		60			al, eth	B10[2], 64
13103	Salicylic acid, butyl ester or Butyl salicylate............ 2-HOC$_6$H$_4$CO$_2$C$_4$H$_9$	194.23	....	270-2, 136-8[10]	-5.9	1.0728[20/4]	1.5115[20]		B10[3], 121
13104	Salicylic acid, isobutyl ester or Iso Butyl salicylate .... 2-HOC$_6$H$_4$CO$_2$-i-C$_4$H$_9$	194.23	....	260-2, 136-8[10]	5.9	1.0639[20/4]	1.5087[20]	al, eth	B10[3], 121
13105	Salicylic acid, 3-chloro 3-Cl-2-HOC$_6$H$_3$CO$_2$H	172.57	nd(w)		180			al, chl, aa	B10[3], 163
13106	Salicylic acid, 3-chloro,ethyl ester or Ethyl 3-chloro salicylate .... 3-Cl-2-HOC$_6$H$_3$CO$_2$C$_2$H$_5$	200.62	nd	269-70, 147[12]	21			al	B10, 101

No.	Name, Synonyms, and Formula	Mol. wt.	Color, crystalline form, specific rotation and λ_{max} (log ε)	b.p. °C	m.p. °C	Density	n_D	Solubility	Ref.
13107	Salicylic acid, 3-chloro,methyl ester or Methyl-3-chloro salicylate. . . . . 3-Cl-2-HOC$_6$H$_3$CO$_2$CH$_3$	186.60	nd (MeOH or al)	260d	38				B10^2, 61
13108	Salicylic acid, 4-chloro or 4-Chloro-2-hydroxy benzoic acid . . . . 4-Cl-2-HOC$_6$H$_3$CO$_2$H	172.57	nd (w)	sub (d)	207			al, bz, chl	B10^3, 164
13109	Salicylic acid, 5-chloro . . . . 5-Cl-2-HOC$_6$H$_3$CO$_2$H	172.57	nd (w, or al)		173-4			al, eth, bz, chl	B10^3, 165
13110	Salicylic acid, 5-chloro,ethyl ester 5-Cl-2-HOC$_6$H$_3$CO$_2$C$_2$H$_5$	200.62	nd (al)		25			al	B10, 103
13111	Salicylic acid, 5-chloro, methyl ester . . . . 5-Cl-2-HOC$_6$H$_3$CO$_2$CH$_3$	186.60	nd (al)	249d	50			al	B10^3, 166
13112	Salicylic acid, 6-chloro 6-Cl-2-HOC$_6$H$_3$CO$_2$H	172.57	nd (w)		166			al, eth, ace, bz	B10, 104
13113	Salicylic acid, 3-diazo or 3-Diazo-2-hydroxy benzoic acid 3-N$_2$-2-HOC$_6$H$_3$CO$_2$H	164.12	ye nd (ace)		155d			ace	B16, 553
13114	Salicylic acid, 3,5-dichloro . . . . 3,5-Cl$_2$-2-HOC$_6$H$_2$CO$_2$H	207.01	nd (dil al), rh pr	sub d	220-1			al, eth	B10^3, 169
13115	Salicylic acid, 3,5-diido 3,5-I$_2$-2-HOC$_6$H$_2$CO$_2$H	389.92	nd (al, aa)		235-6			al, eth	B10^3, 189
13116	Salicylic acid, 3,5-diido, ethyl ester 3,5-I$_2$-2-HOC$_6$H$_2$CO$_2$C$_2$H$_5$	417.98	lf (al)		133			al, eth, bz	B10, 114
13117	Salicylic acid, 3,4-dimethoxy 3,4-(CH$_3$O)$_2$-2-HOC$_6$H$_2$CO$_2$H	198.18	nd (w)		169-72			chl	B10^3, 2058
13118	Salicylic acid, 4,5-dimethoxy 4,5(CH$_3$O)$_2$-2-HOC$_6$H$_2$CO$_2$H	198.19	nd (w)		202d			al	B10^3, 2065
13119	Salicylic acid, 4,5-dimethoxy methyl ester . 4,5-(CH$_3$O)$_2$-2-HOC$_6$H$_2$CO$_2$CH$_3$	212.20	nd (w)		95				B10^3, 2067
13120	Salicylic acid, 4,6-dimethoxy 4,6-(CH$_3$O)$_2$-2-HOC$_6$H$_2$CO$_2$H	198.18	nd (eth-bz)		152-4d			al, eth, ace	B10^3, 2069
13121	Salicylic acid, 2,4-dinitrobenzyl ester . . . . 2-HOC$_6$H$_4$CO$_2$[CH$_2$C$_6$H$_3$(NO$_2$)$_2$-2,4]	318.25	ye pl (aa)		168				B10^2, 52
13122	Salicylic acid, 3,5-dinitro . . . . 3,5-(NO$_2$)$_2$-2-HOC$_6$H$_2$CO$_2$H	228.12	ye nd or pl (+1w)		182			al, eth, bz	B10^3, 207
13123	Salicylic acid, dithio or Dithiosalicylic acid . . . . 2-HOC$_6$H$_4$CS$_2$H	170.25	og-ye nd (peth)		48-50			w, al, eth, bz	B10^2, 78
13124	Salicylic acid, 4-ethoxy . . . . 4-C$_2$H$_5$O-2-HOC$_6$H$_3$CO$_2$H	182.18	nd (w, bz)		154			al, eth, bz	B10, 379
13125	Salicylic acid, ethyl ester or Ethyl salicylate . 2-HOC$_6$H$_4$CO$_2$C$_2$H$_5$	166.18		234, 132.8^{27}	2-3	1.1326$^{20/4}$	1.5296^{20}	al, eth	B10^3, 115
13126	Salicylic acid, ethyl ester, acetate or Ethyl aspirin . . . . 2(CH$_3$CO$_2$)C$_6$H$_4$CO$_2$C$_2$H$_5$	208.22		272, 148-50^{15}		1.1566^{15}		al, eth, ace, bz	B10^2, 48
13127	Salicylic acid, 3-formyl 3-HCO-2-HOC$_6$H$_3$CO$_2$H	166.14	nd (w + 1)		179			al	B10^3, 4207
13128	Salicylic acid, 5-formyl 5-HCO-2-HOC$_6$H$_3$CO$_2$H	166.14	nd (w)		250d			eth	B10^3, 4209
13129	Salicylic acid hydrazide . . . . 2-HOC$_6$H$_4$CONHNH$_2$	152.16	pl (al), pr (w)		147			al, bz	B10^3, 161
13130	Salicylic acid, β-hydroxyethyl ester . . . . 2-HOC$_6$H$_4$CO$_2$CH$_2$CH$_2$OH	182.18		173^{15}	37	1.2537$^{15/15}$		al, eth, bz, chl	B10^3, 138
13131	Salicylic acid, 3-iodo 3-I-2-HOC$_6$H$_3$CO$_2$H	264.02	nd (w)		199			al, eth	B10^3, 186
13132	Salicylic acid, 4-iodo 4-I-2-HOC$_6$H$_3$CO$_2$H	264.02	pr or nd (al)		230d			al, eth	B10^2, 65
13133	Salicylic acid, 5-iodo 5-I-2-HOC$_6$H$_3$CO$_2$H	264.02	nd (w)		197			al, eth	B10^3, 188
13134	Salicylic acid, menthyl ester or Menthyl salicylate . . . . C$_{17}$H$_{24}$O$_3$	276.38		190^{15}, 143$^{0.5}$		1.0467^{20}	1.5198^{26}	al, eth, ace, bz	B10^3, 126
13135	Salicylic acid, 4-methoxy-6-methyl or Everninic acid . . . . . 4-CH$_3$O-6-CH$_3$-2-HOC$_6$H$_2$CO$_2$H	182.18	nd (peth, w)		171-2			al, ace, aa	B10^2, 273
13136	Salicylic acid, 4-methoxy-6-methyl,methyl ester or Sparassol 4-CH$_3$O-6-CH$_3$-2-HOC$_6$H$_2$CO$_2$CH$_3$	196.21	pr (w), lf (MeOH)		67-8			al, eth, ace, chl, peth	B10^2, 273
13137	Salicylic acid, (methoxymethyl) ester or Mesotan . . . . 2-HOC$_6$H$_4$CO$_2$CH$_2$OCH$_3$	182.18		162$^{7.42}$		1.2^{15}		al, eth, ace, bz, chl	B10^3, 143
13138	Salicylic acid, (2-methoxyphenyl) ester 2-HOC$_6$H$_4$CO$_2$(C$_6$H$_4$OCH$_3$-2)	224.25			70			al, eth, chl	B10, 81

No.	Name, Synonyms, and Formula	Mol. wt.	Color, crystalline form, specific rotation and λ_{max} (log ε)	b.p. °C	m.p. °C	Density	n_D	Solubility	Ref.
13139	Salicylic acid, methyl ester .... 2-HOC$_6$H$_4$CO$_2$CH$_3$	152.16	....	223.3	−8	1.1738$^{20/4}$	1.5360^{20}	al, eth	B10^3, 107
13140	Salicylic acid, methyl ester, benzoate .... 2-(C$_6$H$_5$CO$_2$)C$_6$H$_4$CO$_2$CH$_3$	256.26	pr (al, eth)	385d, 270-80^{120}	92	....	....	al, eth, bz, chl	B10^3, 112
13141	Salicylic acid, 3-methyl or 2,3-Cresotic acid............ 3-CH$_3$-2-HOC$_6$H$_3$CO$_2$H	152.16	nd (w, dil al)	....	169-70	....	....	al, eth, chl	B10^3, 505
13142	Salicylic acid, 4-methyl or 2,4-Cresotic acid............ 4-CH$_3$-2-HOC$_6$H$_3$CO$_2$H	152.16	nd (w), pr (al), pl (chl)	sub	177.8	....	....	al, eth, chl	B10^3, 521
13143	Salicylic acid, 5-methyl or 2,5-Cresotic acid............ 5-CH$_3$-2-HOC$_6$H$_3$CO$_2$H	152.16	nd (w, peth)	....	153	....	....	al, eth, bz, chl	B10^3, 516
13144	Salicylic acid, 6-methyl or 2,6-Cresotic acid............ 6-CH$_3$-2-HOC$_6$H$_3$CO$_2$H	152.16	nd (chl)	....	173 (184)	....	....	al, eth, chl	B10^3, 496
13145	Salicylic acid, α-naphthyl ester or Alphol............. 2-HOC$_6$H$_4$CO$_2$-α-C$_{10}$H$_7$	264.28	....	....	83	....	....	eth	B10^2, 52
13146	Salicylic acid, β-naphthyl ester or Betol............. 2HOC$_6$H$_4$CO$_2$-β-C$_{10}$H$_7$	264.28	cr (al)	....	95.5	1.11^{116}	....	eth, bz	B10^3, 136
13147	Salicylic acid, 3-nitro .... 3-NO$_2$-2-HOC$_6$H$_3$CO$_2$H	183.12	yesh nd (aa, w + 1)	....	148-9 (anh), 128-9 (+ w)	....	....	al, eth, ace, bz, chl	B10^3, 190
13148	Salicylic acid, 4-nitro .... 4-NO$_2$-2-HOC$_6$H$_3$CO$_2$H	183.12	ye nd (w, dil al)	....	235	....	....	al, ace, chl, aa	B10^3, 194
13149	Salicylic acid, 5-nitro .... 5-NO$_2$-2-HOC$_6$H$_3$CO$_2$H	183.12	nd (w)	....	229-30	1.650^{20}	....	al, eth, ace, bz	B10^3, 197
13150	Salicylic acid, 6-nitro .... 6-NO$_2$-2-HOC$_6$H$_3$CO$_2$H	183.12	pa ye nd	....	....	....	....	w, al	B10^3, 205
13151	Salicylic acid, 4-nitrobenzyl ester .... 2-HOC$_6$H$_4$CO$_2$(CH$_2$C$_6$H$_4$NO$_2$-4)	273.25	cr (dil al)	....	97-8	....	....	al, ace	B10^2, 52
13152	Salicylic acid, pentyl ester .... 2-HOC$_6$H$_4$CO$_2$C$_5$H$_{11}$	208.24	....	148-54	....	1.065$^{15/15}$	1.506^{20}	**al, eth**	B10^3, 122
13153	Salicylic acid, isopentyl ester or Isopentyl salicylate ...... 2-HOC$_6$H$_4$CO$_2$-i-C$_5$H$_{11}$	208.24	....	276-7^{741}, 151-2^{15}	....	1.0535$^{20/4}$	1.5080^{20}	al, eth, chl	B10^3, 123
13154	Salicylic acid, phenyl ester or Phenyl salicylate Salol ..... 2-HOC$_6$H$_4$CO$_2$C$_6$H$_5$	214.22	pl (MeOH)	173.12	43	1.2614$^{30/4}$	....	al, eth, ace, bz, aa	B10^3, 127
13155	Salicylic acid, propyl ester .... 2-HOC$_6$H$_4$CO$_2$C$_3$H$_7$	180.21	....	239	96-8	1.0979$^{20/4}$	1.5161^{20}	**al, eth**	B10^3, 119
13156	Salicylic acid, iso-propyl ester .... 2-HOC$_6$H$_4$CO$_2$CH(CH$_3$)$_2$	180.21	....	240-2, 118^{17}	....	1.0729$^{20/4}$	1.5065^{20}	**al, eth**	B10^3, 120
13157	Salicylic acid, 3-iso-propyl-6-methyl or o-Thymotinic acid ..... 3-i-C$_3$H$_7$-6-CH$_3$-2-HOC$_6$H$_2$CO$_2$H	194.23	nd (w, bz, lig)	sub	127	....	....	al, eth, bz, aa	B10^3, 629
13158	Salicylic acid, 5-iso-propyl-3-methyl .... 5-i-C$_3$H$_7$-3-CH$_3$-2-HOC$_6$H$_2$CO$_2$H	194.23	nd (w)	....	147	....	....	al	B10, 282
13159	Salicylic acid, 5-iso-propyl-4-methyl .... 5-i-C$_3$H$_7$-4-CH$_3$-2-HOC$_6$H$_2$CO$_2$H	194.23	nd (bz)	....	189-90	....	....	....	B10^2, 171
13160	Salicylic acid, 6-iso-propyl-3-methyl or o-Carvacrotinic acid ..... 6-i-C$_3$H$_7$-3-CH$_3$-2-HOC$_6$H$_2$CO$_2$H	194.23	nd (w)	sub	136	....	....	al, eth	B10, 282
13161	Salicylic acid, 3-sulfo .... 3-SO$_3$H-2-HOC$_6$H$_3$CO$_2$H	218.19	pr (w + 2)	....	120	....	....	w, al, eth, aa	B11^3, 701
13162	Salicylic acid, 5-sulfo .... 5-SO$_3$H-2-HOC$_6$H$_3$CO$_2$H	218.19	hyg nd (w + 2)	....	198d 120 (+2w)	....	....	**w, al**, eth	B11^3, 704
13163	Salicylonitrile .... 2-HOC$_6$H$_4$CN	119.12	pr (bz)	149^{14}	98	1.052$^{100/4}$	1.5372^{100}	al, eth, bz, chl	B10^3, 159
13164	Salicyl chloride .... 2-HOC$_6$H$_4$COCl	156.57	....	92^{15}	19	1.3112^{20}	1.5812^{20}	eth	B10^3, 150
13165	Salicyl chloride, 3-methyl or 3-Methyl salicyl chloride .... 3-CH$_3$-2-HOC$_6$H$_3$COCl	170.60	....	87-9^{16}	27-8	....	....	....	B10^2, 132
13166	Salsoline .... C$_{11}$H$_{15}$NO$_2$	193.26	pw or cr (al), [α]$^{20}_D$ + 34.5 (0.1 NHCl, c=1)	....	221-2	....	....	al, chl	B21^4, 2121

No.	Name, Synonyms, and Formula	Mol. wt.	Color, crystalline form, specific rotation and λ_{max} (log ε)	b.p. °C	m.p. °C	Density	n_D	Solubility	Ref.
13167	Salvarsan or 606. Arspenamine $C_{12}H_{13}N_2As_2O_2 \cdot 2HCl \cdot 2H_2O$	475.03	pw	....	185-95d			w	B16[2], 560
13168	Sambunigrin or d-Mandelonitrile glucoside .... $C_6H_5CH(CN)OC_6H_{11}O_5$	295.30	nd (bz-peth), $[\alpha]^{18}_D$ −75.1		151-2			w, al	B17[4], 3356
13169	Samidin $C_{21}H_{22}O_7$	386.41	cubic, $[\alpha]^{21}_D$ +49.1 (chl, c=1.59)	....	138-9			al, eth	B19[4], 2788
13170	Sanquinarine $C_{20}H_{15}NO_5$	351.36	cr (eth al)	....	266		....	al, eth, ace, bz	B27[2], 614
13171	Santalic acid or Guerbet's acid	234.34	red	....	β:202 γ:189	β:1.5136[20] γ:1.5055[2]			B9[3], 2619
13172	α-Santalol or Arheol .... $C_{15}H_{24}O$	220.36	$[\alpha]^{20}_D$ +17	301-2, 167[14]	....	0.9679[20/4]	1.5023[20]	al	B6[3], 2083
13173	β-Santalol $C_{15}H_{24}O$	220.36	$[\alpha]^{20}_D$ −90.5	167-8[10]		0.9750[20/4]	1.5115[20]	....	B6[3], 2080
13174	Santene or 2,3-Dimethyl-2-norbornene............ C_9H_{14}	122.21	....	140-1, 35[15]		0.8698[17/4]	1.4688[17]	eth, ace, bz	B5[3], 333
13175	Santenic acid (cis, d) or π-Norcamphenic acid .... $C_9H_{14}O_4$	186.21	pl (w), $[\alpha]^{33}_D$ +38.3 (al)	....	170-1			al	B9[3], 3847
13176	β-Santenol (cis, exo) or 2,3-Dimethyl-2-norbornano .... $C_9H_{16}O$	140.23	nd (al), tab (lig)	19[2]	101-2			eth	B6[3], 243
13177	α-Santenone (cis) or 1,7-Dimethyl-1,2-norbarnanone ···· $C_9H_{14}O$	138.21	pl $[s]^{22}_D$ +11.4 (al)	191	55			al, ace	B7[3], 305
13178	Santonin or Santonic acid .... $C_{15}H_{18}O_3$	246.31	orh (w, aa, eth) $[\alpha]^{18}_D$ −173 (al, c=2)	....	174-6	1.187[66/4]	1.590	bz, chl, Py	B17[4], 6232
13179	Sarmentogenin $C_{23}H_{34}O_5$	390.53	pr (95% al, MeOH-eth)	270-5	....			al, Py	B18[4], 2443
13180	Sarpagine or Raupine .... $C_{19}H_{22}N_2O_2$	310.40	nd $[\alpha]^{20}_D$ +54 (Py)		320			al	Am84, 622
13181	Sarsaspogenin or Parigenin $C_{27}H_{44}O_3$	416.65	lo pr nd (ace), $[\alpha]^{25}_D$ −75 (chl, c=0.5)	....	200-1			al, ace, bz, chl	B19[4], 824
13182	Scarlet red .... $C_{24}H_{20}N_4O$	380.46	dr br pw or nd	d 260	186d			chl, peth	B16, 172
13183	Scilliroside (a β-glucoside) .... $C_{32}H_{44}O_{12}$	620.67	lo pr (dil MeOH) d $[\alpha]^{20}_D$ −60 (MeOH)		168-70			al, diox	B18[4], 3178
13184	Scopoline or Oscin .... $C_8H_{13}NO_2$	155.20	hyg nd (lig, eth, chl, peth)	248	108-9	1.0891[134/4]	....	w, al, ace	B27[2], 61
13185	Scyllitol .... $C_6H_{12}O_6$	180.16	pr (+3w)		353d	1.659[19/4]	....		B6[3], 6926
13186	Sebacamide or Decandiamide .... $H_2NOC(CH_2)_8CONH_2$	200.28	pr or pl (aa)		210			aa	B2[4], 2088
13187	Sebacic acid or Decanedioic acid .... $HO_2C(CH_2)_8CO_2H$	202.25	lf	295[100]	134.5	1.2705[20/4]	1.422[133]	al, eth	B2[4], 2078
13188	Sebacic acid, dibutyl ester or di-Butyl sebacate .... $C_4H_9O_2C(CH_2)_8CO_2C_4H_9$	314.47	....	344-5, 227[17]	−10	0.9405[15]	1.4433[15]	eth	B2[4], 2081
13189	Sebacic acid, diethyl ester or Diethyl sebacate .... $C_2H_5O_2C(CH_2)_8CO_2C_2H_5$	258.36	....	306[773], 188[19]	5	0.9646[20/4]	1.4366[20]	al, ace	B2[4], 2080
13190	Sebacic acid, di(2-ethyl butyl) ester .... $(C_2H_5)_2CHCH_2O_2C(CH_2)_8CO_2CH_2CH(C_2H_5)_2$	370.58	....	344-6	−22	0.920[25/4]	....	al, ace, bz	B2[4], 2082
13191	Sebacic acid, di(2-ethyl hexyl) ester .... $C_4H_9CH(C_2H_5)CH_2O_2C(CH_2)_8CO_2CH_2CH(C_2H_5)C_4H_9$	426.69	....	256[5]	−48	0.912[25/4]	1.451[25]	al, ace, bz	B2[4], 2083
13192	Sebacic acid, dimethyl ester or Dimethyl sebacate .... $CH_3O_2C(CH_2)_8CO_2CH_3$	230.31	lo pr	175[20], 144[5]	38	0.9882[28]	1.4355[28]	al, eth, ace	B2[4], 2080
13193	Sebacic acid, monomethyl ester, mononitrile .... $NC(CH_2)_8CO_2CH_3$	197.28	....	178[16]	3-4	0.934[20]	1.4398[25]	eth	B2[4], 2079

No.	Name, Synonyms, and Formula	Mol. wt.	Color, crystalline form, specific rotation and λ_{max} (log ϵ)	b.p. °C	m.p. °C	Density	n_D	Solubility	Ref.
13194	Sebaconitrile NC(CH₂)₈CN	164.25	....	204[16]		0.9313[20/4]	1.4474[20]	....	B2[4], 2089
13195	Sebacyl chloride ClOC(CH₂)₈COCl	239.14	....	220[75], 165[11]	−2.5	1.1212[20/4]	1.4684[18]		B2[4], 2088
13196	Sedarmide H₂NCONHCOCH(CH₂CH=CH₂)CH(CH₃)₂	184.24	nd (al)		194			al, chl	B3[2], 53
13197	Selanthrone C₁₂H₆Se₂	310.13	pr (al), nd (aa)	223[11]	181			CS₂	B19[4], 352
13198	Seleno urea H₂NCSeNH₂	123.02	pr or nd (w)	200d				w	B3[4], 435
13199	Seleno urea, 1-ethyl C₂H₅NHCSeNH₂	151.08	nd (al, peth, w)		ca 125			al	B4[2], 610
13200	α-Selinene C₁₅H₂₄	204.35	[α]'_D +49.5	268-72, 128-32[11]		0.9196[20/15]	1.5048[20]	....	B5[3], 1090
13201	B-Selinene C₁₅H₂₄	204.35	[α]²⁰_D +38.2	121-2[6]		0.9170[18/15]	1.4956[21]		B5[3], 1090
13202	Semicarbazide or Aminourea H₂NCONHNH₂	75.07	pr (al)		96			w, al	B3[4], 177
13203	Semicarbazide, 1,1-diphenyl (C₆H₅)₂NNHCONH₂	227.27	nd (al, bz)		195			al, bz	B15, 304
13204	Semicarbazide, 1,1-diphenyl-3-thio (C₆H₅)₂NNHCSNH₂	243.33	cr (al, bz)		202			al, ace, bz, chl	B15[3], 193
13205	Semicarbazide, 1,4-diphenyl C₆H₅NHNHCONHC₆H₅	227.27	nd or lf (al, bz)		177			al	B15[3], 184
13206	Semicarbazide, 1,4-diphenyl-3-thio C₆H₅NHCSNHNHC₆H₅	243.33			176-7			al, ace, chl	B15[3], 190
13207	Semicarbazide, 2,4-diphenyl H₂NN(C₆H₅)CONHC₆H₅	227.27	lf (al)		165.5			al, eth, bz, chl	B15, 277
13208	Semicarbazide, 2,4-diphenyl-3-thio H₂NN(C₆H₅)CSNHC₆H₅	243.33	lf (al)		139			ace, bz	B15[3], 182
13209	Semicarbazide, 4,4-diphenyl (C₆H₅)₂NCONHNH₂	227.27			154			al, eth, chl	B12[3], 896
13210	Semicarbazide hydrochloride H₂NCONHNH₂.HCl	111.52	pr (dil al)		175-7d			w	B3[4], 177
13211	Semicarbazide, 4-methyl-3-thio CH₃NHCSNHNH₂	105.36			135-8			w, al	B4[4], 220
13212	Semicarbazide, 1-phenyl H₂NCONHNHC₆H₅	151.17			172			al, ace, MeOH	B15[3], 184
13213	Semicarbazide, 1-phenyl-3-thio C₆H₅NHNHCSNH₂	167.23	pr (al)		200-1d				B15[3], 190
13214	Semicarbazide, 2-phenyl H₂NN(C₆H₅)CONH₂	151.17	nd (bz, al)		120			w, al, chl	B15[2], 103
13215	Semicarbazide, 2-phenyl-3-thio H₂NN(C₆H₅)CSNH₂	167.23	w		153			al	B15[1], 70
13216	Semicarbazide, 4-phenyl H₂NNHCONHC₆H₅	151.17	nd (bz), pl (w)		128			al, chl	B12[3], 822
13217	Semicarbazide, 4-phenyl-3-thio H₂NNHCSNHC₆H₅	167.23	pl (al)		140d				B12[2], 232
13218	Semicarbazide, 3-thio or Thiosemicarbazide H₂NNHCSNH₂	91.14	lo nd (w)		183			w, al	B3[4], 374
13219	Semicarbazide, 1-(3-tolyl) or Maretine (3-CH₃C₆H₄)NHNHCONH₂	165.19	lf (w, dil al)		183-4			al	B15, 508
13220	Semicarbazide, 1-(4-tolyl) (4-CH₃C₆H₄)NHNHCONH₂	165.19			190-1			w	B15[2], 239
13221	Semicarbazide, 1,1,4-triphenyl (C₆H₅)₂NNHCONHC₆H₅	303.37	nd (al)		206-7				B15[2], 115
13222	Semicarbazide, 1,4,4,-triphenyl C₆H₅NHNHCON(C₆H₅)₂	303.37	pl (al)		151-2			al	B15[3], 185
13223	Semicarbazide, 2,4,4-triphenyl H₂NN(C₆H₅)CON(C₆H₅)₂	303.37	cr (dil al)		128			al	B15[3], 181
13224	Serine (D) or D-2-Amino-3-hydroxy propionic acid HOCH₂CH(NH₂)CO₂H	105.10	nd or hex pr (w), [α]²⁰'_D +6.9 (w, c=10)	d	228d			w	B4[3], 1572
13225	Serine (DL) or DL-2-Amino-3-hydroxy propionic acid HOCH₂CH(NH₂)CO₂H	105.10	mcl pr or lf (w)	....	246d (sealed tube)	1.603[22 5]	....	w	B4[3], 1573

No.	Name, Synonyms, and Formula	Mol. wt.	Color, crystalline form, specific rotation and λ_{max} (log ε)	b.p. °C	m.p. °C	Density	n_D	Solubility	Ref.
13226	Serine (L) or L-2-Amino-3-hydroxy propionic acid ...... HO-CH₂CH(NH₂)CO₂H	105.10	hex pl or pr (w), $[\alpha]^{20}_D$ −6.8 (w, c=10)	sub150¹⁰⁻⁴	228d	....	....	w	B4³, 1568
13227	Serpentine .... C₂₁H₂₀N₂O₃	348.41	ye rods or lf (al), $[\alpha]^{25}_D$ + 292 (MeOH)	....	175			al, eth, ace, MeOH	Am76 , 2843
13228	Sesamin or Asaranin .... C₂₀H₁₈O₆	354.34	nd (al), $[\alpha]^{20}_D$ + 68.4 (chl, c=24)	....	123-4		....	al, ace, bz, chl	B19⁴, 6236
13229	Shikonine or d-Alkanin .... C₁₆H₁₆O₅	288.30	br-red nd (bz), $[\alpha]^{20}_{644}$ + 135 (bz, c=1.3)		147			al, eth, ace, bz	B8³, 4088
13230	Silane, allyl-trimethyl .... (CH₃)₃SiCH₂CH=CH₂	114.26	....	85.4⁷⁵²		0.7193²⁰/⁴	1.4074²⁰		B4³, 1854
13231	Silane, benzyl-trimethyl............. (CH₃)₃SiCH₂C₆H₅	164.32	....	191-2⁷⁴ᴷ 93¹⁵		0.8933²⁰/⁴	1.5042²⁰		B16³, 1201
13232	Silane, 4-bromophenoxy-trimethyl .... 4-BrC₆H₄OSi(CH₃)₃	245.19	....	126²⁵		1.2619²⁰/⁴	1.5145²⁰		B6⁴, 1056
13233	Silane, butyl-chloro-dimethyl.......... C₄H₉Si(Cl)(CH₃)₂	245.19	....	138.4⁷³⁷		0.8751¹⁰/⁴	1.5143⁴⁴		BOSC 2 , 180
13234	Silane, butyl-trichloro............. C₄H₉SiCl₃	191.56	....	148.9		1.1606²⁰/⁴	1.4363²⁰	bz, eth	B4¹, 582
13235	Silane, isobutyl-trichloro .... i-C₄H₉SiCl₃	191.56	....		141-6	1.154¹⁰/⁴			B4¹, 582
13236	Silane, butyl-trimethyl .... C₄H₉Si(CH₃)₃	130.31	....	115		0.7353²⁰/⁴			B4³, 1851
13237	Silane, isobutyl-trimethyl .... i-C₄H₉Si(CH₃)₃	130.31	....	108-9		0.7330⁰/⁴			B4¹, 580
13238	Silane, chloro-dimethyl-ethyl .... (CH₃)₂Si(Cl)C₂H₅	122.67	....	89-90		0.8675²⁰	1.4105²⁰		B4³, 1865
13239	Silane, chloro-dimethyl-phenyl .... (CH₃)₂Si(Cl)C₆H₅	170.71	....	196		1.0646²⁰	1.5184²⁰		B16³, 1206
13240	Silane, chloro-dimethyl-vinyl .... (CH₃)₂Si(Cl)CH=CH₂	120.65	....	83-4		0.8744²⁰/⁴	1.4141²⁰		KHOC, 240
13241	Silane, chloro-ethyl-methyl-phenyl .... C₆H₅SiCl(CH₃)C₂H₅	184.74	....					CS₂	
13242	Silane, chloro-phenyl .... C₆H₅SiH₂Cl	142.66	....	162-3		1.0683²⁰/⁴	1.5340²⁰		BOSC. 2., 147
13243	Silane, chloro-triethoxy .... (C₂H₅O)₃SiCl	198.72	....	156, 69¹²	−51	1.032²⁰/²⁰	1.3999²⁰	al	B1³, 1336
13244	Silane, chloro-triethyl .... (C₂H₅)₃SiCl	150.72	....	144-5		0.8967²⁰/⁴	1.4314²⁰		B4³, 1867
13245	Silane, chloro-trimethyl .... (CH₃)₃SiCl	108.64	....	57.7, 0⁶³·⁵	−57.7	0.8580²⁰/⁴	1.3885²⁰		B4³, 1857 334, BOSC 2¹,71
13246	Silane, chloro-triphenyl .... (C₆H₅)₃SiCl	294.86	....	240-3¹⁵					B16³, 1208
13247	Silane, chloromethyl-dimethyl-phenyl .... (CH₃)₂Si(C₆H₅)CH₂Cl	184.74	....	225				CS₂	B16³, 1200
13248	Silane, (4-chlorophenoxy)-trimethyl .... 4-ClC₆H₄OSi(CH₃)₃	200.74	....	214⁷⁵ᴷ		1.0320²⁰/⁴	1.4930²⁰		B6⁴, 878
13249	Silane, (2-chlorophenoxy)-trimethyl .... 2-ClC₆H₄OSi(CH₃)₃	200.74	....	95-6¹⁵					B6⁴, 809
13250	Silane, 3-chlorophenyl-trimethyl .... 3-ClC₆H₄Si(CH₃)₃	200.74	....	206-7, 84-5⁹		1.0071²⁰/⁴	1.5108²⁰		B16³, 1200
13251	Silane, 3-chloropropyl-trimethyl .... Cl(CH₂)₃Si(CH₃)₃	150.72	....	148		0.8825²⁰	1.4310²⁰/_D		KHOC, 263
13252	Silane, dibenzyl-diethoxy .... (C₆H₅CH₂)₂Si(OC₂H₅)₂	300.47	....	139⁰·⁵			1.5250²⁰		BOSC 2¹, 674
13253	Silane, dichloro-diisobutyl .... (i-C₄H₉)₂SiCl₂	213.22	....	93¹⁶		1.00²⁰/⁴			B4³, 1893
13254	Silane, dichloro-diethoxy .... (C₂H₅O)₂SiCl₂	189.12	....	135.9, 51.6¹²	−130	1.1290²⁰/⁴			B1⁴, 1364

No.	Name, Synonyms, and Formula	Mol. wt.	Color, crystalline form, specific rotation and λ_{max} (log ϵ)	b.p. °C	m.p. °C	Density	n_D	Solubility	Ref.
13255	Silane, dichloro-diethyl . . . . $(C_2H_5)_2SiCl_2$	157.11	. . . .	d128-30	−96.5	1.0504^{20}	1.4309^{20}		B4³, 1890
13256	Silane, dichloro-dimethyl . . . . $(CH_3)_2SiCl_2$	129.06		70.5	−76	$1.070^{25/25}$	$1.405^{25/}{}_D$	. . . .	B4³, 1877
13257	Silane, dichloro-diphenyl . . . . $(C_6H_5)_2SiCl_2$	253.20		302-5, 163-5¹⁰		$1.2216^{20/4}$	1.5819^{20}	al, eth, ace, bz	B16³, 1214
13258	Silane, dichloro-ethyl-methyl . . . . $C_2H_5SiCl_2CH_3$	143.09		100⁷⁴⁷		$1.0630^{20/4}$	1.4197^{20}		B4³, 1889
13259	Silane, dichloro-methyl-phenyl . . . . $C_6H_5SiCl_2CH_3$	155.68		206-7		1.1866^{20}	1.5180^{20}		B16³, 1211
13260	Silane, dichloro-methyl-isopropyl . . . . i-$C_3H_7SiCl_2CH_3$	157.11	. . . .	121-2		1.0385^{20}	1.4270^{20}		C'52, 6159
13261	Silane, dichloro methyl vinyl . . . . $CH_2=CHSiCl_2CH_3$	141.07		93		$1.085^{27/27}$	$1.4295^{20/}{}_D$		B4³, 1894
13262	Silane, dichloro-phenyl . . . . $C_6H_5SiHCl_2$	177.11		180-2		$1.225^{25/25}$			B16³, 1210
13263	Silane, dichloro-isopropyl-methyl . . . . $CH_3SiCl_2CH(CH_3)_2$	157.11		121-2		1.0385^{20}	1.4270^{20}		C52, 6159
13264	Silane, diethoxy-difluoro . . . . $(C_2H_5O)_2SiF_2$	156.21		83	−122				B1³, 1336
13265	Silane, diethoxy-dimethyl . . . . $(C_2H_5O)_2Si(CH_3)_2$	148.28		113.8, 69.5¹⁷⁹	−87	$0.8395^{20/4}$	$1.3805^{20/}{}_D$		B4³, 1874
13266	Silane, diethoxy-diphenyl . . . . $(C_2H_5O)_2Si(C_6H_5)_2$	272.42		302-4⁷⁶⁷, 151-3⁶		$1.0329^{20/4}$	$1.52695^{20/}{}_D$		B16², 608
13267	Silane, diethoxy-methyl . . . . $(C_2H_5O)_2SiHCH_3$	134.25		135-8⁷¹¹		0.829^{25}	$1.3724^{25/}{}_D$		KHOC, 358
13268	Silane, diethoxy-methyl-vinyl . . . . $(C_2H_5O)_2Si(CH_3)CH=CH_2$	160.29		133		0.8620^{20}	$1.4001^{20/}{}_D$		KHOC, 388
13269	Silane, diethyl . . . . $(C_2H_5)_2SiH_2$	88.22		56	−134.4	$0.6832^{20/4}$	1.3920^{20}		B4³, 1846
13270	Silane, diethyl-difluoro . . . . $(C_2H_5)_2SiF_2$	124.21		60.9⁷⁵⁵	−78.7	$0.9348^{20/4}$	1.3385^{20}		B4³, 1890
13271	Silane, difluoro-diphenyl . . . . $(C_6H_5)_2SiF_2$	220.30		252, 158⁵⁰		$1.145^{17/4}$	1.5221^{25}	bz	B16³, 1213
13272	Silane, dimethoxy-dimethyl . . . . $(CH_3O)_2Si(CH_3)_2$	120.22		80.5-82		0.8535^{20}	$1.3679^{20/}{}_D$		KHOC, 346
13273	Silane, dimethoxy-diphenyl . . . . $(CH_3O)_2Si(C_6H_5)_2$	244.37		90-5¹³		1.0771^{20}	$1.5447^{20/}{}_D$		KHOC, 489
13274	Silane, dimethyl . . . . $(CH_3)_2SiH_2$	60.17		−20.1	−150.2	0.68^{-80}			B4¹, 579
13275	Silane, dimethyl-diphenoxy . . . . $(C_6H_5O)_2Si(CH_3)_2$	244.37		130-2⁵	−23	1.0599^{25}	1.5330^{20}		B6⁴, 764
13276	Silane, dimethyl-diphenyl . . . . $(C_6H_5)_2Si(CH_3)_2$	212.37		276⁷⁴⁰, 173⁴⁵		$0.9867^{20/4}$	1.5644^{20}		B16², 605
13277	Silane, dimethylphenyl . . . . $C_6H_5SiH(CH_3)_2$	136.27		157		0.8891^{20}	1.4995^{20}		KHOC, 396
13278	Silane, dimethylpropyl . . . . $C_3H_7SiH(CH_3)_2$	102.25		73-4					B4³, 1850
13279	Silane, dimethylamino-trimethyl . . . . $(CH_3)_3SiN(CH_3)_2$	117.27		86, 0²².⁵			1.4379^{24}		BOSC 2¹, 138
13280	Silane, diphenyl . . . . $(C_6H_5)_2SiH_2$	184.31		75				CS_2	. . . .
13281	Silane, diphenyl-methyl . . . . $(C_6H_5)_2SiHCH_3$	198.34		266.8, 93.5¹		$0.9973^{20/4}$	1.5694^{20}		BOSC 2², 522
13282	Silane, diphenyl-divinyl . . . . $(C_6H_5)_2Si(CH=CH_2)_2$	236.39		130-1⁰·⁰⁵		$1.0092^{25/4}$	$1.5350^{25/}{}_4$		BOSC 2¹, 614
13283	Silane, ethoxy-trichloro . . . . $C_2H_5OSiCl_3$	179.51		101.9	−135	1.2274^{20}	1.4045^{20}	al	B1⁴, 1364
13284	Silane, ethoxy-triethyl . . . . $(C_2H_5)_3SiOC_2H_5$	160.34		154-5		0.8160^{20}	1.4140^{20}	al, eth	B4³, 1866
13285	Silane, ethoxy-trifluoro . . . . $C_2H_5OSiF_3$	130.15	gas	−7	−122				B1³, 1336
13286	Silane, ethoxy-trimethyl . . . . $C_2H_5OSi(CH_3)_3$	118.25		75⁷⁴⁵		$0.7573^{20/4}$	$1.3741^{20/}{}_D$	al, eth, ace	B4³, 1856
13287	Silane, ethoxy-triphenyl . . . . $(C_6H_5)_3SiOC_2H_5$	304.46	. . . .	344, 207³	65	. . . . . .	. . . .	chl	B16³, 1207

No.	Name, Synonyms, and Formula	Mol. wt.	Color, crystalline form, specific rotation and λ_{max} (log ϵ)	b.p. °C	m.p. °C	Density	n_D	Solubility	Ref.
13288	Silane, ethyl trichloro $C_2H_5SiCl_3$	163.51		97.9	−105.6	1.2381[20/4]	1.4257[20]		B4[3], 1900
13289	Silane, ethyl-triethoxy $(C_2H_5O)_3SiC_2H_5$	192.33		158.9		0.8594[20/4]	1.3955[20]	al, eth	B4[3], 1899
13290	Silane, ethyl-trifluoro $C_2H_5SiF_3$	114.14		−4.4	−105	1.227[−76]			B4[3], 1900
13291	Silane, ethyl-trimethoxy $(CH_3O)_3SiC_2H_5$	150.25		124.3		0.9488[20/4]	1.3838[20]	al	B4[3], 1899
13292	Silane, ethyl-triphenyl $(C_6H_5)_3SiC_2H_5$	288.46			76			chl	
13293	Silane, fluoro-triethoxy $(C_2H_5O)_3SiF$	182.27		134.6					B1[3], 1336
13294	Silane, fluoro-triethyl $(C_2H_5)_3SiF$	134.27		110		0.8354[25/4]	1.3900[25]	peth	B4[3], 1866
13295	Silane, fluoro-triphenyl $(C_6H_5)_3SiF$	278.40		245-52[30], 160-70[0.5]	64				B16[3], 1208
13296	Silane, methyl CH_3SiH_3	46.15		−57	−156.5				B4[1], 579
13297	Silane, methyl-phenyl $C_6H_5SiH_2CH_3$	122.24		140, 46-7[20]		0.8895[20/4]	1.5058[20]		BOSC 2[1], 209
13298	Silane, methyl-tribromo CH_3SiBr_3	282.83		131.3[744]	−28.4	2.2130[25]	1.5152[25]		B4[3], 1897
13299	Silane, methyl-trichloro CH_3SiCl_3	149.48		66.4	−77.8	1.273[25/23]			B4[3], 1896
13300	Silane, methyl-triethoxy $(C_2H_5O)_3SiCH_3$	178.31		143		0.8923[20]	1.3835[20]	al	B4[3], 1895
13301	Silane, methyl-trimethoxy $(CH_3O)_3SiCH_3$	136.22		102		0.951[25/4]	1.3687[25]	chl	BOSC 2[1], 109
13302	Silane, methyl-triphenoxy $(C_6H_5O)_3SiCH_3$	322.44		267-71[100], 210-1[12]		1.135[20]	1.5599[20]		B6[4], 765
13303	Silane, methyl-triphenyl $(C_6H_5)_3SiCH_3$	274.44		196-200[9]	69.5			CS_2	B16[3], 1199
13304	Silane, 2-methylphenoxy-trimethyl $2-CH_3C_6H_4OSi(CH_3)_3$	180.32		192		0.9287[20/4]	1.4830[20]		BOSC 2[1], 386
13305	Silane, 3-methylphenoxy-trimethyl $3-CH_3C_6H_4OSi(CH_3)_3$	180.32		198[3]		0.9186[20/4]	1.4791[20]		BOSC 2[1], 386
13306	Silane, 4-methylphenoxy-trimethyl $4-CH_3C_6H_4OSi(CH_3)_3$	180.32		199		0.9183[20/4]	1.4790[20]		BOSC 2[1], 386
13307	Silane, pentyl-trichloro $C_5H_{11}SiCl_3$	205.59		171[742], 60.5[15]		1.1330[20/4]	1.4503[20]		B4[3], 1905
13308	Silane, pentyl-triethoxy $C_5H_{11}Si(OC_2H_5)_3$	234.41		95[13]		0.8862[20]	1.4059[20]		B4[3], 1905
13309	Silane, phenyl $C_6H_5SiH_3$	108.22		118-20		0.8681[20/4]	1.5125[20]		B16[3], 1198
13310	Silane, phenyl-tri-butyl $(C_4H_9)_3SiC_6H_5$	276.54		140-2[4]		0.8753[20]	1.4915[20]		B16[3], 1199
13311	Silane, phenyl-trichloro $C_6H_5SiCl_3$	211.55		201				chl	B16[3], 1216
13312	Silane, phenyl-triethoxy $C_6H_5Si(OC_6H_5)_3$	240.37		235-8		0.9961[20/4]	1.4718[20]		B16[3], 1215
13313	Silane, phenyl-triethyl $C_6H_5Si(C_2H_5)_3$	192.38		283		0.8915[20/4]	1.4999[20]		B16[3], 1198
13314	Silane, phenyl-trifluoro $C_6H_5SiF_3$	162.19		101.5, 16.5[24]		1.2169[20/4]	1.4110[20]	al, bz	B16[3], 1216
13315	Silane, phenyl-trimethyl $C_6H_5Si(CH_3)_3$	150.30		170				CS_2	B16[3], 1198
13316	Silane, phenyl-tripropyl $(C_3H_7)_3SiC_6H_5$	234.46		146-7[18]		0.8799[20/4]	1.4950[20]		BOSC 2[1], 601
13317	Silane, propyl-trichloro $C_3H_7SiCl_3$	234.46		122.2[740]		1.1851[20/4]	1.4290[20]		B4[3], 1902
13318	Silane, propyl-trichloro $(CH_3)_2CHSiCl_3$	177.53		120.3[748]	−87.7	1.1934[20/4]	1.4319[20/D]		B4[3], 1903
13319	Silane, tetra(allyloxy) $(C_3H_5O)_4Si$	256.37		134-5[34]		0.9824[20/4]	1.4349[20]		B1[3], 1886
13320	Silane, tetraethoxy $(C_2H_5O)_4Si$	208.33		165.5-8	−77	0.9356[20/20]	1.3832[20]		B1[4], 1360
13321	Silane, tetraethyl $(C_2H_5)_4Si$	144.33		153		0.7658[20]	1.4268[20]		B4[3], 1847

No.	Name, Synonyms, and Formula	Mol. wt.	Color, crystalline form, specific rotation and λ_{max} (log ε)	b.p. °C	m.p. °C	Density	n_D	Solubility	Ref.
13322	Silane, tetramethoxy .... $(CH_3O)_4Si$	152.22	....	121-2	......	1.032^{20}	1.3683^{20}	....	B1[4], 1266
13323	Silane, tetramethyl .... $(CH_3)_4Si$	88.22	....	26.5	-102.2	$0.648^{19/4}$	1.3587^{20}	al, eth	B4[3], 1843
13324	Silane, tetraphenyl .... $(C_6H_5)_4Si$	336.51	....	228^5	236-7	$1.078^{20/4}$	....	CS_2	B16[3], 1199
13325	Silane, tetrapropoxy .... $(C_3H_7O)_4Si$	264.44	....	$225-7^{757}$, 94^5	......	$0.9158^{20/4}$	1.4012^{20}	....	B1[4], 1435
13326	Silane, tetravinyl .... $(CH_2=CH)_4Si$	136.27	....	130.2	......	0.7999^{20}	1.4625^{20}	....	KHOC, 278
13327	Silane, 2-tolyl-trichloro .... $(2-CH_3C_6H_4)SiCl_3$	225.58	....	226.7	-26	$1.306^{25/4}$	1.5336^{25}	....	BOSC 2[1], 20-4
13328	Silane, tributyl .... $(c_4H_9)_3SiH$	200.44	....	215-20 $88-9^5$	......	$0.7794^{20/4}$	1.4380^{20}_D	....	BOSC 2[1], 50-4
13329	Silane, triethoxy .... $(C_2H_5O)_3SiH$	164.28	....	132-5	......	$0.8745^{20/4}$	....	....	B1[4], 1359
13330	Silane, triethoxy-vinyl .... $(C_2H_5O)_3SiCH=CH_2$	190.31	....	....	......	....	....	....	BOSC 2[1], 29-4
13331	Silane, triethyl .... $(C_2H_5)_3SiH$	116.28	....	109^{755}	......	0.7302^{20}	1.4117^{20}	....	B4[3], 1847
13332	Silane, trimethyl vinyl .... $(CH_3)_3SiCH=CH_2$	100.24	....	55.5^{767}	......	0.6910^{20}	1.3914^{20}_D	....	KHOC, 354
13333	Silane, trioctyl .... $(C_8H_{17})_3SiH$	368.76	....	$163-5^{0.15}$	......	$0.8207^{20/20}$	1.4545^{20}	....	BOSC 2[2], 81
13334	Silane, triphenyl .... $(C_6H_5)_3SiH$	260.41	....	152-61	36-7	....	....	....	B16[3], 1199
13335	Silane, tripropyl .... $(C_3H_7)_3SiH$	158.36	....	171-3	......	$0.7723^{0/4}$	1.4280^{20}	....	B4[3], 1850
13336	Silane, vinyl-trichloro .... $CH_2=CHSiCl_3$	161.49	....	91-2	-95	$1.2426^{20/4}$	1.4295^{20}	chl	B4[3], 1908
13337	Silanol, dimethyl-ethyl .... $(CH_3)_2Si(OH)C_2H_5$	104.22	....	120^{774}, 58^{50}	......	$0.8332^{20/4}$	1.4070^{20}	....	BOSC 2[1], 10
13338	Silanol, diphenyl-methyl .... $(C_6H_5)_2Si(OH)CH_3$	214.34	....	148^3	167	....	....	CS_2	....
13339	Silanol, triethyl .... $C_{29}H_{50}O$	132.28	....	154, 46.6^9	......	$0.8647^{20/4}$	1.4329^{20}	al, eth	B4[3], 1866
13340	Silanediol, dimethyl, diacetate .... $(CH_3CO_2)_2Si(CH_3)_2$	176.24	....	165^{750}, $44-5^3$	......	$1.0540^{20/4}$	1.4030^{20}	....	B4[3], 1876
13341	Silanetriol, methyl-triacetate .... $(CH_3CO_2)_3SiCH_3$	220.25	....	$110-2^{17}$	40.5	$1.1677^{25/4}$	1.4083^{20}	....	B4[3], 1896
13342	Silin, 1,1-diphenyl-perhydro .... $C_{17}H_{20}Si$	252.43	....	$193-8^5$	......	1.0319^{25}	1.5779^{20}	....	BOSC2[1], 64
13343	Sinapine hydrogen sulfate .... $C_{16}H_{23}NO_5.H_2SO_4$	407.44	lf (al)	....	186-7d	....	....	w	B10[2], 354
13344	Sinapine, thiocyanate, monohydrate .... $C_{16}H_{23}NO_5.HSCN.H_2O$	386.45	ye nd	....	180-1	....	....	....	B10[2], 354
13345	Sinomenine or Cuculine .... $C_{19}H_{23}NO_4$	329.40	nd (bz), $[\alpha]^{26}_D$ -71 (al, c=2.1)	....	162	....	....	al, ace, chl	B21[4], 6670
13346	α-Sitosterol .... $C_{29}H_{48}O$	412.71	nd (al), $[\alpha]^{28}_D$ -1.7 (chl,c=2)	....	166	....	....	al, chl	B6[3], 2876
13347	α₂-Sitosterol .... $C_{30}H_{50}O$	426.74	cr (al-peth), $[\alpha]^{25}_D$ +3.5 (chl,c=2)	....	156	....	....	al	B6[3], 2697
13348	α₃-Sitosterol .... $C_{29}H_{48}O$	412.71	pl (al), $[\alpha]^{20}_D$ +5.2 (chl)	....	142-3	....	....	al	B6[3], 2707
13349	β-Sitosterol or Verosterol .... $C_{29}H_{50}O$	414.72	pl (al), nd (MeOH), $[\alpha]^{25}_D$ -37 (chl,c=2)	....	140	....	....	al, eth, aa	B6[3], 2696
13350	Skimmin or 7-Hydroxycoumarin-β-D-glucose .... $C_{15}H_{16}O_8$	324.29	cr (w + 1), $[\alpha]^{18}_D$ -80	....	219-21	....	....	al	B18[4], 301

No.	Name, Synonyms, and Formula	Mol. wt.	Color, crystalline form, specific rotation and λ_{max} (log ε)	b.p. °C	m.p. °C	Density	n_D	Solubility	Ref.
13351	Smilagenin or Isosapogenin $C_{27}H_{44}O_3$	416.65	silky nd (ace), $[\alpha]^{25}_D$ -69 (chl,c=0.5)		185			al, ace, bz, chl	B19[4], 826
13352	Solanidine-T or Solatubine $C_{27}H_{43}NO$	397.65	lo nd (chl-MeOH)	sub d	218-9			bz, chl	B21[4], 1398
13353	Solanine-S or Purapurine $C_{45}H_{73}NO_{16}$	884.09	nd (al), fl (diox) $[\alpha]^{20}_D$ -69 (al)		ca 190				J1963, 745
13354	Solasodine or Purapuridine $C_{27}H_{43}NO_2$	413.65	hex pl (sub), $[\alpha]^{20}_D$ -92.4 (bz)		202			ace, bz, chl, MeOH, Py	J1942, 13
13355	Sorbaldehyde or 2,4-Hexadienal $CH_3CH=CHCH=CHCHO$	96.14		173-4[756], 76[30]		0.898[20]	1.5384[20]		B1[4], 3545
13356	Sorbamide $CH_3CH=CHCH=CHCONH_2$	111.15	nd (w)		171-2			w, al	B2[4], 1705
13357	Sorbic Acid or 2,4-Hexadienoic acid $CH_3CH=CHCH=CHCO_2H$	112.14	nd (dil al)	228d, 153[50]	134.5	1.204[19/4]		al, eth	B2[4], 1701
13358	Sorbic acid, ethyl ester or Ethyl sorbate	140.19		195.5, 85[20]		0.9506[20/4]	1.4951[20]	al, eth, chl	B2[4], 1703
13359	Sorbic acid, methyl ester or Methyl sorbate $CH_3CH=CHCH=CHCO_2CH_3$	126.16	lf	180, 70[20]	15	0.9777[20/4]	1.5025[22]	al, eth	B2[4], 1703
13360	Sorbylalcohol or 2,4-Hexadien-1-ol $CH_3CH=CHCH=CHCH_2OH$	98.15	nd	76[12]	30-1	0.8967[20/4]	1.4981[20]	al, eth	B1[4], 2240
13361	Sorbyl chloride $CH_3CH=CHCH=CHCOCl$	130.58		78[15]		1.0666[19/4]	1.5545[20]	ace	B2[3], 1460
13362	Sorbierite or L-Iditol $HOCH_2(CHOH)_4CH_2OH$	182.18	pr (al), $[\alpha]^l_D$ -3.5 (w,p=2)		73-4			w	B1[4], 2843
13363	D-Sorbitol or D-Gulcitol $C_6H_{14}O_6$, $HOCH_2(CHOH)_4CH_2OH$	182.18	nd (w + 0.5), $[\alpha]^{25}_D$ -1.98 (w)	295[3.5]	110-2 (anh), 75 (hyd)	1.489[20/4]	1.3330[20]	w, ace, aa	B1[4], 2839
13364	L-Sorbitol or D-Gulitol $C_6H_{14}O_6$, $HOCH_2(CH_2OH)_4CH_2OH$	182.18	nd (w + 0.5), $[\alpha]$ $+1.7$ (w)		77			w, ace, aa	B1[3], 2405
13365	D-Sorbitol, hexaacetate $C_{18}H_{26}O_{12}$	434.38	pr (w), $[\alpha]^{18}_D$ $+6.8$ (aa)		99.5 (120)			al	B2[4], 275
13366	D-Sorbose $C_6H_{12}O_6$	180.16	rh (al), $[\alpha]^{20}_D$ $+42.9$ (w)		165	1.612[17]		w	B1[4], 4411
13367	DL-Sorbose or β-Acrose $C_6H_{12}O_6$	180.16	rh (dil al)		162-3	1.638[17]		w, MeOH	B31, 348
13368	L-Sorbose $C_6H_{12}O_6$	180.16	rh (al), $[\alpha]^{20}_D$ -43.2 (w,c=5)		165	1.612[17]		w	B1[4], 4412
13369	Sparteine (D) or Lupinidine $C_{15}H_{26}N_2$	234.39	$[\alpha]^{20}_D$ -19.5 (al)	325[754], 173[8]	30-1	1.0196[20/4]	1.5312[20]	al, eth, chl	B23[2], 97
13370	Sparteine sulfate, pentahydrate $C_{15}H_{26}N_2 \cdot H_2SO_4 \cdot 5H_2O$	422.53	pr $[\alpha]^{17}_D$ -15.3		242		1.5289	al	B23[2], 99
13371	Spiro [5,5]-hendecane-2,4,8,10-tetraoxa or Pentacrythritol dimethylene ether $C_7H_{12}O_4$	160.17		147[53]	46.5			w, al, eth, ace, bz	B19[4], 5650
13372	Spiropentane C_5H_8	68.12		39-40[746]		0.7266[20/4]	1.4120[20]		B5[4], 218
13373	Squalene or Spinacene $C_{30}H_{50}$	410.74		280[17]	<-20	0.8584[20/4]	1.4990[20]	eth, ace	B1[4], 1146
13374	Squalene, perhydro or 2,6,10,15,19,23-Hexamethyltetracosane $C_{30}H_{62}$	422.83	oil	ca 350, 263[10]	-38	0.8125[15/4]	1.4525[20]	eth, bz, peth, chl	B1[3], 585

No.	Name, Synonyms, and Formula	Mol. wt.	Color, crystalline form, specific rotation and λ_{max} (log ε)	b.p. °C	m.p. °C	Density	n_D	Solubility	Ref.
13375	Stachydrine (L) or Hygric acid methylbetaine .... $C_7H_{13}NO_2$	143.19	cr (w + 1), $[\alpha]_D^{25/}$ −40.2 (w,c=4)	....	235d (anh), 116-8 (+ w)	....	....	w, al	B22^4, 26
13376	Stachydrine-(L)-oxalate .... $C_7H_{13}NO_2.H_2C_2O_4$	233.22	nd	....	105-7				B22^1, 484
13377	Starch or Amylum............................. $(C_6H_{10}O_5)n$	....	amor pw	....	d				C51, 11746
13378	Starch triacetate .... $C_{216}H_{288}O_{144}$	7801.3	pw $[\alpha]_D^{25/}$ + 72.5	....	180d			ace, bz	C46, 8401
13379	Stearamide .... $CH_3(CH_2)_{16}CONH_2$	283.50	lf (al)	250-1^{12}	109			eth, chl	B2^4, 1240
13380	Stearamide, N-phenyl or Stearanilide................. $C_{17}H_{35}CONHC_6H_5$	359.60	nd (al)	153.5^{10}	94			al, eth, ace, bz, chl	B12^3, 486
13381	Stearic acid or Octadecanoic acid .... $CH_3(CH_2)_{16}CO_2H$	284.50	mcl lf (al)	360d, 232^{15}	71.2	0.9408$^{20/4}$	1.4299^{20}	eth, ace, chl	B2^4, 1206
13382	Stearic acid anhydride or Stearic anhydride .... $[C_{17}H_{35}CO]_2O$	550.96	....		72	0.8365$^{82/4}$	1.4362^{80}		B2^4, 1239
13383	Stearic acid, benzyl ester or Benzyl stearate .... $C_{17}H_{35}CO_2CH_2C_6H_5$	374.61	pa ye	....	28 (45)	0.9075$^{50/25}$	1.4663^{50}	....	B6^3, 1481
13384	Stearic acid, butyl ester .... $C_{17}H_{35}CO_2C_4H_9$	340.60	....	343,223^{15}	27.5	0.855$^{20/4}$	1.4328^{50}	al, ace	B2^4, 1219
13385	Stearic acid, isobutyl ester .... $C_{17}H_{35}CO_2$-i-C_4H_9	340.60	wax	199^5	(i)22.5 (ii)28.9	0.8498$^{20/4}$	....	eth	B2^3, 1017
13386	Stearic acid, cyclohexyl ester or Cyclohexyl stearate ...... $C_{17}H_{35}CO_2C_6H_{11}$	366.64	....		44	0.890$^{15/15}$		eth	B6^4, 38
13387	Stearic acid, 9,10-dibromo (trans) .... $CH_3(CH_2)_7CHBrCHBr(CH_2)_7CO_2H$	442.29	col or ye	....	29-30	1.2458$^{30/4}$	1.4893^{42}	eth	B2^4, 1048
13388	Stearic acid, 2,3-dihydroxy .... $C_{15}H_{31}CHOHCHOHCO_2H$	316.49	nd (aa)	....	α: 107 β: 126				B3^4, 1089
11388a	Stearic acid, 9,10-dihydroxy .... $C_8H_{17}CH(OH)CH(OH)(CH_2)_7CO_2H$	316.49	lf	....	95				B2^4, 1218
13389	Stearic acid, ethyl ester or Ethyl stearate............... $C_{17}H_{35}CO_2C_2H_5$	312.54	....	199^{10}	31-3	1.057$^{20/4}$	1.4349^{40}	al, eth, ace	B2^4, 1249
13390	Stearic acid, 9,10,12,13,15,16-hexabromo .... $CH_3[CH_2CHBrCHBr]_3(CH_2)_7CO_2H$	757.89	cr (diox)	....	182				B2^4, 1220
13391	Stearic acid, hexadecyl ester or Cetyl stearate .... $C_{17}H_{35}CO_2C_{16}H_{33}$	508.92	lf or pl (eth, aa)	....	57		1.4410^{70}	eth, ace, chl	B3^4, 934
13392	Stearic acid, 2-hydroxy .... $C_{16}H_{33}CH(OH)CO_2H$	300.49	flat nd (chl, AcOEt)	....	93			al, eth, ace, bz, chl	B3^4, 935
13393	Stearic acid, 3-hydroxy .... $C_{15}H_{31}CH(OH)CH_2CO_2H$	300.49	pl (chl)	....	89-90			al, eth, chl	B3^4, 940
13394	Stearic acid, 10-hydroxy .... $C_8H_{17}CHOH(CH_2)_8CO_2H$	300.49	pl	....	84			al	B3^4, 941
13395	Stearic acid, 11-hydroxy .... $C_7H_{15}CH(OH)(CH_2)_9CO_2H$	300.49	ta or pl (al)	....	81-2				B3^4, 942
13396	Stearic Acid, 12-hydroxy .... $C_6H_{13}CH(OH)(CH_2)_{10}CO_2H$	300.49	(al)	....	82			al, eth, chl	B2^4, 1222
13397	Stearic acid, 2-hydroxyethyl ester or Glycolmonostearate $C_{17}H_{35}CO_2CH_2CH_2OH$	328.54	(peth)	189-91^3	60-1	0.8780$^{60/4}$	1.4310^{60}	....	B2^4, 1216
13398	Stearic acid, methyl ester or Methylstearate .... $C_{17}H_{35}CO_2CH_3$	298.52	....	442-3^{747}, 215^{15}	39.1	0.8498$^{40/4}$	1.4367^{40}	al, eth, ace, chl	B2^4, 1272
13399	Stearic acid, 10-methyl (D) or Tuberculostearic acid ..... $C_8H_{17}CH(CH_3)(CH_2)_8CO_2H$	298.52	$[\alpha]_D^{22/}$ −0.11 (ace)	175-8$^{0.7}$	12-3	0.8771$^{25/4}$	1.4512	....	B3^4, 1093
13401	Stearic acid, 9,10-dioxo or Stearoxylic acid .... $CH_3(CH_2)_7COCO(CH_2)_7CO_2H$	312.46	ye pl (al)	....	86			al, eth, ace, lig	B3^4, 1796
13402	Stearic acid, 9,10,12,13,15,16-hexabromo or α-Linolenic acid hexabromide .... $CH_3[CH_2CHBrCHBr]_3(CH_2)_7CO_2H$	757.89	cr (diox)	....	182				B2^4, 1249
13403	Stearic acid, 3-oxo or Palmitoyl acetic acid .... $C_{15}H_{31}COCH_2CO_2H$	298.47	....	....	102-3			al	B3^4, 1687
13404	Stearic acid, 3-oxo, ethyl ester .... $C_{15}H_{31}COCH_2CO_2C_2H_5$	326.53	(al or peth)	....	37-8			al, peth	B3^4, 1687
13405	Stearic acid, 6-oxo or Lactaric acid .... $C_{12}H_{25}CO(CH_2)_4CO_2H$	298.47	pl (al, peth)	....	87			eth, chl	B3^4, 1688

PHYSICAL CONSTANTS OF ORGANIC COMPOUNDS (Continued)

No.	Name, Synonyms, and Formula	Mol. wt.	Color, crystalline form, specific rotation and λ_{max} (log ε)	b.p. °C	m.p. °C	Density	n_D	Solubility	Ref.
13406	Stearic acid, 6-oxo, ethyl ester $C_{12}H_{23}CO(CH_2)_4CO_2C_2H_5$	326.53	cr	....	47	....	....	al, eth	B3[1], 253
13407	Stearic acid, 10-oxo $C_8H_{17}CO(CH_2)_8CO_2H$	298.47	pl (al)	....	76 (82)	....	....	....	B3[4], 1689
13408	Stearic acid, 10-oxo, ethyl ester $C_8H_{17}CO(CH_2)_8CO_2C_2H_5$	326.53	pl (al)	....	41	....	....	al	B3, 725
13409	Stearic acid, 12-oxo $C_6H_{13}CO(CH_2)_{10}CO_2H$	298.47	lf (aa), cr (lig)	....	82	....	....	al	B3[4], 1690
13410	Stearic acid, 12-oxo, ethyl ester $C_6H_{13}CO(CH_2)_{10}CO_2C_2H_5$	326.53	lf (al)	199-200[3]	38	....	....	al	B3[3], 1294
13411	Stearic acid, pentyl ester or Pentylo, stearate Amylstearate $C_{17}H_{35}CO_2C_5H_{11}$	354.62	pl	....	30	....	1.4342[50]	al, eth	B2[4], 1220
13412	Stearic acid, isopentyl ester or Isoamyl stearate $C_{17}H_{35}CO_2CH_2CH_2CH(CH_3)_2$	354.62	....	192[2]	25.5	0.855[20/4]	1.433[50]	eth, ace	B2[3], 1017
13413	Stearic acid, phenyl ester or Phenyl stearate $C_{17}H_{35}CO_2C_6H_5$	360.59	....	267[15]	51-3	....	....	al, eth	B6[4], 618
13414	Stearic acid, propyl ester or Propyl stearate $C_{17}H_{35}CO_2C_3H_7$	326.57	cr (eth), pr (peth)	186.8[2]	fr 28.9	0.8452[38]	1.4400[30]	al, eth, ace	B2[4], 1219
13415	Stearic acid, isopropyl ester or Isopropyl stearate $C_{17}H_{35}CO_2CH(CH_3)_2$	326.57	....	207[6]	28	0.8403[38]	....	al, eth, ace, chl	B2[4], 1219
13416	Stearic acid, 9,10,11,12-tetrabromo or Linoleic acid tetra-bromide $C_5H_{11}[CHBrCHBrCH_2]_2(CH_2)_6CO_2H$	600.09	pl or lf (aa)	....	114-5	....	....	al, eth, bz, chl, aa	B2[4], 1247
13417	Stearic acid, 9,10,12,13-tetrabromo, ethyl ester $C_5H_{11}[CHBrCHBrCH_2]_2(CH_2)_6CO_2C_2H_5$	628.15	nd	....	63	....	....	....	B2[4], 1940
13418	Stearic acid, 9,10,12,13-tetrabromo, methyl ester $C_5H_{11}[CHBrCHBrCH_2]_2(CH_2)_6CO_2CH_3$	614.12	lf	215[15]	63	....	1.4346[45]	al, eth, chl	B2[4], 1248
13419	Stearic acid, tetrahydrofurfuryl ester $C_{17}H_{35}CO_2CH_2(C_4H_7O)$	368.61	....	22		0.917[88/45]	....	al, chl	Am50, 131
13420	Stearolic acid or 9-Octadecynoic acid $CH_3(CH_2)_7C\equiv C(CH_2)_7CO_2H$	280.46	pr (al, peth), nd (dil al)	260, 189-90[1.8]	48	1.4510[54]	....	eth	B2[4], 1751
13421	Stearolic acid, ethyl ester or Ethylstearolate $CH_3(CH_2)_7C\equiv C(CH_2)_7CO_2C_2H_5$	308.51	....	180[2.5]	....	....	1.4555[20]	....	B2[4], 1752
13422	Stearolic acid, methyl ester or Methylstearolate $CH_3(CH_2)_7C\equiv C(CH_2)_7CO_2CH_3$	294.48	....	175[3]	....	....	1.4562[20]	....	B2[4], 1751
13423	Stearone or 18-Pentatriacontanone $(C_{17}H_{35})_2CO$	506.95	lf (lig)	....	88.4	0.793[95/4]	....	....	B1[4], 3413
13424	Stearonitrile $C_{17}H_{35}CN$	265.49	....	362, 193[10]	41	0.8325[20/4]	1.4389[45]	al, eth, ace, chl	B2[4], 1242
13425	Stearophenone $C_{17}H_{35}COC_6H_5$	330.56	....	....	56	....	....	ace	B7[3], 1317
13426	Stearoxylic acid or 9,10-Dioxostearic acid $C_8H_{17}COCO(CH_2)_7CO_2H$	312.46	yr pl (al)	....	86	....	....	al, eth, ace, lig	B3[4], 1796
13427	Stearoyl chloride $C_{17}H_{35}COCl$	302.93	....	215[15]	23	....	1.4523[24]	al	B2[4], 1240
13428	Stearyl alcohol or 1-Octadecanol $CH_3(CH_2)_{16}CH_2OH$	270.50	lf (al)	210.5[15]	59-60	0.8124[59/4]	....	al, eth, chl	B1[4], 1888
13429	Stearyl sulfate or Octadecyl sulfate $(C_{18}H_{37}O)_2SO_2$	603.06	....	....	70.5	....	....	....	B1[4], 1892
13430	Stibine, bis (trifluoromethyl)-bromo $(F_3C)_2SbBr$	339.67	....	113	....	....	....	....	J1957, 3708
13431	Stibene, bis (trifluoromethyl)-chloro $(F_3C)_2SbCl$	295.22	....	ca 88, 17[20]	....	....	....	....	J1957, 3708
13432	Stibene, bis(trifluoromethyl)-iodo $(F_3C)_2SbI$	386.67	....	ca 129, 16[8]	-42	....	....	....	J1957, 3708
13433	Stibene, dibromo-trifluoromethyl CF_3SbBr_2	350.57	....	ca 157, 34[2.5]	....	....	....	....	J1957, 3708
13434	Stibene, dichloro-phenyl $C_6H_5SbCl_2$	269.71	nd or ta	110-5[10]	69-70	....	....	al, ace, aa	B16[3], 1167
13435	Stibene, diiodo-trifluoromethyl F_3CSbI_2	444.57	bt ye	>200d	4-8	....	....	....	J1957, 3708
13436	Stibene, triethyl $(C_2H_5)_3Sb$	208.94	....	161.4	-98	1.3224[15]	....	al, eth	B4[3], 1834
13437	Stibene, trimethyl $(CH_3)_3Sb$	166.85	....	80.6	-62	1.523[15]	1.42[15]	al, eth	B4[3], 1834

No.	Name, Synonyms, and Formula	Mol. wt.	Color, crystalline form, specific rotation and λ_{max} (log ε)	b.p. °C	m.p. °C	Density	n_D	Solubility	Ref.
13438	Stibene, triphenyl .... $(C_6H_5)_3Sb$	353.07	pr (peth)	>360, >220[1]	53.5	1.4343[25/4]	1.6948[42]	eth, ace, bz, chl, aa	B16[3], 1159
13439	Stibene, tris(trifluoromethyl) .... $(CF_3)_3Sb$	328.77	....	73	−58	....	....		J1957, 3708
13440	Stibene, tri-(2-tolyl) .... $(2-CH_3C_6H_4)_3Sb$	395.15	(al)	....	102			eth, bz, chl, peth	B16, 892
13441	Stibene, tri-(3-tolyl) .... $(3-CH_3C_6H_4)_3Sb$	395.15	(peth)		72	1.3957[16/4]	....	al, eth, bz, chl, aa	B16, 892
13442	Stibene, tri(4-tolyl) .... $(4-CH_3C_6H_4)_3Sb$	395.15	rh (eth, MeOH)		127-8	1.3595[16]		eth, bz, chl	B16[3], 1161
13443	Stigmastanol or β-Sitostanol .... $C_{29}H_{52}O$	416.74	pl (al), cr (+1 w), $[α]^{20}_D$ +24.8 (chl, c=1.1)		144.5			chl	B6[3], 2172
13444	Stigmasterol .... $C_{29}H_{48}O$	412.71	cr (al, +1 w), $[α]^{22}_D$ −51 (chl, c=2)		170			eth, ace, bz, chl	B6[3], 2857
13445	2-Stilbazole (cis) .... $C_{13}H_{11}N$	181.24	....	141[10]	−50			al, eth, bz	J1958, 2202
13446	2-Stilbazole (trans) .... $C_{13}H_{11}N$	181.24		324-5[750], 194[14]	91-2			al, eth, bz, lig	B20[4], 3874
13447	4-Stilbazole -(trans) .... $C_{13}H_{11}N$	181.24		208[35]	131			al, eth, chl	B20[4], 3880
13448	Stilbene - (cis) .... $C_6H_5CH=CHC_6H_5$	180.25		141[12]	5-6	1.0143[20/4]	1.6130[20]	al, eth, ace, bz, chl	B5[3], 1958
13449	Stilbene (trans) or trans-1,2-Diphenylthylene .... $C_6H_5CH=CHC_6H_5$	180.25	cr (al)	305[720], 166-7[12]	124-5	0.9707	1.6264[17]	eth, bz	B5[3], 1953
13450	Stilbene, 2-bromo (cis) .... $2-BrC_6H_4CH=CHC_6H_5$	259.14		121[0.5]	....		1.6404[25]	al, lig	Am78, 475
13451	Stilbene, 2-bromo (trans) .... $2-BrC_6H_4CH=CHC_6H_5$	259.14		145[0.5]	34		1.6822[25]	al, lig	B5[3], 1964
13452	Stilbene, α-chloro (cis) .... $C_6H_5CCl=CHC_6H_5$	214.69		160-2[12]	....		1.6281[19]	al, eth, bz, chl	B5[3], 1961
13453	Stilbene, α-chloro-(trans) .... $C_6H_5CCl=CHC_6H_5$	214.69		320-4	53-4			al, eth, bz, chl	B5[3], 1961
13454	Stilbene, 2-chloro (trans) .... $2-ClC_6H_4CH=CHC_6H_5$	214.09		208-10[30], 138-40[2]	39-40			al, chl	B5[3], 1961
13455	Stilbene, 3-chloro (trans) .... $3-ClC_6H_4CH=CHC_6H_5$	214.09		175-80[0.2]	73-4			al, chl	B5[3], 1961
13456	Stilbene, 2,2'-diamino (cis) .... $2-H_2NC_6H_4CH=CHC_6H_4NH_2-2$	210.28	red nd (w)	....	123			ace	B13[3], 510
13457	Stilbene, 2,2'-diamino (trans) .... $2-H_2NC_6H_4CH=CH(C_6H_4NH_2-2)$	210.28	gold-ye pr (al)	....	176			ace	B13[3], 510
13458	Stilbene, 4,4'-diamino (trans) .... $4-H_2NC_6H_4CH=CH(C_6H_4NH_2-4)$	210.28	ye nd or lf (al)	sub	231			MeOH	B13[3], 513
13459	Stilbene, 2,2'-dibromo (trans) .... $2-BrC_6H_4CH=CH(C_6H_4Br-2)$	338.04		....	215-6			al, bz	B5[3], 1965
13460	Stilbene, α,β-dichloro (cis) .... $C_6H_5CCl=CClC_6H_5$	249.14		180[18]	67.5-8			al, eth	B5[3], 1963
13461	Stilbene, α,β-dichloro-(trans) .... $C_6H_5CCl=CClC_6H_5$	249.14		180[18]	143-4			al, eth	B5[3], 1963
13462	Stilbene, 2,2'-dihydroxy (α-form) .... $2-HOC_6H_4CH=CH(C_6H_4OH-2)$	212.25	nd (al)		95			al, eth	B6, 1022
13463	Stilbene, 2,2'-dihydroxy-(β form) .... $2-HOC_6H_4CH=CH(C_6H_4OH-2)$	212.25	flat nd (al)		197			eth, bz	B6[3], 5574
13464	Stilbene, 3,5-dihydroxy (trans) or Pinosylvin .... $3,5-(HO)_2C_6H_3CH=CHC_6H_5$	212.25	nd (aa)		156			ace, bz, chl, aa	B6[3], 5577
13465	Stilbene, 4,4'-dihydroxy (trans) or Stilbestrol .... $4-HOC_6H_4CH=CH(C_6H_4OH-4)$	212.25	nd (aa), ta (al)		284			ace, bz	B6[3], 5581
13466	Stilbene, 4,4'-dimethoxy or Bianisal .... $4-CH_3OC_6H_4CH=CH(C_6H_4OCH_3-4)$	240.31	lf (bz, aa)	sub	214-5			ace	B6[3], 5582
13467	Stilbene, α,β-dimethyl (cis) .... $C_6H_5C(CH_3)=C(CH_3)C_6H_5$	208.30	....		67-8	0.9537[78/4]	1.5612[78]	al	B5[3], 2009

No.	Name, Synonyms, and Formula	Mol. wt.	Color, crystalline form, specific rotation and λ_{max} (log ε)	b.p. °C	m.p. °C	Density	n_D	Solubility	Ref.
13468	Stilbene, α, β -dimethyl (trans) .... $C_6H_5C(CH_3)=C(CH_3)C_6H_5$	208.30	....	....	107	0.987^{20}	1.6173^{20}	eth	B5³, 2010
13469	Stilbene, α, β -dinitro (cis) .... $C_6H_5C(NO_2)=C(NO_2)C_6H_5$	270.25	ye pym pr (al)	d	108-9	....	....	al, eth, ace, bz, chl	B5³, 1973
13470	Stilbene, α, β dinitro-(trans) .... $C_6H_5C(NO_2)=C(NO_2)C_6H_5$	270.25	pa ye nd or pr (al)	....	187-8	....	....	ace, bz, chl, aa	B5³, 1973
13471	Stilbene, 2,2′-dinitro (cis) .... $2-O_2NC_6H_4CH=CH(C_6H_4NO_2-2)$	270.25	ye nd (aa)	....	126	....	....	....	B5, 637
13472	Stilbene, 2,2′-dinitro (trans) .... $2-O_2NC_6H_4CH=CH(C_6H_4NO_2-2)$	270.25	pa ye nd (chl)	420 exp	199	....	....	....	B5³, 1970
13473	Stilbene, 2,4-dinitro (cis) .... $2,4-(NO_2)_2C_6H_3CH=CHC_6H_5$	270.25	ye pl (aa)	....	127	....	....	bz, chl, aa	B5², 541
13474	Stilbene, 2,4-dinitro (trans) .... $2,4-(NO_2)_2C_6H_3CH=CHC_6H_5$	270.25	pa ye (aa)	412 exp	143-5	....	....	chl, xyl	B5³, 1970
13475	Stilbene, 2,6-dinitro (trans) .... $2,6-(NO_2)_2C_6H_3CH=CHC_6H_5$	270.25	ye nd (bz, aa)	....	114	....	....	chl	B5³, 1970
13476	Stilbene, 3,4′-dinitro (cis) .... $3-O_2NC_6H_4CH=CH(C_6H_4NO_2-4)$	270.25	ye nd (aa)	....	155	....	....	ace, bz, chl	B5², 541
13477	Stilbene, 3,4′-dinitro (trans) .... $3-O_2NC_6H_4CH=CH(C_6H_4NO_2-4)$	270.25	ye nd (aa, Py)	....	220-2	....	....	....	B5³, 1971
13478	Stilbene, 4,4′-dinitro (cis) .... $4-O_2NC_6H_4CH=CH(C_6H_4NO_2-4)$	270.25	ye nd (aa, chl)	....	186	....	....	bz, chl	B5³, 1972
13479	Stilbene, 4,4′-dinitro (trans) .... $4-O_2NC_6H_4CH=CH(C_6H_4NO_2-4)$	270.26	pa ye lf (aa), nd (aa)	....	303-4	....	....	al, eth, ace, bz, chl	B5³, 1971
13480	Stilbene, α-ethyl .... $C_6H_5C(C_2H_5)=CHC_6H_5$	208.30	....	296-7	57	....	....	....	B5³, 2009
13481	Stilbene, 2-methoxy (trans) .... $2-CH_3OC_6H_4CH=CHC_6H_5$	210.27	....	....	59	....	....	....	B6⁰, 3494
13482	Stilbene, 3-methoxy (trans) .... $3-CH_3OC_6H_4CH=CHC_6H_5$	210.27	....	173-4⁴	34	....	....	al, eth, ace	B6³, 3497
13483	Stilbene, 4-methoxy (trans) .... $4-CH_3OC_6H_4CH=CHC_6H_5$	210.27	....	142.5¹⁵	136-6.5	....	....	al, eth, ace, bz	B6³, 3498
13484	Stilbene, α -methyl .... $C_6H_5CH=C(CH_3)C_6H_5$	194.28	....	195-200⁴⁵	82-3	$0.9565^{100/4}$	1.5635^{17}	al, eth, bz	B5³, 1995
13485	Stilbene, 2-nitro (cis) .... $2-O_2NC_6H_4CH=CHC_6H_5$	225.25	....	187¹¹	62.5-3.5	....	....	al, eth	B5³, 1967
13486	Stilbene, 2-nitro (trans) .... $2-O_2NC_6H_4CH=CHC_6H_5$	225.25	....	209¹¹	73	....	....	al, eth	B5³, 1966
13487	Strophanthiden or Corchorgenin .... $C_{23}H_{32}O_6$	404.51	orh ta (MeOH-w), lf (w + 2)	....	235 (anh), 171-5 (+ w)	....	....	al, ace, bz, chl, aa	B18⁴, 3127
13488	g-Strophanthin or Quabain .... $C_{29}H_{44}O_{12}$	584.67	hyg pl (+ aw), $[\alpha]^{25}_D$ −34	....	200	....	....	al	B18⁴, 3554
13489	Strychnine .... $C_{21}H_{22}N_2O_2$	334.42	orh pr (al), $[\alpha]^{18}_D$ −139.3 (chl, c=1)	270⁵	286-8	$1.36^{20/4}$	....	chl	B27², 723
13490	Strychnine hydrochloride .... $C_{21}H_{22}N_2O_2 \cdot HCl.2H_2O$	406.91	nd (w), $[\alpha]_D$ −28.3 (w, c=0.7)	....	....	....	....	w	B27², 730
13491	Strychnine nitrate .... $C_{21}H_{22}N_2O_2 \cdot HNO_3$	397.44	nd (w)	....	280-310	1.627	....	w, MeOH	B27², 732
13492	Strychnine sulfate .... $(C_{21}H_{22}N_2O_2)_2 \cdot H_2SO_4.5H_2O$	857.00	wh mcl $[\alpha]_D$ + 13.2	....	200d	....	....	w, MeOH	B27², 731
13493	Styracin or Cinnamyl cinnamate .... $C_6H_5CH=CHCO_2(CH_2CH=CHC_6H_5)$	264.33	nd	....	44	1.1565^4	....	al, eth	B9³, 2693
13494	Styracitol or 1,5-anhydro-D-mannitol .... $C_6H_{12}O_5$	164.16	pr (90% al), $[\alpha]^{20}_D$ −49.9, −71.7 (w)	....	157	....	....	w	B17⁴, 2579
13495	Styrene or Vinyl benzene .... $C_6H_5CH=CH_2$	104.16	....	145.2, 33.6¹⁰	−30.6	$0.9060^{20/4}$	1.5468^{20}	al, eth, ace, bz, peth	B5⁴, 1334

No.	Name, Synonyms, and Formula	Mol. wt.	Color, crystalline form, specific rotation and λ_{max} (log ε)	b.p. °C	m.p. °C	Density	n_D	Solubility	Ref.
13496	Styrene, 2-amino or 2-Vinyl aniline 2-H$_2$NC$_6$H$_4$CH=CH$_2$	119.17		112[20]		1.0181[20/4]	1.6124[20]	ace, bz	B12[3], 2785
13497	Styrene, 3-amino or 3-Vinyl aniline 3-H$_2$NC$_6$H$_4$CH=CH$_2$	119.17		112-5[12]		1.0216[20/20]	1.6069[26]	ace, bz	B12[3], 2786
13498	Styrene, 4-amino or 4-Vinyl aniline 4-H$_2$NC$_6$H$_4$CH=CH$_2$	119.17		116[9]	23.5	1.012[20/20]	1.6250[22]	ace, bz	B12[3], 2786
13499	Styrene, α-bromo C$_6$H$_5$CBr=CH$_2$	183.05		86-7[14]	−44	1.4025[23]	1.5881[20]		B5[4], 1349
13500	Styrone, β-bromo-(trans) C$_6$H$_5$CH=CHBr	183.05		219d, 108[20]	7	1.4269[16]	1.6093[20]	al, eth	B5[4], 1349
13501	Styrene, 2-bromo 2-BrC$_6$H$_4$CH=CH$_2$	183.05		206.2, 98[20]	fr−52.8	1.4160[20/4]	1.5927[20]		B5[4], 1349
13502	Styrene, 3-bromo 3-BrC$_6$H$_4$CH=CH$_2$	183.05		90-4[20]		1.4059[20/4]	1.5933[20]		B5[3], 1176
13503	Styrone, 4-bromo 4-BrC$_6$H$_4$CH=CH$_2$	183.05		103[20], 50[2.5]	4.5	1.3984[20/4]	1.5947[20]	chl, aa	B5[4], 1349
13504	Styrene, α-chloro C$_6$H$_5$CCl=CH$_2$	138.60		199, 73[16]	−23	1.1016[18/4]	1.5612[20]	al, eth	B5[4], 1345
13505	Styrene, β-chloro-(trans) C$_6$H$_5$CH=CHCl	138.60		199, 90[18]		1.1095[18/4]	1.5648[20]	al, eth, ace	B5[4], 1346
13506	Styrene, 2-chloro 2-ClC$_6$H$_4$CH=CH$_2$	138.60		188.7, 64.6[10]	−63.1	1.1000[20]	1.5649[20]	al, eth, ace, peth, aa	B5[4], 1345
13507	Styrene, 3-chloro 3-ClC$_6$H$_4$CH=CH$_2$	138.60		62-3[6]		1.1168[20/4]	1.5625[20]	al, eth	B5[4], 1345
13508	Styrene, 4-chloro 4-ClC$_6$H$_4$CH=CH$_2$	138.60		192, 66.3[10]	−15.9	1.0868[20/4]	1.5660[20]	al, eth, ace, bz, peth	B5[4], 1345
13509	Styrene, ββ-dichloro C$_6$H$_5$CH=CCl$_2$	173.04		225, 85-6[5]		1.2531[20/4]	1.5852[20]	ace, chl	B5[4], 1348
13510	Styrene, 2,4-dimethyl 2,4-(CH$_3$)$_2$C$_6$H$_3$CH=CH$_2$	132.20		79-80[12]		0.9022[21.5]	1.5214[21.5]		B5[4], 1386
13511	Styrene, 3-ethyl 3-C$_2$H$_5$C$_6$H$_4$CH=CH$_2$	132.20		190	−101	0.8945[20]	1.5351[20]		B5[3], 1217
13512	Styrene, 4-ethyl 4-C$_2$H$_5$C$_6$H$_4$CH=CH$_2$	132.20		192.3	−49.7	0.8925[20]	1.5376[70]		B5[4], 1384
13513	Styrene, 2-fluoro 2-F-C$_6$H$_4$CH=CH$_2$	122.14		46[32]		1.0282[20]	1.5200[20]	al, eth, bz	B5[4], 1342
13514	Styrene, 3-fluoro 3-FC$_6$H$_4$CH=CH$_2$	122.14		30-1[4]		1.0177[20]	1.5170[20]	al, eth, bz	B5[4], 1343
13515	Styrene, 4-fluoro 4-FC$_6$H$_4$CH=CH$_2$	122.14		67.4[50], 29-30[4]	−34.5	1.0220[20/4]	1.5150[20]	al, eth, bz	B5[4], 1343
13516	Styrene, α-methyl C$_6$H$_5$C(CH$_3$)=CH$_2$	118.18		163-4, 60[17]	24.3	0.9082[20/4]	1.5303[20]	eth, bz, chl	B5[3], 1192
13517	Styrene, 2-methyl or 2-Vinyl toluene 2-CH$_3$C$_6$H$_4$CH=CH$_2$	118.18		171		0.9106[20/4]	1.5450[20]	bz, chl	B5[4], 1367
13518	Styrene, 3-methyl or 3-Vinyl toluene 3-CH$_3$C$_6$H$_4$CH=CH$_2$	118.18		168, 61-2[18]	−70	0.9028[20/20]	1.5410[20]	al, eth, bz	B5[4], 1367
13519	Styrene, 4-methyl or 4-Vinyl toluene 4-CH$_3$C$_6$H$_4$CH=CH$_2$	118.18		169, 63[15]	−37.8	0.8760[20/4]	1.5428[20]	bz	B5[4], 1369
13520	Styrene, β-nitro-(trans) C$_6$H$_5$CH=CHNO$_2$	149.15	ye pr (peth, al)	250-60, 150[14]	60			al, eth, ace, chl, peth	B5[4], 1352
13521	Styrene, 3-nitro 3-O$_2$NC$_6$H$_4$CH=CH$_2$	149.15		120-1[11]	−1	1.1552[32/4]	1.5836[20]	al, eth, bz, chl, lig	B5[4], 1351
13522	Styrene, 4-nitro 4-O$_2$NC$_6$H$_4$CH=CH$_2$	149.15	pr (lig)	d	29			al, eth, chl, lig, aa	B5[4], 1351
13523	Styrene oxide or (1,2-Epoxy ethyl) benzene C$_6$H$_5$-CH-CH$_2$	120.16		194.1, 84-5[25]	−35.6	1.0523[16/4]	1.5342[20]	al, eth, bz	B17[4], 398
13524	Styrene, 2,4,6-trimethyl or Vinyl mesitylene 2,4,6-(CH$_3$)$_3$C$_6$H$_2$CH=CH$_2$	146.23		208-10, 83[12]		0.9057[20/4]	1.5296[20]		B5[4], 1408
13525	β-Styrene sulfonyl chloride C$_6$H$_5$CH=CHSO$_2$Cl	202.66	fl (bz)		88				B11[3], 370
13526	Subboric acid, tetramethyl ester (CH$_3$O)$_2$BB(OCH$_3$)$_2$	145.76		93	−24.3				B1[4], 1273
13527	Suberaldehyde or Octanedial OHC(CH$_2$)$_6$CHO	142.20		230-40d, 96-8[3]			1.4439[20]	w, al	B1[4], 3706
13528	Suberaldehyde oxime or Suberaldoxime HON=C(CH$_2$)$_6$C=NOH	172.23	pr (al)		155-6			al	B1[4], 3706
13529	Suberic acid or Octanedioic acid HO$_2$C(CH$_2$)$_6$CO$_2$H	174.20	lo nd or pl (w)	300 sub, 219.5[10]	144			al	B2[4], 2028

No.	Name, Synonyms, and Formula	Mol. wt.	Color, crystalline form, specific rotation and λ_{max} (log ε)	b.p. °C	m.p. °C	Density	n_D	Solubility	Ref.
13530	Suberic acid, diethyl ester or Diethyl suberate . . . . $C_2H_5O_2C(CH_2)_6CO_2C_2H_5$	230.31	. . . .	282-6, 140-1[8]	5.9	0.9811[20/4]	1.4328[20]	al, eth	B2[4], 2029
13531	Suberic acid, dimethyl ester . . . . $CH_3O_2C(CH_2)_6CO_2CH_3$	202.25	. . . .	268, 120[6]	3.1	1.0217[20/4]	1.4341[20]	al, eth, ace	B2[4], 2029
13532	Succinaldehyde or Butanedial . . . . OHCCH₂CH₂CHO	86.09	. . . .	169-70d, 58.5[9]		1.064[20/4]	1.4262[18]	w, al, eth, ace, bz	B1[4], 3642
13533	Succinamicacid or Succinic acid monoamide . . . . HO₂CCH₂CH₂CONH₂	117.11	nd (w, ace)	. . . .	156-8			w	B2[4], 1922
13534	Succinamic acid, N-phenyl or Succinanilic acid . . . . HO₂CCH₂CH₂CONHC₆H₅	193.20	nd (w)	. . . .	148.5			al, eth	B12, 295
13535	Sucinamide . . . . H₂NCOCH₂CH₂CONH₂	116.12	orh nd (w)	125.5 (sub)	268-70d				B2[4], 1922
13536	Succinic acid or Butanedioic acid . . . . HO₂CCH₂CH₂CO₂H	118.09	tcl or mcl pr	235d	188	1.572[25/4]	1.450	al, eth, ace	B2[4], 1908
13537	Succinic acid, acetyl,diethyl ester or Diethyl acetyl succi- nate . . . . $C_2H_5O_2CCH(COCH_3)CH_2CO_2C_2H_5$	216.23	. . . .	254-6, 139[12]	. . . . . .	1.081[20/4]	1.438[16]	al, eth, bz	B3[4], 1826
13538	Succinic acid anhydride or Succinic anhydride . . . . $C_2H_4(CO)_2O$	100.08	nd (al), rh pym (chl)	261, 139[15]	119.6	1.2340[20/4]		al, chl	B17[4], 5820
13539	Sucinic acid, benzyl ester-(mono) . . . . HO₂CCH₂CH₂CO₂CH₂C₆H₅	208.22	cr (bz)		59				B6[4], 2271
13540	Succinic acid, benzyl-(dl) . . . . $C_6H_5CH_2CH(CO_2H)CH_2CO_2H$	208.22	nd or lf	. . . .	163-4			al, eth	B9[3], 4300
13541	Succinic acid, bromo . . . . HO₂CCHBrCH₂CO₂H	196.99	(w)	. . . .	161	2.073	. . . .	w, al	B2[4], 1930
13542	Succinic acid, bromo,dimethyl ester . . . . CH₃O₂CCHBrCH₂CO₂CH₃	225.04	. . . .	132-6[10]	. . . . . .	1.5094[15]	. . . .		B2[4], 1930
13543	Succinic acid, dibenzyl ester . . . . C₆H₅CH₂O₂CCH₂CH₂CO₂CH₂C₆H₅	298.34	. . . .	245[13]	49-50	1.230	1.596	al, eth, bz, chl	B6[4], 2271
13544	Succinic acid, 2,3-dibromo (d) . . . . HO₂CCHBrCHBrCO₂H	275.89	pl (aa-CCl₄), [α][24]D + 147.8 (AcOEt)		157-8			w, al, ace	B2[3], 1679
13545	Succinic acid, 2,3-dibromo(dl) . . . . HO₂CCHBrCHBrCO₂H	275.89	(w or AcOEt)	. . . .	171			w, al, bz	B2[4], 1930
13546	Succinic acid, 2,3-dibromo(l) . . . . HO₂CCHBrCHBrCO₂H	275.89	nd (bz), [α][18]D −148 (AcOEt, c=5.8)		157-8d			w, al, ace	B2[3], 1679
13547	Succinic acid, 2,3-dibromo (meso) . . . . HO₂CCHBrCHBrCO₂H	275.89	. . . .	sub 275	255 (sealed tube)			al, eth	B2[4], 1930
13548	Succinic acid, dibutyl ester or Dibutyl succinate . . . . C₄H₉O₂CCH₂CH₂CO₂C₄H₉	230.31	. . . .	274.5, 145[4]	−29.2	0.9752[20/4]	1.4299[20]	al, eth, bz	B2[4], 1916
13549	Succinic acid, di-sec-butyl ester . . . . sec-C₄H₉O₂CCH₂CH₂CO₂-sec-C₄H₉	230.31	. . . .	256		0.9735[20/4]	1.4238[25]	al, eth, bz	B2[3], 1665
13550	Succinic acid, di(2-carboxy phenyl) ester or Diaspirin. Succinyl salicylic acid	358.29	nd (aa, al)	. . . .	176-8				B10[2], 43
13551	Succinic acid, 2,3-dichloro(d) . . . . HO₂CCHClCHClCO₂H	186.98	mcl, [α][19]D + 3.6 (w, c=3)	. . . .	168	1.820[15]	. . . .	w, eth, ace, chl	B2[4], 1928
13552	Succinic acid, 2,3-dichloro (dl) . . . . HO₂CCHClCHClCO₂H	186.98	pr (w, eth-peth)		175d	1.844[15]		eth, ace, chl	B2[4], 1929
13553	Succinic acid, 2,3-dichloro-(meso) . . . . HO₂CCHClCHClCO₂H	186.98	hex pr (w)		221d			w, al, eth, ace, chl	B2[4], 1928
13554	Succinic acid, 2,3-dichloro,diethyl ester(dl) . . . . $C_2H_5O_2CCHClCHClCO_2C_2H_5$	243.09	. . . .	132[15]	. . . . . .	1.1963[77/4]	1.4512[20]	eth	B2[2], 558
13555	Succinic acid, 2,3-dichloro,diethyl ester(meso) . . . . $C_2H_5O_2CCHClCHClCO_2C_2H_5$	243.09	nd (dil al)	125.5[12.5]	63	1.490[99/4]	1.4266[65]	al, eth	B2[2], 558
13556	Succinic acid, diethyl ester or Diethyl succinate . . . . $C_2H_5O_2CCH_2CH_2CO_2C_2H_5$	174.20	. . . .	216.5, 105[15]	−20.6	1.0402[20/4]	1.4198[20]	**al, eth**, ace	B2[4], 1914
13557	Succinic acid, 2,3-dimethoxy,dimethyl ester . . . . CH₃O₂CCH(OCH₃)CH(OCH₃)CO₂CH₃	208.20	pr, [α][18]D + 79.9 (MeOH, c=2)	130-2[12]	53-4	1.1317[60/4]	1.4340[20]	MeOH	B3[3], 1019
13558	Succinic acid, dimethyl ester or Dimethyl succinate . . . . . CH₃O₂CCH₂CH₂CO₂CH₃	146.14	. . . .	196.4, 80[11]	19	1.1198[20/4]	1.4197[20]	al, eth, ace	B2[4], 1913

No.	Name, Synonyms, and Formula	Mol. wt.	Color, crystalline form, specific rotation and λ_{max} (log ε)	b.p. °C	m.p. °C	Density	n_D	Solubility	Ref.
13559	Succinic acid, diphenyl ester or Diphenyl succinate....... $C_6H_5O_2CCH_2CH_2CO_2C_6H_5$	270.29	lf (al)	222.5[15]	121	....	....	eth, ace, bz	B6[3], 605
13560	Succinic acid, dipropyl ester .... $C_3H_7O_2CCH_2CH_2CO_2C_3H_7$	202.25	....	250.8, 101.5[3]	−5.9	1.0020[20/4]	1.4250[20]	eth, ace, bz	B2[4], 1916
13561	Succinic acid, di-(4-tolyl)ester .... $(4-CH_3C_6H_4)O_2CCH_2CH_2CO_2(C_6H_4CH_3-4)$	298.34	nd or lf (al)		121			al, eth, ace, bz, aa	B6[2], 379
13562	Succinic acid, ethyl ester(mono)dl .... $HO_2CCH_2CH_2CO_2C_2H_5$	146.14	pr or nd	172[42], 119[3]	8	1.1466[20/4]	1.4327[20]	w, al, eth	B2[4], 1915
13563	Succinic acid, ethyl,methyl ester .... $C_2H_5O_2CCH_2CH_2CO_2CH_3$	160.17	....	208.2, 90-5[3]	<−20	1.076[20/4]		al, eth	B2, 609
13564	Succinic acid, ethyl-(d) .... $HO_2CCH(C_2H_5)CH_2CO_2H$	146.14	pr or nd, $[α]^{18}_D$ +20.6 (aa, c=3.7)	180-3	96	1.0017[20/4]		w, al, eth	B2[4], 1995
13565	Succinic acid, ethyl (l) .... $HO_2CCH(C_2H_5)CH_2CO_2H$	146.14	pr or nd, $[α]^{24}_D$ −20.8 (ace, c=4.6)	180-3	96	1.0018[20]		w, al, eth	B2[4], 1995
13566	Succinic acid, formyl,diethyl ester .... $C_2H_5O_2CCH(CHO)CH_2CO_2C_2H_5$	202.21	....	130-4[15]	....		1.4486[25]	**al, eth**	B3[4], 1819
13567	Succinic acid, 2-hydroxy-2-methyl (dl) .... $HO_2C(CH_3)(OH)CH_2CO_2H$	148.13	mcl pr	sub	123			w, al, ace	B3[4], 1149
13568	Succinic acid, 2-hydroxy-3-methyl (d) or d-Citramalic acid $HO_2CCH(CH_3)CH(OH)CH_2CO_2H$	148.13	$[α]^{14}_D$ +34.7 (w)	....	108-9			w, al, ace	B3[4], 1151
13569	Succinic acid, mercapto (d) or d-Thiomalic acid ..... $HO_2CCH(SH)CH_2CO_2H$	150.15	cr (AeOEt-bz), $[α]^{17}_D$ +64.4 (al)		154			w, al, ace	B3[4], 1130
13570	Succinic acid, mercapto (dl) .... $HO_2CCH(SH)CH_2CO_2H$	150.15	cr (eth)		151			w, al, eth, ace	B3[4], 1130
13571	Succinic acid, mercapto (l) .... $HO_2CCH()SH)CH_2CO_2H$	150.15	$[α]^{17}_D$ −64.8 (al)		152-3			w, al, ace	B3[2], 287
13572	Succinic acid, methyl ester (mono) .... $HO_2CCH_2CH_2CO_2CH_3$	132.12		121-3[4]	56			w	B2[4], 1913
13573	Succinic acid, methyl (dl) or Pyrotartaric acid ..... $HO_2CCH(CH_3)CH_2CO_2H$	132.13	pr	d	115		1.4303	w, al, eth	B2[4], 1948
13574	Succinic acid, monochloride,monomethyl ester .... $CH_3O_2CCH_2CH_2COCl$	150.56	....	102[35], 53-4[1]	....		1.4412[20]		B2[4], 1921
13575	Succinic acid, oxo-diethyl ester or Diethyl oxaloacetate ... $C_2H_5O_2CCOCH_2CO_2C_2H_5$	188.18		131-2[24]		1.131[20/4]	1.4561[17]	**al, eth**, ace, bz	B3[4], 1809
13576	Succinic acid, 2-oxo-3-methyl,diethyl ester or Diethyl methyl oxaloacetate ... $C_2H_5O_2CCH(CH_3)COCO_2C_2H_5$	202.21	....	137-8[23], 75-8[2]		1.0970[20/4]	1.4313[20]	**al, eth**	B3[4], 1818
13577	Succinic acid, 2-oxo-3-phenyl-1-ethyl ester-4-nitrile or Ethyl phenyl cyanopyruvate .... $C_2H_5O_2CCOCH(C_6H_5)CN$	217.23	(eth-lig)	206[20]	120			al, chl	B10[3], 3957
13578	Succinic acid, phenyl (d) .... $HO_2CCH(C_6H_5)CH_2CO_2H$	194.19	pr (w), $[α]^{16.5}_D$ +148.3 (al, c=1.5)		173-4			al, eth, ace	B9[3], 4276
13579	Succinic acid, phenyl (dl) .... $HO_2CCH(C_6H_5)CH_2CO_2H$	194.19	lf or nd (w)	d	168			al, eth, ace	B9[3], 4276
13580	Succinic acid, phenyl (l) .... $HO_2CCH(C_6H_5)CH_2CO_2H$	194.19	$[α]^{15}_D$ −173.3 (ace)		173-4			al, eth, ace	B9[3], 4276
13581	Succinic acid, phenyl,anhydride (d) .... $C_{10}H_8O_3$	176.18	nd (bz-peth), $[α]^{15}_D$ +100.9 (bz)		83-4			al, bz, chl	B17[1], 259
13582	Succinic acid, phenyl,anhydride (dl) .... $C_{10}H_8O_3$	176.10	mcl pr or nd (eth)	204-6[22]	54			al, eth, ace, bz	B17[4], 6173
13583	Succinic acid, phenyl,anhydride (l) .... $C_{10}H_8O_3$	176.18	$[α]^{14}_D$ −100.9 (bz)		83-4			bz, chl	B17[4], 6173
13584	Succinic acid, (3-phenyl propenyl) .... $(C_6H_5CH_2CH=CH)CH(CO_2H)CH_2CO_2H$	234.25	lf (eth, bz)		112			eth, ace, bz	B9, 909
13585	Succinic acid, isopropylidene or Tetraconic acid .... $(CH_3)_2C=C(CO_2H)CH_2CO_2H$	158.16	tcl nd (eth)		160-1			al, eth	B2[4], 2251

No.	Name, Synonyms, and Formula	Mol. wt.	Color, crystalline form, specific rotation and λ_{max} (log ε)	b.p. °C	m.p. °C	Density	n_D	Solubility	Ref.
13586	Succinic acid, tetrahydroxy HO$_2$CC(OH)$_2$C(OH)$_2$CO$_2$H	182.09			114-5					B3[4], 1883
13587	Succinic acid, tetramethyl HO$_2$CC(CH$_3$)$_2$C(CH$_3$)$_2$CO$_2$H	174.20	tcl (60% MeOH, lig), mcl (eth, ace)	sub	200	1.30		al, bz, chl	B2[4], 2054
13588	Succinimide C$_4$H$_5$NO$_2$	99.09	pl (+1w, al), rh (ace)	287-8d	126-7	1.418		w	B21[4], 4539
13589	Succinimide, N-benzyl C$_{11}$H$_{11}$NO$_2$	189.22	nd (w), pr (al)	390-400	103			al, eth, bz, chl	B21[4], 4550
13590	Succinimide, N-bromo C$_4$H$_4$BrNO$_2$	177.99	cr (bz)	173.5d		2.098		ace, AcOEt	B21[4], 4575
13591	Succinimide, N-(2-bromophenyl) C$_{10}$H$_8$BrNO$_2$	254.10	(dil al)		91			al	B21[2], 304
13592	Succinimide, N-(3-bromophenyl) C$_{10}$H$_8$BrNO$_2$	254.10	nd (al)		118			al	B21[2], 304
13593	Succinimide, N-(4-bromophenyl) C$_{10}$H$_8$BrNO$_2$	254.10	pr (al)		172			ace, bz	B21[4], 4547
13594	Succinimide, N-chloro C$_4$H$_4$ClNO$_2$	133.53	pl(CCl$_4$)		150	1.65		ace, aa	B21[4], 4575
13595	Succinimide, N-chloromethyl C$_5$H$_6$ClNO$_2$	147.56		158-60[12]	58			w, ace, bz	B21[2], 305
13596	Succinimide, N-(4-chlorophenyl) C$_{10}$H$_8$ClNO$_2$	209.64	nd (dil al)		170			al, ace, bz	B21[4], 4547
13597	Succinimide, N-(ethoxymethyl) C$_7$H$_{11}$NO$_3$	157.17	nd	262, 151-2[14]	31-2			w, al	B21[2], 304
13598	Succinimide, N-(1-ethoxyphenyl) C$_{12}$H$_{13}$NO$_3$	219.24	pr(al)		155				B21[4], 4554
13599	Succinimide, N-ethyl C$_6$H$_9$NO$_2$	127.14	(eth)	236	26			w, al, eth	B21[4], 4544
13600	Succinimide, N-(hydroxymethyl) C$_5$H$_7$NO$_3$	129.12	lf (bz)		66			w	B21[2], 304
13601	Succinimide, N-iodo C$_4$H$_4$INO$_2$	224.99	(ace)	234	135			w, al, eth	B21[4], 4576
13602	Succinimide, N-methyl C$_5$H$_7$NO$_2$	113.12	nd (eth-peth, al, ace)	234	71			w, al, eth	B21[4], 4544
13603	Succinimide, N-α-naphthyl C$_{14}$H$_{11}$NO$_2$	225.25	nd (dil al)		153			al	B21, 376
13604	Succinimide, N-β-naphthyl C$_{14}$H$_{11}$NO$_2$	225.25	nd (al)		183			al, bz	B21[4], 4551
13605	Succinimide, N-phenyl or Succinanil C$_{10}$H$_9$NO$_2$	175.19	mcl pr or nd (w, al)	ca 400	156	1.356		eth	B21[4], 4547
13606	Succinonitrile NCCH$_2$CH$_2$CN	80.09		265-7, 124[5]	57	0.9867[60/4]	1.4173[60]	al, ace, bz, chl	B2[4], 1923
13607	Succinonitrile, tetramethyl NCC(CH$_3$)$_2$C(CH$_3$)$_2$CN	136.20	mcl pl, lf and pr (dil al)		170-1	1.070		al	B2[4], 2054
13608	Succinyl chloride ClCOCH$_2$CH$_2$COCl	154.98	pl or lf	193.3, 88.5[19]	20	1.3748[20/4]	1.4683[20]	eth, ace, bz	B2[4], 1921
13609	Succinyl chloride, 2,3-dichloro-dl) ClCOCHClCHClCOCl	223.87		78.5[7]	39			eth	B2[2], 558
13610	Succinyl chloride, 2,3-dichloride (*meso*) ClCOCHClCHClCOCl	223.87		105-6[45]				CCl$_4$	B2[2], 558
13611	Sucrose or Saccharose	342.30	mcl, [α]$^{20}_D$ +66.37 (w)		185-6	1.5805[17.5]	1.5376	w, Py	B31, 424
13612	Sucrose octaacetate C$_{28}$H$_{38}$O$_{19}$	678.61	nd, (al), [α]$^{20}_D$ +59.6 (chl)	d 285, 260[1]	86-87	1.27[16]	1.4660	eth, ace, bz, chl	B31, 453
13613	Sudan III or Tetrazobenzene-β-naphthol C$_{22}$H$_{16}$N$_4$O	352.40	br lf with gr lustre(aa)		195			al, eth, ace, bz	B16[3], 148
13614	Sulfadiazine or 2-Sulfanilamido pyrimidine C$_{10}$H$_{10}$N$_4$O$_2$S	250.28	cr (w), wh pw		255-6d				C55, 25956
13615	Sulfaguanidine C$_7$H$_{10}$N$_4$O$_2$S	214.25	nd (w)		190-3				B14[3], 1970

No.	Name, Synonyms, and Formula	Mol. wt.	Color, crystalline form, specific rotation and λ_{max} (log ϵ)	b.p. $^\circ$C	m.p. $^\circ$C	Density	n_D	Solubility	Ref.
13616	Sulfaguanidine, monohydrate $C_7H_{10}N_4O_2S.H_2O$	232.27	nd (w)		143 (sealed tube)				B14³, 1971
13617	Sulfamerazine or Sulfamethyldiazine $C_{11}H_{12}N_4O_2S$	264.31	cr		234-8				C55, 5501
13618	Sulfomethazine $C_{12}H_{14}N_4O_2S$	278.34	pa ye(w + ¹/₂) cr (diox-w)		198-9 (205-7)			w	Am64, 567
13619	Sulfamethylthiazole $C_{10}H_{11}N_3O_2S_2$	269.34			236-8			al	Am64, 2905
13620	Sulfamic acid, N-cyclohexyl $C_6H_{11}NHSO_2OH$	179.24	(al)		169-70				B12³, 1594
13621	Sulfamide, tetramethyl $[(CH_3)_2N]_2SO_2$	152.22	pl or nd (dil al)	225	73			al	B4⁴, 270
13622	Sulfanilamide or 4-Amino sulfonamide $4-H_2NC_6H_4SO_2NH_2$	172.22	lf (aq al)		165-6	1.08		al, eth, ace	B14³, 1920
13623	Sulfanilamide, N-acetyl $4-(CH_3CO)NHC_6H_4SO_2NH_2$	214.25	pr (w)		182-4			al, ace	B14³, 2042
13624	Sulfanilic acid or 4-Aminobenzene sulfonic acid $4-H_2NC_6H_4SO_3H$	173.9	rh pl or mcl (w + 2)		288	1.485²⁵/⁴			B14³, 1916
13625	Sulfapyrazine or N-(2-Pyrazinyl)sulfanilamide $C_{10}H_{10}N_4O_2S$	250.28	nd (PhNO₂)		251			Py	Am63, 3153
13626	Sulfapyridine $C_{11}H_{11}N_3O_2S$	249.29	ye og (al)		191-3				C34, 1814
13627	Sulfaquinoxaline $C_{14}H_{12}N_4O_2S$	300.34			247-8				C49, 5521
13628	Sulfathiadiazole $C_8H_9N_5S_2O_2$	256.32			218			al, Py	C40, 5411
13629	Sulfathiazole $C_9H_9N_3O_2S_2$	255.32	br pl, rods or pw (45 al)		(i) 202.5 (ii) 125				C40, 7518
13630	Sulfathiazole, 4-nitro $C_9H_8N_4O_4S_2$	285.30	pa ye pw		258-62				C53, 12281
13631	Sulfathiazole, phthalyl, $C_{17}H_{13}N_3O_5S_2$	403.44			272-4				C51, 9689
13632	Sulfathiazole, succinyl $C_{13}H_{13}N_3O_5S_2$	355.39	cr		192-5				C51, 12148
13633	Sulfoacetic acid or Sulfoethanoic acid $HO_3SCH_2CO_2H$	140.12	hyg ta (w + 1)	245d	84-6			w, al, ace	B4⁴, 102
13634	Sylvestrene (d) or Carvestrene $C_{10}H_{16}$	136.24	$[\alpha]^{18}{}_D$ +83.2 (chl, c=4.3), (undil), +66.3	175⁷⁵¹		0.8479¹⁵/⁴	1.4760¹⁸	al, eth	B14³, 1916
13635	Syringenin or Sinapyl alcohol. 3-(3,5-dimethoxy-4-hydroxyphenyl)-2-propen-1-ol. $[3,5-(CH_3O)_2-4-HOC_6H_2]CH=CHCH_2OH$	210.23	nd (eth-peth)		66-7			eth	B6³, 6690
13636	Syringin or Methoxy coniferine $C_{17}H_{24}O9$	372.38	cr (w), nd (al), $[\alpha]_D$ −17.1		192			al	B31, 222
13637	Syringoyl methyl ketone $C_{11}H_{12}O_5$	224.22	ye nd		80-1			al, bz	Am62, 986

No.	Name, Synonyms, and Formula	Mol. wt.	Color, crystalline form, specific rotation and λ_{max} (log ε)	b.p. °C	m.p. °C	Density	n_D	Solubility	Ref.
13638	Tachysterol . . . . $C_{28}H_{44}O$	396.67	$[\alpha]^{18}{}_D$ −70 (bz)	220 vac	. . . . . .	. . . . . .	. . . . . .	al, eth, ace, bz	B6[3], 3087
13639	D-Tagatose . . . . $C_6H_{12}O_6$	180.16	cr (dil al), $[\alpha]^{20}{}_D$ −5 (w, c=1)		134-5			w	B31, 348
13640	D-Talitol or D-Altritol, D-Talaite . . . . $C_6H_{14}O_6$	182.18	pr (al)$[\alpha]^{20}{}_D$ + 3.1	. . . .	87-8			w, al	B1[4], 2839
13641	D-Talonic acid, hemihydrate. . . . . $C_6H_{12}O_7.1\frac{1}{2}H_2O$	205.17	cr (aq al + $\frac{1}{2}$ w), $[\alpha]^{25}{}_D$ + 16.7→ −21.6 (w, c=4, mut)		138-9			w	B3[3], 1068
13642	D-Talonic acid, γ-lactone. . . . . $C_6H_{10}O_6$	178.14	pr (al), $[\alpha]^{25}{}_D$ −34.6→ −28.4 (w), (mut)		135-7			. . . .	B18[4], 30273
13643	D-Talose . . . . $C_6H_{12}O_6$ HOCH(CHOH)$_4$CHO	180.16	$[\alpha]$cr (al) β:cr (MeOH), $[\alpha]^{27}{}_D$ + 29→ + 19.7 (w, c=1, mut)		α:130-5 β:120-1			w	B31, 283
13644	Tannic acid or Gallotannic acid Tannin . . . . $C_{76}H_{52}O_{46}$	1701.24	pa ye br amor or fl	. . . .	210-5d			w, al, ace	B31, 133
13645	Taraxanthin . . . . $C_{40}H_{56}O_4$	600.89	ye pl (MeOH), $[\alpha]^{20}{}_{Cd}$ + 200 (AcOEt)		181-4			al, ace, petl	B17[], 7239
13646	L-Tartaric acid or L-Threoic acid or L-2,3-Dihydroxy butanedioic acid HO$_2$CCH(OH)CH(OH)CO$_2$H	150.09	mcl (anh), rh pr (w + 1), $[\alpha]^{20}{}_D$ + 12.7 (w, c=17.4)	. . . .	171-4	1.7598[20]	1.4955	w, al, ace	B3[4], 1229
13647	DL-Tartaric acid or Racemic acid. . . . . HO$_2$CCH(OH)CH(OH)CO$_2$H	150.09	mcl pr (w, al + 1w)		206	1.788		w	B3[4], 1229
13648	Tartaric acid (meso) . . . . HO$_2$CCH(OH)CH(OH)CO$_2$H	150.09	tcl pl (w)		146-8	1.666[20/4]	1.5	w, al	B3[4], 1218
13649	Tartaric acid anhydride,diacetate (d) or α,α-Diacetoxy succinic anhydride. . . . . $C_8H_8O_7$	216.16	nd (bz)		135			al, eth, ace	B18[4], 2296
13650	Tartaric acid dibenzyl ester(d) or d-Dibenzyl tartrate . . . . . $C_6H_5CH_2O_2C(CHOH)_2CO_2CH_2C_6H_5$	330.14	$[\alpha]^{15}{}_D$ + 19.3	250-70[4]	50	1.2036[72]		al, Py	B6[3], 1537
13651	Tartaric acid, dibutyl ester (d) or d-Dibutyl tartrate . . . . . . $C_4H_9O_2C(CHOH)_2CO_2C_4H_9$	262.31	pr $[\alpha]^{20}{}_D$ + 11.3 (al)	320, 178[12]	22	1.0909[20/4]	1.4451[20]	w, al, ace	B3[4], 1232
13652	DL-Tartaric acid, dibutyl ester . . . . $C_4H_9O_2C(CHOH)_2CO_2C_4H_9$	262.31	. . . .	320[765], 185[12]		1.0879[25/4]	1.4474[15]	w, al, ace	B3[3], 1032
13653	Tartaric acid, diisobutyl ester-(d) or d-Diisobutyl tartrate i-$C_4H_9O_2C(CHOH)_2CO_2$-i-C_4H_9	262.31	$[\alpha]_D$ + 11.8 (al)	323-5, 183[11]	73	1.0265[81/4]	. . . .	al	B3[3], 1021
13654	DL-Tartaric acid, diiosbutyl ester or DL-Diisobutyl tartrate . . . . . i-$C_4H_9O_2C(CHOH)_2CO_2$-i-C_4H_9	262.31	cr (bz)	311[768], 195[13]	63	1.0386[68/4]		al	B3[4], 1232
13655	Tartaric acid, diethyl ester (d) or d-Diethyl tartrate. . . . . . $C_2H_5O_2C(CHOH)_2CO_2C_2H_5$	206.20	$[\alpha]^{16}{}_D$ + 7.9 (un-dil)	280, 142[8]	18.7	1.2036[20/4]	1.4468[20]	w, al, eth, ace, aa	B3[3], 1025
13656	DL-Tartaric acid, diethyl ester or DL-Diethyl racemate . . . . $C_2H_5O_2C(CHOH)_2CO_2C_2H_5$	206.20	. . . .	281[765], 158[14]	18.7	1.2046[20/4]	1.4438[20]	al, eth, ace	B3[4], 1232
13657	Tartaric acid, diethyl ester-(l) . . . . $C_2H_5O_2C(CHOH)_2CO_2C_2H_5$	206.20	$[\alpha]^{20}{}_D$ −7.55 (un-dil)	280, 162[15]		1.2054[20/4]	1.4468[20]	al, eth, ace	B3[3], 1020
13658	Tartaric acid, diethyl ester-(meso) . . . . $C_2H_5O_2C(CHOH)_2CO_2C_2H_5$	206.20	. . . .	157.5[14]	60	1.1350[99/4]	1.4315[65]	eth, ace	B3[3], 1031
13659	Tartaric acid, diethyl ester, diacetate-(d) . . . . . $C_2H_5O_2C(CHO_2CCH_3)_2CO_2C_2H_5$	290.27	mcl cr (lig), $[\alpha]^{100}{}_D$ + 6.3 (un-dil)	296[764], 163[10]	67	1.1149[66/4]		al, eth	B3[3], 1021

No.	Name, Synonyms, and Formula	Mol. wt.	Color, crystalline form, specific rotation and λ_{max} (log ε)	b.p. °C	m.p. °C	Density	n_D	Solubility	Ref.
13660	Tartaric acid, dimethyl ester (d) .... $CH_3O_2C(CHOH)_2CO_2CH_3$	178.14	(i)cr (bz), (ii)cr (bz), (iii)cr (w), $[\alpha]^{50'}_D$ +6.7 (MeOH, c=16)	280, 166[12]	(i)48(ii) 50(iii) 61	1.306[45]	....	w, al, eth, ace, chl	B3[3], 1018
13661	DL-Tartaric acid, dimethyl ester or DL-Dimethyl tartrate ...... $CH_3O_2C(CHOH)_2CO_2CH_3$	178.14	orh nd (bz), ta (chl)	280, 169[20]	90	1.2604[90]		w, al, eth, ace	B3[4], 1232
13662	Tartaric acid, dimethyl ester-(meso) or meso-Dimethyl tartrate .... $CH_3O_2C(CHOH)_2CO_2CH_3$	178.14	nd (chl), cr (MeOH)	98[0.01] sub	114			w, al, ace	B3[3], 1031
13663	Tartaric acid, di(2-methyl butyl)ester-(d) .... $C_5H_{11}O_2C(CHOH)_2CO_2C_5H_{11}$	290.36	$[\alpha]^{20'}_D$ +14.1	208[20]	....	1.0636[20/4]		al, ace	B3, 519
13664	Tartaric acid, dinitrate (d) $HO_2C(CHONO_2)_2CO_2H$	240.08	nd (eth-bz), $[\alpha]^{20'}_D$ +13.7 (MeOH, p=9)	....	d			al, eth, ace	B3[3], 1018
13665	Tartaric acid, dipropyl ester-(d) or d-Dipropyl tartrate.... $C_3H_7O_2C(CHOH)_2CO_2C_3H_7$	234.25	$[\alpha]^{20'}_D$ +12.4 (w)	303, 181[23]		1.1390[20/4]		w, al, eth, ace	B3[2], 331
13666	DL-Tartaric acid, dipropyl ester .... $C_3H_7O_2C(CHOH)_2CO_2C_3H_7$	234.25	pr (al-eth)	286[765], 167[11]		1.1256[20/4]		w, al, eth, ace	B3[2], 337
13667	Tartaric acid, diisopropyl ester-(d) or d-Diisopropyl tartrate .... i-$C_3H_7O_2C(CHOH)_2CO_2$-i-C_3H_7	234.25	$[\alpha]^{20'}_D$ +14.9	275[765], 152[12]		1.1300[20/4]		al, eth, ace	B3[2], 331
13668	DL-Tartaric acid, diisopropyl ester or DL-Diisopropyl tartrate .... i-$C_3H_7O_2C(CHOH)_2CO_2$-i-C_3H_7	234.25	....	275[765], 154[12]	34	1.1166[20/4]		al, eth, ace	B3[2], 337
13669	Tartaric acid, ethyl ester-(mono) .... $HO_2C(CHOH)_2CO_2C_2H_5$	178.14	$[\alpha]'_D$ +21.8 (w)		90			w, al	B3[3], 1020
13670	Taurine or 2-Amino ethane sulfonic acid .... $H_2NCH_2CH_2SO_3H$	125.15	mcl pr (w)		328			w	B4[3], 1697
13671	Taurine, N,N-dimethyl or 2-(N,N-dimethylamino)ethane sulfonic acid.............. $(CH_3)_2NCH_2CH_2SO_2H$	153.20	pr (MeOH), pl (w + 1)		315-6 (anh), 270-80 d (+ w)			w, aa	B4[3], 1699
13672	Taruine, N-methyl or 2-(N-Methyl)ethane sulfonic acid... $CH_3NHCH_2CH_2SO_3H$	139.18	pr		241-2			w	B4[3], 1699
13673	Taurine, N-methyl-N-phenyl .... $C_6H_5NC(CH_3)CH_2CH_2SO_3H$	215.27	pa vt (al)		239-40			al	B12[2], 285
13674	Taurine, N-phenyl $C_6H_5NHCH_2CH_2SO_3H$	201.25	lf (w), pr (al)		277-80d			w	B12[2], 284
13675	Taurocholic acid or Cholaic acid. Cholyl taurine........ $C_{26}H_{45}NO_7S$	515.72	pr (al-eth), $[\alpha]^{18'}_D$ +38.8 (al, c=2)		ca 125d			w, al	E14, 195
13676	Tephrosin or Hydroxy deguelin .... $C_{23}H_{22}O_7$	410.43	pr (chl-MeOH)		198 (218)			eth, ace, chl	B19[4], 5271
13677	Terebic acid-(dl) or 3,3-Dimethyl paraconic acid........ $C_7H_{10}O_4$	158.16	mcl pr (al)		176	0.8155[24/4]			B18[4], 5287
13678	Terephthaldehyde 1,4-$(OHC)_2C_6H_4$	134.14	nd (w)	247[771]	116			al, eth	B7[3], 3460
13679	Terephthalamide .... 1,4-$C_6H_4(CONH_2)_2$	164.18	nd (w), pl (aa)		>250				B9[3], 4253
13680	Terephthalamide, N,N,N',N'-tetraethyl .... 1,4-$C_6H_4[CON(C_2H_5)_2]_2$	276.38	cr (eth-al)		127			al, bz	B9[3], 4253
13681	Terephthalic acid or 1,4-Benzene dicarboxylic acid....... 1,4-$C_6H_4(CO_2H)_2$	166.14	nd (sub)	>300 (sub without melting)	sub				B9[3], 4249
13682	Terephthalic acid, 2-amino .... 2-$H_2NC_6H_3(CO_2H)_2$-1,4	181.15	ye cr (w)		324-5d				B14[1], 637
13683	Terephthalic acid, 2-amino, dimethyl ester .... 2-$H_2NC_6H_3(CO_2CH_3)_2$-1,4	209.20	nd (bz), cr (al)		134			ace, chl	B14[2], 338
13684	Terephthalic acid, 2-benzyl or Benzophenone-2,5-dicarboxylic acid ...... 2-$(C_6H_5CO)C_6H_3(CO_2H)_2$-1,4	270.25	nd		291-2			al, eth	B10[3], 4008
13685	Terephthalic acid, 2-bromo .... 2-$BrC_6H_3(CO_2H)_2$-1,4	245.04	nd (w, al)		299			al	B9[3], 4258

No.	Name, Synonyms, and Formula	Mol. wt.	Color, crystalline form, specific rotation and λ_{max} (log ϵ)	b.p. °C	m.p. °C	Density	n_D	Solubility	Ref.
13686	Terephthalic acid, 2-chloro .... 2-ClC$_6$H$_3$(CO$_2$H)$_2$-(1,4)	200.58	cr (w)	....	320	....	....	al, eth	B9^3, 4256
13687	Terephthalic acid, 2,5-dichloro .... 2,5-Cl$_2$C$_6$H$_2$(CO$_2$H)$_2$-1,4	235.03	nd (w)	sub	306			al, eth	B9^3, 4257
13688	Terephthalic acid, diethyl ester or Diethyl terephthalate... 1,4-C$_6$H$_4$(CO$_2$C$_2$H$_5$)$_2$	222.24	nd pr (al, peth)	302, 142^2	44	1.1098$^{45/45}$	....	al, eth	B9^3, 4250
13689	Terephthalic acid, 2,5-dihydroxy .... 2,5-(HO)$_2$C$_6$H$_2$(CO$_2$H)$_2$-1,4	198.13	ye cr (al, w)	....	d				B10^3, 2438
13690	Terephthalic acid, dimethyl ester .... 1,4-C$_6$H$_4$(CO$_2$CH$_3$)$_2$	194.19	nd (eth)	sub	141-2			eth	B9^3, 4250
13691	Terephthalic acid, monoamide or Terephthalamic acid ... 4-HO$_2$CC$_6$H$_4$CONH$_2$	165.15	....	sub 250	>300				B9^3, 4253
13692	Terephthalic acid, mononitrile or 4-Cyanobenzoic acid ... 4-NCC$_6$H$_4$CO$_2$H	147.14	pl or lf (w)	....	219			al, eth	B9^3, 4254
13693	Terephthalic acid, 2-nitro .... 2-NO$_2$C$_6$H$_3$(CO$_2$H)$_2$-1,4	211.13	nd (w)	....	270-5			al	B9^3, 4258
13694	Terephthalic acid, tetrabromo .... Br$_4$C$_6$(CO$_2$H)$_2$-1,4	481.72	nd (w)		266 (anh)				B9^1, 367
13695	Terephthalonitrile .... 1,4-C$_6$H$_4$(CN)$_2$	128.14	nd (w, MeOH)	sub	222			bz, aa	B9^3, 4255
13696	Terephthalyl chloride .... 1,4-C$_6$H$_4$(COCl)$_2$	203.03	nd or pl (lig)	125-7^9	83-4			eth	B9^3, 4252
13697	Terephthalyl alcohol or p-Xylylene glycol .... 1,4-(HOCH$_2$)$_2$C$_6$H$_4$	138.17	nd (w)	....	115-6			w, al, eth, ace	B6^3, 4608
13698	Terpenolic acid or Terpenylic acid .... C$_7$H$_8$O$_4$	172.14	lf or pr (w + l)	sub 130-40	90 (anh), 57 (1 w)			w	B18^4, 5302
13699	o-Terphenyl or 1,2-Diphenyl benzene .... 1,2-(C$_6$H$_5$)$_2$C$_6$H$_4$	230.31	mcl pr (MeOH)	332, 160-70^2	58			ace, bz, chl, MeOH	B5^3, 2292
13700	m-Terphenyl or 1-3-Diphenyl benzene .... 1,3-(C$_6$H$_5$)$_2$C$_6$H$_4$	230.31	ye nd (al)	365	89			al, eth, bz, aa	B5^3, 2294
13701	p-Terphenyl .... 1,4-(C$_6$H$_5$)$_2$C$_6$H$_4$	230.31	....	sub 250^{45}	213			eth, bz	B5^3, 2296
13702	α-Terpinene or p-Mentha-1,3-diene .... C$_{10}$H$_{16}$	136.24		177.2, 68-70^{12}		0.8502$^{20/4}$	1.4784^{10}	al, eth	B5^4, 435
13703	β-Terpinene .... C$_{10}$H$_{16}$	136.24		173-4		0.838^{23}	1.4754^{22}	....	B5^4, 437
13704	γ-Terpinene .... C$_{10}$H$_{16}$	136.24		183		0.849$^{20/4}$	1.4765$^{14.5}$		B5^4, 436
13705	α-Terpineol (dl) or dl-p-Menth-1-en-8-ol .... C$_{10}$H$_{18}$O	154.26	cr (peth)	220, 85^2	40-1	0.9337$^{20/4}$	1.4831^{20}	al, eth, ace, bz, chl	B6^4, 251
13706	α-Terpineol acetate .... C$_{12}$H$_{20}$O$_2$	196.29	d: [α]$_D$ +52.5 (undil), l: [α]$_D$ −73 (un-dil)	140^{40}, 104-6^{11}		0.9659$^{20/4}$	1.4689^{21}	al, eth, bz	B6^4, 252
13707	Terpin hydrate (cis) or cis-Terpino hydrate .... C$_{10}$H$_{20}$O$_2$·H$_2$O	214.31	rh cr	sub 100	123d		1.51	al	B6^3, 4113
13708	Terpinolene .... C$_{10}$H$_{16}$	136.24		185, 76^{10}		0.8623$^{20/4}$	1.4883^{20}	al, eth, bz	B5^4, 437
13709	3-Terpinolenone or Piperitenone .... C$_{10}$H$_{14}$O	150.22	[α]$_{546}$−0.1	120-2^{14}		0.9774$^{20/4}$	1.5294^{20}	al, eth	B7^3, 559
13710	Terramycin or Oxytetracycline .... C$_{22}$H$_{24}$N$_2$O$_9$	498.49			184-5	1.634^{20}	....		AM87, 134
13711	Testosterone or 17β-Hydroxy-4-androsten-3-one .... C$_{19}$H$_{28}$O$_2$	288.44	nd (dil ace), [α]$^{24/}_D$ +109 (al, c=4)		155			al, eth, ace	B8^3, 892
13712	Testosterone, 4,5-dihydro-17-methyl .... C$_{20}$H$_{32}$O$_2$	304.48	cr (AcOEt)		192-3				B8^3, 609
13713	Testosterone, 17-ethenyl or 17-Vinyl testosterone .... C$_{21}$H$_{30}$O$_2$	314.47	pr nd (peth-eth), [α]$_D$ +87.6		140-1			eth, ace, chl, MeOH	B8^3, 1067
13714	Testosterone, 17-ethyl .... C$_{21}$H$_{32}$O$_2$	316.49	nd (AcOEt)		143-4				B8^3, 946

No.	Name, Synonyms, and Formula	Mol. wt.	Color, crystalline form, specific rotation and λ_{max} (log ε)	b.p. °C	m.p. °C	Density	n_D	Solubility	Ref.
13715	Testosterone, 17-ethynyl $C_{21}H_{28}O_2$	312.46	cr (chl-MeOH, AcOEt), $[\alpha]^{20}_D$ + 22.5 (diox)	(sub, vac)	270-2	....	....	diox, Py	B8[3], 1206
13716	Testosterone, 17-methyl $C_{20}H_{30}O_2$	302.46	nd (bz), $[\alpha]^{20}_D$ + 82 (al)		165-6			al, eth	B8[3], 939
13717	Testosterone propionate $C_{22}H_{32}O_3$	344.49	$[\alpha]^{25}_D$ + 83-90 (diox, c=1)		118-22			al, eth, Py	B8[3], 897
13718	Tetrabenzotriazaporphyrin $C_{32}H_{18}N_3$	514.55	purple nd and pl (quino-line)					Py	J1939, 1809
13719	Tetrabutyl ammonium iodide $(C_4H_9)_4N^+I^-$	369.38	lf(w bz)		148			al, w	B4[4], 558
13720	Tetracosane $CH_3(CH_2)_{22}CH_3$	338.67	cr (eth)	391.3, 231.3[10]	54	0.7665[70/4], 0.7991[20]	1.4283[70], 1.4480[20]	eth	B1[4], 578
13721	Tetracoscene, 11-decyl $(C_{10}H_{21})_2CHC_{13}H_{27}$	478.93	....	301[10]	10.8	0.8161[20/4]	1.4556[20]	....	B1[4], 598
13722	Tetracosane, 3-ethyl $(C_2H_5)_2CHC_{21}H_{43}$	366.71	....	255.5[10]	30.1	0.7949[40/4]	1.4436[40]	....	B1[4], 584
13723	Tetracyclone or Tetraphenyl cyclopentadienone $C_{29}H_{20}O$	384.48	bk-vt lf, cr (aa, xyl)	....	220-1			al, bz	B7[3], 2997
13724	6,8-Tetradecadiyne $C_5H_{11}C≡CC≡CC_5H_{11}$	190.33	....	118-9[4]	2	0.8699[16/4]	....	eth	B1[3], 1067
13725	Tetradecanal or Myristaldehyde $C_{13}H_{27}CHO$	212.38	lf	166[24]	30			al, eth, ace	B1[4], 3389
13726	Tetradecane $CH_3(CH_2)_{12}CH_3$	198.40	....	253.7, 121.9[10]	5.9	0.7628[20/4]	1.4290[20]	al, eth	B1[4], 520
13727	Tetradecane, 1-amino or Myristyl amine $C_{13}H_{27}CH_2NH_2$	213.41	....	291.2, 162[15]	83.1	0.8079[20/4]	1.4463[20]	al, eth, ace, bz, chl	B4[4], 812
13728	Tetradecane, bromo or Myristyl bromide $C_{13}H_{27}CH_2Br$	277.30	....	307, 181[21]	5.6	1.0170[20/4]	1.4603[20]	al, ace, bz, chl	B1[4], 523
13729	Tetradecane, 1-chloro or Myristyl chloride $C_{13}H_{27}CH_2Br$	232.84	....	292, 153[10]	4.9	0.8665[20/4]	1.4473[20]	al, ace, bz, chl	B1[3], 550
13730	Tetradecane, 1,14-dibromo $Br(CH_2)_{14}Br$	356.20	lf (al-eth), cr (al)	190-2[8]	50.4			al, eth, chl	B1[4], 523
13731	Tetradecane, 1,1-dicycloheryl $C_{13}H_{27}CH(C_6H_{11})_2$	362.68	....	406	37.6	0.8735[20]	1.4799[20]	....	B5[3], 300
13732	Tetradecane, 1-phenyl or Tetradecyl benzene $C_{13}H_{27}CH_2C_6H_5$	274.50	....	358.9, 210[12]	16.1	0.8559[20/4]	1.4818[20]	....	B5[4], 1212
13733	Tetradecanedioic acid, dimethyl ester $CH_3O_2C(CH_2)_{12}CO_2CH_3$	286.42	nd (MeOH)	191-2[10]	43-5	....			B2[4], 2149
13734	1,14-Tetradecanediol or Tetradecamethylene glycol $HO(CH_2)_{14}OH$	230.40	nd (bz)	200[9]	84.5			al, eth	B1[4], 2631
13735	1-Tetradecane sulfonic acid $C_{13}H_{27}CH_2SO_3H$	278.46	....	....	65.5 (anh), 55-6 (+1w)	0.9996[25/4]	....	w	B4[4], 66
13736	1-Tetradecanethiol or Myristyl mercaptan $C_{13}H_{27}CH_2SH$	230.45	....	176-80[22]	....	0.8484[20/20]	1.4597[20]	al, eth	B1[4], 1867
13737	Tetradecanoic acid or Myristic acid $C_{13}H_{27}CO_2H$	228.38	lf (eth 80% aa)	250.5[100], 149.3[2]	58	0.8439[80/4]	1.4305[80]	al, ace, bz, chl	B2[4], 1126
13738	1-Tetradecanol or Myristyl alcohol $C_{13}H_{27}CH_2OH$	214.40	lf	263.2, 167[11]	39-40	0.8236[38/4]		al, eth, ace, bz, chl	B1[4], 1864
13739	2-Tetradecanone or n-Dodecyl methyl ketone $C_{12}H_{25}COCH_3$	212.38	cr (dil al)	205-6[100], 134[13]	33-4			al, ace	B1[4], 3389
13740	3-Tetradecanone or Ethyl undecyl ketone $C_{11}H_{23}COC_2H_5$	212.38	cr (MeOH)	152[16]	34			al, ace	B1[4], 3389
13741	5,9-Tetradecadien-7-yne, 6,9-dimethyl $C_4H_9CH=C(CH_3)C≡CC(CH_3)=CHC_4H_9$	218.39	....	95-8[0.3]		0.8241[20/4]	1.4866[20]	eth, bz	B1[4], 1133
13742	1-Tetradecene $C_{12}H_{25}CH=CH_2$	196.38	....	232-4, 125[15]	-12	0.7745[20/4]	1.4351[20]	al, eth, bz	B1[4], 924
13743	2-Tetradecyne $C_{11}H_{23}C≡CCH_3$	194.36	....	252.5, 134[15]	6.5	0.8000[20/4]	....	al, eth	B1[4], 1070

No.	Name, Synonyms, and Formula	Mol. wt.	Color, crystalline form, specific rotation and λ_{max} (log ε)	b.p. °C	m.p. °C	Density	n_D	Solubility	Ref.
13744	7-Tetradecyne $C_6H_{13}C{\equiv}CC_6H_{13}$	194.36	....	144[30]	.....	0.7991[20/4]	1.4330[25]	al, eth	B1[3], 1067
13745	Tetraethyl ammonium bromide $(C_2H_5)_4N^+Br^-$	210.17	dlq (al)	....	......	1.3970[20/4]	....	w, al, chl	B4[4], 332
13746	Tetraethyl ammonium hydroxide, hydrate $(C_2H_5)_4N^+OH^-.4H_2O$	219.32	nd (w + 4)	d (vac)	49-50	....	....	w, al	B4[4], 331
13747	Tetraethylene glycol $HOCH_2(CH_2OCH_2)_3CH_2OH$ $C_8H_{18}O_5$	194.23	....	328, 198[14]	-6.2	1.1285[15/4]	1.4577[20]	w, al, eth	B1[4], 2403
13748	Tetraethylene glycol, dimethyl ether $CH_3OCH_2(CH_2OCH_2)_3CH_2OCH_3$	222.29	....	275.8		1.0132[20/20]	....	w, al, eth	B1[4], 2404
13749	Tetraethylene glycol, mono stearate $C_{17}H_{35}CO_2CH_2(CH_2OCH_2)_3CH_2OH$	476.72	....	328	40	1.1285[15/4]	1.4593[20]	....	C37, 3202
13750	Tetraethylene pentamine $H_2NCH_2(CH_2NHCH_2)_3CH_2NH_2$	189.31	....	340.3, 186-92[14]			1.5042[20]		B4[1], 1244
13751	Tetraethyl hypophosphate $(C_2H_5O)_2POOP(OC_2H_5)_2$	274.19	....	116-7[2]		1.1283[18]	1.4284[20]	bz	B1[4], 1358
13752	Tetralin or 1,2,3,4-Tetrahydronaphthalene $C_{10}H_{12}$	132.21	....	207.6, 79.4[10]	-35.8	0.9702[20/4]	1.5413[20]	al, eth	B5[4], 1388
13753	Tetralin, 6-acetyl 6-$(CH_3CO)C_{10}H_{11}$	174.25	....	289-91d, 182[20]					B7[3], 1473
13754	Tetralin, 2-amino-(dl) 2-$H_2NC_{10}H_{11}$	147.22	....	249[710], 140[20]	38	1.0295[22/4]	1.5604[22]	al, eth, ace, bz	B12[3], 2811
13755	Tetralin, 5-amino 5-$H_2NC_{10}H_{11}$	147.22	....	276.8, 155[22]		1.0625[16]	1.6050[20]	al, eth	B12[3], 2805
13756	Tetralin, 1,1-dimethyl 1,1-$(CH_3)_2C_{10}H_{10}$	160.26	....	221		0.950[20]	1.5292[20]		B5[1], 1430 (35.5214)
13757	Tetralin, 1-ethyl 1-$C_2H_5C_{10}H_{11}$	160.26	....	239-46		0.95285[20]	1.5318[20]		B5[4], 1429
13758	Tetralin, 2-ethyl 2-$C_2H_5C_{10}H_{11}$	160.26	....	237, 63-5[0.5]		0.9401[15.5]	1.5250[15.5]		B5[3], 1261
13759	Tetralin, 6-ethyl 6-$C_2H_5C_{10}H_{11}$	160.26	....	244		0.9632[17/4]	1.5414[16]		B5[4], 1429
13761	Tetralin, 1-hydroxy 1-$HOC_{10}H_{11}$	148.20	....	102-4[2]			1.5658[20]		B6[3], 2457
13762	Tetralin, 1-methyl 1-$CH_3C_{10}H_{11}$	146.23	....	220.59		0.9583[20]	1.5353[20]		B5[4], 1412 (35.5210)
13763	Tetralin, 2-methyl 2-$CH_3C_{10}H_{11}$	146.23	....	220-2, 99-101[13]					B5[4], 1414
13764	Tetralin, 6-methyl 6-$CH_3C_{10}H_{11}$	146.23	....	220-2		0.9541[15/4]	1.5332[15]		B5[4], 1413
13765	Tetralin, 1,2,3,4-tetrabromo or Naphthalene tetrabromide 1,2,3,4-$Br_4C_{10}H_8$	447.81	mcl pr (chl)	....	111d			CS_2, bz	B5[3], 1229
13766	Tetralin, 5,6,7,8-tetrachloro 5,6,7,8-$Cl_4C_{10}H_8$	269.99	....	180[26]	172				B5[3], 1227
13767	1,4-Tetralin dicarboxylic acid, 1-phenyl-(d,α) or d-α-Isatropic acid $C_{18}H_{16}O_4$	296.33	pr $[\alpha]^{20}_D$ + 9.4 (al, c=12.6)	....	239d				B9[1], 416
13768	5-Tetralin sulfonamide 5-$C_{10}H_{11}SO_2NH_2$	211.29	lf (al, 30% aa)	....	139-40			al	B11[2], 87
13769	5-Tetralin sulfonic acid 5-$C_{10}H_{11}SO_3H$	212.27	cr (chl + 1w)		105-10 (+ 1w)			w	B11[2], 87
13770	5-Tetralin sulfonyl chloride 5-$C_{10}H_{11}SO_2Cl$	230.72	pl (peth)		70.5			eth	B11[2], 87
13771	6-Tetralin sulfonic acid 6-$C_{10}H_{11}SO_3H$	212.26	cr (chl, dil sulf)		75			w, eth, chl	B11[2], 88
13772	6-Tetralin sulfonyl chloride 6-$C_{10}H_{11}SO_2Cl$	230.72	pl (eth)	197-200[18]	58			eth	B11[2], 88
13773	1-Tetralone or 1-oxo-1,2,3,4-tetrahydronaphthalene 1-$C_{10}H_{10}O$	146.19	....	255-7, 129[12]	8	1.0989[16/4]	1.5672[20]		B7[3], 1416
13773a	1-Tetralone, 7-ethyl 7-C_2H_5-(1-$C_{10}H_9O$)	175.25	....	152-3[12]		1.0556[17/4]	1.5599[17]	al, eth, bz	B7[3], 1473
13774	1-Tetralone, 4-methyl 4-$CH_3(C_{10}H_9O)$	160.22	....	133-4[12]		1.0779[19/4]	1.5620[19]		B7[3], 1444
13775	2-Tetralone 2-$C_{10}H_{10}O$	146.19	....	234-40, 138[16]	18	1.1055[27/4]	1.5598[20]	eth, bz	B7[3], 1424

No.	Name, Synonyms, and Formula	Mol. wt.	Color, crystalline form, specific rotation and λ_{max} (log ε)	b.p. °C	m.p. °C	Density	n_D	Solubility	Ref.
13776	Tetramethyl ammonium bromide $(CH_3)_4N^+Br^-$	154.06	dlq ditetr bipym	(360 sub,vac)	230d	1.56	....	w, MeOH	B4[4], 145
13777	Tetramethyl ammonium chloride $(CH_3)_4N^+Cl^-$	109.60	dlq ditetr bipym (dil)	....	420 (sealed tube)	1.169[20/4]	....	w, MeOH	B4[4], 145
13778	Tetramethyl ammonium hydroxide, monohydrate $(CH_3)_4N^+OH^-.H_2O$	109.17		d 130-5				w	B4, 50
13779	Tetramethyl ammonium hydroxide, trihydrate $(CH_3)_4N^+OH^-.3H_2O$	145.20			60			w	B4, 50
13780	Tetramethyl ammonium hydroxide, pentahydrate $(CH_3)_4N^+OH^-.5H_2O$	181.23		d	62-3			w, al	B4[2], 557
13781	Tetramethyl ammonium iodide $(CH_3)_4NI$	201.05			>230d	1.829			B4[4], 146
13782	Tetraphenylene or Tetrabenzocyclooctatetraene $C_{24}H_{16}$	304.40	cr (al, AcOEt)	200[0.2] sub	233			al	B5[3], 2595
13783	Tetrapropyl ammonium bromide $(C_3H_7)_4N^+Br^-$	266.27			252			w, chl	B4[4], 471
13784	Tetrapropyl ammonium iodide $(C_3H_7)_4N^+I^-$	313.27	rh bipym		280d	1.3138[25/4]		w, al, chl, aa	B4[4], 472
13785	Tetrasiloxane, decamethyl $(CH_3)_3Si(OSi(CH_3)_2)_3CH_3$	310.69	....	194, 88[20]	−76	0.8536[20/4]	1.3895[20]	bz, peth	B4[3], 1879
13786	Tetrasiloxane, 1,1,1,3,5,7,7,7,-octamethyl $[(CH_3)_3SiOSi(CH_3)_2]_2O$	310.69	....	170		0.8559[20/4]	1.3854[20]		B4[3], 1874
13787	1-Tetratriacontanol or n-Carnatyl alcohol $CH_3(CH_2)_{33}OH$	494.94	nd (ace)		913			ace	B1[4], 1921
13788	1,2,4,5-Tetrazine or s-Tetrazine $C_2H_2N_4$	82.07	dk red pr	sub	99			w, al, eth	B26[2], 212
13789	1,2,3,4-Tetrazole CH_2N_4	70.05	pl (al)	sub	156			w, al, ace, aa	B26[2], 196
13790	Thebaine or Paramorphine $C_{19}H_{21}NO_3$	311.39	pl (eth), pr (dil al), $[\alpha]^{25}_D$ −218.5 (al, p=2)	sub 91[0.01]	193	1.305[20/4]		al, bz, chl, Py	B27[2], 177
13791	Thebaine, dihydro $C_{19}H_{23}NO_3$	313.40	$[\alpha]^{20}_D$ −267 (bz, c=1.02)		162-3			al, bz, AcOEt	B27[2], 110
13792	Thebaine, hydrochloride monohydrate $C_{19}H_{21}NO_3.HCl.H_2O$	365.86	orh pr (al), $[\alpha]^{20}_D$ −157 (al)					w, al, eth	B27[2], 181
13793	Thebainone A $C_{18}H_{21}NO_3$	299.37	nd (dil al + ½ w), nd or pr (al, aa)		151-2			ace, bz, chl	B21[4], 6548
13794	Theobromine or 3,7-Dimethylxanthine $C_7H_8N_4O_2$	180.18	rh or mcl nd (w)	sub 290	351 (357)				B26[2], 264
13795	Theophylline or 1,3-Dimethylxanthine $C_7H_8N_4O_2$	180.18	nd or pl (w + 1)		272-4			w	B26[2], 263
13796	Thevetin $C_{42}H_{66}O_{18}$	858.99	nd (al), pl or nd (i-proh)		210			al, Py, MeOH	B18[4], 1493
13797	Thiacyclobutane or Trimethylene sulfide C_3H_6S	74.15	....	94.7, 14[30]	−73.2	1.0200[20/4]	1.5102[20]	al, ace, bz	B17[4], 14
13798	1,3,4-Thiadiazole, 2.5-dimethyl $C_4H_6N_2S$	114.16		202-3	64				B27, 565
13799	Thialdine or Thioacetaldehyde ammonia $C_6H_{13}NS_2$	163.31	pl (al-eth)	d	46	1.0632[50/20]		al, eth	B27[2], 525
13800	Thiane $C_5H_{10}S$	102.19	....	141.8, 93[82]	19	0.9861[20/4]	1.5067[20]	al, eth, ace, bz, chl	B17[4], 55
13801	Thianthrene or Dephenylene disulfide $C_{12}H_8S_2$	216.34	mcl pr or pl (al)	336, 204[11]	158-9			eth, bz	B19[4], 347
13802	Thiazole C_3H_3NS	85.13		116.8		1.998[17/4]	1.5969[20]	al, eth, ace	B27[2], 9
13803	Thiazole, 2-acetamido $2-CH_3CONH(C_3H_2NS)$	142.18			208				C35, 5110
13804	Thiazole, 2-amino or Abadol $2-H_2N(C_3H_2NS)$	100.14	ye pl (al)	140[11]	93				B27[2], 205
13805	Thiazole, 2-amino-4-methyl $4-CH_3-2-H_2N(C_3HNS)$	114.17	hyg cr	281-2d, 136[30-40]	45-6			w, al, eth	B27[2], 206

No.	Name, Synonyms, and Formula	Mol. wt.	Color, crystalline form, specific rotation and λ_{max} (log ε)	b.p. °C	m.p. °C	Density	n_D	Solubility	Ref.
13806	Thiazole, 2-amino-5-methyl 5-CH$_3$-2-H$_2$N(C$_3$HNS)	114.17	pl (w)	....	96			al, eth	B27, 162
13807	Thiazole, 2-amino-5-sulfanilyl C$_9$H$_9$N$_3$O$_2$S$_2$	255.32	nd (al)		219-21d			al, eth, ace, diox	C41, 447
13808	Thiazole, 2,4-dimethyl 2,4-(CH$_3$)$_2$(C$_3$HNS)	113.18	....	144-5[719], 70-3[50]		1.0562[15/4]	1.5091[20]	al, eth	B27[2], 10
13809	Thiazole, 4,5-dimethyl 4,5-(CH$_3$)$_2$(C$_3$HNS)	113.18	....	158, 81-3[59]	83-4			al, eth	AM74, 5778
13810	Thiazole, 2,4-diphenyl 2,4-(C$_6$H$_5$)$_2$(C$_3$HNS)	237.33	lf (al)	>360	92-3	1.1554[98/4]		al, eth	B27[2], 43
13811	Thiazole, 5-(2-hydroxyethyl)-4-methyl 4-CH$_3$-5-(HOCH$_2$CH$_2$)(C$_3$HNS)	143.21	col to pa ye	135[7]		1.196[24/4]		w, al, eth, bz, chl	AM71, 2931
13812	Thiazole, 2-mercapto-4-methyl 4-CH$_3$-2-HS(C$_3$HNS)	131.22	ye (dil al)	188[3]	88-9			al	B27[2], 208
13813	Thiazole, 2-methyl 2-CH$_3$(C$_3$H$_2$NS)	99.16		128-9, 65-70[80]			1.510	w, al, ace	B27, 16
13814	Thiazole, 4-methyl 4-CH$_3$(C$_3$H$_2$NS)	99.16		132[745], 70[90]	......	1.112[25]		w, al, eth	B27[2], 9
13815	5-Thiazole carboxylic acid 5-(C$_3$H$_3$NS)-CO$_2$H	129.14	lf ye nd (dil HCl)	....	217-8			eth	C48, 2688
13816	5-Thiazole carboxylic acid, 2-amino,ethyl ester 2-H$_2$N-(C$_3$HNS)CO$_2$C$_2$H$_5$-5	172.22	....	213-5d	163-4				C47, 7453
13817	5-Thiazole carboxylic acid, 4-methyl 4-CH$_3$(C$_3$HNS)CO$_2$H-5	143.17	pr or pl (w), nd (al)	sub >250	280d				B27, 316
13818	5-Thiazole carboxylic acid, 4-methyl,ethyl ester 4-CH$_3$-(C$_3$HNS)CO$_2$C$_2$H$_5$-5	171.22	pr	232-3[715], 110-5[15]	28			al, eth, ace	B27, 316
13819	5-Thiazole carboxylic acid, 4-methyl,ethyl ester hydrochloride C$_7$H$_9$NO$_2$S.HCl	207.68	nd (al)	....	155d			w, al	B19[3], 443
13820	5-Thiazole carboxylic acid, 4-methyl-2-sulfanilamide C$_{11}$H$_{11}$N$_3$O$_4$S$_2$	313.36	cr	....	241-2				C38, 2250
13821	Thiazolidine or Tetrahydrothiazole C$_3$H$_7$NS	89.16	....	164-5		1.131[25/4]	1.551[20]	w, al, eth, ace	AM59, 200
13822	2,4-Thiazolidinedione or 2,4-Dioxothiazolidine C$_3$H$_3$O$_2$NS	117.13	pl (w), pr (al)	179[19]	128			eth	B27[2], 284
13823	2-Thiazolidine thione or 2-thiothiazolidone C$_3$H$_5$NS$_2$	119.21	nd (w, MeOH)		106-7				B27[2], 198
13824	2-Thiazoline or 4,5-Dihydrothiazole C$_3$H$_5$NS	87.14	....	138[750]	......			eth, ace, bz	B27[1], 206
13825	2-Thiazoline, 2-amino C$_3$H$_6$N$_2$S	102.13	nd or lf (bz)	d	84-5			w, al, bz, chl	B27[2], 194
13826	2-Thiazoline-2,5-dimethyl C$_5$H$_9$NS	115.19	....	152					B27, 14
13827	2-Thiazoline, 5-ethyl-2-methyl 5-C$_2$H$_5$-2-CH$_3$(2-C$_3$H$_3$NS)	129.22		62-3[16]					C53, 11368
13828	2-Thiazoline, 4-hexyl-2-methyl 4-C$_6$H$_{11}$-2-CH$_3$(2-C$_3$H$_3$NS)	185.33		92-3[2]					
13829	2-Thiazoline, 5-hexyl-2-methyl 5-C$_6$H$_{11}$-2-CH$_3$(2-C$_3$H$_3$NS)	785.33		82-4[1]					
13830	2-Thiazoline, 2-mercapto or 2-Thiazolinethiol 2HS-(2-C$_3$H$_4$NS)	119.21	nd (w, MeOH)	....	106-7			w, al	B27[2], 198
13831	2-Thiazoline, 2,4,4-trimethyl 2,4,4-(CH$_3$)$_3$(2-C$_3$H$_2$NS)	129.22		146-8		0.969[25]	1.4825[25]		C53, 11368
13832	Thienone or Di-(2-thienyl)ketone (C$_4$H$_3$S)$_2$CO	194.28	nd (al)	326	90			eth, ace	B19[4], 1745
13833	Thiepane C$_6$H$_{12}$S	116.22	....	169-71[747]		0.9883[20/4]	1.5138[20]	eth, ace, chl	B17[4], 81
13834	Thietane C$_3$H$_6$S	74.14	....	94.7, 14[10]	−73.25	1.0200[20/4]	1.5102[20]	al, ace, bz	B17[4], 14
13835	Thietane, 1,1-dioxide C$_3$H$_6$SO$_2$	106.14	....	91.2[14]	75.5-6		1.5156[20]	w, al, eth, lig	B17[4], 16
13836	Thietane, 2-methyl 2-CH$_3$(C$_3$H$_5$S)	88.17	....	105.5-7.5[747]		0.9571[20/4]	1.4852[20]	al, eth, ace, bz, chl	B17[4], 44
13837	Thiirane or Ethylene sulfide C$_2$H$_4$S	60.11	....	55-6d	......	1.0368[0/4]	1.4935[20]	ace, chl	B17[4], 11
13838	Thioacetamide CH$_3$CSNH$_2$	75.13	cr (al), pl (eth)	....	115-6			w, al	B2[4], 565

No.	Name, Synonyms, and Formula	Mol. wt.	Color, crystalline form, specific rotation and λ_{max} (log ε)	b.p. °C	m.p. °C	Density	n_D	Solubility	Ref.
13839	Thioacetamide, N-phenyl or Thioacetanilide .... CH₃CSNHC₆H₅	151.23	nd (w)	d	75-6				B12², 142
13840	Thioacetic acid .... CH₃COSH	76.12	ye	87, 26-7²⁵	<−17	1.064²⁰/⁴	1.4648²⁰	w, al, eth, ace	B2⁴, 542
13841	Thioacetic acid, ethyl ester or Ethyl thioacetate........ CH₃COSC₂H₅	104.18	....	116.4		0.9792²⁰/⁴	1.4583²¹	al, eth	B2⁴, 543
13842	Thioanisole or Methyl phenyl sulfide .... CH₃SC₆H₅	124.22	....	193, 74¹⁰		1.0579²⁰/⁴	1.5868²⁰	al, bz	B6⁴, 1466
13843	Thiobenzophenone (C₆H₅)₂CS	198.29	bl nd (peth)	174¹⁴	53-4			bz, chl	B7³, 2087
13844	Thiobenzophenone, 4,4´ bis(dimethyl amino) .... [4-(CH₃)₂NC₆H₄]₂CS	284.43	pl		204			bz, chl, aa	B14³, 233
13845	Thiocarbamoyl chloride, N,N-diethyl .... (C₂H₅)₂NCSCl	151.66	pr	108¹⁰	48-51				B4⁴, 389
13846	Thiocarbamoylchloride, N,N-dimethyl .... (CH₃)₂NCSCl	123.61	pr	98¹⁰	42-3			eth, chl	B4³, 147
13847	Thiochrome .... C₁₂H₁₄N₄OS	262.34	ye pr (chl)	sub, vac	227-8			w, MeOH	C29, 6242
13848	Thiocyanuric acid or Trithiocyanuric acid .... C₃S₃N₃H₃	177.27	ye pr		200d				B26, 256
13849	Thiodiglycolic acid .... S(CH₂CO₂H)₂	150.15	cr (AcOEt- bz, w)		129			w, al	B3⁴, 612
13850	Thioglycolamide, N-β-naphthyl .... HSCH₂CONH-(β-C₁₀H₇)	217.29	nd		111-2			al	C29, 3330
13851	Thioglycolic acid .... HSCH₂CO₂H	92.12	....	120²⁰	−165	1.3253²⁰	1.5030²⁰	**w, al, eth**	B3⁴, 600
13852	Thioglycolic acid, acetate or Acetylthioglycolic acid..... CH₃COSCH₂CO₂H	134.18	ye	158-9¹⁷				w	B3⁴, 610
13853	Thioindigo .... C₁₆H₈O₂S₂	296.38	br-red nd (xyl), red mcl nd (bz)	sub	359			xyl	B19⁴, 2091
13854	Thiolane or Tetrahydrothiophene .... C₄H₈S	88.17	....	121.1, 14.5¹⁰	−96.2	0.9987²⁰/⁴	1.5048²⁰	al, eth, ace, bz, chl	B17⁴, 34
13855	Thiolane, 1,1-dioxide C₄H₈SO₂	120.17	....	285	27		1.4840²⁰	chl	B17⁴, 37
13856	Thiolane, 2-methyl 2-CH₃(C₄H₇S)	102.19	....	132.5⁷⁵⁰		0.9541²⁰/⁴	1.4900²⁰	al, eth, ace, bz, chl	B17⁴, 62
13857	Thiolane, 3-methyl 3-CH₃(C₄H₇S)	102.19	....	137⁷⁴²		0.9625²⁰/⁴	1.4917²⁰	al, eth, ace, bz, chl	B17⁴, 64
13858	Thiomorpholine or 1,4-Thiazan. C₄H₈S₂	103.19	....	174⁷⁴⁶, 110¹⁰⁰		1.0882²⁰/⁴	1.5386²⁰	w, al, eth, ace, bz	B27², 4
13859	Thionamide, tetramethyl .... [(CH₃)₂N]₂SO	136.22	....	209, 70¹⁶	31			eth	B4⁴, 269
13860	Thiomin hydrochloride or 7-Amino-3-imino-3H-2-phenothiazine hydrochloride Lauth's violet .... C₁₂H₁₀N₃S·HCl	276.77	dk br or gr pl or nd					bz, chl	B27², 447
13861	Thiophene or Thiofuran........... C₄H₄S	84.14	....	84.2	−38.2	1.0649²⁰/⁴	1.5289²⁰	**al, eth, ace, bz, Py**	B17⁴, 234
13862	Thiophene, acetamido .... 2-CH₃CONH(C₄H₃S)	141.19	lf (w)		161-2			al, ace	B17¹, 136
13863	Thiophene, 2-acetyl or Methyl-2-thienyl ketone 2-CH₃CO(C₄H₃S)	126.18	....	213.5, 94-6¹³	10-11	1.1679²⁰/⁴	1.5667²⁰	al, eth	B17⁴, 4507
13864	Thiophene, 2-acetyl-5-bromo 5-Br-2-CH₃CO(C₄H₂S)	205.08	nd (al)	105-7⁴·⁵	94-5				B17⁴, 4512
13865	Thiophene, 2-acetyl-5-chloro 5-Cl-2-CH₃CO(C₄H₂S)	160.62	ta (al, eth)	88-9⁴·⁵	52			w, al, eth	B17⁴, 4510
13866	Thiophene, 2-acetyl-3-hydroxy 2-CH₃CO-3-HO(C₄H₂S)	142.17	....	47-9⁰·²	51.5- 2.5		1.5795²⁰	al	TETRA21, 33
13867	Thiophene, 2-acetyl-5-methyl 2-CH₃CO-5-CH₃(C₄H₂S)	140.20	....	232-3, 98-100⁸	27-8	1.1185²⁵/⁴	1.5604	eth, ace, bz	B17⁴, 4550
13868	Thiophene, 2-amino or Thiophenine........... 2-H₂N(C₄H₃S)	99.16	pa ye tab (al)	77-9¹¹				w, al	B17², 296
13869	Thiophene, 2-benzoyl or Phenyl 2-thienyl ketone (C₄H₃S)COC₆H₅	188.26	nd (dil al)	300	56-7	1.1890⁵⁴/⁴	1.6181⁵⁴	al, eth, ace	B17⁴, 5187
13870	Thiophene, 3-benzoyl or Phenyl 3-thienyl ketone .... 3-C₆H₅CO(C₄H₃S)	188.24	....	129-30³	63-4			al, eth	B17⁴, 5193

No.	Name, Synonyms, and Formula	Mol. wt.	Color, crystalline form, specific rotation and λ_{max} (log ε)	b.p. °C	m.p. °C	Density	n_D	Solubility	Ref.
13871	Thiophene, 2-bromo 2-Br-(C₄H₃S)	163.04	....	149.51, 42-6[13]	......	1.684[20/4]	1.5868[20]	eth, ace	B17[4], 245
13872	Thiophene, 2-bromo-5-chloro 5-Cl-2-Br-(C₄H₂S)	197.49	....	70[18]	−20	1.803[25/25]	1.5925[25]	eth, ace	B17[4], 246
13873	Thiophene, 2-bromo-5-iodo 5-I-2-Br-(C₄H₂O)	288.94	....	116[13]	......	....	....	....	B17[4], 252
13874	Thiophene, 2-bromo-3-methyl 2-Br-3-CH₃(C₄H₂S)	177.07	....	175[729], 27[1.8]	......	1.5844[18/4]	1.5714[20]	eth, bz	B17[4], 279
13875	Thiophene, 2-bromo-5-methyl 2-Br-5-CH₃(C₄H₂S)	177.07	col to pa ye	177[740], 29[1.8]	......	1.5529[20/4]	1.5673[20]	eth, bz	B17[4], 272
13876	Thiophene, 3-bromo 3-Br(C₄H₃S)	163.03	....	159-60, 66-8[31]	......	1.735[20/4]	1.5919[20]	ace, bz	B17[4], 245
13877	Thiophene, 2-butyl 2-C₄H₉(C₄H₃S)	140.24	....	181-2	......	0.9537[20/4]	1.5090[20]	....	B17[4], 305
13878	Thiophene, 2-chloro 2-Cl(C₄H₃S)	118.59	....	128.3	−71.9	1.2863[20/4]	1.5487[20]	al, eth	B17[4], 241
13879	Thiophene, 2-chloro-5-butyl 2-Cl-5-C₄H₉(C₄H₂S)	174.69	....	117-8[88]	......	1.0842[17/4]	1.5162[20]	bz	B17, 44
13880	Thiophene, 2-chloro-5-iodo 2-Cl-5-I(C₄H₂S)	244.48	....	95-6[14]	−25	....	....	....	B17[4], 252
13881	Thiophene, 2-chloro-5-methyl 2-Cl-5-CH₃(C₄H₂S)	132.61	....	154-5[742], 55[19]	......	1.2147[25/4]	1.5372[20]	al, eth, ace, bz	B17[4], 271
13882	Thiophene, 2,5-dibromo 2,5-Br₂(C₄H₂S)	241.94	....	210.3, 76-80[10]	−6	2.147[23/23]	1.6288[20]	al, eth	B17[4], 248
13883	Thiophene, 2,5-dibromo-3,4-dinitro 3,4-(NO₂)₂-2,5-Br₂(C₄S)	331.94	cr (al)	....	139-40	....	....	....	B17[4], 261
13884	Thiophene, 2,3-dichloro 2,3-Cl₂(C₄H₂S)	153.03	....	173-4	−26.2	1.4605[20/4]	1.5651[20]	al, eth, ace, bz	B17[4], 242
13885	Thiophene, 2,4-dichloro 2,4-Cl₂(C₄H₂S)	153.03	....	167.6	−34	1.4553[20/4]	1.5660[20]	al, eth, ace, bz	B17[4], 242
13886	Thiophene, 2,5-dichloro 2,5-Cl₂(C₄H₂S)	153.03	....	162	−40.5	1.4422[20/4]	1.5626[20]	al, eth	B17[4], 243
13887	Thiophene, 2,5-diiodo 2,5-I₂(C₄H₂S)	335.93	lf (al)	139-40[15]	41-2	....	....	al	B17[4], 253
13888	Thiophene, 2,3-dimethyl or 2,3-Thioxene 2,3-(CH₃)₂(C₄H₂S)	112.21	....	141.6, 29.3[10]	−49	1.0021[20/4]	1.5192[20]	al, eth, bz	B17[4], 287
13889	Thiophene, 2,4-dimethyl or 2,4-Thioxene 2,4(CH₃)₂(C₄H₂S)	112.21	....	140.7, 29.9[10]	......	0.9956[20/20]	1.5104[20]	al, eth, bz	B17[4], 288
13890	Thiophene, 2,5-dimethyl or α,α'-Thioxene 2,5-(CH₃)₂(C₄H₂S)	112.21	....	136.7, 26.2[10]	−62.6	0.985[20/4]	1.5129[20]	al, eth, bz	B17[4], 290
13891	Thiophene, 3,4-dimethyl 3,4-(CH₃)₂(C₄H₂S)	112.19	....	144-6[762], 70-1[55]	......	0.994[25/15]	1.5206[20]	al, eth	B17[4], 293
13892	Thiophene, 2,5-dinitro 2,5-(NO₂)₂C₄H₂S)	174.14	(i) ye lf (al) (ii) ye nd (al, w)	290	(i)52 (ii)80-2	....	....	al, eth	B17[4], 260
13893	Thiophene, 2,4-diphenyl 2,4-(C₆H₅)₂(C₄H₂S)	236.33	....	....	119-20	....	....	al, ace, chl	B17[4], 680
13894	Thiophene, 2-ethyl 2-C₂H₅(C₄H₃S)	112.21	....	104, 24[10]	......	0.9930[20/4]	1.5122[20]	al, eth	B17[4], 285
13895	Thiophene, 3-ethyl 3-C₂H₅(C₄H₃S)	112.21	....	136, 26[10]	−89.1	0.9980[20/4]	1.5146[20]	al, eth	B17[4], 286
13896	Thiophene, 2-hydroxy 2-HO(C₄H₃S)	100.13	....	217-9, 91-3[13]	......	1.255[20/4]	1.5644[20]	w, ace, chl	B17[4], 4286
13897	Thiophene, 2-hydroxy-5-methyl or 2,5-Thiotenol 5-CH₃-2-HO(C₄H₂S)	114.17	....	85[40]	......	....	....	al, eth	B17[4], 4301
13898	Thiophene, 2-hydroxymethyl 2-HOCH₂(C₄H₃S)	114.17	....	207, 96[12]	......	1.2053[16/4]	1.5280[20]	al, eth, ace	B17[4], 1242
13899	Thiophene, 2-iodo 2-I-(C₄H₃S)	210.04	....	180-2, 73[15]	−40	....	1.6465[25]	al, eth	B17[4], 250
13900	Thiophene, 2-iodo-5-nitro 5-NO₂-2-I(C₄H₂S)	255.04	ye pr (al)	....	74	....	....	al	B17[4], 259
13901	Thiophene, 2-methyl or α-Thiololene 2-CH₃(C₄H₃S)	98.17	....	112.6, 9.2[10]	−63.4	1.0193[20/4]	1.5203[20]	al, eth, ace, bz	B17[4], 269
13902	Thiophene, 3-methyl or β-Thiololene 3-CH₃(C₄H₃S)	98.17	....	115.4, 11[10]	−69	1.0218[20/4]	1.5204[20]	al, eth, ace, bz, chl	B17[4], 277
13903	Thiophene, 2-methyl-5-phenyl 2-CH₃-5-C₆H₅(C₄H₂S)	174.27	nd	270-2	51	....	....	al, eth, lig	B17[4], 551

No.	Name, Synonyms, and Formula	Mol. wt.	Color, crystalline form, specific rotation and λ_{max} (log ε)	b.p. °C	m.p. °C	Density	n_D	Solubility	Ref.
13904	Thiophene, 2-methylamino or N-Methyl-2-thiophenine 2-CH$_3$NH(C$_4$H$_3$S)	113.10	....	88-92[15]	....	....	....	eth, ace	B17[1], 136
13905	Thiophene, 2-nitro .... 2-O$_2$N(C$_4$H$_3$S)	129.14	lf ye mcl nd (peth)	224-5	46.5	1.3644[43/4]	....	al	B17[4], 255
13906	Thiophene, 3-nitro 3-3NO$_2$(C$_4$H$_3$S)	129.13	....	225, 95[12]	78-9	....	....	al, lig	B17[4], 256
13907	Thiophene, 3-nitro-2,4,5-trichloro .... 2,4,5-Cl$_3$-3-O$_2$N(C$_4$S)	232.47	red-ye nd (al)	....	86	....	....	al, eth, bz	B17, 35
13908	Thiophene, 4-nitro-2,3,5-tribromo 2,3,5-Br$_3$-4-O$_2$N(C$_4$S)	365.84	red-ye nd (al)	....	106	....	....	eth	B17, 35
13909	Thiophene, 2-octyl 2-C$_8$H$_{17}$(C$_4$H$_3$S)	196.35	....	257-9	....	0.8118[20.5 / 20.5]	....	eth	B17[4], 335
13910	Thiophene, 2-isopentyl 2-i-C$_5$H$_{11}$(C$_4$H$_3$S)	154.27	....	74-5[11]	....	0.9481[12/4]	1.5014[12/587.6]	....	B17[4], 316
13911	Thiophene, 2-propyl 2-C$_3$H$_7$(C$_4$H$_3$S)	126.22	....	157-9, 55.5[20]	....	0.9683[20/4]	1.50757[20]	al, eth, bz	B17[4], 297
13912	Thiophene, 3-propyl 3-C$_3$H$_7$(C$_4$H$_3$S)	126.22	....	163.2	....	0.97377[20/4]	1.5057[20]	al, eth, bz	B17[2], 42
13913	Thiophene, tetrabromo C$_4$Br$_4$S	399.74	nd (al)	326, 170-3[14]	117-8	....	....	eth	B17[4], 250
13914	Thiophene, tetrachloro .... C$_4$Cl$_4$S	221.92	nd (dil al)	233.4, 75-7[2]	30-1	1.7036[30/4]	1.5915[30]	al, eth	B17[4], 244
13915	Thiophene, tetrahydro or Thiolane C$_4$H$_8$S	88.17	....	121.1, 14-5[10]	−96.6	0.9987[20/4]	1.5018[20]	al, eth, ace, bz, chl	B17[4], 34
13916	Thiophene, tetraphenyl (C$_6$H$_5$)$_4$(C$_4$S)	388.53	....	400	189-90	....	....	eth, bz	B17[4], 810
13917	Thiophene, 2,3,5-tribromo .... 2,3,5-Br$_3$(C$_4$HS)	320.84	nd (al)	260	29	....	....	al, eth	B17[4], 249
13918	Thiophene, 2,3,5-trichloro 2,3,5-Cl$_3$(C$_4$HS)	187.48	....	198.7	−16.1	1.5856[20/4]	1.5791[20]	al, eth	B17[4], 244
13919	Thiophene, 2,3,5-trimethyl 2,3,5-(CH$_3$)$_3$(C$_4$HS)	126.22	....	164.5, 46.8[10]	....	0.9753[20/4]	1.5112[20]	al, eth, ace, bz	B17[4], 301
13920	2-Thiophenacetic acid 2-HO$_2$C(C$_4$H$_3$S)	142.18	cr (w)	....	76	....	....	al, eth	B18[4], 4062
13921	2-Thiophene carboxaldehyde or 2-Thiophenealdehyde.... 2-OHC(C$_4$H$_3$S)	112.15	pa ye liq	197, 85-6[16]	....	1.215[21/21]	1.5920[20]	al, eth	B17[4], 4477
13922	2-Thiophene carboxaldehyde oxime or 2-Thiophenaldoxime 2-(C$_4$H$_3$S)CH=NOH	127.17	nd	....	133	....	....	eth	B17[4], 4482
13923	2-Thiophene carboxaldehyde phenylhydrazone .... 2-(C$_4$H$_3$S)CH=NNHC$_6$H$_5$	202.28	ye nd (al)	....	134.5	....	....	al	B17[4], 4482
13924	2-Thiophene carboxaldehyde, 5-bromo .... 5-Br-(C$_4$H$_2$S)-CHO-2	191.04	....	105-7[11]	....	....	1.6378[20]	....	B17[4], 4487
13925	2-Thiophene carboxaldehyde, 5-chloro 5-Cl-(C$_4$H$_2$S)CHO-2	146.59	....	77.5[5]	....	....	1.6036[25]	....	B17[4], 4485
13926	2-Thiophene carboxaldehyde, 5-methyl 5-CH$_3$-(C$_4$H$_2$S)CHO-2	126.17	....	114[25]	....	....	1.5825[20]	chl	B17[4], 4529
13927	3-Thiophene carboxaldehyde 3-(C$_4$H$_3$S)CHO	112.15	....	86-7[20]	....	....	1.5855[20]	al, eth	B17[4], 4497
13928	2-Thiophene carboxylic acid or α-Thiophenic acid .... 2-(C$_4$H$_3$S)CO$_2$H	128.15	nd (w)	260d	129-30	....	....	w, al, eth, chl	B18[4], 4011
13929	2-Thiophene carboxylic acid, ethyl ester or Ethyl-α-thiophene carboxylate .... 2-(C$_4$H$_3$S)CO$_2$C$_2$H$_5$	156.22	....	218, 94[10]	....	1.1623[16/4]	1.5248[20]	al, ace	B18[4], 4012
13930	2-Thiophene carboxylic acid hydrazide 2-(H$_2$NNHCO)(C$_4$H$_3$S)	142.18	....	....	136	....	....	....	B18[4], 4024
13931	2-Thiophene carboxylic acid, 3-methyl .... 3-CH$_3$(C$_4$H$_3$S)CO$_2$H-2	142.18	nd (dil al, w)	....	144	....	....	al, eth, ace	B18[4], 4068
13932	2-Thiophene carboxylic acid, 5-methyl 5-CH$_3$(C$_4$H$_3$S)CO$_2$H-2	142.17	....	....	138-9	....	....	al, eth	B18[4], 4086
13933	2-Thiophene (carboxylyl) chloride 2-(C$_4$H$_3$S)COCl	146.60	....	201, 77[10]	....	....	....	....	B18[4], 4016
13934	3-Thiophene carboxylic acid .... 3-HO$_2$C(C$_4$H$_3$S)	128.15	....	....	138.4	....	....	w	B18[4], 4054
13935	2,3-Thiophene dicarboxylic acid .... 2,3-(C$_4$H$_2$S)(CO$_2$H)$_2$	172.16	pr or nd (w)	....	272-4	....	....	eth	B18[4], 4478
13936	2,4-Thiophene dicarboxylic acid 2,4-(C$_4$H$_2$S)(CO$_2$H)$_2$	172.16	cr	sub>200	280d	....	....	....	B18[4], 4480

No.	Name, Synonyms, and Formula	Mol. wt.	Color, crystalline form, specific rotation and λ_{max} (log ε)	b.p. °C	m.p. °C	Density	n_D	Solubility	Ref.
13937	2,5-Thiophene dicarboxylic acid . . . . 2,5-(C₄H₂S)(CO₂H)₂	172.16	. . . .	sub 150-300	359 (sealed tube)	. . . .		al, eth	B18[4], 4495
13938	2,5-Thiophene dicarboxylic acid, diethyl ester or Diethyl-2,5-thiophene dicarboxylate . . . . 2,5-(C₄H₂S)(CO₂C₂H₅)₂	228.27	nd (al)	. . . .	51.5			al	B18, 331
13939	2-Thiophene sulfonamide . . . . 2-(C₄H₃S)SO₂NH₂	163.22	nd (w)		147				B18[4], 6706
13940	2-Thiophene sulfonyl chloride . . . . 2-(C₄H₃S)(SO₂Cl)	182.65	. . . .	99-101[h], sub	28			eth	B18[4], 6706
13941	3-Thiophene sulfonamide . . . . 3-(C₄H₃S)(SO₂NH₂)	163.22	pl (w)		152-3				B18[4], 6714
13942	3-Thiophene sulfonyl chloride . . . . 3-(C₄H₃S)(SO₂Cl)	182.65	cr (eth)	98-9[0.5]	43			eth	B18[4], 6713
13943	Thiophenol or Mercaptobenzene . . . . C₆H₅SH	110.18	. . . .	168.7, 46.4[10]	-14.8	1.0766[20/4]	1.5893[20]	al, eth, bz	B6[4], 1463
13944	Thiophenol, o-amino 2H₂NC₆H₄SH	125.19	nd	234, 125-7[h]	26		1.4606[20]	al, eth	B13[3], 902
13945	Thiophenol, m-amino 3-H₂N-C₆H₄SH	125.19	pa ye oil	180-90[16]				al, aa, eth	B13[3], 982
13946	Thiophenyl, m-amino, hydrochloride 3-H₂NC₆H₄SH.HCl	161.66	. . . .	sub	232			w, al	B13, 425
13947	Thiophenol, p-amino 4-H₂NC₆H₄SH	125.19	cr	140-5[15]	46			w, al	B13[3], 1221
13948	Thiophenol, p-bromo 4-BrC₆H₄SH	189.08	lf (al)	230-1	73			eth, chl	B6[1], 1650
13949	Thiophenol, o-chloro 2-ClC₆H₄SH	144.82	. . . .	205-6, 117[11]	. . . .	1.2752[10]	. . . .	. . . .	B6[4], 1570
13950	Thiophenol, m-chloro 3-ClC₆H₄SH	144.62		205-7		1.2637[13]		al, eth, peth, chl	B6[4], 1576
13951	Thiophenol, p-chloro 4-ClC₆H₄SH	144.62		205.7	61	1.1911[20/4]	1.5480[20]	al, eth	B6[4], 1581
13952	Thiophenyl, m-ethoxy 3-C₂H₅OC₆H₄SH	154.23		238-9				al, eth, ace, bz	B6, 833
13953	Thiophenol, p-ethoxy 4-C₂H₅OC₆H₄SH	154.23		238	1.6			al, eth, ace, bz	B6[3], 4446
13954	Thiophenol, m-methoxy 3-CH₃OC₆H₄SH	140.20		223-6			1.5874[20]	chl	B6[3], 4363
13955	Thiophenol, p-methoxy or 4-Mercaptoanisole 4-CH₃OC₆H₄SH	140.22		227-9, 89-90[5]		1.1313[25/4]	1.5801[25]	al, eth, bz	B6[3], 4445
13956	Thiophenol, o-nitro 2-NO₂C₆H₄SH	155.17		. . . .	58.5			al, eth, bz	B6[4], 1661
13957	Thiophenol, p-nitro 4-NO₂C₆H₄SH	155.18	(eth, chl, ace)		79			al, eth, ace	B6[4], 1687
13958	Thiophenol, 2,4,6-trinitro or Thiopicric acid . . . . 2,4,6-(NO₂)₃C₆H₂SH	245.17	ye nd		114			al, eth, ace, bz, chl	B6[2], 316
13959	Thiophosgene or Thiocarbonyl chloride. . . . . CSCl₂	144.98	red	73		1.508[15]	1.5442[20]	al, eth	B3[4], 281
13960	Thiophthene (solid) or Thieno[3,2-b]thiophene. . . . . C₆H₄S₂	140.23	bipym orh (lig)	221-4	56			eth	B19[4], 189
13961	Thiophthene (liquid) or Thieno[3,2-b]thiophene. . . . . C₆H₄S₂	140.23		224-6, 106[16]	6.5			eth	B19[4], 189
13962	Thiopyran, tetrahydro or Pentamethylene sulfide. . . . . . . C₅H₁₀S	102.20		141.7, 93[82]	19	0.9861[20/4]	1.5067[20]	al, eth, ace, bz	B17[4], 55
13963	Thiopyrine or 1,5-Dimethyl-2-phenyl-3-thio-3-pyrazolone C₁₁H₁₂N₂S	204.30	cr (w)	. . . .	166			al, eth	B24[2], 28
13964	Thiosalicylic acid or 2-Mercaptobenzoic acid. 2-HSC₆H₄CO₂H	154.19	lf or nd (al, w, aa)	sub	168-9			al, eth, aa	B10[3], 265
13965	Thiosemicarbazide or Semicarbazide-3-thio. . . . . H₂NNHCSNH₂	91.4	lo nd (w)	. . . .	183			w, al	B3[4], 374
13966	2-Thiouracil . . . . C₄H₄N₂OS	128.15	pr (w, al)		>340d				B24[2], 171
13967	2-Thiouracil, 6-methyl . . . . C₅H₆N₂OS	142.18	pl (w)		299-303d				B24[2], 183
13968	2-Thiouracil, 6-propyl . . . . C₇H₁₀N₂OS	170.24	pw (w)		219				C43, 674

No.	Name, Synonyms, and Formula	Mol. wt.	Color, crystalline form, specific rotation and λ_{max} (log ϵ)	b.p. °C	m.p. °C	Density	n_D	Solubility	Ref.
13969	4-Thiouracil $C_4H_4N_2OS$	128.15	yesh nd or pr (w)		328d				B24, 323
13970	4-Thiouracil, 6-methyl $C_5H_6N_2OS$	142.18	ye pr (w)		>250d				B24, 352
13971	Thiourea H_2NCSNH_2	76.12	rh (al)		182	1.405		w, al	B3⁴, 342
13972	Thiourea, 1-acetyl $CH_3CONHCSNH_2$	118.16	pr (w), rh (al)		165			al	B3⁴, 354
13973	Thiourea, S-acetyl, hydrochloride $H_2NC(=NH)SCOCH_3.HCl$	154.62	pl		109d			w	B3³, 314
13974	Thiourea, 1-allyl $(CH_2=CHCH_2)NHCSNH_2$	116.20	mcl or rh, pr (w)		(i)71, (ii)78.4	1.219²⁰′²⁰	1.5936⁷⁸	w, al	B4⁴, 1072
13975	Thiourea, 1-allyl-3-phenyl $(CH_2=CHCH_2)NHCSNHC_6H_5$	192.28			98				B12³, 855
13976	Thiourea, 1-(4-aminobenzene sulfonyl) or 1-Sulfanyl thiourea $(4-H_2NC_6H_4SO_2)NHCSNH_2$	231.30	pw, pl (w)		182				B14³, 1975
13977	Thiourea, 1-benzoyl $C_6H_5CONHCSNH_2$	180.24	pr (dil al)		171			al	B9³, 1120
13978	Thiourea, 1-benzyl $C_6H_5CH_2NHCSNH_2$	166.25	pr (w)		164-5				B12³, 2277
13979	Thiourea, 1-benzyl-3-phenyl $C_6H_5CH_2NHCSNHC_6H_5$	242.34			153-4			al, eth	B12², 564
13980	Thiourea, 1-butyl-3-phenyl $C_4H_9NHCONHC_6H_5$	208.33	pr		85			bz	B12³, 853
13981	Thiourea, 1-(2-chlorophenyl) $2-ClC_6H_4NHCSNH_2$	186.66	nd or pl		146			al, bz	B12³, 1296
13982	Thiourea, 1-(4-chlorophenyl) $4-ClC_6H_4NHCSNH_2$	186.66	pl or nd (al)		178			al, bz	B12³, 1368
13983	Thiourea, 1,3-dibenzyl $(C_6H_5CH_2NH)_2CS$	256.36			147-8			al, eth	B12³, 2278
13984	Thiourea, 1,3-diethyl $(C_2H_5NH)_2CS$	132.23		d	144			w, al, eth	B4⁴, 375
13985	Thiourea, 1,3-diethyl-1,3-diphenyl $[C_6H_5N(C_2H_5)]_2CS$	284.43	rh pl (lig), pr (al)		75.5			al	B12³, 884
13986	Thiourea, 1,3-dimethyl $(CH_3NH)_2CS$	104.19	dlq pl		62			w, al, ace, chl	B12⁴, 217
13987	Thiourea, 1,3-di(α-naphthyl) $(α-C_{10}H_7NH)_2CS$	328.44	lf (to)		207.5			to	B12², 696
13988	Thiourea, 1,3-di(β-naphthyl) $(β-C_{10}H_7NH)_2CS$	328.44	lf (aa)		203				B12³, 3048
13989	Thiourea, 1,1-diphenyl $(C_6H_5)_2NCSNH_2$	228.32	pr (al)		210d, (218)			al	B12³, 898
13990	Thiourea, 1,3-diphenyl $(C_6H_5NH)_2CS$	228.32	lf (al)		154, (189)			al, eth, chl	B12³, 858
13991	Thiourea, 1,1-dipropyl $(C_3H_7)_2NCSNH_2$	160.28			67				B4, 144
13992	Thiourea, 1,3-dipropyl $(C_3H_7NH)_2CS$	160.28	lf (w)		71			eth	B4, 143
13993	Thiourea, 1,3-diisopropyl $(i-C_3H_7NH)_2CO$	160.28			141			chl	B4⁴, 526
13994	Thiourea, 1,3-di(2-tolyl) $(2-CH_3C_6H_4NH)_2CS$	256.37	nd (al, sub)	216-8 (sub)	165-6			al, bz, chl, aa	B12³, 1881
13995	Thiourea, 1,3-di(4-tolyl) $(4-CH_3C_6H_4NH)_2CS$	256.37	rh bipym pr		176			eth	B12³, 2099
13996	Thiourea, 1-ethyl-3-phenyl $C_2H_5NHCSNHC_6H_5$	180.27			107			al, bz	B12³, 853
13997	Thiourea, ethylidene $CH_3CH=NCSNH_2$	102.16	(al)		ca 212d				B3¹, 76
13998	Thiourea, S-methyl $HN=C(SCH_3)NH_2$	90.15	lf (ace)		79			w, al, ace	B3⁴, 358
13999	Thiourea, S-methyl, hydroiodide or S-Methyl isothiouronium iodide $HN=C(SCH_3)NH_2.HI$	218.06	pr		117			w, al	B3¹, 78
14000	Thiourea, S-methyl, nitrate or S-Methyl isothiouronium nitrate $HN=C(SCH_3)NH_2.HNO_3$	153.16			109-10			w, al	B3¹, 78

No.	Name, Synonyms, and Formula	Mol. wt.	Color, crystalline form, specific rotation and λ_{max} (log ε)	b.p. °C	m.p. °C	Density	n_D	Solubility	Ref.
14001	Thiourea, S-methyl, sulfate or S-Methyl isothiouronium sulfate. [HN=C(SCH₃)NH₂]₂.H₂SO₄	278.37	nd (al, w)	....	244d			....	B3⁴, 358
14002	Thiourea, S-methyl-1,3-diphenyl C₆H₅N=C(SCH₃)NHC₆H₅	242.35	nd (al)		109-10			al	B12³, 911
14003	Thiourea, 1-methyl CH₃NHCSNH₂	90.15	pr		120-1			w, al, ace	B4⁴, 216
14004	Thiourea, 1-methyl-3-(α-naphthyl) α-C₁₀H₇NHCSNHCH₃	216.31	pl (al)		198			al	B12², 696
14005	Thiourea, 1-methyl-3-phenyl CH₃NHCSNHC₆H₅	166.25	ta, pl		112-3			al	B12³, 853
14006	Thiourea, 1-(α-naphthyl) α-C₁₀H₇NHCSNH₂	202.28	pr (al)		198				B12³, 2941
14007	Thiourea, 1-(α-naphthyl)-3-phenyl α-C₁₀H₇NHCSNHC₆H₅	278.38	lf		162-3				B12, 1241
14008	Thiourea, 1-β-naphthyl β-C₁₀H₇NHCSNH₂	202.28	lf (al)		186			al	B12³, 3045
14009	Thiourea, S-phenyl HN=C(SC₆H₅)NH₂	152.23	nd (bz)		96-7d			bz	B6¹, 146
14010	Thiourea, 1-phenyl C₆H₅NHCSNH₂	152.23	nd (w), pr (al)		154			al	B12³, 852
14011	Thiourea, 1,1,3,3,-tetramethyl [(CH₃)₂N]₂CS	132.23	....	245	78-9			w, al	B4⁴, 232
14012	Thiourea, 1,1,3,3-tetraphenyl [(C₆H₅)₂N]₂CS	300.52	nd (al, MeOH)	....	191 f	....	....	eth, bz	B12⁷, 808
14013	Thiourea, 1-(2-tolyl) (2-CH₃C₆H₄)NHCSNH₂	166.25	nd (dil al, w)		162			w, al	B12³, 1878
14014	Thiourea, 1-(3-tolyl) (3-CH₃C₆H₄)NHCSNH₂	166.25	pr (al)	....	110-1			w	B12⁷, 1915
14015	Thiourea, 1-(4-tolyl) (4-CH₃C₆H₄)NHCSNH₂	166.25	pl (al)		188-9			al	B12³, 2097
14016	Thiourea, 1,1,3-trimethyl (CH₃)₂NCSNHCH₃	118.20	pr (bz-lig)		87-8			w, al, bz, chl	B4⁴, 232
14017	Thiourea, 1,1,3-triphenyl (C₆H₅)₂NCSNHC₆H₅	304.42	nd (al)		152				B12, 432
14018	Thioxanthene or Dibenzthiopyran C₁₃H₁₀S	198.29	nd (al-chl)	340⁷³⁰, sub	128-9			chl, peth	B17⁴, 615
14019	Thioxanthone or 9-Oxothioxanthene C₁₃H₈OS	212.28	ye nd (chl)	371-3⁷¹⁵, sub	209 (212)			bz, aa	B17⁴, 5303
14020	D-Threonic acid or 2,3,4-Trihydroxy butyric acid HOCH₂(CHOH)₂CO₂H	136.12	nd (al), $[\alpha]_D$ −30 (w)		197-8			w	B3⁴, 1112
14021	DL-Threonic acid HOCH₂(CHOH)₂CO₂H	136.12	cr (al, ace)		99			w	B3³, 895
14022	L-Threonic acid HOCH₂(CHOH)₂CO₂H	136.12	nd (al-eth), $[\alpha]_D$ +9.5 (w)		169-70			w, ace	B3⁴, 1112
14023	Threonine (D) or D-2-Amino-3-hydroxy butyric acid CH₃CH(OH)CH(NH₂)CO₂H	119.12	cr (80% al), $[\alpha]_D^{26}$ −28.3 (w, c=1.1)	....	255-7d			w	B4³, 1625
14024	Threonine (DL) or DL-2-Amino-3-hydroxy butyric acid CH₃CH(OH)CH(NH₂)CO₂H	119.12	cr (dil al)	....	234-5d, 229-30 (+ ½ w)			w	B4³, 1625
14025	D-Threose or D-Trihydroxy butyraldehyde HOCH₂(CHOH)₂CHO	120.12	hyg syr or nd(w), $[\alpha]_D^{22}$ +29.1→ 19.6 (w, mut)	....	126-32			w	B1⁴, 4173
14026	Thujane (d) or Sabinane C₁₀H₁₈	138.26	$[\alpha]_D$ +73.1	157⁷⁵⁸	......	0.8139²⁰/⁴	1.4376²⁰	al, eth, ace, bz	B5⁴, 317
14027	3-Thujene C₁₀H₁₆	136.24	$[\alpha]'_D$ −37.20	151			0.8301²⁰/⁴	1.4515²⁰	....	B5⁴, 451
14028	4(10)-Thujene (d) or Sabinene C₁₀H₁₆	136.24	$[\alpha]_D$ +80.2	163-5, 66³⁰		0.842²⁰	1.4678²⁰	al, eth, ace, bz, chl	B5⁴, 451
14029	4(10)Thujene (l) or l-Sabinene C₁₀H₁₆	136.24	$[\alpha]^{15}_D$ −42.5	163-5		0.8464²⁰/⁴	1.4515²⁰	al, eth, ace, bz, chl	B5⁴, 452

No.	Name, Synonyms, and Formula	Mol. wt.	Color, crystalline form, specific rotation and λ_{max} (log ε)	b.p. °C	m.p. °C	Density	n_D	Solubility	Ref.
14030	3-Thujone $C_{10}H_{16}O$	152.24	[α][18/D] −19.9	200-1, 75[9]	......	0.9152[20]	1.4490[25]	**al, eth**, ace	B7[3], 379
14031	Thymidine or Thymine-β-D-2-desoxy-riboside.......... $C_{10}H_{14}N_2O_5$	242.23	nd (AcOEt), [α][25/D] + 30.6 (w, c=1.03)		186-7			w, MeOH, aa, Py	C48, 226
14032	Thymol or 2-Methyl-5-isopropyl phenol............... 2-CH_3-5-i-$C_3H_7C_6H_3OH$	150.22	pl (aa, ace)	233, 92[2]	52	0.925[80/4]	1.5227[20]	al, eth	B6[3], 1893
14033	Thymol acetate .... 5-CH_3-2-$(CH_3)_2CHC_6H_3O_2CCH_3$	192.24	....	245, 131[21]		1.009[9]		**al, eth**, bz, chl	B6[3], 1901
14034	Thymol, 4-amino or 2-Amino-3-methyl-6-iso-propyl phenol.......... 2-H_2N-3-CH_3-6-i-C_3H_7-C_6H_2OH	165.24	nd, sc (bz)	....	178-9			al, bz	B13[3], 1803
14035	Thymol carbonate or Thymatol Tyranol............... 2-CH_3-5-i-$C_3H_7C_6H_3OH.H_2CO_3$	326.44	nd or pr		49			eth, chl	B6[2], 499
14036	Thymol, 6-chloro or 4-Chloro-2-iso-propyl-5-methyl phenol......... 4-Cl-2-i-C_3H_7-5-$CH_3C_6H_2OH$	184.67	nd or pl (lig)	259-63	59-61			w, al, eth, bz, peth	B6[3], 1906
14037	Thymol, 2,6-dinitro or 2,4-Dinitro-3-methyl-6-iso-propyl phenol..... 2,4(NO_2)₂-3-CH_3-6-[($CH_3)_2$CH]C_6HOH	240.22	pr (peth)	....	55.5			al, eth	B6[3], 1911
14038	α-Thymolsulfonic acid or 4-Hydroxy-5-iso-propyl-2 methyl benzene sulfonic acid.......... 4-HO-5$(CH_3)_2$CH-2-$CH_3C_6H_2SO_3H$	230.29	pl (+ 1w)	....	91-2			w	B11[3], 537
14039	Thymol blue or Thymolsulfonephthalein.............. $C_{27}H_{30}O_5S$	466.60	gr-red (al, eth, aa)		221-4d			al, aa	B19[4], 1135
14040	Thymolphthalein .... $C_{28}H_{30}O_4$	430.55	pr or nd (al)	....	253			eth, ace	B18[4], 1955
14041	Thyroxine (d) $C_{15}H_{11}I_4NO_4$	776.88	nd [α][25/546] + 3 (al-NaOH)		237d			....	B14[3], 1566
14042	Thyroxine-(l) $C_{15}H_{11}I_4NO_4$	776.88	nd [α][20/D] −4.4 (al-NaOH)	....	235-6			....	B14[3], 1566
14043	Tiglaldehyde or 2-Methyl-2-butenal .... $CH_3CH=C(CH_3)CHO$	84.13	....	116-7[738], 63-5[119]	......	0.8710[20/4]	1.4475[20]	w, **al, eth**	B1[4], 3464
14044	Tiglic acid or $trans$-2-Methyl-2-butenoic acid.......... $CH_3CH=C(CH_3)CO_2H$	100.13	ta (w)	198.5	64-5	0.9641[20/4]	1.4330[76]	al, eth	B2[4], 1552
14045	Tiglyl chloride or $trans$-2-Methyl-2-butenyl chloride ..... $CH_3CH=C(CH_3)COCl$	118.56	....	146-8, 45[12]				eth	B2[4], 1554
14046	Tigogenin or 5-α,22α-Spirostan-3β-ol................. $C_{27}H_{44}O_3$	416.65	lf (al + 1w), pr (ace), [α][25/D] −49 (Py, c=1)	....	205-6			eth, ace, MeOH, peth	B19[4], 828
14047	α-Tocopherol or Vitamin E. 5,7,8-Trimethyl tocol ...... $C_{29}H_{50}O_2$	430.69	pa ye visc oil [α][25/D] + 0.65 (al)	350d, 140[10-6]	2-3	....	....	al, eth, ace	B17[4], 1436
14048	α-Tocopherol acetate .... $C_{31}H_{52}O_3$	472.75	....	184[0.01]	−27.5	0.9533[21.5]	1.495-1.497[20]	eth, ace, chl	B17[4], 1439
14049	α-Tocopherolquinone or α-tocoquinone.............. $C_{29}H_{50}O_3$	446.72	ye oil	120[0.002]				eth, peth	Am73, 5148
14050	β-Tocopherol or Vitamin E. 5,8-Dimethyl tocol .... $C_{28}H_{48}O_2$	416.70	pa ye visc oil [α][20/D] + 6.4	200-10[0.1]				**al, eth**, ace, chl	B17[4], 1427
14051	γ-Tocopherol or Vitamin E. 7,8-Dimethyltocol......... $C_{28}H_{48}O_2$	416.70	pa ye visc oil [α][20/D] −2.4 (al)	200-10[0.1]	−3	....	....	**al, eth**, ace, chl	B17[4], 1429
14052	γ-Tocopherol or γ-Methyltocol .... $C_{27}H_{46}O_2$	402.67	pa ye visc oil [α][25/546] + 3.4 (al, c=1.5)	150[0.001]	......	....	....	al, eth, ace, chl	B17[4], 1411

No.	Name, Synonyms, and Formula	Mol. wt.	Color, crystalline form, specific rotation and λ_{max} (log ε)	b.p. °C	m.p. °C	Density	n_D	Solubility	Ref.
14053	Tolbutamide or 1-Butyl-3(p-tolylsulfonyl)urea.......... $C_{12}H_{18}N_2O_3S$	270.35	orth cr	....	128-9	1.245[25]	....	al, eth, chl	C53, 13084
14054	o-Tolualdehyde or 2-Methylbenzaldehyde............. 2-$CH_3C_6H_4CHO$	120.16	....	200, 94[10]	......	1.0386[19/4]	1.5481[20]	al, eth, ace, bz, chl	B7³, 1011
14055	m-Tolualdehyde or 3-Methylbenzaldehyde............. 3-$CH_3C_6H_4CHO$	120.16	....	199, 93-4[17]	......	1.0189[21]	....	al, eth, ace, bz, chl	B7³, 1013
14056	p-Tolualdehyde or 4-Methylbenzaldehyde............. 4-$CH_3C_6H_4CHO$	120.16	....	204-5, 106[10]	......	1.0194[17/4]	1.5454[20]	al, eth, ace, chl	B7³, 1016
14057	o-Toluamide or 2-Methylbenzamide .... 2-$CH_3C_6H_4CONH_2$	135.17	pl or nd (w)	....	147	....	....	al	B9³, 2304
14058	m-Toluamide or 3-Methylbanzamide................ 3-$CH_3C_6H_4CONH_2$	135.17	nd (eth)	....	97	....	....	al	B9³, 2322
14059	p-Toluamide or 4-Methylbenzamide .... 4-$CH_3C_6H_4CONH_2$	135.17	nd or pl (w)	....	160	....	....	al, eth	B9³, 2343
14060	Toluene or Methyl benzene...................... $C_6H_5CH_3$	92.15	....	110.6, 14.5[14.5]	−95	0.8669[20/4]	1.4961[20]	al, eth, ace, bz, lig	B5⁴, 766
14061	Toluene, 2-allyl 2-CH_2=$CHCH_2C_6H_4CH_3$	132.20	....	182-3[757], 88-90[25]	......	0.9005[20/4]	1.5187[20]	....	B5³, 1212
14062	Toluene, 2-amino or o-Toluidine .... 2-$CH_3C_6H_4NH_2$	107.16	....	200.2, 80.1[10]	−14.7	0.9984[20/4]	1.5725[20]	al, eth	B12³, 1837
14063	Toluene, 3-amino or m-Toluidine.................. 3-CH_3-$C_6H_4NH_2$	107.16	....	203.3, 82.3[10]	−30.4	0.9889[20/4]	1.5681[20]	al, eth, ace, bz	B12³, 1949
14064	Toluene, 4-amino or p-Toluidine .... 4-$CH_3C_6H_4NH_2$	107.16	lf (w + 1)	200.5, 79.6[10]	44-5	0.9619[20/4]	1.5534[45], 1.5636[20]	al, eth, ace, Py	B12³, 2017
14065	Toluene, 2-bromo or o-Tolylbromide.............. 2-Br-$C_6H_4CH_3$	171.04	....	181.7	−27.8	1.4232[20/4]	1.5565[20]	al, eth, bz	B5⁴, 825
14066	Toluene, 2-bromo-3-nitro .... 2-Br-3-$NO_2C_6H_3CH_3$	216.04	ye pr (al)	157[22]	41[2]	....	....	al	B5⁴, 860
14067	Toluene, 2-bromo-4-nitro .... 2-Br-4-$NO_2C_6H_3CH_3$	216.04	nd (al)	150-1[20]	78	....	....	eth	B5⁴, 861
14068	Toluene, 2-bromo-5-nitro .... 2-Br-5-$NO_2C_6H_3CH_3$	216.04	cr (al)	140-3[17]	78	....	....	eth	B5⁴, 860
14069	Toluene, 2-bromo-6-nitro .... 2-Br-6-$NO_2C_6H_3CH_3$	216.04	pe ye nd (dil al)	143[22]	42	....	....	al	B53, 751
14070	Toluene, 3-bromo or m-Tolylbromide................ 3-$BrC_6H_4CH_3$	171.04	....	183.7	−39.8	1.4099[20/4]	1.5510[20]	al, eth, ace, chl	B5⁴, 827
14071	Toluene, 3-bromo-2-nitro .... 3-Br-2-$NO_2C_6H_3CH_3$	216.04	pa ye nd	129-30[10]	28	....	....	al, eth	B5⁴, 860
14072	Toluene, 3-bromo-4-nitro .... 3-Br-4-$NO_2C_6H_3CH_3$	216.04	pa ye pr or nd (MeOH)	154-5[20]	37	....	....	al, eth	B5⁴, 861
14073	Toluene, 3-bromo-5-nitro .... 3-Br-5-$NO_2C_6H_3CH_3$	216.04	pa ye nd or pr (MeOH)	269-70	84	....	....	al, eth	B5³, 752
14074	Toluene, 4-bromo or p-Tolylbromide .... 4-Br-$C_6H_4CH_3$	171.04	cr (al)	184.3, 61.9[10]	28.5	1.3995[20/4]	1.5477[20]	al, eth, ace, bz, chl	B5⁴, 827
14075	Toluene, 4-bromo-2-nitro .... 4-Br-2-$NO_2C_6H_3CH_3$	216.04	pa ye nd (dil al)	256-7, 130-2	47	....	....	al, eth	B5⁴, 860
14076	Toluene, 4-bromo-3-nitro .... 4-Br-3-$NO_2C_6H_3CH_3$	216.04	pa ye nd (MeOH)	......	35	....	1.5682[20]	....	B5⁴, 860
14077	Toluene, 5-bromo-2-nitro .... 5-Br-2-$NO_2C_6H_3CH_3$	216.04	cr (al)	267, 143[10]	56	....	....	....	B5⁴, 860
14078	Toluene, 2-chloro or o-Tolylchloride................ 2-$ClC_6H_4CH_3$	126.59	....	159.1, 42.6[10]	−35.1	1.0825[20/4]	1.5268[20]	al, eth, ace bz chl	B5⁴, 805
14079	Toluene, 2-chloro-4-nitro .... 2-Cl-4-$NO_2C_6H_3CH_3$	171.58	nd (al)	260	68	....	1.5470[69]	al, eth	B5⁴, 855
14080	Toluene, 2-chloro-6-nitro .... 2-Cl-6-$NO_2C_6H_3CH_3$	171.58	nd (dil al)	238	37-40	....	1.5377[69]	al	B5⁴, 854
14081	Toluene, 3-chloro or m-Tolyl chloride .... 3-Cl-$C_6H_4CH_3$	126.59	....	162	−47.8	1.0722[20/4]	1.5214[19]	al, eth, bz, chl	B5⁴, 806
14082	Toluene, 3-chloro-4-nitro .... 3-Cl-4-$NO_2C_6H_3CH_3$	171.58	pa ye nd	146[19]	24	....	....	al, eth	B5⁴, 856
14083	Toluene, 3-chloro-5-nitro .... 3-Cl-5-$NO_2C_6H_3CH_3$	171.58	ye nd (al)	....	61	....	1.5404[69]	al	B5⁴, 855
14084	Toluene, 4-chloro or p-Tolyl chloride .... 4-$ClC_6H_4CH_3$	126.59	....	162, 44[10]	7.5	1.0697[20/4]	1.5150[20]	al, eth, chl, aa	B5⁴, 806
14085	Toluene, 4-chloro-3-nitro .... 4-Cl-2-$NO_2C_6H_3CH_3$	171.58	mcl nd	240[720], 115.5[11]	38	1.2559[80]	....	eth	B5⁴, 853

No.	Name, Synonyms, and Formula	Mol. wt.	Color, crystalline form, specific rotation and λ_{max} (log ε)	b.p. °C	m.p. °C	Density	n_D	Solubility	Ref.
14086	Toluene, 4-chloro-3-nitro 4-Cl-3-NO$_2$C$_6$H$_3$CH$_3$	171.58	......	260[745], 118[11]	7	......	1.5572[20]	......	B5[4], 855
14087	Toluene, 5-chloro-2-nitro 5-Cl-2-NO$_2$C$_6$H$_3$CH$_3$	171.58	ye	......	24.9		1.5496[65]		B5[4], 854
14088	Toluene, 2,3-diamino 2,3-(NH$_2$)$_2$C$_6$H$_3$CH$_3$	122.17	cr	255	63-4		......	w, al, eth	B13[3], 277
14089	Toluene, 2,4-diamino 2,4-(NH$_2$)$_2$C$_6$H$_3$CH$_3$	122.17	nd (w), cr (al)	29[2], 148-50[8]	99		......	w, al, eth, bz	B13[3], 278
14090	Toluene, 2,5-diamino 2,5-(NH$_2$)$_2$C$_6$H$_3$CH$_3$	122.17	pl (bz)	273-4	64		......	w, al, eth	B13[3], 284
14091	Toluene, 2,6-diamino 2,6-(NH$_2$)$_2$C$_6$H$_3$CH$_3$	122.17	pr (bz, w)		106			w, al	B13[3], 297
14092	Toluene, 3,4-diamino 3,4-(NH$_2$)$_2$C$_6$H$_3$CH$_3$	122.17	lf (lig)	265 sub	89-90			w	B13[3], 292
14093	Toluene, 3,5-diamino 3,5-(NH$_2$)$_2$C$_6$H$_3$CH$_3$	122.17	......	283-5	<0			w, al, eth	B13[3], 299
14094	Toluene, 3,5-diamino, hydrochloride 3,5-(NH$_2$)$_2$C$_6$H$_3$CH$_3$.2HCl	195.09	nd (dil al)		255-60d			w	B13, 164
14095	Toluene, 2,5-dibromo 2,5-Br$_2$C$_6$H$_3$CH$_3$	249.94		236, 135-6[35]	fr 5.6	1.8127[19]	1.5982[18]	......	B5[4], 835
14096	Toluene, 3,5-dibromo 3,5-Br$_2$C$_6$H$_3$CH$_3$	249.94	nd	246	39			......	B5[4], 836
14097	Toluene, 2,3-dichloro 2,3-Cl$_2$C$_6$H$_3$CH$_3$	161.03		207-8, 61-2[3]		......	1.5511[20]	bz	B5[4], 815
14098	Toluene, 2,4-dichloro 2,4-Cl$_2$C$_6$H$_3$CH$_3$	161.03		196-7	−13.5	1.2498[20/20]	1.5511[20]	......	B5[4], 815
14099	Toluene, 2,5-dichloro 2,5-Cl$_2$C$_6$H$_3$CH$_3$	161.03		200[770]	5	1.2535[20]	1.5449[20]	bz	B5[3], 694
14100	Toluene, 2,6-dichloro 2,6-Cl$_2$C$_6$H$_3$CH$_3$	161.03		198			1.5507[20]	chl	B5[4], 815
14101	Toluene, 3,4-dichloro 3,4-Cl$_2$C$_6$H$_3$CH$_3$	161.03		208.9, 81.8[10]	−15.2	1.2564[20/4]	1.5471[20]	al, eth, ace, bz	B5[4], 815
14102	Toluene, 3,5-dichloro 3,5-Cl$_2$C$_6$H$_3$CH$_3$	161.03		201-2	26			......	B5[4], 816
14103	Toluene, 2,5-diethoxy 2,5-(C$_2$H$_5$O)$_2$C$_6$H$_3$CH$_3$	180.25	nd (lig)	247-9	24-5	1.0134[15]	......	al, eth, bz, chl	B6[3], 4499
14104	Toluene, 2,3-dihydroxy or Isohomocatechol 2,3-(HO)$_2$C$_6$H$_3$CH$_3$	124.15	lf (bz)	241, 127[12]	68			w, al, eth, bz, chl	B6[3], 4492
14105	Toluene, 2,4-dihydroxy or Cresorcinol 2,4-(HO)$_2$C$_6$H$_3$CH$_3$	124.15	cr (bz-peth)	267-70	105-7			w, al, eth	B6[3], 4495
14106	Toluene, 2,5-dihydroxy or Toluhydroquinone 2,5-(HO)$_2$C$_6$H$_3$CH$_3$	124.15	rh pl (bz)	163[11] sub	126-7			w, al, eth	B6[3], 4498
14107	Toluene, 2,5-dihydroxy, diacetate or 2.5-Diacetoxy toluene 2.5-(CH$_3$CO$_2$)$_2$C$_6$H$_3$CH$_3$	208.22	nd (aa), pr (lig), nd or pr (w)	......	52			eth, al, aa	B6[3], 4501
14108	Toluene, 3,4-dihydroxy or 4-Homocatechol 3,4-(HO)$_2$C$_6$H$_3$CH$_3$	124.15	lf (bz-lig), pr (bz)	251, 143-6[26] sub	65	1.1287[74/4]	1.5425[74]	w, al, eth, ace	B6[3], 4514
14109	Toluene, 3,5-dihydroxy or Orcinol 3.5-(HO)$_2$C$_6$H$_3$CH$_3$	124.15	pr (w + l), lf (chl)	289-90, 147[5]	107-8 (anh), 58 (hyd)	1.290[4]	......	w, al, eth, bz	B6[3], 4531
14110	Toluene, 2,4-dimercapto or Dithiocresorcinol 2,4-(HS)$_2$C$_6$H$_3$CH$_3$	156.28	......	263	36.7	......	......	......	B6, 873
14111	Toluene, 3,4-dimercapto 3,4-(HS)$_2$C$_6$H$_3$CH$_3$	156.26			28-30			chl	B6[3], 4530
14112	Toluene, 2,3-dimethoxy or Isohomoveratrol 2,3-(CH$_3$O)$_2$C$_6$H$_3$CH$_3$	152.20	......	202-3, 92-3[18]		1.0335[20/4]	1.5121[25]	al, eth	B6[3], 4493
14113	Toluene, 2,4-dimethoxy 2,4-(CH$_3$O)$_2$C$_6$H$_3$CH$_3$	152.20		211, 110-20[30]		......		al, eth	B6[3], 4496
14114	Toluene, 3,4-dimethoxy or Homoveratrol 3,4-(CH$_3$O)$_2$C$_6$H$_3$CH$_3$	152.20	pr (eth)	219-21, 122-4[27]	24	1.0509[25/4]	1.5257[25]		B6[3], 4516
14115	Toluene, 3,5-dimethoxy 3,5-(CH$_3$O)$_2$C$_6$H$_3$CH$_3$	152.20		244, 102[8]		1.0478[15]	1.5234[20]	al, eth, bz, aa	B6[3], 4533
14116	Toluene, 2,4-dinitro 2,4-(NO$_2$)$_2$C$_6$H$_3$CH$_3$	182.14	ye nd or mcl pr	300d	71	1.3208[71]	1.442	al, eth, ace, bz, Py	B5[4], 865
14117	Toluene, 2,5-dinitro 2,5-(NO$_2$)$_2$C$_6$H$_3$CH$_3$	182.14	nd (al)	......	52.5	1.282[111]	......	al, bz	B5[4], 866

No.	Name, Synonyms, and Formula	Mol. wt.	Color, crystalline form, specific rotation and λ_{max} (log ε)	b.p. °C	m.p. °C	Density	n_D	Solubility	Ref.
14118	Toluene, 2,6-dinitro 2,6-$(NO_2)_2C_6H_3CH_3$	182.14	rh nd (al)	....	66	1.2833[111]	1.479	al	B5[4], 866
14119	Toluene, 3,4-dinitro 3,4-$(NO_2)_2C_6H_3CH_3$	182.14	ye nd (CS_2)		58.3	1.2594[111]	....	al	B5[4], 866
14120	Toluene, 3,5-dinitro 3,5-$(NO_2)_2C_6H_3CH_3$	182.14	ye rh nd (aa)	sub	93	1.2772[111]	....	al, eth, bz, chl	B5[4], 867
14121	Toluene, 4,6-Dinitro-3-Fluoro 4,6-$(NO_2)_2$-3-$FC_6H_2CH_3$	200.13	....	129-30[0.5]	78				B5[4], 867
14122	Toluene, 2,4-diisopropyl 2,4-$(i\text{-}C_3H_9)_2C_6H_3CH_3$	176.30		82[7]	....	0.8636[20]	1.4912[20]	CS_2	B5[4], 1152
14123	Toluene, 3,5-diisopropyl 3,5-$(i\text{-}C_3H_7)_2C_6H_3CH_3$	176.30		215-8	....	0.8668[20]	1.4950[20]	....	B5[4], 1153
14124	Toluene, 2-ethoxy or Ethyl o-tolyl ether (2-$CH_3C_6H_4)OC_2H_5$	136.20		184, 70[12]	....	0.9592[13/4]	1.508[13]	al, eth	B6[4], 1944
14125	Toluene, 3-ethoxy or Ethyl m-tolyl ether (3-$CH_3C_6H_4)OC_2H_5$	136.20		192	....	0.949[20]	1.513[20]	al, eth	B6[4], 2039
14126	Toluene, 4-ethoxy or Ethyl p-tolyl ether (4-$CH_3C_6H_4)OC_2H_5$	136.20		188-9	....	0.9509[18/4]	1.5058[18]	al, eth	B6[4], 2099
14127	Toluene, 2-fluoro or o-Tolyl fluoride 2-$FC_6H_4CH_3$	110.13		114, 30[26]	−62	1.0041[13/4]	1.4704[20]	al, eth	B5[4], 799
14128	Toluene, 3-fluoro or m-Tolyl fluoride 3-$FC_6H_4CH_3$	110.13		116	−87.7	0.9986[20]	1.4691[20]	al, eth	B5[4], 799
14129	Toluene, 4-fluoro or 4-Tolyl fluoride 4-$FC_6H_4CH_3$	110.13		116.6	−56.8	1.0007[16/4]	1.4699[20]	al, eth	B5[4], 800
14130	Toluene, 3-(α-hydroxy ethyl) 3-$[CH_3CH(OH)]C_6H_4CH_3$	136.20		112[13]	....	0.9974[15/4]	1.5240[20]	al, eth	B6[3], 1823
14131	Toluene, 4-(α-hydroxy ethyl) 4-$[CH_3CH(OH)]C_6H_4CH_3$	136.20	....	219[756], 120[19]	....	0.9944[20/4]	1.5246[20]	al, eth	B6[3], 1826
14132	Toluene, 2-hydroxylamino or o-Toyl hydroxyl amine 2-$HONHC_6H_4CH_3$	123.16	nd (eth-bz)	....	44			al, eth, bz	B15[3], 15
14133	Toluene, 2-hydroxylamino-6-nitro 2-$HONH$-6-$NO_2C_6H_3CH_3$	168.18	col or ye (bz)		120-1			al, eth	B15[2], 14
14134	Toluene, 3-hydroxylamino or m-Tolyl hydroxyl amine 3-$(HONH)C_6H_4CH_3$	123.16	lf (bz-peth)	....	68.5			al, eth, chl	B15[3], 17
14135	Toluene, 4-hydroxylamino or p-Tolyl hydroxyl amine 4-$(HONH)C_6H_4CH_3$	123.16	lf (bz)	115-20d	98			al, eth, chl	B15[3], 18
14136	Toluene, 4-hydroxylamino-2-nitro 4-$(HONH)$-2-$NO_2C_6H_3CH_3$	168.18	ye (bz)		108-9			bz	B15[2], 16
14137	Toluene, 2-iodo or o-Tolyl iodide 2-$IC_6H_4CH_3$	218.04		211-2, 73-5[7]	....	1.713[20/4]	1.6079[20]	al, eth	B5[4], 838
14138	Toluene, 3-iodo or m-Tolyl iodide 3-$IC_6H_4CH_3$	218.04		213, 80-2[10]	−27.2	1.705[20]	1.6053[20]	al, eth	B5[4], 839
14139	Toluene, 4-iodo or p-Tolyl iodide 4-$IC_6H_4CH_3$	218.04	lf (al)	211 sub	36-7	1.678[20/4]	....	al, eth	B5[4], 840
14140	Toluene, 2-mercapto or o-Thiocresol 2-$CH_3C_6H_4SH$	124.22	pl or lf	194.2, 67.1[10]	15	1.041[20/4]	1.570[20]	al, eth	B6[4], 2014
14141	Toluene, 3-mercapto or m-Thiocresol 3-$CH_3C_6H_4SH$	124.22		195.1, 67.8[10]	<−20	1.044[20/4]	1.572[20]	al, eth	B6[4], 2079
14142	Toluene, 4-mercapto or p-Thiocresol 4-$CH_3C_6H_4SH$	124.22	lf (eth, dil al)	195, 67.6[10]	44	....		al, eth	B6[4], 2153
14143	Toluene, 2-methoxy or Methyl-o-tolyl ether 2-$CH_3OC_6H_4CH_3$	122.17		171.3	....	0.9851[15/15]	1.5161[20]	al, eth, ace	B6[4], 1943
14144	Toluene, 3-methoxy or Methyl-m-tolyl ether 3-$CH_3OC_6H_4CH_3$	122.17		177.2	....	0.9697[25/25]	1.5164[13]	al, eth, ace, bz	B6[4], 2039
14145	Toluene, 4-methoxy or Methyl-p-tolyl ether 4-$CH_3OC_6H_4CH_3$	122.17		176.5	....	0.9689[25/25]	1.5124[19]	al, eth	B6[4], 2098
14146	Toluene, 2-nitro 2-$NO_2C_6H_4CH_3$	137.14	(i) nd, (ii) cr	221.7, 118[16]	(i)-9.5 (ii)-2.9	1.1629[20]	1.5450[20]	al, eth	B5[4], 845
14147	Toluene, 3-nitro 3-$NO_2C_6H_4CH_3$	137.14	pa ye	232.6	16	1.1571[20/4]	1.5466[20]	al, eth, bz	B5[4], 847
14148	Toluene, 3-nitro-4-triazo 4-N_3-3-NO_2-$C_6H_3CH_3$	178.15	ye nd or pl (lig, dil al)	....	38		....	al	B5[3], 773
14149	Toluene, 4-nitro 4-$NO_2C_6H_4CH_3$	137.14	orth cr (al, eth)	238.3, 105[9]	54.5	1.1038[75/4]	....	al, eth, ace, bz	B5[4], 848

No.	Name, Synonyms, and Formula	Mol. wt.	Color, crystalline form, specific rotation and λ_{max} (log ε)	b.p. °C	m.p. °C	Density	n_D	Solubility	Ref.
14150	Toluene, 2-nitroso 2-ONC₆H₄CH₃	121.14	nd or pr	....	72.5	....	....	al, eth, chl	B5³, 728
14151	Toluene, 3-nitroso 3-ONC₆H₄CH₃	121.14	nd	....	53.5			eth, chl	B5⁴, 844
14152	Toluene, 4-nitroso 4-ONC₆H₄CH₃	121.14	nd (lig)	....	48.5			bz, chl	B5⁴, 845
14153	Toluene, 4-isopropenyl 4-[CH₂=C(CH₃)]C₆H₄CH₃	132.20		185.3		0.8936²³ᐟ⁴	1.52832²³	....	B5⁴, 1383
14154	Toluene, 4-isopropyl-2-nitro 4-i-C₃H₇-2-NO₂C₆H₃CH₃	179.22					1.5280²⁰		B5⁴, 1064
14155	Toluene, 2,3,4,5,6-pentabromo C₆Br₅CH₃	486.62			288	2.97¹⁷		bz	B5³, 720
14156	Toluene, 2,3,4,5,6-pentachloro C₆Cl₅CH₃	264.37	nd (bz, peth)	301	224			to	B5⁴, 823
14157	Toluene, 2-phenyl or 2-Methyl biphenyl 2-C₆H₅-C₆H₄CH₃	168.24		255.3, 130-6²⁷	−0.2	1.010²²ᐟ⁴	1.5914²⁰	al, eth	B5⁴, 1855
14158	Toluene, 3-phenyl or 3-Methyl biphenyl 3-C₆H₅-C₆H₄CH₃	168.24		272.7, 148-50²⁰	4.5	1.0182¹⁷ᐟ⁴	1.5972²⁰	al, eth	B5⁴, 1858
14159	Toluene, 4-phenyl 4-C₆H₅C₆H₄CH₃	168.24	pl (lig, MeOH)	267-8, 134-6¹⁵	49-50	1.015²⁷		al, eth	B5⁴, 1860
14160	Toluene, 2-iso-propenyl 2-[CH₂=C(CH₃)]C₆H₄CH₃	132.20		175, 59-62¹¹		0.9181¹⁵ᐟ⁰	1.5112³⁰		B5³, 1214
14161	Toluene, 2-propoxy or Propyl-o-tolyl ether 2-C₃H₇OC₆H₄CH₃	150.22		204.1		0.9517⁰ᐟ⁰			B6³, 1246
14162	Toluene, 3-propoxy or Propyl-m-tolyl ether 3-C₃H₇OC₆H₄CH₃	150.22		210.6		0.9484⁰ᐟ⁰			B6⁴, 2040
14163	Toluene, 4-propoxy or Propyl-p-tolyl ether 4-C₃H₇OC₆H₄CH₃	150.22		210.4		0.9497⁰ᐟ⁰			B6³, 1354
14164	Toluene, 2-propyl or 1-Methyl-2-propyl benzene 2-C₃H₇C₆H₄CH₃	134.22		185, 63.5¹⁰	−60.2	0.8744²⁰	1.4998²⁰		B5⁴, 1056
14165	Toluene, 3-propyl or 1-Methyl-3-propyl benzene 3-C₃H₇C₆H₄CH₃	134.22		182, 61.4¹⁰		0.8610²⁰	1.4936²⁰		B5⁴, 1056
14166	Toluene, 4-propyl or 1-Methyl-4-propyl benzene 4-C₃H₇C₆H₄CH₃	134.22		183, 61.9¹⁰	−63.6	0.8584²⁰	1.4919²⁰	al, eth	B5⁴, 1057
14167	Toluene, 2,3,5,6-tetrabromo 2,3,5,6-Br₄C₆HCH₃	407.75	nd	....	116-7			ace, bz	B5, 310
14168	Toluene, 2,3,4,5-tetrachloro 2,3,4,5-Cl₄C₆HCH₃	229.92	nd (MeOH, dil al)	....	98.1			al, eth, ace	B5², 233
14169	Toluene, 2,3,4,6-tetrachloro 2,3,4,6-Cl₄C₆HCH₃	229.92	nd (al, eth)	276.5	96			al, bz	B5³, 702
14170	Toluene, 2,3,5,6-tetrachloro 2,3,5,6--Cl₄C₆HCH₃	229.92	nd (MeOH)	sub	93-4			al, eth	B5³, 702
14171	Toluene, 2,4,5-triacetoxy 2,4,5-(CH₃CO₂)₃C₆H₂CH₃	266.25	cr (al)	....	114-5			w, al, bz	B6, 1109
14172	Toluene, 2,4,6-triacetoxy 2,4,6-(CH₃CO₂)₃C₆H₂CH₃	266.25	(i) cr (eth-peth) (ii) nd (peth)	....	(i) 76 (ii) 58			al, bz	B6³, 6320
14173	Toluene, 2,4,6-triamino 2,4,6-(NH₂)₃C₆H₂CH₃	137.19	col nd (bz), red in air	....	105			w, al, bz	B13³, 557
14174	Toluene, 2-triazo or o-Tolyl azide 2-N₃C₆H₄CH₃	133.15	pa ye liq	90.5³⁰	<−10			eth	B5⁴, 876
14175	Toluene, 3-triazo or m-Tolyl azide 3-N₃C₆H₄CH₃	133.15		92.5³¹				eth	B5⁴, 876
14176	Toluene, 4-triazo or p-Tolyl azide 4-N₃C₆H₄CH₃	133.15		d 180, 80¹⁰	......	1.0527²³ᐟ⁴		al, eth	B5⁴, 876
14177	Toluene, 2,3,4-tribromo 2,3,4-Br₃C₆H₂CH₃	328.85	rh pl (aa, lig-CS₂)	....	45-6	2.456²⁰			B5¹, 156
14178	Toluene, 2,3,5-tribromo 2,3,5-Br₃C₆H₂CH₃	328.85	mcl pr (eth-to)	....	53-4	2.467¹⁷		eth	B5³, 719
14179	Toluene, 2,3,6-tribromo 2,3,6-Br₃C₆H₂CH₃	328.85	mcl pr or lf (lig, chl)	....	60.5	2.471¹⁷			B5¹, 156
14180	Toluene, 2,4,5-tribromo 2,4,5-Br₃C₆H₂CH₃	328.85	mcl pr (eth-al), nd (al)	....	113.5	2.472¹⁷			B5³, 719
14181	Toluene, 2,4,6-tribromo 2,4,6-Br₃C₆H₂CH₃	328.85	lo nd or mcl pr (eth-AcOEt)	290	70	2.479¹⁷			B5³, 719

No.	Name, Synonyms, and Formula	Mol. wt.	Color, crystalline form, specific rotation and λ_{max} (log ϵ)	b.p. °C	m.p. °C	Density	n_D	Solubility	Ref.
14182	Toluene, 2,3,4-trichloro 2,3,4-Cl$_3$C$_6$H$_2$CH$_3$	195.48	nd (al, MeOH)	244	43-4	....	....	al, eth, ace	B5[4], 819
14183	Toluene, 2,3,5-trichloro 2,3,5-Cl$_3$C$_6$H$_2$CH$_3$	195.48	nd(al)	229 31[757]	45 6			al, ace	B5, 299
14184	Toluene, 2,3,6-trichloro 2,3,6-Cl$_3$C$_6$H$_2$CH$_3$	195.48	nd (al)		45-6			al, ace	B5[4], 819
14185	Toluene, 2,4,5-trichloro 2,4,5-Cl$_3$C$_6$H$_2$CH$_3$	195.48	nd or lf (al)	229-30[716]	82.4			al, ace	B5[4], 819
14186	Toluene, 2,4,6-trichloro 2,4,6-Cl$_3$C$_6$H$_2$CH$_3$	195.48	nd (al)		38			al, ace	B5[4], 819
14187	Toluene, 3,4,5-trichloro 3,4,5-Cl$_3$C$_6$H$_2$CH$_3$	195.48	....	246-7[768]	45-6			al, ace	B5, 299
14188	Toluene, 2,4,6-trihydroxy 2,4,6-(HO)$_3$C$_6$H$_2$CH$_3$	140.14			222-3			w, al, eth	B6[3], 6318
14189	Toluene, 3,4,5-trihydroxy or 5-Methyl pyrogallol 3,4,5-(HO)$_3$C$_6$H$_2$CH$_3$	140.15	pa br nd (bz)	sub	129			bz	B6[3], 6320
14190	Toluene, 2,3,4-triido 2,3,4-I$_3$C$_6$H$_2$CH$_3$	469.83	pa br nd (al)		92			al, bz	B5[1], 157
14191	Toluene, 2,3,5-triido 2,3,5-I$_3$C$_6$H$_2$CH$_3$	469.83	og pl (al)		72-3			....	B5[1], 157
14192	Toluene, 2,3,6-triido 2,3,6-I$_3$C$_6$H$_2$CH$_3$	469.83	nd (al)		80.5			....	B5[1], 158
14193	Toluene, 2,4,5-triido 2,4,5-I$_3$C$_6$H$_2$CH$_3$	469.83	pa br nd (ul)		118-20			....	B5[1], 158
14194	Toluene, 2,4,6-triido 2,4,6-I$_3$C$_6$H$_2$CH$_3$	469.83	nd (al, bz)	300d	118-9			al	B5[1], 158
14195	Toluene, 3,4,5-triido 3,4,5-I$_3$C$_6$H$_2$CH$_3$	469.83	nd(al)	....	122-3			al	B5, 317
14196	Toluene, 2,3,4-trinitro 2,3,4-(NO$_2$)$_3$C$_6$H$_2$CH$_3$	227.13	tcl lf (al), pr (ace)		112	1.62		eth, ace, bz	B5[4], 872
14197	Toluene, 2,4,5-trinitro 2,4,5-(NO$_2$)$_3$C$_6$H$_2$CH$_3$	227.13	yesh pl (ace), pa ye rh bi-pym (al)	exp 290-310	104			eth, ace, bz	B5[4], 872
14198	Toluene, 2,4,6-trinitro or TNT 2,4,6-(NO$_2$)$_3$C$_6$H$_2$CH$_3$	227.13	orh (al)	240 exp	82	1.654		eth, ace, bz, Py	B5[4], 873
14199	Toluene, 2-vinyl 2-(CH$_2$=CH)C$_6$H$_4$CH$_3$	118.18	....	171	....	0.9106[20/4]	1.5450[20]	bz, chl	B5[4], 1367
14200	Toluene, 3-vinyl 3-(CH$_2$=CH)C$_6$H$_4$CH$_3$	118.18	....	168, 61-2[18]	−70	0.9028[20/20]	1.5410[20]	al, eth, bz	B5[4], 1367
14201	Toluene, 4-vinyl 4-(CH$_2$=CH)C$_6$H$_4$CH$_3$	118.18	....	169, 63[15]	−37.8	0.8760[20/4]	1.5428[20]	bz	B5[4], 1369
14202	o-Toluene arsonic acid or o-Tolyl arsinic acid 2-CH$_3$C$_6$H$_4$AsO(OH)$_2$	216.07	nd (w)	....	163-4			al	B16[3], 1060
14203	m-Toluene arsonic acid or m-Tolyl arsinic acid 3-CH$_3$C$_6$H$_4$AsO(OH)$_2$	216.07	nd (w)		150			al	B16[3], 1061
14204	p-Toluene arsonic acid or p-Tolyl arsinic acid 4-CH$_3$C$_6$H$_4$AsO(OH)$_2$	216.07	nd (w)		3 (−w, 105-10)			al	B16[3], 1061
14205	o-Toluene boronic acid or o-Tolyl boronic acid 2-CH$_3$C$_6$H$_4$B(OH)$_2$	135.96	nd or pl (w)	d	165-8			al, eth, bz	B16[3], 1277
14206	m-Toluene boronic acid or m-Tolyl boronic acid 3-CH$_3$C$_6$H$_4$B(OH)$_2$	135.96	cr (w)		137-40			al, eth	B16[3], 1277
14207	p-Toluene boronic acid or p-Tolyl boronic acid 4-CH$_3$C$_6$H$_4$B(OH)$_2$	135.96	nd (w)		245			eth	B16[3], 1277
14208	o-Toluene sulfinic acid or o-Tolyl sulfinic acid 2-CH$_3$C$_6$H$_4$SO$_2$H	156.22	nd (w)	d	80			w, al, eth, ace	B11[3], 6
14209	p-Toluene sulfinic acid or p-Tolyl sulfinic acid 4-CH$_3$C$_6$H$_4$SO$_2$H	156.22	rh pl or lo nd (w)		86-7			al, eth	B11[3], 8
14210	p-Toluene sulfinyl chloride 4-CH$_3$C$_6$H$_4$SOCl	174.65	nd	115-20[4]	54-8			chl	B11[3], 11
14211	o-Toluene sulfonamide or 2-Toluene sulfonamide 2-CH$_3$C$_6$H$_4$SO$_2$NH$_2$	171.22	oct (al), pr (w)		156.3			al	B11[3], 167
14212	o-Toluene sulfonamide, 4-amino 4-NH$_2$-2-CH$_3$C$_6$H$_3$SO$_2$NH$_2$	186.23	nd or pl (w)		164			al	B14[3], 2208
14213	o-Toluene sulfonamide, N-methyl 2-CH$_3$C$_6$H$_4$SO$_2$NHCH$_3$	185.25	pl (bz-lig)		74-5			al, ace, chl	B11, 87

No.	Name, Synonyms, and Formula	Mol. wt.	Color, crystalline form, specific rotation and λ_{max} (log ε)	b.p. °C	m.p. °C	Density	n_D	Solubility	Ref.
14214	o-Toluene sulfonamide, N-phenyl .... 2-CH$_3$C$_6$H$_4$SO$_2$NHC$_6$H$_5$	247.32	pr (dil al)	....	136			al	B12, 566
14215	o-Toluene sulfonamide, 4-propyl .... 4-C$_3$H$_7$-2-CH$_3$C$_6$H$_3$SO$_2$NH$_2$	213.30	pl (bz, dil al)		101-2			w, al	B11, 138
14216	o-Toluene sulfonamide, 5-isopropyl .... 5-i-C$_3$H$_7$-2-CH$_3$C$_6$H$_3$SO$_2$NH$_2$	213.30	pl (al), pr or lf (dil al)		75			lig	B11, 140
14217	o-Toluene sulfonic acid or 2-Toluene sulfonic acid ...... 2-CH$_3$C$_6$H$_4$SO$_3$H	172.21	hyg pl (w + 2)	128.8[25]	67.5			w, al	B11[3], 167
14218	o-Toluene sulfonic acid, 4-amino .... 4-H$_2$N-2-CH$_3$C$_6$H$_3$SO$_3$H	187.22	mcl pr (w + l)		d				B14[3], 2208
14219	o-Toluene sulfonic acid, 5-amino .... 5-H$_2$N-2-CH$_3$C$_6$H$_3$SO$_3$H	187.22	pl (w + l)		>275				B14[3], 2210
14220	o-Toluene sulfonic acid, 6-amino .... 6-H$_2$N-2-CH$_3$C$_6$H$_3$SO$_3$H	187.22	nd (+ ½ w)		130d				B14[1], 723
14221	o-Toluene sulfonic acid, 4-nitro, dihydrate .... 4-NO$_2$-2-CH$_3$C$_6$H$_3$SO$_3$H.2H$_2$O	253.23	pl (w)		133.5			w, al, eth, chl	B11[3], 173
14222	o-Toluene sulfonic acid, 5-nitro .... 5-NO$_2$-2-CH$_3$C$_6$H$_3$SO$_3$H	217.20	pr or pl (w + 2)		130-3			w, al, eth, chl	B11[3], 173
14223	o-Toluene sulfonic acid, 2-tolyl ester or (2-Tolyl)-o-toluene sulfonate ...... 2-CH$_3$C$_6$H$_4$SO$_2$(OC$_6$H$_4$CH$_3$-2)	262.33	cr (al)		50-1			bz, al	B11, 85
14224	2-Toluene sulfonic acid, 3-tolyl ester .... 2-CH$_3$C$_6$H$_4$SO$_2$(OC$_6$H$_4$CH$_3$-3)	262.33	cr (al)		60			al, bz	B11, 85
14225	2-Toluene sulfonic acid, 4-tolyl ester .... 2-CH$_3$C$_6$H$_4$SO$_2$(OC$_6$H$_4$CH$_3$-4)	262.33	cr (al)		70-1			al, bz	B11, 85
14226	o-Toluene sulfonyl bromide or 2-Toluene sulfonyl bromide .... 2-CH$_3$C$_6$H$_4$SO$_2$Br	235.11	cr	138[10]	13				B11, 86
14227	o-Toluene sulfonyl chloride or 2-Toluene sulfonyl chloride .... 2-CH$_3$C$_6$H$_4$SO$_2$Cl	190.65	....	154[36]	10.2	1.3383[20/4]	1.5565[20]	eth, bz	B11[3], 167
14228	m-Toluene sulfonamide or 3-Toluene sulfonamide ...... 3-CH$_3$C$_6$H$_4$SO$_2$NH$_2$	171.22	pl (w), mcl pr		108			al	B11[1], 23
14229	m-Toluene sulfonamide, N-phenyl .... 3-CH$_3$C$_6$H$_4$SO$_2$NHC$_6$H$_5$	247.32	pr (al)		96			al, eth	B11[1], 23
14230	m-Toluene sulfonamide, 4-isopropyl .... 4-i-C$_3$H$_7$-3-CH$_3$C$_6$H$_3$SO$_2$NH$_2$	213.30	fl (dil al)		149.9			al, eth	B11[3], 350
14231	m-Toluene sulfonic acid or 3-Toluene sulfonic acid ...... 3-CH$_3$C$_6$H$_4$SO$_3$H	172.21	oil					w, al, eth	B11[3], 176
14232	m-Toluene sulfonic acid, 4-amino .... 4-NH$_2$-3-CH$_3$C$_6$H$_3$SO$_3$H	187.22	lf ye nd (w + ½)		132d (hyd)			w	B14[3], 2213
14233	m-Toluene sulfonyl chloride-6-acetamido .... 6-(CH$_3$CONH)-3-CH$_3$C$_6$H$_3$SO$_2$Cl	247.70	nd (bz)		159				B14[2], 448
14234	p-Toluene sulfonamide or 4-Toluene sulfonamide .... (4-CH$_3$C$_6$H$_4$)SO$_2$NH$_2$	171.22	mcl pl (w + 2)		138-9 (anh), 105 (hyd)			al	B11[3], 266
14235	p-Toluene sulfonamide, 2-amino .... 2-NH$_2$-4-CH$_3$C$_6$H$_3$SO$_2$NH$_2$	186.23	pr (w)		176				B14[3], 2220
14236	p-Toluene sulfonamide, N,N-dichloro or Dichloramine T .... 4-CH$_3$C$_6$H$_4$SO$_2$NCl$_2$	240.11	pa ye pr (peth-chl)		83			al, eth, bz, aa, chl	B11[3], 301
14237	p-Toluene sulfonamide, N-ethyl .... 4-CH$_3$C$_6$H$_4$SO$_2$NHC$_2$H$_5$	199.27	pl (dil al, lig)		64			al	B11[3], 268
14238	p-Toluene sulfonamide, N-methyl .... 4-CH$_3$C$_6$H$_4$SO$_2$NHCH$_3$	185.25	pl (dil al)		78-9	1.340		al, eth	B11[3], 267
14239	p-Toluene sulfonamide, N-methyl-N-nitroso .... 4-CH$_3$C$_6$H$_4$SO$_2$N(NO)CH$_3$	214.25	cr		60			al, eth	B11[1], 29
14240	p-Toluene sulfonamide, N-methyl-N-phenyl .... 4-CH$_3$C$_6$H$_4$SO$_2$N(CH$_3$)C$_6$H$_5$	261.25	pl or mcl pr (AcOEt)		95			al, eth	B12[3], 1102
14241	p-Toluene sulfonamide, N-methyl-N-(2-tolyl) .... 4-CH$_3$C$_6$H$_4$SO$_2$N(CH$_3$)(C$_6$H$_4$CH$_3$-2)	275.37	pr (al)		119-20			al	B12[1], 388
14242	p-Toluene sulfonamide, N-methyl-N-(4-tolyl) .... 4-CH$_3$C$_6$H$_4$SO$_2$N(CH$_3$)(C$_6$H$_4$CH$_3$-4)	275.37	pr		60				B12[3], 2145
14243	p-Toluene sulfonamide, N-phenyl or 4-Toluene sulfonanilide .... 4-CH$_3$C$_6$H$_4$SO$_2$NHC$_6$H$_5$	247.32	dimorphic α: tcl, β: mcl pr (dil al, bz)		103-4			al, aa	B12[3], 1081

No.	Name, Synonyms, and Formula	Mol. wt.	Color, crystalline form, specific rotation and λ_{max} (log ε)	b.p. °C	m.p. °C	Density	n_D	Solubility	Ref.
14244	p-Toluene sulfonamide, 3-isopropyl 3-i-C₃H₇-4-CH₃C₆H₃SO₂NH₂	213.30	nd (w)	....	162	....		al	B11³, 349
14245	p-Toluene sulfonamide, N-(2-tolyl) .... 4-CH₃C₆H₄SO₂NH(C₆H₄CH₃-2)	261.35	rh bipym (al), nd (dil aa)	....	110			al, eth, acc, bz, chl	B12², 452
14246	p-Toluene sulfonamide, N-(4-tolyl) 4-CH₃C₆H₄SO₂NH(C₆H₄CH₃-4)	261.35	tcl pr or nd (aa)	....	118-9			al, bz, chl	B12³, 2143
14247	p-Toluene sulfonic acid or 4-Toluene sulfonic acid 4-CH₃C₆H₄SO₃H	172.21	hyg pl (w + 1), mcl lf or pl	140²⁰	104-5			w, al, eth	B11³, 183
14248	p-Toluene sulfonic acid, 2-amino 2-NH₂-4-CH₃C₆H₃SO₃H	187.22	pl, nd or pr (w)						B14³, 2219
14249	p-Toluene sulfonic acid, butyl ester .... 4-CH₃C₆H₄SO₂(OC₄H₉)	228.31	....	164-6⁶		1.1319²⁰ᐟ⁴	1.5050²⁰	eth	B11³, 189
14250	p-Toluene sulfonic acid, 2-chloro ethyl ester or 2-Chloro ethyl-p-toluene sulfonate............ 4-CH₃C₆H₄SO₂(OCH₂CH₂Cl)	234.70		210²¹					B11³, 188
14251	p-Toluene sulfonic acid, 2-chloro propyl ester or 2-Chloro propyl-p-toluene sulfonate............ 4-CH₃C₆H₄SO₂(OCH₂CHClCH₃)	248.73	....	216-7¹⁷	....	1.2674²⁰ᐟ⁴	1.5225²¹		B11², 45
14252	P-Toluene sulfonic acid, ethyl ester or Ethyl-p-toluene sulfonate .... 4-CH₃C₆H₄SO₂(OC₂H₅)	200.26	mcl pr (aa, AcOEt)	173¹⁵	34-5	1.166⁴⁸ᐟ⁴	....	al, eth	B11³, 188
14253	4-Toluene sulfonic acid, ethylene ester or Ethylene-p-toluene sulfonate (4-CH₃C₆H₄)SO₂CH₂CH₂(O₃SC₆H₄CH₃-4)	370.44	cr (bz)	....	128				B11³, 225
14254	p-Toluene sulfonic acid, methyl ester or Methyl-p-toluene sulfonate (4-CH₃C₆H₄)SO₂(OCH₃)	186.23	mcl lf or pr (eth-lig)	292, 140²⁰	28-9			al, eth, bz, chl	B11³, 187
14255	p-Toluene sulfonic acid, propyl ester or Propyl-p-toluene sulfonate .... (4-CH₃C₆H₄)SO₂(OC₃H₇)	214.29		189⁹	<−20	1.144²⁰ᐟ⁴	1.4998²⁰		B11³, 188
14256	p-Toluene sulfonic acid, isopropyl ester .... (4-CH₃C₆H₄)SO₂(O-i-C₃H₇)	214.29			21	¹.5065²⁰			B11³, 189
14257	p-Toluene sulfonic acid, 2-tolyl ester or 2-Tolyl-p-toluene sulfonate 4-CH₃C₆H₄SO₂(OC₆H₄CH₃-2)	262.33	nd	....	54-5			al	B11³, 204
14258	p-Toluene sulfonic acid, (2,4,6-tribromo phenyl)ester ..... (2,4,6-Br₃C₆H₂O)SO₂C₆H₄CH₃-4	485.01	cr (al)		113			al	B11², 47
14259	p-Toluene sulfonyl chloride 4-CH₃C₆H₄SO₂Cl	190.65	tcl (eth, peth)	145-6¹⁵	71			al, eth, bz	B11³, 265
14260	o-Toluic acid or 2-Methyl benzoic acid 2-CH₃C₆H₄CO₂H	136.16	pr or nd (w)	258-9⁷⁵¹	107-8	1.062¹¹⁵	1.512¹¹⁵	al, eth, chl	B9³, 2298
14261	o-Toluic acid anhydride .... (2-CH₃C₆H₄CO)₂O	254.29		>325	38-9			eth	B9³, 2303
14262	o-Toluic acid, 3-chloro 3-Cl-2-CH₃C₆H₃CO₂H	170.60	nd (al)		159			al, eth	B9³, 2309
14263	o-Toluic acid, 4-chloro or 4-Chloro-2-methyl benzoic acid 4-Cl-2-CH₃C₆H₃CO₂H	170.60	nd (w, al, bz)		173			w, al, aa	B9², 320
14264	o-Toluic acid, 5-chloro or 5-Chloro-2-methyl benzoic acid 5-Cl-2-CH₃C₆H₃CO₂H	170.60	nd (al)	....	168-9			al	B9², 320
14265	o-Toluic acid, 4,6-dihydroxy or o-Orsellinic acid........ 4,6-(HO)₂-2-CH₃C₆H₂CO₂H	168.16	nd (dil aa, + 1w)		176d			al, eth	B10³, 1479
14266	o-Toluic acid, 4,6-dihydroxy, ethyl ester .... 4,6-(HO)₂-2-HOC₆H₂CO₂C₂H₅	224.24	wh nd		147			al, eth	B10³, 1482
14267	o-Toluic acid, ethyl ester or Ethyl-o-toluate............ 2-CH₃C₆H₄CO₂C₂H₅	164.21		227, 102.5¹²	<−10	1.0325²¹ᐟ⁴	1.507²²	al, eth	B9³, 2301
14268	o-Toluic acid, methyl ester or Methyl-o-toluate........ 2-CH₃C₆H₄CO₂CH₃	150.18		215, 97¹⁵	<−15	1.068²⁰ᐟ⁴		al, eth	B9³, 2301
14269	o-Toluic acid, 6-methyl 6-CH₃-2-CH₃C₆H₃CO₂H	170.60	nd (w)	....	102				B9³, 2310
14270	o-Toluonitrile, or 2-methyl benzonitrile ... 2-CH₃C₆H₄CN	117.15		205, 90¹⁵	−14	0.9955²⁰ᐟ⁴	1.5279²⁰	al, eth	B9³, 2307
14271	o-Toluonitrile, 4-nitro 4-NO₂-2-CH₃C₆H₃CN	162.15	lf (sub)	sub	100 (105)			ace, bz, chl	B9¹, 188
14272	o-Toluonitrile, 5-nitro 5-NO₂-2-CH₃C₆H₃CN	162.15	nd (95% al)	174-5¹⁸	105			al, eth, ace, bz, chl	B9³, 2314

No.	Name, Synonyms, and Formula	Mol. wt.	Color, crystalline form, specific rotation and λ_{max} (log ϵ)	b.p. °C	m.p. °C	Density	n_D	Solubility	Ref.
14273	o-Toluonitrile, 6-nitro or 6-Nitro-o-toluonitrile........ 6-NO$_2$-2-CH$_3$C$_6$H$_3$CN	162.15	pl (bz)	....	109-10	....	....	al, ace, bz	B9[2], 323
14274	o-Toluyl chloride or 2-Methyl benzoyl chloride .... 2-CH$_3$C$_6$H$_4$COCl	154.60	....	213-4, 88-90[12]	......	....	1.5549[20]	eth	B9[3], 2304
14275	m-Toluic acid or 3-Methyl benzoic acid .... 3-CH$_3$C$_6$H$_4$CO$_2$H	136.16	pr (w, al)	263 sub	111-3	1.054[112]	1.509	al, eth	B9[3], 2318
14276	m-Toluic acid, 2-amino or 2-Amino-3-methyl benzoic acid .... 2-H$_2$N-3-CH$_3$C$_6$H$_3$CO$_2$H	151.17	nd (al), pr (w)	....	172	....	....	al, eth	B14[2], 291
14277	m-Toluic acid, 4-amino or 4-Amino-3-methyl benzoic acid .... 4-H$_2$N-3-CH$_3$C$_6$H$_3$CO$_2$H	151.17	nd (w)	....	170	....	....	w	B14[2], 290
14278	m-Toluic acid anhydride .... (3-CH$_3$C$_6$H$_4$CO)$_2$O	254.29	cr (peth)	230[17]	71	....	....	al, eth, ace, bz, chl	B9[3], 2321
14279	m-Toluic acid, 4-chloro or 4-Chloro-3-methyl benzoic acid .... 4-Cl-3-CH$_3$C$_6$H$_3$CO$_2$H	70.60	nd (w)	....	209-10	....	....	....	B9[3], 2327
14280	m-Toluic acid, 5-chloro or 5-Chloro-3-methyl benzoic acid .... 5-Cl-3-CH$_3$C$_6$H$_3$CO$_2$H	170.60	nd (dil al)	....	178	....	....	al	B9, 479
14281	m-Toluic acid, 6-chloro .... 6-CH$_3$-3-CH$_3$C$_6$H$_3$CO$_2$H	170.60	nd (w al)	....	167	....	....	w, al	B9[3], 2327
14282	m-Toluic acid, ethyl ester or Ethyl-m-toluate........ 3-CH$_3$C$_6$H$_4$CO$_2$C$_2$H$_5$	164.21	....	234, 103-5[10]	......	1.0265[21/4]	1.5052[22]	**al, eth**	B9[3], 2320
14283	m-Toluic acid, methyl ester or Methyl-m-toluate....... 3-CH$_3$C$_6$H$_4$CO$_2$CH$_3$	150.18	....	221[758]	......	1.061[20/4]	....	al	B9[3], 2320
14284	m-Toluonitrile 3-CH$_3$C$_6$H$_4$CN	117.15	....	213, 845[10]	−23	1.0316[20/4]	1.5252[20]	al, eth	B9[3], 2324
14285	m-Toluonitrile, 2-nitro 2-NO$_2$-3-CH$_3$C$_6$H$_3$CN	162.15	nd (al)	....	84	....	....	ace, bz, chl	B9[3], 2330
14286	m-Toluonitrile, 4-nitro 4-NO$_2$-3-CH$_3$C$_6$H$_3$CN	162.15	pr (al), nd	....	80	....	....	al	B9[3], 2331
14287	m-Toluonitrile, 5-nitro 5-NO$_2$-3-CH$_3$C$_6$H$_3$CN	162.15	nd (lig)	....	104.5	....	....	al	B9[3], 2331
14288	m-Toluonitrile, 6-nitro 6-NO$_2$-3-CH$_3$C$_6$H$_3$CN	162.15	nd (al)	....	93-4	....	....	al, bz, aa	B9[3], 2332
14289	m-Toluyl chloride or 3-Methyl benzoyl chloride........ 3-CH$_3$C$_6$H$_4$COCl	154.60	....	219.20, 105[20]	−23	1.0265[21/4]	1.505[22]	**al, eth**	B9[3], 2321
14290	p-Toluic acid or 4-Methyl benzoic acid .... 4-CH$_3$C$_6$H$_4$CO$_2$H	136.16	nd (w)	275 (sub)	182	....	....	al, eth	B9[3], 2334
14291	p-Toluic acid anhydride .... (4-CH$_3$C$_6$H$_4$CO)$_2$O	254.29	pl (MeOH), nd (al)	....	95	....	....	eth, ace, bz, chl	B9[3], 2341
14292	p-Toluic acid, 2-chloro or 2-Chloro-4-methyl benzoic acid 2-Cl-4-CH$_3$C$_6$H$_3$CO$_2$H	170.60	nd (al)	....	155.6	....	....	al, eth, bz, chl	B9, 497
14293	p-Toluic acid, 3-chloro .... 3-Cl-4-CH$_3$C$_6$H$_3$CO$_2$H	170.60	nd or lf (dil al)	....	200-2	....	....	al	B9[3], 2355
14294	p-Toluic acid, ethyl ester or Ethyl-p-toluate .... 4-CH$_3$C$_6$H$_4$CO$_2$C$_2$H$_5$	164.21	....	235.7, 110[12]	......	1.0269[18/4]	1.5089[18]	**al, eth**	B9[3], 2337
14295	p-Toluic acid, methyl ester or Methyl-p-toluate........ 4-CH$_3$C$_6$H$_4$CO$_2$CH$_3$	150.18	cr (aq MeOH, peth)	222.5	33.2	....	....	al, eth	B9[3], 2337
14296	p-Toluonitrile or 4-Methyl benzonitrile .... 4-CH$_3$C$_6$H$_4$CN	117.15	nd (al)	217.6, 91[11]	29.5	0.9805[30/30]	....	al, eth	B9[3], 2348
14297	p-Toluonitrile, 2-nitro 2-NO$_2$-4-CH$_3$C$_6$H$_3$CN	162.15	nd (w)	....	101	....	....	al, bz, chl	B9[2], 334
14298	p-Toluonitrile, 3-nitro .... 3-NO$_2$-4-CH$_3$C$_6$H$_3$CN	162.15	pa ye nd (w)	171[12]	107-8	....	....	al, eth, ace, bz, chl	B9[3], 2359
14299	p-Toluyl chloride or 4-Methyl benzoyl chloride .... 4-CH$_3$C$_6$H$_4$COCl	154.60	....	225-7, 102[15]	−2	1.1686[20/4]	1.5547[20]		B9[3], 2342
14300	o-Toluidine or 2-Amino toluene .... 2-H$_2$NC$_6$H$_4$CH$_3$	107.16	....	200.2, 80.1[10]	−14.7	0.9984[20/4]	1.5725[20]	**al, eth**	B12[3], 1837
14301	o-Toluidine, N-acetyl-4-bromo .... 4-Br-2-CH$_3$C$_6$H$_3$(NHCOCH$_3$)	228.10	nd (dil al, lig)	....	159-60	....	....	al	B12[3], 456
14302	o-Toluidine, N-acetyl-5-bromo .... 5-Br-2-CH$_3$·C$_6$H$_3$(NHCOCH$_3$)	228.10	nd (bz)	....	165.5	....	....	al, bz	B12[3], 1923
14303	o-Toluidine, hydrochloride .... 2-CH$_3$C$_6$H$_4$NH$_2$.HCl	143.62	mcl pr (w)	242	215	....	....	w, al	B12[3], 1840

No.	Name, Synonyms, and Formula	Mol. wt.	Color, crystalline form, specific rotation and λ_{max} (log ε)	b.p. °C	m.p. °C	Density	n_D	Solubility	Ref.
14304	o-Toluidine, N-acetyl-6-bromo .... 6-Br-2-CH$_3$C$_6$H$_3$(NHCOCH$_3$)	228.10	nd (bz)	....	166	....	....	al	B12[3], 1922
14305	o-Toluidine, N-benzyl 2-CH$_3$C$_6$H$_4$NHCH$_2$C$_6$H$_5$	197.28	cr (al, eth)	300-5, 176[10]	60	1.0142[65/4]	1.5861[65]	al, ace, chl	B12[3], 2220
14306	o-Toluidine, N-benzyl-N-ethyl 2-CH$_3$C$_6$H$_4$N(C$_2$H$_5$)CH$_2$C$_6$H$_5$	225.34	ye	230[20-5]	....	....	....	....	B12, 1033
14307	o-Toluidine, N-benzyl-N-methyl 2-CH$_3$C$_6$H$_4$N(CH$_3$)CH$_2$C$_6$H$_5$	211.31	ye	167[13]	....	....	....	al	B12, 1033
14308	o-Toluidine, 4-bromo 4-Br-2-CH$_3$-C$_6$H$_3$NH$_2$	186.07	cr (al)	240	59.5	....	....	al, eth, aa	B12[2], 456
14309	o-Toluidine, 5-bromo 5-Br-2-CH$_3$C$_6$H$_3$NH$_2$	186.07	lf	253-7d, 139[17]	33	....	....	al, eth	B12[3], 1923
14310	o-Toluidine, 6-bromo 6-Br-2-CH$_3$C$_6$H$_3$NH$_2$	186.07	....	130[16]	....	....	....	eth	B12[3], 1922
14311	o-Toluidine, 3-chloro 3-Cl-2-CH$_3$C$_6$H$_3$NH$_2$	141.61	....	245, 96-9[10]	0-2	....	1.5880[20]	al	B12[3], 1919
14312	o-Toluidine, 4-chloro 4-Cl-2-CH$_3$C$_6$H$_3$NH$_2$	141.61	lf (al)	241	29-30	....	....	al	B12[3], 1914
14313	o-Toluidine, 5-chloro 5-Cl-2-CH$_3$C$_6$H$_3$NH$_2$	141.61	....	237[722], 140[38]	26	....	....	al	B12[3], 1910
14314	o-Toluidine, N,N-diethyl 2-CH$_3$C$_6$H$_4$N(C$_2$H$_5$)$_2$	163.26	....	208-9[755]	....	....	....	al, eth	B12[3], 1844
14315	o-Toluidine, N,N-dimethyl 2-CH$_3$C$_6$H$_4$N(CH$_3$)$_2$	135.21	....	185.3, 70.3[13]	-60	0.9286[20/4]	1.5152[20]	al, eth	B12[3], 1843
14316	o-Toluidine, N-ethyl 2-CH$_3$C$_6$H$_4$NHC$_2$H$_5$	135.21	....	218, 95.5[10]	< 15	0.948[25/4]	1.5456[20]	al, eth	B12[3], 1843
14317	o-Toluidine, 4-iodo 4-I-2-CH$_3$C$_6$H$_3$NH$_2$	233.06	nd (dil al), pr (lig)	....	91-2	....	....	al, eth, bz, aa, liq	B12[3], 1927
14318	o-Toluidine, 5-iodo 5-I-2-CH$_3$C$_6$H$_3$NH$_2$	233.06	nd (aq al)	273d	48-9	....	....	al, aa	B12[1], 391
14319	o-Toluidine, 4-methoxy or 2-Methyl-p-anisidine .... 4-CH$_3$O-2-CH$_3$C$_6$H$_3$NH$_2$	137.18	cr (lig)	248-9, 146-7[23]	29-30	....	1.5647[20]	al	B13[3], 1560
14320	o-Toluidine, 5-methoxy or 6-Methyl-m-anisidine .... 5-CH$_3$O-2-CH$_3$C$_6$H$_3$NH$_2$	137.18	nd (w)	253, 140[20]	47	....	....	eth	B13[3], 1573
14321	o-Toluidine, 6-methoxy or 6-Methyl-o-anisidine .... 6-CH$_3$O-2-CH$_3$C$_6$H$_3$NH$_2$	137.18	nd (w)	119-21[16]	31	....	....	al	B13[3], 1538
14322	o-Toluidine, N-Methyl .... 2-CH$_3$C$_6$H$_4$NHCH$_3$	121.18	....	207-8, 99[17]	....	0.9769[20/4]	1.5649[20]	al, eth, ace	B12[3], 1842
14323	o-Toluidine, 3-nitro 3-NO$_2$-2-CH$_3$C$_6$H$_3$NH$_2$	152.16	ye rh nd (w), ye lf (al)	305d	92 (97)	....	....	al, eth, bz, chl	B12[3], 1944
14324	o-Toluidine, 4-nitro 4-NO$_2$-2-CH$_3$C$_6$H$_3$NH$_2$	152.16	ye mcl pr or nd (w, al, lig)	....	134-5	1.1586[140/4]	....	al, bz, aa	B12[3], 1938
14325	o-Toluidine, 5-nitro 5-NO$_2$-2-CH$_3$C$_6$H$_3$NH$_2$	152.16	ye mcl pr (al)	....	107-8	....	....	al, eth, ace, bz, chl	B12[3], 1932
14326	o-Toluidine, 6-nitro .... 6-NO$_2$-2-CH$_3$C$_6$H$_3$NH$_2$	152.16	og-ye pr (dil al)	....	97	1.1900[100/4]	....	al, eth, bz, chl	B12[3], 1929
14327	m-Toluidine or 3-Amino toluene .... 3-CH$_3$C$_6$H$_4$NH$_2$	107.16	....	203.3, 82.3[10]	-30.4	0.9889[20/4]	1.5681[20]	al, eth, ace, bz	B12[3], 1949
14328	m-Toluidine, N-benzyl 3-CH$_3$C$_6$H$_4$NHCH$_2$C$_6$H$_5$	197.28	pa ye oil	312, 180[10]	....	1.0083[65/4]	1.5845[65]	eth, ace, bz, chl	B12[2], 552
14329	m-Toluidine, 4-bromo 4-Br-3-CH$_3$C$_6$H$_3$NH$_2$	186.07	pl (50% al), cr (al)	240	81	....	....	al	B12[3], 1999
14330	m-Toluidine, 5-bromo 5-Br-3-CH$_3$C$_6$H$_3$NH$_2$	186.07	....	255-60, 150-1[15]	37-8	1.1422[19]	....	al	B12[2], 474
14331	m-Toluidine, 6-bromo 6-Br-3-CH$_3$-C$_6$H$_3$NH$_2$	186.07	pr	129-30[15]	46	1.474[25/25]	1.5990[25]	al, eth	B12[3], 1998
14332	m-Toluidine, 4-chloro 4-Cl-3-CH$_3$C$_6$H$_3$NH$_2$	141.61	nd (peth)	241	83-4	....	....	al, ace, bz	B12[3], 1993
14333	m-Toluidine, 6-chloro 6-Cl-3-CH$_3$C$_6$H$_3$NH$_2$	141.61	pl	228-30	29-30	....	....	al	B12[3], 1992
14334	m-Toluidine, N,N-dimethyl .... 3-CH$_3$C$_6$H$_4$N(CH$_3$)$_2$	135.21	....	212	....	0.9410[20/4]	1.5492[20]	al, eth	B12[3], 1953
14335	m-Toluidine, N-ethyl 3-CH$_3$C$_6$H$_4$NHC$_2$H$_5$	135.21	....	221, 111-2[20]	....	....	1.5451[20]	al, eth	B12[3], 1954
14336	m-Toluidine hydrochloride .... 3-CH$_3$C$_6$H$_4$NH$_2$.HCl	143.62	lf (w)	250	228	....	....	w, al	B12[3], 1951

No.	Name, Synonyms, and Formula	Mol. wt.	Color, crystalline form, specific rotation and λ_{max} (log ϵ)	b.p. °C	m.p. °C	Density	n_D	Solubility	Ref.
14337	m-Toluidine, 2-iodo 2-I-3-CH$_3$C$_6$H$_3$NH$_2$	233.06	pr	....	41-2	....	....	al, eth, ace	B12[3], 2003
14338	m-Toluidine, 4-iodo 4-I-3-CH$_3$C$_6$H$_3$NH$_2$	233.06	lf or pl (al, peth)		46			al, eth, bz, aa, lig	B12[2], 475
14339	m-Toluidine, 5-iodo 5-I-3-CH$_3$C$_6$H$_3$NH$_2$	233.06	nd (peth)	....	78.5			al, eth, ace	B12[1], 406
14340	m-Toluidine, 6-iodo 6-I-3-CH$_3$C$_6$H$_3$NH$_2$	233.06	nd (dil al), br in air		48			al, eth, chl	B12[2], 475
14341	m-Toluidine, 5-methoxy or 3-Methyl-p-anisidine 5-CH$_3$O-3-CH$_3$C$_6$H$_3$NH$_2$	137.18	cr (dil al)		59-60			al, eth, ace, bz	B13[2], 320
14342	m-Toluidine, 6-methoxy or 5-Methyl-o-anisidine 6-CH$_3$O-3-CH$_3$C$_6$H$_3$NH$_2$	137.18	nd or lf (al, lig, peth)	235	93-4			al, eth, bz	B13[3], 1577
14343	m-Toluidine, N-methyl 3-CH$_3$C$_6$H$_4$NHCH$_3$	121.18	....	206-7, 120-1[40]			1.5557[25]	al, eth, ace	B12[3], 1953
14344	m-Toluidine, 2-nitro 2-NO$_2$-3-CH$_3$C$_6$H$_3$NH$_2$	152.16	ye-og pr or nd (bz-peth)		108			al, eth	B12[3], 2004
14345	m-Toluidine, 4-nitro 4-NO$_2$-3-CH$_3$C$_6$H$_3$NH$_2$	152.16	lt ye nd (w, dil al)		138			al, eth	B12[3], 2008
14346	m-Toluidine, 5-nitro 5-NO$_2$-3-CH$_3$C$_6$H$_3$NH$_2$	152.16	ye-red or red-br nd (al)		98			al, eth, bz	B12[3], 2007
14347	m-Toluidine, 6-nitro 6-NO$_2$-3-CH$_3$C$_6$H$_3$NH$_2$	152.16	ye lf (w), pl (dil al)		112			al, eth, bz, chl	B12[3], 2004
14348	p-Toluidine or 4-Amino toluene 4-CH$_3$C$_6$H$_4$NH$_2$	107.16	lf (w + l)	200.5, 79.6[10]	44-5	0.9619[20/4]	1.5534[45], 1.5636[20]	al, eth, ace, Py	B12[3], 2017
14349	p-Toluidine, N-acetyl-2-bromo 2-Br-4-CH$_3$C$_6$H$_3$NH(COCH$_3$)	228.10	nd (al)	....	118			al	B12[3], 2159
14350	p-Toluidine, N-acetyl-3-bromo 3-Br-4-CH$_3$C$_6$H$_3$NH(COCH$_3$)	228.10	nd (bz, dil al)		113			w, al, bz	B12[3], 2157
14351	p-Toluidine, N-benzyl 4-CH$_3$C$_6$H$_4$NHCH$_2$C$_6$H$_5$	197.28	ye lf	319[765], 181[10]	19-20	1.0064[65/4]	1.5832[65]	al, eth, bz, chl	B12[3], 2220
14352	p-Toluidine, 2-bromo 2-Br-4-CH$_3$C$_6$H$_3$NH$_2$	186.07	lf	240, 120-2[30]	26	1.510[20]	1.5999[20]	al, eth	B12[3], 2158
14353	p-Toluidine, 2-bromo-5-nitro 2-Br-5-NO$_2$-4-CH$_3$C$_6$H$_2$NH$_2$	231.06	br to pa ye nd (al, aa)	....	121				B12[1], 441
14354	p-Toluidine, 3-bromo 3-Br-4-CH$_3$C$_6$H$_3$NH$_2$	186.07		254-7	26			eth	B12[3], 2157
14355	p-Toluidine, 2-chloro 2-Cl-4-CH$_3$C$_6$H$_3$NH$_2$	141.61	....	219[732]	7	1.151[20]	1.5748[22]	....	B12[3], 2152
14356	p-Toluidine, 3-chloro 3-Cl-4-CH$_3$C$_6$H$_3$NH$_2$	141.61	....	242-4, 112-3[3]	26	....		al	B12[3], 2151
14357	p-Toluidine, N,N-diethyl 4-[(C$_2$H$_5$)$_2$N]C$_6$H$_4$CH$_3$	163.26	....	229[770]	....	0.9242[16]	....	al, eth	B12[3], 2028
14358	p-Toluidine, N,N-dimethyl 4-CH$_3$C$_6$H$_4$N(CH$_3$)$_2$	135.21	....	211	......	0.9366[20/4]	1.5366[20]	al, eth	B12[3], 2026
14359	p-Toluidine, 3,5-dinitro 3,5-(NO$_2$)$_2$-4-CH$_3$C$_6$H$_2$NH$_2$	197.45	ye nd (w, aa)	....	171			al, eth, ace, bz, chl	B12[3], 2184
14360	p-Toluidine, N-ethyl 4-CH$_3$C$_6$H$_4$NHC$_2$H$_5$	135.21	....	217		0.9391[16]		al, eth	B12[3], 2027
14361	p-Toluidine hydrochloride 4-CH$_3$C$_6$H$_4$NH$_2$.HCl	143.62	mcl nd (aa-eth)	258	243			w, al, aa	B12[3], 2021
14362	p-Toluidine, 2-iodo 2-I-4-CH$_3$C$_6$H$_3$NH$_2$	233.06	pr	d	40			al, eth, ace, bz, chl, peth	B12[2], 533
14363	p-Toluidine, 3-iodo 3-I-4-CH$_3$C$_6$H$_3$NH$_2$	233.06	nd (dil al, peth)	....	39-40			al, eth, ace, aa	B12[2], 533
14364	p-Toluidine, 2-methoxy or 4-Methyl-o-anisidine 2-CH$_3$O-4-CH$_3$C$_6$H$_3$NH$_2$	137.18	pa ye	237-9, 179-80[46]	....			al, eth, ace	B13[3], 1552
14365	p-Toluidine, 3-methoxy or 4-Methyl-m-anisidine 3-CH$_3$O-4-CH$_3$C$_6$H$_3$NH$_2$	137.18	....	250-2	58			al, eth, bz, lig	B13[1], 213
14366	p-Toluidine, N-methyl 4-CH$_3$C$_6$H$_4$NHCH$_3$	121.18	....	209-11, 102[20]	......	0.9348[55/4]	1.5568[20]	al, eth, ace	B12[3], 2025
14367	p-Toluidine, 2-nitro 2-NO$_2$-4-CH$_3$C$_6$H$_3$NH$_2$	152.16	red lf (dil al), mcl pr (al)	....	117	1.164[121/4]	....	al	B12[3], 2174
14368	p-Toluidine, 3-nitro 3-NO$_2$-4-CH$_3$C$_6$H$_3$NH$_2$	152.16	ye nd (w)	....	78-9			eth, bz	B12[3], 2165

No.	Name, Synonyms, and Formula	Mol. wt.	Color, crystalline form, specific rotation and λ_{max} (log ε)	b.p. °C	m.p. °C	Density	n_D	Solubility	Ref.
14369	o-Tolyl acetic acid or o-Methyl-α-toluic acid 2-CH₃C₆H₄CH₂CO₂H	150.18	nd (w)	. . . .	88-90	. . . .	. . . . . .	w	B9[3], 2426
14370	2-Tolyl acetonitrile or o-Methyl-α-tolunitrile 2-CH₃C₆H₄CH₂CN	131.18	. . . .	244	. . . . . .	1.0156[22]	1.5252[20]	al, eth, bz	B9[3], 2427
14371	3-Tolyl acetic acid or m-Methyl-α-toluic acid 3-CH₃C₆H₄CH₂CO₂H	150.18	nd (w)	120-3[26]	62	. . . .	. . . . . .	.	B9[3], 2429
14372	3-Tolyl acetonitrile or m-Methyl-α-tolunitrile 3-CH₃C₆H₄CH₂CN	131.18	. . . .	245-7[745]/d, 133[15]	. . . . . .	1.0022[22]	1.5233[20]	al, eth, bz	B9[2], 349
14373	4-Tolyl acetic acid or p-Methyl-α-toluic acid 4-CH₃C₆H₄CH₂CO₂H	150.18	nd or pl (al, w)	265-7 sub	91-3	. . . .	. . . . . .	al, eth, bz, chl	B9[3]
14374	2-Tolyl ether or Di-2-Tolyl ether . . . . (2-CH₃C₆H₄)₂O	198.26	. . . .	146-7[17]	. . . . . .	1.047[24]	. . . . . .	al, eth, bz	B6[4], 1947
14375	3-Tolyl ether or Di-3-Tolyl ether . . . . (3-CH₃C₆H₄)₂O	198.26	. . . .	284, 135-7[14]	. . . . . .	1.0323[21]	. . . . . .	al, eth, bz	B6[4], 2043
14376	4-Tolyl ether or Di-4-Tolyl ether . . . . (4-CH₃C₆H₄)₂O	198.26	. . . .	285	51	. . . .	. . . . . .	al, eth, bz	B6[4], 2103
14377	2-Tolyl hydrazine 2-CH₃C₆H₄NHNH₂	122.17	nd (dil al)	. . . .	59	. . . .	. . . . . .	al, eth	B15[3], 654
14378	2-Tolyl isocyanate 2-CH₃C₆H₄NCO	133.15	. . . .	184-6	. . . . . .	. . . .	1.5282[20]	eth, bz	B12[3], 1886
14379	3-Tolyl isocyanate 3-CH₃C₆H₄NCO	133.15	. . . .	195-8	. . . . . .	. . . .	. . . . . .	eth, bz	B12[3], 1979
14380	4-Tolyl isocyanate 4-CH₃C₆H₄NCO	133.15	. . . .	187[751]	. . . . . .	. . . .	. . . . . .	eth, bz	B12[3], 2110
14381	2-Tolyl selenide or Di-(2-Tolyl)selenide . . . . (2-CH₃C₆H₄)₂Se	261.23	pl or lf (al)	106[15]	65	. . . .	. . . . . .	.	B6[3], 343
14382	4-Tolyl selenide or Di-(4-Tolyl)selenide . . . . (4-CH₃C₆H₄)₂Se	261.23	rods or nd (al)	196[16]	69-70	. . . .	. . . . . .	.	B6[4], 2218
14383	2-Tolyl sulfone or 2,2'-Ditolyl sulfone . . . . (2-CH₃C₆H₄)₂SO₂	246.33	nd (al)	. . . .	134-5	. . . .	. . . . . .	al, eth, bz, chl	B6[4], 2020
14384	4-Tolyl sulfone or Di-4-Tolyl sulfone . . . . (4-CH₃C₆H₄)₂SO₂	246.33	pr (bz), nd (w, al), pl (al)	405[714]	159	. . . .	. . . . . .	bz, chl	B6[4], 2174
14385	4-Tolyl sulfoxide or Di-4-Tolyl sulfoxide . . . . (4-CH₃C₆H₄)₂SO	230.33	cr (lig-peth)	. . . .	94	. . . .	. . . . . .	al, eth, bz, chl, aa	B6[4], 2173
14386	4-Tolyl thiocyanate 4-CH₃C₆H₄SCN	149.22	. . . .	240-5, 116-8[10]	. . . . . .	. . . .	. . . . . .	al, bz, chl	B6[3], 1421
14387	Tomatidine C₂₇H₄₅NO₂	415.67	pl, [α][20]/D +5 (MeOH)	. . . .	210-1	. . . .	. . . . . .	eth	Am7[3], 4018
14388	Tomatine or Lycopersicin . . . . C₅₀H₈₃NO₂₁	1034.22	nd (MeOH), [α][20]/D −30 (Py)	. . . .	270	. . . .	. . . . . .	al, diox	C41, 3502
14389	Torularhodin or Torulene . . . . C₃₇H₄₈O₂	524.80	red nd (MeOH-eth), vt-bk (bz-MeOH)	. . . .	201-3d	. . . .	. . . . . .	ace, chl, Py	B9[3], 3661
14390	α-Toxicarol (dl) or Hydroxydequelin C₂₃H₂₂O₇	410.43	gr ye pl (al)	. . . .	219-23 (231)	. . . .	1.580	chl	B19[4], 5270
14391	α-Toxicarol (l) . . . . C₂₃H₂₂O₇	410.43	gr-yesh pl or nd (AcOEt-al)	. . . .	125-7	. . . .	. . . . . .	ace, AcOEt	B19[4], 5270
14392	α-Toxicarol, dihydro- (l) . . . . C₂₃H₂₄O₇	412.45	pa ye pl or rods, [α][20]/D −30 (bz, c=5)	. . . .	179 (206)	. . . .	. . . . . .	.	B19[4], 5262
14393	β-Toxicarol-(dl) . . . . C₂₃H₂₂O₇	410.43	pa ye pl (al)	. . . .	169-70	. . . .	. . . . . .	al	B19[4], 5270
14394	Trasentin hydrochloride or Adephenine. 2-(Diethyl amino)ethyl diphenyl acetate hydrochloride . . . . (C₆H₅)₂CHCO₂CH₂CH₂N(C₂H₅)₂·HCl	347.89	nd	. . . .	114-5	. . . .	. . . . . .	w	B9[3], 3297
14395	α,α-Trehalose or Mycose . . . . C₁₂H₂₂O₁₁	342.30	orh cr, [α][20]/D +199 (w, c=6)	. . . .	214-6 (anh), 97 (hyd)	. . . .	1.58[24/24]	w	B17[4], 3505

No.	Name, Synonyms, and Formula	Mol. wt.	Color, crystalline form, specific rotation and λ_{max} (log ϵ)	b.p. °C	m.p. °C	Density	n_D	Solubility	Ref.
14396	α,σ-Trehalose, dihydrate .... $C_{11}H_{22}O_{11}.2H_2O$	378.33	cr (dil al), $[\alpha]^{20}_D$ +178.3 (w, c=7)	....	103	....	....	w	B17⁴, 3505
14397	Triacetamide or N,N-Diacetyl acetamide .... $(CH_3CO)_2N$	143.14	nd (eth)	....	79	....	....	eth	B2⁴, 416
14398	Triacetin or Glycerol triacetate .... $CH_3CO_2CH(CH_2O_2CCH_3)_2$	218.21	cr (al)	258-60, 130.5⁷	4.1	1.1596²⁰/⁴	1.4301²⁰	al, eth, ace, bz, chl	B2⁴, 253
14399	Triacontane .... $CH_3(CH_2)_{28}CH_3$	422.83	orh (eth bz)	449.7, 304¹⁵	65.8	0.7750⁷⁸/⁴	1.4352⁷⁰, 1.4536²⁰	eth, bz	B1⁴, 592
14400	1-Triacontanol or Myricyl alcohol .... $CH_3(CH_2)_{28}CH_2OH$	438.83	nd (eth), pl (bz)	....	88	0.777⁹⁵	....	al, eth, bz	B1⁴, 1918
14401	Triallylamine .... $(CH_2=CHCH_2)_3N$	137.22	....	155-6	....	0.809²⁰/⁴	1.4502²⁰	al, eth, ace, bz	B4⁴, 1061
14402	1,3,5-Triazine .... $C_3H_3N_3$	81.08	....	114	86	1.38	....	al, eth	Am76, 5646
14403	1,3,5-Triazine, 2,4-diamino or Guanamine .... 2,4-$(NH_2)_2C_3HN_3$	111.11	nd (w)	....	329d	....	....	w	B26¹, 65
14404	1,3,5-Triazine, 2,4-diamino-6-phenyl or Benzoguanamine .... 2,4-$(NH_2)_2$-6-$C_6H_5C_3N_3$	187.21	nd or pr (al)	....	226-8	....	....	al, eth	B26¹, 69
14405	1,3,5-Triazine, perhydro, 1,3,5-tributyl .... 1,3,5-$(C_4H_9)_3C_3H_6N_3, C_{15}H_{33}N_3$	255.45	....	130-2³	....	....	1.4602²⁵	....	B26, 3
14406	1,3,5-Triazine, perhydro, 1,3,5-triethyl .... 1,3,5-$(C_2H_5)_3C_3H_6N_3$	171.28	....	78-9⁶	....	....	1.4580²⁵	....	B26², 3
14407	1,3,5-Triazine, perhydro, 1,3,5-trinitro or Cyclonite . Hexogen .... 1,3,5-$(NO_2)_3C_3H_6N_3$	222.12	orh cr (ace)	....	205-6	1.82²⁰	....	ace, aa	B26², 5
14408	1,3,5-Triazine-perhydro-1,3,5-triphenyl .... 1,3,5-$(C_6H_5)_3C_3H_6N_3$	315.42	nd (lig), pr (eth, chl-al)	185	143	....	....	eth, ace, bz, chl	B26², 3
14409	1,3,5-Triazino-2,4,6-tricarboxylic acid or Cyanuric tricarboxylic acid .... 2,4,6 $C_3N_3(CO_2H)_3$	213.11	pw	....	>250d	....	....	....	B26², 168
14410	1,3,5-Triazino-2,4,6-tricarboxylic acid, triethyl ester ..... 2,4,6-$C_3N_3(CO_2C_2H_5)_3$	297.27	nd (al)	d	168	....	....	....	B26², 168
14411	1,3,5-Triazino-2,4,6-tricarbonitrile .... 2,4,6-$C_3N_3(CN)_3$	156.11	mcl pr (bz)	262⁷⁷¹, 119¹	119	....	....	bz	B26¹, 91
14412	1,2,3-Triazole or Osotriazole .... $C_2N_3H_3$	69.07	hyg cr	203⁷³⁹	23	1.1861²⁵/⁴	1.4854²⁵	w, eth, ace	B26¹, 5
14413	1,2,4-Triazole or Pyrrodiazole .... $C_2N_3H_3$	69.07	pr (w), nd (al, eth, chl, bz)	260	120-1	1.132¹⁵³	1.4854²⁵	w, al	B26², 7
14414	1,2,4-Triazole, 3-amino or Amizol . ATA .... 3-$NH_2C_2N_3H_2$	84.08	cr (w, al, AcOEt)	....	159	....	....	w, al, chl	B26², 76
14415	1,2,4-Triazole, 4-amino .... 4-$NH_2C_2N_3H_2$	84.08	hyg nd (al, chl)	....	82-3	....	....	w, al	B26², 7
14416	1,2,4-Triazole, 3,5-diamino .... $C_2H_5N_5$	99.09	....	....	211-2	....	....	w, al	B26¹, 57
14417	Tribenzylamine .... $(C_6H_5CH_2)_3N$	287.41	pl, (eth), mcl (al)	380-90, 230¹³	91-2	0.9912⁹⁵/⁴	....	eth	B12³, 2226
14418	Triabromo acetamide .... Br_3CCONH_2	295.77	mcl pr (al)	sub	121-2	....	....	w, al, eth	B2³, 485
14419	Tribromoacetic acid .... Br_3CCOOH	296.76	mcl	245	135	....	....	w, al, eth	B2⁴, 534
14420	Tribromoacetic acid, ethyl ester or Ethyl tribromo acetate .... $Br_3CCOOC_2H_5$	324.81	....	225, 148⁷³	....	2.2300²⁰/²⁰	1.5438¹³	al, eth	B2³, 485
14421	Tribromoacetyl bromide .... Br_3CCOBr	359.66	....	210-5	....	....	....	eth, bz, chl	B2³, 485
14422	Bis-(tribromomethyl) trisulfide .... $(Br_3C)_2S_3$	599.67	rh pr (eth)	d	125d	....	....	bz, chl, peth	B3², 107
14423	Tributylamine .... $(C_4H_9)_3N$	185.36	hyg	213, 91-2⁹	....	0.7771²⁰/⁰	1.4297²⁰	al, eth, ace, bz	B4⁴, 554
14424	Tributylamine, perfluoro .... $(C_4F_9)_3N$	671.10	....	179	....	1.873²⁵/⁴	1.291²⁵	ace	B2⁴, 819
14425	Tri-iso-butyl amine .... $(i-C_4H_9)_3N$	185.36	....	191.5, 84¹⁵	-21.8	0.7684²⁰/⁴	1.4252¹⁷	al, eth	B4⁴, 631
14426	Tributyrin or Glycerol tributyrate .... $C_4H_9CO_2CH(CH_2O_2CC_4H_9)_2$	302.37	....	305-10, 190¹⁵	-75	1.0350²⁰/⁴	1.4359²⁰	al, eth, ace, bz	B2⁴, 799

PHYSICAL CONSTANTS OF ORGANIC COMPOUNDS (Continued)

No.	Name, Synonyms, and Formula	Mol. wt.	Color, crystalline form, specific rotation and λ_{max} (log ε)	b.p. °C	m.p. °C	Density	n_D	Solubility	Ref.
14427	Trichloro acetamide … Cl₃CCONH₂	162.40	mcl pr (w)	238-9[746]	142	………	………	al, eth	B2[4], 520
14428	Trichloro acetamide, N,N-diethyl … Cl₃CCON(C₂H₅)₂	218.52	pr	109[9]	27	………	1.4900[24]	………	B4[4], 351
14429	Trichloro acetamide, N,N-dimethyl or N,N-Dimethyl trichloroacetamide … Cl₃CON(CH₃)₂	190.46	………	230-3d, 84[4]	12	1.390[20]	1.5017[25]	bz, chl	B4[4], 182
14430	Trichloro acetamide, N-phenyl or α-Trichloroacetanilide Cl₃CCONHC₆H₅	238.50	lf (dil al)	168-70	95-7	………	………	al	B12[3], 464
14431	Trichloroacetic acid … Cl₃CCO₂H	163.39	dlq cr	197.5, 141-2[25]	α:58 β:49.6	1.62[25/4], 1.6218[64/1]	1.4603[61]	al, eth, w	B2[4], 508
14432	Trichloroacetic acid anhydride or Trichloroacetic anhydride … (Cl₃CCO)₂O	308.76	………	222-4d, 98-100[12]		1.6908[20]	………	eth, aa	B2[4], 518
14433	Trichloroacetic acid butyl ester or Butyl trichloroacetate.. Cl₃CCO₂C₄H₉	219.50	………	203-5, 97-9[19]		1.2778[20/4]	1.4525[25]	al, eth, ace, bz	B2[4], 515
14434	Trichloroacetic acid, iso-butyl ester or iso-Butyl trichloroacetate. Cl₃CCO₂-i-C₄H₉	219.50	………	187-9, 93-4[20]		1.2636[20/4]	1.4483[20]	al, eth, bz	B2[4], 515
14435	Trichloroacetic acid, sec-butyl ester or sec-Butyl trichloroacetate. Cl₃CCO₂ sec-C₄H₉	219.50	………	93-4[24]		1.2636[20/4]	1.4483[20]	al, eth, bz	B2[3], 472
14436	Trichloro acetic acid, tert-butyl ester or tert-Butyl trichloroacetate. Cl₃CCO₂C(CH₃)₃	219.50	cr (MeOH)	54-5[7]	25.5	1.2363[25/4]	1.4398[25]	al, eth	B2[4], 516
14437	Trichloroacetic acid, 2-chloroethyl ester or 2-Chloroethyl trichloro acetate Cl₃CCO₂CH₂CH₂Cl	226.89	………	217, 100[14]		1.5357[20/4]	1.4813[20]	eth	B2[4], 515
14438	Trichloroacetic acid, ethyl ester or Ethyl trichloro acetate Cl₃CCO₂C₂H₅	191.44	………	167-8, 62[18]		1.3836[20/4]	1.4505[20]	al, eth, bz	B2[3], 514
14439	Trichloroacetic acid, 2-hydroxyethyl ester … Cl₃CCO₂CH₂CH₂OH	207.44	………	130-4[12]		1.532[20/4]	1.4775[20]	al	B2[3], 474
14440	Trichloroacetic acid, 2-methoxyethyl ester or 2-Methoxyethyl trichloro acetate. Cl₃CCO₂CH₂CH₂OCH₃	221.47	………	98-9[17]	14.5	1.3826[20/4]	1.4563[20]	al, eth, bz	B2[3], 474
14441	Trichloroacetic acid, methyl ester or Methyl trichloro acetate … Cl₃CCO₂CH₃	177.42	………	153.8, 44.5[12]	−17.5	1.4874[20/4]	1.4572[20]	al, eth	B2[4], 513
14442	Trichloroacetic acid, 2-(2-methyl butyl) ester or 2-(2-Methylbutyl) trichloroacetate Cl₃CCO₂CH₂CH(CH₃)CH₂CH₃	233.52	………	217, 92-5[12]		1.2314[20/4]	1.4521[20]	al, eth	B2[3], 473
14443	Trichloroacetic acid, pentyl ester or Pentyltrichloroacetate Cl₃CCO₂C₅H₁₁	233.52	………	220-2, 118[30]		1.2475[20/20]	………	al, eth	B2[3], 473
14444	Trichloroacetic acid, isopentyl ester or Isopentyl trichloro acetate Cl₃CO₂-i-C₅H₁₁	233.52	………	217, 92-5[11]		1.2314[20/4]	1.4521[20]	al, eth	B2[4], 516
14445	Trichloroacetic acid, propyl ester or Propyl trichloro acetate … Cl₃CCO₂C₃H₇	205.47	………	187, 69[10]		1.3221[20/4]	1.4501[20]	al, eth	B2[4], 515
14446	Trichloroacetic acid, iso-propyl ester or iso-Propyl trichloroacetate Cl₃CCO₂CH(CH₃)₂	205.47	………	173.5[747], 65-7[15]		1.3034[20/4]	1.4428[20]	al, eth, bz	B2[4], 515
14447	Trichloroacetic acid, trichlomethyl ester or Trichloromethyl trichloro acetate. Cl₃CCO₂CCl₃	280.75	………	191-2, 73-4[10]	34	………	………	eth, bz, chl, lig	B3[3], 35
14448	Trichloroacetonitrile Cl₃CCN	144.39	………	84.6[741]	−42	1.4403[25/4]	1.4409[20]	………	B2[4], 524
14449	Trichloroacetyl bromide … Cl₃CCOBr	226.29	………	143		1.900[15/15]	………	eth, bz	B2[3], 476
14450	Trichloroacetyl chloride … Cl₃CCOCl	181.83	………	118		1.6202[20/4]	1.4695[20]	eth	B2[4], 519
14451	Trichloromethyl perchlorate Cl₃COClO₃	217.82	………	exp d 40	−55	………	………	eth	B3[3], 37
14452	Tricosane CH₃(CH₂)₂₁CH₃	324.64	lf (eth-al)	380, 243[15]	47.6	0.7785[48/4] 0.7969[204]	1.4468[20]	eth	B1[4], 576
14453	Tricosane, 2-methyl (CH₃)₂CH(CH₂)₁₉CH₃	338.66	………	205[3]	37.6	0.7539[90/4]	1.4201[90]	………	B1[3], 576
14454	Tricosanoic acid, methyl ester or Methyltricosonate … CH₃(CH₂)₂₁CO₂CH₃	368.64	………		55.6	………	………		B1[3], 1078

No.	Name, Synonyms, and Formula	Mol. wt.	Color, crystalline form, specific rotation and λ_{max} (log ε)	b.p. °C	m.p. °C	Density	n_D	Solubility	Ref.
14455	1-Tricosanol $C_{22}H_{45}CH_2OH$	340.63	....	191-3[0.7]	74	....	....		B1[4], 1908
14456	Tricyclene $C_{10}H_{16}$	136.24	....	153.5	67.5	0.8268[80/4]	1.4389[20]	....	B5[4], 468
14457	1,12-Tridecadiyne $HC\equiv C(CH_2)_9C\equiv CH$	176.31	....	115.5[12]	−3	0.8262[21/4]	1.454[20]		B1[2], 248
14458	Tridecanal $C_{12}H_{25}CHO$	198.35	....	156[13]	14	0.8356[18/4]	1.4384[18]	al	B1[4], 3386
14459	Tridecanal oxime $C_{12}H_{25}CH=NOH$	213.37	nd (dil al)	....	80.5	....	....	eth, chl	B1[2], 769
14460	Tridecane $CH_3(CH_2)_{11}CH_3$	184.37	....	235.4, 107[10]	−5.5	0.7564[20/4]	1.4256[20]	al, eth	B1[4], 512
14461	Tridecane, 1-amino or Tridecyl amine $C_{12}H_{25}CH_2NH_2$	199.38	....	275.8, 140.1[10]	27.4	0.8049[20/4]	1.4443[20]	al, eth	B4[4], 810
14462	Tridecane, 1-bromo or Tridecyl bromide $C_{12}H_{25}CH_2Br$	263.27	....	296, 162[16]	6.2	1.0177[20/4]	1.4593[20]	chl	B1[4], 514
14463	Tridecane, 1,13-dibromo $Br(CH_2)_{13}Br$	342.17	....	188.92[12]	8-10	1.276[15]	1.4880[27]	eth, chl	B1[3], 548
14464	Tridecane, 7-hexyl $(C_6H_{13})_2CH$	268.52	....	170.5[10]	−28.3	0.7877[20/4]	1.4409[20]	....	B1[4], 562
14465	Tridecane, 7-methyl $(C_6H_{13})_2CHCH_3$	198.39	....	115.5[10]	−37.2	0.7634[20/4]	1.4291[20]	....	B1[4], 525
14466	Tridecane, 7-phenyl $(C_6H_{13})_2CHC_6H_5$	260.46	....	183-4[20]	....	0.8723[20/4]	1.49307[18]	....	B5[4], 1210
14467	Tridecanedioic acid $HO_2C(CH_2)_{11}CO_2H$	244.33			114	....	....	al, eth, chl	B2[4], 2141
14468	1,12-Tridecanediol $HO(CH_2)_{11}CH(OH)CH_3$	216.37	cr (dil al)	188-90[8]	60-1	....	....	al	B1[2], 563
14469	1,13-Tridecanediol or Tridecamethylene glycol $HO(CH_2)_{13}OH$	216.37	cr (bz)	195-7[10]	76.5	....	....	al, aa	B1[4], 2630
14470	Tridecan amide $C_{12}H_{25}CONH_2$	213.37	lf (al)	....	100	....	....	al, eth	B2[4], 1119
14471	Tridecanoic acid $C_{12}H_{25}CO_2H$	214.35	cr (peth, ace)	236[100], 140.5[1]	44-5	....	....	al, eth, ace, aa	B2[4], 1117
14472	Tridecanoic acid, methyl ester or Methyl tridecanoate $C_{12}H_{25}CO_2CH_3$	228.38	....	90-5[1]	fr 6.5	....	1.4405[20]	al	B2[4], 1118
14473	Tridecanonitrile $C_{12}H_{25}CN$	195.35	....	293, 142[10]	9.7	0.8257[20/4]	1.4378[20]	al, eth	B2[3], 906
14474	1-Tridecanol or Tridecyl alcohol $C_{12}H_{25}CH_2OH$	200.37	cr (al)	152[14]	32-3	0.8223[31/4]	....	al, eth	B1[4], 1860
14475	2-Tridecanol $C_{11}H_{23}CH(OH)CH_3$	200.36	....	161[30], 95-6[0.5]	23	....	1.4188[70]	al, eth, bz	B1[4], 1861
14476	2-Tridecanone or Undecyl methyl ketone $C_{11}H_{23}COCH_3$	198.35	....	263, 160[16]	30.5	0.8217[30/4]	1.4318[20]	al, eth, ace, bz	B1[4], 3386
14477	3-Tridecanone or Ethyl decyl ketone $C_{10}H_{21}COC_2H_5$	198.35	pl	140[17]	31	....	....	ace	B1[4], 3387
14478	7-Tridecanone or Dihexyl ketone $(C_6H_{13})_2CO$	198.35	lf (al)	261, 138[12]	33	0.825[30]	....	al, eth, chl, lig	B1[4], 3387
14479	1-Tridecene $C_{11}H_{23}CH=CH_2$	182.35	....	232.8, 104[11]	−13	0.7658[20/4]	1.4340[20]	al, eth, bz	B1[4], 92!
14480	1-Tridecyne $C_{11}H_{23}C\equiv CH$	180.34	....	94.5[25]	....	0.7729[20/4]	1.4309[20]	eth, bz	B1[4], 1069
14481	Triethanol amine or Tris (2-hydroxyethyl) amine $(HOCH_2CH_2)_3N$	149.19	hyg cr	277[150]	21-2	1.1242[20/4]	1.4852[20]	w, al, chl	B4[4], 1524
14482	Triethanol amine hydrochloride $(HOCH_2CH_2)_3N.HCl$	185.65	cr (al)	....	179-80	....	....		B4[4], 1525
14483	Triethylamine $(C_2H_5)_3N$	101.19	....	89.3	−114.7	0.7275[20/4]	1.4010[20]	w, al, eth, ace, bz	B4[4], 322
14484	Triethylamine hydrochloride $(C_2H_5)_3N.HCl$	137.65	hex (al)	sub 245	260d	1.0689[21/4]	....	w, al, chl	B4[4], 327
14485	Triethylamine hydroiodide $(C_2H_5)_3N \cdot HI$	229.11	pr (95% al)	....	181	1.924	....	w, al, chl	B4[4], 328
14486	Triethylamine, perfluoro $(C_2F_5)_3N$	371.05	....	70.3	....	1.736[20/4]	1.262[25]	....	B2[4], 471
14487	Triethylene glycol $HO(CH_2CH_2O)_2CH_2CH_2OH$	150.18	hyg liq	278.3, 165[14]	−5	1.1274[15/4]	1.4531[20]	w, al, bz	B1[4], 2400
14488	Triethylene glycol, 3-amino propyl ether $HO(CH_2CH_2O)_2CH_2CH_2OCH_2CH_2CH_2NH_2$	207.27	glassy	184[10]	−50	1.0682[20/20]	1.4668[20]	w, al	C55, 5935

No.	Name, Synonyms, and Formula	Mol. wt.	Color, crystalline form, specific rotation and λ_{max} (log ϵ)	b.p. °C	m.p. °C	Density	n_D	Solubility	Ref.
14489	Triethylene glycol diacette .... $CH_3CO_2(CH_2CH_2O)_2CH_2CH_2O_2CCH_3$	234.25	....	300	....			w, al, eth	B2[4], 215
14490	Triethylene glycol, monobutyl ether .... $HO(CH_2CH_2O)_2CH_2CH_2OC_4H_9$	206.29	....	278	....	$0.9890^{20/4}$	1.4389^{20}	w, al, MeOH	B1[4], 2402
14491	Triethylene tetramine $H_2NCH_2(CH_2NHCH_2)_2CH_2NH_2$	146.24	....	266-7, 157^{20}	12		1.4971^{20}	w, al	B4[4], 1242
14492	Trifluoro acetamide, N-phenyl or Trifluoro acetanilide .. $F_3CCONHC_6H_5$	189.14	(60% al)	....	87.6			al	B12[2], 141
14493	Trifluoroacetic acid F_3CCO_2H	114.02	....	72.4	−15.2	1.5351^0		w, al, eth, ace	B2[4], 458
14494	Trifluoroacetic acid anhydride .... $(F_3CCO)_2O$	210.03	....	39-40	−65	$1.490^{25/4}$	1.269^{26}	eth, aa	B2[4], 469
14495	Trifluoro acetic acid, butyl ester or Butyltrifluoro acetate $F_3CCO_2C_4H_9$	170.13	....	100.2	....	$1.0268^{22/4}$	1.353^{20}	chl	B2[4], 464
14496	Trifluoro acetic acid, tert-butyl ester .... $F_3CCO_2C(CH_3)_3$	170.13	....	83	....		1.3300^{25}	chl	B2[4], 465
14497	Trifluoroacetic acid, ethyl ester or Ethyltrifluoroacetate $F_3CCO_2C_2H_5$	142.08	....	60-2	....	$1.19^{20/4}$	1.308^{20}		B2[4], 463
14498	Trifluoroacetic acid, methyl ester or Methyl trifluoro acetate $F_3CCO_2CH_3$	128.03	....	....	....	$1.28^{20/4}$			B2[4], 463
14499	Trifluoroacetic acid, propyl ester or Propyltrifluoro acetate $F_3CCO_2C_3H_7$	156.10	....	82.5	....	$1.1285^{25/4}$	$1.3233^{22.5}$	chl	B2[4], 464
14500	Trifluoroacetonitrile F_3CCN	95.03	....	−64					B2[4], 472
14501	Trifloromethyl peroxide .... $CF_3-OO-CF_3$	170.01	....	−32					B3[4], 22
14502	Trifluoromethyl sulfide $(CF_3)_2S$	170.08	....	−22.2					B3[4], 278
14503	Trifurfuryl amine .... $(C_5H_3O)_3N$	257.29	....	136-8[1]				eth	B18[4], 7094
14504	Triglycine $H_2N(CH_2CONH)_2CH_2CO_2H$	189.17	nd (dil al)	....	246d			w	B4[3], 1198
14505	Triglycine-N-phthalyl $C_{14}H_{13}N_3O_6$	319.28	nd (al)	....	234-5d			w, al	B21[3], 5186
14506	Tri-heptyl amine .... $(C_7H_{15})_3N$	311.60	....	330^{762}, $151-4^1$	....			al, eth	B4[4], 736
14507	Tri-hexyl amine .... $(C_6H_{13})_3N$	269.52	....	263-4, 119^{15}	....			al, eth	B4[4], 711
14508	Triidoacetic acid I_3CCO_2H	437.74	ye lf	....	150d			w, al, eth	B2[4], 537
14509	Trilaurin or Glycerol trilaurate .... $C_{11}H_{23}CO_2CH(CH_2O_2CC_{11}H_{23})_2$	639.03	nd (al)	....	46.4	0.8986^{55}	1.4404^{60}	al, eth, ace, bz, chl	B2[4], 1098
14510	Trilauryl amine or Tridodecylamine $(_{12}H_{25})_3N$	522.01	....	$220-8^{0.03}$	fr 15.7		1.4567^{25}	eth, bz, chl	Am74, 428[1]
14511	Trimellitic acid or 1,2,4-Benzene tricarboxylic acid....... $1,2,4-C_6H_3(CO_2H)_3$	210.15	nd (w), cr (aa, al)	....	238d			w, al, eth	B9[3], 4792
14512	Trimesic acid or 1,3,5-Benzene tricarboxylic acid .... $1,3,5-C_6H_3(CO_2H)_3$	210.15	pr or nd (w + 1)	....	380 (anh)			al, eth	B9[3], 4793
14513	Trimethadione or 3,5,5-trimethyl 2,4-oxazoldinedione . $C_6H_9NO_3$	143.14	cr (50% MeOH)	$78-80^5$	46			al, eth, ace, bz, chl	B4[4], 801[3]
14514	Trimethyl amine .... $(CH_3)_3N$	59.11	....	2.9	−117.2	$0.6356^{20/4}$	1.3631^0	w, al, eth, bz, chl	B4[4], 134
14515	Trimethyl amine hydrochloride .... $(CH_3)_3N.HCl$	95.57	mcl dlq nd (al)	sub at 200	277-8			w, al, chl	B4[4], 138
14516	Trimethylamine oxide $(CH_3)_3NO$	75.11	hyg nd (w + 2)	....	255-7			w, al	B4[4], 144
14517	Trimethylene oxide, hydrochloride .... $(CH_3)_3NO.HCl$	111.57	nd (al)	....	218-20d			w, al	B4[4], 144
14518	Trimethylamine oxide, hydroiodide .... $(CH_3)_3NO.HI$	203.02	pr (al)	....	130d			w, al	B4, 50
14519	Trimethylamine, perfluoro .... $(CF_3)_3N$	221.03	....	....	−7				B3[4], 79
14520	Tris-(2-methylbutyl) amine .... $[CH_3CH_2CH(CH_3)CH_2]_3N$	227.44	....	230-7, 94^4	....	0.7964^{13}	1.4330^{20}	al, eth, ace, bz	B4, 179
14521	Trimethylene sulfide or Thiacyclobutane .... C_3H_6S	74.15	....	94.7, 14^{30}	−73.2	$1.0200^{20/4}$	1.5102^{20}	al, ace, bz	B17[4], 14

No.	Name, Synonyms, and Formula	Mol. wt.	Color, crystalline form, specific rotation and λ_{max} (log ϵ)	b.p. °C	m.p. °C	Density	n_D	Solubility	Ref.
14522	Trimethylol ethane $CH_3C(CH_2OH)_3$	120.15	wh pw or nd (al)	135-7[15]	204	....		w, al	B1[4], 2780
14523	1,1,1-Trimethylol propane or TMP $CH_3CH_2C(CH_2OH)_3$	134.18	wh pw or pl	160[5]	58			w, al	B1[4], 2786
14524	Trimethyloxonium fluoborate $[(CH_3)_3O]^+BF_4^-$	147.91	hyg nd		148d			ace, chl	B1[4], 1248
14525	Bis-(trimethylsilyl) amine $[(CH_3)_3Si]_2NH$	161.39	....	126.2		0.7741[25]	1.4090[20]		B4[3], 1861
14526	Trinitro acetonitrile $(NO_2)_3CCN$	176.05	wax	220 exp	41.5			eth	B2, 229
14527	Tri-octylamine $(C_8H_{17})_3N$	353.67		365, 182[5]			1.4510[19]		B4[4], 754
14528	Triolein or Glycerol trioleate $C_{17}H_{33}CO_2CH(CH_2O_2CC_{17}H_{33})_2$	885.47	poly-morphic	235-40[18]	-5.5	0.8988[40]	1.4621[40]	eth, chl, peth	B2[4], 1664
14529	1,3,5-Trioxane or Metaformaldehyde $C_3H_6O_3$	90.08	rh nd (eth)	114.5, sub 46[1]	64	1.17[65]		w, al, eth, bz, chl	B19[4], 4710
14530	Tripalmitin or Glycerol tripalmitate $C_{15}H_{31}CO_2CH(CH_2O_2CC_{15}H_{31})_2$	807.35	nd (eth)	310-20	66.4	0.8752[70/4]	1.4381[80]	eth, bz, chl	B2[3], 971
14531	Tripentylamine or Triamylamine $(C_5H_{11})_3N$	227.44		240-5, 130[14]		0.7907[20/4]	1.4366[20]	al, eth	B4[4], 676
14532	Tri-isopentylamine $(i-C_5H_{11})_3N$	227.44		235, 94[4]		0.7848[20/4]	1.4331[20]	al, eth, bz	B4[4], 700
14533	Triphenyl acetic acid $(C_6H_5)_3CCO_2H$	288.35	mcl pr (al), lf (aa)	....	271			al, aa, lig	B9[3], 3585
14534	Triphenyl amine $(C_6H_5)_3N$	245.33	mcl (MeOH, AcOEt, bz)	365	127	0.774[0/0]	1.353[16]	eth, bz	B12[3], 292
14535	Triphenylene $C_{18}H_{12}$	228.29	....	425	199			al, bz, chl	B5[3], 2384
14536	Triphenylethylene, chloro $(C_6H_5)_2C=C(Cl)C_6H_5$	290.80			117			al, eth, ace, bz, chl	B5[3], 2400
14537	Triphenyl methane or Tritan $(C_6H_5)_3CH$	244.34	rh (al)	358-9[754], 190-215[10]	94	1.014[99/4]	1.5839[99]	eth, bz, chl	B5[3], 2307
14538	Triphenyl methane, 3-amino or 3-Amino tritan $3-H_2NC_6H_4CH(C_6H_5)_2$	259.36	nd (eth)		120			al, eth	B12[2], 790
14539	Triphenylmethane, 4-amino or 4-Aminotritan $4-H_2NC_6H_4CH(C_6H_5)_2$	259.36	pr or lf (eth, lig), ta (Peth)	ca 248[12]	84-5			eth, bz, lig	B12[2], 790
14540	Triphenyl methane, α-chloro or Trityl chloride $(C_6H_5)_3CCl$	278.79	nd or pr (bz-peth)	310, 230—5[20]	113-4			eth, ace, bz, chl	B5[3], 2315
14541	Triphenylmethane, 4,4'-diamino $(4-H_2NC_6H_4)_2CHC_6H_5$	274.37	pr (bz, eth)	....	139-40			al, eth, chl, lig	B13[3], 529
14542	Triphenyl methane, 4,4'-dimethyl $(4-CH_3C_6H_4)_2CHC_6H_5$	272.40	nd (MeOH)	218-20[12]	56			al, eth, bz, chl, lig	B13[3], 2342
14543	Triphenylmethane, 2,4-bis (dimethylamino) $2,4-[(CH_3)_2N]_2C_6H_3CH(C_6H_5)_2$	330.48	pl (peth)		122-3			al, eth, ace, bz	B13, 273
14544	Triphenyl methane, 4,4'-bis(dimethyl amino) $(4-(CH_3)_2NC_6H_4)_2CHC_6H_5$	330.48	nd or lf (al, bz)		102			eth, bz	B13[3], 529
14545	Triphenyl methane, 4,4'-bis(dimethyl amino)-4"-amino $[4-(CH_3)_2NC_6H_4]_2CHC_6H_4NH_2-4$	345.49	(al)	....	151-2				B13[3], 566
14546	Triphenyl methane, 4,4'-bis(dimethyl amino)-2"-hydroxy $[4-(CH_3)_2NC_6H_4]_2CH(C_6H_4OH-2)$	346.48	nd (al)		127-8			bz, lig	B13[3], 2063
14547	Triphenyl methane, 4,4'-bis(dimethyl amino)-3"-hydroxy $[4-(CH_3)_2NC_6H_4]_2CH(C_6H_4OH-3)$	346.48	cr (al)		149			bz	B13[2], 440
14548	Triphenyl methane, 4,4'-bis(dimethyl amino)-4"-hydroxy $[4-(CH_3)_2NC_6H_4]_2CH(C_6H_4OH-4)$	346.48	cr (al)		165			al, bz	B13[2], 440
14549	Triphenylmethane, 3-methyl or 3-Methyltritan $(3-CH_3C_6H_4)CH(C_6H_5)_2$	258.37	pr (al)	354[706]	62			eth, bz, chl, aa, lig	B5[3], 2332
14550	Triphenylmethane, 2,4',4"-triamino $(4-H_2NC_6H_4)_2CH(C_6H_4NH_2-2)$	289.38	cr (al)	....	165			al	B13[3], 565
14551	Triphenylmethane, 3,4',4"-triamino $(4-H_2NC_6H_4)_2CH(C_6H_4NH_2-3)$	289.38	nd (eth, eth-lig)		150			al	B13[3], 565
14552	Triphenyl methane, 4,4',4"-triamino or p-Leucaniline $(4-H_2NC_6H_4)_3CH$	289.38	lf (w, al, bz)		208d			al, eth	B13[3], 566
14553	Triphenyl methane, 4,4',4"-tris(dimethylamino) or Leuco crystal violet $[(CH_3)_2NC_6H_4]_3CH$	373.55	lf (al), nd (bz, lig)	....	175			eth, bz, chl, aa	B13[3], 566

No.	Name, Synonyms, and Formula	Mol. wt.	Color, crystalline form, specific rotation and λ_{max} (log ε)	b.p. °C	m.p. °C	Density	n_D	Solubility	Ref.
14554	Triphenyl methane, 4,4′,4″-trihydroxy or Leucoaurin .... (4-HOC₆H₄)₃CH	292.34	pr (aa, dil al)		240			al, chl, aa	B6³, 6578
14555	Triphenylmethane, 2,2′,2″-trimethyl or Tri-2-tolylmethane . [2-CH₃C₆H₄]₃CH	286.42	nd (al)		130-1			al, eth	B5³, 2347
14556	Triphenylmethane, 4,4′,4″-trinitro .... (4-O₂NC₆H₄)₃CH	379.33	sc (bz)		212.5				B5⁴, 2501
14557	Triphenylmethanol or Tritanol .... (C₆H₅)₃COH	260.34	pl (al), trg (bz)	380	164.2	1.199⁰ᐟ⁴		al, eth, ace, bz, aa	B6³, 3640
14558	Triphenylmethanol, 2,2′-bis-(dimethylamino) .... [2-(CH₃)₂NC₆H₄]₂C(OH)C₆H₅	346.48	pr (lig)		105				B13, 741
14559	Triphenylmethanol, 2,3′-bis(dimethylamino) 2,3′[(CH₃)₂NC₆H₄]₂C(OH)C₆H₅	346.48	pl (bz)		183-4			bz, HCl	B13, 742
14560	Triphenylmethanol, 2,4′-bis (dimethylamino) .... C₂₃H₂₆N₂O	346.48	(al)		169-70			bz	B13, 742
14561	Triphenyl methanol, 3,3′-bis(dimethylamino) .... [3-(CH₃)₂NC₆H₄]₂	346.48	cr (eth)		128-9			eth, aa	B13, 742
14562	Triphenylmethanol, 3,4′-bis(dimethyl amino) .... C₂₃H₂₆N₂O	346.48	pr (bz-al)		140			bz	B13, 742
14563	Triphenylmethanol, 4,4′-bis(dimethylamino) or Michler's hydrol .... [4-(CH₃)₂NC₆H₄]₂C(OH)C₆H₅	346.48	cr (eth, bz, lig, MeOH)		121-3			eth, bz, lig	B13³, 2068
14564	Triphenyl methanol, 4,4′-dihydroxy or Benzaurin ...... [4-HOC₆H₄]₂C(OH)C₆H₅	292.34	ye-red pw		110-20				B6³, 6582
14565	Triphenyl methanol, 4,4′,4″-triamino or Pararosaniline ... (4-H₂NC₆H₄)₃COH	305.38	col to red lf		189-205			al	B13³, 2072
14566	Triphenylmethanol, 3,3′,3″-trinitro .... (3-NO₂C₆H₄)₃COH	395.33	rh (MeOH, dil)		167			bz, aa	B6¹, 352
14567	Triphenylmethanol, 4,4′,4″-trinitro .... [-O₂NC₆H₄]₃COH	395.33	mcl pr (bz, aa)		(i) 190, (ii) 167			bz, aa	B6³, 3673
14568	Triphenylselenonium chloride .... (C₆H₅)₃SeCL	345.71	orh (AcOEt)		230d			w, al, chl	B6⁴, 1780
14569	Triphenylselenonium fluoride .... (C₆H₅)₃SeF	329.26	oct deliq		145d			w, al, ace, chl	B6³, 1108
14570	Bis-(2 nitrophenyl) trisulfide .... (2-NO₂C₆H₄)₂S₃	340.41	ye nd (al)		175-6				B6³, 1062
14571	2,3,5-Triphenyl,1,2,3,4-tetrazolium chloride or Tetrazolium salt . T.T.C. . C₁₉H₁₅N₄Cl	334.81	nd (al, chl)		243d			w, al, ace	B26, 363
14572	Trisopropanolamine .... [CH₃CH(OH)CH₂]₃N	227.31		170-80¹⁰	45	1.0²⁰ᐟ⁴		w, al	B4⁴, 1680
14573	Tripropylamine .... (C₃H₇)₃N	143.28		156	−93.5	0.7558²⁰ᐟ⁴	1.4181²⁰	al, eth	B4⁴, 470
14574	Tripropylamine, perfluoro .... (C₃F₇)₃N	521.07		130		1.822⁴ᐟ⁴	1.279²⁵		B2⁴, 742
14575	Triptane or 2,2,3-Trimethyl butane............... (CH₃)₃CHC(CH₃)₃	100.21		80.9	−24.2	0.6901²⁰ᐟ⁴	1.3894²⁰	al, eth, ace, bz	B1⁴, 410
14576	Trisiloxane, 1,5-dichloro-1,1,3,3,5,5-hexamethyl ClSi(CH₃)₂OSi(CH₂)₂OSi(CH₃)₂Cl	277.37		184	−53	1.018²⁰ᐟ⁴			B4³, 1884
14577	Trisiloxane, 1,5-dihydroxy-1,1,3,3,5,5-hexamethyl HOSi(CH₃)₂OSi(CH₂)₂OSi(CH₃)₂OH	240.48			−23	0.9950²⁰	1.4090²⁰ᐟ_D		KHOC, 653
14578	Trisiloxane, 1,1,1,3,5,5,5-heptamethyl (CH₃)₃SiOSiH(CH₃)OSi(CH₃)₃	222.51				0.8194²⁰ᐟ⁴	1.3818²⁰		B4³, 1874
14579	Trisiloxane, octamethyl (CH₃)₃SiOSi(CH₃)₂OSi(CH₃)₃	236.54		153⁷⁷⁴, 50-2¹⁷	−80	0.8200²⁰ᐟ⁴	1.3840²⁰	bz, peth	B4³, 1879
14580	Trisiloxane, 1,1,3,5,5-pentamethyl-1,3,5-triphenyl .... C₆H₅Si(CH₃)₂OSi(CH₃)(C₆H₅)OSi(CH₃)₂C₆H₅	422.75		169⁰·⁷		1.0227²⁰ᐟ⁴	1.5280²⁰	bz, lig	Am70, 1116
14581	Tristearin or Glycerol tristearate .... C₁₇H₃₅CO₂CH(CH₂O₂CC₁₇H₃₅)₂	891.51	cr (eth, peth)		73	0.8559⁹⁰ᐟ⁴	1.4399⁸⁰	ace, chl, CS₂	B2⁴, 1233
14582	Tristearylamine or Trioctadecylamine .... (C₁₈H₃₇)₃N	774.50			54.6			eth, bz, chl	B4⁴, 829
14583	Tritetracontane .... C₄₃H₈₈ (CH₃(CH₂)₄₁CH₃	605.17		332³	85.5	0.7812⁹⁰ᐟ⁴	1.4340⁹⁰		B1³, 593
14584	Tris-(4-tolyl) amine .... (4-CH₃C₆H₄)₃N	287.41	cr (aa)		117			eth, ace, bz, chl	B12³, 2034
14585	3,3′,4″-Tritolylamine .... (3-CH₃C₆H₄)₂N(C₆H₄CH₃-4)	287.41	nd (al)		89.90			al	B12¹, 415

No.	Name, Synonyms, and Formula	Mol. wt.	Color, crystalline form, specific rotation and λ_{max} (log ϵ)	b.p. °C	m.p. °C	Density	n_D	Solubility	Ref.
14586	Bis-(tribromomethyl) trisulfide .... $(CBr_3)_2S_3$	599.67	rh pr (eth)	d	125d	....	....	bz, chl, peth	B3², 107
14587	1,3,5-Trithiane or Trithioformaldehyde .... $C_3H_6S_3$	138.27	hex (bz), pr (w), nd (al)	sub	220	$1.6374^{24/4}$	....	bz	B19⁴, 4711
14588	1,3,5-Trithiane, 2,4,6-trimethyl-(α -form) or Trithioacetaldehyde .... $2,4,6-(CH_3)_3(C_3H_3S_3)$	180.36	mcl (al, ace)	245-8	101	....	....	al, eth, ace, bz, chl	B19⁴, 4719
14589	1,3,5-Trithiane, 2,4,6-trimethyl-(β -form) $1,3,5-(CH_3)_3(C_3H_3S_3)$	180.36	rh nd (ace)	245-8	126-7	....	....	al, eth, ace, bz, chl	B19⁴, 4718
14590	1,3,5-Trithiane-2,4,6-trimethyl-2,4-6-triphenyl .... $C_{24}H_{24}S_3$	408.65	nd (al)		122	....	....	eth, ace, chl	B19⁴, 4775
14591	1,3,5-Trithiane, 2,4,6-triphenyl (α-form) .... $2,4,6(C_6H_5)_3C_3H_3S_3$	366.57	nd (bz-al)	d	167	....	....	bz, chl	B19⁴, 4774
14592	1,3,5-Trithiane, 2,4,6-triphenyl-(β-form) $2,4,6-(C_6H_5)_3C_3H_3S_3$	366.57	cr (bz)	....	229-30	....	....	bz	B19⁴, 4773
14593	α -Tritisterol $C_{30}H_{50}O$	426.74	nd (MeOH -ace), $[\alpha]^{20/}_D$ + 54.3 (al)		114-5	....	....	al, peth, chl	B6³, 2911
14594	β -Tritisterol $C_{30}H_{50}O$	426.74	nd (MeOH), $[\alpha]_D$ + 49.2 (al)		97	....	....	al	H20, 424
14595	Tropacocaine or Benzoyl-ψ -tropeine .... $C_{15}H_{19}NO_2$	245.33	pl or tab	d	49	$1.0426^{100/4}$	1.5080^{100}	al, eth, bz, chl, peth	B21⁴, 174
14596	Tropacocaine hydrochloride $C_{15}H_{19}NO_2 . HCl$	281.79	pl (al)		283d	....	....	w	B21⁴, 174
14597	Tropane or 2,3-Dihydro-8-methylnortropidine .... $C_8H_{15}N$	125.22		167	....	$0.9259^{15/15}$	1.4732^{20}	....	B20⁴, 1963
14598	3-Tropanol $C_8H_{15}NO$	141.21		229	64	....	....	w, al, eth, chl	B21⁴, 168
14599	3-Tropanone $C_8H_{13}NO$	139.20		224-5⁷¹⁴, 113²⁵	42-4	$1.9872^{100/4}$	1.4598^{100}	al, eth, ace, bz	B21⁴, 3299
14600	Tropeine, benzoyl $C_{15}H_{19}NO_2$	245.33	cr (eth)	175-70	41-2	....	....	w, al, eth	B21⁴, 173
14601	Tropeine, benzoyl, dihydrate $C_{15}H_{19}NO_2 . 2H_2O$	281.36	pl (w)		58	....	....	w, al, eth	B21, 19
14602	Tropine or 3-Tropanol $C_8H_{15}NO$	141.22	hyg pl (eth)	229	64	....	....	w, al, eth, chl	B21², 17
14603	Tropinic acid [(d)] $C_8H_{13}NO_4$	187.20	cr (w, dil al), $[\alpha]'_D$ + 14.8 (w)		253d	....	....	w	B22, 123
14604	Tropinic acid [(dl)] $C_8H_{13}NO_4$	187.20	nd (dil al)		251d	....	....	w	B22, 124
14605	Tropinic acid [(l)] $C_8H_{13}NO_4$	187.20	cr (a), $[\alpha]^{20/}_D$ -14.8 (w)		243	....	....	w	B22, 124
14606	Tropinone or 3-Tropanone .... $C_8H_{13}NO$	139.20	lo nd (peth)	224-5⁷¹⁴, 113²⁵	42-4	$1.9872^{100/4}$	1.4598^{100}	al, eth, ace, bz	B21², 225
14607	Truxane or Di-indene .... $C_{18}H_{16}$	232.33	pl (peth)		116	....	....	al, eth, aa	B5³, 2253
14608	Truxane, tetraphenyl or Triphenyl allene dimer .... $(C_6H_5)_4C_{18}H_{12}$	536.73	lf (bz-al)		210	....	....	chl	B5³, 2790
14609	α -Truxilline or Cocamine .... $C_{36}H_{46}N_2O_6$	658.80	amor pw		[ca] 80	....	....	al, eth, bz, chl	B22, 202
14610	Tryptophan [(D)] or α -Amino-β -indolyl propionic acid .. $C_{11}H_{12}N_2O_2$	204.23	pl (dil al), $[\alpha]^{20/}_D$ + 33 (w)		281-2	....	....	al	B22⁴, 6765
14611	Tryptophan [(DL)] $C_{11}H_{12}N_2O_2$	204.23	pl (50% al)		282, (293)	....	....		B22⁴, 6768
14612	Tryptophan [(L)] $C_{11}H_{12}N_2O_2$	204.23	lf or pl (dil al), $[\alpha]^{20/}_D$ + 6.1 (1N NaOH, p=11)		290-2d	....	....	w	B22⁴, 6765
14613	Tryptophan, [N] -acetyl-[(DL)] .... $C_{13}H_{14}N_2O_3$	246.27	pl (dil al)		206-7	....	....	al, eth	B22⁴, 6782

No.	Name, Synonyms, and Formula	Mol. wt.	Color, crystalline form, specific rotation and λ_{max} (log ε)	b.p. °C	m.p. °C	Density	n_D	Solubility	Ref.
14614	Tryptophan, [N]-acetyl-[(L)] .... $C_{13}H_{14}N_2O_3$	246.27	nd (dil MeOH), $[\alpha]^{15}_D$ +25 (95% al, c=1)	....	189-90		....	w, al	B22[4], 6781
14615	Tryptophan, [N]-methyl-[(DL)] .... $C_{12}H_{14}N_2O_2$	218.26	nd (dil al)	....	297d		....	al	B22[4], 6776
14616	Tryptophan, [N]-methyl [(L)] or Abrin .... $C_{12}H_{14}N_2O_2$	218.26	pr (w) $[\alpha]^{21}_D$ +44.4 (dil HCl)		295d			al	B22[4], 6776
14617	Tuduranine .... $C_{18}H_{19}NO_3$	297.36	nd (eth), $[\alpha]^{20}_D$ −127.5 (al)		204			al, eth, ace	B21[4], 2643
14618	Turanose or 3-O-α-D-glucoside-D-fructose .... $C_{12}H_{22}O_{11}$	342.30	pr (w-al, MeOH) $[\alpha]^{20}_D$ +27.2→ +75.8 (w,c=4,mut)		168			w, al	B17[4], 3092
14619	Tyramine or [p]-(β-aminoethyl)phenol .... 4-($H_2NCH_2CH_2$)C_6H_4OH	137.18	pl or nd (bz), cr (al) nd (w)	205-7[25], 165-7[2]	164-5			al, xyl	B13[3], 1637
14620	Tyrosine [(D)] or 2-Amino-3-(4-hydroxyphenyl)propionic acid .... (4-HOC$_6$H$_4$)CH$_2$CH(NH$_2$)CO$_2$H	181.19	nr (n), $[\alpha]^{25}_D$ +10.3 (1N HCl, c=1)	....	318-1d	....		dil HCl	B14[3], 1504
14621	Tyrosine [(DL)] .... (4-HOC$_6$H$_4$)CH$_2$CH(NH$_2$)CO$_2$H	181.19	nd	....	340d				B14[3], 1506
14622	Tyrosine [(L)] .... (4-HOC$_6$H$_4$)CH$_2$CH(NH$_2$)CO$_2$H	181.19	nd (w), $[\alpha]^{22}_D$ −10.6 (1N HCl,c=4)	sub	342-4d (sealed tube)	....			B14[3], 1504
14623	Tyrosine amide [(L)] .... (4-HOC$_6$H$_4$)CH$_2$CH(NH$_2$)CONH$_2$	180.21	pl or pr (al), $[\alpha]^{20}_D$ +19.5 (w)		153-4		....	w, al, MeOH	B14[3], 1511
14624	Tyrosine-3-bromo [(L)] .... (3-Br-4-HOC$_6$H$_3$)CH$_2$CH(NH$_2$)CO$_2$H	260.09	cr (w + 1), nd (w + 2)		246-9d				B14[2], 377
14625	Tyrosine-3,5-dibromo [(L)] .... (3,5-Br$_2$-4-HOC$_6$H$_2$)CH$_2$CH(NH$_2$)CO$_2$H	339.00	nd or pl (w + 2), $[\alpha]^{20}_D$ −5.5 (1N HCl, c=5)		245 (anh)				B14[3], 1561
14626	Tyrosine, ethyl ester [(L)] .... (4-HOC$_6$H$_4$)CH$_2$CH(NH$_2$)CO$_2$C$_2$H$_5$	209.25	pr (AcOEt), $[\alpha]^{20}_D$ +20.4 (MeOH)		108-9			al, bz, AcOEt, MeOH	B14[3], 1510
14627	Tyrosine, [N]-methyl-[(L)] .... (4-HOC$_6$H$_4$)CH$_2$CH(NHCH$_3$)CO$_2$H	195.21	nd $[\alpha]^{21}_D$ +19.8 (dil HCl)		257, (d 280)				B14[3], 1513
14628	Tyrosine, methyl ester .... (4-HOC$_6$H$_4$)CH$_2$CH(NH$_2$)CO$_2$CH$_3$	195.21	pr (AcOEt), $[\alpha]^{20}_D$ +25.7 (MeOH)		136-7			al, MeOH	B14[3], 1510

No.	Name, Synonyms, and Formula	Mol. wt.	Color, crystalline form, specific rotation and λ_{max} (log ε)	b.p. °C	m.p. °C	Density	n_D	Solubility	Ref.
14629	Umbellatine $C_2H_{21}NO_8$	415.71	ye nd	....	206-7	....	....	al	C48, 10034
14630	1,10-Undecadiyne $HC\equiv C(CH_2)_7C\equiv CH$	148.25	....	83[12]	−17	0.8182[21/4]	1.453[21]	ace, bz	B1[2], 248
14631	Undecanal or Undecylaldehyde $CH_3(CH_2)_9CHO$	170.30	....	117[18]	−4	0.8251[23/4]	1.4520[20]	al, eth	B1[4], 3374
14632	Undecanal oxime $CH_3(CH_2)_9CH=NOH$	185.31	wh nd (dil MeOH)	....	72	....	....	w, al, eth	B1[4], 3374
14633	Undecanal, 2-methyl $CH_3(CH_2)_8CH(CH_3)CHO$	184.33	....	114[10]	....	0.830[15/4]	1.4321[20]	al, eth	B1[4], 3383
14634	Undecane $C_{11}H_{23}$	156.32	....	196, 75[10]	−25.6	0.7402[20/4]	1.4398[20]	al	B1[4], 487
14635	Undecane, 1-amino or Undecylamine $CH_3(CH_2)_{10}NH_2$	171.33	cr (eth, al)	242, 112.3[10]	17	0.7979[20/4]	1.4398[20]	al	B4[4], 792
14636	Undecane, 1-bromo-11-fluoro $F(CH_2)_{11}Br$	253.21	....	95[0.6]	....	....	1.4518[25]	al, eth	B1[4], 490
14637	Undecanoic acid or Undecylic acid $CH_3(CH_2)_9CO_2H$	186.30	cr (ace)	280, 164[15]	28.6	0.8907[20/4]	1.4294[45]	al, eth, ace, bz, chl	B2[4], 1068
14638	Undecanoic acid, 2-bromo $CH_3(CH_2)_8CHBrCO_2H$	265.20	....	178-83[14]	10	....	....	....	B2[4], 1073
14639	Undecanoic acid, 10-bromo $CH_3CHBr(CH_2)_8CO_2H$	265.20	cr (peth-eth, bz, ace)	....	35.7	....	....	al	B2[3], 861
14640	Undecanoic acid, 11-bromo $BrCH_2(CH_2)_9CO_2H$	265.20	nd (lig)	188[18]	57	....	....	al, eth, ace, bz	B2[4], 1073
14641	Undecanoic acid, 10,11-dibromo $BrCH_2CH(Br)(CH_2)_8CO_2H$	344.10	cr	....	38	....	....	al	B2[4], 1075
14642	Undecanoic acid, ethyl ester or Ethyl undecanoate $CH_3(CH_2)_9CO_2C_2H_5$	214.35	....	131[14]	−15	0.8633[20/4]	1.4285[20]	al, eth, ace, bz	B2[3], 858
14643	Undecanoic acid, 11-fluoro $F(CH_2)_{10}CO_2H$	204.29	....	113-5[0.25]	36	....	....	al, eth, lig	B2[4], 1071
14644	Undecanoic acid, 4-hydroxy-δ-lactone $C_7H_{15}\underset{\frown}{CHCH_2CH_2}CO$	184.28	....	286, 162[13]	....	0.9494[20/4]	1.4512[20]	al	B17[4], 4260
14645	Undecanoic acid -α-lactone or 2-Undecalactone $C_{11}H_{20}O_2$	184.28	....	286, 162[13]	....	0.9494[20/4]	1.4512[20]	al	B17[2], 284
14646	1-Undecanol or 1-Undecylalcohol $CH_3(CH_2)_9CH_2OH$	172.31	....	243, 131[15]	19	0.8298[20/4]	1.4392[20]	al, eth	B1[4], 1835
14647	1-Undecanol, 2-methyl $CH_3(CH_2)_8CH(CH_3)CH_2OH$	186.34	....	129-31[12]	....	0.8300[15/4]	1.4382[20]	al, eth	B1[4], 1855
14648	2-Undecanol [(d)] $CH_3(CH_2)_8CHOHCH_3$	172.31	(bz), $[\alpha]^{20/}_D$ + 10.3 (bz)	128[20]	12	0.8270[20/4]	1.4369[20]	al, eth	B1[3], 1774
14649	2-Undecanol [(dl)] $CH_3(CH_2)_8CHOHCH_3$	172.31	....	228, 119[12]	0	0.8268[19]	1.4369[20]	al, eth	B1[4], 1838
14650	2-Undecanol -[(l)] $CH_3(CH_2)_8CHOHCH_3$	172.31	$[\alpha]_D$ −0.02	231-3	....	0.8302[20]	1.4381[20]	al, eth	B1[3], 1774
14651	3-Undecanol (l) $CH_3(CH_2)_7CHOHCH_2CH_3$	172.31	$[\alpha]^{20/}_D$ −6.2 (al)	229	17	0.8295[20/4]	1.4367[20]	al, eth, bz	B1[2], 462
14652	5-Undecanol $C_6H_{13}CHOH(C_4H_9)$	172.31	....	229, 107[12]	−3.5	0.8292[20/4]	1.4354[24]	al, eth	B1[4], 1839
14653	6-Undecanol $C_5H_{11}CH(OH)(C_5H_{11})$	172.31	cr (ace)	228, 117-8[16]	25	0.8334[20/4]	1.4374[20]	al, ace	B1[4], 1839
14654	1-Undecanethiol or Undecyl mercaptan $CH_3(CH_2)_9CH_2SH$	188.37	....	257.4	−3	0.8448[20]	1.4585[20]	....	B1[4], 1838 2(1.1000)
14655	2-Undecanone $CH_3(CH_2)_8COCH_3$	170.30	....	231-2, 106[12]	15	0.8250[20/4]	1.4291[20]	al, eth, ace, bz, chl	B1[4], 3374
14656	3-Undecanone $CH_3(CH_2)_7COCH_2CH_3$	170.30	....	227, 104-6[11]	12	0.8272[20/4]	1.4296[20]	al, eth	B1[4], 3375
14657	4-Undecanone $CH_3(CH_2)_6CO(CH_2)_2CH_3$	170.30	....	106[13]	4-5	0.8274[25/4]	1.4248[24]	al, eth	B1[4], 3375
14658	4-Undecanone, 7-ethyl,2-methyl $C_4H_9CH(C_2H_5)CH_2CH_2COCH_2CH(CH_3)_2$	212.37	....	252-3, 101-3[4]	....	0.8362[20/4]	1.4370[20]	al, eth, ace, bz	B1[3], 2920
14659	5-Undecanone $CH_3(CH_2)_5CO(CH_2)_3CH_3$	170.30	....	227, 105-6[12]	2	0.8278[19/4]	1.4275[18]	al, eth	B1[4], 3375
14660	6-Undecanone or n-Caprone $(C_5H_{11})_2CO$	170.30	....	228, 145.7[29]	14-5	0.8308[20/4]	1.4270[20]	al, eth	B1[4], 3376
14661	Undecanonitrile $CH_3(CH_2)_9CN$	167.30	....	253, 143[22]	....	0.8254[30/4]	1.4293[30]	al, eth	B3[2], 860

No.	Name, Synonyms, and Formula	Mol. wt.	Color, crystalline form, specific rotation and λ_{max} (log ϵ)	b.p. °C	m.p. °C	Density	n_D	Solubility	Ref.
14662	Undecasiloxane, tetra cosamethyl $(CH_3)_3SiO[Si(CH_3)_2O]_9Si(CH_3)_3$	829.78	......	322.8, 202[4.7]	......	0.930[20]	1.3994[20]	bz	B4[3], 1881
14663	1-Undecene $CH_3(CH_2)_8CH=CH_2$	154.30	......	192.7, 72[10]	−49.2	0.7503[20/4]	1.4261[20]	eth, chl, lig	B1[4], 910
14664	2-Undecene (trans) $CH_3(CH_2)_7CH=CHCH_3$	154.30	......	192-3, 75[10]	−48.3	0.7528[20/4]	1.4292[20]	....	B1[4], 911
14665	5-Undecene (trans) $CH_3(CH_2)_4CH=CH(CH_2)_3CH_3$	154.30	......	192, 73[10]	−61.1	0.7497[20/4]	1.4285[20]	eth, chl, lig	B1[4], 911
14666	9-Undecenoic acid (trans) $CH_3CH=CH(CH_2)_7CO_2H$	184.28	......	273, 121-3[0.7]	19	0.9119[25/0]	1.4519[20]	eth	B2[4], 1616
14667	10-Undecenoic acid $CH_2=CH(CH_2)_8CO_2H$	184.28	cr	275, 165[13]	24-5	0.9072[24/4]	1.4486[24]	al, eth	B2[4], 1612
14668	10- Undecenoic acid, ethyl ester or Ethyl-10-undecenoate $CH_2=CH(CH_2)_8CO_2C_2H_5$	212.34	......	264-5, 131.5[16]	−38	0.8827[15]	1.4449[23]	al, eth	B2[4], 1614
14669	10- Undecenoic acid, methyl ester $H_2C=CH(CH_2)_8CO_2CH_3$	198.31	......	248, 124[10]	−27.5	0.889[15]	1.4393[20]	al, eth, aa	B2[4], 1613
14670	10 -Undecenoyl chloride $CH_2=CH(CH_2)_8COCl$	202.72	......	120-2[10]	......	0.944[20/4]	1.454[20]		B2[4], 1615
14671	10-Undecen-1-ol or Undecenyl alchohol $CH_2=CH(CH_2)_8CH_2OH$	170.30	......	250, 132-3[15]	−2	0.8495[15/4]	1.4500[20]	al, eth	B1[4], 2194
14672	1-Undecene-3-yne $CH_3(CH_2)_6C≡CCH=CH_2$	150.27	......	74[9]	......	0.7962[20/4]	1.4606[20]	....	B1[3], 1054
14673	1- Undecyne $CH_3(CH_2)_8C≡CH$	152.28	......	196, 73.2[10]	−25	0.7728[20/4]	1.4306[20]	al, eth, ace, bz	B1[4], 1064
14674	5-Undecyne $CH_3(CH_2)_4C≡C(CH_2)_3CH_3$	152.28	......	198, 78.3[10]	−74.1	0.7753[20/4]	1.4369[20]	....	B1[4], 1064
14675	9-Undecynoic acid $CH_3C≡C(CH_2)_6CO_2H$	182.27	pl (dil al)	170[15]	61	....	....	w, al, eth	B2[3], 1471
14676	Uneicosane $CH_3(CH_2)_{19}CH_3$	296.59	cr (w)	358.5, 203[10]	40.5	0.7917[20/4]	1.4441[20]	peth	B1[4], 569
14677	Uneicosane, 1-cyclopentyl $C_5H_9C_{21}H_{43}$	364.70	......	420	42	0.8286[20]	1.4602[20]	....	B5[4], 200 (5.1031)
14678	Uneicosane, 11-decyl $(C_{10}H_{21})_2CH$	436.85	......	282[10]	10.0	0.8116[20/4]	1.4540[20]	....	B1[4], 595
14679	Uneicosane, 11-(2,2-dimethyl propyl) $(CH_3)_3CCH_2CH(C_{10}H_{21})_2$	366.71	......	238[10]	−21	0.8031[20/4]	1.4491[20]	....	B1[4], 585
14680	Uneicosane, 11-phenyl $C_6H_5CH(C_{10}H_{21})_2$	372.68	......	205[1.0]	20.8	0.8531[20/4]	1.4788[20]	....	B5[4], 1226
14681	Uneicosanoic acid $CH_3(CH_2)_{19}CO_2H$	326.57	nd (ace)	....	82	....	....	eth, bz, chl	B2[4], 1285
14682	10-Uneicosene, 11-Phenyl $CH_3(CH_2)_9CH=C(C_6H_5)(CH_2)_9CH_3$	370.66	......	203[1]	48.2	0.8638[20/4]	1.4922[20]	....	B5[4], 2057
14683	Untriacontane $CH_3(CH_2)_{29}CH_3$	436.86	lf (AcOEt)	458, 302[15]	67.9	0.781[68/4]	1.4278[90]	peth	B1[4], 594
14684	Untriacontanoic acid or Melissic acid $CH_3(CH_2)_{29}CO_2H$	466.84	sc or nd (al or ace)	....	93	....	....	bz	B2[2], 382
14685	16-Untriacontanone or Palmitone $(C_{15}H_{31})_2CO$	450.34	lf (al)	....	83	0.7947[91/4]	1.4297[94]	eth	B1[4], 3413
14686	Uracil $C_4H_4N_2O_2$	112.10	nd (w)	....	338	....	....	al, eth	B24[2], 168
14687	Uracil, 5-amino $C_4H_5N_3O_2$	127.11	nd (w)	....	d	....	....	....	B24[2], 266
14688	Uracil, 5,6 dihydro or β -Lactylurea $C_4H_6N_2O_2$	114.10	nd (w)	....	275-6	....	....	w, al, chl	B24[2], 140
14689	Uracil, 5,6-dihydro-5-methyl or Dihydrothymine $C_5H_8N_2O_2$	128.14	(w or al)	....	264-5	....	....	w	B24[1], 306
14690	Urcil, 5-hydroxy or Isobarbituric acid $C_4H_4N_2O_3$	128.09	pr (w)	....	d>300	....	....	w	B24[1], 408
14691	Uracil, 1-methyl $C_5H_6N_2O_2$	126.12	pr (w)	....	179	....	....	w, al	B24[2], 170
14692	Uracil, 3-methyl $C_5H_6N_2O_2$	126.12	pr (al) nd (w)	....	232	....	....	w, al	B24[2], 170
14693	Uracil, 5-methyl or Thymine $C_5H_6N_2O_2$	126.12	nd (al), pl (w)	sub	326	....	....	w, al	B24[2], 183
14694	Uracil, 6-methyl $C_5H_6N_2O_2$	126.12	oct, pr or nd (w, al)	....	270-80d	....	....	w, al	B24[2], 182

No.	Name, Synonyms, and Formula	Mol. wt.	Color, crystalline form, specific rotation and λ_{max} (log ϵ)	b.p. °C	m.p. °C	Density	n_D	Solubility	Ref.
14695	Uracil, 5-nitro .... $C_4H_3N_3O_4$	157.09	gold nd (al)	....	exp> 300	....	....	al	B24², 171
14696	Uracil, 2-thio or 2-Thiouracil........ $C_4H_4N_2OS$	128.15	pr (w, al)		>340d				B24², 171
14697	6-Uracil carboxylic acid or Orotic acid $C_5H_4N_2O_4$	156.10	pr (w + 1)		347d				B25², 249
14698	6-Uracil carboxylic acid, monohydrate .... $C_5H_4N_2O_4 . H_2O$	174.11	pr (w)		125-30d				B25², 249
14699	Uranin or Fluorescein, sodium salt .... $C_{20}H_{10}O_5Na_2$	376.28	ye pw					w, al	B19⁴, 2909
14700	Urea or Carbamide .... H_2NCONH_2	60.06	tetr pr (al)	d	135	1.3230²⁰/⁴	1.484	w, al, Py	B3⁴, 94
14701	Urea, 1-acetyl .... $CH_3CONHCONH_2$	102.09	....	sub 180-190	218			al	B3³, 119
14702	Urea, 1-acetyl-3-methyl .... $CH_3CONHCONHCH_3$	116.13	tcl (w, al), pr (w)	d	180-1			w	B4⁴, 207
14703	Urea, 1-allyl .... $CH_2=CHCH_2NHCONH_2$	100.13	nd (al)		85			w, al	B4⁴, 1070
14704	Urea, 1-allyl-3-phenyl .... $CH_2=CHCH_2NHCONHC_6H_5$	176.22	nd (bz)		115-6			al, bz	B12³, 765
14705	Urea, 1-(4-amino benzene sulfonyl) or 1-Sulfonyl urea.... $4-H_2NC_6H_4CO_2NHCONH_2$	215.23	(w)	sub 320	146-8d				B14³, 1968
14706	Urea, 1-benzoyl .... $C_6H_5CONHCONH_2$	164.18	fl or lf (al)		214-5				B9³, 1115
14707	Urea, 1-benzyl or Phthalamic acid.......... $C_6H_5CH_2NHCONH_2$	150.18	nd (al)	d 200	149			al, ace	B12³, 2271
14708	Urea, 1-(2-bromo-2-ethyl-butanoyl) or Adalin. Uradal $H_2NCONH[COCBr(C_2H_5)_2]$	237.10	rh (dil al)		118	1.544²⁵		ace, bz	B3⁴, 117
14710	Urea, 1-(2-bromo-3-methyl butanoyl) or Adabine Bromural $H_2NCONH[COCHBrCH(CH_3)_2]$	223.08	nd or lf (to)	sub	154 (160)	1.56¹⁵		al, eth, ace, bz	B3³, 123
14711	Urea, 1-(2-bromo phenyl) .... $2-BrC_6H_4NHCONH_2$	215.06	nd (al)		202			bz, chl	B12³, 1419
14712	Urea, 1-(3-bromophenyl) .... $3-BrC_6H_4NHCONH_2$	215.06	nd (bz, al)		164-5			al, eth	B12, 634
14713	Urea, 1-(4-bromophenyl) .... $4-BrC_6H_4NHCONH_2$	215.06	nd (bz, al)	d 260	225-7			al, eth, bz, aa	B12³, 1452
14714	Urea, 1-butyl .... $C_4H_9NHCONH_2$	116.16	ta (w), nd (bz)		96			al, eth	B4⁴, 578
14715	Urea, 1-isobutyl .... $H_2NCONH-i-C_4H_9$	116.16	pr (w), nd (ace)		141				B4⁴, 648
14716	Urea, 1-sec-butyl .... $C_2H_5CH(CH_3)NHCONH_2$	116.16	nd $[\alpha]^{20/}_D$ + 24.1 (al, c=1.5)		166-(d) 169- 70-(dl)			al, eth	B4⁴, 622
14717	Urea, 1-tert-butyl .... $(CH_3)_3CNHCONH_2$	116.16	nd (w dil al)	sub >100	191d			w, al	B4⁴, 665
14718	Urea, 1-(2-chlorophenyl) .... $(2-ClC_6H_4)NHCONH_2$	170.60	pr (w)		152			w, al, ace	B12³, 1293
14719	Urea, 3-(4-chlorophenyl)-1,1-dimethyl or CNV Weed Killer $3-(4-ClC_6H_4)NHCON(CH_3)_2$	198.65	wh pl (MeOH)		170-1				C51, 16534
14720	Urea, 1,3-diacetyl .... $(CH_3CONH)_2CO$	144.14	nd (aa, 50% al)	sub 179-80d	154-5			eth, ace, bz	B3³, 121
14721	Urea, 1,3-dibutyl .... $(C_4H_9NH)_2CO$	172.27	....		72-4				B4⁴, 578
14722	Urea, 1,3-bis-(2,4-dinitro phenyl) .... $(2,4(NO_2)_2C_6H_3NH)_2CO$	392.25	lf or pr (al)		204				B12³, 1690
14723	Urea, 1,3-bis-(2-ethoxyphenyl) .... $(2-C_2H_5OC_6H_4NH)_2CO$	300.36	pr (dil al)		125			al	B13³, 816
14724	Urea, 1,3-bis-(4-ethoxyphenyl) .... $(4-C_2H_5OC_6H_4NH)_2CO$	300.36	nd (aa), pr (abs al)		255-6			al	B13³, 1112
14725	Urea, 1,1-diethyl .... $H_2NCON(C_2H_5)_2$	116.16	pl nd (eth)	94-6⁰·⁰²	75			w, al, eth, bz, lig	B4⁴, 380
14726	Urea, 1,3-diethyl .... $(C_2H_5NH)_2CO$	116.16	ta (lig), dlq nd (al)	263	112.5	1.0415	1.4616⁴⁰	w, al, eth	B4⁴, 370
14727	Urea, 1,3-diethyl-1,3-diphenyl .... $[C_6H_5N(C_2H_5)]_2CO$	268.36	(al, w)		79			al	B12³, 882

No.	Name, Synonyms, and Formula	Mol. wt.	Color, crystalline form, specific rotation and λ_{max} (log ε)	b.p. °C	m.p. °C	Density	n_D	Solubility	Ref.
14728	Urea, 1,3-bis-(hydroxymethyl) or N,N′-Dimethylol urea (HOCH₂NH)₂CO	120.12	pr (abs al), pl (w-al)	d 260	126	1.49²⁵	....	w, al	B3⁴, 107
14729	Urea, 1,3-bis-(1-hydroxy-2,2,2-trichloro ethyl) or Dichloral urea. Cragherbicide [CCl₃CH(OH)NH]₂CO	354.83	....		196		....	al, ace	B3³, 116
14730	Urea, 1,1-dimethyl H₂NCON(CH₃)₂	88.12	mcl pr (al, chl)		182	1.255	....	w	B4⁴, 224
14731	Urea, 1,3-dimethyl (CH₃NH)₂CO	88.12	rh bipym (chl-eth)	268-70	108	1.142	....	w, al	B4⁴, 207
14732	Urea, 1,3-dimethyl-1,3-diphenyl [C₆H₅N(CH₃)]₂CO	240.31	pl (al)	350	121		....	w, al, ace, chl	B12³, 875
14733	Urea, 1,1-di(β-naphthyl) (β-C₁₀H₇)₂NCONH	312.38	nd (al)		192-3		....		B12, 1297
14734	Urea, 1,3-di(α-naphthyl) (α-C₁₀H₇NH)₂CO	312.38	nd (aa, Py)	sub	296		....	Py	B12³, 2926
14735	Urea, 1,3-di(β-naphthyl) (β-C₁₀H₇NH)₂CO	312.38	nd (ace, aa)		310		....	aa	B12³, 3033
14736	Urea, 1,1-diphenyl or Acardite (C₆H₅)₂NCONH₂	212.25	ta (al)	d	189	1.276	....	al, eth, chl	B12³, 893
14737	Urea, 1,3-diphenyl or Carbanilide (C₆H₅NH)₂CO	212.25	rh bipym, pr (al)	262d	238	1.239	....	eth, aa	B12³, 767
14738	Urea, 1,3-diphenyl-1-methyl C₆H₅NHCON(CH₃)C₆H₅	226.28	nd (al, xyl), cr (lig)	203-5d	106		....	eth, bz, chl	B12³ 878
14739	Urea, 1-(3-ethoxyphenyl) (3-C₂H₅OC₆H₄)NHCONH₂	180.21	nd	....	113		....	....	B13¹, 960
14740	Urea, 1-(4-ethoxyphenyl) or Dulcin (4-C₂H₅OC₆H₄)NHCONH₂	180.21	lf (dil al), pl (w)	d	173-4		....	al, AcOEt	B13³, 1109
14741	Urea, 1-ethyl C₂H₅NHCONH₂	88.11	nd (bz, al), eth	d	111		....	w, al, eth, bz, chl	B4⁴, 207
14742	Urea, 1-ethyl-1-phenyl C₆H₅N(C₂H₅)CONH₂	164.21	ta (peth), cr (aa)	....	62-3		....	w, eth, ace	B12³, 882
14743	Urea, 1-ethyl-3-phenyl C₂H₅NHCONHC₆H₅	164.21	nd (dil al)		104		....	al	B12³, 761
14744	Urea hydrochloride (H₂N)₂CO.HCl	96.52	....		145d		....		B3⁴, 102
14745	Urea, 1-hydroxy HONHCONH₂	76.06	nd (al)	d	141		....	w	B3⁴, 170
14746	Urea, 1-hydroxymethyl or Methylol urea HOCH₂NHCONH₂	90.08	pr (al)		111		....	w, al, aa, MeOH	B3⁴, 105
14747	Urea, 1-(2-iodo-3-methyl butanoyl) or α-Iodo isovaleryl urea H₂NCONHCOCHICH(CH₃)₂	270.07	lf (al)		180-1		....	al	B3¹, 29
14748	Urea, 1-(4-methoxyphenyl) (4-CH₃OC₆H₄)NHCONH₂	166.18	pl (w)	....	168		....	eth	B13³, 1096
14749	Urea, 1-methyl CH₃NHCONH₂	74.08	rh pr (w, al)	d	103		....	w, al	B4⁴, 205
14750	Urea, 1-methyl-1-nitroso H₂NCON(NO)CH₃	103.08	col or yesh pl (eth)		123-4d		....	al, eth, ace, bz, chl	B4⁴, 272
14751	Urea, 1-methyl-3-phenyl C₆H₅NHCONHCH₃	150.18	....		151		....	w, al, bz	B12³, 761
14752	Urea, 1-(2-methyl-2-butyl) H₂NCONHC(CH₃)₂C₂H₅	130.19	mcl (w)		162		....		B4⁴, 695
14753	Urea, 1-α-naphthyl α-C₁₀H₇NHCONH₂	186.22	nd (al)	....	219-20		....	al, eth	B12, 1253
14754	Urea, 1-(β-naphthyl) β-C₁₀H₇NHCONH₂	186.22	nd (al)	....	219		....	al	B12³, 3029
14755	Urea nitrate H₂NCONH₂.HNO₃	123.07	mcl lf (w)	....	157d	1.690²⁰/⁴	....	al	B3⁴, 102
14756	Urea, 1-nitro H₂NCONHNO₂	105.05	nd (al-peth), lf or pr(al)	exp	158-9		....	al, eth, ace, aa	B3⁴, 248
14757	Urea oxalate 2(H₂NCONH₂).HO₂CCO₂H	210.15	mcl (w + l)		173d	1.585	....	w	B3², 48
14758	Urea, 1-isopentyl H₂NCONH-(i-C₅H₁₁)	130.19	pl (dil al)	....	96 (150)		....	al	B4⁴, 695
14759	Urea, 1-(2-phenoxy ethyl) (C₆H₅OCH₂CH₂)NHCONH₂	180.21	nd (w + 2), cr (50% al)	....	120-1		....	al	B6¹, 91

No.	Name, Synonyms, and Formula	Mol. wt.	Color, crystalline form, specific rotation and λ_{max} (log ε)	b.p. °C	m.p. °C	Density	n_D	Solubility	Ref.
14760	Urea, 1-phenyl or Phenyl carbamide................ $C_6H_5NHCONH_2$	136.17	nd or pl (w), tab (al)	238	147	....	....	al, AcOEt	B12[3], 760
14761	Urea, 1-phenylacetyl .... $C_6H_5CH_2CONHCONH_2$	178.19	(al)	....	209	....	....	al	B9[3], 2208
14762	Urea, propionyl or Propionyl urea .... $CH_3CH_2CONHCONH_2$	116.13	cr (w)	....	210-11	....	....	w	B3[3], 121
14763	Urea, 1-propyl .... $C_3H_7NHCONH_2$	102.14	pr (al)	....	110	....	....	al	B4[4], 482
14764	Urea, 1,1,3,3-tetraethyl .... $[(C_2H_5)_2N]_2CO$	172.27	....	209, 94-5[12]	....	0.919[20/4]	1.4474[20]	....	B4[4], 380
14765	Urea, 1,1,3,3-tetramethyl .... $[(CH_3)_2N]_2CO$	116.16	....	166-7, 63-4[12]	-1.2	0.9687[20]	1.4496[23]	....	B4[4], 225
14766	Urea, 1,1,3,3-tetraphenyl .... $[(C_6H_5)_2N]_2CO$	364.45	rh (bz)	....	183	1.222	....	....	B12[3], 894
14767	Urea, 1-(2-tolyl) .... $(2-CH_3C_6H_4)NHCONH_2$	150.18	lf (al, w)	....	195-6	....	....	al, eth	B12[3], 1867
14768	Urea, 1-(3-tolyl) .... $(3-CH_3C_6H_4)NHCONH_2$	150.18	lf (w)	....	142	....	....	w, al	B12[3], 1968
14769	Urea, 1-(4-tolyl) .... $(4-CH_3C_6H_4)NHCONH_2$	150.18	nd (w), pl (w-aa)	....	183	....	....	al, ace	B12[3], 2082
14770	Urea, 1,1,3-trimethyl .... $(CH_3)_2NCONHCH_3$	102.14	pr (eth)	232.5[764]	75.5	....	....	w, al	B4[4], 224
14771	1-Urea carboxylic acid, ethyl ester or Allophanic acid, ethyl ester.................... $H_2NCONHCO_2C_2H_5$	132.13	nd (w, bz)	d	195	....	....	....	B3[4], 127
14772	Urethane, N-amino $H_2NNHCO_2C_2H_5$	104.12	cr	198d, 93[9]	46	....	....	al, eth	B3[4], 174
14773	Uric acid or 2,6,8-Purine trione............... $C_5H_4N_4O_3$	168.12	rh pr or pl	d	d	1.89	....	glycerol	B26[2], 293
14774	Uric acid, 1-methyl $C_6H_6N_4O_3$	182.14	nd	....	400	....	....	....	B26[2], 299
14775	Uric acid, 3-methyl $C_6H_6N_4O_3$	182.14	pr (w + l)	....	>350	1.6104[25]	....	....	B26[2], 299
14776	Uric acid, 7-methyl $C_6H_6N_4O_3$	182.14	pl (w)	....	370-80d	1.706[25/4]	....	....	B26[2], 299
14777	Uric acid, 9-methyl $C_6H_6N_4O_3$	182.14	....	....	380-400d	....	....	w	B26[2], 299
14778	Uric acid, 1,3,7-trimethyl $C_8H_{10}N_4O_3$	210.19	....	....	373-5	....	....	....	B26[2], 301
14779	Uridine or 1-β-D-Ribofuranosyluracil..................... $C_9H_{12}N_2O_6$	244.21	nd (aq al), $[\alpha]^{20}_D$ +4 (w)	....	165	....	....	w, Py	J1947, 358
14780	Uridylic acid or Uridine-3′-phosphate.............. $C_9H_{13}N_2O_9P$	324.19	pr (MeOH), $[\alpha]^{20}_D$ + 10.5 (w)	....	202d	....	....	w, MeOH	J1948, 746
14781	Urocanic acid or 4-Imidazolyl acrylic acid .... $C_6H_6N_2O_2$	138.13	pr(w + 2)	....	cis 175-6, trans 218-24	....	....	w, ace	B25[2], 121
14782	Urochloralic acid or 2,2,2-trichloroethyl-β-D-glucuronide.. $C_8H_{11}O_7Cl_3$	325.23	nd	....	142	....	....	w, al	B18[4], 5113
14783	Ursodesoxycholic acid or 3α-7β-Dihydroxy-5β-cholanic acid.................... $C_{24}H_{40}O_4$	392.58	pl (al), $[\alpha]^{20}_D$ + 57 (abs al, c=2)	....	203	....	....	al	B10[3], 1635
14784	Ursolic acid or Malol. Urson .... $C_{30}H_{48}O_3$	456.72	pr (al), $[\alpha]^{21}_D$ + 72.4 (MeOH)	....	284	....	....	eth, ace, chl, MeOH	B10[3], 1038
14785	Usnic acid (d) or Usninic acid.................... $C_{18}H_{16}O_7$	344.33	ye orh pr (ace), $[\alpha]^{20}_D$ + 469 (chl, c=0.7)	....	204	....	....	al, eth	B18[4], 3522
14786	Usnic acid (dl) .... $C_{18}H_{16}O_7$	344.33	....	....	193-4	....	1.710	al, eth, chl	B18[4], 3523

No.	Name, Synonyms, and Formula	Mol. wt.	Color, crystalline form, specific rotation and λ_{max} (log ε)	b.p. °C	m.p. °C	Density	n_D	Solubility	Ref.
14787	Usnic acid (l) .... $C_{18}H_{16}O_7$	344.33	$[\alpha]^{20}_D$ −480	....	203	....	....	al, eth, chl	B18⁴, 3522
14788	Uzarin .... $C_{35}H_{54}O_{14}$	698.82	pr, $[\alpha]_D$ −27 (Py)	....	268-70	....	....	Py	B18⁴, 1497

No.	Name, Synonyms, and Formula	Mol. wt.	Color, crystalline form, specific rotation and λ_{max} (log ε)	b.p. °C	m.p. °C	Density	n_D	Solubility	Ref.
14789	Vacciniin or 6-O-Benzoyl-D-glucose $C_{13}H_{16}O_7$	284.27	amor (aq ace + 1w), $[\alpha]^{21}_D$ + 48 (al)	....	104-6	....	....	w, al, ace	B31, 123
14790	Valeraldehyde or Pentanal $CH_3(CH_2)_3CHO$	86.14	....	103	-91.5	$0.8095^{20/4}$	1.3944^{20}	al, eth	B1[4], 3268
14791	Valeraldehyde, diethyl acetyl or Pentanal diethyl acetal $CH_3(CH_2)_3CH(OC_2H_5)_2$	160.26	....	59^{12}	....	0.829^{22}	1.4029^{22}	....	B1[4], 3269
14792	Valeraldehyde, 3-hydroxy-2-methyl or Propionaldol $C_2H_5CH(OH)CH(CH_3)CHO$	116.16	....	$94-6^{23}$	....	$0.986^{25/4}$	1.4502^{20}	w, al, eth, ace	B1[4], 4022
14793	Valeraldehyde, 3-hydroxy-2,2,4-trimethyl $(CH_3)_2CHCH(OH)C(CH_3)_2CHO$	144.21	....	$118-20^{14}$	....	$0.9482^{20/4}$	1.4501^{20}	....	B1[4], 4049
14794	Valeraldehyde, 4-hydroxy (dl) $CH_3CH(OH)CH_2CH_2CHO$	102.13	....	$63-5^{10}$	....	$1.019^{20/4}$	1.4359^{17}	w, ace	B1[4], 4002
14795	Valeraldehyde, 4-hydroxy (l) $CH_3CH(OH)CH_2CH_2CHO$	102.13	$[\alpha]^{23}_D$ -7.8	$43-6^1$	....	....	....	w, ace	B1[2], 872
14796	Valeraldehyde, 2-methyl $C_3H_7CH(CH_3)CHO$	100.16	....	116^{737}	....	....	....	w, eth, ace	B1[4], 3304
14797	Valeraldehyde, 4-methyl-2-oxo or Formyl isobutyl ketone $(CH_3)_2CHCH_2COCHO$	114.15	ye-gr	$45-6^{12}$	....	....	....	....	B1[1], 406
14798	Valeraldehyde oxime or Valeraldoxime $CH_3(CH_2)_3CH=NOH$	101.15	cr	....	52	....	....	eth	B1[3], 2798
14799	Valeraldehyde, 2-oxo or Butyryl formaldehyde C_3H_7COCHO	100.13	....	$112, 36^{16}$	....	....	1.4043^{25}	w, al, eth	B1[4], 3658
14800	Valeraldehyde, 4-oxo or Levalinaldehyde $CH_3COCH_2CH_2CHO$	100.13	....	186-8d, 70^{12}	<-21	$1.0184^{21/4}$	1.4257^{22}	w, al, eth, ace, bz	B1[4], 3659
14801	Valeramide $CH_3(CH_2)_3CONH_2$	101.15	micalike mcl pl (peth, al)	....	106 (114)	0.8735^{110}	1.4183^{110}	w, al, eth	B2[4], 874
14802	Valeramide, 2-bromo-4-methyl (d) $(CH_3)_2CHCH_2CHBrCONH_2$	194.08	(w), $[\alpha]^{20}_D$ -48.3 (al, c=5.9)	....	118	....	....	al, eth, chl	B2[2], 291
14803	Valeramide, N,N-dimethyl $CH_3(CH_2)_3CON(CH_3)_2$	129.20	....	141^{100}	-51	$0.8962^{25/4}$	1.4419^{25}	w, al, eth	B4[3], 127
14804	Valeramide, 4-methyl $(CH_3)_2CHCH_2CH_2CONH_2$	115.18	nd (al)	....	120-1	....	....	w, al	B2[4], 946
14805	Valeramide, 4-methyl-N-phenyl or Isocaproanilide $(CH_3)_2CHCH_2CH_2CONHC_6H_5$	191.28	nd (bz, dil al)	....	112	....	....	eth	B12[3], 478
14806	Valeramide, N-phenyl $C_4H_9CONHC_6H_5$	177.25	mcl pr (al), cr (peth)	....	63	....	....	eth	B12[3], 476
14807	Valeramide, 2,2,4-trimethyl $(CH_3)_2CHCH_2C(CH_3)_2CONH_2$	143.23	lf (lig)	....	71	....	....	al	B2[2], 305
14808	Valeric acid or Pentanoic acid $CH_3(CH_2)_3CO_2H$	102.13	....	186, 82.7^{10}	-33.8	$0.9391^{20/4}$	$1.4085^{20/4}$	w, al, eth	B2[4], 868
14810	Valeric acid, 2-acetyl, ethyl ester $CH_3COCH(C_3H_7)CO_2C_2H_5$	172.23	....	$210-2^{749}$, 90^{10}	....	$0.9682^{15/4}$	1.4271^{20}	al, eth	B3[4], 1602
14811	Valeric acid, 2-amino (D +) $C_3H_7CH(NH_2)CO_2H$	117.15	lf (w), $[\alpha]^{23}_D$ +32 (6NHCl, c=2)	sub	ca 307	....	....	w	B4[3], 1331
14812	Valeric acid, 2-amino (DL) or Norvaline $C_3H_7CH(NH_2)CO_2H$	117.15	lf (al, w)	sub	303 (sealed tube)	....	....	w	B4[3], 1333
14813	Valeric acid, 2-amino (L-) $C_3H_7CH(NH_2)CO_2H$	117.15	cr (dil al, w), $[\alpha]^{20}_D$ -23 (20% HCl, c=10)	sub	ca-305	....	....	w	B4[3], 1332
14814	Valeric acid, 4-amino (D) $CH_3CH(NH_2)CH_2CH_2CO_2H$	117.15	cr (dil al), $[\alpha]^{20}_D$ +12 (w, p=10)	....	214	....	....	w	B4[3], 1342
14815	Valeric acid, 4-amino (DL) $CH_3CH(NH_2)CH_2CH_2CO_2H$	117.15	cr (w)	d	199 (214)	....	....	w	B4[3], 1342
14816	Valeric acid, 5-amino $H_2N(CH_2)_4CO_2H$	117.15	lf (dil al)	d	157-8d	....	....	w	B4[3], 1343
14817	Valeric acid anhydride $(C_4H_9CO)_2O$	186.25	....	218^{754}, 111^{15}	-56.1	$0.924^{20/4}$	1.4171^{26}	al, eth	B2[4], 874

No.	Name, Synonyms, and Formula	Mol. wt.	Color, crystalline form, specific rotation and λ_{max} (log ε)	b.p. °C	m.p. °C	Density	n_D	Solubility	Ref.
14818	Valeric acid, 2-bromo (dl) .... C₃H₇CHBrCO₂H	181.04	....	118¹²	......	1.381²⁰	....	al, eth	B2⁴, 883
14819	Valeric acid, 2 bromo, ethyl ester .... C₃H₇CHBrCO₂C₂H₅	209.09	....	190 2, 92-4¹⁸		1.226¹⁸/⁴	1.4496²⁰	al, cth	B2³, 681
14820	Valeric acid, 2-bromo-4-methyl (d) .... (CH₃)₂CHCH₂CHBrCO₂H	195.06	oil, [α]²⁰/D +29.8 (eth, c=6)	131-2¹⁸	......	....	....	al, eth	B2³, 747
14821	Valeric acid, 2-bromo-4-methyl (l) .... (CH₃)₂CHCH₂CHBrCO₂H	195.06	oil, [α]²⁰/D −12.1	94⁰·³		....	....	al, eth	B2³, 747
14822	Valeric acid, 2-bromo-4-methyl, ethyl ester .... (CH₃)₂CHCH₂CHBrCO₂C₂H₅	223.12	pa ye	202-4, 86-7¹¹		....	....	....	B2³, 747
14823	Valeric acid, butyl ester or Butyl valarate .... C₄H₉CO₂C₄H₉	158.24	....	185.8, 84-5⁸	−92.8	0.8710¹⁵/⁴	1.4128²⁰	al, eth	B2³, 671
14824	Valeric acid, iso-butyl ester or Isobutyl valarate .... C₄H₉CO₂CH₂CH(CH₃)₂	158.24	....	179		0.8625²⁵/⁴	1.4046²⁰	al, eth, ace	B2³, 671
14825	Valeric acid, sec-butyl ester (d) or sec-Butyl valarate ..... C₄H₉CO₂CH(CH₃)C₂H₅	158.24	[α]²⁰/D + 20.7	174.5, 67¹⁸		0.8605²⁰/⁴	1.4070²⁰	al, eth, bz, Py	B2³, 671
14826	Valeric acid, 2-chloro (dl) .... C₃H₇CHClCO₂H	136.58	....	222⁷⁶³, 133-5³⁹	−15	1.141¹³	1.4481³⁰	al, eth	B2⁴, 878
14827	Valeric acid, 2-chloro, ethyl ester .... C₃H₇CHClCO₂C₂H₅	164.63	....	185⁷⁵²		1.040¹²	1.4307¹¹	eth	B2³, 678
14828	Valeric acid, 3-chloro .... C₂H₅CHClCH₂CO₂H	136.58	cr (bz)	112¹⁰	33	1.1484²⁰/⁴	1.4462²⁰	al, eth	B2⁴, 879
14829	Valeric acid, 3-chloro, ethyl ester .... C₂H₅CHClCH₂CO₂C₂H₅	164.63	....	189, 66-7¹⁰		1.0330²⁰/⁴	1.4278²⁰	al, eth	B2³, 678
14830	Valeric acid, 4-chloro .... CH₃CHClCH₂CH₂CO₂H	136.58	....	116¹⁰		1.1514²⁰	1.4458²⁰	al, eth	B2⁴, 879
14831	Valeric acid, 4-chloro, ethyl ester .... CH₃CHClCH₂CH₂CO₂C₂H₅	164.63	....	196, 70.5⁹		1.0393²⁰/⁴	1.4310²⁰	al, eth	B2⁴, 879
14832	Valeric acid, 5-chloro .... Cl(CH₂)₄CO₂H	136.58	....	230, 141-9¹²	18	1.3416²⁵/⁴	1.4555²⁰	al, eth	B2⁴, 880
14833	Valeric acid, 5-chloro, ethyl ester .... Cl(CH₂)₄CO₂C₂H₅	164.63	....	205-6, 93¹⁶		1.0561²⁰	1.4355²⁰	al, eth	B2⁴, 880
14834	Valeric acid, 2,4-dimethyl (dl) .... (CH₃)₂CHCH₂CH(CH₃)CO₂H	130.17	....	111⁰·⁹	143-5	1.0818²⁷	....	al, eth	B2⁴, 979
14835	Valeric acid, 2,4-dioxo or Acetyl pyruvic acid .... CH₃COCH₂COCO₂H	130.10	nd (bz-chl), pr (bz)	130³⁷ sub	55-63 (+1w) 101 (anh)	....	....	w, al, eth, ace, bz	B3³, 1331
14836	Valeric acid, 2,4-dioxo, ethyl ester or Ethyl acetone oxalate	158.16	....	213-5, 111-2¹⁶	18	1.1251²⁰	1.4757¹⁷	al, eth	B3⁴, 1777
	CH₃COCH₂COCO₂C₂H₅								
14837	Valeric acid, ethyl ester or Ethyl valerate .... C₄H₉CO₂C₂H₅	130.19	....	144.6⁷³⁶, 50.5²⁹	−91.2	0.8770²⁰/⁴	1.4120²⁰	al, eth	B2⁴, 872
14838	Valeric acid, 2-ethyl (dl) or 3-Hexane carboxylic acid ..... C₃H₇CH(C₂H₅)CO₂H	130.19	....	209.2, 105-7¹⁸	......	0.9361³³/³³	....	al, eth	B2⁴, 975
14839	Valeric acid, 3-ethyl .... (C₂H₅)₂CHCH₂CO₂H	130.19	....	212, 104-5¹³		....	1.4250²⁰	....	B2⁴, 976
14840	Valeric acid, 5-fluoro .... F(CH₂)₄CO₂H	120.13	....	83²		....	1.4080²⁵	al, eth	B2⁴, 876
14841	Valeric acid, furfuryl ester or Furfuryl valerate .... C₄H₉CO₂CH₂(C₄H₃O)	182.22	....	228-9⁷⁶⁴, 82-3¹		1.0284²⁰/⁴	....	al, eth	B17², 115
14842	Valeric acid, heptyl ester or Heptyl valerate .... C₄H₉CO₂C₇H₁₅	200.33	....	245.2	−46.4	0.8623²⁰	1.4254¹⁵/ₙₑ	al, eth, ace	B2⁴, 872
14843	Valeric acid, hexyl ester or Hexyl valerate .... C₄H₉CO₂C₆H₁₃	186.30	....	226.3	−63.1	0.8635²⁰	1.4228¹⁵	al, eth, ace	B2³, 671
14844	Valeric acid, 2-hydroxy or Valerolactic acid .... C₃H₇CH(OH)CO₂H	118.13	hyg pl	sub	34	....	....	w, al, eth	B3⁴, 807
14845	Valeric acid, 4-hydroxy, lactone (d) or γ-Valerolactone ... CH₃CHCH₂CH₂CO	100.12	[bα]²⁰/D + 13.5 (undil)	86-90¹⁴		....	....	w, al, ace	B17², 288
14846	Valeric acid, 4-hydroxy-lactone (dl) or γ-Valerolactone ... CH₃CHCH₂CH₂CO	100.12	....	206, 83-4¹³	−31	1.0465²⁵	1.4328²⁰	w, al, ace	B17⁴, 4176
14847	Valeric acid, 4-hydroxy-lactone (l) or γ-Valerolactone ... CH₃CHCH₂CH₂CO	100.12	[α]²⁰/D + 4.6 (eth, c=10)	78-80⁸		....	1.4322²⁰	w, al, ace	B17⁴, 4176
14848	Valeric acid, 5-hydroxy-lactone or γ-Valerolactone ...... CH₂(CH₂)₃CO	100.12	....	218-20, 113-4¹⁴	−12.5	1.0794²⁰	1.4503²⁰	w, al, eth	B17⁴, 4169
14849	Valeric acid, methyl ester or Methyl valerate .... C₄H₉CO₂CH₃	116.16	....	126.5		0.8947²⁰/⁴	1.4003²⁰	al, eth, ace	B2⁴, 871

No.	Name, Synonyms, and Formula	Mol. wt.	Color, crystalline form, specific rotation and λ_{max} (log ε)	b.p. °C	m.p. °C	Density	n_D	Solubility	Ref.
14850	Valeric acid, 2-methyl (d) .... C$_3$H$_7$CH(CH$_3$)CO$_2$H	116.16	$[\alpha]^{25}_D$ + 18.5 (undil)	96[15]	......	0.927[16/4]	1.4112[25]	w, al, eth	B2[4], 942
14851	Valeric acid, 2-methyl (dl) C$_3$H$_7$CH(CH$_3$)CO$_2$H	116.16	....	195-6, 102-5[12]		0.9230[20/4]	1.413[20]	w, al, eth	B2[4], 942
14852	Valeric acid, 2-methyl (l) C$_3$H$_7$CH(CH$_3$)CO$_2$H	116.16	$[\alpha]^{25}_D$ −7.1 (eth)	190-3, 96[15]	......	0.9781[20/4]	1.4117[15]	w, al, eth	B2[3], 740
14853	Valeric acid, 3-methyl (d) C$_2$H$_5$CH(CH$_3$)CH$_2$CO$_2$H	116.16	$[\alpha]^{20}_D$ + 8.5 (un-dil)	199, 92-3[10]		0.9276[20/4]	1.4158[20]	al, eth	B2[4], 948
14854	Valeric acid, 3-methyl (dl) C$_2$H$_5$CH(CH$_3$)CH$_2$CO$_2$H	116.16	....	197-8	−41.6	0.9262[20]	1.4159[20]	al, eth	B2[4], 948
14855	Valeric acid, 3-methyl (l) C$_2$H$_5$CH(CH$_3$)CH$_2$CO$_2$H	116.16	$[\alpha]^{20}_D$ −8.9 (al)	196-7, 105[30]		0.923[25/4]	1.4152[20]	al, eth	B2[4], 948
14856	Valeric acid, 3-methyl-3-phenyl C$_2$H$_5$C(CH$_3$)(C$_6$H$_5$)CH$_2$CO$_2$H	192.26	....	174[14]		1.050[25]	1.5197[25]	....	B9[3], 2548
14857	Valeric acid, 4-methyl or Isocaproic acid .... (CH$_3$)$_2$CHCH$_2$CH$_2$CO$_2$H	116.16	....	200-1, 86-8[11]	−33	0.9225[20/4]	1.4144[20]	al, eth	B2[4], 944
14858	Valeric acid, 4-methyl-2-phenyl or α-Phenyl isocaproic acid .... (CH$_3$)$_2$CHCH$_2$CH(C$_6$H$_5$)CO$_2$H	192.24	pr (peth)	178-80[20]	78-9	....	....	al	B9[3], 2546
14859	Valeric acid, octyl ester or Octyl valerate .... C$_4$H$_9$CO$_2$C$_8$H$_{17}$	214.35	....	261.6	−42.3	0.8615[20]	1.4273[15]	al, eth, ace	B2[3], 672
14860	Valeric acid, 2-oxo C$_3$H$_7$COCO$_2$H	116.12	....	179, 81[12]	6-7	....	....	eth, bz, chl	B3[4], 1558
14861	Valeric acid, 3-oxo, ethyl ester or Ethyl propionyl acetate C$_2$H$_5$COCH$_2$CO$_2$C$_2$H$_5$	144.17	....	191, 90[13]		....	1.4230[20]	al, eth, bz	B3[4], 1559
14862	Valeric acid, 4-oxo or Levulinic acid .... CH$_3$COCH$_2$CH$_2$CO$_2$H	116.13	lf or pl	245-6d, 139-40[8]	37.2	1.1335[20/4]	1.4396[20]	w, al, eth	B3[4], 1560
14863	Valeric acid, pentyl ester or Pentyl valerate .... C$_4$H$_9$CO$_2$C$_5$H$_{11}$	172.27	....	203.7, 103[23]	−78.8	0.8638[20/4]	1.4164[20]	**al, eth**	B2[4], 872
14864	Valeric acid, 2-phenyl (d) C$_3$H$_7$CH(C$_6$H$_5$)CO$_2$H	178.23	$[\alpha]_D$ + 58.8 (chl)	165[14]	......	1.047[20]	....	al, eth, chl	B9[3], 2506
14865	Valeric acid, 2-phenyl (dl) C$_3$H$_7$CH(C$_6$H$_5$)CO$_2$H	178.23	nd (lig)	280	58	....	....	al, eth	B9[3], 2506
14866	Valeric acid, 4-phenyl C$_6$H$_5$CH(CH$_3$)CH$_2$CH$_2$CO$_2$H	178.23	....	210[85], 165[12]	ca 13	1.0554[15/4]	1.5167[20]	al, eth	B9[3], 2505
14867	Valeric acid, 5-phenyl .... C$_6$H$_5$(CH$_2$)$_4$CO$_2$H	178.23	pl (w), pr (peth)	190-3[30]	57-8	....	....	al	B9[3], 2502
14868	Valeric acid, propyl ester or Propyl valerate .... C$_4$H$_9$CO$_2$C$_3$H$_7$	144.22	....	167.5	......	0.8699[20/4]	1.4065[20]	al, eth, chl	B2[4], 872
14869	Valeric acid, isopropyl ester or Isopropyl valerate .... C$_4$H$_9$CO$_2$CH(CH$_3$)$_2$	144.22	....	153.5		0.8579[20/4]	1.4061[20]	al, eth, ace	B2[3], 671
14870	Valeronitrile .... CH$_3$(CH$_2$)$_3$CN	83.13	....	141.3, 30.9[10]	−96	0.8008[20/4]	1.3971[20]	eth, ace, bz	B2[4], 875
14871	Valeronitrile, 2-hydroxy-2-propyl .... (C$_2$H$_7$)$_2$C(OH)CN	141.22	....	119-20[21]		0.9077[18]	1.4337[18]	al	B3[4], 880
14872	Valeronitrile, 4-methyl or Isocapronitrile .... (CH$_3$)$_2$CHCH$_2$CH$_2$CN	97.16	....	156-7, 50[13]	−51	0.8030[20/4]	1.4059[20]	al, eth	B2[4], 946
14873	Valeronitrile, 4-methyl-2-phenyl or Phenyl isocapronitrile (CH$_3$)$_2$CHCH$_2$CH(C$_6$H$_5$)CN	173.26	....	263-6[765], 136-8[15]		0.942[16]	....	al, eth, bz, aa	B9[1], 220
14874	Valeronitrile, 2-phenyl .... C$_3$H$_7$CH(C$_6$H$_5$)CN	159.23	....	254-5[750], 125-8[13]		0.9425	1.5000[20]	**al, bz**	B9[3], 2508
14875	Valerophenone or Butyl phenyl ketone............. C$_4$H$_9$COC$_6$H$_5$	162.23	....	248.5, 131-3[13]		0.988[20/20]	1.5158[20]	al, eth	B7[3], 1114
14876	Valerophenone, γ-oxo or Phenacyl acetone .... C$_6$H$_5$COCH$_2$CH$_2$COCH$_3$	176.22	ye oil	162[12]		....	1.5250[30]	ace	B7[3], 3509
14877	Valeryl chloride .... C$_4$H$_9$COCl	120.58	....	107-10[756]	−110	1.0155[15]	1.4200[20]	....	B2[4], 874
14878	Valeryl chloride, 2-chloro .... C$_3$H$_7$CHClCOCl	155.03	....	155-7[763], 61-2[28]		1.1765[20]	1.4465[20]	eth	B2[4], 879
14879	Valeryl chloride, 4-chloro .... CH$_3$CHClCH$_2$CH$_2$COCl	155.03	....	61[8]		....	....	eth	B2[3], 679
14880	Valeryl chloride, 5-chloro .... Cl(CH$_2$)$_4$COCl	155.03	....	83[12]		1.210[18]	1.4639[20]	eth	B2[4], 881
14881	Valeryl chloride, 4,4-dimethyl or Neopentyl acetyl chloride .... (CH$_3$)$_3$CCH$_2$CH$_2$COCl	148.63	....	150-2, 58[2]		....	1.4294[20]	eth	B2[3], 780

No.	Name, Synonyms, and Formula	Mol. wt.	Color, crystalline form, specific rotation and λ_{max} (log ε)	b.p. °C	m.p. °C	Density	n_D	Solubility	Ref.
14882	Valeryl chloride, 2-ethyl $C_3H_7CH(C_2H_5)COCl$	148.63	....	158-60, 50[11]				eth	B2[4], 975
14883	Valeryl chloride, 2 methyl (dl) $C_3H_7CH(CH_3)COCl$	134.61	....	140-1[745]	....	0.9781[20/4]	1.4330[27]	eth	B2[3], 741
14884	Valeryl chloride, 3-methyl (dl) $C_2H_5CH(CH_3)CH_2COCl$	134.61	....	142-3		0.9781[20/4]		eth	B2[4], 949
14885	Valeryl chloride, 4-methyl $(CH_3)_2CHCH_2CH_2COCl$	134.61	....	144-5[745]		0.9725[20]			B2[4], 945
14886	Valine (D) or l-α-Amino isovaleric acid $(CH_3)_2CHCH(NH_2)CO_2H$	117.15	pl (aq al), $[\alpha]^{20}_D$ −29 (20% al)	....	156-7 (hyd) 293d (sealed tube)			w	B4[3], 1369
14887	Valine (DL) $(CH_3)_2CHCH(NH_2)CO_2H$	117.15		sub	298d (sealed tube)	1.316	....	w	B4[3], 1370
14888	Valine (L) or d-α-Amino isovaleric acid $(CH_3)_2CHCH(NH_2)CO_2H$	117.15	lf (w-al), $[\alpha]^{23}_D$ +22.9 (20%al, c=0.8)		93-6, 315 (sealed tube)	1.230		w	B4[3], 1365
14889	Valine, β-hydroxy-dl $(CH_3)_2C(OH)CH(NH_2)CO_2H$	133.15	pl (dil al)		240d	....		w	B4[3], 1660
14890	Vanillan or 4-Hydroxy-3-methoxy benzaldehyde $4\text{-}HO\text{-}3\text{-}CH_3O\text{-}C_6H_3CHO$	152.16	(i) nd (w, lig), (ii) tetr (w, lig)	285, 170[15]	(i) 77-9, (ii) 81-2	1.056	....	al, eth, ace, bz	B8[3], 2011
14891	Vanillan acetate or 4-Acetoxy-3-methoxy benzaldehyde $4\text{-}CH_3CO_3\text{-}3\text{-}CH_3OC_6H_3CHO$	194.19	nd (eth)	....	102-3			al, eth	B8[3], 2030
14892	Vanillan, 5-bromo $5\text{-}Br\text{-}4\text{-}HO\text{-}3\text{-}CH_3OC_6H_2CHO$	231.05	pl (aa), nd, pl (al)		164-6				B8[3], 2050
14893	Vanillan, 5-chloro $5\text{-}Cl\text{-}4\text{-}HO\text{-}3\text{-}CH_3OC_6H_2CHO$	186.60	tetr		165				B8[3], 2043
14894	Vanillan, 2-iodo $4\text{-}HO\text{-}2\text{-}I\text{-}3\text{-}CH_3OC_6H_2CHO$	278.05	....		155-6			al	B8[3], 2058
14895	Vanillan, 5-iodo $4\text{-}HO\text{-}5\text{-}I\text{-}3\text{-}CH_3OC_6H_2CHO$	278.05	pa ye		180				B8[3], 2059
14896	o-Vanillan or 2-Hydroxy-3-methoxy benzaldehyde $2\text{-}HO\text{-}3\text{-}CH_3OC_6H_3CHO$	152.16	lt ye-lt gr nd (w, lig)	265-6, 128[10]	44-5			al, eth, lig	B8[3], 1979
14897	Vasicine (dl) or Peganine $C_{11}H_{12}N_2O$	188.23	nd (al)	....	209-10			al, ace, chl	B23[2], 342
14898	Vasicine (l) $C_{11}H_{12}N_2O$	188.23	nd (al), $[\alpha]^{14}_D$ −62 (al, c=2.4)		211-2			al, chl	B23[2], 342
14899	Veraridine or 3-Veratroyl veracerine $C_{36}H_{50}NO_{11}$	673.80	yesh amor, $[\alpha]^{22}_D$ +8 (al)		180				B21[4], 6824
14900	Veratramide or 3,4-Dimethoxy benzaldehyde $3,4\text{-}(CH_3O)_2C_6H_3CONH_2$	181.19	cr (w)	....	164			w, eth, bz	B10[3], 1427
14901	Veratramine $C_{27}H_{39}NO_2$	409.62	nd, $[\alpha]^{19}_D$ −70 (MeOH)		209-10 (hyd)			bz, chl	B21[4], 2385
14902	Veratric acid or 3,4-Dimethoxy benzoic acid $3,4\text{-}(CH_3O)_2C_6H_3CO_2H$	182.18	nd (w, aa), rh (sub)	sub	181-2			al, eth	B10[3], 1404
14903	Veratrine or Cevadine $C_{32}H_{49}NO_9$	591.75	rh (+2 al), $[\alpha]^{17}_D$ +12.5 (al)		205d			al, chl, Py	B21[4], 6820
14904	Veratronitrile or 3,4-Dimethoxy benzonitrile $3,4\text{-}(CH_3O)_2C_6H_3CN$	163.18	nd (w)		67-8			bz	B10[2], 264
14905	Veratrosine or Veratramine-β-D-glucoside $C_{33}H_{49}NO_7$	571.76	nd (aq MeOH), $[\alpha]^{24}_D$ −55 (al-chl, c=0.94)		242-3d				B21[4], 2386
14906	Verbenone (d) or d-4-Piperone $C_{10}H_{14}O$	150.22	$[\alpha]^{18}_D$ +249.6	227-8, 103-4[16]	9.8	0.9978[20]	1.4993[18]	w, al, ace, bz	B7[3], 583
14907	Verbenone (l) $C_{10}H_{14}O$	150.22	$[\alpha]_D$ −144	253-5, 100[16]	6.5	0.9731[20]	1.4961[20]	w, al, bz	B7[3], 583

No.	Name, Synonyms, and Formula	Mol. wt.	Color, crystalline form, specific rotation and λ_{max} (log ε)	b.p. °C	m.p. °C	Density	n_D	Solubility	Ref.
14908	α-Vetivone or α-Vetiverone .... $C_{15}H_{22}O$	218.34	(peth), $[\alpha]_D$ +238.2	144[2]	51.5	1.0035[20/4]	1.5370[20]	ace	B7[3], 1265
14909	β-Vetivone or β-Vetiverone............. $C_{15}H_{22}O$	218.34	(pentane), $[\alpha]^{20}_D$ −38.9 (al, c=10.6)	141-2[2]	44.5	1.0001[20/4]	1.5309[20]	ace	B7[3], 1256
14910	Vicianose or 6-O-(α-L-arabinosido)-β-D-glucose .... $C_{11}H_{20}O_{10}$	312.28	nd (dil al), $[\alpha]^{20}_D$ +56.5→ +39.7 (w, mut)	....	ca 210d			w	B17[4], 2447
14911	Vicine (a β-D-glucoside)............. $C_{10}H_{16}N_4O_7$	304.26	nd (w, dil al + l w), $[\alpha]^{25}_D$ −12 (w, c=10)	....	239-42d				B31, 163
14912	Vincamine .... $C_{21}H_{26}N_2O_3$	354.45	$[\alpha]^{23}_D$ +41 (Py)	....	231-2				M, 1107
14913	Violaxanthin or Zeaxanthin diepoxide .... $C_{40}H_{56}O_4$	600.89	red pr (MeOH, al-eth), $[\alpha]^{20}_d$ +35 (chl, c=0.08)		208			al, eth	B19[4], 1139
14914	Violuric acid or Alloxan-5-oxime............. $C_4H_3N_3O_4$	157.09	pa ye rh		203-4d			al	B24[2], 304
14915	Visnadin .... $C_{21}H_{24}O_7$	388.42	nd, $[\alpha]_D$ +9 (al)		85-6			al, eth	B19[4], 2787
14916	Visnagin or 5-Methoxy-2-methyl furanochromone .... $C_{13}H_{10}O_4$	230.22	nd (w, MeOH)	....	144-5			chl	B19[4], 2640
14917	Vitamin A$_1$ or Axerophytol............. $C_{20}H_{30}O$	286.46	ye pr (peth)	137-8[10.6]	63-4			al, eth, ace, bz	B6[3], 2787
14918	Vitamin B$_1$ or Thiamine hydrochloride............. $C_{12}H_{18}Cl_2N_4OS$	337.27	(i) mcl pr (MeOH-al), (ii) pl (MeOH-al, w-al)	....	(i) 233-4, (ii) 250			w	Am74, 2409
14919	Vitamin B$_6$ or Pyroxidin............. $C_8H_{11}NO_3$	169.18	nd (ace)	sub	160			w, al, ace	B21[4], 2509
14920	Vitamin B$_6$ hydrochloride or Pyridoxin hydrochloride.... $C_8H_{11}NO_3.HCl$	205.64	pl (al, ace)	sub	206-8			w	B21[4], 2511
14921	Vitamin K$_1$ .... $C_{31}H_{46}O_2$	450.70	$[\alpha]^{20}_D$ −0.4 (bz 57.5%)	140-5[0.001] dec>120	−20	0.967[25/25]	1.5250[25]	al, eth, ace, bz, peth	B7[3], 3792
14922	Vitamin B$_{12}$ .... $C_{63}H_{88}N_{14}O_{14}COP$	1355.38	....	>300				....	
14923	Vitamin B$_1$-hydrochloride .... $C_{12}H_{17}N_4OSCl.HCl$	337.27			233-4			w	Am74, 2409
14924	Vitamin D$_2$ .... $C_{28}H_{44}O$	396.65	$[\alpha]^{25}_D$ +82.6 (ace, c=3)		115-8				B6[3], 3089
14925	Vitamin D$_3$ .... $C_{27}H_{44}O$	384.64	$[\alpha]^{20}_D$ +84.8 (ace, c=1.6)		84-5				B6[3], 2811
14926	Vomicine............. $C_{22}H_{24}N_2O_4$	380.44	nd (80% al), pr (ace), $[\alpha]^{22}_D$ +80.4 (al, p=0.4)	....	282			chl	B27[2], 795
14927	Vinyl acetate .... $CH_3CO_2CH=CH_2$	86.09	....	72.2	−93.2	0.9317[20/4]	1.3959[20]	al, eth, ace, bz, chl	B2[4], 176
14928	Vinyl acetylene or 1-Buten-3-yne .... $CH_2=CH-C≡CH$	52.08	....	5.1		0.7095[0/0]	1.4161[1]	bz	B1[4], 1083
14929	Vinyl amine .... $CH_2=CHNH_2$	43.07	....	55-6[750]		0.8321[24]	....	w, al, eth	B20[4], 3
14930	Vinyl bromide .... $CH_2=CHBr$	106.96	....	15.8	−139.5	1.4933	1.4410[20]	al, eth, ace, bz, chl	B1[4], 718

No.	Name, Synonyms, and Formula	Mol. wt.	Color, crystalline form, specific rotation and λ_{max} (log ε)	b.p. °C	m.p. °C	Density	n_D	Solubility	Ref.
14931	Vinyl chloride or Chloro ethylene .... $CH_2=CHCl$	62.50	gas	−13.4	−1538	0.9106[20/4]	1.3700[20]	al, eth	B1[4], 700
14932	Vinyl ether or Divinyl ether .... $(CH_2=CH)_2O$	70.09	....	28	−101	0.773[20/4]	1.3989[20]	al, eth, ace	B1[1], 2058
14933	Vinyl ether, hexachloro .... $(Cl_2C=CCl)_2O$	276.76	....	210		1.654[21]	....	....	B1, 725
14934	Vinyl fluoride .... $CH_2=CHF$	46.05	gas	−72.2	−160.5	....	....	al, ace	B1[4], 694
14935	Vinyl iodide .... $CH_2=CHI$	153.95	......	56		2.037[20]	1.5385[20]	al, eth	B1[4], 722
14936	Vinyl propionate .... $CH_3CH_2CO_2CH=CH_2$	100.12	....	91-2		......	....	....	B2[4], 711
14937	2-Vinyl pyridine .... $2-(CH_2=CH)C_5H_4N$	105.14	....	159-60, 50-5[4]		0.9985[20/0]	1.5495[20]	al, eth, ace, chl	B20[4], 2884
14938	3-Vinyl pyridine .... $3-(CH_2=CH)C_5H_4N$	105.14	....	162, 67-8[18]		......	1.5530[20]	al, eth	B20[4], 2886
14939	4-Vinyl pyridine .... $4-(CH_2=CH)C_5H_4N$	105.14	....	65[15]		0.9800[20/4]	1.5449[20]	w, al, chl	B20[4], 2887
14940	Vinyl sulfide .... $(CH_2=CH)_2S$	86.15	....	84	0.9174[1] 5/4	....	....	al, eth, ace	B1[4], 2068
14941	2-Vinyl toluene or 2-Methyl styrene................. $2-CH_3C_6H_4CH=CH_2$	118.18	....	171, 51[9]		0.9106[20/4]	1.5450[20]	bz, chl	B5[4], 1367
14942	m-Vinyl toluene or 3-Methyl styrene .... $3-CH_3C_6H_4CH=CH_2$	118.18	....	168, 61-2[18]	−70	0.9028[20/20]	1.5428[20]	bz	B5[4], 1367
14943	p-Vinyl toluene or 4-Methyl styrene .... $4-CH_3C_6H_4CH=CH_2$	118.18	....	169, 63[15]	−37.8	0.8760[20/4]	1.5428[20]	bz	B5[4], 1369

No.	Name, Synonyms, and Formula	Mol. wt.	Color, crystalline form, specific rotation and λ_{max} (log ε)	b.p. °C	m.p. °C	Density	n_D	Solubility	Ref.
14944	Xanthene or Dibenzo-1,4-Puran .... $C_{13}H_{10}O$	182.23	wh ye lf (al)	310-2	100.5			eth, bz, chl, lig	B17[4], 614
14945	Xanthene, 9-hydroxy or Dibenz-γ-pyranol .... $C_{13}H_{10}O_2$	198.22	nd (aq al)		ca 125			al, eth, chl	B17[4], 1602
14946	Xanthene, 9-phenyl .... $C_{19}H_{14}O$	258.32	(al)		145-6			bz, aa	B17[4], 733
14947	9-Xanthene carboxylic acid or Xanthanoic acid......... $C_{14}H_{10}O_3$	226.24	nd (dil al, MeOH)		223.4			eth	B18[4], 4351
14948	Peri-xanthenoxanthene or 1,1-Binaphthalene-2:8′,2:8-dioxide .... $C_{20}H_{10}O_2$	282.30	ye pr (chl)	400[20.25]	242			bz	B19[4], 448
14949	Xanthine or 2,6-Purinedione .... $C_5H_4N_4O_2$	152.11	yesh pl (w)	sub	d				B26[2], 260
14950	Xanthine, 2-methyl or Heteroxanthine .... $C_6H_6N_4O_2$	166.14	nd (w)		380d				B26[2], 263
14951	Xanthione or 9-Xanthenethione................. $C_{13}H_8OS$	212.28	red nd (al)		156			eth, bz	B17[4], 5301
14952	Xanthogen-diethyl or Auligen $C_2H_5OC(:S)SSC(:S)OC_2H_5$	242.40	ye nd or pl (al)	107-9[0.05]	31-2	1.2604[25/4]		eth, ace, bz, peth, chl	B3[3], 349
14953	Xanthogenic acid or Ethoxy dithioformic acid .... $C_2H_5OCS_2H$	122.21	unst	25d	fr−53				B3[4], 401
14953a	Xanthogenic acid, ethyl ester .... $CH_3CH_2OCS_2C_2H_5$	150.25		200		1.085[19]	1.5370[18.2]		B3[2], 153
14954	Xanthone or Dibenzopyrone .... $C_{13}H_8O_2$	196.22	nd (al)	349-50, 146[3]	174			al, eth, bz, chl	B17[4], 5292
14955	Xanthone, 1,2-dihydroxy .... $C_{13}H_8O_4$	228.22	pa ye nd (aq al + 3 w)		166-7			al, Py	B18[4], 1678
14956	Xanthone, 1,3-dihydroxy .... $C_{13}H_8O_4$	228.22	nd (aq al + l w)	sub	259			al	B18[4], 1680
14957	Xanthone, 1,6-dihydroxy or Isoeuxanthone......... $C_{13}H_8O_4$	228.22	ye nd (dil al)		245-6			al, eth	B18[4], 1682
14958	Xanthone, 1,7-dihydroxy or Euxanthone............... $C_{13}H_8O_4$	228.22	ye nd (to), pl (al)	sub d	240				B18[4], 1682
14959	Xanthone, 1,8-dihydroxy .... $C_{13}H_8O_4$	228.22	ye lf (bz)		187				B18[4], 1683
14960	Xanthone, 2,3-dihydroxy .... $C_{13}H_8O_4$	228.22	ye nd (al)		294			al, ace, bz, aa	B18[4], 1683
14961	Xanthone, 2,7-dihydroxy or β-Isoeuxanthone .... $C_{13}H_8O_4$	228.22	ye nd (al, eth)	sub	>330			al, eth	B18[2], 357
14962	Xanthone, 3,4-dihydroxy .... $C_{13}H_8O_4$	228.22	pa red-ye nd (dil al + 3w)		240			al	B18[4], 1684
14963	Xanthone, 3,6-dihydroxy .... $C_{13}H_8O_4$	228.22	nd (dil al), pr (sub)	sub	300-50d			al, aa	B18[4], 1685
14964	Xanthone, 1,7-dihydroxy-3-methoxy or Gentisin .... $C_{14}H_{10}O_5$	258.23	ye rh	400 sub	266-7			al, Py	B18[4], 2602
14966	Xanthophyll or Lutein. Lateol $C_{40}H_{56}O_2$	568.89	ye or vt pr (eth-MeOH), $[\alpha]_{Ca}$ + 160 (chl)		196			al, eth, bz, peth	B6[3], 5870
14967	Xanthopterin or 2-Amino-4,6-pteridenediol. Uropteim ... $C_6H_5N_5O_2$	179.14	hyg ye amor or og (aa)	98-100[18]	>410d	1.559			B26[2], 313
14968	Xanthotoxin or Ammoidin......................... $C_{12}H_8O_4$	216.21	pr (dil al), nd (peth)		148			al	B19[4], 2633
14969	Xanthotoxol or 8-Hydroxy-4′:5′,6:7-furocoumarin ...... $C_{11}H_6O_4$	202.17			251-2				B19[4], 2633
14970	Xanthoxyletin or Alloxanthylethin .... $C_{15}H_{14}O_4$	258.28	pr (MeOH, peth)		133			ace, bz, chl	B19[4], 2645
14971	Xanthyletin or 2,2-Dimethyl chromeno coumarin........ $C_{14}H_{12}O_3$	228.25	pr	140-5[0.1]	131.5			al	B19[4], 1828
14972	o-Xylene or 1,2-Dimethyl benzene. 1,2-(CH₃)₂C₆H₄	106.17		144.4, 32[10]	−25.2	0.8802[10/4]	1.5055[20]	**al, eth, ace, bz**	B5[4], 917
14973	o-Xylene, 4-chloro or 4-Chloro-1,2-dimethyl benzene .... 1,2-(CH₃)₂C₆H₃Cl-(4)	140.61		194[755]	−6	1.0692[15/15]		ace, bz	B5[3], 816
14974	o-Xylene, 3,4-dinitro or 1,2-Dimethyl-3,4-dinitro benzene 1,2-(CH₃)₂-3,4-(NO₂)₂C₆H₂	196.17	nd (al)		82			eth, bz	B5[3], 822
14975	o-Xylene, 3,5-dinitro 1,2-(CH₃)₂-3,5-(NO₂)₂C₆H₂	196.17	ye nd (al, peth)		77			al, ace, bz, chl	B5[3], 823

No.	Name, Synonyms, and Formula	Mol. wt.	Color, crystalline form, specific rotation and λ_{max} (log ε)	b.p. °C	m.p. °C	Density	n_D	Solubility	Ref.
14976	o-Xylene, 3,6-dinitro 1,2-(CH$_3$)$_2$-3,6-(NO$_2$)$_2$C$_6$H$_2$	196.17	nd (al)		89-90			al, eth, ace, bz, peth	B5[1], 181
14977	o-Xylene, 4,5-dinitro 1,2-(CH$_3$)$_2$-4,5-(NO$_2$)$_2$C$_6$H$_2$	196.17	nd (al, bz, aa)		118			eth, ace, bz	B5[3], 823
14978	o-Xylene, 4-ethyl 1,2-(CH$_3$)$_2$-4-C$_2$H$_5$C$_6$H$_3$	134.22		189.7, 67.8[10]	−67	0.8745[20/4]	1.5031[20]	al, eth, ace, bz, peth	B5[4], 1069
14979	o-Xylene, 3-iodo or 3-Iodo-1,2-dimethyl benzene 3-I-1,2-(CH$_3$)$_2$C$_6$H$_3$	232.07		228-30, 125-6[15]		1.6395[20/4]	1.6074[20]	ace	B5[3], 820
14980	o-Xylene, 4-iodo 4-I-1,2-(CH$_3$O$_2$C$_6$H$_3$	232.07		231-2, 111[11]		1.6334[18/4]	1.6049[18]	ace	B5[3], 821
14981	o-Xylene, 3-methoxy or 2,3-Dimethyl anisole 2,3-(CH$_3$)$_2$C$_6$H$_3$OCH$_3$	136.20		199, 85[18]	29	0.9596[40]	1.5120[40]	al, eth, ace, bz	B6[3], 1723
14982	o-Xylene, 4-methoxy 4-CH$_3$O-1,2-(CH$_3$)$_2$C$_6$H$_3$	136.20		204-5, 96-7[17]		0.9744[14/4]	1.5198[14]	al, eth, bz	B6[3], 1727
14983	o-Xylene, 3-nitro 3-NO$_2$-1,2-(CH$_3$)$_2$C$_6$H$_3$	151.17	nd (al)	240, 131[20]	15	1.1402[/20]	1.5441[20]	al	B5[4], 930
14984	o-Xylene, 4-nitro 4-NO$_2$-1,2-(CH$_3$)$_2$C$_6$H$_3$	151.17	ye pr (al)	254[750], 143[21]	30-1	1.112[15]	1.5202[20]	al	B5[4], 930
14985	o-Xylene, 3,4,5,6-tetrabromo 3,4,5,6-Br$_4$-1,2-(CH$_3$)$_2$C$_6$	421.77	nd (bz)	374-5	262			bz	B5[3], 820
14986	m-Xylene or 1,3-Dimethyl benzene 1,3-(CH$_3$)$_2$C$_6$H$_4$	106.17		139.1, 28.1[10]	−47.9	0.8642[20/4]	1.4972[10]	**al, eth, ace, bz**	B5[4], 932
14987	m-Xylene 4-chloro 1,3-(CH$_3$)$_2$C$_6$H$_3$Cl-(4)	140.61		187-8, 83[11]		1.0598[20/20]	1.5230[25]	ace, bz	B5[3], 834
14988	m-Xylene, 5-chloro 1,3-(CH$_3$)$_2$-C$_6$H$_3$-Cl-(5)	140.61		187-8, 66[12]				ace, bz	B5[3], 834
14989	m-Xylene, 2,5-dinitro 1,3-(CH$_3$)$_2$-2,5-(NO$_2$)$_2$C$_6$H$_2$	196.17	ye (al)		101			al, eth, ace, bz, chl	B5[2], 295
14990	m-Xylene, 4-ethyl 1,3-(CH$_3$)$_2$-4-C$_2$H$_5$C$_6$H$_3$	134.22		188.4, 66.5[10]	−62.9	0.8763[20/4]	1.5038[20]	**al, eth, ace, bz,** peth	B5[4], 1070
14991	m-Xylene, 5-ethyl 1,3-(CH$_3$)$_2$-5-C$_2$H$_5$C$_6$H$_3$	134.22		183.7, 63[10]	−84.3	0.8648[20/4]	1.4981[20]	**al, eth, ace,** bz, peth	B5[4], 1071
14992	m-Xylene, 2-iodo 2-I-1,3-(CH$_3$)$_2$C$_6$H$_3$	232.07	oil	229-30, 102-3[14]	11.2	1.6158[20/4]	1.6035[20]	ace, bz	B5[4], 947
14993	m-Xylene, 4-iodo 4-I-1,3-(CH$_3$)$_2$C$_6$H$_3$	232.07		231d, 111[14]		1.6282[16/4]	1.6008[16]	ace, bz	B5[4], 947
14994	m-Xylene, 5-iodo 5-I-1,3-(CH$_3$)$_2$C$_6$H$_3$	232.07	oil	230-1, 117[27]		1.6085[18.5/4]	1.5967[18.5]	ace	B5[3], 840
14995	m-Xylene, 2-methoxy 2-CH$_3$O-1,3-(CH$_3$)$_2$C$_6$H$_3$	136.20		182-3		0.9619[14/4]	1.5053[14]	al, eth, bz	B6[3], 1737
14996	m-Xylene, 4-methoxy 4-CH$_3$O-1,3-(CH$_3$)$_2$C$_6$H$_3$	136.20		192, 83-4[15]		0.9740[16/4]	1.5190[16]	al, eth, bz	B6[3], 1744
14997	m-Xylene, 5-methoxy 5-CH$_3$O-1,3-(CH$_3$)$_2$C$_6$H$_3$	136.20		194.5, 89[15]		0.9627[15/4]	1.5110[20]	al, eth, bz, aa	B6[3], 1756
14998	m-Xylene, 2-nitro 2-NO$_2$-1,3-(CH$_3$)$_2$C$_6$H$_3$	151.17		222, 84.6[0.05]	13	1.112[15]	1.5202[20]	al	B5[4], 948
14999	m-Xylene, 4-nitro 4-NO$_2$-1,3-(CH$_3$)$_2$C$_6$H$_3$	151.16		246[744], 122[18]	9	1.135[15]	1.5473[25]	eth, ace, bz	B5[4], 948
15000	m-Xylene, 5-nitro 5-NO$_2$-1,3-(CH$_3$)$_2$C$_6$H$_3$	151.17	nd (al)	273[739]	75			al, eth	B5[4], 948
15001	m-Xylene, 2,4,5,6-tetrabromo 2,4,5,6-Br$_4$-1,3-(CH$_3$)$_2$C$_6$	421.77	nd (xyl, al)		248 (252)			bz	B5[3], 839
15002	m-Xylene, 2,4,6-trinitro 2,4,6-(NO$_2$)$_3$-1,3-(CH$_3$)$_2$-C$_6$H	241.16	pa ye pr or lf (al-bz)		184	1.604[19]		bz, chl, Py	B5[4], 950
15003	p-Xylene or 1,4-Dimethyl benxene 1,4-(CH$_3$)$_2$C$_6$H$_4$	106.17	mcl pr (al)	138.3, 27.2[10]	13.3	0.8611[20/4]	1.4958[20]	**al, eth, ace, bz**	B5[4], 951
15004	p-Xylene, 2-chloro 1,4-(CH$_3$)$_2$C$_6$H$_3$Cl-(2)	140.61		187	1.6	1.0589[15/4]		ace, bz	B5[4], 965
15005	p-Xylene, 2,3-dinitro 1,4-(CH$_3$)$_2$-2,3-(NO$_2$)$_2$C$_6$H$_2$	196.17	mcl pr (al)		93			al, eth, ace, bz, chl	B5[4], 972
15006	p-Xylene, 2,5-dinitro 1,4-(CH$_3$)$_2$-2,5-(NO$_2$)$_2$C$_6$H$_2$	196.17	ye nd (al)		147-8			ace, bz, chl	B5[4], 973
15007	p-Xylene, 2,6-dinitro 1,4-(CH$_3$)$_2$-2,6(NO$_2$)$_2$C$_6$H$_2$	196.17	nd (al)		123-4			al, ace, bz, chl	B5[4], 973
15008	p-Xylene, 2-ethyl 1,4-(CH$_3$)$_2$-2-C$_2$H$_5$C$_6$H$_3$	134.22		186.9, 64.9[10]	−53.7	0.8772[20/4]	1.5043[20]	**al, eth, ace, bz,** peth	B5[4], 1070
15009	p-Xylene, 2-iodo 2-I-1,4-(CH$_3$)$_2$C$_6$H$_3$	232.07		227-8d, 106-8[13]		1.6168[17/4]	1.5992[17]	ace, bz	B5[4], 970

No.	Name, Synonyms, and Formula	Mol. wt.	Color, crystalline form, specific rotation and λ_{max} (log ε)	b.p. °C	m.p. °C	Density	n_D	Solubility	Ref.
15010	p-Xylene, 2-methoxy .... 2-CH$_3$O-1,4-(CH$_3$)$_2$C$_6$H$_3$	136.20	....	194[772]	....	0.9693[13/4]	1.5182[13]	al, eth, bz, peth	B6[3], 1772
15011	p-Xylene, 2-nitro .... 2-NO$_2$-1,4-(CH$_3$)$_2$C$_6$H$_3$	151.17	pa ye	240-1, 64-5[0.35]	....	1.132[15]	1.5413[20]	al	B5[4], 971
15012	p-Xylene, 2,3,5-trinitro .... 2,4,5-(NO$_2$)$_3$-1,4-(CH$_3$)$_2$C$_6$H	241.16	mcl nd (al), lf (al-bz)	410 exp	139-40	1.59[19]	....	al	B5[3], 864
15013	2,3-Xylidine or 2,3-Dimethyl aniline.............. 2,3-(CH$_3$)$_2$C$_6$H$_3$NH$_2$	121.18	....	221-2, 106[15]	<-15	0.9931[20]	1.5684[20]	al, eth	B12[3], 2438
15014	2,4-Xylidine or 2,4-Dimethyl aniline.............. 2,4-(CH$_3$)$_2$C$_6$H$_3$NH$_2$	121.18	....	214, 91[10]	-14.3	0.9723[20/4]	1.5569[20]	al, eth, bz	B12[3], 2469
15015	2,5-Xylidine .... 2,5-(CH$_3$)$_2$C$_6$H$_3$NH$_2$	121.18	ye lf (liq)	214, 92-100[10]	15.5	0.9790[21/4]	1.5591[21]	eth	B12[3], 2503
15016	2,6-Xylidine or 2,6-Dimethyl aniline.............. 2,6-(CH$_3$)$_2$C$_6$H$_3$NH$_2$	121.18	....	214[739]	11.2	0.9842[20]	1.5610[20]	al, eth	B12[3], 2462
15017	3,4-Xylidine or 3,4-Dimethyl aniline.............. 3,4-(CH$_3$)$_2$C$_6$H$_3$NH$_2$	121.18	pl or pr (liq)	228	51	1.076[18]	....	eth, lig	B12[3], 2443
15018	3,5-Xylidine or 3,5-Dimethyl aniline.............. 3,5-(CH$_3$)$_2$C$_6$H$_3$NH$_2$	121.18	....	220-1, 99-100[20]	9.8	0.9706[20/4]	1.5581[20]	eth	B12[3], 2495
15019	Xylitol .... C$_5$H$_{12}$O$_5$, HOCH(CHOH)$_3$CH$_2$OH	152.15	(i) rh (al), (ii) mcl (al) (st)	(ii) 215-7	(i) 61, (ii) 93-4	....	....	w, al, Py	B1[4], 2832
15020	D-Xylose or Wood sugar.......................... C$_5$H$_{10}$O$_5$	150.13	mcl nd, [α]$^{20}_D$ + 22.5 (chl)	....	145	1.525[20/4]	....	w	B1[4], 4223
15021	D-Xylose osazone............................... C$_{17}$H$_{20}$N$_4$O$_3$ HOCH$_2$(CHOH)$_2$C(=NNHC$_6$H$_3$)CH=NNHC$_6$H$_5$	328.38	pa ye nd, [α]$_D$ -40.9 (al)	....	159(167 d)	....	....	al, eth, ace	B31, 61

No.	Name, Synonyms, and Formula	Mol. wt.	Color, crystalline form, specific rotation and λ_{max} (log ϵ)	b.p. °C	m.p. °C	Density	n_D	Solubility	Ref.
15022	Yamogenin .... $C_{27}H_{42}O_3$	414.64	pl, $[\alpha]^{25}_D$ −123	....	201	....	....	....	B19[4], 865
15023	Yobyrine .... $C_{19}H_{16}N_2$	272.35	nd(dil al)	150[0.01]	218-9	....	....	al, chl	B23[2], 263
15024	Yohimbine or Corynine .... $C_{21}H_{26}N_2O_3$	354.45	nd (dil al)	159[0.01] sub	241	....	....	al, eth, chl	B25[2], 201
15025	Yohimbine hydrochloride .... $C_{21}H_{26}N_2O_3 \cdot HCl$	390.92	orh nd or pl (w, dil HCl)	....	302	....	....	w	B25[2], 204
15025	Yohimbine-hydrochloride .... $C_{21}H_{26}N_2O_3 \cdot HCl$	390.92	orh nd or pl (w, dil HCl)	....	302	....	....	w	B25[2], 201
15026	Yohimbine-nitrate .... $C_{21}H_{26}N_2O_3 \cdot HNO_3$	417.47	pr (w)	....	276	....	....	w	B25[2], 204
15027	Yuccagenin .... $C_{27}H_{42}O_4$	430.64	nd (al), $[\alpha]^{25}_D$ −113	....	252	....	....	al, eth, diox	B19[4], 1083

No.	Name, Synonyms, and Formula	Mol. wt.	Color, crystalline form, specific rotation and λ_{max} (log ε)	b.p. °C	m.p. °C	Density	n_D	Solubility	Ref.
15028	Zagadinine $C_{27}H_{43}NO_7$	493.65	orh (al), nd (bz), [α] −45 (chl)	....	201-4	....	....	al, chl	Am35, 258
15029	Zeaxanthin or Zeaxanthol $C_{40}H_{56}O_2$	568.89	ye pr (MeOH), rh (chl-eth)	226-9$^{0.06}$	215.5	....	....	eth, ace, bz, chl, Py	B6³, 5865
15030	Zymosterol or 8(14).24(25′) cholestadienol $C_{27}H_{44}O$	384.65	pl (MeOH), nd, $[α]^{20}_D$ + 149 (chl)	160$^{0.001}$ sub	110	....	....	ace, chl	B6³, 2828
15031	Zeaxanthin diacetate $C_{44}H_{60}O_4$	652.96	....	....	154-5	....	....	....	B6³, 5868

STRUCTURAL FORMULAS OF ORGANIC COMPOUNDS

In Numeric as they occur in Organic Compounds Table

1 — $CH(CH_3)_2$, CH_3, CH_3, CO_2H (structure)

3 (structure)

4 (structure, positions 1–8)

16 — HO_2C (structure)

17 (structure with two O)

18 — SO_3H (structure)

19 (structure with O)

20 (structure)

22 — $CH_3CH:N \cdot NH$ — ring with NO_2, NO_2

90 — CH_3CONH — ring with CH_3

92 — CH_3CONH — ring with H_2N

96 — CH_3CONH — ring with Br

100 — CH_3CONH — ring — $(CH_2)_3CH_3$

103 — CH_3CONH — ring with Cl, CH_3

104 — CH_3CONH — ring with Cl, NO_2

111 — CH_3CONH — ring with Cl, NO_2

113 — CH_3CONH — ring with CH_3, Cl

115 — CH_3CONH — ring with NO_2, Cl

118 — CH_3CONH — ring — cyclohexyl

120 — CH_3, CH_3CONH — ring — CH_3, NO_2

121 — CH_3CONH — ring with NO_2, NO_2

128 — C_2H_5O, CH_3CONH — ring

129 — C_2H_5O, CH_3CONH — ring — NO_2

135 — NO_2, CH_3CONH — ring — OC_2H_5

137 — $CH_3CON(C_2H_5)$ — ring — NO_2

139 — HO, CH_3CONH — ring

143 — CH_3, CH_3CONH — ring

146 — HO, $CH_3CON(CH_3)$ — ring

148 — CH_3O, CH_3CONH — ring

150 — CH_3O, CH_3CONH — ring — NO_2

154 — NO_2, OCH_3, CH_3CONH — ring

158 — NO_2, CH_3CONH — ring — OCH_3

162 — $CH_3CON(CH_3)$ — ring — NO_2, OH

164 — CH_3, $CH_3CON(CH_3)$ — ring

167 — NO_2, CH_3CONH — ring — CH_3

168 — NO_2, CH_3CONH — ring

171 — CH_3CONH — ring — $(CH_2)_7CH_3$

CH_3CO_2 — cyclohexyl

196

CH_3CO_2 — cyclopentyl

197

CH_3CO_2 — benzene ring with NO_2 (2,4-dinitro)

201

$CH_3CO_2CH_2$ — furan

210

CH_3CO_2 — furan

211

CH_3CO_2 — cyclohexyl with i-C_3H_7 and CH_3

221

CH_3CO_2 — benzene with CH_3O

224

CH_3CO_2 — naphthalene with CH_3

233

CH_3CO_2 — naphthalene

241

CH_3CO_2 — naphthalene

242

CH_3CO_2 — benzene with NO_2

243

$CH_3CO_2CH_2$ — tetrahydrofuran

264

thiophene — CH_2CO_2H

265

CH_3CO_2 — benzene with CH_3

266

CH_3COCH_2CONH — benzene with H_3C

272

cyclohexylidene $=CHCN$

343

CH_3COCH_3 — benzene (positions 2,3,4,5,6)

351

furan — $CH:CHCOC_6H_5$

395

HO_2C — butenolide, O, $=O$

452

furan — $CH:CHCHO$

483

$CH_2:CHCO_2$ — cyclohexyl

497

furan — $CH:CHCO_2H$

502

521

cyclohexyl — OH $C:C$ HO — cyclohexyl

450

460

CH_3O, HO, OCH_3, $OCOC_6H_5$, OH, CH_3CH_2, HO, $OCOCH_3$, CH_2, OCH_3, OCH_3

acridine (positions 1-9, N-10)

464

HO, HO — $CH(OH)CH_2NHCH_3$

548

HO, HO — $COCH_2NHCH_3$

550

551

CH_3O_2C — with O, CH_3, indole, N—H

552

N—CH_3, OH, CH_2CH_3

553

adenosine with NH_2, $HOCH_2$, O, OH, OH

523

HO—P—OCH_2, OH, adenine, OH, OH

524

522

In Numeric as they occur in Organic Compounds Table

564

592

633

739

574

593

634

744

575

594

636

746

576

595

637

747

578

600

639

748

579

670

749

582

631

673

750

583

632

675

751

586

738

752

823

1139

1313

753

946

1146

1321

754

970

1150

1324

757

1096

1151

1326

758

1111

1194

1327

759

1117

1195

1336

763

1118

1273

1337

765

1127

1292

1338

822

1138

1294

1339

In Numeric as they occur in Organic Compounds Table

1342

1363

1410

1441

1344

1366

1422

1445

1347

1369

1423

1448

1375

1424

1449

1350

1376

1427

1353

1378

1435

1451

1355

1379

1436

1356

1437

1359

1394

1439

1452

1461

1462

1463

1464

1465

1559

1561

1568

1574

1580

1581

1583

1586

1588

1589

1590

1591

1598

1603

1608

1610

1616

1619

1626

1635

1638

1644

1645

1654

1660

1662

1667

1784

1785

1787

1788

1809

1858

1882

2348

2459

1870

2058

2357

2462

1871

2106

2378

2467

1872

2298

2400

2468

2314

2401

2470

1873

2326

2451

2483

2333

2452

2484

1874

2334

2453

2485

2455

2486

1879

2340

2487

2488

1881

2344

2456

2489

2493

2506

2507

2509

2510

2511

2513

2514

2515

2516

2517

2518

2520

2528

2529

2531

2625

2837

2879

2880

2881

2882

3007

3008

3009

3013

3014

3016

3023

3031

3046

3048

3053

3069

3078

3094

3108

3109

3110

3114

3115

3116

3117

3118

3162

3163

3177

3179

3186

3289

3314

3315

3320

3322

3323

3324

3325

3329

3332a

3333

3334

3335

3336

3337

3338

3339

3340

3342

3343

3344

3345

3346

3348

3350

3351

3353

3357

3360

In Numeric as they occur in Organic Compounds Table

3361

3362

3366

3367

3369

3500

3501

3505

3511

3517

3523

3524

3529

3530

3531

3532

3533

3534

3534a

3535

3536

3537

3538

3540

3541

3542

3543

3544

3545

3546

3548

3549

3551

3570

3581

3583

3586

3588

3595

3596

3597

3599

STRUCTURAL FORMULAS OF ORGANIC COMPOUNDS (Continued)

In Numeric as they occur in Organic Compounds Table

3600

$CH_3CHCHCH_3$ (epoxide O)
3757

$CH_3CHCHCH_3$ (epoxide O)
3758

4477

4520

HO_2C—$CONH_2$
4525

3620

$CH_3CH_2CH_2CH$ $\left(\text{—OH}\right)_2$
3765

$CH_3CH_2C(CH_3)$ $\left(\text{—OH}\right)_2$
3766

4478

H_2NOC—CO_2H
4526

3624

CH_3—SO (ring)
3832

CO_2H
4479

OH HO
4531

3628

H_2N—$CO(CH_2)_3CH_3$
4093

4480

4532

3629

$CH_2C(CH_3)_2$ (epoxide O)
4194

4483

CO_3CH_3
4533

$CH_2CHCHCH_2$ (diepoxide O O)
3650

$CH_3C(C_6H_5)CHCO_2C_2H_5$ (epoxide O)
4327

HO
4484

CH_2SO_3H
4534

OH   OH
C:CC:C
3670

CH_2 (γ-butyrolactone O O)
4350

CO_2H
4486

$CH_2CH_2NH_2$
4537

$CH_2CHCHCH_2$ (diepoxide O O)
3738

CH_2CHCH_2CN (epoxide O)
4424

CO_2H
4489

$CH_2CH_2NH_2$
4538

$\left(\text{}\right)_2$ $C(CH_3)CH_2CH_3$
3743

4466

4467

4490

CH_3 OH
H_3C $(CH_2)_4CH_3$ OH
4544

:NOH
4493

4539

O
CO_2H
4514

HO_2C—CO_2H
4518

C-587

STRUCTURAL FORMULAS OF ORGANIC COMPOUNDS (Continued)

In Numeric as they occur in Organic Compounds Table

4545

4734

4754

4769

4546

4743

4772

4744

4756

4777

4575

4779

4746

4760

4576

4780

4577

4747

4762

4588

4781

4764

4597

4749

4765

4782

4637

4751

4784

4649

4766

4785

4697

4752

4768

4811

4722

4753

4812

4814

04824

4825

4826

4836

4838

4843

4845

4848

4860

4862

4863

4864

4865

4889 $ClCH_2CON(C_2H_5)(4-Cl-C_6H_4)$

4986

4987

4989

4998

5003

5007

5008

5009

5010

5011

5019

5020

5025

5026

5028

5032

5033

5039

5043

5047

5052

5053

5054

5055

5056

5059

5064

5065

5066

5067

5070

5081

5084

5128

5167

5168

5169

5171

5179

5187

5188

5193

5194

5195

5200

5201

5202

5203

In Numeric as they occur in Organic Compounds Table

5204

5221

5244

5262

5263

5264

5224

5246

5265

5205

5225

5248

5279

5206

5226

5250

5295

5207

5210

5237

5251

5296

5211

5238

5252

5300

5213

5239

5253

5303

5216

5241

5258

5379

5217

5242

5259

5380

5220

5243

5259

5392

In Numeric as they occur in Organic Compounds Table

5393

5437

5438

5397

5444

5399

5446

5400

5450

5401

5454

5402

5455

5403

5458

5405

5459

5406

5460

5434

5461

5436

5463

5464

5465

5466

5473

5474

5475

5482

5483

5488

5489

5507

5542

5543

5551

5563

5564

5574

5576

5577

5621

5647

5663

5667

5676

5677

In Numeric as they occur in Organic Compounds Table

5678

5697

5698

5699

5702

5703

5704

5707

5713

5717

5725

5726

5732

5738

5739

5740

5742

5744

5745

5746

5748

5749

5752

5754

5756

5756a

5760

5761

5762

5774

5784

5785

5789

5798

5801

5805

5806

5820

5822

5823

5824

5838

5839

5847

STRUCTURAL FORMULAS OF ORGANIC COMPOUNDS (Continued)

In Numeric as they occur in Organic Compounds Table

5849

5856

5858

5859

5860

5861

5862

5863

5877

5878

5879

5881

5887

5893

5895

5897

5947

5948

5949

5950

5951

5953

5954

5955

5956

5957

5960

5966

5982

5983

5985

5986

5987

5989

5990

5991

5992

6006

6009

6010

6011

6012

6016

6017

6021

6023

6024

6025

6026

6028

6029

6072

6073

6074

6075

6076

6077

6079

6082

6083

6084

6086

6129

6130

6131

6132

6133

6134

6136

6140

6145

6152

6155

6168

6181

6182

6196

6207

6215

6220

6223

6224

6231 COCH$(C_6H_5)_2$

6249

6267

6269 H_2N—CH$_2$CH$_2$—NH$_2$

6270

6271 Br—CH$_2$CH$_2$—Br

6274

6314 CH(OH)

6315 $(CH_3$—$)_2$CHOH

6327

6342

6357

6358

6359

6360

6361

6362

6365

6370 CH$_3$——SS——CH$_3$

6372 CH$_3$——S——CH$_3$

6373 CH$_3$——S——CH$_3$

6374 CH$_3$——S——CH$_3$

6376 CH$_3$——SO$_2$——CH$_3$

6379

6423 CH$_3$—N CO$_2$H

6424 CH$_3$—N CO$_2$H, OH

6428 CH$_3$—N CO$_2$H, O$_2$CC$_6$H$_5$

6431

6433

6434

6435

6438

6468

6470

6471

6472

6473

6476

6477

6478

6479

6480 C_6H_5—NH——NH

CH₃O, OCH₃, CH₃O, OCH₃, NH, CH₃CH₂, N — **6481**

H₂NCONHN, HO — **6504**

6521

$C_{12}H_{21}O_{11}$·O, O, OH, OH — **6541**

Br, Br, Br, O, CO₂H, HO, Br — **6485**

O, Br — **6508**

CH₃—C—C—C—C—CH₂OH with OH, OH, H, OH — **6522**

HO, O, OH — **6542**

H₃C CH₃, CH₃, CH₂CH₂NH₂ — **6511**

CH₃—C—C—C—C—CHO with OH, OH, H, OH — **6524**

HO, CH₃, CH₃, CH₂, CH₃ — **6543**

H₂N—CH(OH)CH(NH₂)CH₃·2HCl — **6489**

OH, OH, OH, HO, O, OH — **6512**

HO — **6525**

O, HO — **6526**

O, CH₃, CH₃, CH₂, CH₃ — **6544**

NO₂—CON(CH₃)CH(CH₃)CH(OH)C₆H₅ — **6494**

HO — **6514**

N, CH₂OH — **6528**

O, O, CH₃, CH₃, CH₃, CH₂ — **6545**

HO — **6497 (6496?)**

HO — **6515**

HO, HO, CH₂CH₂NHCH₃ — **6529**

HN, N—CH₃, CONH₂ — **6546**

CH₃CO₂ — **6497**

HO — **6516**

N, CH(OH), CHCH₃, OCH₃ — **6530**

HO — **6499**

HO — **6517**

C₆H₅, CO₂H, HO₂C, C₆H₅ — **6533**

CH(CH₃)₂, CONH, O, OH, N, O, CH(CH₃)₂, CH₃N, N, H — **6547**

CH₃CO₂ — **6500**

HO — **6518**

O, HO — **6536**

HO, O — **6501**

NCH₃, CH₃O, OCH₃, O—CH₂ — **6519**

O, HO — **6539**

CH(CH₃)₂, CONH, O, N, O, CH₃N, N, CH₂C₆H₅, H — **6549**

O — **6502**

HO, OH — **6540**

HON — **6503**

In Numeric as they occur in Organic Compounds Table

6551

6553

6554

6556

6557

6559

6560

06561

6562

6563

6564

6565

6566

6567

6568

6569

6570

6571

6572

6573

6574

6576

6579

R₁ = CH₃ or H
R₂ = H or CH₃

6580

R₁ = CH₃ or H
R₂ = H or CH₃

6581

6582

R₁ = CH₃ or H
R₂ = H or CH₃

6584

6585

6586

6589

6590

6591

6593

6595

6682

6688

6742

6794

6931

6933

6934

6935

6936

6941

6946

6948

6952

6955

6956

6598

6599

6601

6602

6605

6609

6638

6643

6655

6681

6830

6831

6849

6889

6917

6918

6919

6925

6927

6937

6938

6939

6940

6957

6958

6959

6962

6981

6982

6986

6987

7020

7021

7030

7033

7034

7041

7042

7045

7046

7053

7057

7066

7078

7091

7092

7093

7094

7095

7096

7098

7116

7133

7140

7142

7151

7158

7165

7169

7170

7171

7174

7183

7184

7185

7186

7192

7193

7194

7196

7224

In Numeric as they occur in Organic Compounds Table

7225 — furan-2-CN

7227 — furan-CO$_2$H (positions 2,3,4,5)

7240 — furan-COCH(OH)-furan

7241 — furan-CH$_2$CO$_2$H

7242 — furan-CH$_2$CN

7243 — furan-O(CH$_2$)$_7$CH$_3$

7244 — furan-O-phenyl

7245 — furan-CO-phenyl

7246 — furan-S-phenyl

7249 — HOCH$_2$-C(OH)-C(H)-C(H)-C(OH)-CH$_2$OH

7250 — HOCH$_2$-C(OH)-C(H)-C(OH)-C(H)-CO (with O bridge, OH H)

7251 — HOCH$_2$-C(OH)-C(OH)-C(OH)-C(H)-CHO (OH OH H OH)

7252 — CH$_2$-C(OH)-C(OH)-C(OH)-C(H)-CHO (with O bridge)

7253 — CH$_3$OCH$_2$-C(CH$_3$)(CH$_3$)-C(H)-C(O)(OH)-C (with CH$_3$, O bridge)

7254 — HCO$_2$-C(H)(OH)-C(OH)(OH)-C(OH)(OH)-CHO

7256 — quinoline, OCH$_3$, CH$_2$CH$_2$ -C$_6$H$_3$(OCH$_3$)(OCH$_3$)

7257 — xanthone, HO, CO$_2$H

7258 — xanthone, HO, HO, HO, HO, CO$_2$H

7259 — indole alkaloid, NCH$_3$, CH:CH$_2$, N-H

7261 — terpene aldehyde, CHO

7272 — steroid alkaloid, N, OH, OH, HO, OH, OH

7275 — spirostane type, HO, HO, O

7276 — cardenolide, OH, OH, OH, HO, O

7278 — HOCH$_2$-C(H)(OH)-C(H)(OH)-C(H)-C:C-C=O (with O bridge)

7279 — HOCH$_2$-C(H)(OH)-C(H)(OH)-C(H)(OH)-C(H)(OH)-CHO

7280 — CH$_3$-C(H)(OH)-C(H)(OH)-C(H)-C(H)(OH)-CHO

7281 — pyran, HO, HO, CH$_2$OH

7282 — HOCH$_2$-C(H)(OH)-C(H)(OH)-C(H)-C(H)(OH)-CO$_2$H

7283 — HOCH$_2$-C(H)(OH)-C(H)(OH)-C(H)-C(H)-CO (with O bridge)

7284 — HOCH$_2$-C(H)-C(H)(OH)-C(H)(OH)-CO (with O bridge)

7285 — HOCH$_2$-C(H)(OH)-C(H)(OH)-C(H)-C(H)(OH)-CN

7286 — HOCH$_2$-C(H)(OH)-C(H)(OH)-C(H)-C(H)(OH)-CONHNHC$_6$H$_5$

7287 — HOCH$_2$-CO-C(OH)(H)-C(H)(OH)-CO$_2$H

7288 — HOCH$_2$-C(H)(OH)-C(H)(OH)-C(H)-C(H)(OH)-CHO

7289 — HOCH$_2$-C(H)-C(H)(OH)-C(OH)(H)-C(H)-C(H)(OH) (with O bridge)

7291 — HOCH$_2$-C(H)-C(H)(OH)-C(OH)(H)-C(H)-C(H)(OH) (with O bridge)

7295 — HOCH$_2$-C(H)(OH)-C(H)(OH)-C(OH)(H)-C(H)(OH)-CH:NNHC$_6$H$_5$

7297 — HOCH$_2$-C(H)(OH)-C(H)(OH)-C(OH)=NNHC$_6$H$_5$-CH:NNHC$_6$H$_5$

7299 — HOCH$_2$-C(H)(OH)-C(H)(OH)-C(OH)(H)-C(H)(NH$_2$)-CHO

7301 — HOCH$_2$-C(H)-C(OH)(H)-C(OH)(H)-C(H)(NHCH$_3$)-CHO

7302 — HOCH$_2$-C(H)(OH)-C(H)(OH)-C(OH)(H)-C(H)(OH)-CHS

7303 — HOCH$_2$-C(H)-C(H)(OH)-C(OH)(H)-C(H)(OCH$_3$) (with O bridge)

7304 — HOCH$_2$-C(H)-C(H)(OH)-C(OH)(H)-C(H)(OCH$_3$) (with O bridge)

7305 — HO$_2$C-C(H)-C(H)(OH)-C(OH)(H)-C(H)(OH) (with O bridge)

7306 — O-C(H)-C(H)-C(H)(OH)-C(H) (with O bridges, OH)

In Numeric as they occur in Organic Compounds Table

7320

7434

7465

7477

7321

7435

7467

7478

7358

7435a

7468

7371

7443

7470

7479

7375

7453

7471

7480

7389

7406

7454

7472

7481

7414

7474

7482

7419

7455

7475

7484

7420

7462

7476

7516

7421

7464

7476a

7689

In Numeric as they occur in Organic Compounds Table

7690

OH / N—CH$_3$ (O–CH$_2$–O ring)
8052

(methylenedioxy tetrahydroisoquinoline) N—CH$_3$, OCH$_3$
8159

indole CH$_2$CHCO$_2^-$ / N$^+$(CH$_3$)$_3$
8315

N$^+$(CH$_2$)$_{15}$CH$_3$ Cl$^-$
7720

H$_3$C—⬡—NHNHCH$_2$C$_6$H$_5$
8065

C$_2$H$_5$ quinuclidine CH(OH) quinoline
8160

HO–quinoline
CH$_3$—N—N—phenyl pyrazolone
8316

(hexamethylenetetramine structure with N, CH$_2$)
7758

H$_2$NN(—⬡—CH$_3$)$_2$
8093

CH:N—CH—N:CH (furan rings, O)
8162

8317

CH$_2$CHC:CCHCH$_2$ / O O
7997

CH$_3$—⬡—NHNH—⬡—CH$_3$
8096

(methylenedioxy tetrahydroisoquinoline) N—CH$_3$
8163

H OH H OH
HOCH$_2$—C—C—C—C—CH$_2$OH
OH H OH H
8318

p / m / o —CONHCH$_2$CO$_2$H
8003

CH$_3$—⬡—NHNHCH$_3$
8104

OH / CH$_2$CH$_2$CH(CH$_3$)$_2$ (naphthoquinone)
8164

H OH H OH
HOCH$_2$—C—C—C—C—CO$_2$H
OH H OH H
8319

(imidazole) CH$_2$CH$_2$NH$_2$ / NH
8008

CH$_3$—⬡—NHNHCH$_3$
8105

OH H OH H
HOCH$_2$—C—C—C—C—CO / H H OH / O
8320

(imidazole) CH$_2$CH(NH$_2$)CO$_2$H / N NH
8010

4 5 6
3 2 —NHNH$_2$
8113

C$_2$H$_5$ quinuclidine CH(OH) quinoline CH$_3$O
8165

OH H OH H
HOCH$_2$C—C—C—C—CONHNHC$_6$H$_5$
OH H OH H
8321

C$_2$H$_5$O—⬡—NHC:N—⬡—OC$_2$H$_5$ / CH$_3$
8018

CH$_3$—⬡—NHNH—⬡
8118

OH H OH H
HOCH$_2$—C—C—C—C—CHO
H OH H OH
8323

CH$_3$—N—O$_2$CCH(OH)C$_6$H$_5$ (tropane)
8020

CH$_3$—⬡—NHNH—⬡
8134

OH (fluorescein) O OH
8207

NH / N—C=O / H (isatin-like)
8324

HO—⬡—CH$_2$CH$_2$N(CH$_3$)$_2$
8027

4 8 3
5 9 2
6 7 1 (hydrindane)
8138

CH$_3$CH(NH$_2$)CH$_2$—⬡—OH
8217

5 1 NH
4 2
3 (imidazole)
8325

(CH$_3$)$_2$C:CHCH$_2$ HO COCH$_2$CH(CH$_3$)$_2$
HO O
O CH$_2$CH:C(CH$_3$)$_2$
8030

CH$_3$O / CH$_3$O (protoberberine) O—CH$_2$ / O
8142

CH$_3$—N—pyrrolidine CH$_2$COCH$_3$
8297

HN—⬡=O / N—H (imidazole-dione) CO$_2$H CO$_2$H
HN (imidazole)
8331

4 5 O
HN NH
O 2 3 (hydantoin)
8034

CH$_2$OH
CH$_3$—N (tropane) O$_2$CCHC$_6$H$_5$
8298

CH$_3$ / N—CH$_3$ (methylenedioxy...) OCH$_3$ / OCH$_3$ / O / O
8050

C$_2$H$_5$ quinuclidine CH(OH) quinoline
8144

CH$_3$O / HO—⬡—CH$_2$CH$_2$CH$_2$OH
8158

CH$_2$OH
H$_3$C—N (tropane) O$_2$CCHC$_6$H$_5$
8309

HN NH / S (imidazolethione)
8332

In Numeric as they occur in Organic Compounds Table

8333

8334

8336

8338

8365

8366

8367

8375

8379

8381

8382

8383

8394

8395

8403

8404

8405

8406

8407

8408

8409

8410

8411

8421

8425

8426

8428

8429

8430

8431

8432

8433

8434

8435

8442

8445

8446

8447

8448

8452

8454

8455

8456

8457

8459

8466

STRUCTURAL FORMULAS OF ORGANIC COMPOUNDS (Continued)

In Numeric as they occur in Organic Compounds Table

8467

8468

8471

8472

8473

8480

8482

8485

8488

8489

8491

8492

8493

8494

8495

8496

8498

8499

8500

8509

8510

8511

8516

8517

8518

8522

8523

8524

8525

8527

8533

8536

8538

8539

8541

8543

8545

8569

8572

8573

STRUCTURAL FORMULAS OF ORGANIC COMPOUNDS (Continued)

In Numeric as they occur in Organic Compounds Table

8574

8605

8633

8681

8577

8607

8634

8685

8579

8608

8636

8692

8581

8610

8638

8695

8601

8617

8645

8698

8602

8618

8646

8701

8603

8621

8673

8729

8604

8625

8676

8742

8626

8631

8632

8680

8743

8754

8781

HO

8782

$CH(CO_2C_2H_5)_2$

8856

$CH(CO_2C_2H_5)_2$

8857

$C(CO_2C_2H_5)_2$

8858

8928

8930

8763

8785

CH_2OH

$(CH_3)_2C:CHCH_2$ $COCH_2CH(CH_3)_2$

OH

$[CH_2CH:C(CH_3)_2]_2$

8787

8770

$(CH_2)_4CO_2H$

8883

$CO_2C_2H_5$

8898

$N-CH(CO_2C_2H_5)_2$

8932

8937

$HO_2C-C-C-C-CHO$ •H_2O

8771

HO

8796

$[CH:CHC:CH]_2CH:$

CH_3

8908

8772

OH

$CHCH_2$ N CH_2 OH

CH_3

8797

OH OH

8910

Cl^-

HO OH

OH OCH_3

OCH_3

8940

8773

$COCH_2$ N CH_2CO

CH_3

8798

CH_3 CH_3

HO $(CH:CHC:CH)_2CH:$ CH_3 CH_3

CH_3

8941

HO_2C

HO_2C

8774

$COCH_2$ N CH_2CH OH

CH_3

8799

HO_2C $N-H$

N

CH_3

8921

$HOCH_2-C-C-C-CH_2$

8943

OCH_3

CH_3O

8776

8805

S $CH_2CO_2C_2H_5$

$(CH_3O)_2PSCHCO_2C_2H_5$

8922

$HOCH_2-C-C-C-CH_2OH$

8944

HO

HO

CO_2H

CO_2H

8777

C_6H_5 N

C_6H_5 N C_6H_5

H

8820

8926

$HOCH_2-C-C-C-C-CHO$

8780

H_2N NH

NH

8824

N H

8927

$HOCH_2-C-C-C-C-CHO.H_2O$

8947

H_2N NH_2

H_2N

C-607

8950

8951

8953

8954

8956

8957

8961

8966

8969

8970

8986

8989

8997

8998

9000

9001

9019

9069

9074

9075

9076

9078

9103

9126

9129

9159

9166

9179

9195

9196

9209

9210

9211

9212

9213

9215

9216

9225

9239

9242

9243

9258

9262

9267

In Numeric as they occur in Organic Compounds Table

9268

9269

9273

9275

9316

9445

9454

9455

9456

9462

9464

9476

9483

9485

9487

9500

9529

9530

9534

9567

9568

9586

9618

9636

9638

9679

9769

9770

9771

9773

9775

9778

9783

9786

9787

9790

9791

9792

9794

9806

9812

9818

9820

9827

9830

9832

9833

9834

9892

9899

9958

9959

9960

9962

9963

9965

9969

9970

10168

10174

10181

10182

10183

10200

10201

10208

10209

10210

10211

10212

10216

10219

10220

10241

10249

10250

10252

10253

10254

10255

10257

10258

10259

10260

10261

10262

10511

10702

11048

10274

10567

10703

11057

10570

10275

10588

10704

11058

10276

10598

10707

11062

10277

10638

10725

11064

10289

10639

10766

11067

10310

10640

10767

11071

10311

10642

11072

10329

10643

11042

11076

10330

10649

11046

11078

10338

10398

10700

11047

11086

11087

11088

11094

11101

11104

11111

11117

11120

11122

11123

11127

11129

11136

11137

11144

11148

11149

11150

11153

11160

11162

11171

11174

11176

11177

11181

11184

11186

11195

11198

11249

11254

11257

11259

11266

11308

11318

11333

11335

11341

11348

11349

11367

11420

11429

11439

11441

CH_3O —◯— $OC_6H_{11}O_5$ 11538

11585

11702

11446

11539

11447

11554

11587

11705

11448

11556

11591

11707

11449

11560

11593

11708

11594

11504

11561

11614

11712

11615

11714

11528

11563

11616

11715

11532

11565

11617

11717

11533

11570

11669

11534

11695

11734

11536

11581

11696

11537

11584

11701

11736

$C_6H_5CO_2CH_2$ ... O ... CH_2OH
11737

$CH_2CH(CH_3)CH_2Cl$
11815

H_3C — $ClCH_2CH_2CONH$
12063

$CH_3CH_2CO_2CH_2$ ...
12192

11738

11834

CH_3 — CH_3CH_2CONH
12070

$CH_2CH_2CO_2H$
12193

11741

11835

CH_3 — $CH_3CHBrCONH$
12077

CH_2CH_2CN
12212

11755

$CH_3CH CH_2$
11853

H_3C — $CH_3CHClCONH$
12103

$COCH_2CH_3$
12228

11756

$CH_3CH CH_2$
11854

$CH_2CH_2CO_2H$
12111

CH_3O — CH_2COCH_3
12238

11759

CH_2CHCH_2I
11859

$CH_3CH_2CO_2$
12112

H_3C — $COCH_2CH_3$
12241

H_2NSO_2 ... $N:N$ ... $NH_2 \cdot HCl$
11769

CH_2CHCCl_3
11860

CH_2CHCO_2H
12142

$CH_2CBr:CH_2$
12261

$[HO$ ... $CHCH_2CH_3]_2$
11780

CH_3 ... O ... O
11900

$CH_2CH_2CO_2H$
12145

$CH_3C=CH$
12285

$[HO$ ... $C(CH_3)_2]_2$
11781

CH_3 ... SO
11909

$CH_3CH_2CO_2CH_2$
12147

CH_2CHCH_3
12300

CH_2CHCH_2Br
11791

CH_3O — HO — $CH_2CH_2CH_2OH$ CH_3O
11956

12152

$CH_2CH(NHCH_3)CH_3$
12305

CH_2CHCH_2Cl
11813

CH_2CHCH_2OH
11958

OH — $CH_2CH_2CO_2H$
12154

CO_2H — $N:N$ — $N(CH_2CH_2CH_3)_2$
12320

$C:CCO_2H$
12020

HO — $CH_2CH_2CO_2H$ CH_3O
12156

$CH_2C:CH$
12336

O ... O
12041

$CH_2CH_2CO_2H$
12160

$(CH_3)_2NCO_2$ ... $N(CH_3)_3Br$
12345

$CH_2CH_2CO_2H$
12187

12346

12347

12349

12350

12351

12355

12356

12357

12366

12368

12369

12371

12372

12373

12374

12375

12376

12379

12380

12381

12382

12386

12388

12391

12392

12398

12399

12401

12402

12403

In Numeric as they occur in Organic Compounds Table

12419

12421

12424

12429

12430

12438

12439

12440

12441

12544

12547

12548

12564

12565

12566

12567

12569

12575

12576

12577

12578

12602

12603

12611

12612

12613

12617

12618

12623

12649

12650

12652

12656

12660

12662

12671

12675

12679

12685

12686

12687

12688

12689

12690

12691

12692

In Numeric as they occur in Organic Compounds Table

12699

12929

12967

13035

12700

12937

13036

12702

12943

12969

13037

12706

12946

12970

12717

12947

12971

13042

12719

12949

12972

13043

12806

12950

12973

13045

12895

12963

13054

12980

13055

12920

13032

12926

12966

13034

13056

In Numeric as they occur in Organic Compounds Table

13057

13058

13064

13065

13066

13068

CH_3 CH_3 $-(CH:CHCH_2CH)_2CH:CH(CH:CCH:CH)_2-$ 13069

C_6H_5 C_6H_5 C_6H_5 C_6H_5 13070

13072

13073

13074

13075

13076

13080

13081

13082

CS_2H 13123

13166

13167

13169

CH_3 N $OH^\ominus$ 13170

CH_3 CH_3 $CH_2CH_2CHCCO_2H$ 13171

CH_2CH_2 CH_3 C H CH_2OH 13172

CH_2 CH_2CH_2 CH_3 C H CH_2OH 13173

CH_3 CH_3 13174

HO_2C CH_3 CO_2H CH_3 13175

OH 13176

13177

13178

13179

13180

13181

CH_3 CH_3 $N=N$ $N=N$ HO 13182

$C_6H_{11}O_5O$ O_2CCH_3 OH 13183

CH_3N HO 13184

13185

Se Se 13197

CH_3 $C(CH_2)_2CH_3$ CH_3 13200

In Numeric as they occur in Organic Compounds Table

H₂NCONHNH— (aromatic ring with CH₃)
13219

13351

$CH_3(CH_2)_{16}CO_2$—(cyclohexyl)
13386

13227
CH_3O_2C ... CH₃

$CH_3(CH_2)_{16}CO_2CH_2$—(tetrahydrofuranyl)
13419

13352

(aryl-CH₃)₃Sb
13440

13228

13354
HN

13443

13229
OH ... CH(OH)CH₂CH:C(CH₃)₂ ... OH

HOCH₂—C—C—C—C—CH₂OH
OH H OH H
H OH H OH
13362

13444
CH₃ ... H CH₃ ... CH₃ ... CH₂CH₃ ... CH₃ ... CH₃

13342
Si (C₆H₅)₂

HOCH₂—C—C—C—C—CH₂OH
H H OH H
OH OH H OH
13363

13445
N ... CH:CHC₆H₅

13343
CH_3O— ... HO— ... OCH_3 —CH:CHCO₂CH₂CH₂N(CH₃)₂·H₂SO₄

HOCH₂—C—C—C—COCH₂OH
OH H OH
H OH H
13368

13449
3' 2' ... β α ... 2 3
4' —CH:CH— 4
5' 6' ... 6 5

13345
NCH₃ ... OCH₃ ... CH₃O ... OH ... =O

13369

13487
CHO ... HO ... OH ... OH ... O=

13346
HO

13371

13488 } rhamnose
HO ... OHCH₂ ... HO ... OH ... OH ... O=O

13349
HO

13373

13489
O ... N ... O

13350
C₆H₁₁O₅.O— (coumarin ring) =O

13375
H ... ⊖O₂C—N⁺ ... CH₃ ... CH₃

13494
HOCH₂—C—C—C—C—CH₂
H H OH OH
OH H H
O

STRUCTURAL FORMULAS OF ORGANIC COMPOUNDS (Continued)

In Numeric as they occur in Organic Compounds Table

13495

CH₃ —NHSO₂ — NH₂   13619

CH₃O / HO — COCOCH₃ / CH₃O   13637

OCH — CHO   13678

C₆H₅CH—CH₂ (O)   13523

—NHSO₃H   13620

13538

H₂NCO — CONH₂   13679

H₂N — SO₂NH —   13625

CH₃ — O₂CCH₂CH₂CO₂ — CH₃   13561

13638

CON(C₂H₅)₂ / CON(C₂H₅)₂   13680

13581

H₂N — SO₂NH — (pyridine)   13626

HOCH₂—C—C—C—CO—CH₂OH   13639

HO₂C — CO₂H   13681

13588

H₂N — SO₂NH — (quinoxaline)   13627

HOCH₂—C—C—C—C—CH₂OH   13640

CO₂H OH / HO CO₂H   13689

H₂N — SO₂NH — (thiadiazole)   13628

HOCH₂—C—C—C—C—CO₂H · ½H₂O   13641

H₂NCO — CO₂H   13691

13611

H₂N — SO₂NH — (thiazole)   13629

HOCH₂—C—C—C—CO (ring O)   13642

NC — CO₂H   13692

OH / naphthol N=N— —N=N—   13613

O₂N — SO₂NH — (thiazole)   13630

HOCH₂—C—C—C—C—CHO   13643

NC — CN   13695

H₂N — SO₂NH — (pyrimidine)   13614

CO₂H / CONH — SO₂NH — (thiazole)   13631

CH₃CO₂ — O₂CCH₃ (anhydride)   13649

ClCO — COCl   13696

H₂N — SO₂NHC(NH₂):NH   13615

HO₂CCH₂CH₂CONH — SO₂NH — (thiazole)   13632

13675

CH₂CO₂H / O (ring) CH₃ CH₃   13698

13634

13676

13699

CH₃ / H₂N — SO₂NH — (pyrimidine-CH₃)   13617

CH₃O / HO — CH:CHCH₂OH / CH₃O   13635

CO₂H / O CH₃ OCH₃   13677

13700

CH₃ / H₂N — SO₂NH — (pyrimidine-CH₃) / CH₃   13618

OCH₃ / HOCH₂CH:CH— —O C₆H₁₁O₅ / OCH₃   13636

13701

13702

In Numeric as they occur in Organic Compounds Table

13703

13704

13705

13707

13708

13709

13710

13711

13718

13723

13752

13769

13771

13773

13775

13782

13788

13789

13790

13793

13794

13795

13796

thevetose
glucose
glucose

13797

13799

13800

13801

13802

13807

13815

13820

13821

13822

13823

13824

13825

13832

13833

13834

13835

13837

13843

13847

13848

13853

13854

13858

13860

13861

13869

13898

13915

CH_2CO_2H on thiophene
13920

CHO on thiophene
13921

CH:NOH on thiophene
13922

CH:NNH on thiophene with phenyl
13923

CO_2H on thiophene (positions 3,4,5)
13928

CO_2H, CO_2H on thiophene
13935

HO_2C — thiophene — CO_2H
13937

SO_2NH_2 on thiophene
13939

SO_2Cl on thiophene
13940

SH on phenyl
13943

thieno-thiophene
13960

thieno-thiophene
13961

thiopyran
13962

CH_3—N—N—C_6H_5, CH_3, =S
13963

CO_2H, SH on phenyl
13964

NH, =S, O on ring
13966

NH, =O, =S ring (uracil-like)
13969

NHCSNH between two o-tolyl groups
13994

H_2NCSNH — m-tolyl
14013

thioxanthene
14018

thioxanthone (=O, S)
14019

$HOCH_2$—C—C—CO_2H with H, OH, OH, H
14020

OHC—C—C—CH_2OH with H, OH, OH, H
14025

pinane
14026

pinene
14027

pinanone
14028

pinanone
14030

HOH_2C ribose with H, H, H, OH, H — N-ring, =O, NH, =O, CH_3 (thymidine)
14031

CH_3, OH, $CH(CH_3)_2$ on phenol
14032

HO, $C_3H_7^i$, CH_3, CH_3, HO, $C_3H_7^i$ with O—SO_2 (thymol blue type)
14039

HO, $C_3H_7^i$, CH_3, CH_3, HO, $C_3H_7^i$ with O, =O (thymolphthalein)
14040

I, HO, I — O — I, I, $CH_2CH(NH_2)CO_2H$
14041

steroid spiro (sapogenin) H_3C, H_3C, CH_3, CH_3, O, HO
14046

HO, CH_3 chroman with phytyl chain (tocopherol)
14047

=O, ketone with long chain
14049

HO chroman with chain
14050

HO chroman with chain
14051

HO chroman with chain
14052

$CH_3(CH_2)_3NHCONHSO_2$—phenyl—CH_3
14053

$^\alpha CH_3$ on phenyl numbered 1-6
14060

CH_3 phenyl, HO—CH—CH_3
14131

CH_3 phenyl, $AsO(OH)_2$
14202

CH_3 phenyl, $B(OH)_2$
14205

CH_3 phenyl, SO_2H
14208

CH_3—phenyl—$SOCl$
14210

CH_3 phenyl, SO_2NH_2 (positions 4,5,6)
14211

CH_3 phenyl, SO_3H (positions 3,4,5,6)
14217

CH_3—phenyl—$SO_3CH_2CH_2O_3S$—phenyl—CH_3
14253

CH_3 phenyl, $NHCH_2C_6H_5$
14305

CH_3 phenyl, $N(CH_3)CH_2C_6H_5$
14306

CH_3 phenyl, CH_2CO_2H
14369

CH_3 phenyl, CH_2CN
14370

In Numeric as they occur in Organic Compounds Table

14381

14383

14385

14386

14387

14388

xylose
glucose
galactose

14389

14390

14392

14393

14395

14402

14409

14411

14412

14413

14456

14503

14505

14511

14512

14513

14529

14535

14537

14543

14545

14546

14550

14553

14554

14555

14557

14560

14561

14564

14565

14566

14571

14584

14585

14587

14595

14597

14598

14599

STRUCTURAL FORMULAS OF ORGANIC COMPOUNDS (Continued)

In Numeric as they occur in Organic Compounds Table

14600

CH_3-N ... OH
14602

HO_2C ... CH_3-N ... CH_2CO_2H
14603

14606

14607

C_6H_5 C_6H_5 ... C_6H_5 C_6H_5
14608

O_2CCH_3 ... CH_3-N ... O ... $N-CH_3$ ... CH_3CO_2
14609

$CH_2CH(NH_2)CO_2H$
14610

OCH_3 ... CH_3O ... HO ... N ... H
14617

CH_2OH ... CH_2OH ... H ... OH ... CO ... HO ... O ... CH ... H ... OH ... $HCOH$ ... $HCOH$ ... CH_2OH
14618

HO ... $CH_2CH_2NH_2$
14619

HO ... $CH_2CH(NH_2)CO_2H$
14620

$CH_3(CH_2)_6$ ... O
14644

$CH_3(CH_2)_6$ ... O
14645

14686

HN ... O ... NH ... O ... CO_2H
14697

O ... CO_2Na ... NaO
14699

H_2N ... $SO_2NHCONH_2$
14705

Cl ... $NHCON(CH_3)_2$
14719

$NHCONH$
14737

CH_3 ... H_2NCONH
14767

CH_3 ... H_2NCONH
14768

H_2NCONH ... CH_3
14769

OH ... HO ... N ... NH ... O
14773

$HOCH_2$ ... C ... C ... C ... C ... OH OH ... O
14779

$HOCH_2$ ... C ... C ... C ... C ... OH ... PO_3H_2 ... O
14780

N ... $CH:CHCO_2H$ ... N ... H
14781

HO_2C ... OCH_2CCl_3 ... OH ... OH ... O
14782

HO ... OH
14783

HO ... CO_2H
14784

HO ... CH_3 ... $COCH_3$ ... O ... HO ... $COCH_3$
14785

$C_6H_{11}O_5-C_6H_{10}O_6$ ... OH ... O ... O
14788

H H H OH H ... $H-C-C-C-C-C-C=O$ ... C_6H_5COO OH OH H OH
14789

$CH_3(CH_2)_3CO_2CH_2$ ... O
14841

H_3C ... O
14845

14848

CHO ... OH ... OCH_3
14890

CH_3O ... CH_3CO_2 ... CHO
14891

N ... N ... OH
14897

CH_3O ... CH_3O ... CO_2 ... HO ... O ... HO ... OH ... OH ... N ... OH
14899

HO ... OH ... N ... H ... HO
14901

O ... O ... OH ... CH_3 ... CH_3CHCCO_2 ... OH ... N
14903

CH_3 ... O
14906

O
14908

$CH(OH)$ ... $H-C-OH$ ... $HO-C-H$ ... O ... O ... $H-C-OH$ ... HC ... CH_2 ... CH ... $H-C-OH$ ... $HO-C-H$ ... $HO-C-H$ ... CH_2
14910

C-624

STRUCTURAL FORMULAS OF ORGANIC COMPOUNDS (Continued)

In Numeric as they occur in Organic Compounds Table

14911

14912

14913

14914

14915

14916

14917

14918

14919

14921

14923

14924

14925

14926

14944

14947

14948

14949

14951

14954

14966

14967

14968

14969

14970

14971

15019

15020

15021

15022

15023

15024

15027

15029

15030

Temperatures in °C; where values are not precisely known, or where there is a range of melting points, the compound is listed according to the lower temperature.

−197: 9089
−189: 11772
−185: 3942, 12025, 12257
−183: 6603, 11880
−182: 9015
−181: 9047
−169: 6840, 10448
−168: 3971
−165: 13851
−161: 3576
−160: 9100, 14934
−159: 3777, 3808, 10422, 11862, 11872, 12304
−158: 9061
−157: 3963, 6847
−156: 12293, 13296
−153: 10406, 10534, 10535, 14931
−151: 8640, 10539
−150: 9093, 11866, 13274
−148: 10307
−147: 7853
−146: 3661, 8575, 9037
−145: 9051
−144: 6704
−143: 1411, 6646
−142: 5440, 5776, 6876
−141: 7731, 7954, 9079
−140: 3707, 4104, 4189, 10563
−139: 7941, 14930
−138: 3672, 3980, 4742, 5773, 5837, 9155, 10517, 10560
−137: 3970, 7960, 10298, 10356, 11360, 12266, 12277
−136: 580, 3631, 6614, 6775, 7654, 10530, 10540, 11770
−135: 5806, 9064, 10533, 10559, 13283
−134: 3761, 10394, 10527, 10529, 10561, 10562, 12264, 12267, 13269
−133: 1585, 3945, 4013, 7955
−132: 3657, 3782
−131: 3706, 3968, 4101, 7991
−130: 4100, 6158, 6867, 10340, 11937, 13254
−129: 324, 596, 4014
−128: 3715, 3746, 7733
−127: 5815, 5824, 10555
−126: 445, 5545, 7805,

−126: 10114, 10352, 11942, 12244
−125: 4195, 7992, 10308
−124: 12260
−123: 3698, 4099, 5812, 7835, 9124, 10387, 10528
−122: 6863, 9068, 9205, 10392, 11795, 12253, 13264, 13285
−121: 21, 5685, 7571, 7798
−120: 3636, 5772, 7569, 10402
−119: 5832, 7643, 7819, 10391, 12262
−118: 6610, 6770, 7818, 10116, 10399, 10554, 11359
−117: 3993, 4096, 4211, 5780, 5833, 6632, 6708, 6827, 11805, 12254, 14514
−116: 7553, 7832, 7999
−115: 3851, 3964, 4139, 4145, 5687, 5688, 6839, 9023, 10400, 10557
−114: 4741, 12297, 14483
−113: 37, 7552, 7961, 9193, 10056, 10564, 11936, 12259
−112: 435, 3683, 3691, 4095, 4186, 4421, 5937, 6625, 10418, 11385
−111: 3692, 3985, 4097, 4717, 5531, 5781, 6676, 6920
−110: 10115, 10449, 10631, 11782, 12251, 14877
−109: 3979, 4148, 5827, 7566, 7648, 10421, 12278
−108: 3638, 5761, 5767, 6653, 6805, 6871, 11873
−107: 4740, 5562, 10054, 10419
−106: 5551, 6618, 7606, 10565
−105: 3679, 3981, 3994, 4171, 5814, 6619, 6810, 7831, 10363, 10627, 13288, 13290
−104: 3675, 3704, 3771, 4085, 4089, 5777,

−104: 7052, 11843
−103: 3767, 5678, 6102, 6627, 7813, 7995, 9157, 10131, 10469
−102: 7556, 10135, 10417, 12321, 13323
−101: 6046, 6098, 6630, 7836, 9194, 9928, 10105, 10628, 12333, 13511, 14932
−100: 3944, 4328, 4427, 6663, 6799, 9128, 9232, 10109, 10420, 11828
−99: 2236, 3671, 3745, 4222, 4364, 6349, 7083, 10354, 10388, 11831, 12265, 12288
−98: 186, 225, 434, 4106, 4191, 9199, 13436
−97: 4149, 4383, 5756, 6170, 6626, 7054, 8642, 9035
−96: 2237, 5394, 13255, 13854, 13915, 14870
−95: 262, 306, 2140, 2145, 4117, 4125, 5502, 5769, 6220, 6764, 7076, 7760, 8755, 9060, 10350, 11777, 12005, 12250, 13336, 14060
−94: 4123, 5558, 6628, 6651, 6813, 7769, 9183, 12217
−93: 206, 270, 5762, 6156, 7086, 9020, 9127, 9148, 10117, 10538, 11874, 14573, 14927
−92: 261, 3571, 3574, 3930, 4185, 5912, 7047, 7087, 7994, 9192, 12198, 12312, 14823
−91: 4294, 4295, 7075, 7804, 10339, 11829, 14790, 14837
−90: 4460, 4462, 5524, 7522, 7646, 10166, 10401, 10618, 11870, 12307, 12498
−89: 3860, 4069, 4199, 4445, 5483, 5521, 7993, 8954, 11728, 11788, 11971, 12091, 12245,

−89: 12252, 13895
−88: 1982, 4407, 7749, 8812, 11376
−87: 1416, 4617, 5525, 7824, 10110, 10301, 10345, 11721, 12023, 12169, 13265, 13318, 14128
−86: 480, 3918, 6843, 9021, 9161, 9929
−85: 279, 626, 1058, 1985, 3759, 5765, 6837, 6912, 7116, 7515, 10403, 11380, 12302
−84: 3573, 4047, 4351, 4412, 6164, 7765, 9150, 9927, 10502, 14991
−83: 204, 2099, 7601, 7620, 8862, 11773, 12246
−82: 2027, 7625, 9173, 10164, 10630
−81: 4198, 5416, 6754, 7673, 7852, 11771, 12042
−80: 215, 443, 625, 2144, 3733, 3758, 4399, 4604, 4615, 4967, 6720, 6865, 7079, 7865, 8644, 10053, 10512, 14579
−79: 4168, 5474, 5677, 7738, 9814, 10076, 10126, 10455, 11927
−78: 227, 1456, 1584, 4337, 7563, 7642, 12324, 13270, 14863
−77: 185, 8869, 10108, 10495, 13299, 13320
−76: 3649, 4105, 4190, 12497, 13256, 13785
−75: 1987, 1988, 2228, 4278, 5523, 5802, 7392, 7732, 7816, 10447, 10622, 12188, 14426
−74: 2100, 3731, 3935, 4008, 5381, 5501, 5734, 8754, 9926, 14674
−73: 177, 263, 3709, 4372, 5642, 5938, 5946, 6052, 6881, 7085, 7562, 12143, 12177, 13797,

MELTING POINT INDEX OF ORGANIC COMPOUNDS (Continued)

−73: 13834, 14521

−72: 305, 3929, 10378, 11878

−71: 501, 2190, 4433, 4599, 5865, 12092, 13878

−70: 252, 3975, 5742, 6666, 6821, 7615, 13518, 14200, 14942

−69: 7530, 10508, 10589, 10602, 13902

−68: 2192, 4610, 6112, 6118

−67: 534, 4091, 4589, 5867, 6711, 6846, 7572, 7588, 7662, 10089, 14978

−66: 447, 2273, 6355, 6623, 7051, 7590, 9082, 10416, 11540, 12516

−65: 178, 3718, 5861, 14494

−64: 491, 920, 4587, 9050, 9087, 10315, 10398

−63: 914, 977, 2191, 4218, 4648, 5866, 6621, 7599, 9098, 12495, 13506, 13901, 14166, 14843

−62: 2028, 2146, 5771, 7081, 7129, 7825, 12051, 13437, 13890, 14127, 14990

−61: 207, 2132, 6334, 8867, 10129, 13951, 14665

−60: 2230, 3872, 5456, 6051, 6792, 7060, 7393, 7735, 10273, 12029, 14164, 14315

−59: 3566, 4180, 5753, 7930, 11892

−58: 5803, 6893, 7049, 11784, 13439

−57: 53, 933, 1991, 2131, 4336, 4868, 4975, 5555, 5743, 7911, 9024, 9230, 10036, 12149, 13245

−56: 3580, 4564, 4715, 5482, 5495, 5522, 5729, 7525, 7595, 7759, 10027, 11417, 11796, 14129, 14817

−55: 4369, 4593, 6872, 10031, 10341, 10599, 11584, 11821, 14451

−54: 4063, 8576, 8844,

−54: 10047, 12036, 12528

−53: 3756, 3849, 4394, 6861, 11581, 14576, 15008

−52: 1998, 2135, 3265, 8101, 8991, 9053, 10592, 12004, 12999

−51: 179, 1986, 2777, 4232, 5389, 5801, 6065, 8338, 9204, 9918, 10604, 10612, 13243, 14803, 14872

−50: 212, 304, 5438, 5520, 6361, 6656, 6866, 7598, 9950, 12148, 12681, 13445, 14488

−49: 1374, 3673, 3926, 4080, 5486, 5561, 6459, 8657, 10075, 11578, 13512, 13888, 14663

−48: 2164, 3794, 4003, 5911, 6092, 6225, 7564, 7593, 8866, 9009, 11813, 11900, 12010, 12344, 13191, 14664

−47: 3885, 4608, 14081, 14986

−46: 3821, 4392, 4645, 7868, 12170, 14842

−45: 342, 394, 1057, 1455, 2000, 4647, 4716, 6911, 9169, 10021, 10049, 10507, 12074

−44: 310, 2287, 3565, 5550, 5961, 6228, 7787, 8987, 10179, 10491, 11196, 12536, 13499

−43: 919, 2029, 2286, 3177, 3974, 4632, 4727, 5506, 5543, 5887, 6724, 7517, 7579, 9316, 10316, 12358

−42: 1399, 2101, 4039, 4349, 4419, 4631, 5388, 6207, 7165, 10874, 11848, 11886, 12173, 12441, 12474, 13432, 14448, 14859

−41: 2153, 4578, 4976, 5504, 6640, 6690, 7545, 7621, 14854

−40: 425, 1883, 4155, 4246, 4638, 5469, 5643, 7839, 8677,

−40: 10093, 12208, 12322, 12785, 13886, 13899

−39: 292, 1056, 3238, 6126, 6331, 6753, 7084, 7632, 9922, 10373, 10504, 14070

−38: 247, 844, 1403, 4129, 5442, 5805, 7170, 10175, 13374, 13861, 14668

−37: 1025, 1375, 2180, 3729, 5526, 6347, 7972, 11933, 12026, 13519, 14201, 14465, 14943

−36: 309, 2159, 3569, 3668, 5942, 6667, 6674, 6675, 6923, 8996, 10269, 11265

−35: 87, 283, 587, 3259, 4042, 5909, 6403, 6629, 7097, 7580, 7627, 8766, 9012, 10103, 13523, 13752, 14078

−34: 2102, 2690, 4563, 4592, 4642, 7602, 9071, 10272, 13515, 13885

−33: 2235, 5519, 5572, 7592, 7638, 10385, 10454, 11833, 12041, 12152, 14808, 14857

−32: 530, 834, 4202, 4933, 5631, 6113

−31: 260, 2098, 2169, 4329, 5835, 5904, 6886, 7167, 9198, 10085, 10243, 11576, 14846

−30: 1921, 3347, 4635, 4739, 5888, 6310, 6672, 8868, 9999, 10171, 11365, 11935, 12299, 13495, 14063, 14327

−29: 1404, 2139, 3074, 3560, 3932, 4224, 4359, 5899, 5901, 6661, 6671, 7342, 8612, 10743, 13548

−28: 922, 4606, 6309, 9831, 13298, 14464

−27: 299, 2231, 2234, 2260, 5429, 9915, 10757, 12030, 14048, 14065, 14138, 14669

−26: 1033, 1654, 2031,

−26: 4201, 4929, 5473, 5916, 6041, 6624, 6716, 7494, 10141, 10362, 13327, 13884

−25: 61, 446, 2285, 4102, 4465, 6123, 11308, 13880, 14634, 14673, 14972

−24: 1893, 2086, 3136, 3807, 4955, 6306, 6832, 7322, 11572, 13526, 14575

−23: 228, 432, 2147, 2233, 2254, 2876, 5818, 8497, 9092, 10442, 10485, 11169, 12674, 13275, 13504, 14284, 14289, 14577

−22: 1972, 2587, 5425, 5426, 5967, 6782, 6895, 7596, 9391, 9896, 10573, 13190

−21: 1925, 3656, 5263, 5720, 6641, 11283, 11573, 11849, 12249, 14425, 14679

−20: 74, 588, 3681, 4555, 5733, 9803, 10104, 10475, 10615, 11797, 11941, 11980, 12072, 13556, 13872, 14921

−19: 532, 1978, 3333, 3799, 4011, 4156, 4435, 5798, 6416, 6874, 7492, 7761, 7763, 7987, 7989, 8814, 9296, 10042, 10143

−18: 1036, 1151, 1454, 3800, 4557, 4597, 4614, 4641, 6953, 7523, 7666, 10130, 10431, 11546

−17: 2082, 2133, 2143, 3242, 4333, 5494, 5727, 6695, 8724, 11836, 14441, 14630

−16: 947, 3298, 3720, 3739, 4098, 5210, 5621, 5910, 6611, 7706, 7997, 8790, 8973, 9174, 9361, 9797, 10079, 10094, 10187, 10756, 11847, 13918

−15: 198, 336, 533, 536, 3160, 4659, 4914, 6616, 9228, 9794, 11970, 12268,

−15: 13508, 14101, 14493, 14642, 14826

−14: 791, 803, 891, 3903, 4084, 4321, 4738, 5331, 6869, 9025, 9156, 11894, 13943, 14062, 14270, 14300, 15014

−13: 182, 2107, 2848, 2874, 3914, 5639, 6169, 7493, 8161, 9363, 9377, 9850, 14098, 14479

−12: 307, 1924, 2052, 2149, 2740, 5465, 5752, 7585, 7935, 7936, 8140, 10396, 12661, 13742, 14848

−11: 5692, 5925, 6680, 6883, 6918, 8768, 9996

−10: 807, 3708, 4634, 6105, 6111, 10097, 11868, 12306, 12530, 13188

−9: 1726, 1955, 5599, 5757, 5928, 6393, 6395, 6417, 7937, 9043, 9379, 11617

−8: 286, 525, 1952, 3905, 6836, 7837, 9340, 9424, 9792, 10758, 12440, 13139

−7: 325, 1742, 2072, 2137, 2295, 5070, 5475, 5785, 6319, 6931, 7169, 7583, 9378, 9937, 10058, 13083, 14519

−6: 765, 2011, 2156, 2253, 4463, 5431, 5516, 6862, 7753, 8328, 8791, 9358, 10771, 11331, 11754, 13747, 13882, 14973

−5: 350, 996, 3464, 4547, 6838, 7846, 8764, 9032, 9297, 9933, 9976, 10478, 13103, 13560, 14460, 14487, 14528

−4: 921, 1885, 3070, 3140, 3665, 4263, 4283, 4948, 5158, 5509, 5595, 5597, 5749, 9040, 9225, 9298, 9939, 10947, 11665, 12598, 14631

−3: 3226, 5676, 10668,

−3: 13531, 14051, 14457, 14652, 14654

−2: 323, 3077, 3326, 3788, 3848, 3862, 4217, 4571, 5226, 5443, 5924, 7226, 7786, 7849, 7859, 8881, 9299, 10167, 11294, 12878, 13195, 14299, 14671

−1: 537, 804, 931, 1977, 4324, 4598, 4629, 5515, 5598, 8395, 9260, 9357, 9919, 9957, 10145, 13521, 14765

0: 924, 1000, 1605, 1777, 1934, 1937, 3461, 3487, 3645, 3714, 4560, 5330, 6077, 6127, 6399, 6441, 6664, 6665, 6845, 9958, 10028, 10142, 11473, 11555, 11666, 11838, 12247, 12469, 12631, 14157, 14311, 14649, 14940

1: 545, 1402, 1658, 2009, 2697, 2720, 2991, 3206, 3397, 3669, 3802, 4004, 4005, 4144, 4167, 4213, 4757, 4831, 5065, 5066, 5511, 5730, 5927, 7106, 8956, 8960, 10762, 10980, 11174, 11972, 13953, 15004

2: 333, 639, 872, 913, 1030, 3078, 4728, 5455, 5470, 6321, 6801, 7056, 7058, 8805, 10072, 10295, 10740, 10763, 11477, 12525, 13125, 13724, 14047, 14659

3: 1423, 3155, 3255, 8977, 10009, 11548, 12637, 13193, 14204

4: 398, 1745, 1946, 2247, 3340, 3341, 3488, 3650, 3663, 3738, 3742, 4009, 4553, 4582, 4840, 7390, 7715, 9048, 9385, 10746, 10768, 12825, 13435, 13503, 13729, 14158, 14398, 14657

5: 1882, 2163, 2194, 2864, 3567, 3744,

5: 3832, 3904, 5371, 6465, 6811, 6954, 9057, 9375, 10814, 10960, 11580, 12022, 13104, 13189, 13448, 13530, 13726, 13728, 14099

6: 57, 929, 1013, 3405, 3824, 4127, 4325, 5491, 6308, 6952, 9070, 10004, 10146, 10260, 11622, 11722, 11855, 13743, 13961, 14462, 14860, 14907

7: 360, 1011, 1967, 2057, 2068, 3119, 3839, 5164, 5557, 5593, 6288, 8822, 9144, 9359, 12484, 13500, 14084, 14086, 14355

8: 3128, 3432, 4718, 5214, 5329, 6849, 6870, 7072, 8531, 8708, 9097, 9821, 10124, 10775, 11680, 11987, 12033, 12660, 13562, 13773, 14463

9: 539, 748, 902, 1729, 1935, 1954, 2108, 3465, 3809, 4554, 5508, 6402, 6622, 9022, 10281, 10372, 10386, 10843, 11822, 12883, 14473, 14906, 14999, 15018

10: 426, 535, 1065, 1844, 2245, 4138, 4181, 6130, 6727, 6854, 7394, 7699, 7797, 8139, 8593, 9086, 10285, 10296, 11574, 12165, 12359, 13721, 13863, 14227, 14638, 14678

11: 877, 896, 2188, 2244, 3138, 3409, 4437, 5335, 5336, 6215, 7504, 7603, 8557, 11423, 11864, 12463, 13082, 14992, 15016

12: 15, 1679, 2050, 3834, 4438, 5122, 5922, 6346, 6815, 7971, 9254, 9432, 10239, 10257, 11401, 12995, 13399, 14429, 14491, 14648, 14656

13: 486, 1032, 1169, 1737, 5533, 6060, 6104, 7399, 7671, 7693,

13: 7867, 8330, 10236, 12037, 12174, 12444, 12676, 12750, 14226, 14998, 15003

14: 941, 1807, 2431, 2878, 3103, 3468, 5744, 5926, 6108, 6673, 6834, 7374, 7741, 8860, 9095, 9226, 9437, 9793, 10614, 11820, 14440, 14458, 14660

15: 895, 3143, 4017, 7423, 7721, 9932, 10044, 11007, 11332, 11526, 11730, 12121, 12204, 12395, 12517, 12627, 13359, 14140, 14655, 14983, 15015

16: 175, 822, 909, 1741, 3142, 3404, 3886, 4306, 4309, 4624, 7641, 8895, 9004, 10078, 10139, 10228, 10680, 10774, 10835, 10952, 10961, 11889, 12038, 12111, 12555, 12629, 12882, 13732, 14147

17: 801, 1684, 1962, 2266, 5700, 5845, 5858, 5900, 6121, 6447, 7230, 7703, 7704, 9034, 9998, 12166, 14635, 14651

18: 31, 777, 2746, 3891, 4839, 5045, 5513, 5839, 5843, 6175, 6921, 7701, 7707, 7714, 7737, 8424, 8651, 9202, 10291, 10570, 11188, 11239, 11387, 11718, 12013, 12228, 12343, 12678, 13655, 13656, 13775, 14832, 14836

19: 1177, 1923, 3113, 3458, 3840, 6410, 6723, 7793, 8113, 9102, 9255, 9259, 10287, 10338, 10443, 12886, 13164, 13558, 13800, 13962, 14351, 14646, 14666

20: 351, 352, 371, 376, 958, 1733, 1974, 2053, 2216, 2682, 3198, 3335, 3833, 4207, 4305, 5962, 6287, 6744, 7354, 7722,

20: 8979, 9042, 9116, 9253, 9532, 9934, 10235, 11252, 12184, 12907, 13608, 14680

21: 755, 1068, 1666, 1957, 2020, 2556, 2605, 6401, 7224, 7736, 8228, 10749, 13106, 14256, 14481

22: 3240, 3469, 4801, 4942, 5159, 6802, 7488, 7576, 9059, 10007, 10839, 10905, 10924, 11006, 11179, 12328, 12578, 13419, 13651

23: 627, 983, 1791, 2090, 2765, 5626, 8528, 10382, 10684, 10951, 10997, 11928, 13427, 13498, 14412, 14475

24: 67, 99, 835, 1091, 1095, 1178, 2044, 2786, 2872, 3088, 3144, 3149, 3261, 3695, 3908, 4552, 7700, 9114, 9717, 9761, 9817, 9991, 10283, 10568, 11177, 11975, 13183, 12294, 12671, 12758, 13516, 14082, 14087, 14103, 14114, 14667

25: 1886, 2128, 3184, 4010, 4071, 4076, 5269, 5359, 5577, 5592, 5745, 6282, 6635, 9074, 9227, 9343, 10313, 11424, 11794, 11994, 12076, 12594, 13110, 13412, 14436, 14653

26: 3816, 6409, 7365, 7409, 8064, 8581, 8720, 9418, 10608, 11005, 11009, 11168, 11266, 12871, 13098, 13599, 13944, 14102, 14313, 14352, 14354, 14356

27: 295, 2109, 2738, 3309, 3716, 3752, 4616, 5348, 6691, 7205, 7709, 8989, 9434, 10023, 10288, 10845, 10903, 10978, 12819, 13003, 13165, 13384, 13855, 13867, 14428, 14461

28: 79, 418, 419, 1040,

28: 1661, 2021, 2168, 3195, 4805, 5398, 5462, 5615, 5746, 5906, 5907, 6087, 6317, 6372, 6394, 6460, 7140, 7496, 7512, 7544, 7843, 8721, 9935, 9981, 9984, 9985, 10064, 10943, 11178, 11262, 11358, 12950, 13383, 13415, 13818, 13940, 14071, 14074, 14111, 14254, 14637

29: 538, 1041, 1088, 1338, 2877, 3423, 3467, 3485, 3591, 3593, 4813, 5106, 5584, 6339, 6443, 6444, 6464, 7138, 7996, 8131, 10760, 10962, 11040, 11173, 11558, 11856, 11863, 12325, 12752, 13387, 13522, 13917, 14296, 14312, 14319, 14333, 14981

30: 442, 1772, 1904, 2759, 3071, 3192, 3378, 4912, 4980, 4982, 5207, 5301, 5988, 6141, 6216, 6290, 6400, 6446, 6932, 7497, 7746, 9052, 9247, 9760, 9847, 10008, 10233, 10635, 12600, 12935, 13360, 13369, 13411, 13722, 13725, 13914, 14476, 14984

31: 224, 1604, 1606, 2136, 2170, 3118, 3227, 3244, 4366, 4370, 4401, 5564, 5590, 5748, 6405, 6445, 7173, 7587, 8341, 8760, 10090, 10238, 10280, 12779, 13389, 13597, 13859, 14321, 14477, 14952

32: 770, 1553, 2576, 3076, 3110, 3135, 3607, 4178, 4814, 4937, 5464, 6056, 6082, 6357, 6408, 7433, 7456, 7490, 8709, 8842, 8904, 9314, 9553, 9685, 9713, 9904, 11270, 11851, 12506, 12599, 12613, 14474

33: 47, 378, 629, 849, 869, 927, 2171, 3349, 3549, 3749, 4292, 4588, 5103, 5726, 6115, 6387, 6577, 7117, 7718, 8505, 8507, 9269, 9433, 9436, 10687, 10819, 10850, 11092, 11140, 11547, 13739, 14295, 14309, 14478, 14828

34: 246, 409, 603, 768, 963, 1901, 1943, 1966, 2043, 2093, 2478, 2491, 2866, 3212, 3403, 3476, 3838, 3841, 4027, 5160, 5360, 5587, 5630, 5738, 6240, 6283, 6285, 6345, 7204, 7498, 7698, 9282, 9400, 9800, 9987, 10222, 10660, 10915, 11035, 11469, 11565, 12464, 12760, 12772, 13451, 13482, 13668, 13740, 14252, 14447, 14844

35: 320, 508, 880, 1503, 1730, 1773, 1931, 1959, 2113, 3175, 3477, 4391, 4585, 4586, 4633, 5800, 6205, 6212, 7043, 7176, 7178, 7381, 8986, 9329, 9768, 9874, 9880, 10071, 10292, 10476, 10569, 10979, 11138, 11234, 11343, 11403, 11621, 11669, 11725, 11998, 12132, 14076, 14639

36: 374, 658, 948, 1591, 1793, 2340, 2444, 3247, 3302, 3453, 4415, 4891, 5130, 5373, 5375, 5582, 6080, 6340, 6439, 6461, 6467, 7213, 7487, 9170, 9407, 10005, 10227, 10764, 11353, 11445, 12235, 12326, 12417, 12866, 13334, 14110, 14139, 14643

37: 62, 967, 998, 1774, 2960, 3185, 3221, 3246, 3490, 5162, 5619, 5941, 6812, 7807, 8340, 8588,

37: 8746, 9376, 9531, 9705, 9840, 9914, 10286, 11127, 11139, 11714, 11911, 12090, 13049, 13130, 13404, 13731, 14072, 14080, 14330, 14453, 14862

38: 415, 938, 1028, 1066, 2356, 2439, 3611, 4558, 5041, 5213, 5318, 5344, 5667, 5699, 5919, 6369, 7491, 8095, 8623, 9055, 9175, 9279, 9732, 10456, 10486, 10590, 11135, 12048, 12405, 12764, 13107, 13192, 13410, 13754, 14085, 14148, 14186, 14261, 14641

39: 174, 236, 393, 693, 838, 1176, 1597, 1659, 2745, 2853, 2940, 3245, 5094, 5369, 6173, 6224, 7159, 7171, 7716, 9261, 10293, 10844, 10875, 11189, 11660, 11679, 11681, 11702, 12890, 13398, 13454, 13609, 13738, 14096, 14363

40: 243, 433, 1744, 1960, 2110, 2856, 2946, 3231, 3613, 3790, 4210, 4726, 4994, 5044, 5404, 5461, 5840, 6486, 6491, 6633, 6769, 7233, 7360, 7483, 7597, 8587, 8634, 8782, 10866, 11427, 12054, 12658, 13341, 13705, 13749, 14362, 14676

41: 1552, 1710, 1922, 2157, 2523, 2770, 2953, 3228, 3495, 3501, 3927, 4483, 4607, 4685, 4843, 5075, 5215, 5481, 5602, 5606, 5808, 5842, 6213, 6398, 6803, 7757, 7782, 8707, 9423, 9889, 10138, 10636, 10742, 10838, 10858, 10992, 11541, 11726, 12104, 12314, 12461, 12869, 13408, 13424, 13887, 14066, 14337, 14526, 14600

42: 181, 337, 504, 816, 1347, 1405, 1443, 1899, 2446, 2545, 2896, 3139, 3277, 3300, 3459, 3508, 4064, 5081, 5407, 6039, 6106, 6442, 6544, 8224, 8375, 8961, 9081, 9828, 10174, 10208, 10297, 11461, 11527, 11983, 12223, 12773, 12874, 13846, 14069, 14599, 14606, 14677

43: 390, 392, 824, 951, 1522, 1783, 1961, 2085, 2271, 2855, 3288, 3842, 4288, 4800, 4807, 4945, 5408, 5662, 6237, 7469, 7549, 7551, 7840, 7842, 7978, 8571, 9830, 10294, 10446, 10587, 10767, 10852, 10957, 11356, 11914, 13154, 13733, 13942, 14182

44: 966, 1397, 1524, 1702, 1722, 1743, 2218, 2262, 2615, 2954, 3617, 4244a, 4286, 4287, 4338, 4934, 5140, 6329, 6381, 6392, 6407, 7695, 7713, 8281, 8354, 8702, 8706, 8963, 10003, 10069, 10206, 10832, 10876, 11608, 12665, 12776, 13386, 13493, 13688, 14064, 14132, 14142, 14348, 14471, 14896, 14909

45: 314, 756, 850, 1081, 1724, 2379, 2721, 3063, 3251, 3348, 4024, 4038, 4430, 4882, 4985, 5347, 5799, 6286, 6463, 6884, 6905, 8825, 8882, 8897, 9028, 9799, 9931, 10000, 10173, 10662, 10882, 10988, 11426, 11996, 12162, 12475, 12774, 13805, 14177, 14183, 14184, 14187, 14572

46: 319, 339, 767, 839, 1509, 1512, 1694, 2172, 2677, 3233, 3304, 3310, 4381, 4792, 4842, 4923,

46: 6055, 6199, 6284, 7012, 7396, 7702, 8084, 8130, 8718, 9125, 10925, 10983, 11172, 11415, 11484, 12230, 12595, 12649, 12916, 13371, 13799, 13905, 13947, 14331, 14338, 14509, 14513, 14772

47: 46, 1687, 1768, 1917, 2241, 2441, 2692, 2936, 3843, 4379, 5415, 6370, 6766, 7212, 7723, 8073, 8145, 8494, 8770, 9000, 9383, 9867, 10877, 10950, 10986, 11175, 11678, 12058, 12396, 12408, 12985, 13406, 14075, 14320, 14452

48: 672, 680, 2283, 2426, 2623, 3618, 3750, 4162, 4175, 4233, 4343, 4397, 4661, 4818, 4907, 6048, 6335, 6414, 6440, 7502, 8066, 8147, 8191, 8726, 8877, 9099, 9315, 9709, 9846, 10001, 10010, 10221, 10772, 10853, 10953, 11056, 11272, 11457, 11723, 12233, 12284, 12610, 12692, 12787, 13123, 13420, 13845, 14152, 14318, 14340, 14682

49: 173, 241, 257, 397, 956, 1012, 1671, 2032, 2844, 3280, 3489, 3785, 4344, 4556, 4651, 5918, 6474, 7489, 7506, 7581, 7594, 7717, 7734, 8090, 8592, 8647, 8719, 9145, 9276, 9333, 9403, 9995, 10644, 10918, 10933, 11615, 11620, 12234, 12749, 12783, 12784, 12875, 13019, 13543, 13746, 14035, 14159, 14595

50: 358, 361, 752, 837, 1371, 1648, 3173, 3241, 3250, 3496, 3602, 4244, 4711, 4735, 4841, 4946, 5479, 6248, 6383,

50: 6451, 7065, 7712, 8998, 9688, 9829, 9912, 10632, 10646, 10827, 11216, 11377, 11420, 12003, 12125, 12913, 13111, 13650, 13730, 14223

51: 63, 365, 385, 717, 842, 899, 1189, 1527, 1594, 1740, 1912, 2434, 2442, 2849, 2930, 3375, 3392, 3873, 4481, 4499, 4565, 5172, 5314, 5478, 5614, 5969, 8896, 9238, 9570, 9714, 9715, 9745, 9890, 9968, 10232, 10428, 10812, 10840, 10873, 11054, 11895, 12511, 12611, 12746, 13413, 13866, 13903, 13938, 14376, 14908, 15017

52: 9, 403, 404, 813, 833, 1337, 1675, 1688, 1719, 1911, 2347, 2552, 2902, 2903, 3014, 3105, 3666, 3705, 3755, 4382, 4390, 4480, 4482, 4887, 4922, 5430, 5604, 5972, 6268, 6365, 6458, 6473, 6642, 6736, 7174, 7904, 8356, 8411, 9737, 9795, 9798, 10101, 10675, 10887, 11182, 11854, 11957, 12283, 12420, 12754, 12862, 12902, 13865, 14032, 14107, 14117, 14798

53: 52, 402, 703, 827, 875, 1063, 1528, 2089, 2198, 2265, 2850, 2992, 3382, 3427, 3996, 4206, 4643, 4673, 6366, 7015, 7068, 7503, 8519, 8650, 8969, 9252, 10230, 10290, 11097, 11384, 11441, 11719, 11850, 11930, 12150, 12243, 13438, 13453, 13557, 13843, 14151, 14178

54: 483, 769, 1067, 1372, 1649, 1770, 2094, 2173, 2226, 2242, 2284, 2457, 2727,

54: 3389, 3414, 4653, 4668, 5844, 6119, 6217, 6249, 6279, 6384, 6449, 6452, 7082, 7326, 7373, 7379, 7499, 7500, 8372, 9386, 9414, 9989, 10177, 10189, 11357, 11451, 12371, 12392, 12956, 13582, 13720, 14149, 14210, 14257, 14582

55: 136, 164, 244, 823, 878, 1499, 1685, 1900, 2207, 2374, 2438, 2788, 2852, 2998, 3028, 3050, 5321, 5368, 5534, 5620, 6085, 6154, 6292, 7719, 8302, 8303, 9046, 9435, 9676, 9757, 9767, 9979, 10321, 10984, 11076, 11141, 11143, 11170, 11293, 11488, 12770, 13021, 13177, 14037, 14454, 14835

56: 34, 55, 368, 903, 1034, 1554, 1613, 1800, 2091, 2199, 2445, 2755, 3133, 4921, 5343, 5353, 5365, 5366, 5603, 6247, 6660, 6738, 6873, 6892, 7398, 7692, 7705, 8159, 8198, 8306, 8329, 8374, 8386, 8471, 8767, 9078, 9283, 9293, 9295, 9977, 10262, 10647, 10859, 10868, 10871, 12129, 12607, 13425, 13572, 13869, 13960, 14077, 14542

57: 56, 273, 372, 831, 932, 1021, 1024, 1039, 1437, 1914, 2036, 2208, 3026, 3260, 4675, 4809, 4870, 5487, 6373, 6679, 6927, 6928, 6929, 7461, 7519, 8085, 8194, 8404, 8570, 8653, 8994, 9237, 9882, 9891, 9910, 10255, 10633, 10879, 11342, 11921, 12447, 12747, 12958, 13391, 13480, 13606, 14640, 14867

72: 14150, 14191, 14632, 14721

73: 988, 1514, 1976, 2339, 2679, 2712, 2984, 2999, 3180, 3479, 3817, 4062, 4291, 4389, 4691, 4773, 4802, 5141, 5315, 5413, 5586, 6107, 6201, 6204, 6342, 6887, 7061, 8242, 8318, 8815, 9076, 9172, 9390, 9577, 9696, 9901, 10908, 10954, 11281, 11314, 11344, 11456, 11475, 12748, 12821, 12896, 12917, 13362, 13455, 13486, 13621, 13653, 13948, 14581

74: 790, 864, 901, 1730, 2069, 2751, 3145, 3170, 3191, 3480, 3509, 4093, 5131, 5225, 5235, 5345, 5608, 5702, 5970, 6203, 6481, 6853, 8378, 8898, 9774, 10205, 10652, 10913, 11929, 12601, 12788, 12806, 12855, 12948, 13900, 14213, 14455

75: 308, 706, 734, 870, 1940, 2200, 2346, 2492, 2843, 3171, 3248, 3284, 5019, 5234, 5291, 5480, 6241, 6638, 7180, 8648, 9271, 9561, 9738, 9888, 10513, 10572, 10897, 10910, 11215, 11286, 11487, 11489, 11528, 11956, 12065, 12560, 12781, 12786, 12854, 12933, 13771, 13835, 13839, 13985, 14216, 14725, 14770, 15000

76: 265, 379, 499, 820, 894, 898, 1016, 1099, 1116, 1651, 2006, 2121, 2358, 2435, 2521, 3760, 4497, 4500, 6488, 6528, 6959, 6996, 7359, 7371, 7388, 7391, 7475, 9396, 9911, 10137, 10159, 10663, 10896, 12190, 12407, 13292, 13407, 13920, 14469

77: 780, 854, 1080, 1370, 1736, 1895, 2905, 3168, 3328, 3406, 3547, 4310, 4380, 4506, 4545, 4782, 5378, 5759, 6084, 6450, 6702, 7382, 7720, 8296, 8732, 9564, 10982, 11102, 11118, 11195, 11425, 12523, 13364, 14975

78: 990, 1135, 2013, 3018, 3112, 4215, 4235, 4318, 4498, 4501, 4505, 4820, 5010, 5144, 5174, 5674, 5703, 6033, 6196, 7375, 7920, 8083, 8601, 8705, 9671, 9872, 10702, 10828, 10861, 10892, 10999, 11115, 11412, 12063, 12653, 13010, 13906, 14011, 14067, 14068, 14121, 14238, 14339, 14368, 14858

79: 59, 69, 70, 106, 128, 332, 683, 825, 1095a, 1607, 1811, 2255, 2449, 2562, 2859, 2867, 3863, 4579, 4723, 4853, 5067, 5069, 5189, 5374, 6047, 6298, 6367, 6422, 6900, 7370, 9073, 9412, 9482, 9770, 11284, 11514, 12161, 13957, 13998, 14397, 14727

80: 421, 660, 984, 1476, 1647, 1850, 1857, 2081, 2589, 3134, 3167, 4012, 4253, 5008, 5011, 5031, 5312, 6382, 7417, 7518, 7694, 8026, 8209, 8305, 8508, 8523, 8636, 8731, 9275, 10309, 10788, 10970, 11017, 11220, 11381, 11720, 13637, 14192, 14208, 14286, 14459

81: 6, 153, 233, 245, 420, 448, 705, 830, 881, 892, 897, 1133, 1470, 1593, 2479, 2583, 2897, 3413, 3783, 3835, 4783, 5026, 5132, 5242, 5897, 6034, 6722, 7069,

81: 7367, 7387, 7925, 8733, 9136, 9277, 9300, 9388, 9393, 9401, 9586, 9755, 10870, 11280, 11333, 12061, 12071, 13395, 14329

82: 58, 1075, 1532, 1557, 1582, 1695, 1834, 1929, 2335, 2861, 2898, 2904, 4406, 4409, 5120, 5585, 6643, 7172, 7338, 8518, 8520, 9249, 9272, 9360, 9395, 10919, 11105, 11292, 11295, 11508, 11600, 12154, 12500, 12879, 13022, 13396, 13409, 13484, 14185, 14198, 14415, 14681, 14974

83: 166, 771, 1071, 1104, 1124, 2045, 2056, 2621, 2681, 2840, 2870, 3002, 3033, 3161, 3283, 3307, 3540, 3917, 4036, 4729, 4798, 5289, 6198, 6299, 6637, 7329, 7861, 8259, 8440, 8532, 8566, 8927, 9552, 9703, 9864, 9975, 10240, 10333, 10450, 10854, 10883, 11026, 11277, 11741, 12134, 12136, 12904, 13024, 13027, 13145, 13581, 13583, 13696, 13727, 13809, 14236, 14332, 14685

84: 485, 681, 759, 817, 1079, 1510, 1656, 1667, 1683, 1873, 1995, 2073, 2348, 2785, 2945, 2948, 3297, 3306, 4257, 4883, 5073, 5357, 5822, 6018, 6270, 6307, 6518, 7021, 7035, 8952, 9429, 9495, 9601, 9602, 9820, 9902, 10813, 10885, 11190, 11739, 12451, 12940, 13394, 13633, 13734, 13825, 14073, 14285, 14539, 14925

85: 412, 760, 882, 944, 968, 1131, 1346, 1526,

85: 2005, 2070, 2077, 2201, 2425, 2543, 3032, 3061, 3390, 3391, 3426, 3697, 4669, 4689, 5322, 5370, 5993, 5996, 6562, 7324, 8071, 8082, 8417, 8544, 8596, 9324, 9529, 9559, 9687, 10344, 10653, 10796, 10899, 12163, 12554, 12771, 12800, 12805, 13009, 13980, 14583, 14703, 14915

86: 91, 234, 271, 496, 735, 1352, 1801, 2129, 2223, 2851, 3188, 3456, 4255, 4342, 4674, 4795, 4825, 5313, 5323, 5976, 5977, 5992, 6981, 7290, 7419, 8008, 9278, 9584, 9782, 10284, 10430, 10751, 11016, 11159, 12901, 12910, 12923, 13401, 13426, 13612, 13907, 14209, 14402

87: 7, 18, 93, 97, 102, 148, 348, 498, 565, 828, 855, 871, 1590, 2033, 2067, 2075, 2261, 2672, 2928, 3086, 3151, 3178, 3189, 3334, 4046, 4313, 4791, 4803, 4862, 5002, 6194, 6493, 7330, 7334, 8292, 8517, 9302, 9960, 10888, 10895, 10928, 11022, 11222, 12141, 12179, 12593, 12984, 13014, 13405, 13640, 14016, 14492

88: 95, 137, 707, 860, 1076, 1662, 2015, 2422, 2432, 2537, 4690, 4704, 5795, 5895, 6587, 7700, 7851, 9246, 9415, 9661, 10190, 10514, 10698, 10856, 11162, 13423, 13525, 13812, 14369, 14400

89: 180, 317, 344, 1072, 1541, 1775, 2118,

89: 2209, 3256, 3320, 4073, 4722, 5268, 5287, 5308, 6038, 6040, 6421, 6725, 6820, 8542, 8698, 8700, 9335, 10258, 10317, 11053, 11217, 11497, 11964, 12070, 12156, 12431, 12936, 13393, 13700, 14092, 14585, 14976

90: 303, 396, 704, 736, 1486, 1516, 1757, 1784, 1804, 2125, 2332, 3030, 3055, 3109, 3486, 4326, 4479, 4676, 4710, 4793, 4859, 5352, 5559, 5644, 6362, 6588, 6901, 7002, 8108, 8325, 8434, 8826, 9091, 9466, 9974, 10224, 10645, 10676, 10881, 10906, 11446, 12435, 12625, 3413037, 13661, 13669, 13698, 13832, 15020

91: 101, 114, 781, 1078, 1799, 2003, 2281, 2522, 3395, 3400, 3601, 4650, 5288, 6179, 6297, 7131, 7181, 7427, 7435a, 8105, 8118, 8422, 8536, 8905, 9516, 10181, 10733, 10914, 11164, 11267, 12119, 12777, 12865, 12899, 13446, 13591, 14038, 14317, 14373, 14417

92: 20, 356, 573, 722, 886, 1335, 1539, 1956, 2048, 2342, 2550, 2937, 4504, 4678, 4806, 6195, 6289, 6429, 6644, 6999, 7844, 9835, 10829, 10831, 10857, 11390, 12909, 13008, 13039, 13140, 13810, 14190, 14323, 14741

93: 540, 1138, 1533, 1592, 1665, 2480, 2750, 3059, 3286, 3415, 3473, 4503, 4776, 4931, 5001, 5020, 5163, 5794, 5796, 6185, 6462, 8093, 8109, 8666, 8787,

93: 8818, 8824, 8949, 10959, 10985, 11001, 11297, 11428, 11512, 12081, 12088, 12602, 12741, 12798, 12892, 13392, 13804, 14120, 14170, 14288, 14342, 14684, 14888, 15005

94: 168, 171, 541, 1050, 1114, 1190, 1652, 1674, 2443, 2763, 2857, 3013, 3019, 3433, 3507, 4033, 4507, 5995, 6378, 6605, 7970, 8075, 8164, 9103, 9305, 9308, 9769, 10977, 11021, 11378, 11492, 12729, 12918, 13380, 13864, 14385, 14537

95: 80, 81, 135, 274, 414, 728, 737, 858, 1027, 1412, 1704, 1894, 1965, 2204, 2259, 2427, 2482, 2540, 2603, 2894, 2935, 2968, 3024, 3210, 4570, 4709, 5310, 6035, 6083, 6280, 6677, 6678, 7320, 7484, 7998, 8439, 8673, 8675, 8831, 9657, 9771, 10231, 10972, 11023, 11236, 12612, 12708, 12858, 13056, 13119, 13146, 13388a, 13462, 14240, 14291, 14430

96: 4, 167, 363, 399, 989, 1997, 2718, 2860, 2885, 3193, 3232, 3399, 4785, 4787, 4897, 5022, 5325, 7113, 7582, 8287, 8821, 8878, 8899, 9556, 9571, 9721, 9885, 10734, 10752, 10769, 10973, 11038, 11117, 11134, 11155, 11306, 12239, 12695, 13155, 13202, 13564, 13565, 13806, 14009, 14169, 14229, 14714, 14758

97: 24, 42, 131, 380, 542, 1051, 1559, 1999, 2277, 3000, 3004, 3165, 3440, 4028, 4060, 4254, 4789, 5973, 6958, 7014,

97: 7182, 8197, 8294, 8420, 8421, 9567, 9990, 10213, 10335, 10754, 10989, 11507, 12182, 12803, 12863, 12987, 13020, 13151, 14058, 14326, 14594

98: 50, 353, 1037, 1083, 1471, 1535, 1840, 1905, 1964, 2115, 2354, 2504, 2701, 2869, 2944, 3003, 3516, 3584, 4079, 4374, 4699, 5188, 5190, 5209, 5317, 5968, 6234, 6506, 6898, 7430, 8188, 8669, 8723, 8783, 9214, 9317, 9397, 9465, 9651a, 10926, 10971, 11060, 11122, 11242, 11303, 11495, 13004, 13163, 13975, 14135, 14168, 14346

99: 72, 96, 488, 1029, 1603, 1691, 1713, 2197, 2239, 2345, 2352, 2742, 2847, 3235, 3269, 3385, 5025, 5663, 5860, 6012, 6059, 6590, 6705, 6819, 6917, 7318, 7633, 8020, 8773, 9307, 9701, 10266, 10424, 10783, 10920, 11247, 11248, 11408, 12961, 13365, 13788, 14021, 14089

100: 641, 1902, 2813, 2824, 2884, 2929, 3038, 3115, 3172, 3796, 4030, 5021, 5107, 5460, 5576, 5864, 5896, 6962, 8055, 9313, 9439, 9478, 9834, 9852, 10686, 10705, 11108, 11213, 11223, 11549, 13033, 13046, 14271, 14470, 14944

101: 82, 377, 1003, 1164, 1477, 1939, 2758, 2868, 2985, 4568, 4899, 5320, 6648, 8116, 9135, 10649, 10842, 11031, 11059, 11587, 12677, 12829, 12864, 12962, 13176, 14215, 14297, 14588, 14989

102: 161, 357, 1366, 2061, 2315, 2397, 2599, 2810, 3037, 4569, 5240, 5264, 5334, 6295, 7130, 7236, 7460, 8336, 8909, 8943, 9697, 10237, 10694, 10880, 10963, 11211, 11324, 12982, 13403, 13440, 14269, 14544, 14891

103: 14, 149, 172, 502, 634, 815, 2258, 2931, 3130, 3416, 3481, 3583, 3585, 5261, 5294, 5395, 5704, 6304, 6503, 6509, 6585, 7007, 7093, 7108, 7325, 7459, 8547, 9345, 9413, 9420, 9728, 9772, 10204, 10696, 10964, 11082, 11129, 11701, 12546, 12725, 12928, 12983, 13589, 14243, 14396, 14749

104: 115, 133, 401, 547, 574, 775, 1150, 1345, 1518, 3514, 6181, 6564, 6591, 6925, 6998, 7754, 8845, 9464, 9490, 9611, 9750, 10655, 10657, 11352, 12450, 12791, 12797, 12804, 12872, 14197, 14247, 14287, 14743, 14789

105: 100, 108, 391, 843, 1052, 1077, 1677, 2007, 2103, 2279, 2471, 2967, 3029, 3281, 4576, 4612, 4788, 4898, 5447, 5449, 5664, 5671, 6522, 6523, 6592, 7321, 7452, 7458, 8415, 8729, 8833, 9250, 9312, 9365, 9372, 9406, 9549, 9582, 9603, 9754, 10125, 10739, 10965, 11107, 11119, 11354, 11940, 12068, 13023, 13050, 13376, 13769, 14105, 14173, 14272, 14558

106: 354, 514, 638, 1074, 1085, 1086, 1106, 1139, 1365, 1440, 1453, 1505, 1789,

106: 1798, 2074, 2875, 3025, 3156, 4160, 4256, 4262, 4486, 4488, 5024, 5121, 5476, 5623, 6132, 6584, 7017, 7575, 8146, 8309, 8734, 9427, 9473, 9608, 9735, 9930, 10182, 10700, 10956, 11197, 11550, 11571, 11594, 12186, 12356, 12501, 12668, 12955, 13823, 13830, 13908, 14091, 14738, 14801

107: 272, 635, 1097, 1546, 1650, 1907, 1942, 2743, 2756, 5477, 6254, 7328, 8334, 9292, 9332, 9673, 10223, 11553, 11897, 12415, 12478, 13468, 13996, 14109, 14260, 14298, 14325

108: 8, 65, 468, 806, 981, 1536, 1749, 3129, 3282, 3419, 4549, 4910, 5372, 6609, 6709, 7011, 8067, 8173, 8310, 8368, 8398, 8808, 9109, 9599, 9764, 10692, 10939, 11258, 11273, 11310, 11349, 11529, 12472, 12717, 12925, 13184, 13237, 13469, 13568, 14136, 14228, 14344, 14626, 14731

109: 400, 1328, 1332, 1545, 1746, 2889, 2981, 3597, 3764, 4487, 4684, 5293, 5358, 5423, 5947, 6184, 6428, 6756, 7013, 7161, 7215, 9645, 10328, 10598, 10770, 10901, 11913, 12374, 12479, 13379, 13973, 14000, 14002, 14273

110: 143, 375, 865, 1009, 1152, 1193, 1548, 1555, 1682, 1810, 1992, 2360, 2474, 2829, 2883, 3434, 4496, 4822, 5997, 6151, 6174, 6565, 6726, 7299, 8141, 8172, 8599, 8704, 8774, 8971, 9003, 9139, 9201, 9753,

110: 9765, 10949, 11068, 11510, 12056, 12655, 12733, 12905, 13363, 14014, 14245, 14564, 14763, 15030

111: 464, 785, 889, 1111, 2104, 2455, 2890, 2964, 3067, 4251, 4947, 5102, 5566, 6070, 6358, 6689, 6700, 6831, 6858, 8081, 8794, 8972, 9303, 9674, 9851, 10220, 10261, 10275, 10790, 10801, 10847, 11604, 12062, 12103, 12648, 12974, 13765, 13850, 14275, 14746

112: 450, 674, 808, 1523, 1625, 1802, 1852, 1859, 2206, 2670, 2676, 2695, 2892, 2914, 3154, 3557, 4241, 5201, 5510, 5539, 6136, 6153, 6191, 6197, 7163, 7250, 7292, 7418, 8637, 8835, 8925, 9440, 9592, 9610, 9635, 12483, 12766, 13026, 13584, 14005, 14196, 14347, 14726, 14739, 14805

113: 1070, 1501, 1725, 2693, 2704, 2705, 2771, 2841, 2871, 3169, 3285, 3435, 3484, 3500, 3598, 5637, 5665, 6530, 6767, 6997, 8290, 8388, 8843, 9049, 9320, 9347, 9722, 10330, 10741, 10945, 10975, 11012, 12445, 12994, 14180, 14258, 14350, 14540

114: 84, 90, 562, 726, 888, 954, 1327, 1367, 1410, 1507, 1805, 2060, 2351, 2502, 2536, 2729, 2854, 2990, 3062, 3106, 3492, 3534a, 4061, 4240, 4502, 4903, 4904, 5276, 7071, 8365, 8383, 8701, 8871, 9338, 10192, 10816, 10846, 10990, 11032, 11287, 11602, 11734, 12493, 12738, 13416,

114: 13475, 13586, 13662, 13958, 14171, 14394, 14467, 14593

115: 85, 383, 408, 1348, 1513, 1915, 3001, 3021, 3023, 3182, 3466, 3539, 4477, 4493, 4736, 4790, 4943, 5029, 5841, 6177, 6232a, 6271, 7256, 7304, 7335, 8192, 8321, 8699, 8847, 9188, 9270, 9481, 9600, 10738, 10932, 10938, 11015, 11307, 11311, 12180, 12622, 13573, 13697, 13838, 14704, 14924

116: 158, 364, 814, 991, 992, 1439, 1475, 1537, 1751, 1906, 2055, 2280, 2661, 2716, 2916, 2941, 3056, 3159, 3380, 4799, 4871, 4900, 5077, 6560, 6844, 6987, 7193, 8051, 8052, 8276, 8665, 8667, 8891, 8915, 9311, 9318, 9632, 10207, 10679, 11013, 11109, 11124, 11176, 11192, 11296, 11301, 11305, 11506, 11560, 12075, 12477, 12888, 13678, 13802, 14167, 14607

117: 477, 1488, 1497, 1758, 2278, 2782, 3035, 5286, 5458, 5632, 6145, 6517, 6848, 7967, 8018, 8027, 8106, 8162, 8377, 8389, 8997, 9325, 9402, 9643, 10703, 10729, 10936, 11100, 11181, 12409, 12432, 12518, 12654, 12736, 12801, 12989, 12993, 13913, 13999, 14367, 14536, 14584

118: 138, 890, 942, 980, 1015, 1049, 1441, 1484, 1867, 2124, 2256, 2477, 2553, 2557, 2593, 2862, 3358, 3431, 3472, 3791, 4494, 4495, 4508, 4700, 4888, 5307, 5326, 5974, 6236, 6301, 6360,

118: 6438, 7022, 7448, 8212, 8649, 8781, 9428, 9702, 9719, 9873, 10833, 10867, 10958, 11018, 11079, 11098, 11136, 11299, 11603, 12360, 12361, 12362, 12411, 12482, 12686, 12796, 12884, 12966, 13592, 13717, 14193, 14194, 14246, 14349, 14708, 14802, 14977

119: 609, 1047, 1595, 1808, 2202, 3305, 3499, 4161, 5363, 5725, 6250, 6294, 6939, 7157, 7239, 8119, 8811, 8914, 9115, 9356, 9405, 10677, 11037, 11123, 11137, 11391, 12067, 12481, 13011, 13538, 13893, 14241, 14411

120: 725, 1020, 2410, 2451, 3041, 3117, 3394, 3428, 4712, 4743, 4987, 5095, 5971, 5998, 6003, 7830, 8208, 8258, 8413, 8611, 8937, 8992, 9319, 9336, 9471, 9551, 11070, 11530, 11567, 11711, 12794, 13161, 13214, 13577, 14003, 14133, 14413, 14538, 14759, 14804

121: 19, 68, 123, 821, 1531, 1689, 2263, 2469, 2473, 2529, 3114, 3370, 3374, 3844, 4754, 4872, 5207, 5418, 5793, 6589, 8838, 8850, 8912, 9411, 9736, 10848, 10851, 11132, 11246, 11496, 11757, 13034, 13559, 13561, 14353, 14418, 14563, 14732

122: 386, 471, 861, 985, 1333, 1534, 1860, 2531, 2680, 2965, 2987, 3543, 4846, 6013, 6202, 6683, 7155, 7221, 7227, 8176, 8823, 8873, 9441, 10671, 10737, 10777, 10916, 11039, 11166, 11235, 11316,

122: 11559, 11598, 12237, 12352, 12476, 12633, 12857, 12889, 13015, 13040, 14195, 14543, 14590

123: 134, 473, 544, 893, 945, 987, 1344, 1578, 1732, 2820, 3122, 3164, 3442, 4252, 5145, 5236, 7252, 8174, 8271, 8430, 8567, 8584, 9101, 9224, 9408, 9588, 9848, 10065, 10673, 10689, 10791, 11078, 11158, 11269, 11746, 12406, 13036, 13228, 13456, 13567, 13707, 14750, 15007

124: 601, 856, 1018, 1479, 1664, 1672, 1714, 2212, 2341, 2698, 2805, 2891, 2913, 2949, 3046, 3181, 3923, 5175, 5299, 5893, 6260, 8107, 8668, 8670, 8674, 9476, 11229, 11302, 11304, 11494, 12659, 12742, 12757, 12868, 13449

125: 122, 160, 551, 590, 859, 1148, 1181, 1846, 2205, 2316, 2318, 2922, 2986, 3051, 3377, 3407, 3417, 3470, 4855, 5205, 6015, 6448, 7027, 7042, 7287, 8217, 8672, 9457, 9517, 9578, 10811, 11840, 12157, 12520, 12651, 12744, 12998, 13080, 14391, 14422, 14586, 14698, 14723

126: 202, 818, 1863, 2359, 2375, 2918, 2961, 2983, 3047, 4667, 5157, 5701, 5821, 6808, 7710, 9195, 9446, 9501, 9573, 9724, 9747, 10263, 10942, 12885, 13471, 13588, 14025, 14106, 14589, 14728

127: 543, 643, 670, 694, 1474, 1478, 1581, 1601, 1790, 1963, 2291, 2549, 2590, 2659, 2747, 2828, 2900, 3401, 3471,

127: 4259, 4516, 4849, 5027, 5076, 5194, 5200, 6274, 8693, 8864, 9604, 9615, 9656, 10212, 10320, 10855, 10891, 11334, 12423, 12427, 12579, 12990, 13157, 13442, 13473, 13680, 14534, 14546

128: 119, 318, 591, 592, 826, 1192, 1480, 1483, 1547, 1609, 1618, 1692, 1747, 1755, 2472, 2481, 2722, 3108, 3525, 4280, 4514, 4854, 4856, 5016, 5048, 5292, 5705, 5994, 6323, 6404, 7006, 7340, 8193, 8211, 8586, 9579, 9636, 9733, 10996, 11318, 11778, 12513, 12894, 13216, 13223, 13822, 14018, 14053, 14253, 14561

129: 71, 118, 782, 1001, 1023, 1115, 1832, 2162, 2400, 2814, 4581, 5305, 5351, 5819, 6014, 6037, 6681, 7228, 7429, 8535, 9301, 9581, 9605, 9741, 9746, 9816, 10322, 10817, 10824, 11024, 11515, 11756, 12155, 12452, 12508, 12643, 12743, 12893, 13849, 13928, 14189

130: 157, 406, 455, 1100, 1627, 1633, 1707, 1752, 1792, 2112, 2412, 2424, 2790, 2819, 2923, 2963, 2980, 4247, 4810, 5128, 6934, 7301, 7449, 8060, 8177, 8199, 8775, 9106, 9585, 9804, 10241, 10803, 11153, 11298, 11329, 11707, 11780, 11918, 12024, 12027, 12503, 12685, 14220, 14222, 14518, 14555

131: 88, 946, 953, 1490, 1676, 2111, 2117, 2326, 2717, 3420, 3430, 5096, 5397, 5448, 5605, 6566, 6991, 7214, 7282,

131: 8091, 8134, 8283, 8333, 8367, 8549, 9304, 9409, 9665, 10809, 10822, 11025, 11121, 11151, 11214, 11493, 12077, 12559, 12726, 13447, 14971

132: 92, 453, 653, 786, 1504, 1640, 1809, 1824, 1993, 2648, 2660, 3079, 3093, 3097, 3373, 3551, 3596, 4116, 4511, 4539, 5259, 6573, 6701, 8050, 8142, 8888, 8932, 8933, 8934, 9606, 9639, 10210, 10784, 11336, 11569, 11592, 11662, 12066, 12433, 12614, 12898, 14232

133: 103, 819, 904, 1447, 2274, 2436, 2454, 2465, 2837, 2839, 2993, 3614, 3766, 4243, 5104, 6074, 6235, 6432, 6508, 7151, 7197, 8195, 8852, 8911, 9497, 9613, 9684, 10860, 11131, 11288, 12015, 12386, 12442, 12604, 12728, 12919, 13116, 13922, 14221, 14970

134: 367, 478, 610, 668, 1334, 1487, 2906, 2959, 3287, 3294, 3421, 3498, 4248, 4541, 4759, 5152, 5328, 5654, 5656, 5914, 6045, 6375, 6857, 7211, 7283, 7293, 7450, 7745, 8460, 8926, 9623, 9723, 10699, 10941, 11095, 11133, 11517, 11896, 12160, 12833, 13187, 13357, 13639, 13683, 13923, 14324, 14383

135: 710, 1146, 1422, 1508, 1556, 1632, 1660, 1669, 1818, 1994, 2541, 2942, 3332, 3447, 5084, 5098, 5254, 5820, 6269, 6497, 6524, 6546, 7234, 7313, 7336, 8032, 8096, 8133, 8564, 8602, 8830, 8885, 8916, 8917,

135: 9543, 9659, 9679, 10637, 10797, 11125, 11505, 11590, 12172, 12978, 13092, 13211, 13601, 13642, 13649, 14419, 14700

136: 637, 1445, 1588, 2250, 2270, 2476, 3354, 3713, 4515, 4866, 5253, 6031, 6182, 7154, 7209, 8205, 8807, 9310, 9323, 9658, 11083, 11224, 11355, 11413, 11554, 12449, 12541, 12657, 12997, 13087, 13160, 13483, 13930, 14214, 14628

137: 132, 276, 1166, 2213, 2486, 2685, 2838, 2971, 3381, 3418, 3438, 3765, 4045, 4250, 4472, 5361, 5880, 6746, 7847, 8739, 9211, 9580, 9710, 9717, 9743, 10656, 11564, 12020, 12089, 12448, 12881, 14206

138: 209, 275, 407, 1542, 1611, 1868, 2329, 2969, 3092, 3123, 4510, 5005, 5079, 5090, 5113, 5446, 6005, 6520, 6703, 7156, 7238, 7240, 7434, 9350, 9515, 11817, 12723, 13169, 13641, 13932, 13934, 14234, 14345

139: 104, 110, 1014, 1610, 1680, 1686, 1825, 1918, 2243, 2343, 2635, 2767, 3157, 3345, 4509, 5339, 5849, 6180, 6258, 6685, 7280, 8533, 8692, 8806, 12376, 13208, 13768, 13883, 14541, 15012

140: 113, 905, 1118, 1264, 1325, 1612, 2063, 2276, 2586, 2600, 2627, 2760, 2994, 3225, 3425, 4059, 5789, 5847, 7036, 7152, 7203, 7296, 7426, 7481, 8087, 8137, 8293, 8685, 8744, 9122, 9337, 9843, 9855, 9881,

140: 10148, 10337, 10955, 11019, 11071, 11226, 11452, 11535, 12231, 12366, 13057, 13217, 13349, 13713, 14562

141: 503, 776, 908, 1698, 1820, 1856, 1869, 2398, 2419, 2752, 2781, 2909, 3121, 3163, 5015, 5078, 5147, 5540, 5706, 5956, 6516, 6567, 8380, 8414, 9243, 9662, 11165, 11516, 11523, 11703, 12856, 13025, 13235, 13690, 13993, 14715, 14745

142: 86, 1377, 1701, 1817, 1838, 2364, 2594, 2618, 2881, 3084, 3523, 3578, 5570, 6519, 6941, 6990, 7070, 8129, 8326, 8884, 9845, 9853, 9905, 11062, 11397, 11650, 11919, 12467, 12694, 12831, 13086, 13348, 14427, 14768, 14782

143: 1558, 2458, 3045, 3355, 4475, 5000, 5035, 5105, 5450, 5666, 6669, 8127, 8511, 9748, 9752, 10661, 11276, 11289, 11339, 11575, 12634, 12735, 12737, 13461, 13474, 13616, 13714, 14408, 14834

144: 125, 521, 657, 1349, 1413, 1467, 1519, 1549, 1765, 2196, 2466, 2642, 2655, 2825, 2919, 2970, 3279, 3289, 3356, 3537, 5797, 6563, 6759, 8458, 9114, 9352, 9716, 9941, 10061, 10650, 10664, 10681, 10937, 10968, 10969, 11201, 12016, 13088, 13443, 13529, 13931, 13984, 14916

145: 111, 862, 1102, 1845, 1896, 2349, 2417, 2501, 2678, 2772, 2966, 2989, 3057, 3082, 3315, 3343, 3443, 3457, 4836, 4857, 6168, 6200, 6652, 7025, 7346,

145: 7814, 7929, 8045, 8063, 8089, 8249, 8635, 8763, 9507, 9775, 10782, 10795, 10825, 10869, 11094, 11531, 11704, 14569, 14744, 14946

146: 570, 682, 857, 1134, 1153, 1624, 2321, 2544, 2637, 2725, 2792, 2988, 3278, 3408, 5138, 5432, 5949, 6155, 6231, 6938, 7288, 7289, 7451, 8006, 8126, 8180, 8921, 9618, 9884, 11275, 11715, 13648, 13981, 14705

147: 381, 384, 812, 1681, 2610, 2649, 2757, 2915, 5169, 5324, 7000, 7464, 8382, 8840, 9367, 9612, 9667, 9839, 9865, 10250, 10815, 10830, 11072, 11147, 11705, 12976, 12977, 13000, 13129, 13158, 13229, 13939, 13983, 14057, 14266, 14760, 15006

148: 140, 145, 355, 632, 720, 955, 1103, 1173, 1331, 1822, 1870, 1887, 2251, 2373, 2903a, 2933, 3424, 3445, 3545, 4260, 5028, 5080, 5115, 5275, 6137, 6139, 6233, 6568, 6682, 6684, 7019, 7397, 8200, 8384, 8778, 9649, 10183, 10688, 11067, 11232, 11322, 11442, 11480, 12510, 13147, 13534, 13719, 14524, 14968

149: 576, 867, 1140, 1495, 2433, 2631, 2972, 3371, 5250, 7026, 7231, 8122, 8313, 8371, 8453, 9887, 10444, 11003, 11085, 11152, 11312, 11747, 11748, 11924, 12353, 12701, 12709, 13029, 14230, 14547, 14707

150: 116, 583, 959, 1302, 1517, 1521, 1623, 1865, 2561, 2613, 2634, 2921, 4979, 4986, 5009, 5017,

150: 5337, 5655, 5669, 7291, 7442, 8317, 8496, 8772, 9530, 9593, 9751, 10193, 11478, 11597, 12647, 12832, 13016, 13094, 13594, 14203, 14508, 14551

151: 13, 109, 141, 146, 560, 687, 788, 2490, 2657, 2811, 2846, 3293, 3732, 4694, 5459, 6239, 7003, 8062, 8264, 8573, 8730, 8928, 8929, 9640, 9766, 11313, 11509, 12586, 12876, 12953, 13038, 13168, 13222, 13570, 13793, 14545, 14751

152: 330, 940, 964, 1550, 1753, 2363, 2830, 3036, 3652, 5030, 5894, 6559, 8025, 8311, 9421, 10779, 10786, 11120, 11207, 11238, 11749, 11781, 13120, 13571, 13941, 14017, 14718

153: 150, 159, 163, 528, 651, 774, 1715, 1762, 2299, 2320, 2564, 3436, 5036, 5170, 5488, 6027, 7838, 8086, 8136, 8186, 8403, 8919, 8920, 8950, 9511, 9626, 9854, 9868, 10802, 10981, 11144, 11146, 11167, 11338, 12582, 12596, 12616, 12897, 13143, 13215, 13603, 13979, 14623

154: 169, 476, 688, 866, 1168, 2367, 2641, 2689, 2694, 4692, 5277, 5436, 5879, 6243, 6272, 7010, 7341, 7611, 8143, 8272, 8645, 9443, 9474, 9496, 9568, 9620, 9670, 10781, 10808, 10810, 11077, 11149, 11150, 12069, 12731, 12895, 13124, 13209, 13569, 13990, 14010, 14710, 14720, 15031

155: 1084, 1117, 1160, 1324, 1356, 1357, 1540, 1760, 1815,

155: 1837, 2571, 2574, 2597, 2831, 2962, 3317, 3452, 5004, 5032, 5203, 5204, 5255, 5257, 5823, 6044, 6176, 6569, 7004, 7020, 8047, 8813, 9001, 9789, 9838, 12138, 12799, 13054, 13113, 13476, 13528, 13598, 13711, 13819, 14292, 14894

156: 105, 511, 698, 718, 810, 1142, 1262, 1461, 1614, 1630, 1657, 1795, 1829, 1862, 2357, 3158, 3429, 5364, 5444, 5673, 5957, 5979, 7516, 8231, 8385, 8880, 9369, 9682, 11058, 11069, 11382, 12171, 12430, 12712, 12793, 13028, 13347, 13464, 13533, 13605, 13789, 14211, 14886, 14951

157: 83, 107, 1381, 1466, 1629, 1716, 1841, 2514, 2558, 2559, 2633, 2939, 3931, 4713, 5123, 5195, 6252, 8044, 8236, 8288, 9164, 9614, 10244, 10669, 11081, 11087, 11093, 11160, 11328, 11460, 12018, 12920, 13494, 13544, 13546, 14755, 14816

158: 152, 553, 754, 789, 1128, 1628, 1706, 1847, 2298, 2602, 2710, 3020, 3116, 3441, 4478, 4531, 5087, 5142, 5965, 6830, 7016, 7024, 8110, 9634, 10070, 10672, 10728, 11033, 11075, 11084, 11218, 11605, 12456, 12711, 12963, 13801, 14756

159: 1359, 1360, 1668, 2355, 2508, 2579, 2606, 3094, 3376, 4847, 5097, 5435, 5537, 6244, 6377, 6553, 6755, 8033, 8229, 8267, 8918, 9574, 9859, 11145, 11939, 12139, 12385, 12583, 12873, 12951, 13089, 14233, 14262,

MELTING POINT INDEX OF ORGANIC COMPOUNDS (Continued)

159: 14301, 14384, 14414, 15021

160: 758, 1119, 1130, 1626, 1814, 1864, 2222, 2269, 2337, 2350, 2817, 2833, 3538, 4094, 4830, 5285, 7295, 8135, 8233, 8295, 8432, 8908, 9107, 9690, 10780, 10804, 11148, 11315, 11879, 12566, 12707, 12730, 12967, 13069, 13585, 14059, 14919

161: 162, 526, 1543, 1835, 2066, 2554, 2773, 2887, 3039, 3107, 3446, 5279, 5383, 7109, 8178, 8401, 8738, 9349, 9500, 10976, 11063, 12652, 13541, 13862

162: 594, 978, 1217, 1263, 1419, 1491, 1750, 1830, 1858, 1881, 2306, 2791, 2920, 3091, 3439, 4424, 5148, 5341, 5791, 6071, 6296, 6494, 6578, 7815, 8007, 8261, 8569, 9366, 9550, 9616, 9983, 12615, 13345, 13367, 13791, 14007, 14013, 14244, 14752

163: 48, 474, 784, 1112, 1515, 1843, 2257, 2510, 2584, 2723, 2804, 2845, 3081, 3595, 4998, 5256, 7023, 8171, 8924, 9353, 9644, 9842, 10334, 11323, 12900, 12996, 13030, 13540, 13816, 14202

164: 452, 696, 885, 1171, 1172, 1358, 1678, 1756, 1821, 2485, 2591, 2640, 2719, 2807, 3402, 3410, 3422, 3437, 3734, 4371, 4530, 4858, 5133, 5569, 5953, 8182, 8539, 9523, 10800, 11325, 11465, 11500, 11606, 12364, 12834, 13978, 14212, 14557, 14619, 14712, 14892, 14900

165: 94, 155, 559, 709, 1330, 1442, 1693, 1813, 2248, 2275,

165: 2365, 2369, 2572, 2601, 2668, 2996, 3022, 3064, 3497, 3502, 3568, 5018, 5156, 5258, 5267, 5354, 6189, 6363, 6413, 6688, 7005, 7148, 7305, 7414, 8036, 8094, 8227, 8429, 8938, 9154, 9629, 9866, 9964, 11000, 11485, 11922, 12137, 12424, 12453, 12575, 12608, 12880, 13043, 13207, 13366, 13368, 13622, 13716, 13972, 13994, 14205, 14302, 14548, 14550, 14779, 14893

166: 518, 665, 1577, 1833, 2167, 2319, 2368, 2516, 2569, 2662, 3368, 4580, 7691, 9017, 9187, 9878, 10312, 10659, 10935, 11230, 11938, 12679, 12696, 12720, 13112, 13346, 13963, 14304, 14716, 14955

167: 1602, 2215, 2513, 2598, 2651, 2732, 2827, 3223, 5653, 6232, 6571, 7053, 8255, 8697, 9339, 9480, 9871, 10807, 12345, 12524, 12724, 13338, 14281, 14566, 14591

168: 22, 39, 98, 727, 1425, 1427, 1449, 2515, 2530, 2622, 2639, 2736, 2749, 2754, 2806, 2836, 4078, 4714, 5266, 6570, 6582, 7008, 7095, 7303, 8165, 8206, 8798, 8922, 8923, 9587, 9621, 9642, 10318, 10670, 10706, 10865, 11271, 11320, 12732, 12830, 13001, 13101, 13121, 13183, 13551, 13579, 13964, 14264, 14410, 14618, 14748

169: 475, 1019, 1576, 1851, 1872, 2700, 3068, 3344, 3558, 4707, 5393, 5672, 6020, 6251, 6967, 9628, 9730, 10601, 10785, 12740, 13117, 13141,

169: 13620, 14022, 14393, 14560

170: 462, 730, 1194, 1200, 1215, 1218, 1267, 1283, 1493, 1561, 1827, 1828, 1874, 2307, 2493, 2500, 2663, 2924, 3386, 4693, 4696, 4708, 5659, 5747, 6437, 6586, 6936, 8080, 8230, 8268, 8902, 9309, 9470, 9672, 10324, 10864, 11101, 11228, 11619, 11912, 12458, 13045, 13175, 13444, 13596, 13607, 14277, 14719

171: 702, 1434, 2535, 2826, 3083, 3099, 3542, 5013, 5127, 5451, 6016, 6157, 7009, 8166, 8485, 8491, 8530, 8827, 9094, 9514, 9647, 10921, 12387, 12699, 12703, 13135, 13356, 13545, 13646, 13977, 14359

172: 120, 430, 431, 500, 1179, 1380, 1492, 1619, 1754, 1908, 2225, 3372, 4529, 4628, 5051, 5129, 5571, 6178, 6956, 6957, 7479, 7480, 8168, 8181, 8256, 9126, 9213, 9371, 9622, 9624, 9720, 10439, 10678, 11206, 11208, 11750, 13212, 13593, 13766, 14276

173: 1, 577, 733, 957, 2105, 2616, 2907, 2950, 3319, 3454, 5137, 7688, 8169, 8279, 8487, 9222, 9370, 9575, 9651, 9655, 9666, 10067, 10929, 12605, 13109, 13144, 13578, 13580, 14263, 14740, 14757

174: 960, 994, 1620, 1634, 2126, 2175, 2311, 2328, 2632, 2815, 4540, 5003, 5989, 6540, 6758, 8285, 8320, 10776, 10909, 10994, 11471, 11745, 12400, 12702, 13178, 14954

175: 778, 1180, 2303, 2702, 2734, 3104, 3124,

175: 3529, 3582, 5089, 5118, 5252, 5306, 5850, 5851, 5852, 5853, 6086, 6302, 6476, 6505, 6549, 7149, 7273, 7476, 8035, 8046, 8167, 8273, 8324, 8620, 8796, 9105, 9445, 9892, 10196, 10821, 10917, 11008, 11337, 11482, 12133, 13210, 13227, 13552, 14553, 14570

176: 1162, 1631, 1836, 1879, 2294, 2376, 2499, 2647, 2753, 3008, 3009, 3034, 3536, 4476, 4526, 5309, 5955, 6572, 7032, 8495, 8551, 9625, 9779, 9837, 10725, 11237, 12457, 12589, 12721, 13206, 13457, 13550, 13677, 13995, 14235, 14265

177: 89, 723, 2666, 2696, 2794, 2910, 2997, 3007, 5023, 5863, 6187, 6293, 6498, 6940, 6984, 7306, 8265, 8284, 8396, 9503, 10704, 11088, 11453, 12468, 12470, 12706, 12861, 13142, 13205

178: 151, 787, 906, 916, 1144, 1336, 1392, 1551, 1875, 1878, 2214, 2608, 3624, 3820, 4491, 4492, 4753, 5179, 5202, 6583, 6600, 7259, 7317, 7424, 7444, 8038, 8238, 8438, 8585, 8819, 9351, 9510, 9619, 10806, 11321, 11568, 12722, 12775, 13982, 14034, 14280

179: 112, 1175, 1819, 2744, 2895, 4212, 4490, 5338, 5489, 5575, 5650, 6985, 7232, 7443, 11126, 12197, 13031, 13127, 14392, 14482, 14691

180: 312, 579, 907, 969, 1123, 1642, 1761, 1823, 2570, 3096, 3292, 5006, 5112, 5244, 5554, 5792,

180: 6152, 6275, 6607,
7031, 7462, 8031,
8253, 8459, 8638,
9641, 9683, 9781,
9860, 10214, 10902,
11268, 11518, 11737,
12187, 12959, 13105,
13344, 13378, 14702,
14747, 14895, 14899

181: 655, 772, 970, 1204,
1617, 2546, 2619,
2654, 3100, 3383,
3544, 4484, 5790,
6004, 6313, 6499,
7018, 7038, 8392,
8498, 9502, 9650,
9653, 9862, 12019,
12355, 12363, 13197,
14485, 14902

182: 1198, 1282, 1429,
1876, 2240, 2674,
3006, 3581, 6502,
6547, 6561, 6662,
6687, 6982, 9638,
10930, 11200, 11744,
12365, 12688, 13122,
13390, 13402, 13623,
13971, 13976, 14290,
14730

183: 708, 1452, 1560, 1587,
1812, 1866, 3359,
4525, 5116, 5211,
5402, 5421, 6186,
6322, 6993, 8626,
8931, 9196, 12587,
12802, 13218, 13219,
13604, 13965, 14559,
14766, 14769

184: 142, 479, 1185, 1186,
1289, 2282, 3027,
3101, 4517, 4752,
6507, 7217, 8048,
8480, 8695, 8696,
9627, 9786, 9961,
11112, 11447, 11984,
12558, 13099, 13710,
15002

185: 659, 753, 993, 1154,
2059, 2203, 2268,
2289, 2532, 2699,
5284, 6514, 7100,
7315, 7463, 8235,
8252, 8300, 8493,
8607, 8742, 9142,
9534, 9861, 9965,
10323, 10724, 10798,
11111, 11443, 11534,
11740, 12426, 12609,
12691, 12734, 12992,
13167, 13351, 13611,
13645

186: 746, 773, 783, 1468,
2297, 2302, 2327,
2366, 2708, 2893,
3098, 3504, 5054,
5224, 6025, 6650,
8128, 8240, 8286,
8771, 9725, 10074,
10799, 11048, 11459,
11700, 12756, 13063,
13182, 13343, 13478,
14008, 14031

187: 121, 757, 1233, 2726,
2803, 3366, 4705,
4751, 4756, 5114,
5191, 6147, 6264,
6515, 8461, 8492,
10247, 11481, 11499,
12229, 12739, 12941,
13091, 13470, 14959

188: 686, 711, 1212, 2372,
2644, 2938, 4518,
5014, 5298, 5675,
6305, 6529, 6757,
8393, 10639, 11243,
11454, 12454, 13536,
14015

189: 1766, 2518, 5356,
6490, 6542, 7057,
7249, 7850, 8218,
8423, 8538, 9462,
12867, 13073, 13159,
13916, 14565, 14614,
14736

190: 338, 648, 685, 1098,
1197, 1353, 1354,
1586, 2305, 2626,
2951, 3448, 3541,
3552, 5295, 5333,
5377, 5862, 6479,
7229, 8003, 8019,
8580, 8930, 9322,
9355, 9595, 9681,
11064, 11326, 11753,
12763, 13220, 13615

191: 126, 251, 840, 2402,
2568, 3126, 5088,
5091, 5670, 6259,
6963, 7278, 8646,
8686, 9544, 9694,
9857, 12106, 12398,
12809, 13626, 14717

192: 669, 997, 1418, 1424,
1426, 1712, 1816,
2730, 3336, 4163,
4821, 5119, 5296,
5574, 6489, 6877,
7034, 7103, 8005,
8488, 9346, 9596,
9664, 9692, 9777,
9778, 9849, 9963,
10805, 11596, 11751,

192: 12017, 13632, 13636,
13712, 14733

193: 662, 673, 1252, 2296,
2399, 2612, 2799,
3553, 4271, 5046,
5085, 5125, 5453,
7270, 7279, 9675,
9780, 10605, 12399,
12584, 12848, 13790,
14786

194: 646, 1213, 1279, 1444,
2707, 2783, 2834,
3295, 3353, 4536,
5355, 6019, 6988,
8088, 9240, 9430,
9535, 9631, 9966,
11498, 13047, 13052,
13196, 14012

195: 524, 589, 986, 1350,
1705, 1898, 2519,
2800, 4534, 6257,
6427, 6469, 6541,
6554, 7030, 7344,
7584, 8299, 8841,
8936, 9533, 9554,
9609, 9648, 9677,
9718, 9727, 10218,
10246, 11199, 11340,
11464, 11538, 12713,
13203, 13613, 14767,
14771

196: 697, 1107, 1254, 1281,
1386, 1451, 2002,
2973, 3127, 3535,
6555, 8579, 9546,
11198, 11244, 12789,
14729, 14966

197: 124, 747, 2459, 2578,
2798, 3314, 5192,
5260, 5986, 7028,
7153, 7208, 7274,
7482, 8029, 8239,
8470, 8486, 9416,
9879, 10722, 11317,
12590, 12828, 13053,
13133, 13463, 14020

198: 456, 585, 649, 925,
2555, 2687, 2801,
3393, 3555, 4527,
4999, 5033, 5376,
5399, 5951, 6426,
6748, 6875, 7314,
7440, 8004, 8246,
8464, 8779, 9576,
9678, 12028, 13064,
13162, 13618, 13676,
14004, 14006

199: 117, 130, 667, 1155,
1376, 1842, 2880,
3080, 3554, 4532,
5362, 5419, 6000,

199: 7309, 7312, 8042,
8609, 8935, 9327,
9458, 11104, 11219,
12835, 13131, 13472,
14535, 14815

200: 661, 745, 1143, 1247,
1351, 1387, 1390,
1494, 2309, 2607,
2652, 2974, 2975,
2995, 3474, 4273,
5037, 5859, 5876,
6933, 6992, 7938,
8308, 8332, 8409,
8443, 8691, 9217,
9219, 9691, 9698,
11086, 12053, 12176,
12379, 12459, 12568,
12718, 12839, 12851,
13181, 13213, 13488,
13492, 13587, 13848,
14293

201: 156, 1329, 1711, 1831,
1839, 3388, 4258,
5340, 6265, 8232,
8262, 8682, 8801,
9597, 9875, 11128,
12827, 13058, 14389,
15022, 15028

202: 129, 1237, 1385, 1794,
2667, 2711, 2818,
2976, 3044, 3411,
4535, 4828, 6148,
6188, 6368, 8337,
8627, 9668, 9686,
11520, 12561, 13118,
13204, 13354, 14711,
14780

203: 642, 1235, 1636, 2585,
3387, 4275, 5134,
6424, 6521, 8457,
8610, 9858, 11562,
12840, 13067, 13988,
14783, 14787, 14914

204: 382, 460, 1145, 1214,
1363, 1384, 2645,
2665, 2982, 5143,
5283, 6501, 6532,
6579, 7112, 7286,
8170, 8175, 8241,
8274, 8983, 9498,
11479, 11566, 11923,
13844, 14522, 14617,
14722, 14785

205: 633, 1203, 2317, 2323,
2495, 2650, 2669,
3179, 3528, 3546,
5058, 5262, 5982,
6190, 6300, 6425,
6556, 6598, 7092,
8319, 8390, 8426,
9505, 9536, 9598,

205: 9967, 11760, 12588, 12597, 12715, 14046, 14407, 14903

206: 529, 645, 750, 1362, 1481, 1717, 1826, 2377, 2428, 3005, 5063, 5728, 6435, 8312, 8314, 9680, 9869, 11065, 13221, 13647, 14613, 14629, 14920

207: 568, 1136, 1137, 1361, 1496, 1849, 2484, 2809, 3011, 5109, 9264, 9431, 9637, 9660, 9749, 10789, 11073, 13108, 13987

208: 139, 1141, 1182, 1589, 2643, 2768, 2925, 3379, 3588, 3590, 4519, 5099, 5135, 5332, 7455, 8250, 8257, 8391, 8469, 8628, 8629, 8632, 9104, 10186, 10707, 12346, 12650, 13803, 14552, 14913

209: 461, 700, 1132, 1232, 2288, 2503, 2617, 6008, 6096, 8301, 8582, 8583, 8590, 10794, 14019, 14279, 14761, 14897, 14901

210: 513, 691, 1165, 1265, 2177, 2489, 2538, 2539, 2764, 2769, 2808, 2926, 2978, 3589, 4850, 4951, 5012, 5038, 5056, 5178, 6028, 7297, 8028, 8251, 8604, 8945, 9512, 10188, 11449, 11522, 11755, 12227, 12808, 12841, 12947, 13186, 13644, 13796, 13989, 14387, 14608, 14762

211: 512, 549, 1219, 2127, 3043, 5229, 6899, 7441, 9876, 14416, 14898

212: 1110, 1240, 1394, 1643, 2735, 4786, 5400, 5536, 6149, 6551, 7098, 7476a, 8054, 8473, 8474, 8786, 8855, 9263, 9491, 9499, 9521, 9785, 10199, 12140, 14556

213: 465, 467, 1161, 1236,

213: 2300, 2565, 2638, 2731, 3363, 5278, 6192, 6574, 7308, 8179, 8280, 8442, 9562, 9669, 9870, 11202, 11345, 12422, 12556, 12849, 13701

214: 1292, 2714, 3054, 3518, 4276, 4542, 4706, 5991, 6670, 7311, 8234, 8263, 8416, 8475, 9883, 12354, 12581, 12812, 13084, 13100, 13466, 14395, 14706, 14814

215: 548, 1105, 1202, 1274, 1563, 1635, 2573, 2629, 2832, 5126, 5490, 8278, 8366, 8784, 9538, 10149, 10311, 11759, 12421, 12425, 13459, 14303, 15029

216: 170, 1096, 1450, 1498, 2361, 3085, 3125, 3511, 5231, 6256, 6477, 8053, 9537, 9547, 9959, 10216, 10787, 11519, 12382, 12693, 12932

217: 1323, 1382, 1389, 1391, 2517, 2625, 2812, 3506, 3521, 3818, 4274, 5124, 6255, 7164, 7445, 8021, 8185, 9528, 10197, 12402, 12824, 13815

218: 652, 2581, 3600, 4277, 4546, 6487, 6492, 8516, 9442, 9646, 9827, 10731, 11210, 11350, 11483, 11556, 11570, 12480, 12813, 12934, 12960, 13352, 13628, 14517, 14701, 15023

219: 1342, 2362, 3510, 8237, 9368, 13350, 13692, 13807, 13968, 14390, 14753, 14754

220: 12, 16, 527, 1641, 2423, 2709, 3364, 5193, 5198, 5232, 5245, 5246, 5249, 5760, 5856, 6431, 6558, 6599, 7272, 7422, 7435, 8022, 8034, 9508, 9542, 9663, 9689, 9693, 11761, 12351, 12564,

220: 12823, 13114, 13477, 13723, 14587

221: 1260, 1645, 1646, 4521, 4522, 5230, 5251, 7268, 12964, 13166, 13553, 14039

222: 561, 647, 809, 3449, 5649, 6140, 6253, 7029, 8472, 8680, 9328, 9354, 9557, 10638, 11351, 11444, 12196, 12850, 13695, 14188

223: 744, 1253, 1853, 2733, 3451, 3628, 4520, 5392, 8694, 9506, 11735, 14947

224: 1639, 1797, 2494, 2563, 2821, 5280, 6017, 6512, 7310, 8184, 8800, 9726, 10251, 10792, 11346, 12844, 14156

225: 1121, 1748, 1861, 3215, 4755, 5110, 6002, 6423, 6496, 6878, 8465, 8828, 9426, 9526, 9540, 10336, 10778, 11044, 12348, 12576, 14711

226: 763, 1109, 1210, 1277, 1284, 2528, 3012, 3493, 7158, 8245, 8829, 12577, 14404

227: 2264, 2979, 3095, 6093, 6960, 7040, 8727, 8795, 9149, 11045, 12572, 13847

228: 2221, 2334, 2401, 2795, 2835, 3503, 6548, 6557, 7091, 7269, 8207, 10717, 11561, 12434, 12548, 12761, 13224, 13226, 14336

229: 566, 1242, 2560, 3450, 4827, 5050, 5092, 5954, 6022, 6995, 7033, 7438, 8049, 8144, 10710, 10793, 10993, 11046, 11717, 13081, 13149, 14592

230: 1562, 2161, 2488, 3090, 3522, 3548, 3626, 5274, 5282, 5877, 6146, 6262, 8160, 8591, 10766, 10862, 11052, 11524, 12531, 12574, 12592, 12790, 12846, 12927, 13132, 13776, 14568

231: 1159, 6966, 8149, 9223, 9558, 11450, 13458, 14912

232: 962, 1163, 1201, 1288, 1378, 2249, 2301, 3367, 6006, 6303, 6327, 8202, 10690, 12842, 13946, 14692

233: 2308, 3505, 5878, 6011, 8040, 8204, 10708, 13782, 14923

234: 469, 1428, 2636, 5186, 5301, 5990, 6468, 7276, 8148, 8150, 8243, 8402, 9374, 9555, 10150, 12836, 13617, 14024, 14505

235: 523, 550, 690, 701, 805, 1580, 1621, 1644, 2413, 3060, 3290, 5553, 6273, 6597, 8254, 8481, 8802, 9589, 9594, 12714, 13093, 13115, 13148, 13375, 13487, 14042

236: 1022, 1293, 1430, 2178, 3850, 6580, 7634, 8112, 8728, 9266, 9520, 9818, 10776, 12349, 12367, 12585, 13324, 13619

237: 1158, 2437, 2609, 3586, 3587, 5248, 5948, 6504, 6510, 6550, 7179, 8015, 8041, 8270, 9790, 9844, 10319, 11762, 12838, 12979, 14041

238: 578, 1101, 2290, 2460, 2483, 2653, 2927, 3031, 4467, 6281, 6539, 8468, 9186, 10325, 11504, 11764, 12852, 12970, 14511, 14737

239: 731, 1309, 1340, 1759, 3010, 5052, 5247, 7446, 7454, 8467, 11113, 13673, 13767, 14911

240: 764, 766, 1482, 2630, 2802, 3053, 3475, 4834, 5872, 5975, 8097, 8799, 8804, 9456, 10665, 11327, 12118, 14554, 14889, 14958, 14962

241: 470, 2453, 5059, 6581, 7107, 8070, 8527, 8603, 8605, 9527, 9833, 12547, 12971,

241: 13672, 13820, 15024
242: 721, 1266, 2658, 2762,
4695, 5057, 5139,
7271, 8740, 9707,
9877, 13068, 13370,
14905, 14948
243: 1199, 2611, 3291,
3792, 7411, 7473,
8115, 8466, 8633,
9210, 9541, 10176,
10252, 11742, 11758,
12571, 12973, 14361,
14571, 14605
244: 1244, 1383, 1848,
2908, 3396, 6261,
6326, 6434, 8201,
8277, 8560, 11743,
12853, 14001
245: 147, 636, 663, 863,
1108, 1243, 1568,
3327, 5111, 6001,
6167, 6495, 6513,
6552, 7635, 8556,
9513, 10242, 12837,
12860, 14207, 14625,
14957
246: 1268, 2370, 2977,
4750, 4845, 6007,
6415, 6430, 8999,
9451, 12845, 13225,
14504, 14624
247: 3351, 4697, 5873,
6850, 8335, 8437,
8499, 13627
248: 684, 1231, 1436, 1608,
2533, 2675, 2822,
3520, 7046, 7412,
7783, 8043, 8275,
8522, 9633, 9739,
11769, 12486, 12549,
12690, 15001
249: 1315, 1567, 6472,
8009, 8435, 11768,
12822, 12939
250: 1489, 2577, 2706,
3599, 5281, 7045,
7470, 7477, 8394,
9841, 10974, 11491,
11763, 12569, 13128
251: 1321, 3323, 8014,
13625, 14604, 14969
252: 1125, 1174, 1234,
1341, 2547, 3322,
6575, 8016, 9493,
10711, 12552, 13783,
15027
253: 595, 1129, 1196, 2816,
3214, 5535, 6134,
8436, 8681, 12751,
14040, 14603

254: 762, 1598, 2252, 2580,
8463, 9216, 9218,
9742, 9825, 10198,
10326
255: 1295, 3102, 3213,
5047, 5060, 5220,
6029, 6135, 8247,
8315, 9239, 11502,
12373, 12401, 13096,
13547, 13614, 14023,
14094, 14516, 14724
256: 593, 712, 1261, 1287,
1583, 1616, 2534,
2703, 3049, 3329,
3352, 5049, 5136,
8625, 8741, 9522,
10695, 12551, 12843,
13041, 13074
257: 1270, 1565, 2325,
3550, 6010, 6026,
8132, 12372, 12929,
14627
258: 552, 1388, 2415, 6536,
6538, 8037, 8203,
10691, 12704, 13630
259: 1305, 2403, 8017,
13072, 14956
260: 466, 1216, 2385, 4749,
4829, 5108, 5870,
5874, 5875, 6602,
8565, 9487, 10721,
12485, 12820, 12942,
12945, 14484
261: 17, 1291, 5272, 6030,
6965, 7689, 9509,
12429, 12487, 12573
262: 644, 1871, 4864, 6072,
6989, 7410, 11042,
11348, 12553, 14985
263: 3087, 5985
264: 6980, 10712, 12972,
14689
265: 154, 961, 1120, 7472,
8574, 9348, 13071
266: 7160, 8559, 9486,
11490, 12753, 13170,
13694, 14964
267: 1446, 3519, 6961,
8462, 9518
268: 522, 1222, 1246, 5064,
6131, 8803, 10709,
12603, 13535, 14788
269: 1183, 1431, 5983,
6482, 8269, 8502,
9563, 10332, 12759
270: 1248, 1285, 2728,
4519a, 4626, 5882,
6379, 8023, 9485,
9545, 9548, 10217,
10277, 10693, 11766,

270: 12943, 13693, 13715,
14388, 14694
271: 1238, 1311, 7275,
8501, 8503, 9525,
11521, 12957, 14533
272: 3512, 5871, 7471,
12347, 13631, 13795,
13935
273: 1485, 1511, 1520,
6024, 9454, 9969,
11513
274: 3444, 5093, 5273,
9220, 9488, 9823,
11557, 11765, 11767,
12436, 12697
275: 3384, 5100, 6964,
8777, 12926, 14688
276: 1250, 2456, 4627,
5271, 5952, 6537,
6978, 9654, 12780,
15026
277: 3325, 6930, 6975,
10640, 13674, 14515
278: 1306, 2461, 2548,
5199, 6937, 8554,
9373
279: 8606, 11501, 12817
280: 567, 584, 743, 1208,
1249, 2382, 6133,
8013, 8543, 8555,
8797, 11066, 12965,
13055, 13491, 13784,
13817, 13936
281: 1322, 8024, 9130,
14610
282: 1286, 5270, 8688,
9652, 11319, 12369,
12550, 12922, 14611,
14926
283: 1230, 2304, 3040,
8513, 9131, 10719,
11203, 11204, 12930,
13095, 14596
284: 472, 732, 11205,
13465, 14784
285: 564, 631, 2463, 3524,
5379, 6533, 7277,
7758, 8011, 8514,
11227, 12350, 12931
286: 1195, 2333, 2420,
5146, 8039, 8953,
13489
287: 1310, 5197, 8010,
8012, 8552
288: 1147, 1318, 1572,
2391, 5053, 6075,
6601, 8568, 13624,
14155
289: 575, 1241, 1245,
12921

290: 1221, 1225, 1313,
2371, 2395, 3065,
6576, 12428, 14612
291: 716, 1239, 9539,
12938, 13684
292: 1251, 2823, 4267,
4269
293: 1276, 2505, 3321,
8562, 8687, 8689,
8735, 8736, 8737,
10701
294: 1877, 2324, 2394,
2414, 6009, 9267,
12689, 13035, 14960
295: 555, 1002, 1122, 1271,
6436, 6484, 6485,
7467, 8553, 9565,
9566, 11043, 12656,
14616
296: 13048, 14734
297: 1224, 3357, 4573,
14615
298: 557, 699, 1205, 2497,
5171, 8563, 12949,
14887
299: 7039, 13685, 13967
300: 1317, 2383, 3365,
3527, 3682, 6380,
6855, 6968, 7099,
7420, 9321, 9492,
12755, 14963
301: 8510, 9265, 10718
302: 15025
303: 1206, 1220, 5302,
6969, 12727, 13479,
14812
304: 4268, 8690, 9826,
12368
305: 1571, 2392, 7744,
9241, 12349
306: 1280, 2464, 6970,
13066, 13687
308: 1448, 6972, 9450
309: 9447
310: 1273, 2381, 3494,
4784, 8561, 9449,
11736, 14620, 14735
312: 1255, 1272, 5660,
9819
313: 1304, 12384
314: 554, 556, 5538, 7041,
12946
315: 1637, 2389, 3717,
8615, 13671
316: 6971
317: 10727
318: 1301, 2496, 11616
319: 1223, 8525, 12545,
12815
320: 1228, 1599, 5881,

320: 9448, 9483, 12687, 13180, 13686
321: 1316
322: 9268, 12944
323: 8509, 12570
324: 1433, 13682
325: 12557
326: 7260, 14693
327: 2160, 6021, 11503, 12437
328: 3412, 13670, 13969
329: 6977, 7436, 8780, 14403
330: 1307, 1570, 6976, 10641, 10720
331: 1207, 13070
332: 7474
333: 7447, 12563
334: 3515
336: 8548
337: 4395
338: 1432, 7037, 14686
340: 1298, 1438, 1569, 14621
342: 11448, 12937, 14622
345: 8947
347: 6979, 14697
348: 8545
350: 1259, 2166, 5950
351: 13794

352: 8944
353: 1227, 13185
355: 12807
356: 3513, 12698
357: 9262
358: 5101
359: 13853, 13937
360: 1258, 1600, 7453, 12383
361: 6023
367: 11539
370: 14776
373: 14778
374: 1308
375: 12565
380: 2462, 14512, 14777, 14950
385: 5300
390: 8406
395: 362
400: 10714, 14774
410: 12954
415: 5984
419: 10195
420: 13777
422: 1269
438: 5243
450: 6478
470: 8381
490: 5987
762: 12847
913: 13787

BOILING POINT INDEX OF ORGANIC COMPOUNDS

Temperatures in °C; where there is a range of values, the compound is listed according to the lower value.

−164: 9015
−129: 9093
−103: 6840
−88: 6603
−84: 443, 9089
−83: 4741
−82: 9100
−81: 9047
−79: 6651, 6663
−78: 4715, 9079
−77: 4187
−76: 6876
−72: 14934
−64: 14500
−59: 9033
−57: 13296
−56: 8640
−51: 9067
−50: 4742
−48: 12342
−47: 6676, 12025, 12257
−42: 6659, 11772
−40: 9037
−38: 581, 6618
−37: 6159, 6646
−36: 11880
−35: 6762
−34: 580, 11770
−32: 5824, 14501
−31: 1585, 9088
−30: 445
−29: 9061, 12293
−28: 324
−26: 6668, 6847, 11379
−25: 9155
−24: 4211, 6632, 7055, 9035
−23: 12333
−22: 9084, 14502
−21: 5429, 7047, 9123
−20: 9090, 13274
−14: 11374
−13: 14931
−12: 9026, 9174, 11866
−11: 1415, 3808, 11872
−7: 13285
−6: 3942, 4104, 4189, 6158, 9148
−4: 3638, 5442, 13290
−3: 12286
−2: 6170
0: 3672, 3981, 4014, 6657, 9113, 11843
1: 1584, 1863, 11360, 11867
2: 1411, 11491, 14514
3: 3715, 3980, 5456, 6628, 9020

4: 444, 3972, 5833
5: 3151, 4065, 14928
6: 3657, 4740, 6619, 9124
7: 6156, 7096, 11359
8: 4195
9: 9064, 9797, 11847
10: 3631, 3668, 6809
12: 3655, 3814, 5389, 5431, 5438, 6614, 6775, 9205
13: 5980, 6920
14: 9077
15: 14930
16: 6754, 6814
17: 11385
18: 9029
19: 13176
20: 21, 439, 3576, 3971, 5827
21: 3572, 6867, 9068, 11361
22: 12266
23: 4769, 5416, 11386
24: 6225, 9056
25: 5807, 8161, 9038, 10496, 11371, 11373, 11865, 14953
26: 4064, 7514, 10307, 13323
27: 3777, 4202, 9947, 10422
28: 5830, 14932
29: 1465, 4199, 5829, 10537, 14089
30: 6634, 10517, 10520, 13514
31: 3970, 7083, 7116, 9119
32: 448, 3761, 5966, 6627, 9191, 11777, 12250, 12264
33: 1421, 3667, 4225, 6173, 7189
34: 3661, 4068, 5831, 6165, 6791, 8575, 8642, 9041, 11857, 12300
35: 6704, 6839, 11805, 12254
36: 1407, 1409, 5441, 6166, 6804, 6807, 9023, 9906, 10340, 10539, 10540, 11693
37: 5828, 6863, 6864, 9199, 11368, 11380, 12265
38: 4013, 6610, 6770,

38: 7932, 9168, 9190, 10357
39: 178, 3662, 4198, 6349, 13372, 14494
40: 481, 3636, 4450, 5756, 8661, 9060, 10553, 10618
41: 3964, 3977, 7778, 10316, 11842, 11926
42: 3639, 3646, 5733, 7940, 8981, 9082, 10301, 10595, 11721, 12022, 12263
43: 7134, 12222, 14795
44: 4091, 5806, 5832, 10298, 10619, 12225
45: 610, 6673, 7052, 7097, 7934, 7964, 12267, 12318, 14797
46: 616, 4717, 6625, 6630, 7532, 7537, 11795, 12253, 13513
47: 6675, 6866, 10209, 11773, 12246, 12277, 13866
48: 3335, 7982, 9994, 10308, 11362, 11771, 12042, 12260
49: 3745, 5758, 5762, 6343, 7191, 7968, 10315
50: 435, 4227, 5750, 5826, 6793, 6919, 6922, 7423, 7533, 7542, 7933, 9162
51: 197, 493, 8684, 10382, 10534
52: 480, 494, 1416, 3813, 4194, 4995, 5691, 5837, 6780, 8345, 8359, 11937, 12023, 12221
53: 4056, 5800, 6046, 9080, 10535, 11398, 12278
54: 5743, 7079, 7536, 9197, 10626, 14436
55: 3913, 3963, 4066, 4148, 5310, 5936, 6841, 8361, 9118, 10629, 11796, 12331, 13332, 14929
56: 306, 1463, 3759, 4658, 6092, 7406, 7408, 8654, 9034, 10012, 10562, 10628, 12005, 13269, 14935
57: 225, 2219, 4739, 6626,

57: 7739, 7824, 7854, 7889, 7952, 10019, 10303, 10411, 10907, 12259, 12290, 13245
58: 604, 676, 3746, 3875, 3953, 4146, 4228, 4968, 6640, 8658, 8663, 8907, 9948, 10563, 12203, 12258
59: 3643, 3758, 4988, 5640, 6615, 7731, 8364, 9157, 9169, 10593, 11788, 12199, 12252, 12289, 12332, 14791
60: 3247, 3647, 3664, 3978, 4147, 4971, 5538, 5540, 6806, 6865, 7650, 10406, 10533, 10536, 11437, 13270, 14497
61: 1989, 3805, 5430, 6823, 6923, 7897, 7962, 9050, 9098, 10127, 10627, 11265, 12080, 12128, 14879
62: 326, 3994, 4687, 7044, 7528, 7661, 7665, 7848, 7910, 8292, 9081, 10023, 10119, 10302, 10402, 13507, 13827
63: 442, 1462, 3675, 3676, 3677, 3757, 3902, 3949, 4085, 4086, 4087, 4103a, 4154, 4953, 6825, 6826, 7136, 7543, 7941, 7947, 8078, 10187, 10408, 12060, 12285, 12340, 14794
64: 37, 3743, 3955, 3968, 4035, 4050, 4144, 4210, 4236, 7539, 7781, 9058, 9173, 10011
65: 605, 624, 3571, 6821, 7137, 7407, 7728, 8996, 9127, 9128, 11723, 12248, 12256, 12335, 12539, 14939
66: 613, 1673, 3219, 4052, 4662, 4680, 6100, 6364, 6810, 7960, 10597, 11402, 13299
67: 615, 3560, 3899, 4016, 4153, 4966, 6090, 7142, 7202, 7541,

96: 56, 607, 1420, 2053, 3951, 4053, 4121, 4870, 4913, 5454, 6110, 6658, 7050, 8537, 9013, 10356, 10516, 10596, 11699, 11828, 12666, 14850

97: 34, 53, 190, 596, 2770, 3569, 4011, 4029, 4054, 4089, 4868, 7077, 9053, 9895, 10363, 10525, 10554, 10636, 11942, 12198, 12244, 13288

98: 614, 2042, 3648, 3704, 3851, 3948, 4063, 4088, 4139, 4655, 5625, 6872, 6896, 7076, 7522, 7565, 7604, 7647, 7648, 7997, 8076, 8576, 10071, 10130, 10253, 11409, 11946, 12036, 13662, 13846, 13942, 14440, 14967

99: 438, 501, 2184, 2856, 3642, 3703, 3857, 3888, 3889, 3890, 4075, 4221, 4894, 5184, 5745, 5771, 5813, 5845, 6128, 6355, 6935, 7673, 7828, 9180, 10304, 10419, 10611, 12143, 12308, 13940

100: 168, 169, 170, 707, 830, 3566, 3702, 4129, 4313, 4502, 4869, 4972, 5533, 5545, 5889, 7072, 7730, 7737, 7807, 8121, 8433, 8759, 9009, 9071, 9072, 9085, 10259, 11158, 11383, 11697, 11907, 11965, 12316, 12918, 13258, 14495

101: 262, 572, 3683, 4007, 4095, 4485, 4881, 4885, 5467, 5484, 5522, 6215, 6500, 6647, 6818, 7194, 7267, 8098, 8353, 10052, 10504, 10538, 10624, 11040, 11729, 11786, 13283, 13314

102: 211, 3074, 3681, 3712, 3905, 4124, 4351, 4445, 5551, 5688, 6697, 6902, 7119, 7241, 7994, 9788,

102: 9923, 10495, 10586, 10621, 10633, 11868, 12288, 12306, 13301, 13574, 13761

103: 35, 176, 3254, 3982, 4433, 4958, 5773, 5809, 6716, 7120, 7785, 8064, 8219, 8650, 8961, 10157, 10339, 10526, 10550, 11849, 12150, 12215, 13503, 14790

104: 3678, 3731, 3935, 4203, 4220, 4418, 4976, 5381, 5686, 5687, 6108, 6834, 7678, 9032, 9956, 10341, 10565, 10599, 10610, 11394, 12105, 12279, 13894

105: 1402, 1691, 1968, 1974, 3810, 3885, 3946, 4165, 4282, 4600, 4916, 4974, 5580, 5940, 5962, 6207, 6417, 7155, 7207, 7215, 7227, 7244, 7681, 9341, 9836, 9893, 10449, 10888, 11730, 11827, 11996, 12085, 12204, 13039, 13610, 13836, 13864, 13924

106: 612, 3331, 3565, 3796, 3926, 4130, 4201, 5462, 5585, 5854, 5890, 7075, 7798, 8363, 8490, 8810, 9802, 9886, 10083, 11404, 11578, 11617, 11658, 11785, 12033, 12165, 14657

107: 343, 495, 876, 1969, 3641, 3831, 4219, 4317, 4665, 4738, 4763, 4940, 4962, 5469, 6611, 6777, 7139, 7869, 8120, 8226, 9066, 10152, 10354, 10519, 10551, 11663, 11698, 11732, 11890, 14877, 14952

108: 436, 440, 2509, 3224, 3696, 3811, 4070, 4082, 4109, 4112, 4166, 4598, 4691, 4957, 5103, 5185, 5420, 5765, 5812, 6069, 6101, 6621, 6760, 6763, 6862, 7511, 7787, 10202,

108: 10438, 11726, 11830, 11963, 12191, 13845

109: 221, 258, 1953, 2130, 3203, 3632, 3653, 3992, 4167, 4307, 4464, 4879, 4977, 5630, 6164, 6331, 6772, 7056, 7144, 7802, 7804, 8965, 9044, 10546, 11970, 12043, 12189, 13331, 14428

110: 24, 222, 1405, 1752, 1970, 2536, 3146, 3991, 4436, 4657, 4792, 4880, 4938, 4965, 5685, 6351, 6613, 7110, 7248, 7510, 7521, 7663, 7803, 7966, 8260, 8489, 9000, 9027, 10064, 10122, 10335, 10360, 10418, 10542, 10799, 11396, 12087, 12095, 12166, 12315, 12338, 13294, 13341, 13434, 14060

111: 3113, 3658, 3714, 3763, 3990, 4155, 4407, 5581, 5622, 5903, 5963, 6341, 6345, 7058, 7801, 9087, 10361, 11434, 12414, 14834

112: 187, 188, 345, 627, 630, 3708, 3900, 3909, 3910, 3914, 3960, 4150, 4622, 4887, 4954, 5964, 6068, 6130, 6346, 6402, 6664, 6817, 6861, 7172, 7515, 7677, 7805, 7958, 8220, 8476, 9342, 10413, 10448, 10457, 10617, 11254, 11886, 11995, 12051, 12280, 12504, 12612, 12665, 13496, 13497, 13901, 14130, 14799, 14828

113: 3723, 4051, 4315, 4983, 5106, 5475, 5768, 5782, 6674, 6706, 7800, 7944, 9798, 10013, 10421, 11957, 12010, 12344, 13265, 13430, 14643

114: 282, 315, 1089, 2120, 3701, 3901, 4111, 4396, 4457, 5825, 5857, 6211, 6645,

114: 7141, 7176, 7762, 7870, 8357, 9014, 9799, 10069, 10108, 10420, 10509, 10576, 11041, 11395, 11825, 11903, 12039, 12299, 12461, 12639, 13926, 14127, 14402, 14529, 14633

115: 46, 55, 194, 535, 3616, 3700, 3920, 3962, 4823, 4973, 4982, 4984, 5482, 5508, 5689, 5783, 5905, 5932, 6208, 6470, 6656, 6803, 7126, 7568, 7577, 7799, 8477, 8876, 9120, 9159, 9236, 9429, 9611, 10072, 10368, 10400, 10543, 10584, 11549, 12339, 12392, 12441, 12667, 13236, 13902, 14135, 14210, 14457, 14465

116: 189, 506, 2051, 2507, 3249, 3699, 3733, 3940, 3999, 4077, 4306, 4365, 4443, 4458, 4949, 5511, 6138, 6783, 6849, 7206, 9150, 9227, 10369, 10474, 10502, 10564, 11776, 11813, 11993, 12123, 13498, 13751, 13841, 13873, 14043, 14128, 14129, 14796, 14830

117: 175, 186, 1843, 2026, 3245, 3772, 3860, 3921, 3947, 4036, 4069, 4164, 4229, 4452, 5330, 5474, 5664, 6091, 6099, 6109, 7562, 7566, 7570, 7571, 7763, 7806, 9008, 10203, 10350, 10544, 10614, 11664, 11665, 12220, 12952, 13879, 14631

118: 347, 1398, 2523, 3775, 3836, 3852, 3853, 3903, 4401, 4421, 4944, 5465, 5732, 6827, 6854, 7063, 7646, 7808, 10015, 10032, 10352, 10384, 10401, 10466, 10501, 10549, 10579, 10589, 11728, 11783, 11789, 11992, 12032, 12167,

118: 12296, 12502, 12619, 13309, 13724, 14450, 14793, 14818

119: 310, 918, 1909, 3240, 3577, 3579, 3986, 4043, 4047, 4873, 5519, 5526, 5846, 6229, 6400, 6816, 7121, 7246, 7353, 7569, 7965, 9036, 9170, 10581, 10585, 12014, 12670, 14321, 14871

120: 77, 287, 302, 313, 537, 617, 796, 2106, 3044, 3230, 3559, 3690, 3691, 3771, 3904, 3922, 3987, 4118, 4333, 4417, 4906, 4908, 5168, 5523, 5663, 5754, 6471, 6852, 6945, 7384, 7468, 7894, 7945, 7948, 8001, 8114, 8373, 8454, 8925, 8973, 9806, 10469, 10635, 10870, 11163, 11573, 11580, 11593, 11816, 11831, 11874, 11955, 12324, 13318, 13337, 13521, 13709, 13851, 14049, 14371, 14670

121: 191, 1064, 2135, 3689, 3908, 4039, 4048, 4328, 4561, 4602, 5388, 5817, 5935, 6874, 6943, 7581, 7734, 7943, 9062, 10105, 10167, 10518, 10524, 10545, 10580, 10582, 11661, 12153, 12327, 13201, 13260, 13263, 13322, 13450, 13572, 13854, 13915

122: 795, 934, 2049, 2576, 3686, 4067, 4209, 4780, 5396, 5548, 5908, 7205, 7651, 7773, 7957, 8061, 8427, 8619, 8651, 8836, 8882, 9045, 9208, 10114, 10116, 10117, 10416, 10432, 10485, 10591, 10632, 10844, 11432, 11815, 12050, 12059, 12188, 12322, 13317

123: 437, 482, 4149, 4779, 4815, 5521, 6171, 6217, 6695, 6890,

123: 7714, 7750, 7777, 7956, 10059, 10115, 10541, 11678, 11893, 12201, 12506, 12578

124: 226, 235, 396, 3069, 3205, 3239, 3725, 3773, 4152, 4292, 4319, 4341, 5107, 5391, 5524, 5525, 5527, 5764, 6617, 7086, 7831, 9281, 9810, 10507, 10508, 10645, 11639, 11667, 12011, 12073, 13291

125: 517, 1058, 2347, 3744, 3959, 3997, 4006, 4426, 5373, 5461, 5578, 5884, 6585, 6792, 6795, 6912, 7601, 7664, 7918, 7990, 8092, 8982, 9011, 9434, 9800, 9936, 10027, 10109, 10110, 10126, 10370, 11608, 11666, 12499, 12530, 13535, 13555, 13696

126: 185, 546, 2445, 3637, 3687, 3984, 4400, 4461, 4716, 4727, 4896, 4905, 4939, 5781, 5869, 6789, 7218, 7325, 7830, 7835, 8155, 8727, 8816, 9095, 9200, 10447, 10484, 11979, 13232, 14525, 14849

127: 29, 283, 1967, 2030, 3198, 3200, 3570, 3816, 3879, 5939, 6080, 6162, 6511, 6771, 6802, 6932, 7348, 7931, 8446, 9156, 9805, 10358, 11637, 11638, 11644, 11869, 11960, 12008, 12034, 12226, 12270, 12449

128: 63, 1138, 1894, 3155, 3788, 3878, 3882, 4404, 4440, 4455, 4560, 4564, 4639, 4960, 4975, 5481, 5696, 5736, 6047, 6534, 6711, 6712, 6787, 6790, 6846, 7178, 7759, 7911, 8970, 9225, 10257, 10380, 11259, 11540, 11980, 12249, 12507, 12516, 12958, 13813,

128: 13878, 14217, 14648

129: 219, 411, 3227, 3880, 3983, 4000, 4218, 4583, 4807, 4933, 5520, 6176, 6801, 6942, 7243, 7755, 7823, 7988, 8964, 8991, 10345, 10353, 10592, 10683, 11376, 12635, 13870, 14071, 14121, 14331, 14647

130: 183, 253, 666, 1966, 2031, 2065, 2087, 3432, 3610, 3634, 3724, 3767, 3884, 3897, 3998, 4001, 4343, 4384, 4427, 4447, 4764, 4781, 4994, 5499, 5786, 5801, 5885, 6081, 6666, 7684, 7761, 7764, 7855, 7987, 8141, 8376, 8894, 9094, 9167, 9890, 10433, 10772, 10853, 11649, 11873, 11962, 12007, 12045, 12083, 12361, 12623, 12648, 12677, 13275, 13282, 13326, 13557, 13566, 14310, 14405, 14439, 14574, 14835

131: 301, 359, 360, 3837, 3917, 4010, 4318, 4356, 5177, 5493, 5528, 5531, 5780, 5902, 6407, 6622, 6838, 7196, 7829, 7834, 7891, 8447, 8706, 9134, 9147, 10135, 10383, 11601, 11991, 12054, 12218, 13004, 13298, 13575, 14642, 14820

132: 184, 256, 441, 492, 2000, 3216, 3347, 3695, 4072, 4453, 4633, 4637, 5390, 5910, 7085, 7135, 7552, 7924, 8703, 8747, 9065, 9204, 9590, 10173, 10552, 11668, 11948, 12093, 12101, 12131, 13329, 13542, 13554, 13814, 13856

133: 50, 307, 1059, 1399, 2322, 2868, 3726, 3941, 4238, 4393, 4878, 4997, 5437, 5651, 5734, 5784,

133: 6958, 7138, 7554, 7591, 7812, 7942, 8978, 9048, 9153, 10068, 10131, 10388, 10415, 10468, 10471, 10634, 10875, 11260, 11819, 11947, 12100, 13268, 13774

134: 224, 254, 331, 2592, 3688, 3727, 3793, 3812, 3819, 4364, 4876, 4911, 4914, 5468, 5494, 5727, 5861, 6127, 6648, 6732, 6774, 7732, 7741, 7769, 7926, 7927, 7928, 7973, 8504, 8892, 9248, 10395, 10470, 10967, 11791, 11807, 11832, 11997, 12235, 12682, 12796, 13293, 13319

135: 5, 358, 379, 1895, 1995, 3162, 3174, 3645, 3881, 3896, 4151, 4403, 4416, 4454, 4556, 5652, 5684, 5838, 5918, 6053, 6129, 6488, 6910, 7247, 7556, 7560, 7775, 7780, 7892, 7893, 9177, 9889, 10022, 10282, 10365, 10379, 10478, 10506, 10646, 10877, 11673, 11818, 11875, 11923, 13254, 13267, 13811, 14522

136: 309, 366, 598, 1051, 2140, 2152, 2521, 3499, 3869, 4023, 4037, 5387, 5683, 5923, 6764, 6788, 7555, 7559, 7884, 7984, 9444, 10392, 10497, 10515, 11674, 12092, 12159, 12175, 12206, 12397, 12412, 12621, 13890, 13895, 14503

137: 678, 868, 2029, 3346, 3762, 3919, 4411, 5342, 5452, 5512, 5766, 6835, 7557, 7832, 8000, 8225, 8413, 9891, 10184, 10455, 10753, 11577, 11645, 12000, 12012, 12746, 13576, 13857, 14917

138: 1347, 1414, 3685,

138: 3735, 3739, 3774, 3787, 3895, 4591, 4886, 4959, 5369, 5466, 5560, 5561, 5629, 5752, 6270, 6347, 7409, 7645, 7776, 7809, 7882, 7980, 8131, 9030, 9233, 10129, 10480, 10577, 10959, 11159, 11266, 11609, 11640, 11643, 11792, 12212, 12268, 12311, 12598, 12983, 13233, 13824, 14226, 15003

139: 177, 410, 432, 626, 2069, 2216, 3259, 3313, 4434, 4575, 4611, 4793, 5802, 6052, 6350, 7147, 7655, 7833, 8354, 8621, 10080, 10442, 10479, 10571, 10578, 10615, 12416, 12475, 13252, 13887, 14986

140: 759, 761, 1124, 1606, 1741, 1805, 2495, 2540, 3163, 3438, 3463, 3607, 3776, 3782, 3815, 4138, 4181, 4381, 4456, 4553, 4630, 4672, 4875, 5401, 5407, 5409, 5562, 5591, 5602, 5749, 5798, 5960, 5967, 6312, 6616, 7054, 7346, 7500, 7553, 7558, 7649, 7810, 7811, 7836, 7883, 7976, 8452, 8672, 8858, 9152, 9436, 10398, 10414, 10464, 10575, 10583, 10674, 11261, 11641, 11642, 11775, 11821, 11891, 11959, 12082, 12157, 12195, 12509, 12511, 12593, 13174, 13297, 13310, 13706, 13804, 13889, 13947, 14068, 14247, 14477, 14883, 14921, 14971

141: 269, 486, 2374, 2438, 3217, 3459, 3893, 3898, 6079, 7561, 7675, 7768, 7797, 8505, 8507, 8857, 9010, 10057, 10125, 10338, 10390, 10417, 11184, 11800, 11941,

141: 12037, 12072, 12271, 13445, 13448, 13800, 13888, 13962, 14803, 14870, 14909

142: 227, 255, 848, 3870, 4022, 4117, 4266, 4361, 4432, 4567, 4794, 4862, 5682, 6065, 6406, 6473, 6797, 6884, 7183, 7427, 7524, 7853, 7914, 7915, 8725, 9083, 9133, 10048, 10053, 10056, 10347, 10351, 10590, 11196, 11267, 11651, 12274, 12298, 12321, 12514, 12599, 12601, 12641, 13483, 14884

143: 305, 3152, 3740, 3806, 3872, 3929, 4217, 4383, 4442, 4948, 5506, 5730, 5753, 5803, 7886, 7900, 8512, 9968, 10054, 10055, 10138, 10156, 10355, 10366, 10482, 10486, 11652, 11784, 11911, 12080a, 12086, 12219, 12367, 12418, 13300, 14069, 14449

144: 45, 223, 428, 3606, 3650, 3738, 3769, 3871, 4055, 4125, 4448, 4734, 4929, 5071, 5463, 5804, 6851, 6915, 7535, 7540, 7638, 7699, 7767, 7916, 8210, 8341, 8662, 10151, 10404, 11546, 11548, 11779, 12013, 12079, 12124, 12205, 12343, 12419, 13244, 13744, 13808, 13891, 14837, 14885, 14908, 14972

145: 192, 325, 751, 3164, 3307, 3613, 3728, 3824, 3873, 4260, 4298, 4301, 4425, 5161, 5509, 5546, 5695, 5844, 6786, 6837, 6842, 7127, 7187, 7224, 7919, 7937, 8002, 8069, 8724, 8791, 8896, 9183, 9687, 9809, 10362, 10405, 10925, 10979, 11106, 11174, 11595, 11646, 11892, 11977, 12091, 12644,

145: 13451, 13495, 14259
146: 491, 870, 883, 1719, 1854, 3597, 3598, 3694, 3874, 6333, 6667, 6690, 6713, 7143, 7225, 7671, 7939, 8449, 9231, 9435, 9437, 9935, 9942, 10385, 11594, 11915, 11986, 12224, 12519, 13316, 13831, 14045, 14082, 14374

147: 239, 400, 505, 2404, 2960, 3770, 3936, 3938, 4640, 4732, 4924, 5382, 6743, 7150, 7431, 7573, 7632, 7902, 7921, 7979, 8854, 9112, 9256, 9438, 9943, 10101, 10200, 10349, 10381, 10472, 10475, 10492, 10522, 10609, 10770, 12097, 12474, 12640, 13371

148: 26, 237, 238, 390, 656, 1910, 3768, 4120, 4399, 4402, 4449, 4459, 5738, 5744, 7365, 8669, 9031, 10208, 10348, 10374, 10461, 11209, 11303, 11692, 11999, 12460, 12473, 12476, 12495, 12798, 13024, 13152, 13234, 13251, 13338

149: 193, 218, 252, 341, 1686, 2869, 3809, 4316, 6620, 6710, 6794, 7060, 7912, 9097, 9146, 10014, 10313, 10386, 11612, 11687, 11822, 11860, 13163, 13871

150: 2188, 2313, 2721, 3143, 3609, 3894, 4058, 4288, 4410, 4446, 4465, 4936, 5074, 5218, 5358, 5458, 5547, 5742, 5897, 6735, 6905, 6914, 7101, 7486, 7603, 7772, 7888, 7977, 8673, 8675, 8683, 9055, 9428, 9816, 9918, 9950, 9954, 10367, 11002, 11014, 11138, 11194, 11794, 11881, 11987, 12213, 14052, 14067, 14881, 15023

151: 536, 1724, 3961, 4002, 4295, 4599, 4996, 5219, 5661, 6723, 6781, 6946, 6947, 6951, 7188, 7627, 7668, 7852, 7922, 8214, 10463, 10952, 11682, 11913, 12099, 14027

152: 200, 425, 558, 625, 658, 693, 1952, 2237, 3742, 3803, 3932, 4003, 4224, 4279, 4297, 4819, 5217, 5264, 5394, 5603, 5777, 5981, 6311, 7517, 7660, 7695, 8056, 8058, 8412, 8906, 9226, 9399, 10462, 10489, 10493, 10510, 11257, 11653, 11793, 12185, 12216, 12330, 12600, 13334, 13740, 13760, 13826, 14474

153: 412, 597, 963, 1783, 1922, 2435, 3234, 3729, 3781, 4392, 4479, 4617, 5769, 8863, 9953, 10099, 10112, 10291, 10410, 11883, 12170, 12310, 12413, 13321, 13380, 14441, 14456, 14579, 14869

154: 1056, 1057, 1914, 3153, 3206, 3619, 3892, 4103, 4877, 4970, 5075, 5473, 5559, 6098, 6161, 6949, 7012, 7163, 7978, 7986, 8570, 8659, 10107, 10498, 10500, 10587, 11607, 12663, 13009, 13284, 13339, 13881, 14072, 14227

155: 316, 365, 490, 1025, 2180, 2397, 3741, 3956, 4004, 4308, 4367, 5592, 5621, 6063, 6330, 6742, 6948, 7081, 7192, 7378, 7379, 7509, 7567, 7574, 7640, 7644, 7669, 7765, 7779, 7867, 8943, 9007, 9316, 10297, 10465, 10572, 10604, 11056, 11347, 11610, 11654, 11691, 11720,

193: 49, 1035, 1553, 1998, 2183, 2193, 4483, 5616, 5667, 6062, 6084, 6125, 6286, 6336, 6952, 6953, 7108, 7822, 8297, 8401, 9028, 9189, 9894, 9898, 11636, 12112, 12466, 13342, 13608, 13842

194: 30, 61, 249, 279, 350, 872, 1043, 1046, 1981, 2011, 2842, 2953, 3211, 3264, 3268, 4157, 4329, 4537, 4990, 5724, 5746, 5958, 6049, 6060, 6097, 6233, 6335, 6720, 7185, 7743, 7846, 9110, 9212, 10079, 10814, 10989, 13523, 13785, 14140, 14973, 14997, 15010

195: 47, 60, 259, 459, 1055, 1088, 1375, 1552, 1767, 2040, 3267, 3311, 3766, 3849, 4020, 4214, 4648, 4704, 5367, 5615, 5887, 5909, 6048, 6122, 6238, 6526, 6527, 6624, 6714, 7519, 7917, 8328, 8334, 8425, 9475, 9495, 9896, 9937, 10254, 11050, 11092, 11103, 11192, 12147, 13358, 13484, 14141, 14142, 14379, 14469, 14851

196: 248, 250, 394, 933, 1054, 1133, 1925, 1926, 1979, 2255, 2525, 3907, 3911, 3965, 3973, 4309, 4677, 5347, 5556, 5565, 5618, 6203, 6209, 6422, 6447, 6845, 7204, 8715, 8751, 8760, 10090, 11627, 12492, 13239, 13303, 13558, 14098, 14382, 14634, 14673, 14831, 14855

197: 285, 1036, 1742, 1891, 1990, 2025, 2039, 2524, 2526, 3137, 3277, 3720, 4733, 5615, 6446, 6736, 6895, 7062, 7512, 8158, 8820, 8868,

197: 10436, 10616, 10730, 11358, 11418, 11802, 11961, 12631, 13083, 13098, 13772, 13921, 14431, 14854

198: 284, 637, 675, 1033, 1459, 2052, 2157, 2182, 2229, 2254, 3262, 3974, 4040, 4397, 5176, 6680, 6693, 6883, 7084, 7223, 7735, 8130, 8400, 8497, 8757, 9769, 10005, 11542, 11552, 11675, 11877, 12491, 12494, 12629, 13305, 13918, 14044, 14100, 14674, 14772

199: 206, 1041, 1131, 1781, 1935, 1950, 2122, 2186, 2740, 3887, 4278, 6089, 6358, 6445, 6707, 6729, 6730, 8704, 8866, 8883, 10075, 10091, 10239, 10667, 13306, 13309, 13410, 13504, 13505, 14055, 14853, 14981

200: 62, 533, 1003, 1337, 1780, 1793, 2041, 2227, 2486, 3128, 3595, 3801, 3802, 4302, 4843, 4989, 5048, 5275, 5426, 5720, 5892, 6348, 6637, 6699, 7303, 7385, 7698, 9040, 9135, 9381, 10031, 10222, 11659, 11778, 11887, 12439, 12517, 12626, 12786, 12788, 12912, 12968, 12981, 13198, 13734, 13782, 14030, 14050, 14051, 14054, 14062, 14064, 14099, 14300, 14348, 14857, 14953a

201: 300, 504, 2491, 3119, 3226, 4385, 5360, 5619, 5679, 6240, 7181, 8886, 9482, 12900, 13311, 13933, 14102

202: 199, 323, 351, 420, 1948, 3077, 3175, 3846, 4574, 4928, 5336, 5555, 6226, 6444, 7074, 7265, 8664, 8814, 9888, 9919, 9957, 10089,

202: 11253, 11537, 11998, 12057, 12674, 13798, 14112, 14822

203: 740, 741, 879, 917, 974, 1880, 1936, 1947, 2132, 2450, 3996, 4215, 4723, 4923, 5927, 5988, 7789, 7972, 8106, 8985, 9132, 9922, 11251, 11252, 11405, 11724, 11934, 12076, 12529, 12787, 14063, 14327, 14412, 14433, 14682, 14738, 14863

204: 79, 264, 914, 1045, 1090, 1782, 1924, 1938, 1951, 2019, 2054, 2131, 3115, 3509, 3829, 3912, 4405, 4492, 4653, 6443, 7487, 7564, 7705, 7757, 7782, 7841, 7969, 8358, 8966, 9828, 10748, 11393, 11445, 12104, 12192, 12417, 12447, 13194, 13582, 14056, 14161, 14982

205: 34, 76, 288, 621, 1934, 1984, 2100, 2150, 2228, 2253, 2480, 2874, 3160, 3298, 3415, 3482, 4422, 4571, 4584, 4647, 4814, 5153, 5182, 5425, 5552, 6678, 6782, 6822, 7212, 7216, 7456, 7859, 7998, 8352, 8714, 8895, 9326, 9561, 9986, 10004, 10146, 10407, 10734, 10800, 11403, 11670, 11706, 11754, 12305, 12785, 12902, 13739, 13949, 13950, 13951, 14270, 14453, 14619, 14680, 14833

206: 1050, 2158, 3392, 3845, 3865, 4033, 4245, 4349, 4419, 4538, 4801, 5344, 5515, 5644, 5928, 6094, 6332, 6740, 7165, 7167, 7198, 7433, 9232, 9340, 9804, 11270, 11436, 11918, 12116, 12660, 13250, 13259, 13501, 13577, 14343, 14846

207: 829, 1985, 2108, 3596, 3830, 4656, 4724, 4837, 4863, 5180, 5181, 5505, 5542, 6126, 7599, 8872, 9732, 10142, 10944, 12109, 13415, 13752, 13898, 14097, 14322

208: 266, 803, 804, 1525, 1983, 2267, 3602, 3822, 4330, 4337, 4547, 4587, 4632, 5716, 5898, 6391, 6832, 6911, 7491, 7589, 7826, 7971, 8623, 8990, 8993, 10490, 10567, 11541, 12155, 12440, 13078, 13447, 13454, 13524, 13563, 13663, 14101, 14314

209: 75, 127, 278, 540, 919, 1379, 2143, 2634, 3630, 4038, 4174, 4265, 6419, 6819, 7827, 8008, 8327, 8531, 8708, 8753, 8968, 9958, 10220, 10430, 10446, 11812, 11916, 12292, 13486, 13859, 14366, 14764, 14838

210: 247, 281, 1031, 1069, 1978, 2018, 2133, 2194, 2504, 2867, 3287, 3590, 4027, 4156, 4305, 4649, 5166, 5805, 5851, 5925, 5926, 6045, 6370, 6458, 6660, 7125, 7222, 8713, 8967, 9984, 9990, 9993, 10051, 10249, 10425, 10681, 10747, 10882, 10903, 11292, 12069, 12125, 12148, 12379, 12408, 13065, 13428, 13882, 14162, 14163, 14250, 14421, 14810, 14866, 14933

211: 622, 1679, 2141, 2682, 2710, 2777, 3237, 3915, 5151, 5924, 6744, 7881, 8229, 9343, 9380, 9504, 9794, 9795, 9796, 9912, 10179, 10437, 10574, 10755, 10884, 10910, 11288, 12184, 13089, 14113, 14137, 14139, 14358

212: 267, 268, 608, 1958, 2020, 4030, 4312, 4332, 4388, 4409, 4558, 5785, 5919, 6442, 7264, 7266, 8065, 8187, 8577, 8578, 10081, 10082, 10427, 10749, 10918, 12146, 14334, 14839

213: 1429, 1684, 1687, 2096, 2266, 2690, 2876, 3176, 3798, 4484, 4517, 5164, 5365, 6403, 6466, 6499, 6507, 7089, 7492, 8444, 8822, 9230, 9295, 9933, 10271, 10568, 11007, 11901, 11910, 11968, 12061, 12505, 12638, 12678, 13010, 13816, 13863, 14138, 14274, 14284, 14423, 14836

214: 416, 534, 623, 794, 891, 895, 896, 920, 1933, 1937, 2156, 2187, 2912, 4092, 5039, 5424, 5848, 6061, 7106, 7323, 7359, 7723, 7844, 8541, 9317, 9792, 9793, 10850, 10947, 11587, 11588, 11589, 11671, 11975, 12115, 13248, 15014, 15015, 15016

215: 179, 602, 1032, 1151, 1372, 2037, 3043, 3186, 3243, 3611, 3656, 3800, 3817, 3906, 4320, 4360, 4597, 4601, 4614, 4992, 5507, 5568, 6398, 6416, 6452, 7375, 7380, 7864, 8679, 8766, 9379, 9946, 10273, 10288, 11417, 12040, 12209, 12613, 12672, 13328, 13418, 13427, 14123, 14268

216: 336, 844, 1030, 1957, 2273, 2936, 3076, 3797, 4240, 4605, 4674, 5549, 5868, 6393, 7343, 7842, 7866, 8962, 8963, 9568, 10473, 10613, 10960, 10988, 13556, 13994, 14251

217: 606, 921, 1949, 2021,

217: 2272, 2877, 3187, 3242, 3341, 4283, 4604, 4767, 4813, 6104, 6465, 6604, 7321, 7865, 9250, 9921, 10170, 10569, 11672, 12228, 12649, 12999, 13896, 14296, 14360, 14437, 14442, 14444

218: 209, 398, 873, 936, 1726, 1959, 1972, 2072, 2075, 2238, 2265, 2778, 2928, 3120, 3136, 3258, 3567, 4701, 4768, 5149, 5262, 5787, 5879, 6123, 7320, 8140, 8379, 8518, 8519, 8769, 8904, 9078, 9275, 10143, 10768, 10897, 11161, 11256, 11401, 11422, 12415, 12661, 13929, 14316, 14542, 14817, 14848

219: 915, 2058, 3339, 4554, 4572, 4800, 5737, 5779, 5855, 6033, 7109, 7860, 9228, 10852, 10886, 10931, 10950, 11240, 11255, 11657, 11731, 11804, 11912, 11950, 11951, 11985, 12595, 13500, 14114, 14131, 14289, 14355

220: 214, 286, 417, 902, 1540, 1694, 1711, 1727, 1729, 3070, 3310, 3799, 4249, 4650, 4670, 5900, 5953, 5997, 5998, 6234, 6353, 6882, 7149, 7174, 7376, 8090, 8339, 8528, 8758, 9108, 10393, 10409, 10429, 10771, 11004, 11906, 12114, 12409, 12444, 12625, 13195, 13638, 13705, 13762, 13763, 13764, 14443, 14510, 14526, 15018

221: 58, 887, 1887, 3110, 4439, 4603, 4777, 4778, 5371, 7201, 8752, 8900, 10456, 10836, 10900, 11036, 11283, 11547, 12194, 13756, 13960, 14146,

221: 14283, 14335, 15013

222: 246, 971, 1884, 1944, 1945, 2001, 3142, 3194, 3244, 3256, 4766, 6725, 8348, 8718, 8834, 8853, 8881, 9185, 9901, 10373, 10744, 10745, 11558, 11695, 11708, 11889, 12068, 12483, 12908, 13559, 14295, 14432, 14826, 14998

223: 59, 1728, 2857, 3185, 3236, 3591, 3592, 3593, 4090, 4182, 4519a, 5314, 5904, 6875, 7544, 7583, 8145, 8812, 8821, 10144, 10487, 11974, 12058, 13139, 13197, 13384, 13385, 13954

224: 68, 929, 973, 1013, 1649, 2195, 4250, 4557, 4609, 4737, 4872, 5378, 5572, 5726, 6384, 7105, 7367, 8095, 9807, 9821, 10154, 10155, 10272, 10983, 11367, 12381, 13905, 13961, 14599, 14606

225: 2, 793, 941, 943, 977, 1977, 2067, 2068, 2181, 2742, 2826, 2853, 2930, 3141, 3209, 3491, 4336, 4762, 5040, 5313, 5331, 6486, 6491, 6873, 7111, 8139, 8157, 8374, 8478, 8750, 10049, 10923, 11090, 11285, 11419, 13247, 13325, 13509, 13621, 13906, 14299, 14420

226: 418, 972, 2142, 3622, 4608, 4645, 5207, 6054, 6897, 7588, 8815, 9259, 10002, 10904, 11618, 11811, 13059, 13327, 14843, 15029

227: 369, 508, 1081, 2036, 4378, 5529, 7495, 7505, 7825, 8102, 8939, 9234, 9446, 10141, 10212, 10269, 10835, 11130, 11527, 11662, 11845, 13955, 14267, 14656, 14659, 14906, 15009

228: 387, 489, 899, 2147, 3299, 4380, 4429, 4437, 4590, 4816, 5165, 5368, 5375, 6088, 6438, 7425, 7745, 7863, 8428, 8622, 9215, 11005, 11009, 11010, 12173, 13324, 13357, 14333, 14649, 14653, 14660, 14841, 14979, 15017

229: 66, 335, 770, 802, 807, 1960, 3834, 4183, 4184, 5167, 5445, 5921, 6076, 6338, 7171, 7261, 7263, 8764, 8869, 8984, 9253, 9976, 10172, 10481, 10512, 12055, 12145, 12950, 14183, 14185, 14357, 14598, 14602, 14651, 14652, 14992

230: 541, 999, 1455, 1666, 1740, 1776, 1946, 2035, 2115, 3121, 3170, 3556, 3785, 3933, 4173, 4286, 4287, 4651, 4721, 4934, 5329, 5345, 5949, 6144, 6295, 6598, 6768, 6833, 6885, 7151, 7197, 7262, 7351, 7580, 7718, 7843, 8059, 8103, 8397, 8768, 8865, 9934, 9996, 10060, 10612, 10746, 11017, 11245, 11472, 11606, 11871, 12210, 13527, 13948, 14278, 14306, 14429, 14520, 14832, 14994

231: 792, 851, 880, 1699, 1923, 2605, 3078, 3266, 4508, 4770, 4771, 4812, 5586, 6103, 6112, 7493, 7706, 8387, 8749, 14650, 14655, 14980, 14993

232: 260, 388, 811, 911, 996, 1847, 1929, 2226, 2855, 3886, 4563, 4730, 4761, 5564, 7377, 8600, 8832, 8875, 10732, 10894, 11006, 11034, 11341, 11411, 11709, 12606, 13742, 13818, 13867, 14147, 14479, 14770

233: 72, 797, 1670, 1976, 2083, 2723, 3934, 6059, 6352, 6448, 7200, 7394, 8783, 10827, 10890, 11905, 13914, 14032

234: 755, 995, 1068, 1128, 1973, 2034, 2596, 2968, 4804, 5818, 7672, 8860, 9344, 10428, 11169, 11660, 11925, 12182, 12241, 12642, 13082, 13125, 13601, 13602, 13775, 13944, 14282

235: 180, 1040, 1393, 1703, 1796, 1919, 1931, 2044, 2057, 2283, 2336, 2854, 3171, 3833, 3883, 4009, 4760, 5311, 5346, 5349, 5350, 5413, 5557, 6038, 6182, 6474, 6691, 7230, 7400, 8398, 8809, 8846, 8901, 9995, 10162, 10780, 10774, 10834, 10840, 10924, 10978, 11290, 11399, 11570, 11710, 11992, 12594, 12610, 12995, 13312, 13536, 14294, 14342, 14460, 14528, 14532

236: 3218, 3241, 3271, 4387, 6386, 7307, 7322, 7326, 8284, 8613, 10223, 10566, 10819, 10915, 11969, 12356, 13599, 14095, 14471

237: 1063, 1456, 1807, 1920, 1932, 2081, 2092, 2614, 4673, 4757, 5700, 5860, 6913, 7709, 8099, 8748, 9577, 9578, 10140, 10750, 10911, 10980, 10984, 11614, 12673, 13758, 14313, 14364

238: 86, 799, 800, 976, 1975, 2295, 3103, 3140, 3340, 4208, 4281, 4431, 4805, 6174, 6856, 7381, 8153, 8445, 8851, 9201, 9489, 10761, 10762, 10826, 10866, 11439, 12242, 12719, 13952, 13953, 14080,

238: 14149, 14427, 14679, 14760

239: 877, 937, 1061, 1900, 4623, 4624, 5757, 7352, 10078, 10842, 11089, 13155, 13757

240: 414, 427, 801, 1027, 1082, 1490, 1648, 1749, 1902, 3179, 4489, 4582, 4631, 4785, 4930, 5150, 5699, 5901, 5916, 6383, 6449, 6888, 7989, 8100, 8418, 8590, 8893, 8988, 9829, 9989, 10043, 10838, 10841, 11147, 11293, 11900, 11917, 12838, 12853, 12928, 13156, 13246, 14085, 14198, 14308, 14329, 14352, 14386, 14531, 14983, 15011

241: 370, 791, 947, 3246, 4084, 4555, 4718, 4797, 4808, 5348, 6000, 8276, 9400, 12692, 14104, 14312, 14332

242: 1930, 2045, 3148, 2542, 2588, 3144, 3547, 4106, 5731, 6535, 8544, 8581, 9125, 9813, 10171, 11173, 11188, 11236, 14303, 14356, 14635

243: 392, 746, 910, 932, 998, 1021, 1094, 1774, 1928, 2095, 2243, 2595, 2604, 3148, 3263, 4191, 4562, 6665, 8113, 8194, 8859, 11377, 11555, 12181, 13057, 14646

244: 198, 1062, 2567, 3197, 4369, 4496, 4507, 5183, 5723, 7839, 8124, 8125, 9391, 11186, 11425, 11966, 13012, 13759, 14115, 14182, 14370

245: 378, 413, 422, 532, 836, 925, 1005, 1026, 1658, 1734, 1896, 2050, 2103, 2149, 2164, 3073, 3138, 3139, 3578, 4428, 4720, 4758, 4788, 5045, 5339, 5366, 5428, 5676, 5843, 6111, 6113, 6114,

245: 6214, 7598, 7786, 7930, 8542, 8597, 8746, 8849, 9138, 9812, 10759, 11035, 11080, 11187, 11420, 11902, 12232, 12243, 12611, 12869, 12959, 13003, 13295, 13543, 13633, 14011, 14033, 14311, 14372, 14419, 14588, 14589, 14842, 14862

246: 15, 389, 426, 2043, 2242, 3195, 4593, 5237, 5386, 6150, 6401, 6800, 7432, 8183, 8189, 8222, 8223, 8594, 9814, 10932, 11030, 12359, 12371, 13005, 14096, 14187, 14999

247: 368, 1702, 1733, 1745, 3114, 3442, 4852, 4860, 4861, 8152, 8340, 9600, 9705, 10647, 11018, 11110, 11063, 12627, 12077, 12878, 12891, 13090, 13678, 14103

248: 913, 3088, 3479, 4159, 4413, 5201, 6107, 6224, 8593, 10736, 10837, 10839, 10919, 11027, 12614, 12960, 13184, 14319, 14669, 14875

249: 376, 423, 1000, 1737, 1777, 2168, 2531, 3183, 4495, 4985, 5158, 6329, 6957, 10760, 10905, 11864, 11967, 13111, 13754

250: 1, 352, 1901, 2245, 2587, 2940, 3193, 3766, 4375, 4488, 5011, 5195, 5327, 6230, 6570, 7104, 7337, 7358, 7503, 7582, 7586, 7597, 7840, 8068, 8133, 8423, 8437, 8655, 8861, 9251, 9754, 10773, 10951, 10997, 11001, 11074, 11142, 11274, 11366, 11440, 11475, 11781, 12304, 12484, 12671, 13379, 13520, 13560, 13650, 13737, 14336, 14365, 14671

251: 777, 837, 850, 931, 1017, 1514, 2244, 2431, 2684, 2852, 4438, 5692, 6745, 8862, 11177, 12907, 13002, 14108

252: 835, 1008, 1080, 1955, 1956, 3181, 5450, 6213, 6410, 9141, 11314, 11421, 11572, 12455, 13271, 13743, 14658

253: 161, 243, 1522, 3480, 5070, 6354, 6931, 8608, 8844, 10775, 10889, 11139, 11365, 13726, 14309, 14320, 14661, 14907

254: 913, 935, 1093, 1400, 1885, 2241, 2566, 2593, 2737, 3147, 3378, 3433, 3561, 3564, 4578, 5472, 8411, 8776, 10740, 11550, 12994, 13537, 14354, 14874, 14984

255: 361, 903, 1091, 1177, 2076, 2085, 2090, 2218, 2284, 2746, 2072, 3072, 3369, 3487, 4486, 4568, 5318, 6077, 6409, 8372, 8674, 8720, 8848, 9932, 10267, 10874, 10985, 11294, 11357, 12482, 12768, 12819, 13722, 13773, 14088, 14157, 14330

256: 424, 1012, 1092, 1178, 1743, 1963, 2061, 2873, 3288, 3458, 6212, 6906, 7399, 7858, 8595, 9145, 9306, 10820, 10943, 11211, 11364, 11702, 12617, 12774, 13021, 13191, 13549, 14075

257: 603, 948, 1954, 2084, 3173, 3301, 5160, 5162, 6441, 7850, 8325, 8414, 8631, 10290, 10946, 11020, 11234, 11249, 11677, 12890, 13909, 14654

258: 136, 328, 415, 1028, 1066, 1722, 1961, 2093, 2338, 2743, 2858, 2865, 4374, 7390, 7667, 8598, 9299, 9377, 9378, 12238, 12411, 12887,

258: 13062, 14260, 14361, 14398

259: 71, 982, 2691, 3184, 5187, 6037, 6394, 6765, 8721, 9237, 10898, 11656, 12882, 12988, 14036

260: 164, 203, 372, 747, 749, 813, 842, 847, 965, 1007, 1608, 1912, 2256, 2683, 2720, 2864, 3464, 3549, 4642, 5008, 5359, 5439, 5457, 6397, 6609, 7330, 8057, 9140, 9229, 10194, 10431, 11412, 11414, 11424, 11426, 11565, 12021, 12903, 12905, 12916, 13104, 13107, 13420, 13917, 13928, 14079, 14086, 14413

261: 798, 1004, 2575, 3133, 4592, 5130, 7593, 8156, 8386, 8572, 9137, 9979, 12214, 12772, 12810, 13538, 14478, 14859

262: 174, 828, 990, 1095, 2582, 2741, 3149, 4824, 7551, 7576, 8154, 8712, 10977, 11413, 11914, 12607, 12776, 12886, 13597, 14411, 14737

263: 166, 875, 1189, 1735, 2849, 2859, 5159, 5319, 7545, 8173, 8329, 9261, 9358, 9363, 9439, 10042, 10871, 11438, 11714, 12193, 13738, 14110, 14275, 14476, 14507, 14726, 14873

264: 878, 2686, 3071, 4524, 5631, 6282, 6747, 9074, 9361, 9755, 10058, 10912, 11416, 12811, 12818, 12883, 13011, 14668

265: 20, 528, 1226, 1962, 2167, 2217, 3131, 3132, 4501, 4505, 4616, 4719, 5082, 6387, 6577, 7838, 8420, 8421, 8596, 8643, 9284, 9360, 10062, 10878, 10895, 10986, 11102, 11164, 11217, 11899, 12179, 12180, 12764, 12825,

265: 12935, 13023, 13606, 14092, 14373, 14896

266: 173, 1176, 2745, 4338, 5321, 9357, 9910, 11016, 12816, 12854, 13281, 14491

267: 94, 329, 514, 1038, 1440, 2063, 2091, 2659, 3489, 4700, 7670, 8382, 10158, 10178, 10763, 10996, 11197, 11226, 12240, 12779, 12855, 12956, 13302, 13413, 14077, 14105, 14159

268: 852, 874, 1169, 2097, 2660, 5086, 5155, 7585, 8879, 9359, 9365, 9425, 10295, 10296, 11311, 11879, 12826, 13200, 13531, 14731

269: 1011, 8785, 9144, 9171, 9389, 9756, 10159, 10873, 12773, 13106, 14073

270: 257, 377, 397, 458, 849, 975, 1188, 1288, 1603, 2116, 2717, 2738, 3161, 3191, 3460, 3461, 3765, 4379, 4521, 4550, 4818, 4820, 5917, 5947, 7642, 7719, 8283, 8765, 9364, 9970, 10044, 10285, 10644, 11427, 11592, 11834, 12521, 12555, 12752, 13103, 13179, 13489, 13903

271: 2263, 4686, 5122, 6087, 6141, 7065, 7319, 11031, 12643

272: 841, 949, 2704, 2779, 3488, 3530, 4327, 4415, 4871, 5269, 6686, 7575, 7688, 7693, 7899, 8419, 9222, 10876, 11571, 11835, 13126, 14158

273: 371, 671, 771, 3253, 3462, 3469, 4083, 5001, 7361, 8710, 9831, 11719, 11839, 12420, 12522, 14090, 14318, 14666, 15000

274: 834, 1067, 1785, 2661, 3403, 3485, 4466, 4497, 5297, 8495, 8517, 8588, 9172, 9298, 9390, 9424,

274: 9706, 11076, 12750, 13548

275: 457, 839, 1039, 1412, 1980, 2262, 2688, 2759, 3150, 3306, 4606, 5032, 5427, 6749, 7249, 7475, 7607, 9142, 9847, 9850, 9874, 11780, 12111, 13029, 13667, 13668, 13748, 14290, 14461, 14667

276: 748, 846, 1065, 2727, 3476, 3737, 6395, 7592, 8330, 8571, 9258, 9808, 10598, 10892, 12463, 12998, 13153, 13276, 13755, 14169

277: 964, 2223, 2224, 2724, 4635, 6124, 6669, 6870, 8228, 8717, 11461, 12996, 14481

278: 1709, 1899, 7067, 11025, 12730, 12758, 12991, 13060, 14487, 14490

279: 4, 822, 3620, 6115, 6309, 8147, 8870, 10990, 11622

280: 884, 927, 1911, 2174, 2220, 2713, 3308, 3465, 3532, 3736, 4551, 4629, 4690, 4784, 6308, 6798, 6811, 7360, 7587, 8248, 8326, 8388, 8422, 8587, 8634, 9375, 9753, 9817, 9832, 10215, 11263, 11853, 11854, 12283, 12501, 12523, 12754, 12792, 13373, 13655, 13657, 13660, 13661, 14637, 14865

281: 67, 99, 2927, 3467, 5148, 6206, 9291, 11526, 12500, 12814, 12911, 13656, 13805

282: 2062, 2952, 6388, 6812, 7220, 8508, 8558, 9376, 10224, 10872, 11023, 11221, 11570, 11856, 12870, 12875, 13061, 13530, 14678

283: 3389, 3478, 5317, 5808, 7641, 8424, 10765, 10859, 10922, 11473, 12874, 12906, 12975, 12980, 13313, 14093

284: 952, 3404, 3496, 4840, 4981, 7715, 7721, 10764, 14375

285: 538, 1763, 2080, 2105, 2110, 2111, 2624, 2907, 3212, 3390, 5154, 6268, 6371, 6642, 7118, 7245, 8169, 8221, 8282, 8415, 8494, 9334, 9337, 9387, 9704, 10124, 11060, 11514, 11517, 11718, 12909, 12910, 13855, 14376, 14890

286: 385, 503, 703, 2109, 2498, 3752, 4523, 5658, 6310, 6641, 6737, 7342, 8224, 8637, 9331, 10121, 10139, 10998, 11185, 11611, 12464, 12767, 12915, 13666, 14644, 14645

287: 1453, 4430, 4839, 5174, 5992, 7701, 9294, 12423, 12427, 12465, 12913, 13588

288: 853, 2173, 2208, 2850, 3468, 5131, 5537, 5657, 6463, 7066, 8331, 9388, 9571, 10000, 11284, 12782

289: 353, 2175, 3753, 5398, 6036, 6320, 7357, 7707, 9296, 9765, 11410, 12284, 12961, 13753, 14109

290: 1809, 3338, 4382, 4634, 5123, 5748, 5862, 6044, 7354, 7596, 8200, 9564, 10277, 11054, 11415, 11528, 12654, 12760, 12821, 13892, 14181

291: 1454, 1834, 2125, 3406, 3533, 8281, 9314, 9330, 9362, 9559, 10294, 11836, 12653, 13727

292: 988, 5288, 8349, 8523, 8729, 9269, 9297, 13729, 14254

293: 354, 1135, 6319, 6987, 8389, 9422, 9712, 10860, 11372, 11512, 12313, 14473

294: 689, 1338, 1791, 2506, 5173, 9287, 10293, 11495

295: 280, 408, 545, 1348, 1371, 1833, 2712, 3470, 3531, 5914, 6316, 6318, 6451, 8242, 8922, 9254, 9255, 9329, 9588, 9766, 10680, 11233, 11463, 11493, 11838, 13187, 13363

296: 143, 399, 2071, 2128, 2232, 3397, 3481, 6643, 9282, 9867, 10769, 11308, 11837, 12765, 13480, 13659, 14462

297: 1076, 1893, 2468, 2716, 3534, 6205, 6306, 6314, 8276, 9532, 9908

298: 2126, 2162, 3255, 6142, 6288, 6399, 6889, 10662, 11470, 12070

299: 3375, 3398, 9531, 10292

300: 449, 674, 768, 985, 1078, 1152, 1193, 1333, 1810, 2784, 2825, 3180, 3222, 5084, 5660, 6041, 6285, 7504, 8839, 9688, 10061, 11840, 11855, 11935, 12842, 13529, 13869, 14116, 14194, 14305, 14489

301: 2641, 5265, 6321, 7488, 9283, 9384, 12560, 13172, 13721, 14156

302: 3382, 4046, 6249, 7318, 8557, 9385, 10424, 12762, 13257, 13266, 13688

303: 144, 148, 5291, 8163, 8471, 9708, 13665

304: 84, 90, 2478, 6283, 9410, 9570, 11330, 11476, 12743, 12744, 12849

305: 954, 1075, 1111, 1904, 2882, 3477, 3534a, 6291, 7392, 7694, 9063, 9423, 9517, 9737, 10334, 11000, 12314, 12871, 13449, 14323, 14426

306: 319, 1077, 1647, 3221, 4641, 4729, 6382, 6784, 9569, 9722, 13189

307: 145, 966, 2787, 3452, 8356, 9940, 10281, 10286, 10396, 10668, 12503, 13728

308: 7499, 8842, 9386, 9553

309: 1373, 2274, 2956, 6453, 10994, 11331, 12604

310: 722, 769, 958, 1060, 1770, 1838, 2862, 2934, 3400, 3927, 5296, 6177, 6267, 7401, 8151, 9049, 9519, 9853, 11239, 12733, 14530, 14540, 14944

311: 7398, 11057, 13654

312: 845, 4980, 6365, 9412, 10812, 11172, 11220, 14328

313: 10007, 11499

314: 1771, 2774, 2788, 2957, 7713, 11271, 12462

315: 967, 1772, 2014, 2015, 2016, 2293, 3424, 11615

316: 2789, 9740, 9763, 9981

317: 395, 1836, 3405, 3755, 9417, 9744, 9837, 10607, 11264, 11332

318: 1773, 10278, 12435

319: 10, 1327, 2124, 3409, 6366, 8436, 12829, 14351

320: 1113, 1939, 3270, 3275, 3484, 3490, 6093, 7502, 7744, 9241, 9695, 10813, 11387, 11620, 12728, 13097, 13453, 13651, 13652

322: 2161, 3423, 4842, 7702, 7704, 9099, 9760, 10287, 14662

323: 2556, 5539, 9418, 10608, 13653

324: 3015, 12806, 12948, 13446

325: 3374, 3434, 4712, 6307, 6879, 9572, 13369

326: 6279, 6362, 12442, 13832, 13913

327: 2958, 5755, 9914

328: 1370, 3453, 6450, 11861, 13747, 13749

329: 2239, 6589, 9904, 11496

330: 1095a, 2903, 2917, 5859, 6367, 7404, 10145, 10397, 11183, 11281, 11474, 12135, 12422, 14506

331: 320, 955, 2137, 4714

332: 2905, 6012, 13699, 14583

333: 2935, 6467, 9738, 10238, 12732, 13006

334: 3305, 9419

335: 9, 3109, 6317, 9761, 9762, 11262

336: 2258, 7489, 7506, 7703, 13801

337: 1397, 7061, 11447

338: 3014, 3122, 3621, 10204, 11852

339: 1526, 11498

340: 1070, 1096, 1554, 2259, 3492, 6324, 6461, 9151, 10649, 11280, 11335, 11462, 13750, 14018

341: 6460, 7021

342: 2954, 3439, 5464

343: 9500, 6439, 7131

344: 2838, 4841, 7712, 13188, 13190, 13287

345: 464, 1653, 2890, 4848, 9420, 11469

346: 2482, 6280

348: 6679, 9982, 9985

349: 7490, 7498, 10700, 14954

350: 2552, 2898, 2902, 3013, 3652, 5094, 7711, 9289, 9290, 10225, 10729, 14047, 14732

353: 2915

354: 2955, 9075, 11231, 14549

355: 2771, 3370, 3422, 4697

356: 7483

357: 6454, 7708

358: 6670, 9103, 10165, 13732, 14537, 14676

360: 2522, 2545, 3092, 4660, 5028, 9770, 9992, 10205, 10230, 11435, 12693, 12812, 13381

362: 13424

363: 3414, 12901

364: 9441

365: 9773, 13700, 14527, 14534

368: 6381

371: 11048, 14019

374: 14985

375: 6986, 10260

377: 11390

379: 1195, 6323, 11318

380: 8948, 14417, 14452, 14557

383: 9987

385: 13140

390: 6297, 10686, 13033, 13589

391: 13720

393: 12430

395: 2910, 9764

398: 6289

400: 2481, 3417, 9771, 13916, 14948, 14964

401: 479, 10279

405: 6377, 14384

406: 13731

410: 5335, 11423, 15012

412: 7692, 13474

413: 2518

415: 2290, 6878, 10233

418: 2129

419: 10166

420: 12686, 13472, 14677

422: 6440

425: 2880, 14535

428: 12687

430: 575, 1241

431: 9096, 9973

435: 1858

440: 9416, 9900

442: 7485, 13398

445: 9370

446: 1194

448: 5047

449: 14399

452: 3366

458: 14683

459: 1307, 2294

462: 1308

467: 6420

471: 6178

480: 2517

490: 10513

518: 11539

525: 5243

727: 4520

891: 12307

$C_2H_2Cl_2$: 06863, 06865, 06866
$C_2H_2Cl_2FO_2$: 06071
$C_2H_2Cl_2O$: 00032, 04940, 06057
$C_2H_2Cl_2O_2$: 04962, 06060
$C_2H_2Cl_3F_2$: 06630
$C_2H_2Cl_3NO$: 00086, 14427
$C_2H_2Cl_4$: 06666, 06667
$C_2H_2F_2$: 06868
$C_2H_2F_2O_2$: 06127
$C_2H_2F_3NO_2$: 06658
$C_2H_2F_4$: 06668
$C_2H_2I_2$: 06869
$C_2H_2I_2O_2$: 06151
$C_2H_2N_4$: 13788
C_2H_2O: 08640
$C_2H_2O_2$: 11735
$C_2H_2O_3$: 07430
$C_2H_2O_4$: 10169
C_2H_3Br: 14930
C_2H_3BrFNO: 03617
C_2H_3BrO: 00434
$C_2H_3BrO_2$: 03602
$C_2H_3Br_3$: 06671
$C_2H_3Br_3O$: 06722
$C_2H_3Br_3O_2$: 00052
C_2H_3Cl: 14931
C_2H_3ClO: 00027, 00435
$C_2H_3ClO_2$: 04920, 04921, 04922, 04969
$C_2H_3Cl_2F$: 06627, 06631
$C_2H_3Cl_2NO$: 00072, 06059
$C_2H_3Cl_3$: 06672, 06674
$C_2H_3Cl_3O$: 06723
$C_2H_3Cl_3O_2$: 00056, 04870
$C_2H_3Cl_3Si$: 13336
C_2H_3F: 14934
C_2H_3FO: 00439
$C_2H_3FO_2$: 07043
$C_2H_3F_3$: 06676
$C_2H_3F_3O$: 06724
C_2H_3I: 14935
C_2H_3IO: 00440
$C_2H_3IO_2$: 08440
$C_2H_3I_3$: 06677
C_2H_3N: 00342
C_2H_3NO: 08216, 09169
$C_2H_3NO_2$: 06872
$C_2H_3NO_3$: 10188
$C_2H_3NO_4$: 09835
C_2H_3NS: 09170, 09204
$C_2H_3N_3$: 14412, 14413
C_2H_4: 06840
C_2H_4BrCl: 06611
$C_2H_4BrClO_2S$: 06697
C_2H_4BrF: 06612
C_2H_4BrNO: 03601
$C_2H_4Br_2$: 06621, 06622
$C_2H_4Br_2O_5$: 02635

C_2H_4ClF: 06615
C_2H_4ClI: 06616
C_2H_4ClNO: 00068, 04872
$C_2H_4ClNO_2$: 06617
$C_2H_4Cl_2$: 06626, 06629
$C_2H_4Cl_2O$: 04976, 06713
$C_2H_4Cl_2O_2$: 00034
$C_2H_4Cl_2O_2S$: 06698, 06699
$C_2H_4Cl_3NO$: 04869
$C_2H_4F_2$: 06632, 06634
C_2H_4INO: 08439
$C_2H_4I_2$: 06636, 06637
$C_2H_4NO_2$: 00348
$C_2H_4N_2$: 00676
$C_2H_4N_2O_2$: 07424, 08135, 10195
$C_2H_4N_2O_4$: 06639, 06896
$C_2H_4N_2O_6$: 06895
$C_2H_4N_2S_2$: 10191
$C_2H_4N_4$: 07441, 14414, 14415
C_2H_4O: 00021
C_2H_4OS: 13840
$C_2H_4O_2$: 00042, 00175, 07083
$C_2H_4O_2S$: 08973, 13851
$C_2H_4O_3$: 07417, 08209
$C_2H_4O_3S$: 06918
$C_2H_4O_4S$: 06917
$C_2H_4O_5S$: 13633
$C_2H_4S_2$: 06364
$C_2H_5AsF_2$: 01403
C_2H_5Br: 06610, 06770
C_2H_5BrO: 06710
C_2H_5Cl: 06614, 06775
$C_2H_5ClN_2O$: 04931
C_2H_5ClO: 06711, 06804, 06846, 09157
$C_2H_5ClO_2S$: 04995, 06696, 06780
$C_2H_5ClO_3S$: 04996, 06781
$C_2H_5ClO_4$: 06820
C_2H_5ClS: 06706
$C_2H_5Cl_3OSi$: 13283
$C_2H_5Cl_3Si$: 13288
C_2H_5F: 06646
C_2H_5FO: 06716
$C_2H_5F_3OSi$: 13285
$C_2H_5F_3Si$: 13290
C_2H_5I: 06653, 06805
C_2H_5IO: 06717
C_2H_5N: 01463, 06841, 06919, 14929
C_2H_5NO: 00046, 00058, 07064
$C_2H_5NO_2$: 00077, 00644, 04668, 06656, 06814, 07410, 08208
$C_2H_5NO_3$: 06720, 06813
C_2H_5NS: 00085, 13838

$C_2H_5N_3O$: 06721
$C_2H_5N_3O_2$: 03552, 14750
$C_2H_5N_5$: 14416
C_2H_5O: 06920
$C_2H_5O_5P$: 11397
C_2H_6: 06603
C_2H_6AsCl: 04464
$C_2H_6BF_3N$: 03568
$C_2H_6BF_3O$: 09156
$C_2H_6BrNO_2$: 00065
$C_2H_6ClNO_2$: 00661
C_2H_6ClOP: 11393
$C_2H_6Cl_2Si$: 13256
C_2H_6NO: 13746
$C_2H_6N_2$: 00087, 01584
$C_2H_6N_2O$: 00219, 00640, 06161, 14749
$C_2H_6N_2O_2$: 04691, 06160, 14746
$C_2H_6N_2S$: 13998, 14003
$C_2H_6N_4O$: 07452
$C_2H_6N_4O_2$: 06379, 08132, 10176
$C_2H_6N_4S$: 07448
C_2H_6O: 06708, 09155
C_2H_6OS: 06175, 06718, 09202
$C_2H_6O_2$: 06680, 06799, 06883
$C_2H_6O_2S$: 06174, 09201
$C_2H_6O_3S$: 06695, 09200
$C_2H_6O_4S$: 06700, 06798, 09198
$C_2H_6O_6S_2$: 06689
C_2H_6S: 06704, 09199
$C_2H_6S_2$: 06164, 06690
C_2H_6Se: 09197
C_2H_6Te: 09203
C_2H_7As: 01407, 01409
$C_2H_7AsO_3$: 06819
$C_2H_7BO_2$: 06769
C_2H_7BrClN: 06758
C_2H_7BrO: 06169
$C_2H_7ClN_2$: 00089
$C_2H_7ClN_2O$: 00642
C_2H_7ClO: 06170
$C_2H_7Cl_2N$: 06759
$C_2H_7IN_2S$: 13999
C_2H_7N: 06156, 06754
C_2H_7NO: 00024, 06727, 06750, 08291
$C_2H_7NO_3S$: 13670
C_2H_7NS: 06705
$C_2H_7N_3O$: 00660
$C_2H_7N_3O_3S$: 14000
$C_2H_7N_3S$: 13211
$C_2H_7N_5$: 03354
$C_2H_7N_5O_2$: 03554
$C_2H_7O_2P$: 11390
$C_2H_7O_3P$: 06692, 11431

$C_2H_7O_4P$: 11406, 11407
C_2H_7P: 11371, 11373
C_2H_8BrN: 06755
C_2H_8ClN: 06157, 06756
C_2H_8IN: 06757
$C_2H_8N_2$: 06849, 06854, 08078, 08079, 08098
C_2H_8Si: 13274
$C_2H_9NO_2$: 06074
$C_2H_{10}Cl_2N_2$: 06855, 08080
C_2H_{42}: 06439
C_2I_2: 00448
C_2I_4: 06877
C_2N_2: 05429
C_2N_2S: 05433
$C_2N_4O_6$: 14526
C_3AsF_9: 01421
$C_3Cl_2F_4$: 12277
C_3Cl_3: 21229
$C_3Cl_3F_3$: 12297
$C_3Cl_3F_5$: 11878
$C_3Cl_3N_3$: 05436
$C_3Cl_4F_4$: 11886
C_3Cl_4O: 00520
C_3Cl_6: 12292
C_3Cl_6O: 00323, 12223
$C_3Cl_6O_2$: 14447
$C_3Cl_6O_3$: 04723
C_3Cl_8: 11879
C_3Cl_9P: 11384
C_3F_4: 00581
C_3F_6: 12293
C_3F_6NP: 11362
C_3F_6O: 00324
C_3F_7NO: 11866
$C_3F_7NO_2$: 11865
C_3F_8: 11880
C_3F_9N: 14519
C_3F_9OP: 11386
C_3F_9P: 11385
C_3F_9Sb: 13439
C_3HBr_5O: 00332
$C_3HCl_2F_3$: 12278
$C_3HCl_3O_2$: 00515
C_3HCl_5: 12291
$C_3HCl_5F_2$: 11841
C_3HCl_5O: 00333
$C_3HCl_5O_2$: 12176
C_3HCl_7: 11863, 11864
C_3HF_3: 12342
C_3HN: 12022
$C_3H_2AsCl_2I$: 06845
$C_3H_2Br_2$: 12337
$C_3H_2Cl_2O_2$: 00498, 00499, 08907
$C_3H_2Cl_3NO$: 08679, 12209
$C_3H_2Cl_4$: 12295
$C_3H_2Cl_4O$: 00339, 00340
$C_3H_2Cl_4O_2$: 12190
$C_3H_2Cl_6$: 04011

$C_3H_2F_6$: 11867
$C_3H_2N_2$: 08904
$C_3H_2N_2O_3$: 10252
C_3H_2O: 12006, 12332
$C_3H_2O_2$: 12013, 12343
C_3H_3Br: 12009, 12334
$C_3H_3BrF_2$: 12263
$C_3H_3BrO_4$: 08843
C_3H_3Cl: 12335
C_3H_3ClO: 00481, 00519
$C_3H_3ClO_2$: 00495, 00496
$C_3H_3ClO_4$: 08808, 08852
$C_3H_3Cl_2F$: 12273
$C_3H_3Cl_2N$: 12203
$C_3H_3Cl_3$: 12296, 12298, 12299
$C_3H_3Cl_3O$: 00341, 11860, 12060, 12220, 12221, 12222
$C_3H_3Cl_3O_2$: 12195, 14441
$C_3H_3Cl_3O_3$: 08672, 12157
$C_3H_3Cl_5$: 11876, 11877
$C_3H_3F_3O_2$: 14498
C_3H_3I: 12338, 12339
C_3H_3N: 00516
C_3H_3NO: 08618, 10201, 12684
C_3H_3NOS: 00441
$C_3H_3NOS_2$: 13045
$C_3H_3NO_2$: 05420
$C_3H_3NO_2S$: 13822
C_3H_3NS: 13802
$C_3H_3N_3$: 14402
$C_3H_3N_3O_3$: 05406
$C_3H_3N_3S_3$: 13848
$C_3H_3S_3$: 14587
C_3H_4: 00580, 11770, 12333
C_3H_4BrClO: 12218
C_3H_4BrN: 12199, 12200
$C_3H_4Br_2$: 12270, 12271
$C_3H_4Br_2O$: 12047, 12216
$C_3H_4Br_2O_2$: 06050, 12114
C_3H_4ClN: 12201, 12202
$C_3H_4Cl_2$: 12272, 12275, 12276, 12279, 12280, 12281, 12282
$C_3H_4Cl_2O$: 00313, 00314, 12048, 12049, 12095, 12219
$C_3H_4Cl_2O_2$: 04961, 06065, 12122, 12125, 12129
$C_3H_4Cl_3NO_2$: 07063
$C_3H_4Cl_3NO_3$: 11996
$C_3H_4Cl_4$: 11881, 11882, 11883, 11884, 11885
$C_3H_4Cl_4O$: 12001, 12002
$C_3H_4F_3NO_2$: 11875
$C_3H_4F_4O$: 11970
C_3H_4NO: 04435
$C_3H_4N_2$: 08325, 12403

$C_3H_4N_2O$: 05418, 12424
$C_3H_4N_2OS$: 08049, 12373
$C_3H_4N_2OS_2$: 13046
$C_3H_4N_2O_2$: 08034
$C_3H_4N_2O_4$: 10186
$C_3H_4N_2S$: 13804
$C_3H_4N_4O_2$: 00738
C_3H_4O: 00480, 06922, 08644, 09162, 12010, 12023, 12285, 12344
$C_3H_4O_2$: 00486, 07405, 09163, 12037, 12041, 12051, 12152
$C_3H_4O_3$: 06224, 12053, 12142, 12174, 12676, 12968
$C_3H_4O_4$: 08830
$C_3H_4O_5$: 08880
$C_3H_4O_6$: 08992
$C_3H_5BO_3$: 07356
C_3H_5Br: 00609, 12259, 12260, 12262
C_3H_5BrO: 00309, 00597, 11791, 11792, 12043, 12215
$C_3H_5BrO_2$: 00183, 03606, 12076, 12082, 14640
$C_3H_5Br_3$: 11887, 11888, 11889
C_3H_5Cl: 00610, 12264, 12265, 12266, 12267
$C_3H_5ClF_2$: 11796
C_3H_5ClO: 00310, 00598, 00599, 10210, 11813, 11814, 12044, 12045, 12217
$C_3H_5ClO_2$: 00194, 00438, 04933, 04967, 12094, 12104
$C_3H_5Cl_2F$: 11829
$C_3H_5Cl_3$: 11890, 11891, 11892, 11893, 11894
$C_3H_5Cl_3O$: 12003
C_3H_5F: 12286
C_3H_5FO: 00322, 11858, 12225
$C_3H_5FO_2$: 12144
$C_3H_5F_3O$: 12004
$C_3H_5F_3O_3S$: 09120
C_3H_5I: 12288
C_3H_5IO: 00326, 11859, 12226
$C_3H_5IO_2$: 12162, 12163
C_3H_5N: 06654, 12198
C_3H_5NO: 00347, 00485, 05417, 06806, 08677, 12208, 12210
$C_3H_5NO_2$: 09165, 12289, 12290
$C_3H_5NO_4$: 08835

C_3H_5NS: 06837, 06838, 13824
$C_3H_5NS_2$: 13823, 13830
$C_3H_5N_2O_2$: 00579
$C_3H_5N_3O_2$: 07438
$C_3H_5N_3O_9$: 07399
$C_3H_5N_5$: 07436, 14403
$C_3H_5N_5O$: 00739
C_3H_6: 05824, 12025, 12257
C_3H_6BrCl: 11783, 11784, 11789, 11790
C_3H_6BrF: 11786
C_3H_6BrNO: 12062
$C_3H_6BrNO_2$: 11787, 11793
$C_3H_6Br_2$: 11819, 11821, 11823, 11825
$C_3H_6Br_2O$: 11950, 11951, 11985
C_3H_6ClF: 11798
C_3H_6ClI: 11799
$C_3H_6ClNO_2$: 11800, 11801, 11802, 11806, 11807
C_3H_6ClNS: 13846
$C_3H_6Cl_2$: 11826, 11828, 11831, 11833
$C_3H_6Cl_2O$: 11952, 11986, 11988
$C_3H_6Cl_2O_2$: 04928
$C_3H_6Cl_2Si$: 13261
$C_3H_6Cl_3NO$: 14429
$C_3H_6Cl_3P$: 11383
$C_3H_6F_2$: 11842, 11843
C_3H_6INO: 00327
$C_3H_6I_2$: 11844, 11845, 11846
$C_3H_6N_2$: 05412, 12419
$C_3H_6N_2O$: 08333
$C_3H_6N_2OS$: 13972
$C_3H_6N_2O_2$: 08827, 09164, 14701
$C_3H_6N_2O_3$: 08031
$C_3H_6N_2O_4$: 04672, 11848, 11849, 11850
$C_3H_6N_2O_5$: 14757
$C_3H_6N_2O_6$: 11919
$C_3H_6N_2S$: 08332, 13825, 13997
$C_3H_6N_3$: 05434
$C_3H_6N_6$: 08947
C_3H_6O: 00306, 00596, 07406, 07407, 07408, 09205, 10209, 11771, 11857, 12005, 12042, 12300
$C_3H_6OS_2$: 14953
$C_3H_6O_2$: 00044, 00225, 00325, 06220, 07079, 11941, 11958, 12072
$C_3H_6O_2S$: 08981, 09208, 12165, 12166, 13835

$C_3H_6O_3$: 00317, 04728, 07345, 07346, 07347, 08214, 08650, 08651, 08652, 09132, 12150, 12151, 14529
$C_3H_6O_3S$: 11909
$C_3H_6O_4$: 07349, 12130
C_3H_6S: 00615, 09207, 13797, 13834, 14521
$C_3H_6S_2$: 06361
$C_3H_7AsO_3$: 12024
C_3H_7Br: 03944, 11782, 11788, 12251, 12252
C_3H_7BrO: 06613, 06772, 11946, 11977
C_3H_7Cl: 11795, 11805, 12253, 12254
$C_3H_7ClN_2OS$: 13973
C_3H_7ClO: 04952, 04956, 06779, 11947, 11949, 11979
$C_3H_7ClO_2$: 11901, 11915
$C_3H_7ClO_2S$: 11931
C_3H_7ClS: 09152
$C_3H_7Cl_3Si$: 13317, 13318
C_3H_7F: 11862, 12304
C_3H_7FO: 11960
C_3H_7I: 11868, 11870, 12306, 12307
C_3H_7N: 00604, 01462, 05826, 12258
C_3H_7NO: 00079, 00331, 07060, 07062, 12054, 12061
C_3H_7NOS: 04684, 04685
$C_3H_7NO_2$: 00554, 00555, 00556, 00568, 00666, 04661, 08647, 08648, 11873, 11874, 12065, 12317, 12318
$C_3H_7NO_2S$: 05872
$C_3H_7NO_3$: 08609, 11965, 11995, 12315, 12316, 13224, 13225, 13226
$C_3H_7NO_5$: 07378, 07379
$C_3H_7NO_5S$: 05870, 05871
C_3H_7NS: 13821
$C_3H_7N_2O_7$: 07365
$C_3H_7N_3O$: 00048
$C_3H_7N_4O_2$: 07415
C_3H_7O: 04070
C_3H_7S: 04144
C_3H_8: 11772
C_3H_8ClNO: 00308
$C_3H_8ClNO_2$: 00665
$C_3H_8Cl_2N$: 06070
$C_3H_8Cl_2Si$: 13258
C_3H_8N: 00605
C_3H_8NO: 14730
$C_3H_8N_2$: 08061

$C_3H_8N_2O$: 14731, 14741

$C_3H_8N_2O_2$: 08130, 12113, 12249, 14772

$C_3H_8N_2O_3$: 14728

$C_3H_8N_2S$: 13986

$C_3H_8N_2Se$: 13199

$C_3H_8N_6O_6$: 14407

C_3H_8O: 06809, 11942, 11971, 12244, 12245

C_3H_8OS: 06719

$C_3H_8OS_2$: 11955

$C_3H_8O_2$: 06912, 07052, 11898, 11910, 12287

$C_3H_8O_2S$: 11907

$C_3H_8O_3$: 07354

$C_3H_8O_3S$: 11933

C_3H_8S: 06810, 11936, 11937

$C_3H_8S_2$: 11926, 11927

C_3H_9As: 01416

$C_3H_9AsO_3$: 11896

C_3H_9B: 03576

$C_3H_9BF_4O$: 14524

C_3H_9BO: 03572

$C_3H_9BO_2$: 11897

$C_3H_9BO_3$: 03560

C_3H_9BS: 03573

$C_3H_9BS_2$: 03566

$C_3H_9BS_3$: 03567

C_3H_9ClSi: 13245

C_3H_9N: 06807, 09160, 11773, 11777, 12246, 12250, 14514

C_3H_9NO: 06734, 06761, 11943, 11944, 11972, 11973, 14516

$C_3H_9NO_2$: 11899

$C_3H_9NO_2S$: 11930, 11932

$C_3H_9NO_3S$: 13672

$C_3H_9NO_4S$: 12834

$C_3H_9O_3P$: 09111, 11404, 11434, 11928, 11929

$C_3H_9O_4P$: 11418

C_3H_9P: 11380

C_3H_9Sb: 13437

$C_3H_{10}ClN$: 06808, 14515

$C_3H_{10}ClNO$: 14517

$C_3H_{10}Cl_2N_2O$: 00312

$C_3H_{10}INO$: 14518

$C_3H_{10}N_2$: 08119, 08120, 11816, 11818

$C_3H_{10}N_2O$: 11983

$C_3H_{10}O_7$: 00415

$C_3H_{12}ClNO_2$: 00569

$C_3H_{12}Cl_2N_2O$: 11984

$C_3H_{12}N_6O_3$: 07440

C_3N_2O: 08996

C_3O_2: 04740

$C_4As_2F_{12}$: 03331

$C_4Br_2N_2O_4S$: 13883

$C_4Br_3NO_2S$: 13908

C_4Br_4S: 13913

C_4Cl_2: 03669

$C_4Cl_3NO_2S$: 13907

C_4Cl_4S: 13914

C_4Cl_6: 03656

$C_4Cl_6O_3$: 14432

$C_4Cl_8F_2$: 03742

C_4F_6: 03657, 04211, 05456

$C_4F_6O_3$: 14494

C_4F_8: 03972, 04014, 05442

C_4F_{10}: 03715

$C_4F_{12}P_2$: 06325

C_4HBr_3S: 13917

$C_4HCl_2N_3O_2$: 12596

C_4HCl_3O: 07150

$C_4HCl_3O_2$: 07115

C_4HCl_3S: 13918

$C_4HCl_7O_2$: 06217

$C_4HF_7O_2$: 04333

C_4HI_4N: 12647

C_4H_2: 03668

C_4H_2BrClS: 13872

C_4H_2BrIO: 13873

$C_4H_2Br_2N_2O_3$: 01621

$C_4H_2Br_2O$: 07124

$C_4H_2Br_2S$: 13882

C_4H_2ClIS: 13880

$C_4H_2Cl_2O_2$: 07126

$C_4H_2Cl_2O_2$: 07114

$C_4H_2Cl_2O$: 08811

$C_4H_2Cl_2S$: 13884, 13885, 13886

$C_4H_2Cl_4$: 03666

$C_4H_2Cl_4O_2$: 13609, 13610

$C_4H_2Cl_4O_3$: 06061

$C_4H_2Cl_5$: 03665

$C_4H_2Cl_6$: 04012

$C_4H_2Cl_8$: 03783

$C_4H_2Cl_8O$: 04994

$C_4H_2INO_2S$: 13900

$C_4H_2I_2S$: 13887

$C_4H_2N_2$: 07113

$C_4H_2N_2O_4$: 00593

$C_4H_2N_2O_4S$: 13892

$C_4H_2N_2O_5$: 07130

$C_4H_2N_6O_4$: 09836

$C_4H_2O_2$: 08824

$C_4H_2O_3$: 08820

$C_4H_2O_4$: 04212

C_4H_3BrO: 07119, 07120

$C_4H_3BrO_4$: 07100, 08807

C_4H_3BrS: 13871, 13876

C_4H_3Cl: 04066

C_4H_3ClNS: 04941

C_4H_3ClO: 07122, 07123

$C_4H_3ClO_2S_2$: 13940, 13942

$C_4H_3ClO_4$: 07103

C_4H_3ClS: 13878

$C_4H_3Cl_2F_3O_2$: 12123

$C_4H_3Cl_3$: 03667

$C_4H_3Cl_5$: 04015

$C_4H_3Cl_7$: 03762

$C_4H_3F_7O$: 03876

C_4H_3IO: 07134, 07135

C_4H_3IS: 13899

$C_4H_3NO_2S$: 13905, 13906

$C_4H_3NO_3$: 07138

$C_4H_3N_3O_4$: 14695, 14914

$C_4H_3N_3O_5$: 01642

C_4H_4: 04065, 14928

$C_4H_4BrNO_2$: 13590

$C_4H_4Br_2$: 04204

$C_4H_4Br_2O_4$: 13544, 13545, 13546, 13547

$C_4H_4ClNO_2$: 13594

$C_4H_4Cl_2$: 03646, 03647, 03648, 04205

$C_4H_4Cl_2O_2$: 04452, 13608

$C_4H_4Cl_2O_3$: 04923, 06138

$C_4H_4Cl_2O_4$: 13551, 13552, 13553

$C_4H_4Cl_4$: 03976, 03977

$C_4H_4Cl_4O_2$: 14437

$C_4H_4Cl_6$: 03763, 03764

$C_4H_4INO_2$: 13601

$C_4H_4I_2$: 04206

C_4H_4NO: 91288

$C_4H_4NO_2S$: 13815

$C_4H_4N_2$: 12392, 12440, 12570, 13606

$C_4H_4N_2OS$: 13966, 13969, 14696

$C_4H_4N_2O_2$: 08817, 14686

$C_4H_4N_2O_2S$: 01644

$C_4H_4N_2O_3$: 01608, 14690

$C_4H_4N_2O_4$: 01639

$C_4H_4N_2O_5$: 00594

$C_4H_4N_2O_{10}$: 13664

$C_4H_4N_2S_2$: 06901

$C_4H_4N_4O_2$: 12585

C_4H_4O: 04201, 07116

C_4H_4OS: 13896

$C_4H_4O_2$: 04215, 06219

$C_4H_4O_3$: 13538

$C_4H_4O_4$: 07099, 07419, 08806

$C_4H_4O_6$: 08813

C_4H_4S: 13861

C_4H_5: 04195

C_4H_5Br: 03632, 03639

$C_4H_5BrN_2$: 12407

$C_4H_5BrO_2$: 04028

$C_4H_5BrO_4$: 13541

$C_4H_5Br_3O_2$: 14420

C_4H_5Cl: 03633, 03640, 03643, 04196, 04203, 04988

$C_4H_5ClN_2$: 12411

$C_4H_5ClN_2O$: 12597

C_4H_5ClO: 03938, 05391, 09013

$C_4H_5ClO_2$: 00304, 00494, 04018, 04020, 04030, 04033, 04036, 04972

$C_4H_5ClO_3$: 08890, 10184

$C_4H_5Cl_2O$: 04450

$C_4H_5Cl_3$: 03978, 04016, 04115

$C_4H_5Cl_3O$: 04234

$C_4H_5Cl_3O_2$: 04387, 04389, 04390, 04391, 14438

$C_4H_5Cl_3O_3$: 14439

$C_4H_5Cl_5$: 03784, 03785, 03786, 03817, 11811

C_4H_5F: 03655

$C_4H_5F_3O_2$: 14497

C_4H_5I: 03658

C_4H_5N: 00614, 04047, 05389, 05838, 09012, 12623

C_4H_5NO: 00302, 04424, 08619

C_4H_5NOS: 06531

$C_4H_5NOS_2$: 13051

$C_4H_5NO_2$: 05426, 06783, 13588

$C_4H_5NO_3S_2$: 13939, 13941

$C_4H_5NO_3$: 08819

C_4H_5NS: 00625, 00628, 13813, 13814, 13868

$C_4H_5NS_2$: 13812

$C_4H_5N_3$: 12579, 12586

$C_4H_5N_3O$: 05881

$C_4H_5N_3O_2$: 12580, 12591, 14687

$C_4H_5N_3O_3$: 01615

C_4H_6O: 03637

$C_4H_6O_2$: 04366

C_4H_6: 03631, 03638, 04202, 05455, 05837

C_4H_6BrClO: 04446

$C_4H_6BrClO_2$: 03612

$C_4H_6BrFO_2$: 03619

C_4H_6BrN: 04422, 04434

$C_4H_6Br_2$: 03996

$C_4H_6Br_2O$: 04441, 04444, 06860

$C_4H_6Br_2O_2$: 04313, 04314, 06049, 12116

$C_4H_6Br_4$: 03787, 03788, 03789, 03790, 03791, 03792, 03818, 03819, 11794

C_4H_6ClN: 04423

C_4H_6ClNO: 04436, 04938

$C_4H_6Cl_2$: 03959, 03960, 03962, 03997, 03998, 03999, 04000, 04001, 04003, 04004, 04006,

C_5H_7N: 12640, 12641

C_5H_7NO: 07187, 10202, 10203, 10571, 12484

$C_5H_7NOS_2$: 13049, 13050

$C_5H_7NO_2$: 05425, 06782, 08678, 13602

$C_5H_7NO_3$: 12561, 13600

$C_5H_7NO_4S$: 12562

C_5H_7NS: 13808, 13809, 13904

$C_5H_7N_3$: 12476, 12477, 12478, 12479, 12480, 12481, 12583, 12584, 12588, 12589, 12590, 12601

$C_5H_7N_3O$: 05882

$C_5H_7O_4$: 09125

C_5H_8: 03636, 03661, 04199, 05806, 08575, 10298, 10301, 10307, 10308, 10618, 10628, 11721, 13372

$C_5H_8Br_2O_2$: 12115

$C_5H_8Br_4$: 10334

$C_5H_8Cl_2$: 03956, 03961, 04002, 04005, 10553

$C_5H_8Cl_2O$: 04448, 04449, 04462, 11733, 14878, 14879, 14880

$C_5H_8Cl_2O_2$: 00199, 06066, 06067, 12126

$C_5H_8Cl_2O_3$: 04718

$C_5H_8Cl_3$: 04116

$C_5H_8Cl_3NO$: 04901

$C_5H_8Cl_4$: 03712, 10335, 10413

$C_5H_8I_4$: 10336

$C_5H_8N_2$: 08327, 12412, 12413, 12414, 12415, 12416

$C_5H_8N_2O_2$: 08038, 14689

$C_5H_8N_2O_3$: 14720

$C_5H_8N_4O_{12}$: 10337

C_5H_8O: 00624, 03635, 03659, 03660, 03940, 03941, 04063, 04220, 04221, 05801, 05825, 05857, 06828, 06829, 06924, 08576, 10516, 10586, 10591, 12036, 12303, 12389, 14043

$C_5H_8O_2$: 00176, 00261, 00262, 00432, 00756, 04024, 04039, 04040, 05388, 05443, 05846, 07166, 07167, 07168, 07183, 07316, 08745, 09009, 10423, 10438, 10442, 10573, 12661, 14044, 14799, 14800,

$C_5H_8O_2$: 14845, 14846, 14847, 14848, 14936

$C_5H_8O_3$: 00295, 00349, 07224, 08746, 12681, 14860, 14862

$C_5H_8O_4$: 00228, 07318, 08653, 08867, 08871, 10185, 10424, 12622, 13572, 13573

$C_5H_8O_5$: 07331, 07332, 07333, 08882, 13568

$C_5H_8O_7$: 07340, 07341

C_5H_9Br: 03946, 03983, 03986, 10518, 10519, 10520, 10541, 10542, 10543, 10544, 10545

C_5H_9BrO: 04442, 07190

$C_5H_9BrO_2$: 03608, 04285, 04286, 04287, 04288, 04291, 04293, 12078, 12084, 14818

C_5H_9Cl: 03951, 03952, 03957, 03991, 03992, 03995, 05768, 10521, 10523, 10525, 10526, 10546, 10547, 10548, 10550, 10551

C_5H_9ClO: 04457, 04458, 07191, 10498, 10499, 10505, 10506, 11732, 14877

$C_5H_9ClO_2$: 00193, 00195, 04301, 04302, 04305, 04308, 04311, 04403, 04935, 04936, 04959, 04960, 11726, 12097, 12107, 14826, 14828, 14830

$C_5H_9ClO_3$: 04932, 04966

$C_5H_9Cl_2NO$: 04890, 04891

$C_5H_9Cl_3$: 03711, 03779, 03780

$C_5H_9FO_2$: 14840

C_5H_9I: 05775

C_5H_9N: 03775, 04132, 04426, 04427, 11730, 12204, 14870

C_5H_9NO: 04130, 04131, 10446, 11702, 12206, 12207, 12674

$C_5H_9NO_2$: 00061, 05778, 10440, 10441, 11703, 11704, 11759, 11760, 11761

$C_5H_9NO_3$: 11762, 11763, 11764, 11765, 11766, 11767

$C_5H_9NO_4$: 00561, 07308, 07309, 07310

$C_5H_9NO_5$: 07313, 07314

C_5H_9NS: 04133, 04134, 04135, 04136, 04137, 04138, 04179, 04180, 04181, 13826

$C_5H_9N_3$: 08008

C_5H_{10}: 03970, 03971, 04013, 05441, 05762, 05827, 05828, 05829, 05830, 05831, 10517, 10539, 10540

$C_5H_{10}BrF$: 10346

$C_5H_{10}BrNO$: 04243

$C_5H_{10}Br_2$: 10373, 11820, 11824

$C_5H_{10}ClF$: 10355

$C_5H_{10}ClNO$: 04688, 04916, 04917, 11705

$C_5H_{10}ClNO_4$: 07311

$C_5H_{10}ClNS$: 13845

$C_5H_{10}ClO$: 03921

$C_5H_{10}Cl_2$: 03724, 03728, 03730, 10374, 10376, 10377, 10378, 10380, 10381, 10383

$C_5H_{10}F_3NO_2S$: 09118

$C_5H_{10}I_2$: 10386

$C_5H_{10}N_2$: 05411, 12205

$C_5H_{10}N_2O$: 11613, 11672

$C_5H_{10}N_2O_2$: 07317, 10439, 10444

$C_5H_{10}N_2O_3$: 07315, 08032

$C_5H_{10}N_2O_4$: 00667, 04665, 04681, 10395

$C_5H_{10}N_2S_2$: 04743

$C_5H_{10}O$: 00057, 00613, 03930, 04113, 04231, 05798, 07145, 07146, 09192, 10339, 10495, 10504, 10576, 10577, 10578, 10579, 10580, 10581, 10583, 10584, 11722, 12312, 12331, 12390, 14790

$C_5H_{10}OS_2$: 04733

$C_5H_{10}O_2$: 00263, 04351, 04352, 04353, 04354, 04359, 04412, 06211, 06534, 06790, 06904, 07075, 07076, 07077, 07186, 08612, 10454, 10489, 10490, 10492, 10493, 10494, 11725, 11727, 12132, 12143, 14794, 14795

$C_5H_{10}O_2S$: 08980

$C_5H_{10}O_3$: 00223, 04341, 04346, 04347, 04411, 04727, 06210, 06221, 06743, 08215, 08658, 08659, 08660, 09133,

$C_5H_{10}O_3$: 14844

$C_5H_{10}O_4$: 07350, 07351, 07355

$C_5H_{10}O_5$: 01356, 01357, 01358, 01359, 01360, 13056, 15020

$C_5H_{10}O_6$: 01367, 01368, 01369

$C_5H_{10}S$: 10338, 13800, 13856, 13857, 13962

$C_5H_{11}Br$: 03689, 03690, 03691, 03696, 09802, 10345, 10350, 10352, 11785

$C_5H_{11}Cl$: 03702, 03703, 03704, 03709, 03710, 09803, 10354, 10356, 10363, 11797

$C_5H_{11}ClO$: 03894, 03898, 04954, 10457, 10467, 10477

$C_5H_{11}Cl_2N_3$: 08009

$C_5H_{11}Cl_3Si$: 13307

$C_5H_{11}F$: 10402

$C_5H_{11}FO$: 10460

$C_5H_{11}I$: 03768, 03769, 03770, 03773, 03774, 09805, 10403, 10404, 10405, 11869

$C_5H_{11}N$: 05765, 11617, 12669

$C_5H_{11}NO$: 03931, 04250, 07059, 07188, 10503, 11723, 14798, 14801

$C_5H_{11}NO_2$: 03321, 04653, 04656, 04659, 04663, 08614, 10410, 10610, 10611, 14811, 14812, 14813, 14814, 14816, 14886, 14887, 14888

$C_5H_{11}NO_2S$: 09130, 09131

$C_5H_{11}NO_3$: 10609, 14889

$C_5H_{11}NO_4$: 11921

$C_5H_{11}O$: 03905, 04232

C_5H_{12}: 03777, 04008, 09797, 10340, 10422, 11847

$C_5H_{12}ClNO_2$: 00565

$C_5H_{12}Cl_3N$: 04947

$C_5H_{12}N$: 00607

$C_5H_{12}NO$: 04233

$C_5H_{12}NO_2$: 08615, 08616

$C_5H_{12}N_2$: 11609, 11610

$C_5H_{12}N_2O$: 14714, 14715, 14716, 14717, 14725, 14726, 14765

$C_5H_{12}N_2O_2$: 10148

$C_5H_{12}N_2S$: 13984, 14011

$C_5H_{12}N_4O_3$: 04543

$C_5H_{12}O$: 03878, 03879,

C₅H₁₂O: 03880, 03909, 03910, 04145, 04146, 04147, 04148, 06825, 06826, 09798, 10455, 10466, 10474, 11957

C₅H₁₂O₂: 03837, 03845, 03846, 03847, 03877, 04077, 06914, 06915, 07051, 09804, 10425, 10429, 10431, 10434, 11918, 11959, 11991, 11993

C₅H₁₂O₃: 06125, 07363, 10152, 10451, 10452, 10453, 11923, 14522

C₅H₁₂O₄: 10332

C₅H₁₂O₅: 00547, 01364, 01365, 01366, 15019

C₅H₁₂S: 03852, 03853, 03857, 04149, 04150, 06827, 10447, 10448, 10449

C₅H₁₂Si: 13332

C₅H₁₃BO₂: 03558, 10601

C₅H₁₃ClN₂O₂: 10149

C₅H₁₃N: 03679, 03680, 06100, 06823, 06824, 09801, 10341, 10342, 10343, 10599, 10600, 11774, 12248

C₅H₁₃NO: 06645, 06739, 08220, 09811, 10456

C₅H₁₃NO₂: 06150, 08222, 11906, 11911

C₅H₁₃N₂O₂: 14815

C₅H₁₃O: 03882, 03903

C₅H₁₃O₃P: 09110

C₅H₁₄N: 03674

C₅H₁₄N₂: 09016, 10372

C₅H₁₄N₂O₆S: 10150

C₅H₁₄O₂Si: 13267, 13286

C₅H₁₄O₃Si: 13291

C₅H₁₄Si: 13278

C₅H₁₅NO₂: 05034

C₅H₁₅NSi: 13279

C₅H₂₀O₅Si₅: 05761

C₆AsF₁₅: 01420

C₆Br₆: 02160

C₆ClF₅: 02049

C₆Cl₄O₂: 03065

C₆Cl₅NO₂: 02196

C₆Cl₆: 02161

C₆Cl₁₀: 07734

C₆F₆: 02163

C₆F₈: 05485, 05487

C₆F₁₀: 05691

C₆F₁₂: 07952, 07968, 10536

C₆F₁₃N: 05518

C₆F₁₄: 07824

C₆F₁₅N: 14486

C₆HBr₅: 02222

C₆HBr₅O: 10993

C₆HCl₃O₂: 03068

C₆HCl₄NO₂: 02197

C₆HCl₅O: 10994

C₆HF₅O: 10995

C₆HI₅: 02225

C₆H₂Br₂ClNO: 03032, 03033

C₆H₂Br₃NO₂: 02201, 02202, 02203, 02204, 02205, 02206

C₆H₂Br₄: 02239, 02240

C₆H₂Br₄O₂: 04821, 08201

C₆H₂Br₅N: 00961

C₆H₂ClN₃O₆: 02055, 02056

C₆H₂ClO: 05585

C₆H₂Cl₂O₂: 03019, 03038, 03039, 03041

C₆H₂Cl₂O₄: 03040

C₆H₂Cl₃NO: 03042

C₆H₂Cl₃NO₂: 02207, 02208, 02209, 02210, 02211

C₆H₂Cl₄: 02241, 02242, 02243

C₆H₂Cl₄O: 11013, 11014, 11015

C₆H₂Cl₄O₂: 04822, 08202, 13025

C₆H₂Cl₅N: 00962

C₆H₂F₄: 02246, 02247

C₆H₂F₅N: 00963

C₆H₂I₃NO₂: 02212, 02213, 02214, 02215

C₆H₂I₄: 02250, 02251, 02252

C₆H₂I₄O₂: 08203

C₆H₂N₂O₂: 05673

C₆H₂N₂O₈: 09834

C₆H₂N₄O₅: 03116

C₆H₂N₄O₉: 11019

C₆H₂O₆: 13054

C₆H₃BrCl₂: 01928, 01929, 01930, 01931, 01932

C₆H₃BrN₂O₄: 01939, 01940, 01941, 01942, 01943

C₆H₃BrN₂O₅: 10821, 10822, 10828, 10831, 10833

C₆H₃Br₂NO₂: 02070, 02071, 02073, 02074, 02077, 02079

C₆H₃Br₂NO₃: 10867, 10869

C₆H₃Br₃: 02261, 02262, 02263

C₆H₃Br₃O: 11021, 11022, 11023, 11024

C₆H₃Br₃O₂: 08205, 13026

C₆H₃ClNO₂: 02013, 02016

C₆H₃ClN₂O₄: 02014, 02015, 02017

C₆H₃ClN₂O₄S: 02345

C₆H₃ClN₂O₅: 10846, 10855, 10857

C₆H₃ClO₂: 03017, 03018, 03026

C₆H₃Cl₂I: 02090

C₆H₃Cl₂NO₂: 02084, 02085, 02087, 02088, 02091, 02093

C₆H₃Cl₂NO₃: 10891

C₆H₃Cl₂N₃: 02094

C₆H₃Cl₃: 02265, 02266, 02267

C₆H₃Cl₃O: 11026, 11027, 11028, 11029, 11030, 11031

C₆H₃Cl₃O₂: 04823, 13027

C₆H₃Cl₃O₂S: 02439, 02440

C₆H₃Cl₃O₃: 02268, 02269, 02270, 08295, 11355, 12609

C₆H₃Cl₄N: 00980, 00981, 12475

C₆H₃FN₂O₄: 02128

C₆H₃F₂NO₂: 02100

C₆H₃I₃: 02280, 02281, 02282

C₆H₃I₃O: 11032, 11033

C₆H₃N₃O₆: 02291, 02292, 02293, 14409

C₆H₃N₃O₆S: 13958

C₆H₃N₃O₇: 11037, 11038, 11039, 11559

C₆H₃N₃O₈: 13031

C₆H₄BrCl: 01924, 01925, 01926

C₆H₄BrClO: 10820, 10827

C₆H₄BrClO₂S: 02434, 02435

C₆H₄BrCl₂N: 00771, 00780, 00781, 00782

C₆H₄BrF: 01952

C₆H₄BrI: 01954, 01955, 01956

C₆H₄BrNO: 01964, 01965

C₆H₄BrNO₂: 01961, 01962, 01963

C₆H₄BrNO₃: 10815, 10816, 10817, 10818, 10823, 10824, 10825, 10829, 10830, 10832

C₆H₄BrN₂O₂: 00776

C₆H₄BrN₃: 01974

C₆H₄BrN₃O₄: 00772, 00773, 00774, 00783, 00784, 00787, 00789

C₆H₄Br₂: 02068, 02072,

C₆H₄Br₂: 02075

C₆H₄Br₂N₂O₂: 00826, 00829

C₆H₄Br₂O: 10866, 10868, 10870

C₆H₄Br₃N: 00984, 00985, 00987

C₆H₄ClF: 02029, 02030, 02031

C₆H₄ClFO₂S: 02444

C₆H₄ClI: 02034, 02035, 02036

C₆H₄ClNO: 02046, 02047, 02048, 03061

C₆H₄ClNO₂: 02043, 02044, 02045, 03027

C₆H₄ClNO₂S: 02346, 02347

C₆H₄ClNO₃: 10847, 10848, 10849, 10851, 10856, 10858

C₆H₄ClNO₄S: 02447, 02448, 02449

C₆H₄ClN₃: 02053

C₆H₄ClN₃O₄: 00808, 00812

C₆H₄Cl₂: 02082, 02086, 02089, 07932

C₆H₄Cl₂N₂: 03047

C₆H₄Cl₂N₂O₂: 00840

C₆H₄Cl₂O: 10879, 10882, 10884, 10886, 10889, 10890

C₆H₄Cl₂O₂: 04798, 04799, 08180, 08181, 08182, 12994

C₆H₄Cl₂O₂S: 02438

C₆H₄Cl₃N: 00988, 00989, 00990, 12530

C₆H₄Cl₇: 05534

C₆H₄FI: 02154, 02155

C₆H₄FNO: 02156, 02157, 02158

C₆H₄FNO₃: 10954

C₆H₄F₂: 02102, 02107

C₆H₄INO₂: 02173, 02174, 02175

C₆H₄IN₃: 02176

C₆H₄I₂: 02109, 02110, 02111

C₆H₄I₂N₂O₂: 00859, 00862, 00863, 00866, 00867

C₆H₄I₂O: 10893

C₆H₄I₃N: 00991, 00992, 00993, 00994

C₆H₄N₂: 08532, 09829, 11558

C₆H₄N₂O₄: 02124, 02125, 02126, 12399, 12401, 12402, 13001

C₆H₄N₂O₅: 10937, 10938, 10939, 10940, 10941,

C₇H₁₅NO₂: → $C_7H_{15}NO_2$: 04657, 04662, 07584, 07669, 09235

$C_7H_{15}N_2O_3$: 08741

$C_7H_{15}O_4$: 07037

C_7H_{16}: 03807, 07522, 07802, 07803, 07818, 07819, 07820, 07821, 10387, 10389, 10391, 10394, 10399, 14575

$C_7H_{16}BrNO_2$: 05035

$C_7H_{16}ClNO_2$: 05036

$C_7H_{16}ClO$: 04953

$C_7H_{16}N_2$: 11612

$C_7H_{16}N_2O$: 09228, 13993

$C_7H_{16}N_2O_3$: 08740

$C_7H_{16}N_2S$: 13991, 13992

$C_7H_{16}O$: 03917, 04164, 04165, 04166, 06816, 06817, 06818, 07602, 07608, 07609, 07610, 07621, 07873, 07874, 07875, 07876, 07877, 07878, 07879, 07886, 07887, 07888, 07900, 07901, 07902, 07903, 10458, 10459, 10468, 10478, 10479, 10480, 10482

$C_7H_{16}O_2$: 00313, 07054, 07576, 10427, 10433, 11917, 11925, 11978, 11050

$C_7H_{16}O_2Si$: 13268

$C_7H_{16}O_3$: 00045, 07580, 10156

$C_7H_{16}O_4$: 06124, 07348, 08873

$C_7H_{16}O_4S_2$: 11840

$C_7H_{16}O_8$: 08927

$C_7H_{16}S$: 07579

$C_7H_{16}S_3$: 10162

$C_7H_{17}N$: 06171, 07523, 07524, 07666, 07764

$C_7H_{17}NO$: 06788, 11953, 11954, 11990

$C_7H_{17}NO_2$: 06328, 11905

$C_7H_{17}O_3P$: 11402

$C_7H_{18}N_2$: 07544, 11779, 12247

$C_7H_{18}N_2O$: 11974, 11976

$C_7H_{18}O_3Si$: 13300

$C_7H_{18}Si$: 13236, 13237

$C_7H_{20}O_4$: 05855

$C_7H_{22}BrNO_4$: 08303

$C_8Br_4O_3$: 11501

$C_8Br_2S_2$: 03550

$C_8Cl_4O_3$: 11502

C_8F_{16}: 05522

$C_8HF_{15}O_2$: 04643

C_8HN: 05212

$C_8H_2Br_4O_4$: 08568, 11490, 13694

$C_8H_2Cl_2O_3$: 11496, 11497, 11498, 11499

$C_8H_2Cl_4O_4$: 11491

$C_8H_3ClO_3$: 11494, 11495

$C_8H_3NO_5$: 11500

C_8H_4ClO: 08459

$C_8H_4Cl_2N_2$: 12953

$C_8H_4Cl_2O_2$: 08571, 11526, 13696

$C_8H_4Cl_2O_4$: 08555, 11464, 11465, 11466, 11467, 13687

$C_8H_4F_2O_2$: 11527

$C_8H_4F_3N$: 02878

$C_8H_4N_2$: 08569, 11523, 13695

$C_8H_4N_2O_2$: 11524

$C_8H_4N_2O_4$: 08463, 11520

$C_8H_4N_4O_6$: 12697

$C_8H_4O_2S$: 03114

$C_8H_4O_3$: 11493

$C_8H_5BrO_4$: 08552, 11454, 11455, 13685

$C_8H_5Br_3O_3$: 11086

$C_8H_5ClN_2$: 12952

$C_8H_5ClN_2O_5$: 11195

$C_8H_5ClN_6$: 01916

$C_8H_5ClO_3$: 11720

$C_8H_5ClO_4$: 08553, 11458, 11459, 11460, 13686

$C_8H_5Cl_3O$: 00424

$C_8H_5Cl_3O_3$: 11087, 11088

$C_8H_5F_3O$: 00425

C_8H_5I: 02726

C_8H_5NO: 02867, 02868, 11270

$C_8H_5NO_2$: 02625, 08457, 11481, 11504, 13692

$C_8H_5NO_3$: 08466

$C_8H_5NO_4$: 11513, 11531

$C_8H_5NO_5$: 11269

$C_8H_5NO_6$: 08565, 11483, 11485, 12569, 12570, 12571, 12572, 12573, 13693

C_8H_6: 05744, 06923, 11196

C_8H_6BrClO: 00362, 00363, 00364

C_8H_6BrN: 11173, 11174, 11175

C_8H_6BrNO: 01814

$C_8H_6Br_2O$: 00374, 00375

$C_8H_6Br_2O_2$: 03034

$C_8H_6Br_4$: 01906, 01907, 01908, 14985

C_8H_6ClN: 02858, 02859, 11177

$C_8H_6ClNO_4$: 02621, 02623,

$C_8H_6ClNO_4$: 09846, 09864

C_8H_6ClNS: 03089

$C_8H_6ClN_3O_5$: 04904

$C_8H_6Cl_2$: 13509

$C_8H_6Cl_2O$: 00376, 00377, 00378, 00379

$C_8H_6Cl_2O_3$: 11071

$C_8H_6Cl_3NO$: 04899, 04900, 14430

$C_8H_6F_3NO$: 14492

$C_8H_6NO_4$: 00201

$C_8H_6N_2$: 11446, 12692, 12950

$C_8H_6N_2O$: 08324, 11447, 11482, 12693, 12957, 12962

$C_8H_6N_2O_2$: 02875, 08464, 08465, 11189, 11190, 11191, 11192, 11448, 12698, 12954, 14271, 14272, 14273, 14285, 14286, 14287, 14288, 14297, 14298

$C_8H_6N_2O_2S$: 03104

$C_8H_6N_2O_6$: 00202, 02676, 06197, 11126, 11128

$C_8H_6N_2O_7$: 11072, 11073

$C_8H_6N_4O_8$: 00393

C_8H_6O: 02520, 11265

C_8H_6OS: 03111, 03112, 08636

$C_8H_6O_2$: 05264, 07427, 08542, 11267, 11441, 11528, 13678

$C_8H_6O_2S$: 08638

$C_8H_6O_3$: 02701, 02702, 02703, 07431, 11714

$C_8H_6O_4$: 08243, 08277, 08545, 11449, 11717, 13127, 13128, 13681

$C_8H_6O_5$: 08560, 08561, 08562, 11478, 11479

$C_8H_6O_6$: 13689

C_8H_6S: 03110

$C_8H_6S_2$: 03549, 03551

C_8H_7: 02134

C_8H_7Br: 13499, 13500, 13501, 13502, 13503

C_8H_7BrO: 00358, 00359, 00361, 10646

$C_8H_7BrO_2$: 02567, 02576, 02583, 03607, 11105, 11107, 11108, 11109

$C_8H_7BrO_3$: 01678, 11062, 11063, 11136, 14892

$C_8H_7Br_3O$: 10929, 10935

C_8H_7Cl: 05750, 13504, 13505, 13506, 13507, 13508

$C_8H_7ClN_2O_3$: 00104,

$C_8H_7ClN_2O_3$: 00105, 00111, 00115, 00116

C_8H_7ClO: 00368, 00369, 00370, 00371, 10647, 11193, 14274, 14289, 14299

$C_8H_7ClO_2$: 01093, 01094, 01095, 02596, 02597, 02598, 02599, 02605, 02606, 02607, 02608, 02615, 02616, 02617, 02622, 03147, 03148, 03149, 04934, 04958, 11090, 11114, 11115, 11116, 11117, 11118, 11119, 13165, 14262, 14263, 14264, 14279, 14280, 14281, 14292, 14293

$C_8H_7ClO_2S$: 13525

$C_8H_7ClO_3$: 01693, 09136, 11067, 11068, 11069, 11137, 13107, 13111, 14893

$C_8H_7Cl_2N_2O_3$: 04888

$C_8H_7Cl_3O$: 10902, 10909, 10917, 10930, 10936

C_8H_7F: 13513, 13514, 13515

C_8H_7FO: 00391

C_8H_7IO: 00409, 00410, 00411, 00412

$C_8H_7IO_2$: 02724, 02727, 02729

$C_8H_7IO_3$: 01760, 01761, 08917, 14894, 14895

C_8H_7N: 02873, 02874, 02876, 02877, 03262, 08411, 11169, 14270, 14284, 14296

C_8H_7NO: 01091, 01092, 02509, 02872, 03128, 08417, 10212, 11089, 11178, 11179, 11180, 11181, 11182, 11183, 14378, 14379, 14380

C_8H_7NOS: 09147

$C_8H_7NO_2$: 07429, 10214, 13520, 13521, 13522

$C_8H_7NO_3$: 00419, 00420, 00421, 02700, 08456, 10193, 11442, 11480, 11715, 13691

$C_8H_7NO_4$: 00243, 00244, 00245, 01775, 08546, 08550, 09850, 09872, 09885, 11165, 11166, 11167, 11450, 12550, 13682

$C_8H_7NO_5$: 11084, 11148

C_8H_7NS: 03103, 03263, 03288, 14386

$C_8H_7NS_2$: 03098, 03099, 03100, 03101, 03105, 03109

$C_8H_7N_2$: 05408

$C_8H_7N_3$: 12951

$C_8H_7N_3O$: 00590

$C_8H_7N_3O_2$: 08780

$C_8H_7N_3O_5$: 00121, 00122, 00123, 00124, 00125, 00126

$C_8H_7N_3O_6$: 15002, 15012

C_8H_8: 05749, 07097, 13495

C_8H_8BrNO: 00095, 00096, 00097, 00098

$C_8H_8Br_2$: 01894, 01895, 01896, 02069

C_8H_8ClNO: 00101, 00102, 00106, 00112, 00355, 00356, 04690

$C_8H_8ClNO_3S$: 02432, 02433, 14233

C_8H_8ClS: 02026

$C_8H_8Cl_2$: 01900, 01901, 01902, 02083, 02096

$C_8H_8Cl_2O$: 02092, 10750, 10880, 10881, 10883, 10885, 10887, 10888

C_8H_8INO: 00142

$C_8H_8NO_2$: 01178

$C_8H_8N_2$: 00680, 00704, 00706, 00707, 00736, 00737, 01190, 01191, 01192, 02498, 02499, 02501, 02502, 08386, 08387, 08388, 08389, 11170, 11171, 11172

$C_8H_8N_2OS$: 13977

$C_8H_8N_2O_2$: 08543, 11268, 11444, 13058, 13679, 14706

$C_8H_8N_2O_3$: 00168, 00169, 00170

$C_8H_8N_2O_4$: 00669, 14974, 14975, 14976, 14977, 14989, 15005, 15006, 15007

$C_8H_8N_2O_5$: 10751, 10752, 10753, 10754

$C_8H_8N_2O_6$: 02112

$C_8H_8N_2S$: 03082, 03083, 03084, 03108

$C_8H_8N_3O_2$: 00330

$C_8H_8N_4O_4$: 00022, 00039

C_8H_8O: 00047, 00351, 01780, 01781, 01782, 05263, 11040, 11041, 11092, 11347, 13523, 14054, 14055, 14056

$C_8H_8O_2$: 00259, 00396, 00398, 00399, 00400, 00998, 00999, 01000, 01774, 01776, 01777, 02740, 02743, 03050, 03051, 03052, 05481, 07074, 07171, 08416, 10645, 10768, 10769, 10770, 11102, 14260, 14275, 14290

$C_8H_8O_2S$: 11155

$C_8H_8O_3$: 00380, 00381, 00382, 00383, 00384, 00508, 01003, 01006, 01009, 01714, 01743, 01744, 01745, 01751, 01752, 01762, 01763, 02713, 02722, 04789, 04790, 04809, 05702, 05703, 05704, 05705, 05706, 07198, 07428, 08170, 08249, 08250, 08251, 08252, 08283, 08284, 08285, 08911, 08912, 08913, 09135, 09139, 09142, 11060, 11131, 11132, 11133, 11147, 11151, 11152, 11716, 12975, 12976, 12977, 12980, 13139, 13141, 13142, 13143, 13144, 14890, 14896

$C_8H_8O_4$: 02647, 03049, 03056, 05488, 05489, 05490, 05947, 08025, 08247, 08280, 11075, 11077, 11120, 11121, 11351, 12605, 14265

$C_8H_8O_5$: 02811, 02815, 02818, 03044, 07159, 07161, 07163

$C_8H_8O_7$: 13649

C_8H_9: 05758

C_8H_9Br: 01933, 01934, 01935, 01936, 01937, 01938, 01947, 01948, 01949, 01950, 01951, 01957, 01958, 01959

C_8H_9BrO: 01944, 01945, 01946, 01953, 10744, 10745, 10746

$C_8H_9Br_2NO$: 00833

C_8H_9Cl: 02008, 02009, 02010, 02011, 02012, 02022, 02023, 02024, 02025, 02027, 02028, 02039, 02040, 02041, 14973, 14987, 14988, 15004

$C_8H_9ClN_2O_2$: 00881

C_8H_9ClO: 02018, 02019, 02020, 02021, 03249, 03254, 06712, 10747, 10748, 10749, 10845, 10854, 10899, 10900, 10906, 10913, 10926, 10927, 10932, 10933

$C_8H_9ClO_2S$: 02441, 02442, 02443

$C_8H_9Cl_2NO$: 00839, 00843

$C_8H_9Cl_2P$: 11364, 11365, 11366

C_8H_9FO: 10756, 10757, 10758

C_8H_9I: 02142, 02143, 02171, 02172, 14979, 14980, 14992, 14994, 15009

C_8H_9IO: 02170, 10759, 10760

$C_8H_9IO_2$: 02113

C_8H_9N: 00802, 01769, 03312, 08428, 12446, 13496, 13497, 13498

C_8H_9NO: 00083, 00084, 00090, 00352, 00353, 00354, 00422, 01834, 07066, 07067, 07068, 11093, 14057, 14058, 14059

C_8H_9NOS: 08971

$C_8H_9NO_2$: 00082, 00139, 00140, 00141, 00670, 00693, 00694, 00696, 00697, 00726, 00727, 00729, 00730, 01001, 01002, 01175, 01179, 01180, 01181, 01779, 01832, 01833, 02147, 02148, 02149, 02188, 02744, 02747, 02749, 04650, 08527, 09820, 09821, 11059, 11095, 11098, 11104, 11555, 12547, 14276, 14277, 14983, 14984, 14998, 14999, 15000, 15011

$C_8H_9NO_3$: 00663, 00692, 00725, 10763, 10764, 10765, 10901, 10908, 10914, 10915, 10916, 10920, 10921, 10928, 10934

$C_8H_9NO_3S$: 02349, 02350

$C_8H_9NO_4$: 00959, 02114, 02115, 02117, 02118, 02119, 02121

C_8H_9NS: 13839

$C_8H_9N_2$: 07443

$C_8H_9N_3O$: 01797

$C_8H_9N_3O_3$: 14705

$C_8H_9N_4O_2S_2$: 13628

$C_8H_9O_2$: 05480

$C_8H_9S_2$: 14111

C_8H_{10}: 02140, 06764, 07750, 10022, 10023, 10024, 10100, 14972, 14986, 15003

$C_8H_{10}BrN$: 00876, 00877, 00878

$C_8H_{10}ClN$: 00879, 00880, 00915

$C_8H_{10}ClNO$: 00816

$C_8H_{10}ClNO_3$: 12575

$C_8H_{10}Cl_3NO$: 04874

$C_8H_{10}N$: 00606, 00919

$C_8H_{10}N_2$: 00050, 02312, 12961

$C_8H_{10}N_2O$: 00092, 00093, 00094, 00643, 00886, 00918, 08060, 11222, 14707, 14751, 14767, 14768, 14769

$C_8H_{10}N_2O_2$: 00883, 00884, 00885, 00888, 00890, 00892, 00893, 00894, 00897, 00898, 00900, 00901, 00903, 00904, 00905, 11150, 14748

$C_8H_{10}N_2O_3$: 10733, 10734, 10735, 10737, 10738, 10739, 10741, 10742, 12576

$C_8H_{10}N_2O_3S$: 02361, 02362, 13623, 14239

$C_8H_{10}N_2O_4$: 09215

$C_8H_{10}N_2S$: 13978, 14005, 14013, 14014, 14015

$C_8H_{10}N_3O_4$: 00162

$C_8H_{10}N_4O_2$: 04467

$C_8H_{10}N_4O_3$: 14778

$C_8H_{10}N_6O_7$: 09242

$C_8H_{10}O$: 02139, 03185, 03186, 03187, 03190, 03265, 10743, 10897, 10903, 10910, 10918, 10923, 10931, 10944, 10947, 10950, 11250, 11251, 11252, 11253, 11256, 14143, 14144, 14145

$C_8H_{10}OS$: 10761, 10762, 13952, 13953

$C_8H_{10}O_2$: 01913, 01914, 01915, 03183, 03184, 03348, 04801, 04802, 04803, 05371, 06686, 06906, 06913, 08184, 08185, 08186, 08187, 08189, 08570, 10956,

C₁₀H₈Br₄: 13765

C₁₀H₈ClN: 09700, 09701, 09734, 12765, 12766, 12767, 12769, 12770, 12771, 12773, 12777, 12778, 12781, 12783, 12784

C₁₀H₈ClNO₂: 13596

C₁₀H₈Cl₄: 13766

C₁₀H₈FN: 09709

C₁₀H₈IN: 09743

C₁₀H₈N₂: 01903, 01904, 01905, 03530, 03531, 03532, 03533, 03534

C₁₀H₈N₂O: 09751, 12919

C₁₀H₈N₂OS: 08037

C₁₀H₈N₂OS₂: 13047

C₁₀H₈N₂O₂: 09408, 09409, 09414, 09415, 09628, 09660, 09716, 09717, 09718, 09719, 09720, 09721, 09747, 09748, 09749, 09750, 12879, 12880, 12881, 12884, 12885, 12888, 12889, 12892, 12893

C₁₀H₈N₂O₂S: 08036, 08043

C₁₀H₈N₂O₃: 12439, 12867, 12868, 12872, 12873, 12876

C₁₀H₈N₄O₃: 11560

C₁₀H₈O: 09571, 09588

C₁₀H₈OS: 07246

C₁₀H₈O₂: 02521, 05288, 05289, 05290, 05291, 05292, 05293, 07244, 08401, 08402, 09345, 09346, 09348, 09350, 09351, 09352, 09353, 09354, 09355

C₁₀H₈O₂S: 09464, 09465

C₁₀H₈O₃: 00488, 05284, 05286, 05287, 05298, 13582, 13583

C₁₀H₈O₃S: 09466, 09476

C₁₀H₈O₄: 00754, 05099, 05100, 05101, 05270, 05273, 05283, 07205, 07240, 08841, 09427

C₁₀H₈O₄S: 09470, 09471, 09472, 09473, 09479, 09480, 09481

C₁₀H₈O₅: 05274

C₁₀H₈O₆S₂: 09456, 09457, 09458

C₁₀H₈O₇S₂: 09455, 09461

C₁₀H₈O₉S₂: 09460

C₁₀H₈S: 09387, 09388

C₁₀H₈S₂: 09356

C₁₀H₉: 01886

C₁₀H₉BrO: 04058

C₁₀H₉ClN: 09195

C₁₀H₉ClO₂: 05106

C₁₀H₉ClO₃: 11194

C₁₀H₉N: 08593, 08594, 08595, 08596, 08597, 08598, 09688, 09722, 12642, 12643, 12878, 12882, 12883, 12886, 12887, 12890, 12891

C₁₀H₉NO: 00303, 06433, 08412, 08592, 09593, 09594, 09595, 09596, 09597, 09598, 12806, 12836, 12837, 12838, 12840, 12842, 12845, 12846, 12847, 12849, 12850, 12851, 12853, 12855, 12856, 12857, 12858, 12859, 12869, 12870, 12871, 12875, 12948

C₁₀H₉NO₂: 11514, 13605

C₁₀H₉NO₃: 11515

C₁₀H₉NO₃S: 09467, 09469, 09477

C₁₀H₉NO₄: 05133, 05135, 05136, 05137, 05141, 05143, 05148

C₁₀H₉NO₄S: 09468

C₁₀H₉NO₅: 05134

C₁₀H₉NO₆: 08364, 08367, 11486

C₁₀H₉NO₆S₂: 09454

C₁₀H₉NO₇S₂: 09459

C₁₀H₉NO₁₂: 12564

C₁₀H₉N₃O₃: 12426

C₁₀H₉O₃P: 09462

C₁₀H₁₀: 02135, 02136, 03663, 03664, 04200, 08399, 08400, 09340, 09343

C₁₀H₁₀BrNO₂: 00275

C₁₀H₁₀BrNO₃: 00651

C₁₀H₁₀Br₂O: 03923

C₁₀H₁₀ClN: 09711, 09742

C₁₀H₁₀ClNO₂: 00276

C₁₀H₁₀ClNO₃: 00653

C₁₀H₁₀ClO₂: 05103

C₁₀H₁₀Fe: 06957

C₁₀H₁₀N₂: 08106, 08107, 09317, 09319, 09322, 09324, 09325, 09327, 09328, 09830, 09831, 09832, 12728, 12730, 12732, 12736, 12737, 12739, 12740, 12743, 12744, 12747, 12748, 12955, 12956

C₁₀H₁₀N₂O: 12422, 12423, 12427, 12746

C₁₀H₁₀N₂O₄S: 12428

C₁₀H₁₀N₂O₆: 02679, 02680, 06201, 06202, 11127

C₁₀H₁₀N₂O₆S₂: 01576

C₁₀H₁₀N₄O₄S: 13614, 13625

C₁₀H₁₀O: 04064, 05074, 05075, 07194, 13773, 13775

C₁₀H₁₀O₂: 00489, 00601, 01658, 02060, 02542, 03133, 04046, 04059, 04060, 04061, 05130, 05131, 05132, 08608, 09676, 13082

C₁₀H₁₀O₂S₂: 06130

C₁₀H₁₀O₃: 00257, 00397, 02540, 03132, 03159, 04045, 05073, 05129, 10644, 12075, 13090

C₁₀H₁₀O₄: 04794, 05127, 08174, 08558, 08559, 08838, 08943, 09001, 11134, 11135, 11473, 11477, 11718, 12991, 12992, 13578, 13579, 13580, 13690, 14891

C₁₀H₁₀O₅: 08840, 11352

C₁₀H₁₀O₆: 01898, 08556, 11471

C₁₀H₁₁Br: 03947, 03984, 03987

C₁₀H₁₁BrO₂: 11106, 12090

C₁₀H₁₁ClN₂O₂: 02002

C₁₀H₁₁ClO: 00373, 03152, 10648

C₁₀H₁₁ClO₂S: 13770, 13772

C₁₀H₁₁Cl₂NO: 04889

C₁₀H₁₁N: 04431, 04432, 08414, 08415, 09705

C₁₀H₁₁NO: 04429, 04430, 10213

C₁₀H₁₁NO₂: 00062, 00091, 00271, 06129, 08429

C₁₀H₁₁NO₃: 00558, 00559, 00560, 00646, 05085, 10194, 13534

C₁₀H₁₁NO₄: 08547, 08549, 08551, 11451, 13683

C₁₀H₁₁NO₅: 11149

C₁₀H₁₁N₃O₂S₂: 13619

C₁₀H₁₁O₂: 03036, 03037, 08504, 08505

C₁₀H₁₂: 02184, 02189, 03973, 03974, 03975, 05834, 06082, 08350, 08351, 08352, 08353, 13510, 13511, 13512, 13752, 14061, 14153, 14160

C₁₀H₁₂BrNO: 12077

C₁₀H₁₂ClNO: 12063, 12103

C₁₀H₁₂N₂: 00751

C₁₀H₁₂N₂O: 14704

C₁₀H₁₂N₂O₂: 02059, 08828

C₁₀H₁₂N₂O₃: 00120, 00137, 00138, 01620, 08646, 10450

C₁₀H₁₂N₂O₄: 00129, 00130, 00133, 00134, 00135, 02127, 04652

C₁₀H₁₂N₂O₅: 14037

C₁₀H₁₂N₂S: 13975

C₁₀H₁₂N₄: 09339

C₁₀H₁₂N₄O₅: 08434

C₁₀H₁₂O: 00387, 00388, 00389, 00621, 00622, 00623, 00755, 01796, 01807, 03933, 03934, 04437, 04439, 06768, 12059, 12241, 12242, 13761

C₁₀H₁₂O₂: 00260, 00328, 00329, 00392, 00393, 02777, 02778, 02779, 02780, 02781, 02782, 02827, 02828, 02829, 02830, 02831, 02832, 03063, 03067, 04378, 04379, 04381, 04382, 05479, 06213, 06214, 06801, 06802, 06803, 06931, 08153, 08507, 10773, 10775, 10904, 10911, 10924, 10967, 11130, 12238, 12240, 14267, 14282, 14294

C₁₀H₁₂O₃: 00222, 00258, 00385, 00507, 01004, 01007, 01011, 01739, 02718, 02737, 04374, 04376, 05225, 08287, 08914, 08915, 08916, 09137, 09140, 09144, 11074, 11123, 11124, 11125, 11129, 11138, 11139, 11140, 13155, 13156

C₁₀H₁₂O₃S: 13769, 13771

C₁₀H₁₂O₄: 00386, 01805, 01806, 02648, 04546, 08026, 08281, 11076, 11122, 12156, 13136

C₁₀H₁₂O₄S: 13938

C₁₀H₁₂O₅: 02819, 02820, 02824, 02825, 02826, 07371, 13119

C₁₀H₁₃Br: 01923, 01973, 30197

C₁₀H₁₃BrN₂O₃: 01617

C₁₀H₁₃BrO: 01922

C₁₀H₁₃Cl: 02001, 05868

C₁₀H₁₃ClO: 14036

C₁₀H₁₃ClO₂: 06620

FORMULA INDEX FOR ORGANIC COMPOUNDS (Continued)

$C_{14}H_{12}N_4O_2S$: 13627
$C_{14}H_{12}O$: 00040, 01114, 01115, 01116, 02956, 02957, 02958, 03270, 03374
$C_{14}H_{12}OS$: 06045
$C_{14}H_{12}O_2$: 02556, 02557, 02558, 02559, 02787, 02788, 02789, 02790, 02791, 02792, 02794, 02795, 02837, 02838, 02839, 02946, 02947, 02948, 02950, 02953, 02954, 02955, 06233, 08196, 13462, 13463, 13464, 13465
$C_{14}H_{12}O_3$: 02490, 02719, 02739, 02943, 02944, 02945, 06239, 13097, 14971
$C_{14}H_{12}O_4$: 01264, 02920, 02923, 09349, 09490, 13138
$C_{14}H_{12}O_5$: 08645
$C_{14}H_{13}N$: 01771, 01772, 01773, 02960, 04703
$C_{14}H_{13}NO$: 00172, 01845, 01846, 01847, 02885, 02886, 02889, 02890, 02892, 02893, 03370, 03371, 11100
$C_{14}H_{13}NO_2$: 02845, 02846, 02847, 13088
$C_{14}H_{13}NO_4$: 06940
$C_{14}H_{13}NO_4S$: 02560
$C_{14}H_{13}N_3$: 13456
$C_{14}H_{13}N_3O$: 01467
$C_{14}H_{13}N_3O_6$: 14505
$C_{14}H_{14}$: 01139, 03458, 03459, 03460, 03461, 03462, 03463, 03464, 03465, 03467, 03468, 03469, 03470, 03478, 06268, 06309, 06310, 06641, 06642, 10687
$C_{14}H_{14}NO_4PS$: 02340
$C_{14}H_{14}N_2$: 00088, 01499, 01509, 01510, 01518, 01519, 13457
$C_{14}H_{14}N_2O$: 01501, 01504, 01505, 01508, 01513, 01515, 01516, 01517, 01521, 01523, 01524, 01525, 01526, 01555, 01596, 01597, 02843, 14738
$C_{14}H_{14}N_2O_2$: 01543, 01544
$C_{14}H_{14}N_2O_3$: 01593, 01594, 01595
$C_{14}H_{14}N_2S$: 13979, 14002
$C_{14}H_{14}N_3$: 05969

$C_{14}H_{14}N_3NaO_3S$: 09179
$C_{14}H_{14}O$: 03255, 03273, 03274, 03476, 03477, 06276, 06277, 06278, 06714, 14374, 14375, 14376
$C_{14}H_{14}OS$: 03287, 06378, 14385
$C_{14}H_{14}O_2$: 02475, 03435, 03436, 03437, 03441, 03446, 03450, 03452, 03453, 03454, 06682, 06683, 06684, 06685, 06898
$C_{14}H_{14}O_2S$: 03279, 06044, 06376, 06377, 11314, 14383, 14384
$C_{14}H_{14}O_3$: 06279, 08368, 08370, 91095
$C_{14}H_{14}O_3S$: 14223, 14224, 14225, 14257
$C_{14}H_{14}O_4S$: 11328, 11329
$C_{14}H_{14}O_4S_2$: 06607
$C_{14}H_{14}O_6$: 02454, 02455, 02458
$C_{14}H_{14}S$: 03280, 03289, 06371, 06373, 06374
$C_{14}H_{14}S_2$: 06042, 06369, 06370, 06608
$C_{14}H_{14}Se$: 14381, 14382
$C_{14}H_{15}N$: 03222, 06041, 06318, 06365, 06366, 06367, 06368, 14305, 14328, 14351
$C_{14}H_{15}NO_2S$: 03283, 03285, 14240, 14245, 14246
$C_{14}H_{15}NO_3S$: 03284
$C_{14}H_{15}N_2$: 01497
$C_{14}H_{15}N_2O$: 01507
$C_{14}H_{15}N_3$: 01475, 01476, 01477, 01478, 01479, 01480, 05971, 05972, 05973, 05974
$C_{14}H_{15}P$: 11372
$C_{14}H_{16}$: 01127, 09296, 09297, 09298, 09399, 09428, 09429, 10668
$C_{14}H_{16}N_2$: 03332, 03395, 03419, 03420, 06269, 06853, 08064, 08072, 08073, 08093, 08094, 08095, 08096
$C_{14}H_{16}N_2O$: 03421
$C_{14}H_{16}N_2O_2$: 03418
$C_{14}H_{16}N_2O_4$: 00960
$C_{14}H_{16}O_2Si$: 13273, 13275
$C_{14}H_{16}O_4$: 08842
$C_{14}H_{16}Si$: 13276
$C_{14}H_{17}N$: 09704
$C_{14}H_{17}NO_2$: 05268
$C_{14}H_{17}NO_3$: 06939

$C_{14}H_{17}NO_5$: 08837
$C_{14}H_{17}NO_6$: 08404, 12352, 12353, 13168
$C_{14}H_{18}$: 01135, 10680, 10683, 10684
$C_{14}H_{18}N_2O_5$: 00367
$C_{14}H_{18}N_4O_9$: 04469
$C_{14}H_{18}O$: 07518, 07657, 08630
$C_{14}H_{18}O_4$: 08839, 11476
$C_{14}H_{18}O_6$: 11472
$C_{14}H_{18}O_7$: 11538
$C_{14}H_{18}O_8$: 06702
$C_{14}H_{19}NO$: 00118
$C_{14}H_{19}N_3S$: 09129
$C_{14}H_{20}$: 08454
$C_{14}H_{20}N_2O_4$: 01993
$C_{14}H_{20}N_2O_6S$: 10974
$C_{14}H_{20}N_4O_7$: 05234, 05235
$C_{14}H_{20}O_2$: 13171
$C_{14}H_{20}O_8$: 06879
$C_{14}H_{21}NO_4$: 12559, 12658
$C_{14}H_{22}$: 02081, 02244, 02245, 10058, 10059, 10666, 13724
$C_{14}H_{22}O$: 08455, 10124, 10871, 10874, 10875, 10992
$C_{14}H_{22}O_2$: 00450, 04797, 08179
$C_{14}H_{22}O_4$: 10174
$C_{14}H_{23}N$: 00834, 00958
$C_{14}H_{23}N_3O$: 08451, 08453
$C_{14}H_{24}$: 01138, 10685
$C_{14}H_{24}N_2$: 01909, 01910, 01911
$C_{14}H_{24}O_3$: 04551
$C_{14}H_{24}O_4$: 04523
$C_{14}H_{26}$: 03338, 13743, 13744
$C_{14}H_{26}O$: 04825
$C_{14}H_{26}O_2$: 06681
$C_{14}H_{26}O_3$: 06742, 07585
$C_{14}H_{26}O_4$: 00530, 00531, 06405, 13189
$C_{14}H_{26}O_6$: 13663
$C_{14}H_{27}ClO$: 09260
$C_{14}H_{27}N$: 09259
$C_{14}H_{28}$: 10042, 13742
$C_{14}H_{28}Br_2$: 13730
$C_{14}H_{28}ClNO$: 04903
$C_{14}H_{28}O$: 09247, 13725, 13739, 13740, 14658
$C_{14}H_{28}O_2$: 00203, 04606, 07592, 08710, 09251, 13737, 14472
$C_{14}H_{28}O_2S$: 08977
$C_{14}H_{29}Br$: 13728, 13729
$C_{14}H_{29}N$: 11673
$C_{14}H_{29}NO$: 09249, 09250
$C_{14}H_{30}$: 13726, 14465

$C_{14}H_{30}O$: 07667, 09261, 13738
$C_{14}H_{30}O_2$: 08703, 13734
$C_{14}H_{30}O_3S$: 13735
$C_{14}H_{30}O_4S$: 07671
$C_{14}H_{30}S$: 06142, 13736
$C_{14}H_{31}N$: 06206, 13727
$C_{14}H_{31}NO_2$: 08722
$C_{14}H_{32}O_2$: 09254
$C_{14}H_{42}O_5Si_6$: 07930
$C_{14}H_{42}O_7Si_7$: 05473
$C_{15}H_4N_4O_{11}$: 00631
$C_{15}H_7NO_6$: 01318, 01319
$C_{15}H_8O_4$: 01313
$C_{15}H_8O_5$: 05300
$C_{15}H_8O_6$: 01314, 01315, 01316
$C_{15}H_9BrO_2$: 06963
$C_{15}H_9ClO_2$: 01218
$C_{15}H_9N$: 10692, 10694, 10696
$C_{15}H_9NO_4$: 01285
$C_{15}H_{10}BrNO$: 01213
$C_{15}H_{10}BrNO_2$: 01201, 01215
$C_{15}H_{10}BrO_2$: 06035
$C_{15}H_{10}N_2O_5$: 04850
$C_{15}H_{10}N_2O_6$: 06244
$C_{15}H_{10}O$: 01150
$C_{15}H_{10}O_2$: 01282, 01321, 01322, 01323, 02522, 06962, 08371, 10690, 10691, 10693, 10695, 11529
$C_{15}H_{10}O_3$: 01281, 06036, 06320, 06967
$C_{15}H_{10}O_4$: 01243, 01254, 06964, 06969, 08509
$C_{15}H_{10}O_5$: 01253, 01286, 01287, 02911, 06979, 06980, 08510, 11452, 11453, 13684
$C_{15}H_{10}O_6$: 06975, 06976, 06977, 06978
$C_{15}H_{10}O_7$: 06968, 06970, 06971, 06972, 06973, 06974
$C_{15}H_{11}ClO_5$: 10274
$C_{15}H_{11}ClO_7$: 05950
$C_{15}H_{11}I_4NO_4$: 14041, 14042
$C_{15}H_{11}N$: 00517, 12901, 12902, 12903, 12904, 12905, 12906
$C_{15}H_{11}NO$: 10204, 10205, 10206, 12843
$C_{15}H_{11}NO_2$: 01203, 01204, 01283, 01284, 01813, 11506
$C_{15}H_{11}NO_3$: 04855, 04856, 04857, 04858, 04859
$C_{15}H_{11}NO_4$: 00521, 11198,

C-680

$C_{15}H_{11}NO_4$: 11199, 11200, 11201, 11202
$C_{15}H_{11}NS$: 13810
$C_{15}H_{11}O_4$: 02742
$C_{15}H_{12}$: 01131, 01132, 01133, 10673, 10674, 10675, 10676
$C_{15}H_{12}Br_2O$: 04846, 04847, 12237
$C_{15}H_{12}N_2OS$: 08040
$C_{15}H_{12}N_2O_2$: 08039
$C_{15}H_{12}N_2O_3$: 07193, 08162, 08511
$C_{15}H_{12}N_6O_4$: 13042
$C_{15}H_{12}O$: 01653, 04848, 07028
$C_{15}H_{12}O_2$: 00209, 00500, 05152, 06032, 06033, 06034, 06959
$C_{15}H_{12}O_3$: 02537, 02541, 02552, 02553, 02902
$C_{15}H_{12}O_4$: 13140
$C_{15}H_{12}O_6$: 06961, 09210
$C_{15}H_{13}N$: 08422
$C_{15}H_{13}NO$: 04710, 06988, 06989
$C_{15}H_{13}NO_2$: 03129
$C_{15}H_{13}NO_3$: 02333
$C_{15}H_{13}NO_4$: 13091
$C_{15}H_{14}$: 12283, 12284, 13484
$C_{15}H_{14}N_2O$: 07470
$C_{15}H_{14}N_2O_2$: 01469, 01470, 01471, 08829
$C_{15}H_{14}O$: 00319, 00320, 02934, 02935, 13481, 13482, 13483
$C_{15}H_{14}O_2$: 02775, 02793, 02844, 06242, 09211, 12133, 12134, 12135, 12136, 12138
$C_{15}H_{14}O_3$: 02492, 02928, 02929, 02931, 02932, 02933, 04734, 04735, 04736, 06241, 06542, 08692, 12137, 12139, 12140
$C_{15}H_{14}O_4$: 08693, 09347, 10598, 14970
$C_{15}H_{14}O_5$: 04722, 06513, 11348
$C_{15}H_{14}O_6$: 04786, 04787, 06512, 08632
$C_{15}H_{14}O_8$: 04785
$C_{15}H_{15}Cl_3O_2$: 06605
$C_{15}H_{15}N$: 04711
$C_{15}H_{15}NO$: 02936, 02937
$C_{15}H_{15}NO_2$: 04660
$C_{15}H_{15}NO_3$: 14503
$C_{15}H_{15}NO_6$: 08898
$C_{15}H_{15}NS$: 11056

$C_{15}H_{15}N_2O$: 00473
$C_{15}H_{15}N_3O_2$: 01560, 09196
$C_{15}H_{16}$: 01893, 06306, 11853, 11854, 11855, 11856
$C_{15}H_{16}NO$: 01531
$C_{15}H_{16}N_2O$: 01528, 01529, 01530, 14732
$C_{15}H_{16}N_2S$: 13983, 13994, 13995
$C_{15}H_{16}N_4O_4$: 04468
$C_{15}H_{16}N_4O_5$: 04472
$C_{15}H_{16}O$: 02470, 06315, 06655
$C_{15}H_{16}O_2$: 03482, 11780, 11781, 11851, 11852
$C_{15}H_{16}O_3$: 01422, 07367, 08164
$C_{15}H_{16}O_7$: 06432
$C_{15}H_{16}O_8$: 13350
$C_{15}H_{16}O_9$: 06598
$C_{15}H_{17}N$: 03212, 14307
$C_{15}H_{17}NO$: 02471, 02472, 02473
$C_{15}H_{17}NO_2S$: 14241, 14242
$C_{15}H_{17}N_2O$: 01527
$C_{15}H_{17}N_2O_2$: 08315
$C_{15}H_{18}$: 01604, 01605, 01606, 09362
$C_{15}H_{18}ClN$: 03294
$C_{15}H_{18}N_2$: 11778
$C_{15}H_{18}N_4O_4$: 04473
$C_{15}H_{18}O$: 09417, 09418, 09760, 09761, 10607, 10608
$C_{15}H_{18}O_3$: 13178
$C_{15}H_{18}O_6$: 02465
$C_{15}H_{19}NO_2$: 14595, 14600, 14601
$C_{15}H_{19}Si_2$: 06345
$C_{15}H_{20}ClNO_2$: 14596
$C_{15}H_{20}N_2O$: 00749
$C_{15}H_{20}O_2$: 07475
$C_{15}H_{20}O_4$: 08874
$C_{15}H_{21}NO_2$: 06927, 06928, 06929, 08500, 11700
$C_{15}H_{21}N_3O_2$: 11533
$C_{15}H_{22}$: 05401, 08349
$C_{15}H_{22}ClNO_2$: 06930, 08501, 08502, 08503
$C_{15}H_{22}O$: 05988, 06544, 14908, 14909
$C_{15}H_{22}O_2$: 02767, 06545
$C_{15}H_{23}$: 04466
$C_{15}H_{23}ClO_4S$: 01375
$C_{15}H_{24}$: 02220, 02295, 04779, 04780, 04781, 04824, 05187, 05237, 06471, 06942, 06943, 07468, 08776, 10282, 13200, 13201

$C_{15}H_{24}N_2O$: 01337, 08940, 10220
$C_{15}H_{24}O$: 06543, 08729, 10872, 10873, 10878, 13172, 13173
$C_{15}H_{24}O_6$: 00459
$C_{15}H_{24}O_8$: 11934, 11935
$C_{15}H_{26}ClN_4O$: 00709
$C_{15}H_{26}N_2$: 13369
$C_{15}H_{26}O$: 06473, 06944, 06945, 08637, 09808, 09809, 09810, 10262
$C_{15}H_{26}O_6$: 07392
$C_{15}H_{26}O_7$: 05176
$C_{15}H_{26}Si$: 13316
$C_{15}H_{28}$: 10296, 11834, 11835, 11836, 11839
$C_{15}H_{28}N_2O_4S$: 13370
$C_{15}H_{28}O$: 05754
$C_{15}H_{28}O_2$: 08964
$C_{15}H_{28}O_4$: 08863
$C_{15}H_{30}$: 06470, 10295
$C_{15}H_{30}Br_2$: 10288
$C_{15}H_{30}O$: 10283, 10293, 10294
$C_{15}H_{30}O_2$: 04634, 07596, 08714, 08715, 09255, 10290
$C_{15}H_{30}O_4$: 07372, 07373
$C_{15}H_{31}Br$: 10287
$C_{15}H_{31}NO$: 10284
$C_{15}H_{32}$: 10285
$C_{15}H_{32}O$: 10292
$C_{15}H_{33}B$: 03577, 03579
$C_{15}H_{33}BO_3$: 03561, 03564
$C_{15}H_{33}ClN_2O_5$: 10221
$C_{15}H_{33}N$: 10286, 14520, 14531, 14532
$C_{15}H_{33}NO$: 11775
$C_{15}H_{33}N_3$: 14405
$C_{15}H_{33}O_4P$: 11419
$C_{15}H_{34}ClN$: 06415
$C_{16}H_8O_2S_2$: 13853
$C_{16}H_{10}$: 06986, 12430
$C_{16}H_{10}Cl_4O_5$: 06327
$C_{16}H_{10}N_2$: 02881
$C_{16}H_{10}N_2O_2$: 08406, 08410
$C_{16}H_{10}N_2O_8S_2$: 08409
$C_{16}H_{10}O_2$: 09674
$C_{16}H_{10}O_3$: 09667
$C_{16}H_{10}O_7$: 01317
$C_{16}H_{11}N$: 12432
$C_{16}H_{11}NO_2$: 09675, 12927, 12934
$C_{16}H_{11}NO_4$: 01277
$C_{16}H_{11}N_3O_3$: 02324, 02325
$C_{16}H_{11}N_3O_7$: 09421
$C_{16}H_{12}$: 01148, 01149, 09419, 09420, 12433
$C_{16}H_{12}N_2$: 02314, 02326
$C_{16}H_{12}N_2O$: 02322, 02323,

$C_{16}H_{12}N_2O$: 02329
$C_{16}H_{12}N_2O_6$: 06243
$C_{16}H_{12}O$: 01097, 01098, 01099, 07131, 10650, 10651, 10652
$C_{16}H_{12}O_2$: 01262, 01263, 01265, 01266, 01267, 06857, 06858
$C_{16}H_{12}O_3$: 04854, 07035
$C_{16}H_{12}O_4$: 01260, 01261, 07109, 08815
$C_{16}H_{12}O_5$: 03599, 06965, 06966
$C_{16}H_{12}O_6$: 07477, 11711
$C_{16}H_{12}O_7$: 13035
$C_{16}H_{12}S$: 13893
$C_{16}H_{12}S_2$: 06360
$C_{16}H_{13}N$: 09762, 09764, 12633, 12634
$C_{16}H_{13}NO_3$: 05091, 05092, 05093
$C_{16}H_{13}N_3$: 02315, 02316
$C_{16}H_{14}$: 01124, 01125, 01126, 03651, 03652, 10663, 10664, 10667
$C_{16}H_{14}ClN_3$: 02317
$C_{16}H_{14}O$: 09151
$C_{16}H_{14}O_2$: 04851, 04852, 04853, 05094, 10662
$C_{16}H_{14}O_3$: 02551, 02840, 02901, 05128, 11103, 14261, 14278, 14291
$C_{16}H_{14}O_4$: 03509, 03514, 03518, 03522, 06887, 08336, 10181, 10182, 10183, 13098, 13559
$C_{16}H_{14}O_5$: 11061
$C_{16}H_{14}O_6$: 06960, 07481
$C_{16}H_{15}ClO_2$: 04945, 06237
$C_{16}H_{15}Cl_3$: 06644
$C_{16}H_{15}N$: 00344, 06040, 06205
$C_{16}H_{16}$: 01113, 03966, 13467, 13468, 13480
$C_{16}H_{16}ClNO$: 04893
$C_{16}H_{16}N_2O_2$: 03410, 03411, 03412, 06850, 08799
$C_{16}H_{16}N_2O_4S$: 11319
$C_{16}H_{16}O_2$: 06038, 06238, 08151, 11168, 13466
$C_{16}H_{16}O_3$: 02491, 06240
$C_{16}H_{16}O_4$: 02841, 02998, 02999, 03000, 03001, 03002, 03003, 03004, 06476, 09489
$C_{16}H_{16}O_5$: 00576, 13229
$C_{16}H_{16}O_6$: 04838
$C_{16}H_{16}Si$: 13282
$C_{16}H_{17}Br$: 03686, 03687
$C_{16}H_{17}N$: 04701
$C_{16}H_{17}NO$: 00071, 06037

SUBLIMATION DATA FOR ORGANIC COMPOUNDS

Compiled by Mansel Davies

The tables quote the parameters from what appear to be the best data in the literature expressed in the form

$$\log_{10}p(mm) = A - B/T$$

and the temperature range for which they apply. The corresponding heats and entropies (taking the standard state of the vapor to be

$$\Delta H \text{ (sublimation)} = 2.303 \text{ R.B. cal/mol}$$
$$\Delta S \text{ (sublimation)} = 2.303R(A - 2.881) \text{ cal/mol K.}$$

Compound	Temp. range °C	A	B	Ref.
Acenaphthene	18 to 37	11.758	4290.5	1
Acetamide	25 to 77	11.8468	4050.1	2
Acetic acid	−35 to +10	8.502	2177.4	3
Acetic Acid. m-cresyl ester	2 to 44	9.759	3170	14
Acetophenone, 1-chloro-	5 to 50	13.779	4740	14
Acetophenone, 1-chloro-o-nitro	23 to 54	14.24	5413	14
Acetophenone, 1-chloro-m-nitro	26 to 70	14.080	5700	14
Acetophenome p-methoxy	3 to 27	11.367	4056	1
Acetone, benzoyl	5 to 26	12.317	4375	1
Adipic acid	86 to 133	15.463	6757	4
Anthracene	65 to 80	12.638	5320	5
Anthracene	105 to 125	12.002	5102	6
Anthracene, 9,10 diphenyl	208 to 229	16.058	8213	22
Anthraquinone	224 to 286	12.305	5747	3
Arachidic acid	63 to 73	25.453	10,424	23
Arsine, Diphenylcyano	23 to 53	10.724	4420	14
Azobenzene (cis)	30 to 60	9.652	3914	7
Azobenzene (trans)	30 to 60	9.721	3911	7
Behenic acid	71 to 79	23.604	10,100	23
Benzanthrone	—	13.416	6030	8
Benzene	−30 to 5	9.846	2309	3
Benzene	−58 to −30	9.556	2241	3
Benzene, p-chlorobromo	−80 to −40	9.013	3200	15
Benzene, -dichloro	10 to 50	11.985	3570	17
Benzene, α-hexachloro	51 to 71	11.950	4850	14
Benzene, β-hexachloro	95 to 117	11.790	5375	14
Benzene, γ-hexachloro	−10 to 97	15.515	6077	14
Benzene, λ-hexachloro	55 to 75	12.635	5100	14
Benzene, -hexamethyl		11.070	4215	37
Benzene, 1, 2, 3,-trichloro	16 to 30	10.662	3440	6
Benzene, 1, 2, 4-trichloro	6 to 25	10.445	3254	6
Benzene, 1,3,5-trichloro	9 to 28	9.176	2956	6
Benzil	45 to 67	12.708	5140	1
Benzoic Acid	70 to 114	12.870	4776	9
Benzoic acid, p-hydroxy	125 to 160	13.623	6063	9
Benzoic acid, o-methoxy	80 to 95	11.871	4746	9
Benzophenone (stable)	16 to 42	17.46	4966	10
Benzophenone, (meta stable)	11 to 25	17.19	4818	10
Benzoquinone	—	10.00	3280	18
Benzoquinone, 2.6-dichloro	1 to 42	9.85	3670	18
Benzoquinone, trichloro	28 to 54	12.03	4630	18
Benzoquinone, tetrachloro	60 to 83	12.06	5170	18
Benzoquinone, p-xylo	0 to 20	11.53	4030	18
Bibenzyl	13 to 34	12.194	4386	1
Biphenyl	6 to 26	11.168	3959	1
Butyramide	25 to 68	12.739	4546	2
Butyramide	63 to 109	12.594	4513	2
Camphor	0 to 180	8.799	2797	3
Capramide	80 to 97	16.471	6577	2
Capramide, N-methyl	30 to 52	14.594	5371	11
Capric acid	16 to 28	17.130	6119	23
Caproamide	65 to 95	13.328	4968	2
εCaprolactam	21 to 41	11.839	4339	34
Caprylamide	52 to 101	14.920	5783	2
Carbamic acid, n-butyl ester	19 to 43	14.582	4919	11
Carbamic acid, ethyl ester	19 to 43	14.090	4646	11
Carbamic acid, n-hexyl ester	18 to 41	14.748	5018	11
Carbamic acid, methyl ester	14 to 32	11.966	3883	11
Carbon tetrabromide (mono-clinic)	22 to 46	9.3867	2841	12
(cubic)	48 to 56	8.5670	2579	12
Carbon tetrachloride	−64 to −48	9.089	2027	13
o-Cresol, 3, 5-dinitro	17 to 51	14.140	5400	14
Cyclohexane	−5 to +5	8.594	1953	3
Cyclo-trimethylene-trinitramine	110 to 138	11.870	5850	16
Diphenylamine	25 to 51	12.434	4654	21
Dodecanedioic acid	102 to 123	17.728	8006	4
Eicosanedioic acid	107 to 122	18.185	8644	4
Enanthamide	72 to 93	13.617	5182	2
Ethane, 1.1 p dichloro diphenyl tri-chloro	66 to 100	14.191	6160	14
Ethane, hexachloro (cubic)	13 to 174	8.731	2677	26
Ethane, hexachloro (triclinic)	13 to 174	9.890	3077	2b

C-687

Compound	Temp. range °C	A	B	Ref.
Ethylene dibromide	−21 to + 8	9.884	2606	13
Ethylene, *trans* di-iodo	−8 to 20	5.86	2130	19
Fluorene	33 to 49	11.325	4324	5
Formic acid	−5 to + 8	12.486	3160	36
2 Fuoric acid	44 to 55	14.62	5667	25
Hendecanoic acid	20 to 28	16.432	6037	24
Heneicosanoic acid	68 to 73	22.602	9642	24
Heptadecanoic acid	48 to 58	21.836	8769	24
Hydroquinone tetrachloro	77 to 86	10.08	4650	18
Hydroquinone tetrachloro, *p*- xylo	59 to 88	12.36	5280	18
Lauramide	76 to 95	19.169	7980	2
Lauric acid	22 to 41	19.897	7322	23
Methane	−194 to −184	7.651	5169	3
Methane, triphenyl	52 to 76	12.661	5228	1
Myristamide	85 to 100	20.940	8746	2
Myristic acid	38 to 52	18.740	7291	23
Naphthalene	6 to 21	11.597	3783	5
1-Naphthol	25 to 39	13.074	4873	1
	39 to 50	11.526	4389	1
2-Naphthol	25 to 39	13.356	5109	1
	39 to 58	11.660	4579	1
Nonadecanoic acid	58 to 64	35.916	13,815	24
n-Octadecane	15 to 25	22.83	7995	27
Oxalic acid, anhyd.	60 to 105	12.223	4727	29
Oxalic acid, anhyd. (*α*)	38 to 52	13.17	5130	28
Oxalic acid, anhyd. (*β*)	38 to 50	12.57	4875	28
Oxamic acid	82 to 90	12.58	5639	30
Oxamide	80 to 96	12.57	5893	30
Palmitamide	91 to 105	22.690	9489	2
Palmitic acid	46 to 60	20.217	8069	23
Pelargonamide	80 to 97	15.249	5997	2
Pentadecanoic acid	38 to 48	23.110	8813	24
Pentaerythritol (tetrag)	106 to 135	16.17	7528	28
Pentaerythritol, tetranitrate	97 to 138	17.73	7750	16
Phenanthrene	37 to 50	ll.388	4519	5
Phenol	5 to 32	11.421	3540	14
Phenol, *p*-acetyl	47 to 75	12.216	5003	31
Phenol, *p*-benzyl	40 to 62	12.600	5072	31
Phenol, *p-tert* butyl	8 to 30	12.332	4402	31
Phenol, 2-*tert* butyl-4- methyl	2 to 20	11.685	4036	31
Phenol, 4-*tert* butyl-2-methyl	3 to 24	11.199	3952	31
Phenol, *p*- formyl	39 to 63	11.795	4762	31
Phenol, *p*-methoxy	5 to 27	13.132	4624	31
Phenol, *o*-phenyl	19 to 40	11.754	4331	31
Phenol, *p*-phenyl	54 to 74	12.056	5068	31
Phenol, 2:4:6-tri*tert* butyl	18 to 40	11.507	4383	31
Phthalic anhydride	30 to 60	12.249	4632	32
Propionamide	45 to 73	12.041	4139	2
Pyrene	72 to 85	11.270	4904	5
Pyrrole 2-carboxylic acid	77 to 81	16.60	6633	25
Rubeanic acid	87 to 105	12.713	5515	30
Salicylic acid	95 to 134	12.859	4969	9
Sebacic acid	102 to 130	18.911	8395	4
Stearamide	94 to 106	24.449	10,230	2
Steraric acid	57 to 67	21.180	8696	23
Suberic acid	106 to 134	16.937	7472	4
Succinic acid	99 to 128	14.068	6132	4
d-Tartaric acid, dimethyl ester	35 to 44	16.610	5903	20
dl-Tartaric acid, dimenthyl ester	42 to 85	16.127	5941	20
Thapsic acid	104 to 125	17.165	7885	4
2-Thenoic acid	42 to 50	13.53	5065	25
Thymol	0 to 40	14.201	4766	14
Toluene, 2, 4, 6-trinitro	50 to 143	15.34	6180	33
Tridecanoic acid	31 to 39	20.939	7764	24
Valeramide	60 to 101	12.846	4666	2

References

1. Aihara,*Bull. Chem. Soc.*,Japan
2. Davies, Jones, and Thomas,*Trans. Faraday Soc.*, 55, 1100, 1959.
3. *CRC Handbook of Chemistry and Physics*, 41st Ed. p. 2428 et seq.
4. Davies and Thomas, *Trans. Faraday, Soc.*, 56, 185, 1960.
5. Bradley and Cleasby, *J. Chem. Soc.*, 1690, 1953.
6. Sears and Hopke, *J. Am. Chem. Soc.*, 71, 1632, 1949.
7. Bright et al., *Research*, 3, 185, 1950.
8. Inokuchi et al., *Bull. Chem. Soc.*, Japan, 25, 299, 1952.
9. Davies and Jones, *Trans. Soc.*, 50, 1042, 1954.
10. Neumann and Volker, *Z. Physik. Chem.*, 161A, 33, 1932.
11. Davies and Jones, *Trans. Soc.*, 55, 1329, 1959.
12. Bradley and Drury, ibid. , 55, 1844, 1959.
13. Nitta and Seki, *J. Chem. Soc.*, Japan, 69, 85, 1948.
14. Balson, *Trans. Faraday Soc.*, 43, 54, 1947.
15. Ewald, ibid., 49, 1401, 1953.
16. Edwards, ibid., 49, 152, 1953.
17. Darkis et al., *Ind. Eng. Chem.*, 32, 946, 1940.
18. Coolidge and Coolidge, *J. Am. Chem. Soc.* 49, 100, 1927.

19. Broadway and Fraser, *J. Chem. Soc.*, 429, 1933.
20. Crowell and Jones, *J. Phys. Chem.*, 58, 666, 1954.
21. Aihara, *J. Chem. Soc.*, Japan, 74, 437, 1953.
22. Stevens, *J. Chem. Soc.*, 2973, 1953.
23. Davies, Malpass, and Stenhagan, *Ark. Kemi*.
24. Thomas, M. Sc. thesis, Univ. of Wales, 1959.
25. Bradley and Care, *J. Chem. Soc.*, 1688, 1953.
26. Ivin and Dainton, *Trans. Soc.*, 43, 32, 1947.
27. Bradley and Shellard, *Proc. Roy. Soc.*, 198A, 239, 1949.
28. Bradley and Cotson, *J. Chem. Soc.*, 1684, 1953.
29. Noyes and Wobbe, *J. Am. Chem. Soc.*, 48, 1882, 1926.
30. Bradley and Cleasby, *J. Chem. Soc.*, 1681, 1953.
3l. Aihara, *Bull Chem. Soc.*, Japan.
32. Crooks and Feetham, *J. Chem. Soc.*, 899, 1946.
33. Edwards, *Trans. Faraday Soc.*, 46, 423, 1950.
34. Aihara, *J. Chem. Soc.*, Japan, 74, 631, 1953.
35. Seki and Suzuki, *Bull. Chem. Soc.*, Japan, 70, 387, 1949.
36. Coolidge, *J. Am. Chem. Soc.*, 53, 1874, 1930.
37. Nitta et al., *J. Chem. Soc.*, Japan, 70, 387, 1949.

HEAT OF FUSION OF SOME ORGANIC COMPOUNDS

Compiled by R. Loebel

Compounds in the following tables are listed in order of the increasing number of carbon and hydrogen atoms in the molecule. Melting points are listed in degrees Celcius and heats of fusion in calories per gram.

Formula	Name	M.P., °C	H_f, cal/g	Formula	Name	M.P., °C	H_f, cal/g
$CHCl_3$	Trichloromethane	−63.6	17.62	C_5H_8	1,4-Pentadiene	−148.8	21.55
CHN	Hydrogen cyanide	−13.4	74.38	$C_5H_8O_3$	Levulinic acid	33.0	18.97
CH_2Cl_2	Dichloromethane	−95.14	16.89	C_5H_{10}	1-Pentene	−166.2	19.81
CH_2N_2	Cyanamide	44.0	49.81	C_5H_{10}	cis-2-Pentene	−151.4	24.25
CH_2O_2	Formic acid	8.3	66.05	C_5H_{10}	trans-2-Pentene	−140.2	28.48
CH_3D	Monodeuteromethane	−182.7	12.76	C_5H_{10}	Cyclopentane	−93.8	2.07
CH_3Br	Methyl bromide	−93.7	15.05	$C_5H_{10}O_2$	Valeric acid	−59.0	27.83
CH_4	Methane	−182.5	13.96	C_5H_{12}	n-Pentane	−129.7	27.89
CD_4	Deuteromethane	−183.4	10.75	C_5H_{12}	Isopentane	−159.9	17.05
CH_4O	Methanol	−97.8	23.70	C_5H_{12}	2,2-Dimethylpropane	−16.7	10.79
CH_4S	Methyl mercaptan	−121.0	29.35	$C_5H_{12}O$	Amyl alcohol	−78.9	26.65
CH_5N	Methylamine	−93.5	47.20	$C_6H_3Br_3O$	2,4,6-Tribromophenol	93.0	13.38
$CBrCl_3$	Bromotrichloromethane	−5.7	3.05	C_6H_4BrI	o-Bromoiodobenzene	21.0	12.18
CCl_2O	Phosgene	−127.9	13.86	C_6H_4BrI	m-Bromoiodobenzene	9.3	10.27
CCl_3NO_2	Chloropicrin	−64.0	48.16	C_6H_4BrI	p-Bromoiodobenzene	90.1	16.16
CCl_4	Tetrachloromethane	−23.0	5.09	$C_6H_4Br_2$	o-Dibromobenzene	1.8	12.78
CS_2	Carbon disulfide	−111.5	13.80	$C_6H_4Br_2$	m-Dibromobenzene	−6.9	13.38
$C_2HCl_3O_2$	Trichloroacetic acid	57.5	8.60	$C_6H_4Br_2$	p-Dibromobenzene	86.0	20.55
$C_2H_2Br_2Cl_2$	1,2-dibromo-1,1-dichloro-methane	−66.9	7.73	$C_6H_4Br_2O$	2,4-Dibromophenol	12.0	13.97
$C_2H_2Cl_2O_2$	Dichloroacetic acid	10.8	14.21	$C_6H_4ClNO_2$	m-Chloronitrobenzene	44.4	29.38
C_2H_3Br	Bromoethane	−139.5	11.44	$C_6H_4ClNO_2$	p-Chloronitrobenzene	83.5	31.51
$C_2H_3Br_3$	1,1,2-Tribromoethane	−29.2	8.16	$C_6H_4Cl_2$	o-Dichlorobenzene	−16.7	21.02
C_2H_3Cl	Chloroethane	−153.8	18.14	$C_6H_4Cl_2$	m-Dichlorobenzene	−24.8	20.55
$C_2H_3ClO_2$	α-Chloroacetic acid	61.2	31.06	$C_6H_4Cl_2$	p-Dichlorobenzene	53.13	29.07
$C_2H_3ClO_2$	β-Chloroacetic acid	56.0	35.12	$C_6H_4I_2$	o-Diiodobenzene	23.4	10.15
$C_2H_3Cl_3$	Methyl chloroform	−30.4	4.90	$C_6H_4I_2$	m-Diiodobenzene	34.2	11.54
$C_2H_3Cl_3$	1,1,2-Trichloroethane	−36.6	20.68	$C_6H_4I_2$	p-Diiodobenzene	129.0	16.20
$C_2H_3Cl_3$	Chloral hydrate	47.4	33.18	$C_6H_4N_2O_4$	o-Dinitrobenzene	116.93	32.25
$C_2H_3F_3$	1,1,1-Trifluoroethane	−111.3	17.61	$C_6H_4N_2O_4$	m-Dinitrobenzene	89.7	24.70
$C_2H_4Br_2$	1,2-Dibromoethane	9.93	13.79	$C_6H_4N_2O_4$	p-Dinitrobenzene	173.5	39.99
$C_2H_4Cl_2$	1,2-Dichloroethane	−35.5	21.12	$C_6H_4O_2$	Benzoquinone	112.9	40.97
C_2H_4O	Ethylene oxide	−112.5	28.07	C_6H_5Br	Bromobenzene	−30.6	16.17
$C_2H_4O_2$	Acetic acid	16.6	45.91	C_6H_5BrO	p-Bromophenol	63.5	20.50
C_2H_5Cl	Ethyl chloride	−138.3	16.49	C_6H_5Cl	Chlorobenzene	−45.2	20.40
C_2H_6	Ethane	−183.3	22.73	C_6H_5I	Iodobenzene	−31.3	11.43
C_2H_6O	Dimethyl ether	−141.5	25.62	$C_6H_5NO_2$	Nitrobenzene	5.7	22.50
C_2H_6O	Ethanol	−114.5	26.05	$C_6H_5NO_3$	p-Nitrophenol	113.8	41.70
$C_2H_6O_2$	Ethyleneglycol	11.8	43.26	$C_6H_5NO_3$	o-Nitrophenol	45.13	26.76
C_2H_6S	Dimethyl sulfide	−98.3	30.73	C_6H_6	Benzene	5.53	30.45
C_2H_6S	Ethyl mercaptan	−121.0	19.14	$C_6H_6N_2O_2$	o-Nitroaniline	71.2	27.88
$C_2H_6S_2$	Methyl disulfide	−120.5	23.32	$C_6H_6N_2O_2$	p-Nitroaniline	147.0	36.50
C_2H_7N	Dimethylamine	−92.2	31.51	$C_6H_6N_2O_2$	m-Nitroaniline	114.0	40.97
$C_3H_4O_2$	Acrylic acid	12.3	37.03	C_6H_6O	Phenol	40.9	28.67
$C_3H_5Br_3$	1,2,3-Tribromopropane	16.19	20.24	$C_6H_6O_2$	Pyrocatechol	105.0	49.40
$C_3H_5N_3O_9$	Trinitroglycerol	12.3	23.02	$C_6H_6O_2$	Quinol	172.3	58.84
C_3H_6	Cyclopropane	−127.4	30.92	$C_6H_6O_2$	Resorcinol	110.0	46.22
C_3H_6	Propene	−185.3	17.06	C_6H_6S	Thiophenol	−14.9	24.90
C_3H_6BrCl	Trimethylenebromochloride	−58.9	14.0	C_6H_7N	Aniline	−6.3	27.09
$C_3H_6Br_2$	1,3-Dibromopropane	−34.2	16.10	$C_6H_8N_2$	Phenylhydrazine	19.6	36.31
$C_3H_6Cl_2$	1,2-Dichloropropane	−100.5	13.53	$C_6H_8O_4$	Glutaric acid	97.5	37.39
C_3H_6O	Acetone	−94.8	23.42	$C_6H_8O_4$	Methyl fumarate	102.0	57.93
C_3H_7Cl	2-Chloropropane	−117.2	22.48	C_6H_{10}	Cyclohexene	−103.5	9.58
$C_3H_7NO_2$	Urethane	48.7	40.85	$C_6H_{10}O_4$	Methyl succinate	19.5	35.72
C_3H_8	Propane	−181.7	19.11	$C_6H_{10}O_6$	d-Dimethyl tartrate	49.0	21.50
C_3H_8O	Propanol	−126.1	20.66	$C_6H_{10}O_6$	dl-Dimethyl tartrate	87.0	35.12
C_3H_8O	Isopropanol	−89.5	21.37	C_6H_{12}	Methylcyclopentane	−142.5	19.68
$C_3H_8O_3$	Glycerol	18.2	47.95	C_6H_{12}	Cyclohexane	6.6	7.47
C_3H_9N	Trimethylamine	−117.1	26.47	C_6H_{12}	Tetramethylethylene	−74.6	15.51
$C_4H_4N_2$	Succinonitrile	54.5	11.71	$C_6H_{12}O$	Cyclohexanol	25.46	4.19
$C_4H_4O_3$	Succinic anhydride	119.0	48.74	$C_6H_{12}O_3$	Paraldehyde	10.5	25.02
C_4H_4S	Thiophene	−39.4	14.11	C_6H_{14}	2,2-Dimethylbutane	−99.0	1.61
C_4H_6	1,3-Butadiene	−108.9	35.28	C_6H_{14}	2,3-Dimethylbutane	−128.8	2.22
C_4H_6	2-Butyne	−32.26	40.80	C_6H_{14}	n-Hexane	−95.3	36.27
$C_4H_6O_2$	Crotonic acid	72.0	25.32	C_6H_{14}	Isohexane	−153.7	17.41
$C_4H_6O_2$	cis-Crotonic acid	15.5	34.90	C_6H_{14}	2-Methylpentane	−153.7	17.38
$C_4H_6O_4$	Methyl oxalate	54.35	42.64	$C_6H_{14}O$	Isopropyl ether	−86.8	25.79
C_4H_8	Isobutene	−140.4	25.25	$C_6H_{14}O$	n-Propyl ether	−126.1	20.66
C_4H_8	cis-2-Isobutene	−138.9	32.30	$C_7H_5ClO_2$	o-Chlorobenzoic acid	140.2	39.27
$C_4H_8N_2S$	Thiosinamine	77.0	33.45	$C_7H_5ClO_2$	m-Chlorobenzoic acid	154.2	36.39
$C_4H_8O_2$	Ethyl acetate	−83.6	28.43	$C_7H_5ClO_2$	p-Chlorobenzoic acid	239.7	49.23
$C_4H_8O_2$	n-Butyric acid	−5.7	30.04	$C_7H_5NO_4$	o-Nitrobenzoic acid	145.8	40.06
$C_4H_8O_2$	Dioxane	11.0	34.85	$C_7H_5NO_4$	m-Nitrobenzoic acid	141.1	27.59
C_4H_9Br	sec-Butylbromide	−112.7	12.01	$C_7H_5NO_4$	p-Nitrobenzoic acid	239.20	52.80
C_4H_{10}	n-Butane	−138.3	19.18	$C_7H_5N_3O_6$	2,4,6-Trinitrotoluene	80.83	22.34
C_4H_{10}	Isobutane	−159.42	18.96	$C_7H_6N_2O_4$	2,4-Dinitrotoluene	70.14	26.40
$C_4H_{10}O$	n-Butanol	−89.8	29.93	$C_7H_6O_2$	Benzoic acid	122.4	33.89
$C_4H_{10}O$	tert-Butyl alcohol	25.4	21.88	C_7H_7Br	p-Bromotoluene	28.0	20.86
$C_4H_{10}O$	Ethyl ether	−116.3	23.45	C_7H_7I	9-Iodotoluene	34.0	18.75
$C_4H_{12}Si$	Tetramethyl silane	−99.04	18.64	$C_7H_7NO_2$	o-Aminobenzoic acid	145.0	35.98
C_5H_8	Cyclopentene	−135.1	11.80	$C_7H_7NO_2$	m-Aminobenzoic acid	179.5	38.03
C_5H_8	Isoprene	−145.9	16.80	$C_7H_7NO_2$	p-Aminobenzoic acid	188.5	36.46
				C_7H_8	Toluene	−94.99	17.17

Formula	Name	M.P., °C	H_f, cal/g	Formula	Name	M.P., °C	H_f, cal/g
C_7H_8O	Benzyl alcohol	−15.2	19.83	$C_{10}H_8O$	α-Naphthol	95.0	38.94
C_7H_8O	p-Cresol	34.6	26.28	$C_{10}H_8O$	β-Naphthol	120.6	31.30
$C_7H_8O_2$	Dimethylpyrone	132.0	56.14	$C_{10}H_8O_2$	Methylphenyl propionate	18.0	22.86
C_7H_9O	p-Toluidine	43.3	39.90	$C_{10}H_9N$	α-Naphthylamine	50.0	22.34
C_7H_{14}	1-Heptene	−119.7	30.82	$C_{10}H_{10}O_2$	Methyl cinnamate	36.0	26.53
C_7H_{14}	Methylcyclohexane	−126.6	16.43	$C_{10}H_{12}$	Camphene	51.0	57.0
C_7H_{16}	n-Heptane	−90.6	33.78	$C_{10}H_{12}O$	Anethole	22.5	25.80
C_7H_{16}	2-Methylhexane	−118.2	21.16	$C_{10}H_{14}$	Durene	79.3	37.40
C_7H_{16}	2,2-Dimethylpentane	−123.8	13.98	$C_{10}H_{14}$	Prehnitene	−7.7	20.0
C_7H_{16}	2,4-Dimethylpentane	−119.9	15.95	$C_{10}H_{14}$	p-Cymene	−68.9	17.10
C_7H_{16}	3,3-Dimethylpentane	−134.9	16.86	$C_{10}H_{14}O$	Thymol	51.5	27.47
C_7H_{16}	3-Ethylpentane	−118.6	22.78	$C_{10}H_{15}NO$	d-Carvoxime	71.5	23.29
C_7H_{16}	2,2,3-Trimethylbutane	−25.0	5.25	$C_{10}H_{15}NO$	l-Carvoxime	71.0	23.41
$C_8H_6Cl_4$	o-Tetrachloroxylene	86.0	21.02	$C_{10}H_{15}NO$	dl-Carvoxime	91.0	23.61
$C_8H_6Cl_4$	p-Tetrachloroxylene	95.0	22.10	$C_{10}H_{20}O$	1-Menthol	43.5	18.63
C_8H_8	Cyclooctatetraene	−4.7	25.87	$C_{10}H_{20}O_2$	n-Capric acid	31.99	38.87
$C_8H_8Br_2$	o-Xylenedibromide	95.0	24.25	$C_{10}H_{22}$	n-Decane	−29.7	48.34
$C_8H_8Br_2$	m-Xylenedibromide	77.0	21.45	$C_{11}H_{10}$	2-Methylnaphthalene	−34.4	20.11
$C_8H_8Cl_2$	o-Xylenedichloride	55.0	29.03	$C_{11}H_{22}O_2$	n-Undecilic acid	28.25	32.20
$C_8H_8Cl_2$	m-Xylenedichloride	34.0	26.64	$C_{11}H_{24}$	Undecane	−25.6	34.12
$C_8H_8Cl_2$	p-Xylenedichloride	100.0	32.73	$C_{12}H_9N$	Carbazole	243.0	42.05
$C_8H_8O_2$	Phenylacetic acid	76.7	25.44	$C_{12}H_{10}$	Diphenyl	165.5	30.40
$C_8H_8O_2$	o-Toluic acid	103.7	35.40	$C_{12}H_{10}N_2$	Azobenzene	67.1	28.91
$C_8H_8O_2$	m-Toluic acid	108.75	27.59	$C_{12}H_{10}N_2O$	Azoxybenzene	36.0	21.62
$C_8H_8O_2$	p-Toluic acid	179.6	39.90	$C_{12}H_{11}N$	Diphenylamine	52.98	25.23
$C_8H_9NO_2$	Hydroxyacetanilide	91.3	33.59	$C_{12}H_{12}N_2$	Hydrazobenzene	134.0	22.89
C_8H_{10}	o-Xylene	−25.2	30.64	$C_{12}H_{14}O_4$	Apiol	29.5	25.80
C_8H_{10}	m-Xylene	−47.8	26.01	$C_{12}H_{24}O_2$	n-Lauric acid	43.22	43.72
C_8H_{10}	p-Xylene	13.2	37.83	$C_{12}H_{26}$	n-Dodecane	−9.6	51.33
C_8H_{16}	Ethylcyclohexane	−111.3	17.75	$C_{13}H_{10}O$	Benzophenone	47.85	23.53
C_8H_{16}	trans-1,1-dimethylcyclohexane	−33.3	4.38	$C_{13}H_{13}N$	Benzylaniline	32.37	21.86
C_8H_{16}	cis-1,2-dimethylcyclohexane	−49.9	3.50	$C_{14}H_8O_2$	Anthraquinone	284.8	37.48
C_8H_{16}	trans-1,2-dimethylcyclohexane	−88.2	22.35	$C_{14}H_{10}$	Anthracene	216.5	38.70
C_8H_{16}	cis-1,3-dimethylcyclohexane	−75.6	23.05	$C_{14}H_{10}$	Phenanthrene	96.3	25.0
C_8H_{16}	trans-1,3-dimethylcyclohexane	−90.1	21.01	$C_{14}H_{10}O_2$	Benzil	95.2	22.15
C_8H_{16}	cis-1,4-dimethylcyclohexane	−87.4	19.82	$C_{14}H_{12}$	Stilbene	124.0	40.0
C_8H_{16}	trans-1,4-dimethylcyclohexane	−36.9	26.27	$C_{14}H_{28}O_2$	Myristic acid	53.96	47.49
$C_8H_{16}O_2$	Caprylic acid	16.3	35.40	$C_{16}H_{32}O_2$	Palmitic acid	61.82	39.18
C_8H_{18}	n-Octane	−56.8	43.21	$C_{16}H_{34}O$	Cetyl alcohol	49.27	33.80
C_8H_{18}	3-Methylheptane	−120.5	23.81	$C_{18}H_{14}O_3$	Cinnamic anhydride	48.0	28.14
C_8H_{18}	4-Methylheptane	−121.0	22.68	$C_{18}H_{30}O_2$	Stearic acid	68.82	47.54
C_8H_{18}	2,2,4-Trimethylpentane	−107.3	18.92	$C_{18}H_{34}O_2$	Elaidic acid	44.4	52.08
C_9H_7N	Quinoline	−15.6	19.98	$C_{18}H_{38}$	n-Octadiene	28.2	57.65
$C_9H_8O_2$	Cinnamic acid	133.0	36.50	$C_{19}H_{40}$	n-Nonadecane	32.1	40.78
$C_9H_8O_2$	Allocinnamic acid	68.0	27.35	$C_{20}H_{42}$	n-Eicosane	36.8	59.11
$C_9H_{10}O_2$	Hydrocinnamic acid	48.0	28.14	$C_{21}H_{44}$	n-Heneicosane	40.5	38.44
C_9H_{12}	Pseudocumene	−43.8	7.47	$C_{22}H_{46}$	n-Docosane	44.4	37.67
C_9H_{12}	Hemimellitine	−25.4	16.65	$C_{23}H_{48}$	n-Tricosane	47.6	30.74
C_9H_{12}	Mesitylene	−44.7	18.90	$C_{24}H_{50}$	n-Tetracosane	50.9	38.74
$C_9H_{18}O_2$	Pelargonic acid	12.35	30.63	$C_{25}H_{52}$	n-Pentacosane	53.7	39.13
C_9H_{20}	n-Nonane	−53.5	28.83	$C_{27}H_{56}$	n-Heptacosane	59.0	37.93
$C_{10}H_7NO_2$	Nitronaphthalene	56.7	25.44	$C_{28}H_{58}$	n-Octacosane	61.4	39.14
$C_{10}H_8$	Naphthalene	80.2	35.06	$C_{57}H_{110}O_6$	Tristearin	54.5	45.63

HEATS OF VAPORIZATION OF ORGANIC COMPOUNDS

Numerical values in the following table are in the units of gram calories per gram mole. To convert to joules per gram mole, multiply the listed value by 4.184.

Formula	Name	ΔH$_v$
CBrN	Cyanogen bromide	10,882.8
CBr$_4$	Carbon tetrabromide	10,771.4
CBrF$_3$	Bromotrifluoromethane	–
CBr$_2$F$_2$	Dibromodifluoromethane	–
CClF$_3$	Chlorotrifluoromethane	3,996.3
CClN	Cyanogen chloride	5,243.4
CCl$_2$F$_2$	Dichlorodifluoromethane	8,363 1
CCl$_2$O	Phosgene	6,224.3
CCl$_3$F	Trichlorofluoromethane	6,424.1
CCl$_3$NO$_2$	Trichloronitromethane	9,109.7
CCl$_4$	Carbon tetrachloride	8,271.5 / 7,628.8
CFN	Cyanogen fluoride	5,875.3
CF$_4$	Carbon tetrafluoride	3,016.5
CHBr$_3$	Tribromomethane	9,673.3
CHClF$_2$	Chlorodifluoromethane	5,212.9
CHCl$_2$F	Dichlorofluoromethane	6,286.8
CHCl$_3$	Chloroform	7,500.5
CHF$_3$	Trifluoromethane	–
CHN	Hydrogen cyanide	7,338.8
CH$_2$Br$_2$	Dibromomethane	8,722.0
CH$_2$Cl$_2$	Dichloromethane	7,572.3
CH$_2$O	Formaldehyde	5,917.9
CH$_2$O$_2$	Formic acid	9,896.5
CH$_3$AsCl$_2$	Dichloromethylarsine	9,636.8
CH$_3$BO	Borine carbonyl	4,867.6
CH$_3$Br	Methyl bromide	5,925.9
CH$_3$Cl	Methyl chloride	5,375.3
CH$_3$Cl$_3$Si	Trichloromethylsilane	7,450.0
CH$_3$F	Methyl fluoride	3,986.4
CH$_3$I	Methyl iodide	6,616.5
CH$_3$NO	Formamide	15,556.6
CH$_3$NO$_2$	Nitromethane	9,210.9
CH$_4$	Methane	2,128.8
CH$_4$Cl$_2$Si	Dichloromethylsilane	7,011.0
CH$_4$O	Methanol	9,377.2 / 8,978.8
CH$_4$S	Methanethiol	6,331.0
CH$_3$ClSi	Chloromethylsilane	6,349.5
CH$_5$N	Methylamine	6,469.5
CH$_6$Si	Methylsilane	4,683.6
C$_3$H$_8$Si	2-Methylvinylsilane	7,195.6
CIN	Cyanogen iodide	14,065.4
CN$_4$O$_8$	Tetranitromethane	9,848.7
CO	Carbon monoxide	1,613.3
COS	Carbonyl sulfide	4,992.2
COSe	Carbonyl selenide	5,366.5
CO$_2$	Carbon dioxide	5,539.0
CSSe	Carbon selenosulfide	8,003.0
CS$_2$	Carbon disulfide	6,786.8
C$_2$BrCl$_2$O	Trichloroacetyl bromide	9,673.9
C$_2$HClF$_2$	1-Chloro-2,2-difluoroethylene	–
C$_2$ClF$_3$	1-Chloro-1,2,2-trifluoroethylene	5,421.5
C$_2$Cl$_2$F$_2$	1,2-Dichloro-1,2-difluoroethylene	7,185.6 / –
C$_2$F$_4$	Tetrafluoroethylene	–
C$_2$Cl$_2$F$_4$	1,1-Dichloro-1,2,2,2-tetrafluoroethane	–
C$_2$Cl$_2$F$_4$	1,2-Dichloro-1,1,2,2-tetrafluoroethane	6,134.6
C$_2$Cl$_3$F$_3$	1,1,2-Trichloro-1,2,2-trifluoroethane	7,115.4
C$_2$Cl$_4$	Tetrachloroethylene	9,240.5
CCl$_4$F$_2$	Tetrachlorodifluoroethane	–
C$_2$Cl$_4$F$_2$	1,1,2,2-Tetrachloro-1,2-difluoroethane	8,746.2
C$_2$Cl$_3$F$_3$	1,1,2-Trifluoro-1,2,2-trichloroethane	–
C$_2$Cl$_6$	Hexachloroethane	11,711.3
C$_2$ClF$_5$	Chloropentafluoroethane	–
C$_2$F$_6$	Hexafluoroethane	–
C$_2$HBr$_3$O	Tribromoacetaldehyde	11,057.8
C$_2$HCl$_3$	Trichloroethylene	8,314.7
C$_2$HCl$_3$O	Trichloroacetaldehyde	8,469.2
C$_2$HCl$_3$O$_2$	Trichloroacetic acid	13,817.9
C$_2$HCl$_5$	Pentachloroethane	9,800.1
C$_2$H$_2$	Acetylene	4,665.8
C$_2$H$_2$Br$_4$	1,1,2,2-Tetrabromoethane	14,517.3
C$_2$H$_2$Br$_4$	1,1,2,2-Tetrabromoethane	12,911.5
C$_2$H$_2$O$_4$	Oxalic acid	21,630.6
C$_2$H$_2$Cl$_2$	cis-1,2-Dichloroethylene	7,420.6
C$_2$H$_2$Cl$_2$	trans-1,2-Dichloroethylene	7,243.1
C$_2$H$_2$Cl$_2$	1,1-Dichloroethylene	7,211.8
C$_2$H$_2$F$_2$	1,1-Difluoroethylene	–
C$_2$H$_2$Cl$_2$O$_2$	Dichloroacetic acid	12,952.9
C$_2$H$_2$Cl$_4$	1,1,1,2-Tetrachloroethane	9,296.5 / 8,725.6
C$_2$H$_2$Cl$_4$	1,1,2,2-Tetrachloroethane	9,917.1
C$_2$H$_3$Br	1-Bromoethylene	6,076.9
C$_2$H$_3$BrO$_2$	Bromoacetic acid	13,537.8
C$_2$H$_3$Br$_3$	1,1,2-Tribromoethane	11,874.1
C$_2$H$_3$Cl	1-Chloroethylene	6,263.0
C$_2$H$_3$ClO$_2$	Chloroacetic acid	13,134.5
C$_2$H$_3$Cl$_3$	1,1,1-Trichloroethane	8,012.7
C$_2$H$_3$Cl$_3$	1,1,2-Trichloroethane	9,163.2
C$_2$H$_3$ClF$_2$	1-Chloro-1,1-difluoroethane	–
C$_2$H$_3$F$_3$	1,1,1-Trifluoroethane	–
C$_2$H$_3$Cl$_3$O$_2$	Trichloroacetaldehyde-hydrate	12,141.5
C$_2$H$_3$F	1-Fluoroethylene	4,198.1
C$_2$H$_3$N	Acetonitrile	8,173.2
C$_2$H$_3$NS	Methyl thiocyanate	9,424.1
C$_2$H$_3$NS	Methyl isothiocyanate	7,990.1
C$_2$H$_4$	Ethylene	3,453.7
C$_2$H$_4$BrCl	1-Bromo-2-chloroethane	9,314.9
C$_2$H$_4$BrCl	1-Bromo-2-chloroethane	8,995.6
C$_2$H$_4$Br$_2$	1,2-Dibromoethane	9,229.4
C$_2$H$_4$Cl$_2$	1,1-Dichloroethane	7,288.0
C$_2$H$_4$Cl$_2$	1,2-Dichloroethane	7,950.7
C$_2$H$_4$F$_2$	1,1-Difluoroethane	6,068.8
C$_2$H$_4$O	Acetaldehyde	7,267.8 / 6,622.1
C$_2$H$_4$O	Ethylene oxide	6,823.7
C$_2$H$_4$O$_2$	Acetic acid	9,963.9 / 9,486.6
C$_2$H$_4$O$_2$	Methyl formate	7,027.8
C$_2$H$_4$O$_2$S	Mercaptoacetic acid	13,790.7
C$_2$H$_5$Br	Ethyl bromide	6,843.1
C$_2$H$_5$Cl	Ethyl chloride	6,310.6
C$_2$H$_5$ClO	2-Chloroethanol	10,740.6
C$_2$H$_5$Cl$_3$Si	Trichloroethylsilane	9,457.8
C$_2$H$_5$Cl$_3$OSi	Trichloroethoxysilane	8,811.4
C$_2$H$_5$F	Ethyl fluoride	5,519.5
C$_2$H$_5$F$_3$Si	Ethyltrifluorosilane	6,945.7
C$_2$H$_5$I	Ethyl iodide	7,851.8
C$_2$H$_5$NO	Acetamide	14,025.3
C$_2$H$_5$NO	Acetaldoxime	11,317.8
C$_2$H$_5$NO$_2$	Nitroethane	9,531.1
C$_2$H$_6$N$_2$O	Dimethylnitrosamine	10,326.7
C$_2$H$_6$	Ethane	3,739.5
C$_2$H$_6$Cl$_2$Si	Dichlorodimethylsilane	7,995.7
C$_2$H$_6$O	Ethyl alcohol	9,673.9
C$_2$H$_6$O	Dimethyl ether	5,409.8
C$_2$H$_6$O$_2$	1,2-Ethanediol	14,032.4
C$_2$H$_6$S	Dimethyl sulfide	6,742.3
C$_2$H$_6$S	Ethane thiol	6,728.7
C$_2$H$_6$Sb	Dimethylantimony	12,075.7
C$_2$H$_7$N	Ethylamine	6,845.1
C$_2$H$_7$N	Dimethylamine	6,660.0
C$_2$H$_8$N$_2$	1,2-Ethanediamine	10,510.5
C$_2$H$_8$Si$_2$	Dimethylsilane	5,497.8
C$_2$H$_{10}$B$_2$	Dimethyldiborane	5,696.7
C$_2$H$_{11}$NSi$_2$	2-Ethyldisilazane	7,348.3
C$_2$N$_2$	Cyanogen	6,597.3
C$_3$H$_3$N	Acrylonitrile	7,941.4
C$_3$H$_4$	Propadiene	5,141.2
C$_3$H$_4$	Propyne	5,632.4
C$_3$H$_4$Br$_2$	2,3-Dibromopropene	9,886.2
C$_3$H$_4$Cl$_2$O$_2$	Methyl dichloroacetate	10,820.5
C$_3$H$_4$O	2-Propenal	7,628.8
C$_3$H$_4$O$_2$	Acrylic acid	10,955.1
C$_3$H$_4$O$_3$	Pyruvic acid	11,815.7
C$_3$H$_5$Br$_3$	1,2,3-Tribromopropane	12,047.1
C$_3$H$_5$Cl	1-Chloropropene	6,594.3
C$_3$H$_5$Cl	Allyl chloride	7,386.8
C$_3$H$_5$ClO	Epichlorohydrin	9,815.4
C$_3$H$_5$ClO$_2$	Methyl chloroacetate	10,815.0
C$_3$H$_5$Cl$_3$	1,1,1-Trichloropropane	8,933.9
C$_3$H$_5$Cl$_3$	1,2,3-Trichloropropane	10,714.3
C$_3$H$_5$Cl$_3$Si	Allyltrichlorosilane	9,386.1
C$_3$H$_5$N	Propionitrile	8,769.0
C$_3$H$_5$NO	3-Hydroxypropionitrile	13,287.2
C$_3$H$_5$NS	Ethylisothiocyanate	9,574.7
C$_3$H$_5$N$_3$O$_9$	Nitroglycerine	13,753.1
C$_3$H$_6$	Propene	4,697.4
C$_3$H$_6$	Cyclopropane	5,897.7
C$_3$H$_6$BrNO	2-Bromo-2-nitrosopropane	9,619.6
C$_3$H$_6$Br$_2$	1,2-Dibromopropane	9,801.9
C$_3$H$_6$Br$_2$	1,3-Dibromopropane	10,374.4
C$_3$H$_6$Br$_2$O	2,3-Dibromo-1-propanol	13,190.0
C$_3$H$_6$Cl$_2$	1,2-Dichloropropane	8,428.5
C$_3$H$_6$Cl$_2$O	1,3-Dichloro-2-propanol	12,067.6
C$_3$H$_6$O	Acetone	7,641.5
C$_3$H$_6$O	Allyl alcohol	10,577.7
C$_3$H$_6$O	Propylene oxide	7,295.8
C$_3$H$_6$O$_2$	Propanoic acid	12,454.4
C$_3$H$_6$O$_2$	Methyl acetate	7,732.8
C$_3$H$_6$O$_2$	Ethyl formate	7,511.7
C$_3$H$_6$O$_3$	Methyl glycolate	11,105.0
C$_3$H$_6$O$_3$	Methoxyacetic acid	13,451.0
C$_3$H$_7$Br	n-Propyl bromide	8,029.8
C$_3$H$_7$Br	2-Bromopropane	7,591.7
C$_3$H$_7$Cl	n-Propyl chloride	7,485.7 / 6,905.8
C$_3$H$_7$Cl	2-Chloropropane	6,855.2
C$_3$H$_7$Cl$_3$Si	Trichloroisopropylsilane	8,973.3
C$_3$H$_7$I	n-Propyl iodide	8,467 1
C$_3$H$_7$I	2-Iodopropane	8,243.4
C$_3$H$_7$NO	Propionamide	14,554.0
C$_3$H$_7$NO$_2$	1-Nitropropane	9,949.9
C$_3$H$_7$NO$_2$	2-Nitropropane	9,476.9
C$_3$H$_7$NO$_2$	Ethyl carbamate	13,078.6
C$_3$H$_8$	Propane	4,550.0 / 4,811.8
C$_3$H$_8$O	n-Propanol	11,298.8 / 10,421.1
C$_3$H$_8$O	Isopropanol	10,063.5
C$_3$H$_8$O	Ethyl methyl ether	6,388.3
C$_3$H$_8$O$_2$	1,2-Propanediol	13,575.2
C$_3$H$_8$O$_2$	1,3-Propanediol	13,782.3
C$_3$H$_8$O$_2$	2-Methoxyethanol	9,893.8
C$_3$H$_8$O$_3$	Glycerol	18,188.9
C$_3$H$_8$S	Methyl ethyl sulfide	
C$_3$H$_8$S	Propanethiol	7,855.3
C$_3$H$_9$B	Trimethylborine	5,375.4
C$_3$H$_9$ClSi	Chlorotrimethylsilane	7,589.1
C$_3$H$_9$Ga	Trimethylgallium	7,758.8
C$_3$H$_9$N	n-Propylamine	7,408.0
C$_3$H$_9$N	Trimethylamine	6,361.7
C$_3$H$_9$O$_3$P	Trimethyl phosphite	11,010.0
C$_3$H$_{12}$B$_2$	Trimethyldiborane	6,981.8
C$_3$O$_2$	Carbon suboxide	6,446.3
C$_3$S$_2$	Carbon subsulfide	10,466.0
C$_4$Cl$_6$O$_3$	Trichloroacetic anhydride	12,929.0
C$_3$F$_8$	Octafluoropropane	–
C$_4$H$_2$O	Furan	–
C$_4$H$_2$	1,3-Butadiyne	7,761.0
C$_4$H$_2$Br$_2$O$_3$	α,β-Dibromomaleic anhydride	12,579.2
C$_4$H$_2$Cl$_2$O$_2$	trans-Fumaryl chloride	11,251.0
C$_4$H$_2$O$_3$	Maleic anhydride	12,122.3
C$_4$H$_3$NO$_2$S	2-Nitrothiophene	11,926.2
C$_4$H$_4$	Butenyne	6,677.2
C$_4$H$_4$Cl$_2$O$_2$	Succinyl chloride	12,466.1
C$_4$H$_4$O$_3$	Chloromaleic anhydride	14,645.1
C$_4$H$_4$O$_3$	Succinic anhydride	14,726.0
C$_4$H$_4$O$_4$	1,4-Dioxane-2,6-dione	14,013.6
C$_4$H$_4$S	Thiophene	8,748.3
C$_4$H$_4$Se	Selenophene	7,766.1
C$_4$H$_5$ClO$_2$	α-Chlorocrotonic acid	15,440.1
C$_4$H$_5$ClO$_3$	Ethyl chloroglyoxylate	10,268.4
C$_4$H$_5$ClO$_2$	Ethyl chloroacetate	11,625.1
C$_4$H$_5$N	3-Butenenitrile	9,447.8
C$_4$H$_5$N	Methacrylonitrile	8,083.8
C$_4$H$_5$N	cis-Crotononitrile	8,905.4
C$_4$H$_5$N	trans-Crotononitrile	9,277.1
C$_4$H$_5$NO$_2$	Succinimide	16,422.0
C$_4$H$_5$NS	Allylisothiocyanate	9,967.8
C$_4$H$_6$	1,2-Butadiene	6,539.1
C$_4$H$_6$	1,3-Butadiene	5,688.2
C$_4$H$_6$	Cyclobutene	6,167.5
C$_4$H$_6$	1-Butyne	6,596.9
C$_4$H$_6$	2-Butyne	7,868.5
C$_4$H$_6$Cl$_2$O$_2$	Ethyl dichloroacetate	10,842.8
C$_4$H$_6$Cl$_2$O$_2$	2-Chloroethyl chloroacetate	12,588.7
C$_4$H$_6$O$_2$	cis-Crotonic acid	12,964.7
C$_4$H$_6$O$_2$	trans-Crotonic acid	13,252.2
C$_4$H$_6$O$_2$	Methyl acrylate	8,598.0
C$_4$H$_6$O$_2$	Methacrylic acid	12,526.6
C$_4$H$_6$O$_2$	Vinyl acetate	8,470.4
C$_4$H$_6$O$_3$	Acetic anhydride	10,930.4
C$_4$H$_6$O$_4$	Dimethyl oxalate	11,519.4
C$_4$H$_7$Br	cis-1-Bromo-1-butene	8,300.2
C$_4$H$_7$Br	trans-1-Bromo-1-butene	8,515.7
C$_4$H$_7$Br	2-Bromo-1-butene	8,389.7
C$_4$H$_7$Br	cis-2-Bromo-2-butene	8,486.3
C$_4$H$_7$Br	trans-2-Bromo-2-butene	8,238.1
C$_4$H$_7$BrO	1-Bromo-2-butanone	10,980.7
C$_4$H$_7$BrO	2-Methylpropionyl bromide	10,974.6
C$_4$H$_7$Br$_3$	1,1,2-Tribromobutane	11,936.5
C$_4$H$_7$Br$_3$	1,2,2-Tribromobutane	11,622.3

HEATS OF VAPORIZATION OF ORGANIC COMPOUNDS (continued)

Formula	Name	ΔH$_v$
C$_4$H$_7$Br$_3$	1,2,2,-Tribromobutane	11,622.3
C$_4$H$_7$Br$_3$	2,2,3-Tribromobutane	11,664.2
C$_4$H$_7$ClO$_2$	Ethyl chloroacetate	10,522.6
C$_4$H$_7$Cl$_3$	1,1-Dichloro-2-methylpropane	9,111.1
C$_4$H$_7$Cl$_3$	1,2,3-Trichlorobutane	9,447.0
C$_4$H$_7$N	Butyronitrile	9,462.9
C$_4$H$_7$NO$_2$	Diacetamide	14,508.1
C$_4$H$_8$	1-Butene	5,996.7
C$_4$H$_8$	cis-2-Butene	6,401.0
C$_4$H$_8$	trans-2-Butene	6,221.6
C$_4$H$_8$	2-Methylpropene	5,742.9
C$_4$H$_8$	Cyclobutane	6,464.8
C$_4$H$_8$BrClO	2-Bromoethyl-2-chloroethyl ether	12,010.5
C$_4$H$_8$Br$_2$	1,2-Dibromobutane	10,182.1
C$_4$H$_8$Br$_2$	dl-2,3-Dibromobutane	10,136.1
C$_4$H$_8$Br$_2$	meso-2,3-Dibromobutane	9,966.9
C$_4$H$_8$Br$_2$	1,4-Dibromobutane	11,369.3
C$_4$H$_8$Br$_2$O	Di(2-bromoethyl)ether	12,454.4
C$_4$H$_8$Cl$_2$	1,2-Dichlorobutane	8,850.6
C$_4$H$_8$Cl$_2$	2,3-Dichlorobutane	8,975.3
C$_4$H$_8$Cl$_2$	1,1-Dichloro-2-methylpropane	8,795.6
C$_4$H$_8$Cl$_2$	1,2-Dichloro-2-methylpropane	9,260.1
C$_4$H$_8$Cl$_2$	1,3-Dichloro-2-methylpropane	10,519.7
C$_4$H$_8$Cl$_2$O	Di(chloroethyl)ether	11,376.8
C$_4$H$_8$O	1,2-Epoxy-2-methylpropane	7,066.6
C$_4$H$_8$O	Methyl ethyl ketone	8,149.5
C$_4$H$_8$O$_2$	Dioxane	8,546.2
C$_4$H$_8$O$_2$	n-Butyric acid	11,881.2
C$_4$H$_8$O$_2$	Isobutyric acid	11,182.8
C$_4$H$_8$O$_2$	Ethyl acetate	8,301.1
C$_4$H$_8$O$_2$	Methyl propanoate	8,356.2
C$_4$H$_8$O$_2$	n-Propyl formate	8,208.1
C$_4$H$_8$O$_2$	Isopropyl formate	8,230.2
C$_4$H$_8$O$_3$	α-Hydroxyisobutyric acid	15,967.0
C$_4$H$_8$O$_3$	Ethyl glycolate	11,318.1
C$_4$H$_9$Br	n-Butyl bromide	8,789.1
C$_4$H$_9$BrO	1-Bromo-2-butanol	13,473.7
C$_4$H$_9$Cl	n-Butyl chloride	8,144.8
C$_4$H$_9$Cl	sec-Butyl chloride	7,407.9
C$_4$H$_9$Cl	Isobutyl chloride	8,045.1
C$_4$H$_9$Cl	tert-Butyl chloride	6,876.0
C$_4$H$_9$ClO$_2$	2-(2-Chloroethoxy)ethanol	14,082.1
C$_4$H$_9$I	n-Butyl iodide	–
C$_4$H$_9$I	1-Iodo-2-methylpropane	9,650.7
C$_4$H$_9$NO$_2$	Ethyl methylcarbamate	12,161.2
C$_4$H$_9$NO$_2$	Propyl carbamate	14,071.8
C$_4$H$_9$N$_2$O$_2$	Di(nitrosoethyl)amine	10,894.8
C$_4$H$_{10}$	n-Butane	5,801.2
C$_4$H$_{10}$	2-Methylpropane	5,084.4
		5,416.2
C$_4$H$_{10}$Cl$_2$Si	Dichlorodiethylsilane	10,038.6
C$_4$H$_{10}$F$_2$Si	Diethyldifluorosilane	8,214.9
C$_4$H$_{10}$O	n-Butyl alcohol	10,970.5
C$_4$H$_{10}$O	sec-Butyl alcohol	10,712.3
C$_4$H$_{10}$O	Isobutyl alcohol	10,936.0
C$_4$H$_{10}$O	tert-Butyl alcohol	10,413.2
C$_4$H$_{10}$O	Diethyl ether	6,946.2
C$_4$H$_{10}$O	Methyl propyl ether	7,409.7
C$_4$H$_{10}$O$_2$	1,3-Butanediol	10,479.1
C$_4$H$_{10}$O$_2$	2,3-Butanediol	13,708.6
C$_4$H$_{10}$O$_2$	1,2-Dimethoxyethane	7,681.0
C$_4$H$_{10}$O$_2$S	2,2-Thiodiethanol	6,597.0
C$_4$H$_{10}$O$_3$	Diethylene glycol	16,146.7
C$_4$H$_{10}$O$_3$	1,2,3-Butanetriol	16,345.8
C$_4$H$_{10}$O$_3$S	Diethyl sulfite	10,783.0
C$_4$H$_{10}$O$_4$S	Diethyl sulfate	12,518.2
C$_4$H$_{10}$S	n-Butanethiol	–
C$_4$H$_{10}$S	Diethyl sulfide	8,210.8
C$_4$H$_{10}$Se	Diethyl selenide	9,274.7
C$_4$H$_{10}$Zn	Diethyl zinc	9,162.3
C$_4$H$_{11}$N	Diethyl amine	7 307.5
C$_4$H$_{11}$N	Isobutylamine	478.3
C$_4$H$_{12}$Cl$_2$Si$_2$	1,3-Diethoxytetramethyl-disiloxane	9,881.6
C$_4$H$_{12}$Pb	Tetramethyllead	8,843.8
C$_4$H$_{12}$Si	Tetramethylsilane	6,439.2
C$_4$H$_{12}$Sn	Tetramethyl tin	7,897.8
C$_4$H$_{14}$B$_2$	Tetramethyldiborane	7,517.1
C$_4$F$_8$	Octafluorocyclobutane	–
C$_4$F$_{10}$	Perfluoro-n-butane	–
C$_5$H$_4$BrN	3-Bromopyridine	10,863.7
C$_5$H$_4$ClN	2-Chloropyridine	10,614.5
C$_5$H$_4$O$_2$	2-Furaldehyde	11,614.6
C$_5$H$_4$O$_3$	Citraconic anhydride	12,307.8
C$_5$H$_5$N	Pyridine	9,649.4
C$_5$H$_6$Cl$_2$O$_2$	Glutaryl chloride	13,192.1
C$_5$H$_6$N$_2$	Glutaronitrile	13,767.5
C$_5$H$_6$O$_2$	Furfuryl alcohol	12,815.8
C$_5$H$_6$O$_3$	Glutaric anhydride	14,814.1
C$_5$H$_6$O$_3$	Pyrotartaric anhydride	13,251.2
C$_5$H$_6$S	2-Methylthiophene	8,884.2
C$_5$H$_6$S	3-Methylthiophene	9,084.1
C$_5$H$_7$ClO$_2$	Propyl chlorogyloxylate	11,430.0
C$_5$H$_7$N	Tiglonitrile	8,704.6
C$_5$H$_7$N	Angelonitrile	9,707.5
C$_5$H$_7$N	α-Ethylacrylonitrile	8,679.1
C$_5$H$_7$NO$_2$	Ethyl cyanoacetate	15,615.6
C$_5$H$_8$	Cyclopentene	–
C$_5$H$_8$	Isoprene	6,901.8
C$_5$H$_8$	1,3-Pentadiene	7,313.9
C$_5$H$_8$	1,4-Pentadiene	6,826.6
C$_5$H$_8$O	Tiglaldehyde	9,009.2
C$_5$H$_8$O$_2$	Levulinaldehyde	11,483.8
C$_5$H$_8$O$_2$	Tiglic acid	13,756.5
C$_5$H$_8$O$_2$	α-Valerolactone	11,537.0
C$_5$H$_8$O$_2$	α-Ethylacrylic acid	14,417.8
C$_5$H$_8$O$_2$	Ethyl acrylate	9,259.4
C$_5$H$_8$O$_2$	Methyl methacrylate	8,974.9
C$_5$H$_8$O$_3$	Levulinic acid	17,795.0
C$_5$H$_8$O$_4$	Glutaric acid	22,085.2
C$_5$H$_8$O$_4$	Dimethyl malonate	12,608.1
C$_5$H$_9$ClO$_2$	Ethyl α-chloropropionate	11,032.8
C$_5$H$_9$ClO$_2$	Isopropyl chloroacetate	10,575.7
C$_5$H$_9$N	Valeronitrile	9,931.3
C$_5$H$_9$NO	α-Hydroxybutyronitrile	13,577.0
C$_5$H$_{10}$	1-Pentene	6,931.2
C$_5$H$_{10}$	2-Pentene	–
C$_5$H$_{10}$	3-Methyl-2-butene	7,112.8
C$_5$H$_{10}$	2-Methyl-1-butene	6,474.6
C$_5$H$_{10}$	3-Methyl-1-butene	–
C$_5$H$_{10}$	Cyclopentane	7,411.1
C$_5$H$_{10}$	Methylcyclobutane	6,413.2
C$_5$H$_{10}$Br$_2$	1,2-Dibromopentane	11,130.0
C$_5$H$_{10}$Br$_2$	1,2-Dibromo-2-methylbutane	7,616.9
C$_5$H$_{10}$Br$_2$	1,3-Dibromo-3-methylbutane	10,639.6
C$_5$H$_{10}$Cl$_2$Si	Allyldichloroethylsilane	9,833.9
C$_5$H$_{10}$O	Diethyl ketone	11,183.0
C$_5$H$_{10}$O	Methyl n-propyl ketone	11,240.6
C$_5$H$_{10}$O	Methyl isopropyl ketone	11,073.2
C$_5$H$_{10}$Cl$_2$O	2-Chloroethyl 2-chloroisopropyl ether	11,420.8
C$_5$H$_{10}$Cl$_2$O	2-Chloroethyl 2-chloropropyl ether	11,316.9
C$_5$H$_{10}$Cl$_2$O$_2$	Di(2-chloroethoxy)methane	12,908.0
C$_5$H$_{10}$O$_2$	4-Hydroxy-3-methyl-2-butanone	13,639.4
C$_5$H$_{10}$O$_2$	Valeric acid	13,370.3
C$_5$H$_{10}$O$_2$	Isovaleric acid	12,951.1
C$_5$H$_{10}$O$_2$	Ethyl propanoate	8,877.8
C$_5$H$_{10}$O$_2$	n-Propyl acetate	8,921.1
C$_5$H$_{10}$O$_2$	Isopropyl acetate	8,794.8
C$_5$H$_{10}$O$_2$	Methyl butyrate	8,886.0
C$_5$H$_{10}$O$_2$	Methyl isobutyrate	8,593.3
C$_5$H$_{10}$O$_2$	n-Butyl formate	9,285.9
C$_5$H$_{10}$O$_2$	Isobutyl formate	8,678.8
C$_5$H$_{10}$O$_2$	sec-Butyl formate	8,975.7
C$_5$H$_{10}$O$_2$	tert-Butyl formate	8,955.3
C$_5$H$_{10}$O$_2$	Diethyl carbonate	10,159.0
C$_5$H$_{10}$O$_3$	1-Bromopentane	–
C$_5$H$_{11}$Br	1-Bromo-3-methylbutane	9,282.7
C$_5$H$_{11}$Br	1-Chloropentane	–
C$_5$H$_{11}$I	1-Iodopentane	–
C$_5$H$_{11}$I	1-Iodo-3-methylbutane	9,951.6
C$_5$H$_{11}$I	Piperidine	8,911.8
C$_5$H$_{11}$N	Pentanoic acid	–
C$_5$H$_{11}$NO$_2$	Isobutyl carbamate	13,897.1
C$_5$H$_{11}$NO$_3$	Isoamyl nitrate	10,817.2
C$_5$H$_{12}$	n-Pentane	6,595.1
C$_5$H$_{12}$	2-Methylbutane	6,470.8
C$_5$H$_{12}$	2,2-Dimethylpropane	5,648.6
C$_5$H$_{12}$O	Amyl alcohol	12,495.5
C$_5$H$_{12}$O	Isoamyl alcohol	12,497.9
C$_5$H$_{12}$O	2-Pentanol	12,086.2
C$_5$H$_{12}$O	tert-Amyl alcohol	11,239.2
C$_5$H$_{12}$O	Ethyl propyl ether	7,092.7
C$_5$H$_{12}$O	Methyl n-butyl ether	–
C$_5$H$_{12}$O$_3$	2,3,4-Pentanetriol	19,694.4
C$_5$H$_{12}$S	1-Pentanethiol	–
C$_5$H$_{14}$OSi	Ethoxytrimethylsilane	8,030.6
C$_5$H$_{14}$Si	Ethyltrimethylsilane	7,633.4
C$_5$H$_{14}$Sn	Ethyltrimethyltin	8,820.9
C$_6$Cl$_4$O$_2$	Chloranil	21,514.3
C$_6$Cl$_6$	Hexachlorobenzene	15,199.1
C$_6$HCl$_5$	Pentachlorobenzene	15,124.2
C$_6$HCl$_5$O	Pentachlorophenol	16,742.6
C$_6$H$_2$BrCl$_3$O	3-Bromo-2,4,6-trichlorophenol	15,231.9
C$_6$H$_2$Cl$_4$	1,2,3,4-Tetrachlorobenzene	12,872.5
C$_6$H$_2$Cl$_4$	1,2,3,5-Tetrachlorobenzene	11,982.1
C$_6$H$_2$Cl$_4$	1,2,4,5-Tetrachlorobenzene	12,828.8
C$_6$H$_2$Cl$_4$O	2,3,4,6-Tetrachlorophenol	15,362.7
C$_6$H$_3$BrCl$_2$O	2-Bromo-4,6-dichlorophenol	13,829.1
C$_6$H$_3$Cl$_3$	1,2,3-Trichlorobenzene	11,349.5
C$_6$H$_3$Cl$_3$	1,2,4-Trichlorobenzene	11,425.1
C$_6$H$_3$Cl$_3$	1,3,5-Trichlorobenzene	11,211.0
C$_6$H$_3$Cl$_3$O	2,4,5-Trichlorophenol	13,237.0
C$_6$H$_3$Cl$_3$O	2,4,6-Trichlorophenol	14,092.8
C$_6$H$_4$Br$_2$	1,4-Dibromobenzene	13,047.8
C$_6$H$_4$BrCl	1,4-Bromochlorobenzene	16,671.8
		11,451.1
C$_6$H$_4$Cl$_2$	1,2-Dichlorobenzene	10,943.0
C$_6$H$_4$Cl$_2$	1,3-Dichlorobenzene	10,446.8
C$_6$H$_4$Cl$_2$	1,4-Dichlorobenzene	17,260.5
		10,611.0
C$_6$H$_4$Cl$_2$O	2,4-Dichlorophenol	13,230.4
C$_6$H$_4$Cl$_2$O	2,6-Dichlorophenol	13,472.0
C$_6$H$_4$Cl$_3$N	2,4,6-Trichloroaniline	22,297.3
C$_6$H$_5$AsCl$_2$	Dichlorophenylarsine	12,229.5
C$_6$H$_5$Br	Bromobenzene	10,157.7
C$_6$H$_5$Cl	Chlorobenzene	10,098.0
		9,067.3
C$_6$H$_5$ClO	2-Chlorophenol	10,341.1
C$_6$H$_5$ClO	3-Chlorophenol	11,979.7
C$_6$H$_5$ClO	4-Chlorophenol	12,281.6
C$_6$H$_5$ClO$_2$S	Benzenesulfonylchloride	12,621.0
C$_6$H$_5$Cl$_2$O$_2$P	Phenyl dichlorophosphate	13,319.6
C$_6$H$_5$Cl$_3$Si	Trichlorophenylsilane	11,385.9
C$_6$H$_5$F	Fluorobenzene	7,980.4
C$_6$H$_5$F$_3$Si	Trifluorophenylsilane	9,171.6
C$_6$H$_5$I	Iodobenzene	10,277.2
		10,377.8
C$_6$H$_5$NO$_2$	Nitrobenzene	12,168.2
C$_6$H$_5$NO$_3$	2-Nitrophenol	12,497.3
C$_6$H$_6$	1,5-Hexadiene-3-yne	8,288.0
C$_6$H$_6$	Benzene	10,254.3
		8,146.5
C$_6$H$_6$ClN	2-Chloroaniline	12,441.0
C$_6$H$_6$ClN	3-Chloroaniline	13,385.6
C$_6$H$_6$ClN	4-Chloroaniline	12,832.8
C$_6$H$_6$ClO	4-Chlorophenol	12,964.7
C$_6$H$_6$N$_2$O$_2$	2-Nitroaniline	15,284.0
C$_6$H$_6$N$_2$O$_2$	3-Nitroaniline	15,996.3
C$_6$H$_6$N$_2$O$_2$	4-Nitroaniline	17,220.2
C$_6$H$_6$O	Phenol	11,891.5
C$_6$H$_6$O$_2$	Pyrocatechol	13,779.7
C$_6$H$_6$O$_2$	Resorcinol	16,400.8
C$_6$H$_6$O$_2$	Hydroquinone	18,734.0
C$_6$H$_6$O$_3$	Pyrogallol	15,731.8
C$_6$H$_6$S	Benzenethiol	11,320.1
C$_6$H$_7$N	Aniline	11,307.6
C$_6$H$_7$N	2-Picoline	9,933.2
C$_6$H$_7$N	3-Methylpyridine (β-picoline)	–
C$_6$H$_8$	1,3-Cyclohexadiene	–
C$_6$H$_8$Cl$_2$O$_4$	Ethylene-bis-chloroacetate	16,499.1
C$_6$H$_8$N$_2$	1,3-Phenylenediamine	14,761.1
C$_6$H$_8$N$_2$	Phenylhydrazine	13,711.9
C$_6$H$_8$O$_3$	α-Methylglutaric anhydride	14,204.9
C$_6$H$_8$O$_3$	α,α-Dimethylsuccinic anhydride	13,683.1
C$_6$H$_8$O$_4$	Dimethyl maleate	12,615.7
C$_6$H$_{10}$	Cyclohexene	–
C$_6$H$_{10}$	1,5-Hexadiene	–
C$_6$H$_{10}$Cl$_2$O$_2$	Isobutyl dichloroacetate	11,733.1
C$_6$H$_{10}$Cl$_2$Si	Diallyldichlorosilane	10,462.8
C$_6$H$_{10}$O	Cyclohexanone	10,037.6
C$_6$H$_{10}$O	Mesityl oxide	10,109.4
C$_6$H$_{10}$O$_2$	Isocaprolactone	11,685.0
C$_6$H$_{10}$O$_3$	Propionic anhydride	11,572.6
C$_6$H$_{10}$O$_3$	Ethyl acetoacetate	11,842.0
C$_6$H$_{10}$O$_3$	Methyl levulinate	12,249.8
C$_6$H$_{10}$O$_4$	Adipic acid	19,570.2
C$_6$H$_{10}$O$_4$	Diethyl oxalate	14,016.9
C$_6$H$_{10}$O$_4$	Glycol diacetate	12,496.1
C$_6$H$_{10}$O$_5$	Dimethyl-l-malate	14,127.6
C$_6$H$_{10}$O$_6$	Dimethyl-d-tartrate	15,372.6
C$_6$H$_{10}$O$_6$	Dimethyl-dl-tartrate	14,999.1
C$_6$H$_{10}$S	Diallyl sulfide	9,652.6
C$_6$H$_{11}$BrO$_2$	Ethyl α-bromoisobutyrate	10,635.8
C$_6$H$_{11}$ClO$_2$	sec-Butylchloroacetate	11,152.0
C$_6$H$_{11}$N	Capronitrile	10,492.3
C$_6$H$_{12}$	1-Hexene	7,787.6
C$_6$H$_{12}$	2-Hexene	–
C$_6$H$_{12}$	Cyclohexane	7,830.9
C$_6$H$_{12}$	Methylcyclopentane	7,940.0
C$_6$H$_{12}$Cl$_2$O	Dichlorodiisopropyl ether	11,881.1
C$_6$H$_{12}$Cl$_2$O$_2$	bis(2-Chloroethyl)acetal	13,497.1
C$_6$H$_{12}$O	2-Hexanone	12,358.3
C$_6$H$_{12}$O	4-Methyl-2-pentanone	11,669.6
C$_6$H$_{12}$O	Allyl propyl ether	8,621.5
C$_6$H$_{12}$O	Allyl isopropyl ether	8,637.5
C$_6$H$_{12}$O	Cyclohexanol	11,935.8
C$_6$H$_{12}$O$_2$	Caproic acid	16,189.4
C$_6$H$_{12}$O$_2$	Isocaproic acid	14,874.8
C$_6$H$_{12}$O$_2$	4-Hydroxy-4-methyl-2-pentanone	11,718.8

Formula	Name	ΔH_V
$C_6H_{12}O_2$	Methyl pentanoate	–
$C_6H_{12}O_2$	Methyl isovalerate	9,567.5
$C_6H_{12}O_2$	Ethyl n-butyrate	9,468.5
$C_6H_{12}O_2$	Ethyl isobutyrate	8,945.7
$C_6H_{12}O_2$	n-Propyl propanoate	9,857.2
$C_6H_{12}O_2$	n-Butyl acetate	–
$C_6H_{12}O_2$	Isobutyl acetate	9,300.8
$C_6H_{12}O_2$	n-Amyl formate	–
$C_6H_{12}O_2$	Isoamyl formate	9,438.2
$C_6H_{12}O_3$	Paraformaldehyde	10,348.2
C_6H_{14}	Hexane	7,627.2
C_6H_{14}	2-Methylpentane	7,676.6
C_6H_{14}	3-Methylpentane	7,743.9
C_6H_{14}	2,2-Dimethylbutane	7,271.0
C_6H_{14}	2,3-Dimethylbutane	7,120.0
$C_6H_{14}O$	1-Hexanol	12,708.5
$C_6H_{14}O$	2-Hexanol	12,386.5
$C_6H_{14}O$	3-Hexanol	11,157.9
$C_6H_{14}O$	2-Methyl-1-pentanol	12,036.6
$C_6H_{14}O$	2-Methyl-2-pentanol	11,132.0
$C_6H_{14}O$	2-Methyl-4-pentanol	10,985.5
$C_6H_{14}O$	Ethyl butyl ether	–
$C_6H_{14}O$	Di-n-propyl ether	8,229.6
$C_6H_{14}O$	Diisopropyl ether	7,777.3
$C_6H_{14}O_2$	Acetal	9,853.9
$C_6H_{14}O_2$	1,2-Diethoxyethane	8,102.6
$C_6H_{14}O_3$	Di(2-methoxyethyl)ether	11,105.2
$C_6H_{14}O_3$	Diethyleneglycol-diethyl ether	12,669.0
$C_6H_{14}O_3$	Dipropyleneglycol	14,610.4
$C_6H_{14}O_4$	Triethyleneglycol	17,097.1
$C_6H_{15}N$	Dipropyl sulfide	–
$C_6H_{15}N$	Di-n-Propylamine	–
$C_6H_{15}N$	Triethylamine	–
$C_6H_{15}B$	Triethylboron	2,535.0
$C_6H_{15}ClSi$	Chlorotriethylsilane	9,806.9
$C_6H_{15}O_4P$	Triethyl phosphate	11,549.9
$C_6H_{15}Tl$	Triethylthallium	9,458.6
$C_6H_{15}...$	(illegible)	9,768.2
$C_6H_{16}Si$	Trimethylpropylsilane	7,964.6
$C_6H_{16}Sn$	Trimethylpropyltin	9,659.6
$C_6H_{18}Cl_2O_2Si_2$	1,5-Dichlorohexamethyltrisilanane	11,391.5
$C_6H_{18}O_3Si_3$	Hexamethylcyclotrisiloxane	10,503.3
$C_7H_3Cl_2F_3$	3,4-Dichloro-α,α,α-trifluorotoluene	10,253.5
$C_7H_4ClF_3$	2-Chloro-α,α,α-trifluorotoluene	10,016.9
$C_7H_4Cl_4$	2-α,α,α-Tetrachlorotoluene	12,501.3
C_7H_5BrO	Benzoyl bromide	12,070.8
C_7H_5ClO	Benzoyl chloride	11,438.0
$C_7H_5Cl_3$	α,α,α-Trichlorotoluene	12,168.6
$C_7H_5F_3$	α,α,α-Trifluorotoluene	8,869.7
C_7H_5N	Benzonitrile	11,341.0
C_7H_5N	Phenyl isocyanide	10,736.7
C_7H_5NO	Phenyl isocyanate	10,556.7
$C_7H_5NO_3$	2-Nitrobenzaldehyde	13,773.6
$C_7H_5NO_3$	3-Nitrobenzaldehyde	14,726.9
C_7H_5NS	Phenyl isothiocyanate	12,132.7
$C_7H_6Cl_2$	α,α-Dichlorotoluene	11,075.9
C_7H_6O	Benzaldehyde	11,657.8
$C_7H_6O_2$	Benzoic acid	15,253.3
		16,295.1
$C_7H_6O_2$	Salicylaldehyde	11,536.5
$C_7H_6O_2$	4-Hydroxybenzaldehyde	16,043.4
$C_7H_6O_3$	Salicylic acid	18,920.7
C_7H_7Br	α-Bromotoluene	11,360.4
C_7H_7Br	2-Bromotoluene	11,365.0
C_7H_7Br	3-Bromotoluene	10,537.1
C_7H_7Br	4-Bromotoluene	10,076.2
C_7H_7BrO	4-Bromoanisole	12,075.4
C_7H_7Cl	α-Chlorotoluene	11,158.7
C_7H_7Cl	2-Chlorotoluene	10,279.3
C_7H_7Cl	3-Chlorotoluene	10,081.1
C_7H_7Cl	4-Chlorotoluene	10,151.7
C_7H_7F	2-Fluorotoluene	9,164.8
C_7H_7F	3-Fluorotoluene	9,251.8
C_7H_7F	4-Fluorotoluene	9,281.0
C_7H_7I	2-Iodotoluene	11,380.7
$C_7H_7NO_2$	2-Nitrotoluene	12,239.1
$C_7H_7NO_2$	3-Nitrotoluene	11,831.1
$C_7H_7NO_2$	4-Nitrotoluene	11,915.0
C_7H_8	Toluene	9,368.5
		8,580.5
$C_7H_8Cl_2Si$	Benzyldichlorosilane	13,128.7
$C_7H_8Cl_2Si$	Dichloromethylphenylsilane	11,464.7
$C_7H_8Cl_2Si$	Dichloro-4-tolysilane	13,125.7
C_7H_8O	Anisole	10,440.9
C_7H_8O	Benzyl alcohol	14,093.2
C_7H_8O	o-Cresol	12,487.3
C_7H_8O	m-Cresol	13,483.8
C_7H_8O	p-Cresol	13,611.7
$C_7H_8O_2$	3,5-Dimethyl-1,2-pyrone	14,470.6
$C_7H_8O_2$	2-Methoxyphenol	13,425.8
$C_7H_8O_3$	Ethyl 2-furoate	12,144.0
C_7H_9N	2,6-Dimethylpyridine	–
C_7H_9N	Benzylamine	11,703.2
C_7H_9N	N-Methylaniline	11,982.3
C_7H_9N	2-Toluidine	12,663.4
C_7H_9N	3-Toluidine	12,104.1
$C_7H_{10}O_4$	Dimethyl citraconate	12,917.3
$C_7H_{10}O_4$	Dimethyl itaconate	15,613.7
$C_7H_{10}O_4$	trans-Dimethyl mesaconate	12,688.1
$C_7H_{11}NO_2$	2-Cyano-2-butyl acetate	12,720.8
$C_7H_{12}O_2$	Butyl acrylate	10,194.0
C_7H_9N	4-Toluidine	12,428.6
C_7H_9NO	2-Methoxyaniline	13,684.6
$C_7H_{10}N_2$	Toluene-2,4-diamine	15,920.1
$C_7H_{10}N_2$	4-Tolylhydrazine	15,063.1
$C_7H_{10}O_3$	Trimethylsuccinic anhydride	12,196.7
$C_7H_{12}O_3$	Ethyl levulinate	12,733.6
$C_7H_{12}O_4$	Pimelic acid	19,840.8
$C_7H_{12}O_4$	Diethyl malonate	12,227.7
$C_7H_{13}ClO$	Enanthyl chloride	15,242.7
$C_7H_{13}N$	Heptanonitrile	10,830.5
C_7H_{14}	Ethylcyclopentane	8,797.7
C_7H_{14}	2-Heptene	8,643.2
C_7H_{14}	Methylcyclohexane	8,549.2
$C_7H_{14}O$	Enanthaldehyde	11,413.4
$C_7H_{14}O$	2-Heptanone	12,478.9
$C_7H_{14}O$	4-Heptanone	13,451.9
$C_7H_{14}O$	2,5-Dimethyl-3-pentanone	12,266.9
$C_7H_{14}O_2$	Enanthic acid	15,893.8
$C_7H_{14}O_2$	Methyl caproate	10,676.8
$C_7H_{14}O_2$	Ethyl isovalerate	10,183.9
$C_7H_{14}O_2$	Propyl butyrate	10,283.7
$C_7H_{14}O_2$	Propyl isobutyrate	10,259.7
$C_7H_{14}O_2$	Isopropyl isobutyrate	9,717.6
$C_7H_{14}O_2$	Isobutyl propionate	10,495.8
$C_7H_{14}O_2$	Isoamyl acetate	10,494.9
C_7H_{16}	Perfluoro-n-Heptane	–
C_7H_{16}	n-Heptane	8,928.8
		8,409.6
C_7H_{16}	2-Methylhexane	8,538.7
C_7H_{16}	3-Methylhexane	8,598.3
C_7H_{16}	3-Ethylpentane	8,642.8
C_7H_{16}	2,2-Dimethylpentane	8,106.7
C_7H_{16}	2,3-Dimethylpentane	8,390.9
C_7H_{16}	2,4-Dimethylpentane	8,167.1
C_7H_{16}	3,3-Dimethylpentane	8,145.4
C_7H_{16}	2,2,3-Trimethylbutane	7,767.1
$C_7H_{16}O$	n-Heptanol	13,920.9
$C_7H_{16}O_3$	Triethyl orthoformate	10,935.0
$C_7H_{16}O_3Si$	Triethoxymethylsilane	10,306.7
$C_7H_{18}Si$	Butyltrimethylsilane	9,206.0
$C_7H_{18}Si$	Triethylmethylsilane	9,232.5
$C_8H_4Cl_2O_2$	Phthaloyl chloride	13,716.0
$C_8H_4O_3$	Phthalic anhydride	13,919.0
$C_8H_5Cl_3N$	α,α-Dichlorophenylacetonitrile	12,829.9
$C_8H_5Cl_5$	Pentachloroethylbenzene	13,728.7
$C_8H_6Cl_2$	2,3-Dichlorostyrene	12,827.2
$C_8H_6Cl_2$	2,4-Dichlorostyrene	12,511.7
$C_8H_6Cl_2$	2,5-Dichlorostyrene	12,592.5
$C_8H_6Cl_2$	2,6-Dichlorostyrene	12,186.0
$C_8H_6Cl_2$	3,4-Dichlorostyrene	12,626.5
$C_8H_6Cl_2$	3,5-Dichlorostyrene	12,511.7
$C_8H_6Cl_4$	3,4,5,6-Tetrachloro-1,2-xylene	14,763.1
$C_8H_6Cl_4$	1,2,3,5-Tetrachloro-4-ethylbenzene	12,980.3
$C_8H_6O_2$	Phenylglyoxal	13,731.6
$C_8H_6O_2$	Phthalide	14,021.6
$C_8H_6O_3$	Piperonal	14,425.5
C_8H_7Cl	3-Chlorostyrene	10,990.2
C_8H_7ClO	Phenylacetyl chloride	12,627.1
C_8H_7N	2-Tolunitrile	11,557.7
C_8H_7N	4-Tolunitrile	11,562.8
C_8H_7N	Phenylacetonitrile	12,796.2
C_8H_7N	2-Tolyl isocyanide	11,303.3
$C_8H_7NO_2$	2-Nitrophenyl acetate	16,875.3
C_8H_7NS	2-Methylbenzothiazole	14,492.3
C_8H_8	Styrene	9,634.7
$C_8H_8Br_2$	(1,2-Dibromoethyl)benzene	14,874.7
$C_8H_8Cl_2$	1,2-Dichloro-3-ethylbenzene	11,784.3
$C_8H_8Cl_2$	1,2-Dichloro-4-ethylbenzene	11,711.5
$C_8H_8Cl_2$	1,4-Dichloro-2-ethylbenzene	11,262.7
C_8H_8O	Acetophenone	11,731.5
$C_8H_8O_2$	Phenylacetate	12,174.9
$C_8H_8O_2$	Phenylacetic acid	15,668.7
$C_8H_8O_2$	Anisaldehyde	13,581.8
$C_8H_8O_2$	Methyl benzoate	12,077.2
$C_8H_8O_2$	Methyl salicylate	12,658.8
$C_8H_8O_3$	Vanillin	15,703.2
$C_8H_6O_4$	Dihydroacetic acid	14,663.8
C_8H_9Br	2-Bromo-1,4-xylene	11,603.7
C_8H_9Br	1-Bromo-4-ethylbenzene	10,170.0
C_8H_9Br	(2-Bromoethyl)benzene	12,152.5
C_8H_9Cl	1-Chloro-2-ethylbenzene	10,749.7
C_8H_9Cl	1-Chloro-3-ethylbenzene	10,724.1
C_8H_9Cl	1-Chloro-4-ethylbenzene	10,659.9
C_8H_9ClO	1-Chloro-2-ethoxybenzene	12,411.1
C_8H_9ClO	4-Chlorophenylethyl alcohol	14,298.5
$C_8H_9Cl_2Si$	Dichlorophenylethylsilane	11,895.1
C_8H_9NO	Acetanilide	15,474.1
C_8H_9NO	Methyl anthranilate	13,186.3
$C_8H_9NO_2$	4-Nitro-1,3-xylene	12,948.0
C_8H_{10}	Ethylbenzene	9,301.3
C_8H_{10}	o-Xylene	9,998.5
C_8H_{10}	m-Xylene	9,904.2
C_8H_{10}	p-Xylene	9,809.9
$C_8H_{10}Cl_2OSi$	Dichloroethoxyphenylsilane	12,516.5
$C_8H_{10}Cl_2Si$	Dichloroethylphenylsilane	11,721.2
$C_8H_{10}O$	2-Ethylphenol	12,516.7
$C_8H_{10}O$	3-Ethylphenol	13,856.4
$C_8H_{10}O$	4-Ethylphenol	13,437.9
$C_8H_{10}O$	Xylenol	–
$C_8H_{10}O$	2,3-Xylenol	13,106.9
$C_8H_{10}O$	2,4-Xylenol	13,130.2
$C_8H_{10}O$	2,5-Xylenol	13,130.2
$C_8H_{10}O$	3,4-Xylenol	13,991.0
$C_8H_{10}O$	3,5-Xylenol	13,767.7
$C_8H_{10}O$	Phenetole	11,075.8
$C_8H_{10}O$	α-Methyl benzyl alcohol	13,087.4
$C_8H_{10}O$	Phenylethylalcohol	13,307.4
$C_8H_{10}O_2$	4,6-Dimethylresorcinol	12,433.1
$C_8H_{10}O_2$	2-Phenoxyethanol	14,368.3
$C_8H_{10}O_6$	Diethyl dioxosuccinate	13,973.3
$C_8H_{11}ClSi$	Chlorodimethylphenylsilane	11,382.2
$C_8H_{11}N$	N-Ethylaniline	11,817.0
$C_8H_{11}N$	N,N-Dimethylaniline	11,320.4
$C_8H_{11}N$	4-Ethylaniline	12,679.9
$C_8H_{11}N$	2,4-Xylidine	13,099.2
$C_8H_{11}N$	2,6-Xylidine	11,742.6
$C_8H_{11}NO$	2-Phenetidine	13,877.8
$C_8H_{11}NO$	2-Anilinoethanol	15,643.2
$C_8H_{11}AsNO_4$	Dimethyl arsanilate	11,277.7
$C_8H_{11}Cl_3O_5$	Diethyleneglycol bis-chloroacetate	19,000.7
$C_8H_{12}O_4$	Diethyl maleate	12,908.0
$C_8H_{12}O_4$	Diethyl fumarate	12,747.4
$C_8H_{12}Si$	Dimethylphenylsilane	10,274.2
$C_8H_{14}O_3$	Ethyl-α-ethylacetoacetate	12,344.2
$C_8H_{14}O_4$	Propyl levulinate	13,354.4
$C_8H_{14}O_4$	Isopropyl levulinate	12,689.6
$C_8H_{14}O_4$	Dipropyl oxalate	13,056.4
$C_8H_{14}O_4$	Diisopropyl oxalate	12,949.3
$C_8H_{14}O_4$	Diethyl succinate	13,076.1
$C_8H_{14}O_4$	Diethyl isosuccinate	12,087.6
$C_8H_{14}O_4$	Suberic acid	21,089.8
$C_8H_{14}O_4$	Diethyl maleate	14,202.9
$C_8H_{14}O_6$	Diethyl-dl-tartrate	15,150.4
$C_8H_{14}O_6$	Diethyl-d-tartrate	15,517.8
$C_8H_{15}Br$	(2-Bromoethyl)cyclohexane	11,462.7
$C_8H_{15}N$	n-Caprylonitrile	12,221.8
$C_8H_{15}NO_3$	Ethyl-N,N-diethyloxamate	13,758.4
C_8H_{16}	1-Octene	–
C_8H_{16}	2-Octene	–
C_8H_{16}	2-Methyl-2-heptene	9,643.8
C_8H_{16}	1,1-Dimethylcyclohexane	8,949.1
C_8H_{16}	cis-1,2-Dimethylcyclohexane	9,364.9
C_8H_{16}	trans-1,2-Dimethylcyclohexane	9,097.1
C_8H_{16}	cis-1,3-Dimethylcyclohexane	9,232.6
C_8H_{16}	trans-1,3-Dimethylcyclohexane	9,080.3
C_8H_{16}	cis-1,4-Dimethylcyclohexane	9,188.9
C_8H_{16}	trans-1,4-Dimethylcyclohexane	8,951.2
C_8H_{16}	Ethylcyclohexane	9,441.2
$C_8H_{16}O$	Caprylaldehyde	21,201.0
$C_8H_{16}O$	Cyclohexaneethanol	13,152.4
$C_8H_{16}O$	6-Methyl-3-hepten-2-ol	13,864.1
$C_8H_{16}O$	6-Methyl-5-hepten-2-ol	13,999.1
$C_8H_{16}O$	2-Octanone	11,649.2
$C_8H_{16}O$	2,2,4-Trimethyl-3-pentanone	12,854.6
$C_8H_{16}O_2$	Caprylic acid	16,745.7
$C_8H_{16}O_2$	Ethyl isocaproate	10,826.7
$C_8H_{16}O_2$	Propyl isovalerate	10,715.7
$C_8H_{16}O_2$	Isobutyl butyrate	10,283.9
$C_8H_{16}O_2$	Isobutyl isobutyrate	10,706.3
$C_8H_{16}O_3$	Amylisopropionate	10,567.2
$C_8H_{16}ClO_4$	Tetraethyleneglycol-chlorohydrin	16,371.2
$C_8H_{17}I$	1-Iodooctane	11,625.1
$C_8H_{17}NO_2$	Ethyl-l-leucinate	11,383.5
C_8H_{18}	Octane	9,221.0
C_8H_{18}	2-Methylheptane	9,362.0

Formula	Name	ΔH_v	Formula	Name	ΔH_v	Formula	Name	ΔH_v
C_8H_{18}	3-Methylheptane	9,432.0	$C_9H_{16}O_3$	Isobutyl levulinate	13,571.2	$C_{10}H_{16}$	Myrcene	10,704.8
C_8H_{18}	4-Methylheptane	9,404.8	$C_9H_{16}O_4$	Azelaic acid	20,944.2	$C_{10}H_{16}$	α-Phellandrene	11,139.5
C_8H_{18}	2,2-Dimethylhexane	8,927.8	$C_9H_{16}O_4$	Diethyl ethylmalonate	12,842.0	$C_{10}H_{16}$	α-Pinene	9,813.6
C_8H_{18}	2,3-Dimethylhexane	9,224.9	$C_9H_{16}O_4$	Diethyl glutarate	13,261.5	$C_{10}H_{16}$	β-Pinene	10,235.8
C_8H_{18}	2,4-Dimethylhexane	9,086.6	$C_9H_{18}O$	2-Nonanone	11,529.5	$C_{10}H_{16}$	Terpenoline	12,030.8
C_8H_{18}	2,5-Dimethylhexane	9,110.2	$C_9H_{18}O$	Di-isobutyl ketone	–	$C_{10}H_{16}AsNO_3$	Diethyl arsanilinate	12,973.9
C_8H_{18}	3,3-Dimethylhexane	9,065.2	$C_9H_{18}O_2$	Azelaldehyde	12,143.4	$C_{10}H_{16}O$	d-Camphor	12,800.9
C_8H_{18}	3,4-Dimethylhexane	9,239.4	$C_9H_{18}O_2$	Pelargonic acid	17,807.8			11,978.0
C_8H_{18}	3-Ethylhexane	9,416.3	$C_9H_{18}O_2$	Methyl caprylate	11,914.9	$C_{10}H_{16}O$	l-dihydrocarvone	11,825.9
C_8H_{18}	2,2,3-Trimethylpentane	8,861.1	$C_9H_{18}O_2$	Isobutyl isovalerate	10,999.7	$C_{10}H_{16}O$	α-Citral	13,255.5
C_8H_{18}	2,2,4-Trimethylpentane	8,548.0	$C_9H_{18}O_2$	Isoamyl butyrate	11,104.5	$C_{10}H_{16}O$	d-Fenchone	11,273.4
C_8H_{18}	2,3,3-Trimethylpentane	8,960.9	$C_9H_{18}O_2$	Isoamyl isobutyrate	10,870.6	$C_{10}H_{16}O$	Pulegone	13,395.4
C_8H_{18}	2,3,4-Trimethylpentane	8,988.2	$C_9H_{17}I$	1-Iodononane	14,853.0	$C_{10}H_{16}O$	α-Thujone	11,950.8
C_8H_{18}	2-Methyl-3-ethylpentane	9,134.3	C_9H_{20}	n-Nonane	10,456.9	$C_{10}H_{16}OSi$	Ethoxydimethylphenylsilane	11,718.6
C_8H_{18}	3-Methyl-3-ethylpentane	9,208.7	C_9H_{20}	2,6-Dimethylheptane	–	$C_{10}H_{16}O_2$	Campholenic acid	16,324.1
C_8H_{18}	2,2,3,3-Tetramethylbutane	10,351.5	C_9H_{20}	2-Methyloctane	–	$C_{10}H_{16}O_2$	Diosphenol	13,644.0
$C_8H_{18}N_2$	Tetramethylpiperazine	11,187.5	C_9H_{20}	3-Methyloctane	–	$C_{10}H_{16}O_2$	Fencholic acid	16,442.8
$C_8H_{18}O$	n-Octanol	14,262.4	$C_9H_{20}O$	1-Nonanol	13,849.2	$C_{10}H_{18}$	cis-Decalin	10,515.4
$C_8H_{18}O$	2-Octanol	12,468.4	$C_9H_{20}O$	Diisobutyl carbinol	–	$C_{10}H_{18}$	trans-Decalin	8,749.1
$C_8H_{18}O$	Di n-butyl ether	–	$C_9H_{20}O_3$	Dipropyleneglycol isopropyl ether	12,583.8	$C_{10}H_{18}O$	d-Citronellal	12,305.1
$C_8H_{18}O$	Methyl heptyl ether	–				$C_{10}H_{18}O$	Cineol	10,570.8
$C_8H_{18}O_2$	1,2-Dipropoxy ethane	6,370.7	$C_9H_{20}O_4$	Tripropyleneglycol	15,291.4	$C_{10}H_{18}O$	Dihydrocarveol	13,698.5
$C_8H_{18}O_3$	Diethylene glycol butyl ether	14,127.0	$C_9H_{22}Si$	Hexyltrimethylsilane	10,264.9	$C_{10}H_{18}O$	dl-Fenchyl alcohol	12,955.9
$C_8H_{18}O_5$	Tetraethylene glycol	21,296.6	$C_9H_{22}Si$	Triethylpropylsilane	10,709.3	$C_{10}H_{18}O$	Geraniol	14,060.7
$C_8H_{18}S$	Di n-butyl sulfide	11,183.6	$C_{10}H_7Br$	1-Bromonaphthalene	13,274.9	$C_{10}H_{18}O$	d-Linalool	12,269.7
$C_8H_{18}S_2$	Dibutyl disulfide	8,254.1	$C_{10}H_7Cl$	1-Chloronaphthalene	13,570.5	$C_{10}H_{18}O$	Nerol	13,366.1
$C_8H_{19}N$	Diisobutylamine	10,058.3	$C_{10}H_{12}$	Dicyclopentadiene	10,165.9	$C_{10}H_{18}O$	α-Terpineol	12,754.5
$C_8H_{20}O_4Si$	Tetraethoxysilane	10,968.6	$C_{10}H_8$	Naphthalene	17,065.2	$C_{10}H_{18}O_2$	Citronellic acid	16,455.4
$C_8H_{20}Pb$	Tetraethyllead	12,959.7			12,311.6	$C_{10}H_{18}O_3$	Amyl levulinate	14,321.7
$C_8H_{20}Si$	Amyltrimethylsilane	9,659.6	$C_{10}H_8Cl_2Si$	Dichloro-1-naphthylsilane	16,325.3	$C_{10}H_{18}O_3$	Isoamyl levulinate	13,867.9
$C_8H_{20}Si$	Tetraethylsilane	9,893.0	$C_{10}H_8O$	1-Naphthol	14,205.6	$C_{10}H_{18}O_4$	Diethyl ethylmethylmalonate	12,345.6
$C_8H_{20}Sb_2$	Tetraethylbistibine	12,975.4	$C_{10}H_8O$	2-Naphthol	14,138.5	$C_{10}H_{18}O_4$	Diethyl adipate	14,240.6
$C_8H_{22}O_3Si_2$	1,3-Dichlorotetramethyl-disiloxane	11,261.9	$C_{10}H_9N$	1-Naphthylamine	14,529.5	$C_{10}H_{18}O_4$	Diisobutyl oxalate	13,343.1
			$C_{10}H_9N$	2-Naphthylamine	14,679.6	$C_{10}H_{18}O_4$	Dipropyl succinate	13,975.7
$C_8H_{24}Cl_2O_3Si_4$	1,7-Dichlorooctamethyl-tetrasiloxane	12,602.9	$C_{10}H_9N$	2-Methylquinoline	14,154.0	$C_{10}H_{18}O_4$	Sebacic acid	21,978.3
			$C_{10}H_{10}$	1,3-Divinylbenzene	11,384.7	$C_{10}H_{18}O_6$	Dipropyl-d-tartrate	15,754.0
$C_8H_{24}O_2Si_3$	Octamethyltrisiloxane	10,956.0	$C_{10}H_{10}O$	4-Phenyl-3-buten-2-one	13,913.9	$C_{10}H_{18}O_6$	Diisopropyl-d-tartrate	15,836.6
$C_8H_{24}O_4Si_4$	Octamethylcyclotetra siloxane	11,515.0	$C_{10}H_{10}O_2$	α-Methylcinnamic acid	18,149.4	$C_{10}H_{19}N$	Camphylamine	13,224.1
$C_9H_6O_2$	Coumarin	15,202.7	$C_{10}H_{10}O_2$	Methyl cinnamate	13,325.5	$C_{10}H_{20}$	Menthane	10,293.1
C_9H_7N	Quinoline	12,575.4	$C_{10}H_{10}O_2$	Safrole	13,255.8	$C_{10}H_{20}$	1-Decene	10,233.3
C_9H_7N	Isoquinoline	12,847.6	$C_{10}H_{10}O_4$	1,2-Phenylene diacetate	14,986.0	$C_{10}H_{20}$	n-Decane	10,912.0
C_9H_8	Indene	10,496.7	$C_{10}H_{10}O_4$	Dimethylphthalate	14,922.2	$C_{10}H_{20}Br_2$	1,2-Dibromodecane	16,407.7
C_9H_8O	Cinnamylaldehyde	14,048.4	$C_{10}H_{12}$	2,4-Dimethylstyrene	11,454.0	$C_{10}H_{20}O$	Decanol	14,065.1
$C_9H_8O_2$	trans-Cinnamic acid	17,492.9	$C_{10}H_{12}$	2,5-Dimethylstyrene	11,283.5	$C_{10}H_{20}O$	Citronellol	14,214.1
C_9H_9N	Skatole	15,232.7	$C_{10}H_{12}$	3-Ethylstyrene	11,285.7	$C_{10}H_{20}O$	Capraldehyde	13,154.9
$C_9H_9NO_2$	Ethyl 3-nitrobenzoate	15,056.1	$C_{10}H_{12}$	4-Ethylstyrene	11,146.6	$C_{10}H_{20}O$	l-Menthol	13,475.3
C_9H_{10}	α-Methyl styrene	10,214.6	$C_{10}H_{12}$	Tetralin	11,613.0	$C_{10}H_{20}O$	Decan-2-one	12,114.7
C_9H_{10}	β-Methyl styrene	10,701.3	$C_{10}H_{12}O$	Anethole	13,006.8	$C_{10}H_{20}O_2$	Capric acid	19,372.6
C_9H_{10}	2-Methyl styrene	–	$C_{10}H_{12}O$	4-Methylpropiophenone	12,505.0	$C_{10}H_{20}O_2$	Isoamyl isovalerate	11,040.8
C_9H_{10}	3-Methyl styrene	–	$C_{10}H_{12}O$	Estragole	12,879.3	$C_{10}H_{22}$	2,7-Dimethyloctane	10,339.3
C_9H_{10}	4-Methyl styrene	10,724.2	$C_{10}H_{12}O$	Cuminal	12,668.0	$C_{10}H_{22}O$	Diisoamyl ether	11,072.2
$C_9H_{10}O$	2,4-Xylaldehyde	13,618.4	$C_{10}H_{12}O$	4-Vinylphenetole	13,728.7	$C_{10}H_{22}O$	2-Butyl-2-ethylbutane-1,3-diol	15,833.7
$C_9H_{10}O$	Cinnamyl alcohol	13,421.6	$C_{10}H_{12}O_2$	Eugenol	13,907.8	$C_{10}H_{22}O$	Dihydrocitronellol	16,769.8
$C_9H_{10}O$	Propiophenone	12,407.6	$C_{10}H_{12}O_2$	Isoeugenol	14,084.2	$C_{10}H_{22}O_3$	Dipropylene glycol monobutyl ether	13,721.1
$C_9H_{10}O$	3-Vinylanisole	12,756.4	$C_{10}H_{12}O_2$	Chavibetol	14,527.7			
$C_9H_{10}O$	3-Vinylanisole	12,735.8	$C_{10}H_{12}O_2$	Propyl benzoate	12,318.7	$C_{10}H_{22}S$	Diisoamyl sulfide	11,829.9
$C_9H_{10}O$	4-Vinylanisole	12,554.7	$C_{10}H_{12}O_3$	2-Phenoxyethyl acetate	14,070.3	$C_{10}H_{24}Si$	Heptyltrimethylsilane	10,987.3
$C_9H_{10}O_2$	Benzyl acetate	12,107.2	$C_{10}H_{13}ClO$	2-Chloroethyl-α-methylbenzyl ether	12,969.2	$C_{10}H_{24}Si$	Butyltriethylsilane	11,124.0
$C_9H_{10}O_2$	Ethyl benzoate	11,981.5				$C_{10}H_{26}O_4Si_3$	1,5-Diethoxyhexamethyl trisiloxane	12,586.4
$C_9H_{10}O_2$	Hydrocinnamic acid	15,411.9	$C_{10}H_{13}Cl_2O_2P$	4-tert-Butylphenyl dichlorophosphate	13,711.0	$C_{10}H_{30}O_3Si_4$	Decamethyltetrasiloxane	11,981.2
$C_9H_{10}O_2$	Ethyl salicylate	13,030.1				$C_{10}H_{30}O_5Si_5$	Decamethylcyclopenta siloxane	12,272.1
$C_9H_{11}NO$	N-Methylacetanilide	13,235.2	$C_{10}H_{14}$	1,2,3,4-Tetramethylbenzene	12,258.0			
$C_9H_{11}NO_2$	Ethyl carbanilate	19,791.8	$C_{10}H_{14}$	1,2,3,5-Tetramethylbenzene	12,358.4	$C_{11}H_8O_2$	1-Naphthoic acid	22,581.4
C_9H_{12}	1,2,3-Trimethylbenzene	10,781.9	$C_{10}H_{14}$	1,2,4,5-Tetramethylbenzene	12,583.6	$C_{11}H_8O_2$	2-Naphthoic acid	22,630.8
C_9H_{12}	1,2,4-Trimethylbenzene	10,710.2	$C_{10}H_{14}$	4-Ethyl-1,3-xylene	11,070.4	$C_{11}H_{10}$	1-Methylnaphthalene	–
C_9H_{12}	1,3,5-Trimethylbenzene	10,516.8	$C_{10}H_{14}$	5-Ethyl-1,3-xylene	11,045.5	$C_{11}H_{12}O_2$	Ethyl-trans-cinnamate	13,639.9
C_9H_{12}	o-Ethyl toluene	10,488.8	$C_{10}H_{14}$	2-Ethyl-1,4-xylene	11,144.6	$C_{11}H_{12}O_2$	1 Phenyl-1,3-pentanedione	15,033.9
C_9H_{12}	m-Ethyl toluene	10,416.6	$C_{10}H_{14}$	1,2-Diethylbenzene	11,695.5	$C_{11}H_{12}O_3$	Ethyl benzoylacetate	17,115.4
C_9H_{12}	p-Ethyl toluene	10,461.1	$C_{10}H_{14}$	1,3-Diethylbenzene	10,993.9	$C_{11}H_{12}O_3$	Myristicine	14,471.4
C_9H_{12}	Isopropylbenzene	10,335.3	$C_{10}H_{14}$	1,4-Diethylbenzene	10,746.3	$C_{11}H_{14}$	1-Phenylpentane	–
C_9H_{12}	N-Propylbenzene	10,424.1	$C_{10}H_{14}$	1-Methyl-2-isopropylbenzene	–	$C_{11}H_{14}$	2,4,5-Trimethylstyrene	12,076.1
$C_9H_{12}O$	2-Ethylanisole	11,642.8	$C_{10}H_{14}$	1-Methyl-4-isopropylbenzene	11,038.7	$C_{11}H_{14}$	2,4,6-Trimethylstyrene	11,588.8
$C_9H_{12}O$	3-Ethylanisole	11,616.7	$C_{10}H_{14}$	N-Butylbenzene	11,052.1	$C_{11}H_{14}$	4-Isopropylstyrene	11,471.0
$C_9H_{12}O$	4-Ethylanisole	11,625.7	$C_{10}H_{14}$	Isobutylbenzene	8,567.8	$C_{11}H_{14}O$	Isobutyrophenone	12,878.8
$C_9H_{12}O$	3-Phenyl-1-propanol	14,493.9	$C_{10}H_{14}$	sec-Butylbenzene	11,069.3	$C_{11}H_{14}O$	Pivalophenone	13,221.3
$C_9H_{12}O$	2-Isopropylphenol	13,402.3	$C_{10}H_{14}$	tert-Butylbenzene	10,705.5	$C_{11}H_{14}O$	2,3,5-Trimethylacetophenone	14,283.6
$C_9H_{12}O$	3-Isopropylphenol	13,292.2	$C_{10}H_{14}O$	Carvacrol	13,765.7	$C_{11}H_{14}O_2$	Isobutyl benzoate	13,105.8
$C_9H_{12}O$	4-Isopropylphenol	13,878.7	$C_{10}H_{14}O$	Carbone	12,796.2	$C_{11}H_{14}O_2$	4-Allylveratrole	15,027.1
$C_9H_{12}O$	Benzyl ethyl ether	11,315.5	$C_{10}H_{14}O$	Cuminyl alcohol	13,799.2	$C_{11}H_{16}$	Pentamethylbenzene	–
$C_9H_{13}ClOSi$	Chloroethoxymethyl-phenylsilane	12,270.3	$C_{10}H_{14}O$	4-Ethylphenetole	12,766.4	$C_{11}H_{16}$	3,5-Diethyltoluene	11,167.4
			$C_{10}H_{14}O$	2-Isopropyl-5-methylphenol	13,352.8	$C_{11}H_{16}$	1,2,4-Trimethyl-5-ethylbenzene	12,145.3
$C_9H_{13}N$	2,4,5-Trimethylaniline	13,975.0	$C_{10}H_{14}O$	4-Isobutylphenol	14,053.5			
$C_9H_{13}N$	N,N-Dimethyl-O-toluidine	11,648.3	$C_{10}H_{14}O$	4-sec-Butylphenol	13,690.2	$C_{11}H_{16}$	1,3,5-Trimethyl-2-ethylbenzene	11,677.3
$C_9H_{13}N$	N,N-Dimethyl-4-toluidine	12,738.4	$C_{10}H_{14}O$	2-sec-Butylphenol	12,781.3	$C_{11}H_{16}$	3-Ethylcumene	11,233.5
$C_9H_{13}N$	4-Cumidine	13,127.9	$C_{10}H_{14}O$	2-tert-Butylphenol	13,112.3	$C_{11}H_{16}$	4-Ethylcumene	11,425.6
$C_9H_{14}O$	Phorone	12,557.2	$C_{10}H_{14}O$	4-tert-Butylphenol	13,787.7	$C_{11}H_{16}$	sec-Amylbenzene	11,886.0
$C_9H_{14}O$	Isophorone	11,277.6	$C_{10}H_{14}N_2$	Nicotine	12,337.1	$C_{11}H_{16}O$	4-tert-Butyl-2-cresol	13,798.1
$C_9H_{14}O_4$	cis-Diethyl citraconate	12,913.2	$C_{10}H_{15}N$	N-Diethylaniline	12,539.2	$C_{11}H_{16}O$	2-tert-Butyl-4-cresol	14,037.9
$C_9H_{14}O_4$	Diethyl itaconate	12,075.8	$C_{10}H_{15}NO_2$	N-Phenyliminodiethanol	17,482.1	$C_{11}H_{16}O$	4-tert-Amylphenol	13,154.3
$C_9H_{14}O_4$	Diethyl mesaconate	13,326.1	$C_{10}H_{16}$	Camphene	10,505.4	$C_{11}H_{16}O_5$	Ethylcamphoronic anhydride	16,373.6
$C_9H_{14}O_7$	Trimethyl citrate	15,807.7	$C_{10}H_{16}$	Dipentene	10,538.3			
			$C_{10}H_{16}$	d-Limonene	10,508.4			

Formula	Name	Δ H$_v$	Formula	Name	Δ H$_v$	Formula	Name	Δ H$_v$
$C_{11}H_{18}O_2$	Bornyl formate	12,276.0	$C_{12}H_{27}N$	Dodecylamine	14,836.4	$C_{15}H_{32}$	Pentadecane	14,635.9
$C_{11}H_{18}O_2$	Geranyl formate	13,189.7	$C_{12}H_{28}Si$	Triethylhexylsilane	12,119.4	$C_{15}H_{32}O_5$	Tetrapropylene glycol monoisopropyl ether	16,494.6
$C_{11}H_{18}O_2$	Neryl formate	12,959.3	$C_{12}H_{34}O_3Si_4$	1,7-Diethoxyoctamethyl-tetrasiloxane	14,095.9			
$C_{11}H_{18}O_2Si$	Diethoxymethylphenyl silane	13,267.3				$C_{15}H_{34}Si$	Dodecyltrimethylsilane	14,374.6
$C_{11}H_{18}O_5$	Diethyl-gamma-oxoazelate	17,543.6	$C_{12}H_{36}O_4Si_5$	Dodecamethylpentasiloxane	12,942.6	$C_{16}H_{14}O_2$	Benzyl cinnamate	20,840.6
$C_{11}H_{20}O_2$	10-Hendecenoic acid	17,247.5	$C_{12}H_{36}O_6Si_6$	Dodecamethyl-cyclohexasiloxane	13,760.6	$C_{16}H_{16}O$	Di(α-methylbenzyl)ether	14,628.1
$C_{11}H_{20}O_2$	Menthyl formate	12,077.7				$C_{16}H_{20}O_2Si$	Diethoxydiphenylsilane	15,828.8
$C_{11}H_{20}O_2$	2-Ethylhexyl acrylate	12,522.5	$C_{13}H_9N$	Acridine	15,174.6	$C_{16}H_{22}O_4$	Dibutyl phthalate	17,747.0
$C_{11}H_{20}O_2$	Octyl acrylate	12,957.5	$C_{13}H_{10}$	Fluorene	13,682.8	$C_{16}H_{25}Cl$	Pentaethylchlorobenzene	13,707.3
$C_{11}H_{20}O_3$	Hexyl levulinate	14,626.2	$C_{13}H_{10}O$	Benzophenone	14,725.4	$C_{16}H_{26}$	Pentaethylbenzene	13,670.1
$C_{11}H_{22}O$	Hendecan-3-one	14,353.5	$C_{13}H_{10}O_2$	Phenyl benzoate	14,181.7	$C_{16}H_{26}O$	2,6-Di-tert-butyl-4-ethylphenol	14,438.0
$C_{11}H_{22}O_2$	Methyl caprate	13,831.7	$C_{13}H_{10}O_2$	Salol	15,441.6			
$C_{11}H_{22}O_2$	Hendecanoic acid	14,689.9	$C_{13}H_{12}$	Diphenylmethane	13,089.4	$C_{16}H_{26}O$	4,6-Di-tert-butyl-3-ethylphenol	15,954.8
$C_{11}H_{24}$	Undecane	11,481.7	$C_{13}H_{12}O$	Benzhydrol	15,220.2			
$C_{11}H_{24}O$	Hendecan-2-ol	14,216.2	$C_{13}H_{12}O$	Benzyl phenyl ether	14,156.7	$C_{16}H_{30}O$	Muscone	14,722.5
$C_{11}H_{26}Si$	Trimethyloctylsilane	12,285.8	$C_{13}H_{12}O$	1-Proprionaphthone	16,630.8	$C_{16}H_{31}N$	Palmitonitrile	16,433.7
$C_{11}H_{26}Si$	Amyltriethylsilane	11,859.7	$C_{13}H_{13}ClSi$	Chloromethyldiphenylsilane	14,924.6	$C_{16}H_{32}$	1-Hexadecene	15,634.7
$C_{12}H_9Br$	4-Bromobiphenyl	13,493.4	$C_{13}H_{14}Si$	Methyldiphenylsilane	15,396.8	$C_{16}H_{33}$	Tetraisobutylene	12,937.2
$C_{12}H_9BrO$	2-Bromo-4-phenylphenol	13,589.9	$C_{13}H_{14}O$	2-Isopropylnaphthalene	13,036.9	$C_{16}H_{32}O$	2-Hexadecanone	15,194.4
$C_{12}H_9Cl$	2-Chlorobiphenyl	13,925.7	$C_{13}H_{18}O$	Enanthophenone	15,597.7	$C_{16}H_{32}O$	Palmitaldehyde	15,454.2
$C_{12}H_9Cl$	4-Chlorobiphenyl	14,017.4	$C_{13}H_{20}$	Heptylbenzene	13,535.4	$C_{16}H_{32}O_2$	Palmitic acid	17,603.6
$C_{12}H_9ClO$	2-Chloro-3-phenylphenol	15,258.0	$C_{13}H_{20}O$	α-Ionone	14,253.4	$C_{16}H_{34}$	Hexadecane	15,405.5
$C_{12}H_9ClO$	2-Chloro-6-phenylphenol	15,508.4	$C_{13}H_{24}O_2$	Bornyl propionate	13,245.0	$C_{16}H_{34}O$	Cetyl alcohol	14,483.4
$C_{12}H_9Cl_2PO$	2-Xenyl dichlorophosphate	17,127.6	$C_{13}H_{26}O$	2-Tridecanone	14,416.1	$C_{16}H_{35}N$	Cetylamine	15,238.0
$C_{12}H_9N$	Carbazole	15,421.6	$C_{13}H_{26}O_2$	Methyl laurate	14,853.5	$C_{16}H_{36}Si$	Decyltriethylsilane	15,393.7
$C_{12}H_{10}$	Acenaphthene	13,078.5	$C_{13}H_{26}O_2$	Tridecanoic acid	19,214.8	$C_{16}H_{46}O_7Si_6$	1,1,1-Diethoxydodeca methylhexasiloxane	15,945.3
$C_{12}H_{10}$	Diphenyl	12,910.0	$C_{13}H_{28}$	Tridecane	12,991.3			
$C_{12}H_{10}ClPO_3$	Diphenyl chlorophosphate	13,191.2	$C_{13}H_{28}O_4$	Tripropyleneglycol monobutyl ether	15,937.6	$C_{16}H_{48}O_6Si_7$	Hexadecamethylhepta-siloxane	14,841.5
$C_{12}H_{10}Cl_2Si$	Dichlorodiphenylsilane	14,968.5						
$C_{12}H_{10}F_2Si$	Difluorodiphenylsilane	12,913.3	$C_{13}H_{30}Si$	Decyltrimethylsilane	13,311.1	$C_{16}H_{48}O_8Si_8$	Hexadecamethylcyclooctasiloxane	14,986.3
$C_{12}H_{10}N_2$	Azobenzene	14,786.7	$C_{13}H_{30}Si$	Triethylheptylsilane	13,298.3			
$C_{12}H_{10}O$	1-Acetonaphthone	16,095.5	$C_{14}H_8O_2$	Anthraquinone	21,163.1	$C_{17}H_{10}O$	Benzanthrone	18,309.6
$C_{12}H_{10}O$	2-Acetonaphthone	16,496.7	$C_{14}H_8O_4$	1,4-Dihydroxyanthraquinone	17,677.9	$C_{17}H_{18}O_3$	4-tert-Butylphenyl salicylate	16,455.6
$C_{12}H_{10}O$	Diphenyl ether	12,325.5	$C_{14}H_{10}$	Anthracene	16,823.6	$C_{17}H_{26}O_2$	Menthyl benzoate	16,804.5
$C_{12}H_{10}O$	2-Phenylphenol	15,397.8	$C_{14}H_{10}$	Phenanthrene	14,184.0	$C_{17}H_{34}O$	2-Heptadecanone	16,559.8
$C_{12}H_{10}O$	4-Phenylphenol	16,974.3	$C_{14}H_{10}O_2$	Benzil	15,046.4	$C_{17}H_{34}O_2$	Methyl palmitate	17,003.5
$C_{12}H_{10}S$	Diphenyl sulfide	13,974.8	$C_{14}H_{10}O_3$	Benzoic anhydride	16,060.9	$C_{17}H_{36}$	Heptadecane	15,608.5
$C_{12}H_{10}S_2$	Diphenyl disulfide	17,452.0	$C_{14}H_{12}$	1,1-Diphenylethylene	13,778.1	$C_{17}H_{38}Si$	Tetradecyltrimethylsilane	16,439.7
$C_{12}H_{10}Se$	Diphenyl selenide	14,603.4	$C_{14}H_{12}$	trans-Diphenylethylene	15,010.1	$C_{18}H_{12}Cl_3O_3PS$	Tri-2-chlorophenylthiophosphate	24,386.1
$C_{12}H_{11}N$	Diphenylamine	14,920.3	$C_{14}H_{12}O$	Desoxybenzoin	15,642.1			
$C_{12}H_{11}N$	1-Ethylnaphthalene	12,751.3	$C_{14}H_{12}O$	Benzoin	15,957.5	$C_{18}H_{15}O_4P$	Triphenyl phosphate	19,272.3
$C_{12}H_{12}N_2$	1,1-Diphenylhydrazine	15,940.4	$C_{14}H_{14}$	Dibenzyl	13,387.6	$C_{18}H_{30}$	Hexaethylbenzene	14,184.9
$C_{12}H_{14}N_2O_5$	2-Cyclohexyl-4,6-dinitrophenol	19,100.0	$C_{14}H_{16}O$	2-Isobutyronaphthone	17,133.8	$C_{18}H_{30}O$	2,4,6-Tri-tert-butylphenol	14,703.7
			$C_{14}H_{15}O$	Dibenzylamine	16,260.1	$C_{18}H_{34}O_2$	Oleic acid	20,326.7
$C_{12}H_{14}O_2$	Eugenyl acetate	15,120.7	$C_{14}H_{15}N$	Ethyldiphenylamine	14,569.4	$C_{18}H_{34}O_2$	Elaidic acid	19,538.0
$C_{12}H_{14}O_4$	Apiole	16,881.7	$C_{14}H_{18}$	1,2-Dichlorotetraethylbenzene	14,029.0	$C_{18}H_{36}O_2$	Oleamide/dec.acid	16,666.1
$C_{12}H_{14}O_4$	Diethyl phthalate	15,383.0	$C_{14}H_{20}Cl_2$	1,4-Dichlorotetraethylbenzene	13,397.5	$C_{18}H_{36}O_2$	Stearic acid	19,306.6
$C_{12}H_{16}$	2,5-Diethylstyrene	12,150.3	$C_{14}H_{20}O_3$	2-(4-tert-Butylphenoxy) ethyl acetate	16,017.6	$C_{18}H_{38}$	Octadecane	15,447.0
$C_{12}H_{16}$	Phenylcyclohexane	13,345.6				$C_{18}H_{38}$	2-Methylheptadecane	16,095.9
$C_{12}H_{16}O_2$	Isoamyl benzoate	12,782.9	$C_{14}H_{22}$	1,2,3,4-Tetraethylbenzene	12,763.5	$C_{18}H_{38}O$	1-Octadecanol	17,508.0
$C_{12}H_{18}$	Hexamethylbenzene	—	$C_{14}H_{22}O$	2,4-Di-tert-butylphenol	14,237.7	$C_{18}H_{39}N$	Ethylcetylamine	15,718.3
$C_{12}H_{18}$	1,2,4-Triethylbenzene	11,957.9	$C_{14}H_{24}O_2$	Bornyl butyrate	13,746.1	$C_{18}H_{52}O_6Si_7$	1,1,3-Diethoxytetradeca methylheptasiloxane	16,765.8
$C_{12}H_{18}$	1,3,4-Triethylbenzene	12,215.0	$C_{14}H_{24}O_2$	Bornyl isobutyrate	13,501.8			
$C_{12}H_{18}$	1,3,5-Triethylbenzene	—	$C_{14}H_{24}O_2$	Geranyl butyrate	16,086.4	$C_{18}H_{54}O_7Si_8$	Octadecamethyloctasiloxane	15,270.3
$C_{12}H_{18}$	1,2-Diisopropylbenzene	11,751.4	$C_{14}H_{24}O_2$	Geranyl isobutyrate	15,699.5			
$C_{12}H_{18}$	1,3-Diisopropylbenzene	11,498.9	$C_{14}H_{26}O_4$	Diethyl sebacate	16,819.6	$C_{19}H_{16}$	Triphenylmethane	34,470.8
$C_{12}H_{18}O$	2-tert-Butyl-4-ethylphenol	13,994.0	$C_{14}H_{28}O$	2-Tetradecanone	15,102.7	$C_{19}H_{40}$	Nonadecane	16,497.3
$C_{12}H_{18}O$	4-tert-Butyl-2,5-xylenol	14,477.9	$C_{14}H_{28}O$	Myristaldehyde	14,088.9	$C_{20}H_{20}OSi$	Ethoxytriphenylsilane	20,214.2
$C_{12}H_{18}O$	4-tert-Butyl-2,6-xylenol	14,142.5	$C_{14}H_{28}O_2$	Myristic acid	18,380.1	$C_{20}H_{43}N$	Diethylhexadecylamine	15,871.3
$C_{12}H_{18}O$	6-tert-Butyl-2,4-xylenol	13,882.4	$C_{14}H_{29}Cl$	1-Chlorotetradecane	14,083.5	$C_{20}H_{56}O_5Si_8$	1,1,5-Diethoxyhexadeca methyloctasiloxane	17,626.6
$C_{12}H_{18}O$	6-tert-Butyl-3,4-xylenol	14,848.3	$C_{14}H_{30}$	Tetradecane	13,750.0			
$C_{12}H_{20}O_2$	d-Bornyl acetate	11,838.7	$C_{14}H_{31}N$	Tetradecylamine	14,840.8	$C_{20}H_{60}O_8Si_9$	Eicosamethylnonasiloxane	19,522.9
$C_{12}H_{20}O_2$	Geranyl acetate	13,879.9	$C_{14}H_{32}Si$	Triethyloctylsilane	12,954.8	$C_{21}H_{21}O_4P$	Tritolyl phosphate	20,835.9
$C_{12}H_{20}O_2$	Linalyl acetate	12,910.6	$C_{14}H_{40}O_6Si_5$	1,9-Diethoxydeca methylpentasiloxane	15,296.9	$C_{21}H_{44}$	Heneicosane	17,702.2
$C_{12}H_{20}O_3Si$	Triethoxyphenylsilane	14,117.9				$C_{22}H_{42}O_2$	Erucic acid	23,655.2
$C_{12}H_{20}O_7$	Triethyl citrate	14,818.4	$C_{14}H_{42}O_5Si_6$	Tetradecamethylhexasiloxane	13,800.0	$C_{22}H_{42}O_2$	Brassidic acid	24,085.7
$C_{12}H_{21}PO_4$	Trimethallyl phosphate	12,566.1	$C_{14}H_{42}O_7Si_7$	Tetradecamethylcyclo heptasiloxane	14,263.8	$C_{22}H_{46}$	Docosane	16,941.1
$C_{12}H_{22}O_2$	Citronellyl acetate	15,781.3				$C_{22}H_{56}O_9Si_{10}$	Docosamethyldecasiloxane	21,878.6
$C_{12}H_{22}O_2$	Menthyl acetate	12,819.2	$C_{15}H_{14}O$	1,3-Diphenyl-2-propanone	15,429.8	$C_{23}H_{48}$	Tricosane	19,082.1
$C_{12}H_{22}O_4$	Dimethyl sebacate	14,861.3	$C_{15}H_{14}O_2$	1-Biphenyloxy-2,3-epoxypropane	16,160.6	$C_{24}H_{50}$	Tetracosane	19,642.5
$C_{12}H_{22}O_4$	Diisoamyl oxalate	14,123.7				$C_{24}H_{72}O_{10}Si_{11}$	Tetracosamethylhendeca-siloxane	23,941.2
$C_{12}H_{22}O_6$	Diisobutyl-d-tartrate	14,874.9	$C_{15}H_{16}O_2$	4,4-Isopropylidenebisphenol	23,254.0			
$C_{12}H_{24}$	1-Dodecene	12,587.8	$C_{15}H_{18}O$	Isocapronaphthone	17,360.3	$C_{25}H_{52}$	Pentacosane	20,815.9
$C_{12}H_{24}$	Triisobutylene	10,790.4	$C_{15}H_{18}OSi$	Ethoxymethyldiphenylsilane	16,106.4	$C_{26}H_{54}$	Hexacosane	21,605.7
$C_{12}H_{24}O$	Dodecan-2-one	14,138.7	$C_{15}H_{20}O_2$	Helenin	26,532.7	$C_{27}H_{33}O_4P$	Dicarvacryl-2-tolyl phosphate	24,233.3
$C_{12}H_{24}O$	Lauraldehyde	13,644.2	$C_{15}H_{24}$	Cadinene	15,518.3			
$C_{12}H_{24}O_2$	Lauric acid	16,585.3	$C_{15}H_{24}O$	2,6-Di-tert-butyl-4-cresol	14,338.6	$C_{27}H_{56}$	Heptacosane	21,958.1
$C_{12}H_{26}$	n-Dodecane	11,857.7	$C_{15}H_{24}O$	4,6-Di-tert-butyl-2-cresol	14,006.9	$C_{28}H_{58}$	Octacosane	24,144.2
$C_{12}H_{26}O$	Dodecyl alcohol	15,160.0	$C_{15}H_{24}O$	4,6-Di-tert-butyl-3-cresol	15,464.6	$C_{29}H_{60}$	Nonacosane	24,816.8
$C_{12}H_{26}O_4$	Tripropylene glycol monoisopropyl ether	14,171.5	$C_{15}H_{26}O$	Champacol	14,655.9	$C_{32}H_{34}ClO_4P$	Dicarvacryl-mono-(6-chloro-2-xenyl)-phosphate	25,299.5
			$C_{15}H_{26}O_6$	Triethyl camphoronate	16,112.2			
$C_{12}H_{27}N$	Triisobutylamine	12,390.6	$C_{15}H_{30}O_2$	Methyl myristate	16,051.0			

While the solubility of organic compounds in selected solvents may be estimated from the "like dissolves like" rule of thumb, more useful information is available under the "Solubility" heading of the table "Physical Constants of Organic Compounds" in Section C of this book. This table provides qualitative data on solubility in water (w), ethanol (al), ethyl ether (eth), acetone (ace), and benzene (bz).

More quantitative solubility data for nonpolar organic compounds may be calculated from the Hildebrand expression for the square root of the cohesive energy density which is defined as the solubility parameter (δ). As shown by the following expression, δ values may be calculated if information for ΔH_v, Kelvin temperature (T), molecular weight (M), and density (D) is available:

$$\delta = \left(\frac{\Delta E_v}{V}\right)^{1/2} = \left(\frac{D(\Delta H_v - RT)}{M}\right)^{1/2}$$

The value for ΔH_v may be found in the table "Heats of Vaporization of Organic Compounds" immediately following this table. Values for D and M are listed under physical constants in Section C. Thus, the δ value for heptane may be calculated as follows:

$$\delta = \left[\frac{0.684(8670 - 1.99(298))}{100}\right]^{1/2} = 7.4 \text{ H}$$

The dimensions for δ are $(\text{cal cm}^{-3})^{1/2}$ but the Hildebrand unit (H) is used for convenience. It is important to note that the law of mixtures applies for δ values of mixed nonpolar solvents. Thus, the δ value for an equimolar mixture of heptane and carbon disulfide ($\delta = 10.0$ H) is 8.7 H.

When ΔH_v values are not available, Small's molar attraction constants shown in Table I may be used.[a] As illustrated in the following expression which is solved for heptane, the summation of these constants (ΣG) may be used to estimate δ values at 298 K:

$$\delta = \frac{D\Sigma G}{M} = \frac{(2 \times 214) + 5(133)0.684}{100} = 7.5 \text{ H}$$

Solvents such as acetone ($\delta = 9.9$ H) and water ($\delta = 23.4$ H) are completely miscible when a large difference in δ values exists. The critical δ range for solubility of solid solutes and liquid solutes is less than that for liquids, and the critical δ range for nonpolar polymers in nonpolar liquids is less than 2 H at temperatures below 50°C.

These δ values are most useful for nonpolar solvents that are listed in Section A of Table II. Some consideration must be given to the dipole-dipole interactions in more polar solvents that are listed in Section B of Table II. The values shown for hydrogen-bonded solvents that are shown in Section C of Table II must be used with more discretion because of the stronger intermolecular forces present. However, these values are much more useful than the "like dissolves like" rule of thumb and can be used to predict the solubility of most solutes in most solvents.

[a]Small, P. A., *J. Appl. Chem.*, 3, 71, 1953.

Table I
MOLAR-ATTRACTION CONSTANTS AT 298 K

Group	G	Group	G
$-CH_3$	214	CO ketones	275
$-CH_2-$ single bonded	133	COO esters	310
$-CH<$	28	CN	410
$>C<$	-93	Cl (mean)	260
$CH_2=$	190	Cl single	270
$-CH=$ double bonded	111	Cl twinned as in $>CCl_2$	260
$>C=$	19	Cl triple as in $-CCl_3$	250
$CH\equiv C-$	285	Br single	340
$-C\equiv C-$	222	I single	425
Phenyl	735	CF_2 } n-fluorocarbons only	150
Phenylene (o,m,p)	658	CF_3 }	274
Naphthyl	1146	S sulfides	225
Ring, 5 membered	105−115	SH thiols	315
Ring, 6 membered	95−105	ONO_2 nitrates	~440
Conjugation	20−30	NO_2 (aliphatic nitro-compounds)	~440
H (variable)	80−100	PO_4 (organic phosphates)	~500
O ether	70	Si (in silicones)	−38

Table compiled by R. B. Seymour.

Table II
SOLUBILITY PARAMETER VALUES

A. Nonpolar Solvents

Name	δ (H)	Name	δ (H)
Acetic acid nitrile (acetonitrile)	11.9	Benzene, isopropyl (cumene)	8.5
Anthracene	9.9	Benzene, 1-isopropyl-4-methyl (p-cymene)	8.2
Benzene	9.2	Benzene, nitro	10.0
Benzene, chloro	9.5	Benzene, propyl	8.6
Benzene, 1,2-dichloro	10.0	Benzene, 1,3,5-trimethyl (mesitylene)	8.8
Benzene, ethyl	8.8	Benzoic acid nitrile (benzonitrile)	8.4

Table II (continued)
SOLUBILITY PARAMETER VALUES

A. Nonpolar Solvents

Name	δ (H)	Name	δ (H)
Biphenyl, perchloro	8.8	Hexene-1	7.4
1,3-Butadiene	7.1	Malonic acid dinitrile (malononitrile)	15.1
1,3-Butadiene, 2-methyl (isoprene)	7.4	Methane	5.4
Butane	6.8	Methane, bromo	9.6
Butanoic acid nitrile	10.5	Methane, dichloro (methylene chloride)	9.7
Carbon disulfide	10.0	Methane, dichloro-difluoro (Freon 12®)	5.5
Carbon tetrachloride	8.6	Methane, dichloro, manofluoro (Freon 21®)	8.3
Chloroform	9.3	Methane, nitro	12.7
Cyclohexane	8.2	Methane, tetrachloro-difluoro (Freon 112®)	7.8
Cyclohexane, methyl	7.8	Methane, trichloro-monofluoro (Freon 11®)	7.6
Cyclohexane, perfluoro	6.0	Naphthalene	9.9
Cyclopentane	8.7	Nonane	7.8
Decalin	8.8	Octane	7.6
Decane	8.0	Pentane	7.0
Dimethyl sulfide	9.4	Pentane, 1-bromo	7.6
Ethane	6.0	Pentane, 1-chloro	8.3
Ethane, bromo (ethyl bromide)	9.6	Pentanoic acid, nitrile (valeronitrile)	9.6
Ethane, chloro (ethyl chloride)	9.2	Pentene-1	6.9
Ethane, 1,2-dibromo	10.4	Phenanthrene	9.8
Ethane, 1,1-dichloro (ethylidene chloride)	8.9	Propane	6.4
Ethane, difluoro-tetrachloro (Freon 112®)	7.8	Propane, 1-bromo	8.9
Ethane, nitro	11.1	Propane, 2,2-dimethyl (neopentane)	6.3
Ethane, pentachloro	9.4	Propane, 1-nitro	16.3
Ethane, 1,1,2,2-tetrachloro	9.7	Propane-2-nitro	9.9
Ethanethiol (ethyl mercaptan)	9.2	Propene (propylene)	6.5
Ethane, 1,1,2-trichloro	9.6	Propene, 2-methyl (isobutylene)	6.7
Ethane trichloro-trifluoro (Freon 113®)	7.3	Propenoic acid nitrile (acrylonitrile)	10.5
Ethene, (ethylene)	6.1	Propionic acid nitrile	10.8
Ethene, tetrachloro (perchloroethylene)	9.3	Styrene	9.3
Ethene, trichloro	9.2	Terphenyl, hydrogenated	9.0
Heptane	7.4	Tetralin	9.5
Heptane, perfluoro	5.8	Toluene	8.9
Hexane	7.3	Xylene, m-	8.8

B. Moderately Polar Solvents

Name	δ (H)	Name	δ (H)
Acetic acid, butyl ester	8.5	Ethylene glycol, monomethyl ether (methyl Cellosolve)	11.4
Acetic acid, ethyl ester	9.1		
Acetic acid, methyl ester	9.6	Formic acid amide, N,N-diethyl	10.6
Acetic acid, pentyl ester	8.0	Formic acid amide, N,N-dimethyl	12.1
Acetic acid, propyl ester	8.8	Formic acid, ethyl ester	9.4
Acetic acid amide, N,N-diethyl	9.9	Formic acid, methyl ester	10.2
Acetic acid amide, N,N-dimethyl	10.8	Formic acid, 2-methylbutyl ester	8.0
Acrylic acid, butyl ester	8.4	Formic acid, propyl ester	9.2
Acrylic acid, ethyl ester	8.6	Furan	9.4
Acrylic acid, methyl ester	8.9	Furan, tetrahydro	9.1
Adipic acid, dioctyl ester	8.7	Furfural	11.2
Aniline, N,N-dimethyl	9.7	2-Heptanone	8.5
Benzene, 1-methoxy-4-propenyl (anethole)	8.4	Hexanoic acid, 6-aminolactam (ε-caprolactam)	12.7
Benzoic acid, ethyl ester	8.2	Hexanoic acid, 6-hydroxylactone (caprolactone)	10.1
Benzoic acid, methyl ester	10.5	Isophorone	9.1
Butanal	9.0	Lactic acid, butyl ester	9.4
Butane, 1-iodo	8.6	Lactic acid, ethyl ester	10.0
Butanoic acid, 4-hydroxylactone (butyrolactone)	12.6	Methacrylic acid, butyl ester	8.3
2-Butanone	9.3	Methacrylic acid, ethyl ester	8.5
Carbonic acid, diethyl ester	8.8	Methacrylic acid, methyl ester	8.8
Carbonic acid, dimethyl ester	9.9	Oxalic acid, diethyl ester	8.6
Cyclohexanone	9.9	Oxalic acid, dimethyl ester	11.0
Cyclopentanone	10.4	Oxirane (ethylene oxide)	11.1
2-Decanone	7.8	Pentane, 1-iodo	8.4
Diethylene glycol, monobutyl ether (butyl carbitol)	9.5	2-Pentanone	8.7
Diethylene glycol, monoethyl ether (ethyl carbitol)	10.2	Pentanone-2,4-hydroxy,4-methyl (diacetone alcohol)	9.2
Dimethyl sulfoxide	12.0	Pentanone-2,4-methyl (mesityl oxide)	9.0
1,4-Dioxane	10.0	Phosphoric acid, triphenyl ester	8.6
Ethene, chloro (vinyl chloride)	7.8	Phosphoric acid, tri-2-tolyl ester	8.4
Ether, 1,1-dichloroethyl	10.0	Phthalic acid, dibutyl ester	9.3
Ether, diethyl	7.4	Phthalic acid, diethyl ester	10.0
Ether, dimethyl	8.8	Phthalic acid, dihexyl ester	8.9
Ether, dipropyl	7.8	Phthalic acid, dimethyl ester	10.7
Ethylene glycol, monobutyl ether (butyl Cellosolve®)	9.5	Phthalic acid, di-2-methylnonyl ester	7.2
		Phthalic acid dioctyl ester	7.9
Ethylene glycol, monoethyl ether (ethyl Cellosolve)	10.5	Phthalic acid, dipentyl ester	9.1
		Phthalic acid, dipropyl ester	9.7
		Propane, 1,2-epoxy (propylene oxide)	9.2

Table II (continued)
SOLUBILITY PARAMETER VALUES

B. Moderately Polar Solvents

Name	δ (H)	Name	δ (H)
Sebacic acid, dioctyl ester	8.6	Propionic acid, ethyl ester	8.4
Stearic acid, butyl ester	7.5	Propionic acid, methyl ester	8.9
Sulfone, diethyl	12.4	4-Pyrone	13.4
Sulfone, dimethyl	14.5	2-Pyrrolidone, 1-methyl	11.3
Sulfone, dipropyl	11.3	Sebacic acid, dibutyl ester	9.2

C. Hydrogen-bonded Solvents

Name	δ (H)	Name	δ (H)
Acetic acid	10.1	1-Hexanol	10.7
Acetic acid amide, N-ethyl	12.3	1-Hexanol-2-ethyl	9.5
Acetic acid, dichloro	11.0	Maleic acid anhydride	13.6
Acetic acid, anhydride	10.3	Methacrylic acid	11.2
Acrylic acid	12.0	Methacrylic acid amide, N-Methyl	14.6
Amine, diethyl	8.0	Methanol	14.5
Amine, ethyl	10.0	Methanol, 2-furil (furfuryl alcohol)	12.5
Amine, methyl	11.2	1-Nonanol	8.4
Ammonia	16.3	Pentane, 1-amino	8.7
Aniline	10.3	1,3-Pentanediol, 2-methyl	10.3
1,3-Butanediol	10.9	1-Pentanol	11.6
1,4-Butanediol	10.0	2-Pentanol	12.1
2,3-Butanediol	8.7	Piperidine	11.1
1-Butanol	13.6	2-Piperidone	11.4
2-Butanol	12.6	1,2-Propanediol	10.8
1-Butanol, 2-ethyl	11.9	1-Propanol	10.5
1-Butanol, 2-methyl	11.5	2-Propanol	10.0
Butyric acid	10.5	1-Propanol, 2-methyl	10.5
Cyclohexanol	10.6	2-Propanol, 2-methyl	11.4
Diethylene glycol	11.8	2-Propenol (allyl alcohol)	12.1
1-Dodecanol	9.9	Propionic acid	8.1
Ethanol	10.0	Propionic acid anhydride	12.7
Ethanol, 2-chloro (ethylene chlorohydrin)	12.6	1,2-Propanediol	12.2
Ethylene glycol	10.7	Pyridine	14.6
Formic acid	14.7	2-Pyrrolidone	12.1
Formic acid amide, N-ethyl	10.8	Quinoline	13.9
Formic acid amide, N-methyl	15.4	Succinic acid anhydride	16.1
Glycerol	9.9	Tetraethylene glycol	16.5
2,3-Hexanediol	10.2	Toluene, 3-hydroxy (meta cresol)	10.3
1,3-Hexanediol-2-ethyl	23.4	Water	9.4

Table compiled by R. B. Seymour.

MISCIBILITY OF ORGANIC SOLVENT PAIRS

Table A

Doctor J. S. Drury

Industrial and Engineering Chemistry Vol. 44, No. 11, Nov. 1952

(Reprinted by permission)

The classifications were made by shaking together 5 ml. of each of the solvents listed in a test tube for 1 minute, then allowing the mixture to settle. If no interfacial meniscus was observed, the solvent pair was considered miscible. If such a meniscus was present, the solvent pair was regarded as immiscible. The classification of immiscible is a qualitative one since solvent pairs may exhibit some degree of partial miscibility while existing as separate phases. Solvent pairs possessing a pronounced degree of partial miscibility are designated by the symbol Is.

Compound number	Compounds	Acetone	Acetyl acetone	2-Amino-2-methyl-1-propanol	Aniline	Benzaldehyde	Benzene	Benzin	Benzyl alcohol	Butyl acetate	Butyl alcohol	n-Butyl ether	Capryl alcohol	Carbon tetrachloride	Diacetone alcohol	Diethanolamine	Diethyl cellosolve	Diethyl ether	Dimethylaniline	Ethyl alcohol	Ethyl benzoate	Ethylene glycol	2-Ethylhexanol	Formamide	Furfuryl alcohol	Glycerol	Hydroxyethyl-ethylenediamine	Isoamyl alcohol	Methyl isobutyl ketone	Nitromethane	Dibutoxytetra-ethylene glycol	Pyridine	Triethanolamine	Trimethylene glycol
1	Acetone	··	M	M	··	M	M	M	M	M	M	M	M	M	M	M	M	M	M	M	M	M	M	M	M	I	M	M	M	M	M	M	··	M
2	Acetyl acetone	M	··	R	··	M	M	M	M	M	M	M	M	M	M	R	M	M	M	M	M	M	M	M	M	I	R	M	M	M	M	M	··	M
3	Adiponitrile	M	M	M	I	··	M	··	M	M	M	I	··	I	M	··	I	M	M	I	I	M	M	I	M	I	M	I	M	··	M	M	M	··
4	2-Amino-2-methyl-1-propanol	M	R	··	··	M	M	I	··	M	M	M	··	Is	M	M	R	M	M	M	M	M	M	M	M	M	M	M	M	··	M	M	··	M
5	Benzaldehyde	M	M	M	··	··	M	M	M	M	M	M	M	M	M	M	M	M	M	M	M	M	M	Is	M	M	Is	R	M	M	M	M	··	M
6	Benzene	M	M	M	··	M	··	M	M	M	M	M	M	M	M	I	M	M	M	M	M	M	M	I	M	I	I	M	M	M	M	M	··	I
7	Benzin	M	M	I	··	M	M	··	··	I	M	M	M	I	I	M	M	M	M	I	I	M	M	I	I	I	I	M	M	M	··	M	M	I
8	Benzonitrile	M	M	M	M	··	M	··	M	M	M	M	··	M	M	M	M	M	M	M	M	I	M	I	M	I	M	M	··	M	M	M	M	I
9	Benzothiazole	M	M	M	M	M	M	M	M	M	M	M	M	M	M	M	M	M	M	M	M	I	M	I	M	I	M	M	M	M	M	M	M	M
10	Benzyl alcohol	M	M	I	M	M	M	I	··	M	M	M	M	M	M	M	M	M	M	M	M	M	M	I	M	I	M	M	M	M	M	M	R	I
11	Benzyl mercaptan	M	M	I	M	··	M	··	M	M	M	M	M	M	M	M	M	M	M	I	M	I	M	I	M	I	M	M	M	M	M	M	R	I
12	Butyl acetate	M	M	M	··	M	M	M	M	··	M	M	M	M	M	M	M	M	M	M	M	Is	M	I	M	I	I	M	M	M	M	M	··	Is
13	Butyl alcohol	M	M	M	··	M	M	M	M	··	M	M	··	M	M	M	M	M	M	M	M	M	M	I	M	I	M	M	M	M	M	M	··	M
14	n-Butyl ether	M	M	Is	··	M	M	M	M	M	M	··	··	M	M	I	M	M	M	M	M	I	M	I	M	I	I	M	M	M	M	M	··	I
15	Capryl alcohol	M	M	M	··	M	M	M	M	M	M	M	··	M	M	M	M	M	M	M	M	I	M	I	M	I	M	M	M	Is	M	M	··	I
16	Carbon tetrachloride	M	M	M	··	M	M	M	M	M	M	M	M	··	Is	I	M	M	M	M	M	I	M	I	M	I	M	M	M	M	M	M	M	I
17	Diacetone alcohol	M	M	R	··	M	Is	I	M	M	M	M	M	Is	··	M	M	M	M	M	M	M	M	M	M	M	R	M	M	I	I	M	··	M
18	Diethanolamine	M	R	M	I	I	I	I	M	I	M	I	M	I	M	··	I	I	I	Is	M	M	M	M	M	M	M	M	I	I	I	··	I	M
19	Diethyl cellosolve	M	M	M	··	M	M	M	M	M	M	M	M	M	M	I	··	M	M	M	M	M	M	I	M	I	M	M	M	M	M	M	··	M
20	Diethyl ether	M	M	M	··	M	M	M	M	M	M	M	M	M	M	I	M	··	M	M	M	I	M	I	M	I	I	M	M	M	M	M	··	I
21	Dimethylaniline	M	M	M	··	M	M	M	M	M	M	M	M	M	M	Is	M	M	··	M	M	I	M	I	M	I	I	M	M	M	M	M	··	I
22	Di-N-propylaniline	M	M	I	··	M	M	M	M	M	M	M	M	M	M	M	M	M	M	M	M	I	M	I	M	I	I	M	M	M	M	I	I	I
23	Ethyl alcohol	M	M	M	··	M	M	M	M	M	M	M	M	M	M	M	M	M	M	··	M	M	M	M	M	M	M	M	M	M	M	M	M	M
24	Ethyl benzoate	M	M	M	··	M	M	M	M	M	M	M	M	M	M	M	M	M	M	M	··	I	M	I	M	I	M	M	M	M	M	M	M	Is
25	Ethyl isothiocyanate	M	M	R	M	··	M	··	M	M	M	M	M	M	M	M	M	M	M	M	M	I	M	I	M	I	R	M	M	··	M	M	M	I
26	Ethyl thiocyanate	M	M	M	··	M	M	··	M	M	M	M	M	M	M	M	M	M	M	M	M	I	M	I	M	I	M	M	M	M	I	M	M	I
27	Ethylene glycol	M	M	M	··	Is	I	M	M	M	Is	M	I	M	M	I	M	M	I	M	I	··	M	I	M	I	M	M	I	M	M	I	M	M	M
28	2-Ethylhexanol	M	M	M	··	M	M	M	M	M	M	M	M	M	M	I	M	M	M	M	M	M	··	I	M	I	I	M	M	M	M	M	M	M
29	Formamide	M	M	M	··	I	M	I	I	M	I	M	I	I	M	M	M	I	I	M	I	I	I	··	M	M	M	M	I	Is	M	M	M	M
30	Furfuryl alcohol	M	M	M	··	M	M	M	M	M	M	M	M	M	M	M	M	M	M	M	M	I	M	M	··	M	M	M	M	I	M	M	M	M
31	Glycerol	I	I	M	··	I	I	I	M	I	M	I	I	I	M	M	I	I	I	M	I	I	I	M	M	··	M	I	I	I	M	I	M	M
32	Hydroxyethyl-ethylenediamine	M	R	M	··	R	I	Is	M	I	M	I	M	I	R	M	I	I	I	M	M	M	M	M	M	M	··	M	M	M	M	M	M	M
33	Isoamyl alcohol	M	M	M	··	M	M	M	M	M	M	M	M	M	M	M	M	M	M	M	M	I	M	I	M	I	I	··	M	M	M	M	··	M
34	Isoamyl sulfide	M	M	I	··	M	M	M	M	M	M	M	M	M	M	M	M	M	M	I	M	I	M	I	I	I	I	M	M	··	M	M	I	R
35	Isobutyl mercaptan	M	M	M	··	M	M	M	M	M	M	M	M	M	M	M	M	M	M	I	M	I	M	I	M	I	I	M	M	··	M	M	R	R
36	Methyl disulfide	M	M	M	··	M	M	M	M	M	M	M	M	M	M	M	M	M	M	I	M	I	M	I	M	I	I	M	M	··	M	M	I	R
37	Methyl isobutyl ketone	M	M	M	··	M	M	M	M	M	M	M	M	M	M	M	I	M	M	M	M	I	M	Is	M	I	M	M	··	M	M	M	M	I
38	Nitromethane	M	M	M	··	M	M	I	M	M	M	M	I	Is	M	I	M	M	M	M	I	M	I	I	M	I	M	M	M	··	M	M	M	I
39	Dibutoxytetra-ethylene glycol	M	M	M	··	M	M	M	M	M	M	M	M	M	M	I	M	M	M	M	M	I	M	I	M	I	M	M	M	M	··	M	··	M
40	Pyridine	M	M	M	··	M	M	M	M	M	M	M	M	M	M	M	M	M	M	M	M	I	M	M	M	I	M	M	M	M	M	··	M	M
41	Tri-n-butylamine	M	M	I	I	··	M	··	M	M	M	M	··	I	M	I	··	M	M	M	M	I	M	I	··	I	M	M	M	M	M	M	I	I
42	Trimethylene glycol	M	M	M	··	M	I	I	M	I	M	Is	M	I	M	I	M	M	I	I	M	Is	M	I	M	M	M	M	M	I	M	M	··	··

MISCIBILITY OF ORGANIC SOLVENT PAIRS (Continued)

Tables B and C

W. M. Jackson and J. S. Drury

Reprinted from Vol. 51 pp. 1491 to 1493, December 1959. Copyright 1959 by the American Chemical Society and reprinted by permission of the copyright owner.

The classifications were made at 20°C in the following manner. One-milliliter portions of each solvent comprising a pair were shaken together for approximately a minute. If no interfacial meniscus was observed after the contents of the tube were allowed to settle, the solvent pair was considered to be miscible, M. If a meniscus was observed without apparent change in the volume of either solvent, the pair was regarded as immiscible, I. This classification is a qualitative one, since solvent pairs may exhibit various degrees of partial miscibility while existing as separate phases. If an obvious change occurred in the volume of each solvent, but a meniscus was present, the pair was classified as partially miscible, S. The designation R indicates that the two solvents reacted.

Table B

No.	Compounds	1	2	3	4	5	6	7	8	9	10	11	12	13	14	15	16	17	18	19	20	21	22	23	24	25	26	27	28	29	30	31	32	33	No.
1	Acetone	..	M	M	M	M	M	M	M	M	M	M	M	M	M	M	M	M	M	M	M	M	M	M	M	M	M	M	R	M	M	M	M	M	1
2	Isoamyl acetate	M	..	M	M	M	M	M	M	M	M	M	M	M	M	M	M	M	M	M	M	M	M	M	M	M	M	M	R	M	M	I	M	M	2
3	n-Amyl cyanide	M	M	..	M	M	M	M	M	M	M	M	M	M	M	M	M	M	M	M	M	M	M	M	M	M	M	M	R	M	M	M	M	M	3
4	Benzene	M	M	M	..	M	S	M	M	M	M	M	M	M	M	M	M	M	M	M	M	M	M	M	M	M	M	M	R	M	M	S	M	M	4
5	Benzyl ether	M	M	M	M	..	S	M	M	M	M	M	M	M	M	M	M	M	M	M	M	M	M	M	M	M	M	M	R	M	M	I	M	M	5
6	2-Bromoethyl acetate	M	M	M	S	M	..	M	M	R	M	M	R	R	M	M	R	M	M	M	M	M	M	S	M	M	M	M	R	M	M	M	R	M	6
7	Chloroform	M	M	M	M	M	M	..	M	M	M	M	M	M	M	M	M	M	M	M	M	M	M	M	M	M	M	M	R	M	M	M	R	M	7
8	Cinnamaldehyde	M	M	M	M	M	M	M	..	M	M	R	M	M	M	M	M	M	M	M	M	M	M	M	M	M	M	M	R	M	M	M	M	M	8
9	Di-n-amylamine	M	M	M	M	M	M	R	M	..	M	R	I	R	M	M	M	M	M	M	M	M	M	M	M	M	R	R	R	M	M	M	S	M	9
10	Di-n-butyl carbonate	M	M	M	M	M	M	M	M	M	..	M	M	R	M	M	M	M	M	M	M	M	M	M	M	M	M	M	R	M	M	S	M	M	10
11	Diethylacetic acid	M	M	M	M	M	M	M	M	R	M	..	R	M	M	M	R	M	M	M	M	M	M	M	M	M	R	R	R	M	M	I	R	M	11
12	Diethylenetriamine	M	M	M	M	M	M	R	R	I	M	R	..	R	M	R	M	M	M	I	R	R	M	R	M	M	R	R	R	M	M	M	R	I	12
13	Diethyl formamide	M	M	M	M	M	M	M	M	R	M	M	R	..	M	M	R	M	M	M	M	M	M	I	I	R	R	M	R	M	M	M	M	I	13
14	Diisobutyl ketone	M	M	M	M	M	M	M	M	M	M	M	M	M	..	M	M	M	M	M	M	M	M	M	M	M	M	M	R	M	M	I	M	M	14
15	Diisopropylamine	M	M	M	M	M	M	M	M	R	M	R	M	M	M	..	M	M	M	M	M	M	M	M	M	M	R	M	R	M	M	I	M	M	15
16	Di-n-propylaniline	M	M	M	M	M	M	M	M	M	M	M	M	M	M	M	..	M	M	M	M	M	M	M	M	M	M	M	R	M	M	I	M	M	16
17	Ethyl alcohol	M	M	M	M	M	M	M	M	M	M	M	M	M	M	M	M	..	M	M	M	M	M	M	M	M	M	M	R	M	M	M	M	M	17
18	Ethyl benzoate	M	M	M	M	M	M	M	M	M	M	M	M	M	M	M	M	M	..	M	M	M	M	M	M	M	M	M	R	M	M	I	M	M	18
19	Ethyl ether	M	M	M	M	M	M	M	M	M	M	M	M	M	M	M	M	M	M	..	M	M	M	M	M	M	M	M	R	M	M	I	M	M	19
20	Ethyl phenylacetate	M	M	M	M	M	M	M	M	M	M	M	M	M	M	M	M	M	M	M	..	M	M	M	M	M	M	M	R	M	M	I	M	M	20
21	Heptadecanol[a]	M	M	M	M	M	M	M	M	M	M	M	M	M	M	M	M	M	M	M	M	..	M	M	M	M	M	M	R	M	M	I	M	M	21
22	3-Heptanol	M	M	M	M	M	M	M	M	M	M	M	M	M	M	M	M	M	M	M	M	M	..	M	M	M	M	M	R	M	M	I	M	M	22
23	n-Heptyl acetate	M	M	M	M	M	M	M	M	M	M	M	M	I	M	M	M	M	M	M	M	M	M	..	M	M	M	M	R	M	M	I	R	M	23
24	n-Hexyl ether	M	M	M	M	M	S	M	M	M	M	M	M	I	M	M	M	M	M	M	M	M	M	M	..	M	M	M	R	M	M	I	M	M	24
25	Methyl isopropyl ketone	M	M	M	M	M	M	M	M	M	M	M	M	M	M	M	M	M	M	M	M	M	M	M	M	..	M	M	R	M	M	I	R	M	25
26	4-Methyl-n-valeric acid	M	M	M	M	M	M	M	M	R	M	R	R	R	M	R	M	M	M	M	M	M	M	M	M	M	..	M	R	M	M	R	R	M	26
27	o-Phenetidine	M	M	M	M	M	M	M	R	M	M	M	R	M	M	M	M	M	M	M	M	M	M	M	M	M	M	..	R	M	M	R	M	M	27
28	Sulfuric acid (concd.)	R	R	R	I	R	R	R	R	R	R	R	R	R	R	R	R	M	M	M	M	R	R	R	R	R	R	R	..	R	R	M	R	R	28
29	Tetradecanol[a]	M	M	M	M	M	M	M	M	M	M	M	M	M	M	M	M	M	M	M	M	M	M	M	M	M	M	M	R	..	M	I	M	M	29
30	Tri-n-butyl phosphate	M	M	M	M	M	M	M	M	M	M	M	M	M	M	M	M	M	M	M	M	M	M	M	M	M	M	M	R	M	..	I	M	M	30
31	Triethylene glycol	M	I	M	S	I	M	M	M	M	M	M	M	M	M	M	M	I	M	M	M	M	M	I	I	I	M	R	M	I	I	..	M	I	31
32	Triethylenetetramine	M	M	M	M	M	R	M	M	R	I	M	R	R	M	I	M	M	M	M	M	M	M	R	M	M	R	M	R	M	M	M	..	I	32
33	2,6,8-Trimethyl 4-nonanone	M	M	M	M	M	M	M	M	M	M	M	M	I	M	M	M	M	M	M	M	M	M	M	M	M	M	M	R	M	M	I	I	..	33

[a] Union Carbide name.

Column key: 1 Acetone; 2 Isoamyl acetate; 3 n-Amyl cyanide; 4 Benzene; 5 Benzyl ether; 6 2-Bromoethyl acetate; 7 Chloroform; 8 Cinnamaldehyde; 9 Di-n-amylamine; 10 Di-n-butyl carbonate; 11 Diethylacetic acid; 12 Diethylenetriamine; 13 Diethyl formamide; 14 Diisobutyl ketone; 15 Diisopropylamine; 16 Di-n-propylaniline; 17 Ethyl alcohol; 18 Ethyl benzoate; 19 Ethyl ether; 20 Ethyl phenylacetate; 21 Heptadecanol; 22 3-Heptanol; 23 n-Heptyl acetate; 24 n-Hexyl ether; 25 Methyl isopropyl ketone; 26 4-Methyl-n-valeric acid; 27 o-Phenetidine; 28 Sulfuric acid (concd.); 29 Tetradecanol; 30 Tri-n-butyl phosphate; 31 Triethylene glycol; 32 Triethylenetetramine; 33 2,6,8-Trimethyl 4-nonanone.

MISCIBILITY OF ORGANIC SOLVENT PAIRS (Continued)
Table C

Compound number	Compounds	Acetone	Isoamyl acetate	n-Amyl cyanide	Anisaldehyde	Benzene	Benzyl ether	Chloroform	o-Cresol	Diisobutyl ketone	Diethylacetic acid	Diethyl formamide	Di-n-propyl aniline	Ethyl alcohol	Ethyl ether	3-Heptanol	n-Heptyl acetate	n-Hexyl ether	α-Methylbenzylamine	α-Methylbenzyldiethanolamine	α-Methylbenzyldimethylamine	α-Methylbenzylethanolamine	2-Methyl-5-ethylpyridine	Methyl isopropyl ketone	4-Methyl-n-valeric acid	α-Phenetidine	2-Phenylethylamine	isopropanolamine	Pyridine	Salicylaldehyde	Tetradecanol[a]	Tri-n-butyl phosphate	Triethylenetetramine	2,6,8-Trimethyl 4-nonanone
1	1,3-Butylene glycol	M	I	M	I	I	I	M	M	M	I	M	M	M	S	M	I	I	I	M	M	M	M	M	M	M	M	M	M	M	M	M	M	I
2	2,3-Butylene glycol	M	M	M	M	S	I	I	M	M	M	M	M	I	M	S	M	M	I	M	M	M	M	M	M	M	M	M	M	M	M	M	I	I
3	2-Chloroethanol	M	M	M	M	M	M	M	M	M	M	M	M	M	M	M	M	R	M	M	M	R	M	M	M	R	R	M	M	M	M	M	M	M
4	3-Chloro-1,2-propanediol	M	M	M	I		M	M	M	M	M	M	I	M	M	M	M	I	R	M	M	M	M	M	M	R	R	M	M	S	M	R	S	
5	Dibutyl hydrogen phosphite	M	M	M	M		M	M	M	M	M	M	M	M	M	M	M	M	M	M	M	M	M	M	M	M	M	M	M	M	M	M	M	
6	Diethylene glycol dibutyl ether	M	M	M	M		M	M	M	M	M	M	M	M	M	M	M	M	R	M	S	M	M	M	M	M	M	R	R	M	M	M	R	M
7	Diethylene glycol diethyl ether	M	M	M	M		M	M	M	M	M	M	M	M	M	M	M	M	M	M	M	M	M	M	M	M	M	M	M	M	M	M	M	M
8	Diethylene glycol monobutyl ether	M	M	M	M		M	M	M	M	M	M	M	M	M	M	M	M	M	M	M	M	M	M	M	M	M	M	M	M	M	M	M	M
9	Diethylene glycol monoethyl ether	M	M	M	M		M	M	M	M	M	M	M	M	M	M	M	I	M	M	M	M	M	M	M	M	M	M	M	M	M	M	M	M
10	Diethylene glycol monomethyl ether	M	M	M	M		M	M	M	M	M	M	M	M	M	M	M	I	M	M	M	M	M	M	M	M	M	M	M	M	M	M	M	M
11	Dipropylene glycol	M	M	M	M	M	M	M	M	M	M	M	M	M	M	M	M	I	M	M	M	M	M	M	M	M	M	M	M	M	M	M	M	M
12	Ethylene diacetate	M	M	M	M	M	M	M	M	M	M	M	M	M	M	M	M	M	M	M	M	M	M	M	M	M	M	M	M	M	M	M	M	M
13	Ethylene glycol	M	I	I	I	I	S	M	I	M	M	I	M	I	M	I	I	M	M	M	M	I	M	M	M	M	I	M	M	M	M	I	S	M
14	Ethyl glycol ethylbutyl ether	M	M	M	M		M	M	M	M	M	M	M	M	M	M	M	M	M	M	M	M	M	M	M	M	M	M	M	M	M	M	M	M
15	Ethylene glycol monobutyl ether	M	M	M	M		M	M	M	M	M	M	M	M	M	M	M	M	M	M	M	M	M	M	M	M	M	M	M	M	M	M	M	M
16	Ethylene glycol monoethyl ether	M	M	M	M		M	M	M	M	M	M	M	M	M	M	M	M	M	M	M	M	M	M	M	M	M	M	M	M	M	M	M	M
17	Ethylene glycol monomethyl ether	M	M	M	M		M	M	M	M	M	M	M	M	M	M	M	M	M	M	M	M	M	M	M	M	M	M	M	M	M	M	M	M
18	Ethylene glycol monophenyl ether	M	M	M	M		M	M	M	M	M	M	M	M	M	M	M	M	M	M	M	M	M	M	M	M	M	M	M	M	M	M	M	M
19	Glycerol	I	I	I	I	I	I	I	M	I	M	I	M	I	I	I	I	M	M	M	I	I	M	M	M	I	I	M	M	I	I	I	M	I
20	1,2-Propanediol	M	M	M	M	I	I	I	M	M	I	M	M	M	S	M	I	I	I	M	M	M	M	M	M	M	M	M	S	M	M	I	I	
21	1,3-Propanediol	M	I	I	I	I	I	I	M	M	I	M	M	M	M	I	I	I	M	M	M	M	M	M	M	M	M	M	M	M	I	I		
22	Triethylene glycol	M	I	M	M	S	I	I	M	M	I	M	M	M	I	M	I	I	I	M	M	M	M	M	M	M	M	M	I	M	M	I	M	
23	Triethyl phosphate	M	M	M	M		M	M	M	M	M	M	M	M	M	M	M	M	M	M	M	M	M	M	M	M	M	M	M	M	M	M	M	M
24	Trimethylene chlorohydrin	M	M	M	M	M	M	M	M	M	M	M	M	M	M	M	M	R	M	M	M	R	M	M	M	R	R	M	M	M	M	M	R	M

[a] Union Carbide name.

STEROID HORMONES AND OTHER STEROIDAL SYNTHETICS

Compiled by Erwin Di Cyan, Ph.D.

The field of steroids has expanded considerably and rapidly in degree and in kind, because synthetic steroids have been synthesized which though resembling the hormones in the body have no natural counterpart, but exert an effect comparable to those of the natural hormones.

In fact, the term *steroid hormone* thus becomes a misnomer when applied to the newer synthetically prepared steroids which do not have a counterpart in the body of man or other animals—as prednisone. (A hormone, by definition, is a material with certain functions and characteristics, *secreted by the ductless glands.* That part of the definition cannot be met by prednisone or by similar steroids as these are not secreted by the ductless, or endocrine glands.)

All the hormones as well as the synthetic analogues have in common the cyclopentanophenanthrene nucleus. Although chemically very similar, a comparatively slight structural change is in many instances productive of substances which have physiologically dissimilar effects, often acting upon different physiologic systems. But in many cases a small change in structure will result merely in an accentuation of certain effects.

The Cyclopentanophenanthrene Nucleus

Classification. Classification becomes a bizarre problem by reason of the (a) overlapping uses to which these substances are put, and (b) the multiple purposes for which the hormones or synthetic substances are used. Indeed, the steroids may be classified by structure; that however would be uninformative to the student as to their use. Classification by origin, as adrenal, would also be unsuitable because, for example, a number of the adrenal corticosteroids are not found in the adrenal cortex at all, but merely resemble the natural hormones found in the adrenal cortex.

For those reasons the hormonal or hormonelike entries in the tables are classified by-and-large, by their predominant pharmacologic effects. Even that classification has its disparities as for example, the use of male sex hormones, i.e. the androgens, is neither limited to men, nor to uses which entail their effect upon male sex characteristics.

Uses. Originally, the use of steroid hormones was largely based upon one or more of the following predicates:
(a) To supplement the progressively declining secretion of a specific hormone due to natural biologic aging of the organism; in the menopause as an example of such declining secretion, a female sex hormone is used for such supplementation;
(b) To make available to the body a specific hormone, the natural secretion of which is inhibited because of a congenital or developmental anomaly; the underdevelopment of male secondary sex characteristics is an example of such an inhibited secretion, in which a male sex hormone is used—and correspondingly, female sex hormones in underdevelopment in females;
(c) To cause a reversal of hormonal balance in the treatment of diseases peculiar to a sex; for example, in the case of cancer of the female breast, a male sex hormone is administered, and in cancer of the prostate, a female sex hormone is used;
(d) To mimic a natural function, as menstruation, by the administration of estrogens—on withdrawal of which bleeding occurs; or by the alternate use of estrogenic and progestational—both female sex hormones.
(e) To delay a function, as ovulation, as in oral contraceptives, or *birth control pills.*

Since the finding that cortisone ameliorates the symptoms of rheumatoid arthritis (1949) the adrenal corticosteroid hormones and especially the synthetically prepared steroid analogues which have no natural counterpart in the body, have been successfully employed in the treatment of diseases not related to sex or sex function.

Androgens and Anabolic Agents. The agents listed in the tables under this classification have the effect of male sex hormones (androgens) i.e., to stimulate sexual maturation, in the "male climacteric," etc. But all androgens have in greater or lesser degree the ability to stimulate muscle development, i.e., an anabolic effect. Among the synthetically prepared agents which have no counterpart in the body (Methandrostenolone or Oxymetholone) are those which have a lessened androgenic, but a heightened anabolic effect. These qualities are determined by biological tests on animals but principally confirmed by clinical use in man. The anabolic effect includes remineralization of bone, which may be partially demineralized (osteoporosis) by age, or by certain drugs, as the adrenal corticosteroids (q.v.).

Anabolic agents are used for muscle and bone nutrition in men as well as women. The reason for the high interest in synthetic steroidal substances for anabolic use, is based on the need for materials, which within a given effective dose have a greater anabolic-to-androgenic ratio than such androgens as methyl testosterone. Otherwise, the administration of androgens to women produces manifestations of virilism, such as growth of hair on the face, a deepening of the voice, etc. Androgens are also used in the female in the suppression of excessive bleeding and in the treatment of cancer of the breast and cervix. (For other androgen-like agents, see also Progestogens and Progestins.)

Estrogens. Estrogenic agents hasten sexual maturation in the female. Therefore, they are used in underdevelopment in the female. The widest use of estrogens is in the treatment of the menopause, in which they supplement from without, the secretion of natural estrogens by the ovary, which begins to decline at about the 40th year. The menopause is usually a slow process, and the declining secretion gives rise to various symptoms during the time that the secretion declines, until adjustment to the new status takes place. The menopause, a period of physical and psychological stress, is made less precipitous by estrogens.

Frequently, a menopause must be quickly induced, as in cancer of the ovary or in uterine hemorrhage. This is done by radiation or by the removal of the uterus. Severe vasomotor symptoms occur when the menopause is thus suddenly induced. Estrogens—among other drugs—are used in the amelioration of these symptoms.

Estrogens (especially diethylstilbestrol which though not a hormone has an estrogenic effect) are also used in the control of cancer of the prostate in the male. Note the inverse correspondence to the use of male sex hormones in cancer of the breast in the female.

Progestogens and Progestins (Including 19-Norsteroid Compounds). The agents under that listing include progesterone, a female sex hormone, as well as progestins, i.e., synthetic progesterone-like compounds which have no natural counterpart in the body. Their use includes a variety of conditions: functional uterine bleeding, absence of menstruation (amenorrhea) used at times with estrogens, painful menstruation (dysmenorrhea), infertility, habitual abortion in order to maintain pregnancy, and in fact, to suppress ovulation hence their use as antifertility drugs. Certain progestins—as norethindrone combined with

STEROID HORMONES AND OTHER STEROIDAL SYNTHETICS (Continued)

an estrogen, are the principal components of birth control pills—suppressing ovulation, there is no egg to fertilize, hence conception does not take place.

Adrenal Corticosteroids, Including Antiinflammatory, Antiallergic and Antirheumatic Agents. The adrenal cortex secretes a large number of hormones. They usually differ from each other in the accentuation of some phases of their properties. Virtually all of the cortical hormones are catabolic, thus having an effect in this respect, diametrically opposed to the androgens which are anabolic. Nearly all the cortical hormones—differing in degree from each other—cause retention of sodium and water by the body and hasten the excretion of potassium. These effects are utilized in the treatment of adrenal insufficiency or Addison's disease, in which conversely, there is an undue excretion of sodium and a strong retention of potassium. Desoxycorticosterone is used in Addison's disease because it has a particularly strong sodium retaining and potassium excreting effect.

Since the finding in 1949 of the usefulness of cortisone in profoundly reducing the symptoms of rheumatoid arthritis, the adrenal corticosteroids, including hydrocortisone, a natural hormone secreted by the adrenal cortex, and particularly the synthetic analogues not found in the body, as prednisone, have been used in the treatment of a wide variety of inflammatory diseases—especially diseases of collagen tissue. The same antiinflammatory effect is also brought into use in the reduction of inflammations associated with diseases of the skin, allergy, asthma, and in such systematic diseases as disseminated lupus erythematosus, also a collagen disease.

The drawbacks of cortisone, also shared in lesser measure by hydrocortisone, gave the impetus to the synthesis of steroidal substances not native to the body but differing somewhat from cortisone and hydrocortisone, in order to reduce the drawbacks attendant to the use of the latter. The sideeffects—especially those of cortisone—are retention of water and sodium, excretion of potassium, loss of mineral from bone leading to osteoporosis and fractures, hypertension, at times diabetes, personality changes or gastric ulcer. Prednisone and prednisolone among others (see tables) are two such steroidal synthetics which have the effects of cortisone, but fewer or less severe sideeffects. Whereas the synthetic steroidal substances are superior to cortisone with respect to lessened sideeffects, it cannot be said that the sideeffects are absent—they vary in degree from substance to substance.

Diuretic, Antidiuretic and Local Anesthetic Agents. Aldosterone, a natural hormone of the adrenal cortex promotes retention in the body of sodium and water, and facilitates excretion of potassium. Hence its effect is almost diametrically opposed to diuretics—especially the thiazide diuretics. Aldosterone is much more active in this respect than desoxycorticosterone, and is used in the treatment of Addison's disease, a hypofunction of the adrenal glands.

Spironolactone is an antagonist to aldosterone—the latter when elaborated in the body in excessive amounts gives rise to a syndrome called aldosteronism. Spironolactone, a synthetically produced steroid does not have a natural counterpart in the body, is diuretic when mercurial or thiazide diuretics are ineffective; it prevents sodium retention and potassium excretion—effects opposite to aldosterone. Hence spironolactone is used in aldosteronism, against edema, in the treatment of congestive heart failure and in other conditions in which an accumulation of water, and water-retaining salt, is to be corrected.

Doses. The amount of substance which comprises a dose of steroid hormones, or of the steroidal synthetics varies from substance to substance from 0.1 mg for an estradiol ester, to 50 mg for a 19-norsteroid compound. The dose is conditioned upon the order of activity of the substance, the purpose for which it is administered, as well as the patient's response. However, as additional steroids for hormonal use are synthesized—especially those with adrenocortical activity, their average dose is usually smaller than the previously available steroid. The smaller effective dose of the more recent steroid is cited as an advantage over the previously available steroid.

However, a smaller dose cannot be claimed as an inherent advantage of a new steroid in comparison with an existing one, unless the lower dosage exhibits either greater or more prolonged activity or lesser sideeffects. One cannot meaningfully compare a dose, milligram for milligram, without taking into consideration if a heightened effect of the smaller dose produces fewer sideeffects. For example, it does not make any difference if a given effect and the same accompanying sideeffects are produced by a 50 mg or a 5 mg dose.

ADRENAL CORTICOSTEROIDS, INCLUDING ANTIINFLAMMATORY, ANTIALLERGIC AND ANTIRHEUMATIC AGENTS

Names & synonyms:	BETAMETHASONE; 9α-fluoro-16β-methylprednisolone; 16β-methyl-11β,17α,21-trihydroxy-9α-fluoro-1,4-pregnadiene-3,20-dione.	BETAMETHASONE ACETATE; 9α-fluoro-16β-methylprednisolone-21-acetate.	BETAMETHASONE DISODIUM PHOSPHATE; 9α-fluoro-16β-methylprednisolone-21-disodium phosphate.
Formulae:	$C_{22}H_{29}O_5F$	$C_{24}H_{31}O_6F$	$C_{22}H_{28}O_8FNa_2P$
Molecular weight	392.5	434.5	516.4
Melting point (°C)	240 (dec.)	200 to 220 (dec.)	decomposes
Specific rotation	$(\alpha)\frac{25}{D}+112$ to $+120$ (100 mg. in 10 ml. dioxane)	$(\alpha)\frac{25}{D}+120$ to $+128$ (100 mg. in 10 ml. dioxane)	$(\alpha)\frac{25}{D}+99$ to $+105$ (100 mg. in 10 ml. water)
Absorption max.	239 mμ, E(1 %, 1 cm) 390, methanol	239 mμ, methanol	241 mμ, water

Names & synonyms:	CHLOROPREDNISONE ACETATE; 6α-chloroprednisone acetate; 6α-chloro-Δ¹·⁴-pregnadien-17β,21-diol-3,11,20-trione 21-acetate.	CORTICOSTERONE; 11,21-dihydroxyprogesterone; Δ⁴-pregnene-11β,21-diol-3,20-dione; 11β,21-dihydroxy-4-pregnene-3,20-dione; Kendall compound B; Reichstein substance H.	CORTISONE; 17-hydroxy-11-dehydrocorticosterone; 17α,21-dihydroxy-4-pregnene-3,11,20-trione; Δ⁴-pregnene-17α,21-diol-3,11,20-trione; Kendall compound E; Wintersteiner compound F.
Formulae:		$C_{21}H_{30}O_4$	$C_{21}H_{28}O_5$
Molecular weight	436.6	346.40	360.4
Melting point (°C)	207–213	180–182	220–224
Specific rotation	$(\alpha)\frac{25}{D}+137$ to $+142$ (100 mg. in 10 ml. chloroform)	$(\alpha)\frac{15}{D}+222$ (110 mg. in 10 ml. alcohol)	$(\alpha)\frac{25}{D}+209$ (120 mg. in 10 ml. alcohol)
Absorption max.		240 mμ	237 mμ

Names & synonyms:	DESOXYCORTICOSTERONE; deoxycorticosterone; 11-desoxycorticosterone; 21-hydroxyprogesterone; 4-pregnen-21-ol-3,20-dione; Kendall desoxy compound B; Reichstein substance Q.	DESOXYCORTICOSTERONE ACETATE; DCA; 11-desoxycorticosterone acetate.	DESOXYCORTICOSTERONE PIVALATE; desoxycorticosterone trimethylacetate; 21-hydroxy-4-pregnene-3,20-dione pivalate.
Formulae:	$C_{21}H_{30}O_3$	$C_{23}H_{32}O_4$	$C_{26}H_{38}O_4$
Molecular weight	330.2	372.4	414.6
Melting point (°C)	140–142	154–160	198–204
Specific rotation	$(\alpha)\frac{22}{D}+176-+178$ (100 mg. in 10 ml. alcohol)	$(\alpha)\frac{20}{D}+168-+178$ (100 mg. in 10 ml. dioxane)	$(\alpha)\frac{25}{D}+157\pm4$ (1 % in dioxane)
Absorption max.	240 mμ		240 mμ (in ethanol)

Names & synonyms:	DEXAMETHASONE; hexadecadrol; 9α-fluoro-16α-methyl prednisolone; 9α-fluoro-11β,17α-21-trihydroxy-16α-methyl-1,4-pregnadiene-3,20-dione; 16α-methyl-9α-fluoro-1,4-pregnadiene-11β,17α-21-triol-3,20-dione; 16α-methyl-9α-fluoro-Δ¹-hydrocortisone; 1-dehydro-16α-methyl-9α-fluorohydro-cortisone.	DICHLORISONE ACETATE; 9α-11β-dichloro-1,4-pregnadiene-17α,21-diol-3,20-dione-21-acetate	FLUOCINOLONE ACETONIDE; 6α,9α-difluoro-16α hydroxyprednisolone-16,17-acetonide; 6α,9α-difluoro-16α,17α-isopropylidenediosy-1,4-pregnadiene-3,20-dione.
Formulae:	$C_{22}H_{29}FO_5$	$C_{23}H_{28}O_5Cl$	$C_{24}H_{30}O_6F_2$
Molecular weight	392.4	455.3	452.50
Melting point (°C)	262–264	235 (dec.)	255–266
Specific rotation	$(\alpha)\frac{25}{D} + 78$ (100 mg. in 10 ml. dioxane)	$(\alpha)\frac{25}{D} + 160 - 168$ (100 mg. in 10 ml. dioxane)	not less than $+95°$ and not more than $+105°C$ at 25°C.
Absorption max.		$237 m\mu - 316 - 337 (\varepsilon_1^1)$	$237 m\mu \pm 1 m\mu$

Names & synonyms:	FLUOROHYDROCORTISONE; fludrocortisone; 9α-fluorohydrocortisone; 9α-fluorocortisol; fluohydrisone; 9α-fluoro-11β,17α,21-trihydroxy-4-pregnene-3,20-dione; 9α-fluoro-17-hydroxycorticosterone.	FLUOROMETHOLONE; 9α-fluoro-11β,17α-dihydroxy-6α-methyl-1,4-pregnadiene-3,20-dione; 21-desoxy-9α-fluoro-6α-methyl-prednisolone.	FLUPREDNISOLONE; 6α-fluoroprednisolone; 6α-fluoro-1-dehydrohydrocortisone; 6α-fluoro-11β,17α,21-trihydroxy-1,4-pregnadiene-3,20-dione.
Formulae:	$C_{21}H_{29}FO_5$	$C_{22}H_{24}FO_4$	$C_{21}H_2 \cdot FO_5$
Molecular weight	380.4	376.4	378.4
Melting point (°C)	260–262 (dec.)	290 (dec.)	205–210
Specific rotation	$(\alpha)\frac{23}{D} + 139$ (55 mg. in 10 ml. alcohol)	$(\alpha)\frac{25}{D} + 56$ (pyridine)	$(\alpha)_D + 88$ (dioxane)
Absorption max.		239 mu ($a_M = 15,050$) methanol	λ_{max} 241.5 mu (ε 16,000)

Names & synonyms:	FLURANDRENOLONE: 6-fluoro-16α-hydroxyhydrocortisone-16,17-acetonide; 6α-fluoro-11β,21-dehydroxy-16α,17α-isopropylidenedioxy-pregna-4-ene-3,20-dione.	HYDROCORTISONE: cortisol; 17-hydroxycorticosterone; hydrocortisone free alcohol; 11β,17α,21-trihydroxy-4-pregnene-3,20-dione; 4-pregnene-11β,17α,21-triol-3,20-dione; Kendall compound F; Reichstein substance M.	HYDROCORTISONE ACETATE: cortisol acetate; hydrocortisone-21-acetate; 17-hydroxycorticosterone-21-acetate.
Formulae:	$C_{24}H_{33}O_6F$	$C_{21}H_{30}O_5$	$C_{23}H_{32}O_6$
Molecular weight	436.5	362.5	404.5
Melting point (°C)	240–250	215–220 (dec.)	223 (dec.)
Specific rotation	$(\alpha)\frac{25}{D} = +145$ (1 % in $CHCl_3$)	$(\alpha)\frac{25}{D} +150 - +156$ (100 mg. in 10 ml.) dioxane)	$(\alpha)\frac{25}{D} +158 - +165$ (100 mg. in 10 ml. dioxane)
Absorption max.	236 mμ (methanol)	242 mμ	242 mμ (methanol)

Names & synonyms:	HYDROCORTISONE SODIUM SUCCINATE; 11β, 17α, 21-trihydroxy-4-pregnene-3, 20-dione, 21 hydrogen succinate, sodium salt; hydrocortisone, 21 hydrogen succinate, sodium salt.	METHYLPREDNISOLONE; 6α-methylprednisolone; Δ¹-6α-methylhydrocortisone; 1-dehydro-6α-methylhydrocortisone; 11β, 17α,21-trihydroxy-6α-methyl-1,4-pregnadiene-3,20-dione.	METHYLPREDNISOLONE SODIUM SUCCINATE; 1-dehydro-6α-methylhydrocortisone, 21-hydrogen succinate, sodium salt; 6α-methylprednisolone 21-hydrogen succinate, sodium salt; 11β, 17α, 21-trihydroxy-6α-methyl-1, 4-pregnadiene-3, 20-dione, 21-hydrogen succinate, sodium salt.
Formulae:	$C_{25}H_{33}O_8Na$	$C_{22}H_{30}O_5$	$C_{26}H_{33}O_8Na$
Molecular weight	484.5	374.5	496.5
Melting point (°C)	decomposes	230–240 (dec.)	decomposes
Specific rotation	$(\alpha)_D +140 \pm 5$ (alcohol)	$(\alpha)\frac{25}{D} +85$ (dioxane)	$(\alpha)_D +100 \pm 4$ (alcohol)
Absorption max.	λ 242 mμ (ε 15,700)	243 mμ	λ_{max} 242 mμ (ε 14,500)

Names & synonyms:	PARAMETHASONE; 6α-fluoro-16α-methylprednisolone; 6α-fluoro-11β-17α,21-trihydroxy-16α-methyl-1,4-pregnadiene-3,20-dione.	PARAMETHASONE ACETATE; 6α-fluoro-16α-methylprednisolone-21-acetate; 6α-fluoro-16α-methylpregna-1,4-diene-11β,21-diol-3,20-dione-21-acetate; 6α-fluoro-17β,17α,21-trihydroxy-16α-methyl-1,4-pregnadiene-3,20-dione-21-acetate.	PREDNISOLONE; metacortandralone; Δ^1-dehydrocortisol; delta F; Δ^1-hydrocortisone; Δ^1-dehydrohydrocortisone; 1,4-pregnadiene-3,20-dione-11β,17α,21-triol; 11β,17α,21-trihydroxy-1,4-pregnadiene-3,20-dione.
Formulae:	$C_{22}H_{30}O_5$	$C_{24}H_{31}O_6F$	$C_{21}H_{28}O_5$
Molecular weight	392.45	434.5	360.4
Melting point (°C)	228–241	233–246	240 (dec.)
Specific rotation	+59 to +69 at 25°C	$(\alpha)\frac{25}{D} +72$ (1 % in $CHCl_3$)	$(\alpha)\frac{25}{D} +97 - +103$ (100 mg. in 10 ml. dioxane)
Absorption max.	242 mμ	242 mμ (methanol)	242 mμ (ε = 15,000) methanol

Names & synonyms:	PREDNISOLONE PHOSPHATE SODIUM; disodium prednisolone 21-phosphate.	PREDNISOLONE PIVALATE; prednisolone trimethylacetate; 11β,17α,21-trihydroxy-1,4-pregnadiene-3,20-dione 21-pivalate.
Formulae:	$C_{21}H_2 \cdot Na_2O_8P$	$C_{26}H_{36}O_6$
Molecular weight	484.4	444.6
Melting point (°C)		229
Specific rotation	$(\alpha)\frac{25}{D} +102.5$ (100 mg. in 10 ml. H_2O)	+108 ±4 (1 % in dioxane)
Absorption max.	243 mμ	240 and 263 mμ (in absolute ethanol)

Names & synonyms:	PREDNISONE; metacortandricin; Δ^1-dehydrocortisone; delta E; Δ^1-cortisone; 1,4-pregnadiene-17α,21-diol-3,11,20-trione; 17α,21-dihydroxy-1,4-pregnadiene-3,11,20-trione.	TRIAMCINOLONE; 9α-fluoro-16α-hydroxyprednisolone; 9α-fluoro-11β,16α,17α,21-tetrahydroxy-1,4-pregnadiene-3,20-dione.
Formulae:	$C_{21}H_{26}O_5$	$C_{21}H_2\text{-}FO_6$
Molecular weight	358.4	394.4
Melting point (°C)	225 (dec.)	260–262.5 (dec.)
Specific rotation	$(\alpha)\frac{25}{D} + 167 - +175$ (100 mg. in 10 ml. dioxane)	$(\alpha)\frac{25}{D} + 75$ (200 mg. in 100 ml. acetone)
Absorption max.	239 mμ (ε = 15,500) methanol	238 mμ (ε = 15,800)

Names & synonyms:	TRIAMCINOLONE ACETONIDE; 9α-fluoro-11β,21-dihydroxy-16α,17α-isopropylidene-dioxy-1,4-pregnadiene-3,20-dione; 9α-fluoro-16α-hydroxyprednisolone 16,17-acetonide.	TRIAMCINOLONE DIACETATE; 16α,21-diacetoxy-9α-fluoro-11β,17α-dihydroxy-1,4-pregnadiene-3,20-dione; 9α-fluoro-16α-hydroxyprednisolone 16,21-diacetate.
Formulae:	$C_{24}H_{31}FO_6$	$C_{25}H_{31}FO_8$
Molecular weight	434.4	478.49
Melting point (°C)	274–278 (dec.); 292–294	variable: 158–235
Specific rotation	$(\alpha)\frac{25}{D} + 109 - +112$ (53.7 mg. in 10 ml. chloroform)	$(\alpha)\frac{25}{D} + 22$ (78.8 mg. in 10 ml. chloroform)
Absorption max.	238–239 mμ (ε = 14,600)	239 mμ (ε = 15,200)

Names & synonyms:	ANDROSTERONE; cis-androsterone; 3α-hydroxy-17-androstanone; androstane-3α-ol-17-one.	FLUOXYMESTERONE; 9α-fluoro-11β-hydroxy-17α-methyltestosterone 9α-fluoro-11β,17β-dihydroxy-17α-methyl-4-androsten-3-one.	ALDOSTERONE; electrocortin; 18-oxocorticosterone; 18-formyl-11β,21-dihydroxy-4-pregnene-3,20-dione.
Formulae:	$C_{19}H_{30}O_2$	$C_{20}H_{29}FO_3$	$C_{21}H_{28}O_5$
Molecular weight	290.4	336.4	360.4
Melting point (°C)	185–185.5	270 (dec.)	108–112 (hydrate); 164 (anhydrous)
Specific rotation	$(\alpha)\frac{15}{D} +85 - +90$ (150 mg. in 10 ml. dioxane)	$(\alpha)\frac{25}{D} +107 - +109$ (alcohol)	$(\alpha)\frac{25}{D} +161$ (10 mg. in 10 ml. chloroform)
Absorption max.		240 mμ (ε = 16,700) alcohol	240 mμ (log ε = 4.20 monohydr.; ε mol. 15,000 anhydr.)

Names & synonyms:	HYDROXYDIONE SODIUM: 21-hydroxypregnane-3,20-dione-21-sodium hemisuccinate.	SPIRONOLACTONE: 3-(3-oxo-7α-acetylthio-17β-hydroxy-4-androsten-17α-yl)-propionic acid γ lactone.
Formulae:	$C_{25}H_{35}O_6Na$	$C_{24}H_{32}O_4S$
Molecular weight	454.5	416.5
Melting point (°C)	193–203 (dec.)	135 (preliminary)-202 (dec.)
Specific rotation	$(\alpha)\frac{25}{D} +95$ (chloroform) for free acid.	$(\alpha)\frac{25}{D} -34$ (chloroform)
Absorption max.	280 mμ (ε = 93.2)	ε^{238} = 20,200

Names & synonyms:	METHANDROSTENOLONE: 17α-methyl-17β-hydroxy-1,4-androstadien-3-one.	METHYLANDROSTENEDIOL: MAD: methandriol: 17α-methyl-5-androsten-3β,17β-diol.	METHYL TESTOSTERONE: 17-methyl testosterone: 17α-methyl-Δ⁴-androsten-17-β-ol-3-one; 17(β)-hydroxy-17(α-methyl-4-androsten-3-one.
Formulae:	$C_{20}H_{28}O_2$	$C_{20}H_{32}O_2$	$C_{20}H_{30}O_2$
Molecular weight	300.4	304.4	302.4
Melting point (°C)	166–167	205–207	161–166
Specific rotation	$(\alpha)\frac{20}{D}+9 - +17$ (100 mg. in 10 ml. alcohol)	$(\alpha)\frac{20}{D}-73$ (100 mg. in 10 ml. alcohol)	$(\alpha)\frac{25}{D}+69 - +75$ (100 mg. in 10 ml. dioxane)
Absorption max.			

Names & synonyms:	NORETHANDROLONE: 17α-ethyl-19-nortestosterone: 17α-ethyl-17-hydroxy-4-norandrosten-3-one: 17α-ethyl-17-hydroxy-19-norandrost-4-en-3-one.	OXANDROLONE: 17β-hydroxy-17α-methyl-2-oxa-5α-androstane-3-one.
Formulae:	$C_{20}H_{30}O_2$	$C_{19}H_{30}O_3$
Molecular weight	302.4	306.4
Melting point (°C)	130–136	230–233
Specific rotation	$(\alpha)\frac{25}{D}+21$ (dioxane)	$(\alpha)\frac{25}{D}-21$ (1% in chloroform)
Absorption max.	240 mμ (ε = 16,500)	None

Names & synonyms:	OXYMETHOLONE: 17β-hydroxy-2-hydroxymethylene-17α-methyl-3-androstanone: 2-hydroxymethylene-17-α-methyl dihydrotestosterone.	PROMETHOLONE: 2α-methyl-dihydro-testosterone propionate: 2α-methyl-5α-androstane-17β-ol-3-one-propionate.
Formulae:	$C_{21}H_{32}O_3$	
Molecular weight	332.4	360.5
Melting point (°C)	182	124–130
Specific rotation	$(\alpha)\frac{25}{D} = +36$ (200 mg. in 10 ml. dioxane)	$(\alpha)\frac{25}{D} +22 - +29$ (200 mg. in 10 ml. chloroform)
Absorption max.	$E_1^1 = 547$ at 315 mμ (in alkaline methanol made 0.01 N with NaOH)	without significant absorption from 220–300 mμ (methanol)

Names & synonyms:	TESTOSTERONE: trans-testosterone: Δ^4-androsten-17-β-ol-3-one: 17β-hydroxy-4-androsten-3-one.	TESTOSTERONE CYPIONATE: testosterone cyclopentylpropionate: 17β-hydroxy-4-androsten-3-one, cyclopentanepropionate.
Formulae:	$C_{19}H_{28}O_2$	$C_{27}H_{40}O_3$
Molecular weight	288.4	412.6
Melting point (°C)	151–156	100–102
Specific rotation	$(\alpha)\frac{24}{D} +109$ (400 mg. in 10 ml. alcohol)	$(\alpha)_D +88.5 \pm 3.5$ (CHCl$_3$)
Absorption max.	238 mμ	λ_{max} 241 mμ (ϵ 16,125)

Names & synonyms:	TESTOSTERONE ENANTHATE: testosterone heptanoate: 17β-hydroxyandrost-4-en-3-one-17-enanthate.	TESTOSTERONE PHENYLACETATE: 17β-hydroxy-4-androsten-3-one phenyl-acetate; testosterone α-toluate.	TESTOSTERONE PROPIONATE: Δ^4-androstene-17-β-propionate-3-one.
Formulae:	$C_{26}H_{40}O_3$	$C_{27}H_{34}O_3$	$C_{22}H_{32}O_3$
Molecular weight	400.6	406.5	344.4
Melting point (°C)	34–39	129–131	118–122
Specific rotation	$(\alpha)\frac{25}{D}+77 - +82$ (2 % in dioxane)	$(\alpha)\frac{25}{D}+101 \pm 3$ (1 % in chloroform)	$(\alpha)\frac{25}{D}+83 - +90$ (100 mg. in 10 ml. dioxane)
Absorption max.	241 mμ (in ethanol)	241 mμ (in ethanol)	

ESTROGENS

Names & synonyms:	EQUILENIN: 3-hydroxy-17-keto-$\Delta^{1,3,5-10,6,8}$ estrapentaene: 1,3,5–10,6,8-estrapentaen-3-ol-17-one.	EQUILIN: 3-hydroxy-17-keto-$\Delta^{1,3,5-10,7}$ estratetraene: 1,3,5,7-estratetraen-3-ol-17-one.	ESTRADIOL (formerly called α-estradiol): β-estradiol; dihydrofolliculin: dihydroxyestrin; 1,3,5-estratriene-3,17β-diol; 3,17-dihydroxy-$\Delta^{1,3,5-10}$-estratriene: 3,17-epidihydroxyestratriene.
Formulae:	$C_{18}H_{18}O_2$	$C_{18}H_{20}O_2$	$C_{18}H_{24}O_2$
Molecular weight	266.3	268.3	272.3
Melting point (°C)	258–259	236–240	173–179
Specific rotation	$(\alpha)\frac{25}{D}+89$ (dioxane)	$(\alpha)\frac{25}{D}+308$ (200 mg. in 10 ml. dioxane): +325 (200 mg. in 10 ml. alcohol).	$(\alpha)\frac{25}{D}+76 - +83$ (100 mg. in 10 ml. dioxane)
Absorption max.	231, 270, 282, 292, 325, 340 mμ	283–285 mμ	225, 280 mμ

Names & synonyms:	ESTRADIOL BENZOATE; β-estradiol-3-benzoate; estradiol monobenzoate.	ESTRADIOL CYPIONATE; estradiol cyclopentylpropionate; β-estradiol 17-cyclopentanepropionate; 1,3,5(10)-estratriene-3,17β-diol,17-cyclopentanepropionate.
Formulae:	$C_{25}H_{28}O_3$	$C_{26}H_{36}O_3$
Molecular weight	376.4	396.6
Melting point (°C)	191–196	151–154
Specific rotation	$(\alpha)\frac{25}{D} + 58 - +63$ (200 mg. in 10 ml. dioxane)	$(\alpha)_D + 41.5 \pm 3.5$ (dioxane)
Absorption max.		223 mμ

Names & synonyms:	ESTRADIOL DIPROPIONATE; α-estradiol dipropionate; 17β-estradiol dipropionate.	ESTRIOL; trihydroxyestrin; $\Delta^{1,3,5-10}$-estratriene-3-16-cis-17-trans-diol; 1,3,5-estratriene-3,16α,17β-triol.	ESTRONE; folliculin; ketohydroxyestrin; 1,3,5-estratrien-3-ol-17-one.
Formulae:	$C_{24}H_{32}O_4$	$C_{18}H_{24}O_3$	$C_{18}H_{22}O_2$
Molecular weight	384.5	288.3	270.3
Melting point (°C)	104–109	282	258–262
Specific rotation	$(\alpha)\frac{25}{D} + 39 \pm 2$ (1 % in dioxane)	$(\alpha)\frac{25}{D} + 53 - +63$ (40 mg. in 1 ml. dioxane)	$(\alpha)\frac{25}{D} + 158 - +168$ (100 mg. in 10 ml. dioxane)
Absorption max.	268 mμ	280 mμ	283–285 mμ

STEROID HORMONES AND OTHER STEROIDAL SYNTHETICS (Continued)

Names & synonyms:	ESTRONE BENZOATE	ETHYNYL ESTRADIOL; 17-ethinyl estradiol; 17α-ethynyl-1,3,5-estratriene-3,17β-diol.	MESTRANOL; ethynylestradiol 3-methyl ether; 3-methoxy-17α-ethynyl-1,3,5(10)-estratriene-17β-ol; 17α-ethynyl-estradiol-3-methyl ether; 3-methoxy-19-nor-17α-pregna-1,3,5,trien-20-yn-17-ol.
Formulae:	$C_{25}H_{26}O_3$	$C_{20}H_{24}O_2$	$C_{21}H_{26}O_2$
Molecular weight	374.4	296.4	310.4
Melting point (°C)	220	141–146	148–154
Specific rotation	$(\alpha)\frac{25}{D} + 120$ (dioxane)	$(\alpha)\frac{25}{D} + 1 - + 10$ (100 mg. in 10 ml. dioxane)	$(\alpha)\frac{25}{D} + 2$ to $+8$ (200 mg. in 10 ml. dioxane)
Absorption max.		248 mμ	278 to 287 mμ (methanol)

PROGESTOGENS AND PROGESTINS (INCLUDING 19-NORSTEROID COMPOUNDS)

Names & synonyms:	ACETOXYPREGNENOLONE; 21-acetoxypregnenolone; prebediolone acetate; Δ⁵-pregnene-3β,21-diol-20-one-21-monoacetate; 21-acetoxy-5-pregnene-3-ol-20-one; 3-hydroxy-21-acetoxy-5-pregnen-20-one.	ANAGESTONE ACETATE; 6α-methyl-4-pregnen-17α-ol-20-one acetate; 17α-acetoxy-6α-methylpregn-4-en-20-one; 17α-acetoxy-6α-methyl-4-pregnen-20-one.	CHLORMADINONE ACETATE; 6-chloro-Δ⁶-dehydro-17α-acetoxyprogesterone; 6-chloro-Δ⁴·⁶-pregnadiene-17α-ol-3,20-dioneacetate.
Formulae:	$C_{23}H_{34}O_4$	$C_{24}H_{36}O_3$	$C_{23}H_{29}ClO_4$
Molecular weight	374.5	372.6	404.9
Melting point (°C)	184–185	172–178	204–212
Specific rotation	$(\alpha)\frac{20}{D} + 37 - + 43$ (dioxane)	$(\alpha)\frac{25}{D} + 40$ to $+45$ (10 mg. in 10 ml. chloroform)	$(\alpha)\frac{25}{D} 0$ to -6 (200 mg. in 10 ml. chloroform)
Absorption max.			284 mμ (methanol) Log $\varepsilon = 4.34 \pm 0.02$

Names & synonyms:	DIMETHISTERONE; 6α,21-dimethylethisterone; 6α,21-dimethyl-17β-hydroxy-17α-pregn-4-en-20-yn-3-one; 6α-methyl-17α-propynylandrost-4-en-17β-ol-3-one; 17β-hydroxy-6α-methyl-17α-(prop-1-ynyl)-androst-4-ene-3-one.	ETHISTERONE; anhydrohydroxyprogesterone; ethinyl testosterone; pregneninolone; 17α ethynyl testosterone; 17α-ethynyl-17β-hydroxy-4-androsten-3-one.	ETHYNODIOL DIACETATE; 17α-ethynyl-estrene-3β,17β-diol-17-diacetate; 19-nor-17α-pregn-4-en-20-yne-3β,17-diol diacetate.
Formulae:	$C_{23}H_{32}O_2 \cdot H_2O$	$C_{21}H_{28}O_2$	$C_{24}H_{32}O_4$
Molecular weight	358.5	312.4	384.5
Melting point (°C)	App. 100 (dec.)	266–273	126–132
Specific rotation	$(\alpha)\dfrac{20}{D} + 16.5$ to $+18.5$ (2 % solution in chloroform) (calculated to the anhydrous basis)	$(\alpha)\dfrac{25}{D} - 32°$ (100 mg. in 10 ml. pyridine)	$(\alpha)\dfrac{25}{D} - 74$ (1 % in chloroform)
Absorption max.	App. 240 mμ (anhydrous ethanol) $E_1^{1\%}$ cm = 443	241 mμ (methanol)	None

Names & synonyms:	FLUROGESTONE ACETATE; 17α-acetoxy-9α-fluoro-11β-hydroxy-4-pregnene-3,20-dione	HYDROXYMETHYLPRO-GESTERONE; medroxyprogesterone; 17α-hydroxy-6α-methylprogesterone; 17α-hydroxy-6α-methyl-4-pregnene-3,20-dione.	HYDROXYMETHYLPRO-GESTERONE ACETATE; medroxyprogesterone acetate; 17α-hydroxy-6α-methylprogesterone acetate; 17α-hydroxy-6α-methyl-4-pregnene-3,20-dione acetate.
Formulae:	$C_{23}H_{31}O_5F$	$C_{22}H_{32}O_3$	$C_{24}H_{34}O_4$
Molecular weight	406.5	344.5	386.5
Melting point (°C)	250–251	220–223.5	202–207
Specific rotation	$(\alpha)\dfrac{25}{D} + 78$	$(\alpha)\dfrac{25}{D} + 75$	$(\alpha)\dfrac{25}{D} + 51$ (dioxane)
Absorption max.	238 mμ ($\varepsilon = 17,100$)	241 mμ ($\varepsilon = 16,150$)	241 mμ ($\alpha_M = 16,500$) ethanol

Names & synonyms:	HYDROXYPROGESTERONE: 17α-hydroxyprogesterone; 17α-hydroxy-4-pregene-3,20 dione; 4-pregnen-17α-ol-3,20-dione.	HYDROXYPROGESTERONE ACETATE: 17α-acetoxyprogesterone; 17α-hydroxyprogesterone acetate; 17α-hydroxy-4-pregnene-3,20 dione acetate.
Formulae:	 $C_{21}H_{30}O_3$	 $C_{23}H_{32}O_4$
Molecular weight	330.4	372.5
Melting point (°C)	276	249–250
Specific rotation	$(\alpha)\dfrac{17}{D} + 105$ (104 mg. in 10 ml. chloroform)	$(\alpha)\dfrac{25}{D} + 72$ (chloroform)
Absorption max.		240 mμ (a$_M$ = 16,875) ethanol

Names & synonyms:	HYDROXYPROGESTERONE CAPROATE: 17α-hydroxyprogesterone caproate; 17α-hydroxy-4-pregnene-3,20-dione caproate.	MELENGESTROL ACETATE: MGA: 17α-hydroxy-6-methyl-16-methylene-4,6-pregnadiene-3, 20-dione acetate; 6-dehydro-17-hydroxy-6-methyl-16-methylene-progesterone acetate.
Formulae:	 $C_{27}H_{40}O_4$	 $C_{25}H_{32}O_4$
Molecular weight	428.6	396.51
Melting point (°C)	121–123	215–227
Specific rotation	$(\alpha)\dfrac{25}{D} + 57$ (chloroform)	$(\alpha)_D - 127$ to $- 135$ (in CHCl$_3$)
Absorption max.		288 mμ (ε_1^1 = 24,000) (ethanol)

Names & synonyms:	NORETHINDRONE: Norethisterone; 17α-ethynyl-19-nortestosterone; 17α-ethynyl-17-hydroxy-19-nor-17α-4-en-20-yn-3-one.	NORETHINDRONE ACETATE: 17α-ethynyl-19-nortestosterone acetate.
Formulae:	$C_{20}H_{26}O_2$	$C_{22}H_{28}O_3$
Molecular weight	298.4	340.4
Melting point (°C)	202 and 208	157–163
Specific rotation	$(\alpha)\dfrac{25}{D} - 30 - -35$ (200 mg. in 10 ml. dioxane)	$(\alpha)\dfrac{25}{D} - 32 - -35$ (200 mg. in 10 ml. dioxane)
Absorption max.	α (1 %, 1 cm) λ 240 = 535 ± 15	α (1 %, 1 cm.) λ 240 = 490 to 520 (505 ± 15) (ethanol)

Names & synonyms:	NORETHISTERONE: norethindrone; 19-norethisterone; 17α-ethynyl-19-nor-Δ⁴-androstan-17β-ol-3-one; 17α-ethynyl-19-nor-testosterone; 17-hydroxy-3-oxo-19-nor-17α-pregn-4-ene-20-yne; 17-hydroxy-19-nor-17α-pregn-4-en-20-yn-3-one.	NORETHYNODREL: 17α-ethynyl-17β-hydroxy-5(10)-estren-3-one.
Formulae:	$C_{20}H_{26}O_2$	$C_{20}H_{26}O_2$
Molecular weight	298.4	298.4
Melting point (°C)	200–207	174–184
Specific rotation	$(\alpha)\dfrac{25}{D} - 30 - -38$ (200 mg. in 10 ml. dioxane)	$(\alpha)\dfrac{25}{D} + 125$ (dioxane)
Absorption max.	240 mμ $(\varepsilon_1^1 = 576)$	

STEROID HORMONES AND OTHER STEROIDAL SYNTHETICS (Continued)

Names & synonyms:	NORMETHISTERONE; 19-normethisterone; normethandrolone; metalutin; normetandrone; 17α-methyl-19-nor-Δ⁴-androsten-17β-ol-3-one; 17α-methyl-19-nor-testosterone; 17β-hydroxy-3-oxo-17α-methyl-estra-4-ene; 17β-hydroxy-17-methyl-estr-4-en-3-one.	PREGNENOLONE; Δ⁵-pregnenolone; Δ⁵-pregnen-3β-ol-20-one; 17β(1-ketoethyl)-Δ⁵-androstene-3β-ol.	PROGESTERONE; progestin; progestone; pregnendione; Δ⁴-pregnene-3,20-dione.
Formulae:	$C_{19}H_{28}O_2$	$C_{21}H_{32}O_2$	$C_{21}H_{30}O_2$
Molecular weight	288.4	308.4	314.4
Melting point (°C)	153–158	193	(β) isomer 121; (α) isomer 127–131
Specific rotation	$(\alpha)\frac{25}{D}+25$ to $+29$ (200 mg. in 10 ml. chloroform)	$(\alpha)\frac{20}{D}+28 - +30$ (alcohol)	$(\alpha)\frac{20}{D}+172 - +182$ (200 mg. in 10 ml. dioxane)
Absorption max.	241 mμ − 565 ± 15		240 mμ

DIURETIC, ANTIDIURETIC AND LOCAL ANESTHETIC AGENTS

Names & synonyms:	ALDOSTERONE; electrocortin; 18-oxocorticosterone; 18-formyl-11β,21-dihydroxy-4-pregnene-3,20-dione.	HYDROXYDIONE SODIUM; 21-hydroxypregnane-3,20-dione-21-sodium hemisuccinate.	SPIRONOLACTONE; 3-(3-oxo-7α-acetylthio-17β-hydroxy-4-androsten-17α-yl)-propionic acid γ lactone.
Formulae:	$C_{21}H_{28}O_5$	$C_{25}H_{35}O_6Na$	$C_{24}H_{32}O_4S$
Molecular weight	360.4	454.5	416.5
Melting point (°C)	108–112 (hydrate); 164 (anhydrous)	193–203 (dec.)	135; 202 (dec.)
Specific rotation	$(\alpha)\frac{25}{D}+161$ (10 mg. in 10 ml. chloroform)	$(\alpha)\frac{25}{D}+95$ (chloroform) for free acid.	$(\alpha)\frac{25}{D}-34$ (chloroform)
Absorption max.	240 mμ (log ε = 4.20 monohydr.; ε mol. 15,000 anhydr.)	280 mμ (ε = 93.2)	ε^{238} = 20,200

PROPERTIES OF THE AMINO ACIDS
IONIZATION CONSTANTS AND pH VALUES AT THE ISOELECTRIC POINTS OF THE AMINO ACIDS IN WATER AT 25°C

The majority of the recorded values are true thermodynamic constants calculated from electrometric force measurements of cells without liquid junctions. The values for the constants given in the table were derived from the classical, the zwitterionic (Bjerrum), and the acidic (Bronsted) formulations of ionization and the corresponding mass law expressions. pH values at the isoelectric points were calculated from the expression, $pI = \frac{1}{2}(pk_{ai} + pk_w - pk_{bi})$. The error is approximately 0.5% when this expression is used to calculate pI values for cystine, tyrosine, and diiodotyrosine.

Amino acid	Classical				Zwitterionic				Acidic					
	pk_{a1}	pk_{a2}	pk_{b1}	pk_{b2}	pK_{A1}	pK_{A2}	pK_{B1}	pP_{B2}	pK_1	pK_2	pK_3	pK_4	pI	Ref.
DL-Alanine	9.866	—	11.649	—	2.348	—	4.131	—	2.348	9.866	—	—	6.107	1
L-Arginine	12.48	—	4.96	11.99	2.01	—	1.52	4.96	2.01	9.04	12.48	—	10.76	2
L-Aspartic acid	3.86	9.82	11.93	—	2.10	3.86	4.18	—	2.10	3.86	9.82	—	2.98	3
L-Cystine	8.00	10.25	11.95	12.96	1.04	2.05	3.75	6.00	1.04	2.05	8.00	10.25	5.02	4
Diiodo-L-tyrosine	6.48	7.82	11.88	—	2.12	6.48	6.18	—	2.12	6.48	7.82	—	4.29	5,6
L-Glutamic acid	4.07	9.47	11.90	—	2.10	4.07	4.53	—	2.10	4.07	9.47	—	3.08	7
Glycine	9.778	—	11.647	—	2.350	—	4.219	—	2.350	9.778	—	—	6.064	8
L-Histidine	9.18	—	7.90	12.23	1.77	—	4.82	7.90	1.77	6.10	9.18	—	7.64	3
Hydroxy-L-proline	9.73	—	12.08	—	1.92	—	4.27	—	1.92	9.73	—	—	5.82	9
DL-Isoleucine	9.758	—	11.679	—	2.318	—	4.239	—	2.318	9.758	—	—	6.038	1
DL-Leucine	9.744	—	11.669	—	2.328	—	4.253	—	2.328	9.744	—	—	6.036	1
L-Lysine	10.53	—	5.05	11.82	2.18	—	3.47	5.05	2.18	8.95	10.53	—	9.47	2
DL-Methionine	9.21	—	11.72	—	2.28	—	4.79	—	2.28	9.21	—	—	5.74	10
DL-Phenylalanine	9.24	—	11.42	—	2.58	—	4.76	—	2.58	9.24	—	—	5.91	11
L-Proline	10.60	—	12.0	—	2.00	—	3.40	—	2.00	10.60	—	—	6.3	12
DL-Serine	9.15	—	11.79	—	2.21	—	4.85	—	2.21	9.15	—	—	5.68	9
L-Tryptophan	9.39	—	11.62	—	2.38	—	4.61	—	2.38	9.39	—	—	5.88	13
L-Tyrosine	9.11	10.07	11.80	—	2.20	9.11	3.93	—	2.20	9.11	10.07	—	5.63	6
DL-Valine	9.719	—	11.711	—	2.286	—	4.278	—	2.286	9.719	—	—	6.002	1

REFERENCES

1. Smith, P. K., Taylor, A. C., and Smith, E. R. B., *J. Biol. Chem.*, 122, 109, 1937—1938.
2. Schmidt, C. L. A., Kirk, P. L., and Appleman, W. K., *J. Biol. Chem.*, 88, 285, 1930.
3. Greenstein, J. P., *J. Biol. Chem.*, 93, 479, 1931.
4. Borsook, H., Ellis, E. L., and Huffman, H. M., *J. Biol. Chem.*, 117, 281, 1937.
5. Dalton, J. B., Kirk, P. L., and Schmidt, C. L. A., *J. Biol. Chem.*, 88, 589, 1930.
6. Winnek, P. S. and Schmidt, C. L. A., *J. Gen. Physiol.*, 18, 889, 1935.
7. Simms, H. S., *J. Gen. Physiol.*, 11, 629, 1928; 12, 231, 1928.
8. Owen, B. B., *J. Am. Chem. Soc.*, 56, 24, 1934.
9. Kirk, P. L. and Schmidt, C. L. A., *J. Biol. Chem.*, 81, 237, 1929.
10. Emerson, O. H., Kirk, P. L., and Schmidt, C. L. A., *J. Biol. Chem.*, 92, 449, 1931.
11. Miyamoto, S. and Schmidt, C. L. A., *J. Biol. Chem.*, 90, 165, 1931.
12. McCay, C. M. and Schmidt, C. L. A., *J. Gen. Physiol.*, 9, 333, 1926.
13. Schmidt, C. L. A., Appleman, W. K., and Kirk, P. L., *J. Biol. Chem.*, 85, 137, 1929—1930.

IONIZATION CONSTANTS OF THE AMINO ACIDS IN AQUEOUS ETHANOL SOLUTIONS

Amino acid	pK_1	pK_2	pK_3	Volume % ethanol	Temperature (°C)	Ref.	Amino acid	pK_1	pK_2	pK_3	Volume % ethanol	Temperature (°C)	Ref.
Alanine	3.55	10.02	—	72	25	1		3.79	9.99	—	90	19.5	2
Arginine	3.34	9.40	14.1	72	25	1	Histidine	3.00	5.85	9.45	72	25	1
Aspartic acid	2.85	5.20	10.51	72	25	1	Isoleucine	3.69	9.81	—	72	25	1
Glutamic acid	3.16	5.63	10.75	72	25	2	Lysine	2.75	8.95	10.53	48	25	1
Glycine	2.66	9.82	—	10	19.5	2		3.56	8.95	10.49	84	25	1
	2.96	9.76	—	40	19.5	2	Proline	3.04	10.55	—	72	25	1
	3.46	9.82	—	72	25	1	Valine	3.60	9.73	—	72	25	1

REFERENCES

1. Jukes, T. H. and Schmidt, C. L. A., *J. Biol. Chem.*, 105, 359, 1934.
2. Michaelis, L. and Mizutani, M., *Z. Physik. Chem.*, 116, 135, 1925.

IONIZATION CONSTANTS OF THE AMINO ACIDS IN AQUEOUS FORMALDEHYDE SOLUTION[a]

Amino acid	Mole % formaldehyde					Amino acid	Mole % formaldehyde				
	0.99	3.95	5.60	10.0	17.9		0.99	3.95	5.60	10.0	17.9
DL-Alanine	8.36	7.42	6.96[b]	6.56	6.10	DL-Leucine	8.44	7.48	—	6.60	6.20
L-Arginine	—	3.45[c]	3.40[d]	—	—	L-Lysine	—	7.35[c]	7.15[d]	—	—
L-Aspartic acid	—	—	7.21[d]	≤3.8[e]	—	L-Phenylalanine	—	—	6.62[d]	5.9[e]	—
				6.85[f]		DL-Phenylalanine	8.09	7.16	6.80[b]	6.35	6.13
L-Glutamic acid	—	—	6.91[d]	≤4.2[e]	—	L-Proline	—	—	7.78[d]	—	—
				6.8[f]		DL-Serine	6.66	5.74	5.63[b]	—	4.94
Glycine	7.16	6.08	5.92[b]	5.34	5.04	L-Tryptophan	—	—	6.88[d]	—	—
L-Histidine	—	7.90[c]	7.90[d]	—	—	L-Tyrosine	—	—	7.50[d]	6.2[e]	—
Hydroxy-L-proline	—	—	7.19[d]	—	—					>9[f]	
L-Leucine	8.44	7.50	6.92[d]	6.62	6.20	DL-Valine	8.52	7.65	7.47[d]	—	6.52

[a] pK$_2$ at 22°.[1]
[b] pK$_2$ at 22°.[2]
[c] pK$_2$ at 30° for arginine and pK$_3$ at 30° for histidine and lysine.[3]
[d] pK$_2$ at 30°, pK$_3$ at 30° for histidine and lysine.[4]
[e] pK$_2$ at 25° for aspartic acid, glutamic acid, phenylalanine, and tyrosine.[5]
[f] pK$_3$ at 30° for aspartic acid, glutamic acid, and tyrosine.[5]

REFERENCES

1. Dunn, M. S. and Weiner, J. G., *J. Biol. Chem.*, 117, 381, 1937.
2. Dunn, M. S. and Loshakoff, A., *J. Biol. Chem.*, 113, 691, 1936.
3. Levy, M., *J. Biol. Chem.*, 109, 365, 1935.
4. Levy, M. and Silberman, D. E., *J. Biol. Chem.*, 118, 723, 1937.
5. Harris, L. J., *Proc. R. Soc. London, Series B*, 95, 440, 1923—1924.

SPECIFIC ROTATIONS OF THE AMINO ACIDS USING SODIUM LIGHT (5893 Å)

Abbreviations

c —grams of solute per 100 ml of solution.
d —density of the solution.
p —grams of solute per 100 g of solution.
l —length of the tube in decimeters.
α —observed rotation in angular degrees.
[α] —specific rotation in angular degrees calculated from
[α]$_t$ = α × 100/ c × l = α × 100/p × d × l where t is temperature in °C and λ is wave length of the incident light in Ångstroms.

A —prepared from a protein or other naturally occurring material.
B—prepared by resolution of the inactive synthetic form.
C—prepared by resolution of the inactive racemized form.
D—prepared from the inactive synthetic form by a biological method.
E—prepared from the inactive racemized form by a biological method.
?—source not given.

Source	c	Solvent	d	p	Moles acid or base per mole amino acid	l	Temp. (°C)	α	[α]	Ref.
L-Alanine										
A	5.790	0.97 NHCl	1.033	5.605	1.5	2	15	+1.70	+14.7	1
A	10.3	Water	1.03	1.00	0	2	22	+0.55	+2.7	2
A	1.781	3 NNaOH	—	—	15	2	20	—	+3.0	3
D-Alanine										
B	71.344	6 NHCl	—	—	39.4	2	30.4	−0.392	−14.6	4
L-Arginine										
A	1.653	6.0 NHCl	—	—	63	4.001	23.4	+1.777	+26.9	5
A	3.48	Water	—	—	0	2	20	—	+12.5	6
A	0.87	0.50 NNaOH	—	—	10	2	20	—	+11.8	6
L-Aspartic acid										
A	2.002	6.0 NHCl	—	—	39	4.001	24.0	+1.972	+24.6	7
A	1.3300	Water	—	—	0	3	18	—	4.7	3
A	1.3300	3 NNaOH	—	—	30	3	18	—	−1.7	3
D-Aspartic acid										
C	4.289	0.97 NHCl	1.032	4.156	3	1	20	−1.09	−25.5	8
L-Cystine										
A	0.9974	1.02 NHCl	1.0181	0.9797	24.6	2	24.35	−4.277	−214.40	9
A	0.400	0.20 NNaOH	—	—	12	2	18.5	—	−70.0	3
D-Cystine										
C	—	1 NHCl	—	1	24	—	20	—	+223	10
Diiodo-D-tyrosine										
A	5.08	1.1 NHCl	1.05	4.84	9.4	1	20	+0.15	+2.89	11
A	4.41	13.4 NNH$_4$OH	0.9779	4.51	132	1	20	+0.10	+2.27	11
L-Glutamic acid										
A	1.002	6.0 NHCl	—	—	87	4.001	22.4	+1.25	+31.2	12
A	1.471	Water	—	—	0	2	18	—	+11.5	3
A	1.471	1 NNaOH	—	—	10	2	18	—	+10.96	3
D-Glutamic acid										
C	5.425	0.37 NHCl	1.0233	5.3011	1	1	20	−1.63	−30.05	8
L-Histidine										
A	1.480	6.0 NHCl	—	—	63	4.001	22.7	+0.766	+13.0	7
A	1.128	Water	1.0012	1.127	0	4	25.00	−1.714	−39.01	13
A	0.775	0.50 NNaOH	—	—	10	2	20	—	−10.9	6
D-Histidine										
?	4.000	1.0 NHCl	—	—	4	1	20	−0.407	−10.2	14
B	2.66	Water	—	—	0	2	23	+2.11	+39.8	14
Hydroxy-L-proline										
A	1.31	1.0 NHCl	—	—	10	2	20	—	−47.3	6
A	1.001	Water	—	—	0	4.001	22.5	−3.009	−75.2	7
A	0.655	0.50 NNaOH	—	—	10	2	20	—	−70.6	6
Hydroxy-D-proline										
B	4.48	Water	1.03	4.35	0	1	21	+3.37	+75.2	16

Source	c	Solvent	d	p	Moles acid or base per mole amino acid	l	Temp. (°C)	α	[α]	Ref.
Allo-Hydroxy-L-proline										
B	2.617	Water	1.014	2.581	0	1	18	−1.52	−58.1	16
Allo-Hydroxy-D-proline										
B	2.530	Water	1.013	2.998	0	1	17	+1.48	+58.5	16
L-Isoleucine										
B	5.09	6.1 NHCl	1.098	4.64	15	1	20	+2.07	+40.61	17
B	3.10	Water	1.008	3.08	0	2	20	+0.70	+11.29	17
A	3.34	0.33 NNaOH	1.017	3.28	1.3	2	20	+0.74	+11.09	18
D-Isoleucine										
B	4.53	6.1 NHCl	1.083	4.18	17	1	20	−1.85	−40.86	17
B	3.12	Water	1.006	3.10	0	2	20	−0.66	−10.55	17
D-allo-Isoleucine										
D	5.14	6.0 NHCl	1.094	4.70	15.0	2	20	−3.80	−36.95	19
B	2.00	Water	—		0	2	20	−0.285	−14.2	20
L-allo-Isoleucine										
B	3.97	6.0 NHCl	—	—	20	1	20	+1.50	+38.1	20
B	2.00	Water	—	—	0	1	20	+0.28	+14.0	20
L-Leucine										
A	1.999	6.0 NHCl	—	—	38	4.001	25.9	+1.212	+15.1	5
A	2.001	Water	—	—	8	4.001	24.1	0.803	+10.8	3
A	1.31	3.00 NNaOH	—	—	30	2	20		+7.6	3
D-Leucine										
?	4.0	6.0 NHCl	1.1	3.664	19	2	20	+1.26	−15.6	21
?	—	Water	—	2.08	0	2	20	+0.43	+10.34	38
L-Lysine										
A	2.00	6.0 NHCl	—	—	43	4	22.9	+1.652	+25.9	5
A	6.496	Water	—	—	0	2	20	+1.90	+14.6	22
D-Lysine										
B	2.00	0.27 NHCl	—	—	2	2	20	−0.939	−23.48	23
L-Methionine										
B	0.80	Water	—	—	0	2	25	−0.13	−8.11	24
D-Methionine										
B	0.80	0.2001 NHCl	—	—	4	2	25	−0.34	−21.18	24
B	0.80	Water	—	—	0	2	25	+0.13	+8.12	24
B	0.80	0.6 NNaHCO₃	—	—	11	2	25	−0.12	−7.47	24
L-Phenylalanine										
B	1.936	Water	1.0040	1.928	0	2	20	−1.36	−35.14	27
D-Phenylalanine										
B	3.814	5.4 NHCl	1.0895	3.501	23	2	20	+0.54	+7.07	28
B	2.043	Water	1.0045	2.034	0	2	20	+1.43	+35.0	27
L-Proline										
A	0.575	0.50 NHCl	—	—	10	2	20	—	−52.6	6
A	1.001	Water	—	—	0	4.001	23.4	−3.402	−85.0	7
B	2.42	0.6 NKOH	1.031	2.35	3	1	20	−2.25	−93.0	29
D-Proline										
B	3.90	Water	1.01	3.865	0	1	20	+3.18	+81.5	29
L-Serine										
B	9.344	1 NHCl	1.0465	8.929	1	1	25	+1.35	+14.45	30
B	10.414	Water	1.0414	9.997	0	2	20	−1.42	−6.83	30
D-Serine										
B	9.359	1 NHCl	1.0465	8.943	1	1	25	−1.34	−14.32	30
B	10.412	Water	1.0414	9.998	0	2	20	+1.43	+6.87	30
D-Threonine										
B	—	Water	—	1.092	0	2	26	−0.625	−28.3	31
L-Threonine										
B	—	Water	—	1.331	0	2	26	+0.780	+28.4	31
D-allo-Threonine*										
B	—	Water	—	1.634	0	2	26	−0.302	−9.1	31
L-allo-Threonine										
B	—	Water	—	1.643	0	2	26	+0.388	+3.0	31
L-Thyroxine										
A		0.12 NNaOH in 70% EtOH by weight	—	3	1	1	—	−0.141	−4.4	39
L-Tryptophan										
A	1.02	0.50 NHCl	—	—	10	2	20	—	+2.4	6
A	1.004	Water	—	—	0	4.001	22.7	−1.266	−31.5	7
A	2.426	0.5 NNaOH	1.0243	2.368	4.2	1	20	+0.15	+6.17	33
D-Tryptophan										
C	0.5024	Water	—	—	0	2	25	+0.326	+32.45	34
L-Tyrosine										
B	4.40	6.3 NHCl	1.116	3.94	28	2	20	−0.76	−8.64	35
A	0.906	3.0 NNaOH	—	—	60	3	18	—	−13.2	3
D-Tyrosine										
B	5.1484	6.3 NHCl	1.1175	4.6071	24	2	20	+0.89	+8.64	35
L-Valine										
B	3.4	6.0 NHCl	1.1	3.05	20	2	20	+1.93	+28.8	36
B	3.58	Water	1.007	3.56	0	2	20	+0.46	+6.42	36
D-Valine										
B	3.2	6.0 NHCl	1.1	2.91	21	2	20	−1.86	−29.04	36
E	6.24	Water	1.00	6.24	0	1	20	−0.37	−6.06	37

* The levorotatory allothreonine probably belongs to the D family and its enantiomorph to the L family.

REFERENCES

1. Clough, G. W., *J. Chem. Soc.*, 113, 526, 1918.
2. Fischer, E. and Raske, K., *Ber*, 40, 3717, 1907.
3. Lutz, O. and Jirgensons, B., *Ber.*, 63, 448, 1930.
4. Dunn, M. S., Butler, A. W., and Naiditch, M. J., unpublished data.
5. Dunn, M. S. and Courtney, G., unpublished data.
6. Lutz, O. and Jirgensons, B., *Ber.*, 64, 1221, 1931.
7. Dunn, M. S. and Stoddard, M. P., unpublished data.
8. Fischer, E., *Ber.*, 32, 2451, 1899.
9. Toennies, G. and Lavine, T. F., *J. Biol. Chem.*, 89, 153, 1930.
10. Loring, H. S. and du Vigneaud, V., *J. Biol. Chem.*, 107, 267, 1934.
11. Abderhalden, E. and Guggenheim, M., *Ber.*, 41, 1237, 1908.
12. Dunn, M. S. and Sexton, E. L., unpublished data.
13. Dunn, M. S. and Frieden, E. H., unpublished data.
14. Cox, G. J. and Berg, C. P., *J. Biol. Chem.*, 107, 497, 1934.
15. Dakin, H. D., *Biochem. J.*, 13, 398, 1919.
16. Leuchs, H. and Bormann, K., *Ber.*, 52, 2086, 1919.
17. Locquin, R., *Bull. Soc. Chim.*, (4)1, 601, 1907.
18. Ehrlich, F., *Ber.*, 37, 1809, 1904.
19. Ehrlich, F., *Ber.*, 40, 2538, 1907.
20. Abderhalden, E. and Zeisset, W., *Z. Physiol. Chem.*, 196, 121, 1931.
21. Fischer, E., and Warburg, O., *Ber.*, 38, 3997, 1905.
22. Vickery, H. B., private communication, April, 1940.
23. Berg, C. P., *J. Biol. Chem.*, 115, 9, 1936; private communication, June, 1940.
24. Windus, W. and Marvel, C. S., *J. Am. Chem. Soc.*, 53, 3490, 1931.
27. Fischer, E. and Schoeller, W., *Ann.*, 357, 1, 1907.
28. Fischer, E. and Mouneyrat, A., *Ber.*, 33, 2383, 1900.
29. Fischer, E. and Zemplén, G., *Ber.*, 42, 2989, 1909.
30. Fischer, E. and Jacobs, W. A., *Ber.*, 39, 2942, 1906.
31. West, H. D. and Carter, H. E., *J. Biol. Chem.*, 119, 109, 1937; private communication from H. E. Carter, July, 1940.
32. Foster, G. L., Palmer, W. W., and Leland, J. P., *J. Biol. Chem.*, 115, 467, 1936.
33. Abderhalden, E. and Baumann, L., *Z. Physiol. Chem.*, 55, 412, 1908.
34. Berg, C. P., J. Biol. Chem., 100, 79 1933; private communication, July, 1940.
35. Fischer, E., *Ber.*, 32, 3638, 1899.
36. Fischer, E., *Ber.*, 39, 2320, 1906.
37. Ehrlich, F. and Wendel, A., *Biochem. Z.*, 8, 399, 1908.
38. Ehrlich, F., *Biochem. Z.*, 1, 8, 1906.

SOLUBILITIES OF THE AMINO ACIDS IN GRAMS PER 100 GRAMS OF WATER

Amino acid	Temperature (°C)					Ref.	Amino acid	Temperature (°C)					Ref.
	0°	25°	50°	75°	100°			0°	25°	50°	75°	100°	
DL-Alanine	12.11	16.72	23.09	31.89	44.04	1	DL-Leucine	0.797	0.991	1.406	2.276	4.206	1
L-Alanine	12.73	16.65	21.79	28.51	37.30	1	L-Leucine	2.270	2.426^c	2.887^b	3.823	5.638	1
DL-Aspartic acid	0.262	0.778	2.000	4.456	8.594	1	DL-Methionine	1.818	3.381	6.070	10.52	17.60	2
L-Aspartic acid	0.209	0.500	1.199	2.875	6.893	1	DL-Phenylalanine	0.997	1.411	2.187	3.708	6.886	1
L-Cystine^c × 10^2	0.502	1.096	2.394	5.229	11.42	2	L-Phenylalanine	1.983	2.965	4.431	6.624	9.900	2
Diiodo-DL-tyrosine × 10	0.149	0.340	0.773	—	—	3	L-Proline × 10^{-1}	12.74	16.23	20.67	23.90^a	—	3
Diiodo-L-tyrosine × 10	0.204	0.617	1.862	5.62	17.00	1	DL-Serine	2.204	5.023	10.34	19.21	32.24	2
DL-Glutamic acid	0.855	2.054	4.934	11.86	28.49	1	L-Tryptophan	0.823	1.136	1.706	2.795	4.987	2
L-Glutamic acid	0.341	0.864	2.186	5.532	14.00	1	DL-Tyrosine × 10	0.147	0.351	0.836	—	—	3
Glycine	14.18	24.99	39.10	54.39	67.17	1	L-Tyrosine × 10	0.196	0.453	1.052	2.438	5.650	1
L-Histidine	—	4.19	—	—	—	4	D-Tyrosine × 10	0.196	0.453	1.052	—	—	3
Hydroxy-L-proline	28.86	36.11	45.18	51.67^a	—	5	DL-Valine	5.98	7.09	9.11	12.61	18.81	1
DL-Isoleucine	1.826	2.229	3.034	4.607	7.802	1	L-Valine	8.34	8.85	9.62	10.24^a	—	6
L-Isoleucine	3.791	4.117	4.818	6.076	8.255	2							

^{a} Value at 65°.

^{b} Dunn and Stoddard[7] report 2.19 g at 25° for L-leucine rendered methionine-free by repeated recrystallization from 6 *N* HCl. Hlynka[8] found 2.20 g at 25° and 2.66 g at 50° for L-leucine rendered methionine-free (by S. W. Fox[9]) by fractional crystallization of the formyl derivative and identical values for D-leucine obtained by resolution of the DL form.

^{c} The following values were found by Loring and du Vigneaud[10]: DL-cystine (0.0049 g), D-cystine (0.0108 g), and *meso*-cystine (0.0056 g) at 25°.

REFERENCES

1. Dalton, J. B. and Schmidt, C. L. A., *J. Biol. Chem.*, 103, 549, 1933.
2. Dalton, J. B. and Schmidt, C. L. A., *J. Biol. Chem.*, 109, 241, 1935.
3. Winnek, P. S. and Schmidt, C. L. A., *J. Gen. Physiol.*, 18, 889, 1934—1935.
4. Dunn, M. S., Frieden, E. H., and Brown, H. V., unpublished data.
5. Tomiyama, T. and Schmidt, C. L. A., *J. Gen. Physiol.*, 19, 379, 1935—1936.
6. Dalton, J. B. and Schmidt, C. L. A., *J. Gen. Physiol.*, 19, 767, 1935—1936.
7. Dunn, M. S. and Stoddard, M. P., unpublished data.
8. Hlynka, I., Thesis (1939), California Institute of Technology, Pasadena, California.
9. Fox, S. W., *Science*, 84, 163, 1936.
10. Loring, H. S. and du Vigneaud, V., *J. Biol. Chem.*, 107, 270, 1934.

SOLUBILITIES OF THE AMINO ACIDS IN GRAMS PER 100 GRAMS OF WATER-ETHANOL MIXTURES

% ethanol by volume	Temp. °C	Grams amino acid per 100 grams solvent	Ref.	% ethanol by volume	Temp. °C	Grams amino acid per 100 grams solvent	Ref.	% ethanol by volume	Temp. °C	Grams amino acid per 100 grams solvent	Ref.	% ethanol by volume	Temp. °C	Grams amino acid per 100 grams solvent	Ref.
DL-Alanine				74.28	0.03	0.0163	1	50.10	24.97	0.318	1	**L-Tyrosine**			
24.93	0.00	3.84	1	24.56	25.05	0.292	1	74.20	24.97	0.175	1	95	17	0.10	7
50.10	0.00	1.16	1	50.25	25.08	0.131	1	95.14	25.09	0.0258	1	**DL-Tyrosine**			
74.50	0.00	0.305	1	74.35	25.07	0.0370	1	24.93	45.24	0.853	1				
95.14	0.00	0.0167	1	95.14	25.04	0.0044	1	50.10	45.24	0.633	1	95.09	0.00	0.0031	8
10	25	12.25	2	24.55	45.01	0.811	1	74.50	45.18	0.323	1	25.28	24.85	0.0285	8
24.93	24.97	7.09	1	50.18	45.27	0.378	1	95.14	45.18	0.0471	1	50.99	24.75	0.0226	8
50.10	24.97	2.52	1	74.35	44.93	0.0885	1	24.93	65.16	1.45	1	74.63	24.75	0.0117	8
74.20	24.97	0.573	1	95.14	45.20	0.0127	1	50.10	65.20	1.16	1	95.09	25.24	0.0032	8
95.14	25.09	0.0329	1	**Glycine**				74.20	65.15	0.584	1	25.28	45.15	0.0630	8
25.28	45.16	10.6	1					95.09	65.07	0.0844	1	50.99	45.16	0.0513	8
50.10	44.96	4.25	1	24.93	0.02	3.95	1					74.63	44.93	0.0230	8
74.20	44.98	0.949	1	50.10	0.02	1.03	1	**L-Leucine**				95.09	44.98	0.0035	8
95.14	45.19	0.0545	1	74.50	0.02	0.200	1	20	25	1.33	2	95.09	65.06	0.0067	8
24.93	64.96	15.9	1	95.09	0.01	0.0080	1	60	25	0.641	2				
50.10	64.94	6.68	1	10	25	17.13	2	90	25	0.123	2	**DL-Valine**			
74.20	64.94	1.48	1	24.93	24.97	8.72	1	**L-Proline**				24.93	0.02	2.10	1
95.09	65.15	0.0851	1	50.10	24.97	2.47	1					50.10	0.02	0.769	1
DL-Aspartic acid				74.20	24.97	0.448	1	100	19	1.5	5	74.20	0.02	0.269	1
				95.14	25.09	0.0172	1	**DL-Serine**				95.14	0.01	0.0277	1
24.93	0.03	0.0703	1	24.93	44.98	15.0	1					10	25	5.50	2
50.10	0.03	0.0267	1	50.10	44.98	4.62	1	24.93	0.00	0.1530	1	25.28	24.85	3.30	1
74.20	0.02	0.0111	1	74.20	44.97	0.756	1	50.10	0.00	0.146	1	50.99	24.85	1.53	1
24.55	25.06	0.266	1	95.14	45.19	0.0294	1	74.50	0.00	0.0304	1	74.35	24.93	0.570	1
50.25	25.06	0.0992	1	24.93	65.11	24.5	1	95.14	0.00	0.0008	1	95.14	25.04	0.0369	1
74.28	25.14	0.0317	1	50.10	65.10	8.03	1	24.93	25.14	1.54	1	24.55	44.91	5.10	1
95.14	25.07	0.0020	1	74.20	65.07	1.23	1	50.10	25.14	0.461	1	50.25	44.92	2.74	1
24.74	45.25	0.680	1	95.14	65.00	0.0488	1	74.50	25.10	0.0840	1	74.35	44.92	0.999	1
50.18	45.25	0.255	1	**L-Isoleucine**				95.14	25.09	0.0028	1	95.14	45.21	0.0979	1
74.28	45.27	0.0608	1	00	20	0.46	4	24.93	45.15	2.14	1	24.55	65.07	7.44	1
95.14	45.21	0.0042	1	80	78—80	1.16	4	50.10	45.04	0.985	1	50.10	64.94	4.49	1
24.93	64.91	1.53	1	**L-allo-Isoleucine**				74.20	45.04	0.185	1	74.20	64.34	1.62	1
50.10	64.91	0.588	1	80	20	0.81	4	95.14	45.18	0.0058	1	95.09	65.15	0.167	1
74.20	65.07	0.132	1	80	78—80	1.97	4	24.93	65.26	5.99	1	**L-Valine**			
95.14	65.00	0.0129	1	**DL-Leucine**				50.10	65.25	1.88	1				
L-Aspartic acid				24.93	0.00	0.251	1	74.50	65.24	0.318	1	20	25	5.11	2
20	25	0.204	3	50.10	0.00	0.118	1	95.14	65.01	0.0152	1	40	25	2.93	2
50	25	0.0633	3	74.50	0.00	0.0693	1	**DL-Threonine**				60	25	1.61	2
70	25	0.0224	3	95.14	0.00	0.0116	1	95	25	0.07a	6	80	25	0.52	2
90	25	0.0034	3	10	25	0.771	2	**DL-allo-Threonine**							
L-Glutamic acid				24.93	24.97	0.493	1	95	25	0.03a	6				
24.74	0.01	0.0855	1												
50.18	0.01	0.0371	1												

a Grams per 100 ml of solution.

REFERENCES

1. Dunn, M. S. and Ross, F. J., *J. Biol. Chem.*, 125, 309, 1938.
2. Cohn, E. J., McMeekin, T. L., Edsall, J. T., and Weare, J. H., *J. Am. Chem. Soc.*, 56, 2270, 1934.
3. McMeekin, T. L., Cohn, E. J., and Weare, J. H., *J. Am. Chem. Soc.*, 57, 626, 1935.
4. Abderhalden, E. and Zeisset, W., *Z. Physiol. Chem.*, 196, 121, 1931.
5. Kapfhammer, J. and Eck, R., *Z. Physiol. Chem.*, 170, 294, 1927.
6. West, H. D. and Carter, H. E., *J. Biol. Chem.*, 119, 109, 1937.
7. Stutzer, A., *Z. Anal. Chem.*, 31, 501, 1892.
8. Dunn, M. S. and Ross, F. J., unpublished data.

SOLUBILITIES OF THE AMINO ACIDS IN GRAMS PER 100 GRAMS OF ORGANIC SOLVENT

Solvent	Grams amino acid per 100 grams solvent	Temp. (°C)	Ref.	Solvent	Grams amino acid per 100 grams solvent	Temp. (°C)	Ref.
DL-Alanine				Ethanol	0.13	78—80	5
Ethanol	0.0087	25	1	**L-allo-Isoleucine**			
L-Aspartic acid				Ethanol	0.13	20	5
Ethanol	0.000196	25	2	Ethanol	0.19	78—80	5
L-Glutamic acid				**L-Leucine**			
Ethanol	0.000347	25	2	Ethanol	0.0217	25	1
Ethanol	0.0056	44.93	3	**L-Proline**			
Glycine				Ethanol	1.5	19	6
Acetone	0.000291	25	4	**DL-Valine**			
Butanol	0.000892	25	4	Ethanol	0.0136	0.03	3
Ethanol	0.0037	25	1	Ethanol	0.019	25	1
Formamide	0.558	25	4				
Methanol	0.0407	25	4				
L-Isoleucine							
Ethanol	0.09	20	5				

REFERENCES

1. Cohn, E. J., McMeekin, T. L., Edsall, J. T., and Weare, J. H., *J. Am. Chem. Soc.,* 56, 2270, 1934.
2. McMeekin, T. L., Cohn, E. J., and Weare, J. H., *J. Am. Chem. Soc.,* 57, 626, 1935.
3. Dunn, M. S. and Ross, F. J., *J. Biol. Chem.,* 125, 309, 1938.
4. McMeekin, T. L., Cohn, E. J., and Weare, J. H., *J. Am. Chem. Soc.,* 58, 2173, 1936.
5. Abderhalden, E. and Zeisset, W., *Z. Physiol. Chem.,* 196, 121, 1931.
6. Kapfhammer, J. and Eck, R., *Z. Physiol. Chem.,* 170, 294, 1927.

DENSITIES OF CRYSTALLINE AMINO ACIDS

Amino acid	Density	Ref.
DL-Alanine	1.424	1
L-Alanine	1.401	2
β-Alanine	1.404	1
DL-α-Amino-n-butyric acid	1.231	1
α-Aminoisobutyric acid	1.278	1
L-Arginine	1.1	3
L-Aspartic acid	1.66	3
DL-Glutamic acid	1.460	4
L-Glutamic acid	1.538	4
Glycine[a]	1.601	3
	1.607	1
DL-Leucine	1.191	1
L-Leucine	1.165	1
DL-Methionine	1.340	5
DL-Serine	1.537	5
L-Tyrosine	1.456	1
DL-Valine	1.316	1
L-Valine	1.230	1

[a] The density of glycine at 50° is 1.5753 according to Houck[6] who concluded that the figure 1.1607, reported by Curtius[7] and reproduced in chemical handbooks, is a typographical error.

REFERENCES

1. Cohn, E. J., McMeekin, T. L., Edsall, J. T., and Weare, J. H., *J. Am. Chem. Soc.,* 56, 2270, 1934.
2. Dalton, J. B. and Schmidt, C. L. A., *J. Biol. Chem.,* 103, 549, 1933.
3. Huffman, H. M., Ellis, E. L., and Fox, S. W., *J. Am. Chem. Soc.,* 58, 1728, 1936; Huffman, H. M., Fox, S. W., and Ellis, E. L., *J. Am. Chem. Soc.,* 59, 2144, 1937.
4. Schmidt, C. L. A., *Chemistry of the Amino Acids and Proteins,* Charles C Thomas, Springfield, 1938, p. 900.
5. Albrecht, G. and Dunn, M. S., unpublished data.
6. Houck, R. C., *J. Am. Chem. Soc.,* 52, 2420, 1930.
7. Curtius, T., *J. Prakt. Chem.,* 26, 145, 1882.

CARBOHYDRATES

These data for carbohydrates were compiled originally for the Biology Data Book by M. L. Wolfram, G. G. Maher and R. G. Pagnucco (1964). Data are reproduced here by permission of the copyright owners of the above publication, the Federation of American Societies for Experimental Biology, Washington, D.C. pp. 351–359.

All data are for crystalline substances, unless otherwise specified. Selection of substances was restricted to natural carbohydrates found free (or in chemical combination and released on hydrolysis) and to biological oxidation products of the natural carbohydrates. The nomenclature conforms with that of the British-American report as published in the *Journal of Organic Chemistry*, 28:281 (1963). Substances have been arranged alphabetically under the name of the parent sugar within groups formulated according to increasing carbon content (excluding carbon in substituents), with synonymous common names in parentheses. **Melting Point:** b.p. = boiling point; d. = decomposes; s. = sinters. **Specific Rotation** was determined in water at concentrations of 1–5 g per 100 ml. of solution and at 20°–25°C, unless otherwise specified; other temperatures or wavelengths are shown in brackets; c = grams solute per 100 ml of solution.

Part I. NATURAL MONOSACCHARIDES: ALDOSES AND KETOSES

	Substance (Synonym)	Chemical Formula	Melting Point °C	Specific Rotation $[\alpha]_D$
	(A)	(B)	(C)	(D)
	Aldoses			
1	D-Glyceraldehyde	$C_3H_6O_3$	....	+13.5 ± 0.5 (syrup)
2	D-Glyceraldehyde, 3-deoxy-3,3-C-bis-(hydroxymethyl)- (Cordycepose)	$C_5H_{10}O_4$	....	−26 (c 0.6, C_2H_5OH
3	D-Glyceraldehyde, 3,3-bis-(C-hydroxymethyl)- (Apiose)	$C_5H_{10}O_5$	....	+5.6 (c 10) [15°] syrup
4	β-D-Arabinose	$C_5H_{10}O_5$	155	−175 → −103
5	D-Arabinose, 2-O-methyl	$C_6H_{12}O_5$	Syrup	−102
6	α-L-Arabinose	$C_5H_{10}O_5$	158 amorphous	+55.4 → +105
7	β-L-Arabinose	$C_5H_{10}O_5$	160	+190.6 → +104.5
8	DL-Arabinose	$C_5H_{10}O_5$	163.5–164.5	None
9	α-L-Lyxose	$C_5H_{10}O_5$	105	+5.8 → +13.5
10	L-Lyxose, 5-deoxy-3-C-formyl- (Streptose)	$C_6H_{10}O_5$	....	....
11	L-Lyxose, 3-C-formyl- (Hydroxy-streptose)	$C_6H_{10}O_6$	....	....
12	Pentose, 4,5-anhydro-5-deoxy-D-*erythro*-	$C_5H_8O_3$	....	....
13	Pentose, 2-deoxy-D-*erythro*-	$C_5H_{10}O_4$	96–98	−91 → −58
14	D-Ribose	$C_5H_{10}O_5$	87	−23.1 → −23.7
15	D-Ribose, 2-C-hydroxymethyl- (Hamamelose)	$C_6H_{12}O_6$	....	−7.1 [λ578]
16	α-D-Xylose	$C_5H_{10}O_5$	145	+93.6 → +18.8
17	D-Xylose, 5-deoxy-	$C_5H_{10}O_4$	....	+16
18	β-D-Xylose, 2-O-methyl-	$C_6H_{12}O_5$	137–138	−21 → +34
19	α-D-Xylose, 3-O-methyl-	$C_6H_{12}O_5$	95	+45 → +19
20	D-Allose, 6-deoxy-	$C_6H_{12}O_5$	140–143 146–148	+1.6 [18°] (c 0.6) −4.7 → 0
21	D-Allose, 6-deoxy-2,3-di-O-methyl- (Mycinose)	$C_8H_{16}O_5$	102–106	−46 → −29
22	Amicetose (a trideoxy hexose)	$C_6H_{12}O_3$	Oil, b.p. 65–70	+28.6 ($CHCl_3$)
23	Antiarose	$C_6H_{12}O_5$	....	Levo
24	α-D-Galactose	$C_6H_{12}O_6$	167	+150.7 → +80.2
25	β-D-Galactose	$C_6H_{12}O_6$	143–145	+52.8 → +80.2
26	D-Galactose, 3,6-anhydro-	$C_6H_{10}O_5$	....	+21.3 [10°]
27	α-D-Galactose, 6-deoxy- (D-Fucose; Rhodeose)	$C_6H_{12}O_5$	140–145	+127 → +76.3 (c 10)
28	D-Galactose, 6-deoxy-3-O-methyl- (Digitalose)	$C_7H_{14}O_5$	106[1], 119[2]	+106

Substance (Synonym)	Chemical Formula	Melting Point °C	Specific Rotation $[\alpha]_D$
(A)	(B)	(C)	(D)
Aldoses (Con't)			
29 D-Galactose, 6-deoxy-4-O-methyl-	$C_7H_{14}O_5$	131–132	+82
30 D-Galactose, 6-deoxy-2,3-di-O-methyl-	$C_8H_{16}O_5$	....	+73
31 α-D-Galactose, 3-O-methyl-	$C_7H_{14}O_6$	144–147	+150.6 → +108.6
32 α-D-Galactose, 6-O-methyl-	$C_7H_{14}O_6$	122–123	+117 → +77.3
33 L-Galactose	$C_6H_{12}O_6$		See D-Galactose
34 α-L-Galactose, 3,6-anhydro-	$C_6H_{10}O_5$	....	−39.4 → −25.2
35 α-L-Galactose, 6-deoxy- (L-Fucose)	$C_6H_{12}O_5$	145	−124.1 → −76.4
36 L-Galactose, 6-deoxy-2-O-methyl-	$C_7H_{14}O_5$	149–150	−75 ± 4 (c 0.5)
37 L-Galactose, 6-sulfate	$C_6H_{12}O_9S$	....	−47 (c 0.2) (Na salt)
38 DL-Galactose	$C_6H_{12}O_6$	143–144, 163	None (racemic)
39 α-D-Glucose	$C_6H_{12}O_6$	146, 83 (H_2O)	+112 → +52.7
40 β-D-Glucose	$C_6H_{12}O_6$	148–150	+18.7 → +52.7
41 D-Glucose, 6-acetate	$C_7H_{14}O_7$	135	+48
42 D-Glucose, 2,3-di-O-methyl-	$C_8H_{16}O_6$	85–86, 121	+50
43 D-Glucose, 6-O-benzoyl- (Vaccinin)	$C_{13}H_{16}O_7$	Amorphous	+48 (C_2H_5OH)
44 α-D-Glucose, 6-deoxy- (Chinovose; Epirhamnose; Glucomethylose; Isorhamnose; Isorhodeose; Quinovose)	$C_6H_{12}O_5$	139–140	+73.3 → +29.7 (c 8)
45 α-D-Glucose, 6-deoxy-3-O-methyl- (D-Thevetose)	$C_7H_{14}O_5$	116	+84 → +33
46 D-Glucose, 6-sulfonic acid, 6-deoxy- (6-Sulfoquinovose)	$C_6H_{12}O_8S$	173–174	+87[3]
47 D-Glucose, 3-O-methyl-	$C_7H_{14}O_6$	162–167	+98 → +59.5
48 α-L-Glucose	$C_6H_{12}O_6$	141–143	−95.5 → −51.4
49 L-Glucose, 6-deoxy-3-O-methyl- (L-Thevetose)	$C_7H_{14}O_5$	126–129	−36.9 ± 2
50 D-Gulose, 6-deoxy-	$C_6H_{12}O_5$	....	....
51 Hexose, 2-deoxy-D-arabino-[4]	$C_6H_{12}O_5$	148	+46.6 [18°]
52 Hexose, 2,6-dideoxy-3-O-methyl-D-arabino- (D-Oleandrose)	$C_7H_{14}O_4$	....	−11
53 Hexose, 3,6-dideoxy-D-arabino- (Tyvelose)	$C_6H_{12}O_4$	....	+24 ± 2
54 Hexose, 2,6-dideoxy-3-O-methyl-L-arabino- (L-Oleandrose)	$C_7H_{14}O_4$	62–63	+11.9 ± 2.5
55 Hexose, 3,6-dideoxy-L-arabino- (Ascarylose)	$C_6H_{12}O_4$	....	−24 ± 2
56 Hexose, 2,6-dideoxy-3-O-methyl-D-lyxo- (Diginose)	$C_7H_{14}O_4$	90–92	+56 ± 4
57 Hexose, 2,6-dideoxy-L-lyxo- (L-Fucose, 2-deoxy-)	$C_6H_{12}O_4$	103–106	−61.6
58 Hexose, 2,6-dideoxy-3-O-methyl-L-lyxo-	$C_7H_{14}O_4$	78–85	−65
59 Hexose, 2,6-dideoxy-D-ribo- (Digitoxose; D-Altrose, 2,6-dideoxy-)	$C_6H_{12}O_4$	110	+46.4
60 Hexose, 2,6-dideoxy-3-O-methyl-D-ribo- (Cymarose)	$C_7H_{14}O_4$	93	+52
61 Hexose, 3,6-dideoxy-D-ribo- (Paratose)	$C_6H_{12}O_4$	....	+10 ± 2 (c 0.9)
62 Hexose, 4,6-dideoxy-3-O-methyl-D-ribo- (D-Gulose, 4,6-dideoxy-3-O-methyl-; Chalcose)	$C_7H_{14}O_4$	96–99	+120 → +76
63 Hexose, 2,6-dideoxy-D-xylo- (Boivinose)	$C_6H_{12}O_4$	96–98	−3.9 → +3.9

	Substance (Synonym)	Chemical Formula	Melting Point °C	Specific Rotation $[\alpha]_D$
	(A)	(B)	(C)	(D)
	Aldoses (Con't)			
64	Hexose, 2,6-dideoxy-3-O-methyl-D-xylo- (Sarmentose)	$C_7H_{14}O_4$	78–79	$+12 \rightarrow +15.8$
65	Hexose, 3,6-dideoxy-D-xylo- (Abequose)	$C_6H_{12}O_4$	....	-3.2 ± 0.6
66	Hexose, 2,6-dideoxy-3-C-methyl-L-xylo- (Mycarose)	$C_7H_{14}O_4$	129–129	-31.1
67	Hexose, 2,6-dideoxy-3-C-methyl-3-O-methyl-L-xylo-(Cladinose)	$C_8H_{16}O_4$	oil, b.p. 120–132 (0.25 mm)	-23.1
68	Hexose, 3,6-dideoxy-L-xylo- (Colitose)	$C_6H_{12}O_4$	....	$+4$ (H_2O); -51 ± 2 (CH_3OH)
69	D-Idose[5]	$C_6H_{12}O_6$	....	....
70	L-Idose, 1,6-anhydro-	$C_6H_{10}O_5$	....	....
71	α-D-Mannose	$C_6H_{12}O_6$	133	$+29.3 \rightarrow +14.5$
72	β-D-Mannose	$C_6H_{12}O_6$	132	$-16.3 \rightarrow +14.5$
73	D-Mannose, 6-deoxy- (D-Rhamnose)	$C_6H_{12}O_5$	86–90	-7.0
74	α-L-Mannose, 6-deoxy-monohydrate (L-Rhamnose)	$C_6H_{14}O_6$	93–94	$-8.6 \rightarrow +8.2$
75	β-L-Mannose, 6-deoxy-	$C_6H_{12}O_5$	123–125	$+38.4 \rightarrow +8.0$
76	L-Mannose, 6-deoxy-2-O-methyl-	$C_7H_{14}O_5$	....	....
77	L-Mannose, 6-deoxy-3-O-methyl- (L-Acofriose)	$C_7H_{14}O_5$	114–115	$+30$ [18°]
78	L-Mannose, 6-deoxy-2,4-di-O-methyl-	$C_8H_{16}O_5$	82	-19 [16°]
79	L-Mannose, 6-deoxy-5-C-methyl-4-O-methyl-(Noviose)	$C_8H_{16}O_5$	128–130	$+19.9$ (50% C_2H_5OH)
80	Rhodinose (a 2,3,6-trideoxyhexose)	$C_6H_{12}O_3$	....	-11 ± 1.6
81	D-Talose	$C_6H_{12}O_6$	128–132	$+16.9$
82	D-Talose, 6-deoxy- (D-Talomethylose)	$C_6H_{12}O_5$	129–131	$+20.6$
83	L-Talose, 6-deoxy- (L-Talomethylose)	$C_6H_{12}O_5$	116–118	-19.5 ± 2 [18°]
84	L-Talose, 6-deoxy-2-O-methyl- (L-Acovenose)	$C_7H_{14}O_5$	....	-19.4
85	Heptose, D-glycero-D-galacto-	$C_7H_{14}O_7$	139–140	$+47 \rightarrow +64$ (c 0.5)
86	Heptose, D-glycero-D-manno-	$C_7H_{14}O_7$	....	....
87	Heptose, D-glycero-L-manno-	$C_7H_{14}O_7$	....	....
	Ketoses			
88	Dihydroxyacetone	$C_3H_6O_3$	80 (dimer)	None
89	Tetrulose, L-glycero-[8] (L-Erythrulose; Ketoerythritol; L-Threulose)	$C_4H_8O_4$	Syrup	$+12$
90	Pentulose, D-erythro- (Adonose; D-Ribulose)	$C_5H_{10}O_5$	Syrup	$+16.6$ [27°]
91	Pentulose, L-erythro- (L-Ribulose)	$C_5H_{10}O_5$	....	-16.6
92	Pentulose, D-threo- (D-Xylulose)	$C_5H_{10}O_5$	....	-33
93	Pentulose, 5-deoxy-D-threo-	$C_5H_{10}O_4$	....	-5 ± 1 (CH_3OH)
94	Pentulose, L-threo- (L-Xylulose; L-Lyxulose; Xyloketose)	$C_5H_{10}O_5$	Syrup	$+33.1$
95	Hexulose, β-D-arabino-(β-D-Fructose; Levulose)	$C_6H_{12}O_6$	102–104[7]	$-133.5 \rightarrow -92$
96	Hexulose, 6-deoxy-D-arabino- (D-Rhamnulose)	$C_6H_{12}O_5$	....	-13 ± 2
97	Hexulose, D-lyxo- (D-Tagatose)	$C_6H_{12}O_6$	131–132	$+2.7 \rightarrow -4, -5$
98	5-Hexulose, D-lyxo	$C_6H_{12}O_6$	158	-86.6
99	Hexulose, 6-deoxy-L-lyxo- (L-Fuculose)	$C_6H_{12}O_5$	....	....

Substance (Synonym)	Chemical Formula	Melting Point °C	Specific Rotation $[\alpha]_D$
(A)	(B)	(C)	(D)
Ketoses (Con't)			
100 Hexulose, D-*ribo*- (D-Psicose)	$C_6H_{12}O_6$	Amorphous	+4.7
101 Hexulose, L-*xylo*- (L-Sorbose)	$C_6H_{12}O_6$	159–161	−43.1
102 Hexulose, 6-deoxy-L-*xylo*-	$C_6H_{12}O_5$	88	−25 ± 2 (c 0.7)
103 Heptulose, D-*altro*- (Sedoheptulose; Sedoheptose)	$C_7H_{14}O_7$	Amorphous	+2.5 (c 10)
104 Heptulose·hemihydrate, L-*galacto*- (Perseulose)	$C_7H_{14}O_7 \cdot \frac{1}{2}H_2O$	110–115	−90 → −80
105 Heptulose, L-*gulo*-	$C_7H_{14}O_7$	. . . .	−28
106 Heptulose, D-*ido*-	$C_7H_{14}O_7$	172	−34 ± 8 (c 0.3)
107 Heptulose, D-*manno*- (Mannoketoheptose; D-Mannotagatoheptose)	$C_7H_{14}O_7$	152	+29.4
108 Heptulose, D-*talo*-	$C_7H_{14}O_7$	. . . .	. . . .
109 Octulose, D-*glycero*-L-*galacto*-	$C_8H_{16}O_8$	. . . .	−57, −43.4 → −13.4
110 Octulose, D-*glycero*-D-*manno*-	$C_8H_{16}O_8$	. . . .	+20 (CH₃OH)

[1] Original melting point. [2] Melting point after four-months' storage. [3] As a methyl glycoside cyclohexylamine salt. [4] Included because of speculations concerning it in biological processes. [5] Either D-idose or L-altrose is in the polysaccharide varianose. [6] Early literature refers to this as D-erythrose. [7] The $\cdot\frac{1}{2}H_2O$ and $\cdot 2H_2O$ forms also exist.

Part II. NATURAL MONOSACCHARIDES: AMINO SUGARS

Substance (Synonym)	Chemical Formula	Melting Point °C	Specific Rotation $[\alpha]_D$
(A)	(B)	(C)	(D)
Aldosamines			
1 D-Ribose, 3-amino-3-deoxy-	$C_5H_{11}NO_4$	158–158.5 d.	−24.6 (hydrochloride)
2 D-Galactose, 2-amino-2-deoxy- (Galactosamine; Chondrosamine)	$C_6H_{13}NO_5$	185	+121 → +80 (hydrochloride)
3 α-L-Galactose, 2-amino-2,6-dideoxy- (L-Fucosamine)	$C_6H_{13}NO_4$	192–193 d.	−119 → −92 [27°] (hydrochloride)
4 α-D-Glucose, 2-amino-2-deoxy- (Glucosamine; Chitosamine)	$C_6H_{13}NO_5$	88	+100 → +47.5
5 β-D-Glucose, 2-amino-2-deoxy-	$C_6H_{13}NO_5$	110–111	+28 → +47.5
6 D-Glucose, 3-amino-3-deoxy- (Kanosamine)	$C_6H_{13}NO_5$	128 d.	+19 [14°]
7 D-Glucose, 6-amino-6-deoxy-	$C_6H_{13}NO_5$	161–162 d.	+23 → +50.1 (hydrochloride)
8 D-Glucose, 2,6-diamino-2,6-dideoxy- (Neosamine C)	$C_6H_{14}N_2O_4$	>230	+61.5 (dihydrochloride)
9 D-Glucose, 3,6-dideoxy-3-dimethylamino- (Mycaminose)	$C_8H_{17}NO_4$	115–116	+31 (hydrochloride)
10 D-Glucose, 4,6-dideoxy-4-dimethylamino-	$C_8H_{17}NO_4$	192–193	+45.5 (hydrochloride)
11 L-Glucose, 2-deoxy-2-methylamino-	$C_7H_{15}NO_5$	130–132	−64
12 D-Gulose, 2-amino-1,6-anhydro-2-deoxy-	$C_6H_{11}NO_4$	250–260 d.	+41 ± 2 (hydrochloride)
13 D-Gulose, 2-amino-2-deoxy-	$C_6H_{13}NO_5$	152–162 d.	+5.6 → −18.7 (hydrochloride)

	Substance (Synonym)	Chemical Formula	Melting Point °C	Specific Rotation $[\alpha]_D$
	(A)	(B)	(C)	(D)

				Aldosamines (Con't)
14	Hexose, 3,4,6-trideoxy-3-dimethyl-amino-D-*xylo*- (Desosamine; Picrocine)	$C_8H_{17}NO_3$	189–191 d.	+49.5 (c 10) (hydrochloride)
15	Hexose, a 4-acetamido-2-amino-2,4,6-trideoxy-	$C_8H_{16}N_2O_4$	216–219	$+115 \rightarrow +94$ [26°] (c 0.05)
16	Hexose, an amino-deoxy-3-O-carboxyethyl-	$C_9H_{17}NO_7$	. . . .	. . . .
17	Hexose, a 2,6-diamino-2,6-dideoxy- (Neosamine B; Paramose)	$C_6H_{14}N_2O_4$	135–150 d.	+17.5 (c 0.9 (hydrochloride)
18	Hexose, a 3-dimethylamino-2,3,6-trideoxy- (Rhodosamine)	$C_8H_{17}NO_3$	. . . .	. . . .
19	D-Mannose, 2-amino-2-deoxy- (Mannosamine)	$C_6H_{13}NO_5$	142 d.	−4.3 (c 9) (hydrochloride)
20	D-Mannose, 3-amino-3,6-dideoxy- (Mycosamine)	$C_6H_{13}NO_4$	162	−11.5 (hydrochloride)
21	D-Talose, 2-amino-2-deoxy- (Talosamine)	$C_6H_{13}NO_5$	151–153	$+3.4 \rightarrow -5.7$ (c 0.9) (hydrochloride)
22	L-Talose, 2-amino-2,6-dideoxy- (Pneumosamine)	$C_6H_{13}NO_4$	162–163	$+6.9 \rightarrow +10.4$ (hydrochloride)

				Ketosamines
23	Pentulose, 1-(o-carboxyanilino)-1-deoxy-D-*erythro*-	$C_{12}H_{14}NO_6$	. . . .	. . . .
24	Hexulose, 1-(o-carboxyanilino)-1-deoxy-D-*arabino*-	$C_{13}H_{16}NO_7$	. . . .	. . . .
25	Hexulose, 5-amino-5-deoxy-L-*xylo*-	$C_6H_{13}NO_5$	174–176	−62
26	Hexulose, 6-deoxy-6-(N-methyl-acetamido)-L-*xylo*-	$C_9H_{17}NO_6$	. . . .	. . . .

Part III. NATURAL ALDITOLS AND INOSITOLS (with Inososes and Inosamines)

	Substance (Synonym)	Chemical Formula	Melting Point °C	Specific Rotation $[\alpha]_D$
	(A)	(B)	(C)	(D)

				Alditols
1	Glycerol	$C_3H_8O_3$	20	None
2	Glycerol, 1-deoxy- (1,2-Propane-diol)[1]	$C_3H_8O_2$	Oil, b.p. 188–189	None (racemic)
3	Erythritol	$C_4H_{10}O_4$	118–120	None (meso)
4	Erythritol, 1,4-dideoxy- (2,3-Butylene-glycol)	$C_4H_{10}O_2$	25, 34	None (meso)
5	D-Threitol, 1,4-dideoxy-	$C_4H_{10}O_2$	19	−13.0
6	L-Threitol, 1,4-dideoxy-	$C_4H_{10}O_2$	. . . .	+10.2
7	DL-Threitol, 1,4-dideoxy-	$C_4H_{10}O_2$	7.6	None (racemic)
8	D-Arabinitol	$C_5H_{12}O_5$	103	+7.82 (c 8, borax solution)
9	L-Arabinitol	$C_5H_{12}O_5$	101–102	−32 (c 0.4, 5% molybdate)
10	Ribitol (Adomitol)	$C_5H_{12}O_5$	102	None (meso)
11	Galactitol (Dulcitol)	$C_6H_{14}O_6$	186–188	None (meso)

Substance (Synonym)	Chemical Formula	Melting Point °C	Specific Rotation $[\alpha]_D$
(A)	(B)	(C)	(D)
Alditols (Con't)			
12 D-Glucitol (Sorbitol)	$C_6H_{14}O_6$	112	−1.8 [15°]
13 D-Glucitol, 1,5-anhydro- (Polygalitol)	$C_6H_{12}O_5$	140–141	+42.4
14 L-Iditol	$C_6H_{14}O_6$	73.5	−3.5 (c 10)
15 D-Mannitol	$C_6H_{14}O_6$	166	−0.21
16 D-Mannitol, 1,5-anhydro- (Styracitol)	$C_6H_{12}O_5$	157	−49.9
17 Heptitol, D-glycero-D-galacto- (Heptitol, L-glycero-D-manno-; Perseitol)	$C_7H_{16}O_7$	183–185, 188	−1.1
18 Heptitol, D-glycero-D-gluco- (Heptitol, L-glycero-D-talo-; β-Sedoheptitol)	$C_7H_{16}O_7$	131–132	+46 (5% NH₄ molybdate)
19 Heptitol, D-glycero-D-manno- (Heptitol, D-glycero-D-talo-; Volemitol)	$C_7H_{16}O_7$	153	+2.65
20 Octitol, D-erythro-D-galacto-	$C_8H_{18}O_8 \cdot H_2O$	169–170	−11 (5% NH₄ molybdate)
Inositols			
21 Betitol (a dideoxy inositol)	$C_6H_{12}O_4$	224	....
22 Bioinosose (scyllo-Inosose; myo-Inosose-2; a deoxy keto inositol)	$C_6H_{10}O_6$	198–200	None (meso)
23 h-Bornesitol (a myo-inositol monomethyl ether)	$C_7H_{14}O_6$	200	+31.6
24 l-Bornesitol (a myo-inositol monomethyl ether)	$C_7H_{14}O_6$	205–206	−32.1
25 Conduritol (a 2,3-dehydro-2,3-dideoxyinositol)	$C_6H_{10}O_4$	142–143	None (meso)
26 Cordycepic acid (a tetrahydroxycyclohexanecarboxylic acid)²	$C_7H_{12}O_6$	....	....
27 Dambonitol (a myo-inositol dimethyl ether)	$C_8H_{16}O_6$	206	None (meso)
28 DL-Inositol	$C_6H_{12}O_6$	253	None (racemic)
29 d-Inositol	$C_6H_{12}O_6$	....	+60
30 l-Inositol	$C_6H_{12}O_6$	240	−65
31 Laminitol (a C-methyl myo-inositol)	$C_7H_{14}O_6$	266–269	−3
32 Liriodendritol (a myo-inositol dimethyl ether)	$C_8H_{16}O_6$	224	−25
33 muco-Inositol monomethyl ether	$C_7H_{14}O_6$	322–325	....
34 myo-Inositol (meso-Inositol)	$C_6H_{12}O_6$	217–218	None (meso)
35 d-myo-Inosose-1 (a deoxy keto inositol)	$C_6H_{10}O_6$	138–139	+19.6
36 Mytilitol (a C-methyl scyllo-inositol)	$C_7H_{14}O_6$	259	None (meso)
37 neo-Inosamine-2 (a deoxy amino inositol)	$C_6H_{13}O_5N$	239–241 d.	None (meso)
38 d-Ononitol (a myo-inositol monomethyl ether)	$C_7H_{14}O_6$	172	+6.6
39 h-Pinitol (a dextro-inositol monomethyl ether)	$C_7H_{14}O_6$	186	+65.5
40 l-Pinitol (a levo-inositol monomethyl ether)	$C_7H_{14}O_6$	186	−65
41 l-Quebrachitol (a levo-inositol monomethyl ether)	$C_7H_{14}O_6$	190–191	−80.2 [28°]
42 d-Quercitol (a deoxy dextro-inositol)	$C_6H_{12}O_5$	235	+24.2
43 d-Quinic acid (a trideoxy carboxy dextro-inositol)	$C_7H_{12}O_6$	164	+44 (c 10)

CARBOHYDRATES (Continued)
Part III. NATURAL ALDITOLS AND INOSITOLS
(with Inososes and Inosamines) (Continued)

	Substance (Synonym)	Chemical Formula	Melting Point °C	Specific Rotation $[\alpha]_D$
	(A)	(B)	(C)	(D)
	Inositols (Con't)			
44	l-Quinic acid (a trideoxy carboxy levo-inositol)	$C_7H_{12}O_6$	162	−42.1
45	Quinic acid, 5-dehydro-	$C_7H_{10}O_6$	140–142 (138 s.)	−82.4 [28°]
46	Scyllitol (scyllo-Inositol; Cocositol)	$C_6H_{12}O_6$	352–353	None (meso)
47	Sequoyitol (a myo-inositol monomethyl ether)	$C_7H_{14}O_6$	234–235	None (meso)
48	Shikimic acid (a 3,4-anhydro-quinic acid)	$C_7H_{10}O_5$	183–184	−200 [16°]
49	Shikimic acid, 5-dehydro-	$C_7H_8O_5$	150–152	−57.5 [28°] (EtOH)
50	Streptamine (2,4-diaminodideoxy-scyllitol)	$C_6H_{14}O_4N_2$	88, 210–250 d.	None (meso)
51	Streptamine, 2-deoxy-	$C_6H_{14}O_3N_2$	. . . .	None (meso)
52	Streptadine (1,3-Dideoxy-1,3-diguanidino-scyllitol)	$C_8H_{18}N_6O_4$	. . . .	None (meso)
53	Viburnitol (a deoxy levo-inositol)[3]	$C_6H_{12}O_5$	174	−73.9

[1] The 1-phosphate ester of this diol is said to occur in brain tissue and sea-urchin eggs. [2] Strong evidence that cordycepic acid is really D-mannitol. [3] Not an enantiomorph of d-quercitol; other isomeric relationship is involved.

Part IV. NATURAL ALDONIC, URONIC, AND ALDARIC ACIDS

	Substance (Synonym)	Chemical Formula	Melting Point °C	Specific Rotation $[\alpha]_D$
	(A)	(B)	(C)	(D)
	Aldonic Acids			
1	D-Glyceric acid	$C_3H_6O_4$	Gum	Dextro
2	L-Glyceric acid	$C_3H_6O_4$	Gum	Levo
3	D-Arabinonic acid	$C_5H_{10}O_6$	114–116	+10.5 (c 6)
4	L-Arabinonic acid	$C_5H_{10}O_6$	118–119	−9.6 → −41.7[1]
5	L-Arabinonic-1,4-lactone	$C_5H_8O_5$	97–99	−72
6	D-Ribonic acid	$C_5H_{10}O_6$	112–113	−17.0
7	D-Xylonic acid	$C_5H_{10}O_6$	. . . .	−2.9 → +20.1[1]
8	L-Xylonic acid	$C_5H_{10}O_6$	. . . .	−91.8[1]
9	D-Altronic acid	$C_6H_{12}O_7$	. . . .	+11.5 → +24.8[1] (Ca salt, N HCl)
10	D-Galactonic acid	$C_6H_{12}O_7$	122	−11.2 → +57.6[1]
11	D-Gluconic acid	$C_6H_{12}O_7$	130–132 (110–112 s.)	−6.7 → +11.9[1]
12	L-Gulonic acid	$C_6H_{12}O_7$	Exists only in soln.	[ca. 0°]
13	Hexsonic acid, 2-deoxy-D-arabino-	$C_6H_{12}O_6$	93–95	+68 (lactone)
14	2-Hexulosonic acid, D-arabino-	$C_6H_{10}O_7$	. . . .	−81.7 (Na salt)
15	2-Hexulosonic acid, 3-deoxy-D-erythro-	$C_6H_{10}O_6$	. . . .	−29.2 (c 6, Ca salt)
16	2-Hexulosonic acid, D-lyxo-	$C_6H_{10}O_7$	169	−5
17	5-Hexulosonic acid, D-arabino-	$C_6H_{10}O_7$	108–109	. . . .
18	5-Hexulosonic acid, D-xylo-	$C_6H_{10}O_7$	. . . .	−14.5
19	D-Mannonic acid	$C_6H_{12}O_7$	. . . .	−15.6
20	D-Gluconic acid, O-β-D-galactopyranosyl- (1 → 4)- (Lactobionic acid)	$C_{12}H_{22}O_{12}$	. . . .	+25.1 (Ca salt)

Substance (Synonym)	Chemical Formula	Melting Point °C	Specific Rotation $[\alpha]_D$
(A)	(B)	(C)	(D)
Uronic Acids			
21 L-Lyxuronic acid	$C_5H_8O_6$	. . . .	. . . .
22 β-D-Galacturonic acid	$C_6H_{10}O_7$	160	$+27 \rightarrow +55.6$
23 α-D-Galacturonic acid·monohydrate	$C_6H_{12}O_8$	159–160 (110–115 s.)	$+97.9 \rightarrow +50.9$
24 D-Galacturonic acid, 2-amino-2-deoxy-	$C_6H_{11}O_6N$	160 d.	$+84.5$ (pH 2 HCl)
25 β-D-Glucuronic acid	$C_6H_{10}O_7$	156	$+11.7 \rightarrow +36.3$
26 D-Glucuronic acid, 2-amino-2-deoxy-	$C_6H_{11}O_6N$	120–172 d.	$+55$
27 D-Glucuronic acid, 3-O-methyl-	$C_7H_{12}O_7$	Syrup	$+6$
28 L-Guluronic acid	$C_6H_{10}O_7$	. . . .	. . . .
29 L-Iduronic acid	$C_6H_{10}O_7$	. . . .	$+30$
30 β-D-Mannuronic acid	$C_6H_{10}O_7$	165–167	$-47.9 \rightarrow -23.9$
31 α-D-Mannuronic acid·monohydrate	$C_6H_{12}O_8$	110 s., 120–130 d.	$+16 \rightarrow -6.1$ (c 6.8)
Aldaric Acids			
32 D-Tartaric acid	$C_4H_6O_6$	170	-15
33 L-Tartaric acid	$C_4H_6O_6$	170	$+15$ [15°]
34 L-Malic acid	$C_4H_6O_5$	100	-2.3 (c 8.4)

[1] Equilibrates with the lactone.

WAXES

These data for waxes were compiled originally for the Biology Data Book by A. H. Warth. Data are reproduced here by permission of the copyright owners of the above publication, the American Societies for Experimental Biology, Washington, D.C. p. 382.

Specific Gravity (column C) was calculated at the specified temperature, degrees centigrade, and referred to water at the same temperature. **Density.** shown in parentheses (column C), and **Refractive Index** (column D) were measured at the specified temperature, degrees centigrade.

	Wax	Melting Point °C	Specific Gravity or (Density)	Refractive Index $n\frac{°C}{D}$	Iodine Value	Acid Value	Saponification Value
	(A)	(B)	(C)	(D)	(E)	(F)	(G)
1	Bamboo leaf	79–80	$(0.961^{25°})$	....	7.8[1]	14.5	43.4
2	Bayberry (myrtle)	46.7–48.8	$(0.985^{15°})$	$1.436^{80°}$	2.9[2]–3.9[3]	3.5	20.5–21.7
3	Beeswax, crude	62–66	$(0.927–0.970^{15°})$	$1.439–1.483^{40°}$	6.8–16.4[2]	16.8–35.8	89.3–149.0
4	Beeswax, white, U.S.P.	61–69	$(0.959–0.975^{15°})$	$1.447–1.465^{65°}$	7–11[3]	17–24	90–96
5	Beeswax, yellow	62–65	$(0.960–0.964^{15°})$	$1.443–1.449^{65°}$	6–11	18–24	90–97
6	Candelilla, refined	67–69	$(0.982–0.986^{15°})$	$1.454–1.463^{85°}$	14.4–20.4	12.7–18.1	35–86
7	Cape berry[4]	40.5–45.0	$(1.004–1.007^{15°})$	$1.450^{45°}$	0.6–2.4	2.5–3.7	211–215
8	Carandá	79.7–84.5	$(0.990^{25°})$	....	8.0–8.9	5.0–9.5	64.5–78.5
9	Carnauba	83–86	$0.990–1.001^{15°}$	$1.467–1.472^{40°}$	7.2–13.5	2.9–9.7	78–95
10	Castor oil, hydrogenated	83–88	$(0.980–0.990^{20°})$	....	2.5–8.5	1.0–5.0	177–181
11	Chinese insect	81.5–84.0	$0.950–0.970^{15°}$	$1.457^{40°}$	1.4	0.2–1.5	73–93
12	Cotton	68–71	$0.959^{15°}$	....	24.5	32	70.6
13	Cranberry	207–218	$(0.970–0.975^{15°})$	....	44.2–53.2[2]	42.2–59.1	131–134
14	Douglas fir bark	60.0–70.0	$(1.000^{15°})$	$1.468^{80°}$	25.8–62.5	58.6–80.7	112–200
15	Esparto	67.5–78.1	$0.988^{15°}$	....	22–23	22.7–23.9	69.8–79.3
16	Flax	61.5–69.8	$0.908–0.985^{15°}$	....	21.6–28.8	17.5–48.3	77.5–101.5
17	Ghedda, E. Indian beeswax	60.5–66.4	$0.956–0.973^{15°}$	$1.440^{50°}$	5.6–12.6	5.8–7.9	84.5–118.3
18	Indian corn	80–81	....	....	4.2[2]	1.9	120.3
19	Japan wax	48–53	$0.975–0.993^{15°}$	....	4.5–12.5	6–20	206.5–237.5
20	Jojoba	11.2–11.8	$0.864–0.899^{25°}$	$1.465^{25°}$	81.7–88.4[2]	0.2–0.6	92.2–95.0
21	Madagascar	88	....	....	3.2–5.3	17.7–28.0	140.0–159.6
22	Microcrystalline, amber	64–91	$0.913–0.943^{15°}$	$1.424–1.452^{80°}$	0	0	0
23	Microcrystalline, white	71–89	$0.928–0.941^{15°}$	$1.441^{80°}$	0	0	0
24	Montan, crude	76–86	$(1.010–1.020^{25°})$	....	13.9–17.6	22.7–31.0	59.4–92.0
25	Montan, refined	77–84	$(1.010–1.030^{25°})$	....	10–14	24–43	72–103
26	Orange peel	44.0–46.5	$0.985^{15°}$	$1.502^{20°}$	115.7[2]	48.3	120.9
27	Ouricury, refined	79.0–83.8	$1.053^{15°}$	....	6.9–7.8[2]	3.4–21.1	61.8–85.8
28	Ozocerite, refined	74.4–75.0	$0.907–0.920^{15°}$	....	0	0	0
29	Palm	74–86	$(0.991–1.045^{15°})$	....	8.9–16.9[2]	5.0–10.6	64.5–104.0
30	Paraffin, American	49–63	$0.896–0.925^{15°}$	$1.442–1.448^{80°}$	0	0	0
31	Peat wax, natural	73–76	$0.980^{15°}$	....	16–40	60.0–73.3	73.9–136.0
32	Rice bran, refined	75.3–79.9	....	$1.469^{30°}$	11.1–19.4	15–17	56.9–104.4
33	Shellac wax	79–82	$0.971–0.980^{15°}$	....	6.0–8.8[3]	12.1–24.3	63.8–83.0
34	Sisal hemp	74–81	$1.007–1.010^{15°}$	....	28–29[2]	16–19[2]	56–58
35	Sorghum grain	77–82	....	....	15.7–20.9	10.1–16.2	16–44
36	Spanish moss	79–80	....	....	33.0	25.0	120.4
37	Spermaceti	42–50	$0.905–0.945^{15°}$	$1.440^{70°}$	4.8–5.9	2.0–5.2	108–134
38	Sugarcane, crude	52–67	$0.988–0.998^{25°}$	....	32–84	24–57	128–177
39	Sugarcane, double-refined	77–82	$0.961–0.979^{25°}$	$1.510^{25°}$	13–29	8–23	55–95
40	Wool wax, refined	36–43	$0.932–0.945^{15°}$	$1.478–1.482^{40°}$	15.0–46.9	5.6–22.0	80–127

[1] Wijs test. [2] Hanus test. [3] Hubl test. [4] *Myrica cordifolia*.

TRADE NAMES OF DYESTUFF INTERMEDIATES

Trade name	Chemical name
A acid	1,7-Hydroxynaphthalene-3,6-disulfonic acid
Acetyl H acid	N-Acetyl-1-amino-8-naphthol-3,6-disulfonic acid
Alen's acid	1-Naphthylamine-3,6-disulfonic acid (also Freund's ac.)
Alizarin	1,2-Dihydroxyanthraquinone
Amido acid	2-Amino-7-hydroxynaphthalene-5-sulfonic acid
Amido J acid	2-Naphthylamine-5,7-disulfonic acid
Amino G acid	7-Amino-1,3-naphthalene disulfonic acid
	2-Naphthylamine-6,8-disulfonic acid
Amino R acid	3-Amino-2,7-naphthalene disulfonic acid
	2-Naphthylamine-3,6-disulfonic acid
Aminophenolic acid V	1-Amino-3-oxybenzene-5-sulfonic acid
Aminophenol sulfonic acid III	1-Amino-3-oxybenzene-6-sulfonic acid
Andresen's acid	1-Naphthol-3,8-disulfonic acid
Anisidine	o-Aminophenol methylether
Anthrachrysone	1,3,5,7-Tetrahydroanthraquinone
Anthraflavic acid	2,6-Dihydroxyanthraquinone
Anthranilic acid	o-Aminobenzoic acid
Anthrarufin	1,5-Dihydroxyanthraquinone
Anthranol	9-Hydroxyanthracene
α Anthrol	1-Hydroxyanthracene
Armstrong's acid	Naphthalene-1,5-disulfonic acid
Armstrong & Wynne acid	1-Naphthol-3-sulfonic acid
Armstrong & Wynne acid II	2-Naphthylamino-5,7-disulfonic acid
B acid	8-Amino-1-naphthol-4,6-disulfonic acid
Badische acid	2-Naphthylamino-8-sulfonic acid
Bayer's acid	2-Naphthol-8-sulfonic acid
Benzidine	p,p'-Diaminodiphenyl
Bronner's acid	2-Naphthylamino-6-sulfonic acid
β acid	Anthraquinone-2-sulfonic acid
C acid } CLT acid }	6-Chloro-m-toluidine-4-sulfonic acid
Casella's acid	2-Naphthol-7-sulfonic acid (F acid)
Chicago acid	1-Amino-8-naphthyl-2,4-disulfonic acid
Chloro H acid	8-Chloro-1-naphthol-3,6-disulfonic acid
Chromogene I	4,5-Dihydroxy-2,7-naphthalene disulfonic acid
Chromotrope acid	1,8-Dihydronaphthalene-3,6-disulfonic acid
Chromotropic acid	4,5-Dihydroxy-2,7-naphthalene disulfonic acid
Chrysazine	1,8-Dihydroxyanthraquinone
Cleve's acid	1-Naphthylamine-3-sulfonic acid
Cleve's acid	1-Naphthylamine-5-sulfonic acid
Cleve's acid	1-Naphthylamine-6-sulfonic acid
Cleve's acid	1-Naphthylamine-7-sulfonic acid
α Coccinic acid	1-Methyl-5-oxybenzyl-2,4-dicarbonic acid
Cresidine	3-Amino-4-methoxytoluene
Cresotic acid	Cresol carboxylic acid
Croceine acid	2-Naphthol-1-sulfonic acid
DS	4,4'-Diamino-2,2'-stilbene disulfuric acid
DTS	Dihydrothio-p-toluidine sulfonic acid
Dahl's acid	2-Naphthylamine-5-sulfonic acid
Dahl's acid II	1-Naphthylamine-4,6-disulfonic acid
Dahl's acid III	1-Naphthylamine-4,7-disulfonic acid
Dimethyl-γ-acid	7-Dimethylamino-1-naphthol sulfonic acid
Dioxy G acid	1,7-Dihydroxynaphthalene-3-sulfonic acid
Dioxy J acid	1,6-Dihydroxynaphthalene-3-sulfonic acid
Dioxy S acid	1,8-Dihydroxynaphthalene-4-sulfonic acid
Diphenylblack base	p-Aminodiphenylamine
Disulfo acid S	1-Naphthylamine-4,8-disulfonic acid
δ acid	{ 1-Naphthol-4,8-disulfonic acid 1-Naphthylamine-4,8-disulfonic acid

Trade name	Chemical name
Ebert & Merz acid	{ Naphthalene-2,7-disulfonic acid { Naphthalene-2,6-disulfonic acid
Ethyl-γ-acid	7-Ethylamino-1-naphthol-3-sulfonic acid
Ethyl F acid	7-Ethylamino-2-naphthalene sulfonic acid
Ewer & Pick's acid	Naphthalene-1,6-disulfonic acid
ε acid	{ 1-Naphthol-3,8-disulfonic acid { 1-Naphthylamine-3,8-disulfonic acid
F acid	2-Naphthol-7-sulfonic acid (Casella's acid)
Fast Black B base	4,4'-Diamino diphenylamine
Fast Blue base	Dianisidine
Fast Blue Red O base	3-Nitro-p-phenetidine
Fast Bordeaux GP	3-Nitro-p-anisidine
Fast Orange GR	o-Nitroaniline
Fast Orange R	m-Nitroaniline
Fast Red base AL	α-Aminoanthraquinone
Fast Red B base	5-Nitro-o-anisidine
Fast Red GG base	p-Nitroaniline
Fast Red GL base	3-Nitro-p-toluidine
Fast Red 3 GL base	4-Chloro-2-nitroaniline
Fast Red RL base	5-Nitro-o-toluidine
Fast Scarlet G base	4-Nitro-α-toluidine
Fast Scarlet R base	4-Nitro-O-anisidine
Forsling's acid I	2-Naphthylamine-8-sulfonic acid
Forsling's acid II	2-Naphthylamine-5-sulfonic acid
Freund's acid	1-Naphthylamine-3,6-disulfonic acid
G acid	2-Naphthol-6,8-disulfonic acid
GR acid	α-Naphthol-3,6-disulfonic acid
Gallic acid	3,4,5-Trihydroxybenzoic acid
γ acid	2-Amino-8-naphthol-6-sulfonic acid
H acid	1-Amino-8-naphthol-3,6-disulfonic acid
Histazarin	2,3-Dihydroxyanthraquinone
Isoanthraflavic acid	2,7-Dihydroxyanthraquinone
J acid	2-Amino-5-naphthol-7-sulfonic acid
K acid	1-Amino-8-naphthol-4,6-disulfonic acid
Kalle's acid	1-Naphthylamine-2,7-disulfonic acid
Ketone base	Tetramethyl aminobenzophenone
Koch's acid	1-Naphthylamine-3,6,8-trisulfonic acid
L acid	1-Naphthol-5-sulfonic acid
Laurent's acid	1-Naphthylamine-5-sulfonic acid
Lepidine	4-Methylquinoline
Leucotrop	Phenyldimethyl benzylammonium chloride
M acid	1-Amino-5-naphthol-7-sulfonic acid
Mesidine	2,4,6-Thimethylaniline
Metanilic acid	Aniline-m-sulfonic acid
Methyl-γ-acid	7-Methyl-8-naphthol disulfonic acid
Michler's hydrol	Tetramethyl diaminobenzohydrol
Michler's ketone	Tetramethyl diaminobenzophenone
Myrbane oil	Nitrobenzene
Naphthacetol	1-Acetylamino-4-naphthol
Naphthazarin	5,8-Dihydroxy-1,4-naphthoquinone
Naphthionic acid	1-Naphthylamine-4-sulfonic acid
o-Naphthionic acid	1-Naphthylamine-2-sulfonic acid
Naphthol AS	Anilide of hydronaphthoic acid
Naphthoresorcine	1,3-Dihydroxynaphthalene
Nekal BX	Na-salt of 1,4-bis,sec-butylnaphthalene-6-sulfonic acid
Nevile and Winther's acid	1-Naphthol-4-sulfonic acid
Nigrotic acid	1,3,6,7-Dihydroxysulfonaphthoic acid
Nitron 1,2,4-acid	1-Amino-8-nitro-7-naphthol-4-sulfonic acid
Nitroso base	p-Nitrodimethyl aniline

Trade name	Chemical name
NW acid	Nevile and Winther s acid
Oxy L acid	1-Naphthol-5-sulfonic acid
Oxy Tobias acid	β-Naphthol-1-sulfonic acid
Peri acid	1-Naphthylamine-8-sulfonic acid
p-Phenetidine	p-Aminophenol ethylether
Phenyl gamma acid	2-Phenylamine-8-naphthol-6-sulfonic acid
Phenyl Peri acid	Phenyl-1-naphthylamine-8-sulfonic acid
Phosxgene	Carbonyl chloride
Phthalic acid	O-Benzenedicarbolic acid
Picramic acid	2-Amino-4,6-dinitrophenol
Picric acid	2,4,6-Trinitrophenol
Pirio's acid	4-Amino-1-naphthalene sulfonic acid
Primuline base	p-Toluidine heated with sulfur
Purpurine	1,2,4-Trihydroxyanthraquinone
Pyrogallol	1,2,3-Trihydroxybenzene
Quinaldine	2-Methylquinoline
Quinazarin	1,4-Dihydroxyanthraquinone
R acid	2-Naphthol-3,6-disulfonic acid
2 R acid	2-Amino-8-naphthol-3,6-disulfonic acid
Red acid	1,5-Dihydroxynaphthalene-3,7-disulfonic acid
RG acid	1-Naphthol-3,6-disulfonic acid
Resorcinol	1,3-Dihydroxybenzene
Rumpff acid	2-Naphthol-8-sulfonic acid (Croceine acid)
S acid	1-Amino-8-naphthol-4-sulfonic acid
2 S acid	1-Amino-8-naphthol-2,4-disulfonic acid
Salicylic acid	o-Hydroxybenzoic acid
Schäffer's acid	2-Naphthol-6-sulfonic acid
Schäffer and Baum acid	α-Naphthol-2-sulfonic acid
Schollkopf's acid	1-Naphthol-4,8-disulfonic acid 1-Naphthylamine-4,8-disulfonic acid 1-Naphthylamine-8-sulfonic acid
Sulfanilic acid	Aniline-p-sulfonic acid
Thiocarbanilide	Diphenylthiourea
Tobias acid	2-Naphthylamine-1-sulfonic acid
Tolidine	Di-p-aminoditolyl
Toluidine	Aminotoluene
Violet acid	α-Naphthol-3,6-disulfuric acid
Xylidine	Aminoxylene
Y acid	2-Naphthol-6,8-disulfuric acid
Yellow acid	1,3-Dihydroxynaphthalene-5,7-disulfuric acid
1:2:4 acid	1-Amino-2-naphthol-4-sulfonic acid

NOMENCLATURE OF SOME MONOMERS AND POLYMERS

Calvin E. Schildknecht

The small molecules from which industrial polymers are formed are of two general types. The first are the unsaturated or ethylenic compounds. They are often called vinyl-type monomers. Table 1 gives formulas, common names and newer indexing names used by *Chemical Abstracts*. The second type of monomer useful in polymer synthesis are better called intermediates. These are shown in Table 2. Recent names as 2,5-Furandione for Maleic anhydride and 1,3-isobenzofurandione for Phthalic anhydride are employed in *Chemical Abstracts* for indexing, but older names are generally retained in the abstracts. Recent books[1-3] have used capital letters for abbreviations of polymer types, e.g., PVA for polyvinyl alcohols and VC-VAC for vinyl chloride-vinyl acetate copolymers. Preceding small letters may show proved structures, for example, i-PP for substantially isotactic or stereoregular polypropylenes. Experimentally determined physical properties may be added using small numbers and letters following for example, x-i-PP-mp 150-170C indicates a crosslinked isotactic polypropylene of crystal melting range 150-170C. Abbreviations used to indicate the type of polymerization processes are presented in Table 3.

Table 1
NAMES OF SOME ETHYLENIC MONOMERS

Formula	Common Name	New Name
$CH_2=CH_2$	Ethylene	Ethene
$CH_2=CH$	Vinyl (group)	Ethenyl
$CH_2=CHC_6H_5$	Styrene	Ethylbenzene
$CH_2=CHOOCCH_3$	Vinyl acetate	Ethenyl acetate
$CH_2=CHCl$	Vinyl chloride	Chloroethene
$CH_2=CHOCH_3$	Methyl vinyl ether	Methoxyethene
$CH_2=CHOCH=CH_2$	Divinyl ether	1,1-Oxybis (ethene)
$CH_2=CCl_2$	Vinylidene chloride	1,1-Dichloroethene
$CH_2=CHCOOH$	Acrylic acid	2-Propenoic acid
$CH_2=CHCN$	Acrylonitrile	2-Propenenitrile
$CH_2=CHCONH_2$	Acrylamide	2-Propenamide
$CH_2=CHCOOC_2H_5$	Ethyl acrylate	Ethyl 2-Propenoate
$CH_2=C(CH_3)COOH$	Methacrylic acid	2-Methyl-2-propenoic acid
$CH_2=C(CH_3)COOCH_3$	Methyl methacrylate	Ethyl 2-propenoate
$CH_2=CHCH_3$	Propylene	Propene
$CH_2=CHCH_2$	Allyl (group)	2-Propenyl
$CH_2=CHCH_2OH$	Allyl alcohol	2-Propen-1-ol
$CH_2=CHCH_2OOCCH_3$	Allyl acetate	2-Propenyl ethanoate

Table 2

SOME INTERMEDIATES FOR POLYMERS

Formula	Common name	New name
CH_2-CH_2 ring with N-H	Ethylene imine	Aziridine
CH_2-CH_2 ring with O	Ethylene oxide	Oxirane
$CH=CH$ with COOH and HOOC	Fumaric acid	2-Butenedioic acid (Z)
$CH=CH$ with HOOC and COOH	Maleic acid	2-Butenedioic acid (E)
$CH=CH$, OC, CO ring with O	Maleic anhydride	2,5-Furandione
$CH=CH$, OC, CO ring with N-H	Maleimide	1 H-Pyrrole-2,5-dione
CH_2-CH_2, OC, CO ring with O	Succinic anhydride	Dihydro-2,5-furandione
benzene ring fused, OC, CO, O	Phthalic anhydride	1,3-Isobenzofurandione
$HOOC(CH_2)_4COOH$	Adipic acid	Hexanedioic acid
$(CH_2)_5-CO$ ring with N-H	Caprolactam	Hexahydro-2H-azepin-2-one

Table 3
ABBREVIATIONS USED IN POLYMERIZATION PROCESSES

PE	An ethylene high polymer		TMA	Trimellitic anhydride
hd-PE	High density polyethylene		TDI	Toluene diisocyanate
S-AN	Copolymer of styrene and acrylonitrile		PP	A propylene high polymer
			i-PP	Substantially isotactic PP
PIB	An isobutene high polymer		syn-PP	Substantially syndiotactic PP (if it exists)
PMP	Poly-4-methyl-1-pentene			
HD	1,4-Hexadiene		PB	Poly-1-butene
EP	Copolymer of ethylene + propylene		AA	Acrylamide
			DAA	Diacetone acrylamide
EPDM	Ethylene-propylene-difunctional monomer copolymer		PVAC	Vinyl acetate polymer
			PVC	Vinyl chloride polymer
VAC	Vinyl acetate		PVA	Polyvinyl alcohol
VC	Vinyl chloride		PMMA	Methyl methacrylate polymer
AN	Acrylonitrile		PS	A polystyrene
MMA	Methyl methacrylate		DGEBA	Diglycidyl ether of bisphenol A
S	Styrene		UF	Ureo-formaldehyde polymer
ECH	Epichlorohydrin		THF	Tetrahydrofuran
PF	Phenol-formaldehyde polymer		PET	Polyethylene terephthalate
MF	Melamine-formaldehyde polymer		PBT	Polybutylene terephthalate
			PO	1,2-Propylene oxide
DMP	2,6-dimethylphenol		PBI	Polybenzimidazole
DEB	m-Diethynylbenzene		MDI	Methylene bis (4-phenyl isocyanate)
DMT	Dimethyl terephthalate			
BHET	bis(2-Hydroxyethyl)terephtholate		MOCA	Methylene bis (o-chloroaniline)
			PABM	Polyaminobismaleimide
BCMO	3,3-bis (Chloromethyl)oxelane		TAHT	1,3,5-Triacryloyl hexahydrotriazine
PMA	Pyromellitic dionhydride			
BPDA	Benzophenone tetracarboxylate dianhydride		HMM	Hexamethylalmelamine
			HMMM	Hexamethoxymethylolmelamine

REFERENCES

1. **Schildknecht, C. E.**, *Mod. Paint Coat.*, 69 (6), 41—45, 1979.
2. **Schildknecht, C. E.**, *Vinyl and Related Polymers*, (1952); *Polymer Processes*, (editor, 1956); *Allyl Compounds and Their Polymers*, (1973); *Polymerization Processes*, (editor, 1977), Wiley-Interscience, New York.
3. **Billmeyer, F. W.**, *Textbook of Polymer Science*, 2nd ed., Wiley-Interscience, New York, 1971.

IONIC EXCHANGE RESINS

ANION EXCHANGE RESINS

The following table is divided into two parts; the first lists properties of some anionic resins and the second, properties of some cationic resins.

Character S=strong W=weak	Trade name	Manufacturer*	Active group	Matrix	Effective pH	Selectivity	Order of selectivity	Total exchange capacity; meq/ml	Total exchange capacity; meq/gm	Maximum thermal stability; °C	Physical form; s=sphere b=beads	Standard mesh range	Ionic form as shipped	Shipping density; lb./cu. ft.
S	Dowex 1	1	Trimethyl benzyl ammonium	Polystyrene	0–14	Cl/H approx. 25	I, NO₃, Br, Cl, Acetate, OH, F	1.33	3.5	OH⁻ 50 Cl⁻ 150	s	20–50 (wet)	Cl⁻	44
S	Dowex 21 K	1	Trimethyl benzyl ammonium	Polystyrene	0–14	Cl/H approx. 15	I, NO₃, Br, Cl, Acetate, OH, F	1.25	4.5	OH⁻ 50 Cl⁻ 150	s	20–50 (wet)	Cl⁻	43
S	Duolite A-101 D	2	Quaternary ammonium	Polystyrene	0–14	—	—	1.4	4.2	OH⁻ 60 Cl⁻ 100	b	16–50	Cl⁻	—
S	Ionac A-540	3	Quaternary ammonium	Polystyrene	0–14	—	—	1.0	3.6	salt 100 OH⁻ 60	b	16–50	salt	43–66
S	Dowex 2	1	Dimethyl ethanol benzyl ammonium	Polystyrene	0–14	Cl/H approx. 1.5	I, NO₃, Br, Cl, Acetate, OH, F	1.33	3.5	OH⁻ 30 Cl⁻ 150	s	20–50 (wet)	Cl⁻	44
S	Duolite A-102 D	2	Quaternary ammonium	Polystyrene	0–14	—	—	1.4	4.2	OH⁻ 40 Cl⁻ 100	b	16–50	Cl⁻	—

IONIC EXCHANGE RESINS (Continued)

ANION EXCHANGE RESINS (Continued)

Character S=strong W=weak	Trade name	Manu-facturer*	Active group	Matrix	Effective pH	Selectivity	Order of selectivity	Total exchange capacity; meq/ml	Total exchange capacity; meq/gm	Maximum thermal stability; °C	Physical form; s=sphere b=beads	Standard mesh range	Ionic form as shipped	Shipping density; lb./cu. ft.
S	Ionac A-550	3	Dimethyl ethanol benzyl ammonium	Polystyrene	0–14	—	—	1.3	3.5	salt 100 OH⁻ 40	b	16–50	salt	43–46
W	Duolite A-30 B	2	Tertiary amine; Quaternary ammonium	Epoxy polyamines	0–9	—	—	2.6	8.7	80	b	16–50	salt	—
W	Ionac A-300	3	Tertiary amine; Quaternary ammonium	Epoxy amine	0–12	—	—	1.8	5.5	40	g	16–50	salt	19–21
W	Duolite A-6	2	Tertiary amine	Phenolic	0–5	—	—	2.4	7.6	60	g	16–50	salt	—
W	Duolite A-7	2	Secondary amine	Phenolic	0–4	—	—	2.4	9.1	40	g	16–50	salt	—

* 1. Dow
2. Diamond Shamrock
3. Ionac
4. Nalco

CATION EXCHANGE RESINS

Character S=strong W=weak	Trade name	Manu-facturer*	Active group	Matrix	Effective pH	Selectivity	Order of selectivity	Total exchange capacity; meq/ml	Total exchange capacity; meq/mg	Maximum thermal stability; °C	Physical form; s=sphere b=beads	Standard mesh range	Ionic form as shipped	Shipping density; lb./cu. ft.
S	Dowex 50	1	Nuclear sulfonic acid	Polystyrene	0–14	Na/H approx. 1.2	Ag, Cs, Rb, K, NH₄, Na, H, Li, Ba, Sr, Ca, Mg, Be	Na⁺ 1.9 H⁺ 1.7	Na⁺ 4.8 H⁺ 5.0	150	s	20–50 (wet)	H⁺ or Na⁺	H⁺ 50 Na⁺ 53
S	Dowex MPC-1	4	Nuclear sulfonic acid	Polystyrene	0–14	—	—	1.6–1.8 H⁺ form	4.5–4.9 H⁺ form	150	b	20–40 (wet)	Na⁺	50
S	Duolite C-20	2	Nuclear sulfonic acid	Polystyrene	0–14	—	—	2.2	5.1	150	b	16–50	Na⁺	—
S	Ionac 240	3	Nuclear sulfonic acid	Polystyrene	0–14	—	—	1.9	4.6	140 (Na⁺) 130 (H⁺)	b	16–50	Na⁺	50–55
S	Duolite C-3	2	Methylene sulfonic	Phenolic	0–9	—	—	1.1	2.9	60	g	16–50	H⁺	—
W	Dowex CCR-1	4	Carboxylic	Phenolic	0–9	—	—	—	—	38	g	20–50 (wet)	H⁺ (dry)	21
W	Duolite ES-63	2	Phosphonic	Polystyrene	4–14	—	—	3.3	6.5	100	b	16–50	H⁺	—
W	Duolite ES-80	2	Aliphatic	Acrylic	6–14	—	—	3.5	10.2	100	b	16–50	H⁺	—

* 1. Dow
2. Diamond Shamrock
3. Ionac
4. Nalco

AZEOTROPES

Zdzislaw M. Kurtyka

GENERAL CLASSIFICATION OF AZEOTROPES AND THEIR SYMBOLISM

Although the first reported observations of the appearance of minimum vapor pressure in binary mixtures were in 1802 by Dalton,[1] the term "azeotrope" was not introduced until 1911 by Wade and Merriman.[2]

Azeotrope from the Greek "not to boil with change" or "to boil unchanged" means literally the same as the English "constant-boiling", i.e., the vapor boiling from a liquid has the same composition as the liquid.

A liquid mixture of two or more components which can be separated by distillation was termed a "zeotrope" by Swietos-lawski.[3] Such a mixture is also called a "nonazeotrope" or a "non-azeotropic" system in many countries. However, some objection may be placed against the use of the terms "nonazeotrope" and "non-azeotropic" system due to the use of two negatives in the same word. In Greek "a" means "non". Therefore, accuracy indicates that the word nonazeotrope should be replaced by non-non-azeotrope. It is obvious the latter is awkward and that it is reasonable to replace it with Swietos-lawski's suggestion, zeotrope.

Azeotropic systems may be classified broadly in relation to the character of the extremum (maximum or minimum), the number of components in the system and whether they form one or more liquid phases.

Positive and Negative Azeotropes

In 1926 Lecat[4] proposed to divide azeotropes into positive and negative. Positive azeotropes are characterized by a minimum boiling temperature at constant preassure, i.e. a maximum in the vapor pressure at constant temperature. Negative azeotropes, on the other hand, have a maximum boiling temperature and a minimum vapor pressure.

In Anglo-Saxon literature a positive azeotrope is equivalent to a pressure-maximum azeotrope, while a negative one — to a pressure-minimum azeotrope. This description of azeotropes cannot be extended to systems exhibiting neither a minimum nor a maximum in either boiling temperature or vapor pressure (saddle or positive-negative azeotropes).

One may say at this point that neither azeotrope nor zeotrope, thus described, gives any indication whether the vapor phase consists of one, two or more liquid phases. To make this perfectly clear, the terms homo- and heteroazeotrope, and homo- and heterozeotrope were introduced. For practical purposes negative azeotropes are divided into three groups.[5]

In view of the fact that the number of different types of azeotropes continue to increase, symbols are given below for positive and negative azeotropes and zeotropes.

Type of homoazeotrope	Type of zeotrope
1. (A,B) binary positive	(A,B)ₓ binary positive
2. A,B,C) ternary positive	(A,B,C)ₓ ternary positive
3. (A,B,C,D)quaternary positive	(A,B,C,D)ₓquaternary positive
4. [(−)A,B]binary negative	[(−)A,B]ₓ binary negative

Symbols are also used for heteroazeotropes. For example, there is a combination of letters, dots, and dashes. There is one letter for each of the components in the system, one dash for each of the phases in the azeotrope and one less dot than there are dashes. The system may be illustrated by the following two examples. The system benzene-ethanol-water forms a two-phase heteroazeotrope. Symbolically this is written as (B,E,W, -·-). The system nitromethane-water-n-paraffin forms three liquid phases at the boiling temperature. This applies for n-paraffins from heptane to tridecane. If the symbol H_i is assigned to the n-parafffin, the azeotrope will be designated by the symbols (N,W,H$_i$,-·-·-).

Saddle or Positive-Negative Azeotropes

In spite of the fact that a ternary saddle azeotrope was predicted by Ostwald at the end of the last century, the first azeotrope of this type was found in 1945 by Ewell and Welch[7] in the system acetone-chloroform-methanol.

Saddle azeotropic systems, also called positive-negative systems, exhibit a hyperbolic point which is neither a minimum nor a maximum in either boiling temperature or vapor pressure, and are characterized by the presence of a "top-ridge" line. They also exhibit some peculiar properties called distillation anomalies.[7-10]

In general, ternary saddle azeotropes are classified according to the number of binary negative systems forming such an azeotrope. From this point of view, ternary saddle azeotropes may be divided into bipositive-negative and binegative-positive azeotropes. All possible types of ternary bipositive-negative azeotropes were found.[11-14] The bipositive-negative azeotropes may be designated by the symbols [(−)A,B(±)H], where, A, B, and H are the components forming these azeotropes. For the ternary binegative-positive azeotropes with B, E, and C as the components, the symbols are [(+)B,E(−)C] (two binary negative azeotropes [(−)C,B] and [(−)C,E] occur in this system). This type of a ternary saddle azeotrope is rather a rare phenomenon.[15,16]

Although the terms "saddle azeotrope" and "positive-negative" azeotrope are equivalent in the context of ternary systems, the latter is more preferable when discussing multicomponent systems.

THE COMMONNESS OF THE PHENOMENON OF AZEOTROPY

Azeotropes occur in organic and inorganic systems, although most of the known azeotropes are formed by organic compounds. This is because the very large number of organic compounds boil without decomposition in the easily accessible ranges of temperature and pressure.

In the last 2 decades of the 19th century the appearance of an extremum vapor pressure and boiling temperature was believed to be a rare phenomenon. For this reason Ostwald[17] used the term "ausgezeichnete Lösungen" to emphasize the phenomenon was not often encountered. This view survived in certain circles for many years despite convincing evidence the phenomenon of azeotropy is definitely a common one.

The first investigator who provided evidence of the commonness of azeotropy was Ryland.[18] His word dealt with 80 systems, 45 of which were azeotropic; 80 further new azeotropes were described in Lecats doctoral dissertation, published in 1908 to 1909. Lecat was able to list 1000 azeotropes in his monograph 10 years later.[19] Among them were numerous binary and ternary heteroazeotropes. This monograph showed that the appearance of maximum or miminum vapor pressures of binary and ternary mixtures should not be regarded as a rare phenomenon. From then on many physical chemists and technologists began to take an interest in the theoretical and practical application of azeotropy.

In 1949 Lecat[20] published the azeotropic data for 13,290 binary systems. The number of azeotropes reached 6287 or 47% of the systems examined.

In 1973 Horsley[21] published his Azeotropic Data-III. In this volume 15,823 binary, 725 ternary, 21 quaternary, and 2 quinary systems are reported. The number of the azeotropic systems is as follows: binary 7945 (52%), ternary 371 (51%), quaternary 9 (43%), and quinary 1.

It is interesting to see that 119 binary azeotropes (47% of the systems examined), including 32 negative, occur in the systems composed either of two inorganic compounds (elements) or an inorganic-organic compound system; 768 binary systems contain water as one component (among them 665 [86%] are azeotropic). The ternary positive-negative (saddle) azeotropes occur in 40 systems; 267 (72%) ternary azeotropic systems contain water as one component. There are also 4 ternary negative azeotropes. As far as the quaternary systems are concerned, 8 systems form positive azeotropes and one, a positive-negative.

In the last decade, studies of vapor-liquid equilibria were developed rapidly, mostly at a constant temperature, with the aim of correlating and predicting the equilibrium parameters. Unfortunately, a large number of investigators did not report whether an azeotrope or a zeotrope is formed in a particular system. In view of the fact that the number of the systems examined in that period of time is small, compared to that of the systems examined previously, it becomes clear that the current situation regarding the commonness of the phenomenon of azeotropy remains essentially unchanged.

From the existing experimental work one might conclude that the frequency of occurence of azeotropes diminishes as the number of components in the system increases, and that multicomponent azeotropes should be expected to be very rare. Such a point of view, however, would be wrong. Although the conditions of the formation of multicomponent azeotropes are complex, they are not difficult to fulfil in practice.[22-24]

THE AZEOTROPIC RANGE

The idea of relating a certain "azeotropic ability" to the chemical character of a substance is already apparent in the first works of Lecat,[19] where he arranged the experimental data on azeotropes according to the chemical character of the components.

The term "azeotropic range" is due to Swietoslawski.[25] There is also another closely related term known, namely, the "relative azeotropic effect",[26] but its use is very limited. Malesinski[27] developed the concept of Swietoslawski's azeotropic range on a general assumption that the components of the system form a regular solution.

The symmetrical azeotropic range, Z, is given by the formula

$$(\delta_1^{1/2} + \delta_2^{1/2}) = Z_{12} = 0.5Z \tag{1}$$

where z_{12} is the half-value of the symmetrical azeotropic range and the quantities δ_1 and δ_2, known otherwise as azeotropic deviations, for positive azeotropes are defined by $\delta_1 = T_1 - T^{A_1}$ and $\delta_2 = T_2 - T^{A_2}$; T_1, T_2 are the boiling temperatures of the pure components and T^{A_1} is the boiling temperature of the azeotrope. For negative azeotropes we have $\delta_1 = T^{A_1} - T_1$ and $\delta_2 = T^{A_1} - T_2$. The symmetrical azeotropic range means that both parts of the range (the upper and the lower part) are equal.

Equation 1 which is also valid at high pressures[28] enables us to compute z_{12} of any binary azeotropic system, provided that the boiing temperature of the pure components and of the azeotrope are known. It should be remembered that by convention the z_{12} values are positive for positive azeotropic systems, and negative for negative ones.

The knowledge of the azeotropic ranges usually leads to a better understanding of the distillation course of complex azeotropic mixtures, e.g. high- and low-temperature coal tars, petroleum and synthetic gasoline. In addition, the azeotropic ranges appear in the equations for calculating the boiling temperatures and compositions of ternary homoazeotropes. These equations were found useful in predicting the existence of a large number of ternary azeotropes.[24]

THE PREDICTION OF AZEOTROPIC DATA IN BINARY SYSTEMS

Empirical Correlations

Lecat[19] first observed that the composition of a binary azeotrope is related to the difference between the boiling temperatures of the components. He used a power series to relate the above quantities for the systems formed by a common substance with members of a homologous series.

His relation may be written in the form

$$x_1 = A_0 + A_1 \Delta + A_2 \Delta^2 + A_3 \Delta^3 + \cdots \cdots \qquad (2)$$

where x_1 is the weight fraction of component 1 in the azeotrope and Δ is the absolute difference between the boiling temperatures of the components.

Lecat[29] proposed also a relation between Δ and the azeotropic deviation, δ, i.e. the absolute difference between the boiling temperatures of the azeotrope and that of the more volatile component, for positive azeotropes, and the less volatile component, for negative azeotropes. The Δ—δ relation, otherwise known as Lecat's rule is

$$\delta = C_0 + C_1 \Delta + C_2 \Delta^2 + C_3 \Delta^3 + \cdots \cdots \qquad (3)$$

For most cases the terms with Δ higher than Δ^2 may be neglected to give

$$\delta = C_0 + C_1 \Delta + C_2 \Delta^2 \qquad (4)$$

The existence of an approximate relation of this kind was expected for a series of binary regular solutions.[30] For instance, the constants of Equation 4 for the azeotropes of ethanol with aliphatic halogen derivatives are $C_o = 12$, $C_1 = -0.5$ and $C_2 = 0.00526$. The general agreement with Equation 4 for that series is satisfactory, despite that these alcohol solutions show large deviations from regularity.

In the 1940s several graphical correlations of empirical nature for the composition of azeotropes within a series of organic compounds were reported, notably by Mair et al.,[31] Horsley,[32] Meissner and Greenfield,[33] Skolnik,[26] and by Seymour.[34] A decade later Johnson and Madonis[35] described another correlation that included a number of other series of compounds.

The empirical treatment suffered mainly due to many variations in the form of the equations and the number of constants required for their evaluation.

A further attempt in improving the situation in this field was made by Seymour et al.[36] They proposed a correlation between x and Δt in the form of a master equation to correlate, among other things, the data from other series already reported. This equation is

$$\log(10\ \frac{x_1}{x_2} = mf(\tau)\ \Delta t + b \qquad (5)$$

where x_1 and x_2 are the mole fractions of component 1 and 2 in the azeotrope, respectively, Δt is the difference between the boiling temperatures of the two components, m and b are constants and $f(\tau) = \tau_1/\tau_2$. The quantity τ is defined as the ratio of the boiling temperatures (°K) of a compound and a hypothetical n-paraffin of the same molecular weight. The differences between the calculated and observed azeotropic compositions are reported for 15 series of organic compounds.

An approximate linear relation between the logarithm of the mole fraction, $\log x_i$, of the main and the secondary azeotropic agent, and the average condensation temperature, and was found for several ternary homo- and heteropolyazeotropic mixtures.[37] Empirical correlations for the azeotropic composition apply to both homo- and heteroazeotropes.

Composition of a Binary Homoazeotrope

The theory of regular solutions offers methods for calculating the composition of a binary (positive or negative) azeotrope from certain properties of the pure components and of the azeotrope.

Generally speaking there are three known methods serving this purpose. Two of them involve the boiling temperatures of the pure components, T_1, T_2, and that of the azeotrope, T^{Az}, and were developed by Prigogine[30] and Malesinski.[23] The third method is based on the activity coefficients of the components at the azeotropic point, and is due to Kireev.[38]

The Prigogine equation for equal molar vaporization entropies of the components, $\Delta S_1^o = \Delta S_2^o$, may be written in the form

$$x_2 = \alpha(1 + \alpha)^{-1} \qquad (6)$$

were α is the square root of the ratio of azeotropic deviations, i.e., $(\delta_1/\delta_2)^{1/2}$, and x_2 is the mole fraction of components 2 in the azeotrope.

When $\Delta S_1^o \neq \Delta S_2^o$, Equation 6 takes the form

$$x_2 = \alpha'(1 + \alpha')^{-1} \qquad (7)$$

In this case $\alpha' = c\alpha$ and $c = (\Delta S_1^o/\Delta S_2^o)^{1/2}$.

The Malesinski equation for the components having equal molar vaporization entropies is given by

$$x_2 = 0.5 + \frac{T_1 - T_2}{2z_{12}} \tag{8}$$

where z_{12} is the half-value of the symmetrical azeotropic range. For systems with unequal molar vaporization entropies, Malesinski's equation requires evaluation of z_u (the upper part of the azeotropic range), the quantity which is not easily available.

Recently, Equations 6, 7, and 8 were evaluated with regard to their usefulness for the calculation of the composition of binary azeotropes.[39]

The simplest and most suitable form of Kireev's equation for the case in which $\Delta S_1^0 = \Delta S_2^0$, is the expression

$$x_2 = (1 + b)^{-1} \tag{9}$$

where b is the square root of the ratio of the logarithms of the activity coefficients of the components γ_2 and γ_1, i.e. $(\ln \gamma_2 / \ln \gamma_2)^{1/2}$. At the azeotropic point the composition of the liquid and the vapor are equal, $x_2 = y_2$, and in the case of an ideal vapor phase the expressions for γ_1 and γ_2 are given by

$$\gamma_1 = P/p_1^0 \quad \text{and} \quad \alpha_2 = P/p_2^0$$

In these relations p_1^0 and p_2^0 are the vapor pressures of the pure components at the boiling temperature of the aceotrope, and P is the total pressure of the mixture at equilibrium.

Equation 9 for $\Delta S_1^0 \neq \Delta S_2^0$ becomes

$$x_2 = (1 + b')^{-1} \tag{10}$$

where $b' = cb$ and $c = (\Delta S_2^0 / \Delta S_1^0)^{1/2}$.

The possible error in the composition of the binary azeotrope under isobaric conditions is due not only to the deviations of the system from regularity but also to the change of the regular solutions constant, A_{12}, with temperature, and to the differences in the vaporization entropies of the pure components. One of the causes of the deviations of the system from regularity is the nonideality of the vapor phase.

The effect of the nonideality of the vapor phase and the differences in the vaporization entropies of the components on the azeotropic composition were studied by Kurtyka and Kurtyka[40] on the systems of acetic acid with n-paraffins. Acetic acid is associated (dimerized) in the vapor phase and its vaporization entropy is 14.85 cal/g-mole, while those of the hydrocarbons are ~20 cal/g-mole.

To get a general idea about those effects on the azeotropic composition, the results obtained for some systems of acetic acid with n-paraffins are reproduced in part from this paper. The system, $\Delta x_2 = x_{2(calcd.)} - x_{2(obsd.)}$, with x_2 computed by Equation 9; Δx_2, with x_2 computed with the corrections for the dimerization of the acid in the vapor phase, and Δx_2, with x_2 computed by Equation 10 are: acetic acid — n-heptane, 12.3, 1.8, 9.3; acetic acid — n-octane, 12.8, 3.8, 9.2, and acetic acid — n-nonane, 9.8 mole %, 4.3 mole %, and 7.0 mole %, respectively.

These results show that the Δx_2 values are reduced to a reasonable magnitude when the corrections for the dimerization of the acid are taken into account. This simply means that the dimerization of acetic acid in the vapor phase is the dominant factor contributing to the large differences in Δx_2. The effect of the differences in the vaporization entropies of the components on the azeotropic composition as exemplified by the systems containing acetic acid and n-paraffins is small compared to the deviations from regularity caused, among other things, by the dimerization of acetic acid in the vapor phase.

Equation 9 has a built-in unfavorable factor because it involves the vapor pressure, p_i^0, of each pure component at the boiling temperature of the azeotrope. p_i^0 values are almost exclusively computed by means of the Antoine equation. Therefore, the accuracy of the computed p_i^0 is related to the constants of that equation. However, the effect of the differences in p_i^0 values on the azeotropic composition is usually small, except in cases where the boiling temperature of the azeotrope is close to the boiling temperature of one component of the system.

The Azeotropic Data of a Binary Heteroazeotrope

Many systems of two or more components exhibit a limited solubility and most of them form positive heteroazeotropes. In binary systems it is easy to predict from the data on critical solution temperatures (CST) whether an azeotrope that occurs at a certain temperature is a homoazeotrope or heteroazeotrope.[41]

To describe the vapor-liquid equilibrium in a heterogeneous system, it is necessary to introduce certain simplifying assumptions.

The simplest case is obtained upon the assumption that liquids are completely immiscible in each other and that the vapor phase is ideal.

Then the composition of a binary heteroazeotrope, y_1, is given by the formula

$$x_1 = y_1 = \frac{p_1^0 (T^{Az})}{p_1^0 (T^{Az}) + p_2^0 (T^{Az})} \tag{11}$$

where p_1^0 and p_2^0 are the vapor pressures of pure components 1 and 2, respectively, expressed as functions of the boiling temperature of the azeotrope, T^{Az}, or simply, at the boiling temperature of the azeotrope.

And for the boiling temperature of the azeotrope, T^{Az} we have the relation

$$P = p_1^0 (T^{Az}) + p_2^0 (T^{Az}) \qquad (12)$$

where P is the total pressure at equilibrium.

In the cases in which the condition of complete immiscibility is not satisfied, we introduce the correction factors in terms of the activities of the components, α_1 and α_2 in the liquid phase.

Accordingly, the expression for the composition of an azeotrope takes the form

$$y_1 = \frac{p_1^0 (T^{Az}) \alpha_1}{p_1^0 (T^{Az}) \alpha_1 + p_2^0 (T^{Az}) \alpha_2} \qquad (13)$$

And the boiling temperature of the azeotrope may be calculated from

$$P = p_1^0 (T^{Az}) \alpha_1 + p_2^0 (T^{Az}) \alpha_2 \qquad (14)$$

There is also another possibility, namely to assume that the activities of both components in the liquid phase are equal, i.e., $\alpha_1 = \alpha_2 = \alpha$. The systems behavior of which is well described by a common activity, are those of nitromethane with n-paraffins.[42] The computation of the activities, α_1 and α_2 of the components is a straightforward procedure involving the solution of Equations 13 and 14.

Equations 11 and 12 were found to be satisfied for the systems of aromatic and n-paraffin hydrocarbons with water. Examples of the systems in which considerable miscibility of the components occurs are: n-butanol-water, aniline-water, and acetonitrile-n-paraffins. For instance, the solubility of aniline in water at 90°C is 6.4 g/100 ml.

For calculating p_1^0 and p_2^0 at the boiling temperatures of the respective azeotropes, the Antoine equation is recommended.[43] The constants of the Antoine equation, viz. $\log p^0 = A - B/(C + t)$, where p^0 is expressed in millimeters Hg and t is in this case the boiling temperature of the azeotrope (°C), are compiled for quite a large number of organic compounds.[44,45]

For example, the activities of the components, α_1 and α_2 in the systems aniline-water[2] and n-butanol-water were found to be: 0.315, 1.00, and 0.649 and 0.9824, respectively.

The relations, thus described, can be easily extended to ternary and multicomponent heteroazeotropes.

THE PREDICTION OF AZEOTROPIC DATA IN TERNARY SYSTEMS

To separate nonideal liquid mixtures by fractional distillation, it is important to establish, among other things, whether the mixtures to be separated form azeotropes. In view that the experimental methods for determining the azeotropic composition are rather difficult and time-consuming operations, expecially in the case of azeotropes containing three and more components, several computational methods were developed to predict the composition of ternary and multicomponent azeotropes.

In general, two approaches in this area may be distinguished. The first one, originated by Haase[46,47] and developed by Malesinski,[48] is based on the theory of regular solutions and is restricted to homoazeotropes. The Malesinski method makes it possible to predict the appearance of ternary azeotropes of various types and to calculate their composition and boiling temperatures. The second approach, which is not limited only to the azeotropic points, is based on the use of various equations of empirical and semiempirical nautre, that relate the liquid-phase activity coefficients to the composition of the liquid phase.

The Azeotropic Data of a Ternary Homoazeotrope from the Binary Azeotropic Data

The composition of a ternary homoazeotrope (1,2,3) under isobaric conditions, when the vaporization entropies of the components are equal, is related to the two pairs of binary azeotropes by the equations.

Pairs (1,2) and (2,3):

$$x_1 = \frac{x_1^{(1,2)} + ax_3^{(2,3)}}{1 - ab} \qquad (15)$$

$$x_3 = \frac{x_3^{(2,3)} + bx_1^{(1,2)}}{1 - ab} \qquad (16)$$

where

$$a = \frac{z_{13} - z_{23} - z_{12}}{2z_{12}} \text{ and } b = \frac{z_{13} - z_{23} - z_{12}}{2z_{23}}$$

$x_1^{(1,2)}$ and $x_3^{(2,3)}$ are the mole fractions of components 1 and 3 in the binary azeotropes (1,2) and (2,3), respectively; x_1 and x_3 are the mole fractions of the above components in the ternary azeotrope and z_{12}, z_{13}, and z_{23} are the half-values of the symmetrical azeotropic range.

When the vaporization entropies of the components are not equal, the regular solution constants, A_{12}, A_{13}, and A_{23} take the place of z_{12}, z_{13}, and z_{23}, respectively.

Pairs (1,2), and (2,3):

$$x_1 = \frac{x_1^{(1, 3)} + cx_2^{(2, 3)}}{1 - cd} \tag{17}$$

$$x_2 = \frac{x_2^{(2, 3)} + dx_1^{(1, 3)}}{1 - cd} \tag{18}$$

where

$$c = \frac{z_{12} - z_{23} - z_{13}}{2z_{13}} \quad \text{and} \quad d = \frac{z_{12} - z_{23} - z_{13}}{2z_{23}}$$

Pairs (1,2) and (1,3):

$$x_2 = \frac{x_2^{(1, 2)} + ex_3^{(1, 3)}}{1 - ef} \tag{19}$$

$$x_3 = \frac{x_3^{(1, 3)} + fx_2^{(1, 2)}}{1 - ef} \tag{20}$$

where

$$e = \frac{z_{23} - z_{13} - z_{12}}{2z_{12}} \quad \text{and} \quad f = \frac{z_{23} - z_{13} - z_{12}}{2z_{13}}$$

The values of z_{12}, z_{13}, and z_{23} should be computed from Equation 1.

The ternary system is zeotropic if, for example, the composition of one component of the system is zero or takes a negative value.

In the case when one binary system constituting the ternary system is zeotropic, the respective z_{ij} value may be estimated from that of any close member of a homologous series, its isomers or closely related substances.

The sources of error in the calculated composition of the ternary azeotrope are similar to those of the binary azeotrope.

The boiling temperature of a ternary homoazeotrope, T, with component 2 as the reference component, can be computed from the equation

$$T = T_2 - \frac{\delta_2^{(1, 2)} + \left(\frac{\delta_2^{(1, 2)}}{z_{12}} \cdot \frac{\delta_2^{(2, 3)}}{z_{23}}\right)^{1/2} (z_{13} - z_{23} - z_{12}) + \delta_2^{(2, 3)}}{1 - \frac{(z_{13} - z_{23} - z_{12})^2}{4z_{12}z_{23}}} \tag{21}$$

where $\delta_2^{(1, 2)}$ and $\delta_2^{(2, 3)}$ are the azeotropic depressions or elevations in the azeotropes (1,2) and (2,3) in relation to component 2, e.g. $\delta_2^{(1, 2)} = T_2 - T^{(1, 2)}$: $T^{(1, 2)}$ is the boiling temperature of the azeotrope (1,2).

By convention, the $\delta_i^{(i,j)}$ and z_{ij} are positive for a positive azeotrope and negative for a negative one.

By interchanging the components, the boiling temperature of the ternary homoazeotrope may be computed from the two remaining sets of the pairs of the components.

For mixtures which exactly fulfil the requirements for regular solutions, the result is independent of the choice of the reference component. If there are deviations from regularity, then the values of z_{ij} or $\delta_i^{(i,j)}$ depend on which experimentally determined quantity was used in the calculations.

For positive azeotropes Equations 15 to 21 usually give good results for the calculated azeotropic data.[23] In the case of positive-negative (saddle) azeotropes the agreement between the calculated and observed azeotropic data is less satisfactory. But this is understandable in view that these systems are complex mixtures, which often contain polar and associated components, e.g. alcohols and low-molecular fatty acids.

For the series of ternary saddle systems acetic acid-pyridine(2-picoline)-n-paraffins, the results obtained for the calculated azeotropic compositions improved considerably, when the corrections for the association (dimerization) of acetic acid in the vapor phase were taken into account.[49]

Equations 15 to 21 made it possible to predict the existence of a large number of ternary saddle (positive-negative) azeotropes that may appear in the course of fractional distillation of certain fractions of coal tar.[24]

It has been found, for instance, that ternary saddle azeotropes are formed in the series of ternary systems aniline-phenol-n-paraffins (ranging from nonane to tetradecane). Only one saddle azeotrope of this type was examined in the system aniline(1)-phenol(2)-tridecane(3).[50] For this system the differences between the calculated and observed azeotropic compositions, Δx_1, Δx_2, and Δx_3, and that of the boiling temperatures of the azeotrope, ΔT, are: 1.0, −2.1, and 1.0 mole %, and −0.11°C, respectively.

In general, good results for ternary homoazeotropes are obtained for systems which show moderate deviations from regularity, any of the binary azeotropes is close to what is called a tangent azeotrope,[23] and the differences between the boiling temperature of the ternary azeotrope and those of the components are not large.[24]

It should be also mentioned that two empirical correlations for the azeotropic composition in the series of ternary saddle azeotropes were proposed by Zeiborak.[51]

Composition of a Ternary Homoazeotrope from Vapor-Liquid Equilibrium Data of the Binary Systems

A general method for predicting the vapor-liquid equilibrium data in ternary and multicomponent systems can be restricted to the azeotropic point by the use of the relative volatility, α_{ij}, defined as

$$\alpha_{ij} = \frac{y_i x_j}{x_i y_j} = \frac{p_i^0 \gamma_i}{p_j^0 \gamma_j} = 1 \tag{22}$$

In relation (Equation 22) p_i^0 and p_j^0 are the vapor pressures of components i and j at the boiling temperature of the azeotrope, and γ_i and γ_j are the liquid-phase activity coefficients of these components.

The method involves minimization of the function, f, which for a ternary azeotrope becomes

$$f = |\alpha_{13} - 1| + |\alpha_{23} - 1| \tag{23}$$

α_{13} and α_{23} are the relative volatilities of the pairs of components (1,3) and (2,3), respectively.

The value of the function, f, sufficiently close to zero, corresponds to the azeotropic composition.

This method was used by Aristovicz and Stepanova[52] for the calculation of the azeotropic composition in 19 ternary systems and 1 quaternary system. The results obtained were good. In this procedure the liquid-phase activity coefficients were correlated by the Wilson equation.[53,54]

For ternary and multicomponent heteroazeotropes the above procedure remains essentially unchanged but requires the use of an equation that is applicable to partially miscible systems, e.g. the NRTL (Non-Random, Two-Liquid) equation.[55]

The Azeotropic Data of a Ternary Heteroazeotrope

On the basis of the arguments similar to those described previously, which are applicable to heteroazeotropic systems of any number of components, we can obtain the expressions for the composition and the boiling temperature of a ternary heteroazeotrope.

For the case in which the components are immiscible in each other, the expressions for y_1, y_2, and T^{Az} of a ternary heteroazeotrope are

$$y_1 = \frac{p_1^0}{p_1^0 + p_2^0 + p_3^0} \tag{24}$$

$$y_2 = \frac{p_2^0}{p_1^0 + p_2^0 + p_3^0} \tag{25}$$

and

$$P = p_1^0 + p_2^0 + p_3^0 \tag{26}$$

where p_1^0, p_2^0 and p_3^0 are the vapor pressures of the pure components at the boiling temperature of the azeotrope, P is the total pressure of the mixture at equilibrium, and y_1 and y_2 are the mole fractions of component 1 and 2 in the vapor of the azeotrope, respectively.

Other cases, in which the condition of complete immiscibility of the components is not fulfilled, may be described by introducing the activities of the components, α_1, α_2, and α_3.

The procedure, which is straightforward, and involves the multiplication of each p_i^0 by α_i in expressions (24-26), will not be reproduced here.

The activities, α_1, α_2, and α_3 are related to P, p_1^0, p_2^0, and p_3^0 by

$$\alpha_1 = \frac{P - p_2^0 \alpha_2 - p_3^0 \alpha_3}{p_1^0} \tag{27}$$

$$\alpha_2 = \frac{P y_2}{p_2^0} \tag{28}$$

and

$$\alpha_3 = \frac{P y_3}{p_3^0} \tag{29}$$

Only one series of ternary heteroazeotropes with three liquid phases has been investigated to date.[6] The heteroazeotropes occur in the systems nitromethane-water-n-paraffins. For this series satisfactory agreement was obtained between the calculated (using Equations 24 and 25) and observed azeotropic compositions, but the calculated boiling temperatures of the azeotropes by Equation 26 were found to be much lower than those observed. Low calculated boiling temperatures are due to the relatively high miscibility of nitromethane in water at the respective boiling temperatures of the azeotropes. The difference is lower for the systems with higher-boiling hydrocarbons and tends to the difference between the calculated and observed boiling temperatures for the nitromethane-water system.

It is interesting to note that in some cases the behavior of a series of ternary heteroazeotropes, in which two components remain unchanged for the series, may be described by the activities α_1, α_2, and α_3, which are common for each component within the whole series. This case is exemplified by the series of ternary heteroazeotropes water-pyridine-n-paraffins.[56]

Studies of ternary heteroazeotropes based on the theory of regular solutions were made by Malesinska and Malesinski[57] and by Stecki.[58]

All the examined heteroazeotropes were found to be positive. The existence of negative heteroazeotropes is rather doubtful.

REFERENCES

1. Dalton, J., *Mem. Manchester Phil. Soc.*, 5, 585, 1802; *Ann. Phil.*, 9, 186, 1817.
2. Wade, J. and Merriman, R. W., *J. Chem. Soc. Trans.*, 99, 997, 1911.
3. Swietoslawski, W., *Ebulliometric Measurements*, Reinhold, New York, 1945.
4. Lecat, M., *Compt. Rend.*, 183, 880, 1926.
5. Swietoslawski, W., *Rocz. Chem.* 26, 632, 1952; *Bull. Acad. Polon. Sci. Cl. III*, 1, 63, 1953.
6. Malesinska, B. and Malesinski, W., *Bull. Acad. Polon. Sci. Ser. Sci. Chim.*, 11, 475, 1963.
7. Ewell, R. L. and Welch, L. M., *Ind. Eng. Chem.*, 37, 1244, 1945.
8. Lang, H., *Z. Physik. Chem.*, 196, 278, 1950.
9. Swietoslawski, W. and Trabczynski, W., *Bull. Acad. Polon. Sci. Cl. III*, 3, 333, 1955.
10. Galska-Krajewska, A., *Bull. Acad. Polon. Sci. Ser. Sci. Chim.*, 10, 45, 51, 1962.
11. Zieborak, K., *Bull. Acad. Polon. Sci. Cl. III*, 3, 53, 1955.
12. Kurtyka, Z. M., *J. Chem. Eng. Data*, 16, 310, 1971.
13. Kurtyka, Z., *Bull. Acad. Polon. Sci. Ser. Sci. Chim.*, 9, 741, 1961.
14. Zieborak, K. and Wyrzykowska-Stankiewicz, D., *Bull. Acad. Polon. Sci. Ser. Sci. Chim.*, 8, 137, 1960.
15. Orszagh, A., Lelakowska, J., and Beldowicz, M., *Bull. Acad. Polon. Sci. Ser. Sci. Chim. Geol. et Geogr.*, 6, 419, 1958.
16. Orszagh, A., Lelakowska, J., and Radecki, A., *Bull. Acad. Polon. Sci. Ser. Sci. Chim. Geol. et Geogr.*, 6, 605, 1958.
17. Ostwald, W., *Lehrbuch der Allgemeinen Chimie*, Vol. 2, Engelman, Leipzig, 1899.
18. Ryland, G., *Am. Chem. J.*, 22, 384, 1899; *Chem. News*, 81, 15, 42, 50, 1900.
19. Lecat, M., *L'Azeotropisme*, Lamartin, Bruxelles, 1918.
20. Lecat, M., *Tables Azeotropiques*, Vol. 1, L'Auteur, Bruxelles, 1949.
21. Horsley, L. H., *Azeotropic Data-III*, American Chemical Society, Washington, D.C., 1973.
22. Swietoslawski, W., *Azeotropy and Polyazeotrophy*, Pergamon Press, Oxford, London, 1963.
23. Malesinski, W., *Azeotropy and Other Theoretical Problems of Vapour-Liquid Equilibrium*, Wiley-Interscience, New York, 1965.
24. Kurtyka, Z. M., *Azeotropy and Its Applications*, to be published.
25. Swietoslawski, W., *Bull. Acad. Sci. Polon. Ser. A*, 19, 29, 1950, *Przem. Chem.*, 7, 363, 1951.
26. Skolnik, H., *Ind. Eng. Chem.*, 40, 442, 1948.
27. Malesinski, W., *Bull. Acad. Polon. Sci. Cl. III*, 3, 601, 1955; 4, 295, 1956.
28. Zawisza, A., *Bull. Acad. Polon. Sci. Ser. Sci. Chim.*, 9, 141, 1961.
29. Lecat, M., *Azeotropisme et Distillation, Traite de Chimie Organique*, Vol. I, V. Grignard, Ed., Mason et Cie., Paris, 1935.
30. Prigogine, I., and Defay, R., *Chemical Thermodynamics*, translated by D. H. Everett, Longmans, Green, London, 1954.
31. Mair, B. J., Glasgow, A. R., and Rossini, F. D., *J. Res. Bur. Std.*, 27, 39, 1941.
32. Horsley, L. H., *Anal. Chem.*, 19, 508, 1947.
33. Meissner, H. P. and Greenfield, S. H., *Ind. Eng. Chem.*, 40, 438, 1948.
34. Seymour, K. M., Abstracts 50-I, 110th National Meeting of the American Chemical Sciety, Chicago, September 1946.
35. Johnson, A. I. and Madonis, J. A., *Can. J. Chem. Eng.*, 37, 71, 1959.
36. Seymour, K. M., Carmichael, R. H., Carter, J., Ely, J., Isaacs, E., King, J., Taylor, R., and Northern, T., *Ind. Eng. Chem. Fundam.*, 16, 200, 1977.
37. Orszagh, A., *Rocz. Chem.*, 29, 623, 636, 1955.
38. Kireev, V. A. *Acta Physicochim. URSS*, 14, 371, 1941.
39. Kurtyka, Z. M. and Kurtyka, A., *Ind. Eng. Chem. Fundam.*, 19, 225, 1980.
40. Kurtyka, Z. M. and Kurtyka, Z., *Ind. Eng. Chem. Fundam.*, in press.
41. Francis, A. W., *Critical Solution Temperatures*, American Chemical Society, Washington, D.C. 1961.

42. **Malesinska, B . and Malesinski, W.,** *Bull. Acad. Polon. Sci. Ser. Sci. Chim.,* 11, 469, 1963.
43. **Antoine, C.,** *Compt. Rend.,* 107, 681, 836, 1143, 1888.
44. **Hala, E., Wichterle, I., Polak J., and Boublik, T.,** *Vapour-Liquid Equilibrium Data at Normal Pressures,* Pergamon Press, Oxford, London, 1968.
45. **Hirata, M., Ohe, S., and Nagahama, K.,** *Computer-Aided Data Book of Vapour-Liquid Equilibria,* Kodansha-Elsevier, Tokyo, Amsterdam, 1975.
46. **Haase, R.,** *Z. Physik. Chem.,* 195, 362, 1950.
47. **Haase, R.,** *Termodynamik des Mischphasen,* Springer-Verlag, Berlin, 1956.
48. **Malesinski, W.,** *Bull. Acad. Polon. Sci. Cl. III,* 4, 701, 709, 1956; 5, 177, 183, 1957.
49. **Zeiborak, K. and Wyrzykowska-Stankiewicz, D.,** *Bull. Acad. Polon. Sci. Ser. Sci. Chim. Geol. et Geogr.,* 6, 755, 1958.
50. **Stadnicki, J. S.,** *Bull. Acad. Polon. Sci. Ser. Sci. Chim.,* 10, 357, 1962.
51. **Zieborak, K.,** *Bull. Acad. Polon. Sci. Cl. III,* 3, 531, 1955.
52. **Aristovicz, V. Y. and Stepanova, E. I.,** *Zhur. Prikl. Khim.,* 43, 2192, 1970.
53. **Wilson, G. M.,** *J. Am. Chem. Soc.,* 86, 127, 1964.
54. **Prausnitz, J. M.,** *Molecular Thermodynamics of Fluid-Phase Equilibria,* Prentice Hall, Englewood Cliffs, New Jersey, 1969.
55. **Renon, H. and Prausnitz, J. M.,** *A. I. Ch. E. Journal,* 14, 135, 1968.
56. **Trabczynski, W.,** *Bull. Acad. Polon. Sci. Ser. Sci. Chim. Geol. et Geogr.,* 6, 269, 1958.
57. **Malesinska, B. and Malesinski, W.,** *Bull. Acad. Polon. Sci. Ser. Sci. Chim.,* 12, 861, 867, 1964.
58. **Stecki, J.,** *Bull. Acad. Polon. Sci. Cl. III,* 5, 421, 1957; *Ser. Sci. Chim. Geol. et Geogr.,* 6, 47, 1958.

TABLES OF AZEOTROPES AND ZEOTROPES

In Tables 1, 2, and 3 the different types of azeotropes are identified as:

1. Homoazeotrope, positive; no marking
2. Homoazeotrope, negative; N
3. Homoazeotrope, saddle or positive-negative; S
4. Heteroazeotrope, positive, with two phases; H
5. Heteroazeotrope, positive, with three phases; H-3

Throughout Tables 1, 2, and 3 the azeotropic composition is expressed as weight percent (wt %). In Table 1 only the weight percent of component 2, x_2, the variable component, is listed. However, in Tables 2, and 3 the weight percents of all of the components are listed.

Compounds are listed in the following tables according to the empirical formula convention employed by Chemical Abstracts. As further assistance in locating particular systems the following index has been arranged in alphabetical order with respect to component X_1 of each system. The entry number for the particular system is in the second column of the list.

Table 1
BINARY SYSTEMS

Component X_1	Entry No.	Component X_1	Entry No.
Acetal	1364	Benzyl phenyl ether	1743
Acetaldehyde	377	Borneol	1717
Acetic acid	383	Boron fluoride	2
Acetone	544	Boron hydride	10
Acetonitrile	369	Bromoacetic acid	349
Acetophenone	1529	Bromodichloromethane	217
Acrylonitrile	525	Bromoform	224
Allyl alcohol	554	Bromomethane	271
Aluminum chloride	1	1-Butanethiol	1032
n-Amyl alcohol	1143	1-Butanol	908
p-tert-Amyl alcohol	1729	2-Butanol	939
Aniline	1255	2-Butanone	770
o-Anisidine	1511	2-Butoxyethanol	1368
Benzaldehyde	1408	Butyl acetate	1335
Benzene	1180	Butyl alcohol	908
Benzoic acid	1415	Butyl formate	1100
Benzonitrile	1400	Butyl nitrite	893
Benzyl alcohol	1448	Butyraldehyde	782

Table 1 (Continued)
BINARY SYSTEMS

Component X₁	Entry No.	Component X₁	Entry No.
Butyronotile	764	Hydrogen chloride	14
Camphene	1715	Hydrogen cyanide	11
Capric acid	1722	Hydrogen fluoride	15
Caproic acid	1338	Indole	1527
Capronitrile	1325	Iodobenzene	1168
Caprylic acid	1622	Iodoethane	420
Carbon disulfide	204	Iodomethane	272
Carbon tetrachloride	185	Isoamyl alcohol	1148
Carvacol	1696	Isoamyl benzoate	1735
Carvone	1703	Isoamyl formate	1345
Chloroacetic acid	354	Isoamyl oxalate	1739
Chloroform	232	Isobutyl nitrate	902
m-Cresol	1473	Isobutyronitrile	767
o-Cresol	1462	Isobutryic acid	829
p-Cresol	1485	Isopropyl alcohol	663
Cumene	1655	Isopropyl lactate	1354
Cyclohexanol	1327	Isopropyl methyl sulfide	1038
Cyclohexanone	1297	Isovaleric acid	1105
Diethylene glycol	1008	Levulinic acid	1074
Doxane	811	2,4-Lutidine	1501
Dipropylene glycol	1386	Menthol	301
Enanthic acid	1514	Mesitol	1662
Ethanol	460	Mesitylene	1656
2-Ethoxyethanol	994	Methanol	301
Ethyl		Methyl	
acetate	821	acetate	567
acetoacetate	1301	acetoacetate	1085
alcohol	460	acetophenone, para	1643
aniline	1592	alcohol	301
benzene	1551	aniline	1503
bromoacetate	750	anisole, para	1562
carbamate	631	butyrate	1120
chloroacetate	756	chloroacetate	534
formate	563	disulfide	520
fumarate	1613	formate	414
lactate	1136	fumarate	1284
maleate	1616	lactate	876
methyl sulfide	712	maleate	1290
nitrate	455	1-Methylnaphthalene	1724
oxalate	1319	2-Methylnaphthalene	1728
phenol, para	1554	2-Methyl-2-propanol	949
pyruvate	1071	2-Methylthiophene	1063
salicylate	1651	3-Methylthiophene	1067
succinate	1619	2-Octanol	1630
Ethylene glycol	476	sec-Octyl alcohol	1630
Ethylene sulfide	411	Naphthalene	1679
Ethylenediamine	522	Nitrobenzene	1159
Ethylidine diacetate	1313	Nitroethane	453
Formic acid	242	Nitromethane	276
2-Furaldehyde	1044	1-Nitropropane	654
Glycol diacetate	1324	o-Nitrotoluene	1426
Glycol monoacetate	865	p-Nitrotoluene	1437
Glycerol	687	Pelargonic acid	1669
Guaiacol	1497	2-Pentanone	1096
1-Heptanol	1521	Perfluorobutyric acid	724
n-Heptyl alcohol	1521	Phenethyl alcohol	1568
Hexachloroethane	337	o-Phenetidine	1600
1-Hexanol	1357	p-Phenetidine	1607
n-Hexyl alcohol	1357	Phenol	1190

Table 1 (Continued)
BINARY SYSTEMS

Component X_1	Entry No.	Component X_1	Entry No.
Phenyl acetate	1538	Pyridine	1056
Phenyl benzoate	1741	Pyrogallol	1252
Phenyl ether	1732	Pyrrol	734
o-Phenylenediamine	1278	Pyruvic acid	527
3-Phenylpropanol	1664	Qunialdine	1683
Phosphorus oxychloride	16	Quinoline	1635
Phosphorus trichloride	17	Resorcinol	1238
2-Picoline	1273	Silicon tetrachloride	20
3-Picoline	1276	Tetrachlorothylene	327
Pinacol	1377	Tetrahydrothiophene	888
1-Propanethiol	716	Thioacetic acid	380
1-Propanol	673	Thiophene	727
Propioamide	593	Thymol	1707
Propionic acid	574	Tin chloride	26
Propionitrile	541	Toluene	1442
Propiophenone	1648	o-Toluidine	1505
Propyl		Trichloroacetic acid	346
acetate	1123	Trichloroethylene	342
alcohol	673	Trichlorofluoromethane	173
benzene	1659	Trinitromethane	176
benzoate	1690	Triethylene glycol	1391
formate	858	Valeric acid	1128
isovalerate	1626	Water	28
lactate	1350	2,4-Xylenol	1570
nitrite	661	3,4-Xylenol	1582
propionate	1348	2,4-Xylidine	1595
Pseudocumene	1660		

Table 2
TERNARY SYSTEMS

Component X_1	Entry No.	Component X_1	Entry No.
Acetic acid	108	Hydrogen chloride	5
Acetone	141	Hydrogen cyanide	3
Acetonitrile	106	Hydrogen fluoride	7
Aniline	167	Isobutyl alcohol	161
Argon	1	Isobutyl lactate	176
1-Butanol	158	Isopropyl alcohol	151
2-Butanone	154	Methanol	101
Butyric acid	156	Methyl formate	
Carbon tetrachloride	92	Nitromethane	129
Chlorine trifluoride	4	Phenol	163
Chloroform	93	1-Propanol	152
m-Cresol	175	1-Propanol, 2-methyl	161
m,p-Cresol (mixture)	172	2-Propanol	151
p-Dioxane	157	Propionic acid	145
Ethanol	130	Propyl lactate	171
Ethyl benzene	166, 177	Pyridine	162
Ethylene glycol	135	Silicon tetrafluoride	8
Hydrogen bromide	2	Trichloroethylene	107

Table 3
QUATERNARY AND QUINARY SYSTEMS

Component X_1	Entry No.	Component X_1	Entry No.
Acetic acid	18	Hydrogen cyanide	1
Acetone	20	Isopropyl alcohol	21
Chloroform	16	Water	3

BINARY SYSTEMS

No.	System	B.P. (°C)	Azeotropic data Compn. X_2 (wt. %)	Azeotropic data B.P. (°C)
	Aluminum chloride	183		
1	-tantalum chloride (2)	242	90.4	235
	Boron fluoride	−100		
2	-boron hydride	−92	22.8	−106
3	-acetonitrile, N	81.6	38	101
4	-methyl formate, N	31.9	47	91
5	-methyl ether, N	−21	40	127
6	-ethyl formate, N	54.1	52	102
7	-methyl acetate, N	57.1	52	110
8	-ethyl ether, N	34.5	52	125
9	-ethyl propionate, N	99.15	60	116
	Boron hydride	−92.5		
10	-hydrogen chloride	−85	36	−94
	Hydrogen cyanide	26		
11	-methyl alcohol	64.7	Zeotropic	
12	-methyl formate	31.7	48	24
13	-ethyl nitrite	17.4	85	16.5
	Hydrogen chloride	−85		
14	-methyl ether, N	−22	62	−2
	Hydrogen fluoride	19.4		
15	-ethyl ether, N	34.5	60	74
	Phosphorus oxychloride	107.2		
16	-titanium tetrachloride, N	136.5	53.4	143.2
	Phosphorus trichloride	76		
17	-cyclohexane	80.75	Zeotropic	
18	-2,3-dimethylpentane	80.5	27	74.2
19	-2,2,3-trimethylbutane	80.9	23	74.5
	Silicon tetrachloride	56.9		
20	-carbon tetrachloride	76.75	Zeotropic	
21	-chloroform	61.0	30	55.6
22	-nitromethane	101	6	53.8
23	-acetonitrile	82	9.4	49.0
24	-acrylonitrile	79	11	51.2
25	-propionitrile	97	8	55.6
	Tin chloride	113.85		
26	-toluene	110.7	48	109.15
27	-2,5-dimethylhexane	109.4	60	107.5
	Water	100		
28	-hydrogen chloride, N	−85	20.22	108.58
29	-hydrogen bromide, N	−73	47.5	126
30	-hydrogen iodide, N	−34	57	127
31	-hydrogen fluoride, N	19.4	35.6	111.35
32	-nitric acid, N	86	67.4	120.7
33	-hydrogen peroxide	152.1	Zeotropic	
34	-hydrazine, N	113.8	67.7	120
35	-carbon tetrachloride, H	76.75	95.9	66
36	-carbon disulfide, H	46.25	97.2	42.6
37	-chloroform, H	61	97.2	56.1
38	-formic acid, N	100.75	77.4	107.2
39	-nitromethane, H	101.2	76.4	83.59
40	-tetrachloroethylene, H	121	82.8	88.5
41	-trichloroethylene	86.2	83	73.4
42	-acetonitrile, H	80.1	83.7	76.5
43	-acetic acid	118.1	Zeotropic	
44	-acetamide, H	221.2	Zeotropic	
45	-nitroethane, H	114.07	71.5	87.22
46	-ethyl nitrate	87.68	78	74.35
47	-ethyl alcohol	78.32	96	78.17
48	-methyl sulfate	189.1	27	98.6
49	-acrylonitrile, H	77.2	85.7	70.6
50	-acrolein, H	52.8	97.4	52.4
51	-acetone	56.1	Zeotropic	
52	-allyl alcohol	96.9	72.3	88.9
53	-propionaldehyde, H	47.9	98	47.5
54	-ethyl formate	54.2	95	52.6
55	-propionic acid	141.1	17.7	99.9
56	-trioxane	114.5	70	91.4
57	-l-chloropropane	46.6	97.8	44
58	-isopropyl alcohol	82.3	87.4	80.3
59	-propyl alcohol, 740 mm	97.3	71.7	87
60	-perfluorobutyric acid	122.0	29	97
61	-crotonic acid	189	2.2	99.9
62	-methyl acrylate	80	92.8	71
63	-ethyl chloroacetate	143.5	54.9	95.2
64	-butyronitrile, H	117.6	67.5	88.7
65	-isobutyronitrile, H	103	77	82.5
66	-ethyl vinyl ether, H	35.5	98.5	34.6
67	-butyric acid	163.5	3	99.4
68	-ethyl acetate	77.15	91.53	70.38
69	-isopropyl formate	68.8	97	65.0
70	-propyl formate	80.9	97.7	71.6
71	-methyl lactate	144.8	20	99
72	-butyl alcohol, H	117.4	57.5	92.7
73	-sec-butyl alcohol	99.5	73.2	87.0
74	-pyridine	115.5	58.7	93.6
75	-furfuryl alcohol	169.35	20	98.5
76	-furfurylamine	144	26	99
77	-isoprene	34.1	99.86	32.4
78	-cyclopentanone	130.8	57.6	94.6
79	-allyl acetate	104.1	85.3	83
80	-cyclopentanol	140.85	42	96.25
81	-valeraldehyde, H	103.3	81	83
82	-butyl formate	106.6	85.5	83.8
83	-isopropyl acetate	88.6	89.4	76.6
84	-isovaleric acid	176.5	18.4	99.5
85	-methyl butyrate	102.65	88.5	82.7
86	-methyl isobutyrate	92.3	93.2	77.7
87	-valeric acid	188.5	11	99.8
88	-piperidine	105.8	65	92.8
89	-n-pentane, H	36.1	98.6	34.6
90	-n-amyl alcohol, H	137.8	45.6	95.8
91	-tert-amyl alcohol	102.25	72.5	87.35
92	-2-pentanol	119.3	63.5	91.7
93	-N-methylbutylamine	91.1	85	82.7
94	-chlorobenzene	131.8	71.6	90.2
95	-nitrobenzene, H	210.85	—	98.6
96	-benzene, H	80.1	91.17	69.25
97	-phenol	182	9.2	99.52
98	-aniline, H, 742 mm	184.3	19.2	98.6
99	-2-picoline	129.5	52	93.5
100	-3-picoline	144.1	40	97
101	-4-picoline	144.3	37.2	97.35
102	-2,5-dimethylfuran	93.3	88.3	77.0
103	-cyclohexene, H	82.75	91.07	70.8
104	-ethyl crotonate	137.8	62	93.5
105	-ethylene glycol diacetate	190.8	15.4	99.7
106	-butyl chloroacetate	181.9	24.5	98.12
107	-cyclohexane, H	80.8	91.6	69.5
108	-amyl formate	132	71.6	91.6
109	-butyl acetate	126.2	71.3	90.2
110	-ethyl butyrate	120.1	78.5	87.9
111	-isoamyl formate	124.2	79	90.2
112	-isobutyl acetate	117.2	83.5	87.4
113	-isopropyl propionate	110.3	80.1	85.2
114	-propyl propionate	122.1	77	88.9
115	-paraldehyde, H	124	71.5	90
116	-n-hexane, H	68.7	94.4	61.6
117	-butyl ethyl ether, H	92.2	88.1	76.6
118	-n-hexyl alcohol, H	157.1	32.8	97.8
119	-acetal	103.6	85.5	82.6
120	-pinacol	174.35	Zeotropic	
121	-toluene, H	110.7	86.5	84.1
122	-anisole	153.85	59.5	95.5
123	-benzyl alcohol, H	205.2	9	99.9
124	-guaiacol	205.0	12.5	99.5
125	-2,6-lutidine, H	144.0	48.2	96.02
126	-o-toluidine	199.7	84.6	—
127	-p-toluidine, H	200.4	86.2	—
128	-2-heptanone	149	52	95
129	-3-heptanone	147.6	57.8	94.6
130	-4-heptanone	143.7	59.5	94.3
131	-ethyl valerate	145.45	60	94.5
132	-isoamyl acetate	142	63.7	93.8
133	-isobutyl propionate	136.85	47.8	92.75
134	-n-heptane, H	98.4	87.1	79.2
135	-benzyl formate	202.3	20	99.2
136	-methyl benzoate	199.45	20.8	99.08
137	-phenyl acetate	195.7	24.9	98.9
138	-ethylbenzene, H	136.2	67.0	92.0
139	-m-xylene, H	139	64.2	92
140	-N-ethylaniline	204.8	16.1	99.2
141	-l-octene, H	121.28	71.3	88.0
142	-hexyl acetate	171.0	39	97.4
143	-isoamyl propionate	160.3	51.5	96.55
144	-isobutyl butyrate	156.8	54	96.3
145	-isobutyl isobutyrate	147.3	60.6	95.5
146	-propyl isovalerate	155.8	54.8	96.2
147	-n-octane, H	125.7	75.5	89.6
148	-isooctane, H	99.3	88.9	78.8
149	-butyl ether, H	142.6	67	92.9
150	-n-octyl alcohol, H	195.15	10	99.4
151	-dibutylamine, H	159.6	49.5	97

BINARY SYSTEMS (Continued)

No.	System	B.P. (°C)	Compn. X_2 (wt. %)	B.P. (°C)	No.	System	B.P. (°C)	Compn. X_2 (wt. %)	B.P. (°C)
152	-quinoline, H	237.3	3.4	—	227	-phenol	182.2	Zeotropic	
153	-ethyl benzoate	212.4	16.0	99.4	228	-aniline	184.35	Zeotropic	
154	-cumene, H	152.4	56.2	95	229	-toluene	110.65	Zeotropic	
155	-mesitylene, H	164.6	—	96.5	230	-o-cresol	191.1	Zeotropic	
156	-triallylamine, H	151.1	62	95	231	-α-pinene	155.8	25	146.5
157	-isoamyl butyrate	178.5	36.5	98.05		**Chloroform**	61.2		
158	-isobutyl carbonate	190.3	26	98.6	232	-formic acid	100.75	15	59.15
159	-n-nonane, H	150.8	18	94.8	233	-methyl alcohol	64.7	12.6	53.43
160	-naphthalene, H	218.0	16	98.8	234	-ethyl alcohol	78.3	7	59.35
161	methyl phthalate, H	283.2	2.5	99.95	235	-acetone, N	56.5	21.9	64.4
162	-nicotine, H	—	2.5	99.85	236	-propyl alcohol	97.2	Zeotropic	
163	-camphene, H	159.6	—	96.0		**Chloroform**	61.2		
164	-n-decane, H	173.3	—	97.2	237	-p-dioxane	101	Zeotropic	
165	-n-undecane, H	194.5	4.0	98.85	238	-cyclohexane	80.75	Zeotropic	
166	-o-phenyl phenol	—	1.25	99.95	239	-methylcyclopentane	72.0	20	60.5
167	-phenyl ether, H	259.3	3.25	99.33	240	-n-hexane	68.7	16.5	60.4
168	-ethyl phthalate, H	298.5	2.0	99.98	241	-toluene	110.65	Zeotropic	
169	-isoamyl benzoate, H	262.3	4.4	99.9		**Formic acid**	100.75		
170	-n-dodecane, H	214.5	2	99.45	242	-nitromethane	101.22	54.5	97.07
171	-dihexylamine, H	239.8	7.2	99.8	243	-trichloroethylene	86.95	75	74.1
172	-tributylamine, H	213.9	20.3	99.65	244	-tetrachloroethylene	121.1	50	88.15
	Trichlorofluoromethane	24.9			245	-acetic acid	118.1	Zeotropic	
173	-acetaldehyde	20.2	45	15.6	246	-nitroethane	114.2	Zeotropic	
174	-methyl formate	32	18	20.0	247	-ethyl ether	34.6	Zeotropic	
175	-2-methylbutane	27	8	23.16	248	-ethyl sulfide	92.2	65	82.2
	Trichloronitromethane	111.9			249	-pyridine, N	115.5	38.6	127.43
176	-acetic acid	118.1	19.5	107.65	250	-2-methylbutane	27.95	96	27.2
177	-ethyl alcohol	78.3	66	77.5	251	-n-pentane	36.15	80	34.2
178	-isopropyl alcohol	82.4	65	81.95	252	-bromobenzene	156.1	32	98.1
179	-propyl alcohol	97.2	41.5	94.05	253	-chlorobenzene	131.75	41	93.7
180	-isoamyl alcohol	131.9	7	111.15	254	-fluorobenzene	84.9	73	73.0
181	-n-pentanol	119.0	17	100.0	255	-benzene	80.2	69	71.05
182	-toluene	110.75	Zeotropic		256	-aniline	184.35	Zeotropic	
183	-methylcyclohexane	101.15	73	100.8	257	-2-picoline, N	129	75	158.0
184	-n-heptane	98.4	93	98.32	258	-cyclohexane	80.75	30	70.7
	Carbon tetrachloride	76.71			259	-methylcyclopentane	72.0	71	63.3
185	-carbon disulfide	46.25	Zeotropic		260	-n-hexane	68.95	72	60.6
186	-chloroform	62.1	Zeotropic		261	-propyl sulfide	141.5	17	98.0
187	-formic acid	100.7	18.5	66.65	262	-isopropyl sulfide	120.5	38	93.5
188	-nitromethane	101.2	17	71.3	263	-toluene	110.7	50	85.8
189	-methyl alcohol	64.7	20.56	55.7	264	-o-chlorotoluene	159.3	17	100.2
190	-acetonitrile	81.6	17	65.1	265	-methylcyclohexane	101.1	53.5	80.2
191	-acetic acid	118.1	1.54	76	266	-n-heptane	98.45	43.5	78.2
192	-ethyl alcohol	78.3	15.8	65.04	267	-styrene	145.8	27	97.75
193	-acrylonitrile	77.3	21	66.2	268	-o-xylene	143.6	26	95.5
194	-acetone	56.15	88.5	56.08	269	-m-xylene	139.0	28.2	92.8
195	-propyl alcohol	97.25	7.9	73.4	270	-n-octane	125.8	37	90.5
196	-thiophene	84	Zeotropic			**Bromomethane**	3.65		
197	-butyl nitrite	78.2	30	75.3	271	-methyl alcohol	64.7	0.45	3.55
198	-butyl alcohol	117.75	2.4	76.55		**Iodomethane**	42.5		
199	-ethyl ether	34.6	Zeotropic		272	-methyl alcohol	64.7	4.5	37.8
200	-pyridine	115.5	Zeotropic		273	-ethyl alcohol	78.3	3.2	41.2
201	-benzene	80.1	Zeotropic		274	-acetone	56.15	5	42.4
202	-n-heptane	98.4	Zeotropic		275	-n-hexane	68.85	Zeotropic	
203	-o-xylene	143.6	Zeotropic			**Nitromethane**	101.2		
	Carbon disulfide	46.25			276	-methyl alcohol	64.7	90.9	64.4
204	-chloroform	61.2	Zeotropic		277	-acetic acid	118.1	4	101.2
205	-formic acid	100.75	17	42.55	278	-ethyl alcohol	78.3	71	76.05
206	-nitromethane	101.2	81.4	41.2	279	-propyl alcohol	97.15	51.6	89.09
207	-methyl alcohol	64.7	29	39.8	280	-p-dioxane	101.35	43.5	100.55
208	-acetic acid	118.1	Zeotropic		281	-n-butyl alcohol	117.73	28.6	98.0
209	-propyl nitrite	47.75	38	40.15	282	-n-pentane, H	36.07	99	35
210	-ethyl alcohol	78.3	9	42.6	283	-cyclohexane	80.75	73.5	69.5
211	-acetone	56.15	33	39.25	284	-methylcyclopentane	72.0	77	64.2
212	-propyl alcohol	97.1	5.5	45.65	285	-n-hexane, H	68.74	81.5	61.7
213	-ethyl acetate	76.7	3	46.1	286	-toluene	110.75	45	96.5
214	-n-pentane	36.15	89	35.7	287	-n-heptane, 748 mm Hg, H	98.4	64.4	79.7
215	-n-hexane	68.95	Zeotropic		288	-styrene	145.8	Zeotropic	
216	-toluene	110.7	Zeotropic		289	-o-xylene	144.3	Zeotropic	
	Bromodichloromethane	90.2			290	-n-octane, 748 mm; H	125.75	44.8	90.23
217	-nitromethane	101.2	25	87.3	291	-cumene	152.8	Zeotropic	
218	-methyl alchol	64.7	40	63.8	292	-mesitylene	164.6	Zeotropic	
219	-ethyl alcohol	78.3	28	75.5	293	-n-nonane, 748 mm; H	150.85	28.4	96.14
220	-ethyl acetate	77.1	12	90.55	294	-n-decane, 748 mm; H	174.12	16.1	98.81
221	-benzene	80.2	Zeotropic		295	-n-undecane, 748 mm; H	194.5	9.3	100.01
222	-cyclohexane	80.75	Zeotropic		296	-n-dodecane, 748 mm; H	216.0	4.2	100.60
223	-n-hexane	68.8	Zeotropic			**Methyl nitrate**	64.8		
	Bromoform	149.5			297	-methyl alcohol	64.65	27	52.5
224	-formic acid	100.75	48	97.4	298	-cyclohexane	80.75	23	61.0
225	-acetamide	221.15	Zeotropic		299	-n-hexane	68.8	44	56.0
226	-butyric acid	162.45	6.8	146.8	300	-n-heptane	98.4	Zeotropic	

No.	System	B.P. (°C)	Azeotropic data Compn. X₂ (wt. %)	Azeotropic data B.P. (°C)
	Methyl alcohol	64.7		
301	-trichloroethylene	87	62	59.3
302	-bromoethane	38	94.7	34.9
303	-acetic acid	118.1	Zeotropic	
304	-acetone	56.15	88	55.5
305	-methyl acetate	57.1	81	53.5
306	-thiophene	84	83.6	59.71
307	-methyl acrylate	80	46	62.5
308	-p-dioxane	101.05	Zeotropic	
309	-ethyl sulfide	92.2	38	61.2
310	-pyridine	115.4	Zeotropic	
311	-cyclopentane	49.4	86	38.8
312	-isobutyl formate	97.9	5	64.6
313	-piperidine	106.4	Zeotropic	
314	-n-pentane	36.15	93	30.85
315	-chlorobenzene	132.0	Zeotropic	
316	-fluorobenzene	85.15	68	59.7
317	-benzene	80.1	60.9	57.5
318	-cyclohexane	80.7	63.6	53.9
319	-toluene	110.6	27.5	63.5
320	-methylcyclohexane	100.8	46	59.2
321	-n-heptane, H	98.45	48.5	59.1
322	-o-xylene	143.6	Zeotropic	
323	-n-octane	125.75	32.5	62.75
324	-n-nonane	150.7	16.6	64.1
325	-n-decane	173.8	Zeotropic	
326	-methyl tert-butyl ether	55.06	85.7	51.27
	Tetrachloroethylene	121.1		
327	-acetic acid	118.1	38.5	107.35
328	-acetamide	221.2	2.6	120.45
329	-ethylene glycol	197.4	6	119.1
330	-acetone	56.1	Zeotropic	
331	-propionic acid	140.9	8.5	119.1
332	-propyl alcohol	97.25	48	94.05
333	-n-butyl alcohol	117.7	32	110.0
334	-pyridine	115.4	48.5	112.85
335	-n-amyl alcohol	138.2	15	117.0
336	-toluene	110.75	Zeotropic	
	Hexachloroethane	185		
337	-trichloroacetic acid	196	15	181
338	-phenol	182.2	30	173.7
339	-aniline	184.35	34	176.75
340	-benzyl alcohol	205.15	12	182.0
341	-p-cresol	201.7	10	183.0
	Trichloroethylene	86.9		
342	-acetic acid	118.1	3.8	86.5
343	-benzene	80.2	Zeotropic	
344	-cyclohexane	80.7	83.4	80.5
345	-n-heptane	98.45	Zeotropic	
	Trichloroacetic acid	197.55		
346	-pentachloroethane	161.95	96.5	161.8
347	-naphthalene	218.05	Zeotropic	
348	-butylbenzene	183.1	80	181.3
	Bromoacetic acid	205.1		
349	-o-dichlorobenzene	179.5	84	177.0
350	-o-bromotoluene	181.5	82	179.0
351	-acetophenone	202.0	30	206.5
352	-butylbenzene	183.1	75	179.5
353	-cymene	176.7	85	174.7
	Chloroacetic acid	189.35		
354	-bromobenzene	156.1	89	154.3
355	-phenol	181.5	Zeotropic	
356	-m-bromotoluene	183.8	70	174
357	-p-bromotoluene	185.0	66	174.1
358	-styrene	145.8	86	144.8
359	-o-xylene	144.3	88	143.5
360	-m-xylene	139.2	93	139.05
361	-n-octane	125.75	Zeotropic	
362	-cumene	152.8	79	150.8
363	-mesitylene	164.6	83	162.0
364	-pseudocumene	168.2	66	162.8
365	-naphthalene	218.05	22	187.1
366	-cymene	176.7	58	169.0
367	-n-decane	173.3	58	165.2
368	-1,3,5-triethylbenzene	215.5	25	185.5
	Acetonitrile	81.6		
369	-acetic acid	118.1	Zeotropic	
370	-ethyl alcohol	78.3	56	72.5
371	-pyridine	115.5	Zeotropic	
372	-isoprene	34.1	97.6	33.7
373	-isopropyl acetate	89.5	40.0	79.5
374	-toluene	110.7	20	81.4

No.	System	B.P. (°C)	Azeotropic data Compn. X₂ (wt. %)	Azeotropic data B.P. (°C)
375	-ethylbenzene	136.2	Zeotropic	
376	-n-undecane	195.4	Zeotropic	
	Acetaldehyde	20.4		
377	-acetone	56.15	Zeotropic	
378	-ethyl ether	34.5	23.5	18.9
379	-benzene	80.1	Zeotropic	
	Thioacetic acid	89.5		
380	-benzene	80.15	Zeotropic	
381	-cyclohexane	80.75	Zeotropic	
382	-methylcyclopentane	72.0	Zeotropic	
	Acetic acid	118.1		
383	-nitroethane	114.2	70	112.4
384	-dioxane	101.35	23	119.5
385	-acetone	56.1	Zeotropic	
386	-pyridine, N	115.5	48.9	138.1
387	-2-picoline, N	129.3	59.6	144.12
388	-3-picoline, N	144	69.6	152.5
389	-4-picoline, N	144.3	69.7	154.3
390	-benzene	80.2	98.0	80.05
391	-cyclohexane	80.75	90.4	78.8
392	-n-hexane	68.6	94.0	68.25
393	-isopropyl sulfide	120	52	111.5
394	-toluene	110.7	71.9	100.6
395	-triethylamine, N	89	33	163
396	-2,6-lutidine, N	144.0	77.1	148.1
397	-methylcyclohexane	101.1	69	96.3
398	-n-heptane	98.25	67	91.72
399	-styrene	145.2	14.3	116.8
400	-ethylbenzene	136.15	34	114.65
401	-o-xylene	143.6	22	116.6
402	-m-xylene	139.0	27.5	115.35
403	-p-xylene	138.4	28	115.25
404	-ethylcyclohexane	131.8	—	107.9
405	-n-octane	125.75	46.3	105.7
406	-cumene	152.8	16	116.0
407	-mesitylene	164.6	Zeotropic	
408	-n-nonane	150.8	31	112.9
409	-n-decane	173.3	20.5	116.75
410	-n-undecane	194.5	5	117.87
	Ethylene sulfide	55.7		
411	-acetone	56.15	43	51.5
412	-n-hexane	68.8	Zeotropic	
413	-2,3-dimethylbutane	58.0	35	54.0
	Methyl formate	31.7		
414	-ethyl ether	34.6	45	28.4
415	-isoprene	34.1	50	22.5
416	-2-methylbutane	27.95	53	17.05
417	-n-pentane	36.15	47	21.8
418	-n-hexane	69.0	Zeotropic	
419	-2,3-dimethylbutane	58.0	15	30.5
	Iodoethane	72.3		
420	-ethyl alcohol	78.3	14	63
421	-propyl alcohol	97.2	7	70
422	-n-hexane	68.85	24	68.0
	Acetamide	221.2		
423	-benzaldehyde	179.2	93.5	178.6
424	-methylaniline	196.25	86	193.8
425	-m-cresol	202.1	Zeotropic	
426	-styrene	145.8	88	144
427	-o-xylene	144.3	89	142.6
428	-m-xylene	139.0	90	138.4
429	-p-xylene	138.2	92	137.75
430	-2,4-xylenol	210.5	Zeotropic	
431	-3,4-xylenol	226.8	4	221.1
432	-ethylaniline	205.5	82	199.0
433	-quinoline	237.3	Zeotropic	
434	-indene	183.0	82.5	177.2
435	-naphthalene	218.05	73	199.55
436	-safrol	235.9	68	208.8
437	-eugenol	255.0	12	220.8
438	-p-cymene	176.7	81	170.5
439	-diethylaniline	217.05	76	198.05
440	-camphene	159.6	88	155.5
441	-dipentene	177.7	82	169.15
442	-camphor	209.1	77	199.8
443	-isoamyl valerate	192.7	84	184.85
444	-isoamyl sulfide	214.8	83	199.5
445	-1-methylnaphthalene	245.1	56.2	209.8
446	-2-methylnaphthalene	241.15	60	208.25
447	-acenaphthene	277.9	35.8	217.1
448	-biphenyl	255.9	49.5	212.95
449	-phenyl ether	259.3	48	214.55

No.	System	B.P. (°C)	Compn. X_2 (wt. %)	B.P. (°C)
450	-diphenylmethane	265.6	43.5	215.15
451	-1,2-diphenylethane	284	32	218.2
452	-benzyl ether	297	Zeotropic	
	Nitroethane	114.2		
453	-n-hexane, H	68.74	89.4	59.4
454	-toluene	110.75	75	106.2
	Ethyl nitrate	87.68		
455	-thiophene	84.7	Zeotropic	
456	-benzene	80.15	88	80.03
457	-cyclohexane	80.75	64	74.5
458	-n-hexane	68.8	76	66.25
459	-n-heptane	98.4	37	82.6
	Ethyl alcohol	78.3		
460	-acrylonitrile	77.3	59	70.8
461	-acetone	56.1	Zeotropic	
462	-ethyl sulfide	92.2	44	72.6
463	-pyridine	115.4	Zeotropic	
464	-cyclopentane	49.4	92.5	44.7
465	-n-pentane	36.15	95	34.3
466	-fluorobenzene	85.15	25	70.0
467	-benzene	80.1	68.3	67.9
468	-cyclohexane	80.8	70.8	64.8
469	-n-hexane	68.95	79	58.68
470	-propyl ether	90.4	56	74.4
471	-toluene	110.7	32	76.7
472	-ethylbenzene	136.15	Zeotropic	
473	-p-xylene	138.3	Zeotropic	
474	-n-octane	125.6	22	77.0
	Ethylene glycol	197.4		
475	-pyridine	115.5	Zeotropic	
476	-benzene	80.2	Zeotropic	
477	-phenol	182.2	Zeotropic	
478	-aniline	184.35	76	180.55
479	-o-bromotoluene	181.75	75	166.8
480	-o-nitrotoluene	221.75	51.5	188.55
481	-toluene	110.6	97.7	110.1
482	-m-toluidine	200.3	58	188.55
483	-o-cresol	191.1	73	189.6
484	-m-cresol	202.1	40	195.2
485	-2,6-lutidine	144.0	Zeotropic	
486	-n-heptane	98.45	97	97.9
487	-styrene	145.8	83.5	139.5
488	-m-xylene	139.1	93.45	135.1
489	-p-xylene	138.4	93.6	134.5
490	-3,4-xylenol	226.8	11	197.2
491	-2,4,6-collidine	171.3	90.3	170.5
492	-2,4-xylidine	214.0	53	188.6
493	-butyl ether	142.1	93.6	139.5
494	-quinoline	237.3	20.5	196.35
495	-indene	183.0	74	168.4
496	-cumene	152.8	82	147.0
497	-mesitylene	164.6	87	156
498	-propylbenzene	158.8	81	152
499	-cymene	176.7	74.5	163.2
500	-camphene	159.5	80	152.5
501	-camphor	209.1	60	186.15
502	-menthol	216.3	48.5	188.55
503	-n-decane	173.3	77	161.0
504	-naphthalene	218.05	49	183.9
505	-1-methylnaphthalene	245.1	40.0	190.25
506	-2-methylnaphthalene	241.15	42.8	189.1
507	-acenaphthene	277.9	25.8	194.65
508	-biphenyl	256.1	33.5	192.25
509	-fluorene	296.4	18	196.0
510	-diphenylmethane	265.6	31.5	193.3
511	-benzyl phenyl ether	286.5	13	195.5
512	-n-tridecane	234.0	45	188.0
513	-anthracene	340	1.7	197
514	-stilbene	306.4	13	196.8
	Methyl sulfide	37.3		
515	-acetone	56.15	Zeotropic	
516	-isoprene	34.3	65	32.5
517	-cyclopentane	49.35	12.5	37.1
518	-n-pentane	36.15	53.4	31.8
519	-2,2-dimethylbutane	49.7	20.2	36.5
	Methyl disulfide	109.44		
520	-n-heptane	98.4	73.7	96.44
521	-2,3-dimethylhexane	109.15	51.8	102.84
	Ethylenediamine	116.5		
522	-n-butyl alcohol	117.7	64.3	124.7
523	-benzene	80.1	Zeotropic	

No.	System	B.P. (°C)	Compn. X_2 (wt. %)	B.P. (°C)
524	-toluene	110.7	69.2	104
	Acrylonitrile	77.3		
525	-isopropyl alcohol	82.55	44	71.7
526	-benzene	80.2	53	73.3
	Pyruvic acid	166.8		
527	-propionic acid	141.3	Zeotropic	
528	-benzene	80.15	Zeotropic	
529	-toluene	110.75	92.5	110.05
530	-o-xylene	144.3	72	137.0
531	-ethylbenzene	136.15	78	130.5
532	-mesitylene	164.6	60	151.2
533	-propylbenzene	159.3	63	147.6
	Methyl chloroacetate	129.95		
534	-isobutyl alcohol	107.85	88	107.55
535	-cyclopentanol	140.85	23	127.5
536	-amyl alcohol	138.2	30	126.8
537	-isoamyl alcohol	131.3	39.5	124.9
538	-ethylbenzene	136.15	37.5	127.2
539	-m-xylene	139.2	10	128.25
540	-p-xylene	138.45	15	128.3
	Propionitrile	97.2		
541	-propyl alcohol	97.2	50	90.5
542	-n-hexane	68.8	91	63.5
543	-ethylbenzene	136.15	Zeotropic	
	Acetone	56.15		
544	-methyl acetate	57	51.7	55.8
545	-diethylamine	55.5	61.8	51.4
546	-pyridine	115.4	Zeotropic	
547	-cyclopentane	49.3	64	41.0
548	-n-pentane	36.15	80	32.5
549	-benzene	80.1	Zeotropic	
550	-cyclohexane	80.75	32.5	53.8
551	-n-hexane	68.95	41	49.8
552	-isopropyl ether	69.0	39	54.2
553	-n-heptane	98.4	10.5	55.85
	Allyl alcohol	96.15		
554	-ethyl sulfide	92.1	55	85.1
555	-pyridine	115.4	Zeotropic	
556	-benzene	80.2	82.64	76.75
557	-cyclohexane	80.8	42	74.0
558	-n-hexane	68.95	95.5	65.5
559	-methylcyclohexane	101.1	58	85.0
560	-m-xylene	139.0	Zeotropic	
561	-2,5-dimethylhexane	109.4	50	89.3
562	-n-octane	125.75	32	93.4
	Ethyl formate	54.1		
563	-n-pentane	36.2	70	32.5
564	-benzene	80.2	Zeotropic	
565	-methylcyclopentane	72.0	25	51.2
566	-n-hexane	68.95	33	49.0
	Methyl acetate	56.95		
567	-cyclopentane	49.3	62.1	43.2
568	-benzene	80.2	0.3	56.7
569	-cyclohexane	80.7	22.0	55.5
570	-n-hexane	68.95	39.3	51.75
571	-n-heptane	98.45	3.55	56.65
572	-2-methylhexane	90.0	11.4	56.0
573	-2,2,3-trimethylbutane	80.9	25.8	55.1
	Propionic acid	141.0		
574	-pyridine, N	115.5	32.8	148.6
575	-2-picoline, N	129.3	45.0	154.5
576	-chlorobenzene	132.0	82	128.9
577	-benzene	80.15	Zeotropic	
578	-o-xylene	143.6	57	135.4
579	-p-xylene	138.2	66	132.5
580	-n-hexane	68.85	Zeotropic	
581	-n-heptane	98.15	98	97.82
582	-n-octane	125.12	78.5	120.89
583	-n-nonane	150.67	46.0	134.27
584	-n-decane	174.06	19.5	139.76
585	-n-undecane	193.85	Zeotropic	
586	-propyl sulfide	141.5	55	136.5
587	-quinoline	237.5	Zeotropic	
588	-cumene	152.8	35	139.0
589	-mesitylene	164	23	139.3
590	-propylbenzene	158.0	25	139.5
591	-camphene	159.6	35	138.0
592	-α-pinene	155.8	41.5	136.4
	Propionamide	222.2		
593	-p-bromochlorobenzene	196.4	84	189.5
594	-p-dibromobenzene	220.25	78	204.9

No.	System	B.P. (°C)	Compn. X₂ (wt. %)	B.P. (°C)	No.	System	B.P. (°C)	Compn. X₂ (wt. %)	B.P. (°C)
595	-iodobenzene	188.45	90	183.5		**Propyl alcohol**	97.2		
596	-nitrobenzene	210.75	76	205.4	673	-dioxane	101.35	45	95.3
597	-o-nitrophenol	217.25	75.2	211.15	674	-butyl formate	106.8	36	95.5
598	-phenol	182.2	Zeotropic		675	-chlorobenzene	132	20	96.5
599	-p-bromotoluene	185.0	90	181.0	676	-fluorobenzene	85.15	82	80.2
600	-m-nitrotoluene	230.8	56	214.5	677	-benzene	80.2	83.1	77.12
601	-toluene	110.75	Zeotropic		678	-cyclohexane	80.75	81.5	74.69
602	-o-cresol	191.1	Zeotropic		679	-toluene	110.6	48.8	92.5
603	-m-cresol	202.2	Zeotropic		680	-methylcyclohexane	100.8	65.2	87.0
604	-o-toluidine	200.35	97.5	200.25	681	-n-heptane	98.4	65.3	84.6
605	-m-toluidine	203.1	Zeotropic		682	-styrene	145.8	92	97.0
606	-acetophenone	202	85	200.35	683	-o-xylene	143.6	Zeotropic	
607	-methyl salicylate	222.35	66	210.55	684	-m-xylene	139.2	6	97.08
608	-o-xylene	144.3	98	144.0	685	-p-xylene	138.4	7.8	96.88
609	-dimethylaniline	194.15	84.5	190.5	686	-n-octane	125.6	30	93.9
610	-3,4-xylidine	225.5	72	217.2		**Glycerol**	290.5		
611	-quinoline	237.3	Zeotropic		687	-p-chloronitrobenzene	239.1	87	235.6
612	-indene	182.6	88	179.5	688	-triethylene glycol	288.7	63	285.1
613	-ethyl benzoate	212.6	75	205.0	689	-m-nitrotoluene	230.8	87	228.8
614	-cumene	152.8	96	151.8	690	-p-cresol	201.7	Zeotropic	
615	-mesitylene	164.6	90	162.3	691	-methyl salicylate	222.35	92.5	221.4
616	-naphthalene	218.05	68.5	204.65	269	-3,4-xylenol	226.8	Zeotropic	
617	-cymene	176.7	85	172.8	693	-o-xylene	143.6	Zeotropic	
618	-carvone	231.0	52	214.5	694	-quinoline	237.3	Zeotropic	
619	-camphene	159.6	87	156.5	695	-ethyl salicylate	233.7	89.7	230.5
620	-camphor	209.1	83	203.5	696	-naphthalene	218.05	90	215.2
621	-borneol	213.4	78	209.2	697	-safrol	235.9	85.5	231.3
622	-n-decane	173.3	88.2	168	698	-methyl phthalate	283.2	69	271.5
623	-1-methylnaphthalene	245.1	48	213.8	699	-estragol	215.6	92.5	213.5
624	-2-methylnaphthalene	241.15	50	213.0	700	-eugenol	254.5	96	251.3
625	-n-undecane	194.5	79	183	701	-propyl benzoate	230.85	92	228.8
626	-acenaphthene	277.9	25	220.8	702	-carvone	231.0	97	230.85
627	-biphenyl	256.1	45	216.0	703	-2-methylnaphthalene	241.15	83.5	233.7
628	-n-dodecane	216.0	68.4	193.0	704	-acenaphthene	277.9	71	259.1
629	-fluorene	295	10	221.5	705	-biphenyl	254.9	75	246.1
630	-diphenylmethane	265.6	40	218.2	706	-phenyl ether	259.3	78	247.9
	Ethyl carbamate	185.25			707	-1,3,5-triethylbenzene	215.5	92	212.9
631	-bromobenzene	156.1	90.2	153.95	708	-bornyl acetate	227.7	90	226.0
632	-iodobenzene	188.45	67	174.5	709	-diphenylmethane	265.6	73	250.8
633	-nitrobenzene	210.75	12	184.95	710	-benzyl phenyl ether	286.5	70	264.5
634	-phenol	182.2	46.5	190.75	711	-benzyl ether	297.0	64	269.5
635	-benzonitrile	191.1	43	182.1		**Ethyl methyl sulfide**	66.61		
636	-anisole	153.85	95	153.5	712	-cyclohexane	80,75	Zeotropic	
637	-2,4-xylenol	210.5	Zeotropic		713	-methylcyclopentane	71.85	35.9	65.6
638	-n-octyl alcohol	195.2	27.5	183.5	714	-n-hexane	68.75	43.4	63.94
639	-isobutyl sulfide	172.0	77	166.5	715	-2,2-dimethylpentane	79.2	11.8	66.37
640	-indene	182.6	65	172.65		**1-Propanethiol**	67.3		
641	-cumene	152.8	94	151.5	716	-thiophene	84.7	Zeotropic	
642	-mesitylene	164.6	78	159.0	717	-cyclohexane	80.75	2.4	67.77
643	-propylbenzene	159.3	85	157.0	718	-2,3-dimethylbutane	58.0	83.7	57.54
644	-pseudocumene	168.2	75	161.4	719	-n-hexane	68.75	47.4	64.35
645	-naphthalene	218.0	23	184.05	720	-2-methylpentane	60.27	76.1	59.2
646	-butylbenzene	183.1	63	172.0	721	-isopropyl ether	68.3	35	66.0
645	-camphene	159.6	85	157.0	722	-2,2-dimethylpentane	79.2	18.7	67.2
648	-limonene	177.6	68	168.07	723	-2,2,3-trimethylbutane	80.97	12.6	67.57
649	-camphor	209.1	16	184.85		**Perfluorobutyric acid**	122.0		
650	-2-methylnaphthalene	241.15	Zeotropic		724	-ethyl-benzene	136.15	20	115.4
651	-amyl ether	187.4	63	171.0	725	-m-xylene	139.0	17	117.5
652	-isoamyl ether	173.35	73	163.15	726	-p-xylene	138.4	18	117.6
653	-methyl pelargonate	213.8	15	184.3		**Thiophene**	84.7		
	1-Nitropropane	131			727	-benzene	80.15	Zeotropic	
654	-propyl alcohol	97.15	91.2	96.95	728	-cyclohexane	80.8	58.8	77.9
655	-n-butyl alcohol	117.73	67.8	115.3	729	-methylcyclopentane	71.85	86	71.47
656	-isobutyl alcohol	107.89	84.8	105.28	730	-n-hexane	68.75	88.8	68.46
657	-n-heptane	98.43	86.5	96.6	731	-2,3-dimethylpentane	89.9	36	80.9
658	-ethylbenzene	136.19	44.0	129.0	732	-2,4-dimethylpentane	80.55	57.3	76.58
659	-n-octane	125.66	65.8	115.8	733	-n-heptane	98.4	16.8	83.09
660	-n-nonane	150.8	38.4	126.6		**Pyrrol**	129.2		
	Propyl nitrite	47.75			734	-chlorobenzene	131.75	57	124.5
661	-n-pentane	36.15	91	35.8	735	-isopropyl sulfide	120.5	80	117.5
662	-cyclopentane	49.3	46	45.5	736	-propyl sulfide	140.8	35	127.5
	Isopropyl alcohol	82.45			737	-toluene	110.75	Zeotropic	
663	-butylamine	77.8	40	74.7		**Methyl pyruvate**	137.5		
664	-n-pentane	36.15	94	35.5	738	-isoamyl acetate	142.1	35	135.0
665	-fluorobenzene	85.15	70	74.5	739	-m-xylene	139.2	50	130.0
666	-benzene	80.2	66.3	71.74		**Methyl oxalate**	163.3		
667	-cyclohexane	80.7	68	69.4	740	-p-dichlorobenzene	174.35	35	162.05
668	-n-hexane	68.85	77	62.7	741	-pinacol	174.35	19	163.15
669	-toluene	110.6	31	80.6	742	-o-bromotoluene	181.5	2	164.1
670	-n-heptane	98.45	49.5	76.4	743	-butyl butyrate	166.4	42	160.5
671	-o-xylene	144.3	Zeotropic		744	-ethyl caproate	167.7	40	161.0
672	-n-octane	124.75	16	81.6	745	-indene	182.6	17	163.6

No.	System	B.P. (°C)	Azeotropic data Compn. X₂ (wt. %)	B.P. (°C)	No.	System	B.P. (°C)	Azeotropic data Compn. X₂ (wt. %)	B.P. (°C)
746	-mesitylene	164.0	50.2	154.8		**Ethyl acetate**	77.1		
747	-naphthalene	218.0	Zeotropic		821	-butyl nitrite	78.2	29	76.3
748	-2,7-dimethyloctane	160.6	55	147.0	822	-isobutyl nitrite	67.1	Zeotropic	
749	-1,3,5-triethylbenzene	215.5	Zeotropic		823	-tert-butyl alcohol	82.45	27	76.0
	Ethyl bromoacetate	158.8			824	-benzene	80.15	Zeotropic	
750	-butyric acid	164.0	16	157.4	825	-cyclohexane	80.75	44	71.6
751	-isobutyric acid	154.6	60	153.0	826	-methylcyclopentane	72.0	62	67.2
752	-bromobenzene	156.1	72	155.3	827	-n-hexane	68.7	60.1	65.15
753	-cyclohexanol	160.8	35	155.5	828	-methylcyclohexane	101.1	Zeotropic	
754	-o-chlorotoluene	159.3	48	156.2		**Isobutyric acid**	154.6		
755	-propylbenzene	159.3	50	155.8	829	-iodobutane	130.4	93	128.8
	Ethyl chloroacetate	143.55			830	-ethyl pyruvate	155.5	40	153.0
756	-isoamyl acetate	142.1	60	141.7	831	-bromobenzene	156.15	65	148.6
757	-isoamyl alcohol	131.3	77	131.0	832	-chlorobenzene	132.0	92	131.2
758	-allyl sulfide	139.35	78	138.5	833	-phenol	182.2	Zeotropic	
759	-propyl butyrate	142.8	53	141.7	834	-1-bromohexane	156.5	65	148.0
760	-ethylbenzene	136.15	82	135.3	835	-o-bromotoluene	181.5	15	153.9
761	-o-xylene	144.3	42	140.2	836	-toluene	110.75	Zeotropic	
762	-m-xylene	139.0	68	137.45	837	-anisole	153.85	58	149.0
763	-butyl ether	142.4	55	139.8	838	-styrene	145.8	73	142.0
	Butyronitrile	117.9			839	-o-xylene	144.3	78	141.0
764	-n-butyl alcohol	117.8	50	113.0	840	-m-xylene	139.0	85	136.9
765	-toluene	110.75	73	107.0	841	-p-xylene	138.4	87	136.4
766	-methylcyclohexane	101.15	80	90.5	842	-cumene	152.8	65	146.8
	Isobutyronitrile	103.85			843	-propylbenzene	158.9	51	149.3
767	-benzene	80.15	Zeotropic		844	-pseudocumene	168.2	37	152.3
768	-methylcyclohexane	101.15	60	85.5	845	-cymene	176.7	20	153.4
769	-n-heptane	98.4	62	80.5	846	-camphene	159.6	55	148.1
	2-Butanone	79.6			847	-d-limonene	177.8	22	152.5
770	-methyl propionate	79.85	40	79.0	848	-2,7-dimethyloctane	160.2	52	148.55
771	-ethyl acetate	77.1	88.2	77.05	849	-isoamyl ether	173.2	7	154.2
772	-1-chlorobutane	78.5	62	77.0		**Methyl propionate**	79.85		
773	-butyl nitrite	78.2	70	76.7	850	-1-chlorobutane	78.05	62	76.8
774	-tert-butyl alcohol	82.45	31	78.7	851	-butyl nitrite	78.2	88	77.7
775	-butylamine	77.8	65	74.0	852	-n-butyl alcohol	117.8	Zeotropic	
776	-fluorobenzene	84.9	25	79.3	853	-benzene	80.2	48	79.45
777	-benzene	80.1	16	78.33	854	-cyclohexane	80.75	48	75.0
778	-cyclohexane	80.75	60	71.8	855	-methylcyclopentane	72.0	72	69.5
779	-n-hexane	68.8	71.4	64.2	856	-propyl ether	90.5	Zeotropic	
780	-n-heptane	98.5	30	77.0	857	-methylcyclohexane	101.1	11.5	79.3
781	-2,5-dimethylhexane	109.4	5	79.0		**Propyl formate**	80.85		
	Butyraldehyde	74.8			858	-1-chlorobutane	78.5	62	76.1
782	-benzene	80.1	Zeotropic		859	-butyl nitrite	78.2	65	76.8
783	-n-hexane	68.7	74	60.0	860	-tert-butyl alcohol	82.6	60	78.0
	Butyric acid	164.0			861	-benzene	80.2	53	78.5
784	-iodobutane	130.4	97.5	129.8	862	-cyclohexane	80.75	52	75.0
785	-2-furaldehyde	161.45	57.5	159.4	863	-n-hexane	68.95	70.5	63.6
786	-pyridine	115.5	8.0	163.2	864	-n-heptane	98.5	29	78.2
787	-propyl chloroacetate	162.5	60	160.5		**Glycol monoacetate**	190.9		
788	-isoamyl nitrate	149.75	88	147.85	865	-phenol, N	182.2	35	197.5
789	-p-dichlorobenzene	174.4	43	162.0	866	-m-bromotoluene	184.3	68	182.0
790	-chlorobenzene	132.0	97.2	131.75	867	-o-cresol, N	191.1	49	199.45
791	-o-bromotoluene	181.5	28	163.0	868	-m-cresol, N	202.2	69	206.5
792	-m-bromotoluene	184.3	20.5	163.62	869	-p-cresol, N	201.7	67	206.0
793	-p-bromotoluene	185.0	25	161.5	870	-n-octyl alcohol	195.2	29	189.5
794	-anisole	153.85	88	152.85	871	-indene	182.6	80	180.0
795	-n-heptane	98.4	Zeotropic		872	-naphthalene	218.0	Zeotropic	
796	-styrene	145.8	85	143.5	873	-amyl ether	187.5	58	180.8
797	-ethylbenzene	136.15	96	135.8	874	-isoamyl ether	173.2	72	170.2
798	-o-xylene	144.3	90	143.0	875	-1,3,5-triethylbenzene	215.5	Zeotropic	
799	-m-xylene	139.0	94	138.5		**Methyl lactate**	143.8		
800	-p-xylene	138.45	94.5	137.8	876	-phenol	182.2	Zeotropic	
801	-indene	182.6	16	163.65	877	-anisole	153.85	18	142.8
802	-cumene	152.8	80	149.5	878	-4-heptanone	143.55	53	142.7
803	-mesitylene	164.8	62	158.0	879	-ethyl valerate	145.45	42	140.0
804	-propylbenzene	158.9	72	154.5	880	-methyl caproate	149.8	30	141.7
805	-pseudocumene	169	55	159.5	881	-m-xylene	139.0	57.5	131.2
806	-naphthalene	218.1	Zeotropic		882	-p-xylene	138.2	60	130.8
807	-butylbenzene	183.1	25	162.5	883	-n-octane	125.8	70	120.3
808	-cymene	176.7	40	161.0	884	-butyl ether	142.8	58	137.0
809	-camphene	159.6	97.2	152.3	885	-cumene	152.8	38	137.8
810	-n-undecane	194.5	15.5	162.4	886	-camphene	159.6	15	140.0
	Dioxane	101.35			887	-2,7-dimethyloctane	160.1	32	137.8
811	-ethyl acetate	77.1	Zeotropic			**Tetrahydrothiophene**	118.8		
812	-1-bromobutane	101.5	53	98.0	888	-pyridine	115.4	55	113.5
813	-pyridine	115.5	Zeotropic		889	-1-methylpyrrol	112.8	82	111.5
814	-piperidine	106.4	Zeotropic		890	-ethylcyclohexane	131.85	19.3	117.46
815	-tert-amyl alcohol	102.35	20	100.65	891	-2-methylheptane	117.70	61.8	113.96
816	-benzene	80.15	Zeotropic		892	-n-octane	125.7	39.7	117.79
817	-cyclohexane	80.75	75.4	79.5		**Butyl nitrite**	78.2		
818	-ethyl borate	118.6	8	100.7	893	-benzene	80.15	25	77.95
819	-toluene	110.75	Zeotropic		894	-cyclohexane	80.75	37	76.5
820	-n-heptane	98.4	56	91.85					

No.	System	B.P. (°C)	Compn. X₂ (wt. %)	B.P. (°C)	No.	System	B.P. (°C)	Compn. X₂ (wt. %)	B.P. (°C)
895	-n-hexane	68.8	82	68.5	970	-methyl butyrate	102.65	75	101.3
896	-methylcyclohexane	101.15	Zeotropic		971	-propyl acetate	101.6	83	101.0
897	-n-heptane	98.4	Zeotropic		972	-n-pentane	36.15	Zeotropic	
	Isobutyl nitrite	67.1			973	-chlorobenzene	132.0	37	107.1
898	-benzene	80.15	Zeotropic		974	-fluorobenzene	84.9	91	84.0
899	-cyclohexane	80.75	Zeotropic		975	-benzene	80.1	92.6	79.3
900	-methylcyclopentane	72.0	32	65.9	976	-cyclohexene	82.7	85.8	80.5
901	-n-hexane	68.8	46	65.0	977	-cyclohexane	80.75	86	78.3
	Isobutyl nitrate	123.5			978	-methylcyclopentane	72.0	95	71.0
902	-n-butyl alcohol	117.8	55	112.8	979	-isobutyl vinyl ether	83.0	93.8	82.7
903	-isobutyl alcohol	107.85	64	105.6	980	-ethyl isobutyrate	110.1	48	105.5
904	-chlorobenzene	131.75	Zeotropic		981	-n-hexane	68.9	97.5	68.3
905	-propyl sulfide	141.5	Zeotropic		982	-propyl ether	90.55	90	89.5
906	-toluene	110.75	Zeotropic		983	-acetal	103.55	80	98.2
907	-ethylbenzene	136.15	Zeotropic		984	-isopropyl sulfide	100.5	27	105.8
	n-Butyl alcohol	117.75			985	-toluene	110.7	55	101.2
908	-pyridine	115.5	31	118.6	986	-methylcyclohexane	100.8	68	92.6
909	-butyl formate	106.6	76.4	105.8	987	-n-heptane	98.45	73	90.8
910	-ethyl carbonate	125.9	37	116.5	988	-ethylbenzene	136.15	20	107.2
911	-chlorobenzene	132.0	44	115.3	989	-p-xylene	138.4	11.4	107.1
912	-fluorobenzene	84.9	Zeotropic		990	-1,3-dimethylcyclohexane	120.7	44	102.2
913	-benzene	80.1	Zeotropic		991	-2,5-dimethylhexane	109.2	58	98.7
914	-2-picoline	129.4	Zeotropic		992	-2,2,4-trimethylpentane	99.3	73	92.0
915	-cyclohexene	82.7	95	82.0	993	-butyl ether	142.4	Zeotropic	
916	-cyclohexane	80.75	90.5	79.8		**2-Ethoxyethanol**	135.3		
917	-hexaldehyde	128.3	22.9	116.8	994	-toluene	110.75	89.2	110.15
918	-ethyl isobutyrate	110.1	83	109.2	995	-methylcyclohexane	101.15	85	98.6
919	-isoamyl formate	123.8	31	115.9	996	-propyl butyrate	143.7	28	133.5
920	-isobutyl acetate	117.2	50	114.5	997	-n-heptane	98.4	86	96.5
921	-methyl isoyalerate	116.3	60	113.5	998	-styrene	145.8	45	130.0
922	-paraldehyde	123.9	48	115.75	999	-ethylbenzene	136.15	52	127.8
923	-n-hexane	68.95	96.8	68.2	1000	-p-xylene	138.45	50	128.6
924	-acetal	103.55	87	101.0	1001	-n-octane	125.75	62	116.0
925	-isopropyl sulfide	120.5	55	112.0	1002	-cumene	152.8	33	133.2
926	-ethyl borate	118.6	48	113.0	1003	-propylbenzene	159.3	20	134.6
927	-toluene	110.7	72.2	105.5	1004	-camphene	159.6	35	131.0
928	-methylcyclohexane	100.8	80	95.3		**Methyl propyl ether**	38.95		
929	-n-heptane	98.4	42	93.85	1005	-2-methyl-2-butene	37.15	75	36.3
930	-ethylbenzene	136.15	34.9	115.85	1006	-n-pentane	36.2	78	35.6
931	-o-xylene	143.6	25	116.8	1007	-isoprene	34.3	Zeotropic	
932	-m-xylene	139.0	28.5	116.5		**Diethylene glycol**	245.5		
933	-p-xylene	138.3	32	115.7	1008	-p-dibromobenzene	220.25	87	212.85
934	-n-octane	125.75	54.8	108.45	1009	-nitrobenzene	210.75	90	210.0
935	-butyl ether	142.1	17.5	117.65	1010	-o-nitrophenol	217.2	89.5	216.0
936	-isobutyl ether	122.3	52	113.5	1011	-pyrocatechol, N	245.9	54	259.5
937	-n-nonane	150.7	28.5	115.9	1012	-m-nitrotoluene	230.8	75	224.2
938	-2,7-dimethyloctane	160.2	Zeotropic		1013	-methyl salicylate	222.95	85	220.55
	sec-Butyl alcohol	99.5			1014	-p-cresol	202.0	Zeotropic	
939	-butyl formate	106.8	32	98.0	1015	-ethyl fumarate	217.85	90	217.1
940	-ethyl propionate	99.15	53	95.7	1016	-quinoline	237.3	71	233.6
941	-benzene	80.15	84.6	78.5	1017	-benzyl acetate	215.0	93	214.85
942	-cyclohexane	80.75	82	76.0	1018	-naphthalane	218.0	78	212.6
943	-methylcyclopentane	72.0	88.5	69.7	1019	-isosafrol	252.0	54	233.5
944	-propyl ether	90.4	78	87.0	1020	-safrol	235.9	67	225.5
945	-toluene	110.7	45	95.3	1021	-methyl phthalate	283.7	3.7	245.4
946	-methylcyclohexane	101.5	61.8	89.7	1022	-thymol	232.9	87	232.25
947	-n-heptane	98.4	63.3	88.1	1023	-1-methylnaphthalene	244.6	55	277.0
948	-isooctane	99.3	66.2	88.0	1024	-2-methylnaphthalene	241.15	61	225.45
	tert-Butyl alcohol	82.9			1025	-biphenyl	256.1	52	232.65
949	-ethyl sulfide	92.1	30	79.8	1026	-acenaphthene	277.9	38	239.6
950	-flurobenzene	85.15	69	76.0	1027	-1,3,5-triethylbenzene	215.5	78	210.0
951	-benzene	80.2	63.4	73.95	1028	-bornyl acetate	227.6	82	223.0
952	-cyclohexane	80.7	65.8	71.2	1029	-fluorene	295.0	20	243.0
953	-methylcyclopentane	72.0	74	66.6	1030	-diphenylmethane	265.4	48	236.0
954	-n-hexane	68.85	78	63.7	1031	-benzyl phenyl ether	286.5	20	241.5
955	-isopropyl ether	68.3	92.1	67.3		**1-Butanethiol**	97.8		
956	-propyl ether	90.4	48	79.0	1032	-benzene	80.15	Zeotropic	
957	-toluene	110.7	Zeotropic		1033	-pyridine	115.4	Zeotropic	
958	-methylcyclohexane	100.8	34	78.8	1034	-n-heptane	98.42	50.6	95.45
959	-n-heptane	98.45	38	78.0	1035	-2-methylhexane	90.05	84.6	89.74
960	-p-xylene	138.45	Zeotropic		1036	-3-methylhexane	91.95	77.2	91.2
961	-2,5-dimethylhexane	109.2	23	81.5	1037	-2,5-dimethylhexane	109.1	12	98.22
	Ethyl ether	34.6				**Isopropyl methyl sulfide**	84.76		
962	-isoprene	34.3	52	33.2	1038	-cyclohexane	80.85	70	79.76
963	-2-methyl-2-butene	37.1	15	34.2	1039	-3-methylhexane	91.6	17.6	84.38
964	-n-pentane	36.16	44	33.7	1040	-2,4-dimethylpentane	80.55	70.3	79.39
965	-benzene	80.2	Zeotropic			**Methyl propyl sulfide**	95.47		
966	-n-hexane	68.85	Zeotropic		1041	-ethylcyclopentane	103.45	9.3	95.41
	Isobutyl alcohol	108.0			1042	-methylcyclohexane	101.05	22.0	95.06
967	-2-pentanone	102.35	81	101.8	1043	-3-methylhexane	91.6	67.05	90.53
968	-3-pentanone	102.05	80	101.7		**2-Furaldehyde**	161.45		
969	-butyl formate	106.8	60	103.0	1044	-n-heptane	98.4	94.7	98.3

No.	System	B.P. (°C)	Azeotropic data Compn. X₂ (wt. %)	B.P. (°C)
1045	-ethylbenzene	136.15	Zeotropic	
1046	-o-xylene	143.6	87	140.5
1047	-m-xylene	139.0	88	138.4
1048	-p-xylene	138.4	80	138.0
1049	-cumene	152.8	73	148.5
1050	-mesitylene	164.6	40	155.2
1051	-pseudocumene	168.2	33	157.0
1052	-propylbenzene	159.2	58	151.4
1053	-cymene	176.7	32	157.8
1054	-camphene	159.5	60	146.75
1055	-cineol	176.35	41	157.25
	Pyridine	115.4		
1056	-piperidine	105.8	92	106.1
1057	-phenol, N	181.4	86.9	183.1
1058	-toluene	110.75	77.8	110.1
1059	-n-heptane	98.4	74.7	95.6
1060	-n-octane	125.75	43.9	109.5
1061	-n-nonane	150.7	10.1	115.1
1062	-n-decane	173.3	Zeotropic	
	2-Methylthiophene	111.92		
1063	-n-heptane	98.4	97.8	97.77
1064	-2-methylheptane	117.7	32.2	109.77
1065	-2,2-dimethylhexane	106.85	66.8	104.62
1066	-2,5-dimethylhexane	109.15	60.4	106.12
	3-Methylthiophene	114.96		
1067	-ethylcyclopentane	103.45	96.1	102.82
1068	-n-octane	125.75	18.0	114.15
1069	-2-methylheptane	117.7	41.2	111.86
1070	-2,5-dimethylhexane	109.15	68.3	107.12
	Ethyl pyruvate	155.1		
1071	-bromobenzene	156.1	52	149.5
1072	-m-xylene	139.2	70	137.2
1073	-cumene	152.8	55	146.2
	Levulinic acid	252.0		
1074	-m-nitrotoluene	230.8	85	229.5
1075	-p-nitrotoluene	238.9	78	236.4
1076	-methyl salicylate	222.95	94	222.75
1077	-3,4-xylenol	226.8	Zeotropic	
1078	-ethyl salicylate	233.8	82	230.5
1079	-naphthalene	218.0	89	216.7
1080	-safrol	235.9	83	232.5
1081	-1-methylnaphthalene	244.6	64	237.0
1082	-2-methylnaphthalene	241.15	71	234.55
1083	-isobutyl benzoate	241.9	75	238.6
1084	-1,3,5-triethylbenzene	215.5	89	214.0
	Methyl acetoacetate	169.5		
1085	-siobutyl sulfide	172.0	42	166.0
1086	-mesitylene	164.6	57	159.5
1086	-cymene	176.7	44	165.0
1088	-camphene	159.6	60	152.8
1089	-isoamyl ether	173.2	40	160.5
	Methyl malonate	181.4		
1090	-acetophenone	202.0	61	201.0
1091	-naphthalene	218.0	Zeotropic	
1092	-butylbenzene	183.2	48	173.0
1093	-cymene	176.7	60	169.0
1094	-camphene	159.6	74	154.6
1095	-d-limonene	177.8	52	167.3
	2-Pentanone	102.25		
1096	-methyl butyrate	102.65	50	101.9
1097	-toluene	110.7	Zeotropic	
1098	-methylcyclohexane	101.15	60	95.2
1099	-n-heptane	98.4	66	93.2
	Butyl formate	106.8		
1100	-tert-amyl alcohol	102.35	65	101.0
1101	-benzene	80.15	Zeotropic	
1102	-pinacolone	106.2	62	106.0
1103	-methylcyclohexane	101.15	65	96.0
1104	-n-heptane	98.45	60	90.7
	Isovaleric acid	176.5		
1105	-ethyl acetoacetate	180.4	23	176.1
1106	-ethyl oxalate	185.65	16	176.3
1107	-o-xylene	144.3	95	143.8
1108	-butyl sulfide	185.0	27	175.0
1109	-indene	183.0	40	173.0
1110	-cumene	152.8	88	152.0
1111	-mesitylene	164.6	81	162.5
1112	-pseudocumene	168.2	77	165.7
1113	-naphthalene	218.05	Zeotropic	
1114	-butylbenzene	183.1	50	173.0
1115	-cymene	175.3	62	170.8
1116	-camphene	159.6	83	156.5

No.	System	B.P. (°C)	Azeotropic data Compn. X₂ (wt. %)	B.P. (°C)
1117	-cineol	176.3	57.5	175.0
1118	-n-decane	173.3	67	167.0
1119	-n-tridecane	234.0	Zeotropic	
	Methyl butyrate	102.65		
1120	-methylcyclohexane	101.1	55	97.0
1121	-n-heptane	98.45	65	95.1
1122	-n-octane	125.8	Zeotropic	
	Propyl acetate	101.6		
1123	-tert-amyl alcohol	102.0	42	99.5
1124	-benzene	80.2	Zeotropic	
1125	-cyclohexane	80.75	Zeotropic	
1126	-n-hexane	69.0	Zeotropic	
1127	-acetal	103.55	32	101.25
	Valeric acid	186.35		
1128	-phenol	182.2	Zeotropic	
1129	-indene	182.6	70	178.5
1130	-mesitylene	164.6	90	164.0
1131	-naphthalene	218.0	4	186.0
1132	-cymene	176.7	78	176.5
1133	-camphene	159.6	92	158.5
1134	-amyl ether	187.5	55	181.5
1135	-isoamyl ether	173.2	87.5	171.8
	Ethyl lactate	154.1		
1136	-toluene	110.75	Zeotropic	
1137	-o-xylene	144.3	70	140.2
1138	-p-xylene	138.45	83	136.6
1139	-cumene	152.8	52	143.5
1140	-mesitylene	164.9	27	150.05
1141	-pseudocumene	168.2	27	152.4
1142	-camphene	159.5	45	144.95
	n-Amyl alcohol	138.2		
1143	-benzene	80.2	Zeotropic	
1144	-phenol	182.2	Zeotropic	
1145	-amyl formate	132.0	57	131.4
1146	-ethylbenzene	136.15	60	129.8
1147	-p-xylene	138.45	58.1	130.9
	Isoamyl alcohol	131.9		
1148	-bromobenzene	156.15	15	131.65
1149	-butyl acetate	126.0	83.5	125.85
1150	-paraldehyde	124.0	78.0	123.5
1151	-o-fluorotoluene	114.0	86.0	112.1
1152	-toluene	110.7	90	109.7
1153	-n-heptane	98.45	93	97.7
1154	-ethylbenzene	136.15	51	125.7
1155	-n-octane	125.8	70	117.0
1156	-butyl ether	142.1	35	129.8
1157	-cumene	152.8	6	131.6
1158	-camphene	159.6	76	130.9
	Nitrobenzene	210.75		
1159	-aniline	184.35	Zeotropic	
1160	-methyl maleate	204.05	93	203.9
1161	-benzyl alcohol	205.25	62	204.2
1162	-3,4-xylenol	226.8	Zeotropic	
1163	-ethyl benzoate	212.5	19	210.6
1164	-camphor	208.9	65	208.4
1165	-borneol	215.0	42	207.8
1166	-1,3,5-triethylbenzene	215.5	Zeotropic	
1167	-ethyl bornyl ether	204.9	70	203.0
	Iodobenzene	188.55		
1168	-nitrobenzene	210.75	Zeotropic	
1169	-phenol	181.5	47	177.7
1170	-ethyl oxalate	185.65	52	181.0
1171	-caproic acid	205.15	12	186.8
1172	-isocaproic acid	199.5	15	185.5
1173	-benzyl alcohol	205.2	12	187.75
1174	-p-cresol	201.7	10	188.1
1175	-o-toluidine	200.35	Zeotropic	
1176	-isobutyl lactate	182.15	70	180.5
1177	-indene	182.6	Zeotropic	
1178	-isoamyl butyrate	178.5	Zeotropic	
1179	-butylbenzene	183.1	Zeotropic	
	Benzene	80.15		
1180	-aniline	184.35	Zeotropic	
1181	-cyclohexene	82.1	35.3	78.9
1182	-cyclohexane	80.75	48.1	77.56
1183	-methylcyclopentane	71.85	84	71.7
1184	-n-hexane	69.0	95.3	68.5
1185	-2,2-dimethylpentane	79.1	53.7	75.85
1186	-2,3-dimethylpentane	89.79	21.2	79.4
1187	-2,4-dimethylpentane	80.8	51.7	75.2
1188	-n-heptane	98.4	0.7	80.1
1189	-2,2,4-trimethylpentane	99.2	2.3	80.1

No.	System	B.P. (°C)	Azeotropic data Compn. X_2 (wt. %)	B.P. (°C)	No.	System	B.P. (°C)	Azeotropic data Compn. X_2 (wt. %)	B.P. (°C)
	Phenol	182.2			1267	-2-methylnaphthalene	241.15	Zeotropic	
1190	-aniline, N	183.91	58.1	185.84	1268	-n-undecane	194.5	42.5	175.31
1191	-2-picoline, N	129.2	24.6	185.5	1269	-1,3,5-triethylbenzene	215.5	Zeotropic	
1192	-3-picoline, N	143.5	29.8	188.93	1270	-n-dodecane	216.5	28.5	180.37
1194	-4-picoline, N	144.8	32.5	190.0	1271	-n-tridecane	235.4	13.8	182.94
1195	-ethylene diacetate, N	189.86	60.8	195.53	1272	-n-tetradecane	252.5	4.8	183.90
1196	-benzaldehyde	179.2	49.0	175.6		**2-Picoline**	129.3		
1197	-o-cresol	191.1	Zeotropic		1273	-n-octane	125.75	58.0	121.12
1198	-2,4-lutidine, N	159.0	43.0	193.4	1274	-n-nonane	150.7	15.9	129.2
1199	-2,6-lutidine, N	144.0	27.5	185.5	1275	-n-decane	174.6	Zeotropic	
1200	-o-toluidine	200.35	Zeotorpic			**3-Picoline**	144.0		
1201	-2,4,6-collidine, N	171.0	47.7	195.23	1276	-allyl sulfide	139.35	70	135.5
1202	-n-octyl alcohol	195.15	87	195.4	1277	-2,6-lutidine	144.06	27.3	143.5
1203	-sec-octyl alcohol	179.0	50	184.5		**o-Phenylenediamine**	258.6		
1204	-indene	182.2	53	177.8	1278	-isosafrole	252.0	70	249.2
1205	-mesitylene	164.5	79	163.5	1279	-isafrole	235.9	Zeotropic	
1206	-pseudocumene	168.2	75	166.0	1280	-biphenyl	256.1	63	249.7
1207	-naphthalene	218.1	Zeotropic		1281	-phenyl ether	259.0	54	251.2
1208	-butylbenzene	183.1	54	175.0	1282	-diphenylmethane	265.4	30	254.0
1209	-camphene	159.6	78	156.1	1283	-1,2-diphenylethane	284.5	Zeotropic	
1210	-n-decane	173.3	65	168.0		**Methyl fumarate**	193.25		
1211	-2,7-dimethyloctane	160.25	94	159.5	1284	-m-bromotoluene	184.3	84	183.65
1212	-amyl ether	187.5	22	180.2	1285	-o-cresol, N	191.1	40	197.8
1213	-isoamyl ether	173.2	85	172.2	1286	-m-cresol, N	202.2	28	204.3
1214	-isoamyl sulfide	214.8	Zeotropic		1287	-benzyl ethyl ether	185.0	68	183.5
1215	-1,3,5-triethylbenzene	215.5	Zeotropic		1288	-naphthalene	218.0	Zeotropic	
1216	-n-tridecane	235.42	16.9	180.56	1289	-dipentene	177.7	30	172.5
	Pyrocatechol	245.9				**Methyl maleate**	204.05		
1217	-indole	253.5	85.0	255.0	1290	-caproic acid	205.15	37	201.5
1218	-o-phenetidine, N	232.5	8	246.0	1291	-o-cresol, N	191.1	22	204.65
1219	-p-phenetidine, N	249.9	66	253.8	1292	-m-cresol, N	202.2	45	208.75
1220	-quinoline, N	237.4	39	257.9	1293	-p-cresol, N	201.7	44	208.6
1221	-naphthalene	218.05	88.5	217.45	1294	-naphthalene	218.0	13	203.7
1222	-quinaldine, N	246.5	52.0	252.5	1295	-borneol	215.0	22	202.95
1223	-safrole	235.9	77.0	233.55	1296	-isoamyl sulfide	214.8	18	203.0
1224	-isosafrole	252.0	30	243.0		**Cyclohexanone**	155.7		
1225	-eugenol	254.8	1.5	245.85	1297	-n-hexyl alcohol	157.85	6	155.65
1226	-carvone	231.0	29	248.3	1298	-cumene	152.8	35	152.0
1227	-thymol	232.9	83	232.2	1299	-camphene	159.6	42.5	150.55
1228	-1-methylnapthalene	244.9	60	235.1	1300	-2,7-dimethyloctane	160.1	45	151.5
1229	-2-methylnaphthalene	241.15	63	233.25		**Ethyl acetoacetate**	180.4		
1230	-acenaphthene	277.9	16	245.25	1301	-phenetole	170.45	76	169.8
1231	-biphenyl	255.9	43.5	239.85	1302	-isobutyl sulfide	172.0	90	171.0
1232	-phenyl ether	259.3	40.7	242.0	1303	-indene	182.6	32	177.15
1233	-1,3,5-triethylbenzene	215.5	91.1	214.7	1304	-propylbenzene	159.3	76	158.3
1234	-fluorene	295.0	Zeotropic		1305	-pseudocumene	168.2	63	165.2
1235	-diphenyl methane	265.6	35.0	243.05	1306	-butylbenzene	183.1	48	172.0
1236	-n-tridecane	234.0	70.0	229.7	1307	-cymene	176.7	59	170.5
1237	-1,2-diphenylethane	284.9	Zeotropic		1308	-camphene	159.6	70	156.15
	Resorcinol	281.4			1309	-dipentene	177.7	57	169.05
1238	-naphthalene	218.05	Zeotropic		1310	-d-limonene	177.8	57	169.05
1239	-1-naphthol	288.0	30	280.2	1311	-2,7-dimethyloctane	160.1	76	156.0
1240	-2-naphthol	295.0	15	280.8	1312	-amyl ether	187.5	30	174.5
1241	-methyl phthalate	283.7	62	287.5		**Ethylidene diacetate**	168.5		
1242	-1-methylnaphthalene	244.6	85.5	243.1	1313	-phenetole	170.45	44	164.5
1243	-2-methylnaphthalene	241.15	89.5	240.05	1314	-butyl butyrate	166.4	63	163.5
	Resorcinol	281.4			1315	-ethyl caproate	167.7	55	164.0
1244	-p-tert-amylphenol	266.5	85	265.8	1316	-sec-octyl alcohol	180.4	6.5	168.3
1245	-acenaphthene	277.9	59	266.2	1317	-cineole	176.35	34	164.95
1246	-biphenyl	255.9	79	252.15	1318	-isoamyl ether	173.2	43	161.5
1247	-phenyl ether	259.3	77	255.65		**Ethyl oxalate**	185.65		
1248	-fluorene	295.0	52	274.0	1319	-o-cresol, N	191.1	64	194.1
1249	-n-tridecane	234.0	88	233.25	1320	-camphene	159.6	84	158.5
1250	-stilbene	306.5	44	277.5	1321	-2,7-dimethyloctane	160.1	78	159.5
1251	-1,2-diphenylethane	284.9	53	269.7	1322	-amyl ether	187.5	46	177.7
	Pyrogallol	309.0			1323	-isoamyl ether	173.2	71	170.15
1252	-2-naphthol	295.0	22	293.5		**Glycol diacetate**	186.3		
1253	-acenaphthene	277.9	80	272.8	1324	-o-cresol, N	191.1	65	194.5
1254	-biphenyl	256.1	90	253.5		**Capronitrile**	163.9		
	Aniline	184.35			1325	-cumene	152.8	82	150.8
1255	-o-cresol, N	191.1	92	191.25	1326	-camphene	159.6	65	143.0
1256	-n-octyl alcohol	195.2	17	183.95		**Cyclohexanol**	160.8		
1257	-o-xylene	144.3	Zeotropic		1327	-o-xylene	143.6	86	143.0
1258	-indene	182.6	58.5	179.75	1328	-m-xylene	139.0	95	138.9
1259	-mesitylene	164.7	88.0	164.35	1329	-indene	181.7	25	160.0
1260	-pseudocumene	169.35	86.5	168.64	1330	-propylbenzene	158.8	60	153.8
1261	-naphthalene	218.0	Zeotropic		1331	-naphthalene	218.05	Zeotropic	
1262	-butylbenzene	183.1	54	177.8	1332	-cymene	176.7	28	159.5
1263	-n-nonane	150.7	86.5	149.2	1333	-camphene	159.5	59	151.9
1264	-n-decane	174.6	64	167.28	1334	-cineole	176.35	8	160.55
1265	-amyl ether	187.5	45	177.5		**Butyl acetate**	126.0		
1266	-isoamyl ether	173.2	72	169.35	1335	-paraldehyde	124.35	91	124.25

No.	System	B.P. (°C)	Azeotropic data Compn. X₂ (wt. %)	B.P. (°C)	No.	System	B.P. (°C)	Azeotropic data Compn. X₂ (wt. %)	B.P. (°C)
1336	-n-octane	125.8	48	119.0	1406	-amyl ether	187.5	58	180.5
1337	-butyl ether	142.1	5	125.9	1407	-isoamyl ether	173.2	84	171.4
	Caproic acid	205.3				**Benzaldehyde**	179.2		
1338	-m-cresol	202.2	87	201.9	1408	-o-cresol, N	191.1	77	192.0
1339	-guaiacol	205.05	58	200.8	1409	-p-cresol	2017	Zeotropic	
1340	-acetophenone	202.0	68	200.5	1410	-naphthalene	218.0	Zeotropic	
1341	-naphthalene	218.05	39	203.75	1411	-p-cymene	175.3	72	171.0
1342	-1-methylnaphthalene	244.6	Zeotropic		1412	-d-limonene	177.8	57	171.2
1343	-2-methylnaphthalene	241.15	Zeotropic		1413	-camphene	159.6	84.5	158.45
1344	-1,3,5-triethylbenzene	215.5	37	202.0	1414	-isoamyl ether	173.2	62.5	168.6
	Isoamyl formate	123.8				**Benzoic acid**	250.8		
1345	-paraldehyde	124.1	44	123.0	1415	-p-nitrotoluene	238.9	89	237.4
1346	-ethylbenzene	136.15	Zeotropic		1416	-3,4-xylenol	226.8	Zeotropic	
1347	-isobutyl ether	122.3	35	121.5	1417	-propyl succinate	250.5	57	248.0
	Propyl propionate	122.5			1418	-naphthalene	218.05	95	217.7
1348	-toluene	110.75	Zeotropic		1419	-1-methylnaphthalene	244.6	73	239.6
1349	-n-octane	125.8	40	118.2	1420	-2-methylnaphthalene	241.15	75	237.25
	Propyl lactate	171.7			1421	-biphenyl	277.9	49.5	246.05
1350	-o-cresol	191.1	Zeotropic		1422	-phenyl ether	259.3	41	247.3
1351	-isobutyl sulfide	172.0	52	169.0	1423	-fluorene	295.0	Zeotropic	
1352	-mesitylene	164.6	72	160.5	1424	-diphenylmethane	265.6	18	248.95
1353	-isoamyl ether	173.2	47	167.5	1425	-1,2-diphenylethane	284.0	Zeotropic	
	Isopropyl lactate	166.8				**o-Nitrotoluene**	221.75		
1354	-o-cresol	191.1	Zeotropic		1426	-benzyl alcohol	205.2	91	204.75
1355	-mesitylene	164.6	40	159.5	1427	-methyl salicylate	222.95	14	221.65
1356	-camphene	159.6	70	154.2	1428	-3,4-xylenol	226.8	Zeotropic	
	n-Hexyl alcohol	157.8			1429	-2,4-xylidine	214.0	Zeotropic	
1357	-o-cresol	191.1	zeotropic		1430	-naphthalene	218.0	Zeotropic	
1358	-anisole	153.85	63.5	151.0	1431	-diethylaniline	217.05	88	216.85
1359	-m-xylene	139.0	85	138.3	1432	-geraniol	229.6	19	220.7
1360	-cumene	152.8	65	149.5	1433	-menthol	216.3	66	214.65
1361	-mesitylene	164.6	45	153.5	1434	-n-decyl alcohol	232.8	15	221.0
1362	-pseudocumene	168.2	22	166.0	1435	-2-methylnaphthalene	241.15	Zeotropic	
1363	-propylbenzene	158.8	45	152.5	1436	-bornyl acetate	227.6	27	221.15
	Acetal	103.55				**p-Nitrotoluene**	238.9		
1364	-methylcyclohexane	101.15	60	99.65	1437	-quinoline	237.3	92	237.2
1365	-n-heptane	98.47	77	97.15	1438	-safrole	235.9	82	234.5
1366	-2,5-dimethylhexane	109.3	25	103.0	1439	-geraniol	229.6	75	228.8
1367	-n-octane	125.75	Zeotropic		1440	-n-decyl alcohol	232.8	67	231.5
	2-Butoxyethanol	171.15			1441	-bornyl acetate	227.6	90	227.45
1368	-benzaldehyde	179.2	9	170.95		**Toluene**	110.7		
1369	-o-cresol, N	191.1	85	191.55	1442	-2,6-lutidine	144.0	Zeotropic	
1370	-phenetole	170.45	48	167.1	1443	-ethylcyclopentane	103.5	93	103.0
1371	-isobutyl sulfide	172.0	58	163.8	1444	-n-heptane	98.4	Zeotropic	
1372	-mesitylene	164.6	68	162.0	1445	-2,5-dimethylhexane	109.4	65	107.0
1373	-butylbenzene	183.4	26.6	169.6	1446	-2-methylheptane	117.6	18	110.3
1374	-camphene	159.6	70	154.5	1447	-2,3,4-trimethylpentane	113.5	40	109.5
1375	-dipentene	177.7	47	164.0		**Benzyl alcohol**	205.2		
1376	-cineole	176.35	41.5	168.9	1448	-o-cresol	191.1	Zeotropic	
	Pinacol	174.35			1449	-m-cresol	202.2	39	207.1
1377	-o-cresol, N	191.1	92	191.5	1450	-p-cresol, N	201.7	38	206.8
1378	-p-cresol	201.7	Zeotropic		1451	-methylaniline	196.25	70	195.8
1379	-n-octane	125.75	Zeotropic		1452	-o-toluidine	200.35	Zeotropic	
1380	-pseudocumene	168.2	62	162.9	1453	-3,4-xylenol	226.8	Zeotropic	
1381	-propylbenzene	159.3	72	156.3	1454	-dimethylaniline	194.05	93.5	193.9
1382	-naphthalene	218.05	Zeotropic		1455	-ethylaniline	205.5	50.0	202.8
1383	-p-cymene	176.7	50	167.7	1456	-2,4-xylidine	214.0	Zeotropic	
1384	-cineole	176.35	55	168.5	1457	-naphthalene	218.05	40	204.1
1385	-isoamyl ether	173.4	60	167.2	1458	-diethylaniline	217.05	28	20.42
	Dipropylene glycol	229.2			1459	-d-limonene	177.8	89	176.4
1386	-p-cresol	201.7	Zeotropic		1460	-borneol	215.0	14.2	205.07
1387	-methyl salicylate	222.95	65	213.0	1461	-1,3,5-triethylbenzene	215.5	43	203.2
1388	-isosafrole	252.0	40	225.5		**o-Cresol**	191.1		
1389	-safrole	235.9	50	222.0	1462	-benzylamine, N	185.0	33	201.45
1390	-2-methylnaphthalene	241.1	Zeotropic		1463	-phenyl acetate, N	195.7	64	198.5
	Triethylene glycol	288.7			1464	-2,4,6-collidine, N	171.3	37	197.2
1391	-methyl phthalate	283.2	67	277.0	1465	-n-octyl alcohol	195.15	62	196.9
1392	-1-methylnaphthalene	244.6	Zeotropic		1466	-butyl sulfide	185.0	75	183.8
1393	-acenaphthene	277.9	65	271.5	1467	-indene	183.0	91	182.9
1394	-biphenyl	256.1	90	255.3	1468	-naphthalene	218.05	Zeotropic	
1395	-fluorene	294.0	Zeotropic		1469	-terpinene	181.5	72	177.8
1396	-phenyl benzoate	315.0	20	286.0	1470	-terpinolene	184.6	66	179.5
1397	-diphenylmethane	265.4	80	263.0	1471	-thymene	179.7	27	176.6
1398	-stilbene	306.5	40	284.5	1472	-camphor, N	209.1	85	209.85
1399	-1,2-diphenylethane	284.5	58	275.5		**m-Cresol**	202.2		
	Benzonitrile	191.1			1473	-o-toluidine, N	200.35	38.5	203.65
1400	-o-cresol, N	191.1	51	195.95	1474	-m-toluidine, N	203.1	47	205.5
1401	-m-cresol, N	202.2	89	202.5	1475	-p-toluidine, N	200.55	38	204.3
1402	-p-cresol, N	201.7	86	202.1	1476	-phenyl acetate, N	195.7	30	204.4
1403	-o-toluidine	200.35	Zeotropic		1477	-2,4,6-collidine, N	171.3	27	206.2
1404	-isoamyl butyrate	181.05	92	180.85	1478	-isoamyl lactate, N	202.4	50	207.6
1405	-cineole	176.35	86	175.6	1479	-n-octyl alcohol, N	195.15	38	203.3

BINARY SYSTEMS (Continued)

No.	System	B.P. (°C)	Compn. X₂ (wt. %)	B.P. (°C)
1480	-propiophenone, N	217.7	83	218.6
1481	-phorone, N	197.8	45	206.5
1482	-naphthalene	218.05	Zeotropic	
1483	-camphor, N	209.1	63.5	213.35
1484	-1,3,5-triethylbenzene	215.5	Zeotropic	
	p-Cresol	201.6		
1485	-o-toluidine, N	200.35	43	203.5
1486	-m-toluidine, N	203.1	53	204.9
1487	-p-toluidine, N	200.55	43	204.05
1488	-o-anisidine	219.0	Zeotropic	
1489	-acetophenone, N	202.0	53.5	208.4
1490	-benzyl formate, N	202.4	58.0	207.0
1491	-methyl benzoate, N	199.4	60.0	204.35
1492	-phenyl acetate, N	195.7	32	204.3
1493	-isoamyl lactate, N	202.4	52	207.25
1494	-n-octyl alcohol, N	195.2	30	202.25
1495	-camphor, N	209.1	69.5	213.5
1496	-ethyl caprylate, N	208.35	75	209.5
	Guaiacol	205.05		
1497	-acetophenone	202.0	32.5	205.25
1498	-m-toluidine	203.1	Zeotropic	
1499	-ethylaniline	205.5	45	204.4
1500	-ethyl caprylate, N	208.35	85	208.9
	2,4-Lutidine	159.0		
1501	-n-nonane	150.7	67.75	148.3
1502	-n-undecane	195.4	Zeotropic	
	Methylaniline	196.25		
1503	-n-octyl alcohol	195.2	55	193.0
1504	-d-limonene	177.8	97	174.5
	o-Toluidine	200.35		
1505	-acetophenone, N	202.0	68	203.65
1506	-n-octyl alcohol	195.2	77	194.7
1507	-n-decane	174.6	87	173.76
1508	-n-undecane	195.5	60.3	188.25
1509	-n-dodecane	216.5	37	195.75
1510	-n-tridecane	234.6	14.5	199.45
	o-Anisidine	219.0		
1511	-naphthalene	218.0	50	217.0
1512	-2-methylnaphthalene	241.15	Zeotropic	
1513	-1,3,5-triethylbenzene	215.5	65	214.5
	Enanthic acid	222.0		
1514	-ethyl fumarate	217.85	78	216.4
1515	-ethyl maleate	223.3	50	220.0
1516	-ethyl succinate	217.85	80	216.0
1517	-propiophenone	217.7	80	216.5
1518	-naphthalene	218.0	70	214.2
1519	-biphenyl	256.1	Zeotropic	
1520	1,3,5-triethylbenzene	215.5	73	211.0
	n-Heptyl alcohol	176.15		
1521	-benzyl methyl ether	167.8	80	167.0
1522	-p-methylanisole	177.05	48	173.3
1523	-phenetole	170.45	72	169.0
1524	-p-cymene	176.0	52	173.0
1526	-isoamyl ether	173.35	63	170.35
	Indole	253.5		
1527	-carvacrol	237.85	12	254.5
1528	-p-tert-amylphenol	266.5	88	268.0
	Acetophenone	202.		
1529	-p-ethylphenol N	218.8	85	219.5
1530	-2,4-xylenol, N	210.5	70	213.0
1531	-3,4-xylenol	226.8	Zeotropic	
1532	-dimethylaniline	194.15	Zeotropic	
1533	-ethylaniline	205.5	Zeotropic	
1534	-2,4-xylidine	214.0	Zeotropic	
1535	-n-octyl alcohol	195.2	87.5	194.95
1536	-naphthalene	218.0	Zeotropic	
1537	-1,3,5-triethylbenzene	215.5	Zeotropic	
	Phenyl acetate	195.7		
1538	-2,4-xylenol	210.5	Zeotropic	
1539	-n-octyl alcohol	195.15	47	192.4
1540	-indene	182.6	Zeotropic	
1541	-naphthalene	218.05	Zeotropic	
1542	-thymene	179.7	82	179.3
1543	-linalool	198.6	39	193.5
	Methyl salicylate	222.95		
1544	-phenethyl alcohol	219.4	57	218.0
1545	-3,4-xylenol	226.8	Zeotropic	
1546	-ethyl maleate	223.3	40	221.95
1547	-quinoline	237.3	Zeotropic	
1548	-geraniol	229.7	3	222.2
1549	-menthol	216.4	85	216.25
1550	-n-tridecane	234.0	Zeotropic	

No.	System	B.P. (°C)	Compn. X₂ (wt. %)	B.P. (°C)
	Ethylbenzene	136.15		
1551	-ethylcyclohexane	131.8	85	131.2
1552	-n-octane	125.75	Zeotropic	
1553	-n-nonane	150.7	Zeotropic	
	p-Ethylphenol	218.8		
1554	-ethyl fumarate, N	217.85	52	223.0
1555	-ethyl maleate, N	223.3	62	226.3
1556	-p-methylacetophenone, N	226.35	70	229.5
1557	-benzyl acetate, N	215.0	40	221.0
1558	-ethyl benzoate, N	212.5	20	219.8
1559	-naphthalene	218.0	55	215.0
1560	-diethylaniline	217.05	40	214.0
1561	-1,3,5-triethylbenzene	215.5	60	212.0
	p-Methylanisole	177.05		
1562	-sec-octyl alcohol	180.4	21	176.3
1563	-pseudocumene	169.0	Zeotropic	
1564	-butyl isovalerate	177.6	42	176.4
1565	-butylbenzene	183.2	Zeotropic	
1566	-cineole	176.35	65	175.35
1567	-isoamyl ether	173.2	70.5	172.5
	Phenethyl alcohol	219.4		
1568	-3,4-xylenol	226.8	Zeotropic	
1569	-2,4-xylidine	214.0	Zeotropic	
1570	-naphthalene	218.0	56	214.2
1571	-diethylaniline	217.05	60	213.95
1572	-borneol	213.4	80	213.0
1573	-menthol	216.3	70	215.05
1574	-1-methylnaphthalene	244.9	Zeotropic	
1575	-biphenyl	256.1	Zeotropic	
	2,4-Xylenol	210.5		
1576	-ethyl fumarate, N	217.85	68	219.65
1577	-quinoline, N	237.3	92	239.0
1578	-p-methylacetophenone, N	226.35	15	227.0
1579	-propiophenone, N	217.7	35	221.0
1580	-benzyl acetate, N	215.0	64	216.8
1581	-camphor, N	209.1	50	217.0
	3,4-Xylenol	226.8		
1582	-o-phenetidine, N	232.5	92	232.65
1583	-ethyl fumarate, N	217.85	35	228.2
1584	-ethyl maleate, N	223.3	45	230.0
1585	-quinoline, N	237.3	65	241.95
1586	-p-methylacetophenone, N	226.35	49	231.35
1587	-propiophenone, N	217.7	33	228.5
1588	-naphthalene	218.0	84	217.6
1589	-diethylaniline	217.05	92	217.0
1590	-camphor, N	209.1	27	227.55
1591	-n-tridecane	234.0	42	223.5
	Ethylaniline	205.5		
1592	-n-octyl alcohol	195.2	85	194.9
1593	-naphthalene	218.0	Zeotropic	
1594	-camphor	209.1	Zeotropic	
	2,4-Xylidine	217.4		
1595	-menthol	216.3	30	213.5
1596	-n-undecane	195.5	88	194.98
1597	-n-dodecane	216.5	63	209.8
1598	-n-tridecane	234.6	29	215.28
1599	-n-tetradecane	252.0	2.5	217.38
	o-Phenetidine	232.5		
1600	-ethyl salicylate	233.8	18	232.2
1601	-naphthalene	218.0	Zeotropic	
1602	-safrole	235.9	14	232.38
1603	-anethole	235.7	25	232.25
1604	-carvacrol	237.85	87	238.0
1605	-thymol, N	232.9	54.9	234.3
1606	-2-methylnaphthalene	241.15	Zeotropic	
	p-Phenetidine	249.9		
1607	-safrole	235.9	Zeotropic	
1608	-isosafrole	252.0	36	248.8
1609	-1-methylnaphthalene	244.6	73	243.95
1610	-2-methylnaphthalene	241.15	85	240.85
1611	-biphenyl	256.1	10	249.5
1612	-phenyl ether	259.0	15	249.75
	Ethyl fumarate	217.85		
1613	-naphthalene	218.0	42	216.7
1614	-thymol, N	232.9	87.5	233.35
1615	-menthol	216.3	70	216.0
	Ethyl maleate	223.3		
1616	-p-methylacetophenone	226.35	12	223.15
1617	-naphthalene	218.0	77	217.65
1618	-thymol, N	232.9	73	234.9
	Ethyl succinate	217.25		

BINARY SYSTEMS (Continued)

No.	System	B.P. (°C)	Compn. X₂ (wt. %)	B.P. (°C)
1619	-propiophenone	217.7	33	216.7
1620	-naphthalene	218.05	38.5	216.3
1621	-2-methylnaphthalene	241.15	Zeotropic	
	Caprylic acid	238.5		
1622	-naphthalene	218.05	94	216.2
1623	-carvacrol	237.85	75	237.6
1624	-1-methylnaphthalene	244.6	48	233.5
1625	-2-methylnaphthalene	241.15	52	235.0
	Propyl isovalerate	155.7		
1626	-cumene	152.8	Zeotropic	
1627	-propylbenzene	158.9	Zeotropic	
1628	-camphene	159.6	35	145.0
1629	-nopinene	163.8	25	155.0
	***sec*-Octyl alcohol**	179.0		
1630	butylbenzene	183.1	50	178.2
1631	-p-cymene	176.7	56	174.0
1632	-thymene	179.7	48	176.0
1633	-cineole	176.35	73.5	175.85
1634	-amyl ether	187.5	14	178.8
	Quinoline	237.3		
1635	-mesitol, N	220.5	15	240.4
1636	-safrole	235.9	73	235.15
1637	-carvacrol, N	237.85	52	244.3
1638	-thymol, N	232.9	45	243.1
1639	-1-methylnaphthalene	244.6	Zeotropic	
1640	2-methylnaphththalene	241.15	7	237.25
1641	-p-*tert*-amylphenol, N	266.5	94	267.5
1642	-biphenyl	256.1	Zeotropic	
	p-Methylacetophenone	226.35		
1643	-thymol, N	232.9	68	234.9
1644	-geraniol	229.6	5	226.25
1645	-citronellol	224.4	68	223.7
1646	-2-methylnaphthalene	241.15	Zeotropic	
1647	bornyl acetate	227.6	40	225.8
	Propiophenone	217.7		
1648	-benzyl acetate	215.0	Zeotropic	
1649	-borneol	213.0	Zeotropic	
1650	-1,3,5-triethylbenzene	215.5	75	215.4
	Ethylsalicylate	233.8		
1651	-safrole	235.9	12	233.65
1652	-geraniol	229.7	60	228.5
1653	-n-decyl alcohol	232.9	52	230.5
1654	-2-methylnaphthalene	241.15	Zeotropic	
	Cumene	152.8		
1655	-n-nonane	150.75	77	148.0
	Mesitylene	164.6		
1656	-propylbenzene	159.3	Zeotropic	
1657	-camphene	159.6	Zeotropic	
1658	2,7-dimethyloctane	160.1	72	158.6
	Propylbenzene	159.3		
1659	-camphene	159.6	53	158.0
	Pseudocumene	168.2		
1660	-p-cymene	176.7	Zeotropic	
1661	-n-decane	173.3	25	166.5
	Mesitol	230.5		
1662	-naphthalene	218.0	63	215.5
1663	-1,3,5-triethylbenzene	215.5	70	213.0
	3-Phenylpropanol	235.6		
1664	-naphthalene	218.05	~80	217.8
1665	-safrole	235.9	53	233.8
1666	-anethole	235.7	52	234.0
1667	-thymol, N	232.9	38	237.5
1668	-biphenyl	254.9	—	235.4
	Pelargonic acid	254.0		
1669	-naphthalene	218.0	Zeotropic	
1670	-isosafrole	252.0	65	249.5
1671	-eugenol	254.8	48	250.5
1672	-thymol	232.9	Zeotropic	
1673	-1-methylnaphthalene	244.6	82	243.0
1674	-2-methylnaphthalene	241.15	90	240.2
1675	-biphenyl	256.1	55	250.0
1676	-phenyl ether	259.0	45	250.5
1677	-1,3,5-triethylbenzene	215.5	Zeotropic	
1678	-diphenylmethane	265.4	25	252.7
	Naphthalene	218.0		
1679	-borneol	213.4	65	213.0
1680	-citronellol	224.5	30	217.8
1681	-menthol	216.4	74.5	215.5
1682	-n-tridecane	234.0	Zeotropic	
	Quinaldine	246.5		
1683	-safrole	235.9	Zeotropic	
1684	-carvacrol, N	237.85	33	250.8
1685	-thymol, N	232.9	20	250.0
	Methyl phthalate	283.2		
1686	-acenaphthene	277.9	66.5	276.35
1687	-biphenyl	255.9	Zeotropic	
1688	-diphenylmethane	265.6	Zeotropic	
1689	-1,2-diphenylethane	284.0	47	280.5
	Propyl benzoate	230.85		
1690	-carvacrol, N	237.85	82	238.85
1691	-thymol, N	232.8	55	235.5
1692	-2-methylnaphthalene	241.15	Zeotropic	
	p-Cymene	176.7		
1693	-dipentene	177.7	40	175.8
1694	-d-limonene	177.8	25	174.5
1695	-cineoloe	176.35	55	176.2
	Carvacrol	237.85		
1696	-carvenone, N	234.5	45	243.0
1697	-menthenone, N	222.5	25	239.5
1698	-propyl succinate, N	250.5	75	251.5
1699	-n-decyl alcohol	232.8	Zeotropic	
1700	-isobutyl benzoate, N	241.9	67	243.85
1701	-biphenyl	256.1	Zeotropic	
1702	-bornyl acetate, N	227.6	25	238.2
	Carvone	230.95		
1703	-thymol, N	232.9	52	238.65
1704	-geraniol	229.6	60	229.2
1705	-n-decyl alcohol	232.8	19	230.85
1706	-2-methylnaphthalene	241.15	Zeotropic	
	Thymol	232.9		
1707	-carvenone, N	234.5	50	241.0
1708	-pulegone, N	223.8	35	235.3
1709	-geraniol	229.6	42.5	225.6
1710	-menthone, N	209.5	8	233.2
1711	-2-methylnaphthalene	241.15	Zeotropic	
1712	-isobutyl benzoate, N	242.15	80	243.2
1713	-1,3,5-trimethylbenzene	215.5	Zeotropic	
1714	-bornyl acetate, N	227.7	40	235.6
	Camphene	159.6		
1715	-dipentene	177.7	Zeotropic	
1716	-2,7-dimethyloctane	160.25	38	158.0
	Borneol	211.8		
1717	-methol	216.4	Zeotropic	
1718	-1,3,5-triethylbenzene	215.5	38	212.2
	Menthol	216.3		
1719	-2-methylnaphthalene	241.15	Zeotropic	
1720	-terpineol methyl ether	216.2	50	215.3
1721	-1,3,5-triethylbenzene	215.5	~45	214.0
	Capric acid	268.8		
1722	-1-methylnaphthalene	244.6	Zeotropic	
1723	-diphenylmethane	265.4	72	262.5
	1-Methylnaphthalene	244.6		
1724	-2-methylnaphthalene	241.15	Zeotropic	
1725	-biphenyl	256.1	Zeotropic	
1726	-phenyl ether	259.0	Zeotropic	
1727	-diphenylmethane	265.4	Zeotropic	
	2-Methylnaphthalene	241.15		
1728	-isobutyl benzoate	241.9	40	240.8
	p-*tert*-Amylphenol	266.5		
1729	-acenaphthene	277.9	Zeotropic	
1730	-fluorene	295.0	Zeotropic	
1731	-diphenylmethane	265.4	60	263.0
	Phenyl ether	259.0		
1732	-isoamyl benzoate	262.05	10	258.9
1733	-isoamyl oxalate	268.0	Zeotropic	
1734	-diphenylmethane	265.6	Zeotropic	
	Isoamyl benzoate	262.0		
1735	-isoamyl oxalate	268.0	Zeotropic	
1736	-diphenylmethane	265.6	Zeotropic	
	1,3,5-Triethylbenzene	215.5		
1737	-bornyl acetate	227.2	Zeotropic	
1738	-bornyl ethyl ether	204.9	Zeotropic	
	Isoamyl oxalate	268.0		
1739	-diphenylmethane	265.4	86	265.25
1740	-1,2-diphenylethane	284.5	Zeotropic	
	Phenyl benzoate	315.0		
1741	-stilbene	306.5	Zeotropic	
1742	-benzyl ether	297.0	Zeotropic	
	Benzyl phenyl ether	286.5		
1743	-1,2-diphenylethane	284.5	Zeotropic	

TERNARY SYSTEMS

No.	System	B.P. (°C)	Azeotropic data		Type of azeotrope	Ref.
			Compn. (wt. %)	B.P. (°C)		
1	Argon	−186			—	1
	Nitrogen	−195	Zeotropic			
	Oxygen	−183	90—120°K			
2	Hydrogen bromide	−67	10.4		H	2
	Water	100	11.0	105		
	Chlorobenzene	131.8	78.6			
3	Hydrogen cyanide	26			—	3
	Acetonitrile	81.6	Zeotropic			
	Acrolein	52.45				
4	Chlorine trifluoride	—			—	4
	Hydrogen fluoride	19.4	Zeotropic			
	Uranium hexafluoride	56				
5	Hydrogen chloride	−80	15.8		H	5
	Water	100	64.8	107.33		
	Phenol	182	19.4			
6	Hydrogen chloride	−80	5.3		H	5
	Water	100	20.2	96.9		
	Chlorobenzene	131.8	74.5			
7	Hydrogen fluoride	19.4	10		N	6
	Fluosilicic acid	—	36	116.1		
	Water	100	54			
8	Silicon tetrafluoride	—	24.6		—	7
	Hexafluoroethane	−78	32.7	−104		
	Ethane	−88	42.7			
9	Water	100	4.5		H	8
	Carbon tetrachloride	76.75	85.5	62		
	Ethyl alcohol	78.3	10			
10	Water	100	5		H	8
	Carbon tetrachloride	76.75	84	65.15		
	Allyl alcohol	96.95	11			
11	Water	100	4.05		H	9
	Carbon tetrachloride	76.7	91.0	65		
	sec-Butyl alcohol	99	4.95			
12	Water	100	1.6		H	10
	Carbon disulfide	46.25	93.4	41.3		
	Ethyl alcohol	78.3	5.0			
13	Water	100			—	11
	Carbon disulfide	46.25	Zeotropic			
	Dioxane	101.4				
14	Water	100			—	12
	Chloroform	61	Zeotropic			
	Formic acid	100.75				
15	Water	100	1.3		—	13
	Chloroform	61	90.5	52.3		
	Methyl alcohol	64.7	8.2			
16	Water	100	2.3		—	14
	Chloroform	61	94.2	55.3		
	Ethyl alcohol	78.3	3.5			
17	Water	100	18.6		S	15
	Formic acid	100.8	71.9	107.2		
	Propionic acid	140.7	9.5			
18	Water	100	19.5		S	16
	Formic acid	100.8	75.9	107.62		
	Butyric acid	162.4	4.6			
19	Water	100	15.5		S	16
	Formic acid	100.8	66.8	107.02		
	Isobutyric acid	154	17.7			
20	Water	100	21.3		S	16
	Formic acid	100.8	76.3	107.64		
	Isovaleric acid	176.5	2.4			
21	Water	100			—	16
	Formic acid	100.8	Zeotropic			
	Valeric acid	186				
22	Water	100	2.1		H-3	17
	Nitromethane	101.2	6.5	33.1		
	n-Pentane	36.07	91.4			
23	Water	100	7.88		H-3; 748 mm Hg	18
	Nitromethane	101.2	29.73	71.43		
	n-Heptane	98.43	62.39			
24	Water	100	12.4		H-3; 748 mm Hg	18
	Nitromethane	101.2	44.25	77.35		
	n-Octane	125.7	43.35			
25	Water	100	17.4		H-3; 748 mm Hg	18
	Nitromethane	101.2	58.3	80.72		
	n-Nonane	150.8	24.3			
26	Water	100	19.1		H-3; 748 mm Hg	18
	Nitromethane	101.2	68.1	82.35		
	n-Decane	174.12	12.8			

No.	System	B.P. (°C)	Azeotropic data Compn. (wt. %)	Azeotropic data B.P. (°C)	Type of azeotrope	Ref.
27	Water	100	20.6		H-3; 748 mm Hg	18
	Nitromethane	101.2	73.3	82.82		
	n-Undecane	194.5	6.1			
28	Water	100	21.5		H-3; 748 mm Hg	18
	Nitromethane	101.2	75.3	83.13		
	n-Dodecane	214.5	3.2			
29	Water	100	22.8		H-3; 748 mm Hg	18
	Nitromethane	101.2	75.4	83.21		
	n-Tridecane	234	1.8			
30	Water	100			—	19
	Methyl alcohol	64.7	Zeotropic			
	Ethyl alcohol	78.3				
31	Water	100	5.26		—	20
	Methyl alcohol	64.7	81.20	67.85		
	Methyl chloroacetate	131.4	13.54			
32	Water	100			—	21
	Methyl alcohol	64.7	Zeotropic			
	Methyl acetate	57.1				
33	Water	100	0.6		—	22
	Methyl alcohol	64.7	5.4	30.2		
	Isoprene	34.0	94.0			
34	Water	100	12.45		—	23
	Tetrachloroethylene	120.8	66.75	81.18		
	n-Propyl alcohol	97.2	20.8			
35	Water	100	6.4		—	24
	Trichloroethylene	86.95	73.1	67		
	Acetonitrile	81.6	20.5			
36	Water	100	5.5		—	25
	Trichloroethylene	86.95	78.4	67.0		
	Ethyl alcohol	78.3	16.1			
37	Water	100	7		—	26
	Trichloroethylene	86.95	73	69.4		
	Isopropyl alcohol	82.45	20			
38	Water	100			—	24
	Acetonitrile	81.6	Zeotropic			
	Acetone	56.4				
39	Water	100	1		—	26
	Acetonitrile	81.6	44	72.9		
	Ethyl alcohol	78.3	55			
40	Water	100			—	27
	Acetonitrile	81.6	Zeotropic			
	Diethylamine	55.5				
41	Water	100	8.2		—	24
	Acetonitrile	81.6	23.3	66		
	Benzene	80.2	68.5			
42	Water	100	3.5		—	13
	Acetonitrile	81.6	9.6	68.6		
	Triethylamine	89.7	86.9			
43	Water	100			—	8
	Acetic acid	118.1	Zeotropic			
	Toluene	110.7				
44	Water	100	8.4		H	17
	Nitroethane	114.07	9.3	59.5		
	n-Hexane	68.74				
45	Water	100	11.5		H	17
	Nitroethane	114.07	24.5	75.1		
	n-Heptane	98.43	64.0			
46	Water	100	8.7		—	26
	Ethyl alcohol	78.3	20.3	69.5		
	Acrylonitrile	77.2	71.0			
47	Water	100	4.8		—	26
	Ethyl alcohol	78.3	87.9	78.0		
	Crotonaldehyde	102.4	7.3			
48	Water	100	9.0		—	28
	Ethyl alcohol	78.3	8.4	70.23		
	Ethyl acetate	77.05	82.6			
49	Water	100	7.5		—	26
	Ethyl alcohol	78.3	42.5	81.8		
	Butylamine	77.8	50.0			
50	Water	100	6.3		—	26
	Ethyl alcohol	78.3	8.6	62		
	Butyl methyl ether	70.3	85.1			
51	Water	100	7.4		H	29
	Ethyl alcohol	78.3	18.5	64.86		
	Benzene	80.2	74.1			
52	Water	100	4.8		H	30
	Ethyl alcohol	78.3	19.7	62.60		
	Cyclohexane	80.75	75.5			

No.	System	B.P. (°C)	Azeotropic data Compn. (wt. %)	Azeotropic data B.P. (°C)	Type of azeotrope	Ref.
53	Water	100	9		—	31
	Ethyl alcohol	78.3	13	74.7		
	Triethylamine	89.4	78			
54	Water	100	12		H	26
	Ethyl alcohol	78.3	37	74.4		
	Toluene	110.6	51			
55	Water	100	3		H	26
	Ethyl alcohol	78.3	12	56.0		
	n-Hexane	68.7	85			
56	Water	100	6.1		H	26
	Ethyl alcohol	78.3	33.0	68.8		
	n-Heptane	98.45	60.9			
57	Water	100	0.4		—	32
	Acetone	56.7	7.6	32.5		
	Isoprene	34.7	92.0			
58	Water	100	8.5		H	33
	Allyl alcohol	96.95	5.1	59.7		
	n-Hexane	68.95	86.4			
59	Water	100	12.5		H	26
	Isopropyl alcohol	82.3	40.5	83		
	Butylamine	77.8	47.0			
60	Water	100	7.5		H	34
	Isopropyl alcohol	82.45	19.0	66.3		
	Benzene	80.2				
61	Water	100	7.5		H	8
	Isopropyl alcohol	82.45	18.5	64.3		
	Cyclohexane	80.75	74.0			
62	Water	100	13.1		H	26
	Isopropyl alcohol	82.3	38.2	76.3		
	Toluene	110.6	48.7			
63	Water	100	17.0		—	35
	Propyl alcohol	97.3	10.0	82.45		
	Propyl acetate	101.6	73.0			
64	Water	100	7.6	740mm Hg; H	26
	Propyl alcohol	97.2	10.1	67		
	Benzene	80.1	82.3			
65	Water	100	8.5		H	8
	Propyl alcohol	97.2	10.0	66.55		
	Cyclohexane	80.75	81.5			
66	Water	100	19.2		H	36
	n-Butyl alcohol	117.75	2.9	61.5		
	n-Hexane	68.95	77.9			
67	Water	100	41.4		H	36
	n-Butyl alcohol	117.75	7.6	78.1		
	n-Heptane	98.4	51.0			
68	Water	100	60.0		H	36
	n-Butyl alcohol	117.75	14.6	86.1		
	n-Octane	125.75	25.4			
69	Water	100	69.9		H	36
	n-Butyl alcohol	117.75	18.3	90.0		
	n-Nonane	150.7	11.8			
70	Water	100	29.9		H	26
	n-Butyl alcohol	117.75	34.6	90.6		
	Butyl ether	142.1	35.5			
71	Water	100	5		H	37
	2-Butanone	79.6	35	63.6		
	Cyclohexane	80.7	60			
72	Water	100	4		H	37
	Butyraldehyde	74.8	21	55.0		
	n-Hexane	68.7	75			
73	Water	100	8.9		H	14
	sec-Butyl alcohol	99.6	10.8	69.7		
	Cyclohexane	80.75	80.3			
74	Water	100	9		H	13
	sec-Butyl alcohol	99.4	19	76.3		
	Isooctane	99.0	72			
75	Water	100	8.1		H	38
	tert-Butyl alcohol	82.55	21.4	67.3		
	Benzene	80.2	70.5			
76	Water	100			—	22
	tert-Butyl alcohol	82.55	Zeotropic			
	Isoprene	34.0				
77	Water	100	8		H	8
	tert-Butyl alcohol	82.55	21	65.0		
	Cyclohexane	80.75	71			
78	Water	100			H	38
	Isobutyl alcohol	108	Zeotropic			
	Benzene	80.2				
79	Water	100	17.9		H	39
	Isobutyl alcohol	108	16.4	81.3		
	Toluene	110.7	65.7			

TERNARY SYSTEMS (Continued)

No.	System	B.P. (°C)	Azeotropic data Compn. (wt. %)	B.P. (°C)	Type of azeotrope	Ref.
80	Water	100			—	37
	Ethyl acrylate	99.3	Zeotropic			
	Isopropyl ether	68.3				
81	Water	100			—	40
	Toluene	110.7	Zeotropic			
	Benzyl alcohol	204.7				
82	Water	100			—	2
	Pyridine	115.5	Zeotropic			
	Benzene	80.1				
83	Water	100	14.0		H	41
	Pyridine	115.5	15.5	78.6		
	n-Heptane	98.4	70.5			
84	Water	100	22.5		H	41
	Pyridine	115.5	25.5	86.7		
	n-Octane	125.75	52.0			
85	Water	100	30.5		H	41
	Pyridine	115.5	37.0	90.5		
	n-Nonane	150.7	32.5			
86	Water	100	35.5		H	41
	Pyridine	115.5	45.5	92.3		
	N- Decane	173.3	19.0			
87	Water	100	38.5		H	41
	Pyridine	115.5	51.0	93.1		
	n-Undecane	194.5	10.5			
88	Water	100	40.5		H	41
	Pyridine	115.5	54.5	93.5		
	n-Dodecane	216.0	5.0			
89	Water	100	32.4		H	42
	Isoamyl alcohol	131.5	19.6	89.8		
	Isoamyl formate	124.2	48.0			
90	Water	100	44.8		—	43
	Isoamyl alcohol	131.5	31.2	93.6		
	Isoamyl acetate	142.0	24.0			
91	Water	100			—	37
	2-Picoline	129.2	Zeotropic			
	Paraldehyde	124.5				
92	Carbon tetrachloride	76.8				44
	Methyl alcohol	64.7	Zeotropic			
	Benzene	80.1				
93	Chloroform	61			—	45
	Formic Acid	100.75	Zeotropic			
	Acetic acid	118.1				
94	Chloroform	61	47		S	46
	Methyl alcohol	64.7	23	57.5		
	Acetone	56.1	30			
95	Chloroform	61	65.3		S	47
	Ethyl alcohol	78.3	10.4	63.2		
	Acetone	56.1	24.3			
96	Chloroform	61.0	56.1		—	48
	Ethyl alcohol	78.3	9.5	57.3		
	n-Hexane	68.7	34.4			
97	Chloroform	61.2	68.8		—	48
	Acetone	56.5	3.6	60.79		
	n-Hexane	68.7	27.6			
98	Chloroform	61.0			—	49
	Acetone	56.4	Zeotropic			
	Toluene	110.7				
99	Chloroform	61.2	79.70		S	50
	Ethyl formate	54.1	5.3	61.97		
	Isopropyl bromide	59.4	15.7			
100	Chloroform	61.2			—	51
	2-Bromopropane	59.4	Zeotropic			
	Isopropyl formate	68.8				
101	Methyl alcohol	64.7	17.4		—	26
	Acetone	56.1	5.8	53.7		
	Methyl acetate	56.3	76.8			
102	Methyl alcohol	64.7	14.6		—	52
	Acetone	56.25	30.8	47		
	n-Hexane	68.95	59.6			
103	Methyl alcohol	64.7	17.8		—	53
	Methyl acetate	57	48.6	50.8		
	Cyclohexane	80.75	33.6			
104	Methyl alcohol	64.7	14.6		—	14
	Methyl acetate	56.3	36.8	47.4		
	n-Hexane	68.7	48.6			
105	Methyl alcohol	64.7				8
	Benzene	80.1	Zeotropic			
	Cyclohexane	80.75				
106	Acetonitrile	81.6	34		—	26
	Ethyl alcohol	78.3	8	70.1		
	Triethylamine	89.7	58			

TERNARY SYSTEMS (Continued)

No.	System	B.P. (°C)	Azeotropic data Compn. (wt. %)	Azeotropic data B.P. (°C)	Type of azeotrope	Ref.
107	Trichloroethylene	87.2			—	54
	Benzene	80.1	Zeotropic			
	Cyclohexane	80.7				
108	Acetic acid	118.1	23		S	55
	Acetic anhydride	139.6	55	134.4		
	Pyridine	115.5	22			
109	Acetic acid	118.1	3.4		S	56
	Pyridine	115.5	10.6	98.5		
	n-Heptane	98.4	86.0			
110	Acetic acid	118.1	10.4		S	56
	Pyridine	115.5	20.1	115.7		
	n-Octane	125.75	69.5			
111	Acetic acid	118.1	20.7		S	57
	Pyridine	115.5	29.4	128.0		
	n-Nonane	150.7	49.9			
112	Acetic acid	118.1	31.4		S	56
	Pyridine	115.5	38.2	134.1		
	n-Decane	173.3	30.4			
113	Acetic acid	118.1	37.5		S	58
	Pyridine	115.5	43.5	137.1		
	n-Undecane	194.5	19.0			
114	Acetic acid	118.1	13.5		S	59
	Pyridine	115.5	25.2	129.08		
	Ethylbenzene	136.5	61.3			
115	Acetic acid	118.1	17.7		S	57
	Pyridine	115.5	30.5	132.2		
	o-Xylene	143.6	51.8			
116	Acetic acid	118.1	10.2		S	60
	Pyridine	115.5	22.5	129.22		
	p-Xylene	138.4	67.3			
117	Acetic acid	118.1	15		—	61
	Isoamyl alcohol	132	54	132		
	Isoamyl acetate	142	31			
118	Acetic acid	118.1	7.6		—	62
	Benzene	80.1	34.4	77.2		
	Cyclohexane	80.75	58.0			
119	Acetic acid	118.1	3.6		S	63
	2-Picoline	129.45	24.8	121.3		
	n-Octane	125.75	71.6			
120	Acetic acid	118.1	12.8		S	63
	2-Picoline	129.45	38.4	135.0		
	n-Nonane	150.7	48.8			
121	Acetic acid	118.1	19.9		—	63
	2-Picoline	129.45	46.8	141.3		
	n-Decane	173.3	33.3			
122	Acetic acid	118.1	30.5		—	63
	2-Picoline	129.45	55.2	143.4		
	n-Undecane	194.5	14.3			
123	Acetic acid	118.1			—	64
	2,6-Lutidine	144.0	Zeotropic			
	n-Octane	125.75				
124	Acetic acid	118.1	12.6		S	64
	2,6-Lutidine	144.0	74.3	147.0		
	n-Decane	173.3	13.1			
125	Acetic acid	118.1	75.0		S	58
	2,6-Lutidine	144.0	13.8	163.0		
	n-Undecane	194.5	11.2			
126	Acetic acid	118.1			—	65
	Ethylbenzene	136.15	Zeotropic			
	n-Nonane	150.7				
127	Acetic acid	118.1			—	66
	Acetic anhydride	139.6	Zeotropic			
	Methylene diacetate	164.0				
128	Methyl formate	31.9			—	14
	Ethyl ether	34.6	Zeotropic			
	n-Pentane	36.15				
129	Nitroethane	114.2	31.7		S	67
	p-Dioxane	101.3	17.7	102.87		
	Isobutyl alcohol	108	50.6			
130	Ethyl alcohol	78.3	29.6		—	68
	Benzene	80.1	12.8	64.7		
	Cyclohexane	80.75	57.6			
131	Ethyl alcohol	78.3			—	69
	Benzene	80.1	Zeotropic			
	n-Hexane	68.7				
132	Ethyl alcohol	78.3			—	70
	Aniline	184.35	Zeotropic			
	Toluene	110.7				
133	Ethyl alcohol	78.3			—	70
	Aniline	184.35	Zeotropic			
	n-Heptane	98.4				

TERNARY SYSTEMS (Continued)

No.	System	B.P. (°C)	Azeotropic data Compn. (wt. %)	B.P. (°C)	Type of azeotrope	Ref.
134	Ethyl alcohol	78.3			—	70
	Toluene	110.7	Zeotropic			
	n-Heptane	98.4				
135	Ethylene glycol	197.4			—	71
	Pyridine	115.5	Zeotropic			
	Phenol	181.4				
136	Ethylene glycol	197.4	5.9		S	71
	Phenol	181.4	79.1	185.01		
	2-Picoline	128.8	15.0			
137	Ethylene glycol	197.4	15.9		S	71
	Phenol	181.4	67.7	186.41		
	3-Picoline	143.5	16.4			
138	Ethylene glycol	197.4	8.7		S	71
	Phenol	181.4	74.6	185.04		
	2,6-Lutidine	144.0	16.7			
139	Ethylene glycol	197.4	29.5		S	71
	Phenol	181.4	54.8	188.55		
	2,4,6-Collidine	171.0	15.7			
140	Ethylene glycol	197.45	33.6		S	72
	o-Cresol	191.0	62.4	189.65		
	2,4,6-Collidine	171.3	4.0			
141	Acetone	56.1	51.1		—	14
	Methyl acetate	56.3	5.6	49.7		
	n-Hexane	68.7	43.3			
142	Acetone	56.1			—	73
	2-Butanone	79.6	Zeotropic			
	Ethyl acetate	77.0				
143	Acetone	56.4			—	74
	Benzene	80.1	Zeotropic			
	Cyclohexane	80.75				
144	Acetone	56.1			—	75
	Benzene	80.1	Zeotropic			
	Toluene	110.7				
145	Propionic acid	140.7	55.5		S	58
	Pyridine	115.5	26.4	147.1		
	n-Undecane	194.5	18.1			
146	Propionic acid	141.05	4.5		S	76
	2-Picoline	129.3	10.5	123.7		
	n-Octane	125.4	85.0			
147	Propionic acid	141.05	16.5		S	76
	2-Picoline	129.3	21.5	140.1		
	n-Nonane	150.6				
148	Propionic acid	141.05	29.5		S	76
	2-Picoline	129.3	32.0	149.33		
	n-Decane	174.0	38.5			
149	Propionic acid	141.05	43.0		S	76
	2-Picoline	129.3	40.0	153.4		
	n-Undecane	194.8	17.0			
151	Propionic acid	141.05			S	76
	2-Picoline	129.3	Zeotropic			
	n-Dodecane	216.1				
151	Isopropyl alcohol	82.3	31.1		—	76
	Benzene	80.1	15.0	69.1		
	Cyclohexane	80.75	53.9			
152	Propyl alcohol	97.2	15.5		—	77
	Benzene	80.1	30.4	73.81		
	Cyclohexane	80.75	54.2			
153	Propyl alcohol	97.2			—	78
	Benzene	80.1	Zeotropic			
	n-Heptane	98.4				
154	2-Butanone	79.6			—	79
	Ethyl acetate	77.1	Zeotropic			
	n-Hexane	68.7				
155	2-Butanone	79.6			—	80
	Benzene	80.1	Zeotropic			
	Cyclohexane	80.75				
156	Butyric acid	162.45			—	58
	Pyridine	115.5	Zeotropic			
	n-Undecane	194.5				
157	p-Dioxane	101.1	44.3		S	81
	Isobutyl alcohol	107.0	26.7	101.8		
	Toluene	110.7				
158	n-Butyl alcohol	117.75	11.9		—	82
	Pyridine	115.5	20.7	108.7		
	Toluene	110.7	67.4			
159	n-Butyl alcohol	117.75	4		—	83
	Benzene	80.1	48	77.42		
	Cyclohexane	80.75	48			
160	n-Butyl alcohol	117.7			—	84
	Benzene	80.1	Zeotropic			
	n-Heptane	98.4				

TERNARY SYSTEMS (Continued)

No.	System	B.P. (°C)	Azeotropic data Compn. (wt. %)	Azeotropic data B.P. (°C)	Type of azeotrope	Ref.
161	Isobutyl alcohol	107.0	43.2		—	85
	Benzene	80.1	47.0	77.2		
	Cyclohexane	80.75				
162	Pyridine	115.5	8.6		S	86
	Isoamyl alcohol	131.0	4.1	110.79		
	Toluene	110.7				
163	Phenol	181.4	33.5		—	87
	Aniline	183.95	48.5	184.45		
	n-Tridecane	234.0	18.0			
164	Phenol	182.0	26.4		S	88
	Ethylene diacetate	186.0	34.4	194.45		
	Phenyl acetate	195.7	39.2			
165	Phenol	181.4	19.88		S	89
	2,4-Lutidine	159.0	21.52	181.78		
	n-Undecane	194.5	58.60			
166	Ethylbenzene	136.15			—	59
	Pyridine	115.5	Zeotropic			
	n-Nonane	150.7				
167	Aniline	184.35			—	70
	Toluene	110.7	Zeotropic			
	n-Heptane	98.4				
168	Aniline	184.35			—	8
	Benzyl alcohol	205.5	Zeotropic			
	d-Limonene	177.8				
169	Aniline	184.35			—	8
	sec-Octyl alcohol	178.7	Zeotropic			
	d-Limonene	177.8				
170	Aniline	184.35			—	8
	o-Bromotoluene	181.75	Zeotropic			
	sec-Octyl alcohol	178.7				
171	Propyl lactate	171.7	31		—	8
	Phenetole	171.5	33	163.0		
	Menthene	170.8	36			
172	m-,p-Cresol (mixt.)	202	81		S	90
	Pyridine bases (mixt.)	143	9	202.81		
	Naphthalene	218.1	10			
173	m-, p-Cresol (mixt.)	202	65.5		S	90
	Pyridine bases (mixt.)	157	16.5	202.03		
	Naphthalene	218.1				
174	m-, p-Cresol (mixt.)	202	62		S	90
	Pyridine bases (mixt.)	163	17	202.39		
	Naphthalene	218.1	21			
175	m-Cresol	202.8	61.5		S	91
	2,4,6-Collidine	171.3	20.8	205.82		
	Naphthalene	217.9	17.7			
176	Isobutyl lactate	182.15			—	8
	sec-Octyl alcohol	178.7	Zeotropic			
	Terpinene	180.5				
177	Ethylbenzene	136.1			—	92
	Isopropylbenzene	152.8	Zeotropic			
	Butylbenzene	183.1				

REFERENCES

1. Narinskii, *Tr. Vses. Nauch. Issled. Inst. Kisloror. Mashinostr.*, 11, 3, 1967.
2. Dow Chemical Co., unpublished data.
3. Sokolov, Sevryogova, Zhavoronskov, *Rev. Chim. (Bucharest)*, 20, 169, 1969.
4. Ellis, Johnson, *J. Inorg. Nucl. Chem.*, 6, 194, 199, 1958.
5. Prahl, Mathes, *Angew. Chem.*, 47, 11, 1934.
6. Munter, Aepli, Kossatz, *Ind. Eng. Chem.*, 39, 427, 1947.
7. Calfee, Fukuhara, Bigelow, *J. Amer. Chem. Soc.*, 61, 3552, 1939.
8. Lecat, *L'Azeotropisme*, Lamartin, Bruxelles, 1918.
9. Marinichev, Susarev, *Zhur. Fiz. Khim.*, 43, 1132, 1969.
10. Ghysels, *Bull. Soc. Chim. Belges*, 33, 57, 1924.
11. De Mol, *Ingr. Chim.*, 22, 262, 1938.
12. Conti, Othmer, Gilmont, *J. Chem. Eng. Data*, 5, 301, 1960.
13. Kudryavtseva, Susarev, Eisen, *Zhur. Fiz. Khim.*, 43, 437, 1969.
14. Kudryavtseva, Eisen, Susarev, *Zhur. Fiz. Khim.* 40, 1285, 1652, 1966.
15. Kushner, Tatsievskaya, Serafimov, *Zhur. Fiz. Khim.*, 41, 237, 1967.
16. Kushner, Tatsievskaya, Serafimov, *Zhur. Fiz. Khim.*, 42, 2248, 1968.
17. Riddick, Commercial Solvents Corp., unpublished data.
18. Malesinska, Malesinski, *Bull. Acad. Polon. Sci. Ser. Sci. Chim.*, 11, 475, 1963.

19. Delzenne, *J. Chem. Eng. Data*, 3, 224, 1958.
20. Calices, Hannotte, *Ingr. Chim.*, 20, 1, 1936.
21. Balashov, Serafimov, Bessonova, *Zhur. Fiz. Khim.*, 40, 2294, 1966.
22. Lesteva, Kachalova, Morozova, Ogorodnikov, Trenke, *Zhur. Prikl. Khim.*, 40, 1808, 1967.
23. Malesinska, *Bull. Acad. Polon. Sci. Ser. Sci. Chim.*, 12, 853, 1964.
24. Pratt, Preprint, Trans. Inst. Chem. Engrs. (London), March 1947.
25. Licht, Denzler, *Chem. Eng. Progr.*, 44, 627, 1948.
26. Union Carbide Chemicals, *Alcohols*, 1961.
27. Union Carbide Chemicals, unpublished data.
28. Merriman, *J. Chem. Soc. Trans.*, 103, 1790, 1801, 1913.
29. Young, Fortey, *J. Chem. Soc. Trans.*, 81, 717, 1902.
30. Zieborak, Galska, *Bull. Acad. Polon. Sci. Cl. III*, 3, 383, 1955.
31. Tyerman, Br. Pat. 590.713, 1947.
32. Patterson, U.S. Pat. 2,407.997, 1946.
33. Kogan, Tolstova, *Zhur. Fiz. Khim.*, 33, 276, 1959.
34. Yorizane, Yoshimura, *Hiroshima Daigaku Kogakuba Kenkya Hokoku*, 13, 41, 1965.
35. Smirnova, Moraczevskii, Storonkin, *Vest. Leningrad Univ.*, 14, 70, 1959.
36. Kogan, Fridman, Deizenrot, *Zhur. Prikl. Khim.* 30, 1339, 1957.
37. Union Carbide Chemicals Co., unpublished data.
38. Young, Fortey, *J. Chem. Soc. Trans.*, 81, 739, 1902.
39. Frolov, Loginova, Nazarova, *Zhur. Fiz. Khim.* 43, 2632, 1969.
40. Susarev, Gorbunov, *Zhur. Prikl. Khim.*, 36, 459, 1963.
41. Trabczynski, *Bull. Acad. Polon. Sci. Ser. Sci. Chim. Geol. Geogr.*, 6, 269, 1958.
42. Hannotte, *Bull. Soc. Chim. Belges*, 35, 85, 1926.
43. Hyatt, U.S. Patent 2,176.500, 1939.
44. Hirata, Hirose, *Mem. Fac. Technol. Tokyo Metrop. Univ.*, No. 11, 876, 1961.
45. Conti, Othmer, Gilmont, *J. Chem. Eng. Data*, 5, 301, 1960.
46. Ewell, Welch, *Ind. Eng. Chem.*, 37, 1224, 1945.
47. Morachevskii, Leontev, *Zhur. Fiz. Khim.*, 34, 2347, 1960.
48. Kidryavtseva, Susarev, *Zhur. Prikl. Khim.*, 36, 1231, 1471, 1710, 2025, 1963.
49. Satapathy et al., *J. Appl. Chem. (London)*, 6, 261, 1956.
50. Orszagh, Lelakowska, Bełdowica, *Bull. Acad. Polon. Sci. Ser. Sci. Chim. Geol. Geogr.*, 6, 419, 1958.
51. Lelakowska, *Bull. Acad. Polon. Sci. Ser. Sci. Chim. Geol. Geogr.*, 6, 645, 1958.
52. Forman, U.S. Patent 2,581.789, 1952.
53. Fisher, U.S. Patent 2,341.433, 1944.
54. Rao, Dakshinamurty, Rao, *J. Sci. Ind. Res. (India)*, 20B, 218, 1961.
55. Jones, *J. Chem. Eng. Data*, 7, 13, 1962.
56. Zieborak, *Bull. Acad. Polon. Sci. Cl. III*, 3, 531, 1955.
57. Zieborak, Wyrzykowska-Stankiewicz, *Bull. Acad. Polon. Sci. Ser. Sci. Chim. Geol. Geogr.*, 7, 247, 1959.
58. Zieborak, Wyrzykowska-Stankiewicz, *Bull. Acad. Polon. Sci. Ser. Sci. Chim. Geol. Geogr.*, 6, 517, 1958.
59. Galska-Krajewska, Zieborak, *Rocz. Chem.*, 36, 119, 1962.
60. Galska-Krajewska, *Bull. Acad. Polon. Sci. Ser. Sci. Chim.*, 9, 455, 1961.
61. Krokhin, *Zhur. Fiz. Khim.*, 43, 442, 1969.
62. Baradarajan, Satyanarayana, *J. Chem. Eng. Data*, 13, 148, 1968.
63. Zieborak, Wyrzykowska-Stankiewicz, *Bull. Acad. Polon. Sci. Ser. Sci. Chim. Geol. Geogr.*, 6, 377, 1958.
64. Zieborak, Kaczorowana-Badyoczek, Maczynska, *Rocz. Chem.*, 29, 783, 1955.
65. Zieborak, Galska-Krajewska, *Bull. Acad. Polon. Sci. Ser. Sci. Chim. Geol. Geogr.*, 7, 253, 1959.
66. Tatscheff et al., *Z. Phys. Chem. (Leipzig)*, 237, 52, 1968.
67. Malesinska, Malinsinski, *Bull. Acad. Polon. Sci. Ser. Sci. Chim.*, 8, 191, 1960.
68. Morachevskii, Zharov, *Zhur. Prikl. Khim.*, 36, 2771, 1963.
69. Yuan, Ho, Keshpande, Lu, *J. Chem. Eng. Data*, 8, 549, 1963.
70. Hollo, Ember, Lengyel, Weig, *Acta Chim. Acad. Sci. Hung.*, 13, 307, 1957.
71. Razniewska, *Rocz. Cham.*, 38, 851, 1964.
72. Kurtyka, *Bull. Acad. Polon. Sci. Cl. III*, 4, 49, 1956.
73. Babicz, Ivanchikova, Serafimov, *Zhur. Prikl. Khim.*, 42, 1354, 1969.
74. Kurmanadharao, Krishnamurty, Rao, *Rec. Trav. Chim.*, 76, 769, 1957.
75. Vitman, Zharov, *Zhur. Prikl. Khim.*, 42, 2858, 1969.
76. Trabczynski, *Bull. Acad. Polon. Sci. Ser. Sci. Chim.*, 12, 335, 1965.
76. Nagata, *Can. J. Chem. Eng.*, 42, 82, 1964.
77. Moraczevskii, Cheng, *Zhur. Fiz. Khim.*, 35, 2535, 1961.
78. Fu, Lu, *J. Chem. Eng. Data*, 13, 6, 1968.
79. Gorbunova, Lutigina, Malenko, *Zhur. Prikl. Khim.*, 38, 374, 622, 1965.
80. Donald, Ridgway, *J. Appl. Chem. (London)*, 8, 403, 408, 1958.

81. Wyrzykowska-Stankiewicz, Zieborak, *Bull. Acad. Polon. Sci. Ser. Sci. Chim.*, 8, 655, 1960.
82. Hollo, Lengyel, *Ind. Eng. Chem.*, 51, 957, 1959.
83. Zieborak, Galska-Krajewska, *Bull. Acad. Polon. Sci. Ser. Sci. Chim. Geol. Geogr.*, 6, 763, 1958.
84. Vijayaraghavan, Deshpande, Kuloor, *J. Chem. Eng. Data*, 12, 13, 1967.
85. Nataraj, Rao, *Trans. Indian Inst. Chem. Eng.*, 95, 1968.
86. Zieborak, Wyrzykowska-Stankiewicz, *Bull. Acad. Polon. Sci. Ser. Sci. Chim.*, 8, 137, 1960.
87. Stadnicki, *Bull. Acad. Polon. Sci. Ser. Sci. Chim.*, 10, 357, 1962.
88. Orszagh, Lelakowska, Radecki, *Bull. Acad. Polon. Sci. Ser. Sci. Chim. Geol. Geogr.*, 6, 605, 1958.
89. Fahmy, Assal, *Bull. Acad. Polon. Sci. Ser. Sci. Chim.*, 14, 773, 1966.
90. Zieborak, Markowska-Majewska, *Bull. Acad. Polon. Sci. Cl. III*, 2, 341, 1954.
91. Kurtyka, *Bull. Acad. Polon. Sci. Ser. Sci. Chim.*, 9, 741, 1961.
92. Linek, Fried, Pick, *Coll. Czech. Chem. Commun.*, 30, 1358, 1965.

QUATERNARY AND QUINARY SYSTEMS

No.	System	B.P. (°C)	Compn. (wt. %)	B.P. (°C)	Type	Ref.
1	Hydrocyanic acid	26			—	1
	Water	100.0	Zeotropic			
	Acrylonitrile	77.3				
	Acrolein	52.4				
2	Hydrocyanic acid	26			—	1
	Acetonitrile	81.6	Zeotropic			
	Acrylonitrile	77.3				
	Acrolein	52.4				
3	Water	100.0			—	2
	Formic acid	100.8	Zeotropic			
	Acetic acid	118.1				
	Butyric acid	162.4				
4	Water	100.0	7.38		H	3
	Nitromethane	101.2	20.65	76.88		
	Tetrachloroethylene	120.8	59.45			
	n-Propyl alcohol	97.2	12.52			
5	Water	100.0	9.86		H	3
	Nitromethane	101.2	34.40	77.06		
	Tetrachloroethylene	120.8	32.60			
	n-Octane	125.75	23.14			
6	Water	100.0	—		H	3
	Tetrachloroethylene	120.8	—	80.98		
	n-Propyl alcohol	97.2	—			
	n-Octane	125.75	—			
7	Water	100.0	9.98		H	3
	Nitromethane	101.2	41.00	76.34		
	n-Propyl alcohol	97.2	12.42			
	n-Octane	125.75	36.60			
8	Water	100.0			—	4
	Acetonitrile	81.6	Zeotropic			
	Ethyl alcohol	78.3				
	Triethylamine	89.7				
9	Water	100.0	8.7		—	5
	Ethyl alcohol	78.3	11.1	70		
	Crotonaldehyde	102.2	0.1			
	Ethyl acetate	77.1	80.1			
10	Water	100.0	6.1		H	6
	Ethyl alcohol	78.3	19.2	62.14		
	Benzene	80.1	20.4			
	Cyclohexane	80.75	54.3			
11	Water	100.0			—	7
	Ethyl alcohol	78.3	Zeotropic			
	Benzene	80.1				
	n-Hexane	68.95				
12	Water	100.0			—	7
	Ethyl alcohol	78.3	Zeotropic			
	Benzene	80.1				
	Methylcyclohexane	100.88				
13	Water	100.0	6.8		H	8
	Ethyl alcohol	78.3	18.7	64.97		
	Benzene	80.1	62.4			
	n-Heptane	98.4	12.1			
14	Water	100.0	6.7		H	8
	Ethyl alcohol	78.3	17.7	64.69		
	Benzene	80.1	61.4			
	Isooctane	99.3	14.1			
15	Water	100.0			—	9
	1-Chlorobutane	78.44	Zeotropic			
	n-Butyl alcohol	117.73				
	Butyl ether	—				

QUATERNARY AND QUINARY SYSTEMS (Continued)

No.	System	B.P. (°C)	Compn. (wt. %)	B.P. (°C)	Type	Ref.
16	Chloroform	61.2		—		10
	Methyl alcohol	64.6	Zeotropic			
	Methyl acetate	56.9				
	Benzene	80.1				
17	Acetic acid	118.1	17		S	11
	Pyridine	115.4	27	127.9		
	Ethylbenzene	136.4	18			
	n-Nonane	150.8	38			
18	Acetic acid	118.1		—		12
	Pyridine	115.4	Zeotropic			
	p-Xylene	138.4				
	n-Nonane	150.8				
19	Acetone	56.1		—		13
	Isopropyl alcohol	82.3	Zeotropic			
	Benzene	80.1				
	Toluene	110.7				
20	Acetone	56.1		—		14
	Benzene	80.1	Zeotropic			
	Cyclohexane	80.9				
	Toluene	110.7				
21	Isopropyl alcohol	82.3		—		15
	2-Butanone	79.6	Zeotropic			
	Benzene	80.1				
	Cyclohexane	80.9				
22	Water	100.0	9.45		H	3
	Nitromethane	101.2	37.30			
	Tetrachloroethylene	120.8	21.15	76.5		
	n-Propyl alcohol	97.2	10.58			
	n-Octane	125.75	21.52			
23	Chloroform	61.2		—		10
	Methyl alcohol	64.6				
	Acetone	56.15	Zeotropic			
	Methyl acetate	56.9				
	Benzene	80.1				

REFERENCES

1. Sokolov, Sevryoguva, Zhavoronkov, *Teor. Osn. Khim. Tekhnol.*, 3, 288, 1969.
2. Kushner, Lebedeva, Tatsievskaya, Serafimov, *Shur. Prikl. Khim.*, 42, 1104, 1968.
3. Malesinska, *Bull. Acad. Polon. Sci. Ser. Sci. Chim.*, 12, 853, 1964.
4. Union Carbide Chemicals, unpublished data.
5. Eastman Chemical Products, unpublished data.
6. Zieborak, Galska, *Bull. Acad. Polon. Sci. Cl. III*, 3, 383, 1955.
7. Swietoslawski, Zieborak, Galska-Krajewska, *Bull. Acad. Polon. Sci. Ser. Sci. Chim. Geol. Geogr.*, 7, 43, 1959.
8. Swietoslawski, Zieborak, *Bull. Acad. Polon. Sci. Ser. A*, 9, 13, 1950.
9. Riddick, Commercial Solvents, unpublished data.
10. Hudson, van Winkle, *J. Chem. Eng. Data*, 14, 310, 1969.
11. Zieborak, Galska-Krajewska, *Bull. Acad. Polon. Sci. Ser. Sci. Chim. Geol. Geogr.*, 7, 253, 1959.
12. Galska-Krajewska, *Bull. Acad. Polon. Sci. Ser. Sci. Chim.*, 9, 455, 1961.
13. Vitman, Zharov, *Zhur. Prikl. Khim.*, 42, 2858, 1969.
14. Vitman, Markova, *Zhur. Prikl. Khim.*, 42, 2360, 1969.
15. Lutugina, Kolbina, *Zhur. Prikl. Khim.*, 41, 2766, 1968.

HEAT OF FORMATION OF INORGANIC OXIDES

lb/in² × 6894.8 = Pa	lb/in² × 0.070307 = Kg/cm²
lb/in² × 6894.8 = N/m²	lb/in² × 51.7149 = Torr
lb/in² × 0.068046 = Std. atm.	lb/in² × 51.7149 = mmHg at 0°C

The ΔH_o values are given in gram calories per mole. The a, b, and I values listed here make it possible for one to calculate the ΔF and ΔS values by use of the following equations:

$$\Delta F_t = \Delta H_o + 2.303aT \log T + b \times 10^{-3}T^2 + c \times 10^5 T^{-1} + IT$$
$$\Delta S_t = -a - 2.303a \log T - 2b \times 10^{-3}T + c \times 10^5 T^{-2} - I$$

Ref: Bulletin 542, U. S. Bureau of Mines, 1954.

Coefficients in Free-Energy Equations

Reaction and temperature range of validity	ΔH_o	2.303a	b	c	I
2 Ac(c) + 3/2 O₂(g) = Ac₂O₃(c) .... (298.16°–1,000° K.)	−446,090	−16.12	......	.....	+109.89
2 Al(c) + 1/2 O₂(g) = Al₂O(g) .... (298.16°–931.7° K.)	−31,660	+14.97	......	.....	−72.74
2 Al(l) + 1/2 O₂(g) = Al₂O(g) .... (931.7°–2,000° K.)	−38,670	+10.36	......	.....	−51.53
Al(c) + 1/2 O₂(g) = AlO(g) .... (298.16°–931.7° K.)	+10,740	+5.76	......	.....	−37.61
Al(l) + 1/2 O₂(g) = AlO(g) .... (931.7°–2,000° K.)	+8,170	+5.76	......	.....	−34.85
2 Al(c) + 3/2 O₂(g) = Al₂O₃ (corundum). (298.16°–931.7° K.)	−404,080	−15.68	+2.18	+3.935	+123.64

Reaction and temperature range of validity		ΔH_o	2.303a	b	c	I
$2\,Al(l) + 3/2\,O_2(g) = Al_2O_3$ (corundum).	(931.7°–2,000° K.)	−407,950	−6.19	−.78	+3.935	+102.37
$2\,Am(c) + 3/2\,O_2(g) = Am_2O_3(c)$	(298.16°–1,000° K.)	−422,090	−16.12	......	.....	+107.89
$Am(c) + O_2(g) = AmO_2(c)$	(298.16°–1,000° K.)	−240,600	−4.61	......	.....	+55.91
$2\,Sb(c) + 3/2\,O_2(g) = Sb_2O_3$ (cubic)	(298.16°–842° K.)	−169,450	+6.12	−6.01	.....	+52.21
$2\,Sb(c) + 3/2\,O_2(g) = Sb_2O_3$ (orthorhombic).	(298.16°–903° K.)	−168,060	+6.12	−6.01	−.30	+50.56
$2\,Sb(l) + 3/2\,O_2(g) = Sb_2O_3$ (orthorhombic).	(903°–928° K.)	−175,370	+15.29	−7.75	−.30	+33.12
$2\,Sb(l) + 3/2\,O_2(g) = Sb_2O_3(l)$	(928°–1,698° K.)	−173,940	−32.84	+.75	−.30	+166.52
$2\,Sb(l) + 3/2\,O_2(g) = 1/2\,Sb_4O_6(g)$	(1,698°–1,713° K.)	−132,760	+10.91	+.75	−.30	+.96
$2\,Sb(g) + 3/2\,O_2(g) = 1/2\,Sb_4O_6(g)$	(1,713°–2,000° K.)	−234,760	−.74	+.75	−.30	+98.17
$2\,Sb(c) + 2\,O_2(g) = Sb_2O_4(c)$	(298.16°–903° K.)	−208,310	+6.31	−5.36	−.40	+73.02
$2\,Sb(l) + 2\,O_2(g) = Sb_2O_4(c)$	(903°–1,500° K.)	−215,610	+15.47	−7.10	−.40	+55.61
$6\,Sb(c) + 13/2\,O_2(g) = Sb_6O_{13}(c)$	(298.16°–903° K.)	−649,160	+38.46	−25.13	−1.30	+192.54
$6\,Sb(l) + 13/2\,O_2(g) = Sb_6O_{13}(c)$	(903°–1,500° K.)	−691,370	+14.13	−30.35	−1.30	+315.93
$2\,Sb(c) + 5/2\,O_2(g) = Sb_2O_5(c)$	(298.16°–903° K.)	−226,060	+37.12	−22.66	−.50	+18.61
$2\,Sb(l) + 5/2\,O_2(g) = Sb_2O_5(c)$	(903°–1,500° K.)	−240,130	+29.01	−24.40	−.50	+59.74
$2\,As(c) + 3/2\,O_2(g) = As_2O_3$ (orthorhombic).	(298.16°–542° K.)	−154,870	+29.54	−21.33	−.30	−8.83
$2\,As(c) + 3/2\,O_2(g) = As_2O_3$ (monoclinic).	(298.16°–586° K.)	−150,760	+29.54	−21.33	−.30	−16.95
$2\,As(c) + 3/2\,O_2(g) = As_2O_3(l)$	(542°–730.3° K.)	−156,260	−43.29	+2.97	−.30	+180.95
$2\,As(c) + 3/2\,O_2(g) = 1/2\,As_4O_6(g)$	(730.3°–883° K.)	−135,930	+.46	+2.97	−.30	+26.88
$1/2\,As_4(g) + 3/2\,O_2(g) = 1/2\,As_4O_6(g)$	(883°–2,000° K.)	−154,450	−2.90	+.75	−.30	+59.71
$2\,As(c) + 2\,O_2(g) = As_2O_4(c)$	(298.16°–883° K.)	−173,690	+21.52	−13.42	−.40	+34.38
$1/2\,As_4(g) + 2\,O_2(g) = As_2O_4(c)$	(883°–1,500° K.)	−192,210	+18.15	−15.64	−.40	+67.22
$2\,As(c) + 5/2\,O_2(g) = As_2O_5(c)$	(298.16°–883° K.)	−217,080	+12.32	−4.65	−.50	+80.50
$1/2\,As_4(g) + 5/2\,O_2(g) = As_2O_5(c)$	(883°–2,000° K.)	−235,600	+8.96	−6.87	−.50	+113.33
$Ba(\alpha) + 1/2\,O_2(g) = BaO(c)$	(298.16°–648° K.)	−134,590	−7.60	+.87	+.42	+45.76
$Ba(\beta) + 1/2\,O_2(g) = BaO(c)$	(648°–977° K.)	−134,140	−3.34	−.56	+.42	+34.01
$Ba(l) + 1/2\,O_2(g) = BaO(c)$	(977°–1,911° K.)	−135,900	−2.19	−.56	+.42	+32.37
$Ba(g) + 1/2\,O_2(g) = BaO(c)$	(1,911°–2,000° K.)	−176,400	−8.01	−0.56	+0.42	+72.66
$Ba(\alpha) + O_2(g) = BaO_2(c)$	(298.16°–648° K.)	−154,830	−11.05	+.87	+.42	+74.48
$Ba(\beta) + O_2(g) = BaO_2(c)$	(648°–977° K.)	−154,380	−6.79	−.56	+.42	+62.73
$Ba(l) + O_2(g) = BaO_2(c)$	(977°–1,500° K.)	−156,140	−5.64	−.56	+.42	+61.09
$Be(c) + 1/2\,O_2(g) = BeO(c)$	(298.16°–1,556° K.)	−144,220	−1.91	−.46	+1.24	+30.64
$Be(l) + 1/2\,O_2(g) = BeO(c)$	(1,556°–2,000° K.)	−144,300	+6.06	−1.75	+1.485	+7.25
$Bi(c) + 1/2\,O_2(g) = BiO(c)$	(298.16°–544° K.)	−50,450	−4.61	......	.....	+35.51
$Bi(l) + 1/2\,O_2(g) = BiO(c)$	(544°–1,600° K.)	−52,920	−4.61	......	.....	+40.05
$2\,Bi(c) + 3/2\,O_2(g) = Bi_2O_3(c)$	(298.16°–544° K.)	−139,000	−11.56	+2.15	−.30	+96.52
$2\,Bi(l) + 3/2\,O_2(g) = Bi_2O_3(c)$	(544°–1,090° K.)	−142,270	+2.30	−3.25	−.30	+67.55
$2\,Bi(l) + 3/2\,O_2(g) = Bi_2O_3(l)$	(1,090°–1,600° K.)	−147,350	−32.84	+.75	−.30	+174.59
$2\,B(c) + 3/2\,O_2(g) = B_2O_3(c)$	(298.16°–723° K.)	−304,690	+11.72	−7.55	+.355	+34.25
$2\,B(c) + 3/2\,O_2(g) = B_2O_3(gl)$	(298.16°–723° K.)	−298,670	+26.57	−15.90	−.30	−10.40
$2\,B(c) + 3/2\,O_2(g) = B_2O_3(l)$	(723°–2,000° K.)	−308,100	−38.41	+5.15	−.30	+173.24
$Cd(c) + 1/2\,O_2(g) = CdO(c)$	(298.16°–594° K.)	−62,330	−2.05	+.71	−.10	+29.17
$Cd(l) + 1/2\,O_2(g) = CdO(c)$	(594°–1,038° K.)	−63,240	+2.07	−.76	−.10	+20.14
$Cd(g) + 1/2\,O_2(g) = CdO(c)$	(1,038°–2,000° K.)	−89,320	−2.83	−.76	−.10	+60.05
$Ca(\alpha) + 1/2\,O_2(g) = CaO(c)$	(298.16°–673° K.)	−151,850	−6.56	+1.46	+.68	+43.93
$Ca(\beta) + 1/2\,O_2(g) = CaO(c)$	(673°–1,124° K.)	−151,730	−4.14	+.41	+.68	+37.63
$Ca(l) + 1/2\,O_2(g) = CaO(c)$	(1,124°–1,760° K.)	−153,480	−1.36	−.29	+.68	+31.49
$Ca(g) + 1/2\,O_2(g) = CaO(c)$	(1,760°–2,000° K.)	−194,670	−7.18	−.29	+.68	+73.84
$Ca(\alpha) + O_2(g) = CaO_2(c)$	(298.16°–500° K.)	−158,230	−12.32	+1.46	+.68	+78.28
$C(graphite) + 1/2\,O_2(g) = CO(g)$	(298.16°–2,000° K.)	−25,400	+2.05	+.27	−1.095	−28.79
$C(graphite) + O_2(g) = CO_2(g)$	(298.16°–2,000° K.)	−93,690	+1.63	−.07	−.23	−5.64
$2\,Ce(c) + 3/2\,O_2(g) = Ce_2O_3(c)$	(298.16°–1,048° K.)	−435,600	−4.60	......	.....	+92.84
$2\,Ce(l) + 3/2\,O_2(g) = Ce_2O_3(c)$	(1,048°–1,900° K.)	−440,400	−4.60	......	.....	+97.42
$Ce(c) + O_2(g) = CeO_2(c)$	(298.16°–1,048° K.)	−245,490	−6.42	+2.34	−.20	+67.79
$Ce(l) + O_2(g) = CeO_2(c)$	(1,048°–2,000° K.)	−247,930	+.71	−.66	−.20	+51.73
$2\,Cs(c) + 1/2\,O_2(g) = Cs_2O(c)$	(298.16°–301.5°K.)	−75,900	......	......	.....	+36.60
$2\,Cs(l) + 1/2\,O_2(g) = Cs_2O(c)$	(301.5°–763° K.)	−76,900	......	......	.....	+39.92
$2\,Cs(l) + 1/2\,O_2(g) = Cs_2O(l)$	(763°–963° K.)	−75,370	−9.21	......	.....	+64.47
$2\,Cs(g) + 1/2\,O_2(g) = Cs_2O(l)$	(963°–1,500° K.)	−113,790	−23.03	......	.....	+145.60
$2\,Cs(c) + O_2(g) = Cs_2O_2(c)$	(298.16°–301.5° K.)	−96,500	−2.30	......	.....	+62.30
$2\,Cs(l) + O_2(g) = Cs_2O_2(c)$	(301.5°–870° K.)	−97,800	−4.61	......	.....	+72.34
$2\,Cs(l) + O_2(g) = Cs_2O_2(l)$	(870°–963° K.)	−96,060	−18.42	......	.....	+110.94
$2\,Cs(g) + O_2(g) = Cs_2O_2(l)$	(963°–1,500° K.)	−134,000	−31.08	......	.....	+188.11
$2\,Cs(c) + 3/2\,O_2(g) = Cs_2O_3(c)$	(298.16°–301.5° K.)	−112,690	−11.51	......	.....	+110.10
$2\,Cs(l) + 3/2\,O_2(g) = Cs_2O_3(c)$	(301.5°–775° K.)	−113,840	−12.66	......	.....	+116.77
$2\,Cs(l) + 3/2\,O_2(g) = Cs_2O_3(l)$	(775°–963° K.)	−110,740	−26.48	......	.....	+152.70
$2\,Cs(g) + 3/2\,O_2(g) = Cs_2O_3(l)$	(963°–1,500° K.)	−148,680	−39.14	......	.....	+229.87
$Cs(c) + O_2(g) = CsO_2(c)$	(298.16°–301.5° K.)	−63,590	−11.51	......	.....	+72.29
$Cs(l) + O_2(g) = CsO_2(c)$	(301.5°–705° K.)	−64,240	−12.66	......	.....	+77.30
$Cs(l) + O_2(g) = CsO_2(l)$	(705°–963° K.)	−61,770	−18.42	......	.....	+90.20
$Cs(g) + O_2(g) = CsO_2(l)$	(963°–1,500° K.)	−80,500	−24.18	......	.....	+126.83
$Cl_2(g) + 1/2\,O_2(g) = Cl_2O(g)$	(298.16°–2,000° K.)	+17,770	−.71	−.12	+.49	+16.81
$1/2\,Cl_2(g) + 1/2\,O_2(g) = ClO(g)$	(298.16°–1,000° K.)	+33,000	......	......	.....	−.24
$1/2\,Cl_2(g) + O_2(g) = ClO_2(g)$	(298.16°–2,000° K.)	+24,150	−.76	−.105	−.665	+19.08
$1/2\,Cl_2(g) + 3/2\,O_2(g) = ClO_3(g)$	(298.16°–500° K.)	+37,740	+5.76	......	.....	+21.42
$Cl_2(g) + 7/2\,O_2(g) = Cl_2O_7(g)$	(298.16°–500° K.)	+65,040	+12.66	......	.....	+78.01
$2\,Cr(c) + 3/2\,O_2(g) = Cr_2O_3(\beta)$	(298.16°–1,823° K.)	−274,670	−14.07	+2.01	+.69	+105.65
$2\,Cr(l) + 3/2\,O_2(g) = Cr_2O_3(\beta)$	(1,823°–2,000° K.)	−278,030	+2.33	−.35	+1.57	+58.29
$Cr(c) + O_2(g) = CrO_2(c)$	(298.16°–1,000° K.)	−142,500	......	......	.....	+42.00
$Cr(c) + 3/2\,O_2(g) = CrO_3(c)$	(298.16°–471° K.)	−141,590	−13.82	......	.....	+103.90
$Cr(c) + 3/2\,O_2(g) = CrO_3(l)$	(471°–600° K.)	−141,580	−32.24	......	.....	+153.14
$Co(\alpha, \beta) + 1/2\,O_2(g) = CoO(c)$	(298.16°–1,400° K.)	−56,910	+.69	......	.....	+16.03
$Co(\gamma) + 1/2\,O_2(g) = CoO(c)$	(1,400°–1,763° K.)	−58,160	−1.15	......	.....	+22.71
$Co(l) + 1/2\,O_2(g) = CoO(c)$	(1,763°–2,000° K.)	−65,680	−6.22	......	.....	+43.43
$3\,Co(\alpha, \beta, \gamma) + 2\,O_2(g) = Co_3O_4(c)$	(298.16°–1,500° K.)	−207,300	−2.30	......	.....	+90.56
$2\,Cu(c) + 1/2\,O_2(g) = Cu_2O(c)$	(298.16°–1,357° K.)	−40,550	−1.15	−1.10	−.10	+21.92
$2\,Cu(l) + 1/2\,O_2(g) = Cu_2O(c)$	(1,357°–1,502° K.)	−43,880	+8.47	−2.60	−.10	−3.72
$2\,Cu(l) + 1/2\,O_2(g) = Cu_2O(l)$	(1,502°–2,000° K.)	−37,710	−12.48	+.25	−.10	+54.44
$Cu(c) + 1/2\,O_2(g) = CuO(c)$	(298.16°–1,357° K.)	−37,740	−.64	−1.40	−.10	+24.87
$Cu(l) + 1/2\,O_2(g) = CuO(c)$	(1,357°–1,720° K.)	−39,410	+4.17	−2.15	−.10	+12.05
$Cu(l) + 1/2\,O_2(g) = CuO(l)$	(1,720°–2,000° K.)	−41,060	−11.35	+.25	−.10	+59.09
$F_2(g) + 1/2\,O_2(g) = F_2O(g)$	(298.16°–2,000° K.)	+5,070	−.41	−.15	+.535	+16.04
$2\,Ga(c) + 1/2\,O_2(g) = Ga_2O(c)$	(298.16°–302.7° K.)	−81,110	+10.32	−5.75	−.10	−3.66

Reaction and temperature range of validity	ΔH_o	2.303a	b	c	I
2 Ga(l) + 1/2 O_2(g) = Ga_2O(c)........(302.7°–1,000° K.)	−83,360	+13.49	−5.75	−.10	−4.08
2 Ga(c) + 3/2 O_2(g) = Ga_2O_3(c)....(298.16°–302.7° K.)	−256,240	+14.64	−3.75	−.30	+32.23
2 Ga(l) + 3/2 O_2(g) = Ga_2O_3(c)....(302.7°–2,000° K.)	−258,490	+17.82	−3.75	−.30	+31.79
Ge(c) + 1/2 O_2(g) = GeO(c)...........(298.16°–1,200° K.)	−60,900	+1.27	−1.49	−.10	+17.19
Ge(c) + 1/2 O_2(g) = GeO(g)...........(298.16°–1,200° K.)	−21,870	+6.72	−.075	....	−41.25
Ge(c) + O_2(g) = GeO_2(gl)...........(298.16°–1,200° K.)	−127,830	+4.28	−2.52	−0.20	+30.54
2 Au(c) + 3/2 O_2(g) = Au_2O_3(c)......(298.16°–500° K.)	−2,160	−10.36	....	....	+95.14
Hf(c) + O_2(g) = HfO_2(monocl.).......(298.16°–2,000° K.)	−268,380	−9.74	−.28	+1.54	+78.16
H_2(g) + 1/2 O_2(g) = H_2O(l).........(298.16°–373.16° K.)	−70,600	−18.26	+.64	−.04	+91.67
H_2(g) + 1/2 O_2(g) = H_2O(g).........(298.16°–2,000° K.)	−56,930	+6.75	+.64	−.08	−8.74
D_2(g) + 1/2 O_2(g) = D_2O(l).........(298.16°–374.5° K.)	−72,760	−18.10	....	....	+93.59
D_2(g) + 1/2 O_2(g) = D_2O(g).........(298.16°–2,000° K.)	−58,970	+5.50	−.75	+.085	−3.74
1/2 H_2(g) + 1/2 O_2(g) = OH(g)......(298.16°–2,000° K.)	+10,350	+.90	+.005	−.26	−6.69
H_2(g) + O_2(g) = H_2O_2(l)..........(298.16°–425° K.)	−47,140	−13.52	−7.13	....	+99.39
H_2(g) + O_2(g) = H_2O_2(g)..........(298.16°–1,500° K.)	−32,570	+4.77	−.96	+.97	+13.84
2 In(c) + 3/2 O_2(g) = In_2O_3(c).......(298.16°–429.6° K.)	220,410	+5.43	−.50	−.30	+59.49
2 In(l) + 3/2 O_2(g) = In_2O_3(c).......(429.6°–2,000° K.)	−220,970	+13.22	−3.00	−.30	+41.36
I_2(c) + 5/2 O_2(g) = I_2O_5(c).........(298.16°–386.8° K.)	−42,040	+2.30	....	....	+113.71
I_2(l) + 5/2 O_2(g) = I_2O_5(c).....:....(386.8°–456° K.)	−43,490	+16.12	....	....	+81.70
I_2(g) + 5/2 O_2(g) = I_2O_5(c).........(456°–500° K.)	−58,020	−6.91	....	....	+174.79
Ir(c) + O_2(g) = IrO_2(c)............(298.16°–1,200° K.)	−39,480	+8.17	−6.39	−.20	+20.33
0.947 Fe(α) + 1/2 O_2(g) = $Fe_{0.947}O$(c)..(298.16°–1,033° K.)	−65,320	−11.26	+2.61	+0.44	+48.60
0.947 Fe(β) + 1/2 O_2(g) = $Fe_{0.947}O$(c)..(1,033°–1,179° K.)	−62,380	+4.08	−.75	+.235	+3.00
0.947 Fe(γ) + 1/2 O_2(g) = $Fe_{0.947}O$(c)..(1,179°–1,650° K.)	−66,750	−8.04	+.67	−.10	+42.28
0.947 Fe(γ) + 1/2 O_2(g) = $Fe_{0.947}O$(l)..(1,650°–1,674° K.)	−64,200	−18.72	+1.67	−.10	+73.45
0.947 Fe(δ) + 1/2 O_2(g) = $Fe_{0.947}O$(l)..(1,674°–1,803° K.)	−59,650	−6.84	+.25	−.10	+34.81
0.947 Fe(l) + 1/2 O_2(g) = $Fe_{0.947}O$(l)..(1,803°–2,000° K.)	−63,660	−7.48	+.25	−.10	+39.12
3 Fe(α) + 2 O_2(g) = Fe_3O_4(magnetite)..(298.16°–900° K.)	−268,310	+5.87	−12.45	+.245	+73.11
3 Fe(α) + 2 O_2(g) = Fe_3O_4(β)........(900°–1,033° K.)	−272,300	−54.27	+11.65	+.245	+233.52
3 Fe(β) + 2 O_2(g) = Fe_3O_4(β)........(1,033°–1,179° K.)	−262,990	−5.71	+1.00	−.40	+89.19
3 Fe(γ) + 2 O_2(g) = Fe_3O_4(β)........(1,179°–1,674° K.)	−276,990	−44.05	+5.50	−.40	+213.52
3 Fe(δ) + 2 O_2(g) = Fe_3O_4(β)........(1,674°–1,803° K.)	−262,560	−6.40	+1.00	−.40	+91.05
3 Fe(l) + 2 O_2(g) = Fe_3O_4(β)........(1,803°–1,874° K.)	−275,280	−8.74	+1.00	−.40	+104.84
3 Fe(l) + 2 O_2(g) = Fe_3O_4(l)........(1,874°–2,000° K.)	−257,240	−26.89	+1.00	−.40	+155.46
2 Fe(α) + 3/2 O_2(g) = Fe_2O_3(hematite)..(298.16°–950° K.)	−200,000	−13.84	−1.45	+1.905	+108.26
2 Fe(α) + 3/2 O_2(g) = Fe_2O_3(β)......(950°–1,033° K.)	−202,960	−42.64	+7.85	+.13	+188.48
2 Fe(β) + 3/2 O_2(g) = Fe_2O_3(β)......(1,033°–1,050° K.)	−196,740	−10.27	+.75	−.30	+92.26
2 Fe(β) + 3/2 O_2(g) = Fe_2O_3(γ)......(1,050°–1,179° K.)	−193,200	−.39	−.13	−.30	+59.96
2 Fe(γ) + 3/2 O_2(g) = Fe_2O_3(γ)......(1,179°–1,674° K.)	−202,540	−25.95	+2.87	−.30	+142.85
2 Fe(δ) + 3/2 O_2(g) = Fe_2O_3(γ)......(1,674°–1,800° K.)	−192,920	−.85	−.13	−.30	+61.21
2 La(c) + 3/2 O_2(g) = La_2O_3(c)......(298.16°–1,153° K.)	101,120	15.91	+.80	+1.34	+112.36
2 La(l) + 3/2 O_2(g) = La_2O_3(c)......(1,153°–2,000° K.)	434,330	−4.88	+.80	+1.34	+91.17
Pb(c) + 1/2 O_2(g) = PbO (red)........(298.16°–600.5° K.)	−52,800	−2.76	+.80	−.10	+32.49
Pb(l) + 1/2 O_2(g) = PbO (red)........(600.5°–762° K.)	−53,780	−.51	−1.75	−.10	+28.44
Pb(c) + 1/2 O_2(g) = PbO (yellow)......(298.16°–600.5° K.)	−52,040	+.81	−2.00	.10	+29.19
Pb(l) + 1/2 O_2(g) = PbO (yellow)......(600.5°–1,159° K.)	50,020	+3.00	−2.95	−.10	+18.08
Pb(l) + 1/2 O_2(g) = PbO(l)...........(1,159°–1,745° K.)	−53,980	−12.94	+.25	−.10	+64.22
Pb(c) + 1/2 O_2(g) = PbO(g)...........(298.16°–600.5° K.)	+10,270	+1.91	+1.08	+.295	−23.21
Pb(l) + 1/2 O_2(g) = PbO(g)...........(600.5°–2,000° K.)	+9,300	+4.17	+.13	+.295	−27.29
3 Pb(c) + 2 O_2(g) = Pb_3O_4(c)........(298.16°–600.5° K.)	−174,920	+8.82	−8.20	−.40	+72.78
3 Pb(l) + 2 O_2(g) = Pb_3O_4(c)........(600.5°–1,000° K.)	−177,860	+15.59	−11.05	−.40	+60.57
Pb(c) + O_2(g) = PbO_2(c)...........(298.16°–600.5° K.)	−66,120	+.64	−2.45	−.20	+45.58
Pb(l) + O_2(g) = PbO_2(c)...........(600.5°–1,100° K.)	−67,100	+2.90	−3.40	−.20	+41.50
2 Li(c) + 1/2 O_2(g) = Li_2O(c)........(298.16°–452° K.)	−142,220	−3.06	+5.77	−.10	+34.19
2 Li(l) + 1/2 O_2(g) = Li_2O(c)........(452°–1,600° K.)	−141,380	+16.97	−2.63	−.10	−17.05
2 Li(c) + O_2(g) = Li_2O_2(c).........(298.16°–452° K.)	−151,880	−1.38	....	....	+54.83
2 Li(l) + O_2(g) = Li_2O_2(c).........(452°–500° K.)	−153,260	−1.38	....	....	+57.88
Mg(c) + 1/2 O_2(g) = MgO (periclase)..(298.16°–923° K.)	−144,090	−1.06	+.13	+.25	+29.16
Mg(l) + 1/2 O_2(g) = MgO (periclase)..(923°–1,393° K.)	−145,810	+1.84	−.62	+.64	+23.07
Mg(g) + 1/2 O_2(g) = MgO (periclase)..(1,393°–2,000° K.)	−180,700	−3.75	−.62	+.64	+65.69
Mg(c) + O_2(g) = MgO_2(c)...........(298.16°–500° K.)	−150,230	−9.12	+.13	+.25	+70.84
Mn(α) + 1/2 O_2(g) = MnO(c)..........(298.16°–1,000° K.)	−92,600	−4.21	+.97	+.155	+29.66
Mn(β) + 1/2 O_2(g) = MnO(c)..........(1,000°–1,374° K.)	−91,900	+1.84	−.39	+.34	+12.15
Mn(γ) + 1/2 O_2(g) = MnO(c)..........(1,374°–1,410° K.)	−89,810	+7.30	−.72	+.34	−6.05
Mn(δ) + 1/2 O_2(g) = MnO(c)..........(1,410°–1,517° K.)	−89,390	+8.68	−.72	+.34	−10.70
Mn(l) + 1/2 O_2(g) = MnO(c)..........(1,517°–2,000° K.)	−93,350	+7.99	−.72	+.34	−5.90
3 Mn(α) + 2 O_2(g) = Mn_3O_4(α)......(298.16°–1,000° K.)	−332,400	−7.41	+.66	+.145	+106.62
3 Mn(β) + 2 O_2(g) = Mn_3O_4(α)......(1,000°–1,374° K.)	−330,310	+10.75	−3.42	+.70	+54.07
3 Mn(γ) + 2 O_2(g) = Mn_3O_4(α)......(1,374°–1,410° K.)	−324,050	+27.12	−4.41	+.70	−.50
3 Mn(δ) + 2 O_2(g) = Mn_3O_4(α)......(1,410°–1,445° K.)	−322,800	+31.27	−4.41	+.70	−14.46
3 Mn(δ) + 2 O_2(g) = Mn_3O_4(β)......(1,445°–1,517° K.)	−328,870	−4.56	+1.00	−.40	+95.20
3 Mn(l) + 2 O_2(g) = Mn_3O_4(β)......(1,517°–1,800° K.)	−340,730	−6.63	+1.00	−.40	+109.60
2 Mn(α) + 3/2 O_2(g) = Mn_2O_3(c)....(298.16°–1,000° K.)	−230,610	−5.96	−.06	+.945	+80.74
2 Mn(β) + 3/2 O_2(g) = Mn_2O_3(c)....(1,000°–1,374° K.)	−229,210	+6.15	−2.78	+1.315	+45.70
2 Mn(γ) + 3/2 O_2(g) = Mn_2O_3(c)....(1,374°–1,410° K.)	−225,030	+17.06	−3.44	+1.315	+9.33
2 Mn(δ) + 3/2 O_2(g) = Mn_2O_3(c)....(1,410°–1,517° K.)	−224,200	+19.82	−3.44	+1.315	+0.05
2 Mn(l) + 3/2 O_2(g) = Mn_2O_3(c)....(1,517°–1,700° K.)	−232,110	+18.44	−3.44	+1.315	+9.65
Mn(α) + O_2(g) = MnO_2(c)..........(298.16°–629.88° K.)	−126,400	−8.61	+.97	+1.555	+70.14
2 Hg(l) + 1/2 O_2(g) = Hg_2O(c).......(298.16°–629.88° K.)	−22,400	−4.61	....	....	+43.29
2 Hg(g) + 1/2 O_2(g) = Hg_2O(c).......(629.88°–1,000° K.)	−53,800	−16.12	....	....	+125.36
Hg(l) + 1/2 O_2(g) = HgO (red)........(298.16°–629.88° K.)	−21,760	+.85	−2.47	−.10	+24.81
Hg(g) + 1/2 O_2(g) = HgO (red)........(629.88°–1,500° K.)	−36,920	−2.92	−2.47	−.10	+59.42
Mo(c) + O_2(g) = MoO_2(c)...........(298.16°–2,000° K.)	−132,910	−3.91	....	....	+47.42
Mo(c) + 3/2 O_2(g) = MoO_3(c)........(298.16°–1,068° K.)	−182,650	−8.86	−1.55	+1.54	+90.07
Mo(c) + 3/2 O_2(g) = MoO_3(l)........(1,068°–1,500° K.)	−179,770	−36.34	+1.40	−.30	+167.61
2 Nd(c) + 3/2 O_2(g) = Nd_2O_3(hexagonal).(298.16°–1,113° K.)	−435,150	−16.19	+3.21	+1.78	+125.68
2 Nd(c) + 3/2 O_2(g) = Nd_2O_3(hexagonal).(1,113°–1,500° K.)	−437,090	+4.03	−2.13	+1.78	+71.77
Np(c) + O_2(g) = NpO_2(c)...........(298.16°–913° K.)	−246,450	−3.45	....	....	+52.44
Np(l) + O_2(g) = NpO_2(c)...........(913°–1,500° K.)	−249,010	−4.61	....	....	+58.68
Ni(α) + 1/2 O_2(g) = NiO(c)..........(298.16°–633° K.)	−57,640	−4.61	+2.16	−.10	+34.41
Ni(β) + 1/2 O_2(g) = NiO(c)..........(633°–1,725° K.)	−57,460	−.14	−.46	−.10	+23.27
Ni(l) + 1/2 O_2(g) = NiO(c)..........(1,725°–2,000° K.)	−58,830	+7.23	−1.36	−.10	+1.76
2 Nb(c) + 2 O_2(g) = Nb_2O_4(c).......(298.16°–2,000° K.)	−382,050	−9.67	....	....	+116.23
2 Nb(c) + 5/2 O_2(g) = Nb_2O_5(c).....(298.16°–1,785° K.)	−458,640	−16.14	−.56	+1.94	+157.66
2 Nb(c) + 5/2 O_2(g) = Nb_2O_5(l).....(1,785°–2,000° K.)	−463,630	−66.04	+2.21	−.50	+317.84
N_2(g) + 1/2 O_2(g) = N_2O(g)........(298.16°–2,000° K.)	+18,650	−1.57	−.27	+.92	+23.47

Reaction and temperature range of validity	ΔH_o	2.303a	b	c	I
$1/2\ N_2(g) + 1/2\ O_2(g) = NO(g)$ .... (298.16°–2,000° K.)	+21,590	− .28	+ .45	− .03	−2.20
$N_2(g) + 3/2\ O_2(g) = N_2O_3(g)$ .... (298.16°–500° K.)	+17,390	− .35			+54.30
$1/2\ N_2(g) + O_2(g) = NO_2(g)$ .... (298.16°–2,000° K.)	+7,730	+ .53	− .265	+ .605	+13.74
$N_2(g) + 2\ O_2(g) = N_2O_4(g)$ .... (298.16°–1,000° K.)	+1,370	+2.14	−3.24	+1.38	+68.34
$N_2(g) + 5/2\ O_2(g) = N_2O_5(c)$ .... (298.16°–305° K.)	−10,200				+131.70
$N_2(g) + 5/2\ O_2(g) = N_2O_5(g)$ .... (298.16°–500° K.)	+3,600				+86.50
$Os(c) + 2\ O_2(g) = OsO_4$ (yellow) .... (298.16°–329° K.)	−92,260	+14.97			+32.45
$Os(c) + 2\ O_2(g) = OsO_4$ (white) .... (298.16°–315° K.)	−90,560	+14.97			+27.60
$Os(c) + 2\ O_2(g) = OsO_4(l)$ .... (315°–403° K.)	−88,970	+9.67			+35.78
$Os(c) + 2\ O_2(g) = OsO_4(g)$ .... (298.16°–1,000° K.)	−81,200	+8.17	−2.86	+1.94	+20.34
$3/2\ O_2(g) = O_3(g)$ .... (298.16°–2,000° K.)	+33,980	+2.03	− .48	+ .36	+11.45
P (white) $+ 1/2\ O_2(g) = PO(g)$ .... (298.16°–317.4° K.)	−9,370	+2.53			−25.40
$P(l) + 1/2\ O_2(g) = PO(g)$ .... (317.4°–553° K.)	−9,390	+3.45			−27.63
$1/4\ P_4(g) + 1/2\ O_2(g) = PO(g)$ .... (553°–1,500° K.)	−12,640	+2.30			−18.61
$4\ P$ (white) $+ 5\ O_2(g) =$ (298.16°–317.4° K.) P_4O_{10} (hexagonal).	−711,520	+95.67	−51.50	−1.00	−28.24
$4\ P(l) + 5\ O_2(g) = P_4O_{10}$ (hex.) .... (317.4°–553° K.)	−711,800	+97.98	−51.50	−1.00	−33.13
$P_4(l) + 5\ O_2(g) = P_4O_{10}$ (hex.) .... (553°–631° K.)	−725,560	+87.45	−51.07	−2.405	+20.87
$P_4(g) + 5\ O_2(g) = P_4O_{10}(g)$ .... (631°–1,500° K.)	−722,330	−43.45	+2.93	−2.405	+348.20
$Pu(c) + O_2(g) = PuO_2(c)$ .... (298.16°–1,500° K.)	−246,450	−3.45			+52.48
$Po(c) + O_2(g) = PoO_2(c)$ .... (298.16°–900° K.)	−61,510	−9.21			+72.80
$2\ K(c) + 1/2\ O_2(g) = K_2O(c)$ .... (298.16°–336.4° K.)	−86,400				+33.90
$2\ K(l) + 1/2\ O_2(g) = K_2O(c)$ .... (336.4°–1,049° K.)	−87,380	+1.15			+33.90
$2\ K(g) + 1/2\ O_2(g) = K_2O(c)$ .... (1,049°–1,500° K.)	−133,090	−16.12			+129.64
$2\ K(c) + O_2(g) = K_2O_2(c)$ .... (298.16°–336.4° K.)	−118,300	−2.30			+59.60
$2\ K(l) + O_2(g) = K_2O_2(c)$ .... (336.4°–763° K.)	−119,780	−4.61			+69.85
$2\ K(l) + O_2(g) = K_2O_2(l)$ .... (763°–1,049° K.)	−118,250	−18.42			+107.66
$2\ K(g) + O_2(g) = K_2O_2(l)$ .... (1,049°–1,500° K.)	−161,870	−31.08			+187.49
$2\ K(c) + 3/2\ O_2(g) = K_2O_3(c)$ .... (298.16°–336.4° K.)	−126,640	−12.66			+111.75
$2\ K(l) + 3/2\ O_2(g) = K_2O_3(c)$ .... (336.4°–703° K.)	−127,790	−12.66			+115.16
$2\ K(l) + 3/2\ O_2(g) = K_2O_3(l)$ .... (703°–1,000° K.)	−125,330	−27.63			+154.28
$K(c) + O_2(g) = KO_2(c)$ .... (298.16°–336.4° K.)	−68,940	−10.36			+66.45
$K(l) + O_2(g) = KO_2(c)$ .... (336.4°–653° K.)	−69,510	−10.36			+68.15
$K(l) + O_2(g) = KO_2(l)$ .... (653°–1,000° K.)	−67,880	−18.42			+88.34
$K(c) + 3/2\ O_2(g) = KO_3(c)$ .... (298.16°–336.4° K.)	−63,340	−10.36			+85.85
$K(l) + 3/2\ O_2(g) = KO_3(c)$ .... (336.4°–500° K.)	−63,910	−10.36			+87.55
$2\ Pr(c) + 3/2\ O_2(g) = Pr_2O_3(c,\ C'\text{-}type)$ .. (298.16°–1,205° K.)	−440,600	−4.60			+78.38
$2\ Pr(l) + 3/2\ O_2(g) = Pr_2O_3(c,\ C'\text{-}type)$ .. (1,205°–2,000° K.)	−446,100	−4.60			+82.94
$6\ Pr(c) + 11/2\ O_2(g) = Pr_6O_{11}(c)$ ...... (298.16°–1,205° K.)	−1,374,000				+241.04
$6\ Pr(l) + 11/2\ O_2(g) = Pr_6O_{11}(c)$ ...... (1,205°–1,500° K.)	−1,390,500				+254.73
$Pr(c) + O_2(g) = PrO_2(c)$ ...... (298.16°–1,200° K.)	−230,990	−6.42	+2.34	− .20	+61.07
$Ra(c) + 1/2\ O_2(g) = RaO(c)$ ...... (298.16°–1,000° K.)	−130,000				+23.50
$Re(c) + 3/2\ O_2(g) = ReO_3(c)$ ...... (298.16°–433° K.)	−149,090	−16.12			+110.49
$Re(c) + 3/2\ O_2(g) = ReO_3(l)$ ...... (433°–1,000° K.)	−146,750	−31.32			+145.16
$2\ Re(c) + 7/2\ O_7(g) = Re_2O_7(c)$ ...... (298.16°–569° K.)	−301,470	−34.54			+250.57
$2\ Re(c) + 7/2\ O_7(g) = Re_2O_7(l)$ ...... (569°–635.5° K.)	−295,810	−73.68			+348.45
$2\ Re(c) + 7/2\ O_7(g) = Re_2O_7(g)$ ...... (635.5°–1,500° K.)	−256,460	+3.45			+70.33
$2\ Re(c) + 4\ O_2(g) = Re_2O_8(c)$ ...... (298.16°–420° K.)	−313,870	−41.45			+293.57
$2\ Re(c) + 4\ O_2(g) = Re_2O_8(l)$ ...... (420°–600° K.)	−318,470	−87.50			+425.32
$2\ Rh(c) + 1/2\ O_2(g) = Rh_2O(c)$ ...... (298.16°–2,000° K.)	−23,740	−8.06			+35.64
$Rh(c) + 1/2\ O_2(g) = RhO(c)$ ...... (298.16°–1,500° K.)	−22,650	−7.37			+40.54
$2\ Rh(c) + 3/2\ O_2(g) = Rh_2O_3(c)$ ...... (298.16°–1,500° K.)	−70,060	−13.58			+101.72
$2\ Rb(c) + 1/2\ O_2(g) = Rb_2O(c)$ ...... (298.16°–312.2° K.)	−78,900				+32.20
$2\ Rb(l) + 1/2\ O_2(g) = Rb_2O(c)$ ...... (312.2°–750° K.)	−79,950				+35.56
$2\ Rb(l) + 1/2\ O_2(g) = Rb_2O(l)$ ...... (750°–952° K.)	−78,830	−10.36			+63.85
$2\ Rb(g) + 1/2\ O_2(g) = Rb_2O(l)$ ...... (952°–1,500° K.)	−120,290	−23.03			+145.14
$2\ Rb(c) + O_2(g) = Rb_2O_2(c)$ ...... (298.16°–312.2° K.)	−102,000	−2.30			+57.40
$2\ Rb(l) + O_2(g) = Rb_2O_2(c)$ ...... (312.2°–840° K.)	−103,360	−4.61			+67.52
$2\ Rb(l) + O_2(g) = Rb_2O_2(l)$ ...... (840°–952° K.)	−101,680	−18.42			+105.91
$2\ Rb(g) + O_2(g) = Rb_2O_2(l)$ ...... (952°–1,500° K.)	−143,130	−31.08			+187.16
$2\ Rb(c) + 3/2\ O_2(g) = Rb_2O_3(c)$ ...... (298.16°–312.2° K.)	−118,190	−11.51			+104.70
$2\ Rb(l) + 3/2\ O_2(g) = Rb_2O_3(c)$ ...... (312.2°–760° K.)	−119,400	−12.66			+111.43
$2\ Rb(l) + 3/2\ O_2(g) = Rb_2O_3(c)$ ...... (760°–952° K.)	−116,740	−27.63			+151.06
$2\ Rb(g) + 3/2\ O_2(g) = Rb_2O_3(l)$ ...... (952°–1,500° K.)	−157,720	−39.14			+228.39
$Rb(c) + O_2(g) = RbO_2(c)$ ...... (298.16°–312.2° K.)	−52,330	−10.36			+66.25
$Rb(l) + O_2(g) = RbO_2(c)$ ...... (312.2°–685° K.)	−65,120	−11.51			+71.30
$Rb(l) + O_2(g) = RbO_2(l)$ ...... (685°–952° K.)	−63,070	−18.42			+87.89
$Rb(g) + O_2(g) = RbO_2(l)$ ...... (952°–1,500° K.)	−83,560	−24.18			+126.57
$Ru(\alpha,\ \beta,\ \gamma) + O_2(g) = RuO_2(c)$ ...... (298.16°–1,500° K.)	−57,290	−6.91			+62.01
$2\ Sm(c) + 3/2\ O_2(g) = Sm_2O_3(c)$ ...... (298.16°–1,623° K.)	−430,600	−4.60			+78.38
$2\ Sm(l) + 3/2\ O_2(g) = Sm_2O_3(c)$ ...... (1,623°–2,000° K.)	−438,000	−4.60			+82.94
$2\ Sc(c) + 3/2\ O_2(g) = Sc_2O_3(c)$ ...... (298.16°–1,673° K.)	−409,960	+7.78	−1.84	−0.30	+52.73
$2\ Sc(l) + 3/2\ O_2(g) = Sc_2O_3(c)$ ...... (1,673°–2,000° K.)	−412,950	+18.47	−2.93	− .30	+21.88
$Se(c) + 1/2\ O_2(g) = SeO(g)$ ...... (298.16°–490° K.)	+9,280	−3.04	+4.40	+ .30	−14.78
$Se(l) + 1/2\ O_2(g) = SeO(g)$ ...... (490°–1,027° K.)	+9,420	+8.70		+ .30	−44.50
$1/2\ Se_2(g) + 1/2\ O_2(g) = SeO(g)$ ...... (1,027°–2,000° K.)	−7,400	− .37		+ .19	− .80
$Se(c) + O_2(g) = SeO_2(c)$ ...... (298.16°–490° K.)	−53,770	+14.94	−9.41	− .20	+6.94
$Se(l) + O_2(g) = SeO_2(c)$ ...... (490°–595° K.)	−53,640	+27.59	−14.31		−25.05
$Se(l) + O_2(g) = SeO_2(g)$ ...... (595°–1,027° K.)	−32,840	+6.79			−10.80
$1/2\ Se_2(g) + O_2(g) = SeO_2(g)$ ...... (1,027°–2,000° K.)	−49,000	− .74			+27.61
$Si(c) + 1/2\ O_2(g) = SiO(g)$ ...... (298.16°–1,683° K.)	−21,090	+3.84	+ .16	− .295	−33.14
$Si(l) + 1/2\ O_2(g) = SiO(g)$ ...... (1,683°–2,000° K.)	−30,170	−7.78	− .12	+ .25	−40.01
$Si(a) + O_2(g) = SiO_2(\alpha\text{-}quartz)$ ...... (298.16°–848° K.)	−210,070	+3.98	−3.32	+ .605	+34.59
$Si(c) + O_2(g) = SiO_2(\beta\text{-}quartz)$ ...... (848°–1,683° K.)	−209,920	−3.36	− .19	− .745	+53.44
$Si(l) + O_2(g) = SiO_2(\beta\text{-}quartz)$ ...... (1,683°–1,883° K.)	−219,000	+ .58	− .47	− .20	+46.58
$Si(l) + O_2(g) = SiO_2(l)$ ...... (1,883°–2,000° K.)	−228,590	−15.66			+103.97
$Si(c) + O_2(g) = SiO_2(\alpha\text{-}crist.)$ ...... (298.16°–523° K.)	−207,330	+19.96	−9.75	− .745	−9.78
$Si(c) + O_2(g) = SiO_2(\beta\text{-}crist.)$ ...... (523°–1,683° K.)	−209,820	−3.34	− .24	− .745	+53.35
$Si(l) + O_2(g) = SiO_2(\beta\text{-}crist.)$ ...... (1,683°–2,000° K.)	−218,900	+ .60	− .52	− .20	+46.49
$Si(c) + O_2(g) = SiO_2(\alpha\text{-}trid.)$ ...... (298.16°–390° K.)	−207,030	+22.29	−11.62	− .745	−15.64
$Si(c) + O_2(g) = SiO_2(\beta\text{-}trid.)$ ...... (390°–1,683° K.)	−209,350	−1.59	− .54	− .745	+47.86
$Si(l) + O_2(g) = SiO_2(\beta\text{-}trid.)$ ...... (1,683°–1,953° K.)	−218,430	+2.35	− .82	− .20	+41.00
$2\ Ag(c) + 1/2\ O_2(g) = Ag_2O(c)$ ...... (298.16°–1,000° K.)	−7,740	−4.14			+27.84
$2\ Ag(c) + O_2(g) = Ag_2O_2(c)$ ...... (298.16°–500° K.)	−6,620	−3.22			+52.17
$2\ Na(c) + 1/2\ O_2(g) = Na_2O(c)$ ...... (298.16°–371° K.)	−99,820	−7.51	+5.47	− .10	+50.43
$2\ Na(l) + 1/2\ O_2(g) = Na_2O(c)$ ...... (371°–1,187° K.)	−100,150	+4.97	−2.45	− .10	+22.19
$2\ Na(l) + 1/2\ O_2(g) = Na_2O(c)$ ...... (1,187°–1,190° K.)	−156,200	−20.72			+145.48
$2\ Na(g) + 1/2\ O_2(g) = Na_2O(l)$ ...... (1,190°–2,000° K.)	−150,250	−23.03			+147.58
$2\ Na(c) + O_2(g) = Na_2O_2(c)$ ...... (298.16°–371° K.)	−122,500	−2.30			+57.51
$2\ Na(l) + O_2(g) = Na_2O_2(c)$ ...... (371°–733° K.)	−124,320	−5.76			+71.30

Reaction and temperature range of validity	ΔH_0	2.303a	b	c	I
$2\,Na(l) + O_2(g) = Na_2O_2(l)$ .... (733°–1,187° K.)	−123,220	−20.72	.....	.....	+112.66
$2\,Na(g) + O_2(g) = Na_2O_2(l)$ .... (1,187°–1,500° K.)	−174,800	−31.08	.....	.....	+187.97
$Na(c) + O_2(g) = NaO_2(c)$ .... (298.16°–371° K.)	−63,040	−8.06	.....	.....	+56.98
$Na(l) + O_2(g) = NaO_2(c)$ .... (371°–1,000° K.)	−64,220	−11.51	.....	.....	+69.04
$Sr(c) + 1/2\,O_2(g) = SrO(c)$ .... (298.16°–1,043° K.)	−142,410	−6.79	+.305	+.675	+44.33
$Sr(l) + 1/2\,O_2(g) = SrO(c)$ .... (1,043°–1,657° K.)	−143,370	−2.42	−.38	+.675	+32.77
$Sr(g) + 1/2\,O_2(g) = SrO(c)$ .... (1,657°–2,000° K.)	−181,180	−8.24	−.38	+.675	+74.32
$Sr(c) + O_2(g) = SrO_2(c)$ .... (298.16°–1,000° K.)	−155,510	−11.40	+.305	+.675	+75.44
$S(rh) + 1/2\,O_2(g) = SO(g)$ .... (298.16°–368.6° K.)	+19,250	−1.24	+2.95	+.225	−18.84
$S(mon) + 1/2\,O_2(g) = SO(g)$ .... (368.6°–392° K.)	+19,200	−1.29	+3.31	+.225	−18.72
$S(\lambda,\mu) + 1/2\,O_2(g) = SO(g)$ .... (392°–718° K.)	+20,320	+10.22	−.17	+.225	−50.05
$1/2\,S_2(g) + 1/2\,O_2(g) = SO(g)$ .... (298.16°–2,000° K.)	+3,890	+.07	.....	.....	−1.50
$S(rh) + O_2(g) = SO_2(g)$ .... (298.16°–368.6° K.)	−70,980	+.83	+2.35	+.51	−5.85
$S(mon) + O_2(g) = SO_2(g)$ .... (368.6°–392° K.)	−71,020	+.78	+2.71	+.51	−5.74
$S(\lambda,\mu) + O_2(g) = SO_2(g)$ .... (392°–718° K.)	−69,900	+12.30	−.77	+.51	−37.10
$1/2\,S_2(g) + O_2(g) = SO_2(g)$ .... (298.16°–2,000° K.)	−86,330	+2.42	−.70	+.31	+10.71
$S(rh) + 3/2\,O_2(g) = SO_3(c-I)$ .... (298.16°–335.4° K.)	−111,370	6.45	.....	.....	+88.32
$S(rh) + 3/2\,O_2(g) = SO_3(c-II)$ .... (298.16°–305.7° K.)	−108,680	−11.97	.....	.....	+94.95
$S(rh) + 3/2\,O_2(g) = SO_3(l)$ .... (298.16°–335.4° K.)	−107,430	−21.18	.....	.....	+113.76
$S(rh) + 3/2\,O_2(g) = SO_3(g)$ .... (298.16°–368.6° K.)	−95,070	+1.43	+0.66	+1.26	+16.81
$S(mon) + 3/2\,O_2(g) = SO_3(g)$ .... (368.6°–392° K.)	−95,120	+1.38	+1.02	+1.26	+16.93
$S(\lambda,\mu) + 3/2\,O_2(g) = SO_3(g)$ .... (392°–718° K.)	−94,010	+12.89	−2.46	+1.26	−14.40
$1/2\,S_2(g) + 3/2\,O_2(g) = SO_3(g)$ .... (298.16°–1,500° K.)	−110,420	+3.02	−2.39	+1.06	+33.41
$2\,Ta(c) + 5/2\,O_2(g) = Ta_2O_5(c)$ .... (298.16°–2,000° K.)	−492,790	−17.18	−1.25	+2.46	+161.68
$Tc(c) + O_2(g) = TcO_2(c)$ .... (298.16°–500° K.)	−103,400	.....	.....	.....	+41.00
$Tc(c) + 3/2\,O_2(g) = TcO_3(c)$ .... (298.16°–500° K.)	−129,000	.....	.....	.....	+64.50
$2\,Tc(c) + 7/2\,O_2(g) = Tc_2O_7(c)$ .... (298.16°–392.7° K.)	−266,000	.....	.....	.....	+147.00
$2\,Tc(c) + 7/2\,O_2(g) = Tc_2O_7(l)$ .... (392.7°–500° K.)	−258,930	.....	.....	.....	+129.00
$Te(c) + 1/2\,O_2(g) = TeO(g)$ .... (298.16°–723° K.)	+43,110	+1.91	+.84	+.315	−27.22
$Te(l) + 1/2\,O_2(g) = TeO(g)$ .... (723°–1,360° K.)	+39,750	+6.08	+.09	+.315	−33.94
$1/2\,Te_2(g) + 1/2\,O_2(g) = TeO(g)$ .... (1,360°–2,000° K.)	+23,730	−.90	+.09	+.315	−.29
$Te(c) + O_2(g) = TeO_2(c)$ .... (298.16°–723° K.)	−78,090	−2.10	−2.35	−.20	+51.27
$Te(l) + O_2(g) = TeO_2(c)$ .... (723°–1,006° K.)	−81,530	+1.84	−3.10	−.20	+45.30
$Te(l) + O_2(g) = TeO_2(g)$ .... (1,006°–1,300° K.)	−82,090	−21.74	+.50	−.20	+113.04
$2\,Tl(\alpha) + O_2(g) = Tl_2O(c)$ .... (298.16°–505.5° K.)	−44,110	−6.91	.....	.....	+42.30
$2\,Tl(\beta) + O_2(g) = Tl_2O(c)$ .... (505.5°–573° K.)	−44,260	−6.91	.....	.....	+42.60
$2\,Tl(\beta) + 1/2\,O_2(g) = Tl_2O(l)$ .... (573°–576° K.)	−40,880	−13.82	.....	.....	+55.76
$2\,Tl(l) + 1/2\,O_2(g) = Tl_2O(l)$ .... (576°–773° K.)	−42,320	−11.51	.....	.....	+51.89
$2\,Tl(l) + 1/2\,O_2(g) = Tl_2O(l)$ .... (773°–1,730° K.)	−18,400	+11.51	.....	.....	−45.55
$2\,Tl(g) + 1/2\,O_2(g) = Tl_2O(l)$ .... (1,730°–2,000° K.)	−104,670	.....	.....	.....	+41.59
$2\,Tl(\alpha) + 3/2\,O_2(g) = Tl_2O_3(c)$ .... (298.16°–505.5° K.)	−99,410	−16.12	.....	.....	+119.09
$2\,Tl(\beta) + 3/2\,O_2(g) = Tl_2O_3(c)$ .... (505.5°–576° K.)	−99,560	16.10	.....	.....	+119.05
$2\,Tl(l) + 3/2\,O_2(g) = Tl_2O_3(c)$ .... (576°–990° K.)	−101,010	−13.82	.....	.....	+115.55
$2\,Tl(l) + 3/2\,O_2(g) = Tl_2O_3(c)$ .... (990°–1,500° K.)	−94,550	−27.63	.....	.....	+150.39
$2\,Tl(\alpha) + 2\,O_2(g) = Tl_2O_4(c)$ .... (298.16°–505.5° K.)	−117,680	−23.03	.....	.....	+161.19
$2\,Tl(\beta) + 2\,O_2(g) = Tl_2O_4(c)$ .... (505.5°–576° K.)	−117,830	−23.03	.....	.....	+161.49
$2\,Tl(l) + 2\,O_2(g) = Tl_2O_4(c)$ .... (576°–1,000° K.)	−119,970	30.70	.....	.....	+137.63
$Th(c) + O_2(g) = ThO_2(c)$ .... (298.16°–2,000° K.)	−294,350	−5.25	+.59	+.775	+62.81
$Sn(c) + 1/2\,O_2(g) = SnO(c)$ .... (298.16°–505° K.)	−68,600	−3.57	+1.65	−.10	+32.59
$Sn(l) + 1/2\,O_2(g) = SnO(c)$ .... (505°–1,300° K.)	−69,670	+3.06	−1.50	−.10	+18.39
$Sn(c) + 1/2\,O_2(g) = SnO(g)$ .... (298.16°–505° K.)	−1,000	−.97	+3.24	+.32	−17.41
$Sn(l) + 1/2\,O_2(g) = SnO(g)$ .... (505°–2,000° K.)	−2,070	+5.66	+.09	+.32	−31.62
$Sn(c) + O_2(g) = SnO_2(c)$ .... (298.16°–505° K.)	−142,010	−14.00	+2.45	+2.38	+90.74
$Sn(l) + O_2(g) = SnO_2(c)$ .... (505°–1,898° K.)	−143,080	−7.37	−.70	+2.38	+76.53
$Sn(l) + O_2(g) = SnO_2(l)$ .... (1,898°–2,000° K.)	−139,130	−21.97	+.50	−.20	+120.11
$Ti(\alpha) + 1/2\,O_2(g) = TiO(\alpha)$ .... (298.16°–1,150° K.)	−125,010	−4.01	−.29	+.83	+36.28
$Ti(\beta) + 1/2\,O_2(g) = TiO(\alpha)$ .... (1,150°–1,264° K.)	−125,040	+1.17	−1.55	+.83	+21.90
$Ti(\beta) + 1/2\,O_2(g) = TiO(\beta)$ .... (1,264°–2,000° K.)	−125,210	−1.77	−1.25	−.10	+30.83
$Ti(\alpha) + 1/2\,O_2(g) = TiO(g)$ .... (298.16°–1,150° K.)	+11,710	+3.71	+1.07	−.10	−35.50
$Ti(\beta) + 1/2\,O_2(g) = TiO(g)$ .... (1,150°–2,000° K.)	+11,680	+8.89	−.19	−.10	−49.88
$2\,Ti(\alpha) + 3/2\,O_2(g) = Ti_2O_3(\alpha)$ .... (298.16°–473° K.)	−360,660	+32.08	−23.49	−.30	−10.66
$2\,Ti(\alpha) + 3/2\,O_2(g) = Ti_2O_3(\beta)$ .... (473°–1,150° K.)	−369,710	−30.95	+2.62	+4.80	+162.79
$2\,Ti(\beta) + 3/2\,O_2(g) = Ti_2O_3(\beta)$ .... (1,150°–2,000° K.)	−369,760	−20.59	+.10	+4.80	+134.03
$3\,Ti(\alpha) + 5/2\,O_2(g) = Ti_3O_5(\alpha)$ .... (298.16°–450° K.)	−587,980	−4.19	−9.72	−.50	+131.05
$3\,Ti(\alpha) + 5/2\,O_2(g) = Ti_3O_5(\beta)$ .... (450°–1,150° K.)	−586,330	−18.31	+1.03	−.50	+159.98
$3\,Ti(\beta) + 5/2\,O_2(g) = Ti_3O_5(\beta)$ .... (1,150°–2,000° K.)	−586,420	−2.76	−2.75	−.50	+116.81
$Ti(\alpha) + O_2(g) = TiO_2\ (rutile)$ .... (298.16°–1,150° K.)	−228,360	−12.80	+1.62	+1.975	+82.81
$Ti(\beta) + O_2(g) = TiO_2\ (rutile)$ .... (1,150°–2,000° K.)	−228,380	−7.62	+.36	+1.975	+68.43
$W(c) + O_2(g) = WO_2(c)$ .... (298.16°–1,743° K.)	−137,180	−1.38	.....	.....	+45.56
$4\,W(c) + 11/2\,O_2(g) = W_4O_{11}(c)$ .... (298.16°–1,700° K.)	−745,730	−32.70	.....	.....	+321.84
$W(c) + 3/2\,O_2(g) = WO_3(c)$ .... (298.16°–1,743° K.)	−201,180	−2.92	−1.81	−.30	+70.89
$W(c) + 3/2\,O_2(g) = WO_3(l)$ .... (1,743°–2,000° K.)	−203,140	−35.74	+1.13	−.30	+173.27
$U(\alpha) + O_2(g) = UO_2(c)$ .... (298.16°–935° K.)	−262,880	−19.92	+3.70	+2.13	+100.54
$U(\beta) + O_2(g) = UO_2(c)$ .... (935°–1,045° K.)	−260,660	−4.28	−.31	+1.78	+55.50
$U(\gamma) + O_2(g) = UO_2(c)$ .... (1,045°–1,405° K.)	−262,830	−6.54	−.31	+1.78	+64.41
$U(l) + O_2(g) = UO_2(c)$ .... (1,405°–2,000° K.)	−264,790	−5.92	.....	.....	+63.50
$3\,U(\alpha) + 4\,O_2(g) = U_3O_8(c)$ .... (298.16°–935° K.)	−863,370	−56.57	+10.68	+5.20	+330.19
$3\,U(\beta) + 4\,O_2(g) = U_3O_8(c)$ .... (935°–1,045° K.)	−856,720	−9.67	−1.35	+4.15	+195.12
$3\,U(\gamma) + 4\,O_2(g) = U_3O_8(c)$ .... (1,045°–1,405° K.)	−863,230	−16.44	−1.35	+4.15	+221.79
$3\,U(l) + 4\,O_2(g) = U_3O_8(c)$ .... (1,405°–1,500° K.)	−869,460	−10.91	−1.35	+4.15	+208.82
$U(\alpha) + 3/2\,O_2(g) = UO_3\ (hex)$ .... (298.16°–935° K.)	−294,090	−18.33	.....	+1.535	+114.94
$U(\beta) + 3/2\,O_2(g) = UO_3\ (hex)$ .... (935°–1,045° K.)	−291,870	−2.69	−.52	+1.185	+69.90
$U(\gamma) + 3/2\,O_2(g) = UO_3\ (hex)$ .... (1,045°–1,400° K.)	−294,040	−4.95	−.52	+1.185	+78.80
$V(c) + 1/2\,O_2(g) = VO(c)$ .... (298.16°–2,000° K.)	−101,090	−5.39	−.36	+.53	+38.69
$V(c) + 1/2\,O_2(g) = VO(g)$ .... (298.16°–2,000° K.)	+52,090	+1.80	+1.04	+.35	−28.42
$2\,V(c) + 3/2\,O_2(g) = V_2O_3(c)$ .... (298.16°–2,000° K.)	−299,910	−17.98	+.37	+2.41	+118.83
$2\,V(c) + 2\,O_2(g) = V_2O_4(\alpha)$ .... (298.16°–345° K.)	−342,890	−11.03	+3.00	−.40	+117.38
$2\,V(c) + 2\,O_2(g) = V_2O_4(\beta)$ .... (345°–1,818° K.)	−345,330	−24.36	+1.30	+3.545	+155.55
$2\,V(c) + 2\,O_2(g) = V_2O_4(l)$ .... (1,818°–2,000° K.)	−339,880	−59.59	+3.00	−0.40	+264.42
$6\,V(c) + 13/2\,O_2(g) = V_6O_{13}(c)$ .... (298.16°–1,000° K.)	−1,076,340	−95.33	.....	.....	+557.61
$2\,V(c) + 5/2\,O_2(g) = V_2O_5(c)$ .... (298.16°–943° K.)	−381,960	−41.08	+5.20	+6.11	+228.50
$2\,V(c) + 5/2\,O_2(g) = V_2O_5(l)$ .... (943°–2,000° K.)	−365,840	−38.91	+3.25	−.50	+207.54
$2\,Y(c) + 3/2\,O_2(g) = Y_2O_3(c)$ .... (298.16°–1,773° K.)	−419,600	+2.76	−1.73	−.30	+66.36
$2\,Y(l) + 3/2\,O_2(g) = Y_2O_3(c)$ .... (1,773°–2,000° K.)	−422,850	+13.36	−2.75	−.30	+35.56
$Zn(c) + 1/2\,O_2(g) = ZnO(c)$ .... (298.16°–692.7° K.)	−84,670	−6.40	+.84	+.99	+43.25
$Zn(l) + 1/2\,O_2(g) = ZnO(c)$ .... (692.7°–1,180° K.)	−85,520	−1.45	−.36	+.99	+31.25
$Zn(g) + 1/2\,O_2(g) = ZnO(c)$ .... (1,180°–2,000° K.)	−115,940	−7.28	−.36	+.99	+74.94
$Zr(c) + O_2(g) = ZrO_2(\alpha)$ .... (298.16°–1,135° K.)	−262,980	−6.10	+.16	+1.045	+65.00
$Zr(\beta) + O_2(g) = ZrO_2(\alpha)$ .... (1,135°–1,478° K.)	−264,190	−5.09	−.40	+1.48	+63.58
$Zr(\beta) + O_2(g) = ZrO_2(\beta)$ .... (1,478°–2,000° K.)	−262,290	−7.76	+.50	−.20	+69.50

REFRACTORY MATERIALS

Borides

Name	Formula	Molecular weight	Melting point, °C	Crystalline form	Lattice parameter, Å	X-ray density, g/cm³
Chromium boride	CrB_2	73.65	1,850[29]a	Hexagonal A1B₂ type [C 32]	$a = 2.969$, $c = 3.066$[39]	5.16
Hafnium boride	HfB_2	200.14	3,100[31]	Hexagonal A1B₂ type [C 32]	$a = 3.14$[6], $c = 3.47$	10.5
Molybdenum boride	MoB	106.77	2,180[40]	Tetragonal	$a = 3.110$[41], $c = 16.95$	8.77
	MoB_2	117.59	2,100[40]	Hexagonal A1B₂ type [C 32]	$a = 3.05$, $c = 3.113$[42]	7.78
	Mo_2B	202.72	2,000[40] (decomposes)	Tetragonal CuAl₂ type [C 16]	$a = 5.543$[41], $c = 4.735$	9.31
Niobium boride	NbB	103.73	>0.2000[3]	Orthorhombic	$a = 3.298$, $b = 8.724$, $c = 3.137$[35]	
	NbB_2	114.55	2,900[36] (decomposes) >0.2000[3]	Hexagonal AlB₂ type [C 32]	$a = 3.086$[28], $c = 3.306$	7.21
Tantalum boride	TaB	191.77	>0.2000[3]	Orthorhombic	$a = 3.276$, $b = 8.669$, $c = 3.157$[37]	14.29
	TaB_2	202.59	3,000[6]	Hexagonal AlB₂ type [C 32]	$a = 3.088$[28], $c = 3.241$	12.60
Thorium boride	ThB_4	275.53	>0.2500[6]	Tetragonal D_{4k}^5-P4/mbm	$a = 7.256$[43], $c = 4.113$	8.45
Titanium boride	TiB_2	69.54	2,980[6]	Hexagonal AlB₂ type [C 32]	$a = 3.028$[28], $c = 3.228$	4.52
Tungsten boride	WB	194.68	2,860[29]	Tetragonal	$a = 3.115$[6], $c = 16.92$	16.0
	W_2B	378.54	2,770[29]	Tetragonal CuAl₂ type [C 16]	$a = 5.564$[41], $c = 4.740$	16.72
Uranium boride	UB_{12}	367.91	>0.1500[6] (decomposes)	Face-centered cubic	$a = 7.473$[44]	5.82
Vanadium boride	VB_2	72.59	2,100[29]	Hexagonal AlB₂ type [C 32]	$a = 2.998$[28], $c = 3.057$	5.10
Zirconium boride	ZrB_2	112.86	3,040 ± 50[30]	Hexagonal AlB₂ type [C 32]	$a = 3.169$[28], $c = 3.530$	6.09
	ZrB_{12}	221.06	2,680[30]	Face-centered cubic	$a = 7.408$[33]	

Name	Thermal conductivity, cal-sec⁻¹-cm⁻²-cm-°C⁻¹	Electrical resistivity, microhm-cm	Ductility relative scale[b]	Resistance to oxidation[c]	Hardness[d]
Chromium boride		21 at room temperature[38]			1,800 kg/mm²[29]
Hafnium boride		10[31]	3[3]		
Molybdenum boride		α-MoB = 45 at room temperature[40]; β-MoB = 25 at room temperature; 45 at room temperature[40]; 40 at room temperature[40]	3[3]	3[3]	8 Mohs[20]; 1,570 kg/mm²[29]; 1,280 kg/mm²[29]; 8–9 Mohs[20]
Niobium boride	0.040 at 20°C[36]	6.45 at room temperature[36]; 32.0 at room temperature[36]			>0.8 Mohs[36]
Tantalum boride	0.026 at 20°C[6]	100 at room temperature[38]; 68 at room temperature[38]	3[3]	3[3]	
Thorium boride					
Titanium boride	0.0624 at 200°C[28] (15% porosity)	28.4[28]	3[3]	2–3[3]	3,400 kg/mm²[29]
Tungsten boride			3[3]	3[3]	9 Mohs[6]
Uranium boride					
Vanadium boride	16 at 20°C[34]				8–9 Mohs[6]
Zirconium boride	0.0550 at 200°C[28]; 0.029 at room temperature[33]	9.2 at 20°C[31]; 60–80 at room temperature[30]	2[3]	2–3[3]	8 Mohs[32]

Carbides

Name	Formula	Molecular weight	Melting point, °C	Crystalline form	Lattice parameter, Å	X-ray density g/cm³
Boron carbide	B_4C	55.29	2,450	Hexagonal	$a_0 = 5.60$[6], $c_0 = 12.12$	2.52
Chromium carbide	Cr_3C_2	180.05	1,895[15]	Orthorhombic (D5₁₀)	$a = 2.82$, $b = 5.53$, $c = 11.47$[16]	6.7
	Cr_7Cr_3	400.01	1,780[15]	Hexagonal [C_{3v^4}]	$a = 14.01$, $c = 4.532$[17]	6.92
Graphite	C	12.01	3,700 ± 100[3]	Hexagonal	Orthohexagonal axes $a_0 = 2.46$, $b_0 = 4.28$, $c_0 = 6.71$	2.25
Hafnium carbide	HfC	190.51	3,890[9]	Cubic NaCl type (B1)	$a = 4.46$[10]	12.7

Carbides (Continued)

Name	Formula	Molecular weight	Melting point, °C	Crystalline form	Lattice parameter, Å	X-ray density, g/cm³
	MoC	107.96	2,695[9]	Face-centered cubic	a = 4.28[19]	
Molybdenum carbide	Mo$_2$C	203.91	2,690[6]	Hexagonal (L'3)	a = 3.002 c = 4.724[18]	9.2
Niobium carbide	NbC	104.92	3,500[9]	Cubic NaCl type (B1)	a = 4.461	7.85
Silicon carbide						
β	SiC	40.07	Trans. to α at 2,100	Face-centered cubic	a$_0$ = 4.3590[6]	3.22
α	SiC		2,700[6]	Hexagonal (Wurtzite)	a$_0$ = 3.081[6] c$_0$ = 5.0394	
Tantalum carbide	TaC	192.96	3,880[9], 4,730[4]	Cubic NaCl type (B1)	a = 4.455[14]	14.5
Thorium carbide	ThC	244.06	2,625[23]	Cubic NaCl type (B1)	a = 5.34[9]	10.67
	ThC$_2$	256.07	2,655[29]	Monoclinic [C 2/e]	a = 6.53[24] b = 4.24	9.6
Titanium carbide	TiC	59.91	3,160[4]	Cubic NaCl type (B1)	a = 4.32[6]	4.938
Tungsten carbide	W$_2$C	379.73	2,730[21]	Hexagonal (L'3)	a = 2.98[6] c = 4.71	17.34
	WC	195.87	2,630[21] (decomposes)	Hexagonal (L'3)	a = 2.900[22] c = 2.831	15.77
Uranium carbide	UC	250.08	2,450–2,500[25]	Cubic NaCl type (B1)	a = 4.955[26]	13.6
	UC$_2$	262.09	2,350–2,400[25]	Body-centered tetragonal CaCl$_2$ type	a = 3.517[27] c = 5.987	11.68
Vanadium carbide	VC	62.96	2,830[1]	Cubic NaCl type (B1)	a = 4.16[6]	5.8
Zirconium carbide	ZrC	103.23	3,030[4]	Cubic NaCl type (B1)	a = 4.689[5]	6.44

Name	Thermal conductivity, cal-sec⁻¹-cm⁻²-cm-°C⁻¹	Electrical resistivity, microhm-cm	Ductility relative scale[b]	Resistance to oxidation[c]	Hardness[d]
Boron carbide	0.05 at 20–425°C[1]	0.30–0.80[1]	3[3]	3[3]	9.3 Mohs
Chromium carbide			3[3]	3[3]	1,300 kg/mm²[8]
			3[3]	3[3]	
Graphite	0.268–0.451 at room temperature	65 at room temperature[71]	3[3]	4[3]	
Hafnium carbide		109 at room temperature[1]	3[3]	3[3]	
Molybdenum carbide		97[6]	3[3]	5[3]	1,800 kg/mm²[6]
			3[3]	5[3]	>0.7 Mo[b,c][20]
Niobium carbide	0.034 at room temperature[7]	74[6]	2–3[3]	3[3]	2,170 kg/mm²[13]
Silicon carbide					
β	0.10 at 20–425°C[1]	107–200 ohm cm[1] at room temperature	3[3]	2[3]	9.2 Mohs
Tantalum carbide	0.053 at room temperature[6]	30[6]	2–3[3]	3[3]	1,800 kg/mm²[8]
Thorium carbide			3[3]	3[3]	
Titanium carbide	0.049 at room temperature[6]	180–250[4]	3[3]	3[4]	3,200 kg/mm²[8]
Tungsten carbide		80[6]	3[3]	5[3]	3,000 kg/mm²[6]
		53[6]	3[3]	5[3]	2,400 kg/mm²[6]
Uranium carbide			3[3]		
				5[3]	
Vanadium carbide		150[6]	3[3]		2,800 kg/mm²[11]
					2,100 kg/mm²[12]
Zirconium carbide	0.049 at room temperature[6]	70 at room temperature[4]	3[3]	3[3]	2,600 kg/mm²[6]

Nitrides

Name	Formula	Molecular weight	Melting point, °C	Crystalline form	Lattice parameter, Å	X-ray density, g/cm³
Boron nitride	BN	24.83	2,730[6]	Hexagonal (B 12)	a = 2.51 ± .02 c = 6.70 ± .04[52]	2.25
Chromium nitride	CrN	66.02	1,500 (decomposes)	Cubic NaCl type (B1)	a = 4.140[6]	6.14
Hafnium nitride	HfN	192.60	3,310[6]			
Niobium nitride	NbN	106.92	2,050[46]	Cubic NaCl type (B1)	a = 4.41–4.375[6]	7.28
Tantalum nitride	Ta$_2$N	375.77	3,090[46]	Hexagonal	a = 3.05[48] c = 4.95[53]	14.1
Thorium nitride	ThN	246.13	2,630[48]	Cubic NaCl type (B1)	a = 5.2[48]	
Titanium nitride	TiN	61.91	2,930 N$_2$ liberated on melting[46]	Cubic NaCl type (B1)	a = 4.23[53]	5.43
Uranium nitride	UN		2,650	Cubic NaCl type (B1)	a = 4.880[58]	14.32
Vanadium nitride	VN	64.96	2,050[46]	Cubic NaCl type (B1)	a = 4.129[6]	6.102
Zirconium nitride	ZrN	105.22	2,980[47]	Cubic NaCl type (B1)	a = 4.567[53]	7.349

REFRACTORY MATERIALS (Continued)

Nitrides (Continued)

Name	Thermal conductivity, cal-sec^{-1}-cm^{-2}-cm-°C^{-1}	Electrical resistivity, microhm-cm	Ductility relative scale[b]	Resistance to oxidation[c]	Hardness[d]
Boron nitride					
Chromium nitride		1,900 at 2,000°C[1]	3[3]	2[3]	2.0 Mohs
Hafnium nitride					
Niobium nitride		200 at room temperature[46]	5[3]	3[3]	+8 Mohs[46]
Tantalum nitride		135 at room temperature[47]	3[3]	5[3]	+8
Thorium nitride					
Titanium nitride		21.7 at room temperature	3[3]	3[3]	Between 9 and 10 Mohs[46]
Uranium nitride					
Vanadium nitride		200 at room temperature[46]	3[3]		
Zirconium nitride		13.6 at room temperature[46]	3[3]	3[3]	+8 Mohs[6]

Oxides

Name	Formula	Molecular weight	Melting point, °C	Crystalline form	Lattice parameter, A	X-ray density, g/cm^3
Aluminum oxide	Al_2O_3	101.92	2,015[1]	Rhombohedral	a = 5.13 axial angle = 55° 6'	3.965
Beryllium oxide	BeO	25.02	2,550[1]	Hexagonal	a = 2.70 c = 4.39	3.03
Cerium oxide	CeO_2	172.3	2,600[1]	Face-centered cubic	a = 5.41	7.13
Chromic oxide	Cr_2O_3	152.02	2,265[1]	Rhombohedral	a = 5.38 axial angle = 54° 50'	5.21
Hafnium oxide	HfO_2	210.6	2,777[1]	Face-centered cubic	a = 5.11	9.68
Magnesium oxide	MgO	40.32	2,800[1]	Face-centered cubic	a = 4.20	3.58
Silicon oxide	SiO_2	60.06	1,728[1]	Hexagonal	a = 4.90 c = 5.39	2.32 (low cristobalite)
Thorium oxide	ThO_2	264.12	3,300[1]	Face-centered cubic	a = 5.59	9.69
Titanium oxide	TiO_2	79.90	1,840[1]	Tetragonal	a = 4.58 c = 2.98	4.24
Uranium oxide	UO_2	270.07	2,280[1]	Face-centered cubic	a = 5.47	10.96
Zirconium oxide	ZrO_2	123.22	2,677[1]	Monoclinic	a = 5.21 c = 5.37	5.56

Name	Thermal conductivity, cal-sec^{-1}-cm^{-2}-cm-°C^{-1}	Electrical resistivity, microhm-cm	Ductility relative scale[b]	Resistance to oxidation[c]	Hardness[d]
Aluminum oxide	0.0723 at 100°C[2]	1×10^{22} at 14°C[1] 3×10^{19} at 300°C[1] 3.5×10^{14} at 800°C[1]	3[3]	1[3]	9 Mohs[1]
Beryllium oxide	0.525 at 100°C[2]	4×10^{14} at 600°C[1] 5×10^{12} at 1,100°C[1] 8×10^{8} at 2,100°C[1]			9 Mohs[1]
Cerium oxide		6.5×10^{10} at 800°C[1] 3.4×10^{8} at 1,200°C[1]			6 Mohs[1]
Chromium oxide		1.3×10^{9} at 350°C[1] 2.3×10^{7} at 1,200°C[1] 6.8×10^{3} at 600°C[1] 4.5×10^{3} at 1,100°C[1]	3[3]	1[3]	
Hafnium oxide		5×10^{15} at 400°C[1] 1×10^{3} at 1,500°C[1]			
Magnesium oxide	0.0860 at 100°C[2]	2×10^{14} at 850°C[1] 3×10^{13} at 980°C[1] 4.5×10^{8} at 2,100°C[1]			6 Mohs[1]
Silicon oxide		1×10^{21} at 20°C[1] 7×10^{12} at 600°C[1] 4×10^{5} at 1,300°C[1] (vitreous)	3[4] (vitreous)	1[4] (vitreous)	6–7 Mohs[1] (cristobalite)
Thorium oxide	0.0245 at 100°C[2]	2.6×10^{13} at 550°C[1] 8×10^{11} at 800°C[1] 1.5×10^{10} at 1,200°C[1]			6.5 Mohs[1]
Titanium oxide	0.0156 at 100°C[2]	1.2×10^{10} at 800°C[1] 8.5×10^{6} at 1,200°C[1]			5.5–6.0 Mohs[1]
Uranium oxide	0.0234 at 100°C[2]	3.8×10^{10} at 20°C[1] 5×10^{8} at 500°C[1]			
Zirconium oxide	0.00466 at 100°C[2]	1×10^{12} at 385°C[1] 2.2×10^{10} at 700°C[1] 3.6×10^{8} at 1,200°C[1]	3[3]	1[3]	6.5 Mohs[1]

Silicides

Name	Formula	Molecular weight	Melting point, °C	Crystalline form	Lattice parameter, Å	X-ray density g/cm³
Chromium silicide	$CrSi_2$	108.13	1,570[51]	Hexagonal	a = 4.42 c = 6.35[51]	
	CrSi	80.07	1,870 (decomposes in presence of C)	Tetragonal (C 11b)	a = 3.20 c = 7.86[6]	
Molybdenum silicide	$MoSi_2$	152.07	1,870 (decomposes in presence of C)	Tetragonal (C 11b)	a = 3.20 c = 7.86[64][51]	6.24
Niobium silicide	$NbSi_2$		1,950[50]	Hexagonal $CrSi_2$ type	a = 4.785 c = 6.576[51]	5.29
Tantalum silicide	$TaSi_2$	237.00	2,400[50]	Hexagonal $CrSi_2$ type	a = 4.773 c = 6.552[51]	8.83
Titanium silicide	$TiSi_2$	104.02	1,540[49]	Orthorhombic	a = 8.24 b = 4.79 c = 8.52[59]	4.13
	Ti_5Si_3	323.68	2,120[49]	Hexagonal	a = 7.465 c = 5.162[60]	4.32
Tungsten silicide	WSi_2	240.04	2,050[50]	Tetragonal $MoSi_2$ Structure (C 11b)	a = 3.21 c = 7.83[66]	9.3
Uranium silicide	βUSi_2	294.19	1,700[6]	Hexagonal	a = 3.85[6] c = 4.06	9.25
	U_3Si_2	770.33	1,665[6]	Tetragonal	a = 7.3151[6] c = 3.8925	12.20
Vanadium silicide	VSi_2	107.07	1,750[50]	Hexagonal $CrSi_2$ type	a = 4.562 c = 6.359[51]	4.71

Name	Thermal conductivity, cal·sec^{-1}·cm^{-1}·cm^{-1}·°C^{-1}	Electrical resistivity, microhm·cm	Ductility relative scale[b]	Resistance to oxidation[c]	Hardness[d]
Chromium silicide					1,150 kg/mm²[50]
Molybdenum silicide	0.075 at room temperature to 200°C[6]	21.5 at room temperature 18.9 at –80°C[69]	2[3]	1[3]	1,290 kg/mm²[50]
Niobium silicide		6.3[68]	2[3]	4[3]	1,050 kg/mm²[50]
Tantalum silicide		8.5[68]	2[3]	3[3]	1,560 kg/mm²[50]
Titanium silicide		123 (hot pressed)[1]	3[3]	4[3]	870 kg/mm²[70]
			3[3]	4[3]	986 kg/mm²[49]
Tungsten silicide		33.4[68]		1–2[3]	1,310 kg/mm²[68]
					1,090 kg/mm²[68]
Uranium silicide					
Vanadium silicide		9.5[68]			1,090 kg/mm²[50]

[a]Numbers in parentheses refer to references at end of table.

[b]Ductility — 1: capable of being severely drawn, rolled, or otherwise worked without failure; 2: capable of withstanding slight deformation, or consisting of individually ductile crystals fragilely bound together; 3: incapable of being worked, of glasslike brittleness.

[c]Resistance to oxidation — classed according to the temperature range in which the rate of attack by air would cause severe erosion or failure of the coated specimen within a few hours. 1: above 1,700°C; 2: 1,400 to 1,700°C; 3: 1,100 to 1,400°C; 4: 800 to 1,000°C; 5: 500 to 800°C.

[d]Microhardness values taken with 100-g load.

REFERENCES

1. Campbell, I. E., *High Temperature Technology*, Electrochemical Society, John Wiley & Sons New York, 1956.
2. Kingery, W. D., Franch, J., Coble, R. L., and Vasilos, T., Thermal conductivity: X, data for several pure oxide materials corrected to zero porosity, *J. Am. Ceram. Soc.*, 37, 2, 107, 1954.
3. Powel, C. F., Campbell, I. E., and Gonser, B. W., *Vapor Plating*, John Wiley & Sons, New York, 1955.
4. Friederich, E. and Sittig, L., *Z. Anorg. Allg. Chem.*, 144, 169, 1925.
5. Norton, J. T. and Morory, A. L., *Trans. Am. Inst. Min. Metall. Pet. Eng.*, 185, 133, 1949.
6. Schwarzkopf, P. and Kieffer, R., *Refractory Hard Metals*, Macmillan, New York, 1953.
7. Schwarzkopf, P. and Sindeband, S. J., Electrochemical Society, 97th Meeting, Cleveland, Ohio, 1950.
8. Kieffer, R. and Kölbl, F., *Powder Metall. Bull.*, 4, 4, 1949.
9. Agte, C. and Alterthum, H., *Z. Tech. Phys.*, 11, 182, 1930.
10. Becker, K. and Ebert, F., *Z. Phys.*, 31, 268, 1925.
11. Ruff, O. and Martin, W., *Z. Angew. Chem.*, 25, 49, 1912.
12. Hinnuber, J., *Z. VDI*, 92, 111, 1950.

13. Foster, L. S., Forbes, L. W., Jr., Friar, L. B., Moody, L. S., and Smith, W. H., *J. Am. Ceram. Soc.,* 33, 27, 1950.
14. Ellinger, F. H., *Trans. Am. Soc. Met.,* 31, 89, 1943.
15. Bloom, D. S. and Grant, N. J., *Trans. Am. Inst. Min. Metall. Pet. Eng.,* 188, 41, 1950.
16. Hellstrom, K. and Westgren, A., *Kem. Tidskr.,* 45, 141, 1933.
17. Westgren, A., *Jernkontorets Ann.,* 119, 231, 1935.
18. Kuo, K. and Hagg, G., *Nature,* 170, 245, 1952.
19. Nowotny, H. and Kieffer, R., *Z. Anorg. Allg. Chem.,* 267, 261, 1952.
20. Weiss, G., *Ann. Chem.,* 1, 446, 1946.
21. Brewer, L., Bromely, L. A., Gilles, P. W., and Lofgren, N. L., *The Chemistry and Metallurgy of Miscellaneous Materials: Thermodynamics,* McGraw-Hill, New York, 1950.
22. Becker, K., *Z. Phys.,* 51, 481, 1928.
23. Wilhelm, H. A. and Chiotti, P., *Trans. Am. Soc. Met.,* 42, 1295, 1950.
24. Hunt, E. B. and Rundle, R. E., *J. Am. Chem. Soc.,* 73, 4777, 1951.
25. Mallet, W., Gerds, A. F., and Nelson, H. R., *J. Electrochem. Soc.,* 99, 197, 1952.
26. Litz, L. M., Gurrett, A. B., and Croxton, F. C., *J. Am. Chem. Soc.,* 70, 1718, 1948.
27. Rundle, R. E., Baenziger, N. C., Wilson, A. S., and McDonald, R. A., *J. Am. Chem. Soc.,* 70, 99, 1948.
28. Norton, J. T., Blumenthal, H., and Sindeband, S. J., *Trans. Am. Inst. Min. Metall. Pet. Eng.,* 185, 749, 1949.
29. Honak, E. R., Thesis, Tech. Hochsch. Graz, 1951.
30. Glaser, F. W. and Post, B., *J. Met.,* 1953.
31. Moers, K., *Z. Anorg. Allg. Chem.,* 198, 262, 1931.
32. Andrieux, L., *Rev. Mét.,* 45, 49, 1948.
33. Post, B. and Glaser, F. W., *J. Met.,* 4, 631, 1952.
34. Moers, K., *Z. Anorg. Allg. Chem.,* 198, 243, 1931.
35. Anderson, L. H. and Kiessling, R., *Acta Chem. Scand.,* 4, 160, 209, 1950.
36. Glaser, F. W., *J. Met.,* 4, 391, 1952.
37. Kiessling, R., *Acta Chem. Scand.,* 3, 603, 1949.
38. Moers, K., *Z. Anorg. Allg. Chem.,* 198, 262, 1931.
39. Kiessling, R., *Acta Chem. Scand.,* 3, 595, 1949.
40. Steintz, R., *J. Met.,* 4, 148, 1952.
41. Kiessling, R., *Acta Chem. Scand.,* 1, 893, 1947.
42. Bertaut, F. and Blum, P., *Acta Crystallogr.,* 4, 72, 1951.
43. Zalkin, A. and Templeton, D. H., *J. Chem. Phys.,* 18, 391, 1950.
44. Bertaut, F. and Blum, P., *Comptes rendus,* 229, 666, 1949.
45. Baumann, H. N., Jr., Electrochemical Society, 99th Meeting, Washington, D.C., April 1951.
46. Friederich, E. and Sittig, L., *Z. Anorg. Allg. Chem.,* 143, 293, 1925.
47. Agte, C. and Moers, K., *Z. Anorg. Chem.,* 198, 233, 1931.
48. Chiotti, P., *J. Am. Ceram. Soc.,* 35, 123, 1952.
49. Hansen, M., Klasler, H. D., and McPherson, D. J., *Trans. Am. Soc. Met.,* 44, 518, 1952.
50. Cerwenka, E., Thesis, Tech. Hochsch. Graz, 1951.
51. Wallbaum, H. J., *Z. Metallkd.,* 33, 378, 1941.
52. Pease, R. S., *Acta Crystallogr.,* 5, 356, 1952.
53. van Arkel, A. E., *Physica,* 4, 296, 1924.
54. Horn, F. H. and Ziegler, W. T., *J. Am. Chem. Soc.,* 69, 2762, 1947.
55. Pauling, L., Killeffer, D. H., and Linz, A., *Molybdenum Compounds,* Interscience, New York, 1952.
56. Hagg, G., *Z. Phys. Chem. Abt. B,* 7, 339, 1930.
57. Kiessling, R. and Lier, Y. H., *J. Met.,* 3, 639, 1951.
58. Rundle, R. E., Baenziger, N. C., Wilson, A. S., and McDonald, R. A., *J. Am. Chem. Soc.,* 70, 99, 1941.
59. Laues, F. and Wallbaum, H. J., *Z. Kristallogr. A,* 101, 78, 1939.
60. Pietrokowsky, P. and Duwez, P., *J. Met.,* 3, 772, 1951.
61. Naray Szako, S. V., *Z. Kristallogr. A,* 97, 223, 1937.
62. Lundin, C. E., McPherson, D. J., and Hansen, M., American Society for Metals, 34th Ann. Convention, Preprint No. 41, 1952.
63. Wallbaum, H. J., *Z. Metallkd.,* 31, 362, 1939.
64. Zachariasen, W. H., *Z. Phys. Chem.,* 128, 39, 1927.
65. Templeton, D. H. and Dauben, C. H., *Acta Crystallogr.,* 3, 261, 1950.
66. Nowotny, H., Kieffer, R., and Schachner, H., *Mh. Chem.,* 83, 1243, 1952.
67. Brauer, G. and Mitices, A., *Z. Anorg. Allg. Chem.,* 249, 325, 1942.
68. Gallistl, E., Thesis, Tech. Hochsch. Graz, 1951.
69. Glaser, F. W., *J. Appl. Phys.,* 22, 103, 1951.
70. Cerwenka, E., Thesis, Tech. Hochsch. Graz, 1951.
71. Currie, L. M., Hamister, V. C., and MacPherson, H. G., paper presented at the United Nations International Conference on The Peaceful Uses of Atomic Energy, Geneva, 1955.

THERMODYNAMIC PROPERTIES OF ELEMENTS AND OXIDES

Thermodynamic calculations over a wide range of temperatures are generally made with the aid of algebraic equations representing the characteristic properties of the substances being considered. The necessary integrations and differentiations, or other mathematical manipulations, are then most easily effected.

The most convenient starting point in making such calculations for a given substance is the heat capacity at constant pressure. From this quantity and a knowledge of the properties of any phase transitions, the other thermodynamic properties may be computed by the well-known equations given in standard texts on thermodynamics.

Users of the following equations and tables are cautioned that the units for a, b, c, and d are cal/g mole, whereas those for A are Kcal/g mole. The necessary adjustment must be made when the data are substituted into the equations.

Empirical heat capacity equations are generally of the form of a power series with the absolute temperature T as the independent variable:

$$C_p = a' + (b' \times 10^{-3})T + (c' \times 10^{-6})T^2$$

or

$$C_p = a'' + (b'' \times 10^{-3})T + \frac{d \times 10^5}{T^2}$$

Since both forms are used in the ensuing, let

(1)
$$C_p = a + (b \times 10^{-3})T + (c \times 10^{-6})T^2 + \frac{d \times 10^5}{T^2}$$

The constants a, b, c, and d are to be determined either experimentally or by some theoretical or semi-empirical approach.

The heat content or enthalpy H is determined from the heat capacity by a simple integration over the range of temperatures for which (1) is applicable. Thus, if 298°K is taken as a reference temperature,

(2)
$$H_T - H_{298} = \int_{298}^{T} C_p dT$$

$$= a(T - 298) + \frac{1}{2}(b \times 10^{-3})(T^2 - 298^2) + \frac{1}{3}(c \times 10^{-6})(T^3 - 298^3) - (d \times 10^5)\left(\frac{1}{T} - \frac{1}{298}\right)$$

$$= aT + \frac{1}{2}(b \times 10^{-3})T^2 + \frac{1}{3}(c \times 10^{-6})T^3 - \frac{d \times 10^5}{T} - A$$

where all the constants on the right hand side of the equation have been incorporated in the term $-A$.

In general, the enthalpy is given by a sum of terms such as (2) for each phase of the substance involved in the temperature range considered plus terms which represent the heats of transitions:

$$H_T - H_{298} = \sum \int_{T_1}^{T_2} C_p dT + \sum \Delta H_{tr}$$

In a similar manner, the entropy Si is obtained from (1) by performing the integration

(3)
$$S_T - S_{298} = \int_{298}^{T} (C_p/T)dt$$

$$= a \ln(T/298) + (b \times 10^{-3})(T - 298) + \frac{1}{2}(c \times 10^{-6})(T^2 - 298^2) - \frac{1}{2}(d \times 10^5)\left(\frac{1}{T^2} - \frac{1}{298^2}\right)$$

$$= a \ln T + (b \times 10^{-3})T + \frac{1}{2}(c \times 10^{-6})T^2 - \left(\frac{\frac{1}{2}(d \times 10^5)}{T^2}\right) - B'$$

or

(4)
$$S_T = 2.303\,a \log T + (b \times 10^{-3})T + \frac{1}{2}(c \times 10^{-6})T^2 - \frac{1}{2}\left(\frac{d \times 10^{-5}}{T^2}\right) - B$$

where

(5)
$$B = B' - S_{298}$$

From the definition of free energy F:

$$F = H - TS$$

the quantity

$$F_T - H_{298} = (H - H_{298}) - TS_T$$

is obtained from (2) and (4):

(6)
$$F_T - H_{298} = -2.303aT \log T - \frac{1}{2}(b \times 10^{-3})T^2 - \frac{1}{6}(c \times 10^{-6})T^3 - \frac{\frac{1}{2}(d \times 10^5)}{T} + (B + a)T - A$$

and also the free energy function

$$(7) \qquad \frac{F_T - H_{298}}{T} = -2.303a \log T - \frac{1}{2}(b \times 10^{-3})T - \frac{1}{6}(c \times 10^{-6})T^2 - \frac{\frac{1}{2}(d \times 10^5)}{T^2} + (B + a) - \frac{A}{T}$$

In the following 2 tables there has been collected the values of the constants. The first column lists the element or the oxide. The second column gives the phase to which they are applicable. The third, fourth, and fifth columns specify the thermodynamic properties for the transition to the succeeding phase. In column 6, the value of the entropy at 298.15°K, the reference temperature, is given. The remaining columns, except for the last, give the values of the constants a, b, c, d, A, and B required in the thermodynamic equations.

All values throughout the table which represent estimates have been enclosed in parentheses.

The heat capacities at temperatures beyond the range of experimental determination were estimated by extrapolation. Where no experimental values were found, use of analogy with compounds of neighboring elements in the Periodic Table was employed.

THERMODYNAMIC PROPERTIES OF THE ELEMENTS*

Ele-ment	Phase	Temperature of Transition (°K)	Heat of Tran-sition (kcal/g mole)	Entropy of Transition (e.u.)	Entropy at 298°K (e.u.)	cal/g mole a	cal/g mole b	cal/g mole c	cal/g mole d	A (kcal/g mole)	B(e.u.)
Ac	solid	(1090)	(2.5)	(2.3)	(13)	(5.4)	(3.0)	—	—	(1.743)	(18.7)
	liquid	(2750)	(70)	(25)	—	(8)	—	—	—	(0.295)	(31.3)
Ag	solid	1234	2.855	2.313	10.20	5.09	1.02	—	0.36	1.488	19.21
	liquid	2485	60.72	24.43	—	7.30	—	—	—	0.164	30.12
	gas	—	—	—	—	(4.97)	—	—	—	(−66.34)	(−12.52)
Al	solid	931.7	2.57	2.76	6.769	4.94	2.96	—	—	1.604	22.26
	liquid	2600	67.9	26	—	7.0	—	—	—	0.33	30.83
Am	solid	(1200)	(2.4)	(2.0)	(13)	(4.9)	(4.4)	—	—	(1.657)	(16.2)
	liquid	2733	51.7	18.9	—	(8.5)	—	—	—	(0.409)	(34.5)
As	solid	883	3 ¼	35. ¼	8.4	5.17	2.34	—	—	1.646	21.8
Au	solid	1336.16	3.03	2.27	11.32	6.14	−0.175	0.92	—	1.831	23.65
	liquid	2933	74.21	25.30	—	7.00	—	—	—	−0.631	26.99
B	solid	2313	(3.8)	(1.6)	1.42	1.54	4.40	—	—	0.655	8.67
	liquid	2800	75	27	—	(6.0)	—	—	—	(−4.599)	(31.4)
Ba	solid, α	648	0.14	0.22	16	5.55	1.50	—	—	1.722	16.1
	solid, β	977	1.83	1.87	—	5.55	1.50	—	—	1.582	15.9
	liquid	1911	35.665	18.63	—	(7.4)	—	—	—	(0.843)	(25.3)
	gas	—	—	—	—	(4.97)	—	—	—	(−39.65)	(−11.7)
Be	solid	1556	2.919	1.501	2.28	5.07	1.21	—	−1.15	1.951	27.62
	liquid	—	—	—	—	5.27	—	—	—	−1.611	25.68
Bi	solid	544.2	2.63	4.83	13.6	5.38	2.60	—	—	1.720	17.8
	liquid	1900	41.1	21.6	—	7.60	—	—	—	−0.087	25.6
	gas	—	—	—	—	(4.97)	—	—	—	(−46.19)	(−15.9)
C	solid	—	—	—	1.3609	4.10	1.02	—	−2.10	1.972	23.484
Ca	solid, α	723	0.24	0.33	9.95	5.24	3.50	—	—	1.718	20.95
	solid, β	1123	2.2	1.96	—	6.29	1.40	—	—	1.689	26.01
	liquid	1755	38.6	22.0	—	7.4	—	—	—	−0.147	30.28
	gas	—	—	—	—	(4.97)	—	—	—	(−43.015)	(−9.88)
Cd	solid	594.1	1.46	2.46	12.3	5.31	2.94	—	—	1.714	18.8
	liquid	1040	23.86	22.94	—	7.10	—	—	—	0.798	26.1
	gas	—	—	—	—	(4.97)	—	—	—	(−25.28)	(−11.7)
Ce	solid	1048	2.1	2.0	13.8	4.40	6.0	—	—	1.579	13.1
	liquid	2800	73	26	—	(7.9)	—	—	—	(−0.148)	(29.1)
Cl₂	gas	—	—	—	53.286	8.76	0.27	—	−0.65	2.845	−2.929
Co	solid, α	723	0.005	0.007	6.8	4.72	4.30	—	—	1.598	21.4
	solid, β	1398	0.095	0.068	—	3.30	5.86	—	—	0.974	13.1
	solid, γ	1766	3.7	2.1	—	9.60	—	—	—	3.961	50.5
	liquid	3370	93	28	—	8.30	—	—	—	−2.034	38.7
Cr	solid	2173	3.5	1.6	5.68	5.35	2.36	—	−0.44	1.848	25.75
	liquid	2495	72.97	29.25	—	9.40	—	—	—	1.556	50.13
	gas	—	—	—	—	(4.97)	—	—	—	(−82.47)	(−13.8)
Cs	solid	301.9	0.50	1.7	19.8	7.42	—	—	—	2.212	22.5
	liquid	963	16.32	17.0	—	8.00	—	—	—	1.887	24.1
	gas	—	—	—	—	(4.97)	—	—	—	(−17.35)	(−13.6)
Cu	solid	1356.2	3.11	2.29	7.97	5.41	1.50	—	—	1.680	23.30
	liquid	2868	72.8	25.4	—	7.50	—	—	—	0.024	34.05
F₂	gas	—	—	—	48.58	8.29	0.44	—	−0.80	2.760	−0.76
Fe	solid, α	1033	0.410	0.397	6.491	3.37	7.10	—	0.43	1.176	14.59
	solid, β	1180	0.217	0.184	—	10.40	—	—	—	4.281	55.66
	solid, γ	1673	0.15	0.084	—	4.85	3.00	—	—	0.396	19.76
	solid, δ	1808	3.86	2.14	—	10.30	—	—	—	4.382	55.11
	liquid	3008	84.62	28.1	—	10.00	—	—	—	−0.021	50.73
Ga	solid	302.94	1.335	4.407	9.82	5.237	3.03	—	—	1.710	21.01
	liquid	2700	—	—	—	(6.645)	—	—	—	(0.648)	(23.64)
Ge	solid	1232	8.3	6.7	10.1	5.90	1.13	—	—	1.764	23.8
	liquid	2980	68	23	—	(7.3)	—	—	—	(−5.668)	(25.7)
H₂	gas	—	—	—	31.211	6.62	0.81	—	—	2.010	6.75
Hf	solid	(2600)	(6.0)	(2.3)	13.1	(6.00)	(0.52)	—	—	(1.812)	(21.2)
Hg	liquid	629.73	13.985	22.208	18.46	6.61	—	—	—	1.971	19.20
	gas	—	—	—	—	4.969	—	—	—	−13.048	−13.54
In	solid	430	0.775	1.80	13.88	5.81	2.50	—	—	1.844	19.97
	liquid	2440	53.8	22.0	—	7.50	—	—	—	1.564	27.34
	gas	—	—	—	—	(4.97)	—	—	—	(−58.42)	(−14.46)
Ir	solid	2727	6.6	2.4	8.7	5.56	1.42	—	—	1.721	23.4
K	solid	336.4	0.5575	1.657	15.2	1.3264	19.405	—	—	1.258	−1.86
	liquid	1052	18.88	17.95	—	8.8825	−4.565	2.9369	—	1.923	32.55
	gas	—	—	—	—	(4.97)	—	—	—	(−19.689)	(−9.46)

* From the U.S. Atomic Energy Commission Report ANL-5750.

THERMODYNAMIC PROPERTIES OF THE ELEMENTS (Continued)

Element	Phase	Temperature of Transition (°K)	Heat of Transition (kcal/g mole)	Entropy of Transition (e.u.)	Entropy at 298°K (e.u.)	cal/g mole a	cal/g mole b	cal/g mole c	cal/g mole d	A (kcal/g mole)	B (e.u.)
La	solid	1153	(2.3)	(2.0)	13.7	6.17	1.60	—	—	1.911	21.9
	liquid	3000	80	27	—	(7.3)	—	—	—	(−0.15)	(26.0)
Li	solid	459	0.69	1.5	6.70	3.05	8.60	—	—	1.292	12.92
	liquid	1640	32.48	19.81	—	7.0	—	—	—	1.509	32.00
	gas	—	—	—	—	(4.97)	—	—	—	(−34.30)	(−2.84)
Mg	solid	923	2.2	2.4	7.77	5.33	2.45	—	−0.103	1.733	23.39
	liquid	1393	31.5	22.6	—	(8.0)	—	—	—	0.942	30.97
	gas	—	—	—	—	(4.97)	—	—	—	(−34.78)	(−7.60)
Mn	solid, α	1000	0.535	0.535	7.59	5.70	3.38	—	−0.37	1.974	26.11
	solid, β	1374	0.545	0.397	—	8.33	0.66	—	—	2.672	41.02
	solid, γ	1410	0.430	0.305	—	10.70	—	—	—	4.760	56.84
	solid, δ	1517	3.5	2.31	—	11.30	—	—	—	5.176	60.88
	liquid	2368	53.7	22.7	—	11.00	—	—	—	1.221	56.38
	gas	—	—	—	—	6.26	—	—	—	−63.704	−3.13
Mo	solid	2883	(5.8)	(2.0)	6.83	5.48	1.30	—	—	1.692	24.78
N₂	gas	—	—	—	45.767	6.76	0.606	0.13	—	2.044	−7.064
Na	solid	371	0.63	1.7	12.31	5.657	3.252	0.5785	—	1.836	20.92
	liquid	1187	23.4	20.1	—	8.954	−4.577	2.540	—	1.924	36.0
	gas	—	—	—	—	(4.97)	—	—	—	(−24.40)	(−8.7)
Nb	solid	2760	(5.8)	(2.1)	8.3	5.66	0.96	—	—	1.730	24.24
Nd	solid	1297	(2.55)	(1.97)	13.9	5.61	5.34	—	—	1.910	19.7
	liquid	(2750)	(61)	(22)	—	(9.1)	—	—	—	−0.606	35.8
Ni	solid, α	626	0.092	0.15	7.137	4.06	7.04	—	—	1.523	18.095
	solid, β	1728	4.21	2.44	—	6.00	1.80	—	—	1.619	27.16
	liquid	3110	90.48	29.0	—	9.20	—	—	—	0.251	45.47
Np	solid	913	(2.3)	(2.5)	(14)	(5.3)	(3.4)	—	—	(1.731)	(17.9)
	liquid	(2525)	(55)	(22)	—	(9.0)	—	—	—	(1.392)	(37.5)
O₂	gas	—	—	—	49.003	8.27	0.258	—	−1.877	3.007	−0.750
Os	solid	2970	(6.4)	(2.2)	7.8	5.69	0.88	—	—	1.736	24.9
P₄	solid, white	317.4	0.601	1.89	42.4	13.62	28.72	—	—	5.338	43.8
	liquid	553	11.9	21.5	—	19.23	0.51	—	−2.98	6.035	66.7
	gas	—	—	—	—	(19.3)	(0.4)	(1.3)	—	(0.32)	(40.1)
Pa	solid	(1825)	(4.0)	(2.2)	(13.5)	(5.2)	(4.0)	—	—	(1.728)	(17.3)
	liquid	(4500)	(115)	(26)	—	(8.0)	—	—	—	(−3.823)	(28.8)
Pb	solid	600.6	1.141	1.900	15.49	5.64	2.30	—	—	1.784	17.33
	liquid	2023	42.1	21.0	—	7.75	−0.74	—	—	1.862	22.11
	gas	—	—	—	—	(4.97)	—	—	—	(−45.25)	(−13.6)
Pd	solid	1828	4.12	2.25	8.9	5.80	1.38	—	—	1.791	24.6
	liquid	3440	89	26	—	(9.0)	—	—	—	(1.215)	(43.8)
Po	solid	525	(2.4)	(4.6)	13	(5.2)	(3.2)	—	—	(1.693)	(17.6)
	liquid	(1235)	(24.6)	(19.9)	—	(9.0)	—	—	—	(0.847)	(35.2)
	gas	—	—	—	—	(4.97)	—	—	—	(−28.73)	(−13.5)
Pr	solid	1205	(2.5)	(2.1)	(13.5)	(5.0)	(4.6)	—	—	(1.705)	(16.4)
	liquid	3563	—	—	—	(8.0)	—	—	—	(−0.519)	(30.0)
Pt	solid	2042.5	5.2	2.5	10.0	5.74	1.34	—	0.10	1.737	23.0
	liquid	4100	122	29.8	—	(9.0)	—	—	—	(0.406)	(42.6)
Pu	solid	913	(2.26)	(2.48)	(13.0)	(5.2)	(3.6)	—	—	(1.710)	(17.7)
	liquid	—	—	—	—	(8.0)	—	—	—	(0.506)	(31.0)
Ra	solid	1233	(2.3)	(1.9)	(17)	(5.8)	(1.2)	—	—	(1.783)	(16.4)
	liquid	(1700)	(35)	(21)	—	(8.0)	—	—	—	(1.284)	(28.6)
	gas	—	—	—	—	(4.97)	—	—	—	(−38.87)	(−14.5)
Rb	solid	312.0	0.525	1.68	16.6	3.27	13.1	—	—	1.557	5.9
	liquid	952	18.11	19.0	—	7.85	—	—	—	1.814	26.5
	gas	—	—	—	—	(4.97)	—	—	—	(−19.04)	(−12.3)
Re	solid	3440	(7.9)	(2.3)	(8.89)	(5.85)	(0.8)	—	—	(1.780)	(24.7)
Rh	solid	2240	(5.2)	(2.3)	7.6	5.40	2.19	—	—	1.707	23.8
	liquid	4150	127	30.7	—	(9.0)	—	—	—	(−0.923)	(44.4)
Ru	solid, α	1308	0.034	0.026	6.9	5.25	1.50	—	—	1.632	23.5
	solid, β	1473	0	0	—	7.20	—	—	—	2.867	35.5
	solid, γ	1773	0.23	0.13	—	7.20	—	—	—	2.867	35.5
	solid, δ	2700	(6.1)	(2.3)	—	7.50	—	—	—	3.169	37.6
S	solid, α	368.6	0.088	0.24	7.62	3.58	6.24	—	—	1.345	14.64
	solid, β	392	0.293	0.747	—	3.56	6.95	—	—	1.298	14.54
	liquid	717.76	2.5	3.5	—	5.4	5.0	—	—	1.576	24.02
½S₂	gas	—	—	—	—	(4.25)	(0.15)	—	(−1.0)	(−2.859)	(9.57)
Sb	solid (α,β,γ)	903.7	4.8	5.3	10.5	5.51	1.74	—	—	1.720	21.4
	liquid	1713	46.665	27.3	—	7.50	—	—	—	−1.992	28.1
½Sb₂	gas	—	—	—	—	4.47	—	—	−0.11	−53.876	−21.7
Sc	solid	1670	(4.0)	(2.4)	(9.0)	(5.13)	(3.0)	—	—	1.663	21.1
	liquid	3000	80	27	—	(7.50)	—	—	—	(−2.563)	31.3
Se	solid	490.6	1.25	2.55	10.144	3.30	8.80	—	—	1.375	11.28
	liquid	1000	14.27	14.27	—	7.0	—	—	—	0.881	27.34
Si	solid	1683	11.1	6.60	4.50	5.70	1.02	—	−1.06	2.100	28.88
	liquid	2750	71	26	—	7.4	—	—	—	−7.646	33.17
Sm	solid	1623	3.7	2.3	(15)	(6.7)	(3.4)	—	—	(2.149)	(24.2)
	liquid	(2800)	(70)	(25)	—	(9.0)	—	—	—	(−2.296)	(33.4)
Sn	solid (α,β)	505.1	1.69	3.35	12.3	4.42	6.30	—	—	1.598	14.8
	liquid	2473	(55)	(22)	—	7.30	—	—	—	0.559	26.2
	gas	—	—	—	—	(4.97)	—	—	—	(−60.21)	(−14.3)

D-45

THERMODYNAMIC PROPERTIES OF THE ELEMENTS (Continued)

Element	Phase	Temperature of Transition (°K)	Heat of Transition (kcal/g mole)	Entropy of Transition (e.u.)	Entropy at 298°K (e.u.)	cal/g mole a	cal/g mole b	cal/g mole c	cal/g mole d	A (kcal/g mole)	B (e.u.)
Sr	solid	1043	2.2	2.1	13.0	(5.60)	(1.37)	—	—	(1.731)	(19.3)
	liquid	1657	33.61	20.28	—	(7.7)	—	—	—	(0.976)	(30.4)
	gas	—	—	—	—	(4.97)	—	—	—	(−37.16)	(−10.2)
Ta	solid	3250	7.5	2.3	9.9	5.82	0.78	—	—	1.770	23.4
Te	solid	(2400)	(5.5)	(2.3)	(8.0)	(5.6)	(2.0)	—	—	(1.759)	(24.5)
	liquid	(3800)	(120)	(32)	—	(11)	—	—	—	(3.459)	(59.4)
Te	solid, α	621	0.13	0.21	11.88	4.58	5.25	—	—	1.599	15.78
	solid, β	723	4.28	5.92	—	4.58	5.25	—	—	1.469	15.57
	liquid	1360	11.9	8.75	—	9.0	—	—	—	−0.988	34.96
½ Te₂	gas	—	—	—	—	4.47	—	—	−0.10	−19.048	−6.47
Th	solid	2173	(4.6)	(2.1)	12.76	8.2	−0.77	2.04	—	2.591	33.64
	liquid	4500	(130)	(29)	—	(8.0)	—	—	—	(−7.602)	(26.84)
Ti	solid, α	1155	0.950	0.822	7.334	5.25	2.52	—	—	1.677	23.33
	solid, β	2000	(4.6)	(2.3)	—	7.50	—	—	—	1.645	35.46
	liquid	3550	(101)	(28)	—	(7.8)	—	—	—	(−2.355)	(35.45)
Tl	solid, α	508.3	0.082	0.16	15.4	5.26	3.46	—	—	1.722	15.6
	solid, β	576.8	1.03	1.79	—	7.30	—	—	—	2.230	26.4
	liquid	1730	38.81	22.4	—	7.50	—	—	—	1.315	25.9
	gas	—	—	—	—	(4.97)	—	—	—	(−41.88)	(−45.4)
U	solid, α	938	0.665	0.709	12.03	3.25	8.15	—	0.80	1.063	8.47
	solid, β	1049	1.165	1.111	—	10.28	—	—	—	3.493	48.27
	solid, γ	1405	(3.0)	(2.1)	—	9.12	—	—	—	1.110	39.09
	liquid	3800	—	—	—	(8.99)	—	—	—	(−2.073)	36.01
V	solid	2003	(4.0)	(2.0)	7.05	5.57	0.97	—	—	1.704	24.97
	liquid	3800	—	—	—	(8.6)	—	—	—	1.827	44.06
W	solid	3650	8.42	2.3	8.0	5.74	0.76	—	—	1.745	24.9
Y	solid	1750	(4.0)	(2.3)	(11)	(5.6)	(2.2)	—	—	(1.767)	(21.6)
	liquid	3500	(90)	(26)	—	(7.5)	—	—	—	(−2.277)	(29.6)
Zn	solid	692.7	1.595	2.303	9.95	5.35	2.40	—	—	1.702	21.25
	liquid	1180	27.43	23.24	—	7.50	—	—	—	1.020	31.35
	gas	—	—	—	—	—	—	—	—	(−29.407)	(−9.81)
Zr	solid, α	1135	0.920	0.811	9.29	6.83	1.12	—	−0.87	2.378	30.45
	solid, β	2125	(4.9)	(2.3)	—	7.27	—	—	—	1.159	31.43
	liquid	(3900)	(100)	(26)	—	(8.0)	—	—	—	(−2.190)	(34.7)

THERMODYNAMIC PROPERTIES OF THE OXIDES*·**

Oxide	Phase	Temperature of Transition (°K)	Heat of Transition kcal/mole	Entropy of Transition (e.u.)	Entropy at 298°K (e.u.)	cal/g mole a	cal/g mole b	cal/g mole c	cal/g mole d	A (kcal/mole)	B (e.u.)
Ac₂O₃	solid	(2250)	(20)	(8.9)	(36.5)	(20.0)	(20.4)	—	—	(6.870)	(80.9)
	liquid	—	—	—	—	(40)	—	—	—	(−19.767)	(180.5)
Ag₂O	solid	dec. 460	—	—	29.09	13.26	7.04	—	—	4.266	48.56
Ag₂O₂	solid	dec	—	—	(20.4)	(16.4)	(12.2)	—	—	(5.432)	(76.7)
Al₂O₃	solid	2300	26	11	12.186	26.12	4.388	—	−7.269	10.422	142.03
	liquid	dec.	—	—	—	(33)	—	—	—	(−11.655)	(174.1)
Am₂O₃	solid	(2225)	(17)	(7.6)	(37)	(20.0)	(15.6)	—	—	(6.657)	(81.6)
	liquid	(3400)	(85)	(25)	—	(38.5)	—	—	—	(−7.796)	(181.8)
AmO₂	solid	dec.	—	—	(20)	(14.0)	(6.8)	—	—	(4.477)	(61.8)
As₂O₃	solid, α	503	4.1	8.2	25.6	8.37	48.6	—	—	4.656	36.6
	solid, β	586	4.4	7.5	—	8.37	48.6	—	—	0.556	28.4
	liquid	730	7.15	9.79	—	(39)	—	—	—	(5.760)	(187.6)
	gas	—	—	—	—	(21.5)	—	—	—	(−14.164)	(62.5)
AsO₂	solid	(1200)	(9.0)	(7.5)	(13)	(8.5)	(9.4)	—	—	(2.952)	(38.2)
	liquid	(dec.)	—	—	—	(21)	—	—	—	(2.184)	(108.0)
As₂O₅	solid	dec. >1100	—	—	25.2	(31.1)	(16.4)	—	(−5.4)	(11.813)	(159.9)
Au₂O₃	solid	dec.	—	—	30	(23.5)	(4.8)	—	—	(7.220)	(105.3)
B₂O₃	solid	723	5.27	7.29	12.91	8.73	25.40	—	−1.31	4.171	45.04
	liquid	2520	(55)	(22)	—	30.50	—	—	—	7.822	161.59
Ba₂O	solid	(880)	(5.2)	(5.9)	(23.5)	(20.0)	(2.2)	—	—	(6.061)	(91.1)
	liquid	(1040)	(20)	(19)	—	(22)	—	—	—	(1.769)	(96.8)
	gas	—	—	—	—	(15)	—	—	—	(−25.51)	(29.0)
BaO	solid	2196	13.8	6.28	16.8	12.74	1.040	—	−1.984	4.510	57.2
	liquid	3000	(62)	(21)	—	(13.9)	—	—	—	(−9.341)	(57.5)
BaO₂	solid	723	(5.7)	(7.9)	(18.5)	(13.6)	(2.0)	—	—	(4.144)	(59.6)
	liquid	dec.1110	—	—	—	(21)	—	—	—	(3.241)	(99.0)
BeO	solid	dec.	—	—	3.37	8.69	3.65	—	−3.13	3.803	48.99
BiO	solid	(1175)	(3.7)	(3.1)	(15)	(9.7)	(3.0)	—	—	(3.025)	(41.2)
	liquid	(1920)	(54)	(28)	—	(14)	—	—	—	(2.306)	(64.9)
	gas	—	—	—	—	(8.9)	—	—	—	(−61.49)	(−1.8)
Bi₂O₃	solid	1090	6.8	6.2	36.2	23.27	11.05	—	—	7.429	99.7
	liquid	(dec.)	—	—	—	(35.7)	—	—	—	(7.614)	(168.3)
CO	gas	—	—	—	47.30	6.60	1.2	—	—	2.021	−9.34
CO₂	gas	—	—	—	51.06	7.70	5.3	−0.83	—	2.490	−5.64
CaO	solid	2860	(18)	(6.3)	9.5	10.00	4.84	—	−1.08	3.559	49.5
CdO	solid	dec.	—	—	13.1	9.65	2.08	—	—	2.970	42.5
Ce₂O₃	solid	1960	(20)	(10)	(33.5)	(23.0)	(9.0)	—	—	(7.258)	(100.2)
	liquid	(3500)	(80)	(23)	—	(37)	—	—	—	(−2.591)	(178.5)
CeO₂	solid	3000	(19)	(6.3)	17.7	15.0	2.5	—	—	4.579	68.5

*For description of headings and symbols see preceding table.
**From U.S. Atomic Energy Report ANL-5750.

THERMODYNAMIC PROPERTIES OF THE OXIDES (Continued)

Oxide	Phase	Temperature of Transition (°K)	Heat of Transition kcal/mole	Entropy of Transition (e.u.)	Entropy at 298°K (e.u.)	cal/g mole a	cal/g mole b	cal/g mole c	cal/g mole d	A (kcal/mole)	B (e.u.)
CoO	solid	2078	(12)	(5.8)	10.5	(9.8)	(2.2)	—	—	(3.020)	(46.0)
	liquid	(2900)	(61)	(21)	—	(15.5)	—	—	—	(−1.886)	(79.2)
Co₃O₄	solid	dec., 1240	—	—	(35.5)	(29.5)	(17.0)	—	—	(9.551)	(137.6)
Cr₂O₃	solid	2538	(25)	(10)	19.4	28.53	2.20	—	−3.736	9.857	145.9
CrO₂	solid	dec. 700	—	—	(11.5)	(16.1)	(3.0)	—	(−3.0)	(5.946)	(82.8)
CrO₃	solid	460	(6.1)	(13)	(17.5)	(18.1)	(4.0)	—	(−2.0)	(6.245)	(87.9)
	liquid	(1000)	(25)	(25)	—	(27)	—	—	—	(3.381)	(127.0)
	gas	—	—	—	—	(20)	—	—	—	(−28.62)	(53.6)
Cs₂O	solid	763	(4.58)	(6.0)	(23)	(16.5)	(5.4)	—	—	(5.160)	(72.6)
	liquid	dec.	—	—	—	(22)	—	—	—	(3.205)	(99.0)
Cs₂O₂	solid	867	(5.5)	(6.3)	(40)	(21.4)	(11.4)	—	—	(6.887)	(85.3)
	liquid	dec.	—	—	—	(29.5)	—	—	—	(4.125)	(123.8)
Cs₂O₃	solid	775	(7.75)	(10)	(47)	(24.0)	(22.6)	—	—	(8.160)	(96.5)
	liquid	dec.	—	—	—	(35)	—	—	—	(2.148)	(142.2)
Cu₂O	solid	1503	13.4	8.92	22.44	(13.4)	(8.6)	—	—	(4.378)	(54.9)
	liquid	dec.	—	—	—	(21.5)	—	—	—	(3.721)	(96.0)
CuO	solid	1609	(8.9)	(5.5)	10.4	14.34	6.2	—	—	4.551	61.11
	liquid	dec.	—	—	—	(22)	—	—	—	(−4.339)	(98.91)
FeO	solid	1641	7.5	4.6	12.9	9.27	4.80	—	—	(2.977)	(43.8)
	liquid	(2700)	(55)	(20)	—	(14.5)	—	—	—	(−3.721)	(69.2)
Fe₃O₄	solid, α	900	(0)	(0)	35.0	12.38	1.62	—	−0.38	3.826	58.3
	solid, β	dec.	—	—	—	(14.5)	—	—	—	(−2.399)	(66.7)
Fe₂O₃	solid, α	950	0.16	0.17	21.5	21.88	48.20	—	—	8.666	104.0
	solid, β	1050	0	0	—	48.00	—	—	—	12.652	238.3
	solid, γ	dec.	—	—	—	23.49	18.6	—	−3.55	9.021	119.9
Ga₂O	solid	(925)	(8.5)	(9.2)	(22.5)	36.00	—	—	—	11.979	187.6
	liquid	(1000)	(20)	(20.)	—	31.71	1.8	—	—	8.467	159.7
	gas	—	—	—	—	(13.8)	(8.6)	—	—	(4.497)	(58.7)
Ga₂O₃	solid	2013	(22)	(11.)	20.23	(21.5)	—	—	—	(−0.559)	(94.1)
	liquid	(2900)	(75)	(26.)	—	(14.)	—	—	—	(−28.06)	(22.3)
GeO	solid	983	(50)	(51)	(12.5)	11.77	25.2	—	—	(4.630)	(54.35)
	gas	—	—	—	—	(35.5)	—	—	—	(−20.66)	173.2
GeO₂	solid (α,β)	1380	10.6	7.66	(13.6)	(16.4)	(2.0)	—	(0.5)	(3.384)	(47.8)
	liquid	(2625)	(61)	(23.)	—	(8.2)	(0.4)	—	(−0.2)	(−49.67)	(−20.2)
H₂O	liquid	373.16	9.770	26.18	16.716	11.2	7.17	—	—	(3.658)	(53.5)
	gas	—	—	—	—	(11.7)	—	—	—	(1.149)	(111.9)
HfO₂	solid	3063	(17)	(5.6)	14.18	18.03	—	—	—	5.376	86.01
Hg₂O	solid	dec.	—	—	(30.)	7.17	2.56	—	(−0.08)	−8.290	−3.56
HgO	solid	dec.	—	—	16.839	17.39	2.08	—	−3.48	6.445	87.48
In₂O	solid	(600)	(4.5)	(7.5)	(28.)	(14.7)	(7.8)	—	—	(4.730)	(58.1)
	liquid	(800)	(16)	(20.)	—	(22.)	—	—	—	(3.206)	(92.6)
	gas	—	—	—	—	(15.)	—	—	—	(−18.39)	(25.8)
InO	solid	(1325)	(4.0)	(3.0)	(14.5)	(10.0)	(3.2)	—	—	(3.124)	(43.4)
	liquid	(2000)	(60.)	(30)	—	(14.)	—	—	—	(1.615)	(64.9)
	gas	—	—	—	—	(9.0)	—	—	—	(−68.38)	(−3.1)
In₂O₃	solid	(2000)	(20.)	(10.)	30.1	(22.6)	(6.0)	—	—	7.005	(100.5)
	liquid	(3600)	(85)	(24)	—	(35)	—	—	—	(−0.195)	(172.8)
Ir₂O₃	solid	(1450)	(10)	(6.8)	(26.5)	(21.8)	(14.4)	—	—	(7.140)	(102.0)
	liquid	(2250)	(50)	(22)	—	(35)	—	—	—	(0.706)	(170.3)
	gas	—	—	—	—	(20)	(10)	—	—	(−57.73)	(54.8)
IrO₂	solid	dec. 1373	—	—	(15.9)	9.17	15.20	—	—	3.410	40.9
K₂O	solid	(980)	(6.8)	(6.9)	(23)	(15.9)	(6.4)	—	—	(5.025)	(69.5)
	liquid	dec.	—	—	—	(22)	—	—	—	(1.130)	(98.3)
K₂O₂	solid	763	(7.0)	(9.2)	(27)	(20.8)	(5.4)	—	—	(6.442)	(93.1)
	liquid	(1800)	(45)	(25)	—	(29)	—	—	—	(4.127)	(134.2)
	gas	—	—	—	—	(20)	—	—	—	(−57.07)	(41.7)
K₂O₃	solid	703	(6.1)	(8.7)	(33.5)	(19.1)	(23.2)	—	—	(6.750)	(82.2)
	liquid	(975)	(25)	(26)	—	(35.5)	—	—	—	(6.447)	(164.7)
	gas	—	—	—	—	(20)	(5.0)	—	—	(−31.29)	(37.3)
KO₂	solid	653	(4.9)	(7.5)	27.9	(15.0)	(12.0)	—	—	(5.006)	(61.1)
	liquid	dec.	—	—	—	(24)	—	—	—	(3.424)	(105.5)
La₂O₃	solid	2590	(18)	(7)	(36.5)	28.86	3.076	—	−3.275	9.840	(130.7)
Li₂O	solid	2000	(14)	(7)	9.06	(11.4)	(5.4)	—	—	(3.639)	(57.5)
	liquid	2600	(56)	(22)	—	(21)	—	—	—	(−1.961)	(112.7)
Li₂O₂	solid	dec.470	—	—	(16.5)	(17.0)	(5.4)	—	—	(5.309)	(82.0)
MgO	solid	3075	18.5	5.8	6.4	10.86	1.197	—	−2.087	3.991	57.0
MgO₂	solid	dec. 361	—	—	(20.5)	(12.1)	(2.4)	—	—	(3.714)	(49.2)
MnO	solid	2058	13.0	6.32	14.27	11.11	1.94	—	−0.88	3.689	50.10
	liquid	dec.	—	—	—	(13.5)	—	—	—	(−8.543)	(58.02)
Mn₃O₄	solid, α	1445	4.97	3.44	35.5	34.64	10.82	—	−2.20	11.312	166.3
	solid, β	1863	(33)	(18)	—	50.20	—	—	—	17.376	260.4
	liquid	(2900)	(75)	(26)	—	(49)	—	—	—	(−17.86)	(233.4)
Mn₂O₃	solid	dec. 1620	—	—	26.4	24.73	8.38	—	−3.23	8.829	118.8
MnO₂	solid	dec. 1120	—	—	12.7	16.60	2.44	—	−3.88	6.359	84.8
MoO₂	solid	(2200)	(16)	(7.3)	(14.5)	(16.2)	(3.0)	—	(−3.0)	(5.973)	(80.4)
	liquid	dec. 2250	—	—	—	(23)	—	—	—	(−2.463)	(118.4)
MoO₃	solid	1068	12.54	11.74	18.68	13.6	13.5	—	—	4.655	62.83
	liquid	1530	33	22	—	(28.4)	—	—	—	(0.222)	(139.88)
	gas	—	—	—	—	(18.1)	—	—	—	(−48.54)	(42.8)

THERMODYNAMIC PROPERTIES OF THE OXIDES (Continued)

Oxide	Phase	Temperature of Transition (°K)	Heat of Transition kcal/mole	Entropy of Transition (e.u.)	Entropy at 298°K (e.u.)	cal/g mole a	cal/g mole b	cal/g mole c	cal/g mole d	A (kcal/mole)	B (e.u.)
N₂O	gas	—	—	—	52.58	10.92	2.06	—	−2.04	4.032	11.40
Na₂O	solid	1193	(7.1)	(6.0)	17.4	15.70	5.40	—	—	4.921	73.7
	liquid	dec.	—	—	—	(22)	—	—	—	(1.494)	(105.9)
Na₂O₂	solid	dec. 919	—	—	22.6	(20.2)	(3.8)	—	—	(6.192)	(93.6)
NaO₂	solid	(825)	(6.2)	(7.5)	27.7	(16.2)	(3.6)	—	—	(4.990)	(65.7)
	liquid	(1300)	(28)	(22)	—	(23)	—	—	—	3.175	(100.9)
	gas	—	—	—	—	(15)	—	—	—	(−35.22)	(22.0)
NbO	solid	(2650)	(16)	(6.0)	(12)	(9.6)	(4.4)	—	—	(3.058)	(44.0)
NbO₂	solid	(2275)	(16)	(7.0)	(12.7)	(17.1)	(1.6)	—	(−2.8)	(6.109)	(84.6)
	liquid	(3800)	(85)	(22)	—	(24)	—	—	—	(1.033)	(127.2)
Nb₂O₅	solid	1733	(28)	(16)	32.8	21.88	28.2	—	—	7.776	100.3
	liquid	(3200)	(80)	(25)	—	(44.2)	—	—	—	(−24.09)	(201.6)
Nd₂O₃	solid	2545	(22)	(8.8)	(35.3)	28.99	5.760	—	(−4.159)	10.295	(133.9)
NiO	solid	2230	(12.1)	(5.43)	9.22	13.69	0.83	—	−2.915	5.097	70.67
	liquid	dec.	—	—	—	(14.3)	—	—	—	(−7.861)	(67.91)
NpO₂	solid	(2600)	(15)	(5.7)	19.19	(17.7)	(3.2)	—	(−2.6)	(6.292)	(84.08)
Np₂O₅	solid	dec. 800—900°K	—	—	(43)	(32.4)	(12.6)	—	—	(10.22)	(145.4)
OsO₂	solid	dec. 923	—	—	(14.5)	(11.5)	(6.0)	—	—	(3.696)	(52.8)
OsO₄	solid	813.3	3.41	10.9	34.7	(16.4)	(23.1)	—	(−2.4)	(6.726)	(67.0)
	liquid	403	9.45	23.4	—	(33)	—	—	—	(6.612)	(143.0)
	gas	—	—	—	—	16.46	8.60	—	−4.6	(−7.644)	(25.3)
P₂O₃	liquid	448.5	4.5	10	(34)	(34.5)	—	—	—	(10.287)	(162.6)
	gas	—	—	—	—	(15)	(10)	—	—	(−1.953)	(38.0)
PO₂	solid	(350)	(2.7)	(7.7)	(11.5)	(11.3)	(5.0)	—	—	(3.591)	(54.4)
	liquid	(dec)	—	—	—	(20)	—	—	—	(3.640)	(95.9)
P₂O₅	solid	631	8.8	13.9	33.5	8.375	5.40	—	—	4.897	30.3
	gas	—	—	—	—	36.80	—	—	—	3.284	165.6
PaO₂	solid	(2560)	(20)	(7.8)	(17.8)	(14.4)	(2.6)	—	—	(4.409)	(65.0)
Pa₂O₅	solid	(2050)	(26)	(13)	(37.5)	(28.4)	(11.4)	—	—	(8.975)	(127.7)
	liquid	(3350)	(95)	(28)	—	(48)	—	—	—	(−0.800)	(241.1)
PbO	solid, red	762	(0.4)	(0.5)	16.2	10.60	4.00	—	—	3.338	45.4
	solid, yellow	1159	2.8	2.4	—	9.05	6.40	—	—	2.454	36.4
	liquid	1745	51	29	—	(14.6)	—	—	—	1.788	65.7
	gas	—	—	—	—	(8.1)	(0.4)	—	—	(−59.94)	(−11.0)
Pb₃O₄	solid	dec.	—	—	50.5	(31.1)	(17.6)	—	—	(10.055)	(132.0)
PbO₂	solid	dec.	—	—	18.3	12.7	7.80	—	—	4.133	56.4
PdO	solid	dec. 1150	—	—	(9.1)	3.30	14.2	—	—	1.615	(13.9)
PoO₂	solid	(825)	(5.5)	(6.7)	(17)	(14.3)	(5.6)	—	—	(4.513)	(66.1)
	liquid	(dec.)	—	—	—	(22)	—	—	—	(3.460)	(106.5)
Pr₂O₃	solid	(2200)	(22)	(10)	(35.5)	(29.0)	(4.0)	—	(−4.0)	(10.166)	(133.2)
	liquid	(4000)	(90)	(23)	—	(36)	—	—	—	(−6.298)	(168.3)
PrO₂	solid	dec. 700	—	—	(17)	(17.6)	(3.4)	—	(−2.8)	(6.338)	(85.9)
PtO	solid	dec. 780	—	—	(13.5)	(9.0)	(6.4)	—	—	(2.968)	(39.7)
Pt₃O₄	solid	(dec.)	—	—	(41)	(30.8)	17.4	—	—	(9.957)	(139.7)
PtO₂	solid	723	(4.6)	(6.4)	(16.5)	(11.1)	(9.6)	—	—	(3.736)	(49.6)
	liquid	dec. 750	—	—	—	(21)	—	—	—	(3.785)	(101.5)
PuO	solid	(1290)	(7.2)	(5.6)	(20)	(12.0)	(2.4)	—	—	(3.685)	(49.1)
	liquid	(2325)	(47)	(20)	—	(14.5)	—	—	—	(−2.287)	(58.3)
	gas	—	—	—	—	(8.9)	—	—	—	(−62.307)	(−5.3)
Pu₂O₃	solid	(1880)	(16)	(8.5)	(38)	(21.2)	18.2	—	—	(7.130)	(88.2)
	liquid	(3250)	(75)	(23)	—	(40)	—	—	—	(−5.691)	(187.2)
PuO₂	solid	(2400)	(15)	(6.2)	19.7	(17.1)	(3.4)	—	(−2.6)	(6.122)	(80.2)
	liquid	(3500)	(90)	(26)	—	(20.5)	—	—	—	(−10.62)	(92.2)
RaO	solid	(>2500)	—	—	(17)	(10.5)	(2.0)	—	—	(3.220)	(43.4)
Rb₂O	solid	(910)	(5.7)	(6.3)	(27)	(15.4)	(5.8)	—	—	(4.850)	(62.5)
	liquid	dec.	—	—	—	(22)	—	—	—	(2.754)	(95.9)
Rb₂O₂	solid	843	(7.3)	(8.7)	(27.5)	(10.9)	(8.0)	—	—	(6.587)	(94.0)
	liquid	(dec.)	—	—	—	(29)	—	—	—	(3.273)	(133.2)
Rb₂O₃	solid	762	(7.6)	(10)	(32.5)	(20.5)	(13.0)	—	—	(6.690)	(88.2)
	liquid	dec.	—	—	—	(34)	—	—	—	5.603	(157.8)
RbO₂	solid	685	(4.1)	(6.0)	(21.5)	(13.8)	(6.4)	—	—	(4.399)	(59.0)
	liquid	dec.	—	—	—	(21)	—	—	—	(3.720)	(95.7)
ReO₂	solid	(1475)	(12)	(8.1)	(15)	(10.8)	(9.8)	—	—	(3.656)	(49.5)
	liquid	(3250)	(80)	(25)	—	(24.5)	—	—	—	(1.204)	(127.0)
ReO₃	solid	433	5.2	12	19.8	(18.0)	(5.8)	—	—	(5.625)	(84.5)
	liquid	dec.	—	—	—	29	—	—	—	(4.644)	(136.8)
Re₃O₇	solid	569	15.8	27.8	44	(41.8)	(14.8)	—	(−3.0)	(14.127)	(200.3)
	liquid	635.5	17.7	27.9	—	(65.7)	—	—	—	(9.203)	(314.7)
	gas	—	—	—	—	(38.2)	—	—	—	(−25.97)	(109.3)
ReO₄	solid	420	(4.2)	(10)	(34.5)	(21.4)	(10.8)	—	(−2.0)	(7.531)	(91.8)
	liquid	(460)	(9.3)	(20)	—	(33)	—	—	—	(6.775)	(146.7)
	gas	—	—	—	—	(16.5)	(8.6)	—	(−5.0)	(−8.118)	(30.6)
Rh₂O	solid	dec. 1400	—	—	(25.5)	15.59	6.47	—	—	4.936	65.3
RhO	solid	dec. 1394	—	—	(12)	(9.84)	(5.53)	—	—	(3.179)	(45.7)
Rh₂O₃	solid	dec. 1388	—	—	(23)	20.73	13.80	—	—	6.794	99.2
RuO₂	solid	dec. 1400	—	—	(12.5)	(11.4)	(6.0)	—	—	3.666	(54.2)
RuO₄	solid	300	(3.2)	(11)	(32.5)	(20)	—	—	—	(5.963)	(81.5)
	liquid	dec.	—	—	—	(33)	—	—	—	(6.663)	(144.9)
SO₂	gas	—	—	—	59.40	11.4	1.414	—	−2.045	4.148	7.12
Sb₂O₃	solid	928	14.74	15.88	29.4	19.10	17.1	—	—	6.455	84.5
	liquid	1698	8.92	5.25	—	(36)	—	—	—	(0.035)	(168.2)
	gas	—	—	—	—	(20.8)	—	—	—	(−34.70)	(49.9)

Oxide	Phase	Temperature of Transition (°K)	Heat of Transition kcal/mole	Entropy of Transition (e.u.)	Entropy at 298°K (e.u.)	cal/g mole a	cal/g mole b	cal/g mole c	cal/g mole d	A (kcal/mole)	B (e.u.)
SbO₃	solid	dec.	—	—	15.2	11.30	8.1	—	—	3.725	51.6
Sb₂O₃	solid	dec.	—	—	29.9	(22.4)	(23.6)	—	—	(7.728)	(104.8)
Sc₂O₃	solid	(2500)	(23)	(9.3)	24.8	23.17	5.64	—	—	7.159	108.9
SeO	solid	(1375)	(7.6)	(5.5)	(11)	(9.1)	(3.8)	—	—	(2.882)	(42.0)
	liquid	(2075)	(45)	(22)	—	(15.5)	—	—	−0.80	(0.490)	(77.5)
	gas	—	—	—	—	8.20	0.50	—	−0.80	(−58.54)	(0.7)
SeO₂	solid	603	(24.5)	(40.6)	(15)	(12.8)	(6.1)	—	(−0.2)	(4.150)	(59.9)
	gas	—	—	—	—	(14.5)	—	—	—	(−20.45)	(26.4)
SiO	solid	(2550)	(12)	(4.7)	(6.5)	(7.3)	(2.4)	—	—	(2.283)	(35.8)
SiO₂	solid, β	856	0.15	0.18	10.06	11.22	8.20	—	−2.70	4.615	57.83
	solid, α	1883	2.04	1.08	—	14.41	1.94	—	—	4.602	73.67
	liquid	dec. 2250	—	—	—	(20)	—	—	—	(9.649)	(111.08)
Sm₂O₃	solid	(2150)	(20)	(9.3)	(36.5)	(25.9)	(7.0)	—	—	(8.033)	(113.2)
	liquid	(3800)	(80)	(21)	—	(36)	—	—	—	(−6.431)	(166.3)
SnO	solid	(1315)	(6.4)	(4.9)	13.5	9.40	3.62	—	—	2.964	41.1
	liquid	(1800)	(60)	(33)	—	(14.5)	—	—	—	(0.141)	(68.1)
	gas	—	—	—	—	(9.0)	—	—	—	(−69.76)	(−6.4)
SnO₂	solid	1898	(11.39)	(5.95)	12.5	17.66	2.40	—	−5.16	7.103	91.7
	liquid	(3200)	(75)	(23)	—	(22.5)	—	—	—	(0.304)	(117.7)
SrO	solid	2703	16.7	6.2	13.0	12.34	1.120	—	−1.806	4.335	58.7
SrO₂	solid	dec. 488	—	—	(14.8)	(16.8)	(2.2)	—	(−3.0)	(6.113)	(83.3)
Ta₂O₅	solid	2150	(16)	(7.4)	34.2	29.2	10.0	—	—	9.151	135.2
	liquid	—	—	—	—	(46)	—	—	—	(6.158)	(235.1)
TcO₂	solid	(2400)	(18)	(7.5)	(13.5)	(10.4)	(9.2)	—	—	(3.510)	(48.6)
	liquid	(4000)	(105)	(26)	—	(25)	—	—	—	(−5.946)	(132.7)
TcO₃	solid	(dec. <1200)	—	—	(19.5)	(19.4)	(5.2)	—	(−2.0)	(6.686)	(93.7)
Tc₂O₇	solid	392.7	(11)	(28)	(42.5)	(39.1)	(18.6)	—	(−2.4)	(13.29)	(187.2)
	liquid	583.8	(14)	(24)	—	(64)	—	—	—	(10.02)	(299.8)
	gas	—	—	—	—	(25)	(28)	—	—	(−21.98)	(43.8)
TeO	solid	(1020)	(7.1)	(7.0)	(13)	(8.6)	(6.2)	—	—	(2.840)	(37.8)
	liquid	(1775)	(50)	(28)	—	(15.5)	—	—	—	(−0.448)	(72.3)
	gas	—	—	—	—	(8.9)	—	—	—	(−62.16)	(−5.2)
TeO₂	solid	1006	3.2	3.2	16.99	13.85	6.87	—	—	4.435	63.97
	liquid	dec.	—	—	—	(20)	—	—	—	(3.940)	(96.4)
ThO	solid	(2150)	(13)	(6.0)	(16)	(11.0)	(2.4)	—	—	(3.386)	(47.4)
	liquid	(3250)	(65)	(20)	—	(15)	—	—	—	(−6.561)	(66.9)
ThO₂	solid	2225	(19)	(5.6)	15.59	16.45	2.346	—	−2.124	5.721	80.03
TiO	solid, α	1264	0.82	0.65	8.31	10.57	3.60	—	−1.86	3.933	54.03
	solid, β	dec. 2010	—	—	—	11.85	3.00	—	—	4.108	61.71
Ti₂O₂	solid, α	473	0.215	0.455	18.83	7.31	53.52	—	—	4.559	38.78
	solid, β	2400	(24)	(10)	—	34.68	1.30	—	−10.20	13.605	184.48
	liquid	3300	—	—	—	(37.5)	—	—	—	(−7.796)	(193.2)
Ti₅O₅	solid, α	450	2.24	4.98	30.92	35.47	29.50	—	—	11.887	179.98
	solid, β	(2450)	(50)	(20)	—	41.60	8.00	—	—	10.230	202.80
	liquid	(3600)	(85)	(24)	—	(60)	—	—	—	(−18.701)	(306.4)
TiO₂	solid	2128	(16)	(7.5)	12.01	17.97	0.28	—	−4.35	6.829	92.92
	liquid	dec.3200	—	—	—	(21.4)	—	—	—	(−2.610)	(111.08)
Ti₂O	solid	573	(5.0)	(8.7)	23.8	(15.8)	(6.0)	—	(−0.3)	(5.078)	(68.2)
	liquid	773	(17)	(22)	—	(22.1)	—	—	—	(2.651)	(96.0)
	gas	—	—	—	—	(13.7)	—	—	—	(−20.94)	(18.0)
Ti₂O₃	solid	990	(12.4)	(13)	(33.5)	(23.0)	(5.0)	—	—	(7.080)	(99.0)
	liquid	(dec)	—	—	—	(35.5)	—	—	—	(4.604)	(167.8)
UO	solid	(2750)	(14)	(5.1)	(16)	(10.6)	(2.0)	—	—	(3.249)	(45.0)
UO₂	solid	3000	—	—	18.63	19.20	1.62	—	−3.957	7.124	93.37
U₃O₃	solid	dec.	—	—	(66)	(65)	(7.5)	—	(−10.9)	(23.37)	(312.7)
UO₂	solid	dec. 925	—	—	23.57	22.09	2.54	—	−2.973	7.696	104.72
VO	solid	(2350)	(15)	(6.4)	9.3	11.32	1.61	—	−1.26	3.869	56.4
	liquid	(3400)	(70)	(21)	—	(14.5)	—	—	—	(−8.157)	(70.9)
V₂O₃	solid	2240	(24)	(11)	23.58	29.35	4.76	—	−5.42	10.780	148.12
	liquid	dec. 3300	—	—	—	(38)	—	—	—	(−6.028)	(193.4)
V₂O₅	solid	(2100)	(42)	(20)	(32)	(36)	(30)	—	—	(12.07)	(182.1)
	liquid	(dec.)	—	—	—	(55.6)	—	—	—	(−54.72)	(249.1)
VO₂	solid, α	345	1.02	2.96	12.32	14.96	—	—	—	4.460	72.92
	solid, β	1818	13.60	7.48	—	17.85	1.70	—	−3.94	5.680	89.09
	liquid	dec. 3300	—	—	—	25.50	—	—	—	2.962	135.87
V₂O₅	solid	943	15.56	16.50	313	46.54	−3.90	—	−13.22	18.136	240.2
	liquid	(2325)	(63)	(27)	—	45.60	—	—	—	2.122	220.1
	gas	—	—	—	—	(40)	—	—	—	(−73.90)	(149.6)
VO₂	solid	(1543)	(11.5)	(7.45)	(15)	(17.6)	(4.2)	—	(−4.0)	(6.772)	(88.8)
	liquid	dec. 2125	—	—	—	(24)	—	—	—	(−0.112)	(121.8)
WO₃	solid	1743	(17)	(9.8)	19.90	17.33	7.74	—	—	5.511	81.15
	liquid	(2100)	(43)	(20)	—	(30)	—	—	—	(−1.162)	(152.5)
	gas	—	—	—	—	(18)	—	—	—	(−69.36)	(40.2)
Y₂O₃	solid	(2500)	(25)	(10)	(29.5)	(26.0)	(8.2)	—	(−2.2)	(8.846)	(122.3)
ZnO	solid	dec.	—	—	10.4	11.71	1.22	—	−2.18	4.277	57.88
ZrO₂	solid, α	1478	1.420	0.961	12.03	16.64	1.80	—	−3.36	6.168	85.21
	solid, β	2950	20.8	7.0	—	17.80	—	—	—	4.270	89.96

SELECTED VALUES OF CHEMICAL THERMODYNAMIC PROPERTIES

(From National Bureau of Standards Technical Notes 270-3, 270-4, 270-5, 270-6, 270-7 and 270-8

D. D. Wagman, W. H. Evans, V. B. Parker, R. H. Schumm, S. M. Bailey, I. Halow, K. L. Churney, and R. L. Nuttall

The compounds listed in the table represent only a small fraction of those for which data are given in the six Technical Notes referenced above. Copies of these Technical Notes may be purchased from the Superintendent of Documents, U.S. Government Printing Office, Washington, D.C. 20402. Conversion of units used in the table to SI units or other units may be accomplished by employing the following factors.

CONVERSION FACTORS FOR UNITS OF MOLECULAR ENERGY

	J mol⁻¹	cal mol⁻¹	cm³ atm mol⁻¹	kWh mol⁻¹	Btu lb⁻¹ mol⁻¹	cm⁻¹ molecule⁻¹	eV molecule⁻¹
J mol⁻¹ =	1	2.390057×10^{-1}	9.86923	2.77778×10^{-7}	0.429923	8.35940×10^{-2}	1.036409×10^{-5}
cal mol⁻¹ =	4.18400^a	1	41.2929	1.162222×10^{-6}	1.798796	3.49757×10^{-1}	4.33634×10^{-5}
cm³ atm mol⁻¹ =	0.1013250^a	2.42173×10^{-2}	1	2.81458×10^{-8}	4.35619×10^{-2}	8.47016×10^{-3}	1.050141×10^{-6}
kWh mol⁻¹ =	3,600,000	860,421	3.55292×10^{7}	1	1,547,721	300,938	37.3107
Btu lb⁻¹ mol⁻¹ =	2.32600^a	5.55927×10^{-1}	22.9558	6.46111×10^{-7}	1	1.944396×10^{-1}	2.41069×10^{-5}
cm⁻¹ molecule⁻¹ =	11.96258	2.85912	118.0614	3.32294×10^{-6}	5.14299	1	1.239812×10^{-4}
eV molecule⁻¹ =	96487.0^a	23060.9	952,252	2.68019×10^{-2}	41482.0	8065.73	1

^{a} These numbers represent the fundamental values used in deriving the data in the table. The remaining factors were obtained by applying the relationships:

$$n_{ij} = n_{ik} \cdot n_{kj} \qquad n_{ii} = n_{ik} \cdot n_{ki} = 1$$

INTRODUCTION

Substances and Properties Included in the Tables

The tables contain values of the enthalpy and Gibbs energy of formation, enthalpy, entropy and heat capacity at 298.15 K (25°C), and the enthalpy of formation at 0 K, for inorganic substances and organic molecules containing not more than two carbon atoms.

No values are given in these tables for metal alloys or other solid solutions, fused salts, or for substances of undefined chemical composition.

Physical States

The physical state of each substance is indicated in the column headed "State" as crystalline solid (c), liquid (l), glassy or amorphous (amorp), or gaseous (g). For solutions, the physical state is that normal for the indicated solvent at 298.15 K. Isomeric substances or various crystalline modifications of a given substance are designated by a number following the letter designation, as c2, g2, etc.

Definition of Symbols

The symbols used in these tables are defined as follows: P = pressure; V = volume; T = absolute temperature; S = entropy; H = enthalpy (heat content); G = H − TS = Gibbs energy (formerly the free energy); $C_p = (dH/dT)_p$ = heat capacity at constant pressure.

Conventions Regarding Pure Substances

The values of the thermodynamic properties of the pure substances given in these tables are for the substances in their standard states (indicated by the superscript ° on the thermodynamic symbol). These standard states are defined as follows:

1. For a pure solid or liquid, the standard state is the substance in the condensed phase under a pressure of one atmosphere.*
2. For a gas the standard state is the hypothetical ideal gas at unit fugacity, in which state the enthalpy is that of the real gas at the same temperature and zero pressure.

The values of ΔHf° and ΔGf° given in the tables represent the change in the appropriate thermodynamic quantity when one gram-formula weight of the substance in its standard state is formed isothermally at the indicated temperature from the elements, each in its appropriate standard reference state. The standard reference state at 298.15 K for each element except phosphorus has been chosen to be the standard state that is thermodynamically stable at that temperature and one atmosphere pressure. For phosphorus the standard reference state is the crystalline white form; the more stable forms have not been well characterized thermochemically. The same reference states have been maintained for the elements at 0 K except for the liquid elements bromine and mercury, for which the reference states have been chosen as the stable crystalline forms.

The value of H°₂₉₈ − H°₀ represents the enthalpy difference for the given substance between 298.15 K and 0 K. If the indicated standard state at 298.15 K is the gas, the corresponding state at 0 K is the hypothetical ideal gas; if the state at 298.15 K is solid or liquid, the corresponding state at 0 K is the thermodynamically stable crystalline solid, unless otherwise specifically indicated.

The values of S° represent the virtual or "thermal" entropy of the substance in the standard state at 298.15 K, omitting contributions from nuclear spins. Isotope mixing effects, etc., are also excluded except in the case of the hydrogen-deuterium system. Where data have been available only for a particular isotope, they have been corrected when possible to the normal isotopic composition.

The values of the enthalpies of formation of gaseous ionic species are computed on the convention that the value of ΔHf° for the electron is zero. Conversions between 0 and 298.15 K are claculated using the value of H°₂₉₈ − H°₀ = 1.481 kcal per mole of electrons, and assuming that the values of H°₂₉₈ − H°₀ for the ionized and nonionized molecules are the same.

Conventions Regarding Solutions

For all dissolved substances the composition of the solvent is indicated following the chemical formula of the solute. In most instances the number of moles of solvent associated with 1 mole of solute is stated explicitly. In some cases the concentration of the solute cannot be specified. For aqueous solutions this is indicated in the State column by "au" (aqueous, unspecified). Such solutions may be assumed to be "dilute".

*   One standard atmosphere equals 101325 pascal.

The standard state for a strong electrolyte in aqueous solution is the ideal solution at unit mean molality (unit activity). For a non-dissociating solute in aqueous solution the standard state is the ideal solution at unit molality.

The value of $\Delta Hf°$ for a solute in its standard state is equal to the apparent molal enthalpy of formation of the substance in the infinitely dilute solution, since the enthalpy of dilution of an ideal solution is zero. At this dilution the partial molal enthalpy is equal to the apparent molal quantity. At concentrations other than the standard state, the value of $\Delta Hf°$ represents the apparent enthalpy of the reaction of formation of the solution from the elements comprising the solute, each in its standard reference state, and the appropriate total number of moles of solvent. In this representation the value of $\Delta Hf°$ for the solvent is not required. The experimental value for an enthalpy of dilution is obtained directly as the difference between the two values of $\Delta Hf°$ at the corresponding concentrations. At finite concentrations the partial molal enthalpy of formation differs from the apparent enthalpy. In some instances the partial molal enthalpy of formation is given in the Tables. In this case the concentration designation is preceded by "D:".

The values for the thermodynamic properties for an individual ion in aqueous solution are for that undissociated ion in the standard state and are based on the convention that $\Delta Hf°$, $\Delta Gf°$, $S°$, and $Cp°$ for $H^+(a)$ are zero. The properties of the neutral strong electrolyte in aqueous solution in the standard state are equal to the sum of these values for the appropriate number of ions assumed to constitute the molecule of the given electrolyte. By adopting the above convention with respect to $H^+(a)$, it follows that for an individual ionic species the G−H−S relation becomes

$$\Delta Gf° = \Delta Hf° - T\{\Delta Sf° + (n/2)S°[H_2(gs)]\}$$

with n = the algebraic value of the ionic charge. For neutral electrolytes and gaseous ions the normal consistency relation holds (see below).

Unit of Energy and Fundamental Constants

All of the energy values given in these tables are expressed in terms of the thermochemical calorie. This unit, defined as equal to 4.1840 joules, has been generally accepted for presentation of chemical thermodyanmic data. Values reported in other units have been converted to calories by means of the conversion factors for molecular energy given in the table which precedes this discussion.

The following values of the fundamental physical constants have been used in these calculations:

- R = gas constant = 8.3143 ± 0.0012 J/deg mol = 1.98717 ± 0.00029 cal/deg mol
- F = Faraday constant = 96487.0 ± 1.6 coulombs/mol = 23060.9 ± 0.4 cal/V equivalent
- Z = Nhc = 11.96258 ± 0.00107 J/cm^{-1} mol = 2.85912 ± 0.00026 cal/cm^{-1} mol
- c_2 = second radiation constant = hc/k = 1.43879 ± 0.00015 cm deg 0°C = 273.15 K

These constants are consistent with those given in the Table of Physical Constants, recommended by the National Academy of Sciences — National Research Council.* The formula weights listed in the tables have been calculated for the empirical molecular formula given in the Formula and Description column.

Internal Consistency of the Tables

The various aspects of internal consistency are specified below:

1. Subsidiary and auxiliary quantities used. All of the values given in these tables have been calculated from the original articles, using consistent values for all subsidiary and auxiliary quantities. The original data were corrected where possible for differences in energy units, molecular weights, temperature scales, etc. Thus we have sought to maintain a uniform scale of energies for all substances in the tables.

2. Physical and thermodynamic relationships for the tabulated properties of a substance. The tabulated values of the properties of a substance satisfy all the known physical and thermodynamic relationships among these properties. The quantities $\Delta Hf°$, $\Delta Gf°$, and $S°$ at 298.15 K satisfy the relation (within the assumed uncertainty)

$$\Delta Gf° = \Delta Hf° - T\,\Delta Sf°$$

Formula and Description	State	Formula weight	$\Delta Hf_0°$ kcal/mol	$\Delta Hf°$	$\Delta Gf°$ kcal/mol	$H°_{298} - H°_0$	$S°$ cal/deg mol	$Cp°$
ACTINIUM								
Ac	c	227.0280	0	0	0	—	13.5	6.5
	g	227.0280	—	97.0	87.6	1.481	44.92	4.98
ALUMINUM								
Al	c	26.9815	0	0	0	1.094	6.77	5.82
	g	—	77.44	78.0	68.3	1.654	39.30	5.11
Al$^+$	g	—	215.476	217.517	—	—	—	—
Al^{2+}	g	—	649.663	653.185	—	—	—	—
Al^{3+}	g	—	1305.70	1310.70	—	—	—	—
std. state, m = 1	aq	—	—	−127.	−116.	—	−76.9	—
Al$_2$	g	53.9630	116.	116.14	103.57	2.33	55.7	8.7
AlO$_2^-$								
std. state, m = 1	aq	58.9803	—	−219.6	−196.7	—	−5.	—
Al$_2$O$_3$								
α, corundum	c	101.9612	−397.59	−400.5	−378.2	2.394	12.17	18.89
δ	c	—	—	−398.	—	—	—	—
ϱ	c	—	—	−391.	—	—	—	—
χ	c	—	—	−397.	—	—	—	—
γ	c	—	—	−395.	—	—	—	—
	amorp	—	—	−390.	—	—	—	—
Al$_2$O$_3$·H$_2$O								
boehmite	c	119.9765	—	−472.0	−436.3	—	23.15	31.37
diaspore	c	—	—	−478.	−440.	—	16.86	25.22
Al$_2$O$_3$·3H$_2$O								
gibbsite	c	156.0072	—	−612.5	−546.7	—	33.51	44.49
bayerite	c	—	—	−610.1	—	—	—	—
AlH$_3$	c	30.0054	—	−11.	—	—	—	—
AlOH^{2+}								
std. state, m = 1	aq	43.9889	—	—	−165.9	—	—	—
Al(OH)$_3$	amorp	78.0036	—	−305.	—	—	—	—
AlF$_3$	c	83.9767	−358.02	−359.5	−340.6	2.778	15.88	17.95
AlF$_3$	g	—	−287.01	−287.9	−284.0	3.37	66.2	14.97
un-ionized; std. state, m = 1	aq	—	—	−363.	−338.	—	−6.	—
ionized; std. state, m = 1	aq	—	—	−366.	−316.	—	−86.8	—

* NBS Technical News Bulletin, October 1963. See also Report of the CODATA Task Group on Fundamental Constants, CODATA Bulletin 11, December 1973.

Formula and Description	State	Formula weight	0 K ΔHf₀° kcal/mol	298.15 K (25°C) ΔHf°	ΔGf° kcal/mol	H°₂₉₈ − H°₀	S° cal/deg mol	C°ₚ
AlCl₃	c	133.3405	−168.02	−168.3	−150.3	4.104	26.45	21.95
	g	—	—	−139.4	—	—	—	—
std. state, m = 1	aq	—	—	−247.	−210.	—	−36.4	—
AlCl₃·6H₂O	c	241.4325	—	−643.3	—	—	—	—
Al₂Cl₆	g	266.6810	—	−308.5	−291.7	—	117.	—
AlBr₃	c	266.7085	—	−126.0	—	—	—	24.3
	g	—	—	−101.6	—	—	—	—
std. state, m = 1	aq	—	—	−214.	−191.	—	−17.8	—
Al₂Br₆	g	533.4170	—	−232.0	—	—	—	—
AlI₃	c	407.6947	—	−75.0	−71.9	—	38.	23.6
	g	—	—	−49.6	—	—	—	—
	aq	—	—	−165.8	—	—	—	—
Al₂I₆	g	815.3894	—	−123.5	—	—	—	—
Al₂S₃	c	150.1550	—	−173.	—	—	—	—
Al₂(SO₄)₃	c	342.1478	—	−822.38	−740.95	—	57.2	62.00
std. state, m = 1	aq	—	—	−906.	−766.	—	−139.4	—
Al₂(SO₄)₃·6H₂O	c	450.2398	—	−1269.53	−1104.82	—	112.1	117.8
Al₂Se₃	c	290.843	—	−135.	—	—	—	—
Al₂Te₃	c	436.763	—	−78.	—	—	—	—
AlN	c	40.9882	−74.80	−76.0	−68.6	0.925	4.82	7.20
Al(NO₃)₃								
std. state, m = 1	aq	212.9962	—	−276.	−196.	—	28.1	—
Al(NO₃)₃·6H₂O	c	321.0882	—	−681.28	−526.74	—	111.8	103.5
AlCl₃·NH₃	c	150.3711	—	−212.6	—	—	—	—
AlCl₃·3NH₃	c	184.4323	—	−283.0	—	—	—	—
AlCl₃·6NH₃	c	235.5242	—	−363.4	—	—	—	94.
AlBr₃·NH₃	c	283.7391	—	−177.6	—	—	—	—
AlBr₃·3NH₃	c	317.8003	—	−252.7	—	—	—	—
AlBr₃·6NH₃	c	368.8922	—	−343.0	—	—	—	—
AlI₃·NH₃	c	424.7253	—	−119.0	—	—	—	—
AlI₃·3NH₃	c	458.7865	—	−205.9	—	—	—	—
AlI₃·6NH₃	c	509.8784	—	−312.9	—	—	—	—
NH₄Al(SO₄)₂	c	237.1433	—	−562.2	−487.2	—	51.7	54.12
std. state, m = 1	aq	—	—	−593.	−491.	—	−40.2	—
NH₄Al(SO₄)₂·12H₂O	c	453.3274	—	−1420.26	−1180.21	—	166.6	163.3
AlP	c	57.9553	—	−39.8	—	—	—	—
AlPO₄								
berlinite	c	121.9529	−411.40	−414.4	−386.7	3.528	21.70	22.27
AlAs	c	101.9031	—	−27.8	—	—	—	—
Al₄C₃	c	143.9594	−48.71	−49.9	−46.9	3.936	21.26	27.91
Al(CH₃)₃	c	72.0867	−29.76	—	—	—	—	—
	liq	—	—	−32.6	−2.4	8.114	50.05	37.19
	g	—	—	−17.7	—	—	—	—
Al₂(CH₃)₆	g	144.1734	—	−55.19	−2.34	—	125.4	—
Al₂Si₂O₇·2H₂O								
kaolinite	c	258.1615	—	−979.6	−903.0	—	48.5	58.62
halloysite	c	—	—	−975.0	−898.5	—	48.6	58.86
Al₆Si₂O₁₃								
mullite	c	426.0532	—	−1632.8	−1541.2	—	61.	77.94
ANTIMONY								
Sb								
III	c	121.75	0	0	0	1.410	10.92	6.03
IV, explosive	amorp	—	—	2.54	—	—	—	—
	g	—	62.63	62.7	53.1	1.481	43.06	4.97
Sb⁺	g	—	261.91	263.46	—	—	—	—
Sb²⁺	g	—	643.1	646.1	—	—	—	—
Sb³⁺	g	—	1227.1	1231.6	—	—	—	—
Sb⁴⁺	g	—	2245.4	2251.4	—	—	—	—
SbO	g	137.749	48.	47.67	—	2.122	—	—
SbO⁺								
std. state, m = 1	aq	—	—	—	−42.33	—	—	—
SbO₂⁻								
std. state, m = 1	aq	153.749	—	—	−81.32	—	—	—
Sb₂O₃	aq	291.498	—	−164.9	—	—	—	—
Sb₂O₄	c	307.498	—	−216.9	−190.2	—	30.4	27.39
Sb₂O₅	c	323.497	—	−232.3	−198.2	—	29.9	—
SbH₃	g	124.774	36.625	34.681	35.31	2.502	55.61	9.81
HSbO₂								
undissoc.; std. state, m = 1	aq	154.757	—	−116.6	−97.4	—	11.1	—
Sb(OH)₃	c	172.772	—	—	−163.8	—	—	—
undissoc.; std. state, m = 1	aq	—	—	−184.9	−154.1	—	27.8	—
H₃SbO₄	aq	188.772	—	−216.8	—	—	—	—
SbF₃	c	178.745	—	−218.8	—	—	—	—
SbCl₃	c	228.109	—	−91.34	−77.37	—	44.0	25.8
SbCl₃	g	—	−74.57	−75.0	−72.0	4.269	80.71	18.33
SbCl₅	liq	299.015	—	−105.2	−83.7	—	72.	—
SbOCl	c	173.202	—	−89.4	—	—	—	—
	g	—	−25.	−25.5	—	—	—	—
SbBr₃	c	361.477	—	−62.0	−57.2	—	49.5	—
	g	—	−41.03	−46.5	−53.5	4.727	89.09	19.17
in CS₂	—	—	—	−58.4	—	—	—	—
SbI₃	c	502.463	—	−24.0	—	—	—	—
	aq	—	—	−23.6	—	—	—	—
Sb₂S₃								
black	c	339.692	—	−41.8	−41.5	—	43.5	28.65
orange	amorp	—	—	−35.2	—	—	—	—
Sb₂(SO₄)₃	c	531.685	—	−574.2	—	—	—	—
ARGON								
Ar	g	39.948	0	0	0	1.481	36.9822	4.9679
std. state, m = 1	aq	—	—	−2.9	3.9	—	14.2	—
Ar⁺	g	—	363.42	364.90	—	—	—	—
Ar²⁺	g	—	1000.5	1003.5	—	—	—	—
Ar³⁺	g	—	1943.9	1948.3	—	—	—	—
ARSENIC								
As								
α, gray, metallic	c	74.9216	0	0	0	1.226	8.4	5.89
γ, yellow, cubic	c	—	—	3.5	—	—	—	—
β	amorp	—	—	1.0	—	—	—	—
	g	—	72.04	72.3	62.4	1.481	41.61	4.968
As⁺	g	—	298.38	300.12	—	—	—	—
As²⁺	g	—	764.42	767.64	—	—	—	—

SELECTED VALUES OF CHEMICAL THERMODYNAMIC PROPERTIES (Continued)

Formula and Description	State	Formula weight	0 K ΔHf₀° kcal/mol	298.15 K (25°C) ΔHf° kcal/mol	ΔGf° kcal/mol	H°₂₉₈ − H°₀	S° cal/deg mol	Cp° cal/deg mol
As³⁺	g	—	1417.44	1422.14	—	—	—	—
As⁴⁺	g	—	2573.58	2579.76	—	—	—	—
As₂	g	149.8432	53.30	53.1	41.1	2.251	57.2	8.366
AsO	g	90.9210	16.88	16.72	—	2.101	—	—
AsO⁺								
undissoc.; std. state, m = 1	aq	—	—	—	−39.15	—	—	—
AsO₂⁻₃₋								
std. state, m = 1	aq	106.9204	—	−102.54	−83.66	—	9.7	—
AsO₄								
std. state, m = 1	aq	138.9192	—	−212.27	−155.00	—	−38.9	—
As₂O₅	c	229.8402	—	−221.05	−187.0	—	25.2	27.85
As₂O₅·4H₂O	c	301.9016	—	−503.0	—	—	—	—
AsH₃	g	77.9455	17.70	15.88	16.47	2.438	53.22	9.10
HAsO₂								
undissoc.; std. state, m = 1	aq	107.9284	—	−109.1	−96.25	—	30.1	—
HAsO₃²⁻								
undissoc.; std. state, m = 1	aq	139.9272	—	−216.62	−170.82	—	−0.4	—
H₂AsO₃⁻								
undissoc.; std. state, m = 1	aq	124.9357	—	−170.84	−140.35	—	26.4	—
H₂AsO₄⁻								
undissoc.; std. state, m = 1	aq	140.9351	—	−217.39	−180.04	—	28.	—
H₃AsO₄	c	141.9431	—	−216.6	—	—	—	—
	aq	—	—	−216.2	—	—	—	—
AsF₃	liq	131.9168	—	−196.3	−185.04	—	43.31	30.25
	g	—	−186.82	−187.80	−184.22	3.413	69.07	15.68
AsCl₃	liq	181.2806	—	−72.9	−62.0	—	51.7	—
	g	—	−62.12	−62.5	−59.5	4.137	78.17	18.10
AsBr₃	c	314.6486	—	−47.2	—	—	—	—
	g	—	−25.55	−31.	−38.	4.569	86.94	18.92
AsI₃	c	455.6348	−13.91	−13.9	−14.2	5.964	50.92	25.28
	g	—	—	—	—	4.834	92.79	19.27
As₂S₂	c	213.9712	—	−34.1	—	—	—	—
As₂S₃	c	246.0352	—	−40.4	−40.3	—	39.1	27.8
NH₄H₂AsO₃								
std. state, m = 1, (NH₄⁺ + H₂AsO₃⁻)	aq	143.9743	—	−202.51	−159.32	—	53.5	—
(NH₄)₂HAsO₄	c	176.0043	—	−282.4	—	—	—	—
std. state, m = 1, (2NH₄⁺ + HAsO₄²⁻)	aq	—	—	−279.96	−208.76	—	53.8	—
BARIUM								
Ba	c	137.34	0	0	0	1.65	15.0	6.71
	g	—	43.2	43.	35.	1.481	40.663	4.968
Ba⁺	g	—	163.39	164.67	—	—	—	—
Ba²⁺	g	—	394.09	396.86	—	—	—	—
std. state, m = 1	aq	—	—	−128.50	−134.02	—	+ 2.3	—
BaO₂	c	169.339	—	−151.6	—	—	—	16.0
BaO₂·H₂O	c	187.354	—	−222.3	—	—	—	—
BaO₂·8H₂O	c	313.462	—	−718.6	—	—	—	—
BaH	g	138.348	53.6	53.	47.	2.082	52.29	7.19
BaH₂	c	139.356	—	−42.7	—	—	—	—
Ba(OH)₂	c	171.355	—	−225.8	—	—	—	—
Ba(OH)₂·H₂O	c	189.370	—	−298.4	—	—	—	—
Ba(OH)₂·3H₂O	c	225.401	—	−442.0	—	—	—	—
Ba(OH)₂·8H₂O	c	315.477	—	−798.8	−667.6	—	102.	—
BaF₂	c	175.337	−288.19	−288.5	−276.5	3.452	23.03	17.02
	g	—	−195.0	−195.5	−198.0	3.262	71.91	12.85
std. state, m = 1	aq	—	—	−287.50	−267.30	—	−4.3	—
BaCl₂	c	208.246	−205.35	−205.2	−193.7	3.993	29.56	17.96
	g	—	−125.37	−125.7	−128.5	3.511	77.72	13.43
std. state, m = 1	aq	—	—	−208.40	−196.76	—	29.3	—
BaCl₂·H₂O	c	226.261	—	−277.4	−252.32	—	39.9	—
BaCl₂·2H₂O	c	244.277	—	−348.98	−309.86	—	48.5	38.71
Ba(ClO₂)₂	c	272.244	−162.6	−127.0	—	—	47.	—
Ba(ClO₃)₂	c	304.242	—	−184.4	—	—	—	—
Ba(ClO₄)₂	c	336.241	—	−191.2	—	—	—	—
Ba(ClO₄)₂·3H₂O	c	390.287	—	−404.3	−303.7	—	94.	—
BaBr₂	c	297.158	—	−181.0	−176.1	—	35.	—
	g	—	−101.4	−105.	−113.	3.9	79.	14.7
BaBr₂								
std. state, m = 1	aq	—	—	−186.60	−183.72	—	41.7	—
∞ H₂O	aq	—	—	−186.60	—	—	—	—
BaBr₂·H₂O	c	315.173	—	−255.3	—	—	—	—
BaBr₂·2H₂O	c	333.189	—	−326.5	−294.1	—	54.	—
Ba(BrO₃)₂	c	393.154	—	−171.65	−130.1	—	59.	—
Ba(BrO₃)₂								
std. state, m = 1	aq	—	—	−160.56	−125.16	—	79.6	—
Ba(BrO₃)₂·H₂O	c	411.170	−236.99	−243.84	−188.6	9.94	68.9	53.5
BaI₂	c	391.149	—	−143.9	—	—	—	—
std. state, m = 1	aq	—	—	−154.88	−158.68	—	55.5	—
BaI₂·H₂O	c	409.164	—	−219.7	—	—	—	—
BaI₂·2H₂O	c	427.179	—	−290.8	—	—	—	—
BaI₂·7H₂O	c	517.256	—	−639.7	—	—	—	—
Ba(IO₃)₂	c	487.145	−243.31	−245.5	−206.7	8.84	59.6	44.8
std. state, m = 1	aq	—	—	−234.3	−195.2	—	58.9	—
Ba(IO₃)₂·H₂O	c	505.161	—	−316.0	−263.9	—	71.	—
BaS	c	169.404	—	−110.	−109.	—	18.7	11.80
Ba₂S₂	g	338.808	—	−90.	—	—	—	—
BaSO₃	c	217.402	—	−281.9	—	—	—	—
BaSO₄	c	233.402	—	−352.1	−325.6	—	31.6	24.32
std. state, m = 1	aq	—	—	−345.82	−311.99	—	7.1	—
BaS₂O₃	c	249.466	—	—	—	—	—	40.7
BaSi₂O₆	aq	297.464	—	−415.4	—	—	—	—
Ba(HSO₃)₂	aq	299.480	—	−430.0	—	—	—	—
BaI₂·4SO₂	c	647.400	—	−470.2	—	—	—	—
BaSe	c	216.30	—	−89.	—	—	—	—
BaSeO₃	c	264.298	—	−248.7	−231.4	—	40.	—
BaSeO₄	c	280.298	—	−274.0	−249.7	—	42.	—
BaN₂	c	165.353	—	−41.	—	—	—	—
Ba(N₃)₂·H₂O	c	239.396	—	−73.7	−25.1	—	45.	—
Ba(NO₂)₂	c	229.351	—	−183.6	—	—	—	—

Formula and Description	State	Formula weight	ΔHf₀° kcal/mol	ΔHf° kcal/mol	ΔGf° kcal/mol	H°₂₉₈ − H°₀	S° cal/deg mol	C°ₚ cal/deg mol
		0 K		298.15 K (25°C)				
Ba(NO₃)₂	aq	—	—	−178.5		—		
std. state, m = 1	c	261.350	—	−237.11	−190.42	—	51.1	36.18
	aq	—	—	−227.62	−187.24	—	72.3	—
BaHPO₄	c	233.319	—	−433.7		—	—	—
Ba(H₂PO₂)₂	c	267.317	—	−421.2		—	—	—
Ba(H₂PO₂)₂·H₂O	c	285.332	—	−490.5		—	—	—
Ba(H₂PO₄)₂	c	331.315	—	−747.		—	—	—
BaHAsO₄·H₂O	c	295.283	—	−412.6		—	—	—
Ba(H₂AsO₄)₂·2H₂O	c	455.241	—	−696.9		—	—	—
BaC₂	c	161.362	—	−18.		—	—	—
BaCO₃								
witherite	c	197.349	—	−290.7	−271.9	—	26.8	20.40
std. state, m = 1	aq	—	—	−290.34	−260.19	—	−11.3	—
BaC₂O₄	c	225.360	—	−327.1		—	—	—
BaC₂O₄·2H₂O	c	261.391	—	−471.1		—	—	—
Ba(HCO₃)₂								
std. state, m = 1	aq	259.375	—	−459.28	−414.54	—	45.9	
Ba(C₂H₃O₂)₂	c	255.430	—	−354.8		—	—	—
std. state, m = 1	aq	—	—	−360.82	−310.60	—	43.7	
BaCN₂	c	177.365	—	−63.6		—	—	—
Ba(CN)₂	c	189.376	—	−52.2		—	—	—
	aq	—	—	−55.0		—	—	—
BaO·SiO₂	c	213.424	—	−388.05	−368.13	—	26.2	21.51
glassy	amorp	—	—	−376.		—	—	—
BaO·2SiO₂	c	273.509	—	−609.0	−576.2	—	36.6	32.05
2BaO·SiO₂	c	366.764	—	−546.8	−519.8	—	42.1	32.24
2BaO·3SiO₂	c	486.933	—	−1000.2	−947.2	—	61.7	53.68
Ba₂Fe(CN)₆	aq	486.634	—	−145.7		—	—	—
Ba₂Fe(CN)₆·6H₂O	c	594.726	—	−567.0		—	—	—
Ba₃[Fe(CN)₆]₂								
std. state, m = 1	aq	835.928	—	−116.9	−53.5	—	136.1	—
BaMnO₄								
barium manganate	c	256.276	—	—	−267.5		—	—
BaCrO₄	c	253.334	—	−345.6	−321.53	—	379	—
BaMoO₄	c	297.278	—	−370.	−344.1	—	33.	33.6
BaWO₄	c	385.188	—	−407.		—	—	—
BaTiO₃	c	233.238	—	−396.7	−375.8	—	25.8	24.49
BaZrO₃	c	276.558	—	−425.3	−405.0	—	29.8	24.31
BERYLLIUM								
Be	c	9.0122	0	0	0	0.466	2.27	3.93
	g	—	76.49	77.5	68.5	1.481	32.543	4.968
Be⁺	g	—	291.474	293.965		—	—	—
Be²⁺	g	—	711.426	715.398		—	—	—
Std. state, m = 1	aq	—	—	−91.5	−90.75	—	−31.0	—
BeO	c	25.0116	−144.86	−145.7	−138.7	0.669	3.38	6.10
	g	—	—	28.		—	—	—
BeO₂²⁻								
std. state, m = 1	aq	41.0110	—	−189.0	−153.0	—	−38.	—
BeH	g	10.0202	75.	75.6	68.3	2.062	42.21	6.95
BeH₂	c	11.0282	—	−4.60		—	—	—
Be(OH)₂								
α	c	43.0269	—	−215.7	−194.8	—	12.4	—
β	c	—	—	−216.5	−195.4	—	12.	—
fresh precipitated	amorp	—	—	−214.6		—	—	—
Be₃(OH)₃³⁺								
std. state, m = 1	aq	78.0587	—	—	−430.6	—	—	—
BeF₂								
α, quartz	c	47.0090	−244.85	−245.4	−234.1	2.024	12.75	12.39
β, cristobalite	c	—	—	−244.7		—	—	—
glassy	amorp	—	—	−244.3		—	—	—
BeCl₂								
α	c	79.9182	−117.40	−117.2	−106.5	2.863	19.76	15.50
β	c	—	−118.57	−118.5	−107.3	2.729	18.12	14.92
BeCl₂·4H₂O	c	151.9796	—	−432.2		—	—	—
Be₂Cl₄	g	159.8364	—	−202.		—	—	—
BeBr₂	c	168.8302	—	−84.5		—	—	—
BeI₂	c	262.8210	—	−46.0		—	—	—
BeS	c	41.0762	—	−56.0		—	—	—
BeSO₄								
α, tetragonal	c	105.0738	−285.505	−288.05	−261.44	3.125	18.62	20.48
std. state, m = 1	aq	—	—	−308.8	−268.72	—	−26.2	—
BeSO₄·H₂O	c	123.0891	—	−364.2		—	—	—
BeSO₄·2H₂O	c	141.1045	−429.814	−435.74	−381.99	5.865	39.01	36.63
BeSO₄·4H₂O								
tetragonal	c	177.1352	−569.682	−579.29	−497.29	8.306	55.68	51.77
BeSeO₄	c	151.970	—	−212.8		—	—	—
std. state, m = 1	aq	—	—	−234.7	−196.2	—	−18.1	—
BeSeO₄·2H₂O	c	188.000	—	−360.3		—	—	—
BeSeO₄·4H₂O	c	224.031	—	−505.0		—	—	—
Be₃N₂								
α, cubic	c	55.0500	—	−140.6		—	—	—
β, hexagonal	c	—	—	−136.5		—	—	—
Be(NO₃)₂	aq	133.0220	—	−191.0		—	—	—
Be₂C	c	30.0356	—	−28.0		—	—	—
BeCO₃	c	69.0216	—	−245.		—	—	—
Be₂SiO₄	c	110.1080	−510.77	−513.7	−485.8	2.922	15.37	22.84
Be(BO₂)₂	g	94.6318	—	−325.		—	—	—
BeO·Al₂O₃								
chrysoberyl	c	126.9728	−546.32	−549.9	−520.7	3.128	15.84	25.19
BeO·3Al₂O₃	c	330.8952	−1335.65	−1344.9	−1271.6	8.153	42.0	63.38
BeMoO₄	c	168.950	—	−328.		—	—	—
PuBe₁₃	c	356.21	—	−36.		—	—	—
BISMUTH								
Bi	c	208.980	0	0	0	1.536	13.56	6.10
	g	—	49.56	49.5	40.2	1.481	44.669	4.968
Bi⁺	g	—	217.65	219.07		—	—	—
Bi²⁺	g	—	602.5	605.4		—	—	—
Bi³⁺	g	—	1192.0	1196.4		—	—	—
std. state, m = 1	aq	—	—	—	19.8	—	—	—
Bi₂	g	417.960	53.12	52.5	—	2.454		8.83

Formula and Description	State	Formula weight	0 K ΔHf_0° kcal/mol	298.15 K (25°C) ΔHf°	ΔGf° kcal/mol	$H_{298}^\circ - H_0^\circ$	S° cal/deg mol	C_p°
BiO⁺								
std. state, m = 1	aq	—	—	—	−35.0	—	—	—
Bi₂O₃	c	465.9582	—	−137.16	−118.0	—	36.2	27.13
BiO·OH	c	241.9868	—	—	−88.0	—	—	—
Bi(OH)₃	c	260.0021	—	−170.0	—	—	—	—
BiCl₃	c	315.339	—	−90.6	−75.3	—	42.3	25.
BiCl₄⁻								
std. state, m = 1	aq	350.792	—	—	−115.1	—	—	—
BiCl₆³⁻								
std. state, m = 1	aq	421.698	—	—	−178.51	—	—	—
BiOCl	c	260.4324	—	−87.7	−77.0	—	28.8	—
Bi(OH)₂Cl	c	278.4477	—	—	−128.71	—	—	—
BiOBr	c	304.8884	—	—	−71.0	—	—	—
BiI₃	c	589.6932	—	—	−41.9	—	—	—
BiI₄								
std. state, m = 1	aq	716.5976	—	—	−49.9	—	—	—
BiS	g	241.044	—	43.	29.	—	68.	—
Bi₂S₃	c	514.152	—	−34.2	−33.6	—	47.9	29.2
Bi₂(SO₄)₃	c	706.1448	—	−608.1	—	—	—	—
Bi₂Te₃	c	800.760	−18.43	−18.5	−18.4	7.387	62.36	28.8
BiAsO₄	c	347.8992	—	—	−148.	—	—	—
BORON								
B								
β	c	10.811	0	0	0	0.290	1.40	2.65
	amorp	—	—	0.9	—	0.315	1.56 + X*	2.86
	g	—	133.28	134.5	124.0	1.511	36.65	4.971
B⁺	g	—	324.64	327.34	—	—	—	—
B²⁺	g	—	904.74	908.92	—	—	—	—
B³⁺	g	—	1779.43	1785.09	—	—	—	—
B⁵⁺	g	—	15606.2	15614.9	—	—	—	—
BO₂⁻								
std. state, m = 1	aq	—	—	−184.60	−162.27	—	−8.9	—
B₂O₃	c	69.6202	−302.731	−304.20	−285.30	2.223	12.90	15.04
	amorp	—	—	−299.84	−282.6	—	18.6	14.6
	g	—	−201.4	−201.67	−198.85	3.426	66.85	15.98
B₄O₇⁻⁻								
std. state, m = 1	aq	155.2398	—	—	−622.6	—	—	—
BH	g	11.8190	106.7	107.46	100.29	2.065	41.05	6.97
BH₃	g	13.8349	—	24.	—	—	—	—
BH₄⁻								
std. state, m = 1	aq	14.8429	—	11.51	27.31	—	26.4	—
B₂H₆	g	27.6698	12.29	8.3	20.7	2.831	55.45	13.00
		—	—	6.0	23.0	—	39.4	—
B₄H₁₀	g	53.3237	—	15.8	—	—	—	—
H₃BO₃	c	61.8331	−258.312	−261.55	−231.60	—	21.23	19.45
	A	—	—	−277.6	—	—	—	—
un-ionized; std. state, m = 1	aq	—	—	−256.29	−231.56	—	38.8	—
B(OH)₄⁻								
std. state, m = 1	aq	78.8405	—	−321.23	−275.65	—	24.5	—
H₂B₄O₇								
std. state, m = 1, (undissoc.)	aq	157.2557	—	—	−650.1	—	—	—
BF₃	g	67.8062	−271.082	−271.75	−267.77	2.784	60.71	12.06
BF₄⁻								
std. state, m = 1	aq	86.8046	—	−376.4	−355.4	—	43.	—
B₂F₄	g	97.6156	−343.48	−344.2	−337.1	4.08	75.8	18.90
BCl₃	c	117.170	−105.40	—	—	—	—	—
BCl₃	liq	117.170	—	−102.1	−92.6	6.88	49.3	25.5
	g	—	−96.28	−96.50	−92.91	3.362	69.31	14.99
BClF₂	g	84.2608	—	−212.8	−209.4	—	65.	—
BCl₂F	g	100.7154	—	−154.2	−150.9	—	68.	—
BBr₃	liq	250.538	—	−57.3	−57.0	—	54.9	—
	g	—	−43.83	−49.15	−55.56	3.755	77.47	16.20
BBrF₂	g	128.7168	—	—	—	3.054	68.42	13.49
BFBr₂	g	189.6274	—	—	—	3.386	74.06	14.89
BCl₂Br	g	161.626	—	—	—	3.477	74.16	15.39
BClBr₂	g	206.082	—	—	—	3.609	76.90	15.78
BI₃	g	391.5242	18.	17.00	4.96	4.024	83.43	16.92
BS	g	42.875	81.	81.74	69.02	2.085	51.65	7.18
BN	c	24.8177	−60.10	−60.8	−54.6	0.628	3.54	4.71
NH₄BO₂								
std. state, m = 1	aq	60.8484	—	−216.27	−181.24	—	18.2	—
NH₄BO₃	aq	76.8478	—	−195.4	—	—	—	—
B₂C	g	33.6332	181.	—	—	—	—	—
B₄C	c	55.2552	−16.93	−17.	−17.	1.343	6.48	12.62
B(CH₃)₃	liq	55.9162	—	−34.2	−7.7	—	57.1	—
	g	—	−23.38	−29.7	−8.6	3.831	75.2	21.15
B(C₂H₅)₃	liq	97.9974	−42.54	−46.5	2.2	13.02	80.47	57.65
BH(OCH₃)₂								
dimethoxyborane	liq	73.8879	—	−144.7	−113.2	—	57.	—
B(OCH₃)₃								
trimethoxyborane	liq	103.9144	—	−223.2	−178.0	11.45	67.8	45.89
B(OC₂H₅)₃								
triethoxyborane	liq	145.9958	—	−250.8	—	—	—	—
BSi₂	g	66.983	175.	—	—	—	—	—
BSiC	g	50.9082	164.	—	—	—	—	—
BROMINE								
Br	g	79.909	28.189	26.741	19.701	1.481	41.805	4.968
Br⁺	g	—	301.38	301.41	—	—	—	—
Br²⁺	g	—	799.2	800.7	—	—	—	—
Br³⁺	g	—	1627.0	1630.0	—	—	—	—
Br⁻	g	—	−53.0	−55.9	—	—	—	—
std. state, m = 1	aq	—	—	−29.05	−24.85	—	19.7	−33.9
Br₂	c	159.818	0	—	—	—	—	—
	liq	—	—	0	0	5.859	36.384	18.090
	g	—	10.923	7.387	0.751	2.323	58.641	8.61
std. state, m = 1	aq	—	—	−0.62	0.94	—	31.2	—
Br₃⁻								
std. state, m = 1	aq	239.727	—	−31.17	−25.59	—	51.5	—
Br₅⁻								
std. state, m = 1	aq	399.545	—	−34.0	−24.8	—	75.7	—

* X = undetermined residual entropy.

Formula and Description	State	Formula weight	0 K ΔHf_0° kcal/mol	298.15 K (25°C) ΔHf° kcal/mol	ΔGf° kcal/mol	$H_{298}^\circ - H_0^\circ$ cal/deg mol	S° cal/deg mol	C_p° cal/deg mol
BrO⁻								
std. state, m = 1	aq	95.9084	—	−22.5	−8.0	—	10.	—
BrO₂	c	111.9078	—	11.6	—	—	—	—
BrO₃⁻								
std. state, m = 1	aq	127.9072	—	−16.03	4.43	—	38.65	—
BrO₄⁻								
std. state, m = 1	aq	—	—	3.1	28.2	—	47.7	—
HBr	g	80.9170	−6.826	−8.70	−12.77	2.067	47.463	6.965
std. state, m = 1	aq	—	—	−29.05	−24.85	—	19.7	−33.9
HBrO₃								
std. state, m = 1	aq	128.9152	—	−16.03	4.43	—	38.65	—
BrF₃	liq	136.9042	—	−71.9	−57.5	—	42.6	29.78
	g	—	−58.41	−61.09	−54.84	3.416	69.89	15.92
BrCl	g	115.362	5.28	3.50	−0.23	2.245	57.36	8.36
CADMIUM								
Cd								
γ	c	112.40	0	0	0	1.491	12.37	6.21
α	c	—	—	−0.14	−0.14	—	12.37	—
	g	—	26.78	26.77	18.51	1.481	40.066	4.968
in Hg; two-phase amalgam	—	—	—	−5.078	−2.328	—	3.145	—
Cd⁺	g	—	234.18	235.65	—	—	—	—
Cd²⁺	g	—	624.09	627.04	—	—	—	—
std. state, m = 1	aq	—	—	−18.14	−18.542	—	−17.5	—
CdO	c	128.399	—	−61.7	−54.6	—	13.1	10.38
Cd(OH)₂								
precipitated	c	146.415	—	−134.0	−113.2	—	23.	—
std. state, m = 1	aq	—	—	−128.08	−93.73	—	−22.6	—
undissoc.;	aq	—	—	—	−105.8	—	—	—
CdF₂	c	150.397	—	−167.4	−154.8	—	18.5	—
std. state, m = 1	aq	—	—	−177.14	−151.82	—	−24.1	—
CdCl₂	c	183.306	−93.677	−93.57	−82.21	3.791	27.55	17.85
std. state, m = 1	aq	—	—	−98.04	−81.286	—	9.5	—
undissoc.; std. state, m = 1	aq	—	—	−96.8	−85.88	—	29.1	—
CdCl₂·H₂O	c	201.321	—	−164.54	−140.310	—	40.1	—
CdCl₂·5/2H₂O	c	228.344	—	−270.54	−225.644	—	54.3	—
Cd(ClO₄)₂								
std. state, m = 1	aq	311.301	—	−79.96	−22.66	—	69.5	—
Cd(ClO₄)₂·6H₂O	c	419.393	—	−490.6	—	—	—	—
CdCl₂·2HCl·7H₂O	c	382.335	—	−654.8	—	—	—	—
CdBr₂	c	272.218	−72.455	−75.57	−70.82	4.235	32.8	18.32
std. state, m = 1 1	aq	—	—	−76.24	−68.24	—	21.9	—
CdBr₂·4H₂O	c	344.279	—	−356.73	−298.287	—	75.6	—
CdI₂	c	366.209	−48.52	−48.6	−48.13	4.565	38.5	19.11
std. state m = 1	aq	—	—	−44.52	−43.20	—	35.7	—
CdS	c	144.464	—	−38.7	−37.4	—	15.5	—
CdSO₄	c	208.462	−220.720	−223.06	−196.65	4.354	29.407	23.80
std. state, m = 1	aq	—	—	−235.46	−196.51	—	−12.7	—
CdSO₄·H₂O	c	226.477	−292.087	−296.26	−255.46	5.582	36.814	32.16
CdSO₄·8/3H₂O	c	256.502	−406.960	−413.33	−350.224	8.497	54.883	50.97
CdSO₄·2½H₂SO₄	c	453.655	—	−769.6	—	—	—	—
CdSe	c	191.36	—	—	—	—	—	—
CdSeO₃	c	239.358	—	−137.5	−119.0	—	34.0	—
std. state, m = 1	aq	—	—	−139.8	−106.9	—	−14.4	—
CdSeO₄	c	255.358	—	−151.3	−127.1	—	39.3	—
std. state, m = 1	aq	—	—	−161.3	−124.0	—	−4.6	—
CdSeO₄·H₂O	c	273.373	—	−225.2	—	—	—	—
CdTe	c	240.00	—	−22.1	−22.0	—	24.	—
Cd(NO₃)₂	c	236.410	—	−109.06	—	—	—	—
std. state, m = 1	aq	—	—	−117.26	−71.76	—	52.5	—
Cd(NO₃)₂·2H₂O	c	272.440	—	−252.30	—	—	—	—
Cd(NO₃)₂·4H₂O	c	308.471	—	−394.11	—	—	—	—
Cd(NH₃)²⁺								
std. state, m = 1	aq	129.431	—	—	−28.4	—	—	—
Cd(NH₃)₂²⁺								
std. state, m = 1	aq	146.461	—	−63.6	−38.0	—	34.6	—
Cd(NH₃)₄²⁺								
std. state, m = 1	aq	180.522	—	−107.6	−54.1	—	80.4	—
CdCl₂·2NH₃	c	217.367	—	−152.0	−106.1	—	51.	—
CdCl₂·4NH₃	c	251.428	—	−195.4	−116.2	—	79.	—
CdCl₂·6NH₃	c	285.490	—	−237.9	−126.2	—	109.2	—
CdBr₂·NH₃	c	289.249	—	−103.3	—	—	—	—
CdBr₂·2NH₃	c	306.279	—	−131.5	—	—	—	—
CdBr₂·6NH₃	c	374.402	—	−218.4	—	—	—	—
CdI₂·2NH₃	c	400.270	—	−104.0	—	—	—	—
CdI₂·6NH₃	c	468.392	—	−173.0	—	—	—	—
Cd₃(PO₄)₂	c	527.143	—	—	−587.1	—	—	—
CdCO₃	c	172.409	—	−179.4	−160.0	—	22.1	—
CdC₂O₄	c	200.420	—	−218.1	—	—	—	—
std. state, m = 1	aq	—	—	−215.3	−179.6	—	−6.6	—
undissoc.; std. state, m = 1	aq	200.420	—	—	−185.1	—	—	—
CdC₂O₄·3H₂O	c	254.466	—	—	−360.4	—	—	—
CdCN⁺								
std. state, m = 1	aq	138.418	—	—	15.4	—	—	—
Cd(CN)₂	c	164.436	—	38.8	—	—	—	—
std. state, m = 1	aq	—	—	53.9	63.9	—	27.5	—
undissoc.;	aq	—	—	—	49.7	—	—	—
Cd(CN)₃⁻								
std. state, m = 1	aq	190.454	—	—	84.8	—	—	—
Cd(CN)₄²⁻								
std. state, m = 1	aq	216.471	—	102.3	121.3	—	77.	—
Cd(CNS)₂	c	228.564	—	12.43	—	—	—	—
std. state, m = 1	aq	—	—	18.40	25.76	—	51.4	—
undissoc.; std. state, m = 1	aq	—	—	—	23.2	—	—	—
CdSiO₃	c	188.484	—	−284.20	−264.0	—	23.3	21.17
Cd(BO₂)₂	c	198.020	—	—	−354.87	—	—	—
CALCIUM								
Ca	c	40.08	0	0	0	1.364	9.90	6.05
	g	—	42.48	42.6	34.5	1.481	36.992	4.968
Ca⁺	g	—	183.45	185.05	—	—	—	—
Ca²⁺	g	—	457.21	460.29	—	—	—	—

SELECTED VALUES OF CHEMICAL THERMODYNAMIC PROPERTIES (Continued)

Formula and Description	State	Formula weight	0 K ΔHf₀° kcal/mol	298.15 K (25°C) ΔHf° kcal/mol	ΔGf° kcal/mol	H°₂₉₈ − H°₀	S° cal/deg mol	C°ₚ cal/deg mol
std. state, m = 1	aq	—	—	−129.74	−132.30	—	−12.7	—
CaO	c	56.079	—	−151.79	−144.37	—	9.50	10.23
	g	—	+11.	—	—	—	—	—
CaO₂	c	72.079	—	−156.0	—	—	—	—
CaH	g	41.088	55.	54.7	47.9	2.076	48.19	7.11
CaH₂	c	42.096	—	−44.5	−35.2	—	10.	—
CaOH⁺								
std. state, m = 1	aq	—	—	—	−171.7	—	—	—
Ca(OH)₂	c	74.095	—	−235.68	−214.76	—	19.93	20.91
	g	—	—	−130.	—	—	—	—
std. state, m = 1	aq	—	—	−239.68	−207.49	—	−17.8	—
CaF	g	59.078	−64.76	−65.0	−71.2	2.181	54.8	8.03
CaF₂	c	78.077	—	−291.5	−279.0	—	16.46	16.02
	g	—	−186.35	−186.8	−188.9	3.025	65.55	12.25
std. state, m = 1	aq	—	—	−288.74	−265.58	—	−19.3	—
CaCl	g	75.533	−23.23	−23.4	−29.7	2.292	57.70	8.58
CaCl₂	c	110.986	—	−190.2	−178.8	—	25.0	17.35
	g	—	−112.76	−112.7	−114.54	3.613	69.35	14.18
std. state, m = 1	aq	—	—	−209.64	−195.04	—	14.3	—
CaCl₂·H₂O	c	129.001	—	−265.1	—	—	—	—
CaCl₂·2H₂O	c	147.017	—	−335.3	—	—	—	—
CaCl₂·4H₂O	c	183.047	—	−480.3	—	—	—	—
CaCl₂·6H₂O	c	219.078	—	−623.3	—	—	—	—
CaOCl₂	c	126.985	—	−178.4	—	—	—	—
CaOCl₂·H₂O	c	145.001	—	−249.1	—	—	—	—
Ca(OCl)₂	aq	142.985	—	−180.3	—	—	—	—
Ca(ClO₂)₂	c	174.984	—	−162.1	—	—	—	—
Ca(ClO₄)₂	c	238.981	—	−176.09	—	—	—	—
std. state, m = 1	aq	—	—	−191.56	−136.42	—	74.3	—
8 H₂O	aq	—	—	−188.36	—	—	—	—
10 H₂O	aq	—	—	−189.16	—	—	—	—
Ca(ClO₄)₂·4H₂O	c	311.043	—	−465.8	−352.97	—	103.6	—
CaBr₂	c	199.898	—	−163.2	−158.6	—	31.	—
	g	—	—	−95.2	—	—	—	—
std. state, m = 1	aq	—	—	−187.84	−182.00	—	26.7	—
CaBr₂·6H₂O	c	307.990	—	−599.0	−514.6	—	98.	—
Ca(BrO₃)₂	c	295.894	—	−163.9	—	—	—	—
CaI₂	c	293.889	—	−127.5	−126.4	—	34.	—
	g	—	—	−65.	—	—	—	—
std. state, m = 1	aq	—	—	−156.12	−156.96	—	40.5	—
CaI₂·8H₂O	c	438.012	—	−700.2	—	—	—	—
Ca(IO₃)₂	c	389.885	—	−239.6	−200.0	—	55.	—
Ca(IO₃)₂·H₂O	c	407.901	—	−309.1	—	—	—	—
Ca(IO₃)₂·6H₂O	c	497.977	—	−664.6	−542.0	—	108.	—
CaS	c	72.144	—	−115.3	−114.1	—	13.5	11.33
	g	—	—	—	—	—	—	—
CaSO₃	c	120.142	—	—	—	—	24.23	21.92
CaSO₃·2H₂O	c	156.173	—	−418.9	−371.7	—	44.	42.7
CaSO₄								
insol., anhydrite	c	136.142	—	−342.76	−315.93	—	25.5	23.82
sol., α	c	—	—	−340.64	−313.93	—	25.9	23.95
sol., β	c	—	—	−339.58	−312.87	—	25.9	23.67
std. state, m = 1	aq	—	—	−347.06	−310.27	—	−7.9	—
CaSO₄·1/2H₂O	c	145.149	—	−376.85	−343.41	—	31.2	28.54
macro; α								
micro; β	c	—	—	−376.35	−343.18	—	32.1	29.69
CaSO₄·2H₂O								
selenite	c	172.172	—	−483.42	−429.60	—	46.4	44.46
CaSe	c	119.04	—	−88.0	−86.8	—	16.	—
CaSeO₄	c	183.038	—	−265.25	—	—	—	—
CaSeO₄·2H₂O	c	219.068	—	−407.9	−355.4	—	53.	—
Ca(N₃)₂	c	124.120	—	+3.5	—	—	—	—
Ca₃N₂	c	148.253	—	−103.	—	—	—	—
Ca(NO₂)₂	c	132.091	—	−177.2	—	—	—	—
in 800 H₂O	aq	—	—	−179.5	—	—	—	—
Ca(NO₂)₂·4H₂O	c	204.152	—	−450.7	—	—	—	—
Ca(NO₃)₂	c	164.090	—	−224.28	−177.63	—	46.2	35.70
std. state, m = 1	aq	—	—	−228.38	−185.52	—	57.3	—
∞ H₂O	aq	—	—	−228.86	—	—	—	—
Ca(NO₃)₂·2H₂O	c	200.120	—	−368.25	−293.82	—	64.4	—
Ca(NO₃)₂·3H₂O	c	218.136	—	−439.3	−351.8	—	76.3	—
Ca(NO₃)₂·4H₂O	c	236.151	—	−509.64	−409.53	—	89.7	—
Ca(NH₂)₂								
Ca₃P₂	c	182.188	—	−121.	—	—	—	—
Ca(PO₃)₂								
β	c	198.024	—	—	—	5.715	35.05	34.68
glassy	amorp	—	—	−587.0	—	—	—	—
Ca₂P₂O₇								
β	c	254.103	−792.88	−798.0	−748.6	7.430	45.23	44.89
Ca₃(PO₄)₂	c	310.183	—	−984.9	−928.5	—	56.4	54.45
β, low temp. form	c	—	—	−982.3	−926.3	—	57.58	55.35
α, high temp form	c	—	—	−999.8	−883.9	—	−144.	—
std. state, m = 1	aq	—	—	−433.65	−401.83	4.455	26.62	26.30
CaHPO₄	c	136.059	−430.299	−438.57	−392.64	—	−20.7	—
std. state, m = 1	aq	—	—	−574.47	−515.00	7.490	45.28	47.10
CaHPO₄·2H₂O	c	172.090	−568.032	−418.9	—	—	—	—
Ca(H₂PO₂)₂	c	170.057	—	−423.1	—	—	—	—
std. state, m = 1	aq	—	—	−742.04	—	—	—	—
Ca(H₂PO₄)₂	c	234.055	—	−749.38	−672.64	—	30.5	—
std. state, m = 1	aq	—	—	−814.93	−730.98	9.950	62.1	61.86
Ca(H₂PO₄)₂·H₂O	c	252.070	−805.547	−788.4	−732.1	—	54.	—
Ca₃(AsO₄)₂	c	398.074	—	−799.	—	—	—	—
hydrated precipitate	—	—	—	−345.6	—	—	—	—
CaHAsO₄	aq	180.007	—	−563.6	—	—	—	—
Ca(H₂AsO₄)₂	aq	321.950	—	−14.3	−15.5	2.711	16.72	14.99
CaC₂	c	64.102	−15.14					
CaCO₃								
calcite	c	100.089	—	−288.46	−269.80	—	22.2	19.57
aragonite	c	—	—	−288.51	−269.55	—	21.2	19.42
std. state, m = 1	aq	—	—	−291.58	−258.47	—	−26.3	—

Formula and Description	State	Formula weight	0 K ΔHf_0° kcal/mol	298.15 K (25°C) ΔHf° kcal/mol	ΔGf° kcal/mol	$H_{298}^\circ - H_0^\circ$	S° cal/deg mol	C_p° cal/deg mol
CaC_2O_4	c	128.100	—	−325.2	—	—	—	—
std. state, m = 1	aq	—	—	−326.9	−293.37	—	−1.8	—
$CaC_2O_4 \cdot H_2O$	c	146.115	—	−400.30	−361.85	—	37.4	36.52
$CaCN_2$	c	80.105	—	−83.8	—	—	—	—
$Ca(CN)_2$	c	92.116	—	−44.1	—	—	—	—
	aq	—	—	−56.9	—	—	—	—
$CaO \cdot SiO_2$								
wollastonite	c	116.164	—	−390.76	−370.39	—	19.58	20.38
pseudowollastonite	c	—	—	−389.2	−369.2	—	20.88	20.67
glassy	amorp	—	—	−382.65	—	—	—	—
$2CaO \cdot SiO_2$								
β	c	172.244	−548.95	−551.5	−524.1	5.098	30.53	30.78
γ	c	—	−551.25	−554.0	−526.1	4.898	28.87	30.27
$3CaO \cdot SiO_2$	c	228.323	—	−700.1	−665.4	—	40.3	41.08
$3CaO \cdot 2SiO_2$								
rankinite	c	288.408	−942.23	−946.7	−899.0	8.424	50.38	51.24
$CaO \cdot Al_2O_3$	c	158.041	−552.87	−556.0	−527.9	4.569	27.30	28.87
$CaO \cdot 2Al_2O_3$	c	260.002	−944.98	−950.7	−901.2	7.286	42.50	48.00
$2CaO \cdot Al_2O_3$	c	214.120	—	−707	—	—	—	—
$3CaO \cdot Al_2O_3$	c	270.199	−853.21	−857.5	−815.4	8.216	49.2	50.16
$CaO \cdot Al_2O_3 \cdot 2SiO_2$								
anorthite, triclinic	c	278.210	—	−1009.2	−955.5	—	48.4	50.46
anorthite, hexagonal	c	—	—	−1004.3	−949.8	—	45.8	49.76
glassy	amorp	—	—	−991.8	—	—	—	—
$CaO \cdot Fe_2O_3$	c	215.772	−361.787	−363.37	−337.67	6.076	34.74	36.71
$2CaO \cdot Fe_2O_3$	c	271.851	−508.804	−511.30	−478.44	7.564	45.12	46.19
$CaCrO_4$	aq	156.074	—	−340.9	—	—	—	—
$CaMoO_3$	c	184.018	—	−296.	—	—	—	—
$CaMoO_4$	c	200.018	—	−368.4	−342.9	—	29.3	27.32
	g	—	−197.	—	—	—	—	—
std. state, m = 1	aq	—	—	−368.2	−332.2	—	−6.2	—
$CaWO_4$	c	287.928	−391.272	−393.20	−367.71	4.775	30.21	27.28
std. state, m = 1	aq	—	—	−386.8	—	—	—	—
$CaZrO_3$	c	179.298	—	−422.3	−401.8	—	23.92	23.88
$CaHfO_3$	c	266.568	—	−433.0	—	—	—	—
$CaMgC_2O_6$								
dolomite	c	184.411	−552.93	−556.0	−517.1	6.210	37.09	37.65
$2CaO \cdot 5MgO \cdot 8SiO_2 \cdot H_2O$								
tremolite	c	812.410	—	−2954.	−2780.	23.34	131.2	156.7
$CaUO_4$	c	342.107	—	−478.4	—	—	—	—
CARBON								
C								
graphite, Acheson spectroscopic	c	12.0112	0	0	0	0.251	1.372	2.038
diamond	c	—	0.5797	0.4533	0.6930	0.125[a]	0.568	1.4615
C^+	g	—	169.98	171.291	160.442	1.562	37.7597	4.9805
C^{2+}	g	—	429.628	432.420	—	—	—	—
C^{2+}	g	—	991.900	996.173	—	—	—	—
C^{3+}	g	—	2095.98	2101.73	—	—	—	—
CO	g	28.0106	−27.199	−26.416	−32.780	2.0716	47.219	6.959
std. state, m = 1	aq	—	—	−28.91	−28.66	—	25.0	—
in CH_3COOH	—	—	—	−26.416	−28.15	—	31.7	—
CO^+	g	—	295.9	298.16	—	—	—	—
CO^{2+}	g	—	942.	945.5	—	—	—	—
CO_2	g	44.0100	−93.963	−94.051	−94.254	2.2378	51.06	8.87
undissoc.; std. state, m = 1	aq	—	—	−98.90	−92.26	—	28.1	—
CO_2^+	g	—	223.8	225.23	—	—	—	—
CO_3^{2-}								
std. state, m = 1	aq	60.0094	—	−161.84	−126.17	—	−13.6	—
CH	g	13.0191	141.6	142.4	—	—	—	—
CH^+	g	—	398.1	400.4	—	—	—	—
CH_2	g	14.0271	93.9	93.7	—	—	—	—
CH_2^+	g	—	333.6	334.9	—	—	—	—
CH_3	g	15.0351	34.0	33.2	—	—	—	—
CH_3^+	g	—	261.0	261.7	—	—	—	—
CH_4	g	16.0430	−15.970	−17.88	−12.13	2.388	44.492	8.439
std. state, m = 1	aq	—	—	−21.28	−8.22	—	20.0	—
CH_4^+	g	—	277.1	276.7	—	—	—	—
HCO	g	29.0185	−4.2	−4.12	−7.76	2.386	53.68	8.26
$HCOO^-$								
std. state, m = 1	aq	45.0180	—	−101.71	−83.9	—	22.	−21.0
HCO_3^-								
std. state, m = 1	aq	61.0174	—	−165.39	−140.26	—	21.8	—
HCHO	g	30.0265	−27.1	−28.	−27.	2.394	52.26	8.46
unhydrolyzed	aq	—	—	−35.9	—	—	—	—
HCOOH	liq	—	—	−101.51	−86.38	—	30.82	23.67
	g	—	—	−90.48	—	—	—	—
un-ionized; std. state, m = 1	aq	—	—	−101.68	−89.0	—	39.	—
ionized; std. state, m = 1	aq	—	—	−101.71	−83.9	—	22.	−21.0
in ∞ H_2O	aq	—	—	−101.71	—	—	—	—
H_2CO_3								
std. state, m = 1, (undissoc.)	aq	62.0253	—	−167.22	−148.94	—	44.8	—
CH_3OH	liq	32.0424	—	−57.04	−39.76	—	30.3	19.5
	g	—	−45.355	−47.96	−38.72	2.731	57.29	10.49
std. state, m = 1	aq	—	—	−58.779	−41.92	—	31.8	—
CF_3	g	69.0064	−113.	−114.	—	—	—	—
CF_3^+	g	—	119.9	120.4	—	—	—	—
CF_4	g	88.0048	−219.6	−221.	−210.	3.043	62.50	14.60
COF_2	g	66.0074	−150.95	−151.7	−148.0	2.642	61.78	11.19
CH_3F	g	34.0335	—	—	—	2.422	53.25	8.96
CH_2F_2	g	52.0239	−104.97	−106.8	−100.2	2.555	58.94	10.25
CHF_3	g	70.0143	−162.84	−164.5	−156.3	2.764	62.04	12.20
CCl_3	g	118.3702	14.	14.	—	—	—	—
CCl_4	liq	153.8232	—	−32.37	−15.60	—	51.72	31.49
	g	—	−24.08	−24.6	−14.49	4.117	74.03	19.91
$COCl_2$	g	98.9166	−51.89	−52.3	−48.9	3.067	67.74	13.78
CH_3Cl	g	50.4881	−17.426	−19.32	−13.72	2.489	56.04	9.74
std. state, m = 1	aq	—	—	−24.3	−12.3	—	34.6	—

[a] Relative to C, diamond.

Formula and Description	State	Formula weight	0 K ΔHf° kcal/mol	298.15 K (25°C) ΔHf° kcal/mol	ΔGf° kcal/mol	H°298 − H°0	S° cal/deg mol	C°p cal/deg mol
CH₂Cl₂	liq	84.9331	—	−29.03	−16.09	—	42.5	23.9
	g		−20.462	−22.10	−15.75	2.830	64.56	12.18
CF₃Cl	g	104.4594	−164.8	−166.	−156.	3.293	68.16	15.98
CF₂Cl₂	g	120.9140	−113.0	−114.	−105.	3.543	71.86	17.27
CFCl₃	liq	137.3686	—	−72.02	−56.61	—	53.86	29.05
	g		−65.2	−66.	−57.	3.843	74.05	18.66
COFCl	g	86.4620	—	—	—	2.845	66.11	12.52
CH₂ClF	g	68.4785	—	—	—	2.689	63.17	11.24
CHClF₂	g	86.4689	—	—	—	2.955	67.11	13.35
CHCl₂F	g	102.9235	—	—	—	3.170	70.02	14.56
CBr₄								
monoclinic	c	331.6472	—	4.5	11.4	—	50.8	34.5
	g		26.10	19.	16.	4.873	85.55	21.79
COBr₂	liq	187.8286	—	−30.4	—	—	—	—
	g		−19.19	−23.0	−26.5	3.340	73.85	14.78
CH₃Br	g	94.9441	−4.72	−8.4	−6.2	2.536	58.86	10.14
in C₂H₅OH	—		—	−13.08	—	—	—	—
CH₂Br₂	g	173.8451	—	—	—	3.020	70.06	13.07
CHBr₃	liq	252.7461	—	−6.8	−1.2	—	52.8	31.
	g		10.24	4.	2.	3.811	79.07	17.02
CF₃Br	g	148.9154	−150.72	−153.6	−147.3	3.457	71.14	16.57
CCl₃Br	g	198.2792	−8.81	−11.0	−5.1	4.285	79.55	20.38
CH₃I	liq	141.9395	—	−3.7	3.2	—	39.0	30.
	g		5.38	3.1	3.5	2.585	60.71	10.54
CHI₃	c	393.7323	—	33.7	—	—	—	—
	g		—	—	—	4.106	85.1	17.92
CF₃I	g	195.9108	—	—	—	3.579	73.44	16.94
CH₂ClI	g	176.3845	—	—	—	3.002	70.7	13.02
CH₂IBr	g	220.8405	—	—	—	3.102	73.5	13.46
CS₂	liq	76.1392	—	21.44	15.60	—	36.17	18.1
	g		27.86	28.05	16.05	2.547	56.82	10.85
	aq		—	21.3	—	—	—	—
COS	g	60.0746	−33.991	−33.96	−40.47	2.373	55.32	9.92
CH₃SH	liq	48.1070	—	−11.08	−1.85	—	40.44	21.64
	g		−2.885	−5.34	−2.23	2.898	60.96	12.01
CN	g	26.0178	108.	109.	102.	2.07	48.4	6.97
CN⁻								
std. state, m = 1	aq	42.0172	—	−34.9	−23.3	—	25.5	—
HCN	liq	27.0258	—	26.02	29.86	—	26.97	16.88
	g		32.39	32.3	29.8	2.208	48.20	8.57
std. state, m = 1	aq		—	30.8	41.2	—	22.3	—
HCN'								
CH₃NH₂								
methylamine	liq	31.0577	—	−11.3	8.5	—	35.90	—
	g		—	−5.49	7.67	—	58.15	12.7
std. state, m = 1	aq		—	−16.77	4.94	—	29.5	—
CH₂N₂								
diazomethane	g		—	—	—	2.887	58.02	12.55
NH₂CN								
cyanamide	c		—	14.1	—	—	—	—
NH₄CN	c	44.0564	—	0.10	—	—	—	32.
std. state, m = 1	aq		—	4.3	22.2	—	49.6	—
	aq		—	7.7	—	—	—	—
C=NH(NH₂)₂								
guanidine	c	59.0711	—	−18.1	—	—	—	—
HNCO								
isocyanic acid	g	43.0252	—	—	—	2.615	56.85	10.72
HCNO								
cyanic acid, std. state, m = 1, ionized	aq	—	—	−34.9	−23.3	—	25.5	—
un-ionized, std. state, m = 1	aq	—	—	−36.90	−28.0	—	34.6	—
CH₃NO₂								
nitromethane	liq	61.0406	—	−27.03	−3.47	—	41.05	25.33
	g		−14.546	−17.86	−1.65	3.083	65.69	13.70
CH₃ONO								
methyl nitrite	g	61.0406	—	−16.5	—	—	51.9	—
CH₃NO₃	liq	77.0400	−38.82	−38.0	−10.4	8.26	—	37.6
	g		—	−29.8	−9.4	—	76.1	—
NH₄HCO₃	c	79.0559	—	−203.0	−159.2	—	28.9	—
std. state, m = 1	aq		—	−197.06	−159.23	—	48.9	—
NH₄CNO	c	60.0558	—	−72.75	—	—	—	—
std. state, m = 1	aq		—	−66.6	−42.3	—	52.6	—
CO(NH₂)₂								
urea	c	60.0558	—	−79.71	−47.19	—	25.00	22.26
NH₂COONH₄								
ammonium carbamate	c	78.0712	—	−154.17	−107.09	—	31.9	—
(NH₄)₂CO₃								
std. state, m = 1	aq	96.0865	—	−225.18	−164.11	—	40.6	—
CNBr	c	105.9268	—	33.58	—	—	—	—
	g		46.07	44.5	39.5	2.648	59.32	11.22
CNI	c	152.9222	—	39.71	44.22	—	23.0	—
	g		54.04	53.9	47.0	2.724	61.35	11.54
std. state, m = 1	aq		—	42.5	44.95	—	29.9	—
CNS⁻								
thiocyanate ion std. state, m = 1	aq	58.0818	—	18.27	22.15	—	34.5	−9.6
HCNS								
thiocyanic acid, undissoc. std. state, m = 1	aq	—	—	—	23.31	—	—	—
ionized; std. state, m = 1	aq	—	—	18.27	22.15	—	34.5	−9.6
NH₄CNS								
ammonium thiocyanate	c	76.1204	—	−18.8	—	—	—	—
std. state, m = 1	aq		—	−13.40	3.18	—	61.6	9.5
in 200 H₂O	aq		—	−13.4	—	—	—	—
CS(NH₂)₂								
thiourea	c	76.1204	—	−21.1	—	—	—	—
C₂O₄²⁻								
std. state, m = 1	aq	88.0199	—	−197.2	−161.1	—	10.9	—
C₂H₄	g	28.0542	14.515	12.49	16.28	2.525	52.45	10.41
std. state, m = 1	aq		—	8.69	19.43	—	29.2	—

SELECTED VALUES OF CHEMICAL THERMODYNAMIC PROPERTIES (Continued)

Formula and Description	State	Formula weight	ΔHf₀ kcal/mol (0 K)	ΔHf° kcal/mol	ΔGf° kcal/mol	H°₂₉₈ − H°₀	S° cal/deg mol	C°ₚ cal/deg mol
C₂H₅								
ethyl radical	g	29.0622	28.	25.	31.	—	59.2	—
C₂H₅⁺	g	—	222.	220.	—	—	—	—
C₂H₆	g	30.0701	−16.523	−20.24	−7.86	2.856	54.85	12.58
std. state, m = 1	aq	—	—	−24.40	−4.09	—	28.3	—
C₂H₄⁺	g	—	252.2	250.0	—	—	—	—
HC₂O₄⁻								
std. state, m = 1	aq	89.0279	—	−195.6	−166.93	—	35.7	—
CH₂CO								
ketene	g	42.0376	−13.86	−14.6	−14.8	2.819	59.16	12.37
(COOH)₂								
oxalic acid	c	90.0358	—	−197.7	—	—	—	28.
std. state, m = 1	aq	—	—	−197.2	−161.1	—	10.9	—
CH₃COO⁻								
std. state, m = 1	aq	59.0450	—	−116.16	−88.29	—	20.7	−1.5
C₂H₃O₃⁻								
glycolate ion, std. state m = 1	aq	75.0444	—	−155.9	—	—	—	—
C₂H₄O								
ethylene oxide	liq	44.0536	—	−18.60	−2.83	—	36.77	21.02
	g	—	−9.589	−12.58	−3.12	2.596	57.94	11.45
CH₃CHO								
acetaldehyde	liq	—	—	−45.96	−30.64	—	38.3	—
1/3 (CH₃CHO)₃								
paraldehyde	liq	—	—	−54.73	—	—	—	—
1/4 (CH₃CHO)₄								
metaldehyde	c	—	—	−56.2	—	—	—	—
HCOOCH₃								
methyl formate	liq	60.0530	—	−90.60	—	—	—	29.
	g	—	—	−83.7	—	—	—	—
CH₃COOH								
acetic acid	liq	60.0530	—	−115.8	−93.2	—	38.2	29.7
	g	—	−99.972	−103.31	−89.4	3.286	67.5	15.9
ionized; std. state, m = 1	aq	—	—	−116.16	−88.29	—	20.7	−1.5
un-ionized; std. state, m = 1	aq	—	—	−116.10	−94.8	—	42.7	—
CH₂OHCOOH								
hydroxyacetic acid (glycolic acid)	c	76.0524	—	−158.7	—	—	—	—
CH(OH)₂COOH								
dihydroxyacetic acid (glyoxylic acid)	c	92.0518	—	−199.7	—	—	—	—
CH₃CH₂O⁻								
std. state, m = 1	aq	45.0616	—	—	−24.5	—	—	—
C₂H₅OH								
ethanol	liq	46.0695	—	−66.37	−41.80	—	38.4	26.64
	g	—	−51.969	−56.19	−40.29	3.390	67.54	15.64
std. state, m = 1	aq	—	—	−68.9	−43.44	—	35.5	—
C₂H₅OH								
in ∞ H₂O	aq	—	—	−68.9	—	—	—	—
(CH₂OH)₂								
ethylene glycol	liq	—	—	−108.70	−77.25	—	39.9	35.8
C₂F₄								
tetrafluoroethylene	g	100.0159	−154.68	−155.5	−147.2	3.903	71.69	19.23
C₂F₆								
hexafluoroethane	g	138.0127	−308.0	−310.	−290.	4.87	79.4	25.5
CH≡CF	g	44.0287	—	—	—	2.739	55.34	12.52
CH₃CH₂F								
ethyl fluoride	g	48.0606	—	—	—	3.06	63.2	14.0
CH₂=CF₂	g	—	−76.95	−78.6	−73.0	2.980	63.6	14.36
CH₃CHF₂	g	66.0510	−110.98	−114.3	−100.6	3.34	67.5	16.2
CH₃CF₃								
1,1,1-trifluoroethane	g	84.0414	−172.93	−176.0	−159.5	3.631	66.87	18.69
CF₃CH₂F	g	102.0318	—	—	—	4.079	75.85	20.82
C₂Cl₄								
tetrachloroethylene	liq	165.8343	—	−12.5	1.1	—	63.8	33.7
	g	—	−2.70	−2.9	5.4	4.686	81.5	22.69
C₂Cl₆								
I, cubic	c	236.7403	—	−46.0	—	—	—	—
II, monoclinic	c	—	—	−47.9	—	—	—	—
III, triclinic	c	—	—	−48.5	—	—	—	—
CH₂=CHCl								
vinyl chloride	liq	62.4992	—	3.5	—	—	—	—
	g	—	10.31	8.5	12.4	2.825	63.07	12.84
1/n(CH₂=CHCl)ₙ								
polyvinyl chloride	c	—	—	−22.5	—	—	—	14.2
C₂H₅Cl								
ethyl chloride	liq	64.5152	—	−32.63	−14.20	—	45.60	24.94
	g	—	−23.331	−26.81	−14.45	3.179	65.94	15.01
CHCl=CHCl								
cis-1,2-dichloroethylene	liq	—	—	−6.6	5.27	—	47.42	27.
trans-1,2-dichloroethylene	liq	—	—	−5.53	6.52	—	46.81	27.
CH₂ClCH₂Cl								
1,2-dichloroethane	liq	—	—	−39.49	−19.03	—	49.84	30.9
	g	—	−28.357	−31.02	−17.67	4.08	73.68	18.8
CHCl=CCl₂								
trichloroethylene	liq	131.3893	—	−10.1	2.9	—	54.6	28.8
	g	—	−1.032	−1.86	4.31	3.975	77.6	19.18
CH₃COCl								
acetyl chloride	liq	78.4986	—	−65.44	−49.73	—	48.0	28.
	g	—	−56.054	−58.20	−49.20	3.53	70.5	16.2
CH₂ClCH₂OH								
ethylene chlorohydrin	liq	80.5146	—	−70.6	—	—	—	—
CHCl₂COOH								
dichloracetic acid	liq	128.9430	—	−119.0	—	—	—	44.
ionized	aq	—	—	−122.4	—	—	—	—
un-ionized	aq	—	—	−120.4	—	—	—	—
CCl₃CHO								
chloral (trichloroacetaldehyde)	liq	147.3887	—	−56.45	—	—	—	36.
CCl₃COOH								
trichloroacetic acid	c	163.3881	—	−120.7	—	—	—	—
ionized	aq	—	—	−123.4	—	—	—	—

D-60

Formula and Description	State	Formula weight	0 K ΔHf_0° kcal/mol	298.15 K (25°C) ΔHf° kcal/mol	ΔGf° kcal/mol	$H_{298}^\circ - H_0^\circ$	S° cal/deg mol	C_p° cal/deg mol
$CCl_3CH(OH)_2$								
chloral hydrate	c	165.4040	—	−137.7	—	—	—	34.
	g	—	—	−107.2	—	—	—	—
in 150 $CHCl_3$	—	—	—	−131.2	—	—	—	—
$CF_2{=}CFCl$	g	116.4705	−132.04	−132.7	−125.2	4.096	76.96	20.06
CF_3CCl_3	g	187.3765	—	—	—	5.61	88.6	28.8
$CF_2ClCFCl_2$	liq	—	—	−188.37	—	—	—	41.5
	g	—	—	−181.5	—	—	—	—
CCl_3CF_2Cl	g	203.8311	−115.75	−117.1	−97.3	5.65	91.5	29.5
$CF_2{=}CHCl$	g	98.4801	−74.25	−75.4	−69.1	3.570	72.39	17.23
CH_3CF_2Cl	liq	100.4960	—	—	—	—	—	31.4
CBr_3CBr_3								
hexabromoethane	g	503.4763	—	—	—	7.108	105.6	33.30
$CH_2{=}CHBr$								
vinyl bromide	g	106.9552	22.26	18.7	19.3	2.905	65.90	13.27
C_2H_5Br								
ethyl bromide	liq	108.9712	—	−21.99	−6.64	—	47.5	24.1
	g	—	−10.188	−15.42	−6.34	3.259	68.50	15.42
in 2000 CH_3OH	—	—	−21.71	—	—	—	—	—
$CHBr{=}CHBr$								
cis 1,2-dibromoethylene	g	185.8562	—	—	—	3.491	74.38	16.44
trans, 1,2-dibromoethylene	g	—	—	—	—	3.682	74.90	16.79
CH_3CHBr_2	g	187.8722	—	—	—	3.94	78.3	19.3
CH_2BrCH_2Br								
ethylene bromide	liq	—	—	−19.4	−5.0	—	53.37	32.51
	g	—	—	−9.16	−2.47	—	79.1	20.7
CH_3COBr								
acetyl bromide	liq	122.9546	—	−53.39	—	—	—	—
$CF_2{=}CFBr$	g	160.9265	—	—	—	4.238	80.0	20.51
CF_3CF_2Br	g	198.9233	—	—	—	4.98	85.7	26.1
$CF_2{=}CBr_2$	g	221.8371	—	—	—	4.537	83.5	21.58
CF_2BrCF_2Br	g	259.8339	—	−186.5	—	—	—	—
$CF_2{=}CHBr$	g	142.9361	—	—	—	3.682	75.1	17.63
CF_3CH_2Br	g	162.9424	—	—	—	4.299	80.59	21.67
CHF_2CF_2Br	g	180.9329	—	−197.0	—	—	—	—
CH_2ClCH_2Br	liq	143.4162	—	—	—	—	—	31.1
$CHClBrCHClBr$	g	256.7622	—	−8.8	—	—	—	—
$CF_2{=}CBrCl$	g	177.3811	—	—	—	4.385	82.0	21.16
$CF_2BrCHCl_2$	g	213.8421	—	−107.9	—	—	—	—
$CI{=}CI$	g	277.8311	—	—	—	3.901	74.80	16.81
C_2I_2	c	531.6399	—	73.	—	—	—	—
$CH_2{=}CHI$								
vinyl iodide	g	153.9506	—	—	—	3.027	68.1	13.84
C_2H_5I								
ethyl iodide	liq	155.9666	—	−9.6	3.5	—	50.6	27.5
CH_2ICH_2I								
ethylene iodide	c	281.8630	—	0.1	13.8	—	47.	—
	g	—	—	15.9	18.8	—	83.2	19.2
C_2H_5SH								
ethanethiol	liq	—	—	−17.53	−1.28	—	49.48	28.17
	g	—	−6.940	−10.95	−1.05	3.617	70.77	17.37
$(CH_3)_2SO$								
dimethyl sulfoxide	liq	78.1335	—	−48.6	−23.7	—	45.0	35.2
	g	—	−31.427	−35.96	−19.48	4.132	73.20	21.26
$(CH_3)_2SO_2$								
dimethyl sulfone	c	94.1329	—	−107.8	−72.3	—	34.	—
	g	—	−83.3	−88.7	−65.2	4.3	74.2	23.9
$C_2H_5HSO_4$								
ethyl sulfuric acid	aq	126.1317	—	−209.3	—	—	—	—
$N{\equiv}C{-}C{\equiv}N$								
cyanogen	g	52.0357	73.386	73.84	71.07	3.028	57.79	13.58
CH_3CN								
acetonitrile	liq	—	—	12.8	23.7	—	35.76	21.86
	g	—	22.58	20.9	25.0	2.892	58.67	12.48
$C_2H_5NH_2$								
ethylamine	liq	45.0848	—	−17.7	—	—	—	31.
	g	—	—	−11.27	—	—	—	16.7
$(CH_3)_2NH$								
dimethylamine	liq	—	—	−10.5	16.7	—	43.58	32.9
	g	—	—	−4.41	16.35	—	65.24	16.9
std. state, m = 1	aq	—	—	−16.88	13.85	—	31.8	—
$NH_2CH_2CH_2NH_2$								
ethylenediamine	liq	—	—	−5.82	—	—	—	50.
CH_3CONH_2								
acetamide	c	—	—	−76.0	—	—	—	16.
$C_2H_5NO_2$								
nitroethane	liq	75.0676	—	−33.5	—	—	—	33.
	g	—	—	−23.56	—	—	—	—
CH_3CH_2ONO								
ethyl nitrite	liq	—	—	−30.8	—	—	—	—
NH_2CH_2COOH								
glycine	c	—	−121.415	−126.22	−88.09	3.867	24.74	23.71
ionized; std. state, m = 1	aq	—	—	−112.280	−75.278	—	28.54	—
un-ionized, std. state, m = 1	aq	—	—	−122.846	−88.618	—	37.84	—
$CH_3CH_2ONO_2$								
ethyl nitrate	c	—	−45.023	—	—	—	—	—
	liq	—	—	−45.49	−10.29	9.242	59.08	40.7
CH_3COONH_4								
ammonium acetate	c	77.0836	—	−147.26	—	—	—	—
std. state, m = 1	aq	—	—	−147.83	−107.26	—	47.8	17.6
$CH_2OHCOONH_4$								
ammonium glycolate	c	93.0830	—	−190.6	—	—	—	—
$(CH{=}NOH)_2$								
glyoxime	c	—	—	−21.2	—	—	—	—
$(CH_3)_2NH_2NO_3$								
dimethylammonium nitrate	c	108.0977	—	−83.7	—	—	—	—
std. state, m = 1	aq	—	—	−78.30	−27.41	—	76.2	—
$(NH_4)_2C_2O_4$								
ammonium oxalate	c	—	—	−268.4	—	—	—	54.
std. state, m = 1	aq	—	—	−260.5	−199.0	—	65.1	—
$(NH_4)_2C_2O_4 \cdot H_2O$	c	142.1124	—	−340.7	—	—	—	—

Formula and Description	State	Formula weight	0 K ΔHf_0° kcal/mol	298.15 K (25°C) ΔHf° kcal/mol	ΔGf° kcal/mol	$H_{298}^\circ - H_0^\circ$	S° cal/deg mol	C_p°
CH_3NCS								
methyl isothiocyanate	c	73.1169	—	19.0	—	—	—	—
	g	—	33.46	31.3	34.5	3.464	69.29	15.65
CH_3SCN								
methyl thiocyanate	liq	—	—	28.4	—	—	—	—
	g	—	—	38.3	—	—	—	—
CH_3COOCH_3	liq	74.0801	—	−106.42	—	—	—	—
$(CH_3)_3N$								
trimethylamine	liq	59.1119	—	−11.0	24.1	—	49.82	32.31
	g	—	—	−5.81	23.65	—	68.6	—
std. state, m = 1	aq	—	—	−18.17	22.22	—	31.9	—
$(C_2H_5)_2O$								
diethyl ether	liq	74.1237	—	−66.82	—	—	—	—
	g	—	—	−60.26	—	—	—	—
$(C_2H_5)_2S$								
diethyl sulfide	liq	90.1883	—	−28.43	2.81	—	64.36	40.97
	g	—	−13.15	−19.86	4.34	5.467	87.96	27.97
$(C_2H_5)_2NH$								
diethylamine	liq	73.1390	—	−24.7	—	—	—	—
	g	—	—	−17.07	—	—	—	—
CERIUM								
Ce								
γ	c	140.12	0	0	0	1.8	17.2	6.44
	g	—	101.2	101.	92	1.594	45.807	5.515
Ce^+	g	—	227.3	228.5	—	—	—	—
Ce^{2+}	g	—	478.	480.	—	—	—	—
Ce^{3+}	g	—	942.7	947.4	—	—	—	—
std. state, m = 1	aq	—	—	−166.4	−160.6	—	−49.	—
CeO_2	c	172.119	−256.69	−258.80	−244.9	2.478	14.89	14.73
Ce_2O_3	c	328.238	−427.64	−429.3	−407.8	5.13	36.0	27.4
CeH_2	c	142.136	−47.83	−49.	−39.	1.776	13.3	9.78
CeF_3	c	197.115	—	—	—	4.237	27.5	22.3
$CeF_3 \cdot H_2O$	c	215.131	—	−472.4	—	—	—	—
$CeCl_3$	c	246.479	—	−251.8	−233.7	—	36.	20.9
	g	—	—	−174.	—	—	—	—
std. state, m = 1	aq	—	—	−286.2	−254.7	—	−9.	—
$CeCl_3 \cdot 7H_2O$	c	372.587	—	−757.5	—	—	—	—
$CeOCl$	c	191.572	—	−239.	—	—	—	—
$CeClO_4^{2+}$								
std. state, m = 1	aq	239.571	—	−209.1	−165.2	—	−37.	—
$CeBr^{2+}$								
std. state, m = 1	aq	220.029	—	—	−186.3	—	—	—
CeI_3	c	520.833	—	−155.3	—	—	—	—
$Ce(I_3)_3$	c	664.828	—	−332.	—	—	—	—
$Ce(IO_3)_3 \cdot 2H_2O$	c	700.858	—	—	−378.3	—	—	—
CeS_2	c	204.248	—	−146.3	—	—	—	—
Ce_2S	g	312.304	73.	71.6	60.0	3.25	81.	12.9
Ce_2S_3	c	376.432	—	−284.	—	—	—	—
$CeSO_4^+$								
std. state	aq	236.182	—	−380.2	−343.3	—	−17.	—
$Ce(SO_4)_2^-$								
std. state	aq	332.243	—	−595.9	−523.6	—	2.	—
$Ce_2(SO_4)_3$	c	568.425	—	−945.1	—	—	—	—
$Ce_2(SO_4)_3 \cdot 5H_2O$	c	658.502	—	—	—	—	—	—
$Ce_2(SO_4)_3 \cdot 8H_2O$	c	712.548	—	—	−1320.6	—	—	132.
$Ce(NO_3)_3$	c	326.135	—	−293.0	—	—	—	—
$Ce(NO_3)_3 \cdot 3H_2O$	c	380.181	—	−516.	—	—	—	—
$Ce(NO_3)_3 \cdot 4H_2O$	c	398.196	—	−588.9	—	—	—	—
$Ce(NO_3)_3 \cdot 6H_2O$	c	434.227	—	−729.14	—	—	—	—
CeC_2	c	164.142	—	−15.	15.2	—	20.	—
	g	—	136.	136.2	122.9	2.47	64.	10.5
CeC_4	g	188.165	167.1	168.	152.	3.68	73.	17.3
$CeCrO_3$	c	240.114	—	−368.	−347.	—	25.	—
CESIUM								
Cs	c	132.9054	0	0	0	1.843	20.37	7.69
	g	132.9054	18.542	18.180	11.748	1.481	41.942	4.968
Cs^+	g	132.9054	108.337	109.456	—	—	—	—
Cs^{2+}	g	132.9054	686.63	689.23	—	—	—	—
Cs^+	a	132.9054	—	−61.73	−69.79	—	31.80	−2.5
CsO_2	c	164.9042	—	−68.4	—	—	—	—
Cs_2O	c	281.8102	−82.142	−82.64	−73.65	4.225	35.10	18.16
	g	281.8102	—	−37.	—	—	—	—
CsH	c	133.9134	—	−12.950	—	—	—	—
	g	133.9134	28.4	27.7	23.1	2.114	51.40	7.54
$CsOH$	c	149.9128	—	−99.72	—	—	—	—
	g	149.9128	−57.9	−59.	−59.1	2.828	60.88	11.88
	a	149.9128	—	−116.70	−107.38	—	29.23	—
$CsOH \cdot H_2O$	c	167.9282	—	−180.22	—	—	—	—
CsF	c	151.9038	−132.205	−132.3	−125.6	2.802	22.18	12.21
	g	151.9038	−85.21	−85.8	−89.8	2.306	58.11	8.57
$CsCl$	c	168.3584	−105.926	−105.89	−99.08	2.976	24.18	12.54
	g	168.3584	−56.89	−57.41	−61.62	2.42	61.15	8.83
$CsClO$	a	184.3578	—	−87.3	−78.6	—	42.	—
$CsClO_2$	a	200.3572	—	−77.6	−65.7	—	56.0	—
$CsClO_3$	c	216.3566	—	−98.4	−73.6	—	37.3	—
	a	216.3566	—	−86.58	−71.71	—	70.6	—
$CsClO_4$	c	232.3560	−104.136	−105.90	−75.13	5.325	41.84	25.88
	a	232.3560	—	−92.64	−71.85	—	75.3	—
	c	212.8144	−95.358	−96.99	−93.55	3.140	27.02	12.65
	g	212.8144	−47.69	−50.0	−57.6	2.46	63.89	8.86
$CsBrO$	a	228.8138	—	−84.2	−77.8	—	42.	—
$CsBrO_3$	c	260.8126	—	−89.82	−68.11	—	39.1	—
	a	260.8126	—	−77.76	−65.36	—	70.45	—
CsI	c	259.8098	−82.652	−82.84	−81.40	3.232	29.41	12.62
	g	259.8098	−35.40	−36.3	−45.7	2.52	65.77	8.95
$CsIO$	a	275.8092	—	−87.4	−79.0	—	30.5	—
$CsIO_3$	c	307.8080	—	—	−103.7	—	—	—
$CsIO_4$	a	307.8080	—	−114.6	−100.4	—	60.1	—
$CsIO_4$	c	323.8074	—	—	−91.0	—	—	—
Cs_2S	c	297.8748	—	−86.0	—	—	—	—
	a	297.8748	—	−115.6	−119.1	—	60.1	—

Formula and Description	State	Formula weight	ΔHf_0° kcal/mol	ΔHf° kcal/mol	ΔGf° kcal/mol	$H_{298}^\circ - H_0^\circ$	S° cal/deg mol	C_p°
				0 K		298.15 K (25°C)		
Cs₂SO₃	c	345.8730	—	−271.2	—	—	—	—
	a	345.8730	—	−275.4	−255.9	—	56.6	—
Cs₂SO₄	c	361.8724	−342.63	−344.89	−316.36	6.63	50.65	32.24
	g	361.8724	—	−263.6	—	—	—	—
	a	361.8724	—	−340.78	−317.55	—	68.4	—
CsHSO₃								
from HSO₃	a	213.9756	—	−211.40	−195.94	—	65.2	—
CsHSO₄	c	229.9750	—	−276.8	—	—	—	—
from HSO₄⁻	a	229.9750	—	−273.81	−250.48	—	63.3	—
Cs₂Se	a	344.7708	—	—	−108.7	—	—	—
Cs₂SeO₃	a	392.7690	—	−245.2	−228.0	—	67.	—
Cs₂SeO₄	c	408.7684	—	−272.34	—	—	—	—
	a	408.7684	—	−266.7	−245.1	—	76.5	—
CsN₃	c	174.9255	—	−4.7	—	—	—	—
CsNO₂	a	178.9109	—	−86.7	−77.5	—	61.2	—
CsNO₃	c	194.9103	—	−120.93	−97.18	—	37.1	—
	g	194.9103	—	−89.4				
	a	194.9103	—	−111.29	−96.40	—	66.8	−23.7
CsPO₃	c	211.8774	—	−296.7	—	—	42.	—
Cs₃PO₄	a	493.6876	—	−490.5	−452.9	—	42.	—
Cs₄P₂O₇	a	705.5650	—	−789.7	−737.9	—	99.2	—
CsH₂PO₄	c	229.8928	—	−374.0	—	—	—	—
CsAsO₂	a	239.8258	—	−164.27	−153.45	—	41.5	—
Cs₃AsO₄	a	537.6354	—	−397.46	−364.37	—	56.5	—
Cs₂C₂O₄								
oxalate	c	353.8308	—	—	—	7.45	56.92	—
	a	353.8308	—	−320.7	−300.7	—	74.5	—
CsHCO₃	c	193.9228	—	−230.9	—	—	—	—
from HCO₃⁻	a	193.9228	—	−227.12	−210.05	—	53.6	—
CsHC₂O₄								
from HC₂O₄⁻	a	221.9334	—	−257.3	−236.72	—	67.5	—
CH₃COOCs								
acetate	a	191.9506	—	−177.89	−158.08	—	52.5	—
CsCN	c	158.9233	—	—	—	4.330	33.40	15.70
	a	158.9233	—	−25.7	−28.6	—	54.3	—
CsCNS								
thiocyanate	a	190.9873	—	−43.46	−47.64	—	66.3	—
CsMnO₄	a	251.8410	—	−191.1	−176.7	—	77.5	—
	a	664.7288	—	−305.7	−280.5	—	125.0	—
Cs₂CrO₄	c	381.8044	—	−341.57	—	—	—	—
Cs₂Cr₂O₇	c	481.7986	−497.023	−499.24	−456.09	10.671	78.89	55.34
	a	481.7986	—	−479.9	−450.6	—	126.2	—
Cs₂WO₄	a	513.6584	—	−380.6	—	—	—	—
CsVO₃	a	231.8456	—	−274.0	−257.1	—	44.	—
Cs₃VO₄	a	513.6558	—	—	−424.3	—	—	—
Cs₂UO₄	c	567.0374	−457.01	−461.0	−431.7	7.562	52.50	36.31
CHLORINE								
Cl	g	35.453	28.68	29.082	25.262	1.499	39.457	5.220
Cl⁺	g	—	328.86	330.74	—	—	—	—
Cl²⁺	g	—	877.81	881.17	—	—	—	—
Cl³⁺	g	—	1798.26	1803.10	—	—	—	—
Cl⁻	g	—	−57.7	−58.8	—	—	—	—
std. state, m = 1	aq	—	—	−39.952	−31.372	—	13.5	−32.6
Cl₂	g	70.906	0	0	0	2.193	53.288	8.104
std. state, m = 1	aq	—	—	−5.6	1.65	—	29.	—
Cl₃⁻								
std. state, m = 1	aq	106.359	—	—	−28.8	—	—	—
ClO	g	51.4524	24.36	24.34	23.45	2.114	54.14	7.52
ClO⁻								
std. state, m = 1	aq	—	—	−25.6	−8.8	—	10.	—
ClO₂	g	67.4518	25.09	24.5	28.8	2.580	61.36	10.03
std. state, m = 1	aq	—	—	17.9	28.7	—	39.4	—
ClO₂⁻								
std. state, m = 1	aq	—	—	−15.9	4.1	—	24.2	—
ClO₃	g	83.4512	—	37.	—	—	—	—
ClO₃⁻								
std. state, m = 1	aq	—	—	−23.7	−0.8	—	38.8	—
ClO₄								
std. state, m = 1	aq	99.4506	—	−30.91	−2.06	—	43.5	—
Cl₂O	g	86.9054	19.71	19.2	23.4	2.719	63.60	10.85
HCl	g	36.4610	−22.020	−22.062	−22.777	2.066	44.646	6.96
std state, m = 1	aq	—	—	−39.952	−31.372	—	13.5	−32.6
HCl								
∞ H₂O	aq	—	—	−39.952	—	—	—	—
HClO	g	52.4604	—	—	—	2.440	56.54	8.88
undissoc.; std. state, m = 1	aq	—	—	−28.9	−19.1	—	34.	—
HClO₂								
undissoc.; std. state, m = 1	aq	68.4598	—	−12.4	1.4	—	45.0	—
HClO₃								
std. state, m = 1	aq	84.4592	—	24.85	−1.92	—	38.8	—
HClO₄	liq	100.4586	—	−9.70	—	—	—	—
std. state, m = 1	aq	—	—	−30.91	−2.06	—	43.5	—
HClO₄								
∞ H₂O	aq	—	—	−30.91	—	—	—	—
ClF	g	54.4514	−13.0	−13.02	−13.37	2.127	52.05	7.66
ClF₃	liq	92.4482	—	−45.3	—	—	—	—
CHROMIUM								
Cr	c	51.996	0	0	0	0.970	5.68	5.58
	g	—	94.29	94.8	84.1	1.481	41.68	4.97
Cr⁺	g	—	250.30	252.29	—	—	—	—
Cr²⁺	g	—	630.7	634.2	—	—	—	—
std. state, m = 1	aq	—	—	−34.3	—	—	—	—
Cr³⁺	g	—	1345.	1350.	—	—	—	—
CrO₃	c	99.9942	—	−140.9	—	—	—	—
	g	—	—	−92.2	—	—	—	—
CrO₄²⁻								
std. state, m = 1	aq	115.9936	—	−210.60	−173.96	—	12.00	—
Cr₂O₃	c	151.9902	—	−272.4	−252.9	—	19.4	28.38
Cr₂O₇²⁻								
std. state, m = 1	aq	215.9878	—	−356.2	−311.0	—	62.6	—

Formula and Description	State	Formula weight	0 K ΔHf$_0^\circ$ kcal/mol	298.15 K (25°C) ΔHf° kcal/mol	ΔGf° kcal/mol	H$_{298}^\circ$ − H$_0^\circ$	S° cal/deg mol	C$_p^\circ$
ErI$_3$	c	547.973	—	−146.5	—	—	—	—
Er(IO$_3$)$_3$	c	691.968	—	−330.	—	—	—	—
ErC$_2$	g	191.282	138.	138.2	125.4	2.47	63.	10.5
EUROPIUM								
Eu	c	151.96	0	0	0	1.913	18.59	6.61
	g	—	42.232	41.9	34.0	1.481	45.097	4.968
Eu$^+$	g	—	172.93	174.08	—	—	—	—
Eu^{2+}	g	—	432.	435.	—	—	—	—
std. state, m = 1	aq	—	—	−126.	−129.1	—	−2.	—
Eu^{3+}								
std. state, m = 1	aq	—	—	−144.6	−137.2	—	−53.	2.
EuO	c	167.959	—	−141.5	−133.1	—	15.	—
Eu$_2$O$_3$								
cubic	c	351.918	—	−397.4	—	—	—	29.6
monoclinic	c	—	—	−394.7	−372.1	—	35.	29.2
Eu$_3$O$_4$	c	519.878	—	−543.	−512.	—	49.	—
Eu(OH)$_3$	c	202.982	—	—	−285.5	—	—	—
EuF	g	170.958	—	−70.	—	—	—	—
EuF$_3$	c	208.955	—	—	—	—	—	—
EuCl$_2$	c	222.866	—	−197.	—	—	—	—
	g	—	—	−110.	—	—	—	—
EuCl$_3$	c	258.319	—	−223.7	—	—	—	—
std. state, m = 1	aq	—	—	−264.4	−231.3	—	−13.	−96.
EuCl$_3$·6H$_2$O	c	366.411	—	−665.6	−565.5	—	97.3	87.7
Eu(BrO$_3$)$_3$·9H$_2$O	c	697.820	—	−823.4	—	—	—	—
Eu(IO$_3$)$_3$	c	676.668	—	−308.4	—	—	—	—
EuS	g	184.024	27.	—	—	—	—	—
Eu$_2$(SO$_4$)$_3$·8H$_2$O	c	736.228	—	—	—	—	160.6	146.0
EuC$_2$	c	175.982	—	−15.	−16.	—	24.	—
FLUORINE								
F	g	18.9984	18.38	18.88	14.80	1.558	37.917	5.436
F$^+$	g	—	420.16	422.14	—	—	—	—
F^{2+}	g	—	1226.98	1230.44	—	—	—	—
F^{3+}	g	—	2672.0	2676.9	—	—	—	—
F^{4+}	g	—	4684.2	4690.6	—	—	—	—
F$^-$	g	—	−63.7	−64.7	—	—	—	—
std. state, m = 1	aq	—	—	−79.50	−66.64	—	−3.3	−25.5
F$_2$	g	37.9968	0	0	0	2.108	48.44	7.48
F$_2^+$	g	—	365.1	366.6	—	—	—	—
FO	g	34.9978	41.	41.	—	—	—	—
F$_2$O	g	53.9962	−4.7	−5.2	−1.1	2.604	59.11	10.35
HF	liq	20.0064	—	−71.65	—	—	18.02 + x°	12.35
	g	—	−64.789	−64.8	−65.3	2.055	41.508	6.963
ionized; std. state, m = 1	aq	—	—	−79.50	−66.64	—	−3.3	−25.5
HF								
∞ H$_2$O	aq	—	—	−79.50	—	—	—	—
HF$_2^-$								
std. state, m = 1	aq	39.0048	—	−155.34	−138.18	—	22.1	—
XeF$_4$	c	207.294	—	−62.5	—	—	—	—
FRANCIUM								
Fr	cs	223.0000	0	0	0	—	22.8	—
FrF	c	241.9984	—	—	—	2.80	26.0	12.6
FrCl	c	258.4530	—	—	—	3.10	27.0	12.80
FrBr	c	302.9040	—	—	—	3.30	31.0	12.90
FrI	c	349.9045	—	—	—	3.40	33.0	12.90
GADOLINIUM								
Gd	c	157.25	0	0	0	2.178	16.27	8.85
	g	—	95.353	95.0	86.0	1.825	46.416	6.584
Gd$^+$	g	—	237.0	238.1	—	—	—	—
Gd^{2+}	g	—	517.	519.	—	—	—	—
Gd^{3+}								
std. state, m = 1	aq	—	—	−164.	−158.	—	−49.2	0.
GdO	g	173.249	−17.	—	—	—	—	—
Gd$_2$O$_3$								
monoclinic	c	362.498	—	−434.9	—	—	—	25.5
cubic	c	—	—	—	—	4.45	36.0	25.22
GdH$_2$	c	159.266	—	−45.5	—	—	—	—
GdF	g	176.248	—	−41.	—	—	—	—
GdF$_2$	g	195.247	—	−169.	—	—	—	—
GdF$_3$	g	214.245	—	−310.	—	—	—	—
GdCl$_3$	c	263.609	—	−241.	—	—	—	21.
std. state, m = 1	aq	—	—	−284.	−253.	—	−8.8	−98.
GdCl$_3$·6H$_2$O	c	371.701	−675.66	−685.	−586.	14.43	97.56	83.0
Gd(BrO$_3$)$_3$·9H$_2$O	c	703.110	—	−842.9	—	—	—	—
GdI$_3$	c	537.963	—	−142.	—	—	—	—
Gd(IO$_3$)$_3$	c	681.958	—	−327.	—	—	—	—
GdS	g	189.314	38.	—	—	—	—	—
Gd$_2$(SO$_4$)$_3$·8H$_2$O	c	746.808	—	−1513.	−1322.	—	155.8	140.5
Gd(NO$_3$)$_3$·6H$_2$O	c	451.357	—	—	—	19.16	133.2	106.2
GdPO$_4$·H$_2$O	c	270.237	—	—	−490.	—	—	—
GdC$_2$	c	181.272	—	−25.	—	—	—	—
	g	—	128.2	128.2	—	2.5	—	—
GALLIUM								
Ga	c	69.72	—	—	—	1.331	9.77	6.18
	liq	—	—	1.33	—	—	—	—
	g	—	66.	66.2	57.1	1.566	40.38	6.06
Ga$^+$	g	—	204.32	206.00	—	—	—	—
Ga^{2+}	g	—	677.38	680.54	—	—	—	—
std. state, m = 1	aq	—	—	−21.	—	—	—	—
Ga^{3+}	g	—	1385.	1390.	—	—	—	—
std. state, m = 1	aq	—	—	−50.6	−38.0	—	−79.	—
Ga^{4+}	g	—	2865.	2871.	—	—	—	—
GaO	g	85.719	67.	66.8	60.6	2.127	55.2	7.66
Ga$_2$O	c	155.439	—	−85.	—	—	—	—
Ga$_2$O$_3$								
β, rhombic	c	187.438	—	−260.3	−238.6	—	20.31	22.00
GaH	g	70.728	53.	52.7	46.3	2.07	46.69	7.00
Ga(OH)$_3$	c	120.742	—	−230.5	−198.7	—	24.	—
GaF	g	88.718	−60.	−60.2	—	2.167	—	7.95
GaCl	g	105.173	−19.	−19.1	−25.4	2.29	57.4	8.50

° x = Undetermined residual entropy

SELECTED VALUES OF CHEMICAL THERMODYNAMIC PROPERTIES (Continued)

Formula and Description	State	Formula weight	ΔHf₀ kcal/mol	ΔHf° kcal/mol	ΔGf° kcal/mol	H°₂₉₈ − H°₀	S° cal/deg mol	C°p cal/deg mol
				0 K	298.15 K (25°C)			
Cr₃O₄	c	219.9856	—	−366.	—	—	—	—
HCrO₄⁻								
std. state, m = 1	aq	117.0016	—	−209.9	−182.8	—	44.0	—
Cr(OH)₃								
precipitated	c	103.0181	—	−254.3	—	—	—	—
CrF₂	c	89.9928	—	−186.	—	—	—	—
	g	—	—	−99.	—	—	—	—
CrF₃	c	108.9912	−276.22	−277.	−260.	3.357	22.44	18.82
CrCl₂	c	122.902	−94.93	−94.5	−85.1	3.593	27.56	17.01
	g	—	—	−30.7	—	—	—	—
	aq	—	—	−114.2	—	—	—	—
CrCl₂·2H₂O								
light green	c	158.9327	—	−237.1	—	—	—	—
CrCl₂·3H₂O								
pale blue	c	176.9480	—	−308.9	—	—	—	—
CrCl₂·4H₂O								
dark blue	c	194.9634	—	−384.4	—	—	—	—
CrCl₃	c	158.355	−132.96	−133.0	−116.2	4.22	29.4	21.94
CrO₂Cl₂	liq	154.9008	—	−138.5	−122.1	—	53.0	—
	g	—	−127.68	−128.6	−119.9	4.32	78.8	20.2
CrBr₂	c	211.814	—	−72.2	—	—	—	—
	g	—	—	−17.	—	—	—	—
(CrBr₂)₂	g	423.628	—	−84.	—	—	—	—
CrI₂	c	305.8048	—	−37.5	—	—	—	—
	g	—	—	24.	—	—	—	—
	aq	—	—	−60.1	—	—	—	—
CrI₃	c	432.7092	—	−49.0	—	—	—	—
CrICl₂	c	249.8064	—	−100.	—	—	—	—
CrIBr₂	c	338.7184	—	−79.	—	—	—	—
Cr₂(SO₄)₃	c	392.1768	—	—	—	—	—	67.5
CrN	c	66.0027	—	−29.8	—	—	—	11.0
Cr₂N	c	117.9987	—	−30.5	—	—	—	—
NH₄HCrO₄								
std. state, m = 1	aq	135.0402	—	−241.6	−201.8	—	71.1	—
(NH₄)₂CrO₄	c	152.0708	—	−279.0	—	—	—	—
std. state, m = 1	aq	—	—	−273.94	−211.90	—	66.2	—
in 300 H₂O	aq	—	—	−273.5	—	—	—	—
(NH₄)₂Cr₂O₇	c	252.0650	—	−431.8	—	—	—	—
std. state, m = 1	aq	—	—	−419.5	−348.9	—	116.8	—
Cr₃C₂	c	180.0103	−19.51	−19.3	−19.5	3.621	20.42	23.53
Cr₇C₃	c	400.0054	—	−38.7	−39.9	—	48.0	49.92
Cr₂₃C₆	c	1267.9749	—	−87.2	−90.0	—	145.0	149.2
Cr(CO)₆	c	220.0593	—	−257.4	—	—	—	—
	g	—	—	−240.4	—	—	—	—
TlClO₄	c	323.184	—	−222.5	—	—	—	—
Tl₂CrO₄	c	524.734	—	−225.8	−205.9	—	67.5	—
AgₓCrO₄	c	331.7330	—	−174.89	−153.40	—	52.0	34.00
std. state, m = 1	aq	—	—	−160.13	−137.09	—	46.8	—
FeCr₂O₄	c	223.8366	—	−345.3	−321.2	—	34.9	31.94
COBALT								
Co								
α, hexagonal	c	58.9332	0	0	0	1.139	7.18	5.93
β, f. c. cubic	c	—	—	0.11	0.06	—	7.34	—
	g	—	101.119	101.5	90.9	1.520	42.879	5.502
Co⁺	g	—	282.486	284.348	—	—	—	—
Co²⁺	g	—	675.82	679.17	—	—	—	—
std. state, m = 1	aq	—	—	−13.9	−13.0	—	−27.	—
Co³⁺	g	—	1448.35	1453.18	—	—	−73.	—
std. state, m = 1	aq	—	—	22.	32.	—	—	—
Co₂CoO	c	74.9326	—	−56.87	−51.20	—	12.66	13.20
Co₃O₄	c	240.7972	—	−213.	−185.	—	24.5	29.5
Co(OH)₂								
blue, precipitated	c	92.9479	—	—	−107.6	—	—	—
pink, precipitated	c	—	—	−129.0	−108.6	—	19.	—
pink, precipitated, aged	c	—	—	—	−109.5	—	—	—
std. state, m = 1	aq	—	—	−123.8	−88.2	—	−32.	—
undissoc.; std. state, m = 1	aq	—	—	—	−100.8	—	—	—
Co(OH)₃								
precipitated	c	109.9553	—	−171.3	—	—	—	—
CoF₃	c	115.9284	—	−193.8	—	—	—	—
CoCl₂	c	129.8392	−74.74	−74.7	−64.5	3.375	26.09	18.76
std. state, m = 1	aq	—	—	−93.8	−75.7	—	0.	—
CoCl₂								
in ∞ H₂O	aq	—	—	−93.8	—	—	—	—
CoCl₂·H₂O	c	147.8545	—	−147.	—	—	—	—
CoCl₂·2H₂O	c	165.8699	—	−220.6	−182.8	—	45.	—
CoCl₂·6H₂O	c	237.9312	—	−505.6	−412.4	—	82.	—
Co(ClO₄)₂								
std. state, m = 1	aq	257.8344	—	−75.7	−17.1	—	60.	—
in ∞ H₂O	aq	—	—	−75.7	—	—	—	19.0
CoBr₂	c	218.7512	—	−52.8	—	—	—	—
std. state, m = 1	aq	—	—	−72.0	−62.7	—	12.	—
CoBr₂·6H₂O	c	326.8432	—	−482.8	—	—	—	—
CoI₂	c	312.7420	—	−21.2	—	—	—	—
std. state, m = 1	aq	—	—	−40.3	−37.7	—	26.	—
Co(IO₃)₂								
std. state, m = 1	aq	408.7384	—	−119.7	−74.2	—	30.	—
Co(IO₃)₂·2H₂O	c	444.7691	—	−258.6	−190.2	—	64.	—
CoS	c	90.9972	—	−19.8	—	—	—	—
Co₂S₃								
precipitated	c	214.0584	—	−35.2	—	—	—	—
CoSO₄	c	154.9948	—	−212.3	−187.0	—	28.2	—
std. state, m = 1	aq	—	—	−231.2	−191.0	—	−22.	—
CoSO₄·6H₂O	c	263.0868	−630.257	−641.4	−534.35	13.525	87.86	84.46
CoSO₄·7H₂O	c	281.1022	−699.547	−712.22	−591.26	15.097	97.05	93.33
CoSe	c	137.893	—	−14.6	—	—	—	—
CoTe₂	c	314.133	—	−31.	—	—	—	—
Co(NO₃)₂	c	182.9430	—	−100.5	—	—	—	—
std. state, m = 1	aq	—	—	−113.0	−66.2	—	43.	—
[Co(NH₃)]²⁺								
std. state, m = 1	aq	75.9638	—	−34.7	−22.1	—	3.	—

SELECTED VALUES OF CHEMICAL THERMODYNAMIC PROPERTIES (Continued)

Formula and Description	State	Formula weight	0 K $\Delta Hf_0°$ kcal/mol	298.15 K (25°C) $\Delta Hf°$ kcal/mol	$\Delta Gf°$ kcal/mol	$H°_{298} - H°_0$	S° cal/deg mol	$C_p°$
$[Co(NH_3)_2]^{2+}$								
std. state, m = 1	aq	92.9944	—	—	−30.5	—	—	—
$[Co(NH_3)_3]^{2+}$								
std. state, m = 1	aq	110.0250	—	—	−38.1	—	—	—
$[Co(NH_3)_4]^{2+}$								
std. state, m = 1	aq	127.0556	—	—	−45.3	—	—	—
$[Co(NH_3)_6]^{3+}$								
std. state, m = 1	aq	161.1169	—	−139.8	−38.9	—	40.0	
$[Co(NH_3)_6]N_3^{3+}$								
std. state, m = 1	aq	203.1370	—	—	42.9	—	60	
$[Co(NH_3)_6](NO_3)_3$	c	347.1316	—	−306.4	−125.5	—	107.	
std. state, m = 1	aq	—	—	−288.5	−117.4	—	140.	
$CoBr_2 \cdot NH_3$	c	235.7818	—	−85.1	—	—	—	—
$CoBr_2 \cdot 2NH_3$								
rose	c	252.8124	—	−116.0	—	—	—	—
$[Co(NH_3)_5]Br^{2+}$								
std. state, m = 1	aq	241.0259	—	−171.7	−60.5	—	39.	
$[Co(NH_3)_6]Br_2$	c	320.9349	—	−216.4	—	—	—	—
$[Co(NH_3)_6]Br_3$	c	400.8439	−217.52	−239.7	−119.8	12.18	77.7	78.1
std. state, m = 1	aq	—	—	−227.0	−112.2	—	94.	
Co_2P	c	148.8402	—	−45.	—	—	—	—
$Co_3(PO_4)_2$	c	366.7424	—	—	−573.3	—	—	—
$CoHPO_4$	c	154.9126	—	—	−282.5	—	—	—
$CoAs$	c	133.8548	—	−9.7	—	—	—	—
$CoAs_2$	c	208.7764	—	−14.7	—	—	—	—
Co_2As	c	192.7880	—	−9.5	—	—	—	—
Co_2As_3	c	342.6312	—	−23.3	—	—	—	—
Co_3As_2	c	326.6428	—	−19.4	—	—	—	—
Co_5As_2	c	444.5092	—	−19.0	—	—	—	—
$Co_3(AsO_4)_2$	c	454.6380	—	—	−387.4	—	—	—
$CoSb$	c	180.683	—	−10.	—	—	—	—
$CoSb_2$	c	302.433	—	−13.	—	—	—	—
$CoSb_3$	c	424.183	—	−16.	—	—	—	—
Co_2C	c	129.8776	—	−10.	—	—	—	—
$CoCo_3$	c	118.9426	—	−170.4	—	—	—	—
CoC_2O_4	c	146.9531	—	−203.5	—	—	—	—
std. state, m = 1	aq	—	—	−211.1	−174.1	—	−16.	
$Co(C_2O_4)_2^{2-}$								
std. state, m = 1	aq	234.9730	—	−408.5	−344.8	—	26.	
$CoSi$	c	87.0192	−23.87	−24.0	−23.6	1.78	10.3	10.6
$CoSi_2$	c	115.1052	—	−24.6	—	—	—	—
$CoSi_3$	c	143.1912	—	−25.6	—	—	—	—
Co_2SiO_4	c	209.9500	—	−353.	—	—	—	—
COPPER								
Cu	c	63.54	0	0	0	1.196	7.923	5.840
	g	—	80.58	80.86	71.37	1.481	39.74	4.968
Cu^+	g	—	258.752	260.513	—	—	—	—
std. state, m = 1	aq	—	—	17.13	11.95	—	9.7	—
Cu^{2+}	g	—	726.69	729.93	—	—	—	—
std. state, m = 1	aq	—	—	15.48	15.66	—	−23.8	—
Cu^{3+}	g	—	1576.1	1580.8	—	—	—	—
Cu_2	g	127.08	115.7	115.72	103.24	2.370	57.71	8.75
CuO	c	79.539	—	−37.6	−31.0	—	10.19	10.11
Cu_2O	c	143.079	—	−40.3	−34.9	—	22.26	15.21
CuH	c	64.548	—	5.1	—	—	—	—
$Cu(OH)_2$	c	97.555	—	−107.5	—	—	—	—
std. state, m = 1	aq	—	—	−94.46	−59.53	—	−28.9	—
CuF^+								
std. state, m = 1	aq	82.538	—	−62.4	−52.7	—	−16.	—
CuF_2	c	101.537	—	−129.7	—	—	—	—
$CuF_2 \cdot 2H_2O$	c	137.567	—	—	−234.6	—	—	—
$CuCl$	c	98.993	—	−32.8	−28.65	—	20.6	11.6
$CuCl^+$								
std. state, m = 1	aq	—	—	—	−16.3	—	—	—
$CuCl_2$	c	134.446	−52.79	−52.6	−42.0	3.581	25.83	17.18
$CuCl_2 \cdot 2H_2O$	c	170.477	—	−196.3	−156.8	—	40.	—
$CuCl_2$								
std. state, m = 1	aq	134.446	—	—	−57.4	—	—	—
$Cu(ClO_4)_2$								
std. state, m = 1	aq	262.441	—	−46.34	11.54	—	63.2	—
$Cu(ClO_4)_2 \cdot 6H_2O$	c	370.533	—	−460.9	—	—	—	—
Cu_2OCl_2	c	213.985	—	−90.	—	—	—	—
$CuBr$	c	143.449	−23.76	−25.0	−24.1	2.893	22.97	13.08
$CuBr^+$								
std. state, m = 1	aq	—	—	—	−11.9	—	—	—
$CuBr_2$	c	223.358	—	−33.9	—	—	—	—
$CuBr_2 \cdot 4H_2O$	c	295.419	—	−317.0	—	—	—	—
$CuBr_2 \cdot 3Cu(OH)_2$	c	516.022	—	−378.1	−306.2	—	68.	—
$Cu(BrO_3)_2 \cdot 3Cu(OH)_2$	c	612.019	—	—	−244.2	—	—	—
CuI	c	190.444	—	−16.2	−16.6	—	23.1	12.92
$Cu(IO_3)_2$								
std. state, m = 1	aq	413.345	—	−90.3	−45.5	—	32.8	—
$Cu(IO_3)_2 \cdot H_2O$	c	431.361	—	−165.4	−112.0	—	59.1	—
$Cu(IO_3)_2 \cdot 3Cu(OH)_2$	c	706.009	—	—	−160.0	—	—	—
CuS	c	95.604	—	−12.7	−12.8	—	15.9	11.43
Cu_2S								
α	c	159.144	—	−19.0	−20.6	—	28.9	18.24
$CuSO_4$	c	159.602	—	−184.36	−158.2	—	26.	23.9
std. state, m = 1	aq	—	—	−201.84	−162.31	—	−19.0	—
$CuSO_4$								
in ∞ H_2O	aq	—	—	−201.84	—	—	—	—
$CuSO_4 \cdot H_2O$	c	177.617	—	−259.52	−219.46	—	34.9	32.
$CuSO_4 \cdot 3H_2O$	c	213.648	—	−402.56	−334.65	—	52.9	49.
$CuSO_4 \cdot 5H_2O$	c	249.678	—	−544.85	−449.344	—	71.8	67.
Cu_2SO_4	c	223.142	—	−179.6	—	—	—	—
$CuO \cdot CuSO_4$	c	239.141	—	−223.8	—	—	—	—
$CuSO_4 \cdot 2Cu(OH)_2$								
antlerite	c	354.711	—	—	−345.8	—	—	—
$CuSO_4 \cdot 3Cu(OH)_2$								
brochantite	c	452.266	—	—	−434.5	—	—	—

Formula and Description	State	Formula weight	0 K ΔHf° kcal/mol	298.15 K (25°C) ΔHf° kcal/mol	ΔGf° kcal/mol	H°₂₉₈ − H°₀	S° cal/deg mol	C°ₚ
CuSO₄·3Cu(OH)₂·H₂O								
langite	c	470.281	—	−594.	−488.6	—	80.	—
CuSe	c	142.50	—	−9.45	—	—	—	—
CuSe₂	c	221.46	—	−10.3	—	—	—	—
Cu₂Se	c	206.04	—	−14.2	—	—	—	—
Cu₂Te	c	254.68	—	5.	—	—	—	—
CuN₃	c	105.560	—	66.7	82.4	—	24.	—
Cu(N₃)₂	c	147.580	—	143.0	—	—	—	—
Cu₃N	c	204.627	—	17.8	—	—	—	22.
Cu(NO₃)₂	c	187.550	—	−72.4	—	—	—	—
std. state, m = 1	aq	—	—	−83.64	−37.56	—	46.2	—
Cu(NO₃)₂								
in ∞ H₂O	aq	—	—	−83.64	—	—	—	—
Cu(NO₃)₂·3H₂O	c	241.596	—	−290.9	—	—	—	—
Cu(NO₃)₂·6H₂O	c	295.642	—	−504.5	—	—	—	—
Cu(NH₃)²⁺								
std. state, m = 1	aq	80.571	—	−9.3	3.72	—	2.9	—
Cu(NH₃)₂²⁺								
std. state, m = 1	aq	97.601	—	−34.0	−7.28	—	26.6	—
Cu(NH₃)₃²⁺								
std. state, m = 1	aq	114.632	—	−58.7	−17.48	—	47.7	—
Cu(NH₃)₄²⁺								
std. state, m = 1	aq	131.662	—	−83.3	−26.60	—	65.4	—
Cu(NH₃)₅²⁺								
std. state, m = 1	aq	148.693	—	—	−32.13	—	—	—
CuSO₄·NH₃	c	176.632	—	−218.0	—	—	—	—
CuSO₄·2NH₃	c	193.663	—	−248.2	—	—	—	—
CuSO₄·5NH₃	c	244.755	—	−328.1	—	—	—	—
CuP₂	c	125.488	—	−29.	—	—	—	—
Cu₃(PO₄)₂	c	380.563	—	—	−490.3	—	—	—
Cu₃As	c	265.542	—	−2.8	—	—	—	—
Cu₃(AsO₄)₂	c	468.458	—	—	−310.9	—	—	—
std. state, m = 1	aq	—	—	−378.10	−263.02	—	−149.2	—
CuC₂O₄	c	151.560	—	—	−158.2	—	—	—
std. state, m = 1	aq	—	—	−181.7	−145.5	—	−12.9	—
undissoc.; std. state, m = 1	aq	—	—	—	−153.1	—	—	—
CuCO₃·Cu(OH)₂								
malachite	c	221.104	—	−251.3	−213.6	—	44.5	—
2CuCO₃·Cu(OH)₂								
azurite	c	344.653	—	−390.1	—	—	—	—
CuCN	c	89.550	—	25.6	26.0	—	20.1	—
Cu(CN)₂⁻								
std. state, m = 1	aq	115.576	—	—	61.6	—	—	—
Cu(CN)₃²⁻								
std. state, m = 1	aq	141.594	—	—	96.5	—	—	—
Cu(CN)₄³⁻								
std. state, m = 1	aq	167.611	—	—	135.4	—	—	—
CuONC								
cuprous fulminate	c	105.557	—	26.3	—	—	—	—
CuCNS	c	121.622	—	—	16.7	—	—	—
std. state, m = 1	aq	—	—	35.40	34.10	—	44.2	—
(Cu(CNS)₂								
std. state, m = 1	aq	179.704	—	52.02	59.96	—	45.2	—
CuAl	c	90.522	—	−9.8	—	—	—	—
CuAl₂	c	117.503	—	−9.75	—	—	—	—
Cu₂Al	c	154.062	—	−16.5	—	—	—	—
Cu₃Al	c	217.602	—	−16.8	—	—	—	—
Cu₃Al₂	c	244.583	—	−26.2	—	—	—	—
DYSPROSIUM								
Dy	c	162.50	0	0	0	2.116	17.87	6.73
	g	—	70.04	69.4	60.8	1.48	46.97	4.97
Dy⁺	g	—	206.8	207.8	—	—	—	—
Dy²⁺	g	—	476.	477.0	—	—	—	—
Dy³⁺								
std. state, m = 1	aq	—	—	−167.	−159.	—	−55.2	5.
DyO	g	178.499	−19.	—	—	—	—	—
Dy₂O₃	c	372.998	−443.02	−445.3	−423.4	5.04	35.8	27.79
DyF	g	181.498	—	−43.	—	—	—	—
DyF²⁺								
std. state, m = 1	aq	—	—	—	−231.7	—	—	—
DyCl₃								
β	c	268.859	—	−239.	—	—	—	—
γ	c	—	—	−236.	—	—	—	—
std. state, m = 1	aq	—	—	−286.	−253.	—	−14.8	−93.
DyCl₃								
in ∞ H₂O	aq	—	—	−286.	—	—	—	—
DyCl₃·6H₂O	c	376.951	−676.83	−686.	−586.	14.42	96.00	82.7
DyI₃	c	543.213	—	−145.	—	—	—	—
Dy(IO₃)₃	c	687.208	—	−329.	—	—	—	—
DyC₂	g	186.522	206.	206.1	193.2	2.47	64.	10.5
ERBIUM								
Er	c	167.26	0	0	0	1.765	17.49	6.72
	g	—	76.08	75.8	67.1	1.48	46.72	4.97
Er⁺	g	—	216.8	218.0	—	—	—	—
Er²⁺	g	—	492.	495.	—	—	—	—
Er³⁺								
std. state, m = 1	aq	—	—	−168.6	−159.9	—	−58.4	5.
ErO	g	183.259	−13.	—	—	—	—	—
Er₂O₃	c	382.518	−451.69	−453.6	−432.3	4.78	37.2	25.93
ErH₂	c	169.276	—	−49.0	—	—	—	—
Er²H₂	c	171.288	—	−49.3	—	—	—	—
ErH₃	c	170.284	—	−58.	—	—	—	—
ErF	g	186.258	—	−45.	—	—	—	—
ErF₂	g	205.257	—	−164.	—	—	—	—
ErF₃	c	224.255	—	−409.	—	—	—	—
	g	—	—	−294.	—	—	—	—
ErCl₃	c	273.619	—	−238.7	—	—	—	24.
ErCl₃	g	—	—	−167.	—	—	—	—
ErCl₃·6H₂O	c	381.711	−677.92	−687.0	−586.6	14.39	95.3	82.0
Er(BrO₃)₃·9H₂O	c	713.120	—	−844.9	—	—	—	—

SELECTED VALUES OF CHEMICAL THERMODYNAMIC PROPERTIES (Continued)

Formula and Description	State	Formula weight	0 K ΔHf₀° kcal/mol	298.15 K (25°C) ΔHf° kcal/mol	ΔGf° kcal/mol	H°₂₉₈ − H°₀	S° cal/deg mol	C°ₚ cal/deg mol
GaCl₃	c	176.079	—	−125.4	−108.7	—	34.	
	g		—	−107.0	—	—		
GaBr	g	149.629	−10.	−11.9	−21.5	2.37	60.2	8.70
GaBr₃	c	309.447	—	−92.4	−86.0	—	43.	
GaI	g	196.624	7.4	6.9	—	2.41	—	8.76
GaI₃	c	450.433	—	−57.1	—	—	—	
	g	—	—	−34.0	—	—	—	
Ga₂(SO₄)₃	c	427.625	—	—	—	—	—	62.4
GaN	c	83.727	—	−26.4	—	—	—	—
GaP	c	100.694	—	−21.	—	—	—	—
GaPO₄	c	164.691	—	—	−310.1	—	—	—
GaAs	c	144.642	—	−17.	−16.2	—	15.34	11.05
GaSb	c	191.47	—	−10.	−9.3	—	18.18	11.60
Ga₂C₂	g	163.462	—	134.	—	—	—	—
GERMANIUM								
Ge	c	72.59	0	0	0	1.105	7.43	5.580
	g	—	89.34	90.0	80.3	1.768	40.103	7.345
Ge⁺	g	—	271.18	273.32	—	—	—	—
Ge²⁺	g	—	638.63	642.25	—	—	—	—
Ge³⁺	g	—	1427.85	1432.95	—	—	—	—
GeO								
brown	c	88.589	—	−62.6	−56.7	—	12.	—
yellow	c	—	—	—	−49.5	—	—	—
	g	—	−11.	−11.04	−17.49	2.102	53.58	7.39
GeO₂								
hexagonal	c	104.589	—	−131.7	−118.8	—	13.21	12.45
amorp	amorp	—	—	−128.4	—	—	—	—
Ge₂O₂	g	177.179	—	−112.	—	—	—	—
Ge₃O₃	g	265.768	—	−212.	—	—	—	—
GeH₄	g	76.622	24.29	21.7	27.1	2.567	51.87	10.76
Ge₂H₆	liq	151.228	—	32.82	—	—	—	—
	g	—	—	38.8	—	—	—	—
Ge₃H₈	liq	225.834	—	46.3	—	—	—	—
	g	—	—	54.2	—	—	—	—
H₂GeO₃	aq	122.604	—	−195.73	—	—	—	—
GeF	g	91.588	−8.	−7.97	—	2.185	—	8.30
GeF₂	g	110.587	−121.	—	—	—	—	—
GeF₄	g	148.584	—	—	—	4.163	72.36	19.56
GeCl	g	108.043	37.	37.09	29.7	2.290	59.	8.81
GeCl₄	liq	214.402	—	−127.1	−110.6	—	58.7	—
GeH₃Cl	g	111.067	—	—	—	2.865	63.00	13.08
GeHCl₃	g		—	—	—	4.192	79.06	19.40
GeBr	g	152.499	58.	56.32	—	2.355	—	8.87
GeBr₂	g	232.408	—	−15.0	−25.5	—	79.1	—
GeBr₄	liq	392.226	—	−83.1	−79.2	—	67.1	—
	g	—	−64.61	−71.7	−76.0	5.736	94.66	24.34
GeI₂	c	326.399	—	−21.	−20.	—	32.	—
	g	—	—	11.2	−1.0	—	76.	—
GeI₄	c	580.208	—	−33.9	−34.5	—	64.8	—
	g	—	−12.31	−13.6	−25.4	6.12	102.49	24.89
GeS	c	104.654	—	−16.5	−17.1	—	17.	—
	g	—	22.0	22.	10.	2.185	56.	8.05
GeS₂	c	136.718	—	−45.3	—	—	—	—
GeSe	c	151.55	—	−22.0	—	—	—	—
GeTe	c	200.19	—	−6.	—	—	—	—
Ge₃N₄	c	273.797	—	−15.1	—	—	—	—
GeP	c	103.564	—	−5.	−4.	—	15.	—
GeC	g	84.601	150.	151.	—	—	—	—
GeC₂	g	96.612	142.	143.	—	—	—	—
Ge₂C	g	157.191	130.	131.	—	—	—	—
GeSi	g	100.676	126.	127.	—	—	—	—
Ge₂Si	g	173.266	123.	124.	—	—	—	—
Ge₃Si	g	245.856	121.	122.	—	—	—	—
GOLD								
Au	c	196.967	—	—	—	1.436	11.33	6.075
	g	—	87.46	87.5	78.0	1.481	43.115	4.968
Au⁺	g	—	300.20	301.73	—	—	—	—
Au²⁺	g	—	773.0	776.0	—	—	—	—
AuO₃³⁻								
std. state, m = 1	aq	244.9652	—	—	−12.4	—	—	—
AuH	g	197.9750	70.9	70.5	63.5	2.068	50.441	6.968
HAuO₃²⁻								
std. state, m = 1	aq	245.9732	—	—	−34.0	—	—	—
H₂AuO₃⁻								
std. state, m = 1	aq	246.9811	—	—	−52.2	—	—	—
Au(OH)₃								
precipitated	c	247.9891	—	−101.5	−75.77	—	45.3	—
AuF₃	c	253.9622	—	−86.9	—	—	—	—
AuCl	c	232.420	—	−8.3	—	—	—	—
AuCl₂⁻								
std. state, m = 1	aq	267.873	—	—	−36.13	—	—	—
AuCl₃	c	303.326	—	−28.1	—	—	—	—
AuCl₃·2H₂O	c	339.357	—	−170.9	—	—	—	—
AuCl₄⁻								
std. state, m = 1	aq	338.779	—	−77.0	−56.22	—	63.8	—
HAuCl₄								
std. state, m = 1	aq	339.787	—	−77.0	−56.22	—	63.8	—
HAuCl₄·3H₂O	c	393.833	—	−284.9	—	—	—	—
HAuCl₄·4H₂O	c	411.848	—	−355.9	—	—	—	—
AuBr	c	276.876	—	−3.34	—	—	—	—
AuBr₂⁻								
std. state, m = 1	aq	356.785	—	−30.7	−27.49	—	52.5	—
AuBr₃	c	436.695	—	−12.73	—	—	—	—
AuBr₄⁻								
std. state, m = 1	aq	516.603	—	−45.8	−40.0	—	80.3	—
HAuBr₄·5H₂O	c	607.688	—	−398.7	—	—	—	—
AuI	c	323.8714	—	0.	—	—	—	—
AuSb₂	c	440.467	—	−2.6	—	—	—	—
Au(CN)₂⁻								
std. state, m = 1	aq	249.003	—	57.9	68.3	—	41.	—

Formula and Description	State	Formula weight	0 K ΔHf° kcal/mol	298.15 K (25°C) ΔHf° kcal/mol	ΔGf° kcal/mol	$H^\circ_{298} - H^\circ_0$	S° cal/deg mol	C°$_p$ cal/deg mol
Au(SCN)$_2^-$								
std. state, m = 1	aq	313.131	—	—	60.2	—	—	—
Au(SCN)$_4^-$								
std. state, m = 1	aq	429.294	—	—	134.2	—	—	—
Au(SCN)$_5^{2-}$								
std. state, m = 1	aq	487.376	—	—	156.4	—	—	—
Au(SCN)$_6^{3-}$								
std. state, m = 1	aq	545.458	—	—	178.5	—	—	—
AuSn	c	315.657	—	−7.28	−7.15	—	23.22	12.06
AuSn$_2$	c	434.347	—	−10.14	−9.07	—	32.4	—
AuSn$_4$	c	671.727	—	−9.25	−9.04	—	59.9	—
AuPb$_2$	c	611.347	—	−1.5	—	—	—	—
AuIn	c	311.787	—	−10.8	—	—	—	—
AuIn$_2$	c	426.607	—	−18.0	—	—	—	—
AuCd	c	309.367	—	−9.28	−9.34	—	23.9	—
AuCu	c	260.507	—	−4.46	−4.40	—	19.1	11.9
AuCu$_3$	c	387.587	—	−6.84	−6.97	—	35.7	24.0
HAFNIUM								
Hf								
α, hexagonal	c	178.49	0	0	0	1.397	10.41	6.15
	g	—	147.92	148.0	137.8	1.481	44.642	4.972
Hf$^+$	g	—	309.	311.	—	—	—	—
Hf^{2+}	g	—	653.	656.	—	—	—	—
Hf^{3+}	g	—	1190.	1195.	—	—	—	—
HfO	g	194.489	—	12.	—	—	—	—
HfO$_2$	c	210.489	—	−273.6	−260.1	—	14.18	14.40
HfF$_4$								
monoclinic	c	254.484	—	−461.4	−437.5	—	27.	—
HfCl$_4$	c	320.302	—	−236.70	−215.42	—	45.6	28.80
HfN	c	192.497	—	−88.3	—	—	—	—
HfC	c	190.501	—	−60.1	—	—	—	—
HfB	c	189.301	—	−47.	—	—	—	—
HfB$_2$	c	200.112	−80.09	−80.3	−79.4	1.77	10.2	11.89
HELIUM								
He	g	4.0026	0	0	0	1.481	30.1244	4.9679
std. state, m = 1	aq	—	—	−0.4	4.6	—	13.3	—
He$^+$	g	—	566.978	568.459	—	—	—	—
HOLMIUM								
Ho	c	164.930	0	0	0	1.91	18.0	6.49
	g	—	72.33	71.9	63.3	1.48	46.72	4.97
Ho$^+$	g	—	111.1	111.1	—	—	—	—
Ho^{2+}	g	—	483.	486.	—	—	—	—
Ho^{3+}								
std. state, m = 1	aq	—	—	−168.5	−161.0	—	−54.2	4.
HoO	g	180.930	−22.					
Ho$_2$O$_3$	c	377.858	−447.59	−449.5	−428.1	5.02	37.8	27.48
HoH$_2$	c	166.946	—	−51.7	—	—	—	—
HoF	g	183.928	—	−43.	—	—	—	—
HoF$_2$	g	202.927	—	−163.	—	—	—	—
HoF$_3$	c	221.925	—	−408.	—	—	—	—
	g	—	—	−294.	—	—	—	—
HoCl$_3$	c	271.289	—	−240.3	—	—	—	21.
	g	—	—	−168.	—	—	—	—
HoCl$_3$								
∞ H$_2$O	aq	—	—	−288.4	—	—	—	—
HoCl$_3 \cdot$6H$_2$O	c	379.381	−678.80	−687.9	−588.0	14.49	97.08	83.0
HoOCl	c	216.382	—	−239.2	—	—	—	—
HoI$_3$	c	545.643	—	−149.0	—	—	—	—
HoS	g	196.994	43.	—	—	—	—	—
HoAs	c	239.852	—	−72	—	—	—	—
HoC$_2$	c	188.952	—	−26.	−26.7	—	23.	—
	g	—	135.	135.1	—	2.47	—	10.5
HoC$_4$	g	212.975	165.	—	—	—	—	—
Ho$_2$C$_3$	c	365.893	—	−56.	—	—	—	—
HoAu	g	361.897	100.	99.06	—	2.41	—	8.77
HYDROGEN								
H	g	1.0080	51.626	52.095	48.581	1.481	27.391	4.9679
^{1}H	g	1.0078	51.626	52.095	48.581	1.481	27.391	4.9679
^{2}H	g	2.0141	52.524	52.981	49.360	1.481	29.455	4.9679
H$^+$	g	1.0080	365.211	367.161	—	—	—	—
std. state, m = 1	aq	—	—	0	0	—	0	0
H$^-$	g	—	34.40	33.39	—	—	—	—
H$_2$	g	2.0159	0	0	0	2.0238	31.208	6.889
^{1}H$_2$	g	2.0156	0	0	0	2.0238	31.208	6.889
^{2}H$_2$	g	4.0282	0	0	0	2.0481	34.620	6.978
^{1}H^{2}H	g	3.0219	0.079	0.076	−0.350	2.0328	34.343	6.978
H$_2$								
std. state, m = 1	aq	2.0159	—	−1.0	4.2	—	13.8	—
H$_2^+$	g	—	355.74	357.23	—	—	—	—
OH	g	17.0074	9.25	9.31	8.18	2.1070	43.890	7.143
OH$^-$	g	—	−32.3	−33.67	—	—	—	—
std. state, m = 1	aq	—	—	−54.970	−37.594	—	−2.57	−35.5
HO$_2$	g	33.0068	6.	5.	—	—	—	—
H$_2$O	liq	18.0153	—	−68.315	−56.687	—	16.71	17.995
^{2}H$_2$O	liq	20.0276	—	−70.411	−58.195	—	18.15	20.16
^{1}H^{2}HO	liq	19.0213	—	−69.285	−57.817	—	18.95	—
H$_2$O	g	18.0153	−57.102	−57.796	−54.634	2.3667	45.104	8.025
^{1}H$_2$O	g	18.0150	−57.102	−57.796	−54.634	2.3667	45.103	8.025
^{2}H$_2$O	g	20.0276	−58.855	−59.560	−56.059	2.3801	47.378	8.19
^{1}H^{2}HO	g	19.0213	−57.927	−58.628	−55.719	2.3721	47.658	8.08
H$_2$O$_2$	liq	34.0147	—	−44.88	−28.78	−26.2	—	21.3
	g	—	−31.08	−32.58	−25.24	2.594	55.6	10.3
undissoc.; std. state, m = 1	aq	—	—	−45.69	−32.05	—	34.4	—
H$_2$O$_2$								
∞ H$_2$O	aq	—	—	−45.69	—	—	—	—
INDIUM								
In	c	114.82	0	0	0	1.578	13.82	6.39
	g	—	58.25	58.15	49.89	1.48	41.51	4.98
In$^+$	g	—	191.68	193.06	—	—	—	—
std. state, m = 1	aq	—	—	—	−2.9	—	—	—

		0 K		298.15 K (25°C)			
Formula and Description	State	Formula weight	ΔHf₀ kcal/mol	ΔHf° kcal/mol	ΔGf° kcal/mol	H°₂₉₈ − H°₀	S°	C°ₚ cal/deg mol

Formula and Description	State	Formula weight	ΔHf₀ kcal/mol	ΔHf° kcal/mol	ΔGf° kcal/mol	H°₂₉₈ − H°₀	S° cal/deg mol	C°ₚ cal/deg mol
In²⁺	g	—	626.82	629.68	—	—	—	—
std. state, m = 1	aq	—	—	—	−12.1	—	—	—
In³⁺	g	—	1273.2	1277.56	—	—	—	—
std. state, m = 1	aq	—	—	−25.	−23.4	—	−36.	—
InO	g	130.819	92.	93.	87.1	2.14	56.5	7.78
In₂O₃	c	277.638	—	−221.27	−198.55	—	24.9	22.
InH	g	115.828	52.	51.5	45.49	2.075	49.60	7.07
InOH	g	131.827	−18.	−19.	—	—	—	—
InF	g	133.818	−48.2	−48.61	—	2.198	—	—
InCl₃	c	221.179	—	−128.4	—	—	—	—
In₂Cl₃	g	335.999	—	−103.6	—	—	—	—
InBr	c	194.729	—	−41.9	−40.4	—	27.	—
	g	—	−11.5	−13.6	−22.54	2.406	61.99	8.76
InBr₃	c	354.547	—	−102.5	—	—	—	—
	g	—	—	−67.4	—	—	31.	—
InI	c	241.724	—	−27.8	−28.8	—	31.	—
	g	—	2.52	1.8	−9.0	2.437	63.87	8.80
InI₃	c	495.533	—	−57.	—	—	—	—
	g	—	—	−28.8	—	—	—	—
InS	c	146.884	—	−33.0	−31.5	—	16.	—
	g	—	—	57.	—	—	—	—
In₂S	g	261.704	15.	2.9	—	76.	—	—
In₂S₃	c	325.832	—	−102.	−98.6	—	39.1	28.20
In₂(SO₄)₃	c	517.825	—	−666.	−583.	—	65.	67.
InN	c	128.827	—	−4.2	—	—	—	—
InP	c	145.794	—	−21.2	−18.4	—	14.3	10.86
InAs	c	189.742	—	−14.0	−12.8	—	18.1	11.42
InSb	c	236.57	—	−7.3	−6.1	—	20.6	11.82
InSb₂	g	358.32	—	75.	—	—	—	—
IODINE								
I	g	126.9044	25.631	25.535	16.798	1.481	43.184	4.968
I⁺	g	—	266.77	268.16	—	—	—	—
I²⁺	g	—	707.22	710.09	—	—	—	—
I⁻	g	—	−45.4	−47.0	—	—	—	—
std. state, m = 1	aq	—	—	−13.19	−12.33	—	26.6	−34.0
I₂	c	253.8088	0	0	0	3.154	27.757	13.011
	g	—	15.659	14.923	4.627	2.418	62.28	8.82
std. state, m = 1	aq	—	—	5.4	3.92	—	32.8	—
I₃⁻								
std. state, m = 1	aq	380.7132	—	−12.3	−12.3	—	57.2	—
IO⁻								
std. state, m = 1	aq	—	—	−25.7	−9.2	—	−1.3	—
IO₃⁻								
std. state, m = 1	aq	174.9026	—	−52.9	−30.6	—	28.3	—
IO₄⁻								
std. state, m = 1	aq	190.9020	—	−36.2	−14.0	—	−53.	—
I₂O₅	c	333.8058	—	−37.78	—	—	—	—
HI	g	127.9124	6.850	6.33	0.41	2.069	49.351	6.969
std. state, m = 1	aq	—	—	−13.19	−12.33	—	26.6	−34.0
HI								
in ∞ H₂O	aq	—	—	−13.19	—	—	—	—
HIO								
undissoc.; std. state, m = 1	aq	143.9118	—	−33.0	−23.7	—	22.8	—
HIO₃	c	175.9106	—	−55.0	—	—	—	—
undissoc.; std. state, m = 1	aq	—	—	−50.5	−31.7	—	39.9	—
IF	g	145.9028	−22.40	−22.86	−28.32	2.174	56.42	7.99
IF₅	liq	221.8964	—	−206.7	—	—	—	—
ICl								
α	c	162.3574	—	−8.4	—	—	—	—
	liq	—	—	−5.71	−3.25	—	32.3	—
	g	—	4.64	4.25	−1.30	2.282	59.140	8.50
std. state, m = 1	aq	—	—	—	−4.1	—	—	—
ICl₃	c	233.2634	—	−21.4	−5.34	—	40.0	—
IBr	c	206.8134	—	−2.5	—	—	—	—
	g	—	11.90	9.76	0.89	2.367	61.822	8.71
std. state, m = 1	aq	—	—	—	−1.0	—	—	—
IRIDIUM								
Ir	c	192.2	0	0	0	1.260	8.48	6.00
	g	—	158.78	159.0	147.7	1.481	46.240	4.968
Ir⁺	g	—	373.	375.	—	—	—	—
IrO₂	c	224.20	—	−65.5	—	—	—	13.7
IrO₃	g	240.20	—	1.9	—	—	—	—
IrF₆	c	306.19	—	−138.54	−110.34	—	59.2	—
IrCl	c	227.65	—	−19.5	—	—	—	—
IrCl₃	c	298.56	—	−58.7	—	—	—	—
IrS₂	c	256.33	—	−33.	—	—	—	—
Ir₂S₃	c	480.59	—	−56.	—	—	—	—
IRON								
Fe								
α	c	55.847	0	0	0	1.073	6.52	6.00
	g	—	98.94	99.5	88.6	1.6374	43.112	6.137
Fe⁺	g	—	281.12	283.16	—	—	—	—
Fe²⁺	g	—	654.2	657.8	—	—	—	—
std. state, m = 1	aq	—	—	−21.3	−18.85	—	−32.9	—
Fe³⁺	g	—	1360.8	1365.9	—	—	—	—
std. state, m = 1	aq	—	—	−11.6	−1.1	—	−75.5	—
Fe₀.₉₄₇O								
wustite	c	68.8865	−63.85	−63.64	−58.89	2.26	13.74	11.50
FeO	c	71.8464	—	−65.0	—	—	—	—
Fe₂O₃								
hematite	c	159.6922	−195.46	−197.0	−177.4	3.719	20.89	24.82
Fe₃O₄								
magnetite	c	231.5386	−265.80	−267.3	−242.7	5.87	35.0	34.28
FeO(OH)								
goethite	c	88.8538	—	−133.6	—	—	—	—
Fe(OH)₂								
precipitated	c	89.8617	—	−136.0	−116.3	—	21.	—
Fe(OH)₃								
precipitated	c	106.8691	—	−196.7	−166.5	—	25.5	—
undissoc.; std. state, m = 1	aq	—	—	—	−157.6	—	—	—

Formula and Description	State	Formula weight	ΔHf_0° kcal/mol (0 K)	ΔHf° kcal/mol	ΔGf° kcal/mol	$H_{298}^\circ - H_0^\circ$	S° cal/deg mol	C_p° cal/deg mol
FeF₂	c	93.8438	—	—	—	3.049	20.79	16.28
FeF₂⁺								
std. state, m = 1	aq	—	—	−166.4	−150.2	—	−15.	—
FeF₃	aq	112.8422	—	−242.9	—	—	—	—
FeCl₂	c	126.753	−82.313	−81.69	−72.26	3.889	28.19	18.32
std. state, m = 1	aq	—	—	−101.2	−81.59	—	−5.9	—
FeCl₂·2H₂O	c	162.7837	—	−227.8	—	—	—	—
FeCl₂·4H₂O	c	198.8143	—	−370.3	—	—	—	—
FeCl₃	c	162.206	−95.828	−95.48	−79.84	4.710	34.0	23.10
std. state, m = 1	aq	—	—	−131.5	−95.2	—	−35.0	—
FeCl₃·6H₂O	c	270.2980	—	−531.5	—	—	—	—
FeOCl	c	107.2994	—	−90.1	—	—	—	—
Fe(ClO₄)₂								
std. state, m = 1	aq	254.7482	—	−83.1	−22.97	—	54.1	—
FeBr₂	c	215.665	—	−59.7	—	—	—	—
std. state, m = 1	aq	—	—	−79.4	−68.55	—	6.5	—
FeBr₃	c	295.574	—	−64.1	—	—	—	—
std. state, m = 1	aq	—	—	−98.8	−75.7	—	−16.4	—
FeI₂	c	309.6558	—	−27.0	—	—	—	—
std. state, m = 1	aq	—	—	−47.7	−43.51	—	20.3	—
FeI₃	g	436.5602	—	17.	—	—	—	—
std. state, m = 1	aq	—	—	−51.2	−38.1	—	4.3	—
Fe₁.₀₀₀S								
Iron-rich pyrrhotite, α	c	87.911	−24.01	−23.9	−24.0	2.235	14.41	12.08
FeS₂								
pyrite	c	119.975	−41.72	−42.6	−39.9	2.302	12.65	14.86
markasite	c	—	—	−37.0	—	—	—	—
FeSO₄	c	151.9086	—	−221.9	−196.2	—	25.7	24.04
std. state, m = 1	aq	—	—	−238.6	−196.82	—	−28.1	—
FeSO₄·H₂O	c	169.9239	—	−297.25	—	—	—	—
FeSO₄·4H₂O	c	223.9700	—	−508.9	—	—	—	—
FeSO₄·7H₂O	c	278.0160	—	−720.50	−599.97	—	97.8	94.28
FeSe	c	134.807	—	−18.0	—	—	—	—
FeTe	c	183.447	—	−15.0	—	—	—	—
Fe₄N	c	237.3947	—	−2.5	0.9	—	37.	—
Fe(NO₃)₃	aq	241.8617	—	−161.3	—	—	—	—
std. state, m = 1	aq	—	—	−160.3	−80.9	—	29.5	—
FeP	c	86.8208	—	−30.	—	—	—	—
FeP₂	c	117.7946	—	−46.	—	—	—	—
Fe₂P	c	142.6678	—	−39.	—	—	—	—
Fe₃P	c	198.5148	—	−39	—	—	—	—
Fe₃PO₄	c	150.8184	—	−310.1	—	—	—	—
FePO₄·2H₂O								
strengite	c	186.8491	−445.28	−451.3	−396.2	6.607	40.93	43.15
Fe₃C								
α, cementite	c	179.5522	—	6.0	4.8	—	25.0	25.3
FeCO₃								
siderite	c	115.8564	—	−177.00	−159.35	—	22.2	19.63
Fe(CO)₅	liq	195.8998	—	−185.0	−168.6	—	80.8	57.5
	g	—	—	−175.4	−166.65	—	106.4	—
HFe(CN)₆³⁻								
std. state, m = 1	aq	212.9621	—	108.9	160.40	—	42.	—
H₂Fe(CN)₆²⁻								
std.state, m = 1	aq	213.9700	—	108.9	157.37	—	52.	—
H₃Fe(CN)₆⁻								
std. state, m = 1	aq	—	—	108.9	—	—	—	—
Fe(SCN)²⁺								
thiocyanate; std. state, m = 1	aq	113.9288	—	5.6	17.0	—	−31.	—
FeSi	c	83.933	−17.67	−17.6	−17.6	1.91	11.0	11.4
FeSi₂								
β-lebeanite	c	112.019	−19.19	−19.4	−18.7	2.40	13.3	15.79
FeSi₂.₃₃								
α-lebeanite	c	121.287	−14.03	−14.	−14.	2.89	16.6	17.62
Fe₃Si	c	195.627	−22.55	−22.4	−22.6	4.14	24.8	23.50
Fe₂SiO₄								
fayalite	c	203.7776	—	−353.7	−329.6	—	34.7	31.76
FeAl₂O₄	c	173.8076	—	−470.	−442.	—	25.4	29.53
KRYPTON								
Kr	g	83.80	0	0	0	1.481	39.1905	4.9679
std. state, m = 1	aq	—	—	−3.7	3.6	—	14.7	—
Kr⁺	g	—	322.84	324.32	—	—	—	—
Kr²⁺	g	—	889.47	892.43	—	—	—	—
Kr³⁺	g	—	1741.6	1746.0	—	—	—	—
LANTHANUM								
La								
α	c	138.91	0	0	0	1.593	13.6	6.48
	g	—	103.084	103.0	94.07	1.509	43.563	5.438
La⁺	g	—	231.690	233.087	—	—	—	—
La²⁺	g	—	487.	490.	—	—	—	—
La³⁺	g	—	929.2	933.2	—	—	—	—
std. state, m = 1	aq	—	—	−169.0	−163.4	—	−52.0	−3.
LaO	g	154.909	−28.5	−29.01	−34.72	2.121	57.27	7.59
La₂O	g	293.819	−2.	−3.2	—	3.00	—	12.0
La₂O₂	g	309.819	−145.	−146.6	—	3.67	—	16.2
La₂O₃	c	325.818	−427.19	−428.7	−407.7	4.742	30.43	26.00
LaH₂	c	140.926	—	−48.3	—	—	—	—
La(OH)₃	c	189.932	—	−337.0	—	—	—	—
LaF	g	157.908	—	—	—	2.173	56.9	7.98
LaF₃	g	195.905	—	—	—	4.14	78.7	17.4
LaCl₃	c	245.269	—	−256.0	—	—	—	26.0
std. state, m = 1	aq	—	—	−288.9	−257.5	—	−12.	−101.
LaCl₃								
in ∞ H₂O	aq	—	—	−288.9	—	—	—	—
LaCl₃·7H₂O	c	371.376	—	−759.7	−648.5	17.12	110.6	103.0
La(BrO₃)₃·9H₂O	c	684.770	—	−846.0	—	—	—	—
LaI₃	c	519.623	—	−159.4	—	—	—	—
La(IO₃)₃	c	663.618	—	−334.	−270.4	—	62.	—
LaS	c	170.974	−108.9	−109.	−107.9	2.6	17.5	14.
La₂S₃	c	374.012	—	−289	—	—	—	—
La₂(SO₄)₃	c	566.005	—	−942.0	—	—	—	67.
La₂(SO₄)₃·9H₂O	c	728.143	—	−1589.	—	—	—	152.

Formula and Description	State	Formula weight	0 K ΔHf° kcal/mol	298.15 K (25°C) ΔHf° kcal/mol	ΔGf° kcal/mol	H°₂₉₈ − H°₀	S° cal/deg mol	C°ₚ cal/deg mol
LaN	c	152.917	—	−72.5	—	—	—	—
La(NO₃)₃	c	324.925	—	−299.8	—	—	—	—
La(NO₃)₃ in ∞ H₂O	aq	—	—	−317.7	—	—	—	—
La(NO₃)₃·3H₂O	c	378.971	—	−520.0	—	—	—	—
La(NO₃)₃·4H₂O	c	396.986	—	−592.3	—	—	—	—
La(NO₃)₃·6H₂O	c	433.017	—	−732.23	—	—	—	—
LaC₂	c	162.932	—	−17.	−17.3	—	17.	—
La₂(CO₃)₃	c	457.848	—	—	−750.9	—	—	—
LaB₆	c	203.776	—	−31.	—	—	—	—
LaAu	g	335.877	111.4	110.8	98.2	2.46	67.	8.8
LEAD								
Pb	c	207.19	0	0	0	1.644	15.49	6.32
	g	—	46.76	46.6	38.7	1.481	41.889	4.968
Pb⁺	g	—	217.795	219.116	—	—	—	—
Pb²⁺	g	—	564.44	567.25	—	—	—	—
std. state, m = 1	aq	—	—	−0.4	−5.83	—	2.5	—
Pb³⁺	g	—	1300.93	1305.22	—	—	—	—
Pb⁴⁺	g	—	2276.9	2282.7	—	—	—	—
PbO yellow	c	223.189	−51.766	−51.466	−44.91	2.207	16.42	10.94
red	c	—	—	−52.34	−45.16	—	15.9	10.95
PbO₂	c	239.189	—	−66.3	−51.95	—	16.4	15.45
Pb₂O₃	c	462.378	—	—	—	—	36.3	25.74
Pb₃O₄	c	685.568	—	−171.7	−143.7	—	50.5	35.1
Pb(OH)₂	c	241.205	—	—	−108.1	—	—	—
precipitated	c	—	—	−123.3	—	—	—	—
PbF₂	c	245.187	—	−158.7	−147.5	—	26.4	—
std. state, m = 1	aq	—	—	−159.4	−139.11	—	−4.1	—
PbF₄	c	283.184	—	−225.1	—	—	—	—
PbCl₂	c	278.096	—	−85.90	−75.98	—	32.5	—
ionized; std. state, m = 1	aq	—	—	−80.3	−68.57	—	29.5	—
PbCl₄	liq	349.002	—	−78.7	—	—	—	—
PbBr₂	c	367.008	—	−66.6	−62.60	—	38.6	19.15
ionized; std. state, m = 1	aq	—	—	−58.5	−55.53	—	41.9	—
PbI₂	c	460.999	−41.81	−41.94	−41.50	4.666	41.79	18.49
ionized; std. state, m = 1	aq	—	—	−26.8	−30.49	—	55.7	—
PbS	c	239.254	—	−24.0	−23.6	—	21.8	11.83
PbSO₃	c	287.252	—	−160.1	—	—	—	—
PbSO₄	c	303.252	−217.82	−219.87	−194.36	4.795	35.51	24.667
PbS₂O₃	c	319.316	—	−161.1	—	—	—	—
PbSe	c	286.15	—	−24.6	−24.3	—	24.5	12.0
PbSeO₄	c	350.148	—	−145.6	−120.7	—	40.1	—
PbTeO₃·0.667H₂O (amorp)	—	394.804	—	−185.5	—	—	—	—
Pb₃(PO₄)₂	c	811.513	—	—	—	—	84.4	61.25
PbCO₃	c	267.199	—	−167.1	−149.5	—	31.3	20.89
PbC₂O₄ std. state, m = 1	c	295.210	—	−197.6	−166.9	—	34.9	25.2
Pb(CH₃)₄	liq	267.330	—	23.4	—	—	—	—
Pb(C₂H₅)₄	liq	323.439	—	12.6	—	—	—	—
	g	—	—	26.19	—	—	—	—
PSiO₃	c	283.274	—	−273.83	−253.86	—	26.2	21.52
Pb₂SiO₄	c	506.464	—	−325.8	−299.4	—	44.6	32.78
LITHIUM								
Li	c	6.941	0	0	0	1.106	6.96	5.92
	g	6.941	37.715	38.09	30.28	1.481	33.14	4.968
Li⁺	g	6.941	162.045	162.42	—	—	—	—
Li²⁺	g	6.941	—	1936.7	—	—	—	—
Li³⁺	g	6.941	—	4760.5	—	—	—	—
Li⁺	a	6.941	—	−66.56	−70.10	—	3.2	16.4
Li in D:99Hg		6.941	—	−19.60	—	—	—	—
Li Hg:x		6.941	—	—	−19.5	—	—	—
LiO	g	22.9404	18.10	18.1	12.5	2.14	50.40	7.75
LiO₂	g	38.9398	—	—	—	2.60	58.27	10.34
Li₂O	c	29.8814	−141.393	−142.91	−134.13	1.732	8.98	12.93
	g	29.8814	−38.18	−38.4	−43.4	3.03	55.30	11.91
Li₂O₂	c	45.8808	—	−151.6	—	—	—	—
	a	45.8808	—	−158.8	—	—	—	—
LiH	c	7.9490	−20.425	−21.64	−16.34	0.903	4.782	6.66
LiD	c	8.9551	−20.684	−21.73	−16.18	1.085	5.640	8.20
LiOH	c	23.9484	−114.517	−115.90	−104.92	1.772	10.23	11.87
LiOD	c	24.9545	—	−116.8	—	—	—	—
LiF	c	25.9394	−146.657	−147.22	−140.47	1.547	8.52	9.94
Li₂F₂	g	51.8788	−223.67	−224.8	−224.7	3.19	61.79	15.09
Li₃F₃	g	77.8182	−360.30	−361.9	−357.	4.88	76.	24.5
LiHF₂	c	45.9458	−223.66	−225.22	−209.11	2.826	17.0	16.77
LiCl	c	42.394	−97.68	−97.66	−91.87	2.224	14.18	11.47
	g	42.394	−46.663	−46.7	−51.8	2.165	50.840	7.94
	a	42.394	—	−106.51	−101.48	—	16.7	−16.2
LiCl·H₂O	c	60.4094	—	−170.31	−151.01	—	24.58	—
LiCl·2H₂O	c	78.4248	—	−242.03	—	—	—	—
LiCl·3H₂O	c	96.4402	—	−313.4	—	—	—	—
Li₂Cl₂	g	84.788	−141.20	−141.9	−142.5	3.70	69.0	17.26
LiClO₃	c	90.3922	—	−88.2	—	—	—	—
LiClO₄	c	106.3916	—	−91.06	—	—	—	—
	a	106.3916	—	−97.472	−72.2	—	46.7	−1.8
LiClO₄·H₂O	c	124.4070	—	−166.6	−121.8	—	37.1	—
LiClO₄·3H₂O	c	160.4378	—	−310.22	−239.30	—	60.9	—
LiBr	c	86.850	—	−83.942	−81.74	—	17.75	—
	a	86.850	—	−95.612	−94.95	—	22.9	−17.5
LiBr·H₂O	c	104.8654	—	−158.36	−142.05	—	26.2	—
LiBr·2H₂O	c	122.8808	—	−230.1	−200.9	—	38.8	—
LiBrO₃	c	134.8482	—	−82.93	—	—	—	—
LiI	c	133.8454	−64.663	−64.63	−64.60	2.716	20.74	12.20
	a	133.8454	—	−79.75	−82.4	—	29.8	−17.6
LiI·H₂O	c	151.8608	—	−141.09	−127.0	—	29.4	—
LiI·2H₂O	c	169.8762	—	−212.81	−186.5	—	44.	—
LiI·3H₂O	c	187.8916	—	−284.93	—	—	—	—
LiIO₃	c	181.8436	—	−120.31	—	—	—	—

SELECTED VALUES OF CHEMICAL THERMODYNAMIC PROPERTIES (Continued)

Formula and Description	State	Formula weight (0 K)	ΔHf_0° kcal/mol (0 K)	ΔHf° kcal/mol	ΔGf° kcal/mol	$H_{298}^\circ - H_0^\circ$	S° cal/deg mol	C_p° cal/deg mol
	a	181.8436	—	−119.46	−100.70	—	31.4	−13.2
Li₂S	c	45.946	—	−105.5	—	—	—	—
Li₂S₂	c	78.010	—	−104.7	—	—	—	—
Li₂SO₃	c	93.9442	—	−281.3	—	—	—	—
Li₂SO₄	c	109.9436	−340.367	−343.33	−315.91	4.452	27.5	28.10
	a	109.9436	—	−350.44	−318.18	—	11.3	−37.2
LiSO₄·H₂O	c	127.9590	−410.02	−414.8	−374.2	5.697	39.1	36.11
Li₂SO₄·D₂O	c	129.9712	—	−416.9	−375.6	—	40.	—
LiHS	c	40.0130	—	−60.1	—	—	—	—
LiNO₂	c	52.9465	—	−89.0	−72.2	—	23.	—
	a	52.9465	—	−91.56	−77.8	—	32.7	−6.9
LiNO₃	c	68.9459	—	−115.47	−91.1	—	21.5	—
	a	68.9459	—	−116.12	−96.7	—	38.3	−4.3
LiNO₃·3H₂O	c	122.9921	—	−328.5	−263.8	—	53.4	—
Li₃PO₄	c	115.7944	—	−500.9	—	—	—	—
Li₂C₂	c	37.9044	—	−14.2	—	—	—	—
Li₂CO₃	c	73.8914	−288.652	−290.6	−270.58	3.627	21.60	23.69
Li₂SiO₃	c	89.9662	−391.26	−393.9	−372.2	3.453	19.08	23.68
Li₂SiF₆	c	155.9584	—	−704.4	—	—	—	—
LiBO₂	c	49.7508	−245.37	−246.7	−233.3	2.144	12.3	14.3
LiBH₄	c	21.7840	−43.20	−45.6	−29.9	3.049	18.13	19.73
	a	21.7840	—	−55.05	−42.8	—	29.7	—
LiAlSi₂O₆								
α spodumene	c	186.0909	—	−730.1	−688.71	—	30.90	38.0
β spodumene	c	186.0909	—	−723.4	−683.79	—	36.90	38.9
Li₂CrO₄	c	129.8756	—	−331.9	—	—	—	—
	a	129.8756	—	−343.7	−314.2	—	18.5	—
Li₂MoO₄	c	173.8196	—	−363.36	−336.9	—	30.	—
Li₂WO₄	c	261.7296	—	−335.	—	—	—	—
Li₂TiO₃	c	109.7802	−396.780	−399.3	−377.6	3.953	21.93	26.54
Li₂ZrO₃	c	153.1002	—	−420.7	—	—	—	—
Li₂ZrO₄	c	182.9816	—	−567.	—	—	—	—
Li₄PuF₈	c	418.8012	—	−1033.	—	—	—	—
LUTETIUM								
Lu	c	174.97	0	0	0	1.524	12.18	6.42
	g	—	102.242	102.2	96.7	1.482	44.142	4.986
Lu⁺	g	—	227.36	228.80	—	—	—	—
Lu²⁺	g	—	547.	549.9	—	—	—	—
Lu³⁺								
std. state, m = 1	aq	—	—	−159.	−150.	—	−63.	6.
Lu₂O₃	c	397.938	−446.93	−448.9	−427.6	4.192	26.28	24.32
LuCl₃	c	281.329	—	−276.0	—	—	—	—
std. state, m = 1	aq	—	—	−279.	−244.	—	−23.	−92.
LuCl₃·6H₂O	c	389.421	−667.41	−676.6	−576.3	13.96	89.9	82.0
LuOCl	c	226.422	—	−227	—	—	—	—
Lu(BrO₃)₃·9H₂O	c	720.830	—	−833.8	—	—	—	—
LuI₃	c	555.683	—	−131	—	—	—	—
Lu(IO₃)₃	c	699.678	—	320.	—	—	—	—
LuS	g	207.034	48.	—	—	—	—	—
MAGNESIUM								
Mg	c	24.312	0	0	0	1.195	7.81	5.95
	g	—	35.014	35.30	27.04	1.481	35.502	4.968
Mg⁺	g	—	211.333	213.100	—	—	—	—
Mg²⁺	g	—	558.052	561.299	—	—	—	—
std. state, m = 1	aq	—	—	−111.58	−108.7	—	−33.0	—
Mg³⁺	g	—	2406.08	2410.81	—	—	—	—
Mg⁴⁺	g	—	4927.14	4933.35	—	—	—	—
MgO								
macrocrystal (periclase)	c	40.3114	−142.813	−143.81	−136.10	1.235	6.44	8.88
microcrystal	c	—	−141.954	−142.92	−135.27	1.266	6.67	9.00
MgH₂	c	26.3279	−16.05	−18.0	−8.6	1.270	7.43	8.45
Mg(OH)₂	c	58.3267	−218.402	−220.97	−199.23	2.725	15.10	18.41
precipitate	amorp	—	—	−220.0	—	—	—	—
std. state, m = 1	aq	—	—	−221.52	−183.9	—	−38.1	—
MgF	g	43.3104	−52.89	−53.0	−59.2	2.143	52.79	7.78
MgF₂	c	62.3088	−267.57	−268.5	−255.8	2.370	13.68	14.72
MgCl₂	c	95.218	−153.180	−153.28	−141.45	3.288	21.42	17.06
MgCl₂								
in ∞ H₂O	aq	—	—	−191.48	—	—	—	—
MgCl₂·H₂O	c	113.2333	—	−231.03	−205.98	—	32.8	27.55
MgCl₂·2H₂O	c	131.2487	—	−305.86	−267.24	—	43.0	38.05
MgCl₂·4H₂O	c	167.2793	—	−453.87	−388.03	—	63.1	57.70
MgCl₂·6H₂O	c	203.3100	—	−597.28	−505.49	—	87.5	75.30
Mg(ClO₄)₂	c	223.2132	—	−135.97	—	—	—	—
std. state, m = 1	aq	—	—	−173.40	−112.8	—	54.0	—
Mg(ClO₄)₂								
in ∞ H₂O	aq	—	—	−173.40	—	—	—	—
Mg(ClO₄)₂·2H₂O	c	259.2439	—	−291.3	—	—	—	—
Mg(ClO₄)₂·4H₂O	c	295.2746	—	−439.1	—	—	—	—
Mg(ClO₄)₂·6H₂O	c	331.3052	—	−584.5	−445.3	—	124.5	—
MgBr₂	c	184.130	—	−125.3	−120.4	—	28.0	—
	g	—	—	−74.0	—	—	—	—
std. state, m = 1	aq	—	—	−169.68	−158.4	—	6.4	—
MgBr₂								
in ∞ H₂O	aq	—	—	−169.68	—	—	—	—
MgI₂	c	278.1208	—	−87.0	−85.6	—	31.0	—
	g	—	—	−41.	—	—	—	—
std. state, m = 1	aq	—	—	−137.96	−133.4	—	20.2	—
in ∞ H₂O	aq	—	—	−137.96	—	—	—	—
MgS	c	56.376	−82.44	−82.7	−81.7	1.992	12.03	10.89
MgSO₃	c	104.3742	—	−241.0	—	—	—	—
MgSO₄	c	120.3736	—	−307.1	−279.8	—	21.9	23.06
undissociated; std. state,	aq	—	—	−324.1	−289.74	—	−1.7	—
m = 1	aq	—	—	−328.90	−286.7	—	−28.2	—
MgSO₄								
in ∞ H₂O	aq	—	—	−328.90	—	—	—	—
MgSO₄·2H₂O	c	156.4043	—	−453.2	—	—	—	—
MgSO₄·4H₂O	c	192.4350	—	−596.7	—	—	—	—
MgSO₄·6H₂O	c	228.4656	—	−737.8	−629.1	—	83.2	83.20
MgSO₄·7H₂O	c	246.4810	—	−809.92	−686.4	—	89.	—

SELECTED VALUES OF CHEMICAL THERMODYNAMIC PROPERTIES (Continued)

				0 K		298.15 K (25°C)		
Formula and Description	State	Formula weight	ΔHf₀° kcal/mol	ΔHf° kcal/mol	ΔGf° kcal/mol	H°₂₉₈ − H°₀	S° cal/deg mol	C°p cal/deg mol
Mg(NO₃)₂	c	148.3218	—	−188.97	−140.9	—	39.2	33.92
std. state, m = 1	aq	—	—	−210.70	−161.9	—	37.0	—
Mg(NO₃)₂								
in ∞ H₂O	aq	—	—	−210.70	—	—	—	—
Mg(NO₃)₂·2H₂O	c	184.3525	—	−336.8	—	—	—	—
Mg(NO₃)₂·6H₂O	c	256.4138	—	−624.59	−497.3	—	108.	—
Mg₂P₂O₇	c	222.5674	—	—	—	6.467	37.02	42.53
Mg₃(PO₄)₂	c	262.8788	−896.98	−903.6	−845.8	7.825	45.22	51.02
MgC₂	c	48.3343	—	+20.	—	—	—	—
Mg₂C₃	c	84.6574	—	+17	—	—	—	—
MgCO₃								
magnesite	c	84.3214	—	−261.9	−241.9	—	15.7	18.05
MgCO₃·3H₂O								
nesquehonite	c	138.3674	—	—	−412.6	—	—	—
MgCO₃·5H₂O								
lansfordite	c	174.3980	—	—	−525.7	—	—	—
MgC₂O₄	c	112.3319	—	−303.3	—	—	—	—
std. state, m = 1	aq	—	—	−308.8	−269.8	—	−22.1	—
Mg₂Si	c	76.710	—	−18.6	−18.0	—	18.	17.6
MgSiO₃								
clinoenstatite	c	100.3962	−368.039	−370.22	−349.46	2.895	16.19	19.45
Mg₂SiO₄								
forsterite	c	140.7076	−516.42	−519.6	−491.2	4.129	22.74	28.32
Mg₃Si₂O₅(OH)₄								
chrysotile	c	277.1345	—	−1043.4	−965.1	—	52.9	65.41
antigorite	c	—	—	—	—	—	53.2	65.47
Mg₃Si₄O₁₀(OH)₂								
talc	c	379.2887	−1405.57	−1415.5	−1324.8	11.20	62.3	76.9
Mg₂Al₄Si₅O₁₈								
cordierite	c	584.9692	—	−2177.	−2055.	—	97.3	108.1
MgCrO₄	c	140.3056	—	−321.1	—	—	—	—
MgCr₂O₄	c	192.3016	—	−426.3	−398.9	—	25.34	30.30
MgMoO₄	c	184.250	—	−334.81	−309.69	—	28.4	26.57
MgWO₄	c	272.160	−363.88	−366.3	−339.6	4.115	24.18	26.14
MgV₂O₆								
metavanadate	c	222.1924	—	−526.19	−487.43	—	38.4	39.47
Mg₂V₂O₇								
pyrovanadate	c	262.5038	—	−677.80	−632.24	—	47.9	48.63
MgTiO₃								
metatitanate	c	120.210	−373.69	−375.9	−354.8	3.240	17.82	21.96
Mg₂TiO₄								
orthotitanate	c	160.522	−514.32	−517.5	−489.2	4.502	26.13	30.75
MgUO₄	c	326.339	—	−443.9	−418.2	—	31.5	30.6
MgU₃O₁₀	c	898.393	—	—	—	—	80.9	73.
MANGANESE								
Mn								
α	c	54.9380	0	0	0	1.194	7.65	6.29
β	c					1.234	8.22	6.34
γ	c	—	.343	.37	.34	1.221	7.75	6.59
	g	—	66.77	67.1	57.0	1.481	41.49	4.97
Mn⁺	g	—	238.20	240.0	—	—	—	—
Mn⁺⁺	g	—	598.87	602.1	—	—	—	—
std. state, m = 1	aq	—	—	−52.76	−54.5	—	−17.6	12.
MnO	c	70.9374	—	−92.07	−86.74	—	14.27	10.86
MnO₂	c	86.9368	—	−124.29	−111.18	—	12.68	12.94
precipitated	amorp	—	—	−120.1	—	—	—	—
Mn₂O₃	c	157.8742	—	−229.2	−210.6	—	26.4	25.73
Mn₃O₄	c	228.8116	—	−331.7	−306.7	—	37.2	33.38
MnH	g	55.9460	—	—	—	2.077	51.03	7.035
Mn(OH)₂								
precipitated	amorp	88.9527	—	−166.2	−147.0	—	23.7	—
MnF	g	73.9364	−5.	−5.2	—	—	—	—
MnF₂	c	92.9348	—	—	—	3.11	22.05	15.96
MnCl₂	c	125.8440	−115.245	−115.03	−105.29	3.602	28.26	17.43
std. state, m = 1	aq	—	—	−132.66	−117.3	—	9.3	−53.
MnCl₂·H₂O	c	143.8593	—	−188.8	−166.4	—	41.6	—
MnCl₂·2H₂O	c	161.8747	—	−261.0	−225.2	—	52.3	—
MnCl₂·4H₂O	c	197.9054	—	−403.3	−340.3	—	72.5	—
MnBr	g	134.8470	20.8	19.1	—	—	—	—
MnBr₂	c	214.7560	—	−92.0	—	—	—	—
std. state, m = 1	aq	—	—	−110.9	—	—	—	—
MnBr₂·H₂O	c	232.7713	—	−168.5	—	—	—	—
MnBr₂·4H₂O	c	286.8174	—	−380.1	—	—	—	—
MnI	g	181.8424	25.7	25.5	—	—	—	—
MnI₂								
std. state, m = 1	aq	308.7468	—	−79.1	—	—	—	—
MnI₂·2H₂O	c	344.7748	—	−201.4	—	—	—	—
MnI₂·4H₂O	c	380.8082	—	−343.9	—	—	—	—
Mn(IO₃)₂	c	404.7432	—	−160.	−124.4	—	63.	—
MnS								
green	c	87.0020	—	−51.2	−52.2	—	18.7	11.94
precipitated, pink	amorp	—	—	−51.1	—	—	—	—
MnSO₄	c	150.9996	—	−254.60	−228.83	—	26.8	24.02
std. state, m = 1	aq	—	—	−270.1	−232.5	—	−12.8	−58.
MnSO₄								
in ∞ H₂O	aq	—	—	−270.1	—	—	—	—
MnSO₄·H₂O								
α	c	169.0149	—	−329.0	—	—	—	—
β	c	—	—	−322.2	—	—	—	—
MnSO₄·4H₂O	c	223.0610	—	−539.7	—	—	—	—
MnSO₄·5H₂O	c	241.0763	—	−610.2	—	—	—	78.
MnSO₄·7H₂O	c	277.1070	—	−750.3	—	—	—	—
Mn(N₃)₂								
manganese azide	c	138.9782	—	92.2	—	—	—	—
Mn₅N₂	c	302.7034	—	−48.8	—	—	—	—
Mn(NO₃)₂	c	178.9478	—	−137.73	—	—	—	—
std. state, m = 1	aq	—	—	−151.9	−107.8	—	52.	−29.
MnP	c	85.9118	—	−27.	—	—	—	—
MnP₃	c	147.8594	—	−51.	—	—	—	—
Mn₃(PO₄)₂	c	354.7568	—	−744.9	—	—	—	—
MnHPO₄	c	150.9174	—	—	−332.5	—	—	—

Formula and Description	State	Formula weight	0 K ΔHf° kcal/mol	298.15 K (25°C) ΔHf° kcal/mol	ΔGf° kcal/mol	H°₂₉₈ − H°₀	S° cal/deg mol	C°p cal/deg mol
Mn₃C	c	176.8252	—	1.1	1.3	—	23.6	22.33
Mn₇C₃	c	420.5995	—	−10.	—	—	—	—
MnCO₃								
natural	c	114.9474	—	−213.7	−195.2	—	20.5	19.48
precipitated	c	—	—	−211.1	—	—	—	—
MnC₂O₄	c	142.9579	—	−245.9	—	—	—	—
undissoc.; std. state, m = 1	aq	—	—	−248.5	−221.0	—	16.1	—
MnC₂O₄·2H₂O	c	178.9886	—	−389.2	−338.2	—	48.	—
MnC₂O₄·3H₂O	c	197.0039	—	−459.1	—	—	—	—
MnSiO₃	c	131.0222	—	−315.7	−296.5	—	21.3	20.66
glassy	amorp	—	—	−307.2	—	—	—	—
Mn₂SiO₄	c	201.9596	—	−413.6	−390.1	—	39.0	31.04
MERCURY								
Hg	c	200.59	0	—	—	—	—	—
	liq	—	—	0	0	2.233	18.17	6.688
	g	—	15.407	14.655	7.613	1.481	41.79	4.968
std. state, m = 1	aq	—	—	9.0	9.4	—	17.	—
Hg⁺	g	—	256.10	256.82	—	—	—	—
Hg²⁺	g	—	688.63	690.83	—	—	—	—
std. state, m = 1	aq	—	—	40.9	39.30	—	−7.7	—
Hg³⁺	g	—	1478.	1480.	—	—	—	—
Hg₂²⁺								
std. state, m = 1	aq	—	—	41.2	36.70	—	20.2	—
HgO								
red, orthorhombic	c	216.589	—	−21.71	−13.995	—	16.80	10.53
yellow	c	—	—	−21.62	−13.964	—	17.0	—
hexagonal	c	—	—	−21.4	−13.92	—	17.6	—
HgH	g	201.598	58.36	57.20	51.63	2.078	52.46	7.16
Hg(OH)₂								
undissoc.; std. state, m = 1	aq	234.605	—	−84.9	−65.7	—	34.	—
Hg₂F₂	c	439.177	—	—	−104.1	—	—	—
HgCl⁺								
std. state, m = 1	aq	—	—	−4.5	−1.3	—	18.	—
HgCl₂	c	271.496	—	−53.6	−42.7	—	34.9	—
undissoc.; std. state, m = 1	aq	—	—	−51.7	−41.4	—	37.	—
Hg₂Cl₂	c	472.086	—	−63.39	−50.377	—	46.0	—
HgBr₂	c	360.408	—	−40.8	−36.6	—	41.	—
undissoc.; std. state, m = 1	aq	—	—	−38.4	−34.2	—	41.	—
Hg₂Br₂	c	560.998	—	−49.45	−43.278	—	52.	—
HgI	g	327.494	32.9	31.64	21.14	2.546	67.26	8.99
HgI₂								
red	c	454.399	—	−25.2	−24.3	—	43.	—
yellow	c	—	—	−24.6	—	—	—	—
Hg₂I₂	c	654.989	—	−29.00	−26.53	—	55.8	—
HgS								
red	c	232.654	—	−13.9	−12.1	—	19.7	11.57
black	c	—	—	−12.8	−11.4	—	19.1	—
HgSO₄	c	437.142	—	−177.61	−149.589	—	47.96	31.54
Hg₂(N₃)₂	c	485.220	—	142.0	178.4	—	49.	—
HgC₂O₄	c	288.610	—	−162.1	—	—	—	—
Hg₂CO₃	c	461.189	—	−132.3	−111.9	—	43.	—
Hg₂C₂O₄	c	489.200	—	—	−141.8	—	—	—
HgCH₃	g	215.625	—	40.	—	—	—	—
Hg(CH₃)₂	liq	230.660	—	14.3	33.5	—	50.	—
Hg(C₂H₅)₂	liq	258.715	—	7.2	—	—	—	—
	g	—	—	18.0	—	—	—	—
Hg(CN)₂	c	252.626	—	63.0	—	—	—	—
undissoc.; std. state, m = 1	aq	—	—	66.5	74.6	—	39.5	—
Hg(CN)₃⁻								
std. state, m = 1	aq	278.644	—	94.9	110.7	—	53.8	—
Hg(CN)₄²⁻								
std. state, m = 1	aq	304.662	—	125.8	147.8	—	73.	—
Hg(ONC)₂								
mercuric fulminate	c	284.624	—	64.	—	—	—	—
MOLYBDENUM								
Mo	c	95.94	0	0	0	1.098	6.85	5.75
	g	—	156.92	157.3	146.4	1.4812	43.461	4.968
Mo⁺	g	—	320.6	322.5	—	—	—	—
Mo²⁺	g	—	693.2	696.5	—	—	—	—
Mo³⁺	g	—	1319.0	1323.8	—	—	—	—
Mo⁴⁺	g	—	2388.9	2395.2	—	—	—	—
Mo⁵⁺	g	—	3799.4	3807.2	—	—	—	—
MoO	g	111.939	101.	—	—	—	—	—
MoO₂	c	127.939	—	−140.76	−127.40	—	11.06	13.38
	g	—	—	3.	—	—	—	—
MoO₃	c	143.939	—	−178.08	−159.66	—	18.58	17.92
	g	—	—	−78.	—	—	—	—
	aq	—	—	−172.5	—	—	—	—
MoO₄	aq	159.938	—	−158.0	—	—	—	—
MoO₄²⁻								
std. state, m = 1	aq	—	—	−238.5	−199.9	—	6.5	—
MoO₅	aq	175.937	—	−139.6	—	—	—	—
H₂MoO₄								
white	c	161.954	—	−250.0	—	—	—	—
H₂MoO₄·H₂O								
yellow	c	179.969	—	−325.	—	—	—	—
MoF₆	liq	209.930	−381.733	−378.95	−352.08	10.205	62.06	40.58
	g	—	−370.608	−372.29	−351.88	5.74	83.75	28.82
MoCl₂	c	166.846	—	−67.4	—	—	—	—
MoCl₃	c	202.299	—	−92.5	—	—	—	—
MoCl₄	c	237.752	—	−114.8	—	—	—	—
MoCl₅	c	273.205	—	−126.0	—	—	—	—
MoO₂Cl₂	c	198.845	—	−171.4	—	—	—	—
MoO₂Cl₂·H₂O	c	216.860	—	−245.4	—	—	—	—
MoOCl₄	c	253.751	—	−153.0	—	—	—	—
MoBr₂	c	255.758	—	−62.4	—	—	—	18.3
MoBr₄	c	415.576	—	−76.8	—	—	—	—
MoO₂Br₂	c	287.757	—	−150.4	—	—	—	—
MoS₂	c	160.068	−55.52	−56.2	−54.0	2.528	14.96	15.19
Mo₂S₃	c	288.072	—	−87.	—	—	—	—

Formula and Description	State	Formula weight	ΔHf₀ kcal/mol	ΔHf° kcal/mol	ΔGf° kcal/mol	H°₂₉₈ − H°₀	S°	C°ₚ
							cal/deg mol	
Mo₂N	c	205.887	—	−19.50	—	—	—	—
MoC	c	107.951	—	−2.4	—	—	—	—
Mo₂C	c	203.891	—	−10.9	—	—	—	—
Mo(CO)₆	c	264.003	—	−234.9	−209.8	—	77.9	57.90
FeMoO₄	c	215.785	—	−257.	−233.	—	30.9	28.31
NEODYMIUM								
Nd	c	144.24	0	0	0	1.73	17.1	6.56
	g	—	78.53	78.3	69.9	1.498	45.243	5.280
Nd⁺	g	—	205.1	206.3	—	—	—	—
Nd²⁺	g	—	452.	455.	—	—	—	—
Nd³⁺								
std. state, m = 1	aq	—	—	−166.4	−160.5	—	−49.4	−5.
NdO	g	160.239	−30.2	—	—	—	—	—
Nd₂O₃								
hexagonal	c	336.478	−430.50	−432.1	−411.3	5.00	37.9	26.60
NdH₂	c	146.256	—	−46.	—	—	—	—
NdF	g	163.238	—	−38.	—	—	—	—
NdF₂	g	182.237	—	−165.	—	—	—	—
NdF₃	c	201.235	—	−396.	—	—	—	—
NdCl₂	c	215.146	—	−163.	—	—	—	—
NdCl₃	c	250.599	—	−248.8	—	—	—	27.
NdCl₃								
∞ H₂O	aq	—	—	−286.3	—	—	—	—
NdCl₃·6H₂O	c	358.691	−678.72	−687.0	−588.1	15.14	99.7	86.25
NdI₃	c	524.953	—	−152.8	—	—	—	—
Nd(IO₃)₃	c	668.948	—	−332.	—	—	—	—
NdS	g	176.304	33.	—	—	—	—	—
Nd₂S₃	c	384.672	—	−284.	−280.2	6.160	44.28	29.28
Nd₂(SO₄)₃·8H₂O	c	720.788	—	−1513.1	—	—	160.9	144.9
Nd₂Se₃	c	525.36	—	—	—	7.13	53.6	31.1
Nd₂(SeO₃)₃·8H₂O								
amorp.	c	813.477	—	−1230.3	—	—	—	—
Nd₂(SeO₃)₃·5H₂O	c	807.430	—	−1100.4	—	—	—	—
Nd(NO₃)₃	c	330.255	—	−294.2	—	—	—	—
Nd(NO₃)₃·3H₂O	c	384.301	—	−515.	—	—	—	—
Nd(NO₃)₃·4H₂O	c	402.316	—	−588.6	—	—	—	—
Nd(NO₃)₃·6H₂O	c	438.347	—	−728.39	—	—	—	—
NdC₂	g	168.262	130.5	130.75	117.9	2.48	6.3	10.6
Nd₂(CO₃)₃	c	468.508	—	—	−744.5	—	—	—
NEON								
Ne	g	20.183	0	0	0	1.481	34.9471	4.9679
std. state, m = 1	aq	—	—	−1.1	4.6	—	15.8	—
Ne⁺	g	—	497.29	498.77	—	—	—	—
Ne²⁺	g	1444.7	1447.6	—	—	—	—	—
Ne³⁺	g	—	2915.	2919.	—	—	—	—
Ne⁴⁺	g	—	5156	5162	—	—	—	—
Ne⁵⁺	g	—	8072.	8079.	—	—	—	—
NICKEL								
Ni	c	58.71	0	0	0	1.144	7.14	6.23
	g	—	102.213	102.7	91.9	1.631	43.519	5.583
Ni⁺	g	—	278.275	280.243	—	—	—	—
Ni²⁺	g	—	696.87	700.32	—	—	—	—
std. state, m = 1	aq	—	—	−12.9	−10.9	—	−30.8	—
Ni³⁺	g	—	1508.0	1512.9	—	—	—	—
NiO	c	74.709	−56.7	−57.3	−50.6	1.6	9.08	10.59
Ni₂O₃	c	165.418	—	−117.0	—	—	—	—
Ni(OH)₂	c	92.725	—	−126.6	−106.9	—	21.	—
std. state, m = 1	aq	—	—	−122.8	−86.1	—	−35.9	—
Ni(OH)₃								
precipitated	c	109.732	—	−160.	—	—	—	—
NiF₂	c	96.707	−155.18	−155.7	−144.4	2.729	17.59	15.31
NiCl₂	c	129.616	−73.077	−72.976	−61.918	3.438	23.34	17.13
NiCl₂								
in ∞H₂O	aq	—	—	−92.8	—	—	—	—
NiCl₂·2H₂O	c	165.647	—	−220.4	−181.7	—	42.	—
NiCl₂·4H₂O	c	201.677	—	−362.5	−295.2	—	58.	—
NiCl₂·6H₂O	c	237.708	—	−502.67	−409.54	—	82.3	—
Ni(ClO₄)₂								
std. state, m = 1	aq	257.611	—	−74.7	−15.0	—	56.2	—
Ni(ClO₄)₂·6H₂O	c	365.703	—	−486.6	—	—	—	—
NiBr₂	c	218.528	—	−50.7	—	—	—	—
std. state, m = 1	aq	—	—	−71.0	−60.6	—	8.6	—
NiBr₂·3H₂O	c	272.574	—	−274.0	—	—	—	—
NiI₂	c	312.519	—	−18.7	—	—	—	—
std. state, m = 1	aq	—	—	−39.3	−35.6	—	22.4	—
Ni(IO₃)₂	c	408.515	—	−116.9	−78.0	—	51.	—
NiS	c	90.774	—	−19.6	−19.0	—	12.66	11.26
precipitated	c	—	—	−18.5	—	—	—	—
Ni₃S₂	c	240.258	—	−48.5	−47.1	—	32.0	28.12
NiSO₄	c	154.772	—	−208.63	−181.6	—	22.	33.
std. state, m = 1	aq	—	—	−230.2	−188.9	—	−26.0	—
NiSO₄								
∞ H₂O	aq	—	—	−230.2	—	—	—	—
NiSO₄·4H₂O	c	226.833	—	−502.9	—	—	—	—
NiSO₄·6H₂O								
α, tetragonal, green	c	262.864	−628.887	−641.21	−531.78	12.391	79.94	78.36
β, monoclinic, blue	c	—	—	−638.7	—	—	—	—
NiSO₄·7H₂O	c	280.879	−697.670	−711.36	−588.49	14.085	90.57	87.14
Ni₃N	c	190.137	—	0.2	—	—	—	—
Ni(NO₃)₂	c	182.720	—	−99.2	—	—	—	—
std. state, m = 1	aq	—	—	−112.0	−64.2	—	39.2	—
Ni(NO₃)₂·3H₂O	c	236.766	—	−317.0	—	—	—	—
Ni(NO₃)₂·6H₂O	c	290.812	—	−528.6	—	—	—	111.
Ni₃C	c	188.141	—	16.1	—	—	—	—
NiCO₃	c	118.719	—	—	−146.4	—	—	—
Ni(CO)₄	liq	170.752	—	−151.3	−140.6	—	74.9	48.9
	g	—	−144.877	−144.10	−140.36	7.074	98.1	34.70
Ni(HCO₂)₂	c	148.746	—	−208.4	—	—	—	—
Ni(CN)₂								
precipitated	c	110.746	—	30.5	—	—	—	—
std. state, m = 1	aq	—	—	59.1	71.5	—	14.2	—

Formula and Description	State	Formula weight	ΔHf°_0 kcal/mol	ΔHf° kcal/mol	ΔGf° kcal/mol	$H^\circ_{298} - H^\circ_0$	S° cal/deg mol	C°_p cal/deg mol
				298.15 K (25°C)				
Ni(CNS)₂	c	174.874	—	22.8	—	—	—	—
NIOBIUM								
Nb	c	92.906	0	0	0	1.255	8.70	5.88
	g	—	172.758	173.5	162.8	1.997	44.490	7.208
Nb⁺	g	—	330.32	332.54	—	—	—	—
Nb²⁺	g	—	660.55	664.25	—	—	—	—
Nb³⁺	g	—	1238.0	1243.1	—	—	—	—
Nb⁴⁺	g	—	2121.3	2127.9	—	—	—	—
NbO	c	108.9054	—	−97.0	−90.5	—	11.5	9.86
	g	—	51.2	51.	44.	2.099	57.09	7.36
NbO₂	c	124.9048	−189.19	−190.3	−177.0	2.222	13.03	13.74
	g	—	—	−51.3	−52.3	—	61.0	—
Nb₂O₅								
(high temp. form)	c	265.809	−451.63	−454.0	−422.1	5.325	32.80	31.57
NbF₅	c	187.898	−432.68	−433.5	−406.1	5.707	38.3	32.2
NbCl₅	c	270.171	—	−190.6	−163.3	—	50.3	35.4
NbOCl₃	c	179.8114	—	−185.1	—	—	—	—
NbOCl₃	g	215.2644	—	−210.2	−187.	—	34.	—
NbBr₅	c	492.451	—	−132.9	—	—	—	—
NbOBr₃	c	348.6174	—	−179.3	—	—	—	—
NbI₅	c	727.428	—	−64.2	—	—	—	—
NbN	c	106.9127	−55.35	−56.2	−49.2	1.439	8.25	9.32
Nb₂N	c	199.8187	—	−59.9	—	—	—	—
NbC	c	104.9172	−33.12	−33.2	−32.7	1.422	8.46	8.81
Nb₂C	c	197.8232	—	−45.4	−44.4	—	15.3	14.48
NITROGEN								
N	g	14.0067	112.534	112.979	108.886	1.481	36.613	4.968
N⁺	g	—	447.663	449.589	—	—	—	—
N²⁺	g	—	1130.55	1133.96	—	—	—	—
N³⁺	g	—	2224.52	2229.41	—	—	—	—
N⁴⁺	g	—	4011.04	4017.41	—	—	—	—
N⁵⁺	g	—	6268.41	6276.26	—	—	—	—
N₂	g	28.0134	0	0	0	2.072	45.77	6.961
N₃	g	—	45.	43.2	—	—	—	—
std. state, m = 1	aq	—	—	65.76	83.2	—	25.8	—
NO	g	30.0061	21.45	21.57	20.69	2.199	50.347	7.133
NO₂	g	46.0055	8.60	7.93	12.26	2.438	57.35	8.89
NO₂⁻								
std. state, m = 1	aq	—	—	−25.0	−7.7	—	29.4	−23.3
NO₃⁻								
nitrate; std. state, m = 1	aq	62.0049	—	−49.56	−26.61	—	35.0	−20.7
N₂O	g	44.0128	20.102	15.01	24.90	2.284	52.52	9.19
N₂O₃	liq	76.0116	—	12.02	—	—	—	—
	g	—	21.628	20.01	33.32	3.566	74.61	15.68
N₂O₄	liq	92.0110	—	−4.66	23.29	—	50.0	34.1
	g	—	4.49	2.19	23.38	7.018	72.70	18.47
N₄O₉	c	109.0101	—	10.3	27.2	—	42.6	34.2
	g	—	5.7	2.7	27.5	4.237	85.0	20.2
NH₃	g	17.0306	−9.34	−11.02	−3.94	2.388	45.97	8.38
undissoc.; std. state, m = 1	aq	—	—	−19.19	−6.35	—	26.6	—
NH₄⁺								
std. state, m = 1	aq	18.0386	—	−31.67	−18.97	—	27.1	19.1
N₂H₄	liq	32.0453	—	12.10	35.67	—	28.97	23.63
	g	—	26.18	22.80	38.07	2.743	56.97	11.85
undissoc.; std. state, m = 1	aq	—	—	8.20	30.6	—	33.	—
HN₃	liq	43.0281	—	63.1	78.2	—	33.6	—
	g	—	71.82	70.3	78.4	2.599	57.09	10.44
undissoc.; std. state, m = 1	aq	—	—	62.16	76.9	—	34.9	—
HNO₂								
cis	g	47.0135	−17.12	−18.64	−10.27	2.608	59.43	10.70
trans	g	—	−17.68	−19.15	−10.82	2.652	59.54	11.01
cis-trans mixture, equil.	g	—	—	−19.0	−11.0	—	60.7	10.9
undissoc.; std. state, m = 1	aq	—	—	−28.5	−12.1	—	32.4	—
HNO₃	liq	63.0129	—	−41.61	−19.31	—	37.19	26.26
	g	—	−29.94	−32.28	−17.87	2.815	63.64	12.75
std. state, m = 1		—	—	−49.56	−26.61	—	35.0	−20.7
HNO₃								
∞ H₂O	aq	—	—	−49.56	—	—	—	—
NH₄OH	liq	35.0460	—	−86.33	−60.74	—	39.57	37.02
undissoc.; std. state, m = 1	aq	—	—	−87.505	−63.04	—	43.3	—
ionized; std. state, m = 1	aq	—	—	−86.64	−56.56	—	24.5	−16.4
NH₄OH								
∞ H₂O	aq	—	—	−86.64	—	—	—	—
NH₄NO₂	c	64.0441	—	−61.3	—	—	—	—
std. state, m = 1	aq	—	—	−56.7	−26.7	—	56.5	−4.2
NH₄NO₃	c	80.0435	—	−87.37	−43.98	—	36.11	33.3
std. state, m = 1	aq	—	—	−81.23	−45.58	—	62.1	−1.6
NH₄NO₃								
∞ H₂O	aq	—	—	−81.23	—	—	—	—
(NH₄)₂O	liq	52.0766	—	−102.94	−63.84	—	63.94	59.08
NF₃	g	71.0019	−28.43	−29.8	−19.9	2.827	62.29	12.7
NH₄F	c	37.0370	−107.41	−110.89	−83.36	2.655	17.20	15.60
std. state, m = 1	aq	—	—	−111.17	−85.61	—	23.8	−6.4
NCl₃	liq	120.3657	—	55.	—	—	—	—
NOCl	g	65.4591	12.81	12.36	15.79	2.716	62.52	10.68
NH₄Cl	c	53.4916	—	−75.15	−48.51	—	22.6	20.1
std. state, m = 1	aq	—	—	−71.62	−50.34	—	40.6	−13.5
NH₄Cl								
∞ H₂O	aq	—	—	−71.62	—	—	—	—
NH₂OH·HCl	c	69.4910	—	−75.9	—	—	—	22.2
NH₄ClO								
std. state, m = 1	aq	69.4910	—	−57.3	−27.8	—	37.	—
NH₄ClO₂								
std. state, m = 1	aq	85.4904	—	−47.6	−14.9	—	51.3	—
std. state, m = 1	aq	101.4898	—	−56.52	−20.89	—	65.9	—
NH₄ClO₄	c	117.4892	—	−70.58	−21.25	—	44.5	—
std. state, m = 1	aq	—	—	−62.58	−21.03	—	70.6	—
NH₄Br	c	94.9477	—	−64.73	−41.9	—	27.	23.
std. state, m = 1	aq	—	—	−60.72	−43.82	—	46.8	−14.8
NH₄Br								
∞ H₂O	aq	—	—	−60.72	—	—	—	—

Formula and Description	State	Formula weight	0 K ΔHf_0° kcal/mol	298.15 K (25°C) ΔHf° kcal/mol	ΔGf° kcal/mol	$H_{298}^\circ - H_0^\circ$	S° cal/deg mol	C_p°
NH₄BrO								
std. state, m = 1	aq	113.9470	—	−54.2	−27.0	—	37.	—
NH₄BrO₃								
std. state, m = 1	aq	145.9458	—	−47.70	−14.54	—	65.75	—
NH₄Br₂Cl								
std. state, m = −72.4	aq	213.3096	—	−72.4	−49.7	—	72.2	—
NH₄I	c	144.9430	—	−48.14	−26.9	—	28.	—
std. state, m = 1	aq	—	—	−44.86	−31.30	—	53.7	−14.9
NH₄I								
∞ H₂O	aq	—	—	−44.86	—	—	—	—
NH₄IO								
std. state, m = 1	aq	160.9424	—	−57.4	−28.2	—	25.8	—
NH₄IO₃	c	192.9412	—	−92.2	—	—	—	—
std. state, m = 1	aq	—	—	−84.6	−49.6	—	55.4	—
NH₄IO₄	aq	208.9406	—	−67.9	—	—	—	—
NH₄HS	c	51.1106	—	−37.5	−12.1	—	23.3	—
std. state, m = 1 (NH₄⁺ + HS⁻)	aq	—	—	−35.9	−16.09	—	42.1	—
(NH₄)₂S								
std. state, m = 1	aq	68.1412	—	−55.4	−17.4	—	50.7	—
(NH₄)₂S₂								
std. state, m = 1	aq	100.2052	—	−56.1	−18.9	—	61.0	—
(NH₄)₂S₃								
std. state, m = 1	aq	132.2692	—	−57.1	−20.3	—	70.0	—
H₂NSO₃H								
sulfamic acid	c	97.0928	—	−161.3	—	—	—	—
	aq	—	—	−156.3	—	—	—	—
NH₄HSO₃	c	99.1088	—	−183.7	—	—	—	—
NH₄HSO₄	c	115.1082	—	−245.45	—	—	—	—
SO₂(NH₂)₂								
sulfamide	c	96.1081	—	−129.3	—	—	—	—
(NH₄)₂SO₃	c	116.1394	—	−211.6	—	—	—	—
std. state, m = 1	aq	—	—	−215.2	−154.2	—	47.2	—
(NH₄)₂SO₄	c	132.1388	−282.23	−215.56	—	52.6	44.81	—
std. state, m = 1, (2NH₄⁺ + SO₄²⁻)	aq	—	—	−280.66	−215.91	—	59.0	−31.8
OSMIUM								
Os	c	190.2	0	0	0	—	7.8	5.9
	g	—	—	189.	178.	1.481	46.000	4.968
Os⁺	g	—	—	391.91	—	—	—	—
Os²⁺	g	—	—	785.	—	—	—	—
OsO₃	g	238.20	—	−67.8	—	—	—	—
OsO₄								
yellow	c	254.20	—	−94.2	−72.9	—	34.4	—
white	c	—	—	−92.2	−72.6	—	40.1	—
Os(OH)₄	amorp	258.23	—	—	−161.0	—	—	—
OsCl₃	c	296.56	—	−45.5	—	—	—	—
OsCl₄	c	332.01	—	−60.9	—	—	—	—
OsS₂	c	254.33	—	−34.9	—	—	—	—
O	g	15.9994	58.983	59.553	55.389	1.607	38.467	5.237
O⁺	g	—	373.019	375.070	—	—	—	—
O²⁺	g	—	1183.73	1187.26	—	—	—	—
O³⁺	g	—	2450.87	2455.88	—	—	—	—
O⁴⁺	g	—	4236.1	4242.6	—	—	—	—
O⁵⁺	g	—	6862.8	6870.8	—	—	—	—
O₂	g	31.9988	0	0	0	2.0746	49.003	7.016
std. state, m = 1	aq	—	—	−2.8	3.9	—	26.5	—
O₃	g	47.9982	34.74	34.1	39.0	2.4736	57.08	9.37
	aq	—	—	30.1	—	—	—	—
PALLADIUM								
Pd	c	106.4	0	0	0	1.299	8.98	6.21
	g	—	90.2	90.4	81.2	1.481	39.90	4.968
Pd⁺	g	—	282.436	284.117	—	—	—	—
Pd²⁺	g	—	730.5	733.6	—	—	—	—
std. state, m = 1	aq	—	—	35.6	42.2	—	−44.	—
Pd³⁺	g	—	1490.	1494.	—	—	—	—
PdO	c	122.40	—	−20.4	—	—	—	7.5
Pd₂H	c	213.81	—	−4.7	—	—	—	—
Pd(OH)₂								
precipitated	c	140.41	—	−94.4	—	—	—	—
Pd(OH)₄								
precipitated	c	174.43	—	−171.1	—	—	—	—
PdCl₂	c	177.31	—	−47.5	—	—	—	—
PdBr₂	c	266.22	—	−24.9	—	—	—	—
PdI₂	c	360.21	—	−15.1	−17.1	—	43.	—
PdS	c	138.46	—	−18.	−16.	—	11.	—
PdS₂	c	170.53	—	−19.4	−17.8	—	19.	—
Pd₄S	c	457.66	—	−16.	−16.	—	43.	—
Pd(CN)₂	c	158.44	—	49.1	—	—	—	—
Pd(CNS)₂	c	222.56	—	—	56.0	—	—	—
PHOSPHORUS								
P								
α, white	c	30.9738	0	0	0	1.281*	9.82	5.698
red, triclinic	c	—	−3.78	−4.2	−2.9	0.862	5.45	5.07
black	c	—	—	−9.4	—	—	—	—
red	amorp	—	—	−1.8	—	—	—	—
	g	—	75.	75.20	66.51	1.481	38.978	4.968
in CS₂	—	—	—	0.5	—	—	—	—
P⁺	g	—	328.20	329.88	—	—	—	—
P²⁺	g	—	781.51	784.67	—	—	—	—
P³⁺	g	—	1477.11	1481.75	—	—	—	—
P⁴⁺	g	—	2661.67	2667.79	—	—	—	—
P⁵⁺	g	—	4161.2	4168.8	—	—	—	—
P₂	g	61.9476	34.94	34.5	24.8	2.126	52.108	7.66
P₄	g	123.8952	15.83	14.08	5.85	3.378	66.89	16.05
PO	g	46.9732	−6.7	−6.8	−12.4	2.245	53.22	7.59
P₄O₆	c	219.8916	—	−392.0	—	—	—	—
P₄O₁₀								
hexagonal	c	283.8892	−705.82	−713.2	−644.8	8.117	54.70	50.60
	amorp	—	—	−727.	—	—	—	—
PH₃	g	33.9977	3.20	1.3	3.2	2.420	50.22	8.87

Formula and Description	State	Formula weight	0 K ΔHf$_0^\circ$ kcal/mol	298.15 K (25°C) ΔHf° kcal/mol	ΔGf° kcal/mol	H$_{298}^\circ$ − H$_0^\circ$	S° cal/deg mol	C$_p^\circ$
std. state, m = 1	aq	—	—	−2.27	6.05	—	28.7	—
HPO₃	c	79.9800	—	−226.7	—	—	—	—
H₃PO₃	c	81.9959	—	−230.5	—	—	—	—
H₃PO₄	c	97.9953	−301.29	−305.7	−267.5	4.059	26.41	25.35
ionized, std. state, m = 1	aq	—	—	−305.3	−243.5	—	−53.	—
PF	g	49.9722	—	—	—	2.117	53.74	7.56
PF₃	g	87.9690	−218.25	−219.6	−214.5	3.092	65.28	14.03
PF₅	g	125.9658	—	−381.4	—	—	—	—
PCl₃	liq	137.3328	—	−76.4	−65.1	—	51.9	—
	g	—	−67.85	−68.6	−64.0	3.817	74.49	17.17
PCl₅	c	208.2388	—	−106.0	—	—	—	—
POCl₃	c	153.3322	−145.81	—	—	—	—	—
PBr₃	liq	270.7008	—	−44.1	−42.0	—	57.4	—
	g	—	−27.47	−33.3	−38.9	4.240	83.17	18.16
PBr₅	c	430.5188	—	−64.5	—	—	—	—
POBr₃	c	286.7002	—	−109.6	—	—	—	—
PI₃	c	411.6870	—	−10.9	—	—	—	—
	g	—	—	—	—	4.542	89.45	18.73
PH₄I	c	161.9101	—	−16.7	0.2	—	29.4	26.2
P₂S₃	c	158.1396	—	−19.2	—	—	—	—
PN	g	44.9805	26.5	26.26	20.97	2.080	50.45	7.10
P₃N₅	c	162.9549	—	−71.4	—	—	—	36.
NH₄H₂PO₄	c	115.0259	—	−345.38	−289.33	—	36.32	34.00
(NH₄)₂HPO₃	aq	116.0571	—	−294.9	—	—	—	—
(NH₄)₂HPO₄	c	132.0565	—	−374.50	—	—	—	45.
(NH₄)₃PO₄	c	149.0871	—	−399.6	—	—	—	—
std. state, m = 1	aq	—	—	−400.3	−300.4	—	28.	—
PLATINUM								
Pt	c	195.09	0	0	0	1.372	9.95	6.18
	g	—	134.90	135.1	124.4	1.572	45.960	6.102
Pt⁺	g	—	341.6	343.3	—	—	—	—
std. state, m = 1	aq	—	—	—	60.9	—	—	—
PtO₂	g	227.089	—	41.0	40.1	62.	—	—
Pt₃O₄	c	649.268	—	−39.	—	—	—	—
Pt(OH)₂	c	229.105	—	−84.1	—	—	—	—
PtCl	c	230.543	—	−13.5	—	—	—	—
PtCl₂	c	265.996	—	−29.5	—	—	—	—
std. state, m = 1	aq	—	—	—	—	—	—	—
PtCl₃	c	301.449	—	−43.5	—	—	—	—
PtCl₄	c	336.902	—	−55.4	—	—	—	—
HPtCl₅·2H₂O	c	409.394	—	−242.0	—	—	—	—
H₂PtCl₆·6H₂O	c	511.910	—	500.1	—	—	—	—
PtBr	c	274.999	—	−9.2	—	—	—	—
PtBr₂	c	354.908	—	−19.6	—	—	—	—
PtBr₃	c	434.817	—	−28.9	—	—	—	—
PtBr₄	c	514.726	—	−37.4	—	—	—	—
PtI₄	c	702.708	—	−17.4	—	—	—	—
PtS	c	227.154	−19.020	−19.5	−18.2	1.946	13.16	10.37
PtS₂	c	259.218	−25.333	−26.0	−23.8	2.813	17.85	15.75
POLONIUM								
Po	c	210.	0	0	0	—	—	—
Po²⁺								
std. state, m = 1	aq	—	—	—	17.	—	—	—
Po⁴⁺								
std. state, m = 1	aq	—	—	—	70.	—	—	—
Po(OH)₄	c	278.0	—	—	−130.	—	—	—
PoS	c	242.1	—	—	−1.	—	—	—
POTASSIUM								
K	c	39.1020	0	0	0	1.695	15.34	7.07
	g	39.1020	21.544	21.33	14.49	1.481	38.295	4.968
K⁺	g	39.1020	—	122.92	—	—	—	—
K²⁺	g	39.1020	—	853.70	—	—	—	—
K³⁺	g	39.1020	—	1909.5	—	—	—	—
K⁴⁺	g	39.1020	—	3315.6	—	—	—	—
K⁺	a	39.1020	—	−60.32	−67.70	—	24.5	5.2
K in 88.81 Hg		39.1020	—	—	−24.41	—	—	—
KO₂	c	71.1008	—	−68.10	−57.23	—	27.9	18.53
KO₃	c	87.1002	—	−62.2	—	—	—	—
K₂O	c	94.2034	—	−86.4	—	—	—	—
	g	94.2034	—	−15.	—	—	—	—
K₂O₂	c	110.2028	—	−118.1	−101.6	—	24.4	—
	g	110.2028	—	−38.	—	—	—	—
KH	c	40.1100	—	−13.80	—	—	—	—
KD	c	41.1160	—	−13.21	—	—	—	—
KOH	c	56.1094	−100.681	−101.521	−90.61	−2.904	18.85	15.51
	a	56.1094	—	−115.29	−105.29	—	21.9	−30.3
in ∞ H₂O	a	56.1094	—	−115.29	—	—	—	—
KF	c	58.1004	−135.223	−135.58	−128.53	2.392	15.91	11.72
KF in ∞ H₂O	a	58.1004	—	−139.82	—	—	—	—
KF·2H₂O	c	94.1312	—	−278.112	−244.17	—	37.1	—
KHF₂	cα	78.1068	−220.560	−221.72	−205.48	3.655	24.92	18.39
KCl	c	74.5550	−104.310	−104.385	−97.79	2.717	19.74	12.26
	a	74.5550	—	−100.27	−99.07	—	38.0	−27.4
KClO	a	90.5544	—	−85.9	−76.5	—	35.	—
KClO₂	a	106.5538	—	−76.2	−63.6	—	48.7	—
KClO₃	c	122.5532	—	−95.06	−70.82	—	34.2	23.96
	a	122.5532	—	−85.17	−69.62	—	63.3	—
KClO₄	c	138.5526	−101.525	−103.43	−72.46	5.036	36.1	26.86
	a	138.5526	—	−91.23	−69.76	—	68.0	—
KBr	c	119.0110	−92.414	−94.120	−90.98	2.919	22.92	12.50
	g	119.0110	−40.832	−43.04	−50.89	2.416	59.85	8.824
	a	119.0110	—	−89.37	−92.55	—	44.2	−28.7
KBr₃	a	278.8290	—	−91.49	−93.29	—	76.0	—
KBrO	a	135.0104	—	−82.8	−75.7	—	34.	—
KBrO₃	c	167.0092	−83.957	−86.10	−64.82	5.593	35.65	28.72
	a	167.0092	—	−76.35	−63.27	—	63.15	—
KBrO₄	c	183.0086	−65.619	−68.80	−41.70	5.593	40.65	28.72
	a	183.0086	—	−57.2	−39.5	—	72.2	—
KI	c	166.0065	−78.137	−78.370	−77.651	3.039	25.41	12.65

Formula and Description	State	Formula weight	0 K ΔHf° kcal/mol	298.15 K (25°C) ΔHf° kcal/mol	ΔGf° kcal/mol	$H°_{298} - H°_0$	S° cal/deg mol	C°p cal/deg mol
KI₃	c	419.8152	—	−78.4	—	—	—	—
	a	419.8152	—	−72.6	−80.0	—	81.7	—
KIO	a	182.0059	—	−86.0	−76.9	—	23.2	—
KIO₃	c	214.0047	−118.54	−119.83	−100.00	5.09	36.20	25.45
	a	214.0047	—	−113.2	−98.3	—	52.8	—
KIO₄	c	230.0041	—	−111.67	−86.38	—	42.	—
	a	230.0041	—	−96.5	−81.7	—	77.	—
K₂S	c	110.2680	—	−91.0	−87.0	—	25.0	—
	a	110.2680	—	−112.7	−114.9	—	45.5	—
K₂SO₃	c	158.2662	—	−269.0	—	—	—	—
K₂SO₃	a	158.2662	—	−272.5	−251.7	—	42.	—
K₂SO₄	c	174.2656	−341.126	−343.64	−315.83	6.079	41.96	31.42
	a	174.2656	—	−337.96	−313.37	—	53.8	−60.
K₂S₂O₃	c	190.3302	—	−280.5	—	—	—	—
	a	190.3302	—	−276.5	−260.3	—	65.	—
K₂S₂O₇	c	254.3278	—	−474.8	−428.2	—	61.	—
	a	254.3278	—	−455.5	—	—	—	—
KHSO₄	c	136.1716	—	−277.4	−246.5	—	33.0	—
	a	136.1716	—	272.40	−248.39	—	56.0	−15.
KNO₂ rhombic	c	85.1075	−88.45	−88.39	−73.28	4.871	36.35	25.67
	a	85.1075	—	−85.3	−75.4	—	53.9	—
KNO₃	c	101.1069	−116.86	−118.22	−94.39	4.488	31.80	23.04
KNO₃	a	101.1069	—	−109.88	−94.31	—	59.5	−15.5
KPO₃	c	118.0740	—	—	—	3.886	25.93	21.56
	a	118.0740	—	−293.8	—	—	—	—
K₃PO₄	c	212.2774	—	−466.1	—	—	—	—
	a	212.2774	—	−486.3	−446.6	—	21.0	—
K₃AsO₄	a	256.2252	—	−393.23	−358.10	—	34.6	—
KH₂AsO₄	c	180.0372	−278.60	−282.2	−247.6	5.490	37.05	30.29
K₂CO₃	c	138.2134	−273.76	−275.1	−254.2	5.417	37.17	27.35
K₂C₂O₄ oxalate	c	166.2240	—	−321.9	—	—	—	—
HCOOK formate	c	84.1200	—	−162.46	—	—	—	—
KHCO₃	c	100.1194	—	−230.2	−206.4	—	27.6	—
CH₃COOK acetate	c	98.1472	—	−172.8	—	—	—	—
	a	98.1472	—	−176.48	−155.99	—	45.2	3.7
KCN	c	65.1199	−28.174	−27.0	−24.35	4.157	30.71	15.84
	g	65.1199	21.494	21.7	15.34	3.188	62.57	12.51
KCNO cyanate	c	81.1193	—	−100.06	—	—	—	—
	a	81.1193	—	−95.2	−91.0	—	50.0	—
KCNS	c	97.1839	−47.980	−47.84	−42.62	4.176	29.70	21.16
	a	97.1839	—	−42.05	−45.55	—	59.0	−4.4
K₂SiO₃	c	154.2882	—	—	—	5.230	34.9	28.3
K₂SiF₆	c	220.2804	—	−706.5	−668.9	—	54.0	—
KAl(SO₄)₂	c	258.2067	—	−590.4	−535.4	—	48.9	46.12
KAl(SO₄)₂·3H₂O	c	312.2529	—	−808.1	−711.0	—	75.0	—
KAl(SO₄)₂·12H₂O	c	474.3915	—	−1448.8	−1228.9	—	164.3	155.6
KAlSi₃O₁₀(OH)₂ muscovite	c	398.3133	—	−1430.3	−1340.5	—	73.2	—
K₃Fe(CN)₆	c	329.2604	—	−59.7	−31.0	—	101.83	—
	a	329.2604	—	−46.7	−28.8	—	138.1	—
K₄Fe(CN)₆	c	368.3624	−141.30	−142.0	−108.3	14.871	100.1	79.40
K₄Fe(CN)₆	a	368.3624	—	−132.4	−104.71	—	120.7	—
KMnO₄	c	158.0376	—	−200.1	−176.3	—	41.04	28.10
	a	158.0376	—	—	—	—	—	−14.4
K₂CrO₄	c	194.1976	−333.796	−335.5	−309.7	6.805	47.83	34.89
	a	194.1976	—	−331.24	−309.36	—	61.0	—
C₂Cr₂O₇	c	294.1918	—	−492.7	−449.8	—	69	52.4
	a	294.1918	—	−476.8	−446.4	—	111.6	—
NaK	l	62.0918	—	1.5	—	—	—	—
Na₂K	l	85.0816	—	2.0	—	—	—	—
NaK₂	l	101.1938	—	2.5	—	—	—	—
PRASEODYMIUM								
Pr	c	140.907	0	0	0	1.74	17.5	6.50
	g	—	85.25	85.0	76.7	1.487	45.339	5.105
Pr⁺	g	—	210.3	211.5	—	—	—	—
Pr²⁺	g	—	453.6	456.3	—	—	—	—
Pr³⁺	g	—	952.3	956.5	—	—	—	—
std. state, m = 1	aq	—	—	−168.4	−162.3	—	−50.	−7.
Pr⁴⁺	g	—	1851.	1856.	—	—	—	—
Pr⁵⁺	g	—	3176.	3183.	—	—	—	—
PrO	g	156.906	−38.	—	—	—	—	—
PrO₂	c	172.906	—	−226.9	—	—	—	28.06
Pr₂O₃ hexagonal	c	329.812	—	−432.5	—	—	—	28.06
cubic	c	—	—	−432.5	—	—	—	—
PrH₂	c	142.923	−45.47	−47.4	−36.9	1.83	13.6	9.8
PrOH²⁺ std. state, m = 1	aq	157.914	—	—	−206.	—	—	—
Pr(OH)₂⁺ std. state, m = 1	aq	174.922	—	—	−257.	—	—	—
Pr(OH)₃	c	191.929	—	—	−307.1	—	—	—
PrCl₃	c	247.266	—	−252.6	—	—	—	24.
	g	—	—	−187.	—	—	—	—
std. state, m = 1	aq	—	—	−288.3	−256.4	—	−10.	−105.
PrOCl	c	192.359	—	−242.	—	—	—	—
PrI₃	c	521.620	—	−156.4	—	—	—	—
Pr(IO₃)₃	c	665.615	—	−333.8	—	—	—	—
PrSO₄⁺ std. state, m = 1	aq	236.969	—	−382.3	−345.1	—	−17.	—
Pr(SO₄)₂⁻ std. state, m = 1	aq	333.030	—	−598.4	−525.6	—	0.	—
Pr(NO₃)₃	c	326.922	—	−293.8	—	—	—	—
in HNO₃(aq)	aq	—	—	−316.7	—	—	—	—

Formula and Description	State	Formula weight	0 K ΔHf_0° kcal/mol	298.15 K (25°C) ΔHf° kcal/mol	ΔGf° kcal/mol	$H_{298}^\circ - H_0^\circ$	S° cal/deg mol	C_p° cal/deg mol
$Pr(NO_3)_3 \cdot 6H_2O$	c	435.014	—	−731.05	—	—	—	—
PrC	c	152.918	—	−13.0	—	—	—	—
PrC_2	g	164.929	131.	131.3	118.7	2.48	62.6	10.6
$Pr_2(CO_3)_3$	c	461.842	—	−768.	—	—	—	—
PROMETHIUM								
147_{Pm}	g	146.915	—	—	—	1.545	44.692	5.797
PROTACTINIUM								
Pa	c	231.0359	0	0	0	—	12.4	—
	g	231.0359	—	145.	134.6	1.518	47.31	5.48
Pa^{4+}	a	231.0359	—	−148.0	—	—	—	—
in HCl + 3.43 H₂O:A		231.0359	—	−159.6	—	—	—	—
Pa^{5+}								
in HCl + 3.43 H₂O:A		231.0359	—	−161.8	—	—	—	—
Pa^{4+}								
in HCl + 8.16 H₂O:A		231.0359	—	−144.8	—	—	—	—
in HCl + 54.4 H₂O:A		231.0359	—	−147,7	—	—	—	—
PaO_2	c	263.0347	—	—	—	—	17.8	—
$PaCl_4$	c	372.8479	—	−249.3	−227.7	—	46.0	—
in HCl + 3.43 H₂O:Au		372.8479	—	−290.1	—	—	—	—
in HCl + 8.16 H₂O:Au		372.8479	—	−291.6	—	—	—	—
in HCl + 54.4 H₂O:Au		372.8479	—	−304.9	—	—	—	—
$PaCl_5$	c	408.3009	—	−273.6	−247.2	—	57.	—
	g	408.3009	—	−251.	−236.	—	94.	—
in HCl + 3.43 H₂O:Au		408.3009	—	−322.2	—	—	—	—
$PaBr_4$	c	550.6719	—	−197.0	−188.3	—	56.0	—
in HCl + 8.16 H₂O:Au		550.6719	—	−248.70	—	—	—	—
in HCl + 54.4 H₂O:Au		550.6719	—	−262.44	—	—	—	—
$PaBr_5$	c	630.5809	—	−206.	−196.	—	69.	—
	g	630.5809	—	−180.	−182.	—	111.	—
in HCl + 3.43 H₂O:Au		630.5809	—	−264.8	—	—	—	—
$PaOBr_2$	c	406.8533	—	−239.	—	—	—	—
PaI_4	c	738.6535	—	−123.2	—	—	—	—
in HCl + 54.4 H₂O:Au		738.6535	—	−199.1	—	—	—	—
RADIUM								
Ra	c	226.025	0	0	0	—	17.	—
	g	—	—	38.	31.	1.481	42.15	4.97
Ra^+	g	—	—	161.22	—	—	—	—
Ra^{2+}	g	—	—	396.70	—	—	—	—
std. state, m = 1	aq	—	—	−126.1	−134.2	—	13.	—
RaO	c	242.0244	—	−125.	—	—	32.	—
$RaCl_2$	c	296.931	—	—	—	—	32.	—
std. state, m = 1	aq	—	—	−206.0	−196.0	—	101.	—
$RaCl_2 \cdot 2H_2O$	c	332.9617	—	−350.	−311.4	—	51.	—
$Ra(IO_3)_2$	c	575.8302	—	−245.4	−207.6	—	65.	—
$RaSO_4$	c	322.0866	—	−351.6	−326.4	—	33.	—
std. state, m = 1	aq	—	—	−343.4	−312.2	—	18.	—
$Ra(OH)_2$	c	260.0410	—	−237.	−190.3	—	53.	—
std. state, m = 1	aq	—	—	−225.2	−187.4	—	83.	—
RADON								
Rn	g	222.	0	0	0	1.481	42.09	4.968
Rn^+	g	—	247.86	249.34	—	—	—	—
RHENIUM								
Re	c	186.2	0	0	0	1.296	8.81	6.09
	g	—	183.8	184.0	173.2	1.481	45.131	4.968
Re^+	g	—	365.5	367.1	—	—	—	—
std. state, m = 1	aq	—	—	—	−8.	—	—	—
ReO_2	c	218.20	—	—	−88.	—	—	—
ReO_3	c	234.20	—	−144.6	—	—	—	—
$HReO_4$	c	251.21	—	−182.2	−156.9	—	37.8	—
std. state, m = 1	aq	—	—	−188.2	−166.0	—	48.1	−3.2
$HReO_4$								
∞ H₂O	aq	—	—	−188.2	—	—	—	—
$ReCl_3$	c	292.56	−62.73	−63.	−45.	4.319	29.6	22.08
$ReCl_5$	c	363.47	—	−89.	—	—	—	—
H_2ReCl_4	c	330.03	—	−152.	—	—	—	—
$ReBr_3$	c	425.93	—	−40.	—	—	—	—
ReS_2	c	250.33	—	−43.	—	—	—	—
RHODIUM								
Rh	c	102.905	0	0	0	1.174	7.53	5.97
	g	—	132.79	133.1	122.1	1.483	44.383	5.022
Rh^+	g	—	304.90	306.69	—	—	—	—
Rh^{2+}	g	—	721.8	725.0	—	—	—	—
Rh^{3+}	g	—	1438.	1443.	—	—	—	—
RhO	g	118.9044	—	92.	—	—	—	—
RhO^+	g	—	—	308.	—	—	—	—
RhO_2	g	134.9038	—	44.	—	—	—	—
Rh_2O_3	c	253.8082	—	−82.	—	—	—	24.8
$RhCl_2$	g	173.811	—	30.3	—	—	—	—
$RhCl_3$	c	209.264	—	−71.5	—	—	—	—
$RhCl_6^{3-}$	aq	315.623	—	−202.8	—	—	—	—
$RhCl_3 \cdot 3(C_2H_5)_2S$	c	479.8289	—	−166.	—	—	—	—
RUBIDIUM								
Rb	c	85.4678	0	0	0	1.790	18.35	7.424
	g	85.4678	19.639	19.330	12.690	1.481	40.626	4.968
Rb^+	g	85.4678	115.965	117.137	—	—	—	—
Rb^{2+}	g	85.4678	745.10	747.76	—	—	—	—
Rb^{3+}	g	85.4678	1660.0	1664.1	—	—	—	—
Rb^+	a	85.4678	—	−60.03	−67.87	—	29.04	—
Rb								
in 185 Hg		85.4678	—	—	−25.25	—	—	—
RbO_2	c	117.4666	—	−66.6	—	—	—	—
Rb_2O	c	186.9350	—	−81.	—	—	—	—
Rb_2O_2	c	202.9344	—	−112.8	—	—	—	—
RbH	c	86.4758	—	−12.5	—	—	—	—
RbOH	c	102.4752	—	−99.95	—	—	—	—
$RbOH \cdot H_2O$	c	120.4906	—	−178.98	—	—	—	—
$RbOH \cdot 2H_2O$	c	138.5060	—	−251.73	—	—	—	—
RbF	c	104.4662	—	−133.3	—	—	—	—
	a	104.4662	—	−139.53	−134.510	—	25.70	—
$RbHF_2$	c	124.4726	−219.52	−220.5	−204.5	3.932	28.70	18.97

Formula and Description	State	Formula weight	0 K ΔHf° kcal/mol	298.15 K (25°C) ΔHf° kcal/mol	ΔGf° kcal/mol	H°298 − H°0	S° cal/deg mol	C°p cal/deg mol
RbCl	c	120.9208	−104.080	−104.05	−97.47	2.917	22.92	12.52
RbCl₃	a	191.8268	—	—	−96.7	—	—	—
RbClO	a	136.9202	—	−85.6	−76.7	—	39.	—
RbClO₂	a	152.9196	—	−75.9	−63.8	—	53.2	—
RbClO₃	c	168.9190	—	−96.3	−71.8	—	36.3	24.66
	a	168.9190	—	−84.88	−69.77	—	67.80	—
RbClO₄	c	184.9184	—	−104.50	−73.54	—	39.2	—
	a	184.9184	—	−90.94	−69.93	—	72.5	—
RbBr	c	165.3768	−92.714	−94.31	−91.25	3.124	26.28	12.63
	a	165.3768	—	−89.08	−92.72	—	48.74	—
RbBr₃	c	325.1948	—	−100.0	—	—	—	—
	a	325.1948	—	−91.20	−93.46	—	80.5	—
RbBr₅	a	485.0128	—	−94.0	−92.7	—	104.7	—
RbBrO₃	c	213.3750	—	−87.78	−66.47	—	38.5	—
RbBrO₃	a	213.3750	—	−76.06	−63.44	—	67.69	—
RbBrO₄	a	229.3744	—	−56.9	−39.7	—	76.7	—
RbBrCl₂	c	236.2828	—	−116.2	—	—	—	—
RbI	c	212.3722	−79.603	−79.78	−78.60	3.190	28.30	12.71
	a	212.3722	—	−73.22	−80.20	—	55.6	—
RbI₃	c	466.1810	—	−82.8	−81.0	—	53.9	—
	a	466.1810	—	−72.3	−80.2	—	86.2	—
RbIO₃	c	260.3704	—	—	−101.90	—	—	—
	a	260.3704	—	−112.90	−98.50	—	57.30	—
Rb₂S	c	202.9996	—	−86.2	—	—	—	—
	a	202.9996	—	−112.2	−115.2	—	54.6	—
Rb₂S₂	a	235.0636	—	−112.9	−116.7	—	64.9	—
Rb₂SO₃	a	250.9978	—	−272.0	−252.0	—	51.	—
Rb₂SO₄	c	266.9972	−340.795	−343.12	−314.76	6.458	47.19	32.04
Rb₂SO₄	a	266.9972	—	−337.38	−313.71	—	62.9	—
RbNO₃	c	147.4727	—	−118.32	−94.61	—	35.2	24.4
	a	147.4727	—	−109.59	−94.48	—	64.0	—
Rb₃PO₄	a	351.3748	—	−485.4	−447.1	—	34.	—
Rb₄P₂O₇	a	515.8146	—	−782.9	−730.2	—	88.	—
RbAsO₂	a	192.3882	—	−162.57	−151.53	—	38.7	—
Rb₃AsO₄	a	395.3226	—	−392.36	−358.61	—	48.2	—
Rb₂CO₃	c	230.9450	−270.41	−271.5	−251.2	5.851	43.34	28.11
	a	230.9450	—	−281.90	−261.91	—	44.5	—
RbHCO₃	c	146.4852	—	−230.2	−206.4	—	29.0	—
	a	146.4852	—	−225.42	−208.13	—	50.84	—
RbCN	c	111.4857	—	—	—	4.159	33.67	16.20
	a	111.4857	—	−24.0	−26.7	—	51.5	—
RbCNO cyanate	a	127.4851	—	−94.9	−91.2	—	54.5	—
RbCNS thiocyanate	a	143.5497	—	−41.76	−45.72	—	63.5	—
RbBO₂	c	128.2776	−231.0	−232.0	−218.2	3.181	22.54	17.7
RbBH₄	a	100.3108	—	−48.52	−40.56	—	55.4	—
RbMnO₄	a	204.4034	—	−189.4	−174.8	—	74.7	—
Rb₂MnO₄	a	289.8712	—	−276.	−255.4	—	72.	—
Rb₂CrO₄	c	286.9292	—	−338.0	—	—	—	—
	a	286.9292	—	−330.66	−309.70	—	70.08	—
Rb₂Cr₂O₇	a	386.9234	—	−476.3	−446.7	—	120.7	—
RUTHENIUM								
Ru	c	101.07	0	0	0	1.100	6.82	5.75
	g	—	153.210	153.6	142.4	1.490	44.550	5.144
Ru⁺	g	—	323.07	324.94	—	—	—	—
Ru²⁺	g	—	709.6	713.0	—	—	—	—
Ru³⁺	g	—	1366.0	1370.9	—	—	—	—
RuO₄	c	133.069	—	−72.9	—	—	—	—
RuCl₃ black	c	207.429	—	−49.	—	—	—	—
RuCl₄	g	242.882	—	−12.4	—	—	—	—
RuBr₃	c	340.797	—	−33.	—	—	—	—
RuI₃	c	481.783	—	−15.7	—	—	—	—
RuS₂	c	165.198	—	−47.	—	—	—	—
SAMARIUM								
Sm	c	150.35	0	0	0	1.81	16.63	7.06
	g	—	49.26	49.4	41.3	1.953	43.722	7.255
Sm⁺	g	—	179.1	180.7	—	—	—	—
Sm²⁺	g	—	434.	438.	—	—	—	—
std. state, m = 1	aq	—	—	—	−118.9	—	—	—
Sm³⁺								
std. state, m = 1	aq	—	—	−165.3	−159.3	—	−50.6	−5.
SmO	g	166.349	−31.	—	—	—	—	—
Sm₂O₃								
monoclinic	c	348.698	−433.89	−435.7	−414.6	5.02	36.1	27.37
cubic	c	—	—	—	—	5.0	—	26.86
SmF	g	169.348	—	−63.	—	—	—	—
SmF₃	c	207.345	—	−425	—	—	—	—
SmF₃·H₂O	c	216.353	—	−436.2	—	—	—	—
SmCl₃	c	256.709	—	−245.2	—	—	—	—
∞ H₂O	aq	—	—	−285.2	—	—	—	—
SmCl₃·6H₂O	c	364.801	—	−686.0	−587.1	—	99.	86.4
SmI₃	c	531.063	—	−148.2	—	—	—	—
Sm(IO₃)₃	c	675.058	—	−330.	—	—	—	—
Sm₂(SO₃)₃	c	540.887	—	—	−697.4	—	—	—
Sm₂(SO₄)₃	c	588.885	—	−931.9	—	—	—	—
Sm(NO₃)₃	c	336.365	—	−289.7	—	—	—	—
SmC₂	c	174.372	—	−17.	−18.1	—	23.	—
Sm₂(CO₃)₃	c	480.728	—	—	−741.4	—	—	—
SCANDIUM								
Sc	c	44.956	0	0	0	1.247	8.28	6.10
	g	—	89.87	90.3	80.32	1.674	41.75	5.28
Sc⁺	g	—	240.69	—	—	—	—	—
Sc²⁺	g	—	535.87	—	—	—	—	—
Sc³⁺	g	—	1106.86	—	—	—	—	—
std. state, m = 1	aq	—	—	−146.8	−140.2	—	−61.	—
Sc⁴⁺	g	—	2801.1	—	—	—	—	—

Formula and Description	State	Formula weight	0 K ΔHf_0° kcal/mol	ΔHf° kcal/mol	298.15 K (25°C) ΔGf° kcal/mol	$H_{298}^\circ - H_0^\circ$	S° cal/deg mol	C_p°
Sc^{3+}	g	—	4914.9	—	—	—	—	—
ScO	g	60.955	−13.5	−13.68	−19.90	2.100	53.65	7.38
Sc_2O	g	105.911	−6.	−6.9	—	2.7	—	11.2
Sc_2O_3	c	137.910	−453.95	−456.22	−434.85	3.34	18.4	22.52
$Sc(OH)_3$	c	95.978	—	−325.9	−294.8	—	24.	—
ScF	g	63.9544	−33.	−33.2	−39.3	2.138	53.11	7.74
ScF_2	g	82.9528	−153.	−153.5	−156.6	2.84	67.0	11.5
ScF_3	c	101.9512	—	−389.4	−371.8	—	22.	—
$ScCl_2$	g	115.862	—	—	—	3.1	72.5	12.6
$ScCl_3$	c	151.315	—	−221.1	—	—	—	—
in 5500 H_2O	aq	—	—	−268.8	—	—	—	—
$ScCl_3 \cdot 6H_2O$	c	259.411	—	−666.6	—	—	—	—
$Sc(OH)_2Cl$	c	114.424	—	−303.	−276.3	—	26.	—
$ScBr_3$	c	284.683	—	−177.6	—	—	—	—
ScI_3	g	298.765	—	—	—	3.4	81.6	13.3
ScS	g	77.020	41.9	41.8	29.7	2.2	56.3	8.0
$Sc_2(SO_4)_3$	c	378.097	—	—	—	—	—	62.0
ScC_2	g	68.978	143.	143.6	—	—	—	—
$Sc(HCO_2)_3$	c	180.009	—	—	—	—	—	54.
$Sc_2(C_2O_4)_3$	c	353.971	—	—	—	—	—	124.
SELENIUM								
Se								
hexagonal, black	c	78.96	0	0	0	1.319	10.144	6.062
monoclinic, red	c	—	—	1.6	—	—	—	—
	g	—	54.11	54.27	44.71	1.4815	42.21	4.978
glassy	amorp	—	—	1.2	—	—	—	—
Se^{2-}								
std. state, m = 1	aq	—	—	—	30.9	—	—	—
Se_2	g	157.92	35.26	34.9	23.0	2.275	60.2	8.46
Se_6	g	473.76	—	39.2	—	—	—	—
SeO	g	94.959	13.0	12.75	6.41	2.108	55.9	7.47
SeO_2	c	110.959	—	−53.86	—	—	—	—
	aq	—	—	−52.97	—	—	—	—
SeO_3	c	126.958	—	−39.9	—	—	—	—
Se_2O_5	c	237.917	—	−97.6	—	—	—	—
H_2SeO_3	c	128.974	—	−125.35	—	—	—	—
undissoc.; std. state, m = 1	aq	—	—	−121.29	−101.87	—	49.7	—
	aq	—	—	−121.24	—	—	—	—
H_2SeO_4	c	144.974	—	−126.7	—	—	—	—
$H_2SeO_4 \cdot H_2O$	c	162.989	—	−200.9	—	—	—	—
	liq	—	—	−196.1	—	—	—	—
SeF_6	g	192.950	−264.1	−267.	−743	4.740	74.00	26.1
Se_2Cl_2	g	149.866	—	−7.6	—	—	—	—
$SeCl_4$	c	220.772	—	−43.8	—	—	—	—
SILICON								
Si	c	28.086	0	0	0	0.769	4.50	4.78
	amorp	—	—	1.0	—	—	—	—
	g	—	107.86	108.9	98.3	1.805	40.12	5.318
Si^+	g	—	295.83	298.35	—	—	—	—
Si^{2+}	g	—	672.71	676.71	—	—	—	—
Si^{3+}	g	—	1444.50	1449.98	—	—	—	—
Si^{4+}	g	—	2485.50	2492.46	—	—	—	—
Si^{5+}	g	—	6331.3	6339.8	—	—	—	—
Si^{6+}	g	—	11062.6	11072.6	—	—	—	—
Si_2	g	56.172	141.32	142.	128.	2.22	54.92	8.22
Si_3	g	84.258	146.4	147.	—	2.9	—	12.9
SiO	g	44.0854	−24.08	−23.8	−30.2	2.082	50.55	7.15
SiO_2								
α, quartz	c	60.0848	—	−217.72	−204.75	1.657	10.00	10.62
α, cristobalite	c	—	—	−217.37	−204.46	1.671	10.20	10.56
SiO_2								
α, tridymite	c	—	—	−217.27	−204.42	1.693	10.4	10.66
	amorp	—	—	−215.94	−203.33	—	11.2	10.6
	g	—	—	−77.	—	—	—	—
	aq	—	—	−214.4	—	—	—	—
SiH	g	29.0940	86.	86.28	—	2.069	—	—
SiH_4	g	32.1179	10.30	10.5	13.6	2.517	48.88	10.24
Si_2H_6	g	62.2198	23.04	19.2	30.4	3.768	65.14	19.31
Si_3H_8	liq	92.3218	—	22.1	—	—	—	—
	g	—	—	28.9	—	—	—	—
H_2SiO_3	c	78.1001	—	−284.1	−261.1	—	32.	—
undissoc.; std. state, m = 1	aq	—	—	−282.7	−258.0	—	26.	—
H_4SiO_4	c	96.1155	—	−354.0	−318.6	—	46.	—
SiF	g	47.0844	1.	1.7	−5.8	2.260	53.94	7.80
SiF_2	g	66.0828	−147.75	−148.	−150.	2.630	60.38	10.49
SiF_4	g	104.0796	−384.66	−385.98	−375.88	3.663	67.49	17.60
std. state, m = 1	aq	—	—	—	−384.2	—	—	—
SiF_6^{2-}								
std. state, m = 1	aq	142.0764	—	−571.0	−525.7	—	29.2	—
H_2SiF_6	aq	144.0923	—	—	—	—	—	—
SiCl	g	63.539	45.	45.39	—	2.267	—	8.81
$SiCl_2$	g	98.992	−39.61	−39.59	−42.35	2.98	67.0	12.16
$SiCl_4$	liq	169.898	—	−164.2	−148.16	—	57.3	34.73
	g	—	−156.508	−157.03	−147.47	4.633	79.02	21.57
SiBr	g	107.995	51.30	50.	—	2.40	—	9.23
$SiBr_4$	liq	347.722	—	−109.3	−106.1	—	66.4	—
	g	—	—	−99.3	−103.2	5.317	90.29	23.21
SiI_4	c	535.7036	—	−45.3	—	—	—	—
SiH_3I	g	158.0143	—	—	—	2.886	64.73	13.00
SiS	g	60.150	26.6	26.88	14.56	2.135	53.43	7.71
SiS_2	c	92.214	—	−49.5	—	—	—	—
Si_3N_4								
α	c	140.2848	—	−177.7	−153.6	—	24.2	—
SiC								
β, cubic	c	40.0972	−15.36	−15.6	−15.0	0.781	3.97	6.42
α, hexagonal	c	—	—	−15.0	−14.4	—	3.94	6.38
	g	—	175.6	177.	—	2.4	—	—
SiC_2	g	52.1083	145.7	147.	—	2.5	—	—
Si_2C	g	68.1832	131.1	132.	—	2.7	—	—
Si_2C_2	g	80.1943	167.	—	—	—	—	—

Formula and Description	State	Formula weight	ΔHf₀ kcal/mol (0 K)	ΔHf° kcal/mol (298.15 K (25°C))	ΔGf° kcal/mol	H°₂₉₈ − H°₀	S° cal/deg mol	C°p cal/deg mol
Si_2C_3	g	92.2054	176.	—	—	—	—	—
Si_3C	g	96.2692	161.	—	—	—	—	—
SILVER								
Ag	c	107.870	0	0	0	1.373	10.17	6.059
	g	—	67.90	68.01	58.72	1.481	41.321	4.9679
Ag⁺	g	—	242.00	243.59				
std. state, m = 1	aq	—	—	25.234	18.433	—	17.37	5.2
Ag_2O	c	231.7394	−7.034	−7.42	−2.68	3.397	29.0	15.74
Ag_2O_2	c	247.7388	—	−5.8	6.6	—	28.	21.
Ag_2O_3	c	263.7382	—	8.1	29.0	—	24.	—
AgF	c	126.8684	—	−48.9	—	—	—	—
std. state, m = 1	aq	—	—	−5·.27	−48.21	—	14.1	−20.3
in ∞ H_2O	aq	—	—	−54.27	—	—	—	—
AgF_2	c	145.8668	—	−86.	—	—	—	—
AgCl	c	143.323	—	−30.370	−26.244	—	23.0	12.14
std. state, m = 1	aq	—	—	−14.718	−12.939	—	30.9	−27.4
$AgClO_2$	c	175.3218	—	2.10	18.1	—	32.16	20.87
$AgClO_3$	c	191.3212	—	−7.24	15.4	—	34	—
std. state, m = 1	aq	—	—	0.38	16.51	—	56.2	—
$AgClO_4$	c	207.3206	—	−7.44	—	—	—	—
std. state, m = 1	aq	—	−5.68	16.37	—	60.9	—	—
AgBr	c	187.779	—	−23.99	−23.16	—	25.6	12.52
std. state, m = 1	aq	—	—	−3.82	−6.42	—	37.1	−28.7
$AgBrO_3$	c	235.7772	—	−2.5	−17.04	—	36.3	—
std. state, m = 1	aq	—	—	9.20	22.86	—	56.02	—
AgI	c	234.7744	—	−14.78	−15.82	—	27.6	13.58
std. state, m = 1	aq	—	—	12.04	6.10	—	44.0	−28.8
AgI_2^-								
std. state, m = 1	aq	361.6788	—	—	−20.8	—	—	—
AgI_2^{2-}								
std. state, m = 1	aq	488.5832	—	−43.5	−36.8	—	60.5	—
AgI_4^{3-}								
std. state, m = 1	aq	615.4876	—	—	−50.1	—	—	—
$AgIO_3$	c	282.7726	—	−40.9	−22.4	—	35.7	24.60
std. state, m = 1	aq	—	—	−27.7	−12.2	—	45.7	—
Ag_2S								
α, orthorhombic	c	247.804	−8.126	−7.79	−9.72	4.136	34.42	18.29
β	c	—	—	−7.03	−9.43	—	36.0	—
Ag_2SO_3	c	295.8022	—	−117.3	−98.3	—	37.8	—
std. state, m = 1	aq	—	—	−101.4	−79.4	—	27.8	—
Ag_2SO_4	c	311.8016	—	−171.10	−147.82	—	47.9	31.40
std. state, m = 1	aq	—	—	−166.85	−141.10	—	39.6	−60.
Ag_2Se	c	294.70	−9.42	−9.	−10.6	4.48	36.02	19.54
AgN_3	c	149.8901	—	73.8	89.9	—	24.9	—
std. state, m = 1	aq	—	—	90.99	101.6	—	43.2	—
Ag_3N	c	337.6167	—	47.6	—	—	—	—
$AgNO_2$	c	153.8755	—	−10.77	4.56	—	30.64	19.17
std. state, m = 1	aq	—	—	10.7	9.5	—	46.8	−18.1
$AgNO_3$	c	169.8749	—	−29.73	−8.00	—	33.68	22.24
nitrate; std. state, m = 1	aq	—	—	−24.33	−8.18	—	52.4	−15.5
$Ag(NH_3)_2^+$								
std. state, m = 1	aq	141.9312	—	−26.60	−4.12	—	58.6	—
$Ag(NH_3)_2NO_3$	c	203.9361	—	−85.0	—	—	—	—
std. state, m = 1	aq	—	—	−76.16	−30.73	—	93.6	—
$Ag(NH_3)_2Cl$								
std. state, m = 1	aq	177.3842	—	−66.55	−35.49	—	72.1	—
undissoc.; std. state, m = 1	aq	—	—	—	−35.01	—	—	—
$Ag(NH_3)_2Br$								
std. state, m = 1	aq	221.8402	—	−55.65	−28.97	—	78.3	—
Ag_3PO_4	c	418.5814	—	—	−210.	—	—	—
AgCN	c	133.8878	34.354	34.9	37.5	3.206	25.62	15.95
std. state, m = 1	aq	—	—	61.2	59.6	—	39.9	—
$Ag(CN)_2$								
std. state, m = 1	aq	159.9057	—	64.6	73.0	—	46.	—
AgONC								
silver fulminate	c	149.8872	—	43.	—	—	—	—
AgSCN	c	165.9518	—	21.0	24.23	—	31.3	15.
std. state, m = 1	aq	—	—	43.50	40.58	—	51.9	−4.4
SODIUM								
Na	c	22.9898	0	0	0	1.54	12.24	6.75
	g	22.9898	25.709	25.65	18.354	1.481	36.712	4.968
Na⁺	g	22.9898	—	145.55	—	—	—	—
Na²⁺	g	22.9898	—	1237.48	—	—	—	—
Na³⁺	g	22.9898	—	2891.0	—	—	—	—
Na⁴⁺	g	22.9898	—	5173.5	—	—	—	—
Na⁵⁺	g	22.9898	—	8366.	—	—	—	—
Na⁺	a	22.9898	—	−57.39	−62.593	—	14.1	11.1
NaO	g	38.9892	25.4	25.	19.7	2.22	54.6	8.3
NaO_2	c	54.9886	−62.96	−62.2	−52.2	4.37	27.7	17.24
Na_2O	c	61.9790	−97.85	−99.00	−89.74	2.964	17.94	16.52
Na_2O_2	c	77.9784	−120.70	−122.10	−107.00	3.75	22.70	21.33
NaOH	c	39.9972	−100.641	−101.723	−90.709	2.507	15.405	14.23
	g	39.9972	−48.6	−49.5	−50.2	2.72	54.57	11.56
in ∞ H_2O	a	39.9972	—	−112.36	—	—	—	—
NaF	c	41.9882	−136.542	−137.105	−129.902	2.031	12.30	11.20
	a	41.9882	—	−136.89	−129.23	—	10.8	−14.4
NaCl	c	58.4428	−98.168	−98.268	−91.815	2.536	17.24	12.07
	g	58.4428	−41.9	−42.22	−47.00	2.298	54.90	8.55
	a	58.4428	—	−97.34	−93.965	—	27.6	−21.5
in ∞ H_2O	a	58.4428	—	−97.34	—	—	—	—
NaClO	a	74.4422	—	−83.0	−71.4	—	24.	—
in 400 H_2O		74.4422	—	−82.8	—	—	—	—
$NaClO_2$	c	90.4416	—	−73.38	—	—	—	—
	a	90.4416	—	−73.3	−58.5	—	38.3	—
$NaClO_3$	c	106.4410	—	−87.422	−62.697	—	29.5	—
	a	106.4410	—	−82.24	−64.51	—	52.9	—
$NaClO_4$	c	122.4404	—	−91.61	−60.93	—	34.0	—
	a	122.4404	—	−88.30	−64.65	—	57.6	—

Formula and Description	State	Formula weight	0 K ΔHf° kcal/mol	298.15 K (25°C) ΔHf° kcal/mol	ΔGf° kcal/mol	H°$_{298}$ − H°$_0$	S° cal/deg mol	C°$_p$
NaBr	c	102.8988	−84.596	−86.296	−83.409	2.770	20.75	12.28
	g	102.8988	−32.08	−34.2	−42.31	2.346	57.62	8.68
	a	102.8988	—	−86.440	−87.440	—	33.8	−22.8
NaBrO	a	118.8982	—	−79.9	−70.6	—	24.	—
NaBrO$_3$	c	150.8970	—	−79.85	−58.04	—	30.8	—
	a	150.8970	—	−73.42	−58.16	—	52.8	—
NaBrO$_4$	a	166.8964	—	−54.30	−34.40	—	61.80	—
NaI	c	149.8942	−68.593	−68.78	−68.37	2.93	23.55	12.45
	a	149.8942	—	−70.58	−74.92	—	40.7	−22.9
NaI$_3$	c	403.7030	—	−56.2	—	—	—	—
NaI$_3$	a	407.7030	—	−69.7	−74.9	—	71.3	—
NaIO$_3$	c	197.8924	—	−115.150	—	—	—	22.0
	a	197.8924	—	−110.30	−93.20	—	42.4	—
NaIO$_3$·H$_2$O	c	215.9078	—	−186.30	−151.56	—	38.8	—
NaIO$_3$·5H$_2$O	c	287.9694	—	−466.60	—	—	—	—
NaIO$_4$	c	213.8918	—	−102.60	−77.22	—	39.0	—
	a	213.8918	—	−93.60	−76.60	—	67.	—
in 2,000 H$_2$O		213.8918	—	−93.808	—	—	—	—
NaIO$_3$·3H$_2$O	c	267.9380	—	—	−247.8	—	—	—
NaH$_4$IO$_6$ from H$_4$IO$_6^-$	au	249.9226	—	−237.8	—	—	—	—
Na$_2$H$_3$IO$_6$ from H$_3$IO$_6^{2-}$	au	271.9044	—	−294.4	—	—	—	—
NaIO$_2$F$_2$	c	219.8898	—	−202.50	—	—	—	—
NaICl$_2$	c	220.8002	—	−96.0	—	—	—	—
	a	220.8002	—	—	−101.1	—	—	—
NaICl$_4$	c	291.7062	—	−112.0	—	—	—	—
NaIBr$_2$	c	309.7122	—	−83.0	—	—	—	—
	a	309.7122	—	—	−92.0	—	—	—
NaBrI$_2$	a	356.7076	—	−88.0	−88.9	—	61.3	—
Na$_2$S	c	78.0436	—	−87.2	−83.6	—	20.0	—
	a	78.0436	—	−106.9	−104.7	—	24.7	—
Na$_2$S$_2$	c	110.1076	—	−94.9	−90.5	—	25.	—
Na$_2$S$_2$	a	110.1076	—	−107.6	−106.2	—	35.0	—
Na$_2$SO$_3$	c	126.0418	−261.21	−263.1	−242.0	5.36	34.88	28.74
	a	126.0418	—	−266.70	−241.50	—	21.0	—
Na$_2$SO$_4$	c	142.0412	−328.789	−331.52	−303.59	5.551	35.75	30.64
	a	142.0412	—	−332.10	−303.16	—	33.0	−48.
Na$_2$SO$_4$·10H$_2$O	c	322.1952	—	−1034.24	−871.75	—	141.5	—
Na$_2$S$_2$O$_3$	c	158.1058	—	−268.4	−245.7	—	37.	—
	a	158.1058	—	−270.10	−247.0	—	44.0	—
Na$_2$S$_2$O$_3$·5H$_2$O	c	248.1828	—	−623.31	−533.0	—	89.0	—
Na$_2$S$_2$O$_4$	c	174.1052	—	−294.5	—	—	—	—
NaHSO$_4$	c	120.0594	—	−269.0	−237.3	—	27.0	—
from HSO$_4^-$	a	120.0594	—	−269.47	−243.28	—	45.6	−9.
Na$_2$SeO$_4$	c	188.9372	—	−266.8	—	—	—	—
	a	188.9372	—	258.0	−230.7	—	41.1	—
NaN$_3$	c	65.0099	6.316	5.19	22.41	3.522	23.15	18.31
	a	65.0099	—	8.37	20.6	—	39.9	—
NaNO$_2$	c	68.9953	—	−85.72	−68.02	—	24.8	—
	a	68.9953	—	−82.4	−70.3	—	43.5	−12.2
NaNO$_3$	c	84.9947	−110.248	−111.82	−87.73	4.115	27.85	22.20
	a	84.9947	—	−106.95	−89.20	—	49.1	−9.6
NaNH$_2$	c	39.0125	−27.84	−29.6	−15.3	2.842	18.38	15.81
NaNH$_3$	c	40.0205	—	−16.27	−2.94	—	37.2	—
Na$_3$PO$_4$	c	163.9408	−454.79	−458.27	−427.55	6.566	41.54	36.68
	a	163.9408	—	−477.5	−431.3	—	−11.	—
(NaPO$_3$)$_3$	c	305.8854	−865.9	−873.	−808.	10.720	68.47	62.00
Na$_4$P$_2$O$_7$	c	265.9026	−756.2	−762.	−709.7	10.180	64.60	57.63
Na$_5$P$_3$O$_{10}$								
form I, quenched	c	367.8644	−1043.57	−1051.4	−978.5	14.070	91.25	78.16
form II	c2	367.8644	−1045.76	−1054.0	−980.0	13.67	87.37	77.72
in 5,200 H$_2$O		367.8644	—	−1068.1	—	—	—	—
NaH$_2$PO$_2$	c	87.9784	—	−200.5	—	—	—	—
NaH$_2$PO$_3$	c	103.9778	—	−288.0	—	—	—	—
NaH$_2$PO$_4$	c	119.9772	−363.06	−367.3	−331.3	4.75	30.47	27.93
NaH$_2$PO$_4$·H$_2$O	c	137.9926	—	−438.1	—	—	—	—
Na$_2$HPO$_3$	c	125.9596	—	−336.8	—	—	—	—
Na$_2$HPO$_4$	c	141.9590	−413.92	−417.8	−384.4	5.646	35.97	32.34
Na$_2$HPO$_4$·2H$_2$O	c	177.9898	—	−560.7	−499.2	—	52.9	—
Na$_2$HPO$_4$·7H$_2$O	c	268.0668	—	−913.4	−784.0	—	103.87	—
Na$_2$HPO$_4$·12H$_2$O	c	358.1438	—	−1266.2	−1068.0	—	151.49	—
Na$_2$H$_2$P$_2$O$_7$	c	221.9390	−654.06	−660.8	−602.9	8.184	52.63	47.36
NaAsO$_2$	c	129.9102	—	−157.87	—	—	—	—
	a	129.9102	—	−159.93	−146.25	—	23.8	—
Na$_3$AsO$_4$	c	207.8886	—	−368.	—	—	—	—
	a	207.8886	—	−384.44	−342.78	—	3.4	—
Na$_3$BiO$_4$	c	341.9470	—	−291.	—	—	—	—
Na$_2$CO$_3$	c	105.9890	−268.76	−270.24	−249.64	4.959	32.26	26.84
	a	105.9890	—	−276.62	−251.36	—	14.6	—
Na$_2$CO$_3$·H$_2$O	c	124.0044	−338.87	−342.08	−307.22	6.296	40.18	34.80
Na$_2$CO$_3$·7H$_2$O	c	232.0968	—	−764.81	−648.8	—	102.	—
Na$_2$CO$_3$·10H$_2$O	c	286.1430	−959.62	−975.46	−819.36	21.21	134.8	131.53
Na$_2$C$_2$O$_4$	c	133.9996	—	−315.0	—	—	—	34.
HCOONa	c	68.0078	−158.19	−159.3	−143.4	3.767	24.80	19.76
NaHCO$_3$	c	84.0072	−225.15	−227.25	−203.4	3.81	24.3	20.94
NaOCH$_3$	c	54.0244	−85.41	−87.9	−70.46	3.374	26.43	16.60
NaC$_2$H$_3$O$_2$	c	82.0350	—	−169.41	−145.14	—	29.4	19.1
	a	82.0350	—	−173.55	−150.88	—	34.8	9.6
NaC$_2$H$_3$O$_2$·3H$_2$O	c	136.0812	—	−383.2	−317.6	—	58.	—
NaOC$_2$H$_5$	c	68.0516	—	−98.90	—	—	—	—
	a	68.0516	—	−113.	—	—	—	—
CCl$_3$COONa sodium trichloroacetate	c	185.3700	—	−178.9	—	—	—	—
NaCN								
(C,I,cubic)	c	49.0077	−22.56	−20.91	−18.27	4.480	27.63	16.82
(c,II,orthorhombic)	c2	49.0077	—	−21.69	—	—	—	—

Formula and Description	State	Formula weight	0 K ΔHf_0° kcal/mol	298.15 K (25°C) ΔHf° kcal/mol	ΔGf° kcal/mol	$H_{298}^\circ - H_0^\circ$	S° cal/deg mol	C_p° cal/deg mol
$NaCN \cdot 1/2H_2O$	c	58.0154	—	−56.35	—	—	—	—
$NaCN \cdot 2H_2O$	c	85.0385	—	−162.47	—	—	—	—
$NaCNO$	c	65.0071	—	−96.89	−85.6	—	23.1	20.7
	a	65.0071	—	−92.3	−85.9	—	39.6	—
$NaCNS$	c	81.0717	—	−40.75	—	—	—	—
	a	81.0717	—	−39.12	−40.44	—	48.6	1.5
Na_2SiO_3								
sodium metasilicate	c	122.0638	—	−371.63	−349.19	—	27.21	—
	gl	122.0638	—	−368.1	—	—	—	—
$Na_2SiO_3 \cdot 5H_2O$	c	212.1408	—	−728.6	—	—	—	—
$Na_2SiO_3 \cdot 9H_2O$	c	284.2024	—	−1010.7	—	—	—	—
$Na_2Si_2O_5$	c	182.1486	—	−589.8	−555.0	—	39.21	—
sodium disilicate, stable up to 951K [formerly β]								
	c2	182.1486	—	−589.22	—	—	—	—
stable 951K to m.pt.(1147K)[formerly α]								
unstable	c3	182.1486	—	−587.28	—	—	—	—
	gl	122.0638	—	−368.1	—	—	—	—
Na_4SiO_4	c	184.0428	—	—	—	—	46.76	—
$NaBO_2$	c	65.7996	−232.38	−233.5	−220.06	2.780	17.57	15.76
$NaBO_2 \cdot 2H_2O$	c	101.8304	—	−378.	−337.0	—	37.	—
$NaBO_2 \cdot 4H_2O$	c	137.8612	—	−520.	−451.3	—	55.	—
$NaBO_3 \cdot 4H_2O$	c	153.8606	—	−505.3	—	—	—	—
$Na_2B_4O_7$	c	201.2194	−782.36	−786.6	−740.0	7.262	45.30	44.64
	am	201.2194	—	−781.8	−735.4	—	46.1	—
	a	201.2194	—	—	−747.8	—	—	—
$Na_2B_4O_7 \cdot 4H_2O$	c	273.2810	—	−1077.3	—	—	—	—
$Na_2B_4O_7 \cdot 5H_2O$	c	291.2964	—	−1147.8	—	—	—	—
$Na_2B_4O_7 \cdot 10H_2O$								
borax	c	381.3734	—	−1503.0	−1318.5	—	140.	147.
$NaBH_4$	c	37.8328	−43.100	−45.08	−29.62	3.890	24.21	20.74
	a	37.8328	—	−45.88	−35.28	—	40.5	—
$NaBF_4$	c	109.7944	−440.04	−440.9	−418.30	5.191	34.73	28.74
$NaBF_4$	a	109.7944	—	−433.8	−418.0	—	58.	—
$NaAlO_2$	c	81.9701	—	−271.30	−256.06	—	16.90	17.52
	a	81.9701	—	−277.0	−259.4	—	9.	—
$NaAl(SO_4)_2 \cdot 12H_2O$ alum	c	458.2793	—	−1434.69	—	—	—	—
$NaAlSiO_4$								
nepheline, nephelite	c	142.0549	—	−500.2	−472.8	—	29.7	—
$NaAlSi_2O_6$								
jadeite	c	202.1397	—	−724.4	−681.7	—	31.9	—
dehydrated analcite	c2	202.1397	—	−713.5	−673.8	—	41.9	39.30
$NaHg$	c	223.5798	—	−11.3	−9.76	—	25.	—
$NaHg_2$	c	424.1698	—	−18.2	−16.2	—	42.	—
$NaHg_4$	c	825.3498	—	−21.2	−17.7	—	73.	—
Na_3Hg	c	269.5594	—	−11.0	−10.84	—	54.	—
Na_3Hg_2	c	470.1494	—	−21.9	−20.7	—	69.	—
Na_5Hg_2	c	516.1290	—	—	−21.7	—	—	—
Na_7Hg_8	c	1765.6486	—	−84.4	−76.0	—	203.	—
$Na_3Fe(CN)_6$	a	280.9238	—	−37.9	−13.5	—	106.9	—
$Na_4Fe(CN)_6$	a	303.9136	—	−120.7	−84.28	—	79.1	—
$NaMnO_4$	a	141.9254	—	−186.8	−169.5	—	59.8	—
Na_2MnO_4	c	164.9152	—	−276.3	—	—	—	—
	a	164.9152	—	−271.	−244.9	—	42.	—
Na_2CrO_4	c	161.9732	−318.2	−320.8	−295.17	6.323	42.21	33.97
	a	161.9732	—	−325.38	−299.15	—	40.2	—
$Na_2Cr_2O_7$	c	261.9674	—	−472.9	—	—	—	—
	a	216.9674	—	−471.0	−436.2	—	90.8	—
Na_2MoO_4	c	205.9172	−348.63	−350.89	−323.71	6.070	38.17	33.87
$Na_2MoO_4 \cdot 2H_2O$	c	241.9480	—	−492.1	−437.46	—	57.5	—
Na_2WO_4	c	293.8272	−367.83	−370.2	−342.86	6.05	38.6	33.41
$Na_2W_2O_7$	c	525.6754	−571.44	−574.8	−529.84	9.36	60.8	51.36
$NaVO_3$	c	121.9300	−272.31	−273.85	−254.33	4.217	27.17	23.32
Na_3VO_4	c	183.9090	−417.36	−420.14	−391.45	7.093	45.4	39.40
$Na_4V_2O_7$	c	305.8390	−693.73	−697.62	−650.46	11.75	76.1	64.47
$NaNbO_3$	c	163.8940	—	−314.5	−294.7	—	28.	—
Na_2TiO_3	c	141.8778	−377.88	−380.3	−357.6	4.917	29.08	30.03
$Na_2Ti_2O_5$	c	221.7766	—	—	—	6.910	41.56	41.68
$NaUO_3$	c	309.0170	—	−360.	—	—	—	—
Na_2UO_4								
α form	c	348.0062	−450.02	−452.5	−424.90	6.268	39.68	35.05
β form	c2	348.0062	—	−450.2	—	—	—	—
Na_3UO_4	c	370.9960	−481.15	−484.0	−454.4	7.435	47.37	41.35
STRONTIUM								
Sr	c	87.62	0	0	0	—	12.5	6.3
	g	—	—	39.3	31.3	1.481	39.32	4.968
Sr^+	g	—	—	172.11	—	—	—	—
Sr^{2+}	g	—	—	427.96	—	—	—	—
std. state, m = 1	aq	—	—	−130.45	−133.71	—	−7.8	—
Sr^{3+}	g	—	—	1435.	—	—	—	—
Sr^{4+}	g	—	—	2751.	—	—	—	—
SrO	c	103.619	—	−141.5	−134.3	—	13.0	10.76
	g	—	—	−2.	—	—	—	—
SrO_2	c	119.619	—	−151.4	—	—	—	—
$SrO_2 \cdot 8H_2O$	c	263.742	—	−722.6	—	—	—	—
Sr_2O	c	191.239	—	−154.7	—	—	—	—
SrH	g	88.628	—	+52.	45.	2.080	50.80	7.17
SrH_2	c	89.636	—	−43.1	—	—	—	—
$SrOH$	g	104.627	—	−41.	—	—	—	—
$Sr(OH)_2$	c	121.635	—	−229.2	—	—	—	—
	g	—	—	−135.	—	—	—	—
in 800 H_2O	aq	—	—	−240.1	—	—	—	—
SrF	g	106.618	—	−69.0	−75.1	2.219	57.31	8.27
SrF_2	c	125.617	—	−290.7	−278.4	3.125	19.63	16.73
	g	—	—	−182.7	−185.3	3.192	69.54	12.66
$SrCl$	g	123.073	—	−30.2	−36.5	2.338	60.2	8.73
$SrCl_2$								
α	c	158.526	—	−198.1	−186.7	3.880	27.45	18.07
	g	—	—	−116.1	−118.6	3.454	74.26	13.33
$SrCl_2 \cdot H_2O$	c	176.541	—	−271.7	−247.7	—	41.	28.7
$SrCl_2 \cdot 2H_2O$	c	194.557	—	−343.7	−306.4	—	52.	38.3

Formula and Description	State	Formula weight	0 K ΔHf° kcal/mol	298.15 K (25°C) ΔHf° kcal/mol	ΔGf° kcal/mol	H°₂₉₈ − H°₀	S° cal/deg mol	C°ₚ
SrCl₂·6H₂O	c	266.618	—	−627.1	−535.67	—	93.4	—
Sr(ClO₄)₂	c	286.521	—	−182.31	—	—	—	—
std. state, m = 1	aq	—	—	−192.27	−137.83	—	79.2	—
∞ H₂O	aq	—	—	−192.27	—	—	—	—
SrBr₂	c	247.438	—	−171.5	−166.6	4.261	32.29	18.01
	g	—	—	−98.	−106.	3.8	77.3	14.5
std. state, m = 1	aq	—	—	−188.55	−183.41	—	31.6	—
∞ H₂O	aq	—	—	−188.55	—	—	—	—
SrBr₂	c	247.438	—	−171.5	−166.6	4.261	32.29	18.01
	g	—	—	−98.	−106	3.8	77.3	14.5
std. state, m = 1	aq	—	—	−188.55	−183.41	—	31.6	—
Sr(IO₃)₂	c	437.425	—	−243.6	−204.4	—	56.	—
Sr(IO₃)₂·H₂O	c	455.441	—	−313.2	−260.4	—	66.	—
Sr(IO₃)₂·6H₂O	c	545.517	—	−666.8	−543.7	—	109.	—
SrS	c	119.684	—	−112.9	−111.8	—	16.3	11.64
SrSO₃	c	167.682	—	−281.3	—	—	—	—
SrSO₄	c	183.682	—	−347.3	−320.5	—	28.	—
precipitate	c	—	—	−346.5	—	—	—	—
std. state, m = 1	aq	—	—	−347.77	−311.68	—	−3.0	—
SrSeO₃	c	214.578	—	−250.4	—	—	—	—
SrSeO₄	c	230.578	—	−273.1	—	—	—	—
Sr(N₃)₂	c	171.660	—	+2.1	—	—	—	—
Sr₃N₂	c	290.873	—	−93.5	—	—	—	—
Sr(NO₂)₂	c	179.631	—	−182.3	—	—	—	—
In 800 H₂O	aq	—	—	−180.5	—	—	—	—
Sr(NO₃)₂	c	211.630	—	−233.80	−186.46	6.854	46.50	35.83
std. state, m = 1	aq	—	—	−229.57	−186.93	—	62.2	—
Sr(NO₃)₂ in ∞ H₂O	aq	—	—	−229.57	—	—	—	—
Sr₃P₂	c	324.808	—	−152.	—	—	—	—
Sr₃(PO₄)₂	c	452.803	—	−985.4	—	—	—	—
SrHPO₄	c	183.599	—	−435.4	−403.6	—	29.	—
Sr(H₂PO₄)₂	c	281.595	—	−749.2	—	—	—	—
Sr₃(AsO₄)₂	c	540.698	—	−792.8	−736.2	—	61.	—
hydrated precipitate	—	—	—	−803.	—	—	—	—
SrHAsO₄	aq	227.547	—	−345.7	—	—	—	—
Sr(H₂AsO₄)₂	aq	369.490	—	−563.2	—	—	—	—
SrC₂	c	111.642	—	−18.	—	—	—	—
SrCO₃								
strontianite	c	147.629	—	−291.6	−272.5	—	23.2	19.46
std. state, m = 1	aq	—	—	−292.79	−259.88	—	31.4	—
SrC₂O₄	c	175.640	—	−327.6	—	—	—	—
std. state, m = 1	aq	—	—	−327.6	−294.8	—	+2.1	—
SrCN₂	c	127.645	—	−72.5	—	—	—	—
Sr(CN)₂	aq	139.656	—	−57.0	—	—	—	—
Sr(CN)₂·4H₂O	c	211.717	—	−333.1	—	—	—	—
SrSiO₃	c	163.704	—	−390.5	−370.4	—	23.1	21.16
Sr₂SiO₄	c	267.324	—	−550.8	−523.7	—	36.6	32.09
SrMoO₃	c	231.558	—	−306.	—	—	—	—
SrMoO₄	c	247.558	—	−370.	—	—	—	—
SrWO₄	c	335.468	—	−391.9	−366.	—	33.	—
SrTiO₃	c	183.518	—	−399.71	−379.64	—	26.0	23.51
Sr₂TiO₄	c	287.138	—	−546.7	−520.7	—	38.0	34.34
SrZrO₃	c	226.838	—	−422.4	−402.2	—	27.5	24.71
SULFUR								
S								
rhombic	c	32.064	0	0	0	1.054	7.60	5.41
monoclinic	c	—	—	0.08	—	—	—	—
	g	—	66.1	66.636	56.951	1.591	40.084	5.658
S⁺	g	—	304.95	306.97	—	—	—	—
S²⁺	g	—	844.8	848.2	—	—	—	—
S³⁺	g	—	1653.0	1657.9	—	—	—	—
S⁴⁺	g	—	2743.6	2750.0	—	—	—	—
S⁵⁺	g	—	4415.	4422.	—	—	—	—
S⁶⁺	g	—	6445.	6654.	—	—	—	—
S₂	g	64.128	30.647	30.68	18.96	2.141	54.51	7.76
S₂²⁻								
std. state, m = 1	aq	—	—	7.2	19.0	—	6.8	—
S₃	g	96.192	—	31.7	—	—	—	—
S₄	g	128.256	—	32.7	—	—	—	—
S₅	g	160.320	—	29.6	—	—	—	—
S₆	g	192.384	—	24.5	—	—	—	—
SO	g	48.0634	1.5	1.496	−4.741	2.087	53.02	7.21
SO₂	liq	64.0628	—	−76.6	—	—	—	—
	g	—	−70.336	−70.944	−71.748	2.521	59.30	9.53
undissoc.; std. state, m = 1	aq	—	—	−77.194	−71.871	—	38.7	—
SO₂ in 10,000 H₂O	aq	—	—	−80.584	—	—	—	—
SO₃								
I, β	c	80.0622	—	−108.63	−88.19	—	12.5	—
	liq	—	—	−105.41	−88.04	—	22.85	—
	g	—	−93.21	−94.58	−88.69	2.796	61.34	12.11
SO₃²⁻								
std. state, m = 1	aq	—	—	−151.9	−116.3	—	−7.	—
SO₄²⁻								
std. state, m = 1	aq	96.0616	—	−217.32	−177.97	—	4.8	−70.
S₂O₃²⁻	aq	112.1262	—	−155.9	—	—	—	—
S₂O₄²⁻								
std. state, m = 1	aq	128.1256	—	−180.1	−143.5	—	22.	—
S₂O₇²⁻	aq	176.1238	—	−334.9	—	—	—	—
S₂O₈²⁻								
std. state, m = 1	aq	192.1232	—	−320.0	−265.4	—	59.3	—
H₂S	g	34.0799	−4.232	−4.93	−8.02	2.379	49.16	8.18
std. state, m = 1	aq	—	—	−9.5	−6.66	—	29.	—
HSO₃								
std. state, m = 1	aq	81.0702	—	−149.67	−126.15	—	33.4	—
HSO₄⁻								
std. state, m = 1	aq	97.0696	—	−212.08	−180.69	—	31.5	−20.
H₂SO₃								
undissoc.; std. state, m = 1	aq	82.0781	—	−145.51	−128.56	—	55.5	—
H₂SO₄	c	98.0775	−194.069	—	—	—	—	—

Formula and Description	State	Formula weight	ΔHf₀° kcal/mol (0 K)	ΔHf° kcal/mol	ΔGf° kcal/mol	H°₂₉₈ − H°₀	S° cal/deg mol	Cₚ°
std. state, m = 1	liq	—	—	−194.548	−164.938	6.748	37.501	33.20
	aq	—	—	−217.32	−177.97	—	4.8	−70.
H₂SO₄								
in ∞ H₂O	aq	—	—	−217.32	—	—	—	—
H₂SO₄·1H₂O	liq	116.0929	—	−269.508	−227.182	—	50.56	51.35
H₂SO₄·2H₂O	liq	134.1082	—	−341.085	−286.770	—	66.06	62.34
H₂SO₄·3H₂O	liq	152.1236	—	−411.186	−345.178	—	82.55	76.23
H₂SO₄·4H₂O	liq	170.1389	—	−480.688	−403.001	—	99.09	91.35
H₂SO₄·6.5 H₂O	liq	215.1772	—	−653.264	−546.403	—	140.51	136.30
H₂S₂O₄								
std. state, m = 1	aq	130.1415	—	—	−147.4	—	—	—
undissoc.								
H₂S₂O₆	aq	162.1403	—	−286.4	—	—	—	—
H₂S₂O₇	c	178.1397	—	−304.4	—	—	—	27.
H₂S₂O₈								
std. state, m = 1	aq	194.1391	—	−320.0	−265.4	—	59.3	—
SF₄	g	108.0576	−183.4	−185.2	−174.8	3.482	69.77	17.45
SF₆	g	146.0544	−285.7	−289.	−264.2	4.056	69.72	23.25
std. state, m = 1	aq	—	—	−293.0	−259.3	—	39.8	—
SOCl₂	liq	118.9694	—	−58.7	—	—	—	29.
	g	—	−50.07	−50.8	−47.4	3.559	74.01	15.9
in C₆H₆	—	—	—	−59.7	—	—	—	—
SO₂Cl₂	liq	134.9688	—	−94.2	—	—	—	32.
	g	—	−85.50	−87.0	−76.5	3.825	74.53	18.4
in C₆H₆	—	—	—	−94.2	—	—	—	—
TANTALUM								
Ta	c	180.948	0	0	0	1.347	9.92	6.06
	g	—	186.765	186.9	176.7	1.482	44.241	4.985
Ta⁺	g	—	368.72	370.33	—	—	—	—
TaO	g	196.9474	60.3	60.	53.	2.10	57.6	7.31
TaO₂	g	212.9468	−40.7	−41.	−43.	3.1	64.	12.5
Ta₂O₅								
β	c	441.8930	—	−489.0	−456.8	—	34.2	32.30
Ta₂H	c	362.9040	−6.93	−7.8	−16.5	2.84	18.9	21.7
TaF₅	c	275.9400	—	−454.97	—	—	—	—
TaCl₃	c	287.307	—	−132.2	—	—	—	—
TaCl₄	c	322.760	—	−167.7	—	—	—	—
TaCl₅	c	358.213	—	−205.3	—	—	—	—
TaBr₅	c	580.493	—	−143.0	—	—	—	—
TaS₂	c	245.076	—	−111.	—	—	—	—
TaN	c	194.9547	—	−60.1	—	—	—	9.7
Ta₂N	c	375.9027	—	−65.	—	—	—	—
TaC	c	192.9592	−34.96	−35.0	−34.6	1.56	10.11	8.79
Ta₂C	c	373.9072	—	−51.0	−50.8	—	20.7	—
TaSi₂	c	237.120	—	−28.	—	—	—	—
Ta₅Si₃	c	988.998	—	−76.	—	—	—	—
TaB₂	c	202.570	—	−46.	—	—	—	—
TECHNETIUM								
Tc	c	98.906	0	0	0	—	—	—
	g	—	—	162.	—	1.481	43.25	4.97
Tc⁺	g	—	—	331.	—	—	—	—
Tc₂O₇	c	309.8078	—	−266.	—	—	—	—
HTcO₄	c	163.9115	—	−167.	—	—	—	—
std. state, m = 1	aq	—	—	−173.	—	—	—	—
TELLURIUM								
Te	c	127.60	0	0	0	1.463	11.88	6.15
	g	—	47.	47.02	37.55	1.481	43.65	4.968
	amorp	—	—	2.7	—	—	—	—
Te⁺	g	—	254.76	256.26	—	—	—	—
Te²⁺	g	—	684.	687.	—	—	—	—
Te³⁺	g	—	1390.	1394.	—	—	—	—
Te⁴⁺	g	—	2262.	2268.	—	—	—	—
Te⁵⁺	g	—	3652.	3659.	—	—	—	—
TeO	g	143.599	16.	15.6	9.2	2.093	57.7	7.19
TeO₂	c	159.599	—	−77.1	−64.6	—	19.0	—
H₂Te	g	129.616	—	23.8	—	—	—	—
H₂TeO₃								
std. state, m = 1	aq	177.614	—	—	−76.2	—	—	—
H₆TeO₆	c	229.644	—	−310.4	—	—	—	—
TeF₆	g	241.590	—	−315.	—	—	—	—
TeCl₄	c	269.412	—	−78.0	—	—	—	33.1
TeBr₄	c	447.236	—	−45.5	—	—	—	—
TERBIUM								
Tb	c	158.924	0	0	0	2.250	17.50	6.91
	g	—	93.36	92.9	83.6	1.79	48.63	5.87
Tb⁺	g	—	228.3	229.3	—	—	—	—
Tb²⁺	g	—	494.	496.	—	—	—	—
Tb³⁺								
std. state, m = 1	aq	—	—	−163.2	−155.8	—	−54.	4.
TbO	g	174.923	−19.	—	—	—	—	—
TbO₂	c	190.923	—	−232.2	—	—	—	—
Tb₂O₃	c	365.846	—	−445.8	—	—	—	27.7
TbCl₃	c	265.283	—	−238.3	—	—	—	—
std. state, m = 1	aq	—	—	−283.0	−249.9	—	−14.	−94.
TbCl₃								
in ∞ H₂O	aq	—	—	−283.0	—	—	—	—
TbCl₃·6H₂O	c	373.375	—	−683.4	−583.4	—	96.4	—
Tb(BrO₃)₃								
in 39.1 H₂O (satd)	aq	542.646	—	−214.1	—	—	—	—
Tb(IO₃)₃	c	683.632	—	−326.	—	—	—	—
TbC₂	g	182.946	212.	211.7	198.7	2.47	64.	10.5
Tb₂(CO₃)₃	c	497.876	—	−795.7	—	—	—	—
THALLIUM								
Tl	c	204.37	0	0	0	1.632	15.45	6.29
	g	—	43.701	43.55	35.24	1.481	43.225	4.968
in Hg	liq	—	—	0.076	−0.062	—	15.80	—
Tl⁺	g	—	184.553	185.883	—	—	—	—
std. state, m = 1	aq	—	—	1.28	−7.74	—	30.0	—
Tl²⁺	g	—	655.636	658.447	—	—	—	—
Tl³⁺	g	—	1343.5	1347.8	—	—	—	—
std. state, m = 1	aq	—	—	47.0	51.3	—	−46.	—

Formula and Description	State	Formula weight	$\Delta H f_0^\circ$ kcal/mol	$\Delta H f^\circ$ kcal/mol	$\Delta G f^\circ$ kcal/mol	$H_{298}^\circ - H_0^\circ$	S° cal/deg mol	C_p° cal/deg mol
			0 K	**298.15 K (25°C)**				
Tl₂O	c	424.7394	—	−42.7	−35.2	—	30.	—
Tl₂O₃	c	456.7382	—	−74.5	—	—	—	—
Tl₂O₄	c	472.7376	—	—	−83.0	—	—	—
TlOH	c	221.3774	—	−57.1	−46.8	—	21.	—
TlOH								
std. state, m = 1	aq	—	—	−53.69	−45.33	—	27.4	—
Tl(OH)₃	c	255.3922	—	—	−121.2	—	—	—
TlF	c	223.3684	—	−77.6	—	—	—	—
std. state, m = 1	aq	—	—	−78.22	−74.38	—	26.7	—
TlCl	c	239.8230	−49.091	−48.79	−44.20	3.030	26.59	12.17
std. state, m = 1	aq	—	—	−38.67	−39.11	—	43.5	—
TlCl₃	c	310.7290	—	−75.3	—	—	—	—
TlCl₃								
std. state, m = 1, Tl³⁺ + 3Cl⁻	aq	—	—	−72.9	42.8	—	−5.5	—
TlClO₃								
std. state, m = 1	aq	287.8212	—	−22.4	−8.5	—	68.8	—
TlBr	c	284.2790	—	−41.4	−40.00	—	28.8	—
std. state, m = 1	aq	—	—	−27.77	−32.59	—	49.7	—
TlBr₃	aq	444.0970	—	−59.7	—	—	—	—
std. state, m = 1	aq	—	—	−40.2	−23.2	—	13.	—
TlBrO₃	c	332.2772	—	−32.6	−12.70	—	40.3	—
std. state, m = 1	aq	—	—	−18.7	−7.3	—	69.0	—
TlI	c	331.2744	—	−29.6	−29.97	—	30.5	—
std. state, m = 1	aq	—	—	−11.91	−20.07	—	56.6	—
TlIO₃	c	379.2726	—	−63.9	−45.86	—	42.2	—
std. state, m = 1	aq	—	—	−51.6	−38.3	—	58.3	—
Tl₂S	c	440.8040	—	−23.2	−22.4	—	36.	—
Tl₂SO₄	c	504.8016	—	−222.7	−198.49	—	55.1	—
std. state, m = 1	aq	—	—	−214.76	−193.45	—	64.8	—
TlN₃	c	246.3901	—	55.8	70.38	—	35.1	—
TlNO₃	c	266.3749	—	−58.30	−36.44	—	38.4	23.78
undissoc.; std. state, m = 1	aq	—	—	−48.93	−34.85	—	64.5	—
Tl₂CO₃	c	468.7493	—	−167.3	−146.9	—	37.1	—
TlCNS	c	262.4518	—	6.8	9.21	—	39.	—
undissoc.; std. state, m = 1	aq	—	—	16.59	13.32	—	58.2	—
THORIUM								
Th	c	232.0381	0	0	0	1.556	12.76	6.53
	g	232.0381	143.08	143.0	133.26	1.481	45.420	4.97
Th⁺	g	232.0381	—	283.1				
Th²⁺	g	232.0381	—	550.6	—	—	—	—
Th³⁺	g	232.0381	—	1012.	—	—	—	—
Th⁴⁺	g	232.0381	—	1677.	—	—	—	—
	a	232.0381	—	−183.8	−168.5	—	−101.0	—
ThO	g	248.0375	6.5	0.0	−1.11	1.100	57.330	7.47
ThH	c	264.0369	−292.01	−293.12	−279.35	2.523	15.590	14.76
ThH₂	c	234.0541	−31.4	−33.4	−23.9	1.616	12.120	8.77
ThF	g	251.0365	—	—	—	2.23	61.50	8.28
ThF³⁺	a	251.0365	—	−264.5	−246.1	—	−71.7	—
ThF₄	c	308.0317	−499.24	−499.90	−477.30	5.114	33.950	26.420
	g	308.0317	−417.1	−418.0	−409.7	4.90	81.7	22.2
ThCl	g	267.4911	—	—	—	2.35	64.3	8.71
ThCl₂	g	302.9441	—	—	—	3.36	75.8	13.2
ThCl₄	c	373.8501	—	−283.6	−261.6	—	45.5	—
ThCl₄·2H₂O	c	409.8809	—	−436.0	—	—	—	—
ThCl₄·4H₂O	c	445.9117	—	−587.8	—	—	—	—
ThCl₄·7H₂O	c	499.9579	—	−804.4	—	—	—	—
ThOCl₂	c	318.9435	—	−294.5	−276.3	—	29.5	—
ThBr₄	c	551.6741	—	−230.7	−221.6	—	55.	—
	g	551.6741	−175.26	−182.3	−187.5	6.23	103.	25.1
ThBr₄·7H₂O	c	677.7819	—	−757.2	—	—	—	—
ThBr₃·10H₂O	c	731.8281	—	−975.1	—	—	—	—
ThOBr₂	c	407.8555	—	−283.	—	—	—	—
ThI	g	358.9425	—	—	—	2.49	68.9	8.98
ThI₂	g	485.8469	—	—	—	3.69	85.0	13.7
ThI₃	g	612.7513	—	—	—	5.12	102.7	19.6
ThI₄	c	739.6557	—	−158.9	−156.6	—	61.	—
	g	739.6557	−108.89	−110.1	−123.1	6.65	112.	25.4
ThS	c	264.1021	—	−94.5	−93.4	—	16.68	—
ThS₂	c	296.1661	—	−149.7	−148.2	—	23.0	—
Th₂S₃	c	560.2682	—	−259.0	−257.4	—	43.	—
ThN	c	246.0448	−92.93	−93.5	−86.9	2.020	13.40	10.8
	g	246.0448	—	118.	—	—	—	—
Th₃N₄	c	752.1411	—	−314.3	−289.9	—	48.	—
Th(NO₃)₄	c	480.0577	—	−344.5	—	—	—	—
ThP	c	263.0119	—	−83.2	−81.54	—	17.0	—
	g	263.0119	128.5	128.	116.	2.29	63.9	8.5
Th₃P₄	c	820.0095	—	−273.0	−266.0	—	53.0	—
ThC	c	244.0493	—	−29.60	—	—	—	—
ThC₁.₉₄	c	255.3398	−35.4	−35.	−35.3	2.447	16.37	13.55
ThC₂	g	256.0605	—	173.	—	—	—	—
THULIUM								
Tm	c	168.934	0	0	0	1.767	17.69	6.46
	g	—	55.786	55.5	47.2	1.481	45.412	4.968
Tm⁺	g	—	198.3	199.5				
Tm²⁺	g	—	476.	479.	—	—	—	—
Tm³⁺	g	—	1023.	1027.	—	—	—	—
std. state, m = 1	aq	—	—	−166.8	−158.2	—	−58.	6.
TmO	g	184.933	−19.	—	—	—	—	—
Tm₂O₃	c	385.866	−449.75	−451.4	−428.9	4.99	33.4	27.9
TmCl₃	c	275.293	—	−235.8	—	—	—	—
	g	—	—	−166.	—	—	—	—
std. state, m = 1	aq	—	—	−286.6	−252.3	—	−18.	−92.
TmCl₃								
in ∞ H₂O	aq	—	—	−286.6	—	—	—	—
TmCl₃·6H₂O	c	383.385	—	—	—	—	95.5	—
TmOCl	c	220.386	—	−236.0	—	—	—	—
TmI₃	c	549.647	—	−143.8	—	—	—	—
Tm(IO₃)₃	c	693.642	—	−328.	—	—	—	—
TmC₂	c	192.956	—	−22.	—	—	—	—

Formula and Description	State	Formula weight	ΔHf_0° kcal/mol	ΔHf° kcal/mol	ΔGf° kcal/mol	$H_{298}^\circ - H_0^\circ$	S° cal/deg mol	C_p°
			0 K	298.15 K (25°C)				
TIN								
Sn								
I, white	c	118.69	0	0	0	1.505	12.32	6.45
II, gray	c	—	−0.371	−0.50	0.03	1.376	10.55	6.16
	g	—	72.18	72.2	63.9	1.485	40.243	5.081
Sn^+	g	—	241.54	243.04	—	—	—	—
Sn^{2+}	g	—	578.97	581.95	—	—	—	—
in aq HCl, std. state, m = 1	aq	—	—	−2.1	−6.5	—	−4.	—
Sn^{3+}	g	—	1282.9	1287.4	—	—	—	—
SnO	c	134.689	−68.3	−68.3	−61.4	—	13.5	10.59
	g	—	—	—	—	2.117	55.45	7.55
SnO_2	c	150.689	—	−138.8	−124.2	—	12.5	12.57
SnH_4	g	122.722	41.78	38.9	45.0	2.669	54.39	11.70
$Sn(OH)_2$								
precipitated	c	152.705	—	−134.1	−117.5	—	37.	—
$Sn(OH)_4$								
precipitated	c	186.719	—	−265.3	—	—	—	—
$SnCl_2$	c	189.596	—	−77.7	—	—	—	—
un-ionized, in aq HCl; std. state, m = 1	aq	—	—	−78.8	−71.6	—	41.	—
$SnCl_2 \cdot 2H_2O$	c	225.627	—	−220.2	—	—	—	—
$SnCl_4$	liq	260.502	—	−122.2	−105.2	—	61.8	39.5
in aq HCl; std. state, m = 1	aq	—	—	−152.5	−124.9	—	26.	—
$SnBr_2$	c	278.508	—	−58.2	—	—	—	—
	aq	—	—	−56.8	—	—	—	—
$SnBr_4$	c	438.326	—	−90.2	−83.7	—	63.2	—
SnI_2	c	372.499	—	−34.3	—	—	—	—
SnI_4	c	626.308	—	—	—	—	—	20.3
SnS	c	150.754	—	−24.	−23.5	—	18.4	11.77
SnS_2	c	182.818	—	—	—	—	20.9	16.76
TITANIUM								
Ti	c	47.90	0	0	0	1.149	7.32	5.98
	g	—	111.65	112.3	101.6	1.802	43.066	5.839
Ti^+	g	—	268.9	271.1	—	—	—	—
Ti^{2+}	g	—	582.1	585.7	—	—	—	—
Ti^{3+}	g	—	1216.0	1221.1	—	—	—	—
Ti^{4+}	g	—	2213.8	2220.4	—	—	—	—
TiO								
α	c	63.899	−123.49	−124.2	−118.3	1.48	8.31	12.
TiO_2								
anatase	c	79.899	−223.44	−224.6	−211.4	2.062	11.93	13.26
Brookite	c	—	—	−225.1	—	—	—	—
rutile	c	—	−224.6	−225.8	−212.6	2.065	12.03	13.15
	amorp	—	—	−210.	—	—	—	—
Ti_2O_3	c	143.798	−361.52	−363.5	−342.8	3.431	18.83	23.27
TiH_2	c	49.916	−26.61	−28.6	−19.2	1.18	7.1	7.2
$TiCl_2$	c	118.806	−122.64	−122.8	−111.0	3.18	20.9	16.69
$TiCl_3$	c	154.259	−172.86	−172.3	−156.2	5.00	33.4	23.22
$TiCl_4$	c	189.712	−195.75	—	—	—	—	—
	liq	—	—	−192.2	−176.2	—	60.31	34.70
$TiBr_2$	c	207.718	—	−96.	—	—	—	—
$TiBr_3$	c	287.627	−126.65	−131.1	−125.2	5.49	42.2	24.31
$TiBr_4$	c	367.536	−141.36	−147.4	−140.9	6.825	58.2	31.43
	g	—	−124.14	−131.3	−135.8	5.71	95.2	24.1
TiI_2	c	301.709	—	−63.	—	—	—	—
TiI_4	c	555.518	—	−89.8	−88.8	—	59.6	30.03
	g	—	—	−66.4	—	—	—	—
TiS	c	79.964	—	−57.	—	—	—	—
TiS_2	c	112.028	—	—	—	—	18.73	16.23
TiN	c	61.907	−79.93	−80.8	−74.0	1.311	7.23	8.86
TiP	c	78.874	—	−67.6	—	—	—	—
TiC	c	59.911	−43.80	−44.1	−43.2	1.101	5.79	8.04
TiB_2	c	69.522	−77.0	−77.4	−76.4	1.333	6.81	10.58
TUNGSTEN								
W	c	183.85	0	0	0	1.190	7.80	5.80
	g	—	202.70	203.0	192.9	1.486	41.549	5.093
W^+	g	—	386.8	388.6	—	—	—	—
WO	g	199.849	—	108.	—	—	—	—
WO_2	c	215.849	−139.762	−140.94	−127.61	2.087	12.08	13.41
WO_3	c	231.848	−200.100	−201.45	−182.62	2.952	18.14	17.63
WO_4^{2-}								
std. state, m = 1	aq	247.848	—	−257.1	—	—	—	—
H_2WO_4	c	249.864	—	−270.5	—	—	—	—
WS_2	c	247.978	—	−50.	—	—	—	—
WC	c	195.861	—	−9.69	—	—	—	—
W_2C	c	379.711	—	−6.3	—	—	—	—
$W(CO)_6$	c	351.913	—	−227.9	—	—	—	—
Fe_3W_6	c	1494.029	—	—	—	—	—	77.7
$FeWO_4$	c	303.695	—	−276.	−252.	—	31.5	27.39
$MnWO_4$	c	302.786	—	−311.9	—	—	—	29.7
URANIUM								
U	c	238.0290	0	0	0	1.521	12.00	6.612
	g	238.0290	127.97	128.0	117.4	1.553	47.72	5.663
U^+	g	238.0290	270.81	272.32	—	—	—	—
U^{2+}	g	238.0290	520.	—	—	—	—	—
U^{3+}	g	238.0290	960.	—	—	—	—	—
	g	238.0290	—	−116.9	−113.6	—	−46.	—
U^{4+}	a	238.0290	—	−141.3	−126.9	—	−98.	—
UO	g	254.0284	—	5.	—	—	—	—
UO_2	c	270.0278	−258.40	−259.3	−246.6	2.696	18.41	15.20
	g	270.0278	−110.85	−111.3	−112.7	3.15	65.6	12.28
UO_2	a	270.0278	—	−230.1	—	—	—	—
UO_2^{2+}	a	270.0278	—	−243.7	−227.9	—	−23.3	—
UO_3								
γ, orthorhombic	c	286.0272	−291.35	−292.5	−273.9	3.486	22.97	19.52
ϵ form, triclinic, red	c2	286.0272	—	−291.0	—	—	—	—
	c3	286.0272	−289.96	−291.0	−272.6	3.596	23.76	19.57
α, orthorhombic, prev. described as hexagonal								
	c4	286.0272	−290.526	−291.65	−273.02	3.509	23.02	19.44

SELECTED VALUES OF CHEMICAL THERMODYNAMIC PROPERTIES (Continued)

Formula and Description	State	Formula weight	0 K ΔHf° kcal/mol	298.15 K (25°C) ΔHf° kcal/mol	ΔGf° kcal/mol	H°298 − H°0	S° cal/deg mol	C°p cal/deg mol
β, orthorhombic, orange-red; δ, cubic, dark red	c5	286.0272	—	−291.5	—	—	—	—
amorphous, orange	am	286.0272	—	−288.8	—	—	—	—
	g	286.0272	−195.	—	—	—	—	—
UO₃·H₂O								
β, orthorhombic	c	304.0426	—	−366.6	−333.4	—	30.	—
ε form, monoclinic	c2	304.0426	—	−366.0	—	—	—	—
α, transition to β 278.3k	c3	304.0426	—	−365.2	—	—	—	—
U₃O₇								
β, tetragonal	c	826.0828	−816.38	−819.1	−775.1	9.108	59.88	51.51
α, tetragonal	c2	826.0828	—	—	—	9.009	59.19	51.09
U₃O₈								
α, orthorhombic	c	842.0822	−851.75	−854.4	−805.4	10.216	67.54	56.97
UF	g	257.0274	−5.7	−6.	−13.	2.28	60	9.04
UF₂	g	276.0258	−134.6	−135.	−138.	3.28	71.	16.0
UF₃	c	295.0242	−360.3	−360.6	−344.2	4.392	29.50	22.73
UF₄								
monoclinic	c	314.0226	−458.75	−459.1	−437.4	5.390	36.25	27.73
UF₄	g	314.0226	−382.72	−383.7	−377.5	4.76	88.	21.8
	g	333.0210	−462.8	−464.	−452.	5.6	93.	26.2
UF₆	c	352.0194	−524.80	−525.1	−494.4	7.545	54.4	39.86
UOF₂	c	292.0252	—	−358.3	−341.5	—	28.5	—
UCl₃	c	344.3880	−207.61	−207.1	−191.	5.318	38.0	24.5
UCl₄	c	379.8410	−243.97	−243.6	−222.3	6.28	47.1	29.16
UCl₆	c	450.7470	−261.8	−261.	−230.	8.90	68.3	42.0
UOCl₂	c	324.9344	−254.83	−255.0	−238.1	4.586	33.06	22.72
UO₂Cl₂	c	340.9338	−296.67	−297.3	−274.0	5.157	35.98	25.78
UBr₃	c	477.7560	—	−167.1	−161.0	—	46.	26.0
UBr₄	c	557.6650	—	−191.8	−183.5	—	57.0	30.6
UOBr₂	c	413.8464	−229.27	−232.7	−222.2	4.989	37.66	23.42
UO₂Br₂	c	429.8458	—	−271.9	−254.9	—	40.5	—
UI₃	c	618.7422	—	−110.1	−109.9	—	53.	26.8
UI₄	c	745.6466	—	−122.4	−121.1	—	63.	32.1
US₂								
β	c	302.1570	−126.1	−126.	−125.8	3.698	26.39	17.84
UO₂SO₄								
β	c	366.0894	—	−441.0	−402.4	—	37.0	34.7
U(SO₄)₂	c	430.1522	—	−554.	—	—	—	—
UN	c	252.0357	−69.12	−69.5	−63.5	2.173	14.92	11.37
UO₂(NO₃)₂	c	394.0376	—	−322.5	−264.1	—	58.	—
UP	c	260.9988	03.0	04.	−63.	2.58	18.7	11.9
UP₂	c	299.9766	−72.6	−73.	−71.	3.679	24.3	19.12
U₃P₄	c	837.9822	−199.2	−200.	−196.	8.87	61.8	41.8
UC	c	250.0402	−23.90	−23.5	−23.7	2.176	14.15	11.98
U₂C₃	c	512.0916	−44.43	−43.4	−41.0	4.029	32.43	15.44
UO₂CO₃	c	330.0372	—	−404.2	−373.5	—	33.	—
USi	c	266.1150	—	−19.2	—	—	—	—
USi₂	c	294.2010	—	−31.2	—	—	—	—
USi₃	c	322.2870	—	−31.6	—	—	—	—
VANADIUM								
V	c	50.942	0	0	0	1.109	6.91	5.95
	g	—	122.12	122.90	180.32	1.8898	43.544	6.217
V⁺	g	—	277.550	279.811	—	—	—	—
V²⁺	g	—	615.39	619.13	—	—	—	—
V³⁺	g	—	1292.69	1297.91	—	—	—	—
V⁴⁺	g	—	2369.8	2376.5	—	—	—	—
VO	c	66.9414	—	−103.2	−96.6	—	9.3	10.86
VO²⁺								
std. state	aq	—	—	−116.3	−106.7	—	−32.0	—
VO₂	g	82.9408	—	−57.	—	—	—	—
V₂O₃	c	149.8822	—	−291.3	−272.3	—	23.5	24.67
V₂O₄								
α	c	165.8816	—	−341.1	−315.1	—	24.5	27.96
V₂O₅	c	181.8810	—	−370.6	−339.3	—	31.3	30.51
V₃O₅	c	232.8230	—	−462.	−434.	—	39.	—
V₄O₇	c	315.7638	—	−631.	−587.	—	52.	—
V₆O₁₃	c	513.6442	—	−1062.	—	—	—	—
HVO₄²⁻								
std. state	aq	115.9476	—	−277.0	−233.0	—	4.	—
H₂VO₄⁻								
std. state	aq	116.9555	—	−280.6	−244.0	—	29.	—
HV₂O₇³⁻								
std. state	aq	214.8878	—	—	−428.4	—	—	—
H₃V₂O₇⁻								
std. state	aq	216.9037	—	—	−445.5	—	—	—
VF₃	c	107.9372	—	—	—	—	23.18	21.62
VF₄	c	126.9356	—	−335.4	—	—	—	—
VF₅	liq	145.9340	—	−353.8	−328.2	—	42.0	—
VCl₂	c	121.848	—	−108.	−97.	—	23.2	17.26
VCl₃	c	157.301	—	−138.8	−122.2	—	31.3	22.27
VCl₄	liq	192.754	—	−136.1	−120.4	—	61.	—
VOCl	c	102.3944	—	−145.	−133.	—	18.	—
VBr₂	c	210.760	—	−87.3	—	—	—	—
VBr₃	c	290.669	—	−103.6	—	—	—	—
VI₂	c	304.7508	—	−60.1	—	—	—	—
VI₃	c	431.6552	—	−64.7	—	—	—	—
VI₄	g	558.5596	—	−29.3	—	—	—	—
V₂S₃	c	198.076	—	−227.	—	—	—	—
VN	c	64.9487	—	−51.9	−45.7	—	8.91	9.08
XENON								
Xe	g	131.30	0	0	0	1.481	40.5290	4.9679
std. state, m = 1	aq	—	—	−4.2	3.2	—	15.7	—
Xe⁺	g	—	279.72	281.20	—	—	—	—
Xe²⁺	g	—	768.8	771.8	—	—	—	—
Xe³⁺	g	—	1509.6	1514.1	—	—	—	—
YTTERBIUM								
Yb	c	173.04	0	0	0	1.604	14.31	6.39
	g	—	36.479	36.4	28.3	1.481	41.352	4.968
Yb⁺	g	—	180.70	182.05	—	—	—	—
Yb²⁺	g	—	462.	464.8	—	—	—	—
std. state, m = 1	aq	—	—	—	−126.	—	—	—

Substance			0 K	298.15 K (25°C)				
Formula and Description	State	Formula weight	$\Delta H f_0^\circ$ kcal/mol	$\Delta H f^\circ$ kcal/mol	$\Delta G f^\circ$ kcal/mol	$H_{298}^\circ - H_0^\circ$	S° cal/deg mol	C_p°
Yb³⁺	g	—	1043.	1047.3	—	—	—	—
std. state, m = 1	aq	—	—	−161.2	−153.9	—	−57.	6.
Yb₂O₃	c	394.078	−432.08	−433.7	−412.7	4.69	31.8	27.57
YbH	g	174.048	—	—	—	2.083	52.66	7.178
YbCl₂	c	243.946	—	−191.1	—	—	—	—
YbCl₃	c	279.399	—	−229.4	—	—	—	—
std. state, m = 1	aq	—	—	−281.1	−248.0	—	−17.	−92.
YbCl₃ in ∞ H₂O	aq	—	—	−281.1	—	—	—	—
YbCl₃·6H₂O	c	387.491	—	−680.2	−580.6	—	94.6	81.6
YbOCl	c	224.492	—	−229.9	—	—	—	—
Yb(IO₃)₃	c	697.748	—	−322.	—	—	—	—
Yb(NO₃)₃ in 200 H₂O	aq	359.055	—	−308.997	—	—	—	—
YbC₂	c	197.062	—	−17.9	−18.5	—	19.	—
YTTRIUM								
Y	c	88.905	0	0	0	1.426	10.62	6.34
	g	—	100.49	100.7	91.1	1.639	42.87	6.18
Y⁺	g	—	247.62	—	—	—	—	—
Y²⁺	g	—	529.88	—	—	—	—	—
Y³⁺	g	—	1003.1	—	—	—	—	—
std. state, m = 1	aq	—	—	−172.9	−165.8	—	−60.	—
Y⁴⁺	g	—	2428.	—	—	—	—	—
YO	g	104.9044	−9.	−9.3	−15.5	2.115	55.88	7.53
Y₂O	g	193.8094	2.	1.0	—	2.9	—	11.7
Y₂O₂	g	209.8088	−126.	−127.4	—	3.5	—	15.8
Y₂O₃	c	225.8082	−453.40	−455.38	−434.19	3.983	23.68	24.50
YH₂	c	90.9209	−51.95	−54.0	−44.3	1.403	9.18	8.24
YH₃	c	91.9289	−61.15	−64.0	−49.9	1.613	10.02	10.36
YF	g	107.9034	−32.7	−33.	−39.	2.163	55.38	7.92
YF₂	g	126.9018	—	—	—	2.9	69.3	11.7
YF₃	c	145.900	—	−410.8	−393.1	—	24.	—
YCl²⁺	g	124.358	48.	47.8	41.5	2.286	58.33	8.56
YCl²⁺ std. state, m = 1	aq	—	—	−214.0	−198.7	—	−46.	—
YCl₃	c	195.264	—	−239.0	—	—	—	—
YCl₃·6H₂O	c	303.356	—	−691.3	−592.1	—	92.	—
YBr₂	g	248.723	—	—	—	3.4	80.0	13.2
YI₂	g	342.7138	—	—	—	3.5	84.0	13.5
YI₃	c	469.6182	—	−147.4	—	—	—	—
Y(IO₃)₃	c	613.6128	—	—	−271.2	—	—	—
YS	g	120.969	42.	41.7	29.7	2.2	58.	8.2
Y₂(SO₄)₃	c	465.995	—	—	−866.8	—	—	138.
YC₂	c	112.9274	—	−26.	−26.	—	13.	—
Y₂(CO₃)₃	c	357.8379	—	—	−752.4	—	—	—
ZINC								
Zn	c	65.37	0	0	0	1.350	9.95	6.07
	g	—	31.114	31.245	22.748	1.481	38.450	4.968
Zn⁺	g	—	247.740	249.352	—	—	—	—
Zn²⁺	g	—	662.00	665.09	—	—	—	—
std. state, m = 1	aq	—	—	−36.78	−35.14	—	−26.8	11.
ZnO	c	81.369	—	−83.24	−76.08	—	10.43	9.62
ZnO₂²⁻ std. state, m = 1	aq	97.369	—	—	−91.85	—	—	—
HZnO₂⁻ std. state, m = 1	aq	98.377	—	—	−109.26	—	—	—
Zn(OH)₂								
γ	c	99.385	—	—	−132.38	—	—	—
β	c	—	—	−153.42	−132.31	—	19.4	—
e	c	—	—	−153.74	−132.68	—	19.5	17.3
precipitated	—	—	—	−153.5	—	—	—	—
Zn(OH)₂ std. state, m = 1	aq	—	—	−146.72	−110.33	—	−31.9	−60.
Zn(OH)₄²⁻ std. state, m = 1	aq	133.399	—	—	−205.23	—	—	—
ZnF₂	c	103.367	−182.06	−182.7	−170.5	2.821	17.61	15.69
ZnCl₂	c	136.276	−99.255	−99.20	−88.296	3.598	26.64	17.05
std. state, m = 1	aq	—	—	−116.68	−97.88	—	0.2	−54.
in ∞ H₂O	aq	—	—	−116.68	—	—	—	—
Zn(ClO₄)₂ std. state, m = 1	aq	264.271	—	−98.60	−39.26	—	60.2	—
Zn(ClO₄)₂·6H₂O	c	372.363	—	−509.89	—	—	—	—
ZnBr₂	c	225.188	—	−78.55	−74.60	—	33.1	—
std. state, m = 1	aq	—	—	−94.88	−84.84	—	12.6	−57.
ZnBr₂·2H₂O	c	261.219	—	−224.0	−191.1	—	47.5	—
ZnI₂	c	319.179	—	−49.72	−49.94	—	38.5	—
std. state, m = 1	aq	—	—	−63.16	−59.80	—	25.2	−57.
Zn(IO₃)₂	c	415.175	—	—	−103.68	—	—	—
std. state, m = 1	aq	—	—	−142.6	−96.3	—	29.8	—
ZnS								
wurtzite	c	97.434	—	−46.04	—	—	—	—
sphalerite	c	—	—	−49.23	−48.11	—	13.8	11.0
ZnSO₄	c	161.432	—	−234.9	−209.0	—	28.6	—
ZnSO₄ std. state, m = 1	aq	—	—	−254.10	−213.11	—	−22.0	−59.
in ∞ H₂O	aq	—	—	−254.10	—	—	—	—
ZnSO₄·H₂O	c	179.447	—	−311.78	−270.58	—	33.1	—
ZnSO₄·6H₂O	c	269.524	—	−663.83	−555.64	—	86.9	85.49
ZnSO₄·7H₂O	c	287.539	—	−735.60	−612.59	—	92.9	91.64
Zn(N₃)₂	c	149.410	—	52.	—	—	—	—
undissoc.; std. state, m = 1	aq	—	—	—	129.6	—	—	—
Zn₃N₂	c	224.123	—	−5.4	—	—	—	26.
Zn(NO₃)₂	c	189.380	—	−115.6	—	—	—	—
ionized; std. state, m = 1	aq	—	—	−135.90	−88.36	—	43.2	−30.
Zn(NO₃)₂ in ∞ H₂O	aq	—	—	−135.90	—	—	—	—
Zn(NO₃)₂·H₂O	c	207.395	—	−192.4	—	—	—	—
Zn(NO₃)₂·2H₂O	c	225.410	—	−265.36	—	—	—	—
Zn(NO₃)₂·4H₂O	c	261.441	—	−406.10	—	—	—	—
Zn(NO₃)₂·6H₂O	c	279.472	—	−551.30	−423.79	—	109.2	77.2

Formula and Description	State	Formula weight	0 K ΔHf° kcal/mol	298.15 K (25°C) ΔHf° kcal/mol	ΔGf° kcal/mol	$H^{\circ}_{298} - H^{\circ}_0$	S° cal/deg mol	C°_p
Zn_3P_2	c	258.058	—	−113.	—	—	—	—
$Zn(PO_3)_2$	c	223.314	—	−497.9	—	—	—	—
$Zn_2(P_2O_7)$	c	304.683	—	−600.0	—	—	—	—
$Z_3(PO_4)_2$	c	386.053	—	−691.3	—	—	—	—
$ZnCO_3$	c	125.379	—	−194.26	−174.85	—	19.7	19.05
$ZnCO_3 \cdot H_2O$	c	143.395	—	—	−232.0	—	—	—
ZnC_2O_4								
std. state, m = 1	aq	153.390	—	−234.0	−196.2	—	−15.9	—
$Zn(CH_3)_2$	liq	95.440	—	5.6	—	—	—	—
$Zn(C_2H_5)_2$	liq	123.494	—	2.5	—	—	—	—
$Zn(CN)_2$	c	117.406	—	22.9	—	—	—	—
$ZnSiO_3$	c	141.454	—	−301.2	—	—	—	—
Zn_2SiO_4	c	222.824	—	−391.19	−364.06	—	31.4	29.48
ZIRCONIUM								
Zr								
α, hexagonal	c	91.22	0	0	0	1.322	9.32	6.06
	g	—	145.19	145.5	135.4	1.629	43.32	6.37
Zr^+	g	—	302.9	304.7	—	—	—	—
Zr^{2+}	g	—	605.7	609.0	—	—	—	—
Zr^{3+}	g	—	1135.9	1140.6	—	—	—	—
Zr^{4+}	g	—	1927.8	1934.0	—	—	—	—
ZrO	g	107.219	—	15.	—	—	—	—
ZrO_2								
α, monoclinic	c	123.219	−261.734	−263.04	−249.24	2.091	12.04	13.43
ZrO_2								
hydrated ppt	—	—	—	−260.4	—	—	—	—
ZrO_3								
ppt	c	139.218	—	−241.	—	—	—	—
ZrH_2								
zirconium hydride	c	93.236	−38.34	−40.4	−30.8	1.284	8.37	7.40
ZrF_4								
β, monoclinic	c	167.214	−455.44	−456.8	−432.6	4.183	25.00	24.79
ZrCl	c	126.673	—	−63.	—	—	—	—
$ZrCl_2$	c	162.126	—	−120.	—	—	—	—
$ZrCl_3$	c	197.579	—	−179.	—	—	—	—
$ZrCl_4$	c	233.032	−234.60	−234.35	−212.7	5.957	43.4	28.63
$ZrOCl_2$	aq	178.125	—	−280.3	—	—	—	—
$ZrBr_4$	c	410.856	—	−181.8	—	—	—	—
$ZrOBr_2$	aq	267.037	—	−259.9	—	—	—	—
ZrI_4	c	598.838	—	−115.1	—	—	—	—
ZrS_2	c	155.348	—	−135.3	—	—	—	—
$Zr(SO_4)_2$	c	283.343	—	−529.9	—	—	—	41.
$Zr(SO_4)_2 \cdot H_2O$	c	301.359	—	−610.4	—	—	—	—
$Zr(SO_4)_2 \cdot 4H_2O$	c	355.405	—	−825.6	—	—	—	—
ZrN	c	105.227	−86.42	−87.2	−80.4	1.575	9.29	9.66
ZrC	c	103.231	40.11	−46.3	−41.7	1.401	7.96	9.00
ZrB	c	119.306	—	−37.	—	—	—	—
$ZrSi_2$	c	147.392	—	−38.	—	—	—	—
Zr_2Si	c	210.526	—	−50.	—	—	—	—
Zr_3Si	c	301.746	—	−52.	—	—	—	—
Zr_3Si_2	c	329.832	—	−92.	—	—	—	—
$ZrSiO_4$	c	183.304	−483.32	−486.0	−458.7	3.562	20.1	23.58
ZrB_2	c	112.842	−77.69	−78.0	−77.0	1.590	8.59	11.53

T °K	C_P° J/deg-mol			$H_T^\circ - H_0^\circ$ J/mol			$(H_T^\circ - H_0^\circ)/T$ J/deg-mol		
	Cu	Ag	Au	Cu	Ag	Au	Cu	Ag	Au
1.00	0.000743	0.000818	0.00118	0.000359	0.000367	0.000478	0.000359	0.000367	0.000478
2.00	0.00177	0.00265	0.00504	0.00158	0.00197	0.00326	0.000790	0.000987	0.00163
3.00	0.00337	0.00650	0.0141	0.00409	0.00633	0.0123	0.00136	0.00211	0.00410
4.00	0.00582	0.0134	0.0306	0.00860	0.0160	0.0340	0.00215	0.00399	0.00849
5.00	0.00943	0.0243	0.0570	0.0161	0.0344	0.0768	0.00322	0.00689	0.0154
6.00	0.0145	0.0403	0.0955	0.0279	0.0663	0.152	0.00466	0.0110	0.0253
7.00	0.0213	0.0626	0.149	0.0456	0.117	0.273	0.00652	0.0167	0.0390
8.00	0.0301	0.0927	0.220	0.0712	0.194	0.456	0.00889	0.0243	0.0570
9.00	0.0414	0.132	0.313	0.107	0.306	0.720	0.0119	0.0340	0.0800
10.00	0.0555	0.183	0.431	0.155	0.462	1.090	0.0155	0.0462	0.109
11.00	0.0727	0.247	0.577	0.219	0.676	1.592	0.0199	0.0614	0.145
12.00	0.0936	0.325	0.755	0.302	0.961	2.255	0.0251	0.0801	0.188
13.00	0.119	0.421	0.963	0.407	1.332	3.112	0.0313	0.102	0.239
14.00	0.149	0.535	1.203	0.541	1.809	4.193	0.0386	0.129	0.299
15.00	0.184	0.670	1.474	0.706	2.409	5.529	0.0471	0.161	0.369
16.00	0.225	0.826	1.772	0.910	3.155	7.149	0.0569	0.197	0.447
17.00	0.273	1.002	2.096	1.158	4.067	9.081	0.0681	0.239	0.534
18.00	0.328	1.199	2.442	1.458	5.166	11.35	0.0810	0.287	0.630
19.00	0.390	1.414	2.807	1.816	6.471	13.97	0.0956	0.341	0.735
20.00	0.462	1.647	3.187	2.242	8.001	16.97	0.112	0.400	0.848
25.00	0.963	3.066	5.245	5.703	19.62	37.97	0.228	0.785	1.519
30.00	1.693	4.774	7.375	12.25	39.14	69.53	0.408	1.305	2.318
35.00	2.638	6.612	9.395	22.99	67.58	111.5	0.657	1.931	3.186
40.00	3.740	8.419	11.22	38.89	105.2	163.2	0.972	2.630	4.079
45.00	4.928	10.11	12.86	60.54	151.6	223.4	1.345	3.368	4.965
50.00	6.154	11.66	14.29	88.23	206.1	291.4	1.765	4.121	5.828
55.00	7.385	13.04	15.52	122.1	267.9	366.0	2.220	4.871	6.654
60.00	8.595	14.27	16.59	162.0	336.2	446.3	2.701	5.604	7.438
65.00	9.759	15.35	17.51	208.0	410.4	531.6	3.199	6.313	8.179
70.00	10.86	16.30	18.31	259.5	489.5	621.2	3.708	6.993	8.874
75.00	11.89	17.14	19.01	316.4	573.2	714.6	4.219	7.642	9.528
80.00	12.85	17.87	19.63	378.4	660.7	811.2	4.729	8.259	10.14
85.00	13.74	18.53	20.17	444.9	751.8	910.7	5.234	8.844	10.71
90.00	14.56	19.11	20.64	515.7	845.9	1013.	5.730	9.399	11.25
95.00	15.31	19.63	21.06	590.4	942.8	1117.	6.215	9.924	11.76
100.00	16.01	20.10	21.44	668.7	1042.	1223.	6.687	10.42	12.23
105.00	16.64	20.52	21.77	750.3	1144.	1331.	7.146	10.89	12.68
110.00	17.22	20.89	22.06	835.0	1247.	1441.	7.591	11.34	13.10
115.00	17.76	21.23	22.33	922.5	1353.	1552.	8.021	11.76	13.49
120.00	18.25	21.54	22.56	1013.	1460.	1664.	8.438	12.16	13.87
125.00	18.70	21.82	22.78	1105.	1568.	1777.	8.839	12.54	14.22
130.00	19.12	22.07	22.97	1199.	1678.	1892.	9.227	12.91	14.55
135.00	19.51	22.31	23.15	1296.	1789.	2007.	9.601	13.25	14.87
140.00	19.87	22.52	23.31	1395.	1901.	2123.	9.961	13.58	15.17
145.00	20.20	22.72	23.45	1495.	2014.	2240.	10.31	13.89	15.45
150.00	20.51	22.90	23.59	1597.	2128.	2358.	10.64	14.19	15.72
155.00	20.79	23.07	23.70	1700.	2243.	2476.	10.97	14.47	15.97
160.00	21.05	23.22	23.81	1804.	2358.	2595.	11.28	14.74	16.22
165.00	21.30	23.37	23.91	1910.	2475.	2714.	11.58	15.00	16.45
170.00	21.53	23.50	24.00	2017.	2592.	2834.	11.87	15.25	16.67
175.00	21.74	23.63	24.08	2125.	2710.	2954.	12.15	15.49	16.88
180.00	21.94	23.75	24.15	2235.	2828.	3075.	12.42	15.71	17.08
185.00	22.13	23.86	24.22	2345.	2947.	3196.	12.68	15.93	17.27
190.00	22.31	23.96	24.29	2456.	3067.	3317.	12.93	16.14	17.46
195.00	22.47	24.06	24.35	2568.	3187.	3438.	13.17	16.34	17.63
200.00	22.63	24.16	24.41	2681.	3308.	3650.	13.40	16.54	17.80
205.00	22.77	24.24	24.48	2794.	3429.	3683.	13.63	16.72	17.96
210.00	22.91	24.33	24.54	2908.	3550.	3805.	13.85	16.90	18.12
215.00	23.04	24.41	24.60	3023.	3672.	3928.	14.06	17.08	18.27
220.00	23.17	24.49	24.65	3139.	3794.	4051.	14.27	17.25	18.41
225.00	23.28	24.56	24.71	3255.	3917.	4174.	14.47	17.41	18.55
230.00	23.39	24.63	24.76	3372.	4040.	4298.	14.66	17.56	18.69
235.00	23.50	24.69	24.82	3489.	4163.	4422.	14.85	17.71	18.82
240.00	23.60	24.76	24.87	3607.	4287.	4546.	15.03	17.86	18.94
245.00	23.69	24.82	24.92	3725.	4411.	4671.	15.20	18.00	19.06
250.00	23.78	24.88	24.97	3844.	4535.	4796.	15.37	18.14	19.18
255.00	23.86	24.93	25.02	3963.	4659.	4921.	15.54	18.27	19.30
260.00	23.94	24.99	25.07	4082.	4784.	5046.	15.70	18.40	19.41
265.00	24.02	25.04	25.12	4202.	4909.	5171.	15.86	18.53	19.51
270.00	24.09	25.09	25.17	4322.	5035.	5297.	16.01	18.65	19.62
273.15	24.13	25.12	25.20	4398.	5114.	5376.	16.10	18.72	19.68
275.00	24.15	25.14	25.21	4443.	5160.	5423.	16.16	18.76	19.72
280.00	24.22	25.19	25.26	4564.	5286.	5549.	16.30	18.88	19.82
285.00	24.28	25.24	25.31	4685.	5412.	5676.	16.44	18.99	19.91
290.00	24.34	25.28	25.35	4807.	5538.	5802.	16.57	19.10	20.01
295.00	24.40	25.32	25.39	4929.	5665.	5929.	16.71	19.20	20.10
298.15	24.44	25.35	25.42	5005.	5745.	6009.	16.79	19.27	20.15
300.00	24.46	25.37	25.43	5051.	5792.	6056.	16.84	19.31	20.19

T °K	S°_T J/deg-mol			$-(G^\circ_T - H^\circ_0)$ J/mol			$-(G^\circ_T - H^\circ_0)/T$ J/deg-mol		
	Cu	Ag	Au	Cu	Ag	Au	Cu	Ag	Au
1.00	0.000711	0.000706	0.000880	0.000351	0.000339	0.000402	0.000351	0.000339	0.000402
2.00	0.00152	0.00175	0.00266	0.00145	0.00152	0.00206	0.000727	0.000762	0.00103
3.00	0.00251	0.00347	0.00620	0.00345	0.00406	0.00631	0.00115	0.00135	0.00210
4.00	0.00379	0.00619	0.0123	0.00657	0.00879	0.0153	0.00164	0.00220	0.00383
5.00	0.00546	0.0103	0.0218	0.0112	0.0169	0.0321	0.00223	0.00338	0.00641
6.00	0.00760	0.0160	0.0354	0.0176	0.0299	0.0603	0.00294	0.00498	0.0100
7.00	0.0103	0.0238	0.0539	0.0265	0.0496	0.104	0.00379	0.00709	0.0149
8.00	0.0137	0.0341	0.0782	0.0385	0.0783	0.170	0.00481	0.00979	0.0212
9.00	0.0179	0.0472	0.109	0.0542	0.119	0.263	0.00602	0.0132	0.0292
10.00	0.0229	0.0636	0.148	0.0746	0.174	0.391	0.00746	0.0174	0.0391
11.00	0.0290	0.0839	0.196	0.100	0.247	0.562	0.00913	0.0225	0.0511
12.00	0.0362	0.109	0.253	0.133	0.343	0.786	0.0111	0.0286	0.0655
13.00	0.0447	0.138	0.322	0.173	0.466	1.073	0.0133	0.0359	0.0825
14.00	0.0545	0.174	0.402	0.223	0.622	1.434	0.0159	0.0444	0.102
15.00	0.0660	0.215	0.494	0.283	0.815	1.880	0.0189	0.0544	0.125
16.00	0.0791	0.263	0.598	0.355	1.054	2.426	0.0222	0.0659	0.152
17.00	0.0941	0.318	0.715	0.442	1.344	3.081	0.0260	0.0790	0.181
18.00	0.111	0.381	0.845	0.544	1.693	3.861	0.0302	0.0940	0.214
19.00	0.131	0.452	0.987	0.665	2.109	4.775	0.0350	0.111	0.251
20.00	0.152	0.530	1.140	0.806	2.599	5.838	0.0403	0.130	0.292
25.00	0.305	1.043	2.069	1.917	6.446	13.76	0.0767	0.258	0.550
30.00	0.541	1.750	3.214	3.995	13.35	26.89	0.133	0.445	0.896
35.00	0.871	2.623	4.505	7.487	24.22	46.14	0.214	0.692	1.318
40.00	1.294	3.625	5.881	12.86	39.79	72.08	0.322	0.995	1.802
45.00	1.802	4.715	7.299	20.57	60.61	105.0	0.457	1.347	2.334
50.00	2.385	5.862	8.729	31.01	87.04	145.1	0.620	1.741	2.902
55.00	3.029	7.040	10.15	44.52	119.3	192.3	0.809	2.169	3.496
60.00	3.724	8.228	11.55	61.38	157.5	246.5	1.023	2.624	4.109
65.00	4.458	9.414	12.91	81.82	201.6	307.7	1.259	3.101	4.734
70.00	5.222	10.59	14.24	106.0	251.6	375.6	1.514	3.594	5.366
75.00	6.007	11.74	15.53	134.1	307.4	450.0	1.788	4.099	6.001
80.00	6.806	12.87	16.78	166.1	368.9	530.8	2.076	4.612	6.635
85.00	7.612	13.97	17.98	202.1	436.1	617.7	2.378	5.130	7.267
90.00	8.421	15.05	19.15	242.2	508.6	710.6	2.691	5.651	7.895
95.00	9.229	16.10	20.28	286.4	586.5	809.1	3.014	6.174	8.517
100.00	10.03	17.12	21.37	334.5	669.6	913.3	3.345	6.696	9.133
105.00	10.83	18.11	22.42	386.7	757.7	1023.	3.683	7.216	9.740
110.00	11.62	19.07	23.44	442.8	850.6	1137.	4.025	7.733	10.34
115.00	12.39	20.01	24.43	502.8	948.3	1257.	4.372	8.246	10.93
120.00	13.16	20.92	25.38	566.7	1051.	1382.	4.723	8.755	11.51
125.00	13.91	21.80	26.31	634.4	1157.	1511.	5.075	9.260	12.09
130.00	14.66	22.66	27.20	705.8	1269.	1645.	5.429	9.759	12.65
135.00	15.39	23.50	28.07	780.9	1384.	1783.	5.785	10.25	13.21
140.00	16.10	24.32	28.92	859.7	1504.	1925.	6.140	10.74	13.75
145.00	16.80	25.11	29.74	941.9	1627.	2072.	6.496	11.22	14.29
150.00	17.49	25.88	30.54	1028.	1755.	2223.	6.851	11.70	14.82
155.00	18.17	26.64	31.31	1117.	1886.	2377.	7.206	12.17	15.34
160.00	18.84	27.37	32.07	1209.	2021.	2536.	7.559	12.63	15.85
165.00	19.49	28.09	32.80	1305.	2160.	2698.	7.910	13.09	16.35
170.00	20.13	28.79	33.52	1404.	2302.	2864.	8.260	13.54	16.85
175.00	20.75	29.47	34.21	1506.	2448.	3033.	8.608	13.99	17.33
180.00	21.37	30.14	34.89	1612.	2597.	3206.	8.954	14.43	17.81
185.00	21.97	30.79	35.55	1720.	2749.	3382.	9.298	14.86	18.28
190.00	22.57	31.43	36.20	1831.	2904.	3561.	9.639	15.29	18.74
195.00	23.15	32.05	36.83	1946.	3063.	3744.	9.978	15.71	19.20
200.00	23.72	32.66	37.45	2063.	3225.	3930.	10.31	16.12	19.65
205.00	24.28	33.26	38.05	2183.	3390.	4118.	10.65	16.54	20.09
210.00	24.83	33.85	38.64	2306.	3558.	4310.	10.98	16.94	20.52
215.00	25.37	34.42	39.22	2431.	3728.	4505.	11.31	17.34	20.95
220.00	25.90	34.98	39.79	2559.	3902.	4702.	11.63	17.74	21.37
225.00	26.42	35.53	40.34	2690.	4078.	4903.	11.96	18.12	21.79
230.00	26.94	36.07	40.89	2824.	4257.	5106.	12.28	18.51	22.20
235.00	27.44	36.60	41.42	2960.	4439.	5312.	12.59	18.89	22.60
240.00	27.94	37.12	41.94	3098.	4623.	5520.	12.91	19.26	23.00
245.00	28.42	37.63	42.46	3239.	4810.	5731.	13.22	19.63	23.39
250.00	28.90	38.14	42.96	3382.	4999.	5945.	13.53	20.00	23.78
255.00	29.37	38.63	43.46	3528.	5191.	6161.	13.83	20.36	24.16
260.00	29.84	39.11	43.94	3676.	5386.	6379.	14.14	20.71	24.53
265.00	30.30	39.59	44.42	3826.	5582.	6600.	14.44	21.07	24.91
270.00	30.75	40.06	44.89	3979.	5782.	6823.	14.74	21.41	25.27
273.15	31.02	40.35	45.18	4076.	5908.	6965.	14.92	21.63	25.50
275.00	31.19	40.52	45.35	4134.	5983.	7049.	15.03	21.76	25.63
280.00	31.62	40.97	45.81	4291.	6187.	7277.	15.32	22.10	25.99
285.00	32.05	41.42	46.25	4450.	6393.	7507.	15.61	22.43	26.34
290.00	32.48	41.86	46.69	4611.	6601.	7739.	15.90	22.76	26.69
295.00	32.89	42.29	47.13	4775.	6811.	7974.	16.19	23.09	27.03
298.15	33.15	42.56	47.40	4879.	6945.	8123.	16.36	23.29	27.24
300.00	33.30	42.72	47.56	4940.	7024.	8211.	16.47	23.41	27.37

VALUES OF CHEMICAL THERMODYNAMIC PROPERTIES OF HYDROCARBONS

The values in this table are for the ideal gas state at 298.15 K. The units for $\Delta Hf°$, $\Delta Ff°$, and $\text{Log}_{10} Kf$ are Kcal/g mol. The units for absolute entropy, $S°$, are cal/$°K$ g mol.

It is frequently possible to calculate values for compounds not listed since the following increments are known for an addition of a methylene group, CH_2, to the following types of compounds:

> Normal alkyl cyclohexanes
> Normal alkyl benzenes
> Normal alkyl cyclopentanes
> Normal monoolefins (1-alkenes)
> Normal acetylenes (1-alkynes)

For each of the above types of compounds the increments per CH_2 group are

$\Delta Hf°$:	−4.926 kcal/g mol
$\Delta Ff°$:	−2.048 kcal/g mol
$\text{Log}_{10} Kf$:	−1.5012 kcal/g mol
$S°$:	−9.183 cal/deg g mol

Relationships to SI units — The symbols cal mole^{-1} deg^{-1} and gibbs/mol are identical and refer to units of calories per degree-mole. These units can be converted to SI units of joules per degree-mole by multiplying the tabulated values by 4.184. Similarly, values in kilocalories per mole can be converted to joules per mole by multiplying with the factor 4184. For further discussions of the SI system and for conversions from other units, see *Pure and Applied Chemistry*, 21, 1, 1970.

Formula	Compound	$\Delta Hf°$	$\Delta Ff°$	$\text{Log}_{10} Kf$	$S°$
CH_4	Methane	−17.889	−12.140	8.8985	44.50
C_2H_2	Ethyne (acetylene)	54.194	50.000	−36.6490	47.997
C_2H_4	Ethene (ethylene)	12.496	16.282	−11.9345	52.54
C_2H_6	Ethane	−20.236	− 7.860	5.7613	54.85
C_3H_4	Propadiene (allene)	45.92	48.37	−35.4519	58.30
C_3H_4	Propyne (methyl-acetylene)	44.319	46.313	−33.9469	59.30
C_3H_6	Propene (propylene)	4.879	14.990	−10.9875	63.80
C_3H_8	Propane	−24.820	− 5.614	4.1150	64.51
C_4H_6	1,2-Butadiene	39.55	48.21	−35.3377	70.03
C_4H_6	1,3-Butadiene	26.75	36.43	−26.7004	66.62
C_4H_6	1-Butyne (ethyl acetylene)	39.70	48.52	−35.5616	69.51
C_4H_6	2-Butyne (dimethylacetylene)	35.374	44.725	−32.7823	67.71
C_4H_8	1-Butene	0.280	17.217	−12.6199	73.48
C_4H_8	cis-2-Butene	−1.362	16.046	−11.7618	71.90
C_4H_8	trans-2-Butene	−2.405	15.315	−11.2255	70.86
C_4H_8	2-Methylpropane (isobutene)	− 3.343	14.582	−10.6888	70.17
C_4H_{10}	n-Butane	−29.812	− 3.754	2.7516	74.10
C_4H_{10}	2-Methylpropane (isobutane)	−31.452	− 4.296	3.1489	70.42
C_5H_8	1-Pentyne	34.50	50.17	−36.7712	79.10
C_5H_8	2-Pentyne	30.80	46.41	−34.0177	79.30
C_5H_8	3-Methyl-1-butyne	32.60	49.12	−36.0061	76.23
C_5H_8	1,2-Pentadiene	34.80	50.29	−36.861	79.7
C_5H_8	cis-1,3-Pentadiene (cis-piperylene)	18.70	34.88	−25.563	77.5
C_5H_8	trans-1,3-Pentadiene (trans-piperylene)	18.60	35.07	−25.707	76.4
C_5H_8	1,4-Pentadiene	25.20	40.69	−29.824	79.7
C_5H_8	2,3-Pentadiene	33.10	49.22	−36.074	77.6
C_5H_8	3-Methyl-1,2-butadiene	31.00	47.47	−34.657	76.4
C_5H_8	2-Methyl-1,3-butadiene (isoprene)	18.10	34.87	−25.560	75.44
C_5H_{10}	1-Pentene	− 5.000	18.787	−13.7704	83.08
C_5H_{10}	cis-2-Pentene	− 6.710	17.173	−12.5874	82.76

VALUES OF CHEMICAL THERMODYNAMIC PROPERTIES OF HYDROCARBONS (Continued)

Formula	Compound	$\Delta Hf°$	$\Delta Ff°$	$Log_{10}Kf$	$S°$
C_5H_{10}	trans-2-Pentene	− 7.590	16.575	−12.1495	81.81
C_5H_{10}	2-Methyl-1-butene	− 8.680	15.509	−11.3680	81.73
C_5H_{10}	3-Methyl-1-butene	− 6.920	17.874	−13.1017	79.70
C_5H_{10}	2-Methyl-2-butene	−10.170	14.267	−10.4572	80.90
C_5H_{10}	Cyclopentane	−18.46	9.23	− 6.7643	70.00
C_5H_{12}	n-Pentane	−35.00	− 1.96	1.4366	83.27
C_6H_6	Benzene	19.820	30.989	−22.7143	64.34
C_6H_{10}	1-Hexyne	29.55	52.19	−38.258	88.27
C_6H_{12}	1-Hexene	− 9.96	20.80	−15.2491	92.25
C_6H_{12}	cis-2-Hexene	−11.56	19.18	−14.0549	92.35
C_6H_{12}	trans-2-Hexene	−12.56	18.46	−13.5291	91.40
C_6H_{12}	cis-3-Hexene	−11.56	19.66	−14.4094	90.73
C_6H_{12}	trans-3-Hexene	−12.56	18.86	−13.8262	90.04
C_6H_{12}	2-Methyl-1-pentene	−13.56	17.48	−12.8135	91.32
C_6H_{12}	3-Methyl-1-pentene	−11.02	20.28	−14.8655	90.45
C_6H_{12}	4-Methyl-1-pentene	−11.66	19.90	−14.5865	89.58
C_6H_{12}	2-Methyl-2-pentene	−14.96	16.34	−11.9780	90.45
C_6H_{12}	cis-3-Methyl-2-pentene	−14.32	16.98	−12.4471	90.45
C_6H_{12}	trans-3-Methyl-2-pentene	−14.32	16.74	−12.2697	91.26
C_6H_{12}	cis-4-Methyl-2-pentene	−13.26	18.40	−13.4903	89.23
C_6H_{12}	trans-4-Methyl-2-pentene	−14.26	17.77	−13.0216	88.02
C_6H_{12}	2-Ethyl-1-butene	−12.92	18.51	−13.5690	90.01
C_6H_{12}	2,3-Dimethyl-1-butene	−14.70	17.43	−12.7702	89.39
C_6H_{12}	3,3-Dimethyl-1-butene	−14.25	19.04	−13.9578	83.79
C_6H_{12}	2,3-Dimethyl-2-butene	−15.91	16.52	−12.1073	86.67
C_6H_{12}	Methylcyclopentane	−25.50	8.55	− 6.2649	81.24
C_6H_{12}	Cyclohexane	−29.43	7.59	− 5.5605	71.28
C_6H_{14}	n-Hexane	−39.96	0.05	0.037	92.45
C_7H_8	Methylbenzene (toluene)	11.950	29.228	−21.4236	76.42
C_7H_{12}	1-Heptyne	24.62	54.24	−39.759	97.25
C_7H_{12}	1-Heptene	−14.85	22.84	−16.742	101.43
C_7H_{14}	Ethylcyclopentane	−30.37	10.59	− 7.7632	90.62
C_7H_{14}	1,1-Dimethylcyclopentane	−33.05	9.33	− 6.8372	85.87
C_7H_{14}	1,cis-2-Dimethylcyclopentane	−30.96	10.93	− 8.0107	87.51
C_7H_{14}	1,trans-2-Dimethylcyclopentane	−32.67	9.17	− 6.7224	87.67
C_7H_{14}	1,cis-3-Dimethylcyclopentane	−31.93	9.91	− 7.2648	87.67
C_7H_{14}	1,trans-3-Dimethylcyclopentane	−32.47	9.37	− 6.8690	87.67
C_7H_{14}	Methylcyclohexane	−36.99	6.52	− 4.7819	82.06
C_7H_{16}	n-Heptane	−44.89	2.09	− 1.532	101.64
C_8H_8	Ethenylbenzene (styrene)	35.32	51.10	−37.4532	82.48
C_8H_{10}	Ethylbenzene	7.120	31.208	−22.8750	86.15
C_8H_{10}	1,2-Dimethylbenzene (o-xylene)	4.540	29.177	−21.3860	84.31
C_8H_{10}	1,3-Dimethylbenzene (m-xylene)	4.120	28.405	−20.8202	85.49
C_8H_{10}	1,4-Dimethylbenzene (p-xylene)	4.290	28.952	−21.2214	84.23
C_8H_{14}	1-Octyne	19.70	56.29	−41.260	106.63
C_8H_{16}	1-Octene	−19.82	24.89	−18.244	110.61
C_8H_{16}	n-Propylcyclopentane	−35.39	12.54	− 9.195	99.80
C_8H_{16}	1,1-Dimethylcyclohexane	−43.26	8.42	− 6.174	87.24
C_8H_{16}	cis-1,2-Dimethylcyclohexane	−41.15	9.85	− 7.225	89.51
C_8H_{16}	trans-1,2-Dimethylcyclohexane	−43.02	8.24	− 6.038	88.65

VALUES OF CHEMICAL THERMODYNAMIC PROPERTIES OF HYDROCARBONS (Continued)

Formula	Compound	$\Delta Hf°$	$\Delta Ff°$	$Log_{10}Kf$	$S°$
C_8H_{16}	cis-1,3-Dimethylcyclohexane	−44.16	7.13	− 5.228	88.54
C_8H_{16}	trans-1,3-Dimethylcyclohexane	−42.20	8.68	− 6.363	89.92
C_8H_{16}	cis-1,4-Dimethylcyclohexane	−42.22	9.07	− 6.650	88.54
C_8H_{16}	trans-1,4-Dimethylcyclohexane	−44.12	7.58	− 5.552	87.19
C_8H_{18}	n-Octane	−49.82	4.14	− 3.035	110.82
C_8H_{18}	2-Methylheptane	−51.50	3.06	− 2.243	108.81
C_8H_{18}	3-Methylheptane	−50.82	3.29	− 2.412	110.32
C_8H_{18}	4-Methylheptane	−50.69	4.00	− 2.932	108.35
C_8H_{18}	3-Ethylhexane	−50.40	3.95	− 2.895	109.51
C_8H_{18}	2,2-Dimethylhexane	−53.71	2.56	− 1.876	103.06
C_8H_{18}	2,3-Dimethylhexane	−51.13	4.23	− 3.101	106.11
C_8H_{18}	2,4-Dimethylhexane	−52.44	2.80	− 2.052	106.51
C_8H_{18}	2,5-Dimethylhexane	−53.21	2.50	− 1.832	104.93
C_8H_{18}	3,3-Dimethylhexane	−52.61	3.17	− 2.324	104.70
C_8H_{18}	3,4-Dimethylhexane	−50.91	4.14	− 3.035	107.15
C_8H_{18}	2-Methyl-3-ethylpentane	−50.48	5.08	− 3.724	105.43
C_8H_{18}	3-Methyl-3-ethylpentane	−51.38	4.76	− 3.489	103.48
C_8H_{18}	2,2,3-Trimethylpentane	−52.61	4.09	− 2.998	101.62
C_8H_{18}	2,2,4-Trimethylpentane	−53.57	3.13	− 2.294	101.62
C_8H_{18}	2,3,3-Trimethylpentane	−51.73	4.52	− 3.313	103.14
C_8H_{18}	2,3,4-Trimethylpentane	−51.97	4.32	− 3.167	102.99
C_8H_{18}	2,2,3,3-Tetramethylbutane	−53.99	4.88	− 3.577	94.34
C_9H_{10}	Isopropenylbenzene (α-methylstyrene; 2-phenyl-1-propene)	27.00	49.84	−36.531	91.70
C_9H_{10}	1-Methyl-2-ethenylbenzene (o-Methylstyrene)	28.30	51.14	−37.484	91.70
C_9H_{10}	1-Methyl-3-ethenylbenzene (m-methylstyrene)	27.60	50.02	−36.665	93.1
C_9H_{10}	1-Methyl-4-ethenylbenzene (p-methylstyrene)	27.40	50.24	−36.825	91.7
C_9H_{12}	n-Propylbenzene	1.870	32.810	−24.049	95.74
C_9H_{12}	Isopropylbenzene (Cumene)	0.940	32.738	−23.996	92.87
C_9H_{12}	1,3,5-Trimethylbenzene (Mesitylene)	− 3.840	28.172	−20.6497	92.15
C_9H_{16}	1-Nonyne	14.77	58.34	−42.761	115.82
C_9H_{18}	1-Nonene	−24.74	26.94	−19.747	119.80
C_9H_{18}	n-Butylcyclopentane	−40.22	14.69	−10.768	108.99
C_9H_{20}	n-Nonane	−54.74	6.18	− 4.536	120.00
$C_{10}H_{14}$	n-Butylbenzene	− 3.30	34.62	−25.374	104.91
$C_{10}H_{18}$	1-Decyne	9.85	60.39	−44.262	125.00
$C_{10}H_{10}$	1-Decene	−29.67	28.99	−21.249	128.98
$C_{10}H_{22}$	n-Decane	−59.67	8.23	6.037	129.19
$C_{11}H_{22}$	1-Undecene	−34.60	31.03	−22.745	138.16
$C_{11}H_{24}$	n-Undecane	−64.60	10.28	− 7.539	138.37
$C_{12}H_{24}$	1-Dodecene	−39.52	33.08	−24.297	147.34

From Rossini, F. D., Pitzer, K. S., Arnett, R. L., Braun, R. M., and Pimentel, G. C., *Selected Values of Physical and Thermodynamic Properties of Hydrocarbons and Related Compounds,* Carnegie Press, Pittsburgh, 1953.

KEY VALUES FOR THERMODYNAMICS

The following table was prepared from data published in CODATA Bulletin 28 (April, 1978) which is a report of the CODATA Task Group on Key Values of Thermodynamics. A bibliography, references citing methods for calculations and measurements, bases for selection of data and limitations of the data are presented in Bulletin 28. One may contact CODATA by writing to CODATA Secretariat, 51 Boulevard de Montmorency, 75016 Paris, France.

The recommended values were derived from a reconsideration of all the tentative data published in CODATA Special Report 4, March 1977 ("Tentative set of key values for thermodynamics. Part VI"), which took place at a meeting of the Task Group held in Lund, Sweden, in August 1977. It was decided that values of $\Delta_f H^\circ$ (298.15 K) for NO^-_3, (aq) and all values for Cd(g), Ge(g) and U(g) should remain tentative for the present, that P_4O_{10}(cr) and PbO(cr, yellow) should be dropped from the program, and that the remaining values given in CODATA Special Report 4 should advance to recommended status, albeit with some minor numerical adjustments.

In the following tables, the species are listed in the "Standard Order of Arrangement", as used in most modern compendia of thermodynamic data. The reference state for each element at 298.15 K is the thermodynamically stable standard state except for phosphorus, for which the "white" crystal modification has been selected, as it is the most reproducible; there is a subtlety concerning the reference state for tin, where the "white" form (thermodynamcally stable at 298.15 K) is taken as the reference state at all temperatures down to zero, even though that form is known to be metastable below 286 K.

The usual definitions of standard states have been adopted. For crystalline solids (cr) and liquids (l) the standard state is that of the pure substance (in a stated crystallographic modification, where appropriate) under a pressure of 101 325 Pa. For gases (g), the standard state is that of the ideal gas at a pressure of 101 325 Pa. For species in aqueous solution (aq), the standard state is the hypothetical ideal solution at unit activity (molality scale); the properties of ideal aqueous ionic solutions are taken equal to the sum of the properties of the individual ions. It should be noted that values of H° (298.15 K) − H° (0) for gases relate to the hypothetical ideal-gas state at zero temperature, while values of H° (298.15 K) − H° (0) for both liquids and crystalline solids relate to crystalline solids at zero temperature.

Relative atomic masses were taken from the recommendations of the IUPAC Commission on Atomic Weights, 1970. Data in the tables relate to the natural mixture of isotopic species; nuclear spin contributions have been ignored. The joule table is the primary table; values in the calorie table were derived from the corresponding values in joules by dividing by 4.184, with retention of sufficient decimal places to assure accurate reconversion.

For consistency with earlier work, the following values of the fundamental constants were employed in the calculations: gas constant, R = (8.314 33 ± 0.000 80) $J \cdot K^{-1} \cdot mol^{-1}$; Faraday constant, F = (96 487.0 ± 1.0) $J \cdot V^{-1} \cdot mol^{-1}$; constant relating wave number and energy, $N_A hc$ = (0.119 625 6 ± 0.000 002 6) $J \cdot m \cdot mol^{-1}$. These values differ very slightly from those recommended by CODATA in 1973 (CODATA Bulletin 11, December 1973). However, adoption of the 1973 values of the fundamental constants would change the values in the present tables by far less than their assigned uncertainties, so the thermodynamic data where reported may be said to be consistent with the 1973 set of fundamental constants.

CODATA RECOMMENDED KEY VALUES FOR THERMODYNAMICS, 1977
(To convert joules to calories multiply by 0.239006)

Substance	State	$\Delta_f H^\circ$ (298.15 K) kJ·mol⁻¹	S° (298.15 K) J·K⁻¹·mol⁻¹	H° (298.15 K) − H° (0) kJ·mol⁻¹
O	g	249.17 ± 0.10	160.946 ± 0.020	6.728 ± 0.003
O_2	g	0	205.037 ± 0.033	8.682 ± 0.004
H	g	217.997 ± 0.006	114.604 ± 0.015	6.197 ± 0.002
H^+	aq	0	0	—
H_2	g	0	130.570 ± 0.033	8.468 ± 0.003
OH^-	aq	−230.025 ± 0.045	−10.71 ± 0.20	—
H_2O	l	−285.830 ± 0.042	69.950 ± 0.080	13.293 ± 0.020
H_2O	g	−241.814 ± 0.042	188.724 ± 0.040	9.908 ± 0.008
He	g	0	126.039 ± 0.012	6.197 ± 0.002
Ne	g	0	146.214 ± 0.016	6.197 ± 0.002
Ar	g	0	154.732 ± 0.020	6.197 ± 0.002
Kr	g	0	163.971 ± 0.020	6.197 ± 0.002
Xe	g	0	169.573 ± 0.020	6.197 ± 0.002
F	g	79.39 ± 0.30	158.640 ± 0.020	6.518 ± 0.004
F^-	aq	−335.35 ± 0.65	−13.18 ± 0.54	—
F_2	g	0	202.685 ± 0.040	8.825 ± 0.004
HF	g	−273.30 ± 0.70	173.655 ± 0.035	8.599 ± 0.004
Cl	g	121.302 ± 0.008	165.076 ± 0.020	6.272 ± 0.003
Cl^-	aq	−167.080 ± 0.088	56.73 ± 0.16	—
Cl_2	g	0	222.965 ± 0.040	9.180 ± 0.008
HCl	g	−92.31 ± 0.13	186.786 ± 0.033	8.640 ± 0.004
Br	g	111.86 ± 0.12	174.904 ± 0.020	6.197 ± 0.020
Br^-	aq	−121.50 ± 0.15	82.84 ± 0.20	—
Br_2	l	0	152.210 ± 0.040	24.52 ± 0.13
Br_2	g	30.91 ± 0.11	245.350 ± 0.054	9.724 ± 0.012
HBr	g	−36.38 ± 0.17	198.585 ± 0.033	8.648 ± 0.004
I	g	106.762 ± 0.040	180.673 ± 0.020	6.197 ± 0.002
I^-	aq	−56.90 ± 0.84	106.70 ± 0.20	—
I_2	cr	0	116.139 ± 0.080	13.196 ± 0.040
I_2	g	62.421 ± 0.080	260.567 ± 0.063	10.117 ± 0.021
HI	g	26.36 ± 0.80	206.480 ± 0.040	8.657 ± 0.006
S	cr, rhombic	0	32.054 ± 0.050	4.412 ± 0.060
S	g	276.98 ± 0.25	167.715 ± 0.035	6.657 ± 0.004
S_2	g	128.49 ± 0.30	228.055 ± 0.050	9.131 ± 0.008
SO_2	g	−296.81 ± 0	248.11 ± 0.06	10.548 ± 0.013
SO_4^{-2}	aq	−909.60 ± 0.40	18.83 ± 0.50	—
N	g	472.68 ± 0.40	153.189 ± 0.020	6.197 ± 0.002
N_2	g	0	191.502 ± 0.025	8.669 ± 0.003
NO_3^-	aq	—	146.94 ± 0.85	—
NH_3	g	−45.94 ± 0.35	192.67 ± 0.08	10.046 ± 0.008

CODATA RECOMMENDED KEY VALUES FOR THERMODYNAMICS, 1977
(To convert joules to calories multiply by 0.239006) (continued)

Substance	State	$\Delta_f H°$ (298.15 K) kJ·mol⁻¹	S° (298.15 K) J·K⁻¹·mol⁻¹	H° (298.15 K) − H° (0) kJ·mol⁻¹
NH₄⁺	aq	−133.26 ± 0.25	111.17 ± 0.75	—
P	cr, white	0	41.09 ± 0.25	5.360 ± 0.015
P	g	316.5 ± 1.0	163.085 ± 0.020	6.197 ± 0.002
P₂	g	144.0 ± 2.0	218.01 ± 0.04	8.903 ± 0.0008
P₄	g	58.9 ± 0.3	279.9 ± 0.5	14.10 ± 0.24
C	cr	0	5.74 ± 0.12	1.050 ± 0.020
C	g	716.67 ± 0.44	157.988 ± 0.020	6.535 ± 0.006
CO	g	−110.53 ± 0.17	197.556 ± 0.032	8.673 ± 0.008
CO₂	g	−393.51 ± 0.13	213.677 ± 0.040	9.364 ± 0.008
Si	cr	0	18.81 ± 0.08	3.217 ± 0.008
Si	g	450 ± 8	167.870 ± 0.035	7.550 ± 0.004
SiO₂	cr, α-quartz	−910.7 ± 1.0	41.46 ± 0.20	6.916 ± 0.020
SiF₄	g	−1614.95 ± 0.85	282.65 ± 0.40	15.36 ± 0.05
Ge	Cr, cubic	0	31.09 ± 0.13	4.636 ± 0.015
GeO₂	cr, tetrag	−580.2 ± 1.2	39.71 ± 0.15	7.230 ± 0.020
GeF₄	g	−1190.15 ± 0.50	301.8 ± 1.0	17.30 ± 0.080
Sn	cr, white	0	51.18 ± 0.08	6.323 ± 0.008
Sn	g	301.2 ± 1.7	168.380 ± 0.020	6.215 ± 0.002
Sn⁺²	aq	−8.9 ± 0.8	−15.8 ± 4.0	—
SnO	cr	−285.93 ± 0.70	57.17 ± 0.30	8.736 ± 0.022
SnO₂	cr	−580.78 ± 0.40	52.3 ± 1.2	8.76 ± 0.08
Pb	cr	0	64.80 ± 0.30	6.870 ± 0.020
Pb	g	195.20 ± 0.80	175.270 ± 0.020	6.197 ± 0.002
Pb⁺²	aq	0.92 ± 0.25	17.7 ± 0.8	—
PbSO₄	cr	−919.94 ± 0.90	148.49 ± 0.40	20.050 ± 0.040
B	cr	0	5.90 ± 0.08	1.222 ± 0.008
B	g	560 ± 12	153.325 ± 0.035	6.315 ± 0.004
B₂O₃	cr	−1273.5 ± 1.4	53.97 ± 0.30	9.301 ± 0.040
BF₃	g	−1135.95 ± 0.80	254.31 ± 0.10	11.650 ± 0.020
Al	cr	0	28.35 ± 0.08	4.565 ± 0.010
Al	g	329.7 ± 4.0	164.440 ± 0.030	6.919 ± 0.004
Al₂O₃	cr, α-corundum	−1675.7 ± 1.3	50.92 ± 0.10	10.016 ± 0.020
AlF₃	cr	−1510.4 ± 1.3	66.5 ± 0.4	11.62 ± 0.04
Zn	cr	0	41.63 ± 0.13	5.657 ± 0.020
Zn	g	130.42 ± 0.20	160.875 ± 0.025	6.197 ± 0.002
Zn⁺²	aq	−153.39 ± 0.20	−109.6 ± 0.7	—
ZnO	cr	−350.46 ± 0.27	43.64 ± 0.40	6.933 ± 0.040
Cd	cr	0	51.80 ± 0.15	6.247 ± 0.004
Cd⁺²	aq	−75.88 ± 0.60	−72.8 ± 1.2	—
CdO	cr	−258.1 ± 0.8	54.8 ± 1.7	8.41 ± 0.08
CdSO₄·8/3H₂O	cr	−1729.55 ± 0.80	229.66 ± 0.40	35.56 ± 0.04
Hg	l	0	75.90 ± 0.12	9.342 ± 0.008
Hg	g	61.38 ± 0.04	174.860 ± 0.020	6.197 ± 0.002
Hg⁺²	aq	170.16 ± 0.20	−36.32 ± 0.80	—
Hg₂⁺²	aq	166.82 ± 0.20	65.52 ± 0.80	—
HgO	cr, red	−90.83 ± 0.12	70.25 ± 0.30	9.117 ± 0.025
Hg₂Cl₂	cr	−265.45 ± 0.30	191.6 ± 1.5	23.25 ± 0.20
Hg₂SO₄	cr	−743.41 ± 0.50	200.71 ± 0.20	26.070 ± 0.030
Cu	cr	0	33.15 ± 0.08	5.004 ± 0.008
Cu	g	337.6 ± 1.2	166.285 ± 0.025	6.197 ± 0.002
Cu⁺²	aq	65.69 ± 0.80	−97.1 ± 1.2	—
CuSO₄	cr	−771.1 ± 1.2	109.2 ± 0.4	16.86 ± 0.08
Ag	cr	0	42.55 ± 0.21	5.745 ± 0.020
Ag	g	284.9 ± 0.8	172.883 ± 0.025	6.197 ± 0.002
Ag⁺	aq	105.750 ± 0.085	73.38 ± 0.40	—
AgCl	cr	−127.070 ± 0.085	96.23 ± 0.20	12.033 ± 0.040
Th	cr	0	53.39 ± 0.40	6.510 ± 0.020
Th	g	598 ± 6	190.06 ± 0.40	6.197 ± 0.002
ThO₂	cr	−1226.4 ± 3.5	65.23 ± 0.20	10.560 ± 0.020
U	cr	0	50.20 ± 0.20	6.364 ± 0.020
UO₂	cr	−1085.0 ± 1.0	77.03 ± 0.20	11.280 ± 0.020
UO₂⁺²	aq	−1019.2 ± 2.5	−98.3 ± 4.0	—
UO₃	cr, gamma	−1223.8 ± 2.0	96.11 ± 0.40	14.585 ± 0.080
U₃O₈	cr	−3574.8 ± 2.5	282.55 ± 0.50	42.74 ± 0.10
Be	cr	0	9.50 ± 0.08	1.950 ± 0.020
Be	g	324 ± 5	136.165 ± 0.020	6.197 ± 0.002
BeO	cr	−609.4 ± 2.5	13.77 ± 0.04	2.837 ± 0.008
Mg	cr	0	32.68 ± 0.10	5.000 ± 0.030
Mg	g	147.10 ± 0.80	148.535 ± 0.020	6.197 ± 0.002
MgO	cr	−601.5 ± 0.3	26.95 ± 0.15	5.160 ± 0.020
MgF₂	cr	−1124.2 ± 1.2	57.2 ± 0.4	9.92 ± 0.06

Substance	State	$\Delta_f H°$ (298.15 K) kJ·mol^{-1}	S° (298.15 K) J·K^{-1}·mol^{-1}	H° (298.15 K) − H° (0) kJ·mol^{-1}
Ca	cr	0	41.6 ± 0.4	5.73 ± 0.04
Ca	g	177.8 ± 0.8	154.775 ± 0.020	6.197 ± 0.002
Ca^{+2}	aq	−543.10 ± 0.80	−56.4 ± 0.4	—
CaO	cr	−635.09 ± 0.90	38.1 ± 0.4	6.75 ± 0.06
Li	cr	0	29.12 ± 0.20	4.632 ± 0.040
Li$^+$	aq	−278.455 ± 0.090	11.30 ± 0.35	—
Na	cr	0	51.30 ± 0.20	6.460 ± 0.020
Na$^+$	aq	−240.300 ± 0.065	58.41 ± 0.20	—
K	cr	0	64.68 ± 0.20	7.088 ± 0.020
K$^+$	aq	−252.17 ± 0.10	101.04 ± 0.25	—
Rb	cr	0	76.78 ± 0.30	7.489 ± 0.020
Rb$^+$	aq	−251.12 ± 0.13	120.46 ± 0.40	—
Cs	cr	0	85.23 ± 0.40	7.711 ± 0.020
Cs$^+$	aq	−258.04 ± 0.13	132.84 ± 0.40	—

LATTICE ENERGIES

H. D. B. Jenkins

Table 1 contains calculated values of lattice energies, U_{Pot}, of crystalline salts M_aX_b. U_{Pot} is expressed in the units of kilojoules per mole, KJmol^{-1}. M and X can be complex or simple ions. Also cited is the lattice energy obtained from the Born-Fajans-Haber Cycle, U^{BFHC}_{Pot} using thermochemical data published in U.S. Government publications plus certain other data which are located at the end of this table. The values quoted are of variable reliability and a full discussion of the values is to appear in a review by Jenkins and Waddington currently (1978) nearing completion.

$$M_aX_{b(c)} \xrightarrow{\Delta H_L} aM^{b+}{}_{(g)} + bX^{a-}{}_{(g)}$$

$$\Delta H_f^\theta (M_aX_g)_{(c)} \qquad a\Delta H_f^\theta (M^{b+})_{(g)} + b\Delta H_f^\theta (X^{a-})_{(g)}$$

$$aM_{(ss)} + bX_{(ss)}$$

where, (ss) is the standard state of the ion or element

$$\Delta H_L = U_{Pot}(M_aX_b) + \left[a\left(\frac{n_{M^{b+}}}{2} - 2\right) + b\left(\frac{n_{X^{a-}}}{2} - 2\right) \right] RT$$

$$\Delta H_L = a\Delta H_f^\theta (M^{b+})(g) + b\Delta H_f^\theta (X^{a-})(g) - \Delta H_f^\theta (M_aX_b)(c)$$

Where n_{M^b+}, n_{X^a-} is equal to 3 for monatomic ions, 5 for linear polyatomic ions, and 6 for polyatomic nonlinear ions.

The data listed in Table 2 were employed in the calculation of the Born-Haber Cycle values in Table 1 and are not listed in Technical Notes 270 of the National Bureau of Standards.

Table 1
LATTICE ENERGIES

Salt	Calculated lattice energy (kJmol⁻¹)	Literature source	Thermochemical cycle lattice energy (kJmol⁻¹)
Acetates			
Li(CH₃COO)	—	—	881
Na(CH₃COO)	761	Morris (1959)	763
K(CH₃COO)	686	Morris (1959)	682
Rb(CH₃COO)	715	Yatsimirskii (1956)	656
Cs(CH₃COO)	682	Yatsimirskii (1956)	—
NH₄(CH₃COO)	725	Yatsimirskii (1956)	695
Ag(CH₃COO)	—	—	863
Tl(CH₃COO)	—	—	750
Ca(CH₃COO)₂	2431	Yatsimirskii (1956)	2294
Sr(CH₃COO)₂	2280	Yatsimirskii (1956)	2166
Ba(CH₃COO)₂	2180	Yatsimirskii (1956)	2033
Mn(CH₃COO)₂	2548	Yatsimirskii (1956)	2616
Cu(CH₃COO)₂	—	—	2835
Zn(CH₃COO)₂	2615	Yatsimirskii (1956)	2750
Hg(CH₃COO)₂	2368	Yatsimirskii (1961)	2595
Pb(CH₃COO)₂	2247	Yatsimirskii (1961)	2225
Acetylides			
CaC₂	2911	Vinek, Neckel, Nowotny (1967)	2904
SrC₂	2788	Vinek, Neckel, Nowotny (1967)	2784
BaC₂	2647	Vinek, Neckel, Nowotny (1967)	2654
Ammonium salts			
NH₄(CH₃COO)	725	Yatsimirskii (1956)	656
NH₄HF₂	705	Jenkins, Pratt (1977)	705
NH₄HCO₃	—	—	741
NH₄HSO₄	640	Yatsimirskii (1956)	(645)
NH₄BF₄	582	Ladd, Lee (1961)	—
NH₄NCO	724	Waddington (1959)	—
NH₄CN	617	Ladd (1970)	670
NH₄HCO₂	715	Morris (1959)	—
NH₄	650	Bernard, Businot, Decker (1974)	—
(NH₄CH₃CO₂)	—	—	—
NH₄IO	—	—	808
NH₄IO₂F₂	678	Finch, Gates, Jenkinson (1972)	685
NH₄HS	661	Yatsimirskii (1956)	666
NH₄NO₂	661	Morris (1958)	676
NH₄ClO₂	583	Jenkins, Pratt (1978)	580
(NH₄)₂S	2008	Karapet'yants (1954)	(2026)
(NH₄)₂SO₄	1766	Jenkins, Smith (1975)	1777
(NH₄)₂GeF₆	1657	Jenkins, Pratt (1978)	—
(NH₄)₂IrCl₆	1442	Jenkins, Pratt (1978)	1440
(NH₄)₂OsCl₆	1433	Jenkins, Pratt (1978)	—
(NH₄)₂PdCl₆	1481	Jenkins, Pratt (1978)	—
(NH₄)₂PbCl₆	1355	Jenkins, Pratt (1978)	—
(NH₄)₂PbCl₆	1468	Jenkins, Pratt (1978)	1390
(NH₄)₂ReCl₆	1402	Jenkins, Pratt (1978)	1727
(NH₄)₂SiF₆	1657	Jenkins, Pratt (1978)	1334
(NH₄)₂SnCl₆	1370	Jenkins, Pratt (1978)	—
(NH₄)₂SnBr₆	1319	Jenkins, Pratt (1978)	—
(NH₄)₂TeCl₆	1318	Jenkins, Pratt (1978)	—
(NH₄)₂TeBr₆	1294	Jenkins, Pratt (1978)	—
(NH₄)₂TiCl₆	1413	Jenkins, Pratt (1978)	—
NH₄CNS	605	Gill, Singla, Paul, Narula (1972)	611
(NH₄)₂SeCl₆	1420	Jenkins, Pratt (1978)	—
(NH₄)₂SeBr₆	1380	Jenkins, Pratt (1978)	—
Arsenates			
Mg₃(AsO₄)₂	10669	Grekenschikov (1967)	10716
Ca₃(AsO₄)₂	9749	Grekenschikov (1967)	9653
Sr₃(AsO₄)₂	9330	Grekenschikov (1967)	9266
Ba₃(AsO₄)₂	8870	Grekenschikov (1967)	8985
AlAsO₄	7255	Grekenschikov (1967)	—
GaAsO₄	7243	Grekenschikov (1967)	—
Astatides			
LiAt	720	Ladd, Lee (1961)	—
NaAt	657	Ladd, Lee (1961)	—
KAt	615	Ladd, Lee (1961)	—
RbAt	594	Ladd, Lee (1961)	—
CsAt	586	Ladd, Lee (1961)	—
FrAt	573	Ladd, Lee (1961)	—
Azides			
LiN₃	812	Jenkins, Pratt (1977)	818
NaN₃	732	Jenkins, Pratt (1977)	731
KN₃	659	Jenkins, Pratt (1977)	658
RbN₃	637	Jenkins, Pratt (1977)	632
CsN₃	612	Jenkins, Pratt (1977)	604
AgN₃	854	Jenkins, Pratt (1977)	—
TlN₃	689	Jenkins, Pratt (1977)	686
Ca(N₃)₂	2186	Jenkins, Pratt (1977)	2162
Sr(N₃)₂	2056	Jenkins, Pratt (1977)	2066
Ba(N₃)₂	2021	Gora (1971)	1965
Mn(N₃)₂	2408	Jenkins, Pratt (1977)	2416
Cu(N₃)₂	2730	Jenkins, Pratt (1977)	2738
Zn(N₃)₂	2840	Jenkins, Pratt (1977)	2848
Cd(N₃)₂	2446	Jenkins, Pratt (1977)	2454
Pb(N₃)₂	—	Jenkins, Pratt (1977)	2173
Bihalide salts			
LiHF₂	893	Jenkins, Pratt (1977)	866
NaHF₂	788	Jenkins, Pratt (1977)	788

Table 1 (continued)
LATTICE ENERGIES

Salt	Calculated lattice energy (kJmol⁻¹)	Thermochemical cycle lattice energy (kJmol⁻¹)	Literature source
KHF₂	703	698	Jenkins, Pratt (1977)
RbHF₂	674	676	Jenkins, Pratt (1977)
CsHF₂	646	628	Jenkins, Pratt (1977)
NH₄HF₂	705	705	Jenkins, Pratt (1977)
CsHCl₂	601	—	Thomson, Clark, Waddington, Jenkins (1975)
Me₄NHCl₂	427	—	Thompson, Clark, Waddington, Jenkins (1975)
Et₄NHCl₂	346	—	Thompson, Clark, Waddington, Jenkins (1975)
Bu₄NHCl₂	290	—	Thompson, Clark, Waddington, Jenkins (1975)
Bicarbonates			
NaHCO₃	820	818	Yatsimirskii (1956)
KHCO₃	741	736	Yatsimirskii (1956)
RbHCO₃	707	714	Yatsimirskii (1956)
CsHCO₃	678	709	Yatsimirskii (1956)
NH₄HCO₃	—	741	—
Ca(HCO₃)₂	2402	(2403)	Yatsimirskii (1956)
Sr(HCO₃)₂	2255	(2272)	Yatsimirskii (1956)
Ba(HCO₃)₂	2155	(2159)	Yatsimirskii (1956)
Bisulphates			
NH₄HSO₄	640	(645)	Yatsimirskii (1956)
Borides			
CaB₆	5146	—	Samsanov, Shulishova (1956)
SrB₆	5104	—	Samsanov, Shulishova (1956)
BaB₆	5021	—	Samsanov, Shulishova (1956)
YB₆	7447	—	Samsanov, Shulishova (1956)
LaB₆	7406	—	Samsanov, Shulishova (1956)
CeB₆	10083	—	Samsanov, Shulishova (1956)
PrB₆	7447	—	Samsanov, Shulishova (1956)
NdB₆	7447	—	Samsanov, Shulishova (1956)
PmB₆	7406	—	Samsanov, Shulishova (1956)
SmB₆	7447	—	Samsanov, Shulishova (1956)
EuB₆	5104	—	Samsanov, Shulishova (1956)
GdB₆	7489	—	Samsanov, Shulishova (1956)
TbB₆	7489	—	Samsanov, Shulishova (1956)
DyB₆	7489	—	Samsanov, Shulishova (1956)
HoB₆	7489	—	Samsanov, Shulishova (1956)
ErB₆	7489	—	Samsanov, Shulishova (1956)
TmB₆	7489	—	Samsanov, Shulishova (1956)
YbB₆	5146	—	Samsanov, Shulishova (1956)
LuB₆	7489	—	Samsanov, Shulishova (1956)
ThB₆	10167	—	Samsanov, Shulishova (1956)
Borohydrides			
LiBH₄	778	—	Altschuller (1955)
NaBH₄	703	—	Altschuller (1955)
KBH₄	665	—	Altschuller (1955)
RbBH₄	648	—	Altschuller (1955)
CsBH₄	628	—	Altschuller (1955)
Borohalide			
LiBF₄	699	—	Kapustinskii, Yatsimirskii (1949)
NaBF₄	657	—	Kapustinskii, Yatsimirskii (1949)
KBF₄	611	619	Ladd, Lee (1961)
RbBF₄	577	631	Kapustinskii, Yatsimirskii (1949)
CsBF₄	556	605	Kapustinskii, Yatsimirskii (1949)
NH₄BF₄	582	—	Ladd, Lee (1961)
Co(BF₄)₂	2127	—	Bhattacharya, Das Gupta (1970)
Ni(BF₄)₂	2136	—	Bhattacharya, Das Gupta (1970)
Zn(BF₄)₂	2063	—	Bhattacharya, Das Gupta (1970)
Cd(BF₄)₂	1937	—	Bhattacharya, Das Gupta (1970)
KBCl₄	506	—	Krivtsov, Titova, Rosolovskii (1973)
RbBCl₄	489	—	Krivtsov, Titova, Rosolovskii (1973)
CsBCl₄	473	—	Krivtsov, Titova, Rosolovskii (1973)
Carbonates			
Li₂CO₃	2523	2269	Yatsimirskii (1956)
Na₂CO₃	2301	2030	Yatsimirskii (1956)
K₂CO₃	2084	1858	Yatsimirskii (1956)
Rb₂CO₃	2000	1795	Yatsimirskii (1956)
Cs₂CO₃	1920	1702	Yatsimirskii (1956)
MgCO₃	3180	3122	Waddington (1959)
CaCO₃	2804	2810	Jenkins, Waddington, Pratt (1976)
SrCO₃	2720	2688	Waddington (1959)
BaCO₃	2615	2554	Waddington (1959)
MnCO₃	3046	3151	Waddington (1959)
FeCO₃	3121	3171	Waddington (1959)
CoCO₃	3443	3232	Yatsimirskii (1958)
NiCO₃	—	3297	
CuCO₃	3494	3327	Yatsimirskii (1958)
ZnCO₃	3121	3273	Waddington (1959)
CdCO₃	2929	3052	Waddington (1959)
SnCO₃	2904	(2853)	Yatsimirskii (1961)
PbCO₃	2728	2750	Yatsimirskii (1961)
Cyanates			
LiNCO	770	—	Waddington (1959)
NaNCO	736	—	Waddington (1959)
KNCO	653	—	Waddington (1959)
RbNCO	615	—	Waddington (1959)
CsNCO	586	—	Waddington (1959)
NH₄NCO	724	—	Waddington (1959)
Cyanides			
LiCN	—	849	—
NaCN	738	739	Jenkins, Pratt (1977)
KCN	674	669	Jenkins, Pratt (1977)

Table 1 (continued)
LATTICE ENERGIES

Salt	Calculated lattice energy (kJ mol⁻¹)	Literature source	Thermochemical cycle lattice energy (kJ mol⁻¹)
RbCN	646	Jenkins, Pratt (1977)	(640)
CsCN	612	Jenkins, Pratt (1977)	602
NH₄CN	617	Ladd (1970)	670
Ca(CN)₂	2268	Yatsimirskii (1956)	2191
Sr(CN)₂	2138	Yatsimirskii (1956)	(2076)
Ba(CN)₂	2046	Yatsimirskii (1956)	1960
CuCN	—	—	1035
AgCN	741	Yatsimirskii (1961)	914
Zn(CN)₂	2431	Ladd, Lee (1960)	2768
Cd(CN)₂	2284	Ladd, Lee (1960)	2542
Formates			
Li(HCO₂)	865	Morris (1959)	872
Na(HCO₂)	791	Morris (1959)	811
K(HCO₂)	713	Morris (1959)	729
Rb(HCO₂)	685	Morris (1959)	682
Cs(HCO₂)	651	Morris (1959)	644
NH₄HCO₂	715	Morris (1959)	—
Mg(HCO₂)₂	2674	Morris (1959)	2390
Ca(HCO₂)₂	2360	Morris (1959)	2261
Sr(HCO₂)₂	2221	Morris (1959)	2134
Ba(HCO₂)₂	2092	Morris (1959)	2701
Mn(HCO₂)₂	2598	Morris (1959)	2792
Co(HCO₂)₂	—	—	2880
Ni(HCO₂)₂	—	Morris (1959)	2913
Cu(HCO₂)₂	2870	Morris (1959)	2847
Zn(HCO₂)₂	2791	Morris (1959)	—
Cd(HCO₂)₂	2556	Morris (1959)	2330
Pb(HCO₂)₂	2276	Morris (1959)	—
Germanates			
Mg₂GeO₄	7991	Grekenschikov (1967)	—
Ca₂GeO₄	7301	Grekenschikov (1967)	7306
Sr₂GeO₄	6987	Grekenschikov (1967)	—
Ba₂GeO₄	6653	Grekenschikov (1967)	6643
Glycinates			
Na(NH₃CH₂COO)	739	Bernard, Businot, Decker (1974)	—
K(NH₃CH₂COO)	668	Bernard, Businot, Decker (1974)	—
Rb(NH₃CH₂COO)	648	Bernard, Businot, Decker (1974)	—
NH₄(NH₃CH₂COO)	650	Bernard, Businot, Decker (1974)	—
Cu(NH₃CH₂COO)₂	2694	Bernard, Businot, Decker (1974)	—
Halates			
LiBrO₃	883	Finch, Gardner (1965)	897
NaBrO₃	803	Finch, Gardner (1965)	814
KBrO₃	740	Finch, Gardner (1965)	745
RbBrO₃	720	Finch, Gardner (1965)	742
CsBrO₃	694	Finch, Gardner (1965)	681
NaClO₃	770	Morris (1958)	770
KClO₃	711	Morris (1958)	706

Salt	Calculated lattice energy (kJ mol⁻¹)	Literature source	Thermochemical cycle lattice energy (kJ mol⁻¹)
RbClO₃	690	Morris (1958)	687
CsClO₃	—	—	647
LiIO₃	975	Finch, Gardner (1965)	—
NaIO₃	883	Finch, Gardner (1965)	882
KIO₃	820	Finch, Gardner (1965)	806
RbIO₃	791	Finch, Gardner (1965)	—
CsIO₃	761	Finch, Gardner (1965)	769
NH₄IO₃	—	—	808
Mg(ClO₃)₂	2535	Finch, Gardner (1965)	(2475)
Ca(ClO₃)₂	2259	Finch, Gardner (1965)	2286
Sr(ClO₃)₂	2138	Finch, Gardner (1965)	2155
Ba(ClO₃)₂	2021	Finch, Gardner (1965)	2027
Halides			
LiF	1030	Jenkins, Pratt (1977)	1036
LiCL	834	Jenkins, Pratt (1977)	853
LiBr	788	Jenkins, Pratt (1977)	807
LiI	730	Jenkins, Pratt (1977)	757
NaF	910	Jenkins, Pratt (1977)	923
NaCl	769	Jenkins, Pratt (1977)	786
NaBr	732	Jenkins, Pratt (1977)	747
NaI	682	Jenkins, Pratt (1977)	704
KF	808	Jenkins, Pratt (1977)	821
KCl	701	Jenkins, Pratt (1977)	715
KBr	671	Jenkins, Pratt (1977)	682
KI	632	Jenkins, Pratt (1977)	649
RbF	774	Jenkins, Pratt (1977)	785
RbCl	680	Jenkins, Pratt (1977)	689
RbBr	651	Jenkins, Pratt (1977)	689
RbI	617	Jenkins, Pratt (1977)	660
CsF	744	Jenkins, Pratt (1977)	630
CsCl	657	Jenkins, Pratt (1977)	740
CsBr	632	Jenkins, Pratt (1977)	659
CsI	600	Jenkins, Pratt (1977)	631
FrF	715	Ladd, Lee (1961)	604
FrCl	632	Ladd, Lee (1961)	—
FrBr	611	Ladd, Lee (1961)	—
FrI	582	Ladd, Lee (1961)	—
CuCl	921	Sharma, Madan (1964)	996
CuBr	879	Sharma, Madan (1964)	979
CuI	835	Sharma, Madan (1964)	966
AgF	953	Sharma, Madan (1964)	967
AgCl	864	Sharma, Madan (1964)	915
AgBr	830	Sharma, Madan (1964)	904
AgI	808	Sharma, Madan (1964)	889
AuCl	1013	Mamulov (1961)	1066
AuBr	1015	Mamulov (1961)	1061
AuI	1015	Mamulov (1961)	1070

Table 1 (continued)
LATTICE ENERGIES

Salt	Calculated lattice energy (kJmol⁻¹)	Literature source	Thermochemical cycle lattice energy (kJmol⁻¹)
InCl	—	—	763
InBr	—	—	767
InI	—	—	732
TlF	782	Jenkins, Alcock (1975)	845
TlCl	713	Sharma, Madan (1964)	751
TlBr	687	Sharma, Madan (1964)	735
TlI	—	Krasnov (1961)	709
BeF₂	3150	Krasnov (1961)	3505
BeCl₂	3004	Brackett, Brackett (1965)	3020
BeBr₂	2950	Krasnov (1961)	2914
BeI₂	2653	Brackett, Brackett (1965)	2800
MgF₂	2913	Brackett, Brackett (1965)	2957
MgCl₂	2326	Brackett, Brackett (1965)	2526
MgBr₂	2097	Brackett, Brackett (1965)	2440
MgI₂	1944	Brackett, Brackett (1965)	2327
CaF₂	2609	Brackett, Brackett (1965)	2630
CaCl₂	2223	Brackett, Brackett (1965)	2258
CaBr₂	2132	Brackett, Brackett (1965)	2176
CaI₂	1905	Brackett, Brackett (1965)	2074
SrF₂	2476	Brackett, Brackett (1965)	2492
SrCl₂	2127	Brackett, Brackett (1965)	2156
SrBr₂	2008	Brackett, Brackett (1965)	2075
SrI₂	1937	Karapet'yants (1954)	1963
BaF₂	2341	Brackett, Brackett (1965)	2352
BaCl₂	2033	Brackett, Brackett (1965)	2056
BaBr₂	1950	Brackett, Brackett (1965)	1985
BaI₂	1831	Brackett, Brackett (1965)	1877
RaF₂	2284	Yatsimirskii, Krestov (1960)	—
RaCl₂	2004	Yatsimirskii, Krestov (1960)	—
RaBr₂	1929	Yatsimirskii, Krestov (1960)	—
RaI₂	1803	Yatsimirskii, Krestov (1960)	—
ScCl₃	2380	Nelson, Sharpe (1966)	—
ScBr₃	2291	Nelson, Sharpe (1966)	—
ScI₃	2201	Nelson, Sharpe (1966)	—
TiF₂	2724	Mamulov (1961)	2501
TiCl₂	2431	Mamulov (1961)	2419
TiBr₂	2360	Mamulov (1961)	2329
TiI₂	2259	Mamulov (1961)	2579
VCl₂	2607	Mamulov (1961)	2523
VBr₂	—	Mamulov (1961)	2456
VI₂	2778	Mamulov (1961)	2917
CrF₂	2455	Mamulov (1961)	2586
CrCl₂	2377	Mamulov (1961)	2523
CrBr₂	2269	Mamulov (1961)	2425
CrI₂	—	Mamulov (1961)	—
MoCl₂	2485	Mamulov (1961)	2733
MoBr₂	2448	Mamulov (1961)	2742
MoI₂	2422	Mamulov (1961)	2630
MnF₂	2644	Mamulov (1961)	—
MnCl₂	2368	Mamulov (1961)	2537
MnBr₂	2304	Mamulov (1961)	2471
MnI₂	2212	Mamulov (1961)	—
FeF₂	2769	Mamulov (1961)	2631
FeCl₂	2525	Mamulov (1961)	2569
FeBr₂	2464	Mamulov (1961)	2480
FeI₂	2382	Mamulov (1961)	3018
CoF₂	2878	Morris (1957)	2691
CoCl₂	2709	Mamulov (1961)	2629
CoBr₂	2648	Yatsimirskii (1958)	2545
CoI₂	2569	Yatsimirskii (1958)	3066
NiF₂	2845	VonBaur (1961)	2772
NiCl₂	2753	Yatsimirskii (1958)	2709
NiBr₂	2699	Yatsimirskii (1958)	2623
NiI₂	2607	Yatsimirskii (1958)	2778
PdCl₂	2766	Yatsimirskii (1958)	2623
PdBr₂	—	—	2741
PdI₂	—	—	2748
CuF₂	3046	Perret (1961)	3082
CuCl₂	2774	Perret (1961)	2811
CuBr₂	2711	Perret (1961)	2763
CuI₂	2640	Perret (1961)	—
AgF₂	2919	Mamulov (1961)	2942
ZnF₂	2930	Mamulov (1961)	3032
ZnCl₂	2690	Mamulov (1961)	2734
ZnBr₂	2632	Mamulov (1961)	2678
ZnI₂	2549	Mamulov (1961)	2605
CdF₂	2740	Karapet'yants (1954)	2809
CdCl₂	2226	Krasnov (1961)	2552
CdBr₂	2468	Karapet'yants (1954)	2507
CdI₂	2406	Karapet'yants (1954)	2441
HgF₂	2757	Karapet'yants (1954)	(2798)
HgCl₂	2569	Yatsimirskii (1961)	2651
HgBr₂	2598	Karapet'yants (1954)	2628
HgI₂	2569	Karapet'yants (1954)	2610
SnF₂	2551	Mamulov (1961)	2297
SnCl₂	2276	Mamulov (1961)	2245
SnBr₂	2211	Mamulov (1961)	2193
SnI₂	2123	Mamulov (1961)	—
PbF₂	2460	Mamulov (1961)	2522
PbCl₂	2229	Mamulov (1961)	2269
PbBr₂	2169	Mamulov (1961)	2219
PbI₂	2086	Mamulov (1961)	2163
ScF₃	5096	Mamulov (1961)	5492
ScCl₃	4874	Kapustinskii, Yatsimirskii (1956)	4866

Table 1 (continued)
LATTICE ENERGIES

Salt	Calculated lattice energy (kJmol⁻¹)	Literature source	Thermochemical cycle lattice energy (kJmol⁻¹)
$ScBr_3$	4711	Kapustinskii, Yatsimirskii (1956)	4729
ScI_3	4640	Mamulov (1961)	—
YF_3	4983	Mamulov (1961)	—
YCl_3	4447	Mamulov (1961)	4506
YBr_3	4410	Wen, Sho (1975)	—
YI_3	4125	Mamulov (1961)	4240
TiF_3	5644	Mamulov (1961)	—
$TiCl_3$	5134	Mamulov (1961)	5134
$TiBr_3$	5012	Mamulov (1961)	5007
TiI_3	4845	Mamulov (1961)	—
ZrF_3	—	—	(5400)
$ZrCl_3$	—	—	4791
$ZrBr_3$	5895	Cavell, Clark (1965)	4758
ZrI_3	5322	Mamulov (1961)	(4591)
VF_3	5192	Nelson, Sharpe (1966)	5315
VCl_3	5058	Nelson, Sharpe (1966)	5214
VBr_3	5062	Jenkins, Waddington*	5121
VI_3	4980	Jenkins, Waddington*	—
$NbCl_3$	4860	Jenkins, Waddington*	—
$NbBr_3$	5958	Mamulov (1961)	6033
NbI_3	5473	Mamulov (1961)	5509
CrF_3	5355	Mamulov (1961)	—
$CrCl_3$	5201	Mamulov (1961)	5274
$CrBr_3$	(6459)	Van Gool, Picken (1964)	5230
CrI_3	—	Van Gool, Picken (1964)	5230
MoF_3	—	Van Gool, Picken (1964)	5156
$MoCl_3$	6017	Cavell, Clark (1965)	5073
$MoBr_3$	5544	Nelson, Sharpe (1966)	—
MoI_3	5448	Nelson, Sharpe (1966)	—
MnF_3	5330	Jenkins, Waddington*	—
$MnCl_3$	5270	Jenkins, Waddington*	—
$MnBr_3$	5215	Jenkins, Waddington*	—
MnI_3	5188	Jenkins, Waddington*	—
$TcCl_3$	5870	Perret (1961)	5359
$TcBr_3$	5364	Perret (1961)	5333
TcI_3	5268	Perret (1961)	—
FeF_3	5117	Perret (1961)	5245
$FeCl_3$	—	—	5223
$FeBr_3$	—	—	5222
FeI_3	—	—	6118
$RuCl_3$	5991	Cavell, Clark (1965)	5641
$RuBr_3$	(6112)	Hoppe (1970)	—
RuI_3	(4794)	Hoppe (1970)	—
CoF_3	5991	Cavell, Clark (1965)	—
$RhCl_3$	—	—	—
IrF_3	(6112)	Hoppe (1970)	—
$IrBr_3$	(4794)	Hoppe (1970)	—

Salt	Calculated lattice energy (kJmol⁻¹)	Literature source	Thermochemical cycle lattice energy (kJmol⁻¹)
NiF_3	(6111)	Van Gool, Picken (1969)	—
AuF_3	(5777)	Hoppe (1970)	—
$AuCl_3$	(4605)	Hoppe (1970)	—
$ZnCl_3$	5832	Nelson, Sharpe (1963)	—
$ZnBr_3$	5732	Nelson, Sharpe (1963)	—
ZnI_3	5636	Nelson, Sharpe (1963)	5215
AlF_3	5924	Mamulov (1961)	5492
$AlCl_3$	5376	Mamulov (1961)	5361
$AlBr_3$	5247	Mamulov (1961)	5218
AlI_3	5070	Mamulov (1961)	6205
GaF_3	—	—	5645
$GaCl_3$	5217	Mamulov (1961)	5552
$GaBr_3$	4966	Mamulov (1961)	5476
GaI_3	4611	Mamulov (1961)	5187
$InCl_3$	4736	Yatsimirskii (1961)	5124
$InBr_3$	4535	Yatsimirskii (1961)	5005
InI_3	4234	Yatsimirskii (1961)	—
TlF_3	5493	Mamulov (1961)	5258
$TlCl_3$	5252	Mamulov (1961)	—
$TlBr_3$	5171	Mamulov (1961)	—
TlI_3	5088	Mamulov (1961)	—
$AsBr_3$	—	—	5497
AsI_3	(3758)	Van Gool, Picken (1969)	4824
SbF_3	—	—	5295
$SbCl_3$	—	—	5032
$SbBr_3$	—	—	4954
SbI_3	—	—	4867
$BiCl_3$	—	—	4689
$BiBr_3$	—	—	—
BiI_3	(3774)	Van Gool, Picken (1969)	—
LaF_3	4682	Mamulov (1961)	—
$LaCl_3$	4343	Yatsimirskii (1961)	4242
$LaBr_3$	4209	Wen, Sho (1975)	—
LaI_3	3916	Yatsimirskii (1961)	3986
$CeCl_3$	4297	Ladd, Lee (1961)	4284
CeI_3	—	Ladd, Lee (1961)	4029
$PrCl_3$	4322	Ladd, Lee (1961)	4326
PrI_3	—	Ladd, Lee (1961)	4071
$NdCl_3$	4343	Ladd, Lee (1961)	—
$SmCl_3$	4376	Ladd, Lee (1961)	—
$EuCl_3$	4393	Ladd, Lee (1961)	—
$GdCl_3$	4406	Ladd, Lee (1961)	—
$DyCl_3$	4481	Ladd, Lee (1961)	—
$HoCl_3$	4501	Ladd, Lee (1961)	—
$ErCl_3$	4527	Ladd, Lee (1961)	—
$TmCl_3$	4548	Ladd, Lee (1961)	4550
TmI_3	—	—	4314

Table 1 (continued)
LATTICE ENERGIES

Salt	Calculated lattice energy (kJmol⁻¹)	Literature source	Thermochemical cycle lattice energy (kJmol⁻¹)
YbCl₃	—	Ladd, Lee (1961)	4546
AcCl₃	4096	Ladd, Lee (1961)	—
UCl₃	4243	Ladd, Lee (1961)	—
NpCl₃	4268	Ladd, Lee (1961)	—
PuCl₃	4289	Ladd, Lee (1961)	—
PuBr₃	(3959)	Van Gool, Picken (1969)	—
AmCl₃	4293	Van Gool, Picken (1969)	—
TiF₃	10012	Mamulov (1961)	9908
TiCl₃	9431	Mamulov (1961)	—
TiBr₃	9288	Mamulov (1961)	9039
TiI₃	9108	Mamulov (1961)	8893
ZrF₄	8853	Mamulov (1961)	8971
ZrCl₄	8096	Mamulov (1961)	8144
ZrBr₄	7916	Mamulov (1961)	7984
ZrI₄	7661	Mamulov (1961)	7801
MoF₄	8795	Mamulov (1961)	—
MoCl₄	8556	Mamulov (1961)	9573
MoBr₄	8510	Mamulov (1961)	9475
MoI₄	8427	Mamulov (1961)	—
SnCl₄	8355	Mamulov (1961)	—
SnBr₄	7970	Mamulov (1961)	8833
PbF₄	—	Perret (1961)	9461
CrF₂Cl	5795	Perret (1961)	—
CrF₂Br	5753	Perret (1961)	—
CrF₂I	5669	Perret (1961)	—
CrCl₂Br	5448	Perret (1961)	—
CrCl₂I	5381	Perret (1961)	—
CrBr₂I	5330	Perret (1961)	—
CuFCl	2891	Perret (1961)	—
CuFBr	2853	Perret (1961)	—
CuFI	2803	Perret (1961)	—
CuClBr	2753	Perret (1961)	—
CuClI	2694	Perret (1961)	—
CuBrI	2669	Perret (1961)	—
FeF₂Cl	5711	Perret (1961)	—
FeF₂Br	5653	Perret (1961)	—
FeF₂I	5569	Perret (1961)	—
FeCl₂Br	5339	Perret (1961)	—
FeCl₂I	5272	Perret (1961)	—
FeBr₂I	5209	Perret (1961)	—
LiIO₂F₂	845	Finch, Gates, Jenkinson (1972)	764
NaIO₂F₂	766	Finch, Gates, Jenkinson (1972)	697
KIO₂F₂	699	Finch, Gates, Jenkinson (1972)	671
RbIO₂F₂	674	Finch, Gates, Jenkinson (1972)	—
CsIO₂F₂	636	Finch, Gates, Jenkinson (1972)	—
NH₄IO₂F₂	678	Finch, Gates, Jenkinson (1972)	—
AgIO₂F₂	736	Finch, Gates, Jenkinson (1972)	685

Salt	Calculated lattice energy (kJmol⁻¹)	Literature source	Thermochemical cycle lattice energy (kJmol⁻¹)
Hydrides			
LiH	858	Tsai (1964)	920
NaH	782	Tsai (1964)	808
KH	699	Tsai (1964)	714
RbH	674	Tsai (1964)	685
CsH	648	Tsai (1964)	644
TiH	—	—	1407
ZrH	—	—	1590
VH	—	—	(1344)
NbH	—	—	(1633)
PdH	—	—	(1368)
CuH	996	Gibb (1962)	1254
TiH	916	Gibb (1962)	1407
ZrH	904	Gibb (1962)	1590
HfH	828	Gibb (1962)	—
LaH	1184	Gibb (1962)	(1344)
VH	1163	Gibb (1962)	(1633)
NbH	1021	Gibb (1962)	—
TaH	1050	Gibb (1962)	—
CrH	929	Gibb (1962)	1368
NiH	979	Gibb (1962)	—
PdH	937	Gibb (1962)	1254
PtH	828	Gibb (1962)	—
CuH	941	Gibb (1962)	—
AgH	1033	Karapet'yants (1954)	—
AuH	745	Karapet'yants (1954)	—
TlH	950	Karapet'yants (1954)	—
GeH	778	Gibb (1962)	—
PbH	(2598)	Gibb (1962)	—
BeH₂	3205	Gibb (1962)	3295
MgH₂	2791	Gibb (1962)	2706
CaH₂	2410	Karapet'yants (1954)	2394
SrH₂	2250	Karapet'yants (1954)	2253
BaH₂	2121	Karapet'yants (1954)	2121
ScH₂	2711	Gibb (1962)	2659
TiH₂	2380	Gibb (1962)	2670
VH₂	2414	Gibb (1962)	2522
FeH₂	2448	Gibb (1962)	2484
MnH₂	2464	Gibb (1962)	2398
CdH₂	2519	Gibb (1962)	2367
ZnH₂	2510	Gibb (1962)	—
CuH₂	2494	Gibb (1962)	2389
AgH₂	2372	Gibb (1962)	—
TlH₂	2711	Gibb (1962)	2706
PbH₂	2519	Gibb (1962)	—
AuH₂	2544	Gibb (1962)	—

Table 1 (continued)
LATTICE ENERGIES

Salt	Calculated lattice energy (kJmol⁻¹)	Literature source	Thermochemical cycle lattice energy (kJmol⁻¹)
TiH_3	2866	Gibb (1962)	2845
ZrH_3	2711	Gibb (1962)	2999
CuH_3	2941	Karapet'yants (1954)	—
ZnH_3	2870	Karapet'yants (1954)	—
HgH_3	2707	Gibb (1962)	—
AlH_3	5924	Gibb (1962)	—
FeH_3	5724	Gibb (1962)	—
ScH_3	5439	Gibb (1962)	—
YH_3	5063	Gibb (1962)	4493
LaH_3	4895	Gibb (1962)	—
FeH_3	5724	Gibb (1962)	—
GaH_3	5690	Gibb (1962)	—
InH_3	5092	Gibb (1962)	—
TlH_3	5092	Gibb (1962)	—
Hydroselenides			
NaHSe	703	Yatsimirskii (1956)	—
KHSe	644	Yatsimirskii (1956)	—
RbHSe	623	Yatsimirskii (1956)	—
CsHSe	598	Yatsimirskii (1956)	—
Hydrosulphides			
LiHS	759	Waddington (1959)	821
NaHS	704	Waddington (1959)	747
KHS	650	Waddington (1959)	659
RbHS	623	Waddington (1959)	637
CsHS	582	Waddington (1959)	595
NH_4HS	661	Yatsimirskii (1956)	666
$Ca(HS)_2$	2184	Yatsimirskii (1956)	(2171)
$Sr(HS)_2$	2063	Yatsimirskii (1956)	—
$Ba(HS)_2$	1979	Yatsimirskii (1956)	(1956)
Hydroxides			
LiOH	1021	Saloman (1970)	1039
NaOH	887	Saloman (1970)	900
KOH	789	Saloman (1970)	804
RbOH	766	Yatsimirskii (1956)	773
CsOH	721	Yatsimirskii (1956)	724
$Be(OH)_2$	3477	Finch, Gardner (1965)	3629
$Mg(OH)_2$	2870	Finch, Gardner (1965)	3006
$Ca(OH)_2$	2506	Finch, Gardner (1965)	2645
$Sr(OH)_2$	2330	Finch, Gardner (1965)	2483
$Ba(OH)_2$	2141	Finch, Gardner (1965)	2339
$Ti(OH)_2$	—	—	2962
$Mn(OH)_2$	2909	Wen, Sho (1975)	3008
$Fe(OH)_2$	2653	Karapet'yants (1954)	3055
$Co(OH)_2$	2786	Karapet'yants (1954)	3115
$Ni(OH)_2$	2832	Karapet'yants (1954)	3193
$Pd(OH)_2$	—	—	3175
$Cu(OH)_2$	2870	Karapet'yants (1954)	3237
CuOH	1006	Karapet'yants (1954)	—
AgOH	918	Karapet'yants (1954)	—
AuOH	1033	Karapet'yants (1954)	—
TlOH	705	Karapet'yants (1954)	3158
$Zn(OH)_2$	2795	Yatsimirskii (1961)	2918
$Cd(OH)_2$	2607	Yatsimirskii (1961)	—
$Hg(OH)_2$	2669	Karapet'yants (1954)	2729
$Sn(OH)_2$	2489	Yatsimirskii (1961)	2623
$Pb(OH)_2$	2376	Yatsimirskii (1961)	—
$Sc(OH)_3$	5063	Karapet'yants (1954)	—
$Y(OH)_3$	4707	Karapet'yants (1954)	—
$La(OH)_3$	4443	Karapet'yants (1954)	—
$Cr(OH)_3$	5556	Karapet'yants (1954)	—
$Mn(OH)_3$	6213	Wen, Sho (1975)	—
$Al(OH)_3$	5627	Karapet'yants (1954)	—
$Ga(OH)_3$	5732	Karapet'yants (1954)	—
$In(OH)_3$	5280	Karapet'yants (1954)	—
$Tl(OH)_3$	5314	Karapet'yants (1954)	—
$Ti(OH)_3$	9456	Karapet'yants (1954)	—
$Zr(OH)_4$	8619	Karapet'yants (1954)	—
$Mn(OH)_4$	10933	Brenet, Goelfier, Cabano (1963)	—
$Sn(OH)_4$	9188	Karapet'yants (1954)	—
Imides			
CaNH	3293	Altschuller (1955)	—
SrNH	3146	Altschuller (1955)	—
BaNH	2975	Altschuller (1955)	—
Metaniobates			
$NaNbO_3$	789	Lapitskii, Nebyhtsyn (1961)	—
$Ca(NbO_3)_2$	2315	Lapitskii, Nebyhtsyn (1961)	—
$Fe(NbO_3)_2$	2502	Lapitskii, Nebyhtsyn (1961)	—
Metatantalates			
$NaTaO_3$	789	Lapitskii, Nebyhtsyn (1961)	—
$Ca(TaO_3)_2$	2315	Lapitskii, Nebyhtsyn (1961)	—
$Fe(TaO_3)_2$	2502	Lapitskii, Nebyhtsyn (1961)	—
Metavanadates			
Li_3VO_4	3945	Golvkin, Fotiev (1970)	—
Na_3VO_4	3766	Golvkin, Fotiev (1970)	—
K_3VO_4	3376	Golvkin, Fotiev (1970)	—
Rb_3VO_4	3243	Golvkin, Fotiev (1970)	—
Cs_3VO_4	3137	Golvkin, Fotiev (1970)	—
Nitrates			
$LiNO_3$	848	Jenkins, Morris (1977)	848
$NaNO_3$	755	Jenkins, Morris (1977)	756
KNO_3	685	Jenkins, Morris (1977)	687
$RbNO_3$	662	Jenkins, Morris (1977)	658
$CsNO_3$	648	Jenkins, Morris (1977)	625
NH_4NO_3	661	Morris (1958)	676

Table 1 (continued)
LATTICE ENERGIES

Salt	Calculated lattice energy (kJmol⁻¹)	Literature source	Thermochemical cycle lattice energy (kJmol⁻¹)	Salt	Calculated lattice energy (kJmol⁻¹)	Literature source	Thermochemical cycle lattice energy (kJmol⁻¹)
$AgNO_3$	820	Morris (1958)	822	K_2O_2	2114	Wood, D'Orazio (1965)	2078
$TlNO_3$	690	Morris (1958)	700	Rb_2O_2	2025	Wood, D'Orazio (1965)	2006
$Mg(NO_3)_2$	2468	Finch, Gardner (1965)	2503	Cs_2O_2	1948	Wood, D'Orazio (1965)	1861
$Ca(NO_3)_2$	2209	Finch, Gardner (1965)	2228	MgO_2	3356	Wood, D'Orazio (1965)	3526
$Sr(NO_3)_2$	2092	Finch, Gardner (1965)	2132	CaO_2	3144	Wood, D'Orazio (1965)	3133
$Ba(NO_3)_2$	1975	Finch, Gardner (1965)	2016	SrO_2	3037	Wood, D'Orazio (1965)	2849
$Mn(NO_3)_2$	2318	Yatsimirskii (1961)	2519	KO_3	697	Wood, D'Orazio (1965)	—
$Fe(NO_3)_2$	—		(2563)	BeO	4293	Huggins, Sakamoto (1957)	4443
$CO(NO_3)_2$	2560	Wen,Sho (1975)	2626	MgO	3795	Huggins, Sakamoto (1957)	3791
$Ni(NO_3)_2$	—		2709	CaO	3414	Huggins, Sakamoto (1957)	3401
$Cu(NO_3)_2$	—		2720	SrO	3217	Huggins, Sakamoto (1957)	3223
$Zn(NO_3)_2$	2376	Yatsimirskii (1961)	2628	BaO	3029	Huggins, Sakamoto (1957)	3054
$Cd(NO_3)_2$	2238	Yatsimirskii (1961)	2443	TiO	3832	Huggins, Sakamoto (1957)	3811
$Ha(NO_3)_2$	2255	Yatsimirskii (1961)	—	VO	3932	Ladd, Lee (1961)	3863
$Sn(NO_3)_2$	2155	Yatsimirskii (1961)	2254	MnO	3724	Ladd, Lee (1961)	3745
$Pb(NO_3)_2$	2067	Yatsimirskii (1961)	2189	FeO	3795	Ladd, Lee (1961)	3865
Nitrides				CoO	3837	Ladd, Lee (1961)	3910
ScN	7547	Baughan (1959)	7506	NiO	3908	Ladd, Lee (1961)	4010
LaN	6876	Baughan (1959)	6793	PdO	3736	Ladd, Lee (1961)	—
TiN	8130	Baughan (1959)	8033	CuO	4135	Mamulov (1961)	4050
ZrN	7633	Baughan (1959)	7723	ZnO	4142	Ladd, Lee (1961)	3971
VN	8283	Baughan (1959)	8233	CdO	3806	Ladd, Lee (1961)	—
NbN	7939	Baughan (1959)	8022	HgO	3907	Ladd, Lee (1961)	—
CrN	8269	Baughan (1959)	8358	GeO	3919	Ladd, Lee (1961)	—
Nitrites				SnO	3652	Ladd, Lee (1961)	—
$NaNO_2$	774	Morris (1958)	748	PbO	3520	Ladd, Lee (1961)	—
KNO_2	660	Jenkins (1977)	664	Sc_2O_3	13557	Gasharov, Sovers (1970)	13708
$RbNO_2$	638	Jenkins (1977)	765	Y_2O_3	12705	Gasharov, Sovers (1970)	—
$CsNO_2$	598	Jenkins (1977)	—	La_2O_3	12452	Johnson (1969)	—
$Ca(NO_2)_2$	2460	Yatsimirskii (1956)	2225	Ce_2O_3	12661	Johnson (1969)	—
$Sr(NO_2)_2$	2305	Yatsimirskii (1956)	2111	Pr_2O_3	12703	Johnson (1969)	—
$Ba(NO_2)_2$	2205	Yatsimirskii (1956)	1987	Nd_2O_3	12736	Johnson (1969)	—
Oxides				Pm_2O_3	12811	Johnson (1969)	—
Li_2O	2799	Baughan (1959)		Sm_2O_3	12878	Johnson (1969)	—
Na_2O	2481	Baughan (1959)		Eu_2O_3	12945	Johnson (1969)	—
K_2O	2238	Baughan (1959)		Gd_2O_3	12996	Johnson (1969)	—
Rb_2O	2163	Baughan (1959)		Tb_2O_3	13071	Johnson (1969)	—
Cu_2O	3273	Mamulov (1961)		Dy_2O_3	13138	Johnson (1969)	—
Ag_2O	3002	Mamulov (1961)		Ho_2O_3	13180	Johnson (1969)	—
Tl_2O	2659	Mamulov (1961)		Er_2O_3	13263	Johnson (1969)	—
LiO_2	(878)	D'Orazio, Wood (1965)	(872)	Tm_2O_3	13322	Johnson (1969)	—
NaO_2	799	Yatsimirskii (1959)	796	Yb_2O_3	13380	Johnson (1969)	—
KO_2	741	D'Orazio, Wood (1965)	725	Lu_2O_3	13665	Ladd, Lee (1961)	—
RbO_2	706	D'Orazio, Wood (1965)	695	Ac_2O_3	12573	Krestov, Krestova (1969)	—
CsO_2	679	D'Orazio, Wood (1965)	668	Ti_2O_3	—		—
Li_2O_2	2592	Wood, D'Orazio (1965)	256	V_2O_3	15096	Mamulov (1961)	14149
Na_2O_2	2309	Wood, D'Orazio (1965)	2305	Cr_2O_3	15276	Mamulov (1961)	14520
							114957

Table 1 (continued)
LATTICE ENERGIES

Salt	Calculated lattice energy ($kJmol^{-1}$)	Literature source	Thermochemical cycle lattice energy ($kJmol^{-1}$)
Mn_3O_4	15146	Mamulov (1961)	15035
Fe_3O_4	14309	Mamulov (1961)	14774
Al_2O_3	15916	Yatsimirskii (1961)	—
Ga_2O_3	15590	Yatsimirskii (1961)	15220
In_2O_3	13928	Yatsimirskii (1961)	—
Ti_2O_3	14702	Mamulov (1961)	—
Pb_2O_3	(14841)	Van Gool, Picken (1969)	—
CeO_2	9627	VanBaur (1961)	—
ThO_2	10397	Ladd, Lee (1961)	—
PaO_2	10573	Ladd, Lee (1961)	—
VO_2	10644	Ladd, Lee (1961)	—
NpO_2	10707	Ladd, Lee (1961)	—
PuO_2	10786	Ladd, Lee (1961)	—
AmO_2	10799	Ladd, Lee (1961)	—
CmO_2	10832	Ladd, Lee (1961)	—
TiO_2	12150	Ladd, Lee (1961)	—
ZrO_2	11188	Ladd, Lee (1961)	—
MoO_2	11648	Ladd, Lee (1961)	—
MnO_2	12970	Ladd, Lee (1961)	—
SiO_2	13125	Ladd, Lee (1961)	—
GeO_2	12828	Ladd, Lee (1961)	—
SnO_2	11807	Ladd, Lee (1961)	—
PbO_2	11217	—	—
Perchlorates			
$LiClO_4$	709	Gill, Singla, Paul, Narula (1972)	723
$NaClO_4$	643	Jenkins, Pratt (1978)	648
$KClO_4$	599	Jenkins, Pratt (1978)	602
$RbClO_4$	564	Gill, Singla, Paul, Narula (1972)	582
$CsClO_4$	636	Morris (1958)	(542)
NH_4ClO_4	583	Jenkins, Pratt (1978)	580
$Ca(ClO_4)_2$	1958	Yatsimirskii (1956)	1971
$Sr(ClO_4)_2$	1862	Yatsimirskii (1956)	1862
$Ba(ClO_4)_2$	1795	Yatsimirskii (1958)	1769
$NaMnO_4$	661	Yatsimirskii (1961)	—
$KMnO_4$	607	Yatsimirskii (1956)	—
$RbMnO_4$	586	Yatsimirskii (1956)	—
$CsMnO_4$	565	Yatsimirskii (1956)	—
$Ca(MnO_4)_2$	1937	Yatsimirskii (1956)	—
$Sr(MnO_4)_2$	1845	Yatsimirskii (1956)	—
$Ba(MnO_4)_2$	1778	Yatsimirskii (1956)	—
Phosphates			
$Mg_3(PO_4)_2$	11632	Grekenschikov (1967)	11407
$Ca_3(PO_4)_2$	10602	Grekenschikov (1967)	10479
$Sr_3(PO_4)_2$	10125	Grekenschikov (1967)	10075
$Ba_3(PO_4)_2$	9652	Grekenschikov (1967)	9654
$MnPO_4$	7397	Grekenschikov (1967)	—
$FePO_4$	7251	Grekenschikov (1967)	7303
BPO_4	8201	Grekenschikov (1967)	—
$AlPO_4$	7427	Grekenschikov (1967)	7509
$GaPO_4$	7381	Grekenschikov (1967)	—
Phosphonium salts			
PH_4Br	616	Waddington (1965)	—
PH_4I	590	Waddington (1965)	—
Selenides			
Li_2Se	2364	Bevan, Morris (1960)	—
Na_2Se	2130	Bevan, Morris (1960)	—
K_2Se	1933	Bevan, Morris (1960)	—
Rb_2Se	1837	Bevan, Morris (1960)	—
Cs_2Se	1745	Bevan, Morris (1960)	—
Ag_2Se	2686	Mamulov (1961)	—
Tl_2Se	2209	Mamulov (1961)	—
$BeSe$	3431	Huggins, Sakamoto (1957)	—
$MgSe$	3071	Huggins, Sakamoto (1957)	—
$CaSe$	2858	Huggins, Sakamoto (1957)	2862
$SrSe$	2736	Huggins, Sakamoto (1957)	—
$BaSe$	2611	Huggins, Sakamoto (1957)	—
$MnSe$	3176	Mamulov (1961)	3194
$FeSe$	3499	Mamulov (1961)	3396
$CoSe$	3554	Mamulov (1961)	3471
$NiSe$	3658	Mamulov (1961)	3558
$CuSe$	3736	Mamulov (1961)	3662
$ZnSe$	3502	Lambardi (1969)	3514
$CdSe$	3330	Mamulov (1961)	—
$HgSe$	3501	Mamulov (1961)	—
$SnSe$	3058	Mamulov (1961)	—
$PbSe$	3050	Mamulov (1961)	—
Selenites			
Li_2SeO_3	2171	Klushina, Selivanova (1972)	—
Na_2SeO_3	1950	Klushina, Selivanova (1972)	1931
K_2SeO_3	1774	Klushina, Selivanova (1972)	—
Rb_2SeO_3	1715	Klushina, Selivanova (1972)	1675
Cs_2SeO_3	1640	Klushina, Selivanova (1972)	—
Tl_2SeO_3	1879	Klushina, Selivanova (1972)	—
Ag_2SeO_3	2113	Klushina, Selivanova (1972)	—
$BeSeO_3$	3322	Klushina, Selivanova (1972)	—
$MgSeO_3$	3012	Klushina, Selivanova (1972)	2996
$CaSeO_3$	2732	Klushina, Selivanova (1972)	—
$SrSeO_3$	2586	Klushina, Selivanova (1972)	2586
$BaSeO_3$	2460	Klushina, Selivanova (1972)	2448
$RaSeO_3$	2456	Klushina, Selivanova (1972)	—
$MnSeO_3$	2975	Klushina, Selivanova (1972)	—
$FeSeO_3$	2895	Klushina, Selivanova (1972)	—
$CoSeO_3$	3155	Klushina, Selivanova (1972)	—
$NiSeO_3$	2945	Klushina, Selivanova (1972)	—

Table 1 (continued)
LATTICE ENERGIES

Salt	Calculated lattice energy (kJmol⁻¹)	Literature source	Thermochemical cycle lattice energy (kJmol⁻¹)
$CuSeO_4$	3209	Klushina, Selivanova (1972)	—
$ZnSeO_4$	3167	Klushina, Selivanova (1972)	—
$CdSeO_4$	2962	Klushina, Selivanova (1972)	—
$PbSeO_4$	2669	Klushina, Selivanova (1972)	—
Selenates			
Li_2SeO_4	2054	Selivanova, Karapet'yants (1963)	—
Na_2SeO_4	1879	Selivanova, Karapet'yants (1963)	—
K_2SeO_4	1732	Selivanova, Karapet'yants (1963)	—
Rb_2SeO_4	1686	Selivanova, Karapet'yants (1963)	—
Cs_2SeO_4	1615	Selivanova, Karapet'yants (1963)	—
Cu_2SeO_4	2201	Selivanova, Karapet'yants (1963)	—
Ag_2SeO_4	2033	Selivanova, Karapet'yants (1963)	—
Tl_2SeO_4	1766	Selivanova, Karapet'yants (1963)	—
Hg_2SeO_4	2163	Selivanova, Karapet'yants (1963)	—
$BeSeO_4$	3448	Selivanova, Karapet'yants (1963)	—
$MgSeO_4$	2895	Selivanova, Karapet'yants (1963)	—
$CaSeO_4$	2632	Selivanova, Karapet'yants (1963)	—
$SrSeO_4$	2489	Selivanova, Karapet'yants (1963)	—
$BaSeO_4$	2385	Selivanova, Karapet'yants (1963)	—
$RaSeO_4$	2364	Selivanova, Karapet'yants (1963)	—
$GdSeO_4$	2753	Selivanova, Karapet'yants (1963)	—
$MnSeO_4$	2837	Selivanova, Karapet'yants (1963)	—
$FeSeO_4$	3008	Selivanova, Karapet'yants (1963)	—
$CoSeO_4$	2912	Selivanova, Karapet'yants (1963)	—
$NiSeO_4$	3079	Selivanova, Karapet'yants (1963)	—
$CuSeO_4$	3104	Selivanova, Karapet'yants (1963)	—
$ZnSeO_4$	3021	Selivanova, Karapet'yants (1963)	—
$CdSeO_4$	2833	Selivanova, Karapet'yants (1963)	—
$HgSeO_4$	2845	Selivanova, Karapet'yants (1963)	—
$PbSeO_4$	2561	Selivanova, Karapet'yants (1963)	—
Sulphides			
Li_2S	2464	Morris (1958)	2472
Na_2S	2192	Morris (1958)	2203
K_2S	1979	Morris (1958)	(2052)
Rb_2S	1929	Morris (1958)	1949
Cs_2S	1892	Karapet'yants (1954)	1850
$(NH_4)_2S$	2008	Karapet'yants (1954)	(2026)
Cu_2S	2786	Karapet'yants (1954)	2865
Ag_2S	2606	Karapet'yants (1954)	2677
Au_2S	2908	Mamulov (1961)	—
Tl_2S	2298		2258
Sulphates			
Li_2SO_4	2229	Jenkins (1975)	2142
Na_2SO_4	1827	Jenkins (1975)	1938
K_2SO_4	1700	Jenkins (1975)	1796
Rb_2SO_4	1636	Jenkins (1975)	1748
Cs_2SO_4	1596	Jenkins (1975)	1658
$(NH_4)_2SO_4$	1766	Jenkins, Smith (1975)	1777
Cu_2SO_4	2276	Selivanova, Karapet'yants (1963)	2166
Ag_2SO_4	2104	Selivanova, Karapet'yants (1963)	1989
Tl_2SO_4	1828	Selivanova, Karapet'yants (1963)	1722
Hg_2SO_4	—	Selivanova, Karapet'yants (1963)	2127
$BeSO_4$	2489	Ladd, Lee (1968)	2480
$MgSO_4$	2577	Selivanova, Karapet'yants (1963)	2484
$CaSO_4$	2469	Selivanova, Karapet'yants (1963)	2374
$SrSO_4$	2920	Selivanova, Karapet'yants (1963)	2825
$BaSO_4$	2983	Selivanova, Karapet'yants (1963)	2921
$MnSO_4$	3088	Selivanova, Karapet'yants (1963)	2917
$FeSO_4$	3167	Selivanova, Karapet'yants (1963)	3044
$CoSO_4$	3167	Selivanova, Karapet'yants (1963)	3066
$NiSO_4$	3100	Selivanova, Karapet'yants (1963)	3006
$CuSO_4$	2891	Selivanova, Karapet'yants (1963)	—
$CdSO_4$	2635	Selivanova, Karapet'yants (1963)	2534
$Sr(SO_4)_2$	—	—	9616
Ternary salts			
Cs_2CuCl_4	1393	Blake, Cotton (1964)	—
Rb_2ZnCl_4	1529	Paoletti (1965)	—
Cs_2ZnCl_4	1492	Paoletti (1965)	—
Rb_2ZnBr_4	1498	Paoletti (1965)	—
Cs_2ZnBr_4	1454	Paoletti (1965)	—
$(Me_4N)_2ZnBr_4$	1364	Paoletti (1965)	—
Cs_2ZnI_4	1386	Paoletti (1965)	—
Cs_2CaCl_4	494	Jenkins (1976)	—
Na_2AlCl_4	556	Jenkins (1976)	—
Cs_2AlCl_4	486	Jenkins (1976)	—
$NaFeCl_4$	492	Blake, Cotton (1963)	—
Rb_2CoCl_4	1447	Lister, Nyburg, Poyntz (1974)	—
Cs_2CoCl_4	1391	Lister, Nyburg, Poyntz (1974)	—
K_2PtCl_4	1574	Hartley (1972)	1550
Cs_2GeF_6	1573	Jenkins, Pratt (1978)	—
$(NH_4)_2GeF_6$	1657	Jenkins, Pratt (1978)	—
K_2PtCl_6	1375	Jenkins, Pratt (1078)	1375
K_2IrCl_6	1345	Jenkins, Pratt (1978)	1461
K_2IrCl_6	1442	Jenkins, Pratt (1978)	1440
$(NH_4)_2IrCl_6$	1442	Jenkins, Pratt (1978)	—
Na_2IrCl_6	1526	Jenkins, Pratt (1978)	—
K_2MnCl_6	1418	Jenkins, Pratt (1978)	1433
Rb_2PtCl_6	1399	Jenkins, Pratt (1978)	1399
Cs_2PtCl_6	1347	Jenkins, Pratt (1978)	1347
K_2ReCl_6	1375	Jenkins, Pratt (1978)	1398
Rb_2ReCl_6	1371	Jenkins, Pratt (1978)	1385
Cs_2ReCl_6	1381	Jenkins, Pratt (1978)	1344

Table 1 (continued)
LATTICE ENERGIES

Salt	Calculated lattice energy ($kJmol^{-1}$)	Literature source	Thermochemical cycle lattice energy ($kJmol^{-1}$)
K_2OsCl_6	1447	Jenkins, Pratt (1978)	1447
Cs_2OsCl_6	1409	Jenkins, Pratt (1978)	—
$(NH_4)_2OsCl_6$	1433	Jenkins, Pratt (1978)	—
K_2OsBr_6	1396	Jenkins, Pratt (1978)	—
K_2PdCl_6	1450	Jenkins, Pratt (1978)	1466
Rb_2PdCl_6	1449	Jenkins, Pratt (1978)	—
Cs_2PdCl_6	1426	Jenkins, Pratt (1978)	—
$(NH_4)_2PdCl_6$	1481	Jenkins, Pratt (1978)	—
Rb_2PbCl_6	1343	Jenkins, Pratt (1978)	1343
Cs_2PbCl_6	1344	Jenkins, Pratt (1978)	—
$(NH_4)_2PbCl_6$	1355	Jenkins, Pratt (1978)	—
K_2PtCl_6	1468	Jenkins, Pratt (1978)	—
Rb_2PtCl_6	1464	Jenkins, Pratt (1978)	—
Cs_2PtCl_6	1444	Jenkins, Pratt (1978)	—
$(NH_4)_2PtCl_6$	1468	Jenkins, Pratt (1978)	—
Tl_2PtCl_6	1546	Jenkins, Pratt (1978)	—
Ag_2PtCl_6	1773	Jenkins, Pratt (1978)	—
$BaPtCl_6$	2047	Jenkins, Pratt (1978)	—
K_2PtBr_6	1423	Jenkins, Pratt (1978)	—
Ag_2PtBr_6	1791	Jenkins, Pratt (1978)	—
K_2PtI_6	1421	Lister, Nyburg, Poyntz (1974)	1442
K_2ReCl_6	1416	Jenkins, Pratt (1978)	—
Rb_2ReCl_6	1414	Jenkins, Pratt (1978)	—
Cs_2ReCl_6	1398	Jenkins, Pratt (1978)	1390
$(NH_4)_2ReCl_6$	1402	Jenkins, Pratt (1978)	1375
K_2ReBr_6	1375	Jenkins, Pratt (1978)	—
K_2SiF_6	1670	Jenkins, Pratt (1978)	1628
Rb_2SiF_6	1639	Jenkins, Pratt (1978)	1621
Cs_2SiF_6	1604	Jenkins, Pratt (1978)	1498
Tl_2SiF_6	1675	Jenkins, Pratt (1978)	—
$(NH_4)_2SiF_6$	1657	Jenkins, Pratt (1978)	1727
K_2SnCl_6	1363	Jenkins, Pratt (1978)	1390
Rb_2SnCl_6	1361	Jenkins, Pratt (1978)	1363
Cs_2SnCl_6	1358	Jenkins, Pratt (1978)	—
Tl_2SnCl_6	1437	Jenkins, Pratt (1978)	1334
$(NH_4)_2SnCl_6$	1370	Jenkins, Pratt (1978)	—
Rb_2SnBr_6	1309	Jenkins, Pratt (1978)	—
Cl_2SnBr_6	1306	Jenkins, Pratt (1978)	—
$(NH_4)_2SnBr_6$	1319	Jenkins, Pratt (1978)	—
Rb_2SnI_6	1226	Jenkins, Pratt (1978)	—
Cs_2SnI_6	1243	Jenkins, Pratt (1978)	—
K_2TeCl_6	1318	Jenkins, Pratt (1978)	1320
Rb_2TeCl_6	1321	Jenkins, Pratt (1978)	—
Cs_2TeCl_6	1323	Jenkins, Pratt (1978)	—
Tl_2TeCl_6	1392	Jenkins, Pratt (1978)	—
$(NH_4)_2TeCl_6$	1318	Jenkins, Pratt (1978)	—
K_2RuCl_6	1451	Jenkins, Pratt (1978)	—
Rb_2CoF_6	1688	Jenkins, Pratt (1978)	—
Cs_2CoF_6	1632	Jenkins, Pratt (1978)	—
K_2NiF_6	1721	Jenkins, Pratt (1978)	—
Rb_2NiF_6	1688	Jenkins, Pratt (1978)	—
Rb_2SbCl_6	1357	Jenkins, Pratt (1978)	—
Rb_2SeCl_6	1409	Jenkins, Pratt (1978)	—
Cs_2SeCl_6	1397	Jenkins, Pratt (1978)	—
$(NH_4)_2SeCl_6$	1420	Jenkins, Pratt (1978)	—
K_2SeBr_6	1379	Jenkins, Pratt (1978)	—
$(NH_4)_2SeBr_6$	1380	Jenkins, Pratt (1978)	—
$(NH_4)_2PoCl_6$	1338	Jenkins, Pratt (1978)	—
Cs_2PoBr_6	1286	Jenkins, Pratt (1978)	—
$(NH_4)_2PoBr_6$	1292	Jenkins, Pratt (1978)	—
Cs_2CrF_6	1603	Jenkins, Pratt (1978)	—
Rb_2MnF_6	1688	Jenkins, Pratt (1978)	—
Cs_2MnF_6	1620	Jenkins, Pratt (1978)	—
K_2MnCl_6	1462	Jenkins, Pratt (1978)	—
Rb_2MnCl_6	1451	Jenkins, Pratt (1978)	—
$(NH_4)_2MnCl_6$	1464	Jenkins, Pratt (1978)	—
Cs_2TeBr_6	1306	Jenkins, Pratt (1978)	—
$(NH_4)_2TeBr_6$	1294	Jenkins, Pratt (1978)	—
Cs_2TeI_6	1246	Jenkins, Pratt (1978)	1443
K_2TiCl_6	1412	Jenkins, Pratt (1978)	1425
Rb_2TiCl_6	1415	Jenkins, Pratt (1978)	1370
Cs_2TiCl_6	1402	Jenkins, Pratt (1978)	1553
Tl_2TiCl_6	1560	Jenkins, Pratt (1978)	—
$(NH_4)_2TiCl_6$	1413	Jenkins, Pratt (1978)	1379
K_2TiBr_6	1379	Jenkins, Pratt (1978)	1365
Rb_2TiBr_6	1341	Jenkins, Pratt (1978)	1316
Cs_2TiBr_6	1339	Jenkins, Pratt (1978)	—
Na_2UBr_6	1504	Vdorenko, Suglobova, Chirkst (1974)	—
K_2UBr_6	1484	Vdorenko, Suglobova, Chirkst (1974)	—
Rb_2UBr_6	1473	Vdorenko, Suglobova, Chirkst (1974)	—
Cs_2UBr_6	1459	Vdorenko, Suglobova, Chirkst (1974)	—
K_2WCl_6	1398	Jenkins, Pratt (1978)	1423
Rb_2WCl_6	1397	Jenkins, Pratt (1978)	1434
Cs_2WCl_6	1392	Jenkins, Pratt (1978)	1366
K_2WBr_6	1408	Jenkins, Pratt (1978)	1408
Rb_2WBr_6	1361	Jenkins, Pratt (1978)	1391
Cs_2WBr_6	1362	Jenkins, Pratt (1978)	1332
K_2ZrCl_6	1339	Jenkins, Pratt (1978)	1371
Rb_2ZrCl_6	1341	Jenkins, Pratt (1978)	—
Cs_2ZrCl_6	1339	Jenkins, Pratt (1978)	1308
Tetraalkyl ammonium salts			
Me_4NCl	566	Wilson (1976)	—

Table 1 (continued)
LATTICE ENERGIES

Salt	Calculated lattice energy (kJmol⁻¹)	Literature source	Thermochemical cycle lattice energy (kJmol⁻¹)
Me₄NBr	553	Wilson (1976)	—
Me₄NI	544	Wilson (1976)	—
Tellurides			
Li₂Te	2212	Bevan, Morris (1960)	—
Na₂Te	1997	Bevan, Morris (1960)	2095
K₂Te	1830	Bevans, Morris (1960)	—
Rb₂Te	1837	Bevan, Morris (1960)	—
Cs₂Te	1745	Bevan, Morris (1960)	—
Cu₂Te	2706	Mamulov (1961)	2683
Ag₂Te	2607	Mamulov (1961)	2600
Tl₂Te	2084	Mamulov (1961)	2172
BeTe	3319	Das, Keer, Rao (1963)	3081
MgTe	2878	Das, Keer, Rao (1963)	—
CaTe	2721	Das, Keer, Rao (1963)	—
SrTe	2599	Das, Keer, Rao (1963)	—
BaTe	2473	Das, Keer, Rao (1963)	—
RaTe	2481	Yatsimirskii, Krestov (1960)	—
MnTe	3041	Mamulov (1961)	—
FeTe	3399	Mamulov (1961)	—
CoTe	3429	Mamulov (1961)	—
NiTe	3534	Mamulov (1961)	—
CuTe	3639	Mamulov (1961)	—
ZnTe	3416	Mamulov (1961)	—
CdTe	3212	Mamulov (1961)	—
PbTe	2930	Mamulov (1961)	—
Thiocyanates			
LiCNS	764	Gill, Singla, Paul, Narula (1972)	(765)
NaCNS	682	Morris (1958)	682
KCNS	623	Morris (1958)	616
RbCNS	623	Morris (1958)	619
CsCNS	623	Yatsimirskii (1956)	568
NH₄CNS	605	Gill, Singla, Paul, Narula (1972)	611
Ca(CNS)₂	2184	Yatsimirskii (1956)	2118
Sr(CNS)₂	2063	Yatsimirskii (1956)	1957
Ba(CNS)₂	1979	Yatsimirskii (1956)	1852
Mn(CNS)₂	2280	Yatsimirskii (1961)	2351
Zn(CNS)₂	2335	Yatsimirskii (1961)	2560
Cd(CNS)₂	2201	Yatsimirskii (1961)	2374
Hg(CNS)₂	2146	Yatsimirskii (1961)	2492
Sn(CNS)₂	2117	Yatsimirskii (1961)	2142
Pb(CNS)₂	2058	Morris (1958)	—
Vanadates			
LiVO₃	810	Golovkin, Fotiev (1970)	—
NaVO₃	761	Golovkin, Fotiev (1970)	—
KVO₃	686	Golovkin, Fotiev (1970)	—
RbVO₃	657	Golovkin, Fotiev (1970)	—
CsVO₃	628	Golovkin, Fotiev (1970)	—

* Jenkins and Waddington refers to a forthcoming review article: "The Use of Lattice Energy in Inorganic Chemistry"; publication details can be obtained by writing to H. D. B. Jenkins, Department of Chemistry, University of Warwick, Coventry CV 4 7AL, Warwickshire.

Table 2
ANCILLARY THERMODYNAMIC DATA

Salt or ion	State	Source	ΔH_f^{θ} (kJmol^{-1})	Salt or ion	State	Source	ΔH_f^{θ} (kJmol^{-1})
Li$^+$	g	JANAF	687.16	LiClO$_4$	c	JANAF	-380.7
Na$^+$	g	JANAF	609.84	CsClO$_4$	c	Wilcox, Bromley (1963)	(-435.1)
K$^+$	g	JANAF	514.19	LiH	c	JANAF	-90.62
Cs$^+$	g	JANAF	452.3	NaH	c	JANAF	-56.4
O$_2^-$	g	Jenkins, Waddington*	-74	KH	c	JANAF	-57.8
CN$^-$	g	Jenkins, Pratt, Waddington (1977)	41	CsH	c	Smith, Bass (1963)	-49.9
NH$_4^+$	g	Jenkins, Morris (1977)	630.2	LiN$_3$	c	Gray, Waddington (1956)	10.8
C$_2^{-2}$	g	Jenkins, Waddington*	918	NaN$_3$	c	Gray, Waddington (1956)	21.25
O$_2^{-2}$	g	Jenkins, Waddington*	553	KN$_3$	c	Gray, Waddington (1956)	-1.38
NO$_3^-$	g	Jenkins, Morris (1977)	-320.1	RbN$_3$	c	Gray, Waddington (1956)	-0.29
IO$_2$F$_2^-$	g	Jenkins, Waddington*	-693	CsN$_3$	c	Gray, Waddington (1956)	-9.92
IO$_3^-$	g	Jenkins, Waddington*	-208	TlN$_3$	c	Gray, Waddington (1956)	233.4
ClO$_3^-$	g	Jenkins, Waddington*	-200	AgN$_3$	c	Gray, Waddington (1956)	310.3
BrO$_3^-$	g	Jenkins, Waddington*	-145	CsHS	c	NBS Circular 500	-263.2
HSO$_4^-$	g	Jenkins, Waddington*	(-1012)	Ca(HS)$_2$	c	Wilcox, Bromley (1963)	(-481)
HCO$_3^-$	g	Jenkins, Waddington*	-738	Ba(HS)$_2$	c	Wilcox, Bromley (1956)	(-531.4)
NH$_2$CH$_2$CO$_2^-$	g	Bernard	-564	LiHS	c	Juza, Laurer (1953)	-254.4
N^{-3}	g	Jenkins, Waddington*	(2588)	K$_2$TiCl$_6$	c	Jenkins, Pratt (1978)[b]	-1747
CO$_3^{-2}$	g	Jenkins, Waddington, Pratt	-321	Rb$_2$TiCl$_6$	c	Jenkins, Pratt (1978)[b]	-1767
PO$_4^{-3}$	g	Jenkins, Waddington*	(291)	Cs$_2$TiCl$_6$	c	Jenkins, Pratt (1978)[b]	-1797
AsO$_4^{-3}$	g	Jenkins, Waddington*	(289)	Tl$_2$TiCl$_6$	c	Jenkins, Pratt (1978)[b]	-1330
GeO$_4^{-4}$	g	Jenkins, Waddington*	(1460)	K$_2$TiBr$_6$	c	Jenkins, Pratt (1978)[b]	-1493
CH$_3$COO$^-$	g	Jenkins, Waddington*	-554	Rb$_2$TiBr$_6$	c	Jenkins, Pratt (1978)[b]	-1517
ClO$_4^-$	g	Jenkins, Pratt (1978)	-344	Cs$_2$TiBr$_6$	c	Jenkins, Pratt (1978)[b]	-1553
NO$_2^-$	g	Jenkins (1977)	-219	K$_2$ZrCl$_6$	c	Jenkins, Pratt (1978)[b]	-1932
SeO$_3^{-2}$	g	Jenkins, Waddington*	(-249)	Cs$_2$ZrCl$_6$	c	Jenkins, Pratt (1978)[b]	-1992
HCO$_2^-$	g	Jenkins, Waddington*	-463	K$_2$HfCl$_6$	c	Jenkins, Pratt (1978)[b]	-1957
BH$_4^-$	g	Waddington (1959)	-96	K$_2$NbCl$_6$	c	Jenkins, Pratt (1978)[b]	-1594
HS$^-$	g	Jenkins, Waddington*	-120	Rb$_2$NbCl$_6$	c	Jenkins, Pratt (1978)[b]	-1619
TiCl$_6^{-2}$	g	Jenkins, Pratt (1978)	-1330	Cs$_2$NbCl$_6$	c	Jenkins, Pratt (1978)[b]	-1663
TiBr$_6^{-2}$	g	Jenkins, Pratt (1978)	-1142	K$_2$ReCl$_6$	c	Jenkins, Pratt (1978)[b]	-1333
ZrCl$_6^{-2}$	g	Jenkins, Pratt (1978)	-1526	(NH$_4$)$_2$ReCl$_6$	c	Jenkins, Pratt (1978)[b]	-1056
HfCl$_6^{-2}$	g	Jenkins, Pratt (1978)	-1640	K$_2$ReBr$_6$	c	Jenkins, Pratt (1978)[b]	-1036
NbCl$_6^{-2}$	g	Jenkins, Pratt (1978)	-1224	K$_2$OsCl$_6$	c	Jenkins, Pratt (1978)[b]	-1171
TaCl$_6^{-2}$	g	Jenkins, Pratt (1978)	-1275	K$_2$IrCl$_6$	c	Jenkins, Pratt (1978)[b]	-1197
MoCl$_6^{-2}$	g	Jenkins, Pratt (1978)	-1070	K$_2$SiF$_6$	c	Jenkins, Pratt (1978)[b]	-2807
WCl$_6^{-2}$	g	Jenkins, Pratt (1978)	-985	Rb$_2$SiF$_6$	c	Jenkins, Pratt (1978)[b]	-2838
WBr$_6^{-2}$	g	Jenkins, Pratt (1978)	-705	Cs$_2$SiF$_6$	c	Jenkins, Pratt (1978)[b]	-2801
ReCl$_6^{-2}$	g	Jenkins, Pratt (1978)	-919	(NH$_4$)$_2$SiF$_6$	c	Jenkins, Pratt (1978)[b]	-2681
ReBr$_6^{-2}$	g	Jenkins, Pratt (1978)	-689	Rb$_2$PbCl$_6$	c	Jenkins, Pratt (1978)[b]	-1293
OsCl$_6^{-2}$	g	Jenkins, Pratt (1978)	-752	Mn(CN)$_2$	c	Karapet'yants, Karapet'yants	60.7
IrCl$_6^{-2}$	g	Jenkins, Pratt (1978)	-785	RbCN	c	Pritchard (1953)	(-108.8)
PdCl$_6^{-2}$	g	Jenkins, Pratt (1978)	-749	CsCN	c	Pritchard (1953)	(-108.7)
PtCl$_6^{-2}$	g	Jenkins, Pratt (1978)	-774	LiCN	c	Jenkins, Pratt, Waddington (1977)	-121
PtBr$_6^{-2}$	g	Jenkins, Pratt (1978)	-645	Sr(CN)$_2$	c	Wilcox, Bromley (1963)	(-205)
SiF$_6^{-2}$	g	Jenkins, Pratt (1978)	-2207	LiNo$_3$	c	NBS Circular 500	-482.3
GeCl$_6^{-2}$	g	Jenkins, Pratt (1978)	-981	RbNO$_3$	c	NBS Circular 500	-489.7
SuCl$_6^{-2}$	g	Jenkins, Pratt (1978)	-1156	CsNO$_3$	c	NBS Circular 500	-494.2
PbCl$_6^{-2}$	g	Jenkins, Pratt (1978)	-940	Fe(NO$_3$)$_2$	c	Wilcox, Bromley (1963)	(-447.7)
TeCl$_6^{-2}$	g	Jenkins, Pratt (1978)	-902	Sn(NO$_3$)$_2$	c	Yatsimirskii (1961)	-456.1
NO$_2^-$	g	Jenkins, Pratt (1978)	-202	NaBF$_4$	c	Altschuller (1955)	-1774
LiO$_2$	c	D'Orazio, Wood (1965)	-259.4	CsIO$_3$	c	Shedlovskii, Voskresenskii (1966)	-526.3
NaO$_2$	c	JANAF	-260.7	CsClO$_3$	c	Volodina, Shidlovskii, Voskresenskii (1966)	-395.8
KO$_2$	c	JANAF 1974 Suppl.	-284.5	Mg(ClO$_3$)$_2$	c	Wilcox, Bromley (1963)	(-523)
CsO$_2$	c	D'Orazio, Wood (1965)	-289.5	Ca(ClO$_3$)$_2$	c	Rudnitskii (1961)	-757
Li$_2$O$_2$	c	JANAF	-632.6	Sr(ClO$_3$)$_2$	c	Rudnitskii (1961)	-766
Na$_2$O$_2$	c	JANAF	-531.2	ScN	c	Neuman, Kroger, Kunz	-284.5
K$_2$O$_2$	c	JANAF	-495.8	Li$_2$CO$_3$	c	JANAF	-1216.0
Cs$_2$O$_2$	c	NBS Circular 500	-402.5	Cs$_2$CO$_3$	c	NBS Circular 500	-1118.8
ThO$_2$	c	Smirnov, Ivanovskii (1957)	-1230.5	NiCO$_3$	c	Karapet'yants (1955)	-689.1
MgO$_2$	c	NBS Circular 500	-623.0	CuCO$_3$	c	NBS Circular 500	-595.0
NaIO$_2$F$_2$	c	Finch, Gates, Jenkinson	-848.7	SnCO$_3$	c	Wilcox, Bromley (1963)	(-740.6)
KIO$_2$F$_2$	c	Finch, Gates, Jenkinson	-876.6	HCO$_3$Li	c	Morris (1959)	-649.3
RbIO$_2$F$_2$	c	Finch, Gates, Jenkinson	-875.3	HCO$_3$Rb	c	Morris(1959)	-656.0
NH$_4$IO$_2$F$_2$	c	Finch, Gates, Jenkinson	-747.7	TiH	c	Gibb (1962)	-130.5
LiBrO$_3$	c	Boyd, Vaslow (1962)	-356.4	ZrH	c	Gibb (1962)	-173.5
CsBrO$_3$	c	Boyd, Vaslow (1962)	-374.5	VH	c	Gibb (1962)	(-32)
CsHCO$_3$	c	NBS Circular 500	-995.6	NbH	c	Gibb (1962)	(-100)
Ca(HCO$_3$)$_2$	c	Wilcox, Bromley (1963)	(-1950)	PdH	c	Gibb (1962)	(-37)
Sr(HCO$_3$)$_2$	c	Wilcox, Bromley (1963)	(-1954)	CuH	c	Gibb (1962)	(-21.4)
Ba(HCO$_3$)$_2$	c	Wilcox, Bromley (1963)	(-1971)	LaH$_2$	c	Gibb (1962)	-230.1
CH$_3$COONH$_4$	c	NBS Circular 500	-618.4	CeH$_2$	c	Gibb (1962)	-141.8
CH$_3$COOLi	c	Rudnitskii (1961)	-748.9				
CH$_3$COORb	c	Morris (1959)	-720.9				

Table 2 (continued)
ANCILLARY THERMODYNAMIC DATA

Salt or ion	State	Source	ΔH_f^{θ} (kJmol^{-1})	Salt or ion	State	Source	ΔH_f^{θ} (kJmol^{-1})
PrH$_2$	c	Gibb (1962)	−200.0	Rb$_2$WCl$_6$	c	Jenkins, Pratt (1978)[b]	−1429
NdH$_2$	c	Gibb (1962)	−187.4	Cs$_2$WCl$_6$	c	Jenkins, Pratt (1978)[b]	−1446
SmH$_2$	c	Gibb (1962)	−187.4	K$_2$WBr$_6$	c	Jenkins, Pratt (1978)[b]	−1085
GdH$_2$	c	Gibb (1962)	−196.2	Rb$_2$WBr$_6$	c	Jenkins, Pratt (1978)[b]	−1106
CsOH	c	NBS Circular 500	−406.7	Cs$_2$WBr$_6$	c	Jenkins, Pratt (1978)[b]	−1132
Ti(OH)$_2$	c	Wilcox, Bromley (1963)	(−778.0)	K$_2$PdCl$_6$	c	Jenkins, Pratt (1978)[b]	−1187
LiOH	c	NBS Circular 500	−487.2	Ag$_2$PtCl$_6$	c	Jenkins, Pratt (1978)[b]	−527
K$_2$TaCl$_6$	c	Jenkins, Pratt (1978)[b]	−1648	BaPtCl$_6$	c	Jenkins, Pratt (1978)[b]	−1180
Rb$_2$TaCl$_6$	c	Jenkins, Pratt (1978)[b]	−1669	K$_2$PtBr$_6$	c	Jenkins, Pratt (1978)[b]	−1040
Cs$_2$TaCl$_6$	c	Jenkins, Pratt (1978)[b]	−1711	Ag$_2$PtBr$_6$	c	Jenkins, Pratt (1978)[b]	−398
Na$_2$MoCl$_6$	c	Jenkins, Pratt (1978)[b]	−1376	Cs$_2$GeCl$_6$	c	Jenkins, Pratt (1978)[b]	−1451
K$_2$MoCl$_6$	c	Jenkins, Pratt (1978)[b]	−1475	K$_2$SnCl$_6$	c	Jenkins, Pratt (1978)[b]	−1518
Rb$_2$MoCl$_6$	c	Jenkins, Pratt (1978)[b]	−1479	Rb$_2$SnCl$_6$	c	Jenkins, Pratt (1978)[b]	−1529
Cs$_2$MoCl$_6$	c	Jenkins, Pratt (1978)[b]	−1512	(NH$_4$)$_2$SnCl$_6$	c	Jenkins, Pratt (1978)[b]	−1237
K$_2$WCl$_6$	c	Jenkins, Pratt (1978)[b]	−1380	Rb$_2$TeCl$_6$	c	Jenkins, Pratt (1978)[b]	−1237

[a] Jenkins and Waddington refers to a forthcoming review article: "The Use of Lattice Energy in Inorganic Chemistry"; publication details can be obtained by writing to H. D. B. Jenkins, Department of Chemistry, University of Warwick, Coventry CV 4 7AL, Warwickshire.
[b] For sources of data used in this ancillary table, refer to *Adv. Inorg. Chem. Radiochem.*, 22, 1, 1978.

THERMODYNAMIC PROPERTIES OF ALKANE HYDROCARBONS[a]

Symbols and Units

$(G° - H°_0)/T$	=	Gibbs energy function, cal K^{-1} mol^{-1}
$(H° - H°_0)/T$	=	enthalpy function, cal K^{-1} mol^{-1}
$S°$	=	entropy, cal K^{-1} mol^{-1}
C_p	=	heat capacity at constant pressure, cal K^{-1} mol^{-1}
$\Delta Hf°$	=	enthalpy of formation, kcal mol^{-1}
$\Delta Gf°$	=	Gibbs energy of formation, kcal mol^{-1}
$\log_{10} Kf$	=	common logarithm of the equilibrium constant of formation
T	=	temperature, kelvins
$H° - H°_0$	=	enthalpy, kcal mol^{-1}

Conversion Factors

The tabulated values may be converted to SI units by use of the relations:

calories × 4.184 = joules
kilocalories × 4184 = joules

T, K	$(G° - H°_0)/T$, cal K^{-1} mol^{-1}	$(H° - H°_0)/T$, cal K^{-1} mol^{-1}	$H° - H°_0$, kcal mol^{-1}	$S°$, cal K^{-1} mol^{-1}	$C_p°$, cal K^{-1} mol^{-1}	$\Delta H_f°$, kcal mol^{-1}	$\Delta G_f°$, kcal mol^{-1}	$\log_{10} K_f$
				Methane (CH$_4$)				
0	0	0	0	0	0	−15.9$_5$	−15.9$_5$	∞
200	−33.3$_1$	7.9$_4$	1.58$_8$	41.2$_5$	8.0$_1$	−17.1$_8$	−13.8$_8$	15.1$_7$
273.15	−35.7$_9$	7.9$_9$	2.18$_4$	43.7$_8$	8.3$_3$	−17.6$_8$	−12.6$_0$	10.0$_8$
298.15	−36.4$_6$	8.0$_3$	2.39	44.5$_2$	8.5$_3$	−17.8$_9$	−12.1	8.8$_8$
400	−38.8$_8$	8.3$_0$	3.32	47.1$_8$	9.7$_1$	−18.6$_0$	−10.0$_4$	5.4$_9$
600	−42.4	9.2	5.5$_4$	51.6	12.5	−19.9	−5.5	1.9$_9$
800	−45.2	10.4	8.3	55.6	15.2	−20.8	−0.5	0.1$_4$
1,000	−47.7	11.6	11.6	59.3	17.$_4$	−21.4	4.6	−1.0$_0$
1,200	−49.$_9$	12.$_7$	15.$_3$	62.$_6$	19.$_2$	−21.$_7$	9.$_4$	−1.7$_2$
1,500	−52.$_9$	14.$_2$	21.$_3$	67.$_1$	21.	−21.$_8$	17.$_7$	−2.5$_9$
				Ethane (C$_2$H$_6$)				
0	0	0	0	0	0	−16.2$_6$	−16.2$_6$	∞
200	−41.6$_3$	8.6$_7$	1.73$_5$	50.3$_0$	10.1$_0$	−18.8$_0$	−11.4$_4$	12.5$_2$
273.15	−44.4$_2$	9.2$_8$	2.53$_4$	53.7$_0$	11.8$_5$	−19.7$_0$	−8.6$_3$	6.9$_1$
298.15	−45.2$_4$	9.5$_2$	2.84	54.7$_6$	12.5$_7$	−20.0$_0$	−7.6$_0$	5.5$_7$
400	−48.2$_0$	10.7$_1$	4.28	58.9$_1$	15.8$_3$	−21.1$_8$	−3.1$_8$	1.7$_4$
600	−53.1	13.5	8.0$_9$	66.6	22.2	−23.0	6.2	−2.2$_7$
800	−57.3	16.7	13.3	74.0	27.4	−23.7	16.1	−4.4$_1$
1,000	−61.3	19.0	19.0	80.3	31.$_7$	−24.3	26.2	−5.7$_3$
1,200	−65.$_0$	21.$_4$	25.$_7$	86.$_4$	35.$_1$	−24.$_1$	36.$_3$	−6.6$_1$
1,500	−70.$_1$	24.$_6$	36.$_9$	94.$_7$	39.	−23.$_1$	51.$_3$	−7.4$_7$
				Propane (C$_3$H$_8$)				
0	0	0	0	0	0	−19.7$_4$	−19.7$_4$	∞
200	−48.3$_6$	10.2$_1$	2.04$_1$	58.5$_7$	12.6$_4$	−23.4$_2$	−11.8$_8$	12.9$_8$
273.15	−51.7$_0$	11.3$_3$	3.09$_6$	63.0$_3$	16.2$_2$	−24.6$_6$	−7.4$_7$	5.9$_7$

THERMODYNAMIC PROPERTIES OF ALKANE HYDROCARBONS (Continued)

T, K	$(G° - H°_0)/T$, cal K⁻¹ mol⁻¹	$(H° - H°_0)/T$, cal K⁻¹ mol⁻¹	$H° - H°_0$, kcal mol⁻¹	$S°$, cal K⁻¹ mol⁻¹	$C_p°$, cal K⁻¹ mol⁻¹	$\Delta H_f°$, kcal mol⁻¹	$\Delta G_f°$, kcal mol⁻¹	$\log_{10} K_f$
298.15	-52.7_1	11.8_0	3.52	64.5_1	17.4_4	-25.0_7	-5.8_7	4.3_0
400	-56.4_6	13.8_8	5.55	70.3_4	22.4_6	-26.6_1	0.9_4	-0.5_1
600	-62.9	18.2	10.9_5	81.1	31.2	-28.9	15.2	-5.5_5
800	-68.7	22.4	17.9	91.1	38.0	-30.2	30.2	-8.2_5
1,000	-74.1	26.1	26.1	100.2	$43._4$	-30.8	45.4	-9.9_2
1,200	$-79._2$	$29._3$	$35._2$	$108._5$	$47._6$	$-30._7$	$60._6$	-11.0_4
1,500	$-86._2$	$33._5$	$50._2$	$119._7$	$52.$	$-29._7$	$83._3$	-12.1_4

n-Butane (C₄H₁₀)

T, K	$(G° - H°_0)/T$	$(H° - H°_0)/T$	$H° - H°_0$	$S°$	$C_p°$	$\Delta H_f°$	$\Delta G_f°$	$\log_{10} K_f$
0	0	0	0	0	0	-23.6_3	-23.6_3	∞
200	-52.5_7	13.4_1	2.68_1	65.9_8	18.3_6	-28.1_3	-12.2_3	13.3_6
273.15	-57.0_1	15.1_8	4.14_5	72.1_9	21.8_9	-29.5_7	-6.1_9	4.9_5
298.15	-58.3_7	15.8_0	4.71	74.1_7	23.3_0	-30.0_5	-4.0_2	2.9_5
400	-63.3_8	18.4_7	7.40	81.8_7	29.4_5	-31.9_0	5.1_7	-2.8_2
600	-71.9	24.0	14.4_2	95.9	40.4	-34.7	24.3	-8.8_7
800	-79.6	29.2	23.4	108.8	48.8	-36.3	44.3	-12.1_0
1,000	-86.6	33.8	33.8	120.4	$55._2$	-37.0	64.5	-14.1_0
1,200	$-93._1$	$37._8$	$45._4$	$130._4$	$60._1$	$-37._0$	$84._1$	-15.4_5
1,500	$-102._1$	$42._8$	$64._3$	$144._9$	$66.$	$-36._1$	$115._2$	-16.7_8

2-Methylpropane (C₄H₁₀)

T, K	$(G° - H°_0)/T$	$(H° - H°_0)/T$	$H° - H°_0$	$S°$	$C_p°$	$\Delta H_f°$	$\Delta G_f°$	$\log_{10} K_f$
0	0	0	0	0	0	-25.3_4	-25.3_4	∞
200	-50.7_9	11.7_2	2.34_4	62.5_1	16.8_5	-30.1_7	-13.5_4	14.8_4
273.15	-54.7_3	13.7_1	3.74_4	68.4_4	21.5_2	-31.6_7	-7.2_7	5.8_2
298.15	-55.9_7	14.4_4	4.30	70.4_1	23.1_3	-32.1_6	-5.0_0	3.6_7
400	-60.6_3	17.4_9	6.99	78.1_2	29.6_2	-34.0_0	4.5_6	-2.4_9
600	-68.9	23.4	14.0_1	92.3	40.8	-36.7	24.5	-8.9_2
800	-76.4	28.9	23.1	105.3	49.5	-38.2	45.1	-12.3_3
1,000	-83.3	33.8	33.8	117.1	$56._4$	-38.7	66.1	-14.4_4
1,200	$-89._9$	$38._0$	$45._6$	$127._9$	$61._2$	$-38._4$	$87._0$	-15.8_0
1,500	$-99._0$	$43._3$	$65._0$	$142._3$	$68.$	$-37._0$	$118._2$	-17.2_2

n-Pentane (C₅H₁₂)

T, K	$(G° - H°_0)/T$	$(H° - H°_0)/T$	$H° - H°_0$	$S°$	$C_p°$	$\Delta H_f°$	$\Delta G_f°$	$\log_{10} K_f$
0	0	0	0	0	0	-27.3_5	-27.3_5	∞
200	-57.0_3	16.4_6	3.29_1	73.4_9	22.3_6	-32.7_0	-12.4_5	13.6_0
273.15	-62.4_7	18.6_1	5.08_1	81.0_6	26.9_0	-34.4_1	-4.7_7	3.8_2
298.15	-64.1_4	19.3_7	5.78	83.5_2	28.6_9	-34.9_4	-2.0_3	1.4_9
400	-70.2_9	22.7_4	9.10	93.0_3	36.4_6	-37.1_6	9.5_8	-5.2_3
600	-80.8	29.7	17.7_9	110.5	49.9	-40.3	33.7	-12.2_7
800	-90.3	36.0	28.8	126.3	59.9	-42.2	58.7	-16.0_3
1,000	-98.9	41.6	41.6	140.5	$67._3$	-43.0	84.0	-18.3_3
1,200	$-106._9$	$46._3$	$55._6$	$153._2$	$72._8$	$-43._0$	$109._4$	-19.9_2
1,500	$-117._9$	$52._3$	$78._4$	$170._2$	$79.$	$-42._1$	$147._4$	-21.4_6

2-Methybutane (C₅H₁₂)

T, K	$(G° - H°_0)/T$	$(H° - H°_0)/T$	$H° - H°_0$	$S°$	$C_p°$	$\Delta H_f°$	$\Delta G_f°$	$\log_{10} K_f$
0	0	0	0	0	0	-28.4_4	-28.4_4	∞
200	-58.1_6	14.3_4	2.86_7	72.5_0	20.3_0	-34.2_0	-13.7_6	15.0_3
273.15	-62.9_6	16.7_4	4.57_3	79.7_2	26.3_8	-36.0_0	-6.0_0	4.8_0
298.15	-64.4_9	17.6_4	5.26	82.1_3	28.4_1	-36.5_4	-3.2_1	2.3_6
400	-70.2_0	21.4_3	8.57	91.6_3	36.5_4	-38.7_6	8.5_3	-4.6_6
600	-80.3	28.8	17.3_0	109.1	50.2	-41.9	32.9	-11.9_9
800	-89.5	35.5	28.4	125.0	60.5	-43.7	58.2	-15.8_9
1,000	-98.1	41.3	41.3	139.4	$68._4$	-44.3	83.7	-18.3_0
1,200	$-106._1$	$46._3$	$55._6$	$152._4$	$74._4$	$-44._1$	$109._3$	-19.9_1
1,500	$-117._1$	$52._7$	$79._0$	$169._8$	$81.$	$-42._6$	$147._5$	-21.4_9

2,2-Dimethylpropane (C₅H₁₂)

T, K	$(G° - H°_0)/T$	$(H° - H°_0)/T$	$H° - H°_0$	$S°$	$C_p°$	$\Delta H_f°$	$\Delta G_f°$	$\log_{10} K_f$
0	0	0	0	0	0	-32.3_8	-32.3_8	∞
200	-48.3_2	13.4_3	2.68_5	61.7_5	19.2_5	-38.3_3	-15.7_4	17.2_0
273.15	-53.0_1	17.4_6	4.77_0	70.4_7	26.6_6	-39.7_6	-7.2_2	5.7_8
298.15	-54.5_3	18.5_8	5.54	73.1_1	28.8_8	-40.2_5	-4.1_9	3.0_7
400	-60.3_3	22.7_2	9.09	83.0_5	37.2_8	-42.2_0	8.5_3	-4.6_6
600	-70.7	30.2	18.1_1	100.9	51.3	-45.0	34.7	-12.6_5
800	-80.2	36.9	29.6	117.1	62.4	-46.5	61.7	-16.8_5
1,000	-88.9	43.0	43.0	131.9	$71._2$	-46.6	88.9	-19.4_3
1,200	$-97._2$	$48._3$	$57._9$	$145._5$	$78._2$	$-45._7$	$116._0$	-21.1_3
1,500	$-108._7$	$55._1$	$82._7$	$163._8$	$86.$	$-42._9$	$156._2$	-22.7_6

n-Hexane (C₆H₁₄)

T, K	$(G° - H°_0)/T$	$(H° - H°_0)/T$	$H° - H°_0$	$S°$	$C_p°$	$\Delta H_f°$	$\Delta G_f°$	$\log_{10} K_f$
0	0	0	0	0	0	-31.0_9	-31.0_9	∞
200	-61.4_5	19.5_6	3.91_1	81.0_1	26.4_3	-37.2_6	-12.6_8	13.8_5
273.15	-67.9_2	22.1_0	6.03_1	90.0_2	31.9_2	-39.2_2	-3.3_5	2.7_0
298.15	-69.9_0	23.0_1	6.86	92.9_6	34.0_6	-39.9_1	-0.0_3	0.0_3
400	-77.2_1	27.0_1	10.80	104.2_2	43.3_9	-42.4_4	13.9_7	-7.6_3

T, K	$(G° - H°_0)/T$, cal K⁻¹ mol⁻¹	$(H° - H°_0)/T$, cal K⁻¹ mol⁻¹	$H° - H°_0$, kcal mol⁻¹	$S°$, cal K⁻¹ mol⁻¹	$C_p°$, cal K⁻¹ mol⁻¹	$\Delta H_f°$, kcal mol⁻¹	$\Delta G_f°$, kcal mol⁻¹	$\log_{10} K_f$
600	-89.7	35.3	21.1_5	125.0	59.3	-46.0	43.0	-15.6_7
800	-100.9	42.8	34.2	143.7	70.8	-48.1	73.1	-19.9_6
1,000	-111.2	49.3	49.3	160.5	$79._2$	-49.0	103.5	-22.6_2
1,200	$-120._7$	$54._8$	$65._8$	$175._5$	$85._4$	$-49._1$	$134._0$	-24.4_0
1,500	$-133._7$	$61._7$	$92._5$	$195._4$	$93.$	$-48._2$	$179._7$	-26.1_8

2-Methylpentane (C_6H_{14})

T, K	$(G° - H°_0)/T$	$(H° - H°_0)/T$	$H° - H°_0$	$S°$	$C_p°$	$\Delta H_f°$	$\Delta G_f°$	$\log_{10} K_f$
0	0	0	0	0	0	-32.1_6	-32.1_6	∞
200	-62.3_5	17.1_8	3.43_6	79.5_3	24.2_2	-38.8_1	-13.9_3	15.2_1
273.15	-68.1_2	20.0_4	5.47_3	88.1_6	31.5_2	-40.8_9	-4.5_0	3.6_0
298.15	-69.9_2	21.1_1	6.29	91.0_5	33.9_7	-41.5_7	-1.1_3	0.8_3
400	-76.7_5	25.6_7	10.27	102.4_7	43.8_0	-44.0_3	13.0_8	-7.1_5
600	-88.9	34.5	20.7_1	123.4	60.0	-47.5	42.5	-15.4_7
800	-99.9	42.5	34.0	142.4	71.8	-49.5	72.8	-19.8_5
1,000	-110.1	49.3	49.3	159.4	$80._6$	-50.1	103.5	-22.6_1
1,200	$-119._6$	$55._1$	$66._1$	$174._7$	$87._2$	$-49._9$	$134._2$	-24.4_3
1,500	$-132._7$	$62._3$	$93._4$	$195._0$	$95.$	$-48._4$	$180._0$	-26.2_3

3-Methylpentane (C_6H_{14})

T, K	$(G° - H°_0)/T$	$(H° - H°_0)/T$	$H° - H°_0$	$S°$	$C_p°$	$\Delta H_f°$	$\Delta G_f°$	$\log_{10} K_f$
0	0	0	0	0	0	-31.4_8	-31.4_8	∞
200	-63.1_3	17.1_0	3.41_9	80.2_3	23.6_4	-38.1_4	-13.4_0	14.6_4
273.15	-68.8_5	19.8_4	5.41_9	88.6_9	31.0_3	-40.2_6	-4.0_1	3.2_1
298.15	-70.6_3	20.8_9	6.23	91.5_2	33.4_9	-40.9_3	-0.6_5	0.4_8
400	-77.3_9	25.3_6	10.14	102.7_5	43.3_0	-43.4_6	13.5_1	-7.3_8
600	-89.3	34.2	20.5_0	123.5	59.5	-47.1	42.9	-15.6_1
800	-100.3	42.1	33.6	142.4	71.4	-49.1	73.2	-19.9_9
1,000	-110.4	48.9	48.9	159.3	$80._3$	-49.8	103.9	-22.7_0
1,200	$-119._9$	$54._7$	$65._6$	$174._6$	$87._0$	$-49._7$	$134._1$	-24.5_1
1,500	$-132._9$	$61._9$	$92._9$	$194._8$	$95.$	$-48._2$	$180._5$	-26.3_0

2,2-Dimethylbutane (C_6H_{14})

T, K	$(G° - H°_0)/T$	$(H° - H°_0)/T$	$H° - H°_0$	$S°$	$C_p°$	$\Delta H_f°$	$\Delta G_f°$	$\log_{10} K_f$
0	0	0	0	0	0	-34.2_8	-34.2_8	∞
200	-58.3_5	15.8_2	3.16_3	74.1_7	24.2_5	-41.2_0	-15.2_4	16.6_5
273.15	-63.7_5	19.0_0	5.19_1	82.7_7	31.2_4	$-43.2_?$	$-1.4_?$	$4._?$
298.15	-65.4_7	20.1_5	6.01	85.6_2	33.8_1	-43.9_5	-1.9_1	1.4_0
400	-72.0_5	24.9_1	9.96	96.9_2	43.7_7	-46.4_4	12.8_5	-7.0_2
600	-83.9	34.1	20.4_6	118.0	60.5	-49.9	43.3	-15.7_8
800	-94.9	42.3	33.9	137.2	73.3	-51.7	74.7	-20.4_1
1,000	-105.1	49.6	49.6	154.7	$83._2$	-51.9	106.4	-23.2_5
1,200	$-114._7$	$55._8$	$67._0$	$170._5$	$90._9$	$-51._1$	$137._9$	-25.1_2
1,500	$-128._1$	$63._8$	$95._7$	$191._9$	$100.$	$-48._3$	$184._9$	-26.9_9

2,3-Dimethylbutane (C_6H_{14})

T, K	$(G° - H°_0)/T$	$(H° - H°_0)/T$	$H° - H°_0$	$S°$	$C_p°$	$\Delta H_f°$	$\Delta G_f°$	$\log_{10} K_f$
0	0	0	0	0	0	-32.2_9	-32.2_9	∞
200	-60.8_6	15.5_6	3.11_1	76.4_2	22.2_6	-39.2_7	-13.7_6	15.0_4
273.15	-66.1_2	18.5_0	5.05_1	84.6_2	30.6_7	-41.4_4	-4.0_8	3.2_7
298.15	-67.8_0	19.6_3	5.85	87.4_3	33.3_2	-42.1_2	-0.6_2	0.4_6
400	-74.2_3	24.4_3	9.77	98.6_6	43.4_3	-44.6_5	13.9_6	-7.6_3
600	-85.9	33.6	20.1_7	119.5	59.8	-48.2	44.1	-16.0_7
800	-96.7	41.8	33.4	138.5	72.1	-50.1	75.2	-20.5_5
1,000	-106.8	48.8	48.8	155.6	$81._4$	-50.7	106.7	-23.3_1
1,200	$-116._3$	$54._8$	$65._8$	$171._1$	$88._6$	$-50._3$	$138._1$	-25.1_5
1,500	$-129._4$	$62._4$	$93._7$	$191._8$	$97.$	$-48._3$	$185._1$	-26.9_5

n-Heptane (C_7H_{16})

T, K	$(G° - H°_0)/T$	$(H° - H°_0)/T$	$H° - H°_0$	$S°$	$C_p°$	$\Delta H_f°$	$\Delta G_f°$	$\log_{10} K_f$
0	0	0	0	0	0	-34.8_3	-34.8_3	∞
200	-65.8_6	22.6_7	4.53_4	88.5_3	30.5_1	-41.8_3	-12.9_0	14.1_0
273.15	-73.3_5	25.5_8	6.98_8	98.9_4	36.9_6	-44.0_9	-1.9_4	1.5_7
298.15	-75.6_4	26.6_5	7.94	102.2_9	39.4_8	-44.8_4	1.9_4	-1.4_2
400	-84.1_1	31.3_0	12.52	115.4_1	50.3_5	-47.6_7	18.3_7	-10.0_4
600	-98.6	40.9	24.5_2	139.5	68.7	-51.7	52.4	-19.0_7
800	-111.6	49.6	39.6	161.2	81.8	-54.1	87.4	-23.8_6
1,000	-123.5	57.0	57.0	180.5	$91._2$	-55.1	123.0	-26.8_7
1,200	$-134._4$	$63._3$	$76._0$	$197._7$	$98._1$	$-55._2$	$158._6$	-28.8_8
1,500	$-149._4$	$71._1$	$106._6$	$220._5$	$106.$	$-54._4$	$212._0$	-30.8_8

2-Methylhexane (C_7H_{16})

T, K	$(G° - H°_0)/T$	$(H° - H°_0)/T$	$H° - H°_0$	$S°$	$C_p°$	$\Delta H_f°$	$\Delta G_f°$	$\log_{10} K_f$
0	0	0	0	0	0	-35.9_1	-35.9_1	∞
200	-66.7_7	20.3_5	4.07_0	87.1_2	28.3_4	-43.3_8	-14.1_7	15.4_8
273.15	-73.5_8	23.5_8	6.43_9	97.1_6	36.5_1	-45.7_3	-3.1_1	2.4_9
298.15	-75.7_0	24.7_8	7.39	100.4_8	39.3_2	-46.4_8	0.8_4	-0.6_1
400	-83.6_9	29.9_4	11.98	113.6_3	50.6_6	-49.3_0	17.4_5	-9.5_4
600	-97.8	40.1	24.0_6	137.9	69.2	-53.3	51.8	-18.8_6
800	-110.6	49.1	39.3	159.7	82.6	-55.5	87.2	-23.8_4
1,000	-122.4	56.9	56.9	179.3	$92._4$	-56.3	123.0	-26.8_7

T, K	$(G° - H°_0)/T$, cal K⁻¹ mol⁻¹	$(H° - H°_0)/T$, cal K⁻¹ mol⁻¹	$H° - H°_0$, kcal mol⁻¹	$S°$, cal K⁻¹ mol⁻¹	$C_p°$, cal K⁻¹ mol⁻¹	$\Delta H_f°$, kcal mol⁻¹	$\Delta G_f°$, kcal mol⁻¹	$\log_{10} K_f$
1,200	-133.4	63.4	76.1	196.8	99.8	-56.2	158.8	-28.92
1,500	-148.4	71.6	107.4	220.0	108.	-54.7	212.4	-30.95

3-Methylhexane (C₇H₁₆)

T, K	$(G° - H°_0)/T$	$(H° - H°_0)/T$	$H° - H°_0$	$S°$	$C_p°$	$\Delta H_f°$	$\Delta G_f°$	$\log_{10} K_f$
0	-1.38	0	0	1.38	0	-35.22	-35.22	∞
200	-68.68	19.95	3.989	88.63	27.58	-42.77	-13.85	15.14
273.15	-75.36	23.14	6.322	98.50	36.20	-45.15	-2.90	2.32
298.15	-77.44	24.36	7.26	101.80	39.10	-45.91	1.01	-0.74
400	-85.32	29.60	11.84	114.92	50.60	-48.74	17.50	-9.56
600	-99.3	39.9	23.92	139.2	69.2	-52.7	51.6	-18.73
800	-112.0	49.0	39.2	161.0	82.6	-54.9	86.7	-23.68
1,000	-123.8	56.7	56.7	180.5	92.4	-55.8	122.2	-26.71
1,200	-134.4	63.3	76.0	198.1	99.8	-55.6	157.8	-28.74
1,500	-149.8	71.5	107.2	221.3	108.	-54.1	211.0	-30.75

3-Ethylpentane (C₇H₁₆)

T, K	$(G° - H°_0)/T$	$(H° - H°_0)/T$	$H° - H°_0$	$S°$	$C_p°$	$\Delta H_f°$	$\Delta G_f°$	$\log_{10} K_f$
0	0	0	0	0	0	-34.84	-34.84	∞
200	-64.14	20.50	4.099	84.64	29.96	-42.28	-12.56	13.73
273.15	-71.04	23.95	6.541	94.99	37.06	-44.55	-1.34	1.07
298.15	-73.20	25.16	7.50	98.35	39.67	-45.29	2.66	-1.95
400	-81.29	30.25	12.10	111.54	50.56	-48.10	19.49	-10.65
600	-95.5	40.2	24.13	135.7	68.9	-52.1	54.2	-19.76
800	-108.3	49.1	39.3	157.4	82.2	-54.4	90.1	-24.61
1,000	-120.1	56.8	56.8	176.9	92.1	-55.3	126.3	-27.61
1,200	-131.0	63.3	76.0	194.3	99.5	-55.2	162.6	-29.61
1,500	-146.1	71.4	107.2	217.5	108.	-53.8	217.0	-31.62

2,2-Dimethylpentane (C₇H₁₆)

T, K	$(G° - H°_0)/T$	$(H° - H°_0)/T$	$H° - H°_0$	$S°$	$C_p°$	$\Delta H_f°$	$\Delta G_f°$	$\log_{10} K_f$
0	0	0	0	0	0	-38.02	-38.02	∞
200	-62.19	18.28	3.655	80.47	27.93	-45.90	-15.36	16.78
273.15	-68.43	22.03	6.017	90.46	36.79	-48.26	-3.81	3.06
298.15	-70.43	23.40	6.98	93.83	39.84	-49.00	0.30	-0.22
400	-78.10	29.14	11.66	107.24	51.81	-51.73	17.58	-9.61
600	-92.0	40.0	24.02	132.0	70.9	-55.4	53.1	-19.35
800	-104.9	49.6	39.7	154.5	84.9	-57.2	89.6	-24.45
1,000	-116.8	57.8	57.8	174.6	95.6	-57.5	126.4	-27.63
1,200	-128.0	64.8	77.8	192.8	103.9	-56.6	163.1	-29.71
1,500	-143.5	73.6	110.4	217.1	114.	-53.8	217.8	-31.73

2,3-Dimethylpentane (C₇H₁₆)

T, K	$(G° - H°_0)/T$	$(H° - H°_0)/T$	$H° - H°_0$	$S°$	$C_p°$	$\Delta H_f°$	$\Delta G_f°$	$\log_{10} K_f$
0	-1.38	0	0	1.38	0	-35.20	-35.20	∞
200	-68.32	18.16	3.631	86.48	25.09	-43.11	-13.77	15.05
273.15	-74.44	21.40	5.846	95.84	35.28	-45.61	-2.64	2.11
298.15	-76.37	22.71	6.77	99.08	38.44	-46.39	1.34	-0.97
400	-83.81	28.28	11.31	112.09	50.44	-49.26	18.11	-9.90
600	-97.3	39.0	23.38	136.3	69.3	-53.3	52.7	-19.21
800	-109.9	48.4	38.7	158.3	83.1	-55.4	88.4	-24.14
1,000	-121.5	56.4	56.4	177.9	93.5	-56.1	124.5	-27.2
1,200	-132.4	63.3	75.9	195.7	101.3	-55.7	160.6	-29.2
1,500	-147.5	71.8	107.7	219.3	110.	-53.6	214.4	-31.24

2,4-Dimethylpentane (C₇H₁₆)

T, K	$(G° - H°_0)/T$	$(H° - H°_0)/T$	$H° - H°_0$	$S°$	$C_p°$	$\Delta H_f°$	$\Delta G_f°$	$\log_{10} K_f$
0	0	0	0	0	0	-37.21	-37.21	∞
200	-63.18	18.12	3.623	81.30	27.59	-45.13	-14.75	16.12
273.15	-69.39	22.02	6.014	91.41	37.64	-47.45	-3.27	2.61
298.15	-71.37	23.47	7.00	94.86	40.81	-48.17	0.82	-0.60
400	-79.11	29.43	11.77	108.54	52.60	-50.86	17.94	-9.83
600	-93.2	40.3	24.21	133.5	70.8	-54.4	53.2	-19.39
800	-106.1	49.7	39.8	155.8	84.2	-56.3	89.5	-24.49
1,000	-118.0	57.7	57.7	175.7	94.2	-56.8	126.0	-27.54
1,200	-129.2	64.4	77.3	193.6	101.9	-56.3	162.5	-29.60
1,500	-144.5	72.8	109.2	217.3	111.	-54.1	217.0	-31.61

3,3-Dimethylpentane (C₇H₁₆)

T, K	$(G° - H°_0)/T$	$(H° - H°_0)/T$	$H° - H°_0$	$S°$	$C_p°$	$\Delta H_f°$	$\Delta G_f°$	$\log_{10} K_f$
0	0	0	0	0	0	-36.64	-36.64	∞
200	-63.70	18.24	3.648	81.94	27.18	-44.53	-14.28	15.60
273.15	-69.91	21.90	5.982	91.81	36.55	-46.91	-2.84	2.27
298.15	-71.85	23.27	6.94	95.16	39.62	-47.66	1.24	-0.91
400	-79.51	28.97	11.59	108.48	51.47	-50.42	18.39	-10.05
600	-93.3	39.8	23.89	133.1	70.7	-54.2	53.7	-19.56
800	-106.1	49.4	39.5	155.5	84.9	-56.0	90.0	-24.58
1,000	-118.1	57.6	57.6	175.7	95.7	-56.3	126.6	-27.66
1,200	-129.2	64.7	77.6	193.9	104.1	-55.4	163.0	-29.69
1,500	-144.7	73.6	110.4	218.3	114.	-52.4	217.3	-31.66

T, K	$(G° - H°_0)/T$, cal K^{-1} mol^{-1}	$(H° - H°_0)/T$, cal K^{-1} mol^{-1}	$H° - H°_0$, kcal mol^{-1}	$S°$, cal K^{-1} mol^{-1}	$C_p°$, cal K^{-1} mol^{-1}	$\Delta H_f°$, kcal mol^{-1}	$\Delta G_f°$, kcal mol^{-1}	$\log_{10} K_f$
			2,2,3-Trimethylbutane (C$_7$H$_{16}$)					
0	0	0	0	0	0	-37.4_2	-37.4_2	∞
200	-61.3_4	17.3_0	3.45_9	78.6_4	26.6_4	-45.5_0	-14.5_7	15.9_4
273.15	-67.2_7	21.0_5	5.74_9	88.3_2	35.9_6	-47.9_3	-2.9_0	2.3_2
298.15	-69.1_8	22.4_4	6.69	91.6_2	39.0_2	-48.6_9	1.2_7	-0.9_3
400	-76.5_7	28.2_0	11.28	104.7_7	50.8_7	-51.5_1	18.7_5	-10.2_7
600	-90.1	39.1	23.4_7	129.2	70.2	-55.4	54.9	-19.9_8
800	-102.7	48.8	39.0	151.5	84.9	-57.3	91.9	-25.1_2
1,000	-114.5	57.2	57.2	171.7	$96._3$	-57.5	129.3	-28.2_6
1,200	$-125._6$	$64._5$	$77._4$	$190._1$	$105._2$	$-56._4$	$166._6$	-30.3_4
1,500	$-141._0$	$73._7$	$110._6$	$214._7$	115.	$-53._0$	$222._0$	-32.3_4
			n-Octane (C$_8$H$_{18}$)					
0	0	0	0	0	0	-38.5_7	-38.5_7	∞
200	-70.2_6	25.7_9	5.15_8	96.0_5	34.6_0	-46.4_0	-13.1_2	14.3_4
273.15	-78.7_7	29.0_9	7.94_5	107.8_6	41.9_0	-48.9_4	-0.5_6	0.4_5
298.15	-81.3_8	30.2_9	9.03	111.6_7	44.8_6	-49.7_7	3.9_2	-2.8_7
400	-91.0_0	35.5_9	14.24	126.5_9	57.3_0	-52.9_3	22.7_7	-12.4_4
600	-107.5	46.5	27.8_8	154.0	78.1	-57.4	61.7	-22.4_7
800	-122.3	56.3	45.1	178.6	92.8	-60.0	101.8	-27.8_2
1,000	-135.7	64.7	64.7	200.4	$103._1$	-61.2	142.4	-31.1_3
1,200	$-148._2$	$71._8$	$86._1$	$220._0$	$110._7$	$-61._4$	$183._2$	-33.3_4
1,500	$-165._2$	$80._5$	$120._7$	$245._7$	119.	$-60._5$	$244._3$	-35.5_9
			2-Methylheptane (C$_8$H$_{18}$)					
0	0	0	0	0	0	-39.6_5	-39.6_5	∞
200	-71.1_5	23.4_7	4.69_3	94.6_2	32.4_2	-47.9_5	-14.3_9	15.7_2
273.15	-78.9_9	27.0_7	7.39_4	106.0_6	41.5_6	-50.5_7	-1.7_0	1.3_6
298.15	-81.4_2	28.4_2	8.47	109.8_4	44.7_5	-51.4_1	2.8_2	-2.0_7
400	-90.5_7	34.2_4	13.70	124.8_1	57.6_6	-54.5_5	21.8_6	-11.9_4
600	-106.7	45.7	27.4_4	152.4	78.7	-59.0	61.1	-22.2_6
800	-121.3	55.9	44.8	177.2	93.6	-61.4	101.5	-27.7_4
1,000	-134.7	64.6	64.6	199.3	$104._1$	-61.7	142.4	-31.1_3
1,200	$-147._1$	$72._0$	$86._3$	$219._0$	$112._5$	$-62._2$	$183._4$	-33.4_0
1,500	$-164._2$	$81._0$	$121._5$	$245._2$	122.	$-60._8$	$244._6$	-35.6_4
			3-Methylheptane (C$_8$H$_{18}$)					
0	-1.38	0	0	1.38	0	-38.9_7	-38.9_7	∞
200	-73.1_0	23.1_3	4.62_5	96.2_3	31.7_1	-47.3_4	-14.0_9	15.4_0
273.15	-80.8_2	26.6_9	7.28_9	107.5_1	41.1_8	-50.0_0	-1.5_2	1.2_1
298.15	-83.2_2	28.0_5	8.36	111.2_6	44.4_1	-50.8_4	2.9_7	-2.1_8
400	-92.2_6	33.8_8	13.55	126.1_4	57.4_0	-54.0_1	21.8_6	-11.9_5
600	-108.2	45.4	27.2_5	153.6	78.4	-58.5	60.9	-22.1_7
800	-122.7	55.7	44.5	178.4	93.5	-60.9	101.0	-27.6_1
1,000	-136.1	64.4	64.4	200.5	$104._3$	-61.9	141.7	-30.9_7
1,200	$-148._5$	$71._7$	$86._1$	$220._2$	$112._4$	$-61._8$	$182._4$	-33.2_2
1,500	$-165._5$	$80._8$	$121._2$	$246._3$	122.	$-60._4$	$243._4$	-35.4_6
			4-Methylheptane (C$_8$H$_{18}$)					
0	0	0	0	0	0	-38.9_6	-38.9_6	∞
200	-71.4_7	22.8_0	4.56_0	94.2_7	31.5_2	-47.3_9	-13.7_6	15.0_4
273.15	-79.1_0	26.4_5	7.22_5	105.5_5	41.3_7	-50.0_5	-1.0_4	0.8_3
298.15	-81.4_8	27.8_4	8.30	109.3_2	44.7_0	-50.8_9	3.5_0	-2.5_7
400	-90.4_8	33.8_5	13.54	124.3_3	57.9_2	-54.0_2	22.5_8	-12.3_4
600	-106.4	45.6	27.3_4	152.0	79.0	-58.4	61.9	-22.5_5
800	-121.0	55.9	44.7^-	176.9	93.9	-60.8	102.4	-27.9_6
1,000	-134.5	64.6	64.6	199.1	$104._6$	-61.7	143.4	-31.3_3
1,200	$-146._9$	$72._0$	$86._4$	$218._9$	$112._6$	$-61._5$	$184._3$	-33.5_7
1,500	$-164._0$	$81._0$	$121._6$	$245._0$	122.	$-60._0$	$245._7$	-35.7_9
			2,2-Dimethylhexane (C$_8$H$_{18}$)					
0	0	0	0	0	0	-41.7_7	-41.7_7	∞
200	-66.6_3	21.5_8	4.31_5	88.2_1	32.2_0	-50.4_5	-15.6_1	17.0_5
273.15	-73.9_5	25.6_7	7.01_0	99.6_2	41.6_8	-53.0_8	-2.4_4	1.9_6
298.15	-76.2_7	27.1_5	8.10	103.4_2	45.0_0	-53.9_1	2.2_4	-1.6_4
400	-85.1_1	33.4_2	13.37	118.5_3	58.2_8	-57.0_0	21.9_2	-11.9_6
600	-101.0	45.5	27.2_8	146.5	79.8	-61.2	62.4	-22.7_3
800	-115.6	56.1	44.9	171.7	95.6	-63.4	104.0	-28.4_0
1,000	-129.1	65.3	65.3	194.4	$107._4$	-63.8	145.9	-31.8_9
1,200	$-141._7$	$73._1$	$87._7$	$214._6$	$116._0$	$-63._0$	$187._5$	-34.2_0
1,500	$-159._1$	$82._8$	$124._2$	$241._9$	127.	$-60._2$	$250._2$	-36.4_5
			3,3-Dimethylhexane (C$_8$H$_{18}$)					
0	0	0	0	0	0	-40.3_8	-40.3_8	∞
200	-68.9_2	20.7_0	4.14_0	89.6_2	30.8_4	-49.2_4	-14.6_7	16.0_3
273.15	-75.9_7	24.9_3	6.80_7	100.9_0	42.0_0	-51.8_9	-1.6_1	1.2_8
298.15	-78.2_3	26.5_1	7.90	104.7_4	45.6_2	-52.7_1	3.0_5	-2.2_3
400	-86.9_4	33.1_8	13.27	120.1_2	59.4_4	-55.7_1	22.5_8	-12.3_4
600	-102.8	45.7	27.4_3	143.5	81.0	-59.7	62.7	-22.8_3
800	-117.5	56.6	45.2	174.1	96.5	-61.6	103.8	-28.3_6
1,000	-131.1	65.8	65.8	196.9	$108._1$	-61.9	145.3	-31.7_5

T, K	$(G° - H°_0)/T$, cal K^{-1} mol^{-1}	$(H° - H°_0)/T$, cal K^{-1} mol^{-1}	$H° - H°_0$, kcal mol^{-1}	$S°$, cal K^{-1} mol^{-1}	$C_p°$, cal K^{-1} mol^{-1}	$\Delta H_f°$, kcal mol^{-1}	$\Delta G_f°$, kcal mol^{-1}	$\log_{10} K_f$
1,200	-143.6	73.6	88.3	217.4	117.0	-61.0	186.6	-33.95
1,500	-161.3	83.4	125.0	244.7	127.	-58.0	248.2	-36.16

2,2,4-Trimethylpentane (Isooctane) (C$_8$H$_{18}$)

T, K	$(G° - H°_0)/T$	$(H° - H°_0)/T$	$H° - H°_0$	$S°$	$C_p°$	$\Delta H_f°$	$\Delta G_f°$	$\log_{10} K_f$
0	0	0	0	0	0	-41.23	-41.23	∞
200	-66.30	19.73	3.945	86.03	31.21	-50.25	-15.00	16.39
273.15	-73.10	24.15	6.606	97.28	41.58	-52.94	-1.67	1.35
298.15	-75.29	25.80	7.69	101.09	45.03	-53.77	3.07	-2.25
400	-83.79	32.4	12.98	116.23	58.46	-56.85	22.99	-12.56
600	-99.3	44.9	26.94	144.2	80.2	-61.0	63.9	-23.26
800	-113.8	55.8	44.7	169.6	96.4	-63.1	105.9	-28.94
1,000	-127.3	65.2	65.2	192.5	108.7	-63.3	148.2	-32.40
1,200	-139.9	73.3	88.0	213.2	118.2	-62.2	190.4	-34.68
1,500	-157.4	83.4	125.1	240.8	129.	-58.7	253.2	-36.90

n-Nonane (C$_9$H$_{20}$)

T, K	$(G° - H°_0)/T$	$(H° - H°_0)/T$	$H° - H°_0$	$S°$	$C_p°$	$\Delta H_f°$	$\Delta G_f°$	$\log_{10} K_f$
0	0	0	0	0	0	-42.32	-42.32	∞
200	-74.64	28.9	5.783	103.56	38.70	-50.98	-13.36	14.60
273.15	-84.20	32.59	8.902	116.77	47.03	-53.79	0.8	-0.67
298.15	-87.10	33.94	10.12	121.04	50.29	-54.71	5.90	-4.32
400	-97.89	39.88	15.95	137.77	64.25	-58.20	27.16	-14.84
600	-116.4	52.1	31.25	168.5	87.5	-63.2	71.0	-25.86
800	-132.9	63.1	50.5	196.0	103.7	-66.0	116.2	-31.74
1,000	-148.0	72.4	72.4	220.4	115.1	-67.3	161.9	-35.33
1,200	-161.5	80.3	96.3	242.2	123.4	-67.5	207.8	-37.84
1,500	-180.9	89.9	134.8	270.8	133.	-66.7	276.5	-40.29

2-Methyloctane (C$_9$H$_{20}$)

T, K	$(G° - H°_0)/T$	$(H° - H°_0)/T$	$H° - H°_0$	$S°$	$C_p°$	$\Delta H_f°$	$\Delta G_f°$	$\log_{10} K_f$
0	0	0	0	0	0	-43.40	-43.40	∞
200	-75.54	26.59	5.318	102.14	36.52	-52.53	-14.62	15.98
273.15	-84.40	30.58	8.352	114.98	46.60	-55.42	-0.30	0.24
298.15	-87.15	32.0	9.56	119.22	50.15	-56.35	4.80	-3.52
400	-97.45	38.54	15.41	135.94	64.61	-59.82	26.25	-14.34
600	-115.5	51.4	30.81	166.9	88.1	-64.7	70.4	-25.66
800	-131.9	62.7	50.2	194.6	104.6	-67.4	115.9	-31.67
1,000	-146.9	72.4	72.4	219.3	116.	-68.4	161.9	-35.3
1,200	-160.9	80.4	96.5	241.3	125.1	-68.4	208.0	-37.89
1,500	-180.0	90.4	135.6	270.4	135.	-66.9	276.9	-40.35

3-Methyloctane (C$_9$H$_{20}$)

T, K	$(G° - H°_0)/T$	$(H° - H°_0)/T$	$H° - H°_0$	$S°$	$C_p°$	$\Delta H_f°$	$\Delta G_f°$	$\log_{10} K_f$
0	-1.38	0	0	1.38	0	-42.71	-42.71	∞
200	-77.48	26.24	5.248	103.72	35.71	-51.91	-14.32	15.64
273.15	-86.23	30.18	8.246	116.41	46.24	-54.83	-0.10	0.08
298.15	-88.93	31.69	9.45	120.62	49.85	-55.77	4.96	-3.64
400	-99.14	38.18	5.00	137.32	64.40	-59.26	26.27	-14.35
600	-117.1	51.0	30.63	168.1	87.9	-64.2	70.2	-25.57
800	-133.4	62.4	50.0	195.8	104.5	-66.9	115.4	-31.54
1,000	-148.4	72.1	72.1	220.5	116.3	-68.0	161.1	-35.24
1,200	-162.3	80.2	96.3	242.5	125.0	-67.9	207.0	-37.70
1,500	-181.3	90.2	135.4	271.1	135.	-66.5	275.6	-40.16

4-Methyloctane (C$_9$H$_{20}$)

T, K	$(G° - H°_0)/T$	$(H° - H°_0)/T$	$H° - H°_0$	$S°$	$C_p°$	$\Delta H_f°$	$\Delta G_f°$	$\log_{10} K_f$
0	-1.38	0	0	1.38	0	-42.70	-42.70	∞
200	-77.27	25.97	5.194	103.24	35.64	-51.95	-14.26	15.59
273.15	-85.94	29.99	8.190	115.93	46.35	-54.88	-0.02	0.01
298.15	-88.63	31.52	9.40	120.15	50.02	-55.81	5.06	-3.71
400	-98.80	38.12	15.25	136.92	64.71	-59.28	26.42	-14.44
600	-116.8	51.1	30.67	167.9	88.2	-64.1	70.4	-25.65
800	-133.1	62.5	50.1	195.6	104.7	-66.8	115.7	-31.66
1,000	-148.1	72.2	72.2	220.3	116.2	-67.8	161.5	-35.24
1,200	-162.0	80.4	96.4	242.4	125.2	-67.8	207.3	-37.79
1,500	-181.0	90.4	135.5	271.4	135.	-66.3	276.0	-40.21

3-Ethylheptane (C$_9$H$_{20}$)

T, K	$(G° - H°_0)/T$	$(H° - H°_0)/T$	$H° - H°_0$	$S°$	$C_p°$	$\Delta H_f°$	$\Delta G_f°$	$\log_{10} K_f$
0	0	0	0	0	0	-42.34	-42.34	∞
200	-74.90	26.50	5.300	101.40	38.01	-51.48	-13.41	14.67
273.15	-83.80	30.79	8.410	114.59	47.27	-54.30	0.93	-0.74
298.15	-86.56	32.32	9.64	118.87	50.64	-55.21	6.04	-4.43
400	-96.94	38.78	15.51	135.72	64.62	-58.66	27.52	-15.04
600	-115.1	51.4	30.86	166.5	87.7	-63.6	71.8	-26.14
800	-131.5	62.6	50.1	194.1	104.2	-66.3	117.3	-32.05
1,000	-146.5	72.2	72.2	218.7	116.0	-67.5	163.4	-35.72
1,200	-160.4	80.3	96.3	240.7	124.1	-67.5	209.6	-38.17
1,500	-179.5	90.2	135.4	269.7	135.	-66.1	278.8	-40.61

THERMODYNAMIC PROPERTIES OF ALKANE HYDROCARBONS (Continued)

T, K	$(G° - H°_0)/T$, cal K⁻¹ mol⁻¹	$(H° - H°_0)/T$, cal K⁻¹ mol⁻¹	$H° - H°_0$, kcal mol⁻¹	$S°$, cal K⁻¹ mol⁻¹	$C_p°$, cal K⁻¹ mol⁻¹	$\Delta H_f°$, kcal mol⁻¹	$\Delta G_f°$, kcal mol⁻¹	$\log_{10} K_f$

4-Ethylheptane (C₉H₂₀)

T, K	$(G°-H°_0)/T$	$(H°-H°_0)/T$	$H°-H°_0$	$S°$	$C_p°$	$\Delta H_f°$	$\Delta G_f°$	$\log_{10}K_f$
0	0	0	0	0	0	-42.3_4	-42.3_4	∞
200	-74.6_3	26.1_7	5.23_3	100.8_0	37.9_3	-51.5_6	-13.3_8	14.6_2
273.15	-83.4_3	30.5_7	8.34_3	114.0_0	47.5_2	-54.3_6	1.0_2	-0.8_2
298.15	-86.1_8	32.1_3	9.58	118.3_1	50.9_8	-55.2_7	6.1_5	-4.5_1
400	-96.5_3	38.7_5	15.50	135.2_8	65.0_9	-58.6_7	27.6_8	-15.1_2
600	-114.7	51.6	30.9_4	166.3	88.1	-63.5	72.0	-26.2_2
800	-131.1	62.9	50.3	194.0	104.5	-66.2	117.6	-32.1_3
1,000	-146.2	72.4	72.4	218.6	$116._2$	-67.3	163.7	-35.7_8
1,200	$-160._1$	$80._5$	$96._6$	$240._6$	$124._9$	$-67._3$	$209._9$	-38.2_3
1,500	$-179._2$	$90._4$	$135._6$	$269._6$	135.	$-65._9$	$279._1$	-40.6_6

n-Decane

T, K	$(G°-H°_0)/T$	$(H°-H°_0)/T$	$H°-H°_0$	$S°$	$C_p°$	$\Delta H_f°$	$\Delta G_f°$	$\log_{10}K_f$
0	0	0	0	0	0	-46.0_6	-46.0_6	∞
200	-79.0_2	32.0_5	6.40_8	111.0_7	42.8_0	-55.5_6	-13.5_4	14.8_4
273.15	-89.6_1	36.1_0	9.86_1	125.7_1	52.0_5	-58.6_3	2.2_5	-1.8_0
298.15	-92.8_3	37.5_5	11.21	130.4_2	55.7_0	-59.6_4	7.8_3	-5.7_5
400	-104.7_7	44.1_8	17.67	148.9_5	71.2_4	-63.4_7	31.5_6	-17.2_4
600	-125.3	57.7	34.6_3	183.0	97.0	-68.8	80.3	-29.2_6
800	-143.6	69.9	55.9	213.5	114.7	-71.9	130.6	-35.6_4
1,000	-160.3	80.2	80.2	240.5	$127._1$	-73.3	181.4	-39.6_4
1,200	$-175._7$	$88._8$	$106._5$	$264._5$	$136._1$	$-73._6$	$232._4$	-42.3_2
1,500	$-196._7$	$99._3$	$149._0$	$296._0$	146.	$-72._8$	$308._8$	-44.9_9

2-Methylnonane (C₁₀H₂₂)

T, K	$(G°-H°_0)/T$	$(H°-H°_0)/T$	$H°-H°_0$	$S°$	$C_p°$	$\Delta H_f°$	$\Delta G_f°$	$\log_{10}K_f$
0	0	0	0	0	0	-47.1_5	-47.1_5	∞
200	-79.9_2	29.7_2	5.94_4	109.6_4	40.6_2	-57.1_0	-14.8_5	16.2_2
273.15	-89.8_1	34.0_9	9.31_0	123.9_0	51.6_5	-60.2_6	1.1_5	-0.8_8
298.15	-92.8_7	35.7_2	10.65	128.5_9	55.5_6	-61.2_8	6.7_8	-4.9_7
400	-101.0_4	42.8_4	17.13	147.1_8	71.5_8	-65.0_6	30.6	-16.7_1
600	-124.4	57.0	34.1_9	181.4	97.5	-70.4	79.8	-29.0_5
800	-142.6	69.5	55.6	212.1	115.6	-73.3	130.3	-35.6_0
1,000	-159.2	80.1	80.1	239.3	$128._4$	-74.5	181.4	-39.6_4
1,200	$-174._7$	$88._9$	$106._8$	$263._6$	$137._8$	$-74._5$	$232._5$	-42.3_5
1,500	$-195._7$	$99._9$	$149._8$	$295._6$	149.	$-73._0$	$309._2$	-45.0_4

3-Methylnonane (C₁₀H₂₂)

T, K	$(G°-H°_0)/T$	$(H°-H°_0)/T$	$H°-H°_0$	$S°$	$C_p°$	$\Delta H_f°$	$\Delta G_f°$	$\log_{10}K_f$
0	-1.38	0	0	1.38	0	-46.4_5	-46.4_5	∞
200	-81.8_6	29.3_7	5.87_3	111.2_3	39.8_8	-56.4_8	-14.5_4	15.8_9
273.15	-91.6_4	33.6_9	9.20_2	125.3_3	51.2_8	-59.6_7	1.3_0	-1.0_4
298.15	-94.6_6	35.3_3	10.54	129.9_9	55.2_5	-60.7_0	6.9_5	-5.0_9
400	-106.0_2	42.4_8	16.99	148.5_0	71.3_5	-64.5_2	30.6_7	-16.7_6
600	-125.9	56.7	34.0_0	182.6	97.3	-69.9	79.5	-28.9_7
800	-144.1	69.2	55.4	213.3	115.4	-72.8	129.8	-35.4_4
1,000	-160.7	79.8	79.8	240.5	$128._3$	-74.0	180.7	-39.4_7
1,200	$-176._0$	$88._7$	$106._5$	$264._7$	$137._7$	$-74._0$	$231._6$	-42.1_8
1,500	$-197._1$	$99._6$	$149._5$	$296._7$	149.	$-72._6$	$307._9$	-44.8_6

4-Methylnonane (C₁₀H₂₂)

T, K	$(G°-H°_0)/T$	$(H°-H°_0)/T$	$H°-H°_0$	$S°$	$C_p°$	$\Delta H_f°$	$\Delta G_f°$	$\log_{10}K_f$
0	-1.38	0	0	1.38	0	-46.4_5	-46.4_5	∞
200	-81.6_5	29.0_9	5.81_7	110.7_4	39.7_2	-56.5_3	-14.5_0	15.8_4
273.15	-91.3_5	33.4_8	9.14_6	124.8_3	51.4_1	-59.7_3	1.3_5	-1.1_1
298.15	-94.3_5	35.1_6	10.48	129.5_1	55.4_6	-60.7_7	7.0_4	-5.1_6
400	-105.6_5	42.4_2	16.97	148.1_0	71.7_1	-64.5_4	30.8_1	-16.8_3
600	-125.6	56.8	34.0_5	182.4	97.7	-69.8	79.7	-29.0_4
800	-143.7	69.4	55.5	213.1	115.7	-72.7	130.1	-35.5_3
1,000	-160.4	80.0	80.0	240.4	$128._5$	-73.9	180.9	-39.5_4
1,200	$-175._8$	$88._9$	$106._7$	$264._7$	$137._8$	$-73._8$	$231._9$	-42.2_3
1,500	$-196._8$	$99._8$	$149._7$	$296._6$	149.	$-72._4$	$308._2$	-44.9_0

5-Methylnonane (C₁₀H₂₂)

T, K	$(G°-H°_0)/T$	$(H°-H°_0)/T$	$H°-H°_0$	$S°$	$C_p°$	$\Delta H_f°$	$\Delta G_f°$	$\log_{10}K_f$
0	0	0	0	0	0	-46.4_3	-46.4_3	∞
200	-80.3_3	29.1_5	5.82_9	109.4_8	39.7_5	-56.5_1	-14.2_2	15.5_4
273.15	-90.0_4	33.5_2	9.15_5	123.5_6	51.3_1	-59.7_0	1.7_6	-1.4_0
298.15	-93.0_5	35.1_8	10.49	128.2_3	55.3_2	-60.7_3	7.4_4	-5.4_6
400	-104.3_8	42.3_9	16.96	146.7_7	71.5_2	-64.5_4	31.3_4	-17.1_3
600	-124.3	56.7	34.0_0	181.0	97.5	-69.8	80.5	-29.3_4
800	-142.4	69.3	55.4	211.7	115.6	-72.8	131.2	-35.8_3
1,000	-159.0	79.9	79.9	238.9	$128._4$	-73.9	182.3	-39.8_4
1,200	$-174._4$	$88._8$	$106._6$	$263._2$	$137._8$	$-73._9$	$233._6$	-42.5_4
1,500	$-195._4$	$99._7$	$149._6$	$295._1$	149.	$-72._5$	$310._3$	-45.2_1

Data from Scott, D. W., U.S. Bureau of Mines Bulletin 666 (stock number 2404-01547), U.S. Government Printing Office, Washington, D.C., 1974.

[a]The original source of the data in this table contains the thermodynamic properties of all the alkane hydrocarbons through decane. Including the *meso*, *(d, l)*, *cis*, and *trans* isomers, there are 182 compounds in the original source.

HEAT OF DILUTION OF ACIDS

From National Standards Reference Data Systems NSRDS-NBS 2

Vivian B. Parker

ΔH_{diln}, the integral heat of dilution, is the change in enthalpy, per mole of solute, when a solution of concentration m_1 is diluted to a final finite concentration m_2. When the dilution is carried out by addition of an infinite amount of solvent, so the final solution is infinitely dilute, the enthalpy change is the integral heat of dilution to infinite dilution. Since Φ_L, the relative apparent molal enthalpy, is equal to and opposite in sign to this, only Φ_L is referred to here.

Φ_L, cal/mole, at 25°C

n	m	HF	HCl	HClO$_4$	HBr	HI	HNO$_3$	CH$_2$O$_2$	C$_2$H$_4$O$_2$
∞	0.00	0	0	0	0	0	0	0	0
500,000	.000111	300	5	5	5	5	5	9	40
100,000	.000555	900	10	10	9	9	11	13	50
50,000	.00111	1,300	16	14	13	12	15	20	53
20,000	.00278	1,800	25	22	22	20	23	23	55
10,000	.00555	2,130	34	30	31	29	31	25	58
7,000	.00793	2,250	40	35	37	34	36	26	59
5,000	.01110	2,360	47	40	44	41	42	26	61
4,000	.01388	2,450	54	43	49	46	46	27	62
3,000	.01850	2,550	60	47	56	52	51	28	62
2,000	.02775	2,700	74	54	68	63	59	28	63
1,500	.03700	2,812	85	58	77	71	65	29	64
1,110	.05000	2,927	97	62	89	81	73	29	65
1,000	.05551	2,969	102	62	92	84	76	29	65
900	.0617	2,989	107	63	97	88	78	30	66
800	.0694	3,015	113	64	102	92	81	31	67
700	.0793	3,037	120	65	108	96	84	32	68
600	.0925	3,057	129	65	115	102	88	32	68
555.1	.1000	3,060	133	65	119	105	89	32	69
500	.1110	3,077	140	65	124	108	92	32	70
400	.1388	3,097	156	64	135	116	97	33	72
300	.1850	3,126	176	61	150	125	103	34	76
277.5	.2000	3,129	182	59	155	128	105	35	79
200	.2775	3,142	212	50	176	140	117	36	82
150	.3700	3,148	242	36	197	154	118	39	88
111.0	.5000	3,156	280	18	225	170	119	42	97
100	.5551	3,160	295	+12	235	176	120	44	101
75	.7401	3,167	343	−14	270	194	121	49	113
55.51	1.0000	3,179	405	−48	314	223	121	54	130
50	1.1101	3,184	431	−61	331	234	121	56	147
40	1.3877	3,192	493	−91	379	260	121	60	155
37.00	1.5000	3,194	518	−103	398	269	121	62	162
30	1.8502	3,200	595	−138	455	301	124	65	183
27.75	2.0000	3,203	627	−149	477	315	126	66	192
25	2.2202	3,208	674	−162	510	336	130	67	204
22.20	2.5000	3,211	732	−173	550	365	139	68	218
20	2.7753	3,214	792	−182	590	396	149	69	233
18.50	3.0000	3,216	838	−187	624	427	159	69	245
15.86	3.500	3,221	946	−196	709	503	189	69	268
15	3.7004	3,227	988	−195	743	536	203	69	277
13.88	4.0000	3,234	1,052	−188	796	588	229	69	291
12.33	4.5000	3,246	1,171	−175	887	676	265	69	313
12	4.6255	3,249	1,190	−170	911	700	277	69	318
11.10	5.0000	3,256	1,271	−150	983	764	313	69	333
10	5.5506	3,265	1,396	−117	1,097	855	368	68	353
9.5	5.8427	3,269	1,462	−97	1,156	920	400	68	363
9.251	6.0000	3,272	1,498	−84	1,196	950	418	67	368
9.0	6.1674	3,274	1,535	−72	1,230	980	437	67	373
8.5	6.5301	3,278	1,618	−40	1,313	1,050	480	66	383
8.0	6.9383	3,282	1,710	+4	1,401	1,115	530	65	392
7.929	7.0000	3,283	1,725	11	1,416	1,130	538	65	394

n	m	HF	HCl	HClO$_4$	HBr	HI	HNO$_3$	CH$_2$O$_2$	C$_2$H$_4$O$_2$
7.5	7.4008	3,286	1,820	61	1,497	1,210	595	63	402
7.0	7.9295	3,290	1,942	135	1,608	1,325	661	61	411
6.938	8.0000	3,291	1,960	146	1,622	1,340	667	61	412
6.5	8.5394	3,296	2,090	229	1,738	1,450	745	58	420
6.167	9.0000	3,302	2,202	306	1,845	1,570	805	55	426
6.0	9.2510	3,305	2,265	348	1,903	1,630	840	53	429
5.551	10.0000	3,316	2,447	481	2,078	1,820	940	49	436
5.5	10.0920	3,317	2,472	499	2,102	1,850	950	49	437
5.0	11.1012	3,335	2,721	730	2,344	2,100	1,008	43	445
4.5	12.3346	3,362	3,025	1,144	2,655	2,460	1,270	37	453
4.0	13.8765	3,400	3,404	1,574	3,089	2,960	1,495	29	462
3.700	15.0000	3,428	3,680	1,893	3,415	3,350	1,645	26	469
3.5	15.8589	3,450	3,882	2,150	3,668	3,660	1,770	21	473
3.25	17.0788	3,483	4,160	2,460	4,005	4,110	1,920	17	481
3.0	18.5020	3,520	4,460	2,880	4,370	4,630	2,101	13	488
2.775	20.0000	3,557	4,750	3,300	4,760	5,190	2,270	9	496
2.5	22.2024	3,607	5,180	4,000	5,300	6,000	2,520	+4	506
2.0	27.7530	3,712	6,260	5,500	6,650	.....	3,060	−5	528
1.5	37.0040	.....	8,240	.....	8,530	.....	3,770	−13	532
1.0	55.506	.....	10,900	.....	11,670	.....	4,715	+11	518
0.5	111.012	.....	.....	.....	.....	.....	.....	77	495
0.25	222.02	.....	.....	.....	.....	.....	.....	129	.....

HEATS OF SOLUTION

From National Standards Reference Data Systems NSRDS–NBS 2

Vivian B. Parker

ΔH°_∞ 25°C for uni-univalent electrolytes in H$_2$O

Substance	State	ΔH°_∞	Substance	State	ΔH°_∞	Substance	State	ΔH°_∞
		cal/mole			cal/mole			cal/mole
HF	g	−14,700	LiBr·2H$_2$O	c	−2,250	KCl	c	4,115
HCl	g	−17,888	LiBrO$_3$	c	340	KClO$_3$	c	9,890
HClO$_4$	l	−21,215	LiI	c	−15,130	KClO$_4$	c	12,200
HClO$_4$·H$_2$O	c	−7,875	LiI·H$_2$O	c	−7,090	KBr	c	4,750
HBr	g	−20,350	LiI·2H$_2$O	c	−3,530	KBrO$_3$	c	9,830
HI	g	−19,520	LiI·3H$_2$O	c	140	KI	c	4,860
HIO$_3$	c	2,100	LiNO$_2$	c	−2,630	KIO$_3$	c	6,630
HNO$_3$	l	−7,954	LiNO$_2$·H$_2$O	c	1,680	KNO$_2$	c	3,190
HCOOH	l	−205	LiNO$_3$	c	−600	KNO$_3$	c	8,340
CH$_3$COOH	l	−360				KC$_2$H$_3$O$_2$	c	−3,665
			NaOH	c	−10,637	KCN	c	2,800
NH$_3$	g	−7,290	NaOH·H$_2$O	c	−5,118	KCNO	c	4,840
NH$_4$Cl	c	3,533	NaF	c	218	KCNS	c	5,790
NH$_4$ClO$_4$	c	8,000	NaCl	c	928	KMnO$_4$	c	10,410
NH$_4$Br	c	4,010	NaClO$_2$	c	80			
NH$_4$I	c	3,280	NaClO$_2$·3H$_2$O	c	6,830	RbOH	c	−14,900
NH$_4$IO$_3$	c	7,600	NaClO$_3$	c	5,191	RbOH·H$_2$O	c	−4,310
NH$_4$NO$_2$	c	4,600	NaClO$_4$	c	3,317	RbF	c	210
NH$_4$NO$_3$	c	6,140	NaClO$_4$·H$_2$O	c	5,380	RbF·H$_2$O	c	−6,240
NH$_4$C$_2$H$_3$O$_2$	c	−570	NaBr	c	−144	RbF·1½H$_2$O	c	−100
NH$_4$CN	c	4,200	NaBr·2H$_2$O	c	4,454	RbCl	c	320
NH$_4$CNS	c	5,400	NaBrO$_3$	c	6,430	RbCl	c	4,130
CH$_3$NH$_3$Cl	c	1,378	NaI	c	−1,800	RbClO$_3$	c	11,410
(CH$_3$)$_2$NHCl	c	350	NaI·2H$_2$O	c	3,855	RbClO$_4$	c	13,560
N(CH$_3$)$_4$Cl	c	975	NaIO$_3$	c	4,850	RbBr	c	5,230
N(CH$_3$)$_4$Br	c	5,800	NaNO$_2$	c	3,320	RbBrO$_3$	c	11,700
N(CH$_3$)$_4$I	c	10,055	NaNO$_3$	c	4,900	RbI	c	6,000
			NaC$_2$H$_3$O$_2$	c	−4,140	RbNO$_3$	c	8,720
AgClO$_4$	c	1,760	NaC$_2$H$_3$O$_2$·3H$_2$O	c	4,700			
AgNO$_2$	c	8,830	NaCN	c	290	CsOH	c	−17,100
AgNO$_3$	c	5,400	NaCN·½H$_2$O	c	790	CsOH·H$_2$O	c	−4,900
			NaCN·2H$_2$O	c	4,440	CsF	c	−8,810
LiOH	c	−5,632	NaCNO	c	4,590	CsF·H$_2$O	c	−2,500
LiOH·H$_2$O	c	−1,600	NaCNS	c	1,632	CsF·1½H$_2$O	c	−1,300
LiF	c	1,130				CsCl	c	4,250
LiCl	c	−8,850	KOH	c	−13,769	CsClO$_4$	c	13,250
LiCl·H$_2$O	c	−4,560	KOH·H$_2$O	c	−3,500	CsBr	c	6,210
LiClO$_4$	c	−6,345	KOH·1½H$_2$O	c	−2,500	CsBrO$_3$	c	12,060
LiClO$_4$·3H$_2$O	c	7,795	KF	c	−4,238	CsI	c	7,970
LiBr	c	−11,670	KF·2H$_2$O	c	1,666	CsNO$_3$	c	9,560
LiBr·H$_2$O	c	−5,560						

HEAT CAPACITY OF AQUEOUS SOLUTIONS OF VARIOUS ACIDS

From National Standards Reference Data Systems NSRDS-NBS 2

Vivian B. Parker

Φ_C is the apparent molal heat capacity of the solute, equal to $[(1000 + mM_2)C - 1000C°]/m$ where C and C° are the specific heats (per unit mass) of the solution and pure solvent, respectively, m is the molality, and M_2 is the molecular weight of the solute.

Φ_C, cal/deg mole, at 25°C

n	m	HF	HCl	HBr	HI	HIO$_3$	HNO$_3$	CH$_2$O$_2$	C$_2$H$_4$O$_2$	C$_3$H$_6$O$_2$
∞	0.00	−25.5	−32.6	−33.9	−34.0	−29.6	−20.7	−21.0	−1.5	+26.7
500,000	.000111	−23.0	.....	.....	.....	.....	.....	−9.8	+25.8	38.0
100,000	.000555	−18.8	−32.4	−33.8	−33.9	−29.4	−20.6	−1.2	32.6	45.1
50,000	.00111	−16.6	−32.4	−33.7	−33.8	−29.3	−20.5	+3.1	34.0	48.3
20,000	.00278	−12.4	−32.3	−33.6	−33.7	−29.1	−20.4	10.2	35.7	52.2
10,000	.00555	−8.6	−32.2	−33.4	−33.5	−28.5	−20.3	12.7	36.9	54.1
7,000	.00793	−6.7	−32.1	−33.4	−33.4	−28.4	−20.2	13.7	37.4	54.8
5,000	.01110	−4.9	−32.0	−33.3	−33.3	−28.1	−20.1	14.3	37.8	55.3
4,000	.01388	−3.6	−31.9	−33.2	−33.2	−27.7	−20.1	14.7	37.9	55.6
3,000	.01850	−2.2	−31.8	−33.1	−33.1	−27.2	−20.0	15.1	38.1	56.1
2,000	.02775	−0.5	−31.6	−32.9	−32.9	−25.8	−19.8	15.8	38.5	56.7
1,500	.03700	+0.4	−31.4	−32.7	−32.7	−24.9	−19.6	16.4	38.7	57.1
1,000	.05551	1.6	−31.2	−32.5	−32.5	−23.0	−19.3	16.8	39.0	57.7
900	.0617	1.8	−31.2	−32.4	−32.4	−22.3	−19.2	17.0	39.0	57.8
800	.0694	2.0	−31.1	−32.3	−32.3	−21.5	−19.1	17.2	39.1	57.9
700	.0793	2.2	−30.9	−32.2	−32.1	−20.2	−19.0	17.3	39.2	58.0
600	.0925	2.4	−30.8	−32.1	−32.0	−18.7	−18.9	17.5	39.3	58.3
500	.1110	2.7	−30.6	−31.9	−31.7	−16.7	−18.7	17.7	39.4	58.4
400	.1388	2.8	−30.3	−31.6	−31.4	−13.9	−18.4	18.0	39.4	58.6
300	.1850	3.2	−30.0	−31.2	−30.9	−10.1	−17.9	18.3	39.4	58.8
200	.2775	3.3	−29.3	−30.6	−30.2	−4.7	−17.2	18.7	39.3	58.9
150	.3700	3.5	−28.8	−30.1	−29.6	−0.7	−16.4	18.9	39.2	58.8
100	.5551	3.8	−27.8	−29.1	−28.4	+4.8	−15.0	19.0	39.1	58.7
75	.7401	4.1	−27.0	−28.5	−27.5	9.2	−13.7	19.2	38.9	58.6
50	1.1101	4.5	−25.6	−26.8	−25.9	15.7	−11.5	19.8	38.6	58.3
40	1.3877	4.9	−24.8	−25.9	−24.8	19.0	−9.9	19.8	38.4	58.0
30	1.8502	5.3	−23.6	−24.5	−23.3	23.1	−7.3	20.0	38.0	57.1
25	2.2202	5.6	−22.7	−23.6	−22.2	25.7	−5.4	20.1	37.7	56.2
20	2.7753	5.7	−21.5	−22.2	−20.8	.....	−2.7	20.2	37.1	55.0
15	3.7004	6.0	−19.8	−20.3	.....	.....	+1.4	20.4	36.3	53.1
12	4.6255	6.2	−18.2	−18.5	.....	.....	5.0	20.4	35.4	51.3
10	5.5506	6.3	−16.8	−16.8	.....	.....	8.5	20.6	34.8	49.7
9.5	5.8427	6.4	−16.4	−16.3	.....	.....	9.2	20.6	34.7	.....
9.0	6.1674	6.4	−15.8	−15.7	.....	.....	10.3	20.6	34.5	.....
8.5	6.5301	6.5	−15.4	−15.1	.....	.....	11.4	20.7	34.4	.....
8.0	6.9383	6.6	−14.8	−14.4	.....	.....	12.5	20.7	34.2	.....
7.5	7.4008	6.7	−14.2	−13.2	.....	.....	13.7	20.8	34.0	.....
7.0	7.9295	6.9	−13.5	−12.7	.....	.....	14.9	20.8	33.8	.....
6.5	8.5394	7.0	−12.7	−11.9	.....	.....	16.1	20.8	33.5	.....
6.0	9.2510	7.1	−11.8	−10.8	.....	.....	17.1	20.9	33.3	.....
5.5	10.0920	7.2	−10.8	−9.6	.....	.....	18.3	20.9	33.0	.....
5.0	11.1012	7.3	−9.6	−8.2	.....	.....	19.3	21.0	32.8	.....
4.5	12.3346	7.4	−8.7	−6.8	.....	.....	20.4	21.1	32.5	.....
4.0	13.8765	7.5	−6.6	−5.5	.....	.....	21.3	21.2	32.2	.....
3.5	15.8589	7.6	−4.7	−4.0	.....	.....	22.1	21.3	31.8	.....
3.25	17.0788	7.6	.....	−3.2	.....	.....	22.6	21.4	31.7	.....
3.0	18.5020	7.7	.....	−2.3	.....	.....	23.0	21.5	31.5	.....
2.5	22.2024	7.8	.....	−0.4	.....	.....	23.8	21.6	31.2	.....
2.0	27.7530	7.9	.....	.....	.....	.....	24.6	21.7	30.8	.....
1.5	37.0040	.....	.....	.....	.....	.....	25.2	21.8	30.4	.....
1.0	55.506	.....	.....	.....	.....	.....	25.7	22.0	30.1	.....

THERMODYNAMIC FORMULAS

Compiled by Doctor E. A. Coomes

Legend: p = Pressure
V = Volume
T = Temperature
n = Number of mols
S = Entropy
U = Internal energy (some books use E)
Cp = Molal specific heat at constant pressure
β = Coefficient volume expansion
K = Compressibility
H = U + pV = Total heat or enthalpy
A = U − TS = Helmholtz free energy
G = H − TS = Gibbs' free energy (some books use F).

Use of Table — Partial derivatives of the first order for the eight fundamental thermodynamic variables, namely, p, V, T, U, S, H, A, G, may be obtained in terms of $(\partial V/\partial T)_p$, $(\partial V/\partial p)_T$, and $(\partial H/\partial T)_p$; the latter three are connected to measurable quantities as follows:

$$\frac{1}{V}\left(\frac{\partial V}{\partial T}\right)_p = \beta; \quad -\frac{1}{V}\left(\frac{\partial V}{\partial p}\right)_T = K; \quad \left(\frac{\partial H}{\partial T}\right)_p = nC_p$$

Computation by the Table — The method of using the table will become apparent in working several examples. Suppose it is desired to know $(\partial H/\partial p)$, in terms of p, V, and T. Under caption "Constant" move horizontally to column marked "S"; across from "H" beside caption "differential" find "$-V nC_p/T$." Across from "p" in column "S" find "$-nC_p/T$"; $(\partial H/\partial p)$, is found by taking the ratio of the two:

$$(\partial H/\partial p)_S = (-VnC_p/T)/(-nC_p/T) = V$$

To find $(\partial S/\partial V)_T$ in terms of p, V, T, move to column "T" under "constant." Opposite "S" beside "differential" find "$(\partial V/\partial T)_p$"; opposite "V" find "$-(\partial V/\partial p)_T$". Taking the ratio:

$$(\partial S/\partial V)_T = (\partial V/\partial T)_p/ - (\partial V/\partial p)_T = (\partial p/\partial T)_V$$

THERMODYNAMIC FORMULAS (Continued)

Constant

Differential	T	p	V	S
T	0	1	$\left(\dfrac{\partial V}{\partial p}\right)_T$	$-\left(\dfrac{\partial V}{\partial T}\right)_p$
p	-1	0	$-\left(\dfrac{\partial V}{\partial T}\right)_p$	$-\dfrac{nC_p}{T}$
V	$-\left(\dfrac{\partial V}{\partial p}\right)_T$	$\left(\dfrac{\partial V}{\partial T}\right)_p$	0	$\left(-\dfrac{1}{T}\right)\left[nC_p\left(\dfrac{\partial V}{\partial p}\right)_T + T\left(\dfrac{\partial V}{\partial T}\right)_p^2\right]$
S	$\left(\dfrac{\partial V}{\partial T}\right)_p$	$\dfrac{nC_p}{T}$	$\left(\dfrac{1}{T}\right)\left[nC_p\left(\dfrac{\partial V}{\partial p}\right)_T + T\left(\dfrac{\partial V}{\partial T}\right)_p^2\right]$	0
U	$T\left(\dfrac{\partial V}{\partial T}\right)_p + p\left(\dfrac{\partial V}{\partial p}\right)_T$	$nC_p - p\left(\dfrac{\partial V}{\partial T}\right)_p$	$nC_p\left(\dfrac{\partial V}{\partial p}\right)_T + T\left(\dfrac{\partial V}{\partial T}\right)_p^2$	$\left(\dfrac{p}{T}\right)\left[nC_p\left(\dfrac{\partial V}{\partial p}\right)_T + T\left(\dfrac{\partial V}{\partial T}\right)_p^2\right]$
H	$-V + T\left(\dfrac{\partial V}{\partial T}\right)_p$	nC_p	$nC_p\left(\dfrac{\partial V}{\partial p}\right)_T + T\left(\dfrac{\partial V}{\partial T}\right)_p^2 - V\left(\dfrac{\partial V}{\partial T}\right)$	$-\dfrac{VnC_p}{T}$
A	$p\left(\dfrac{\partial V}{\partial p}\right)_T$	$-S - p\left(\dfrac{\partial V}{\partial T}\right)_p$	$-S\left(\dfrac{\partial V}{\partial p}\right)_T$	$\left(\dfrac{1}{T}\right)\left[pnC_p\left(\dfrac{\partial V}{\partial p}\right)_T + pT\left(\dfrac{\partial V}{\partial T}\right)_p^2 + TS\left(\dfrac{\partial V}{\partial T}\right)_p\right]$
G	$-V$	$-S$	$-V\left(\dfrac{\partial V}{\partial T}\right)_p - S\left(\dfrac{\partial V}{\partial p}\right)_T$	$\left(-\dfrac{1}{T}\right)\left[nC_pV - TS\left(\dfrac{\partial V}{\partial T}\right)_p\right]$

Constant

Differential	U	H	A	G
T	$-T\left(\dfrac{\partial V}{\partial T}\right)_p - p\left(\dfrac{\partial V}{\partial p}\right)_T$	$V - T\left(\dfrac{\partial V}{\partial T}\right)_p$	$-p\left(\dfrac{\partial V}{\partial p}\right)_T$	V
p	$-nC_p + p\left(\dfrac{\partial V}{\partial T}\right)_p$	$-nC_p$	$S + p\left(\dfrac{\partial V}{\partial T}\right)_p$	S
V	$-nC_p\left(\dfrac{\partial V}{\partial p}\right)_T - T\left(\dfrac{\partial V}{\partial T}\right)_p^2$	$-nC_p\left(\dfrac{\partial V}{\partial p}\right)_T - T\left(\dfrac{\partial V}{\partial T}\right)_p^2 + V\left(\dfrac{\partial V}{\partial T}\right)_p$	$S\left(\dfrac{\partial V}{\partial p}\right)_T$	$V\left(\dfrac{\partial V}{\partial T}\right)_p + S\left(\dfrac{\partial V}{\partial p}\right)_T$
S	$\left(-\dfrac{p}{T}\right)\left[nC_p\left(\dfrac{\partial V}{\partial p}\right)_T + T\left(\dfrac{\partial V}{\partial T}\right)_p^2\right]$	$\dfrac{VnC_p}{T}$	$\left(-\dfrac{1}{T}\right)\left[pnC_p\left(\dfrac{\partial V}{\partial p}\right)_T + pT\left(\dfrac{\partial V}{\partial T}\right)_p^2 + TS\left(\dfrac{\partial V}{\partial T}\right)_p\right]$	$\left(\dfrac{1}{T}\right)\left[nC_pV - TS\left(\dfrac{\partial V}{\partial T}\right)_p\right]$
U	0	$V\left[nC_p - p\left(\dfrac{\partial V}{\partial T}\right)_p\right] + p\left[nC_p\left(\dfrac{\partial V}{\partial p}\right)_T + T\left(\dfrac{\partial V}{\partial T}\right)_p^2\right]$	$-p\left[nC_p\left(\dfrac{\partial V}{\partial p}\right)_T + T\left(\dfrac{\partial V}{\partial T}\right)_p^2\right] - S\left[T\left(\dfrac{\partial V}{\partial T}\right)_p + p\left(\dfrac{\partial V}{\partial p}\right)_T\right]$	$V\left[nC_p - p\left(\dfrac{\partial V}{\partial T}\right)_p\right] - S\left[T\left(\dfrac{\partial V}{\partial T}\right)_p + p\left(\dfrac{\partial V}{\partial p}\right)_T\right]$
H	$-V\left[nC_p - p\left(\dfrac{\partial V}{\partial T}\right)_p\right] - p\left[nC_p\left(\dfrac{\partial V}{\partial p}\right)_T + T\left(\dfrac{\partial V}{\partial T}\right)_p^2\right]$	0	$-\left[S + p\left(\dfrac{\partial V}{\partial T}\right)_p\right] \times \left[V - T\left(\dfrac{\partial V}{\partial T}\right)_p - pnC_p\left(\dfrac{\partial V}{\partial p}\right)_T\right]$	$VnC_p + VS - TS\left(\dfrac{\partial V}{\partial T}\right)_p$
A	$p\left[nC_p\left(\dfrac{\partial V}{\partial p}\right)_T + T\left(\dfrac{\partial V}{\partial T}\right)_p^2\right] + S\left[T\left(\dfrac{\partial V}{\partial T}\right)_p + p\left(\dfrac{\partial V}{\partial p}\right)_T\right]$	$-\left[S + p\left(\dfrac{\partial V}{\partial T}\right)_p\right] \times \left[V - T\left(\dfrac{\partial V}{\partial T}\right)_p - pnC_p\left(\dfrac{\partial V}{\partial p}\right)_T\right]$	0	$-S\left[V + p\left(\dfrac{\partial V}{\partial p}\right)_T\right] + pV\left(\dfrac{\partial V}{\partial T}\right)_p$
G	$-V\left[nC_p - p\left(\dfrac{\partial V}{\partial T}\right)_p\right] + S\left[T\left(\dfrac{\partial V}{\partial T}\right)_p + p\left(\dfrac{\partial V}{\partial p}\right)_T\right]$	$-VnC_p - VS + TS\left(\dfrac{\partial V}{\partial T}\right)_p$	$S\left[V + p\left(\dfrac{\partial V}{\partial p}\right)_T\right] + pV\left(\dfrac{\partial V}{\partial T}\right)_p$	0

LIMITS OF INFLAMMABILITY

Reprinted from ''Combustion Flame and Explosions of Gases'', B. Lewis and G. von Elbe, authors, Academic Press (1951), publishers, by special permission.

The limits of inflammability given in the following tables were all determined at atmospheric pressure and room* temperature for upward propagation in a tube or bomb 2 inches or more in diameter. Values are on a percentage-by-volume basis.

LIMITS OF INFLAMMABILITY OF GASES AND VAPORS IN AIR

Compound	Empirical formula	Limits of inflammability		Compound	Empirical formula	Limits of inflammability	
		Lower	Upper			Lower	Upper
Paraffin hydrocarbons				Methyl propyl ketone	$C_5H_{10}O$	1.55	8.15
Methane	CH_4	5.00	15.00	Methylbutyl ketone	$C_6H_{12}O$	1.35	7.60
Ethane	C_2H_6	3.00	12.50	Acids			
Propane	C_3H_8	2.12	9.35	Acetic acid	$C_2H_4O_2$	5.40	—
Butane	C_4H_{10}	1.86	8.41	Hydrocyanic acid	HCN	5.60	40.00
Isobutane	C_4H_{10}	1.80	8.44	Esters			
Pentane	C_5H_{12}	1.40	7.80	Methyl formate	$C_2H_4O_2$	5.05	22.70
Isopentane	C_5H_{12}	1.32	—	Ethyl formate	$C_3H_6O_2$	2.75	16.40
2,2-Dimethylpropane	C_5H_{12}	1.38	7.50	Methyl acetate	$C_3H_6O_2$	3.15	15.60
Hexane	C_6H_{14}	1.18	7.40	Ethyl acetate	$C_4H_8O_2$	2.18	11.40
Heptane	C_7H_{16}	1.10	6.70	Propyl acetate	$C_5H_{10}O_2$	1.77	8.00
2,3-Dimethylpentane	C_7H_{16}	1.12	6.75	Isopropyl acetate	$C_6H_{10}O_2$	1.78	7.80
Octane	C_8H_{18}	0.95	—	Butyl acetate	$C_6H_{12}O_2$	1.39	7.55
Nonane	C_9H_{20}	0.83	—	Amyl acetate	$C_7H_{14}O_2$	1.10	—
Decane	$C_{10}H_{22}$	0.77	5.35	Hydrogen			
Olefins				Hydrogen	H_2	4.00	74.20
Ethylene	C_2H_4	2.75	28.60	Nitrogen compounds			
Propylene	C_3H_6	2.00	11.10	Ammonia	NH_3	15.50	27.00
Butene-1	C_4H_8	1.65	9.95	Cyanogen	C_2N_2	6.60	42.60
Butene-2	C_4H_8	1.75	9.70	Pyridine	C_5H_5N	1.81	12.40
Amylene	C_5H_{10}	1.42	8.70	Ethyl nitrate	$C_2H_5NO_3$	3.80	—
Acetylenes				Ethyl nitrite	$C_2H_5NO_2$	3.01	50.00
Acetylene	C_2H_2	2.50	80.00	Oxides			
Aromatics				Carbon monoxide	CO	12.50	74.20
Benzene	C_6H_6	1.40	7.10	Ethylene oxide	C_2H_4O	3.00	80.00
Toluene	C_7H_8	1.27	6.75	Propylene oxide	C_3H_6O	2.00	22.00
o-Xylene	C_8H_{10}	1.00	6.00	Dioxan	$C_4H_8O_2$	1.97	22.25
Cyclic hydrocarbons				Diethyl peroxide	$C_4H_{10}O_2$	2.34	—
Cyclopropane	C_3H_6	2.40	10.40	Sulfides			
Cyclohexane	C_6H_{12}	1.26	7.75	Carbon disulfide	CS_2	1.25	50.00
Methylcyclohexane	C_7H_{14}	1.15	—	Hydrogen sulfide	H_2S	4.30	45.50
Terpenes				Carbon oxysulfide	COS	11.90	28.50
Turpentine	$C_{10}H_{16}$	0.80	—	Chlorides			
Alcohols				Methyl chloride	CH_3Cl	8.25	18.70
Methyl alcohol	CH_4O	6.72	36.50	Ethyl chloride	C_2H_5Cl	4.00	14.80
Ethyl alcohol	C_2H_6O	3.28	18.95	Propyl chloride	C_3H_7Cl	2.60	11.10
Allyl alcohol	C_3H_6O	2.50	18.00	Butyl chloride	C_4H_9Cl	1.85	10.10
η-Propyl alcohol	C_3H_8O	2.15	13.50	Isobutyl chloride	C_4H_9Cl	2.05	8.75
Isopropyl alcohol	C_3H_8O	2.02	11.80	Allyl chloride	C_3H_5Cl	3.28	11.15
η-Butyl alcohol	$C_4H_{10}O$	1.45	11.25	Amyl chloride	$C_5H_{11}Cl$	1.60	8.63
Isobutyl alcohol	$C_4H_{10}O$	1.68	—	Vinyl chloride	C_2H_3Cl	4.00	21.70
η-Amyl alcohol	$C_5H_{12}O$	1.19	—	Ethylene dichloride	$C_2H_4Cl_2$	6.20	15.90
Isoamyl alcohol	$C_5H_{12}O$	1.20	—	Propylene dichloride	$C_3H_6Cl_2$	3.40	14.50
Aldehydes				Bromides			
Acetaldehyde	C_2H_4O	3.97	57.00	Methyl bromide	CH_3Br	13.50	14.50
Crotonic aldehyde	C_4H_6O	2.12	15.50	Ethyl bromide	C_2H_5Br	6.75	11.25
Furfural	$C_5H_4O_2$	2.10	—	Allyl bromide	C_3H_5Br	4.36	7.25
Paraldehyde	$C_6H_{12}O_3$	1.30	—	Amines			
Ethers				Methyl amine	CH_5N	4.95	20.75
Methylethyl ether	C_3H_8O	2.00	10.00	Ethyl amine	C_2H_7N	3.55	13.95
Diethyl ether	$C_4H_{10}O$	1.85	36.50	Dimethyl amine	C_2H_7N	2.80	14.40
Divinyl ether	C_4H_6O	1.70	27.00	Propyl amine	C_3H_9N	2.01	10.35
Ketones				Diethyl amine	$C_4H_{11}N$	1.77	10.10
Acetone	C_3H_6O	2.55	12.80	Trimethyl amine	C_3H_9N	2.00	11.60
Methylethyl ketone	C_4H_8O	1.81	9.50	Triethyl amine	$C_6H_{16}N$	1.25	7.90

* The upper limits of some vapors were determined at somewhat higher temperatures because of their low vapor pressures.

LIMITS OF INFLAMMABILITY OF GASES AND VAPORS IN OXYGEN

Compound	Formula	Limits of inflammability		Compound	Formula	Limits of inflammability	
		Lower	Upper			Lower	Upper
Hydrogen	H_2	4.65	93.9	Propylene	C_3H_6	2.10	52.8
Deuterium	D_2	5.00	95.0	Cyclopropane	C_3H_6	2.45	63.1
Carbon monoxide	CO	15.50	93.9	Ammonia	NH_3	13.50	79.0
Methane	CH_4	5.40	59.2	Diethyl ether	$C_4H_{10}O$	2.10	82.0
Ethane	C_2H_6	4.10	50.5	Divinyl ether	C_4H_6O	1.85	85.5
Ethylene	C_2H_4	2.90	79.9				

LIMITS FOR HUMAN EXPOSURE TO AIR CONTAMINANTS

From United States Federal Register
Volume 36, Number 105

Exposures by inhalation, ingestion, skin absorption, or contact to any material or substance (1) at a concentration above those specified in the "Threshold Limit Values of Airborne Contaminants for 1970" of the American Conference of Governmental Industrial Hygienists, listed in Table 1, except for the American National Standards listed in Table 2 of this section and except for values of mineral dusts listed in Table 3 of this section, and (2) concentrations above those specified in Table 1, 2, and 3 of this section, shall be avoided, or protective equipment shall be provided and used.

Table 1

Substance	ppm[a]	mg/m³[b]	Substance	ppm[a]	mg/m³[b]
Abate		15	2-Butanone	200	590
Acetaldehyde	200	360	2-Butoxy ethanol (Butyl Cel-		
Acetic acid	10	25	losolve)–Skin	50	240
Acetic anhydride	5	20	Butyl acetate (n -butyl acetate)	150	710
Acetone	1,000	2,400	sec-Butyl acetate	200	950
Acetonitrile	40	70	tert-Butyl acetate	200	950
Acetylene dichloride, see 1,2-			Butyl alcohol	100	300
Dichloroethylene			sec-Butyl alcohol	150	450
Acetylene tetrabromide	1	14	tert-Butyl alcohol	100	300
Acrolein	0.1	0.25	C Butylamine–Skin	5	15
Acrylamide–Skin		0.3	C tert-Butyl chromate (as CrO₃)--		
Acrylonitrile–Skin	20	45	Skin		0.1
Aldrin–Skin		0.25	n-Butyl glycidyl ether (BGE)	50	270
Allyl alcohol–Skin	2	5	*Butyl mercaptan	0.5	1.5
Allyl chloride	1	3	p-tert-Butyltoluene	10	60
**C Allyl glycidyl ether (AGE)	10	45	Calcium arsenate		1
Allyl propyl disulfide	2	12	Calcium oxide		5
2-Aminoethanol, see Ethanol-			**Camphor (Synthetic)	2	
amine			Carbaryl (Sevin®)		5
2-Aminopyridine	0.5	2	Carbon black		3.5
**Ammonia	50	35	Carbon dioxide	5,000	9,000
Ammonium sulfamate (Ammate)		15	Carbon monoxide	50	55
n-Amyl acetate	100	525	Chlordane–Skin		0.5
sec-Amyl acetate	125	650	Chlorinated camphene–Skin		0.5
Aniline–Skin	5	19	Chlorinated diphenyl oxide		0.5
Anisidine (o,p-isomers)–Skin		0.5	*Chlorine	1	3
Antimony and compounds (as Sb)		0.5	Chlorine dioxide	0.1	0.3
ANTU (alpha naphthyl thiourea)		0.3	C Chlorine trifluoride	0.1	0.4
Arsenic and compounds (as As)		0.5	C Chloroacetaldehyde	1	3
Arsine	0.05	0.2	α-Chloroacetophenone (phenacyl-		
Azinphos-methyl–Skin		0.2	chloride)	0.05	0.3
Barium (soluble compounds)		0.5	Chlorobenzene (monochloroben-		
p-Benzoquinone, see Quinone			zene)	75	350
Benzoyl peroxide		5	o-Chlorobenzylidene malononi-		
Benzyl chloride	1	5	trile (OCBM)	0.05	0.4
Biphenyl, see Diphenyl			Chlorobromomethane	200	1,050
Bisphenol A, see Diglycidyl ether			2-Chloro-1,3-butadiene, see		
Boron oxide		15	Chloroprene		
Boron tribromide	1	10	Chlorodiphenyl (42 percent		
C Boron trifluoride	1	3	Chlorine)–Skin		1
Bromine	0.1	0.7	Chlorodiphenyl (54 percent		
*Bromine pentafluoride	0.1	0.7	Chlorine)–Skin		0.5
Bromoform–Skin	0.5	5	1-Chloro-2,3-epoxypropane, see		
Butadiene (1,3-butadiene)	1,000	2,200	Epichlorhydrin		
Butanethiol, see Butyl mercaptan					

Table 1 (continued)

Substance	ppm[a]	mg/m³[b]	Substance	ppm[a]	mg/m³[b]
2-Chloroethanol, see Ethylene chlorohydrin			Dieldrin–Skin		0.25
Chloroethylene, see Vinyl chloride			Diethylamine	25	75
			Diethylamino ethanol–Skin	10	50
C Chloroform (trichloromethane)	50	240	**C Diethylene triamine–Skin	10	42
1-Chloro-1-nitropropane	20	100	Diethylether, see Ethyl ether		
Chloropicrin	0.1	0.7	Difluorodibromomethane	100	860
Chloroprene (2-chloro-1,3-butadiene)–Skin	25	90	C Diglycidyl ether (DGE)	0.5	2.8
Chromium, sol. chromic, chromous salts as Cr		0.5	Dihydroxybenzene, see Hydroquinone		
Metal and insol. salts		1	Diisobutyl ketone	50	290
Coal tar pitch volatiles (benzene soluble fraction) anthracene, BaP, phenanthrene, acridine, chrysene, pyrene		0.2	Diisopropylamine–Skin	5	20
			Dimethoxymethane, see Methylal		
			Dimethyl acetamide–Skin	10	35
			Dimethylamine	10	18
Cobalt, metal fume and dust		0.1	Dimethylaminobenzene, see Xylidene		
Copper fume		0.1	Dimethylaniline (*N*-dimethyl-aniline)–Skin	5	25
Dusts and Mists		1	Dimethylbenzene, see Xylene		
Cotton dust (raw)		1	Dimethyl 1,2-dibromo-2,2-di-chloroethyl phosphate, (Dibrom)		3
Crag® herbicide		15			
Cresol (all isomers)–Skin	5	22	Dimethylformamide–Skin	10	30
Crotonaldehyde	2	6	2,6-Dimethylheptanone, see Diisobutyl ketone		
Cumene–Skin	50	245	1,1-Dimethylhydrazine–Skin	0.5	1
Cyanide (as CN)–Skin		5	Dimethylphthalate		5
*Cyanogen	100		Dimethylsulfate–Skin	1	5
Cyclohexane	300	1,050	Dinitrobenzene (all isomers)–Skin		1
Cyclohexanol	50	200	Dinitro-*o*-cresol–Skin		0.2
Cyclohexanone	50	200	Dinitrotoluene–Skin		1.5
Cyclohexene	300	1,015	Dioxane (Diethylene dioxide)–Skin	100	360
Cyclopentadiene	75	200	Diphenyl	0.2	1
2,4-D		10	Diphenylamine		10
DDT–Skin		1	Diphenylmethane diisocyanate (see Methylene bisphenyl isocyanate (MDI)		
DDVP, see Dichlorvos					
Decaborane–Skin	0.05	0.3	Dipropylene glycol methyl ether–Skin	100	600
Demeton®–Skin		0.1			
Diacetone alcohol (4-hydroxy-4-methyl-2-pentanone)	50	240	Di-*sec*, octyl phthalate (Di-2-ethylhexylphthalate)		5
1,2-Diaminoethane, see Ethylenediamine			*Endosulfan (Thiodan®)–Skin		0.1
Diazomethane	0.2	0.4	Endrin–Skin		0.1
Diborane	0.1	0.1	Epichlorhydrin–Skin	5	19
Dibutyl phosphate	1	5	EPN–Skin		0.5
Dibutylphthalate		5	1,2-Epoxypropane, see Propyleneoxide		
*C Dichloroacetylene	0.1	0.4	2,3-Epoxy-1-propanol, see Glycidol		
C *o*-Dichlorobenzene	50	300	Ethanethiol, see Ethylmercaptan		
p-Dichlorobenzene	75	450	Ethanolamine	3	6
Dichlorodifluoromethane	1,000	4,950	2-Ethoxyethanol–Skin	200	740
1,3-Dichloro-5,5-dimethyl hydantoin		0.2	2-Ethoxyethylacetate (Cellosolve acetate)–Skin	100	540
1,1-Dichloroethane	100	400	Ethyl acetate	400	1,400
1,2-Dichloroethylene	200	790	Ethyl acrylate–Skin	25	100
C Dichloroethyl ether–Skin	15	90	Ethyl alcohol (ethanol)	1,000	1,900
Dichloromethane, see Methylenechloride			Ethylamine	10	18
Dichloromonofluoromethane	1,000	4,200			
C 1,1-Dichloro-1-nitroethane	10	60			
1,2-Dichloropropane, see Propylenedichloride					
Dichlorotetrafluoroethane	1,000	7,000			
Dichlorvos (DDVP)–Skin		1			

Table 1 (continued)

Substance	ppm[a]	mg/m³[b]	Substance	ppm[a]	mg/m³[b]
Ethyl sec-amyl ketone (5-methyl-3-heptanone)	25	130	Isoamyl alcohol	100	360
Ethyl benzene	100	435	Isobutyl acetate	150	700
Ethyl bromide	200	890	Isobutyl alcohol	100	300
Ethyl butyl ketone (3-Heptanone)	50	230	Isophorone	25	140
Ethyl chloride	1,000	2,600	Isopropyl acetate	250	950
Ethyl ether	400	1,200	Isopropyl alcohol	400	980
Ethyl formate	100	300	Isopropylamine	5	12
Ethyl mercaptan	0.5	1	Isopropylether	500	2,100
Ethyl silicate	100	850	Isopropyl glycidyl ether (IGE)	50	240
Ethylene chlorohydrin–Skin	5	16	Ketene	0.5	0.9
Ethylenediamine	10	25	Lead arsenate		0.15
Ethylene dibromide, see 1,2-Dibromoethane			Lindane–Skin		0.5
			Lithium hydride		0.025
Ethylene dichloride, see 1,2-Dichloroethane			L.P.G. (liquefied petroleum gas)	1,000	1,800
			Magnesium oxide fume		15
C Ethylene glycol dinitrate and/or Nitroglycerin–Skin	[d]0.2		Malathion–Skin		15
			Maleic anhydride	0.25	1
Ethylene glycol monomethyl ether acetate, see Methyl cellosolve acetate			C Manganese and compounds, as Mn		5
			Mesityl oxide	25	100
Ethylene imine–Skin	0.5	1	Methanethiol, see Methyl mercaptan		
Ethylene oxide	50	90	Methoxychlor		15
Ethylidine chloride, see 1,1-Dichloroethane			2-Methoxyethanol, see Methyl cellosolve		
N-Ethylmorpholine–Skin	20	94	Methyl acetate	200	610
Ferbam		15	Methyl acetylene (propyne)	1,000	1,650
Ferrovanadium dust		1	Methyl acetylene-propadiene mixture (MAPP)	1,000	1,800
Fluoride (as F)		2.5			
Fluorine	0.1	0.2	Methyl acrylate–Skin	10	35
Fluorotrichloromethane	1,000	5,600	Methylal (dimethoxymethane)	1,000	3,100
Formic acid	5	9	Methyl alcohol (methanol)	200	260
Furfural–Skin	5	20	Methylamine	10	12
Furfuryl alcohol	50	200	Methyl amyl alcohol, see Methyl isobutyl carbinol		
Glycidol (2,3-Epoxy-1-propanol)	50	150	*Methyl isoamyl ketone	100	475
Glycol monoethyl ether, see 2-Ethoxyethanol			Methyl (n-amyl) ketone (2-Heptanone)	100	465
Guthion®, see Azinphosmethyl					
Hafnium		0.5	C Methyl bromide–Skin	20	80
Heptachlor–Skin		0.5	Methyl butyl ketone, see 2-Hexanone		
Heptane (n-heptane)	500	2,000			
Hexachloroethane–Skin	1	10	Methyl cellosolve–Skin	25	80
Hexachloronaphthalene–Skin		0.2	Methyl cellosolve acetate–Skin	25	120
Hexane (n-hexane)	500	1,800	Methyl chloroform	350	1,900
2-Hexanone	100	410	Methylcyclohexane	500	2,000
Hexone (Methyl isobutyl ketone)	100	410	Methylcyclohexanol	100	470
sec-Hexyl acetate	50	300	o-Methylcyclohexanone–Skin	100	460
Hydrazine–Skin	1	1.3	Methyl ethyl ketone (MEK), see 2-Butanone		
Hydrogen bromide	3	10			
C Hydrogen chloride	5	7	Methyl formate	100	250
Hydrogen cyanide–Skin	10	11	Methyl iodide–Skin	5	28
Hydrogen peroxide	1	1.4	Methyl isobutyl carbinol–Skin	25	100
Hydrogen selenide	0.05	0.2	Methyl isobutyl ketone, see Hexone		
Hydroquinone		2			
*Indene	10	45	Methyl isocyanate–Skin	0.02	0.05
Indium and compounds, as In		0.1	*Methyl mercaptan	0.5	1
C Iodine	0.1	1	Methyl methacrylate	100	410
Iron oxide fume		10	Methyl propyl ketone, see 2-Pentanone		
Iron salts, soluble, as Fe		1			
Isoamyl acetate	100	525	C Methyl silicate	5	30

Table 1 (continued)

Substance	ppm[a]	mg/M³ [b]	Substance	ppm[a]	mg/M³ [b]
C α-Methyl styrene	100	480	TEPP–Skin		0.05
C Methylene bisphenyl isocyanate			C Terphenyls	1	9
(MDI)	0.02	0.2	1,1,1,2-Tetrachloro-2,2-difluoro-		
Molybdenum:			ethane	500	4,170
Soluble compounds		5	1,1,2,2-Tetrachloro-1,2-difluoro-		
Insoluble compounds		15	ethane	500	4,170
Monomethyl aniline–Skin	2	9	1,1,2,2-Tetrachloroethane–Skin	5	35
C Monomethyl hydrazine–Skin	0.2	0.35	Tetrachloroethylene, see Per-		
Morpholine–Skin	20	70	chloroethylene		
Naphtha (coaltar)	100	400	Tetrachloromethane, see Carbon		
Naphthalene	10	50	tetrachloride		
Nickel carbonyl	0.001	0.007	Tetrachloronaphthalene–Skin		2
Nickel, metal and soluble cmpds,			Tetraethyl lead (as Pb)–Skin		ƒ0.100
as Ni		1	Tetrahydrofuran	200	590
Nicotine–Skin		0.5	Tetramethyl lead (as Pb)–Skin		ƒ0.150
Nitric acid	2	5	Tetramethyl succinonitrile–Skin	0.5	3
Nitric oxide	25	30	Tetranitromethane	1	8
p-Nitroaniline–Skin	1	6	Tetryl (2,4,6-trinitrophenyl-		
Nitrobenzene–Skin	1	5	methylnitramine)–Skin		1.5
p-Nitrochlorobenzene–Skin		1	Thallium (soluble compounds)–		
Nitroethane	100	310	Skin as T1		0.1
Nitrogen dioxide	5	9	Thiram		5
Nitrogen trifluoride	10	29	Tin (inorganic cmpds, except		
Nitroglycerin–Skin	0.2	2	SnH₄ and SnO₂)		2
Nitromethane	100	250	Tin (organic cmpds)		0.1
1-Nitropropane	25	90	C Toluene 2,4-diisocyanate	0.02	0.14
2-Nitropropane	25	90	o-Toluidine–Skin	5	22
Nitrotoluene–Skin	5	30	Toxaphene, see Chlorinated		
Nitrotrichloromethane, see			camphene		
Chloropicrin			Tributyl phosphate		5
Octachloronaphthalene–Skin		0.1	1,1,1-Trichloroethane (see		
*Octane	400	1,900	Methyl chloroform		
*Oil mist, particulate		5	1,1,2-Trichloroethane–Skin	10	45
Osmium tetroxide		0.002	Parathion–Skin		0.1
Oxalic acid		1	Pentaborane	0.005	0.01
Oxygen difluoride	0.05	0.1	Pentachloronaphthalene–Skin		0.5
Ozone	0.1	0.2	Pentachlorophenol–Skin		0.5
Paraquat–Skin		0.5	*Pentane	500	1,500
Silver, metal and soluble com-			2-Pentanone	200	700
pounds		0.01	Perchloromethyl mercaptan	0.1	0.8
Sodium fluoroacetate (1080)–			Perchloryl fluoride	3	13.5
Skin		0.05	Phenol–Skin	5	19
Sodium hydroxide		2	p-Phenylene diamine–Skin		0.1
Stibine	0.1	0.5	Phenyl ether (vapor)	1	7
*Stoddard solvent	200	1,150	Phenyl ether-biphenyl mixture		
Strychnine		0.15	(vapor)	1	7
Sulfur dioxide	5	13	Phenylethylene, see Styrene		
Sulfur hexafluoride	1,000	6,000	Phenyl glycidyl ether (PGE)	10	60
Sulfuric acid		1	Phenylhydrazine–Skin	5	22
Sulfur monochloride	1	6	Phosdrin (Mevinphos®)–Skin		0.1
Sulfur pentafluoride	0.025	0.25	Phosgene (carbonyl chloride)	0.1	0.4
Sulfuryl fluoride	5	20	Phosphine	0.3	0.4
Systox, see Demeton®			Phosphoric acid		1
2,4,5T		10	Phosphorus (yellow)		0.1
Tantalum		5	Phosphorus pentachloride		1
TEDP–Skin		0.2	Phosphorus pentasulfide		1
Tellurium		0.1	Phosphorus trichloride	0.5	3
Tellurium hexafluoride	0.02	0.2	Phthalic anhydride	2	12

Table 1 (continued)

Substance	ppm[a]	mg/M³ [b]
Picric acid–Skin		0.1
Pival® (2-Pivalyl-1,3-indandione)		0.1
Platinum (Soluble Salts) as Pt		0.002
Propargyl alcohol–Skin	1	
n-Propyl acetate	200	840
Propyl alcohol	200	500
n-Propyl nitrate	25	110
Propylene dichloride	75	350
Propylene imine–Skin	2	5
Propylene oxide	100	240
Propyne, see Methylacetylene		
Pyrethrum		5
Pyridine	5	15
Quinone	0.1	0.4
RDX–Skin		1.5
Rhodium, Metal fume and dusts, as Rh		0.1
Soluble salts		0.001
Ronnel		10
Rotenone (commercial)		5
Selenium compounds (as Se)		0.2
Selenium hexafluoride	0.05	0.4
Trichloromethane, see Chloroform		
Trichloronaphthalene–Skin		5
1,2,3-Trichloropropane	50	300
1,1,2-Trichloro 1,2,2-trifluoroethane	1,000	7,600
Triethylamine	25	100
Trifluoromonobromomethane	1,000	6,100
*Trimethyl benzene	25	120
2,4,6-Trinitrophenol, see Picric acid		
2,4,6-Trinitrophenylmethylnitramine, see Tetryl		
Trinitrotoluene–Skin		1.5
Triorthocresyl phosphate		0.1
Triphenyl phosphate		3
Tungsten and compounds, as W:		
Soluble		1
Insoluble		5
Turpentine	100	560
Uranium (natural) sol. and insol. compounds as U		0.2
C Vanadium:		
V₂O₅ dust		0.5
V₂O₅ fume		0.1
Vinyl benzene, see Styrene		
**C Vinyl chloride	500	1,300
Vinylcyanide, see Acrylonitrile		
Vinyl toluene	100	480
Warfarin		0.1
Xylene (xylol)	100	435
Xylidine–Skin	5	25
Yttrium		1
Zinc chloride fume		1
Zinc oxide fume		5
Zirconium compounds (as Zr)		5

*1970 Addition.

[a] Parts of vapor or gas per million parts of contaminated air by volume at 25°C and 760 mm Hg pressure.

[b] Approximate milligrams of particulate per cubic meter of air.

(No footnote "c" is used to avoid confusion with ceiling value notations.)

[d] An atmospheric concentration of not more than 0.02 ppm, or personal protection may be necessary to avoid headache.

[e] As sampled method that does not collect vapor.

[f] For control of general room air, biologic monitoring is essential for personnel control.

Table 2

	8-hour time weighted average
Benzene (Z37.4–1969)	10 ppm
Beryllium and beryllium compounds (Z37.29–1970)	0.002 mg/M³
Cadmium dust (as Cd) (Z37.5–1970)	0.2 mg/M³
Cadmium fume (as Cd) (Z37.5–1970)	0.1 mg/M³
Carbon disulfide (Z37.3–1968)	20 ppm
Carbon tetrachloride (Z37.17–1967)	10 ppm
Ethylene dibromide (Z37.31–1970)	20 ppm
Ethylene dichloride (Z37.21–1969)	50 ppm
Formaldehyde (Z37.16–1967)	3 ppm
Hydrogen fluoride (Z37.28–1969)	3 ppm
Fluoride as dust (Z37.28–1966)	2.5 mg/M³
Lead and its inorganic compounds (Z37.11–1969)	0.2 mg/M³
Methyl chloride (Z37.18–1969)	100 ppm
Methylene chloride (Z37.23–1969)	500 ppm
Organo (alkyl) mercury (Z37.30–1969)	0.01 mg/M³
Styrene (Z37.12–1969)	100 ppm
Tetrachloroethylene (Z37.22–1967)	100 ppm
Toluene (Z37.12–1967)	200 ppm

	Acceptable ceiling concentration
Hydrogen sulfide (Z37.2–1966)	20 ppm
Chromic acid and chromates (Z37.3–1971)	1 mg/10M³
Mercury (Z37.8–1971)	1 mg/10M³

Table 3

Substance	Mppcf[a]	Mg/M³	Substance	Mppcf[a]	Mg/M³
Silica:			Inert or Nuisance Dust:		
Crystalline:			Respirable fraction	15	5 mg/M³
Quartz (respirable)	250[f]	10 mg/M³ [m]	Total dust	50	15 mg/M³
	$\frac{\%SiO_2 + 5}{}$	$\frac{\%SiO_2 + 2}{30 mg/M^3}$			
Quartz (total dust)					
		$\%SiO_2 + 2$			

NOTE: Conversion factors—

mppcf x 35.3 = million particles per cubic meter

              = particles per cc

[a]Millions of particles per cubic foot of air, based on impinger samples counted by light-field technics.

[f]The percentage of crystalline silica in the formula is the amount determined from air-borne samples, except in those instances in which other methods have been shown to be applicable.

[i]As determined by the membrane filter method at 430 x phase contrast magnification.

[m]Both concentration and percent quartz for the application of this limit are to be determined from the fraction passing a size-selector with the following characteristics:

Cristobalite: Use ½ the value calculated from the count or mass formulae for quartz.

Trioymite: Use ½ the value calculated from the formulae for quartz.

Substance	Mppcf[a]	Mg/M³
Amorphous, including natural diatomaceous earth	20	80 mg/M³
Tremolite	5	$\frac{\%SiO_2}{20 mg/M^3}$
		$\%SiO_2$

Aerodynamic diameter (unit density sphere)	Percent passing selector
2	90
2.5	75
3.5	50
5.0	25
10	0

Substance	Mppcf[a]	Mg/M³
Silicates (less than 1% crystalline silica):		
Asbestos 12 fibers per milliliter greater than 5 microns in length,[i] or	2	
Mica	20	
Soapstone	20	
Talc	20	
Portland cement	50	
Graphite (natural)	15	
Coal dust (respirable fraction less than 5% SiO₂)		2.4 mg/M³ or
For more than 5% SiO₂		10 mg/M³
		$\%SiO_2 + 2$

The measurements under this note refer to the use of an AEC instrument. If the respirable fraction of coal dust is determined with a MRE the figure corresponding to that of 2.4 Mg/M³ in the table for coal dust is 4.5 Mg/M³.

FLAME AND BEAD TESTS

Flame Colorations

Violet

Potassium compounds. Purple red through blue glass. Easily obscured by sodium flame. Bluish green through green glass. Rubidium and Cesium compounds impart same flame as potassium compounds

Blues

Azure.—Copper chloride. Copper bromide gives azure blue followed by green. Other copper compounds give same coloration when moistened with hydrochloric acid.

Light Blue.—Lead, Arsenic, Selenium.

Greens

Emerald.—Copper compounds except the halides, and when not moistened with hydrochloric acid.

Pure Green.—Compounds of thallium and tellurium.

Yellowish.—Barium compounds. Some molybdenum compounds. Borates, especially when treated with sulphuric acid or when burned with alcohol.

Bluish.—Phosphates with sulphuric acid.

Feeble.—Antimony compounds. Ammonium compounds.

Whitish.—Zinc.

Reds

Carmine.—Lithium compounds. Violet through blue glass. Invisible through green glass. Masked by barium flame.

Scarlet.—Strontium compounds. Violet through blue glass. Yellowish through green glass. Masked by barium flame.

Yellowish.—Calcium compounds. Greenish through blue glass. Green through green glass. Masked by barium flame.

Yellow

Yellow.—All sodium compounds. Invisible with blue glass.

Borax Beads

Abbreviations employed: s., saturated; s.s., supersaturated; n.s.; not saturated; h., hot; c., cold.

Substance	Oxidizing flame	Reducing flame
Aluminum	Colorless (h.c., n.s.); opaque (s.s.)	Colorless; opaque (s.)
Antimony	Colorless; yellow or brownish (h., s.s.)	Gray and opaque
Barium	Colorless (n.s.)	
Bismuth	Colorless; yellow or brownish (h., s.s.)	Gray and opaque
Cadmium	Colorless	Gray and opaque
Calcium	Colorless (n.s.)	
Cerium	Red (h.)	Colorless (h.c.)
Chromium	Green (c.)	Green
Cobalt	Blue (h.c.)	Blue (h.c.)
Copper	Green (h.); blue (c.)	Red (c.): opaque (s.s.): colorless (h.)
Iron	Yellow or brownish red (h., n.s.)	Green (s.s.)
Lead	Colorless; yellow or brownish (h., s.s.)	Gray and opaque
Magnesium	Colorless (n.s.)	
Manganese	Violet (h.c.)	Colorless (h.c.)
Molybdenum	Colorless	Yellow or brown (h.)
Nickel	Brown; red (c.)	Gray and opaque
Silicon	Colorless (h.c.); opaque (s.s.)	Colorless; opaque (s.)
Silver	Colorless (n.s.)	Gray and opaque
Strontium	Colorless (n.s.)	
Tin	Colorless (h.c.); opaque (s.s.)	Colorless; opaque (s.)
Titanium	Colorless	Yellow (h.); violet (c.)
Tungsten	Colorless	Brown
Uranium	Yellow or brownish (h., n.s.)	Green
Vanadium	Colorless	Green

Beads of Microcosmic Salt
NaNH₄HPO₄

Substance	Oxidizing flame	Reducing flame
Aluminum	Colorless; opaque (s.)	Colorless; not clear (s.s)
Antimony	Colorless (n.s.)	Gray and opaque
Barium	Colorless; opaque (s.)	Colorless; not clear (s.s.)
Bismuth	Colorless (n.s.)	Gray and opaque
Cadmium	Colorless (n.s.)	Gray and opaque
Calcium	Colorless; opaque (s.)	Colorless; not clear (s.s.)
Cerium	Yellow or brownish red (h., s.)	Colorless
Chromium	Red (h., s.); green (c.)	Green (c.)
Cobalt	Blue (h.c.)	Blue (h.c.)
Copper	Blue (c.); green (h.)	Red and opaque (c.)
Iron	Yellow or brown (h., s.)	Colorless; yellow or brownish (h.)
Lead	Colorless (n.s.)	Gray and opaque
Magnesium	Colorless; opaque (s.)	Colorless; not clear (s.s.)
Manganese	Violet (h.c.)	Colorless
Molybdenum	Colorless; green (h.)	Green (h.)
Nickel	Yellow (c.); red (h., s.)	Yellow (c.); red (h.); gray and opaque
Silicon	(Swims undissolved)	(Swims undissolved)
Silver		Gray and opaque
Strontium	Colorless; opaque (s.)	Colorless; not clear (s.s.)
Tin	Colorless; opaque (s.)	Colorless
Titanium	Colorless (n.s.)	Violet (c.); yellow or brownish (h.)
Uranium	Green; yellow or brownish (h., s.)	Green (h.)
Vanadium	Yellow	Green
Zinc	Colorless (n.s.)	Gray and opaque

Sodium Carbonate Bead

Substance	Oxidizing flame	Reducing flame
Manganese	Green	Colorless

PREPARATION OF REAGENTS

The following pages present directions for the preparation of various reagents. The collection has been prepared with the active collaboration of W. D. Bonner, R. K. Carleton, L. L. Carrick, Giles B. Cooke, E. J. Cragoe, Thos. De Vries, James L. Kassner, Thos. W. Mason, F. C. Mathers, M. G. Mellon, W. C. Pierce, J. H. Reedy, Arthur A. Vernon and S. R. Wood. Many others have contributed valuable suggestions.

Volumes have been stated in milliliters (ml) and liters (l). One milliliter is equivalent to one cubic centimeter (cm^3 or cc.). Masses are indicated in grams (g).

The relation to molar solution *(M)* or normal solution *(N)* is indicated in many cases.

Distilled water should be used.

LABORATORY REAGENTS FOR GENERAL USE

DILUTE ACIDS, 3 molar. Use the amount of concentrated acid indicated and dilute to one liter.

Acetic acid, 3 *N.* Use 172 ml of 17.4 *M* acid (99-100%).

Hydrochloric acid, 3 *N.* Use 258 ml of 11.6 *M* acid (36% HCl).

Nitric acid, 3 *N.* Use 195 ml of 15.4 *M* acid (69% HNO_3).

Phosphoric acid, 9 *N.* Use 205 ml of 14.6 *M* acid (85% H_3PO_4)

Sulfuric acid, 6 *N.* Use 168 ml of 17.8 *M* acid (95% H_2SO_4).

DILUTE BASES.

Ammonium hydroxide, 3 *M,* 3 *N.* Dilute 200 ml of concentrated solution (14.8 *M,* 28% NH_3) to 1 liter.

Barium hydroxide, 0.2 *M,* 0.4 *N.* Saturated solution, 63 g per liter of $Ba(OH)_2 \cdot 8H_2O$. Use some excess, filter off $BaCO_3$ and protect from CO_2 of the air with soda lime or ascarite in a guard tube.

Calcium hydroxide, 0.02 *M,* 0.04 *N.* Saturated solution, 1.5 g per liter of $Ca(OH)_2$. Use some excess, filter off $CaCO_3$ and protect from CO_2 of the air.

Potassium hydroxide, 3 *M,* 3 *N.* Dissolve 176 g of the sticks (95%) in water and dilute to 1 liter.

Sodium hydroxide, 3 *M,* 3 *N.* Dissolve 126 g of the sticks (95%) in water and dilute to 1 liter.

GENERAL REAGENTS (See also Decinormal Solutions of Salts and Other Reagents.)

Aluminum chloride, 0.167 *M,* 0.5 *N.* Dissolve 22 g of $AlCl_3$ in 1 liter of water.

Aluminum nitrate, 0.167 *M,* 0.5 *N.* Dissolve 58 g of $Al(NO_3)_3 \cdot 7.5H_2O$ in 1 liter of water.

Aluminum sulfate, 0.083 *M,* 0.5 *N.* Dissolve 56 g of $Al_2(SO_4)_3 \cdot 18H_2O$ in 1 liter of water.

Ammonium acetate, 3 *M,* 3 *N.* Dissolve 230 g of $NH_4C_2H_3O_2$ in water and dilute to 1 liter.

Ammonium carbonate, 1.5 *M* Dissolve 144 g of the commercial salt (mixture of $(NH_4)_2CO_3 \cdot H_2O$ and $NH_4CO_2NH_2$) in 500 ml of 3 *N* NH_4OH and dilute to 1 liter.

Ammonium chloride, 3 *M,* 3 *N.* Dissolve 160 g of NH_4Cl in water. Dilute to 1 liter.

Ammonium molybdate.

1. 0.5 *M,* 1 *N.* Mix well 72 g of pure MoO_3 (or 81 g of H_2MoO_4) with 200 ml of water, and add 60 ml of conc. ammonium hydroxide. When solution is complete, filter and pour filtrate, very slowly and with rapid stirring, into a mixture of 270 ml of conc. HNO_3 and 400 ml of water. Allow to stand over night, filter and dilute to 1 liter.

2. The reagent is prepared as two solutions which are mixed as needed, thus always providing fresh reagent of proper strength and composition. Since ammonium molybdate is an expensive reagent, and since an acid solution of this reagent as usually prepared keeps for only a few days, the method proposed will avoid loss of reagent and provide more certain results for quantitative work.

Solution 1. Dissolve 100 g of ammonium molybdate (C.P. grade) in 400 ml of water and 80 ml of 15 *M* NH_4OH. Filter if necessary, though this seldom has to be done.

Solution 2. Mix 400 ml of 16 *M* nitric acid with 600 ml of water.

For use, mix the calculated amount of solution 1 with twice its volume of solution 2, adding solution 1 to solution 2 slowly with vigorous stirring. Thus, for amounts of phosphorus up to 20 mg, 10 ml of solution 1 to 20 ml of solution 2 is adequate. Increase amount as needed.

Ammonium nitrate, 1 *M,* 1 *N.* Dissolve 80 g of NH_4NO_3 in 1 liter of water.

Ammonium oxalate, 0.25 *M,* 0.5 *N.* Dissolve 35.5 g of $(NH_4)_2C_2O_4 \cdot H_2O$ in water. Dilute to 1 liter.

Ammonium sulfate, 0.25 *M,* 0.5 *N,* Dissolve 33 g of $(NH_4)_2SO_4$ in 1 liter of water.

Ammonium sulfide, colorless.

1. 3 *M.* Treat 200 ml of conc. NH_4OH with H_2S until saturated, keeping the solution cold. Add 200 ml of conc. NH_4OH and dilute of 1 liter.

2. 6 *N.* Saturate 6 *N* ammonium hydroxide (40 ml conc. ammonia solution + 60 ml H_2O) with washed H_2S gas. The ammonium hydroxide bottle must be completely full and must be kept surrounded by ice while being saturated (about 48 hours for two liters). The reagent is best preserved in brown, completely filled, glass-stoppered bottles.

Ammonium sulfide, yellow, Treat 150 ml of conc. NH_4OH with H_2S until saturated, keeping the solution cool. Add 250 ml of conc. NH_4OH and 10 g of powdered sulfur. Shake the mixture until the sulfur is dissolved and dilute to 1 liter with water. In the solution the concentration of $(NH_4)_2S_2$, $(NH_4)_2S$ and NH_4OH are 0.625, 0.4 and 1.5 normal respectively. On standing, the concentration of $(NH_4)_2S_2$ increases and that of $(NH_4)_2S$ and NH_4OH decreases.

Antimony pentachloride, 0.1 *M,* 0.5 *N.* Dissolve 30 g of $SbCl_5$ in 1 liter of water.

Antimony trichloride, 0.167 *M,* 0.5 *N.* Dissolve 38 g of $SbCl_3$ in 1 liter of water.

Aqua regia. Mix 1 part concentrated HNO_3 with 3 parts of concentrated HCl. This formula should include one volume of water if the aqua regia is to be stored for any length of time. Without water, objectionable quantities of chlorine and other gases are evolved.

Barium chloride, 0.25 *M,* 0.5 *N.* Dissolve 61 g of $BaCl_2 \cdot 2H_2O$ in water. Dilute to 1 liter.

Barium hydroxide, 0.1 *M,* about 0.2 *N.* Dissolve 32 g of $Ba(OH)_2 8H_2O$ in 1 liter of water.

Barium nitrate, 0.25 *M,* 0.5 *N.* Dissolve 65 g of $Ba(NO_3)_2$ in 1 liter of water.

Bismuth cloride, 0.167 *M,* 0.5 *N.* Dissolve 53 g of $BiCl_3$ in 1 liter of dilute HCl. Use 1 part HCl to 5 parts water.

Bismuth nitrate, 0.083 *M,* 0.25 *N.* Dissolve 40 g of $Bi(NO_3)_3 \cdot 5H_2O$ in 1 liter of dilute HNO_3. U se 1 part of HNO_3 to 5 parts of water.

Cadmium chloride, 0.25 *M,* 0.5 *N.* Dissolve 46 g of $CdCl_2$ in 1 liter of water.

Cadmium nitrate, 0.25 *M,* - 0.5 *N.* Dissolve 77 g of $Cd(NO_3)_2 \cdot 4H_2O$ in 1 liter of water.

Cadmium sulfate, 0.25 *M,* 0.5 *N.* Dissolve 70 g of $CdSO_4 \cdot 4H_2O$ in 1 liter of water.

Calcium chloride, 0.25 *M,* 0.5 *N.* Dissolve 55 g of $CaCl_2 \cdot 6H_2O$ in water. Dilute to 1 liter.

Calcium nitrate, 0.25 *M,* 0.5 *N.* Dissolve 41 g of $Ca(NO_3)_2$ in 1 liter of water.

Chloroplatinic acid.

1. 0.0512 *M,* 0.102 *N.* Dissolve 26.53 g of $H_2PtCl_6 \cdot 6H_2O$ in water. Dilute to 100 ml. Contains 0.100 g Pt per ml.

2. Make a 10% solution by dissolving 1 g of $H_2PtCl_6 \cdot 6H_2O$ in 9 ml of water. Shake thoroughly to insure complete mixing. Keep in a dropping bottle.

Chromic chloride, 0.167 *M,* 0.5 *N.* Dissolve 26 g of $CrCl_3$ in 1 liter of water.

Chromic nitrate, 0.167 *M,* 0.5 *N.* Dissolve 40 g of $Cr(NO_3)_3$ in 1 liter of water.

Chromic sulfate, 0.083 *M,* -.5 *N.* Dissolve 60 g of $Cr_2(SO_4)_3 \cdot 18H_2O$ in 1 liter of water.

Cobaltous nitrate, 0.25 *M,* 0.5 *N.* Dissolve 73 g of $Co(NO_3)_2 \cdot 6H_2O$ in 1 liter of water.

Cobaltous sulfate, 0.25 *M,* 0.5 *N.* Dissolve 70 g of $CoSO_4 \cdot 7H_2O$ in 1 liter of water.

Cupric chloride, 0.25 *M,* 0.5 *N.* Dissolve 43 g of $CuCl_2 \cdot 2H_2O$ in 1 liter of water.

Cupric nitrate, 0.25 *M,* 0.5 *N.* Dissolve 74 g of $Cu(NO_3)_2 \cdot 6H_2O$ in 1 liter of water.

Cupric sulfate, 0.5 *M,* 1 *N.* Dissolve 124.8 g of $CuSO_4 \cdot 5H_2O$ in water to which 5 ml of H_2SO_4 has been added. Dilute to 1 liter.

Ferric chloride, 0.5 *M,* 1.5 *N.* Dissolve 135.2 g of $FeCl_3 \cdot 6H_2O$ in water containing 20 ml of conc. HCl. Dilute to 1 liter.

Ferric nitrate, 0.167 *M,* 0.5 *N.* Dissolve 67 g of $Fe(NO_3)_3 \cdot 9H_2O$ in 1 liter of water.

LABORATORY REAGENTS FOR GENERAL USE (Continued)

Ferric sulfate, 0.25 M, 0.5 N. Dissolve 140.5 g of $Fe_2(SO_4)_3 \cdot 9H_2O$ in water containing 100 ml of conc. H_2SO_4. Dilute to 1 liter.

Ferrous ammonium sulfate, 0.5 M, 1 N. Dissolve 196 g of $Fe(NH_4SO_4)_2 \cdot 6H_2O$ in water containing 10 ml of conc. H_2SO_4. Dilute to 1 liter. Prepare fresh solutions for best results.

Ferrous sulfate, 0.5 M, 1 N. Dissolve 139 g of $FeSO_4 \cdot 7H_2O$ in water containing 10 ml of conc. H_2SO_4. Dilute to 1 liter. Solution does not keep well.

Lead acetate, 0.5 M, 1 N. Dissolve 190 g of $Pb(C_2H_3O_2)_2 \cdot 3H_2O$ in water. Dilute to 1 liter.

Lead nitrate, 0.25 M, 0.5 N. Dissolve 83 g of $Pb(NO_3)_2$ in water. Dilute to one liter.

Lime water, See Calcium hydroxide.

Magnesium chloride, 0.25 M, 0.5 N. Dissolve 51 g of $MgCl_2 \cdot 6H_2O$ in 1 liter of water.

Magnesium chloride reagent. Dissolve 50 g of $MgCl_2 \cdot 6H_2O$ and 100 g of NH_4Cl in 500 ml of water. Add 10 ml of conc. NH_4OH, allow to stand over night and filter if a precipitate has formed. Make acid to methyl red with dilute HCl. Dilute to 1 liter. Solution contains 0.25 M $MgCl_2$ and 2 M NH_4Cl. Solution may also be diluted with 133 ml of conc. NH_4OH and water to make 1 liter. Such a solution will contain 2 M NH_4OH.

Magnesium nitrate, 0.25 M, 0.5 N. Dissolve 64 g of $Mg(NO_3)_2 \cdot 6H_2O$ in 1 liter of water.

Magnesium sulfate, 0.25 M, 0.5 N. Dissolve 62 g of $MgSO_4 \cdot 7H_2O$ in 1 liter of water.

Manganous chloride, 0.25 M, 0.05 N. Dissolve 50 g of $MnCl_2 \cdot 4H_2O$ in 1 liter of water.

Manganous nitrate, 0.25 M, 0.5 N. Dissolve 72 g of $Mn(NO_3)_2 \cdot 6H_2O$ in 1 liter of water.

Manganous sulfate, 0.25 M, 0.5 N. Dissolve 69 g of $MnSO_4 \cdot 7H_2O$ in 1 liter of water.

Mercuric chloride, 0.25 M, 0.5 N. Dissolve 68 g of $HgCl_2$ in water. Dilute to 1 liter.

Mercuric nitrate, 0.25 M, 0.5 N. Dissolve 81 g of $Hg(NO_3)_2$ in 1 liter of water.

Mercuric sulfate, 0.25 M, 0.5 N. Dissolve 74 g of $HgSO_4$ in 1 liter of water.

Mercurous nitrate. Use 1 part $HgNO_3$, 20 parts water and 1 part HNO_3.

Nickel chloride, 0.25 M, 0.5 N. Dissolve 59 g of $NiCl_2 \cdot 6H_2O$ in 1 liter of water.

Nickel nitrate, 0.25 M, 0.5 N. Dissolve 73 g of $Ni(NO_3)_2 \cdot 6H_2O$ in 1 liter of water.

Nickel sulfate, 0.25 M, 0.5 N. Dissolve 66 g of $NiSO_4 \cdot 6H_2O$ in 1 liter of water.

Potassium bromide, 0.5 M, 0.5 N. Dissolve 60 g of KBr in 1 liter of water.

Potassium carbonate, 1.5 M, 3 N. Dissolve 207 g of K_2CO_3 in 1 liter of water.

Potassium chloride, 0.5 M, 0.5 N. Dissolve 37 g of KCl in 1 liter of water.

Potassium chromate, 0.25 M, 0.5 N. Dissolve 49 g of K_2CrO_4 in 1 liter of water.

Potassium cyanide, 0.5 M, 0.5 N. Dissolve 33 g of KCN in 1 liter of water.

Potassium dichromate, 0.125 M. Dissolve 37 g of $K_2Cr_2O_7$ in 1 liter of water.

Potassium ferricyanide, 0.167 M, 0.5 N. Dissolve 55 g of $K_3Fe(CN)_6$ in 1 liter of water.

Potassium ferrocyanide, 0.5 M, 2 N. Dissolve 211 g of $K_4Fe(CN)_6 \cdot 3H_2O$ in water. Dilute to 1 liter.

Potassium iodide, 0.5 M, 0.5 N. Dissolve 83 g of KI in 1 liter of water.

Potassium nitrate, 0.5 M, 0.5 N. Dissolve 51 g of KNO_3 in 1 liter of water.

Potassium sulfate, 0.25 M, 0.5 N. Dissolve 44 g of K_2SO_4 in 1 liter of water.

Silver nitrate, 0.5 M, 0.5 N. Dissolve 85 g of $AgNO_3$ in water. Dilute to 1 liter.

Sodium acetate, 3 M, 3 N. Dissolve 408 g of $NaC_2H_3O_2 \cdot 3H_2O$ in water. Dilute to 1 liter.

Sodium carbonate, 1.5 M, 3 N. Dissolve 159 g of Na_2CO_3, or 430 g of $Na_2CO_3 \cdot 10H_2O$ in water. Dilute to 1 liter.

Sodium chloride, 0.5 M, 0.5 N. Dissolve 29 g of NaCl in 1 liter of water.

Sodium cobaltinitrite, 0.08 M (reagent for potassium). Dissolve 25 g of $NaNO_2$ in 75 ml of water, add 2 ml of glacial acetic acid and then 2.5 g of $Co(NO_3)_2 \cdot 6H_2O$. Allow to stand for several days, filter and dilute to 100 ml. Reagent is somewhat unstable.

Sodium hydrogen phosphate, 0.167 M, 0.5 N. Dissolve 60 g of $Na_2HPO_4 \cdot 12H_2O$ in 1 liter of water.

Sodium nitrate, 0.5 M, 0.5 N. Dissolve 43 g of $NaNO_3$ in 1 liter of water.

SPECIAL SOLUTIONS AND REAGENTS

Sodium Sulfate, 0.25 M, 0.5 N. Dissolve 36 g of Na_2SO_4 in 1 liter of water.

Sodium sulfide, 0.5 M, 1 N. Dissolve 120 g of $Na_2S \cdot 9H_2O$ in water and dilute to 1 liter. Or, saturate 500 ml of 1 M NaOH (21 g of 95% NaOH sticks) with H_2S, keeping the solution cool, and dilute with 500 ml of 1 M NaOH.

Stannic chloride, 0.125 M, 0.5 N. Dissolve 33 g of $SnCl_4$ in 1 liter of water.

Stannous chloride, 0.5 M, 1 N. Dissolve 113 g of $SnCl_2 \cdot 2H_2O$ in 170 ml of conc. HCl, using heat if necessary. Dilute with water to 1 liter. Add a few pieces of tin foil. Prepare solution fresh at frequent intervals.

Stannous chloride (for Bettendorf test). Dissolve 113 g of $SnCl_2 \cdot 2H_2O$ in 75 ml of conc. HCl. Add a few pieces of tin foil.

Strontium chloride, 0.25 M, 0.5 N. Dissolve 67 g of $SrCl_2 \cdot 6H_2O$ in 1 liter of water.

Zinc nitrate, 0.25 M, 0.5 N. Dissolve 74 g of $Zn(NO_3)_2 \cdot 6H_2O$ in 1 liter of water.

Zinc sulfate, 0.25 M, 0.5 N. Dissolve 72 g of $ZnSO_4 \cdot 7H_2O$ in 1 liter of water.

SPECIAL SOLUTIONS AND REAGENTS

Aluminon (qualitative test for aluminum). Aluminon is a trade name for the ammonium salt of aurin tricarboxylic acid. Dissolve 1 g of the salt in 1 liter of distilled water. Shake the solution well to insure thorough mixing.

Bang's reagent (for glucose estimation). Dissolve 100 g of K_2CO_3, 66 g of KCl and 160 g of $KHCO_3$ in the order given in about 700 ml of water at 30° C. Add 4.4 g of $CuSO_4$ and dilute to 1 liter after the CO_2 is evolved. This solution should be shaken only in such a manner as not to allow entry of air. After 24 hours 300 ml are diluted to 1 liter with saturated KCl solution, shaken gently and used after 24 hours; 50 ml equivalent to 10 mg glucose.

Barfoed's reagent (test for glucose). See Cupric acetate.

Baudisch's reagent. See Cupferron.

Benedict's solution (qualitative reagent for glucose). With the aid of heat, dissolve 173 g of sodium citrate and 100 g of Na_2CO_3 in 800 ml of water. Filter, if necessary, and dilute to 850 ml. Dissolve 17.3 g of $CuSO_4 \cdot 5H_2O$ in 100 ml of water. Pour the latter solution, with constant stirring, into the arbonate-citrate solution, and make up to 1 liter.

Benzidine hydrochloride solution (for sulfite determination). Make a paste of 8 g of benzidine hydrochloride ($C_{12}H_8(NH_2)_2 \cdot 2HCl$) and 20 ml of water, add 20 ml of HCl (sp. gr. 1.12) and dilute to 1 liter with water. Each ml of this solution is equivalent to 0.00357 g of H_2SO_4.

Bertrand's reagent (glucose estimation). Consists of the following solutions:

(a) Dissolve 200 g of Rochelle salts and 150 g of NaOH in sufficient water to make 1 liter of solution.

(b) Dissolve 40 g of $CuSO_4$ in enough water to make 1 liter of solution.

(c) Dissolve 50 g of $Fe_2(SO_4)_3$ and 200 g of H_2SO_4 (sp. gr. 1.84) in sufficient water to make 1 liter of solution.

(d) Dissolve 5 g of $KMnO_4$ in sufficient water to make 1 liter of solution.

Bial's reagent (for pentose). Dissolve 1 g of orcinol ($CH_3 \cdot C_6H_3(OH)_2$) in 500 ml of 30% HCl to which 30 drops of a 10% solution of $FeCl_3$ has been added.

SPECIAL SOLUTIONS AND REAGENTS (Continued)

Boutron-Boudet soap solution.
(a) Dissolve 100 g of pure castile soap in about 2500 ml of 56% ethyl alcohol.
(b) Dissolve 0.59 g of Ba(NO₃)₂ in 1 liter of water.
Adjust the castile soap solution so that 2.4 ml of it will give a permanent lather with 40 ml of solution (b). When adjusted, 2.4 ml of soap solution is equivalent to 220 parts per million of hardness (as CaCO₃) for a 40 ml sample.
See also Soap solution.

Brucke's reagent (protein precipitation). See Potassium iodide-mercuric iodide.

Clarke's soap solution (or A.P.H.A., standard method). Estimation of hardness in water.
(a) Dissolve 100 g of pure powdered castile soap in 1 liter of 80% ethyl alcohol and allow to stand over night.
(b) Prepare a standard solution of CaCl₂ by dissolving 0.5 g of CaCO₃ in HCl (sp. gr. 1.19), neutralize with NH₄OH and make slightly alkaline to litmus, and dilute to 500 ml. One ml is equivalent to 1 mg of CaCO₃.
Titrate (a) against (b) and dilute (a) with 80% ethyl alcohol until 1 ml of the resulting solution is equivalent to 1 ml of (b) after making allowance for the lather factor (the amount of standard soap solution required to produce a permanent lather in 50 ml of distilled water). One ml of the adjusted solution after subtracting the lather factor is equivalent to 1 mg of CaCO₃.
See also Soap solution.

Cobalticyanide paper (Rinnmann's test for Zn). Dissolve 4 g of K₃Co(CN)₆ and 1 g of KClO₃ in 100 ml of water. Soak filter paper in solution and dry at 100°C. Apply drop of zinc solution and burn in an evaporating dish. A green disk is obtained if zinc is present.

Cochineal. Extract 1 g of cochineal for four days with 20 ml of alcohol and 60 ml of distilled water. Filter.

Congo red. Dissolve 0.5 g of congo red in 90 ml of distilled water and 10 ml of alcohol.

Cupferron (Baudisch's reagent for iron analysis). Dissolve 6 g of the ammonium salt of nitroso-phenyl-hydroxyl-amine (cupferron) in 100 ml of H₂O. Reagent good for one week only and must be kept in the dark.

Cupric acetate (Barfoed's reagent for reducing monosaccharides). Dissolve 66 g of cupric acetate and 10 ml of glacial acetic acid in water and dilute to 1 liter.

Cupric oxide, ammoniacal; Schweitzer's reagent (dissolves cotton, linen and silk, but not wool).
1. Dissolve 5 g of cupric sulfate in 100 ml of boiling water, and add sodium hydroxide until precipitation is complete. Wash the precipitate well, and dissolve it in a minimum quantity of ammonium hydroxide.
2. Bubble a slow stream of air through 300 ml of strong ammonium hydroxide containing 50 g of fine copper turnings. Continue for one hour.

Cupric sulfate in glycerin-potassium hydroxide (reagent for silk). Dissolve 10 g of cupric sulfate, CuSO₄·5H₂O, in 100 ml of water and add 5 g of glucerin. Add KOH solution slowly until a deep blue solution is obtained.

Cupron (benzoin oxime). Dissolve 5 g in 100 ml of 95% alcohol.

Cuprous chloride, acidic (reagent for CO in gas analysis).
1. Cover the bottom of a two-liter flask with a layer of cupric oxide about one-half inch deep, suspend a bunch of copper wire so as to reach from the bottom to the top of the solution, and fill the flask with hydrochloric acid (sp. gr. 1.10). Shake occasionally. When the solution becomes nearly colorless, transfer to reagent bottles, which should also contain copper wire. The stock bottle may be refilled with dilute hydrochloric acid until either the cupric oxide or the copper wire is used up.
Copper sulfate may be substituted for copper oxide in the above procedure.
2. Dissolve 340 g of CuCl₂·2H₂O in 600 ml of conc. HCl and reduce the cupric chloride by adding 190 ml of a saturated solution of stannous chloride or until the solution is colorless. The stannous chloride is prepared by treating 300 g of metallic tin in a 500 ml flask with conc. HCl until no more tin goes into solution.
3 (Winkler method). Add a mixture of 86 g of CuO and 17 g of finely divided metallic Cu, made by the reduction of CuO with hydrogen, to a solution of HCl, made by diluting 650 ml of conc. HCl with 325 ml of water. After the mixture has been added slowly and with frequent stirring, a spiral of copper wire is suspended in the bottle, reaching all the way to the bottom. Shake occasionally, and when the solution becomes colorless, it is ready for use.

Cuprous chloride, ammoniacal (reagent for CO in gas analysis).
1. The acid solution of cuprous chloride as prepared above is neutralized with ammonium hydroxide until an ammonia odor persists. An excess of metallic copper must be kept in the solution.
2. Pour 800 ml of acidic cuprous chloride, prepared by the Winkler method, into about 4 liters of water. Transfer the precipitate to a 250 ml graduate. After several hours, siphon off the liquid above the 50 ml mark and refill with 7.5% NH₄OH solution which may be prepared by diluting 50 ml of conc. NH₄OH with 150 ml of water. The solution is well shaken and allowed to stand for several hours. It should have a faint odor of ammonia.

Dichlorofluorescein indicator. Dissolve 1 g in 1 liter of 70% alcohol or 1 g of the sodium salt in 1 liter of water.

Dimethylglyoxime (diacetyl dioxime), 0.01 N. Dissolve 0.6 g of dimethylglyoxime, (CH₃CNOH)₂, in 500 ml of 95% ethyl alcohol. This is an especially sensitive test for nickel, a very definite crimson color being produced.

Diphenylamine (reagent for rayon). Dissolve 0.2 g in 100 ml of concentrated sulfuric acid.

Diphenylamine sulfonate (for titration of iron with K₂Cr₂O₇). Dissolve 0.32 g of the barium salt of diphenylamine sulfonic acid in 100 ml of water, add 0.5 g of sodium sulfate and filter off the precipitate of BaSO₄.

Diphenylcarbazide. Dissolve 0.2 g of diphenylcarbazide in 10 ml of glacial acetic acid and dilute to 100 ml with 95% ethyl alcohol.

Esbach's reagent (estimation of protein). To a water solution of 10 g of picric acid and 20 g of citric acid, add sufficient water to make one liter of solution.

Eschka's compound. Two parts of calcined ("light") magnesia are thoroughly mixed with one part of anhydrous sodium carbonate.

Fehling's solution (reagent for reducing sugars.)
(a) Copper sulfate solution. Dissolve 34.66 g of CuSO₄·5H₂O in water and dilute to 500 ml.
(b) Alkaline tartrate solution. Dissolve 173 g of potassium sodium tartrate (Rochelle salts, KNaC₄H₄O₆·4H₂O) and 50 g of NaOH in water and dilute when cold to 500 ml.
For use, mix equal volumes of the two solutions at the time of using.

Ferric-alum indicator. Dissolve 140 g of ferric-ammonium sulfate cyrstals in 400 ml of hot water. When cool, filter, and make up to a volume of 500 ml with dilute (6 N) nitric acid.

Folin's mixture (for uric acid). To 650 ml of water add 500 g of (NH₄)₂SO₄, 5 g of uranium acetate and 6 g of glacial acetic acid. Dilute to 1 liter.

Formaldehyde-sulfuric acid (Marquis' reagent for alkaloids). Add 10 ml of formaldehyde solution to 50 ml of sulfuric acid.

Froehde's reagent. See Sulfomolybdic acid.

Fuchsin (reagent for linen). Dissolve 1 g of fuchsin in 100 ml of alcohol.

Fuchsin-sulfurous acid (Schiff's reagent for aldehydes). Dissolve 0.5 g of fuchsin and 9 g of sodium bisulfite in 500 ml of water, and add 10 ml of HCl. Keep in well-stoppered bottles and protect from light.

Gunzberg's reagent (detection of HCl in gastric juice). Prepare as needed a solution containing 4 g of phloroglucinol and 2 g of vanillin in 100 ml of absolute ethyl alcohol.

Hager's reagent. See Picric acid.

Hanus solution (for iodine number). Dissolve 13.2 g of resublimed iodine in one liter of glacial acetic acid which will pass the dichromate test for reducible matter. Add sufficient bromine to double the halogen content, determined by titration (3 ml is about the proper amount). The iodine may be dissolved by the aid of heat, but the solution should be cold when the bromine is added.

Iodine, tincture of. To 50 ml of water add 70 g of I₂ and 50 g of KI. Dilute to 1 liter with alcohol.

Iodo-potassium iodide (Wagner's reagent for alkaloids). Dissolve 2 g of iodine and 6 g of KI in 100 ml of water.

Litmus (indicator). Extract litmus powder three times with boiling alcohol, each treatment consuming an hour. Reject the alcoholic extract. Treat residue with an equal weight of cold water and filter; then exhaust with five times its weight of boiling water, cool and filter. Combine the aqueous extracts.

Magnesia mixture (reagent for phosphates and arsenates). Dissolve 55 g of magnesium chloride and 105 g of ammonium chloride in water, barely acidify with hydrochloric acid, and dilute to 1 liter. The ammonium hydroxide may be omitted until just previous to use. The reagent, if completely mixed and stored for any period of time, becomes turbid.

Magnesium reagent. See S and O reagent.

Magnesium uranyl acetate. Dissolve 100 g of $UO_2(C_2H_3O_2)_2 \cdot 2H_2O$ in 60 ml of glacial acetic acid and dilute to 500 ml. Dissolve 330 g of $Mg(C_2H_3O_2)_2 \cdot 4H_2O$ in 60 ml of glacial acetic acid and dilute to 200 ml. Heat solutions to te boiling point until clear, pour the magnesium solution into the uranyl solution, cool and dilute to 1 liter. Let stand over night and filter if necessary.

Marme's reagent. See Potassium-cadmium iodide.

Marquis' reagent. See Formaldehyde-sulfuric acid.

Mayer's reagent (white precipitate with most alkaloids in slightly acid solutions). Dissolve 1.358 g of $HgCl_2$ in 60 ml of water and pour into a solution of 5 g of KI in 10 ml of H_2O. Add sufficient water to make 100 ml.

Methyl orange indicator. Dissolve 1 g of methyl orange in 1 liter of water. Filter, if necessary.

Methyl orange, modified. Dissolve 2 g of methyl orange and 2.8 g of xylene cyanole FF in 1 liter of 50% alcohol.

Methyl red indicator. Dissolve 1 g of methyl red in 600 ml of alcohol and dilute with 400 ml of water.

Methyl red, modified. Dissolve 0.50 g of methyl red and 1.25 g of xylene cyanole FF in 1 liter of 90% alcohol. Or, dissolve 1.25 g of methyl red and 0.825 g of methylene blue in 1 liter of 90% alcohol.

Millon's reagent (for albumins and phenols). Dissolve 1 part of mercury in 1 part of cold fuming nitric acid. Dilute with twice the volume of water and decant the clear solution after several hours.

Mixed indicator. Prepared by adding about 1.4 g of xylene cyanole FF to 1 g of methyl orange. The dye is seldom pure enough for these proportions to be satisfactory. Each new lot of dye should be tsted by adding additional amounts of the dye until a test portion gives the proper color change. The acid color of this indicator is like that of permanganate; the neutral color is gray; and the alkaline color is green. Described by Hickman and Linstead, J. Chem. Soc. (Lon.), 121, 2502 (1922).

Molisch's reagent. See α-Naphthol.

α-Naphthol (Molisch's reagent for wool). Dissolve 15 g of α-naphthol in 100 ml of alcohol or chloroform.

Nessler's reagent (for ammonia). Dissolve 50 g of KI the smallest possible quantity of cold water (50 ml). Add a saturated solution of mercuric chloride (about 22 g in 350 ml of water will be needed) until an excess is indicated by the formation of a precipitate. Then add 200 ml of 5 N NaOH and dilute to 1 liter. Let settle, and draw off the clear liquid.

Nickel oxide, ammoniacal (reagent for silk). Dissolve 5 g of nickel sulfate in 100 ml of water, and add sodium hydroxide solution until nickel hydroxide is completely precipitated. Wash the precipitate well and dissolve in 25 ml of concentrated ammonium hydroxide and 25 ml of water.

ϱ-Nitrobenzene-azo-resorcinol (reagent for magnesium). Dissolve 1 g of the dye in 10 ml of NNaOH and dilute to 1 liter.

Nitron (detection of nitrate radical). Dissolve 10 g of nitron ($C_{20}H_{16}N_4$, 4, 5-dihydro-1, 4-diphenyl-3, 5-phenylimino-1, 2, 4-triazole) in 5 ml of glacial acetic acid and 95 ml of water. The solution may be filtered with slight suction through an alumdum crucible and kept in a dark bottle.

α-Nitroso-β-naphthol. Make a saturated solution in 50% acetic acid (1 part of glacial acetic acid with 1 part of water). Does not keep well.

Nylander's solution (carbohydrates). Dissolve 20 g of bismuth subnitrate and 40 g of Rochelle salts in 1 liter of 8% NaOH solution. Cool and filter.

Obermayer's reagent (for indoxyl in urine). Dissolve 4 g of $FeCl_3$ in one liter of HCl (sp. gr. 1.19).

Oxine. Dissolve 14 g of HC_9H_6ON in 30 ml of glacial acetic acid. Warm slightly, if necessary. Dilute to 1 liter.

Oxygen absorbent. Dissolve 300 g of ammonium chloride in one liter of water and add one liter of concentrated ammonium hydroxide solution. Shake the solution thorughly. For use as n oxygen absorbent, a bottle half full of copper turnings is filled nearly full with the NH_4-Cl-NH_4OH solution and the gas passed through.

Pasteur's salt solution. To one liter of distilled water add 2.5 g of potassium phosphate, 0.25 g of calcium phosphate, 0.25 g of magnesium sulfate and 12.00 g of ammonium tartrate.

Pavy's solution (glucose reagent). To 120 ml of Fehling's solution, add 300 ml of NH_4OH (sp. gr. 0.88) and dilute to 1 liter with water.

Phenathroline ferrous ion indicator. Dissolve 1.485 g of phenanthroline monohydrate in 100 ml of 0.025 Mferrous sulfate solution.

Phenolphthalein. Dissolve 1 g of phenolphthalein in 50 ml of alcohol and add 50 ml of water.

Phenolsulfonic acid (determination of nitrogen as nitrate). Dissolve 25 g of phenol in 150 ml of conc. H_2SO_4, add 75 ml of fuming H_2SO_4 (15% SO_3), stir well and heat for two hours at 100° C.

Phloroglucinol solution (pentosans). Make a 3% phloroglucinol solution in alcohol. Keep in a dark bottle.

Phosphomolybdic acid (Sonnenschein's reagent for alkaloids).

1. Prepare ammonium phosphomolybdate and after washing with water, boil with nitric acid and expel NH_3; evaporate to dryness and dissolve in 2 Nnitric acid.

2. Dissolve ammonium molybdate in HNO_3 and treat with phosphoric acid. Filter, wash the precipitate, and boil with aqua regia until the ammonium salt is decomposed. Evaporate to dryness. The residue dissolved in 10% HNO_3 constitutes Sonnenschein's reagent.

Phosphoric acid—sulfuric acid mixture. Dilute 150 ml of conc. H_2SO_4 and 100 ml of conc. H_3PO_4 (85%) with water to a volume of 1 liter.

Phosphotungstic acid (Scheibler's reagent for alkaloids).

1. Dissolve 20 g of sodium tungstate and 15 g of sodium phosphate in 100 ml of water containing a little nitric acid.

2. The reagent is a 10% solution of phosphotungstic acid in water. The phosphotungstic acid is prepared by evaporating a mixture of 10 g of sodium tungstate dissolved in 5 g of phosphoric acid (sp. gr. 1.13) and enough boiling water to effect solution. Crystals of phosphotungstic acid separate.

Picric acid (Hager's reagent for alkaloids, wool and silk). Dissolve 1 g of picric acid in 100 ml of water.

Potassium antimonate (reagent for sodium). Boil 22 g of potassium antimonate with 1 liter of water until nearly all of the salt has dissolved, cool quickly, and add 35 ml of 10% potassium hydroxide. Filter after standing over night.

Potassium-cadmium iodide (Marme's reagent for alkaloids). Add 2 g of CdI_2 to a boiling solution of 4 g of KI in 12 ml of water, and then mix with 12 ml of saturated KI solution.

Potassium hydroxide (for CO_2 absorption). Dissolve 360 gof KOH in water and dilute to 1 liter.

Potassium iodide-mercric iodide (Brucke's reagent for proteins). Dissolve 50 g of KI in 500 ml of water, and saturate with mercuric iodide (about 120 g). Dilute to 1 liter.

Potassium pyrogallate (for oxygen absorption). For mixtures of gases containing less than 28% oxygen, add 100 ml of KOH solution (50 g of KOH to 100 ml of water) to 5 g of pyrogallol. For mixtures containing more than 28% orygen the KOH solution should contain 120 g of KOH to 100 ml of water.

Pyrogallol, alkaline.

(a) Dissolve 75 g of pyrogallic acid in 75 ml of water.

(b) Dissolve 500 g of KOH in 250 ml of water. When cool, adjust until sp. gr. is 1.55.

For use, add 270 ml of solution (b) to 30 ml of solution (a).

Rosolic acid (indicator). Dissolve 1 g of rosolic acid in 10 ml of alcohol and add 100 ml of water.

S and O reagent (Suitsu and Okuma's test for Mg). Dissolve 0.5 g of the dye (o-p-dihydroxy-monazo-p-nitrobenzene) in 100 ml of 0.25 N NaOH.

Scheibler's reagent. See Phosphotungstic acid.

Schiff's reagent. See Fuchsin-sulfurous acid.

Schweitzer's reagent. See Cupric oxide, ammoniacal.

Soap solution (reagent for hardness in water). D issolve 100 g of dry castile soap in 1 liter of 80% alcohol (5 parts alcohol to 1 part water). Allow to stand several days and dilute with 70% to 80% alcohol until 6.4 ml produces a permanent lather with 20 ml of standard calcium solution. The latter solution is made by dissolving 0.2 g of $CaCO_3$ in a small amount of dilute HCl, evaporating to dryness and making up to 1 liter.

Sodium bismuthate (oxidation of manganese). Heat 20 parts of NaOH nearly to redness in an iron or nickel crucible and add slowly 10 parts of basic bismuth nitrate which has been previously dried. Add two parts of sodium peroxide, and pour the brownish-yellow fused mass on an iron plate to cool. When cold, break up in a mortar, extract with water, and collect on an asbestos filter.

Sodium hydroxide (for CO_2 absorption). Dissolve 330 g of NaOH in water and dilute to 1 liter.

Sodium nitroprusside (reagent for hydrogen sulfide and wool). Use a freshly prepared solution of 1 g of sodium nitroprusside in 10 ml of water.

Sodium oxalate, according to Sorensen (primary standard). Dissolve 30 g of the commercial salt in 1 liter of water, make slightly alkaline with sodium hydroxide, and let stand until perfectly clear. Filter and evaporate the filtrate to 100 ml. Cool and filter. Pulverize the residue and wash it several times with small volumes of water. The procedure is repeated until the mother liquor is free from sulfate and is neutral to phenolphthalein.

Sodium plumbite (reagent for wool). Dissolve 5 g of sodium hydroxide in 100 ml of water. Add 5 g of litharge and boil until dissolved.

Sodium polysulfide. Dissolve 480 g of $Na_2S \cdot 9H_2O$ in 50G ml of water, add 40 g of NaOH and 18 g of sulfur. Stir thoroughly and dilute to 1 liter with water.

Sonnenschein's reagent. See Phosphomolybdic acid.

Starch solution.

1. Make a paste with 2 g of soluble starch and 0.01 g of HgI_2 with a small amount of water. Add the mixture slowly to 1 liter of boiling water and boil for a few minutes. Keep in a glass stoppered bottle. If other than soluble starch is used, the solution will not clear on boiling; it should be allowed to stand and the clear liquid decanted.

2. A solution of starch which keeps indefinitely is made as follows: Mix 500 ml of saturated NaCl solution (filtered), 80 ml of glacial acetic acid, 20 ml of water and 3 g of starch. Bring slowly to a boil and boil for two minutes.

3. Make a paste with 1 g of soluble starch and 5 mg of HgI_2, using as little cold water as possible. Then pour about 200 ml of boiling water on the paste and stir immediately. This will give a clear solution if the paste is prepared correctly and the water actually boiling. Cool and add 4 g of KI. Starch solution decomposes on standing due to bacterial action, but this solution will keep a long time if stored under a layer of toluene.

Stoke's reagent. Dissolve 30 g of $FeSO_4$ and 20 g of tartaric acid in water and dilute to 1 liter. Just before using, add concentrated NH_4OH until the precipitate first formed is redissolved.

Sulfanilic acid (reagent for nitrites). Dissolve 0.5 g of sulfanilic acid in a mixture of 15 ml of glacial acetic acid and 135 ml of recently boiled water.

Sulfomolybdic acid (Froehde's reagent for alkaloids and glucosides). Dissolve 10 g of molybdic acid or sodium molybdate in 100 ml of conc. H_2SO_4.

Tannic acid (reagent for albumen, alkaloids and gelatin). Dissolve 10 g of tannic acid in 10 ml of alcohol and dilute with water to 100 ml.

Titration mixture. (residual chlorine in water analysis). Prepare 1 liter of dilute HCl (100 ml of HCl (sp. gr. 1.19) in sufficient water to make 1 liter). Dissolve 1 g of o-tolidine in 100 ml of the dilute HCl and dilute to 1 liter with dilute HCl solution.

Trinitrophenol solution. See Picric acid.

Turmeric paper. Impregnant white, unsized paper with the tincture, and dry.

Turmeric tincture (reagent for borates). Digest ground turmeric root with several quantities of water which are discarded. Dry the residue and digest it several days with six times its weight of alcohol. Filter.

Uffelmann's reagent (turns yellow in presence of a lactic acid). To a 2% solution of pure phenol in water, add a water solution of $FeCl_3$ until the phenol solution becomes a violet in color.

Wagner's reagent. See Iodo-potassium iodide.

Wagner's solution (used in phosphate rock analysis to prevent precipitation of iron and aluminum). Dissolve 25 g of citric acid and 1 g of salicylic acid in water and dilute to 1 liter. Use 50 ml of the reagent.

Wij's iodine monochloride solution (for iodine number). Dissolve 13 g of resublimed iodine in 1 liter of glacial acetic acid which will pass the dichromate test for reducible matter. Set aside 25 ml of this solution. Pass into the remainder of the solution dry chlorine gas (dried and washed by passing through H_2SO_4 (sp. gr. 1.84)) until the characteristic color of free iodine has been discharged. Now add the iodine solution which was reserved, until all free chlorine has been destroyed. A slight excess of iodine does little or no harm, but an excess of chlorine must be avoided. Preserve in well stoppered, amber colored bottles. Avoid use of solutions which have been prepared for more than 30 days.

Wij's special solution (for iodine number—Analyst 58, 523-7, 1933). To 200 ml of glacial acetic acid that will pass the dichromate test for reducible matter, add 12 g of dichloroamine T (paratoluene-sulfonedichloroamide), and 16.6 g of dry KI (in small quantities with continual shaking until all the KI has dissolved). Make up to 1 liter with the same quality of acetic acid used above and preserve in a dark colored bottle.

Zimmermann Reinhardt reagent (determination of iron). Dissolve 70 g of $MnSO_4 \cdot 4H_2O$ in 500 ml of water, add 125 ml of conc. H_2SO_4 and 125 ml of 85% H_3PO_4, and dilute to 1 liter.

Zinc chloride solution, basic (reagent for silk). Dissolve 1000 g of zinc chloride in 850 ml of water, and add 40 g of zinc oxide. Heat until solution is complete.

Zinc uranyl acetate (reagent for sodium). Dissolve 10 g of $UO_2(C_2H_3O_2)_2 \cdot 2H_2O$ in 6 g of 30% acetic acid with heat, if necessary, and dilute to 50 ml. Dissolve 30 g of $Zn(C_2H_3O_2)_2 \cdot 2H_2O$ in 3 g of 30% acetic acid and dilute to 50 ml. Mix the two solutions, add 50 mg of NcCl, allow to stand over night and filter.

DECI-NORMAL SOLUTIONS OF SALTS AND OTHER REAGENTS

Atomic and molecular weights in the following table are based upon the 1965 atomic weight scale and the isotope C-12. The weight in grams of the compound in 1 cc of the following deci-normal solutions is found by dividing the H equivalent in the last column by 1000.

Name	Formula	Atomic or molecular weight	Hydrogen equivalent	0.1 Hydrogen equivalent in g
Acetic acid	$HC_2H_3O_2$	60.0530	$HC_2H_3O_2$	6.0053
Ammonia	NH_3	17.0306	NH_3	1.7031
Ammonium ion	NH_4^+	18.0386	NH_4	1.8039
Ammonium chloride	NH_4Cl	53.4916	NH_4Cl	5.3492
Ammonium sulfate	$(NH_4)_2SO_4$	132.1388	$\frac{1}{2}(NH_4)_2SO_4$	6.6069
Ammonium thiocyanate	NH_4CNS	76.1204	NH_4CNS	7.6120
Barium	Ba	137.34	$\frac{1}{2}Ba$	6.867
Barium carbonate	$BaCO_3$	197.3494	$\frac{1}{2}BaCO_3$	9.8675
Barium chloride hydrate	$BaCl_2 \cdot 2H_2O$	244.2767	$\frac{1}{2}BaCl_2 \cdot 2H_2O$	12.2138
Barium hydroxide	$Ba(OH)_2$	171.3547	$\frac{1}{2}Ba(OH)_2$	8.5677
Barium oxide	BaO	153.3394	$\frac{1}{2}BaO$	7.6670
Bromine	Br	79.909	Br	7.9909
Calcium	Ca	40.08	$\frac{1}{2}Ca$	2.004
Calcium carbonate	$CaCO_3$	100.0894	$\frac{1}{2}CaCO_3$	5.0045
Calcium chloride	$CaCl_2$	110.9860	$\frac{1}{2}CaCl_2$	5.5493
Calcium chloride hydrate	$CaCl_2 \cdot 6H_2O$	219.0150	$\frac{1}{2}CaCl_2 \cdot 6H_2O$	10.9508
Calcium hydroxide	$Ca(OH)_2$	74.0947	$\frac{1}{2}Ca(OH)_2$	3.7047
Calcium oxide	CaO	56.0794	$\frac{1}{2}CaO$	2.8040
Chlorine	Cl	35.453	Cl	3.5453
Citric acid	$C_6H_8O_7 \cdot H_2O$	210.1418	$\frac{1}{3}C_6H_8O_7 \cdot H_2O$	7.0047
Cobalt	Co	58.9332	$\frac{1}{2}Co$	2.9466
Copper	Cu	63.54	$\frac{1}{2}Cu$	3.177
Copper oxide (cupric)	CuO	79.5394	$\frac{1}{2}CuO$	3.9770
Copper sulfate hydrate	$CuSO_4 \cdot 5H_2O$	249.6783	$\frac{1}{2}CuSO_4 \cdot 5H_2O$	12.4839
Cyanogen	$(CN)_2$	26.0179	CN	2.6018
Hydrochloric acid	HCl	36.4610	HCl	3.6461
Hydrocyanic acid	HCN	27.0258	HCN	2.7026
Iodine	I	126.9044	I	12.6904
Lactic acid	$C_3H_6O_3$	90.0795	$C_3H_6O_3$	9.0080
Malic acid	$C_4H_6O_5$	134.0894	$\frac{1}{2}C_4H_6O_5$	6.7045
Magnesium	Mg	24.312	$\frac{1}{2}Mg$	1.2156
Magnesium carbonate	$MgCO_3$	84.3214	$\frac{1}{2}MgCO_3$	4.2161
Magnesium chloride	$MgCl_2$	95.2180	$\frac{1}{2}MgCl_2$	4.7609
Magnesium chloride hydrate	$MgCl_2 \cdot 6H_2O$	203.2370	$\frac{1}{2}MgCl_2 \cdot 6H_2O$	10.1623
Magnesium oxide	MgO	40.3114	$\frac{1}{2}MgO$	2.0156
Manganese	Mn	54.938	$\frac{1}{2}Mn$	2.7469
Manganese sulfate	$MnSO_4$	150.9996	$\frac{1}{2}MnSO_4$	7.5500
Mercuric chloride	$HgCl_2$	271.4960	$\frac{1}{2}HgCl_2$	13.5748
Nickel	Ni	58.71	$\frac{1}{2}Ni$	2.9356
Nitric acid	HNO_3	63.0129	HNO_3	6.3013
Nitrogen	N	14.0067	N	1.4007
Nitrogen pentoxide	N_2O_5	108.0104	$\frac{1}{2}N_2O_5$	5.4005
Oxalic acid	$H_2C_2O_4$	90.0358	$\frac{1}{2}H_2C_2O_4$	4.5018
Oxalic acid hydrate	$H_2C_2O_4 \cdot 2H_2O$	126.0665	$\frac{1}{2}H_2C_2O_4 \cdot 2H_2O$	6.3033
Oxalic acid anhydride	C_2O_3	72.0205	$\frac{1}{2}C_2O_3$	3.6010
Phosphoric acid	H_3PO_4	97.9953	$\frac{1}{3}H_3PO_4$	3.2665
Potassium	K	39.102	K	3.9102
Potassium bicarbonate	$KHCO_3$	100.1193	$KHCO_3$	10.0119
Potassium carbonate	K_2CO_3	138.2134	$\frac{1}{2}K_2CO_3$	6.9106
Potassium chloride	KCl	74.5550	KCl	7.4555
Potassium cyanide	KCN	65.1199	KCN	6.5120
Potassium hydroxide	KOH	56.1094	KOH	5.6109
Potassium oxide	K_2O	94.2034	$\frac{1}{2}K_2O$	4.7102
Potassium permanganate for Co estimation	$KMnO_4$	158.0376	$\frac{1}{6}KMnO_4$	2.6339
Potassium permanganate for Mn estimation	$KMnO_4$	158.0376	$\frac{1}{3}KMnO_4$	5.2678
Potassium tartrate	$K_2H_4C_4O_6$	226.2769	$\frac{1}{2}K_2H_4C_4O_6$	11.3139
Silver	Ag	107.87	Ag	10.787
Silver nitrate	$AgNO_3$	169.8749	$AgNO_3$	16.9875
Sodium	Na	22.9898	Na	2.2990
Sodium bicarbonate	$NaHCO_3$	84.0071	$NaHCO_3$	8.4007
Sodium carbonate	Na_2CO_3	105.9890	$\frac{1}{2}Na_2CO_3$	5.2995
Sodium chloride	$NaCl$	58.4428	$NaCl$	5.8443
Sodium hydroxide	$NaOH$	39.9972	$NaOH$	3.9997
Sodium oxide	Na_2O	61.9790	$\frac{1}{2}Na_2O$	3.0990
Sodium sulfide	Na_2S	78.0436	$\frac{1}{2}Na_2S$	3.9022
Succinic acid	$H_2C_4H_4O_4$	118.0900	$\frac{1}{2}H_2C_4H_4O_4$	5.9045
Sulfuric acid	H_2SO_4	98.0775	$\frac{1}{2}H_2SO_4$	4.9039
Sulfur trioxide	SO_3	80.0622	$\frac{1}{2}SO_3$	4.0031
Tartaric acid	$C_4H_6O_6$	150.0888	$\frac{1}{2}C_4H_6O_6$	7.5044
Zinc	Zn	65.37	$\frac{1}{2}Zn$	3.269
Zinc sulfate	$ZnSO_4 \cdot 7H_2O$	287.5390	$\frac{1}{2}ZnSO_4 \cdot 7H_2O$	14.3769

DECI–NORMAL SOLUTIONS OF OXIDATION AND REDUCTION REAGENTS

Atomic and molecular weights in the following table are based upon the 1965 atomic weight scale and the isotope C-12. The weight in grams of the compound in 1 cc of the following deci-normal solutions is found by dividing the H equivalent in the last column by 1000.

Name	Formula	Atomic or molecular weight	Hydrogen equivalent	0.1 Hydrogen equivalent in g
Antimony	Sb	121.75	$\frac{1}{2}Sb$	6.0875
Arsenic	As	74.9216	$\frac{1}{2}As$	3.7461
Arsenic trisulfide	As_2S_3	246.0352	$\frac{1}{4}As_2S_3$	6.1509
Arsenous oxide	As_2O_3	197.8414	$\frac{1}{4}As_2O_3$	4.9460
Barium peroxide	BaO_2	169.3388	$\frac{1}{2}BaO_2$	8.4669
Barium peroxide hydrate	$BaO_2 \cdot 8H_2O$	313.4615	$\frac{1}{2}BaO_2 \cdot 8H_2O$	15.6730
Calcium	Ca	40.08	$\frac{1}{2}Ca$	2.004
Calcium carbonate	$CaCO_3$	100.0894	$\frac{1}{2}CaCO_3$	5.0045
Calcium hypochlorite	$Ca(OCl)_2$	142.9848	$\frac{1}{4}Ca(OCl)_2$	3.5746
Calcium oxide	CaO	56.0794	$\frac{1}{2}CaO$	2.8040
Chlorine	Cl	35.453	Cl	3.5453
Chromium trioxide	CrO_3	99.9942	$\frac{1}{3}CrO_3$	3.3331
Ferrous ammonium sulfate	$FeSO_4(NH_4)SO_4 \cdot 6H_2O$	392.0764	$FeSO_4(NH_4)_2SO_4 \cdot 6H_2O$	39.2076
Hydroferrocyanic acid	$H_4Fe(CN)_6$	215.9860	$H_4Fe(CN)_6$	21.5986
Hydrogen peroxide	H_2O_2	34.0147	$\frac{1}{2}H_2O_2$	1.7007
Hydrogen sulfide	H_2S	34.0799	$\frac{1}{2}H_2S$	1.7040
Iodine	I	126.9044	I	12.6904
Iron	Fe	55.847	Fe	5.5847
Iron oxide (ferrous)	FeO	71.8464	FeO	7.1846
Iron oxide (ferric)	Fe_2O_3	159.6922	$\frac{1}{2}Fe_2O_3$	7.9846
Lead peroxide	PbO_2	239.1888	$\frac{1}{2}PbO_2$	11.9594
Manganese dioxide	MnO_2	86.9368	$\frac{1}{2}MnO_2$	4.3468
Nitric acid	HNO_3	63.0129	$\frac{1}{3}HNO_3$	2.1004
Nitrogen trioxide	N_2O_3	76.0116	$\frac{1}{4}N_2O_3$	1.9002
Nitrogen pentoxide	N_2O_5	108.0104	$\frac{1}{5}N_2O_5$	1.8001
Oxalic acid	$C_2H_2O_4$	90.0358	$\frac{1}{2}C_2H_2O_4$	4.5018
Oxalic acid hydrate	$C_2H_2O_4 \cdot 2H_2O$	126.0665	$\frac{1}{2}C_2H_2O_4 \cdot 2H_2O$	6.3033
Oxygen	O	15.9994	$\frac{1}{2}O$	0.8000
Potassium dichromate	$K_2Cr_2O_7$	294.1918	$\frac{1}{6}K_2Cr_2O_7$	4.9032
Potassium chlorate	$KClO_3$	122.5532	$\frac{1}{6}KClO_3$	2.0425
Potassium chromate	K_2CrO_4	194.1076	$\frac{1}{3}K_2CrO_4$	6.4733
Potassium ferrocyanide	$K_4Fe(CN)_6$	368.3621	$K_4Fe(CN)_6$	36.8362
Potassium ferrocyanide	$K_4Fe(CN)_6 \cdot 3H_2O$	422.4081	$K_4Fe(CN)_6 \cdot 3H_2O$	42.2408
Potassium iodide	KI	166.0064	KI	16.6006
Potassium nitrate	KNO_3	101.1069	$\frac{1}{3}KNO_3$	3.3702
Potassium perchlorate	$KClO_4$	138.5526	$\frac{1}{8}KClO_4$	1.7319
Potassium permanganate	$KMnO_4$	158.0376	$\frac{1}{5}KMnO_4$	3.1608
Sodium chlorate	$NaClO_3$	106.4410	$\frac{1}{6}NaClO_3$	1.7740
Sodium nitrate	$NaNO_3$	84.9947	$\frac{1}{3}NaNO_3$	2.8332
Sodium thiosulfate	$Na_2S_2O_3 \cdot 5H_2O$	248.1825	$Na_2S_2O_3 \cdot 5H_2O$	24.8183
Stannous chloride	$SnCl_2$	189.5960	$\frac{1}{2}SnCl_2$	9.4798
Stannous oxide	SnO	134.6894	$\frac{1}{2}SnO$	6.7345
Sulfur dioxide	SO_2	64.0628	$\frac{1}{2}SO_2$	3.2031
Tin	Sn	118.69	$\frac{1}{2}Sn$	5.935

ORGANIC ANALYTICAL REAGENTS

Compiled by John H. Yoe

Determination	Reagent	Reference
Acetate	o-Nitrobenzaldehyde	Feigl, p. 342 (3)
Aldehydes	Dimethyl-dihydro-resorcin (Dimedon)	Ind. Eng. Chem., Anal. Ed. **3**, 365 (1931)
Aluminum	Alizarin S	J. Am. Chem. Soc. **50**, 748 (1928)
	Ammonium salt of aurin tricarboxylic acid ("Aluminon")	J. Am. Chem. Soc. **49**, 2395 (1927) J. Am. Chem. Soc. **55**, 2437 (1933)
	Ammonium salt of nitrosophenyl hydroxylamine ("Cupferron")	Bull. soc. chim. Belg. **36**, 288 (1927)
	Eriochrome cyanine	Z. anal. Chem. **96**, 91 (1934)
	Hematoxylin	Ind. Eng. Chem. **16**, 233 (1924)
	8-Hydroxyquinoline	J. Am. Chem. Soc. **50**, 1900 (1928)
	Morin	Feigl, p. 182 (2) Ind. Eng. Chem., Anal. Ed. **12**, 229 (1940)
	Quinalizarine	J. Am. Pharm. Assoc. **17**, 260 (1928)
	Sodium or zinc salt of 4-sulfo-2 hydroxy-α-naphthalene-azo-β-naphthol (Ponta-chrome Blue Black R)	Sandell, p. 241 (6)
	Urea	Ind. Eng. Chem., Anal. Ed. **9**, 357 (1937)
Ammonia	Hematoxylin	Helv. Chim. Acta **12**, 730 (1929)
	p-Nitrobenzenediazonium chloride	Feigl, p. 235 (2)
	Phenol and sodium hypochloride	Snell, Vol. II, p. 818 (8)
	Tannin—AgNO₃	Snell, Vol. II, p. 819 (8)
	Thymol	J. Biol. Chem. **131**, 309 (1939)
	Zinc oxinate	Feigl, p. 674 (9)
Antimony	Hexamethylenetetramine	Z. anal. Chem. **67**, 298 (1925)
	9-Methyl-2,3,7-tryhydroxyfluorone	Helv. Chim. Acta **20**, 1427 (1937)
	Phenylthiohydantoic acid	Compt. rend. **176**, 1221 (1923)
	Pyridine	Analyst **53**, 373 (1928)
	Pyrogallol	Z. anal. Chem. **64**, 44 (1924)
	Rhodamine B	Z. anal. Chem. **70**, 400 (1927)
Arsenic	Cocaine-molybdate	Biochem. Z. **185**, 14 (1927)
	N-Ethyl-8-hydroxy-tetrahydroquinoline hydrochloride	Z. anal. Chem. **99**, 180 (1934)
	Quinine arsenomolybdate	Analyst **47**, 317 (1922)
	Strychnine-molybdate	Ann. Chim. applicata **23**, 517 (1933)
Barium	Sodium rhodizonate	Feigl, p. 216 (2)
	Tetrahydroxyquinone	Welcher, Vol. I, p. 225 (9)
Beryllium	1-Amino-4-hydroxy-anthraquinone	Ind. Eng. Chem., Anal. **13**, 809 (1941)
	Aurin trycarboxylic	J. Am. Chem. Soc. **48**, 2125 (1926); ibid., **50**, 353 (1928)
	Curcumin	J. Am. Chem. Soc. **50**, 393 (1928)
	1,4-Dihydroxyanthraquinone (Quinizarin)	Ind. Eng. Chem. Anal. Ed., **18**, 179 (1946)
	1,4-Dihydroxyanthraquinone-2-sulfonic acid (quinizarine-2-sulfonic acid)	Sandell, p. 318 (6)
	8-Hydroxyquinoline	Bur. Standards J. Research **3**, 91 (1929)
	Morin	Ind. Eng. Chem., Anal. Ed. **12**, 674, 762 (1940)
	Naphthochrome Azurine 2B	Sandell, p. 318 (6)
	Naphthochrome Green G	Sandell, p. 318 (6)
	p-Nitrobenzeneazo-orcinol	Mikrochemie **14**, 315 (1934)
	1,2,5,8-Tetrahydroxy-anthraquinone (Quinalizarin)	Siemens-Konzerns, Beryllium, p. 25 (1932)
Bismuth	Caffeine (as sulfate or nitrate)	Ind. Eng. Chem., Anal. Ed. **14**, 43 (1942)
	Cinchonine	Scott, p. 158 (7)
	Diethyldithiocarbamate	Sandell, p. 338 (6)
	Dimercaptothiodiazole	Z. anal. Chem. **98**, 184 (1934); ibid., **100**, 408 (1935)
	Dimethylglyoxime	Z. anal. Chem. **72**, 11 (1927)
	Diphenylthiocarbazone	Z. Angew. Chem. **47**, 685 (1934); Ind. Eng. Chem., Anal. Ed. **7**, 285 (1935)
	8-Hydroxyquinoline	Z. anal. Chem. **72**, 177 (1927)
	Phenyldithiobiazolon-ethiol	J. Indian Chem. Soc., **21**, 240, 347 (1944)
	Pyrogallol	Z. anal. Chem. **65**, 448 (1925)
	Thiourea	Z. anal. Chem. **94**, 161 (1933)

Determination	Reagent	Reference
Boron	Curcumin	Chem. News **87**, 27 (1903)
	1,1-Dianthramide	Boltz, p. 346 (1)
	Diaminochrysazin	Anal. Chem. **29**, 1251 (1957)
	Diaminoanthrarufin	Ibid.
	Tribromoanthrarufin	Ibid.
	Mannitol	Scott, p. 168 (7)
	Methyl alcohol	J. Am. Chem. Soc. **50**, 1385 (1928)
	p-Nitrobenzeneazo-chromotropic acid (Chromotrope 2B)	Feigl, p. 341 (2)
	Quinalizarin	Ind. Eng. Chem., Anal. Ed. **11**, 540 (1939)
	Titan Yellow (Clayton Yellow)	Compt. rend. **138**, 1046 (1904)
	Turmeric	Ind. Eng. Chem., Anal. Ed. **4**, 180 (1932)
Bromine	Fluorescein	Snell, Vol. II, p. 725 (8)
	Fuchsin	Snell, Vol. II, p. 724 (8)
	Phenol red	Snell, Vol. II, p. 725 (8)
Cadmium	Allythiourea	Helvetica Chim. Acta. **12**, 718 (1929)
	Di-β-naphthylthiocarbazide	Ind. Eng. Chem., Anal. Ed., **16**, 333, (1944)
	Dinitrodiphenyl-carbazide	Feigl, p. 96 (2)
	Diphenylcarbazide	Feigl, p. 99 (2)
	Diphenylthiocarbazone	Z. angew. Chem. **47**, 685 (1934); Ind. Eng. Chem., Anal. Ed. **11**, 364 (1939)
	Ethylenediamine	Anal. Chem. **77**, 340 (1929)
	Hexamethylenetetramine alliodide	C. A., **24**, 311 (1930)
	β-Naphthoquinoline	Analyst, **58**, 667 (1933)
	Nitrophenolarsinic acid	Mikrochemie **8**, 277 (1930)
	4-Nitrophthalene-diazo-aminobenzene-4-azobenzene (Cadion 2B)	C. A. **32**, 2871 (1938)
	Phenyl-trimethyl-ammonium iodide	Analyst **58**, 667 (1933)
	Pyridine	Z. Anal. Chem. **73**, 279 (1928)
Calcium	Alizarin	Biochem. Z. **16**, 494 (1922); Yoe, Vol. I, p. 139 (2)
	1-amino-2-naphthol-4-sulfonic acid	J. Biol. Chem. **81**, 1 (1929)
	Ammonium oxalate	Snell, Vol. II, p. 592 (8)
	Ammonium stearate	J. Biol. Chem. **29**, 169 (1917); Yoe, Vol. II, p. 119 (3)
	2,5-Dichloro-3,6-dihydroxyquinone (Chloranilic acid)	Anal. Chem., **20**, 76 (1948)
	Dihydroxytartaric acid osazone (sodium salt)	Feigl, p. 221 (2)
	Picrolonic acid	Biochem. Z. **265**, 85 (1933)
	Potassium oleate	Sandell, p. 377 (6)
	Pyrogallol carboxylic acid	Sandell, p. 380 (6)
	Sodium sulforicinate	Biochem. Z. **137**, 157 (1923); Yoe, Vol. II, p. 125 (3)
Cerium	Benzidine	Feigl, p. 211 (2)
	Brucine	Sandell, p. 386 (6)
	Gallic acid	Snell, Vol. II, p. 608 (8)
	8-Hydroxyquinoline	Analyst, **75**, 275 (1948)
	Malachite Green	C. A. **30**, 5143 (1936), **31**, 626 (1937)
	Morphine	Sandell, p. 386 (6)
	Sulfanilic acid	Sandell, p. 386 (6)
Cesium	Dipicrylamine (Hexanitrodiphenylamine)	Mikrochemie **18**, 175 (1935)
Chlorate	Aniline hydrochloride	Snell, Vol. II, p. 717 (8)
Chlorine	Benzidine hydrochloride	Ind. Eng. Chem., Anal. Ed. **4**, 2 (1932)
	Dimethyl-p-phenylene-diamine	Chem. Weekblad. **23**, 203 (1926)
	Oleic acid	J. Soc. Chem. Ind. **42**, 427A (1923)
	Sodium sulforicinate	Biochem. Z., **137**, 157 (1923); Yoe, Vol. II, p. 125 (3)
	Thymolphthalein	Ind. Eng. Chem. **19**, 112 (1927)
	o-Tolidine	Yoe, Vol. I, p. 157 (2)
Columbium	Benzidine	Feigl, p. 171 (2)
	1,8-Dihydroxynaphthalene-3,6-Disulfonate	Ind. Eng. Chem. **5**, 298 (1913)
	s-Diphenylcarbazide	J. Am. Chem. Soc. **50**, 2363 (1928)
	Pyrogallol dimethyl ether	C. A. **4**, 3178 (1910)
	Serichrome Blue R	Ind. Eng. Chem., Anal. Ed., **4**, 245 (1932)

Determination	Reagent	Reference
Cobalt........	Anthranilic acid (o-aminobenzoic acid)	Z. anal. Chem. 93, 241 (1933)
	Cysteine hydrochloride	J. Biol. Chem. 83, 367 (1929)
	Dimethylglyoxime	J. Am. Chem. Soc. 43, 482 (1921)
	3,5-Dimethylpyrazole	Ind. Eng. Chem., Anal. Ed. 2, 38 (1930)
	Dinitrosoresorcinol	J. Am. Chem. Soc. 45, 1439 (1923)
	Formaldoxime reagent	C. A. 27, 927 (1933)
	o-Nitrosocresol	Sandell, p. 422 (6)
	α-Nitroso-β-naphthol	Chem. Zeit. 46, 430 (1922)
	Nitrose-R-salt	J. Am. Chem. Soc. 43, 746 (1921)
	β-Nitroso-α-naphthol	Ind. Eng. Chem., Anal. Ed. 12, 405 (1940)
	o-Nitrosoresorcinol	Ind. Eng. Chem., Anal. Ed. 15, 310 (1943)
	o-Nitrosophenol	Sandell, p. 422 (6)
	Phenylthiohydantoic acid	J. Am. Chem. Soc. 44, 2219 (1922)
	Rubeanic acid (Dithio-oxamide)	Anal. Chim. Acta 20, 332, 435 (1959)
	2,2',2''-Terpyridyl	Ind. Eng. Chem., Anal. Ed., 15, 74 (1943)
Columbium ... Niobium	See Niobium	
Copper.......	Ammonium salt of nitrosophenyl-hydroxylamine ("Cupferron")	Ind. Eng. Chem. 3, 629 (1911)
	m-Benzamino-semicarbazide	Snell, Vol. II, p. 129 (6)
	Benzidine	Z. anal. Chem. 67, 31 (1925)
	α-Benzionoxime (Cupron)	Ber. 56, 2083 (1923)
	Benzotriazole	Ind. Eng. Chem., Anal. Ed. 13, 349 (1941)
	2-Carboxy-a'-hydroxy-5' sulfoformazylbenzene (Zincon)	Anal. Chem. 26, 1345 (1954)
	Diacetyl-dioxime	Analyst 54, 333 (1929)
	Dibenzyldithiocarbamate	Sandell, p. 443 (6)
	Dihydroxypropyl thiocarbamic acid	Sandell, p. 443 (6)
	p-Dimethylamino-benzalrhodanine	J. Am. Chem. Soc. 52, 2222 (1930) J. Am. Chem. Soc. 47, 1268 (1925)
	s-Diphenylcarbazide	Chem. Weekblad. 21, 20 (1924)
	Diphenylthiocarbazone (Dithizone)	J. Assoc. Official Agr. Chem. 18, 192 (1935)
	Hydroquinone	Bull. soc. chim. 31, 1176 (1922)
	Isatin	Rec trav chim. 42, 199 (1923)
	Mercaptobenzothiazole	Z. anal. Chem. 102, 24, 108 (1935)
	α-Naphthol	Bull. soc. chim. 31, 1176 (1922)
	β-Naphthol	Am. J. Pharm. 105, 62 (1933)
	Phenolphthalein	Compt. rend. 173, 1082 (1921)
	Phenylthiohydantoic acid	J. Am. Chem. Soc. 44, 225 (1922)
	Piperidinium piperidyl-dithioformate	Analyst 56, 736 (1931)
	Potassium ethyl xanthate	Yoe, Vol. I, p. 184 (2)
	Pyridine	Z. anal. Chem. 67, 27 (1925)
	Rubeanic acid (Dithio-oxamide)	Anal. Chim. Acta 20, 332, 435 (1959)
	Salicylaldoxime	J. Chem. Soc. (1933) 314; Feigl, p. 40 (8)
	Salicylic acid	Yoe, Vol. I, p. 183 (2)
	Sodium diethyldithiocarbamate	Analyst 54, 650 (1929)
	o-Toluidine	Z. anal. Chem. 67, 31 (1925)
	Urobilin	Chem. Weekblad. 27, 552 (1930)
Cyanide.....	Benzidine	Feigl, p. 276 (2)
	Phenolphthalin	Analyst 60, 294 (1935)
	Picric acid	Helv. Chim. Acta 12, 713 (1929); J. Am. Chem. Soc. 51, 1171 (1929)
Fluoride......	Acetylacetone	Ind. Eng. Chem., Anal. Ed. 5, 300 (1933)
	Alizarin sodium sulfonate—Zr(NO3)4	Ind. Eng. Chem., Anal. Ed. 7, 23 (1935)
	p-Dimethylaminoazophenylarsonic acid	Feigl, p. 271 (2)
	Triphenyltin chloride	J. Am. Chem. Soc. 54, 4625 (1932)
	Quinalizarine—Zr(NO3)4	Ind. Eng. Chem., Anal. Ed. 6, 61 (1934)
Gallium......	Camphoric acid	Welcher, Vol. II, p. 31 (9)
	8-Hydroxyquinoline	Ind. Eng. Chem., Anal. Ed. 13, 844 (1941)
	Morin	Mikrochemie 20, 194 (1936)
	Quinalizarin	J. Am. Chem. Soc. 59, 40 (1937)
Germanium ...	Benzidine	Mikrochemie 18, 66 (1935)
	Quinalizarin	Mikrochemie 18, 48 (1935)
	Tannin	Welcher, Vol. II, p. 160 (9)

Determination	Reagent	Reference
Gold.......	Benzidine	Bull. Chim. Farm. 52, 461 (1912)
	o-Dianisidine	Sandell, p. 507 (6)
	Dimethylaminobenzylidene rhodanine	Feigl, p. 127 (2)
	Formaldehyde	Bull. soc. chim. 31, 717 (1922)
	p-Fuchsine	Ind. Eng. Chem., Anal. Ed., 18, 400 (1946)
	Malachite Green	Sandell, p. 506 (6)
	m-Phenylenediamine sulfate	Chem. Zeit. 36, 934 (1912)
	Phenylhydrazine	Ann. chim. anal. 12, 90 (1907)
	o-Toluidine	Analyst 44, 94 (1919)
Hafnium...	2-Hydroxy-5-methyl-azobenzene-4-sulfonic acid	Feigl, p. 288 (4)
Hydrogen sulfide	p-phenylenedimethyl-diamine sulfate	Yoe, Vol. I, p. 375 (2)
Indium....	Diphenylthiocarbazone (Dithizone)	Sandell, p. 516 (6)
	8-Hydroxyquinoline	Z. anorg. allgem. Chem. 209, 129 (1932)
	Morin	Mikrochim. Acta 2, 287 (1937)
Iodine....	Starch	Snell, Vol. II, p. 740 (8)
	o-Toluidine	J. Am. Chem. Soc. 47, 1000 (1925)
Iridium....	Benzidine	Snell, Vol. II, p. 526 (8)
	Malachite Green, leuco base	C. A. 24, 2689 (1930)
Iron........	Acetylacetone	J. Am. Chem. Soc. 26, 967 (1904)
	Alloxantin	Compt. rend. 180, 519 (1925)
	Ammonium salt of nitrosophenyl hydroxylamine ("Cupferron")	Ind. Eng. Chem. 3, 629 (1911)
	Bis-p-chlorophenyl-phosphoric acid	Feigl, p. 294 (4)
	Cysteine	Biochem. Z. 187, 255 (1927)
	Dimethyl glyoxime	Z. anorg. Chemie 80, 401 (1911)
	Dinitrosoresorcinol	J. Am. Chem. Soc. 47, 1268 (1925)
	Diisonitrosoacetone	Chem. Listy 23, 496 (1929); C. A. 24, 801 (1930)
	Dioximes (various)	Anal. Chem. 19, 1017 (1947)
	Diphenylamine	J. Am. Chem. Soc. 46, 968 (1924)
	2,2-Bipyridyl	Snell, Vol. II, p. 316 (8)
	Disodium-1,2-dihydroxybenzene-3,5-disulfonate ("Tiron")	Ind. Eng. Chem., Anal. Ed. 16, 111 (1944)
	Hexamethylenetetramine	Bull soc. chim. Rom. 2, 89 (1921)
	4 Hydroxybiphenyl-3-carboxylic acid	J. Am. Chem. Soc. 70, 648 (1948)
	8-Hydroxyquinoline	Bull. soc. chim. biol., 17, 432 (1935)
	4-Hydroxybiphenyl-3-carboxylic acid	J. Am. Chem. Soc. 70, 648 (1948)
	7-Iodo-8-hydroxy-quinoline-5-sulfonic acid (Ferron)	J. Am. Chem. Soc. 59, 872 (1937)
	Isonitrosoacetophenone	Ber. 60, 527 (1927)
	Isonitrosodimethyl-dihydroresorcinol	Anal. Chem., 20, 1205 (1948)
	Kojic acid	Ind. Eng. Chem., Anal. Ed., 13, 612 (1941)
	α-Nitroso-β-naphthol	Bull. soc. chim. 35, 641 (1924)
	Nitroso-R salt	Ind. Eng. Chem., Anal. Ed., 14, 756 (1942); 16, 276 (1944)
	o-Phenanthroline	Ind. Eng. Chem., Anal. Ed. 9, 67 (1937)
	Protacatechvic acid	J. Biol. Chem., 137, 417 (1941)
	Pyramidone	Pharm. Weekblad, 63, 1121 (1926)
	Pyrocatechol	Helv. chim. Acta 9, 835 (1926)
	Salicylaldoxime	Ind Eng. Chem., Anal. Ed., 12, 448 (1940)
	Salicylic acid	J. Chem. Soc. 93, 93 (1908)
	Salicylsulfonic acid	Snell, Vol. II, p. 321 (8)
	Sulfosalicylic acid	Biochem. Z. 181, 391 (1927)
	Thioglycollic acid	J. Am. Chem. Soc. 49, 1916 (1927)
	8-Hydroxyquinoline	Z. anal. Chem. 107, 191 (1936)
Lanthanum	Sodium alizarinesulfonate	J. Am. Chem. Soc., 53, 1217 (1931)
Lead......	Ammonium thiocyanate and pyridine	Z. anal. Chem. 72, 289 (1927)
	Aniline	Ind. Eng. Chem. 11, 1055 (1919); Yoe, Vol. I, p. 257 (2)
	Anthranilic acid	Z. anal. Chem. 101, 85 (1935)
	Carminic acid	Mikrochem 7, 301 (1929)
	s-Diphenylcarbazide	Yoe, Vol. I, p. 255 (2)
	Diphenylthiocarbazone (Dithizone)	Snell, Vol. IIA, p. 10 (8)
	Hematein	Yoe, Vol. I, p. 257 (2)
	Salicylaldoxime	Ind. Eng. Chem., Anal. Ed. 14, 359 (1942)
	Tetramethyldiamido-diphenylmethane	Snell, Vol. II, p. 43 (8)
	Thiourea	Z. anorg. Chem., 234, 224 (1937)

Determination	Reagent	Reference
Lithium......	Ammonium stearate	J. Am. Chem. Soc. **52**, 2754 (1930)
	Hexamethylenetetr-amine	Welcher, Vol. III, p. 131 (9)
Magnesium..	Pyridine	Welcher, Vol. III, p. 33 (9)
	Brilliant yellow	Sandell, p. 598 (6)
	Curcumin	Ind. Eng. Chem., Anal. Ed. **4**, 426 (1932)
	Dimethylamine	Z. anorg. Chem. **26**, 347 (1901)
	Hydroquinone	Yoe, Vol. I, p. 264 (2)
	8-Hydroxyquinoline	Z. anal. Chem. **71**, 122 (1927)
	p-Nitrobenzeneazo-α-naphthol	Feigl, p. 225 (2)
	p-Nitrobenzeneazo-resorcinol	J. Am. Chem. Soc. **51**, 1456 (1929)
	Oleic acid	Yoe, Vol. I, p. 270 (2)
	Quinalizarin	Feigl, p. 224 (2)
	Sodium 1-azo-2-hydroxy-3-(2,4-dimethylcarbox-aniliodonaphthalene-1'(2-hydroxyben-zene-5-sulfonate)	Anal. Chem. **28**, 202 (1956); cf. Anal. Chim. Acta **16**, 155 (1957)
	Thiazole yellow	Sandell, p. 598 (6)
	Titan yellow	Sandell, p. 591 (6)
	Tropeoline OO	Rec. trav. chim., **61**, 849 (1942)
Manganese..	Benzidine	Snell, Vol. II, p. 397 (8)
	Formaldoxime	Ind. Eng. Chem., Anal. Ed. **9**, 445 (1937); **12**, 307 (1940)
	Tetramethyldiamino-diphenylmethane	Feigl, p. 174 (2); Snell, Vol. IIA, p. 311 (8)
	4,4-Tetramethyl-diaminotriphenyl-methane	J. Biol. Chem., **168**, 537 (1947)
Mercury....	Anthranilic acid	Z. anal. Chem. **101**, 88 (1935)
	p-Dimethylamino-benzalrhodamine	J. Am. Chem. Soc. **52**, 2222 (1930)
	Di-β-naphthylthio-carbazone	Sandell, p. 630 (6)
	s-Diphenylcarbazide	Z. angew. Chem. **39**, 791 (1926)
	Diphenylthiocarba-zone (Dithizone)	Z. anal. Chem. **103**, 241 (1935)
	Potassium s-diphenyl-carbazone	Snell, Vol. II 78 (8)
	Strychnine	" II 76 (8)
Molybdenum	α-Benzoin-oxime (Cupron)	B. S. J. Research **9**, 1 (1932)
	Disodium-1,2-dihy-droxy-benzene-3,5-disulfonate (Tiron)	Anal. Chim. Acta 8, 546 (1953)
	4-Methyl-1,2-dimer-captobenzene (Dithiol)	J. Am. Pharm. Assoc., **37**, 255 (1948)
	Phenylhydrazine	Ber. **36**, 512 (1903)
	Potassium ethyl xan-thate and chloroform	J. Am. Chem. Soc. **44**, 1462 (1922)
	Tannic acid	Chem. Eng. Mining Rev. **11**, 258 (1919)
Nickel......	α-Benzil-dioxime	Analyst **38**, 316 (1913)
	Cyclohexanedione-dioxime	Ann. **437**, 148 (1925)
	Dicyandiamidine sulfate	Chem. Zeit. **31**, 335, 911 (1907)
	Diethyldithiocar-bamate	Ind. Eng. Chem., Anal. Ed., **18**, 206 (1946)
	Dimethylglyoxime	Chem. Weekblad. **21**, 358 (1924)
	Formaldoxime	Snell, Vol. IIA, p. 265 (8)
	α-Furildioxime	J. Am. Chem. Soc. **47**, 918 (1925)
	Potassium dithiooxa-late	J. Am. Chem. Soc. **54**, 1866 (1932)
	Rubianic acid (Dithio-oxamide)	Anal. Chim. Acta **20**, 332 435 (1959)
Niobium (Columbium)	Ammonium salt of ni-trosophenyl hydrox-ylamine ("Cupfer-ron")	Hillebrand et al., p. 120 (5)
	8 Hydroxyquinolin	Wechler, Vol. I, p. 302 (2)
Nitrate.....	Brucine	Yoe, Vol. I, p. 318 (2)
	Diphenylamine sulfonic acid	J. Am. Chem. Soc. **55**, 1448 (1933)
	Diphenylbenzidine	Yoe, Vol. I, p. 316 (2)
	Diphenyl-endo-anilo-hydrotriazole ("Nitron")	Fales and Kenny, Inorg. Quant. Anal., p. 347 (1938)
	Phenoldisulfonic acid	Yoe, Vol. I, p. 313 (2)
	Pyrogallol	" p. 319 (2)
	Strychnine sulfate	" p. 320 (2)
	2:4-Xylenol	J. Assoc. Off. Agri. Chem. **18**, 459 (1935)

Determination	Reagent	Reference
Nitrite......	Antipyrin	Yoe, Vol. I, p. 311 (2)
	Dimethylaniline	" p. 311 (2)
	Dimethyl-α-Naph-thylamine	Ind. Eng. Chem., Anal. Ed. **1**, 28 (1929)
	Diphenylamine sulfate	Yoe, Vol. I, p. 654 (2)
	α-Naphthylamine and β-Naphthylamine-6,8-Disulfonic acid	J. Pharmacol. **51**, 398 (1934)
	α-Naphthylamine hydrochloride	Yoe, Vol. I, p. 309 (2)
	m-Phenylenediamine	Yoe, Vol. I, p. 310 (2)
	Sulfanilic acid and α-naphthylamine	Yoe, Vol. I, p. 308 (2)
Osmium..	s-Diphenylthiourea	Yoe and Sarver, p. 155 (12)
	1-Naphthylamine-4,6,8-trisulfonic acid	Anal. Chim. Acta **20**, 205 (1959)
	Strychnine	Welcher, Vol. IV, p. 272 (9)
	Thiourea	Compt. rend. **167**, 235 (1918)
Oxygen.....	Indigo carmine	Snell, Vol. IIA, p. 726 (8)
	Pyrogallol	Dennis, Gas Analysis, p. 174 (1929)
Palladium..	Dimethylglyoxime	J. Am. Chem. Soc. **57**, 2565 (1935)
	p-Fuchsine	Ind. Eng. Chem., Anal. Ed., **18**, 400 (1946)
	β-Furfuraldoxime	Ind. Eng. Chem., Anal. Ed. **14**, 491 (1942)
	6-Nitroquinoline	J. Am. Chem. Soc. **50**, 3018 (1928)
	p-Nitrosodiphenyl-amine	J. Am. Chem. Soc. **61**, 2058 (1939)
	p-Nitrosodimethyl-aniline	J. Am. Chem. Soc. **63**, 3224 (1941)
	p-Nitrosodiethyl-aniline	Ibid.
	Thiomalic acid	Talanta **2**, 223 (1959)
Phosphate..	1,2,4-Aminonaphtho-sulfonic acid	Yoe, Vol. I, p. 348 (2)
	Hydroquinone	Yoe, Vol. I, pp. 346 and 353 (2)
	Quinine-molybdate	Yoe, Vol. I, p. 343 (2)
	Strychnine-molybdate	Yoe, Vol. II, p. 142 (3)
Phosphorus.	Hydrazine sulfate	Yoe, Vol. I, p. 341 (2)
Platinum...	p-Nitrosodimethyl-aniline	Anal. Chem. **26**, 1335, 1340 (1954)
Potassium..	6-Chloro-5-nitrotol-uene-3-sulfonic acid	Mikrochem. **14**, 368 (1934)
	Dipicrylamine	Z. angew. Chem. **49**, 827 (1936)
	Picric acid	J. Am. Chem. Soc. **53**, 539 (1931)
Rhodium...	Thiomalic acid	Talanta **2**, 239 (1959)
Ruthenium..	s-Diphenylthiourea	Yoe and Sarver, p. 155 (12)
	5-Hydroxyquinoline-8-carboxylic acid	Canadian J. Research B25, **49**, (1945)
	1-Naphthylamine-3,5,7-trisulfonic acid	Anal. Chim Acta **20**, 211 (1959)
	Rubianic acid (Dithio-oxamide)	Mikrochemie **15**, 295 (1934)
	Thioglycolyl-β-amido-naphthalide (Thion-alid)	Ind. Eng. Chem., Anal. Ed. **12**, 5611 (1940)
	Thiourea	Sandell, p. 781 (6)
Scandium..	Morin	Mikrochem. Acta **2**, 9, 287 (1937)
Selenium...	Codeine phosphate	Arch Pharm. **252**, 161 (1914)
	Hydrazine	Boltz, p. 321 (1)
	Hydroquinone	Am. J. Sci. **15**, 253 (1928)
	Hydroxylamine hydrochloride	J. Am. Chem. Soc. **47**, 2456 (1925)
	Pyrrol	Snell, Vol. II, p. 779 (8)
	Thiourea	Ann. chim. applicata **17**, 357 (1927); Feigl, p. 231 (8)
Silver......	Chromotropic acid	Helvetica Chim. Acta **12**, 714 (1929)
	Dichlorofluorescein	J. Am. Chem. Soc. **51**, 3273 (1929)
	p-Dimethylamino-benzalrhodamine	J. Am. Chem. Soc. **52**, 2222 (1930)
	Diphenylthiocar-bazone (Dithizone)	Z. anal. Chem. **101**, 1 (1935)
	Formazylcarboxylic acid	J. Anal. Chem. Russ., **2**, 131 (1934); abs. in Analyst, **73**, 352 (1948)
	Methylamine	Mikrochemie **7**, 233 (1929)
	2-Thio-5-keto-4-car-bethoxy-1,3-dihydro-pyrimidine	Ind. Eng. Chem., Anal. Ed. **14**, 148 (1942)
Sodium....	6,8-Dichlorobenzoyl-urea	J. Org. Chem. **3**, 414 (1938)
	Dihydroxy-tartaric acid	J. Russ. Phys. Chem. Soc. **60**, 661 (1928)
	Uranyl zinc acetate	J. Am. Chem. Soc. **51**, 1664 (1929)
Strontium.	Sodium rhodizonate	Mikrochemie **2**, 187 (1924)
Sulfide....	p-Aminodimethyl-aniline	Snell, Vol. IIA, p. 658 (8)
Sulfur....	p-Phenylenedimethyl-diamine-hydro-chloride	Yoe, Vol. I, p. 373 (2)

Determination	Reagent	Reference	Determination	Reagent	Reference
Tantalum.....	Ammonium salt of nitrosophenyl hydroxylamine ("Cupferron")	Hillebrand et. al., p. 120 (5)	Tungsten (Wolfram) (Cont.)	Toluene-3,4-dithiol	Analyst, **69**, 109 (1944); J. Am. Pharm. Assoc., **37**, 255 (1948)
	Pyrogallol	Sandell, p. 695 (6)		Uric acid	Ann. chim. anal. **9**, 371 (1904)
Tellurium.....	Hydrazine hydrochloride	J. Am. Chem. Soc. **47**, 2456 (1925)		α-Benzoinoxime ("Cupron")	Bur. Std. J. Research **9**, 1 (1932)
	Hydroquinone	Am. J. Sci. **15**, 253 (1928)		Cinchonine	Hillebrand et al., p. 689 (5)
	Thiourea	Ann. chim. applicata **17**, 359 (1927)		Hydroquinone	Z. angew. Chem. **44**, 237 (1931)
Thallium......	Diphenylthiocarbazone (Dithizone)	Analyst **60**, 394 (1935)		Phenylhydrazine	Bull. soc. chim. Belg. **38**, 385 (1929)
	Thionalid (Thioglycolyl-β-amidonaphthalide)	Z. angew. Chem. **48**, 430, 597 (1935)		Rhodamine B.	Snell, Vol. II, p. 469 (8)
Thorium......	1-Amino-4-hydroxyanthraquinone	Ind. Eng. Chem., Anal. Ed. **13**, 809 (1941)	Uranium....	Toluene-3,4-dithiol	Analyst, **69**, 109 (1944); J. Am. Pharm. Assoc., **37**, 255 (1948)
	Cupferron (Ammonium nitrosophenylhydroxylamine)	Chem. Ztg. **33**, 1298 (1908)		Uric acid	Ann. chim. anal. **9**, 371 (1904)
	8-Hydroxyquinoline	Z. anal. Chem. **100**, 98 (1935)		Dibenzoylmethane	Anal. Chem. **25**, 1200 (1953)
	Phenylarsonic acid	J. Am. Chem. Soc. **48**, 895 (1926)		o-Hydroxybenzoic acid	Snell, Vol. II, p. 492 (8)
Tin..........	Ammonium salt of nitrosophenyl hydroxylamine ("Cupferron")	Hillebrand et al., p. 120 (5)		α-Quinaldinic acid	Z. anal. Chem. **95**, 400 (1933)
	Cacotheline	Ind. Eng. Chem., Anal. Ed. **7**, 26 (1935)		Sodium diethyldithiocarbamate	Sandell, p. 921 (6)
	4-Chloro-1,2-dimercaptobenzene (1-chloro-benzene-3,4-dithiol)	J. Chem. Soc. **149**, 175 (1936)		Sodium salicylate	Chem. Zeit. **43**, 739 (1919)
	Hematoxylin	Sandell, p. 866 (6)		Thioglycolic acid (Mercapto acetic acid)	Anl. Chem., **21**, 1093 (1949)!
	Quinalizarin	Sandell, p. 866 (6)	Urea......	Xanthydrol	Mikrochem. **14**, 132 (1934)
	Toluene-3,4-dithiol (4-methyl-1,2-dimercaptobenzene)	Analyst **61**, 242 (1936); Ibid., **62**, 661 (1937)	Vanadium..	Ammonium benzoate	C. A. **29**, 2880 (1935)
Titanium.....	Ammonium salt of nitrosophenyl hydroxylamine ("Cupferron")	Hillebrand et al., p. 119 (5); Z. anal. Chem. **83**, 345 (1931)		Aniline	C. A. **24**, 567 (1930)
	Chromotropic acid	Feigl, p. 107 (2)		Benzidine	J. Applied Chem. (U.S.S.R.) **17**, 83 (1944); C. A. **39**, 1115 (1945)
	5,7-Dibromo-8-hydroxyquinoline	Z. anorg. Chem. **204**, 215 (1932)		Diphenylamine	Yoe, Vol. I, p. 715 (2)
	Dihydroxymaleic acid	Snell, Vol. II, p. 445 (8)		Diphenylbenzidine	Ind. Eng. Chem. **20**, 764 (1928)
	Disodium 1,2-dihydroxybenzene-3,5-disulfonate ("Tiron")	Ind. Eng. Chem., Anal. Ed. **19**, 100 (1947)		8-Hydroxyquinoline	Feigl, p. 125 (2)
	Gallic acid	Snell, Vol. II, p. 444 (8)		Safranine	Vol. Anal., Vol. II, p. 326 (6)
	p-Hydroxyphenylarsonic acid	Ind. Eng. Chem., Anal. Ed. **10**, 642 (1938)		Strychnine	Yoe, Vol. I, p. 393 (2)
	8-Hydroxyquinoline	Z. anal. Chem. **81**, 1 (1930)	Wolfram....	See Tungsten	
	Resoflavine (Color Index 1015)	Anal. Chim. Acta, **1**, 244 (1947)	Zinc........	Anthranilic acid	Z. anal. Chem. **91**, 332 (1933)
	Tannic acid	Analyst **55**, 605 (1930)		2 Carboxy-2'hydroxy-5'-sulfoformazybenzene (Zincon)	Anal. Chem. **26**, 1345 (1954)
	Thymol	Yoe, Vol. I, p. 381 (2)		Di-β-naphthylthiocarbazone	Sandell, pp. 945, 962 (6)
Tungsten (Wolfram)	Anti-1,5-di-(p-methoxyphenyl)-1-hydroxylamino-3-oximino-4-pentene ("Wolfron")	Ind. Eng. Chem., Anal. Ed. **16**, 45 (1944)		Diphenylamine	J. Am. Chem. Soc. **49**, 2214 (1927)
	Benzidine	Ber. **38**, 783 (1905)		Diphenylbenzidine	J. Am. Chem. Soc. **49**, 356 (1927)
	α-Benzoinoxime ("Cupron")	Bur. Std. J. Research **9**, 1 (1932)		Diphenylthiocarbazone (Dithizone)	Ind. Eng. Chem., Anal. Ed. **9**, 127 (1937)
	Cinchonine	Hillebrand et al., p. 689 (5)		8-Hydroxyquinoline	Z. anal. Chem. **71**, 171 (1927)
	Hydroquinone	Z. angew. Chem. **44**, 237 (1931)		5-Nitroquinaldic acid	Ind. Eng. Chem., Anal. Ed. **10**, 335 (1938)
	Phenylhydrazine	Bull. soc. chim. Belg. **38**, 385 (1929)		Pyridine	Z. anal. Chem. **73**, 356 (1928)
	Rhodamine B.	Snell, Vol. II, p. 469 (8)		α-Quinaldinic acid	Z. anal. Chem. **100**, 324 (1935)
				Resorcinol	Yoe, Vol. I, p. 396 (2)
				Urobilin	J. Ind. Hyg. **7**, 273 (1925)
			Zirconium...	Ammonium salt of nitrosophenyl hydroxylamine ("Cupferron")	Hillebrand et al., p. 572 (5)
				5-Chlorobromamine acid	Ind. Eng. Chem., Anal. Ed. **15**, 73 (1943)
				p-Dimethylaminoazophenylarsenic acid	Ind. Eng. Chem., Anal. Ed., **13**, 603 (1941)
				2-Hydroxy-5-methylazobenzene-4-sulfonic acid	Feigl, p. 288 (4)
				Morin	Sandell, p. 971 (6)
				Phenylarsenic acid	J. Am. Chem. Soc. **48**, 895 (1926)
				Sodium alizarin sulfonate (Alizarin S)	Chem. Weekblad **20**, 404 (1924); Feigl, p. 200 (2)

(1) Colorimetric Determination of Nonmetals, Boltz, Edition 1958.
(2) Feigl, Spot Tests in Inorganic Analysis, 5th English Ed., translated by Oesper, 1958.
(3) Feigl, Spot Tests in Organic Analysis, 5th English Ed., translated by Oesper, 1956.
(4) Feigl, Chemistry of Specific, Selective and Sensitive Reactions translated by Oesper, 1949.
(5) Hillebrand, Lundell, Bright and Hoffman, Applied Inorganic Analysis, 2nd Edition, 1953.
(6) Sandell, Colorimetric Determination of Traces of Metals, 3rd Edition, 1959.
(7) Scott's Standard Methods of Chemical Analysis, Furman, 5th Edition 1939.
(8) Snell, Snell and Snell, Colorimetric Methods of Analysis, Volume II (1949), II A (1959).
(9) Welcher, Organic Analytical Reagents, Vols. I–IV, 1947–1948.
(10) Yoe, Photometric Chemical Analysis, Vol. I, Colorimetry, 1928.
(11) Yoe, Photometric Chemical Analysis Vol. II, Nephelometry, 1929
(12) Yoe and Sarver, Organic Analytical Reagents, 1941

CALIBRATION OF VOLUMETRIC GLASSWARE FROM THE WEIGHT OF THE CONTAINED WATER OR MERCURY WHEN WEIGHED IN AIR WITH BRASS WEIGHTS

D. F. SWINEHART

A borosilicate glass vessel containing g_t grams of water at a temperature of $t°C$ has, at the same temperature, a volume $V_t = W_t \times g_t$ cubic centimeters. Similarly when filled with G_t grams of mercury at a temperature of $t°C$ the volume at the same temperature is given by $V_t = M_t \times G_t$ cubic centimeters.

When filled with g_t grams of water at a temperature of $t°C$ the volume of the vessel at 18°C is given by $V_{18} = W_{18} \times g_t$ cubic centimeters and the true volume at 25°C is given by $V_{25} = W_{25} \times g_t$. The volumes at 18°C and 25°C are given similarly when using mercury by using the values under M_{18} and M_{25}, respectively.

The data on water are adapted from the data of G. S. Kell, *Journal of Chemical and Engineering Data*, 12, 67–68 (1967) (Table on p. F5, 52nd Edition, this handbook) and the data on mercury are adapted from *Smithsonian Tables*, Ninth Revised Edition, Volume 120, Publication No. 4169. The coefficient of linear expansion for borosilicate glass used here is 32.5×10^{-7} deg^{-1} and the volume coefficient of expansion is 97.5×10^{-7} deg^{-1}.

$t°C$	W_t	W_{18}	W_{25}	M_t	M_{18}	M_{25}
0	1.001 220	1.001 396	1.001 466	0.0735 519	0.0735 648	0.0735 698
1	1.001 161	1.001 327	1.001 395	0.0735 653	0.0735 775	0.0735 825
2	1.001 120	1.001 276	1.001 345	0.0735 787	0.0735 902	0.0735 952
3	1.001 096	1.001 242	1.001 311	0.0735 920	0.0736 028	0.0736 078
4	1.001 088	1.001 225	1.001 293	0.0736 054	0.0736 154	0.0736 205
5	1.001 096	1.001 223	1.001 291	0.0736 188	0.0736 281	0.0736 332
6	1.001 120	1.001 237	1.001 306	0.0736 322	0.0736 408	0.0736 458
7	1.001 158	1.001 265	1.001 334	0.0736 456	0.0736 535	0.0736 585
8	1.001 211	1.001 309	1.001 377	0.0736 590	0.0736 662	0.0736 712
9	1.001 279	1.001 367	1.001 435	0.0736 724	0.0736 789	0.0736 839
10	1.001 360	1.001 438	1.001 506	0.0736 858	0.0736 915	0.0736 966
11	1.001 455	1.001 523	1.001 592	0.0736 992	0.0737 042	0.0737 093
12	1.001 563	1.001 622	1.001 690	0.0737 125	0.0737 168	0.0737 218
13	1.001 684	1.001 733	1.001 801	0.0737 259	0.0737 295	0.0737 345
14	1.001 816	1.001 855	1.001 923	0.0737 393	0.0737 422	0.0737 472
15	1.001 961	1.001 990	1.002 059	0.0737 526	0.0737 548	0.0737 598
16	1.002 118	1.002 138	1.002 206	0.0737 660	0.0737 674	0.0737 725
17	1.002 286	1.002 296	1.002 364	0.0737 794	0.0737 801	0.0737 852
18	1.002 466	1.002 466	1.002 534	0.0737 928	0.0737 928	0.0737 978
19	1.002 658	1.002 648	1.002 717	0.0738 062	0.0738 055	0.0738 105
20	1.002 859	1.002 839	1.002 908	0.0738 196	0.0738 182	0.0738 232
21	1.003 072	1.003 043	1.003 111	0.0738 330	0.0738 308	0.0738 359
22	1.003 294	1.003 255	1.003 323	0.0738 463	0.0738 434	0.0738 485
23	1.003 528	1.003 479	1.003 548	0.0738 597	0.0738 561	0.0738 611
24	1.003 771	1.003 712	1.003 781	0.0738 731	0.0738 688	0.0738 738
25	1.004 024	1.003 955	1.004 024	0.0738 864	0.0738 814	0.0738 864
26	1.004 287	1.004 209	1.004 277	0.0738 998	0.0738 940	0.0738 991
27	1.004 560	1.004 472	1.004 540	0.0739 132	0.0739 067	0.0739 118
28	1.004 842	1.004 744	1.004 813	0.0739 266	0.0739 194	0.0739 244
29	1.005 133	1.005 025	1.005 094	0.0739 400	0.0739 321	0.0739 371
30	1.005 434	1.005 316	1.005 385	0.0739 534	0.0739 447	0.0739 498
31	1.005 743	1.005 615	1.005 684	0.0739 669	0.0739 575	0.0739 626
32	1.006 060	1.005 923	1.005 991	0.0739 801	0.0739 700	0.0739 750
33	1.006 388	1.006 241	1.006 310	0.0739 934	0.0739 826	0.0739 876
34	1.006 723	1.006 566	1.006 635	0.0740 068	0.0739 953	0.0740 003
35	1.007 066	1.006 899	1.006 968	0.0740 202	0.0740 079	0.0740 130
36	1.007 418	1.007 242	1.007 311	0.0740 335	0.0740 205	0.0740 256
37	1.007 780	1.007 593	1.007 669	0.0740 469	0.0740 332	0.0740 382
38	1.008 149	1.007 952	1.008 021	0.0740 603	0.0740 459	0.0740 509
39	1.008 525	1.008 318	1.008 387	0.0740 737	0.0740 585	0.0740 636
40	1.008 910	1.008 694	1.008 762	0.0740 871	0.0740 712	0.0740 763
41	1.009 303	1.009 077	1.009 146	0.0741 007	0.0740 841	0.0740 891
42	1.009 703	1.009 467	1.009 536	0.0741 139	0.0740 966	0.0741 016
43	1.010 112	1.009 866	1.009 935	0.0741 273	0.0741 092	0.0741 143
44	1.010 528	1.010 272	1.010 341	0.0741 407	0.0741 219	0.0741 270
45	1.010 951	1.010 685	1.010 754	0.0741 541	0.0741 346	0.0741 396
46	1.011 382	1.011 106	1.011 175	0.0741 675	0.0741 473	0.0741 523
47	1.011 820	1.011 534	1.011 603	0.0741 810	0.0741 600	0.0741 651
48	1.012 266	1.011 970	1.012 039	0.0741 944	0.0741 727	0.0741 778
49	1.012 719	1.012 413	1.012 482	0.0742 078	0.0741 854	0.0741 904
50	1.013 180	1.012 864	1.012 933	0.0742 213	0.0741 981	0.0742 031

SOLUBILITY CHART

Abbreviations: W, soluble in water; A, insoluble in water but soluble in acids; w, sparingly soluble in water but soluble in acids; a, insoluble in water and only sparingly soluble in acids; I, insoluble in both water and acids; d, decomposes in water. * Certain salts occur in two modifications.

No.		Al	NH₄	Sb	Ba	Bi	Cd	Ca	Cr	Co	Cu	Au'	Au'''	H	Fe''	Fe'''
1	Acetates —(C₂H₃O₂)	W Al(—)₃	W NH₄(—)	......	W Ba(—)₂	W Bi(—)₃	W Cd(—)₂	W Ca(—)₂	W Cr(—)₃	W Co(—)₂	W Cu(—)₂	......	......	W C₂H₄O₂	W Fe(—)₂	W Fe₂(—)₆
2	Arsenate —(AsO₄)	a Al(—)	W (NH₄)₃(—)	A Sb(—)	W Ba₃(—)₂	A Bi(—)	A Cd₃(—)₂	w Ca₃(—)₂		A Co₃(—)₂	A Cu₃(—)₂			W H₃AsO₄	A Fe₃(—)₂	A Fe(—)
3	Arsenite —(AsO₂)	......	W NH₄AsO₂	A Sb(—)				W Ca₃(—)₂		A Co₃H₆( )₄	A CuH(—)			......		
4	Benzoate —(C₇H₅O₂)	......	W NH₄(—)		W Ba(—)₂	A Bi(—)₃	W Cd(—)₂	W C₃(—)₂		W Co(—)₂	w Cu(—)₂			W C₇H₆O₂	W Fe(—)₂	W Fe₂(—)₆
5	Bromide	W AlBr₃	W NH₄B₁	d SbBr₃	W BaBr₂	d BiBr₃	W CdBr₂	W CaBr₂	W(I)* CrBr₃	W CoBr₂	W CuBr₂	w AuBr	W AuBr₃	W HBr	W FeBr₂	W FeBr₃
6	Carbonate	......	W (NH₄)₂CO₃		A BaCO₃		A CdCO₃	A CaCO₃	W CrCO₃	A CoCO₃					w FeCO₃	
7	Chlorate —(ClO₃)	W Al(—)₃	W NH₄(—)		W Ba(—)₂	W Bi(—)₃	W Cd(—)₂	W Ca(—)₂		W Co(—)₂	W Cu(—)₂			W HClO₃	W Fe(—)₂	W Fe(—)₃
8	Chloride	W AlCl₃	W NH₄Cl	W SbCl₃	W BaCl₂	d BiCl₃	W CdCl₂	W CaCl₂	I CrCl₃	W CoCl₂	W CuCl₂	W AuCl	W AuCl₃	W HCl	W FeCl₂	W FeCl₃
9	Chromate —(CrO₄)		W (NH₄)₂(—)		A Ba(—)		W Cd(—)	W Ca(—)		W Co(—)						A Fe₂(—)₃
10	Citrate —(C₆H₅O₇)	W Al(—)	W (NH₄)₃(—)		w Ba₃(—)₂	A Bi(—)	A Cd₃(—)₂	A Ca₃(—)₂		W Co₃(—)₂				W C₆H₈O₇		W Fe(—)
11	Cyanide	......	W NH₄CN		W Ba(CN)₂	w Bi(CN)₃	W Cd(CN)₂	W Ca(CN)₂	A Cr(CN)₃	W Co(CN)₂	A Cu(CN)₂	w AuCN	W Au(CN)₃	W HCN	a Fe(CN)₂	
12	Ferricy'de —(Fe(CN)₆)	......	W (NH₄)₃(—)		W Ba₃(—)₂		W Cd₃(—)₂	W Ca₃(—)₂		I Co₃(—)₂	A Cu₃(—)₂			W H₃(—)	I Fe₃(—)₂	
13	Ferrocy'de —(Fe(CN)₆)	w Al₄(—)₃	W (NH₄)₄(—)		W Ba₂(—)		A Cd₂(—)	W Ca₂(—)		I Co₂(—)	I Cu₂(—)			W H₄(—)	I Fe₂(—)	a Fe₄(—)₃
14	Fluoride	W AlF₃	W NH₄F	W SbF₃	W BaF₂	W BiF₃	W CdF₂	w CaF₂	W(a)* CrF₃	W CoF₂	W CuF₂			W HF	w FeF₂	w FeF₃
15	Formate —(CHO₂)	W Al(—)₃	W NH₄(—)		w Ba(—)₂	w Bi(—)₃	W Cd(—)₂	W Ca(—)₂		W Co(—)₂	W Cu(—)₂			W CH₂O₂	W Fe(—)₂	W Fe(—)₃
16	Hydroxide	A Al(OH)₃	W NH₄OH		W Ba(OH)₂	A Bi(OH)₃	A Cd(OH)₂	W Ca(OH)₂	A Cr(OH)₃	A Co(OH)₂	A Cu(OH)₂	W AuOH	A Au(OH)₃	......	A Fe(OH)₂	A Fe(OH)₃
17	Iodide	W AlI₃	W NH₄I	d SbI₃	W BaI₂	A BiI₃	W CdI₂	W CaI₂	W CrI₂	W CoI₂	a CuI	a AuI	a AuI₃	W HI	W FeI₂	W FeI₃
18	Nitrate	W Al(NO₃)₃	W NH₄NO₃		W Ba(NO₃)₂	W Bi(NO₃)₃	W Cd(NO₃)₂	W Ca(NO₃)₂	W Cr(NO₃)₃	W Co(NO₃)₂	W Cu(NO₃)₂			W HNO₃	W Fe(NO₃)₂	W Fe(NO₃)₃
19	Oxalate —(C₂O₄)	A Al₂(—)₃	W (NH₄)₂(—)		A Ba(—)	A Bi₂(—)₃	A Cd(—)	A Ca(—)	W Cr(—)	a Co(—)	A Cu(—)		A	W C₂H₂O₄	A Fe(—)	W Fe₂(—)₃
20	Oxide	A Al₂O₃	......	w Sb₂O₃	W BaO	A Bi₂O₃	A CdO	W CaO	A Cr₂O₃	A CoO	A CuO	A Au₂O	A Au₂O₃	W H₂O₂	A FeO	A Fe₂O₃
21	Phosphate	A AlPO₄	W NH₄H₂PO₄		A Ba₃(PO₄)₂	A BiPO₄	A Cd₃(PO₄)₂	A Ca₃(PO₄)₂	A Cr₂(PO₄)₂	A Co₃(PO₄)₂	A Cu₃(PO₄)₂			W H₃PO₄	A Fe₃(PO₄)₂	A FePO₄
22	Silicate, —(SiO₃)	I Al₂(—)₃			W Ba(—)		W Cd(—)	W Ca(—)		A Co₂SiO₄	W Cu(—)			I H₂SiO₃		
23	Sulfate	W Al₂(SO₄)₃	W (NH₄)₂SO₄	A Sb₂(SO₄)₃	A BaSO₄	d Bi₂(SO₄)₃	W CdSO₄	W CaSO₄	W(I)* Cr₂(SO₄)₃	W CoSO₄	W CuSO₄			W H₂SO₄	W FeSO₄	W Fe₂(SO₄)₃
24	Sulfide	d Al₂S₃	W (NH₄)₂S	A Sb₂S₃	W BaS	A Bi₂S₃	A CdS	W CaS	A Cr₂S₃	A CoS	A CuS	I Au₂S	I Au₂S₃	W H₂S	A FeS	d Fe₂S₃
25	Tartrate —(C₄H₄O₆)	w Al₂(—)₃	W (NH₄)₂(—)	W Sb₂(—)₃	w Ba(—)	A Bi₂(—)₃	W Cd(—)	A Ca(—)		w Co(—)	w Cu(—)			W C₄H₆O₆	W Fe(—)	W Fe₂(—)₃
26	Thiocy'te	......	W NH₄CNS		W Ba(CNS)₂			W Ca(CNS)		W Co(CNS)₂	d CuCNS			W CNSH	W Fe(CNS)₂	W Fe(CNS)₃

No.		Pb	Mg	Mn	Hg'	Hg''	Ni	K	Ag	Na	Sn''''	Sn''	Sr	Zn	Pt
1	Acetate —(C₂H₃O₂)	W Pb(—)₂	W Mg(—)₂	W Mn(—)₂	w Hg(—)	W Hg(—)₂	W Ni(—)₂	W K(—)	w Ag(—)	W Na(—)	W Sn(—)₄	d Sn(—)₂	W Sr(—)₂	W Zn(—)₂	......
2	Arsenate —(AsO₄)	A PbH(—)	W Mg₃(—)	W MnH(—)	A Hg₃(—)	A Hg₃(—)₂	A Ni₃(—)₂	W K₃(—)	A Ag₃(—)	W Na₃(—)			A SrH(—)	A Zn₃(—)₂	
3	Arsenite —(AsO₂)	......	W Mg₃(—)	A Mn₃H₆(—)₄	A Hg₃(—)	A Hg₃(—)	A Ni₃H₆(—)₄	W K₃AsO₃	A Ag₃(—)	W Na₂H(—)		A Sn₃(—)	w Sr₃(—)₂		
4	Benzoate —(C₇H₅O₂)	w Pb(—)₂	W Mg(—)₂	W Mn(—)₂	A Hg₂(—)₂	W Hg(—)₂	W Ni(—)₂	W K(—)	W Ag(—)	W Na(—)				W Zn(—)₂	
5	Bromide	W PbBr₂	W MgBr₂	W MnBr₂	A HgBr	W HgBr₂	W NiBr₂	W KBr	A AgBr	W NaBr	W SnBr₄	W SnBr₂	W SrBr₂	W ZnBr₂	W PtBr₄
6	Carbonate	A PbCO₃	w MgCO₃	A MnCO₃	A Hg₂CO₃	......	A NiCO₃	W K₂CO₃	A Ag₂CO₃	W Na₂CO₃			A SrCO₃	A ZnCO₃	
7	Chlorate —(ClO₃)	W Pb(—)₂	W Mg(—)₂	W Mn(—)₂	w Hg(—)	W Hg(—)₂	W Ni(—)₂	W K(—)	W Ag(—)	W Na(—)		W Sn(—)₂	W Sr(—)₂	W Zn(—)₂	
8	Chloride	W PbCl₂	W MgCl₂	W MnCl₂	A HgCl	W HgCl₂	W NiCl₂	W KCl	A AgCl	W NaCl	W SnCl₄	W SnCl₂	W SrCl₂	W ZnCl₂	W PtCl₄
9	Chromate —(CrO₄)	A Pb(—)	W Mg(—)		w Hg₂(—)	w Hg(—)	A Ni(—)	W K₂(—)	A Ag₂(—)	W Na₂(—)	W Sn(—)₂	A Sn(—)	w Sr(—)	w Zn(—)	......
10	Citrate —(C₆H₅O₇)	W Pb₃(—)₂	W Mg₃(—)₂	w MnH(—)	w Hg₃(—)		W Ni₃(—)₂	W K₃(—)₂	W Ag₃(—)	W Na₃(—)			A SrH(—)	w Zn₃(—)₂	

SOLUBILITY CHART (Continued)

No.		Pb	Mg	Mn	Hg'	Hg''	Ni	K	Ag	Na	Sn''''	Sn''	Sr	Zn	Pt
11	Cyanide	w $Pb(CN)_2$	W $Mg(CN)_2$	....	A $HgCN$	W $Hg(CN)_2$	a $Ni(CN)_2$	W KCN	a $AgCN$	W $NaCN$	....	....	W $Sr(CN)_2$	A $Zn(CN)_2$	I $Pt(CN)_2$
12	Ferricy'de —$Fe(CN)_6$	w $Pb_3(—)_2$	W $Mg_3(—)_2$	....	....	A $Hg_3(—)_2$	I $Ni_3(—)_2$	W $K_3(—)$	I $Ag_3(—)$	W $Na_3(—)$	....	A $Sn_3(—)_2$	W $Sr_3(—)_2$	A $Zn_3(—)_2$	....
13	Ferrocy'de —$Fe(CN)_6$	a $Pb_2(—)$	W $Mg_2(—)$	A $Mn_2(—)$	.....	I $Hg_2(—)$	I $Ni_2(—)$	W $K_4(—)$	I $Ag_4(—)$	W $Na_4(—)$	....	a $Sn_2(—)$	W $Sr_2(—)$	I $Zn_2(—)$	....
14	Fluoride	w PbF_2	w MgF_2	A MnF_2	d HgF	d HgF_2	w NiF_2	W KF	W AgF	W NaF	W SnF_4	W SnF_2	W SrF_2	w ZnF_2	W PtF_4
15	Formate —(CHO_2)	W $Pb(—)_2$	W $Mg(—)_2$	W $Mn(—)_2$	w $Hg(—)$	W $Hg(—)_2$	W $Ni(—)_2$	W $K(—)$	W $Ag(—)$	W $Na(—)$	....	....	W $Sr(—)_2$	W $Zn(—)_2$	....
16	Hydroxide	w $Pb(OH)_2$	A $Mg(OH)_2$	A $Mn(OH)_2$	.....	W $Hg(OH)_2$	A $Ni(OH)_2$	W KOH	....	W $NaOH$	A $Sn(OH)_4$	A $Sn(OH)_2$	W $Sr(OH)_2$	A $Zn(OH)_2$	A $Pt(OH)_4$
17	Iodide	w PbI_2	W MgI_2	W MnI_2	A HgI	w HgI_2	W NiI_2	W KI	I AgI	W NaI	d SnI_4	W SnI_2	W SrI_2	W ZnI_2	I PtI_2
18	Nitrate	W $Pb(NO_3)_2$	W $Mg(NO_3)_2$	W $Mn(NO_3)_2$	W $HgNO_3$	W $Hg(NO_3)_2$	W $Ni(NO_3)_2$	W KNO_3	W $AgNO_3$	W $NaNO_3$	....	d $Sn(NO_3)_2$	W $Sr(NO_3)_2$	W $Zn(NO_3)_2$	W $Pt(NO_3)_4$
19	Oxalate —(C_2O_4)	A $Pb(—)$	w $Mg(—)$	w $Mn(—)$	a $Hg_2(—)$	A $Hg(—)$	A $Ni(—)$	W $K_2(—)$	a $Ag_2(—)$	W $Na_2(—)$	....	A $Sn(—)$	A $Sr(—)$	A $Zn(—)$	....
20	Oxide	w PbO	A MgO	A MnO	A Hg_2O	w HgO	A NiO	W K_2O	w Ag_2O	d Na_2O	A SnO_2	A SnO	A SrO	w ZnO	A PtO
21	Phosphate	A $Pb_3(PO_4)_2$	A $Mg_3(PO_4)_2$	A $Mn_3(PO_4)_2$	A Hg_3PO_4	A $Hg_3(PO_4)_2$	A $Ni_3(PO_4)_2$	W K_3PO_4	A Ag_3PO_4	W Na_3PO_4	....	A $Sn_3(PO_4)_2$	A $Sr_3(PO_4)_2$	A $Zn_3(PO_4)_2$	....
22	Silicate --(SiO_3)	A $Pb(—)$	A $Mg(—)$	I $Mn(—)$	....	....	....	W $K_2(—)$	....	W $Na_2(—)$	....	....	A $Sr(—)$	A $Zn(—)$	....
23	Sulfate	w $PbSO_4$	W $MgSO_4$	W $MnSO_4$	w Hg_2SO_4	d $HgSO_4$	W $NiSO_4$	W K_2SO_4	w Ag_2SO_4	W Na_2SO_4	W $Sn(SO_4)_2$	W $SnSO_4$	w $SrSO_4$	W $ZnSO_4$	W $Pt(SO_4)_2$
24	Sulfide	A PbS	d MgS	A MnS	I Hg_2S	I HgS	A NiS	W K_2S	A Ag_2S	W Na_2S	A SnS_2	A SnS	W SrS	A ZnS	I PtS
25	Tartrate —$(C_4H_4O_6)$	A $Pb(—)$	w $Mg(—)$	w $Mn(—)$	I $Hg_2(—)$	....	w $Ni(—)$	W $K_2(—)$	w $Ag_2(—)$	W $Na_2(—)$	....	W $Sn(—)$	w $Sr(—)$	w $Zn(—)$	....
26	Thiocy'te	w $Pb(CNS)_2$	W $Mg(CNS)_2$	W $Mn(CNS)_2$	A Hg_2CNS	W $Hg(CNS)_2$	....	W $KCNS$	I $AgCNS$	W $NaCNS$	....	....	W $Sr(CNS)_2$	W $Zn(CNS)_2$	....

REDUCTIONS OF WEIGHINGS IN AIR TO VACUO

When the weight M in grams of a body is determined in air, a correction is necessary for the buoyancy of the air. The following table is computed for an air density of 0.0012. The corrected weight $= M + kM/1000$. values of k being found in the table.

Density of body weighed	Correction factor, k.			Density of body weighed	Correction factor, k.		
	Pt Ir weights	Brass weights	Quartz or Al weights		Pt Ir weights	Brass weights	Quartz or Al weights
.5	+2.34	+2.26	+1.95	1.6	+ .69	+ .61	+ .30
.6	+1.94	+1.86	+1.55	1.7	+ .65	+ .56	+ .25
.7	+1.66	+1.57	+1.26	1.8	+ .62	+ .52	+ .21
.75	+1.55	+1.46	+1.15	1.9	+ .58	+ .49	+ .18
.80	+1.44	+1.36	+1.05	2.0	+ .54	+ .46	+ .15
.85	+1.36	+1.27	+0.96	2.5	+ .43	+ .34	+ .03
.90	+1.28	+1.19	+ .88	3.0	+ .34	+ .26	— .05
.95	+1.21	+1.12	+ .81	4.0	+ .24	+ .16	— .15
1.00	+1.14	+1.06	+ .75	6.0	+ .14	+ .06	— .25
1.1	+1.04	+0.95	+ .64	8.0	+ .09	+ .01	— .30
1.2	+0.94	+ .86	+ .55	10.0	+ .06	— .02	— .33
1.3	+ .87	+ .78	+ .47	15.0	+ .03	— .06	— .37
1.4	+ .80	+ .71	+ .40	20.0	+ .004	— .08	— .39
1.5	+ .75	+ .66	+ .35	22.0	— .001	— .09	— .40

BUFFER SOLUTIONS
OPERATIONAL DEFINITIONS OF pH

Prepared by R. A. Robinson

The operational definition of pH is:

$$pH = pH(s) + E/k$$

where E is the e.m.f. of the cell:

$$H_2|Solution, pH|Saturated\ KCl|Solution, pH(s)|H_2$$

the half cell on the left containing the solution whose pH is being measured and that on the right a standard buffer mixture of known pH; $k = 2.303RT/F$, where R is the gas constant, T the temperature in degrees Kelvin and F the value of the faraday.

Alternatively, the cell:

$$Glass\ electrode|Solution, pH|Saturated\ calomel\ electrode$$

can be used, the glass electrode being calibrated using a standard buffer mixture or, if possible, two standard buffer mixtures whose pH values lie on either side of that of the solution which is being measured. Suitable standard buffer mixtures are:

0.05 M potassium hydrogen phthalate (pH = 4.008 at 25°C)
0.025 M potassium dihydrogen phosphate
0.025 M disodium hydrogen phosphate (pH = 6.865 at 25°C)
0.01 M borax (pH = 9.180 at 25°C)

For most purposes pH can be equated to $-\log_{10} \gamma_{H^+}m_{H^+}$, i.e., to the negative logarithm of the hydrogen ion activity. There is a small difference between those two quantities if pH > 9.2 or pH < 4.0, given by:

$$-\log \gamma_{H^+}m_{H^+} = pH + 0.014(pH - 9.2)\ for\ pH > 9.2$$
$$= pH + 0.009(4.0 - pH)\ for\ pH < 4.0$$

It should be noted that in the table titled "Solutions giving Round Values of pH at 25°C" it is $-\log \gamma_{H^+}m_{H^+}$ and not pH which is quoted when there is a difference between them.

References:
R. G. Bates, "Electrometric pH Determinations: Theory and Practice" Wiley, New York, 1954.
R. A. Robinson and R. H. Stokes, "Electrolyte Solutions," 2nd edition, Butterworths, London; Academic Press, Inc. New York, 1959. R. C. Bates, J. Res. of N.B.S. 66 A, 179 (1962).

National Bureau of Standards
R. G. Bates and S. F. Acree, Res. **34**, 373 (1945); W. J. Hammer, C. D. Pinching and S. F. Acree, ibid. **36**, 47 (1946); G. G. Manor, N. J. DeLollis, P. W. Lindwall and S. F. Acree, ibid., **36**, 543 (1946); R. G. Bates, ibid., **39**, 411 (1947); R. G. Bates, V. E. Bower, R. G. Miller and E. R. Smith, ibid., **47**, 433 (1951); V. E. Bower, R. G. Bates and E. R. Smith, ibid., **51**, 189 (1953); V. E. Bower and R. G. Bates, ibid., **55**, 197 (1955); R. G. Bates, V. E. Bower and E. R. Smith, ibid., **56**, 305 (1956); V. E. Bower and R. G. Bates, ibid., **59**, 261; R. G. Bates and V. E. Bower, Anal. Chem., **28**, 1322 (1956).

PROPERTIES OF STANDARD AQUEOUS BUFFER SOLUTIONS AT 25°C

Solution	Buffer substance	Molality m	Weight of salt in air per liter solution	Density g/ml	Molarity M	Dilution value $\Delta pH_{\frac{1}{2}}$	ΔpH_s[a]	Buffer value, equiv. per pH	Temp coeff., dpH_s/dt. Units per °C
Tetroxalate	$KH_3(C_2O_4)_2 \cdot 2H_2O$	0.05	12.61	1.0032	0.04962	+0.186	−0.0028	0.070	+0.001
Tartrate	$KHC_4H_4O_6$, sat. sol'n. at 25°C	0.0341	......	1.0036	0.034	+0.049	−0.0003	0.027	−0.0014
Phthalate	$KHC_8O_4H_4$	0.05	10.12	1.0017	0.04958	+0.052	−0.0009	0.016	+0.0012
Phosphate	$KH_2PO_4 +$ Na_2HPO_4	0.025[b]	3.39 3.53	1.0028	0.0249[b]	+0.080	−0.0006	0.029	−0.0028
Phosphate	$KH_2PO_4 +$ Na_2HPO_4	0.008695[c] 0.03043[d]	1.179 4.30	1.0020	0.008665[c] 0.03032[d]	+0.07[e]	−0.0005	0.016	−0.0028
Borax	$Na_2B_4O_7 \cdot 10H_2O$	0.01	3.80	0.9996	0.009971	+0.01	−0.0001	0.020	−0.0082
Calcium hydroxide	$Ca(OH)_2$, sat. sol'n. at 25°C	0.0203	......	0.9991	0.02025	−0.28	+0.0014	0.09	−0.033

[a] ΔpH_s = pHs (M Molar solution) − pHs (m molal solution).
[b] Concentration of each phosphate salt.
[c] KH_2PO_4.
[d] Na_2HPO_4.
[e] Calculated value.

SOLUTIONS GIVING ROUND VALUES OF pH AT 25°C

Reproduced from "Electrolyte Solutions" by permission from
Robinson and Stokes, authors, and Butterworth's Scientific
Publications.

A*		B*		C*		D*		E*	
pH	x	pH	x	pH	x	pH	x	pH	x
1.00	67.0	2.20	49.5	4.10	1.3	5.80	3.6	7.00	46.6
1.10	52.8	2.30	45.8	4.20	3.0	5.90	4.6	7.10	45.7
1.20	42.5	2.40	42.2	4.30	4.7	6.00	5.6	7.20	44.7
1.30	33.6	2.50	38.8	4.40	6.6	6.10	6.8	7.30	43.4
1.40	26.6	2.60	35.4	4.50	8.7	6.20	8.1	7.40	42.0
1.50	20.7	2.70	32.1	4.60	11.1	6.30	9.7	7.50	40.3
1.60	16.2	2.80	28.9	4.70	13.6	6.40	11.6	7.60	38.5
1.70	13.0	2.90	25.7	4.80	16.5	6.50	13.9	7.70	36.6
1.80	10.2	3.00	22.3	4.90	19.4	6.60	16.4	7.80	34.5
1.90	8.1	3.10	18.8	5.00	22.6	6.70	19.3	7.90	32.0
2.00	6.5	3.20	15.7	5.10	25.5	6.80	22.4	8.00	29.2
2.10	5.1	3.30	12.9	5.20	28.8	6.90	25.9	8.10	26.2
2.20	3.9	3.40	10.4	5.30	31.6	7.00	29.1	8.20	22.9
		3.50	8.2	5.40	34.1	7.10	32.1	8.30	19.9
		3.60	6.3	5.50	36.6	7.20	34.7	8.40	17.2
		3.70	4.5	5.60	38.8	7.30	37.0	8.50	14.7
		3.80	2.9	5.70	40.6	7.40	39.1	8.60	12.2
		3.90	1.4	5.80	42.3	7.50	40.9	8.70	10.3
		4.00	0.1	5.90	43.7	7.60	42.4	8.80	8.5
						7.70	43.5	8.90	7.0
						7.80	44.5	9.00	5.7
						7.90	45.3		
						8.00	46.1		

F*		G*		H*		I*		J*	
pH	x	pH	x	pH	x	pH	x	pH	x
8.00	20.5	9.20	0.9	9.60	5.0	10.90	3.3	12.00	6.0
8.10	19.7	9.30	3.6	9.70	6.2	11.00	4.1	12.10	8.0
8.20	18.8	9.40	6.2	9.80	7.6	11.10	5.1	12.20	10.2
8.30	17.7	9.50	8.8	9.90	9.1	11.20	6.3	12.30	12.8
8.40	16.6	9.60	11.1	10.00	10.7	11.30	7.6	12.40	16.2
8.50	15.2	9.70	13.1	10.10	12.2	11.40	9.1	12.50	20.4
8.60	13.5	9.80	15.0	10.20	13.8	11.50	11.1	12.60	25.6
8.70	11.6	9.90	16.7	10.30	15.2	11.60	13.5	12.70	32.2
8.80	9.6	10.00	18.3	10.40	16.5	11.70	16.2	12.80	41.2
8.90	7.1	10.10	19.5	10.50	17.8	11.80	19.4	12.90	53.0
9.00	4.6	10.20	20.5	10.60	19.1	11.90	23.0	13.00	66.0
9.10	2.0	10.30	21.3	10.70	20.2	12.00	26.9		
		10.40	22.1	10.80	21.2				
		10.50	22.7	10.90	22.0				
		10.60	23.3	11.00	22.7				
		10.70	23.8						
		10.80	24.25						

*A. 25 ml of 0.2 molar KCl + x ml of 0.2 molar HCl.
*B. 50 ml of 0.1 molar potassium hydrogen phthalate + x ml of 0.1 molar HCl.
*C. 50 ml of 0.1 molar potassium hydrogen phthalate + x ml of 0.1 molar NaOH.
*D. 50 ml of 0.1 molar potassium dihydrogen phosphate + x ml 0.1 molar NaOH.
*E. 50 ml of 0.1 molar tris(hydroxymethyl) aminomethane + x ml of 0.1 M HCl.
*F. 50 ml of 0.025 molar borax + x ml of 0.1 molar HCl.
*G. 50 ml of 0.025 molar borax + x ml of 0.1 molar NaOH.
*H. 50 ml of 0.05 molar sodium bicarbonate + x ml of 0.1 molar NaOH.
*I. 50 ml of 0.05 molar disodium hydrogen phosphate + x ml of 0.1 molar NaOH.
*J. 25 ml of 0.2 molar KCl + x ml of 0.2 molar NaOH.
Final Volume of Mixtures = 100 ml

STANDARD VALUES OF pH AT TEMPERATURE 0–95°C

Temperature	Tetroxalate 0.05 molal	Tartrate 0.0341 molal (sat'd at 25°C)	Phthalate 0.05 molal	Phosphate[a]	Phosphate[b]	Borax 0.01 molal	Calcium hydroxide (sat'd at 25°C)
0	1.666	....	4.003	6.984	7.534	9.464	13.423
5	1.668	....	3.999	6.951	7.500	9.395	13.207
10	1.670	....	3.998	6.923	7.472	9.332	13.003
15	1.672	....	3.999	6.900	7.448	9.276	12.810
20	1.675	....	4.002	6.881	7.429	9.225	12.627
25	1.679	3.557	4.008	6.865	7.413	9.180	12.454
30	1.683	3.552	4.015	6.853	7.400	9.139	12.289
35	1.688	3.549	4.024	6.844	7.389	9.102	12.133
38	1.691	3.548	4.030	6.840	7.384	9.081	12.043
40	1.694	3.547	4.035	6.838	7.380	9.068	11.984
45	1.700	3.547	4.047	6.834	7.373	9.038	11.841
50	1.707	3.549	4.060	6.833	7.367	9.011	11.705
55	1.715	3.554	4.075	6.834	....	8.985	11.574
60	1.723	3.560	4.091	6.836	....	8.962	11.449
70	1.743	3.580	4.126	6.845	....	8.921	....
80	1.766	3.609	4.164	6.859	....	8.885	....
90	1.792	3.650	4.205	6.877	....	8.850	....
95	1.806	3.674	4.227	6.886	....	8.833	....

[a] Solution 0.025 m KH_2PO_4 and 0.025 m Na_2HPO_4.
[b] Solution 0.008695 m KH_2PO_4 and 0.03043 m Na_2HPO_4.

APPROXIMATE pH VALUES

The following tables give approximate pH values for a number of substances such as acids, bases, foods, biological fluids, etc. All values are rounded off to the nearest tenth and are based on measurements made at 25° C. A few buffer systems with their pH values are also given.

From Modern pH and Chlorine Control, W. A. Taylor & Co., by permission

ACIDS

Hydrochloric, N	0.1	Oxalic, 0.1N	1.6	Acetic, 0.01N	3.4
Hydrochloric, 0.1N	1.1	Tartaric, 0.1N	2.2	Benzoic, 0.01N	3.1
Hydrochloric, 0.01N	2.0	Malic, 0.1N	2.2	Alum, 0.1N	3.2
Sulfuric, N	0.3	Citric, 0.1N	2.2	Carbonic (saturated)	3.8
Sulfuric, 0.1N	1.2	Formic, 0.1N	2.3	Hydrogen sulfide, 0.1N	4.1
Sulfuric, 0.01N	2.1	Lactic, 0.1N	2.4	Arsenious (saturated)	5.0
Orthophosphoric, 0.1N	1.5	Acetic, N	2.4	Hydrocyanic, 0.1N	5.1
Sulfurous, 0.1N	1.5	Acetic, 0.1N	2.9	Boric, 0.1N	5.2

BASES

Sodium hydroxide, N	14.0	Lime (saturated)	12.4	Magnesia (saturated)	10.5
Sodium hydroxide, 0.1N	13.0	Trisodium phosphate, 0.1N	12.0	Sodium sesquicarbonate, 0.1M	10.1
Sodium hydroxide, 0.01N	12.0	Sodium carbonate, 0.1N	11.6	Ferrous hydroxide (saturated)	9.5
Potassium hydroxide, N	14.0	Ammonia, N	11.6	Calcium carbonate (saturated)	9.4
Potassium hydroxide, 0.1N	13.0	Ammonia, 0.1N	11.1	Borax, 0.1N	9.2
Potassium hydroxide, 0.01N	12.0	Ammonia, 0.01N	10.6	Sodium bicarbonate, 0.1N	8.4
Sodium metasilicate, 0.1N	12.6	Potassium cyanide, 0.1N	11.0		

BIOLOGIC MATERIALS

Blood, plasma, human	7.3–7.5	Gastric contents, human	1.0–3.0	Milk, human	6.6–7.6
Spinal fluid, human	7.3–7.5	Duodenal contents, human	4.8–8.2	Bile, human	6.8–7.0
Blood, whole, dog	6.9–7.2	Feces, human	4.6–8.4		
Saliva, human	6.5–7.5	Urine, human	4.8–8.4		

FOODS

Apples	2.9–3.3	Gooseberries	2.8–3.0	Potatoes	5.6–6.0
Apricots	3.6–4.0	Grapefruit	3.0–3.3	Pumpkin	4.8–5.2
Asparagus	5.4–5.8	Grapes	3.5–4.5	Raspberries	3.2–3.6
Bananas	4.5–4.7	Hominy (lye)	6.8–8.0	Rhubarb	3.1–3.2
Beans	5.0–6.0	Jams, fruit	3.5–4.0	Salmon	6.1–6.3
Beers	4.0–5.0	Jellies, fruit	2.8–3.4	Sauerkraut	3.4–3.6
Beets	4.9–5.5	Lemons	2.2–2.4	Shrimp	6.8–7.0
Blackberries	3.2–3.6	Limes	1.8–2.0	Soft drinks	2.0–4.0
Bread, white	5.0–6.0	Maple syrup	6.5–7.0	Spinach	5.1–5.7
Butter	6.1–6.4	Milk, cows	6.3–6.6	Squash	5.0–5.4
Cabbage	5.2–5.4	Olives	3.6–3.8	Strawberries	3.0–3.5
Carrots	4.9–5.3	Oranges	3.0–4.0	Sweet potatoes	5.3–5.6
Cheese	4.8–6.4	Oysters	6.1–6.6	Tomatoes	4.0–4.4
Cherries	3.2–4.0	Peaches	3.4–3.6	Tuna	5.9–6.1
Cider	2.9–3.3	Pears	3.6–4.0	Turnips	5.2–5.6
Corn	6.0–6.5	Peas	5.8–6.4	Vinegar	2.4–3.4
Crackers	6.5–8.5	Pickles, dill	3.2–3.6	Water, drinking	6.5–8.0
Dates	6.2–6.4	Pickles, sour	3.0–3.4	Wines	2.8–3.8
Eggs, fresh white	7.6–8.0	Pimento	4.6–5.2		
Flour, wheat	5.5–6.5	Plums	2.8–3.0		

Indicator	Approximate pH range	Color-change	Preparation
Methyl Violet	0.0–1.6	yel to bl	0.01–0.05% in water
Crystal Violet	0.0–1.8	yel to bl	0.02% in water
Ethyl Violet	0.0–2.4	yel to bl	0.1 g in 50 ml of MeOH + 50 ml of water
Malachite Green	0.2–1.8	yel to bl grn	water
Methyl Green	0.2–1.8	yel to bl	0.1% in water
2-(p-dimethylaminophenylazo)pyridine	0.2–1.8	yel to bl	0.1% in EtOH
	4.4–5.6	red to yel	
o-Cresolsulfonephthalein (Cresol Red)	0.4–1.8	yel to red	0.1 g in 26.2 ml 0.01N
	7.0–8.8	yel to red	NaOH + 223.8 ml water
Quinaldine Red	1.0–2.2	col to red	1% in EtOH
p-(p-dimethylaminophenylazo)-benzoic acid, Na-salt (Paramethyl Red)	1.0–3.0	red to yel	EtOH
m-(p-anilnophenylazo)benzene sulfonic acid, Na-salt (Metanil Yellow)	1.2–2.4	red to yel	0.01% in water
4-Phenylazodiphenylamine	1.2–2.6	red to yel	0.01 g in 1 ml 1N HCl + 50 ml EtOH + 49 ml water
Thymolsulfonephthalein (Thymol Blue)	1.2–2.8	red to yel	0.1 g in 21.5 ml
	8.0–9.6	yel to bl	0.01N NaOH + 229.5 ml water
m-Cresolsulfonephthalein (Metacresol Purple)	1.2–2.8	red to yel	0.1 g in 26.2 ml
	7.4–9.0	yel to purp	0.01N NaOH + 223.8 ml water
p-(p-anilinophenylazo)benzenesulfonic acid, Na-salt (Orange IV)	1.4–2.8	red to yel	0.01% in water
4-o-Tolylazo-o-toluidine	1.4–2.8	or to yel	water
Erythrosine, disodium salt	2.2–3.6	or to red	0.1% in water
Benzopurpurine 48	2.2–4.2	vt to red	0.1% in water
N,N-dimethyl-p-(m-tolylazo)aniline	2.6–4.8	red to yel	0.1% in water
4,4'-Bix(2-amino-1-naphthylazo)2,2'-stil-benedisulfonic acid	3.0–4.0	purp to red	0.1 g in 5.9 ml 0.05N NaOH + 94.1 ml water
Tetrabromophenolphthaleinethyl ester, K-salt	3.0–4.2	yel to bl	0.1% in EtOH
3',3'',5',5''-tetrabromophenol-sulfonephthalein (Bromophenol Blue)	3.0–4.6	yel to bl	0.1 g in 14.9 ml 0.01N NaOH + 235.1 ml water
2,4-Dinitrophenol	2.8–4.0	col to yel	saturated water solution
N,N-Dimethyl-p-phenylazoaniline (p-Dimethylaminoazobenzene)	2.8–4.4	red to yel	0.1 g in 90 ml in EtOH + 10 ml water
Congo Red	3.0–5.0	blue to red	0.1% in water
Methyl Orange-Xylene Cyanole solution	3.2–4.2	purp to grn	ready solution
Methyl Orange	3.2–4.4	red to yel	0.01% in water
Ethyl Orange	3.4–4.8	red to yel	0.05–0.2% in water or aqueous EtOH
4-(4-Dimethylamino-1-naphthylazo)-3-methoxybenzenesulfonic acid	3.5–4.8	vt to yel	0.1% in 60% EtOH
3',3'',5',5''-Tetrabromo-m-cresol-sulfonephthalein (Bromocresol Green)	3.8–5.4	yel to blue	0.1 g in 14.3 ml 0.01N NaOH + 235.7 ml water
Resazurin	3.8–6.4	or to vt	water
4-Phenylazo-1-naphthylamine	4.0–5.6	red to yel	0.1% in EtOH
Ethyl Red	4.0–5.8	col to red	0.1 g in 50 ml MeOH + 50 ml water
2-(p-Dimethylaminophenylazo)-pyridine	0.2–1.8	yel to red	0.1% in EtOH
	4.4–5.6	red to yel	
4-(p-ethoxyphenylazo)-m-phenylene-diamine monohydrochloride	4.4–5.8	or to yel	0.1% in water
Lacmoid	4.4–6.2	red to bl	0.2% in EtOH
Alizarin Red S	4.6–6.0	yel to red	dilute solution in water
Methyl Red	4.8–6.0	red to yel	0.02 g in 60 ml EtOH + 40 ml water

ACID BASE INDICATORS (Continued)

Indicator	Approximate pH range	Color-change	Preparation
Propyl Red	4.8–6.6	red to yel	EtOH
5',5''-Dibromo-o-cresolsulfone-phthalein (Bromocresol Purple)	5.2–6.8	yel to purp	0.1 g in 18.5 ml 0.01N NaOH + 231.5 ml water
3',3''-Dichlorophenolsulfonephthalein (Chlorophenol Red)	5.2–6.8	yel to red	0 1 g in 23.6 ml 0.01N NaOH + 226.4 ml water
p-Nitrophenol	5.4–6.6	col to yel	0.1% in water
Alizarin	5.6–7.2	yel to red	0.1% in MeOH
	11.0–12.4	red to purp	
2-(2,4-Dinitrophenylazo)-1-naphthol-3, 6-disulfonic acid, di-Na salt	6.0–7.0	yel to bl	0.1% in water
3',3''-Dibromothymolsulfonephthalein (Bromothymol Blue)	6.0–7.6	yel to bl	0.1 g in 16 ml 0.01N NaOH + 234 ml water
6,8-Dinitro-2,4-(1H)quinazolinedione (m-Dinitrobenzoylene urea)	6.4–8.0	col to yel	25 g in 115 ml M NaOH + 50 ml boiling water 0.292 g of NaCl in 100 ml water
Brilliant Yellow	6.6–7.8	yel to or	1% in water
Phenolsulfonephthalein (Phenol Red)	6.6–8.0	yel to red	0.1 g in 28.2 ml 0.01N NaOH + 221.8 ml water
Neutral Red	6.8–8.0	red to amb	0.01 g in 50 ml EtOH + 50 ml water
m-Nitrophenol	6.8–8.6	col to yel	0.3% in water
o-Cresolsulfonephthalein (Cresol Red)	0.0–1.0	red to yel	0.1 g in 26.2 ml 0.01N NaOH + 223.8 ml water
	7.0–8.8	yel to red	
Curcumin	7.4–8.6	yel to red	EtOH
	10.2–11.8		
m-Cresolsulfonephthalein (Metacresol Purple)	1.2–2.8	red to yel	0.1 g in 26.2 ml 0.01N NaOH + 223.8 ml water
	7.4–9.0	yel to purp	
4,4'-Bis(4-amino-1-naphthylazo) 2,2'stilbene disulfonic acid	8.0–9.0	bl to red	0.1 g in 5.9 ml 0.05N NaOH + 94.1 ml water
Thymolsulfonephthalein (Thymol Blue)	1.2–2.8	red to yel	0.1 g in 21.5 ml 0.01N NaOH + 228.5 ml water
	8.0–9.6		
o-Cresolphthalein	8.2–9.8	col to red	0.04% in EtOH
p-Naphtholbenzene	8.2–10.0	or to bl	1% in dil. alkali
Phenolphthalein	8.2–10.0	col to pink	0.05 g in 50 ml EtOH + 50 ml water
Ethyl-bis(2,4-dimethylphenyl)acetate	8.4–9.6	col to bl	saturated solution in 50% acetone alcohol
Thymolphthalein	9.4–10.6	col to bl	0.04 g in 50 ml EtOH + 50 ml water
5-(p-Nitrophenylazo)salicylic acid, Na-salt (Alizarin Yellow R)	10.1–12.0	yel to red	0.01% in water
p-(2,4-Dihydroxyphenylazo)benzene-sulfonic acid, Na-salt	11.4–12.6	yel to or	0.1% in water
5,5'-Indigodisulfonic acid, di-Na-salt	11.4–13.0	bl to yel	water
2,4,6-Trinitrotoluene	11.5–13.0	col to or	0.1–0.5% in EtOH
1,3,5-Trinitrobenzene	12.0–14.0	col to or	0.1–0.5% in EtOH
Clayton Yellow	12.2–13.2	yel to amb	0.1% in water

FLUORESCENT INDICATORS

Jack DeMent

Fluorescent indicators are substances which show definite changes in fluorescence with change in pH. Some fluorescent materials are not suitable for indicators since their change in fluorescence is too gradual. Fluorescent indicators find greatest utility in the titration of opaque, highly turbid or deeply colored solutions. A long wavelength ultraviolet ("black light") lamp in a dimly lighted room provides the best environment for titrations involving fluorescent indicators, although bright daylight is sometimes sufficient to evoke a response in the bright green, yellow and orange fluorescent indicators. Titrations are carried out in non-fluorescent glassware. One should check the glassware prior to use to make certain that it does not fluoresce due to the wavelengths of light involved in the titration. The meniscus of the liquid in the burette can be followed when a few particles of an insoluble fluorescent solid are dropped onto its surface.

In this table the indicators are arranged by approximate pH range covered. In the case of some of the dyestuffs the end point may vary slightly with the source or manufacturer.

pH 0 to 2

Indicator	C.I.	From pH	To pH
Benzoflavine	—	0.3, yellow fl.	1.7, green fl.
3,6-Dioxyphthalimide	—	0, blue fl.	2.4, green fl.
Eosine YS	768	0, yellow colored	3.0, yellow fl.
Erythrosine	772	0, yellow colored	3.6, yellow fl.
Esculin	—	1.5, colorless	2, blue fl.
4-Ethoxyacridone	—	1.2, green fl.	3.2, blue fl.
3,6-Tetramethyldiaminooxanthone	—	1.2, green fl.	3.4, blue fl.

pH 2 to 4

Indicator	C.I.	From pH	To pH
Chromotropic acid	—	3.5, colorless	4.5, blue fl.
Fluorescein	766	4, colorless	4.5, green fl.
Magdala Red	—	3.0, purple colored	4.0, fl.
α-Naphthylamine	—	3.4, colorless	4.8, blue fl.
β-Naphthylamine	—	2.8, colorless	4.4, violet fl.
Phloxine	774	3.4, colorless	5.0, bright yellow fl.
Salicylic acid	—	2.5, colorless	3.5, blue fl.

pH 4 to 6

Indicator	C.I.	From pH	To pH
Acridine	788	4.9, green fl.	5.1, violet colored
Dichlorofluorescein	—	4.0, colorless	5.0, green fl.
3,6-Dioxyxanthone	—	5.4, colorless	7.6, blue-violet fl.
Erythrosine	772	4.0, colorless	4.5, yellow-green fl.
β-Methylesculetin	—	4.0, colorless	6.2, blue fl.
Neville-Winther acid	—	6.0, colorless	6.5, blue fl.
Resorufin	—	4.4, yellow fl.	6.4, weak orange fl.
Quininic acid	—	4.0, yellow colored	5.0, blue fl.
Quinine [first end point]	—	5.0, blue fl.	6.1, violet fl.

pH 6 to 8

Indicator	C.I.	From pH	To pH
Acid R Phosphine	—	(claimed for range pH 6.0–7.0)	
Brilliant Diazol Yellow	—	6.5, colorless	7.5, violet fl.
Cleves acid	—	6.5, colorless	7.5, green fl.
Coumaric acid	—	7.2, colorless	9.0, green fl.
3,6-Dioxyphthalic dinitrile	—	5.8, blue fl.	8.2, green fl.
Magnesium 8-hydroxyquinolinate	—	6.5, colorless	7.5, golden fl.
β-Methylumbelliferone	—	7.0, colorless	7.5, blue fl.
1-Naphthol-4-sulfonic acid	—	6.0, colorless	6.5, blue fl.
Orcinaurine	—	6.5, colorless	8.0, green fl.
Patent Phosphine	789	(for the range pH 6.0–7.0, green-yellow fl.)	
Thioflavine	816	(for the region pH 6.5–7.0, yellow fl.)	
Umbelliferone	—	6.5, colorless	7.6, blue fl.

FLUORESCENT INDICATORS (Continued)

pH 8 to 10

Indicator	C.I.	From pH	To pH
Acridine Orange	788	8.4, orange colored	10.4, green fl.
Ethoxyphenylnaphthostilbazonium chloride	—	9, green fl.	11, non-fl.
G Salt	—	9.0, dull blue fl.	9.5, bright blue fl.
Naphthazol derivatives	—	8.2, colorless	10.0, yellow or green fl.
α-Naphthionic acid	—	9, blue fl.	11, green fl.
2-Naphthol-3,6-disulfonic acid	—	9.5, dark blue fl.	Light blue fl. at higher pH
β-Naphthol	—	8.6, colorless	Blue fl. at higher pH
α-Naphtholsulfonic acid	—	8.0, dark blue fl.	9.0, bright violet fl.
1,4-Naphtholsulfonic acid	—	8.2, dark blue fl.	Light blue fl. at higher pH
Orcinsulfonphthalein	—	8.6, yellow colored	10.0 fl.
Quinine [second end point]	—	9.5, violet fl.	10.0, colorless
R-Salt	—	9.0, dull blue fl.	9.5, bright blue fl.
Sodium 1-naphthol-2-sulfonate	—	9.0, dark blue fl.	10.0, bright violet fl.

pH 10 to 12

Indicator	C.I.	From pH	To pH
Coumarin	—	9.8, deep green fl.	12, light green fl.
Eosine BN	771	10.5, colorless	14.0, yellow fl.
Papaverine (permanganate oxidized)	—	9.5, yellow fl.	11.0, blue fl.
Schaffers Salt	—	5.0, violet fl.	11.0, green-blue fl.
SS-Acid (sodium salt)	—	10.0, violet fl.	12.0, yellow colored

pH 12 to 14

Indicator	C.I.	From pH	To pH
Cotarnine	—	12.0, yellow fl.	13.0, white fl.
α-Naphthionic acid	—	12, blue fl.	13, green fl.
β-Naphthionic acid	—	12, blue fl.	13, violet fl.

FORMULAS FOR CALCULATING TITRATION DATA, pH VS. ML. OF REAGENT

A Substance Titrated V_0 ml. of solution M_0 its molarity	B Initial $[H^+]$ or $[OH^-]$	C Intermediate Points 10, 50, 99, etc., per cent neutralized, V_1 ml. of reagent of M_1 molarity added	D Equivalence Point	E Excess of Reagent V_1 volume reagent M_1 its molarity V_T total volume
(1) Strong Acid	$[H^+] = M_0$	$[H^+] = \dfrac{V_0 M_0 - V_1 M_1}{V_0 + V_1}$	$\sqrt{K_w}$	$[OH^-] = \dfrac{V_1 M_1}{V_T}$
(2) Strong Base	$[OH^-] = M_0$	$[OH^-] = \dfrac{V_0 M_0 - V_1 M_1}{V_0 + V_1}$	$\sqrt{K_w}$	$[H^+] = \dfrac{V_1 M_1}{V_T}$
(3) Weak Acid ($K_a = 10^{-5}$ to 10^{-8})	$[H^+] = \sqrt{M_0 K_a}$	$[H^+] = \dfrac{[Acid]}{[Salt]} K_a$	$[OH^-] = \sqrt{\dfrac{K_w}{K_a} c}$	$[OH^-] = \dfrac{V_1 M_1}{V_T}$ (Value in column D to be added)
(4) Weak Base ($K_b = 10^{-5}$ to 10^{-8})	$[OH^-] = \sqrt{M_0 K_b}$	$[OH^-] = \dfrac{[Base]}{[Salt]} K_b$	$[H^+] = \sqrt{\dfrac{K_w}{K_b} c}$	$[H^+] = \dfrac{V_1 M_1}{V_T}$ (Value in column D to be added)
(5) Salt of a Very Weak Acid (e.g. KCN)	$[OH^-] = \sqrt{\dfrac{K_w}{K_a} c}$	$[H^+] = \dfrac{[Acid]}{[Salt]} K_a$	$[H^+] = \sqrt{[Acid] K_a}$	$[H^+] = \dfrac{V_1 M_1}{V_T}$ (Correct for value in column D)
(6) Salt of a Very Weak Base	$[H^+] = \sqrt{\dfrac{K_w}{K_b} c}$	$[OH^-] = \dfrac{[Base]}{[Salt]} K_b$	$[OH^-] = \sqrt{[Base] K_b}$	$[OH^-] = \dfrac{V_1 M_1}{V_T}$ (Add to $[OH^-]$ found in column D)

A = Weight per cent of solute   G = Molality
B = Molecular weight of solvent   M = Molarity
E = Molecular weight of solute   N = Mole fraction
F = Grams of solute per liter of solution   R = Density of solution grams per cc

Concentration of solute— SOUGHT	Concentration of solute—GIVEN				
	A	N	G	M	F
A	—	$\dfrac{100N \times E}{N \times E + (1-N)B}$	$\dfrac{100G \times E}{1000 + G \times E}$	$\dfrac{M \times E}{10R}$	$\dfrac{F}{10R}$
N	$\dfrac{\frac{A}{E}}{\frac{A}{E} + \frac{100 - A}{B}}$	—	$\dfrac{B \times G}{B \times G + 1000}$	$\dfrac{B \times M}{M(B-E) + 1000R}$	$\dfrac{B \times F}{F(B-E) + 1000R \times E}$
G	$\dfrac{1000A}{E(100-A)}$	$\dfrac{1000N}{B - N \times B}$	—	$\dfrac{1000M}{1000R - (M \times E)}$	$\dfrac{1000F}{E(1000R - F)}$
M	$\dfrac{10R \times A}{E}$	$\dfrac{1000R \times N}{N \times E + (1-N)B}$	$\dfrac{1000R \times G}{1000 + E \times G}$	—	$\dfrac{F}{E}$
F	$10AR$	$\dfrac{1000R \times N \times E}{N \times E + (1-N)B}$	$\dfrac{1000R \times G \times E}{1000 + G \times E}$	$M \times E$	—

ELECTROCHEMICAL SERIES

Petr Vanýsek

There are three tables for this Electrochemical Series. Each table lists standard reduction potentials, E° values, at 298.15 K (25°C), and at a pressure of 101.325 kPa (1 atm.). Table 1 is an alphabetical listing of the elements according to the symbols for the elements. Thus, data for Silver (Ag) precedes those for Aluminum (Al). Table 2 lists only those reduction reactions which have E° values positive to the potential of the Standard Hydrogen Electrode. In Table 2, the reactions are listed in the order of increasing positive potential and range from 0.000 V to +3.053 V. Table 3 lists only those reduction reactions which have E° values negative to the potential of the Standard Hydrogen Electrode. In Table 3, reactions are listed in the order of increasing negative potential and range from −0.017 to −4.10 V.

Table 1
ALPHABETICAL LISTING

Reaction	E°, V	Reaction	E°, V
$Ag^+ + e \rightleftharpoons Ag$	0.7996	$Ag_2WO_4 + 2\,e \rightleftharpoons 2\,Ag + WO_4^{2-}$	0.4660
$Ag^{2+} + e \rightleftharpoons Ag^+$	1.980	$Al^{3+} + 3\,e \rightleftharpoons Al$	−1.662
$Ag(ac) + e \rightleftharpoons Ag + (ac)^-$	0.643	$H_2AlO_3^- + H_2O + 3\,e \rightleftharpoons Al + 4\,OH^-$	−2.33
$AgBr + e \rightleftharpoons Ag + Br^-$	0.07133	$AlF_6^{3-} + 3\,e \rightleftharpoons Al + 6\,F^-$	−2.069
$AgBrO_3 + e \rightleftharpoons Ag + BrO_3^-$	0.546	$As + 3\,H^+ + 3\,e \rightleftharpoons AsH_3$	−0.608
$Ag_2C_2O_4 + 2\,e \rightleftharpoons 2Ag + C_2O_{44}^{2-}$	0.4647	$As_2O_3 + 6\,H^+ + 6\,e \rightleftharpoons 2\,As + 3\,H_2O$	0.234
$gCl + e \rightleftharpoons Ag + Cl^-$	0.22233	$HAsO_2 + 3\,H^+ + 3\,e \rightleftharpoons As + 2\,H_2O$	0.248
$AgCN + e \rightleftharpoons Ag + CN^-$	−0.017	$AsO_2^- + 2\,H_2O + 3\,e \rightleftharpoons As + 4\,OH^-$	−0.68
$Ag_2CO_3 + 2\,e \rightleftharpoons 2\,Ag + CO_3^{2-}$	0.47	$H_3AsO_4 + 2\,H^+ + 2\,e^- \rightleftharpoons HAsO_2 + 2\,H_2O$	0.560
$Ag_2CrO_4 + 2\,e \rightleftharpoons 2\,Ag + CrO_4^{2-}$	0.4470	$AsO_4^{3-} + 2\,H_2O + 2\,e \rightleftharpoons AsO_2^- + 4\,OH^-$	−0.71
$AgF + e \rightleftharpoons Ag + F^-$	0.779	$Au^+ + e \rightleftharpoons Au$	1.692
$Ag_4[Fe(CN)_6] + 4\,e \rightleftharpoons 4\,Ag + [Fe(CN)_6]^{4-}$	0.1478	$Au^{3+} + 2\,e \rightleftharpoons Au^+$	1.401
$AgI + e \rightleftharpoons Ag + I^-$	−0.15224	$Au^{3+} + 3\,e \rightleftharpoons Au$	1.498
$AgIO_3 + e \rightleftharpoons Ag + IO_3^-$	0.354	$AuBr_2^- + e \rightleftharpoons Au + 2\,Br^-$	0.959
$Ag_2MoO_4 + 2\,e \rightleftharpoons 2\,Ag + MoO_4^{2-}$	0.4573	$AuBr_4^- + 3\,e \rightleftharpoons Au + 4\,Br^-$	0.854
$AgNO_2 + e \rightleftharpoons Ag + NO_2^-$	0.564	$AuCl_4^- + 3\,e \rightleftharpoons Au + 4\,Cl^-$	1.002
$Ag_2O + H_2O + 2\,e \rightleftharpoons 2\,Ag + 2\,OH^-$	0.342	$Au(OH)_3 + 3\,H^+ + 3\,e \rightleftharpoons Au + 3\,H_2O$	1.45
$Ag_2O_3 + H_2O + 2\,e \rightleftharpoons 2\,AgO + 2\,OH^-$	0.739	$H_2BO_3^- + 5\,H_2O + 8\,e \rightleftharpoons BH_4^- + 8\,OH^-$	−1.24
$2\,AgO + H_2O + 2\,e \rightleftharpoons Ag_2O + 2\,OH^-$	0.607	$H_2BO_3^- + H_2O + 3\,e \rightleftharpoons B + 4\,OH^-$	−1.79
$AgOCN + e \rightleftharpoons Ag + OCN^-$	0.41	$H_3BO_3 + 3\,H^+ + 3\,e \rightleftharpoons B + 3\,H_2O$	−0.8698
$Ag_2S + 2\,e \rightleftharpoons 2\,Ag + S^{2-}$	−0.691	$Ba^{2+} + 2\,e \rightleftharpoons Ba$	−2.912
$Ag_2S + 2\,H^+ + 2\,e \rightleftharpoons 2\,Ag + H_2S$	−0.0366	$Ba^{2+} + 2\,e \rightleftharpoons Ba(Hg)$	−1.570
$AgSCN + e \rightleftharpoons Ag + SCN^-$	0.08951	$Ba(OH)_2 + 2\,e \rightleftharpoons Ba + 2\,OH^-$	−2.99
$Ag_2SeO_3 + 2\,e \rightleftharpoons 2\,Ag + SeO_3^{2-}$	0.3629	$Be^{2+} + 2\,e \rightleftharpoons Be$	−1.847
$Ag_2SO_4 + 2\,e \rightleftharpoons 2\,Ag + SO_4^{2-}$	0.654	$Be_2O_3^{2-} + 3\,H_2O + 4\,e \rightleftharpoons 2\,Be + 6\,OH^-$	−2.63

Table 1 (continued)
ALPHABETICAL LISTING

Reaction	E°, V	Reaction	E°, V
p-benzoquinone + 2 H^+ + 2 e $\rightleftharpoons$ hydroquinone	0.6992	Co^{3+} + e $\rightleftharpoons$ Co^{2+} (2 mol/ℓ H_2SO_4)	1.83
$BiCl_4^-$ + 3 e $\rightleftharpoons$ Bi + 4 Cl^-	0.16	$[Co(NH_3)_6]^{3+}$ + e $\rightleftharpoons$ $[Co(NH_3)_6]^{2+}$	0.108
Bi_2O_3 + 3 H_2O + 6 e $\rightleftharpoons$ 2 Bi + 6 OH^-	−0.46	$Co(OH)_2$ + 2 e $\rightleftharpoons$ Co + 2 OH^-	−0.73
Bi_2O_4 + 4 H^+ + 2 e $\rightleftharpoons$ 2 BiO^+ + 2 H_2O	1.593	$Co(OH)_3$ + e $\rightleftharpoons$ $Co(OH)_2$ + OH^-	0.17
BiO^+ + 2 H^+ + 3 e $\rightleftharpoons$ Bi + H_2O	0.320	CO_2 + 2 H^+ + 2 e $\rightleftharpoons$ HCOOH	−0.199
BiOCl + 2 H^+ + 3 e $\rightleftharpoons$ Bi + Cl^- + H_2O	0.1583	Cr^{2+} + 2 e $\rightleftharpoons$ Cr	−0.913
Br_2(aq) + 2 e $\rightleftharpoons$ 2 Br^-	1.0873	Cr^{3+} + e $\rightleftharpoons$ Cr^{2+}	−0.407
Br_2(1) + 2 e $\rightleftharpoons$ 2 Br^-	1.066	Cr^{3+} + 3 e $\rightleftharpoons$ Cr	−0.744
HBrO + H^+ + 2 e $\rightleftharpoons$ Br^- + H_2O	1.331	$Cr_2O_7^{2-}$ + 14 H^+ + 6 e $\rightleftharpoons$ 2 Cr^{3+} + 7 H_2O	1.232
HBrO + H^+ + e $\rightleftharpoons$ 1/2Br_2(aq) + H_2O	1.574	CrO_2^- + 2 H_2O + 3 e $\rightleftharpoons$ Cr + 4 OH^-	−1.2
HBrO + H^+ + e $\rightleftharpoons$ 1/2Br_2(ℓ) + H_2O	1.596	$HCrO_4^-$ + 7 H^+ + 3 e $\rightleftharpoons$ Cr^{3+} + 4 H_2O	1.350
BrO^- + H_2O + 2 e $\rightleftharpoons$ Br^- + 2 OH^-	0.761	CrO_4^- + 4 H_2O + 3 e $\rightleftharpoons$ $Cr(OH)_3$ + 5 OH^-	−0.13
BrO_3^- + 6 H^+ + 5 e $\rightleftharpoons$ 1/2Br_2 + 3 H_2O	1.482	$Cr(OH)_3$ + 3 e $\rightleftharpoons$ Cr + 3 OH^-	−1.48
BrO_3^- + 6 H^+ + 6 e $\rightleftharpoons$ Br^- + 3 H_2O	1.423	Cs^+ + e $\rightleftharpoons$ Cs	−2.92
BrO_3^- + 3 H_2O + 6 e $\rightleftharpoons$ Br^- + 6 OH^-	0.61	Cu^+ + e $\rightleftharpoons$ Cu	0.521
Ca^+ + e $\rightleftharpoons$ Ca	−3.80	Cu^{2+} + e $\rightleftharpoons$ Cu^+	0.153
Ca^{2+} + 2 e $\rightleftharpoons$ Ca	−2.868	Cu^{2+} + 2 e $\rightleftharpoons$ Cu	0.3419
Calomel electrode, 1 mol/ℓ KCl	0.2800	Cu^{2+} + 2 e $\rightleftharpoons$ Cu(Hg)	0.345
Calomel electrode, 1 mol/ℓ KCl (NCE)	0.2801	Cu^{2+} + 2 CN^- + e $\rightleftharpoons$ $[Cu(CN)_2]^-$	1.103
Calomel electrode, 0.1 mol/ℓ KCl	0.3337	CuI_2^- + e $\rightleftharpoons$ Cu + 2 I^-	0.00
Calomel electrode, saturated KCl (SCE)	0.2412	Cu_2O + H_2O + 2 e $\rightleftharpoons$ 2 Cu + 2 OH^-	−0.360
Calomel electrode, saturated NaCl (SSCE)	0.2360	$Cu(OH)_2$ + 2 e $\rightleftharpoons$ Cu + 2 OH^-	−0.222
$Ca(OH)_2$ + 2 e $\rightleftharpoons$ Ca + 2 OH^-	−3.02	2 $Cu(OH)_2$ + 2 e $\rightleftharpoons$ Cu_2O + 2 OH^- + H_2O	−0.080
Cd^{2+} + 2 e $\rightleftharpoons$ Cd	−0.4030	D^+ + e $\rightleftharpoons$ 1/2D_2	−0.0034
Cd^{2+} + 2 e $\rightleftharpoons$ Cd(Hg)	−0.3521	2 D^+ + 2 e $\rightleftharpoons$ D_2	−0.044
$Cd(OH)_2$ + 2 e $\rightleftharpoons$ Cd(Hg) + 2 OH^-	−0.809	Eu^{2+} + 2 e $\rightleftharpoons$ Eu	−3.395
$CdSO_4$ + 2 e $\rightleftharpoons$ Cu + SO_4^{2-}	−0.246	Eu^{3+} + 3 e $\rightleftharpoons$ Eu	−2.407
Ce^{3+} + 3 e $\rightleftharpoons$ Ce	−2.483	Eu^{3+} + e $\rightleftharpoons$ Eu^{2+}	−0.36
Ce^{3+} + 3 e $\rightleftharpoons$ Ce(Hg)	−1.4373	F_2 + 2 H^+ + 2 e $\rightleftharpoons$ 2 HF	3.053
Ce^{4+} + e $\rightleftharpoons$ Ce^{3+}	1.61	F_2 + 2 e $\rightleftharpoons$ 2 F^-	2.866
$CeOH^{3+}$ + H^+ + e $\rightleftharpoons$ Ce^{3+} + H_2O	1.715	F_2O + 2 H^+ + 4 e $\rightleftharpoons$ H_2O + 2 F^-	2.153
Cl_2(g) + 2 e $\rightleftharpoons$ Cl^-	1.35827	Fe^{2+} + 2 e $\rightleftharpoons$ Fe	−0.447
HClO + H^+ + e $\rightleftharpoons$ 1/2Cl_2 + H_2O	1.611	Fe^{3+} + 3 e $\rightleftharpoons$ Fe	−0.037
HClO + H^+ 2 e $\rightleftharpoons$ Cl^- + H_2O	1.482	Fe^{3+} + e $\rightleftharpoons$ Fe^{2+}	0.771
ClO^- + H_2O + 2 e $\rightleftharpoons$ Cl^- + 2 OH^-	0.81	$[Fe(CN)_6]^{3-}$ + e $\rightleftharpoons$ $[Fe(CN)_6]^{4-}$	0.358
ClO_2 + H^+ + e $\rightleftharpoons$ $HClO_2$	1.277	FeO_4^{2-} + 8 H^+ + 3 e $\rightleftharpoons$ Fe^{3+} + 4 H_2O	2.20
$HClO_2$ + 2 H^+ + 2 e $\rightleftharpoons$ HClO + H_2O	1.645	$Fe(OH)_3$ + e $\rightleftharpoons$ $Fe(OH)_2$ + OH^-	−0.56
$HClO_2$ + 3 H^+ + 3 e $\rightleftharpoons$ 1/2Cl_2 + 2 H_2O	1.628	$[Fe(phenanthroline)_3]^{3+}$ + e $\rightleftharpoons$ $[Fe(phen)_3]^{2+}$	1.147
$HClO_2$ + 3 H^+ + 4 e $\rightleftharpoons$ Cl^- + 2 H_2O	1.570	$[Fe(phen)_3]^{3+}$ + e $\rightleftharpoons$ $[Fe(phen)_3]^{2+}$ (1 mol/ℓ H_2SO_4)	1.06
ClO_2^- + H_2O + 2 e $\rightleftharpoons$ ClO^- + 2 OH^-	0.66	$[Ferricinium]^+$ + e $\rightleftharpoons$ ferrocene	0.400
ClO_2^- + 2 H_2O + 4 e $\rightleftharpoons$ Cl^- + 4 OH^-	0.76	Ga^{3+} + 3 e $\rightleftharpoons$ Ga	−0.560
ClO_2(aq) + e $\rightleftharpoons$ ClO_2^-	0.954	$H_2GaO_3^-$ + H_2O + 3 e $\rightleftharpoons$ Ga + 4 OH^-	−1.219
ClO_3^- + 2 H^+ + e $\rightleftharpoons$ ClO_2 + H_2O	1.152	Ge^{2+} + 2 e $\rightleftharpoons$ Ge	0.24
ClO_3^- + 3 H^+ + 2 e $\rightleftharpoons$ $HClO_2$ + H_2O	1.214	Ge^{4+} + 4 e $\rightleftharpoons$ Ge	0.124
ClO_3^- + 6 H^+ + 5 e $\rightleftharpoons$ 1/2Cl_2 + 3 H_2O	1.47	Ge^{4+} + 2 e $\rightleftharpoons$ Ge^{2+}	0.00
ClO_3^- + 6 H^+ + 6 e $\rightleftharpoons$ Cl^- + 3 H_2O	1.451	GeO_2 + 2 H^+ + 2 e $\rightleftharpoons$ GeO + H_2O	−0.118
ClO_3^- + H_2O + 2 e $\rightleftharpoons$ ClO_2^- + 2 OH^-	0.33	H_2GeO_3 + 4 H^+ + 4 e $\rightleftharpoons$ Ge + 3 H_2O	−0.182
ClO_3^- + 3 H_2O + 6 e $\rightleftharpoons$ Cl^- + 6 OH^-	0.62	2 H^+ + 2 e $\rightleftharpoons$ H_2	0.00000
ClO_4^- + 2 H^+ + 2 e $\rightleftharpoons$ ClO_3^- + H_2O	1.189	H_2 + 2 e $\rightleftharpoons$ 2 H^-	−2.23
ClO_4^- + 8 H^+ + 7 e $\rightleftharpoons$ 1/2Cl_2 + 4 H_2O	1.39	HO_2 + H^+ + e $\rightleftharpoons$ H_2O_2	1.495
ClO_4^- + 8 H^+ + 8 e $\rightleftharpoons$ Cl^- + 4 H_2O	1.389	2 H_2O + 2 e $\rightleftharpoons$ H_2 + 2 OH^-	−0.8277
ClO_4^- + H_2O + 2 e $\rightleftharpoons$ ClO_3^- + 2 OH^-	0.36	H_2O_2 + 2 H^+ + 2 e $\rightleftharpoons$ 2 H_2O	1.776
$(CN)_2$ + 2 H^+ + 2 e $\rightleftharpoons$ 2 HCN	0.373	HfO^{2+} + 2 H^+ + 4 e $\rightleftharpoons$ Hf + H_2O	−1.724
2 HCNO + 2 H^+ + 2 e $\rightleftharpoons$ $(CN)_2$ + 2 H_2O	−0.330	HfO_2 + 4 H^+ + 4 e $\rightleftharpoons$ Hf + 2 H_2O	−1.505
$(CNS)_2$ + 2 e $\rightleftharpoons$ 2 CNS^-	0.77	$HfO(OH)_2$ + H_2O + 4 e $\rightleftharpoons$ Hf + 4 OH^-	−2.50
Co^{2+} + 2 e $\rightleftharpoons$ Co	−0.28	Hg^{2+} + 2 e $\rightleftharpoons$ Hg	0.851
		2 Hg^{2+} + 2 e $\rightleftharpoons$ Hg_2^{2+}	0.920
		Hg_2^{2+} + 2 e $\rightleftharpoons$ 2 Hg	0.7973

Table 1 (continued)
ALPHABETICAL LISTING

Reaction	E°, V	Reaction	E°, V
$Hg_2(ac)_2 + 2\,e \rightleftharpoons 2\,Hg + 2\,(ac)^-$	0.51163	$HNO_2 + H^+ + e \rightleftharpoons NO + H_2O$	0.983
$Hg_2Br_2 + 2\,e \rightleftharpoons 2\,Hg + 2\,Br^-$	0.13923	$2\,HNO_2 + 4\,H^+ + 4\,e \rightleftharpoons H_2N_2O_2 + 2\,H_2O$	0.86
$Hg_2Cl_2 + 2\,e \rightleftharpoons 2\,Hg + 2\,Cl^-$	0.26808	$2\,HNO_2 + 4\,H^+ + 4\,e \rightleftharpoons N_2O + 3\,H_2O$	1.297
$Hg_2HPO_4 + 2\,e \rightleftharpoons 2\,Hg + HPO_4^{2-}$	0.6359	$NO_2^- + H_2O + 3\,e \rightleftharpoons NO + 2\,OH^-$	−0.46
$Hg_2I_2 + 2\,e \rightleftharpoons 2\,Hg + 2\,I^-$	−0.0405	$2\,NO_2^- + 2\,H_2O + 4\,e \rightleftharpoons N_2^{2-} + 4\,OH^-$	−0.18
$Hg_2O + H_2O + 2\,e \rightleftharpoons 2\,Hg + 2\,OH^-$	0.123	$2\,NO_2^- + 3\,H_2O + 4\,e \rightleftharpoons N_2O + 6\,OH^-$	0.15
$HgO + H_2O + 2\,e \rightleftharpoons Hg + 2\,OH^-$	0.0977	$NO_3^- + 3\,H^+ + 2\,e \rightleftharpoons HNO_2 + H_2O$	0.934
$Hg_2SO_4 + 2\,e \rightleftharpoons 2\,Hg + SO_4^{2-}$	0.6125	$NO_3^- + 4\,H^+ + 3\,e \rightleftharpoons NO + 2\,H_2O$	0.957
$I_2 + 2\,e \rightleftharpoons 2\,I^-$	0.5355	$2\,NO_3^- + 4\,H^+ + 2\,e \rightleftharpoons N_2O_4 + 2\,H_2O$	0.803
$I_3^- + 2\,e \rightleftharpoons 3\,I^-$	0.536	$NO_3^- + H_2O + 2\,e \rightleftharpoons NO_2^- + 2\,OH^-$	0.01
$H_3IO_6 + 2\,e \rightleftharpoons IO_3^- + 3\,OH^-$	0.7	$2\,NO_3^- + 2\,H_2O + 2\,e \rightleftharpoons N_2O_4 + 4\,OH^-$	−0.85
$H_5IO_6 + H^+ + 2\,e \rightleftharpoons IO_3^- + 3\,H_2O$	1.601	$Na^+ + e \rightleftharpoons Na$	−2.71
$2\,HIO + 2\,H^+ + 2\,e \rightleftharpoons I_2 + 2\,H_2O$	1.439	$Nb^{3+} + 3\,e \rightleftharpoons Nb$	−1.099
$HIO + H^+ + 2\,e \rightleftharpoons I^- + H_2O$	0.987	$Nb_2O_5 + 10\,H^+ + 10\,e \rightleftharpoons 2\,Nb + 5\,H_2O$	−0.644
$IO^- + H_2O + 2\,e \rightleftharpoons I^- + 2\,OH^-$	0.485	$Nd^{3+} + 3\,e \rightleftharpoons Nd$	−2.431
$2\,IO_3^- + 12\,H^+ + 10\,e \rightleftharpoons I_2 + 6\,H_2O$	1.195	$Ni^{2+} + 2\,e \rightleftharpoons Ni$	−0.257
$IO_3^- + 6\,H^+ + 6\,e \rightleftharpoons I^- + 3\,H_2O$	1.085	$Ni(OH)_2 + 2\,e \rightleftharpoons Ni + 2\,OH^-$	−0.72
$IO_3^- + 2\,H_2O + 4\,e \rightleftharpoons IO^- + 4\,OH^-$	0.56	$NiO_2 + 4\,H^+ + 2\,e \rightleftharpoons Ni^{2+} + 2\,H_2O$	1.678
$IO_3^- + 3\,H_2O + 6\,e \rightleftharpoons I^- + 6\,OH^-$	0.26	$NiO_2 + 2\,H_2O + 2\,e \rightleftharpoons Ni(OH)_2 + 2\,OH^-$	−0.490
$In^+ + e \rightleftharpoons In$	−0.14	$Np^{3+} + 3\,e \rightleftharpoons Np$	−1.856
$In^{2+} + e \rightleftharpoons In^+$	−0.40	$Np^{4+} + e \rightleftharpoons Np^{3+}$	0.147
$In^{3+} + e \rightleftharpoons In^{2+}$	−0.49	$NpO_2 + H_2O + H^+ + e \rightleftharpoons Np(OH)_3$	−0.962
$In^{3+} + 2\,e \rightleftharpoons In^+$	−0.443	$O_2 + 2\,H^+ + 2\,e \rightleftharpoons H_2O_2$	0.695
$In^{3+} + 3\,e \rightleftharpoons In$	−0.3382	$O_2 + 4\,H^+ + 4\,e \rightleftharpoons 2\,H_2O$	1.229
$Ir^{3+} + 3\,e \rightleftharpoons Ir$	1.156	$O_2 + H_2O + 2\,e \rightleftharpoons HO_2^- + OH^-$	−0.076
$[IrCl_6]^{2-} + e \rightleftharpoons [IrCl_6]^{3-}$	0.8665	$O_2 + 2\,H_2O + 2\,e \rightleftharpoons H_2O_2 + 2\,OH^-$	−0.146
$[IrCl_6]^{3-} + 3\,e \rightleftharpoons Ir + 6\,Cl^-$	0.77	$O_2 + 2\,H_2O + 4\,e \rightleftharpoons 4\,OH^-$	0.401
$Ir_2O_3 + 3\,H_2O + 6\,e \rightleftharpoons 2\,Ir + 6\,OH^-$	0.098	$O_3 + 2\,H^+ + 2\,e \rightleftharpoons O_2 + H_2O$	2.076
$K^+ + e \rightleftharpoons K$	−2.931	$O_3 + H_2O + 2\,e \rightleftharpoons O_2 + 2\,OH^-$	1.24
$La^{3+} + 3\,e \rightleftharpoons La$	−2.522	$O(g) + 2\,H^+ + 2\,e \rightleftharpoons H_2O$	2.421
$La(OH)_3 + 3\,e \rightleftharpoons La + 3\,OH^-$	−2.90	$OH + e \rightleftharpoons OH^-$	2.02
$Li^+ + e \rightleftharpoons Li$	−3.0401	$HO_2^- + H_2O + 2\,e \rightleftharpoons 3\,OH^-$	0.878
$Mg^+ + e \rightleftharpoons Mg$	−2.70	$OsO_4 + 8\,H^+ + 8\,e \rightleftharpoons Os + 4\,H_2O$	0.85
$Mg^{2+} + 2\,e \rightleftharpoons Mg$	−2.372	$P(red) + 3\,H^+ + 3\,e \rightleftharpoons PH_3(g)$	−0.111
$Mg(OH)_2 + 2\,e \rightleftharpoons Mg + 2\,OH^-$	−2.690	$P(white) + 3\,H^+ + 3\,e \rightleftharpoons PH_3(g)$	−0.063
$Mn^{2+} + 2\,e \rightleftharpoons Mn$	−1.185	$P + 3\,H_2O + 3\,e \rightleftharpoons PH_3(g) + 3\,OH^-$	−0.87
$Mn^{3+} + 3 \rightleftharpoons Mn^{2+}$	1.5415	$H_2PO_2^- + e \rightleftharpoons P + 2\,OH^-$	−1.82
$MnO_2 + 4\,H^+ + 2\,e \rightleftharpoons Mn^{2+} + 2\,H_2O$	1.224	$H_3PO_2 + H^+ + 3\,e \rightleftharpoons P + 2\,H_2O$	−0.508
$MnO_4^- + e \rightleftharpoons MnO_4^{2-}$	0.558	$H_3PO_3 + 2\,H^+ + 2\,e \rightleftharpoons H_3PO_2 + H_2O$	−0.499
$MnO_4^- + 4\,H^+ + 3\,e \rightleftharpoons MnO_2 + 2\,H_2O$	1.679	$H_3PO_3 + 3\,H^+ + 3\,e \rightleftharpoons P + 3\,H_2O$	−0.454
$MnO_4^- + 8\,H^+ + 5\,e \rightleftharpoons Mn^{2+} + 4\,H_2O$	1.507	$HPO_3^{2-} + 2\,H_2O + 2\,e \rightleftharpoons H_2PO_2^- + 3\,OH^-$	−1.65
$MnO_4^- + 2\,H_2O + 3\,e \rightleftharpoons MnO_2 + 4\,OH^-$	0.595	$HPO_3^{2-} + 2\,H_2O + 3\,e \rightleftharpoons P + 5\,OH^-$	−1.71
$MnO_4^{2-} + 2\,H_2O + 2\,e \rightleftharpoons MnO_2 + 4\,OH^-$	0.60	$H_3PO_4 + 2\,H^+ + 2\,e \rightleftharpoons H_3PO_3 + H_2O$	−0.276
$Mn(OH)_2 + 2\,e \rightleftharpoons Mn + 2\,OH^-$	−1.56	$PO_4^{3-} + 2\,H_2O + 2\,e \rightleftharpoons HPO_3^{2-} + 3\,OH^-$	−1.05
$Mn(OH)_3 + e \rightleftharpoons Mn(OH)_2 + OH^-$	0.15	$Pb^{2+} + 2\,e \rightleftharpoons Pb$	−0.1262
$Mo^{3+} + 3\,e \rightleftharpoons Mo$	−0.200	$Pb^{2+} + 2\,e \rightleftharpoons Pb(Hg)$	−0.1205
$N_2 + 2\,H_2O + 6\,H^+ + 6\,e \rightleftharpoons 2\,NH_4OH$	0.092	$PbBr_2 + 2\,e \rightleftharpoons Pb + 2\,Br^-$	−0.284
$3\,N_2 + 2\,H^+ + 2\,e \rightleftharpoons 2\,NH_3$	−3.09	$PbCl_2 + 2\,e \rightleftharpoons Pb + 2\,Cl^-$	−0.2675
$N_5^+ + 3\,H^+ + 2\,e \rightleftharpoons 2\,NH_4^+$	1.275	$PbF_2 + 2\,e \rightleftharpoons Pb + 2\,F^-$	−0.3444
$N_2O + 2\,H^+ + 2\,e \rightleftharpoons N_2 + H_2O$	1.766	$PbHPO_4 + 2\,e \rightleftharpoons Pb + HPO_4^{2-}$	−0.465
$H_2N_2O_2 + 2\,H^+ + 2\,e \rightleftharpoons N_2 + 2\,H_2O$	2.65	$PbI_2 + 2\,e \rightleftharpoons Pb + 2\,I^-$	−0.365
$N_2O_4 + 2\,e \rightleftharpoons 2\,NO_2^-$	0.867	$PbO + H_2O + 2\,e \rightleftharpoons Pb + 2\,OH^-$	−0.580
$N_2O_4 + 2\,H^+ + 2\,e \rightleftharpoons 2\,HNO_2$	1.065	$PbO_2 + 4\,H^+ + 2\,e \rightleftharpoons Pb^{2+} + 2\,H_2O$	1.455
$N_2O_4 + 4\,H^+ + 4\,e \rightleftharpoons 2\,NO + 2\,H_2O$	1.035	$HPbO_2^- + H_2O + 2\,e \rightleftharpoons Pb + 3\,OH^-$	−0.537
$2\,NH_3OH^+ + H^+ + 2\,e \rightleftharpoons N_2H_5^+ + 2\,H_2O$	1.42	$PbO_2 + H_2O + 2\,e \rightleftharpoons PbO + 2\,OH^-$	0.247
$2\,NO + 2\,e \rightleftharpoons N_2O_2^{2-}$	0.10	$PbO_2 + SO_4^{2-} + 4\,H^+ + 2\,e \rightleftharpoons PbSO_4 + 2\,H_2O$	1.6913
$2\,NO + 2\,H^+ + 2\,e \rightleftharpoons N_2O + H_2O$	1.591	$PbSO_4 + 2\,e \rightleftharpoons Pb + SO_4^{2-}$	−0.3588
$2\,NO + H_2O + 2\,e \rightleftharpoons N_2O + 2\,OH^-$	0.76		

Table 1 (continued)
ALPHABETICAL LISTING

Reaction	$E°$, V	Reaction	$E°$, V
$PbSO_4 + 2\,e \rightleftharpoons Pb(Hg) + SO_4^{2-}$	−0.3505	$Se + 2\,H^+ + 2\,e \rightleftharpoons H_2Se(aq)$	−0.399
$Pd^{2+} + 2\,e \rightleftharpoons Pd$	0.951	$H_2SeO_3 + 4\,H^+ + 4\,e \rightleftharpoons Se + 3\,H_2O$	−0.74
$[PdCl_4]^{2-} + 2\,e \rightleftharpoons Pd + 4\,Cl^-$	0.591	$SeO_3^{2-} + 3\,H_2O + 4\,e \rightleftharpoons Se + 6\,OH^-$	−0.366
$[PdCl_6]^{2-} + 2\,e \rightleftharpoons [PdCl_4]^{2-} + 2\,Cl^-$	1.288	$SeO_4^{2-} + 4\,H^+ + 2\,e \rightleftharpoons H_2SeO_3 + H_2O$	1.151
$Pd(OH)_2 + 2\,e \rightleftharpoons Pd + 2\,OH^-$	0.07	$SeO_4^{2-} + H_2O + 2\,e \rightleftharpoons SeO_3^{2-} + 2\,OH^-$	0.05
$Pt^{2+} + 2\,e \rightleftharpoons Pt$	1.118	$SiF_6^{2-} + 4\,e \rightleftharpoons Si + 6\,F^-$	−1.24
$[PtCl_4]^{2-} + 2\,e \rightleftharpoons Pt + 4\,Cl^-$	0.755	$SiO_2 \text{ (quartz)} + 4\,H^+ + 4\,e \rightleftharpoons Si + 2\,H_2O$	0.857
$[PtCl_6]^{2-} + 2\,e \rightleftharpoons [PtCl_4]^{2-} + 2\,Cl^-$	0.68	$SiO_3^{2-} + 3\,H_2O + 4\,e \rightleftharpoons Si + 6\,OH^-$	−1.697
$Pt(OH)_2 + 2\,e \rightleftharpoons Pt + 2\,OH^-$	0.14	$Sn^{2+} + 2\,e \rightleftharpoons Sn$	−1.375
$Pu^{3+} + 3\,e \rightleftharpoons Pu$	2.031	$Sn^{4+} + 2\,e \rightleftharpoons Sn^{2+}$	0.151
$Pu^{4+} + e \rightleftharpoons Pu^{3+}$	1.006	$HSnO_2^- + H_2O + 2\,e \rightleftharpoons Sn + 3\,OH^-$	−0.909
$Pu^{5+} + e \rightleftharpoons Pu^{4+}$	1.099	$Sn(OH)_6^{2-} + 2\,e \rightleftharpoons HSnO_2^- + 3\,OH^- + H_2O$	−0.93
$PuO_2(OH)_2 + 2\,H^+ + 2\,e \rightleftharpoons Pu(OH)_4$	1.325	$Sr^+ + e \rightleftharpoons Sr$	−4.10
$PuO_2(OH)_2 + H^+ + e \rightleftharpoons PuO_2OH + H_2O$	1.062	$Sr^{2+} + 2\,e \rightleftharpoons Sr$	−2.89
$Rb^+ + e \rightleftharpoons Rb$	−2.98	$Sr^{2+} + 2\,e \rightleftharpoons Sr(Hg)$	−1.793
$Re^{3+} + 3\,e \rightleftharpoons Re$	0.300	$Sr(OH)_2 + 2\,e \rightleftharpoons Sr + 2\,OH^-$	−2.88
$ReO_4^- + 4\,H^+ + 3\,e \rightleftharpoons ReO_2 + 2\,H_2O$	0.510	$Ta_2O_5 + 10\,H^+ + 10\,e \rightleftharpoons 2\,Ta + 5\,H_2O$	−0.750
$ReO_2 + 4\,H^+ + 4\,e \rightleftharpoons Re + 2\,H_2O$	0.2513	$Tc^{2+} + 2\,e \rightleftharpoons Tc$	0.400
$ReO_4^- + 2\,H^+ + e \rightleftharpoons ReO_3 + H_2O$	0.768	$TcO_4^- + 4\,H^+ + 3\,e \rightleftharpoons TcO_2 + 2\,H_2O$	0.782
$ReO_4^- + 4\,H_2O + 7\,e \rightleftharpoons Re + 8\,OH^-$	−0.584	$Te + 2\,e \rightleftharpoons Te^{2-}$	−1.143
$ReO_4^- + 8\,H^+ + 7\,e \rightleftharpoons Re + 4\,H_2O$	0.368	$Te + 2\,H^+ + 2\,e \rightleftharpoons H_2Te$	−0.793
$Rh^+ + e \rightleftharpoons Rh$	0.600	$Te^{4+} + 4\,e \rightleftharpoons Te$	0.568
$Rh^{2+} + 2\,e \rightleftharpoons Rh$	0.600	$TeO_2 + 4\,H^+ + 4\,e \rightleftharpoons Te + 2\,H_2O$	0.593
$Rh^{3+} + 3\,e \rightleftharpoons Rh$	0.758	$TeO_3^{2-} + 3\,H_2O + 4\,e \rightleftharpoons Te + 6\,OH^-$	−0.57
$[RhCl_6]^{3-} + 3\,e \rightleftharpoons Rh + 6\,Cl^-$	0.431	$TeO_4^- + 8\,H^+ + 7\,e \rightleftharpoons Te + 4\,H_2O$	0.472
$Ru^{2+} + 2\,e \rightleftharpoons Ru$	0.455	$H_6TeO_6 + 2\,H^+ + 2\,e \rightleftharpoons TeO_2 + 4\,H_2O$	1.02
$Ru^{3+} + e \rightleftharpoons Ru^{2+}$	0.2487	$Th^{4+} + 4\,e \rightleftharpoons Th$	−1.899
$RuO_2 + 4\,H^+ + 2\,e \rightleftharpoons Ru^{2+} + 2\,H_2O$	1.120	$ThO_2 + 4\,H^+ + 4\,e \rightleftharpoons Th + 2\,H_2O$	−1.789
$RuO_4^- + e \rightleftharpoons RuO_4^{2-}$	0.59	$Th(OH)_4 + 4\,e \rightleftharpoons Th + 4\,OH^-$	−2.48
$RuO_4 + e \rightleftharpoons RuO_4^-$	1.00	$Ti^{2+} + 2\,e \rightleftharpoons Ti$	−1.630
$S + 2\,e \rightleftharpoons S^{2-}$	−0.47627	$Ti^{3+} + e \rightleftharpoons Ti^{2+}$	−0.368
$S + 2\,H^+ + 2\,e \rightleftharpoons H_2S(aq)$	0.142	$TiO_2 + 4\,H^+ + 2\,e \rightleftharpoons Ti^{2+} + 2\,H_2O$	−0.502
$S + H_2O + 2\,e \rightleftharpoons HS^- + OH^-$	−0.478	$TiOH^{3+} + H^+ + e \rightleftharpoons Ti^{3+} + H_2O$	−0.055
$2\,S + 2\,e \rightleftharpoons S_2^{2-}$	−0.42836	$Tl^+ + e \rightleftharpoons Tl$	−0.336
$S_2O_6^{2-} + 4\,H^+ + 2\,e \rightleftharpoons 2\,H_2SO_3$	0.564	$Tl^+ + e \rightleftharpoons Tl(Hg)$	−0.3338
$S_2O_8^{2-} + 2\,e \rightleftharpoons 2\,SO_4^{2-}$	2.010	$Tl^{3+} + 2\,e \rightleftharpoons Tl^+$	1.252
$S_2O_8^{2-} + 2\,H^+ + 2\,e \rightleftharpoons 2\,HSO_4^-$	2.123	$TlBr + e \rightleftharpoons Tl + Br^-$	−0.658
$S_4O_6^{2-} + 2\,e \rightleftharpoons 2\,S_2O_3^{2-}$	0.08	$TlCl + e \rightleftharpoons Tl + Cl^-$	−0.5568
$2\,H_2SO_3 + H^+ + 2\,e \rightleftharpoons HS_2O_4^- + 2\,H_2O$	−0.056	$TlI + e \rightleftharpoons Tl + I^-$	−0.752
$H_2SO_3 + 4\,H^+ + 4\,e \rightleftharpoons S + 3\,H_2O$	0.449	$Tl_2O_3 + 3\,H_2O + 4\,e \rightleftharpoons 2\,Tl^+ + 6\,OH^-$	0.02
$2\,SO_3^{2-} + 2\,H_2O + 2\,e \rightleftharpoons S_2O_4^{2-} + 4\,OH^-$	−1.12	$TlOH + e \rightleftharpoons Tl + OH^-$	−0.34
$2\,SO_3^{2-} + 3\,H_2O + 4\,e \rightleftharpoons S_2O_3^{2-} + 6\,OH^-$	−0.571	$Tl(OH)_3 + 2\,e \rightleftharpoons TlOH + 2\,OH^-$	−0.05
$SO_4^{2-} + 4\,H^+ + 2\,e \rightleftharpoons H_2SO_3 + H_2O$	0.172	$Tl_2SO_4 + 2\,e \rightleftharpoons Tl + SO_4^{2-}$	−0.4360
$2\,SO_4^{2-} + 4\,H^+ + 2\,e \rightleftharpoons S_2O_6^{2-} + H_2O$	−0.22	$U^{3+} + 3\,e \rightleftharpoons U$	−1.798
$SO_4^{2-} + H_2O + 2\,e \rightleftharpoons SO_3^{2-} + 2\,OH^-$	−0.93	$U^{4+} + e \rightleftharpoons U^{3+}$	−0.607
$Sb + 3\,H^+ + 3\,e \rightleftharpoons SbH_3$	−0.510	$UO_2^+ + 4\,H^+ + e \rightleftharpoons U^{4+} + 2\,H_2O$	0.612
$Sb_2O_3 + 6\,H^+ + 6\,e \rightleftharpoons 2\,Sb + 3\,H_2O$	0.152	$UO_2^{2+} + e \rightleftharpoons UO_2^+$	0.062
$Sb_2O_5 \text{ (senarmontite)} + 4\,H^+ + 4\,e \rightleftharpoons Sb_2O_3 + 2\,H_2O$	0.671	$UO_2^{2+} + 4\,H^+ + 2\,e \rightleftharpoons U^{4+} + 2\,H_2O$	0.327
$Sb_2O_5 \text{ (valentinite)} + 4\,H^+ + 4\,e \rightleftharpoons Sb_2O_3 + 2\,H_2O$	0.649	$UO_2^{2+} + 4\,H^+ + 6\,e \rightleftharpoons U + 2\,H_2O$	−1.444
$Sb_2O_5 + 6\,H^+ + 4\,e \rightleftharpoons 2\,SbO^+ + 3\,H_2O$	0.581	$V^{2+} + 2\,e \rightleftharpoons V$	−1.175
$SbO^+ + 2\,H^+ + 3\,e \rightleftharpoons Sb + 2\,H_2O$	0.212	$V^{3+} + e \rightleftharpoons V^{2+}$	−0.255
$SbO_2^- + 2\,H_2O + 3\,e \rightleftharpoons Sb + 4\,OH^-$	−0.66	$VO^{2+} + 2\,H^+ + e \rightleftharpoons V^{3+} + H_2O$	0.337
$SbO_3^- + H_2O + 2\,e \rightleftharpoons SbO_2^- + 2\,OH^-$	−0.59	$VO_2^+ + 2\,H^+ + e \rightleftharpoons VO^{2+} + H_2O$	0.991
$Sc^{3+} + 3\,e \rightleftharpoons Sc$	−2.077	$V(OH)_4^+ + 2\,H^+ + e \rightleftharpoons VO^{2+} + 3\,H_2O$	1.00
$Se + 2\,e \rightleftharpoons Se^{2-}$	−0.924	$V(OH)_4^+ + 4\,H^+ + 5\,e \rightleftharpoons V + 4\,H_2O$	−0.254
		$W_2O_5 + 2\,H^+ + 2\,e \rightleftharpoons 2\,WO_2 + H_2O$	−0.031

Table 1 (continued)
ALPHABETICAL LISTING

Reaction	E°, V	Reaction	E°, V
$WO_2 + 4\,H^+ + 4\,e \rightleftharpoons W + 2\,H_2O$	−0.119	$ZnO_2^{2-} + 2\,H_2O + 2\,e \rightleftharpoons Zn + 4\,OH^-$	−1.215
$WO_3 + 6\,H^+ + 6\,e \rightleftharpoons W + 3\,H_2O$	−0.090	$ZnSO_4 \cdot 7H_2O + 2\,e \rightleftharpoons Zn(Hg) + SO_4^{2-}$	−0.7993
$2\,WO_3 + 2\,H^+ + 2\,e \rightleftharpoons W_2O_5 + H_2O$	−0.029	(Sat'd $ZnSO_4$)	
$Y^{3+} + 3\,e \rightleftharpoons Y$	−2.372	$ZrO_2 + 4\,H^+ + 4\,e \rightleftharpoons Zr + 2\,H_2O$	−1.553
$Zn^{2+} + 2\,e \rightleftharpoons Zn$	−0.7618	$ZrO(OH)_2 + H_2O + 4\,e \rightleftharpoons Zr + 4\,OH^-$	−2.36
$Zn^{2+} + 2\,e \rightleftharpoons Zn(Hg)$	−0.7628		

Table 2
REDUCTION REACTIONS HAVING E° VALUES MORE POSITIVE THAN THAT OF THE STANDARD HYDROGEN ELECTRODE

Reaction	E°, V	Reaction	E°, V
$2\,H^+ + 2\,e \rightleftharpoons H_2$	0.00000	$Re^{3+} + 3\,e \rightleftharpoons Re$	0.300
$CuI_2^- + e \rightleftharpoons Cu + 2\,I^-$	0.00	$BiO^+ + 2\,H^+ + 3\,e \rightleftharpoons Bi + H_2O$	0.320
$Ge^{4+} + 2\,e \rightleftharpoons Ge^{2+}$	0.00	$UO_2^{2+} + 4\,H^+ + 2\,e \rightleftharpoons U^{4+} + 2\,H_2O$	0.327
$NO_3^- + H_2O + 2\,e \rightleftharpoons NO_2^- + 2\,OH^-$	0.01	$ClO_3^- + H_2O + 2\,e \rightleftharpoons ClO_2^- + 2\,OH^-$	0.33
$Tl_2O_3 + 3\,H_2O + 4\,e \rightleftharpoons 2\,Tl^+ + 6\,OH^-$	0.02	Calomel electrode, 0.1 mol/1 KCl	0.3337
$SeO_4^{2-} + H_2O + 2\,e \rightleftharpoons SeO_3^{2-} + 2\,OH^-$	0.05	$VO^{2+} + 2\,H^+ + e \rightleftharpoons V^{3+} + H_2O$	0.337
$UO_2^{2+} + e \rightleftharpoons UO_2^+$	0.062	$Cu^{2+} + 2\,e \rightleftharpoons Cu$	0.3419
$Pd(OH)_2 + 2\,e \rightleftharpoons Pd + 2\,OH^-$	0.07	$Ag_2O + H_2O + 2\,e \rightleftharpoons 2\,Ag + 2\,OH^-$	0.342
$AgBr + e \rightleftharpoons Ag + Br^-$	0.07133	$Cu^{2+} + 2\,e \rightleftharpoons Cu(Hg)$	0.345
$S_4O_6^{2-} + 2\,e \rightleftharpoons 2\,S_2O_3^{2-}$	0.08	$AgIO_3 + e \rightleftharpoons Ag + IO_3^-$	0.354
$AgSCN + e \rightleftharpoons Ag + SCN^-$	0.8951	$[Fe(CN)_6]^{3-} + e \rightleftharpoons [Fe(CN)_6]^{4-}$	0.358
$N_2 + 2\,H_2O + 6\,H^+ + 6\,e \rightleftharpoons 2\,NH_4OH$	0.092	$ClO^- + H_2O + 2\,e \rightleftharpoons ClO_3^- + 2\,OH^-$	0.36
$HgO + H_2O + 2\,e \rightleftharpoons Hg + 2\,OH^-$	0.0977	$Ag_2SeO_3 + 2\,e \rightleftharpoons 2\,Ag + SeO_3^{2-}$	0.3629
$Ir_2O_3 + 3\,H_2O + 6\,e \rightleftharpoons 2\,Ir + 6\,OH^-$	0.098	$ReO_4^- + 8\,H^+ + 7\,e \rightleftharpoons Re + 4\,H_2O$	0.368
$2\,NO + 2\,e \rightleftharpoons N_2O_2^{2-}$	0.10	$(CN)_2 + 2\,H^+ + 2\,e \rightleftharpoons 2\,HCN$	0.373
$[Co(NH_3)_6]^{3+} + e \rightleftharpoons [Co(NH_3)_6]^{2+}$	0.108	$[Ferricinium]^+ + e \rightleftharpoons ferrocene$	0.400
$Hg_2O + H_2O + 2\,e \rightleftharpoons 2\,Hg + 2\,OH^-$	0.123	$Tc^{2+} + 2\,e \rightleftharpoons Tc$	0.400
$Ge^{4+} + 4\,e \rightleftharpoons Ge$	0.124	$O_2 + 2\,H_2O + 4\,e \rightleftharpoons 4\,OH^-$	0.401
$Hg_2Br_2 + 2\,e \rightleftharpoons 2\,Hg + 2\,Br^-$	0.13923	$AgOCN + e \rightleftharpoons Ag + OCN^-$	0.41
$Pt(OH)_2 + 2\,e \rightleftharpoons Pt + 2\,OH^-$	0.14	$[RhCl_6]^{3-} + 3\,e \rightleftharpoons Rh + 6\,Cl^-$	0.431
$S + 2\,H^+ + 2\,e \rightleftharpoons H_2S(aq)$	0.142	$Ag_2CrO_4 + 2\,e \rightleftharpoons 2\,Ag + CrO_4^{2-}$	0.4470
$Np^{4+} + e \rightleftharpoons Np^{3+}$	0.147	$H_2SO_3 + 4\,H^+ + 4\,e \rightleftharpoons S + 3\,H_2O$	0.449
$Ag_4[Fe(CN)_6] + 4\,e \rightleftharpoons 4\,Ag + [Fe(CN)_6]^{4-}$	0.1478	$Ru^{2+} + 2\,e \rightleftharpoons Ru$	0.455
$Mn(OH)_3 + e \rightleftharpoons Mn(OH)_2 + OH^-$	0.15	$Ag_2MoO_4 + 2\,e \rightleftharpoons 2\,Ag + MoO_4^{2-}$	0.4573
$2\,NO_2^- + 3\,H_2O + 4\,e \rightleftharpoons N_2O + 6\,OH^-$	0.15	$Ag_2C_2O_4 + 2\,e \rightleftharpoons 2\,Ag + C_2O_4^{2-}$	0.4647
$Sn^{4+} + 2\,e \rightleftharpoons Sn^{2+}$	0.151	$Ag_2WO_4 + 2\,e \rightleftharpoons 2\,Ag + WO_4^{2-}$	0.4660
$Sb_2O_3 + 6\,H^+ + 6\,e \rightleftharpoons 2\,Sb + 3\,H_2O$	0.152	$Ag_2CO_3 + 2\,e \rightleftharpoons 2\,Ag + CO_3^{2-}$	0.47
$Cu^{2+} + e \rightleftharpoons Cu^+$	0.153	$TeO_4^- + 8\,H^+ + 7\,e \rightleftharpoons Te + 4\,H_2O$	0.472
$BiOCl + 2\,H^+ + 3\,e \rightleftharpoons Bi + Cl^- + H_2O$	0.1583	$IO^- + H_2O + 2\,e \rightleftharpoons I^- + 2\,OH^-$	0.485
$Bi(Cl)_4^- + 3\,e \rightleftharpoons Bi + 4\,Cl^-$	0.16	$ReO_4^- + 4\,H^+ + 3\,e \rightleftharpoons ReO_2 + 2\,H_2O$	0.510
$Co(OH)_3 + e \rightleftharpoons Co(OH)_2 + OH^-$	0.17	$Hg_2(ac)_2 + 2\,e \rightleftharpoons 2\,Hg + 2\,(ac)^-$	0.51163
$SO_4^{2-} + 4\,H^+ + 2\,e \rightleftharpoons H_2SO_3 + H_2O$	0.172	$Cu^+ + e \rightleftharpoons Cu$	0.521
$SbO^+ + 2\,H^+ + 3\,e \rightleftharpoons Sb + 2\,H_2O$	0.212	$I_2 + 2\,e \rightleftharpoons 2\,I^-$	0.5355
$AgCl + e \rightleftharpoons Ag + Cl^-$	0.22233	$I_3^- + 2\,e \rightleftharpoons 3\,I^-$	0.536
$As_2O_3 + 6\,H^+ + 6\,e \rightleftharpoons 2\,As + 3\,H_2O$	0.234	$AgBrO_3 + e \rightleftharpoons Ag + BrO_3^-$	0.546
Calomel electrode, saturated NaCl (SSCE)	0.2360	$MnO_4^- + e \rightleftharpoons MnO_4^{2-}$	0.558
$Ge^{2+} + 2\,e \rightleftharpoons Ge$	0.24	$H_3AsO_4 + 2\,H^+ + 2\,e^- \rightleftharpoons HAsO_2 + 2\,H_2O$	0.560
Calomel electrode, saturated KCl	0.2412	$IO_3^- + 2\,H_2O + 4\,e \rightleftharpoons IO^- + 4\,OH^-$	0.56
$PbO_2 + H_2O + 2\,e \rightleftharpoons PbO + 2\,OH^-$	0.247	$S_2O_6^{2-} + 4\,H^+ + 2\,e \rightleftharpoons 2\,H_2SO_3$	0.564
$HAsO_2 + 3\,H^+ + 3_e \rightleftharpoons As + 2\,H_2O$	0.248	$AgNO_2 + e \rightleftharpoons Ag + NO_2^-$	0.564
$Ru^{3+} + e \rightleftharpoons Ru^{2+}$	0.2487	$Te^{4+} + 4\,e \rightleftharpoons Te$	0.568
$ReO_2 + 4\,H^+ + 4\,e \rightleftharpoons Re + 2\,H_2O$	0.2513	$Sb_2O_5 + 6\,H^+ + 4\,e \rightleftharpoons 2\,SbO^+ + 3\,H_2O$	0.581
$IO_3^- + 3\,H_2O + 6\,e \rightleftharpoons I^- + OH^-$	0.26	$RuO_4^- + e \rightleftharpoons RuO_4^{2-}$	0.59
$Hg_2Cl_2 + 2\,e \rightleftharpoons 2\,Hg + 2\,Cl^-$	0.26808	$[PdCl_4]^{2-} + 2\,e \rightleftharpoons Pd + 4\,Cl^-$	0.591
Calomel electrode, molal KCl	0.2800		
Calomel electroce, 1 mol/1 KCl (NCE)	0.2801		

Table 2 (continued)
REDUCTION REACTIONS HAVING E° VALUES MORE POSITIVE THAN THAT OF THE STANDARD HYDROGEN ELECTRODE

Reaction	E°, V
$TeO_2 + 4H^+ + 4e \rightleftharpoons Te + 2H_2O$	0.593
$MnO_4^- + 2H_2O + 3e \rightleftharpoons MnO_2 + 4OH^-$	0.595
$Rh^{2+} + 2e \rightleftharpoons Rh$	0.600
$Rh^+ + e \rightleftharpoons Rh$	0.600
$MnO_4^{2-} + 2H_2O + 2e \rightleftharpoons MnO_2 + 4OH^-$	0.60
$2AgO + H_2O + 2e \rightleftharpoons Ag_2O + 2OH^-$	0.607
$BrO_3^- + 3H_2O + 6e \rightleftharpoons Br^- + 6OH^-$	0.61
$UO_2^+ + 4H^+ + e \rightleftharpoons U^{4+} + 2H_2O$	0.612
$Hg_2SO_4 + 2e \rightleftharpoons 2Hg + SO_4^{2-}$	0.6125
$ClO_3^- + 3H_2O + 6e \rightleftharpoons Cl^- + 6OH^-$	0.62
$Hg_2HPO_4 + 2e \rightleftharpoons 2Hg + HPO_4^{2-}$	0.6359
$Ag(ac) + e \rightleftharpoons Ag + (ac)^-$	0.643
$Sb_2O_5(valentinite) + 4H^+ + 4e \rightleftharpoons Sb_2O_3 + 2H_2O$	0.649
$Ag_2SO_4 + 2e \rightleftharpoons 2Ag + SO_4^{2-}$	0.654
$ClO_2^- + H_2O + 2e \rightleftharpoons ClO^- + 2OH^-$	0.66
$Sb_2O_5(senarmontite) + 4H^+ + 4e \rightleftharpoons Sb_2O_3 + 2H_2O$	0.671
$[PtCl_6]^{2-} + 2e \rightleftharpoons [PtCl_4]^{2-} + 2Cl^-$	0.68
$O_2 + 2H^+ + 2e \rightleftharpoons H_2O_2$	0.695
p-benzoquinone $+ 2H^+ + 2e \rightleftharpoons$ hydroquinone	0.6992
$H_5IO_6 + 2e \rightleftharpoons IO_3^- + 3OH^-$	0.7
$Ag_2O_3 + H_2O + 2e \rightleftharpoons 2AgO + 2OH^-$	0.739
$[PtCl_4]^{2-} + 2e \rightleftharpoons Pt + 4Cl^-$	0.755
$Rh^{3+} + 3e \rightleftharpoons Rh$	0.758
$ClO_2^- + 2H_2O + 4e \rightleftharpoons Cl^- + 4OH^-$	0.76
$2NO + H_2O + 2e \rightleftharpoons N_2O + 2OH^-$	0.76
$BrO^- + H_2O + 2e \rightleftharpoons Br^- + 2OH^-$	0.761
$ReO_4^- + 2H^+ + e \rightleftharpoons ReO_3 + H_2O$	0.768
$(CNS)_2 + 2e \rightleftharpoons 2CNS^-$	0.77
$[IrCl_6]^{3-} + 3e \rightleftharpoons Ir + 6Cl^-$	0.77
$Fe^{3+} + e \rightleftharpoons Fe^{2+}$	0.771
$Ag(F) + e \rightleftharpoons Ag + F^-$	0.779
$TcO_4^- + 4H^+ + 3e \rightleftharpoons TcO_2 + 2H_2O$	0.782
$Hg_2^{2+} + 2e \rightleftharpoons 2Hg$	0.7973
$Ag^+ + e \rightleftharpoons Ag$	0.7996
$2NO_3^- + 4H^+ + 2e \rightleftharpoons N_2O_4 + 2H_2O$	0.803
$ClO^- + H_2O + 2e \rightleftharpoons Cl^- + 2OH^-$	0.841
$OsO_4 + 8H^+ + 8e \rightleftharpoons Os + 4H_2O$	0.85
$Hg^{2+} + 2e \rightleftharpoons Hg$	0.851
$AuBr_4^- + 3e \rightleftharpoons Au + 4Br^-$	0.854
$SiO_2(quartz) + 4H^+ + 4e \rightleftharpoons Si + 2H_2O$	0.857
$2HNO_2 + 4H^+ + 4e \rightleftharpoons H_2N_2O_2 + H_2O$	0.86
$[IrCl_6]^{2-} + e \rightleftharpoons [IrCl_6]^{3-}$	0.8665
$N_2O_4 + 2e \rightleftharpoons 2NO_2^-$	0.867
$HO_2^- + H_2O + 2e \rightleftharpoons 3OH^-$	0.878
$2Hg^{2+} + 2e \rightleftharpoons Hg_2^{2+}$	0.920
$NO_3^- + 3H^+ + 2e \rightleftharpoons HNO_2 + H_2O$	0.934
$Pd^{2+} + 2e \rightleftharpoons Pd$	0.951
$ClO_2(aq) + e \rightleftharpoons ClO_2^-$	0.954
$NO_3^- + 4H^+ + 3e \rightleftharpoons NO + 2H_2O$	0.957
$AuBr_2^- + e \rightleftharpoons Au + 2Br^-$	0.959
$HNO_2 + H^+ + e \rightleftharpoons NO + H_2O$	0.983
$HIO + H^+ + 2e \rightleftharpoons I^- + H_2O$	0.987

Reaction	E°, V
$VO_2^+ + 2H^+ + e \rightleftharpoons VO^{2+} + H_2O$	0.991
$RuO_4 + e \rightleftharpoons RuO_4^-$	1.00
$V(OH)_4^+ + 2H^+ + e \rightleftharpoons VO^{2+} + 3H_2O$	1.00
$AuCl_4^- + 3e \rightleftharpoons Au + 4Cl^-$	1.002
$Pu^{4+} + e \rightleftharpoons Pu^{3+}$	1.006
$H_6TeO_6 + 2H^+ + 2e \rightleftharpoons TeO_2 + 4H_2O$	1.02
$N_2O_4 + 4H^+ + 4e \rightleftharpoons 2NO + 2H_2O$	1.035
$[Fe(phen)_3]^{3+} + e \rightleftharpoons [Fe(phen)_3]^{2+}$ (1 (Mol/l H_2SO_4)	1.06
$PuO_2(OH)_2 + H^+ + e \rightleftharpoons PuO_2OH + H_2O$	1.062
$N_2O_4 + 2H^+ + 2e \rightleftharpoons 2HNO_2$	1.065
$Br_2(l) + 2e \rightleftharpoons 2Br^-$	1.066
$IO_3^- + 6H^+ + 6e \rightleftharpoons I^- + 3H_2O$	1.085
$Br_2(aq) + 2e \rightleftharpoons 2Br^-$	1.0873
$Pu^{5+} + e \rightleftharpoons Pu^{4+}$	1.099
$Cu^{2+} + 2CN^- + e \rightleftharpoons [Cu(CN)_2]^-$	
$Pt^{2+} + 2e \rightleftharpoons Pt$	1.118
$RuO_2 + 4H^+ + 2e \rightleftharpoons Ru^{2+} + 2H_2O$	1.120
$[Fe(phenanthroline)_3]^{3+} + e \rightleftharpoons [Fe(phen)_3]^{2+}$	1.147
$SeO_4^{2-} + 4H^+ + 2e \rightleftharpoons H_2SeO_3 + H_2O$	1.151
$ClO_3^- + 2H^+ + e \rightleftharpoons ClO_2 + H_2O$	1.152
$Ir^{3+} + 3e \rightleftharpoons Ir$	1.156
$ClO_4^- + 2H^+ + 2e \rightleftharpoons ClO_3^- + H_2O$	1.189
$2IO_3^- + 12H^+ + 10e \rightleftharpoons I_2 + 6H_2O$	1.195
$ClO_3^- + 3H^+ + 2e \rightleftharpoons HClO_2 + H_2O$	1.214
$MnO_2 + 4H^+ + 2e \rightleftharpoons Mn^{2+} + 2H_2O$	1.224
$O_2 + 4H^+ + 4e \rightleftharpoons 2H_2O$	1.229
$Cr_2O_7^{2-} + 14H^+ + 6e \rightleftharpoons 2Cr^{3+} + 7H_2O$	1.232
$O_3 + H_2O + 2e \rightleftharpoons O_2 + 2OH^-$	1.24
$Tl^{3+} + 2e \rightleftharpoons Tl^+$	1.252
$N_2H_5^+ + 3H^+ + 2e \rightleftharpoons 2NH_4^+$	1.275
$ClO_2 + H^+ + e \rightleftharpoons HClO_2$	1.277
$[PdCl_6]^{2-} + 2e \rightleftharpoons [PdCl_4]^{2-} + 2Cl^-$	1.288
$2HNO_2 + 4H^+ + 4e \rightleftharpoons N_2O + 3H_2O$	1.297
$PuO_2(OH)_2 + 2H^+ + 2e \rightleftharpoons Pu(OH)_4$	1.325
$HBrO + H^+ + 2e \rightleftharpoons Br^- + H_2O$	1.331
$HCrO_4^- + 7H^+ + 3e \rightleftharpoons Cr^{3+} + 4H_2O$	1.350
$Cl_2(g) + 2e \rightleftharpoons Cl^-$	1.35827
$ClO_4^- + 8H^+ + 8e \rightleftharpoons Cl^- + 4H_2O$	1.389
$ClO_4^- + 8H^+ + 7e \rightleftharpoons \frac{1}{2}Cl_2 + 4H_2O$	1.39
$Au^{3+} + 2e \rightleftharpoons Au^+$	1.401
$2NH_3OH^+ + H^+ + 2e \rightleftharpoons N_2H_5^+ + 2H_2O$	1.42
$BrO_3^- + 6H^+ + 6e \rightleftharpoons Br^- + 3H_2O$	1.423
$2HIO + 2H^+ + 2e \rightleftharpoons I_2 + 2H_2O$	1.439
$Au(OH)_3 + 3H^+ + 3e \rightleftharpoons Au^- + 3H_2O$	1.45
$3IO_3^- + 6H^+ + 6e \rightleftharpoons Cl^- + 3H_2O$	1.451
$PbO_2 + 4H^+ + 2e \rightleftharpoons Pb^{2+} + 2H_2O$	1.455
$ClO_3^- + 6H^+ + 5e \rightleftharpoons \frac{1}{2}Cl_2 + 3H_2O$	1.47
$BrO_3^- + 6H^+ + 5e \rightleftharpoons \frac{1}{2}Br_2 + 3H_2O$	1.482
$HClO + H^+ + 2e \rightleftharpoons Cl^- + H_2O$	1.482
$HO_2 + H^+ + e \rightleftharpoons H_2O_2$	1.495
$Au^{3+} + 3e \rightleftharpoons Au$	1.498
$MnO_4^- + 8H^+ + 5e \rightleftharpoons Mn^{2+} + 4H_2O$	1.507
$Mn^{3+} + e \rightleftharpoons Mn^{2+}$	1.5415
$HClO_2 + 3H^+ + 4e \rightleftharpoons Cl^- + 2H_2O$	1.570
$HBrO + H^+ + e \rightleftharpoons \frac{1}{2}Br_2(aq) + H_2O$	1.574

Table 2 (continued)
REDUCTION REACTIONS HAVING E° VALUES MORE POSITIVE THAN THAT OF THE STANDARD HYDROGEN ELECTRODE

Reaction	E°, V	Reaction	E°, V
$2\ NO + 2\ H^+ + 2\ e \rightleftharpoons N_2O + H_2O$	1.591	$N_2O + 2\ H^+ + 2\ e \rightleftharpoons N_2 + H_2O$	1.766
$Bi_2O_4 + 4\ H^+ + 2\ e \rightleftharpoons 2\ BiO^+ + 2\ H_2O$	1.593	$H_2O_2 + 2\ H^+ + 2\ e \rightleftharpoons 2\ H_2O$	1.776
$HBrO + H^+ + e \rightleftharpoons 1/2\ Br_2(\ell) + H_2O$	1.596	$Co^{3+} + e \rightleftharpoons Co^{2+} (2\ mol/\ell\ H_2SO_4)$	1.83
$H_5IO_6 + H^+ + 2\ e \rightleftharpoons IO_3^- + 3\ H_2O$	1.601	$Ag^{2+} + e \rightleftharpoons Ag^+$	1.980
$Ce^{4+} + e \rightleftharpoons Ce^{3+}$	1.61	$S_2O_8^{2-} + 2\ e \rightleftharpoons 2\ SO_4^{2-}$	2.010
$HClO + H^+ + e \rightleftharpoons 1/2Cl_2 + H_2O$	1.611	$OH + e \rightleftharpoons OH^-$	2.02
$HClO_2 + 3\ H^+ + 3\ e \rightleftharpoons 1/2Cl_2 + 2\ H_2O$	1.628	$O_3 + 2\ H^+ + 2\ e \rightleftharpoons O_2 + H_2O$	2.076
$HClO_2 + 2\ H^+ + 2\ e \rightleftharpoons HClO + H_2O$	1.645	$S_2O_8^2 + 2\ H^+ + 2\ e \rightleftharpoons 2\ HSO_4^-$	2.123
$NiO_2 + 4\ H^+ + 2\ e \rightleftharpoons Ni^{2+} + 2\ H_2O$	1.678	$F_2O + 2\ H^+ + 4\ e \rightleftharpoons H_2O + 2\ F^-$	2.153
$MnO_4^- + 4\ H^+ + 3\ e \rightleftharpoons MnO_2 + 2\ H_2O$	1.679	$FeO_4^{2-} + 8\ H^+ + 3\ e \rightleftharpoons Fe^{3+} + 4\ H_2O$	2.20
$PbO_2 + SO_4^{2-} + 4\ H^+ + 2\ e \rightleftharpoons PbSO_4 + 2\ H_2O$	1.6913	$O(g) + 2\ H^+ + 2\ e \rightleftharpoons H_2O$	2.421
		$H_2N_2O_2 + 2\ H^+ + 2\ e \rightleftharpoons N_2 + 2\ H_2O$	2.65
$Au^+ + e \rightleftharpoons Au$	1.692	$F_2 + 2\ e \rightleftharpoons 2\ F^-$	2.866
$CeOH^{3+} + H^+ + e \rightleftharpoons Ce^{3+} + H_2O$	1.715	$F_2 + 2\ H^+ + 2\ e \rightleftharpoons 2\ HF$	3.053

Table 3
REDUCTION REACTIONS HAVING E° VALUES MORE NEGATIVE THAN THAT OF THE STANDARD HYDROGEN ELECTRODE

Reaction	E°, V	Reaction	E°, V
$2\ H^+ + 2\ e \rightleftharpoons H_2$	0.00000	$H_3PO_4 + 2\ H^+ + 2\ e \rightleftharpoons H_3PO_3 + H_2O$	−0.276
$AgCN + e \rightleftharpoons Ag + CN^-$	−0.017	$Co^{2+} + 2\ e \rightleftharpoons Co$	−0.28
$2\ WO_3 + 2\ H^+ + 2\ e \rightleftharpoons W_2O_5 + H_2O$	−0.029	$PbBr_2 + 2\ e \rightleftharpoons Pb + 2\ Br^-$	−0.284
$W_2O_5 + 2\ H^+ + 2\ e \rightleftharpoons 2\ WO_2 + H_2O$	−0.031	$2\ HCNO + 2\ H^+ + 2\ e \rightleftharpoons (CN)_2 + 2\ H_2O$	−0.330
$D^+ + e \rightleftharpoons 1/2D_2$	−0.0034	$Tl^+ + e \rightleftharpoons Tl(Hg)$	−0.3338
$Ag_2S + 2\ H^+ + 2\ e \rightleftharpoons 2\ Ag + H_2S$	−0.0366	$Tl^+ + e \rightleftharpoons Tl$	−0.336
$Fe^{3+} + 3\ e \rightleftharpoons Fe$	−0.037	$In^{3+} + 3\ e \rightleftharpoons In$	−0.3382
$Hg_2I_2 + 2\ e \rightleftharpoons 2\ Hg + 2\ I^-$	−0.0405	$TlOH + e \rightleftharpoons Tl + OH^-$	−0.34
$2\ D^+ + 2\ e \rightleftharpoons D_2$	−0.044	$PbF_2 + 2\ e \rightleftharpoons Pb + 2\ F^-$	−0.3444
$Tl(OH)_3 + 2\ e \rightleftharpoons TlOH + 2\ OH^-$	−0.05	$PbSO_4 + 2\ e \rightleftharpoons Pb(Hg) + SO_4^{2-}$	−0.3505
$TiOH^{3+} + H^+ + e \rightleftharpoons Ti^{3+} + H_2O$	−0.055	$Cd^{2+} + 2\ e \rightleftharpoons Cd(Hg)$	−0.3521
$2\ H_2SO_3 + H^+ + 2\ e \rightleftharpoons HS_2O_4^- + 2\ H_2O$	−0.056	$PbSO_4 + 2\ e \rightleftharpoons Pb + SO_4^{2-}$	−0.3588
$P(white) + 3\ H^+ + 3\ e \rightleftharpoons PH_3(g)$	−0.063	$Cu_2O + H_2O + 2\ e \rightleftharpoons 2\ Cu + 2\ OH^-$	−0.360
$O_2^- + H_2O + 2\ e \rightleftharpoons HO_2^- + OH^-$	−0.076	$Eu^{3+} + e \rightleftharpoons Eu^{2+}$	−0.36
$2\ Cu(OH)_2 + 2\ e \rightleftharpoons Cu_2O + 2\ OH^- + H_2O$	−0.080	$PbI_2 + 2\ e \rightleftharpoons Pb + 2\ I^-$	−0.365
$WO_3 + 6\ H^+ + 6\ e \rightleftharpoons W + 3\ H_2O$	−0.090	$SeO_3^{2-} + 3\ H_2O + 4\ e \rightleftharpoons Se + 6\ OH^-$	−0.366
$P(red) + 3\ H^+ + 3\ e \rightleftharpoons PH_3(g)$	−0.111	$Ti^{3+} + e \rightleftharpoons Ti^{2+}$	−0.368
$GeO_2 + 2\ H^+ + 2\ e \rightleftharpoons GeO + H_2O$	−0.118	$Se + 2\ H^+ + 2\ e \rightleftharpoons H_2Se(aq)$	−0.399
$WO_2 + 4\ H^+ + 4\ e \rightleftharpoons W + 2\ H_2O$	−0.119	$In^{2+} + e \rightleftharpoons In^+$	−0.40
$Pb^{2+} + 2\ e \rightleftharpoons Pb(Hg)$	−0.1205	$Cd^{2+} + 2\ e \rightleftharpoons Cd$	−0.4030
$Pb^{2+} + 2\ e \rightleftharpoons Pb$	−0.1262	$Cr^{3+} + e \rightleftharpoons Cr^{2+}$	−0.407
$CrO_4^{2-} + 4\ H_2O + 3\ e \rightleftharpoons Cr(OH)_3 + 5\ OH^-$	−0.13	$2\ S + 2\ e \rightleftharpoons S_2^{2-}$	−0.42836
$In^+ + e \rightleftharpoons In$	−0.14	$Tl_2SO_4 + 2\ e \rightleftharpoons Tl + SO_4^{2-}$	−0.4360
$O_2 + 2\ H_2O + 2\ e \rightleftharpoons H_2O_2 + 2\ OH^-$	−0.146	$In^{3+} + 2\ e \rightleftharpoons In^+$	−0.443
$AgI + e \rightleftharpoons Ag + I^-$	−0.15224	$Fe^{2+} + 2\ e \rightleftharpoons Fe$	−0.447
$2\ NO_2^- + 2\ H_2O + 4\ e \rightleftharpoons N_2O_2^{2-} + 4\ OH^-$	−0.18	$H_3PO_2 + 3\ H^+ + 3\ e \rightleftharpoons P + 3\ H_2O$	−0.454
$H_2GeO_3 + 4\ H^+ + 4\ e \rightleftharpoons Ge + 3\ H_2O$	−0.182	$Bi_2O_3 + 3\ H_2O + 6\ e \rightleftharpoons 2\ Bi + 6\ OH^-$	−0.46
$CO_2 + 2\ H^+ + 2\ e \rightleftharpoons HCOOH$	−0.199	$NO_2^- + H_2O + e \rightleftharpoons NO + 2\ OH$	−0.46
$Mo^{3+} + 3\ e \rightleftharpoons Mo$	−0.200	$PbHPO_4 + 2\ e \rightleftharpoons Pb + HPO_4^{2-}$	−0.465
$2\ SO_3^{2-} + 4\ H^+ + 2\ e \rightleftharpoons S_2O_6^{2-} + H_2O$	−0.22	$S + 2\ e \rightleftharpoons S^{2-}$	−0.47627
$Cu(OH)_2 + 2\ e \rightleftharpoons Cu + 2\ OH^-$	−0.222	$S + H_2O + 2\ e \rightleftharpoons HS^- + OH^-$	−0.478
$CdSO_4 + 2\ e \rightleftharpoons Cd + SO_4^{2-}$	−0.246	$NiO_2 + 2\ H_2O + 2\ e \rightleftharpoons Ni\ (OH)_2 + 2\ OH^-$	−0.490
$V(OH)_4^+ + 4\ H^+ + 5\ e \rightleftharpoons V + 4\ H_2O$	−0.254	$In^{3+} + e \rightleftharpoons In^{2+}$	−0.49
$V^{3+} + e \rightleftharpoons V^{2+}$	−0.255	$H_3PO_3 + 2\ H^+ + 2\ e \rightleftharpoons H_3PO_2 + H_2O$	−0.499
$Ni^{2+} + 2\ e \rightleftharpoons Ni$	−0.257	$TiO_2 + 4\ H^+ + 2\ e \rightleftharpoons Ti^{2+} + 2\ H_2O$	−0.502
$PbCl_2 + 2\ e \rightleftharpoons Pb + 2\ Cl^-$	−0.2675	$H_3PO_2 + H^+ + e \rightleftharpoons P + 2\ H_2O$	−0.508

Table 3 (continued)

REDUCTION REACTIONS HAVING E° VALUES MORE NEGATIVE THAN THAT OF THE STANDARD HYDROGEN ELECTRODE

Reaction	E°, V	Reaction	E°, V
$Sb + 3 H^+ + 3 e \rightleftharpoons SbH_3$	−0.510	$UO_2^{2+} + 4 H^+ + 6 e \rightleftharpoons U + 2 H_2O$	−1.444
$HPbO_2^- + H_2O + 2 e \rightleftharpoons Pb + 3 OH^-$	−0.537	$Cr(OH)_3 + 3 e \rightleftharpoons Cr + 3 OH^-$	−1.48
$TlCl + e \rightleftharpoons Tl + Cl^-$	−0.5568	$HfO_2 + 4 H^+ + 4 e \rightleftharpoons Hf + 2 H_2O$	−1.505
$Ga^{3+} + 3 e \rightleftharpoons Ga$	−0.560	$ZrO_2 + 4 H^+ + 4 e \rightleftharpoons Zr + 2 H_2O$	−1.553
$Fe(OH)_3 + e \rightleftharpoons Fe(OH)_2 + OH^-$	−0.56	$Mn(OH)_2 + 2 e \rightleftharpoons Mn + 2 OH^-$	−1.56
$TeO_3^{2-} + 3 H_2O + 4 e \rightleftharpoons Te + 6 OH^-$	−0.57	$Ba^{2+} + 2 e \rightleftharpoons Ba(Hg)$	−1.570
$2 SO_3^{2-} + 3 H_2O + 4 e \rightleftharpoons S_2O_3^{2-} + 6 OH^-$	−0.571	$Ti^{2+} + 2 e \rightleftharpoons Ti$	−1.630
$PbO + H_2O + 2 e \rightleftharpoons Pb + 2 OH^-$	−0.580	$HPO_3^{2-} + 2 H_2O + 2 e \rightleftharpoons H_2PO_2^- + 3 OH^-$	−1.65
$ReO_2^- + 4 H_2O + 7 e \rightleftharpoons Re + 8 OH^-$	−0.584	$Al^{3+} + 3 e \rightleftharpoons Al$	−1.662
$SbO_3^- + H_2O + 2 e \rightleftharpoons SbO_2^- + 2 OH^-$	−0.59	$SiO_3^{2-} + H_2O + 4 e \rightleftharpoons Si + 6 OH^-$	−1.697
$U^{4+} + e \rightleftharpoons U^{3+}$	−0.607	$HPO_3^{2-} + 2 H_2O + 3 e \rightleftharpoons P + 5 OH^-$	−1.71
$As + 3 H^+ + 3 e \rightleftharpoons AsH_3$	−0.608	$HfO^{2+} + 2 H^+ + 4 e \rightleftharpoons Hf + H_2O$	−1.724
$Nb_2O_5 + 10 H^+ + 10 e \rightleftharpoons 2 Nb + 5 H_2O$	−0.644	$ThO_2 + 4 H^+ + 4 e \rightleftharpoons Th + 2 H_2O$	−1.789
$TlBr + e \rightleftharpoons Tl + Br^-$	−0.658	$H_2BO_3^- + H_2O + 3 e \rightleftharpoons B + 4 OH^-$	−1.79
$SbO_2^- + 2 H_2O + 3 e \rightleftharpoons Sb + 4 OH^-$	−0.66	$Sr^{2+} + 2 e \rightleftharpoons Sr(Hg)$	−1.793
$AsO_2^- + 2 H_2O + 3 e \rightleftharpoons As + 4 OH^-$	−0.68	$U^{3+} + 3 e \rightleftharpoons U$	−1.798
$Ag_2S + 2 e \rightleftharpoons 2 Ag + S^{2-}$	−0.691	$H_2PO_2^- + e \rightleftharpoons P + 2 OH^-$	−1.82
$AsO_4^{3-} + 2 H_2O + 2 e \rightleftharpoons AsO_2^- + 4 OH^-$	−0.71	$Be^{2+} + 2 e \rightleftharpoons Be$	−1.847
$Ni(OH)_2 + 2 e \rightleftharpoons Ni + 2 OH^-$	−0.72	$Np^{3+} + 3 e \rightleftharpoons Np$	−1.856
$Co(OH)_2 + 2 e \rightleftharpoons Co + 2 OH^-$	−0.73	$Th^{4+} + 4 e \rightleftharpoons Th$	−1.899
$H_2SeO_3 + 4 H^+ + 4 e \rightleftharpoons Se + 3 H_2O$	−0.74	$Pu^{3+} + 3 e \rightleftharpoons Pu$	−2.031
$Cr^{3+} + 3 e \rightleftharpoons Cr$	−0.744	$AlF_6^{3-} + 3 e \rightleftharpoons Al + 6 F^-$	−2.069
$Ta_2O_5 + 10 H^+ + 10 e \rightleftharpoons 2 Ta + 5 H_2O$	−0.750	$Sc^{3+} + 3 e \rightleftharpoons Sc$	−2.077
$TlI + e \rightleftharpoons Tl + I^-$	−0.752	$H_2 + 2 e \rightleftharpoons 2 H^-$	−2.23
$Zn^{2+} + 2 e \rightleftharpoons Zn$	−0.7618	$H_2AlO_3^- + H_2O + 3 e \rightleftharpoons Al + 4 OH^-$	−2.33
$Zn^{2+} + 2 e \rightleftharpoons Zn(Hg)$	−0.7628	$ZrO(OH)_2 + H_2O + 4 e \rightleftharpoons Zr + 4 OH^-$	−2.36
$Te + 2 H^+ + 2 e \rightleftharpoons H_2Te$	−0.793	$Mg^{2+} + 2 e \rightleftharpoons Mg$	−2.372
$ZnSO_4 \cdot 7H_2O + 2 e \rightleftharpoons Zn(Hg) + SO_4^{2-}$	−0.7993	$Y^{3+} + 3 e \rightleftharpoons Y$	−2.372
(Sat'd $ZnSO_4$)		$Eu^{3+} + 3 e \rightleftharpoons Eu$	−2.407
$Cd(OH)_2 + 2 e \rightleftharpoons Cd(Hg) + 2 OH^-$	−0.809	$Nd^{3+} + 3 e \rightleftharpoons Nd$	−2.431
$2 H_2O + 2 e \rightleftharpoons H_2 + 2 OH^-$	−0.8277	$Th(OH)_4 + 4 e \rightleftharpoons Th + 4 OH^-$	−2.48
$2 NO_3^- + 2 H_2O + 2 e \rightleftharpoons N_2O_4 + 4 OH^-$	−0.85	$Ce^{3+} + 3 e \rightleftharpoons Ce$	−2.483
$H_3BO_3 + 3 H^+ + 3 e \rightleftharpoons B + 3 H_2O$	−0.8698	$HfO(OH)_2 + H_2O + 4 e \rightleftharpoons Hf + 4 OH^-$	−2.50
$P + 3 H_2O + 3 e \rightleftharpoons PH_3(g) + 3 OH^-$	−0.87	$La^{3+} + 3 e \rightleftharpoons La$	−2.522
$HSnO_2^- + H_2O + 2 e \rightleftharpoons Sn + 3 OH^-$	−0.909	$Be_2O_3^{2-} + 3 H_2O + 4 e \rightleftharpoons 2 Be + 6 OH^-$	−2.63
$Cr^{2+} + 2 e \rightleftharpoons Cr$	−0.913	$Mg(OH)_2 + 2 e \rightleftharpoons Mg + 2 OH^-$	−2.690
$Se + 2 e \rightleftharpoons Se^{2-}$	−0.924	$Mg^+ + e \rightleftharpoons Mg$	−2.70
$SO_4^{2-} + H_2O + 2 e \rightleftharpoons SO_3^{2-} + 2 OH^-$	−0.93	$Na^+ + e \rightleftharpoons Na$	−2.71
$Sn(OH)_6^{2-} + 2 e \rightleftharpoons HSnO_2^- + 3 OH^- + H_2O$	−0.93	$Ca^{2+} + 2 e \rightleftharpoons Ca$	−2.868
$NpO_2 + H_2O + H^+ + e \rightleftharpoons Np(OH)_3$	−0.962	$Sr(OH)_2 + 2 e \rightleftharpoons Sr + 2 OH^-$	−2.88
$PO_4^{3-} + 2 H_2O + 2 e \rightleftharpoons HPO_3^{2-} + 3 OH^-$	−1.05	$Sr^{2+} + 2 e \rightleftharpoons Sr$	−2.89
$Nb^{3+} + 3 e \rightleftharpoons Nb$	−1.099	$La(OH)_3 + 3 e \rightleftharpoons La + 3 OH^-$	−2.90
$2 SO_3^{2-} + 2 H_2O + 2 e \rightleftharpoons S_2O_4^{2-} + 4 OH^-$	−1.12	$Ba^{2+} + 2 e \rightleftharpoons Ba$	−2.912
$Te + 2 e \rightleftharpoons Te^{2-}$	−1.143	$Cs^+ + e \rightleftharpoons Cs$	−2.92
$V^{2+} + 2 e \rightleftharpoons V$	−1.175	$K^+ + e \rightleftharpoons K$	−2.931
$Mn^{2+} + 2 e \rightleftharpoons Mn$	−1.185	$Rb^+ + e \rightleftharpoons Rb$	−2.98
$CrO_2^- + 2 H_2O + 3 e \rightleftharpoons Cr + 4 OH^-$	−1.2	$Ba(OH)_2 + 2 e \rightleftharpoons Ba + 2 OH^-$	−2.99
$ZnO_2^- + 2 H_2O + 2 e \rightleftharpoons Zn + 4 OH^-$	−1.215	$Ca(OH)_2 + 2 e \rightleftharpoons Ca + 2 OH^-$	−3.02
$H_2GaO_3^- + H_2O + 3 e \rightleftharpoons Ga + 4 OH^-$	−1.219	$Li^+ + e \rightleftharpoons Li$	−3.0401
$H_2BO_3^- + 5 H_2O + 8 e \rightleftharpoons BH_4^- + 8 OH^-$	−1.24	$3 N_2 + 2 H^+ + 2 e \rightleftharpoons 2 NH_3$	−3.09
$SiF_6^{2-} + 4 e \rightleftharpoons Si + 6 F^-$	−1.24	$Eu^{2+} + 2 e \rightleftharpoons Eu$	−3.395
$Sn^{2+} + 2 e \rightleftharpoons Sn$	−1.375	$Ca^+ + e \rightleftharpoons Ca$	−3.80
$Ce^{3+} + 3 e \rightleftharpoons Ce(Hg)$	−1.4373	$Sr^+ + e \rightleftharpoons Sr$	−4.10

DISSOCIATION CONSTANTS OF ORGANIC BASES IN AQUEOUS SOLUTION

No.	Compound	Temp. °C	Step	pK_a	K_a
a1	Acetamide............	25		0.63	2.34×10^{-1}
a2	Acridine.............	20		5.58	2.63×10^{-6}
a3	α-Alanine	25		2.345	4.52×10^{-3}
a4	Alanine, glycyl-......	25		3.153	7.03×10^{-4}
a5	Alanine, methoxy-(DL)	25		2.037	9.18×10^{-3}
a6	Alanine, phenyl......	25	2	9.18	6.61×10^{-10}
a7	Allothreonine........	25	1	2.108	7.80×10^{-3}
	Allothreonine........	25	2	9.096	8.02×10^{-10}
a8	n-Amylamine........	25		10.63	2.34×10^{-11}
a9	Aniline.............	25		4.63	2.34×10^{-5}
a10	Aniline, n-allyl......	25		4.17	6.76×10^{-5}
a11	Aniline, 4-(p-aminobenzoyl)...	25	1	2.932	1.17×10^{-3}
a12	Aniline, 4-benzyl.....	25		2.17	6.76×10^{-3}
a13	Aniline, 2-bromo.....	25		2.53	2.95×10^{-3}
a14	Aniline, 3-bromo.....	25		3.58	2.63×10^{-4}
a15	Aniline, 4-bromo.....	25		3.86	1.38×10^{-4}
a16	Aniline, 4-bromo-N,N-dimethyl.....	25		4.232	5.86×10^{-5}
a17	Aniline, o-chloro.....	25		2.65	2.24×10^{-3}
a18	Aniline, m-chloro.....	25		3.46	3.47×10^{-4}
a19	Aniline, p-chloro.....	25		4.15	7.08×10^{-5}
a20	Aniline, 3-chloro-N,N-dimethyl.....	20		3.837	1.46×10^{-4}
a21	Aniline, 4-chloro-N,N-dimethyl.....	20		4.395	4.03×10^{-5}
a22	Aniline, 3,5-dibromo...	25		2.34	4.57×10^{-3}
a23	Aniline, 2,4-dichloro...	22		2.05	8.91×10^{-3}
a24	Aniline, N,N-diethyl...	22		6.61	2.46×10^{-7}
a25	Aniline, N,N-dimethyl..	25		5.15	7.08×10^{-6}
a26	Aniline, N,N-dimethyl-3-nitro....	25		2.626	2.37×10^{-3}
a27	Aniline, N-ethyl......	24		5.12	7.59×10^{-6}
a28	Aniline, 2-fluoro.....	25		3.20	6.31×10^{-4}
a29	Aniline, 3-fluoro.....	25		3.50	3.16×10^{-4}
a30	Aniline, 4-fluoro.....	25		4.65	2.24×10^{-5}
a31	Aniline, 2-iodo.......	25		2.60	2.51×10^{-3}
a32	Aniline, N-methyl.....	25		4.848	1.41×10^{-5}
a33	Aniline, 4-methylthio..	25		4.35	4.46×10^{-5}
a34	Aniline, 3-nitro......	25		2.466	3.42×10^{-3}
a35	Aniline, 4-nitro......	25		1.0	1.00×10^{-1}
a36	Aniline, 2-sulfonic acid	25	2	2.459	3.47×10^{-3}
a37	Aniline, 3-sulfonic acid	25	2	3.738	1.82×10^{-4}
a38	Aniline, 4-sulfonic acid	25	2	3.227	5.92×10^{-4}
a39	o-Anisidine.........	25		4.52	3.02×10^{-5}
a40	m-Anisidine.........	25		4.23	5.89×10^{-5}
a41	p-Anisidine.........	25		5.34	4.57×10^{-6}
a42	Arginine	25	1	1.8217	1.51×10^{-2}
		25	2	8.9936	1.01×10^{-9}
a43	Asparagine..........	20	1	2.213	6.12×10^{-3}
		20	2	8.85	1.41×10^{-9}
a44	Asparagine, glycyl.....	25	1	2.942	1.14×10^{-3}
		18	2	8.44	3.63×10^{-9}
a45	DL-Aspartic acid......	1	1	2.122	7.55×10^{-3}
		1	2	4.006	1.00×10^{-4}
a46	Azetidine (Trimethylimidine)..	25		11.29	5.12×10^{-12}
a47	Aziridine............	25		8.01	9.77×10^{-9}
b1	Benzene, 4-aminoazo..	25		2.82	1.51×10^{-3}
b2	Benzene, 2-aminoethyl (β-Phenylamine)....	25		9.84	1.45×10^{-10}
b3	Benzene, 4-dimethylaminoazo.	25		3.226	5.94×10^{-4}
b4	Benzidine............	30	1	4.66	2.19×10^{-5}
		30	2	3.57	2.69×10^{-4}
b5	Benzimidazole.......	25		5.532	2.94×10^{-6}
b6	Benzimidazole, 2-ethyl.	25		6.18	6.61×10^{-7}
b7	Benzimidazole, 2-methyl.	25		6.19	6.46×10^{-7}
b8	Benzimidazole, 2-phenyl.	25	1	5.23	5.89×10^{-6}
		25	2	11.91	1.23×10^{-12}
b9	Benzoic acid, 2-amino (Anthranilic acid)...	25	1	2.108	7.80×10^{-3}
		25	2	4.946	1.13×10^{-5}
b10	Benzoic acid, 4-amino..	25	1	2.501	3.15×10^{-3}
		25	2	4.874	1.33×10^{-5}
b11	Benzylamine.........	25		9.33	4.67×10^{-10}
b12	Betaine.............	0		1.83	1.48×10^{-2}
b13	Biphenyl, 2-amino.....	22		3.82	1.51×10^{-4}
b14	Bornylamine(trans-)...	25		10.17	6.76×10^{-11}
b15	Brucine.............	25	1	8.28	5.24×10^{-9}
b16	Butane, 1-amino-3-methyl............	25		10.60	2.51×10^{-11}
b17	Butane, 2-amino-2-methyl............	19		10.85	1.41×10^{-11}
b18	Butane, 1,4-diamino (Putrescine)........	10	1	11.15	7.08×10^{-12}
		10	2	9.71	1.95×10^{-10}
b19	n-Butylamine........	20		10.77	1.69×10^{-11}
b20	t-Butylamine........	18		10.83	1.48×10^{-11}
b21	Butyric acid, 4-amino..	25	1	4.0312	9.31×10^{-5}
		25	2	10.5557	2.78×10^{-11}
b22	n-Butyric acid, glycyl-2-amino......	25	1	3.1546	7.01×10^{-4}
c1	Cacodylic acid........	25	1	1.57	2.69×10^{-2}
			2	6.27	5.37×10^{-7}
c2	β-Chlortriethyl-ammonium.........	25		8.80	1.59×10^{-9}
c3	Cinnoline............	20		2.37	4.27×10^{-3}
c4	Codeine.............	25		8.21	6.15×10^{-9}
c5	Cyclohexaneamine, n-butyl............	25		11.23	5.89×10^{-12}
c6	Cyclohexylamine......	24		10.66	2.19×10^{-11}
c7	Cystine.............	30	1	1.90	1.25×10^{-2}
		30	2	8.24	5.76×10^{-9}
d1	n-Decylamine........	25		10.64	2.29×10^{-11}
d2	Diethylamine........	40		10.489	3.24×10^{-11}
d3	Diisobutylamine......	21		10.91	1.23×10^{-11}
d4	Diisopropylamine.....	28.5		10.96	1.09×10^{-11}
d5	Dimethylamine.......	25		10.732	1.85×10^{-11}
d6	n-Diphenylamine.....	25		0.79	1.62×10^{-1}
d7	n-Dodecaneamine (Laurylamine)......	25		10.63	2.35×10^{-11}
e1	d-Ephedrine.........	10		10.139	7.26×10^{-11}
e2	l-Ephedrine.........	10		9.958	1.10×10^{-10}
e3	Ethane, 1-amino-3-methoxy.........	10		9.89	1.29×10^{-10}
e4	Ethane, 1,2-bismethylamino..	25	1	10.40	3.98×10^{-11}
		25	2	8.26	5.50×10^{-9}
e5	Ethanol, 2-amino.....	25		9.50	3.16×10^{-11}
e6	Ethylamine..........	20		10.807	1.56×10^{-11}
e7	Ethylenediamine.....	0	1	10.712	1.94×10^{-11}
		0	2	7.564	2.73×10^{-8}
g1	l-Glutamic acid......	25	1	2.13	7.41×10^{-3}
		25	2	4.31	4.90×10^{-5}
e2	Glutamic acid, α-monoethyl.......	25	1	3.846	1.42×10^{-4}
		25	2	7.838	1.45×10^{-8}
e3	l-Glutamine.........	—		9.28	5.25×10^{-10}
e4	l-Glutathione........	25	2	3.59	2.57×10^{-4}
e5	Glycine.............	25	1	2.3503	4.46×10^{-3}
		25	2	9.7796	1.68×10^{-10}
e6	Glycine, n-acetyl.....	25		3.6698	2.14×10^{-4}
e7	Glycine, dimethyl.....	5		10.3371	4.60×10^{-11}
e8	Glycine, glycyl......	25		3.1397	7.25×10^{-4}
e9	Glycine, glycylglycyl...	25	1	3.225	5.96×10^{-4}
		25	2	8.090	8.13×10^{-9}
e10	Glycine, leucyl.......	25	1	3.25	5.62×10^{-4}
		25	2	8.28	5.25×10^{-9}
e11	Glycine, methyl (Sarcosine)........	25	1	2.21	6.16×10^{-3}
		25	2	10.12	7.58×10^{-11}
g12	Glycine, phenyl......	25	1	1.83	1.48×10^{-2}
		25	2	4.39	4.07×10^{-5}
g13	Glycine, N,n-propyl...	25	1	2.35	4.46×10^{-3}
		25	2	10.19	6.46×10^{-11}
g14	Glycine, tetraglycyl...	20	1	3.10	7.94×10^{-4}
		20	2	8.02	9.55×10^{-9}
g15	Glycylserine.........	25	1	2.9808	1.04×10^{-3}
		25	2	8.38	4.17×10^{-9}
h1	Hexadecaneamine.....	25		10.63	2.35×10^{-11}
h2	Heptane, 1-amino.....	25		10.66	2.19×10^{-11}
h3	Heptane, 2-amino.....	19		10.88	1.58×10^{-11}
h4	Heptane, 2-methylamino......	17		10.99	1.02×10^{-11}
h5	Hexadecaneamine.....	25		10.61	2.46×10^{-11}

No.	Compound	Temp. °C	Step	pK_a	K_a
h6	Hexamethylene-diamine	0	1	11.857	1.39×10^{-12}
		0	2	10.762	1.73×10^{-11}
h7	Hexanoic acid, 6-amino	25	1	4.373	4.23×10^{-5}
		25	2	10.804	1.57×10^{-11}
h8	n-Hexylamine	25		10.56	2.75×10^{-11}
h9	dl-Histidine	25	1	1.80	1.58×10^{-2}
		25	2	6.04	9.12×10^{-7}
		25	3	9.33	4.67×10^{-10}
h10	Histidine, β-alanyl (Carnosine)	20	1	2.73	1.86×10^{-3}
		20	2	6.87	1.35×10^{-7}
		20	3	9.73	1.48×10^{-10}
i1	Imidazol	25		6.953	1.11×10^{-7}
i2	Imidazol, 2,4-dimethyl	25		8.359	5.50×10^{-9}
i3	Imidazol, 1-methyl (Oxalmethyline)	25		6.95	1.12×10^{-7}
i4	Indane, 1-amino (d-1-Hydrindamine)	22.5		9.21	6.17×10^{-10}
i5	Isobutyric acid, 2-amino	25	1	2.357	4.30×10^{-3}
		25	2	10.205	6.23×10^{-11}
i6	Isoleucine	25	1	2.318	4.81×10^{-3}
		25	2	9.758	1.74×10^{-10}
i7	Isoquinoline (Leucoline)	20		5.42	3.80×10^{-6}
i8	Isoquinoline, 1-amino	20		7.59	2.57×10^{-8}
i9	Isoquinoline, 7-hydroxy	20	1	5.68	2.09×10^{-6}
		20	2	8.90	1.26×10^{-9}
l1	L-Leucine	25	1	2.328	4.70×10^{-3}
		25	2	9.744	1.80×10^{-10}
l2	Leucine, glycyl	25		3.18	6.61×10^{-4}
m1	Methionine	25	1	2.22	6.02×10^{-3}
		25	2	9.27	5.37×10^{-10}
m2	Methylamine	25		10.657	2.70×10^{-11}
m3	Morphine	25		8.21	6.16×10^{-9}
m4	Morpholine	25		8.33	4.67×10^{-9}
n1	Naphthalene, 1-amino-6-hydroxy	25		3.97	1.07×10^{-4}
n2	Naphthalene, dimethylamino	25		4.560	2.72×10^{-5}
n3	α-Naphthylamine	25		3.92	1.20×10^{-4}
n4	β-Naphthylamine	25		4.16	6.92×10^{-5}
n5	α-Naphthylamine, n-methyl	27		3.67	2.13×10^{-4}
n6	Neobornylamine(cis-)	25		10.01	9.77×10^{-11}
n7	Nicotine	25	1	8.02	9.55×10^{-9}
		25	2	3.12	7.59×10^{-4}
n8	n-Nonylamine	25		10.64	2.29×10^{-11}
n9	Norleucine	25		2.335	4.62×10^{-3}
o1	Octadecaneamine	25		10.60	2.51×10^{-11}
o2	Octylamine	25		10.65	2.24×10^{-11}
o3	Ornithine	25	1	1.705	1.97×10^{-2}
		25	2	8.690	2.04×10^{-9}
p1	Papaverine	25		6.40	3.98×10^{-7}
p2	Pentane, 3-amino	17		10.59	2.57×10^{-11}
p3	Pentane, 3-amino-3-methyl	16		11.01	9.77×10^{-12}
p4	n-Pentadecylamine	25		10.61	2.46×10^{-11}
p5	Pentanoic acid, 5-amino (Valeric acid)	25	1	4.270	5.37×10^{-5}
		25	2	10.766	1.71×10^{-11}
p6	Perimidine	20		6.35	4.47×10^{-7}
p7	Phenanthridine	20		5.58	2.63×10^{-6}
p8	1,10-Phenanthroline	25		4.84	1.44×10^{-5}
p9	o-Phenetidine (2-Ethoxyaniline)	28		4.43	3.72×10^{-5}
p10	m-Phenetidine (3-Ethoxyaniline)	25		4.18	6.60×10^{-5}
p11	p-Phenetidine (4-Ethoxyaniline)	28		5.20	6.31×10^{-6}
p12	α-Picoline	20		5.97	1.07×10^{-6}
p13	β-Picoline	20		5.68	2.09×10^{-6}
p14	γ-Picoline	20		6.02	9.55×10^{-7}
p15	Pilocarpine	30		6.87	1.35×10^{-7}
p16	Piperazine	23.5	1	9.83	1.48×10^{-10}
		23.5	2	5.56	2.76×10^{-6}
p17	Piperazine, 2,5-dimethyl(trans-)	25	1	9.66	2.19×10^{-10}
		25	2	5.20	6.31×10^{-6}
p18	Piperidine	25		11.123	7.53×10^{-12}
p19	Piperidine, 3-acetyl	25		3.18	6.61×10^{-4}
p20	Piperidine, 1-n-butyl	23		10.47	3.39×10^{-11}
p21	Piperidine, 1,2-dimethyl	25		10.22	6.03×10^{-11}
p22	Piperidine, 1-ethyl	23		10.45	3.55×10^{-11}
p23	Piperidine, 1-methyl	25		10.08	8.32×10^{-11}
p24	Piperidine, 2,2,6,6-tetramethyl	25		11.07	8.51×10^{-12}
p25	Piperidine, 2,2,4-trimethyl	30		11.04	9.12×10^{-12}
p26	Proline	25	1	1.952	1.11×10^{-2}
		25	2	10.640	2.29×10^{-11}
p27	Proline, hydroxy	25	1	1.818	1.52×10^{-2}
		25	2	9.662	2.18×10^{-10}
p28	Propane, 1-amino-2,2-dimethyl	25		10.15	7.08×10^{-11}
p29	Propane, 1,2-diamino	25	1	9.82	1.52×10^{-10}
		25	2	6.61	2.46×10^{-7}
p30	Propane, 1,3-diamino	10	1	10.94	1.15×10^{-11}
		10	2	9.03	9.33×10^{-10}
p31	Propane, 1,2,3-triamino	20	1	9.59	2.57×10^{-10}
		20	2	7.95	1.12×10^{-8}
p32	Propanoic acid, 3-amino (β-Alanine)	25	1	3.551	2.81×10^{-4}
		25	2	10.238	5.78×10^{-11}
p33	Propylamine	20		10.708	1.96×10^{-11}
p34	Pteridine	20		4.05	8.91×10^{-5}
p35	Pteridine, 2-amino-4,6-dihydroxy	20	2	6.59	2.57×10^{-7}
		20	3	9.31	4.90×10^{-10}
p36	Pteridine, 2-amino-4-hydroxy	20	1	8.07	5.01×10^{-8}
		20	2	7.96	1.10×10^{-8}
p37	Pteridine, 6-chloro	20		3.68	2.09×10^{-4}
p38	Pteridine, 6-hydroxy-4-methyl	20	1	4.08	8.32×10^{-5}
		20	2	8.11	8.09×10^{-9}
p39	Purine	20	1	2.30	5.01×10^{-3}
		20	2	8.96	1.10×10^{-9}
p40	Purine, 6-amino (Adenine)	25	1	4.12	7.59×10^{-5}
		25	2	9.83	1.48×10^{-10}
p41	Purine, 2-dimethylamino	20	1	4.00	1.00×10^{-4}
		20	2	10.24	5.75×10^{-11}
p42	Purine, 8-hydroxy	20	1	2.56	2.75×10^{-3}
		20	2	8.26	9.49×10^{-9}
p43	Pyrazine	27		0.65	2.24×10^{-1}
p44	Pyrazine, 2-methyl	27		1.45	3.54×10^{-2}
p45	Pyrazine, methylamino	25		3.39	4.07×10^{-4}
p46	Pyridazine	20		2.24	5.76×10^{-3}
p47	Pyrimidine, 2-amino	20		3.45	3.54×10^{-4}
p48	Pyrimidine, 2-amino-4,6-dimethyl	20		4.82	1.51×10^{-5}
p49	Pyrimidine, 2-amino-5-nitro	20		0.35	4.46×10^{-1}
p50	Pyridine	25		5.25	5.62×10^{-6}
p51	Pyridine, 2-aldoxime	20	1	3.59	2.57×10^{-4}
		20	2	10.18	6.61×10^{-11}
p52	Pyridine, 2-amino	20		6.82	1.51×10^{-7}
p53	Pyridine, 4-amino	25		9.1141	7.69×10^{-10}
p54	Pyridine, 2-benzyl	25		5.13	7.41×10^{-6}
p55	Pyridine, 3-bromo	25		2.84	1.45×10^{-3}
p56	Pyridine, 3-chloro	25		2.84	1.45×10^{-3}
p57	Pyridine, 2,5-diamino	20		6.48	3.31×10^{-7}
p58	Pyridine, 2,3-dimethyl (2,3-Lutidine)	25		6.57	2.69×10^{-7}
p59	Pyridine, 2,4-dimethyl (2,4-Lutidine)	25		6.99	1.02×10^{-7}
p60	Pyridine, 3,5-dimethyl (3,5-Lutidine)	25		6.15	7.08×10^{-7}
p61	Pyridine, 2-ethyl	25		5.89	1.28×10^{-6}
p62	Pyridine, 2-formyl	20		3.80	1.59×10^{-4}
p63	Pyridine, 2-hydroxy (2-Pyridol)	20	1	0.75	9.82×10^{-1}
		20	2	11.65	2.24×10^{-12}

DISSOCIATION CONSTANTS OF ORGANIC BASES IN AQUEOUS SOLUTION (*Continued*)

No.	Compound	Temp. °C	Step	pK_a	K_a	No.	Compound	Temp. °C	Step	pK_a	K_a
p64	Pyridine, 4-hydroxy...	20	1	3.20	6.31×10^{-4}	q15	Quinoxaline (Quinazine)........	20		0.56	3.63×10^{-1}
		20	2	11.12	7.59×10^{-12}	s1	Serine (2-amino-3-hydroxypropanoic acid)	25	1	2.186	5.49×10^{-3}
p65	Pyridine, methoxy....	25		6.47	3.30×10^{-7}			25	2	9.208	6.19×10^{-10}
p66	Pyridine, 4-methylamino......	20		9.65	2.24×10^{-10}	s2	Strychnine..........	25		8.26	5.49×10^{-9}
p67	Pyridine, 2,4,6-trimethyl......	25		7.43	3.72×10^{-8}	t1	Taurine (2-Aminoethane sulfonic acid).......	25	2	9.0614	8.69×10^{-10}
p68	Pyrrolidine..........	25		11.27	5.37×10^{-12}	t2	Tetradecaneamine (Myristilamine).....	25		10.62	2.40×10^{-11}
p69	Pyrrolidine, 1,2-dimethyl.......	26		10.20	6.31×10^{-11}	t3	Thiazole............	20		2.44	3.63×10^{-3}
p70	Pyrrolidine, n-methyl..	25		10.32	4.79×10^{-11}	t4	Thiazole, 2-amino.....	20		5.36	4.36×10^{-5}
q1	Quinazoline..........	20		3.43	3.72×10^{-4}	t5	Threonine...........	25	1	2.088	8.16×10^{-3}
q2	Quinazoline, 5-hydroxy.	20	1	3.62	2.40×10^{-4}			25	2	9.10	7.94×10^{-10}
		20	2	7.41	3.89×10^{-8}	t6	o-Toluidine.........	25		4.44	3.63×10^{-5}
q3	Quinine.............	25	1	8.52	3.02×10^{-9}	t7	m-Toluidine.........	25		4.73	1.86×10^{-5}
		25	2	4.13	7.41×10^{-5}	t8	p-Toluidine.........	25		5.08	8.32×10^{-6}
q4	Quinoline............	20		4.90	1.25×10^{-5}	t9	1,3,5-Triazine, 2,4,6-triamino.......	25		5.00	1.00×10^{-5}
q5	Quinoline, 3-amino....	20		4.91	1.23×10^{-5}	t10	Tridecaneamine.......	25		10.63	2.35×10^{-11}
q6	Quinoline, 3-bromo....	25		2.69	2.04×10^{-3}	t11	Triethylamine........	18		11.01	9.77×10^{-12}
q7	Quinoline, 8-carboxy...	25		1.82	1.51×10^{-2}	t12	Trimethylamine.......	25		9.81	1.55×10^{-10}
q8	Quinoline, 3-hydroxy (3-Quinolinol)......	20	1	4.28	5.25×10^{-5}	t13	Tryptophan..........	25	1	2.43	3.72×10^{-3}
		20	2	8.08	8.32×10^{-9}			25	2	9.44	3.63×10^{-10}
q9	Quinoline, 8-hydroxy (8-Quinolinol)......	20	1	5.017	1.21×10^{-6}	t14	Tyrosine............	25	2	9.11	7.76×10^{-10}
		25	2	9.812	1.54×10^{-10}			25	3	10.13	7.41×10^{-11}
q10	Quinoline, 8-hydroxy-5-sulfo............	25	1	4.112	7.73×10^{-5}	t15	Tyrosineamide........	25		7.33	4.68×10^{-8}
		25	2	8.757	1.75×10^{-9}	u1	Urea................	21		0.10	7.94×10^{-1}
q11	Quinoline, 6-methoxy..	20		5.03	9.33×10^{-6}	v1	Valine..............	25	1	2.286	5.17×10^{-3}
q12	Quinoline, 2-methyl (Quinaldine)........	20		5.83	1.48×10^{-6}			25	2	9.719	1.91×10^{-10}
q13	Quinoline, 4-methyl (Lepidine)...........	20		5.67	2.14×10^{-6}						
q14	Quinoline, 5-methyl...	20		5.20	6.31×10^{-6}						

DISSOCIATION CONSTANTS OF INORGANIC BASES IN AQUEOUS SOLUTIONS

(Approximately 0.1–0.01 N)

Compound	T°C	Step	K_b	pK_b
Ammonium hydroxide.............	25		1.79×10^{-5}	4.75
Arsenous oxide....................	25		1.1×10^{-4}	3.96
Beryllium hydroxide................	25	2	5×10^{-11}	10.30
Calcium hydroxide................	25	1	3.74×10^{-3}	2.43
Calcium hydroxide................	30	2	4.0×10^{-2}	1.40
Deuteroammonium hydroxide.......	25		1.1×10^{-5}	4.96
Hydrazine.......................	20		1.7×10^{-6}	5.77
Hydroxylamine...................	20		1.07×10^{-8}	7.97
Lead Hydroxide...................	25		9.6×10^{-4}	3.02
Silver Hydroxide..................	25		1.1×10^{-4}	3.96
Zinc Hydroxide...................	25		9.6×10^{-4}	3.02

DISSOCIATION CONSTANTS OF ORGANIC ACIDS IN AQUEOUS SOLUTIONS

Compound	T°C	Step	K	pK	Compound	T°C	Step	K	pK
Acetic	25		1.76×10^{-5}	4.75	Cyclopropane-1:1-dicarboxylic	25	1	1.51×10^{-2}	1.82
Acetoacetic	18		2.62×10^{-4}	3.58	Cyclopropane-1:1-dicarboxylic	25	2	3.72×10^{-8}	7.43
Acrylic	25		5.6×10^{-5}	4.25	DL-Cysteine	30	1	7.25×10^{-9}	8.14
Adipamic	25		2.35×10^{-5}	4.63	DL-Cysteine	30	2	4.6×10^{-11}	10.34
Adipic	25	1	3.71×10^{-5}	4.43	L-Cystine	25	1	1.4×10^{-8}	7.85
Adipic	25	2	3.87×10^{-6}	5.41	L-Cystine	25	2	1.4×10^{-10}	9.85
d-Alanine	25		1.35×10^{-10}	9.87					
Allantoin	25		1.10×10^{-9}	8.96	Deuteroacetic (in D_2O)	25		5.5×10^{-6}	5.25
Alloxanic	25		2.3×1^{-7}	6.64	Dichloroacetic	25		3.32×0^{-2}	1.48
a-Aminoacetic (glycine)	25		1.67×10^{-10}	9.78	Dichloroacetylacetic	?		7.8×10^{-3}	2.11
o-Aminobenzoic	25		1.07×10^{-7}	6.97	Dichlorophenol (2,3-)	25		3.6×10^{-8}	7.44
m-Aminobenzoic	25		1.67×10^{-5}	4.78	Dihydroxybenzoic (2,2-)	25		1.14×10^{-3}	2.94
p-Aminobenzoic	25		1.2×10^{-5}	4.92	Dihydroxybenzoic (2,5-)	25		1.08×10^{-3}	2.97
o-Aminobenzosulfonic	25		3.3×10^{-3}	2.48	Dihydroxybenzoic (3,4-)	25		3.3×10^{-5}	4.48
m-Aminobenzosulfonic	25		1.85×10^{-4}	3.73	Dihydroxybenzoic (3,5-)	25		9.1×10^{-5}	4.04
p-Aminobenzosulfonic	25		5.81×10^{-4}	3.24	Dihydroxymalic	25		1.12×10^{-2}	1.92
Anisic	25		3.38×10^{-5}	4.47	Dihydroxytartaric	25		1.2×10^{-2}	1.92
o-β-Anisylpropionic	25		1.59×10^{-5}	4.80	Dimethylglycine	25		1.3×10^{-10}	9.89
m-β-Anisylpropionic	25		2.24×10^{-5}	4.65	Dimethylmalic	25	1	6.83×10^{-4}	3.17
p-β-Anisylpropionic	25		2.04×10^{-5}	4.69	Dimethylmalic	25	2	8.72×10^{-7}	6.06
Ascorbic	24	1	7.94×10^{-5}	4.10	Dimethylmalonic	25		7.08×10^{-4}	3.15
Ascorbic	16	2	1.62×10^{-12}	11.79					
DL-Aspartic	25	1	1.38×10^{-4}	3.86	Dinicotinic	25		1.6×10^{-3}	2.80
DL-Aspartic	25	2	1.51×10^{-10}	9.82	Dinitrophenol (2,4-)	15		1.1×10^{-4}	3.96
					Dinitrophenol (3,6-)	15		7.1×10^{-6}	5.15
Barbituric	25		9.8×10^{-5}	4.01	Diphenylacetic	25		1.15×10^{-4}	3.94
Benzoic	25		6.46×10^{-5}	4.19					
Benzosulfonic	25		2×10^{-1}	0.70	Ethylbenzoic	25		4.47×10^{-5}	4.35
Bromoacetic	25		2.05×10^{-3}	2.69	Ethylphenylacetic	25		4.27×10^{-5}	4.37
o-Bromobenzoic	25		1.45×10^{-3}	2.84					
m-Bromobenzoic	25		1.37×10^{-4}	3.86	Fluorobenzoic	17		1.25×10^{-3}	2.90
n-Butyric	20		1.54×10^{-5}	4.81	Formic	20		1.77×10^{-4}	3.75
iso-Butyric	18		1.44×10^{-5}	4.84	Fumaric (trans-)	18	1	9.30×10^{-4}	3.03
					Fumaric (trans-)	18	2	3.62×10^{-5}	4.44
Cacodylic	25		6.4×10^{-7}	6.19	Furancarboxylic	25		7.1×10^{-4}	3.15
n-Caproic	18		1.43×10^{-5}	4.83	Furoic	25		6.76×10^{-4}	3.17
iso-Caproic	18		1.46×10^{-5}	4.84					
Chloroacetic	25		1.40×10^{-3}	2.85	Gallic	25		3.9×10^{-5}	4.41
o-Chlorobenzoic	25		1.20×10^{-3}	2.92	Glutaramic	25		3.98×10^{-5}	4.60
m-Chlorobenzoic	25		1.51×10^{-4}	3.82	Glutaric	25	1	4.58×10^{-5}	4.31
p-Chlorobenzoic	25		1.04×10^{-4}	3.98	Glutaric	25	2	3.89×10^{-6}	5.41
α-Chlorobutyric	R.T.		1.39×10^{-3}	2.86	Glycerol	25		7×10^{-15}	14.15
β-Chlorobutyric	R.T.		8.9×10^{-5}	4.05	Glycine	25		1.67×10^{-10}	9.78
γ-Chlorobutyric	R.T.		3.0×10^{-5}	4.52	Glycol	25		6×10^{-15}	14.22
o-Chlorocinnamic	25		5.89×10^{-5}	4.23	Glycolic	25		1.48×10^{-4}	3.83
m-Chlorocinnamic	25		5.13×10^{-5}	4.29					
p-Chlorocinnamic	25		3.89×10^{-5}	4.41	Heptanoic	25		1.28×10^{-5}	4.89
o-Chlorophenoxyacetic	25		8.91×10^{-4}	3.05	Hexahydrobenzoic	25		1.26×10^{-5}	4.90
m-Chlorophenoxyacetic	25		7.94×10^{-4}	3.10	Hexanoic	25		1.31×10^{-5}	4.88
o-Chlorophenylacetic	25		1.18×10^{-5}	4.07	Hippuric	25		1.57×10^{-4}	3.80
m-Chlorophenylacetic	25		7.25×10^{-5}	4.14	Histidine	25		6.7×10^{-10}	9.17
p-Chlorophenylacetic	25		6.46×10^{-5}	4.19	Hydroquinone	20		4.5×10^{-11}	10.35
β-(o-Chlorophenyl) propionic	25		2.63×10^{-5}	4.58	o-Hydroxybenzoic	19	1	1.07×10^{-3}	2.97
β-(m-Chlorophenyl) propionic	25		2.57×10^{-5}	4.59	o-Hydroxybenzoic	18	2	4×10^{-14}	13.40
β-(p-Chlorophenyl) propionic	25		2.46×10^{-5}	4.61	m-Hydroxybenzoic	19	1	8.7×10^{-5}	4.06
α-Chloropropinic	25		1.47×10^{-3}	2.83	m-Hydroxybenzoic	19	2	1.2×10^{-10}	9.92
β-Chloropropionic	25		1.04×10^{-4}	3.98	p-Hydroxybenzoic	19	1	3.3×10^{-5}	4.48
cis-Cinnamic	25		1.3×10^{-4}	3.89	p-Hydroxybenzoic	19	2	4.8×10^{-10}	9.32
trans-Cinnamic	25		3.65×10^{-5}	4.44	β-Hydroxybutyric	25		2×10^{-5}	4.70
Citric	20	1	7.10×10^{-4}	3.14	γ-Hydroxybutyric	25		1.9×10^{-5}	4.72
Citric	20	2	1.68×10^{-5}	4.77	β-Hydroxypropionic	25		3.1×10^{-5}	4.51
Citric	20	3	6.4×10^{-6}	6.39	γ-Hydroxyquinoline	20		3.1×10^{-10}	9.51
o-Cresol	25		6.3×10^{-11}	10.20					
m-Cresol	25		9.8×10^{-11}	10.01	Iodoacetic	25		7.5×10^{-4}	3.12
p-Cresol	25		6.7×10^{-11}	10.17	o-Iodobenzoic	25		1.4×10^{-3}	2.85
Crotonic (trans-)	25		2.03×10^{-5}	4.69	m-Iodobenzoic	25		1.6×10^{-4}	3.80
Cyanoacetic	25		3.65×10^{-3}	2.45	Itaconic	25	1	1.40×10^{-4}	3.85
γ-Cyanobutyric	25		3.80×10^{-3}	2.42	Itaconic	25	2	3.56×10^{-6}	5.45
o-Cyanophenoxyacetic	25		1.05×10^{-3}	2.98					
m-Cyanophenoxyacetic	25		9.33×10^{-4}	3.03	Lactic	100		8.4×10^{-4}	3.08
p-Cyanophenoxyacetic	25		1.18×10^{-3}	2.93	Lutidinic	25		7.0×10^{-3}	2.15
Cyanopropionic	25		3.6×10^{-3}	2.44	Lysine	25		2.95×10^{-11}	10.53
Cyclohexane-1:1-dicarboxylic	25	1	3.55×10^{-4}	3.45					
Cyclohexane-1:1-dicarboxylic	25	2	7.76×10^{-7}	6.11	Maleic	25	1	1.42×10^{-2}	1.83

Compound	T°C	Step	K	pK	Compound	T°C	Step	K	pK
Maleic	25	2	8.57×10^{-7}	6.07	o-Phthalic	25	1	1.3×10^{-3}	2.89
Malic	25	1	3.9×10^{-4}	3.40	o-Phthalic	25	2	3.9×10^{-8}	5.51
Malic	25	2	7.8×10^{-6}	5.11	m-Phthalic	25	1	2.9×10^{-4}	3.54
Malonic	25	1	1.49×10^{-2}	2.83	m-Phthalic	18	2	2.5×10^{-5}	4.60
Malonic	25	2	2.03×10^{-6}	5.69	p-Phthalic	25	1	3.1×10^{-4}	3.51
DL-Mandelic	25		1.4×10^{-4}	3.85	p-Phthalic	16	2	1.5×10^{-5}	4.82
Mesaconic	25	1	8.22×10^{-4}	3.09	Picric	25		4.2×10^{-1}	0.38
Mesaconic	25	2	1.78×10^{-5}	4.75	Pimelic	25		3.09×10^{-5}	4.71
Mesitylenic	25		4.8×10^{-5}	4.32	Propionic	25		1.34×10^{-5}	4.87
Methyl-o-aminobenzoic	25		4.6×10^{-5}	5.34	iso-Propylbenzoic	25		3.98×10^{-5}	4.40
Methyl-m-aminobenzoic	25		8×10^{-6}	5.10	2-Pyridinecarboxylic	25		3×10^{-6}	5.52
Methyl-p-aminobenzoic	25		9.2×10^{-6}	5.04	3-Pyridinecarboxylic	25		1.4×10^{-5}	4.85
o-Methylcinnamic	25		3.16×10^{-5}	4.50	4-Pyridinecariboxylic	25		1.1×10^{-5}	4.96
m-Methylcinnamic	25		3.63×10^{-5}	4.44	Pyrocatcchol	20		1.4×10^{-10}	9.85
p-Methylcinnamic	25		2.76×10^{-5}	4.56					
β-Methylglutaric	25		5.75×10^{-5}	4.24	Quinolinic	25		3×10^{-3}	2.52
n-Methylglycine	18		1.2×10^{-10}	9.92					
Methylmalonic	25		1.17×10^{-4}	3.07	Resorcinol	25		1.55×10^{-10}	9.81
Methylsuccinic	25	1	7.4×10^{-5}	4.13					
Methylsuccinic	25	2	2.3×10^{-6}	5.64	Saccharin	18		2.1×10^{-12}	11.68
o-Monochlorophenol	25		3.2×10^{-9}	8.49	Suberic	25		2.99×10^{-5}	4.52
m-Monochlorophenol	25		1.4×10^{-9}	8.85	Succinic	25	1	6.89×10^{-5}	4.16
p-Monochlorophenol	25		6.6×10^{-10}	9.18	Succinic	25	2	2.47×10^{-6}	5.61
					Sulfanilic	25		5.9×10^{-4}	3.23
Naphthalenesulfonic	25		2.7×10^{-1}	0.57					
α-Naphthoic	25		2×10^{-4}	3.70	α-Tataric	25	1	1.04×10^{-3}	2.98
β-Naphthoic	25		6.8×10^{-5}	4.17	α-Tartaric	25	2	4.55×10^{-5}	4.34
α-Naphthol	25		4.6×10^{-10}	9.34	meso-Tartaric	25	1	6×10^{-4}	3.22
β-Naphthol	25		3.1×10^{-10}	9.51	meso-Tartaric	25	2	1.53×10^{-5}	4.82
Nitrobenzene	0		1.05×10^{-4}	3.98	Theobromine	18		1.3×10^{-8}	7.89
o-Nitrobenzoic	18		6.95×10^{-3}	2.16	Terephthalic	25		3.1×10^{-4}	3.51
m-Nitrobenzoic	25		3.4×10^{-4}	3.47	Thioacetic	25		4.7×10^{-4}	3.33
p-Nitrobenzoic	25		3.93×10^{-4}	3.41	Thiophenecarboxylic	25		3.3×10^{-4}	3.48
o-Nitrophenol	25		6.8×10^{-8}	7.17	o-Toluic	25		1.22×10^{-4}	3.91
m-Nitrophenol	25		5.3×10^{-9}	8.28	m-Toluic	25		5.32×10^{-5}	4.27
p-Nitrophenol	25		7×10^{-8}	7.15	p-Toluic	25		4.33×10^{-5}	4.36
o-Nitrophenylacetic	25		1.00×10^{-4}	4.00	Trichloroacetic	25		2×10^{-1}	0.70
m-Nitrophenylacetic	25		1.07×10^{-4}	3.97	Trichlorophenol	25		1×10^{-6}	6.00
p-Nitrophenylacetic	25		1.41×10^{-4}	3.85	Trihydroxybenzoic (2,4,6-)	25		2.1×10^{-2}	1.68
o-β-Nitrophenylpropionic	25		3.16×10^{-5}	4.50	Trimethylacetic	18		9.4×10^{-6}	5.03
p-β-Nitrophenylopropionic	25		3.39×10^{-5}	4.47	Trinitrophenol (2,4,6-)	25		4.2×10^{-1}	0.38
					Tryptophan	25		4.2×10^{-10}	9.38
Nonanic	25		1.09×10^{-5}	4.96	Tyrosine	17		3.98×10^{-9}	8.40
Octanoic	25		1.28×10^{-5}	4.89					
Oxalic	25	1	5.90×10^{-2}	1.23	Uric	12		1.3×10^{-4}	3.89
Oxalic	25	2	6.40×10^{-5}	4.19					
Phenol	20		1.28×10^{-10}	9.89	n-Valeric	18		1.51×10^{-5}	4.82
Phenylacetic	18		5.2×10^{-5}	4.28	iso-Valeric	25		1.7×10^{-5}	4.77
o-Phenylbenzoic	25		3.47×10^{-4}	3.46	Veronal	25		3.7×10^{-8}	7.43
γ-Phenylbutyric	25		1.74×10^{-5}	4.76	Vinylacetic	25		4.57×10^{-5}	4.34
α-Phenylpropionic	25		2.27×10^{-5}	4.64					
β-Phenylpropionic	25		4.25×10^{-5}	4.37	Xanthine	40		1.24×10^{-10}	9.91

DISSOCIATION CONTANTS OF INORGANIC ACIDS IN AQUEOUS SOLUTIONS

(Approximately 0.1—0.01 N)

Compound	T°C	Step	K	pK	Compound	T°C	Step	K	pK
Arsenic	18	1	5.62×10^{-3}	2.25	o-Phosphoric	25	2	6.23×10^{-8}	7.21
Arsenic	18	2	1.70×10^{-7}	6.77	o-Phosphoric	18	3	2.2×10^{-13}	12.67
Arsenic	18	3	3.95×10^{-12}	11.60	Phosphorous	18	1	1.0×10^{-2}	2.00
Arsenious	25		6×10^{-10}	9.23	Phosphorous	18	2	2.6×10^{-7}	6.59
o-Boric	20	1	7.3×10^{-10}	9.14	Pyrophosphoric	18	1	1.4×10^{-1}	0.85
o-Boric	20	2	1.8×10^{-13}	12.74	Pyrophosphoric	18	2	3.2×10^{-2}	1.49
o-Boric	20	3	1.6×10^{-14}	13.80	Pyrophosphoric	18	3	1.7×10^{-6}	5.77
Carbonic	25	1	4.30×10^{-7}	6.37	Pyrophosphoric	18	4	6×10^{-9}	8.22
Carbonic	25	2	5.61×10^{-11}	10.25	Selenic	25	2	1.2×10^{-2}	1.92
Chromic	25	1	1.8×10^{-1}	0.74	Selenious	25	1	3.5×10^{-2}	2.46
Chromic	25	2	3.20×10^{-7}	6.49	Selenious	25	2	5×10^{-8}	7.31
Germanic	25	1	2.6×10^{-9}	8.59	m-Silicic	R.T.	1	2×10^{-10}	9.70
Germanic	25	2	1.9×10^{-13}	12.72	m-Silicic	R.T.	2	1×10^{-12}	12.00
Hyrocyanic	25		4.93×10^{-10}	9.31	o-Silicic	30	1	2.2×10^{-10}	9.66
Hydrofluoric	25		3.53×10^{-4}	3.45	o-Silicic	30	2	2×10^{-12}	11.70
Hydrogen sulfide	18	1	9.1×10^{-8}	7.04	o-Silicic	30	3	1×10^{-12}	12.00
Hydrogen sulfide	18	2	1.1×10^{-12}	11.96	o-Silicic	30	4	1×10^{-12}	12.00
Hydrogen peroide	25		2.4×10^{-12}	11.62	Sulfuric	25	2	1.20×10^{-2}	1.92
Hypobromous	25		2.06×10^{-9}	8.69	Sulfurous	18	1	1.54×10^{-2}	1.81
Hypochlorous	18		2.95×10^{-8}	7.53	Sulfurous	18	2	1.02×10^{-7}	6.91
Hypoiodous	25		2.3×10^{-11}	10.64	Telluric	18	1	2.09×10^{-8}	7.68
Iodic	25		1.69×10^{-1}	0.77	Telluric	18	2	6.46×10^{-12}	11.29
Nitrous	12.5		4.6×10^{-4}	3.37	Tellurous	25	1	3×10^{-3}	2.48
Periodic	25		2.3×10^{-2}	1.64	Tellurous	25	2	2×10^{-8}	7.70
o-Phosphoric	25	1	7.52×10^{-3}	2.12	Tetraboric	25	1	$\sim 10^{-4}$	~ 4.00
					Tetraboric	25	2	$\sim 10^{-9}$	~ 9.00

DISSOCIATION CONSTANTS (K_b) OF AQUEOUS AMMONIA FROM 0 TO 50°C

Temperature (°C)	pK_b	K_b
0	4.862	1.374×10^{-5}
5	4.830	1.479×10^{-5}
10	4.804	1.570×10^{-5}
15	4.782	1.652×10^{-5}
20	4.767	1.710×10^{-5}
25	4.751	1.774×10^{-5}
30	4.740	1.820×10^{-5}
35	4.733	1.849×10^{-5}
40	4.730	1.862×10^{-5}
45	4.726	1.879×10^{-5}
50	4.723	1.892×10^{-5}

Values of K_b accurate to ±0.005; determined by e.m.f. method by: Bates, R. G. and Pinching, G. D., *J. Am. Chem. Soc.*, 72, 1393, 1950.

ION PRODUCT OF WATER SUBSTANCE

William L. Marshall and E. U. Franck

Pressure (bars)	Temperature (°C)																	
	0	25	50	75	100	150	200	250	300	350	400	450	500	600	700	800	900	1000
Sat'd Vapor	14.938	13.995	13.275	12.712	12.265	11.638	11.289	11.191	11.406	12.30	—	—	—	—	—	—	—	—
250	14.83	13.90	13.19	12.63	12.18	11.54	11.16	11.01	11.14	11.77	19.43	21.59	22.40	23.27	23.81	24.23	24.59	24.93
500	14.72	13.82	13.11	12.55	12.10	11.45	11.05	10.85	10.86	11.14	11.88	13.74	16.13	18.30	19.29	19.92	20.39	20.80
750	14.62	13.73	13.04	12.48	12.03	11.36	10.95	10.72	10.66	10.79	11.17	11.89	13.01	15.25	16.55	17.35	17.93	18.39
1,000	14.53	13.66	12.96	12.41	11.96	11.29	10.86	10.60	10.50	10.54	10.77	11.19	11.81	13.40	14.70	15.58	16.22	16.72
1,500	14.34	13.53	12.85	12.29	11.84	11.16	10.71	10.43	10.26	10.22	10.29	10.48	10.77	11.59	12.50	13.30	13.97	14.50
2,000	14.21	13.40	12.73	12.18	11.72	11.04	10.57	10.27	10.08	9.98	9.98	10.07	10.23	10.73	11.36	11.98	12.54	12.97
2,500	14.08	13.28	12.62	12.07	11.61	10.92	10.45	10.12	9.91	9.79	9.74	9.77	9.86	10.18	10.63	11.11	11.59	12.02
3,000	13.97	13.18	12.53	11.98	11.53	10.83	10.34	9.99	9.76	9.61	9.54	9.53	9.57	9.78	10.11	10.49	10.89	11.24
3,500	13.87	13.09	12.44	11.90	11.44	10.74	10.24	9.88	9.63	9.47	9.37	9.33	9.34	9.48	9.71	10.02	10.35	10.62
4,000	13.77	13.00	12.35	11.82	11.37	10.66	10.16	9.79	9.52	9.34	9.22	9.16	9.15	9.23	9.41	9.65	9.93	10.13
5,000	13.60	12.83	12.19	11.66	11.22	10.52	10.00	9.62	9.34	9.13	8.99	8.90	8.85	8.85	8.95	9.11	9.30	9.42
6,000	13.44	12.68	12.05	11.53	11.09	10.39	9.87	9.48	9.18	8.96	8.80	8.69	8.62	8.57	8.61	8.72	8.86	8.97
7,000	13.31	12.55	11.93	11.41	10.97	10.27	9.75	9.35	9.04	8.81	8.64	8.51	8.42	8.34	8.34	8.40	8.51	8.64
8,000	13.18	12.43	11.82	11.30	10.86	10.17	9.64	9.24	8.93	8.68	8.50	8.36	8.25	8.13	8.10	8.13	8.21	8.38
9,000	13.04	12.31	11.71	11.20	10.77	10.07	9.54	9.14	8.82	8.57	8.37	8.22	8.10	7.95	7.89	7.89	7.95	8.12
10,000	12.91	12.21	11.62	11.11	10.68	9.98	9.45	9.04	8.71	8.46	8.25	8.09	7.96	7.78	7.70	7.68	7.70	7.85

Note: Data in the following table were calculated from the equation, $\log_{10} K^*_w = A + B/T + C/T^2 + D/T^3 + (E + F/T + G/T^2) \log_{10} \varrho^*_w$, where, $K^*_w = K_w/(\text{mol kg}^{-1})$, and $\varrho^*_w = \varrho_w/(\text{g cm}^{-3})$. The parameters are:

$A = -4.098$ $C = +2.2362 \times 10^5 K^2$ $E = +13.957$ $G = +8.5641 \times 10^5 K^2$
$B = -3245.2K$ $D = -3.984 \times 10^7 K^3$ $F = -1262.3K$

Users of this table may wish to refer to the reference cited for the background for this international formulation of the ion product of water substance.

From *J. Phys. Chem. Ref. Data*, 10, 295, 1981. With permission.

IONIZATION CONSTANT FOR WATER (K_w)

$-\log_{10} K_w$	Temperature °C.	$-\log_{10} K_w$	Temperature °C.
14.9435	0	13.8330	30
14.7338	5	13.6801	35
14.5346	10	13.5348	40
14.3463	15	13.3960	45
14.1669	20	13.2617	50
14.0000	24	13.1369	55
13.9965	25	13.0171	60

IONIZATION CONSTANTS OF ACIDS IN WATER AT VARIOUS TEMPERATURES

Acids		0°	5°	10°	15°	20°	25°	30°	35°	40°	45°	50°
Formic	$K_A \cdot 10^4$	1.638	1.691	1.728	1.749	1.765	1.772	1.768	1.747	1.716	1.685	1.650
Acetic	$K_A \cdot 10^5$	1.657	1.700	1.729	1.745	1.753	1.754	1.750	1.728	1.703	1.670	1.633
Propionic	$K_A \cdot 10^5$	1.274	1.305	1.326	1.336	1.338	1.336	1.326	1.310	1.280	1.257	1.229
n-Butyric	$K_A \cdot 10^5$	1.563	1.574	1.576	1.569	1.542	1.515	1.484	1.439	1.395	1.347	1.302
Chloracetic	$K_A \cdot 10^3$	1.528	....	1.488	....	....	1.379	....	....	1.230	....	....
Lactic	$K_A \cdot 10^4$	1.287	....	....	....	....	1.374	....	....	....	....	1.270
Glycollic	$K_A \cdot 10^4$	1.334	....	....	....	....	1.475	....	....	....	....	1.415
Oxalic	$K_{2A} \cdot 10^5$	5.91	5.82	5.70	5.55	5.40	5.18	4.92	4.67	4.41	4.09	3.83
Malonic	$K_{2A} \cdot 10^6$	2.140	2.165	2.152	2.124	2.076	2.014	1.948	1.863	1.768	1.670	1.575
Phosphoric	$K_A \cdot 10^3$	8.968	....	....	....	....	7.516	....	....	....	....	5.495
Phosphoric	$K_{2A} \cdot 10^8$	4.85	5.24	5.57	5.89	6.12	6.34	6.46	6.53	6.58	6.59	6.55
Boric	$K_A \cdot 10^{10}$	....	3.63	4.17	4.72	5.26	5.79	6.34	6.86	7.38		8.32
Carbonic	$K_{1A} \cdot 10^7$	2.64	3.04	3.44	3.81	4.16	4.45	4.71	4.90	5.04	5.13	5.19
Phenol-sulfonic	$K_{2A} \cdot 10^{10}$	4.45	5.20	6.03	6.92	7.85	8.85	9.89	10.94	12.00	13.09	14.16
Glycine	$K_{1A} \cdot 10^7$	....	3.82	3.99	4.17	4.32	4.46	4.57	4.66	4.73	4.77	4.79
Citric	$K_{1A} \cdot 10^4$	6.03	6.31	6.69	6.92	7.21	7.45	7.66	7.78	7.96	7.99	8.04
	$K_{2A} \cdot 10^5$	1.45	1.54	1.60	1.65	1.70	1.73	1.76	1.77	1.78	1.76	1.75
	$K_{3A} \cdot 10^7$	4.05	4.11	4.14	4.13	4.09	4.02	3.99	3.78	3.69	3.45	3.28

Reproducibility between various workers is about $\pm (0.01 - 0.02) \cdot 10^5$.
All values are on the m-scale.

COMPOSITION OF SOME INORGANIC ACIDS AND BASES

The following acids and bases are frequently supplied as concentrated aqueous solutions. This table presents certain data concerning these solutions.

	Formula weight	Molarity	Specific gravity	Weight percent	
Acetic acid	60.05	17.5	1.05	99–100	CH_3COOH
Ammonium hydroxide	35.05	14.8	0.90	28–30	NH_3
Hydriodic acid	127.91	5.5	1.5	47–47.5	HI
Hydrobromic acid	80.93	9.0	1.5	47–49	HBr
Hydrochloric acid	36.46	12.0	1.18	36.5–38	HCl
Hydrofluoric acid	20.01	28.9	1.17	48–51	HF
Phosphoric acid	98.00	14.7	1.7	85	H_3PO_4
Sulfuric acid	98.08	18.0	1.84	95–98	H_2SO_4

IONIZATION CONSTANTS FOR DEUTERIUM OXIDE FROM 10 TO 50°C.

From NBS Technical Note 400

The subscript m indicates values on the molal scale, whereas the subscript c indicates values on the molar scale.

t, °C	pK_m	pK_c
10	15.526	15.439
20	15.136	15.049
25	14.955	14.869
30	14.784	14.699
40	14.468	14.385
50	14.182	14.103

EQUIVALENT CONDUCTANCES OF SOME ELECTROLYTES IN AQUEOUS SOLUTIONS AT 25°C

Compound	Concentration in gram equivalents per 1000 cm³								Compound	Concentration in gram equivalents per 1000 cm³							
	Infinite dilution	0.0005	0.001	0.005	0.01	0.02	0.05	0.1		Infinite dilution	0.0005	0.001	0.005	0.01	0.02	0.05	0.1
AgNO₃	133.36	131.36	130.51	127.20	124.76	121.41	115.24	109.14	LaCl₃	145.8	139.6	137.0	127.5	121.8	115.3	106.2	99.1
BaCl₂	139.98	135.96	134.34	128.02	123.94	119.09	111.48	105.19	LiCl	115.03	113.15	112.40	109.40	107.40	104.65	100.11	95.86
CaCl₂	135.84	131.93	130.36	124.25	120.36	115.65	108.47	102.4	LiClO₄	105.98	104.18	103.44	100.57	98.61	96.18	92.20	88.56
Ca(OH)₂	257.9			232.9	225.9	213.9			MgCl₂	129.40	125.61	124.11	118.31	114.55	110.04	103.08	97.10
CuSO₄	133.6	121.6	115.26	94.07	83.12	72.20	59.05	50.58	NH₄Cl	149.7		146.8	143.5	141.28	138.33	133.29	128.75
HCl	426.16	422.74	421.36	415.80	412.00	407.24	399.09	391.32	NaCl	126.45	124.50	123.74	120.65	118.51	115.51	111.06	106.74
KBr	151.9			146.09	143.43	140.48	135.68	131.39	NaClO₄	117.48	115.64	114.87	111.75	109.59	106.90	102.40	98.43
KCl	149.86	147.81	146.95	143.35	141.27	138.34	133.37	128.96	NaI	126.94	125.36	124.25	121.25	119.24	116.70	112.70	108.78
KClO₄	140.04	138.76	137.87	134.16	131.40	127.92	121.62	115.20	NaOOCCH₃	91.0	89.2	88.5	85.72	83.76	81.24	76.92	72.80
K₃Fe(CN)₆	174.5	166.4	163.1	150.7					NaOOCC₂H₅	85.9		83.5	80.9	79.1	76.6		
K₄Fe(CN)₆	184.5		167.24	146.09	134.83	122.82	107.70	97.87	NaOOCC₃H₇	82.70	81.04	80.31	77.58	75.76	73.39	69.32	65.27
KHCO₃	118.0	116.10	115.34	112.24	110.08	107.22			NaOH	247.8	245.6	244.7	240.8	238.0			
KI	150.38			144.37	142.18	139.45	134.97	131.11	Na₂SO₄	129.9	125.74	124.15	117.15	112.44	106.78	97.75	89.98
KIO₄	127.92	125.80	124.94	121.24	118.51	114.14	106.72	98.12	SrCl₂	135.80	131.90	130.33	124.24	120.24	115.54	108.25	102.19
KNO₃	144.96	142.77	141.84	138.48	132.82	132.41	126.31	120.40	ZnSO₄	132.8	121.4	114.53	95.49	84.91	74.24	61.20	52.64
KReO₄	128.20	126.03	125.12	121.31	118.49	114.49	106.40	97.40									

THE EQUIVALENT CONDUCTANCE OF THE SEPARATE IONS

(From Smithsonian Physical Tables)

Ion,	0°	18°	25°	50°	75°	100°	100°	150°	Ion.	0°	18°	25°	50°	75°	100°	128°	156°
K	40.4	64.6	74.5	115	159	206	263	317	½SO₄	41.	68.	79.	125	177	234	303	370
Na	26.	43.5	50.9	82	116	155	203	249	½C₂O₄	39.	63.	73.	115	163	213	271	336
NH₄	40.2	64.5	74.5	115	159	207	264	319									
½Mg	02.9	54.5	05.5	100	147	190	215	209	⅓C₆H₅O₇	36.	60.	70.	113	161	214		
½Ba	33.	55.	65.	104	149	200	262	322	¼Fe(CN)₆	58.	95.	111.	173	244	321		
½Ca	30.	51.	60.	98	142	191	252	312									
⅓La	35.	61.	72.	119	173	235	312	388	H	240.	314.	350.	465	565	644	722	777
Cl	41.1	65.5	75.5	116	160	207	264	318	OH	105.	172.	192.	284	360	439	525	592
NO₃	40.4	61.7	70.6	104	140	178	222	263									
C₂H₃O₂	20.3	34.6	40.8	67	96	130	171	211									

ACTIVITY COEFFICIENTS OF ACIDS, BASES AND SALTS

The following coefficients are valid only at 25°C. The concentrations are expressed as molalities.

Name	0.1	0.2	0.3	0.4	0.5	0.6	0.7	0.8	0.9	1.0	Name	0.1	0.2	0.3	0.4	0.5	0.6	0.7	0.8	0.9	1.0
AgNO₃	0.734	0.657	0.606	0.567	0.536	0.509	0.485	0.464	0.446	0.429	KOH	0.798	0.760	0.742	0.734	0.732	0.733	0.736	0.742	0.749	0.756
AlCl₃	(0.337)	0.305	0.302	0.313	0.331	0.356	0.388	0.429	0.479	0.539	LiBr	0.796	0.766	0.756	0.752	0.753	0.758	0.767	0.777	0.789	0.803
Al₂(SO₄)₃	(0.0350)	0.0225	0.0176	0.0153	0.0143	0.0142	0.0142	0.0149	0.0159	0.0175	LiCl	0.790	0.757	0.744	0.740	0.739	0.743	0.748	0.755	0.764	0.774
CdSO₄	(0.150)	0.102	0.082	0.069	0.061	0.055	0.050	0.046	0.043	0.041	LiClO₄	0.812	0.794	0.792	0.798	0.808	0.820	0.834	0.852	0.869	0.887
CrCl₃	(0.331)	0.298	0.294	0.300	0.314	0.335	0.362	0.397	0.436	0.481	LiI	0.815	0.802	0.804	0.813	0.824	0.838	0.852	0.870	0.888	0.910
Cr(NO₃)₃	(0.319)	0.285	0.279	0.281	0.291	0.304	0.322	0.344	0.371	0.401	LiNO₃	0.788	0.752	0.736	0.728	0.726	0.727	0.729	0.733	0.737	0.743
Cr₂(SO₄)₃	(0.0458)	0.0300	0.0238	0.0207	0.0190	0.0182	0.0181	0.0185	0.0194	0.0208	LiOAc	0.784	0.742	0.721	0.709	0.700	0.691	0.689	0.688	0.688	0.689
CsBr	0.754	0.694	0.654	0.626	0.603	0.586	0.571	0.558	0.547	0.538	MgSO₄	(0.150)	0.108	0.088	0.076	0.068	0.062	0.057	0.054	0.051	0.049
CsCl	0.756	0.694	0.656	0.628	0.606	0.589	0.575	0.563	0.553	0.544	MnSO₄	(0.150)	0.106	0.085	0.073	0.064	0.058	0.053	0.049	0.046	0.044
CsI	0.754	0.692	0.651	0.621	0.599	0.581	0.567	0.554	0.543	0.533	NaBr	0.782	0.741	0.719	0.704	0.697	0.692	0.689	0.687	0.687	0.687
CsNO₃	0.733	0.655	0.602	0.561	0.528	0.501	0.478	0.458	0.439	0.422	NaCl	0.778	0.735	0.710	0.693	0.681	0.673	0.667	0.662	0.659	0.657
CsOAc	0.799	0.771	0.761	0.759	0.762	0.768	0.776	0.783	0.792	0.802	NaClO₄	0.775	0.729	0.701	0.683	0.668	0.656	0.648	0.641	0.635	0.629
CuSO₄	(0.150)	0.104	0.083	0.071	0.062	0.056	0.052	0.048	0.045	0.043	NaCNS	0.787	0.750	0.731	0.720	0.715	0.712	0.710	0.710	0.711	0.712
HBr	0.805	0.782	0.777	0.781	0.789	0.801	0.815	0.832	0.850	0.871	NaF	0.765	0.710	0.676	0.651	0.632	0.616	0.603	0.592	0.582	0.573
HCl	0.796	0.767	0.756	0.755	0.757	0.763	0.772	0.783	0.795	0.809	NaH₂PO₄	0.744	0.675	0.629	0.593	0.563	0.539	0.517	0.499	0.483	0.468
HClO₄	0.803	0.778	0.768	0.766	0.769	0.776	0.785	0.795	0.808	0.823	NaI	0.787	0.751	0.735	0.727	0.723	0.723	0.724	0.727	0.731	0.736
HI	0.818	0.807	0.811	0.823	0.839	0.860	0.883	0.908	0.935	0.963	NaNO₃	0.762	0.703	0.666	0.638	0.617	0.599	0.583	0.570	0.558	0.548
HNO₃	0.791	0.754	0.735	0.725	0.720	0.717	0.717	0.718	0.721	0.724	NaOAc	0.791	0.757	0.744	0.737	0.735	0.736	0.740	0.745	0.752	0.757
KBr	0.772	0.722	0.693	0.673	0.657	0.646	0.636	0.629	0.622	0.617	NaOH	0.766	0.727	0.708	0.697	0.690	0.685	0.681	0.679	0.678	0.678
KCl	0.770	0.718	0.688	0.666	0.649	0.637	0.626	0.618	0.610	0.604	NiSO₄	(0.150)	0.105	0.084	0.071	0.063	0.056	0.052	0.047	0.044	0.042
KCNS	0.769	0.716	0.685	0.663	0.646	0.633	0.623	0.614	0.606	0.599	RbBr	0.763	0.706	0.673	0.650	0.632	0.617	0.605	0.595	0.586	0.578
KF	0.775	0.727	0.700	0.682	0.670	0.661	0.654	0.650	0.646	0.645	RbCl	0.764	0.709	0.675	0.652	0.634	0.620	0.608	0.599	0.590	0.583
KI	0.778	0.733	0.707	0.689	0.676	0.667	0.660	0.654	0.649	0.645	RbI	0.762	0.705	0.671	0.647	0.629	0.614	0.602	0.591	0.583	0.575
KNO₃	0.739	0.663	0.614	0.576	0.545	0.519	0.496	0.476	0.459	0.443	RbNO₃	0.734	0.658	0.606	0.565	0.534	0.508	0.485	0.465	0.446	0.430
KOAc	0.796	0.766	0.754	0.750	0.751	0.754	0.759	0.766	0.774	0.783	RbOAc	0.796	0.767	0.756	0.753	0.755	0.759	0.766	0.773	0.782	0.792
											TlNO₃	0.702	0.606	0.545	0.500	....	....	....	....	....	....
											ZnSO₄	(0.150)	0.104	0.083	0.071	0.063	0.057	0.052	0.048	0.046	0.043

EQUIVALENT CONDUCTANCES OF AQUEOUS SOLUTIONS OF HYDROHALOGEN ACIDS

From NSRDS-NBS 33
Walter J. Hamer and Harold J. DeWane

One may wish to refer to the above document which defines terms relating to conductance of electrolytic solutions and also discusses some general considerations of the migration of ions under applied potential gradients. In NSRDS-NBS 33 conductance equations are given and some treatment of the Debye–Hückel–Onsager–Fuoss theories is presented.

Equivalent conductances (Ω^{-1} cm^2 equiv^{-1}) of aqueous solutions of HF at 0, 16, 18, 20, and 25°C

c	0°C	16°C	18°C	20°C	25°C
mol l^{-1}					
0.004	106.7	—	—	—	140.5
0.005	97.7	—	—	—	128.1
0.006	90.9	112.9	114.6	116.3	118.8
0.007	85.5	105.7	107.3	108.9	111.4
0.008	81.1	99.9	101.3	102.9	105.4
0.009	77.3	95.0	96.4	97.8	100.4
0.01	74.2	90.8	92.1	93.5	96.1
0.02	56.2	67.7	68.8	69.8	72.2
0.05	39.3	46.8	47.6	48.3	50.1
0.07	34.7	41.3	41.9	42.6	44.3
0.10	30.7	36.4	37.0	37.7	39.1
0.20	24.8	29.6	30.1	30.7	31.7
0.50	20.4	—	—	—	26.3
0.70	19.5	—	—	—	25.1
1.0	18.8	—	—	—	24.3

Equivalent conductances (Ω^{-1} cm^2 equiv^{-1}) of aqueous solutions of HCl from −20 to 65°C

c	−20°C	−10°C	0°C	5°C	10°C	15°C	20°C	25°C	30°C	35°C	40°C	45°C	50°C	55°C	65°C
mol l^{-1}															
0.0001	—	—	—	296.4	—	360.8	—	424.5	—	487.0	—	547.9	577.7	606.6	662.9
0.0005	—	—	—	295.2	—	359.2	—	422.6	—	484.7	—	545.2	575.1	603.5	660.0
0.001	—	—	—	294.3	—	358.0	—	421.2	—	483.1	—	543.2	573.1	601.3	657.8
0.005	—	—	—	291.0	—	353.5	—	415.7	—	476.7	—	535.5	564.4	592.6	647.3
0.01	—	—	—	288.6	—	350.3	—	411.9	—	472.2	—	530.3	558.7	586.5	641.2
0.05	—	—	—	280.3	—	339.9	—	398.9	—	456.7	—	512.4	—	565.6	616.9
0.1	—	—	—	275.0	—	333.3	—	391.1	—	446.8	—	501.1	—	552.8	602.8
0.5	—	—	228.7	254.8	283.0	308.1	336.4	360.7	386.8	411.9	436.9	461.1	482.4	508.0	552.3
1.0	—	—	211.7	235.2	261.6	283.9	312.2	332.2	359.0	379.4	402.9	424.8	445.3	468.1	509.3
1.5	—	—	196.2	216.9	241.5	261.5	287.5	305.8	331.1	349.4	371.6	391.5	410.8	431.7	469.9
2.0	—	—	182.0	199.9	222.7	240.7	262.9	281.4	303.3	321.6	342.4	360.6	378.2	398.0	433.6
2.5	—	131.7	168.5	184.3	205.1	221.4	239.8	258.9	277.0	295.8	315.2	332.0	347.6	366.7	399.9
3.0	—	120.8	154.6	169.5	188.5	203.4	219.3	237.6	253.3	271.5	289.3	304.8	319.0	336.9	368.0
3.5	85.5	111.3	139.6	155.6	172.2	186.5	201.6	218.3	232.9	248.6	263.9	279.4	292.1	308.6	337.2
4.0	79.3	102.7	129.2	143.4	158.1	171.5	185.6	200.0	214.2	228.4	242.2	256.6	268.2	283.6	310.1
4.5	73.7	94.9	119.5	132.0	145.4	157.4	170.6	183.1	196.6	209.5	222.5	235.2	246.7	260.2	284.7
5.0	68.5	87.8	110.3	121.3	133.5	144.4	156.6	167.4	180.2	191.9	204.1	215.4	226.5	238.4	261.0
5.5	63.6	81.1	101.7	111.4	122.5	132.3	143.6	152.9	165.0	175.6	187.1	197.1	207.7	218.3	239.1
6.0	58.9	74.9	93.7	102.2	112.3	121.2	131.5	139.7	151.0	160.6	171.3	180.2	190.3	199.7	218.9
6.5	54.4	69.1	86.2	93.8	103.0	111.0	120.4	127.7	138.2	146.8	156.9	164.8	174.3	182.7	200.4
7.0	50.2	63.7	79.3	86.0	94.4	101.7	110.2	116.9	126.4	134.3	143.3	150.8	159.7	167.2	183.5
7.5	46.3	58.6	73.0	78.9	86.5	93.3	100.9	107.0	115.7	122.9	131.6	138.1	146.2	153.1	168.1
8.0	42.7	54.0	67.1	72.4	79.4	85.6	92.4	98.2	106.1	112.6	120.6	126.7	134.0	140.4	154.1
8.5	39.4	49.8	61.7	66.5	72.9	78.7	84.7	90.3	97.3	103.2	110.7	116.4	123.0	128.8	141.5
9.0	36.4	45.9	56.8	61.2	67.1	72.5	77.8	83.1	89.4	94.8	101.7	107.1	112.9	118.3	130.1
9.5	33.6	42.3	52.3	56.4	61.8	66.8	71.5	76.6	82.3	87.3	93.6	98.7	103.9	108.8	119.7
10.0	31.2	39.1	48.2	52.0	57.0	61.6	65.8	70.7	75.9	80.5	86.3	91.1	95.7	—	—
10.5	28.9	36.1	44.5	48.0	52.7	56.9	60.7	65.3	70.1	74.3	79.6	84.1	88.4	—	—
11.0	26.8	33.4	41.1	44.4	48.8	52.6	56.1	60.2	64.9	68.7	73.6	77.7	81.7	—	—
11.5	24.9	31.0	38.0	41.1	45.3	48.5	51.9	55.3	60.1	63.6	68.0	71.7	75.6	—	—
12.0	23.1	28.7	35.3	38.0	42.0	—	48.0	—	55.6	—	62.8	—	70.0	—	—
12.5	21.4	26.7	32.7	—	39.0	—	44.4	—	51.4	—	57.9	—	64.8	—	—
13.0	20.0	24.9	—	—	—	—	—	—	—	—	—	—	—	—	—

Equivalent conductances (Ω^{-1} cm^2 equiv^{-1}) of aqueous solutions of HBr at 25°C

c	Λ	c	Λ	c	Λ	c	Λ	c	Λ	c	Λ
mol l^{-1}		mol l^{-1}		mol l^{-1}		mol l^{-1}		mol l^{-1}		mol l^{-1}	
0.0002	425.5	0.003	419.7	0.04	402.7	0.30	374.0	0.80	345.3	4.0	199.4
0.0003	425.0	0.004	418.5	0.05	400.4	0.35	370.8	0.85	342.6	4.5	182.4
0.0004	424.6	0.005	417.6	0.06	398.4	0.40	367.7	0.90	339.9	5.0	166.5
0.0005	424.3	0.006	416.7	0.07	396.5	0.45	364.7	0.95	337.2	5.5	151.8
0.0006	424.0	0.007	415.8	0.08	394.9	0.50	361.9	1.0	334.5	6.0	138.2
0.0007	423.7	0.008	415.1	0.09	393.4	0.55	359.0	1.5	307.6	6.5	125.7
0.0008	423.4	0.009	414.4	0.10	391.9	0.60	356.2	2.0	281.7	7.0	114.2
0.0009	423.2	0.01	413.7	0.15	386.0	0.65	353.5	2.5	257.8	7.5	103.8
0.001	422.9	0.02	408.9	0.20	381.4	0.70	350.8	3.0	236.8	8.0	94.4
0.002	421.1	0.03	405.4	0.25	377.5	0.75	348.0	3.5	217.5	8.5	85.8

EQUIVALENT CONDUCTANCES OF AQUEOUS SOLUTIONS OF HYDROHALOGEN ACIDS (Continued)

Equivalent conductances (Ω^{-1} cm^2 equiv^{-1}) of aqueous solutions of HBr from -20 to 50°C

c	-20°C	-10°C	0°C	10°C	20°C	30°C	40°C	50°C
mol l^{-1}								
0.50	—	—	240.9	295.9	347.0	398.9	453.6	496.8
0.75	—	—	234.7	284.9	339.0	387.2	433.8	480.6
1.00	—	—	229.6	276.0	329.0	380.4	418.6	465.2
1.25	—	—	221.7	265.8	314.9	362.8	401.8	442.9
1.50	—	—	209.5	254.9	298.9	340.6	381.8	421.4
1.75	—	—	198.3	243.1	284.6	327.3	366.2	404.8
2.00	—	150.8	188.6	231.3	271.8	314.1	350.5	387.4
2.25	—	143.4	180.1	219.6	258.3	296.9	332.3	367.0
2.50	—	136.8	171.7	208.3	244.8	281.7	316.0	349.1
2.75	—	131.1	164.1	198.3	232.9	267.6	301.2	333.2
3.00	—	125.7	157.2	189.5	222.2	255.0	287.8	318.6
3.25	—	120.6	150.8	181.7	212.4	244.3	275.4	304.9
3.50	—	116.1	144.1	174.6	203.2	234.4	263.7	291.9
3.75	87.1	112.0	137.6	167.4	194.7	224.2	252.2	279.4
4.00	84.0	107.5	132.3	160.2	186.8	214.2	239.7	266.9
4.25	80.9	103.1	127.7	153.2	179.0	204.5	228.8	254.6
4.50	78.0	99.0	123.0	146.4	171.2	195.1	218.8	242.6
4.75	75.1	95.1	117.7	139.9	163.1	186.1	209.3	231.3
5.00	72.3	91.4	112.6	134.0	155.7	178.2	199.6	221.3
5.25	69.6	87.8	107.8	128.3	148.7	170.5	190.0	211.4
5.50	67.0	84.2	103.1	122.7	142.1	162.8	181.4	201.8
5.75	64.4	80.6	98.6	117.3	135.7	155.3	173.2	192.4
6.00	61.8	77.2	94.3	112.0	129.6	148.0	165.4	183.4
6.25	59.3	73.8	90.1	106.9	123.7	140.9	157.8	174.7
6.50	56.8	70.7	86.0	102.0	118.0	134.1	150.5	166.3
6.75	54.4	67.7	82.1	97.2	112.5	127.6	143.3	158.4
7.00	51.9	64.6	78.4	92.6	107.1	121.4	136.3	150.8

Equivalent conductances (Ω^{-1} cm^2 equiv^{-1}) of aqueous solutions of HI at 25°C

c	Λ	c	Λ	c	Λ	c	Λ	c	Λ
mol l^{-1}		mol l^{-1}		mol l^{-1}		mol l^{-1}		mol l^{-1}	
0.00045	423.2								
0.00050	423.0	0.0035	417.9	0.025	406.6	0.10	394.0	0.90	349.2
0.00055	422.9	0.0040	417.3	0.030	405.2	0.15	389.5	0.95	346.6
0.00060	422.7	0.0045	416.8	0.035	403.9	0.20	385.9	1.0	343.9
0.00065	422.6	0.0050	416.4	0.040	402.8	0.25	382.9	1.5	316.4
0.00070	422.4	0.0055	415.9	0.045	401.7	0.30	380.1	2.0	288.9
0.00075	422.3	0.0060	415.5	0.050	400.8	0.40	374.9	2.5	262.5
0.00080	422.2	0.0065	415.1	0.055	399.9	0.45	372.3	3.0	237.9
0.00085	422.0	0.0070	414.8	0.060	399.1	0.50	369.8	3.5	215.4
0.00090	421.9	0.0075	414.4	0.065	398.3	0.55	367.3	4.0	195.1
0.00095	421.8	0.0080	414.1	0.070	397.6	0.60	364.8	4.5	176.8
0.0010	421.7	0.0090	413.4	0.075	396.9	0.65	362.2	5.0	160.4
0.0015	420.7	0.0095	413.1	0.080	396.3	0.70	359.7	5.5	145.5
0.0020	419.8	0.010	412.8	0.085	395.6	0.75	357.1	6.0	131.7
0.0025	419.1	0.015	410.3	0.090	395.1	0.80	354.5	6.5	118.6
0.0030	418.5	0.020	408.3	0.095	394.5	0.85	351.9	7.0	105.7

Equivalent conductances (Ω^{-1} cm^2 equiv^{-1}) of aqueous solutions of HI from -20 to 50°C

c	-20°C	-10°C	0°C	10°C	20°C	30°C	40°C	50°C
mol l^{-1}								
0.4	—	—	253.9	300.7	354.5	411.9	453.6	501.8
0.6	—	—	249.1	293.6	344.6	401.1	441.5	488.4
0.8	—	—	242.4	285.5	333.9	389.1	429.3	474.8
1.0	—	—	234.8	276.9	322.7	376.3	416.5	460.8
1.2	—	—	226.6	267.9	311.4	363.1	403.4	446.3
1.4	—	—	218.2	258.8	300.2	349.8	390.0	431.4
1.6	—	—	209.9	249.7	289.1	336.6	376.5	416.3
1.8	—	—	201.8	240.7	278.4	323.7	363.0	401.2
2.0	—	—	194.0	231.9	268.1	311.1	349.6	386.1
2.2	—	147.7	186.6	223.4	258.2	299.0	336.3	371.2
2.4	—	143.5	179.5	215.1	248.8	287.3	323.3	356.5
2.6	—	138.8	172.8	207.2	239.7	276.1	310.5	342.2
2.8	—	133.9	166.5	199.5	231.0	265.4	298.1	328.2
3.0	99.9	129.0	160.5	192.0	222.7	255.1	286.1	314.7
3.2	97.4	124.1	154.7	184.8	214.6	245.2	274.5	301.7
3.4	94.2	119.5	149.1	177.8	206.7	235.7	263.4	289.2
3.6	90.6	115.0	143.7	171.0	198.9	226.4	252.6	277.3
3.8	86.9	110.9	138.3	164.3	191.2	217.4	242.3	266.0
4.0	83.6	106.9	132.8	157.6	183.5	208.4	232.5	255.3
4.2	80.9	103.0	127.3	151.0	175.6	199.6	223.1	245.3
4.4	79.3	99.2	121.6	144.3	167.5	190.7	214.1	235.9

SPECIFIC HEAT OF WATER

Heat Capacity of Air-free Water 0°—100°C at 1 Atmosphere Pressure

The heat capacity of air-free water is given in international steam table calories per gram and in absolute joules per gram. (1 absolute joule — 0.238846 I.T. Cal.).

The enthalpy or heat content is given for air-free water in I.T. Cal. per gram and in absolute joules per gram.

From Osborne, Stimson and Ginnings; B. of S. Jour. Res. 23, 238, 1939.

Temp. °C.	Thermal Capacity		Enthalpy		Temp. °C	Thermal Capacity		Enthalpy	
	Cal./g/°C	Joules/g/°C	Cal./g	Joules/g		Cal/g/°C	Joules/g/°C	Cal/g	Joules/g
0	1.00738	4.2177	0.0245	0.1026	50	.99854	4.1807	50.0079	209.3729
1	1.00652	4.2141	1.0314	4.3184	51	.99862	4.1810	51.0065	213.5538
2	1.00571	4.2107	2.0376	8.5308	52	.99871	4.1814	52.0051	217.7350
3	1.00499	4.2077	3.0429	12.7400	53	.99878	4.1817	53 0039	221.9166
4	1.00430	4.2048	4.0475	16.9462	54	.99885	4.1820	54.0027	226.0984
5	1.00368	4.2022	5.0515	21.1498	55	.99895	4.1824	55.0016	230.2806
6	1.00313	4.1999	6.0549	25.3508	56	.99905	4.1828	56.0006	234.4632
7	1.00260	4.1977	7.0578	29.5496	57	.99914	4.1832	56.9997	238.6462
8	1.00213	4.1957	8.0602	33.7463	58	.99924	4.1836	57.9989	242.8296
9	1.00170	4.1939	9.0621	37.9410	59	.99933	4.1840	58.9982	247.0134
10	1.00129	4.1922	10.0636	42.1341	60	.99943	4.1844	59.9975	251.1976
11	1.00093	4.1907	11.0647	46.3255	61	.99955	4.1849	60.9970	255.3822
12	1.00060	4.1893	12.0654	50.5155	62	.99964	4.1853	61.9966	259.5673
13	1.00029	4.1880	13.0659	54.7041	63	.99976	4.1858	62.9963	263.7529
14	1.00002	4.1869	14.0660	58.8916	64	.99988	4.1863	63.9962	267.9390
15	.99976	4.1858	15.0659	63 0779	65	1.00000	4.1868	64.9961	272.1256
16	.99955	4.1849	16.0655	67.2632	66	1.00014	4.1874	65.9962	276.3127
17	.99933	4.1840	17.0650	71.4476	67	1.00026	4.1879	66.9964	280.5003
18	.99914	4.1832	18.0642	75.6312	68	1.00041	4.1885	67.9967	284.6885
19	.99897	4.1825	19.0633	79.8141	69	1.00053	4.1890	68.9972	288.8772
20	.99883	4.1819	20.0622	83.9963	70	1.00067	4.1896	69.9977	293.0665
21	.99869	4.1813	21.0609	88.1778	71	1.00081	4.1902	70.9985	297.2564
22	.99857	4.1808	22.0596	92.3589	72	1.00096	4.1908	71.9994	301.4469
23	.99847	4.1804	23.0581	96.5395	73	1.00112	4.1915	73.0004	305.6381
24	.99838	4.1800	24.0565	100.7196	74	1.00127	4.1921	74.0016	309.8299
25	.99828	4.1796	25.0548	104.8994	75	1.00143	4.1928	75.0030	314.0224
26	.99821	4.1793	26.0530	109.0788	76	1.00160	4.1935	76.0045	318.2155
27	.99814	4.1790	27.0512	113.2580	77	1.00177	4.1942	77.0062	322.4094
28	.99809	4.1788	28.0493	117.4369	78	1.00194	4.1949	78.0080	326.6039
29	.99804	4.1786	29.0474	121.6157	79	1.00213	4.1957	79.0101	330.7992
30	.99802	4 1785	30.0455	125 7943	80	1.00229	4.1964	80.0123	334.9952
31	.99799	4.1784	31.0435	129.9727	81	1.00248	4.1972	81.0147	339.1920
32	.99797	4.1783	32.0414	134.1510	82	1.00268	4.1980	82.0172	343.3897
33	.99797	4.1783	33.0394	138.3293	83	1.00287	4.1988	83.0200	347.5881
34	.99795	4.1782	34.0374	142.5076	84	1.00308	4.1997	84.0230	351.7873
35	.99795	4.1782	35.0353	146.6858	85	1.00327	4.2005	85.0262	355.9874
36	.99797	4.1783	36.0333	150.8641	86	1.00349	4.2014	86.0295	360.1883
37	.99797	4.1783	37.0312	155.0423	87	1.00370	4.2023	87.0331	364.3902
38	.99799	4.1784	38.0292	159.2207	88	1.00392	4.2032	88.0369	368.5929
39	.99802	4.1785	39.0272	163.3991	89	1.00416	4.2042	89.0410	372.7966
40	.99804	4.1786	40.0253	167.5777	90	1.00437	4.2051	90.0452	377.0012
41	.99807	4.1787	41.0233	171.7563	91	1.00461	4.2061	91.0497	381.2068
42	.99811	4.1789	42.0214	175.9351	92	1.00485	4.2071	92.0545	385.4135
43	.99816	4.1791	43.0195	180.1141	93	1.00509	4.2081	93.0594	389.6211
44	.99819	4.1792	44.0177	184.2933	94	1.00535	4.2092	94.0647	393.8297
45	.99826	4.1795	45.0159	188.4726	95	1.00561	4.2103	95.0701	398.0395
46	.99830	4.1797	46.0142	192.6522	96	1.00588	4.2114	96.0759	402.2503
47	.99835	4.1799	47.0125	196.8320	97	1.00614	4.2125	97.0819	406.4622
48	.99842	4.1802	48.0109	201.0120	98	1.00640	4.2136	98.0882	410.6753
49	.99847	4.1804	49.0094	205.1923	99	1.00669	4.2148	99.0947	414.8895
					100	1.00697	4.2160	100.1015	419.1049

Enthalpy of Air-saturated Water
1 Atmosphere Pressure 0-100°C

Temp. °C	Enthalpy		Temp. °C	Enthalpy	
	Cal/g	Joules/g		Cal/g	Joules/g
0	0	0	50	49.9896	209.2964
5	5.0276	21.0496	55	54.9842	230.2077
10	10.0402	42.0363	60	59.9811	251.1289
15	15.0431	62.9826	65	64.9808	272.0619
20	20.0400	83.9034	70	69.9839	293.0087
25	25.0332	104.8089	75	74.9907	313.9712
30	30.0244	125.7063	80	80.0019	334.9519
35	35.0149	146.6003	85	85.0180	355.9532
40	40.0055	167.4949	90	90.0395	376.9773
45	44.9968	188.3928	95	95.0671	398.0270
			100	100.1016	419.1053

SPECIFIC HEAT OF WATER (Continued)

Specific Heat of Super-heated Steam

Specific heat of steam under constant pressure given in atmospheres and at temperatures above saturation in Cal./g/°C

Temp. °C	Pressure in atmospheres 1	2	4	6	8	10	12	Temp. °C	Pressure in atmospheres 1	2	4	6	8	10	12
110	0.481							310	0.478	0.482	0.408	0.495	0.502	0.510	0.518
120	0.477	0.498						320	0.480	0.483	0.489	0.496	0.502	0.509	0.516
130	0.475	0.494						330	0.482	0.484	0.490	0.496	0.502	0.508	0.515
140	0.473	0.489						340	0.483	0.485	0.491	0.496	0.502	0.507	0.513
150	0.472	0.486	0.519					350	0.484	0.486	0.492	0.497	0.502	0.507	0.512
160	0.471	0.483	0.512	0.549				360	0.485	0.487	0.492	0.497	0.502	0.507	0.511
170	0.470	0.481	0.507	0.538				370	0.486	0.488	0.493	0.498	0.503	0.507	0.511
180	0.469	0.479	0.502	0.528	0.561	0.602		380	0.488	0.490	0.494	0.498	0.503	0.507	0.511
190	0.469	0.478	0.498	0.522	0.549	0.583	0.625	390	0.489	0.491	0.495	0.499	0.503		
200	0.469	0.478	0.495	0.515	0.539	0.567	0.601	400	0.490	0.492	0.496	0.500	0.504		
210	0.470	0.477	0.493	0.510	0.531	0.555	0.584	410	0.492	0.494	0.497	0.501	0.505		
220	0.470	0.477	0.491	0.506	0.524	0.545	0.569	420	0.494	0.496	0.498	0.502	0.506		
230	0.471	0.477	0.489	0.504	0.519	0.537	0.557	430	0.495	0.497	0.500	0.504	0.507		
240	0.472	0.477	0.488	0.501	0.515	0.530	0.548	440	0.497	0.499	0.501	0.505	0.508		
250	0.473	0.477	0.488	0.499	0.512	0.525	0.540	450	0.498	0.500	0.503	0.506	0.509		
260	0.474	0.478	0.587	0.498	0.509	0.521	0.534	460	0.500	0.501	0.505	0.507	0.510		
270	0.474	0.478	0.487	0.497	0.507	0.518	0.529	470	0.502	0.503	0.506	0.508	0.512		
280	0.475	0.479	0.487	0.496	0.505	0.515	0.525	480	0.504	0.505	0.507	0.509	0.513		
290	0.476	0.480	0.487	0.495	0.504	0.513	0.523	490	0.505	0.506	0.509	0.511	0.514		
300	0.477	0.481	0.488	0.495	0.503	0.511	0.519	500	0.506	0.508	0.510	0.512	0.515		

Specific Heat of Water Above 100°C

Mean specific heat of water in 15°C calories between 0°C and the temperature stated.
Heat content (Enthalpy) in joules per gram between 0°C and the temperature stated.
From data by Osborne, Stimson and Fiock, B of S Jour. Res. 5, 411, 1930.

Temp. °C	Specific heat mean 0-t°C	Heat content 0-t joules/g	Temp. °C	Specific heat mean 0-t°C	Heat content 0-t joules/g	Temp. °C	Specific heat mean 0-t°C	Heat content 0-t joules/g	Temp. °C	Specific heat mean 0-t°C	Heat content 0-t joules/g
100	1.0008	418.75	190	1.0153	807.15	150	1.0067	631.75	240	1.0326	1036.97
110	1.0015	460.97	200	1.0181	852.02	160	1.0083	675.06	250	1.0376	1084.97
120	1.0025	503.36	210	1.0212	897.55	170	1.0103	718.66	260	1.0423	1133.87
130	1.0037	545.93	220	1.0247	943.24	180	1.0127	762.72	270	1.0483	1184.32
140	1.0050	588.71	230	1.0285	989.75						

Specific Heat of Ice — Cal./g/°C

Temp. °C	Specific heat	Observer	Temp. °C.	Specific heat	Observer	Temp. °C	Specific heat	Observer	Temp. °C.	Specific heat	Observer
−252 to −188	.146	Dieterici, 1903	−31.8	.4454	Dickinson-Osborne, 1915	− 60	.392		− 4.9	.4932	Dickinson-Osborne, 1915
−250	.0361		−23.7	.4599	Dickinson-Osborne, 1915	− 38.3	.4346	Dickinson-Osborne, 1915	− 2.6	.5003	Dickinson-Osborne, 1915
−200	.162	Mean	−24.5	.4605	Dickinson-Osborne, 1915	− 34.3	.4411	Dickinson-Osborne, 1915	− 2.2	.5018	Dickinson-Osborne, 1915
−188 to −78	.285	Dieterici, 1903	−20.8	.4668	Dickinson-Osborne, 1915	− 30.6	.4488	Dickinson-Osborne, 1915			
−180	.199	Nernst, 1910	−14.8	.4782	Dickinson-Osborne, 1915						
−160	.230	Nernst, 1910	−14.6	.4779	Dickinson-Osborne, 1915				Water Below 0°C		
−150	.246		−11.0	.4861	Dickinson-Osborne, 1915						
−140	.262	Nernst, 1910	− 8.1	.4896	Dickinson-Osborne, 1915	− 6	1.0119	Martinetti, 1890	− 3	1.0102	Martinetti, 1890
−100	.329	Mean	− 4.3	.4989	Dickinson-Osborne, 1915	− 5	1.0155	Barnes, 1902	− 2	1.0097	Martinetti, 1890
− 78 to −18	.463	Dieterici, 1903	− 4.5	.4984	Dickinson-Osborne, 1915	− 5	1.0113	Martinetti, 1890	− 1	1.0092	Martinetti, 1890
						− 4	1.0105	Martinetti, 1890			

HEAT CAPACITY OF MERCURY

The specific heat of solid mercury is given in relation to water at 15°C. The values are from Carpenter and Stoodley, Phil. Mag. 10, 249, 1930. Heat capacity is given in calories per gram and in calories per gram atom (1 cal = 4.1840 absolute joules and the atomic weight of mercury 200.61). Values for the liquid and vapor are from Douglas, Ball and Ginnings, Jour. of Res. Bureau of Standards 46, 334, 1951.

Temp. °C	Specific heat	Heat capacity cal/g-atom	Temp. °C	Heat capacity cal/g	Heat capacity cal/g-atom	Temp. °C	Heat capacity cal/g	Heat capacity cal/g-atom	Temp. °C	Heat capacity cal/g	Heat capacity cal/g-atom
	Solid	Solid		Liquid	Liquid		Liquid	Liquid		Vapor	Vapor
−75.6	.0319	6.3995	−38.88	.033686	6.7578	240	.032356	6.4910	0	.02476	4.968
−72.9	.0324	6.4998	−20	.033534	6.7272	260	.032336	6.4869	100	.02476	4.968
−65.4	.0324	6.4998	0	.033382	6.6967	280	.032325	6.4847	200	.02477	4.969
−59.5	.0324	6.4998	20	.033240	6.6683	300	.032322	6.4843	300	.02480	4.975
−44.9	.0336	6.7405	25	.033206	6.6615	320	.032330	6.4858	400	.02489	4.993
−42.2	.0336	6.7405	40	.033109	6.6419	340	.032346	6.4890	500	.02507	5.030
−40.0	.0337	6.7606	60	.032987	6.6176	356.58	.032366	6.4930			
			80	.032877	6.5954	360	.032371	6.4940			
			100	.032776	6.5752	380	.032404	6.5005			
			120	.032686	6.5571	400	.032445	6.5087			
			140	.032606	6.5410	420	.032494	6.5186			
			160	.032535	6.5270	440	.032550	6.5298			
			180	.032476	6.5150	460	.032614	6.5426			
			200	.032426	6.5050	480	.032684	6.5567			
			220	.032386	6.4970	500	.032762	6.5723			

HEAT CAPACITY (C_p) OF ORGANIC LIQUIDS AND VAPORS AT 25C

The values in this table are expressed as cal/(g mole) (°K) at 1 atm pressure. Calories, g × 4.184 = joules.

Robert Shaw

Reproduced from *Chemical and Engineering Data*, *14*, 461 (1969) with permission of the copyright owner, the American Chemical Society.

	$C_p(l)$	$C_p(g)$
Alkanes		
2-Methyl butane	39.4	28.5
n-Hexane	46.6	34.2
2-Methyl pentane	46.2	34.5
3-Methyl pentane	45.6	34.2
2,3-Dimethyl butane	45.0	33.6
2,2-Dimethyl butane	45.0	33.9
n-Heptane	53.7	39.7
2-Methyl hexane	53.1	
3-Ethyl pentane	52.5	
2,2-Dimethyl pentane	52.8	
2,3-Dimethyl pentane	51.8	
2,4-Dimethyl pentane	53.5	
3,3-Dimethyl pentane	51.4	
2,2,3-Trimethyl butane	51.0	39.3
n-Octane	60.7	45.1
2,2,4-Trimethyl pentane	56.5	
2,2,3,3-Tetramethyl butane	55.9	46.0
n-Nonane	68.0	50.6
n-Decane	75.2	56.1
2-Methyl nonane	75.6	
n-Undecane	82.5	61.5
n-Dodecane	89.7	67.0
n-Tridecane	97.2	72.5
n-Tetradecane	104.8	77.9
n-Pentadecane	112.2	83.4
n-Hexadecane	119.9	88.9
Olefins		
Pentene-1	37.1	26.2
cis-Pentene-2	36.3	24.3
trans-Pentene-2	37.6	25.9
2-Methyl butene-1	37.5	26.7
3-Methyl butene-1	37.3	28.4
2-Methyl butene-2	36.4	25.1
Hexene-1	43.8	31.6
2,3-Dimethyl butene-2	41.7	30.5
Heptene-1	50.5	37.1
Octene-1	57.7	42.6
Decene-1	71.8	53.5
Undecene-1	78.9	59.0
Dodecene-1	86.2	64.4
Hexadecene-1	115.9	86.3
Butadiene-1,3	29.5	19.0
Pentadiene-1,4	34.8	25.1
2-Methyl butadiene-1,3	36.7	25.0
Butyne-2	29.4	18.6
Aromatics		
Benzene	32.4	19.5
Toluene	37.3	24.8
Ethyl benzene	44.5	30.7
p-Xylene	43.6	30.3
1,2,3-Trimethyl benzene	51.8	36.9
1,2,4-Trimethyl benzene	51.6	37.1
1,3,5-Trimethyl benzene	50.1	35.9
Oxygen-Containing		
Methanol	19.5	10.5
Ethanol	27.0	15.7
Propanol-1	34.3	20.9
Propanol-2	37.1	21.3
2-Methyl propanol-2	53.2	27.2
estimated alternative	42.2	
Di-n-butyl sulfide	68.0	49.7
Methyl isopropyl sulfide	41.2	28.8
Ethyl mercaptan	28.2	17.4
N-butyl mercaptan	41.3	28.4
Iso-butyl mercaptan	41.1	28.4
Amyl mercaptan	48.2	33.9
Iso-propyl mercaptan	34.7	23.1
Sec-butyl mercaptan	40.9	28.6
Cyclo-pentyl mercaptan	39.5	25.9
Tert-butyl mercaptan	41.8	29.0
Dimethyl disulfide	34.9	22.6

	$C_p(l)$	$C_p(g)$
Diethyl ether	41.1	25.8
Acetone	30.2	18.0
Formic acid	23.6	10.8
Acetic acid	29.4	16.0
Ethyl acetate	40.6	
Benzyl alcohol	52.1	
Diphenyl ether	64.2	
Furan	27.4	
Nitrogen-Containing		
Hydrogen cyanide	16.9	8.6
Hydrazine	22.2	12.2
Unsym. dimethyl hydrazine	39.2	
Trimethyl hydrazine	44.5	
Aniline	46.2	26.1
Quinoline	47.6	
Acrylonitrile	26.5	14.9
Perfluoro piperidene	71.0	
Methyl nitrate	37.5	
Tetrahydropyrole	37.4	
Nitromethane	25.4	13.8
Ring Compounds		
Cyclopentane	30.4	19.8
Methyl cyclopentane	38.0	26.2
Ethyl cyclopentane	44.5	31.5
1,1-Dimethyl cyclopentane	44.6	31.9
cis-1,2-Dimethyl cyclopentane	45.1	32.1
trans-1,2-Dimethyl cyclopentane	45.1	32.1
trans-1,3-Dimethyl cyclopentane	45.1	32.1
Cyclopentane	29.4	18.0
Cyclohexane	36.4	25.4
Methyl cyclohexane	44.2	32.3
Ethyl cyclohexane	50.0	38.0
1,2-Dimethyl cyclohexane	50.4	37.4
cis-1,3-Dimethyl cyclohexane	50.0	37.6
trans-1,3-Dimethyl cyclohexane	50.8	37.6
cis-1,4-Dimethyl cyclohexane	50.5	37.6
trans-1,4-Dimethyl cyclohexane	50.3	37.7
Cyclohexane	35.6	25.1
Cyclooctatetraene	44	
Tetralin $C_6H_4(CH_2)_4$	52	
cis-Decalin	55.4	
trans-Decalin	54.5	
Halogen-Containing		
Chloroform	27.6	15.7
Carbon tetrachloride	31.7	19.9
1,2-Dibromoethane	32.2	18.2
1,1,1-Trichloro ethane	34.4	22.4
1,1-Dichloro ethylene	26.6	16.1
Fluoro benzene	35.1	22.7
Chloro benzene	34.9	23.6
Bromo benzene	37.1	24.2
Sulfur-Containing		
Carbon disulfide[d]	18.2	10.9
Dimethyl sulfide	28.3	17.8
Methylethyl sulfide	34.6	22.8
Methyl n-propyl sulfide	41.0	28.2
Methyl n-butyl sulfide	48.0	33.8
Diethyl sulfide	40.9	28.1
Ethyl n-propyl sulfide	47.4	33.4
Di-n-propyl sulfide	54.0	38.7
Diethyl disulfide	48.7	33.9
$(CH_2)_3S$	27.1	16.7
$(CH_2)_4S$	33.5	21.9
$(CH_2)_5S$	39.0	26.0
Phenyl mercaptan	39.1	25.2
Silicon-Containing		
Tetramethylsilane	49.8	33.5
Hexamethyldisiloxane	74.4	

HEAT CAPACITY (C_p) OF ORGANIC GASES AT 300 and 800 K

These data are a portion of those contained in "Additivity Rules for the Estimation of Thermochemical Properties", authored by S. W. Benson, F. R. Cruickshank, D. M. Golden, G. R. Haugen, H. E. O'Neal, A. S. Rodgers, R. Shaw and R. Walsh in *Chemical Reviews*, 69, 279 (1969). These data are reproduced here with permission of the copyright owner, the American Chemical Society.

	C_p			C_p	
	300 K	800 K		300 K	800 K
Alkanes			1,2,4-Trimethylbenzene	36.99	76.93
Ethane	12.65	25.83	1,3,5-Trimethylbenzene	36.10	76.84
Propane	17.66	37.08	1,2,3,4-Tetramethylbenzene	45.50	89.42
n-Butane	23.40	48.23	1,2,3,5-Tetramethylbenzene	44.57	88.79
n-Hexane	34.37	70.36	1,2,4,5-Tetramethylbenzene	44.77	88.41
n-Octane	43.35	92.50	Pentamethylbenzene	51.99	101.29
n-Decane	(56.34)	(114.63)	Hexamethylbenzene	59.73	113.51
n-Dodecane	(67.33)	(136.76)	*Unsaturated Benzenes*		
2-Methylpropane	23.25	48.49	Styrene	29.35	61.40
2-Methylbutane	28.54	59.71	S-Methylstyrene	(34.9)	(71.8)
2,2-Dimethylpropane	29.21	60.78	*cis*-β-Methylstyrene	(34.9)	(71.8)
2,3-Dimethylbutane	33.76	70.7	*trans*-β-Methylstyrene	(35.1)	(72.2)
2,2,3-Trimethylbutane	39.54	82.73	*o*-Methylstyrene	(34.9)	(71.8)
2,2,3,3-Tetramethylbutane	46.29	96.18	*m*-Methylstyrene	(34.9)	(71.8)
Alkenes			*p*-Methylstyrene	(34.9)	(71.8)
Ethene	10.45	20.20	Ethynylbenzene	27.63	55.79
Propene	15.34	30.68	*Polyaromatics*		
But-1-ene	20.57	41.80	Biphenyl	39.05	86.92
cis-But-2-ene	18.96	40.87			
trans-But-2-ene	21.08	41.50	*Nonaromatic Rings: Ring Corrections*		
2-Methylpropene	21.39	41.86	Cyclopropane	13.44	31.44
Pent-1-ene	26.31	52.95	Cyclobutane	17.37	42.42
cis-Pent-2-ene	(21.45)	(52.29)	Cyclobutene	16.03	32.26
trans-Pent-2-ene	(26.04)	(52.45)	Cyclopentane	19.98	52.44
2-Methylbut-1-ene	26.41	53.15	Cyclopentene	18.08	45.78
3-Methylbut-1-ene	(28.47)	(53.85)	Methylcyclopentane	26.43	63.8
2-Methylbut-2-ene	25.22	52.05	1-Methylcyclopentene	(24.3)	(57.0)
Allenic Dienes			1,1-Dimethylcyclopentane	(32.16)	(76.18)
Allene	14.16	25.42	*cis*-1,2-Dimethylcyclopentane	(32.34)	74.98
1,2-Butadiene	19.23	36.01	*trans*-1,2-Dimethylcyclopentane	(32.44)	(75.84)
1,2-Pentadiene	(25.3)	(47.7)	*cis*-1,3-Dimethylcyclopentane	32.72	75.68
2,3-Pentadiene	(24.3)	(46.6)	*trans*-1,3-Dimethylcyclopentane	(32.72)	(75.68)
3-Methyl-1,2-butadiene	(25.3)	(47.2)	Cyclohexane	25.58	66.76
Nonallenic Dienes			Cyclohexene	25.28	59.49
1,3-Butadiene	19.11	36.84	Methylcyclohexane	32.51	78.74
cis-1,3-Pentadiene	(22.7)	(47.0)	Ethylcyclohexane	38.23	90.1
trans-1,3-Pentadiene	(24.9)	(47.7)	1,1-Dimethylcyclohexane	(37.2)	(90.7)
1,4-Pentadiene	(25.2)	(47.6)	*cis*-1,2-Dimethylcyclohexane	(37.7)	(90.1)
2-Methyl-1,3-butadiene (isoprene)	(25.2)	(48.0)	*trans*-1,2-Dimethylcyclohexane	(38.3)	(90.5)
Alkynes			*cis*-1,3-Dimethylcyclohexane	(37.9)	(90.5)
Ethyne	10.53	14.93	*trans*-1,3-Dimethylcyclohexane	(37.9)	(89.8)
Propyne	14.55	25.14	*cis*-1,4-Dimethylcyclohexane	(37.9)	(89.8)
But-1-yne	19.54	35.95	*trans*-1,4-Dimethylcyclohexane	(38.0)	(90.6)
But-2-yne	18.70	35.14	Spiropentane	21.19	47.91
Pent-1-yne	(25.65)	(47.1)			
Pent-2-yne	(23.69)	(45.9)	*Alcohols and Phenols*		
Aromatics			Methanol, CH_3OH	10.5	19.0
Benzene	19.65	45.06	Ethanol, C_2H_5OH	15.7	30.3
Toluene	24.94	56.61	1-Propanol, C_3H_7OH	20.9	41.0
Ethylbenzene	30.88	67.15	2-Propanol $(CH_3)_2CHOH$	21.3	42.1
1,2-Dimethylbenzene	32.10	66.50	2-Butanol $(CH_3CH_2)(CH_3)CHOH$	27.2	52.7
1,3-Dimethylbenzene	30.66	66.41	2-Methyl-2-propanol,		
1,4-Dimethylbenzene	30.49	66.14	$(CH_3)_3COH$	27.2	53.3
n-Propylbenzene	(36.99)	(78.30)	Phenol, C_6H_5OH	24.8	50.7
n-Butylbenzene	(42.09)	(89.37)	*o*-Cresol, $CH_3C_6H_4OH$	31.2	61.6
1-Propylbenzene	(36.47)	(78.6)	*m*-Cresol, $CH_3C_6H_4OH$	29.8	61.3
1-Methyl-2-ethylbenzene	(37.94)	(78.1)	*p*-Cresol, $CH_3C_6H_4OH$	29.8	61.1
1-Methyl-3-ethylbenzene	(36.59)	(77.8)			
1-Methyl-4-ethylbenzene	(36.42)	(77.6)	*Ethers*		
1,2,3-Trimethylbenzene	37.82	77.56	Dimethyl ether, CH_3OCH_3	15.8	30.4

	C_p			C_p	
	300 K	800 K		300 K	800 K
Peroxides and Hydroperoxides			$HONO_2$	12.80	20.33
Hydrogen peroxide, HOOH	10.3	14.3	CH_2NO_2	13.76	25.56
			N_2	6.96	7.51
Cyclics			NO	7.13	7.83
Ethylene oxide	11.5	24.6	NO_2	8.85	11.88
Propylene oxide	17.4	35.7	NO_3 sym	11.26	17.51
Trimethylene oxide	14.3	35.1	N_2O	9.25	12.49
Aldehydes and Ketones			N_2O_3	15.72	21.39
Acetaldehyde, CH_3CHO	13.2	24.2	N_2O_4	18.52	27.11
Propionaldehyde, C_2H_5CHO	18.8	35.2	N_2O_5	23.09	32.74
Acetone, CH_3COCH_3	18.0	34.9	$CH_2=CHNO_2$	17.43	31.7
Methyl ethyl ketone,					
$CH_3CH_2COCH_3$	24.7	46.1	*Haloalkanes*		
			CF_3CCl_3	28.1	39.6
Acids, Esters, and Anhydrides			CF_2ClCF_2Cl	28.9	...
Formic acid, HCOOH	10.8	18.3	CF_3CF_3	25.5	38.3
Acetic acid, CH_3COOH	16.0	29.1	CCl_3CCl_3	32.7	41.3
			CBr_3CBr_3	33.4	39.9
Aliphatic and Aromatic Amines			$CF_3CHClBr$	25.1	37.2
CH_3NH_2	11.91	22.43	CF_3CHCl_2	24.4	37.3
$(CH_3)_2NH$	16.58	33.94	CF_3CH_2Cl	21.4	34.8
$(CH_3)_3N$	22.05	45.62	CF_3CH_2Br	21.8	35.1
$C_6H_5NH_2$	26.07	53.79	CF_3CH_2I	21.9	35.1
			CF_3CH_2F	20.9	34.5
Cyanides			CF_3CH_3	18.9	...
C_2H_5CN	17.30	32.7	$CHCl_2CH_2Cl$	21.3	33.3
C_6H_5CN	26.2	52.1	CH_3CCl_3	22.4	33.3
c-C_3H_5CN	18.95	37.68	CH_3CHF_2	16.1	...
$CH_2=CHCN$	15.30	26.43	CH_3CHCl_2	18.2	30.9
trans-$CH_3CH=CHCN$	19.71	36.97	CH_3CH_2F	14.1	...
$C_2(CN)_2$	20.58	27.26	CH_3CH_2Cl	15.1	28.3
			CH_3CH_2Br	15.5	28.6
Nitrites			$CF_2ClCF_2CF_3$	37.4	...
CH_3ONO	15.36	26.27	$CF_3CF_2CH_3$	28.8	...
			$CH_3CH_2CH_2Cl$	20.3	39.2
Hydrazines, Azo Compounds, and Tetrazines			$CH_3CHClCH_3$	21.3	...
N_2H_4	12.19	21.08	$CH_3CH_2CH_2Br$	20.7	39.4
CH_3NHNH_2	17.1	31.3	$(CH_3)_3CCl$	27.9	...
$NH=NH$	8.75	13.55	CH_2ICH_22I	19.2	
			CH_3CHICH_2I	(24.1)	
Amides					
$HCONH_2$	11.10	21.12	*Haloalkenes, Alkynes, and Arenes*		
CH_3CONH_2	15.63	32.60	$CFCl=CCl_2$	21.8	28.8
			$CF_2=CClBr$	21.8	28.5
Heterocyclic N-Atom-Containing Ring Compounds in the Ideal Gas State			$CFCl=CFCl$	21.0	28.4
Pyrrole	17.16	...	$CF_2=CBr_2$	22.2	...
Pyridine	18.80	...	$CF_2=CFCl$	20.0	28.0
Picolines, α	24.05	...	$CF_2=CFBr$	20.5	28.1
β	23.94	...	$CF_2=CF_2$	19.3	27.6
			$CCl_2=CCl_2$	22.7	29.2
Unique Groups			$CBr_2=CBr_2$	24.5	29.7
NH_3	8.53	12.23	$CHF=CCl$	18.3	26.5
HCN	8.61	11.45	$CF_2=CHBr$	17.6	26.2
CH_3CN	12.55	21.32	$CF_2=CHF$	16.5	25.6
$(CN)_2$	13.59	17.42	$CHCl=CCl_2$	19.3	26.9
$CH_2(CN)_2$	17.38	27.72	$CHBr=CBr_2$	20.5	...
HNC	9.33	11.49	$CH_2=CFCl$	15.3	24.4
CH_3NC	12.83	21.34	$CH_2=CCl_2$	16.1	24.7
HN_3	10.46	15.10	$CHCl=CHCl$ (av)	15.8	24.6
CH_2N_2	12.58	18.73	$CHBr=CHBr$ *cis*	16.5	...
$CH_2\begin{smallmatrix}N\\ \\N\end{smallmatrix}$	10.23	18.35	$CH_2=CHCl$	12.9	22.4
			$CH_2=CHBr$	13.4	22.5
			$CH_2=CHI$	13.9	22.7
HCNO	10.57	15.19	$CH_3C\equiv CCl$	17.2	27.0
HNO	8.29	10.77	$CH_3C\equiv CBr$	17.6	27.1

	C_p			C_p	
	300 K	800 K		300 K	800 K
Haloalkenes, Alkynes, and Arenes (Continued)			(thiirane ring)	12.90	25.61
$CH_3C{\equiv}CI$	17.8	27.2			
C_6H_5F	22.7	47.6	(thietane ring)	16.66	36.40
$o\text{-}C_6H_4F_2$	25.6	49.5			
$m\text{-}C_6H_4F_2$	25.5	49.7	(thiolane ring)	21.85	47.66
$p\text{-}C_6H_4F_2$	25.7	49.7			
$p\text{-}FC_6H_4CH_3$	27.9	58.6			
$C_6H_5CF_3$	31.4	62.8	(thiane ring)	26.03	64.00
C_6H_5Cl	23.6	47.9			
C_6H_5Br	24.0	48.0			
C_6H_5I	24.2	48.0	(thiophene ring)	17.52	36.01
Organosulfur Compounds					
CH_3SH	12.05	20.32			
CH_3CH_2SH	17.44	31.83	(2-methylthiophene)	22.92	46.43
$CH_3(CH_2)_2SH$	22.75	43.60			
$CH_3(CH_2)_3SH$	28.37	55.68			
$CH_3(CH_2)_4SH$	33.91	66.78	(3-methylthiophene)	22.80	45.95
$((CH_3)_2CH)SH$	23.04	43.26			
$(CH_3)_2CHCH_2SH$	28.41	53.77			
$(CH_3)(CH_3CH_2)CHSH$	28.64	54.29			
$(CH_3)_3CSH$	29.04	55.53			
$(CH_3)_2(CH_3CH_2)CSH$	34.46	66.28	*Compounds Containing Unique Groups*		
			SOF	9.97	12.79
			$SOCl_2$	15.94	18.88
(cyclopentyl)SH	25.94	58.61	SOCl	10.88	13.03
			S_2F_2	15.33	18.94
			SF_2	10.44	13.14
(cyclohexyl)SH	32.13	75.19	SF	7.57	8.68
			S_2Cl_2	17.44	19.46
			SCl_2	12.19	13.61
(phenyl)SH	25.22	51.59	SCl	8.20	8.95
			$CSCl_2$	15.34	18.78
$(CH_3)_2S$	17.77	31.59	NSF_3	17.23	23.56
$CH_3CH_2S_3CH$	22.82	42.93	BO_2	10.08	13.45
$CH_3S(CH_2)_2CH_3$	28.17	54.45	H_3SiBr	12.69	20.02
$(CH_3)_2CHSCH_3$	28.82	54.89	H_3SiI	12.86	
$CH_3S(CH_2)_3CH_3$	33.79	66.53	H_2SiBr_2	15.68	21.73
$CH_3SC(CH_3)_3$	34.66	66.74	$SiBr_4$	23.21	25.39
$CH_3CH_2SCH_2CH_3$	28.09	54.91	$ClSiBr_3$	22.78	25.31
$CH_3CH_2SCH_2CH_2CH_3$	33.40	66.68	Cl_2SiBr_2	22.39	25.23
$CH_3(CH_2)_2S(CH_2)_2CH_3$	38.71	78.45	Cl_3SiBr	21.72	25.13
$(CH_3)_2CHSCH(CH_3)_2$	40.65	77.12	$ISiBr_3$	23.39	25.42
$CH_3(CH_2)_3S(CH_2)_3CH_3$	49.70	100.58	$Si(CH_3)_4$	33.52	60.55
CH_3SSCH_3	22.61	37.66	$(CH_3)_3SiCl$	31.36	52.15
$CH_3CH_2SSCH_2CH_3$	33.91	60.19	$(CH_3)_2SiCl_2$	26.17	42.54
$CH_3(CH_2)_2SS(CH_2)_2CH_3$	44.50	83.70	$(CH_3)_3SiF$	29.42	51.30
$CH_3(SO)CH_3$	21.26	37.02	$(CH_3)_2SiF_2$	25.26	42.16
$CH_3(SO_2)CH_3$	23.9		CH_3SiF_3	22.73	
NH_2CSNH_2	17.72	29.06	$[(CH_3)_3Si]_2O$	57.22	99.95
CH_3NCS	15.70	25.87	$(CH_3)_3SiH$	28.36	50.50
S_8	37.25	42.62	$(CH_3)_2SiH_2$	22.03	39.79

SPECIFIC HEAT OF THE ELEMENTS AT 25°C

$$C_p = cal\ g^{-1}\ °K^{-1}$$

Element	Kelly: Bureau of Mines Bulletin 592 (1961)	Hultgren: Selected values of Thermodynamic properties of Metals and Alloys (1963)	N.B.S. Circular #500 Part 1 (1952)
Aluminum	0.215	0.215	0.2154
Antimony	0.049	0.0495	0.0501
Argon	0.124	......	0.124
Arsenic	0.0785	......	0.0796
Barium	0.046	0.0362	0.0458
Beryllium	0.436	0.436	0.4733
Bismuth	0.0296	0.0238	0.0292
Boron	0.245	......	0.2463
Bromine (Br$_2$)	0.113	0.0537	
Cadmium	0.0555	0.0552	0.0554
Calcium	0.156	0.155	0.1566
Carbon (Diamond)	0.124	......	0.120
" (Graphite)	0.170	......	0.172
Cerium	0.049	0.0459	0.0442
Cesium	0.057	0.0575	0.0558
Chlorine (Cl$_2$)	0.114	......	0.114
Chromium	......	0.107	0.1073
Cobalt	0.109	0.107	0.1037
Columbium		See Niobium	
Copper	0.092	0.0924	0.0920
Dysprosium	0.0414	0.0414	
Erbium	0.0401	0.0401	
Europium	0.0421	0.0326	
Fluorine (F$_2$)	0.197	0.197	
Gadolinium	0.055	0.056	
Gallium	0.089	0.088	0.0911
Germanium	0.077		
Gold	0.0308	0.0308	0.0305
Hafnium	0.035	0.028	0.0344
Helium	1.24	......	1.242
Holmium	0.0393	0.0394	
Hydrogen (H$_2$)	3.41	......	3.42
Indium	0.056	0.0556	0.0570
Iodine (I$_2$)	0.102	......	0.034
Iridium	0.0317	0.0312	0.0305
Iron (α)	0.106	0.1075	0.1078
Krypton	0.059	......	0.059
Lanthanum	0.047	0.0479	0.0475
Lead	0.038	0.0305	0.0308
Lithium	0.85	0.834	0.814
Lutetium	0.037	0.0285	
Magnesium	0.243	0.245	0.235
Manganese (α)	0.114	0.114	0.1147
" (β)	0.119	......	0.1120
Mercury	0.0331	0.0333	0.0331
Molybdenum	0.0599	0.0597	0.0584
Neodymium	0.049	0.0453	0.0499
Neon	0.246	......	0.246
Nickel	0.106	0.1061	0.1057
Niobium	0.064	0.0633	
Nitrogen (N$_2$)	0.249	......	0.249
Osmium	0.03127	0.0310	0.0310
Oxygen (O$_2$)	0.219	0.219	
Palladium	0.0584	0.0583	0.0590

SPECIFIC HEAT OF THE ELEMENTS AT 25°C (Continued)

Element	Kelly: Bureau of Mines Bulletin 592 (1961)	Hultgren: Selected values of Thermodynamic properties of Metals and Alloys (1963)	N.B.S. Circular #500 Part 1 (1952)
Phosphorus, white	0.181	. . . . . .	0.178
" red, triclinic	0.160		
Platinum	0.0317	0.0317	0.0325
Polonium	0.030		
Potassium	0.180	0.180	0.1787
Praseodymium	0.046	(0.0467)	0.0482
Promethium	0.0442		
Protactinium	0.029		
Radium	0.0288		
Radon	0.0224	. . . . . .	0.0224
Rhenium	0.0329	0.0330	0.0327
Rhodium	0.0583	0.0580	0.0592
Rubidium	0.0861	0.0860	0.0850
Ruthenium	0.057	0.0569	
Samarium	0.043	0.0469	
Scandium	0.133	0.1173	
Selenium (Se$_2$)	0.0767	. . . . . .	0.0535
Silicon	0.168	. . . . . .	0.169
Silver	0.0566	0.0562	0.0564
Sodium	0.293	0.292	0.2952
Strontium	0.0719	0.0719	0.0684
Sulfur, yellow	0.175	. . . . .	0.177
Tantalum	0.0334	0.0334	0.0335
Technetium	0.058		
Tellurium	0.0481	. . . . . .	0.482
Terbium	0.0437	0.0435	
Thallium	0.0307	0.0307	0.0310
Thorium	0.0271	0.0281	0.0331
Thulium	0.0382		
Tin (α)	0.0510	0.0519	0.0518
Tin (β)	0.0530	0.0543	0.0530
Titanium	0.125	0.1248	0.1231
Tungsten	0.0317	0.0322	0.0324
Uranium	0.0276	0.0278	0.0276
Vanadium	0.116	0.116	0.1147
Xenon	0.0378	. . . . . .	0.0379
Ytterbium	0.0346	0.0287	
Yttrium	0.068	0.0713	
Zinc	0.0928	0.0922	0.0916
Zirconium	0.0671	0.0660	

SPECIFIC HEAT AND ENTHALPY OF SOME SOLIDS AT LOW TEMPERATURES

R. J. Corruccini and J. J. Gniewek

For a more extensive listing of data one is referred to N.B.S. Monograph 21 (1960)

Joules/gm × 453.6 = joules/lb × 0.239 = cal/gm × 0.4299 = Btu/lb

METALS

T	Aluminum		Beryllium		Bismuth		Cadmium	
	C_p	$H - H_0$	C_p	$H - H_0$	C_p	$H - H_0$	C_p	$H - H_0$
°K	$jg^{-1} deg^{-1} K$	jg^{-1}	$jg^{-1} deg^{-1} K$	jg^{-1}	$jg^{-1} deg^{-1} K$	jg^{-1}	$jg^{-1} deg^{-1} K$	jg^{-1}
1	0.000 10[a]	....						
1	.000 051	0.000 025	0.000 025	0.000 013	0.000 00598	0.000 00158	0.000 008	0.000 003
2	.000 108	.000 105	.000 051	.000 051	.000 0461	.000 0233	.000 033	.000 022
3	.000 176	.000 246	.000 079	.000 116	.000 170	.000 123	.000 090	.000 082
4	.000 261	.000 463	.000 109	.000 209	.000 493	.000 432	.000 21	.000 22
6	.000 50	.001 21	.000 180	.000 496	.002 14	.002 88	.001 30	.001 5
8	.000 88	.002 6	.000 271	.000 944	.005 47	.010 2	.004 3	.007 0
10	.001 4	.004 9	.000 389	.001 60	.010 4	.025 9	.008 0	.019
15	.004 0	.018	.000 842	.004 57	.023 8	.111	.025	.102
20	.008 9	.048	.001 61	.010 5	.036 3	.262	.046	.28
25	.017 5	.112	.002 79	.021 2	.047 7	.472	.066	.56
30	.031 5	.232	.004 50	.039 2	.057 5	.734	.086	.94
35	.051 5	.436						
40	.077 5	.755	.009 96	.109	.072 7	1.38	.117	1.96
50	.142	1.85	.019 2	.253	.084 6	2.17	.141	3.26
60	.214	3.64	.034 1	.523	.093 5	3.06	.159	4.76
70	.287	6.15	.056 2	.971	.100	4.03	.172	6.43
80	.357	9.37	.090 6	1.69	.105	5.05	.182	8.20
90	.422	13.25	.139	2.82	.108	6.12	.190	10.1
100	.481	17.76	.199	4.51	.111	7.21	.196	12.0

T	Chromium		Copper		Germanium[b]		Gold	
	C_p	$H - H_0$	C_p	$H - H_0$	C_p	$H - H_0$	C_p	$H - H_0$
°K	$jg^{-1} deg^{-1} K$	jg^{-1}	$jg^{-1} deg^{-1} K$	jg^{-1}	$g^{-1} deg^{-1} K$	jg^{-1}	$jg^{-1} deg^{-1} K$	jg^{-1}
1	0.000 0285	0.000 0142	0.000 012	0.000 006	0.000 000 528	0.000 000 132	0.000 006	0.000 002
2	.000 058	.000 0573	.000 028	.000 025	.000 004 23	.000 002 11	.000 025	.000 016
3	.000 089	.000 131	.000 053	.000 064	.000 014 4	.000 010 7	.000 070	.000 061
4	.000 124	.000 237	.000 091	.000 13	.000 034 4	.000 034 3	.000 16	.000 17
6	.000 206	.000 567	.000 23	.000 44	.000 125	.000 179	.000 50	.000 78
8	.000 312	.001 07	.000 47	.001 12	.000 335	.000 612	.001 2	.002 4
10	.000 451	.001 82	.000 86	.002 4	.000 813	.001 69	.002 2	.005 6
15	.001 02	.005 28	.002 7	.010 7	.004 45	.013 6	.007 4	.028
20	.002 10	.012 8	.007 7	.034	.012 5	.054 0	.015 9	.086
25	.003 92	.027 4	.016	.090	.024 0	.145	.026 3	.191
30	.006 83	.053 2	.027	.195	.036 6	.296	.037 1	.349
40	.017 1	.163	.060	.61	.061 7	.786	.057 2	.821
50	.035 8	.421	.099	1.40	.085 8	1.52	.072 6	1.47
60	.062 1	.904	.137	2.58	.108	2.50	.084 2	2.25
70	.093	1.68	.173	4.13	.131	3.70	.092 8	3.14
80	.127	2.77	.205	6.02	.153	5.12	.099 2	4.10
90	.161	4.21	.232	8.22	.173	6.74	.104 3	5.12
100	.193	5.98	.254	10.6	.191	8.55	.108 3	6.18

T	Indium		α-Iron[c]		γ-Iron[d]		Lead	
	C_p	$H - H_0$	C_p	$H - H_0$	C_p	$H - H_0$	C_p	$H - H_0$
°K	$jg^{-1} deg^{-1} K$	jg^{-1}	$jg^{-1} deg^{-1} K$	jg^{-1}	$jg^{-1} deg^{-1} K$	jg^{-1}	$jg^{-1} deg^{-1} K$	jg^{-1}
1	0.000 029	0.000 011	0.000 090	0.000 045	....	....	0.000 026	0.000 010
1	[a].000 019	[a].000 006	....	....			[a].000 012	[a].000 003
2	.000 138	.000 085	.000 183	.000 181	....	....	.000 12	.000 07
2	[a].000 141	[a].000 073	....				[a].000 09	.000 05
								.000 28
3	.000 410	.000 341	.000 279	.000 412	....	....	.000 33	.000 23
3	[a].000 464	[a].000 357	....				[a].000 31	
[e]3.40	.000 584	.000 537	....					
3.40	[a].000 669	[a].000 581	....				...........•	
4	.000 95	.000 99	.000 382	.000 742	....	....	.000 7	.000 8
4	....	....					[a].000 7	[a].000 7
5	....	....					.001 5	.001 8
5	....	....					[a].001 5	[a].001 8
6	.003 59	.005 20	.000 615	.001 73	....	....	.002 9	.003 9
6	....	....					[a].003 0	[a].004 0
7	....	....					.004 8	.008
7	....	....					[a].005 0	[a].008
8	.008 55	.017 0	.000 90	.003 23	....	....	.007 3	.014
10	.015 5	.040 8	.001 24	.005 37	....	....	.013 7	.034
15	.036 7	.170	.002 49	.014 5	....	....	.033 5	.150
20	.060 8	.413	.004 5	.031 6	0.007	0	.053 1	.368
25	.085 7	.778	.007 5	.061	....	....	.068 1	.672
30	.108	1.265	.012 4	.110	.016	.11	.079 6	1.042
40	.141	2.52	.029	.31	.041	.39	.094 4	1.920
50	.162	4.04	.055	.73	.090	1.0$_2$	.103	2.91
60	.176	5.73	.087	1.43	.13$_7$	2.1$_6$	.108	3.97
70	.186	7.53	.121	2.46	.18$_0$	3.7$_5$	.112	5.07
80	.193	9.42	.154	3.84	.21$_8$	5.7$_4$	.114	6.20
90	.198	11.38	.186	5.55	.25$_5$	8.1$_1$	.116	7.35
100	.203	13.39	.216	7.56	.28$_8$	10.$_8$	.118	8.53

T	Molybdenum C_p	Molybdenum $H - H_0$	Nickel C_p	Nickel $H - H_0$	Palladium C_p	Palladium $H - H_0$
°K	$jg^{-1}deg^{-1}K$	jg^{-1}	$jg^{-1}deg^{-1}K$	jg^{-1}	$jg^{-1}deg^{-1}K$	jg^{-1}
1	0.000 0229	0.000 0105	0.000 120	0.000 060	0.000 099	0.000 0493
2	.000 0472	.000 0445	.000 242	.000 241	.000 203	.000 200
3	.000 0745	.000 105	.000 369	.000 546	.000 318	.000 459
4	.000 106	.000 194	.000 503	.000 98	.000 447	.000 840
5						
6	.000 191	.000 484	.000 82	.002 28	.000 891	.002 31
7						
8	.000 317	.000 981	.001 19	.004 28	.001 41	.004 60
9						
10	.000 498	.001 78	.001 62	.007 1	.002 10	.008 07
15	.001 31	.006 10	.003 1	.018 5	.004 71	.024 5
20	.002 87	.016 1	.005 8	.041	.009 22	.058 6
25	.005 77	.037 4	.010 1	.079	.016 0	.120
30	.009 60	.072 9	.016 7	.145	.025 8	.223
40	.023 6	.232	.038 1	.413	.050 7	.600
50	.041 0	.554	.068 2	.937	.077 7	1.24
60	.061 9	1.07	.103	1.79	.101	2.14
70	.083 8	1.80	.139	3.00	.122	3.26
80	.104	2.74	.173	4.56	.139	4.56
90	.123	3.88	.204	6.45	.154	6.03
100	.139	5.20	.232	8.63	.167	7.63

T	Platinum C_p	Platinum $H - H_0$	Rhodium C_p	Rhodium $H - H_0$	Silicon[l] C_p	Silicon[l] $H - H_0$	Silver C_p	Silver $H - H_0$
°K	$jg^{-1}deg^{-1}K$	jg^{-1}	$jg^{-1}deg^{-1}K$	jg^{-1}	$jg^{-1}deg^{-1}K$	jg^{-1}	$jg^{-1}deg^{-1}K$	jg^{-1}
1	0.000 035	0.000 0175	0.000 048	0.000 024	0.000 000 263	0.000 000 0658	0.000 0072	0.000 0032
2	.000 074	.000 071	.000 097	.000 096	.000 002 10	.000 001 05	.000 0239	.000 0176
3	.000 122	.000 168	.000 147	.000 218	.000 007 09	.000 005 32	.000 0595	.000 0574
4	.000 186	.000 320	.000 204	.000 303	.000 016	.000 010	.000 131	.000 146
6	.000 37	.000 85	.000 32	.000 91	.000 059 6	.000 085 3	.000 39	.000 62
8	.000 67	.001 88	.000 47	.001 70	.000 140	.000 279	.000 91	.001 87
10	.001 12	.003 65	.000 65	.002 81	.000 275	.000 670	.001 8	.004 52
15	.003 3	.013 5	.001 35	.007 65	.001 09	.003 74	.006 4	.023 3
20	.007 4	.039 5	.002 71	.017 4	.003 37	.013 8	.015 5	.076
25	.013 7	.092	.005 61	.037 3	.008 49	.042 3	.028 7	.185
30	.021 2	.182	.010 6	.077 1	.017 1	.105	.044 2	.368
40	.038	.48	.026 6	.256	.044 0	.400	.078	.979
50	.055	.95	.048 9	.633	.078 5	1.00	.108	1.91
60	.068	1.56	.072 4	1.238	.115	1.97	.133	3.12
70	.079	2.29	.094	2.07	.152	3.31	.151	4.54
80	.088	3.12	.114	3.11	.188	5.01	.166	6.13
90	.094	4.02	.132	4.34	.224	7.06	.177	7.85
100	.100	5.01	.147	5.74	.259	9.47	.187	9.67

T	Sodium[m] C_p	Sodium[m] $H - H_0$	Tantalum C_p	Tantalum $H - H_0$	Tin (white) C_p	Tin (white) $H - H_0$	Titanium C_p	Titanium $H - H_0$
°K	$jg^{-1}deg^{-1}K$	jg^{-1}	$jg^{-1}deg^{-1}K$	jg^{-1}	$jg^{-1}deg^{-1}K$	jg^{-1}	$jg^{-1}deg^{-1}K$	jg^{-1}
1	0.000 081	0.000 035	0.000 032	0.000 016	0.000 0170	0.000 0079	0.000 071	0.000 035
1			a.000 0063	a.000 0021	a.000 0041	a.000 0009		
2	.000 289	.000 204	.000 068	.000 065	.000 047	.000 0383	.000 146	.000 143
2			a.000 054	a.000 026	a.000 048	a.000 0228		
3	.000 76	.000 70	.000 112	.000 155	.000 109	.000 113	.000 226	.000 329
3			a.000 178	a.000 138	a.000 151	a.000 116		
n3.72					.000 198	.000 221		
3.72					a.000 285	a.000 270		
4	.001 60	.001 84	.000 171	.000 295	.000 245	.000 283	.000 317	.000 599
4			a.000 352	a.000 400				
o4.39			.000 201	.000 368				
4.39			a.000 433	a.000 553				
5	.002 98	.004 08			.000 54	.000 65		
6	.005 1	.008 1	.000 333	.000 776	.001 27	.001 51	.000 54	.001 45
8	.012 2	.024 7	.000 648	.001 73	.004 2	.006 8	.000 84	.002 81
10	.023 8	.060 2	.001 17	.003 52	.008 1	.019 0	.001 26	.004 89
12	.039 7	.123						
14	.063	.225						
15			.003 60	.014 5	.022 6	.093	.003 3	.015 6
16	.093	.380						
18	.124	.597						
20	.155	.875	.008 23	.043 2	.040	.251	.007 0	.040
25	.259	1.90	.015 3	.102	.058	.498	.013 4	.090
30	.364	3.45	.024 0	.202	.076	.834	.024 5	.182
40	.544	8.03	.043 0	.540	.106	1.75	.057 1	.581
50	.695	14.2	.060 4	1.06	.130	2.93	.099 2	1.358
60	.793	21.7	.075 4	1.74	.148	4.33	.146 7	2.592
70	.86	30.0	.087 9	2.56	.162	5.88	.189	4.27
80	.91	38.9	.097 6	3.49	.173	7.55	.230	6.37
90	.95	48.2	.105	4.50	.182	9.33	.267	8.86
100	.98	57.9	.111	5.58	.189	11.18	.300	11.69

METALS (*Continued*)

T	Tungsten		Zinc	
	C_p	$H - H_0$	C_p	$H - H_0$
°K	$jg^{-1} deg^{-1} K$	jg^{-1}	$jg^{-1} deg^{-1} K$	jg^{-1}
1	0.000 0074	0.000 0037	0.000 011	0.000 005
2	.000 0158	.000 0152	.000 028	.000 023
3	.000 0262	.000 0360	.000 058	.000 065
4	.000 0393	.000 0685	.000 11	.000 14
6	.000 0783	.000 182	.000 29	.000 53
8	.000 141	.000 396	.000 96	.001 6
10	.000 234	.000 765	.002 5	.005 0
15	.000 725	.002 97	.011	.034
20	.001 89	.009 27	.026	.125
25	.004 21	.023 7	.049	.31
30	.007 83	.053 4	.076	.62
40	.018 4	.181	.125	1.62
50	.033 2	.436	.171	3.11
60	.048 3	.843	.208	5.01
70	.060 5	1.39	.236	7.23
80	.071 5	2.05	.258	9.70
90	.081 0	2.81	.277	12.38
100	.088 8	3.66	.293	15.24

a Superconducting.
b In germanium the electronic specific heat is markedly dependent on impurities. The values given are for pure germanium (negligible electronic specific heat).
c α-Iron is the form that is thermodynamically stable at low temperatures. It has the body-centered cubic lattice which is the basis of the ferritic steels.
d γ-Iron is stable between 910° and 1,400°C. It has the face-centered cubic structure which is the basis of the austenitic steels. Since pure γ-iron is not stable at low temperatures the above values were calculated by application of the Kopp-Neumann rule to experimental data on two austenitic Fe-Mn alloys and are of uncertain accuracy.
e Superconducting transition temperature.
i Superconducting transition temperature of mercury.

j Melting temperature of mercury.
l In silicon the electronic specific heat, γT, is markedly dependent on impurities. Values of the coefficient, γ, from zero to 2.4×10^{-6} jg^{-1} deg^{-2} K have been reported. The values in the above table are for pure silicon ($\gamma = 0$).
m It has been shown (Barrett 1956, Hull & Rosenberg 1959) that sodium partially transforms at low temperatures from the normal body-centered cubic structure to close-packed hexagonal. The transformation is of the martensitic type and is promoted by cold-working at the low temperatures. Inasmuch as none of the calorimetric measurements on sodium were accompanied by crystallographic analysis, the tabulated data below 100°K are to some degree ambiguous.
n Superconducting transition temperature of tin.
o Superconducting transition temperature of tantalum.

CONSTANTS OF DEBYE-SOMMERFELD EQUATION

$C_v = \gamma T + \alpha T^3; \alpha = 12\pi^4 R/5\theta_0^3; 0 \lesssim T \lesssim T_{max}; T_{max}$ = maximum temperature to which the equation can be used with the limiting value of θ.

Substance	$10^6 \gamma$	γ	$10^6 \alpha$	θ_0	T_{max}
Metals:	$jg^{-1} deg^{-2} K$	mjg-$atom^{-1} deg^{-2} K$	$jg^{-1} deg^{-4} K$	$deg K$	$deg K$
Aluminum	50.4	1.36	0.93	426	4
Beryllium	25	0.226	.138	1160	20
Bismuth	0.32	.067	5.66	118	2
Cadmium	5.6	.63	2.69	186	3
Chromium	28.3	1.47	0.165	610	4
Copper	10.81	0.687	.746	344.5	10
Germanium	(a)	(a)	.528	370	2
Gold	3.75	0.74	2.19	165	15
Indium	15.8	1.81	13.1	109	2
α-Iron	90	5.0	0.349	464	10
Lead	15.1	3.1	10.6	96	4
Magnesium	54	1.32	1.19	406	4
α-Manganese	251	13.8	0.328	476	12
Molybdenum	23	2.18	.238	440	4
Nickel	120	7.0	.39	440	4
Niobium	85	7.9	.64	320	1
Palladium	98	10.5	.89	274	4
Platinum	34.1	6.7	.72	240	3
Rhodium	48	4.9	.173	478	4
Silicon	(a)	(a)	.263	640	4
Silver	5.65	0.610	1.58	225	4
Sodium a	60	1.37	21.4	158	4
Tantalum	31.5	5.7	0.69	250	4
Tin (white)	14.7	1.75	2.21	195	2
Titanium	71	3.4	0.54	420	10
Tungsten	7	1.3	.16	405	4
Zinc	9.6	0.63	1.10	300	4
Alloys:					
Constantan a	113	6.9	0.56	384	15
Monel a	108	6.5	.62	374	20
Other inorg. subs.:					
Diamond	....	....	0.0152	2200	50
Ice	....	....	15.2	192	10
Pyrex	....	....	3.14	......	5
Organic subs.:					
Glyptal	....	....	27	......	4
Lucite	....	....	35	......	4
Polystyrene	....	....	63	......	4

(a) Superconducting.

BOILING POINT OF WATER

(Hydrogen Scale)

Pressure mm	Tenths of millimeters										Pressure mm	Tenths of millimeters									
	.0	.1	.2	.3	.4	.5	.6	.7	.8	.9		.0	.1	.2	.3	.4	.5	.6	.7	.8	.9
700	97.714	718	722	725	729	733	737	741	745	749	750	99.630	633	637	641	645	648	652	656	659	663
701	753	757	761	765	769	773	777	781	785	789	751	667	671	674	678	682	686	689	693	697	700
702	792	796	800	804	808	812	816	820	824	828	752	704	708	712	715	719	723	726	730	734	738
703	832	836	840	844	847	851	855	859	863	867	753	741	745	749	752	756	760	764	767	771	775
704	871	875	879	883	887	891	895	899	902	906	754	778	782	786	790	793	797	801	804	808	812
705	97.910	914	918	922	926	930	934	938	942	946	755	99.815	819	823	827	830	834	838	841	845	849
706	949	953	957	961	965	969	973	977	981	985	756	852	856	860	863	867	871	875	878	882	886
707	989	993	996	*000	*004	*008	*012	*016	*020	*024	757	889	893	897	900	904	908	911	915	919	923
708	98.028	032	036	040	043	047	051	055	059	063	758	926	930	934	937	941	945	948	952	956	959
709	067	071	075	079	082	086	090	094	098	102	759	963	967	970	974	978	982	985	989	993	996
710	98.106	110	114	118	121	125	129	133	137	141	760	100.000	004	007	011	015	018	022	026	029	033
711	145	149	153	157	160	164	168	172	176	180	761	037	040	044	048	052	055	059	063	066	070
712	184	188	192	195	199	203	207	211	215	219	762	074	077	081	085	088	092	096	099	103	107
713	223	227	230	234	238	242	246	250	254	258	763	110	114	118	121	125	129	132	136	140	143
714	261	265	269	273	277	281	285	289	292	296	764	147	151	154	158	162	165	169	173	176	180
715	98.300	304	308	312	316	320	323	327	331	335	765	100.184	187	191	195	198	202	206	209	213	216
716	339	343	347	351	355	358	362	366	370	374	766	220	224	227	231	235	238	242	246	249	253
717	378	382	385	389	393	397	401	405	409	412	767	257	260	264	268	271	275	279	283	286	290
718	416	420	424	428	432	436	440	443	447	451	768	293	297	300	304	308	311	315	319	322	326
719	455	459	463	467	470	474	478	482	486	490	769	330	333	337	341	344	348	352	355	359	363
720	98.493	497	501	505	509	513	517	52C	524	528	770	100.366	370	373	377	381	384	388	392	395	399
721	532	536	540	544	547	551	555	559	563	567	771	403	406	410	414	417	421	424	428	432	435
722	570	574	578	582	586	590	593	597	601	605	772	439	442	446	450	453	457	461	464	468	472
723	609	613	617	620	624	628	632	636	640	643	773	475	479	483	486	490	493	497	501	504	508
724	647	651	655	659	662	666	670	674	678	682	774	511	515	519	522	526	530	533	537	540	544
725	98.686	689	693	697	701	705	709	712	716	720	775	100.548	551	555	559	562	566	569	573	577	580
726	724	728	732	735	739	743	747	751	755	758	776	584	588	591	595	598	602	606	609	613	616
727	762	766	770	774	777	781	785	789	793	797	777	620	624	627	631	634	638	642	645	649	653
728	800	804	808	812	816	819	823	827	831	835	778	656	660	663	667	671	674	678	681	685	689
729	838	842	846	850	854	858	861	865	869	873	779	692	696	699	703	707	710	714	718	721	725
730	98.877	880	881	888	892	896	899	903	907	911	780	100.728	732	735	739	743	746	750	753	757	761
731	915	918	922	926	030	934	937	941	945	949	781	764	768	772	775	779	782	786	790	793	797
732	953	956	960	964	968	972	975	979	983	987	782	800	804	807	811	815	818	822	825	829	833
733	991	994	998	*002	*006	*010	*013	*017	*021	*025	783	836	840	843	847	851	854	858	861	865	869
734	99.029	032	036	040	044	048	051	055	059	000	784	872	876	879	883	886	890	894	897	901	904
735	99.067	070	074	078	082	085	089	093	097	101	785	100.908	912	915	919	922	926	929	933	937	940
736	104	108	112	116	119	123	127	131	135	138	786	944	947	951	954	958	962	965	969	972	976
737	142	146	150	153	157	161	165	169	172	176	787	979	983	987	990	994	997	*001	*005	*008	*012
738	180	184	187	191	195	199	203	206	210	214	788	101.015	019	022	026	029	033	037	040	044	047
739	218	221	225	229	233	236	240	244	248	252	789	051	054	058	062	065	069	072	076	079	083
740	99.255	259	263	267	270	274	278	282	285	289	790	101.087	090	094	097	101	104	108	112	115	119
741	293	297	300	304	308	312	316	319	323	327	791	122	126	129	133	136	140	144	147	151	154
742	331	334	338	342	346	349	353	357	361	364	792	158	161	165	168	172	176	179	183	186	190
743	368	372	376	379	383	387	391	394	398	402	793	193	197	200	204	207	211	215	218	222	225
744	406	409	413	417	421	424	428	432	436	439	794	229	232	236	239	243	246	250	254	257	261
745	99.443	447	451	454	458	462	466	469	473	477	795	101.264	268	271	275	278	282	286	289	293	296
746	481	484	488	492	495	499	503	507	510	514	796	300	303	307	310	314	317	321	324	328	332
747	518	522	525	529	533	537	540	544	548	551	797	335	339	342	346	349	353	356	360	363	367
748	555	559	563	566	570	574	578	581	585	589	798	370	374	377	381	385	388	392	395	399	402
749	592	596	600	604	607	611	615	619	622	626	799	406	409	413	416	420	423	427	430	434	437
											800	101.441	....	....	....	....	....	....	....	....	....

For lower pressures see under Vapor Tension of Water

MELTING POINTS OF MIXTURES OF METALS

(*Smithsonian Physical Tables*)

Melting-points, °C.

Metals	Percentage of metal in second column.										
	0%	10%	20%	30%	40%	50%	60%	70%	80%	90%	100%
Pb. Sn.	* 326	295	276	262	240	220	190	185	200	216	232
Bi.	322	290	...	179	145	126	168	205	...		268
Te.	322	710	790	880	917	760	600	480	410	425	446
Ag.	328	460	545	590	620	650	705	775	840	905	959
Na.	...	360	420	400	370	330	290	250	200	130	96
Cu.	326	870	920	925	945	950	955	985	1005	1020	1084
Sb.	326	250	275	330	395	440	490	525	560	600	632
Al. Sb.	650	750	840	925	945	950	970	1000	1040	1010	632
Cu.	650	630	600	560	540	580	610	755	930	1055	1084
Au.	655	675	740	800	855	915	970	1025	1055	675	1062
Ag.	650	625	615	600	590	580	575	570	650	750	954
Zn.	654	640	620	600	580	560	530	510	475	425	419
Fe.	653	860	1015	1110	1145	1145	1220	1315	1425	1500	1515
Sb. Bi.	632	610	590	575	555	540	520	470	405	330	268
Ag.	630	595	570	545	520	500	505	545	680	850	959
Sn.	622	600	570	525	480	430	395	350	310	255	232
Zn.	632	555	510	540	570	565	540	525	510	470	419
Ni. Sn.	* 1455	1380	1290	1200	1235	1290	1305	1230	1060	800	232
Na. Bi.	96	425	520	590	645	690	720	730	715	570	268
Cd.	96	125	185	245	285	325	330	340	360	390	322
Cd. Ag.	322	420	520	610	700	760	805	850	895	940	954
Tl.	321	300	285	270	262	258	245	230	210	235	302
Zn.	322	280	270	295	313	327	340	355	370	390	419
Au. Cu.	1063	910	890	895	905	925	975	1000	1025	1060	1084
Ag.	1064	1062	1061	1058	1054	1049	1039	1025	1006	982	963
Pt.	1075	1125	1190	1250	1320	1380	1455	1530	1610	1685	1775
K. Na.	62	17.5	−10	−3.5	5	11	26	41	58	77	97.5
Hg.	...	...	...	...	...	90	110	135	162	265	...
Tl.	62.5	133	165	188	205	215	220	240	280	305	301
Cu. Ni.	1080	1180	1240	1290	1320	1355	1380	1410	1430	1440	1455
Ag.	1082	1035	990	945	910	870	830	788	814	875	960
Sn.	1084	1005	890	755	725	680	630	580	530	440	232
Zn.	1084	1040	995	930	900	880	820	780	700	580	419
Ag. Zn.	959	850	755	705	690	660	630	610	570	505	419
Sn.	959	870	750	630	550	495	450	420	375	300	232
Na. Hg.	96.5	90	80	70	60	45	22	55	95	215	...

* The data in this table are compiled from various sources,— hence variations in the melting point of the metals as shown in this column.

The triple point of water, 0.01°C (273.16°K) is the thermodynamic point for temperature measurements.

Miscellaneous Properties
Properties (Typical Only)

Common name and classification	Thermal conductivity			Specific gravity	Coeff. of linear expansion, μ in./in. °F	Electrical resistivity, microhm-cm	Modulus of elasticity, millions of psi	Approximate melting point	
	J/sec cm °K	Btu/hr ft °F	kcal/sec cm °C					°F	°C
Ingot iron (included for comparison)	1.3	77	0.32	7.86	6.8	9	30	2800	1538
Plain carbon steel AISI–SAE 1020	1.0	56	0.23	7.86	6.7	10	30	2760	1515
Stainless steel type 304	0.3	19	0.08	8.02	9.6	72	28	2600	1427
Cast gray iron ASTM A48–48. Class 25	0.8	48	0.20	7.2	6.7	67	13	2150	1177
Malleable iron ASTM A47				7.32	6.6	30	25	2250	1232
Ductile cast iron ASTM A339, A395	0.6	34	0.14	7.2	7.5	60	25	2100	1149
Ni-resist cast iron, type 2	0.7	41	0.17	7.3	9.6	170	15.6	2250	1232
Cast 28-7 alloy (IID) ASTM A297–63T	0.04	2	0.01	7.6	9.2	41	27	2700	1482
Hastelloy C	0.2	10	0.04	3.94	6.3	139	30	2350	1288
Inconel X, annealed	0.3	17	0.07	8.25	6.7	122	31	2550	1399
Haynes Stellite alloy 25 (L605)	0.2	10	0.04	9.15	7.61	88	34	2500	1371
Aluminum alloy 3003, rolled ASTM B221	2.8	164	0.68	2.73	12.9	4	10	1200	649
Aluminum alloy 2017, annealed ASTM B221	3.0	174	0.72	2.8	12.7	4	10.5	1185	641
Aluminum alloy 380 ASTM SC84B	1.8	102	0.42	2.7	11.6	7.5	10.3	1050	566
Copper ASTM B152, B124, B133, B1, B2, B3	7.1	411	1.70	8.91	9.3	1.7	17	1980	1082
Yellow brass (high brass) ASTM B36, B134, B135	2.2	126	0.52	8.47	10.5	7	15	1710	932
Aluminum bronze ASTM B169, alloy A; ASTM B124, B150	1.3	75	0.31	7.8	9.2	12	17	1900	1038
Beryllium copper 25 ASTM B194	0.2	12	0.05	8.25	9.3	–	19	1700	927
Nickel silver 18% alloy A (wrought) ASTM B122, No. 2	0.6	34	0.14	8.8	9.0	29	18	2030	1110
Cupronickel 30%	0.5	31	0.13	8.95	8.5	35	22	2240	1227
Red brass (cast) ASTM B30, No. 4A	1.3	77	0.32	8.7	10	11	13	1825	996
Chemical lead	0.6	36	0.15	11.35	16.4	21	2	621	327
Antimonial lead (hard lead)	0.5	31	0.13	10.9	15.1	23	3	554	290
Solder 50–50	0.8	48	0.20	8.89	13.1	15	–	420	216
Magnesium alloy AZ31B	1.4	82	0.34	1.77	14.5	9	6.5	1160	627
K Monel	0.3	19	0.08	8.47	7.4	58	26	2430	1332
Nickel ASTM B160, B161, B162	1.1	63	0.26	8.89	6.6	10	30	2625	1441
Cupronickel 55–45 (Constantan)	0.4	24	0.10	8.9	8.1	49	24	2300	1260
Commercial titanium	0.3	19	0.08	5	4.9	80	16.5	3300	1816
Zinc ASTM B69	2.0	114	0.47	7.14	18	6	–	785	418
Zirconium, commercial	0.3	19	0.08	6.5	2.9	41	12	3350	1843

THERMAL PROPERTIES OF PURE METALS

From Handbook of Tables for Applied Engineering Science by R. E. Bolz and G. L. Tuve, The Chemical Rubber Co., 1970

		At Atmospheric Pressure					Liquid Metal			
		At 100°K		At 25°C (77°F)				Vapor pressure		
Metal	Latent heat of fusion, cal/g	Thermal conductivity, watts cm°C	Specific heat, cal/g°C	Specific heat, cal/g°C	Coeff. of linear expansion, $(\times 10^6)$ $(°C)^{-1}$	Thermal conductivity, watts/cm°C*	Specific heat (liquid) at 2000°K cal/g°C	10^{-3} atm	10^{-6} atm	10^{-9} atm
								Boiling point temperatures, °K		
Aluminum	95	3.00	0.115	0.215	25	2.37	0.26	1,782	1,333	1,063
Antimony	38.5	—	0.040	0.050	9	0.185	0.062	1,007	741	612
Beryllium	324	—	0.049	0.436	12	2.18	0.78	1,793	1,347	1,085
Bismuth	12.4	—	0.026	0.030	13	0.084	0.036	1,155	851	677
Cadmium	13.2	1.03	0.047	0.055	30	0.93	0.063	655	486	388
Chromium	79	1.58	0.046	0.110	6	0.91	0.224	1,992	1,530	1,247
Cobalt	66	—	0.057	0.10	12	0.69	0.164	2,167	1,652	1,345
Copper	49	4.83	0.061	0.092	16.6	3.98	0.118	1,862	1,391	1,120
Gold	15	3.45	0.026	0.031	14.2	3.15	0.0355	2,023	1,510	1,211
Iridium	33	—	0.022	0.031	6	1.47	0.0434	3,253	2,515	2,062
Iron	65	1.32	0.052	0.108	12	0.803	0.197	2,093	1,594	1,297
Lead	5.5	0.396	0.028	0.031	29	0.346	0.033	1,230	889	698
Magnesium	88.0	1.69	0.016	0.243	25	1.59	0.32	857	638	509
Manganese	64	—	0.064	0.114	22	—	0.20	1,495	1,131	913
Mercury	2.7	—	0.029	0.033	—	0.0839	—	393	287	227
Molybdenum	69	1.79	0.033	0.060	5	1.4	0.089	3,344	2,558	2,079
Nickel	71	1.58	0.055	0.106	13	0.899	0.175	2,156	1,646	1,343
Niobium (Columbium)	68	0.552	0.045	0.064	7	0.52	0.083	3,523	2,721	2,232
Osmium	34	—		0.031	5	0.61	0.039	—	—	—
Platinum	24	0.79	0.024	0.032	9	0.73	0.043	2,817	2,155	1,757
Plutonium	3	—	0.019	0.032	54	0.08	0.041	2,200	1,596	1,252
Potassium	14.5	—	0.150	0.180	83	0.99	—	606	430	335
Rhodium	50	—	—	0.058	8	1.50	0.092	—	—	—
Selenium	16	—	—	0.077	37	0.005	—	—	—	—
Silicon	430	—	0.062	0.17	3	0.835	0.217	2,340	1,749	1,427
Silver	26.5	4.50	0.045	0.057	19	4.27	0.068	1,582	1,179	952
Sodium	27	—	0.234	0.293	70	1.34	—	701	504	394
Tantalum	41	0.592	0.026	0.034	6.5	0.54	0.040	3,959	3,052	2,495
Thorium	17	—	0.024	0.03	12	0.41	0.047	3,251	2,407	1,919
Tin	14.1	0.85	0.039	0.054	20	0.64	0.058	1,857	1,366	1,080
Titanium	100	0.312	0.072	0.125	8.5	0.2	0.188	2,405	1,827	1,484
Tungsten	46	2.35	0.021	0.032	4.5	1.78	0.040	4,139	3,228	2,656
Uranium	12	—	0.022	0.028	13.4	0.25	0.048	2,861	2,128	1,699
Vanadium	98	—	0.061	0.116	8	0.60	0.207	2,525	1,948	1,591
Zinc	27	1.32	0.063	0.093	35	1.15	—	752	559	449

* $(watts/cm°C) \times 860.421 = Cal(gm)hr^{-1}cm^{-1}°C^{-1}$
$(watts/cm°C) \times 57.818 = Btu\ hr^{-1}\ ft^{-1}\ °F.$

BOILING POINTS AND TRIPLE POINTS OF SOME LOW BOILING ELEMENTS

Element	Boiling point °K	Triple point °K	λ point °K
He	4.216		2.174
p-H$_2$	20.27	13.81	
n-H$_2$	20.379	13.95	
N$_2$	77.35		
O$_2$	90.188	54.34	
A	87.45		

MOLECULAR ELEVATION OF THE BOILING POINT*

Molecular elevation of the boiling point showing the elevation of the boiling point in degrees C due to the addition of one gram molecular weight of the dissolved substance to 1000 g of any one of the solvents below. The correction in the last column gives the number of degrees to be subtracted for each mm. of difference between the barometric reading and 760 mm.

Solvent	K_θ	Barometric correction per mm.
Acetic acid	3.07	0.0008
Acetone	1.71	0.0004
Aniline	3.52	0.0009
Benzene	2.53	0.0007
Bromobenzene	6.26	0.0016
Carbon bisulfide	2.34	0.0006
Carbon tetrachloride	5.03	0.0013
Chloroform	3.63	0.0009
Cyclohexane	2.79	0.0007
Ethanol (ethyl alcohol)	1.22	0.0003
Ethyl acetate	2.77	0.0007
Ethyl ether	2.02	0.0005
n-Hexane	2.75	0.0007
Methanol (methyl alcohol)	0.83	0.0002
Methyl acetate	2.15	0.0005
Nitrobenzene	5.24	0.0013
n-Octane	4.02	0.0010
Phenol	3.56	0.0009
Toluene	3.33	0.0008
Water	0.512	0.0001

* Most values from Hoyt, C. S. and Fink, C. K., Journal of Physical Chemistry, Vol. 41, No. 3., March, 1937.

MOLECULAR DEPRESSION OF THE FREEZING POINT

Showing the depression of the freezing point due to the addition of one gram molecular weight of dissolved substance, for various solvents.

Solvent	Depression for one gram molecular weight dissolved in 100 g °C
Acetic acid	39.0
Benzene	49.0
Benzophenone	98.0
Diphenyl	80.0
Diphenylamine	86.0
Ethylene dibromide	118.0
Formic acid	27.7
Naphthalene	68—69
Nitrobenzene	70.0
Phenol	74.0
Stearic acid	45.0
Triphenyl methane	124.5
Urethane	51.4
Water	18.5—18.7

CORRECTION OF BOILING POINTS TO STANDARD PRESSURE

By H. B. Hass and R. F. Newton

This correction may be made by using the equation:

$$\Delta t = \frac{(273.1 + t)(2.8808 - \log p)}{\phi + .15(2.8808 - \log p)} \qquad (1)$$

where Δt = degrees C to be added to the observed boiling point.

t = the observed boiling point.

$\log p$ = the logarithm of the observed pressure in millimeters of mercury.

$$\phi = \frac{\Delta H_{vap}}{2.303RT_b} = \frac{\Delta S_{vap}}{2.303R}$$

The value of ϕ may be estimated from the graph and the table. Substances not included in the table may be classified by grouping them with compounds which bear a close physical or structural resemblance to them.

Example 1. Benzene boils at 20°C. at 75 mm pressure. What is its normal boiling point? We do not find benzene in the table but we find hydrocarbons in group 2, and a group 2 compound with a boiling point of 20° has a ϕ of 4.6.

Substituting in the equation

$$\Delta t = \frac{(273.1 + 20)(2.8808 - 1.8751)}{4.60 + .15(2.8808 - 1.8751)} = 62°$$

Adding this to 20° gives 82° as a first approximation.

The graph shows that the ϕ for a compound of group 2 boiling at 82° is 4.72 instead of 4.60 which we originally used. Since ϕ is in the denominator, this increase will lower our Δt by the ratio, 4.60/4.72, or the corrected Δt is 62 × 4.60/4.72 = 60.4. Adding Δt to t, gives 80.4° as a second approximation.

The formula can best be used in a slightly different form when the reverse calculation is desired, i.e., when one calculates the vapor pressure at a given temperature, lower than the normal boiling point.

$$2.8808 - \log p = \frac{\phi \Delta t}{273.1 + t - .15\Delta t} \qquad (2)$$

Example 2. Alcohol boils at 78.4°C. What is its vapor pressure at 20°C.? Substituting in equation 2:

$$2.8808 - \log p = \frac{6.06 \times 58.4}{293.1 - (.15 \times 58.4)} = 1.245$$

$$\log p = 2.8808 - 1.245 = 1.6358$$

$$p = 43.2 \text{ mm.}$$

Here no second approximation is necessary, since the correct value of ϕ was taken immediately, the normal boiling point having been known.

Compound	Group	Compound	Group
Acetaldehyde	3	Amines	3
Acetic acid	4	n-Amyl alcohol	8
Acetic anhydride	6	Anthracene	1
Acetone	3	Anthraquinone	1
Acetophenone	4	Benzaldehyde	2

CORRECTION OF BOILING POINTS TO STANDARD PRESSURE (Continued)

Compound	Group	Compound	Group
Benzoic acid	5	Hydrogen cyanide	3
Benzonitrile	2	Isoamyl alcohol	7
Benzophenone	2	Isobutyl alcohol	8
Benzyl alcohol	5	Isobutyric acid	6
Butylethylene	1	Isocaproic acid	7
Butyric acid	7	Methane	1
Camphor	2	Methanol	7
Carbon monoxide	1	Methyl amine	5
Carbon oxysulfide	2	Methyl benzoate	3
Carbon suboxide	2	Methyl ether	3
Carbon sulfoselenide	2	Methyl ethyl ether	3
m.p. Chloroanilines	3	Methyl ethyl ketone	2
Chlorinated derivatives	Same group as though Cl were H	Methyl fluoride	3
		Methyl formate	4
o.m.p. Cresols	4	Methyl salicylate	2
Cyanogen	4	Methyl silicane	1
Cyanogen chloride	3	α, β Naphthols	3
Dibenzyl ketone	2	Nitrobenzene	3
Dimethyl amine	4	Nitromethane	3
Dimethyl oxalate	4	o.m.p. Nitrotoluenes	2
Dimethyl silicane	2	o.m.p. Nitrotoluidines	2
Esters	3	Phenanthrene	1
Ethanol	8	Phenol	5
Ethers	2	Phosgene	2
Ethylamine	4	Phthalic anhydride	2
Ethylene glycol	7	Propionic acid	5
Ethylene oxide	3	n-Propyl alcohol	8
Formic acid	3	Quinoline	2
Glycol diacetate	4	Sulfides	2
Halogen derivatives	Same group as though halogen were hydrogen.	Tetranitromethane	3
		Trichloroethylene	1
		Valeric acid	7
Heptylic acid	7	Water	6
Hydrocarbons	2		

BOILING POINT AT STANDARD PRESSURE

VAN DER WAALS' CONSTANTS FOR GASES

(Calculated from Amagat units in Landolt-Bornstein Physical Chemical Tables)

Van der Waals' equation is an equation of state for real gases. It may be written

$$\left(P + \frac{a}{V^2}\right)(V - b) = RT \text{ for one mole.} \qquad \text{or} \qquad \left(P + \frac{n^2 a}{V^2}\right)(V - nb) = nRT \text{ for } n \text{ moles.}$$

The term a is a measure of the attractive force between the molecules. The term b is due to the finite volume of the molecules and to their general incompressibility. It is known that a and b vary to some extent with temperature.

The values for a and b in the following table are those to be used when the pressure is in atmospheres and the volume is in liters. Thus R in the above equation will be 0.08206 liter atmospheres per mole per degree. T is degrees Kelvin.

Name	Formula	a (liters)² × atm. (mole)²	b liters mole	Name	Formula	a (liters)² × atm. (mole)²	b liters mole
Acetic acid	CH_3CO_2H	17.59	0.1068	n-Hexane	C_6H_{14}	24.39	0.1735
Acetic anhydride	$(CH_3CO)_2O$	19.90	0.1263	Hydrogen	H_2	0.2444	0.02661
Acetone	$(CH_3)_2CO$	13.91	0.0994	Hydrogen bromide	HBr	4.451	0.04431
Acetonitrile	CH_3CN	17.58	0.1168	Hydrogen chloride	HCl	3.667	0.04081
Acetylene	C_2H_2	4.390	0.05136	Hydrogen selenide	H_2Se	5.268	0.04637
Ammonia	NH_3	4.170	0.03707	Hydrogen sulfide	H_2S	4.431	0.04287
Amyl formate	$HCO_2C_5H_{11}$	27.58	0.1730	Iodobenzene	C_6H_5I	33.08	0.1656
Amylene	C_5H_{10}	15.90	0.1207	Krypton	Kr	2.318	0.03978
Isoamylene	C_5H_{10}	18.08	0.1405	Mercury	Hg	8.093	0.01696
Aniline	$C_6H_5NH_2$	26.50	0.1369	Mesitylene	$(CH_3)_3C_6H_3$	34.32	0.1979
Argon	A	1.345	0.03219	Methane	CH_4	2.253	0.04278
Benzene	C_6H_6	18.00	0.1154	Methyl acetate	$CH_3CO_2CH_3$	15.29	0.1091
Benzonitrile	C_6H_5CN	33.39	0.1724	Methyl alcohol	CH_3OH	9.523	0.06702
Bromobenzene	C_6H_5Br	28.56	0.1539	Methylamine	CH_3NH_2	7.130	0.05992
n-Butane	C_4H_{10}	14.47	0.1226	Methyl butyrate	$C_3H_7CO_2CH_3$	23.94	0.1569
iso-Butane	C_4H_{10}	12.87	0.1142	Methyl isobutyrate	$C_3H_7CO_2CH_3$	24.50	0.1637
iso-Butyl acetate	$CH_3CO_2C_4H_9$	28.50	0.1833	Methyl chloride	CH_3Cl	7.471	0.06483
iso-Butyl alcohol	C_4H_9OH	17.03	0.1143	Methyl ether	$(CH_3)_2O$	8.073	0.07246
iso-Butyl benzene	$C_6H_5C_4H_9$	38.59	0.2144	Methyl ethyl ether	$CH_3OC_2H_5$	11.95	0.09775
iso-Butyl formate	$HCO_2C_4H_9$	22.54	0.1476	Methyl ethyl sulfide	$CH_3SC_2H_5$	19.00	0.1304
Butyronitrile	C_3H_7CN	25.72	0.1596	Methyl fluoride	CH_3F	4.631	0.05264
Capronitrile	$C_5H_{11}CN$	34.16	0.1984	Methyl formate	HCO_2CH_3	10.84	0.08068
Carbon dioxide	CO_2	3.592	0.04267	Methyl propionate	$C_2H_5CO_2CH_3$	19.91	0.1360
Carbon disulfide	CS_2	11.62	0.07685	Methyl sulfide	$(CH_3)_2S$	12.87	0.09213
Carbon monoxide	CO	1.485	0.03985	Methyl valerate	$C_4H_9CO_2CH_3$	30.00	0.1848
Carbon oxysulfide	COS	0.955	0.05817	Naphthalene	$C_{10}H_8$	39.74	0.1937
Carbon tetrachloride	CCl_4	20.39	0.1383	Neon	Ne	0.2107	0.01709
Chlorine	Cl_2	6.493	0.05622	Nitric oxide	NO	1.340	0.02789
Chlorobenzene	C_6H_5Cl	25.43	0.1453	Nitrogen	N_2	1.390	0.03913
Chloroform	$CHCl_3$	15.17	0.1022	Nitrogen dioxide	NO_2	5.284	0.04424
m-Cresol	$CH_3C_6H_4OH$	31.38	0.1607	Nitrous oxide	N_2O	3.782	0.04415
Cyanogen	C_2N_2	7.667	0.06901	n-Octane	C_8H_{18}	37.32	0.2368
Cyclohexane	C_6H_{12}	22.81	0.1424	Oxygen	O_2	1.360	0.03183
Cymene	$C_{10}H_{14}$	42.16	0.2336	n-Pentane	C_5H_{12}	19.01	0.1460
Decane	$C_{10}H_{12}$	48.55	0.2905	iso-Pentane	C_5H_{12}	18.05	0.1417
Di-isobutyl	C_8H_{18}	34.97	0.2296	Phenetole	$C_6H_5OC_2H_5$	35.16	0.1963
Diethylamine	$(C_2H_5)_2NH$	19.15	0.1392	Phosphine	PH_3	4.631	0.05156
Dimethylamine	$(CH_3)_2NH$	10.38	0.08570	Phosphonium chloride	PH_4Cl	4.054	0.04545
Dimethylaniline	$C_6H_5N(CH_3)_2$	37.49	0.1970	Phosphorus	P	52.94	0.1566
Diphenyl	$(C_6H_5)_2$	52.79	0.2480	Propane	C_3H_8	8.664	0.08445
Diphenyl methane	$(C_6H_5)_2CH_2$	38.20	0.2240	Propionic acid	$C_2H_5CO_2H$	20.11	0.1187
Dipropylamine	$(C_2H_7)_2NH$	27.72	0.1820	Propionitrile	C_2H_5CN	16.44	0.1064
Di-isopropyl	$(C_3H_7)_2$	23.13	0.1669	Propyl acetate	$CH_3CO_2C_3H_7$	24.63	0.1619
Durene	$C_{10}H_{14}$	45.32	0.2424	Propyl alcohol	C_3H_7OH	14.92	0.1019
Ethane	C_2H_6	5.489	0.06380	Propylamine	$C_3H_7NH_2$	14.99	0.1090
Ethyl acetate	$CH_3CO_2C_2H_5$	20.45	0.1412	Propyl benzene	$C_6H_5C_3H_7$	35.85	0.2028
Ethyl alcohol	C_2H_5OH	12.02	0.08407	iso-Propyl benzene	$C_6H_5C_3H_7$	35.64	0.2025
Ethylamine	$C_2H_5NH_2$	10.60	0.08409	Propyl chloride	C_3H_7Cl	15.91	0.1141
Ethyl benzene	$C_6H_5C_2H_5$	28.60	0.1667	Propyl formate	$HCO_2C_3H_7$	18.95	0.1280
Ethyl butyrate	$C_3H_7CO_2C_2H_5$	30.07	0.1919	Propylene	C_2H_6	8.379	0.08272
Ethyl isobutyrate	$C_3H_7CO_2C_2H_5$	28.87	0.1994	Pseudo-cumene	$C_6H_3(CH_3)_3$	36.61	0.2021
Ethyl chloride	C_2H_5Cl	10.91	0.08651	Silicon fluoride	SiF_4	4.195	0.05571
Ethyl ether	$(C_2H_5)_2O$	17.38	0.1344	Silicon tetrahydride	SiH_4	4.320	0.05786
Ethyl formate	$HCO_2C_2H_5$	14.80	0.1056	Stannic chloride	$SnCl_4$	26.91	0.1642
Ethyl mercaptan	C_2H_5SH	11.24	0.08098	Sulfur dioxide	SO_2	6.714	0.05636
Ethyl propionate	$C_2H_5CO_2C_2H_5$	24.39	0.1615	Thiophene	C_4H_4S	20.72	0.1270
Ethyl sulfide	$(C_2H_5)_2S$	18.75	0.1214	Toluene	$C_6H_5CH_3$	24.06	0.1463
Ethylene	C_2H_4	4.471	0.05714	Triethylamine	$(C_2H_5)_3N$	27.17	0.1831
Ethylene bromide	$(CH_2Br)_2$	13.98	0.08664	Trimethylamine	$(CH_3)_3N$	13.02	0.1084
Ethylene chloride	$(CH_2Cl)_2$	16.91	0.1086	Xenon	Xe	4.194	0.05105
Ethylidene chloride	CH_3CHCl_2	15.50	0.1073	m-Xylene	$C_6H_4(CH_3)_2$	30.36	0.1772
Fluorobenzene	C_6H_5F	19.93	0.1286	o-Xylene	$C_6H_4(CH_3)_2$	29.98	0.1755
Germanium tetrachloride	$GeCl_4$	22.60	0.1485	p-Xylene	$C_6H_4(CH_3)_2$	30.93	0.1809
Helium	He	0.03412	0.02370	Water	H_2O	5.464	0.03049
n-Heptane	C_7H_{16}	31.51	0.2065				

VAN DER WAAL'S RADII IN Å

N	1.5	Te	2.20
P	1.9	H	1.2
As	2.0	F	1.35
Sb	2.2	Cl	1.80
O	1.40	Br	1.95
S	1.85	I	2.15
Se	2.00		

EMERGENT STEM CORRECTION FOR LIQUID-IN-GLASS THERMOMETERS

Accurate thermometers are calibrated with the entire stem immersed in the bath which determines the temperature of the thermometer bulb. However, for reasons of convenience it is common practice when using a thermometer to permit its stem to extend out of the apparatus. Under these conditions both the stem and the mercury in the exposed stem are at a temperature different from that of the bulb. This introduces an error into the observed temperature. Since the coefficient of thermal expansion of glass is less than that of mercury, the observed temperature will be less than the true temperature if the bulb is hotter than the stem and greater than the true temperature, providing the thermal gradient is reversed. For exact work the magnitude of this error can only be determined by experiment. However, for most purposes it is sufficiently accurate to apply the following equation which takes into account the difference of the thermal expansion of glass and mercury:

$$T_c = T_o + F \times L(T_o - T_m)$$

Where

T_c = corrected temperature
T_o = observed temperature
T_m = mean temperature of exposed stem. The mean temperature of the exposed stem may be determined by fastening the bulb of a second thermometer against the midpoint of the exposed liquid column.
L = the length of the exposed column in degrees above the surface of the substance whose temperature is being determined.
F = correction factor. For approximate work and when the liquid in the thermometer is mercury a value for F of 0.00016 is generally used. For more accurate work with mercury filled thermometers values as given in the following table are used. For thermometers filled with organic liquids it is customary to use 0.001 for the value of F.

Values of F for various glasses

Tm°C.	Corning 0041	Corning 8800	Corning 8810	Jena 16 III	Jena 59 III
50	0.000157	0.000166	0.000156	0.000158	0.000164
150	0.000159	0.000167	0.000157	0.000158	0.000165
250	0.000163	0.000168	0.000161	0.000161	0.000170
350	0.000168	0.000173	0.000166	—	0.000177

PRESSURE OF AQUEOUS VAPOR

Vapor Pressure of Ice

Pressure of aqueous vapor over ice in mm of Hg for temperatures from −98 to 0°C.

Temp. °C	0.0	0.2	0.4	0.6	0.8	Temp. °C	0.0	0.2	0.4	0.6	0.8
−90	0.000070	0.000048	0.000033	0.000022	0.000015	−17	1.031	1.012	0.993	0.975	0.956
−80	0.00040	0.00029	0.00020	0.00014	0.00010	−16	1.132	1.111	1.091	1.070	1.051
−70	0.00194	0.00143	0.00105	0.00077	0.00056	−15	1.241	1.219	1.196	1.175	1.153
−60	0.00808	0.00614	0.00464	0.00349	0.00261	−14	1.361	1.336	1.312	1.288	1.264
−50	0.02955	0.0230	0.0178	0.0138	0.0106	−13	1.490	1.464	1.437	1.411	1.386
−40	0.0966	0.0768	0.0609	0.0481	0.0378	−12	1.632	1.602	1.574	1.546	1.518
−30	0.2859	0.2318	0.1873	0.1507	0.1209	−11	1.785	1.753	1.722	1.691	1.661
−29	0.317	0.311	0.304	0.298	0.292	−10	1.950	1.916	1.883	1.849	1.817
−28	0.351	0.344	0.337	0.330	0.324	−9	2.131	2.093	2.057	2.021	1.985
−27	0.389	0.381	0.374	0.366	0.359	−8	2.326	2.285	2.246	2.207	2.168
−26	0.430	0.422	0.414	0.405	0.397	−7	2.537	2.493	2.450	2.408	2.367
−25	0.476	0.467	0.457	0.448	0.439	−6	2.765	2.718	2.672	2.626	2.581
−24	0.526	0.515	0.505	0.495	0.486	−5	3.013	2.962	2.912	2.862	2.813
−23	0.580	0.569	0.558	0.547	0.536	−4	3.280	3.225	3.171	3.117	3.065
−22	0.640	0.627	0.615	0.603	0.592	−3	3.568	3.509	3.451	3.393	3.336
−21	0.705	0.691	0.678	0.665	0.652	−2	3.880	3.816	3.753	3.691	3.630
−20	0.776	0.761	0.747	0.733	0.719	−1	4.217	4.147	4.079	4.012	3.946
−19	0.854	0.838	0.822	0.806	0.791	−0	4.579	4.504	4.431	4.359	4.287
−18	0.939	0.921	0.904	0.887	0.870						

VAPOR PRESSURE OF WATER BELOW 100°C

Pressure of aqueous vapor over water in mm of Hg for temperatures from −15.8 to 100°C. Values for fractional degrees between 50 and 89 were obtained by interpolation.

Temp. °C	0.0	0.2	0.4	0.6	0.8	Temp. °C	0.0	0.2	0.4	0.6	0.8
−15	1.436	1.414	1.390	1.368	1.345	−5	3.163	3.115	3.069	3.022	2.976
−14	1.560	1.534	1.511	1.485	1.460	−4	3.410	3.359	3.309	3.259	3.211
−13	1.691	1.665	1.637	1.611	1.585	−3	3.673	3.620	3.567	3.514	3.461
−12	1.834	1.804	1.776	1.748	1.720	−2	3.956	3.898	3.841	3.785	3.730
−11	1.987	1.955	1.924	1.893	1.863	−1	4.258	4.196	4.135	4.075	4.016
−10	2.149	2.116	2.084	2.050	2.018	−0	4.579	4.513	4.448	4.385	4.320
−9	2.326	2.289	2.254	2.219	2.184	0	4.579	4.647	4.715	4.785	4.855
−8	2.514	2.475	2.437	2.399	2.362	1	4.926	4.998	5.070	5.144	5.219
−7	2.715	2.674	2.633	2.593	2.553	2	5.294	5.370	5.447	5.525	5.605
−6	2.931	2.887	2.843	2.800	2.757	3	5.685	5.766	5.848	5.931	6.015

VAPOR PRESSURE OF WATER BELOW 100°C (continued)

Temp. °C	0.0	0.2	0.4	0.6	0.8	Temp. °C	0.0	0.2	0.4	0.6	0.8
4	6.101	6.187	6.274	6.363	6.453	53	107.20	108.2	109.3	110.4	111.4
5	6.543	6.635	6.728	6.822	6.917	54	112.51	113.6	114.7	115.8	116.9
6	7.013	7.111	7.209	7.309	7.411	55	118.04	119.1	120.3	121.5	122.6
7	7.513	7.617	7.722	7.828	7.936	56	123.80	125.0	126.2	127.4	128.6
8	8.045	8.155	8.267	8.380	8.494	57	129.82	131.0	132.3	133.5	134.7
9	8.609	8.727	8.845	8.965	9.086	58	136.08	137.3	138.5	139.9	141.2
10	9.209	9.333	9.458	9.585	9.714	59	142.60	143.9	145.2	146.6	148.0
11	9.844	9.976	10.109	10.244	10.380	60	149.38	150.7	152.1	153.5	155.0
12	10.518	10.658	10.799	10.941	11.085	61	156.43	157.8	159.3	160.8	162.3
13	11.231	11.379	11.528	11.680	11.833	62	163.77	165.2	166.8	168.3	169.8
14	11.987	12.144	12.302	12.462	12.624	63	171.38	172.9	174.5	176.1	177.7
15	12.788	12.953	13.121	13.290	13.461	64	179.31	180.9	182.5	184.2	185.8
16	13.634	13.809	13.987	14.166	14.347	65	187.54	189.2	190.9	192.6	194.3
17	14.530	14.715	14.903	15.092	15.284	66	196.09	197.8	199.5	201.3	203.1
18	15.477	15.673	15.871	16.071	16.272	67	204.96	206.8	208.6	210.5	212.3
19	16.477	16.685	16.894	17.105	17.319	68	214.17	216.0	218.0	219.9	221.8
20	17.535	17.753	17.974	18.197	18.422	69	223.73	225.7	227.7	229.7	231.7
21	18.650	18.880	19.113	19.349	19.587	70	233.7	235.7	237.7	239.7	241.8
22	19.827	20.070	20.316	20.565	20.815	71	243.9	246.0	248.2	250.3	252.4
23	21.068	21.324	21.583	21.845	22.110	72	254.6	256.8	259.0	261.2	263.4
24	22.377	22.648	22.922	23.198	23.476	73	265.7	268.0	270.2	272.6	274.8
25	23.756	24.039	24.326	24.617	24.912	74	277.2	279.4	281.8	284.2	286.6
26	25.209	25.509	25.812	26.117	26.426	75	289.1	291.5	294.0	296.4	298.8
27	26.739	27.055	27.374	27.696	28.021	76	301.4	303.8	306.4	308.9	311.4
28	28.349	28.680	29.015	29.354	29.697	77	314.1	316.6	319.2	322.0	324.6
29	30.043	30.392	30.745	31.102	31.461	78	327.3	330.0	332.8	335.6	338.2
30	31.824	32.191	32.561	32.934	33.312	79	341.0	343.8	346.6	349.4	352.2
31	33.695	34.082	34.471	34.864	35.261	80	355.1	358.0	361.0	363.8	366.8
32	35.663	36.068	36.477	36.891	37.308	81	369.7	372.6	375.6	378.8	381.8
33	37.729	38.155	38.584	39.018	39.457	82	384.9	388.0	391.2	394.4	397.4
34	39.898	40.344	40.796	41.251	41.710	83	400.6	403.8	407.0	410.2	413.6
35	41.175	42.644	43.117	43.595	44.078	84	416.8	420.2	423.6	426.8	430.2
36	44.563	45.054	45.549	46.050	46.556	85	433.6	437.0	440.4	444.0	447.5
37	47.067	47.582	48.102	48.627	49.157	86	450.9	454.4	458.0	461.6	465.2
38	49.692	50.231	50.774	51.323	51.879	87	468.7	472.4	476.0	479.8	483.4
39	52.442	53.009	53.580	54.156	54.737	88	487.1	491.0	494.7	498.5	502.2
40	55.324	55.91	56.51	57.11	57.72	89	506.1	510.0	513.9	517.8	521.8
41	58.34	58.96	59.58	60.22	60.86	90	525.76	529.77	533.80	537.86	541.95
42	61.50	62.14	62.80	63.46	64.13	91	546.05	550.18	554.35	558.53	562.75
43	64.80	65.48	66.16	66.86	67.56	92	566.99	571.26	575.55	579.87	584.22
44	68.26	68.97	69.69	70.41	71.14	93	588.60	593.00	597.43	601.89	606.38
45	71.88	72.62	73.36	74.12	74.88	94	610.90	615.44	620.01	624.61	629.24
46	75.65	76.43	77.21	78.00	78.80	95	633.90	638.59	643.30	648.05	652.82
47	79.60	80.41	81.23	82.05	82.87	96	657.62	662.45	667.31	672.20	677.12
48	83.71	84.56	85.42	86.28	87.14	97	682.07	687.04	692.05	697.10	702.17
49	88.02	88.90	89.79	90.69	91.59	98	707.27	712.40	717.56	722.75	727.98
50	92.51	93.5	94.4	95.3	96.3	99	733.24	738.53	743.85	749.20	754.58
51	97.20	98.2	99.1	100.1	101.1	100	760.00	765.45	770.93	776.44	782.00
52	102.09	103.1	104.1	105.1	106.2	101	787.57	793.18	798.82	804.50	810.21

VAPOR PRESSURE OF WATER ABOVE 100°C.

Based on values given by Keyes in the International Critical Tables.

Temp. °C	Pressure mm	Pressure Pounds per sq. in.	Temp. °F	Temp. °C	Pressure mm	Pressure Pounds per sq. in.	Temp. °F	Temp. °C	Pressure mm	Pressure Pounds per sq. in.	Temp. °F	Temp. °C	Pressure mm	Pressure Pounds per sq. in.	Temp. °F
100	760.	14.696	212.0	127	1850.83	35.789	260.6	154	3970.24	76.772	309.2	181	7694.24	148.782	357.8
101	787.51	15.228	213.8	128	1907.83	36.891	262.4	155	4075.88	78.815	311.0	182	7872.08	152.221	359.6
102	815.86	15.776	215.6	129	1966.35	38.023	264.2	156	4183.80	80.901	312.8	183	8052.96	155.719	361.4
103	845.12	16.342	217.4	130	2026.10	39.180	266.0	157	4293.24	83.018	314.6	184	8236.88	159.275	363.2
104	875.06	16.921	219.2	131	2087.42	40.364	267.8	158	4404.96	85.178	316.4	185	8423.84	162.890	365.0
105	906.07	17.521	221.0	132	2150.42	41.582	269.6	159	4519.72	87.397	318.2	186	8616.12	166.609	366.8
106	937.92	18.136	222.8	133	2214.64	42.824	271.4	160	4636.00	89.646	320.0	187	8809.92	170.356	368.6
107	970.60	18.768	224.6	134	2280.76	44.103	273.2	161	4755.32	91.953	321.8	188	9007.52	174.177	370.4
108	1004.42	19.422	226.4	135	2347.26	45.389	275.0	162	4876.92	94.304	323.6	189	9208.16	178.057	372.2
109	1038.92	20.089	228.2	136	2416.34	46.724	276.8	163	5000.04	96.685	325.4	190	9413.36	182.025	374.0
110	1074.56	20.779	230.0	137	2488.16	48.113	278.6	164	5126.96	99.139	327.2	191	9620.08	186.022	375.8
111	1111.20	21.487	231.8	138	2560.67	49.515	280.4	165	5256.16	101.638	329.0	192	9831.36	190.107	377.6
112	1148.74	22.213	233.6	139	2634.45	50.950	282.2	166	5386.88	104.165	330.8	193	10047.20	194.281	379.4
113	1187.42	22.961	235.4	140	2710.92	52.421	284.0	167	5521.40	106.766	332.6	194	10265.32	198.499	381.2
114	1227.25	23.731	237.2	141	2788.44	53.920	285.8	168	5658.20	109.412	334.4	195	10488.76	202.819	383.0
115	1267.98	24.519	239.0	142	2867.48	55.448	287.6	169	5798.04	112.116	336.2	196	10715.24	207.199	384.8
116	1309.94	25.330	240.8	143	2948.80	57.020	289.4	170	5940.92	114.879	338.0	197	10944.76	211.637	386.6
117	1352.95	26.162	242.6	144	3031.64	58.622	291.2	171	6085.32	117.671	339.8	198	11179.60	216.178	388.4
118	1397.18	27.017	244.4	145	3116.76	60.268	293.0	172	6233.52	120.537	341.6	199	11417.48	220.778	390.2
119	1442.63	27.896	246.2	146	3203.40	61.944	294.8	173	6383.24	123.432	343.4	200	11659.16	225.451	392.0
120	1489.14	28.795	248.0	147	3292.32	63.663	296.6	174	6538.28	126.430	345.2	201	11905.40	230.213	393.8
121	1536.80	29.717	249.8	148	3382.76	65.412	298.4	175	6694.08	129.442	347.0	202	12155.44	235.048	395.6
122	1586.04	30.669	251.6	149	3476.24	67.220	300.2	176	6852.92	132.514	348.8	203	12408.52	239.942	397.4
123	1636.36	31.642	253.4	150	3570.48	69.042	302.0	177	7015.56	135.659	350.6	204	12666.16	244.924	399.2
124	1687.81	32.637	255.2	151	3667.00	70.908	303.8	178	7180.48	138.848	352.4	205	12929.12	250.008	401.0
125	1740.93	33.664	257.0	152	3766.56	72.833	305.6	179	7349.20	142.110	354.2	206	13197.40	255.196	402.8
126	1795.12	34.712	258.8	153	3866.88	74.773	307.4	180	7520.20	145.417	356.0	207	13467.96	260.428	404.6

VAPOR PRESSURE OF WATER ABOVE 100°C (continued)

Temp. °C	mm	Pounds per sq. in.	Temp. °F	Temp. °C	mm	Pounds per sq. in.	Temp. °F	Temp. °C	mm	Pounds per sq. in.	Temp. °F
208	13742.32	265.733	406.4	264	37529.56	725.703	507.2	320	84686.80	1637.575	608.0
209	14022.76	271.156	408.2	265	38133.00	737.372	509.0	321	85819.20	1659.472	609.8
210	14305.48	276.623	410.0	266	38742.52	749.158	510.8	322	86959.20	1681.516	611.6
211	14595.04	282.222	411.8	267	39361.92	761.135	512.6	323	88114.40	1703.854	613.4
212	14888.40	287.895	413.6	268	39986.64	773.215	514.4	324	89277.20	1726.339	615.2
213	15184.80	293.626	415.4	269	40619.72	785.457	516.2	325	90447.60	1748.971	617.0
214	15488.04	299.490	417.2	270	41261.16	797.861	518.0	326	91633.20	1771.897	618.8
215	15792.80	305.383	419.0	271	41910.20	810.411	519.8	327	92826.40	1794.969	620.6
216	16104.40	311.408	420.8	272	42566.08	823.094	521.6	328	94042.40	1818.483	622.4
217	16420.56	317.522	422.6	273	43229.56	835.923	523.4	329	95273.60	1842.291	624.2
218	16742.04	323.738	424.4	274	43902.16	848.929	525.2	330	96512.40	1866.245	626.0
219	17067.32	330.028	426.2	275	44580.84	862.053	527.0	331	97758.80	1890.346	627.8
220	17395.64	336.377	428.0	276	45269.40	875.367	528.8	332	99020.40	1914.742	629.6
221	17731.56	342.872	429.8	277	45964.04	888.799	530.6	333	100297.20	1939.431	631.4
222	18072.80	349.471	431.6	278	46669.32	902.437	532.4	334	101581.60	1964.267	633.2
223	18417.84	356.143	433.4	279	47382.20	916.222	534.2	335	102881.20	1989.398	635.0
224	18766.68	362.888	435.2	280	48104.20	930.183	536.0	336	104196.00	2014.822	636.8
225	19123.12	369.781	437.0	281	48833.80	944.291	537.8	337	105526.00	2040.540	638.6
226	19482.60	376.732	438.8	282	49570.24	958.532	539.6	338	106871.20	2066.552	640.4
227	19848.92	383.815	440.6	283	50316.56	972.963	541.4	339	108224.00	2092.710	642.2
228	20219.80	390.987	442.4	284	51072.76	987.586	543.2	340	109592.00	2119.163	644.0
229	20596.76	398.276	444.2	285	51838.08	1002.385	545.0	341	110967.60	2145.763	645.8
230	20978.28	405.654	446.0	286	52611.76	1017.345	546.8	342	112358.40	2172.657	647.6
231	21365.12	413.134	447.8	287	53395.32	1032.497	548.6	343	113749.20	2199.550	649.4
232	21757.28	420.717	449.6	288	54187.24	1047.810	550.4	344	115178.00	2227.179	651.2
233	22154.00	428.388	451.4	289	54989.04	1063.314	552.2	345	116614.40	2254.954	653.0
234	22558.32	436.207	453.2	290	55799.20	1078.980	554.0	346	118073.60	2283.171	654.8
235	22967.96	444.128	455.0	291	56612.40	1094.705	555.8	347	119532.80	2311.387	656.6
236	23382.92	452.152	456.8	292	57448.40	1110.871	557.6	348	121014.80	2340.044	658.4
237	23802.44	460.264	458.6	293	58284.40	1127.036	559.4	349	122504.40	2368.848	660.2
238	24229.56	468.523	460.4	294	59135.60	1143.496	561.2	350	124001.60	2397.799	662.0
239	24661.24	476.871	462.2	295	59994.40	1160.102	563.0	351	125521.60	2427.191	663.8
240	25100.52	485.365	464.0	296	60860.80	1176.856	564.8	352	127049.20	2456.730	665.6
241	25543.60	493.933	465.8	297	61742.40	1193.903	566.6	353	128599.60	2486.710	667.4
242	25994.28	502.647	467.6	298	62624.00	1210.950	568.4	354	130157.60	2516.837	669.2
243	26449.52	511.450	469.4	299	63528.40	1228.439	570.2	355	131730.80	2547.258	671.0
244	26912.36	520.400	471.2	300	64432.80	1245.927	572.0	356	133326.80	2578.119	672.8
245	27381.28	529.467	473.0	301	65352.40	1263.709	573.8	357	134945.60	2609.422	674.6
246	27855.52	538.638	474.8	302	66279.60	1281.638	575.6	358	136579.60	2641.018	676.4
247	28335.84	547.926	476.6	303	67214.40	1299.714	577.4	359	138228.80	2672.908	678.2
248	28823.76	557.360	478.4	304	68156.80	1317.937	579.2	360	139893.20	2705.093	680.0
249	29317.00	566.898	480.2	305	69114.40	1336.454	581.0	361	141572.80	2737.571	681.8
250	29817.84	576.583	482.0	306	70072.00	1354.971	582.8	362	143275.20	2770.490	683.6
251	30324.00	586.370	483.8	307	71052.40	1373.929	584.6	363	144992.80	2803.703	685.4
252	30837.76	596.305	485.6	308	72048.00	1393.181	586.4	364	146733.20	2837.357	687.2
253	31356.84	606.342	487.4	309	73028.40	1412.139	588.2	365	148519.20	2871.892	689.0
254	31885.04	616.556	489.2	310	74024.00	1431.390	590.0	366	150320.40	2906.722	690.8
255	32417.80	626.858	491.0	311	75042.40	1451.083	591.8	367	152129.20	2941.698	692.6
256	32957.40	637.292	492.8	312	76076.00	1471.070	593.6	368	153960.80	2977.116	694.4
257	33505.36	647.888	494.6	313	77117.20	1491.203	595.4	369	155815.20	3012.974	696.2
258	34059.40	658.601	496.4	314	78166.00	1511.484	597.2	370	157692.40	3049.273	698.0
259	34618.76	669.417	498.2	315	79230.00	1532.058	599.0	371	159584.80	3085.866	699.8
260	35188.00	680.425	500.0	316	80294.00	1552.632	600.8	372	161507.60	3123.047	701.6
261	35761.80	691.520	501.8	317	81373.20	1573.501	602.6	373	163468.40	3160.963	703.4
262	36343.20	702.763	503.6	318	82467.60	1594.663	604.4	374	165467.20	3199.613	705.2
263	36932.20	714.152	505.4	319	83569.60	1615.972	606.2				

VAPOR PRESSURE

The following table is an abridged form of the very extensive compilation by Daniel R. Stull and published in *Industrial and Engineering Chemistry*, 39, 517, 1947.

The table gives the temperatures in degrees centigrade at which the vapor of the compound listed at the left has the pressure indicated at the top of the column. Organic compounds are listed in the order of their empirical formulae and inorganic compounds in the alphabetic order of their names. Pressures greater than one atmosphere are listed in separate tables.

Abbreviations: d = decomposes; *d* = dextrorotatory; *dl* = inactive (50% *d* and 50% *l*); e = explodes; *l* = levorotatory; M.P. = melting point; p = polymerizes; s = solid.

INORGANIC COMPOUNDS
Pressures Less than 1 Atmosphere

Name	Formula	Temperature (°C)						M.P.
		1 mm	10 mm	40 mm	100 mm	400 mm	760 mm	
Aluminum	Al	1284	1487	1635	1749	1947	2056	660
Aluminum borohydride	AlB$_3$H$_{12}$	s	−42.9	−20.9	−3.9	+28.1	45.9	−64.5
Aluminum bromide	AlBr$_3$	81.3s	118.0	150.6	176.1	227.0	256.3	97.5
Aluminum chloride	AlCl$_3$	100.0s	123.8s	139.9s	152.0s	171.6s	180.2s	192.4
Aluminum fluoride	AlF$_3$	1238	1324	1378	1422	1496	1537	1040
Aluminum iodide	AlI$_3$	178.0s	225.8	265.0	294.5	354.0	385.5	—
Aluminum oxide	Al$_2$O$_3$	2148	2385	2549	2665	2874	2977	2050
Ammonia	NH$_3$	−109.1s	−91.9s	−79.2s	−68.4	−45.4	−33.6	−77.7
Deutero ammonia	ND$_3$	s	s	s	−67.4	−45.4	−33.4	−74.0
Ammonium axide	NH$_4$N$_3$	29.2s	59.2s	80.1s	95.2s	120.4s	133.8s	—
Ammonium bromide	NH$_4$Br	198.3s	252.0s	290.0s	320.0s	370.9s	396.0s	—
Ammonium carbamate	NH$_4$CO$_2$NH$_2$	−26.1s	−2.9s	+14.0s	26.7s	48.0s	58.3s	—
Ammonium chloride	NH$_4$Cl	160.4s	209.8s	245.0s	271.5s	316.5s	337.8s	520
Ammonium hydrogen sulfide	NH$_4$HS	−51.1	−28.7	−12.3	0.0	+21.8	33.3	—
Ammonium iodide	NH$_4$I	210.9s	263.5s	302.8s	331.8s	381.0s	404.9s	—
Ammonium cyanide	NH$_4$CN	−50.6s	−28.6s	−12.6s	−0.5s	+20.5s	31.7s	36
Antimony	Sb	886	1033	1141	1223	1364	1440	630.5
Antimony tribromide	SbBr$_3$	93.9	142.7	177.4	203.5	250.2	275.0	96.0
Antimony trichloride	SbCl$_3$	49.2s	83.3	114.0	143.3	192.2	219.0	73.4
Antimony pentachloride	SbCl$_5$	22.7	61.8	91.0	114.1	—	—	2.8
Antimony triiodide	SbI$_3$	163.6s	223.5	267.8	303.5	368.5	401.0	167
Antimony trioxide	Sb$_2$O$_3$	574s	666	812	957	1242	1425	656
Argon	A	−218.2s	−210.9s	−204.9s	−200.5s	−190.6s	−185.6	−189.2
Arsenic (metallic)	As	372s	437s	483s	518s	579s	610s	814
Arsenic tribromide	AsBr$_3$	41.8	85.2	118.7	145.2	193.6	220.0	—
Arsenic trichloride	AsCl$_3$	−11.4	+23.5	50.0	70.9	109.7	130.4	−18
Arsenic trifluoride	AsF$_3$	s	s	−2.5	+13.2	41.5	56.3	−5.9
Arsenic pentafluoride	AsF$_5$	−117.9s	−103.1s	−92.4s	−84.3s	−64.0	−52.8	−79.8
Arsenic hydride (arsine)	AsH$_3$	−142.6s	−124.7s	−110.2	−98.0	−75.2	−62.1	−116.3
Arsenic trioxide	As$_2$O$_3$	212.5s	259.7s	299.2s	332.5	412.2	457.2	312.8
Barium	Ba	s	1049	1195	1301	1518	1638	850
Beryllium borohydride	BeB$_2$H$_8$	+1.0s	28.1s	46.2s	58.6s	79.7s	90.0s	123
Beryllium bromide	BeBr$_2$	289s	342s	379s	405s	451s	474s	490
Beryllium chloride	BeCl$_2$	291s	346s	384s	411	461	487	405
Beryllium iodide	BeI$_2$	283s	341s	382s	411s	461s	487s	488
Bismuth	Bi	1021	1136	1217	1271	1370	1420	271
Bismuth tribromide	BiBr$_3$	s	282	327	360	425	461	218
Bismuth trichloride	BiCl$_3$	s	264	311	343	405	441	230
Borine carbonyl	BH$_3$CO	−139.2	−121.1	−106.6	−95.3	−74.8	−64.0	−137.0
Boron tribromide	BBr$_3$	−41.4	−10.1	+14.0	33.5	70.0	91.7	−45
Boron trichloride	BCl$_3$	−91.5	−66.9	−47.8	−32.4	−3.6	+12.7	−107
Boron trifluoride	BF$_3$	−154.6	−141.3s	−131.0s	−123.0	−108.3	−110.7	−126.8
Dihydrodiborane	B$_2$H$_6$	−159.7	−144.3	−131.6	−120.9	−99.6	−86.5	−169
Diborane hydrobromide	B$_2$BrH$_5$	−93.3	−66.3	−45.4	−29.0	0.0	+16.3	−104.2
Triborine triamine	B$_3$H$_6$N$_3$	−63.0s	−35.3	−13.2	+4.0	34.3	50.6	−58.2
Tetrahydrotetraborane	B$_4$H$_{10}$	−90.9	−64.3	−44.3	−28.1	+0.8	16.1	−119.9
Dihydropentaborane	B$_5$H$_9$	s	−30.7	−8.0	+9.6	40.8	58.1	−47.0
Tetrahydropentaborane	B$_5$H$_{11}$	−50.2	−19.9	+2.7	20.1	51.2	67.0	—
Dihydrodecaborane	B$_{10}$H$_{14}$	60.0s	90.2s	117.4	142.3	d	d	99.6
Bromine	Br$_2$	−48.7s	−25.0s	−8.0s	+9.3	41.0	58.2	−7.3
Bromine pentafluoride	BrF$_5$	−69.3s	−41.9	−21.0	−4.5	+25.7	40.0	−61.4
Cadmium	Cd	394	484	553	611	711	765	320.9
Cadmium chloride	CdCl$_2$	s	656	736	797	908	967	568
Cadmium fluoride	CdF$_2$	1112	1286	1400	1486	1651	1751	520
Cadmium iodide	CdI$_2$	416	512	584	640	742	796	385
Cadmium oxide	CdO	1000s	1149s	1257s	1341s	1484s	1559s	—
Calcium	Ca	s	983	1111	1207	1388	1487	851
Carbon	C	3586s	3946s	4196s	4373s	4660s	4827s	—
Carbon tetrabromide	CBr$_4$	s	s	96.3	119.7	163.5	189.5	90.1
Carbon tetrachloride	CCl$_4$	−50.0s	−19.6	+4.3	23.0	57.8	76.7	−22.6
Carbon tetrafluoride	CF$_4$	−184.6s	−169.3	−158.8	−150.7	−135.5	−127.7	−183.7
Carbon dioxide	CO$_2$	−134.3s	−119.5s	−108.6s	−100.2s	−85.7s	−78.2s	−57.5
Carbon suboxide	C$_3$O$_2$	−94.8	−71.0	−52.0	−36.9	−8.9	+6.3	−107
Carbon disulfide	CS$_2$	−73.8	−44.7	−22.5	−5.1	+28.0	46.5	−110.8
Carbon subsulfide	C$_3$S$_2$	14.0	54.9	85.6	109.9	p	—	+0.4
Carbon selenosulfide	CSSe	−47.3	−16.0	+8.6	28.3	65.2	85.6	−75.2
Carbon monoxide	CO	−222.0s	−215.0s	−210.0s	−205.7s	−196.3	−191.3	−205.0

Name	Formula	Temperature (°C)						M.P.
		1 mm	10 mm	40 mm	100 mm	400 mm	760 mm	
Carbonyl chloride	$COCl_2$	−92.9	−69.3	−50.3	−35.6	−7.6	+ 8.3	−104
Carbonyl selenide	COSe	−117.1	−95.0	−76.4	−61.7	−35.6	−21.9	—
Carbonyl sulfide	COS	−132.4	−113.3	−98.3	−85.9	−62.7	−49.9	−138.8
Chloropierin	CCl_3NO_2	−25.5	+ 7.8	33.8	53.8	91.8	111.9	−64
Chlorotrifluoromethane	$CClF_3$	−149.5	−134.1	−121.9	−111.7	−92.7	−81.2	—
Cyanogen	C_2N_2	−95.8s	−76.8s	−62.7s	−51.8s	−33.0	−21.0	−34.4
Cyanogen bromide	CBrN	−35.7s	−10.0s	+ 8.6s	22.6s	46.0s	61.5	58
Cyanogen chloride	CCIN	−76.7s	−53.8s	−37.5s	−24.9s	−2.3	+ 13.1	−6.5
Cyanogen fluoride	CFN	−134.4s	−118.5s	−106.4s	−97.0s	−80.5s	−72.6s	—
Cyanogen iodide	CIN	25.2s	57.7s	80.3s	97.6s	126.1s	141.1s	—
Deuterocyanic acid	CDN	−68.9s	−46.7s	−30.1s	−17.5s	+ 10.0	26.2	−12
Dichlorodifluoromethane	CCl_2F_2	−118.5	−97.8	−81.6	−68.6	−43.9	−29.8	—
Dichlorofluoromethane	$CHCl_2F$	−91.3	−67.5	−48.8	−33.9	−6.2	+ 8.9	−135
Chlorodifluoromethane	$CHClF_2$	−122.8	−103.7	−88.6	−76.4	−53.6	−40.8	−160
Trichlorofluoromethane	CCl_3F	−84.3	−59.0	−39.0	−23.0	+ 6.8	23.7	—
Cesium	Cs	279	375	449	509	624	690	28.5
Cesium bromide	CsBr	748	887	993	1072	1221	1300	636
Cesium chloride	CsCl	744	884	989	1069	1217	1300	646
Cesium fluoride	CsF	712	844	947	1025	1170	1251	683
Cesium iodide	CsI	738	873	976	1055	1200	1280	621
Chlorine	Cl_2	−118.0s	−101.6s	−84.5	−71.7	−47.3	−33.8	−100.7
Chlorine fluoride	ClF	s	−139.0	−128.8	−120.8	−107.0	−100.5	−145
Chlorine trifluoride	ClF_3	s	−71.8	−51.3	−34.7	−4.9	+ 11.5	−83
Chlorine monoxide	Cl_2O	−98.5	−73.1	−54.3	−39.4	−12.5	+ 2.2	−116
Chlorine dioxide	ClO_2	s	−59.0	−42.8	−29.4	−4.0	+ 11.1	−59
Dichlorine hexoxide	Cl_2O_6	+ 7.5	42.0	68.0	87.7	123.8	142.0	3.5
Chlorine heptoxide	Cl_2O_7	−45.3	−13.2	+ 10.3	29.1	62.2	78.8	−91
Chlorosulfonic acid	HSO_3Cl	32.0	64.0	87.6	105.3	136.1	151.0d	−80
Chromium	Cr	1616	1845	2013	2139	2361	2482	1615
Chromium carbonyl	$Cr(CO)_6$	36.0	68.3	91.2	108.0	137.2	151.0	—
Chromyl chloride	CrO_2Cl_2	−18.4	+ 13.8	38.5	58.0	95.2	117.1	—
Cobaltous chloride	$COCl_2$	s	s	770	843	974	1050	735
Cobalt nitrosyl tricarbonyl	$Co(CO)_2NO$	s	s	+ 11.0	29.0	62.0	80.0	−11
Columbium pentafluoride	CbF_5	s	86.3	121.5	148.5	198.0	225.0	75.5
Copper	Cu	1628	1879	2067	2207	2465	2595	1083
Cuprous bromide	Cu_2Br_2	572	718	844	951	1189	1355	504
Cuprous chloride	Cu_2Cl_2	546	702	838	960	1249	1490	422
Cuprous iodide	Cu_2L_2	s	656	786	907	1158	1336	605
Ferric chloride	$FeCl_3$	194.0s	235.5s	256.8s	272.5s	298.0s	319.0	304
Ferrous chloride	$FeCl_2$		700	779	842	961	1026	—
Fluorine	F_2	−223.0	−214.1	−207.7	−202.7	−193.2	−187.9	−223
Fluorine monoxide	F_2O	−196.1	−182.3	−173.0	−165.8	−151.9	−144.6	−223.9
Gallium	Ga	1349	1541	1680	1784	1974	2071	30
Gallium trichloride	$GaCl_3$	48.0s	76.5s	107.5	132.0	176.3	200.0	77.0
Germanium hydride	GeH_4	−163.0	−145.3	−131.6	−120.4	−100.2	−88.9	−165
Germanium bromide	$GeBr_4$	s	56.8	88.1	113.2	161.6	189.0	26.1
Germanium chloride	$GeCl_4$	−45.0	−15.0	+ 8.0	27.5	63.8	84.0	−49.5
Trichlorogermane	$GeHCl_3$	−41.3	−13.0	+ 8.8	26.5	58.3d	75.0d	−71.1
Tetramethylgermanium	$Ge(CH_3)_4$	−73.2	−45.2	−23.4	−6.3	+ 26.0	44.0	−88
Digermane	Ge_2H_6	−88.7	−60.1	−38.2	−20.3	+ 13.3	31.5	−109
Trigermane	Ge_3H_8	−36.9	−0.9	+ 26.3	47.9	88.6	110.8	−105.6
Gold	Au	1869	2154	2363	2521	2807	2966	1063
Helium	He	−271.7	−271.3	−270.7	−270.3	−269.3	−268.6	—
Hydrogen	H_2	−263.3s	−261.3	−259.6s	−257.9	−254.5	−252.5	−259.1
Hydrogen deuteride	HD		−259.8	−258.2	−256.6	−253.0	−251.0	—
Hydrogen bromide	HBr	−138.8s	−121.8s	−108.3s	−97.7s	−78.0	−66.5	−87.0
Hydrogen chloride	HCl	−150.8s	−135.6s	−123.8s	−114.0	−95.3	−84.8	−114.3
Hydrogen cyanide	HCN	−71.0s	−47.7s	−30.9s	−17.8s	+ 10.2	25.9	−13.2
Hydrogen fluoride	HF	s	−65.8	−45.0	−28.2	+ 2.5	19.7	−83.7
Hydrogen iodide	HI	−123.3s	−102.3s	−85.6s	−72.1s	−48.3	−35.1	−50.9
Hydrogen peroxide	H_2O_2	15.3	50.4	77.0	97.9	137.4d	158.0d	−0.9
Hydrogen selenide	H_2Se	−115.3s	−97.9s	−84.7s	−74.2s	−53.6	−41.1	−64
Hydrogen sulfide	H_2S	−134.3s	−116.3s	−102.3s	−91.6s	−71.8	−60.4	−85.5
Hydrogen disulfide	H_2S_2	−43.2	−15.2	+ 6.0	22.0	49.6	64.0	−89.7
Hydrogen teluride	H_2Te	−96.4s	−75.4s	−59.1s	−45.7	−17.2	−2.0	−49.0
Hydroxylamine	NH_2OH	s	47.2	64.6	77.5	99.2	110.0	34.0
Iodine	I_2	38.7s	73.2s	97.5s	116.5	159.8	183.0	112.9
Iodine pentafluoride	IF_5	−15.2	+ 8.5	32.3	50.0	81.2	97.0	8.0
Iodine heptafluoride	IF_7	−87.0s	−63.0s	−45.3s	31.9s	−8.3s	+ 4.0s	5.5
Iron	Fe	1787	2039	2224	2360	2605	2735	1535
Iron pentacarbonyl	$Fe(CO)_5$	—	+ 4.6	30.3	50.3	86.1	105.0	−21
Krypton	Kr	−199.3s	−187.2s	−178.4s	−171.8s	−159.0s	−152.0	−1156.7
Lead	Pb	973	1162	1309	1421	1630	1744	327.5
Lead bromide	$PbBr_2$	513	610	686	745	856	914	373
Lead chloride	$PbCl_2$	547	648	725	784	893	954	501
Lead fluoride	PbF_2	s	904	1003	1080	1219	1293	855
Lead iodide	PbI	479	571	644	701	807	872	402
Lead oxide	PbO	943	1085	1189	1265	1402	1472	890
Lead sulfide	PbS	852s	975s	1048s	1108s	1221	1281	1114
Lithium	Li	723	881	1003	1097	1273	1372	186
Lithium bromide	LiBr	748	888	994	1076	1226	1310	547

Name	Formula	Temperature (°C)						M.P.
		1 mm	10 mm	40 mm	100 mm	400 mm	760 mm	
Lithium chloride	LiCl	783	932	1045	1129	1290	1382	614
Lithium fluoride	LiF	1047	1211	1333	1425	1591	1681	870
Lithium iodide	LiI	723	841	927	993	1110	1171	446
Magnesium	Mg	621s	743	838	909	1034	1107	651
Magnesium chloride	MgCl₂	778	930	1050	1142	1316	1418	712
Manganese	Mn	1292	1505	1666	1792	2029	2151	1260
Manganous chloride	MnCl₂	s	778	879	960	1108	1190	650
Mercury	Hg	126.2	184.0	228.8	261.7	323.0	357.0	−38.9
Mercuric bromide	HgBr₂	136.5s	179.8s	211.5s	237.8	290.0	319.0	237
Mercuric chloride	HgCl₂	136.2s	180.2s	212.5s	237.0s	275.5s	304.0	277
Mercuric iodide	HgI₂	157.5s	204.5s	238.2s	261.8	324.2	354.0	259
Molybdenum	Mo	3102	3535	3859	4109	4553	5560	2622
Molybdenum hexafluoride	MoF₆	−65.5s	−40.8s	−22.1s	−8.0s	+17.2	36.0	17
Molybdenum trioxide	MoO₂	734s	814	892	955	1082	1151	795
Neon	Ne	−257.3s	−254.6s	−252.6s	−251.0s	−248.1	−246.0	−248.7
Nickel	Ni	1810	2057	2234	2364	2603	2732	1452
Nickel chloride	NiCl₂	671s	759s	821s	866s	945s	987s	1001
Nickel carbonyl	Ni(CO)₄	s	s	−23.0	−6.0	+25.8	42.5	−25
Nitrogen	N₂	−226.1s	−219.1s	−214.0s	−209.7	−200.9	−195.8	−210.0
Nitrogen trifluoride	NF₃	s	−170.7	−160.2	−152.3	−137.4	−129.0	−183.7
Nitric oxide	NO	−184.5s	−178.2s	−171.7s	−166.0s	−156.8s	−151.7	−161
Nitrous oxide	N₂O	−143.4s	−128.7s	−118.3s	−110.3s	−96.2s	−88.5	−90.9
Nitrogen tetroxide	N₂O₄	−55.6s	−36.7s	−23.9s	−14.7s	+8.0	21.0	−9.3
Nitrogen pentoxide	N₂O₅	−36.8s	−16.7s	−2.9s	+7.4s	24.4s	32.4	30
Nitrosyl chloride	NOCl	s	s	−60.2	−46.3	−20.3	−6.4	−64.5
Nitrosyl fluoride	NOF	−132.0	−114.3	−100.3	−88.8	−68.2	−56.0	−134
Nitroxyl fluoride	NO₂F	−143.7s	−126.2	−112.8	−102.3	−83.2	−72.0	−139
Osmium tetroxide (white)	O₅O₄	−5.6s	+26.0s	50.5	71.5	109.3	130.0	42*
Osmium tetroxide (yellow)	O₅O₄	3.2s	31.3s	51.7s	71.6	109.3	130.0	56
Oxygen	O₂	−219.1s	−210.6	−204.1	−198.8	−188.8	−182.96	−218.4
Ozone	O₃	−180.4	−163.2	−150.7	−141.0	−122.5	−111.9	−192.1 ±a1
Phosphorus (yellow)	P	76.6	128.0	166.7	197.3	251.0	280.0	44.1
Phosphorus (violet)	P	237s	287s	323s	349s	391s	417s	590
Phosphorus (black)	P	290s	338s	371s	393s	432s	453s	—
Phosphorus tribromide	PBr₃	7.8	47.8	79.0	103.6	149.7	175.3	−40
Phosphorus trichloride	PCl₃	−51.6	−21.3	+2.3	21.0	56.9	74.2	−111.8
Phosphorus pentachloride	PCl₅	55.5s	83.2s	102.5s	117.0s	147.2s	162.0s	—
Phosphorus hydride (Phosphene)	PH₃	s	s	−129.4	−118.8	−98.3	−87.5	−132.5
Phosphonium bromide	PH₄Br	−43.7s	−21.2s	−5.0s	+7.4s	28.0s	38.3d	
Phosphonium chloride	PH₄Cl	−91.0s	−74.0s	−61.5s	−52.0s	−35.4s	−27.0s	−28.5
Phosphonium iodide	PH₄I	−25.2s	−1.1s	16.1s	29.3s	51.6s	62.3s	—
Phosphorus trioxide	P₂O₃	s	53.0	84.0	108.3	150.3	173.1	22.5
Phosphorus oxychloride	POCl₃	s	2.0	27.3	47.4	84.3	105.1	2
Phosphorus pentoxide (stable form)	P₂O₅	384s	442s	481s	510s	556s	591	569
Phosphorus pentoxide (Metastable form)	P₂O₅	189	236	270	294	336	358	—
Phosphorus thiobromide	PSBr₃	50.0	83.6	108.0	126.3	157.8	175.0d	38
Phosphorus thiochloride	PSCl₃	−18.3	+16.1	42.7	63.8	102.3	124.0	−36.2
Platinum	Pt	2730	3146	3469	3714	4169	4407	1755
Potassium	K	341	443	524	586	708	774	62.3
Potassium bromide	KBr	795	982	1050	1137	1297	1383	730
Potassium chloride	KCl	821	968	1078	1164	1322	1407	790
Potassium fluoride	KF	885	1039	1156	1245	1411	1502	880
Potassium hydroxide	KOH	719	863	976	1064	1233	1327	380
Potassium iodide	KI	745	887	995	1080	1238	1324	723
Radon	Rn	−144.2s	−126.3s	−111.3s	−99.0s	−75.0s	−61.8	−71
Rhenium heptoxide	Re₂O₇	212.5s	248.0s	272.0s	289.0s	336.0	362.4	296
Rubidium	Rb	297	389	459	514	620	679	38.5
Rubidium bromide	RbBr	781	923	1031	1114	1267	1352	682
Rubidium chloride	RbCl	792	937	1047	1133	1294	1381	715
Rubidium fluoride	RbF	921	1016	1096	1168	1322	1408	760
Rubidium iodide	RbI	748	884	991	1072	1223	1304	642
Selenium	Se	356	442	506	554	637	680	217
Selenium dioxide	SeO₂	157.0s	202.5s	234.1s	258.0s	297.7s	317.0s	340
Selenium hexafluoride	SeF₆	−118.6s	−98.9s	−84.7	−73.9s	−55.2s	−45.8s	−34.7
Selenium oxychloride	SeOCl₂	34.8	71.9	99.8	118.0	151.7	168.0	8.5
Selenium tetrachloride	SeCl₄	74.0s	107.4s	130.1s	147.5s	176.4s	191.5d	—
Silane	SiH₄	−179.3	−163.0	−150.3	−140.5	−122.0	−111.5	−185
Silicon	Si	1724	1888	2000	2083	2220	2287	1420
Silicon dioxide	SiO₂	s	1732	1867	1969	2141	2227	1710
Silicon tetrachloride	SiCl₄	−63.4	−34.4	−12.1	+5.4	38.4	56.8	−68.8
Silicon tetrafluoride	SiF₄	−144.0s	−130.4s	−120.8s	−113.3s	−100.7s	−94.8s	−90
Bromosilane	SiH₃Br	s	−77.3	−57.8	−42.3	−13.3	+2.4	−93.9
Chlorosilane	SiH₃Cl	−117.8	−97.7	−81.8	−68.5	−44.5	−30.4	—
Fluorosilane	SiH₃F	−153.0	−141.2	−130.8	−122.4	−106.8	−98.0	—
Iodosilane	SiH₃I	s	−43.7	−21.8	−4.4	+27.9	45.4	−57.0
Bromodichlorofluorosilane	SiBrCl₂F	−86.5	−59.0	−37.0	−19.5	+15.4	35.4	−112.3
Bromotrifluorosilane	SiBrF₃	s	s	s	s	−55.9	−41.7	−70.5
Chlorotrifluorosilane	SiClF₃	−144.0s	−127.0	−112.8	−101.7	−81.0	−70.0	−142
Dibromochlorofluorosilane	SiBr₂ClF	−65.2	−35.6	−12.0	+6.3	43.0	59.5	−99.3
Dibromodifluorosilane	SiBr₂F₂	s	−66.8	−47.4	−31.9	−2.6	+13.7	−66.9

Name	Formula	Temperature (°C) 1 mm	10 mm	40 mm	100 mm	400 mm	760 mm	M.P.
Dibromosilane	SiH₂Br₂	−60.9	−29.4	−5.2	+14.1	50.7	70.5	−70.2
Dichlorodifluorosilane	SiCl₂F₂	−124.7	−102.9	−85.0	−70.3	−45.0	−31.8	−139.7
Difluorosilane	SiH₂F₂	−146.7	−130.4	−117.6	−107.3	−87.6	−77.8	
Diiodosilane	SiH₂I₂	s	18.0	52.6	79.4	125.5	149.5	−1.0
Disilane	Si₂H₆	−114.8	−91.4	−72.8	−57.5	−29.0	−14.3	−132.6
Disiloxane	(SiH₃)₂O	−112.5	−88.2	−70.4	−55.9	−29.3	−15.4	−144.2
Fluorotrichlorosilane	SiCl₃F	−92.6	−68.3	−48.8	−33.2	−4.0	+12.2	−120.8
Hexachlorodisilane	Si₂Cl₆	+4.0	38.8	65.3	85.4	120.6	139.0	−1.2
Hexachlorodisiloxane	(SiCl₃)₂O	−5.0	+29.4	55.2	75.4	113.6	135.6	−33.2
Hexafluorodisilane	Si₂F₆	−81.0s	−63.1s	−50.6s	−41.7s	−26.4s	−18.9s	−18.6
Octachlorotrisilane	Si₃Cl₈	46.3	89.3	121.5	146.0	189.5	211.4	—
Tetrasilane	Si₄H₁₀	−27.7	+4.3	28.4	47.4	81.7	100.0	−93.6
Tribromofluorosilane	SiBr₃F	−46.1	−15.1	+9.2	28.6	64.6	83.8	−82.5
Tribromosilane	SiHBr₃	−30.5	+3.4	30.0	51.6	90.2	111.8	−73.5
Trichlorosilane	SiHCl₃	−80.7	−53.4	−32.9	−16.4	+14.5	31.8	−126.6
Trifluorosilane	SiHF₃	−152.0s	−138.2s	−127.3	−118.7	−102.8	−95.0	−131.4
Trisilane	Si₃H₈	−68.9	−40.0	−16.9	+1.6	35.5	53.1	−117.2
Disilazane	(SiH₃)₃N	−68.7	1575	−18.5	−1.1	+31.0	48.7	−105.7
Silver	Ag	1357	1074	1743	1865	2090	2212	960.5
Silver chloride	AgCl	912	983	1200	1297	1467	1564	455
Silver iodide	AgI	820	549	1111	1210	1400	1506	552
Sodium	Na	439	952	633	701	823	892	97.5
Sodium bromide	NaBr	806	1017	1063	1148	1304	1392	755
Sodium chloride	NaCl	865	983	1131	1220	1379	1465	800
Sodium cyanide	NaCN	817	1240	1115	1214	1401	1497	564
Sodium fluoride	NaF	1077	897	1363	1455	1617	1704	992
Sodium hydroxide	NaOH	739	903	1017	1111	1286	1378	318
Sodium iodide	NaI	767	72.7	1005	1083	1225	1304	651
Stannic bromide	SnBr₄	s	+10.0	105.5	131.0	177.7	204.7	31.0
Stannic chloride	SnCl₄	−22.7	−118.5	35.2	54.7	92.1	113.0	−30.2
Stannic hydride	SnH₄	−140.0	175.8	−102.3	−80.2	−65.2	−52.3	−149.9
Stannic iodide	SnI₄	s	391	218.8	254.2	315.5	348.0	144.5
Stannous chloride	SnCl₂	316	898	450	493	577	623	246.8
Strontium	Sr	s	2262s	1018	1111	1285	1384	800
Strontium oxide	SrO	2068s	243.8	2410s	—	—	—	2430
Sulfur	S	183.8	−114.7s	288.3	327.2	399.6	444.6	112.8
Sulfur hexafluoride	SF₆	−132.7s	−76.8s	−101.5s	−90.9s	−72.6s	−63.5s	−50.2
Sulfur dioxide	SO₂	−95.5s	+27.5	−60.5	−46.9	−23.0	−10.0	−73.2
Sulfur monochloride	S₂Cl₂	−7.4	−24.8	54.1	75.3	115.4	138.0	−80
Sulfuryl chloride	SO₂Cl₂	s	−16.5s	−1.0	+17.8	51.3	69.2	−54.1
Sulfur trioxide (α)	SO₃	−39.0s	−12.3s	−1.0s	+10.5s	32.6	44.8	16.8
Sulfur trioxide (β)	SO₃	−34.0s	+4.3s	+3.2s	14.3s	32.6	44.8	32.3
Sulfur trioxide (γ)	SO₃	−15.3s	194.2	17.9s	28.0s	44.0s	51.6s	62.1
Sulfuric acid	H₂SO₄	145.8	+31.0	229.7	257.0	305.0	330.0d	10.5
Thionyl bromide	SOBr₂	−6.7	−21.9	58.8	80.6	119.2	139.5	−52.2
Thionyl chloride	SOCl₂	−52.9	s	+2.2	21.4	56.5	75.4	−104.5
Tantalum pentafluoride	TaF₅	s	650	s	130.0	194.0	230.0	96.8
Tellurium	Te	520	233	753	838	997	1087	452
Tellurium tetrachloride	TeCl₄	s	−92.4s	273	304	360	392	224
Tellurium hexafluoride	TeF₆	−111.3s	983	−78.4s	−67.9s	−48.2s	−38.6s	−37.8
Thallium	Tl	825	522	1103	1196	1364	1457	303.5
Thallium bromide	TlBr	s	517	598	653	759	819	460
Thallium chloride	TlCl	s	531	589	645	748	807	430
Thallium iodide	TlI	440	1703	607	663	763	823	440
Tin	Sn	1492	+21.3	1855	1968	2169	2270	231.9
Titanium tetrachloride	TiCl₄	−13.9	4507	48.4	71.0	112.7	136.0	−30
Tungsten	W	3990	−49.2s	4886	5168	5666	5927	3370
Tungsten hexafluoride	WF₆	−71.4s	−13.8s	−33.0s	−20.3s	+1.2	17.3	−0.5
Uranium hexafluoride	UF₆	−38.8s	+12.2	+4.4s	18.2s	42.7s	55.7s	69.2
Vanadyl trichloride	VOCl₃	−23.2	+11.3	40.0	62.5	103.5	127.2 —	
Water	H₂O	−17.3s	−152.8s	34.1	51.6	83.0	100.0	0.0
Xenon	Xe	−168.5s	593	−141.2s	−132.8s	−117.1s	−108.0	−111.6
Zinc	Zn	487	508	673	736	844	907	419.4
Zinc chloride	ZnCl₂	428	1086	566	610	689	732	365
Zinc fluoride	ZnF₂	970	250s	1175	1254	1417	1497	872
Zirconium tetrabromide	ZrBr₄	207s	230s	281s	301s	337s	357s	450
Zirconium tetrachloride	ZrCl₄	190s	311s	259s	279s	312s	331s	437
Zirconium tetraiodide	ZrI₄	264s		344s	369s	409s	431s	499

Pressures Greater than 1 Atmosphere

Name	Formula	Temperature, °C 1 atm.	2 atm.	5 Atm.	10 atm.	20 atm.	40 atm.	60 atm.
Ammonia	NH₃	−33.6	−18.7	+4.7	25.7	50.1	78.9	98.3
Argon	A	−185.6	−179.0	−166.7	−154.9	−141.3	−124.9	—
Boron trichloride	BCl₃	12.7	33.2	66.0	96.7	135.4	—	—
Boron trifluoride	BF₃	−100.7	−89.4	−72.6	−57.7	−40.0	−19.0	—
Bromine	Br₂	58.2	78.8	110.3	139.8	174.0	216.0	243.5

INORGANIC COMPOUNDS
Pressures Greater than One Atmosphere (Continued)

Name	Formula	Temperature, °C						
		1 atm.	2 atm.	5 atm.	10 atm.	20 atm.	40 atm.	60 atm.
Carbontetrachloride	CCl_4	76.7	102.0	141.7	178.0	222.0	276.0	—
Carbon dioxide	CO_2	−78.2s	−69.1s	−56.7	−39.5	−18.9	+5.9	22.4
Carbon disulfide	CS_2	46.5	69.1	104.8	136.3	175.5	222.8	256.0
Carbon monoxide	CO	−191.3	−183.5	−170.7	−161.0	−149.7	—	—
Carbonyl chloride	$COCl_2$	+8.3	27.3	57.2	85.0	119.0	159.8	—
Chlorotrifluoromethane	$CClF_3$	−81.2	−66.7	−42.7	−18.5	+12.0	52.8	—
Cyanogen	C_2N_2	−21.0	−4.4	+21.4	44.6	72.6	106.5	—
Dichlorodifluoromethane	CCl_2F_2	−29.8	−12.2	+16.1	42.4	74.0	—	—
Dichlorofluoromethane	$CHCl_2F$	8.9	28.4	59.0	87.0	121.2	162.6	—
Chlorodifluoromethane	$CHClF_2$	−40.8	−24.7	+0.3	24.0	52.0	85.3	—
Trichlorofluoromethane	CCl_3F	23.7	44.1	77.3	108.2	146.7	194.0	—
Chlorine	Cl_2	−33.8	−16.9	+10.3	35.6	65.0	101.6	127.1
Helium	He	−268.6	−268.0	—	—	—	—	—
Hydrogen	H_2	−252.5	−250.2	−246.0	−241.8	—	—	—
Hydrogen bromide	HBr	−66.5	−51.5	−29.1	−8.4	+16.8	48.1	70.6
Hydrogen chloride	HCl	−84.8	−71.4	−50.5	−31.7	−8.8	+17.8	36.2
Hydrogen cyanide	HCN	25.9	45.8	75.8	102.7	135.0	169.9	—
Hydrogen iodide	HI	−35.1	−18.9	+7.3	32.0	62.2	100.7	127.5
Hydrogen sulfide	H_2S	−60.4	−45.9	−22.3	−0.4	+25.5	55.8	76.3
Hydrogen selenide	H_2Se	−41.1	−25.2	0.0	+23.4	50.8	84.6	108.7
Krypton	Kr	−152.0	−143.5	−130.0	−118.0	−101.7	−78.4	—
Neon	Ne	−246.0	−243.8	−239.9	−236.0	−230.8	—	—
Nitrogen	N_2	−195.8	−189.2	−179.1	−169.8	−157.6	—	—
Nitric oxide	NO	−151.7	−145.1	−135.7	−127.3	−116.8	−103.2	−94.8
Nitrous oxide	N_2O	−88.5	−76.8	−58.0	−40.7	−18.8	+8.0	27.4
Nitrogen tetroxide	N_2O_4	21.0	37.3	59.8	79.4	100.3	121.4	132.2
Oxygen	O_2	−183.1	−176.0	−164.5	−153.2	−140.0	−124.1	—
Silicon tetrafluoride	SiF_4	−94.8s	−84.4	−67.9	−52.6	−33.4	—	—
Chlorotrifluorosilane	$SiClF_3$	−70.0	−57.3	−37.2	−18.6	+4.1	—	—
Dichlorodifluorosilane	$SiCl_2F_2$	−31.8	−15.1	+11.6	30.0	66.2	—	—
Fluorotrichlorosilane	$SiCl_3F$	12.2	32.4	64.6	94.2	131.8	—	—
Stannic chloride	$SnCl_4$	113.0	141.3	184.3	223.0	270.0	—	—
Sulfur dioxide	SO_2	−10.0	+6.3	32.1	55.5	83.8	118.0	141.7
Sulfur trioxide	SO_3	44.8	60.0	82.1	100.0	130.0	173.0	198.0
Water	H_2O	100.0	120.1	152.4	180.5	213.1	251.1	276.5

ORGANIC COMPOUNDS
Pressures Less than 1 Atmosphere

Name	Formula	Temperature °C						M.P.
		1 mm	10 mm	40 mm	100 mm	400 mm	760 mm	
Cyanogen bromide	$CBrN$	−35.7s	−10.0s	+8.6s	22.6s	46.0s	61.5	58
Carbon tetrabromide	CBr_4	s	s	96.3	119.7	163.5	189.5	90.1
Chlorotrifluoromethane	$CClF_3$	−149.5	−134.1	−121.9	−111.7	−92.7	−81.2	—
Cyanogen chloride	$CClN$	−76.7s	−53.8s	−37.5s	−24.9s	−2.3	+13.1	−6.5
Dichlorodifluoromethane	CCl_2F_2	−118.5	−97.8	−81.6	−68.6	−43.9	−29.8	—
Carbonyl chloride	CCl_2O	−92.9	−69.3	−50.3	−35.6	−7.6	+8.3	−104
Trichlorofluoromethane	CCl_3F	−84.3	−59.0	−39.0	−23.0	+6.8	23.7	—
Trichloronitromethane	CCl_3NO_2	−25.5	+7.8	33.8	53.8	91.8	111.9	−64
Carbontetrachloride	CCl_4	−50.0s	−19.6	+4.3	23.0	57.8	76.7	−22.6
Cyanogen fluoride	CFN	−134.4s	−118.5s	−106.4s	−97.0s	−80.5s	−72.6s	—
Carbontetrafluoride	CF_4	−184.6s	−169.3	−158.8	−150.7	−135.5	−127.7	—
Tribromomethane	$CHBr_3$	s	34.0	63.6	85.9	127.9	150.5	8.5
Chlorodifluoromethane	$CHClF_2$	−122.8	−103.7	−88.6	−76.4	−53.6	−40.8	−160
Dichlorofluoromethane	$CHCl_2F$	−91.3	−67.5	−48.8	−33.9	−6.2	+8.9	−135
Trichloromethane	$CHCl_3$	−58.0	−29.7	−7.1	+10.4	42.7	61.3	−63.5
Hydrocyanic acid	CHN	−70.8s	−48.2s	−31.3s	−18.8s	+9.8	25.8	−14
Dibromomethane	CH_2Br_2	−35.1	−2.4	+23.3	42.3	79.0	98.6	−52.8
Dichloromethane	CH_2Cl_2	−70.0	−43.3	−22.3	−6.3	24.1	40.7	−96.7
Formaldehyde	CH_2O	s	−88.0	−70.6	−57.3	−33.0	−19.5	−92
Formic acid	CH_2O_2	−20.0s	+2.1s	24.0	43.8	80.3	100.6	8.2
Dichloromethylarsine	CH_3AsCl_2	−11.1	+24.3	51.5	73.0	112.7	134.5	−59
Borine carbonyl	CH_3BO	−139.2	−121.1	−106.6	−95.3	−74.8	−64.0	—
Methyl bromide	CH_3Br	−96.3s	−72.8	−54.2	−39.4	−11.9	+3.6	−93
Methyl chloride	CH_3Cl	s	−92.4	−76.0	−63.0	−38.0	−24.0	−97.7
Trichloromethylsilane	CH_3Cl_3Si	−60.8	−30.7	−7.0	+12.1	47.0	66.4	−90
Methyl fluoride	CH_3F	−147.3	−131.6	−119.1	−109.0	−89.5	78.2	—
Methyl iodide	CH_3I	s	−45.8	−24.2	−7.0	+25.3	42.4	−64.4
Formamide	CH_3NO	70.5	109.5	137.5	157.5	193.5	210.5d	—
Nitromethane	CH_3NO_2	−29.0	+2.8	27.5	46.6	82.0	101.2	−29
Methane	CH_4	−205.9s	−195.5s	−187.7s	−181.4	−168.8	−161.5	−182.5
Dichloromethylsilane	CH_4Cl_2Si	−75.0	−47.8	−26.2	−9.0	+23.7	41.9	—
Methanol	CH_4O	−44.0	−16.2	+5.0	21.2	49.9	64.7	−97.8
Methanethiol	CH_4S	−90.7	−67.5	−49.2	−34.8	−7.9	+6.8	−121
Chloromethylsilane	CH_5ClSi	−95.0	−71.0	−51.7	−36.4	−7.8	+8.7	—
Methylamine	CH_5N	−95.8s	−73.8	−56.9	−43.7	−19.7	−6.3	−93.5
Methylsilane	CH_6Si	−138.5	−120.0	−104.8	−93.0	−70.3	−56.9	—

Name	Formula	Temperature °C						M.P.
		1 mm	10 mm	40 mm	100 mm	400 mm	760 mm	
2-Methyldisilazane	CH_9NSi_2	−76.3	−50.1	−29.6	−13.1	+ 17.2	34.0	—
Cyanogen iodide	CIN	25.2s	57.7s	80.3s	97.6s	126.1s	141.1s	—
Tetranitromethane	CN_4O_8	s	22.7	48.4	68.9	105.9	125.7d	13
Carbon monoxide	CO	−222.0s	−215.0s	−210.0s	−205.7s	−196.3	−191.3	−205.0
Carbonyl sulfide	COS	−132.4	−113.3	−98.3	−85.9	−62.7	−49.9	−138.8
Carbonyl selenide	COSe	−117.1	−95.0	−76.4	−61.7	−35.6	−21.9	—
Carbon dioxide	CO_2	−134.3s	−119.5s	−108.6s	−100.2s	−85.7s	−78.2s	−57.5
Carbon Selenosulfide	CSSe	−47.3	−16.0	+ 8.6	28.3	65.2	85.6	−75.2
Carbon disulfide	CS_2	−73.8	−44.7	−22.5	−5.1	+ 28.0	46.5	−110.8
Trichloroacetyl bromide	C_2BrCl_3O	−7.4	+ 29.3	57.2	79.5	120.2	143.0	—
1-Chloro-1,2,2-trifluoroethylene	C_2ClF_3	−116.0	−95.9	−79.7	−66.7	−41.7	−27.9	−157.5
1,2-Dichloro-1,2-difluoroethylene	$C_2Cl_2F_2$	−82.0	−57.3	−38.2	−23.0	+ 5.0	20.9	−112
1,2-Dichloro-1,1,2,2-tetrafluoroethane	$C_2Cl_2F_4$	−95.4	−72.3	−53.7	−39.1	−12.0	+ 3.5	−94
1,1,2-Trichloro-1,2,2-trifluoroethane	$C_2Cl_3F_2$	−68.0s	40.3s	−18.5	−1.7	+ 30.2	47.6	−35
Tetrachloroethylene	C_2Cl_4	−20.6s	+ 13.8	40.1	61.3	100.0	120.8	−19.0
1,1,2,2-Tetrachloro-1,2-difluoroethane	$C_2Cl_4F_2$	−37.5s	−5.0s	+ 19.8s	38.6	73.1	92.0	26.5
Hexachloroethane	C_2Cl_6	32.7s	73.5s	102.3s	124.2s	163.8s	185.6s	186.6
Tribromoacetaldehyde	C_2HBr_2O	18.5	58.0	87.8	110.2	151.6	174.0d	—
Trichloroethylene	C_2HCl_3	−43.8	−12.4	+ 11.9	31.4	67.0	86.7	−73
Trichloroacetaldehyde	C_2HCl_3O	−37.8	−5.0	20.2	40.2	77.5	97.7	−57
Trichloroacetic acid	$C_2HCl_3O_2$	51.0s	88.2	116.3	137.8	175.2	195.6	57
Pentachloroethane	C_2HCl_5	+ 1.0	39.8	69.9	93.5	137.2	160.5	−22
Acetylene	C_2H_2	−142.9s	−128.2s	−116.7s	−2107.9s	−92.0s	−84.0s	−81.5
1,1,1,2-Tetrabomoethane	$C_2H_2Br_4$	58.0	95.7	123.2	144.0	181.0	200.0d	—
1,1,2,2-Tetrabromoethane	$C_2H_2Br_4$	65.0	110.0	144.0	170.0	217.5	243.5	—
cis-1,2-Dichloroethylene	$C_2H_2Cl_2$	−58.4	−29.9	−7.9	+ 9.5	41.0	59.0	−80.5
trans-1,2-Dichloroethylene	$C_3H_2Cl_2$	−65.4s	−38.0	−17.0	−0.2	+ 30.8	47.8	−50.0
1,1-Dichloroethane	$C_3H_2Cl_2$	−77.2	−51.2	−31.1	−15.0	+ 14.8	31.7	−122.5
Dichloroacetic acid	$C_2H_2Cl_2O_2$	44.0	82.6	111.8	134.0	173.7	194.4	9.7
1,1,1,2-Tetrachloroethane	$C_2H_2Cl_4$	−16.3	+ 19.3	46.7	68.0	108.2	130.5	−68.7
1,1,2,2-Tetrachloroethane	$C_2H_2 Cl_4$	−3.8	+ 33.0	60.8	83.2	124.0	145.9	−36
1-Bromoethylene	C_2H_3Br	−95.4	−68.8	−48.1	−31.9	−1.1	+ 15.8	−138
Bromoacetic acid	$C_2H_3BrO_2$	54.7	94.1	124.0	146.3	186.7	208.0	49.5
1,1,2-Tribromoethane	$C_2H_3Br_3$	32.6	70.6	100.0	123.5	165.4	188.4	−26
1-Chloroethylene	C_2H_3Cl	−105.6	−83.7	−66.8	−53.2	−28.0	−13.8	−153.7
Chloroacetic acid	$C_2H_3ClO_2$	43.0s	81.0	109.2	130.7	169.0	189.5	61.2
1,1,1-Trichloroethane	$C_2H_3Cl_3$	−52.0	−21.9	+ 1.6	20.0	54.6	74.1	−30.6
1,1,2-Trichloroethane	$C_2H_3Cl_3$	−24.0	+ 8.3	35.2	55.7	93.0	113.9	−36.7
Trichloroacetaldehyde hydrate	$C_2H_3Cl_3O_2$	−9.8s	+ 19.5s	39.7s	55.0	82.1	96.2d	51.7
1-Fluoroethylene	C_2H_3F	−149.3	−132.2	−118.0	−106.2	−84.0	−72.2	−160.5
Acetonitrile	C_2H_3N	−47.0s	−16.3	+ 7.7	27.0	62.5	81.8	−41
Methyl thiocyanate	C_2H_3NS	−14.0	+ 21.6	49.0	70.4	110.8	132.9	−51
Methyl isothiocyanate	C_2H_3NS	−34.7s	+ 5.4s	38.2	59.3	97.8	119.0	35.5
Ethylene	C_2H_4	−168.3	−153.2	−141.3	−131.8	−113.9	−103.7	−169
1-Bromo-1-chloroethane	C_2H_4BrCl	−36.0s	−9.4s	+ 10.4s	28.0	63.4	82.7	16.6
1-Bromo-2-chloroethane	C_2H_4BrCl	−28.8s	+ 4.1	29.7	49.5	86.0	106.7	−16.6
1,2-Dibromoethane	$C_2H_4Br_2$	−27.0s	+ 18.6	48.0	70.4	110.1	131.5	10
1,1-Dichloroethane	$C_2H_4Cl_2$	−60.7	−32.3	−10.2	+ 7.2	39.8	57.4	−96.7
1,2-Dichloroethane	$C_2H_4Cl_2$	−44.5s	−13.6	+ 10.0	29.4	64.0	82.4	−35.3
1,1-Difluoroethane	$C_2H_4F_2$	−112.5	−91.7	−75.8	−63.2	−39.5	−26.5	−117
Acetaldehyde	C_2H_4O	−81.5	−56.8	−37.8	−22.6	+ 4.9	20.2	−123.5
Ethylene oxide	C_2H_4O	−89.7	−65.7	−46.9	−32.1	−4.9	+ 10.7	−111.3
Acetic acid	$C_2H_4O_2$	−17.2s	+ 17.5	43.0	63.0	99.0	118.1	16.7
Methyl formate	$C_2H_4O_2$	−74.2	−48.6	−28.7	−12.9	16.0	32.0	−99.8
Mercaptoacetic acid	$C_2H_4O_2S$	−60.0	101.5	131.8	154.0d	—	—	−16.5
Ethyl bromide	C_2H_5Br	−74.3	−47.5	−26.7	−10.0	+ 21.0	38.4	−117.8
Ethyl chloride	C_2H_5Cl	−89.8	−65.8	−47.0	−32.0	−3.9	+ 12.3	−139
2-Chloroethanol	C_2H_5ClO	−4.0	+ 30.3	56.0	75.0	110.0	128.8	−69
Trichloroethylsilane	$C_2H_5Cl_3Si$	−27.9	+ 3.6	27.9	46.3	80.3	99.5	−40
Trichloroethyoxysilane	$C_2H_5Cl_2OSi$	−32.4	0.0	+ 25.3	45.2	82.2	102.4	—
Ethyl fluoride	C_2H_5F	−117.0	−97.7	−81.8	−69.3	−45.5	−32.0	—
Ethyltrifluorosilane	$C_2H_5F_3Si$	−95.4	−73.7	−56.8	−43.6	−19.1	−5.4	—
Ethyl Iodide	C_2H_5I	−54.4	−24.3	−0.9	+ 18.0	52.3	72.4	−105
Acetamide	C_2H_5NO	65.0s	105.0	135.8	158.0	200.0	222.0	81
Acetaldoxime	C_2H_5NO	−5.8s	+ 25.8	48.6	66.2	98.0	115.0	47
Nitroethane	$C_2H_5NO_2$	−21.0	+ 12.5	38.0	57.8	94.0	114.0	−90
Di(nitrosomethyl)amine	$C_2H_5N_3O_2$	+ 3.2	40.0	68.2	90.3	131.3	153.0	—
Ethane	C_2H_6	−159.5	−142.9	−129.8	−119.3	−99.7	−88.6	−183.2
Dichlorodimethylsilane	$C_2H_6Cl_2Si$	−53.5	−23.8	−0.4	+ 17.5	51.9	70.3	−86.0
Ethanol	C_2H_6O	−31.3	−2.3	+ 19.0	34.9	63.5	78.4	−112
Dimethyl ether	C_2H_6O	−115.7	−93.3	−76.2	−62.7	−37.8	−23.7	−138.5
1,2-Ethanediol	$C_2H_6O_2$	53.0	92.1	120.0	141.8	178.5	197.3	−15.6
Dimethyl sulfide	C_2H_6S	−75.6	−49.2	−28.4	−12.0	+ 18.7	36.0	−83.2
Ethanethiol	C_2H_6S	−76.7	−50.2	−29.8	−13.0	+ 17.7	35.5	−121
Dimethylantimony	C_2H_6Sb	44.0	86.0	118.3	143.5	187.2	211.0	—
Ethylamine	C_2H_7N	−82.3s	−58.3	−39.8	−25.1	+ 2.0	16.6	−80.6
Dimethylamine	C_2H_7N	−87.7	−64.6	−46.7	−32.6	−7.1	+ 7.4	−96
1,2-Ethanediamine	$C_2H_8N_2$	−11.0s	+ 21.5	45.8	62.5	99.0	117.2	8.5
Dimethylsilane	C_2H_8Si	−115.0	−93.1	−75.7	−61.4	−35.0	−20.1	—
Dimethyldiborane	$C_2H_{10}B_2$	−106.5	−82.1	−62.4	−47.0	−18.8	−2.6	−150.2
2-Ethyldisilazane	$C_2H_{11}NSi_2$	−62.0	−32.2	−8.3	+ 10.4	45.9	65.9	−127

Name	Formula	Temperature °C						M.P.
		1 mm	10 mm	40 mm	100 mm	400 mm	760 mm	
Cyanogen	C_2N_2	−95.8s	−76.8s	−62.7s	−51.8s	−33.0	−21.0	−34.4
Acrylonitrile	C_3H_3N	−51.0	−20.3	+3.8	22.8	58.3	78.5	−82
Propadiene	C_3H_4	−120.6	−101.0	−85.2	−72.5	−48.5	−35.0	−136
Propyne	C_3H_4	−111.0s	−90.5	−74.3	−61.3	−37.2	−23.3	−102.7
2,3-Dibromopropene	$C_3H_4Br_2$	−6.0	+30.0	57.8	79.5	119.5	141.2	—
Methyl dichloroacetate	$C_3H_4Cl_2O_2$	3.2	38.1	64.7	85.4	122.6	143.0	—
2-Propenal	C_3H_4O	−64.5	−36.7	−15.0	+2.5	34.5	52.5	−87.7
Acrylic acid	$C_3H_4O_2$	+3.5s	39.0	66.2	86.1	122.0	141.0	14
Pyruvic acid	$C_3H_4O_3$	21.4	57.9	85.3	106.5	144.7	165.0d	13.6
1,2,3-Tribromopropane	$C_3H_5Br_3$	47.5	90.0	122.8	148.0	195.0	220.0	16.5
1-Chloropropene	C_3H_5Cl	−81.3	−54.1	−32.7	−15.1	+18.0	37.0	−99.0
3-Chloropropene	C_3H_5Cl	−70.0	−42.9	−21.2	−4.5	+27.5	44.6	−136.4
Epichlorohydrine	C_3H_5ClO	−16.5	+16.6	42.0	62.0	98.0	117.9	−25.6
Methyl chloroacetate	$C_3H_5ClO_2$	−2.9	+30.0	54.5	73.5	109.5	130.3	−31.9
1,1,1-Trichloropropane	$C_3H_5Cl_3$	−28.8	+4.2	29.9	50.0	87.5	108.2	−77.7
1,2,3-Trichloropropane	$C_3H_5Cl_3$	+9.0	46.0	74.0	96.1	137.0	158.0	−14.7
Allyltrichlorosilane	$C_3H_5Cl_3Si$	−20.7	+13.2	39.2	59.3	97.1	118.0	—
Propionitrile	C_3H_5N	−35.0	−3.0	+22.0	41.4	77.7	97.1	−91.9
3-Hydroxypropionitrile	C_3H_5NO	58.7	102.0	134.1	157.7	200.0	221.0	—
Ethylisothiocyanate	C_3H_5NS	−13.2s	+22.8	50.8	71.9	110.1	131.0	−5.9
Nitroglycerine	$C_3H_5N_3O_9$	127	188	235	e	—	—	11
Propylene	C_3H_6	−131.9	−112.1	−96.5	−84.1	−60.9	−47.7	−185
Cyclopropane	C_3H_6	−116.8	−97.5	−82.3	−70.0	−46.9	−33.5	−126.6
2-Bromo-2-nitrosopropane	C_3H_6BrNO	−33.5	−4.3	+17.9	35.2	66.2	83.0	—
1,2-Dibromopropane	$C_3H_6Br_2$	−7.0	+29.4	57.2	78.7	118.5	141.6	−55.5
1,3-Dibromopropane	$C_3H_6Br_2$	+9.7	48.0	77.8	101.3	144.1	167.5	−34.4
2,3-Dibromo-1-propanol	$C_3G_6Br_2O$	57.0	98.2	129.8	153.0	196.0	219.0	—
1,2-Dichloropropane	$C_3H_6Cl_2$	−38.5	−6.1	+19.4	39.4	76.0	96.8	—
1,3-Dichloro-2-propanol	$C_3H_6Cl_2O$	28.0	64.7	93.0	114.8	153.5	174.3	—
Acetone	C_3H_6O	−59.4	−31.1	−9.4	+7.7	39.5	56.5	−94.6
Allyl alcohol	C_3H_6O	20.0	+10.5	33.4	50.0	80.2	96.6	−129
Propylene oxide	C_3H_6O	−75.0	−49.0	−28.4	−12.0	+17.8	34.5	−112.1
Propionic acid	$C_3H_6O_2$	4.6	39.7	65.8	85.8	122.0	141.1	−22
Methyl acetate	$C_3H_6O_2$	−57.2	−29.3	−7.9	+9.4	40.0	57.8	98.7
Ethyl formate	$C_3H_6O_2$	−60.5	−33.0	−11.5	+5.4	37.1	54.3	−79
Methyl glycolate	$C_3H_6O_3$	+9.6	45.3	72.3	93.7	131.7	151.5	—
Methoxyacetic acid	$C_3H_6O_3$	52.5	92.0	122.0	144.5	184.2	204.0	—
1-Bromopropane	C_3H_7Br	−53.0	−23.3	−0.3	+18.0	52.0	71.0	−109.9
2-Bromopropane	C_3H_7Br	−61.8	−32.8	−10.1	+8.0	41.5	60.0	−89.0
1-Chloropropane	C_3H_7Cl	−68.3	−41.0	−19.5	−2.5	+29.4	46.4	−122.8
2-Chloropropane	C_3H_7Cl	−78.8	−52.0	−31.0	−13.7	+18.0	36.5	−117
Trichloroisopropylsilane	$C_3H_7Cl_3Si$	−24.3	+9.9	36.5	57.8	96.8	118.5	—
1-Iodopropane	C_3H_7I	−36.0	−2.4	+23.6	43.8	81.8	102.5	−98.8
2-Iodopropane	C_3H_7I	−43.3	−11.7	+13.2	32.8	69.5	89.5	−90
Propionamide	C_3H_7NO	65.0s	105.0	134.8	156.0	194.0	213.0	79
1-Nitropropane	$C_3H_7NO_2$	−9.6	+25.3	51.8	72.3	110.6	131.6	−108
2-Nitropropane	$C_3H_7NO_2$	−18.8	+15.8	41.8	62.0	99.8	120.3	−93
Ethyl Carbamate	$C_3H_7NO_2$	s	77.8	105.6	126.2	164.0	184.0	49
Propane	C_3H_8	−128.9	−108.5	−92.4	−79.6	−55.6	−42.1	−187.1
Dichloroethoxymethylsilane	$C_3H_8Cl_2OSi$	−33.8	−1.3	+24.4	44.1	80.3	100.6	—
1-Propanol	C_3H_8O	−15.0	+14.7	36.4	52.8	82.0	97.8	−127
2-Propanol	C_3H_8O	−26.1	+2.4	23.8	39.5	67.8	82.5	−85.8
Ethyl methyl ether	C_3H_8O	−91.0	−67.8	−49.4	−34.8	−7.8	+7.5	—
1,2-Propanediol	$C_3H_8O_2$	45.5	83.2	111.2	132.0	168.1	188.2	—
1,3-Propanediol	$C_3H_8O_2$	59.4	100.6	131.0	153.4	193.8	214.2	—
2-Methoxyethanol	$C_3H_8O_2$	−13.5	+22.0	47.8	68.0	104.3	124.4	—
Glycerol	$C_3H_8O_3$	125.5	167.2	198.0	220.1	263.0	290.0	17.9
1-Propanethiol	C_3H_8S	−56.0	−26.3	−3.2	+15.3	49.2	67.4	−112
Trimethylborine	C_3H_9B	−118.0	−92.4	−74.7	−60.8	−34.7	−20.1	—
Chlorotrimethylsilane	C_3H_9ClSi	−62.8	−34.0	−11.4	+6.0	39.4	57.9	—
Trimethylgallium	C_3H_9Ga	−62.3s	−31.7s	−9.0	+8.0	39.0	55.6	−19
Propylamine	C_3H_9N	−64.4	−37.2	−16.0	+0.5	31.5	48.5	−83
Trimethylamine	C_3H_9N	−97.1	−73.8	−55.2	−40.3	−12.5	+2.9	−117.1
Trimethyl phosphate	$C_3H_9O_4P$	26.0	67.8	100.0	124.0	167.8	192.7	—
Trimethyldiborane	$C_3H_{12}B_2$	−74.0	−44.8	−22.0	−4.4	+27.8	45.5	−122.9
Carbon suboxide	C_3O_2	−94.8	−71.0	−52.0	−36.9	−8.9	+6.3	−107
Carbon subsulfide	C_3S_2	14.0	54.9	85.6	109.9	—	—	+0.4
Trichloroacetic anhydride	$C_4Cl_6O_3$	56.2	99.6	131.2	155.2	199.8	223.0	—
1,3-Butadiyne	C_4H_2	−82.5s	−61.2s	−45.9s	−34.0	−6.1	+9.7	−34.9
α,β-Dibromomaleic achydride	$C_4H_2Br_2O_3$	50.0	92.0	123.5	147.7	192.0	214.0	—
trans-Fumaryl chloride	$C_4H_2Cl_2O_2$	+15.0	51.8	79.5	101.0	140.0	100.0	—
Maleic anhydride	$C_4H_2O_3$	44.0s	78.7	111.8	135.8	179.5	202.0	58
2-Nitrothiophene	$C_4H_3NO_2S$	48.2	92.0	125.8	151.5	199.6	224.5	46
Butenyne	C_4H_4	−93.2	−70.0	−51.7	−37.1	−10.1	+5.3	—
Succinyl chloride	$C_4H_4Cl_2O_2$	39.0	78.0	107.5	130.0	170.0	192.5	17
Chloroaccetic anhydride	$C_4H_4Cl_2O_3$	67.2	108.0	138.2	159.8	197.0	217.0	46
Succinic anhydride	$C_4H_4O_3$	92.0s	128.2	163.0	189.0	237.0	261.0	119.6
1,4-Dioxane-2,6-dione	$C_4H_4O_4$	s	116.6	148.6	173.2	217.0	240.0	97
Thiophene	C_4H_4S	−40.7s	−10.9	+12.5	30.5	64.7	84.4	−38.3
Selenophene	C_4H_4Se	−39.0	−4.0	+24.1	47.0	89.8	114.3	—
α-Chlorocrotonic acid	$C_4H_5ClO_2$	70.0	108.0	135.6	155.9	193.2	212.0	—

Name	Formula	Temperature °C						M.P.
		1 mm	10 mm	40 mm	100 mm	400 mm	760 mm	
Ethyl chloroglyoxylate	C₄H₅ClO₃	−5.1	+ 29.9	56.0	76.6	114.7	135.0	—
Ethyl trichloroacetate	C₄H₅Cl₃O₂	20.7	57.7	85.5	107.4	146.0	67.0	—
3-Butenenitrile	C₄H₅N	−19.6	+ 14.1	40.0	60.2	98.0	119.0	—
Methacrylonitrile	C₄H₅N	−44.5	−12.5	+ 12.8	32.8	32.8	90.3	—
cis-Crotononitrile	C₄H₅N	−29.0	+ 4.0	30.0	50.0	88.0	108.0	—
trans-Crotononitrile	C₄H₅N	−19.5	+ 15.0	41.8	62.8	101.5	122.8	—
Succinimide	C₄H₅NO₂	115.0s	157.0	192.0	217.4	263.5	287.5	125.5
Allylisothiocyanate	C₄H₅NS	−2.0	+ 38.3	67.4	89.5	129.8	156.7	−80
1,2-Butadiene	C₄H₆	−89.0	−64.2	−44.3	−28.3	+ 1.8	18.5	—
1,3-Butadiene	C₄H₆	−102.8	−79.7	−61.3	−46.8	−19.3	−4.5	−108.9
Cyclobutene	C₄H₆	−99.1	−75.4	−56.4	−41.2	−12.2	+ 2.4	—
1 Butyne	C₄H₆	−92.5	−68.7	−50.0	−34.9	−6.9	+ 8.7	−130
2-Butyne	C₄H₆	−73.0s	−50.5s	−33.9s	−18.8	+ 10.6	27.2	−32.5
Ethyl dichloroacctate	C₄H₆Cl₂O₂	9.6	46.3	74.0	96.1	135.9	156.5	—
2-Chloroethyl chloroacetate	C₄H₆Cl₂O₂	46.0	86.0	116.0	140.0	182.2	205.5	—
cis-Crotonic acid	C₄H₆O₂	33.5	69.0	96.0	116.3	152.2	171.9d	15.5
trans-Crotonic acid	C₄H₆O₂	s	80.0	107.8	128.0	165.5	185.0	72
Methyl acrylate	C₄H₆O₂	−43.7	−13.5	+ 9.2	28.0	61.8	80.2p	—
Methacrylic acid	C₄H₆O₂	25.5	60.0	86.4	106.6	142.5	161.0	15
Vinyl acetate	C₄H₆O₂	−48.0	−18.9	+ 5.3	23.3	55.5	72.5	—
Acetic anhydride	C₄H₆O₃	1.7	36.0	62.1	82.2	119.8	139.6	−73
Dimethyl oxalate	C₄H₆O₄	20.0	56.0	83.6	104.8	143.3	163.3	—
cis-1 Bromo-1-butene	C₄H₇Br	−44.0	−12.8	+ 11.5	30.8	66.8	86.2	—
trans-1-Bromo-1-butene	C₄H₇Br	−38.4	−6.4	+ 18.4	38.1	75.0	94.7	−100.3
2-Bromo-1-butene	C₄H₇Br	−47.3	−16.8	+ 7.2	26.3	61.9	81.0	−133.4
cis-2-Bromo-2-butene	C₄H₇Br	−39.0	−7.2	+ 17.7	37.5	74.0	93.0	−111.2
trans-2-Bromo-2-butene	C₄H₇Br	−45.0	−13.8	+ 10.5	29.0	66.0	85.5	−114.6
1-Bromo-2-Butanone	C₄H₇BrO	+ 6.2	41.8	68.2	89.2	126.3	147.0	—
2-Methylpropionyl bromide	C₄H₇BrO	13.5	50.6	79.4	101.6	141.7	163.0	—
1,1,2-Tribromobutane	C₄H₇Br₃	45.0	87.8	120.2	146.0	192.0	216.2	—
1,2,2-Tribromobutane	C₄H₇Br₃	41.0	83.2	116.0	141.8	188.0	213.8	—
2,2,3-Tribromobutane	C₄H₇Br₃	38.2	79.8	111.8	136.3	182.2	206.5	—
Ethyl chloroacetate	C₄H₇ClO₂	+ 1.0	37.5	65.2	86.0	123.8	144.2	−26
1,2,3-Trichlorobutane	C₄H₇Cl₃	+ 0.5	40.0	71.5	96.2	143.0	169.0	—
Butyronitrile	C₄H₇N	−20.0	+ 13.4	38.4	59.0	96.8	117.5	—
Diacetamide	C₄H₇ NO₂	70.0s	108.0	138.2	160.6	202.0	223.0	78.5
1-Butene	C₄H₈	−104.8	−81.6	−63.4	−48.9	−21.7	−6.3	−130
cis-2-Butene	C₄H₈	−96.4	−73.4	−54.7	−39.8	−12.0	+ 3.7	−138.9
trans-2-Butene	C₄H₈	−99.4	−76.3	−57.6	−42.9	−14.8	+ 0.9	−105.4
2-Methylpropene	C₄H₈	−105.1	−81.9	−63.8	−49.3	−22.2	−6.9	−140.3
Cyclobutane	C₄H₈	−92.0s	−67.9s	−48.4	−32.8	−3.4	+ 12.9	−50
Methylcyclopropane	C₄H₈	−96.0	−72.8	−54.2	−39.3	−11 .3	+ 4.5	—
2-Bromoethyl 2-chloroethyl ether	C₄H₈BrClO	36.5	76.3	106.6	129.8	172.3	195.8	—
1,2-Dibromobutane	C₄H₈Br₂	7.5	46.1	76.0	99.8	143.5	166.3	−64.5
dl-2,3-Dibromobutane	C₄H₈Br₂	+ 5.0	41.6	72.0	95.3	138.0	160.5	—
meso-2,3-Dibromobutane	C₄H₈Br₂	+ 1.5	39.3	68.2	91.7	134.2	157.3	−34.5
1,4-Dibromobutane	C₄H₈Br₂	32.0	72.4	104.0	128.7	173.8	197.5	−20
1,2-Dibromo-2-methylpropane	C₄H₈Br₂	−28.8	+ 10.5	42.3	68.8	119.8	149.0	−70.3
1,3-Dibromo-2-methylpropane	C₄H₈Br₂	14.0	53.0	83.5	107.4	150.6	174.6	—
Di(2-bromoethyl) ether	C₄H₈Br₂O	47.7	88.5	119.8	144.0	188.0	212.5	—
1,2-Dichlorobutane	C₄H₈Cl₂	−23.6	+ 11.5	37.7	60.2	100.8	123.5	—
2,3-Dichlorobutane	C₄H₈Cl₂	−25.2	+ 8.5	35.0	56.0	94.2	116.0	−80.4
1,1-Dichloro-2-methylpropane	C₄H₈Cl₂	−31.0	+ 2.6	28.2	48.2	85.4	−106.0d	—
1,2-Dichloro-2-methylpropane	C₄H₈Cl₂	−25.8	+ 6.7	32.0	51.7	87.8	108.0	—
1,3-Dichloro-2-methylpropane	C₄H₈Cl₂	−3.0	+ 32.0	58.6	78.8	115.4	135.0	—
di(2-chloroethyl) ether	C₄H₈Cl₂O	23.5	62.0	91.5	114.5	155.4	178.5	—
1,2-Epoxy-2-methylpropane	C₄H₈O	−69.0	−40.3	−17.3	+ 1.2	36.0	55.5	—
2-Butanone	C₄H₈O	−48.3	−17.7	+ 6.0	25.0	60.0	79.6	−85.9
1,4-Dioxane	C₄H₈O₂	−35.8s	−1.2s	+ 25.2	45.1	81.8	101.1	10
Butyric acid	C₄H₈O₂	25.5	61.5	88.0	108.0	144.5	163.5	−4.7
Isobutyric acid	C₄H₈O₂	14.7	51.2	77.8	98.0	134.5	154.5	−47
Ethyl acetate	C₄H₈O₂	−43.4	−13.5	+ 9.1	27.0	59.3	77.1	−82.4
Methyl propionate	C₄H₈O₂	−42.0	−11.8	+ 11.0	29.0	61.8	79.8	−87.5
Propyl formate	C₄H₈O₂	−43.0	−12.6	+ 10.8	29.5	62.6	81.3	−92.9
Isopropyl formate	C₄H₈O₂	−52.0	−22.7	0.2	+ 17.8	50.5	68.3	—
α-Hydroxyisobutyric acid	C₄H₈O₃	73.5s	110.5	138.0	157.7	193.8	212.0	79
Ethyl glycolate	C₄H₈O₃	14.3	50.5	78.1	99.8	138.0	158.2	—
1-Bromobutane	C₄H₉Br	−33.0	−0.3	+ 24.8	44.9	81.7	101.6	−112.4
1-Bromo-2-butanol	C₄H₉BrO	23.7	55.8	79.5	97.6	128.3	145.0	—
1-Chlorobutane	C₄H₉Cl	−49.0	−18.6	+ 5.0	24.0	58.8	77.8	−123.1
sec-Butyl chloride	C₄H₉Cl	−60.2	−29.2	−5.0	+ 14.2	50.0	68.0	−131.3
Isobutyl chloride	C₄H₉Cl	−53.8	−24.5	−1.0	+ 16.0	50.0	68.9	−131.2
tert-Butyl chloride	C₄H₉Cl	s	s	19.0	−1.0	+ 32.6	51.0	−26.5
2-(2-Chloroethoxy) ethanol	C₄H₉ClO₂	53.0	90.7	118.4	139.5	176.5	196.0	—
1-Iodo-2-methylpropane	C₄H₉I	−17.0	+ 17.0	42.8	63.5	100.3	120.4	−90.7
Ethyl methylcarbamate	C₄H₉NO₂	26.5	63.2	91.0	112.0	149.8	170.0	—
Propyl carbamate	C₄H₉NO₂	52.4	90.0	117.7	138.3	175.8	195.0	—
Di(nitrosoethyl)amine	C₄H₉N₂O₂	18.5	57.7	87.6	111.0	153.5	176.9	—
Butane	C₄H₁₀	−101.5	−77.8	−59.1	−44.2	−16.3	−0.5	−135
2-Methylpropane	C₄H₁₀	−109.2	−86.4	−68.4	−54.1	−27.1	−11.7	−145
Dichlorodiethylsilane	C₄H₁₀Cl₂Si	−9.2	+ 25.4	51.6	71.8	110.0	130.4	—

Name	Formula	Temperature °C						M.P.
		1 mm	10 mm	40 mm	100 mm	400 mm	760 mm	
Diethyldifluorosilane	$C_4H_{10}F_2Si$	−56.8	−28.8	−7.3	+9.8	40.5	58.0	—
Butyl alcohol	$C_4H_{10}O$	−1.2	+30.2	53.4	70.1	100.8	117.5	−79.9
sec-Butyl alcohol	$C_4H_{10}O$	−12.2	+16.9	38.1	54.1	83.9	99.5	−114.7
Isobutyl alcohol	$C_4H_{10}O$	−9.0	+21.7	44.1	61.5	91.4	108.0	−108
tert-Butyl alcohol	$C_4H_{10}O$	−20.4s	+5.5s	24.5s	39.8	68.0	82.9	25.3
Diethyl ether	$C_4H_{10}O$	−74.3	−48.1	−27.7	−11.5	+17.9	34.6	−116.3
Methyl propyl ether	$C_4H_{10}O$	−72.2	−45.4	−24.3	−8.1	+22.5	39.1	—
1,3-Butanediol	$C_4H_{10}O_2$	22.2s	85.3	117.4	141.2	183.8	206.5	77
2,3-Butanediol	$C_4H_{10}O_2$	44.0	80.3	107.8	127.8	164.0	182.0	22.5
1,2-Dimethoxyethane	$C_4H_{10}O_2$	−48.0	−15.3	+10.7	31.8	70.8	93.0	—
2,2′-Thiodiethanol	$C_4H_{10}O_2S$	42.0	128.0	210.0d	285d	—	—	—
Diethylene glycol	$C_4H_{10}O_3$	91.8	133.8	164.3	187.5	226.5	244.8	—
1,2,3-Butanetriol	$C_4H_{10}O_3$	102.0	146.0	178.0	202.5	243.5	264.0	—
Diethyl sulfite	$C_4H_{10}O_3S$	10.0	46.4	74.2	96.3	137.0	159.0	—
Diethyl sulfate	$C_4H_{10}O_4S$	47.0	87.7	118.0	142.5	185.5	209.5d	−25.0
Diethl sulfide	$C_4H_{10}S$	−39.6	−8.0	+16.1	35.0	69.7	88.0	−99.5
Diethyl selenide	$C_4H_{10}Se$	−25.7	+7.0	31.2	51.8	88.0	108.0	—
Diethylzinc	$C_4H_{10}Zn$	−22.4	+11.7	38.0	59.1	97.3	118.0	−28
Diethylamine	$C_4H_{11}N$	s	−33.0	−11.3	+6.0	38.0	55.5	−38.9
Isobutylamine	$C_4H_{11}N$	−50.0	−21.0	+1.3	18.8	50.7	68.6	−85.0
1,3-Dichlorotetramethyldisiloxane	$C_4H_{12}Cl_2Si_2$	−7.4	+28.3	55.7	76.9	116.3	138.0	−37
Tetramethyllead	$C_4H_{12}Pb$	−29.0s	+4.4	30.3	50.8	89.0	110.0	−27.5
Tetramethylsilane	$C_4H_{12}Si$	−83.8	−58.0	−37.4	−20.9	+10.0	27.0	−102.1
Tetramethyltin	$C_4H_{12}Sn$	−51.3	−20.6	+3.5	22.8	58.5	78.0	—
Tetramethyldiborane	$C_4H_{14}B_2$	−59.6	−27.4	−3.4	+15.3	49.8	68.6	−72.5
3-Bromopyridine	C_5H_4BrN	16.8	55.2	84.1	107.8	150.0	173.4	—
2-Chloropyridine	C_5H_4ClN	13.3	41.7	81.7	104.6	147.7	170.2	—
2-Furaldehyde	$C_5H_4O_2$	18.5	54.8	82.1	103.4	141.8	151.8	−36.5
Citraconic anhydride	$C_5H_4O_2$	47.1	88.9	120.3	145.4	189.8	213.5	—
Pyridine	C_5H_5N	−18.9	+13.2	38.0	57.8	95.6	115.4	42
Glutaryl chloride	$C_5H_6Cl_2O_2$	56.1	97.8	128.3	141.8	195.3	217.0	—
Glutaronitrile	$C_5H_6N_2$	91.3	140.0	176.4	205.5	257.3	286.2	—
Furfuryl alcohol	$C_5H_6O_2$	31.8	68.0	95.7	115.9	151.8	170.0	—
Glutaric anhydride	$C_5H_6O_3$	100.8	149.5	185.5	212.5	261.0	207.0	—
Pyrotartaric anhydride	$C_5H_6O_3$	69.7	114.2	147.8	173.8	221.0	247.4	—
2-Methylthiophene	C_5H_6S	−27.4	+6.0	32.3	53.1	91.8	112.5	−63.5
3-Methylthiophene	C_5H_6S	−24.5	+9.1	35.4	55.8	93.8	115.4	68.9
Propyl chloroglyoxylate	$C_5H_7ClO_2$	9.7	43.5	68.8	88.0	123.0	150.0	—
Tiglonitrile	C_5H_7N	−25.5	+9.4	36.7	58.2	99.7	122.0	—
Angelonitrile	C_5H_7N	−8.0	+28.0	55.8	77.5	117.7	140.0	—
α-Ethylacrylonitrile	C_5H_7N	−29.0	+5.0	31.8	53.0	92.2	114.0	—
Ethyl cyanoacetate	$C_5H_7NO_2$	67.8	106.0	133.8	152.8	187.8	206.0	—
Isoprene	C_5H_8	−79.8	−53.3	−32.6	−16.0	+15.4	32.6	−146.7
1,3-Pentadiene	C_5H_8	−71.8	−45.0	−23.4	−6.7	+24.7	42.0	—
1,4-Pentadiene	C_5H_8	−83.5	−57.1	−37.0	−20.6	8.3	26.1	—
Tiglaldehyde	C_5H_8O	−25.0	+10.0	37.0	57.7	95.5	116.4	—
Levulinaldehyde	$C_5H_8O_2$	28.1	68.0	98.3	121.8	164.0	187.0	—
Tiglic acid	$C_5H_8O_2$	52.0s	90.2	119.0	140.5	179.2	198.5	64.5
α-Valerolactone	$C_5H_8O_2$	37.5	79.8	101.9	136.5	182.3	207.5	—
α-Ethylacrylic acid	$C_5H_8O_2$	47.0	82.0	108.1	127.5	160.7	179.2	—
Ethyl acrylate	$C_5H_8O_2$	−29.5	+2.0	26.0	44.5	80.0	99.5	−71.2
Methyl methacrylate	$C_5H_8O_2$	−30.5	+1.0	25.5	47.0	82.0	101.0	—
Levulinic acid	$C_5H_8O_3$	102.0	141.8	169.5	190.2	227.4	256.8d	33.5
Glutaric acid	$C_5H_8O_4$	155.5	196.0	226.3	247.0	283.5	303.0	97.5
Dimethyl malonate	$C_5H_8O_4$	35.0	72.0	100.0	121.9	159.8	180.7	−62
Ethyl α-chloropropionate	$C_5H_9ClO_2$	+6.6	41.0	68.2	89.3	126.2	146.5	—
Isopropyl chloroacetate	$C_5H_9ClO_2$	+3.8	40.2	68.7	90.3	128.0	148.6	—
Valeronitrile	C_5H_9N	−6.0	+30.0	57.8	78.6	118.7	140.8	—
α-Hydroxybutyronitrile	C_5H_9NO	41.0	77.8	104.8	125.0	159.8	178.8	—
1-Pentene	C_5H_{10}	−80.4	−54.5	−34.1	−17.7	+12.8	30.1	—
2-Methyl-2-butene	C_5H_{10}	−75.4	−47.9	−26.7	−9.9	+21.6	38.5	−133
2-Methyl-1-butene	C_5H_{10}	−89.1	−64.3	−44.1	−28.0	+2.5	20.2	−135
Cyclopentane	C_5H_{10}	−68.0	−40.4	−18.6	−1.3	+31.0	49.3	−93.7
1,2-Dibromopentane	$C_5H_{10}Br_2$	19.8	58.0	87.4	110.1	151.8	175.0	—
2-Chloroethyl 2-chloroisopropyl ether	$C_5H_{10}Cl_2O$	24.7	63.0	92.4	115.8	156.5	180.0	—
2-Chloroethyl 2-chloropropyl ether	$C_5H_{10}Cl_2O$	29.8	70.0	101.5	125.6	169.8	194.1	—
Di(2-chloroethoxy)methane	$C_5H_{10}Cl_2O_2$	53.0	94.0	125.5	149.6	192.0	215.0	—
Allyldichloroethylsilane	$C_5H_{10}Cl_2Si$	−3.0	34.2	62.7	85.2	127.0	150.3	—
3-Pentanone	$C_5H_{10}O$	−12.7	+17.2	39.4	56.2	86.3	102.7	−42
2-Pentanone	$C_5H_{10}O$	−12.0	+17.9	39.8	56.8	86.8	103.3	−77.8
3-Methyl-2-butanone	$C_5H_{10}O$	−19.9	+8.3	29.6	45.5	73.8	88.9	−92
4-Hydroxy-3-methyl-2-butanone	$C_5H_{10}O_2$	44.6	81.0	108.2	129.0	165.5	185.0	—
Valeric acid	$C_5H_{10}O_2$	42.2	79.8	107.8	128.3	165.0	184.4	−34.5
Isovaleric acid	$C_5H_{10}O_2$	34.5	91.3	98.0	118.9	155.2	175.1	−37.6
Ethyl propionate	$C_5H_{10}O_2$	−28.0	+3.4	27.2	45.2	79.8	99.1	−72.6
Propyl acetate	$C_5H_{10}O_2$	−26.7	+5.0	28.8	47.8	82.0	101.8	−92.5
Isopropyl acetate	$C_5H_{10}O_2$	−38.3	−7.2	+17.0	35.7	69.8	89.0	—
Methyl butyrate	$C_5H_{10}O_2$	−26.8	+5.0	29.6	48.0	83.1	102.3	—
Methyl isobutyrate	$C_5H_{10}O_2$	−34.1	−2.9	+21.0	39.6	73.6	92.6	−84.7
Butyl formate	$C_5H_{10}O_2$	−26.4	+6.1	31.6	51.0	86.2	106.0	—
Isobutyl formate	$C_5H_{10}O_2$	−32.7	−0.8	+24.1	43.4	79.0	98.2	−95.3

Name	Formula	Temperature °C						M.P.
		1 mm	10 mm	40 mm	100 mm	400 mm	760 mm	
sec-Butyl formate	$C_5H_{10}O_2$	−34.4	−3.1	+21.3	40.2	75.2	93.6	—
Diethyl carbonate	$C_5H_{10}O_3$	−10.1	+23.8	49.5	69.7	105.8	125.8	−43
1-Bromo-3-methylbutane	$C_5H_{11}Br$	−20.4	+13.6	39.8	60.4	99.4	120.4	—
1-Iodo-3-methylbutane	$C_5H_{11}I$	−2.5	+34.1	62.3	84.4	125.8	148.2	—
Piperidine	$C_5H_{11}N$	s	+3.9	29.2	49.0	85.7	106.0	−9
Isobutyl carbamate	$C_5H_{11}NO_2$	s	96.4	125.3	147.2	186.0	206.5	65
Isoamyl nitrate	$C_5H_{11}NO_3$	+5.2	40.3	67.6	88.6	126.5	147.5	—
Pentane	C_5H_{12}	−76.6	−50.1	−29.2	−12.6	+18.5	36.1	−129.7
2-Methylbutane	C_5H_{12}	−82.9	−57.0	−36.5	−20.2	+10.5	27.8	−159.7
2,2-Dimethylpropane	C_5H_{12}	−102.0	−76.7s	−56.1s	−39.1s	−7.1	+9.5	−16.6
Amyl alcohol	$C_5H_{12}O$	+13.6	44.9	68.0	85.8	119.8	137.8	—
Isoamyl alcohol	$C_5H_{12}O$	+10.0	40.8	63.4	80.7	113.7	130.6	−17.2
2-Pentanol	$C_5H_{12}O$	+1.5	32.2	54.1	70.7	102.3	119.7	—
tert-Amyl alcohol	$C_5H_{12}O$	−12.9s	+17.2	38.8	55.3	85.7	101.7	−11.9
Ethyl propyl ether	$C_5H_{12}O$	−64.3	−35.0	−12.0	+6.8	41.6	61.7	—
2,3,4-Pentanetriol	$C_5H_{12}O_3$	155.0	204.5	239.6	263.5	307.0	327.2	—
Ethoxytrimethylsilane	$C_5H_{14}OSi$	−50.9	−20.7	+3.7	22.1	56.3	75.7	—
Ethyltrimethylsilane	$C_5H_{14}Si$	−60.6	−31.8	−9.0	+9.2	42.8	62.0	—
Ethyltrimethyltin	$C_5H_{14}Sn$	−30.0	+3.8	30.0	50.0	87.6	108.8	—
Chloranil	$C_6Cl_4O_2$	70.7s	97.8s	116.1s	129.5s	151.3s	162.6s	290
Hexachlorobenzene	C_6Cl_6	114.4s	166.4s	206.0s	235.5	283.5	309.4	230
Pentachlorobenzene	C_6HCl_5	98.6	144.3	178.5	205.5	251.6	276.0	85.5
Pentachlorophenol	C_6HCl_5O	s	s	211.2	239.6	285.0	309.3d	188.5
3-Bromo-2,4,6-trichlorophenol	C_6HBrCl_3O	112.4	163.2	200.5	229.3	278.0	305.8	—
1,2,3,4-Tetrachlorobenzene	$C_6H_2Cl_4$	68.5	114.7	149.2	175.7	225.5	254.0	46.5
1,2,3,5-Tetrachlorobenzene	$C_6H_2Cl_4$	58.2	104.1	140.0	168.0	220.0	246.0	54.5
1,2,4,5-Tetrachlorobenzene	$C_6H_2Cl_4$	s	s	146.0	173.5	220.5	256.0	139
2,3,4,6-Tetrachlorophenol	$C_6H_2Cl_4O$	100.0	145.3	179.1	205.2	250.4	275.0	69.5
2-Bromo-4,6-dichlorophenol	$C_6H_3BrCl_2O$	84.0	130.8	165.8	193.2	242.0	268.0	68
1,2,3-Trichlorobenzene	$C_6H_3Cl_3$	40.0s	85.6	119.8	146.0	193.5	218.5	52.5
1,2,4-Trichlorobenzene	$C_6H_3Cl_3$	38.4	81.7	114.8	140.0	187.7	213.0	17
1,3,5-Trichlorobenzene	$C_6H_3Cl_3$	s	78.0	110.8	136.0	183.0	208.4	63.5
2,4,5-Trichlorophenol	$C_6H_3Cl_3O$	72.0	117.3	151.5	178.0	226.5	251.8	62
2,4,6-Trichlorophenol	$C_6H_3Cl_3O$	76.5	120.2	152.2	177.8	222.5	246.0	68.5
1,4-Dibromobenzene	$C_6H_4Br_2$	61.0s	87.7	120.8	146.5	192.5	218.6	87.5
1,4-Bromochlorobenzene	C_6H_4BrCl	32.0	72.7	103.8	128.0	172.6	196.9	—
1,2-Dichlorobenzene	$C_6H_4Cl_2$	20.0	59.1	89.4	112.9	155.8	179.0	−17.6
1,3-Dichlorobenzene	$C_6H_4Cl_2$	12.1	52.0	82.0	105.0	149.0	173.0	−24.2
1,4-Dichlorobenzene	$C_6H_4Cl_2$	s	54.8	84.8	108.4	150.2	173.9	53.0
2,4-Dichlorophenol	$C_6H_4Cl_2O$	53.0	92.8	123.4	146.0	187.5	210.0	45.0
2,6-Dichlorophenol	$C_6H_4Cl_2O$	59.5	101.0	131.6	154.6	197.7	220.0	—
2,4,6-Trielfloroaniline	$C_6H_4Cl_3N$	134.0	170.0	195.8	214.6	246.4	262.0	78
Dichlorophenylarsine	$C_6H_5AsCl_2$	61.8	116.0	151.0	178.9	228.8	256.5	—
Bromobenzene	C_6H_5Br	+2.9	40.0	68.6	90.8	132.3	156.2	30.7
Chlorobenzene	C_6H_5Cl	−13.0	+22.2	49.7	70.7	110.0	132.2	−45.2
2-Chlorophenol	C_6H_5ClO	12.1	51.2	82.0	106.0	149.8	174.5	7.0
3-Chlorophenol	C_6H_5ClO	44.2	86.1	118.0	143.0	188.7	214.0	32.5
4-Chlorophenol	C_6H_5ClO	49.8	92.2	125.0	150.0	196.0	220.0	42
Benzenesulfonylchloride	$C_6H_5ClO_2S$	65.9	112.0	147.7	174.5	224.0	251.5d	14.5
Phenyl dichlorophosphate	$C_6H_5Cl_2O_2P$	66.7	110.0	143.4	168.0	213.0	239.5	—
Trichlorophenylsilane	$C_6H_5Cl_3Si$	33.0	74.2	105.8	130.5	175.7	201.0	—
Fluorobenzene	C_6H_5F	−43.4s	−12.4	+11.5	30.4	65.7	84.7	−42.1
Trifluorophenylsilane	$C_6H_5F_3Si$	−31.0	+0.8	25.4	44.2	78.7	98.3	—
Iodobenzene	C_6H_5I	24.1	64.0	94.4	118.3	163.9	188.6	−28.5
Nitrobenzene	$C_6H_5NO_2$	44.4	84.9	115.4	139.9	185.8	210.6	+5.7
2-Nitrophenol	$C_6H_5NO_3$	49.3	90.4	122.1	146.4	191.0	214.5	45
1,5-Hexadiene-3-yne	C_6H_6	−45.1	−14.0	+10.0	29.5	64.4	84.0	—
Benzene	C_6H_6	−36.7	−11.5s	+7.6	26.1	60.6	80.1	+5.5
2-Chloroaniline	C_6H_6ClN	46.3	84.8	115.6	139.5	183.7	208.8	—
3-Chloroaniline	C_6H_6ClN	63.5	102.0	133.6	158.0	203.5	228.5	−10.4
4-Chloroaniline	C_6H_6ClN	59.3s	102.1	135.0	159.9	2.6.6	230.5	70.5
4-Chlorophenol	C_6H_6ClO	54.3	95.8	125.8	150.6	194.3	217.0	42
2-Nitroaniline	$C_6H_6N_2O_2$	104.0	150.4	186.0	213.0	260.0	284.5d	71.5
3-Nitroaniline	$C_6H_6N_2O_2$	119.3	167.8	204.2	232.1	280.2	305.7d	114
4-Nitroaniline	$C_6H_6N_2O_2$	142.4s	194.4	234.2	261.8	310.2d	336.0d	156.5
Phenol	C_6H_6O	40.1s	73.8	100.1	121.4	160.0	181.9	40.6
Pyorcatechol	$C_6H_6O_2$	s	118.3	150.6	176.0	221.5	245.5	105
Resorcinol	$C_6H_6O_2$	108.4s	152.1	185.3	209.8	253.4	276.5	110.7
Hydroquinone	$C_6H_6O_2$	132.4s	163.5s	192.0	216.5	262.5	286.2	170.3
Pyrogallol	$C_6H_6O_3$	s	167.7	204.2	232.0	281.5	309.9d	133
Benzenethiol	C_6H_6S	18.6	56.0	84.2	106.6	146.7	168.0	—
Aniline	C_6H_7N	34.8	69.4	96.7	119.9	161.9	184.4	−6.2
2-Picoline	C_6H_7N	−11.1	+24.4	51.2	71.4	108.4	128.8	−70
Ethylene-bis-(chloroacetate)	$C_6H_8Cl_2O_4$	112.0	158.0	191.0	215.0	259.5	283.5	—
1,3-Phenylenediamine	$C_6H_8N_2$	99.8	147.0	182.5	209.9	259.0	285.5	62.8
Phenylhydrazine	$C_6H_8N_2$	71.8	115.8	148.2	173.5	218.2	243.5d	19.6
α-Methylglutaric anhydride	$C_6H_8O_3$	93.8	141.8	177.5	205.0	255.5	282.5	—
α,α-Dimethylsuccinic anhydride	$C_6H_8O_3$	61.4	102.0	132.3	155.3	197.5	219.5	—
Dimethyl maleate	$C_6H_8O_4$	45.7	86.4	117.2	140.3	182.2	205.0	—
Isobutyl dichloroacetate	$C_6H_{10}Cl_2O_2$	28.6	67.5	96.7	119.8	160.0	183.0	—
Diallyldichlorosilane	$C_6H_{10}Cl_2Si$	+9.5	47.4	76.4	99.7	142.0	165.3	—

Name	Formula	Temperature °C						M.P.
		1 mm	10 mm	40 mm	100 mm	400 mm	760 mm	
Cyclohexanone	C₆H₁₀O	+ 1.4	38.7	67.8	90.4	132.5	155.6	−45.0
Mesityl oxide	C₆H₁₀O	−8.7	+ 26.0	51.7	72.1	109.8	130.0	−59
Isocaprolactone	C₆H₁₀O₂	38.3	80.3	112.3	137.2	182.1	207.0	—
Propionic anhydride	C₆H₁₀O₃	20.6	57.7	85.6	107.2	146.0	167.0	−45
Ethyl acetoacetate	C₆H₁₀O₃	28.5	67.3	96.2	118.5	158.2	180.8	−45
Methyl levulinate	C₆H₁₀O₃	39.8	79.7	109.5	133.0	175.8	197.7	—
Adipic acid	C₆H₁₀O₄	159.5	205.5	240.5	265.0	312.5	337.5	152
Diethyl oxalate	C₆H₁₀O₄	47.4	83.8	110.6	130.8	166.2	185.7	−40.6
Glycol diacetate	C₆H₁₀O₄	38.3	77.1	106.1	128.0	108.3	190.5	−31
Dimethyl-l-malate	C₆H₁₀O₅	75.4	118.3	150.1	175.1	219.5	242.6	—
Dimethyl-d-tartrate	C₆H₁₀O₆	102.1	148.2	182.4	208.8	255.0	280.0	61.5
Dimethyl-dl-tartrate	C₆H₁₀O₆	100.4	147.5	182.4	209.5	257.4	282.0	89
Diallyl sulfide	C₆H₁₀S	−9.5	+ 26.6	54.2	75.8	116.1	138.6	−83
Ethyl α-bromoisobutyrate	C₆H₁₁BrO₂	10.6	48.0	77.0	99.8	141.2	163.6	—
sec-Butyl chloroacetate	C₆H₁₁ClO₂	17.0	54.6	83.6	105.5	147.0	167.8	—
Capronitrile	C₆H₁₁N	+ 9.2	47.5	76.9	99.8	141.0	163.7	—
1-Hexene	C₆H₁₂	−57.5	−28.1	−5.0	+ 13.0	46.8	66.0	−98.5
Cyclohexane	C₆H₁₂	−45.3s	−15.9s	+ 6.7	25.5	60.8	80.7	+ 6.6
Methylcyclopentane	C₆H₁₂	−53.7	−23.7	−0.6	+ 17.9	42.3	71.8	−142.4
Dichlorodiisopropyl ether	C₆H₁₂Cl₂O	29.6	68.2	97.3	119.7	159.8	182.7	—
Bis-2-chloroethyl acetal	C₆H₁₂Cl₂O₂	56.2	97.6	127.8	150.7	190.5	212.6	—
2-Hexanone	C₆H₁₂O	+ 7.7	38.8	62.0	79.8	111.0	127.5	−56.9
4-Methyl-2-pentanone	C₆H₁₂O	−1.4	+ 30.0	52.8	70.4	102.0	119.0	−84.7
Allyl propyl ether	C₆H₁₂O	−39.0	−7.9	+ 16.4	35.8	71.4	90.5	—
Allyl isopropyl ether	C₆H₁₂O	−43.7	−12.9	+ 10.9	29.0	61.7	79.5	—
Cyclohexanol	C₆H₁₂O	21.0s	56.0	83.0	103.7	141.4	161.0	23.9
Caproic acid	C₆H₁₂O₂	71.4	99.5	125.0	144.0	181.0	202.0	−1.5
Isocaproic acid	C₆H₁₂O₂	66.2	94.0	120.4	141.4	181.0	207.7	−35
4-Hydroxy-4-methyl-2-pentanone	C₆H₁₂O₂	22.0	58.8	86.7	108.2	147.5	167.9	−47
Methyl isovalerate	C₆H₁₂O₂	−19.2	+ 14.0	39.8	59.8	96.7	116.7	—
Ethyl butyrate	C₆H₁₂O₂	−16.4	+ 15.3	41.5	62.0	100.0	121.0	−93.3
Ethyl isobutyrate	C₆H₁₂O₂	−24.3	+ 8.4	33.8	53.5	90.0	110.1	−88.2
Propyl propionate	C₆H₁₂O₂	−14.2	+ 19.4	45.0	64.2	102.0	122.4	−76
Isobutyl acetate	C₆H₁₂O₂	−21.2	+ 12.8	39.2	59.7	97.5	118.0	−98.9
Isoamyl formate	C₆H₁₂O₂	17.6	+ 17.1	44.0	63.4	102.7	123.3	—
Paraformaldehyde	C₆H₁₂O₃	−9.4s	+ 24.1s	49.5s	69.0s	104.3s	124.0s	155 ± 5
sec-Butyl glycolate	C₆H₁₂O₃	28.3	66.0	94.2	116.4	155.6	177.5	—
Hexane	C₆H₁₄	−53.9	−25.0	−2.3	+ 15.8	49.6	68.7	−95.3
2-Methylpentane	C₆H₁₄	−60.9	−32.1	−9.7	+ 8.1	41.6	60.3	−154
3-Methylpentane	C₆H₁₄	−59.0	−30.1	−7.3	+ 10.5	44.2	63.3	−118
2,2-Dimethylbutane	C₆H₁₄	−69.3	−41.5	−19.5	−2.0	+ 31.0	49.7	−99.8
2,3-Dimethylbutane	C₆H₁₄	−63.6	−34.9	−12.4	+ 5.4	39.0	58.0	−128.2
1-Hexanol	C₆H₁₄O	24.4	58.2	83.7	102.8	138.8	157.0	−51.6
2-Hexanol	C₆H₁₄O	14.6	45.0	67.9	87.3	121.8	139.9	—
3-Hexanol	C₆H₁₄O	+ 2.5	36.7	62.2	81.8	117.0	135.5	—
2-Methyl-1-pentanol	C₆H₁₄O	15.4	49.6	74.7	94.2	129.8	147.9	—
2-Methyl-2-pentanol	C₆H₁₄O	−4.5	+ 27.6	51.3	69.2	102.6	121.1	−103
2-Methyl-4-pentanol	C₆H₁₄O	−0.3	+ 33.3	58.2	78.0	113.5	131.7	—
Dipropyl ether	C₆H₁₄O	−43.3	−11.8	+ 13.2	33.0	69.5	89.5	−122
Diisopropyl ether	C₆H₁₄O	−57.0	−27.4	−4.5	+ 13.7	48.2	67.5	−60
Acetal	C₆H₁₄O₂	−23.0	+ 8.0	31.9	50.1	84.0	102.0	—
1,2-Diethoxyethane	C₆H₁₄O₂	−33.5	+ 1.6	29.7	51.8	92.1	119.5	—
Di(2-methoxyethyl) ether	C₆H₁₄O₃	13.0	50.0	77.5	99.5	138.5	159.8	—
Diethylene glycol, ethyl ether	C₆H₁₄O₃	45.3	85.8	116.7	140.3	180.3	201.9	—
Dipropyleneglycol	C₆H₁₄O₃	73.8	116.2	147.4	169.9	210.5	231.8	—
Triethyleneglycol	C₆H₁₄O₄	114.0	158.1	191.3	214.6	256.6	278.3	—
Triethylboron	C₆H₁₅B	—	−148.0	−131.4	−116.0	−81.0	−56.2	—
Chlorotriethylsilane	C₆H₁₅ClSi	−4.9	+ 32.0	60.2	82.3	123.6	146.3	—
Triethyl phosphate	C₆H₁₅O₄P	39.6	82.1	115.7	141.6	187.0	211.0	—
Triethylthallium	C₆H₁₅Tl	+ 9.3	51.7	85.4	112.1	163.5	192.9d	−63.0
Diethoxydimethylsilane	C₆H₁₆O₂Si	−19.1	+ 13.3	38.0	57.6	93.2	113.5	—
Trimethylpropylsilane	C₆H₁₆Si	−46.0	−13.9	+ 11.3	31.6	69.2	90.0	—
Trimethylpropyltin	C₆H₁₆Sn	−12.0	+ 21.8	48.5	69.8	109.6	131.7	—
1,5-Dichlorohexamethyltrisiloxane	C₆H₁₈Cl₂O₂Si₃	26.0	65.1	94.8	118.2	160.2	184.0	−53
Hexamethylcyclotrisiloxane	C₆H₁₈O₃Si₃	s	s	s	78.7	114.7	134.0	64
Hexamethyldisiloxane	C₆H₁₈OSi₂	−29.0	+ 2.8	26.7	45.6	80.0	99.2	—
3,4-Dichloro-α,α,α-trifluorotoluene	C₇H₃Cl₂F₃	11.0	52.2	84.0	109.2	150.5	172.8	−12.1
2-Chloro-α,α,α-trifluorotoluene	C₇H₄ClF₃	0.0	37.1	65.9	88.3	130.0	152.2	−6.0
2-α,α,α-tetrachlorotoluene	C₇H₄Cl₄	69.0	117.9	155.0	185.0	233.0	262.1	28.7
Benzoyl bromide	C₇H₅BrO	47.0	89.8	122.6	147.7	193.7	218.5	0
Benzoyl chloride	C₇H₅ClO	32.1	73.0	103.8	128.0	172.8	197.2	−0.5
α,α,α-Trichlorotoluene	C₇H₅Cl₃	45.8	87.6	119.8	144.3	189.2	213.5	−21.2
α,α,α-Trifluorotoluene	C₇H₅F₃	−32.0s	+ 0.4	25.7	45.3	82.0	102.2	−29.3
Benzonitrile	C₇H₅N	28.2	69.2	99.6	123.5	166.7	109.6	−12.9
Phenyl isocyanide	C₇H₅N	12.0	49.7	78.3	101.0	142.3	165.0d	—
Phenyl isocyanate	C₇H₅NO	10.6	48.5	77.7	100.6	142.7	165.6	—
2-Nitrobenzaldehyde	C₇H₅NO₃	85.8	133.4	168.8	196.2	246.8	273.5	40.9
3-Nitrobenzaldehyde	C₇H₅NO₃	96.2	142.8	177.7	204.3	252.1	278.3	58
Phenyl isothiocyanate	C₇H₅NS	47.2	89.8	122.5	147.7	194.0	218.5	−21.0
α,α-dichlorotoluene	C₇H₆Cl₂	35.4	78.7	112.1	138.3	187.0	214.0	−16.1
Benzaldehyde	C₇H₆O	26.2	62.0	90.1	112.5	154.1	179.0	−26

Name	Formula	Temperature °C						M.P.
		1 mm	10 mm	40 mm	100 mm	400 mm	760 mm	
Benzoic acid	C₇H₆O₂	96.0s	132.1	162.6	186.2	227.0	249.2	121.7
Salicylaldehyde	C₇H₆O₂	33.0	73.8	105.2	129.4	173.7	196.5	−7
4-Hydroxybenzaldihyde	C₇H₆O₂	121.2	169.7	206.0	233.5	282.6	310.0	115.5
Salicylic acid	C₇H₆O₃	113.7s	146.2s	172.2	193.4	230.5	256.0	159
α-Bromotoluene	C₇H₇Br	32.2	73.4	104.8	129.8	175.2	198.5	−4
2-Bromotoluene	C₇H₇Br	24.4	62.3	91.0	112.0	157.3	181.8	−28
3-Bromotoluene	C₇H₇Br	14.8	64.0	93.9	117.8	160.0	183.7	−39.8
4-Bromotoluene	C₇H₇Br	10.3	61.1	91.8	116.4	160.2	184.5	28.5
4-Bromoanisole	C₇H₇BrO	48.8	91.9	125.0	150.1	197.5	223.0	12.5
α-Chlorotoluene	C₇H₇Cl	22.0	60.8	90.7	114.2	155.8	179.4	−39
2-Chlorotoluene	C₇H₇Cl	+ 5.4	43.2	72.0	94.7	137.1	159.3	—
3-Chlorotoluene	C₇H₇Cl	+ 4.8	43.2	73.0	96.3	139.7	162.3	—
4-Chlorotoluene	C₇H₇Cl	+ 5.5	43.8	73.5	96.6	139.8	162.3	+ 7.3
2-Fluorotoluene	C₇H₇F	−24.2	+ 8.9	34.7	55.3	92.8	114.0	−80
30Fluorotoluene	C₇H₇F	−22.4	+ 11.0	37.0	57.5	95.4	116.0	−110.8
4-Fluorotoluene	C₇H₇F	−21.8	+ 11.8	37.8	58.1	96.1	117.0	—
2-Iodotoluene	C₇H₇I	37.2	79.8	112.4	138.1	185.7	211.0	—
2-Nitrotoluene	C₇H₇NO₂	50.0	93.8	126.3	151.5	197.7	222.3	−4.1
3-Nitrotoluene	C₇H₇NO₂	50.2	96.0	130.7	156.9	206.8	231.9	15.1
4-Nitrotoluene	C₇H₇NO₂	53.7	100.5	136.0	163.0	212.5	238.3	51.9
Toluene	C₇H₈	−26.7	+ 6.4	31.8	51.9	89.5	110.6	−95.0
Benzyldichlorosilane	C₇H₈Cl₂Si	45.3	83.2	111.8	133.5	173.0	194.3	—
Dichloromethylphenylsilane	C₇H₈Cl₂Si	35.7	77.4	109.5	134.2	180.2	205.5	—
Dichloro-4-tolysilane	C₇H₈Cl₂Si	46.2	84.2	113.2	135.5	175.2	196.3	—
Anisole	C₇H₈O	+ 5.4	42.2	70.7	93.0	133.8	155.5	−37.3
Benzyl alcohol	C₇H₈O	58.0	92.6	119.8	141.9	183.0	204.7	−15.3
2-Cresol	C₇H₈O	38.2	76.7	105.8	127.4	168.4	190.8	30.8
3-Cresol	C₇H₈O	52.0	87.8	116.0	138.0	179.0	202.8	10.9
4-Cresol	C₇H₈O	53.0	88.6	117.7	140.0	179.4	201.8	35.5
3,5-Dimethyl-1,2-pyrone	C₇H₈O₂	78.6	122.0	152.7	177.5	221.0	245.0	51.5
2-Methoxyphenol	C₇H₈O₂	52.4	92.0	121.6	144.0	184.1	205.0	28.3
Ethyl 2-furoate	C₇H₈O₃	37.6	77.1	107.5	130.4	172.5	195.0	34
Benzylamine	C₇H₉N	29.0	67.7	97.3	120.0	161.3	184.5	—
N-Methylaniline	C₇H₉N	36.0	76.2	106.0	129.8	172.0	195.5	−57
2-Toluidine	C₇H₉N	44.0	81.4	110.0	133.0	176.2	199.7	−16.3
3-Toluidine	C₇H₉N	41.0	82.0	113.5	136.7	180.6	203.3	−31.5
4-Toluidine	C₇H₉N	42.0	81.8	111.5	133.7	176.9	200.4	44.5
2-Methoxyaniline	C₇H₉NO	61.0	101.7	132.0	155.2	197.3	218.5	5.2
Toluene-2,4-diamine	C₇H₁₀N₂	106.5	151.7	185.7	211.5	256.0	280.0	99
4-Tolylhydrazine	C₇H₁₀N₂	82.0	123.8	154.1	178.0	219.5	242.0d	65.5
Trimethysuccinic anhydride	C₇H₁₀O₃	53.5	97.4	131.0	145.4	205.5	231.0	—
Dimethyl citraconate	C₇H₁₀O₄	50.8	91.8	122.6	145.8	188.0	210.5	—
Dimethyl itaconate	C₇H₁₀O₄	69.3	106.6	133.7	153.7	189.8	208.0	38
trans-Dimethyl mesaconate	C₇H₁₀O₄	46.8	87.8	118.0	141.5	183.5	206.0	—
2-Cyano-2-butyl acetate	C₇H₁₁NO₂	42.0	82.0	111.8	133.8	173.4	195.2	—
Butyl acrylate	C₇H₁₂O₂	−0.5	+ 35.5	63.4	85.1	125.2	147.4	−64.6
Ethyl levulinate	C₇H₁₂O₃	47.3	87.3	117.7	141.3	183.0	206.2	—
Pimelic acid	C₇H₁₂O₄	163.4	212.0	247.0	272.0	318.5	342.1	103
Diethyl malonate	C₇H₁₂O₄	40.0	81.3	113.3	136.2	176.8	198.9	−49.8
Enanthyl chloride	C₇H₁₃ClO	34.2	64.6	86.4	102.7	130.7	145.0	—
Enanthonitrile	C₇H₁₃N	21.0	61.6	92.6	116.8	160.0	184.6	—
Ethylcyclopentane	C₇H₁₄	−32.2	−0.1	+ 25.0	45.0	82.3	103.4	−138.6
Methylcyclohexane	C₇H₁₄	−35.9	−3.2	+ 22.0	42.1	79.6	100.9	−126.4
2-Heptene	C₇H₁₄	−35.8	−3.5	+ 21.5	41.3	78.1	98.5	—
Enanthaldehyde	C₇H₁₄O	12.0	43.0	66.3	84.0	125.5	155.0	−42
2-Heptanone	C₇H₁₄O	19.3	55.5	81.2	100.0	133.2	150.2	—
4-Heptanone	C₇H₁₄O	23.0	55.0	78.1	96.0	127.3	143.7	−32.6
2,4-Dimethyl-3-pentanone	C₇H₁₄O	+ 5.2	36.7	59.6	77.0	108.0	123.7	—
Enanthic acid	C₇H₁₄O₂	78.0	113.2	139.5	160.0	199.6	221.5	−10
Methyl caproate	C₇H₁₄O₂	+ 5.0	42.0	70.0	91.4	129.8	150.0d	—
Ethyl isovalerate	C₇H₁₄O₂	−6.1	+ 28.7	55.2	7.9	114.0	134.3	−99.3
Propyl butyrate	C₇H₁₄O₂	−1.6	+ 34.0	61.5	82.6	121.7	142.7	−95.2
Propyl isobutyrate	C₇H₁₄O₂	−6.2	+ 28.3	54.3	73.9	112.0	133.9	—
Isopropyl isobutyrate	C₇H₁₄O₂	−16.3	+ 17.0	42.4	62.3	100.0	120.5	—
Isobutyl propionate	C₇H₁₄O₂	−2.3	+ 32.3	58.5	79.5	116.4	136.8	−71
Isoamyl acetate	C₇H₁₄O₂	0.0	+ 35.2	62.1	83.2	121.5	142.0	—
Heptane	C₇H₁₆	−34.0	−2.1	+ 22.3	41.8	78.0	98.4	−90.6
2-Methylhexane	C₇H₁₆	−40.4	−9.1	+ 14.9	34.1	69.8	90.0	−118.2
3-Methylhexane	C₇H₁₆	−39.0	−7.8	+ 16.4	35.6	71.6	91.9	−119.4
3-Ethylpentane	C₇H₁₆	−37.8	−6.8	+ 17.5	36.9	73.0	93.5	−118.6
2,2-Dimethylpentane	C₇H₁₆	−49.0	−18.7	+ 5.0	23.9	59.2	79.2	−123.7
2,3-Dimethylpentane	C₇H₁₆	−42.0	−10.3	+ 13.9	33.3	69.4	89.8	−135
2,4-Dimethylpentane	C₇H₁₆	−48.0	−17.1	+ 6.5	25.4	60.6	80.5	−119.5
3,3-Dimethylpentane	C₇H₁₆	−45.9	−14.4	+ 9.9	29.3	65.5	86.1	−135.0
2,3,3-Trimethylbutane	C₇H₁₆	s	−18.8	+ 5.2	24.4	60.4	80.9	−25.0
1-Heptanol	C₇H₁₆O	42.4	74.7	99.8	119.5	155.6	175.8	34.6
Triethyl orthoformate	C₇H₁₆O₃	+ 5.5	40.5	67.5	88.0	125.7	146.0	—
Triethoxymethylsilane	C₇H₁₈O₃Si	−1.5	+ 34.6	61.7	82.7	121.8	143.5	—
Butyltrimethylsilane	C₇H₈Si	−23.4	+ 9.9	35.9	56.3	93.8	115.0	—
Triethylmethylsilane	C₇H₁₈Si	−18.2	+ 16.6	44.0	65.6	105.3	127.0	—
Phthaloyl chloride	C₈H₄Cl₂O₂	86.3s	134.2	170.0	197.8	248.3	275.8	88.5

Name	Formula	Temperature °C						M.P.
		1 mm	10 mm	40 mm	100 mm	400 mm	760 mm	
Phthalic anhydride	$C_8H_4O_3$	96.5s	134.0	172.0	202.3	256.8	284.5	130.8
α,α-Dichlorophenylacetonitrile	$C_8H_5Cl_2N$	56.0	98.1	130.0	154.5	199.5	223.5	—
Pentachloroethylbenzene	$C_8H_5Cl_5$	96.2	148.0	186.2	216.0	269.3	299.0	—
Phenylglyoxylonitrile	C_8H_5NO	44.5	85.5	116.6	141.0	185.0	208.0	33.5
2,3-Dichlorostyrene	$C_8H_6Cl_2$	61.0	104.6	137.8	163.5	210.0p	235.0p	—
2,4-Dichlorostyrene	$C_8H_6Cl_2$	53.5	97.4	129.2	153.8	200.0p	225.0p	—
2,5-Dichlorostyrene	$C_8H_6Cl_2$	55.5	98.2	131.0	155.8	202.5p	227.0p	—
2,6-Dichlorostyrene	$C_8H_6Cl_2$	47.8	90.0	122.4	147.6	193.5p	217.0p	—
3,4-Dichlorostyrene	$C_8H_6Cl_2$	57.2	100.4	133.7	158.2	205.7p	230.0p	—
3,5-Dichlorostyrene	$C_8H_6Cl_2$	53.5	97.4	129.2	153.8	200.0p	225.0p	—
3,4,5,6-Tetrachloro-1,2-xylene	$C_8H_6Cl_4$	94.4	140.3	174.2	200.5	248.3	273.5	—
1,2,3,5-Tetrachloro-4-ethylbenzene	$C_8H_6Cl_4$	77.0	126.0	162.1	191.6	243.0	270.0	—
Phenylglyoxal	$C_8H_6O_2$	s	87.8	115.5	136.2	173.5	193.5	73
Phthalide	$C_8H_6O_2$	95.5	144.0	181.0	210.0	261.8	290.0	73
Piperonal	$C_8H_6O_3$	87.0	132.0	165.7	191.7	238.5	263.0	37
3-Chlorostyrene	C_8H_7Cl	25.3	65.2	96.5	121.2	165.7p	190.0p	—
4-Chlorostyrene	C_8H_7Cl	28.0	67.5	98.0	122.0	166.0p	191.0p	−15.0
Phenylacetyl chloride	C_8H_7ClO	48.0	89.0	119.8	143.5	186.0	210.0	—
2-Tolunitrile	C_8H_7N	36.7	77.9	110.0	135.0	180.0	205.2	−13
4-Tolunitrile	C_8H_7N	42.5	85.8	109.5	145.2	193.0	217.6	29.5
Phenylacetonitrile	C_8H_7N	60.0	103.5	136.3	161.8	208.5	233.5	−23.8
2-Tolyl isocyanide	C_8H_7N	25.2	64.0	94.0	117.7	159.9	183.5	—
2-Nitrophenyl acetate	$C_8H_7NO_4$	100.0	142.0	172.8	194.1	233.5	253.0d	—
2-Methylbenzothiazole	C_8H_7NS	70.0	111.2	141.2	163.9	204.5	225.5	15.4
Benzylisothiocyanate	C_8H_7NS	79.5	121.8	153.0	177.7	220.4	243.0	—
Styrene	C_8H_8	−7.0	+30.8	59.8	82.0p	122.5p	145.2p	−30.6
(1,2-Dibromoethyl) benzene	$C_8H_8Br_2$	86.0	129.8	161.8	186.3	230.0	254.0	—
1,2-Dichloro-3-ethylbenzene	$C_8H_8Cl_2$	46.0	90.0	123.8	149.8	197.0	222.1	−40.8
1,2-Dichloro-4-ethylbenzene	$C_8H_8Cl_2$	47.0	92.3	127.5	153.3	201.7	226.6	−76.4
1,4-Dichloro-2-ethylbenzene	$C_8H_8Cl_2$	38.5	83.2	118.0	144.0	191.3	216.3	−61.2
Acetophenone	C_8H_8O	37.1	78.0	109.4	133.6	178.0	202.4	20.5
Phenyl acetate	$C_8H_8O_2$	38.2	78.0	108.1	131.6	173.5	195.9	—
Phenylacetic acid	$C_8H_8O_2$	97.0	141.3	173.6	198.2	243.0	265.5	76.5
Anisaldehyde	$C_8H_8O_2$	73.2	117.8	150.5	176.7	223.0	248.0	2.5
Methyl benzoate	$C_8H_8O_2$	39.0	77.5	107.8	130.8	174.7	199.5	−12.5
Methyl salicylate	$C_8H_8O_3$	54.0	95.3	126.2	150.0	197.5	223.2	−8.3
Vanillin	$C_8H_8O_3$	107.0	154.0	188.7	214.5	260.0	285.0	81.5
Dehydroacetic acid	$C_8H_8O_4$	91.7	137.3	171.0	197.5	244.5	269.0	—
2-Bromo-1,4-xylene	C_8H_9Br	37.5	78.8	110.6	135.7	181.0	206.7	+9.5
1-Bromo-4-ethylbenzene	C_8H_9Br	30.4	74.0	108.5	135.5	182.0	206.0	−45.0
(2-Bromoethyl) benzene	C_8H_9Br	48.0	90.5	123.2	148.2	194.0	219.0	—
1-Chloro-2-ethylbenzene	C_9H_9Cl	17.2	56.1	86.2	110.0	152.2	177.6	−80.2
1-Chloro-3-ethylbenzene	C_8H_9Cl	18.6	58.2	89.2	113.6	156.7	181.1	−53.3
1-Chloro-4-ethylbenzene	C_8H_9Cl	19.2	60.0	91.8	116.0	159.8	184.3	−62.6
1-Chloro-2-ethoxybenzene	C_8H_9ClO	45.8	86.5	117.8	141.8	185.5	208.0	—
4-Chlorophenethyl alcohol	C_8H_9ClO	84.0	129.0	162.0	188.1	234.5	259.3	—
Acetanilide	C_8H_9NO	114.0	162.0	199.6	227.2	277.0	303.8	113.5
Methyl anthranilate	$C_8H_9NO_2$	77.6	124.2	159.7	187.8	238.5	266.5	24
4-Nitro-1,3-xylene	$C_9H_9NO_2$	65.6	109.8	143.3	168.5	217.5	244.0	+2
Ethylbenzene	C_8H_{10}	−9.8	+25.9	52.8	74.1	113.8	136.2	−94.0
2-Xylene	C_8H_{10}	−3.8	+32.1	59.5	81.3	121.7	144.4	−25.2
3-Xylene	C_8H_{10}	−6.9	+28.3	55.3	76.8	116.7	139.1	−47.9
4-Xylene	C_8H_{10}	−8.1	+27.3	54.4	75.9	115.9	138.3	+13.3
Dichloroethoxyphenylsilane	$C_8H_{10}Cl_2OSi$	52.4	94.6	126.2	151.4	197.2	222.2	—
Dichloroethylpenylsilane	$C_8H_{10}Cl_2Si$	48.5	92.4	126.7	153.3	203.5	230.0	—
2-Ethylphenol	$C_8H_{10}O$	46.2	87.0	117.9	141.8	184.5	207.5	−5
3-Ethylphenol	$C_8H_{10}O$	60.0	100.2	130.0	152.0	193.3	214.0	−4
4-Ethylphenol	$C_8H_{10}O$	59.3	100.2	131.3	154.2	197.4	219.0	46.5
2,3-Xylenol	$C_8H_{10}O$	56.0s	97.6	129.2	152.2	196.0	218.0	75
2,4-Xylenol	$C_8H_{10}O$	51.8	92.3	121.5	143.0	184.2	211.5	25.5
2,5-Xylenol	$C_8H_{10}O$	51.8s	91.3	121.5	143.0	184.2	211.5	74.5
3,4-Xylenol	$C_8H_{10}O$	66.2	107.7	138.0	161.0	203.6	225.2	62.5
3,5-Xylenol	$C_8H_{10}O$	62.0s	102.4	133.3	156.0	197.8	219.5	68
Phenetole	$C_8H_{10}O$	18.1	56.4	86.6	108.4	149.8	172.0	−30.2
α-Methyl benzyl alcohol	$C_8H_{10}O$	49.0	88.0	117.8	140.3	180.7	204.0	—
Phenethylacohol	$C_8H_{10}O$	58.2	100.0	130.5	154.0	197.5	219.5	—
4,6-Dimethylresorcinol	$C_8H_{10}O_2$	49.0	90.7	122.5	147.3	193.0	215.0	—
2-Phenoxyethanol	$C_8H_{10}O_2$	78.0	121.2	152.2	176.5	221.0	245.3	11.6
Diethyl dioxosuccinate	$C_8H_{10}O_6$	70.0	112.0	143.8	167.7	210.8	233.5	—
Chlorodimethylphenylsilane	$C_8H_{11}ClSi$	29.8	70.0	101.2	124.7	168.6	193.5	—
N-Ethylaniline	$C_8H_{11}N$	38.5	80.6	113.2	137.3	180.8	204.0	−63.5
N,N-Dimethylaniline	$C_8H_{11}N$	29.5	70.0	101.6	125.8	169.2	193.1	+2.5
4-Ethylaniline	$C_8H_{11}N$	52.0	93.8	125.7	149.8	194.2	217.4	−4
2,4-Xylidine	$C_8H_{11}N$	52.6	93.0	123.8	146.8	188.3	211.5	—
2,6-Xylidine	$C_8H_{11}N$	44.0	87.0	120.2	146.0	193.7	217.9	—
2-Phenetidine	$C_8H_{11}NO$	67.0	108.6	139.9	163.5	207.0	228.0	—
2-Anilinoethanol	$C_8H_{11}NO$	104.0	149.6	183.7	209.5	254.5	279.6	—
Dimethyl arsanilate	$C_8H_{12}AsNO_2$	15.0	51.8	79.7	101.0	140.3	160.5	—
Diethyleneglycol-bis-chloroacetate	$C_8H_{12}Cl_2O_5$	148.3	195.8	229.0	252.0	291.8	313.0	—
Diethyl maleate	$C_8H_{12}O_4$	57.3	100.0	131.8	156.0	201.7	225.0	—
Diethyl fumarate	$C_8H_{12}O_4$	53.2	95.3	126.7	151.1	195.8	218.5	+0.6

Name	Formula	Temperature °C						M.P.
		1 mm	10 mm	40 mm	100 mm	400 mm	760 mm	
Dimethylphenylsilane	$C_8H_{12}Si$	+ 5.3	42.6	71.4	94.2	136.4	159.3	—
Ethyl-α-ethylacetoacetate	$C_8H_{14}O_3$	40.5	80.2	110.3	133.8	175.6	198.0	—
Propyl levulinate	$C_8H_{14}O_3$	59.7	99.9	130.1	154.0	198.0	221.2	—
Isopropyl levulinate	$C_8H_{14}O_3$	48.0	88.0	118.1	141.8	185.2	208.2	—
Dipropyl oxalate	$C_8H_{14}O_4$	53.4	93.9	124.6	148.1	190.3	213.5	—
Diisopropyl oxalate	$C_8H_{14}O_4$	43.2	81.9	110.5	132.6	171.8	193.5	—
Diethyl succinate	$C_8H_{14}O_4$	54.6	96.6	127.8	151.1	193.8	216.5	−20.8
Diethyl isosuccinate	$C_8H_{14}O_4$	39.8	80.0	111.0	134.8	177.7	201.3	—
Suberic acid	$C_8H_{14}O_4$	172.8	219.5	254.6	279.0	322.8	345.5	142
Diethyl malate	$C_8H_{14}O_5$	80.7	125.6	157.8	183.9	229.5	253.4	—
Diethyl-dl-tartrate	$C_8H_{14}O_6$	100.0	147.2	181.7	208.0	254.3	280.0	—
Diethyl-d-tartrate	$C_8H_{14}O_6$	102.0	148.0	182.3	208.5	254.8	280.0	—
(2-Bromoethyl) cyclohexane	$C_8H_{15}Br$	38.7	80.5	113.0	138.0	186.2	213.0	—
Caprylonitrile	$C_8H_{15}N$	43.0	80.4	110.6	134.8	179.5	204.5	—
Ethyl N,N-diethyloxamate	$C_8H_{15}NO_3$	76.0	121.7	154.4	180.3	226.5	252.0	—
2-Methyl-2-heptene	C_8H_{16}	−16.1	+ 17.8	44.0	64.6	102.2	122.5	—
1,1-Dimethylcyclohexane	C_8H_{16}	−24.4	+ 10.3	37.3	57.9	97.2	119.5	−34
cis-1,2-Dimethylcyclohexane	C_8H_{16}	−15.9	+ 18.4	45.3	66.8	107.0	129.7	−50.0
trans-1,2-Dimethylcyclohexane	C_8H_{16}	−21.1	+ 13.0	39.7	61.0	100.9	123.4	−88.0
cis-1,3-Dimethylcyclohexane	C_8H_{16}	−22.7	+ 11.2	37.5	58.5	97.8	120.1	−76.2
trans-1,3-Dimethylcyclohexane	C_8H_{16}	−19.4	+ 14.9	41.4	62.5	102.1	124.4	−92.0
cis-1,4-Dimethylcyclohexane	C_8H_{16}	−20.0	+ 14.5	41.1	62.3	101.9	124.3	−87.4
trans-1,4-Dimethylcyclohexane	C_8H_{16}	−24.3	+ 10.1	36.5	57.6	97.0	119.3	−36.9
Ethylcyclohexane	C_8H_{16}	−14.5	+ 20.6	47.6	69.0	109.1	131.8	−111.3
Caprylaldehyde	$C_8H_{16}O$	73.4	101.2	120.0	133.9	156.5	168.5	—
Cyclohexaneethanol	$C_8H_{16}O$	50.4	90.0	119.8	142.7	183.5	205.4	—
6-Methyl-3-hepten-2-ol	$C_8H_{16}O$	41.6	76.7	102.7	122.6	156.6	175.5	—
6-Methyl-5-hepten-2-ol	$C_8H_{16}O$	41.9	77.8	104.1	123.8	156.6	174.3	—
2-Octanone	$C_8H_{16}O$	23.6	60.9	89.8	111.7	151.0	172.9	−16
2,2,4-Trimethyl-3-pentanone	$C_8H_{16}O$	14.7	46.4	69.8	87.6	118.4	135.0	—
Caprylic acid	$C_8H_{16}O_2$	92.3	124.0	150.6	172.2	213.9	237.5	16
Ethyl isocaproate	$C_8H_{16}O_2$	11.0	48.0	76.3	98.4	139.2	160.4	—
Propyl isovalerate	$C_8H_{16}O$	+ 8.0	45.1	72.8	95.0	135.0	155.9	—
Isobutyl butyrate	$C_8H_{16}O$	+ 4.6	42.2	71.7	94.0	135.7	156.9	—
Isobutyl isobutyrate	$C_8H_{16}O_2$	+ 4.1	39.9	67.2	88.0	126.3	147.5	−80.7
Amyl propionate	$C_8H_{16}O_2$	+ 8.5	46.3	75.5	97.6	138.4	160.2	—
Tetraethyleneglycol cholorhydrin	$C_8H_{16}ClO_4$	10.1	156.1	190.0	214.7	258.2	281.5	—
1-Iodooctane	$C_8H_{17}I$	45.8	90.0	123.3	150.0	199.3	225.5	−45.9
Ethyl-l-leucinate	$C_8H_{17}NO_2$	27.8	72.1	106.0	131.8	167.3	184.0	—
Octane	C_8H_{18}	−14.0	+ 19.2	45.1	65.7	104.0	125.0	−56.8
2-Methylheptane	C_8H_{18}	−21.0	+ 12.3	37.9	58.3	96.2	117.6	−109.5
3-Methylheptane	C_8H_{18}	−19.8	+ 13.3	38.9	59.4	97.4	118.9	−120.8
4-Methylheptane	C_8H_{18}	−20.4	+ 12.4	38.0	58.3	96.3	117.7	−121.1
2,2-Dimethylhexane	C_8H_{18}	−29.7	+ 3.1	28.2	48.2	85.6	106.8	—
2,3-Dimethylhexane	C_8H_{18}	−23.0	+ 9.9	35.6	56.0	94.1	115.6	—
2,4-Dimethylhexane	C_8H_{18}	−26.9	+ 5.2	30.5	50.6	88.2	109.4	—
2,5-Dimethylhexane	C_8H_{18}	−26.7	+ 5.3	30.4	50.5	97.9	109.1	−90.7
3,3-Dimethylhexane	C_8H_{18}	−25.8	+ 6.1	31.7	52.5	90.4	112.0	—
3,4-Dimethylhexane	C_8H_{18}	−22.1	+ 11.3	37.1	57.7	96.0	117.7	—
3-Ethylhexane	C_8H_{18}	−20.0	+ 12.8	38.5	58.9	97.0	118.5	—
2,2,3-Trimethylpentane	C_8H_{18}	−29.0	+ 3.9	29.5	49.9	88.2	109.8	−112.3
2,2,4-Trimethylpentane	C_8H_{18}	−36.5	−4.3	+ 20.7	40.7	78.0	99.2	−107.3
2,3,3-Trimethylpentane	C_8H_{18}	−25.8	+ 6.9	33.0	53.8	92.7	114.8	−101.5
2,3,4-Trimethylpentane	C_8H_{18}	−26.3	7.1	32.0	53.4	91.8	113.5	−109.2
2-Methyl-3-ethylpentane	C_8H_{18}	−24.0	+ 9.5	35.2	55.7	94.0	115.6	−114.5
3-Methyl-3-ethylpentane	C_8H_{18}	−23.9	+ 9.9	36.2	57.1	96.2	118.3	−90
2,2,3,3-Tetramethylbutane	C_8H_{18}	−17.4	+ 13.5	36.8	54.8	87.4	106.3	+ 100.7
Tetramethylpiperazine	$C_8H_{18}N_2$	23.7	61.7	90.0	113.8	157.8	183.5	—
1-Octanol	$C_8H_{18}O$	54.0	88.3	115.2	135.2	173.8	195.2	−15.4
2-Octanol	$C_8H_{18}O$	32.8	70.0	98.0	119.8	157.5	178.5	−38.6
1,2-Dipropoxy ethane	$C_8H_{18}O_2$	−38.8	+ 5.0	42.3	74.2	140.0	180.0	—
Diethylene glycol butyl ether	$C_8H_{18}O_3$	70.0	107.8	135.5	159.8	205.0	231.2	—
Tetraethylene glycol	$C_8H_{18}O_5$	153.9	197.1	228.0	250.0	228.0	307.8	—
Dibutyl sulfide	$C_8H_{18}S$	+ 21.7	66.4	96.0	118.6	159.0	182.0	−79.7
Dibutyl disulfide	$C_8H_{18}S_2$	+ 34.6	94.0	145.1	188.0	275.5	330.5	—
Diisobutylamine	$C_8H_{19}N$	−5.1	+ 30.6	57.8	79.2	118.0	139.5	−70
Tetraethoxysilane	$C_8H_{20}O_4Si$	16.0	52.6	81.4	103.6	146.2	168.5	—
Tetraethyllead	$C_8H_{20}Pb$	38.4	74.8	102.4	123.8	161.8	183.0	−136
Amyltrimethylsilane	$C_8H_{20}Si$	−9.2	+ 26.7	54.4	76.2	116.6	139.0	—
Tetraethylsilane	$C_8H_{20}Si$	−1.0	+ 36.3	65.3	88.0	130.2	153.0	—
Tetraethyl bistibine	$C_8H_{20}Sb_2$	97.0	151.2	193.2	225.6	286.2	320.3	—
1,3-Diethoxytetramethyldisiloxane	$C_8H_{22}O_3Si_2$	14.8	51.2	78.7	100.3	139.8	160.7	—
1,7-Dichlorooctamethyltetrasiloxane	$C_8H_{24}Cl_2O_3Si_4$	53.3	95.8	127.8	152.7	197.8	222.0	−62
Octamethyltrisiloxane	$C_8H_{24}O_2Si_2$	7.4	43.1	70.0	91.1	129.4	150.2	—
Octamethylcyclotetrasiloxane	$C_8H_{24}O_4Si_4$	21.7	59.0	87.4	110.0	149.6	171.2	17.4
Coumarin	$C_9H_6O_2$	106.0	153.4	189.0	216.5	264.7	291.0	70
Quinoline	C_9H_7N	59.7	103.8	136.7	163.2	212.3	237.7	−15
Isoquinoline	C_9H_7N	63.5	107.8	141.6	167.6	214.5	240.5	24.6
Indene	C_9H_3	16.4	58.5	90.7	114.7	157.8	181.6	−2
Cinnamylaldehyde	C_9H_8O	76.1	120.0	152.2	177.7	222.4	246.0	−7.5
trans-Cinnamic acid	$C_9H_8O_2$	127.5s	173.0	207.1	232.4	276.7	300.0	133

ORGANIC COMPOUNDS
Pressures Less than One Atmosphere (Continued)

Name	Formula	Temperature °C						M.P.
		1 mm	10 mm	40 mm	100 mm	400 mm	760 mm	
Skatole	C₉H₉N	95.0	139.6	171.9	197.4	242.5	266.2	95
Ethyl 3-nitrobenzoate	C₉H₉NO₄	108.1	155.0	192.6	220.3	270.6	298.0	47
α-Methyl styrene	C₉H₁₀	7.4	47.1	77.8	102.2	143.0	165.4p	−23.2
β-Methyl styrene	C₉H₁₀	17.5	57.0	87.7	111.7	154.7	179.0	−30.1
4-Methyl styrene	C₉H₁₀	16.0	55.1	85.0	108.6	151.2	175.0p	—
Propenylbenzene	C₉H₁₀	17.5	57.0	87.7	111.7	154.7	179.0	−30.1
2,4-Xylaldehyde	C₉H₁₀O	59.9s	99.0	129.7	152.2	194.1	215.5	75
Cinnamyl alcohol	C₉H₁₀O	72.6	117.8	151.0	177.8	224.6	250.0	33
Propiophenone	C₉H₁₀O	50.0	92.2	124.3	149.3	194.2	218.0	21
2-Vinylanisole	C₉H₁₀O	41.9	81.0	110.0	132.3	172.1	194.0p	—
3-Vinylanisole	C₉H₁₀O	43.4	83.0	112.5	135.3	175.8	197.5p	—
4-Vinylanisole	C₉H₁₀O	45.2	85.7	116.0	139.7	182.0	204.5p	—
Benzyl acetate	C₉H₁₀O₂	45.0	87.6	119.6	144.0	189.0	213.5	−51.5
Ethyl Benzoate	C₉H₁₀O₂	44.0	86.0	118.2	143.2	188.4	213.4	−34.6
Hydrocinnamic acid	C₉H₁₀O₂	102.2	148.7	183.3	209.0	255.0	279.8	48.5
Ethyl salicylate	C₉H₁₀O₃	61.2	104.2	136.7	161.5	207.0	231.5	1.3
N-Methylacetanilide	C₉H₁₁NO	s	118.6	152.2	179.8	227.4	253.0	102
Ethyl carbanilate	C₉H₁₁NO₂	107.8	143.7	168.8	187.9	220.0	237.0	52.5
1,2,3-Trimethylbenzene	C₉H₁₂	16.8	55.9	85.4	108.8	152.0	176.1	−25.5
1,2,4-Trimethylbenzene	C₉H₁₂	13.6	50.7	79.8	102.8	145.4	169.2	−44.1
1,3,5-Trimethylbenzene	C₉H₁₂	9.6	47.4	76.1	98.9	141.0	164.7	−44.8
2-Ethyltoluene	C₉H₁₂	9.4	47.6	76.4	99.0	141.4	165.1	—
3-Ethyltoluene	C₉H₁₂	7.2	44.7	73.3	95.9	137.8	161.3	−95.5
4-Ethyltoluene	C₉H₁₂	7.6	44.9	73.6	96.3	136.4	162.0	—
Cumene	C₉H₁₂	2.9	38.3	66.1	88.1	129.2	152.4	−96.0
Propylbenzene	C₉H₁₂	6.3	43.4	71.6	94.0	135.7	159.2	−99.5
2-Ethylanisole	C₉H₁₂O	29.7	69.0	98.8	122.3	164.2	187.1	—
3-Ethylanisole	C₉H₁₂O	33.7	73.9	104.8	129.2	172.8	196.5	—
4-Ethylanisole	C₉H₁₂O	33.5	73.9	104.7	128.4	172.3	196.5	—
3-Phenyl-1-propanol	C₉H₁₂O	74.7	116.0	147.4	170.3	212.8	235.0	—
2-Isopropylphenol	C₉H₁₂O	56.6	97.0	127.5	150.3	192.6	214.5	15.5
3-Isopropylphenol	C₉H₁₂O	62.0	104.1	136.2	160.2	205.0	228.0	26
4-Isopropylphenol	C₉H₁₂O	67.0	108.0	139.8	163.3	206.1	228.2	61
Benzyl ethyl ether	C₉H₁₂O	26.0	65.0	95.4	118.9	161.5	185.0	—
Chloroethoxymethylphenylsilane	C₉H₁₃ClOSi	44.8	84.6	117.8	142.6	187.7	212.0	—
2,4,5-Trimethylaniline	C₉H₁₃N	68.4	109.0	139.8	162.0	203.7	234.5	67
N,N-Dimethyl-2-toluidine	C₉H₁₃N	28.8	66.2	95.0	118.1	161.5	184.8	−61
N,N-Dimethyl-4-toluidine	C₉H₁₃N	50.1	86.7	116.3	140.3	185.4	209.5	—
4-Cumidine	C₉H₁₃N	60.0	102.2	134.2	158.0	203.2	227.0	—
Phorone	C₉H₁₄O	42.0	81.5	111.3	134.0	175.3	197.2	28
Isophorone	C₉H₁₄O	38.0	81.2	114.5	140.6	188.7	215.2	—
cis-Diethyl citraconate	C₉H₁₄O₄	59.8	103.0	135.7	160.0	206.5	230.3	—
Diethyl itaconate	C₉H₁₄O₄	51.3	95.2	128.2	154.3	203.1	227.9	—
Diethyl mesaconate	C₉H₁₄O₄	62.8	105.3	137.3	161.6	205.8	229.0	—
Trimethyl citrate	C₉H₁₄O₇	106.2	160.4	194.2	219.6	264.2	287.0d	78.5
Isobutyl levulinate	C₉H₁₆O₃	65.0	105.9	136.2	160.2	205.5	229.9	—
Azelaic acid	C₉H₁₆O₄	178.3	225.5	260.0	286.5	332.8	356.5d	106.5
Diethyl ethylmalonate	C₉H₁₆O₄	50.8	91.6	122.4	146.0	188.7	211.5	—
Diethyl glutarate	C₉H₁₆O₄	65.6	109.7	142.8	167.8	212.8	237.0	—
2-Nonanone	C₉H₁₈O	32.1	72.3	103.4	127.4	171.2	195.0	−19
Azelaldehyde	C₉H₁₈O	33.3	71.6	100.2	123.0	163.4	185.0	—
Pelargonic acid	C₉H₁₈O₂	108.2	137.4	163.7	184.4	227.5	253.5	12.5
Methyl caprylate	C₉H₁₈O₂	34.2	74.9	105.3	128.0	170.0	193.0d	−40
Isobutyl isovalerate	C₉H₁₈O₂	16.0	53.8	82.7	105.2	146.4	168.7	—
Isoamyl butyrate	C₉H₁₈O₂	21.2	59.9	90.0	113.1	155.3	178.6	—
Isoamyl isobutyrate	C₉H₁₈O₂	14.8	52.8	81.8	104.4	146.0	168.8	—
Iodononane	C₉H₁₉I	70.0	109.0	138.1	159.8	199.3	219.5	—
Nonane	C₉H₂₀	+1.4	38.0	66.0	88.1	128.2	150.8	−53.7
1-Nonanol	C₉H₂₀O	59.5	99.7	129.0	151.3	192.1	213.5	−5
Dipropyleneglycol, isopropyl ether	C₉H₂₀O₃	46.0	86.2	117.0	140.3	183.1	205.6	—
Tripropyleneglycol	C₉H₂₀O₄	96.0	140.5	173.7	199.0	244.3	267.2	—
Hexyltrimethylsilane	C₉H₂₂Si	+6.7	44.8	74.0	97.2	139.9	163.0	—
Triethylpropylsilane	C₉H₂₂Si	15.2	54.0	83.7	107.4	149.8	173.0	—
1-Bromonaphthalene	C₁₀H₇Br	84.2	133.6	170.2	198.8	252.0	281.1	5.5
1-Chloronaphthalene	C₁₀H₇Cl	80.6	118.6	153.2	180.4	230.8	259.3	−20
Dicyclopentadiene	C₁₀H₁₂	s	47.6	77.9	101.7	144.2	166.6d	32.9
Naphthalene	C₁₀H₈	52.6s	85.8	119.3	145.5	193.2	217.9	80.2
Dichloro-1-naphthylsilane	C₁₀H₈Cl₂Si	106.2	149.2	181.7	205.9	249.7	273.3	—
1-Naphthol	C₁₀H₈O	94.0s	142.0	177.8	206.0	255.8	288.0	96
2-Naphthol	C₁₀H₈O	s	145.5	181.7	209.8	260.6	300.8	122.5
1-Naphthylamine	C₁₀H₉N	104.3	153.8	191.5	220.0	272.2	306.1	50
2-Naphthylamine	C₁₀H₉N	108.0	157.6	195.7	224.3	277.4	246.5	111.5
2-Methylquinoline	C₁₀H₉N	75.3	119.0	150.8	176.2	211.7	199.5p	−1
1,3-Divinylbenzene	C₁₀H₁₀	32.7	73.8	105.5	130.3	175.2	261.0	−66.9
4-Phenyl-3-buten-2-one	C₁₀H₁₀O	81.7	127.4	161.3	187.8	235.4	288.0	41.5
α-Methylcinnamic acid	C₁₀H₁₀O₂	125.7	169.8	201.8	224.8	266.8	263.0	—
Methyl cinnamate	C₁₀H₁₀O₂	77.4	123.0	157.9	185.8	235.0	263.0	33.4
Safrole	C₁₀H₁₀O₂	63.8	107.6	140.1	165.1	210.0	233.0	11.2
1,2-Phenylene diacetate	C₁₀H₁₀O₄	98.0	145.7	179.8	206.5	253.3	278.0	—
Dimethyl phthalate	C₁₀H₁₀O₄	100.3	147.6	182.8	210.0	257.8	283.7	—
2,4-Dimethylstyrene	C₁₀H₁₂	34.2	75.8	107.7	132.3	177.5	202.0p	—

Name	Formula	Temperature °C						M.P.
		1 mm	10 mm	40 mm	100 mm	400 mm	760 mm	
2,5-Dimethylstyrene	C₁₀H₁₂	29.0	69.0	100.2	124.7	168.7	193.0p	—
3-Ethylstyrene	C₁₀H₁₂	28.3	68.3	99.2	123.2	167.2	191.5p	—
4-Ethylstyrene	C₁₀H₁₂	26.0	66.3	97.3	121.5	165.0	189.0p	—
Tetralin	C₁₀H₁₂	38.0	79.0	110.4	135.3	181.8	207.2	−31.0
Anethole	C₁₀H₁₂O	62.6	106.0	139.3	164.2	210.5	235.3	22.5
4-Methylpropiophenone	C₁₀H₁₂O	59.6	103.8	138.0	164.	212.7	238.5	—
Estragole	C₁₀H₁₂O	52.6	93.7	124.6	148.5	192.0	215.0	—
Cuminal	C₁₀H₁₂O	58.0	102.0	135.2	160.0	206.7	232.0	—
4-Vinylphenetole	C₁₀H₁₂O	64.0	105.6	136.3	159.8	202.8	225.0p	—
Eugenol	C₁₀H₁₂O₂	78.4	123.0	155.8	182.2	228.3	253.5	—
Isoeugenol	C₁₀H₁₂O₂	86.3	132.4	167.0	194.0	242.3	267.5	−10
Chavibetol	C₁₀H₁₂O₂	83.6	127.0	159.8	185.5	229.8	254.0	—
Propyl benzoate	C₁₀H₁₂O₂	54.6	98.0	131.8	157.4	205.2	231.0	−51.6
2-Phenoxyethyl acetate	C₁₀H₁₂O₃	82.6	128.0	162.3	189.2	235.0	259.7	−6.7
2-Chloroethyl α-methylbenzyl ether	C₁₀H₁₃ClO	62.3	106.0	139.6	164.8	210.8	235.0	—
4-tert-Butylphenyl dichlorophosphate	C₁₀H₁₃Cl₂O₂P	96.0	146.0	184.3	214.3	268.2	299.0	—
1,2,3,4-Tetramethylbenzene	C₁₀H₁₄	42.6	81.8	111.5	135.7	180.0	204.4	−6.2
1,2,3,5-Tetramethylbenzene	C₁₀H₁₄	40.6	77.8	105.8	128.3	173.7	197.9	−24.0
1,2,4,5-Tetramethylbenzene	C₁₀H₁₄	45.0s	74.6s	104.2	128.1	172.1	195.9	79.5
4-Ethyl-1,3-xylene	C₁₀H₁₄	26.3	66.4	97.2	121.2	164.4	188.4	—
5-Ethyl-1,3-xylene	C₁₀H₁₄	22.1	62.1	92.6	116.5	159.6	183.7	—
2-Ethyl-1,4-xylene	C₁₀H₁₄	25.7	65.6	96.0	120.0	163.1	186.9	—
1,2-Diethylbenzene	C₁₀H₁₄	22.3	62.0	92.5	116.2	159.0	183.5	−31.4
1,3-Diethylbenzene	C₁₀H₁₄	20.7	59.9	90.4	114.4	156.9	181.1	−83.9
1,4-Diethylbenzene	C₁₀H₁₄	20.7	60.3	91.1	115.3	159.0	183.8	−43.2
Cymene	C₁₀H₁₄	17.3	57.0	87.0	110.8	153.5	177.2	−68.2
Butylbenzene	C₁₀H₁₄	22.7	62.0	92.4	116.2	159.2	183.1	−88.0
Isobutylbenzene	C₁₀H₁₄	14.1	53.7	83.3	107.0	149.6	172.8	−51.5
sec-Butylbenzene	C₁₀H₁₄	18.6	57.0	86.2	109.5	150.3	173.5	−75.5
tert-Butylbenzene	C₁₀H₁₄	13.0	51.7	80.8	103.8	145.8	168.5	−58
Carvacrol	C₁₀H₁₄O	70.0	113.2	145.2	109.7	213.8	237.0	+0.5
Carbone	C₁₀H₁₄O	57.4	100.4	133.0	157.3	203.5	227.5	—
Cuminyl alcohol	C₁₀H₁₄O	74.2	118.0	150.3	176.2	221.7	246.6	—
4-Ethylphenetole	C₁₀H₁₄O	48.5	89.5	119.8	143.5	185.7	208.0	—
Thymol	C₁₀H₁₄O	64.3	107.4	139.8	164.1	209.2	231.8	51.5
4-Isobutylphenol	C₁₀H₁₄O	72.1	115.5	147.2	171.2	214.7	237.0	—
4-sec-Butylphenol	C₁₀H₁₄O	71.4	114.8	147.8	172.4	217.6	242.1	—
2-sec-Butylphenol	C₁₀H₁₄O	57.4	100.8	133.4	157.3	203.8	228.0	—
2-tert-Butylphenol	C₁₀H₁₄O	56.6	98.1	129.2	153.3	196.3	219.5	—
4-tert-Butylphenol	C₁₀H₁₄O	70.0	114.0	146.0	170.2	214.0	238.0	99
Nicotine	C₁₀H₁₄N₂	61.8	107.2	142.1	169.5	219.8	247.3	—
N-Diethylaniline	C₁₀N₁₅N	49.7	91.9	123.6	147.3	192.4	215.5	−34.4
N-Phenyliminodiethano	C₁₀H₁₅NO₂	145.0	195.8	233.0	260.6	311.3	337.8	—
Camphene	C₁₀H₁₆	s	47.2s	75.7	97.9	138.7	160.5	50
Dipentene	C₁₀H₁₆	14.0	53.8	84.3	108.3	150.5	174.6	—
d-Limonene	C₁₀H₁₆	14.0	53.8	84.3	108.3	151.4	175.0	−96.9
Myrcene	C₁₀H₁₆	14.5	53.2	82.6	106.0	148.3	171.5	—
α-Phellandrene	C₁₀H₁₆	20.0	58.0	87.8	110.6	152.0	175.0	—
α-Pinene	C₁₀H₁₆	−1.0	+37.3	66.8	90.1	132.3	155.0	−55
β-Pinene	C₁₀H₁₆	+4.2	42.3	71.5	94.0	136.1	158.3	—
Terpenoline	C₁₀H₁₆	32.3	70.6	100.0	122.7	163.5	185.0	—
Diethyl arsanilate	C₁₀H₁₆AsNO₂	38.0	74.8	102.6	123.8	161.0	181.0	—
d-Camphor	C₁₀H₁₆O	41.5s	82.3s	114.0s	138.0s	182.0	209.2	178.5
l-Dihydrocarvone	C₁₀H₁₆O	46.6	90.0	123.7	149.7	197.0	223.0	—
α-Citral	C₁₀H₁₆O	61.7	103.9	135.9	160.0	205.0	228.0d	—
d-Fenchone	C₁₀H₁₆O	28.0	68.3	99.5	123.6	166.8	191.0	5
Pulegone	C₁₀H₁₆O	58.3	94.0	121.7	143.1	189.8	221.0	—
α-Thyjone	C₁₀H₁₆O	38.3	79.3	110.0	134.0	177.8	201	—
Ethoxydimethylphenylsilane	C₁₀H₁₆OSi	36.3	76.2	107.2	131.4	175.0	199.5	—
Campholenic acid	C₁₀H₁₆O₂	97.6	139.8	170.0	193.7	234.0	256.0	—
Diosphenol	C₁₀H₁₆O₂	66.7	109.0	141.2	165.6	209.5	232.0	—
Fencholic acid	C₁₀H₁₆O₂	101.7	142.3	171.8	194.0	237.8	264.1	19
cis-Decalin	C₁₀H₁₈	22.5	64.2	97.2	123.2	169.9	194.6	−43.3
trans-Decalin	C₁₀H₁₈	−0.8	+47.2	85.7	114.6	160.1	186.7	−30.7
d-Citronellal	C₁₀H₁₈O	44.0	84.8	116.1	140.1	183.8	206.5	—
Cineol	C₁₀H₁₈O	15.0	54.1	84.2	108.2	151.6	176.0	−1
Dihydrocarveol	C₁₀H₁₈O	63.9	105.0	136.1	159.8	202.8	225.0	—
dl-Fenchyl alcohol	C₁₀H₁₈O	45.8	82.1	110.8	132.3	173.2	202.0	35
Geraniol	C₁₀H₁₈O	69.2	110.0	141.8	165.3	207.8	230.0	—
d-Linalool	C₁₀H₁₈O	40.0	79.8	109.9	133.8	175.6	198.0	—
Nerol	C₁₀H₁₈O	61.7	104.0	136.1	159.8	203.5	226.0	—
α-Terpineol	C₁₀H₁₈O	52.8	94.3	126.0	150.1	194.3	217.5	35
Citronellic acid	C₁₀H₁₈O₂	99.5	141.4	171.9	195.4	236.6	257.0	—
Amyl levulinate	C₁₀H₁₈O₃	81.3	124.0	155.8	180.5	227.4	253.2	—
Isoamyl levulinate	C₁₀H₁₈O₃	75.6	118.8	151.7	177.0	222.7	247.9	—
Diethyl ethylmethylmalonate	C₁₀H₁₈O₄	44.7	85.7	116.7	140.8	184.1	207.5	—
Diethyl adipate	C₁₀H₁₈O₄	74.0	123.0	154.6	179.0	219.1	240.0	—
Diisobutyl oxalate	C₁₀H₁₈O₄	63.2	105.3	137.5	161.8	205.8	229.5	−21
Dipropyl succinate	C₁₀H₁₈O₄	77.5	122.2	154.8	180.3	226.5	250.8	—
Sebacic acid	C₁₀H₁₈O₄	183.0	232.0	268.2	294.5	332.8	352.3d	—
Dipropyl-d-tartrate	C₁₀H₁₈O₆	115.6	163.5	199.7	227.0	275.6	303.0	134.5

Name	Formula	Temperature °C						M.P.
		1 mm	10 mm	40 mm	100 mm	400 mm	760 mm	
Diisopropyl-*d*-Tartrate	$C_{10}H_{18}O_6$	103.7	148.2	181.8	207.3	251.8	275.0	—
Camphylamine	$C_{10}H_{19}N$	45.3	83.7	112.5	134.6	173.8	195.0	—
Menthane	$C_{10}H_{20}$	+9.7	48.3	78.3	102.1	146.0	169.0	—
1-Decene	$C_{10}H_{20}$	14.7	53.7	83.3	106.5	149.2	172.0	—
1,2-Dibromodecane	$C_{10}H_{20}Br_2$	95.7	137.3	167.4	190.2	229.8	250.4	—
Citronellol	$C_{10}H_{20}O$	66.4	107.0	137.2	159.8	201.0	221.5	—
Capraldehyde	$C_{10}H_{20}O$	51.9	92.0	122.2	145.3	186.3	208.5	—
l-Menthol	$C_{10}H_{20}O$	56.0	96.0	126.1	149.4	190.2	212.0	42.5
Decan-2-one	$C_{10}H_{20}O$	44.2	85.8	117.1	142.4	186.7	211.0	+3.5
Capric acid	$C_{10}H_{20}O_2$	125.0	152.2	179.9	200.0	240.3	208.4	31.5
Isoamyl isovalerate	$C_{10}H_{20}O_2$	27.0	68.6	100.6	125.1	169.5	194.0	—
Decane	$C_{10}H_{22}$	16.5	55.7	85.5	108.6	150.6	174.1	−29.7
2,7-Dimethyloctane	$C_{10}H_{22}$	+6.3	42.3	71.2	93.9	136.0	159.7	−52.8
Decyl alcohol	$C_{10}H_{22}O$	69.5	111.3	142.1	105.8	208.8	231.0	+7
Diisoamyl ether	$C_{10}H_{22}O$	18.6	57.0	86.3	109.6	150.3	173.4	—
2-Butyl-2-ethylbutane-1,3-diol	$C_{10}H_{22}O_2$	94.1	136.8	167.8	191.9	233.5	255.0	—
Dihydrocitronellol	$C_{10}H_{22}O$	68.0	103.0	127.6	145.9	176.8	193.5	—
Dipropylene glycol monobutyl ether	$C_{10}H_{22}O_3$	64.7	106.0	136.3	159.8	203.8	227.0	—
Diisoamyl sulfide	$C_{10}H_{22}S$	43.0	87.6	120.0	145.3	191.0	216.0	—
Heptyltrimethylsilane	$C_{10}H_{24}Si$	22.3	62.1	92.4	116.5	159.8	184.0	—
Butyltriethylsilane	$C_{10}H_{24}Si$	27.1	67.5	98.3	123.2	167.5	192.0	—
1,5-Diethoxyhexamethyltrisiloxane	$C_{10}H_{28}O_4Si_3$	41.8	80.7	110.0	133.2	174.0	196.6	—
Decamethyltetrasiloxane	$C_{10}H_{30}O_3Si_4$	35.3	74.3	104.0	127.3	169.8	193.5	—
Decamethylcyclopentasiloxane	$C_{10}H_{30}O_5Si_5$	45.2	86.2	117.7	142.0	186.0	210.0	−38.0
1-Naphthoic acid	$C_{11}H_8O_2$	156.0	196.8	225.0	245.8	281.4	300.0	160.5
2-Naphthoic acid	$C_{11}H_8O_2$	160.8	202.8	231.5	252.7	289.5	308.5	184
Ethyl-*trans*-cinnamate	$C_{11}H_{12}O_2$	87.6	123.0	169.1	196.0	245.0	271.0	12
1-Phenyl-1,3-pentanedione	$C_{11}H_{12}O_2$	98.0	144.0	178.0	204.5	251.2	276.5	—
Ethyl benzoylacetate	$C_{11}H_{12}O_3$	107.6	150.3	181.8	205.0	244.7	265.0d	—
Myristicine	$C_{11}H_{12}O_3$	95.2	142.0	177.7	206.0	253.5	200.0	—
2,4,5-Trimethylstyrene	$C_{11}H_{14}$	48.1	91.6	124.2	149.8	196.1	221.2p	
2,4,6-Trimethylstyrene	$C_{11}H_{14}$	37.5	79.7	111.8	136.8	182.3	207.0p	
4-Isopropylstyrene	$C_{11}H_{14}$	34.7	76.0	108.0	132.8	178.0	202.5p	
Isovalerophenone	$C_{11}H_{14}O$	58.3	101.4	133.0	150.0	204.7	228.0	—
Pivalophenone	$C_{11}H_{14}O$	57.8	99.0	13.4	154.0	197.7	220.0	—
2,3,5-Trimethylacetophenone	$C_{11}H_{14}O$	79.0	122.3	154.2	179.7	224.3	247.5	—
Isobutyl benzoate	$C_{11}H_{14}O_2$	64.0	108.6	141.8	166.4	212.8	237.0	—
4-Allylveratrole	$C_{11}H_{14}O_2$	85.0	127.0	158.3	183.7	226.2	248.0	—
3,5-Diethyltoluene	$C_{11}H_{16}$	34.0	75.3	107.0	131.7	176.5	200.7	—
1,2,4-Trimethyl-5-ethylbenzene	$C_{11}H_{16}$	43.7	84.6	106.0	140.3	184.5	208.1	—
1,3,5-Trimethyl-2-ethylbenzene	$C_{11}H_{16}$	38.8	80.5	113.2	137.9	183.5	208.0	—
3-Ethylcumene	$C_{11}H_{16}$	28.3	68.8	99.9	124.3	168.2	193.0	—
4-Ethylcumene	$C_{11}H_{16}$	31.5	72.0	103.3	127.2	171.8	195.8	—
sec-Amylbenzene	$C_{11}H_{16}$	29.0	69.2	100.0	124.1	168.0	193.0	—
4-*tert*-Butyl-2-cresol	$C_{11}H_{16}O$	74.3	118.0	150.8	176.2	221.8	247.0	—
2-*tert*-Butyl-4-cresol	$C_{11}H_{16}O$	70.0	112.0	143.9	167.0	210.0	232.6	—
4-*tert*-Amylphenol	$C_{11}H_{16}O$	s	125.5	160.3	189.0	239.5	266.0	93
Ethylcamphoronic anhydride	$C_{11}H_{16}O_5$	118.2	165.0	199.8	226.6	272.8	298.0	—
Bornyl formate	$C_{11}H_{18}O_2$	47.0	89.3	121.2	145.8	190.2	214.0	—
Geranyl formate	$C_{11}H_{18}O_2$	61.8	104.3	136.2	160.7	205.8	230.0	—
Neryl formate	$C_{11}H_{18}O_2$	57.3	99.7	131.5	155.6	200.0	224.5	—
Diethoxymethylphenylsilane	$C_{11}H_{18}O_2Si$	56.5	97.2	127.5	151.2	193.8	216.5	—
Diethyl-γ-oxoazelate	$C_{11}H_{18}O_5$	121.0	165.7	197.7	221.6	264.5	286.0	—
10-Hendecenoic acid	$C_{11}H_{20}O_2$	114.0	156.3	188.7	213.5	254.0	275.0	24.5
Menthyl formate	$C_{11}H_{20}O_2$	47.3	90.0	123.0	148.0	194.2	219.0	—
2-Ethylhexyl acrylate	$C_{11}H_{20}O_2$	50.0	91.8	123.7	147.9	192.2	216.0	—
Octyl acrylate	$C_{11}H_{20}O_2$	58.5	102.0	135.6	159.1	204.0	227.0	—
Hexyl levulinate	$C_{11}H_{20}O_2$	90.0	134.7	107.8	193.6	241.0	266.8	—
Hendecan-2-one	$C_{11}H_{22}O$	68.2	108.9	139.0	161.0	202.3	224.0	15
Methyl caprate	$C_{11}H_{22}O_2$	63.7	108.0	139.0	161.5	202.9	224.0d	−18
Hendecanoic acid	$C_{11}H_{22}O_2$	101.4	149.0	185.6	212.5	262.8	290.0	29.5
Hendecane	$C_{11}H_{24}$	32.7	73.9	104.4	128.1	171.9	195.8	−25.6
Nendecan-2-ol	$C_{11}H_{24}O$	71.1	112.8	143.7	167.2	209.8	232.0	—
Trimethyloctylsilane	$C_{11}H_{26}Si$	41.8	82.3	113.0	136.5	179.5	202.0	—
Amyltriethylsilane	$C_{11}H_{26}Si$	41.8	83.8	110.0	141.2	186.3	211.0	—
1-Bromobiphenyl	$C_{12}H_9Br$	98.0	150.6	190.8	221.8	277.7	310.0	90.5
2-Bromo-4-Phenylphenol	$C_{12}H_9BrO$	100.0	152.3	193.8	224.5	280.2	311.0	95
2-Chlorobiphenyl	$C_{12}H_9Cl$	89.3	134.7	109.9	197.0	243.8	267.5	34
4-Chlorobiphenyl	$C_{12}H_9Cl$	96.4	146.0	183.8	212.5	204.5	292.5	75.5
2-Chloro-3-phenylphenol	$C_{12}H_9ClO$	118.0	169.7	207.4	237.0	289.4	317.5	+6
2-Chloro-6-phenylphenol	$C_{12}H_9ClO$	119.8	170.7	208.2	237.1	289.5	317.0	—
2-Xenyl dichlorophosphate	$C_{12}H_9Cl_2PO$	138.2	187.0	223.8	251.5	301.5	328.5	—
Carbazole	$C_{12}H_9N$	s	s	s	265.0	323.0	354.8	244.8
Acenaphthene	$C_{12}H_{10}$	s	131.2	168.2	197.5	250.0	277.5	95
Biphenyl	$C_{12}H_{10}$	70.6	117.0	152.5	180.7	229.4	254.9	69.5
Diphenyl chlorophosphate	$C_{12}H_{10}ClPO_3$	121.5	182.0	227.9	265.0	337.2	278.0	—
Dichlorodiphenylsilane	$C_{12}H_{10}Cl_2Si$	109.6	158.0	195.5	223.8	275.5	304.0	—
Difluourodiphenylsilane	$C_{12}H_{10}F_2Si$	68.4	115.5	149.8	176.3	225.4	252.5	—
Azobenzene	$C_{12}H_{10}N_2$	103.5	151.5	187.9	216.0	266.1	293.0	68
1-Acetonaphthone	$C_{12}H_{10}O$	115.6	161.5	196.8	223.8	270.5	295.5	—
Diphenyl ether	$C_{12}H_{10}O$	120.2	168.5	203.8	229.8	275.8	301.0	55.5

Name	Formula	Temperature °C						M.P.
		1 mm	10 mm	40 mm	100 mm	400 mm	760 mm	
2-Phenylphenol	$C_{12}H_{10}O$	66.1	114.0	150.0	178.8	230.7	258.5	27
4-Phenylphenol	$C_{12}H_{10}O$	100.0	146.2	180.3	205.9	251.8	275.0	56.5
Diphenyl sulfide	$C_{12}H_{10}S$	s	176.2	213.0	240.9	285.5	308.0	164.5
Diphenyl disulfide	$C_{12}H_{10}S$	96.1	145.0	182.8	211.8	263.9	292.5	—
Diphenyl selenide	$C_{12}H_{10}Se$	131.6	180.0	214.8	241.3	285.8	310.0	61
Diphenylamine	$C_{12}H_{11}N$	105.7	154.4	192.2	220.8	273.2	301.5	+ 2.5
1-Ethylnaphthalene	$C_{12}H_{12}$	108.3	'57.0	194.3	222.8	274.1	302.0	52.9
1,1-Diphenylhydrazine	$C_{12}H_{12}$	70.0	₁16.8	152.0	180.0	230.8	258.1d	−27
1,1-Diphenylhydrazine	$C_1;2H_{12}N_2$	126.0	176.1	213.5	242.5	294.0	322.2	44
2-Cyclohexyl-4,6-dinitrophenol	$C_{12}H_{14}N_2O_5$	132.8	175.9	206.7	229.0	269.8	291.5	—
Eugenyl acetate	$C_{12}H_{14}O_3$	101.6	148.0	183.0	209.7	257.4	282.0	29.5
Apiole	$C_{12}H_{14}O_4$	116.0	160.2	193.7	218.0	262.1	285.0	30
Diethyl phthalate	$C_{12}H_{14}O_4$	108.8	156.0	192.1	219.5	267.5	294.0	—
2,5-Diethylstyrene	$C_{12}H_{16}$	49.7	92.6	125.8	151.0	198.0	223.0p	—
Phenylcyclohexane	$C_{12}H_{16}$	67.5	111.3	144.0	169.3	214.6	240.0	+ 7.5
Isoamyl benzoate	$C_{12}H_{16}O_2$	72.0	121.6	158.3	186.8	235.8	262.0	—
1,2,4-Triethylbenzene	$C_{12}H_{18}$	46.0	88.5	121.7	146.8	193.7	218.0	—
1,4-Diisopropylbenzene	$C_{12}H_{18}$	40.0	81.8	114.0	138.7	184.3	209.0	—
1,3-Diisopropylbenzene	$C_{12}H_{18}$	34.7	76.0	107.9	132.3	177.6	202.0	−105
2-tert-Butyl-4-ethylphenol	$C_{12}H_{18}O$	76.3	121.0	154.0	179.0	223.8	247.8	—
4-tert-Butyl-2,5-xylenol	$C_{12}H_{18}O$	88.2	135.0	169.8	195.0	241.8	265.3	—
4-tert-Butyl-2,6-xylenol	$C_{12}H_{18}O$	74.0	119.0	152.2	176.0	217.8	239.8	—
6-tert-Butyl-2,4-xylenol	$C_{12}H_{18}O$	70.3	115.0	148.5	172.0	214.2	236.5	—
6-tert-Butyl-3,4-xylenol	$C_{12}H_{13}O$	83.9	127.0	159.7	184.0	226.7	249.5	—
d-Bornyl acetate	$C_{12}H_{20}O_2$	46.9	90.2	123.7	149.8	197.5	223.0	29.
Geranyl acetate	$C_{12}H_{20}O_2$	73.5	117.9	150.0	175.2	219.8	243.3d	—
Linalyl acetate	$C_{12}H_{20}O_2$	55.4	96.0	127.7	151.8	196.2	220.0d	—
Triethoxyphenylsilane	$C_{12}H_{20}O_3Si$	71.0	112.6	143.5	167.5	210.5	233.5	—
Triethyl citrate	$C_{12}H_{20}O_7$	107.0	144.0	190.4	217.8	267.5	294.0d	—
Trimethallyl phosphate	$C_{12}H_{21}PO_4$	93.7	149.8	192.0	225.7	288.5	324.0	—
Citronellyl acetate	$C_{12}H_{22}O_2$	74.7	113.0	140.5	161.0	197.8	217.0	—
Menthyl acetate	$C_{12}H_{22}O_2$	57.4	100.0	132.1	156.7	202.8	227.0	—
Dimethyl sebacate	$C_{12}H_{22}O_4$	104.0	156.2	196.0	222.6	269.6	293.5	38
Diisoamyl oxalate	$C_{12}H_{22}O_4$	85.4	131.4	165.7	192.2	240.0	165.0	—
Diisobutyl-d-tartrate	$C_{12}H_{22}O_6$	117.8	169.0	208.5	239.5	294.0	324.0	73.5
1-Dodecene	$C_{12}H_{24}$	47.2	87.8	118.6	142.3	185.5	208.0	−31.5
Triisobutylene	$C_{12}H_{24}$	18.0	56.5	86.7	110.0	153.0	179.0	—
Dodecan-2-one	$C_{12}H_{24}O$	77.1	120.4	152.4	177.5	222.5	246.5	—
Lauraldehyde	$C_{12}H_{24}O$	77.7	123.7	157.8	184.5	231.8	257.0	44.5
Lauric acid	$C_{12}H_{24}O_2$	121.0	166.0	201.4	227.5	273.8	299.2	48.
Dodecane	$C_{12}H_{26}$	47.8	90.0	121.7	146.2	191.0	216.2	−9.6
Dodecyl alcohol	$C_{12}H_{26}O$	91.0	134.7	167.2	192.0	235.7	259.0	24
Tripropylene glycol monoisopropyl ether	$C_{12}H_{26}O_4$	82.4	127.3	161.4	187.8	232.8	256.6	—
Triisobutylamine	$C_{12}H_{27}N$	32.3	69.8	97.8	119.7	157.8	179.0	−22
Dodecylamine	$C_{12}H_{27}N$	82.8	127.8	157.4	182.1	225.0	248.0	—
Triethylhexylsilane	$C_{12}H_{28}Si$	52.4	96.4	130.0	156.0	204.6	230.0	—
1,7-Diethoxyoctamethyltetrasiloxane	$C_{12}H_{34}O_3Si_4$	67.7	108.6	319.0	162.0	204.0	227.5	—
Dodecamethylpentasiloxane	$C_{12}H_{36}O_4Si_5$	56.6	98.0	128.8	162.8	196.5	220.5	—
Dodecamethylcyclohexasiloxane	$C_{12}H_{36}O_6Si_6$	67.3	110.0	141.8	166.3	210.6	236.0	−3.0
Acridine	$C_{13}H_9N$	129.4	184.0	224.2	256.0	314.3	346.0	110.5
Fluorene	$C_{13}H_{10}$	s	146.0	185.2	214.7	268.6	295.0	113
Benzophenone	$C_{13}H_{10}O$	108.2	157.6	195.7	224.4	276.8	305.4	48.5
Phenyl benzoate	$C_{13}H_{10}O_2$	106.8	157.8	197.6	227.8	283.5	314.0	70.5
Salol	$C_{13}H_{10}O_3$	117.8	167.0	205.0	233.8	284.8	313.0	42.5
Diphenylmethane	$C_{13}H_{12}$	76.0	122.8	157.8	186.3	237.5	264.0	26.5
Benzhydrol	$C_{13}H_{12}O$	110.0	162.0	200.0	227.5	275.6	301.0	68.5
Benzyl phenyl ether	$C_{13}H_{12}O$	95.4	144.0	180.1	209.2	259.8	287.0	—
1-Propionaphthone	$C_{13}H_{12}O$	124.0	171.0	206.9	233.5	280.2	306.0	—
Chloromethyldiphenylsilane	$C_{13}H_{13}ClSi$	105.0	152.7	189.2	216.0	266.5	295.0	—
Methyldiphenylamine	$C_{13}H_{13}N$	103.5	149.7	184.0	210.1	257.0	282.0	−7.6
2-Isopropylnaphthalene	$C_{13}H_{14}$	76.0	123.4	159.0	187.6	238.5	266.0	—
Methyldiphenylsilane	$C_{13}H_{14}Si$	88.0	132.8	164.4	193.7	241.5	266.8	—
Enanthophenone	$C_{13}H_{18}O$	100.0	145.5	178.9	204.2	248.3	271.3	—
Heptylbenzene	$C_{13}H_{20}$	64.0	110.0	144.0	170.2	217.8	244.0	—
a-Ionone	$C_{13}H_{20}O$	79.5	123.0	155.6	181.2	225.2	250.0	—
Bornyl propionate	$C_{13}H_{22}O_2$	64.6	108.0	140.4	165.7	211.2	235.0	—
2-Tridecanone	$C_{13}H_{26}O$	86.8	131.8	165.7	191.5	238.3	262.5	28.5
Methyl laurate	$C_{13}H_{26}O_2$	87.8	133.2	166.0	190.8	d	d	5
Tridecanoic acid	$C_{13}H_{26}O_2$	137.8	181.0	212.4	236.0	276.5	299.0	41
Tridecane	$C_{13}H_{28}$	59.4	104.0	137.7	162.5	209.4	234.0	−6.2
Tripropyleneglycol, monobutyl ether	$C_{13}H_{28}O_4$	101.5	147.0	179.8	204.4	247.0	269.5	—
Decyltrimethylsilane	$C_{13}H_{30}Si$	67.4	111.0	144.0	169.5	215.5	240.0	—
Triethylheptylsilane	$C_{13}H_{30}Si$	70.0	114.6	148.0	174.0	221.0	247.0	—
Anthraquinone	$C_{14}H_8O_2$	190.0s	234.2s	264.3s	285.0s	346.2	379.9	286
1,4-Dihydroxyanthraquinone	$C_{14}H_8O_4$	196.7	259.8	307.4	344.5	413.0d	450.0d	194
Anthracene	$C_{14}H_{10}$	145.0s	187.2s	217.5s	250.0	310.2	342.0	217.5
Phenanthrene	$C_{14}H_{10}$	118.2	173.0	215.9	249.0	308.0	340.2	99.5
Benzil	$C_{14}H_{10}O_2$	128.4	183.0	224.5	255.8	314.3	347.0	95
Benzoic anhydride	$C_{14}H_{10}O_3$	143.8	198.0	239.8	270.4	328.8	360.0	42
1,1-Diphenylethylene	$C_{14}H_{12}$	87.4	135.0	170.8	198.6	249.8	277.0	—
trans-Diphenylethylene	$C_{14}H_{12}$	113.2	161.0	199.0	227.4	287.3	306.5	124
Desoxybenzoin	$C_{14}H_{12}O$	123.3	173.5	212.0	241.3	293.0	321.2	60

ORGANIC COMPOUNDS
Pressures Less than One Atmosphere (Continued)

Name	Formula	1 mm	10 mm	40 mm	100 mm	400 mm	760 mm	M.P.
		Temperature °C						
Benzoin	$C_{14}H_{12}O$	135.6	188.0	227.6	258.0	313.5	343.0	132
Dibenzyl	$C_{14}H_{14}$	86.8	136.0	173.7	202.8	255.0	284.0	51.5
2-Isobutyronaphthone	$C_{14}H_{14}O$	133.2	181.0	215.6	242.3	288.2	313.0	—
Dibenzylamine	$C_{14}H_{15}O$	118.3	165.6	200.2	227.3	274.3	300.0	—
Ethyldiphenylamine	$C_{14}H_{15}N$	98.3	146.0	182.0	209.8	258.8	286.0	—
1,2-Dichlorotetraethylbenzene	$C_{14}H_{20}Cl_2$	105.6	155.0	192.2	220.7	272.8	302.0	—
1,4-Dichlorotetraethylbenzene	$C_{14}H_{20}Cl_2$	91.7	143.8	183.2	212.0	265.8	296.5	—
2-(4-tert-Butylphenoxy) ethyl acetate	$C_{14}H_{20}O_3$	118.0	165.8	201.5	228.0	277.6	304.4	—
1,2,4,5-Tetraethylbenzene	$C_{14}H_{22}$	65.7	111.6	145.8	172.4	221.4	248.0	11.6
2,4-Di-tert-butylphenol	$C_{14}H_{22}O$	84.5	130.0	164.3	190.0	237.0	260.8	—
Bornyl butyrate	$C_{14}H_{24}O_2$	74.0	118.0	150.7	176.4	222.2	247.0	—
Bornyl isobutyrate	$C_{14}H_{24}O_2$	70.0	114.0	147.2	172.2	218.2	243.0	—
Geranyl butyrate	$C_{14}H_{24}O_2$	96.8	139.0	170.1	193.8	235.0	257.4	—
Geranyl isobutyrate	$C_{14}H_{24}O_2$	90.7	133.0	164.0	187.7	228.5	251.0	—
Diethyl sebacate	$C_{14}H_{26}O_4$	125.3	172.1	207.5	234.4	280.3	305.5	1.3
2-Tetradecanone	$C_{14}H_{28}O$	99.3	145.5	179.8	206.0	253.0	278.0	—
Myristaldehyde	$C_{14}H_{28}O$	99.0	148.3	186.0	214.5	267.9	297.8	23.5
Myristic acid	$C_{14}H_{28}O_2$	142.0	190.8	223.5	250.5	294.6	318.0	57.5
1-Chlorotetradecane	$C_{14}H_{29}Cl$	98.5	148.2	187.0	215.5	267.5	296.0	+0.9
Tetradecane	$C_{14}H_{30}$	76.4	120.7	152.7	178.5	226.8	252.5	5.5
Tetradecylamine	$C_{14}H_{31}N$	102.6	152.0	189.0	215.7	264.6	291.2	—
Triethyloctylsilane	$C_{14}H_{32}Si$	73.7	120.6	155.7	184.3	235.0	262.0	—
1,9-Diethoxydecamethylpentasiloxane	$C_{14}H_{40}O_6Si_5$	89.0	131.5	162.2	187.0	230.0	253.3	—
Tetradecamethylhexasiloxane	$C_{14}H_{42}O_5Si_6$	73.7	117.6	149.8	175.2	220.5	245.5	—
Tetradecamethylcycloheptasiloxane	$C_{14}H_{42}O_7Si_7$	86.3	131.5	165.3	191.8	239.2	264.0	−32
1,3-Diphenyl-2-propanone	$C_{15}H_{14}O$	125.5	177.6	216.6	246.6	301.7	330.5	34.5
1-Biphenyloxy-2,3-epoxypropane	$C_{15}H_{14}O_2$	135.3	187.2	226.3	255.0	309.8	340.0	—
1-Isovaleronaphthone	$C_{15}H_{16}O$	136.0	184.0	219.7	246.7	294.0	320.0	—
4,4-Isopropylidenebisphenol	$C_{15}H_{16}O_2$	193.0	240.8	273.0	297.0	339.0	360.5	—
Ethoxymethyldiphenylsilane	$C_{15}H_{18}OSi$	109.0	152.7	186.0	211.8	256.8	282.0	—
Helenin	$C_{15}H_{20}O_2$	157.7	192.1	215.2	232.6	260.6	275.0	76
Cadinene	$C_{15}H_{24}$	101.3	146.0	179.8	205.6	250.7	275.0	—
2,6-Di-tert-butyl-4-cresol	$C_{15}H_{24}O$	85.8	131.0	164.1	190.0	237.6	262.5	—
4,6-Di-tert-butyl-2-cresol	$C_{15}H_{24}O$	86.2	132.4	167.4	194.0	243.4	260.2	—
4,6-Di-tert-butyl-3-cresol	$C_{15}H_{24}O$	103.7	150.0	183.3	211.0	257.1	282.0	—
Champacol	$C_{15}H_{24}O$	100.0	148.0	184.0	211.9	261.2	288.0	91
Trimethyl camphoronate	$C_{15}H_{26}O_6$		166.0	201.8	228.6	276.0	301.0	135
Methyl myristate	$C_{15}H_{30}O_2$	115.0	160.8	195.8	222.6	269.8	295.8	18.5
Pentadecane	$C_{15}H_{32}$	91.6	135.4	167.7	194.0	242.8	270.5	10
Tetrapropylene glycol monoisopropyl ether	$C_{15}H_{32}O_5$	116.6	163.0	197.7	223.3	268.3	292.7	—
Dodecyltrimethylsilane	$C_{15}H_{34}Si$	91.2	137.7	172.1	199.5	248.0	273.0	—
Benzyl cinnamate	$C_{16}H_{14}O_2$	173.8	221.5	255.8	281.5	326.7	350.0	39
Di(α-methylbenzyl) ether	$C_{16}H_{18}O$	96.7	144.0	179.6	206.8	254.8	281.0	—
Diethoxydiphenylsilane	$C_{16}H_{20}O_2Si$	111.5	157.6	193.2	220.0	259.7	296.0	—
Dibutyl phthalate	$C_{16}H_{22}O_4$	148.2	198.2	235.8	263.7	313.5	340.0	—
Pentaethylchlorobenzene	$C_{16}H_{25}Cl$	90.0	140.7	178.2	208.0	257.2	285.0	—
Pentaethylbenzene	$C_{16}H_{26}$	86.0	135.8	171.9	200.0	250.2	277.0	—
2,6-Di-tert-butyl-4-ethylphenol	$C_{16}H_{26}O$	89.1	137.0	172.1	198.0	244.0	268.6	—
4,6-Di-tert-butyl-3-ethylphenol	$C_{16}H_{26}O$	111.5	157.4	192.3	218.0	264.6	290.0	—
Muscone	$C_{16}H_{30}O$	118.0	170.0	210.0	241.5	297.2	328.0	—
Palmitonitrile	$C_{16}H_{31}N$	134.3	185.8	223.8	251.5	304.5	332.0	31
1-Hexadecene	$C_{16}H_{32}$	101.6	146.2	178.8	205.3	250.0	274.0	4
Tetraisobutylene	$C_{16}H_{32}$	63.8	108.5	142.2	167.5	214.6	240.0	—
2-Hexadecanone	$C_{16}H_{32}O$	109.8	167.3	203.7	230.5	279.8	307.0	—
Palmitaldehyde	$C_{16}H_{32}O$	121.6	171.8	210.0	239.5	292.3	321.0	34
Palmitic acid	$C_{16}H_{32}O_2$	153.6	205.8	244.4	271.5	326.0	353.8	64.0
Hexadecane	$C_{16}H_{34}$	105.3	149.8	181.3	208.5	258.3	287.5	18.5
Cetyl alcohol	$C_{16}H_{34}O$	122.7	177.8	219.8	251.7	312.7	344.0	49.3
Cetylamine	$C_{16}H_{36}N$	123.6	176.0	215.7	245.8	300.4	330.0	—
Decyltriethylsilane	$C_{16}H_{36}Si$	108.5	155.6	191.7	218.3	267.5	293.0	—
1,1,1-Diethoxydodecamethylhexasiloxane	$C_{16}H_{46}O_7Si_6$	103.6	147.5	180.0	205.5	250.0	273.5	—
Hexadecamethylheptasiloxane	$C_{16}H_{48}O_6Si_7$	93.2	138.5	171.8	198.0	244.7	270.0	—
Hexadecamethylcyclooctasiloxane	$C_{16}H_{48}O_8Si_8$	103.5	150.5	186.3	213.8	263.0	290.0	31.5
Benzanthrone	$C_{17}H_{10}O$	225.0	297.2	350.0	390.0			174
4-tert-Butylphenyl salicylate	$C_{17}H_{18}O_3$	166.2	225.0	270.7	305.8	370.6	404.0d	—
Menthyl benzoate	$C_{17}H_{24}O_2$	123.2	170.0	204.3	230.4	277.1	301.0	54.5
2-Heptadecanone	$C_{17}H_{34}O$	129.6	178.0	214.3	242.0	291.7	319.5	—
Methyl palmitate	$C_{17}H_{34}O_2$	134.3	184.3	d	—	—	—	30
Heptadecane	$C_{17}H_{36}$	115.0	160.0	195.8	223.0	274.5	303.0	22.5
Tetradecyltrimethylsilane	$C_{17}H_{38}Si$	120.0	166.2	201.5	227.8	275.0	300.0	—
Tri-2-chlorophenylthiophosphate	$C_{18}H_{12}Cl_3O_3PS$	188.2	231.2	261.7	283.8	322.0	341.3	—
Triphenyl phosphate	$C_{18}H_{15}O_4P$	193.5	249.8	290.3	322.5	379.2	413.5	49.4
Hexaethylbenzene	$C_{18}H_{30}$	s	150.3	187.7	216.0	268.5	298.3	130
2,4,6-Tri-tert-butylphenol	$C_{18}H_{30}O$	95.2	142.0	177.4	203.0	250.6	276.3	—
Oleic acid	$C_{18}H_{34}O_2$	176.5	223.0	257.2	286.0	334.7	360.0d	14
Elaidic acid	$C_{18}H_{34}O_2$	171.3	223.5	260.8	288.0	337.0	362.0	51.5
Stearaldehyde	$C_{18}H_{36}O$	140.0	192.1	230.8	260.0	313.8	342.5	63.5
Stearic acid	$C_{18}H_{36}O_2$	173.7	225.0	263.3	291.0	343.0	370.0d	69.3
Octadecane	$C_{18}H_{38}$	119.6	169.6	207.4	236.0	288.0	317.0	28
2-Methylheptadecane	$C_{18}H_{38}$	119.8	168.7	204.8	231.5	279.8	306.5	—
1-Octadecanol	$C_{18}H_{38}O$	150.3	202.0	240.4	269.4	320.3	349.5	58.5

Name	Formula	Temperature °C						M.P.
		1 mm	10 mm	40 mm	100 mm	400 mm	760 mm	
Ethylcetylamine	$C_{18}H_{39}N$	133.2	186.0	226.5	256.8	313.0	342.0d	—
1,1,3-Diethoxytetradecamethylheptasiloxane	$C_{18}H_{52}O_8Si_7$	119.0	163.5	197.0	223.2	268.3	293.5	—
Octadecamethyloctasiloxane	$C_{18}H_{54}O_7Si_9$	105.8	152.3	187.5	214.5	263.5	290.0	—
Triphenylmethane	$C_{19}H_{16}$	169.7	197.0	215.5	228.4	249.8	259.2	93.4
Diphenyl-2-tolyl thiophosphate	$C_{19}H_{17}O_3PS$	159.7	201.6	230.6	252.5	290.0	310.0	—
Nonadecane	$C_{19}H_{40}$	133.2	183.5	220.0	248.0	299.8	330.0	32
Ethoxytriphenylsilane	$C_{20}H_{20}OSi$	167.0	213.5	247.0	273.5	319.5	344.0	—
Diethylhexadecylamine	$C_{20}H_{43}N$	139.8	194.0	235.0	265.5	324.6	355.0	—
1,1,5-Diethoxyhexadecamethyloctasiloxane	$C_{20}H_{58}O_9Si_8$	133.7	179.7	213.8	240.0	286.0	311.5	—
Eicosamethylnonasiloxane	$C_{20}H_{60}O_8Si_9$	144.0	189.0	220.5	244.3	286.0	307.5	—
Tritolyl phosphate	$C_{21}H_{21}O_4P$	154.6	198.0	229.7	252.2	292.7	313.0	—
Heneicosane	$C_{21}H_{44}$	152.6	205.4	243.4	272.0	323.8	350.5	40.4
Erucic acid	$C_{22}H_{42}O_2$	206.7	254.5	289.1	314.4	358.8	381.5d	33.5
Brassidic acid	$C_{22}H_{42}O_2$	209.6	256.0	290.0	316.2	359.6	382.5d	61.5
Docosane	$C_{22}H_{46}$	157.8	213.0	254.5	286.0	343.5	376.0	44.5
Docosamethyldecasiloxane	$C_{22}H_{56}O_9Si_{10}$	160.3	202.8	233.8	255.0	293.8	314.0	—
Tricosane	$C_{23}H_{48}$	170.0	223.0	261.3	289.8	339.8	366.5	47.7
Tetracosane	$C_{24}H_{50}$	183.8	237.6	276.3	305.2	358.0	386.4	51.1
Tetracosamethylhendecasiloxane	$C_{24}H_{72}O_{10}Si_{11}$	175.2	216.7	246.2	266.3	303.7	322.8	—
Pentacosane	$C_{25}H_{52}$	194.2	248.2	285.6	314.0	365.4	390.3	53.3
Hexacosane	$C_{26}H_{54}$	204.0	257.4	295.2	323.2	374.6	399.8	56.6
Dicarvacryl-2-tolyl phosphate	$C_{27}H_{33}O_4P$	180.2	221.8	251.5	272.5	309.8	330.0	—
Heptacosane	$C_{27}H_{56}$	211.7	266.8	305.7	333.5	385.0	410.6	59.5
Octacosane	$C_{28}H_{58}$	226.5	277.4	314.2	341.8	388.9	412.5	61.6
Nonacosane	$C_{29}H_{60}$	234.2	286.4	323.2	350.0	397.2	421.8	63.8
Dicarvacryl-mono-(6-chloro-2-xenyl) phosphate	$C_{32}H_{34}ClO_4P$	204.2	249.3	280.5	304.9	342.0	361.0	—

Pressures Greater than One Atmosphere

Name	Formula	Temperature °C						
		1 atm.	2 atm.	5 atm.	10 atm.	20 atm.	40 atm.	60 atm.
Chlorotrifluoromethane	$CClF_3$	−81.2	−66.7	−42.7	−18.5	+ 12.0	52.8	—
Dichlorodifluoromethane	CCl_2F_2	−29.8	−12.2	+ 16.1	42.4	74.0	—	—
Carbonyl chloride	CCl_2O	8.3	27.3	57.2	85.0	119.0	159.8	—
Trichlorofluoromethane	CCl_3F	23.7	44.1	77.3	108.2	146.7	194.0	—
Carbontetrachloride	CCl_4	76.7	102.0	141.7	178.0	222.0	276.0	—
Chlorodifluoromethane	$CHClF_2$	−40.8	−24.7	+ 0.3	24.0	52.0	85.3	—
Dichlorofluoromethane	$CHCl_2F$	8.9	28.4	59.0	87.0	121.2	162.6	—
Trichloromethane	$CHCl_3$	61.3	83.9	120.0	152.3	191.8	237.5	—
Hydrocyanic acid	CHN	25.8	45.5	75.5	103.5	134.2	170.2	—
Methyl bromide	CH_3Br	3.6	23.3	54.8	84.0	121.7	170.2	—
Methyl chloride	CH_3Cl	−24.0	−6.4	+ 22.0	47.3	77.3	113.8	137.5
Methyl fluoride	CH_3F	−78.2	−64.5	−42.0	−21.0	+ 2.6	26.5	43.5
Methyl iodide	CH_3I	42.4	65.5	101.8	138.0	176.5	228.5	—
Methane	CH_4	−161.5	−152.3	−138.3	−124.8	−108.5	−80.3	—
Methanol	CH_4O	64.7	84.0	112.5	138.0	167.8	203.5	224.0
Methanethiol	CH_4S	6.8	26.1	55.9	83.4	117.5	157.7	185.0
Methylamine	CH_5N	−6.3	+ 10.1	36.0	59.5	87.8	121.8	144.6
Carbon monoxide	CO	−191.3	−183.5	−170.7	−161.0	−149.7	—	—
Carbon dioxide	CO_2	−78.2	−69.1s	−56.7	−39.5	−18.9	+ 5.9	22.4
Carbon disulfide	CS_2	46.5	69.1	104.8	136.3	175.5	222.8	256.0
1-Chloro-1,2,2-trifluoroethylene	C_2ClF_3	−27.9	−11.1	+ 15.5	40.0	71.1	—	—
1,2-Dichloro-1,1,2,2-tetrafluoroethane	$C_2Cl_2F_4$	3.5	22.8	54.0	82.3	117.5	—	—
1,1,2-Trichloro-1,2,2-trifluoroethane	$C_2Cl_3F_3$	47.6	70.0	105.5	138.0	177.7	—	—
Acetylene	C_2H_2	−84.0s	−71.6	−50.2	−32.7	−10.0	+ 16.8	34.8
cis-1,2-Dichloroethylene	$C_2H_2Cl_2$	59.0	82.1	119.3	152.3	194.0	244.5	—
trans-1,2-Dichloroethylene	$C_2H_2Cl_2$	47.8	69.8	104.0	135.7	174.0	220.0	—
Ethylene	C_2H_4	−103.7	−90.8	−71.1	−52.8	−29.1	−1.5	—
1,2-Dibromoethane	$C_2H_4Br_2$	131.5	157.7	200.0	237.0	269.0	295.0	304.5
1,1-Dichloroethane	$C_2H_4Cl_2$	57.3	80.2	117.3	150.3	192.7	243.0	—
1,2-Dichloroethane	$C_2H_4Cl_2$	83.7	108.1	147.8	183.5	226.5	272.0	—
Acetic acid	$C_2H_4O_2$	118.1	143.5	180.3	214.0	252.0	297.0	—
Methyl formate	$C_2H_4O_2$	32.0	51.9	83.5	112.0	147.2	188.5	—
Ethyl bromide	C_2H_5Br	38.4	60.2	95.0	126.8	164.3	206.5	229.5
Ethyl chloride	C_2H_5Cl	12.3	32.5	64.0	92.6	127.3	167.0	—
Ethyl fluoride	C_2H_5F	−32.0	−16.7	+ 7.7	30.2	57.5	90.0	—
Ethane	C_2H_6	−88.6	−75.0	−52.8	−32.0	−6.4	+ 23.6	—
Ethanol	C_2H_6O	78.4	97.5	126.0	151.8	183.0	218.0	242.0
Dimethyl ether	C_2H_6O	−23.7	−6.4	+ 20.8	45.5	75.7	112.1	—
Ethanethiol	C_2H_6S	35.0	56.6	90.7	121.9	159.5	204.7	—
Dimethyl sulfide	C_2H_6S	36.0	57.8	92.3	124.5	163.8	209.0	—
Ethylamine	C_2H_7N	16.6	35.7	65.3	91.8	124.0	163.0	—
Dimethylamine	C_2H_7N	7.4	25.0	53.9	80.0	111.7	149.8	—
Cyanogen	C_2N_2	−21.0	−4.4	+ 21.4	44.6	72.6	106.5	—
Propadiene	C_3H_4	−35.0	−18.4	+ 8.0	33.2	64.5	103.5	—
Propyne	C_3H_4	−23.3	−7.1	+ 19.5	43.8	74.0	111.5	—
Propylene	C_3H_6	−47.7	−31.4	−4.8	+ 19.8	49.5	85.0	—
Acetone	C_3H_6O	56.5	78.6	113.0	144.5	181.0	214.5	—
Propionic acid	$C_3H_6O_2$	141.1	160.0	186.0	203.5	220.0	233.0	—

Name	Formula	Temperature °C						
		1 atm.	2 atm.	5 atm.	10 atm.	20 atm.	40 atm.	60 atm.
Methyl acetate	$C_3H_6O_2$	57.8	79.5	113.1	144.2	181.0	225.0	—
Ethyl formate	$C_3H_6O_2$	54.3	76.0	110.5	142.2	180.0	225.0	—
Propane	C_3H_8	−42.1	−25.6	+ 1.4	26.9	58.1	94.8	—
1-Propanol	C_3H_8O	97.8	117.0	149.0	177.0	210.8	250.0	—
2-Propanol	C_3H_8O	82.5	101.3	130.2	155.7	186.0	220.2	—
Ethyl methyl ether	C_3H_8O	7.5	26.5	56.4	84.0	108.0	160.0	—
Propylamine	C_3H_9N	48.5	69.8	102.8	133.4	170.0	214.5	—
1,3-Butadiene	C_4H_6	−4.5	+ 15.3	47.0	76.0	114.0	158.0	—
Acetic anhydride	$C_4H_6O_3$	139.6	162.0	194.0	221.5	253.0	288.5	—
Dimethyl oxalate	$C_4H_6O_4$	163.3	189.6	228.7	—	—	—	—
Butyric acid	$C_4H_8O_2$	163.5	188.3	225.0	257.0	295.0	338.0	—
Isobutyric acid	$C_4H_8O_2$	154.4	179.8	217.0	250.0	289.0	336.0	—
Ethyl acetate	$C_4H_8O_2$	77.1	100.6	136.6	169.7	209.5	—	—
Methyl propionate	$C_4H_8O_2$	79.8	103.0	139.8	172.6	212.5	—	—
Propyl formate	$C_4H_8O_2$	81.3	104.3	142.0	176.4	217.5	—	—
Butane	C_4H_{10}	−0.5	+ 18.8	50.0	79.5	116.0	—	—
2-Methylpropane	C_4H_{10}	−11.7	+ 7.5	39.0	66.8	99.5	—	—
Butyl alcohol	$C_4H_{10}O$	117.5	139.8	172.5	203.0	237.0	277.0	—
sec-Butyl alcohol	$C_4H_{10}O$	99.5	118.2	147.5	172.0	204.0	251.0	—
Isobutyl alcohol	$C_4H_{10}O$	108.0	127.3	156.2	182.0	212.5	251.0	—
tert-Butyl alcohol	$C_4H_{10}O$	82.9	102.0	130.0	154.2	184.5	222.5	—
Diethyl ether	$C_4H_{10}O$	34.6	56.0	90.0	122.0	159.0	—	—
Diethyl sulfide	$C_4H_{10}S$	88.0	112.0	153.8	190.2	234.0	—	—
Diethylamine	$C_4H_{11}N$	55.5	77.8	113.0	145.3	184.5	—	—
Tetramethylsilane	$C_4H_{12}Si$	27.0	48.0	82.0	113.0	152.0	—	—
Ethyl propionate	$C_5H_{10}O_2$	99.1	123.8	162.7	197.8	240.0	—	—
Propyl acetate	$C_5H_{10}O_2$	101.8	126.8	165.7	200.5	242.8	—	—
Isobutyl formate	$C_5H_{10}O_2$	98.2	121.8	157.8	192.4	234.0	—	—
Methyl butyrate	$C_5H_{10}O_2$	102.3	127.5	166.7	203.0	244.5	—	—
Methyl isobutyrate	$C_5H_{10}O_2$	92.6	116.7	155.2	190.2	232.0	—	—
Pentane	C_5H_{12}	36.1	58.0	93.1	124.7	164.3	—	—
2-Methylbutane	C_5H_{12}	27.8	48.8	82.8	114.5	154.0	—	—
2,2-Dimethylpropane	C_5H_{12}	+ 9.5	29.5	61.1	90.7	127.6	—	—
Ethyl propyl ether	$C_5H_{12}O$	61.7	85.3	123.1	156.2	197.2	—	—
Bromobenzene	C_6H_5Br	156.2	186.2	232.1	270.6	327.0	387.5	—
Chlorobenzene	C_6H_5Cl	132.2	160.2	205.0	245.3	292.8	349.8	—
Fluorobenzene	C_6H_5F	84.7	109.9	148.5	184.4	227.6	279.3	—
Iodobenzene	C_6H_5I	188.6	220.0	270.0	315.7	371.5	437.2	—
Benzene	C_6H_6	80.1	103.8	142.5	178.8	221.5	—	—
Phenol	C_6H_6O	181.9	208.0	248.2	283.8	328.7	272.3	418.7
Aniline	C_6H_7N	184.4	212.8	254.8	292.7	342.0	382.1	—
Cyclohexane	C_6H_{12}	80.7	106.0	146.4	184.0	228.4	400.0	—
Ethyl isobutyrate	$C_6H_{12}O_2$	110.1	135.5	174.2	210.0	253.0	—	—
Hexane	C_6H_{14}	68.7	93.0	131.7	166.6	209.4	—	—
2,3-Dimethylbutane	C_6H_{14}	58.0	82.0	120.3	155.7	198.7	—	—
Toluene	C_7H_8	110.6	136.5	178.0	215.8	262.5	—	—
Heptane	C_7H_{16}	98.4	124.8	165.7	202.8	247.5	319.0	—
Ethylbenzene	C_8H_{10}	136.2	163.5	207.5	246.3	294.5	—	—
Octane	C_8H_{18}	125.6	152.7	196.2	235.8	181.4	—	—
Dodecane	$C_{12}H_{26}$	216.2	249.2	300.0	345.8	—	—	—

VAPOR PRESSURE
Variation with Temperature

The following table gives the value of the constants a and b in the following equation:

$$\log_{10} p = -\frac{0.05223a}{T} + b$$

where p is the pressure in mm of mercury of the saturated vapor at the absolute temperature T. (T = t°C + 273.1). The values obtained by the use of the equation given above are valid within the temperature ranges indicated for each of the compounds.

Elements and Inorganic Compounds

Compound	Formula	Temp. range °C	a	b
Aluminum oxide	Al_2O_3	1840 to 2200 liq.	540,000	14.22
Ammonia	NH_3	−127 to −78 sol.	31,211	9.9974
Ammonium bromide	NH_4Br	250 to 400 sol.	90,208	9.9404
Ammonium chloride	NH_4Cl	100 to 400 sol.	83,486	10.0164
Ammonium cyanide	NH_4CN	7 to 17 sol.	41,484	9.978
Ammonium iodide	NH_4I	300 to 400 sol.	95,730	10.2700
Ammonium sulfhydrate	NH_4HS	6 to 40 sol.	46,025	10.7500
Antimony	Sb	1070 to 1325 liq.	189,000	9.051

VAPOR PRESSURE (Continued)

Compound	Formula	Temp. range (°C)	a	b
Argon	A	−208 to −189 sol.	7,814.5	7.5741
		−189 to −183 liq.	6,826.0	6.9605
Arsenic	As	800 to 860 liq.	47,100	6.692
Arsenous oxide	As₂O₃	440 to 815 sol.	133,000	10.800
		100 to 310 sol.	111,350	12.127
		315 to 490 liq.	52,120	6.513
Barium	Ba	930 to 1130 liq.	350,000	15.765
Bismuth	Bi	1210 to 1420 liq.	200,000	8.876
Bismuth trichloride	BiCl₃	91 to 213 sol.	13,125	2.681
Cadmium	Cd	150 to 320.9 sol.	109,000	8.564
		500 to 840 liq.	99,900	7.897
Cadmium iodide	CdI₂	385 to 450 liq.	122,200	9.269
Cesium	Cs	200 to 350 liq.	73,400	6.949
Cesium chloride	CsCl	986 to 1295 liq.	163,200	8.340
Calcium	Ca	960 to 1110 liq.	370,000	16.240
Carbon	C	3880 to 4430 liq.	540,000*	9.596*
Carbon dioxide	CO₂	−135 to −56.7 sol.	26,179.3	9.9082
Carbon monoxide	CO	−220 to −206 liq.	6,354	6.976
Chlorine	Cl	−154 to −103 sol.	29,293	9.950
Cobalt	Co	2375 liq.	309,000	7.571
Copper	Cu	2100 to 2310 liq.	468,000	12.344
Cuprous chloride	Cu₂Cl₂	878 to 1369 liq.	80,700	5.454
Cyanogen	(CN)₂	−72 to −28 sol.	32,437	9.6539
		−32 to −6 liq.	23,750	7.808
Ferrous chloride	FeCl₂	700 to 930 sol.	135,200	8.33
Gold	Au	2315 to 2500 liq.	385,000	9.853
Hydriodic acid	HI	−97 to −51 sol	24,160	8.259
		−50 to −34 liq.	21,580	7.630
Hydrobromic acid	HBr	−114 to −86 sol.	22,420	8.734
		−86 to −66 liq.	17,960	7.427
Hydrochloric acid	HCl	−158 to −110 sol.	19,588	8.4430
Hydrocyanic acid	HCN	−8 to +27 liq.	27,830	7.7446
Hydrofluoric acid	HF	−83 to +48 liq.	25,180	7.370
Hydrogen peroxide	H₂O₂	10 to 90 liq.	48,530	8.853
Hydrogen sulfide	H₂S	−110 to −83 sol.	20,690	7.880
Iron	Fe	2220 to 2450 liq.	309,000	7.482
Krypton	Kr	−189 to −169 sol.	10,065	7.1770
		−169 to −150 liq.	9,377.0	6.92387
Lead	Pb	525 to 1325 liq.	188,500	7.827
Lead bromide	PbBr₂	735 to 918 liq.	118,000	8.064
Lead chloride	PbCl₂	500 to 950 liq.	141,900	8.961
Lithium bromide	LiBr	1010 to 1265 liq.	152,700	8.068
Lithium chloride	LiCl	1045 to 1325 liq.	155,900	7.939
Lithium fluoride	LiF	1398 to 1666 liq.	218,400	8.753
Lithium iodide	LiI	940 to 1140 liq.	143,600	8.011
Magnesium	Mg	900 to 1070 liq.	260,000	12.993
Manganese	Mn	1510 to 1900 liq.	267,000	9.300
Mercuric bromide	HgBr₂	111 to 235 sol.	79,800	10.181
		238 to 331 liq.	61,250	8.284
Mercuric chloride	HgCl₂	60 to 130 sol.	85,030	10.888
		130 to 270 sol.	78,850	10.094
		275 to 309 liq.	61,020	8.409
Mercuric iodide	HgI₂	100 to 250 sol.	82,340	10.057
		266 to 360 liq.	62,770	8.115
Mercury	Hg	−80 to −38.87 sol.	73,000	10.383
		400 to 1300 liq.	58,700	7.752
Molybdenum	Mo	1800 to 2240 sol.	680,000	10.844
Nitrogen	N₂	−215 to −210 sol.	6,881.3	7.66558
Nitrogen dioxide	NO	−200 to −161 sol.	16,423	10.048
		−163.7 to −148 liq.	13,040	8.440
Nitrogen monoxide	N₂O	−144 to −90 sol.	23,590	9.579
		−90.1 to −88.7 liq.	16,440	7.535
Nitrogen pentoxide	N₂O₅	−30 to +30 sol.	57,180	12.647
Nitrogen tetroxide	N₂O₄	−100 to −40 sol.	55,160	13.400
		−40 to −10 sol.	45,440	11.214
		−8 to +43.2 liq.	33,430	8.814
Nitrogen trioxide	N₂O₃	−25 to 0 liq.	39,400	10.30
Phosphorus (white)	P	20 ro 44.1 sol.	63,123	9.6511
Phosphorus (violet)	P	380 to 590 sol.	108,510	11.0842
Platinum	Pt	1425 to 1765 sol.	486,000	7.786
Potassium	K	260 to 760 liq.	84,900	7.183
Potassium bromide	KBr	906 to 1063 liq.	168,100	8.2470
		1095 to 1375 liq.	163,800	7.936
Potassium chloride	KCl	906 to 1105 liq.	174,500	8.3526
		1116 to 1418 liq.	169,700	8.130
Potassium fluoride	KF	1278 to 1500 liq.	207,500	9.000
Potassium hydroxide	KOH	1170 to 1327 liq.	136,000	7.330
Potassium iodide	KI	843 to 1028 liq.	157,600	8.0957
		1063 to 1333 liq.	155,700	7.949
Rubidium	Rb	250 to 370 liq.	76,000	6.976
Rubidium chloride	RbCl	1142 to 1395 liq.	198,600	9.111
Silicon	Si	1200 to 1320 liq.	170,000	5.950
Silicon dioxide	SiO₂	1860 to 2230 liq.	506,000	13.43
Silver	Ag	1650 to 1950 liq.	250,000	8.762

* Based on boiling point of 3927°C or 4200° absolute.

VAPOR PRESSURE (Continued)

Compound	Formula	Temp. range °C	a	b
Silver chloride	AgCl	1255 to 1442 liq.	185,500	8.179
Sodium	Na	180 to 883 liq.	103,300	7.553
Sodium bromide	NaBr	1138 to 1394 liq.	161,600	7.948
Sodium chloride	NaCl	976 to 1155 liq.	180,300	8.3297
		1156 to 1430 liq.	185,800	8.548
Sodium cyanide	NaCN	800 to 1360 liq.	155,520	7.472
Sodium fluoride	NaF	1562 to 1701 liq.	218,200	8.640
Sodium hydroxide	NaOH	1010 to 1402 liq.	132,000	7.030
Sodium iodide	NaI	1063 to 1307 liq.	165,100	8.371
Stannic chloride	SnCl₄	−52 to −38 sol.	46,740	9.824
Strontium	Sr	940 to 1140 liq.	360,000	16.056
Sulfur dioxide	SO₂	95 to −75 sol.	35,827	10.5916
Sulfur trioxide	SO₃	24 to 48 liq.	43,450	10.022
Thallium	Tl	950 to 1200 liq.	120,000	6.140
Thallium chloride	TlCl	665 to 807 liq.	105,200	7.974
Tin	Sn	1950 to 2270 liq.	328,000	9.643
Tungsten	W	2230 to 2770 sol.	897,000	9.920
Zinc	Zn	250 to 419.4 sol.	133,000	9.200
		600 to 985 liq.	118,000	8.108

Organic Compounds

Compound	Formula	Temp. range °C	a	b
Acetaldehyde	C₂H₄O	−24.3 to +27.5 liq.	27,707	7.8206
Acetic acid	C₂H₄O₂	−35 to 10 sol.	41,689	8.502
Acetic anhydride	C₄H₆O₃	100 to 140 liq.	45,585	8.688
Acetylene	C₂H₂	−140 to −82 sol.	21,914	8.933
Aniline	C₆H₇N	145 to 185 liq.	45,951.6	8.1278
Anthracene	C₁₄H₁₀	100 to 600 liq.	70,390	8.706
		100 to 160 liq.	72,000	8.91
		223 to 342 liq.	59,219	7.910
Anthraquinone	C₁₄H₈O₂	224 to 286 sol.	110,040	12.305
Benzene	C₆H₆	−58 to −30 sol.	42,904	9.556
		−30 to +5 sol.	44,222	9.846
		0 to 42 liq.	34,172	7.9622
		42 to 100 liq.	32,295	7.6546
Benzoic acid	C₇H₆O₂	60 to 110 sol.	67,930	9.033
Benzophenone	C₁₃H₁₀O	200 to 308 liq.	58,221	8.137
Benzoyl chloride	C₇H₅ClO	140 to 200 liq.	45,416	7.9245
Benzyl alcohol	C₇H₈O	100 to 135 liq.	59,491	9.5152
		135 to 205 liq.	53,118	8.6977
Butane	C₄H₁₀	−100 to +12 liq.	23,450	7.395
iso-Butane	C₄H₁₀	−115 to −34 liq.	21,273	7.25
n-Butyl alcohol	C₄H₁₀O	75 to 117.5 liq.	46,774	9.1362
Butyric acid	C₄H₈O₂	80 to 165 liq.	51,103	9.010
Bromobenzene	C₆H₅Br	−26 to −15 liq.	42,500	8.075
p-Bromochlorobenzene	C₆H₄BrCl	23 to 63 sol.	69,755	11.629
Camphor	C₁₉H₁₆O	0 to 180 sol.	53,559	8.799
Carbon tetrachloride	CCl₄	−70 to −50 sol.	34,608	8.05
		−19 to +20 liq.	33,914	8.004
Chlorobenzene	C₆H₅Cl	−35 to −15 liq.	42,250	8.500
Cyclohexane	C₆H₁₂	−5 to +5 sol.	37,394	8.594
p-Dichlorobenzene	C₆H₄Cl₂	30 to 50 sol.	72,218	12.480
Dichloroethane-1,1	C₂H₄Cl₂	0 to 30 liq.	31,706	7.909
Dichloroethane-1,2	C₂H₄Cl₂	0 to 30 liq.	35,598	8.126
Diphenylamine	C₁₂H₁₁N	278 to 284 liq.	57,350	8.088
Ethyl chloride	C₂H₅Cl	−30 to +30 liq.	26,319	7.691
Ethylene	C₂H₄	−160 to −104 liq.	14,396	7.330
Ethylene bromide	C₂H₄Br₂	10 to 150 liq.	38,082	7.792
Heptane	C₇H₁₆	−63 to −40 liq.	37,358	8.2585
Hexane	C₆H₁₄	−10 to +90 liq.	31,679	7.724
Iodobenzene	C₆H₅I	−30 to +18 liq.	43,000	7.500
Methane	CH₄	−194 to −184 sol.	9,896.2	7.6509
		−174 to −163 liq.	8,516.9	6.8626
Methylalcohol	CH₄O	−62 to −44 liq.	39,234	8.9547
		−10 to +80 liq.	38,324	8.8017
Methyl chloride	CH₃Cl	−47 to −10 liq.	21,988	7.481
Methyl ether	C₂H₆O	−70 to −20 liq.	23,025	7.720
Methyl fluoride	CH₃F	−102 to −76 liq.	17,053	7.445
Methyl salicylate	C₈H₈O₃	175 to 215 liq.	48,670	8.008
Naphthalene	C₁₀H₈	0 to 80 sol.	71,401	11.450
		120 to 200 liq.	47,362	7.927
o-Nitroaniline	C₆H₆N₂O₂	150 to 260 liq.	63,881	8.8684
m-Nitroaniline	C₆H₆N₂O₂	170 to 260 liq.	65,880	8.8188
p-Nitroaniline	C₆H₆N₂O₂	190 to 260 liq.	77,345	9.5595
Nitrobenzene	C₆H₅NO₂	112 to 209 liq.	48,955	8.192
Nitromethane	CH₃NO₂	47 to 100 liq.	36,914	8.033
Oxalic acid	C₂H₂O₄	55 to 105 sol.	90,502.6	12.2229
n-Pentane	C₅H₁₂	−20 to +50 liq.	27,691	7.558
Phenol	C₆H₆O	116 to 180 liq.	49,644	8.587
Phthalic anhydride	C₈H₄O₃	160 to 285 liq.	54,920	8.022
Propane	C₃H₈	−136 to −40 liq.	19,037	7.217
Propionic acid	C₃H₆O₂	20 to 140 liq.	46,150	8.715
n-Propyl alcohol	C₃H₈O	−45 to −10 liq.	47,274	9.5180
Propyl bromide	C₃H₇Br	0 to 30 liq.	32,430	7.821
Propyl chloride	C₃H₇Cl	0 to 50 liq.	28,894	7.593

VAPOR PRESSURE (Continued)

Compound	Formula	Temp. range (°C)	a	b
Propylene	C₃H₆	−95 to −48 liq.	19,693	7.4463
Quinoline	C₉H₇N	180 to 240 liq.	49,720	7.969
Tetrachloroethane-1,1,1,2	C₂H₂Cl₄	105 to 145 liq.	36,508	7.605
Tetrachloroethane-1,1,2,2	C₂H₂Cl₄	26 to 145 liq.	39,729	7.846
Toluene	C₇H₈	−92 to +15 liq.	39,198	8.330

Compound	Formula	Temp. range (°C)	a	b
Propylene	C_3H_6	−95 to −48 liq.	19,693	7.4463
Quinoline	C_9H_7N	180 to 240 liq.	49,720	7.969
Tetrachloroethane-1,1,1,2	$C_2H_2Cl_4$	105 to 145 liq.	36,508	7.605
Tetrachloroethane-1,1,2,2	$C_2H_2Cl_4$	26 to 145 liq.	39,729	7.846
Toluene	C_7H_8	−92 to +15 liq.	39,198	8.330

VAPOR PRESSURE OF NITRIC ACID

Temperature (°C)	Vapor Pressure, mm. of Hg	
	100% HNO_3	90% of HNO_3
0	14.4	5.5
10	26.6	11.
20	47.9	20.
30	81.3	37.3
40	133.	64.4
50	208.	107.
70	467.	242.
80	670.	352.
90	937.	504.
100	1282.	710.

VAPOR PRESSURE OF THE ELEMENTS

Rudolf Loebel

This table lists the temperature in degrees Celsius at which an element has a vapor pressure indicated by the headings of the columns. For pressures of one atmosphere and lower, the pressures are given in millimeters of mercury; for pressures above one atmosphere, the pressures are given in atmospheres.

Element		mm Hg					atm.				
		1	10	100	400	760	2	5	10	20	40
Aluminum	Al	1540	1780	2080	2320	2467	2610	2850	3050	3270	3530
Antimony	Sb		960	1280	1570	1750	1960	2490			
Arsenic	As	380	440	510	580	610					
Barium	Ba	860	1050	1300	1520	1640	1790	2030	2230		
Beryllium	Be	1520	1860	2230	2770	2970	3240	3730	4110	4720	5610
Bismuth	Bi		1060	1280	1450	1560	1660	1850	2000	2180	
Boron	Б	2660	3030	3460	3810	4000					
Bromine	Br	−60	−30	+9	39	59	78	110			
Cadmium	Cd	393	486	610	710	765	830	930	1030	1120	1240
Calcium	Ca	800	970	1200	1390	1490	1630	1850	2020	2290	
Cesium	Cs		373	513	624	690					
Chlorine	Cl	−123	−101	−71	−46	−34	−17	+9	30	55	97
Chromium	Cr	1610	1840	2140	2360	2480	2630	2850	3010	3180	
Cobalt	Co	1910	2170	2500	2760	2870	3040	3270			
Copper	Cu		1870	2190	2440	2600	2760	3010	3500	3460	3740
Fluorine	F			−203	−193	−188	−180.7	−169.1	−159.6		
Gallium	Ga	1350	1570	1850	2060	2180	2320	2560	2730		
Germanium	Ge		2080	2440	2710	2830	2970	3200	3430		
Gold	Au	1880	2160	2520	2800	2940	3120	3490	3630	3890	
Indium	In			1960	2080	2230	2440	2600			
Iridium	Ir	2830	3170	3630	3960	4130	4310	4650			
Iron	Fe	1780	2040	2370	2620	2750	2900	3150	3360	3570	
Iodine	I	40	72	115	160	185	216	265			
Lanthanum	La				3230	3420	3620	3960	4270		
Lead	Pb	970	1160	1420	1630	1740	1880	2140	2320	2620	
Lithium	Li	750	890	1080	1240	1310	1420	1518			
Magnesium	Mg	620	740	900	1040	1110	1190	1330	1430	1560	
Manganese	Mn		1510	1810	2050	2100	2360	2580	2850		
Mercury	Hg			260	330	356.9	398	461	517	581	657
Molybdenum	Mo	3300	3770	4200	4580	4830	5050	5340	5680	5980	
Neodymium	Nd			2870	3100	3300	3680	3990			
Nickel	Ni	1800	2090	2370	2620	2730	2880	3120	3300	3310	
Palladium	Pd	1470	2290	2670	2950	3140	3270	3560	3840		
Phosphorus	P		127	199	253	283	319				
Platinum	Pt	2600	2940	3360	3650	3830	4000	4310	4570	4860	
Polonium	Po	472	587	752	890	960	1060	1200	1340		
Potassium	K			590	710	770	850	950	1110	1240	1420
Rhodium	Rh	2530	2850	3260	3590	3760	3930	4230	4440		
Rubidium	Rb		390	527	640	700					
Selenium	Se		429	547	640	685	750	850	920	1010	1120
Silver	Ag	1310	1540	1850	2060	2210	2360	2600	2850	3050	3300
Sodium	Na	440	546	700	830	890	980	1120	1230	1370	
Strontium	Sr	740	900	1100	1280	1380	1480	1670	1850	2030	
Sulfur	S		246	333	407	445	493	574	640	720	
Tellurium	Te	520	633	792	900	962	1030	1160	1250		
Thallium	Tl		1000	1210	1370	1470	1560	1750	1900	2050	2260
Tin	Sn	1610	1890	2270	2580	2750	2950	3270	3540	3890	
Titanium	Ti	2180	2480	2860	3100	3260	3400	3800			
Tungsten	W	3980	4490	5160	5470	5940	6260	6670	7250	7670	
Uranium	U	2450	2800	3270	3620	3800	4040	4420			
Vanadium	V	2290	2570	2950	3220	3380	3540	3800			
Zinc	Zn		590	730	840	907	970	1090	1180	1290	

VAPOR PRESSURE OF ELEMENTS THAT ARE GASEOUS AT STANDARD CONDITIONS

The following tables contain vapor pressure data for helium, hydrogen, neon, nitrogen, and oxygen.

Vapor Pressure (atm) vs. Temperature (K) for Helium-4

Temperature, K	Pressure atm
2.177	0.04969
2.20	0.05256
2.25	0.05916
2.30	0.06629
2.35	0.07399
2.40	0.08228
2.45	0.09120
2.50	0.1008
2.55	0.1110
2.60	0.1219
2.65	0.1336
2.70	0.1460
2.75	0.1591
2.80	0.1730
2.85	0.1878
2.90	0.2033
2.95	0.2198
3.00	0.2371
3.05	0.2553
3.10	0.2744
3.15	0.2945
3.20	0.3156
3.25	0.3376
3.30	0.3607
3.35	0.3848
3.40	0.4100
3.45	0.4363
3.50	0.4637
3.55	0.4923
3.60	0.5220
3.65	0.5528
3.70	0.5849
3.75	0.6182
3.80	0.6528
3.85	0.6886
3.90	0.7257
3.95	0.7642
4.00	0.8040
4.05	0.8452
4.10	0.8878
4.15	0.9318
4.20	0.9772
4.224	1.000
4.25	1.024
4.30	1.073
4.35	1.123
4.40	1.174
4.45	1.227
4.50	1.282
4.55	1.339
4.60	1.397
4.65	1.457
4.70	1.519
4.75	1.582
4.80	1.648
4.85	1.715
4.90	1.784
4.95	1.856
5.00	1.929
5.05	2.004
5.10	2.082
5.201	2.245

Vapor Pressure (atm) vs. Temperature (K) for Equilibrium Hydrogen

Temperature, K	Pressure, atm
13.803	0.069$_5$
14	0.077$_8$
15	0.133
16	0.213
17	0.325
18	0.476
19	0.673
20	0.923
20.268	1.000
21	1.233
22	1.613
23	2.069
24	2.611
25	3.245
26	3.982
27	4.829
28	5.794
29	6.887
30	8.118
31	9.501
32	11.051
32.976	12.759

Vapor Pressure (atm) vs. Temperature (K) for Neon

Temperature, K	Pressure, atm
25	0.50366
26	0.70902
27	0.97255
28	1.3037
29	1.7124
30	2.2088
31	2.8031
32	3.5061
33	4.3286
34	5.2818
35	6.3773
36	7.6271
37	9.0439
38	10.641
39	12.432
40	14.434
41	16.661
42	19.133
43	21.867
44	24.887
44.4	26.19

Vapor Pressure (atm) vs. Temperature (K) for Nitrogen

Temperature, K	Pressure, atm
63.148	0.1237
64	0.1443
66	0.2037
68	0.2813
70	0.3807
72	0.5059
74	0.6610
76	0.8506
77.347	1.0000
78	1.0793
80	1.3520
82	1.6739
84	2.0503
86	2.4865
88	2.9882
90	3.5607
92	4.2099
94	4.9415
98	6.6748
100	7.6885
102	8.8083
104	10.041
106	11.392
108	12.870
110	14.481
112	16.233
114	18.133
116	20.190
118	22.411
120	24.806
122	27.386
124	30.174
126	33.227
126.200	33.555

Vapor Pressure (atm) vs. Temperature (K) for Oxygen

Temperature, K	Pressure, atm
54.351	0.001
56	0.002
58	0.004
60	0.007
62	0.012
64	0.018
66	0.028
68	0.042
70	0.061
72	0.087
74	0.122
76	0.167
78	0.224
80	0.297
82	0.387
84	0.497
86	0.631
88	0.791
90	0.981
90.180	1.000
92	1.205
94	1.466
96	1.768
98	2.114
100	2.509
102	2.957
104	3.462
106	4.029
108	4.661
110	5.363

Vapor Pressure (atm) vs. Temperature (K) for Oxygen (Continued)

Temperature, K	Pressure atm	Temperature, K	Pressure, atm
112	6.139	136	22.986
114	6.995	138	25.170
116	7.934	140	27.501
118	8.961		
120	10.082	142	29.986
		144	32.631
122	11.300	146	35.448
124	12.621	148	38.446
126	14.049	150	41.638
128	15.591		
130	17.249	152	45.041
		154	48.675
132	19.031	154.576	49.767
134	20.942		

Data from Sparks, L. L., ASRDI Oxygen Technology Survey, Vol. IV: Low Temperature Measurement, NASA-3073, 1974.

VAPOR PRESSURE OF NITRIC ACID

Temperature °C	Vapor Pressure, mm. of Hg	
	100% HNO$_3$	90% of HNO$_3$
0	14.4	5.5
10	26.6	11.
20	47.9	20.
30	81.3	37.3
40	133.	64.4
50	208.	107.
70	467.	242.
80	670.	352.
90	937.	504.
100	1282.	710.

LOW TEMPERATURE LIQUID BATHS

Liquid thermostat baths suitable for many physical measurements can be produced by using a stirred solid–liquid mixture at its melting point. Dry-ice or liquid air can be used to produce the solid. A Dewar flask is preferable as a container and, with adequate insulation or immersion in another somewhat colder bath, good temperature constancy can be maintained over several hours. Such baths are especially useful over the temperature range between dry-ice (−78C) and liquid air (−190). The following table gives the melting and normal boiling points of some readily available organic liquids suitable for this purpose. The compounds are listed in order of their increasing melting points. Temperatures are in degrees Celcius.

Compound	M.P.	B.P.	Compound	M.P.	B.P.
Isopentane (2-Methyl butane)	−159.9	27.85	Ethyl acetate	−84	77
Methyl cyclopentane	−142.4	71.8	(Dry-ice + acetone)	−78	——
Allyl chloride	−134.5	45	p-Cymene	−67.9	177.1
n-Pentane	−129.7	36.1	Chloroform	−63.5	61.7
Allyl alcohol	−129	97	N-Methyl aniline	−57	196
Ethyl alcohol	−117.3	78.5	Chlorobenzene	−45.6	132
Carbon disulfide	−110.8	46.3	Anisole	−37.5	155
Isobutyl alcohol	−108	108.1	Bromobenzene	−30.8	156
Acetone	−95.4	56.2	Carbon tetrachloride	−23	76.5
Toluene	−95	110.6	Benzonitrile	−13	205

FATS AND OILS

These data for fats and oils were compiled originally for the Biology Data Book by H. J. Harwood, and R. P. Geyer. 1964. Data are reproduced here by permission of the copyright owners of the above publication, the Federation of American Societies for Experimental Biology, Washington, D.C. pp. 380–382.

Values are typical rather than average, and frequently were derived from specific analyses for particular samples (especially the constituent fatty acids). Extreme variations may occur, depending on a number

	Fat or Oil	Source	Melting (or Solidification) Point, °C	Specific Gravity (or Density)	Refractive Index 40°C $n\frac{}{D}$	Iodine Value	Saponification Value
	(A)	(B)	(C)	(D)	(E)	(F)	(G)
	Land Animals						
1	Butterfat	*Bos taurus*	32.2	$0.911^{40°/15°}$	1.4548	36.1	227
2	Depot fat	*Homo sapiens*	(15)	$0.918^{15°}$	1.4602	67.6	196.2
3	Lard oil	*Sus scrofa*	(30.5)	$0.919^{15°}$	1.4615	58.6	194.6
4	Neat's-foot oil	*B. taurus*	....	$0.910^{25°}$	$1.464^{25°}$	69–76	190–199
5	Tallow, beef	*B. taurus*	....			49.5	197
6	Tallow, mutton	*Ovis aries*	(42.0)	$0.945^{15°}$	1.4565	40	194
	Marine Animals						
7	Cod-liver oil	*Gadus morhua*	....	0.925^{25}	$1.481^{25°}$	165	186
8	Herring oil	*Clupea harengus*	....	$0.900^{60°}$	$1.4610^{60°}$	140	192
9	Menhaden oil	*Brevoortia tyrannus*	....	$0.903^{60°}$	$1.4645^{60°}$	170	191
10	Sardine oil	*Sardinops caerulea*	....	$0.905^{60°}$	$1.4660^{60°}$	185	191
11	Sperm oil, body	*Physeter macrocephalus*	....	....	....	76–88	122–130
12	Sperm oil, head	*P. macrocephalus*	....	....	....	70	140–144
13	Whale oil	*Balaena mysticetus*	....	0.892^{60}	$1.460^{60°}$	120	195
	Plants						
14	Babassu oil	*Attalea funifera*	22–26	$(0.893^{60°})$	$1.443^{60°}$	15.5	247
15	Castor oil	*Ricinus communis*	(−18.0)	$0.961^{15°}$	1.4770	85.5	180.3
16	Cocoa butter	*Theobroma cacao*	34.1	$0.964^{15°}$	1.4568	36.5	193.8
17	Coconut oil	*Cocos nucifera*	25.1	$0.924^{15°}$	1.4493	10.4	268
18	Corn oil	*Zea mays*	(−20.0)	$0.922^{15°}$	1.4734	122.6	192.0
19	Cotton seed oil	*Gossypium hirsutum*	(−1.0)	$0.917^{25°}$	1.4735	105.7	194.3
20	Linseed oil	*Linum usitatissimum*	(−24.0)	$0.938^{15°}$	$1.4782^{25°}$	178.7	190.3
21	Mustard oil	*Brassica hirta*	....	$0.9145^{15°}$	1.475	102	174
22	Neem oil	*Melia azadirachta*	−3	$0.917^{15°}$	1.4615	71	194.5
23	Niger-seed oil	*Guizotia abyssinica*	....	$0.925^{15°}$	1.471	128.5	190
24	Oiticica oil	*Licania rigida*	....	$0.974^{25°}$	....	140–180	....
25	Olive oil	*Olea europaea sativa*	(−6.0)	$0.918^{15°}$	1.4679	81.1	189.7
26	Palm oil	*Elaeis guineensis*	35.0	0.915^{15c}	1.4578	54.2	199.1
27	Palm-kernel oil	*E. guineensis*	24.1	$0.923^{15°}$	1.4569	37.0	219.9
28	Peanut oil	*Arachis hypogaea*	(3.0)	$0.914^{15°}$	1.4691	93.4	192.1
29	Perilla oil	*Perilla frutescens*	....	$(0.935^{15°})$	$1.481^{25°}$	195	192
30	Poppy-seed oil	*Papaver somniferum*	(−15)	$0.925^{15°}$	1.4685	135	194
31	Rapeseed oil	*Brassica campestris*	(−10)	$0.915^{15°}$	1.4706	98.6	174.7
32	Safflower oil	*Carthamus tinctorius*	....	$(0.9000^{60°})$	$1.462^{60°}$	145	192
33	Sesame oil	*Sesamum indicum*	(−6.0)	$0.919^{25°}$	1.4646	106.6	187.9
34	Soybean oil	*Glycine soja*	(−16.0)	$0.927^{15°}$	1.4729	130.0	190.6
35	Sunflower-seed oil	*Helianthus annuus*	(−17.0)	$0.923^{15°}$	1.4694	125.5	188.7
36	Tung oil	*Aleurites fordi*	(−2.5)	$0.934^{15°}$	$1.5174^{25°}$	168.2	193.1
37	Wheat-germ oil	*Triticum aestivum*	....	....	....	125	....

[1] Caproic. [2] Capryli. [3] Capric. [4] Butyric. [5] Decenoic. [6] C_{12} monoethenoic. [7] C_{14} monoethenoic. [8] Gadoleic plus erucic. [9] C_{12} n-pentadecanoic. [10] C_{17} margaric. [11] 12-Methyl tetradecanoic. [12] C_{20} polyethenoic.

of variables such as source, treatment, and age of a fat or oil. **Specific Gravity** (column D) was calculated at the specified temperature (degrees centigrade) and referred to water at the same temperature, unless otherwise specified. **Density,** shown in parentheses (column D), was measured at the specified temperature (degrees centigrade). **Refractive Index** (column E) was measured at 50°C, unless otherwise specified.

Constituent Fatty Acids, g/100 g total fatty acids

	Saturated						Unsaturated				
	Lauric	Myristic	Palmitic	Stearic	Arachidic	Other	Palmitoleic	Oleic	Linoleic	Linolenic	Other
	(H)	(I)	(J)	(K)	(L)	(M)	(N)	(O)	(P)	(Q)	(R)
1	2.5	11.1	29.0	9.2	2.4	2.0[1]; 0.5[2]; 2.3[3]	4.6	26.7	3.6	....	3.6[4]; 0.1[5]; 0.1[6]; 0.9[7] 1.4[8]; 1.0[9]; 1.0[10]; 0.4[11]
2	....	2.7	24.0	8.4	....	....	5	46.9	10.2	....	2.5[8]
3	....	1.3	28.3	11.9	....	....	2.7	47.5	6	....	0.2[7]; 2.1[8]
4	....	....	17–18	2–3	....	....	....	74–76	....	....	....
5	....	6.3	27.4	14.1	....	....	....	49.6	2.5	....	....
6	....	4.6	24.6	30.5	....	....	....	36.0	4.3	....	....
7	....	5.8	8.4	0.6	....	....	20.0	←—29.1—→		....	25.4[12]; 9.6[13]
8	....	7.3	13.0	Trace	....	....	4.9	....	....	20.7	30.1[12]; 23.2[13]
9	....	5.9	16.3	0.6	0.6	....	15.5	....	....	29.6	19.0[12]; 11.7[13]; 0.8[14]
10	....	5.1	14.6	3.2	....	....	11.8	←—17.8—→		....	18.1[12]; 14.0[13]; trace[7]; 15.4[15]
11	1	5	6.5	....	....	....	26.5	37	19	....	1[13]; 4[7]; 19[16]
12	16	14	8	2	....	3.5[3]	15	17	6.5	....	4[6]; 14[7]; 6.5[16]
13	0.2	9.3	15.6	2.8	....	....	14.4	35.2	....	....	13.6[12]; 5.9[13]; 2.5[7]; 0.2[17]
14	44.1	15.4	8.5	2.7	0.2	0.2[1]; 4.8[2]; 6.6[3]	....	16.1	1.4	....	....
15	←———— 2.4 ————→						....	7.4	3.1	....	87[18]
16	....	....	24.4	35.4	....	....	....	38.1	2.1	....	....
17	45.4	18.0	10.5	2.3	0.4[19]	0.8[1]; 5.4[2]; 8.4[3]	0.4	7.5	Trace	....	....
18	....	1.4	10.2	3.0	....	....	1.5	49.6	34.3	....	....
19	....	1.4	23.4	1.1	1.3	....	2.0	22.9	47.8	....	....
20	....	....	6.3	2.5	0.5	....	....	19.0	24.1	47.4	0.2[14]
21	....	1.3[20]	....	....	....	....	....	27.2[20]	16.6[20]	1.8[20]	1.1[14]; 1.0[21]; 51.0[22]
22	....	2.6[20]	14.1[20]	24.0[20]	0.8[20]	....	....	58.5[20]	....	....	....
23	....	3.3[20]	8.2[20]	4.8[20]	0.5[20]	....	....	30.3[20]	57.3[20]	....	....
24	←———— 11.3[23] ————→						....	6.2	....	....	82.5[24]
25	....	Trace	6.9	2.3	0.1	....	....	84.4	4.6	....	....
26	....	1.4	40.1	5.5	....	....	....	42.7	10.3	....	....
27	46.9	14.1	8.8	1.3	....	2.7[2]; 7.0[3]	....	18.5	0.7	....	....
28	....	....	8.3	3.1	2.4	....	....	56.0	26.0	....	3.1[14]; 1.1[21]
29	←———— 9.6[23] ————→						....	17.8	....	17.5	....
30	....	....	4.8[20]	2.9[20]	....	....	....	30.1[20]	62.2[20]	....	....
31	....	....	1	....	....	....	....	32	15	1	50[22]
32	←———— 6.8[23] ————→						....	18.6	70.1	3.4	....
33	....	....	9.1	4.3	0.8	....	....	45.4	40.4	....	....
34	0.2	0.1	9.8	2.4	0.9	....	0.4	28.9	50.7	6.5	0.1[7]
35	....	....	5.6	2.2	0.9	....	....	25.1	66.2	....	....
36	←———— 4.6[23] ————→						....	4.1	0.6	....	90.7[25]
37	←———— 16.0[23] ————→						....	28.1	52.3	3.6	....

[13] C_{22} polyethenoic. [14] Behenic. [15] C_{14} polyethenoic. [16] Gadoleic. [17] C_{24} polyethenoic. [18] Ricinoleic. [19] Includes behenic and lignoceric. [20] Percent by weight. [21] Lignoceric. [22] Erucic. [23] Includes behenic. [24] Licanic. [25] Eleostearic.

CONCENTRATIVE PROPERTIES OF AQUEOUS SOLUTIONS: CONVERSION TABLES

A. V. Wolf, Morden G. Brown and Phoebe G. Prentiss

The table columns are:

$A\%$ = anhydrous solute weight per cent, g solute/100 g solution.

$H\%$ = hydrated solute weight per cent, g solute/100 g solution.

ρ or D_4^{20} = relative density at 20°C, kg/l.

D_{20}^{20} = specific gravity at 20°C.

C_s = anhydrous solute concentration, g/l.

M = molar concentration, g-mol/l.

C_w = total water concentration, g/l.

$(C_0 - C_w)$ = water displaced by anhydrous solute, g/l.

$(n - n_0) \times 10^4$ = index of refraction increment above index of refraction of pure water $\times 10^4$; refraction at 20°C.

n = index of refraction at 20°C relative to air for sodium yellow light.

Δ = freezing point depression, °C.

O = osmolality, Os/kg water.

S = osmosity, molar concentration of NaCl solution having same freezing point or osmotic pressure as given solution, g-mol/l.

η/η_0 = relative viscosity, ratio of the absolute viscosity of a solution at 20°C to the absolute viscosity of water at 20°C.

η/ρ = kinematic viscosity, ratio of absolute viscosity at 20°C (centipoise, cP) to relative density, (centistokes, cS).

ϕ = fluidity, reciprocal of absolute viscosity at 20°C, (poise^{-1}), rhe.

γ = specific conductance (electrical) at 20°C, mmho/cm.

T = condosity, molar concentration of NaCl solution having same specific conductance (electrical) at 20°C as given solution, g-mol/l.

Some related measures are:

Specific gravity, $D_{20}^{20} = D_4^{20}/.99823$

Relative density, $D_4^{20} = (C_s + C_w)/1000$

Molality, g-mol/kg water $= 1000 \times M/C_w$

Osmolality, $O = \Delta/1.86$

Absolute viscosity (cP) = kinematic viscosity (cS) $\times D_4^{20}$

Specific viscosity $= \eta/\eta_0 - 1$

Specific refraction $= (n - 1)/\rho$

Specific refractive increment $= 1000(n - n_0)/C_w$

Molar conductance (ohm^{-1} cm^2 g-mol^{-1}), $\Lambda_M = \lambda/M$

Relative salinity, $\Sigma = T/S$

Ratios such as S/M, T/M, r/A%, $(D_{20}^{20} - 1)/S$, etc.

G water displaced/g solute $= (C_0 - C_w)/C_s$

Ml water displaced/g solute $= (C_0 - C_w)/(.99823 \times C_s)$

G water displaced/mol solute $= (C_0 - C_w)/M$

Ml water displaced/mol solute $= (C_0 - C_w)/(.99823 \times M)$

Mols water displaced/mol solute $= (C_0 - C_w)/(18.015 \times M)$

Concentration of solute relative to water, g/kg water $= 1000 \times C_s/C_w$

Relative specific refractivity = contribution of unit weight of solute relative to unit weight of water to the imaginary concentration of water with same refractive index as solution, equivalent concentration of water (ECW) $= 3694.1788 \times n - 300 \times n^2 - 3394.1788$. The refractive index for this equation is relative to a vacuum whereas the tabulated values are relative to air. The above equation for the equivalent concentration of water provides better constancy of relative specific refractivity than Lorentz and Lorenz relation, ECW $= 4848.3431 - 14545.029/(n^2 + 2)$.

Tables are based upon *Handbook of Chemistry and Physics,* 51st Edition (1970); G. F. Hewitt, Tables of the resistivity of aqueous chloride solutions, Chem. Eng. Div., U.K.A.E.A. Res. Group, Harwell, Gr. Brit. Oct. 1960, HL 60/5450 (S.C. 2), AERE-R 3497; A. V. Wolf, *Aqueous Solutions and Body Fluids,* Hoeber, 1966; and new data.

LIST OF TABLES

1 ACETIC ACID, CH₃COOH

MOLECULAR WEIGHT = 60.05
RELATIVE SPECIFIC REFRACTIVITY = 1.065

0.00 % by wt. data are the same for all compounds.
For Values of 0.00 wt. % solutions see Table 1, Acetic Acid.

A% by wt.	ρ D_4^{20}	D_{20}^{20}	C_s g/l	M g-mol/l	C_w g/l	$(C_o - C_w)$ g/l	$(n - n_o)$ × 10⁴	n	Δ °C	O Os/kg	S g-mol/l	η/η_o	η/ρ cS	φ rhe	γ mmho/cm	T g-mol/l
0.00	0.9982	1.0000	0.0	0.000	998.2	0.0	0	1.3330	0.000	0.000	0.000	1.000	1.004	99.80	0.0	0.000
0.50	0.9989	1.0007	5.0	0.083	993.9	4.3	4	1.3334	0.159	0.086	0.045	1.010	1.013	98.81	0.3	0.003
1.00	0.9996	1.0014	10.0	0.166	989.6	8.6	7	1.3337	0.317	0.170	0.091	1.020	1.022	97.84	0.6	0.006
1.50	1.0003	1.0021	15.0	0.250	985.3	12.9	11	1.3341	0.474	0.255	0.137	1.030	1.032	96.90	0.7	0.007
2.00	1.0011	1.0028	20.0	0.333	981.0	17.2	15	1.3345	0.630	0.339	0.183	1.040	1.041	95.96	0.8	0.008
2.50	1.0018	1.0035	25.0	0.417	976.7	21.5	18	1.3348	0.786	0.423	0.229	1.050	1.051	95.02	0.9	0.009
3.00	1.0025	1.0042	30.1	0.501	972.4	25.9	22	1.3352	0.942	0.507	0.275	1.061	1.060	94.07	1.0	0.010
3.50	1.0031	1.0049	35.1	0.585	968.0	30.2	26	1.3355	1.099	0.591	0.321	1.072	1.070	93.14	1.1	0.010
4.00	1.0038	1.0056	40.2	0.669	963.7	34.5	29	1.3359	1.256	0.675	0.367	1.082	1.080	92.24	1.1	0.011
4.50	1.0045	1.0063	45.2	0.753	959.3	38.9	33	1.3363	1.415	0.761	0.414	1.092	1.090	91.36	1.2	0.011
5.00	1.0052	1.0070	50.3	0.837	955.0	43.3	36	1.3366	1.576	0.847	0.461	1.103	1.099	90.50	1.2	0.012
5.50	1.0059	1.0077	55.3	0.921	950.6	47.6	40	1.3370	1.737	0.934	0.508	1.113	1.109	89.66	1.3	0.012
6.00	1.0066	1.0084	60.4	1.006	946.2	52.0	44	1.3373	1.899	1.021	0.555	1.123	1.118	88.87	1.3	0.013
6.50	1.0073	1.0091	65.5	1.090	941.8	56.4	47	1.3377	2.062	1.108	0.602	1.132	1.126	88.14	1.4	0.013
7.00	1.0080	1.0098	70.6	1.175	937.4	60.8	51	1.3381	2.225	1.196	0.650	1.141	1.134	87.46	1.4	0.013
7.50	1.0087	1.0105	75.7	1.260	933.0	65.2	54	1.3384	2.390	1.285	0.697	1.150	1.142	86.78	1.4	0.014
8.00	1.0093	1.0111	80.7	1.345	928.6	69.6	58	1.3388	2.555	1.374	0.745	1.160	1.152	86.03	1.4	0.014
8.50	1.0100	1.0118	85.9	1.430	924.2	74.1	62	1.3391	2.722	1.463	0.792	1.171	1.162	85.20	1.5	0.014
9.00	1.0107	1.0125	91.0	1.515	919.7	78.5	65	1.3395	2.889	1.553	0.840	1.184	1.173	84.31	1.5	0.014
9.50	1.0114	1.0132	96.1	1.600	915.3	82.9	69	1.3399	3.057	1.644	0.888	1.196	1.185	83.43	1.5	0.015
10.00	1.0121	1.0138	101.2	1.685	910.8	87.4	72	1.3402	3.226	1.734	0.935	1.208	1.196	82.62	1.5	0.015
11.00	1.0134	1.0152	111.5	1.856	901.9	96.3	79	1.3409	3.567	1.918	1.031	1.229	1.215	81.19	1.6	0.015
12.00	1.0147	1.0165	121.8	2.028	893.0	105.3	86	1.3416	3.911	2.103	1.127	1.250	1.234	79.84	1.6	0.016
13.00	1.0161	1.0178	132.1	2.200	884.0	114.3	93	1.3423	4.259	2.290	1.223	1.272	1.255	78.43	1.6	0.016
14.00	1.0174	1.0192	142.4	2.372	874.9	123.3	100	1.3430	4.611	2.479	1.320	1.295	1.275	77.07	1.7	0.016
15.00	1.0187	1.0205	152.8	2.545	865.9	132.4	107	1.3437	4.967	2.670	1.415	1.317	1.295	75.79	1.7	0.016
16.00	1.0200	1.0218	163.2	2.718	856.8	141.5	114	1.3444	5.33	2.86	1.511	1.338	1.314	74.59	1.7	0.016
17.00	1.0213	1.0231	173.6	2.891	847.6	150.6	121	1.3451	5.69	3.06	1.607	1.357	1.331	73.54	1.7	0.017
18.00	1.0225	1.0243	184.1	3.065	838.5	159.8	128	1.3458	6.06	3.26	1.703	1.377	1.349	72.48	1.7	0.017
19.00	1.0238	1.0256	194.5	3.239	829.3	169.0	135	1.3465	6.43	3.46	1.798	1.402	1.372	71.18	1.7	0.017
20.00	1.0250	1.0269	205.0	3.414	820.0	178.2	142	1.3472	6.81	3.66	1.894	1.428	1.396	69.89	1.7	0.017
22.00	1.0275	1.0293	226.1	3.764	801.5	196.8	155	1.3485	7.57	4.07	2.084	1.475	1.438	67.66	1.7	0.017
24.00	1.0299	1.0318	247.2	4.116	782.8	215.6	169	1.3498	8.36	4.49	2.273	1.522	1.481	65.57	1.7	0.016
26.00	1.0323	1.0341	268.4	4.470	763.9	234.3	182	1.3512	9.17	4.93	2.462	1.569	1.523	63.61	1.6	0.015
28.00	1.0346	1.0365	289.7	4.824	744.9	253.3	195	1.3525	10.00	5.38	2.650	1.610	1.559	61.99	1.5	0.015
30.00	1.0369	1.0388	311.1	5.180	725.8	272.4	207	1.3537	10.84	5.83	2.835	1.666	1.610	59.90	1.4	0.014
32.00	1.0391	1.0410	332.5	5.537	706.6	291.6	220	1.3550	11.70	6.29	3.017	1.712	1.651	58.29	1.4	0.013
34.00	1.0413	1.0431	354.0	5.896	687.3	311.0	232	1.3562	12.55	6.75	3.192	1.758	1.692	56.77	1.3	0.012
36.00	1.0434	1.0452	375.6	6.255	667.8	330.5	245	1.3574	13.38	7.20	3.359	1.808	1.736	55.20	1.2	0.011
38.00	1.0454	1.0473	397.3	6.615	648.2	350.1	257	1.3586				1.848	1.771	54.00	1.1	0.011
40.00	1.0474	1.0492	419.0	6.977	628.4	369.8	268	1.3598				1.908	1.825	52.31	1.1	0.010
42.00	1.0493	1.0511	440.7	7.339	608.6	389.7	280	1.3610				1.956	1.868	51.02	1.0	0.010
44.00	1.0510	1.0529	462.5	7.701	588.6	409.7	291	1.3621				2.003	1.910	49.83	1.0	0.009
46.00	1.0528	1.0547	484.3	8.065	568.5	429.7	302	1.3632				2.048	1.949	48.73	0.9	0.009
48.00	1.0545	1.0564	506.2	8.429	548.4	449.9	312	1.3642				2.106	2.001	47.39	0.8	0.008
50.00	1.0562	1.0581	528.1	8.794	528.1	470.1	323	1.3653				2.154	2.043	46.33	0.8	0.007
52.00	1.0577	1.0596	550.0	9.159	507.7	490.5	333	1.3663				2.208	2.092	45.20	0.7	0.007
54.00	1.0592	1.0611	572.0	9.525	487.2	511.0	343	1.3673				2.260	2.138	44.16	0.7	0.006
56.00	1.0605	1.0624	593.9	9.890	466.6	531.6	352	1.3682				2.303	2.176	43.33	0.5	0.005
58.00	1.0618	1.0636	615.8	10.255	445.9	552.3	361	1.3691				2.355	2.222	42.38	0.3	0.003
60.00	1.0629	1.0648	637.7	10.620	425.2	573.1	370	1.3700				2.404	2.266	41.51	0.4	0.004
62.00	1.0640	1.0659	659.7	10.985	404.3	593.9	378	1.3708				2.451	2.308	40.72		
64.00	1.0650	1.0668	681.6	11.350	383.4	614.8	386	1.3716				2.497	2.349	39.97		
66.00	1.0659	1.0678	703.5	11.715	362.4	635.8	394	1.3724				2.548	2.395	39.17		
68.00	1.0668	1.0687	725.4	12.080	341.4	656.9	402	1.3732				2.589	2.432	38.55		
70.00	1.0673	1.0692	747.1	12.441	320.2	678.0	409	1.3738				2.624	2.463	38.03		
72.00	1.0676	1.0695	768.7	12.800	298.9	699.3	415	1.3745				2.657	2.494	37.56		
74.00	1.0678	1.0697	790.2	13.158	277.6	720.6	421	1.3751				2.682	2.517	37.21		
76.00	1.0680	1.0699	811.7	13.516	256.3	741.9	427	1.3757				2.709	2.542	36.84		
78.00	1.0681	1.0700	833.1	13.956	235.0	763.2	432	1.3762				2.715	2.547	36.76		
80.00	1.0680	1.0699	854.4	14.227	213.6	784.6	437	1.3767				2.715	2.547	36.76		
82.00	1.0677	1.0696	875.5	14.579	192.2	806.0	441	1.3770				2.691	2.525	37.09		
84.00	1.0673	1.0692	896.5	14.928	170.8	827.4	443	1.3773				2.653	2.491	37.62		
86.00	1.0666	1.0685	917.3	15.275	149.3	848.9	444	1.3774				2.591	2.434	38.52		
88.00	1.0658	1.0677	937.9	15.618	127.9	870.3	444	1.3774				2.506	2.356	39.82		
90.00	1.0644	1.0663	958.0	15.953	106.4	891.8	441	1.3771				2.381	2.241	41.92		
92.00	1.0629	1.0648	977.9	16.284	85.0	913.2	436	1.3766				2.236	2.108	44.63		
94.00	1.0606	1.0625	997.0	16.602	63.6	934.6	429	1.3759				2.032	1.920	49.11		
96.00	1.0578	1.0597	1015.5	16.912	42.3	955.9	418	1.3748				1.809	1.714	55.17		
98.00	1.0538	1.0557	1032.7	17.196	21.1	977.1	404	1.3734				1.532	1.457	65.14		
100.00	1.0477	1.0496	1047.7	17.446	0.0	998.2	386	1.3716				1.221	1.168	81.73		

2 ACETONE, CH₃COCH₃

MOLECULAR WEIGHT = 58.05
RELATIVE SPECIFIC REFRACTIVITY = 1.349

0.00 % by wt. data are the same for all compounds.
For Values of 0.00 wt. % solutions see Table 1, Acetic Acid.

A% by wt.	ρ D_4^{20}	D_{20}^{20}	C_s g/l	M g-mol/l	C_w g/l	$(C_o - C_w)$ g/l	$(n - n_o)$ × 10⁴	n	Δ °C	O Os/kg	S g-mol/l	η/η_o	η/ρ cS	φ rhe	γ mmho/cm	T g-mol/l
0.50	0.9975	0.9993	5.0	0.086	992.5	5.7	4	1.3334	0.160	0.086	0.045	1.011	1.015	98.73		
1.00	0.9968	0.9985	10.0	0.172	986.8	11.4	7	1.3337	0.321	0.173	0.092	1.022	1.027	97.65		
1.50	0.9961	0.9978	14.9	0.257	981.1	17.1	11	1.3341	0.483	0.260	0.140	1.033	1.039	96.58		

2 ACETONE, CH₃COCH₃—(Continued)

A% by wt.	ρ D_4^{20}	D_{20}^{20}	C_s g/l	M g-mol/l	C_w g/l	$(C_o - C_w)$ g/l	$(n - n_o)$ $\times 10^4$	n	Δ °C	O Os/kg	S g-mol/l	η/η_o	η/ρ cS	ϕ rhe	γ mmho/cm	T g-mol/l
2.00	0.9954	0.9971	19.9	0.343	975.5	22.8	15	1.3344	0.645	0.347	0.187	1.045	1.052	95.50		
2.50	0.9947	0.9964	24.9	0.428	969.8	28.4	18	1.3348	0.807	0.434	0.235	1.057	1.065	94.39		
3.00	0.9940	0.9957	29.8	0.514	964.1	34.1	22	1.3352	0.970	0.521	0.283	1.070	1.079	93.25		
3.50	0.9933	0.9950	34.8	0.599	958.5	39.7	25	1.3355	1.133	0.609	0.331	1.083	1.093	92.11		
4.00	0.9926	0.9943	39.7	0.684	952.9	45.4	29	1.3359	1.296	0.697	0.379	1.097	1.107	90.99		
4.50	0.9919	0.9937	44.6	0.769	947.3	51.0	33	1.3363	1.460	0.785	0.427	1.110	1.121	89.90		
5.00	0.9912	0.9930	49.6	0.854	941.7	56.6	36	1.3366	1.625	0.874	0.475	1.123	1.135	88.87		
5.50	0.9906	0.9923	54.5	0.939	936.1	62.1	40	1.3370	1.790	0.962	0.523	1.136	1.149	87.89		
6.00	0.9899	0.9917	59.4	1.023	930.5	67.7	44	1.3373	1.955	1.051	0.571	1.148	1.162	86.94		
6.50	0.9893	0.9910	64.3	1.108	925.0	73.3	47	1.3377	2.120	1.140	0.619	1.160	1.175	86.03		
7.00	0.9886	0.9904	69.2	1.192	919.4	78.8	51	1.3381	2.286	1.229	0.667	1.172	1.188	85.14		
7.50	0.9880	0.9897	74.1	1.276	913.9	84.4	54	1.3384	2.452	1.318	0.715	1.184	1.201	84.28		
8.00	0.9874	0.9891	79.0	1.361	908.4	89.9	58	1.3388	2.619	1.408	0.763	1.196	1.214	83.44		
8.50	0.9867	0.9885	83.9	1.445	902.9	95.4	62	1.3392	2.785	1.498	0.811	1.208	1.226	82.64		
9.00	0.9861	0.9879	88.8	1.529	897.4	100.9	65	1.3395	2.952	1.587	0.858	1.219	1.239	81.85		
9.50	0.9855	0.9873	93.6	1.613	891.9	106.3	69	1.3399	3.119	1.677	0.905	1.231	1.251	81.09		
10.00	0.9849	0.9867	98.5	1.697	886.4	111.8	73	1.3402	3.287	1.767	0.953	1.242	1.264	80.35		

3 AMMONIA, NH₃, AND AMMONIUM HYDROXIDE, NH₄OH

MOLECULAR WEIGHT, NH₃ = 17.03   FORMULA WEIGHT, NH₄OH = 35.05
RELATIVE SPECIFIC REFRACTIVITY = 1.582

0.00 % by wt. data are the same for all compounds.
For Values of 0.00 wt. % solutions see Table 1, Acetic Acid.

NH₃ % by wt.	NH₄OH % by wt.	ρ D_4^{20}	D_{20}^{20}	C_s (NH₃) g/l	M g-mol/l	C_w g/l	$(C_o - C_w)$ g/l	$(n - n_o)$ $\times 10^4$	n	Δ °C	O Os/kg	S g-mol/l	η/η_o	η/ρ cS	ϕ rhe	γ mmho/cm	T g-mol/l
0.50	1.03	0.9960	0.9978	5.0	0.292	991.0	7.2	2	1.3332	0.550	0.296	0.160	1.007	1.013	99.14	0.5	0.005
1.00	2.06	0.9938	0.9956	9.9	0.584	983.9	14.3	5	1.3335	1.135	0.610	0.332	1.013	1.022	98.48	0.7	0.007
1.50	3.09	0.9917	0.9934	14.9	0.873	976.8	21.5	7	1.3337	1.725	0.927	0.504	1.020	1.031	97.82	0.9	0.008
2.00	4.12	0.9895	0.9913	19.8	1.162	969.7	28.5	9	1.3339	2.319	1.247	0.677	1.027	1.040	97.17	1.0	0.009
2.50	5.15	0.9874	0.9891	24.7	1.449	962.7	35.5	12	1.3342	2.921	1.570	0.849	1.034	1.049	96.52	1.0	0.010
3.00	6.17	0.9853	0.9870	29.6	1.736	955.7	42.5	14	1.3344	3.531	1.898	1.021	1.041	1.059	95.88	1.1	0.011
3.50	7.20	0.9832	0.9849	34.4	2.021	948.8	49.5	17	1.3347	4.150	2.231	1.193	1.048	1.068	95.24	1.1	0.011
4.00	8.23	0.9811	0.9828	39.2	2.304	941.8	56.4	19	1.3349	4.781	2.570	1.365	1.055	1.077	94.60	1.1	0.011
4.50	9.26	0.9790	0.9808	44.1	2.587	935.0	63.3	22	1.3352	5.42	2.92	1.537	1.062	1.087	93.98	1.1	0.011
5.00	10.29	0.9770	0.9787	48.8	2.868	928.1	70.1	24	1.3354	6.08	3.27	1.708	1.069	1.096	93.36	1.1	0.011
5.50	11.32	0.9750	0.9767	53.6	3.149	921.3	76.9	27	1.3357	6.75	3.63	1.879	1.076	1.106	92.75	1.1	0.011
6.00	12.35	0.9730	0.9747	58.4	3.428	914.6	83.6	30	1.3359	7.43	4.00	2.049	1.083	1.115	92.15	1.1	0.011
6.50	13.38	0.9710	0.9727	63.1	3.706	907.9	90.4	32	1.3362	8.13	4.37	2.219	1.090	1.125	91.55	1.1	0.011
7.00	14.41	0.9690	0.9707	67.8	3.983	901.2	97.1	35	1.3365	8.85	4.76	2.388	1.097	1.134	90.96	1.1	0.011
7.50	15.44	0.9671	0.9688	72.5	4.259	894.5	103.7	38	1.3367	9.58	5.15	2.557	1.104	1.144	90.38	1.1	0.011
8.00	16.47	0.9651	0.9668	77.2	4.534	887.9	110.3	40	1.3370	10.34	5.56	2.725	1.111	1.154	89.81	1.1	0.011
8.50	17.49	0.9632	0.9649	81.9	4.807	881.3	116.9	43	1.3373	11.11	5.97	2.892	1.118	1.163	89.25	1.1	0.010
9.00	18.52	0.9613	0.9630	86.5	5.080	874.8	123.5	46	1.3376	11.90	6.40	3.059	1.125	1.173	88.70	1.1	0.010
9.50	19.55	0.9594	0.9611	91.1	5.352	868.3	130.0	48	1.3378	12.71	6.84	3.225	1.132	1.182	88.16	1.0	0.010
10.00	20.58	0.9575	0.9592	95.8	5.623	861.8	136.5	51	1.3381	13.55	7.28	3.391	1.139	1.192	87.63	1.0	0.010
11.00	22.64	0.9538	0.9555	104.9	6.161	848.9	149.3	57	1.3387	15.29	8.22	3.719	1.153	1.211	86.57	1.0	0.009
12.00	24.70	0.9502	0.9519	114.0	6.695	836.2	162.1	63	1.3393	17.13	9.21	4.044	1.167	1.230	85.53	0.9	0.009
13.00	26.76	0.9466	0.9483	123.1	7.226	823.5	174.7	68	1.3398	19.07	10.25	4.365	1.180	1.250	84.55	0.8	0.008
14.00	28.81	0.9431	0.9447	132.0	7.753	811.0	187.2	74	1.3404	21.13	11.36		1.193	1.268	83.64	0.8	0.007
15.00	30.87	0.9396	0.9412	140.9	8.276	798.6	199.6	80	1.3410	23.32	12.54		1.205	1.285	82.84	0.7	0.007
16.00	32.93	0.9361	0.9378	149.8	8.795	786.4	211.9	86	1.3416	25.63	13.78		1.216	1.301	82.10	0.7	0.006
17.00	34.99	0.9327	0.9344	158.6	9.311	774.2	224.1	92	1.3422	28.09	15.10		1.226	1.317	81.43	0.6	0.006
18.00	37.05	0.9294	0.9310	167.3	9.823	762.1	236.1	98	1.3428	30.70	16.51		1.235	1.331	80.82	0.6	0.005
19.00	39.10	0.9261	0.9277	176.0	10.332	750.1	248.1	104	1.3434	33.47	18.00		1.243	1.345	80.26	0.5	0.005
20.00	41.16	0.9228	0.9245	184.6	10.838	738.3	260.0	110	1.3440	36.42	19.58		1.251	1.359	79.75	0.5	0.005
22.00	45.28	0.9164	0.9181	201.6	11.839	714.8	283.4	123	1.3453	43.36	23.31		1.265	1.383	78.88	0.4	0.004
24.00	49.40	0.9102	0.9118	218.4	12.827	691.7	306.5	135	1.3465	51.38	27.62		1.277	1.405	78.18	0.4	0.004
26.00	53.51	0.9040	0.9056	235.0	13.802	669.0	329.3	148	1.3477	60.77	32.67		1.285	1.425	77.64	0.4	0.004
28.00	57.63	0.8980	0.8996	251.4	14.764	646.5	351.7	160	1.3490	71.66	38.53						
30.00	61.74	0.8920	0.8936	267.6	15.713	624.4	373.8	172	1.3502	84.06	45.19						

4 AMMONIUM CHLORIDE, NH₄Cl

MOLECULAR WEIGHT = 53.50
RELATIVE SPECIFIC REFRACTIVITY = 1.241

0.00 % by wt. data are the same for all compounds.
For Values of 0.00 wt. % solutions see Table 1, Acetic Acid.

A% by wt.	ρ D_4^{20}	D_{20}^{20}	C_s g/l	M g-mol/l	C_w g/l	$(C_o - C_w)$ g/l	$(n - n_o)$ $\times 10^4$	n	Δ °C	O Os/kg	S g-mol/l	η/η_o	η/ρ cS	ϕ rhe	γ mmho/cm	T g-mol/l
0.50	0.9998	1.0016	5.0	0.093	994.8	3.4	10	1.3340	0.322	0.173	0.092	0.997	0.999	100.13	10.5	0.110
1.00	1.0014	1.0032	10.0	0.187	991.4	6.9	19	1.3349	0.637	0.343	0.185	0.994	0.995	100.40	20.4	0.225
1.50	1.0030	1.0047	15.0	0.281	987.9	10.3	29	1.3359	0.953	0.512	0.278	0.992	0.991	100.62	30.3	0.347
2.00	1.0045	1.0063	20.1	0.376	984.4	13.8	39	1.3369	1.270	0.683	0.371	0.990	0.988	100.81	40.3	0.474
2.50	1.0061	1.0079	25.2	0.470	980.9	17.3	48	1.3378	1.590	0.855	0.465	0.988	0.984	101.00	49.9	0.601
3.00	1.0076	1.0094	30.2	0.565	977.4	20.8	58	1.3388	1.913	1.029	0.559	0.986	0.981	101.18	59.2	0.728
3.50	1.0092	1.0110	35.3	0.660	973.8	24.4	68	1.3397	2.240	1.205	0.654	0.985	0.978	101.35	68.3	0.858
4.00	1.0107	1.0125	40.4	0.756	970.3	28.0	77	1.3407	2.571	1.382	0.749	0.983	0.975	101.52	77.3	0.992
4.50	1.0122	1.0140	45.6	0.851	966.7	31.6	87	1.3417	2.907	1.563	0.845	0.982	0.972	101.68	86.3	1.13
5.00	1.0138	1.0155	50.7	0.947	963.1	35.2	96	1.3426	3.246	1.745	0.941	0.980	0.969	101.84	95.3	1.27
5.50	1.0153	1.0171	55.8	1.044	959.4	38.8	106	1.3436	3.591	1.931	1.038	0.979	0.966	101.99	104.	1.42
6.00	1.0168	1.0186	61.0	1.140	955.8	42.5	115	1.3445	3.941	2.119	1.135	0.977	0.963	102.15	113.	1.58
6.50	1.0183	1.0201	66.2	1.237	952.1	46.1	125	1.3455	4.296	2.310	1.233	0.976	0.960	102.30	122.	1.74

4 AMMONIUM CHLORIDE, NH₄Cl—(Continued)

A % by wt.	ρ D_4^{20}	D_{20}^{20}	C_s g/l	M g-mol/l	C_w g/l	$(C_o - C_w)$ g/l	$(n - n_o)$ × 10⁴	n	Δ °C	O Os/kg	S g-mol/l	η/η_o	η/ρ cS	ϕ rhe	γ mmho/cm	T g-mol/l
7.00	1.0198	1.0216	71.4	1.334	948.4	49.8	135	1.3464	4.657	2.504	1.332	0.974	0.957	102.44	131.	1.91
7.50	1.0213	1.0231	76.6	1.432	944.7	53.6	144	1.3474	5.02	2.70	1.431	0.973	0.955	102.58	139.	2.09
8.00	1.0227	1.0246	81.8	1.529	940.9	57.3	154	1.3483	5.40	2.90	1.530	0.972	0.952	102.71	147.	2.27
8.50	1.0242	1.0260	87.1	1.627	937.2	61.1	163	1.3493	5.77	3.10	1.629	0.971	0.950	102.83	156.	2.46
9.00	1.0257	1.0275	92.3	1.725	933.4	64.9	172	1.3502	6.16	3.31	1.728	0.970	0.947	102.93	164.	2.66
9.50	1.0272	1.0290	97.6	1.824	929.6	68.7	182	1.3512	6.55	3.52	1.828	0.969	0.945	103.02	172.	2.86
10.00	1.0286	1.0304	102.9	1.923	925.8	72.5	191	1.3521	6.95	3.73	1.928	0.968	0.943	103.10	180.	3.07
11.00	1.0315	1.0333	113.5	2.121	918.1	80.2	210	1.3540	7.76	4.17	2.129	0.967	0.939	103.20	200.	3.71
12.00	1.0344	1.0362	124.1	2.320	910.3	88.0	229	1.3559	8.60	4.62	2.330	0.967	0.936	103.23	220.	4.84
13.00	1.0373	1.0391	134.8	2.520	902.4	95.8	248	1.3578	9.47	5.09	2.531	0.967	0.934	103.22		
14.00	1.0401	1.0419	145.6	2.722	894.5	103.8	267	1.3596				0.967	0.932	103.17		
15.00	1.0429	1.0447	156.4	2.924	886.5	111.8	285	1.3615				0.968	0.930	103.10		
16.00	1.0457	1.0475	167.3	3.127	878.4	119.8	304	1.3634				0.969	0.928	103.01		
17.00	1.0485	1.0503	178.2	3.332	870.2	128.0	322	1.3652				0.970	0.927	102.90		
18.00	1.0512	1.0531	189.2	3.537	862.0	136.2	341	1.3671				0.971	0.926	102.75		
19.00	1.0540	1.0558	200.3	3.743	853.7	144.5	360	1.3689				0.973	0.925	102.54		
20.00	1.0567	1.0586	211.3	3.950	845.3	152.9	378	1.3708				0.976	0.925	102.25		
22.00	1.0621	1.0640	233.7	4.367	828.4	169.8	415	1.3745				0.984	0.928	101.46		
24.00	1.0674	1.0693	256.2	4.788	811.2	187.0	452	1.3782				0.994	0.933	100.40		

5 AMMONIUM SULFATE, (NH₄)₂SO₄

MOLECULAR WEIGHT = 132.14
RELATIVE SPECIFIC REFRACTIVITY = 0.886

0.00 % by wt. data are the same for all compounds.
For Values of 0.00 wt. % solutions see Table 1, Acetic Acid.

A % by wt.	ρ D_4^{20}	D_{20}^{20}	C_s g/l	M g-mol/l	C_w g/l	$(C_o - C_w)$ g/l	$(n - n_o)$ × 10⁴	n	Δ °C	O Os/kg	S g-mol/l	η/η_o	η/ρ cS	ϕ rhe	γ mmho/cm	T g-mol/l
0.50	1.0012	1.0030	5.0	0.038	996.2	2.1	8	1.3338	0.173	0.093	0.049	1.006	1.007	99.22	7.4	0.077
1.00	1.0042	1.0059	10.0	0.076	994.1	4.1	16	1.3346	0.331	0.178	0.095	1.012	1.010	98.62	14.2	0.151
2.00	1.0101	1.0119	20.2	0.153	989.9	8.3	33	1.3363	0.631	0.339	0.183	1.025	1.017	97.37	25.7	0.290
3.00	1.0160	1.0178	30.5	0.231	985.6	12.7	49	1.3379	0.921	0.495	0.269	1.039	1.025	96.03	36.7	0.428
4.00	1.0220	1.0238	40.9	0.309	981.1	17.1	65	1.3395	1.208	0.649	0.353	1.055	1.034	94.63	47.2	0.565
5.00	1.0279	1.0297	51.4	0.390	976.5	21.7	82	1.3411	1.489	0.800	0.435	1.071	1.044	93.18	57.4	0.704
6.00	1.0338	1.0356	62.0	0.469	971.8	26.5	98	1.3428	1.768	0.951	0.517	1.088	1.055	91.71	67.4	0.845
7.00	1.0397	1.0416	72.8	0.551	966.9	31.3	114	1.3444	2.047	1.101	0.598	1.106	1.066	90.21	77.0	0.987
8.00	1.0456	1.0475	83.6	0.633	962.0	36.3	130	1.3460	2.327	1.251	0.679	1.125	1.079	88.81	86.7	1.13
9.00	1.0515	1.0534	94.6	0.716	956.9	41.4	146	1.3476	2.608	1.402	0.760	1.145	1.091	87.13	95.5	1.28
10.00	1.0574	1.0593	105.7	0.800	951.6	46.6	162	1.3492	2.892	1.555	0.841	1.166	1.105	85.59	105.	1.43
12.00	1.0691	1.0710	128.3	0.971	940.8	57.4	193	1.3523	3.473	1.867	1.005	1.208	1.133	82.58	122.	1.75
14.00	1.0808	1.0827	151.3	1.145	929.5	68.8	225	1.3555	4.073	2.190	1.171	1.253	1.162	79.63	139.	2.09
16.00	1.0924	1.0943	174.8	1.323	917.6	80.6	256	1.3586	4.689	2.521	1.341	1.302	1.194	76.66	156.	2.44
18.00	1.1039	1.1059	198.7	1.504	905.2	93.0	287	1.3616				1.356	1.231	73.59	171.	2.82
20.00	1.1154	1.1174	223.1	1.688	892.3	105.9	317	1.3647				1.418	1.274	70.38	185.	3.21
22.00	1.1269	1.1289	247.9	1.876	879.0	119.3	347	1.3677				1.487	1.323	67.10	198.	3.67
24.00	1.1383	1.1403	273.2	2.067	865.1	133.2	377	1.3707				1.563	1.376	63.84	210.	4.18
26.00	1.1496	1.1516	298.9	2.262	850.7	147.5	407	1.3737				1.647	1.435	60.60	220.	4.84
28.00	1.1609	1.1629	325.0	2.460	835.8	162.4	436	1.3766				1.740	1.502	57.37		
30.00	1.1721	1.1742	351.6	2.661	820.5	177.7	465	1.3795				1.843	1.576	54.15		
32.00	1.1833	1.1854	378.7	2.866	804.7	193.6	494	1.3824				1.957	1.657	50.99		
34.00	1.1945	1.1966	406.1	3.073	788.3	209.9	523	1.3853				2.082	1.747	47.93		
36.00	1.2056	1.2077	434.0	3.284	771.6	226.7	551	1.3881				2.218	1.844	44.99		
38.00	1.2166	1.2188	462.3	3.499	754.3	243.9	580	1.3909				2.366	1.948	42.19		
40.00	1.2277	1.2298	491.1	3.716	736.6	261.6	608	1.3938				2.525	2.061	39.52		

6 BARIUM CHLORIDE, BaCl₂ · 2H₂O

MOLECULAR WEIGHT = 208.27   FORMULA WEIGHT, HYDRATE = 244.31
RELATIVE SPECIFIC REFRACTIVITY = 0.552

0.00 % by wt. data are the same for all compounds.
For Values of 0.00 wt. % solutions see Table 1, Acetic Acid.

A % by wt.	H % by wt.	ρ D_4^{20}	D_{20}^{20}	C_s g/l	M g-mol/l	C_w g/l	$(C_o - C_w)$ g/l	$(n - n_o)$ × 10⁴	n	Δ °C	O Os/kg	S g-mol/l	η/η_o	η/ρ cS	ϕ rhe	γ mmho/cm	T g-mol/l
0.50	0.59	1.0026	1.0044	5.0	0.024	997.6	0.6	7	1.3337	0.119	0.064	0.033	1.007	1.007	99.06	4.7	0.047
1.00	1.17	1.0070	1.0088	10.1	0.048	996.9	1.3	15	1.3345	0.233	0.125	0.066	1.014	1.009	98.42	9.1	0.094
2.00	2.35	1.0159	1.0177	20.3	0.098	995.6	2.6	30	1.3360	0.461	0.248	0.133	1.024	1.010	97.46	17.4	0.189
3.00	3.52	1.0249	1.0267	30.7	0.148	994.2	4.0	45	1.3375	0.691	0.372	0.201	1.035	1.012	96.41	25.2	0.284
4.00	4.69	1.0341	1.0359	41.4	0.199	992.7	5.5	61	1.3391	0.928	0.499	0.271	1.047	1.015	95.28	32.8	0.379
5.00	5.87	1.0434	1.0452	52.2	0.250	991.2	7.0	76	1.3406	1.177	0.633	0.344	1.060	1.018	94.15	40.4	0.476
6.00	7.04	1.0528	1.0547	63.2	0.303	989.6	8.6	92	1.3422	1.435	0.772	0.420	1.073	1.021	93.05	47.9	0.575
7.00	8.21	1.0624	1.0643	74.4	0.357	988.0	10.2	108	1.3438	1.704	0.916	0.498	1.085	1.024	91.95	55.3	0.675
8.00	9.38	1.0721	1.0740	85.8	0.412	986.3	11.9	124	1.3454	1.984	1.067	0.580	1.099	1.027	90.85	62.6	0.777
9.00	10.56	1.0820	1.0839	97.4	0.468	984.6	13.6	140	1.3470	2.274	1.222	0.664	1.112	1.030	89.72	69.8	0.880
10.00	11.73	1.0921	1.0940	109.2	0.524	982.9	15.4	157	1.3487	2.575	1.385	0.751	1.127	1.034	88.55	76.7	0.983
12.00	14.08	1.1128	1.1148	133.5	0.641	979.3	19.0	191	1.3520	3.218	1.730	0.933	1.159	1.043	86.14	90.1	1.19
14.00	16.42	1.1342	1.1362	158.8	0.762	975.4	22.8	225	1.3555	3.921	2.108	1.129	1.193	1.054	83.63	103.0	1.40
16.00	18.77	1.1564	1.1584	185.0	0.888	971.4	26.9	261	1.3591	4.689	2.521	1.341	1.232	1.067	81.03	115.0	1.61
18.00	21.11	1.1793	1.1814	212.3	1.019	967.0	31.2	297	1.3627				1.274	1.083	78.32	126.0	1.82
20.00	23.46	1.2031	1.2052	240.6	1.155	962.5	35.8	334	1.3664				1.322	1.101	75.49	137.0	2.04
22.00	25.81	1.2277	1.2299	270.1	1.297	957.6	40.6	373	1.3703				1.375	1.122	72.57	147.0	2.26
24.00	28.15	1.2531	1.2553	300.7	1.444	952.4	45.9	411	1.3741				1.434	1.147	69.57	157.0	2.48
26.00	30.50	1.2793	1.2816	332.6	1.597	946.7	51.6	451	1.3781				1.500	1.175	66.53	166.0	2.70

7 CADMIUM CHLORIDE, CdCl$_2$

MOLECULAR WEIGHT = 183.32

A% by wt.	ρ D_4^{20}	D_{20}^{20}	C_s g/l	M g·mol/l	C_W g/l	$(C_0 - C_W)$ g/l	$(n - n_0)$ × 10⁴	n	Δ °C	O Os/kg	S g·mol/l	η/η_0	η/ρ cS	ϕ rhe	γ mmho/cm	T g·mol/l
1.00	1.0062	1.0080	10.1	0.055	996.1	2.1	16.	1.3346	0.186	0.100	0.053	1.012	1.008	98.60	5.3	0.053
2.00	1.0148	1.0166	20.3	0.111	994.5	3.7	31.	1.3361	0.368	0.198	0.106	1.028	1.015	97.09	9.0	0.093
3.00	1.0237	1.0255	30.7	0.168	993.0	5.2	47.	1.3377	0.555	0.299	0.161	1.044	1.022	95.59	11.8	0.124
4.00	1.0325	1.0344	41.3	0.225	991.2	7.0	62.	1.3392	0.715	0.384	0.208	1.059	1.028	94.21	14.1	0.150
5.00	1.0415	1.0434	52.1	0.284	989.4	8.8	78.	1.3408	0.878	0.472	0.256	1.076	1.035	92.77	16.0	0.172
6.00	1.0508	1.0527	63.0	0.344	987.8	10.4	94.	1.3424	1.035	0.557	0.302	1.092	1.041	91.42	17.5	0.190
7.00	1.0600	1.0619	74.2	0.405	985.8	12.4	110.	1.3440	1.191	0.640	0.348	1.109	1.048	90.02	18.9	0.207
8.00	1.0692	1.0711	85.5	0.467	983.7	14.5	127.	1.3457	1.351	0.726	0.395	1.125	1.054	88.74	20.2	0.222
9.00	1.0794	1.0813	97.1	0.530	982.3	15.9	143.	1.3473	1.491	0.802	0.436	1.146	1.064	87.08	21.2	0.235
10.00	1.0897	1.0917	109.0	0.594	980.7	17.5	159.	1.3489	1.634	0.878	0.478	1.166	1.072	85.60	22.1	0.246
12.00	1.1103	1.1123	133.2	0.727	977.1	21.1	193.	1.3523	1.936	1.041	0.566	1.208	1.090	82.63	23.7	0.265
14.00	1.1308	1.1328	158.3	0.864	972.5	25.7	228.	1.3558	2.236	1.203	0.653	1.264	1.120	78.96	24.7	0.278
16.00	1.1531	1.1552	184.5	1.006	968.6	29.6	264.	1.3594	2.564	1.378	0.747	1.322	1.149	75.48	25.5	0.287
18.00	1.1749	1.1770	211.5	1.154	963.4	34.8	301.	1.3631	2.889	1.553	0.840	1.386	1.182	72.01	25.9	0.293
20.00	1.1988	1.2010	239.8	1.308	959.0	39.2	338.	1.3668	3.246	1.745	0.941	1.455	1.216	68.60	26.3	0.297
24.00	1.2476	1.2498	299.4	1.633	948.2	50.0	420.	1.3750	3.984	2.142	1.147	1.596	1.282	62.52	25.8	0.291
28.00	1.3003	1.3026	364.1	1.986	936.2	62.0	506.	1.3836	4.750	2.554	1.357	1.832	1.412	54.47	24.5	0.275
32.00	1.3558	1.3582	433.8	2.367	922.0	76.2	597.	1.3927				2.135	1.578	46.74	22.5	0.251
36.00	1.4178	1.4204	510.4	2.784	907.4	90.8	696.	1.4026				2.556	1.806	39.05	20.2	0.222
40.00	1.4843	1.4870	593.7	3.239	890.6	107.6	803.	1.4133				3.259	2.200	30.62	17.2	0.187
44.00	1.5563	1.5591	684.8	3.735	871.5	126.7	918.	1.4248				4.155	2.675	24.02	14.3	0.152
48.00	1.6345	1.6374	784.5	4.280	850.0	148.2	1044.	1.4374				5.740	3.519	17.39	11.3	0.119
52.00	1.7206	1.7237	894.7	4.881	825.9	172.3	1180.	1.4510				8.811	5.131	11.33	8.4	0.087
56.00	1.8131	1.8164	1015.4	5.539	797.7	200.5	1326.	1.4656				14.67	8.109	6.802	5.9	0.060

8 CADMIUM SULFATE, CdSO$_4$

MOLECULAR WEIGHT = 208.46

A% by wt.	ρ D_4^{20}	D_{20}^{20}	C_s g/l	M g·mol/l	C_W g/l	$(C_0 - C_W)$ g/l	$(n - n_0)$ × 10⁴	n	Δ °C	O Os/kg	S g·mol/l	η/η_0	η/ρ cS	ϕ rhe	γ mmho/cm	T g·mol/l
1.00	1.0078	1.0096	10.1	0.048	997.7	0.5	14.	1.3344	0.135	0.053	0.028	1.029	1.023	97.00	4.2	0.042
2.00	1.0175	1.0193	20.3	0.098	997.2	1.0	28.	1.3358	0.194	0.104	0.055	1.059	1.043	94.23	7.4	0.075
3.00	1.0271	1.0290	30.8	0.148	996.3	1.9	42.	1.3372	0.290	0.156	0.083	1.089	1.062	91.67	10.2	0.105
4.00	1.0375	1.0394	41.5	0.199	996.0	2.2	56.	1.3386	0.382	0.205	0.110	1.125	1.086	88.75	12.6	0.133
5.00	1.0478	1.0497	52.4	0.251	995.4	2.8	69.	1.3399	0.477	0.256	0.138	1.162	1.111	85.90	14.9	0.160
6.00	1.0579	1.0598	63.5	0.304	994.4	3.8	84.	1.3414	0.569	0.306	0.165	1.197	1.134	83.36	17.1	0.186
7.00	1.0684	1.0703	74.8	0.359	993.6	4.6	98.	1.3428	0.657	0.354	0.191	1.238	1.161	80.62	20.1	0.211
8.00	1.0789	1.0808	86.3	0.414	992.6	5.6	114.	1.3444	0.742	0.399	0.216	1.280	1.189	77.96	21.2	0.235
9.00	1.0894	1.0914	98.0	0.470	991.4	6.8	129.	1.3459	0.831	0.447	0.242	1.325	1.219	75.30	23.2	0.259
10.00	1.1004	1.1024	110.0	0.528	990.4	7.8	143.	1.3473	0.933	0.502	0.272	1.375	1.252	72.58	25.1	0.282
12.00	1.1233	1.1253	134.8	0.647	988.5	9.7	173.	1.3503	1.137	0.611	0.332	1.480	1.320	67.44	28.7	0.327
14.00	1.1470	1.1491	160.6	0.770	986.4	11.8	203.	1.3533	1.354	0.728	0.396	1.603	1.400	62.27	32.0	0.369
16.00	1.1715	1.1736	187.4	0.899	984.1	14.1	234.	1.3564	1.604	0.862	0.469	1.742	1.490	57.29	35.6	0.405
18.00	1.1967	1.1989	215.4	1.033	981.3	16.9	266.	1.3596	1.854	0.997	0.542	1.909	1.598	52.29	37.5	0.439
20.00	1.2229	1.2251	244.6	1.173	978.3	19.9	299.	1.3629	2.140	1.150	0.625	2.081	1.705	47.96	39.8	0.469
24.00	1.2784	1.2807	306.8	1.472	971.6	26.6	369.	1.3699	2.864	1.540	0.833	2.515	1.971	39.69	43.2	0.513
28.00	1.3385	1.3409	374.8	1.798	963.7	34.5	445.	1.3775	3.959	2.129	1.140	3.150	2.358	31.68	45.2	0.539
32.00	1.3996	1.4021	447.9	2.148	951.7	46.5	520.	1.3850				4.083	2.923	24.44	45.5	0.542
36.00	1.4655	1.4681	527.6	2.531	937.9	60.3	597.	1.3927				5.334	3.647	18.71	44.1	0.524
40.00	1.5456	1.5484	618.2	2.966	927.4	70.8	676.	1.4006				7.381	4.785	13.52	40.3	0.475

9 CALCIUM CHLORIDE, CaCl₂ · 2H₂O

MOLECULAR WEIGHT = 110.99   FORMULA WEIGHT, HYDRATE = 147.03

RELATIVE SPECIFIC REFRACTIVITY = 0.863

0.00 % by wt. data are the same for all compounds.
For Values of 0.00 wt. % solutions see Table 1, Acetic Acid.

A% by wt.	H% by wt.	ρ D_4^{20}	D_{20}^{20}	C_s g/l	M g-mol/l	C_w g/l	$(C_o - C_w)$ g/l	$(n - n_o)$ × 10⁴	n	Δ °C	O Os/kg	S g-mol/l	η/η_o	η/ρ cS	ϕ rhe	γ mmho/cm	T g-mol/l
0.50	0.66	1.0024	1.0041	5.0	0.045	997.3	0.9	12	1.3342	0.222	0.119	0.063	1.013	1.013	98.47	8.1	0.083
1.00	1.32	1.0065	1.0083	10.1	0.091	996.4	1.8	24	1.3354	0.440	0.237	0.127	1.026	1.021	97.27	15.7	0.169
1.50	1.99	1.0106	1.0124	15.2	0.137	995.5	2.7	36	1.3366	0.661	0.355	0.192	1.037	1.028	96.24	22.7	0.254
2.00	2.65	1.0148	1.0166	20.3	0.183	994.5	3.7	48	1.3378	0.880	0.473	0.257	1.048	1.035	95.23	29.4	0.337
2.50	3.31	1.0190	1.0208	25.5	0.230	993.5	4.7	60	1.3390	1.102	0.593	0.322	1.061	1.043	94.06	36.1	0.421
3.00	3.97	1.0232	1.0250	30.7	0.277	992.5	5.8	72	1.3402	1.330	0.715	0.389	1.076	1.053	92.79	42.6	0.504
3.50	4.64	1.0274	1.0292	36.0	0.324	991.4	6.8	84	1.3414	1.567	0.843	0.458	1.091	1.064	91.45	48.9	0.588
4.00	5.30	1.0316	1.0334	41.3	0.372	990.3	7.9	96	1.3426	1.815	0.976	0.531	1.108	1.076	90.09	55.1	0.672
4.50	5.96	1.0358	1.0377	46.6	0.420	989.2	9.0	109	1.3438	2.074	1.115	0.606	1.125	1.088	88.75	61.1	0.756
5.00	6.62	1.0401	1.0419	52.0	0.469	988.1	10.2	121	1.3451	2.345	1.261	0.684	1.141	1.099	87.47	67.0	0.840
5.50	7.29	1.0443	1.0462	57.4	0.518	986.9	11.3	133	1.3463	2.630	1.414	0.766	1.157	1.110	86.25	72.8	0.924
6.00	7.95	1.0486	1.0505	62.9	0.567	985.7	12.5	145	1.3475	2.930	1.575	0.852	1.173	1.121	85.07	78.3	1.01
6.50	8.61	1.0529	1.0548	68.4	0.617	984.5	13.8	158	1.3487	3.244	1.744	0.941	1.189	1.132	83.91	83.6	1.09
7.00	9.27	1.0572	1.0591	74.0	0.667	983.2	15.0	170	1.3500	3.573	1.921	1.033	1.206	1.143	82.77	88.7	1.17
7.50	9.94	1.0615	1.0634	79.6	0.717	981.9	16.3	182	1.3512	3.917	2.106	1.128	1.222	1.154	81.64	93.6	1.25
8.00	10.60	1.0659	1.0678	85.3	0.768	980.6	17.6	195	1.3525	4.275	2.299	1.227	1.240	1.165	80.51	98.4	1.33
8.50	11.26	1.0703	1.0722	91.0	0.820	979.3	18.9	207	1.3537	4.649	2.499	1.330	1.257	1.177	79.38	103.	1.41
9.00	11.92	1.0747	1.0766	96.7	0.871	977.9	20.3	220	1.3549	5.04	2.71	1.434	1.276	1.190	78.22	108.	1.49
9.50	12.58	1.0791	1.0810	102.5	0.924	976.6	21.7	232	1.3562	5.44	2.92	1.541	1.295	1.203	77.04	112.	1.57
10.00	13.25	1.0835	1.0854	108.3	0.976	975.1	23.1	245	1.3575	5.86	3.15	1.651	1.316	1.217	75.84	117.	1.65
11.00	14.57	1.0923	1.0943	120.2	1.083	972.2	26.1	270	1.3600	6.74	3.63	1.877	1.359	1.247	73.41	125.	1.81
12.00	15.90	1.1014	1.1033	132.2	1.191	969.2	29.0	295	1.3625	7.70	4.14	2.113	1.405	1.278	71.03	133.	1.97
13.00	17.22	1.1105	1.1125	144.4	1.301	966.2	32.1	321	1.3651	8.72	4.69	2.358	1.454	1.311	68.66	141.	2.13
14.00	18.55	1.1198	1.1218	156.8	1.412	963.0	35.2	347	1.3677	9.83	5.28	2.611	1.505	1.347	66.30	148.	2.28
15.00	19.87	1.1292	1.1312	169.4	1.526	959.8	38.4	374	1.3704	11.01	5.92	2.872	1.561	1.385	63.93	154.	2.42
16.00	21.20	1.1386	1.1407	182.2	1.641	956.5	41.8	400	1.3730	12.28	6.60	3.138	1.622	1.427	61.54	160.	2.55
17.00	22.52	1.1482	1.1502	195.2	1.759	953.0	45.2	427	1.3757	13.65	7.34	3.410	1.688	1.473	59.14	165.	2.67
18.00	23.84	1.1579	1.1599	208.4	1.878	949.5	48.8	454	1.3784	15.11	8.12	3.686	1.760	1.523	56.72	169.	2.78
19.00	25.17	1.1677	1.1697	221.9	1.999	945.8	52.4	482	1.3812	16.7	8.98	3.97	1.839	1.578	54.27	173.	2.88
20.00	26.49	1.1775	1.1796	235.5	2.122	942.0	56.2	509	1.3839	18.3	9.84	4.24	1.926	1.639	51.82	177.	2.97
22.00	29.14	1.1976	1.1997	263.5	2.374	934.1	64.1	565	1.3895	21.7	11.67		2.123	1.777	47.00	182.	3.11
24.00	31.79	1.2180	1.2201	292.3	2.634	925.7	72.6	621	1.3951	25.3	13.60		2.351	1.934	42.45	183.	3.16
26.00	34.44	1.2388	1.2410	322.1	2.902	916.7	81.5	678	1.4008	29.7	15.97		2.640	2.136	32.81	182.	3.12
28.00	37.09	1.2600	1.2622	352.8	3.179	907.2	91.0	736	1.4066	34.7	18.66		2.994	2.381	33.33	179.	3.03
30.00	39.74	1.2816	1.2838	384.6	3.464	897.1	101.1	794	1.4124	41.0	22.04		3.460	2.705	28.85	172.	2.85
32.00	42.39	1.3036	1.3059	417.1	3.758	886.4	111.8	853	1.4183	49.7	26.72		4.027	3.095	24.79	162.	2.61
34.00	45.04	1.3260	1.3283	450.8	4.062	875.1	123.1	912	1.4242				4.810	3.611	20.45	150.	2.34
36.00	47.69	1.3488	1.3512	485.6	4.375	863.2	135.0	971	1.4301				5.795	4.305	17.22	137.	2.05
38.00	50.34	1.3770	1.3745	521.4	4.697	850.7	147.6	1031	1.4361				7.306	5.336	13.66	123.	1.76
40.00	52.99	1.3957	1.3982	558.3	5.030	837.4	160.8	1090	1.4420				8.979	6.446	11.12	106.	1.45

10 CESIUM CHLORIDE, CsCl

MOLECULAR WEIGHT = 168.37

RELATIVE SPECIFIC REFRACTIVITY = 0.464

0.00 % by wt. data are the same for all compounds.
For Values of 0.00 wt. % solutions see Table 1, Acetic Acid.

A% by wt.	ρ D_4^{20}	D_{20}^{20}	C_s g/l	M g-mol/l	C_w g/l	$(C_o - C_w)$ g/l	$(n - n_o)$ × 10⁴	n	Δ °C	O Os/kg	S g-mol/l	η/η_o	η/ρ cS	ϕ rhe	γ mmho/cm	T g-mol/l
0.50	1.0020	1.0038	5.0	0.030	997.0	1.2	4	1.3334	0.100	0.054	0.028	0.998	0.997	100.05	3.8	0.038
1.00	1.0058	1.0076	10.1	0.060	995.8	2.5	8	1.3337	0.201	0.108	0.057	0.995	0.991	100.30	7.4	0.075
1.50	1.0097	1.0114	15.1	0.090	994.5	3.7	11	1.3341	0.302	0.162	0.087	0.992	0.985	100.56	10.6	0.111
2.00	1.0135	1.0153	20.3	0.120	993.3	5.0	15	1.3345	0.403	0.217	0.116	0.990	0.979	100.81	13.8	0.147
2.50	1.0174	1.0192	25.4	0.151	992.0	6.2	19	1.3349	0.505	0.272	0.146	0.988	0.973	101.04	17.0	0.184
3.00	1.0214	1.0232	30.6	0.182	990.7	7.5	23	1.3353	0.607	0.326	0.176	0.986	0.967	101.26	20.2	0.222
3.50	1.0253	1.0271	35.9	0.213	989.4	8.8	27	1.3357	0.709	0.381	0.206	0.984	0.961	101.46	23.3	0.260
4.00	1.0293	1.0311	41.2	0.245	988.2	10.1	31	1.3361	0.812	0.436	0.236	0.982	0.956	101.65	26.5	0.300
4.50	1.0334	1.0352	46.5	0.276	986.9	11.4	35	1.3365	0.915	0.492	0.267	0.980	0.950	101.84	29.7	0.340
5.00	1.0374	1.0392	51.9	0.308	985.5	12.7	39	1.3369	1.018	0.547	0.297	0.978	0.945	102.02	32.9	0.380
5.50	1.0415	1.0433	57.3	0.340	984.2	14.0	43	1.3373	1.121	0.602	0.327	0.976	0.939	102.20	36.1	0.421
6.00	1.0456	1.0475	62.7	0.373	982.9	15.3	47	1.3377	1.223	0.658	0.358	0.975	0.934	102.37	39.3	0.462
6.50	1.0498	1.0516	68.2	0.405	981.6	16.7	52	1.3382	1.326	0.713	0.388	0.973	0.929	102.53	42.5	0.504
7.00	1.0540	1.0558	73.8	0.438	980.2	18.0	56	1.3386	1.428	0.768	0.418	0.972	0.924	102.67	45.8	0.546
7.50	1.0582	1.0601	79.4	0.471	978.8	19.4	60	1.3390	1.531	0.823	0.448	0.971	0.919	102.82	49.0	0.589
8.00	1.0625	1.0643	85.0	0.505	977.5	20.8	64	1.3394	1.635	0.879	0.478	0.969	0.914	102.95	52.3	0.633
8.50	1.0668	1.0686	90.7	0.539	976.1	22.2	69	1.3399	1.739	0.935	0.509	0.968	0.909	103.09	55.6	0.678
9.00	1.0711	1.0730	96.4	0.573	974.7	23.5	73	1.3403	1.845	0.992	0.539	0.967	0.904	103.22	59.0	0.724
9.50	1.0754	1.0773	102.2	0.607	973.3	25.0	77	1.3407	1.952	1.049	0.571	0.966	0.900	103.36	62.4	0.772
10.00	1.0798	1.0818	108.0	0.641	971.9	26.4	82	1.3412	2.060	1.108	0.602	0.964	0.895	103.51	65.8	0.822
11.00	1.0887	1.0907	119.8	0.711	969.0	29.3	91	1.3421	2.281	1.227	0.666	0.961	0.885	103.80	72.8	0.926
12.00	1.0978	1.0997	131.7	0.782	966.1	32.2	100	1.3430	2.507	1.348	0.731	0.959	0.875	104.09	80.0	1.03
13.00	1.1070	1.1089	143.9	0.855	963.1	35.2	109	1.3439	2.737	1.472	0.797	0.956	0.865	104.38	87.3	1.15
14.00	1.1163	1.1183	156.3	0.928	960.0	38.2	118	1.3448	2.972	1.598	0.864	0.953	0.856	104.67	94.8	1.27
15.00	1.1258	1.1278	168.9	1.003	956.9	41.3	128	1.3458	3.212	1.727	0.932	0.951	0.846	104.96	102.	1.40
16.00	1.1355	1.1375	181.7	1.079	953.8	44.5	138	1.3468	3.457	1.859	1.000	0.948	0.837	105.27	110.	1.54
17.00	1.1453	1.1473	194.7	1.156	950.6	47.7	147	1.3477	3.708	1.993	1.070	0.945	0.827	105.58	118.	1.68
18.00	1.1552	1.1573	207.9	1.235	947.3	51.0	157	1.3487	3.963	2.131	1.141	0.943	0.817	105.89	126.	1.83
19.00	1.1653	1.1674	221.4	1.315	943.9	54.3	167	1.3497	4.224	2.271	1.213	0.940	0.808	106.19	134.	1.99
20.00	1.1756	1.1777	235.1	1.396	940.5	57.7	178	1.3507	4.491	2.415	1.287	0.937	0.799	106.49	142.	2.17
22.00	1.1967	1.1989	263.3	1.564	933.5	64.8	198	1.3528				0.932	0.781	107.05	159.	2.53
24.00	1.2185	1.2207	292.4	1.737	926.1	72.1	220	1.3550				0.928	0.763	107.56	176.	2.95
26.00	1.2411	1.2433	322.7	1.916	918.4	79.8	242	1.3572				0.924	0.746	107.97	192.	3.48

10 CESIUM CHLORIDE, CsCl—(Continued)

A% by wt.	ρ D_4^{20}	D_{20}^{20}	C_s g/l	M g-mol/l	C_w g/l	$(C_o - C_w)$ g/l	$(n - n_o)$ $\times 10^4$	n	Δ °C	O Os/kg	S g-mol/l	η/η_o	η/ρ cS	ϕ rhe	γ mmho/cm	T g-mol/l
28.00	1.2644	1.2666	354.0	2.103	910.4	87.9	264	1.3594				0.922	0.730	108.27	207.	4.15
30.00	1.2885	1.2908	386.6	2.296	902.0	96.3	288	1.3617				0.920	0.716	108.44	221.	4.95
32.00	1.3135	1.3158	420.3	2.496	893.1	105.1	311	1.3641				0.920	0.702	108.43		
34.00	1.3393	1.3417	455.4	2.705	883.9	114.3	336	1.3666				0.922	0.690	108.25		
36.00	1.3661	1.3685	491.8	2.921	874.3	124.0	361	1.3691				0.924	0.678	107.97		
38.00	1.3938	1.3963	529.6	3.146	864.2	134.1	387	1.3717				0.928	0.667	107.56		
40.00	1.4226	1.4251	569.0	3.380	853.5	144.7	414	1.3744				0.932	0.657	107.02		
42.00	1.4525	1.4550	610.0	3.623	842.4	155.8	442	1.3771				0.938	0.647	106.40		
44.00	1.4835	1.4861	652.7	3.877	830.8	167.5	470	1.3800				0.945	0.638	105.63		
46.00	1.5158	1.5185	697.3	4.141	818.5	179.7	500	1.3829				0.954	0.630	104.64		
48.00	1.5495	1.5522	743.7	4.417	805.7	192.5	530	1.3860				0.965	0.624	103.45		
50.00	1.5846	1.5874	792.3	4.706	792.3	206.0	562	1.3892				0.979	0.619	101.94		
52.00	1.6212	1.6241	843.0	5.007	778.2	220.0	595	1.3925				0.998	0.617	100.02		
54.00	1.6596	1.6625	896.2	5.323	763.4	234.8	630	1.3960				1.021	0.616	97.76		
56.00	1.6999	1.7029	951.9	5.654	747.9	250.3	666	1.3996				1.048	0.618	95.25		
58.00	1.7422	1.7453	1010.5	6.001	731.7	266.5	705	1.4035				1.078	0.620	92.54		
60.00	1.7868	1.7900	1072.1	6.367	714.7	283.5	746	1.4076				1.118	0.627	89.27		
62.00	1.8340	1.8373	1137.1	6.754	696.9	301.3	790	1.4120				1.170	0.639	85.28		
64.00	1.8842	1.8875	1205.9	7.162	678.3	319.9	837	1.4167				1.236	0.657	80.74		

11 CITRIC ACID, (COOH)CH₂C(OH)(COOH)CH₂COOH · 1H₂O

MOLECULAR WEIGHT = 192.12   **FORMULA WEIGHT, HYDRATE** = 210.14

RELATIVE SPECIFIC REFRACTIVITY = 0.951

0.00 % by wt. data are the same for all compounds.
For Values of 0.00 wt. % solutions see Table 1, Acetic Acid.

A% by wt.	H% by wt.	ρ D_4^{20}	D_{20}^{20}	C_s g/l	M g-mol/l	C_w g/l	$(C_o - C_w)$ g/l	$(n - n_o)$ $\times 10^4$	n	Δ °C	O Os/kg	S g-mol/l	η/η_o	η/ρ cS	ϕ rhe	γ mmho/cm	T g-mol/l
0.50	0.55	1.0002	1.0020	5.0	0.026	995.2	3.0	7	1.3336	0.052	0.028	0.014	1.011	1.013	98.74	1.2	0.012
1.00	1.09	1.0022	1.0040	10.0	0.052	992.2	6.0	13	1.3343	0.105	0.056	0.029	1.022	1.022	97.65	2.1	0.021
2.00	2.19	1.0063	1.0081	20.1	0.105	986.2	12.0	26	1.3356	0.211	0.113	0.060	1.046	1.041	95.41	3.0	0.030
3.00	3.28	1.0105	1.0123	30.3	0.158	980.2	18.1	38	1.3368	0.318	0.171	0.091	1.071	1.062	93.21	3.7	0.037
4.00	4.38	1.0147	1.0165	40.6	0.211	974.1	24.1	51	1.3381	0.428	0.230	0.123	1.096	1.082	91.05	4.2	0.042
5.00	5.47	1.0189	1.0207	50.9	0.265	968.0	30.3	64	1.3394	0.539	0.290	0.156	1.123	1.104	88.87	4.7	0.047
6.00	6.56	1.0232	1.0250	61.4	0.320	961.8	36.5	77	1.3407	0.651	0.350	0.189	1.151	1.128	86.68	5.1	0.051
7.00	7.66	1.0274	1.0292	71.9	0.374	955.5	42.8	90	1.3420	0.764	0.411	0.222	1.181	1.152	84.50	5.5	0.055
8.00	8.75	1.0316	1.0335	82.5	0.430	949.1	49.1	103	1.3433	0.879	0.473	0.256	1.212	1.177	82.32	5.7	0.058
9.00	9.84	1.0359	1.0377	93.2	0.485	942.7	55.6	116	1.3446	0.997	0.536	0.291	1.245	1.204	80.15	6.0	0.060
10.00	10.94	1.0402	1.0420	104.0	0.541	936.1	62.1	129	1.3459	1.120	0.602	0.327	1.280	1.233	77.97	6.2	0.063
12.00	13.13	1.0490	1.0509	125.9	0.655	923.2	75.1	156	1.3486	1.379	0.741	0.403	1.354	1.293	73.72	6.6	0.067
14.00	15.31	1.0580	1.0599	148.1	0.771	909.9	88.3	184	1.3514	1.656	0.890	0.484	1.433	1.358	69.62	6.9	0.070
16.00	17.50	1.0672	1.0690	170.7	0.889	896.4	101.8	211	1.3541	1.950	1.048	0.570	1.522	1.429	65.58	7.1	0.072
18.00	19.69	1.0764	1.0783	193.8	1.009	882.7	115.6	239	1.3569	2.255	1.213	0.658	1.622	1.510	61.53	7.1	0.072
20.00	21.88	1.0858	1.0877	217.2	1.130	868.6	129.6	268	1.3598	2.567	1.380	0.748	1.737	1.603	57.46	7.2	0.073
22.00	24.06	1.0953	1.0972	241.0	1.254	854.3	143.9	296	1.3626	2.884	1.550	0.839	1.868	1.709	53.43	7.2	0.073
24.00	26.25	1.1049	1.1069	265.2	1.380	839.8	158.5	325	1.3655	3.210	1.726	0.931	2.013	1.826	49.57	7.2	0.073
26.00	28.44	1.1147	1.1167	289.8	1.509	824.9	173.4	355	1.3684	3.547	1.907	1.025	2.174	1.954	45.91	7.1	0.072
28.00	30.63	1.1246	1.1266	314.9	1.639	809.7	188.5	384	1.3714	3.893	2.093	1.122	2.351	2.094	42.46	6.8	0.069
30.00	32.81	1.1346	1.1366	340.4	1.772	794.2	204.0	414	1.3744	4.250	2.285	1.220	2.544	2.247	39.23	6.5	0.066

12 COBALTOUS CHLORIDE, CoCl₂ · 6H₂O

MOLECULAR WEIGHT = 129.85   **FORMULA WEIGHT, HYDRATE** = 237.95

RELATIVE SPECIFIC REFRACTIVITY = 0.733

0.00 % by wt. data are the same for all compounds.
For Values of 0.00 wt. % solutions see Table 1, Acetic Acid.

A% by wt.	H% by wt.	ρ D_4^{20}	D_{20}^{20}	C_s g/l	M g-mol/l	C_w g/l	$(C_o - C_w)$ g/l	$(n - n_o)$ $\times 10^4$	n	Δ °C	O Os/kg	S g-mol/l	η/η_o	η/ρ cS	ϕ rhe	γ mmho/cm	T g-mol/l
0.50	0.92	1.0027	1.0045	5.0	0.039	997.7	0.5	11	1.3341	0.191	0.103	0.054	1.017	1.017	98.09	6.7	0.068
1.00	1.83	1.0073	1.0090	10.1	0.078	997.2	1.1	23	1.3352	0.380	0.204	0.110	1.034	1.029	96.52	12.7	0.134
2.00	3.66	1.0164	1.0182	20.3	0.157	996.0	2.2	45	1.3375	0.771	0.415	0.224	1.064	1.049	93.80	23.4	0.262
3.00	5.50	1.0256	1.0274	30.8	0.237	993.5	3.4	68	1.3398	1.187	0.638	0.347	1.096	1.071	91.05	33.6	0.389
4.00	7.33	1.0349	1.0368	41.4	0.319	993.5	4.7	91	1.3421	1.627	0.875	0.476	1.130	1.094	88.30	43.3	0.514
5.00	9.16	1.0444	1.0462	52.2	0.402	992.2	6.1	114	1.3444	2.099	1.128	0.613	1.166	1.119	85.59	52.5	0.636
6.00	10.99	1.0540	1.0559	63.2	0.487	990.8	7.5	137	1.3467	2.611	1.404	0.761	1.203	1.143	82.97	61.0	0.754
7.00	12.83	1.0637	1.0656	74.5	0.573	989.3	9.0	161	1.3491	3.164	1.701	0.918	1.241	1.169	80.43	69.0	0.868
8.00	14.66	1.0736	1.0755	85.9	0.661	987.7	10.5	185	1.3515	3.761	2.022	1.085	1.281	1.195	77.91	76.4	0.979
9.00	16.49	1.0837	1.0856	97.5	0.751	986.2	12.1	209	1.3539				1.324	1.224	75.40	83.4	1.08
10.00	18.32	1.0939	1.0958	109.4	0.842	984.5	13.7	234	1.3563				1.370	1.255	72.85	89.9	1.18
12.00	21.99	1.1148	1.1168	133.8	1.030	981.1	17.2	283	1.3613				1.474	1.324	67.73	102.	1.38
14.00	25.65	1.1364	1.1384	159.1	1.225	977.3	20.9	334	1.3664				1.592	1.403	62.71	112.	1.55
16.00	29.32	1.1587	1.1607	185.4	1.428	973.3	24.9	386	1.3716				1.725	1.492	57.85	120.	1.70
18.00	32.98	1.1816	1.1836	212.7	1.638	968.9	29.4	439	1.3769				1.876	1.591	53.19	126.	1.82
20.00	36.65	1.2050	1.2071	241.0	1.856	964.0	34.3	492	1.3822				2.046	1.701	48.78	130.	1.90

13 CREATININE, CH₃NC(:NH)NHCOCH₂

$CH_3NC(:NH)NHCOCH_2$ (bracket joining)

MOLECULAR WEIGHT = 113.12
RELATIVE SPECIFIC REFRACTIVITY = 1.301

0.00 % by wt. data are the same for all compounds.
For Values of 0.00 wt. % solutions see Table 1, Acetic Acid.

A % by wt.	D_4^{20}	D_{20}^{20}	C_s g/l	M g-mol/l	C_w g/l	$(C_o - C_w)$ g/l	$(n-n_o) \times 10^4$	n	Δ °C	O Os/kg	S g-mol/l	η/η_o	η/ρ cS	φ rhe	γ mmho/cm	T g-mol/l
0.50	0.9995	1.0012	5.0	0.044	994.5	3.8	9	1.3339	0.080	0.043	0.022	1.009	1.011	98.95		
1.00	1.0007	1.0025	10.0	0.088	990.7	7.5	19	1.3349	0.150	0.081	0.042	1.019	1.020	97.97		
1.50	1.0019	1.0037	15.0	0.133	986.9	11.3	28	1.3358	0.220	0.118	0.063	1.030	1.030	96.89		
2.00	1.0032	1.0050	20.1	0.177	983.1	15.1	38	1.3368	0.300	0.161	0.086	1.043	1.041	95.73		
2.50	1.0045	1.0063	25.1	0.222	979.4	18.9	48	1.3378	0.380	0.204	0.110	1.056	1.053	94.52		
3.00	1.0058	1.0075	30.2	0.267	975.6	22.6	57	1.3387	0.460	0.247	0.133	1.070	1.066	93.29		
3.50	1.0071	1.0088	35.2	0.312	971.8	26.4	67	1.3397				1.084	1.079	92.05		
4.00	1.0084	1.0102	40.3	0.357	968.0	30.2	77	1.3407				1.099	1.092	90.83		
4.50	1.0097	1.0115	45.4	0.402	964.3	34.0	87	1.3417				1.113	1.105	89.65		
5.00	1.0110	1.0128	50.6	0.447	960.5	37.7	97	1.3427				1.128	1.117	88.51		
5.50	1.0124	1.0142	55.7	0.492	956.7	41.5	107	1.3437				1.141	1.130	87.45		
6.00	1.0138	1.0156	60.8	0.538	952.9	45.3	117	1.3447				1.154	1.141	86.46		
6.50	1.0151	1.0169	66.0	0.583	949.2	49.1	127	1.3457				1.166	1.151	85.57		
7.00	1.0165	1.0183	71.2	0.629	945.4	52.8	137	1.3467				1.177	1.160	84.78		
7.50	1.0180	1.0198	76.3	0.675	941.6	56.6	148	1.3478				1.186	1.168	84.11		
8.00	1.0194	1.0212	81.6	0.721	937.8	60.4	158	1.3488				1.194	1.174	83.58		

14 CUPRIC SULFATE, CuSO₄ · 5H₂O

MOLECULAR WEIGHT = 159.61   FORMULA WEIGHT, HYDRATE = 249.69
RELATIVE SPECIFIC REFRACTIVITY = 0.517

0.00 % by wt. data are the same for all compounds.
For Values of 0.00 wt. % solutions see Table 1, Acetic Acid.

A % by wt.	H % by wt.	ρ D_4^{20}	D_{20}^{20}	C_s g/l	M g-mol/l	C_w g/l	$(C_o - C_w)$ g/l	$(n-n_o) \times 10^4$	n	Δ °C	O Os/kg	S g-mol/l	η/η_o	η/ρ cS	φ rhe	γ mmho/cm	T g-mol/l
0.50	0.78	1.0033	1.0051	5.0	0.031	998.3	−0.1	9	1.3339	0.075	0.040	0.021	1.015	1.014	98.30	2.9	0.028
1.00	1.56	1.0085	1.0103	10.1	0.063	998.4	−0.2	18	1.3348	0.140	0.075	0.039	1.034	1.027	96.52	5.4	0.054
1.50	2.35	1.0137	1.0155	15.2	0.095	998.5	−0.3	28	1.3358	0.200	0.108	0.057	1.057	1.045	94.40	7.4	0.076
2.00	3.13	1.0190	1.0208	20.4	0.128	998.6	−0.4	37	1.3367	0.259	0.139	0.074	1.082	1.064	92.24	9.3	0.096
2.50	3.91	1.0243	1.0261	25.6	0.160	998.7	−0.4	47	1.3377	0.316	0.170	0.091	1.105	1.081	90.31	11.1	0.116
3.00	4.69	1.0296	1.0314	30.9	0.194	998.7	−0.5	56	1.3386	0.372	0.200	0.107	1.127	1.097	88.54	12.8	0.135
3.50	5.48	1.0349	1.0368	36.2	0.227	998.7	−0.5	66	1.3396	0.478	0.231	0.124	1.149	1.112	86.86	14.6	0.154
4.00	6.26	1.0403	1.0421	41.6	0.261	998.7	−0.5	75	1.3405	0.484	0.260	0.140	1.171	1.128	85.23	16.0	0.172
4.50	7.04	1.0457	1.0475	47.1	0.295	998.6	−0.4	85	1.3414	0.539	0.290	0.156	1.194	1.144	83.58	17.5	0.190
5.00	7.82	1.0511	1.0530	52.6	0.329	998.6	−0.3	94	1.3424	0.594	0.319	0.172	1.219	1.162	81.87	19.0	0.208
5.50	8.60	1.0565	1.0584	58.1	0.364	998.4	−0.2	104	1.3433	0.648	0.349	0.188	1.246	1.181	80.13	20.5	0.226
6.00	9.39	1.0620	1.0639	63.7	0.399	998.3	0.0	113	1.3443	0.703	0.378	0.205	1.273	1.201	78.37	21.9	0.243
6.50	10.17	1.0675	1.0694	69.4	0.435	998.1	0.1	123	1.3452	0.759	0.408	0.221	1.302	1.222	76.63	23.3	0.260
7.00	10.95	1.0730	1.0749	75.1	0.471	997.9	0.3	132	1.3462	0.816	0.439	0.238	1.333	1.244	74.90	24.6	0.277
7.50	11.73	1.0786	1.0805	80.9	0.507	997.7	0.5	142	1.3472	0.874	0.470	0.255	1.364	1.267	73.17	26.0	0.293
8.00	12.52	1.0842	1.0861	86.7	0.543	997.5	0.8	151	1.3481	0.933	0.501	0.272	1.397	1.291	71.46	27.2	0.309
8.50	13.30	1.0898	1.0918	92.6	0.580	997.2	1.0	161	1.3491	0.992	0.533	0.289	1.431	1.315	69.77	28.5	0.325
9.00	14.08	1.0955	1.0975	98.6	0.618	996.9	1.3	171	1.3501	1.052	0.566	0.307	1.466	1.341	68.09	29.7	0.341
9.50	14.86	1.1012	1.1032	104.6	0.655	996.6	1.6	181	1.3511	1.113	0.599	0.325	1.502	1.367	66.44	31.0	0.356
10.00	15.64	1.1070	1.1090	110.7	0.694	996.3	1.9	191	1.3520	1.177	0.633	0.344	1.540	1.394	64.81	32.2	0.371
11.00	17.21	1.1186	1.1206	123.0	0.771	995.6	2.7	210	1.3540	1.308	0.703	0.382	1.617	1.449	61.70	34.4	0.400
12.00	18.77	1.1304	1.1324	135.6	0.850	994.7	3.5	230	1.3560	1.447	0.778	0.423	1.698	1.505	58.77	36.6	0.427
13.00	20.34	1.1424	1.1444	148.5	0.930	993.8	4.4	251	1.3581	1.596	0.858	0.467	1.786	1.566	55.88	38.6	0.453
14.00	21.90	1.1545	1.1566	161.6	1.013	992.9	5.3	271	1.3601	1.753	0.942	0.513	1.885	1.636	52.95	40.5	0.478
15.00	23.47	1.1669	1.1690	175.0	1.097	991.9	6.4	292	1.3622				2.000	1.717	49.90	42.3	0.501
16.00	25.03	1.1796	1.1817	188.7	1.182	990.9	7.4	314	1.3644				2.132	1.811	46.81	44.0	0.523
17.00	26.59	1.1926	1.1947	202.7	1.270	989.8	8.4	336	1.3666				2.280	1.916	43.77	45.6	0.543
18.00	28.16	1.2059	1.2080	217.1	1.360	988.8	9.4	359	1.3689				2.444	2.031	40.83	47.0	0.562

15 DEXTRAN, (C₆H₁₀O₅)ₓ

$(C_6H_{10}O_5)_x$

MOLECULAR WEIGHT, AVERAGE = 72,000
RELATIVE SPECIFIC REFRACTIVITY = 1.045

0.00 % by wt. data are the same for all compounds.
For Values of 0.00 wt. % solutions see Table 1, Acetic Acid.

A % by wt.	ρ D_4^{20}	d D_{20}^{20}	C_s g/l	M g-mol/l	C_w g/l	$(C_o - C_w)$ g/l	$(n-n_o) \times 10^4$	n	Δ °C	O Os/kg	S g-mol/l	η/η_o	η/ρ cS	φ rhe	γ mmho/cm	T g-mol/l
0.50	1.0001	1.0019	5.0		995.1	3.1	7	1.3337				1.129	1.131	88.38		
1.00	1.0020	1.0037	10.0		992.0	6.3	15	1.3345				1.280	1.280	77.97		
1.50	1.0039	1.0056	15.1		988.8	9.4	22	1.3352				1.453	1.450	68.69		
2.00	1.0057	1.0075	20.1		985.6	12.6	29	1.3359				1.647	1.641	60.60		
2.50	1.0076	1.0094	25.2		982.5	15.8	37	1.3367				1.858	1.848	53.71		
3.00	1.0096	1.0113	30.3		979.3	19.0	44	1.3374				2.086	2.071	47.83		
3.50	1.0115	1.0133	35.4		976.1	22.2	52	1.3382				2.336	2.314	42.73		
4.00	1.0134	1.0152	40.5		972.9	25.4	59	1.3389				2.611	2.581	38.23		
4.50	1.0153	1.0171	45.7		969.6	28.6	67	1.3397				2.915	2.877	34.24		
5.00	1.0173	1.0191	50.9		966.4	31.8	74	1.3404				3.254	3.205	30.67		
5.50	1.0192	1.0210	56.1		963.2	35.1	82	1.3412				3.627	3.566	27.52		
6.00	1.0212	1.0230	61.3		959.9	38.3	89	1.3419				4.031	3.955	24.76		
6.50	1.0232	1.0250	66.5		956.7	41.6	97	1.3427				4.466	4.373	22.35		
7.00	1.0251	1.0270	71.8		953.4	44.8	105	1.3435				4.933	4.821	20.23		
7.50	1.0271	1.0290	77.0		950.1	48.1	112	1.3442				5.431	5.299	18.37		
8.00	1.0291	1.0310	82.3		946.8	51.4	120	1.3450				5.963	5.805	16.74		
8.50	1.0312	1.0330	87.6		943.5	54.7	128	1.3458				6.527	6.342	15.29		
9.00	1.0332	1.0350	93.0		940.2	58.0	135	1.3465				7.124	6.909	14.01		
9.50	1.0352	1.0371	98.3		936.9	61.4	143	1.3473				7.754	7.506	12.87		
10.00	1.0373	1.0391	103.7		933.5	64.7	151	1.3481				8.419	8.133	11.85		

16 ETHANOL, CH₃CH₂OH

MOLECULAR WEIGHT = 46.07
RELATIVE SPECIFIC REFRACTIVITY = 1.363

0.00 % by wt. data are the same for all compounds.
For Values of 0.00 wt. %, solutions see Table 1, Acetic Acid.

A % by wt.	ρ D_4^{20}	D_{20}^{20}	C_s g/l	M g-mol/l	C_w g/l	$(C_o - C_w)$ g/l	$(n - n_o)$ × 10⁴	n	Δ °C	O Os/kg	S g-mol/l	η/η_o	η/ρ cS	ϕ rhe	γ mmho/cm	T g-mol/l
0.50	0.9973	0.9991	5.0	0.108	992.3	5.9	3	1.3333	0.20	0.11	0.057	1.021	1.026	97.71		
1.00	0.9963	0.9981	10.0	0.216	986.4	11.8	6	1.3336	0.40	0.22	0.116	1.044	1.050	95.59		
1.50	0.9954	0.9972	14.9	0.324	980.5	17.8	9	1.3339	0.60	0.32	0.175	1.068	1.075	93.42		
2.00	0.9945	0.9962	19.9	0.432	974.6	23.6	12	1.3342	0.81	0.44	0.236	1.093	1.101	91.31		
2.50	0.9936	0.9953	24.8	0.539	968.7	29.5	15	1.3345	1.02	0.55	0.297	1.116	1.126	89.42		
3.00	0.9927	0.9945	29.8	0.646	962.9	35.3	18	1.3348	1.23	0.66	0.359	1.138	1.149	87.70		
3.50	0.9918	0.9936	34.7	0.754	957.1	41.1	21	1.3351	1.44	0.77	0.421	1.159	1.171	86.08		
4.00	0.9910	0.9927	39.6	0.860	951.4	46.9	24	1.3354	1.65	0.89	0.484	1.181	1.194	84.52		
4.50	0.9902	0.9919	44.6	0.967	945.6	52.6	27	1.3357	1.87	1.01	0.547	1.203	1.217	82.98		
5.00	0.9893	0.9911	49.5	1.074	939.9	58.4	31	1.3360	2.09	1.12	0.611	1.226	1.242	81.40		
5.50	0.9885	0.9903	54.4	1.180	934.2	64.1	34	1.3364	2.31	1.24	0.675	1.250	1.267	79.81		
6.00	0.9878	0.9895	59.3	1.286	928.5	69.7	37	1.3367	2.54	1.36	0.739	1.276	1.294	78.24		
6.50	0.9870	0.9887	64.2	1.393	922.8	75.4	41	1.3370	2.76	1.49	0.804	1.301	1.321	76.69		
7.00	0.9862	0.9880	69.0	1.498	917.2	81.0	44	1.3374	2.99	1.61	0.870	1.328	1.349	75.16		
7.50	0.9855	0.9872	73.9	1.604	911.6	86.7	48	1.3377	3.23	1.74	0.936	1.355	1.378	73.66		
8.00	0.9847	0.9865	78.8	1.710	906.0	92.3	51	1.3381	3.47	1.86	1.003	1.382	1.407	72.19		
8.50	0.9840	0.9857	83.6	1.816	900.4	97.9	55	1.3384	3.71	2.00	1.071	1.411	1.436	70.75		
9.00	0.9833	0.9850	88.5	1.921	894.8	103.4	58	1.3388	3.96	2.13	1.140	1.439	1.467	69.34		
9.50	0.9826	0.9843	93.3	2.026	889.2	109.0	62	1.3392	4.21	2.27	1.210	1.468	1.497	67.97		
10.00	0.9819	0.9836	98.2	2.131	883.7	114.5	65	1.3395	4.47	2.40	1.282	1.498	1.529	66.62		
11.00	0.9805	0.9822	107.9	2.341	872.6	125.6	73	1.3403	5.00	2.69	1.426	1.560	1.594	63.99		
12.00	0.9792	0.9809	117.5	2.550	861.7	136.6	80	1.3410	5.56	2.99	1.572	1.624	1.662	61.44		
13.00	0.9778	0.9796	127.1	2.759	850.7	147.5	87	1.3417	6.13	3.30	1.722	1.691	1.732	59.03		
14.00	0.9765	0.9782	136.7	2.967	839.8	158.4	95	1.3425	6.73	3.62	1.874	1.757	1.803	56.80		
15.00	0.9752	0.9769	146.3	3.175	828.9	169.3	102	1.3432	7.36	3.96	2.030	1.822	1.872	54.78		
16.00	0.9739	0.9756	155.8	3.382	818.1	180.2	110	1.3440	8.01	4.31	2.189	1.886	1.941	52.91		
17.00	0.9726	0.9743	165.3	3.589	807.3	191.0	117	1.3447	8.69	4.67	2.351	1.951	2.010	51.16		
18.00	0.9713	0.9730	174.8	3.795	796.5	201.8	125	1.3455	9.40	5.05	2.516	2.015	2.078	49.54		
19.00	0.9700	0.9717	184.3	4.000	785.7	212.5	132	1.3462	10.14	5.45	2.683	2.077	2.146	48.04		
20.00	0.9687	0.9704	193.7	4.205	774.9	223.3	140	1.3469	10.92	5.87	2.853	2.138	2.212	46.68		
22.00	0.9660	0.9677	212.5	4.613	753.5	244.8	154	1.3484	12.60	6.78	3.203	2.254	2.338	44.27		
24.00	0.9632	0.9649	231.2	5.018	732.0	266.2	168	1.3498	14.47	7.78	3.568	2.365	2.460	42.20		
26.00	0.9602	0.9619	249.7	5.419	710.6	287.7	181	1.3511	16.41	8.82	3.920	2.471	2.579	40.39		
28.00	0.9571	0.9588	268.0	5.817	689.1	309.1	194	1.3524	18.43	9.91	4.262	2.576	2.696	38.75		
30.00	0.9539	0.9556	286.2	6.211	667.7	330.5	205	1.3535	20.47	11.01	4.583	2.662	2.796	37.49		
32.00	0.9504	0.9521	304.1	6.601	646.3	352.0	216	1.3546	22.44	12.07		2.721	2.869	36.68		
34.00	0.9468	0.9485	321.9	6.988	624.9	373.3	227	1.3557	24.27	13.05		2.762	2.923	36.13		
36.00	0.9431	0.9447	339.5	7.369	603.6	394.7	236	1.3566	25.98	13.97		2.797	2.971	35.69		
38.00	0.9392	0.9408	356.9	7.747	582.3	415.9	245	1.3575	27.62	14.85		2.823	3.012	35.35		
40.00	0.9352	0.9369	374.1	8.120	561.1	437.1	253	1.3583	29.26	15.73		2.840	3.043	35.14		
42.00	0.9311	0.9328	391.1	8.488	540.0	458.2	260	1.3590	30.98	16.66		2.846	3.063	35.06		
44.00	0.9269	0.9286	407.8	8.853	519.1	479.2	268	1.3598	32.68	17.57		2.844	3.074	35.09		
46.00	0.9227	0.9243	424.4	9.213	498.2	500.0	274	1.3604	34.36	18.47		2.837	3.081	35.18		
48.00	0.9183	0.9199	440.8	9.568	477.5	520.7	280	1.3610	36.04	19.38		2.826	3.083	35.32		
50.00	0.9139	0.9155	457.0	9.919	457.0	541.3	286	1.3616	37.67	20.25		2.807	3.078	35.55		
52.00	0.9095	0.9111	472.9	10.265	436.5	561.7	291	1.3621	39.20	21.08		2.783	3.066	35.87		
54.00	0.9049	0.9065	488.7	10.607	416.3	582.0	296	1.3626	40.65	21.85		2.749	3.044	36.31		
56.00	0.9004	0.9020	504.2	10.944	396.2	602.1	300	1.3630	42.06	22.61		2.696	3.000	37.02		
58.00	0.8958	0.8974	519.5	11.277	376.2	622.0	304	1.3634	43.49	23.38		2.627	2.938	38.00		
60.00	0.8911	0.8927	534.7	11.606	356.5	641.8	308	1.3638	44.93	24.15		2.542	2.858	39.26		
62.00	0.8865	0.8880	549.6	11.930	336.9	661.4	311	1.3641	46.28	24.88		2.474	2.796	40.34		
64.00	0.8818	0.8833	564.3	12.250	317.4	680.8	315	1.3644	47.52	25.55		2.410	2.739	41.41		
66.00	0.8771	0.8786	578.9	12.565	298.2	700.0	317	1.3647	48.64	26.15		2.342	2.676	42.61		
68.00	0.8724	0.8739	593.2	12.876	279.2	719.1	320	1.3650	49.52	26.62		2.276	2.614	43.85		
70.00	0.8676	0.8692	607.3	13.183	260.3	737.9	322	1.3652				2.210	2.552	45.17		
72.00	0.8629	0.8644	621.3	13.486	241.6	756.6	324	1.3654				2.144	2.490	46.55		
74.00	0.8581	0.8596	635.0	13.784	223.1	775.1	325	1.3655				2.078	2.426	48.03		
76.00	0.8533	0.8549	648.5	14.077	204.8	793.4	327	1.3657				2.011	2.361	49.63		
78.00	0.8485	0.8500	661.8	14.366	186.7	811.6	327	1.3657				1.944	2.296	51.33		
80.00	0.8436	0.8451	674.9	14.650	168.7	829.5	328	1.3658				1.877	2.229	53.15		
82.00	0.8387	0.8401	687.7	14.927	151.0	847.3	328	1.3657				1.804	2.155	55.31		
84.00	0.8335	0.8350	700.2	15.198	133.4	864.9	326	1.3656				1.738	2.089	57.42		
86.00	0.8284	0.8299	712.4	15.464	116.0	882.3	325	1.3655				1.671	2.021	59.74		
88.00	0.8232	0.8247	724.4	15.725	98.8	899.4	323	1.3653				1.603	1.951	62.26		
90.00	0.8180	0.8194	736.2	15.979	81.8	916.4	320	1.3650				1.539	1.885	64.85		
92.00	0.8125	0.8140	747.5	16.226	65.0	933.2	316	1.3646				1.472	1.815	67.80		
94.00	0.8070	0.8084	758.6	16.466	48.4	949.8	312	1.3642				1.404	1.743	71.07		
96.00	0.8013	0.8027	769.2	16.697	32.1	966.2	306	1.3636				1.339	1.674	74.53		
98.00	0.7954	0.7968	779.5	16.920	15.9	982.3	300	1.3630				1.270	1.600	78.59		
100.00	0.7893	0.7907	789.3	17.133	0.0	998.2	284	1.3614				1.201	1.525	83.10		

17 (ETHYLENEDINITRILO)TETRAACETIC ACID DISODIUM SALT, EDTA DISODIUM*, $Na_2C_{10}H_{14}O_8N_2 \cdot 2H_2O$

MOLECULAR WEIGHT = 336.21 FORMULA WEIGHT, HYDRATE = 372.24
RELATIVE SPECIFIC REFRACTIVITY = 0.977

0.00 % by wt. data are the same for all compounds.
For Values of 0.00 wt. % solutions see Table 1, Acetic Acid.

A% by wt.	H% by wt.	ρ D_4^{20}	D_{20}^{20}	C_s g/l	M g-mol/l	C_w g/l	$(C_o - C_w)$ g/l	$(n - n_o)$ ×10⁴	n	Δ °C	O Os/kg	S g-mol/l	η/η_o	η/ρ cS	ϕ rhe	γ mmho/cm	T g-mol/l
0.50	0.55	1.0009	1.0027	5.0	0.015	995.9	2.3	9	1.3339	0.07	0.04	0.020	1.015	1.016	98.35	1.9	0.018
1.00	1.11	1.0036	1.0053	10.0	0.030	993.5	4.7	18	1.3348	0.14	0.07	0.039	1.030	1.028	96.94	3.5	0.034
1.50	1.66	1.0062	1.0080	15.1	0.045	991.1	7.1	26	1.3356	0.21	0.11	0.059	1.044	1.040	95.55	5.0	0.049
2.00	2.21	1.0089	1.0107	20.2	0.060	988.7	9.5	35	1.3365	0.27	0.15	0.078	1.060	1.052	94.19	6.3	0.064
2.50	2.77	1.0115	1.0133	25.3	0.075	986.2	12.0	44	1.3374	0.33	0.18	0.096	1.075	1.065	92.84	7.6	0.077
3.00	3.32	1.0142	1.0160	30.4	0.090	983.8	14.5	53	1.3383	0.40	0.21	0.115	1.091	1.078	91.51	8.8	0.091
3.50	3.88	1.0169	1.0187	35.6	0.106	981.3	16.9	62	1.3392	0.46	0.25	0.133	1.107	1.091	90.18	10.1	0.105
4.00	4.43	1.0196	1.0214	40.8	0.121	978.8	19.5	70	1.3400	0.52	0.28	0.151	1.123	1.104	88.85	11.3	0.119
4.50	4.98	1.0223	1.0241	46.0	0.137	976.3	22.0	79	1.3409	0.58	0.31	0.169	1.140	1.118	87.53	12.5	0.133
5.00	5.54	1.0250	1.0268	51.2	0.152	973.7	24.5	88	1.3418	0.65	0.35	0.187	1.158	1.132	86.20	13.7	0.145
5.50	6.09	1.0277	1.0295	56.5	0.168	971.2	27.0	97	1.3427	0.71	0.38	0.206	1.176	1.147	84.86	14.6	0.156
6.00	6.64	1.0305	1.0323	61.8	0.184	968.6	29.6	106	1.3436	0.77	0.42	0.225	1.195	1.162	83.51	15.3	0.164

* Sodium(Di)Ethylenediamine Tetraacetate.

18 ETHYLENE GLYCOL, CH_2OHCH_2OH

MOLECULAR WEIGHT = 62.07
RELATIVE SPECIFIC REFRACTIVITY = 1.147

0.00 % by wt. data are the same for all compounds.
For Values of 0.00 wt. % solutions see Table 1, Acetic Acid.

A% by wt.	ρ D_4^{20}	D_{20}^{20}	C_s g/l	M g-mol/l	C_w g/l	$(C_o - C_w)$ g/l	$(n - n_o)$ ×10⁴	n	Δ °C	O Os/kg	S g-mol/l	η/η_o	η/ρ cS	ϕ rhe	γ mmho/cm	T g-mol/l
0.50	0.9988	1.0006	5.0	0.080	993.8	4.4	5	1.3335	0.15	0.08	0.042	1.008	1.011	99.03		
1.00	0.9995	1.0012	10.0	0.161	989.5	8.8	9	1.3339	0.30	0.16	0.086	1.018	1.021	98.04		
2.00	1.0007	1.0025	20.0	0.322	980.7	17.6	19	1.3348	0.61	0.33	0.176	1.046	1.047	95.41		
3.00	1.0019	1.0037	30.1	0.484	971.9	26.4	28	1.3358	0.92	0.50	0.269	1.072	1.072	93.09		
4.00	1.0032	1.0049	40.1	0.646	963.0	35.2	37	1.3367	1.24	0.67	0.364	1.097	1.096	90.98		
5.00	1.0044	1.0062	50.2	0.809	954.2	44.0	47	1.3377	1.58	0.85	0.461	1.123	1.120	88.87		
6.00	1.0057	1.0075	60.3	0.972	945.3	52.9	56	1.3386	1.91	1.03	0.560	1.151	1.147	86.77		
7.00	1.0070	1.0087	70.5	1.136	936.5	61.8	66	1.3396	2.26	1.22	0.660	1.180	1.174	84.60		
8.00	1.0082	1.0100	80.7	1.299	927.6	70.6	75	1.3405	2.62	1.41	0.763	1.210	1.202	82.49		
9.00	1.0095	1.0113	90.9	1.464	918.7	79.6	85	1.3415	2.99	1.61	0.868	1.241	1.232	80.41		
10.00	1.0108	1.0126	101.1	1.629	909.8	88.5	95	1.3425	3.37	1.81	0.975	1.274	1.263	78.31		
12.00	1.0134	1.0152	121.6	1.959	891.8	106.4	114	1.3444	4.16	2.24	1.196	1.343	1.330	74.21		
14.00	1.0161	1.0179	142.4	2.292	873.8	124.4	134	1.3464	5.01	2.69	1.426	1.421	1.401	70.25		
16.00	1.0188	1.0206	163.0	2.626	855.8	142.5	154	1.3484	5.91	3.18	1.665	1.497	1.473	66.65		
18.00	1.0214	1.0232	183.9	2.962	837.6	160.7	174	1.3503	6.89	3.70	1.913	1.575	1.545	63.35		
20.00	1.0241	1.0259	204.8	3.300	819.3	178.9	194	1.3523	7.93	4.27	2.171	1.658	1.622	60.19		
24.00	1.0296	1.0314	247.1	3.981	782.5	215.8	234	1.3564	10.28	5.53	2.714	1.839	1.790	54.26		
28.00	1.0350	1.0369	289.8	4.669	745.2	253.0	275	1.3605	13.03	7.01	3.289	2.043	1.978	48.84		
32.00	1.0405	1.0424	333.0	5.364	707.6	290.7	316	1.3646	16.23	8.73	3.889	2.275	2.191	43.87		
36.00	1.0460	1.0478	376.6	6.067	669.4	328.8	357	1.3687	19.82	10.66	4.483	2.532	2.425	39.42		
40.00	1.0514	1.0532	420.6	6.775	630.8	367.4	398	1.3728	23.84	12.82		2.826	2.693	35.32		
44.00	1.0567	1.0586	465.0	7.491	591.8	406.5	440	1.3769	28.32	15.23		3.160	2.996	31.59		
48.00	1.0619	1.0638	509.7	8.212	552.2	446.0	481	1.3811	33.30	17.90		3.537	3.337	28.22		
52.00	1.0670	1.0689	554.8	8.939	512.2	486.1	521	1.3851	38.81	20.87		3.973	3.731	25.12		
56.00	1.0719	1.0738	600.3	9.671	471.6	526.6	562	1.3892	44.83	24.10		4.466	4.175	22.35		
60.00	1.0765	1.0784	645.9	10.406	430.6	567.6	602	1.3931	51.23	27.54		5.016	4.669	19.90		

19 FERRIC CHLORIDE, $FeCl_3 \cdot 6H_2O$

MOLECULAR WEIGHT = 162.22 FORMULA WEIGHT, HYDRATE = 270.32
RELATIVE SPECIFIC REFRACTIVITY = 0.946

0.00 % by wt. data are the same for all compounds.
For Values of 0.00 wt. % solutions see Table 1, Acetic Acid.

A% by wt.	H% by wt.	ρ D_4^{20}	D_{20}^{20}	C_s g/l	M g-mol/l	C_w g/l	$(C_o - C_w)$ g/l	$(n - n_o)$ ×10⁴	n	Δ °C	O Os/kg	S g-mol/l	η/η_o	η/ρ cS	ϕ rhe	γ mmho/cm	T g-mol/l
0.50	0.83	1.0025	1.0043	5.0	0.031	997.5	0.7	14	1.3344	0.206	0.111	0.059	1.022	1.022	97.61		
1.00	1.67	1.0068	1.0086	10.1	0.062	996.7	1.5	28	1.3358	0.385	0.207	0.111	1.045	1.040	95.49		
2.00	3.33	1.0153	1.0171	20.3	0.125	995.0	3.3	56	1.3386	0.753	0.405	0.219	1.091	1.077	91.48		
3.00	5.00	1.0238	1.0256	30.7	0.189	993.1	5.2	83	1.3413	1.146	0.616	0.335	1.137	1.113	87.79		
4.00	6.67	1.0323	1.0341	41.3	0.255	991.0	7.3	111	1.3441	1.562	0.840	0.457	1.185	1.150	84.22		
5.00	8.33	1.0408	1.0426	52.0	0.321	988.7	9.5	138	1.3468	2.002	1.076	0.585	1.236	1.190	80.76		
6.00	10.00	1.0493	1.0512	63.0	0.388	986.4	11.9	166	1.3496	2.475	1.331	0.722	1.289	1.231	77.41		
7.00	11.66	1.0580	1.0599	74.1	0.457	984.0	14.3	194	1.3524	2.994	1.609	0.870	1.347	1.275	74.11		
8.00	13.33	1.0668	1.0687	85.3	0.526	981.5	16.7	222	1.3552	3.566	1.917	1.031	1.409	1.324	70.82		
9.00	15.00	1.0760	1.0779	96.8	0.597	979.2	19.0	251	1.3581	4.185	2.250	1.202	1.477	1.376	67.55		
10.00	16.66	1.0853	1.0872	108.5	0.669	976.8	21.5	281	1.3611	4.854	2.610	1.385	1.550	1.431	64.39		
12.00	20.00	1.1040	1.1059	132.5	0.817	971.5	26.8	340	1.3670	6.38	3.43	1.785	1.704	1.547	58.57		
14.00	23.33	1.1228	1.1248	157.2	0.969	965.6	32.6	400	1.3730	8.22	4.42	2.240	1.875	1.673	53.22		
16.00	26.66	1.1420	1.1440	182.7	1.126	959.3	39.0			10.45	5.62	2.750	2.076	1.822	48.07		
18.00	29.99	1.1615	1.1636	209.1	1.289	952.5	45.8			13.08	7.03	3.299	2.306	1.990	43.27		
20.00	33.33	1.1816	1.1837	236.3	1.457	945.3	53.0			16.14	8.68	3.873	2.565	2.175	38.91		
24.00	39.99	1.2234	1.2256	293.6	1.810	929.8	68.4			23.79	12.79		3.172	2.598	31.47		
28.00	46.66	1.2679	1.2702	355.0	2.189	912.9	85.3			33.61	18.07		4.030	3.185	24.76		
32.00	53.32	1.3153	1.3176	420.9	2.595	894.4	103.8			49.16	26.43		5.263	4.010	18.96		
36.00	59.99	1.3654	1.3678	491.5	3.030	873.9	124.4						7.116	5.222	14.03		
40.00	66.66	1.4176	1.4201	567.0	3.496	850.6	147.7						9.655	6.824	10.34		

20 FORMIC ACID, HCOOH

MOLECULAR WEIGHT = 46.03
RELATIVE SPECIFIC REFRACTIVITY = 0.916

A% by wt.	ρ D_4^{20}	D_{20}^{20}	C_s g/l	M g-mol/l	C_w g/l	$(C_o - C_w)$ g/l	$(n - n_o)$ $\times 10^4$	n	Δ °C	O Os/kg	S g-mol/l	η/η_o	η/ρ cS	ϕ rhe	γ mmho/cm	T g-mol/l
0.50	0.9994	1.0012	5.0	0.109	994.4	3.8	3	1.3333	0.210	0.113	0.060	1.004	1.007	99.36	1.4	0.013
1.00	1.0006	1.0023	10.0	0.217	990.6	7.7	6	1.3336	0.418	0.225	0.121	1.009	1.010	98.96	2.4	0.024
2.00	1.0029	1.0047	20.1	0.436	982.9	15.3	12	1.3342	0.824	0.443	0.240	1.015	1.015	98.28	3.5	0.034
3.00	1.0053	1.0071	30.2	0.655	975.2	23.1	18	1.3348	1.243	0.668	0.363	1.023	1.019	97.59	4.3	0.043
4.00	1.0077	1.0095	40.3	0.876	967.4	30.8	24	1.3354	1.672	0.899	0.489	1.030	1.024	96.90	5.0	0.050
5.00	1.0102	1.0119	50.5	1.097	959.6	38.6	29	1.3359	2.103	1.131	0.615	1.037	1.029	96.24	5.6	0.056
6.00	1.0126	1.0144	60.8	1.320	951.8	46.4	35	1.3365	2.534	1.363	0.739	1.044	1.033	95.63	6.1	0.062
7.00	1.0150	1.0168	71.1	1.544	944.0	54.3	41	1.3371	2.967	1.595	0.862	1.050	1.036	95.06	6.6	0.067
8.00	1.0175	1.0193	81.4	1.768	936.1	62.2	46	1.3376	3.402	1.829	0.985	1.056	1.040	94.51	7.0	0.071
9.00	1.0199	1.0217	91.8	1.994	928.1	70.1	52	1.3382	3.838	2.063	1.106	1.062	1.043	93.98	7.4	0.075
10.00	1.0224	1.0242	102.2	2.221	920.1	78.1	57	1.3387	4.267	2.294	1.225	1.068	1.047	93.45	7.8	0.079
12.00	1.0273	1.0291	123.3	2.678	904.0	94.2	67	1.3397	5.19	2.79	1.475	1.080	1.053	92.41	8.4	0.086
14.00	1.0322	1.0340	144.5	3.139	887.7	110.5	78	1.3408	6.11	3.28	1.715	1.092	1.060	91.40	8.8	0.091
16.00	1.0371	1.0389	165.9	3.605	871.1	127.1	88	1.3418	7.06	3.80	1.957	1.104	1.067	90.38	9.3	0.096
18.00	1.0419	1.0438	187.5	4.074	854.4	143.9	98	1.3428	8.08	4.35	2.207	1.117	1.074	89.35	9.6	0.100
20.00	1.0467	1.0486	209.3	4.548	837.4	160.9	107	1.3437	9.11	4.90	2.448	1.130	1.081	88.36	9.9	0.103
24.00	1.0562	1.0580	253.5	5.507	802.7	195.5	126	1.3456	11.10	5.97	2.891	1.154	1.094	86.52	10.3	0.108
28.00	1.0654	1.0673	298.3	6.481	767.1	231.1	145	1.3475	13.10	7.04	3.303	1.177	1.107	84.79	10.5	0.109
32.00	1.0746	1.0765	343.9	7.471	730.7	267.5	163	1.3493	15.28	8.21	3.717	1.201	1.120	83.09	10.5	0.109
36.00	1.0839	1.0858	390.2	8.477	693.7	304.6	181	1.3511	17.65	9.49	4.132	1.225	1.133	81.44	10.3	0.107
40.00	1.0935	1.0955	437.4	9.503	656.1	342.1	199	1.3529	20.18	10.85	4.539	1.251	1.146	79.78	9.9	0.103
44.00	1.1015	1.1035	484.7	10.529	616.9	381.4	217	1.3547	22.93	12.33		1.278	1.162	78.11	9.5	0.098
48.00	1.1097	1.1117	532.7	11.572	577.1	421.2	235	1.3565	26.06	14.01		1.306	1.179	76.42	8.9	0.092
52.00	1.1183	1.1203	581.5	12.634	536.8	461.4	251	1.3581	29.69	15.96		1.337	1.198	74.65	8.3	0.085
56.00	1.1273	1.1293	631.3	13.714	496.0	502.2	267	1.3597	33.81	18.18		1.371	1.218	72.81	7.6	0.078
60.00	1.1364	1.1385	681.9	14.813	454.6	543.7	282	1.3612	38.26	20.57		1.407	1.241	70.93	6.9	0.070
64.00	1.1456	1.1476	733.2	15.928	412.4	585.8	296	1.3626	43.02	23.13		1.446	1.265	69.03	6.2	0.062
68.00	1.1544	1.1565	785.0	17.055	369.4	628.8	311	1.3641				1.487	1.291	67.11	5.4	0.054
70.00	1.1586	1.1607	811.1	17.620	347.6	650.6	318	1.3648				1.508	1.305	66.16	5.1	0.051

21 D-FRUCTOSE (LEVULOSE), C6H12O6

MOLECULAR WEIGHT = 180.16
RELATIVE SPECIFIC REFRACTIVITY = 1.021

A% by wt.	ρ D_4^{20}	D_{20}^{20}	C_s g/l	M g-mol/l	C_w g/l	$(C_o - C_w)$ g/l	$(n - n_o)$ $\times 10^4$	n	Δ °C	O Os/kg	S g-mol/l	η/η_o	η/ρ cS	ϕ rhe	γ mmho/cm	T g-mol/l
0.50	1.0002	1.0020	5.0	0.028	995.2	3.0	7	1.3337	0.052	0.028	0.014	1.013	1.015	98.53		
1.00	1.0021	1.0039	10.0	0.056	992.1	6.1	14	1.3344	0.104	0.056	0.029	1.026	1.026	97.28		
1.50	1.0041	1.0059	15.1	0.084	989.1	9.2	21	1.3351	0.157	0.085	0.045	1.039	1.037	96.06		
2.00	1.0061	1.0079	20.1	0.112	986.0	12.3	28	1.3358	0.211	0.113	0.060	1.052	1.048	94.87		
2.50	1.0081	1.0098	25.2	0.140	982.9	15.4	36	1.3366	0.265	0.142	0.076	1.065	1.059	93.71		
3.00	1.0101	1.0118	30.3	0.168	979.7	18.5	43	1.3373	0.319	0.172	0.092	1.078	1.069	92.59		
3.50	1.0120	1.0138	35.4	0.197	976.6	21.6	50	1.3380	0.374	0.201	0.108	1.091	1.080	91.50		
4.00	1.0140	1.0158	40.6	0.225	973.5	24.8	57	1.3387	0.430	0.231	0.124	1.104	1.091	90.40		
4.50	1.0160	1.0178	45.7	0.254	970.3	27.9	65	1.3395	0.487	0.262	0.141	1.118	1.102	89.30		
5.00	1.0181	1.0199	50.9	0.283	967.2	31.1	72	1.3402	0.544	0.292	0.158	1.132	1.114	88.16		
5.50	1.0201	1.0219	56.1	0.311	964.0	34.3	79	1.3409	0.602	0.324	0.175	1.147	1.127	87.00		
6.00	1.0221	1.0239	61.3	0.340	960.8	37.5	87	1.3417	0.660	0.355	0.192	1.163	1.140	85.83		
6.50	1.0241	1.0259	66.6	0.369	957.6	40.7	94	1.3424	0.720	0.387	0.209	1.179	1.153	84.66		
7.00	1.0262	1.0280	71.8	0.399	954.3	43.9	101	1.3431	0.780	0.419	0.227	1.196	1.167	83.48		
7.50	1.0282	1.0300	77.1	0.428	951.1	47.1	109	1.3439	0.841	0.452	0.245	1.213	1.182	82.29		
8.00	1.0303	1.0321	82.4	0.457	947.8	50.4	116	1.3446	0.902	0.485	0.263	1.230	1.197	81.11		
8.50	1.0323	1.0342	87.7	0.487	944.6	53.7	124	1.3454	0.965	0.519	0.282	1.249	1.212	79.93		
9.00	1.0344	1.0362	93.1	0.517	941.3	56.9	131	1.3461	1.028	0.553	0.300	1.267	1.228	78.75		
9.50	1.0365	1.0383	98.5	0.547	938.0	60.2	139	1.3469	1.092	0.587	0.319	1.286	1.244	77.58		
10.00	1.0385	1.0404	103.9	0.576	934.7	63.5	147	1.3476	1.158	0.622	0.338	1.306	1.260	76.42		
11.00	1.0427	1.0446	114.7	0.637	928.0	70.2	162	1.3492	1.290	0.694	0.377	1.346	1.294	74.13		
12.00	1.0469	1.0488	125.6	0.697	921.3	76.9	177	1.3507	1.427	0.767	0.417	1.388	1.329	71.89		
13.00	1.0512	1.0530	136.7	0.758	914.5	83.7	193	1.3522	1.567	0.842	0.458	1.432	1.365	69.67		
14.00	1.0554	1.0573	147.8	0.820	907.7	90.6	208	1.3538	1.710	0.920	0.500	1.480	1.405	67.45		
15.00	1.0597	1.0616	159.0	0.882	900.8	97.5	224	1.3554	1.858	0.999	0.543	1.530	1.447	65.23		
16.00	1.0640	1.0659	170.2	0.945	893.8	104.5	240	1.3569	2.009	1.080	0.587	1.584	1.491	63.02		
17.00	1.0684	1.0703	181.6	1.008	886.8	111.5	255	1.3585	2.162	1.163	0.632	1.640	1.538	60.84		
18.00	1.0728	1.0747	193.1	1.072	879.7	118.6	272	1.3601	2.319	1.247	0.677	1.700	1.588	58.69		
19.00	1.0772	1.0791	204.7	1.136	872.5	125.7	288	1.3618	2.479	1.333	0.723	1.764	1.641	56.56		
20.00	1.0816	1.0835	216.3	1.201	865.3	133.0	304	1.3634	2.640	1.419	0.769	1.833	1.698	54.45		
22.00	1.0906	1.0925	239.9	1.332	850.6	147.6	337	1.3667	3.054	1.642	0.887	1.982	1.821	50.35		
24.00	1.0996	1.1016	263.9	1.465	835.7	162.5	370	1.3700	3.431	1.845	0.993	2.150	1.959	46.41		
26.00	1.1089	1.1108	288.3	1.600	820.5	177.7	404	1.3734	3.815	2.051	1.100	2.343	2.117	42.60		
28.00	1.1182	1.1202	313.1	1.738	805.1	193.1	438	1.3768	4.196	2.256	1.205	2.557	2.291	39.03		
30.00	1.1276	1.1296	338.3	1.878	789.3	208.9	473	1.3803				2.811	2.498	35.50		
32.00	1.1372	1.1392	363.9	2.020	773.3	224.9	509	1.3839				3.106	2.737	32.13		
34.00	1.1469	1.1490	390.0	2.165	757.0	241.3	545	1.3874				3.455	3.018	28.89		
36.00	1.1568	1.1588	416.4	2.312	740.3	257.9	581	1.3911				3.891	3.370	25.65		
38.00	1.1668	1.1688	443.4	2.461	723.4	274.8	618	1.3948				4.409	3.786	22.64		
40.00	1.1769	1.1790	470.8	2.613	706.1	292.1	655	1.3985				5.036	4.288	19.82		
42.00	1.1871	1.1892	498.6	2.767	688.5	309.7	694	1.4023				5.761	4.863	17.32		
44.00	1.1975	1.1996	526.9	2.925	670.6	327.6	732	1.4062				6.631	5.548	15.05		
46.00	1.2080	1.2102	555.7	3.084	652.3	345.9	771	1.4101				7.738	6.419	12.90		
48.00	1.2187	1.2208	585.0	3.247	633.7	364.5	811	1.4141				9.042	7.434	11.04		

21 D-FRUCTOSE (LEVULOSE), $C_6H_{12}O_6$—(Continued)

A% by wt.	ρ D_4^{20}	D_{20}^{20}	C_s g/l	M g-mol/l	C_w g/l	(C_o-C_w) g/l	$(n-n_o)$ ×10^4	n	Δ °C	O Os/kg	S g-mol/l	η/η_o	η/ρ cS	ϕ rhe	γ mmho/cm	T g-mol/l
50.00	1.2295	1.2316	614.7	3.412	614.7	383.5	851	1.4181					10.80	8.80	9.24	
52.00	1.2404	1.2426	645.0	3.580	595.4	402.9	892	1.4222					12.99	10.49	7.69	
54.00	1.2514	1.2536	675.8	3.751	575.7	422.6	934	1.4264					15.84	12.68	6.30	
56.00	1.2626	1.2649	707.1	3.925	555.6	442.7	976	1.4306					20.05	15.91	4.98	
58.00	1.2739	1.2762	738.9	4.101	535.1	463.2	1019	1.4348					25.60	20.13	3.90	
60.00	1.2854	1.2877	771.2	4.281	514.2	484.1	1062	1.4392					32.47	25.31	3.07	
62.00	1.2970	1.2992	804.1	4.463	492.8	505.4	1105	1.4435					42.97	33.20	2.32	
64.00	1.3086	1.3110	837.5	4.649	471.1	527.1	1150	1.4480					58.17	44.54	1.72	
66.00	1.3204	1.3228	871.5	4.837	448.9	549.3	1194	1.4524					82.19	62.37	1.21	
68.00	1.3323	1.3347	906.0	5.029	426.3	571.9	1240	1.4570					118.4	89.08	0.84	
70.00	1.3443	1.3467	941.0	5.223	403.3	594.9	1285	1.4615					177.8	132.5	0.56	

22 D-GLUCOSE (DEXTROSE), $C_6H_{12}O_6 \cdot H_2O$

MOLECULAR WEIGHT = 180.16   FORMULA WEIGHT, HYDRATE = 198.17
RELATIVE SPECIFIC REFRACTIVITY = 1.031

0.00 % by wt. data are the same for all compounds.
For Values of 0.00 wt. % solutions see Table 1, Acetic Acid.

A% by wt.	H% by wt.	ρ D_4^{20}	D_{20}^{20}	C_s g/l	M g-mol/l	C_w g/l	(C_o-C_w) g/l	$(n-n_o)$ ×10^4	n	Δ °C	O Os/kg	S g-mol/l	η/η_o	η/ρ cS	ϕ rhe	γ mmho/cm	T g-mol/l
0.50	0.55	1.0001	1.0019	5.0	0.028	995.1	3.1	7	1.3337	0.047	0.025	0.013	1.008	1.010	99.01		
1.00	1.10	1.0020	1.0038	10.0	0.056	992.0	6.2	14	1.3344	0.107	0.057	0.030	1.019	1.019	97.94		
1.50	1.65	1.0039	1.0057	15.1	0.084	988.9	9.4	21	1.3351	0.158	0.085	0.045	1.034	1.032	96.54		
2.00	2.20	1.0058	1.0076	20.1	0.112	985.7	12.5	28	1.3358	0.214	0.115	0.061	1.050	1.046	95.05		
2.50	2.75	1.0078	1.0095	25.2	0.140	982.6	15.7	36	1.3366	0.270	0.145	0.077	1.065	1.059	93.68		
3.00	3.30	1.0097	1.0115	30.3	0.168	979.4	18.8	43	1.3373	0.323	0.174	0.093	1.081	1.072	92.36		
3.50	3.85	1.0116	1.0134	35.4	0.197	976.2	22.0	50	1.3380	0.377	0.202	0.108	1.096	1.085	91.08		
4.00	4.40	1.0136	1.0154	40.5	0.225	973.0	25.2	57	1.3387	0.433	0.233	0.125	1.111	1.098	89.82		
4.50	4.95	1.0155	1.0173	45.7	0.254	969.8	28.4	65	1.3395	0.490	0.264	0.142	1.127	1.112	88.57		
5.00	5.50	1.0175	1.0193	50.9	0.282	966.6	31.6	72	1.3402	0.549	0.295	0.159	1.143	1.126	87.31		
5.50	6.05	1.0194	1.0212	56.1	0.311	963.4	34.9	79	1.3409	0.608	0.327	0.176	1.160	1.140	86.06		
6.00	6.60	1.0214	1.0232	61.3	0.340	960.1	38.1	87	1.3417	0.668	0.359	0.194	1.177	1.154	84.82		
6.50	7.15	1.0234	1.0252	66.5	0.369	956.9	41.4	94	1.3424	0.729	0.392	0.212	1.194	1.169	83.59		
7.00	7.70	1.0254	1.0272	71.8	0.398	953.6	44.6	102	1.3432	0.790	0.425	0.230	1.212	1.184	82.37		
7.50	8.25	1.0274	1.0292	77.1	0.428	950.3	47.3	109	1.3439	0.851	0.458	0.248	1.230	1.199	81.17		
8.00	8.80	1.0294	1.0312	82.4	0.457	947.0	51.2	117	1.3447	0.913	0.491	0.266	1.248	1.215	79.97		
8.50	9.35	1.0314	1.0332	87.7	0.487	943.7	54.5	124	1.3454	0.975	0.524	0.284	1.267	1.231	78.77		
9.00	9.90	1.0334	1.0352	93.0	0.516	940.4	57.8	132	1.3462	1.038	0.558	0.303	1.286	1.247	77.60		
9.50	10.45	1.0354	1.0373	98.4	0.546	937.1	61.1	139	1.3469	1.102	0.592	0.322	1.306	1.264	76.39		
10.00	11.00	1.0375	1.0393	103.7	0.576	933.7	64.5	147	1.3477	1.167	0.627	0.341	1.327	1.282	75.21		
11.00	12.10	1.0416	1.0434	114.6	0.636	927.0	71.2	162	1.3492	1.303	0.700	0.381	1.369	1.317	72.89		
12.00	13.20	1.0457	1.0475	125.5	0.697	920.2	78.0	178	1.3508	1.443	0.776	0.422	1.413	1.354	70.63		
13.00	14.30	1.0498	1.0517	136.5	0.758	913.4	84.9	193	1.3523	1.586	0.853	0.464	1.459	1.393	68.40		
14.00	15.40	1.0540	1.0559	147.6	0.819	906.4	91.8	209	1.3539	1.731	0.931	0.506	1.509	1.434	66.15		
15.00	16.50	1.0582	1.0601	158.7	0.881	899.5	98.8	225	1.3555	1.880	1.011	0.550	1.563	1.480	63.85		
16.00	17.60	1.0624	1.0643	170.0	0.944	892.5	105.8	241	1.3571	2.033	1.093	0.594	1.622	1.530	61.53		
17.00	18.70	1.0667	1.0686	181.3	1.007	885.4	112.9	257	1.3587	2.190	1.178	0.640	1.685	1.583	59.21		
18.00	19.80	1.0710	1.0729	192.8	1.070	878.2	120.0	273	1.3603	2.353	1.265	0.687	1.753	1.640	56.93		
19.00	20.90	1.0753	1.0772	204.3	1.134	871.0	127.2	289	1.3619	2.521	1.355	0.735	1.825	1.700	54.70		
20.00	22.00	1.0797	1.0816	215.9	1.199	863.7	134.5	306	1.3635	2.696	1.449	0.785	1.900	1.763	52.53		
22.00	24.20	1.0884	1.0904	239.5	1.329	849.0	149.3	339	1.3668	3.073	1.652	0.892	2.059	1.896	48.47		
24.00	26.40	1.0973	1.0993	263.4	1.462	834.0	164.3	372	1.3702	3.480	1.871	1.007	2.238	2.044	44.59		
26.00	28.60	1.1063	1.1083	287.6	1.597	818.7	179.6	406	1.3736	3.900	2.097	1.124	2.453	2.222	40.69		
28.00	30.80	1.1154	1.1174	312.3	1.734	803.1	195.1	440	1.3770	4.337	2.332	1.244	2.702	2.427	36.94		
30.00	33.00	1.1246	1.1266	337.4	1.873	787.2	211.0	475	1.3805	4.794	2.577	1.369	2.992	2.666	33.36		
32.00	35.20	1.1340	1.1360	362.9	2.014	771.1	227.1	511	1.3840				3.317	2.931	30.09		
34.00	37.40	1.1434	1.1454	388.8	2.158	754.6	243.6	546	1.3876				3.697	3.240	26.99		
36.00	39.60	1.1529	1.1550	415.1	2.304	737.9	260.3	582	1.3912				4.185	3.637	23.85		
38.00	41.80	1.1626	1.1647	441.8	2.452	720.8	277.4	619	1.3949				4.776	4.116	20.90		
40.00	44.00	1.1724	1.1745	469.0	2.603	703.5	294.8	656	1.3986				5.482	4.685	18.21		
42.00	46.20	1.1823	1.1844	496.6	2.756	685.8	312.5	694	1.4024				6.275	5.318	15.90		
44.00	48.40	1.1924	1.1945	524.7	2.912	667.7	330.5	732	1.4062				7.221	6.068	13.82		
46.00	50.60	1.2026	1.2047	553.2	3.071	649.4	348.8	771	1.4101				8.437	7.029	11.83		
48.00	52.80	1.2130	1.2151	582.2	3.232	630.7	367.5	811	1.4141				9.863	8.148	10.12		
50.00	55.00	1.2235	1.2257	611.7	3.396	611.7	386.5	851	1.4181				11.86	9.713	8.41		
52.00	57.20	1.2342	1.2364	641.8	3.562	592.4	405.8	892	1.4222				14.46	11.74	6.90		
54.00	59.40	1.2451	1.2473	672.4	3.732	572.8	425.5	933	1.4263				17.88	14.39	5.58		
56.00	61.60	1.2562	1.2585	703.5	3.905	552.7	445.5	976	1.4306				22.84	18.22	4.37		
58.00	63.80	1.2676	1.2699	735.2	4.081	532.4	465.8	1020	1.4349				29.33	23.18	3.40		
60.00	66.00	1.2793	1.2816	767.6	4.261	511.7	486.5	1064	1.4394				37.37	29.27	2.67		

23 GLYCEROL, $CH_2OHCHOHCH_2OH$

MOLECULAR WEIGHT = 92.09
RELATIVE SPECIFIC REFRACTIVITY = 1.109

0.00 % by wt. data are the same for all compounds.
For Values of 0.00 wt. % solutions see Table 1, Acetic Acid.

A% by wt.	ρ D_4^{20}	D_{20}^{20}	C_s g/l	M g-mol/l	C_w g/l	(C_o-C_w) g/l	$(n-n_o)$ ×10^4	n	Δ °C	O Os/kg	S g-mol/l	η/η_o	η/ρ cS	ϕ rhe	γ mmho/cm	T g-mol/l
0.50	0.9994	1.0011	5.0	0.054	994.4	3.9	6	1.3336	0.072	0.039	0.020	1.009	1.012	98.89		
1.00	1.0005	1.0023	10.0	0.109	990.5	7.7	12	1.3342	0.180	0.097	0.051	1.020	1.022	97.84		
2.00	1.0028	1.0046	20.1	0.218	982.7	15.5	23	1.3353	0.411	0.221	0.119	1.046	1.045	95.41		
3.00	1.0051	1.0069	30.2	0.327	974.9	23.3	35	1.3365	0.627	0.337	0.182	1.072	1.068	93.12		
4.00	1.0074	1.0092	40.3	0.438	967.1	31.2	46	1.3376	0.849	0.456	0.247	1.098	1.092	90.93		

23 GLYCEROL, $CH_2OHCHOHCH_2OH$—(Continued)

A% by wt.	ρ D_4^{20}	D_{20}^{20}	C_s g/l	M g-mol/l	C_w g/l	$(C_o - C_w)$ g/l	$(n - n_o)$ $\times 10^4$	n	Δ °C	O Os/kg	S g-mol/l	η/η_o	η/ρ cS	ϕ rhe	γ mmho/cm	T g-mol/l
5.00	1.0097	1.0115	50.5	0.548	959.2	39.0	58	1.3388	1.078	0.580	0.315	1.125	1.116	88.71		
6.00	1.0120	1.0138	60.7	0.659	951.3	46.9	70	1.3400	1.316	0.708	0.385	1.155	1.143	86.44		
7.00	1.0144	1.0162	71.0	0.771	943.4	54.9	82	1.3412	1.561	0.839	0.457	1.186	1.171	84.17		
8.00	1.0167	1.0185	81.3	0.883	935.4	62.9	94	1.3424	1.811	0.974	0.530	1.218	1.201	81.90		
9.00	1.0191	1.0209	91.7	0.996	927.4	70.9	106	1.3436	2.064	1.110	0.603	1.253	1.232	79.67		
10.00	1.0215	1.0233	102.1	1.109	919.3	78.9	118	1.3448	2.323	1.249	0.678	1.288	1.263	77.48		
12.00	1.0262	1.0281	123.1	1.337	903.1	95.1	142	1.3472	2.880	1.548	0.837	1.362	1.330	73.28		
14.00	1.0311	1.0329	144.4	1.568	886.7	111.5	167	1.3496	3.469	1.865	1.004	1.442	1.401	69.22		
16.00	1.0360	1.0378	165.8	1.800	870.2	128.0	191	1.3521	4.094	2.201	1.177	1.530	1.480	65.22		
18.00	1.0409	1.0428	187.4	2.035	853.6	144.7	217	1.3547	4.756	2.557	1.359	1.627	1.566	61.34		
20.00	1.0459	1.0478	209.2	2.272	836.8	161.5	242	1.3572	5.46	2.93	1.546	1.734	1.661	57.56		
24.00	1.0561	1.0580	253.5	2.752	802.6	195.6	294	1.3624	7.01	3.77	1.944	1.984	1.882	50.31		
28.00	1.0664	1.0683	298.6	3.243	767.8	230.4	347	1.3676	8.77	4.71	2.370	2.274	2.136	43.89		
32.00	1.0770	1.0789	344.6	3.742	732.3	265.9	400	1.3730	10.74	5.78	2.814	2.632	2.449	37.91		
36.00	1.0876	1.0896	391.5	4.252	696.1	302.2	455	1.3785	12.96	6.97	3.276	3.082	2.839	32.38		
40.00	1.0984	1.1003	439.4	4.771	659.0	339.2	511	1.3841	15.50	8.33	3.757	3.646	3.326	27.37		
44.00	1.1092	1.1112	488.1	5.300	621.2	377.1	567	1.3897				4.434	4.005	22.51		
48.00	1.1200	1.1220	537.6	5.838	582.4	415.8	624	1.3954				5.402	4.833	18.47		
52.00	1.1308	1.1328	588.0	6.385	542.8	455.4	681	1.4011				6.653	5.895	15.00		
56.00	1.1419	1.1439	639.4	6.944	502.4	495.8	739	1.4069				8.332	7.311	11.98		
60.00	1.1530	1.1551	691.8	7.513	461.2	537.0	799	1.4129				10.66	9.264	9.36		
64.00	1.1643	1.1663	745.1	8.091	419.1	579.1	859	1.4189				13.63	11.73	7.32		
68.00	1.1755	1.1775	799.3	8.681	376.1	622.1	919	1.4249				18.42	15.70	5.42		
72.00	1.1866	1.1887	854.3	9.277	332.2	666.0	980	1.4310				27.57	23.28	3.62		
76.00	1.1976	1.1997	910.2	9.883	287.4	710.8	1040	1.4370				40.49	33.88	2.46		
80.00	1.2085	1.2106	966.8	10.498	241.7	756.5	1101	1.4431				59.78	49.57	1.67		
84.00	1.2192	1.2214	1024.2	11.121	195.1	803.2	1162	1.4492				84.17	69.18	1.19		
88.00	1.2299	1.2320	1082.3	11.752	147.6	850.7	1223	1.4553				147.2	119.9	0.68		
92.00	1.2404	1.2426	1141.1	12.392	99.2	899.0	1284	1.4613				383.7	310.0	0.26		
96.00	1.2508	1.2530	1200.7	13.039	50.0	948.2	1344	1.4674				778.9	624.0	0.13		
100.00	1.2611	1.2633	1261.1	13.694	0.0	998.2	1405	1.4735				1487.0	1181.3	0.06		

24 HYDROCHLORIC ACID, HCl

MOLECULAR WEIGHT = 36.47
RELATIVE SPECIFIC REFRACTIVITY = 1.152

0.00 % by wt. data are the same for all compounds.
For Values of 0.00 wt. % solutions see Table 1, Acetic Acid.

A% by wt.	ρ D_4^{20}	D_{20}^{20}	C_s g/l	M g-mol/l	C_w g/l	$(C_o - C_w)$ g/l	$(n - n_o)$ $\times 10^4$	n	Δ °C	O Os/kg	S g-mol/l	η/η_o	η/ρ cS	ϕ rhe	γ mmho/cm	T g-mol/l
0.50	1.0007	1.0025	5.0	0.137	995.7	2.5	12	1.3341	0.486	0.261	0.141	1.006	1.008	99.16	45.1	0.537
1.00	1.0031	1.0049	10.0	0.275	993.1	5.1	23	1.3353	0.989	0.532	0.289	1.013	1.012	98.50	92.9	1.23
1.50	1.0056	1.0074	15.1	0.414	990.5	7.7	35	1.3365	1.519	0.817	0.444	1.020	1.016	97.84	140.	2.10
2.00	1.0081	1.0098	20.2	0.553	987.9	10.3	46	1.3376	2.076	1.116	0.607	1.027	1.021	97.17	183.	3.14
2.50	1.0105	1.0123	25.3	0.693	985.3	13.0	58	1.3388	2.662	1.431	0.775	1.034	1.026	96.49	220.	4.85
3.00	1.0130	1.0148	30.4	0.833	982.6	15.6	69	1.3399	3.276	1.761	0.949	1.042	1.030	95.81		
3.50	1.0154	1.0172	35.5	0.975	979.9	18.3	81	1.3411	3.916	2.105	1.128	1.049	1.035	95.12		
4.00	1.0179	1.0197	40.7	1.116	977.2	21.0	92	1.3422	4.579	2.462	1.311	1.057	1.040	94.43		
4.50	1.0204	1.0222	45.9	1.259	974.4	23.8	104	1.3434	5.27	2.83	1.496	1.065	1.046	93.73		
5.00	1.0228	1.0246	51.1	1.402	971.7	26.5	115	1.3445	5.98	3.22	1.683	1.073	1.051	93.04		
5.50	1.0253	1.0271	56.4	1.546	968.9	29.3	127	1.3457	6.73	3.62	1.874	1.081	1.056	92.34		
6.00	1.0278	1.0296	61.7	1.691	966.1	32.1	138	1.3468	7.52	4.04	2.070	1.089	1.062	91.64		
6.50	1.0302	1.0321	67.0	1.836	963.3	35.0	150	1.3480	8.34	4.49	2.269	1.097	1.067	90.94		
7.00	1.0327	1.0345	72.3	1.982	960.4	37.8	162	1.3491	9.22	4.96	2.474	1.106	1.073	90.24		
7.50	1.0352	1.0370	77.6	2.129	957.5	40.7	173	1.3503	10.14	5.45	2.681	1.115	1.079	89.54		
8.00	1.0377	1.0395	83.0	2.276	954.6	43.6	185	1.3515	11.10	5.97	2.890	1.123	1.085	88.85		
8.50	1.0401	1.0420	88.4	2.424	951.7	46.5	196	1.3526	12.10	6.51	3.101	1.132	1.091	88.15		
9.00	1.0426	1.0445	93.8	2.573	948.8	49.5	208	1.3538	13.15	7.07	3.313	1.141	1.097	87.46		
9.50	1.0451	1.0469	99.3	2.722	945.8	52.4	219	1.3549	14.25	7.66	3.526	1.150	1.103	86.77		
10.00	1.0476	1.0494	104.8	2.872	942.8	55.4	231	1.3561	15.40	8.28	3.740	1.159	1.109	86.08		
11.00	1.0526	1.0544	115.8	3.175	936.8	61.4	254	1.3584	17.85	9.60	4.166	1.178	1.122	84.71		
12.00	1.0576	1.0594	126.9	3.480	930.7	67.6	277	1.3607	20.51	11.03	4.590	1.197	1.134	83.36		
13.00	1.0626	1.0645	138.1	3.788	924.4	73.8	300	1.3630				1.217	1.148	82.01		
14.00	1.0676	1.0695	149.5	4.098	918.1	80.1	323	1.3653				1.237	1.161	80.68		
15.00	1.0726	1.0745	160.9	4.412	911.8	86.5	347	1.3676				1.258	1.175	79.35		
16.00	1.0777	1.0796	172.4	4.728	905.3	93.0	370	1.3700				1.279	1.189	78.03		
17.00	1.0828	1.0847	184.1	5.047	898.7	99.5	393	1.3723				1.301	1.204	76.72		
18.00	1.0878	1.0898	195.8	5.369	892.0	106.2	416	1.3746				1.323	1.219	75.42		
19.00	1.0929	1.0949	207.7	5.694	885.3	113.0	439	1.3769				1.347	1.235	74.11		
20.00	1.0980	1.1000	219.6	6.022	878.4	119.8	462	1.3792				1.371	1.251	72.80		
22.00	1.1083	1.1102	243.8	6.686	864.5	133.8	509	1.3838				1.423	1.286	70.15		
24.00	1.1185	1.1205	268.4	7.361	850.1	148.1	555	1.3884				1.480	1.326	67.44		
26.00	1.1288	1.1308	293.5	8.047	835.3	162.9	600	1.3930				1.544	1.371	64.63		
28.00	1.1391	1.1411	318.9	8.745	820.1	178.1	646	1.3976				1.617	1.423	61.71		
30.00	1.1492	1.1513	344.8	9.454	804.5	193.8	691	1.4020				1.702	1.484	58.64		
32.00	1.1594	1.1614	371.0	10.173	788.4	209.9	736	1.4066				1.795	1.551	55.61		
34.00	1.1693	1.1714	397.6	10.901	771.8	226.5	782	1.4112				1.896	1.625	52.63		
36.00	1.1791	1.1812	424.5	11.639	754.6	243.6	828	1.4158				1.998	1.698	49.95		
38.00	1.1886	1.1907	451.7	12.385	736.9	261.3	874	1.4204				2.101	1.771	47.51		
40.00	1.1977	1.1999	479.1	13.137	718.6	279.6	920	1.4250								

25 INULIN, $(C_6H_{10}O_5)_x$*

MOLECULAR WEIGHT, AVERAGE = 5200
RELATIVE SPECIFIC REFRACTIVITY = 1.034

0.00 % by wt. data are the same for all compounds.
For Values of 0.00 wt. % solutions see Table 1, Acetic Acid.

A % by wt.	ρ D_4^{20}	D_{20}^{20}	C_s g/l	M g-mol/l	C_w g/l	$(C_o - C_w)$ g/l	$(n - n_o)$ $\times 10^4$	n	Δ °C	O Os/kg	S g-mol/l	η/η_o	η/ρ cS	ϕ rhe	γ mmho/cm	T g-mol/l
0.50	1.0001	1.0019	5.0	0.001	995.1	3.1	7	1.3337	0.003	0.001	0.001	1.030	1.032	96.93		
1.00	1.0020	1.0038	10.0	0.002	992.0	6.3	14	1.3344	0.005	0.003	0.001	1.059	1.059	94.24		
1.50	1.0038	1.0056	15.1	0.003	988.8	9.5	22	1.3351	0.007	0.004	0.002	1.088	1.086	91.74		
2.00	1.0057	1.0074	20.1	0.004	985.6	12.7	29	1.3359	0.010	0.005	0.003	1.117	1.113	89.35		
2.50	1.0075	1.0093	25.2	0.005	982.3	15.9	36	1.3366	0.012	0.007	0.003	1.147	1.140	87.03		
3.00	1.0093	1.0111	30.3	0.006	979.0	19.2	43	1.3373	0.015	0.008	0.004	1.177	1.168	84.81		
3.50	1.0112	1.0129	35.4	0.007	975.8	22.5	51	1.3381	0.017	0.009	0.005	1.208	1.197	82.65		
4.00	1.0130	1.0148	40.5	0.008	972.5	25.8	58	1.3388	0.019	0.010	0.005	1.240	1.226	80.51		
4.50	1.0148	1.0166	45.7	0.009	969.2	29.1	65	1.3395	0.022	0.012	0.006	1.274	1.258	78.36		
5.00	1.0167	1.0185	50.8	0.010	965.9	32.4	73	1.3403	0.024	0.013	0.007	1.310	1.291	76.18		
5.50	1.0186	1.0204	56.0	0.011	962.6	35.7	80	1.3410	0.027	0.015	0.007	1.349	1.327	74.00		
6.00	1.0205	1.0223	61.2	0.012	959.3	38.9	88	1.3418	0.030	0.016	0.008	1.389	1.364	71.83		
6.50	1.0225	1.0243	66.5	0.013	956.0	42.2	95	1.3425	0.033	0.018	0.009	1.432	1.403	69.69		
7.00	1.0243	1.0263	71.7	0.014	952.7	45.5	103	1.3433	0.036	0.019	0.010	1.477	1.444	67.59		
7.50	1.0265	1.0283	77.0	0.015	949.5	48.7	110	1.3440	0.039	0.021	0.011	1.523	1.487	65.52		
8.00	1.0286	1.0304	82.3	0.016	946.3	52.0	118	1.3448	0.043	0.023	0.012	1.572	1.531	63.49		
8.50	1.0307	1.0325	87.6	0.017	943.1	55.2	125	1.3455	0.047	0.025	0.013	1.623	1.578	61.50		
9.00	1.0329	1.0347	93.0	0.018	939.9	58.3	133	1.3463	0.051	0.027	0.014	1.676	1.626	59.56		
9.50	1.0351	1.0370	98.3	0.019	936.8	61.4	141	1.3471	0.055	0.030	0.015	1.731	1.675	57.67		
10.00	1.0374	1.0393	103.7	0.020	933.7	64.5	148	1.3478	0.060	0.032	0.017	1.788	1.727	55.82		

* Supersaturated solutions.

26 LACTIC ACID, $CH_3CHOHCOOH$, (D-PROPANOIC ACID, 2-HYDROXY)

MOLECULAR WEIGHT = 90.08
RELATIVE SPECIFIC REFRACTIVITY = 1.058

0.00 % by wt. data are the same for all compounds.
For Values of 0.00 wt. % solutions see Table 1, Acetic Acid.

A % by wt.	ρ D_4^{20}	D_{20}^{20}	C_s g/l	M g-mol/l	C_w g/l	$(C_o - C_w)$ g/l	$(n - n_o)$ $\times 10^4$	n	Δ °C	O Os/kg	S g-mol/l	η/η_o	η/ρ cS	ϕ rhe	γ mmho/cm	T g-mol/l
0.50	0.9992	1.0010	05.0	0.055	994.2	4.0	5	1.3335	0.095	0.051	0.027	1.013	1.013	98.62	1.0	0.010
1.00	1.0002	1.0020	10.0	0.111	990.2	8.0	10	1.3340	0.190	0.102	0.054	1.025	1.027	97.37	1.8	0.018
2.00	1.0023	1.0040	20.0	0.223	982.2	16.0	20	1.3350	0.378	0.203	0.109	1.054	1.053	94.69	2.6	0.026
3.00	1.0043	1.0061	30.1	0.334	974.2	24.0	30	1.3360	0.567	0.305	0.165	1.082	1.078	92.27	3.2	0.031
4.00	1.0065	1.0083	40.3	0.447	966.2	32.0	40	1.3370	0.758	0.407	0.221	1.108	1.104	90.03	3.5	0.035
5.00	1.0086	1.0104	50.4	0.560	958.2	40.0	50	1.3380	0.953	0.512	0.278	1.136	1.127	87.85	3.8	0.038
6.00	1.0108	1.0126	60.4	0.670	950.2	48.1	60	1.3390	1.155	0.621	0.337	1.165	1.154	85.64	4.1	0.041
7.00	1.0131	1.0149	70.9	0.787	942.1	56.1	70	1.3400	1.361	0.732	0.398	1.196	1.181	83.47	4.4	0.043
8.00	1.0153	1.0171	81.2	0.902	934.1	64.1	80	1.3410	1.573	0.846	0.460	1.227	1.210	81.34	4.6	0.045
9.00	1.0176	1.0194	91.6	1.017	926.0	72.2	90	1.3420	1.791	0.963	0.524	1.259	1.239	79.24	4.7	0.047
10.00	1.0199	1.0217	102.0	1.132	917.9	80.3	100	1.3430	2.015	1.084	0.589	1.293	1.269	77.19	4.9	0.049
12.00	1.0246	1.0264	123.0	1.365	901.6	96.6	120	1.3450	2.491	1.339	0.726	1.363	1.332	73.20	5.1	0.051
14.00	1.0294	1.0312	144.1	1.600	885.2	113.0	140	1.3470	2.988	1.607	0.868	1.438	1.399	69.38	5.2	0.052
16.00	1.0342	1.0360	165.5	1.837	868.7	129.5	161	1.3491	3.478	1.870	1.006	1.519	1.471	65.71	5.2	0.052
18.00	1.0390	1.0408	187.0	2.076	852.0	146.3	181	1.3511	3.962	2.130	1.141	1.604	1.546	62.22	5.2	0.052
20.00	1.0439	1.0457	208.8	2.318	835.1	163.1	202	1.3532	4.440	2.387	1.273	1.696	1.628	58.84	5.1	0.051
24.00	1.0536	1.0554	252.9	2.807	800.7	197.5	243	1.3573				1.898	1.806	52.57	4.9	0.048
28.00	1.0632	1.0651	297.7	3.305	765.5	232.7	285	1.3615				2.132	2.010	46.80	4.5	0.045
32.00	1.0728	1.0747	343.3	3.811	729.5	268.8	327	1.3657				2.409	2.251	41.43	4.1	0.041
36.00	1.0822	1.0841	389.6	4.325	692.6	305.7	370	1.3700				2.725	2.524	36.63	3.7	0.036
40.00	1.0915	1.0934	436.6	4.847	654.9	343.4	413	1.3743				3.108	2.853	32.11	3.2	0.032
44.00	1.1008	1.1028	484.4	5.377	616.5	381.8	456	1.3786				3.559	3.239	28.04	2.8	0.027
48.00	1.1105	1.1125	533.0	5.917	577.5	420.8	498	1.3828				4.098	3.699	24.35	2.3	0.022
52.00	1.1201	1.1221	582.4	6.466	537.6	460.6	541	1.3871				4.779	4.276	20.89	1.8	0.018
56.00	1.1297	1.1317	632.6	7.023	497.0	501.2	584	1.3914				5.568	4.940	17.93	1.5	0.014
60.00	1.1392	1.1412	683.5	7.588	455.7	542.6	628	1.3958				6.666	5.865	14.97	1.1	0.011
64.00	1.1486	1.1506	735.1	8.160	413.5	584.7	671	1.4001				8.008	6.988	12.46	0.8	0.008
68.00	1.1579	1.1599	787.4	8.741	370.5	627.7	715	1.4045				9.843	8.519	10.14	0.5	0.005
72.00	1.1670	1.1691	840.3	9.328	326.8	671.5	758	1.4088				12.84	11.02	7.77	0.3	0.003
76.00	1.1760	1.1781	893.8	9.922	282.2	716.0	801	1.4131				16.94	14.43	5.89	0.1	0.001
80.00	1.1848	1.1869	947.9	10.523	237.0	761.3	843	1.4173				22.12	18.70	4.51	0.1	0.001

27 LACTOSE, $C_{12}H_{22}O_{11} \cdot 1H_2O$

MOLECULAR WEIGHT = 342.30   FORMULA WEIGHT, HYDRATE = 360.31
RELATIVE SPECIFIC REFRACTIVITY = 1.036

0.00 % by wt. data are the same for all compounds.
For Values of 0.00 wt. % solutions see Table 1, Acetic Acid.

A % by wt.	H % by wt.	ρ D_4^{20}	D_{20}^{20}	C_s g/l	M g-mol/l	C_w g/l	$(C_o - C_w)$ g/l	$(n - n_o)$ $\times 10^4$	n	Δ °C	O Os/kg	S g-mol/l	η/η_o	η/ρ cS	ϕ rhe	γ mmho/cm	T g-mol/l
0.50	0.53	1.0002	1.0019	5.0	0.015	995.2	3.1	7	1.3337	0.027	0.015	0.007	1.011	1.013	98.71		
1.00	1.05	1.0021	1.0039	10.0	0.029	992.1	6.1	15	1.3345	0.055	0.030	0.015	1.024	1.024	97.46		
1.50	1.58	1.0041	1.0059	15.1	0.044	989.0	9.2	22	1.3352	0.083	0.045	0.023	1.039	1.037	96.01		
2.00	2.11	1.0061	1.0079	20.1	0.059	986.0	12.3	30	1.3359	0.112	0.060	0.031	1.056	1.052	94.51		
2.50	2.63	1.0081	1.0099	25.2	0.074	982.9	15.3	37	1.3367	0.140	0.075	0.040	1.072	1.065	93.12		
3.00	3.16	1.0102	1.0119	30.3	0.089	979.8	18.4	45	1.3375	0.169	0.091	0.048	1.087	1.078	91.80		
3.50	3.68	1.0122	1.0140	35.4	0.103	976.8	21.5	53	1.3382	0.198	0.107	0.056	1.103	1.091	90.52		
4.00	4.21	1.0143	1.0160	40.6	0.119	973.7	24.6	60	1.3390	0.228	0.122	0.065	1.118	1.105	89.25		
4.50	4.74	1.0163	1.0181	45.7	0.134	970.6	27.7	68	1.3398	0.258	0.139	0.074	1.135	1.119	87.96		
5.00	5.26	1.0184	1.0202	50.9	0.149	967.5	30.8	76	1.3406	0.288	0.155	0.083	1.152	1.133	86.63		
5.50	5.79	1.0204	1.0223	56.1	0.164	964.3	33.9	84	1.3413	0.319	0.172	0.092	1.170	1.149	85.27		

27 LACTOSE, $C_{12}H_{22}O_{11} \cdot 1H_2O$—(Continued)

A % by wt.	H % by wt.	ρ D_4^{20}	D_{20}^{20}	C_s g/l	M g·mol/l	C_w g/l	$(C_o - C_w)$ g/l	$(n - n_o)$ × 10^4	n	Δ °C	O Os/kg	S g·mol/l	η/η_o	η/ρ cS	ϕ rhe	γ mmho/cm	T g·mol/l
6.00	6.32	1.0225	1.0243	61.4	0.179	961.2	37.1	91	1.3421	0.351	0.189	0.101	1.189	1.166	83.91		
6.50	6.84	1.0246	1.0264	66.6	0.195	958.0	40.2	99	1.3429	0.385	0.207	0.111	1.209	1.183	82.54		
7.00	7.37	1.0267	1.0285	71.9	0.210	954.8	43.4	107	1.3437	0.420	0.226	0.121	1.230	1.200	81.16		
7.50	7.89	1.0287	1.0305	77.2	0.225	951.6	46.7	115	1.3445	0.456	0.245	0.132	1.251	1.218	79.79		
8.00	8.42	1.0308	1.0326	82.5	0.241	948.3	49.9	123	1.3453	0.495	0.266	0.143	1.273	1.237	78.42		
8.50	8.95	1.0329	1.0347	87.8	0.256	945.1	53.2	130	1.3460				1.295	1.257	77.05		
9.00	9.47	1.0349	1.0367	93.1	0.272	941.8	56.5	138	1.3468				1.318	1.277	75.69		
9.50	10.00	1.0370	1.0388	98.5	0.288	938.5	59.8	146	1.3476				1.342	1.297	74.34		
10.00	10.53	1.0390	1.0409	103.9	0.304	935.1	63.1	154	1.3484				1.367	1.318	73.01		
11.00	11.58	1.0432	1.0450	114.7	0.335	928.4	69.8	170	1.3500				1.418	1.362	70.37		
12.00	12.63	1.0473	1.0492	125.7	0.367	921.7	76.6	185	1.3515				1.473	1.409	67.77		
13.00	13.68	1.0515	1.0534	136.7	0.399	914.8	83.4	201	1.3531				1.530	1.458	65.24		
14.00	14.74	1.0558	1.0577	147.8	0.432	908.0	90.2	218	1.3548				1.590	1.509	62.76		
15.00	15.79	1.0602	1.0621	159.0	0.465	901.2	97.0	234	1.3564				1.654	1.563	60.35		
16.00	16.84	1.0648	1.0667	170.4	0.498	894.4	103.8	252	1.3582				1.721	1.619	58.01		
17.00	17.89	1.0696	1.0714	181.8	0.531	887.7	110.5	270	1.3600				1.791	1.678	55.73		
18.00	18.95	1.0746	1.0765	193.4	0.565	881.2	117.1	289	1.3619				1.865	1.739	53.51		

28 LANTHANUM NITRATE, $La(NO_3)_3 \cdot 6H_2O$

MOLECULAR WEIGHT = 324.93 FORMULA WEIGHT, HYDRATE = 433.02

A% by wt.	H% by wt.	ρ D_4^{20}	D_{20}^{20}	C_s g/l	M g·mol/l	C_w g/l	$(C_o - C_w)$ g/l	$(n - n_o)$ × 10^4	n	Δ °C	O Os/kg	S g·mol/l	η/η_o	η/ρ cS	ϕ rhe	γ mmho/cm	T g·mol/l
1.00	1.33	1.0062	1.0080	10.1	0.031	996.1	2.1	14.	1.3344	0.180	0.097	0.051	1.013	1.009	98.50	7.8	0.080
2.00	2.67	1.0142	1.0160	20.3	0.062	993.9	4.3	28.	1.3358	0.358	0.192	0.103	1.025	1.013	97.33	14.2	0.151
3.00	4.00	1.0226	1.0244	30.7	0.094	991.9	6.3	42.	1.3372	0.525	0.283	0.152	1.040	1.019	95.97	20.1	0.221
4.00	5.33	1.0309	1.0328	41.2	0.127	989.7	8.5	57.	1.3387	0.698	0.375	0.203	1.056	1.026	94.54	25.6	0.289
5.00	6.66	1.0394	1.0413	52.0	0.160	987.4	10.8	71.	1.3401	0.889	0.478	0.259	1.074	1.035	92.96	31.1	0.358
6.00	8.00	1.0483	1.0502	62.9	0.194	985.4	12.8	86.	1.3416	1.079	0.580	0.315	1.091	1.043	91.46	36.0	0.420
7.00	9.33	1.0575	1.0594	74.0	0.228	983.5	14.7	101.	1.3431	1.266	0.681	0.370	1.110	1.052	89.89	40.6	0.479
8.00	10.66	1.0667	1.0686	85.3	0.263	981.4	16.8	115.	1.3445	1.463	0.787	0.428	1.130	1.061	88.36	45.0	0.535
9.00	11.99	1.0761	1.0780	96.8	0.298	979.3	18.9	130.	1.3460	1.676	0.901	0.490	1.149	1.070	86.85	49.5	0.596
10.00	13.33	1.0853	1.0873	108.5	0.334	976.8	21.4	145.	1.3475	1.874	1.007	0.548	1.170	1.080	85.32	53.4	0.648
12.00	15.99	1.1055	1.1075	132.7	0.408	972.8	25.4	176.	1.3506	2.340	1.258	0.683	1.214	1.100	82.23	60.7	0.750
14.00	18.66	1.1251	1.1271	157.5	0.485	967.6	30.6	208.	1.3538	2.822	1.517	0.821	1.262	1.124	79.08	67.4	0.845
16.00	21.32	1.1456	1.1477	183.3	0.564	962.3	35.9	241.	1.3571	3.331	1.791	0.965	1.322	1.156	75.51	73.2	0.930
18.00	23.99	1.1667	1.1688	210.0	0.646	956.7	41.5	274.	1.3604	3.869	2.080	1.115	1.390	1.194	71.79	77.7	0.999
20.00	26.65	1.1892	1.1914	237.9	0.732	951.3	46.9	308.	1.3638	4.418	2.375	1.267	1.463	1.233	68.20	81.4	1.055
24.00	31.98	1.2337	1.2359	296.1	0.911	937.6	60.6	379.	1.3709				1.641	1.333	60.81	86.7	1.137
28.00	37.31	1.2829	1.2852	359.2	1.105	923.7	74.5	453.	1.3783				1.854	1.448	53.83	86.8	1.138
32.00	42.64	1.3319	1.3343	426.2	1.312	905.7	92.5	530.	1.3860				2.156	1.622	46.29	84.0	1.095
36.00	47.98	1.3874	1.3899	499.5	1.537	887.9	110.3	613.	1.3943				2.580	1.863	38.69	81.4	1.055
40.00	53.31	1.4477	1.4503	579.1	1.782	868.6	129.6	700.	1.4030				3.205	2.218	31.14	71.9	0.911
44.00	58.64	1.5084	1.5111	663.7	2.043	844.7	153.5	788.	1.4118				4.105	2.727	24.31	61.9	0.767

29 LEAD NITRATE, $Pb(NO_3)_2$

MOLECULAR WEIGHT = 331.23
RELATIVE SPECIFIC REFRACTIVITY = 0.479

0.00 % by wt. data are the same for all compounds.
For Values of 0.00 wt. % solutions see Table 1, Acetic Acid.

A% by wt.	ρ D_4^{20}	D_{20}^{20}	C_s g/l	M g-mol/l	C_w g/l	$(C_o - C_w)$ g/l	$(n - n_o)$ ×10⁴	n	Δ °C	O Os/kg	S g-mol/l	η/η_o	η/ρ cS	ϕ rhe	γ mmho/cm	T g-mol/l
0.50	1.0025	1.0043	5.0	0.015	997.5	0.8	6	1.3336	0.072	0.039	0.020	1.003	1.003	99.49	2.8	0.028
1.00	1.0068	1.0086	10.1	0.030	996.7	1.5	12	1.3342	0.137	0.074	0.039	1.006	1.001	99.21	5.4	0.054
2.00	1.0155	1.0173	20.3	0.061	995.2	3.1	24	1.3354	0.258	0.139	0.074	1.011	0.998	98.71	9.6	0.099
3.00	1.0243	1.0261	30.7	0.093	993.6	4.6	36	1.3366	0.372	0.200	0.107	1.016	0.993	98.27	13.3	0.141
4.00	1.0333	1.0352	41.3	0.125	992.0	6.2	49	1.3379	0.482	0.259	0.140	1.020	0.989	97.85	16.7	0.181
5.00	1.0425	1.0444	52.1	0.157	990.4	7.8	61	1.3391	0.587	0.316	0.170	1.025	0.985	97.37	19.9	0.219
6.00	1.0519	1.0537	63.1	0.191	988.7	9.5	73	1.3403	0.690	0.371	0.201	1.031	0.982	96.78	23.0	0.256
7.00	1.0614	1.0632	74.3	0.224	987.1	11.2	86	1.3415	0.792	0.426	0.231	1.038	0.980	96.12	25.9	0.292
8.00	1.0710	1.0729	85.7	0.259	985.4	12.9	98	1.3428	0.891	0.479	0.260	1.046	0.979	95.40	28.6	0.326
9.00	1.0809	1.0828	97.3	0.294	983.6	14.6	110	1.3440	0.987	0.531	0.288	1.054	0.978	94.64	31.3	0.360
10.00	1.0909	1.0929	109.1	0.329	981.8	16.4	124	1.3454	1.081	0.581	0.316	1.063	0.976	93.89	33.9	0.393
12.00	1.1115	1.1135	133.4	0.403	978.1	20.1	151	1.3481	1.265	0.680	0.370	1.081	0.974	92.33	38.9	0.457
14.00	1.1329	1.1349	158.6	0.479	974.3	24.0	180	1.3510	1.450	0.780	0.424	1.101	0.973	90.69	43.4	0.515
16.00	1.1550	1.1571	184.8	0.558	970.2	28.0	209	1.3539	1.639	0.881	0.479	1.122	0.973	88.94	47.5	0.570
18.00	1.1779	1.1800	212.0	0.640	965.9	32.3	239	1.3569	1.825	0.981	0.534	1.146	0.975	87.10	51.4	0.621
20.00	1.2017	1.2038	240.3	0.726	961.4	36.9	271	1.3601	2.008	1.080	0.587	1.172	0.977	85.15	55.2	0.673
24.00	1.2519	1.2541	300.5	0.907	951.5	46.8	337	1.3667	2.361	1.269	0.689	1.233	0.987	80.97	62.5	0.772
28.00	1.3059	1.3082	365.7	1.104	940.3	58.0	407	1.3737	2.714	1.459	0.790	1.306	1.002	76.43	68.4	0.858
32.00	1.3640	1.3664	436.5	1.318	927.5	70.7	481	1.3811				1.398	1.027	71.39	72.5	0.920
34.00	1.3947	1.3972	474.2	1.432	920.5	77.7	520	1.3850				1.459	1.048	68.42	74.8	0.954

30 LITHIUM CHLORIDE, LiCl

MOLECULAR WEIGHT = 42.40
RELATIVE SPECIFIC REFRACTIVITY = 1.023

0.00 % by wt. data are the same for all compounds.
For Values of 0.00 wt. % solutions see Table 1, Acetic Acid.

A% by wt.	ρ D_4^{20}	D_{20}^{20}	C_s g/l	M g-mol/l	C_w g/l	$(C_o - C_w)$ g/l	$(n - n_o)$ ×10⁴	n	Δ °C	O Os/kg	S g-mol/l	η/η_o	η/ρ cS	ϕ rhe	γ mmho/cm	T g-mol/l
0.50	1.0012	1.0029	5.0	0.118	996.2	2.1	11	1.3341	0.415	0.223	0.120	1.017	1.018	98.08	10.1	0.105
1.00	1.0041	1.0059	10.0	0.237	994.0	4.2	21	1.3351	0.836	0.450	0.244	1.035	1.033	96.43	19.0	0.208
2.00	1.0099	1.0117	20.2	0.476	989.7	8.5	43	1.3373	1.724	0.927	0.504	1.070	1.062	93.27	34.9	0.406
3.00	1.0157	1.0175	30.5	0.719	985.2	13.0	64	1.3394	2.684	1.443	0.782	1.106	1.091	90.22	49.8	0.600
4.00	1.0215	1.0233	40.9	0.964	980.6	17.6	85	1.3415	3.727	2.004	1.075	1.144	1.122	87.26	63.6	0.791
5.00	1.0272	1.0290	51.4	1.211	975.9	22.4	106	1.3436	4.859	2.612	1.386	1.183	1.154	84.36	76.4	0.978
6.00	1.0330	1.0348	62.0	1.462	971.0	27.2	127	1.3457	6.14	3.30	1.723	1.224	1.187	81.55	88.3	1.16
7.00	1.0387	1.0405	72.7	1.715	966.0	32.2	148	1.3478	7.56	4.07	2.081	1.266	1.221	78.83	99.5	1.34
8.00	1.0444	1.0463	83.6	1.971	960.9	37.3	169	1.3499	9.11	4.90	2.449	1.310	1.257	76.16	110.	1.52
9.00	1.0502	1.0521	94.5	2.229	955.7	42.6	190	1.3520	10.79	5.80	2.825	1.357	1.295	73.52	119.	1.69
10.00	1.0560	1.0578	105.6	2.490	950.4	47.9	211	1.3541	12.61	6.78	3.205	1.408	1.336	70.88	127.	1.84
12.00	1.0675	1.0694	128.1	3.021	939.4	58.8	253	1.3583	16.59	8.92	3.951	1.519	1.426	65.68	140.	2.12
14.00	1.0792	1.0811	151.1	3.563	928.1	70.1	295	1.3625	21.04	11.31	4.670	1.644	1.527	60.70	151.	2.34
16.00	1.0910	1.0929	174.6	4.117	916.4	81.8	338	1.3668				1.783	1.637	55.99	160.	2.56
18.00	1.1029	1.1048	198.5	4.682	904.4	93.9	381	1.3711				1.938	1.761	51.50	167.	2.73
20.00	1.1150	1.1170	223.0	5.260	892.0	106.2	425	1.3755				2.124	1.909	46.99	170.	2.80
22.00	1.1274	1.1294	248.0	5.850	879.3	118.9	469	1.3799				2.336	2.076	42.72	170.	2.79
24.00	1.1399	1.1419	273.6	6.452	866.3	131.9	514	1.3844				2.595	2.281	38.46	167.	2.73
26.00	1.1527	1.1548	299.7	7.069	853.0	145.2	560	1.3890				2.919	2.537	34.19	163.	2.63
28.00	1.1658	1.1678	326.4	7.699	839.4	158.9	606	1.3936				3.311	2.846	30.14	156.	2.47
30.00	1.1791	1.1812	353.7	8.343	825.4	172.9	653	1.3983				3.777	3.210	26.42	146.	2.25

31 MAGNESIUM CHLORIDE, MgCl₂ · 6H₂O

MOLECULAR WEIGHT = 95.23   FORMULA WEIGHT, HYDRATE = 203.33
RELATIVE SPECIFIC REFRACTIVITY = 0.893

0.00 % by wt. data are the same for all compounds.
For Values of 0.00 wt. % solutions see Table 1, Acetic Acid.

A % by wt.	H % by wt.	ρ D₄²⁰	D₂₀²⁰	C_s g/l	M g-mol/l	C_w g/l	(C_o − C_w) g/l	(n − n_o) × 10⁴	n	Δ °C	O Os/kg	S g-mol/l	η/η_o	η/ρ cS	φ rhe	γ mmho/cm	T g-mol/l
0.50	1.07	1.0022	1.0040	5.0	0.053	997.2	1.0	13	1.3343	0.255	0.137	0.073	1.022	1.022	97.66	8.6	0.092
1.00	2.14	1.0062	1.0080	10.1	0.106	996.2	2.0	26	1.3356	0.517	0.278	0.150	1.044	1.040	95.59	16.6	0.183
2.00	4.27	1.0144	1.0162	20.3	0.213	994.1	4.1	51	1.3381	1.063	0.572	0.311	1.089	1.076	91.60	31.2	0.356
3.00	6.41	1.0226	1.0244	30.7	0.322	991.9	6.3	76	1.3406	1.653	0.889	0.483	1.137	1.114	87.80	44.5	0.530
4.00	8.54	1.0309	1.0328	41.2	0.433	989.7	8.5	102	1.3432	2.297	1.235	0.671	1.186	1.153	84.13	56.4	0.690
5.00	10.68	1.0394	1.0412	52.0	0.546	987.4	10.8	127	1.3457	3.012	1.619	0.875	1.239	1.194	80.57	66.9	0.840
6.00	12.81	1.0479	1.0497	62.9	0.660	985.0	13.2	153	1.3483				1.295	1.238	77.09	77.2	0.990
7.00	14.95	1.0564	1.0583	74.0	0.777	982.5	15.7	178	1.3508				1.355	1.285	73.67	86.4	1.13
8.00	17.08	1.0651	1.0670	85.2	0.895	979.9	18.3	204	1.3534				1.420	1.336	70.30	94.5	1.26
9.00	19.22	1.0738	1.0757	96.6	1.015	977.2	21.1	231	1.3560				1.490	1.390	66.98	102.0	1.38
10.00	21.35	1.0826	1.0845	108.3	1.137	974.4	23.9	257	1.3587				1.567	1.450	63.70	108.0	1.49
12.00	25.62	1.1005	1.1025	132.1	1.387	968.5	29.8	311	1.3641				1.742	1.586	57.30	119.0	1.69
14.00	29.89	1.1189	1.1209	156.6	1.645	962.3	36.0	365	1.3695				1.952	1.748	51.13	127.0	1.84
16.00	34.16	1.1372	1.1392	181.9	1.911	955.2	43.0	419	1.3749				2.203	1.941	45.30	132.0	1.94
18.00	38.43	1.1553	1.1574	208.0	2.184	947.4	50.8	474	1.3804				2.502	2.170	39.89	134.0	1.99
20.00	42.70	1.1742	1.1763	234.8	2.466	939.4	58.9	529	1.3859				2.861	2.441	34.89	134.0	1.99
22.00	46.97	1.1938	1.1959	262.6	2.758	931.2	67.1	585	1.3915				3.316	2.783	30.10	131.0	1.93
24.00	51.24	1.2140	1.2162	291.4	3.060	922.6	75.6	642	1.3972				3.909	3.226	25.53	126.0	1.81
26.00	55.51	1.2346	1.2368	321.0	3.371	913.6	84.6	700	1.4030				4.685	3.802	21.30	118.0	1.67
28.00	59.78	1.2555	1.2577	351.5	3.691	903.9	94.3	759	1.4089				5.698	4.547	17.52	109.0	1.50
30.00	64.05	1.2763	1.2786	382.9	4.021	893.4	104.8	818	1.4148				7.003	5.498	14.25	97.9	1.32

32 MAGNESIUM SULFATE, MgSO₄ · 7H₂O

MOLECULAR WEIGHT = 120.37   FORMULA WEIGHT, HYDRATE = 246.48
RELATIVE SPECIFIC REFRACTIVITY = 0.572

0.00 % by wt. data are the same for all compounds.
For Values of 0.00 wt. % solutions see Table 1, Acetic Acid.

A % by wt.	H % by wt.	ρ D₄²⁰	D₂₀²⁰	C_s g/l	M g-mol/l	C_w g/l	(C_o − C_w) g/l	(n − n_o) × 10⁴	n	Δ °C	O Os/kg	S g-mol/l	η/η_o	η/ρ cS	φ rhe	γ mmho/cm	T g-mol/l
0.50	1.02	1.0033	1.0051	5.0	0.042	998.3	0.0	10	1.3340	0.103	0.055	0.029	1.025	1.024	97.35	4.1	0.041
1.00	2.05	1.0084	1.0102	10.1	0.084	998.3	−0.1	20	1.3350	0.192	0.103	0.055	1.052	1.045	94.88	7.6	0.078
2.00	4.10	1.0186	1.0204	20.4	0.169	998.2	0.0	41	1.3371	0.361	0.194	0.104	1.110	1.092	89.91	13.3	0.141
3.00	6.15	1.0289	1.0307	30.9	0.256	998.0	0.2	61	1.3391	0.522	0.281	0.151	1.175	1.144	84.94	18.4	0.201
4.00	8.19	1.0392	1.0411	41.6	0.345	997.6	0.6	81	1.3411	0.692	0.372	0.201	1.247	1.202	80.05	23.1	0.258
5.00	10.24	1.0497	1.0515	52.5	0.436	997.2	1.1	101	1.3431	0.868	0.467	0.253	1.325	1.265	75.32	27.4	0.311
6.00	12.29	1.0602	1.0621	63.6	0.528	996.6	1.6	121	1.3451	1.048	0.564	0.306	1.408	1.330	70.90	31.1	0.358
7.00	14.34	1.0708	1.0727	75.0	0.623	995.9	2.3	142	1.3471	1.235	0.664	0.361	1.495	1.399	66.77	34.4	0.399
8.00	16.39	1.0816	1.0835	86.5	0.719	995.1	3.2	162	1.3492	1.431	0.770	0.419	1.590	1.473	62.75	37.3	0.436
9.00	18.44	1.0924	1.0944	98.3	0.817	994.1	4.1	182	1.3512	1.635	0.879	0.478	1.699	1.558	58.75	40.0	0.471
10.00	20.48	1.1034	1.1053	110.3	0.917	993.0	5.2	202	1.3532	1.849	0.994	0.541	1.825	1.657	54.69	42.7	0.506
12.00	24.57	1.1257	1.1276	135.1	1.122	990.6	7.7	242	1.3572	2.313	1.243	0.675	2.100	1.869	47.53	48.4	0.581
14.00	28.67	1.1484	1.1504	160.8	1.335	987.6	10.6	283	1.3613	2.861	1.538	0.832	2.407	2.100	41.46	53.3	0.647
16.00	32.76	1.1717	1.1737	187.5	1.557	984.2	14.0	324	1.3654	3.673	1.974	1.061	2.803	2.397	35.61	55.2	0.673
18.00	36.86	1.1955	1.1976	215.2	1.787	980.3	17.9	364	1.3694				3.353	2.810	29.77	53.7	0.652
20.00	40.95	1.2198	1.2220	244.0	2.026	975.9	22.4	405	1.3735				4.139	3.400	24.11	51.1	0.617
22.00	45.05	1.2447	1.2469	273.8	2.275	970.9	27.4	446	1.3776				5.189	4.177	19.23	48.8	0.586
24.00	49.14	1.2701	1.2724	304.8	2.532	965.3	32.9	487	1.3817				6.485	5.116	15.39	45.9	0.547
26.00	53.24	1.2961	1.2984	337.0	2.799	959.1	39.1	528	1.3858				8.050	6.223	12.40	42.3	0.500

33 MALTOSE, C₁₂H₂₂O₁₁ · 1H₂O

MOLECULAR WEIGHT = 342.29   FORMULA WEIGHT, HYDRATE = 360.31
RELATIVE SPECIFIC REFRACTIVITY = 1.038

0.00 % by wt. data are the same for all compounds.
For Values of 0.00 wt. % solutions see Table 1, Acetic Acid.

A % by wt.	H % by wt.	ρ D₄²⁰	D₂₀²⁰	C_s g/l	M g-mol/l	C_w g/l	(C_o − C_w) g/l	(n − n_o) × 10⁴	n	Δ °C	O Os/kg	S g-mol/l	η/η_o	η/ρ cS	φ rhe	γ mmho/cm	T g-mol/l
0.50	0.53	1.0003	1.0020	5.0	0.015	995.3	3.0	7	1.3337	0.027	0.015	0.007	1.014	1.016	98.45		
1.00	1.05	1.0023	1.0041	10.0	0.029	992.3	6.0	15	1.3345	0.055	0.030	0.015	1.028	1.028	97.08		
1.50	1.58	1.0043	1.0061	15.1	0.044	989.2	9.0	22	1.3352	0.083	0.045	0.023	1.043	1.040	95.71		
2.00	2.11	1.0063	1.0081	20.1	0.059	986.2	12.0	29	1.3359	0.112	0.060	0.031	1.058	1.053	94.33		
2.50	2.63	1.0083	1.0101	25.2	0.074	983.1	15.1	37	1.3367	0.140	0.075	0.040	1.074	1.067	92.95		
3.00	3.16	1.0104	1.0121	30.3	0.089	980.0	18.2	44	1.3374	0.169	0.091	0.048	1.090	1.081	91.56		
3.50	3.68	1.0124	1.0142	35.4	0.104	976.9	21.3	52	1.3382	0.199	0.107	0.057	1.107	1.095	90.17		
4.00	4.21	1.0144	1.0162	40.6	0.119	973.8	24.4	59	1.3389	0.229	0.123	0.065	1.124	1.110	88.78		
4.50	4.74	1.0164	1.0182	45.7	0.134	970.7	27.6	67	1.3397	0.259	0.139	0.074	1.142	1.126	87.41		
5.00	5.26	1.0184	1.0202	50.9	0.149	967.5	30.7	74	1.3404	0.290	0.156	0.083	1.160	1.141	86.03		
5.50	5.79	1.0204	1.0222	56.1	0.164	964.3	33.9	82	1.3412	0.321	0.172	0.092	1.179	1.157	84.68		
6.00	6.32	1.0224	1.0243	61.3	0.179	961.1	37.1	90	1.3420	0.352	0.189	0.101	1.198	1.174	83.34		
6.50	6.84	1.0245	1.0263	66.6	0.195	957.9	40.4	97	1.3427	0.384	0.206	0.111	1.217	1.190	82.01		
7.00	7.37	1.0265	1.0283	71.9	0.210	954.6	43.6	105	1.3435	0.416	0.224	0.120	1.237	1.207	80.69		
7.50	7.89	1.0285	1.0303	77.1	0.225	951.3	46.9	113	1.3443	0.449	0.241	0.130	1.257	1.225	79.38		
8.00	8.42	1.0305	1.0323	82.4	0.241	948.0	50.2	120	1.3450	0.482	0.259	0.139	1.278	1.243	78.07		
8.50	8.95	1.0325	1.0343	87.8	0.256	944.7	53.5	128	1.3458	0.516	0.277	0.149	1.300	1.262	76.77		
9.00	9.47	1.0345	1.0363	93.1	0.272	941.4	56.8	136	1.3466	0.550	0.296	0.159	1.322	1.281	75.48		
9.50	10.00	1.0365	1.0383	98.5	0.288	938.0	60.2	144	1.3474	0.584	0.314	0.169	1.345	1.300	74.19		
10.00	10.53	1.0385	1.0403	103.8	0.303	934.6	63.6	152	1.3482	0.619	0.333	0.180	1.369	1.321	72.90		
11.00	11.58	1.0425	1.0443	114.7	0.335	927.8	70.4	167	1.3497	0.691	0.371	0.201	1.419	1.364	70.34		
12.00	12.63	1.0465	1.0483	125.6	0.367	920.9	77.3	183	1.3513	0.765	0.411	0.223	1.471	1.409	67.83		

33 MALTOSE, $C_{12}H_{22}O_{11} \cdot 1H_2O$—(Continued)

A % by wt.	H % by wt.	ρ D_4^{20}	D_{20}^{20}	C_s g/l	M g-mol/l	C_w g/l	$(C_o - C_w)$ g/l	$(n - n_0)$ $\times 10^4$	n	Δ °C	O Os/kg	S g-mol/l	η/η_o	η/ρ cS	ϕ rhe	γ mmho/cm	T g-mol/l
13.00	13.68	1.0505	1.0523	136.6	0.399	913.9	84.3	199	1.3529	0.841	0.452	0.245	1.527	1.456	65.37		
14.00	14.74	1.0545	1.0563	147.6	0.431	906.8	91.4	216	1.3546	0.919	0.494	0.268	1.585	1.506	62.97		
15.00	15.79	1.0585	1.0603	158.8	0.464	899.7	98.6	232	1.3562	1.000	0.538	0.292	1.646	1.558	60.63		
16.00	16.84	1.0629	1.0648	170.1	0.497	892.8	105.4	248	1.3578	1.080	0.581	0.315	1.712	1.614	58.31		
17.00	17.89	1.0672	1.0691	181.4	0.530	885.8	112.4	265	1.3595	1.165	0.626	0.340	1.780	1.672	56.05		
18.00	18.95	1.0716	1.0735	192.9	0.563	878.7	119.5	282	1.3612	1.250	0.672	0.365	1.855	1.735	53.79		
19.00	20.00	1.0759	1.0778	204.4	0.597	871.5	126.7	298	1.3628	1.338	0.719	0.391	1.936	1.803	51.55		
20.00	21.05	1.0801	1.0820	216.0	0.631	864.1	134.1	314	1.3644	1.430	0.769	0.418	2.017	1.871	49.48		
22.00	23.16	1.0894	1.0913	239.7	0.700	849.7	148.5	348	1.3678	1.635	0.879	0.478	2.212	2.034	45.13		
24.00	25.26	1.0984	1.1004	263.6	0.770	834.8	163.4	384	1.3714	1.848	0.993	0.540	2.458	2.242	40.60		
26.00	27.37	1.1080	1.1100	288.1	0.841	819.9	178.3	419	1.3749	2.080	1.118	0.608	2.748	2.485	36.32		
28.00	29.47	1.1171	1.1191	312.8	0.913	804.3	193.9	455	1.3785	2.335	1.255	0.681	3.060	2.745	32.62		
30.00	31.58	1.1269	1.1289	338.1	0.987	788.8	209.4	491	1.3821	2.615	1.406	0.762	3.420	3.041	29.18		
32.00	33.68	1.1367	1.1387	363.7	1.062	773.0	225.2	528	1.3858	2.925	1.572	0.850	3.910	3.447	25.52		
34.00	35.79	1.1463	1.1483	389.7	1.138	756.6	241.6	566	1.3896	3.252	1.748	0.943	4.438	3.879	22.49		
36.00	37.89	1.1561	1.1582	416.2	1.215	739.9	258.3	605	1.3935	3.600	1.935	1.040	5.040	4.368	19.80		
38.00	40.00	1.1663	1.1684	443.2	1.294	723.1	275.1	644	1.3974	3.990	2.145	1.148	5.820	5.000	17.15		
40.00	42.10	1.1769	1.1790	470.8	1.375	706.1	292.1	683	1.4013	4.408	2.370	1.264	6.912	5.885	14.44		
42.00	44.21	1.1878	1.1899	498.9	1.457	688.9	309.3	721	1.4051	4.878	2.622	1.390	8.175	6.896	12.21		
44.00	46.31	1.1979	1.2000	527.1	1.539	670.8	327.4	764	1.4094	5.352	2.877	1.518	9.630	8.055	10.36		
46.00	48.42	1.2084	1.2105	555.9	1.623	652.5	345.7	806	1.4136				11.45	9.492	8.72		
48.00	50.52	1.2194	1.2216	585.3	1.709	634.1	364.1	847	1.4177				14.12	11.60	7.07		
50.00	52.63	1.2304	1.2326	615.2	1.796	615.2	383.0	887	1.4217				17.75	14.46	5.62		
52.00	54.74	1.2416	1.2438	645.6	1.885	596.0	402.2	930	1.4260				21.99	17.74	4.54		
54.00	56.84	1.2528	1.2550	676.5	1.975	576.3	421.9	978	1.4308				28.70	22.96	3.48		
56.00	58.94	1.2638	1.2660	707.7	2.066	556.1	442.1	1020	1.4350				38.15	30.25	2.62		
58.00	61.05	1.2746	1.2769	739.3	2.159	535.3	462.9	1064	1.4394				49.20	38.68	2.03		
60.00	63.16	1.2855	1.2878	771.3	2.252	514.2	484.0	1110	1.4440								

34 MANGANOUS SULFATE, $MnSO_4 \cdot 1H_2O$

MOLECULAR WEIGHT = 151.00   FORMULA WEIGHT, HYDRATE = 169.01
RELATIVE SPECIFIC REFRACTIVITY = 0.541

0.00 % by wt. data are the same for all compounds.
For Values of 0.00 wt. % solutions see Table 1, Acetic Acid

A % by wt.	H % by wt.	ρ D_4^{20}	D_{20}^{20}	C_s g/l	M g-mol/l	C_w g/l	$(C_o - C_w)$ g/l	$(n - n_0)$ $\times 10^4$	n	Δ °C	O Os/kg	S g-mol/l	η/η_o	η/ρ cS	ϕ rhe	γ mmho/cm	T g-mol/l
1.00	1.12	1.0080	1.0098	10.1	0.067	997.9	0.3	18	1.3348	0.160	0.086	0.045	1.044	1.030	95.55	8.2	0.063
2.00	2.24	1.0178	1.0196	20.4	0.135	997.3	0.8	36	1.3366	0.306	0.164	0.088	1.088	1.071	91.73	10.6	0.111
3.00	3.36	1.0277	1.0296	30.8	0.204	996.9	1.3	54	1.3384	0.437	0.235	0.126	1.135	1.107	87.93	14.6	0.157
4.00	4.48	1.0378	1.0396	41.5	0.275	996.3	2.0	72	1.3402	0.569	0.306	0.165	1.185	1.145	84.19	18.3	0.199
5.00	5.60	1.0480	1.0498	52.4	0.347	995.6	2.7	90	1.3420	0.701	0.377	0.204	1.240	1.186	80.48	21.6	0.240
6.00	6.72	1.0583	1.0602	63.5	0.421	994.8	3.4	108	1.3438	0.838	0.451	0.244	1.298	1.229	76.87	24.7	0.278
7.00	7.83	1.0688	1.0707	74.8	0.495	994.0	4.3	127	1.3457	0.978	0.526	0.285	1.360	1.275	73.36	27.4	0.312
8.00	8.95	1.0794	1.0813	86.4	0.572	993.1	5.2	145	1.3475	1.122	0.603	0.328	1.428	1.325	69.90	29.9	0.343
9.00	10.07	1.0902	1.0921	98.1	0.650	992.1	6.1	164	1.3494	1.275	0.685	0.373	1.502	1.380	66.46	32.3	0.372
10.00	11.19	1.1012	1.1031	110.1	0.729	991.1	7.2	183	1.3513	1.439	0.773	0.421	1.584	1.441	63.01	34.5	0.401
11.00	12.31	1.1123	1.1143	122.4	0.810	990.0	8.3	202	1.3532	1.612	0.867	0.472	1.675	1.509	59.57	36.7	0.428
12.00	13.43	1.1236	1.1256	134.8	0.893	988.8	9.5	221	1.3551	1.797	0.966	0.526	1.775	1.583	56.23	38.7	0.454
13.00	14.55	1.1351	1.1371	147.6	0.977	987.5	10.7	240	1.3570	1.995	1.073	0.583	1.883	1.663	52.99	40.5	0.478
14.00	15.67	1.1467	1.1487	160.5	1.063	986.2	12.1	259	1.3589	2.207	1.187	0.645	2.001	1.749	49.87	42.2	0.499
15.00	16.79	1.1585	1.1606	173.8	1.151	984.7	13.5	279	1.3609	2.431	1.307	0.709	2.129	1.841	46.88	43.7	0.518
16.00	17.91	1.1705	1.1726	187.3	1.240	983.2	15.0	299	1.3629	2.669	1.435	0.777	2.267	1.940	44.03	44.9	0.534
17.00	19.03	1.1827	1.1847	201.1	1.331	981.6	16.6	318	1.3648	2.922	1.571	0.850	2.415	2.046	41.32	45.9	0.548
18.00	20.15	1.1950	1.1971	215.1	1.425	979.9	18.3	338	1.3668	3.194	1.717	0.926	2.575	2.160	38.75	46.7	0.559
19.00	21.27	1.2075	1.2097	229.4	1.519	978.1	20.1	358	1.3688	3.486	1.874	1.008	2.747	2.280	36.32	47.3	0.566
20.00	22.39	1.2203	1.2224	244.1	1.616	976.2	22.0	378	1.3708	3.801	2.043	1.096	2.932	2.408	34.04	47.6	0.570

35 D-MANNITOL, $CH_2OH(CHOH)_4CH_2OH$

MOLECULAR WEIGHT = 182.17
RELATIVE SPECIFIC REFRACTIVITY = 1.063

0.00 % by wt. data are the same for all compounds.
For Values of 0.00 wt. % solutions see Table 1, Acetic Acid.

A % by wt.	ρ D_4^{20}	D_{20}^{20}	C_s g/l	M g-mol/l	C_w g/l	$(C_o - C_w)$ g/l	$(n - n_0)$ $\times 10^4$	n	Δ °C	O Os/kg	S g-mol/l	η/η_o	η/ρ cS	ϕ rhe	γ mmho/cm	T g-mol/l
0.50	1.0000	1.0018	5.0	0.027	995.0	3.2	7	1.3337	0.051	0.028	0.014	1.015	1.017	98.30		
1.00	1.0017	1.0035	10.0	0.055	991.7	6.5	15	1.3345	0.103	0.055	0.029	1.030	1.030	96.89		
1.50	1.0035	1.0053	15.1	0.083	988.5	9.8	22	1.3352	0.156	0.084	0.044	1.044	1.042	95.59		
2.00	1.0053	1.0071	20.1	0.110	985.2	13.1	29	1.3359	0.209	0.112	0.059	1.058	1.055	94.33		
2.50	1.0070	1.0088	25.2	0.138	981.9	16.4	37	1.3367	0.262	0.141	0.075	1.072	1.067	93.07		
3.00	1.0088	1.0106	30.3	0.166	978.7	19.7	44	1.3374	0.317	0.170	0.091	1.087	1.079	91.84		
3.50	1.0106	1.0124	35.4	0.194	975.2	23.0	52	1.3381	0.371	0.200	0.107	1.101	1.092	90.61		
4.00	1.0124	1.0141	40.5	0.222	971.9	26.4	59	1.3389	0.427	0.230	0.123	1.117	1.105	89.38		
4.50	1.0141	1.0159	45.6	0.251	968.5	29.7	66	1.3396	0.483	0.260	0.140	1.132	1.119	88.13		
5.00	1.0159	1.0177	50.8	0.279	965.1	33.1	73	1.3403	0.540	0.290	0.156	1.149	1.133	86.86		
5.50	1.0177	1.0195	56.0	0.307	961.7	36.5	81	1.3411	0.598	0.321	0.173	1.166	1.148	85.56		
6.00	1.0195	1.0213	61.2	0.336	958.3	39.9	88	1.3418	0.656	0.353	0.191	1.185	1.164	84.24		
6.50	1.0212	1.0230	66.4	0.364	954.9	43.4	95	1.3425	0.715	0.384	0.208	1.204	1.181	82.92		
7.00	1.0230	1.0248	71.6	0.393	951.4	46.8	103	1.3433	0.774	0.416	0.225	1.223	1.198	81.59		
7.50	1.0248	1.0266	76.9	0.422	947.9	50.3	110	1.3440	0.835	0.449	0.243	1.244	1.216	80.26		
8.00	1.0266	1.0284	82.1	0.451	944.5	53.8	117	1.3447	0.896	0.482	0.261	1.264	1.234	78.93		
8.50	1.0284	1.0302	87.4	0.480	941.0	57.2	125	1.3455	0.958	0.515	0.279	1.286	1.253	77.62		
9.00	1.0302	1.0320	92.7	0.509	937.5	60.7	132	1.3462	1.020	0.549	0.298	1.308	1.272	76.31		

35 D-MANNITOL, CH₂OH(CHOH)₄CH₂OH—(Continued)

A% by wt.	ρ D₄²⁰	D₂₀²⁰	C_s g/l	M g-mol/l	C_w g/l	(C_o − C_w) g/l	(n − n_o) × 10⁴	n	Δ °C	O Os/kg	S g-mol/l	η/η_o	η/ρ cS	φ rhe	γ mmho/cm	T g-mol/l
9.50	1.0320	1.0338	98.0	0.538	934.0	64.3	139	1.3469	1.084	0.583	0.316	1.330	1.291	75.03		
10.00	1.0338	1.0357	103.4	0.568	930.5	67.8	147	1.3477	1.148	0.617	0.335	1.353	1.311	73.76		
11.00	1.0375	1.0393	114.1	0.626	923.4	74.8	161	1.3491	1.279	0.688	0.374	1.400	1.352	71.28		
12.00	1.0412	1.0431	124.9	0.686	916.3	81.9	176	1.3506	1.413	0.760	0.413	1.450	1.395	68.85		
13.00	1.0450	1.0469	135.9	0.746	909.2	89.0	191	1.3521	1.551	0.834	0.454	1.501	1.440	66.47		
14.00	1.0489	1.0508	146.9	0.806	902.1	96.1	207	1.3536	1.692	0.910	0.495	1.555	1.486	64.16		
15.00	1.0529	1.0548	157.9	0.867	895.0	103.2	222	1.3552	1.837	0.988	0.537	1.612	1.534	61.91		

36 METHANOL, CH₃OH

MOLECULAR WEIGHT = 32.03
RELATIVE SPECIFIC REFRACTIVITY = 1.247

0.00 % by wt. data are the same for all compounds.
For Values of 0.00 wt. % solutions see Table 1, Acetic Acid.

A% by wt.	ρ D₄²⁰	D₂₀²⁰	C_s g/l	M g-mol/l	C_w g/l	(C_o − C_w) g/l	(n − n_o) × 10⁴	n	Δ °C	O Os/kg	S g-mol/l	η/η_o	η/ρ cS	φ rhe	γ mmho/cm	T g-mol/l
0.50	0.9973	0.9991	5.0	0.156	992.4	5.9	1	1.3331	0.278	0.149	0.080	1.020	1.025	97.84		
1.00	0.9964	0.9982	10.0	0.311	986.5	11.7	2	1.3332	0.560	0.301	0.162	1.038	1.044	96.15		
1.50	0.9956	0.9973	14.9	0.466	980.6	17.6	3	1.3333	0.847	0.455	0.247	1.053	1.060	94.74		
2.00	0.9947	0.9965	19.9	0.621	974.8	23.4	4	1.3334	1.140	0.613	0.333	1.068	1.076	93.45		
2.50	0.9938	0.9956	24.8	0.776	969.0	29.2	5	1.3335	1.442	0.776	0.422	1.083	1.092	92.14		
3.00	0.9930	0.9947	29.8	0.930	963.2	35.0	6	1.3336	1.750	0.941	0.512	1.098	1.108	90.85		
3.50	0.9921	0.9939	34.7	1.084	957.4	40.8	8	1.3337	2.058	1.106	0.601	1.114	1.125	89.59		
4.00	0.9913	0.9930	39.7	1.238	951.6	46.6	9	1.3339	2.370	1.274	0.692	1.129	1.142	88.36		
4.50	0.9904	0.9922	44.6	1.392	945.9	52.4	10	1.3340	2.691	1.447	0.784	1.145	1.159	87.15		
5.00	0.9896	0.9914	49.5	1.545	940.1	58.1	11	1.3341	3.020	1.624	0.877	1.161	1.176	85.96		
5.50	0.9888	0.9905	54.4	1.698	934.4	63.8	12	1.3342	3.362	1.807	0.974	1.177	1.193	84.79		
6.00	0.9880	0.9897	59.3	1.851	928.7	69.5	14	1.3343	3.710	1.995	1.071	1.194	1.210	83.62		
6.50	0.9872	0.9889	64.2	2.003	923.0	75.2	15	1.3345	4.058	2.182	1.167	1.210	1.228	82.46		
7.00	0.9864	0.9881	69.0	2.156	917.3	80.9	16	1.3346	4.410	2.371	1.265	1.227	1.247	81.33		
7.50	0.9855	0.9873	73.9	2.308	911.6	86.6	17	1.3347	4.769	2.564	1.362	1.244	1.265	80.22		
8.00	0.9848	0.9865	78.8	2.460	906.0	92.3	19	1.3348	5.13	2.76	1.459	1.261	1.283	79.15		
8.50	0.9840	0.9857	83.6	2.611	900.3	97.9	20	1.3350	5.49	2.95	1.554	1.278	1.301	78.11		
9.00	0.9832	0.9849	88.5	2.763	894.7	103.5	21	1.3351	5.85	3.15	1.649	1.294	1.319	77.12		
9.50	0.9824	0.9841	93.3	2.914	889.1	109.2	23	1.3352	6.22	3.35	1.745	1.310	1.336	76.17		
10.00	0.9816	0.9833	98.2	3.065	883.5	114.8	24	1.3354	6.60	3.55	1.841	1.326	1.354	75.26		
11.00	0.9801	0.9818	107.8	3.366	872.3	126.0	27	1.3356	7.36	3.96	2.032	1.357	1.387	73.57		
12.00	0.9785	0.9803	117.4	3.666	861.1	137.1	29	1.3359	8.14	4.38	2.221	1.386	1.419	72.00		
13.00	0.9770	0.9787	127.0	3.965	850.0	148.2	32	1.3362	8.93	4.80	2.407	1.415	1.451	70.53		
14.00	0.9755	0.9772	136.6	4.264	838.9	159.3	35	1.3365	9.72	5.23	2.588	1.443	1.482	69.15		
15.00	0.9740	0.9757	146.1	4.561	827.9	170.3	38	1.3367	10.53	5.66	2.767	1.471	1.513	67.85		
16.00	0.9725	0.9742	155.6	4.858	816.9	181.3	40	1.3370	11.36	6.11	2.946	1.498	1.544	66.61		
17.00	0.9710	0.9727	165.1	5.154	805.9	192.3	43	1.3373	12.23	6.57	3.126	1.525	1.574	65.45		
18.00	0.9695	0.9712	174.5	5.449	795.0	203.2	46	1.3376	13.13	7.06	3.308	1.551	1.603	64.35		
19.00	0.9680	0.9698	183.9	5.742	784.1	214.1	49	1.3379	14.06	7.56	3.490	1.576	1.632	63.31		
20.00	0.9666	0.9683	193.3	6.035	773.2	225.0	51	1.3381	15.02	8.07	3.669	1.601	1.660	62.34		
22.00	0.9636	0.9653	212.0	6.618	751.6	246.6	57	1.3387	16.98	9.13	4.019	1.649	1.715	60.52		
24.00	0.9606	0.9623	230.5	7.198	730.1	268.2	62	1.3392	19.04	10.23	4.359	1.694	1.767	58.91		
26.00	0.9576	0.9593	249.0	7.773	708.6	289.6	67	1.3397	21.23	11.42		1.732	1.813	57.61		
28.00	0.9545	0.9562	267.3	8.344	687.2	311.0	72	1.3402	23.59	12.68		1.765	1.853	56.55		
30.00	0.9514	0.9531	285.4	8.911	666.0	332.3	77	1.3407	25.91	13.93		1.791	1.886	55.72		
32.00	0.9482	0.9499	303.4	9.473	644.8	353.5	81	1.3411	28.15	15.13		1.810	1.913	55.13		
34.00	0.9450	0.9466	321.3	10.031	623.7	374.6	85	1.3415	30.48	16.39		1.823	1.933	54.74		
36.00	0.9416	0.9433	339.0	10.583	602.6	395.6	89	1.3419	32.97	17.73		1.831	1.949	54.49		
38.00	0.9382	0.9399	356.5	11.131	581.7	416.6	92	1.3422	35.60	19.14		1.835	1.959	54.40		
40.00	0.9347	0.9363	373.9	11.672	560.8	437.4	95	1.3425	38.6	20.8		1.833	1.965	54.45		
42.00	0.9311	0.9327	391.0	12.209	540.0	458.2	97	1.3427	41.5	22.3		1.827	1.966	54.62		
44.00	0.9273	0.9290	408.0	12.739	519.3	478.9	99	1.3429	44.5	23.9		1.817	1.963	54.93		
46.00	0.9235	0.9251	424.8	13.263	498.7	499.5	100	1.3430	47.8	25.7		1.801	1.955	55.40		
48.00	0.9196	0.9212	441.4	13.781	478.2	520.1	101	1.3431	51.2	27.5		1.781	1.940	56.05		
50.00	0.9156	0.9172	457.8	14.292	457.8	540.5	101	1.3431	54.5	29.3		1.757	1.923	56.80		
52.00	0.9114	0.9130	473.9	14.797	437.5	560.7	101	1.3431	58.1	31.2		1.732	1.905	57.61		
54.00	0.9072	0.9088	489.9	15.295	417.3	580.9	100	1.3430	62.0	33.3		1.705	1.883	58.52		
56.00	0.9030	0.9046	505.7	15.787	397.3	600.9	99	1.3429	66.0	35.5		1.673	1.857	59.64		
58.00	0.8987	0.9003	521.2	16.274	377.5	620.8	98	1.3428	70.0	37.6		1.638	1.826	60.94		
60.00	0.8944	0.8960	536.6	16.754	357.8	640.5	96	1.3426	74.5	40.0		1.597	1.789	62.49		
62.00	0.8901	0.8917	551.9	17.230	338.3	660.0	95	1.3425	79.3	42.6		1.550	1.744	64.40		
64.00	0.8856	0.8872	566.8	17.695	318.8	679.4	92	1.3422	84.4	45.4		1.500	1.697	66.54		
66.00	0.8810	0.8825	581.4	18.153	299.5	698.7	89	1.3419	89.6	48.2		1.453	1.653	68.66		
68.00	0.8763	0.8778	595.9	18.603	280.4	717.8	85	1.3415	96.3	51.8		1.410	1.612	70.78		
70.00	0.8715	0.8730	610.0	19.046	261.4	736.8	81	1.3411				1.365	1.569	73.11		
72.00	0.8667	0.8682	624.0	19.482	242.7	755.6	77	1.3407				1.315	1.521	75.87		
74.00	0.8618	0.8633	637.7	19.910	224.1	774.2	72	1.3402				1.264	1.470	78.93		
76.00	0.8568	0.8583	651.2	20.331	205.6	792.6	67	1.3397				1.216	1.422	82.05		
78.00	0.8518	0.8533	664.4	20.744	187.4	810.8	61	1.3391				1.170	1.377	85.28		
80.00	0.8468	0.8483	677.4	21.149	169.4	828.9	55	1.3385				1.126	1.332	88.63		
82.00	0.8416	0.8431	690.2	21.547	151.5	846.7	49	1.3379				1.072	1.276	93.11		
84.00	0.8365	0.8380	702.6	21.937	133.8	864.4	42	1.3372				1.021	1.223	97.75		
86.00	0.8312	0.8327	714.9	22.318	116.4	881.9	35	1.3365				0.968	1.167	103.1		
88.00	0.8259	0.8274	726.8	22.691	99.1	899.1	27	1.3357				0.914	1.109	109.2		
90.00	0.8204	0.8219	738.4	23.053	82.0	916.2	19	1.3348				0.859	1.049	116.2		
92.00	0.8148	0.8163	749.6	23.404	65.2	933.0	9	1.3339				0.804	0.989	124.1		
94.00	0.8089	0.8103	760.3	23.738	48.5	949.7	−2	1.3328				0.748	0.927	133.4		
96.00	0.8034	0.8048	771.3	24.081	32.1	966.1	−14	1.3316				0.694	0.866	143.8		
98.00	0.7976	0.7990	781.6	24.402	16.0	982.2	−26	1.3304				0.638	0.801	156.4		
100.0	0.7917	0.7931	791.7	24.717	0.0	998.2	−40	1.3290				0.585	0.740	170.6		

37 NICKEL SULFATE, NiSO$_4$ · 6H$_2$O

MOLECULAR WEIGHT = 154.75   FORMULA WEIGHT, HYDRATE = 262.85
RELATIVE SPECIFIC REFRACTIVITY = 0.524

0.00 % by wt. data are the same for all compounds.
For Values of 0.00 wt. % solutions see Table 1, Acetic Acid.

A % by wt.	H % by wt.	D_4^{20}	D_{20}^{20}	C, g/l	M g-mol/l	C_w g/l	$(C_o - C_w)$ g/l	$(n - n_o)$ × 10^4	n	Δ °C	O Os/kg	S g-mol/l	η/η_o	η/ρ cS	φ rhe	γ mmho/cm	T g-mol/l
0.50	0.85	1.0035	1.0053	5.0	0.032	998.5	− 0.3	10	1.3340	0.081	0.044	0.023	1.022	1.020	97.66	3.1	0.031
1.00	1.70	1.0089	1.0107	10.1	0.065	998.8	− 0.6	20	1.3349	0.150	0.081	0.043	1.044	1.037	95.59	5.8	0.058
1.50	2.55	1.0142	1.0160	15.2	0.098	999.0	− 0.8	29	1.3359	0.217	0.117	0.062	1.066	1.053	93.60	8.0	0.081
2.00	3.40	1.0196	1.0214	20.4	0.132	999.2	− 1.0	39	1.3369	0.282	0.152	0.081	1.089	1.070	91.64	9.9	0.103
2.50	4.25	1.0250	1.0268	25.6	0.166	999.4	− 1.2	49	1.3379	0.346	0.186	0.099	1.113	1.088	89.67	11.9	0.125
3.00	5.10	1.0304	1.0323	30.9	0.200	999.5	− 1.3	60	1.3389	0.410	0.220	0.118	1.138	1.107	87.70	13.7	0.146
3.50	5.94	1.0359	1.0377	36.3	0.234	999.6	− 1.4	70	1.3400	0.475	0.255	0.137	1.164	1.126	85.75	15.5	0.167
4.00	6.79	1.0413	1.0431	41.7	0.269	999.6	1.4	80	1.3410	0.540	0.290	0.156	1.191	1.146	83.80	17.3	0.187
4.50	7.64	1.0467	1.0486	47.1	0.304	999.6	1.4	90	1.3420	0.605	0.325	0.176	1.219	1.167	81.87	18.9	0.207
5.00	8.49	1.0521	1.0540	52.6	0.340	999.5	− 1.3	100	1.3430	0.671	0.361	0.195	1.248	1.189	79.95	20.5	0.226
5.50	9.34	1.0575	1.0594	58.2	0.376	999.4	− 1.1	110	1.3440	0.736	0.396	0.214	1.279	1.211	78.06	22.0	0.245
6.00	10.19	1.0629	1.0648	63.8	0.412	999.1	− 0.9	120	1.3450	0.803	0.432	0.234	1.310	1.235	76.18	23.5	0.263

38 NITRIC ACID, HNO$_3$

MOLECULAR WEIGHT = 63.02
RELATIVE SPECIFIC REFRACTIVITY = 0.818

0.00 % by wt. data are the same for all compounds.
For Values of 0.00 wt. % solutions see Table 1, Acetic Acid.

A % by wt.	ρ D_4^{20}	d D_{20}^{20}	C, g/l	M g-mol/l	C_w g/l	$(C_o - C_w)$ g/l	$(n - n_o)$ × 10^4	n	Δ °C	O Os/kg	S g-mol/l	η/η_o	η/ρ cS	φ rhe	γ mmho/cm	T g-mol/l
0.50	1.0009	1.0027	5.0	0.079	995.9	2.3	6	1.3336	0.281	0.151	0.080	1.002	1.003	99.64	28.4	0.323
1.00	1.0037	1.0054	10.0	0.159	993.6	4.6	13	1.3343	0.558	0.300	0.162	1.003	1.001	99.50	56.1	0.686
1.50	1.0064	1.0082	15.1	0.240	991.3	6.9	19	1.3349	0.837	0.450	0.244	1.004	1.000	99.40	84.7	1.10
2.00	1.0091	1.0109	20.2	0.320	988.9	9.3	26	1.3356	1.120	0.602	0.327	1.005	0.998	99.30	108.	1.50
2.50	1.0119	1.0137	25.3	0.401	986.6	11.7	32	1.3362	1.408	0.757	0.412	1.006	0.997	99.17	138.	1.97
3.00	1.0146	1.0164	30.4	0.483	984.2	14.0	39	1.3368	1.704	0.916	0.498	1.008	0.995	99.01	160.	2.57
3.50	1.0174	1.0192	35.6	0.565	981.8	16.5	45	1.3375	2.006	1.078	0.586	1.010	0.995	98.83	184.	3.18
4.00	1.0202	1.0220	40.0	0.648	979.4	18.9	51	1.3381	2.315	1.245	0.676	1.012	0.994	98.64	213.	4.31
4.50	1.0230	1.0248	46.0	0.730	976.9	21.3	58	1.3388	2.632	1.415	0.767	1.014	0.993	98.43		
5.00	1.0257	1.0276	51.3	0.814	974.5	23.8	64	1.3394	2.958	1.590	0.859	1.016	0.992	98.23		
5.50	1.0286	1.0304	56.6	0.898	972.0	26.3	71	1.3401	3.290	1.769	0.951	1.018	0.992	98.03		
6.00	1.0311	1.0332	61.9	0.982	969.6	28.7	78	1.3407	3.629	1.951	1.048	1.020	0.991	97.81		
6.50	1.0342	1.0360	67.2	1.067	967.0	31.3	84	1.3414	3.974	2.137	1.144	1.023	0.991	97.59		
7.00	1.0370	1.0389	72.6	1.152	964.4	33.8	91	1.3421	4.327	2.326	1.241	1.025	0.990	97.36		
7.50	1.0399	1.0417	78.0	1.238	961.9	36.3	97	1.3427	4.687	2.520	1.340	1.028	0.990	97.12		
8.00	1.0427	1.0446	83.4	1.324	959.3	38.9	104	1.3434	5.05	2.72	1.439	1.030	0.990	96.88		
8.50	1.0456	1.0475	88.9	1.410	956.7	41.5	110	1.3440	5.43	2.92	1.538	1.033	0.990	96.62		
9.00	1.0485	1.0504	94.4	1.497	954.1	44.1	117	1.3447	5.81	3.12	1.639	1.036	0.990	96.35		
9.50	1.0514	1.0533	99.9	1.585	951.5	46.7	124	1.3454	6.20	3.33	1.740	1.039	0.990	96.07		
10.00	1.0543	1.0562	105.4	1.673	948.9	49.4	130	1.3460	6.60	3.55	1.841	1.042	0.990	95.78		
11.00	1.0602	1.0620	116.6	1.850	943.5	54.7	144	1.3474	7.42	3.99	2.045	1.049	0.991	95.15		
12.00	1.0660	1.0679	127.9	2.030	938.1	60.1	157	1.3487	8.27	4.45	2.251	1.056	0.993	94.48		
13.00	1.0720	1.0739	139.4	2.211	932.6	65.6	170	1.3500	9.15	4.92	2.459	1.064	0.995	93.76		
14.00	1.0780	1.0799	150.9	2.395	927.1	71.2	184	1.3514	10.08	5.42	2.667	1.073	0.997	93.00		
15.00	1.0840	1.0859	162.6	2.580	921.4	76.8	198	1.3527	11.04	5.93	2.877	1.082	1.001	92.20		
16.00	1.0901	1.0921	174.4	2.768	915.7	82.5	211	1.3541	12.04	6.47	3.087	1.092	1.004	91.35		
17.00	1.0963	1.0982	186.4	2.957	909.9	88.3	225	1.3555	13.08	7.03	3.298	1.103	1.008	90.47		
18.00	1.1025	1.1044	198.4	3.149	904.0	94.2	239	1.3569	14.16	7.61	3.509	1.114	1.013	89.55		
19.00	1.1087	1.1107	210.7	3.343	898.0	100.2	253	1.3582	15.30	8.22	3.720	1.126	1.018	88.60		
20.00	1.1150	1.1170	223.0	3.538	892.0	106.2	266	1.3596				1.139	1.024	87.62		
22.00	1.1277	1.1297	248.1	3.937	879.6	118.6	294	1.3624				1.167	1.037	85.55		
24.00	1.1406	1.1426	273.7	4.344	866.8	131.4	322	1.3652				1.197	1.052	83.36		
26.00	1.1536	1.1557	299.9	4.759	853.7	144.6	350	1.3680				1.231	1.069	81.06		
28.00	1.1668	1.1688	326.7	5.184	840.1	158.1	378	1.3708				1.268	1.089	78.70		
30.00	1.1801	1.1822	354.0	5.618	826.0	172.2	406	1.3736				1.308	1.111	76.30		
32.00	1.1934	1.1955	381.9	6.060	811.5	186.7	433	1.3763				1.351	1.134	73.87		
34.00	1.2068	1.2090	410.3	6.511	796.5	201.7	460	1.3790				1.397	1.160	71.42		
36.00	1.2202	1.2224	439.3	6.970	780.9	217.3	487	1.3817				1.447	1.188	68.96		
38.00	1.2335	1.2357	468.7	7.438	764.8	233.5	513	1.3842				1.501	1.219	66.50		
40.00	1.2466	1.2489	498.7	7.913	748.0	250.2	537	1.3867				1.558	1.252	64.06		

39 OXALIC ACID, HO_2CCO_2H

MOLECULAR WEIGHT = 90.04

A% by wt.	ρ D_4^{20}	D_{20}^{20}	C_s g/l	M g-mol/l	C_w g/l	(C_0-C_w) g/l	$(n-n_0)$ $\times 10^4$	n	Δ °C	O Os/kg	S g-mol/l	η/η_0	η/ρ cS	ϕ rhe	γ mmho/cm	T g-mol/l
0.50	1.0006	1.0024	5.0	0.056	995.6	2.6	6.	1.3336	0.159	0.086	0.045	1.011	1.012	98.76	14.0	0.149
1.00	1.0030	1.0048	10.0	0.111	993.0	5.2	12.	1.3342	0.304	0.163	0.087	1.021	1.020	97.75	21.8	0.242
1.50	1.0054	1.0072	15.1	0.167	990.3	7.9	17.	1.3347	0.440	0.237	0.127	1.031	1.028	96.75	29.3	0.335
2.00	1.0079	1.0097	20.2	0.224	987.7	10.5	23.	1.3353	0.572	0.308	0.166	1.042	1.036	95.77	35.3	0.411
2.50	1.0103	1.0121	25.3	0.281	985.0	13.2	29.	1.3359	0.705	0.379	0.205	1.053	1.044	94.81	41.2	0.486
3.00	1.0126	1.0144	30.4	0.337	982.2	16.0	34.	1.3364	0.838	0.450	0.244	1.063	1.052	93.87	46.9	0.561
3.50	1.0150	1.0168	35.5	0.395	979.5	18.7	40.	1.3370	0.967	0.520	0.282	1.074	1.060	92.95	51.7	0.626
4.00	1.0174	1.0192	40.7	0.452	976.7	21.5	45.	1.3375	1.093	0.587	0.319	1.084	1.068	92.03	57.1	0.700
4.50	1.0197	1.0215	45.9	0.510	973.8	24.4	51.	1.3381				1.095	1.076	91.14	61.2	0.757
5.00	1.0220	1.0238	51.1	0.568	970.9	27.3	56.	1.3386				1.106	1.084	90.27	65.6	0.819
5.50	1.0244	1.0262	56.3	0.626	968.1	30.1	62.	1.3392				1.116	1.092	89.39	69.7	0.879
6.00	1.0265	1.0284	61.6	0.684	964.9	33.3	67.	1.3397				1.127	1.100	88.56	73.7	0.937
6.50	1.0288	1.0307	66.9	0.743	961.9	36.3	72.	1.3402				1.138	1.108	87.73	77.2	0.990
7.00	1.0310	1.0329	72.2	0.802	958.8	39.4	77.	1.3407				1.148	1.116	86.91	80.4	1.038
7.50	1.0332	1.0351	77.5	0.861	955.7	42.5	83.	1.3413				1.159	1.124	86.11	83.4	1.085
8.00	1.0355	1.0374	82.8	0.920	952.7	45.5	88.	1.3418				1.170	1.132	85.31	86.1	1.128

40 PHOSPHORIC ACID, H_3PO_4

MOLECULAR WEIGHT = 98.00
RELATIVE SPECIFIC REFRACTIVITY = 0.727

0.00 % by wt. data are the same for all compounds.
For Values of 0.00 wt. % solutions see Table 1, Acetic Acid.

A% by wt.	ρ D_4^{20}	D_{20}^{20}	C_s g/l	M g-mol/l	C_w g/l	(C_0-C_w) g/l	$(n-n_0)$ $\times 10^4$	n	Δ °C	O Os/kg	S g-mol/l	η/η_0	η/ρ cS	ϕ rhe	γ mmho/cm	T g-mol/l
0.50	1.0010	1.0028	5.0	0.051	996.0	2.2	5	1.3335	0.124	0.067	0.035	1.008	1.009	99.06	5.5	0.057
1.00	1.0038	1.0056	10.0	0.102	993.7	4.5	10	1.3340	0.243	0.131	0.069	1.018	1.016	98.04	10.1	0.105
1.50	1.0065	1.0083	15.1	0.154	991.4	6.8	15	1.3345	0.354	0.190	0.102	1.032	1.028	96.68	13.4	0.142
2.00	1.0092	1.0110	20.2	0.206	989.0	9.2	19	1.3349	0.464	0.249	0.134	1.048	1.041	95.23	16.2	0.175
2.50	1.0119	1.0137	25.3	0.258	986.6	11.6	24	1.3354	0.578	0.311	0.168	1.063	1.052	93.91	19.0	0.208
3.00	1.0146	1.0164	30.4	0.311	984.2	14.1	28	1.3358	0.694	0.373	0.202	1.077	1.064	92.64	21.6	0.239
3.50	1.0173	1.0191	35.6	0.363	981.7	16.5	33	1.3363	0.811	0.436	0.236	1.092	1.075	91.41	24.1	0.270
4.00	1.0200	1.0218	40.8	0.416	979.2	19.0	37	1.3367	0.929	0.499	0.271	1.106	1.087	90.22	26.5	0.300
4.50	1.0227	1.0245	46.0	0.470	976.7	21.6	42	1.3372	1.045	0.562	0.305	1.121	1.098	89.03	29.0	0.331
5.00	1.0254	1.0272	51.3	0.523	974.1	24.1	46	1.3376	1.160	0.624	0.339	1.136	1.110	87.85	31.5	0.363
5.50	1.0281	1.0299	56.5	0.577	971.6	26.7	51	1.3381	1.271	0.683	0.372	1.151	1.122	86.69	34.2	0.397

40 PHOSPHORIC ACID, H_3PO_4—(Continued)

A% by wt.	ρ D_4^{20}	D_{20}^{20}	C_s g/l	M g-mol/l	C_w g/l	$(C_o - C_w)$ g/l	$(n - n_o)$ × 10^4	n	Δ °C	O Os/kg	S g-mol/l	η/η_o	η/ρ cS	ϕ rhe	γ mmho/cm	T g-mol/l
6.00	1.0309	1.0327	61.9	0.631	969.0	29.2	55	1.3385	1.381	0.742	0.404	1.167	1.134	85.54	36.9	0.431
6.50	1.0336	1.0354	67.2	0.686	966.4	31.9	60	1.3390	1.496	0.804	0.438	1.182	1.146	84.42	39.6	0.466
7.00	1.0363	1.0381	72.5	0.740	963.8	34.5	64	1.3394	1.619	0.871	0.474	1.198	1.158	83.32	42.4	0.502
7.50	1.0391	1.0409	77.9	0.795	961.1	37.1	69	1.3399	1.748	0.940	0.511	1.214	1.170	82.23	45.2	0.539
8.00	1.0418	1.0437	83.3	0.850	958.5	39.7	73	1.3403	1.882	1.012	0.550	1.230	1.183	81.14	48.0	0.576
8.50	1.0446	1.0465	88.8	0.906	955.8	42.4	78	1.3408	2.020	1.086	0.590	1.247	1.196	80.05	50.8	0.614
9.00	1.0474	1.0493	94.3	0.962	953.2	45.1	83	1.3413	2.160	1.161	0.631	1.264	1.209	78.96	53.7	0.653
9.50	1.0503	1.0521	99.8	1.018	950.5	47.7	87	1.3417	2.305	1.239	0.673	1.282	1.223	77.87	56.5	0.692
10.00	1.0531	1.0550	105.3	1.075	947.8	50.4	92	1.3422	2.450	1.317	0.715	1.300	1.237	76.77	59.4	0.731
11.00	1.0589	1.0607	116.5	1.189	942.4	55.9	101	1.3431	2.721	1.463	0.792	1.338	1.267	74.56	65.1	0.812
12.00	1.0647	1.0665	127.8	1.304	936.9	61.3	111	1.3441	3.010	1.618	0.874	1.379	1.298	72.38	70.9	0.896
13.00	1.0705	1.0724	139.2	1.420	931.4	66.9	120	1.3450	3.382	1.818	0.979	1.421	1.330	70.22	76.7	0.983
14.00	1.0765	1.0784	150.7	1.538	925.8	72.5	130	1.3460	3.760	2.022	1.085	1.466	1.364	68.08	82.5	1.07
15.00	1.0825	1.0844	162.4	1.657	920.1	78.1	140	1.3470	4.075	2.191	1.172	1.513	1.400	65.98	88.4	1.17
16.00	1.0885	1.0905	174.2	1.777	914.4	83.9	150	1.3480	4.450	2.392	1.276	1.562	1.438	63.90	94.3	1.26
17.00	1.0947	1.0966	186.1	1.899	908.6	89.7	160	1.3489	4.820	2.591	1.376	1.613	1.477	61.86	100.	1.36
18.00	1.1009	1.1028	198.2	2.022	902.7	95.5	170	1.3500	5.25	2.82	1.491	1.668	1.518	59.85	106.	1.46
19.00	1.1071	1.1091	210.4	2.146	896.8	101.5	180	1.3510	5.72	3.07	1.614	1.724	1.561	57.88	112.	1.56
20.00	1.1135	1.1154	222.7	2.272	890.8	107.5	190	1.3520	6.23	3.35	1.747	1.784	1.605	55.94	118.	1.67
22.00	1.1263	1.1283	247.8	2.528	878.5	119.7	211	1.3540	7.38	3.97	2.037	1.910	1.699	52.26	129.	1.89
24.00	1.1395	1.1415	273.5	2.790	866.0	132.2	231	1.3561	8.69	4.67	2.350	2.045	1.798	48.80	141.	2.12
26.00	1.1528	1.1549	299.7	3.059	853.1	145.1	253	1.3582	10.12	5.44	2.677	2.194	1.907	45.50	152.	2.37
28.00	1.1665	1.1685	326.6	3.333	839.9	158.4	274	1.3604	11.64	6.26	3.004	2.360	2.027	42.29	163.	2.63
30.00	1.1804	1.1825	354.1	3.613	826.3	172.0	295	1.3625	13.23	7.11	3.328	2.548	2.163	39.17	173.	2.88
32.00	1.1945	1.1966	382.2	3.900	812.3	186.0	317	1.3647	14.94	8.03	3.656	2.760	2.315	36.16	183.	3.13
34.00	1.2089	1.2111	411.0	4.194	797.9	200.3	339	1.3669	16.81	9.04	3.989	2.995	2.482	33.33	191.	3.38
36.00	1.2236	1.2257	440.5	4.495	783.1	215.1	361	1.3691	18.85	10.13	4.329	3.253	2.664	30.68	199.	3.63
38.00	1.2385	1.2407	470.6	4.802	767.8	230.4	383	1.3713	21.09	11.34	4.677	3.537	2.862	28.21	205.	3.88
40.00	1.2536	1.2558	501.4	5.117	752.2	246.1	405	1.3735	23.58	12.68		3.848	3.076	25.94	209.	4.13

41 POTASSIUM BICARBONATE, $KHCO_3$

MOLECULAR WEIGHT = 100.12
RELATIVE SPECIFIC REFRACTIVITY = 0.667

0.00 % by wt. data are the same for all compounds.
For Values of 0.00 wt. % solutions see Table 1, Acetic Acid.

A% by wt.	ρ D_4^{20}	D_{20}^{20}	C_s g/l	M g-mol/l	C_w g/l	$(C_o - C_w)$ g/l	$(n - n_o)$ × 10^4	n	Δ °C	O Os/kg	S g-mol/l	η/η_o	η/ρ cS	ϕ rhe	γ mmho/cm	T g-mol/l
0.50	1.0014	1.0031	5.0	0.050	996.4	1.9	5	1.3335	0.175	0.094	0.050	1.007	1.007	99.14	4.6	0.046
1.00	1.0046	1.0064	10.0	0.100	994.6	3.7	11	1.3341	0.344	0.185	0.099	1.013	1.010	98.52	8.9	0.092
2.00	1.0114	1.0132	20.2	0.202	991.1	7.1	23	1.3353	0.668	0.359	0.194	1.025	1.016	97.37	17.0	0.184
3.00	1.0181	1.0199	30.5	0.305	987.6	10.6	35	1.3365	0.983	0.528	0.287	1.038	1.021	96.17	24.6	0.275
4.00	1.0247	1.0265	41.0	0.409	983.7	14.5	46	1.3376	1.291	0.694	0.378	1.051	1.028	94.95	31.7	0.365
5.00	1.0310	1.0329	51.6	0.515	979.5	18.8	56	1.3386	1.600	0.860	0.468	1.065	1.035	93.71	38.8	0.455
6.00	1.0379	1.0397	62.3	0.622	975.6	22.6	68	1.3397	1.911	1.028	0.559	1.079	1.042	92.46	45.8	0.547
7.00	1.0446	1.0465	73.1	0.730	971.5	26.7	79	1.3409	2.222	1.195	0.649	1.094	1.049	91.21	52.7	0.639
8.00	1.0514	1.0532	84.1	0.840	967.3	31.0	90	1.3419	2.533	1.362	0.738	1.110	1.057	89.95	59.4	0.732
9.00	1.0581	1.0600	95.2	0.951	962.9	35.3	100	1.3430	2.844	1.529	0.827	1.126	1.066	88.65	65.9	0.825
10.00	1.0650	1.0668	106.5	1.064	958.5	39.8	111	1.3441	3.157	1.697	0.916	1.143	1.075	87.31	72.4	0.918
12.00	1.0788	1.0807	129.5	1.293	949.3	48.9	132	1.3462	3.785	2.035	1.092	1.181	1.097	84.54	84.5	1.10
14.00	1.0929	1.0948	153.0	1.528	939.9	58.4	154	1.3484	4.414	2.373	1.266	1.222	1.120	81.68	95.9	1.29
16.00	1.1073	1.1093	177.2	1.770	930.1	68.1	176	1.3506				1.267	1.147	78.77	107.0	1.47
18.00	1.1221	1.1241	202.0	2.017	920.1	78.1	198	1.3528				1.316	1.175	75.83	118.0	1.67
20.00	1.1372	1.1392	227.4	2.272	909.7	88.5	220	1.3550				1.370	1.207	72.85	128.0	1.86
22.00	1.1527	1.1547	253.6	2.533	899.1	99.2	242	1.3572				1.429	1.242	69.83	137.0	2.05
24.00	1.1685	1.1706	280.4	2.801	888.1	110.2	265	1.3595				1.494	1.281	66.80	145.0	2.23

42 POTASSIUM BIPHTHALATE, $KHC_8H_4O_4$

MOLECULAR WEIGHT = 204.23
RELATIVE SPECIFIC REFRACTIVITY = 1.097

0.00 % by wt. data are the same for all compounds.
For Values of 0.00 wt. % solutions see Table 1, Acetic Acid.

A% by wt.	ρ D_4^{20}	D_{20}^{20}	C_s g/l	M g-mol/l	C_w g/l	$(C_o - C_w)$ g/l	$(n - n_o)$ × 10^4	n	Δ °C	O Os/kg	S g-mol/l	η/η_o	η/ρ cS	ϕ rhe	γ mmho/cm	T g-mol/l
0.50	1.0004	1.0021	5.0	0.024	995.4	2.9	9	1.3339	0.085	0.046	0.024	1.012	1.013	98.67	2.0	0.020
1.00	1.0025	1.0043	10.0	0.049	992.5	5.7	19	1.3349	0.169	0.091	0.048	1.023	1.022	97.56	4.0	0.040
1.50	1.0047	1.0065	15.1	0.074	989.6	8.6	28	1.3358	0.254	0.136	0.073	1.034	1.032	96.48	5.9	0.059
2.00	1.0069	1.0087	20.1	0.099	986.8	11.5	38	1.3368	0.337	0.181	0.097	1.046	1.041	95.41	7.7	0.079
2.50	1.0092	1.0109	25.2	0.124	983.9	14.3	47	1.3377	0.421	0.226	0.121	1.058	1.050	94.35	9.4	0.098
3.00	1.0114	1.0132	30.3	0.149	981.1	17.2	57	1.3387	0.503	0.271	0.146	1.070	1.060	93.29	11.0	0.115
3.50	1.0137	1.0155	35.5	0.174	978.2	20.0	66	1.3396	0.585	0.315	0.170	1.082	1.070	92.24	12.5	0.132
4.00	1.0160	1.0178	40.6	0.199	975.3	22.9	76	1.3406	0.667	0.359	0.194	1.094	1.079	91.19	14.0	0.149
4.50	1.0183	1.0201	45.8	0.224	972.5	25.8	86	1.3416	0.749	0.403	0.218	1.107	1.089	90.15	15.3	0.165
5.00	1.0206	1.0224	51.0	0.250	969.6	28.7	95	1.3425	0.831	0.447	0.242	1.120	1.100	89.11	16.7	0.181
5.50	1.0229	1.0247	56.3	0.275	966.7	31.6	105	1.3435	0.913	0.491	0.266	1.133	1.110	88.07	18.1	0.197
6.00	1.0252	1.0270	61.5	0.301	963.7	34.5	115	1.3444	0.995	0.535	0.290	1.147	1.121	87.04	19.4	0.213
6.50	1.0275	1.0294	66.8	0.327	960.8	37.5	124	1.3454	1.076	0.579	0.314	1.160	1.132	86.01	20.7	0.229
7.00	1.0298	1.0317	72.1	0.353	957.8	40.5	133	1.3463	1.158	0.622	0.338	1.174	1.143	84.99	21.9	0.244
7.50	1.0321	1.0339	77.4	0.379	954.7	43.5	143	1.3473	1.239	0.666	0.362	1.189	1.154	83.97	23.1	0.258
8.00	1.0344	1.0362	82.7	0.405	951.6	46.6	152	1.3482	1.320	0.710	0.386	1.203	1.165	82.96	24.2	0.272

43 POTASSIUM BROMIDE, KBr

MOLECULAR WEIGHT = 119.01
RELATIVE SPECIFIC REFRACTIVITY = 0.627

0.00 % by wt. data are the same for all compounds.
For Values of 0.00 wt. % solutions see Table 1, Acetic Acid.

A % by wt.	ρ D_4^{20}	D_{20}^{20}	C_s g/l	M g-mol/l	C_w g/l	$(C_o - C_w)$ g/l	$(n - n_o)$ × 10⁴	n	Δ °C	O Os/kg	S g-mol/l	η/η_o	η/ρ cS	ϕ rhe	γ mmho/cm	T g-mol/l
0.50	1.0018	1.0036	5.0	0.042	996.8	1.4	6	1.3336	0.148	0.080	0.042	0.998	0.998	100.00	5.2	0.053
1.00	1.0054	1.0072	10.1	0.084	995.4	2.9	12	1.3342	0.294	0.158	0.084	0.996	0.993	100.20	10.2	0.106
1.50	1.0090	1.0108	15.1	0.127	993.9	4.3	18	1.3348	0.439	0.236	0.127	0.994	0.987	100.40	14.9	0.160
2.00	1.0127	1.0145	20.3	0.170	992.4	5.8	24	1.3354	0.585	0.315	0.170	0.992	0.982	100.61	19.5	0.214
2.50	1.0164	1.0182	25.4	0.214	990.9	7.3	30	1.3360	0.731	0.393	0.213	0.990	0.976	100.82	24.2	0.270
3.00	1.0200	1.0218	30.6	0.257	989.4	8.8	36	1.3366	0.878	0.472	0.256	0.988	0.970	101.04	28.8	0.328
3.50	1.0238	1.0256	35.8	0.301	987.9	10.3	42	1.3372	1.026	0.552	0.300	0.985	0.965	101.27	33.5	0.387
4.00	1.0275	1.0293	41.1	0.345	986.4	11.9	49	1.3379	1.175	0.632	0.343	0.983	0.959	101.50	38.3	0.447
4.50	1.0312	1.0331	46.4	0.390	984.8	13.4	55	1.3385	1.325	0.712	0.387	0.981	0.953	101.73	43.0	0.509
5.00	1.0350	1.0368	51.8	0.435	983.3	15.0	61	1.3391	1.476	0.794	0.432	0.979	0.948	101.94	47.7	0.572
5.50	1.0388	1.0406	57.1	0.480	981.7	16.6	67	1.3397	1.629	0.876	0.477	0.977	0.942	102.15	52.5	0.636
6.00	1.0426	1.0445	62.6	0.526	980.1	18.2	74	1.3403	1.784	0.959	0.522	0.975	0.937	102.34	57.2	0.701
6.50	1.0465	1.0483	68.0	0.572	978.4	19.8	80	1.3410	1.939	1.043	0.567	0.973	0.932	102.54	62.0	0.767
7.00	1.0503	1.0522	73.5	0.618	976.8	21.4	86	1.3416	2.097	1.127	0.613	0.972	0.927	102.72	66.7	0.834
7.50	1.0542	1.0561	79.1	0.664	975.1	23.1	92	1.3422	2.255	1.213	0.658	0.970	0.922	102.90	71.5	0.903
8.00	1.0581	1.0600	84.6	0.711	973.5	24.8	99	1.3429	2.416	1.299	0.705	0.968	0.917	103.08	76.3	0.974
8.50	1.0620	1.0639	90.3	0.759	971.8	26.5	105	1.3435	2.578	1.386	0.751	0.967	0.912	103.25	81.1	1.05
9.00	1.0660	1.0679	95.9	0.806	970.1	28.2	112	1.3441	2.741	1.474	0.798	0.965	0.907	103.41	85.9	1.12
9.50	1.0700	1.0719	101.6	0.854	968.3	29.9	118	1.3448	2.907	1.563	0.845	0.964	0.902	103.58	90.7	1.20
10.00	1.0740	1.0759	107.4	0.902	966.6	31.6	124	1.3454	3.074	1.653	0.893	0.962	0.898	103.74	95.6	1.28
11.00	1.0821	1.0840	119.0	1.000	963.1	35.2	137	1.3467	3.415	1.836	0.989	0.959	0.888	104.06	105.	1.45
12.00	1.0903	1.0922	130.8	1.099	959.5	38.8	151	1.3481	3.764	2.024	1.086	0.956	0.879	104.36	115.	1.63
13.00	1.0986	1.1005	142.8	1.200	955.8	42.5	164	1.3494	4.122	2.216	1.185	0.954	0.870	104.64	125.	1.82
14.00	1.1070	1.1090	155.0	1.302	952.0	46.2	177	1.3507	4.487	2.412	1.286	0.951	0.861	104.90	134.	2.02
15.00	1.1155	1.1175	167.3	1.406	948.2	50.0	191	1.3521	4.862	2.614	1.387	0.949	0.853	105.13	144.	2.23
16.00	1.1242	1.1262	179.9	1.511	944.3	53.9	205	1.3535	5.25	2.82	1.490	0.947	0.844	105.35	154.	2.46
17.00	1.1330	1.1350	192.6	1.618	940.4	57.9	218	1.3548	5.64	3.03	1.593	0.946	0.836	105.54	164.	2.69
18.00	1.1419	1.1439	205.5	1.727	936.4	61.9	232	1.3562	6.04	3.25	1.698	0.944	0.829	105.70	174.	2.95
19.00	1.1509	1.1530	218.7	1.837	932.2	66.0	247	1.3577	6.46	3.47	1.804	0.943	0.821	105.84	184.	3.22
20.00	1.1601	1.1621	232.0	1.950	928.1	70.2	261	1.3591	6.88	3.70	1.912	0.942	0.814	105.95	194.	3.50
22.00	1.1788	1.1809	259.3	2.179	919.4	78.8	290	1.3620	7.76	4.17	2.130	0.941	0.800	106.07	216.	4.55
24.00	1.1980	1.2002	287.5	2.416	910.5	87.7	320	1.3650	8.70	4.68	2.352	0.941	0.787	106.07		
26.00	1.2179	1.2200	316.6	2.661	901.2	97.0	350	1.3680	9.68	5.20	2.579	0.942	0.775	105.92		
28.00	1.2383	1.2405	346.7	2.913	891.5	106.7	382	1.3711	10.72	5.76	2.809	0.945	0.765	105.59		
30.00	1.2593	1.2615	377.8	3.174	881.5	116.7	413	1.3743	11.82	6.35	3.042	0.950	0.756	105.05		
32.00	1.2810	1.2832	409.9	3.444	871.0	127.2	446	1.3776	12.98	6.98	3.279	0.957	0.748	104.30		
34.00	1.3033	1.3056	443.1	3.723	860.2	138.1	479	1.3809				0.966	0.743	103.33		
36.00	1.3263	1.3287	477.5	4.012	848.8	149.4	513	1.3843				0.977	0.738	102.13		
38.00	1.3501	1.3525	513.0	4.311	837.0	161.2	548	1.3878				0.991	0.736	100.69		
40.00	1.3746	1.3770	549.8	4.620	824.8	173.5	584	1.3914				1.008	0.735	99.01		

44 POTASSIUM CARBONATE, K₂CO₃ · 1½H₂O

MOLECULAR WEIGHT = 138.20   FORMULA WEIGHT, HYDRATE = 165.23
RELATIVE SPECIFIC REFRACTIVITY = 0.608

0.00 % by wt. data are the same for all compounds.
For Values of 0.00 wt. % solutions see Table 1, Acetic Acid.

A % by wt.	H % by wt.	ρ D_4^{20}	D_{20}^{20}	C_s g/l	M g-mol/l	C_w g/l	$(C_o - C_w)$ g/l	$(n - n_o)$ × 10⁴	n	Δ °C	O Os/kg	S g-mol/l	η/η_o	η/ρ cS	ϕ rhe	γ mmho/cm	T g-mol/l
0.50	0.60	1.0027	1.0045	5.0	0.036	997.7	0.5	9	1.3339	0.176	0.095	0.050	1.011	1.011	98.68	7.0	0.072
1.00	1.20	1.0072	1.0090	10.1	0.073	997.2	1.1	17	1.3347	0.339	0.182	0.097	1.023	1.018	97.57	13.6	0.144
2.00	2.39	1.0163	1.0181	20.3	0.147	995.9	2.3	35	1.3365	0.661	0.355	0.192	1.046	1.031	95.41	25.4	0.287
3.00	3.59	1.0254	1.0272	30.8	0.223	994.6	3.6	52	1.3382	0.986	0.530	0.288	1.069	1.045	93.36	36.7	0.428
4.00	4.78	1.0345	1.0363	41.4	0.299	993.1	5.1	69	1.3399	1.322	0.711	0.387	1.092	1.058	91.37	47.4	0.568
5.00	5.98	1.0437	1.0455	52.2	0.378	991.5	6.7	86	1.3416	1.668	0.897	0.488	1.117	1.072	89.35	58.0	0.712
6.00	7.17	1.0529	1.0548	63.2	0.457	989.7	8.5	103	1.3433	2.025	1.089	0.592	1.144	1.088	87.27	68.5	0.862
7.00	8.37	1.0622	1.0640	74.4	0.538	987.8	10.4	120	1.3450	2.395	1.287	0.699	1.172	1.105	85.17	78.8	1.02
8.00	9.56	1.0715	1.0734	85.7	0.620	985.8	12.5	137	1.3467	2.774	1.492	0.807	1.202	1.124	83.06	88.9	1.17
9.00	10.76	1.0809	1.0828	97.3	0.704	983.6	14.6	154	1.3484	3.167	1.702	0.919	1.233	1.143	80.94	98.8	1.34
10.00	11.96	1.0904	1.0923	109.0	0.789	981.3	16.9	171	1.3501	3.574	1.921	1.033	1.266	1.163	78.83	109.	1.50
12.00	14.35	1.1095	1.1115	133.1	0.963	976.4	21.8	205	1.3535	4.445	2.390	1.274	1.336	1.206	74.73	127.	1.83
14.00	16.74	1.1291	1.1311	158.1	1.144	971.0	27.2	239	1.3569	5.39	2.90	1.529	1.411	1.252	70.75	144.	2.17
16.00	19.13	1.1490	1.1510	183.8	1.330	965.2	33.1	273	1.3603	6.42	3.45	1.794	1.494	1.303	66.79	160.	2.53
18.00	21.52	1.1692	1.1713	210.5	1.523	958.8	39.5	307	1.3637	7.55	4.06	2.077	1.591	1.363	62.74	175.	2.90
20.00	23.91	1.1898	1.1919	238.0	1.722	951.9	46.4	341	1.3671	8.82	4.74	2.381	1.704	1.435	58.67	188.	3.31
24.00	28.69	1.2320	1.2342	295.7	2.140	936.3	61.9	409	1.3739	11.96	6.43	3.071	1.974	1.605	50.57	211.	4.24
28.00	33.48	1.2755	1.2778	357.1	2.584	918.4	79.9	477	1.3807	16.01	8.61	3.849	2.326	1.828	42.90		
32.00	38.26	1.3204	1.3227	422.5	3.057	897.8	100.4	544	1.3874	21.46	11.54		2.828	2.146	35.29		
36.00	43.04	1.3665	1.3690	492.0	3.560	874.6	123.6	610	1.3940	28.58	15.37		3.496	2.563	28.55		
40.00	47.82	1.4142	1.4167	565.7	4.093	848.5	149.7	676	1.4006	37.55	20.19		4.351	3.083	22.94		
44.00	52.61	1.4633	1.4659	643.9	4.659	819.5	178.8	741	1.4071				5.709	3.909	17.48		
48.00	57.39	1.5142	1.5169	726.8	5.259	787.4	210.8	806	1.4136				7.749	5.128	12.88		
50.00	59.78	1.5404	1.5431	770.2	5.573	770.2	228.0	838	1.4168				9.350	6.082	10.67		

45 POTASSIUM CHLORIDE, KCl

MOLECULAR WEIGHT = 74.55
RELATIVE SPECIFIC REFRACTIVITY = 0.758

0.00 % by wt. data are the same for all compounds.
For Values of 0.00 wt. % solutions see Table 1, Acetic Acid.

A % by wt.	D_4^{20}	D_{20}^{20}	C_s g/l	M g-mol/l	C_w g/l	$(C_o - C_w)$ g/l	$(n - n_o)$ ×10⁴	n	Δ °C	O Os/kg	S g-mol/l	η/η_o	η/ρ cS	ϕ rhe	γ mmho/cm	T g-mol/l
0.50	1.0014	1.0032	5.0	0.067	996.4	1.8	7	1.3337	0.234	0.126	0.067	0.998	0.999	99.97	8.2	0.084
1.00	1.0046	1.0064	10.0	0.135	994.6	3.7	14	1.3343	0.463	0.249	0.134	0.997	0.995	100.08	15.7	0.169
1.50	1.0078	1.0096	15.1	0.203	992.7	5.5	20	1.3350	0.691	0.372	0.201	0.997	0.991	100.11	22.7	0.253
2.00	1.0110	1.0128	20.2	0.271	990.8	7.4	27	1.3357	0.920	0.495	0.268	0.997	0.988	100.12	29.5	0.338
2.50	1.0142	1.0160	25.4	0.340	988.9	9.4	34	1.3364	1.149	0.618	0.336	0.996	0.984	100.16	36.5	0.426
3.00	1.0174	1.0192	30.5	0.409	986.9	11.3	41	1.3371	1.380	0.742	0.404	0.996	0.981	100.20	43.6	0.517
3.50	1.0207	1.0225	35.7	0.479	984.9	13.3	47	1.3377	1.613	0.867	0.472	0.995	0.977	100.26	50.6	0.610
4.00	1.0239	1.0257	41.0	0.549	982.9	15.3	54	1.3384	1.847	0.993	0.540	0.995	0.974	100.32	57.6	0.707
4.50	1.0271	1.0289	46.2	0.620	980.9	17.3	61	1.3391	2.083	1.120	0.609	0.994	0.970	100.38	64.8	0.807
5.00	1.0304	1.0322	51.5	0.691	978.8	19.4	68	1.3398	2.322	1.248	0.678	0.994	0.966	100.44	71.9	0.911
5.50	1.0336	1.0354	56.8	0.763	976.8	21.5	74	1.3404	2.561	1.377	0.747	0.993	0.963	100.51	79.1	1.01
6.00	1.0369	1.0387	62.2	0.835	974.7	23.6	81	1.3411	2.803	1.507	0.816	0.992	0.959	100.60	86.2	1.12
6.50	1.0402	1.0420	67.6	0.907	972.5	25.7	88	1.3418	3.047	1.638	0.885	0.991	0.955	100.69	93.3	1.24
7.00	1.0434	1.0453	73.0	0.980	970.4	27.8	95	1.3425	3.294	1.771	0.955	0.990	0.951	100.79	100.	1.36
7.50	1.0467	1.0486	78.5	1.053	968.2	30.0	102	1.3431	3.544	1.905	1.025	0.989	0.947	100.89	108.	1.48
8.00	1.0500	1.0519	84.0	1.127	966.0	32.2	108	1.3438	3.797	2.041	1.095	0.988	0.943	100.99	115.	1.61
8.50	1.0533	1.0552	89.5	1.201	963.8	34.4	115	1.3445	4.050	2.177	1.165	0.987	0.939	101.07	122.	1.74
9.00	1.0566	1.0585	95.1	1.276	961.6	36.7	122	1.3452	4.300	2.312	1.234	0.987	0.936	101.14	129.	1.88
9.50	1.0600	1.0619	100.7	1.351	959.3	39.0	129	1.3459	4.550	2.446	1.303	0.986	0.932	101.18	136.	2.02
10.00	1.0633	1.0652	106.3	1.426	957.0	41.2	136	1.3466	4.805	2.583	1.372	0.986	0.929	101.20	143.	2.17
11.00	1.0700	1.0719	117.7	1.579	952.3	45.9	149	1.3479	5.33	2.87	1.513	0.987	0.924	101.15	157.	2.49
12.00	1.0768	1.0787	129.2	1.733	947.6	50.7	163	1.3493	5.88	3.16	1.657	0.988	0.919	101.01	172.	2.84
13.00	1.0836	1.0855	140.9	1.890	942.7	55.5	177	1.3507	6.45	3.47	1.804	0.990	0.915	100.82	185.	3.21
14.00	1.0905	1.0924	152.7	2.048	937.8	60.4	191	1.3521				0.992	0.912	100.58	197.	3.61
15.00	1.0974	1.0993	164.6	2.208	932.8	65.5	205	1.3535				0.995	0.908	100.34	208.	4.05
16.00	1.1043	1.1063	176.7	2.370	927.7	70.6	219	1.3549				0.997	0.905	100.11	220.	4.85
17.00	1.1114	1.1133	188.9	2.534	922.4	75.8	233	1.3563				0.999	0.901	99.87		
18.00	1.1185	1.1204	201.3	2.701	917.1	81.1	247	1.3577				1.002	0.898	99.60		
19.00	1.1256	1.1276	213.9	2.869	911.7	86.5	262	1.3592				1.005	0.895	99.26		
20.00	1.1328	1.1348	226.6	3.039	906.2	92.0	276	1.3606				1.010	0.893	98.81		
22.00	1.1474	1.1494	252.4	3.386	895.0	103.2	305	1.3635				1.022	0.893	97.64		
24.00	1.1623	1.1643	278.9	3.742	883.3	114.9	335	1.3665				1.038	0.895	96.15		

46 POTASSIUM CHROMATE, K₂CrO₄

MOLECULAR WEIGHT = 194.20
RELATIVE SPECIFIC REFRACTIVITY = 0.820

0.00 % by wt. data are the same for all compounds.
For Values of 0.00 wt. % solutions see Table 1, Acetic Acid.

A % by wt.	ρ D_4^{20}	D_{20}^{20}	C_s g/l	M g-mol/l	C_w g/l	$(C_o - C_w)$ g/l	$(n - n_o)$ ×10⁴	n	Δ °C	O Os/kg	S g-mol/l	η/η_o	η/ρ cS	ϕ rhe	γ mmho/cm	T g-mol/l
0.50	1.0023	1.0040	5.0	0.026	997.3	1.0	11	1.3341	0.118	0.063	0.033	1.004	1.004	99.35	5.6	0.056
1.00	1.0063	1.0081	10.1	0.052	996.2	2.0	22	1.3352	0.235	0.126	0.067	1.009	1.005	98.91	10.8	0.113
2.00	1.0144	1.0162	20.3	0.104	994.1	4.2	44	1.3374	0.464	0.249	0.134	1.018	1.006	98.04	20.5	0.226
3.00	1.0224	1.0242	30.7	0.158	991.7	6.5	66	1.3396	0.682	0.367	0.198	1.027	1.006	97.21	29.8	0.340
4.00	1.0305	1.0323	41.2	0.212	989.2	9.0	87	1.3417	0.892	0.480	0.260	1.035	1.007	96.41	38.8	0.455
5.00	1.0386	1.0404	51.9	0.267	986.6	11.6	109	1.3439	1.103	0.593	0.322	1.044	1.007	95.59	47.8	0.573
6.00	1.0467	1.0486	62.8	0.323	983.9	14.3	131	1.3461	1.314	0.707	0.384	1.053	1.008	94.73	56.7	0.695
7.00	1.0550	1.0568	73.8	0.380	981.1	17.1	153	1.3483	1.524	0.819	0.446	1.063	1.010	93.85	65.6	0.819
8.00	1.0633	1.0652	85.1	0.438	978.2	20.0	175	1.3505	1.734	0.932	0.507	1.074	1.012	92.95	74.3	0.947
9.00	1.0718	1.0736	96.5	0.497	975.3	22.9	197	1.3527	1.946	1.047	0.569	1.085	1.014	92.02	83.0	1.08
10.00	1.0803	1.0823	108.0	0.556	972.3	25.9	220	1.3550	2.164	1.163	0.632	1.096	1.017	91.06	91.6	1.22
12.00	1.0980	1.0999	131.8	0.678	966.2	32.0	267	1.3597	2.613	1.405	0.761	1.120	1.022	89.07	109.	1.50
14.00	1.1163	1.1183	156.3	0.805	960.0	38.2	315	1.3645	3.076	1.654	0.893	1.147	1.030	87.00	125.	1.81
16.00	1.1344	1.1364	181.5	0.935	952.9	45.3	363	1.3693	3.555	1.911	1.028	1.176	1.039	84.84	141.	2.14
18.00	1.1527	1.1547	207.5	1.068	945.2	53.0	410	1.3740	4.050	2.177	1.165	1.209	1.051	82.56	157.	2.49
20.00	1.1720	1.1740	234.4	1.207	937.6	60.7	459	1.3789	4.561	2.452	1.306	1.245	1.064	80.16	172.	2.86
22.00	1.1920	1.1941	262.2	1.350	929.7	68.5	509	1.3839				1.285	1.081	77.64	189.	3.32
24.00	1.2125	1.2146	291.0	1.498	921.5	76.7	561	1.3891				1.330	1.099	75.04	201.	3.78
26.00	1.2335	1.2356	320.7	1.651	912.8	85.5	614	1.3944				1.379	1.120	72.37	214.	4.40
28.00	1.2547	1.2570	351.3	1.809	903.4	94.8	668	1.3998				1.433	1.144	69.64		
30.00	1.2763	1.2785	382.9	1.972	893.4	104.8	724	1.4054				1.492	1.171	66.89		

47 POTASSIUM DICHROMATE, K₂Cr₂O₇

MOLECULAR WEIGHT = 294.21
RELATIVE SPECIFIC REFRACTIVITY = 0.835

0.00 % by wt. data are the same for all compounds.
For Values of 0.00 wt. % solutions see Table 1, Acetic Acid.

A % by wt.	ρ D_4^{20}	D_{20}^{20}	C_s g/l	M g-mol/l	C_w g/l	$(C_o - C_w)$ g/l	$(n - n_o)$ ×10⁴	n	Δ °C	O Os/kg	S g-mol/l	η/η_o	η/ρ cS	ϕ rhe	γ mmho/cm	T g-mol/l
0.50	1.0017	1.0035	5.0	0.017	996.7	1.5	9	1.3339	0.101	0.054	0.028	1.000	1.001	99.75	3.7	0.037
1.00	1.0052	1.0070	10.1	0.034	995.1	3.1	18	1.3348	0.192	0.103	0.055	1.001	0.998	99.70	7.3	0.074
1.50	1.0087	1.0105	15.1	0.051	993.6	4.7	27	1.3357	0.278	0.149	0.080	1.001	0.995	99.65	10.6	0.111
2.00	1.0122	1.0140	20.2	0.069	992.0	6.3	36	1.3366	0.359	0.193	0.103	1.002	0.992	99.60	13.8	0.147
2.50	1.0157	1.0175	25.4	0.086	990.3	7.9	46	1.3375	0.435	0.234	0.126	1.003	0.989	99.55	17.0	0.184
3.00	1.0193	1.0211	30.6	0.104	988.7	9.5	55	1.3385				1.003	0.986	99.50	20.1	0.221
3.50	1.0228	1.0246	35.8	0.122	987.0	11.2	64	1.3394				1.003	0.983	99.45	23.1	0.258
4.00	1.0264	1.0282	41.1	0.140	985.3	12.9	73	1.3403				1.004	0.980	99.40	26.2	0.296
4.50	1.0300	1.0318	46.3	0.158	983.6	14.6	83	1.3413				1.004	0.977	99.35	29.2	0.334
5.00	1.0336	1.0354	51.7	0.176	981.9	16.3	92	1.3422				1.005	0.974	99.30	32.2	0.372

47 POTASSIUM DICHROMATE, K₂Cr₂O₇—(Continued)

A% by wt.	D_4^{20}	D_{20}^{20}	C_s g/l	M g-mol/l	C_w g/l	$(C_o - C_w)$ g/l	$(n - n_o)$ × 10⁴	n	Δ °C	O Os/kg	S g-mol/l	η/η_o	η/ρ cS	φ rhe	γ mmho/cm	T g-mol/l
5.50	1.0372	1.0390	57.0	0.194	980.1	18.1	102	1.3431				1.006	0.971	99.25	35.3	0.410
6.00	1.0408	1.0426	62.4	0.212	978.3	19.9	111	1.3441				1.006	0.969	99.21	38.3	0.449
6.50	1.0444	1.0463	67.9	0.231	976.5	21.7	121	1.3450				1.007	0.966	99.16	41.4	0.488
7.00	1.0481	1.0499	73.4	0.249	974.7	23.5	130	1.3460				1.007	0.963	99.11	44.4	0.528
7.50	1.0517	1.0536	78.9	0.268	972.8	25.4	140	1.3470				1.008	0.960	99.06	47.4	0.567
8.00	1.0554	1.0573	84.4	0.287	971.0	27.3	149	1.3479				1.008	0.957	99.01	50.3	0.607
8.50	1.0591	1.0610	90.0	0.306	969.1	29.2	159	1.3489				1.009	0.954	98.96	53.3	0.647
9.00	1.0628	1.0647	95.7	0.325	967.2	31.1	169	1.3499				1.009	0.951	98.91	56.3	0.688
9.50	1.0665	1.0684	101.3	0.344	965.2	33.0	179	1.3509				1.010	0.948	98.86	59.2	0.729
10.00	1.0703	1.0722	107.0	0.364	963.3	35.0	189	1.3519				1.010	0.946	98.81	62.1	0.770

48 POTASSIUM FERRICYANIDE, K₃Fe(CN)₆

MOLECULAR WEIGHT = 329.26
RELATIVE SPECIFIC REFRACTIVITY = 0.929

0.00 % by wt. data are the same for all compounds.
For Values of 0.00 wt. % solutions see Table 1, Acetic Acid.

A% by wt.	D_4^{20}	D_{20}^{20}	C_s g/l	M g-mol/l	C_w g/l	$(C_o - C_w)$ g/l	$(n - n_o)$ × 10⁴	n	Δ °C	O Os/kg	S g-mol/l	η/η_o	η/ρ cS	φ rhe	γ mmho/cm	T g-mol/l
0.50	1.0008	1.0026	5.0	0.015	995.8	2.4	8	1.3338	0.093	0.050	0.026	1.003	1.004	99.49	5.0	0.050
1.00	1.0034	1.0052	10.0	0.030	993.4	4.8	17	1.3347	0.180	0.097	0.051	1.006	1.005	99.21	9.7	0.100
1.50	1.0061	1.0079	15.1	0.046	991.0	7.3	25	1.3355	0.258	0.139	0.074	1.009	1.005	98.95	14.0	0.150
2.00	1.0087	1.0105	20.2	0.061	988.5	9.7	34	1.3364	0.334	0.180	0.096	1.011	1.004	98.71	18.2	0.199
2.50	1.0114	1.0132	25.3	0.077	986.1	12.1	42	1.3372	0.412	0.222	0.119	1.013	1.004	98.51	22.4	0.249
3.00	1.0140	1.0158	30.4	0.092	983.6	14.6	50	1.3380	0.491	0.264	0.142	1.015	1.003	98.33	26.5	0.298
3.50	1.0167	1.0185	35.6	0.108	981.1	17.1	59	1.3389	0.570	0.306	0.165	1.017	1.002	98.17	30.5	0.348
4.00	1.0194	1.0212	40.8	0.124	978.6	19.6	67	1.3397	0.649	0.349	0.189	1.018	1.001	98.02	34.5	0.399
4.50	1.0221	1.0239	46.0	0.140	976.1	22.1	75	1.3405	0.728	0.392	0.212	1.020	1.000	97.85	38.4	0.450
5.00	1.0249	1.0267	51.2	0.156	973.6	24.6	84	1.3414	0.807	0.434	0.235	1.022	0.999	97.65	42.4	0.502
5.50	1.0276	1.0294	56.5	0.172	971.1	27.2	92	1.3422	0.885	0.476	0.258	1.024	0.999	97.43	46.4	0.555
6.00	1.0304	1.0322	61.8	0.188	968.5	29.7	100	1.3430	0.962	0.517	0.281	1.027	0.999	97.19	50.5	0.609
6.50	1.0331	1.0350	67.2	0.204	966.0	32.3	109	1.3438	1.039	0.559	0.303	1.030	0.999	96.94	54.5	0.663
7.00	1.0359	1.0377	72.5	0.220	963.4	34.8	117	1.3447	1.115	0.600	0.326	1.032	0.999	96.67	58.5	0.719
7.50	1.0387	1.0406	77.9	0.237	960.8	37.4	125	1.3455	1.192	0.641	0.348	1.035	0.999	96.39	62.5	0.775
8.00	1.0415	1.0434	83.3	0.253	958.2	40.0	134	1.3464	1.268	0.682	0.371	1.039	0.999	96.10	66.5	0.832
8.50	1.0444	1.0462	88.8	0.270	955.6	42.6	142	1.3472	1.345	0.723	0.393	1.042	0.999	95.80	70.4	0.890
9.00	1.0472	1.0491	94.2	0.286	953.0	45.3	150	1.3480	1.422	0.764	0.416	1.045	1.000	95.49	74.4	0.949
9.50	1.0501	1.0519	99.8	0.303	950.3	47.9	159	1.3489	1.500	0.807	0.439	1.049	1.000	95.18	78.4	1.01
10.00	1.0530	1.0548	105.3	0.320	947.7	50.6	167	1.3497	1.580	0.849	0.462	1.052	1.001	94.87	82.4	1.07
11.00	1.0588	1.0607	116.5	0.354	942.3	55.9	184	1.3514	1.740	0.935	0.509	1.059	1.002	94.22	90.4	1.20
12.00	1.0647	1.0665	127.8	0.388	936.9	61.3	201	1.3531	1.901	1.022	0.556	1.067	1.004	93.55	98.4	1.33
13.00	1.0706	1.0725	139.2	0.423	931.4	66.8	219	1.3549	2.066	1.110	0.604	1.075	1.006	92.85	106.0	1.46
14.00	1.0766	1.0785	150.7	0.458	925.9	72.4	236	1.3566	2.235	1.202	0.653	1.084	1.008	92.11	114.0	1.60
15.00	1.0826	1.0845	162.4	0.493	920.2	78.0	254	1.3583	2.413	1.298	0.704	1.093	1.012	91.31	122.0	1.74
16.00	1.0887	1.0907	174.2	0.529	914.5	83.7	271	1.3601	2.600	1.398	0.758	1.103	1.015	90.46	129.0	1.88
17.00	1.0949	1.0968	186.1	0.565	908.8	89.5	289	1.3619	2.795	1.503	0.813	1.114	1.020	89.58	136.0	2.03
18.00	1.1011	1.1031	198.2	0.602	902.9	95.3	307	1.3637	2.997	1.611	0.871	1.126	1.024	88.65	143.0	2.17
19.00	1.1074	1.1093	210.4	0.639	897.0	101.3	325	1.3655	3.208	1.725	0.930	1.138	1.030	87.67	150.0	2.33
20.00	1.1137	1.1156	222.7	0.676	890.9	107.3	344	1.3674	3.427	1.843	0.992	1.152	1.036	86.63	157.0	2.50
22.00	1.1263	1.1283	247.8	0.753	878.6	119.7	381	1.3711				1.182	1.052	84.41	171.0	2.88
24.00	1.1391	1.1411	273.4	0.830	865.7	132.5	419	1.3749				1.217	1.070	82.02	185.0	3.31
26.00	1.1519	1.1540	299.5	0.910	852.4	145.8	457	1.3787				1.256	1.092	79.49	198.0	3.79
28.00	1.1647	1.1668	326.1	0.990	838.6	159.6	495	1.3825				1.299	1.117	76.83	210.0	4.34
30.00	1.1773	1.1794	353.2	1.073	824.1	174.1	533	1.3863				1.347	1.146	74.09	222.0	4.96

49 POTASSIUM FERROCYANIDE, K₄Fe(CN)₆ · 3H₂O

MOLECULAR WEIGHT = 368.36   FORMULA WEIGHT, HYDRATE = 422.41
RELATIVE SPECIFIC REFRACTIVITY = 0.921

0.00 % by wt. data are the same for all compounds.
For Values of 0.00 wt. % solutions see Table 1, Acetic Acid.

A% by wt.	H% by wt.	D_4^{20}	D_{20}^{20}	C_s g/l	M g-mol/l	C_w g/l	$(C_o - C_w)$ g/l	$(n - n_o)$ × 10⁴	n	Δ °C	O Os/kg	S g-mol/l	η/η_o	η/ρ cS	φ rhe	γ mmho/cm	T g-mol/l
0.50	0.57	1.0016	1.0034	5.0	0.014	996.6	1.6	10	1.3340	0.083	0.044	0.023	1.007	1.008	99.07	4.8	0.048
1.00	1.15	1.0049	1.0067	10.0	0.027	994.9	3.3	21	1.3351	0.162	0.087	0.046	1.014	1.011	98.42	9.3	0.096
1.50	1.72	1.0083	1.0101	15.1	0.041	993.3	5.1	31	1.3361	0.239	0.129	0.068	1.020	1.013	97.87	13.4	0.142
2.00	2.29	1.0116	1.0134	20.2	0.055	991.4	6.9	41	1.3371	0.314	0.169	0.090	1.025	1.015	97.37	17.3	0.188
2.50	2.87	1.0149	1.0167	25.4	0.069	989.5	8.7	51	1.3381	0.385	0.207	0.111	1.031	1.017	96.84	21.3	0.235
3.00	3.44	1.0182	1.0200	30.5	0.083	987.7	10.6	62	1.3391	0.453	0.243	0.131	1.036	1.020	96.33	25.2	0.282
3.50	4.01	1.0215	1.0233	35.8	0.097	985.8	12.5	72	1.3402	0.518	0.278	0.150	1.042	1.022	95.81	29.0	0.331
4.00	4.59	1.0248	1.0266	41.0	0.111	983.8	14.4	82	1.3412	0.582	0.313	0.169	1.047	1.024	95.30	32.9	0.379
4.50	5.16	1.0281	1.0300	46.3	0.126	981.9	16.4	92	1.3422	0.645	0.347	0.187	1.053	1.026	94.77	36.7	0.428
5.00	5.73	1.0315	1.0333	51.6	0.140	979.9	18.4	102	1.3432	0.709	0.381	0.206	1.059	1.029	94.24	40.6	0.478
5.50	6.31	1.0348	1.0366	56.9	0.155	977.9	20.4	112	1.3442	0.773	0.415	0.225	1.065	1.031	93.70	44.4	0.528
6.00	6.88	1.0381	1.0400	62.3	0.169	975.8	22.4	122	1.3452	0.836	0.450	0.244	1.071	1.034	93.15	48.2	0.578
6.50	7.45	1.0415	1.0433	67.7	0.184	973.8	24.5	132	1.3462	0.899	0.483	0.262	1.078	1.037	92.60	51.9	0.629
7.00	8.03	1.0449	1.0467	73.1	0.199	971.7	26.5	143	1.3473	0.961	0.517	0.281	1.084	1.040	92.04	55.7	0.680
7.50	8.60	1.0483	1.0501	78.6	0.213	969.6	28.6	153	1.3483	1.023	0.550	0.299	1.091	1.043	91.47	59.4	0.731
8.00	9.17	1.0517	1.0535	84.1	0.228	967.5	30.7	163	1.3493	1.084	0.583	0.317	1.098	1.046	90.90	63.1	0.784
8.50	9.75	1.0553	1.0572	89.7	0.244	965.6	32.6	175	1.3505	1.144	0.615	0.334	1.105	1.049	90.32	66.8	0.837
9.00	10.32	1.0588	1.0607	95.3	0.259	963.5	34.7	185	1.3515	1.204	0.647	0.352	1.112	1.052	89.74	70.6	0.891
9.50	10.89	1.0623	1.0641	100.9	0.274	961.3	36.9	196	1.3525	1.262	0.679	0.369	1.119	1.056	89.15	74.3	0.947
10.00	11.47	1.0657	1.0676	106.6	0.289	959.2	39.1	206	1.3536	1.320	0.710	0.386	1.127	1.060	88.55	78.1	1.00
11.00	12.61	1.0727	1.0746	118.0	0.320	954.7	43.5	227	1.3557	1.433	0.771	0.419	1.143	1.067	87.35	85.8	1.12

49 POTASSIUM FERROCYANIDE, $K_4Fe(CN)_6 \cdot 3H_2O$—(Continued)

A% by wt.	H% by wt.	ρ D_4^{20}	D_{20}^{20}	C_s g/l	M g-mol/l	C_w g/l	$(C_o - C_w)$ g/l	$(n - n_o)$ $\times 10^4$	n	Δ °C	O Os/kg	S g-mol/l	η/η_o	η/ρ cS	ϕ rhe	γ mmho/cm	T g-mol/l
12.00	13.76	1.0798	1.0817	129.6	0.352	950.2	48.0	248	1.3578				1.159	1.075	86.13	93.7	1.25
13.00	14.91	1.0869	1.0889	141.3	0.384	945.6	52.6	270	1.3600				1.176	1.084	84.89	101.	1.38
14.00	16.05	1.0941	1.0961	153.2	0.416	941.0	57.3	291	1.3621				1.193	1.093	83.63	109.	1.51
15.00	17.20	1.1014	1.1034	165.2	0.449	936.2	62.0	313	1.3643				1.212	1.103	82.34	117.	1.64
16.00	18.35	1.1088	1.1107	177.4	0.482	931.4	66.9	335	1.3665				1.232	1.113	81.03	124.	1.78
17.00	19.49	1.1162	1.1182	189.8	0.515	926.4	71.8	357	1.3687				1.252	1.124	79.70	131.	1.91
18.00	20.64	1.1237	1.1257	202.3	0.549	921.4	76.8	379	1.3709				1.274	1.136	78.35	137.	2.05
19.00	21.79	1.1312	1.1332	214.9	0.583	916.3	81.9	401	1.3731				1.296	1.148	76.98	143.	2.18
20.00	22.93	1.1389	1.1409	227.8	0.618	911.1	87.1	424	1.3754				1.320	1.161	75.61	149.	2.32

50 POTASSIUM HYDROXIDE, KOH

MOLECULAR WEIGHT = 56.11
RELATIVE SPECIFIC REFRACTIVITY = 0.680

0.00 % by wt. data are the same for all compounds.
For Values of 0.00 wt. % solutions see Table 1, Acetic Acid.

A% by wt.	ρ D_4^{20}	D_{20}^{20}	C_s g/l	M g-mol/l	C_w g/l	$(C_o - C_w)$ g/l	$(n - n_o)$ $\times 10^4$	n	Δ °C	O Os/kg	S g-mol/l	η/η_o	η/ρ cS	ϕ rhe	γ mmho/cm	T g-mol/l
0.50	1.0025	1.0043	5.0	0.089	997.5	0.7	10	1.3340	0.299	0.161	0.086	1.008	1.008	98.98	20.0	0.220
1.00	1.0068	1.0086	10.1	0.179	996.7	1.5	20	1.3350	0.609	0.327	0.177	1.017	1.012	98.13	38.5	0.452
1.50	1.0111	1.0129	15.2	0.270	995.9	2.3	29	1.3359	0.924	0.497	0.269	1.026	1.017	97.25	56.9	0.697
2.00	1.0155	1.0172	20.3	0.362	995.1	3.1	39	1.3369	1.242	0.668	0.363	1.036	1.022	96.36	75.0	0.957
2.50	1.0198	1.0216	25.5	0.454	994.3	3.9	49	1.3379	1.562	0.840	0.457	1.046	1.027	95.45	92.8	1.23
3.00	1.0242	1.0260	30.7	0.548	993.5	4.8	58	1.3388	1.886	1.014	0.551	1.056	1.033	94.52	110.	1.53
3.50	1.0286	1.0304	36.0	0.642	992.6	5.6	68	1.3398	2.217	1.192	0.648	1.066	1.039	93.58	128.	1.85
4.00	1.0330	1.0348	41.3	0.736	991.7	6.5	78	1.3408	2.565	1.379	0.748	1.077	1.045	92.63	144.	2.20
4.50	1.0374	1.0393	46.7	0.832	990.8	7.5	87	1.3417	2.940	1.581	0.854	1.089	1.051	91.68	161.	2.58
5.00	1.0419	1.0437	52.1	0.928	989.8	8.4	97	1.3427	3.356	1.804	0.972	1.100	1.058	90.71	178.	3.00
5.50	1.0464	1.0482	57.6	1.026	988.8	9.4	106	1.3436	3.747	2.014	1.081	1.112	1.065	89.75	193.	3.47
6.00	1.0509	1.0527	63.1	1.124	987.8	10.4	116	1.3445	4.144	2.228	1.191	1.124	1.072	88.78	206.	3.97
6.50	1.0554	1.0572	68.6	1.223	986.8	11.5	125	1.3455	4.521	2.431	1.295	1.137	1.079	87.81	216.	4.52
7.00	1.0599	1.0618	74.2	1.322	985.7	12.5	134	1.3464	4.921	2.649	1.403	1.149	1.086	86.84		
7.50	1.0644	1.0663	79.8	1.423	984.6	13.6	144	1.3474				1.162	1.094	85.87		
8.00	1.0690	1.0709	85.5	1.524	983.5	14.8	153	1.3483				1.175	1.102	84.91		
8.50	1.0736	1.0755	91.3	1.626	982.3	16.0	162	1.3492				1.189	1.110	83.95		
9.00	1.0781	1.0801	97.0	1.729	981.1	17.1	172	1.3502				1.203	1.118	82.98		
9.50	1.0827	1.0847	102.9	1.833	979.9	18.4	181	1.3511				1.217	1.126	82.03		
10.00	1.0873	1.0893	108.7	1.938	978.6	19.6	190	1.3520				1.231	1.134	81.07		
11.00	1.0966	1.0985	120.6	2.150	976.0	22.1	210	1.3539				1.261	1.152	79.17		
12.00	1.1060	1.1079	132.7	2.365	973.2	25.0	228	1.3558				1.291	1.170	77.28		
13.00	1.1153	1.1172	145.0	2.584	970.3	28.0	246	1.3576				1.324	1.189	75.39		
14.00	1.1246	1.1266	157.5	2.806	967.2	31.0	265	1.3595				1.358	1.210	73.50		
15.00	1.1341	1.1361	170.1	3.032	964.0	34.3	284	1.3614				1.394	1.232	71.59		
16.00	1.1435	1.1456	183.0	3.261	960.6	37.7	302	1.3632				1.433	1.255	69.66		
17.00	1.1531	1.1551	196.0	3.493	957.0	41.2	321	1.3651				1.474	1.281	67.69		
18.00	1.1626	1.1647	209.3	3.730	953.3	44.9	340	1.3670				1.518	1.308	65.75		
19.00	1.1722	1.1743	222.7	3.969	949.5	48.8	358	1.3688				1.565	1.338	63.77		
20.00	1.1818	1.1839	236.4	4.212	945.4	52.8	377	1.3707				1.616	1.370	61.76		
22.00	1.2014	1.2035	264.3	4.710	937.1	61.2	414	1.3744				1.729	1.442	57.72		
24.00	1.2210	1.2231	293.0	5.223	927.9	70.3	451	1.3781				1.857	1.524	53.74		
26.00	1.2408	1.2430	322.6	5.750	918.2	80.0	488	1.3818				2.002	1.617	49.85		
28.00	1.2609	1.2632	353.1	6.292	907.9	90.4	524	1.3854				2.166	1.721	46.09		
30.00	1.2813	1.2836	384.4	6.851	896.9	101.3	559	1.3889				2.352	1.839	42.44		
32.00	1.3020	1.3043	416.6	7.425	885.4	112.9	593	1.3923				2.565	1.974	38.90		
34.00	1.3230	1.3253	449.8	8.017	873.2	125.0	628	1.3957				2.814	2.131	35.47		
36.00	1.3444	1.3468	484.0	8.626	860.4	137.8	664	1.3993				3.105	2.315	32.14		
38.00	1.3661	1.3685	519.1	9.252	847.0	151.3	700	1.4030				3.453	2.533	28.90		
40.00	1.3881	1.3906	555.2	9.896	832.9	165.4	738	1.4068				3.871	2.795	25.78		
42.00	1.4104	1.4129	592.4	10.558	818.1	180.2	776	1.4106				4.380	3.112	22.78		
44.00	1.4331	1.4356	630.6	11.238	802.5	195.7	813	1.4143				5.003	3.498	19.95		
46.00	1.4560	1.4586	669.8	11.936	786.2	212.0	849	1.4179				5.769	3.970	17.30		
48.00	1.4791	1.4817	710.0	12.653	769.1	229.1	884	1.4214				6.713	4.547	14.87		
50.00	1.5024	1.5050	751.2	13.388	751.2	247.0	917	1.4247				7.876	5.253	12.67		

51 POTASSIUM IODIDE, KI

MOLECULAR WEIGHT = 166.03
RELATIVE SPECIFIC REFRACTIVITY = 0.647

0.00 % by wt. data are the same for all compounds.
For Values of 0.00 wt. % solutions see Table 1, Acetic Acid.

A% by wt.	ρ D_4^{20}	D_{20}^{20}	C_s g/l	M g-mol/l	C_w g/l	$(C_o - C_w)$ g/l	$(n - n_o)$ $\times 10^4$	n	Δ °C	O Os/kg	S g-mol/l	η/η_o	η/ρ cS	ϕ rhe	γ mmho/cm	T g-mol/l
0.50	1.0019	1.0037	5.0	0.030	996.9	1.3	7	1.3337	0.111	0.060	0.031	0.9972	0.9973	100.1	3.8	0.038
1.00	1.0056	1.0074	10.1	0.061	995.5	2.7	13	1.3343	0.220	0.118	0.063	0.9945	0.9910	100.3	7.5	0.076
2.00	1.0131	1.0149	20.3	0.122	992.8	5.4	27	1.3357	0.431	0.232	0.124	0.9892	0.9784	100.5	14.2	0.152
3.00	1.0206	1.0224	30.6	0.184	990.0	8.2	40	1.3370	0.642	0.345	0.187	0.9840	0.9661	101.4	21.1	0.233
4.00	1.0282	1.0301	41.1	0.248	987.1	11.1	54	1.3384	0.858	0.461	0.250	0.9788	0.9538	102.0	28.1	0.320
5.00	1.0360	1.0378	51.8	0.312	984.2	14.1	67	1.3397	1.079	0.580	0.315	0.9736	0.9417	102.5	35.2	0.409
6.00	1.0438	1.0457	62.6	0.377	981.2	17.1	81	1.3411	1.304	0.701	0.381	0.9670	0.9283	103.2	42.3	0.501
7.00	1.0517	1.0536	73.6	0.443	978.1	20.1	95	1.3425	1.534	0.825	0.449	0.9608	0.9153	103.9	49.6	0.597
8.00	1.0598	1.0617	84.8	0.511	975.0	23.2	110	1.3440	1.769	0.951	0.517	0.9549	0.9028	104.5	56.9	0.697
9.00	1.0679	1.0698	96.1	0.579	971.8	26.4	124	1.3454	2.010	1.081	0.588	0.9494	0.8908	105.1	64.3	0.801
10.00	1.0762	1.0781	107.6	0.648	968.6	29.6	139	1.3469	2.257	1.213	0.659	0.9442	0.8791	105.7	71.8	0.909
12.00	1.0931	1.0950	131.2	0.790	961.9	36.3	168	1.3498	2.767	1.488	0.805	0.9348	0.8569	106.8	86.8	1.14

51 POTASSIUM IODIDE, KI—(Continued)

A % by wt.	ρ D_4^{20}	D_{20}^{20}	C_s g/l	M g-mol/l	C_w g/l	$(C_o - C_w)$ g/l	$(n - n_o)$ $\times 10^4$	n	Δ °C	O Os/kg	S g-mol/l	η/η_o	η/ρ cS	ϕ rhe	γ mmho/cm	T g-mol/l
14.00	1.1105	1.1125	155.5	0.936	955.0	43.2	199	1.3529	3.304	1.776	0.957	0.9265	0.8360	107.7	102.	1.39
16.00	1.1284	1.1304	180.5	1.087	947.9	50.4	231	1.3560	3.869	2.080	1.115	0.9193	0.8163	108.6	118.	1.66
18.00	1.1469	1.1489	206.4	1.243	940.4	57.8	263	1.3593	4.464	2.400	1.279	0.9131	0.7977	109.3	133.	1.96
20.00	1.1659	1.1680	233.2	1.405	932.8	65.5	296	1.3626	5.09	2.74	1.449	0.9077	0.7801	109.9	149.	2.30
22.00	1.1856	1.1877	260.8	1.571	924.8	73.4	331	1.3661	5.76	3.09	1.624	0.9031	0.7632	110.5	164.	2.66
24.00	1.2060	1.2081	289.4	1.743	916.5	81.7	366	1.3696	6.46	3.47	1.805	0.8991	0.7470	111.0	180.	3.05
26.00	1.2270	1.2291	319.0	1.921	908.0	90.3	403	1.3733	7.21	3.87	1.993	0.8956	0.7314	111.4	195.	3.53
28.00	1.2487	1.2509	349.6	2.106	899.1	99.2	441	1.3771	8.01	4.30	2.188	0.8928	0.7164	111.8	210.	4.16
30.00	1.2712	1.2734	381.4	2.297	889.8	108.4	480	1.3810	8.86	4.76	2.390	0.8904	0.7019	112.1	224.	5.20
32.00	1.2944	1.2967	414.2	2.495	880.2	118.0	521	1.3851	9.76	5.25	2.597	0.8888	0.6880	112.3		
34.00	1.3185	1.3208	448.3	2.700	870.2	128.0	563	1.3893	10.72	5.76	2.808	0.8880	0.6748	112.4		
36.00	1.3434	1.3458	483.6	2.913	859.8	138.5	606	1.3936	11.73	6.31	3.024	0.8884	0.6626	112.3		
38.00	1.3692	1.3716	520.3	3.134	848.9	149.3	651	1.3981	12.81	6.89	3.246	0.8905	0.6517	112.1		
40.00	1.3959	1.3984	558.4	3.363	837.6	160.7	697	1.4027	13.97	7.51	3.472	0.8952	0.6426	111.5		

52 POTASSIUM NITRATE, KNO₃

MOLECULAR WEIGHT = 101.10
RELATIVE SPECIFIC REFRACTIVITY = 0.651

0.00 % by wt. data are the same for all compounds.
For Values of 0.00 wt. % solutions see Table 1, Acetic Acid.

A % by wt.	ρ D_4^{20}	D_{20}^{20}	C_s g/l	M g-mol/l	C_w g/l	$(C_o - C_w)$ g/l	$(n - n_o)$ $\times 10^4$	n	Δ °C	O Os/kg	S g-mol/l	η/η_o	η/ρ cS	ϕ rhe	γ mmho/cm	T g-mol/l
0.50	1.0014	1.0031	5.0	0.050	996.4	1.9	5	1.3335	0.171	0.092	0.049	0.997	0.998	100.10	5.5	0.056
1.00	1.0045	1.0063	10.0	0.099	994.5	3.8	10	1.3339	0.333	0.179	0.096	0.994	0.992	100.40	10.7	0.112
2.00	1.0108	1.0126	20.2	0.200	990.6	7.7	19	1.3349	0.642	0.345	0.187	0.988	0.979	101.01	20.1	0.222
3.00	1.0171	1.0189	30.5	0.302	986.6	11.7	28	1.3358	0.938	0.504	0.274	0.984	0.969	101.46	29.3	0.334
4.00	1.0234	1.0252	40.9	0.405	982.5	15.7	38	1.3368	1.223	0.657	0.357	0.981	0.960	101.77	38.3	0.448
5.00	1.0298	1.0317	51.5	0.509	978.3	19.9	47	1.3377	1.498	0.805	0.438	0.978	0.952	102.05	47.0	0.563
6.00	1.0363	1.0381	62.2	0.615	974.1	24.1	57	1.3386	1.763	0.948	0.516	0.975	0.943	102.33	55.5	0.678
7.00	1.0428	1.0447	73.0	0.722	969.8	28.4	66	1.3396	2.022	1.087	0.591	0.973	0.935	102.59	63.7	0.793
8.00	1.0494	1.0512	84.0	0.830	965.4	32.8	76	1.3405	2.273	1.222	0.664	0.971	0.927	102.81	71.7	0.908
9.00	1.0560	1.0579	95.0	0.940	961.0	37.3	85	1.3415	2.517	1.353	0.734	0.969	0.920	102.98	79.5	1.03
10.00	1.0627	1.0646	106.3	1.051	956.4	41.8	95	1.3425	2.754	1.481	0.802	0.968	0.913	103.10	87.3	1.15
12.00	1.0762	1.0781	129.1	1.277	947.0	51.2	114	1.3444				0.968	0.901	103.13	103.	1.40
14.00	1.0899	1.0918	152.6	1.509	937.3	60.9	133	1.3463				0.970	0.892	102.90	117.	1.66
16.00	1.1039	1.1058	176.6	1.747	927.3	71.0	152	1.3482				0.974	0.884	102.44	131.	1.94
18.00	1.1181	1.1201	201.3	1.991	916.9	81.4	172	1.3502				0.980	0.879	101.79	145.	2.22
20.00	1.1326	1.1346	226.5	2.241	906.1	92.2	191	1.3521				0.988	0.874	101.01	157.	2.49
22.00	1.1473	1.1493	252.4	2.497	894.9	103.3	211	1.3541				0.997	0.871	100.10	168.	2.75
24.00	1.1623	1.1644	279.0	2.759	883.3	114.9	231	1.3561				1.008	0.869	99.01	178.	3.00

53 POTASSIUM OXALATE, K₂C₂O₄ · 1H₂O

MOLECULAR WEIGHT = 166.21   FORMULA WEIGHT, HYDRATE = 184.23
RELATIVE SPECIFIC REFRACTIVITY = 0.676

0.00 % by wt. data are the same for all compounds.
For Values of 0.00 wt. % solutions see Table 1, Acetic Acid.

A % by wt.	H % by wt.	ρ D_4^{20}	D_{20}^{20}	C_s g/l	M g-mol/l	C_w g/l	$(C_o - C_w)$ g/l	$(n - n_o)$ $\times 10^4$	n	Δ °C	O Os/kg	S g-mol/l	η/η_o	η/ρ cS	ϕ rhe	γ mmho/cm	T g-mol/l
0.50	0.55	1.0018	1.0035	5.0	0.030	996.8	1.5	8	1.3337	0.137	0.074	0.039	1.005	1.005	99.33	5.8	0.059
1.00	1.11	1.0053	1.0071	10.1	0.060	995.3	3.0	15	1.3345	0.273	0.147	0.078	1.010	1.007	98.81	11.4	0.119
1.50	1.66	1.0089	1.0107	15.1	0.091	993.8	4.5	22	1.3352	0.407	0.219	0.118	1.016	1.009	98.25	16.6	0.180
2.00	2.22	1.0125	1.0143	20.3	0.122	992.3	6.0	29	1.3359	0.540	0.290	0.156	1.022	1.011	97.65	21.7	0.241
2.50	2.77	1.0161	1.0179	25.4	0.153	990.7	7.5	36	1.3366	0.669	0.360	0.194	1.028	1.014	97.04	26.8	0.302
3.00	3.33	1.0198	1.0216	30.6	0.184	989.2	9.0	43	1.3373	0.797	0.429	0.232	1.035	1.017	96.41	31.7	0.364
3.50	3.88	1.0235	1.0253	35.8	0.216	987.6	10.6	50	1.3380	0.925	0.497	0.270	1.042	1.020	95.77	36.6	0.426
4.00	4.43	1.0272	1.0290	41.1	0.247	986.1	12.2	56	1.3386	1.055	0.567	0.308	1.049	1.024	95.11	41.4	0.489
4.50	4.99	1.0309	1.0327	46.4	0.279	984.5	13.8	63	1.3393	1.186	0.638	0.347	1.057	1.027	94.45	46.2	0.552
5.00	5.54	1.0346	1.0364	51.7	0.311	982.9	15.4	70	1.3400	1.318	0.709	0.385	1.064	1.030	93.80	50.9	0.615
5.50	6.10	1.0383	1.0402	57.1	0.344	981.2	17.0	77	1.3407	1.451	0.780	0.425	1.071	1.034	93.14	55.6	0.679
6.00	6.65	1.0421	1.0439	62.5	0.376	979.6	18.7	84	1.3414	1.585	0.852	0.464	1.079	1.038	92.48	60.3	0.743
6.50	7.20	1.0459	1.0477	68.0	0.409	977.9	20.3	90	1.3420	1.720	0.925	0.503	1.087	1.041	91.82	64.8	0.807
7.00	7.76	1.0497	1.0516	73.5	0.442	976.2	22.0	97	1.3427	1.856	0.998	0.543	1.095	1.045	91.15	69.3	0.872
7.50	8.31	1.0535	1.0554	79.0	0.475	974.5	23.7	104	1.3434	1.994	1.072	0.583	1.103	1.049	90.48	73.8	0.937
8.00	8.87	1.0574	1.0592	84.6	0.509	972.8	25.5	111	1.3441	2.132	1.146	0.623	1.111	1.053	89.81	78.1	1.00
8.50	9.42	1.0612	1.0631	90.2	0.543	971.0	27.2	118	1.3448	2.273	1.222	0.663	1.120	1.057	89.13	82.4	1.07
9.00	9.98	1.0651	1.0670	95.9	0.577	969.3	29.0	126	1.3456	2.414	1.298	0.704	1.128	1.061	88.45	86.6	1.13
9.50	10.53	1.0691	1.0709	101.6	0.611	967.5	30.7	133	1.3463	2.557	1.375	0.745	1.137	1.065	87.77	90.8	1.20
10.00	11.08	1.0730	1.0749	107.3	0.646	965.7	32.5	140	1.3470	2.701	1.452	0.786	1.146	1.070	87.09	94.8	1.27
11.00	12.19	1.0810	1.0829	118.9	0.715	962.1	36.2	154	1.3484	2.993	1.609	0.870	1.166	1.081	85.58	104.	1.42
12.00	13.30	1.0891	1.0910	130.7	0.786	958.4	39.8	168	1.3498	3.291	1.769	0.954	1.188	1.093	84.01	112.	1.57
13.00	14.41	1.0973	1.0993	142.7	0.858	954.7	43.6	182	1.3512	3.594	1.932	1.039	1.211	1.106	82.40	121.	1.72
14.00	15.52	1.1057	1.1077	154.8	0.931	950.9	47.3	196	1.3526	3.903	2.098	1.125	1.234	1.118	80.89	129.	1.87

54 POTASSIUM PERMANGANATE, KMnO₄

MOLECULAR WEIGHT = 158.04
RELATIVE SPECIFIC REFRACTIVITY = 0.811

0.00 % by wt. data are the same for all compounds.
For Values of 0.00 wt. % solutions see Table 1, Acetic Acid.

A % by wt.	ρ D_4^{20}	D_{20}^{20}	C_s g/l	M g-mol/l	C_w g/l	$(C_o - C_w)$ g/l	$(n - n_o)$ × 10⁴	n	Δ °C	O Os/kg	S g-mol/l	η/η_o	η/ρ cS	φ rhe	γ mmho/cm	T g-mol/l
0.50	1.0017	1.0034	5.0	0.032	996.7	1.6			0.110	0.059	0.031	0.999	0.999	99.93	3.5	0.035
1.00	1.0051	1.0068	10.1	0.064	995.0	3.2			0.217	0.117	0.062	0.998	0.995	100.04	6.9	0.070
1.50	1.0085	1.0102	15.1	0.096	993.3	4.9			0.323	0.174	0.093	0.997	0.990	100.14	10.0	0.104
2.00	1.0118	1.0136	20.2	0.128	991.6	6.6			0.426	0.229	0.123	0.996	0.986	100.24	13.0	0.138
2.50	1.0152	1.0170	25.4	0.161	989.9	8.4						0.994	0.981	100.36	16.0	0.173
3.00	1.0186	1.0204	30.6	0.193	988.1	10.2						0.993	0.977	100.49	19.0	0.209
3.50	1.0220	1.0238	35.8	0.226	986.3	12.0						0.992	0.972	100.63	22.0	0.244
4.00	1.0254	1.0272	41.0	0.260	984.4	13.8						0.990	0.968	100.78	24.9	0.280
4.50	1.0288	1.0306	46.3	0.293	982.5	15.7						0.989	0.963	100.94	27.7	0.315
5.00	1.0322	1.0340	51.6	0.327	980.6	17.7						0.987	0.958	101.12	30.5	0.350
5.50	1.0356	1.0374	57.0	0.360	978.6	19.6						0.985	0.953	101.30	33.3	0.385
6.00	1.0390	1.0408	62.3	0.394	976.6	21.6						0.983	0.948	101.51	36.0	0.420

55 POTASSIUM PHOSPHATE, DIHYDROGEN (MONOBASIC), KH₂PO₄

MOLECULAR WEIGHT = 136.13
RELATIVE SPECIFIC REFRACTIVITY = 0.632

0.00 % by wt. data are the same for all compounds.
For Values of 0.00 wt. % solutions see Table 1, Acetic Acid.

A % by wt.	ρ D_4^{20}	D_{20}^{20}	C_s g/l	M g-mol/l	C_w g/l	$(C_o - C_w)$ g/l	$(n - n_o)$ × 10⁴	n	Δ °C	O Os/kg	S g-mol/l	η/η_o	η/ρ cS	φ rhe	γ mmho/cm	T g-mol/l
0.50	1.0018	1.0035	5.0	0.037	996.8	1.5	6	1.3336	0.130	0.070	0.037	1.008	1.008	98.98	3.0	0.030
1.00	1.0053	1.0071	10.1	0.074	995.3	3.0	12	1.3342	0.252	0.136	0.072	1.017	1.014	98.13	5.9	0.059
1.50	1.0089	1.0107	15.1	0.111	993.8	4.5	18	1.3348	0.371	0.200	0.107	1.026	1.019	97.25	8.5	0.087
2.00	1.0125	1.0143	20.2	0.149	992.2	6.0	24	1.3354	0.488	0.262	0.141	1.036	1.025	96.33	11.0	0.115
2.50	1.0161	1.0179	25.4	0.187	990.7	7.5	29	1.3359	0.603	0.324	0.175	1.046	1.032	95.37	13.5	0.143
3.00	1.0197	1.0215	30.6	0.225	989.1	9.1	35	1.3365	0.720	0.387	0.209	1.058	1.039	94.37	15.9	0.171
3.50	1.0233	1.0251	35.8	0.263	987.5	10.7	41	1.3371	0.838	0.451	0.244	1.069	1.047	93.34	18.3	0.199
4.00	1.0269	1.0288	41.0	0.302	985.9	12.4	47	1.3377	0.957	0.515	0.279	1.081	1.055	92.30	20.6	0.227
4.50	1.0306	1.0324	46.4	0.341	984.2	14.0	52	1.3382	1.075	0.578	0.314	1.094	1.063	91.26	22.8	0.255
5.00	1.0342	1.0360	51.7	0.380	982.5	15.8	58	1.3388	1.190	0.640	0.348	1.106	1.072	90.24	25.0	0.282
5.50	1.0378	1.0396	57.1	0.419	980.7	17.5	64	1.3394	1.303	0.700	0.381	1.119	1.080	89.23	27.2	0.309
6.00	1.0414	1.0432	62.5	0.459	978.9	19.3	70	1.3400	1.414	0.760	0.414	1.131	1.088	88.22	29.4	0.335
6.50	1.0450	1.0469	67.9	0.499	977.1	21.2	75	1.3405	1.523	0.819	0.446	1.144	1.097	87.21	31.5	0.361
7.00	1.0486	1.0505	73.4	0.539	975.2	23.0	81	1.3411	1.631	0.877	0.477	1.158	1.106	86.21	33.5	0.387
7.50	1.0522	1.0541	78.9	0.580	973.3	24.9	87	1.3417	1.736	0.933	0.508	1.171	1.114	85.11	35.5	0.412
8.00	1.0558	1.0577	84.5	0.620	971.4	26.9	91	1.3422	1.839	0.989	0.538	1.185	1.125	84.22	37.5	0.437
8.50	1.0594	1.0613	90.1	0.662	969.4	28.9	98	1.3428	1.941	1.043	0.567	1.199	1.134	83.23	39.3	0.461
9.00	1.0630	1.0649	95.7	0.703	967.4	30.9	104	1.3434	2.039	1.096	0.596	1.213	1.144	82.25	41.2	0.485
9.50	1.0667	1.0686	101.3	0.744	965.3	32.9	109	1.3439	2.136	1.148	0.624	1.228	1.154	81.27	42.9	0.508
10.00	1.0703	1.0722	107.0	0.786	963.3	35.0	115	1.3445	2.230	1.199	0.651	1.243	1.164	80.29	44.6	0.531

56 POTASSIUM PHOSPHATE, MONOHYDROGEN (DIBASIC), K₂HPO₄ · 3H₂O

MOLECULAR WEIGHT = 174.18   FORMULA WEIGHT, HYDRATE = 228.23
RELATIVE SPECIFIC REFRACTIVITY = 0.589

0.00 % by wt. data are the same for all compounds.
For Values of 0.00 wt. % solutions see Table 1, Acetic Acid.

A % by wt.	H % by wt.	ρ D_4^{20}	D_{20}^{20}	C_s g/l	M g.mol/l	C_w g/l	$(C_o - C_w)$ g/l	$(n - n_o)$ × 10⁴	n	Δ °C	O Os/kg	S g-mol/l	η/η_o	η/ρ cS	φ rhe	γ mmho/cm	T g-mol/l
0.50	0.66	1.0025	1.0043	5.0	0.029	997.5	0.7	8.	1.3338	0.126	0.068	0.036	1.011	1.010	98.76	5.2	0.052
1.00	1.31	1.0068	1.0086	10.1	0.058	996.7	1.5	15.	1.3345	0.250	0.134	0.071	1.021	1.016	97.76	9.9	0.102
1.50	1.97	1.0110	1.0128	15.2	0.087	995.8	2.4	23.	1.3353	0.371	0.199	0.107	1.032	1.023	96.69	14.2	0.151
2.00	2.62	1.0153	1.0171	20.3	0.117	994.9	3.3	31.	1.3361	0.490	0.263	0.142	1.044	1.030	95.62	18.3	0.200
2.50	3.28	1.0195	1.0213	25.5	0.146	994.0	4.2	38.	1.3368	0.613	0.329	0.178	1.055	1.037	94.59	23.0	0.248
3.00	3.93	1.0238	1.0256	30.7	0.176	993.1	5.2	46.	1.3376	0.732	0.394	0.213	1.067	1.044	93.56	26.8	0.294
3.50	4.59	1.0281	1.0299	36.0	0.207	992.1	6.1	54.	1.3384	0.855	0.457	0.249	1.079	1.052	92.46	29.7	0.340
4.00	5.24	1.0324	1.0342	41.3	0.237	991.1	7.1	62.	1.3392	0.974	0.523	0.284	1.092	1.060	92.18	33.3	0.385
4.50	5.90	1.0368	1.0386	46.7	0.268	990.1	8.1	69.	1.3399	1.096	0.589	0.320	1.105	1.068	90.31	36.8	0.430
5.00	6.55	1.0412	1.0430	52.1	0.299	989.1	9.1	77.	1.3407	1.218	0.655	0.356	1.118	1.076	89.26	40.3	0.475
5.50	7.21	1.0456	1.0474	57.5	0.330	988.1	10.2	85.	1.3415	1.337	0.719	0.391	1.131	1.084	88.23	43.8	0.520
6.00	7.86	1.0500	1.0519	63.0	0.362	987.0	11.2	92.	1.3422	1.460	0.785	0.427	1.145	1.093	87.13	47.2	0.565
6.50	8.52	1.0545	1.0564	68.5	0.394	986.0	12.3	100.	1.3430	1.580	0.850	0.462	1.160	1.102	86.05	50.6	0.610
7.00	9.17	1.0590	1.0609	74.1	0.426	984.9	13.3	108.	1.3438	1.703	0.915	0.498	1.175	1.112	84.92	53.8	0.654
7.50	9.83	1.0635	1.0654	79.8	0.458	983.8	14.5	115.	1.3445	1.823	0.980	0.533	1.191	1.122	83.80	57.0	0.698
8.00	10.48	1.0680	1.0699	85.4	0.491	982.6	15.7	123.	1.3453	1.947	1.046	0.569	1.207	1.132	82.71	60.0	0.740

57 POTASSIUM SULFATE, K₂SO₄

MOLECULAR WEIGHT = 174.26
RELATIVE SPECIFIC REFRACTIVITY = 0.565

0.00 % by wt. data are the same for all compounds.
For Values of 0.00 wt. % solutions see Table 1, Acetic Acid.

A % by wt.	D_4^{20}	D_{20}^{20}	C_s g/l	M g-mol/l	C_w g/l	$(C_o - C_w)$ g/l	$(n - n_o)$ × 10⁴	n	Δ °C	O Os/kg	S g-mol/l	η/η_o	η/ρ cS	φ rhe	γ mmho/cm	T g-mol/l
0.50	1.0022	1.0040	5.0	0.029	997.2	1.0	6	1.3336	0.136	0.073	0.039	1.004	1.004	99.37	5.8	0.059
1.00	1.0062	1.0080	10.1	0.058	996.2	2.1	13	1.3343	0.262	0.141	0.075	1.009	1.005	98.91	11.2	0.117
1.50	1.0102	1.0120	15.2	0.087	995.1	3.1	19	1.3349	0.382	0.205	0.110	1.014	1.006	98.42	16.2	0.175
2.00	1.0143	1.0161	20.3	0.116	994.0	4.2	25	1.3355	0.499	0.268	0.144	1.019	1.007	97.92	21.0	0.232
2.50	1.0183	1.0201	25.5	0.146	992.9	5.4	32	1.3362	0.614	0.330	0.178	1.025	1.008	97.39	25.6	0.289
3.00	1.0224	1.0242	30.7	0.176	991.7	6.5	38	1.3368	0.726	0.390	0.211	1.031	1.010	96.84	30.2	0.347

57 POTASSIUM SULFATE, K_2SO_4—(Continued)

A% by wt.	D_4^{20}	D_{20}^{20}	C_s g/l	M g-mol/l	C_w g/l	$(C_o - C_w)$ g/l	$(n - n_o) \times 10^4$	n	Δ °C	O Os/kg	S g-mol/l	η/η_o	η/ρ cS	ϕ rhe	γ mmho/cm	T g-mol/l
3.50	1.0265	1.0283	35.9	0.206	990.6	7.7	44	1.3374	0.839	0.451	0.244	1.037	1.012	96.27	34.7	0.404
4.00	1.0306	1.0324	41.2	0.237	989.4	8.9	50	1.3380	0.950	0.511	0.277	1.043	1.014	95.69	39.2	0.461
4.50	1.0347	1.0365	46.6	0.267	988.1	10.1	57	1.3386	1.061	0.571	0.310	1.049	1.016	95.10	43.7	0.518
5.00	1.0388	1.0406	51.9	0.298	986.8	11.4	63	1.3393	1.172	0.630	0.342	1.056	1.019	94.50	48.0	0.576
5.50	1.0429	1.0447	57.4	0.329	985.5	12.7	69	1.3399				1.063	1.021	93.89	52.3	0.634
6.00	1.0470	1.0489	62.8	0.361	984.2	14.0	75	1.3405				1.070	1.024	93.27	56.6	0.692
6.50	1.0512	1.0530	68.3	0.392	982.8	15.4	81	1.3411				1.077	1.027	92.64	60.8	0.751
7.00	1.0553	1.0572	73.9	0.424	981.4	16.8	87	1.3417				1.085	1.030	92.02	64.9	0.810
7.50	1.0595	1.0614	79.5	0.456	980.0	18.2	93	1.3422				1.092	1.033	91.39	69.0	0.869
8.00	1.0637	1.0655	85.1	0.488	978.6	19.7	98	1.3428				1.100	1.036	90.77	73.1	0.928
8.50	1.0679	1.0697	90.8	0.521	977.1	21.1	104	1.3434				1.107	1.039	90.14	77.0	0.988
9.00	1.0721	1.0740	96.5	0.554	975.6	22.6	110	1.3440				1.115	1.042	89.53	80.9	1.04
9.50	1.0763	1.0782	102.3	0.587	974.1	24.2	116	1.3446				1.122	1.045	88.92	84.8	1.10
10.00	1.0806	1.0825	108.1	0.620	972.5	25.7	122	1.3452				1.130	1.048	88.32	88.6	1.16

58 POTASSIUM THIOCYANATE, KSCN

MOLECULAR WEIGHT = 97.18
RELATIVE SPECIFIC REFRACTIVITY = 1.025

0.00 % by wt. data are the same for all compounds.
For Values of 0.00 wt. % solutions see Table 1, Acetic Acid.

A% by wt.	D_4^{20}	D_{20}^{20}	C_s g/l	M g-mol/l	C_w g/l	$(C_o - C_w)$ g/l	$(n - n_o) \times 10^4$	n	Δ °C	O Os/kg	S g-mol/l	η/η_o	η/ρ cS	ϕ rhe	γ mmho/cm	T g-mol/l
0.50	1.0007	1.0025	5.0	0.051	995.7	2.5	9	1.3339	0.177	0.095	0.050	0.9974	0.9987	100.1	5.8	0.058
1.00	1.0032	1.0049	10.0	0.103	993.1	5.1	19	1.3349	0.355	0.191	0.102	0.9947	0.9936	100.3	11.4	0.119
2.00	1.0081	1.0099	20.2	0.207	987.9	10.3	37	1.3367	0.709	0.381	0.206	0.9887	0.9827	100.9	21.9	0.243
3.00	1.0130	1.0148	30.4	0.313	982.6	15.6	55	1.3385	1.061	0.570	0.310	0.9833	0.9726	101.5	32.1	0.370
4.00	1.0180	1.0198	40.7	0.419	977.3	21.0	74	1.3404	1.414	0.760	0.413	0.9785	0.9632	102.0	42.1	0.499
5.00	1.0229	1.0248	51.1	0.526	971.8	26.4	93	1.3423	1.771	0.952	0.518	0.9741	0.9542	102.5	52.2	0.632
6.00	1.0279	1.0298	61.7	0.635	966.3	32.0	111	1.3441	2.136	1.148	0.624	0.9700	0.9456	102.9	62.4	0.773
7.00	1.0330	1.0348	72.3	0.744	960.7	37.6	130	1.3460	2.506	1.347	0.731	0.9664	0.9374	103.3	72.7	0.922
8.00	1.0380	1.0398	83.0	0.854	955.0	43.3	149	1.3479	2.882	1.549	0.838	0.9632	0.9298	103.6	82.9	1.08
9.00	1.0431	1.0449	93.9	0.966	949.2	49.0	168	1.3498	3.264	1.755	0.946	0.9604	0.9225	103.9	93.	1.24
10.00	1.0482	1.0500	104.8	1.079	943.3	54.9	187	1.3517	3.653	1.964	1.055	0.9579	0.9157	104.2	103.	1.40
12.00	1.0585	1.0603	127.0	1.307	931.4	66.8	226	1.3556	4.451	2.393	1.276	0.9541	0.9032	104.6	121.	1.72
14.00	1.0689	1.0708	149.6	1.540	919.2	79.0	265	1.3595	5.28	2.84	1.498	0.9521	0.8925	104.8	138.	2.06
16.00	1.0795	1.0814	172.7	1.777	906.8	91.5	304	1.3634				0.9523	0.8840	104.8	156.	2.47
18.00	1.0902	1.0922	196.2	2.019	894.0	104.3	345	1.3674				0.9550	0.8777	104.5	176.	2.95
20.00	1.1011	1.1031	220.2	2.266	880.9	117.3	385	1.3715				0.9597	0.8733	104.0	197.	3.60
24.00	1.1236	1.1255	269.7	2.775	853.9	144.3	469	1.3798				0.9755	0.8700	102.3		
28.00	1.1467	1.1488	321.1	3.304	825.6	172.6	554	1.3884				1.001	0.8747	99.69		
32.00	1.1706	1.1727	374.6	3.855	796.0	202.2	643	1.3972				1.037	0.8874	96.27		
36.00	1.1949	1.1970	430.2	4.427	764.7	233.5	732	1.4062				1.081	0.9069	92.28		
40.00	1.2199	1.2220	488.0	5.021	731.9	266.3	824	1.4154				1.150	0.9446	86.78		
44.00	1.2455	1.2477	548.0	5.639	697.5	300.7	918	1.4248				1.253	1.008	79.65		
48.00	1.2719	1.2741	610.5	6.282	661.4	336.9	1014	1.4343				1.389	1.094	71.86		
52.00	1.2990	1.3013	675.5	6.951	623.5	374.7	1112	1.4442				1.552	1.197	64.31		
56.00	1.3269	1.3292	743.1	7.646	583.8	414.4	1212	1.4542				1.735	1.310	57.52		
60.00	1.3556	1.3580	813.4	8.370	542.2	456.0	1315	1.4645				2.013	1.488	49.58		
64.00	1.3853	1.3877	886.6	9.123	498.7	499.5	1420	1.4750				2.434	1.761	41.00		

59 PROCAINE HYDROCHLORIDE, $C_6H_4[COOCH_2CH_2N(C_2H_5)_2][NH_2] \cdot HCl\text{-}1,4$

MOLECULAR WEIGHT = 272.78
RELATIVE SPECIFIC REFRACTIVITY = 1.460

0.00 % by wt. data are the same for all compounds.
For Values of 0.00 wt. % solutions see Table 1, Acetic Acid.

A% by wt.	D_4^{20}	D_{20}^{20}	C_s g/l	M g-mol/l	C_w g/l	$(C_o - C_w)$ g/l	$(n - n_o) \times 10^4$	n	Δ °C	O Os/kg	S g-mol/l	η/η_o	η/ρ cS	ϕ rhe	γ mmho/cm	T g-mol/l
0.50	0.9991	1.0008	5.0	0.018	994.1	4.2	11	1.3341	0.062	0.033	0.017	1.014	1.017	98.42	1.4	0.014
1.00	0.9999	1.0017	10.0	0.037	989.9	8.3	22	1.3352	0.126	0.068	0.035	1.028	1.030	97.08	2.7	0.027
2.00	1.0016	1.0034	20.0	0.073	981.6	16.7	44	1.3374	0.237	0.127	0.068	1.056	1.056	94.51	5.2	0.052
3.00	1.0033	1.0051	30.1	0.110	973.2	25.0	66	1.3396	0.343	0.184	0.099	1.083	1.082	92.16	7.4	0.076
4.00	1.0050	1.0068	40.2	0.147	964.8	33.4	89	1.3419	0.433	0.233	0.125	1.110	1.106	89.95	9.4	0.098
5.00	1.0067	1.0085	50.3	0.185	956.4	41.8	111	1.3441	0.519	0.279	0.150	1.138	1.133	87.70	11.4	0.119
6.00	1.0085	1.0103	60.5	0.222	948.0	50.2	133	1.3463	0.593	0.319	0.172	1.169	1.161	85.38	13.1	0.139
7.00	1.0103	1.0120	70.7	0.259	939.5	58.7	156	1.3486	0.675	0.363	0.196	1.202	1.192	83.06	14.7	0.157
8.00	1.0120	1.0138	81.0	0.297	931.1	67.2	179	1.3508	0.756	0.406	0.220	1.236	1.224	80.73	16.2	0.174
9.00	1.0138	1.0156	91.2	0.334	922.6	75.6	201	1.3531	0.824	0.443	0.240	1.273	1.258	78.40	17.5	0.191
10.00	1.0156	1.0174	101.6	0.372	914.1	84.2	224	1.3554	0.889	0.478	0.259	1.312	1.294	76.07	18.8	0.206
12.00	1.0193	1.0211	122.3	0.448	897.0	101.3	270	1.3600	1.026	0.552	0.300	1.397	1.373	71.45	21.1	0.233
14.00	1.0230	1.0248	143.2	0.525	879.8	118.5	316	1.3646	1.164	0.626	0.340	1.491	1.460	66.94	23.0	0.256
16.00	1.0267	1.0285	164.3	0.602	862.4	135.8	362	1.3692	1.297	0.698	0.379	1.595	1.557	62.57	24.6	0.277
18.00	1.0305	1.0323	185.5	0.680	845.0	153.2	409	1.3739	1.425	0.766	0.417	1.708	1.661	58.44	26.0	0.294
20.00	1.0343	1.0361	206.9	0.758	827.4	170.8	456	1.3786	1.545	0.831	0.452	1.838	1.781	54.30	27.2	0.309
24.00	1.0420	1.0439	250.1	0.917	791.9	206.3	552	1.3881				2.143	2.061	46.57	29.0	0.331
28.00	1.0499	1.0517	294.0	1.078	755.9	242.3	649	1.3978				2.542	2.426	39.26	29.8	0.342
32.00	1.0578	1.0596	338.5	1.241	719.3	278.9	747	1.4077				3.089	2.927	32.30	30.2	0.346
36.00	1.0658	1.0676	383.7	1.407	682.1	316.2	848	1.4178				3.746	3.522	26.64	29.5	0.338
40.00	1.0738	1.0757	429.5	1.575	644.3	354.0	951	1.4280				4.757	4.439	20.98	28.3	0.323
44.00	1.0819	1.0838	476.0	1.745	605.9	392.4	1055	1.4385				6.028	5.583	16.56	26.8	0.304
48.00	1.0900	1.0920	523.2	1.918	566.8	431.4	1162	1.4491				7.888	7.251	12.65	24.7	0.278
52.00	1.0982	1.1002	571.1	2.094	527.2	471.1	1270	1.4600				11.22	10.24	9.89	22.3	0.248
56.00	1.1066	1.1085	619.7	2.272	486.9	511.3	1381	1.4711				15.97	14.46	6.25	19.5	0.214
60.00	1.1151	1.1171	669.1	2.453	446.0	552.2	1493	1.4823				22.12	19.88	4.51	16.1	0.174

61 2-PROPANOL, CH₃CHOHCH₃

61 2-PROPANOL, $CH_3CHOHCH_3$

MOLECULAR WEIGHT = 60.09

A% by wt.	ρ D_4^{20}	D_{20}^{20}	C_s g/l	M g·mol/l	C_w g/l	$(C_0 - C_w)$ g/l	$(n - n_0)$ × 10⁴	n	Δ °C	O Os/kg	S g·mol/l	η/η_0	η/ρ cS	ϕ rhe	γ mmho/cm	T g·mol/l
1.00	0.9960	0.9978	10.0	0.166	986.0	12.2	8.	1.3338	0.304	0.163	0.087	1.054	1.060	94.72	0.0	0.000
2.00	0.9939	0.9957	19.9	0.331	974.0	24.2	16.	1.3346	0.603	0.324	0.175	1.110	1.119	89.91		
3.00	0.9920	0.9938	29.8	0.495	962.0	36.0	25.	1.3355	0.926	0.498	0.270	1.164	1.176	85.72		
4.00	0.9902	0.9920	39.6	0.659	950.6	47.6	34.	1.3364	1.259	0.677	0.368	1.223	1.238	81.57		
5.00	0.9884	0.9902	49.4	0.822	939.0	59.2	43.	1.3373	1.610	0.866	0.471	1.284	1.302	77.71		
6.00	0.9871	0.9889	59.2	0.986	927.9	70.3	52.	1.3382	1.960	1.054	0.573	1.349	1.369	74.00		
7.00	0.9855	0.9873	69.0	1.148	916.5	81.7	62.	1.3392	2.320	1.247	0.677	1.414	1.438	70.36		
8.00	0.9843	0.9861	78.7	1.310	905.6	92.6	71.	1.3401	2.679	1.440	0.780	1.482	1.509	67.33		
9.00	0.9831	0.9849	88.5	1.472	894.6	103.6	80.	1.3410	3.062	1.646	0.889	1.550	1.580	64.38		
10.00	0.9816	0.9834	98.2	1.634	883.4	114.8	90.	1.3420	3.481	1.871	1.007	1.626	1.660	61.37		
12.00	0.9793	0.9811	117.5	1.956	861.8	136.4	109.	1.3439	4.433	2.383	1.271	1.790	1.831	55.77		
14.00	0.9772	0.9790	136.8	2.277	840.4	157.8	129.	1.3459	5.29	2.85	1.503	1.966	2.016	50.76		
16.00	0.9751	0.9769	156.0	2.596	819.1	179.1	148.	1.3478	6.36	3.42	1.78	2.156	2.215	46.30		
18.00	0.9725	0.9743	175.1	2.913	797.4	200.8	166.	1.3496	7.40	3.98	2.04	2.347	2.418	42.53		
20.00	0.9696	0.9713	193.9	3.227	775.7	222.5	184.	1.3514	8.52	4.58	2.31	2.545	2.630	39.21		
24.00	0.9630	0.9647	231.1	3.846	731.9	266.3	217.	1.3547	10.91	5.87	2.85	2.851	2.966	35.01		
28.00	0.9555	0.9572	267.5	4.452	688.0	310.2	245.	1.3575				3.115	3.267	32.03		
32.00	0.9478	0.9495	303.3	5.047	644.5	353.7	270.	1.3600				3.330	3.520	29.97		
36.00	0.9393	0.9410	338.2	5.627	601.1	397.1	292.	1.3622				3.499	3.733	28.52		
40.00	0.9302	0.9319	372.1	6.192	558.1	440.1	312.	1.3642				3.630	3.910	27.49		
44.00	0.9207	0.9224	405.1	6.742	515.6	482.6	330.	1.3660				3.732	4.061	26.75		
48.00	0.9113	0.9129	437.4	7.279	473.9	524.3	346.	1.3676				3.772	4.147	26.46		
52.00	0.9019	0.9035	469.0	7.805	432.9	565.3	362.	1.3692				3.789	4.210	26.34		
56.00	0.8920	0.8936	499.5	8.313	392.5	605.7	375.	1.3705				3.781	4.247	26.40		
60.00	0.8824	0.8840	529.4	8.811	353.0	645.2	387.	1.3717				3.719	4.223	26.84		
64.00	0.8729	0.8745	558.7	9.297	314.2	684.0	400.	1.3730				3.636	4.174	27.45		
68.00	0.8632	0.8648	587.0	9.769	276.2	722.0	411.	1.3741				3.528	4.095	28.29		
72.00	0.8535	0.8550	614.5	10.226	239.0	759.2	421.	1.3751				3.391	3.981	29.43		
76.00	0.8438	0.8453	641.3	10.672	202.5	795.7	428.	1.3758				3.235	3.841	30.85		
80.00	0.8341	0.8356	667.3	11.105	166.8	831.4	435.	1.3765				3.072	3.690	32.49		
84.00	0.8243	0.8258	692.4	11.523	131.9	866.3	440.	1.3770				2.896	3.488	34.58		
88.00	0.8145	0.8160	716.8	11.929	97.7	900.5	445.	1.3773				2.684	3.302	37.18		
92.00	0.8046	0.8061	738.4	12.319	64.3	933.9	447.	1.3777				2.530	3.151	39.44		
96.00	0.7949	0.7963	763.1	12.699	31.8	966.4	446.	1.3776				2.424	3.055	41.18		
100.00	0.7848	0.7862	784.8	13.060	0.0	998.2	442.	1.3742				2.428	3.100	41.10		

60 1-PROPANOL, $CH_3CH_2CH_2OH$

MOLECULAR WEIGHT = 60.09

A% by wt.	ρ D_4^{20}	D_{20}^{20}	C_s g/l	M g·mol/l	C_w g/l	$(C_0 - C_w)$ g/l	$(n - n_0)$ × 10⁴	n	Δ °C	O Os/kg	S g·mol/l	η/η_0	η/ρ cS	ϕ rhe	γ mmho/cm	T g·mol/l
1.00	0.9963	0.9981	10.0	0.166	986.3	11.9	9.	1.3339	0.307	0.165	0.088	1.049	1.055	95.14	0.0	0.000
2.00	0.9946	0.9964	19.9	0.331	974.7	23.5	18.	1.3348	0.613	0.329	0.178	1.098	1.106	90.91		
3.00	0.9928	0.9946	29.8	0.496	963.0	35.2	27.	1.3357	0.933	0.502	0.272	1.150	1.161	86.76		
4.00	0.9911	0.9929	39.6	0.660	951.5	46.7	36.	1.3366	1.239	0.666	0.362	1.206	1.219	82.77		
5.00	0.9896	0.9914	49.5	0.823	940.1	58.1	46.	1.3376	1.573	0.846	0.460	1.264	1.280	78.94		
6.00	0.9882	0.9900	59.3	0.987	928.9	69.3	55.	1.3385	1.906	1.024	0.557	1.322	1.340	75.52		
7.00	0.9868	0.9886	69.1	1.150	917.7	80.5	64.	1.3394	2.258	1.214	0.659	1.384	1.405	72.12		
8.00	0.9855	0.9873	78.8	1.312	906.7	91.5	74.	1.3404	2.609	1.403	0.760	1.446	1.470	69.03		
9.00	0.9842	0.9860	88.6	1.474	895.6	102.6	84.	1.3414	2.988	1.606	0.868	1.511	1.538	66.06		
10.00	0.9829	0.9847	98.3	1.636	884.6	113.6	93.	1.3423	3.356	1.805	0.972	1.574	1.605	63.39		
12.00	0.9804	0.9822	117.6	1.958	862.8	135.4	112.	1.3442	3.477	2.014	1.181	1.707	1.745	58.45		
14.00	0.9779	0.9797	136.9	2.278	841.0	157.2	130.	1.3460	4.914	2.642	1.401	1.845	1.890	54.10		
16.00	0.9749	0.9767	156.0	2.596	818.9	179.3	147.	1.3477	5.776	3.108	1.629	1.982	2.037	50.35		
18.00	0.9719	0.9737	174.9	2.911	797.0	201.2	164.	1.3494	6.667	3.588	1.859	2.102	2.167	47.48		
20.00	0.9686	0.9703	193.7	3.224	774.9	223.3	180.	1.3510	7.555	4.062	2.079	2.214	2.290	45.09		
24.00	0.9612	0.9629	230.7	3.839	730.5	267.7	209.	1.3539	9.115	4.901	2.450	2.427	2.530	41.12		
28.00	0.9533	0.9550	266.9	4.442	686.4	311.8	236.	1.3566	10.17	5.47	2.688	2.607	2.740	38.28		
32.00	0.9452	0.9469	302.5	5.034	642.7	355.5	262.	1.3592	10.66	5.73	2.797	2.759	2.925	36.17		
36.00	0.9370	0.9387	337.3	5.614	599.7	398.5	284.	1.3614				2.894	3.095	34.48		
40.00	0.9288	0.9305	371.5	6.183	557.3	440.9	305.	1.3635				3.004	3.241	33.22		
44.00	0.9206	0.9223	405.1	6.741	515.5	482.7	328.	1.3658				3.096	3.370	32.23		
48.00	0.9127	0.9143	438.1	7.291	483.6	514.6	348.	1.3678				3.162	3.472	31.56		
52.00	0.9043	0.9059	470.2	7.826	434.1	564.1	367.	1.3697				3.195	3.540	31.24		
56.00	0.8959	0.8975	501.7	8.349	394.2	604.0	385.	1.3715				3.197	3.576	31.21		
60.00	0.8875	0.8891	532.5	8.862	355.0	643.2	404.	1.3734				3.180	3.590	31.39		
64.00	0.8790	0.8806	562.6	9.362	316.4	681.8	422.	1.3752				3.135	3.574	31.83		
68.00	0.8706	0.8722	592.0	9.852	278.6	719.6	437.	1.3767				3.075	3.539	32.46		
72.00	0.8623	0.8639	620.9	10.332	241.4	756.8	453.	1.3783				2.997	3.482	33.30		
76.00	0.8549	0.8564	649.7	10.813	205.2	793.0	467.	1.3797				2.909	3.410	34.30		
80.00	0.8470	0.8485	677.6	11.276	169.4	828.8	482.	1.3812				2.816	3.332	35.43		
84.00	0.8390	0.8405	704.8	11.728	134.2	864.0	495.	1.3825				2.720	3.249	36.69		
88.00	0.8306	0.8321	730.9	12.164	99.7	898.5	505.	1.3835				2.590	3.125	38.53		
92.00	0.8218	0.8233	756.1	12.582	65.7	932.5	513.	1.3843				2.467	3.008	40.45		
96.00	0.8130	0.8145	780.5	12.989	32.5	965.7	518.	1.3848				2.345	2.890	42.56		
100.00	0.8034	0.8048	803.4	13.370	0.0	998.2	522.	1.3852				2.223	2.773	44.89		

62 PROPYLENE GLYCOL, CH₂OHCHOHCH₃

$CH_2OHCHOHCH_3$

MOLECULAR WEIGHT = 76.09
RELATIVE SPECIFIC REFRACTIVITY = 1.236

0.00 % by wt. data are the same for all compounds.
For Values of 0.00 wt. % solutions see Table 1, Acetic Acid.

A% by wt.	ρ D_4^{20}	D_{20}^{20}	C, g/l	M g-mol/l	C_w g/l	$(C_o - C_w)$ g/l	$(n - n_o) \times 10^4$	n	Δ °C	O Os/kg	S g-mol/l	η/η_o	η/ρ cS	ϕ rhe	γ mmho/cm	T g-mol/l
0.50	0.9985	1.0003	5.0	0.066	993.5	4.7	5	1.3335	0.131	0.071	0.037	1.013	1.017	98.52		
1.00	0.9988	1.0006	10.0	0.131	988.8	9.4	10	1.3340	0.259	0.139	0.074	1.028	1.031	97.08		
2.00	0.9994	1.0012	20.0	0.263	979.4	18.8	20	1.3350	0.501	0.269	0.145	1.064	1.067	93.80		
3.00	1.0001	1.0019	30.0	0.394	970.1	28.2	31	1.3361	0.758	0.407	0.221	1.101	1.104	90.61		
4.00	1.0008	1.0025	40.0	0.526	960.7	37.5	41	1.3371	1.027	0.552	0.300	1.140	1.142	87.52		
5.00	1.0015	1.0032	50.1	0.658	951.4	46.8	52	1.3382	1.303	0.701	0.381	1.181	1.182	84.50		
6.00	1.0022	1.0040	60.1	0.790	942.1	56.2	63	1.3393	1.583	0.851	0.463	1.223	1.223	81.59		
7.00	1.0030	1.0047	70.2	0.923	932.8	65.5	74	1.3403	1.870	1.005	0.547	1.267	1.266	78.77		
8.00	1.0037	1.0055	80.3	1.055	923.4	74.8	84	1.3414	2.164	1.163	0.632	1.313	1.310	76.03		
9.00	1.0045	1.0063	90.4	1.188	914.1	84.1	95	1.3425	2.465	1.325	0.719	1.361	1.357	73.34		
10.00	1.0054	1.0071	100.5	1.321	904.8	93.4	107	1.3436	2.773	1.491	0.807	1.412	1.407	70.68		
12.00	1.0070	1.0088	120.8	1.588	886.2	112.0	129	1.3459	3.534	1.900	1.022	1.524	1.516	65.50		
14.00	1.0088	1.0106	141.2	1.856	867.6	130.7	152	1.3482	4.393	2.362	1.260	1.646	1.635	60.65		
16.00	1.0106	1.0124	161.7	2.125	848.9	149.4	175	1.3504	5.44	2.92	1.540	1.776	1.761	56.19		
18.00	1.0124	1.0142	182.2	2.395	830.2	168.1	198	1.3528				1.915	1.895	52.12		
20.00	1.0142	1.0160	202.8	2.666	811.4	186.9	221	1.3551				2.066	2.041	48.31		
24.00	1.0178	1.0196	244.3	3.210	773.6	224.7	267	1.3597				2.410	2.372	41.42		
28.00	1.0214	1.0232	286.0	3.759	735.4	262.8	314	1.3644				2.811	2.758	35.50		
32.00	1.0248	1.0266	327.9	4.310	696.8	301.4	360	1.3690				3.280	3.207	30.43		
36.00	1.0279	1.0297	370.1	4.863	657.9	340.4	406	1.3736				3.811	3.715	26.19		
40.00	1.0308	1.0326	412.3	5.419	618.5	379.8	451	1.3780				4.432	4.308	22.52		
44.00	1.0333	1.0352	454.7	5.975	578.7	419.6	494	1.3824				5.144	4.988	19.40		
48.00	1.0356	1.0374	497.1	6.533	538.5	459.7	537	1.3867				5.963	5.769	16.74		
52.00	1.0377	1.0395	539.6	7.091	498.1	500.2	580	1.3910				6.939	6.700	14.38		
56.00	1.0396	1.0415	582.2	7.651	457.4	540.8	622	1.3952				8.067	7.775	12.37		
60.00	1.0418	1.0436	625.1	8.215	416.7	581.5	665	1.3995				9.348	8.991	10.68		

63 SEA WATER*

RELATIVE SPECIFIC REFRACTIVITY = 0.789

0.00 % by wt. data are the same for all compounds.
For Values of 0.00 wt. % solutions see Table 1, Acetic Acid.

A% by wt.	Salinity ‰	Chlorinity ‰	ρ D_4^{20}	D_{20}^{20}	C, g/l	C_w g/l	$(C_o - C_w)$ g/l	$(n - n_o) \times 10^4$	n	Δ °C	O Os/kg	S g-mol/l	η/η_o	η/ρ cS	ϕ rhe	γ mmho/cm	T g-mol/l
0.50	4.94	2.72	1.0019	1.0037	5.0	996.9	1.3	9	1.3339	0.269	0.144	0.077	1.008	1.008	99.02	7.8	0.082
1.00	9.92	5.48	1.0057	1.0075	10.1	995.5	2.7	18	1.3348	0.537	0.289	0.156	1.016	1.012	98.23	15.2	0.163
1.50	14.91	8.24	1.0094	1.0112	15.1	994.1	4.1	27	1.3357	0.806	0.433	0.235	1.024	1.017	97.45	22.0	0.244
2.00	19.89	11.00	1.0132	1.0150	20.3	992.7	5.6	36	1.3366	1.075	0.578	0.314	1.033	1.022	96.61	28.4	0.324
2.50	24.87	13.76	1.0170	1.0188	25.4	991.2	7.0	45	1.3375	1.346	0.724	0.394	1.044	1.029	95.64	34.5	0.401
3.00	29.86	16.53	1.0207	1.0225	30.6	989.7	8.5	54	1.3384	1.616	0.869	0.473	1.055	1.036	94.58	40.4	0.476
3.50	34.84	19.29	1.0245	1.0263	35.8	988.1	10.1	63	1.3393	1.877	1.009	0.549	1.067	1.044	93.55	46.3	0.552
4.00	39.82	22.05	1.0283	1.0301	41.1	986.6	11.7	72	1.3402	2.129	1.145	0.622	1.077	1.050	92.67	52.4	0.635
4.50	44.81	24.81	1.0320	1.0339	46.4	985.0	13.3	81	1.3411	2.374	1.277	0.693	1.087	1.056	91.83	58.6	0.721
5.00	49.79	27.57	1.0358	1.0376	51.8	983.3	14.9	90	1.3420	2.612	1.404	0.761	1.100	1.065	90.76	64.2	0.799
5.50	54.78	30.33	1.0396	1.0414	57.1	981.6	16.6	99	1.3429	2.842	1.528	0.827	1.117	1.077	89.38	68.7	0.865
6.00	59.76	33.09	1.0434	1.0452	62.5	979.9	18.3	108	1.3438	3.064	1.647	0.890	1.137	1.093	87.80	72.6	0.922
6.50	64.74	35.85	1.0472	1.0490	68.0	978.1	20.1	118	1.3448								
7.00	69.73	38.61	1.0509	1.0528	73.5	976.3	21.9	127	1.3457								
7.50	74.71	41.37	1.0547	1.0566	79.0	974.6	23.6	136	1.3466								
8.00	79.69	44.13	1.0585	1.0604	84.6	972.9	25.3	146	1.3476								
8.50	84.68	46.90	1.0623	1.0642	90.2	971.2	27.1	155	1.3485								
9.00	89.66	49.66	1.0662	1.0680	95.9	969.3	28.9	164	1.3494								
9.50	94.64	52.42	1.0700	1.0719	101.6	967.5	30.8	173	1.3503								
10.00	99.63	55.18	1.0738	1.0757	107.3	965.6	32.6	182	1.3512								
11.00	109.6	60.70	1.0814	1.0833	118.9	961.7	36.5	200	1.3530								
12.00	119.6	66.22	1.0891	1.0910	130.6	957.8	40.5	219	1.3548								
13.00	129.5	71.74	1.0968	1.0987	142.5	953.7	44.5	237	1.3567								
14.00	139.5	77.27	1.1045	1.1065	154.6	949.5	48.7	255	1.3585								
15.00	149.5	82.79	1.1122	1.1142	166.8	945.2	53.0	273	1.3602								

* Values corresponding to A% from 0.00 6.00 are based upon measurements of artificial sea water as formulated by Lyman and Fleming, *J. Marine Res.* **3**, 134, 194
Values corresponding to A% above 6.00 are based upon natural sea water from the Gulf of Mexico at Aransas Pass, Texas, concentrated by evaporation, according
Behrens, ibid. **23**: 165, 1965.

64 SERUM OR PLASMA, HUMAN

0.00 % by wt. data are the same for all compounds.
For Values of 0.00 wt. % solutions see Table 1, Acetic Acid.

A% by wt.	ρ D_4^{20}	D_{20}^{20}	C, g/l	C_{Pr}, g 100 ml	C_{NPr}, g 100 ml	C_w g/l	$(C_o - C_w)$ g/l	$(n - n_o) \times 10^4$	n
1.00	1.0029	1.0047	10.0	0.0	1.00	993.	5.0	16	1.3346
1.50	1.0051	1.0069	15.0	0.0	1.50	990.	8.0	24	1.3354
2.00	1.0069	1.0087	20.0	0.3	1.69	987.	11.0	33	1.3363
2.50	1.0082	1.0100	25.0	0.8	1.68	983.	15.0	42	1.3372
3.00	1.0096	1.0114	30.0	1.4	1.68	979.	19.0	52	1.3382
3.50	1.0110	1.0128	35.0	1.9	1.67	976.	22.0	61	1.3391
4.00	1.0123	1.0141	40.0	2.4	1.66	972.	26.0	71	1.3401
4.50	1.0137	1.0155	46.0	2.9	1.66	968.	30.0	80	1.3410
5.00	1.0150	1.0168	51.0	3.4	1.65	964.	34.0	90	1.3420
5.50	1.0164	1.0182	56.0	3.9	1.64	960.	38.0	99	1.3429

A% by wt.	ρ D_4^{20}	D_{20}^{20}	C_s g/l	C_{Pr} g/100 ml	C_{NPr} g/100 ml	C_w g/l	$(C_o - C_w)$ g/l	$(n - n_o)$ × 10^4	n
6.00	1.0177	1.0195	61.	4.5	1.64	957.	41.	108	1.3438
6.50	1.0190	1.0208	66.	5.0	1.63	953.	45.	118	1.3448
7.00	1.0203	1.0221	71.	5.5	1.62	949.	49.	127	1.3457
7.50	1.0216	1.0234	77.	6.0	1.62	945.	53.	137	1.3467
8.00	1.0229	1.0247	82.	6.6	1.61	941.	57.	146	1.3476
8.50	1.0242	1.0260	87.	7.1	1.60	937.	61.	156	1.3486
9.00	1.0255	1.0273	92.	7.6	1.60	933.	65.	165	1.3495
9.50	1.0267	1.0286	98.	8.2	1.59	929.	69.	175	1.3505
10.00	1.0279	1.0298	103.	8.7	1.58	925.	73.	184	1.3514
11.00	1.0305	1.0324	113.	9.8	1.57	917.	81.	203	1.3533
12.00	1.0329	1.0348	124.	10.8	1.55	909.	89.	222	1.3552
13.00	1.0354	1.0373	135.	11.9	1.54	901.	97.	241	1.3571
14.00	1.0378	1.0397	145.	13.0	1.53	893.	105.	260	1.3590
15.00	1.0401	1.0420	156.	14.1	1.51	884.	114.	279	1.3609

Notes: C_{Pr} = g protein/100 ml; C_{NPr} = g nonprotein solute/100 ml.

65 SERUM OR PLASMA, RABBIT OR GUINEA PIG

0.00 % by wt. data are the same for all compounds.
For Values of 0.00 wt. % solutions see Table 1, Acetic Acid.

A% by wt.	ρ D_4^{20}	D_{20}^{20}	C_s g/l	C_{Pr} g/100 ml	C_w g/l	$(C_o - C_w)$ g/l	$(n - n_o)$ × 10^4	n
4.0	1.0125	1.0143	41.	3.3	972.	28.	71	1.3401
4.5	1.0139	1.0156	46.	3.8	968.	30.	80	1.3410
5.0	1.0152	1.0170	51.	4.2	964.	34.	89	1.3419
5.5	1.0165	1.0183	56.	4.6	961.	38.	98	1.3428
6.0	1.0179	1.0197	61.	5.0	957.	41.	107	1.3437
6.5	1.0192	1.0210	66.	5.1	953.	45.	116	1.3446
7.0	1.0206	1.0224	71.	5.8	949.	49.	125	1.3455
7.5	1.0219	1.0238	77.	6.2	945.	53.	134	1.3464
8.0	1.0233	1.0251	82.	6.6	941.	57.	143	1.3473
8.5	1.0247	1.0265	87.	7.0	937.	61.	152	1.3482
9.0	1.0260	1.0278	92.	7.4	934.	65.	161	1.3491
9.5	1.0274	1.0292	98.	7.8	930.	68.	170	1.3500
10.0	1.0287	1.0306	103.	8.2	926.	72.	179	1.3509
11.0	1.0315	1.0333	113.	9.0	918.	80.	197	1.3527
12.0	1.0342	1.0360	124.	9.8	910.	88.	215	1.3545

Note: C_{Pr} = g protein/100 ml.

66 SILVER NITRATE, AgNO₃

MOLECULAR WEIGHT = 169.89
RELATIVE SPECIFIC REFRACTIVITY = 0.475

0.00 % by wt. data are the same for all compounds.
For Values of 0.00 wt. % solutions see Table 1, Acetic Acid.

A% by wt.	ρ D_4^{20}	D_{20}^{20}	C_s g/l	M g-mol/l	C_w g/l	$(C_o - C_w)$ g/l	$(n - n_o)$ × 10^4	n	* Δ °C	O Os/kg	S g-mol/l	η/η_o	η/ρ cS	ϕ rhe	γ mmho/cm	T g-mol/l
0.50	1.0027	1.0045	5.0	0.030	997.7	0.5	6	1.3336	0.103	0.056	0.029	1.001	1.001	99.66	3.1	0.031
1.00	1.0070	1.0088	10.1	0.059	996.9	1.3	12	1.3342	0.202	0.109	0.058	1.003	0.998	99.50	6.1	0.062
2.00	1.0154	1.0172	20.3	0.120	995.1	3.1	22	1.3352	0.395	0.213	0.114	1.007	0.994	99.11	12.0	0.126
3.00	1.0239	1.0257	30.7	0.181	993.2	5.1	33	1.3363	0.586	0.315	0.170	1.011	0.989	98.74	17.2	0.187
4.00	1.0327	1.0345	41.3	0.243	991.4	6.8	44	1.3374	0.775	0.416	0.226	1.014	0.984	98.39	22.0	0.245
5.00	1.0417	1.0436	52.1	0.307	989.6	8.6	55	1.3385	0.961	0.517	0.281	1.018	0.979	98.04	26.7	0.303
6.00	1.0506	1.0524	63.0	0.371	987.5	10.7	66	1.3396	1.146	0.616	0.335	1.022	0.974	97.69	31.4	0.362
7.00	1.0597	1.0616	74.2	0.437	985.5	12.7	78	1.3407	1.329	0.715	0.389	1.025	0.969	97.34	36.1	0.421
8.00	1.0690	1.0709	85.5	0.503	983.5	14.7	89	1.3419	1.511	0.812	0.442	1.029	0.964	97.00	40.7	0.480
9.00	1.0785	1.0804	97.1	0.571	981.4	16.8	101	1.3431	1.690	0.909	0.494	1.033	0.959	96.63	45.3	0.540
10.00	1.0882	1.0901	108.8	0.641	979.3	18.9	113	1.3443	1.868	1.004	0.546	1.037	0.955	96.24	49.8	0.600
12.00	1.1079	1.1099	133.0	0.783	975.0	23.2	138	1.3467	2.213	1.190	0.646	1.047	0.947	95.35	58.8	0.723
14.00	1.1284	1.1304	158.0	0.930	970.4	27.8	163	1.3493	2.546	1.369	0.742	1.058	0.939	94.35	67.6	0.849
16.00	1.1496	1.1516	183.9	1.083	965.7	32.6	189	1.3519	2.862	1.539	0.832	1.070	0.933	93.25	76.3	0.976
18.00	1.1715	1.1736	210.9	1.241	960.6	37.6	216	1.3546				1.084	0.927	92.07	84.7	1.10
20.00	1.1942	1.1963	238.8	1.406	955.4	42.9	244	1.3574				1.099	0.922	90.81	92.8	1.23
22.00	1.2177	1.2198	267.9	1.577	949.8	48.5	273	1.3602				1.115	0.918	89.49	101.0	1.36
24.00	1.2420	1.2442	298.1	1.755	943.9	54.3	302	1.3632				1.133	0.914	88.09	108.0	1.49
26.00	1.2672	1.2694	329.5	1.939	937.7	60.5	332	1.3662				1.152	0.911	86.60	115.0	1.61
28.00	1.2933	1.2956	362.1	2.132	931.2	67.1	364	1.3694				1.174	0.910	85.01	122.0	1.74
30.00	1.3204	1.3228	396.1	2.332	924.3	73.9	397	1.3726				1.198	0.909	83.31	129.0	1.88
32.00	1.3487	1.3510	431.6	2.540	917.1	81.1	430	1.3760				1.225	0.910	81.49	136.0	2.02
34.00	1.3780	1.3805	468.5	2.758	909.5	88.7	465	1.3795				1.254	0.912	79.58	142.0	2.16
36.00	1.4087	1.4112	507.1	2.985	901.6	96.7	502	1.3832				1.287	0.915	77.56	149.0	2.31
38.00	1.4407	1.4433	547.5	3.223	893.2	105.0	541	1.3871				1.323	0.920	75.44	156.0	2.46
40.00	1.4743	1.4769	589.7	3.471	884.6	113.7	581	1.3911				1.363	0.926	73.22	162.0	2.61

67 SODIUM ACETATE, CH₃COONa

MOLECULAR WEIGHT = 82.04
RELATIVE SPECIFIC REFRACTIVITY = 0.884

0.00 % by wt. data are the same for all compounds.
For Values of 0.00 wt. % solutions see Table 1, Acetic Acid.

A% by wt.	ρ D_4^{20}	D_{20}^{20}	C_s g/l	M g-mol/l	C_w g/l	$(C_o - C_w)$ g/l	$(n - n_o)$ $\times 10^4$	n	Δ °C	O Os/kg	S g-mol/l	η/η_o	η/ρ cS	ϕ rhe	γ mmho/cm	T g-mol/l
0.50	1.0008	1.0026	5.0	0.061	995.8	2.4	7	1.3337	0.217	0.116	0.062	1.019	1.020	97.96	3.9	0.039
1.00	1.0034	1.0052	10.0	0.122	993.4	4.9	14	1.3344	0.433	0.233	0.125	1.038	1.037	96.15	7.6	0.078
2.00	1.0085	1.0103	20.2	0.246	988.3	9.9	28	1.3358	0.877	0.471	0.256	1.078	1.071	92.58	14.4	0.154
3.00	1.0135	1.0153	30.4	0.371	983.1	15.2	42	1.3372	1.339	0.720	0.391	1.122	1.109	88.98	20.4	0.225
4.00	1.0184	1.0202	40.7	0.497	977.7	20.6	56	1.3386	1.820	0.978	0.532	1.169	1.150	85.38	25.7	0.291
5.00	1.0234	1.0252	51.2	0.624	972.2	26.0	70	1.3400	2.323	1.249	0.678	1.220	1.195	81.80	30.9	0.355
6.00	1.0283	1.0302	61.7	0.752	966.6	31.6	85	1.3414	2.848	1.531	0.828	1.275	1.242	78.27	35.9	0.419
7.00	1.0334	1.0352	72.3	0.882	961.1	37.2	99	1.3428	3.398	1.827	0.984	1.334	1.294	74.79	40.8	0.482
8.00	1.0386	1.0404	83.1	1.013	955.5	42.7	112	1.3442	3.975	2.137	1.144	1.398	1.348	71.41	45.4	0.542
9.00	1.0440	1.0458	94.0	1.145	950.0	48.2	126	1.3456	4.566	2.455	1.307	1.465	1.406	68.13	49.6	0.598
10.00	1.0495	1.0514	105.0	1.279	944.6	53.6	140	1.3470				1.536	1.466	64.97	53.4	0.648
12.00	1.0607	1.0625	127.3	1.551	933.4	64.9	168	1.3498				1.685	1.592	59.22	59.0	0.725
14.00	1.0718	1.0737	150.1	1.829	921.8	76.5	196	1.3526				1.851	1.731	53.90	62.7	0.777
16.00	1.0830	1.0849	173.3	2.112	909.7	88.5	224	1.3554				2.050	1.897	48.68	65.5	0.818
18.00	1.0940	1.0960	196.9	2.400	897.1	101.1	253	1.3583				2.279	2.087	43.79	67.8	0.852
20.00	1.1050	1.1070	221.0	2.694	884.0	114.2	281	1.3611				2.562	2.323	38.95	69.3	0.873
22.00	1.1159	1.1179	245.5	2.993	870.4	127.8	309	1.3639				2.942	2.642	33.92	69.9	0.882
24.00	1.1268	1.1288	270.4	3.296	856.4	141.9	336	1.3666				3.393	3.017	29.41	69.7	0.878
26.00	1.1377	1.1397	295.8	3.606	841.9	156.3	363	1.3693				3.869	3.408	25.79	68.8	0.865
28.00	1.1488	1.1508	321.7	3.921	827.1	171.1	390	1.3720				4.379	3.819	22.79	67.0	0.840
30.00	1.1602	1.1623	348.1	4.243	812.2	186.1	418	1.3748				4.930	4.258	20.24	64.3	0.800

68 SODIUM BICARBONATE, NaHCO₃

MOLECULAR WEIGHT = 84.01
RELATIVE SPECIFIC REFRACTIVITY = 0.687

0.00 % by wt. data are the same for all compounds.
For Values of 0.00 wt. % solutions see Table 1, Acetic Acid.

A% by wt.	ρ D_4^{20}	D_{20}^{20}	C_s g/l	M g-mol/l	C_w g/l	$(C_o - C_w)$ g/l	$(n - n_o)$ $\times 10^4$	n	Δ °C	O Os/kg	S g-mol/l	η/η_o	η/ρ cS	ϕ rhe	γ mmho/cm	T g-mol/l
0.50	1.0018	1.0036	5.0	0.060	996.8	1.4	7	1.3337	0.199	0.107	0.057	1.013	1.013	98.52	4.2	0.041
1.00	1.0054	1.0072	10.1	0.120	995.3	2.9	14	1.3344	0.396	0.213	0.114	1.026	1.023	97.23	8.2	0.083
1.50	1.0089	1.0107	15.1	0.180	993.8	4.4	21	1.3351	0.591	0.318	0.172	1.040	1.033	95.94	11.7	0.123
2.00	1.0125	1.0143	20.2	0.241	992.2	6.0	27	1.3357	0.784	0.422	0.228	1.055	1.044	94.64	15.0	0.161
2.50	1.0160	1.0178	25.4	0.302	990.6	7.6	34	1.3364	0.975	0.524	0.284	1.069	1.055	93.33	18.0	0.197
3.00	1.0196	1.0214	30.6	0.364	989.0	9.3	41	1.3370	1.164	0.626	0.340	1.084	1.066	92.03	20.8	0.230
3.50	1.0231	1.0249	35.8	0.426	987.3	11.0	47	1.3377	1.350	0.726	0.395	1.100	1.077	90.73	23.5	0.263
4.00	1.0266	1.0284	41.1	0.489	985.5	12.7	54	1.3383	1.535	0.825	0.449	1.116	1.089	89.44	26.1	0.295
4.50	1.0301	1.0320	46.4	0.552	983.8	14.5	60	1.3390	1.718	0.924	0.502	1.132	1.101	88.16	28.7	0.327
5.00	1.0337	1.0355	51.7	0.615	982.0	16.2	67	1.3396	1.899	1.021	0.555	1.149	1.114	86.88	31.4	0.361
5.50	1.0372	1.0391	57.0	0.679	980.2	18.1	73	1.3403	2.078	1.117	0.607	1.166	1.126	85.61	34.2	0.397
6.00	1.0408	1.0426	62.4	0.743	978.4	19.9	79	1.3409	2.256	1.213	0.659	1.183	1.139	84.36	37.3	0.437

69 SODIUM BROMIDE, NaBr

MOLECULAR WEIGHT = 102.91
RELATIVE SPECIFIC REFRACTIVITY = 0.626

0.00 % by wt. data are the same for all compounds.
For Values of 0.00 wt. % solutions see Table 1, Acetic Acid.

A% by wt.	ρ D_4^{20}	D_{20}^{20}	C_s g/l	M g-mol/l	C_w g/l	$(C_o - C_w)$ g/l	$(n - n_o)$ $\times 10^4$	n	Δ °C	O Os/kg	S g-mol/l	η/η_o	η/ρ cS	ϕ rhe	γ mmho/cm	T g-mol/l
0.50	1.0021	1.0039	5.0	0.049	997.1	1.1	7	1.3337	0.173	0.093	0.049	1.002	1.002	99.55	5.0	0.050
1.00	1.0060	1.0078	10.1	0.098	995.9	2.3	14	1.3344	0.344	0.185	0.099	1.005	1.001	99.30	9.7	0.100
1.50	1.0099	1.0117	15.1	0.147	994.8	3.5	21	1.3351	0.515	0.277	0.149	1.007	1.000	99.06	14.1	0.150
2.00	1.0139	1.0157	20.3	0.197	993.6	4.7	28	1.3358	0.687	0.369	0.200	1.010	0.998	98.81	18.4	0.201
2.50	1.0178	1.0196	25.4	0.247	992.4	5.9	35	1.3365	0.860	0.463	0.251	1.013	0.997	98.56	22.7	0.253
3.00	1.0218	1.0236	30.7	0.298	991.1	7.1	42	1.3372	1.036	0.557	0.303	1.015	0.995	98.31	26.9	0.305
3.50	1.0258	1.0276	35.9	0.349	989.9	8.3	49	1.3379	1.214	0.653	0.355	1.018	0.994	98.06	31.2	0.359
4.00	1.0298	1.0317	41.2	0.400	988.6	9.6	57	1.3386	1.394	0.750	0.408	1.020	0.993	97.80	35.5	0.413
4.50	1.0339	1.0357	46.5	0.452	987.4	10.9	64	1.3394	1.577	0.848	0.461	1.023	0.992	97.54	39.8	0.468
5.00	1.0380	1.0398	51.9	0.504	986.1	12.2	71	1.3401	1.761	0.947	0.515	1.026	0.990	97.27	44.0	0.523
5.50	1.0421	1.0439	57.3	0.557	984.8	13.5	78	1.3408	1.949	1.048	0.570	1.029	0.989	97.00	48.2	0.578
6.00	1.0462	1.0481	62.8	0.610	983.5	14.8	86	1.3415	2.139	1.150	0.625	1.032	0.988	96.73	52.4	0.634
6.50	1.0504	1.0522	68.3	0.663	982.1	16.1	93	1.3423	2.332	1.254	0.681	1.035	0.987	96.46	56.5	0.691
7.00	1.0546	1.0564	73.8	0.717	980.8	17.5	100	1.3430	2.528	1.359	0.737	1.038	0.986	96.18	60.6	0.748
7.50	1.0588	1.0607	79.4	0.772	979.4	18.8	108	1.3438	2.728	1.466	0.794	1.041	0.985	95.90	64.7	0.806
8.00	1.0630	1.0649	85.0	0.826	978.0	20.2	115	1.3445	2.929	1.575	0.851	1.044	0.984	95.61	68.7	0.864
8.50	1.0673	1.0692	90.7	0.882	976.6	21.6	123	1.3453	3.133	1.685	0.909	1.047	0.983	95.31	72.7	0.923
9.00	1.0716	1.0735	96.4	0.937	975.2	23.1	130	1.3460	3.342	1.797	0.968	1.051	0.982	95.00	76.7	0.983
9.50	1.0760	1.0779	102.2	0.993	973.7	24.5	138	1.3468	3.554	1.911	1.027	1.054	0.982	94.67	80.6	1.04
10.00	1.0803	1.0822	108.0	1.050	972.3	25.9	145	1.3475	3.768	2.026	1.087	1.058	0.981	94.33	84.6	1.10
11.00	1.0892	1.0911	119.8	1.164	969.3	28.9	161	1.3491	4.206	2.261	1.208	1.066	0.981	93.61	92.4	1.22
12.00	1.0981	1.1000	131.8	1.280	966.3	31.9	176	1.3506	4.665	2.508	1.334	1.075	0.981	92.85	100.	1.35
13.00	1.1072	1.1091	143.9	1.399	963.2	35.0	192	1.3522	5.150	2.770	1.463	1.084	0.981	92.05	108.	1.48
14.00	1.1164	1.1184	156.3	1.519	960.1	38.1	208	1.3538	5.650	3.040	1.597	1.094	0.982	91.21	115.	1.61
15.00	1.1257	1.1277	168.9	1.641	956.9	41.4	224	1.3554	6.180	3.320	1.734	1.105	0.984	90.32	122.	1.75
16.00	1.1352	1.1372	181.6	1.765	953.6	44.7	241	1.3570	6.740	3.620	1.875	1.117	0.986	89.38	129.	1.89
17.00	1.1448	1.1469	194.6	1.891	950.2	48.0	257	1.3587	7.320	3.930	2.021	1.129	0.988	88.40	136.	2.03
18.00	1.1546	1.1566	207.8	2.019	946.8	51.5	274	1.3604				1.142	0.991	87.37	143.	2.18
19.00	1.1645	1.1665	221.3	2.150	943.2	55.0	291	1.3621				1.157	0.995	86.25	150.	2.33
20.00	1.1745	1.1766	234.9	2.283	939.6	58.6	308	1.3638				1.172	1.000	85.15	157.	2.50
22.00	1.1951	1.1972	262.9	2.555	932.2	66.1	343	1.3673				1.205	1.010	82.81	172.	2.84

69 SODIUM BROMIDE, NaBr—(Continued)

A% by wt.	ρ D_4^{20}	D_{20}^{20}	C_s g/l	M g-mol/l	C_w g/l	$(C_o - C_w)$ g/l	$(n - n_o)$ $\times 10^4$	n	Δ °C	O Os/kg	S g-mol/l	η/η_o	η/ρ cS	ϕ rhe	γ mmho/cm	T g-mol/l
24.00	1.2163	1.2184	291.9	2.837	924.4	73.9	378	1.3708				1.242	1.023	80.37	185.0	3.21
26.00	1.2382	1.2404	321.9	3.128	916.2	82.0	415	1.3745				1.284	1.039	77.75	197.0	3.60
28.00	1.2608	1.2630	353.0	3.430	907.8	90.5	453	1.3783				1.333	1.059	74.88	207.0	4.03
30.00	1.2842	1.2864	385.2	3.744	889.9	99.3	492	1.3822				1.392	1.086	71.70	216.0	4.50
32.00	1.3083	1.3106	418.7	4.068	889.7	108.6	532	1.3862				1.462	1.120	68.26		
34.00	1.3333	1.3357	453.3	4.405	880.0	118.2	573	1.3903				1.543	1.160	64.68		
36.00	1.3592	1.3616	489.3	4.755	869.9	128.3	616	1.3946				1.636	1.206	61.01		
38.00	1.3860	1.3885	526.7	5.118	859.3	138.9	660	1.3990				1.742	1.259	57.30		
40.00	1.4138	1.4163	565.5	5.495	848.3	150.0	705	1.4035				1.862	1.320	53.60		

70 SODIUM CARBONATE, Na$_2$CO$_3$ · 10H$_2$O

MOLECULAR WEIGHT = 106.00   FORMULA WEIGHT, HYDRATE = 286.16
RELATIVE SPECIFIC REFRACTIVITY = 0.619

0.00 % by wt. data are the same for all compounds.
For Values of 0.00 wt. % solutions see Table 1, Acetic Acid.

A% by wt.	H% by wt.	ρ D_4^{20}	D_{20}^{20}	C_s g/l	M g-mol/l	C_w g/l	$(C_o - C_w)$ g/l	$(n - n_o)$ $\times 10^4$	n	Δ °C	O Os/kg	S g-mol/l	η/η_o	η/ρ cS	ϕ rhe	γ mmho/cm	T g-mol/l
0.50	1.35	1.0034	1.0052	5.0	0.047	998.4	−0.2	11	1.3341	0.222	0.119	0.063	1.023	1.021	97.58	7.0	0.072
1.00	2.70	1.0086	1.0104	10.1	0.095	998.5	−0.3	23	1.3352	0.425	0.229	0.123	1.047	1.040	95.32	13.1	0.139
1.50	4.05	1.0138	1.0156	15.2	0.143	998.6	−0.4	34	1.3364	0.598	0.322	0.174	1.073	1.060	93.03	18.4	0.201
2.00	5.40	1.0190	1.0208	20.4	0.192	998.6	−0.4	45	1.3375	0.751	0.404	0.219	1.100	1.082	90.73	23.3	0.260
2.50	6.75	1.0242	1.0260	25.6	0.242	998.6	−0.4	56	1.3386	0.915	0.492	0.267	1.128	1.104	88.46	27.8	0.316
3.00	8.10	1.0294	1.0312	30.9	0.291	998.5	−0.3	67	1.3397	1.082	0.582	0.316	1.157	1.127	86.23	32.0	0.369
3.50	9.45	1.0346	1.0364	36.2	0.342	998.4	−0.2	78	1.3408	1.250	0.672	0.365	1.188	1.150	84.01	36.0	0.420
4.00	10.80	1.0398	1.0416	41.6	0.392	998.2	0.0	89	1.3419	1.421	0.764	0.416	1.220	1.176	81.81	39.8	0.468
4.50	12.15	1.0450	1.0468	47.0	0.444	998.0	0.3	100	1.3430	1.593	0.857	0.466	1.253	1.202	79.62	43.5	0.515
5.00	13.50	1.0502	1.0521	52.5	0.495	997.7	0.5	110	1.3440	1.768	0.951	0.517	1.289	1.230	77.42	47.0	0.562
5.50	14.85	1.0554	1.0573	58.0	0.548	997.3	0.9	121	1.3451	1.945	1.046	0.569	1.326	1.259	75.26	50.4	0.608
6.00	16.20	1.0606	1.0625	63.6	0.600	997.0	1.3	132	1.3462	2.125	1.142	0.621	1.364	1.289	73.16	53.6	0.652
6.50	17.55	1.0658	1.0677	69.3	0.654	996.5	1.7	142	1.3472				1.404	1.320	71.09	56.7	0.694
7.00	18.90	1.0711	1.0730	75.0	0.707	996.1	2.1	153	1.3483				1.445	1.352	69.05	59.7	0.735
7.50	20.25	1.0763	1.0782	80.7	0.762	995.6	2.6	164	1.3494				1.489	1.386	67.04	62.4	0.774
8.00	21.60	1.0816	1.0835	86.5	0.816	995.1	3.1	174	1.3504				1.535	1.422	65.03	65.1	0.812
8.50	22.95	1.0869	1.0888	92.4	0.872	994.6	3.7	185	1.3515				1.583	1.460	63.03	67.6	0.848
9.00	24.30	1.0922	1.0942	98.3	0.927	993.9	4.3	196	1.3525				1.635	1.500	61.03	70.0	0.883
9.50	25.65	1.0975	1.0995	104.3	0.984	993.3	5.0	206	1.3536				1.691	1.543	59.03	72.2	0.916
10.00	27.00	1.1029	1.1048	110.3	1.040	992.6	5.6	217	1.3547				1.750	1.590	57.03	74.4	0.948
11.00	29.70	1.1136	1.1156	122.5	1.156	991.1	7.1	230	1.3568				1.880	1.691	53.09	78.3	1.00
12.00	32.40	1.1244	1.1264	134.9	1.273	989.5	8.8	259	1.3589				2.024	1.803	49.32	81.7	1.05
13.00	35.10	1.1353	1.1373	147.6	1.392	987.7	10.5	280	1.3610				2.182	1.926	45.74	84.6	1.10
14.00	37.79	1.1463	1.1483	160.5	1.514	985.8	12.4	301	1.3631				2.356	2.059	42.36	86.9	1.14
15.00	40.49	1.1574	1.1595	173.6	1.638	983.8	14.4	322	1.3652				2.546	2.204	39.20	88.6	1.16

71 SODIUM CHLORIDE, NaCl

MOLECULAR WEIGHT = 58.44
RELATIVE SPECIFIC REFRACTIVITY = 0.794

0.00 % by wt. data are the same for all compounds.
For Values of 0.00 wt. % solutions see Table 1, Acetic Acid.

A% by wt.	ρ D_4^{20}	D_{20}^{20}	C_s g/l	M g-mol/l	C_w g/l	$(C_o - C_w)$ g/l	$(n - n_o)$ $\times 10^4$	n	Δ °C	O Os/kg	S g-mol/l	η/η_o	η/ρ cS	ϕ rhe	γ mmho/cm	T g-mol/l
0.10	0.9989	1.0007	1.0	0.017	997.9	0.3	2	1.3332	0.062	0.033	0.017	1.002	1.005	99.61	1.7	0.017
0.20	0.9997	1.0014	2.0	0.034	997.7	0.6	4	1.3333	0.121	0.065	0.034	1.004	1.006	99.43	3.3	0.034
0.30	1.0004	1.0021	3.0	0.051	997.4	0.9	5	1.3335	0.181	0.097	0.051	1.006	1.007	99.24	5.0	0.051
0.40	1.0011	1.0028	4.0	0.069	997.1	1.2	7	1.3337	0.240	0.129	0.069	1.007	1.008	99.06	6.6	0.069
0.50	1.0018	1.0036	5.0	0.086	996.8	1.5	9	1.3339	0.299	0.161	0.086	1.009	1.009	98.89	8.2	0.086
0.60	1.0025	1.0043	6.0	0.103	996.5	1.8	11	1.3340	0.358	0.192	0.103	1.011	1.011	98.71	9.8	0.103
0.70	1.0032	1.0050	7.0	0.120	996.2	2.1	12	1.3342	0.417	0.224	0.120	1.013	1.012	98.54	11.4	0.120
0.80	1.0039	1.0057	8.0	0.137	995.9	2.4	14	1.3344	0.475	0.256	0.137	1.015	1.013	98.37	12.9	0.137
0.90	1.0046	1.0064	9.0	0.155	995.6	2.7	16	1.3346	0.534	0.287	0.155	1.016	1.014	98.20	14.4	0.155
1.00	1.0053	1.0071	10.1	0.172	995.3	3.0	18	1.3347	0.593	0.319	0.172	1.018	1.015	98.04	16.0	0.172
1.10	1.0060	1.0078	11.1	0.189	995.0	3.3	19	1.3349	0.652	0.351	0.189	1.020	1.016	97.87	17.5	0.189
1.20	1.0068	1.0085	12.1	0.207	994.7	3.6	21	1.3351	0.711	0.382	0.207	1.021	1.017	97.72	18.9	0.207
1.30	1.0075	1.0093	13.1	0.224	994.4	3.9	23	1.3353	0.770	0.414	0.224	1.023	1.017	97.56	20.4	0.224
1.40	1.0082	1.0100	14.1	0.241	994.1	4.2	25	1.3354	0.829	0.446	0.241	1.025	1.018	97.41	21.8	0.241
1.50	1.0089	1.0107	15.1	0.259	993.8	4.5	26	1.3356	0.888	0.478	0.259	1.026	1.019	97.26	23.2	0.259
1.60	1.0096	1.0114	16.2	0.276	993.5	4.8	28	1.3358	0.948	0.509	0.276	1.028	1.020	97.11	24.6	0.276
1.70	1.0103	1.0121	17.2	0.294	993.1	5.1	30	1.3360	1.007	0.541	0.294	1.029	1.021	96.96	26.0	0.294
1.80	1.0110	1.0128	18.2	0.311	992.8	5.4	32	1.3362	1.067	0.573	0.311	1.031	1.022	96.82	27.4	0.311
1.90	1.0117	1.0135	19.2	0.329	992.5	5.7	33	1.3363	1.126	0.605	0.329	1.032	1.022	96.67	28.8	0.329
2.00	1.0125	1.0143	20.2	0.346	992.2	6.0	35	1.3365	1.186	0.638	0.346	1.034	1.023	96.52	30.2	0.346
2.10	1.0132	1.0150	21.3	0.364	991.9	6.3	37	1.3367	1.246	0.670	0.364	1.036	1.024	96.37	31.6	0.364
2.20	1.0139	1.0157	22.3	0.382	991.6	6.6	39	1.3369	1.306	0.702	0.382	1.037	1.025	96.22	33.0	0.382
2.30	1.0146	1.0164	23.3	0.399	991.3	7.0	40	1.3370	1.366	0.734	0.399	1.039	1.026	96.07	34.3	0.399
2.40	1.0153	1.0171	24.4	0.418	991.0	7.3	42	1.3372	1.426	0.767	0.418	1.040	1.027	95.93	35.7	0.418
2.50	1.0160	1.0178	25.4	0.435	990.6	7.6	44	1.3374	1.486	0.799	0.435	1.042	1.028	95.78	37.1	0.435
2.60	1.0168	1.0185	26.4	0.452	990.3	7.9	46	1.3376	1.547	0.832	0.452	1.044	1.028	95.64	38.5	0.452
2.70	1.0175	1.0193	27.5	0.470	990.0	8.2	47	1.3377	1.607	0.864	0.470	1.045	1.029	95.49	39.9	0.470
2.80	1.0182	1.0200	28.5	0.488	989.7	8.6	49	1.3379	1.668	0.897	0.488	1.047	1.030	95.35	41.2	0.488
2.90	1.0189	1.0207	29.5	0.505	989.4	8.9	51	1.3381	1.729	0.930	0.505	1.048	1.031	95.21	42.6	0.505
3.00	1.0196	1.0214	30.6	0.523	989.0	9.2	53	1.3383	1.790	0.962	0.523	1.050	1.032	95.06	44.0	0.523
3.10	1.0203	1.0221	31.6	0.541	988.7	9.5	54	1.3384	1.851	0.995	0.541	1.051	1.033	94.92	45.3	0.541
3.20	1.0211	1.0229	32.7	0.559	988.4	9.9	56	1.3386	1.913	1.028	0.559	1.053	1.033	94.78	46.7	0.559

A % by wt.	ρ D_4^{20}	D_{20}^{20}	C, g/l	M g-mol/l	C_w g/l	$(C_o - C_w)$ g/l	$(n - n_o)$ $\times 10^4$	n	Δ °C	O Os/kg	S g-mol/l	η/η_o	η/ρ cS	ϕ rhe	γ mmho/cm	T g-mol/l
3.30	1.0218	1.0236	33.7	0.577	988.0	10.2	58	1.3388	1.974	1.061	0.577	1.055	1.034	94.64	48.0	0.577
3.40	1.0225	1.0243	34.8	0.595	987.7	10.5	60	1.3390	2.036	1.094	0.595	1.056	1.035	94.50	49.4	0.595
3.50	1.0232	1.0250	35.8	0.613	987.4	10.8	61	1.3391	2.098	1.128	0.613	1.058	1.036	94.35	50.7	0.613
3.60	1.0239	1.0257	36.9	0.631	987.1	11.2	63	1.3393	2.160	1.161	0.631	1.059	1.037	94.21	52.0	0.631
3.70	1.0246	1.0265	37.9	0.649	986.7	11.5	65	1.3395	2.222	1.194	0.649	1.061	1.037	94.07	53.3	0.649
3.80	1.0254	1.0272	39.0	0.667	986.4	11.8	67	1.3397	2.284	1.228	0.667	1.063	1.038	93.93	54.7	0.667
3.90	1.0261	1.0279	40.0	0.685	986.1	12.2	68	1.3398	2.347	1.262	0.685	1.064	1.039	93.78	56.0	0.685
4.00	1.0268	1.0286	41.1	0.703	985.7	12.5	70	1.3400	2.409	1.295	0.703	1.066	1.040	93.64	57.3	0.703
4.10	1.0275	1.0293	42.1	0.721	985.4	12.8	72	1.3402	2.472	1.329	0.721	1.067	1.041	93.49	58.6	0.721
4.20	1.0282	1.0301	43.2	0.739	985.1	13.2	74	1.3404	2.535	1.363	0.739	1.069	1.042	93.35	59.9	0.739
4.30	1.0290	1.0308	44.2	0.757	984.7	13.5	75	1.3405	2.598	1.397	0.757	1.071	1.043	93.20	61.2	0.757
4.40	1.0297	1.0315	45.3	0.775	984.4	13.9	77	1.3407	2.662	1.431	0.775	1.072	1.044	93.06	62.5	0.775
4.50	1.0304	1.0322	46.4	0.794	984.0	14.2	79	1.3409	2.725	1.465	0.794	1.074	1.045	92.91	63.8	0.794
4.60	1.0311	1.0330	47.4	0.812	983.7	14.5	81	1.3411	2.789	1.500	0.812	1.076	1.046	92.76	65.1	0.812
4.70	1.0318	1.0337	48.5	0.830	983.4	14.9	82	1.3412	2.853	1.534	0.830	1.078	1.046	92.61	66.3	0.830
4.80	1.0326	1.0344	49.6	0.848	983.0	15.2	84	1.3414	2.917	1.569	0.848	1.079	1.047	92.46	67.6	0.848
4.90	1.0333	1.0351	50.6	0.866	982.7	15.6	86	1.3416	2.982	1.603	0.866	1.081	1.048	92.31	68.9	0.866
5.00	1.0340	1.0358	51.7	0.885	982.3	15.9	88	1.3418	3.046	1.638	0.885	1.083	1.049	92.15	70.1	0.885
5.20	1.0355	1.0373	53.8	0.921	981.6	16.6	91	1.3421	3.176	1.708	0.921	1.087	1.052	91.84	72.6	0.921
5.40	1.0369	1.0387	56.0	0.958	980.9	17.3	95	1.3425	3.307	1.778	0.958	1.090	1.054	91.53	75.1	0.958
5.60	1.0384	1.0402	58.1	0.995	980.2	18.0	98	1.3428	3.438	1.848	0.995	1.094	1.056	91.22	77.5	0.995
5.80	1.0398	1.0417	60.3	1.032	979.5	18.7	102	1.3432	3.570	1.919	1.032	1.098	1.058	90.90	80.0	1.032
6.00	1.0413	1.0431	62.5	1.069	978.8	19.4	105	1.3435	3.703	1.991	1.069	1.102	1.060	90.59	82.4	1.069
6.20	1.0427	1.0446	64.6	1.106	978.1	20.2	109	1.3439	3.837	2.063	1.106	1.106	1.062	90.27	84.7	1.106
6.40	1.0442	1.0460	66.8	1.144	977.4	20.9	112	1.3442	3.972	2.135	1.144	1.109	1.065	89.95	87.1	1.144
6.60	1.0456	1.0475	69.0	1.181	976.6	21.6	116	1.3446	4.107	2.208	1.181	1.113	1.067	89.63	89.4	1.181
6.80	1.0471	1.0490	71.2	1.218	975.9	22.3	119	1.3449	4.244	2.282	1.218	1.117	1.069	89.31	91.8	1.218
7.00	1.0486	1.0504	73.4	1.256	975.2	23.1	123	1.3453	4.378	2.354	1.256	1.122	1.072	88.99	94.1	1.256
7.20	1.0500	1.0519	75.6	1.294	974.4	23.8	126	1.3456	4.516	2.428	1.294	1.126	1.074	88.66	96.4	1.294
7.40	1.0515	1.0533	77.8	1.331	973.7	24.6	130	1.3460	4.655	2.503	1.331	1.130	1.077	88.33	98.7	1.331
7.60	1.0530	1.0548	80.0	1.369	972.9	25.3	133	1.3463	4.795	2.578	1.369	1.134	1.079	88.00	101.	1.369
7.80	1.0544	1.0563	82.2	1.407	972.2	26.1	137	1.3467	4.937	2.654	1.407	1.138	1.082	87.67	103.	1.407
8.00	1.0559	1.0578	84.5	1.445	971.4	26.8	140	1.3470	5.079	2.731	1.445	1.143	1.084	87.33	105.	1.445
8.20	1.0574	1.0592	86.7	1.484	970.7	27.6	144	1.3474	5.222	2.808	1.484	1.147	1.087	86.99	107.	1.484
8.40	1.0588	1.0607	88.9	1.522	969.9	28.3	147	1.3477	5.367	2.885	1.522	1.152	1.090	86.65	109.	1.522
8.60	1.0603	1.0622	91.2	1.560	969.1	29.1	151	1.3481	5.512	2.964	1.560	1.156	1.093	86.30	112.	1.560
8.80	1.0618	1.0637	93.4	1.599	968.3	29.9	154	1.3484	5.659	3.043	1.599	1.161	1.096	85.95	114.	1.599
9.00	1.0633	1.0651	95.7	1.637	967.6	30.7	158	1.3488	5.807	3.122	1.637	1.166	1.099	85.60	116.	1.637
9.20	1.0647	1.0666	98.0	1.676	966.8	31.4	161	1.3491	5.956	3.202	1.676	1.171	1.102	85.25	118.	1.676
9.40	1.0662	1.0681	100.2	1.715	966.0	32.2	165	1.3495	6.106	3.283	1.715	1.176	1.105	84.89	120.	1.715
9.60	1.0677	1.0696	102.5	1.754	965.2	33.0	168	1.3498	6.258	3.364	1.754	1.181	1.108	84.53	122.	1.754
9.80	1.0692	1.0711	104.8	1.793	964.4	33.8	172	1.3502	6.410	3.446	1.793	1.186	1.111	84.16	124.	1.793
10.00	1.0707	1.0726	107.1	1.832	963.6	34.6	175	1.3505	6.564	3.529	1.832	1.191	1.115	83.80	126.	1.832
10.50	1.0744	1.0763	112.8	1.930	961.6	36.6	184	1.3514	6.954	3.739	1.930	1.204	1.123	82.86	131.	1.930
11.00	1.0781	1.0801	118.6	2.029	959.5	38.7	193	1.3523	7.353	3.953	2.029	1.218	1.132	81.90	136.	2.029
11.50	1.0819	1.0838	124.4	2.129	957.5	40.8	202	1.3532	7.760	4.172	2.129	1.233	1.142	80.94	140.	2.129
12.00	1.0857	1.0876	130.3	2.229	955.4	42.9	211	1.3541	8.176	4.396	2.229	1.248	1.152	79.95	145.	2.229
12.50	1.0894	1.0914	136.2	2.330	953.3	45.0	220	1.3549	8.602	4.625	2.330	1.264	1.162	78.96	149.	2.330
13.00	1.0932	1.0952	142.1	2.432	951.1	47.1	228	1.3558	9.038	4.859	2.432	1.280	1.173	77.97	154.	2.432
13.50	1.0970	1.0990	148.1	2.534	948.9	49.3	237	1.3567	9.484	5.099	2.534	1.297	1.184	76.97	158.	2.534
14.00	1.1008	1.1028	154.1	2.637	946.7	51.5	246	1.3576	9.940	5.344	2.637	1.314	1.196	75.97	163.	2.637
14.50	1.1047	1.1066	160.2	2.741	944.5	53.7	255	1.3585	10.408	5.596	2.741	1.331	1.207	74.97	167.	2.741
15.00	1.1085	1.1105	166.3	2.845	942.2	56.0	264	1.3594	10.888	5.854	2.845	1.349	1.219	73.98	171.	2.845
16.00	1.1162	1.1182	178.6	3.056	937.6	60.6	282	1.3612	11.885	6.390	3.056	1.385	1.243	72.07	179.	3.056
17.00	1.1240	1.1260	191.1	3.270	932.9	65.3	300	1.3630	12.935	6.954	3.270	1.421	1.267	70.21	186.	3.270
18.00	1.1319	1.1339	203.7	3.486	928.1	70.1	318	1.3648	14.044	7.550	3.486	1.460	1.293	68.34	193.	3.486
19.00	1.1398	1.1418	216.6	3.706	923.2	75.0	336	1.3666	15.216	8.181	3.706	1.504	1.322	66.36	199.	3.705
20.00	1.1478	1.1498	229.6	3.928	918.2	80.0	354	1.3684	16.458	8.849	3.928	1.554	1.357	64.22	204.	3.928
21.00	1.1558	1.1579	242.7	4.153	913.1	85.1	372	1.3702	17.776	9.557	4.153	1.611	1.396	61.95	209.	4.153
22.00	1.1640	1.1660	256.1	4.382	907.9	90.4	391	1.3721	19.176	10.310	4.382	1.673	1.441	59.64	213.	4.382
23.00	1.1721	1.1742	269.6	4.613	902.6	95.7	409	1.3739	20.667	11.111	4.613	1.742	1.489	57.29	217.	4.613
24.00	1.1804	1.1825	283.3	4.848	897.1	101.1	428	1.3757			4.848	1.817	1.542	54.93	220.	4.848
25.00	1.1887	1.1909	297.2	5.085	891.6	106.7	446	1.3776			5.085	1.898	1.600	52.58	222.	5.085
26.00	1.1972	1.1993	311.3	5.326	885.9	112.3	465	1.3795			5.326	1.986	1.662	50.26	225.	5.326

72 SODIUM CITRATE, Na₃C₆H₅O₇ · 2H₂O

MOLECULAR WEIGHT = 258.068   FORMULA WEIGHT, HYDRATE = 294.10

A% by wt.	H% by wt.	ρ D_4^{20}	D_{20}^{20}	C_s g/l	M g-mol/l	C_w g/l	$(C_o - C_w)$ g/l	$(n - n_o)$ ×10⁴	n	Δ °C	O Os/kg	S g-mol/l	η/η_o	η/ρ cS	ϕ rhe	γ mmho/cm	T g-mol/l
1.00	1.14	1.0049	1.0067	10.0	0.039	994.9	3.3	18.	1.3348	0.197	0.106	0.056	1.041	1.038	95.87	7.4	0.075
2.00	2.28	1.0120	1.0138	20.2	0.078	991.8	6.4	36.	1.3366	0.389	0.209	0.112	1.079	1.068	92.52	12.8	0.136
3.00	3.42	1.0186	1.0204	30.6	0.118	988.0	10.2	53.	1.3383	0.586	0.315	0.170	1.120	1.102	89.09	17.9	0.195
4.00	4.56	1.0260	1.0278	41.0	0.159	985.0	13.2	71.	1.3401	0.787	0.423	0.229	1.164	1.137	85.72	22.1	0.246
5.00	5.70	1.0331	1.0350	51.7	0.200	981.4	16.8	89.	1.3419	0.974	0.523	0.284	1.208	1.172	82.59	26.2	0.296
6.00	6.84	1.0405	1.0424	62.4	0.242	978.1	20.1	107.	1.3437	1.171	0.630	0.342	1.260	1.213	79.23	29.9	0.343
7.00	7.98	1.0482	1.0501	73.4	0.284	974.8	23.4	125.	1.3455	1.361	0.732	0.398	1.311	1.253	76.14	33.2	0.384
8.00	9.12	1.0557	1.0576	84.5	0.327	971.2	27.0	143.	1.3473	1.566	0.842	0.458	1.368	1.290	71.98	36.5	0.426
9.00	10.26	1.0632	1.0651	95.7	0.371	967.5	30.7	161.	1.3491	1.768	0.951	0.517	1.424	1.342	70.09	39.4	0.463
10.00	11.40	1.0708	1.0727	107.1	0.415	963.7	34.5	179.	1.3509	1.957	1.052	0.572	1.496	1.400	66.71	42.1	0.497
12.00	13.68	1.0861	1.0881	130.3	0.505	955.8	42.4	216.	1.3546	2.376	1.277	0.693	1.646	1.519	60.61	48.9	0.588
14.00	15.95	1.1019	1.1039	154.3	0.598	947.6	50.6	253.	1.3583	2.819	1.515	0.820	1.828	1.662	54.60	50.4	0.608
16.00	18.23	1.1173	1.1193	178.8	0.693	938.5	59.7	288.	1.3618	3.274	1.760	0.949	2.041	1.830	48.91	53.5	0.650
18.00	20.51	1.1327	1.1347	203.9	0.790	928.8	69.4	326.	1.3656	3.815	2.051	1.100	2.285	2.021	43.68	55.7	0.680
20.00	22.79	1.1492	1.1513	229.8	0.891	919.4	78.8	363.	1.3693	4.393	2.362	1.260	2.591	2.259	38.52	57.1	0.700
24.00	27.35	1.1813	1.1834	283.5	1.099	897.8	100.4	437.	1.3767				3.402	2.886	29.33	57.7	0.708
28.00	31.91	1.2151	1.2173	340.2	1.318	874.9	123.3	515.	1.3845				4.577	3.774	21.81	55.9	0.683
32.00	36.47	1.2487	1.2510	399.6	1.548	849.1	149.1	593.	1.3923				6.528	5.238	15.29	51.0	0.616
36.00	41.03	1.2843	1.2866	462.3	1.792	822.0	176.2	671.	1.4001				9.768	7.621	10.22	44.5	0.529

73 SODIUM DIATRIZOATE, (CH₃CONH)₂C₆I₃COONa, HYPAQUE®

MOLECULAR WEIGHT = 635.92
RELATIVE SPECIFIC REFRACTIVITY = 0.823

0.00 % by wt. data are the same for all compounds.
For Values of 0.00 wt. % solutions see Table 1, Acetic Acid.

A% by wt.	ρ D_4^{20}	D_{20}^{20}	C_s g/l	M g-mol/l	C_w g/l	$(C_o - C_w)$ g/l	$(n - n_o)$ ×10⁴	n	Δ °C	O Os/kg	S g-mol/l	η/η_o	η/ρ cS	ϕ rhe	γ mmho/cm	T g-mol/l
1.00	1.0041	1.0059	10.0	0.016	994.1	4.1	16	1.3346	0.058	0.031	0.016	1.021	1.019	97.75	0.8	0.008
2.00	1.0102	1.0120	20.2	0.032	990.0	8.2	31	1.3361	0.110	0.059	0.031	1.041	1.033	95.87	1.6	0.016
3.00	1.0164	1.0182	30.5	0.048	985.9	12.3	48	1.3377	0.164	0.088	0.047	1.061	1.046	94.05	2.4	0.024
4.00	1.0227	1.0245	40.9	0.064	981.8	16.4	64	1.3394	0.220	0.118	0.063	1.082	1.060	92.26	3.1	0.031
5.00	1.0292	1.0310	51.5	0.081	977.7	20.5	80	1.3410	0.276	0.148	0.079	1.104	1.075	90.40	3.8	0.038
6.00	1.0357	1.0376	62.1	0.098	973.6	24.6	97	1.3427	0.332	0.178	0.095	1.128	1.092	88.44	4.5	0.045
7.00	1.0424	1.0443	73.0	0.115	969.5	28.8	114	1.3444	0.389	0.209	0.112	1.155	1.110	86.42	5.2	0.052
8.00	1.0493	1.0511	83.9	0.132	965.3	32.9	131	1.3461	0.446	0.240	0.129	1.183	1.130	84.38	5.9	0.059
9.00	1.0562	1.0581	95.1	0.150	961.1	37.1	148	1.3478	0.504	0.271	0.146	1.212	1.150	82.32	6.5	0.066
10.00	1.0632	1.0651	106.3	0.167	956.9	41.3	165	1.3495	0.562	0.302	0.163	1.243	1.171	80.29	7.2	0.073

®Registered trademark, sodium 3.5-diacetamido-2,4,6 tri-iodo benzoate.

73 SODIUM DIATRIZOATE, (CH₃CONH)₂C₆I₃COONa, HYPAQUE®—(Continued)

A% by wt.	ρ D₄²⁰	D₂₀²⁰	C_s g/l	M g-mol/l	C_w g/l	(C₀−C_w) g/l	(n−n₀) ×10⁴	n	Δ °C	O Os/kg	S g-mol/l	η/η₀	η/ρ cS	φ rhe	γ mmho/cm	T g-mol/l
12.00	1.0776	1.0796	129.3	0.203	948.3	49.9	201	1.3530	0.683	0.367	0.199	1.305	1.213	76.50	8.4	0.086
14.00	1.0925	1.0944	152.9	0.240	939.5	58.7	237	1.3567	0.808	0.434	0.235	1.375	1.261	72.60	9.6	0.099
16.00	1.1077	1.1097	177.2	0.279	930.5	67.7	274	1.3603	0.936	0.503	0.273	1.470	1.329	67.91	10.6	0.110
18.00	1.1234	1.1254	202.2	0.318	921.2	77.0	313	1.3643	1.067	0.574	0.312	1.592	1.420	62.70	11.5	0.120
20.00	1.1395	1.1415	227.9	0.358	911.6	86.6	353	1.3683	1.205	0.648	0.352	1.724	1.516	57.89	12.2	0.129
22.00	1.1560	1.1581	254.3	0.400	901.7	96.5	394	1.3724	1.351	0.726	0.395	1.847	1.601	54.03	13.0	0.138
24.00	1.1730	1.1751	281.5	0.443	891.5	106.7	436	1.3766	1.506	0.810	0.441	1.965	1.678	50.79	13.7	0.146
26.00	1.1906	1.1927	309.5	0.487	881.0	117.2	480	1.3810	1.671	0.898	0.489	2.094	1.762	47.66	14.4	0.153
28.00	1.2086	1.2108	338.4	0.532	870.2	128.0	525	1.3854	1.846	0.993	0.540	2.254	1.869	44.27	15.0	0.160
30.00	1.2273	1.2295	368.2	0.579	859.1	139.1	571	1.3901	2.033	1.093	0.594	2.469	2.016	40.42	15.5	0.167
32.00	1.2467	1.2489	399.0	0.627	847.8	150.5	619	1.3949	2.232	1.200	0.652	2.742	2.204	36.39	16.2	0.174
34.00	1.2670	1.2692	430.8	0.677	836.2	162.0	669	1.3999	2.444	1.314	0.713	3.066	2.425	32.55	16.8	0.182
36.00	1.2882	1.2904	463.7	0.729	824.4	173.8	722	1.4052	2.672	1.437	0.778	3.449	2.683	28.93	17.2	0.187
38.00	1.3105	1.3128	498.0	0.783	812.5	185.8	778	1.4108	2.917	1.568	0.848	3.902	2.984	25.58	17.2	0.187
40.00	1.3286	1.3310	531.5	0.836	797.2	201.0	834	1.4164				4.400	3.318	22.68	16.6	0.180

®Registered trademark, sodium 3,5-diacetamido-2,4,6 tri-iodo benzoate.

74 SODIUM DICHROMATE, Na₂Cr₂O₇ · 2H₂O

MOLECULAR WEIGHT = 261.97   FORMULA WEIGHT, HYDRATE = 298.00
RELATIVE SPECIFIC REFRACTIVITY = 0.859

0.00 % by wt. data are the same for all compounds.
For Values of 0.00 wt. % solutions see Table 1, Acetic Acid.

A% by wt.	H% by wt.	ρ D₄²⁰	D₂₀²⁰	C_s g/l	M g-mol/l	C_w g/l	(C₀−C_w) g/l	(n−n₀) ×10⁴	C_s n	Δ °C	O Os/kg	S g-mol/l	η/η₀	η/ρ cS	φ rhe	γ mmho/cm	T g-mol/l
0.50	0.57	1.0019	1.0037	5.0	0.019	996.9	1.4	10	1.3340	0.103	0.056	0.029	1.005	1.005	99.33	3.5	0.035
1.00	1.14	1.0056	1.0073	10.1	0.038	995.5	2.7	21	1.3351	0.204	0.110	0.058	1.009	1.005	98.91	6.9	0.070
2.00	2.28	1.0130	1.0148	20.3	0.077	992.7	5.5	42	1.3371	0.396	0.213	0.114	1.016	1.005	98.23	13.1	0.139
3.00	3.41	1.0204	1.0222	30.6	0.117	989.8	8.4	63	1.3392	0.581	0.312	0.169	1.024	1.005	97.47	19.1	0.209
4.00	4.55	1.0280	1.0298	41.1	0.157	986.9	11.4	84	1.3414	0.762	0.410	0.222	1.033	1.007	96.64	24.9	0.280
5.00	5.69	1.0356	1.0374	51.8	0.198	983.8	14.4	105	1.3435	0.946	0.509	0.276	1.042	1.008	95.78	30.5	0.350
6.00	6.83	1.0433	1.0452	62.6	0.239	980.7	17.5	127	1.3457	1.135	0.610	0.331	1.052	1.010	94.91	36.0	0.419
7.00	7.96	1.0511	1.0530	73.6	0.281	977.5	20.7	149	1.3479	1.325	0.712	0.387	1.061	1.012	94.02	41.2	0.486
8.00	9.10	1.0590	1.0609	84.7	0.323	974.3	23.9	171	1.3501	1.519	0.817	0.444	1.072	1.014	93.10	46.3	0.553
9.00	10.24	1.0670	1.0689	96.0	0.367	971.0	27.2	193	1.3523	1.720	0.924	0.503	1.083	1.017	92.15	51.5	0.622
10.00	11.38	1.0751	1.0770	107.5	0.410	967.6	30.6	216	1.3546	1.929	1.037	0.564	1.095	1.021	91.14	56.7	0.694
12.00	13.65	1.0916	1.0936	131.0	0.500	960.6	37.6	262	1.3592	2.388	1.284	0.697	1.121	1.029	89.01	67.8	0.851
14.00	15.93	1.1086	1.1105	155.2	0.592	953.4	44.8	310	1.3640	2.876	1.546	0.836	1.151	1.040	86.72	78.9	1.02
16.00	18.20	1.1260	1.1280	180.2	0.688	945.9	52.4	358	1.3688	3.356	1.804	0.972	1.184	1.054	84.28	89.2	1.18
18.00	20.48	1.1439	1.1460	205.9	0.786	938.0	60.2	408	1.3738	3.830	2.059	1.104	1.221	1.070	81.71	98.6	1.33
20.00	22.75	1.1624	1.1644	232.5	0.887	929.9	68.3	460	1.3790	4.301	2.312	1.234	1.264	1.090	78.96	107.	1.48
24.00	27.30	1.2009	1.2031	288.2	1.100	912.7	85.5	567	1.3897				1.364	1.138	73.18	121.	1.73
28.00	31.85	1.2418	1.2440	347.7	1.327	894.1	104.1	681	1.4011				1.494	1.206	66.79	131.	1.93
32.00	36.40	1.2851	1.2874	411.2	1.570	873.9	124.4	801	1.4131				1.677	1.308	59.51	140.	2.10
36.00	40.95	1.3302	1.3326	478.9	1.828	851.4	146.9	925	1.4255				1.914	1.442	52.13	145.	2.22
40.00	45.50	1.3786	1.3810	551.4	2.105	827.1	171.1	1058	1.4388				2.246	1.632	44.43	148.	2.29
44.00	50.05	1.4304	1.4330	629.4	2.403	801.0	197.2	1200	1.4530				2.646	1.853	37.72	147.	2.26
48.00	54.60	1.4860	1.4886	713.3	2.723	772.7	225.5	1353	1.4683				3.225	2.174	30.95	141.	2.12
52.00	59.15	1.5454	1.5482	803.6	3.068	741.8	256.4						4.297	2.786	23.23	131.	1.94
56.00	63.70	1.6088	1.6117	900.9	3.439	707.9	290.3						5.933	3.695	16.82	117.	1.67
60.00	68.25	1.6763	1.6793	1005.8	3.839	670.5	327.7						8.221	4.914	12.14	93.9	1.25

75 SODIUM FERROCYANIDE, Na₄Fe(CN)₆ · 10H₂O

MOLECULAR WEIGHT = 303.91   FORMULA WEIGHT, HYDRATE = 484.07
RELATIVE SPECIFIC REFRACTIVITY = 0.991

0.00 % by wt. data are the same for all compounds.
For Values of 0.00 wt. % solutions see Table 1, Acetic Acid.

A% by wt.	H% by wt.	ρ D₄²⁰	D₂₀²⁰	C_s g/l	M g-mol/l	C_w g/l	(C₀−C_w) g/l	(n−n₀) ×10⁴	n	Δ °C	O Os/kg	S g-mol/l	η/η₀	η/ρ cS	φ rhe	γ mmho/cm	T g-mol/l
0.50	0.80	1.0018	1.0036	5.0	0.016	996.6	1.4	13	1.3342	0.091	0.049	0.026	1.016	1.016	98.23	5.0	0.050
1.00	1.59	1.0055	1.0073	10.1	0.033	995.4	2.8	25	1.3355	0.192	0.103	0.055	1.032	1.028	96.73	9.8	0.102
1.50	2.39	1.0092	1.0110	15.1	0.050	994.0	4.2	38	1.3368	0.302	0.162	0.087	1.047	1.040	95.28	14.3	0.153
2.00	3.19	1.0129	1.0147	20.3	0.067	992.6	5.6	51	1.3381	0.394	0.212	0.114	1.063	1.052	93.87	18.7	0.204
2.50	3.98	1.0166	1.0184	25.4	0.084	991.2	7.0	64	1.3394	0.486	0.261	0.141	1.079	1.063	92.50	22.8	0.254
3.00	4.78	1.0204	1.0222	30.6	0.101	989.8	8.5	77	1.3407	0.576	0.310	0.167	1.095	1.075	91.16	26.7	0.303
3.50	5.57	1.0239	1.0258	35.8	0.118	988.1	10.1	89	1.3419	0.666	0.358	0.194	1.111	1.087	89.83	30.5	0.350
4.00	6.37	1.0276	1.0294	41.1	0.135	986.5	11.7	102	1.3432	0.755	0.406	0.220	1.127	1.099	88.52	34.1	0.396
4.50	7.17	1.0313	1.0331	46.4	0.153	984.9	13.3	115	1.3445	0.842	0.453	0.245	1.144	1.112	87.22	37.7	0.441
5.00	7.96	1.0350	1.0368	51.8	0.170	983.3	15.0	128	1.3457	0.929	0.499	0.271	1.161	1.124	85.93	41.1	0.485
5.50	8.76	1.0387	1.0406	57.1	0.188	981.6	16.6	140	1.3470	1.014	0.545	0.296	1.179	1.138	84.63	44.5	0.529
6.00	9.56	1.0425	1.0443	62.5	0.206	979.9	18.3	153	1.3483	1.098	0.590	0.321	1.198	1.151	83.33	47.7	0.572
6.50	10.35	1.0462	1.0481	68.0	0.224	978.2	20.0	166	1.3496	1.181	0.635	0.345	1.217	1.165	82.02	51.0	0.616
7.00	11.15	1.0500	1.0519	73.5	0.242	976.5	21.7	179	1.3509	1.262	0.679	0.369	1.237	1.180	80.70	54.2	0.659
7.50	11.95	1.0539	1.0557	79.0	0.260	974.8	23.4	192	1.3522	1.342	0.722	0.393	1.257	1.195	79.38	57.3	0.702
8.00	12.74	1.0577	1.0596	84.6	0.278	973.1	25.1	205	1.3535	1.421	0.764	0.416	1.279	1.211	78.04	60.5	0.746
8.50	13.54	1.0616	1.0634	90.2	0.297	971.3	26.9	218	1.3548	1.498	0.806	0.438	1.301	1.228	76.73	63.6	0.790
9.00	14.34	1.0655	1.0673	95.9	0.316	969.6	28.7	232	1.3562	1.574	0.846	0.460	1.325	1.246	75.34	66.7	0.835
9.50	15.13	1.0694	1.0713	101.6	0.334	967.8	30.4	245	1.3575	1.648	0.886	0.482	1.349	1.264	73.98	69.7	0.879
10.00	15.93	1.0733	1.0752	107.3	0.353	966.0	32.2	258	1.3588	1.720	0.925	0.503	1.375	1.283	72.61	72.8	0.925
11.00	17.52	1.0813	1.0832	118.9	0.391	962.4	35.9	285	1.3615				1.429	1.324	69.85	78.8	1.02
12.00	19.11	1.0894	1.0913	130.7	0.430	958.6	39.6	312	1.3642				1.488	1.368	67.09	84.7	1.10
13.00	20.71	1.0976	1.0995	142.7	0.469	954.9	43.4	340	1.3670				1.551	1.416	64.34	90.0	1.19
14.00	22.30	1.1058	1.1078	154.8	0.509	951.0	47.2	367	1.3697				1.619	1.467	61.63	94.7	1.27
15.00	23.89	1.1142	1.1162	167.1	0.550	947.1	51.1	395	1.3725				1.692	1.522	58.98	98.1	1.32

76 SODIUM HYDROXIDE, NaOH

MOLECULAR WEIGHT = 40.01
RELATIVE SPECIFIC REFRACTIVITY = 0.692

A % by wt.	ρ D_4^{20}	D_{20}^{20}	C_s g/l	M g-mol/l	C_w g/l	$(C_o - C_w)$ g/l	$(n - n_o)$ × 10^4	n	Δ °C	O Os/kg	S g-mol/l	η/η_o	η/ρ cS	ϕ rhe	γ mmho/cm	T g-mol/l
0.50	1.0039	1.0057	5.0	0.125	998.9	− 0.6	14	1.3344	0.429	0.231	0.124	1.025	1.023	97.34	24.8	0.283
1.00	1.0095	1.0113	10.1	0.252	999.4	− 1.2	29	1.3358	0.860	0.462	0.251	1.052	1.044	94.87	48.6	0.584
1.50	1.0151	1.0169	15.2	0.381	999.9	− 1.7	43	1.3373	1.294	0.695	0.378	1.080	1.066	92.39	71.3	0.900
2.00	1.0207	1.0225	20.4	0.510	1000.3	− 2.0	56	1.3386	1.735	0.933	0.507	1.110	1.090	89.91	93.1	1.24
2.50	1.0262	1.0281	25.7	0.641	1000.6	− 2.4	70	1.3400	2.184	1.174	0.638	1.141	1.114	87.44	114.	1.61
3.00	1.0318	1.0336	31.0	0.774	1000.8	− 2.6	84	1.3414	2.642	1.420	0.770	1.174	1.140	84.98	134.	2.02
3.50	1.0373	1.0391	36.3	0.907	1001.0	− 2.8	97	1.3427	3.109	1.671	0.902	1.209	1.168	82.54	153.	2.45
4.00	1.0428	1.0446	41.7	1.043	1001.1	− 2.9	111	1.3441	3.587	1.929	1.037	1.246	1.197	80.11	171.	2.92
4.50	1.0483	1.0502	47.2	1.179	1001.1	− 2.9	124	1.3454	4.074	2.190	1.172	1.285	1.228	77.69	189.	3.42
5.00	1.0538	1.0557	52.7	1.317	1001.1	− 2.9	138	1.3467	4.569	2.457	1.308	1.326	1.261	75.26	206.	3.96
5.50	1.0593	1.0612	58.3	1.456	1001.0	− 2.8	151	1.3481	5.08	2.73	1.445	1.369	1.295	72.90	222.	4.54
6.00	1.0648	1.0667	63.9	1.597	1000.9	− 2.7	164	1.3494	5.60	3.01	1.587	1.413	1.330	70.62		
6.50	1.0703	1.0722	69.6	1.739	1000.7	− 2.5	177	1.3507	6.13	3.30	1.722	1.459	1.366	68.40		
7.00	1.0758	1.0777	75.3	1.882	1000.5	− 2.3	190	1.3520	6.69	3.60	1.864	1.507	1.404	66.21		
7.50	1.0813	1.0833	81.1	2.027	1000.2	− 2.0	203	1.3533	7.27	3.91	2.008	1.558	1.444	64.05		
8.00	1.0869	1.0888	86.9	2.173	999.9	− 1.7	216	1.3546	7.87	4.23	2.155	1.613	1.487	61.89		
8.50	1.0924	1.0943	92.9	2.321	999.5	− 1.3	229	1.3559	8.48	4.56	2.303	1.671	1.533	59.72		
9.00	1.0979	1.0998	98.8	2.470	999.1	− 0.9	242	1.3572	9.12	4.91	2.452	1.734	1.583	57.54		
9.50	1.1034	1.1054	104.8	2.620	998.6	− 0.3	255	1.3585	9.78	5.26	2.602	1.803	1.637	55.35		
10.00	1.1089	1.1109	110.9	2.772	998.0	0.2	268	1.3597	10.47	5.63	2.753	1.878	1.697	53.14		
11.00	1.1199	1.1219	123.2	3.079	996.7	1.5	293	1.3623	11.89	6.39	3.058	2.035	1.821	49.04		
12.00	1.1309	1.1329	135.7	3.392	995.2	3.0	318	1.3648	13.42	7.21	3.365	2.197	1.947	45.42		
13.00	1.1419	1.1440	148.5	3.710	993.5	4.8	343	1.3673	15.04	8.08	3.673	2.371	2.080	42.09		
14.00	1.1530	1.1550	161.4	4.034	991.6	6.7	367	1.3697	16.76	9.01	3.981	2.563	2.228	38.93		
15.00	1.1640	1.1661	174.6	4.364	989.4	8.8	392	1.3722				2.783	2.395	35.87		
16.00	1.1751	1.1771	188.0	4.699	987.0	11.2	416	1.3746				3.037	2.590	32.86		
17.00	1.1861	1.1882	201.6	5.040	984.5	13.8	440	1.3770				3.337	2.819	29.91		
18.00	1.1971	1.1993	215.5	5.386	981.7	16.6	463	1.3793				3.691	3.089	27.04		
19.00	1.2082	1.2103	229.6	5.737	978.6	19.6	487	1.3817				4.111	3.410	24.27		
20.00	1.2192	1.2214	243.8	6.094	975.4	22.9	510	1.3840				4.610	3.789	21.65		
22.00	1.2412	1.2434	273.1	6.825	968.1	30.1	555	1.3885				5.753	4.644	17.35		
24.00	1.2631	1.2653	303.1	7.576	959.9	38.3	599	1.3929				7.086	5.621	14.08		
26.00	1.2848	1.2871	334.0	8.349	950.8	47.5	641	1.3971				8.727	6.806	11.44		
28.00	1.3064	1.3087	365.8	9.142	940.6	57.6	682	1.4012				10.81	8.29	9.23		
30.00	1.3277	1.3301	398.3	9.956	929.4	68.8	721	1.4051				13.49	10.18	7.40		
32.00	1.3488	1.3512	431.6	10.788	917.2	81.0	758	1.4088				16.81	12.49	5.94		
34.00	1.3697	1.3721	465.7	11.639	904.0	94.3	793	1.4123				20.71	15.15	4.82		
36.00	1.3901	1.3926	500.5	12.508	889.7	108.5	826	1.4155				25.18	18.19	3.95		
38.00	1.4102	1.4127	535.9	13.394	874.3	123.9	856	1.4186				30.40	21.60	3.28		
40.00	1.4299	1.4324	571.9	14.295	857.9	140.3	885	1.4215				36.24	25.40	2.75		

77 SODIUM MOLYBDATE, Na₂MoO₄ · 2H₂O

MOLECULAR WEIGHT = 205.94   FORMULA WEIGHT, HYDRATE = 241.98
RELATIVE SPECIFIC REFRACTIVITY = 0.642

A % by wt.	H % by wt.	ρ D_4^{20}	D_{20}^{20}	C_s g/l	M g-mol/l	C_w g/l	$(C_o - C_w)$ g/l	$(n - n_o)$ × 10^4	n	Δ °C	O Os/kg	S g-mol/l	η/η_o	η/ρ cS	ϕ rhe	γ mmho/cm	T g-mol/l
0.50	0.59	1.0024	1.0042	5.0	0.024	997.4	0.8	8	1.3338	0.096	0.052	0.027	1.013	1.012	98.56	4.1	0.041
1.00	1.18	1.0067	1.0084	10.1	0.049	996.6	1.6	17	1.3347	0.203	0.109	0.058	1.025	1.020	97.37	7.8	0.080
1.50	1.76	1.0110	1.0128	15.2	0.074	995.8	2.4	25	1.3355	0.322	0.173	0.092	1.037	1.028	96.24	11.2	0.118
2.00	2.35	1.0153	1.0171	20.3	0.099	995.0	3.2	34	1.3364	0.440	0.237	0.127	1.049	1.035	95.14	14.4	0.154
2.50	2.94	1.0197	1.0215	25.5	0.124	994.2	4.0	43	1.3373	0.558	0.300	0.162	1.061	1.043	94.05	17.5	0.190
3.00	3.53	1.0242	1.0260	30.7	0.149	993.4	4.8	52	1.3382	0.670	0.360	0.195	1.073	1.050	92.98	20.5	0.226
3.50	4.11	1.0286	1.0305	36.0	0.175	992.6	5.6	61	1.3391	0.782	0.420	0.228	1.086	1.058	91.91	23.3	0.261
4.00	4.70	1.0332	1.0350	41.3	0.201	991.8	6.4	70	1.3400	0.900	0.484	0.263	1.098	1.065	90.85	26.1	0.296
4.50	5.29	1.0377	1.0395	46.7	0.227	991.0	7.2	79	1.3409	1.015	0.546	0.296	1.112	1.073	89.79	28.9	0.330
5.00	5.88	1.0423	1.0441	52.1	0.253	990.1	8.1	88	1.3418	1.126	0.606	0.329	1.125	1.082	88.71	31.6	0.364
5.50	6.46	1.0468	1.0487	57.6	0.280	989.3	9.0	97	1.3427	1.238	0.666	0.362	1.139	1.090	87.62	34.3	0.398
6.00	7.05	1.0514	1.0533	63.1	0.306	988.4	9.9	106	1.3436	1.350	0.726	0.395	1.153	1.099	86.53	36.9	0.431
6.50	7.64	1.0561	1.0579	68.6	0.333	987.4	10.8	115	1.3445				1.168	1.108	85.44	39.5	0.465
7.00	8.23	1.0607	1.0626	74.2	0.361	986.4	11.8	124	1.3454				1.183	1.118	84.35	42.1	0.497
7.50	8.81	1.0653	1.0672	79.9	0.388	985.4	12.8	133	1.3463				1.199	1.127	83.25	44.5	0.530
8.00	9.40	1.0700	1.0719	85.6	0.416	984.4	13.8	142	1.3472				1.215	1.138	82.16	46.9	0.562
8.50	9.99	1.0747	1.0766	91.3	0.444	983.3	14.9	151	1.3481				1.231	1.148	81.06	49.3	0.593
9.00	10.58	1.0794	1.0813	97.1	0.472	982.2	16.0	160	1.3490				1.248	1.159	79.97	51.6	0.624

78 SODIUM NITRATE, NaNO₃

MOLECULAR WEIGHT = 85.01
RELATIVE SPECIFIC REFRACTIVITY = 0.657

A % by wt.	ρ D_4^{20}	D_{20}^{20}	C_s g/l	M g-mol/l	C_w g/l	$(C_o - C_w)$ g/l	$(n - n_o)$ × 10^4	n	Δ °C	O Os/kg	S g-mol/l	η/η_o	η/ρ cS	ϕ rhe	γ mmho/cm	T g-mol/l
0.50	1.0016	1.0034	5.0	0.059	996.6	1.6	6	1.3336	0.204	0.110	0.058	1.002	1.003	99.55	5.4	0.055
1.00	1.0050	1.0067	10.0	0.118	994.9	3.3	11	1.3341	0.403	0.216	0.116	1.005	1.005	99.30	10.6	0.111
2.00	1.0117	1.0135	20.2	0.238	991.5	6.7	23	1.3353	0.793	0.426	0.231	1.010	1.000	98.81	20.4	0.225
3.00	1.0185	1.0203	30.6	0.359	988.0	10.3	34	1.3364	1.177	0.633	0.344	1.016	0.999	98.24	29.5	0.336
4.00	1.0254	1.0272	41.0	0.482	984.4	13.9	46	1.3375	1.561	0.839	0.456	1.023	0.999	97.60	38.0	0.445
5.00	1.0322	1.0341	51.6	0.607	980.6	17.6	57	1.3387	1.942	1.044	0.568	1.030	1.000	96.89	46.2	0.552
6.00	1.0392	1.0410	62.4	0.733	976.8	21.4	68	1.3398	2.323	1.249	0.678	1.038	1.001	96.14	54.2	0.659
7.00	1.0462	1.0480	73.2	0.861	972.9	25.3	79	1.3409	2.703	1.453	0.787	1.047	1.003	95.32	61.7	0.764

78 SODIUM NITRATE, NaNO₃—(Continued)

A% by wt.	ρ D_4^{20}	D_{20}^{20}	C_s g/l	M g-mol/l	C_w g/l	$(C_o - C_w)$ g/l	$(n - n_o)$ $\times 10^4$	n	Δ °C	O Os/kg	S g-mol/l	η/η_o	η/ρ cS	ϕ rhe	γ mmho/cm	T g-mol/l
8.00	1.0532	1.0550	84.3	0.991	968.9	29.3	91	1.3421	3.083	1.657	0.895	1.057	1.005	94.45	68.9	0.868
9.00	1.0603	1.0622	95.4	1.123	964.8	33.4	102	1.3432	3.462	1.862	1.002	1.067	1.009	93.51	75.8	0.971
10.00	1.0674	1.0693	106.7	1.256	960.7	37.6	113	1.3443	3.841	2.065	1.107	1.079	1.013	92.49	82.6	1.07
12.00	1.0819	1.0838	129.8	1.527	952.1	46.1	136	1.3466	4.599	2.472	1.316	1.105	1.024	90.28	95.3	1.27
14.00	1.0967	1.0986	153.5	1.806	943.2	55.1	159	1.3489	5.37	2.89	1.523	1.136	1.038	87.85	106.	1.46
18.00	1.1272	1.1292	202.9	2.387	924.3	73.9	206	1.3536	6.98	3.75	1.937	1.213	1.078	82.30	125.	1.82
20.00	1.1429	1.1449	228.6	2.689	914.3	83.9	230	1.3559	7.81	4.20	2.142	1.260	1.105	79.21	134.	1.99
24.00	1.1752	1.1772	282.0	3.318	893.1	105.1	277	1.3607	9.52	5.12	2.542	1.374	1.172	72.64	149.	2.30
28.00	1.2085	1.2106	338.4	3.981	870.1	128.1	325	1.3654	11.28	6.07	2.929	1.519	1.259	65.71	160.	2.56
30.00	1.2256	1.2278	367.7	4.325	858.0	140.3	348	1.3678				1.606	1.313	62.14	165.	2.68
34.00	1.2610	1.2632	428.7	5.043	832.2	166.0	397	1.3726				1.814	1.442	55.01	173.	2.87
40.00	1.3175	1.3198	527.0	6.199	790.5	207.7	472	1.3802				2.222	1.690	44.91	178.	3.00

79 SODIUM PHOSPHATE (TRIBASIC), Na₃PO₄ · 12H₂O

MOLECULAR WEIGHT = 163.96   FORMULA WEIGHT, HYDRATE = 380.14
RELATIVE SPECIFIC REFRACTIVITY = 0.555

0.00 % by wt. data are the same for all compounds.
For Values of 0.00 wt. % solutions see Table 1, Acetic Acid.

A% by wt.	H% by wt.	ρ D_4^{20}	D_{20}^{20}	C_s g/l	M g-mol/l	C_w g/l	$(C_o - C_w)$ g/l	$(n - n_o)$ $\times 10^4$	n	Δ °C	O Os/kg	S g-mol/l	η/η_o	η/ρ cS	ϕ rhe	γ mmho/cm	T g-mol/l
0.50	1.16	1.0042	1.0059	5.0	0.031	999.1	-0.9	13	1.3343	0.192	0.103	0.055	1.031	1.029	96.80	7.3	0.075
1.00	2.32	1.0100	1.0118	10.1	0.062	999.9	-1.7	26	1.3356	0.368	0.198	0.106	1.062	1.054	93.97	14.1	0.150
1.50	3.48	1.0158	1.0176	15.2	0.093	1000.6	-2.3	39	1.3369	0.527	0.283	0.153	1.092	1.078	91.36	18.8	0.206
2.00	4.64	1.0216	1.0234	20.4	0.125	1001.2	-2.9	51	1.3381	0.668	0.359	0.194	1.124	1.102	88.79	22.7	0.253
2.50	5.80	1.0275	1.0293	25.7	0.157	1001.8	-3.6	64	1.3394	0.790	0.425	0.230	1.159	1.130	86.13	26.7	0.302
3.00	6.96	1.0335	1.0353	31.0	0.189	1002.5	-4.2	76	1.3406				1.196	1.160	83.45	30.4	0.349
3.50	8.11	1.0395	1.0413	36.4	0.222	1003.1	-4.9	89	1.3419				1.236	1.191	80.76	34.0	0.394
4.00	9.27	1.0456	1.0474	41.8	0.255	1003.8	-5.5	102	1.3432				1.278	1.225	78.06	37.3	0.436
4.50	10.43	1.0517	1.0536	47.3	0.289	1004.4	-6.2	114	1.3444				1.324	1.261	75.39	40.5	0.477
5.00	11.59	1.0579	1.0598	52.9	0.323	1005.0	-6.8	127	1.3457				1.372	1.299	72.74	43.5	0.516
5.50	12.75	1.0642	1.0661	58.5	0.357	1005.7	-7.4	140	1.3470				1.423	1.340	70.13	46.4	0.554
6.00	13.91	1.0705	1.0724	64.2	0.392	1006.3	-8.0	152	1.3482				1.477	1.383	67.56	49.0	0.589
6.50	15.07	1.0768	1.0787	70.0	0.427	1006.8	-8.6	165	1.3495				1.535	1.428	65.03	51.5	0.622
7.00	16.23	1.0832	1.0851	75.8	0.462	1007.4	-9.1	177	1.3507				1.595	1.475	62.57	53.7	0.653
7.50	17.39	1.0896	1.0915	81.7	0.498	1007.9	-9.7	189	1.3519				1.659	1.525	60.16	55.8	0.681
8.00	18.55	1.0961	1.0980	87.7	0.535	1008.4	-10.1	202	1.3532				1.726	1.578	57.82	57.6	0.706

80 SODIUM PHOSPHATE, DIHYDROGEN (MONOBASIC), NaH₂PO₄ · 1H₂O

MOLECULAR WEIGHT = 119.97   FORMULA WEIGHT, HYDRATE = 137.99
RELATIVE SPECIFIC REFRACTIVITY = 0.637

0.00 % by wt. data are the same for all compounds.
For Values of 0.00 wt. % solutions see Table 1, Acetic Acid.

A% by wt.	H% by wt.	ρ D_4^{20}	D_{20}^{20}	C_s g/l	M g-mol/l	C_w g/l	$(C_o - C_w)$ g/l	$(n - n_o)$ $\times 10^4$	n	Δ °C	O Os/kg	S g-mol/l	η/η_o	η/ρ cS	ϕ rhe	γ mmho/cm	T g-mol/l
0.50	0.58	1.0019	1.0037	5.0	0.042	996.9	1.3	7	1.3336	0.140	0.075	0.040	1.016	1.017	98.18	2.2	0.021
1.00	1.15	1.0056	1.0074	10.1	0.084	995.6	2.7	13	1.3343	0.279	0.150	0.080	1.033	1.029	96.61	4.4	0.044
1.50	1.73	1.0094	1.0111	15.1	0.126	994.2	4.0	20	1.3349	0.418	0.225	0.120	1.049	1.042	95.10	6.8	0.068
2.00	2.30	1.0131	1.0149	20.3	0.169	992.8	5.4	26	1.3356	0.556	0.299	0.161	1.066	1.054	93.62	9.1	0.094
2.50	2.88	1.0168	1.0186	25.4	0.212	991.4	6.8	33	1.3362	0.695	0.374	0.202	1.083	1.067	92.13	11.3	0.119
3.00	3.45	1.0206	1.0224	30.6	0.255	990.0	8.3	39	1.3369	0.835	0.449	0.243	1.101	1.081	90.65	13.4	0.143
3.50	4.03	1.0244	1.0262	35.9	0.299	988.5	9.7	45	1.3375	0.976	0.525	0.285	1.119	1.095	89.17	15.4	0.166
4.00	4.60	1.0281	1.0300	41.1	0.343	987.0	11.2	52	1.3382	1.116	0.600	0.326	1.138	1.109	87.69	17.4	0.189
4.50	5.18	1.0319	1.0338	46.4	0.387	985.5	12.7	58	1.3388	1.254	0.674	0.367	1.158	1.124	86.21	19.2	0.211
5.00	5.75	1.0358	1.0376	51.8	0.432	984.0	14.3	65	1.3395	1.388	0.746	0.406	1.178	1.140	84.72	21.0	0.232
5.50	6.33	1.0396	1.0414	57.2	0.477	982.4	15.8	71	1.3401	1.519	0.817	0.444	1.199	1.156	83.23	22.6	0.252
6.00	6.90	1.0434	1.0453	62.6	0.522	980.8	17.4	78	1.3408	1.645	0.884	0.481	1.221	1.172	81.74	24.1	0.270
6.50	7.48	1.0473	1.0491	68.1	0.567	979.2	19.0	84	1.3414	1.767	0.950	0.517	1.243	1.190	80.26	25.5	0.287
7.00	8.05	1.0511	1.0530	73.6	0.613	977.6	20.7	91	1.3421	1.887	1.014	0.552	1.267	1.208	78.78	26.7	0.303
7.50	8.63	1.0550	1.0569	79.1	0.660	975.9	22.3	97	1.3427	2.004	1.077	0.586	1.291	1.226	77.31	27.9	0.317
8.00	9.20	1.0589	1.0608	84.7	0.706	974.2	24.0	104	1.3434	2.119	1.139	0.619	1.316	1.245	75.85	29.0	0.330
8.50	9.78	1.0628	1.0647	90.3	0.753	972.5	25.7	110	1.3440	2.234	1.201	0.652	1.342	1.265	74.39	30.0	0.344
9.00	10.35	1.0668	1.0686	96.0	0.800	970.8	27.5	117	1.3447	2.350	1.263	0.686	1.368	1.285	72.93	31.0	0.357
9.50	10.93	1.0707	1.0726	101.7	0.848	969.0	29.2	123	1.3453	2.466	1.326	0.719	1.396	1.307	71.48	32.1	0.370
10.00	11.50	1.0747	1.0766	107.5	0.896	967.2	31.0	130	1.3460	2.584	1.389	0.753	1.425	1.329	70.04	33.2	0.384
11.00	12.65	1.0826	1.0846	119.1	0.993	963.6	34.7	143	1.3473	2.821	1.517	0.821	1.485	1.374	67.20	35.4	0.412
12.00	13.80	1.0907	1.0926	130.9	1.091	959.8	38.4	156	1.3486	3.056	1.643	0.887	1.549	1.423	64.44	37.5	0.439
13.00	14.95	1.0988	1.1007	142.8	1.191	955.9	42.3	169	1.3499	3.291	1.769	0.954	1.617	1.474	61.73	39.6	0.465
14.00	16.10	1.1070	1.1089	155.0	1.292	952.0	46.2	182	1.3512	3.529	1.897	1.021	1.691	1.530	59.04	41.5	0.490
15.00	17.25	1.1152	1.1172	167.3	1.394	947.9	50.3	195	1.3525	3.775	2.029	1.089	1.771	1.591	56.35	43.3	0.513
16.00	18.40	1.1236	1.1255	179.8	1.498	943.8	54.4	209	1.3538	4.028	2.166	1.159	1.857	1.656	53.74	44.8	0.534
17.00	19.55	1.1320	1.1340	192.4	1.604	939.5	58.7	222	1.3552	4.287	2.305	1.230	1.948	1.724	51.24	46.2	0.552
18.00	20.70	1.1404	1.1425	205.3	1.711	935.2	63.1	235	1.3565	4.551	2.447	1.303	2.046	1.798	48.77	47.4	0.568
19.00	21.85	1.1490	1.1510	218.3	1.820	930.7	67.5	248	1.3578	4.820	2.592	1.376	2.155	1.880	46.30	48.5	0.583
20.00	23.00	1.1576	1.1597	231.5	1.930	926.1	72.1	262	1.3592	5.10	2.74	1.450	2.278	1.972	43.81	49.6	0.597
22.00	25.30	1.1752	1.1773	258.5	2.155	916.6	81.6	289	1.3618				2.545	2.170	39.21	51.4	0.622
24.00	27.60	1.1931	1.1952	286.3	2.387	906.7	91.5	316	1.3646				2.844	2.389	35.09	52.7	0.639
26.00	29.91	1.2113	1.2134	314.9	2.625	896.3	101.9	343	1.3673				3.208	2.654	31.11	53.5	0.650
28.00	32.21	1.2299	1.2320	344.4	2.870	885.5	112.7	370	1.3700				3.675	2.994	27.16	53.9	0.655
30.00	34.51	1.2488	1.2510	374.6	3.123	874.2	124.1	398	1.3728				4.291	3.443	23.26	54.0	0.656
32.00	36.81	1.2682	1.2704	405.8	3.383	862.4	135.9	426	1.3756				5.069	4.005	19.69	53.7	0.653
34.00	39.11	1.2879	1.2902	437.9	3.650	850.0	148.2	454	1.3784				5.996	4.665	16.64	52.9	0.642
36.00	41.41	1.3080	1.3103	470.9	3.925	837.1	161.1	482	1.3812				7.084	5.427	14.09	51.5	0.622
38.00	43.71	1.3285	1.3308	504.8	4.208	823.6	174.6	511	1.3840				8.346	6.295	11.96	49.3	0.593
40.00	46.01	1.3493	1.3517	539.7	4.499	809.6	188.6	539	1.3869				9.794	7.273	10.19	46.1	0.551

81 SODIUM PHOSPHATE, MONOHYDROGEN (DIBASIC), Na$_2$HPO$_4 \cdot$ 7H$_2$O

MOLECULAR WEIGHT = 141.97   FORMULA WEIGHT, HYDRATE = 268.09   0.00 % by wt. data are the same for all compounds.
RELATIVE SPECIFIC REFRACTIVITY = 0.576   For Values of 0.00 wt. % solutions see Table 1, Acetic Acid.

A% by wt.	H% by wt.	ρ D_4^{20}	D_{20}^{20}	C_s g/l	M g-mol/l	C_w g/l	$(C_o - C_w)$ g/l	$(n - n_o)$ ×10^4	n	Δ °C	O Os/kg	S g-mol/l	η/η_o	η/ρ cS	φ rhe	γ mmho/cm	T g-mol/l
0.50	0.94	1.0032	1.0050	5.0	0.035	998.2	0.0	10	1.3340	0.167	0.090	0.047	1.019	1.018	97.91	4.6	0.046
1.00	1.89	1.0082	1.0100	10.1	0.071	998.1	0.1	19	1.3349	0.319	0.172	0.092	1.040	1.034	95.96	8.7	0.090
1.50	2.83	1.0131	1.0149	15.2	0.107	997.9	0.3	28	1.3358	0.462	0.248	0.133	1.062	1.051	93.94	12.3	0.130
2.00	3.78	1.0180	1.0198	20.4	0.143	997.7	0.6	38	1.3368				1.086	1.069	91.90	15.6	0.168
2.50	4.72	1.0229	1.0247	25.6	0.180	997.4	0.9	47	1.3377				1.111	1.088	89.86	18.7	0.205
3.00	5.67	1.0279	1.0297	30.8	0.217	997.0	1.2	56	1.3386				1.136	1.108	87.82	21.6	0.240
3.50	6.61	1.0328	1.0346	36.1	0.255	996.7	1.6	66	1.3396				1.163	1.129	85.80	24.3	0.273
4.00	7.55	1.0378	1.0396	41.5	0.292	996.3	1.9	75	1.3405				1.191	1.150	83.78	26.8	0.304
4.50	8.50	1.0428	1.0446	46.9	0.331	995.9	2.4	84	1.3414				1.221	1.173	81.77	29.2	0.334
5.00	9.44	1.0478	1.0496	52.4	0.369	995.4	2.8	94	1.3424				1.251	1.196	79.78	31.4	0.362
5.50	10.39	1.0528	1.0546	57.9	0.408	994.9	3.4	103	1.3433				1.283	1.221	77.80	33.5	0.388

82 SODIUM SULFATE, Na$_2$SO$_4 \cdot$ 10H$_2$O

MOLECULAR WEIGHT = 142.06   FORMULA WEIGHT, HYDRATE = 322.22   0.00 % by wt. data are the same for all compounds.
RELATIVE SPECIFIC REFRACTIVITY = 0.548   For Values of 0.00 wt. % solutions see Table 1, Acetic Acid.

A% by wt.	H% by wt.	ρ D_4^{20}	D_{20}^{20}	C_s g/l	M g-mol/l	C_w g/l	$(C_o - C_w)$ g/l	$(n - n_o)$ ×10^4	n	Δ °C	O Os/kg	S g-mol/l	η/η_o	η/ρ cS	φ rhe	γ mmho/cm	T g-mol/l
0.50	1.13	1.0027	1.0044	5.0	0.035	997.7	0.6	8	1.3338	0.165	0.089	0.047	1.011	1.010	98.71	5.9	0.060
1.00	2.27	1.0071	1.0089	10.1	0.071	997.1	1.2	15	1.3345	0.320	0.172	0.092	1.024	1.019	97.46	11.2	0.117
1.50	3.40	1.0116	1.0134	15.2	0.107	996.4	1.8	23	1.3353	0.466	0.251	0.135	1.039	1.030	96.02	15.7	0.169
2.00	4.54	1.0161	1.0179	20.3	0.143	995.8	2.4	30	1.3360	0.606	0.326	0.176	1.056	1.041	94.51	19.8	0.218
2.50	5.67	1.0206	1.0225	25.5	0.180	995.1	3.1	38	1.3368	0.742	0.399	0.216	1.072	1.053	93.06	23.9	0.268
3.00	6.80	1.0252	1.0270	30.8	0.216	994.4	3.8	46	1.3376	0.873	0.469	0.254	1.089	1.065	91.63	27.9	0.317
3.50	7.94	1.0298	1.0316	36.0	0.254	993.7	4.5	53	1.3383	1.001	0.538	0.292	1.106	1.077	90.21	31.8	0.365
4.00	9.07	1.0343	1.0362	41.4	0.291	993.0	5.3	61	1.3391	1.125	0.605	0.329	1.124	1.089	88.80	35.5	0.413
4.50	10.21	1.0389	1.0408	46.8	0.329	992.2	6.0	68	1.3398	1.245	0.669	0.364	1.142	1.101	87.38	39.2	0.460
5.00	11.34	1.0436	1.0454	52.2	0.367	991.4	6.8	76	1.3406	1.359	0.731	0.397	1.161	1.115	85.96	42.7	0.506
5.50	12.48	1.0481	1.0499	57.6	0.406	990.4	7.8	83	1.3413	1.465	0.788	0.429	1.180	1.129	84.54	46.1	0.551
6.00	13.61	1.0528	1.0545	63.2	0.445	989.5	8.8	90	1.3420	1.560	0.839	0.456	1.200	1.143	83.14	49.4	0.594
6.50	14.74	1.0572	1.0591	68.7	0.484	988.5	9.7	97	1.3427				1.221	1.157	81.75	52.5	0.636
7.00	15.88	1.0619	1.0638	74.3	0.523	987.5	10.7	105	1.3435				1.242	1.172	80.36	55.5	0.676
7.50	17.01	1.0666	1.0684	80.0	0.563	986.6	11.7	112	1.3442				1.264	1.187	78.98	58.5	0.716
8.00	18.15	1.0712	1.0732	85.7	0.603	985.6	12.7	120	1.3449				1.286	1.203	77.59	61.1	0.755
8.50	19.28	1.0760	1.0779	91.5	0.644	984.6	13.7	127	1.3457				1.310	1.220	76.20	63.7	0.793
9.00	20.41	1.0808	1.0827	97.3	0.685	983.5	14.7	134	1.3464				1.334	1.237	74.80	66.3	0.830
9.50	21.55	1.0856	1.0875	103.1	0.726	982.5	15.7	142	1.3472				1.360	1.255	73.38	68.8	0.866
10.00	22.68	1.0905	1.0924	109.0	0.768	981.4	16.8	149	1.3479				1.387	1.274	71.95	71.3	0.902
11.00	24.95	1.1002	1.1022	121.0	0.852	979.2	19.0	164	1.3494				1.444	1.315	69.11	75.9	0.970
12.00	27.22	1.1101	1.1121	133.2	0.938	976.9	21.3	179	1.3509				1.505	1.359	66.30	80.1	1.03
13.00	29.49	1.1201	1.1221	145.6	1.025	974.5	23.8	194	1.3524				1.571	1.406	63.52	83.9	1.09
14.00	31.75	1.1301	1.1321	158.2	1.114	971.9	26.3	209	1.3539				1.643	1.457	60.74	87.5	1.15
15.00	34.02	1.1402	1.1422	171.0	1.204	969.2	29.1	223	1.3553				1.722	1.513	57.96	91.1	1.21
16.00	36.29	1.1503	1.1523	184.0	1.296	966.2	32.0	237	1.3567				1.808	1.575	55.20	94.9	1.27
17.00	38.56	1.1604	1.1625	197.3	1.389	963.1	35.1	251	1.3581				1.901	1.641	52.50	98.5	1.33
18.00	40.83	1.1705	1.1726	210.7	1.483	959.8	38.4	265	1.3595				2.001	1.713	49.88	102.0	1.39
19.00	43.10	1.1806	1.1827	224.3	1.579	956.3	41.9	278	1.3608				2.108	1.789	47.34	105.0	1.44
20.00	45.36	1.1907	1.1928	238.1	1.676	952.6	45.7	290	1.3620				2.223	1.871	44.89	109.0	1.50
22.00	49.90	1.2106	1.2127	266.3	1.875	944.2	54.0	313	1.3643				2.476	2.050	40.30	114.0	1.61

83 SODIUM TARTRATE, NaOOC(CHOH)$_2$COONa $\cdot$ 2H$_2$O

MOLECULAR WEIGHT = 194.06   FORMULA WEIGHT, HYDRATE = 230.10   0.00 % by wt. data are the same for all compounds.
RELATIVE SPECIFIC REFRACTIVITY = 0.781   For Values of 0.00 wt. % solutions see Table 1, Acetic Acid.

A% by wt.	H% by wt.	ρ D_4^{20}	D_{20}^{20}	C_s g/l	M g-mol/l	C_w g/l	$(C_o - C_w)$ g/l	$(n - n_o)$ ×10^4	n	Δ °C	O Os/kg	S g-mol/l	η/η_o	η/ρ cS	φ rhe	γ mmho/cm	T g-mol/l
0.50	0.59	1.0017	1.0035	5.0	0.026	996.7	1.5	8	1.3338	0.111	0.060	0.031	1.015	1.016	98.28	3.8	0.037
1.00	1.19	1.0052	1.0070	10.1	0.052	995.2	3.1	17	1.3347	0.224	0.121	0.064	1.031	1.028	96.80	7.1	0.072
2.00	2.37	1.0123	1.0141	20.2	0.104	992.0	6.2	33	1.3363	0.457	0.246	0.132	1.062	1.051	93.97	12.6	0.133
3.00	3.56	1.0194	1.0212	30.6	0.158	988.8	9.4	50	1.3380	0.678	0.365	0.197	1.094	1.075	91.21	17.5	0.191
4.00	4.74	1.0266	1.0284	41.1	0.212	985.5	12.7	67	1.3397	0.891	0.479	0.260	1.128	1.101	88.49	22.1	0.245
5.00	5.93	1.0338	1.0356	51.7	0.266	982.1	16.1	84	1.3414	1.103	0.593	0.322	1.164	1.128	85.74	26.2	0.297
6.00	7.11	1.0410	1.0428	62.5	0.322	978.5	19.7	101	1.3431	1.317	0.708	0.385	1.203	1.157	82.99	30.1	0.346
7.00	8.30	1.0482	1.0501	73.4	0.378	974.9	23.4	118	1.3448	1.529	0.822	0.447	1.243	1.188	80.28	33.7	0.390
8.00	9.49	1.0555	1.0574	84.4	0.435	971.1	27.2	135	1.3465	1.742	0.937	0.509	1.287	1.222	77.56	37.0	0.432
9.00	10.67	1.0628	1.0647	95.7	0.493	967.2	31.1	152	1.3482	1.955	1.051	0.572	1.334	1.258	74.81	40.0	0.471
10.00	11.86	1.0702	1.0721	107.0	0.551	963.2	35.1	169	1.3499	2.171	1.167	0.634	1.386	1.298	72.01	43.0	0.509
12.00	14.23	1.0851	1.0870	130.2	0.671	954.9	43.4	203	1.3533	2.607	1.402	0.760	1.499	1.384	66.58	48.1	0.577
14.00	16.60	1.1002	1.1022	154.0	0.794	946.2	52.0	237	1.3567	3.048	1.639	0.885	1.625	1.480	61.43	52.2	0.632
15.00	17.79	1.1079	1.1098	166.2	0.856	941.7	56.5	254	1.3584	3.269	1.758	0.948	1.695	1.533	58.88	53.9	0.655
18.00	21.34	1.1313	1.1333	203.6	1.049	927.6	70.6	307	1.3637	3.938	2.117	1.134	1.951	1.728	51.14	57.9	0.710
20.00	23.71	1.1471	1.1492	229.4	1.182	917.7	80.5	342	1.3672	4.385	2.358	1.258	2.174	1.899	45.99	59.9	0.738
22.00	26.09	1.1633	1.1653	255.9	1.319	907.4	90.9	378	1.3708				2.444	2.105	40.84	61.4	0.759
24.00	28.46	1.1797	1.1818	283.1	1.459	896.6	101.7	414	1.3744				2.757	2.342	36.20	62.0	0.768
26.00	30.83	1.1963	1.1984	311.0	1.603	885.3	113.0	450	1.3780				3.117	2.611	32.01	61.6	0.762
28.00	33.20	1.2132	1.2153	339.7	1.750	873.5	124.7	487	1.3817				3.528	2.914	28.29	60.1	0.741

84 SODIUM THIOCYANATE, NaCNS

MOLECULAR WEIGHT = 81.07

A% by wt.	ρ D_4^{20}	D_{20}^{20}	C_s g/l	M g-mol/l	C_w g/l	$(C_0 - C_w)$ g/l	$(n - n_0)$ ×10⁴	n	Δ °C	O Os/kg	S g-mol/l	η/η_0	η/ρ cS	ϕ rhe	γ mmho/cm	T g-mol/l
1.00	1.0030	1.0048	10.0	0.124	993.0	5.2	22.	1.3352	0.426	0.229	0.123	1.006	1.005	99.21	10.0	0.113
2.00	1.0078	1.0096	20.2	0.249	987.6	10.6	44.	1.3374	0.848	0.456	0.247	1.012	1.006	98.64	20.5	0.226
3.00	1.0127	1.0145	30.4	0.375	982.3	15.9	66.	1.3396	1.280	0.688	0.374	1.019	1.008	97.96	29.8	0.341
4.00	1.0176	1.0194	40.7	0.502	976.9	21.3	88.	1.3418	1.734	0.932	0.507	1.025	1.009	97.40	39.0	0.458
5.00	1.0226	1.0244	51.1	0.631	971.5	26.7	110.	1.3440	2.202	1.184	0.643	1.032	1.011	96.73	48.0	0.576
6.00	1.0275	1.0294	61.7	0.760	965.8	32.4	132.	1.3462	2.679	1.440	0.780	1.039	1.013	96.07	56.7	0.694
7.00	1.0327	1.0346	72.3	0.892	960.4	37.8	155.	1.3485	3.168	1.703	0.919	1.047	1.016	95.31	65.1	0.812
8.00	1.0381	1.0400	83.1	1.024	955.0	43.2	178.	1.3508	3.671	1.974	1.060	1.057	1.020	94.44	73.3	0.932
9.00	1.0435	1.0454	93.9	1.158	949.6	48.6	201.	1.3531	4.184	2.250	1.202	1.067	1.025	93.49	81.2	1.052
10.00	1.0491	1.0510	104.9	1.294	944.2	54.0	224.	1.3554	4.706	2.530	1.345	1.079	1.031	92.45	89.0	1.174
12.00	1.0603	1.0622	127.2	1.569	933.1	65.1	270.	1.3600				1.106	1.045	90.25	104.	1.418
14.00	1.0715	1.0734	150.0	1.850	921.5	76.7	318.	1.3648				1.138	1.064	87.72	118.	1.664
16.00	1.0828	1.0848	173.3	2.137	909.5	88.7	366.	1.3696				1.175	1.087	84.96	130.	1.912
18.00	1.0942	1.0962	197.0	2.430	897.2	101.0	414.	1.3744				1.217	1.114	82.04	142.	2.162
20.00	1.1056	1.1076	221.1	2.728	884.5	113.7	464.	1.3794				1.268	1.149	78.72	154.	2.412
24.00	1.1284	1.1304	270.8	3.340	857.6	140.6	564.	1.3894				1.396	1.240	71.47	170.	2.786
28.00	1.1515	1.1536	322.4	3.977	829.1	169.1	666.	1.3996				1.546	1.345	64.57	178.	3.016
32.00	1.1763	1.1784	376.4	4.643	799.9	198.3	774.	1.4104				1.764	1.503	56.56	184.	3.194
36.00	1.2018	1.2040	432.7	5.337	769.1	229.1	882.	1.4212				2.101	1.752	47.49	185.	3.227
40.00	1.2282	1.2304	491.3	6.060	736.9	261.3	990.	1.4320				2.483	2.026	40.19	177.	2.986
44.00	1.2549	1.2572	552.2	6.811	702.7	295.5	1105.	1.4435				3.151	2.516	31.67	165.	2.678
48.00	1.2821	1.2844	615.4	7.591	666.7	331.5	1223.	1.4553				3.936	3.076	25.36	148.	2.290
52.00	1.3100	1.3124	681.2	8.403	628.8	369.4	1342.	1.4672				5.279	4.038	18.90	130.	1.900
56.00	1.3388	1.3412	749.7	9.248	589.1	409.1	1461.	1.4791				7.554	5.654	13.21	108.	1.493

85 SODIUM THIOSULFATE, Na₂S₂O₃ · 5H₂O

MOLECULAR WEIGHT = 158.13   FORMULA WEIGHT, HYDRATE = 248.21
RELATIVE SPECIFIC REFRACTIVITY = 0.773

0.00 % by wt. data are the same for all compounds.
For Values of 0.00 wt. % solutions see Table 1, Acetic Acid.

A% by wt.	H% by wt.	ρ D_4^{20}	D_{20}^{20}	C_s g/l	M g-mol/l	C_w g/l	$(C_0 - C_w)$ g/l	$(n - n_0)$ ×10⁴	n	Δ °C	O Os/kg	S g-mol/l	η/η_0	η/ρ cS	ϕ rhe	γ mmho/cm	T g-mol/l
0.50	0.78	1.0024	1.0041	5.0	0.032	997.4	0.9	10	1.3340	0.139	0.075	0.039	1.010	1.010	98.76	5.7	0.057
1.00	1.57	1.0065	1.0083	10.1	0.064	996.4	1.8	21	1.3351	0.279	0.150	0.080	1.021	1.016	97.75	10.7	0.112
2.00	3.14	1.0148	1.0166	20.3	0.128	994.5	3.7	42	1.3371	0.566	0.304	0.164	1.042	1.029	95.78	19.5	0.214
3.00	4.71	1.0231	1.0249	30.7	0.194	992.4	5.8	62	1.3392	0.835	0.449	0.243	1.064	1.042	93.77	27.7	0.315
4.00	6.28	1.0315	1.0333	41.3	0.261	990.2	8.0	83	1.3413	1.090	0.586	0.318	1.088	1.057	91.74	35.6	0.415
5.00	7.85	1.0399	1.0417	52.0	0.329	987.9	10.3	104	1.3434	1.341	0.721	0.392	1.113	1.072	89.67	43.3	0.513
6.00	9.42	1.0483	1.0502	62.9	0.398	985.4	12.8	124	1.3454	1.588	0.854	0.464	1.139	1.089	87.59	50.6	0.610
7.00	10.99	1.0568	1.0587	74.0	0.468	982.8	15.4	145	1.3475	1.826	0.982	0.534	1.167	1.107	85.50	57.5	0.705
8.00	12.56	1.0654	1.0673	85.2	0.539	980.1	18.1	166	1.3496	2.062	1.108	0.602	1.197	1.126	83.38	64.2	0.799
9.00	14.13	1.0740	1.0759	96.7	0.611	977.3	20.9	187	1.3517	2.299	1.236	0.671	1.229	1.147	81.21	70.5	0.891
10.00	15.70	1.0827	1.0846	108.3	0.685	974.4	23.8	208	1.3538	2.546	1.369	0.742	1.264	1.170	78.96	76.7	0.982
12.00	18.84	1.1003	1.1023	132.0	0.835	968.3	29.9	251	1.3581	3.064	1.647	0.890	1.342	1.222	74.35	88.2	1.16
14.00	21.98	1.1182	1.1202	156.6	0.990	961.7	36.5	294	1.3624	3.604	1.938	1.041	1.432	1.283	69.71	98.8	1.33
16.00	25.11	1.1365	1.1385	181.8	1.150	954.6	43.6	337	1.3667	4.168	2.241	1.197	1.534	1.353	65.04	108.	1.49
18.00	28.25	1.1551	1.1571	207.9	1.315	947.1	51.1	381	1.3711	4.758	2.558	1.359	1.654	1.435	60.34	117.	1.64
20.00	31.39	1.1740	1.1760	234.8	1.485	939.2	59.1	426	1.3756	5.37	2.89	1.524	1.794	1.531	55.63	123.	1.77
22.00	34.53	1.1932	1.1953	262.5	1.660	930.7	67.5	471	1.3801				1.954	1.641	51.08	129.	1.87
24.00	37.67	1.2128	1.2150	291.1	1.841	921.8	76.5	517	1.3847				2.137	1.765	46.70	132.	1.94
26.00	40.81	1.2328	1.2350	320.5	2.027	912.3	85.9	563	1.3893				2.351	1.911	42.45	135.	2.00
28.00	43.95	1.2532	1.2554	350.9	2.219	902.3	95.9	610	1.3940				2.591	2.072	38.52	136.	2.02
30.00	47.09	1.2739	1.2762	382.2	2.417	891.7	106.5	657	1.3987				2.897	2.279	34.45	136.	2.03
32.00	50.23	1.2950	1.2973	414.4	2.621	880.6	117.6	705	1.4035				3.298	2.552	30.26	135.	1.99
34.00	53.37	1.3164	1.3188	447.6	2.831	868.8	129.4	754	1.4084				3.784	2.881	26.37	132.	1.93
36.00	56.51	1.3382	1.3406	481.8	3.047	856.5	141.8	802	1.4132				4.350	3.257	22.95	128.	1.86
38.00	59.65	1.3603	1.3627	516.9	3.269	843.4	154.8	851	1.4181				5.001	3.684	19.96	124.	1.77
40.00	62.79	1.3827	1.3852	553.1	3.498	829.6	168.6	899	1.4229				5.747	4.165	17.37	118.	1.66

86 SODIUM TUNGSTATE, Na$_2$WO$_4 \cdot$ 2H$_2$O

MOLECULAR WEIGHT = 293.91 FORMULA WEIGHT, HYDRATE = 329.95
RELATIVE SPECIFIC REFRACTIVITY = 0.414

0.00 % by wt. data are the same for all compounds.
For Values of 0.00 wt. % solutions see Table 1, Acetic Acid.

A% by wt.	H% by wt.	ρ D$_4^{20}$	D$_{20}^{20}$	C$_s$ g/l	M g-mol/l	C$_w$ g/l	(C$_o$ - C$_w$) g/l	(n - n$_o$) ×10^4	n	Δ °C	O Os/kg	S g-mol/l	η/η_o	η/ρ cS	ϕ rhe	γ mmho/cm	T g-mol/l
0.50	0.56	1.0028	1.0046	5.0	0.017	997.8	0.4	5	1.3335	0.081	0.044	0.023	1.007	1.006	99.13	2.8	0.028
1.00	1.12	1.0074	1.0091	10.1	0.034	997.3	0.9	11	1.3341	0.161	0.086	0.046	1.014	1.009	98.42	5.5	0.055
1.50	1.68	1.0119	1.0137	15.2	0.052	996.8	1.5	16	1.3346	0.239	0.129	0.068	1.022	1.012	97.67	7.9	0.081
2.00	2.25	1.0166	1.0184	20.3	0.069	996.2	2.0	21	1.3351	0.317	0.171	0.091	1.030	1.015	96.89	10.3	0.107
2.50	2.81	1.0212	1.0230	25.5	0.087	995.7	2.5	27	1.3357	0.396	0.213	0.114	1.039	1.019	96.08	12.5	0.132
3.00	3.37	1.0259	1.0278	30.8	0.105	995.2	3.1	33	1.3362	0.474	0.255	0.137	1.048	1.024	95.23	14.7	0.157
3.50	3.93	1.0307	1.0326	36.1	0.123	994.7	3.6	38	1.3368	0.553	0.297	0.160	1.058	1.028	94.35	16.8	0.182
4.00	4.49	1.0356	1.0374	41.4	0.141	994.1	4.1	44	1.3374	0.633	0.340	0.184	1.068	1.033	93.46	18.8	0.206
4.50	5.05	1.0405	1.0423	46.8	0.159	993.6	4.6	50	1.3380	0.714	0.384	0.208	1.078	1.038	92.58	20.8	0.230
5.00	5.61	1.0453	1.0472	52.3	0.178	993.1	5.2	56	1.3386	0.796	0.428	0.232	1.088	1.043	91.73	22.8	0.254
5.50	6.17	1.0502	1.0520	57.8	0.197	992.4	5.8	62	1.3392	0.879	0.472	0.256	1.098	1.048	90.89	24.7	0.278
6.00	6.74	1.0550	1.0569	63.3	0.215	991.7	6.5	67	1.3397	0.962	0.517	0.281	1.108	1.052	90.06	26.7	0.302
6.50	7.30	1.0599	1.0618	68.9	0.234	991.0	7.2	73	1.3403	1.046	0.562	0.305	1.118	1.057	89.24	28.6	0.326
7.00	7.86	1.0648	1.0667	74.5	0.254	990.3	8.0	79	1.3409	1.130	0.607	0.330	1.129	1.062	88.43	30.5	0.350
7.50	8.42	1.0698	1.0717	80.2	0.273	989.5	8.7	85	1.3414	1.212	0.652	0.354	1.139	1.067	87.62	32.4	0.374
8.00	8.98	1.0748	1.0767	86.0	0.293	988.8	9.4	91	1.3420	1.293	0.695	0.378	1.150	1.072	86.81	34.3	0.398
8.50	9.54	1.0799	1.0818	91.8	0.312	988.1	10.1	97	1.3427	1.371	0.737	0.401	1.160	1.077	86.02	36.1	0.422
9.00	10.10	1.0852	1.0871	97.7	0.332	987.5	10.7	103	1.3433	1.445	0.777	0.423	1.171	1.081	85.23	38.0	0.445

87 STRONTIUM CHLORIDE, SrCl$_2 \cdot$ 6H$_2$O

MOLECULAR WEIGHT = 158.54 FORMULA WEIGHT, HYDRATE = 266.64
RELATIVE SPECIFIC REFRACTIVITY = 0.635

0.00 % by wt. data are the same for all compounds.
For Values of 0.00 wt. % solutions see Table 1, Acetic Acid.

A% by wt.	H% by wt.	ρ D$_4^{20}$	D$_{20}^{20}$	C$_s$ g/l	M g-mol/l	C$_w$ g/l	(C$_o$ - C$_w$) g/l	(n - n$_o$) ×10^4	n	Δ °C	O Os/kg	S g-mol/l	η/η_o	η/ρ cS	ϕ rhe	γ mmho/cm	T g-mol/l
0.50	0.84	1.0027	1.0044	5.0	0.032	997.6	0.6	9	1.3339	0.158	0.085	0.045	1.010	1.009	99.45	5.9	0.060
1.00	1.68	1.0071	1.0089	10.1	0.064	997.0	1.2	18	1.3348	0.167	0.089	0.089	1.019	1.014	97.94	11.4	0.120
2.00	3.36	1.0161	1.0179	20.3	0.128	995.8	2.3	36	1.3366	0.619	0.333	0.180	1.037	1.023	96.24	22.0	0.244
3.00	5.05	1.0252	1.0270	30.8	0.194	994.4	3.8	54	1.3384	0.934	0.502	0.272	1.055	1.032	94.56	31.4	0.362
4.00	6.73	1.0344	1.0362	41.4	0.261	993.0	5.2	73	1.3402	1.261	0.678	0.369	1.074	1.041	92.89	40.4	0.475
5.00	8.41	1.0437	1.0456	52.2	0.329	991.5	6.7	91	1.3421	1.607	0.864	0.470	1.094	1.050	91.23	49.1	0.591
6.00	10.09	1.0532	1.0551	63.2	0.399	990.0	8.2	110	1.3440	1.978	1.064	0.578	1.114	1.060	89.59	58.0	0.712
7.00	11.77	1.0628	1.0647	74.4	0.469	988.4	9.8	129	1.3459	2.377	1.278	0.693	1.134	1.069	87.98	66.8	0.836
8.00	13.45	1.0726	1.0745	85.8	0.541	986.8	11.4	148	1.3478	2.800	1.505	0.815	1.155	1.079	86.37	75.3	0.962
9.00	15.14	1.0825	1.0844	97.4	0.614	985.0	13.2	168	1.3498	3.252	1.748	0.943	1.178	1.090	84.73	83.6	1.08
10.00	16.82	1.0925	1.0944	109.3	0.689	983.3	15.0	188	1.3518	3.736	2.009	1.078	1.202	1.102	83.03	91.5	1.21
12.00	20.18	1.1131	1.1150	133.6	0.842	979.5	18.7	228	1.3558	4.811	2.587	1.374	1.255	1.130	79.53	107.	1.46
14.00	23.55	1.1342	1.1362	158.8	1.002	975.4	22.9	269	1.3599	6.03	3.24	1.696	1.314	1.160	75.98	120.	1.71
16.00	26.91	1.1558	1.1579	184.9	1.166	970.9	27.3	311	1.3641	7.41	3.99	2.044	1.380	1.196	72.32	133.	1.96
18.00	30.27	1.1780	1.1801	212.0	1.337	966.0	32.3	354	1.3684	8.98	4.83	2.418	1.457	1.239	68.52	144.	2.19
20.00	33.64	1.2008	1.2029	240.2	1.515	960.6	37.6	398	1.3728	10.74	5.78	2.814	1.546	1.290	64.55	153.	2.40
22.00	37.00	1.2241	1.2263	269.3	1.699	954.8	43.4	442	1.3772	12.74	6.85	3.230	1.647	1.348	60.58	160.	2.56
24.00	40.36	1.2481	1.2503	299.5	1.889	948.6	49.7	487	1.3817	14.99	8.06	3.664	1.761	1.414	56.67	165.	2.68
26.00	43.73	1.2728	1.2751	330.9	2.087	941.9	56.3	534	1.3864				1.893	1.491	52.71	170.	2.79
28.00	47.09	1.2983	1.3006	363.5	2.293	934.8	63.4	582	1.3911				2.052	1.584	48.64	174.	2.90

A% by wt.	H% by wt.	ρ D$_4^{20}$	D$_{20}^{20}$	C$_x$ g/l	M g-mol/l	C$_w$ g/l	(C$_o$ − C$_w$) g/l	(n − n$_o$) × 10^4	n	Δ °C	O Os/kg	S g-mol/l	η/η_o	η/ρ cS	ϕ rhe	γ mmho/cm	T g-mol/l
30.00	50.46	1.3248	1.3271	397.4	2.507	927.3	70.9	631	1.3961			2.241	1.695	44.53	178.0	3.02
32.00	53.82	1.3523	1.3547	432.7	2.729	919.6	78.7	683	1.4013			2.522	1.869	39.57	178.0	3.02
34.00	57.18	1.3811	1.3835	469.6	2.962	911.5	86.7	737	1.4067			2.840	2.060	35.14	175.0	2.92
36.00	60.55	1.4114	1.4139	508.1	3.205	903.3	94.9	794	1.4124			3.200	2.272	31.19	166.0	2.70

88 SUCROSE, C$_{12}$H$_{22}$O$_{11}$

MOLECULAR WEIGHT = 342.30
RELATIVE SPECIFIC REFRACTIVITY = 1.031

0.00 % by wt. data are the same for all compounds.
For Values of 0.00 wt. % solutions see Table 1, Acetic Acid.

A% by wt.	ρ D$_4^{20}$	D$_{20}^{20}$	C$_s$ g/l	M g-mol/l	C$_w$ g/l	(C$_o$ − C$_w$) g/l	(n − n$_o$) × 10^4	n	Δ °C	O Os/kg	S g-mol/l	η/η_o	η/ρ cS	ϕ rhe	γ mmho/cm	T g-mol/l
0.50	1.0002	1.0019	5.0	0.015	995.2	3.1	7	1.3337	0.027	0.015	0.007	1.013	1.015	98.53		
1.00	1.0021	1.0039	10.0	0.029	992.1	6.2	14	1.3344	0.055	0.030	0.015	1.026	1.026	97.27		
1.50	1.0040	1.0058	15.1	0.044	989.0	9.3	22	1.3351	0.083	0.045	0.023	1.039	1.037	96.03		
2.00	1.0060	1.0078	20.1	0.059	985.9	12.4	29	1.3359	0.112	0.060	0.031	1.053	1.049	94.78		
2.50	1.0079	1.0097	25.2	0.074	982.7	15.5	36	1.3366	0.140	0.076	0.040	1.067	1.061	93.51		
3.00	1.0099	1.0117	30.3	0.089	979.6	18.6	43	1.3373	0.170	0.091	0.048	1.082	1.074	92.24		
3.50	1.0119	1.0137	35.4	0.103	976.5	21.8	51	1.3381	0.199	0.107	0.057	1.097	1.086	90.99		
4.00	1.0139	1.0156	40.6	0.118	973.3	24.9	58	1.3388	0.229	0.123	0.066	1.112	1.099	89.75		
4.50	1.0158	1.0176	45.7	0.134	970.1	28.1	65	1.3395	0.260	0.140	0.074	1.128	1.112	88.49		
5.00	1.0178	1.0196	50.9	0.149	966.9	31.3	73	1.3403	0.291	0.156	0.083	1.144	1.126	87.24		
5.50	1.0198	1.0216	56.1	0.164	963.7	34.5	80	1.3410	0.322	0.173	0.093	1.160	1.140	86.02		
6.00	1.0218	1.0236	61.3	0.179	960.5	37.7	88	1.3418	0.354	0.190	0.102	1.177	1.154	84.79		
6.50	1.0238	1.0257	66.5	0.194	957.3	40.9	95	1.3425	0.386	0.208	0.111	1.195	1.169	83.54		
7.00	1.0259	1.0277	71.8	0.210	954.1	44.2	103	1.3433	0.419	0.225	0.121	1.213	1.185	82.28		
7.50	1.0279	1.0297	77.1	0.225	950.8	47.4	110	1.3440	0.452	0.243	0.131	1.232	1.201	81.02		
8.00	1.0299	1.0317	82.4	0.241	947.5	50.7	118	1.3448	0.485	0.261	0.140	1.251	1.217	79.78		
8.50	1.0320	1.0338	87.7	0.256	944.2	54.0	125	1.3455	0.520	0.279	0.150	1.271	1.234	78.54		
9.00	1.0340	1.0358	93.1	0.272	940.9	57.3	133	1.3463	0.554	0.298	0.161	1.291	1.251	77.30		
9.50	1.0361	1.0379	98.4	0.288	937.6	60.6	141	1.3471	0.589	0.317	0.171	1.312	1.269	76.09		
10.00	1.0381	1.0400	103.8	0.303	934.3	63.9	148	1.3478	0.625	0.336	0.181	1.333	1.287	74.87		
11.00	1.0423	1.0441	114.7	0.335	927.6	70.6	164	1.3494	0.698	0.375	0.203	1.378	1.325	72.42		
12.00	1.0465	1.0483	125.6	0.367	920.9	77.4	180	1.3509	0.773	0.415	0.225	1.426	1.365	69.99		
13.00	1.0507	1.0525	136.6	0.399	914.1	84.2	195	1.3525	0.850	0.457	0.248	1.477	1.409	67.57		
14.00	1.0549	1.0568	147.7	0.431	907.2	91.0	211	1.3541	0.930	0.500	0.271	1.531	1.454	65.19		
15.00	1.0592	1.0610	158.9	0.464	900.3	97.9	227	1.3557	1.012	0.544	0.295	1.589	1.503	62.81		
16.00	1.0635	1.0653	170.2	0.497	893.3	104.9	243	1.3573	1.097	0.590	0.320	1.650	1.555	60.49		
17.00	1.0678	1.0697	181.5	0.530	886.3	112.0	259	1.3589	1.185	0.637	0.346	1.716	1.610	58.16		
18.00	1.0722	1.0741	193.0	0.564	879.2	119.1	276	1.3606	1.275	0.686	0.373	1.786	1.669	55.88		
19.00	1.0766	1.0785	204.5	0.598	872.0	126.2	292	1.3622	1.369	0.736	0.400	1.861	1.732	53.63		
20.00	1.0810	1.0829	216.2	0.632	864.8	133.5	309	1.3639	1.465	0.788	0.429	1.941	1.799	51.42		
22.00	1.0899	1.0918	239.8	0.701	850.1	148.1	342	1.3672	1.668	0.897	0.488	2.120	1.949	47.08		
24.00	1.0990	1.1009	263.8	0.771	835.2	163.0	376	1.3706	1.886	1.014	0.551	2.326	2.121	42.91		
26.00	1.1082	1.1102	288.1	0.842	820.1	178.2	411	1.3741	2.120	1.140	0.619	2.568	2.322	38.86		
28.00	1.1175	1.1195	312.9	0.914	804.6	193.6	446	1.3776	2.371	1.275	0.692	2.849	2.554	35.03		
30.00	1.1270	1.1290	338.1	0.988	788.9	209.3	482	1.3812	2.644	1.421	0.770	3.181	2.828	31.37		
32.00	1.1366	1.1386	363.7	1.063	772.9	225.3	518	1.3848	2.942	1.582	0.855	3.754	3.309	26.59		
34.00	1.1464	1.1484	389.8	1.139	756.6	241.6	555	1.3885	3.268	1.757	0.947	4.044	3.535	24.68		
36.00	1.1562	1.1583	416.2	1.216	740.0	258.2	592	1.3922	3.625	1.949	1.047	4.612	3.997	21.64		
38.00	1.1663	1.1683	443.2	1.295	723.1	275.1	630	1.3960	4.018	2.160	1.156	5.304	4.557	18.82		
40.00	1.1765	1.1785	470.6	1.375	705.9	292.4	669	1.3999	4.452	2.394	1.276	6.150	5.238	16.23		
42.00	1.1868	1.1889	498.4	1.456	688.3	309.9	708	1.4038	4.932	2.652	1.406	7.220	6.096	13.82		
44.00	1.1972	1.1994	526.8	1.539	670.5	327.8	748	1.4078				8.579	7.180	11.63		
46.00	1.2079	1.2100	555.6	1.623	652.2	346.0	788	1.4118				10.28	8.53	9.71		
48.00	1.2186	1.2208	584.9	1.709	633.7	364.6	829	1.4159				12.49	10.27	7.99		
50.00	1.2295	1.2317	614.8	1.796	614.8	383.5	871	1.4201				15.40	12.55	6.48		
52.00	1.2406	1.2428	645.1	1.885	595.5	402.7	913	1.4243				19.30	15.59	5.17		
54.00	1.2518	1.2540	676.0	1.975	575.8	422.4	956	1.4286				24.63	19.71	4.05		
56.00	1.2632	1.2654	707.4	2.067	555.8	442.4	1000	1.4330				32.06	25.43	3.11		
58.00	1.2747	1.2770	739.3	2.160	535.4	462.8	1044	1.4374				42.69	33.56	2.34		
60.00	1.2864	1.2887	771.9	2.255	514.6	483.7	1089	1.4419				58.37	45.46	1.71		
62.00	1.2983	1.3006	804.9	2.352	493.3	504.9	1135	1.4465				82.26	63.49	1.21		
64.00	1.3103	1.3126	838.6	2.450	471.7	526.5	1181	1.4511				119.9	91.69	0.83		
66.00	1.3224	1.3248	872.8	2.550	449.6	548.6	1228	1.4558				181.7	137.6	0.55		
68.00	1.3348	1.3371	907.6	2.652	427.1	571.1	1276	1.4606				287.9	216.1	0.35		
70.00	1.3472	1.3496	943.1	2.755	404.2	594.1	1324	1.4654				480.6	357.4	0.21		
72.00	1.3599	1.3623	979.1	2.860	380.8	617.5	1373	1.4703				853.2	628.6	0.12		
74.00	1.3726	1.3751	1015.7	2.967	356.9	641.4	1423	1.4753				1628.	1188.	0.06		
76.00	1.3855	1.3880	1053.0	3.076	332.5	665.7	1473	1.4803								
78.00	1.3986	1.4011	1090.9	3.187	307.7	690.5	1524	1.4854								
80.00	1.4117	1.4142	1129.4	3.299	282.3	715.9	1576	1.4906								
82.00	1.4250	1.4275	1168.5	3.414	256.5	741.7	1628	1.4958								
84.00	1.4383	1.4409	1208.2	3.530	230.1	768.1	1681	1.5010								

89 SULFURIC ACID, H₂SO₄

MOLECULAR WEIGHT = 98.08
RELATIVE SPECIFIC REFRACTIVITY = 0.685

0.00 % by wt. data are the same for all compounds.
For Values of 0.00 wt. % solutions see Table 1, Acetic Acid.

A % by wt.	ρ D_4^{20}	D_{20}^{20}	C, g/l	M g-mol/l	C_w g/l	$(C_o - C_w)$ g/l	$(n - n_o)$ × 10⁴	n	Δ °C	O Os/kg	S g-mol/l	η/η_o	η/ρ cS	φ rhe	γ mmho/cm	T g-mol/l
0.50	1.0016	1.0034	5.0	0.051	996.6	1.7	6	1.3336	0.210	0.113	0.060	1.008	1.009	98.96	24.3	0.277
1.00	1.0049	1.0067	10.0	0.102	994.9	3.3	13	1.3342	0.423	0.227	0.122	1.017	1.014	98.13	47.8	0.573
1.50	1.0083	1.0101	15.1	0.154	993.2	5.1	19	1.3349	0.662	0.356	0.192	1.025	1.019	97.34	70.3	0.886
2.00	1.0116	1.0134	20.2	0.206	991.4	6.8	25	1.3355	0.796	0.428	0.232	1.034	1.024	96.52	92.	1.22
2.50	1.0150	1.0168	25.4	0.259	989.6	8.6	31	1.3361	1.004	0.540	0.293	1.044	1.031	95.55	113.	1.58
3.00	1.0183	1.0201	30.6	0.311	987.8	10.4	37	1.3367	1.172	0.630	0.343	1.057	1.040	94.45	134.	1.98
3.50	1.0217	1.0235	35.8	0.365	985.9	12.3	43	1.3373	1.354	0.728	0.396	1.070	1.049	93.29	155.	2.42
4.00	1.0250	1.0269	41.0	0.418	984.0	14.2	49	1.3379	1.599	0.860	0.468	1.083	1.059	92.11	173.	2.93
4.50	1.0284	1.0302	46.3	0.472	982.1	16.1	55	1.3385	1.855	0.998	0.543	1.097	1.069	90.97	194.	3.57
5.00	1.0318	1.0336	51.6	0.526	980.2	18.0	61	1.3391	2.047	1.101	0.598	1.110	1.078	89.91	211.	4.25
5.50	1.0352	1.0370	56.9	0.580	978.2	20.0	67	1.3397	2.259	1.214	0.659	1.122	1.086	88.93		
6.00	1.0385	1.0404	62.3	0.635	976.2	22.0	73	1.3403	2.495	1.341	0.727	1.134	1.094	88.00		
6.50	1.0419	1.0438	67.7	0.691	974.2	24.0	79	1.3409	2.730	1.468	0.795	1.146	1.102	87.10		
7.00	1.0453	1.0472	73.2	0.746	972.2	26.1	85	1.3415	2.952	1.587	0.858	1.157	1.109	86.24		
7.50	1.0488	1.0506	78.7	0.802	970.1	28.1	91	1.3421	3.197	1.719	0.927	1.169	1.117	85.39		
8.00	1.0522	1.0541	84.2	0.858	968.0	30.2	97	1.3427	3.493	1.878	1.010	1.180	1.124	84.56		
8.50	1.0556	1.0575	89.7	0.915	965.9	32.3	103	1.3433	3.801	2.043	1.096	1.192	1.131	83.74		
9.00	1.0591	1.0610	95.3	0.972	963.8	34.4	109	1.3439	4.083	2.195	1.174	1.204	1.139	82.92		
9.50	1.0626	1.0645	100.9	1.029	961.6	36.6	115	1.3445	4.360	2.344	1.250	1.216	1.146	82.10		
10.00	1.0661	1.0680	106.6	1.087	959.5	38.8	121	1.3451	4.644	2.497	1.328	1.228	1.154	81.27		
11.00	1.0731	1.0750	118.0	1.204	955.1	43.2	133	1.3463	5.25	2.82	1.490	1.253	1.170	79.63		
12.00	1.0802	1.0821	129.6	1.322	950.6	47.6	145	1.3475	5.93	3.19	1.669	1.279	1.187	78.02		
13.00	1.0874	1.0893	141.4	1.441	946.0	52.2	158	1.3488	6.67	3.59	1.859	1.306	1.203	76.43		
14.00	1.0947	1.0966	153.3	1.563	941.4	56.8	170	1.3500	7.49	4.03	2.063	1.334	1.221	74.82		
15.00	1.1020	1.1039	165.3	1.685	936.7	61.5	183	1.3513	8.35	4.49	2.270	1.364	1.240	73.17		
16.00	1.1094	1.1114	177.5	1.810	931.9	66.3	195	1.3525	9.26	4.98	2.483	1.396	1.261	71.47		
17.00	1.1169	1.1189	189.9	1.936	927.0	71.2	208	1.3538	10.23	5.50	2.702	1.431	1.284	69.74		
18.00	1.1245	1.1265	202.4	2.064	922.1	76.2	221	1.3551	11.29	6.07	2.932	1.467	1.308	68.01		
19.00	1.1321	1.1341	215.1	2.193	917.0	81.2	233	1.3563	12.43	6.68	3.169	1.505	1.337	66.32		
20.00	1.1398	1.1418	228.0	2.324	911.6	86.4	246	1.3576	13.64	7.33	3.409	1.543	1.356	64.68		
22.00	1.1554	1.1575	254.2	2.592	901.2	97.0	272	1.3602	16.48	8.86	3.932	1.621	1.405	61.58		
24.00	1.1714	1.1735	281.1	2.866	890.3	108.0	298	1.3628	19.85	10.67	4.488	1.703	1.457	58.60		
26.00	1.1872	1.1893	308.7	3.147	878.5	119.7	323	1.3653	24.29	13.06		1.793	1.513	55.61		
28.00	1.2031	1.2052	336.9	3.435	866.2	132.0	347	1.3677	29.65	15.94		1.890	1.574	52.81		
30.00	1.2191	1.2213	365.7	3.729	853.4	144.9	371	1.3701	36.21	19.47		1.997	1.641	49.98		
32.00	1.2353	1.2375	395.4	4.030	840.0	158.2	395	1.3725	44.76	24.07		2.118	1.718	47.12		
34.00	1.2518	1.2540	425.6	4.339	826.2	172.1	419	1.3749	55.28	29.72		2.250	1.801	44.36		
36.00	1.2685	1.2707	456.7	4.656	811.8	186.4	443	1.3773				2.387	1.885	41.82		
38.00	1.2855	1.2878	488.5	4.981	797.0	201.2	467	1.3797				2.528	1.970	39.48		
40.00	1.3028	1.3051	521.1	5.313	781.7	216.5	491	1.3821				2.685	2.065	37.17		
42.00	1.3205	1.3229	554.6	5.655	765.9	232.3	516	1.3846				2.866	2.174	34.83		
44.00	1.3386	1.3410	589.0	6.005	749.6	248.6	540	1.3870				3.067	2.296	32.53		
46.00	1.3570	1.3594	624.2	6.365	732.8	265.4	565	1.3895				3.292	2.431	30.32		
48.00	1.3759	1.3783	660.4	6.734	715.5	282.8	590	1.3920				3.539	2.577	28.20		
50.00	1.3952	1.3977	697.6	7.113	697.6	300.6	616	1.3945				3.818	2.742	26.14		
52.00	1.4149	1.4174	735.8	7.502	679.2	319.1	641	1.3971				4.134	2.927	24.14		
54.00	1.4351	1.4377	775.0	7.901	660.2	338.1	667	1.3997				4.490	3.135	22.23		
56.00	1.4558	1.4584	815.3	8.312	640.6	357.7	694	1.4024				4.896	3.370	20.38		
58.00	1.4770	1.4796	856.7	8.734	620.3	377.9	720	1.4050				5.343	3.625	18.68		
60.00	1.4987	1.5013	899.2	9.168	599.5	398.8	747	1.4077				5.905	3.948	16.90		
62.00	1.5200	1.5227	942.4	9.608	577.6	420.6										
64.00	1.5421	1.5448	986.9	10.062	555.2	443.0										
66.00	1.5646	1.5674	1032.6	10.528	532.0	466.2										
68.00	1.5874	1.5902	1079.4	11.005	508.0	490.2										
70.00	1.6105	1.6134	1127.4	11.495	483.1	515.1										
72.00	1.6338	1.6367	1176.3	11.993	457.5	540.7										
74.00	1.6574	1.6603	1226.5	12.505	430.9	567.3										
76.00	1.6810	1.6840	1277.6	13.026	403.4	594.8										
78.00	1.7043	1.7073	1329.4	13.554	374.9	623.3										
80.00	1.7272	1.7303	1381.8	14.088	345.4	652.8										
82.00	1.7491	1.7522	1434.3	14.624	314.8	683.4										
84.00	1.7693	1.7724	1486.2	15.153	283.1	715.1										
86.00	1.7872	1.7904	1537.0	15.671	250.2	748.0										
88.00	1.8022	1.8054	1585.9	16.169	216.3	781.9										
90.00	1.8144	1.8176	1633.0	16.650	181.4	816.8										
92.00	1.8240	1.8272	1678.1	17.110	145.9	852.3										
94.00	1.8312	1.8344	1721.3	17.550	109.9	888.3										
96.00	1.8355	1.8388	1762.1	17.966	73.4	924.8										
98.00	1.8361	1.8394	1799.4	18.346	36.7	961.5										
100.00	1.8305	1.8337	1830.5	18.663	0.0	998.2										

90 TARTARIC ACID, $HO_2C(CHOH)_2CO_2H$

MOLECULAR WEIGHT = 150.09

A% by wt.	ρ D_4^{20}	D_{20}^{20}	C_s g/l	M g·mol/l	C_w g/l	$(C_0 - C_w)$ g/l	$(n - n_0)$ ×10⁴	n	Δ °C	O Os/kg	S g·mol/l	η/η_0	η/ρ cS	ϕ rhe	γ mmho/cm	T g·mol/l
1.00	1.0029	1.0047	10.0	0.067	992.9	5.3	12.	1.3342	0.132	0.070	0.037	1.021	1.020	97.76	2.5	0.025
2.00	1.0072	1.0090	20.1	0.134	987.1	11.1	25.	1.3355	0.262	0.141	0.075	1.044	1.039	95.56	3.9	0.039
3.00	1.0121	1.0139	30.4	0.202	981.7	16.5	37.	1.3367	0.402	0.216	0.116	1.069	1.058	93.39	4.9	0.049
4.00	1.0163	1.0181	40.7	0.271	975.6	22.6	49.	1.3379	0.538	0.290	0.156	1.092	1.077	91.36	5.5	0.056
5.00	1.0208	1.0226	51.0	0.340	969.8	28.4	62.	1.3392	0.681	0.366	0.198	1.119	1.098	89.22	6.2	0.063
6.00	1.0255	1.0273	61.5	0.410	964.0	34.2	73.	1.3403	0.824	0.443	0.240	1.147	1.121	86.99	6.9	0.070
7.00	1.0310	1.0329	72.2	0.481	958.8	39.4	88.	1.3418	0.974	0.523	0.284	1.176	1.143	84.86	7.3	0.075
8.00	1.0348	1.0367	82.8	0.552	952.0	46.2	99.	1.3429	1.130	0.607	0.330	1.207	1.169	82.67	7.8	0.080
9.00	1.0393	1.0412	93.5	0.623	945.8	52.4	111.	1.3441	1.283	0.690	0.375	1.236	1.192	80.72	8.2	0.084
10.00	1.0443	1.0462	104.4	0.696	939.9	58.3	124.	1.3454	1.443	0.776	0.422	1.272	1.220	78.49	8.5	0.088
12.00	1.0540	1.0559	126.5	0.843	927.5	70.7	151.	1.3481	1.768	0.951	0.517	1.343	1.277	74.30	9.2	0.095
14.00	1.0639	1.0658	148.9	0.992	915.0	83.2	178.	1.3508	2.130	1.145	0.622	1.413	1.331	70.62	9.7	0.100
16.00	1.0738	1.0757	171.8	1.145	902.0	96.2	205.	1.3535	2.504	1.346	0.730	1.498	1.398	66.61	10.1	0.104
18.00	1.0839	1.0859	195.1	1.300	888.8	109.4	231.	1.3561	2.917	1.568	0.848	1.591	1.471	62.72	10.3	0.107
20.00	1.0942	1.0962	218.8	1.458	875.4	122.8	260.	1.3590	3.360	1.807	0.973	1.701	1.558	58.66	10.5	0.109
24.00	1.1152	1.1172	267.6	1.783	847.6	150.6	316.	1.3646	4.309	2.316	1.236	1.957	1.758	51.01	10.6	0.110
28.00	1.1366	1.1387	318.3	2.120	818.3	179.9	375.	1.3705	5.46	2.93	1.546	2.296	2.024	43.47	10.4	0.108
32.00	1.1588	1.1609	370.8	2.471	788.0	210.2	436.	1.3766	6.73	3.62	1.874	2.695	2.330	37.04	10.1	0.104
36.00	1.1813	1.1834	425.3	2.833	756.0	242.2	499.	1.3829	8.30	4.47	2.259	3.242	2.750	30.78	9.3	0.096
40.00	1.2049	1.2071	482.0	3.211	722.9	275.3	561.	1.3891	10.20	5.48	2.695	4.002	3.328	24.94	8.2	0.085
44.00	1.2290	1.2312	540.8	3.603	688.2	310.0	627.	1.3957				5.088	4.148	19.62	7.3	0.074
48.00	1.2535	1.2558	601.7	4.009	651.8	346.4	692.	1.4022				6.605	5.280	15.11	6.2	0.063
52.00	1.2787	1.2810	664.9	4.430	613.8	384.4	762.	1.4092				8.961	7.022	11.14	5.0	0.050
56.00	1.3058	1.3082	731.3	4.872	574.5	423.7	834.	1.4164				12.60	9.670	7.92	3.7	0.037
58.00	1.3196	1.3220	765.4	5.099	554.2	444.0	869.	1.4199				15.64	11.87	6.38	3.0	0.030

91 TETRACAINE HYDROCHLORIDE, $C_{15}H_{24}N_2O_2 \cdot HCl$

MOLECULAR WEIGHT = 300.84

A% by wt.	ρ D_4^{20}	D_{20}^{20}	C_s g/l	M g·mol/l	C_w g/l	$(C_0 - C_w)$ g/l	$(n - n_0)$ ×10⁴	n	Δ °C	O Os/kg	S g·mol/l	η/η_0	η/ρ cS	ϕ rhe	γ mmho/cm	T g·mol/l
0.50	0.9988	1.0006	5.0	0.017	993.8	4.4	11.	1.3341	0.058	0.031	0.016	1.013	1.016	98.54	1.2	0.012
1.00	0.9995	1.0013	10.0	0.033	989.5	8.7	23.	1.3353	0.110	0.059	0.031	1.029	1.032	96.95	2.4	0.024
1.50	1.0001	1.0019	15.0	0.050	985.1	13.1	34.	1.3364	0.159	0.086	0.045	1.045	1.047	95.50	3.6	0.036
2.00	1.0008	1.0026	20.0	0.067	980.8	17.4	46.	1.3376	0.204	0.109	0.058	1.063	1.064	93.91	4.7	0.047
2.50	1.0014	1.0032	25.0	0.083	976.4	21.8	57.	1.3387				1.081	1.082	92.29	5.6	0.057
3.00	1.0021	1.0039	30.1	0.100	972.0	26.2	69.	1.3399				1.100	1.100	90.72	6.6	0.067
3.50	1.0027	1.0045	35.1	0.117	967.6	30.6	80.	1.3410				1.121	1.120	89.05	7.4	0.076
4.00	1.0034	1.0052	40.1	0.133	963.3	34.9	92.	1.3422				1.142	1.140	87.42	8.2	0.084
4.50	1.0040	1.0058	45.2	0.150	958.8	39.4	103.	1.3433				1.163	1.161	85.79	8.9	0.092
5.00	1.0046	1.0064	50.2	0.167	954.4	43.8	115.	1.3445				1.187	1.184	84.07	9.6	0.099
5.50	1.0053	1.0071	55.3	0.184	950.0	48.2	126.	1.3456				1.211	1.207	82.41	10.2	0.106
6.00	1.0059	1.0077	60.4	0.201	945.5	52.7	138.	1.3468				1.237	1.232	80.69	10.8	0.113
6.50	1.0066	1.0084	65.4	0.217	941.2	57.0	149.	1.3479				1.263	1.257	79.03	11.4	0.120
7.00	1.0072	1.0090	70.5	0.234	936.7	61.5	160.	1.3490				1.291	1.284	77.33	11.9	0.126
7.50	1.0079	1.0097	75.6	0.251	932.3	65.9	172.	1.3502				1.319	1.311	75.68	12.5	0.132
8.00	1.0085	1.0103	80.7	0.268	927.8	70.4	183.	1.3513				1.348	1.339	74.05	13.0	0.138
8.50	1.0092	1.0110	85.8	0.285	923.4	74.8	195.	1.3525				1.379	1.369	72.38	13.6	0.144
9.00	1.0098	1.0116	90.9	0.302	918.9	79.3	206.	1.3536				1.410	1.399	70.79	14.0	0.149
9.50	1.0105	1.0123	96.0	0.319	914.5	83.7	218.	1.3548				1.441	1.429	69.25	14.4	0.154
10.00	1.0111	1.0129	101.1	0.336	910.0	88.2	229.	1.3559				1.477	1.464	67.56	14.7	0.158
11.00	1.0124	1.0142	111.4	0.370	901.0	97.2	252.	1.3582				1.548	1.532	64.48	15.8	0.170
12.00	1.0137	1.0155	121.6	0.404	892.1	106.1	275.	1.3605				1.623	1.604	61.50	16.8	0.182

92 TRICHLOROACETIC ACID, CCl_3COOH

MOLECULAR WEIGHT = 163.38
RELATIVE SPECIFIC REFRACTIVITY = 0.865

0.00 % by wt. data are the same for all compounds.
For Values of 0.00 wt. % solutions see Table 1, Acetic Acid.

A% by wt.	ρ D_4^{20}	D_{20}^{20}	C_s g/l	M g·mol/l	C_w g/l	$(C_0 - C_w)$ g/l	$(n - n_0)$ ×10⁴	n	Δ °C	O Os/kg	S g·mol/l	$(n - n_0)$ η/η_0	η/ρ cS	ϕ rhe	γ mmho/cm	T g·mol/l
0.50	1.0008	1.0026	5.0	0.031	995.8	2.4	7	1.3337	0.105	0.057	0.030	1.009	1.010	98.91	10.3	0.107
1.00	1.0034	1.0051	10.0	0.061	993.3	4.9	13	1.3343	0.211	0.113	0.060	1.019	1.018	97.94	19.6	0.215
2.00	1.0083	1.0101	20.2	0.123	988.2	10.1	26	1.3356	0.423	0.227	0.122	1.042	1.035	95.78	37.2	0.435
3.00	1.0133	1.0151	30.4	0.186	982.9	15.4	39	1.3369	0.638	0.343	0.185	1.067	1.055	93.53	54.0	0.656

92 TRICHLOROACETIC ACID, CCl₃COOH—(Continued)

A % by wt.	ρ D_4^{20}	D_{20}^{20}	C_s g/l	M g-mol/l	C_w g/l	$(C_o - C_w)$ g/l	$(n - n_o)$ × 10⁴	n	Δ °C	O Os/kg	S g-mol/l	η/η_o	η/ρ cS	ϕ rhe	γ mmho/cm	T g-mol/l
4.00	1.0182	1.0200	40.7	0.249	977.4	20.8	52	1.3381	0.857	0.461	0.250	1.094	1.077	91.24	69.8	0.880
5.00	1.0230	1.0248	51.2	0.313	971.9	26.4	64	1.3394	1.079	0.580	0.315	1.121	1.098	89.03	84.7	1.10
6.00	1.0279	1.0297	61.7	0.377	966.2	32.0	76	1.3406	1.303	0.700	0.381	1.148	1.119	86.95	99.	1.33
7.00	1.0328	1.0346	72.3	0.443	960.5	37.7	89	1.3418	1.529	0.822	0.447	1.175	1.140	84.95	113.	1.57
8.00	1.0378	1.0396	83.0	0.508	954.8	43.5	101	1.3431	1.758	0.945	0.514	1.202	1.161	83.01	125.	1.80
9.00	1.0428	1.0446	93.9	0.574	948.9	49.3	114	1.3444	1.992	1.071	0.582	1.231	1.182	81.10	137.	2.04
10.00	1.0479	1.0497	104.8	0.641	943.1	55.1	126	1.3456	2.233	1.201	0.652	1.260	1.205	79.21	148.	2.28
12.00	1.0583	1.0602	127.0	0.777	931.3	66.9	153	1.3483	2.734	1.470	0.796	1.323	1.252	75.44	169.	2.76
14.00	1.0692	1.0710	149.7	0.916	919.5	78.8	180	1.3510	3.261	1.753	0.945	1.390	1.302	71.81	186.	3.24
16.00	1.0806	1.0825	172.9	1.058	907.7	90.6	209	1.3539	3.815	2.051	1.100	1.459	1.353	68.41	200.	3.73
18.00	1.0921	1.0941	196.6	1.203	895.5	102.7	239	1.3568				1.530	1.404	65.21	211.	4.22
20.00	1.1035	1.1055	220.7	1.351	882.8	115.4	267	1.3597				1.605	1.457	62.18	221.	4.90
24.00	1.1260	1.1280	270.2	1.654	855.8	142.5	322	1.3652				1.764	1.570	56.58		
28.00	1.1485	1.1505	321.6	1.968	826.9	171.3	376	1.3705				1.931	1.685	51.68		
32.00	1.1713	1.1734	374.8	2.294	796.5	201.8	429	1.3759				2.114	1.808	47.22		
36.00	1.1947	1.1968	430.1	2.633	764.6	233.6	483	1.3813				2.315	1.942	43.11		
40.00	1.2188	1.2210	487.5	2.984	731.3	266.9	538	1.3868				2.540	2.088	39.29		
44.00	1.2435	1.2457	547.1	3.349	696.3	301.9	593	1.3923				2.791	2.249	35.75		
48.00	1.2682	1.2704	608.7	3.726	659.5	338.8	647	1.3977				3.070	2.425	32.51		
50.00	1.2803	1.2826	640.2	3.918	640.2	358.1	673	1.4003				3.219	2.519	31.00		

93 TRIS(HYDROXYMETHYL)AMINOMETHANE,* THAM,® H₂NC(CH₂OH)₃

MOLECULAR WEIGHT = 121.14
RELATIVE SPECIFIC REFRACTIVITY = 1.169

0.00 % by wt. data are the same for all compounds.
For Values of 0.00 wt. % solutions see Table 1, Acetic Acid.

A % by wt.	ρ D_4^{20}	D_{20}^{20}	C_s g/l	M g-mol/l	C_w g/l	$(C_o - C_w)$ g/l	$(n - n_o)$ × 10⁴	n	Δ °C	O Os/kg	S g-mol/l	η/η_o	η/ρ cS	ϕ rhe	γ mmho/cm	T g-mol/l
0.50	0.9994	1.0012	5.0	0.041	994.4	3.8	7	1.3337	0.080	0.043	0.022	1.012	1.015	98.59		
1.00	1.0006	1.0024	10.0	0.083	990.6	7.6	15	1.3344	0.159	0.085	0.045	1.025	1.026	97.37		
2.00	1.0030	1.0048	20.1	0.166	982.9	15.3	29	1.3359	0.314	0.169	0.090	1.052	1.051	94.87		
3.00	1.0054	1.0072	30.2	0.249	975.2	23.0	44	1.3374	0.473	0.254	0.137	1.081	1.078	92.79		
4.00	1.0078	1.0096	40.3	0.333	967.5	30.7	58	1.3388	0.636	0.342	0.185	1.113	1.106	89.67		
5.00	1.0103	1.0121	50.5	0.417	959.8	38.4	73	1.3403	0.804	0.432	0.234	1.146	1.137	87.09		
6.00	1.0128	1.0146	60.8	0.502	952.0	46.2	88	1.3418	0.974	0.524	0.284	1.180	1.168	84.57		
7.00	1.0153	1.0171	71.1	0.587	944.3	54.0	103	1.3433	1.149	0.618	0.336	1.216	1.200	82.10		
8.00	1.0179	1.0197	81.4	0.672	936.4	61.8	118	1.3448	1.328	0.714	0.388	1.253	1.233	79.66		
9.00	1.0204	1.0222	91.8	0.758	928.6	69.6	133	1.3463	1.510	0.812	0.442	1.292	1.269	77.24		
10.00	1.0230	1.0248	102.3	0.844	920.7	77.5	148	1.3478	1.696	0.912	0.496	1.334	1.307	74.81		
12.00	1.0282	1.0301	123.4	1.019	904.8	93.4	178	1.3508	2.076	1.116	0.606	1.424	1.388	70.09		
14.00	1.0335	1.0354	144.7	1.194	888.8	109.4	209	1.3539	2.473	1.329	0.721	1.524	1.478	65.48		
16.00	1.0389	1.0407	166.2	1.372	872.7	125.6	240	1.3570	2.899	1.559	0.843	1.639	1.581	60.88		
18.00	1.0443	1.0462	188.0	1.552	856.4	141.9	271	1.3601	3.357	1.805	0.972	1.768	1.696	56.45		
20.00	1.0498	1.0517	210.0	1.733	839.9	158.4	303	1.3633	3.847	2.068	1.109	1.916	1.829	52.09		
22.00	1.0554	1.0572	232.2	1.917	823.2	175.0	335	1.3665				2.079	1.974	48.00		
24.00	1.0610	1.0628	254.6	2.102	806.3	191.9	367	1.3697				2.256	2.131	44.23		
26.00	1.0666	1.0685	277.3	2.289	789.3	208.9	400	1.3730				2.459	2.310	40.59		
28.00	1.0723	1.0742	300.2	2.479	772.1	226.2	433	1.3763				2.700	2.523	36.97		
30.00	1.0781	1.0800	323.4	2.670	754.6	243.6	467	1.3797				2.992	2.781	33.36		
32.00	1.0839	1.0858	346.8	2.863	737.0	261.2	501	1.3831				3.337	3.085	29.91		
34.00	1.0897	1.0916	370.5	3.058	719.2	279.0	535	1.3865				3.729	3.429	26.76		
36.00	1.0956	1.0975	394.4	3.256	701.2	297.0	570	1.3900				4.169	3.812	23.94		
38.00	1.1016	1.1035	418.6	3.456	683.0	315.2	605	1.3935				4.658	4.237	21.43		
40.00	1.1076	1.1096	443.1	3.657	664.6	333.6	640	1.3970				5.198	4.702	19.20		

* 1,2-Propanediol, 2-amino-2(hydroxymethyl).   ® Registered trademark.

94 UREA, NH₂CONH₂

MOLECULAR WEIGHT = 60.06
RELATIVE SPECIFIC REFRACTIVITY = 1.147

0.00 % by wt. data are the same for all compounds.
For Values of 0.00 wt. % solutions see Table 1, Acetic Acid.

A % by wt.	ρ D_4^{20}	D_{20}^{20}	C_s g/l	M g-mol/l	C_w g/l	$(C_o - C_w)$ g/l	$(n - n_o)$ × 10⁴	n	Δ °C	O Os/kg	S g-mol/l	η/η_o	η/ρ cS	ϕ rhe	γ mmho/cm	T g-mol/l
0.50	0.9995	1.0012	5.0	0.083	994.5	3.8	7	1.3337	0.155	0.083	0.044	1.005	1.007	99.33		
1.00	1.0007	1.0025	10.0	0.167	990.7	7.5	14	1.3344	0.310	0.167	0.089	1.008	1.009	99.01		
1.50	1.0020	1.0038	15.0	0.250	987.0	11.3	21	1.3351	0.465	0.250	0.134	1.009	1.009	98.89		
2.00	1.0033	1.0050	20.1	0.334	983.2	15.0	28	1.3358	0.620	0.333	0.180	1.010	1.009	98.81		
2.50	1.0045	1.0063	25.1	0.418	979.4	18.8	35	1.3365	0.775	0.417	0.226	1.012	1.010	98.60		
3.00	1.0058	1.0076	30.2	0.502	975.7	22.6	42	1.3372	0.928	0.499	0.271	1.015	1.011	98.30		
3.50	1.0071	1.0089	35.3	0.587	971.9	26.3	50	1.3379	1.082	0.582	0.316	1.019	1.014	97.95		
4.00	1.0085	1.0102	40.3	0.672	968.1	30.1	57	1.3387	1.237	0.665	0.362	1.023	1.016	97.56		
4.50	1.0098	1.0116	45.4	0.757	964.3	33.9	64	1.3394	1.393	0.749	0.407	1.027	1.019	97.17		
5.00	1.0111	1.0129	50.6	0.842	960.6	37.7	71	1.3401	1.552	0.835	0.454	1.031	1.022	96.80		
5.50	1.0125	1.0142	55.7	0.927	956.8	41.5	79	1.3409	1.715	0.922	0.501	1.035	1.024	96.45		
6.00	1.0138	1.0156	60.8	1.013	953.0	45.3	86	1.3416	1.880	1.011	0.550	1.039	1.027	96.09		
6.50	1.0151	1.0169	66.0	1.099	949.2	49.1	93	1.3423	2.048	1.101	0.598	1.043	1.029	95.72		
7.00	1.0165	1.0183	71.2	1.185	945.4	52.9	101	1.3431	2.218	1.192	0.648	1.047	1.032	95.35		
7.50	1.0179	1.0197	76.3	1.271	941.5	56.7	108	1.3438	2.389	1.285	0.697	1.051	1.034	94.98		
8.00	1.0192	1.0210	81.5	1.358	937.7	60.5	116	1.3446	2.562	1.378	0.747	1.055	1.037	94.60		
8.50	1.0206	1.0224	86.8	1.444	933.9	64.4	123	1.3453	2.737	1.471	0.797	1.059	1.040	94.23		

94 UREA, NH$_2$CONH$_2$—(Continued)

A% by wt.	ρ D$_4^{20}$	D$_{20}^{20}$	C$_s$ g/l	M g-mol/l	C$_w$ g/l	(C$_o$ − C$_w$) g/l	(n − n$_o$) × 10^4	n	Δ °C	O Os/kg	S g-mol/l	η/η_o	η/ρ cS	ϕ rhe	γ mmho/cm	T g-mol/l
9.00	1.0220	1.0238	92.0	1.531	930.0	68.2	131	1.3461	2.911	1.565	0.846	1.063	1.043	93.85		
9.50	1.0234	1.0252	97.2	1.619	926.2	72.1	138	1.3468	3.086	1.659	0.896	1.068	1.045	93.47		
10.00	1.0248	1.0266	102.5	1.706	922.3	75.9	146	1.3476	3.260	1.753	0.945	1.072	1.048	93.10		
11.00	1.0276	1.0294	113.0	1.882	914.5	83.7	161	1.3491	3.606	1.939	1.042	1.081	1.054	92.36		
12.00	1.0304	1.0322	123.6	2.059	906.7	91.5	176	1.3506	3.952	2.125	1.138	1.089	1.059	91.64		
13.00	1.0332	1.0350	134.3	2.236	898.9	99.4	192	1.3521	4.301	2.312	1.234	1.098	1.065	90.91		
14.00	1.0360	1.0378	145.0	2.415	891.0	107.3	207	1.3537	4.656	2.503	1.332	1.107	1.071	90.15		
15.00	1.0388	1.0407	155.8	2.594	883.0	115.2	222	1.3552	5.02	2.70	1.430	1.117	1.077	89.35		
16.00	1.0417	1.0435	166.7	2.775	875.0	123.2	238	1.3568	5.40	2.90	1.531	1.128	1.085	88.50		
17.00	1.0445	1.0463	177.6	2.956	866.9	131.3	253	1.3583	5.79	3.11	1.633	1.139	1.093	87.62		
18.00	1.0473	1.0492	188.5	3.139	858.8	139.4	269	1.3599	6.19	3.33	1.736	1.151	1.101	86.72		
19.00	1.0502	1.0520	199.5	3.322	850.6	147.6	284	1.3614	6.59	3.54	1.839	1.163	1.110	85.80		
20.00	1.0530	1.0549	210.6	3.506	842.4	155.8	300	1.3629	7.00	3.76	1.941	1.176	1.119	84.86		
22.00	1.0586	1.0605	232.9	3.878	825.7	172.5	331	1.3661	7.81	4.20	2.141	1.203	1.139	82.93		
24.00	1.0643	1.0662	255.4	4.253	808.9	189.4	362	1.3692	8.64	4.65	2.339	1.233	1.161	80.95		
26.00	1.0699	1.0718	278.2	4.632	791.7	206.5	393	1.3723	9.52	5.12	2.543	1.263	1.183	79.00		
28.00	1.0756	1.0775	301.2	5.014	774.4	223.8	424	1.3754	10.45	5.62	2.751	1.295	1.206	77.09		
30.00	1.0812	1.0831	324.4	5.401	756.8	241.4	455	1.3785	11.40	6.13	2.955	1.329	1.232	75.09		
32.00	1.0869	1.0888	347.8	5.791	739.1	259.2	487	1.3817	12.34	6.63	3.150	1.368	1.261	72.97		
34.00	1.0926	1.0945	371.5	6.185	721.1	277.1	518	1.3848	13.27	7.14	3.337	1.410	1.293	70.77		
36.00	1.0984	1.1004	395.4	6.584	703.0	295.2	551	1.3881	14.20	7.63	3.516	1.456	1.329	68.52		
38.00	1.1044	1.1064	419.7	6.988	684.7	313.5	583	1.3913	15.11	8.12	3.686	1.506	1.367	66.26		
40.00	1.1106	1.1126	444.2	7.397	666.4	331.9	617	1.3947	15.99	8.60	3.845	1.562	1.409	63.89		
42.00	1.1171	1.1191	469.2	7.812	647.9	350.3	652	1.3982	16.83	9.05	3.993	1.626	1.458	61.40		
44.00	1.1239	1.1259	494.5	8.234	629.4	368.8	688	1.4018	17.62	9.47	4.127	1.697	1.512	58.83		
46.00	1.1313	1.1333	520.4	8.665	610.9	387.3	726	1.4056				1.776	1.573	56.21		

95 URINE, CAT

RELATIVE SPECIFIC REFRACTIVITY = 1.08

0.00 % by wt. data are the same for all compounds.
For Values of 0.00 wt. % solutions see Table 1, Acetic Acid.

A% by wt.	ρ D$_4^{20}$	D$_{20}^{20}$	C$_s$ g/l	C$_w$ g/l	(C$_o$ − C$_w$) g/l	(n − n$_o$) × 10^4	n
0.50	1.000	1.002	5	995	3	8	1.3338
1.00	1.002	1.004	10	992	6	16	1.3346
1.50	1.004	1.006	15	989	9	24	1.3354
2.00	1.006	1.008	20	986	12	33	1.3363
2.50	1.008	1.010	25	983	15	41	1.3371
3.00	1.010	1.012	30	979	19	49	1.3379
3.50	1.011	1.013	35	976	22	57	1.3387
4.00	1.013	1.015	41	973	25	65	1.3395
4.50	1.015	1.017	46	970	28	73	1.3403
5.00	1.017	1.019	51	966	32	81	1.3411
5.50	1.019	1.021	56	963	35	89	1.3419
6.00	1.021	1.023	61	960	38	97	1.3427
6.50	1.023	1.025	67	956	42	105	1.3435
7.00	1.025	1.027	72	953	45	113	1.3443
7.50	1.026	1.028	77	950	48	121	1.3451
8.00	1.028	1.030	82	946	52	129	1.3459
8.50	1.030	1.032	88	943	55	136	1.3466
9.00	1.032	1.034	93	939	59	144	1.3474
9.50	1.034	1.036	98	936	62	152	1.3482
10.00	1.036	1.038	104	932	66	160	1.3490
11.00	1.039	1.041	114	925	73	175	1.3505
12.00	1.043	1.045	125	918	80	191	1.3521
13.00	1.047	1.049	136	911	87	206	1.3536
14.00	1.050	1.052	147	903	95	222	1.3552
15.00	1.054	1.056	158	896	102	237	1.3567
16.00	1.058	1.060	169	888	110	252	1.3582
17.00	1.061	1.063	180	881	117	267	1.3597
18.00	1.065	1.067	192	873	125	282	1.3612
19.00	1.068	1.070	203	865	133	296	1.3626
20.00	1.072	1.074	214	857	141	311	1.3641

96 URINE, GUINEA PIG

RELATIVE SPECIFIC REFRACTIVITIES: CLEAR* = 0.984; TURBID = 0.955

0.00 % by wt. data are the same for all compounds.
For Values of 0.00 wt. % solutions see Table 1, Acetic Acid.

	CLEAR							TURBID				
A% by wt.	ρ D$_4^{20}$	D$_{20}^{20}$	C$_s$ g/l	C$_w$ g/l	(C$_o$ − C$_w$) g/l	(n − n$_o$) × 10^4	n	A% by wt.	ρ D$_4^{20}$	D$_{20}^{20}$	C$_s$ g/l	C$_w$ g/l	(C$_o$ − C$_w$) g/l
---	---	---	---	---	---	---	---	---	---	---	---	---	---
0.5	1.001	1.002	5	996	2	8	1.3338	0.5	1.001	1.002	5	996	2
1.0	1.003	1.005	10	993	5	16	1.3346	1.0	1.003	1.005	10	993	5
1.5	1.005	1.007	15	990	8	23	1.3353	1.5	1.005	1.007	15	990	8
2.0	1.008	1.009	20	988	10	31	1.3361	2.1	1.008	1.010	21	987	11
2.5	1.010	1.012	25	985	13	39	1.3369	2.6	1.010	1.012	26	984	14
3.0	1.012	1.014	30	982	16	47	1.3377	3.1	1.013	1.015	31	982	16

* Cleared by centrifuging 15 minutes at 25,000 G.

96 URINE, GUINEA PIG—(Continued)

	CLEAR							TURBID				
A % by wt.	ρ D_4^{20}	D_{20}^{20}	C_s g/l	C_w g/l	$(C_o - C_w)$ g/l	$(n - n_o)$ $\times 10^4$	n	A % by wt.	ρ D_4^{20}	D_{20}^{20}	C_s g/l	C_w g/l	$(C_o - C_w)$ g/l
3.5	1.015	1.016	36	979	19	55	1.3385	3.6	1.015	1.017	37	978	20
4.0	1.017	1.019	41	976	22	63	1.3393	4.1	1.018	1.020	42	976	22
4.5	1.019	1.021	46	973	25	70	1.3400	4.6	1.020	1.022	47	973	25
5.0	1.022	1.024	51	971	27	78	1.3408	5.2	1.023	1.025	53	970	28
5.5	1.024	1.026	56	968	30	86	1.3416	5.7	1.026	1.027	58	968	30
6.0	1.026	1.028	62	964	34	94	1.3424	6.2	1.029	1.030	64	965	33
6.5	1.029	1.031	67	962	36	102	1.3432	6.7	1.031	1.032	69	962	36
7.0	1.031	1.033	72	959	39	110	1.3440	7.2	1.034	1.035	74	960	38
7.5	1.033	1.035	77	956	42	117	1.3447	7.8	1.036	1.038	81	955	43
8.0	1.036	1.038	83	953	45	125	1.3455	8.3	1.038	1.040	86	952	46
8.5	1.038	1.040	88	950	48	133	1.3463	8.8	1.040	1.042	92	948	50
9.0	1.040	1.042	94	946	52	141	1.3471	9.3	1.043	1.045	97	946	52
9.5	1.043	1.045	99	944	54	149	1.3479	9.8	1.046	1.047	103	943	55
10.0	1.045	1.047	104	941	57	157	1.3487	10.3	1.049	1.050	108	941	57
11.0	1.050	1.052	116	934	64	172	1.3502	11.4	1.054	1.055	120	934	64
12.0	1.054	1.056	126	928	70	188	1.3518	12.4	1.058	1.060	131	927	71
13.0	1.059	1.061	138	921	77	204	1.3534	13.4	1.064	1.066	143	921	77
14.0	1.064	1.066	149	915	83	220	1.3550	14.5	1.069	1.071	155	914	84
15.0	1.069	1.070	160	909	89	235	1.3565	15.5	1.074	1.076	166	908	90
16.0	1.073	1.075	172	901	97	251	1.3581	16.5	1.078	1.080	178	900	98
17.0	1.078	1.080	183	895	103	266	1.3596	17.6	1.084	1.086	191	893	105
18.0	1.083	1.085	195	888	110	282	1.3612	18.6	1.089	1.091	203	886	112
19.0	1.087	1.089	207	880	118	298	1.3628	19.6	1.094	1.096	214	880	118
20.0	1.092	1.094	218	874	124	314	1.3644	20.7	1.099	1.101	227	872	126

97 URINE, HUMAN

RELATIVE SPECIFIC REFRACTIVITY = 1.02

0.00 % by wt. data are the same for all compounds.
For Values of 0.00 wt. % solutions see Table 1, Acetic Acid.

A % by wt.	ρ D_4^{20}	D_{20}^{20}	C_s g/l	C_w g/l	$(C_o - C_w)$ g/l	$(n - n_o)$ $\times 10^4$	n
0.50	1.000	1.002	5	996	2	8	1.3338
1.00	1.003	1.003	10	993	5	16	1.3346
1.50	1.005	1.007	15	990	8	24	1.3354
2.00	1.007	1.009	20	987	11	32	1.3362
2.50	1.009	1.011	24	984	14	40	1.3370
3.00	1.012	1.014	31	981	17	47	1.3377
3.50	1.014	1.016	36	978	20	55	1.3385
4.00	1.016	1.018	41	975	23	63	1.3393
4.50	1.018	1.020	46	972	26	71	1.3401
5.00	1.020	1.022	51	969	29	78	1.3408
5.50	1.022	1.024	56	966	32	86	1.3416
6.00	1.024	1.026	61	962	36	94	1.3424
6.50	1.026	1.028	67	959	39	101	1.3431
7.00	1.027	1.029	72	956	42	109	1.3439
7.50	1.029	1.031	77	952	46	116	1.3446
8.00	1.031	1.033	83	949	49	123	1.3453
8.50	1.033	1.035	88	945	53	131	1.3461
9.00	1.034	1.036	93	941	57	138	1.3468
9.50	1.036	1.038	98	937	61	145	1.3475
10.00	1.037	1.039	104	934	64	152	1.3482

98 URINE, RABBIT

RELATIVE SPECIFIC REFRACTIVITIES: CLEAR* = 0.997; TURBID = 0.958

0.00 % by wt. data are the same for all compounds.
For Values of 0.00 wt. % solutions see Table 1, Acetic Acid.

	CLEAR							TURBID				
A % by wt.	ρ D_4^{20}	D_{20}^{20}	C_s g/l	C_w g/l	$(C_o - C_w)$ g/l	$(n - n_o)$ $\times 10^4$	n	A % by wt.	ρ D_4^{20}	D_{20}^{20}	C_s g/l	C_w g/l	$(C_o - C_w)$ g/l
0.5	1.000	1.002	5	995	3	7	1.3337	0.6	1.002	1.003	6	996	2
1.0	1.003	1.004	10	993	5	15	1.3345	1.1	1.003	1.005	11	992	6
1.5	1.005	1.007	15	990	8	22	1.3352	1.7	1.006	1.008	17	989	9
2.0	1.007	1.009	20	987	11	30	1.3360	2.2	1.008	1.010	22	986	12
2.5	1.009	1.011	25	984	14	38	1.3368	2.7	1.011	1.012	27	984	14
3.0	1.011	1.013	30	981	17	45	1.3375	3.2	1.013	1.014	32	981	17
3.5	1.014	1.015	35	979	19	53	1.3383	3.8	1.016	1.017	39	977	21
4.0	1.016	1.018	41	975	23	60	1.3390	4.3	1.018	1.019	44	974	24
4.5	1.018	1.020	46	972	26	68	1.3398	4.9	1.020	1.021	50	970	28
5.0	1.020	1.022	51	969	29	76	1.3406	5.4	1.022	1.024	55	967	31
5.5	1.022	1.024	56	966	32	83	1.3413	5.9	1.024	1.026	60	964	34
6.0	1.025	1.026	62	963	35	91	1.3421	6.5	1.027	1.029	67	960	38
6.5	1.027	1.029	67	960	38	98	1.3428	7.0	1.029	1.031	72	957	41

* Cleared by centrifuging 15 minutes at 25,000 G.

98 URINE, RABBIT—(Continued)

CLEAR								TURBID					
A% by wt.	ρ D_4^{20}	D_{20}^{20}	C, g/l	C_w g/l	$(C_o - C_w)$ g/l	$(n - n_o)$ $\times 10^4$	n	A% by wt.	ρ D_4^{20}	D_{20}^{20}	C, g/l	C_w g/l	$(C_o - C_w)$ g/l
7.0	1.029	1.031	72	957	41	106	1.3436	7.5	1.032	1.034	77	955	43
7.5	1.031	1.033	77	954	44	113	1.3443	8.0	1.034	1.036	83	951	47
8.0	1.033	1.035	83	950	48	121	1.3451	8.6	1.037	1.039	89	948	50
8.5	1.036	1.037	88	948	50	128	1.3458	9.1	1.039	1.041	95	944	54
9.0	1.038	1.040	93	945	53	136	1.3466	9.7	1.042	1.044	101	941	57
9.5	1.040	1.042	99	941	57	143	1.3473	10.2	1.044	1.046	106	938	60
10.0	1.042	1.044	104	938	60	151	1.3481	10.8	1.047	1.049	113	934	64
11.0	1.046	1.048	115	931	67	166	1.3496	11.8	1.051	1.053	124	927	71
12.0	1.051	1.053	126	925	73	181	1.3511	12.9	1.056	1.058	136	920	78
13.0	1.055	1.057	137	918	80	196	1.3526	14.0	1.061	1.063	149	912	86
14.0	1.060	1.062	148	912	86	211	1.3541	15.1	1.066	1.068	161	905	93
15.0	1.064	1.066	160	904	94	226	1.3556	16.2	1.071	1.073	174	897	101
16.0	1.068	1.070	171	897	101	242	1.3572	17.2	1.076	1.077	185	891	107
17.0	1.073	1.075	182	891	107	257	1.3587	18.3	1.080	1.082	198	882	116

99 ZINC SULFATE, $ZnSO_4 \cdot 7H_2O$

MOLECULAR WEIGHT = 161.44   FORMULA WEIGHT, HYDRATE = 287.56
RELATIVE SPECIFIC REFRACTIVITY = 0.493

0.00 % by wt. data are the same for all compounds.
For Values of 0.00 wt. % solutions see Table 1, Acetic Acid.

A% by wt.	H% by wt.	ρ D_4^{20}	D_{20}^{20}	C, g/l	M g-mol/l	C_w g/l	$(C_o - C_w)$ g/l	$(n - n_o)$ $\times 10^4$	n	Δ °C	O Os/kg	S g-mol/l	η/η_o	η/ρ cS	ϕ rhe	γ mmho/cm	T g-mol/l
0.50	0.89	1.0034	1.0051	5.0	0.031	998.3	−0.1	9	1.3339	0.079	0.043	0.022	1.019	1.017	97.97	2.8	0.028
1.00	1.78	1.0085	1.0103	10.1	0.062	998.4	−0.2	18	1.3348	0.150	0.081	0.042	1.038	1.031	96.15	5.4	0.054
1.50	2.67	1.0137	1.0155	15.2	0.094	998.5	−0.3	27	1.3357	0.217	0.117	0.062	1.058	1.046	94.32	7.8	0.080
2.00	3.56	1.0190	1.0208	20.4	0.126	998.6	−0.4	36	1.3366	0.282	0.152	0.081	1.079	1.061	92.49	10.0	0.104
2.50	4.45	1.0243	1.0261	25.6	0.159	998.7	−0.4	45	1.3375	0.346	0.186	0.099	1.101	1.077	90.65	12.1	0.127
3.00	5.34	1.0296	1.0314	30.9	0.191	998.7	−0.5	54	1.3384	0.408	0.219	0.118	1.124	1.094	88.80	13.9	0.148
3.50	6.23	1.0349	1.0368	36.2	0.224	998.7	−0.5	64	1.3393	0.470	0.253	0.136	1.148	1.111	86.94	15.7	0.169
4.00	7.12	1.0403	1.0421	41.6	0.258	998.7	−0.5	73	1.3403	0.531	0.286	0.154	1.173	1.130	85.10	17.3	0.188
4.50	8.02	1.0457	1.0475	47.1	0.291	998.6	−0.4	82	1.3412	0.591	0.318	0.172	1.198	1.148	83.27	18.9	0.207
5.00	8.91	1.0511	1.0530	52.6	0.326	998.5	−0.3	91	1.3421	0.651	0.350	0.189	1.225	1.168	81.47	20.5	0.226
5.50	9.80	1.0565	1.0584	58.1	0.360	998.4	−0.2	100	1.3430	0.711	0.382	0.207	1.252	1.188	79.70	22.0	0.245
6.00	10.69	1.0620	1.0639	63.7	0.395	998.3	0.0	109	1.3439	0.772	0.415	0.225	1.280	1.208	77.98	23.5	0.263
6.50	11.58	1.0675	1.0694	69.4	0.430	998.1	0.1	118	1.3448	0.833	0.448	0.243	1.308	1.228	76.28	24.9	0.280
7.00	12.47	1.0730	1.0749	75.1	0.465	997.9	0.3	127	1.3457	0.893	0.480	0.260	1.338	1.249	74.60	26.3	0.297
7.50	13.36	1.0786	1.0805	80.9	0.501	997.7	0.5	136	1.3466	0.953	0.512	0.278	1.368	1.271	72.95	27.6	0.314
8.00	14.25	1.0842	1.0861	86.7	0.537	997.5	0.8	146	1.3475	1.013	0.544	0.296	1.400	1.294	71.30	28.9	0.330
8.50	15.14	1.0899	1.0918	92.6	0.574	997.2	1.0	155	1.3485	1.073	0.577	0.313	1.433	1.317	69.65	30.1	0.345
9.00	16.03	1.0956	1.0975	98.6	0.611	997.0	1.3	164	1.3494	1.136	0.611	0.332	1.467	1.342	68.01	31.3	0.360
9.50	16.92	1.1013	1.1032	104.6	0.648	996.7	1.6	173	1.3503	1.200	0.645	0.351	1.504	1.368	66.37	32.5	0.375
10.00	17.81	1.1071	1.1091	110.7	0.686	996.4	1.8	183	1.3513	1.267	0.681	0.370	1.542	1.396	64.72	33.7	0.390
11.00	19.59	1.1188	1.1208	123.1	0.762	995.8	2.5	202	1.3532	1.407	0.756	0.411	1.624	1.454	61.45	35.9	0.418
12.00	21.37	1.1308	1.1328	135.7	0.841	995.1	3.1	221	1.3551	1.553	0.835	0.454	1.713	1.518	58.25	38.0	0.445
13.00	23.16	1.1429	1.1450	148.5	0.920	994.4	3.9	241	1.3570	1.711	0.920	0.500	1.810	1.587	55.14	39.9	0.470
14.00	24.94	1.1553	1.1573	161.7	1.002	993.6	4.7	260	1.3590	1.889	1.015	0.552	1.914	1.660	52.14	41.7	0.492
15.00	26.72	1.1679	1.1699	175.2	1.085	992.7	5.6	280	1.3610	2.087	1.122	0.610	2.027	1.739	49.24	43.3	0.513
16.00	28.50	1.1806	1.1827	188.9	1.170	991.7	6.5	300	1.3630	2.306	1.240	0.673	2.148	1.823	46.46	44.6	0.531

OSMOTIC PARAMETERS AND ELECTRICAL CONDUCTIVITIES OF AQUEOUS SOLUTIONS

For any aqueous solutions within the range, the two right hand columns provide values for either condosity (T) or specific conductance (γ) where the other value is given; the three left hand columns provide values for freezing point depression (Δ), osmolality (O) or osmosity (S) where any one of these values is given. Values in all four columns are interconvertible *only* for solutions of sodium chloride, in which case, by definition, osmosity, molarity (M) and condosity are always equal. Other concentrative properties of specific solutions may be found in conversion tables on preceding pages.

Table columns are:

Δ = freezing point depression, °C

O = osmolality ($=\Delta/1.86$, Os/kg

S = osmosity, molar concentration of NaCl solution having same freezing point or osmotic pressure as given solution, g-mol/l

M = molarity of NaCl solution, g-mol/l

T = condosity, molar concentration of NaCl solution having same specific conductance at 20°C as given solution, g-mol/l

γ = specific conductance (electrical) at 20°C, mmho/cm.

Δ °C	O Os/Kg	S,M,T g-mol/l	γ mmho/cm	Δ °C	O Os/Kg	S,M,T g-mol/l	γ mmho/cm	Δ °C	O Os/Kg	S,M,T g-mol/l	γ mmho/cm
0.000	0.000	0.000	0.0	1.881	1.011	0.550	46.1	3.815	2.051	1.100	84.3
0.037	0.020	0.010	1.0	1.916	1.030	0.560	46.8	3.851	2.070	1.110	85.0
0.072	0.039	0.020	2.0	1.950	1.048	0.570	47.6	3.887	2.090	1.120	85.6
0.107	0.057	0.030	3.0	1.985	1.067	0.580	48.3	3.923	2.109	1.130	86.2
0.142	0.076	0.040	4.0	2.019	1.086	0.590	49.1	3.959	2.129	1.140	86.9
0.176	0.095	0.050	5.0	2.054	1.104	0.600	49.8	3.995	2.148	1.150	87.5
0.211	0.113	0.060	5.9	2.088	1.123	0.610	50.6	4.032	2.168	1.160	88.1
0.245	0.132	0.070	6.9	2.123	1.141	0.620	51.3	4.068	2.187	1.170	88.8
0.279	0.150	0.080	7.8	2.157	1.160	0.630	52.0	4.104	2.207	1.180	89.4
0.314	0.169	0.090	8.7	2.192	1.178	0.640	52.8	4.141	2.226	1.190	90.0
0.348	0.187	0.100	9.7	2.226	1.197	0.650	53.5	4.177	2.246	1.200	90.6
0.382	0.205	0.110	10.6	2.261	1.216	0.660	54.2	4.214	2.265	1.210	91.3
0.416	0.224	0.120	11.4	2.296	1.234	0.670	55.0	4.250	2.285	1.220	91.9
0.450	0.242	0.130	12.3	2.330	1.253	0.680	55.7	4.287	2.305	1.230	92.5
0.484	0.260	0.140	13.2	2.365	1.271	0.690	56.4	4.323	2.324	1.240	93.1
0.518	0.279	0.150	14.1	2.400	1.290	0.700	57.1	4.360	2.344	1.250	93.7
0.552	0.297	0.160	14.9	2.434	1.309	0.710	57.9	4.393	2.362	1.260	94.3
0.586	0.315	0.170	15.8	2.469	1.328	0.720	58.6	4.429	2.381	1.270	94.9
0.620	0.333	0.180	16.6	2.504	1.346	0.730	59.3	4.466	2.401	1.280	95.6
0.654	0.352	0.190	17.5	2.539	1.365	0.740	60.0	4.503	2.421	1.290	96.2
0.688	0.370	0.200	18.3	2.574	1.384	0.750	60.7	4.539	2.441	1.300	96.8
0.722	0.388	0.210	19.2	2.609	1.403	0.760	61.4	4.576	2.460	1.310	97.4
0.756	0.407	0.220	20.0	2.644	1.421	0.770	62.1	4.613	2.480	1.320	98.0
0.790	0.425	0.230	20.8	2.679	1.440	0.780	62.9	4.650	2.500	1.330	98.6
0.824	0.443	0.240	21.6	2.714	1.459	0.790	63.6	4.687	2.520	1.340	99.2
0.858	0.461	0.250	22.4	2.749	1.478	0.800	64.3	4.724	2.540	1.350	99.8
0.892	0.480	0.260	23.3	2.784	1.497	0.810	65.0	4.761	2.560	1.360	100
0.926	0.498	0.270	24.1	2.819	1.515	0.820	65.7	4.798	2.580	1.370	101
0.960	0.516	0.280	24.9	2.854	1.534	0.830	66.3	4.835	2.600	1.380	102
0.994	0.534	0.290	25.7	2.889	1.553	0.840	67.0	4.872	2.620	1.390	102
1.028	0.553	0.300	26.5	2.924	1.572	0.850	67.7	4.910	2.640	1.400	103
1.062	0.571	0.310	27.3	2.959	1.591	0.860	68.4	4.947	2.660	1.410	103
1.096	0.589	0.320	28.1	2.995	1.610	0.870	69.1	4.984	2.680	1.420	104
1.130	0.607	0.330	28.9	3.030	1.629	0.880	69.8	5.02	2.70	1.430	104
1.164	0.626	0.340	29.7	3.065	1.648	0.890	70.5	5.06	2.72	1.440	105
1.198	0.644	0.350	30.5	3.101	1.667	0.900	71.2	5.10	2.74	1.450	106
1.232	0.662	0.360	31.3	3.136	1.686	0.910	71.8	5.13	2.76	1.460	106
1.266	0.681	0.370	32.1	3.171	1.705	0.920	72.5	5.17	2.78	1.470	107
1.300	0.699	0.380	32.9	3.207	1.724	0.930	73.2	5.21	2.80	1.480	107
1.334	0.717	0.390	33.7	3.242	1.743	0.940	73.9	5.25	2.82	1.490	108
1.368	0.736	0.400	34.4	3.278	1.762	0.950	74.5	5.28	2.84	1.500	109
1.402	0.754	0.410	35.2	3.313	1.781	0.960	75.2	5.32	2.86	1.510	109
1.436	0.772	0.420	36.0	3.349	1.801	0.970	75.9	5.36	2.88	1.520	110
1.470	0.791	0.430	36.8	3.385	1.820	0.980	76.5	5.40	2.90	1.530	110
1.505	0.809	0.440	37.6	3.420	1.839	0.990	77.2	5.44	2.92	1.540	111
1.539	0.827	0.450	38.4	3.456	1.858	1.000	77.8	5.47	2.94	1.550	111
1.573	0.846	0.460	39.2	3.492	1.877	1.010	78.5	5.51	2.96	1.560	112
1.607	0.864	0.470	39.9	3.527	1.896	1.020	79.2	5.55	2.98	1.570	112
1.641	0.882	0.480	40.7	3.563	1.916	1.030	79.8	5.59	3.00	1.580	113
1.676	0.901	0.490	41.5	3.599	1.935	1.040	80.5	5.63	3.02	1.590	114
1.710	0.919	0.500	42.3	3.635	1.954	1.050	81.1	5.66	3.05	1.600	114
1.744	0.938	0.510	43.0	3.671	1.974	1.060	81.8	5.70	3.07	1.610	115
1.778	0.956	0.520	43.8	3.707	1.993	1.070	82.4	5.74	3.09	1.620	115
1.813	0.975	0.530	44.6	3.743	2.012	1.080	83.1	5.78	3.11	1.630	116
1.847	0.993	0.540	45.3	3.779	2.032	1.090	83.7	5.82	3.13	1.640	116

Δ °C	O Os/Kg	S,M,T. g-mol/l	γ mmho/cm
5.86	3.15	1.650	117
5.89	3.17	1.660	117
5.93	3.19	1.670	118
5.97	3.21	1.680	119
6.01	3.23	1.690	119
6.05	3.25	1.700	120
6.09	3.27	1.710	120
6.13	3.29	1.720	121
6.16	3.31	1.730	121
6.20	3.34	1.740	122
6.24	3.36	1.750	122
6.28	3.38	1.760	123
6.32	3.40	1.770	123
6.36	3.42	1.780	124
6.40	3.44	1.790	125
6.44	3.46	1.800	125
6.48	3.48	1.810	126
6.52	3.50	1.820	126
6.56	3.52	1.830	127
6.60	3.55	1.840	127
6.64	3.57	1.850	128
6.67	3.59	1.860	128
6.71	3.61	1.870	129
6.75	3.63	1.880	129
6.79	3.65	1.890	129
6.83	3.67	1.900	130
6.87	3.70	1.910	130
6.91	3.72	1.920	131
6.95	3.74	1.930	131
6.99	3.76	1.940	132
7.03	3.78	1.950	132
7.07	3.80	1.960	133
7.11	3.82	1.970	133
7.15	3.85	1.980	134
7.19	3.87	1.990	134
7.23	3.89	2.000	135
7.27	3.91	2.010	135
7.32	3.93	2.020	136
7.36	3.95	2.030	136
7.40	3.98	2.040	137
7.44	4.00	2.050	137
7.48	4.02	2.060	137
7.52	4.04	2.070	138
7.56	4.06	2.080	138
7.60	4.09	2.090	139
7.64	4.11	2.100	139
7.68	4.13	2.110	140
7.72	4.15	2.120	140
7.76	4.17	2.130	141
7.81	4.20	2.140	141
7.85	4.22	2.150	142
7.89	4.24	2.160	142
7.93	4.26	2.170	142
7.97	4.29	2.180	143
8.01	4.31	2.190	143
8.05	4.33	2.200	144
8.10	4.35	2.210	144
8.14	4.38	2.220	145
8.18	4.40	2.230	145
8.22	4.42	2.240	146
8.26	4.44	2.250	146
8.31	4.47	2.260	147
8.35	4.49	2.270	147
8.39	4.51	2.280	148
8.43	4.53	2.290	148
8.47	4.56	2.300	148
8.52	4.58	2.310	149
8.56	4.60	2.320	149
8.60	4.62	2.330	150
8.64	4.65	2.340	150
8.69	4.67	2.350	151
8.73	4.69	2.360	151
8.77	4.72	2.370	152
8.82	4.74	2.380	152
8.86	4.76	2.390	153
8.90	4.79	2.400	153
8.94	4.81	2.410	153
8.99	4.83	2.420	154

Δ °C	O Os/Kg	S,M,T. g-mol/l	γ mmho/cm
9.03	4.85	2.430	154
9.07	4.88	2.440	155
9.12	4.90	2.450	155
9.16	4.92	2.460	156
9.20	4.95	2.470	156
9.25	4.97	2.480	156
9.29	5.00	2.490	157
9.33	5.02	2.500	157
9.38	5.04	2.510	158
9.42	5.07	2.520	158
9.47	5.09	2.530	159
9.51	5.11	2.540	159
9.55	5.14	2.550	160
9.60	5.16	2.560	160
9.64	5.18	2.570	160
9.69	5.21	2.580	161
9.73	5.23	2.590	161
9.78	5.26	2.600	162
9.82	5.28	2.610	162
9.86	5.30	2.620	163
9.91	5.33	2.630	163
9.95	5.35	2.640	163
10.00	5.38	2.650	164
10.04	5.40	2.660	164
10.09	5.42	2.670	165
10.13	5.45	2.680	165
10.18	5.47	2.690	165
10.22	5.50	2.700	166
10.27	5.52	2.710	166
10.31	5.55	2.720	167
10.36	5.57	2.730	167
10.41	5.59	2.740	168
10.45	5.62	2.750	168
10.50	5.64	2.760	168
10.54	5.67	2.770	169
10.59	5.69	2.780	169
10.63	5.72	2.790	170
10.68	5.74	2.800	170
10.73	5.77	2.810	170
10.77	5.79	2.820	171
10.82	5.82	2.830	171
10.86	5.84	2.840	172
10.91	5.87	2.850	172
10.96	5.89	2.860	172
11.00	5.92	2.870	173
11.05	5.94	2.880	173
11.10	5.97	2.890	173
11.14	5.99	2.900	174
11.19	6.02	2.910	174
11.24	6.04	2.920	175
11.28	6.07	2.930	175
11.33	6.09	2.940	175
11.38	6.12	2.950	176
11.43	6.14	2.960	176
11.47	6.17	2.970	176
11.52	6.19	2.980	177
11.57	6.22	2.990	177
11.62	6.25	3.000	178
11.71	6.30	3.020	178
11.81	6.35	3.040	179
11.90	6.40	3.060	180
12.00	6.45	3.080	181
12.10	6.50	3.100	181
12.20	6.56	3.120	182
12.29	6.61	3.140	183
12.39	6.66	3.160	183
12.49	6.71	3.180	184
12.59	6.77	3.200	185
12.69	6.82	3.220	185
12.79	6.87	3.240	186
12.89	6.93	3.260	187
12.99	6.98	3.280	187
13.09	7.04	3.300	188
13.19	7.09	3.320	189
13.29	7.15	3.340	189
13.39	7.20	3.360	190
13.49	7.26	3.380	190
13.60	7.31	3.400	191

Δ °C	O Os/Kg	S,M,T. g-mol/l	γ mmho/cm
13.70	7.37	3.420	192
13.80	7.42	3.440	192
13.91	7.48	3.460	193
14.01	7.53	3.480	193
14.12	7.59	3.500	194
14.22	7.65	3.520	195
14.33	7.70	3.540	195
14.43	7.76	3.560	196
14.54	7.82	3.580	196
14.65	7.87	3.600	197
14.75	7.93	3.620	197
14.86	7.99	3.640	198
14.97	8.05	3.660	198
15.08	8.11	3.680	199
15.19	8.16	3.700	199
15.30	8.22	3.720	200
15.41	8.28	3.740	200
15.52	8.34	3.760	201
15.63	8.40	3.780	201
15.74	8.46	3.800	202
15.85	8.52	3.820	202
15.96	8.58	3.840	203
16.07	8.64	3.860	203
16.19	8.70	3.880	204
16.30	8.76	3.900	204
16.41	8.82	3.920	205
16.53	8.89	3.940	205
16.64	8.95	3.960	206
16.76	9.01	3.980	206
16.87	9.07	4.000	206
16.99	9.13	4.020	207
17.11	9.20	4.040	207
17.22	9.26	4.060	208
17.34	9.32	4.080	208
17.46	9.39	4.100	209
17.58	9.45	4.120	209
17.70	9.51	4.140	209
17.82	9.58	4.160	210
17.94	9.64	4.180	210
18.06	9.71	4.200	210
18.18	9.77	4.220	211
18.30	9.84	4.240	211
18.42	9.90	4.260	212
18.55	9.97	4.280	212
18.67	10.04	4.300	212
18.79	10.10	4.320	213
18.92	10.17	4.340	213
19.04	10.24	4.360	213
19.17	10.30	4.380	214
19.29	10.37	4.400	214
19.42	10.44	4.420	214
19.55	10.51	4.440	215
19.67	10.58	4.460	215
19.80	10.65	4.480	215
19.93	10.72	4.500	216
20.06	10.78	4.520	216
20.19	10.85	4.540	216
20.32	10.92	4.560	216
20.45	10.99	4.580	217
20.58	11.07	4.600	217
20.71	11.14	4.620	217
20.85	11.21	4.640	218
20.98	11.28	4.660	218
21.11	11.35	4.680	218
		4.700	218
		4.750	219
		4.800	220
		4.850	220
		4.900	221
		4.950	221
		5.000	222
		5.050	222
		5.100	223
		5.150	223
		5.200	224
		5.250	224
		5.300	225

SPECIFIC GRAVITY OF AQUEOUS INVERT SUGAR SOLUTIONS

DENSITY OF AQUEOUS INVERT SUGAR SOLUTIONS

(From NBS Monograph 64, June 7, 1963) To convert density to specific gravity multiply listed values by 0.998234.

$t = 20°C$

Invert sugar is the equimolar mixture of fructose and glucose obtained by hydrolyzing sucrose.

Wt % (Vacuum)	Specific Gravity	G solute per liter of solution	lb solute per cu ft of solution	lb solute per gallon (U.S.) of solution
1	1.00211	10.021	0.62561	0.08363
2	1.00601	20.120	1.25609	0.16790
3	1.00994	30.298	1.89150	0.25284
4	1.01389	40.556	2.53191	0.33844
5	1.01786	50.893	3.17725	0.42470
6	1.02186	61.312	3.82771	0.51165
7	1.02589	71.812	4.48322	0.59927
8	1.02995	82.396	5.14398	0.68759
9	1.03403	93.063	5.80992	0.77661
10	1.03814	103.814	6.48111	0.86633
11	1.04227	114.649	7.15754	0.95675
12	1.04644	125.573	7.83952	1.04791
13	1.05063	136.582	8.52681	1.13978
14	1.05484	147.678	9.21954	1.23237
15	1.05909	158.864	9.91788	1.32572
16	1.06337	170.139	10.62178	1.41981
17	1.06767	181.504	11.33129	1.51465
18	1.07200	192.960	12.04649	1.61025
19	1.07637	204.510	12.76756	1.70664
20	1.08076	216.152	13.49437	1.80379
21	1.08518	227.888	14.22705	1.90173
22	1.08963	239.719	14.96566	2.00046
23	1.09411	251.645	15.71020	2.09998
24	1.09862	263.669	16.46086	2.20032
25	1.10316	275.790	17.21757	2.30147
26	1.10773	288.010	17.98046	2.40344
27	1.11233	300.329	18.74954	2.50625
28	1.11697	312.752	19.52511	2.60992
29	1.12163	325.273	20.30679	2.71440
30	1.12633	337.899	21.09503	2.81977
31	1.13105	350.626	21.88958	2.92597
32	1.13581	363.459	22.69075	3.03307
33	1.14060	376.398	23.49853	3.14104
34	1.14542	389.443	24.31293	3.24990
35	1.15027	402.595	25.13401	3.35966
36	1.15516	415.858	25.96201	3.47034
37	1.16007	429.226	26.79658	3.58199
38	1.16502	442.708	27.63826	3.69440
39	1.17000	456.300	28.48681	3.80728
40	1.17502	470.008	29.34260	3.92222
41	1.18006	483.825	30.20519	4.03752
42	1.18514	497.759	31.07509	4.15380
43	1.19025	511.808	31.95217	4.27104
44	1.19539	525.972	32.83643	4.38924
45	1.20057	540.257	33.72824	4.50844
46	1.20577	554.654	34.62705	4.62859
47	1.21101	569.175	35.53360	4.74977
48	1.21629	583.819	36.44782	4.87197
49	1.22159	598.579	37.36929	4.99514
50	1.22693	613.465	38.29862	5.11937
51	1.23230	628.473	39.23557	5.24461
52	1.23771	643.609	40.18051	5.37092
53	1.24315	658.870	41.13325	5.49827
54	1.24861	674.249	42.09337	5.62661
55	1.25412	689.766	43.06209	5.75610
56	1.25965	705.404	44.03837	5.88660
57	1.26522	721.175	45.02296	6.01821
58	1.27082	737.076	46.01565	6.15090
59	1.27646	753.111	47.01672	6.28471
60	1.28112	768.672	47.98819	6.41457
61	1.28782	785.570	49.04310	6.55558
62	1.29355	802.001	50.06892	6.69270
63	1.29932	818.572	51.10345	6.83098
64	1.30511	835.270	52.14591	6.97033
65	1.31094	852.111	53.19729	7.11087
66	1.31680	869.088	54.25716	7.25254
67	1.32269	886.202	55.32559	7.39536
68	1.32862	903.462	56.40313	7.53939
69	1.33458	920.860	57.48929	7.68458
70	1.34057	938.399	58.58425	7.83094
71	1.34659	956.079	59.68801	7.79848
72	1.35264	973.901	60.80064	8.12720
73	1.35872	991.866	61.92219	8.27712
74	1.36484	1009.982	63.05318	8.42830
75	1.37099	1028.243	64.19321	8.59069
76	1.37717	1046.649	65.34230	8.73429
77	1.38337	1065.195	66.50012	8.88905
78	1.38962	1083.904	67.66813	9.04518
79	1.39589	1102.753	68.84487	9.20247
80	1.40219	1121.752	70.03098	9.36102
81	1.40852	1140.901	71.22645	9.52082
82	1.41488	1160.202	72.43141	9.68189
83	1.42128	1179.662	73.64630	9.84428
84	1.42770	1199.268	74.87030	10.00790
85	1.43415	1219.028	76.10392	10.17279

HEATS OF COMBUSTION
For Organic Compounds

The heat of combustion is given in kilogram calores per gram molecular weight of the substance when combustion takes place at atmospheric pressure and at either 20°C or 25°C. If the data are for 20°C there is no asterisk for the numerical value of the heat of combustion. If the numerical value is for 25°C there is an asterisk marking the numerical value of the heat of combustion. The final products of combustion are gaseous carbon dioxide, liquid water and nitrogen gas for C, H, N compounds. For method of computing heats of formation see statement following this table.

Name	Formula	Physical state	Heat of combustion, kg. calories
Acetaldehyde	CH_3CHO	liquid	278.77*
Acetamide	CH_3CONH_2	solid	282.6
Acetanilide	$CH_3CONHC_6H_5$	solid	1010.4
Acetic acid	CH_3CO_2H	liquid	209.02*
Acetic anhydride	$(CH_3CO)_2O$	liquid	431.70*
Acetone	$(CH_3)_2CO$	liquid	427.92*
Acetonitrile	CH_3CN	liquid	302.4
Acetophenone	$C_6H_5COCH_3$	liquid	991.60*
Acetylacetone	$CH_3COCH_2COCH_3$	liquid	615.9
Acetylene	$(CH)_2$	gas	310.61*
Acrolein	CH_2CHO	liquid	389.6
Acrylic acid	CH_2CO_2H	liquid	327.0*
Adipic acid	$(CH_2)_4(CH_2H)_2$	solid	668.29*
Alanine	$CH_3CH(NH_2)CO_2H$	solid	387.7
Aldol, see β-hydroxybutyr-aldehyde			
Alizarin, see Dihydroxyanthraquinone			
Allyl alcohol	CH_2CH_2OH	liquid	442.4
Allylene	CH_3C	gas	465.1
p-Aminoazobenzene	$H_2NC_6H_4N_2C_6H_5$	solid	1574.0
p-Aminophenol	$HOC_6H_4NH_2$	solid	760.0
Amygdalin	$C_{20}H_{27}O_{11}N$	solid	2348.4
Amyl acetate	$C_4H_9CO_2C_2H_5$	liquid	1042.5
Amyl aocohol	$(CH_3)_2CHCH_2CH_2OH$	liquid	793.7
Amylene	C_5H_{10}	liquid	803.4
Anethole	$C_{10}H_{12}O$	solid	1324.4
Aniline	$C_6H_5NH_2$	liquid	811.7
p-Anisidine	$CH_3OC_6H_4NH_2$	solid	924.0
Anisole	$C_6H_5OCH_3$	liquid	905.1
Anthracene	$C_{14}H_{10}$	solid	1712.0*
Anthraquinone	$C_{14}H_8O_2$	solid	1544.5
Arabinose	$C_5H_{10}O_5$	solid	559.9
Arabitol	$C_5H_{12}O_5$	solid	661.2
Arachidic acid	$C_{20}H_{40}O_2$	solid	3025.9
Azelaic acid	$(CH_2)_7(CO_2H)_2$	solid	1141.7
Azobenzene	$(C_6H_5N)_2$	solid	1545.9
Azoxybenzene	$(C_6H_5N)_2O$	solid	1534.5
Behenic acid	$C_{22}H_{44}O_2$	solid	3338.4
Benzalacetone	$C_6H_5CHCOCH_3$	solid	1257.4
Benzaldehyde	C_6H_5CHO	liquid	843.2*
Benzamide	$C_6H_5CONH_2$	solid	847.6
Benzanilide	$C_6H_5CONHC_6H_5$	solid	1575.5
Benzene	C_6H_6	liquid	780.96*
Benzenediazonium nitrate	$C_6H_5N_2NO_3$	solid	782.6
Benzidine	$(C_6H_4NH_2)_2$	solid	1560.9
Benzil	$(C_6H_5CO)_2$	solid	1624.6
Benzoic acid*	$C_6H_5CO_2H$	solid	771.24*
Benzoic anhydride	$(C_6H_5CO)_2O$	solid	1555.1
Benzoin	$C_6H_5.CHOH.COC_6H_5$	solid	1671.4
Benzonitrile	C_6H_5CN	liquid	865.5
Benzophenone	$(C_6H_5)_2CO$	solid	1556.5
Benzoyl chloride	C_6H_5COCl	liquid	782.8
Benzoyl peroxide	$(C_6H_5CO)_2O_2$	solid	1551.7
Benzyl alcohol	$C_6H_5CH_2OH$	liquid	894.3
Benzylamine	$C_6H_5CH_2NH_2$	liquid	969.4
Benzyl carbylamine	$C_6H_5CH_2NC$	liquid	1046.5
Benzyl chloride	$C_6H_5CH_2Cl$	liquid	886.4
Benzyl cyanide	$C_6H_5CH_2CN$	liquid	1023.5
Borneol	$C_{10}H_{18}O$	liquid	1469.6
Brucine	$C_{23}H_{26}O_4N_2$	gas	687.68*
n-Butyl alcohol	C_4H_9OH	liquid	639.53*
tert-Butyl alcohol, see Trimethyl carbinol			
n-Butylamine	$C_4H_9NH_2$	liquid	710.6
sec-Butylamine	$(CH_3)(C_2H_5)NH_2$	liquid	713.0
tert-Butylamine	$(CH_3)_3CNH_2$	liquid	716.0
tert-Butylbenzene	$C_6H_5C(CH_3)_3$	liquid	1400.4
n-Butyramide	$C_3H_7CONH_2$	solid	596.0
n-Butyric acid	$C_3H_7CO_2H$	liquid	521.87*
n-Butyronitrile	C_3H_7CN	liquid	613.3
Caffeine	$C_8H_{10}O_2N_4$	solid	1014.2
Camphene	$C_{10}H_{16}$	solid	1468.8
Camphor	$C_{10}H_{16}O$	solid	1411.0
Cane sugar, see Sucrose			
Capric acid	$C_9H_{18}O_2$	solid	1453.07*
Caproic acid	$C_5H_{11}CO_2H$	liquid	834.49*
Carbon disulfide	CS_2	liquid	246.6
Carbon subnitride	$(C.CN)_2$	solid	514.8

* 25°C

Name	Formula	Physical state	Heat of combustion. kg. calories
Carbon tetrachloride	CCl_4	liquid	37.3
Carbonyl sulfide	COS	gas	130.5
Carvacrol	$C_{10}H_{14}O$	liquid	1354.5
Cetyl alcohol	$C_{16}H_{34}O$	solid	2504.4
Cetyl palmitate	$C_{32}H_{64}O$	solid	4872.8
Chloracetic acid	$ClCH_2CO_2H$	solid	171.0
o-Chlorobenzoic acid	$ClC_6H_4CO_2H$	solid	734.5
Chloroform	$CHCl_3$	liquid	89.2
Chrysene	$C_{18}H_{12}$	solid	2139.1
Cinnamic acid *(trans)*	$C_6H_5CH{:}CHCO_2H$	solid	1040.2
Cinnamic aldehyde	$C_6H_5CH{:}CHCHO$	liquid	1112.3
Cinnamic anhydride	$C_{18}H_{14}O_3$	solid	2091.3
d-Citrene	$C_{10}H_{16}$	liquid	1483.0
Citric acid (anhydr)	$C_6H_8O_7$	solid	468.6*
Codeine	$C_{18}H_{21}O_3N.H_2O$	solid	2327.6
Coniine	$C_8H_{17}N$	liquid	1275.5
Creatine (anhydr)	$C_4H_9O_2N_3$	solid	559.8
Creatinine	$C_4H_7ON_3$	solid	563.4
o-Cresol	$CH_3C_6H_4OH$	liquid	882.6
o-Cresol	$CH_3C_6H_4OH$	solid	882.72*
m-Cresol	$CH_3C_6H_4OH$	liquid	880.5
p-Cresol	$CH_3C_6H_4OH$	liquid	882.5
p-Cresol	$CH_3C_6H_4OH$	solid	883.99*
m-Cresolmethyl ether	$CH_3C_6H_4OCH_3$	liquid	1057.0
Crotonaldehyde	C_3H_5CHO	liquid	542.1
Cyanoacetic acid	$NCCH_2CO_2H$	solid	298.8
Cyanogen	$(CN)_2$	gas	258.3
Cyclobutane	C_4H_8	liquid	650.22*
Cycloheptane	$(CH_2)_7$	liquid	1087.2
Cycloheptanol	$CH_2(CH_2)_5CHOH$	liquid	1050.2
Cycloheptene	C_7H_{12}	liquid	1099.09*
Cyclohexane	$(CH_2)_6$	liquid	936.87*
Cyclohexanol	$CH_2(CH_2)_4CHOH$	liquid	890.7
Cyclohexene, see Tetrahydrobenzene			
Cyclopentane	$(CH_2)_5$	liquid	786.55*
Cyclopropane, see Trimethylene			
Cymene	$C_6H_4(CH_3)(CH_3COCO_3){-}(1,4)$	liquid	1402.8
Decahydronaphthalene *(cis)*	$C_{10}H_{18}$	liquid	1502.5
Decahydronaphthalene *(trons)*	$C_{10}H_{18}$	liquid	1499.5
Decane	$C_{10}H_{22}$	liquid	1610.2
Dextrose, see Glucose			
Diallyl	$(CH_2CHCH_2)_2$	vapor	903.4
Diamyl ether	$(C_5H_{11})_2O$	liquid	1609.3
Diamylene	$C_{10}H_{20}$	liquid	1582.2
Dibenzyl	$(C_6H_5CH_2)_2$	solid	1810.6
Dibenzyl amine	$(C_6H_5CH_2)_2NH$	solid	1853.0
o-Dichlorobenzene	$C_6H_4Cl_2$	liquid	671.8
Diethylacetic acid	$(C_2H_5)_2CHCO_2H$	liquid	830.8
Diethyl amine	$(C_2H_5)_2NH$	liquid	716.9
Diethylaniline	$C_6H_5(C_2H_5)_2$	liquid	1451.6
Diethyl carbonate	$CO(OC_2H_5)_2$	liquid	647.9
Diethyl ether	$(C_2H_5)_2O$	liquid	657.52*
Diethyl ketone	$(C_2H_5)_2CO$	liquid	735.6
Diethyl malonate	$CH_2(CO_2C_2H_5)_2$	liquid	860.4
Diethyl oxalate	$(CO_2C_2H_5)_2$	liquid	716.0
Diethyl succintae	$(CH_2CO_2C_2H_5)$	liquid	1007.3
Dihydrobenzene	C_6H_8	liquid	847.8
δ₁-Dihydronaphthalene	$C_{10}H_{10}$	liquid	1296.3
δ₁-Dihydronaphthalene	$C_{10}H_{10}$	solid	1298.3
Dihydroxyanthraquinone	$C_{14}H_6O_2(OH)_2{-}(1,2)$	solid	1448.9
Diisoamyl	$[(CH_3)_2CHCH_2CH_2]_2$	liquid	1615.8
Diisobutylene	$[(CH_3)_2CHCH_2]_2$	liquid	1252.4
Diisopropyl	$[(CH_3)_2CH]_2$	vapor	993.9
Diisopropyl ketone	$[(CH_3)_2CH]_2CO$	liquid	1045.5
Dimethyl amine	$(CH_3)_2NH$	liquid	416.7
Dimethylaniline	$C_6H_5N(CH_3)_2$	liquid	1142.7
Dimethyl carbonate	$CO(OCH_3)_2$	liquid	340.8
Dimethyl ether	$(CH_3)_2O$	gas	
Dimethylethyl carbinol	$C_2H_5(CH_3)_2CHOH$	liquid	784.6
Dimethyl fumarate	$(CHCO_2CH_3)_2$	solid	663.3*
2,5-Dimethylhexane	$(CH_3)_2CH.C_2H_4CH(CH_3)_2$	liquid	1303.3
3,4-Dimethylhexane	$[(C_2H_5)(CH_3)CH]_2$	liquid	1303.7
Dimethyl maleate	$(CHCO_2CH_3)_2$	liquid	669.4*
Dimethyl oxalate	$(CO_2CH_3)_2$	liquid	400.2*
2,2-Dimethylpentane	$(CH_3)_3C.C_3H_7$	liquid	1148.9
2,3-Dimethylpentane	$(CH_3)_2CHCH(CH_3)C_2H_5$	liquid	1148.9
2,4-Dimethylpentane	$(CH_3)_2CHCH_2CH(CH_3)_2$	liquid	1148.9
3,3-Dimethylpentane	$(CH_3)_2C(C_2H_5)_2$	liquid	1147.9
Dimethyl phthalate	$C_6H_4(CO_2CH_3)_2$	liquid	1119.7
Dimethyl succinate	$(CH_2CO_2CH_3)_2$	solid	706.3*
m-Dinitrobenzene	$C_6H_4(NO_2)_2$	solid	696.8

Name	Formula	Physical state	Heat of combustion. kg. calories
Dinitrophenol	$C_6H_3(OH)(NO_2)_2$—(1, 2,4)	solid	648.0
Dinitrotoluene	$C_6H_3)(NO_2)_2$—(1,2,4)	solid	852.8
Diphenyl	$(C_6H_5)_2$	solid	1493.6
Diphenyl amine	$(C_6H_5)_2NH$	solid	1536.2
Diphenyl carbinol	$(C_6H_5)_2CHOH$	solid	1615.4
Diphenylmethane	$(C_6H_5)_2CH_2$	solid	1655.0
Diphenylnitrosamine	$(C_6H_5)_2N.NO$	solid	1532.6
Dipropargyl	$(CH: C.CH_2)_2$	vapor	882.9
Dipropyl ketone	$(C_3H_7)_2CO$	liquid	1050.5
Dulcitol	$C_6H_{14}O_6$	solid	729.1
Durene	$C_6H_2(CH_3)_4$—(1,2,4,5)	solid	1393.6
Eicosane	$C_{20}H_{42}$	solid	3183.1
Erythritol	$C_4H_{10}O_4$	solid	504.1
Ethane	C_2H_6	gas	372.81*
Ethine, see Acetylene			
Ethyl acetate	$CH_3CO_7C_2H_5$	liquid	536.9
Ethyl acetoacetate	$CH_3COCH_2CO_2C_2H_5$	liquid	690.8
Ethyl alcohol	C_2H_6OH	liquid	326.68*
Ethyl amine	$C_2H_5NH_2$	liquid	409.5*
Ethylaniline	$C_6H_5NHC_2H_5$	liquid	1121.5
Ethylbenzene	$C_2H_5C_6H_5$	liquid	1091.2
Ethyl benzoate	$C_6H_5CO_2C_2H_5$	liquid	1098.7
Ethyl bromide	C_2H_5Br	vapor	340.5
Ethyl n-butyrate	$C_3H_7CO_2C_2H_5$	liquid	851.2
Ethyl carbylamine	C_2H_5NC	liquid	477.1
Ethyl chloride	C_2H_5Cl	vapor	316.7
Ethylcycloheptane	$C_2H_5C_7H_{13}$	liquid	1406.8
Ethyl formate	$HCO_2C_2H_5$	liquid	391.7
3-Ethylhexane	$(C_2H_5)_2CH.C_3H_7$	liquid	1302.3
Ethyl iodide	C_2H_5I	liquid	356.0
Ethyl isobutyrate	$(CH_3)_2CHCH_2CO_2C_2H_5$	liquid	845.7
Ethyl isocyanate	C_2H_5NCO	liquid	424.5
Ethyl nitrate	$C_2H_5ONO_2$	vapor	322.4
Ethyl nitrite	C_2H_5ONO	vapor	332.6
3-Ethylpentane	$(C_2H_5)_3CH$	liquid	1149.9
Ethyl propionate	$C_2H_5CO_2C_2H_5$	liquid	690.8
Ethyl salicylate	$HOC_6H_4CO_2C_2H_5$	liquid	1051.2
Ethyl valerate	$C_4H_9CO_2C_2H_5$	liquid	1017.5
Ethylene	CH_2CH_2	gas	337.23*
Ethylene chloride	$(CH_7Cl)_2$	vapor	271.0
Ethylene diamine	$(CH_2NH_2)_3$	liquid	452.6
Ethylene glycol	$(CH_2OH)_2$	liquid	281.9
Ethylene iodide	$(CH_2I)_2$	solid	324.8
Ethylene oxide	CH_2CH_2O	liquid	302.1
Ethylidene chloride	CH_3CHCl_2	liquid	267.1
Eugenol	$C_{10}H_{12}O_2$	liquid	1286.6
Fenchane	$C_{10}H_{18}$	liquid	1502.6
Fluorene	$(C_6H_4)_2: CH_2$	solid	1584.9
Fluorobenzene	C_6H_5F	liquid	747.2
Formaldehyde	CH_2O	gas	136.42*
Formamide	$HCONH_2$	solid	134.9
Formic acid	HCO_2H	liquid	60.86*
β-D-Fructose	$C_6H_{12}O_6$	solid	672.0*
Fumaric acid (trans)	$(CHCO_2H)_2$	solid	318.99*
Furfural	C_4H_3OCHO	liquid	559.5
a-D-Galactose	$C_6H_{12}O_6$	solid	670.1*
Gallic acid	$C_6H_2(OH)_3CO_2H$—(1,3,5,6)	solid	633.7
a-D-Glucose	$C_6H_{12}O_6$	solid	669.94*
Glutaric acid	$(CH_2)_3(CO_2H)_2$	solid	514.08*
Glycerol	$(CH_2OH)_2CHOH$	liquid	397.0
Glyceryl tributyrate	$C_{15}H_{26}O_6$	liquid	1941.1
Glycine	$H_2NCH_2CO_2H$	solid	232.67*
Glycogen	$(C_6H_{10}O_5)x$ per kg	solid	4186.8
Glycollic acid	CH_7OHCO_2H	solid	166.1*
Glycylglycine	$C_4H_8O_3N_2$	solid	470.7
n-Heptaldehyde	$CH_3(CH_2)_5CHO$	liquid	1062.2*
n-Heptane	C_7H_{16}	liquid	1149.9
Heptine-l	$CH: C(CH_2)_4CH_3$	liquid	1091.2
n-Heptyl alcohol	$CH_3(CH_2)_5CH_2OH$	liquid	1104.9
Heptyl amine	$C_7H_{15}NH_2$	liquid	1178.9
Heptylic acid	$C_7H_{14}O_2$	liquid	986.1
n-Hexane	C_6H_{14}	liquid	995.01*
Hexachlorbenzene	C_6Cl_6	solid	509.0
Hexachlorethane	C_2Cl_6	solid	110.0
Hexadecane	$C_{16}II_{34}$	solid	2559.1
Hexahydronaphthalene	$C_{10}H_{14}$	liquid	1419.3
Hexamethylbenzene	$C_6(CH_3)_6$	solid	1711.9
Hexamethylenetetramine	$(CH_2)_6N_4$	solid	1006.7
Hexamethylethane	$(CH_3)_3Cl_2$	solid	1301.8
Hexyl amine	$C_6H_{13}NH_2$	liquid	1022.2
Hexylene	C_6H_{12}	liquid	952.6

Name	Formula	Physical state	Heat of combustion. kg. calories
Hippuric acid	$C_6H_5CONHCH_2CO_2H$	solid	1012.4
Hydantoic acid	$C_3H_6O_3N_2$	solid	308.6
Hydrazobenzene	$(C_6H_5NH)_2$	solid	1597.3
Hydroquinol	$C_6H_4(OH)_2$	solid	681.78*
Hydroquinoldimethyl ether	$(CH_3O)_2C_6H_4$	solid	1014.7
p-Hydroxyazobenzene	$HOC_6H_4N_2C_6H_5$	solid	1502.0
o-Hydroxybenzaldehyde	$C_6H_4(OH)CHO$	liquid	796.4*
m-Hydroxybenzaldehyde	$C_6H_4(OH)CHO$	solid	788.7
p-Hydroxybenzaldehyde	$C_6H_4(OH)CHO$	solid	792.7
m-Hydroxybenzoic acid	$HOC_6H_4CO_2H$	solid	726.1
p-Hydroxybenzoic acid	$HOC_6H_4CO_2H$	solid	725.4
β-Hydroxybutyraldehyde	$CH_3CHOHCH_2CHO$	liquid	546.6
Indigo	$C_{16}H_{10}O_2N_2$	solid	1815.0
Indole	C_8H_7N	solid	1022.0
Inositol	$C_6H_{12}O_6$	solid	662.1
Iodoform	CHI_3	solid	161.9
Isoamyl amine	$(CH_3)_2CHC_2H_4NH_2$	liquid	866.8
Isobutane	$(CH_3)_2CH$	gas	683.4
Isobutyl alcohol	$(CH_3)_2CHCH_2OH$	liquid	638.2
Isobutyl amine	$C_4H_9NH_2$	liquid	713.6
Isobutylene	$(CH_3)_2C:CH_2$	gas	647.2
Isobutyraldehyde	$(CH_3)_2CHCHO$	vapor	596.8
Isobutyramide	$(CH_3)_2CHCONH_2$	solid	595.9
Isobutyric acid	$(CH_3)_2CHCO_2H$	liquid	517.4
Isoeugenol	$C_{10}H_{12}O_2$	liquid	1277.6
Isopentane	C_5H_{12}	gas	843.5(?)
Isopentane	C_5H_{12}	liquid	838.3(?)
Isophthalic acid	$C_6H_4(CO_2H)_2$	solid	768.3
Isopropyl alcohol	$(CH_3)_2CHOH$	liquid	474.0
Isopropylbenzene	$(CH_3)_2CHC_6H_5$	liquid	1247.3
Isopropyltoluene	$C_6H_4(CH_3)(CH_3CHC_3)—(1,3)$	liquid	1409.5
Isopropyltoluene, see Cymene			
Isosafrole	$C_{10}H_{10}O_2$	liquid	1733.9
Lactic acid, DL	$CH_3CHOHCO_2H$	liquid	326.8*
Lactose (anhydr.)	$C_{12}H_{22}O_{11}$	solid	1350.0*
Lauric acid	$C_{12}H_{24}O_2$	solid	1763.25*
Leucine	$C_6H_{13}O_2N$	solid	855.6
d-Limonene	$C_{10}H_{16}$	liquid	1471.2
Maleic acid (cis)	$(CHCO_2H)_2$	solid	323.89*
Maleic anhydride	$(CHCO)_2O$	solid	332.10*
l-Malic acid	$(CHOHCH_2):(CO_2H)_2$	solid	317.37*
Malonic acid	$CH_2(CO_2H)_2$	solid	205.82*
Maltose	$C_{12}H_{22}O_{11}$	solid	1349.3*
Mandelic acid	$C_6H_5CHOHCO_2H$	solid	890.3
d-Mannitol	$C_6H_{14}O_6$	solid	727.6
Menthene	$C_{10}H_{18}$	liquid	1523.2
Menthol	$C_{10}H_{20}O$	solid	1508.8
Mesitylene	$(CH_3)_3C_6H_3—(1,3,5)$	liquid	1243.6
Mesityl oxide	$(CH_3)_2C:CHCOCH_3$	liquid	846.7
Mesotartaric acid	$(CHOH)_2(CO_2H)_2$	solid	276.0
Methane	CH_4	gas	212.79*
Methyl acetate	$CH_3CO_2CH_3$	liquid	381.2
Methyl alcohol	CH_3OH	liquid	173.64*
Methyl amine	CH_3NH_2	liquid	253.5*
Methylaniline	$C_6H_5NHCH_3$	liquid	973.5
Methyl benzoate	$C_6H_5CO_2CH_3$	liquid	945.9*
Methyl bromide	CH_3Br	vapor	184.0
Methyl butyl ketone	$CH_3COC_4H_9$	liquid	895.2
Methyl tert-butyl ketone, see Pinacoline			
Methyl butyrate	$C_3H_7COOCH_3$	liquid	692.8
Methyl carbylamine	CH_3NC	liquid	320.1
Methyl chloride	CH_3Cl	gas	164.2
Methyl cinnamate	$C_{10}H_{10}O_2$	solid	1213.0
Methylcyclobutane	$CH_3CHCH_2CH_2CH_2$	liquid	784.2
Methylcycloheptane	$CH_3C_7H_{13}$	liquid	1244.5
Methylcyclohexane	$CH_3C_6H_{11}$	liquid	1091.8
Methylcyclopentane	$CH_3CH.C_5H_6CH_2$	liquid	937.9
Methyldiethyl carbinol	$CH_3(C_2H_5)_2CHOH$	liquid	927.0
Methylene chloride	CH_2Cl_2	vapor	106.8
Methylene iodide	CH_2I_2	liquid	178.4
Methylethyl ether	$CH_3OC_2H_5$	vapor	503.69*
Methylethyl ketone	$CH_3COC_2H_5$	liquid	584.17*
Methyl formate	HCO_2CH_3	liquid	234.1*
2-Methylheptane	$(CH_3)_2CH.C_5H_{11}$	liquid	1306.1
2-Methylhexane	$(CH_3)_2CHC_4H_9$	liquid	1148.9
3-Methylhexane	$(C_2H_5)(CH_3)CHC_3H_7$	liquid	1148.9
Methylhexyl ketone	$CH_3COC_6H_{13}$	liquid	1205.1
Methyl iodide	CH_3I	liquid	194.7
Methyl isobutyrate	$(CH_3)_2CHCO_2CH_3$	liquid	694.2
Methyl isocyanate	CH_3NCO	liquid	269.4
Methylisopropyl ketone	$CH_3COCH(CH_3)_2$	liquid	733.9
Methyl lactate	$CH_3CHOHCO_2CH_3$	liquid	497.2

Name	Formula	Physical state	Heat of combustion. kg. calories
Methyl propionate	$C_2H_5CO_2CH_3$	vapor	552.3
Methylpropyl ketone	$CH_3COC_3H_7$	liquid	740.78*
Methyl salicylate	$HOC_6H_4CO_2CH_3$	liquid	898.6*
Milk sugar, see Lactose			
Morphine	$C_{17}H_{19}O_3N.H_2O$	solid	2146.3
Mucic acid	$C_6H_{10}O_8$	solid	483.0*
Myristic acid	$C_{14}H_{28}O_2$	solid	2073.91*
Naphthalene	$C_{10}H_8$	solid	1231.8*
α-Naphthoic acid	$C_{10}H_7CO_2H$	solid	1231.8
β-Naphthoic acid	$C_{10}H_7CO_2H$	solid	1227.6
α-Naphthol	$C_{10}H_7OH$	solid	1185.4
β-Naphthol	$C_{10}H_7OH$	solid	1187.2
α-Naphthonitrile	$C_{10}H_7CN$	solid	1326.2
β-Naphthonitrile	$C_{10}H_7CN$	solid	1321.0
α-Naphthoquinone	$C_{10}H_6O_2$	solid	1100.8
β-Naphthoquinone	$C_{10}H_6O_2$	solid	1106.4
α-Naphthyl amine	$C_{10}H_7NH_2$	solid	1263.5
β-Naphthyl amine	$C_{10}H_7NH_2$	solid	1261.0
Narceine	$C_{23}H_{27}O_8N.2H_2O$	solid	2802.9
Narcotine	$C_{22}H_{23}O_7N$	solid	2644.5
Nicotine	$C_{10}H_{14}N_2$	liquid	1427.7
o-Nitraniline	$C_6H_4(NH_2)(NO_2)$	solid	765.8
m-Nitraniline	$C_6H_4(NH_2)(NO_2)$	solid	765.2
p-Nitraniline	$C_6H_4(NH_2)(NO_2)$	solid	761.0
m-Nitrobenzaldehyde	$O_2NC_6H_4CHO$	solid	800.4
Nitrobenzene	$C_6H_5NO_2$	liquid	739.2
m-Nitrobenzoic acid	$O_2NC_6H_4CO_2H$	solid	729.1
Nitroethane	$C_2H_5NO_2$	liquid	322.2
Nitroglycerine, see Trinitroglycerol			
Nitromethane	CH_3NO_2	liquid	169.4
o-Nitrophenol	$HOC_6H_6NO_2$	solid	689.1
m-Nitrophenol	$HOC_5H_4NO_2$	solid	684.4
p-Nitrophenol	$HOC_6H_4NO_2$	solid	688.8
Nitropropane	$C_3H_7NO_2$	liquid	477.9
p-Nitrotoluene	$CH_3C_6H_4NO_2$	liquid	897.0
p-Nitrotoluene	$CH_3C_6H_6NO_2$	solid	888.6
Octahydronaphthalene	$C_{10}H_{16}$	liquid	1461.7
n-Octane	C_8H_{18}	liquid	1302.7
Octyl alcohol	$C_8H_{18}O$	liquid	1262.0
Oleic acid	$C_{18}H_{34}O_2$	liquid	2657.4*
Oxalic acid, a	$(CO_2H)_2$	solid	58.7*
Oxamide	$(CONH_2)_2$	solid	203.2
Palmitic acid	$C_{16}H_{32}O_2$	solid	2384.76*
Papaverine	$C_{20}H_{21}O_4N$	solid	2478.1
Pentamethylbenzene	$C_6H(CH_3)_5$	solid	1554.0
n-Pentane	C_5H_{12}	gas	845.16*
n-Pentane	C_8H_{12}	liquid	838.78*
Phenacetin	$C_{10}H_{13}O_2N$	solid	1285.2
Phenanthraquinone	$C_{14}H_8O_2$	solid	1544.0
Phenanthrene	$C_{14}H_{10}$	solid	1685.6*
Phenetole	$C_6H_5OC_2H_5$	liquid	1060.3
Phenol	C_6H_5OH	solid	729.80*
Phenylacetic acid	$C_6H_5CH_2CO_2H$	solid	930.4*
Phenylacetylene	C_6H_5C	liquid	1024.2
Phenylalanine	$C_9H_{11}O_2N$	solid	1111.3
p-Phenylenediamine	$C_6H_4(NH_2)_2$	solid	843.4
Phenylethylene, see Styrene			
Phenylglycine	$C_2H_5NHCH_2CO_2H$	solid	955.1
Phenylhydrazine	$C_6H_5N_2H_3$	solid	875.4
Phenylhydroxylamine	C_6H_5NHOH	liquid	803.7
Phenyl iodide	C_6H_5I	liquid	770.7
Phloroglucinol	$C_6H_3(OH)_3$	solid	635.7
o-Phthalic acid	$C_6H_4(CO_2H)_2$	solid	770.44*
Phthalic anhydride	$C_6H_4(CO_2O$	solid	783.4
Phthalimide	$C_8H_5O_2N$	solid	849.5
Pieric acid	$C_6H_2(OH)(NO_2)_3$—(1,2,4,6)	solid	611.8
Pinacoline	$CH_3COC(CH_3)_3$	solid	891.8
Piperidine	$C_5H_{11}N$	liquid	826.6
Piperonal	$C_8H_6O_3$	solid	870.7
Propane	C_3H_8	gas	530.57*
Propine, see Allylene			
Propionaodehyde	C_2H_5CHO	liquid	434.1*
Propionamide	$C_2H_5CONH_2$	solid	439.9
Propionic acid	$C_2H_5CO_2H$	liquid	365.03*
Propionic anhydride	$(C_2H_5CO)_2O$	liquid	746.6
Propionitrile	C_2H_5CN	liquid	456.4
n-Propyl alcohol	C_3H_7OH	liquid	482.75*
Propyl amine	$C_3H_7NH_2$	liquid	565.3*
n-Propylbenzene	$C_3H_7C_6H_5$	liquid	1246.4
Propyl bromide	C_3H_7Br	vapor	497.3
Propyl carbylamine	C_3H_7NC	liquid	639.6

89 SULFURIC ACID, H₂SO₄

MOLECULAR WEIGHT = 98.08
RELATIVE SPECIFIC REFRACTIVITY = 0.685

0.00 % by wt. data are the same for all compounds.
For Values of 0.00 wt. % solutions see Table 1, Acetic Acid.

A % by wt.	ρ D_4^{20}	D_{20}^{20}	C, g/l	M g-mol/l	C_w g/l	$(C_o - C_w)$ g/l	$(n - n_o)$ × 10⁴	n	Δ °C	O Os/kg	S g-mol/l	η/η_a	η/ρ cS	ϕ rhe	γ mmho/cm	T g-mol/l
0.50	1.0016	1.0034	5.0	0.051	996.6	1.7	6	1.3336	0.210	0.113	0.060	1.008	1.009	98.96	24.3	0.277
1.00	1.0049	1.0067	10.0	0.102	994.9	3.3	13	1.3342	0.423	0.227	0.122	1.017	1.014	98.13	47.8	0.573
1.50	1.0083	1.0101	15.1	0.154	993.2	5.1	19	1.3349	0.662	0.356	0.192	1.025	1.019	97.34	70.3	0.886
2.00	1.0116	1.0134	20.2	0.206	991.4	6.8	25	1.3355	0.796	0.428	0.232	1.034	1.024	96.52	92.	1.22
2.50	1.0150	1.0168	25.4	0.259	989.6	8.6	31	1.3361	1.004	0.540	0.293	1.044	1.031	95.55	113.	1.58
3.00	1.0183	1.0201	30.6	0.311	987.8	10.4	37	1.3367	1.172	0.630	0.343	1.057	1.040	94.45	134.	1.98
3.50	1.0217	1.0235	35.8	0.365	985.9	12.3	43	1.3373	1.354	0.728	0.396	1.070	1.049	93.29	155.	2.42
4.00	1.0250	1.0269	41.0	0.418	984.0	14.2	49	1.3379	1.599	0.860	0.468	1.083	1.059	92.11	175.	2.93
4.50	1.0284	1.0302	46.3	0.472	982.1	16.1	55	1.3385	1.855	0.998	0.543	1.097	1.069	90.97	194.	3.57
5.00	1.0318	1.0336	51.6	0.526	980.2	18.0	61	1.3391	2.047	1.101	0.598	1.110	1.078	89.91	211.	4.25
5.50	1.0352	1.0370	56.9	0.580	978.2	20.0	67	1.3397	2.259	1.214	0.659	1.122	1.086	88.93		
6.00	1.0385	1.0404	62.3	0.635	976.2	22.0	73	1.3403	2.495	1.341	0.727	1.134	1.094	88.00		
6.50	1.0419	1.0438	67.7	0.691	974.2	24.0	79	1.3409	2.730	1.468	0.795	1.146	1.102	87.10		
7.00	1.0453	1.0472	73.2	0.746	972.2	26.1	85	1.3415	2.952	1.587	0.858	1.157	1.109	86.24		
7.50	1.0488	1.0506	78.7	0.802	970.1	28.1	91	1.3421	3.197	1.719	0.927	1.169	1.117	85.39		
8.00	1.0522	1.0541	84.2	0.858	968.0	30.2	97	1.3427	3.493	1.878	1.010	1.180	1.124	84.56		
8.50	1.0556	1.0575	89.7	0.915	965.9	32.3	103	1.3433	3.801	2.043	1.096	1.192	1.131	83.74		
9.00	1.0591	1.0610	95.3	0.972	963.8	34.4	109	1.3439	4.083	2.195	1.174	1.204	1.139	82.92		
9.50	1.0626	1.0645	100.9	1.029	961.6	36.6	115	1.3445	4.360	2.344	1.250	1.216	1.146	82.10		
10.00	1.0661	1.0680	106.6	1.087	959.5	38.8	121	1.3451	4.644	2.497	1.328	1.228	1.154	81.27		
11.00	1.0731	1.0750	118.0	1.204	955.1	43.2	133	1.3463	5.25	2.82	1.490	1.253	1.170	79.63		
12.00	1.0802	1.0821	129.6	1.322	950.6	47.6	145	1.3475	5.93	3.19	1.669	1.279	1.187	78.02		
13.00	1.0874	1.0893	141.4	1.441	946.0	52.2	158	1.3488	6.67	3.59	1.859	1.306	1.203	76.43		
14.00	1.0947	1.0966	153.3	1.563	941.4	56.8	170	1.3500	7.49	4.03	2.063	1.334	1.221	74.82		
15.00	1.1020	1.1039	165.3	1.685	936.7	61.5	183	1.3513	8.35	4.49	2.270	1.364	1.240	73.17		
16.00	1.1094	1.1114	177.5	1.810	931.9	66.3	195	1.3525	9.26	4.98	2.483	1.396	1.261	71.47		
17.00	1.1169	1.1189	189.9	1.936	927.0	71.2	208	1.3538	10.23	5.50	2.702	1.431	1.284	69.74		
18.00	1.1245	1.1265	202.4	2.064	922.1	76.2	221	1.3551	11.29	6.07	2.932	1.467	1.308	68.01		
19.00	1.1321	1.1341	215.1	2.193	917.0	81.2	233	1.3563	12.43	6.68	3.169	1.505	1.332	66.32		
20.00	1.1398	1.1418	228.0	2.324	911.9	86.4	246	1.3576	13.64	7.33	3.409	1.543	1.356	64.68		
22.00	1.1554	1.1575	254.2	2.592	901.2	97.0	272	1.3602	16.48	8.86	3.933	1.631	1.408	61.38		
24.00	1.1714	1.1735	281.1	2.866	890.3	108.0	298	1.3628	19.85	10.67	4.488	1.703	1.457	58.60		
26.00	1.1872	1.1893	308.7	3.147	878.5	119.7	323	1.3653	24.29	13.06		1.793	1.513	55.67		
28.00	1.2031	1.2052	336.9	3.435	866.2	132.0	347	1.3677	29.65	15.94		1.890	1.574	52.81		
30.00	1.2191	1.2213	365.7	3.729	853.4	144.9	371	1.3701	36.21	19.47		1.997	1.641	49.98		
32.00	1.2353	1.2373	395.3	4.030	840.0	158.2	395	1.3725	44.76	24.07		2.118	1.718	47.12		
34.00	1.2518	1.2540	425.6	4.339	826.2	172.1	419	1.3749	55.28	29.72		2.250	1.801	44.36		
36.00	1.2685	1.2707	456.7	4.656	811.8	186.4	443	1.3773				2.387	1.885	41.82		
38.00	1.2855	1.2878	488.5	4.981	797.0	201.2	467	1.3797				2.528	1.970	39.48		
40.00	1.3028	1.3051	521.1	5.313	781.7	216.5	491	1.3821				2.685	2.065	37.17		
42.00	1.3205	1.3229	554.6	5.655	765.9	232.3	516	1.3846				2.866	2.174	34.83		
44.00	1.3386	1.3410	589.0	6.005	749.6	248.6	540	1.3870				3.067	2.296	32.53		
46.00	1.3570	1.3594	624.2	6.365	732.8	265.4	565	1.3895				3.292	2.431	30.32		
48.00	1.3759	1.3783	660.4	6.734	715.5	282.8	590	1.3920				3.539	2.577	28.20		
50.00	1.3952	1.3977	697.6	7.113	697.6	300.6	616	1.3945				3.818	2.742	26.14		
52.00	1.4149	1.4174	735.8	7.502	679.2	319.1	641	1.3971				4.134	2.927	24.14		
54.00	1.4351	1.4377	775.0	7.901	660.2	338.1	667	1.3997				4.490	3.135	22.23		
56.00	1.4558	1.4584	815.3	8.312	640.6	357.7	694	1.4024				4.896	3.370	20.38		
58.00	1.4770	1.4796	856.7	8.734	620.3	377.9	720	1.4050				5.343	3.625	18.68		
60.00	1.4987	1.5013	899.2	9.168	599.5	398.8	747	1.4077				5.905	3.948	16.90		
62.00	1.5200	1.5227	942.4	9.608	577.6	420.6										
64.00	1.5421	1.5448	986.9	10.062	555.2	443.0										
66.00	1.5646	1.5674	1032.6	10.528	532.0	466.2										
68.00	1.5874	1.5902	1079.4	11.005	508.0	490.2										
70.00	1.6105	1.6134	1127.4	11.495	483.1	515.1										
72.00	1.6338	1.6367	1176.3	11.993	457.5	540.7										
74.00	1.6574	1.6603	1226.5	12.505	430.9	567.3										
76.00	1.6810	1.6840	1277.6	13.026	403.4	594.8										
78.00	1.7043	1.7073	1329.4	13.554	374.9	623.3										
80.00	1.7272	1.7303	1381.8	14.088	345.4	652.8										
82.00	1.7491	1.7522	1434.3	14.624	314.8	683.4										
84.00	1.7693	1.7724	1486.2	15.153	283.1	715.1										
86.00	1.7872	1.7904	1537.0	15.671	250.2	748.0										
88.00	1.8022	1.8054	1585.9	16.169	216.3	781.9										
90.00	1.8144	1.8176	1633.0	16.650	181.4	816.8										
92.00	1.8240	1.8272	1678.1	17.110	145.9	852.3										
94.00	1.8312	1.8344	1721.3	17.550	109.9	888.3										
96.00	1.8355	1.8388	1762.1	17.966	73.4	924.8										
98.00	1.8361	1.8394	1799.4	18.346	36.7	961.5										
100.00	1.8305	1.8337	1830.5	18.663	0.0	998.2										

90 TARTARIC ACID, $HO_2C(CHOH)_2CO_2H$

MOLECULAR WEIGHT = 150.09

A% by wt.	ρ D_4^{20}	D_{20}^{20}	C_s g/l	M g-mol/l	C_w g/l	$(C_o - C_w)$ g/l	$(n - n_0)$ $\times 10^4$	n	Δ °C	O Os/kg	S g-mol/l	η/η_0	η/ρ cS	ϕ rhe	γ mmho/cm	T g-mol/l
1.00	1.0029	1.0047	10.0	0.067	992.9	5.3	12.	1.3342	0.132	0.070	0.037	1.021	1.020	97.76	2.5	0.025
2.00	1.0072	1.0090	20.1	0.134	987.1	11.1	25.	1.3355	0.262	0.141	0.075	1.044	1.039	95.56	3.9	0.039
3.00	1.0121	1.0139	30.4	0.202	981.7	16.5	37.	1.3367	0.402	0.216	0.116	1.069	1.058	93.39	4.9	0.049
4.00	1.0163	1.0181	40.7	0.271	975.6	22.6	49.	1.3379	0.538	0.290	0.156	1.092	1.077	91.36	5.5	0.056
5.00	1.0208	1.0226	51.0	0.340	969.8	28.4	62.	1.3392	0.681	0.366	0.198	1.119	1.098	89.22	6.2	0.063
6.00	1.0255	1.0273	61.5	0.410	964.0	34.2	73.	1.3403	0.824	0.443	0.240	1.147	1.121	86.99	6.9	0.070
7.00	1.0310	1.0329	72.2	0.481	958.8	39.4	88.	1.3418	0.974	0.523	0.284	1.176	1.143	84.86	7.3	0.075
8.00	1.0348	1.0367	82.8	0.552	952.0	46.2	99.	1.3429	1.130	0.607	0.330	1.207	1.169	82.67	7.8	0.080
9.00	1.0393	1.0412	93.5	0.623	945.8	52.4	111.	1.3441	1.283	0.690	0.375	1.236	1.192	80.72	8.2	0.084
10.00	1.0443	1.0462	104.4	0.696	939.9	58.3	124.	1.3454	1.443	0.776	0.422	1.272	1.220	78.49	8.5	0.088
12.00	1.0540	1.0559	126.5	0.843	927.5	70.7	151.	1.3481	1.768	0.951	0.517	1.343	1.277	74.30	9.2	0.095
14.00	1.0639	1.0658	148.9	0.992	915.0	83.2	178.	1.3508	2.130	1.145	0.622	1.413	1.331	70.62	9.7	0.100
16.00	1.0738	1.0757	171.8	1.145	902.0	96.2	205.	1.3535	2.504	1.346	0.730	1.498	1.398	66.61	10.1	0.104
18.00	1.0839	1.0859	195.1	1.300	888.8	109.4	231.	1.3561	2.917	1.568	0.848	1.591	1.471	62.72	10.3	0.107
20.00	1.0942	1.0962	218.8	1.458	875.4	122.8	260.	1.3590	3.360	1.807	0.973	1.701	1.558	58.66	10.5	0.109
24.00	1.1152	1.1172	267.6	1.783	847.6	150.6	316.	1.3646	4.309	2.316	1.236	1.957	1.758	51.01	10.6	0.110
28.00	1.1366	1.1387	318.3	2.120	818.3	179.9	375.	1.3705	5.46	2.93	1.546	2.296	2.024	43.47	10.4	0.108
32.00	1.1588	1.1609	370.8	2.471	788.0	210.2	436.	1.3766	6.73	3.62	1.874	2.695	2.330	37.04	10.1	0.104
36.00	1.1813	1.1834	425.3	2.833	756.0	242.2	499.	1.3829	8.30	4.47	2.259	3.242	2.750	30.78	9.3	0.096
40.00	1.2049	1.2071	482.0	3.211	722.9	275.3	561.	1.3891	10.20	5.48	2.695	4.002	3.328	24.94	8.2	0.085
44.00	1.2290	1.2312	540.8	3.603	688.2	310.0	627.	1.3957				5.088	4.148	19.62	7.3	0.074
48.00	1.2535	1.2558	601.7	4.009	651.8	346.4	692.	1.4022				6.605	5.280	15.11	6.2	0.063
52.00	1.2787	1.2810	664.9	4.430	613.8	384.4	762.	1.4092				8.961	7.022	11.14	5.0	0.050
56.00	1.3058	1.3082	731.3	4.872	574.5	423.7	834.	1.4164				12.60	9.670	7.92	3.7	0.037
58.00	1.3196	1.3220	765.4	5.099	554.2	444.0	869.	1.4199				15.64	11.87	6.38	3.0	0.030

91 TETRACAINE HYDROCHLORIDE, $C_{15}H_{24}N_2O_2 \cdot HCl$

MOLECULAR WEIGHT = 300.84

A% by wt.	ρ D_4^{20}	D_{20}^{20}	C_s g/l	M g-mol/l	C_w g/l	$(C_o - C_w)$ g/l	$(n - n_0)$ $\times 10^4$	n	Δ °C	O Os/kg	S g-mol/l	η/η_0	η/ρ cS	ϕ rhe	γ mmho/cm	T g-mol/l
0.50	0.9988	1.0006	5.0	0.017	993.8	4.4	11.	1.3341	0.058	0.031	0.016	1.013	1.016	98.54	1.2	0.012
1.00	0.9995	1.0013	10.0	0.033	989.5	8.7	23.	1.3353	0.110	0.059	0.031	1.029	1.032	96.95	2.4	0.024
1.50	1.0001	1.0019	15.0	0.050	985.1	13.1	34.	1.3364	0.159	0.086	0.045	1.045	1.047	95.50	3.6	0.036
2.00	1.0008	1.0026	20.0	0.067	980.8	17.4	46.	1.3376	0.204	0.109	0.058	1.063	1.064	93.91	4.7	0.047
2.50	1.0014	1.0032	25.0	0.083	976.4	21.8	57.	1.3387				1.081	1.082	92.29	5.6	0.057
3.00	1.0021	1.0039	30.1	0.100	972.0	26.2	69.	1.3399				1.100	1.100	90.72	6.6	0.067
3.50	1.0027	1.0045	35.1	0.117	967.6	30.6	80.	1.3410				1.121	1.120	89.05	7.4	0.076
4.00	1.0034	1.0052	40.1	0.133	963.3	34.9	92.	1.3422				1.142	1.140	87.42	8.2	0.084
4.50	1.0040	1.0058	45.2	0.150	958.8	39.4	103.	1.3433				1.163	1.161	85.79	8.9	0.092
5.00	1.0046	1.0064	50.2	0.167	954.4	43.8	115.	1.3445				1.187	1.184	84.07	9.6	0.099
5.50	1.0053	1.0071	55.3	0.184	950.0	48.2	126.	1.3456				1.211	1.207	82.41	10.2	0.106
6.00	1.0059	1.0077	60.4	0.201	945.5	52.7	138.	1.3468				1.237	1.232	80.69	10.8	0.113
6.50	1.0066	1.0084	65.4	0.217	941.2	57.0	149.	1.3479				1.263	1.257	79.03	11.4	0.120
7.00	1.0072	1.0090	70.5	0.234	936.7	61.5	160.	1.3490				1.291	1.284	77.33	11.9	0.126
7.50	1.0079	1.0097	75.6	0.251	932.3	65.9	172.	1.3502				1.319	1.311	75.68	12.5	0.132
8.00	1.0085	1.0103	80.7	0.268	927.8	70.4	183.	1.3513				1.348	1.339	74.05	13.0	0.138
8.50	1.0092	1.0110	85.8	0.285	923.4	74.8	195.	1.3525				1.379	1.369	72.38	13.6	0.144
9.00	1.0098	1.0116	90.9	0.302	918.9	79.3	206.	1.3536				1.410	1.399	70.79	14.0	0.149
9.50	1.0105	1.0123	96.0	0.319	914.5	83.7	218.	1.3548				1.441	1.429	69.25	14.4	0.154
10.00	1.0111	1.0129	101.1	0.336	910.0	88.2	229.	1.3559				1.477	1.464	67.56	14.7	0.158
11.00	1.0124	1.0142	111.4	0.370	901.0	97.2	252.	1.3582				1.548	1.532	64.48	15.8	0.170
12.00	1.0137	1.0155	121.6	0.404	892.1	106.1	275.	1.3605				1.623	1.604	61.50	16.8	0.182

92 TRICHLOROACETIC ACID, CCl_3COOH

MOLECULAR WEIGHT = 163.38
RELATIVE SPECIFIC REFRACTIVITY = 0.865

0.00 % by wt. data are the same for all compounds.
For Values of 0.00 wt. % solutions see Table 1. Acetic Acid.

A% by wt.	ρ D_4^{20}	D_{20}^{20}	C_v g/l	M g-mol/l	C_w g/l	$(C_o - C_w)$ g/l	$(n - n_0)$ $\times 10^4$	n	Δ °C	O Os/kg	S g-mol/l	$(n - n_0)$ η/η_0	η/ρ cS	ϕ rhe	γ mmho/cm	T g-mol/l
0.50	1.0008	1.0026	5.0	0.031	995.8	2.4	7	1.3337	0.105	0.057	0.030	1.009	1.010	98.91	10.3	0.107
1.00	1.0034	1.0051	10.0	0.061	993.3	4.9	13	1.3343	0.211	0.113	0.060	1.019	1.018	97.94	19.6	0.215
2.00	1.0083	1.0101	20.2	0.123	988.2	10.1	26	1.3356	0.423	0.227	0.122	1.042	1.035	95.78	37.2	0.435
3.00	1.0133	1.0151	30.4	0.186	982.9	15.4	39	1.3369	0.638	0.343	0.185	1.067	1.055	93.53	54.0	0.656

HEAT OF COMBUSTION
For Organic Compounds (Continued)

Name	Formula	Physical state	Heat of combustion. kg. calories
Propyl chloride	C_3H_7Cl	vapor	478.3
Propylene	CH_3CH_2	gas	490.2
Propylene glycol	$CH_3CHOHCH_2OH$	liquid	431.0
n-Propyl iodide	C_3H_7I	liquid	514.3
n-Propyltoluene	$C_6H_4(CH_3)(C_3H_7)—(1,3)$	liquid	1405.4
Pseudocumene	$C_6H_3(CH_3)_3—(1,2,4)$	liquid	1241.7
Pyridine	C_5H_5N	liquid	665.0*
Pyrocatechol	$C_6H_4(OH)_2$	solid	683.0*
Pyrogallol	$C_6H_3(OH)_3$	solid	638.7
Pyrrole	C_4H_5N	liquid	567.7
Quercitol	$C_6H_{12}O_5$	solid	704.2
Quinoline	C_9H_7N	liquid	1123.5
Quinone	$O:C_6H_4:O$	solid	656.6
Raffinose	$C_{18}H_{32}O_{16}$	solid	2025.5
Retene	$C_{18}H_{18}$	solid	2306.8
Resorcinol	$C_6H_4(OH)_2$	solid	681.30*
Resorcinoldimethyl ether	$(CH_3O)_2C_6H_4$	liquid	1022.6
Rhamnose	$C_6H_{12}O_6$	solid	718.3
Safrole	$C_{10}H_{10}O_2$	liquid	1244.1
Salicylaldehyde, see o-Hydroxybenzaldehyde			
Salicylic acid*	$HOC_6H_4CO_2H—(1,2)$	solid	722.4**
Sarcosine	$CH_3NHCH_2CO_2H$	solid	401.1
Sebacic acid	$(CH_2)_8(CO_2H)_2$	solid	1297.3
Skatole	C_9H_9N	liquid	1170.5
d-Sorbose	$C_6H_{12}O_6$	solid	668.3
Starch	$(C_6H_{10}O_5)x$ per kg	solid	4178.8
Stearic acid	$C_{18}H_{36}O_2$	solid	2696.12*
Strychnine	$C_{21}H_{22}O_2N_2$	solid	2685.7
Styrene	$C_6H_5CH_2$	liquid	1047.1
Suberic acid	$(CH_2)_6(CO_2H)_2$	solid	985.2
Succinic acid	$(CH_2CO_2H)_2$	solid	356.36*
Succinic acid nitrile	$(CH_2CN)_2$	liquid	545.7
Succinic anhydride	$(CH_2CO)_2O$	solid	369.0*
Succinimide	$C_4H_5O_2N$	solid	437.9
Sucrose	$C_{12}H_{22}O_{11}$	solid	1348.2*
Sylvestrene	$C_{10}H_{15}$	liquid	1464.7
l-Tartaric acid	$(CHOH)_2(CO_2H)_2$	solid	274.7*
d,l-Tartaric acid (anhydr)	$(CHOH)_2(CO_2H)_2$	solid	272.6*
Terephthalic acid	$C_6H_4(CO_2H)_2$	solid	770.4
Terpin hydrate	$C_{10}H_{22}O_3$	solid	1451.0
Terpineol	$C_{10}H_{18}O$	solid	1469.5
Tetrahydrobenzene	C_6H_{10}	liquid	891.9
Tetrahydronaphthalene	$C_{10}H_{12}$	liquid	1352.4
Tetramethylmethane	$(CH_3)_4C$	gas	842.6
Tetraphenylmethane	$(C_6H_5)_4C$	solid	3102.4
Tetryl	$C_6H_5N_5O_8$	solid	842.3
Thebaine	$C_{19}H_{21}O_3N$	solid	2441.3
Thiophene	C_4H_4S	liquid	670.5
Thujane	$C_{10}H_{18}$	liquid	1506.4
Thymol	$C_{10}H_{14}O$	liquid	1353.4
Thymol	$C_{10}H_{14}O$	solid	1349.7
Thymoquinone	$C_{10}H_{12}O_2$	solid	1271.3
Toluene	$CH_3C_6H_5$	liquid	934.2
o-Toluic acid	$CH_3C_6H_4CO_2H$	solid	928.9
m-Toluic acid	$CH_3C_6H_4CO_2H$	solid	928.6
p-Toluic acid	$CH_3C_6H_4CO_2H$	solid	926.9
o-Toluidine	$CH_3C_6H_4NH_2$	liquid	964.3
m-Toluidine	$CH_3C_6H_4NH_2$	liquid	965.3
p-Toluidine	$Ch_3C_6H_4NH_2$	solid	958.4
o-Tolunitrile	$CH_3C_6H_4CN$	liquid	1030.3
Toluquinone	$C_7H_6O_3$	solid	803.2
Triaminotriphenyl carbinol	$(C_6H_4NH_2)_3COH$	solid	2483.5
Tribenzyl amine	$(C_6H_5CH_2)_3N$	solid	2762.1
Trichloracetic acid	$Cl_2C.CO_2H$	solid	92.8
Triethyl amine	$(C_2H_5)_3N$	liquid	1036.8
Triethyl carbinol	$(C_2H_5)_3CHOH$	liquid	1080.0
Triisoamyl amine	$[(CH_3)_2CHCH_2CH_2]_3N$	liquid	2459.3
Triisobutyl amine	$[(CH_3)_2CHCH_2]_3N$	liquid	1973.6
Trimethyl amine	$(CH_3)_3N$	liquid	578.6
2,2,3-Trimethylbutane	$(CH_3)_3C.CH(CH_3)_2$	liquid	1147.9
Trimethyl carbinol	$(CH_3)_3COH$	liquid	629.3
Trimethylene	$CH_2CH_2CH_2$	gas	499.89*
Trimethylethylene	$(CH_3)_2CCH_2$	liquid	796.0
Trimethylethylene	$(CH_3)_2CCH_3$	vapor	803.6
2,2,4-Trimethylpentane	$(CH_3)_3C.CH_2CH(CH_3)_2$	liquid	1303.9
Trinitrobenzene	$C_6H_2(NO_2)_3—(1,3,5)$	solid	663.7
Trinitroglycerol	$C_3H_5(NO_3)_3$	liquid	368.4
Trinitrotoluene	$C_5H_2(CH_3)(NO_2)_3—(1,2,4,6)$	solid	820.7
Triphenyl amine	$(C_6H_5)_3N$	solid	2267.8
Triphenylbenzene	$C_6H_3(C_6H_5)_3—(1,3,5)$	solid	2936.7
Triphenyl carbinol	$(C_6H_5)_3CHOH$	sed	2340.8

** Recommended as a secondary thermochemical standard.

HEAT OF COMBUSTION
For Organic Compounds (Continued)

Name	Formula	Physical state	Heat of combustion. kg. calories
Triphenylmethane	$(C_6H_5)_2CH$	solid	2388.7
Triphenyl methyl	$(C_6H_5)_3C$	solid	2378.5
Tyrosine	$C_9H_{11}O_3N$	solid	1070.2
Undecyclic acid	$C_{11}H_{22}O_2$	solid	1615.9
Urea	$(NH_2)_2CO$	solid	150.97*
Urethane	$NH_2CO_2C_2H_5$	solid	397.2
Uric acid	$C_5H_4O_3N_4$	solid	460.2
n-Valeric acid	$C_4H_9CO_2H$	liquid	678.12*
Vanillin	$C_6H_3(OH)(OCH_3)CHO$ (1,2,4)	solid	914.1
o-Xylene	$(CH_3)_2C_6H_4$	liquid	1091.7
m-Xylene	$(CH_3)_2C_6H_4$	liquid	1088.4
p-Xylene	$(CH_3)_2C_6H_4$	liquid	1089.1
Xylose	$C_5H_{10}O_5$	solid	559.0*

HEAT OF FORMATION

The thermal change involved in the formation of 1 mol of substance from its elements is the heat of formation, ΔH, of the substance. If all of the substances involved in the reaction are each in their standard states and each substance is at unit activity, the thermal change is the standard heat of formation, $\Delta H°$. By definition, all elements in their standard state have a heat of formation of zero. By further definition, the sign of ΔH is negative if heat is evolved and is positive if heat is absorbed. Thus, for the thermochemical reaction:

$$7C \text{ (s)} + 3H_2(g) + O_2(g) = C_6H_5COOH \text{ (s)} \qquad \Delta H° = -771.24 \text{ kcal}$$
$$\text{(benzoic acid)}$$

Heats of formation may be calculated from heats of combustion. The heat of formation of compound "A" is equal to the sum of the heats of formation of the products of combustion of compound "A" minus the heat of combustion of compound "A". Some heats of formation of some products of combustion of organic compounds are:

Substance	ΔH, heat of formation (kcal/ g mole)
Free Elements	0
CO	− 26.416
CO_2	− 93.963
½ H_2O (1), from 1 H	− 34.158
H_2O (1)	− 68.317
HF (Dilute aqueous solution)	− 76.531
HCl (Dilute aqueous solution)	− 39.850
HBr (Dilute aqueous solution)	− 28.958
HI (Dilute aqueous solution)	− 13.106
HNO_3 (Dilute aqueous solution)	− 48.484
H_2SO_4 (Dilute aqueous solution)	−213.552
SO_2 (g)	− 70.336

Two examples of calculations are:

Example I

Calculate the heat of formation of methane (CH_4) where

$$\text{Heat of combustion of } CH_4 = -210.8 \text{ kcal/g mol}$$
$$\text{Heat of formation of } CO_2 = -93.963 \text{ kcal/g mol}$$
$$\text{Heat of formation of } H_2O = -68.317 \text{ kcal/g mol}$$

and where the combustion reaction occurs according to the following equation:

$$CH_4(gas) + 2O_2(gas) = CO_2(gas) + 2H_2O \text{ (liquid)}$$

Heat of formation of CH_4 equals:

	Heat of formation of CO_2	=	−93.963 kcal
+	Two times heat of formation of H_2O	=	2(−68.317)kcal
−	Heat of combustion of CH_4	=	−(−210.8) kcal
			−19.8 kcal/g mol

Example II

Calculate the heat of formation of ethylene (C_2H_4) where

$$\text{Heat of combustion of } C_2H_4 = -337.23 \text{ kcal/g mol}$$

and where the combustion reaction is as follows:

$$C_2H_4(gas) + 3O_2(gas) = 2CO_2(gas) + 2H_2O \text{ (liquid)}$$

	Two times heat of formation of CO_2	=	2(−93.963)kcal
+	Two times heat of formation of H_2O	=	2(−68.317)kcal
−	Heat of combustion of C_2H_4	=	−(−337.23) kcal
			+ 12.67 kcal/g mol

LOWERING OF VAPOR PRESSURE BY SALTS IN AQUEOUS SOLUTIONS

The table gives the reduction of the vapor pressure in millimeters due to the presence of the number of grammolecules of salt per liter of water given at the head of the columns, at the temperature 100° C, at which temperature the vapor pressure of pure water is 760 millimeters.
(From Smithsonian Tables.)

Substance	0.5	1.0	2.0	3.0	4.0	5.0	6.0	8.0	10.0
$Al_2(SO_4)_3$	12.8	36.5							
$AlCl_3$	22.5	61.0	179.0	318.0					
BaS_2O_6	6.6	15.4	34.4						
$Ba(OH)_2$	12.3	22.5	39.0						
$Ba(NO_3)_2$	13.5	27.0							
$Ba(ClO_3)_2$	15.8	33.3	70.5	108.2					
$BaCl_2$	16.4	36.7	77.6						
$BaBr_2$	16.8	38.8	91.4	150.0	204.7				
CaS_2O_8	9.9	23.0	56.0	106.0					
$Ca(NO_3)_2$	16.4	34.8	74.6	139.3	161.7	205.4			
$CaCl_2$	17.0	39.8	95.3	166.6	241.5	319.5			
$CaBr_2$	17.7	44.2	105.8	191.0	283.3	368.5			
$CdSO_4$	4.1	8.9	18.1						
CdI_2	7.6	14.8	33.5	52.7					
$CdBr_2$	8.6	17.8	36.7	55.7	80.0				
$CdCl_2$	9.6	18.8	36.7	57.0	77.3	99.0			
$Cd(NO_3)_2$	15.9	36.1	78.0	122.2					
$Cd(ClO_3)_2$	17.5								
$CoSO_4$	5.5	10.7	22.9	45.5					
$CoCl_2$	15.0	34.8	83.0	150.0	180.4				
$Co(NO_3)_2$	17.3	39.2	89.0	152.0	218.7	282.0	332.0		
$FeSO_4$	5.8	10.7	24.0	42.4					
H_3BO_3	6.0	12.3	25.1	38.0	51.0				
H_3PO_4	6.6	14.0	28.6	45.2	62.0	81.5	103.0	146.9	189.5
H_3AsO_4	7.3	15.0	30.2	46.4	64.9				
H_2SO_4	12.9	26.5	62.8	104.0	148.0	198.4	247.0	343.2	
KH_2PO_4	10.2	19.5	33.3	47.8	60.5	73.1	85.2		
KNO_3	10.3	21.1	40.1	57.6	74.5	88.2	102.1	126.3	148.0
$KClO_3$	10.6	21.6	42.8	62.1	80.0				
$KBrO_3$	10.9	22.4	45.0						
$KHSO_4$	10.9	21.9	43.3	65.3	85.5	107.8	129.9	170.0	
KNO_2	11.1	22.8	44.8	67.0	90.0	110.5	130.7	167.0	198.8
$KClO_4$	11.5	22.3							
KCl	12.2	24.4	48.8	74.1	100.9	128.5	152.2		
$KHCO_3$	11.6	23.6	59.0	77.6	104.2	132.0	160.0	210.0	255.0
KI	12.5	25.3	52.2	82.6	112.2	141.5	171.8	225.5	278.5
$K_2C_2O_4$	13.9	28.3	59.8	94.2	131.0				
K_2WO_4	13.9	33.0	75.0	123.8	175.4	226.4			
K_2CO_3	14.4	31.0	68.3	105.5	152.0	209.0	258.5	350.0	
KOH	15.0	29.5	64.0	99.2	140.0	181.8	223.0	309.5	387.8
K_2CrO_4	16.2	29.5	60.0						
$LiNO_3$	12.2	25.9	55.7	88.9	122.2	155.1	188.0	253.4	309.2
$LiCl$	12.1	25.5	57.1	95.0	132.5	175.5	219.5	311.5	393.5
$LiBr$	12.2	26.2	60.0	97.0	140.0	186.3	241.5	341.5	438.0
Li_2SO_4	13.3	28.1	56.8	89.0					
$LiHSO_4$	12.8	27.0	57.0	93.0	130.0	168.0			
LiI	13.6	28.6	64.7	105.2	154.5	206.0	264.0	357.0	445.0
Li_2SiFl_6	15.4	34.0	70.0	106.0					

Substance	0.5	1.0	2.0	3.0	4.0	5.0	6.0	8.0	10.0
$LiOH$	15.9	37.4	78.1						
Li_2CrO_4	16.4	32.6	74.0	120.0	171.0				
$MgSO_4$	6.5	12.0	24.5	47.5					
$MgCl_2$	16.8	39.0	100.5	183.3	277.0	377.0			
$Mg(NO_3)_2$	17.6	42.0	101.0	174.8					
$MgBr_2$	17.9	44.0	115.8	205.3	298.5				
$MgH_2(SO_4)_2$	18.3	46.0	116.0						
$MnSO_4$	6.0	10.5	21.0						
$MnCl_2$	15.0	34.0	76.0	122.3	167.0	209.0			
NaH_2PO_4	10.5	20.0	36.5	51.7	66.8	82.0	96.5	126.7	157.1
$NaHSO_4$	10.9	22.1	47.3	75.0	100.2	126.1	148.5	189.7	231.4
$NaNO_3$	10.6	22.5	46.2	68.1	90.3	111.5	131.7	167.8	198.8
$NaClO_3$	10.5	23.0	48.4	73.5	98.5	123.3	147.5	196.5	223.5
$(NaPO_3)_6$	11.6								
$NaOH$	11.8	22.8	48.2	77.3	107.5	139.1	172.5	243.3	314.0
$NaNO_2$	11.6	24.4	50.0	75.0	98.2	122.5	146.5	189.0	226.2
Na_2HPO_4	12.1	23.5	43.0	60.0	78.7	99.8	122.1		
$NaHCO_3$	12.9	24.1	48.2	77.6	102.2	127.8	152.0	198.0	239.4
Na_2SO_4	12.6	25.0	48.9	74.2					
$NaCl$	12.0	25.9	52.1	80.0	111.0	142.0	176.5		
$NaBrO_3$	12.1	25.0	54.1	81.3	108.8	136.0			
$NaBr$	12.6	25.9	57.0	89.2	124.2	159.5	197.5	268.0	
NaI	12.1	25.6	60.2	99.5	136.7	177.5	221.0	301.5	370.0
$Na_4P_2O_7$	13.2	22.0							
Na_2CO_3	14.3	27.3	53.5	80.2	111.0				
$Na_2C_2O_4$	14.5	30.0	65.8	105.8	146.0				
Na_2WO_4	14.8	33.6	71.6	115.7	162.6				
Na_3PO_4	16.5	30.0	52.5						
$(NaPO_3)_3$	17.1	36.5							
NH_4NO_3	12.8	22.0	42.1	62.7	82.9	103.8	121.0	152.2	180.0
$(NH_4)_2SiFl_6$	11.5	25.0	44.5						
NH_4Cl	12.0	23.7	45.1	69.3	94.2	118.5	138.2	179.0	213.8
NH_4HSO_4	11.5	22.0	46.8	71.0	94.5	118.	139.0	181.2	218.0
$(NH_4)_2SO_4$	11.0	24.0	46.5	69.5	93.0	117.0	141.8		
NH_4Br	11.9	23.9	48.8	74.1	99.4	121.5	145.5	190.2	228.5
NH_4I	12.9	25.1	49.8	78.5	104.5	132.3	156.0	200.0	243.5
$NiSO_4$	5.0	10.2	21.5						
$NiCl_2$	16.1	37.0	86.7	147.0	212.8				
$Ni(NO_3)_2$	16.1	37.3	91.3	156.2	235.0				
$Pb(NO_3)_2$	12.3	23.5	45.0	63.0					
$Sr(SO_3)_2$	7.2	20.3	47.0						
$Sr(NO_3)_2$	15.8	31.0	64.0	97.4	131.4				
$SrCl_2$	16.8	38.8	91.4	156.8	223.3	281.5			
$SrBr_2$	17.8	42.0	101.1	179.0	267.0				
$ZnSO_4$	4.9	10.4	21.5	42.1	66.2				
$ZnCl_2$	9.2	18.7	46.2	75.0	107.0	153.0	195.0		
$Zn(NO_3)_2$	16.6	39.0	93.5	157.5	223.8				

THERMAL CONDUCTIVITY OF GASES

The values in this table are given as cal/(sec)(cm²)(°C/cm) $\times 10^{-6}$. To convert these values to Btu/(hr)(ft²)(°F/ft) $\times 10^{-6}$ multiply by 241.909.

Gas °F °C	−400 −240	−300 −184.4	−200 −128.9	−100 −73.3	−40 −40	−20 −28.9	0 −17.8	20 −6.7	40 4.4	60 15.6	80 26.7	100 37.8	120 48.9	200 93.3
Acetylene	.....	.....	.....	28.10	34.71	37.19	39.67	42.15	45.04	47.94	50.83	53.72	56.62	69.43
Air	.....	.....	.....	.....	50.09	52.15	54.22	56.24	58.31	60.34	62.20	64.22	66.04	.....
Ammonia	.....	.....	.....	.....	43.39	45.87	48.35	50.83	53.31	55.79	58.68	61.58	64.47	.....
Argon	.....	.....	.....	.....	34.30	35.95	37.19	38.85	40.09	41.33	42.57	44.22	45.46	.....
Bromine	.....	.....	.....	.....	.....	.....	9.09	.....	.....	.....	.....	11.57	.....	.....
n-Butane	.....	.....	.....	.....	.....	.....	.....	30.99	33.06	35.54	38.02	40.91	43.39	54.14
i-Butane	.....	.....	.....	.....	.....	.....	.....	32.65	33.89	36.37	38.85	41.74	44.22	55.79
Carbon dioxide	.....	.....	.....	27.90	.....	29.75	31.70	33.68	35.62	37.61	39.67	41.74	43.81	.....
Carbon disulfide	.....	.....	.....	.....	.....	.....	14.05	15.29	16.53	17.77	19.01	19.84	.....	.....
Carbon monoxide	.....	.....	.....	.....	47.94	50.00	51.95	53.85	55.87	57.86	59.92	61.99	63.89	.....
Chlorine	.....	.....	.....	.....	15.29	16.53	17.36	18.18	19.01	20.25	21.08	21.90	23.14	.....
Deuterium	.....	.....	.....	.....	274.82	285.15	295.07	305.81	309.95	322.34	334.74	343.01	355.40	.....
Ethane	.....	.....	.....	23.97	32.65	35.54	38.43	41.33	44.63	47.94	51.24	54.55	58.27	74.39
Ethanol	.....	.....	.....	.....	.....	.....	.....	29.34	30.99	32.65	34.71	36.78	.....	.....
Ethylamine	.....	.....	.....	.....	.....	.....	.....	31.41	33.47	35.54	37.61	39.67	42.15	.....
Ethylene	.....	.....	26.86	.....	33.06	35.54	38.02	40.50	43.39	46.29	49.18	52.07	54.96	68.19
Fluorine	.....	18.18	30.58	43.39	50.83	52.90	55.38	57.86	59.92	61.99	64.06	66.12	68.19	76.04
Helium	84.31	163.24	221.51	274.8	304.99	314.49	324.00	333.50	343.42	352.10	360.36	368.63	376.07	.....
Hydrogen	59.92	142.57	227.29	308.7	357.47	371.93	388.46	405.00	417.39	433.92	446.32	458.72	471.11	.....
Hydrogen bromide	.....	.....	.....	.....	15.29	16.11	17.77	18.60	19.84	20.66	21.49	.....	.....	.....
Hydrogen chloride	.....	.....	.....	.....	25.62	26.86	28.51	29.75	30.99	32.23	33.89	35.12	.....	.....
Hydrogen cyanide	.....	.....	.....	.....	.....	.....	23.97	25.62	26.86	28.10	29.75	30.99	32.65	.....
Hydrogen sulfide	.....	.....	.....	.....	.....	.....	28.10	29.75	31.41	33.47	.....	36.78	.....	.....
Krypton	.....	.....	.....	.....	.....	.....	19.84	.....	.....	.....	.....	23.56	.....	.....
Methane	.....	22.32	36.86	52.07	61.37	64.55	67.86	71.08	74.39	78.11	81.83	85.54	89.26	106.62
Neon	.....	.....	.....	.....	97.94	100.84	104.14	107.03	109.93	112.82	115.71	118.19	121.09	.....
Nitric oxide	.....	.....	30.91	42.40	49.01	51.24	53.39	55.54	57.65	59.76	61.99	64.06	66.12	74.39
Nitrogen	.....	20.25	33.06	44.22	50.42	52.48	54.55	56.20	58.27	60.34	62.40	64.06	65.71	.....
Nitrous oxide	.....	.....	.....	.....	28.93	30.91	32.90	35.04	37.15	39.30	41.45	43.81	46.08	.....
Oxygen	.....	18.84	31.66	43.72	50.54	52.81	54.96	57.24	59.43	61.58	63.64	65.91	68.19	76.87
n-Propane	.....	.....	.....	27.69	.....	29.75	32.23	34.71	37.19	39.67	42.47	45.46	48.35	60.75
R-11(CCl₃F)	.....	.....	.....	.....	.....	.....	12.81	13.64	14.88	15.70	16.53	17.77	18.60	.....
R-12(CCl₂F₂)	.....	.....	.....	.....	.....	17.36	18.60	19.42	20.66	21.49	22.73	23.56	.....	.....
R-21(CHCl₂F)	.....	.....	.....	.....	.....	.....	21.90	22.32	22.73	23.14	23.56	23.97	.....	.....
R-22(CHClF₂)	.....	.....	.....	.....	.....	.....	24.80	25.62	26.45	27.28	28.10	28.93	.....	.....
Water	.....	.....	.....	.....	.....	.....	34.71	36.78	38.85	40.50	42.57	44.63	46.70	54.96

THERMAL CONDUCTIVITY OF GASEOUS HELIUM, NITROGEN AND WATER

From NSRDS-NBS 8

R. W. Powell, C. Y. Ho, and P. E. Liley

The thermal conductivity, k, is given in the units Milliwatt cm^{-1} °K^{-1}. To convert to Cal(gm) hr^{-1} cm^{-1} °K^{-1} multiply the values listed in the table by 0.860421

T (K)	k He	k N$_2$	T (K)	k He	k N$_2$	k H$_2$O	T (K)	k He	k N$_2$	k H$_2$O	T (K)	k He	k N$_2$
0.08	0.00044		150	0.950	0.1385		650	2.64	0.467	0.518	1750	5.57	0.981
0.09	0.00053		160	0.992	0.1474		660	2.67	0.472	0.529	1800	5.70	1.013
0.10	0.00064		170	1.033	0.1562		670	2.69	0.478	0.540	1850	5.83	1.046
0.15	0.00130		180	1.072	0.1651		680	2.72	0.483	0.551	1900	5.96	1.080
0.20	0.00231		190	1.112	0.1739		690	2.75	0.488	0.562	1950	6.08	1.113
0.25	0.0030		200	1.151	0.1826		700	2.78	0.493	0.572	2000	6.20	1.146
0.30	0.0062		210	1.190	0.1908		710	2.81	0.498	0.58	2100	6.44	1.207
0.35	0.0089		220	1.228	0.1989		720	2.84	0.503	0.59	2200	6.69	1.263
0.40	0.0120		230	1.266	0.2067		730	2.87	0.508	0.60	2300	6.93	1.314
0.45	0.0154		240	1.304	0.2145		740	2.90	0.513	0.62	2400	7.16	1.361
0.5	0.0187		250	1.338	0.2222	(0.140)*	750	2.92	0.517	0.63	2500	7.39	1.406
0.6	0.0231		260	1.372	0.2298	(0.148)*	760	2.95	0.522	0.64	2600	7.62	1.449
0.7	0.0252		270	1.405	0.2374	(0.156)*	770	2.98	0.526	0.65	2700	7.85	1.494
0.8	0.0262		280	1.437	0.2449	0.164	780	3.01	0.531	0.66	2800	8.07	1.542
0.9	0.0266		290	1.468	0.2524	0.172	790	3.04	0.536	0.67	2900	8.29	1.590
1.0	0.0269		300	1.499	0.2598	0.181	800	3.07	0.541	0.68	3000	8.51	1.640
1.25	0.0281		310	1.530	0.2671	0.189	810	3.09	0.546	0.69	3100	8.72	1.691
1.5	0.0306		320	1.560	0.2741	0.197	820	3.12	0.551	0.70	3200	8.95	1.743
2.0	0.0393		330	1.590	0.2808	0.205	830	3.15	0.555	0.71	3300	9.16	1.795
2.5	0.0502		340	1.619	0.2874	0.214	840	3.18	0.559	0.72	3400	9.37	1.853
3.0	0.0607		350	1.649	0.2939	0.222	850	3.21	0.564	0.73	3500	9.58	1.915
3.5	0.0732		360	1.678	0.3002	0.231	860	3.23	0.569	0.74	3600	9.79	
4.0	0.0803		370	1.708	0.3065	0.239	870	3.26	0.574	0.75	3700	10.00	
4.5	0.0879		380	1.737	0.3127	0.248	880	3.29	0.578	0.76	3800	10.22	
5.0	0.0982		390	1.766	0.3190	0.256	890	3.32	0.583	0.77	3900	10.43	
6	0.1113		400	1.795	0.3252	0.264	900	3.35	0.587	0.78	4000	10.64	
7	0.1247		410	1.824	0.3314	0.273	910	3.37	0.592		4100	10.85	
8	0.1090		420	1.853	0.3376	0.282	920	3.40	0.596		4200	11.06	
9	0.1523		430	1.882	0.3438	0.291	930	3.43	0.600		4300	11.27	
10	0.1640		440	1.914	0.3501	0.300	940	3.46	0.605		4400	11.48	
12	0.1866		450	1.947	0.3564	0.307	950	3.49	0.609		4500	11.69	
14	0.2067		460	1.980	0.3626	0.317	960	3.52	0.613		4600	11.90	
16	0.2259		470	2.013	0.3688	0.327	970	3.54	0.618		4700	12.11	
18	0.2435		480	2.046	0.3749	0.337	980	3.57	0.622		4800	12.31	
20	0.2582		490	2.080	0.3808	0.347	990	3.60	0.626		4900	12.51	
25	0.2962		500	2.114	0.3864	0.357	1000	3.63	0.631		5000	12.71	
30	0.3330		510	2.15	0.392	0.368	1050	3.76	0.651				
35	0.3669		520	2.18	0.398	0.378	1100	3.89	0.672				
40	0.4000		530	2.22	0.403	0.389	1150	4.03	0.693				
45	0.4314		540	2.25	0.408	0.400	1200	4.16	0.713				
50	0.4623	(0.0485)*	550	2.29	0.414	0.411	1250	4.29	0.733				
60	0.521	(0.0578)*	560	2.33	0.420	0.422	1300	4.43	0.754				
70	0.578	(0.0670)*	570	2.36	0.425	0.432	1350	4.55	0.775				
80	0.631	0.0762	580	2.40	0.431	0.443	1400	4.69	0.797				
90	0.679	0.0852	590	2.43	0.436	0.454	1450	4.82	0.819				
100	0.730	0.0941	600	2.47	0.441	0.464	1500	4.94	0.842				
110	0.776	0.1030	610	2.51	0.446	0.475	1550	5.07	0.867				
120	0.819	0.1119	620	2.54	0.452	0.486	1600	5.21	0.893				
130	0.863	0.1208	630	2.58	0.457	0.497	1650	5.33	0.921				
140	0.907	0.1296	640	2.61	0.462	0.508	1700	5.45	0.950				

THERMAL CONDUCTIVITY OF DIELECTRIC CRYSTALS

Name	Remarks	Conductivity nw/cm deg K 83°K	273°K	Name	Remarks	Conductivity nw/cm deg K 83°K	273°K
Marble	Small crystals, 99.9% CaCO₃	42	33	90% KBr, 10% KCl	Do	50	29
Do	99.99% CaCO₃	54	38	75% KBr, 25% KCl	Do	29	21
Do	Large crystals	50	33	50% KBr, 50% KCl	Do	25	25
Calcite	Main crystal axis perpendicular to rod axis	180	46	25% KBr, 75% KCl	Pressed at 8000 atm	46	33
				10% KBr, 90% KCl	Do	80	50
Do	Main crystal axis parallel to rod axis	293	54	50% KCl, 50% NaCl	Do	188	71
				KNO₂	Do	17	21
Sylvite	Natural crystal	159	75	Mercuric chloride	Do	17	13
KCl	Pressed at 8000 atm	314	88	NH₄Cl	Do	109	25
KCl	From a melt	402	92	NH₂Br	Do	67	25
NaCl	Do	343	92	Ba(NO₃)₂	Do	33	13
NaCl	Pressed at 8000 atm	251	71	Copper sulfate		29	21
Rock salt	Do	180	63	Magnesium sulfate		25	25
Sylvite	Do	343	84	K₄Fe(CN)₆		17	17
KCl	Pressed at 1250 atm	243	75	Chrom alum		13	21
KCl	Pressed at 2500 atm	368	92	Potassium alum		13	21
KCl	Pressed at 8900 atm	402	96	Potassium bichromate	Main crystal axis perpendicular to rod axis	17	21
KBr	Pressed at 8000 atm	92	38				
NaBr	Do	50	25	Do	Main crystal axis parallel to rod axis	17	17
KI	Do	121	29				
KF	Do	234	71	Topaz	Mineral		234
NaF	Do	519	105	Zincblend	Do	63	264
RbI	Do	59	33	Beryll	Do	88	84
RbCl	Do	29	21	Tourmaline	Do	38	46

THERMAL CONDUCTIVITY OF ORGANIC COMPOUNDS[a]

Substance	k	t, °C	t, °F	Substance	k	t, °C	t, °F
Acetaldehyde	0.0004089	21	69.8	Freon-22 (CHClF₂)	0.0002309	40	104
Acetic acid	0.0004109	20	68	Freon-113 (CCl₂FCCCl₂F)	0.0002379	0—80	32—176
Acetic anhydride	0.0005286	21	69.8	Freon-114 (C₂H₂F₄)	0.0002127	0—75	32—167
Acetone	0.0004750	−80	−112	Glycerol	0.000703	20	68
	0.0004543	16	61	Heptane (n)	0.0003354	30	86
	0.0004031	75	167	Heptyl alcohol	0.0003882	70—100	86—212
Allyl alcohol	0.0004295	30	86	Hexane (n)	0.0003287	30—100	86—212
Amyl acetate(n)	0.0003085	20	68	Hexyl alcohol (n)	0.0003857	30—100	86—212
(iso)	0.000310	20	68	Iodobenzene	0.0002874	30—100	86—212
Amyl alcohol(n)	0.0003874	30—100	86—212	Mesitylene	0.0003246	20	68
(iso)	0.0003531	30	86	Methyl alcohol	0.0004832	20	68
Amyl bromide (n)	0.0002350	18.64.4		Methyl aniline	0.0004419	21.5	70.5
Aniline	0.0004237	16.5	61.5	Methyl chloride	0.0004597	−15(−)+30	5—86
Benzene	0.0003780	22.5	72.5	Methyl cyclohexane	0.0003052	30	86
	0.0003275	50	122	Methylene chloride	0.0002908	0	32
	0.0003630	60	140	Nitrobenzene	0.0003907	30—100	86—212
	0.0002870	140	284	Nitromethane	0.0005142	30	86
Bromobenzene	0.0002664	20	68	Nonane (n)	0.0003374	30—100	86—212
Butyl acetate (n)	0.000327	20	68	Nonyl alcohol (n)	0.0004014	30—100	86-212
Butyl alcohol (n)	0.0003663	20	68	Octane (n)	0.0003469	30	86
Carbon tetrachloride	0.0002470	20	68	Octyl alcohol (n)	0.0003973	30—100	86—212
	0.0002333	50	122	Oleic acid	0.0005514	26.5	79.7
Chlorobenzene	0.0003457	30—100	86—212	Palmitic acid	0.0004097	72.5	162.5
Chlorotoluene (p)	0.000310	20	68	Pentachloroethane	0.0002994	20	68
Chloroform	0.0002891	16	61	Pentane (n)	0.0003221	30	86
	0.000246	20	68	Phenetole	0.0003577	−20	−4
Cresol(m)	0.0003581	20	68	Phenyl hydrazine	0.0004121	25	69.8
(p)	0.000345	20.1	68.2	Propyl acetate (sio)	0.000321	20	68
Cumene	0.000298	20	68	Propyl alcohol (iso)	0.0003362	20	68
Cymene (p)	0.0003217	30	86	Propylene chloride	0.0002994	20—50	68—122
Decane	0.0003349	30	86	Propylene glycol (1—2)	0.0004799	20—80	68—176
Diethyl ether	0.0003283	30	86	Stearic acid	0.0003824	72.5	162.5
Dichloroethane, 1—2	0.000302	20	68	Tetrachloroethane (sym)	0.000272	20	68
Di-isopropyl ether	0.000262	20	68	Tetrachloroethylene	0.0003866	20	68
Ethyl acetate	0.0003560	16	60.8	Toluene	0.0003804	−80	−112
Ethyl alcohol	0.0003995	20	68		0.0003221	20	68
Ethyl benzene	0.0003160	20	68		0.0002808	80	176
Ethyl bromide	0.0002862	30	86	Trichloroethylene	0.0003246	−60	−76
Ethyl ether	0.0003283	30	86		0.0002775	20	68
Ethyl iodide	0.0002651	30	86	Triethylamine	0.0003498	−80	−112
Ethylene glycol	0.0006236	20	68		0.0002891	20	68
	0.0006323	15	122		0.0002664	44.4	112
	0.0006443	80	176	Xylene (o)	0.0003411	−20(−)+80	(−4)−176
Freon-12 (CCl₂F₂)	0.0002310	0—75	32—167	Xylene (m)	0.0003767	25	77
Freon-21 (CHCl₂F)	0.0003180	0—75	32—167				

[a] The values in this table are given as cal/(sec)(cm²)(°C/cm). To convert these values to Btu/(hr)(ft₂)(°F/ft) multiply by 242.08 .

THERMAL CONDUCTIVITY OF INORGANIC COMPOUNDS

Ammonia	0.0001198	−15(−)+30	5—86		
Argon	0.0002895	−183	−297		
	0.0001677	−133	−207		
	0.000553	−105	−157.5		
	0.0000409	−75	−102.5		
Carbon dioxide	0.0002040	−50	−58		
	0.0002412	−40	−40		
	0.0002664	−30	−22		
	0.0002746	−20	−4		
	0.0002495	0	32		
	0.0001677	30	86		
Nitrogen	0.0003400	−196	−321.5		
	0.0002028	−158	−253		
	0.0000640	−105	−155		
Oxygen	0.0000500	(−207)—(−191)	(−340)—(−312)		
	0.0000504	(−178)—(−182)	(−288)—(−295)		
Water	0.001326	−3	27		
	0.001372	+7	45		
	0.001456	27	81		
	0.001522	47	117		
	0.001575	67	153		
	0.001625	97	207		
	0.001635	107	225		
	0.001628	157	315		
	0.001580	197	387		
	0.001463	247	441		
	0.001288	297	567		
	0.001004	347	657		

THERMAL CONDUCTIVITY OF MISCELLANEOUS SUBSTANCES

Chlorinated diphenyl 1242	0.0002936	30—100	86—212	Petroleum ether	0.0003118	30	86
Chlorinated diphenyl 1248	0.0002808	30—100	86—212	Red oil	0.0003366	30	86
Kerosene	0.0003572	30	86	Transformer oil	0.0004242	70—100	86—212
Light heat transfer oil	0.0003159	30—100	86—212				

THERMAL CONDUCTIVITY OF MATERIALS
(Bureau of Standards Letter Circular No. 227)

D = Density in pound per cubic foot.
K = Thermal conductivity in B.T.U. per hour, square foot, and temperature gradient of 1 degree Fahrenheit per inch thickness. The lower the conductivity, the greater the insulating values.

Soft Flexible Materials in Sheet Form

		D	K
Dry zero	Kapok between burlap or paper	1.0	0.24
		2.0	0.25
Cabots quilt	Eel grass between kraft paper	3.4	8.25
		4.6	0.26
Hair felt	Felted cattle hair	11.0	0.26
		13.0	0.26
Balsam wool	Chemically treated wood fibre	2.2	0.27
Hairinsul	75% hair 25% jute	6.3	0.27
	50% hair 50% jute	6.1	0.26
Linofelt	Flax fibers between paper	4.9	0.28
Thermofelt	Jute and asbestos fibers, felted	10.0	0.37
	Hair and asbestos fibers, felted	7.8	0.28

Loose Materials

		D	K
Rock wool	Fibrous material made from rock also made in sheet form, felted and confined with wire netting	6.0	0.26
		10.0	0.27
		14.0	0.28
		18.0	0.29
Glass wool	Pyrex glass, curled	4.0	0.29
		10.0	0.29
Sil-O-Cel	Powdered diatomaceous earth	10.6	0.31
Regranulated cork	Fine particles	9.4	0.30
	about 3/16 inch particles	8.1	0.31
Thermofill	Gypsum in powdered form	26	0.52
		34	0.60
Sawdust	Various	12.0	0.41
	redwood	10.9	0.42
Savings	Various, from planer	8.8	0.41
Charcoal	From maple, beech and birch, coarse	13.2	0.36
	6 mesh	15.2	0.37
	20 mesh	19.2	0.39

Semiflexible Materials in Sheet Form

		D	K
Flaxlinum	Flax fiber	13.0	0.31
Fibrofelt	Flax and rye fiber	13.6	0.32

Semiflexible Materials in Sheet Form

		D	K
Flaxlinum	Flax fiber	13.0	0.31
Fibrofelt	Flax and rye fiber	13.6	0.32

Semirigid Materials in Board Form

		D	K
Corkboard	No added binder; very low density	5.4	0.25
Corkboard	No added binder; low density	7.0	0.27
Corkboard	No added binder; medium density	10.6	0.30

		D	K
Corkboard	No added binder; High density	14.0	0.34
Eureka	Corkboard with asphaltic binder	14.5	0.32
Rock Cork	Rock wood block with binder	14.5	0.326
	Also called "Tucork"		
Lith	Board containing rock wool, flax and straw pulp	11.2	0.40

Stiff Fibrous Materials in Sheet Form

		D	K
Insulite	Wood pulp	16.2	0.34
		16.9	0.34
Celotex	Sugar cane fiber	13.2	0.34
		14.8	0.34
*Masonite		K =	0.33
*Inso-board			0.33
*Malzewood			0.33 to 0.39
*Cornstalk Pith Board			0.24 to 0.30
*Maftex			0.34

Cellular Gypsum

		D	K
Insulex or Pyrocell		8	0.35
		12	0.44
		18	0.59
		24	0.77
		30	1.00

Woods (Across Grain)

	D	K
Balsa	7.3	0.33
	8.8	0.38
	20	0.58
Cypress	29	0.67
White pine	32	0.78
Mahogany	34	0.90
Virginia pine	34	0.98
Oak	38	1.02
Maple	44	1.10

Miscellaneous Bulding Materials
(Data taken from various sources)

	K		K
Cinder concrete	2 to 3	Limestone	4 to 9
Building gypsum	About 3	Concrete	6 to 9
Plaster	2 to 5	Sandstone	8 to 16
Building brick	3 to 6	Marble	14 to 20
Glass	5 to 6	Granite	13 to 28

* From various commercial laboratories and the work of O. R. Sweeney at Iowa State College.

THERMAL CONDUCTIVITY DATA ON CERAMIC MATERIALS

Description[a]	Class[b]	Water Abs. %	Bulk Density g/cc	Thermal[c] Conductivity			Description[a]	Class[b]	Water Abs. %	Bulk Density g/cc	Thermal[c] Conductivity		
				100°F	200°F	300°F					100°F	200°F	300°F
Single Crystals							Beryl, aquamarine, a-axis	4	—	—	2.52	2.52	2.52
Silicon carbide	5	—	—	52.0	50.0	49.0	Zircon, a-axis	4	—	—	2.45	2.45	2.45
Periclase	5	—	—	26.7	22.5	19.5	Zircon, c-axis	4	—	—	2.34	2.34	2.35
Sapphire, c-axis	5	—	—	20.2	16.0	14.0	Polycrystalline Single Oxide Ceramics						
Sapphire, a-axis	5	—	—	18.7	15.0	12.9	Pure BeO, hot pressed	2	0.03	2.97	125.0	104.0	92.0
Topaz, a-axis	5	—	—	10.8	9.4	7.9	MgO (spec. pure)	1	0.83	3.21	21.2	18.4	16.0
Kyanite, c-axis	5	—	—	10.00	8.6	7.4	SnO₂ 98%	1	0.03	6.62	17.5	15.0	12.7
Kyanite, b-axis	5	—	—	9.6	8.3	7.1	ZnO (yellow)	1	0.00	5.28	16.8	14.6	12.5
Spinel, MgO·Al₂O₂	5	—	—	6.80	6.20	5.50	ZnO gray)	1	0.03	5.20	13.6	11.8	10.2
Quartz, c-axis	4	—	—	6.40	5.40	5.20	CuO (100%)	1	0.04	6.76	10.2	9.00	7.80
Quartz, A-axis	4	—	—	3.40	3.00	2.60	ThO₂, hot pressed	2	—	9.58	8.00	7.02	6.50
Rutile, c-axis	5	—	—	5.60	4.80	4.40	CeO₂	1	0.00	6.20	6.63	6.29	5.20
Rutile, a-axis	5	—	—	3.20	3.20	3.20	Mn₂O₄	1	0.02	4.21	4.18	3.80	3.41
Fluorite	5	—	—	5.30	4.37	3.45	PbO (100%)	1	0.38	7.98	1.6	1.25	0.98
Beryl, aquamarine, c-axis	4	—	—	3.18	3.15	3.12							

[a] Composition: 90%, MgO, 10%, Al₂O₂ designates weight percent. Li₂O:4B₂O₃ designates mole composition, does not indicate compound formation.

[b] Classification: 1 = research body; 2 = industrial research body; 3 = commercial body; 4 = natural mineral; 5 = synthetic mineral.

[c] Thermal conductivity: Units in Btu/(hr) (sq ft) (°F/ft); to convert to cal/(sec) (sq cm)(°C/cm) multiply by 0.00413. (I) = determination made with high vacuum apparatus, inconel thermodes. No letters following value, determination made with high vacuum thermal conductivity apparatus, co-per thermodes.

By permission from Engineering Research Bulletin No. 40 Rutgers University, 1958.

THERMAL CONDUCTIVITIES OF GLASSES BETWEEN −150 AND +100°C
E. H. Ratcliffe

Type of glass	Approximate silica contents (wt. %)	Approximate contents other oxides normally present in quantity (wt. %)		Estimated approximate thermal conductivity at various temperatures	
				Temperature (°C)	Thermal conductivity $\left(\dfrac{cal\ cm}{cm^2\ s\ deg\ C}\right) \times 10^4$
(a) Vitreous silica	100	. . . .		−150	20.0
				−100	25.0
				− 50	28.8
				0	31.5
				50	33.7
				100	35.4
(b) 'Vycor' glass	96	B₂O₃	3	−100	24
				0	30
				100	34
(c) General information 'Crown' glasses	50–75	Various		−100	12–20.5
				30	19–26
				100	21–29
'Flint' glasses	20–55	Various		−100	9–15
				30	13–21
				100	15–23
(d) Pyrex type chemically-resistant borosilicate glasses	80–81	B₂O₃	12–13	−100	21
		Na₂O	4	0	26
		Al	2	100	30
(e) Borosilicate crown glasses	60–65	B₂O₃	15–20	−100	16–17.5
				0	21–22.5
				100	24–25.5
	65–70	B₂O₃	10–15	−100	17.5–19
				0	22.5–24
				100	25.5–27
	70–75	B₂O₃	5–10	−100	19–20.5
				0	24.5–26
				100	27.5–29
(f) (i) Zinc crown glasses	55–65	ZnO	5–15	−100	21–22
		Remainder		0	26–27
		B₂O₃, Al₂O₃		100	28–30
		ZnO	5–15	−100	14–17
		Remainder		0	17–21
		Na₂O, K₂O		100	20–23
		ZnO	15–25	−100	21–22
		Remainder		0	26–27
		B₂O₃, Al₂O₃		100	27–29
		ZnO	15–25	−100	16–19
		Remainder		0	20–23
		Na₂O, K₂O		100	22–25

THERMAL CONDUCTIVITIES OF GLASSES BETWEEN −150 AND +100°C (Continued)

Type of glass	Approximate silica contents (wt. %)	Approximate contents other oxides normally present in quantity (wt. %)		Estimated approximate thermal conductivity at various temperatures	
				Temperature (°C)	Thermal conductivity $\left(\dfrac{\text{cal cm}}{\text{cm}^2\,\text{s deg C}}\right) \times 10^4$
(f) (ii) Zinc crown glasses	65–75	ZnO	5–15	−100	21–22
		Remainder		0	27–28
		B$_2$O$_3$, Al$_2$O$_3$		100	29–31
		ZnO	5–15	−100	17–20
		Remainder		0	21–25
		Na$_2$O, K$_2$O		100	24–27
		ZnO	15–25	−100	21–23
		Remainder		0	27–28
		B$_2$O$_3$, Al$_2$O$_3$		100	29–30
		ZnO	15–25	−100	16–20
		Remainder		0	20–24
		Na$_2$O, K$_2$O		100	25–29
(g) Barium crown glasses	31	B$_2$O$_3$	12	−100	13
		Al$_2$O$_3$	8	0	17
		BaO	48	100	19
	41	B$_2$O$_3$	6	−100	14
		Al$_2$O$_3$	2	0	18
		ZnO	8	100	20
		BaO	43		
	47	B$_2$O$_3$	4	−100	15
		Na$_2$O	1	0	18
		K$_2$O	7	100	21
		ZnO	8		
		BaO	32		
	65	B$_2$O$_3$	2	−100	17
		Na$_2$O	5	0	21
		K$_2$O	15	100	24
		ZnO	2		
		BaO	10		
(h) Borate glasses					
Borate flint glass	9	B$_2$O$_3$	36	−100	13
		Na$_2$O	1	0	16
		K$_2$O	2	100	19
		PbO	36		
		Al$_2$O$_3$	10		
		ZnO	6		
Borate flint glass	.....	B$_2$O$_3$	56	−100	12
		Al$_2$O$_3$	12	0	16
		PbO	32	100	20
Borate flint glass	.....	B$_2$O$_3$	43	100	0
		Al$_2$O$_3$	5	0	13
		PbO	52	100	17
Borate glass	4	B$_2$O$_3$	55	−100	15
		Al$_2$O$_3$	14	0	19
		PbO	11	100	21
		K$_2$O	4		
		ZnO	12		
Borate crown glass	.....	B$_2$O$_3$	64	−100	12
		Na$_2$O	8	0	16
		K$_2$O	3	100	20
		BaO	4		
		PbO	3		
		Al$_2$O$_3$	18		
Light borate crown glass	.....	B$_2$O$_3$	69	−100	13
		Na$_2$O	8	0	17
		BaO	5	100	21
		Al$_2$O$_3$	18		
Zinc borate glass	.....	B$_2$O$_3$	40	−100	16
		ZnO	60	0	18
				100	20
(i) Phosphate crown glasses					
Potash phosphate glass	.....	P$_2$O$_5$	70	0	18
		B$_2$O$_3$	3	100	20
		K$_2$O	12		
	.....	Al$_2$O$_3$	10		
		MgO	4		
Baryta phosphate glass	.....	P$_2$O$_5$	60	45	18
		B$_2$O$_3$	3		
		Al$_2$O$_3$	8		
		BaO	28		

E-7

Type of glass	Approximate silica contents (wt. %)	Approximate contents other oxides normally present in quantity (wt. %)		Estimated approximate thermal conductivity at various temperatures	
				Temperature (°C)	Thermal conductivity $\left(\dfrac{\text{cal cm}}{\text{cm}^2 \text{ s deg C}}\right) \times 10^4$
(j) Soda-lime glasses	75	Na$_2$O	17	−100	18
		CaO	8	0	23
				100	26
	75	Na$_2$O	12	−100	21
		CaO	13	0	26
				100	28
	72	Na$_2$O	15	−100	19
		CaO	11	0	24
		Al$_2$O$_3$	2	100	27
	65	Na$_2$O	25	−100	16
		CaO	10	0	20
				100	23
	65	Na$_2$O	15	−100	20
		CaO	20	0	24
				100	26
	60	Na$_2$O	20	−100	18
		CaO	20	0	22
				100	24
(k) Other crown glasses					
Crown glass	75	Na$_2$O	9	−100	19
		K$_2$O	11	0	24
		CaO	5	100	26
High dispersion crown glass	68	Na$_2$O	16	−100	16
		ZnO	3	0	20
		PbO	13	100	24
(l) Miscellaneous flint glasses					
(i) Silicate flint glasses					
Light flint glasses	65	PbO	25	−100	16–17
		Others	10	0	21–22
				100	24–25
	55	PbO	35	−100	14–16
		Others	10	0	18–20
				100	21–22
Ordinary flint glass	45	PbO	45	−100	12–14
		Others	10	0	16–18
				100	19–20
Heavy flint glass	35	PbO	60	−100	11–12
		Others	5	0	14–15
				100	17–18
Very heavy flint glasses	25	PbO	73	−100	10–11
		Others	2	0	13–14
				100	15–16
	20	PbO	80	−100	10
				0	12
				100	14
(ii) Borosilicate flint glass	33	B$_2$O$_3$	31	−100	15
		PbO	25	0	20
		Al$_2$O$_3$	7	100	23
		K$_2$O	3		
		Na$_2$O	1		
(iii) Barium flint glass	50	BaO	24	−100	14
		PbO	6	0	17
		K$_2$O	8	100	20
		Na$_2$O	3		
		ZnO	8		
		Sb$_2$O$_3$	1		
(m) Other glasses					
(i) Potassium glass	59	K$_2$O	33	50	21–22
		CaO	8		
(ii) Iron glasses	63	Fe$_2$O$_3$	10	−100	19
		Na$_2$O	17	0	23
		MgO	4	100	25
		CaO	3		
		Al$_2$O$_3$	2		
	67	Fe$_2$O$_3$	15	0	21–22
		Na$_2$O$_3$	18	100	24–25
	62	Fe$_2$O$_3$	20	0	20.5–21.5
		Na$_2$O	18	100	23–24
(ii) Rock glasses					
Obsidian				0	32
				100	35
Artificial diabase				0	27
				100	30

THERMAL CONDUCTIVITY OF CERTAIN METALS

From NSRDS-NBS 8

R. W. Powell, C. Y. Ho, and P. E. Liley

The thermal conductivity, k, is given in the units Watt cm^{-1} °K^{-1}.

To convert to Cal(gm) hr^{-1} cm^{-1} °C^{-1} multiply the values listed in the tables by 860.421

To convert to Btu hr^{-1} ft^{-1} °F^{-1} multiply the values listed in the tables by 57.818.

ρ_0 is the residual electrical resistivity and the value of ρ at 4.2°K is used approximately as ρ_0.

T,K	Aluminum 99.996+% $\rho_0 = 0.00315$ μohm cm	Copper 99.999+% $\rho_0 = 0.000851$ μohm cm	Gold 99.999+% $\rho_0 = 0.0055$ μohm cm	Iron 99.998+% $\rho_0 = 0.0327$ μohm cm	Manganin	Platinum 99.999% $\rho_0 = 0.0106$ μohm cm	Silver 99.999+% $\rho_0 = 0.00062$ μohm cm	Tungsten 99.99+% $\rho_0 = 0.0017$ μohm cm
0	0	0	0	0	0	0	0	0
1	7.8	28.7	4.4	0.75	0.0007	2.31	39.4	14.4
2	15.5	57.3	8.9	1.49	0.0018	4.60	78.3	28.7
3	23.2	85.5	13.1	2.24	0.0031	6.79	115	42.6
4	30.8	113	17.1	2.97	0.0046	8.8	147	55.6
5	38.1	138	20.7	3.71	0.0062	10.5	172	67.1
6	45.1	159	23.7	4.42	0.0078	11.8	187	76.2
7	51.5	177	26.0	5.13	0.0095	12.6	193	82.4
8	57.3	189	27.5	5.80	0.0111	12.9	190	85.3
9	62.2	195	28.2	6.45	0.0128	12.8	181	85.1
10	66.1	196	28.2	7.05	0.0145	12.3	168	82.4
11	69.0	193	27.7	7.62	0.0162	11.7	154	77.9
12	70.8	185	26.7	8.13	0.0180	10.9	139	72.4
13	71.5	176	25.5	8.58	0.0197	10.1	124	66.4
14	71.3	166	24.1	8.97	0.0215	9.3	109	60.4
15	70.2	156	22.6	9.30	0.0232	8.4	96	54.8
16	68.4	145	20.9	9.56	0.0250	7.6	85	49.3
18	63.5	124	17.7	9.88	0.0285	6.1	66	40.0
20	56.5	105	15.0	9.97	0.0322	4.9	51	32.6
25	40.0	68	10.2	9.36	0.0410	3.15	29.5	20.4
30	28.5	43	7.6	8.14	0.0497	2.28	19.3	13.1
35	21.0	29	6.1	6.84	0.0585	1.80	13.7	8.9
40	16.0	20.5	5.2	5.55	0.067	1.51	10.5	6.5
45	12.5	15.3	4.6	4.50	0.075	1.32	8.4	5.07
50	10.0	12.2	4.2	3.72	0.082	1.18	7.0	4.17
60	6.7	8.5	3.8	2.65	0.097	1.01	5.5	3.18
70	5.0	6.7	3.58	2.04	0.110	0.90	4.97	2.76
80	4.0	5.7	3.52	1.68	0.120	0.84	4.71	2.56
90	3.4	5.14	3.48	1.46	0.127	0.81	4.60	2.44
100	3.0	4.83	3.45	1.32	0.133	0.79	4.50	2.35
150	2.47	4.28	3.35	1.04	0.156	0.762	4.32	2.10
200	2.37	4.13	3.27	0.94	0.172	0.748	4.30	1.97
250	2.35	4.04	3.20	0.865	0.193	0.737	4.28	1.86
273	2.36	4.01	3.18	0.835	0.206	0.734	4.28	1.82
300	2.37	3.98	3.15	0.803	0.222	0.730	4.27	1.78
350	2.40	3.94	3.13	0.744	0.250	0.726	4.24	1.70
400	2.40	3.92	3.12	0.694	(0.279)	0.722	4.20	1.62
500	2.37	3.88	3.09	0.613	(0.338)	0.719	4.13	1.49
600	2.32	3.83	3.04	0.547	(0.397)	0.720	4.05	1.39
700	2.26	3.77	2.98	0.487		0.723	3.97	1.33
800	2.20	3.71	2.92	0.433		0.729	3.89	1.28
900	2.13	3.64	2.85	0.380		0.737	3.82	1.24
1000	[0.93]**	3.57	(2.78)	0.326		0.748	(3.74)	1.21
1100	[0.96]	3.50	(2.71)	0.297		0.760	(3.66)	1.18
1200	[0.99]	3.42	(2.62)	0.282		0.775	(3.58)	1.15
1300	[1.02]	(3.34)†	(2.51)	0.299		0.791		1.13
1400				0.309		0.807		1.11
1500				0.318		0.824		1.09
1600				(0.327)		0.842		1.07
						0.860		1.05
						0.877		1.03
						(0.895)		1.02
						(0.913)		1.00
								0.98
								0.96
								0.94
								0.925
								0.915
								0.905
								0.900
								(0.895)

* In the table the third significant figure is given only for the purpose of comparison and for smoothness and is not indicative of the degree of accuracy.

** Values in square brackets are for liquid state.

† Values in parentheses are extrapolated.

‡ Estimated.

THERMAL CONDUCTIVITY OF CERTAIN LIQUIDS

From NSRDS-NBS 8
R. W. Powell, C. Y. Ho, and P. E. Liley

The thermal conductivity, k, is given in the units Milliwatt cm^{-1} °K^{-1}. To convert to Cal(gm) hr^{-1} cm^{-1} °K^{-1} multiply the values listed in the table by 0.860421

T (K)	Helium	Nitrogen	Argon	Carbon tetra-chloride	Diphenyl	m-Terphenyl	Toluene	Water
2.4	0.192							
2.6	0.193							
2.8	0.197							
3.0	0.204							
3.2	0.214							
3.4	0.227							
3.6	0.241							
3.8	0.260							
4.0	0.282							
4.2	0.307							
4.4	(0.335)‡							
4.6	(0.366)‡							
4.8	(0.400)‡							
5.0	(0.437)‡							
5.2	(0.477)‡							
60		1.692†						
65		1.598						
70		1.504						
75		1.411						
80		1.320‡	1.315†					
85		1.229‡	1.258					
90		1.140‡	1.200‡					
95		1.051‡	1.141‡					
100		0.965‡	1.082‡					
105		0.879‡	1.023‡					
110		0.794‡	0.963‡					
115		0.710‡	0.903‡					
120		0.627‡	0.842‡					
125		0.544‡	0.780‡					
130			0.717‡					
135			0.654‡					
140			0.591‡					
145			0.527‡					
150			0.463‡				(1.719)†	
160							(1.694)†	
170							(1.669)†	
180							1.644	
190							1.619	
200							1.594	
210							1.569	
220							1.543	
230				(1.169)†			1.518	
240				(1.150)†			1.492	
250				1.131			1.467	5.22†
260				1.112			1.442	5.39†
270				1.093			1.416	5.55†
280				1.074			1.391	5.74
290				1.055			1.365	5.92
300				1.036			1.340	6.09
310				1.017			1.315	6.23
320				0.997			1.289	6.37
330				0.978	(1.402)†		1.264	6.48
340				0.959	(1.387)†		1.238	6.59
350				0.940	1.373	(1.361)†	1.213	6.68
360				(0.921)	1.359	(1.356)†	1.188	6.75
370				(0.902)	1.345	1.351	1.162	6.80
380				(0.882)	1.331	1.346	1.137	6.84‡
390				(0.863)	1.316	1.341	(1.112)‡	6.86‡
400				(0.844)	1.302	1.335	(1.086)‡	6.86‡
410				(0.825)	1.288	1.329	(1.061)‡	6.86‡
420				(0.806)	1.274	1.323	(1.036)‡	6.84‡
430				(0.787)	1.259	1.317	(1.013)‡	6.81‡
440				(0.768)	1.245	1.310	(0.985)‡	6.78‡
450				(0.749)	1.231	1.304	(0.959)‡	6.73‡
460					1.217	1.297	(0.933)‡	6.67‡
470					1.202	1.290	(0.908)‡	6.61‡
480					1.188	1.283	(0.885)‡	6.53‡
490					1.174	1.276	(0.862)‡	6.45‡
500					1.160	1.268	(0.839)‡	6.35‡
510					1.146	1.261		6.24‡
520					1.131	1.254		6.12‡
530					1.117‡	1.246		5.99‡
540					1.103‡	1.238		5.86‡
550					1.089‡	1.230		5.71‡
560					1.074‡	1.222		5.55‡
570					1.060‡	1.213		5.39‡
580					1.046‡	1.205		5.20‡
590					1.032‡	1.197		5.01‡
600					1.018‡	1.188		4.81‡
610						1.180		4.60‡
620						1.172		4.40‡
630						1.163		(4.20)‡
640						1.155‡		(4.01)‡
650						1.146‡		

† Extrapolated for the supercooled liquid. [Approximate n.m.p. in K: N$_2$, 63; A, 84; CCl$_4$, 250; C$_{12}$H$_{10}$, 342; m-C$_{18}$H$_{14}$, 361; p-C$_{18}$H$_{14}$, 486; C$_7$H$_{10}$, 178; H$_2$O, 273.1].

‡ Under saturation vapor pressure [Approximate n.b.p. in K: He, 4.3; N$_2$, 77; A, 88; CCl$_4$, 350; C$_{12}$H$_{10}$, 528; m-C$_{18}$H$_{14}$, 637; p-C$_{18}$H$_{14}$, 658; C$_7$H$_{10}$, 384; H$_2$O, 373].

THERMAL CONDUCTIVITY OF THE ELEMENTS

Data contained in the following table were extracted from the extensive compilation prepared by C. Y. Ho, R. W. Powell, and P. E. Liley under the National Standard Reference Data System (NSRDS) of the National Bureau of Standards project at the Thermophysical Properties Research Center (TPRC) at Purdue University and published in the Journal of Physical and Chemical Reference Data, *1*, 279-421 (1972). The data in the table below are used with the permission of the authors and the copyright owners, the American Institute of Physics and the American Chemical Society. Users are referred to their more extensive compilation for conductivities at temperatures other than those listed in the table below, and also to obtain an understanding of the basis of selection of recommended and provisional values. Temperatures are in kelvins (K) and conductivities, k, in watt per centimeter kelvin (W cm^{-1} K^{-1}), except as noted. If the numerical value of k has a superscript m, the units of k are milliwatt per centimeter kelvin (mW cm^{-1} K^{-1}).To convert the listed conductivities to units other than those in the tables, one should make use of the conversion factors listed in the table "Conversion Factors for Units of Thermal Conductivity", which follows this table. Conductivity values listed with an asterisk*, are provisional values.

Element	State or Condition	Conductivity at 273.2K	298.2K	373.2K
Aluminum	Solid	2.36	2.37	2.40
Antimony	Polycrystalline	0.255	0.244	0.219
Argon	Gas at 1 atm.	0.1619^{m} (270K)	0.1772^{m} (300K)	0.2103^{m} (370K)
Arsenic	Solid, Gray, Polycrystalline	0.539*	0.502*	0.427*
Barium	Solid	0.185*	0.184* (295K)	
Beryllium	Polycrystalline	2.18*	2.01	1.68
Bismuth	Solid			
	∥ to triagonal axis	0.0554	0.0530	0.0481
	⊥ to triagonal axis	0.0953	0.0913	0.0844
	Polycrystalline	0.0822	0.0792	0.0722
Boron	Solid	0.318	0.274	0.188
Bromine	Saturated liquid	1.30^{m}* (270K)	1.22^{m}* (300K)	1.06^{m}* (370K)
	Saturated vapor		0.048^{m}* (300K)	
	Gas	0.042^{m}* (270K)	0.048^{m}* (300K)	0.057^{m}* (350K)
Cadmium	Solid			
	∥ to c-axis	0.835	0.830	0.816
	⊥ to c-axis	1.04	1.04	1.02
	Polycrystalline	0.975	0.969	0.953
Calcium	Solid	2.06*	2.01*	1.92*
Carbon	Solid, Amorphous	0.0150	0.0159	0.0182
	Solid, Type I (Diamond)	9.94	9.90	7.03*
	Solid, Type IIa (Diamond)	26.2	23.2	17.0*
	Solid, Type IIb (Diamond)	15.2	13.6	10.2*
	Solid, Acheson graphite			
	∥ to axis of extrusion	1.69	1.65	1.50
	⊥ to axis of extrusion	1.21	1.19	1.11
	Solid, AGOT graphite			
	∥ to axis of extrusion	2.28	2.21	1.96
	⊥ to axis of extrusion	1.41	1.38	1.22
	Solid, ATJ graphite			
	∥ to molding pressure	0.984	0.982	0.933
	⊥ to molding pressure	1.31	1.29	1.21
	Solid, AWG graphite			
	∥ to molding pressure	0.807	0.796	0.733
	⊥ to molding pressure	1.32	1.28	1.16
	Solid, Pyrolytic graphite			
	∥ to layer planes	21.3	19.6	15.1
	⊥ to layer planes	0.0636	0.0573	0.0442
	Solid, 875S graphite			
	∥ to axis of extrusion	1.97*	1.92*	1.75*
	⊥ to axis of extrusion	1.49*	1.46*	1.34*
	Solid, 890S graphite			
	∥ to axis of extrusion	1.87*	1.83*	1.66*
	⊥ to axis of extrusion	1.51*	1.48*	1.36*
Cerium	Solid, Polycrystalline	0.108*	0.113	0.128*
Cesium	Solid	0.361*	0.359 (301.9K)	
	Liquid		0.197 (301.9K)	0.201
Chlorine	Saturated liquid	1.49^{m}* (270K)	1.34^{m}* (300K)	0.95^{m}* (370K)
	Saturated vapor	0.082^{m}* (270K)	0.097^{m}* (300K)	0.155^{m}* (370K)

Element	State or Condition	Conductivity at 273.2K	298.2K	373.2K
	Gas, 1 atm.	0.078^m (270K)	0.089^m (300K)	0.114^m (370K)
Chromium	Solid, Polycrystalline	0.965	0.939	0.921
Cobalt	Solid, Polycrystalline	1.05	1.00	0.890
Copper	Solid	4.03	4.01	3.95
Dysprosium	Solid			
	// to c-axis	0.114*	0.117*	
	⊥ to c-axis	0.101*	0.103*	
	Polycrystalline	0.105*	0.107*	0.108*
Erbium	Solid			
	// to c-axis	0.187*	0.184*	
	⊥ to c-axis	0.127*	0.126*	
	Polycrystalline	0.147*	0.145*	0.140*
Europium	Solid	0.140*	0.139*	
Fluorine	Gas, 1 atm.	0.251^m (270K)	0.279^m (300K)	0.344^m (370K)
Gadolinium	Solid			
	// to c-axis	0.104*	0.108*	
	⊥ to c-axis	0.103*	0.103*	
	Polycrystalline	0.103*	0.105*	
Gallium	Solid			
	// to a-axis	0.410	0.408	
	// to b-axis	0.884	0.883	
	// to c-axis	0.160	0.159	
	Liquid		0.281 (302.93K)	0.328
Germanium	Solid	0.667	0.602	0.465
Gold	Solid	3.19	3.18	3.13
Hafnium	Solid, Polycrystalline	0.233*	0.230	0.224
Helium	Solid, ^{3}He	0.033 (0.9K)	0.020 (1K)	0.0021 (2K)
	Solid, ^{4}He	0.650 (0.9K)	0.245 (1K)	0.0018 (2K)
	Liquid, saturated; He-I	0.191^m (2.5K)	0.232^m (3.5K)	0.434^m (5K)
	Gas, 1 atm.	1.411^m	1.520^m	1.766^m
Holmium	Solid			
	// to c-axis	0.215*	0.222*	
	⊥ to c-axis	0.136*	0.138*	
	Polycrystalline	0.159*	0.162*	0.170*
Hydrogen	Solid, Normal Hydrogen	2.30 (4K)	0.0158 (10K)	0.0090 (15K)
	Liquid, saturated; Normal Hydrogen	1.022^m (15K)	1.269^m (25K)	0.60^{m*} (33K)
	Gas, 1 atm. Normal Hydrogen	1.665^m	1.815^m	2.106^m
	Liquid, saturated; para-Hydrogen	0.824^m (14K)	0.998^m (25K)	0.58^m (32K)
	Vapor, saturated; para-Hydrogen	0.081^{m*} (10K)	0.242^{m*} (25K)	0.58^m (32K)
	Gas, 1 atm.; para-Hydrogen	1.768^{m*}	1.880^{m*}	2.126^{m*}
	Deuterium:			
	Liquid, saturated	1.26^m (20K)	1.37^m (30K)	0.83^{m*} (38K)
	Vapor, saturated	0.084^{m*} (20K)	0.26^{m*} (30K)	0.83^{m*} (38K)
	Gas, 1 atm.	1.294^{m*} (270K)	1.406^{m*} (300K)	1.66^{m*} (370K)
	Tritium:			
	Liquid, saturated	1.25 (21K)	1.34 (30K)	0.68 (44K)
Indium	Solid, Polycrystalline	0.837	0.818	0.762
Iodine	Solid	4.81^{m*}	4.49^{m*} (300K)	3.75^{m*} (386.8K)
	Liquid, saturated			1.16^{m*} (386.8K)
Iridium	Solid	1.48	1.47	1.45
Iron	Solid	0.865	0.804	0.720
	Armco Iron	0.747	0.728	0.676
Krypton	Solid	0.4^{m*} (1K)	17^m (10K)	2.5^m (116K)
	Liquid, saturated			0.931^m (116K)
	Vapor, saturated	0.0406^{m*} (120K)	0.0554^{m*} (150K)	0.21^{m*} (210K)
	Gas	0.0860^m (270K)	0.0949^m (300K)	0.1145^m (370K)

Element	State or Condition	Conductivity at		
		273.2K	298.2K	373.2K
Lanthanum	Solid, Polycrystalline	0.131	0.134	0.145
Lead	Solid	0.356	0.353	0.344
Lithium	Solid	0.859	0.848	0.818
Lutetium	Solid			
	// to c-axis	0.236*	0.232*	
	⊥ to c-axis	0.140*	0.138*	
	Polycrystalline	0.167*	0.164*	
Magnesium	Solid, Polycrystalline	1.57	1.56	1.54
Manganese	Solid	0.0768*	0.0781*	
Mercury	Liquid	0.0782	0.0830	0.0947
Molybdenum	Solid	1.39*	1.38*	1.33*
Neodymium	Solid, Polycrystalline	0.165*	0.165*	0.167*
Neon	Gas	0.461^{m*}	0.493^{m*}	0.563^{m*}
		(270K)	(300K)	(370K)
Neptunium	Solid		0.063*	
			(300K)	
Nickel	Solid	0.941	0.909	0.827
Niobium	Solid	0.533	0.537	0.548
Nitrogen	Solid	56^m	17^m	3.2^m
		(4K)	(10K)	(25K)
	Liquid, saturated	1.60^m	0.966^m	0.37^{m*}
		(65K)	(100K)	(126K)
	Vapor, saturated	0.061^{m*}	0.111^{m*}	0.37^{m*}
		(65K)	(100K)	(126K)
	Gas, 1 atm.	0.2374^m	0.2598^m	0.3065^m
		(270K)	(300K)	(370K)
Cesium	Solid			
	// to c-axis	2.93	14.3	15.4
		(2K)	(10K)	(30K)
	⊥ to c-axis	1.76	8.65	11.1
		(2K)	(10K)	(30K)
	Polycrystalline	2.09	10.2	12.4
		(2K)	(10K)	(30K)
	Polycrystalline	0.880*	0.876*	0.870*
Oxygen	Liquid, saturated	1.501^m	1.023^m	0.41^m
		(90K)	(125K)	(155K)
	Vapor, saturated	0.081^{m*}	0.135^{m*}	0.41^{m*}
		(90K)	(125K)	(155K)
	Gas, 1 atm.	0.2424	0.2674	0.3204
		(270K)	(300K)	(370K)
Palladium	Solid	0.716*	0.718	0.730
Phosphorus	Solid			
	Black (Polycrystalline)	0.132	0.121	
	White	0.00250*	0.00236*	
	Liquid, White			0.00181
Platinum	Solid	0.717	0.716	0.717
Plutonium	Solid, polycrystalline	0.0616*	0.0670*	0.0790*
				(350K)
Potassium	Solid	1.036*	1.025	
	Liquid			0.532
Praeseodymium	Solid, polycrystalline	0.120	0.125	0.134
Promethium	Solid, polycrystalline		0.179*	0.184*
Radium	Solid		0.186	
			(293.2K)	
Radon	Gas, 1 atm.	0.0327^{m*}	0.0364^{m*}	0.0445^{m*}
		(270K)	(300K)	(370K)
Rhenium	Solid, polycrystalline	0.486	0.480	0.466
Rhodium	Solid	1.51	1.50	1.47
Rubidium	Solid	0.583*	0.582	0.581
				(312.04K)
	Liquid			0.333
				(312.04K)
Ruthenium	Solid, polycrystalline	1.17	1.17	1.15
Samarium	Solid, polycrystalline	0.133*	0.133*	0.133*
Scandium	Solid, polycrystalline	0.157	0.158	
Selenium	Solid			
	// to c-axis	0.0481	0.0452	0.0483
	⊥ to c-axis	0.0137	0.0131	0.0139
	Amorphous	0.00428	0.00519	0.00818
				(323.2K)
Silicon	Solid	1.68	1.49	1.08
Silver	Solid	4.29	4.29	4.26
Sodium	Solid	1.42	1.42	1.32
				(371K)
Strontium	Solid, polycrystalline	0.364*	0.354*	0.325*

Element	State or Condition	Conductivity at		
		273.2K	298.2K	373.2K
Sulfur	Solid, polycrystalline	0.00287	0.00270	0.00154
	Solid, amorphous	0.00200	0.00205	0.00216*
				(350K)
	Liquid			0.00129
				(392.2K)
Tantalum	Solid	0.574	0.575	0.577
Technetium	Solid, polycrystalline	0.509*	0.506	0.501
Tellurium	Solid			
	// to c-axis	0.0360	0.0338	0.0292
	⊥ to c-axis	0.0208	0.197	0.173
Terbium	Solid			
	// to c-axis	0.138*	0.147*	
	⊥ to c-axis	0.0900*	0.0956*	
	Polycrystalline	0.104*	0.111*	
Thallium	Solid, polycrystalline	0.469	0.461	0.443
Thorium	Solid	0.540*	0.540*	0.543*
Thulium	Solid			
	// to c-axis	0.242*	0.242*	
	⊥ to c-axis	0.140*	0.141*	
	Polycrystalline	0.168*	0.169*	
Tin	Solid			
	// to c-axis	0.527	0.516	0.489
	⊥ to c-axis	0.759	0.743	0.704
	Polycrystalline	0.682	0.668	0.632
Titanium	Solid, polycrystalline	0.224	0.219	0.207
Tungsten	Solid	1.77	1.73	1.63
Uranium	Solid, polycrystalline	0.270	0.275	0.291
Vanadium	Solid	0.307*	0.307	0.310
Xenon	Liquid, saturated	0.31^m	0.16^{m*}	
		(270K)	(290K)	
	Vapor, saturated	0.084^{m*}	0.16^{m*}	
	Gas, 1 atm.	0.0514^m	0.0569^m	0.0695^m
		(270K)	(300K)	(370K)
Ytterbium	Solid	0.354*	0.349*	0.343*
Yttrium	Solid, polycrystalline	0.170*	0.172*	0.177*
Zinc	Solid, polycrystalline	1.17	1.16	1.12
Zirconium	Solid, polycrystalline	0.232*	0.227	0.218
			(300K)	

THERMAL CONDUCTIVITY OF ROCKS

Rock	Temperature, °C	Conductivity, Kcal m^{-1} hr^{-1} deg^{-1}	Heat Capacity, cal g^{-1} deg^{-1}
Granite	0	3.02	0.192
	50	2.81	–
	100	2.59	–
	200	2.34	0.228
	300	2.12	–
	400	–	0.258
Marble	118	1.44	0.21
	196	1.29	0.24
	245	1.19	–
	360	0.95	0.271
Dolomitic	130	1.41	–
limestone	181	1.37	–
	268	1.29	–
	377	1.15	–
	0	1.65	0.17
Shale	100	1.51	–
	120	1.33	–
	188	1.41	0.24
	304	1.26	0.245
	0	4.9	–
Sandstone	100	3.82	0.26
(quartzitic)	200	3.24	–

CONVERSION FACTORS FOR UNITS OF THERMAL CONDUCTIVITY

MULTIPLY

by appropriate factor to	$Btu_{IT} \ h^{-1}$ $ft^{-1} \ F^{-1}$	Btu_{IT} in. $h^{-1} \ ft^{-2} \ F^{-1}$	$Btu_{th} \ h^{-1}$ $ft^{-1} \ F^{-1}$	Btu_{th} in. $h^{-1} \ ft^{-2} \ F^{-1}$	$cal_{IT} \ s^{-1}$ $cm^{-1} \ C^{-1}$	$cal_{th} \ s^{-1}$ $cm^{-1} \ C^{-1}$
OBTAIN→						
$Btu_{IT} \ h^{-1} \ ft^{-1} \ F^{-1}$	1	12	1.00067	12.0080	4.13379×10^{-3}	4.13656×10^{-3}
Btu_{IT} in. $h^{-1} \ ft^{-2} \ F^{-1}$	8.33333×10^{-2}	1	8.33891×10^{-2}	1.00067	3.44482×10^{-4}	3.44713×10^{-4}
$Btu_{th} \ h^{-1} \ ft^{-1} \ F^{-1}$	0.999331	11.9920	1	12	4.13102×10^{-3}	4.13379×10^{-3}
Btu_{th} in. $h^{-1} \ ft^{-2} \ F^{-1}$	8.32776×10^{-2}	0.999331	8.33333×10^{-2}	1	3.44252×10^{-4}	3.44482×10^{-4}
$cal_{IT} \ s^{-1} \ cm^{-1} \ C^{-1}$	2.41909×10^{2}	2.90291×10^{3}	2.42071×10^{2}	2.90485×10^{3}	1	1.00067
$cal_{th} \ s^{-1} \ cm^{-1} \ C^{-1}$	2.41747×10^{2}	2.90096×10^{3}	2.41909×10^{2}	2.90291×10^{3}	0.999331	1
$kcal_{th} \ h^{-1} \ m^{-1} \ C^{-1}$	0.671520	8.05824	0.671969	8.06363	2.77592×10^{-3}	2.77778×10^{-3}
$J \ s^{-1} \ cm^{-1} \ K^{-1}$	57.7789	6.93347×10^{2}	57.8176	6.93811×10^{2}	0.238846	0.239006
$W \ cm^{-1} \ K^{-1}$	57.7789	6.93347×10^{2}	57.8176	6.93811×10^{2}	0.238846	0.239006
$W \ m^{-1} \ K^{-1}$	0.577789	6.93347	0.578176	6.93811	2.38846×10^{-3}	2.39006×10^{-3}
$mW \ cm^{-1} \ K^{-1}$	5.77789×10^{-2}	0.693347	5.78176×10^{-2}	0.693811	2.38846×10^{-4}	2.39006×10^{-4}

MULTIPLY

by appropriate factor to	$kcal_{th} \ h^{-1}$ $m^{-1} \ C^{-1}$	$J \ s^{-1}$ $cm^{-1} \ K^{-1}$	$W \ cm^{-1} \ K^{-1}$	$W \ m^{-1} \ K^{-1}$	$mW \ cm^{-1} \ K^{-1}$
OBTAIN→					
$Btu_{IT} \ h^{-1} \ ft^{-1} \ F^{-1}$	1.48916	1.73073×10^{-2}	1.73073×10^{-2}	1.73073	17.3073
Btu_{IT} in. $h^{-1} \ ft^{-2} \ F^{-1}$	0.124097	1.44228×10^{-3}	1.44228×10^{-3}	0.144228	1.44228
$Btu_{th} \ h^{-1} \ ft^{-1} \ F^{-1}$	1.48816	1.72958×10^{-2}	1.72958×10^{-2}	1.72958	17.2958
Btu_{th} in. $h^{-1} \ ft^{-2} \ F^{-1}$	0.124014	1.44131×10^{-3}	1.44131×10^{-3}	0.144131	1.44131
$cal_{IT} \ s^{-1} \ cm^{-1} \ C^{-1}$	3.60241×10^{2}	4.1868	4.1868	4.1868×10^{2}	4.1868×10^{3}
$cal_{th} \ s^{-1} \ cm^{-1} \ C^{-1}$	3.6×10^{2}	4.184	4.184	4.184×10^{2}	4.184×10^{3}
$kcal_{th} \ h^{-1} \ m^{-1} \ C^{-1}$	1	1.16222×10^{-2}	1.16222×10^{-2}	1.16222	11.6222
$J \ s^{-1} \ cm^{-1} \ K^{-1}$	86.0421	1	1	1×10^{2}	1×10^{3}
$W \ cm^{-1} \ K^{-1}$	86.0421	1	1	1×10^{2}	1×10^{3}
$W \ m^{-1} \ K^{-1}$	0.860421	1×10^{-2}	1×10^{-2}	1	10
$mW \ cm^{-1} \ K^{-1}$	8.60421×10^{-2}	1×10^{-3}	1×10^{-3}	0.1	1

STEAM TABLES

Reproduced by permission of the publishers and copyright owners of the 1967 ASME Steam Tables. Further data and information on the thermodynamic and transport properties of steam and water are contained in the above ASME publication. It is obtainable from The American Society of Mechanical Engineers, United Engineering Center, 345 East 47th Street, New York, New York 10017.

Properties of Saturated Steam and Saturated Water

Temp. F	Press. psia	Volume, ft³/lbm Water v_f	Volume, ft³/lbm Evap. v_{fg}	Volume, ft³/lbm Steam v_g	Enthalpy, Btu/lbm Water h_f	Enthalpy, Btu/lbm Evap. h_{fg}	Enthalpy, Btu/lbm Steam h_g	Entropy, Btu/lbm×F Water s_f	Entropy, Btu/lbm×F Evap. s_{fg}	Entropy, Btu/lbm×F Steam s_g	Temp. F
705.47	3208.2	0.05078	0.00000	0.05078	906.0	0.0	906.0	1.0612	0.0000	1.0612	705.47
705.0	3198.3	0.04427	0.01304	0.05730	873.0	61.4	934.4	1.0329	0.0527	1.0856	705.0
704.5	3187.8	0.04233	0.01822	0.06055	861.9	85.3	947.2	1.0234	0.0732	1.0967	704.5
704.0	3177.2	0.04108	0.02192	0.06300	854.2	102.0	956.2	1.0169	0.0876	1.1046	704.0
703.5	3166.8	0.04015	0.02489	0.06504	848.2	115.2	963.5	1.0118	0.0991	1.1109	703.5
703.0	3156.3	0.03940	0.02744	0.06684	843.2	126.4	969.6	1.0076	0.1087	1.1163	703.0
702.5	3145.9	0.03878	0.02969	0.06847	838.9	136.1	974.9	1.0039	0.1171	1.1210	702.5
702.0	3135.5	0.03824	0.03173	0.06997	835.0	144.7	979.7	1.0006	0.1246	1.1252	702.0
701.5	3125.2	0.03777	0.03361	0.07138	831.5	152.6	984.0	0.9977	0.1314	1.1291	701.5
701.0	3114.9	0.03735	0.03536	0.07271	828.2	159.8	988.0	0.9949	0.1377	1.1326	701.0
700.5	3104.6	0.03697	0.03701	0.07397	825.2	166.5	991.7	0.9924	0.1435	1.1359	700.5
700.0	3094.3	0.03662	0.03857	0.07519	822.4	172.7	995.2	0.9901	0.1490	1.1390	700.0
699.0	3073.9	0.03600	0.04149	0.07749	817.3	184.2	1001.5	0.9858	0.1590	1.1447	699.0
698.0	3053.6	0.03546	0.04420	0.07966	812.6	194.6	1007.2	0.9818	0.1681	1.1499	698.0
697.0	3033.5	0.03498	0.04674	0.08172	808.4	204.0	1012.4	0.9783	0.1764	1.1547	697.0
696.0	3013.4	0.03455	0.04916	0.08371	804.4	212.8	1017.2	0.9749	0.1841	1.1591	696.0
695.0	2993.5	0.03415	0.05147	0.08563	800.6	221.0	1021.7	0.9718	0.1914	1.1632	695.0
694.0	2973.7	0.03379	0.05370	0.08749	797.1	228.8	1025.9	0.9689	0.1983	1.1671	694.0
693.0	2954.0	0.03345	0.05587	0.08931	793.8	236.1	1029.9	0.9660	0.2048	1.1708	693.0
692.0	2934.5	0.03313	0.05797	0.09110	790.5	243.1	1033.6	0.9634	0.2110	1.1744	692.0
690.0	2895.7	0.03256	0.06203	0.09459	784.5	256.1	1040.6	0.9583	0.2227	1.1810	690.0
688.0	2857.4	0.03204	0.06595	0.09799	778.8	268.2	1047.0	0.9535	0.2337	1.1872	688.0
686.0	2819.5	0.03157	0.06976	0.10133	773.4	279.5	1052.9	0.9490	0.2439	1.1930	686.0
684.0	2782.1	0.03114	0.07349	0.10463	768.2	290.2	1058.4	0.9447	0.2537	1.1984	684.0
682.0	2745.1	0.03074	0.07716	0.10790	763.3	300.4	1063.6	0.9406	0.2631	1.2036	682.0
680.0	2708.6	0.03037	0.08080	0.11117	758.5	310.1	1068.5	0.9365	0.2720	1.2086	680.0
678.0	2672.5	0.03002	0.08440	0.11442	753.8	319.4	1073.2	0.9326	0.2807	1.2133	678.0
676.0	2636.8	0.02970	0.08799	0.11769	749.2	328.5	1077.6	0.9287	0.2892	1.2179	676.0
674.0	2601.5	0.02939	0.09156	0.12096	744.7	337.2	1081.9	0.9249	0.2974	1.2223	674.0
672.0	2566.6	0.02911	0.09514	0.12424	740.2	345.7	1085.9	0.9212	0.3054	1.2266	672.0
670.0	2532.2	0.02884	0.09871	0.12755	735.8	354.0	1089.8	0.9174	0.3133	1.2307	670.0
668.0	2498.1	0.02858	0.10229	0.13087	731.5	362.1	1093.5	0.9137	0.3210	1.2347	668.0
666.0	2464.4	0.02834	0.10588	0.13421	727.1	370.0	1097.1	0.9100	0.3286	1.2387	666.0
664.0	2431.1	0.02811	0.10947	0.13757	722.9	377.7	1100.6	0.9064	0.3361	1.2425	664.0
662.0	2398.2	0.02789	0.11306	0.14095	718.8	385.1	1103.9	0.9028	0.3434	1.2462	662.0
660.0	2365.7	0.02768	0.11663	0.14431	714.9	392.1	1107.0	0.8995	0.3502	1.2498	660.0
658.0	2333.5	0.02748	0.12023	0.14771	711.1	399.0	1110.1	0.8963	0.3570	1.2533	658.0
656.0	2301.7	0.02728	0.12387	0.15115	707.4	405.7	1113.1	0.8931	0.3637	1.2567	656.0
654.0	2270.3	0.02709	0.12754	0.15463	703.7	412.2	1115.9	0.8899	0.3702	1.2601	654.0
652.0	2239.2	0.02691	0.13124	0.15816	700.0	418.7	1118.7	0.8868	0.3767	1.2634	652.0
650.0	2208.4	0.02674	0.13499	0.16173	696.4	425.0	1121.4	0.8837	0.3830	1.2667	650.0
648.0	2178.1	0.02657	0.13876	0.16534	692.9	431.1	1124.0	0.8806	0.3893	1.2699	648.0
646.0	2148.0	0.02641	0.14258	0.16899	689.4	437.2	1126.6	0.8776	0.3954	1.2730	646.0
644.0	2118.3	0.02625	0.14644	0.17269	685.9	443.1	1129.0	0.8746	0.4015	1.2761	644.0
642.0	2088.9	0.02610	0.15033	0.17643	682.5	448.9	1131.4	0.8716	0.4075	1.2791	642.0
640.0	2059.9	0.02595	0.15427	0.18021	679.1	454.6	1133.7	0.8686	0.4134	1.2821	640.0
638.0	2031.2	0.02580	0.15824	0.18405	675.8	460.2	1136.0	0.8657	0.4193	1.2850	638.0
636.0	2002.8	0.02566	0.16226	0.18792	672.4	465.7	1138.1	0.8628	0.4251	1.2879	636.0
634.0	1974.7	0.02553	0.16633	0.19185	669.1	471.1	1140.2	0.8599	0.4307	1.2907	634.0
632.0	1947.0	0.02539	0.17044	0.19583	665.9	476.4	1142.2	0.8571	0.4364	1.2934	632.0
630.0	1919.5	0.02526	0.17459	0.19986	662.7	481.6	1144.2	0.8542	0.4419	1.2962	630.0
628.0	1892.4	0.02514	0.17880	0.20394	659.5	486.7	1146.1	0.8514	0.4474	1.2988	628.0
626.0	1865.6	0.02501	0.18306	0.20807	656.3	491.7	1148.0	0.8486	0.4529	1.3015	626.0
624.0	1839.0	0.02489	0.18737	0.21226	653.1	496.6	1149.8	0.8458	0.4583	1.3041	624.0
622.0	1812.8	0.02477	0.19173	0.21650	650.0	501.5	1151.5	0.8430	0.4636	1.3066	622.0
620.0	1786.9	0.02466	0.19615	0.22081	646.9	506.3	1153.2	0.8403	0.4689	1.3092	620.0
618.0	1761.2	0.02455	0.20063	0.22517	643.8	511.0	1154.8	0.8375	0.4742	1.3117	618.0
616.0	1735.9	0.02444	0.20516	0.22960	640.8	515.6	1156.4	0.8348	0.4794	1.3141	616.0
614.0	1710.8	0.02433	0.20976	0.23409	637.8	520.2	1158.0	0.8321	0.4845	1.3166	614.0
612.0	1686.1	0.02422	0.21442	0.23865	634.8	524.7	1159.5	0.8294	0.4896	1.3190	612.0
610.0	1661.6	0.02412	0.21915	0.24327	631.8	529.2	1160.9	0.8267	0.4947	1.3214	610.0
608.0	1637.3	0.02402	0.22394	0.24796	628.8	533.6	1162.4	0.8240	0.4997	1.3238	608.0
606.0	1613.4	0.02392	0.22881	0.25273	625.9	537.9	1163.8	0.8214	0.5048	1.3261	606.0
604.0	1589.7	0.02382	0.23374	0.25757	622.9	542.2	1165.1	0.8187	0.5097	1.3284	604.0
602.0	1566.3	0.02373	0.23875	0.26248	620.0	546.4	1166.4	0.8161	0.5147	1.3307	602.0
600.0	1543.2	0.02364	0.24384	0.26747	617.1	550.6	1167.7	0.8134	0.5196	1.3330	600.0
598.0	1520.4	0.02354	0.24900	0.27255	614.3	554.7	1169.0	0.8108	0.5245	1.3353	598.0
596.0	1497.8	0.02345	0.25425	0.27770	611.4	558.8	1170.2	0.8082	0.5293	1.3375	596.0
594.0	1475.4	0.02337	0.25958	0.28294	608.6	562.8	1171.4	0.8056	0.5342	1.3398	594.0
592.0	1453.3	0.02328	0.26499	0.28827	605.7	566.8	1172.6	0.8030	0.5390	1.3420	592.0
590.0	1431.5	0.02319	0.27049	0.29368	602.9	570.8	1173.7	0.8004	0.5437	1.3442	590.0
588.0	1410.0	0.02311	0.27608	0.29919	600.1	574.7	1174.8	0.7978	0.5485	1.3464	588.0
586.0	1388.6	0.02303	0.28176	0.30478	597.3	578.5	1175.9	0.7953	0.5532	1.3485	586.0
584.0	1367.6	0.02295	0.28753	0.31048	594.6	582.4	1176.9	0.7927	0.5580	1.3507	584.0
582.0	1346.7	0.02287	0.29340	0.31627	591.8	586.1	1178.0	0.7902	0.5627	1.3528	582.0
580.0	1326.2	0.02279	0.29937	0.32216	589.1	589.9	1179.0	0.7876	0.5673	1.3550	580.0

Quantities for saturated liquid v_f h_f s_f

Quantities for saturated vapor v_g h_g s_g

Increment for evaporation v_{fg} h_{fg} s_{fg}

Temp. F	Press. psia	Volume, ft³/lbm			Enthalpy, Btu/lbm			Entropy, Btu/lbm ×F			Temp. F
		Water v_f	Evap. v_{fg}	Steam v_g	Water h_f	Evap. h_{fg}	Steam h_g	Water s_f	Evap. s_{fg}	Steam s_g	
580.0	1326.17	0.02279	0.29937	0.32216	589.1	589.9	1179.0	0.7876	0.5673	1.3550	580.0
578.0	1305.84	0.02271	0.30544	0.32816	586.4	593.6	1179.9	0.7851	0.5720	1.3571	578.0
576.0	1285.74	0.02264	0.31162	0.33426	583.7	597.2	1180.9	0.7825	0.5766	1.3592	576.0
574.0	1265.89	0.02256	0.31790	0.34046	581.0	600.9	1181.8	0.7800	0.5813	1.3613	574.0
572.0	1246.26	0.02249	0.32429	0.34678	578.3	604.5	1182.7	0.7775	0.5859	1.3634	572.0
570.0	1226.88	0.02242	0.33079	0.35321	575.6	608.0	1183.6	0.7750	0.5905	1.3654	570.0
568.0	1207.72	0.02235	0.33741	0.35975	572.9	611.5	1184.5	0.7725	0.5950	1.3675	568.0
566.0	1188.80	0.02228	0.34414	0.36642	570.3	615.0	1185.3	0.7699	0.5996	1.3696	566.0
564.0	1170.10	0.02221	0.35099	0.37320	567.6	618.5	1186.1	0.7674	0.6041	1.3716	564.0
562.0	1151.63	0.02214	0.35797	0.38011	565.0	621.9	1186.9	0.7650	0.6087	1.3736	562.0
560.0	1133.38	0.02207	0.36507	0.38714	562.4	625.3	1187.7	0.7625	0.6132	1.3757	560.0
558.0	1115.36	0.02201	0.37230	0.39431	559.8	628.6	1188.4	0.7600	0.6177	1.3777	558.0
556.0	1097.55	0.02194	0.37966	0.40160	557.2	632.0	1189.2	0.7575	0.6222	1.3797	556.0
554.0	1079.96	0.02188	0.38715	0.40903	554.6	635.3	1189.9	0.7550	0.6267	1.3817	554.0
552.0	1062.59	0.02182	0.39479	0.41660	552.0	638.5	1190.6	0.7525	0.6311	1.3837	552.0
550.0	1045.43	0.02176	0.40256	0.42432	549.5	641.8	1191.2	0.7501	0.6356	1.3856	550.0
548.0	1028.49	0.02169	0.41048	0.43217	546.9	645.0	1191.9	0.7476	0.6400	1.3876	548.0
546.0	1011.75	0.02163	0.41855	0.44018	544.4	648.1	1192.5	0.7451	0.6445	1.3896	546.0
544.0	995.22	0.02157	0.42677	0.44834	541.8	651.3	1193.1	0.7427	0.6489	1.3915	544.0
542.0	978.90	0.02151	0.43514	0.45665	539.3	654.4	1193.7	0.7402	0.6533	1.3935	542.0
540.0	962.79	0.02146	0.44367	0.46513	536.8	657.5	1194.3	0.7378	0.6577	1.3954	540.0
538.0	946.88	0.02140	0.45237	0.47377	534.2	660.6	1194.8	0.7353	0.6621	1.3974	538.0
536.0	931.17	0.02134	0.46123	0.48257	531.7	663.6	1195.4	0.7329	0.6665	1.3993	536.0
534.0	915.66	0.02129	0.47026	0.49155	529.2	666.6	1195.9	0.7304	0.6708	1.4012	534.0
532.0	900.34	0.02123	0.47947	0.50070	526.8	669.6	1196.4	0.7280	0.6752	1.4032	532.0
530.0	885.23	0.02118	0.48886	0.51004	524.3	672.6	1196.9	0.7255	0.6796	1.4051	530.0
528.0	870.31	0.02112	0.49843	0.51955	521.8	675.5	1197.3	0.7231	0.6839	1.4070	528.0
526.0	855.58	0.02107	0.50819	0.52926	519.3	678.4	1197.8	0.7206	0.6883	1.4089	526.0
524.0	841.04	0.02102	0.51814	0.53916	516.9	681.3	1198.2	0.7182	0.6926	1.4108	524.0
522.0	826.69	0.02097	0.52829	0.54926	514.4	684.2	1198.6	0.7158	0.6969	1.4127	522.0
520.0	812.53	0.02091	0.53864	0.55956	512.0	687.0	1199.0	0.7133	0.7013	1.4146	520.0
518.0	798.55	0.02086	0.54920	0.57006	509.6	689.9	1199.4	0.7109	0.7056	1.4165	518.0
516.0	784.76	0.02081	0.55997	0.58079	507.1	692.7	1199.8	0.7085	0.7099	1.4183	516.0
514.0	771.15	0.02076	0.57096	0.59173	504.7	695.4	1200.2	0.7060	0.7142	1.4202	514.0
512.0	757.72	0.02072	0.58218	0.60289	502.3	698.2	1200.5	0.7036	0.7185	1.4221	512.0
510.0	744.47	0.02067	0.59362	0.61429	499.9	700.9	1200.8	0.7012	0.7228	1.4240	510.0
508.0	731.40	0.02062	0.60530	0.62592	497.5	703.7	1201.1	0.6987	0.7271	1.4258	508.0
506.0	718.50	0.02057	0.61722	0.63779	495.1	706.3	1201.4	0.6963	0.7314	1.4277	506.0
504.0	705.78	0.02053	0.62938	0.64991	492.7	709.0	1201.7	0.6939	0.7357	1.4296	504.0
502.0	693.23	0.02048	0.64180	0.66228	490.3	711.7	1202.0	0.6915	0.7400	1.4314	502.0
500.0	680.86	0.02043	0.65448	0.67492	487.9	714.3	1202.2	0.6890	0.7443	1.4333	500.0
498.0	668.65	0.02039	0.66743	0.68782	485.6	716.9	1202.5	0.6866	0.7486	1.4352	498.0
496.0	656.61	0.02034	0.68065	0.70100	483.2	719.5	1202.7	0.6842	0.7528	1.4370	496.0
494.0	644.73	0.02030	0.69415	0.71445	480.8	722.1	1202.9	0.6818	0.7571	1.4389	494.0
492.0	633.03	0.02026	0.70794	0.72820	478.5	724.6	1203.1	0.6793	0.7614	1.4407	492.0
490.0	621.48	0.02021	0.72203	0.74224	476.1	727.2	1203.3	0.6769	0.7657	1.4426	490.0
488.0	610.10	0.02017	0.73641	0.75658	473.8	729.7	1203.5	0.6745	0.7700	1.4444	488.0
486.0	598.87	0.02013	0.75111	0.77124	471.5	732.2	1203.7	0.6721	0.7742	1.4463	486.0
484.0	587.81	0.02009	0.76613	0.78622	469.1	734.7	1203.8	0.6696	0.7785	1.4481	484.0
482.0	576.90	0.02004	0.78148	0.80152	466.8	737.2	1204.0	0.6672	0.7828	1.4500	482.0
480.0	566.15	0.02000	0.79716	0.81717	464.5	739.6	1204.1	0.6648	0.7871	1.4518	480.0
478.0	555.55	0.01996	0.81319	0.83315	462.2	742.1	1204.2	0.6624	0.7913	1.4537	478.0
476.0	545.11	0.01992	0.82958	0.84950	459.9	744.5	1204.3	0.6599	0.7956	1.4555	476.0
474.0	534.81	0.01988	0.84632	0.86621	457.5	746.9	1204.4	0.6575	0.7999	1.4574	474.0
472.0	524.67	0.01984	0.86345	0.88329	455.2	749.3	1204.5	0.6551	0.8042	1.4592	472.0
470.0	514.67	0.01980	0.88095	0.90076	452.9	751.6	1204.6	0.6527	0.8084	1.4611	470.0
468.0	504.83	0.01976	0.89885	0.91862	450.7	754.0	1204.6	0.6502	0.8127	1.4629	468.0
466.0	495.12	0.01973	0.91716	0.93689	448.4	756.3	1204.7	0.6478	0.8170	1.4648	466.0
464.0	485.56	0.01969	0.93588	0.95557	446.1	758.6	1204.7	0.6454	0.8213	1.4667	464.0
462.0	476.14	0.01965	0.95504	0.97469	443.8	761.0	1204.8	0.6429	0.8256	1.4685	462.0
460.0	466.87	0.01961	0.97463	0.99424	441.5	763.2	1204.8	0.6405	0.8299	1.4704	460.0
458.0	457.73	0.01958	0.99467	1.01425	439.3	765.5	1204.8	0.6381	0.8342	1.4722	458.0
456.0	448.73	0.01954	1.01518	1.03472	437.0	767.8	1204.8	0.6356	0.8385	1.4741	456.0
454.0	439.87	0.01950	1.03616	1.05567	434.7	770.0	1204.8	0.6332	0.8428	1.4759	454.0
452.0	431.14	0.01947	1.05764	1.07711	432.5	772.3	1204.8	0.6308	0.8471	1.4778	452.0
450.0	422.55	0.01943	1.07962	1.09905	430.2	774.5	1204.7	0.6283	0.8514	1.4797	450.0
448.0	414.09	0.01940	1.10212	1.12152	428.0	776.7	1204.7	0.6259	0.8557	1.4815	448.0
446.0	405.76	0.01936	1.12515	1.14452	425.7	778.9	1204.6	0.6234	0.8600	1.4834	446.0
444.0	397.56	0.01933	1.14874	1.16806	423.5	781.1	1204.6	0.6210	0.8643	1.4853	444.0
442.0	389.49	0.01929	1.17288	1.19217	421.3	783.2	1204.5	0.6185	0.8686	1.4872	442.0
440.0	381.54	0.01926	1.19761	1.21687	419.0	785.4	1204.4	0.6161	0.8729	1.4890	440.0
438.0	373.72	0.01923	1.22293	1.24216	416.8	787.5	1204.3	0.6136	0.8773	1.4909	438.0
436.0	366.03	0.01919	1.24887	1.26806	414.6	789.7	1204.2	0.6112	0.8816	1.4928	436.0
434.0	358.46	0.01916	1.27544	1.29460	412.4	791.8	1204.1	0.6087	0.8859	1.4947	434.0
432.0	351.00	0.01913	1.30266	1.32179	410.1	793.9	1204.0	0.6063	0.8903	1.4966	432.0
430.0	343.67	0.01909	1.33055	1.34965	407.9	796.0	1203.9	0.6038	0.8946	1.4985	430.0

Properties of Saturated Steam and Saturated Water

Temp. F	Press. psia	Volume, ft³/lbm			Enthalpy, Btu/lbm			Entropy, Btu/lbm ×F			Temp. F
		Water v_f	Evap. v_{fg}	Steam v_g	Water h_f	Evap. h_{fg}	Steam h_g	Water s_f	Evap. s_{fg}	Steam s_g	
430.0	343.674	0.01909	1.3306	1.3496	407.9	796.0	1203.9	0.6038	0.8946	1.4985	430.0
428.0	336.463	0.01906	1.3591	1.3782	405.7	798.0	1203.7	0.6014	0.8990	1.5004	428.0
426.0	329.369	0.01903	1.3884	1.4075	403.5	800.1	1203.6	0.5989	0.9034	1.5023	426.0
424.0	322.391	0.01900	1.4184	1.4374	401.3	802.2	1203.5	0.5964	0.9077	1.5042	424.0
422.0	315.529	0.01897	1.4492	1.4682	399.1	804.2	1203.3	0.5940	0.9121	1.5061	422.0
420.0	308.780	0.01894	1.4808	1.4997	396.9	806.2	1203.1	0.5915	0.9165	1.5080	420.0
418.0	302.143	0.01890	1.5131	1.5320	394.7	808.2	1202.9	0.5890	0.9209	1.5099	418.0
416.0	295.617	0.01887	1.5463	1.5651	392.5	810.2	1202.8	0.5866	0.9253	1.5118	416.0
414.0	289.201	0.01884	1.5803	1.5991	390.3	812.2	1202.6	0.5841	0.9297	1.5137	414.0
412.0	282.894	0.01881	1.6152	1.6340	388.1	814.2	1202.4	0.5816	0.9341	1.5157	412.0
410.0	276.694	0.01878	1.6510	1.6697	386.0	816.2	1202.1	0.5791	0.9385	1.5176	410.0
408.0	270.600	0.01875	1.6877	1.7064	383.8	818.2	1201.9	0.5766	0.9429	1.5195	408.0
406.0	264.611	0.01872	1.7253	1.7441	381.6	820.1	1201.7	0.5742	0.9473	1.5215	406.0
404.0	258.725	0.01870	1.7640	1.7827	379.4	822.0	1201.5	0.5717	0.9518	1.5234	404.0
402.0	252.942	0.01867	1.8037	1.8223	377.3	824.0	1201.2	0.5692	0.9562	1.5254	402.0
400.0	247.259	0.01864	1.8444	1.8630	375.1	825.9	1201.0	0.5667	0.9607	1.5274	400.0
398.0	241.677	0.01861	1.8862	1.9048	372.9	827.8	1200.7	0.5642	0.9651	1.5293	398.0
396.0	236.193	0.01858	1.9291	1.9477	370.8	829.7	1200.4	0.5617	0.9696	1.5313	396.0
394.0	230.807	0.01855	1.9731	1.9917	368.6	831.6	1200.2	0.5592	0.9741	1.5333	394.0
392.0	225.516	0.01853	2.0184	2.0369	366.5	833.4	1199.9	0.5567	0.9786	1.5352	392.0
390.0	220.321	0.01850	2.0649	2.0833	364.3	835.3	1199.6	0.5542	0.9831	1.5372	390.0
388.0	215.220	0.01847	2.1126	2.1311	362.2	837.2	1199.3	0.5516	0.9876	1.5392	388.0
386.0	210.211	0.01844	2.1616	2.1801	360.0	839.0	1199.0	0.5491	0.9921	1.5412	386.0
384.0	205.294	0.01842	2.2120	2.2304	357.9	840.8	1198.7	0.5466	0.9966	1.5432	384.0
382.0	200.467	0.01839	2.2638	2.2821	355.7	842.7	1198.4	0.5441	1.0012	1.5452	382.0
380.0	195.729	0.01836	2.3170	2.3353	353.6	844.5	1198.0	0.5416	1.0057	1.5473	380.0
378.0	191.080	0.01834	2.3716	2.3900	351.4	846.3	1197.7	0.5390	1.0103	1.5493	378.0
376.0	186.517	0.01831	2.4279	2.4462	349.3	848.1	1197.4	0.5365	1.0148	1.5513	376.0
374.0	182.040	0.01829	2.4857	2.5039	347.2	849.8	1197.0	0.5340	1.0194	1.5534	374.0
372.0	177.648	0.01826	2.5451	2.5633	345.0	851.6	1196.7	0.5314	1.0240	1.5554	372.0
370.0	173.339	0.01823	2.6062	2.6244	342.9	853.4	1196.3	0.5289	1.0286	1.5575	370.0
368.0	169.113	0.01821	2.6691	2.6873	340.8	855.1	1195.9	0.5263	1.0332	1.5595	368.0
366.0	164.968	0.01818	2.7337	2.7519	338.7	856.9	1195.6	0.5238	1.0378	1.5616	366.0
364.0	160.903	0.01816	2.8002	2.8184	336.5	858.6	1195.2	0.5212	1.0424	1.5637	364.0
362.0	156.917	0.01813	2.8687	2.8868	334.4	860.4	1194.8	0.5187	1.0471	1.5658	362.0
360.0	153.010	0.01811	2.9392	2.9573	332.3	862.1	1194.4	0.5161	1.0517	1.5678	360.0
358.0	149.179	0.01809	3.0117	3.0298	330.2	863.8	1194.0	0.5135	1.0564	1.5699	358.0
356.0	145.424	0.01806	3.0863	3.1044	328.1	865.5	1193.6	0.5110	1.0611	1.5721	356.0
354.0	141.744	0.01804	3.1632	3.1812	326.0	867.2	1193.2	0.5084	1.0658	1.5742	354.0
352.0	138.138	0.01801	3.2423	3.2603	323.9	868.9	1192.7	0.5058	1.0705	1.5763	352.0
350.0	134.604	0.01799	3.3238	3.3418	321.8	870.6	1192.3	0.5032	1.0752	1.5784	350.0
348.0	131.142	0.01797	3.4078	3.4258	319.7	872.2	1191.9	0.5006	1.0799	1.5806	348.0
346.0	127.751	0.01794	3.4943	3.5122	317.6	873.9	1191.4	0.4980	1.0847	1.5827	346.0
344.0	124.430	0.01792	3.5834	3.6013	315.5	875.5	1191.0	0.4954	1.0894	1.5849	344.0
342.0	121.177	0.01790	3.6752	3.6931	313.4	877.2	1190.5	0.4928	1.0942	1.5871	342.0
340.0	117.992	0.01787	3.7699	3.7878	311.3	878.8	1190.1	0.4902	1.0990	1.5892	340.0
338.0	114.873	0.01785	3.8675	3.8853	309.2	880.5	1189.6	0.4876	1.1038	1.5914	338.0
336.0	111.820	0.01783	3.9681	3.9859	307.1	882.1	1189.1	0.4850	1.1086	1.5936	336.0
334.0	108.832	0.01781	4.0718	4.0896	305.0	883.7	1188.7	0.4824	1.1134	1.5958	334.0
332.0	105.907	0.01779	4.1788	4.1966	302.9	885.3	1188.2	0.4798	1.1183	1.5981	332.0
330.0	103.045	0.01776	4.2892	4.3069	300.8	886.9	1187.7	0.4772	1.1231	1.6003	330.0
328.0	100.245	0.01774	4.4030	4.4208	298.7	888.5	1187.2	0.4745	1.1280	1.6025	328.0
326.0	97.506	0.01772	4.5205	4.5382	296.6	890.1	1186.7	0.4719	1.1329	1.6048	326.0
324.0	94.826	0.01770	4.6418	4.6595	294.6	891.6	1186.2	0.4692	1.1378	1.6071	324.0
322.0	92.205	0.01768	4.7669	4.7846	292.5	893.2	1185.7	0.4666	1.1427	1.6093	322.0
320.0	89.643	0.01766	4.8961	4.9138	290.4	894.8	1185.2	0.4640	1.1477	1.6116	320.0
318.0	87.137	0.01764	5.0295	5.0471	288.3	896.3	1184.7	0.4613	1.1526	1.6139	318.0
316.0	84.688	0.01761	5.1673	5.1849	286.3	897.9	1184.1	0.4586	1.1576	1.6162	316.0
314.0	82.293	0.01759	5.3096	5.3272	284.2	899.4	1183.6	0.4560	1.1626	1.6185	314.0
312.0	79.953	0.01757	5.4566	5.4742	282.1	901.0	1183.1	0.4533	1.1676	1.6209	312.0
310.0	77.667	0.01755	5.6085	5.6260	280.0	902.5	1182.5	0.4506	1.1726	1.6232	310.0
308.0	75.433	0.01753	5.7655	5.7830	278.0	904.0	1182.0	0.4479	1.1776	1.6256	308.0
306.0	73.251	0.01751	5.9277	5.9452	275.9	905.5	1181.4	0.4453	1.1827	1.6279	306.0
304.0	71.119	0.01749	6.0955	6.1130	273.8	907.0	1180.9	0.4426	1.1877	1.6303	304.0
302.0	69.038	0.01747	6.2689	6.2864	271.8	908.5	1180.3	0.4399	1.1928	1.6327	302.0
300.0	67.005	0.01745	6.4483	6.4658	269.7	910.0	1179.7	0.4372	1.1979	1.6351	300.0
298.0	65.021	0.01743	6.6339	6.6513	267.7	911.5	1179.2	0.4345	1.2031	1.6375	298.0
296.0	63.084	0.01741	6.8259	6.8433	265.6	913.0	1178.6	0.4317	1.2082	1.6400	296.0
294.0	61.194	0.01739	7.0245	7.0419	263.5	914.5	1178.0	0.4290	1.2134	1.6424	294.0
292.0	59.350	0.01738	7.2301	7.2475	261.5	915.9	1177.4	0.4263	1.2186	1.6449	292.0
290.0	57.550	0.01736	7.4430	7.4603	259.4	917.4	1176.8	0.4236	1.2238	1.6473	290.0
288.0	55.795	0.01734	7.6634	7.6807	257.4	918.8	1176.2	0.4208	1.2290	1.6498	288.0
286.0	54.083	0.01732	7.8916	7.9089	255.3	920.3	1175.6	0.4181	1.2342	1.6523	286.0
284.0	52.414	0.01730	8.1280	8.1453	253.3	921.7	1175.0	0.4154	1.2395	1.6548	284.0
282.0	50.786	0.01728	8.3729	8.3902	251.2	923.2	1174.4	0.4126	1.2448	1.6574	282.0
280.0	49.200	0.01726	8.6267	8.6439	249.2	924.6	1173.8	0.4098	1.2501	1.6599	280.0

Properties of Saturated Steam and Saturated Water

Temp. F	Press. psia	Volume, ft³/lbm			Enthalpy, Btu/lbm			Entropy, Btu/lbm×F			Temp. F
		Water v_f	Evap. v_{fg}	Steam v_g	Water h_f	Evap. h_{fg}	Steam h_g	Water s_f	Evap. s_{fg}	Steam s_g	
280.0	49.200	0.017264	8.627	8.644	249.17	924.6	1173.8	0.4098	1.2501	1.6599	280.0
278.0	47.653	0.017246	8.890	8.907	247.13	926.0	1173.2	0.4071	1.2554	1.6625	278.0
276.0	46.147	0.017228	9.162	9.180	245.08	927.5	1172.5	0.4043	1.2607	1.6650	276.0
274.0	44.678	0.017210	9.445	9.462	243.03	928.9	1171.9	0.4015	1.2661	1.6676	274.0
272.0	43.249	0.017193	9.738	9.755	240.99	930.3	1171.3	0.3987	1.2715	1.6702	272.0
270.0	41.856	0.017175	10.042	10.060	238.95	931.7	1170.6	0.3960	1.2769	1.6729	270.0
268.0	40.500	0.017157	10.358	10.375	236.91	933.1	1170.0	0.3932	1.2823	1.6755	268.0
266.0	39.179	0.017140	10.685	10.703	234.87	934.5	1169.3	0.3904	1.2878	1.6781	266.0
264.0	37.894	0.017123	11.025	11.042	232.83	935.9	1168.7	0.3876	1.2933	1.6808	264.0
262.0	36.644	0.017106	11.378	11.395	230.79	937.3	1168.0	0.3847	1.2988	1.6835	262.0
260.0	35.427	0.017089	11.745	11.762	228.76	938.6	1167.4	0.3819	1.3043	1.6862	260.0
258.0	34.243	0.017072	12.125	12.142	226.72	940.0	1166.7	0.3791	1.3098	1.6889	258.0
256.0	33.091	0.017055	12.520	12.538	224.69	941.4	1166.1	0.3763	1.3154	1.6917	256.0
254.0	31.972	0.017039	12.931	12.948	222.65	942.7	1165.4	0.3734	1.3210	1.6944	254.0
252.0	30.883	0.017022	13.358	13.375	220.62	944.1	1164.7	0.3706	1.3266	1.6972	252.0
250.0	29.825	0.017006	13.802	13.819	218.59	945.4	1164.0	0.3677	1.3323	1.7000	250.0
248.0	28.796	0.016990	14.264	14.281	216.56	946.8	1163.4	0.3649	1.3379	1.7028	248.0
246.0	27.797	0.016974	14.744	14.761	214.53	948.1	1162.7	0.3620	1.3436	1.7056	246.0
244.0	26.826	0.016958	15.243	15.260	212.50	949.5	1162.0	0.3591	1.3494	1.7085	244.0
242.0	25.883	0.016942	15.763	15.780	210.48	950.8	1161.3	0.3562	1.3551	1.7113	242.0
240.0	24.968	0.016926	16.304	16.321	208.45	952.1	1160.6	0.3533	1.3609	1.7142	240.0
238.0	24.079	0.016910	16.867	16.884	206.42	953.5	1159.9	0.3505	1.3667	1.7171	238.0
236.0	23.216	0.016895	17.454	17.471	204.40	954.8	1159.2	0.3476	1.3725	1.7201	236.0
234.0	22.379	0.016880	18.065	18.082	202.38	956.1	1158.5	0.3446	1.3784	1.7230	234.0
232.0	21.567	0.016864	18.701	18.718	200.35	957.4	1157.8	0.3417	1.3842	1.7260	232.0
230.0	20.779	0.016849	19.364	19.381	198.33	958.7	1157.1	0.3388	1.3902	1.7290	230.0
229.0	20.394	0.016842	19.707	19.723	197.32	959.4	1156.7	0.3373	1.3931	1.7305	229.0
228.0	20.015	0.016834	20.056	20.073	196.31	960.0	1156.3	0.3359	1.3961	1.7320	228.0
227.0	19.642	0.016827	20.413	20.429	195.30	960.7	1156.0	0.3344	1.3991	1.7335	227.0
226.0	19.274	0.016819	20.777	20.794	194.29	961.3	1155.6	0.3329	1.4021	1.7350	226.0
225.0	18.912	0.016812	21.149	21.166	193.28	962.0	1155.3	0.3315	1.4051	1.7365	225.0
224.0	18.556	0.016805	21.529	21.545	192.27	962.6	1154.9	0.3300	1.4081	1.7380	224.0
223.0	18.206	0.016797	21.917	21.933	191.26	963.3	1154.5	0.3285	1.4111	1.7396	223.0
222.0	17.860	0.016790	22.313	22.330	190.25	963.9	1154.2	0.3270	1.4141	1.7411	222.0
221.0	17.521	0.016783	22.718	22.735	189.24	964.6	1153.8	0.3255	1.4171	1.7427	221.0
220.0	17.186	0.016775	23.131	23.148	188.23	965.2	1153.4	0.3241	1.4201	1.7442	220.0
219.0	16.857	0.016768	23.554	23.571	187.22	965.8	1153.1	0.3226	1.4232	1.7458	219.0
218.0	16.533	0.016761	23.986	24.002	186.21	966.5	1152.7	0.3211	1.4262	1.7473	218.0
217.0	16.214	0.016754	24.427	24.444	185.21	967.1	1152.3	0.3196	1.4293	1.7489	217.0
216.0	15.901	0.016747	24.878	24.894	184.20	967.8	1152.0	0.3181	1.4323	1.7505	216.0
215.0	15.592	0.016740	25.338	25.355	183.19	968.4	1151.6	0.3166	1.4354	1.7520	215.0
214.0	15.280	0.016733	25.809	25.820	182.18	969.0	1151.2	0.3151	1.4385	1.7536	214.0
213.0	14.990	0.016726	26.290	26.307	181.17	969.7	1150.8	0.3136	1.4416	1.7552	213.0
212.0	14.696	0.016719	26.782	26.799	180.17	970.3	1150.5	0.3121	1.4447	1.7568	212.0
211.0	14.407	0.016712	27.285	27.302	179.16	970.9	1150.1	0.3106	1.4478	1.7584	211.0
210.0	14.123	0.016705	27.799	27.816	178.15	971.6	1149.7	0.3091	1.4509	1.7600	210.0
209.0	13.843	0.016698	28.324	28.341	177.14	972.2	1149.4	0.3076	1.4540	1.7616	209.0
208.0	13.568	0.016691	28.862	28.878	176.14	972.8	1149.0	0.3061	1.4571	1.7632	208.0
207.0	13.297	0.016684	29.411	29.428	175.13	973.5	1148.6	0.3046	1.4602	1.7649	207.0
206.0	13.031	0.016677	29.973	29.989	174.12	974.1	1148.2	0.3031	1.4634	1.7665	206.0
205.0	12.770	0.016670	30.547	30.564	173.12	974.7	1147.9	0.3016	1.4665	1.7681	205.0
204.0	12.512	0.016664	31.135	31.151	172.11	975.4	1147.5	0.3001	1.4697	1.7698	204.0
203.0	12.259	0.016657	31.736	31.752	171.10	976.0	1147.1	0.2986	1.4728	1.7714	203.0
202.0	12.011	0.016650	32.350	32.367	170.10	976.6	1146.7	0.2971	1.4760	1.7731	202.0
201.0	11.766	0.016643	32.979	32.996	169.09	977.2	1146.3	0.2955	1.4792	1.7747	201.0
200.0	11.526	0.016637	33.622	33.639	168.09	977.9	1146.0	0.2940	1.4824	1.7764	200.0
199.0	11.290	0.016630	34.280	34.297	167.08	978.5	1145.6	0.2925	1.4856	1.7781	199.0
198.0	11.058	0.016624	34.954	34.970	166.08	979.1	1145.2	0.2910	1.4888	1.7798	198.0
197.0	10.830	0.016617	35.643	35.659	165.07	979.7	1144.8	0.2894	1.4920	1.7814	197.0
196.0	10.605	0.016611	36.348	36.364	164.06	980.4	1144.4	0.2879	1.4952	1.7831	196.0
195.0	10.385	0.016604	37.069	37.086	163.06	981.0	1144.0	0.2864	1.4985	1.7848	195.0
194.0	10.168	0.016598	37.808	37.824	162.05	981.6	1143.7	0.2848	1.5017	1.7865	194.0
193.0	9.956	0.016591	38.564	38.580	161.05	982.2	1143.3	0.2833	1.5050	1.7882	193.0
192.0	9.747	0.016585	39.337	39.354	160.05	982.8	1142.9	0.2818	1.5082	1.7900	192.0
191.0	9.541	0.016578	40.130	40.146	159.04	983.5	1142.5	0.2802	1.5115	1.7917	191.0
190.0	9.340	0.016572	40.941	40.957	158.04	984.1	1142.1	0.2787	1.5148	1.7934	190.0
189.0	9.141	0.016566	41.771	41.787	157.03	984.7	1141.7	0.2771	1.5180	1.7952	189.0
188.0	8.947	0.016559	42.621	42.638	156.03	985.3	1141.3	0.2756	1.5213	1.7969	188.0
187.0	8.756	0.016553	43.492	43.508	155.02	985.9	1140.9	0.2740	1.5246	1.7987	187.0
186.0	8.568	0.016546	44.383	44.400	154.02	986.5	1140.5	0.2725	1.5279	1.8004	186.0
185.0	8.384	0.016541	45.297	45.313	153.02	987.1	1140.2	0.2709	1.5313	1.8022	185.0
184.0	8.203	0.016534	46.232	46.249	152.01	987.8	1139.8	0.2694	1.5346	1.8040	184.0
183.0	8.025	0.016528	47.190	47.207	151.01	988.4	1139.4	0.2678	1.5379	1.8057	183.0
182.0	7.850	0.016522	48.172	48.189	150.01	989.0	1139.0	0.2662	1.5413	1.8075	182.0
181.0	7.679	0.016516	49.178	49.194	149.00	989.6	1138.6	0.2647	1.5446	1.8093	181.0
180.0	7.511	0.016510	50.208	50.225	148.00	990.2	1138.2	0.2631	1.5480	1.8111	180.0

Properties of Saturated Steam and Saturated Water

Temp. F	Press. psia	Volume, ft³/lbm			Enthalpy, Btu/lbm			Entropy, Btu/lbm×F			Temp. F
		Water v_f	Evap. v_{fg}	Steam v_g	Water h_f	Evap. h_{fg}	Steam h_g	Water s_f	Evap. s_{fg}	Steam s_g	
180.0	7.5110	0.016510	50.21	50.22	148.00	990.2	1138.2	0.2631	1.5480	1.8111	180.0
179.0	7.3460	0.016504	51.26	51.28	147.00	990.8	1137.8	0.2615	1.5514	1.8129	179.0
178.0	7.1840	0.016498	52.35	52.36	145.99	991.4	1137.4	0.2600	1.5548	1.8147	178.0
177.0	7.0250	0.016492	53.46	53.47	144.99	992.0	1137.0	0.2584	1.5582	1.8166	177.0
176.0	6.8690	0.016486	54.59	54.61	143.99	992.6	1136.6	0.2568	1.5616	1.8184	176.0
175.0	6.7159	0.016480	55.76	55.77	142.99	993.2	1136.2	0.2552	1.5650	1.8202	175.0
174.0	6.5656	0.016474	56.95	56.97	141.98	993.8	1135.8	0.2537	1.5684	1.8221	174.0
173.0	6.4182	0.016468	58.18	58.19	140.98	994.4	1135.4	0.2521	1.5718	1.8239	173.0
172.0	6.2736	0.016463	59.43	59.45	139.98	995.0	1135.0	0.2505	1.5753	1.8258	172.0
171.0	6.1318	0.016457	60.72	60.74	138.98	995.6	1134.6	0.2489	1.5787	1.8276	171.0
170.0	5.9926	0.016451	62.04	62.06	137.97	996.2	1134.2	0.2473	1.5822	1.8295	170.0
169.0	5.8562	0.016445	63.39	63.41	136.97	996.8	1133.8	0.2457	1.5857	1.8314	169.0
168.0	5.7223	0.016440	64.78	64.80	135.97	997.4	1133.4	0.2441	1.5892	1.8333	168.0
167.0	5.5911	0.016434	66.21	66.22	134.97	998.0	1133.0	0.2425	1.5926	1.8352	167.0
166.0	5.4623	0.016428	67.67	67.68	133.97	998.6	1132.6	0.2409	1.5961	1.8371	166.0
165.0	5.3361	0.016423	69.17	69.18	132.96	999.2	1132.2	0.2393	1.5997	1.8390	165.0
164.0	5.2124	0.016417	70.70	70.72	131.96	999.8	1131.8	0.2377	1.6032	1.8409	164.0
163.0	5.0911	0.016412	72.28	72.30	130.96	1000.4	1131.4	0.2361	1.6067	1.8428	163.0
162.0	4.9722	0.016406	73.90	73.92	129.96	1001.0	1131.0	0.2345	1.6103	1.8448	162.0
161.0	4.8556	0.016401	75.56	75.58	128.96	1001.6	1130.6	0.2329	1.6138	1.8467	161.0
160.0	4.7414	0.016395	77.27	77.29	127.96	1002.2	1130.2	0.2313	1.6174	1.8487	160.0
159.0	4.6294	0.016390	79.02	79.04	126.96	1002.8	1129.8	0.2297	1.6210	1.8506	159.0
158.0	4.5197	0.016384	80.82	80.83	125.96	1003.4	1129.4	0.2281	1.6245	1.8526	158.0
157.0	4.4122	0.016379	82.66	82.68	124.95	1004.0	1129.0	0.2264	1.6281	1.8546	157.0
156.0	4.3068	0.016374	84.56	84.57	123.95	1004.6	1128.6	0.2248	1.6318	1.8566	156.0
155.0	4.2036	0.016369	86.50	86.52	122.95	1005.2	1128.2	0.2232	1.6354	1.8586	155.0
154.0	4.1025	0.016363	88.50	88.52	121.95	1005.8	1127.7	0.2216	1.6390	1.8606	154.0
153.0	4.0035	0.016358	90.55	90.57	120.95	1006.4	1127.3	0.2199	1.6426	1.8626	153.0
152.0	3.9065	0.016353	92.66	92.68	119.95	1007.0	1126.9	0.2183	1.6463	1.8646	152.0
151.0	3.8114	0.016348	94.83	94.84	118.95	1007.6	1126.5	0.2167	1.6500	1.8666	151.0
150.0	3.7184	0.016343	97.05	97.07	117.95	1008.2	1126.1	0.2150	1.6536	1.8686	150.0
149.0	3.6273	0.016337	99.33	99.35	116.95	1008.7	1125.7	0.2134	1.6573	1.8707	149.0
148.0	3.5381	0.016332	101.68	101.70	115.95	1009.3	1125.3	0.2117	1.6610	1.8727	148.0
147.0	3.4508	0.016327	104.10	104.11	114.95	1009.9	1124.9	0.2101	1.6647	1.8748	147.0
146.0	3.3653	0.016322	106.58	106.59	113.95	1010.5	1124.5	0.2084	1.6684	1.8769	146.0
145.0	3.2816	0.016317	109.12	109.14	112.95	1011.1	1124.0	0.2068	1.6722	1.8789	145.0
144.0	3.1997	0.016312	111.74	111.76	111.95	1011.7	1123.6	0.2051	1.6759	1.8810	144.0
143.0	3.1195	0.016308	114.44	114.45	110.95	1012.3	1123.2	0.2035	1.6797	1.8831	143.0
142.0	3.0411	0.016303	117.21	117.22	109.95	1012.9	1122.8	0.2018	1.6834	1.8852	142.0
141.0	2.9643	0.016298	120.05	120.07	108.95	1013.4	1122.4	0.2001	1.6872	1.8873	141.0
140.0	2.8892	0.016293	122.98	123.00	107.95	1014.0	1122.0	0.1985	1.6910	1.8895	140.0
139.0	2.8157	0.016288	125.99	126.01	106.95	1014.6	1121.6	0.1968	1.6948	1.8916	139.0
138.0	2.7438	0.016284	129.09	129.11	105.95	1015.2	1121.1	0.1951	1.6986	1.8937	138.0
137.0	2.6735	0.016279	132.28	132.29	104.95	1015.8	1120.7	0.1935	1.7024	1.8959	137.0
136.0	2.6047	0.016274	135.55	135.57	103.95	1016.4	1120.3	0.1918	1.7063	1.8980	136.0
135.0	2.5375	0.016270	138.93	138.94	102.95	1016.9	1119.9	0.1901	1.7101	1.9002	135.0
134.0	2.4717	0.016265	142.40	142.41	101.95	1017.5	1119.5	0.1884	1.7140	1.9024	134.0
133.0	2.4074	0.016260	145.97	145.98	100.95	1018.1	1119.1	0.1867	1.7178	1.9046	133.0
132.0	2.3445	0.016256	149.64	149.66	99.95	1018.7	1118.6	0.1851	1.7217	1.9068	132.0
131.0	2.2830	0.016251	153.42	153.44	98.95	1019.3	1118.2	0.1834	1.7256	1.9090	131.0
130.0	2.2230	0.016247	157.32	157.33	97.96	1019.8	1117.8	0.1817	1.7295	1.9112	130.0
129.0	2.1642	0.016243	161.32	161.34	96.96	1020.4	1117.4	0.1800	1.7335	1.9134	129.0
128.0	2.1068	0.016238	165.45	165.47	95.96	1021.0	1117.0	0.1783	1.7374	1.9157	128.0
127.0	2.0507	0.016234	169.70	169.72	94.96	1021.6	1116.5	0.1766	1.7413	1.9179	127.0
126.0	1.9959	0.016229	174.08	174.09	93.96	1022.2	1116.1	0.1749	1.7453	1.9202	126.0
125.0	1.9424	0.016225	178.58	178.60	92.96	1022.7	1115.7	0.1732	1.7493	1.9224	125.0
124.0	1.8901	0.016221	183.23	183.24	91.96	1023.3	1115.3	0.1715	1.7533	1.9247	124.0
123.0	1.8390	0.016217	188.01	188.03	90.96	1023.9	1114.9	0.1697	1.7573	1.9270	123.0
122.0	1.7891	0.016213	192.94	192.95	89.96	1024.5	1114.4	0.1680	1.7613	1.9293	122.0
121.0	1.7403	0.016208	198.01	198.03	88.96	1025.0	1114.0	0.1663	1.7653	1.9316	121.0
120.0	1.6927	0.016204	203.25	203.26	87.97	1025.6	1113.6	0.1646	1.7693	1.9339	120.0
119.0	1.6463	0.016200	208.64	208.66	86.97	1026.2	1113.2	0.1629	1.7734	1.9362	119.0
118.0	1.6009	0.016196	214.20	214.21	85.97	1026.8	1112.7	0.1611	1.7774	1.9386	118.0
117.0	1.5566	0.016192	219.93	219.94	84.97	1027.3	1112.3	0.1594	1.7815	1.9409	117.0
116.0	1.5133	0.016188	225.84	225.85	83.97	1027.9	1111.9	0.1577	1.7856	1.9433	116.0
115.0	1.4711	0.016184	231.93	231.94	82.97	1028.5	1111.5	0.1559	1.7897	1.9457	115.0
114.0	1.4299	0.016180	238.21	238.22	81.97	1029.1	1111.0	0.1542	1.7938	1.9480	114.0
113.0	1.3898	0.016177	244.69	244.70	80.98	1029.6	1110.6	0.1525	1.7980	1.9504	113.0
112.0	1.3505	0.016173	251.37	251.38	79.98	1030.2	1110.2	0.1507	1.8021	1.9528	112.0
111.0	1.3123	0.016169	258.26	258.28	78.98	1030.8	1109.8	0.1490	1.8063	1.9552	111.0
110.0	1.2750	0.016165	265.37	265.39	77.98	1031.4	1109.3	0.1472	1.8105	1.9577	110.0
109.0	1.2385	0.016162	272.71	272.72	76.98	1031.9	1108.9	0.1455	1.8146	1.9601	109.0
108.0	1.2030	0.016158	280.28	280.30	75.98	1032.5	1108.5	0.1437	1.8188	1.9626	108.0
107.0	1.1684	0.016154	288.09	288.11	74.99	1033.1	1108.1	0.1419	1.8231	1.9650	107.0
106.0	1.1347	0.016151	296.16	296.18	73.99	1033.6	1107.6	0.1402	1.8273	1.9675	106.0
105.0	1.1017	0.016147	304.49	304.50	72.99	1034.2	1107.2	0.1384	1.8315	1.9700	105.0

Properties of Saturated Steam and Saturated Water

Temp. F	Press. psia	Volume, ft³/lbm Water v_f	Volume, ft³/lbm Evap. v_{fg}	Volume, ft³/lbm Steam v_g	Enthalpy, Btu/lbm Water h_f	Enthalpy, Btu/lbm Evap. h_{fg}	Enthalpy, Btu/lbm Steam h_g	Entropy, Btu/lbm×F Water s_f	Entropy, Btu/lbm×F Evap. s_{fg}	Entropy, Btu/lbm×F Steam s_g	Temp. F
105.0	1.10174	0.016147	304.5	304.5	72.990	1034.2	1107.2	0.1384	1.8315	1.9700	105.0
104.0	1.06965	0.016144	313.1	313.1	71.992	1034.8	1106.8	0.1366	1.8358	1.9725	104.0
103.0	1.03838	0.016140	322.0	322.0	70.993	1035.4	1106.3	0.1349	1.8401	1.9750	103.0
102.0	1.00789	0.016137	331.1	331.1	69.995	1035.9	1105.9	0.1331	1.8444	1.9775	102.0
101.0	0.97818	0.016133	340.6	340.6	68.997	1036.5	1105.5	0.1313	1.8487	1.9800	101.0
100.0	0.94924	0.016130	350.4	350.4	67.999	1037.1	1105.1	0.1295	1.8530	1.9825	100.0
99.0	0.92103	0.016127	360.5	360.5	67.001	1037.6	1104.6	0.1278	1.8573	1.9851	99.0
98.0	0.89356	0.016123	370.9	370.9	66.003	1038.2	1104.2	0.1260	1.8617	1.9876	98.0
97.0	0.86679	0.016120	381.7	381.7	65.005	1038.8	1103.8	0.1242	1.8660	1.9902	97.0
96.0	0.84072	0.016117	392.8	392.9	64.006	1039.3	1103.3	0.1224	1.8704	1.9928	96.0
95.0	0.81534	0.016114	404.4	404.4	63.008	1039.9	1102.9	0.1206	1.8748	1.9954	95.0
94.0	0.79062	0.016111	416.3	416.3	62.010	1040.5	1102.5	0.1188	1.8792	1.9980	94.0
93.0	0.76655	0.016108	428.6	428.6	61.012	1041.0	1102.1	0.1170	1.8837	2.0006	93.0
92.0	0.74313	0.016105	441.3	441.3	60.014	1041.6	1101.6	0.1152	1.8881	2.0033	92.0
91.0	0.72032	0.016102	454.5	454.5	59.016	1042.2	1101.2	0.1134	1.8926	2.0059	91.0
90.0	0.69813	0.016099	468.1	468.1	58.018	1042.7	1100.8	0.1115	1.8970	2.0086	90.0
89.0	0.67653	0.016096	482.2	482.2	57.020	1043.3	1100.3	0.1097	1.9015	2.0112	89.0
88.0	0.65551	0.016093	496.8	496.8	56.022	1043.9	1099.9	0.1079	1.9060	2.0139	88.0
87.0	0.63507	0.016090	511.9	511.9	55.024	1044.4	1099.5	0.1061	1.9105	2.0166	87.0
86.0	0.61518	0.016087	527.5	527.5	54.026	1045.0	1099.0	0.1043	1.9151	2.0193	86.0
85.0	0.59583	0.016085	543.6	543.6	53.027	1045.6	1098.6	0.1024	1.9196	2.0221	85.0
84.0	0.57702	0.016082	560.3	560.3	52.029	1046.1	1098.2	0.1006	1.9242	2.0248	84.0
83.0	0.55872	0.016079	577.6	577.6	51.031	1046.7	1097.7	0.0988	1.9288	2.0275	83.0
82.0	0.54093	0.016077	595.5	595.6	50.033	1047.3	1097.3	0.0969	1.9334	2.0303	82.0
81.0	0.52364	0.016074	614.1	614.1	49.035	1047.8	1096.9	0.0951	1.9380	2.0331	81.0
80.0	0.050683	0.016072	633.3	633.3	48.037	1048.4	1096.4	0.0932	1.9426	2.0359	80.0
79.0	0.49049	0.016070	653.2	653.2	47.038	1049.0	1096.0	0.0914	1.9473	2.0387	79.0
78.0	0.47461	0.016067	673.8	673.9	46.040	1049.5	1095.6	0.0895	1.9520	2.0415	78.0
77.0	0.45919	0.016065	695.2	695.2	45.042	1050.1	1095.1	0.0877	1.9567	2.0443	77.0
76.0	0.44190	0.016060	717.1	717.4	44.043	1050.7	1094.7	0.0858	1.9614	2.0472	76.0
75.0	0.42964	0.016060	740.3	740.3	43.045	1051.2	1094.3	0.0839	1.9661	2.0500	75.0
74.0	0.41550	0.016058	764.1	764.1	42.046	1051.8	1093.8	0.0821	1.9708	2.0529	74.0
73.0	0.40177	0.016056	788.8	788.8	41.048	1052.4	1093.4	0.0802	1.9756	2.0558	73.0
72.0	0.38844	0.016054	814.3	814.3	40.049	1052.9	1093.0	0.0783	1.9804	2.0587	72.0
71.0	0.37549	0.016052	840.8	840.9	39.050	1053.5	1092.5	0.0764	1.9852	2.0616	71.0
70.0	0.36292	0.016050	868.3	868.4	38.052	1054.0	1092.1	0.0745	1.9900	2.0645	70.0
69.0	0.35073	0.016048	896.9	896.9	37.053	1054.6	1091.7	0.0727	1.9948	2.0675	69.0
68.0	0.33889	0.016046	926.5	926.5	36.054	1055.2	1091.2	0.0708	1.9996	2.0704	68.0
67.0	0.32740	0.016044	957.2	957.2	35.055	1055.7	1090.8	0.0689	2.0045	2.0734	67.0
66.0	0.31626	0.016043	989.0	989.1	34.056	1056.3	1090.4	0.0670	2.0094	2.0764	66.0
65.0	0.30545	0.016041	1022.1	1022.1	33.057	1056.9	1089.9	0.0651	2.0143	2.0794	65.0
64.0	0.29497	0.016039	1056.5	1056.5	32.058	1057.4	1089.5	0.0632	2.0192	2.0824	64.0
63.0	0.28480	0.016038	1092.1	1092.1	31.058	1058.0	1089.0	0.0613	2.0242	2.0854	63.0
62.0	0.27494	0.016036	1129.2	1129.2	30.059	1058.5	1088.6	0.0593	2.0291	2.0885	62.0
61.0	0.26538	0.016035	1167.6	1167.6	29.059	1059.1	1088.2	0.0574	2.0341	2.0915	61.0
60.0	0.25611	0.016033	1207.6	1207.6	28.060	1059.7	1087.7	0.0555	2.0391	2.0946	60.0
59.0	0.24713	0.016032	1249.1	1249.1	27.060	1060.2	1087.3	0.0536	2.0441	2.0977	59.0
58.0	0.23843	0.016031	1292.2	1292.2	26.060	1060.8	1086.9	0.0516	2.0491	2.1008	58.0
57.0	0.23000	0.016029	1337.0	1337.0	25.060	1061.4	1086.4	0.0497	2.0542	2.1039	57.0
56.0	0.22183	0.016028	1383.6	1383.6	24.059	1061.9	1086.0	0.0478	2.0593	2.1070	56.0
55.0	0.21392	0.016027	1432.0	1432.0	23.059	1062.5	1085.6	0.0458	2.0644	2.1102	55.0
54.0	0.20625	0.016026	1482.4	1482.4	22.058	1063.1	1085.1	0.0439	2.0695	2.1134	54.0
53.0	0.19883	0.016025	1534.7	1534.8	21.058	1063.6	1084.7	0.0419	2.0746	2.1165	53.0
52.0	0.19165	0.016024	1589.2	1589.2	20.057	1064.2	1084.2	0.0400	2.0798	2.1197	52.0
51.0	0.18469	0.016023	1645.9	1645.9	19.056	1064.7	1083.8	0.0380	2.0849	2.1230	51.0
50.0	0.17796	0.016023	1704.8	1704.8	18.054	1065.3	1083.4	0.0361	2.0901	2.1262	50.0
49.0	0.17144	0.016022	1766.2	1766.2	17.053	1065.9	1082.9	0.0341	2.0953	2.1294	49.0
48.0	0.16514	0.016021	1830.0	1830.0	16.051	1066.4	1082.5	0.0321	2.1006	2.1327	48.0
47.0	0.15904	0.016021	1896.5	1896.5	15.049	1067.0	1082.1	0.0301	2.1058	2.1360	47.0
46.0	0.15314	0.016020	1965.7	1965.7	14.047	1067.6	1081.6	0.0282	2.1111	2.1393	46.0
45.0	0.14744	0.016020	2037.7	2037.8	13.044	1068.1	1081.2	0.0262	2.1164	2.1426	45.0
44.0	0.14192	0.016019	2112.8	2112.8	12.041	1068.7	1080.7	0.0242	2.1217	2.1459	44.0
43.0	0.13659	0.016019	2191.0	2191.0	11.038	1069.3	1080.3	0.0222	2.1271	2.1493	43.0
42.0	0.13143	0.016019	2272.4	2272.4	10.035	1069.8	1079.9	0.0202	2.1325	2.1527	42.0
41.0	0.12645	0.016019	2357.3	2357.3	9.031	1070.4	1079.4	0.0182	2.1378	2.1560	41.0
40.0	0.12163	0.016019	2445.8	2445.8	8.027	1071.0	1079.0	0.0162	2.1432	2.1594	40.0
39.0	0.11698	0.016019	2538.0	2538.0	7.023	1071.5	1078.5	0.0142	2.1487	2.1629	39.0
38.0	0.11249	0.016019	2634.1	2634.2	6.018	1072.1	1078.1	0.0122	2.1541	2.1663	38.0
37.0	0.10815	0.016019	2734.4	2734.4	5.013	1072.7	1077.7	0.0101	2.1596	2.1697	37.0
36.0	0.10395	0.016020	2839.0	2839.0	4.008	1073.2	1077.2	0.0081	2.1651	2.1732	36.0
35.0	0.09991	0.016020	2948.1	2948.1	3.002	1073.8	1076.8	0.0061	2.1706	2.1767	35.0
34.0	0.09600	0.016021	3061.9	3061.9	1.996	1074.4	1076.4	0.0041	2.1762	2.1802	34.0
33.0	0.09223	0.016021	3180.7	3180.7	0.989	1074.9	1075.9	0.0020	2.1817	2.1837	33.0
32.018	0.08865	0.016022	3302.4	3302.4	0.0003	1075.5	1075.5	0.0000	2.1872	2.1872	32.018
*32.0	0.08859	0.016022	3304.7	3304.7	−0.0179	1075.5	1075.5	−0.0000	2.1873	2.1873	32.0

*The states here shown are metastable

E-21

STEAM TABLES (Continued)

Specific Heat at constant pressure of Steam and of Water

Temp. F							c_p, Btu/lbm $\times$ F									Temp. F
Press., psia	1	1.5	2	3	4	6	8	10	15	20	30	40	60	80	100	Press., psia
Sat. Water	0.998	0.998	0.999	1.000	1.000	1.002	1.003	1.004	1.007	1.010	1.014	1.019	1.026	1.033	1.039	Sat. Water
Sat. Steam	0.450	0.452	0.454	0.458	0.461	0.466	0.471	0.475	0.485	0.493	0.508	0.521	0.543	0.564	0.582	Sat. Steam
1500	0.559	0.559	0.559	0.559	0.559	0.559	0.559	0.559	0.559	0.559	0.560	0.560	0.560	0.561	0.561	1500
1480	0.557	0.557	0.557	0.557	0.557	0.557	0.557	0.558	0.558	0.558	0.558	0.558	0.559	0.559	0.559	1480
1460	0.556	0.556	0.556	0.556	0.556	0.556	0.556	0.556	0.556	0.556	0.556	0.557	0.557	0.557	0.558	1460
1440	0.554	0.554	0.554	0.554	0.554	0.554	0.554	0.554	0.554	0.554	0.555	0.555	0.555	0.556	0.556	1440
1420	0.552	0.552	0.552	0.552	0.552	0.552	0.553	0.553	0.553	0.553	0.553	0.553	0.554	0.554	0.555	1420
1400	0.551	0.551	0.551	0.551	0.551	0.551	0.551	0.551	0.551	0.551	0.551	0.552	0.552	0.553	0.553	1400
1380	0.549	0.549	0.549	0.549	0.549	0.549	0.549	0.549	0.549	0.549	0.550	0.550	0.550	0.551	0.551	1380
1360	0.547	0.547	0.547	0.547	0.547	0.547	0.547	0.547	0.548	0.548	0.548	0.548	0.549	0.549	0.550	1360
1340	0.546	0.546	0.546	0.546	0.546	0.546	0.546	0.546	0.546	0.546	0.546	0.547	0.547	0.548	0.548	1340
1320	0.544	0.544	0.544	0.544	0.544	0.544	0.544	0.544	0.544	0.544	0.545	0.545	0.545	0.546	0.546	1320
1300	0.542	0.542	0.542	0.542	0.542	0.542	0.542	0.542	0.542	0.543	0.543	0.543	0.544	0.544	0.545	1300
1280	0.540	0.540	0.540	0.540	0.540	0.540	0.540	0.541	0.541	0.541	0.541	0.541	0.542	0.543	0.543	1280
1260	0.538	0.539	0.539	0.539	0.539	0.539	0.539	0.539	0.539	0.539	0.539	0.540	0.540	0.541	0.541	1260
1240	0.537	0.537	0.537	0.537	0.537	0.537	0.537	0.537	0.537	0.537	0.538	0.538	0.539	0.539	0.540	1240
1220	0.535	0.535	0.535	0.535	0.535	0.535	0.535	0.535	0.535	0.536	0.536	0.536	0.537	0.537	0.538	1220
1200	0.533	0.533	0.533	0.533	0.533	0.533	0.533	0.533	0.534	0.534	0.534	0.534	0.535	0.536	0.536	1200
1180	0.531	0.531	0.531	0.531	0.531	0.531	0.532	0.532	0.532	0.532	0.532	0.533	0.533	0.534	0.535	1180
1160	0.529	0.529	0.530	0.530	0.530	0.530	0.530	0.530	0.530	0.530	0.530	0.531	0.532	0.532	0.533	1160
1140	0.528	0.528	0.528	0.528	0.528	0.528	0.528	0.528	0.528	0.528	0.529	0.529	0.530	0.531	0.531	1140
1120	0.526	0.526	0.526	0.526	0.526	0.526	0.526	0.526	0.526	0.527	0.527	0.527	0.528	0.529	0.530	1120
1100	0.524	0.524	0.524	0.524	0.524	0.524	0.524	0.524	0.525	0.525	0.525	0.526	0.526	0.527	0.528	1100
1080	0.522	0.522	0.522	0.522	0.522	0.522	0.522	0.523	0.523	0.523	0.523	0.524	0.525	0.525	0.526	1080
1060	0.520	0.520	0.520	0.520	0.520	0.521	0.521	0.521	0.521	0.521	0.522	0.522	0.523	0.524	0.524	1060
1040	0.518	0.519	0.519	0.519	0.519	0.519	0.519	0.519	0.519	0.519	0.520	0.520	0.521	0.522	0.523	1040
1020	0.517	0.517	0.517	0.517	0.517	0.517	0.517	0.517	0.517	0.518	0.518	0.518	0.519	0.520	0.521	1020
1000	0.515	0.515	0.515	0.515	0.515	0.515	0.515	0.515	0.515	0.516	0.516	0.517	0.518	0.519	0.519	1000
980	0.513	0.513	0.513	0.513	0.513	0.513	0.513	0.513	0.514	0.514	0.514	0.515	0.516	0.517	0.518	980
960	0.511	0.511	0.511	0.511	0.511	0.511	0.512	0.512	0.512	0.512	0.513	0.513	0.514	0.515	0.516	960
940	0.509	0.509	0.509	0.509	0.509	0.510	0.510	0.510	0.510	0.510	0.511	0.511	0.512	0.514	0.515	940
920	0.507	0.508	0.508	0.508	0.508	0.508	0.508	0.508	0.508	0.509	0.509	0.510	0.511	0.512	0.513	920
900	0.506	0.506	0.506	0.506	0.506	0.506	0.506	0.506	0.506	0.507	0.507	0.508	0.509	0.510	0.512	900
880	0.504	0.504	0.504	0.504	0.504	0.504	0.504	0.504	0.505	0.505	0.506	0.506	0.508	0.509	0.510	880
860	0.502	0.502	0.502	0.502	0.502	0.502	0.503	0.503	0.503	0.503	0.504	0.505	0.506	0.507	0.509	860
840	0.500	0.500	0.500	0.500	0.500	0.501	0.501	0.501	0.501	0.502	0.502	0.503	0.504	0.506	0.507	840
820	0.498	0.498	0.499	0.499	0.499	0.499	0.499	0.499	0.499	0.500	0.501	0.501	0.503	0.504	0.506	820
800	0.497	0.497	0.497	0.497	0.497	0.497	0.497	0.497	0.498	0.498	0.499	0.500	0.501	0.503	0.505	800
780	0.495	0.495	0.495	0.495	0.495	0.495	0.495	0.496	0.496	0.496	0.497	0.498	0.500	0.502	0.503	780
760	0.493	0.493	0.493	0.493	0.493	0.494	0.494	0.494	0.494	0.495	0.496	0.497	0.499	0.500	0.502	760
740	0.491	0.491	0.491	0.492	0.492	0.492	0.492	0.492	0.493	0.493	0.494	0.495	0.497	0.499	0.501	740
720	0.490	0.490	0.490	0.490	0.490	0.490	0.490	0.491	0.491	0.492	0.493	0.494	0.496	0.498	0.500	720
700	0.488	0.488	0.488	0.488	0.488	0.488	0.489	0.489	0.490	0.490	0.491	0.492	0.495	0.497	0.500	700
680	0.486	0.486	0.486	0.486	0.487	0.487	0.487	0.487	0.488	0.489	0.490	0.491	0.494	0.496	0.499	680
660	0.484	0.485	0.485	0.485	0.485	0.485	0.485	0.486	0.486	0.487	0.489	0.490	0.493	0.496	0.499	660
640	0.483	0.483	0.483	0.483	0.483	0.484	0.484	0.484	0.485	0.486	0.487	0.489	0.492	0.495	0.499	640
620	0.481	0.481	0.481	0.481	0.482	0.482	0.482	0.483	0.483	0.484	0.486	0.488	0.491	0.495	0.499	620
600	0.479	0.480	0.480	0.480	0.480	0.480	0.481	0.481	0.482	0.483	0.485	0.487	0.491	0.495	0.499	600
580	0.478	0.478	0.478	0.478	0.478	0.479	0.479	0.480	0.481	0.482	0.484	0.486	0.491	0.495	0.500	580
560	0.476	0.476	0.476	0.477	0.477	0.477	0.478	0.478	0.479	0.481	0.483	0.485	0.490	0.496	0.501	560
540	0.475	0.475	0.475	0.475	0.475	0.476	0.476	0.477	0.478	0.480	0.482	0.485	0.491	0.497	0.503	540
520	0.473	0.473	0.473	0.474	0.474	0.475	0.475	0.476	0.477	0.479	0.482	0.485	0.491	0.498	0.505	520
500	0.472	0.472	0.472	0.472	0.473	0.473	0.474	0.475	0.476	0.478	0.481	0.485	0.492	0.500	0.508	500
480	0.470	0.470	0.470	0.471	0.471	0.472	0.473	0.473	0.475	0.477	0.481	0.485	0.493	0.502	0.511	480
460	0.469	0.469	0.469	0.469	0.470	0.471	0.472	0.472	0.475	0.477	0.481	0.486	0.495	0.505	0.516	460
440	0.467	0.467	0.468	0.468	0.469	0.470	0.470	0.471	0.474	0.476	0.481	0.487	0.498	0.509	0.522	440
420	0.466	0.466	0.466	0.467	0.467	0.468	0.470	0.471	0.473	0.476	0.482	0.488	0.501	0.514	0.528	420
400	0.464	0.465	0.465	0.466	0.466	0.467	0.469	0.470	0.473	0.476	0.483	0.490	0.504	0.520	0.536	400
380	0.463	0.463	0.464	0.464	0.465	0.466	0.468	0.469	0.473	0.477	0.484	0.492	0.509	0.527	0.546	380
360	0.462	0.462	0.462	0.463	0.464	0.466	0.467	0.469	0.473	0.477	0.486	0.495	0.515	0.536	0.558	360
340	0.460	0.461	0.461	0.462	0.463	0.465	0.467	0.469	0.473	0.478	0.488	0.499	0.521	0.546	_0.572_	340
320	0.459	0.460	0.460	0.461	0.462	0.464	0.467	0.469	0.474	0.480	0.491	0.504	0.530	_0.558_	_1.036_	320
300	0.458	0.459	0.459	0.460	0.462	0.464	0.466	0.469	0.475	0.482	0.495	0.509	_0.539_	1.029	1.029	300
280	0.457	0.458	0.458	0.460	0.461	0.464	0.467	0.469	0.477	0.484	0.499	_0.516_	1.022	1.022	1.022	280
260	0.456	0.457	0.457	0.459	0.461	0.464	0.467	0.470	0.478	0.487	_0.505_	1.017	1.017	1.017	1.016	260
240	0.455	0.456	0.457	0.458	0.460	0.464	0.468	0.471	0.481	_0.491_	1.012	1.012	1.012	1.012	1.012	240
220	0.454	0.455	0.456	0.458	0.460	0.464	0.468	0.473	_0.484_	1.008	1.008	1.008	1.008	1.008	1.008	220
200	0.453	0.454	0.455	0.458	0.460	0.465	_0.470_	0.475	1.005	1.005	1.005	1.005	1.005	1.005	1.005	200
180	0.452	0.454	0.455	0.458	0.460	_0.466_	_1.003_	1.003	1.003	1.003	1.003	1.003	1.003	1.002	1.002	180
160	0.451	0.453	0.455	_0.458_	_0.461_	1.001	1.001	1.001	1.001	1.001	1.001	1.001	1.001	1.001	1.001	160
140	0.451	0.453	_0.454_	1.000	1.000	1.000	1.000	1.000	0.999	0.999	0.999	0.999	0.999	0.999	0.999	140
120	_0.450_	_0.452_	0.999	0.999	0.999	0.999	0.999	0.999	0.999	0.999	0.999	0.998	0.998	0.998	0.998	120
100	0.998	0.998	0.998	0.998	0.998	0.998	0.998	0.998	0.998	0.998	0.998	0.998	0.998	0.998	0.998	100
80	0.998	0.998	0.998	0.998	0.998	0.998	0.998	0.998	0.998	0.998	0.998	0.998	0.998	0.998	0.998	80
60	1.000	1.000	1.000	1.000	1.000	1.000	1.000	1.000	1.000	1.000	1.000	1.000	0.999	0.999	0.999	60
40	1.004	1.004	1.004	1.004	1.004	1.004	1.004	1.004	1.004	1.004	1.004	1.004	1.004	1.004	1.003	40
32	1.007	1.007	1.007	1.007	1.007	1.007	1.007	1.007	1.007	1.007	1.007	1.007	1.007	1.007	1.006	32

STEAM TABLES (Continued)

Specific Heat at constant pressure of Steam and of Water

Temp. F	150	200	300	400	600	800	1000	1500	2000	3000	4000	6000	8000	10000	15000	Temp. F
Press., psia																Press., psia
Sat. Water	1.054	1.067	1.093	1.118	1.168	1.224	1.286	1.492	1.841	7.646	—	—	—	—	—	Sat. Water
Sat. Steam	0.624	0.661	0.729	0.792	0.915	1.046	1.191	1.667	2.557	13.66	—	—	—	—	—	Sat. Steam
1500	0.562	0.563	0.565	0.567	0.571	0.576	0.580	0.590	0.601	0.623	0.645	0.691	0.737	0.780	0.868	1500
1480	0.561	0.562	0.564	0.566	0.570	0.575	0.579	0.590	0.601	0.623	0.647	0.694	0.742	0.786	0.878	1480
1460	0.559	0.560	0.562	0.565	0.569	0.573	0.578	0.589	0.601	0.624	0.648	0.698	0.747	0.793	0.888	1460
1440	0.557	0.559	0.561	0.563	0.568	0.572	0.577	0.589	0.600	0.625	0.650	0.701	0.753	0.800	0.900	1440
1420	0.556	0.557	0.559	0.562	0.566	0.571	0.576	0.588	0.600	0.625	0.651	0.705	0.759	0.808	0.909	1420
1400	0.554	0.555	0.558	0.560	0.565	0.570	0.575	0.587	0.600	0.626	0.653	0.709	0.765	0.817	0.926	1400
1380	0.553	0.554	0.556	0.559	0.564	0.569	0.574	0.587	0.600	0.627	0.655	0.714	0.773	0.827	0.939	1380
1360	0.551	0.552	0.555	0.558	0.563	0.568	0.573	0.586	0.600	0.628	0.657	0.719	0.781	0.838	0.953	1360
1340	0.549	0.551	0.553	0.556	0.561	0.567	0.572	0.586	0.600	0.629	0.660	0.725	0.790	0.850	0.968	1340
1320	0.548	0.549	0.552	0.555	0.560	0.566	0.571	0.585	0.600	0.630	0.663	0.731	0.800	0.864	0.983	1320
1300	0.546	0.548	0.550	0.553	0.559	0.565	0.570	0.585	0.600	0.632	0.666	0.738	0.811	0.879	0.998	1300
1280	0.545	0.546	0.549	0.552	0.558	0.564	0.570	0.585	0.600	0.634	0.669	0.746	0.824	0.897	1.014	1280
1260	0.543	0.544	0.547	0.550	0.556	0.563	0.569	0.585	0.601	0.636	0.673	0.755	0.838	0.918	1.033	1260
1240	0.541	0.543	0.546	0.549	0.555	0.562	0.568	0.584	0.601	0.638	0.678	0.765	0.855	0.942	1.053	1240
1220	0.540	0.541	0.544	0.548	0.554	0.561	0.567	0.584	0.602	0.641	0.683	0.777	0.875	0.969	1.072	1220
1200	0.538	0.540	0.543	0.546	0.553	0.560	0.567	0.584	0.603	0.644	0.689	0.790	0.897	1.000	1.095	1200
1180	0.536	0.538	0.541	0.545	0.552	0.559	0.566	0.584	0.604	0.647	0.696	0.805	0.922	1.033	1.117	1180
1160	0.535	0.536	0.540	0.544	0.551	0.558	0.565	0.585	0.606	0.652	0.704	0.823	0.952	1.070	1.143	1160
1140	0.533	0.535	0.539	0.542	0.550	0.557	0.565	0.585	0.607	0.656	0.713	0.843	0.986	1.107	1.167	1140
1120	0.531	0.533	0.537	0.541	0.549	0.557	0.565	0.586	0.609	0.662	0.723	0.866	1.025	1.149	1.190	1120
1100	0.530	0.532	0.536	0.540	0.548	0.556	0.564	0.587	0.612	0.668	0.735	0.893	1.070	1.193	1.220	1100
1080	0.528	0.530	0.534	0.538	0.547	0.555	0.564	0.588	0.615	0.676	0.749	0.924	1.120	1.242	1.240	1080
1060	0.527	0.529	0.533	0.537	0.546	0.555	0.564	0.590	0.618	0.685	0.765	0.960	1.176	1.295	1.260	1060
1040	0.525	0.527	0.532	0.536	0.545	0.555	0.565	0.592	0.622	0.695	0.783	1.002	1.238	1.351	1.282	1040
1020	0.523	0.526	0.530	0.535	0.545	0.555	0.565	0.594	0.627	0.707	0.804	1.051	1.306	1.399	1.298	1020
1000	0.522	0.524	0.529	0.534	0.544	0.555	0.566	0.597	0.633	0.721	0.829	1.110	1.382	1.471	1.306	1000
980	0.520	0.523	0.528	0.533	0.544	0.555	0.567	0.601	0.640	0.737	0.855	1.160	1.473	1.531	1.312	980
960	0.519	0.521	0.527	0.533	0.543	0.556	0.568	0.605	0.648	0.756	0.893	1.267	1.598	1.595	1.310	960
940	0.517	0.520	0.526	0.531	0.543	0.556	0.570	0.610	0.658	0.778	0.934	1.376	1.708	1.639	1.299	940
920	0.516	0.519	0.525	0.531	0.544	0.558	0.573	0.617	0.669	0.803	0.984	1.520	1.819	1.667	1.281	920
900	0.515	0.518	0.524	0.530	0.544	0.559	0.576	0.624	0.682	0.834	1.040	1.710	1.932	1.666	1.259	900
880	0.513	0.516	0.523	0.530	0.545	0.561	0.580	0.633	0.699	0.872	1.130	1.993	2.000	1.633	1.232	880
860	0.512	0.515	0.523	0.530	0.546	0.564	0.584	0.644	0.718	0.918	1.240	2.316	2.019	1.593	1.212	860
840	0.511	0.514	0.522	0.530	0.548	0.568	0.590	0.657	0.740	0.977	1.395	2.653	1.978	1.547	1.192	840
820	0.510	0.514	0.522	0.531	0.550	0.572	0.597	0.672	0.767	1.054	1.620	2.886	1.888	1.503	1.175	820
800	0.509	0.513	0.522	0.532	0.553	0.577	0.605	0.690	0.800	1.160	1.967	2.872	1.768	1.459	1.157	800
780	0.508	0.513	0.522	0.533	0.557	0.584	0.615	0.712	0.840	1.312	2.550	2.547	1.670	1.416	1.142	780
760	0.507	0.512	0.523	0.535	0.561	0.592	0.628	0.738	0.892	1.542	4.462	2.156	1.576	1.370	1.126	760
740	0.507	0.512	0.524	0.537	0.567	0.602	0.642	0.770	0.960	1.913	8.119	1.886	1.493	1.332	1.114	740
720	0.506	0.512	0.525	0.540	0.574	0.613	0.660	0.811	1.052	2.584	3.458	1.696	1.421	1.290	1.100	720
700	0.506	0.513	0.528	0.544	0.582	0.627	0.681	0.861	1.181	6.145⊙	2.237	1.557	1.358	1.250	1.089	700
680	0.506	0.514	0.530	0.549	0.592	0.644	0.707	0.927	1.365	2.469	1.789	1.450	1.303	1.217	1.079	680
660	0.507	0.515	0.534	0.555	0.604	0.665	0.738	1.015	1.639	1.851	1.587	1.369	1.256	1.187	1.071	660
640	0.507	0.517	0.538	0.562	0.619	0.690	0.777	1.135	2.219	1.601	1.454	1.303	1.216	1.157	1.063	640
620	0.509	0.519	0.543	0.571	0.637	0.720	0.826	1.308	1.614	1.455	1.362	1.252	1.184	1.136	1.056	620
600	0.510	0.522	0.550	0.582	0.659	0.757	0.888	1.586	1.453	1.358	1.295	1.211	1.157	1.118	1.052	600
580	0.513	0.526	0.558	0.595	0.685	0.804	0.969	1.393	1.351	1.289	1.243	1.178	1.134	1.102	1.046	580
560	0.516	0.531	0.568	0.611	0.717	0.862	1.079	1.309	1.281	1.237	1.202	1.151	1.115	1.087	1.039	560
540	0.519	0.538	0.580	0.630	0.756	0.937	1.272	1.249	1.229	1.196	1.169	1.128	1.098	1.074	1.031	540
520	0.524	0.545	0.594	0.653	0.804	1.035	1.221	1.204	1.189	1.164	1.142	1.109	1.083	1.062	1.024	520
500	0.530	0.554	0.611	0.680	0.865	1.187	1.181	1.169	1.157	1.137	1.120	1.092	1.069	1.051	1.017	500
480	0.537	0.565	0.632	0.714	1.159	1.154	1.150	1.140	1.131	1.115	1.101	1.077	1.057	1.041	1.010	480
460	0.545	0.578	0.657	0.755	1.132	1.128	1.125	1.117	1.110	1.096	1.084	1.064	1.047	1.033	1.004	460
440	0.556	0.594	0.687	1.113	1.110	1.107	1.104	1.098	1.092	1.080	1.070	1.052	1.038	1.025	0.999	440
420	0.568	0.614	0.724	1.094	1.091	1.089	1.087	1.081	1.076	1.067	1.058	1.042	1.029	1.018	0.994	420
400	0.583	0.636	1.079	1.078	1.076	1.074	1.072	1.067	1.063	1.055	1.047	1.034	1.022	1.011	0.990	400
380	0.601	1.066	1.065	1.065	1.063	1.061	1.059	1.056	1.052	1.044	1.038	1.026	1.015	1.006	0.986	380
360	0.622	1.054	1.054	1.053	1.052	1.050	1.049	1.045	1.042	1.036	1.030	1.019	1.009	1.001	0.982	360
340	1.045	1.044	1.044	1.043	1.042	1.040	1.039	1.036	1.033	1.028	1.022	1.013	1.004	0.996	0.979	340
320	1.036	1.036	1.035	1.034	1.033	1.032	1.031	1.028	1.026	1.021	1.016	1.007	0.999	0.992	0.976	320
300	1.028	1.028	1.028	1.027	1.026	1.025	1.024	1.022	1.019	1.015	1.010	1.002	0.995	0.988	0.973	300
280	1.022	1.022	1.021	1.021	1.020	1.019	1.018	1.016	1.014	1.009	1.005	0.998	0.991	0.985	0.971	280
260	1.016	1.016	1.016	1.015	1.014	1.013	1.013	1.011	1.009	1.005	1.001	0.994	0.988	0.982	0.968	260
240	1.012	1.011	1.011	1.011	1.010	1.009	1.008	1.006	1.004	1.001	0.997	0.991	0.985	0.979	0.966	240
220	1.008	1.008	1.007	1.007	1.006	1.005	1.005	1.003	1.001	0.998	0.994	0.988	0.982	0.977	0.964	220
200	1.005	1.004	1.004	1.004	1.003	1.002	1.002	1.000	0.998	0.995	0.992	0.986	0.980	0.975	0.963	200
180	1.002	1.002	1.002	1.001	1.001	1.000	0.999	0.998	0.996	0.993	0.989	0.983	0.978	0.973	0.961	180
160	1.000	1.000	1.000	0.999	0.999	0.998	0.997	0.996	0.994	0.991	0.987	0.981	0.976	0.971	0.959	160
140	0.999	0.999	0.998	0.998	0.997	0.997	0.996	0.994	0.992	0.989	0.986	0.980	0.974	0.969	0.958	140
120	0.998	0.998	0.997	0.997	0.996	0.996	0.995	0.993	0.991	0.988	0.984	0.978	0.972	0.967	0.957	120
100	0.997	0.997	0.997	0.996	0.996	0.995	0.994	0.992	0.990	0.986	0.983	0.976	0.970	0.965	0.955	100
80	0.998	0.997	0.997	0.996	0.995	0.994	0.994	0.991	0.989	0.985	0.981	0.974	0.968	0.962	0.951	80
60	0.999	0.999	0.998	0.997	0.996	0.995	0.994	0.991	0.989	0.984	0.979	0.970	0.963	0.956	0.942	60
40	1.003	1.003	1.002	1.001	1.000	0.998	0.997	0.993	0.989	0.983	0.976	0.965	0.954	0.945	0.920	40
32	1.006	1.006	1.005	1.004	1.002	1.000	0.999	0.994	0.990	0.983	0.975	0.962	0.949	0.937	0.904	32

⊙Critical point.

Thermal Conductivity of Steam and Water

Temp. F Press., psia	1	2	5	10	20	50	100	200	500	1000	2000	5000	7500
Sat. Water	364.0	373.1	383.8	390.4	395.2	397.4	394.7	386.2	361.7	327.6	271.8	—	—
Sat. Steam	11.6	12.2	13.0	13.8	14.8	16.6	18.4	21.1	27.2	36.5	61.3	—	—
1500	63.7	63.7	63.7	63.7	63.7	63.8	64.0	64.3	65.4	67.1	70.7	82.0	92.2
1450	61.4	61.4	61.5	61.5	61.5	61.6	61.8	62.1	63.2	64.9	68.5	80.1	90.6
1400	59.2	59.2	59.2	59.2	59.3	59.4	59.6	59.9	60.9	62.7	66.3	78.2	89.2
1350	57.0	57.0	57.0	57.0	57.1	57.2	57.3	57.7	58.7	60.5	64.2	76.3	87.9
1300	54.8	54.8	54.8	54.8	54.8	54.9	55.1	55.5	56.5	58.3	62.0	74.6	86.9
1250	52.6	52.6	52.6	52.6	52.6	52.7	52.9	53.2	54.3	56.1	59.9	73.0	86.3
1200	50.4	50.4	50.4	50.4	50.4	50.5	50.7	51.0	52.1	53.9	57.8	71.6	86.2
1150	48.2	48.2	48.2	48.2	48.2	48.3	48.5	48.9	49.9	51.8	55.7	70.5	87.0
1100	46.0	46.0	46.0	46.0	46.1	46.2	46.3	46.7	47.8	49.6	53.7	69.8	89.0
1050	43.9	43.9	43.9	43.9	43.9	44.0	44.2	44.6	45.6	47.5	51.8	69.7	93.4
1000	41.7	41.7	41.8	41.8	41.8	41.9	42.1	42.4	43.5	45.5	50.0	70.7	102.9
950	39.6	39.6	39.7	39.7	39.7	39.8	40.0	40.3	41.4	43.5	48.3	73.5	115.5
900	37.6	37.6	37.6	37.6	37.6	37.7	37.9	38.3	39.4	41.5	46.8	80.2	138.7
850	35.5	35.6	35.6	35.6	35.6	35.7	35.9	36.3	37.4	39.7	45.6	96.7	178.8
800	33.6	33.6	33.6	33.6	33.6	33.7	33.9	34.3	35.5	37.9	44.9	129.6	223.2
750	31.6	31.6	31.6	31.6	31.7	31.8	32.0	32.3	33.6	36.3	45.2	202.5	258.3
700	29.7	29.7	29.7	29.7	29.8	29.9	30.1	30.4	31.8	35.0	47.5⊙	262.8	295.1
650	27.8	27.8	27.9	27.9	27.9	28.0	28.2	28.6	30.1	34.1	55.7	304.3	326.7
600	26.0	26.0	26.1	26.1	26.1	26.2	26.4	26.9	28.7	34.1	301.9	333.7	349.3
550	24.3	24.3	24.3	24.3	24.4	24.5	24.7	25.2	27.5	36.1	333.7	356.1	368.0
500	22.6	22.6	22.6	22.6	22.7	22.8	23.0	23.6	26.9	350.8	357.4	373.8	383.6
450	21.0	21.0	21.0	21.0	21.0	21.2	21.4	22.3	368.1	370.6	375.3	387.9	396.5
400	19.4	19.4	19.4	19.4	19.5	19.6	20.0	21.3	383.0	384.9	388.5	398.6	406.4
350	17.9	17.9	17.9	17.9	18.0	18.2	18.8	392.0	392.9	394.4	397.4	406.1	413.2
300	16.5	16.5	16.5	16.5	16.6	16.9	396.9	397.2	398.0	399.3	402.0	409.9	416.4
250	15.1	15.1	15.1	15.2	15.3	396.9	397.0	397.3	398.1	399.4	402.1	409.7	415.8
200	13.8	13.8	13.9	14.0	391.6	391.6	391.8	392.1	393.0	394.4	397.2	404.9	410.6
150	12.7	12.7	380.5	380.5	380.6	380.7	380.8	381.1	382.1	383.7	386.7	394.7	400.3
100	363.3	363.3	363.3	363.3	363.3	363.4	363.6	363.9	365.0	366.6	369.8	378.3	384.1
50	339.1	339.1	339.1	339.1	339.2	339.3	339.4	339.8	340.8	342.5	345.7	354.6	361.0
32	328.6	328.6	328.6	328.6	328.6	328.7	328.9	329.2	330.3	331.9	335.1	344.1	350.8

⊙ Critical point.

THERMODYNAMIC PROPERTIES

Ammonia (NH₃)

Temp. (°F)	Abs. press. sat. vap lb/in.²	Abs. press. sat. vap kg/cm²	Heat content abv. −40°F BTU/lb Liq.	Heat content abv. −40°F BTU/lb Vap.	Ht. of vaporiz. BTU/lb	Heat content abv. −40°C g-cal/g Liq.	Heat content abv. −40°C g-cal/g Vap.	Ht. of vaporiz. g-cal/g	Spec. vol. sat. vap. ft³/lb	Spec. vol. sat. vap. m³/kg	Density sat. vap. lb/ft³	Density sat. vap. kg/m³	Dens. liq. lb/ft³	Entropy from −40°F BTU/lb/°F Liq.	Entropy from −40°F BTU/lb/°F Vap.	Temp. (°C)
−60	5.55	0.390	−21.2	589.6	610.8	−11.8	327.6	339.3	44.73	2.792	0.02235	0.3580	43.91	−0.0517	1.4769	−51.11
−58	5.93	0.417	−19.1	590.4	609.5	−10.6	328.0	338.6	42.05	2.625	0.02378	.3809		−0.0464	1.4713	−50.00
−56	6.33	0.445	−17.0	591.2	608.2	−9.44	328.4	337.9	39.56	2.470	0.02528	.4049		−0.0412	1.4658	−48.89
−54	6.75	0.475	−14.8	592.1	606.9	−8.22	328.9	337.2	37.24	2.325	0.02685	.4301		−0.0360	1.4604	−47.78
−52	7.20	0.506	−12.7	592.9	605.6	−7.06	329.4	336.4	35.09	2.191	0.02850	.4565		−0.0307	1.4551	−46.67
−50	7.67	0.539	−10.6	593.7	604.3	−5.89	329.8	335.7	33.08	2.065	0.03023	0.4842	43.49	−0.0256	1.4497	−45.56
−48	8.16	0.574	−8.5	594.4	602.9	−4.7	330.2	334.9	31.20	1.948	0.03205	.5134		−0.0204	1.4445	−44.44
−46	8.68	0.610	−6.4	595.2	601.6	−3.6	330.7	334.2	29.45	1.839	0.03395	.5438		−0.0153	1.4393	−43.33
−44	9.23	0.649	−4.3	596.0	600.3	−2.4	331.1	333.5	27.82	1.737	0.03595	.5758		−0.0102	1.4342	−42.22
−42	9.81	0.690	−2.1	596.8	598.9	−1.2	331.6	332.7	26.29	1.641	0.03804	.6093		−0.0051	1.4292	−41.11
−40	10.41	0.7319	0.0	597.6	597.6	0.0	332.0	332.0	24.86	1.552	0.04022	0.6442	43.08	0.0000	1.4242	−40.00
−38	11.04	0.7762	2.1	598.3	596.2	1.2	332.4	331.2	23.53	1.469	0.04251	.6809		0.0051	1.4193	−38.89
−36	11.71	0.8233	4.3	599.1	594.8	2.4	332.8	330.4	22.27	1.390	0.04489	.7190		0.0101	1.4144	−37.78
−34	12.41	0.8725	6.4	599.9	593.5	3.6	333.3	329.7	21.10	1.317	0.04739	.7591		0.0151	1.4096	−36.67
−32	13.14	0.9238	8.5	600.6	592.1	4.7	333.7	328.9	20.00	1.249	0.04999	.8007		0.0201	1.4048	−35.56
−30	13.90	0.9773	10.7	601.4	590.7	5.94	334.1	328.2	18.97	1.184	0.05271	0.8443	42.65	0.0250	1.4001	−34.44
−28	14.71	1.034	12.8	602.1	589.3	7.11	334.5	327.4	18.00	1.124	0.05555	.8898		0.0300	1.3955	−33.33
−26	15.55	1.093	14.9	602.8	587.9	8.28	334.9	326.6	17.09	1.067	0.05850	.9371		0.0350	1.3909	−32.22
−24	16.42	1.154	17.1	603.6	586.5	9.50	335.3	325.8	16.24	1.014	0.06158	.9864		0.0399	1.3863	−31.11
−22	17.34	1.219	19.2	604.3	585.1	10.7	335.7	325.1	15.43	0.9633	0.06479	1.038		0.0448	1.3818	−30.00
−20	18.30	1.287	21.4	605.0	583.6	11.9	336.1	324.2	14.68	0.9164	0.06813	1.091	42.22	0.0497	1.3774	−28.89
−18	19.30	1.357	23.5	605.7	582.2	13.1	336.5	323.4	13.97	0.8721	0.07161	1.147		0.0545	1.3729	−27.78
−16	20.34	1.430	25.6	606.4	580.8	14.2	336.9	322.7	13.29	0.8297	0.07522	1.205		0.0594	1.3686	−26.67
−14	21.43	1.507	27.8	607.1	579.3	15.4	337.3	321.8	12.66	0.7903	0.07898	1.265		0.0642	1.3643	−25.56
−12	22.56	1.586	30.0	607.8	577.8	16.7	337.7	321.0	12.06	0.7529	0.08289	1.328		0.0690	1.3600	−24.44
−10	23.74	1.669	32.1	608.5	576.4	17.8	338.1	320.2	11.50	0.7179	0.08695	1.393	41.78	0.0738	1.3558	−23.33
−8	24.97	1.756	34.3	609.2	574.9	19.1	338.4	319.4	10.97	0.6848	0.09117	1.460		0.0786	1.3516	−22.22
−6	26.26	1.846	36.4	609.8	573.4	20.2	338.8	318.6	10.47	0.6536	0.09555	1.531		0.0833	1.3474	−21.11
−4	27.59	1.940	38.6	610.5	571.9	21.4	339.2	317.7	9.991	0.6237	0.1001	1.603		0.0880	1.3433	−20.00
−2	28.98	2.037	40.7	611.1	570.4	22.6	339.5	316.9	9.541	0.5956	0.1048	1.679		0.0928	1.3393	−18.89
0	30.42	2.139	42.9	611.8	568.9	23.8	339.9	316.1	9.116	0.5691	0.1097	1.757	41.34	0.0975	1.3352	−17.78
2	31.92	2.244	45.1	612.4	567.3	25.1	340.2	315.2	8.714	0.5440	0.1148	1.839		0.1022	1.3312	−16.67
4	33.47	2.353	47.2	613.0	565.8	26.2	340.6	314.3	8.333	0.5202	0.1200	1.922		0.1069	1.3273	−15.56
6	35.00	2.467	49.4	613.6	564.2	27.4	340.9	313.4	7.971	0.4976	0.1254	2.009		0.1115	1.3234	−14.44

THERMODYNAMIC PROPERTIES (continued)

Temp. (°F)	Abs. press. sat. vap lb/in.²	kg/cm²	Heat content abv. −40°F BTU/lb Liq.	Vap.	Ht. of vaporiz. BTU/lb	Heat content abv. −40°C g-cal/g Liq.	Vap.	Ht. of vaporiz. g-cal/g	Spec. vol. sat. vap. ft³/lb	m³/kg	Density sat. vap. lb/ft³	kg/m³	Dens. liq. lb/ft³	Entropy from −40°F BTU/lb/°F Liq.	Vap.	Temp. (°C)
8	36.77	2.585	51.6	614.3	562.7	28.7	341.3	312.6	7.629	0.4763	0.1311	2.100		0.1162	1.3195	−13.33
10	38.51	2.708	53.8	614.9	561.1	29.9	341.6	311.7	7.304	0.4560	0.1369	2.193	40.89	0.1208	1.3157	−12.22
12	40.31	2.834	56.0	615.5	559.5	31.1	341.9	310.8	6.996	0.4367	0.1429	2.289		0.1254	1.3118	−11.11
14	42.18	2.966	58.2	616.1	557.9	32.3	342.3	309.9	6.703	0.4185	0.1492	2.390		0.1300	1.3081	−10.00
16	44.12	3.102	60.3	616.6	556.3	33.5	342.6	309.1	6.425	0.4011	0.1556	2.492		0.1346	1.3043	−8.89
18	46.13	3.243	62.5	617.2	554.7	34.7	342.9	308.2	6.161	0.3846	0.1623	2.600		0.1392	1.3006	−7.78
20	48.21	3.390	64.7	617.8	553.1	35.9	343.2	307.3	5.910	0.3690	0.1692	2.710	40.43	0.1437	1.2969	−6.67
22	50.36	3.541	66.9	618.3	551.4	37.2	343.5	306.3	5.671	0.3540	0.1763	2.824		0.1483	1.2933	−5.56
24	52.59	3.697	69.1	618.9	549.8	38.4	343.8	305.4	5.443	0.3398	0.1837	2.943		0.1528	1.2897	−4.44
26	54.90	3.860	71.3	619.4	548.1	39.6	344.1	304.5	5.227	0.3263	0.1913	3.064		0.1573	1.2861	−3.33
28	57.28	4.027	73.5	619.9	546.4	40.8	344.4	303.6	5.021	0.3135	0.1992	3.191		0.1618	1.2825	−2.22
30	59.74	4.200	75.7	620.5	544.8	42.1	344.7	302.7	4.825	0.3012	0.2073	3.321	39.96	0.1663	1.2790	−1.11
32	62.29	4.379	77.9	621.0	543.1	43.3	345.0	301.7	4.637	0.2895	0.2156	3.453		0.1708	1.2755	0.00
34	64.91	4.564	80.1	621.5	541.4	44.5	345.3	300.8	4.459	0.2784	0.2243	3.593		0.1753	1.2721	1.11
36	67.63	4.755	82.3	622.0	539.7	45.7	345.6	299.8	4.289	0.2678	0.2332	3.735		0.1797	1.2686	2.22
38	70.43	4.952	84.6	622.5	537.9	47.0	345.8	298.8	4.126	0.2576	0.2423	3.881		0.1841	1.2652	3.33
40	73.32	5.155	86.8	623.0	536.2	48.2	346.1	297.9	3.971	0.2479	0.2518	4.003	39.49	0.1885	1.2618	4.44
42	76.31	5.365	89.0	623.4	534.4	49.4	346.3	296.9	3.823	0.2387	0.2616	4.190		0.1930	1.2585	5.56
44	79.38	5.581	91.2	623.9	532.7	50.7	346.6	295.9	3.682	0.2299	0.2716	4.350		0.1974	1.2552	6.67
46	82.55	5.804	93.5	624.4	530.9	51.9	346.9	294.9	3.547	0.2214	0.2819	4.515		0.2018	1.2519	7.78
48	85.82	6.034	95.7	624.8	529.1	53.2	347.1	293.9	3.418	0.2134	0.2926	4.687		0.2062	1.2486	8.89
50	89.19	6.271	97.9	625.2	527.3	54.4	347.3	292.9	3.294	0.2056	0.3036	4.863	39.00	0.2105	1.2453	10.00
52	92.66	6.515	100.2	625.7	525.5	55.67	347.6	291.9	3.176	0.1983	0.3149	5.044		0.2149	1.2421	11.11
54	96.23	6.766	102.4	626.1	523.7	56.89	347.8	290.9	3.063	0.1912	0.3265	5.230		0.2192	1.2389	12.22
56	99.91	7.024	104.7	626.5	521.8	58.17	348.1	289.9	2.954	0.1844	0.3385	5.422		0.2236	1.2357	13.33
58	103.7	7.291	106.9	626.9	520.0	59.39	348.3	288.9	2.851	0.1780	0.3508	5.619		0.2279	1.2325	14.44
60	107.6	7.565	109.2	627.3	518.1	60.67	348.5	287.8	2.751	0.1717	0.3635	5.823	38.50	0.2322	1.2294	15.56
62	111.6	7.840	111.3	627.7	316.2	61.94	348.7	286.8	2.656	0.1658	0.3765	6.031		0.2365	1.2262	16.67
64	115.7	8.135	113.7	628.0	514.3	63.17	348.9	285.7	2.565	0.1601	0.3899	6.245		0.2408	1.2231	17.78
66	120.0	8.437	116.0	628.4	512.4	64.44	349.1	284.7	2.477	0.1546	0.4037	6.466		0.2451	1.2201	18.89
68	124.3	8.739	118.3	628.8	510.5	65.72	349.3	283.6	2.393	0.1494	0.4179	6.694		0.2494	1.2170	20.00
70	128.8	9.056	120.5	629.1	508.6	66.94	349.5	282.6	2.312	0.1443	0.4325	6.928	38.00	0.2537	1.2140	21.11
72	133.4	9.379	122.8	629.4	506.6	68.22	349.7	281.4	2.235	0.1395	0.4474	7.166		0.2579	1.2110	22.22
74	138.1	9.709	125.1	629.8	504.7	69.50	349.9	280.4	2.161	0.1349	0.4628	7.413		0.2622	1.2080	23.33
76	143.0	10.05	127.4	630.1	502.7	70.78	350.1	279.3	2.089	0.1304	0.4786	7.666		0.2664	1.2050	24.44
78	147.9	10.40	129.7	630.4	500.7	72.06	350.2	278.2	2.021	0.1262	0.4949	7.927		0.2706	1.2020	25.56
80	153.0	10.76	132.0	630.7	498.7	73.33	350.4	277.1	1.955	0.1220	0.5115	8.193	37.48	0.2749	1.1991	26.67
82	158.3	11.13	134.3	631.0	496.7	74.61	350.6	275.9	1.892	0.1181	0.5287	8.469		0.2791	1.1962	27.78
84	163.7	11.51	136.6	631.3	494.7	75.89	350.7	274.8	1.831	0.1143	0.5462	8.749		0.2833	1.1933	28.89
86	169.2	11.90	138.9	631.5	492.6	77.17	350.8	273.7	1.772	0.1106	0.5643	9.039		0.2875	1.1904	30.00
88	174.8	12.29	141.2	631.8	490.6	78.44	351.0	272.6	1.716	0.1071	0.5828	9.335		0.2917	1.1875	31.11
90	180.6	12.70	143.5	632.0	488.5	79.72	351.1	271.4	1.661	0.1037	0.6019	9.641	36.95	0.2958	1.1846	32.22
92	186.6	13.12	145.8	632.2	486.4	81.00	351.2	270.2	1.609	0.1004	0.6214	9.954		0.3000	1.1818	33.33
94	192.7	13.55	148.2	632.5	484.3	82.33	351.4	269.1	1.559	0.09733	0.6415	10.28		0.3041	1.1789	34.44
96	198.9	13.98	150.5	632.6	482.1	83.61	351.4	267.8	1.510	0.09427	0.6620	10.60		0.3083	1.1761	35.56
98	205.3	14.43	152.9	632.9	480.0	84.94	351.6	266.7	1.464	0.09140	0.6832	10.94		0.3125	1.1733	36.67
100	211.9	14.90	155.2	633.0	477.8	86.22	351.7	265.4	1.419	0.08859	0.7048	11.29	36.40	0.3166	1.1705	37.78
102	218.6	15.37	157.6	633.2	475.6	87.56	351.8	264.2	1.375	0.08584	0.7270	11.65		0.3207	1.1677	38.89
104	225.4	15.85	159.9	633.4	473.5	88.83	351.9	263.1	1.334	0.08328	0.7498	12.01		0.3248	1.1649	40.00
106	232.5	16.35	162.3	633.5	471.2	90.17	351.9	261.8	1.293	0.08072	0.7732	12.39		0.3289	1.1621	41.11
108	239.7	16.85	164.6	633.6	469.0	91.44	352.0	260.6	1.254	0.07829	0.7972	12.77		0.3330	1.1593	42.22
110	247.0	17.37	167.0	633.7	466.7	92.78	352.1	259.3	1.217	0.07598	0.8219	13.17	35.84	0.3372	1.1566	43.33
112	254.5	17.89	169.4	633.8	464.4	94.11	352.1	258.0	1.180	0.07367	0.8471	13.57		0.3413	1.1538	44.44
114	262.2	18.43	171.8	633.9	462.1	95.44	352.2	256.7	1.145	0.07148	0.8730	13.98		0.3453	1.1510	45.56
116	270.1	18.99	174.2	634.0	459.8	96.78	352.2	255.4	1.112	0.06942	0.8996	14.41		0.3495	1.1483	46.67
118	278.2	19.56	176.6	634.0	457.4	98.11	352.2	254.1	1.079	0.06736	0.9269	14.85		0.3535	1.1455	47.78
120	286.4	20.14	179.0	634.0	455.0	99.45	352.2	252.8	1.047	0.06536	0.9549	15.30	35.26	0.3576	1.1427	48.89
122	294.8	20.73	181.4	634.0	452.6	100.8	352.2	251.4	1.017	0.06349	0.9837	15.76		0.3618	1.1400	50.00
124	303.4	21.33	183.9	634.0	450.1	102.2	352.2	250.1	0.987	0.0616	1.0132	16.229		0.3659	1.1372	51.11

Carbon Dioxide (CO₂)

Temp. (°F)	Abs. press sat. vap lb/in.²	kg/cm²	Heat content abv. 32°F BTU/lb Liq.	Vap	Heat of vaporiz. BTU/lb	Heat content abv. 0°C g-cal/g Liq.	Vap	Heat of vaporiz. g-cal/g	Spec. vol. sat. vap. ft³/lb	m³/kg	Density sat. vap. lb/ft³	kg/m³	Dens. liq. lb/ft³	Entropy from 32°F BTU/lb/°F Liq.	Vap.	Temp. (°C)
−20	220.6	15.51	−23.96	102.0	126.0	−13.31	56.67	70.00	0.4166	0.02601	2.401	38.46	64.34	−0.0514	0.2353	−28.89
−18	228.4	16.06	−23.13	102.1	125.2	−12.85	56.72	69.56	0.4018	0.02508	2.489	39.87	64.15	−0.0495	0.2342	−27.78
−16	236.4	16.62	−22.30	102.2	124.5	−12.39	56.78	69.17	0.3876	0.02420	2.580	41.33	63.94	−0.0476	0.2331	−26.67
−14	244.6	17.20	−21.46	102.2	123.7	−11.92	56.78	68.72	0.3739	0.02334	2.674	42.83	63.73	−0.0458	0.2319	−25.56

THERMODYNAMIC PROPERTIES (continued)

Temp. (°F)	Abs. press sat. vap lb/in.²	kg/cm²	Heat content abv. 32°F BTU/lb Liq.	Vap	Heat of vaporiz. BTU/lb	Heat content abv. 0°C g-cal/g Liq.	Vap	Heat of vaporiz. g-cal/g	Spec. vol. sat. vap. ft³/lb	m³/kg	Density sat. vap. lb/ft³	kg/m³	Dens. liq. lb/ft³	Entropy from 32°F BTU/lb/°F Liq.	Vap.	Temp. (°C)
−12	253.0	17.79	−20.61	102.3	122.9	−11.45	56.83	68.28	0.3608	0.02252	2.772	44.40	63.49	−0.0439	0.2307	−24.44
−10	261.7	18.40	−19.76	102.3	122.0	−10.98	56.83	67.78	0.3482	0.02174	2.872	46.00	63.25	−0.0420	0.2296	−23.33
−8	270.6	19.03	−18.90	102.3	121.2	−10.50	56.83	67.33	0.3360	0.02098	2.976	47.67	63.01	−0.0401	0.2284	−22.22
−6	279.7	19.66	−18.04	102.3	120.3	−10.02	56.83	66.83	0.3243	0.02025	3.083	49.38	62.76	−0.0382	0.2273	−21.11
−4	289.1	20.33	−17.17	102.3	119.5	−9.539	56.83	66.39	0.3131	0.01955	3.194	51.16	62.50	−0.0362	0.2261	−20.00
−2	298.7	21.00	−16.29	102.3	118.6	−9.050	56.83	65.89	0.3022	0.01887	3.309	53.00	62.23	−0.0343	0.2249	−18.89
0	308.6	21.70	−15.41	102.2	117.7	−8.561	56.78	65.39	0.2918	0.01822	3.427	54.89	61.95	−0.0324	0.2237	−17.78
2	318.7	22.41	−14.51	102.2	116.7	−8.061	56.78	64.83	0.2817	0.01759	3.550	56.86	61.65	−0.0304	0.2225	−16.67
4	329.1	23.14	−13.61	102.1	115.8	−7.561	56.72	64.33	0.2720	0.01698	3.676	58.88	61.36	−0.0285	0.2213	−15.56
6	339.8	23.89	−12.71	102.1	114.8	−7.061	56.72	63.78	0.2627	0.01640	3.807	60.98	61.07	−0.0266	0.2201	−14.44
8	350.7	24.66	−11.79	102.0	113.8	−6.550	56.67	63.22	0.2537	0.01584	3.942	63.14	60.77	−0.0246	0.2189	−13.33
10	361.8	25.44	−10.87	101.9	112.8	−6.039	56.61	62.67	0.2450	0.01529	4.082	65.39	60.48	−0.0226	0.2176	−12.22
12	373.3	26.25	−9.934	101.8	111.7	−5.519	56.56	62.06	0.2366	0.01477	4.227	67.71	60.18	−0.0206	0.2164	−11.11
14	385.0	27.07	−8.992	101.7	110.7	−4.996	56.50	61.50	0.2285	0.01426	4.377	70.11	59.88	−0.0186	0.2151	−10.00
16	397.1	27.92	−8.038	101.5	109.6	−4.466	56.39	60.89	0.2207	0.01378	4.532	72.59	59.58	−0.0166	0.2139	−8.89
18	409.4	28.78	−7.076	101.4	108.5	−3.931	56.33	60.28	0.2131	0.01330	4.692	75.16	59.27	−0.0146	0.2126	−7.78
20	422.0	29.67	−6.102	101.2	107.3	−3.390	56.22	59.61	0.2058	0.01285	4.859	77.83	58.95	−0.0126	0.2113	−6.67
22	434.9	30.58	−5.117	101.0	106.1	−2.843	56.11	58.94	0.1987	0.01240	5.031	80.59	58.64	−0.0105	0.2100	−5.56
24	448.1	31.50	−4.121	100.8	104.9	−2.289	56.00	58.28	0.1919	0.01198	5.211	83.47	58.31	−0.0085	0.2087	−4.44
25	454.8	31.98	−3.618	100.7	104.3	−2.010	55.94	57.94	0.1886	0.01177	5.303	84.94	58.14	−0.0074	0.2080	−3.89
27	468.5	32.94	−2.601	100.5	103.1	−1.445	55.83	57.28	0.1821	0.01137	5.492	87.97	57.81	−0.0053	0.2066	−2.78
29	482.5	33.92	−1.570	100.2	101.8	−0.8722	55.67	56.56	0.1758	0.01097	5.688	91.11	57.47	−0.0032	0.2053	−1.67
31	496.8	34.93	−0.525	99.98	100.5	−0.292	55.54	55.83	0.1697	0.01059	5.892	94.38	57.12	−0.0011	0.2039	−0.56
33	511.4	35.95	0.531	99.69	99.16	0.295	55.38	55.09	0.1639	0.01023	6.103	97.76	56.77	0.0011	0.2025	0.56
35	526.4	37.01	1.604	99.38	97.77	0.8911	55.21	54.32	0.1581	0.009870	6.323	101.3	56.41	0.0033	0.2010	1.67
37	541.7	38.09	2.697	99.05	96.35	1.498	55.03	53.53	0.1526	0.009527	6.553	105.0	56.03	0.0055	0.1996	2.78
39	557.4	39.19	3.806	98.69	94.88	2.114	54.83	52.71	0.1472	0.009189	6.792	108.8	55.65	0.0077	0.1981	3.89
41	573.4	40.31	4.932	98.31	93.37	2.740	54.62	51.87	0.1420	0.008865	7.040	112.8	55.25	0.0099	0.1965	5.00
43	589.8	41.47	6.080	97.90	91.82	3.378	54.39	51.01	0.1370	0.008553	7.300	116.9	54.84	0.0122	0.1950	6.11
45	606.5	42.64	7.251	97.46	90.21	4.028	54.14	50.12	0.1321	0.008247	7.571	121.3	54.41	0.0146	0.1934	7.22
47	623.6	43.84	8.443	96.99	88.55	4.691	53.88	49.19	0.1273	0.007947	7.854	125.8	53.97	0.0169	0.1918	8.33
49	641.1	45.07	9.664	95.50	86.83	5.369	53.61	48.24	0.1227	0.007660	8.151	130.6	53.51	0.0193	0.1901	9.44
51	659.0	46.33	10.91	95.97	85.06	6.061	53.32	47.26	0.1182	0.007379	8.461	135.5	53.04	0.0218	0.1884	10.56
53	677.3	47.62	12.19	95.40	83.21	6.772	53.00	46.23	0.1138	0.007104	8.787	140.8	52.55	0.0243	0.1867	11.67
55	695.9	48.93	13.49	94.78	81.29	7.494	52.66	45.16	0.1095	0.006836	9.132	146.3	52.05	0.0268	0.1849	12.78
57	714.9	50.26	14.84	94.13	79.30	8.244	52.29	44.06	0.1053	0.006574	9.497	152.1	51.53	0.0294	0.1830	13.89
59	734.3	51.63	16.22	93.44	77.22	9.011	51.91	42.90	0.1012	0.006318	9.880	158.3	50.99	0.0321	0.1811	15.00
61	754.2	53.03	17.65	92.69	75.04	9.806	51.49	41.69	0.0972	0.00607	10.29	164.8	50.42	0.0348	0.1790	16.11
63	774.5	54.45	19.13	91.88	72.75	10.63	51.04	40.42	0.0933	0.00582	10.72	171.7	49.80	0.0377	0.1770	17.22
65	795.1	55.90	20.66	91.01	70.35	11.48	50.56	39.08	0.0894	0.00558	11.18	179.1	49.14	0.0406	0.1748	18.33
67	816.2	57.38	22.25	90.07	67.81	12.36	50.04	37.67	0.0856	0.00534	11.67	186.9	48.44	0.0436	0.1725	19.44
69	837.8	58.90	23.92	89.04	65.12	13.29	49.47	36.18	0.0819	0.00511	12.21	195.6	47.69	0.0468	0.1701	20.56
71	859.8	60.45	25.67	87.92	62.25	14.26	48.84	34.58	0.0782	0.00488	12.82	205.4	46.87	0.0501	0.1676	21.67
73	882.2	62.02	27.52	86.69	59.17	15.29	48.16	32.87	0.0745	0.00465	13.43	215.1	45.99	0.0535	0.1647	22.78
75	905.1	63.63	29.50	85.33	55.83	16.39	47.41	31.02	0.0708	0.00442	14.13	226.3	45.05	0.0573	0.1618	23.89
77	928.4	65.27	31.62	83.80	52.17	17.57	46.56	28.98	0.0671	0.00419	14.90	238.7	44.06	0.0613	0.1585	25.00
79	952.2	66.95	33.95	82.06	48.11	18.86	45.59	26.73	0.0633	0.00395	15.81	253.2	43.04	0.0656	0.1550	26.11
81	976.5	68.65	36.54	80.03	43.49	20.30	44.46	24.16	0.0592	0.00370	16.90	270.7	41.95	0.0704	0.1509	27.22
83	1001.0	70.377	39.53	77.60	38.07	21.96	43.11	21.15	0.0548	0.00342	18.25	292.3	40.62	0.0759	0.1461	28.33
85	1027.0	72.205	43.18	74.47	31.29	23.99	41.37	17.38	0.0500	0.00312	20.00	320.4	38.76	0.0826	0.1401	29.44
86	1039.0	73.049	45.45	72.46	27.00	25.25	40.26	15.00	0.0474	0.00296	21.09	337.8	37.41	0.0868	0.1363	30.00
87	1052.0	73.963	48.32	69.84	21.52	26.84	38.80	11.96	0.0446	0.00278	22.42	359.1	35.34	0.0921	0.1314	30.56
88	1065.0	74.877	52.78	65.62	12.84	29.32	36.46	7.133	0.0401	0.00250	24.95	399.6	32.79	0.1002	0.1237	31.11

Sulfur Dioxide (SO₂)

Temp. (°F)	Abs. press. sat. vap. lb/in.²	kg/cm²	Heat content abv. −40°F BTU/lb Liq.	Vap.	Ht. of vaporiz BTU/lb	Heat content abv. −40°C g-cal/g Liq.	Vap.	Ht. of vaporiz. g-cal/g	Spec. vol. sat. vap. ft³/lb	m³/kg	Density sat. vap. lb/ft³	kg/m³	Dens. liq. lb/ft³	Entropy from −40°F BTU/lb/°F Liq.	Vap.	Temp. (°C)
−40	3.136	0.2205	0.00	178.61	178.61	0.00	99.228	99.228	22.42	1.400	0.04460	0.7144	95.79	0.00000	0.42562	−40.00
−30	4.331	0.3045	2.93	179.90	176.97	1.63	99.945	98.317	16.56	1.034	0.06039	0.9673	94.94	0.00674	0.41864	−34.44
−20	5.883	0.4136	5.98	181.07	175.09	3.32	100.59	97.272	12.42	0.7754	0.08119	1.301	94.10	0.01366	0.41192	−28.89
−10	7.863	0.5528	9.16	182.13	172.97	5.09	101.18	96.095	9.44	0.5893	0.1025	1.642	93.27	0.02075	0.40544	−23.33
0	10.35	0.7277	12.44	183.07	170.63	6.911	101.71	94.795	7.280	0.4545	0.1374	2.201	92.42	0.02795	0.39917	−17.78
2	10.91	0.7670	13.12	183.25	170.13	7.289	101.81	94.517	6.923	0.4322	0.1444	2.313	92.25	0.02941	0.39794	−16.67
4	11.50	0.8085	13.78	183.41	169.63	7.656	101.89	94.239	6.584	0.4110	0.1501	2.404	92.08	0.03084	0.39670	−15.56
5	11.81	0.8303	14.11	183.49	169.38	7.839	101.94	94.100	6.421	0.4009	0.1558	2.496	92.00	0.03155	0.39609	−15.00
6	12.12	0.8521	14.45	183.57	169.12	8.028	101.98	93.956	6.266	0.3912	0.1596	2.556	91.91	0.03228	0.39547	−14.44
8	12.75	0.8964	15.13	183.73	168.60	8.406	102.07	93.667	5.967	0.3725	0.1676	2.685	91.74	0.03373	0.39426	−13.33

THERMODYNAMIC PROPERTIES (continued)

Temp. (°F)	Abs. press. sat. vap.		Heat content abv. −40°F BTU/lb		Ht. of vaporiz. BTU/lb	Heat content abv. −40°C g-cal/g		Ht. of vaporiz. g-cal/g	Spec. vol. sat. vap.		Density sat. vap.		Dens. liq. lb/ft³	Entropy from −40°F BTU/lb/°F		Temp. (°C)
	lb/in.²	kg/cm²	Liq.	Vap.	BTU/lb	Liq.	Vap.	g-cal/g	ft³/lb	m³/kg	lb/ft³	kg/m³		Liq.	Vap.	
10	13.42	0.9435	15.80	183.87	168.07	8.778	102.15	93.372	5.682	0.3547	0.1760	2.819	91.58	0.03519	0.39306	−12.22
12	14.12	0.9927	16.48	184.01	167.53	9.156	102.23	93.072	5.417	0.3382	0.1846	2.957	91.41	0.03664	0.39185	−11.11
14	14.84	1.043	17.15	184.14	166.97	9.528	102.30	92.761	5.164	0.3224	0.1936	3.101	91.24	0.03808	0.39065	−10.00
16	15.59	1.096	17.84	184.28	166.44	9.911	102.38	92.467	4.926	0.3075	0.2030	3.252	91.07	0.03953	0.38946	−8.89
18	16.37	1.1509	18.52	184.40	165.88	10.29	102.44	92.156	4.701	0.2935	0.2127	3.407	90.89	0.04098	0.38827	−7.78
20	17.18	1.208	19.20	184.52	165.32	10.67	102.51	91.845	4.487	0.2801	0.2228	3.569	90.71	0.04241	0.38707	−6.67
22	18.03	1.268	19.90	184.64	164.74	11.06	102.58	91.522	4.287	0.2676	0.2332	3.735	90.53	0.04385	0.38589	−5.56
24	18.89	1.328	20.58	184.74	164.16	11.43	102.63	91.200	4.096	0.2557	0.2441	3.910	90.33	0.04528	0.38471	−4.44
26	19.80	1.392	21.26	184.84	163.58	11.81	102.69	90.878	3.915	0.2444	0.2559	4.099	90.15	0.04671	0.38354	−3.33
28	20.73	1.457	21.96	184.94	162.98	12.20	102.74	90.545	3.744	0.2337	0.2671	4.278	89.96	0.04814	0.38237	−2.22
30	21.70	1.526	22.64	185.02	162.38	12.58	102.79	90.211	3.581	0.2236	0.2800	4.485	89.76	0.04956	0.38119	−1.11
32	22.71	1.597	23.33	185.10	161.77	12.96	102.83	89.872	3.437	0.2146	0.2909	4.660	89.58	0.05099	0.38003	0.00
34	23.75	1.670	24.03	185.18	161.15	13.35	102.88	89.528	3.283	0.2050	0.3046	4.879	89.39	0.05242	0.37887	1.11
36	24.82	1.745	24.72	185.25	160.53	13.73	102.92	89.183	3.144	0.1963	0.3181	5.095	89.18	0.05384	0.37772	2.22
38	25.95	1.824	25.42	185.31	159.89	14.12	102.95	88.828	3.013	0.1881	0.3319	5.316	89.00	0.05527	0.37657	3.33
40	27.10	1.905	26.12	185.37	159.25	14.51	102.98	88.472	2.887	0.1802	0.3464	5.549	88.81	0.05668	0.37541	4.44
42	28.29	1.989	26.81	185.42	158.61	14.89	103.01	88.117	2.769	0.1729	0.3611	5.784	88.62	0.05809	0.37425	5.56
44	29.52	2.075	27.51	185.46	157.95	15.28	103.03	87.750	2.656	0.1658	0.3765	6.031	88.43	0.05949	0.37311	6.67
46	30.79	2.165	28.21	185.50	157.29	15.67	103.06	87.383	2.548	0.1591	0.3925	6.287	88.24	0.06090	0.37197	7.78
48	32.10	2.257	28.92	185.54	156.62	16.07	103.08	87.011	2.446	0.1527	0.4088	6.548	88.05	0.06231	0.37083	8.89
50	33.45	2.352	29.61	185.56	155.95	16.45	103.09	86.639	2.348	0.1466	0.4259	6.822	87.87	0.06370	0.36969	10.00
52	34.86	2.451	30.31	185.58	155.27	16.84	103.10	86.261	2.256	0.1408	0.4433	7.101	87.67	0.06509	0.36857	11.11
54	36.31	2.553	31.00	185.59	154.59	17.22	103.11	85.883	2.167	0.1353	0.4615	7.392	87.51	0.06646	0.36743	12.22
56	37.80	2.658	31.72	185.61	153.89	17.62	103.12	85.495	2.083	0.1300	0.4801	7.690	87.31	0.06785	0.36629	13.33
58	39.33	2.765	32.42	185.61	153.19	18.01	103.12	85.106	2.003	0.1250	0.4992	7.996	87.13	0.06923	0.36517	14.44
60	40.93	2.878	33.10	185.59	152.49	18.39	103.11	84.717	1.926	0.1202	0.5194	8.320	86.95	0.07066	0.36405	15.56
62	42.58	2.994	33.79	185.57	151.78	18.77	103.09	84.322	1.853	0.1157	0.5396	8.643	86.77	0.07196	0.36293	16.67
64	44.27	3.112	34.49	185.55	151.06	19.16	103.08	83.922	1.783	0.1113	0.5609	8.984	86.59	0.07333	0.36181	17.78
66	46.00	3.234	35.19	185.53	150.34	19.55	103.07	83.522	1.716	0.1071	0.5827	9.334	86.41	0.07469	0.36070	18.89
68	47.78	3.359	35.88	185.50	149.62	19.93	103.06	83.122	1.652	0.1031	0.6054	9.697	86.22	0.07602	0.35958	20.00
70	49.62	3.489	36.58	185.46	148.88	20.32	103.03	82.711	1.590	0.09926	0.6290	10.08	86.02	0.07736	0.35846	21.11
72	51.54	3.624	37.28	185.42	148.14	20.71	103.01	82.300	1.533	0.09564	0.6527	10.46	85.82	0.07871	0.35736	22.22
74	53.48	3.760	37.97	185.37	147.40	21.09	102.98	81.889	1.476	0.09214	0.6777	10.86	85.62	0.08003	0.35624	23.33
76	55.48	3.901	38.67	185.31	146.64	21.48	102.95	81.467	1.422	0.08877	0.7030	11.26	85.42	0.08135	0.35512	24.44
78	57.56	4.047	39.36	185.24	145.88	21.87	102.91	81.045	1.371	0.08559	0.7295	11.69	85.23	0.08268	0.35401	25.56
80	59.68	4.196	40.05	185.17	145.12	22.25	102.87	80.622	1.321	0.08247	0.7570	12.13	85.03	0.8399	0.35291	26.67
82	61.88	4.351	40.73	185.09	144.36	22.63	102.83	80.200	1.274	0.07953	0.7850	12.57	84.84	0.08525	0.35177	27.78
84	64.14	4.509	41.43	185.01	143.58	23.02	102.78	79.767	1.229	0.07672	0.8140	13.04	84.64	0.08653	0.35065	28.89
86	66.45	4.672	42.12	184.92	142.80	23.40	102.73	79.333	1.185	0.07398	0.8440	13.52	84.44	0.08783	0.34954	30.00
88	68.84	4.840	42.80	184.82	142.02	23.78	102.68	78.900	1.144	0.07142	0.8740	14.00	84.25	0.08910	0.34843	31.11
90	71.25	5.009	43.50	184.72	141.22	24.17	102.62	78.456	1.104	0.06892	0.9058	14.51	84.05	0.09038	0.34731	32.22
92	73.70	5.182	44.19	184.61	140.42	24.55	102.56	78.011	1.065	0.06649	0.9390	15.04	83.86	0.09165	0.34620	33.33
94	76.30	5.364	44.86	184.49	139.63	24.92	102.49	77.567	1.028	0.06418	0.9730	15.59	83.67	0.09389	0.34508	34.44
96	79.03	5.556	45.54	184.37	138.83	25.30	102.43	77.128	0.9931	0.06200	1.007	16.13	83.47	0.09411	0.34397	35.56
98	81.77	5.749	46.22	184.25	138.03	25.68	102.36	76.683	0.9591	0.05987	1.043	16.71	83.27	0.09532	0.34285	36.67
100	84.52	5.942	46.90	184.10	137.20	26.06	102.28	76.222	0.9262	0.05782	1.080	17.30	83.07	0.09657	0.34173	37.78

Butane ($CH_3(CH_2)_2CH_3$)

Temp. (°F)	Abs. press. sat. vap.		Heat content abv. 32°F BTU/lb		Ht. of vaporiz. BTU/lb	Heat content abv. 0°C g-cal/g		Ht. of vaporiz. g-cal/g	Spec. vol. sat. vap.		Density of sat. vap.		Density of liq.		Temp. (°C)
	lb/in.²	kg/cm²	Liq.	Vap.	BTU/lb	Liq.	Vap.	g-cal/g	ft³/lb	m³/kg	lb/ft³	kg/m³	lb/ft³	kg/m³	
0	7.3	0.51	−17.2	153.3	170.5	−9.56	85.17	94.72	11.1	0.693	0.0901	1.44	38.59	618.1	−17.78
10	9.2	0.65	−11.7	156.8	168.5	−6.50	87.11	93.61	8.95	0.559	0.112	1.79	38.24	612.5	−12.22
20	11.6	0.816	−6.7	160.3	167.0	−3.7	89.06	92.78	7.23	0.451	0.138	2.21	37.89	606.9	−6.67
30	14.4	1.01	−1.2	164.3	165.5	−0.67	91.28	91.94	5.90	0.368	0.169	2.71	37.54	601.3	−1.11
40	17.7	1.24	4.3	167.8	163.5	2.4	93.22	90.83	4.88	0.305	0.205	3.28	37.19	595.7	4.44
50	21.6	1.52	9.8	171.3	161.5	5.4	95.17	89.72	4.07	0.254	0.246	3.94	36.82	589.8	10.00
60	26.3	1.85	15.8	175.3	159.5	8.78	97.39	88.61	3.40	0.212	0.294	4.71	36.45	583.9	15.56
70	31.6	2.22	21.3	178.8	157.5	11.8	99.33	87.50	2.88	0.180	0.347	5.56	36.06	577.6	21.11
80	37.6	2.64	27.3	182.3	155.0	15.2	101.3	86.11	2.46	0.154	0.407	6.52	35.65	571.0	26.67
90	44.5	3.13	33.8	185.8	152.0	18.8	103.2	84.44	2.10	0.131	0.476	7.62	35.24	564.5	32.22
100	52.2	3.67	39.8	189.3	149.5	22.1	105.2	83.06	1.81	0.113	0.552	8.84	34.84	558.1	37.78
110	60.8	4.27	46.3	193.3	147.0	25.7	107.4	81.67	1.58	0.0986	0.633	10.1	34.41	551.2	43.33
120	70.8	4.98	52.8	196.3	143.5	29.3	109.1	79.72	1.38	0.0862	0.725	11.6	33.96	544.0	48.89
130	81.4	5.72	59.3	199.8	140.5	32.9	111.0	78.06	1.21	0.0755	0.826	13.2	33.49	536.4	54.44
140	92.6	6.51	66.3	203.8	137.5	36.8	113.2	76.39	1.07	0.0668	0.934	15.0	32.98	528.3	60.00

THERMODYNAMIC PROPERTIES (continued)

Isobutane (CH₃)₃CH

Temp. (°F)	Abs. press. sat. vap.		Heat content abv. 32°F BTU/lb		Ht. of vaporiz. BTU/lb	Heat content abv. 0°C g-cal/g		Ht. of vaporiz. g-cal/g	Spec. vol. sat. vap.		Density of sat. vap.		Density of liq.		Temp. (°C)
	lb/in.²	kg/cm²	Liq.	Vap.		Liq.	Vap.		ft³/lb	m³/kg	lb/ft³	kg/m³	lb/ft³	kg/m³	
−20	7.50	0.527	−25.5	140.0	165.5	−14.2	77.78	91.94	10.5	0.655	0.0952	1.52	38.35	614.3	−28.89
−10	9.28	0.652	−21.0	142.0	163.0	−11.7	78.89	90.56	8.91	0.556	0.112	1.79	37.95	607.9	−23.33
0	11.6	0.816	−16.5	144.0	160.5	−9.17	80.00	89.17	7.17	0.448	0.139	2.23	37.60	602.3	−17.78
10	14.6	1.03	−11.5	147.0	158.5	−6.39	81.67	88.06	5.75	0.359	0.174	2.79	37.20	595.9	−12.22
20	18.2	1.28	−6.5	149.5	156.0	−3.6	83.06	86.67	4.68	0.292	0.214	3.43	36.80	589.5	−6.67
30	22.3	1.57	−1.0	152.5	153.5	−0.56	84.72	85.28	3.86	0.241	0.259	4.15	36.40	583.1	−1.11
40	26.9	1.89	4.5	155.5	151.0	2.5	86.39	83.89	3.22	0.201	0.311	4.98	36.00	576.6	4.44
50	32.5	2.28	10.5	159.0	148.5	5.83	88.33	82.50	2.71	0.169	0.369	5.91	35.60	570.2	10.00
60	38.7	2.72	16.5	162.5	146.0	9.17	90.28	81.11	2.28	0.142	0.439	7.03	35.20	563.8	15.56
70	45.8	3.22	23.0	166.5	143.5	12.8	92.50	79.72	1.94	0.121	0.515	8.25	34.80	557.4	21.11
80	53.9	3.79	30.0	170.5	140.5	16.7	94.72	78.06	1.66	0.104	0.602	9.64	34.35	550.2	26.67
90	63.3	4.45	37.0	174.5	137.5	20.6	96.94	76.39	1.42	0.0886	0.704	11.3	33.90	543.0	32.22
100	73.7	5.18	44.5	179.0	134.5	24.7	99.44	74.72	1.23	0.0768	0.813	13.0	33.35	535.8	37.78
110	85.1	5.98	52.5	183.5	131.0	29.2	101.9	72.78	1.07	0.0668	0.935	15.0	33.00	528.6	43.33
120	98.0	6.89	60.5	188.0	127.5	33.6	104.4	70.83	0.926	0.0578	1.08	17.3	32.50	520.6	48.89
130	112.0	7.87	69.5	193.5	124.0	38.6	107.5	68.89	0.811	0.0506	1.23	19.7	32.00	512.6	54.44
140	126.8	8.915	78.5	199.0	120.5	43.6	110.6	66.94	0.716	0.0447	1.32	21.1	31.80	509.4	60.00

Propane (C₃H₈)

Temp. (°F)	Abs. press. sat. vap.		Heat content abv. 32°F BTU/lb		Ht. of vaporiz. BTU/lb	Heat content abv. 0°C g-cal/g		Ht. of vaporiz. g-cal/g	Spec. vol. sat. vap.		Density of sat. vap.		Density of liq.		Temp. (°C)
	lb/in.²	kg/cm²	Liq.	Vap.		Liq.	Vap.		ft³/lb	m³/kg	lb/ft³	kg/m³	lb/ft³	kg/m³	
−70	7.37	0.518	−55.2	134.3	189.5	−30.7	74.61	105.3	12.9	0.805	0.0775	1.24	37.40	599.1	−56.67
−60	9.72	0.683	−50.2	136.8	187.0	−27.9	76.00	103.9	9.93	0.620	0.111	1.78	37.00	592.7	−51.11
−50	12.6	0.886	−44.7	139.8	184.5	−24.8	77.67	102.5	7.74	0.483	0.129	2.07	36.60	586.3	−45.56
−40	16.2	1.14	−39.7	141.8	181.5	−22.1	78.78	100.8	6.13	0.383	0.163	2.61	36.19	579.7	−40.00
−30	20.3	1.43	−34.2	144.8	179.0	−19.0	80.44	99.44	4.93	0.308	0.203	3.25	35.78	573.1	−34.44
−20	25.4	1.79	−29.2	146.8	176.0	−16.2	81.56	97.78	4.00	0.250	0.250	4.00	35.37	566.6	−28.89
−10	31.4	2.21	−23.7	149.8	173.5	−13.2	83.22	96.39	3.26	0.204	0.307	4.92	34.96	560.0	−23.33
0	38.2	2.69	−18.2	152.3	170.5	−10.1	84.61	94.72	2.71	0.169	0.369	5.91	34.54	553.3	−17.78
10	46.0	3.23	−12.7	155.3	168.0	−7.06	86.28	93.33	2.27	0.142	0.441	7.06	34.12	546.5	−12.22
20	55.5	3.90	−7.2	157.8	165.0	−4.0	87.67	91.67	1.90	0.119	0.526	8.43	33.67	539.3	−6.67
30	66.3	4.66	−1.2	160.8	162.0	−0.67	89.33	90.00	1.60	0.0999	0.625	10.0	33.20	531.8	−1.11
40	78.0	5.48	4.8	163.8	159.0	2.7	91.00	88.33	1.37	0.0855	0.730	11.7	32.73	524.3	4.44
50	91.8	6.45	10.8	166.8	156.0	6.00	92.67	86.67	1.18	0.0737	0.847	13.6	32.24	516.4	10.00
60	107.1	7.530	16.8	169.8	153.0	9.33	94.33	85.00	1.01	0.0631	0.990	15.9	31.75	508.6	15.56
70	124.0	8.718	22.8	172.3	149.5	12.7	95.72	83.06	0.883	0.0551	1.13	18.1	31.24	500.4	21.11
80	142.8	10.04	29.3	175.3	146.0	16.3	97.39	81.11	0.770	0.0481	1.30	20.8	30.70	491.8	26.67
90	164.0	11.53	35.8	178.3	142.5	19.9	99.06	79.17	0.673	0.0420	1.49	23.9	30.15	482.9	32.22
100	187.0	13.15	42.3	180.8	138.5	23.5	100.4	76.94	0.591	0.0369	1.69	27.1	29.58	473.8	37.78
110	213.0	14.98	48.8	182.8	134.0	27.1	101.6	74.44	0.519	0.0324	1.96	31.4	28.85	462.1	43.33
120	240.0	16.87	55.3	184.3	129.0	30.7	102.4	71.67	0.459	0.0287	2.18	34.9	28.30	453.3	48.89

Difluorodichloromethane (CCl₂F₂ ("F-12"))

Temp. (°F)	Abs. press. sat. vap.		Heat content abv. −40°F BTU/lb		Ht. of vaporiz. BTU/lb	Heat content abv. −40°C g-cal/g		Ht. of vaporiz. g-cal/g	Spec. vol sat. vap.		Density of vap.		Dens. liq. lb/ft³	Entropy from −40°F BTU/lb/°F		Temp. (°C)
	lb/in.²	kg/cm²	Liq.	Vap.		Liq.	Vap.		ft³/lb	m³/kg	lb/ft³	kg/m³		Liq.	Vap.	
−40	9.32	0.655	0	73.50	73.50	0	40.83	40.83	3.911	0.2442	0.2557	4.096	94.58	0	0.17517	−40.00
−30	12.02	0.845	2.03	74.70	72.67	1.13	41.50	40.37	3.088	0.1928	0.3238	5.187	93.59	0.00471	0.17387	−34.44
−20	15.28	1.074	4.07	75.87	71.80	2.26	42.15	39.89	2.474	0.1544	0.4042	6.474	92.58	0.00940	0.17275	−28.89
−10	19.20	1.350	6.14	77.05	70.91	3.41	42.81	39.39	2.003	0.1250	0.4993	7.998	91.57	0.01403	0.17175	−23.33
0	23.87	1.678	8.25	78.21	69.96	4.58	43.45	38.87	1.637	0.1022	0.6109	9.785	90.52	0.01869	0.17091	−17.78
10	29.35	2.064	10.39	79.36	68.97	5.771	44.09	38.32	1.351	0.08434	0.7402	11.86	89.45	0.02328	0.17015	−12.22
20	35.75	2.513	12.55	80.49	67.94	6.972	44.72	37.74	1.121	0.06998	0.8921	14.29	88.37	0.02783	0.16949	−6.67
30	43.16	3.034	14.76	81.61	66.85	8.200	45.34	37.14	0.939	0.0586	1.065	17.06	87.24	0.03233	0.16887	−1.11
40	51.68	3.633	17.00	82.71	65.71	9.444	45.95	36.51	0.792	0.0494	1.263	20.23	86.10	0.03680	0.16833	4.44
50	61.39	4.316	19.27	83.78	64.51	10.71	46.54	35.84	0.673	0.0420	1.485	23.79	84.94	0.04126	0.16785	10.00
60	72.41	5.091	21.57	84.82	63.25	11.98	47.12	35.14	0.575	0.0359	1.740	27.87	83.78	0.04568	0.16741	15.56
70	84.82	5.963	23.90	85.82	61.92	13.28	47.68	34.40	0.493	0.0308	2.028	32.48	82.60	0.05009	0.16701	21.11
80	98.76	6.944	26.28	86.80	60.52	14.60	48.22	33.62	0.425	0.0265	2.353	37.69	81.39	0.05446	0.16662	26.67
90	114.3	8.036	28.70	87.74	59.04	15.94	48.74	32.80	0.368	0.0230	2.721	43.58	80.11	0.05882	0.16624	32.22
100	131.6	9.252	31.16	88.62	57.46	17.31	49.23	31.92	0.319	0.0199	3.135	50.22	78.80	0.06316	0.16584	37.78
110	150.7	10.60	33.65	89.43	55.78	18.69	49.68	30.99	0.277	0.0173	3.610	57.82	77.46	0.06749	0.16542	43.33
120	171.8	12.08	36.16	90.15	53.99	20.09	50.08	29.99	0.240	0.0150	4.167	66.75	76.02	0.07180	0.16495	48.89

Carbon Disulfide (CS₂)

Temp. (°F)	Abs. press. sat. vap.		Heat content abv. 32°F BTU/lb		Ht. of vaporiz. BTU/lb	Heat content abv. 0°C g-cal/g		Ht. of vaporiz. g-cal/g	Spec. vol. sat. vap.		Density sat. vap.		Temp. (°C)
	lb/in.²	kg/cm²	Liq.	Vap.	BTU/lb	Liq.	Vap.	g	ft³/lb	m³/kg	lb/ft³	kg/m³	
0	1.10	0.0773	−8.60	156.90	165.5	−4.78	87.167	91.94	53.76	3.356	0.0186	0.2979	−17.78
10	1.46	0.103	−5.60	158.90	164.5	−3.11	88.278	91.39	43.47	2.714	0.0230	0.3684	−12.22
20	1.89	0.133	−3.00	160.20	163.2	−1.67	89.000	90.67	34.84	2.175	0.0287	0.4597	−6.67
30	2.36	0.166	−0.50	161.70	162.2	−0.28	89.833	90.11	29.49	1.841	0.0339	0.5430	−1.11
40	3.03	0.213	2.05	163.25	161.2	1.14	90.695	89.56	23.52	1.468	0.0425	0.6808	4.44
50	3.90	0.274	4.24	164.24	160.0	2.36	91.245	88.89	20.60	1.286	0.0482	0.7721	10.00
60	4.95	0.348	7.20	166.40	159.2	4.00	92.445	88.44	18.00	1.124	0.0555	0.8890	15.56
70	5.85	0.411	9.80	167.90	158.1	5.44	92.278	87.83	13.20	0.824	0.0758	1.214	21.11
80	7.30	0.513	11.70	168.60	156.9	6.500	93.667	87.17	10.40	0.649	0.0961	1.539	26.67
90	9.15	0.643	13.80	169.40	155.6	7.667	94.111	86.44	8.30	0.518	0.1204	1.920	32.22
100	11.08	0.7790	16.15	170.50	154.4	8.972	94.722	85.78	7.03	0.439	0.1369	2.193	37.78
110	13.50	0.9491	18.30	171.50	153.2	10.17	95.278	85.11	5.80	0.362	0.1724	2.762	43.33
120	16.10	1.132	20.01	172.01	152.0	11.12	95.561	84.44	5.10	0.318	0.1960	3.140	48.89

Carbon Tetrachloride (CCl₄)

Temp. (°F)	Abs. press. sat. vap.		Heat content abv. 32°F BTU/lb		Ht. of vaporiz. BTU/lb	Heat content abv. 0°C g-cal/g		Ht. of vaporiz. g-cal/g	Spec. vol. sat. vap.		Density sat. vap.		Temp. (°C)
	lb/in.²	kg/cm²	Liq.	Vap.	BTU/lb	Liq.	Vap.	g	ft³/lb	m³/kg	lb/ft³	kg/m³	
20	0.40	0.028	−2.00	92.45	94.45	−1.11	51.36	52.47	69.5	4.34	0.01438	0.2303	−6.67
30	0.60	0.042	−0.25	93.45	93.70	−0.14	51.92	52.06	53.0	3.31	0.01886	0.3021	−1.11
40	0.84	0.059	1.60	94.80	93.20	0.889	52.67	51.79	40.0	2.50	0.02500	0.4005	4.44
60	1.42	0.100	5.95	98.15	92.20	3.31	54.33	51.22	24.0	1.50	0.04166	0.6673	15.56
70	1.85	0.130	8.20	99.53	91.40	4.56	55.29	50.78	19.5	1.22	0.05128	0.8214	21.11
80	2.40	0.169	9.80	99.87	90.07	5.44	55.48	50.04	16.0	0.000	0.06245	1.016	26.67
90	3.12	0.219	11.60	101.62	90.02	6.444	56.46	50.01	13.0	0.812	0.07692	1.232	32.22
100	4.00	0.281	13.40	102.80	89.40	7.444	57.11	49.67	10.0	0.624	0.1000	1.602	37.78
110	4.89	0.344	15.80	104.50	88.70	8.778	58.06	49.28	8.5	0.53	0.1176	1.884	43.33
120	5.95	0.418	18.06	105.90	87.90	10.03	58.83	48.83	7.5	0.47	0.1333	2.135	48.89

Ethyl Ether ((C₂H₅)₂O)

Temp. (°F)	Abs. press. sat. vap.		Heat content abv. 32°F BTU/lb		Ht. of vaporiz. BTU/lb	Heat content abv. 0°C g-cal/g		Ht. of vaporiz. g-cal/g	Spec. vol. sat. vap.		Density sat. vap.		Temp. (°C)
	lb/in.²	kg/cm²	Liq.	Vap.	BTU/lb	Liq.	Vap.	g	ft³/lb	m³/kg	lb/ft³	kg/m³	
0	1.3	0.091	−18.00	153.00	171.0	−10.00	85.000	95.00	38.0	2.37	0.0263	0.4213	−17.78
10	1.8	0.13	−12.0	158.43	170.4	−6.67	88.017	94.67	32.5	2.03	0.0332	0.5318	−12.22
20	2.5	0.18	−6.50	163.50	170.0	−3.61	90.833	94.44	27.0	1.69	0.0372	0.5959	−6.67
30	3.4	0.24	−1.50	167.90	169.4	−0.833	93.278	94.11	21.4	1.34	0.0468	0.7496	−1.11
40	4.4	0.31	4.00	172.40	168.4	2.22	95.778	93.56	17.0	1.06	0.0588	0.9419	4.44
50	5.5	0.39	9.57	177.17	167.6	5.32	98.428	93.11	13.2	0.824	0.0757	1.213	10.00
70	8.8	0.62	20.04	185.44	165.4	11.13	103.02	91.89	7.8	0.49	0.1280	2.050	21.11
80	10.9	0.766	26.40	190.60	164.2	14.67	105.89	91.22	6.2	0.39	0.1620	2.595	26.67
90	13.4	0.942	31.50	194.50	163.0	17.50	108.60	90.56	5.1	0.32	0.1960	3.140	32.22
100	16.0	1.12	36.50	197.50	161.5	20.28	109.72	89.72	4.5	0.28	0.2220	3.556	37.78

Methyl Chloride (CH₃Cl)

Methyl Chloride (CH_3Cl)

Temp. (°F)	Abs. press. sat. vap.		Heat content abv. 32°F BTU/lb		Ht. of vaporiz. BTU/lb	Heat content abv. 0°C g-cal/g		Ht. of vaporiz. g-cal/g	Spec. vol. sat. vap.		Spec. vol. liq.		Density sat. vap.		Density of liq.		Temp. (°C)
	lb/in.²	kg/cm²	Liq.	Vap.		Liq.	Vap.		ft³/lb	m³/kg	ft³/lb	m³/lb	lb/ft³	kg/m³	lb/ft³	kg/m³	
−20	11.75	0.8261	−19.0	167.36	186.36	−10.6	92.978	103.53	8.09	0.505	0.015827	0.0009880	0.124	1.98	63.185	1012.1	−28.89
−10	15.0	1.055	−15.38	168.83	184.21	−8.544	93.795	102.34	6.46	0.403	0.015985	0.0009979	0.155	2.48	62.560	1002.1	−23.33
−5	16.79	1.180	−13.58	169.54	183.12	−7.544	94.189	101.73	5.80	0.362	0.016013	0.0009997	0.172	2.76	62.450	1000.3	−20.56
0	18.8	1.32	−11.75	170.23	181.98	−6.528	94.572	101.10	5.18	0.323	0.016146	0.001008	0.193	3.09	61.936	992.09	−17.78
5	21.0	1.48	−9.93	170.96	180.84	−5.517	94.978	100.47	4.68	0.292	0.016228	0.001013	0.214	3.42	61.623	987.08	−15.00
10	23.3	1.64	−8.06	171.58	179.65	−4.478	95.322	99.806	4.18	0.261	0.016310	0.001018	0.239	3.83	61.311	982.08	−12.22
15	25.9	1.82	−6.74	172.24	178.47	−3.744	95.689	99.150	3.88	0.242	0.016388	0.001023	0.258	4.13	61.022	977.45	−9.44
20	28.8	2.02	−4.32	172.95	177.27	−2.400	96.083	98.483	3.41	0.213	0.016474	0.001028	0.293	4.70	60.702	972.32	−6.67
25	31.8	2.24	−2.48	173.63	176.10	−1.378	96.461	97.833	3.09	0.193	0.016552	0.001033	0.324	5.18	60.415	967.73	−3.89
30	35.2	2.47	−0.62	174.28	174.90	−0.344	96.822	97.167	2.81	0.175	0.016645	0.001039	0.356	5.70	60.077	962.31	−1.11
35	38.7	2.72	1.75	174.92	173.77	0.972	97.178	96.539	2.50	0.156	0.016746	0.001045	0.400	6.41	59.715	956.51	1.67
40	42.6	3.00	3.15	175.57	172.42	1.75	97.539	95.789	2.31	0.144	0.016809	0.001049	0.433	6.93	59.492	952.94	4.44
45	46.9	3.30	5.04	176.20	171.16	2.80	97.889	95.089	2.10	0.131	0.016929	0.001057	0.476	7.63	59.069	946.17	7.22
50	51.5	3.62	6.88	176.78	169.90	3.82	98.211	94.389	1.93	0.120	0.017023	0.001063	0.518	8.30	58.745	940.98	10.00
55	56.4	3.97	8.80	177.45	168.65	4.89	98.583	93.695	1.75	0.109	0.017118	0.001069	0.571	9.16	58.419	935.76	12.78
60	61.6	4.33	10.70	178.05	167.35	5.944	98.917	92.972	1.61	0.101	0.017219	0.001075	0.621	9.95	58.077	930.28	15.56
65	67.3	4.73	12.62	178.64	166.02	7.011	99.245	92.233	1.47	0.0918	0.017318	0.001081	0.680	10.9	57.742	924.91	18.33
70	73.3	5.15	14.52	179.17	164.65	8.067	99.539	91.472	1.34	0.0837	0.017421	0.001088	0.746	12.0	57.403	919.48	21.11
75	79.2	5.57	16.46	179.78	163.28	9.144	99.878	90.711	1.24	0.0774	0.017526	0.001094	0.806	12.9	57.058	913.96	23.89
80	85.3	6.00	18.36	180.24	161.88	10.20	100.13	89.933	1.14	0.0712	0.017632	0.001101	0.877	14.1	56.714	908.44	26.67
85	94.1	6.62	20.12	180.74	160.48	11.18	100.41	89.156	1.05	0.0655	0.017740	0.001108	0.952	15.3	56.369	902.92	29.44
90	102.1	7.178	22.13	181.22	159.09	12.29	100.68	88.383	0.98	0.061	0.017850	0.001114	1.02	16.3	56.022	897.36	32.22
95	110.3	7.755	24.07	181.76	157.70	13.37	100.98	87.611	0.91	0.057	0.017961	0.001121	1.10	17.6	55.675	891.80	35.00
100	118.8	8.352	26.06	182.36	156.30	14.48	101.31	86.833	0.85	0.053	0.018074	0.001128	1.18	18.8	55.327	886.23	37.78

Mercury (Hg)

Temp. (°F)	Abs. press. sat. vap.		Heat content abv. 32°F BTU/lb		Ht. of vaporiz. BTU/lb	Heat conent abv. 0°C g-cal/g		Ht. of vaporix. g-cal/g	Spec. vol. sat. vap.		Density of sat. vap.		Entropy above 32°F BTU/lb/°F			Temp (°C)
	lb/in.²	kg/cm²	Liq.	Vap.		Liq.	Vap.		ft³/lb	m³/kg	lb/ft³	kg/m³	Liq.	Vap.	Evap.	
402	0.4	0.03	13.81	141.96	128.15	7.672	78.867	71.195	114.50	7.1480	0.008733	0.1399	0.0209	0.1696	0.1487	205.56
444	0.8	0.06	15.36	142.60	127.24	8.533	79.222	70.689	59.72	3.728	0.016745	0.268220	0.0227	0.1635	0.1408	228.89
458	1.0	0.07	15.89	142.81	126.92	8.828	79.339	70.511	48.45	3.025	0.02064	0.3306	0.0233	0.1616	0.1383	236.67
485	1.5	0.11	16.90	143.23	126.33	9.389	79.572	70.183	33.14	2.069	0.03017	0.4833	0.0244	0.1581	0.1337	251.67
505	2.0	0.14	17.65	143.54	125.89	9.806	79.745	69.939	25.32	1.581	0.03948	0.6324	0.0251	0.1556	0.1305	262.78
558	4.0	0.28	19.62	144.34	124.72	10.90	80.189	69.289	13.26	0.8278	0.07540	1.208	0.0271	0.1497	0.1226	292.22
591	6.0	0.42	20.87	144.86	123.99	11.59	80.478	68.883	9.096	0.5678	0.10993	1.7609	0.0283	0.1462	0.1179	310.56
617	8.0	0.56	21.81	145.24	123.43	12.12	80.689	68.572	6.9630	0.43469	0.14361	2.3003	0.0292	0.1439	0.1147	325.00
637	10.0	0.703	22.58	145.56	122.98	12.54	80.867	68.322	5.6610	0.35341	0.17664	2.8294	0.0299	0.1420	0.1121	336.11
676	15.0	1.05	24.04	146.16	122.12	13.36	81.200	67.844	3.8923	0.24299	0.25691	4.1152	0.0312	0.1387	0.1075	357.78
706	20.0	1.41	25.15	146.61	121.46	13.97	81.450	67.478	2.983	0.1862	0.3352	5.369	0.0322	0.1364	0.1042	374.44
730	25.0	1.76	26.05	146.98	120.93	14.47	81.656	67.183	2.429	0.1516	0.4117	6.595	0.0330	0.1346	0.1016	387.78
751	30.0	2.11	26.81	147.29	120.48	14.89	81.828	66.933	2.053	0.1282	0.4871	7.802	0.0336	0.1331	0.0995	399.44
769	35.0	2.46	27.49	147.57	120.08	15.27	81.983	66.711	1.7815	0.11122	0.5613	8.991	0.0342	0.1319	0.0977	409.44
785	40.0	2.81	28.08	147.81	119.73	15.60	82.117	66.517	1.5762	0.098399	0.6344	10.16	0.0346	0.1308	0.0962	418.33
799	45.0	3.16	28.62	148.04	119.42	15.90	82.245	66.344	1.4147	0.088317	0.7069	11.32	0.0351	0.1300	0.0949	426.11
812	50	3.5	29.11	148.24	119.13	16.17	82.356	66.183	1.284	0.08016	0.7788	12.47	0.0355	0.1291	0.0936	433.33
836	60	4.2	29.99	148.60	118.61	16.66	82.556	65.894	1.086	0.06780	0.9204	14.74	0.0361	0.1276	0.0915	446.67
857	70	4.9	30.75	148.90	118.15	17.08	82.722	65.639	0.9436	0.05891	1.0597	16.974	0.0367	0.1265	0.0898	458.33
875	80	5.6	31.44	149.19	117.75	17.47	82.883	65.417	0.8349	0.05212	1.1977	19.185	0.0372	0.1254	0.0882	468.33
892	90	6.3	32.06	149.44	117.38	17.81	83.022	65.211	0.7497	0.04680	1.3338	21.365	0.0377	0.1247	0.0870	477.78
907	100	7.03	32.63	149.68	117.05	18.13	83.156	65.028	0.6811	0.04252	1.4682	23.518	0.0381	0.1237	0.0856	486.11
921	110	7.73	33.16	149.90	116.74	18.42	83.278	64.856	0.6242	0.03897	1.6020	25.661	0.0385	0.1230	0.0845	493.89
934	120	8.44	33.66	150.10	116.44	18.70	83.389	64.689	0.5767	0.03600	1.7340	27.775	0.0389	0.1224	0.0835	501.11
947	130	9.14	34.12	150.29	116.17	18.96	82.495	64.539	0.5360	0.03346	1.8656	29.883	0.0392	0.1218	0.0826	508.33
958	140	9.84	34.55	150.47	115.92	19.19	82.595	64.400	0.5012	0.03129	1.9952	31.959	0.0395	0.1213	0.0818	514.44
969	150	10.5	34.96	150.63	115.67	19.42	83.683	64.261	0.4706	0.02938	2.125	34.04	0.0398	0.1207	0.0809	520.56
1000	180	12.7	36.09	151.10	115.01	20.05	83.945	63.894	0.3990	0.02491	2.506	40.14	0.0406	0.1194	0.0788	537.78

PHYSICAL PROPERTIES OF FLUOROCARBON REFRIGERANTS

Property No.	Refrigerant name		11	12	13	13B1	14	21
1	Formula		CCl_3F	CCl_2F_2	$CClF_3$	$CBrF_3$	CF_4	$CHCl_2F$
2	Molecular weight		137.37	120.91	104.46	148.92	88.01	102.92
3	Normal boiling point; °C		23.82	-29.79	-81.4	-57.75	-127.96	8.92
4	Normal freezing point; °C		-111	-158	-181	-168	-184	-135
5	Critical temperature; °C		198	112	28.9	67	-45.67	178.5
6	Critical pressure; atm		43.5	40.6	38.2	39.1	36.96	51
7	Critical volume; cm/mol		247	217	181	200	141	197
8	Critical density; g/cm		0.554	0.558	0.578	0.745	0.626	0.522
9	Density of liquid at 25°C; g/cm		1.467	1.311	$1.298^{-30°C}$	1.538	$1.317^{-80°C}$	1.366
10	Density of saturated vapor at B.P.; g/liter		5.86	6.33	7.01	8.71	7.62	4.57
11	Specific heat of liquid at 25°C; cal/g		0.208	0.232	$0.247^{-30°C}$	0.208	$0.294^{-80°C}$	0.256
12	Specific heat of vapor at 25°C and 1 atm; cal/g		$0.142^{38°C}$	0.145	0.158	0.112	0.169	0.140
13	Heat of vaporization at B.P.; cal/g		43.10	39.47	35.47	28.38	32.49	57.86
14	Thermal conductivity at 25°C; Btu/(hr)(ft)(°F)	liquid	0.050	0.041	0.020	0.025	$0.040^{-100°F}$	0.063
		vapor; 1 atm	0.00484	0.00557	—	—	—	0.0569
15	Viscosity at 25°C; centipoise	liquid	0.42	0.26	0.016	0.15	0.020	0.34
		vapor; 1 atm	0.011	0.013		0.016	—	0.011
16	Surface tension at 25°C; dyne cm		18	9	$14^{-73.3°C}$	4	$14^{-73.3°C}$	18
17	Refractive index of liquid at 25°C		1.374	1.287	$1.199^{-73.3°C}$	1.238	$1.151^{-73.3°C}$	1.354
18	Dielectric constant	liquid	$2.28^{29°C}$	$2.13^{29°C}$	—	—	—	$5.34^{28°C}$
		vapor at 0.5 atm	$1.10019^{29°C}$	$1.0016^{29°C}$	$1.0013^{29°C}$	—	$1.0006^{24.5°C}$	$1.0035^{30°C}$
19	Solubility in water at 25°C and 1 atm; wt %		0.011	0.028	0.009	0.03	0.0015	0.95
20	Solubility of water in compound at 25°C; wt %		0.011	0.009	—	$0.0095^{21.1°C}$	—	0.13
21	Toxicity; Group number (See separate table for definition of group number.)		5a	6	probably 6	6	probably 6	<4 >5

PHYSICAL PROPERTIES OF FLUOROCARBON REFRIGERANTS (Continued)

Property No.	22	23	112	113	114	114B2	115	116	500	502
1	$CHClF_2$	CHF_3	$C_2Cl_4F_2$	$C_2Cl_3F_3$	$C_2Cl_2F_4$	$C_2Br_2F_4$	C_2ClF_5	C_2F_6	*	**
2	80.47	70.01	203.83	187.38	170.91	259.83	154.47	138.01	105.5	120.7
3	−40.75	−82.03	92.8	47.57	3.77	47.26	−38.7	−78.2	−33.5	−45.6
4	−160	−155.2	26	−35	−94	−110.5	−106	−100.6	−158.9	—
5	96	25.9	278	214.1	145.7	214.5	80	19.7	105.4	179.89
6	49.12	47.7	34	33.7	32.2	34	30.8	29.4	43.7	590.3
7	165	133	370	325	293	329	259	225	—	290
8	0.525	0.525	0.55	0.576	0.582	0.790	0.596	0.612	0.497	0.56
9	1.194	0.670	$1.634^{30°C}$	1.565	1.456	2.163	1.291	$1.587^{-73.3°C}$	$1.138^{30°C}$	1.242
10	4.72	4.66	7.02	7.38	7.83	—	8.37	9.01	—	6.05
11	0.300	1.553	—	0.218	0.243	0.166	0.285	$0.232^{-73.3°C}$	$0.161^{30°C}$	—
12	0.157	0.176	—	$0.161^{60°C}$	0.170	—	0.164	0.182^{0mm}	—	—
13	55.81	57.23	37 (est)	35.07	32.51	25 (est)	30.11	27.97	—	42.48
14	0.052	0.008	0.040	0.038	0.034	0.027	0.026	$0.045^{-100°F}$	—	0.038
	0.00678	—	—	$0.0045^{0.5\ atm}$	0.00646	—	0.00803	0.0098 (est)	—	—
15	0.23	0.016 (est)	1.21	0.68	0.38	0.72	0.26	—	$0.292^{-15°C}$	0.25
	0.013	—	—	$0.01^{0.1\ atm}$	0.011	—	0.013	—	—	—
16	8	$15^{-73.3°C}$	$23^{30°C}$	19	12	18	5	$16^{-73.3°C}$	—	8
17	1.256	$1.251^{-73.3°C}$	1.413	1.354	1.288	1.367	1.214	$1.206^{-73.3°C}$	—	1.235
18	$6.11^{24°C}$	—	$2.52^{25°C}$	$2.41^{25°C}$	$2.26^{25°C}$	$2.34^{25°C}$	—	—	—	—
	$1.0035^{25.4°C}$	—	—	—	$1.0021^{26.5°C}$	—	$1.0018^{27.4°C}$	—	—	—
19	0.30	0.10	0.12 (sat. pr)	0.017 (sat. pr)	0.013	—	0.006	—	—	—
20	0.13	—	—	0.011	0.009	—	—	—	—	0.560
21	5a	probably 6	<4 >5	<4 >5	6	5a	6	probably 6	5a	5a

* Azeotrope of CCl_2F_2 (73.8 wt %) and $C_2H_4F_2$ (26.2 wt %) ** Azeotrope of $CHClF_2$ (48.8 wt %) and C_2ClF_5 (51.2 wt %).

UNDERWRITERS' LABORATORIES' CLASSIFICATION OF COMPARATIVE LIFE HAZARD OF GASES AND VAPORS

(Group number definition)

Group	Definition	Examples
1	Gases or vapors which in concentrations of the order of $\frac{1}{2}$ to 1 percent for durations of exposure of the order of 5 minutes are lethal or produce serious injury.	Sulfur dioxide
2	Gases or vapors which in concentrations of the order of $\frac{1}{2}$ to 1 percent for durations of exposure of the order of $\frac{1}{2}$ hour are lethal or produce serious injury.	Ammonia, Methyl bromide
3	Gases or vapors which in concentrations of the order of 2 to $2\frac{1}{2}$ percent for durations of exposure of the order of 1 hour are lethal or produce serious injury.	Bromochloromethane, Carbon tetrachloride, Chloroform, Methyl formate
4	Gases or vapors which in concentrations of the order of 2 to $2\frac{1}{2}$ percent for durations of exposure of the order of 2 hours are lethal or produce serious injury.	Dichloroethylene, Methyl chloride, Ethyl bromide
Between 4 and 5	Appear to classify as somewhat less toxic than group 4. Much less toxic than group 4 but somewhat more toxic than group 5.	Methylene chloride, Ethyl chloride, Refrigerant 112*, Refrigerant 113, Refrigerant 21
5a	Gases or vapors much less toxic than group 4 but more toxic than group 6.	Refrigerant 11, Refrigerant 22, Refrigerant 114B2, Refrigerant 502, Carbon dioxide
5b	Gases or vapors which available data indicate would classify as either group 5a or group 6.	Ethane, Propane, Butane
6	Gases or vapors which in concentrations up to at least about 20 percent by volume for durations of exposure of the order of 2 hours do not appear to produce injury.	Refrigerant 13B1, Refrigerant 12, Refrigerant 114, Refrigerant 115, Refrigerant 13*, Refrigerant 14*, Refrigerant 23*, Refrigerant 116*, Refrigerant C318*

* Not tested by U.L. but estimated to belong in group indicated

THERMAL CONDUCTIVITY OF LIQUID FLUOROCARBONS

To convert from W/(m)(K) to Btu/(hr)(ft)($^\circ$F) divide by 1.7296.
To convert from W/(m)(K) to cal/(sec)(cm)($^\circ$C) divide by 418.4.

Fluorocarbon	Formula	Temperature, K	Conductivity, W/(m)(K)
12	CCl_2F_2	277.2	94.14
		298.1	97.49
		303.8	103.76
		317.9	110.04
		329.9	117.15
		346.8	126.36
22	$CHClF_2$	289.5	107.11
		302.8	114.64
		327.8	126.76
		346.8	140.16
114	$C_2Cl_2F_4$	303.6	113.39
		316.8	122.59
		328.7	130.96
		343.0	140.58
13B1	$CBrF_3$	277.3	91.21
		282.3	93.30
		303.5	103.34
		318.7	111.71
		331.9	118.41
		342.8	123.01
		346.9	128.03
C-318	C_4F_8	280.1	112.97
		287.3	117.57
		298.0	130.96
		310.7	141.42
		323.2	148.11
		332.2	156.48
		342.9	158.99
		348.4	165.69
		350.8	166.94

MISCELLANEOUS PROPERTIES OF COMMON REFRIGERANTS

Refrigerant	Formula	Flash Point °F.	Ignition Temp. °F.	Explosive Limits % by Volume Lower	Upper	Vapor Density (Air = 1)	Boiling Point °F	Threshold Limit Value* Parts per Million in Air	Water Soluble	Odor
Ammonia	NH₃	—	1204	16	25	0.59	−28	100	yes	yes
Bromotrifluoromethane (Kulene-131)	CF₃Br	nonflammable				5.25	−73.6	—	no	yes
Butane	C₄H₁₀	−76	806	1.8	8.4	2.04	33	—	no	no
Carbon dioxide	CO₂	nonflammable				1.53	−108	5000	yes	no
Carbon tetrachloride	CCL₄	nonflammable				5.32	170	25	no	yes
Dichlorodifluoromethane (Freon-12)	CCl₂F₂	nonflammable				4.17	−21.6	—	no	no
Dichlorodifluoromethane, 73.8% Ethylidene fluoride, 26.2% (Carrene-7)	CCl₂F₂ CH₃CHF₂	nonflammable				3.24	−28.0		no	yes
Dichloromonofluoromethane (Freon-21)	CHCl₂F	practically nonflammable				3.55	48		no	yes
Dichlorotetrafluoroethane (Freon-114)	C₂Cl₂F₄	practically nonflammable				5.89	38		no	no
Ethane	C₂H₆	<20	950	3.0	12.5	1.04	−128	—	no	no
Ethylene	C₂H₄	<20	842	3.1	32	0.972	−155	—	yes	yes
Isobutane	(CH₃)₃CH	<20	1010	1.8	8.4	2.01	14	—	no	no
Methyl chloride	CH₃Cl	632	1170	10.7	11.4	1.78	−11	100	yes	yes
Monochlorodifluoromethane (Freon-22)	CHClF₂	practically nonflammable				2.9	−41	—	yes	no
Monochlorotrifluoromethane (Freon-13)	CClF₃	nonflammable				3.6	−112	—	—	no
Propane	C₃H₈	<20	871	2.2	9.5	1.56	−45	—	no	no
Propylene	C₃H₆	<20	927	2.4	10.3	1.49	−53	—	yes	yes
Sulfur dioxide	SO₂	nonflammable				2.2	14	10	yes	yes
Tetrafluoromethane (Freon-14)	CF₄	nonflammable				3.0	−198	—	no	no
Trichloroethylene	C₂HCl₃	nonflammable at normal temperature				4.53	189	200	no	yes
Trichloromonofluoromethane (Freon-11)(Carrene-2)	CCL₃F	nonflammable				4.7	75.3	—	no	no
Trichlorotrifluoroethane (Freon-113)	C₂Cl₃F₃	practically nonflammable				6.4	118	—	no	no

* Maximum average atmospheric concentration of contaminants to which workers may be exposed for an eight-hour work day without injury to health. (American Conference of Governmental Industrial Hygienists: "Threshold Limit Values for 1954.") Where blanks appear in this column no published information was available on threshold limit values.

HYGROMETRIC AND BAROMETRIC TABLES

CONVERSION TABLE FOR BAROMETRIC READINGS

U.S. inches to cm.

Inches	.00	.01	.02	.03	.04	.05	.06	.07	.08	.09	Inches	.00	.01	.02	.03	.04	.05	.06	.07	.08	.09
27.0	68.580	.606	.631	.656	.682	.707	.733	.758	.783	.809	29.0	.660	.686	.711	.736	.762	.787	.813	.838	.863	.889
27.1	.834	.860	.885	.910	.936	.961	.987	*.012	*.037	*.063	29.1	.914	.940	.965	.990	*.016	*.041	*.067	*.092	*.117	*.143
27.2	69.088	.114	.139	.164	.190	.215	.241	.266	.291	.317	29.2	74.168	.194	.219	.244	.270	.295	.321	.346	.371	.397
27.3	.342	.368	.393	.418	.444	.469	.495	.520	.545	.571	29.3	.422	.448	.473	.498	.524	.549	.575	.600	.625	.651
27.4	.596	.622	.647	.672	.698	.723	.749	.774	.799	.825	29.4	.676	.702	.727	.752	.778	.803	.829	.854	.879	.905
27.5	.850	.876	.901	.926	.952	.977	*.002	*.028	*.053	*.079	29.5	.930	.956	.981	*.006	*.032	*.057	*.083	*.108	*.133	*.159
27.6	70.104	.130	.155	.180	.206	.231	.257	.282	.307	.333	29.6	75.184	.210	.235	.260	.286	.311	.337	.362	.387	.413
27.7	.358	.384	.409	.434	.460	.485	.511	.536	.561	.587	29.7	.438	.464	.489	.514	.540	.565	.591	.616	.641	.667
27.8	.612	.638	.663	.688	.714	.739	.765	.790	.815	.841	29.8	.692	.718	.743	.768	.794	.819	.845	.870	.895	.921
27.9	.866	.892	.917	.942	.968	.993	*.018	*.044	*.069	*.095	29.9	.946	.972	.997	*.022	*.048	*.073	*.099	*.124	*.149	*.175
28.0	71.120	.146	.171	.196	.222	.247	.273	.298	.323	.349	30.0	76.200	.226	.251	.277	.302	.327	.353	.378	.404	.429
28.1	.374	.400	.425	.450	.476	.501	.527	.552	.577	.603	30.1	.454	.480	.505	.531	.556	.581	.607	.632	.658	.683
28.2	.628	.654	.679	.704	.730	.755	.781	.806	.831	.857	30.2	.708	.734	.759	.785	.810	.835	.861	.886	.912	.937
28.3	.882	.908	.933	.958	.984	*.009	*.035	*.060	*.085	*.111	30.3	.962	.988	*.013	*.039	*.064	*.089	*.115	*.140	*.166	*.191
28.4	72.136	.162	.187	.212	.238	.263	.289	.314	.339	.365	30.4	77.216	.242	.267	.293	.318	.343	.369	.394	.420	.445
28.5	.390	.416	.441	.466	.492	.517	.543	.568	.593	.619	30.5	.470	.496	.521	.547	.572	.597	.623	.648	.674	.699
28.6	.644	.670	.695	.720	.746	.771	.797	.822	.847	.873	30.6	.724	.750	.775	.801	.826	.851	.877	.902	.928	.953
28.7	.898	.924	.949	.974	*.000	*.025	*.051	*.076	*.101	*.127	30.7	.978	*.004	*.029	*.055	*.080	*.105	*.131	*.156	*.182	*.207
28.8	73.152	.178	.203	.228	.254	.279	.305	.330	.355	.381	30.8	78.232	.258	.283	.309	.334	.359	.385	.410	.436	.461
28.9	.406	.432	.457	.482	.508	.533	.559	.584	.609	.635	30.9	.486	.512	.537	.563	.588	.613	.639	.664	.690	.715

U.S. Inches to Millibars

Based on the relation 1 inch of mercury at 32°F represents a pressure of 33.8639 millibars.
Note: Figures in last nine columns to be preceded by 7, 8, 9 or 10 as indicated in column 2.

Inches	.00	.01	.02	.03	.04	.05	.06	.07	.08	.09	Inches	.00	.01	.02	.03	.04	.05	.06	.07	.08	.09
23.0	7 78.87	79.21	79.55	79.89	80.22	80.56	80.90	81.24	81.58	81.92	23.7	8 02.57	02.91	03.25	03.59	03.93	04.27	04.61	04.94	05.28	05.62
23.1	7 82.26	82.59	82.93	83.27	83.61	83.95	84.29	84.63	84.97	85.30	23.8	8 05.96	06.30	06.64	06.98	07.32	07.65	07.99	08.33	08.67	09.01
23.2	7 85.64	85.98	86.32	86.66	87.00	87.34	87.67	88.01	88.35	88.69	23.9	8 09.35	09.69	10.02	10.36	10.70	11.04	11.38	11.72	12.06	12.39
23.3	7 89.03	89.37	89.71	90.04	90.38	90.72	91.06	91.40	91.74	92.08	24.0	8 12.73	13.07	13.41	13.75	14.09	14.43	14.77	15.10	15.44	15.78
23.4	7 92.42	92.75	93.09	93.43	93.77	94.11	94.45	94.79	95.12	95.46	24.1	8 16.12	16.46	16.80	17.14	17.47	17.81	18.15	18.49	18.83	19.17
23.5	7 95.80	96.14	96.48	96.82	97.16	97.49	97.83	98.17	98.51	98.85	24.2	8 19.51	19.85	20.18	20.52	20.86	21.20	21.54	21.88	22.22	22.55
23.6	7 99.19	99.53	99.87	*00.20	*00.54	*00.88	*01.22	*01.56	*01.90	*02.24	24.3	8 22.89	23.23	23.57	23.91	24.25	24.59	24.92	25.26	25.60	25.94

Inches	.00	.01	.02	.03	.04	.05	.06	.07	.08	.09
24.4	8 26.28	26.62	26.96	27.30	27.63	27.97	28.31	28.65	28.99	29.33
24.5	8 29.67	30.00	30.34	30.68	31.02	31.36	31.70	32.04	32.37	32.71
24.6	8 33.05	33.39	33.73	34.07	34.41	34.75	35.08	35.42	35.76	36.10
24.7	8 36.44	36.78	37.12	37.45	37.79	38.13	38.47	38.81	39.15	39.49
24.8	8 39.82	40.16	40.50	40.84	41.18	41.52	41.86	42.20	42.53	42.87
24.9	8 43.21	43.55	43.89	44.23	44.57	44.90	45.24	45.58	45.92	46.26
25.0	8 46.60	46.94	47.27	47.61	47.95	48.29	48.63	48.97	49.31	49.65
25.1	8 49.98	50.32	50.66	51.00	51.34	51.68	52.02	52.35	52.69	53.03
25.2	8 53.37	53.71	54.05	54.39	54.72	55.06	55.40	55.74	56.08	56.42
25.3	8 56.76	57.10	57.43	57.77	58.11	58.45	58.79	59.13	59.47	59.80
25.4	8 60.14	60.48	60.82	61.16	61.50	61.84	62.17	62.51	62.85	63.19
25.5	8 63.53	63.87	64.21	64.55	64.88	65.22	65.56	65.90	66.24	66.58
25.6	8 66.92	67.25	67.59	67.93	68.27	68.61	68.95	69.29	69.62	69.96
25.7	8 70.30	70.64	70.98	71.32	71.66	72.00	72.33	72.67	73.01	73.35
25.8	8 73.69	74.03	74.37	74.70	75.04	75.38	75.72	76.06	76.40	76.74
25.9	8 77.08	77.41	77.75	78.09	78.43	78.77	79.11	79.45	79.78	80.12
26.0	8 80.46	80.80	81.14	81.48	81.82	82.15	82.49	82.83	83.17	83.51
26.1	8 83.85	84.19	84.53	84.86	85.20	85.54	85.88	86.22	86.56	86.90
26.2	8 87.23	87.57	87.91	88.25	88.59	88.93	89.27	89.60	89.94	90.28
26.3	8 90.62	90.96	91.30	91.64	91.98	92.31	92.65	92.99	93.33	93.67
26.4	8 94.01	94.35	94.68	95.02	95.36	95.70	96.04	96.38	96.72	97.05
26.5	8 97.39	97.73	98.07	98.41	98.75	99.09	99.43	99.76	*00.10	*00.44
26.6	9 00.78	01.12	01.46	01.80	02.13	02.47	02.81	03.15	03.49	03.83
26.7	9 04.17	04.50	04.84	05.18	05.52	05.86	06.20	06.54	06.88	07.21
26.8	9 07.55	07.89	08.23	08.57	08.91	09.25	09.58	09.92	10.26	10.60
26.9	9 10.94	11.28	11.62	11.95	12.29	12.63	12.97	13.31	13.65	13.99
27.0	9 14.33	14.66	15.00	15.34	15.68	16.02	16.36	16.70	17.03	17.37
27.1	9 17.71	18.05	18.39	18.73	19.07	19.40	19.74	20.08	20.42	20.76
27.2	9 21.10	21.44	21.78	22.11	22.45	22.79	23.13	23.47	23.81	24.15
27.3	9 24.48	24.82	25.16	25.50	25.84	26.18	26.52	26.85	27.19	27.53
27.4	9 27.87	28.21	28.55	28.89	29.23	29.56	29.90	30.24	30.58	30.03
27.5	9 31.26	31.60	31.93	32.27	32.61	32.95	33.29	33.63	33.97	34.31
27.6	9 34.64	34.98	35.32	35.66	36.00	36.34	36.68	37.01	37.35	37.69
27.7	9 38.03	38.37	38.71	39.05	39.38	39.72	40.06	40.40	40.74	41.08
27.8	9 41.42	41.76	42.09	42.43	42.77	43.11	43.45	43.79	44.13	44.46
27.9	9 44.80	45.14	45.48	45.82	46.16	46.50	46.83	47.17	47.51	47.85
28.0	9 48.19	48.53	48.87	49.21	49.54	49.88	50.22	50.56	50.90	51.24
28.1	9 51.58	51.91	52.25	52.59	52.93	53.27	53.61	53.95	54.28	54.62
28.2	9 54.96	55.30	55.64	55.98	56.32	56.66	56.99	57.33	57.67	58.01
28.3	9 58.35	58.69	59.03	59.36	59.70	60.04	60.38	60.72	61.06	61.40
28.4	9 61.73	62.07	62.41	62.75	63.09	63.43	63.77	64.11	64.44	64.78
28.5	9 65.12	65.46	65.80	66.14	66.48	66.81	67.15	67.49	67.83	68.17
28.6	9 68.51	68.85	69.18	69.52	69.86	70.20	70.54	70.88	71.22	71.56
28.7	9 71.89	72.23	72.57	72.91	73.25	73.59	73.93	74.26	74.60	74.94
28.8	9 75.28	75.62	75.96	76.30	76.63	76.97	77.31	77.65	77.99	78.33
28.9	9 78.67	79.01	79.34	79.68	80.02	80.36	80.70	81.04	81.38	81.71
29.0	9 82.05	82.39	82.73	83.07	83.41	83.75	84.08	84.42	84.76	85.10
29.1	9 85.44	85.78	86.12	86.46	86.79	87.13	87.47	87.81	88.15	88.49
29.2	9 88.83	89.16	89.50	89.84	90.18	90.52	90.86	91.20	91.53	91.87
29.3	9 92.21	92.55	92.89	93.23	93.57	93.91	94.24	94.58	94.92	95.26
29.4	9 95.60	95.94	96.28	96.61	96.95	97.29	97.63	97.97	98.31	98.65
29.5	9 98.99	99.32	99.66	00.00	*00.34	*00.68	*01.02	*01.36	*01.69	*02.03
29.6	10 02.37	02.71	03.05	03.39	03.73	04.06	04.40	04.74	05.08	05.42
29.7	10 05.76	06.10	06.44	06.77	07.11	07.45	07.79	08.13	08.47	08.81
29.8	10 09.14	09.48	09.82	10.16	10.50	10.84	11.18	11.51	11.85	12.19
29.9	10 12.53	12.87	13.21	13.55	13.89	14.22	14.56	14.90	15.24	15.58
30.0	10 15.92	16.26	16.59	16.93	17.27	17.61	17.95	18.29	18.63	18.96
30.1	10 19.30	19.64	19.98	20.32	20.66	21.00	21.34	21.67	22.01	22.35
30.2	10 22.69	23.03	23.37	23.71	24.04	24.38	24.72	25.06	25.40	25.74
30.3	10 26.08	26.41	26.75	27.09	27.43	27.77	28.11	28.45	28.79	29.12
30.4	10 29.46	29.80	30.14	30.48	30.82	31.16	31.49	31.83	32.17	32.51
30.5	10 32.85	33.19	33.53	33.86	34.20	34.54	34.88	35.22	35.56	35.90
30.6	10 36.24	36.57	36.91	37.25	37.59	37.93	38.27	38.61	38.94	39.28
30.7	10 39.62	39.96	40.30	40.64	40.98	41.31	41.65	41.99	42.33	42.67
30.8	10 43.01	43.35	43.69	44.02	44.36	44.70	45.04	45.38	45.72	46.06
30.9	10 46.39	46.73	47.07	47.41	47.75	48.09	48.43	48.76	49.10	49.44
31.0	10 49.78	50.12	50.46	50.80	51.14	51.47	51.81	52.15	52.49	52.83
31.1	10 53.17	53.51	53.84	54.18	54.52	54.86	55.20	55.54	55.88	56.22
31.2	10 56.55	56.89	57.23	57.57	57.91	58.25	58.59	58.92	59.26	59.60
31.3	10 59.94	60.28	60.62	60.96	61.29	61.63	61.97	62.31	62.65	62.99
31.4	10 63.33	63.67	64.00	64.34	64.68	65.02	65.36	65.70	66.04	66.37
31.5	10 66.71	67.05	67.39	67.73	68.07	68.41	68.74	69.08	69.42	69.76
31.6	10 70.10	70.44	70.78	71.12	71.45	71.79	72.13	72.47	72.81	73.15
31.7	10 73.49	73.82	74.16	74.50	74.84	75.18	75.52	75.86	76.19	76.53
31.8	10 76.87	77.21	77.55	77.89	78.23	78.57	78.90	79.24	79.58	79.92
31.9	10 80.26	80.60	80.94	81.27	81.61	81.95	82.29	82.63	82.97	83.31

Centimeters to Millibars

Based on the relation 1 centimeter of mercury at 0°C represents a pressure of 13.3322 millibars.

Note: Figures in last nine columns to be preceded by 9.

Centimeters	.00	.01	.02	.03	.04	.05	.06	.07	.08	.09
68.0	9 06.59	06.72	06.86	06.99	07.12	07.26	07.39	07.52	07.66	07.79
68.1	9 07.92	08.06	08.19	08.32	08.46	08.59	08.72	08.86	08.99	09.12
68.2	9 09.26	09.39	09.52	09.66	09.79	09.92	10.06	10.19	10.32	10.46
68.3	9 10.59	10.72	10.86	10.99	11.12	11.26	11.39	11.52	11.66	11.79
68.4	9 11.92	12.06	12.19	12.32	12.46	12.59	12.72	12.86	12.99	13.12
68.5	9 13.26	13.39	13.52	13.66	13.79	13.92	14.06	14.19	14.32	14.46
68.6	9 14.59	14.72	14.86	14.99	15.12	15.26	15.39	15.52	15.66	15.79
68.7	9 15.92	16.06	16.19	16.32	16.46	16.59	16.72	16.86	16.99	17.12
68.8	9 17.26	17.39	17.52	17.66	17.79	17.92	18.06	18.19	18.32	18.46
68.9	9 18.59	18.72	18.86	18.99	19.12	19.26	19.39	19.52	19.66	19.79
69.0	9 19.92	20.06	20.19	20.32	20.46	20.59	20.72	20.86	20.99	21.12
69.1	9 21.26	21.39	21.52	21.65	21.79	21.92	22.05	22.19	22.32	22.45
69.2	9 22.59	22.72	22.85	22.99	23.12	23.25	23.39	23.52	23.65	23.79
69.3	9 23.92	24.05	24.19	24.32	24.45	24.59	24.72	24.85	24.99	25.12
69.4	9 25.25	25.39	25.52	25.65	25.79	25.92	26.05	26.19	26.32	26.45
69.5	9 26.59	26.72	26.85	26.99	27.12	27.25	27.39	27.52	27.65	27.79
69.6	9 27.92	28.05	28.19	28.32	28.45	28.59	28.72	28.85	28.99	29.12
69.7	9 29.25	29.39	29.52	29.65	29.79	29.92	30.05	30.19	30.32	30.45
69.8	9 30.59	30.72	30.85	30.99	31.12	31.25	31.39	31.52	31.65	31.79
69.9	9 31.92	32.05	32.19	32.32	32.45	32.59	32.72	32.85	32.99	33.12
70.0	9 33.25	33.39	33.52	33.65	33.79	33.92	34.05	34.19	34.32	34.45
70.1	9 34.59	34.72	34.85	34.99	35.12	35.25	35.39	35.52	35.65	35.79
70.2	9 35.92	36.05	36.19	36.32	36.45	36.59	36.72	36.85	36.99	37.12
70.3	9 37.25	37.39	37.52	37.65	37.79	37.92	38.05	38.19	38.32	38.45
70.4	9 38.59	38.72	38.85	38.99	39.12	39.25	39.39	39.52	39.65	39.79
70.5	9 39.92	40.05	40.19	40.32	40.45	40.59	40.72	40.85	40.99	41.12
70.6	9 41.25	41.39	41.52	41.65	41.79	41.92	42.05	42.19	42.32	42.45
70.7	9 42.59	42.72	42.85	42.99	43.12	43.25	43.39	43.52	43.65	43.79
70.8	9 43.92	44.05	44.19	44.32	44.45	44.59	44.72	44.85	44.99	45.12
70.9	9 45.25	45.39	45.52	45.65	45.79	45.92	46.05	46.19	46.32	46.45
71.0	9 46.59	46.72	46.85	46.99	47.12	47.25	47.39	47.52	47.65	47.79
71.1	9 47.92	48.05	48.19	48.32	48.45	48.59	48.72	48.85	48.99	49.12
71.2	9 49.25	49.39	49.52	49.65	49.79	49.92	50.05	50.19	50.32	50.45
71.3	9 50.59	50.72	50.85	50.99	51.12	51.25	51.39	51.52	51.65	51.79
71.4	9 51.92	52.05	52.19	52.32	52.45	52.59	52.72	52.85	52.99	53.12
71.5	9 53.25	53.39	53.52	53.65	53.79	53.92	54.05	54.19	54.32	54.45
71.6	9 54.59	54.72	54.85	54.99	55.12	55.25	55.39	55.52	55.65	55.79
71.7	9 55.92	56.05	56.19	56.32	56.45	56.59	56.72	56.85	56.99	57.12
71.8	9 57.25	57.39	57.52	57.65	57.79	57.92	58.05	58.19	58.32	58.45
71.9	9 58.59	58.72	58.85	58.99	59.12	59.25	59.39	59.52	59.65	59.79
72.0	9 59.92	60.05	60.19	60.32	60.45	60.59	60.72	60.85	60.98	61.12
72.1	9 61.25	61.38	61.52	61.65	61.78	61.92	62.05	62.18	62.32	62.45
72.2	9 62.58	62.72	62.85	62.98	63.12	63.25	63.38	63.52	63.65	63.78
72.3	9 63.92	64.05	64.18	64.32	64.45	64.58	64.72	64.85	64.98	65.12
72.4	9 65.25	65.38	65.52	65.65	65.78	65.92	66.05	66.18	66.32	66.45
72.5	9 66.58	66.72	66.85	66.98	67.12	67.25	67.38	67.52	67.65	67.78
72.6	9 67.92	68.05	68.18	68.32	68.45	68.58	68.72	68.85	68.98	69.12
72.7	9 69.25	69.38	69.52	69.65	69.78	69.92	70.05	70.18	70.32	70.45
72.8	9 70.58	70.72	70.85	70.98	71.12	71.25	71.38	71.52	71.65	71.78
72.9	9 71.92	72.05	72.18	72.32	72.45	72.58	72.72	72.85	72.98	73.12
73.0	9 73.25	73.38	73.52	73.65	73.78	73.92	74.05	74.18	74.32	74.45
73.1	9 74.58	74.72	74.85	74.98	75.12	75.25	75.38	75.52	75.65	75.78
73.2	9 75.92	76.05	76.18	76.32	76.45	76.58	76.72	76.85	76.98	77.12
73.3	9 77.25	77.38	77.52	77.65	77.78	77.92	78.05	78.18	78.32	78.45
73.4	9 78.58	78.72	78.85	78.98	79.12	79.25	79.38	79.52	79.65	79.78
73.5	9 79.92	80.05	80.18	80.32	80.45	80.58	80.72	80.85	80.98	81.12
73.6	9 81.25	81.38	81.52	81.65	81.78	81.92	82.05	82.18	82.32	82.45
73.7	9 82.58	82.72	82.85	82.98	83.12	83.25	83.38	83.52	83.65	83.78
73.8	9 83.92	84.05	84.18	84.32	84.45	84.58	84.72	84.85	84.98	85.12
73.9	9 85.25	85.38	85.52	85.65	85.78	85.92	86.05	86.18	86.32	86.45
74.0	9 86.58	86.72	86.85	86.98	87.12	87.25	87.38	87.52	87.65	87.78
74.1	9 87.92	88.05	88.18	88.32	88.45	88.58	88.72	88.85	88.98	89.12
74.2	9 89.25	89.38	89.52	89.65	89.78	89.92	90.05	90.18	90.32	90.45
74.3	9 90.58	90.72	90.85	90.98	91.12	91.25	91.38	91.52	91.65	91.78
74.4	9 91.92	92.05	92.18	92.32	92.45	92.58	92.72	92.85	92.98	93.12
74.5	9 93.25	93.38	93.52	93.65	93.78	93.92	94.05	94.18	94.32	94.45
74.6	9 94.58	94.72	94.85	94.98	95.12	95.25	95.38	95.52	95.65	95.78
74.7	9 95.92	96.05	96.18	96.32	96.45	96.58	96.72	96.85	96.98	97.12
74.8	9 97.25	97.38	97.52	97.65	97.78	97.92	98.05	98.18	98.32	98.45
74.9	9 98.58	98.72	98.85	98.98	99.12	99.25	99.38	99.52	99.65	99.78
75.0	9 99.92	*00.05	*00.18	*00.31	*00.45	*00.58	*00.71	*00.85	*00.98	*01.11
75.1	10 01.25	01.38	01.51	01.65	01.78	01.91	02.05	02.18	02.31	02.45
75.2	10 02.58	02.71	02.85	02.98	03.11	03.25	03.38	03.51	03.65	03.78
75.3	10 03.91	04.05	04.18	04.31	04.45	04.58	04.71	04.85	04.98	05.11
75.4	10 05.25	05.38	05.51	05.65	05.78	05.91	06.05	06.18	06.31	06.45
75.5	10 06.58	06.71	06.85	06.98	07.11	07.25	07.38	07.51	07.65	07.78
75.6	10 07.91	08.05	08.18	08.31	08.45	08.58	08.71	08.85	08.98	09.11
75.7	10 09.25	09.38	09.51	09.65	09.78	09.91	10.05	10.18	10.31	10.45
75.8	10 10.58	10.71	10.85	10.98	11.11	11.25	11.38	11.51	11.65	11.78
75.9	10 11.91	12.05	12.18	12.31	12.45	12.58	12.71	12.85	12.98	13.11
76.0	10 13.25	13.38	13.51	13.65	13.78	13.91	14.05	14.18	14.31	14.45
76.1	10 14.58	14.71	14.85	14.98	15.11	15.25	15.38	15.51	15.65	15.78
76.2	10 15.91	16.05	16.18	16.31	16.45	16.58	16.71	16.85	16.98	17.11
76.3	10 17.25	17.38	17.51	17.65	17.78	17.91	18.05	18.18	18.31	18.45
76.4	10 18.58	18.71	18.85	18.98	19.11	19.25	19.38	19.51	19.65	19.78
76.5	10 19.91	20.05	20.18	20.31	20.45	20.58	20.71	20.85	20.98	21.11
76.6	10 21.25	21.38	21.51	21.65	21.78	21.91	22.05	22.18	22.31	22.45
76.7	10 22.58	22.71	22.85	22.98	23.11	23.25	23.38	23.51	23.65	23.78
76.8	10 23.91	24.05	24.18	24.31	24.45	24.58	24.71	24.85	24.98	25.11
76.9	10 25.25	25.38	25.51	25.65	25.78	25.91	26.05	26.18	26.31	26.45
77.0	10 26.58	26.71	26.85	26.98	27.11	27.25	27.38	27.51	27.65	27.78
77.1	10 27.91	28.05	28.18	28.31	28.45	28.58	28.71	28.85	28.98	29.11
77.2	10 29.25	29.38	29.51	29.65	29.78	29.91	30.05	30.18	30.31	30.45
77.3	10 30.58	30.71	30.85	30.98	31.11	31.25	31.38	31.51	31.65	31.78
77.4	10 31.91	32.05	32.18	32.31	32.45	32.58	32.71	32.85	32.98	33.11
77.5	10 33.25	33.38	33.51	33.65	33.78	33.91	34.05	34.18	34.31	34.45
77.6	10 34.58	34.71	34.85	34.98	35.11	35.25	35.38	35.51	35.65	35.78
77.7	10 35.91	36.05	36.18	36.31	36.45	36.58	36.71	36.85	36.98	37.11
77.8	10 37.25	37.38	37.51	37.65	37.78	37.91	38.05	38.18	38.31	38.45
77.9	10 38.58	38.71	38.85	38.98	39.11	39.24	39.38	39.51	39.64	39.78

TEMPERATURE CORRECTION FOR BAROMETER READINGS

Brass Scale — Metric Units

To reduce readings of a mercurial barometer with a brass scale to 0°C subtract the appropriate quantity as found in the table. These values are based on the coefficient of expansion of mercury $(181792 + 0.175t + 0.035116t^2) \times 10^{-9}$, and of brass 0.0000184 per °C. Corrections are in millimeters.

Temp. °C								Observed height in millimeters										
	620	630	640	650	660	670	680	690	700	710	720	730	740	750	760	770	780	790
0	0.00	0.00	0.00	0.00	0.00	0.00	0.00	0.00	0.00	0.00	0.00	0.00	0.00	0.00	0.00	0.00	0.00	0.00
1	.10	.10	.10	.11	.11	.11	.11	.11	.11	.12	.12	.12	.12	.12	.12	.13	.13	.13
2	.20	.21	.21	.21	.22	.22	.22	.23	.23	.23	.24	.24	.24	.25	.25	.25	.25	.26
3	.30	.31	.31	.32	.32	.33	.33	.34	.34	.35	.35	.36	.36	.37	.37	.38	.38	.39
4	.40	.41	.42	.42	.43	.44	.44	.45	.46	.46	.47	.48	.48	.49	.50	.50	.51	.52
5	0.51	0.51	0.52	0.53	0.54	0.55	0.56	0.56	0.57	0.58	0.59	0.60	0.60	0.61	0.62	0.63	0.64	0.64
6	.61	.62	.63	.64	.65	.66	.67	.68	.69	.70	.71	.71	.72	.73	.74	.75	.76	.77
7	.71	.72	.73	.74	.75	.77	.78	.79	.80	.81	.82	.83	.85	.86	.87	.88	.89	.90
8	.81	.82	.84	.85	.86	.87	.89	.90	.91	.93	.94	.95	.97	.98	.99	1.01	1.02	1.03
9	.91	.92	.94	.95	.97	.98	1.00	1.01	1.03	1.04	1.06	1.07	1.09	1.10	1.12	1.13	1.15	1.16
10	1.01	1.03	1.04	1.06	1.08	1.09	1.11	1.13	1.14	1.16	1.17	1.19	1.21	1.22	1.24	1.26	1.27	1.29
11	1.11	1.13	1.15	1.17	1.18	1.20	1.22	1.24	1.26	1.27	1.29	1.31	1.33	1.35	1.36	1.38	1.40	1.42
12	1.21	1.23	1.25	1.27	1.29	1.31	1.33	1.35	1.37	1.39	1.41	1.43	1.45	1.47	1.49	1.51	1.53	1.55
13	1.31	1.34	1.36	1.38	1.40	1.42	1.44	1.46	1.48	1.50	1.53	1.55	1.57	1.59	1.61	1.63	1.65	1.67
14	1.41	1.44	1.46	1.48	1.51	1.53	1.55	1.57	1.60	1.62	1.64	1.67	1.69	1.71	1.73	1.76	1.78	1.80
15	1.52	1.54	1.56	1.59	1.61	1.64	1.66	1.69	1.71	1.74	1.76	1.78	1.81	1.83	1.86	1.88	1.91	1.93
16	1.62	1.64	1.67	1.69	1.72	1.75	1.77	1.80	1.82	1.85	1.88	1.90	1.93	1.96	1.98	2.01	2.03	2.06
17	1.72	1.74	1.77	1.80	1.83	1.86	1.88	1.91	1.94	1.97	1.99	2.02	2.05	2.08	2.10	2.13	2.16	2.19
18	1.82	1.85	1.88	1.91	1.93	1.96	1.99	2.02	2.05	2.08	2.11	2.14	2.17	2.20	2.23	2.26	2.29	2.32
19	1.92	1.95	1.98	2.01	2.04	2.07	2.10	2.13	2.17	2.20	2.23	2.26	2.29	2.32	2.35	2.38	2.41	2.44
20	2.02	2.05	2.08	2.12	2.15	2.18	2.21	2.25	2.28	2.31	2.34	2.38	2.41	2.44	2.47	2.51	2.54	2.57
21	2.12	2.15	2.19	2.22	2.26	2.29	2.32	2.36	2.39	2.43	2.46	2.50	2.53	2.56	2.60	2.63	2.67	2.70
22	2.22	2.26	2.29	2.33	2.36	2.40	2.43	2.47	2.51	2.54	2.58	2.61	2.65	2.69	2.72	2.76	2.79	2.83
23	2.32	2.36	2.40	2.43	2.47	2.51	2.54	2.58	2.62	2.66	2.69	2.73	2.77	2.81	2.84	2.88	2.92	2.96
24	2.42	2.46	2.50	2.54	2.58	2.62	2.66	2.69	2.73	2.77	2.81	2.85	2.89	2.93	2.97	3.01	3.05	3.08
25	2.52	2.56	2.60	2.64	2.68	2.72	2.77	2.81	2.85	2.89	2.93	2.97	3.01	3.05	3.09	3.13	3.17	3.21
26	2.62	2.66	2.71	2.75	2.79	2.83	2.88	2.92	2.96	3.00	3.04	3.09	3.13	3.17	3.21	3.26	3.30	3.34
27	2.72	2.77	2.81	2.85	2.90	2.94	2.99	3.03	3.07	3.12	3.16	3.20	3.25	3.29	3.34	3.38	3.42	3.47
28	2.82	2.87	2.91	2.96	3.00	3.05	3.10	3.14	3.19	3.23	3.28	3.32	3.37	3.41	3.46	3.51	3.55	3.60
29	2.92	2.97	3.02	3.06	3.11	3.16	3.21	3.25	3.30	3.35	3.39	3.44	3.49	3.54	3.58	3.63	3.68	3.72
30	3.02	3.07	3.12	3.17	3.22	3.27	3.32	3.36	3.41	3.46	3.51	3.56	3.61	3.66	3.71	3.75	3.80	3.85
31	3.12	3.17	3.22	3.27	3.32	3.37	3.43	3.48	3.53	3.58	3.63	3.68	3.73	3.78	3.83	3.88	3.93	3.98
32	3.22	3.28	3.33	3.38	3.43	3.48	3.54	3.59	3.64	3.69	3.74	3.79	3.85	3.90	3.95	4.00	4.05	4.11
33	3.32	3.38	3.43	3.48	3.54	3.59	3.64	3.70	3.75	3.81	3.86	3.91	3.97	4.02	4.07	4.13	4.18	4.23
34	3.42	3.48	3.53	3.59	3.64	3.70	3.75	3.81	3.87	3.92	3.98	4.03	4.09	4.14	4.20	4.25	4.31	4.36
35	3.52	3.58	3.64	3.69	3.75	3.81	3.86	3.92	3.98	4.03	4.09	4.15	4.21	4.26	4.32	4.38	4.43	4.49

Brass Scale — English Units

Standard Temperature of scale 62°F; of mercury, 32°F. Zero correction at 28.5°F; subtract corrections above, add below. Owing to the difference in the standard temperature of English and metric scales, readings taken in inches to be reduced to centimeters should first be corrected for temperature.

Temp. °F	23.0 in.	23.5 in.	24.0 in.	24.5 in.	25.0 in.	25.5 in.	26.0 in.	26.5 in.	27.0 in.	27.5 in.	28.0 in.	28.5 in.	29.0 in.	29.5 in.	30.0 in.	30.5 in.	31.0 in.	31.5 in.
0	+.060	+.061	+.063	+.064	+.065	+.067	+.068	+.069	+.070	.072	.073	.075	.076	.077	.078	.080	.081	.082
2	.056	.057	.058	.060	.061	.062	.063	.065	.065	.067	.068	.069	.070	.072	.073	.074	.075	.077
4	.052	.053	.054	.055	.056	.057	.058	.060	.061	.062	.063	.064	.065	.066	.067	.069	.070	.071
6	.047	.048	.049	.051	.052	.053	.054	.055	.056	.057	.058	.059	.060	.061	.062	.063	.064	.065
8	.043	.044	.045	.046	.047	.048	.049	.050	.051	.052	.053	.054	.054	.056	.056	.057	.058	.059
10	.039	.040	.041	.042	.042	.043	.044	.045	.046	.047	.047	.048	.049	.050	.051	.052	.053	.054
12	.035	.036	.036	.037	.038	.039	.039	.040	.041	.042	.042	.043	.044	.045	.045	.046	.047	.048
14	.031	.031	.032	.033	.033	.034	.035	.035	.036	.037	.037	.038	.039	.039	.040	.041	.041	.042
16	.026	.027	.028	.028	.029	.029	.030	.031	.031	.032	.032	.033	.033	.034	.034	.035	.036	.036
18	.022	.023	.023	.024	.024	.025	.025	.026	.026	.027	.027	.028	.028	.029	.029	.030	.030	.031
20	.018	.018	.019	.019	.020	.020	.020	.021	.021	.022	.022	.022	.023	.023	.024	.024	.024	.025
22	.014	.014	.014	.015	.015	.015	.016	.016	.016	.017	.017	.017	.017	.018	.018	.018	.019	.019
24	.010	.010	.010	.010	.011	.011	.011	.011	.011	.012	.012	.012	.012	.013	.013	.013	.013	.013
26	.005	.006	.006	.006	.006	.006	.006	.006	.006	.007	.007	.007	.007	.007	.007	.007	.007	.003
28	+.001	+.001	+.001	+.001	+.001	+.001	+.001	+.002	+.002	+.002	+.002	+.002	+.002	+.002	+.002	+.002	+.002	+.002
30	−.003	−.003	−.003	−.003	−.003	−.003	−.003	−.003	−.003	−.003	−.003	−.004	−.004	−.004	−.004	−.004	−.004	−.004
32	.007	.007	.007	.008	.008	.008	.008	.008	.008	.008	.009	.009	.009	.009	.009	.009	.009	.010
34	.011	.011	.012	.012	.012	.012	.013	.013	.013	.013	.014	.014	.014	.014	.015	.015	.015	.015
36	.015	.016	.016	.016	.017	.017	.017	.018	.018	.018	.019	.019	.019	.020	.020	.020	.021	.021
38	.020	.020	.020	.021	.021	.022	.022	.023	.023	.023	.024	.024	.025	.025	.026	.026	.026	.027
40	.024	.024	.025	.025	.026	.026	.027	.027	.028	.028	.029	.030	.030	.031	.031	.032	.032	.033
42	.028	.029	.029	.030	.030	.031	.032	.032	.033	.033	.034	.035	.035	.036	.036	.037	.038	.038
44	.032	.033	.033	.034	.035	.036	.036	.037	.038	.038	.039	.040	.040	.041	.042	.043	.043	.044
46	.036	.037	.038	.039	.039	.040	.041	.042	.043	.043	.044	.045	.046	.047	.047	.048	.049	.050
48	.040	.041	.042	.043	.044	.045	.046	.047	.048	.049	.050	.051	.052	.053	.053	.054	.054	.055
50	.045	.046	.046	.048	.048	.050	.050	.052	.052	.053	.054	.055	.056	.057	.058	.059	.060	.061
52	.049	.050	.051	.052	.053	.054	.055	.056	.057	.058	.059	.061	.061	.063	.064	.065	.066	.067
54	.053	.054	.055	.057	.057	.059	.060	.061	.062	.063	.064	.066	.067	.068	.069	.070	.071	.073
56	.057	.058	.060	.061	.062	.063	.064	.066	.067	.068	.069	.071	.072	.073	.074	.076	.077	.078
58	.061	.063	.064	.065	.066	.068	.069	.071	.072	.073	.074	.076	.077	.079	.080	.081	.082	.084
60	.065	.067	.068	.070	.071	.073	.074	.076	.077	.078	.080	.081	.082	.084	.085	.087	.088	.090
62	.069	.071	.073	.074	.076	.077	.079	.080	.082	.083	.085	.086	.088	.089	.091	.092	.094	.095
64	.074	.075	.077	.079	.080	.082	.083	.085	.086	.088	.090	.092	.093	.095	.096	.098	.099	.101
66	.078	.079	.081	.083	.085	.087	.088	.090	.091	.093	.095	.097	.098	.100	.101	.103	.105	.107
68	.082	.084	.085	.088	.089	.091	.093	.095	.096	.098	.100	.102	.103	.105	.107	.109	.110	.113
70	.086	.088	.090	.092	.094	.096	.097	.100	.101	.103	.105	.107	.109	.111	.112	.115	.116	.118

Observed height in inches

Temp. °F	23.0 in.	23.5 in.	24.0 in.	24.5 in.	25.0 in.	25.5 in.	26.0 in.	26.5 in.	27.0 in.	27.5 in.	28.0 in.	28.5 in.	29.0 in.	29.5 in.	30.0 in.	30.5 in.	31.0 in.	31.5 in.
72	.090	.092	.094	.096	.098	.100	.102	.104	.106	.108	.110	.112	.114	.116	.118	.120	.122	.124
74	.094	.096	.098	.101	.103	.105	.107	.109	.111	.113	.115	.117	.119	.121	.123	.126	.127	.130
76	.098	.101	.103	.105	.107	.110	.111	.114	.116	.118	.120	.122	.124	.127	.128	.131	.133	.135
78	.103	.105	.107	.110	.112	.114	.116	.119	.120	.123	.125	.128	.129	.132	.134	.137	.138	.141
80	.107	.109	.111	.114	.116	.119	.121	.123	.125	.128	.130	.133	.135	.137	.139	.142	.144	.147
82	.111	.113	.116	.119	.121	.123	.125	.128	.130	.133	.135	.138	.140	.143	.145	.148	.149	.152
84	.115	.118	.120	.123	.125	.128	.130	.133	.135	.138	.140	.143	.145	.148	.150	.153	.155	.158
86	.119	.122	.124	.127	.130	.133	.135	.138	.140	.143	.145	.148	.150	153	.155	.159	.161	.164
88	.123	.126	.129	.132	.134	.137	.139	.143	.145	.148	.150	.153	.155	.159	.161	.164	.166	.169
90	.127	.130	.133	.136	.138	.142	.144	.147	.150	.153	.155	.158	.161	.164	.166	.170	.172	.175
92	.132	.134	.137	.141	.143	.146	.149	.152	.154	.158	.160	.163	.166	.169	.172	.175	.177	.181
94	.136	.139	.142	.145	.147	.151	.153	.157	.159	.163	.165	.169	.171	.175	.177	.180	.183	.186
96	.140	.143	.146	.150	.152	.155	.158	.161	.164	.168	.170	.174	.176	.180	.182	.186	.188	.192
98	.144	.147	.150	.154	.156	.160	.163	.166	.169	.172	.175	.179	.181	.185	.188	.191	.194	.197
100	.148	.151	.154	.158	.161	.164	.167	.171	.174	.177	.180	.184	.187	.190	.193	.197	.200	.203

TEMPERATURE CORRECTION, GLASS SCALE

Metric

To reduce readings of a mercurial barometer with a glass scale to 0°C. subtract the appropriate quantity as found in table.

Temp °C.	Observed height in centimeters. 70 cm.	71 cm.	72 cm.	73 cm.	74 cm.	75 cm.	76 cm.	77 cm.	78 cm.	Temp. °C.	Observed height in centimeters. 70 cm.	71 cm.	72 cm.	73 cm.	74 cm.	75 cm.	76 cm.	77 cm.	78 cm.
0	0.000	0.000	0.000	0.000	0.000	0.000	0.000	0.000	0.000	15	0.181	0.184	0.186	0.189	0.191	0.193	0.196	0.198	0.201
1	.012	.012	.013	.013	.013	.013	.013	.013	.014	16	.194	.196	.199	.201	.204	.207	.209	.212	.214
2	.025	.025	.025	.026	.026	.026	.026	.027	.027	17	.205	.208	.210	.213	.216	.219	.221	.224	.227
3	.036	.036	.037	.037	.038	.038	.039	.039	.040	18	.217	.220	.223	.226	.229	.232	.235	.238	.241
4	.048	.049	.049	.050	.051	.051	.052	.053	.053	19	.230	.233	.236	.239	.242	.245	.248	.251	.254
5	.060	.061	.062	.063	.064	.064	.065	.066	.067	20	.242	.245	.248	.252	.255	.258	.261	.264	.268
6	.073	.074	.074	.076	.077	.077	.078	.079	.080	21	.254	.258	.261	.264	.268	.271	.275	.278	.281
7	.085	.086	.087	.088	.089	.091	.092	.093	.094	22	.266	.269	.273	.276	.280	.283	.287	.290	.294
8	.096	.098	.099	.100	.101	.103	.104	.105	.107	23	.278	.282	.285	.289	.293	.296	.300	.304	.308
9	.109	.110	.111	.113	.114	.116	.117	.119	.120	24	.290	.294	.298	.302	.306	.310	.313	.317	.321
10	.121	.122	.124	.126	.127	.129	.130	.132	.134	25	.303	.307	.311	.315	.319	.323	.327	.331	.335
11	.133	.135	.137	.138	.140	.142	.143	.145	.147	26	.313	.319	.323	.327	.332	.336	.340	.344	.348
12	.144	.146	.148	.150	.152	.154	.156	.158	.160	27	.326	.331	.335	.339	.344	.348	.352	.357	.361
13	.157	.159	.161	.163	.165	.167	.169	.171	.174	28	.339	.343	.348	.352	.357	.361	.366	.370	.375
14	.169	.171	.174	.176	.178	.180	.183	.185	.187	29	.351	.356	.360	.365	.370	.374	.379	.384	.388
										30	.363	.368	.373	.378	.383	.387	.392	.397	.402

WEIGHT IN GRAMS OF A CUBIC METER OF SATURATED AQUEOUS VAPOR

(From Smithsonian Tables)

Mass in grams per cubic meter.

Temp. °C	0.0	1.0	2.0	3.0	4.0	5.0	6.0	7.0	8.0	9.0
−20	1.074	.988	.909	.836	.768	.705	.646	.592	.542	.496
−10	2.358	2.186	2.026	1.876	1.736	1.605	1.483	1.369	1.264	1.165
− 0	4.847	4.523	4.217	3.930	3.660	3.407	3.169	2.946	2.737	2.541
+ 0	4.847	5.192	5.559	5.947	6.360	6.797	7.260	7.750	8.270	8.819
+ 10	9.399	10.01	10.66	11.35	12.07	12.83	13.63	14.84	15.37	16.21
+ 20	17.30	18.34	19.43	20.58	21.78	23.05	24.38	25.78	27.24	28.78
+ 30	30.38	32.07	33.83	35.68	37.61	39.63	41.75	43.96	46.26	28.67

EFFICIENCY OF DRYING AGENTS

Compiled by John H. Yoe

A. Drying agents depending upon chemical action (absorption) for their efficiency:*

Substance	Residual water, mg per liter of dry air**	Reference
P_2O_5	<1 mg in 40,000 l,	Morley, Am. J. Sci., 34, 199 (1887); J.A.C.S., 26, 1171 (1904).
$Mg(ClO_4)_2$ anhyd.	"Unweighable" in 2101,	Willard and Smith, J.A.C.S., 44, 2255 (1922).
BaO	0.00065	Bower, Bur. Std. J. Res., 12, 241 (1934).
KOH fused	0.002	Baxter and Starkweather, J.A.C.S., 38, 2038 (1916).
CaO	0.003	Bower, loc. cit.
H_2SO_4	0.003	Baxter and Starkweather, loc. cit.
$CaSO_4$ anhyd.	0.005	Bower loc. cit.
Al_2O_3	0.005	Ibid.
KOH sticks	0.014	Ibid.
NaOH fused	0.16	Baxter and Starkweather, loc. cit.
$CaBr_2$	0.18	Baxter and Warren, J.A.C.S., 33, 340 (1911).
$CaCl_2$ fused	0.34	Baxter and Starkweather, loc. cit.
NaOH sticks	0.80	Bower loc. cit.
$Ba(ClO_4)_2$	0.82	Ibid.
$ZnCl_2$	0.85	Baxter and Warren, loc. cit.
$ZnBr_2$	1.16	Ibid.
$CaCl_2$ granular	1.5	Bower, loc. cit.
$CuSO_4$ anhyd.	2.8	Ibid.

B. Drying agents depending upon physical action (adsorption) for their efficiency:* Alumina (low temperature fired), asbestos, charcoal, clay and porcelain (low temperature fired), glass wool, kieselguhr, silica gel, refrigeration.

* It should be noted that the efficiency of some drying agents (e.g. Al_2O_3 and anhydrous $CaCl_2$, and probably also BaO, anhydrous $Mg(ClO_4)_2$, $Mg(ClO_4)_2 \cdot 3H_2O$, anhydrous $Ba(ClO_4)_2$, and $CaSO_4$) depends upon both adsorption and absorption.

** 30°C. for Bower's values; others 25°C. or room temp.

REDUCTION OF BAROMETER TO SEA LEVEL

The correction to be added to reduce barometric readings to "sea level" values depends principally on three factors: The temperature of the air column (assumed) from the station to sea level, the altitude of the station, and the value of the reading itself. Two tables are provided. Table I is entered with the altitude and assumed temperature and is a fator "2000 m" taken out. Table II is entered with the above factor and the approximate barometer reading and the final correction taken out.

The correction is to be added. If B_0 is the corrected or sea level value; B the barometer reading at the station; C the correction, —

$$C = B_0 - B = B(10^m - 1)$$

The actual barometer reading at the station should be corrected for temperature of the mercury column by the usual methods before entering the tables or applying the sea level correction.

A complete explanation of the theory of the corrections and a more extended set of tables will be found in the Smithsonian Meterological Tables.

Latitude Factor

The influence of the latitude on the value of the correction is usually negligible, being overshadowed by uncertainties in the assumed temperature of the air column. For cases where this correction is desirable the table below is provided. The value of the temperature-altitude factor "2000 m" obtained in Table I is corrected for latitude by subtracting for latitudes 0-45° and adding for latitudes from 45-90° the values found. With this corrected value of "2000 m" Table II is entered for the value of the correction.

LATITUDE FACTOR

To be used in connection with Tables I and II, either English or metric units, to obtain latitude corrections to temperature-altitude factor. For latitudes 0-45° subtract the correction. For latitudes 45-90° add the correction.

Temp.-Alt. from Table I	Latitude			
	0°	15°	30°	45°
100	0.3	0.2	0.1	0.0
200	0.5	0.5	0.3	0.0
300	0.8	0.7	0.4	0.0
	90°	75°	60°	45°

METRIC UNITS—TABLE I

Values of the temperature-altitude factor (2000 m.) for entering table II.

Altitude in meters	Assumed temperature of air column °C									
	−16°	−8°	0°	+4°	+8°	+12°	+16°	+20°	+24°	+28°
10	1.2	1.1	1.1	1.1	1.0	1.0	1.0	1.0	1.0	1.0
50	5.8	5.6	5.4	5.3	5.2	5.2	5.1	5.0	4.9	4.9
100	11.5	11.2	10.8	10.7	10.5	10.3	10.2	10.0	9.9	9.7
150	17.3	16.7	16.2	16.0	15.7	15.5	15.3	15.0	14.8	14.6
200	23.0	22.3	21.6	21.3	21.0	20.7	20.3	20.0	19.7	19.5
250	28.8	27.9	27.0	26.6	26.2	25.8	25.4	25.0	24.7	24.3
300	34.5	33.5	32.5	32.0	31.5	31.0	30.5	30.1	29.6	29.2
350	40.3	39.0	37.9	37.3	36.7	36.2	35.6	35.1	34.6	34.0
400	46.0	44.6	43.3	42.6	42.0	41.3	40.7	40.1	39.5	38.9
450	51.8	50.2	48.7	47.9	47.2	46.5	45.8	45.1	44.4	43.8
500	57.5	55.8	54.1	53.3	52.4	51.6	50.9	50.1	49.4	48.6
550	63.3	61.4	59.5	58.6	57.7	56.8	55.9	55.1	54.3	53.5
600	69.0	66.9	64.9	63.9	62.9	62.0	61.0	60.1	59.2	58.3
650	74.8	72.5	70.3	69.2	68.2	67.1	66.1	65.1	64.2	63.2
700	80.6	78.1	75.7	74.6	73.4	72.3	71.2	70.1	69.1	68.1
750	86.3	83.7	81.1	79.9	78.7	77.5	76.3	75.1	74.0	72.9
800	92.1	89.2	86.5	85.2	83.9	82.6	81.4	80.1	79.0	77.8
850	97.8	94.8	92.0	90.5	89.2	87.8	86.4	85.2	83.9	82.7
900	103.6	100.4	97.4	95.9	94.4	93.0	91.5	90.2	88.8	87.5
950	109.3	106.0	102.8	101.2	99.6	98.1	96.6	95.2	93.8	92.4
1000	115.1	111.5	108.2	106.5	104.9	103.3	101.7	100.2	98.7	97.3
1050	120.8	117.1	113.6	111.8	110.1	108.4	106.8	105.2	103.6	102.1
1100	126.6	122.7	119.0	117.2	115.4	113.6	111.9	110.2	108.6	107.0
1150	132.3	128.3	124.4	122.5	120.6	118.8	117.0	115.2	113.5	111.8
1200	138.1	133.8	129.8	127.8	125.9	123.9	122.0	120.2	118.4	116.7
1250	143.8	139.4	135.2	133.1	131.1	129.1	127.1	125.2	123.4	121.6
1300	149.6	145.0	140.6	138.5	136.3	134.3	132.2	130.2	128.3	126.4
1350	155.3	150.6	146.0	143.8	141.6	139.4	137.3	135.2	133.2	131.3
1400	161.1	156.2	151.4	149.1	146.8	144.6	142.4	140.2	138.2	136.2
1450	166.8	161.7	156.8	154.5	152.1	149.7	147.5	145.3	143.1	141.0
1500	172.6	167.3	162.3	159.8	157.3	154.9	152.5	150.3	148.0	145.9
1550	178.3	172.9	167.7	165.1	162.6	160.1	157.6	155.3	153.0	150.7
1600	184.1	178.5	173.1	170.4	167.8	165.2	162.7	160.3	157.9	155.6
1650	189.8	184.0	178.5	175.7	173.0	170.4	167.8	165.3	162.8	160.5
1700	195.6	189.6	183.9	181.1	178.3	175.6	172.9	170.3	167.8	165.3
1750	201.4	195.2	189.3	186.4	183.5	180.7	178.0	175.3	172.7	170.2
1800	207.1	200.8	194.7	191.7	188.8	185.9	183.1	180.3	177.6	175.0
1850	212.9	206.3	200.1	197.0	194.0	191.0	188.1	185.3	182.6	179.9
1900	218.6	211.9	205.5	202.4	199.3	196.2	193.2	190.3	187.5	184.8
1950	224.4	217.5	210.9	207.7	204.5	201.4	198.3	195.3	192.4	189.6
2000	230.1	223.0	216.3	213.0	209.7	206.5	203.4	200.3	197.4	194.5
2050	235.9	228.6	221.7	218.3	215.0	211.7	208.5	205.3	202.3	199.3
2100	241.6	234.2	227.1	223.7	220.2	216.8	213.5	210.4	207.2	204.2
2150	247.4	239.8	232.5	229.0	225.5	222.0	218.6	215.4	212.2	209.1
2200	253.1	245.4	237.9	234.3	230.7	227.2	223.7	220.4	217.1	213.9
2250	258.9	250.9	243.4	239.6	235.9	232.3	228.8	225.4	222.0	218.8
2300	264.6	256.5	248.8	245.0	241.2	237.5	233.9	230.4	227.0	223.6
2350	270.4	262.1	254.2	250.3	246.4	242.7	239.0	235.4	231.9	228.5
2400	276.1	267.7	259.6	255.6	251.7	247.8	244.0	240.4	236.8	233.4
2450	281.9	273.2	265.0	260.9	256.9	253.0	249.1	245.4	241.8	238.2
2500	287.6	278.8	270.4	266.2	262.2	258.1	254.2	250.4	246.7	243.1
2550	293.4	284.4	275.8	271.6	267.4	263.3	259.3	255.4	251.6	247.9
2600	299.1	290.0	281.2	276.9	272.6	268.5	264.4	260.4	256.6	252.8
2650	304.9	295.5	286.6	282.2	277.9	273.6	269.5	265.4	261.5	257.7
2700	310.6	301.1	292.0	287.5	283.1	278.8	274.5	270.4	266.4	262.5
2750	316.4	306.7	297.4	292.9	288.4	283.9	279.6	275.4	271.4	267.4
2800	322.1	312.3	302.8	298.2	293.6	289.1	284.7	280.4	276.3	272.2
2850	327.9	317.8	308.2	303.5	298.8	294.3	289.8	285.4	281.2	277.1
2900	333.6	323.4	313.6	308.8	304.1	299.4	294.9	290.4	286.2	282.0
2950	339.4	329.0	319.0	314.2	309.3	304.6	299.9	295.5	291.1	286.8
3000	345.1	334.5	324.4	319.5	314.6	309.7	305.0	300.5	296.0	291.7

METRIC UNITS — TABLE II
Values of Correction to be Added

Barometer reading

Temp.-alt.-factor	780 mm	760 mm	740 mm	720 mm	700 mm
1	0.9	0.9	0.9	0.8	0.8
5	4.5	4.4	4.3	4.2	4.0
10	9.0	8.8	8.6	8.3	8.1
15	13.6	13.2	12.9	12.5	12.2
20	18.2	17.7	17.2	16.8	16.3
25	22.8	22.2	21.6	21.0	20.4
30	27.4	26.7	26.0	25.3	24.6
35	—	31.2	30.4	29.6	28.8

Temp.-alt.-factor	760 mm	740 mm	720 mm	700 mm	680 mm	660 mm
40	35.8	34.9	33.9	33.0	32.0	31.1
45	40.4	39.3	38.3	37.2	36.2	35.1
50	45.0	43.8	42.7	41.5	40.3	39.1
55	49.7	48.4	47.1	45.8	44.5	43.1
60	—	52.9	51.5	50.1	48.6	47.2
65	—	57.5	55.9	54.4	52.8	51.3
70	—	62.1	60.4	58.7	57.1	55.4
75	—	66.7	64.9	63.1	61.3	59.5

Temp.-alt.-factor	720 mm	700 mm	680 mm	660 mm	640 mm
80	69.5	67.5	65.6	63.7	61.7
85	74.0	72.0	69.9	67.9	65.8
90	78.6	76.4	74.2	72.1	69.9
95	83.2	80.9	78.6	76.3	74.0
100	87.9	85.4	83.0	80.5	78.1
105	—	89.9	87.4	84.8	82.2
110	—	94.5	91.8	89.1	86.4
115	—	99.1	96.3	93.4	90.6
120	—	103.7	100.7	97.8	94.8
125	—	108.3	105.3	102.2	99.1

Temp.-alt.-factor	680 mm	660 mm	640 mm	620 mm	600 mm
125	105.3	102.2	99.1	96.0	92.9
130	109.8	106.6	103.3	100.1	96.9
135	114.3	111.0	107.6	104.3	100.9
140	118.9	115.4	111.9	108.4	104.9
145	123.5	119.9	116.3	112.6	109.0
150	128.2	124.4	120.6	116.9	113.1
155	—	128.9	125.0	121.1	117.2
160	—	133.5	129.4	125.4	121.4
165	—	138.1	133.9	129.7	125.5
170	—	142.7	138.4	134.0	129.7

Barometer reading

Temp.-alt.-factor	640 mm	620 mm	600 mm	580 mm	560 mm
170	138.4	134.0	129.7	125.4	121.1
175	142.9	138.4	133.9	129.5	125.0
180	147.4	142.8	138.2	133.6	129.0
185	151.9	147.2	142.4	137.7	132.9
190	156.5	151.6	146.7	141.8	136.9
195	161.1	156.1	151.0	146.0	141.0
200	165.7	160.5	155.4	150.2	145.0
205	170.4	165.0	159.7	154.4	149.1
210	—	169.6	164.1	158.6	153.2
215	—	174.1	168.5	162.9	157.3

Temp.-alt.-factor	620 mm	600 mm	580 mm	560 mm	540 mm
215	174.1	168.5	162.9	157.3	151.7
220	178.7	172.9	167.2	161.4	155.7
225	183.3	177.4	171.5	165.6	159.7
230	188.0	181.9	175.8	169.8	163.7
235	192.6	186.4	180.2	174.0	167.8
240	—	191.0	184.6	178.2	171.9
245	—	195.5	189.0	182.5	176.0
250	—	200.1	193.4	186.8	180.1
255	—	204.7	197.9	191.1	184.3
260	—	209.4	202.4	195.4	188.4

Temp.-alt.-factor	580 mm	560 mm	540 mm	520 mm
260	202.4	195.4	188.4	181.5
265	206.9	199.8	192.6	185.5
270	211.3	204.2	196.9	189.6
275	216.0	208.6	201.1	193.7
280	220.6	213.0	205.4	197.8
285	225.2	217.5	209.7	201.9
290	229.9	222.0	214.0	206.1
295	—	226.5	218.4	210.3
300	—	231.0	222.8	214.5

Temp.-alt.-factor	560 mm	540 mm	520 mm	500 mm	480 mm
305	235.6	227.2	218.8	210.3	201.9
310	240.2	231.6	223.0	214.4	205.9
315	244.8	236.0	227.3	218.6	209.8
320	249.4	240.5	231.6	222.7	213.8
325	254.1	245.0	236.0	226.9	217.8
330	—	249.6	240.3	231.1	221.8
335	—	254.1	244.7	235.3	225.9
340	—	258.7	249.1	239.6	230.0
345	—	263.3	253.6	243.8	234.1

ENGLISH UNITS — TABLE I
Values of the temperature-altitude factor (2000 m.) for entering table II

Altitude feet	−20	0	+10	+20	+30	+40	+50	+60	+70	+80
200	7.4	7.1	6.9	6.8	6.6	6.5	6.3	6.2	6.1	6.0
400	14.8	14.1	13.8	13.5	13.2	13.0	12.7	12.4	12.2	11.9
600	22.2	21.2	20.7	20.3	19.9	19.5	19.0	18.6	18.2	17.9
800	29.6	28.3	27.7	27.1	26.5	25.9	25.4	24.8	24.3	23.8
1000	37.0	35.3	34.6	33.8	33.1	32.4	31.7	31.1	30.4	29.8
1200	44.3	42.4	41.5	40.6	39.7	38.9	38.1	37.3	36.5	35.8
1400	51.7	49.5	48.4	47.4	46.4	45.4	44.4	43.5	42.6	41.7
1600	59.1	56.5	55.3	54.1	53.0	51.9	50.8	49.7	48.7	47.7
1800	66.5	63.6	62.2	60.9	59.6	58.4	57.1	55.9	54.7	53.6
2000	73.9	70.6	69.1	67.7	66.2	64.8	63.4	62.1	60.8	59.6
2200	81.3	77.7	76.0	74.4	72.9	71.3	69.8	68.3	66.9	65.5
2400	88.7	84.8	82.9	81.2	79.5	77.8	76.1	74.5	73.0	71.5
2600	96.1	91.8	89.9	87.9	86.1	84.3	82.5	80.7	79.1	77.5
2800	103.5	98.9	96.8	94.7	92.7	90.8	88.8	87.0	85.1	83.4
3000	110.9	106.0	103.7	101.5	99.3	97.2	95.2	93.2	91.2	89.4
3200	118.2	113.0	110.6	108.2	106.0	103.7	101.5	99.4	97.3	95.3
3400	125.6	120.1	117.5	115.0	112.6	110.2	107.9	105.6	103.4	101.3
3600	133.0	127.2	124.4	121.8	119.2	116.7	114.2	111.8	109.5	107.2
3800	140.4	134.2	131.3	128.5	125.8	123.2	120.5	118.0	115.5	113.2
4000	147.8	141.3	138.2	135.3	132.4	129.6	126.9	124.2	121.6	119.2
4200	155.2	148.3	145.1	142.1	139.1	136.1	133.2	130.4	127.7	125.1
4400	162.6	155.4	152.0	148.8	145.7	142.6	139.6	136.6	133.8	131.1
4600	170.0	162.5	159.0	155.6	152.3	149.1	145.9	142.8	139.9	137.0
4800	177.3	169.5	165.9	162.3	158.9	155.6	152.2	149.0	145.9	143.0
5000	184.7	176.6	172.8	169.1	165.6	162.0	158.6	155.2	152.0	148.9
5200	192.1	183.7	179.7	175.9	172.2	168.5	164.9	161.5	158.1	154.9
5400	199.5	190.7	186.6	182.6	178.8	175.0	171.3	167.7	164.2	160.8
5600	206.9	197.8	193.5	189.4	185.4	181.5	177.6	173.9	170.3	166.8
5800	214.3	204.8	200.4	196.2	192.0	188.0	184.0	180.1	176.3	172.8
6000	221.7	211.9	207.3	202.9	198.7	194.4	190.3	186.3	182.4	178.7
6200	229.1	219.0	214.2	209.7	205.3	200.9	196.6	192.5	188.5	184.7
6400	236.4	226.0	221.1	216.4	211.9	207.4	203.0	198.7	194.6	190.6
6600	243.8	233.1	228.0	223.2	218.5	213.0	209.3	204.9	200.7	196.6
6800	251.2	240.1	235.0	230.0	225.1	220.4	215.7	211.1	206.7	202.5
7000	258.6	247.2	241.9	236.7	231.8	226.8	222.0	217.3	212.8	208.5
7200	266.0	254.3	248.8	243.5	238.4	233.3	228.4	223.5	218.9	214.4
7400	273.4	261.3	255.7	250.2	245.0	239.8	234.7	229.7	225.0	220.4
7600	280.8	268.4	262.6	257.0	251.6	246.3	241.0	235.9	231.1	226.4
7800	288.1	275.4	269.5	263.8	258.2	252.8	247.4	242.2	237.1	232.3
8000	295.5	282.5	276.4	270.5	264.8	259.2	253.7	248.4	243.2	238.3
8200	302.9	289.6	283.3	277.3	271.5	265.7	260.1	254.6	249.3	244.2
8400	310.3	296.6	290.2	284.0	278.1	272.2	266.4	260.8	255.4	250.2
8600	317.7	303.7	297.1	290.8	284.7	278.7	272.7	267.0	261.4	256.1
8800	325.1	310.7	304.0	297.6	291.3	285.2	279.1	273.2	267.5	262.1
9000	332.5	317.8	310.9	304.3	297.9	291.6	285.4	279.4	273.6	268.0

ENGLISH UNITS — TABLE II

Value of Correction to be Added.

Temp. alt. factor	31	30	29	28	27
	in.	in.	in.	in.	in.
1	0.04	0.03	0.03	—	—
5	0.18	0.17	0.17	—	—
10	0.36	0.35	0.34	0.32	—
15	0.54	0.52	0.51	0.49	—
20	0.72	0.70	0.68	0.65	—
25	—	0.88	0.85	0.82	—
30	—	1.05	1.02	0.98	—
35	—	1.23	1.19	1.15	—
40	—	1.41	1.37	1.32	1.27
45	—	1.60	1.54	1.49	1.44
50	—	—	1.72	1.66	1.60
55	—	—	1.90	1.83	1.76
60	—	—	2.07	2.00	1.93
65	—	—	2.25	2.18	2.10
70	—	—	2.43	2.35	2.27
75	—	—	—	2.53	2.43
80	—	—	—	2.70	2.60

Temp. alt. factor	26	25	24	23	22
	in.	in.	in.	in.	in.
165	5.44	5.23	5.02	—	—
170	5.62	5.40	5.19	—	—
175	—	5.58	5.36	—	—
180	—	5.76	5.53	5.30	—
185	—	5.93	5.70	5.46	—
190	—	6.11	5.87	5.62	—
195	—	6.29	6.04	5.79	—
200	—	—	6.21	5.96	—
205	—	—	6.39	6.12	—
210	—	—	6.56	6.29	—
215	—	—	6.74	6.46	—
220	—	—	6.92	6.63	6.34
225	—	—	7.10	6.80	6.51
230	—	—	7.28	6.97	6.67
235	—	—	7.46	7.15	6.84
240	—	—	—	7.32	7.00
245	—	—	—	7.49	7.17

Temp. alt. factor	28	27	26	25	24
	in.	in.	in.	in.	in.
75	2.53	2.43	2.34	—	—
80	2.70	2.60	2.51	—	—
85	2.88	2.78	2.67	—	—
90	3.06	2.95	2.84	—	—
95	3.24	3.12	3.01	—	—
100	3.42	3.29	3.17	—	—
105	3.60	3.47	3.34	3.21	—
110	—	3.85	3.51	3.38	—
115	—	3.82	3.68	3.54	—
120	—	4.00	3.85	3.70	—
125	—	4.18	4.02	3.87	—
130	—	4.36	4.20	4.04	—
135	—	4.54	4.37	4.20	—
140	—	—	4.55	4.37	4.20
145	—	—	4.72	4.54	4.36
150	—	—	4.90	4.71	4.52
155	—	—	5.08	4.88	4.69
160	—	—	5.26	5.06	4.85

Temp. alt. factor	23	22	21	20
	in.	in.	in.	in.
250	7.67	7.34	—	—
255	7.85	7.51	—	—
260	8.03	7.68	7.33	—
265	8.21	7.85	7.49	—
270	8.39	8.02	7.66	—
275	8.57	8.19	7.82	—
280	—	8.37	7.99	—
285	—	8.54	8.16	—
290	—	8.72	8.32	—
295	—	8.90	8.49	8.09
300	—	9.08	8.66	8.25
305	—	9.26	8.83	8.41
310	—	9.44	9.01	8.58
315	—	9.62	9.18	8.74
320	—	9.80	9.35	8.91
325	—	—	9.53	9.08
330	—	—	9.71	9.24

REDUCTION OF BAROMETER TO GRAVITY AT SEA LEVEL

Metric Units
Correction to be subtracted given in millimeters
(From Smithsonian Physical Tables)

Height above sea level in meters	Observed Height of Barometer in Millimeters						
	500	550	600	650	700	750	800
100	—	—	—	—	.02	.02	.02
200	—	—	—	—	.04	.05	.05
300	—	—	—	—	.07	.07	.07
400	—	—	—	—	.09	.10	.10
500	—	—	—	—	.11	.12	.13
600	—	—	—	.12	.13	.14	—
700	—	—	—	.14	.15	.16	—
800	—	—	—	.16	.18	.19	—
900	—	—	—	.18	.20	.22	—
1000	—	.18	.19	.20	.22	.24	—
1100	—	.19	.21	.22	.24	—	—
1200	—	.21	.23	.24	.26	—	—
1300	—	.22	.24	.26	.29	—	—
1400	—	.24	.26	.28	.31	—	—
1500	.24	.26	.28	.30	.33	—	—
1600	.25	.28	.30	.32	—	—	—
1700	.27	.30	.32	.34	—	—	—
1800	.28	.31	.34	.36	—	—	—
1900	.30	.33	.36	.39	—	—	—
2000	.31	.34	.38	.41	—	—	—
2100	.33	.36	.40	—	—	—	—
2200	.35	.38	.41	—	—	—	—
2300	.36	.40	.43	—	—	—	—

Height above sea level in meters	Observed Height of Barometer in Millimeters						
	500	550	600	650	700	750	800
2400	.38	.42	.45	—	—	—	—
2500	.39	.43	.47	—	—	—	—

English Units

Height above sea level in feet	Observed Height in Inches						
	18	20	22	24	26	28	30
1000	—	—	—	—	.003	.003	.003
2000	—	—	—	.004	.005	.005	.006
3000	—	—	.007	.007	.008	.008	—
4000	—	—	.009	.009	.010	—	—
4500	—	—	0.10	0.10	0.11	—	—
5000	—	0.10	0.11	0.11	0.12	—	—
5500	—	0.11	0.12	0.13	—	—	—
6000	—	0.11	0.13	0.14	—	—	—
6500	0.11	0.12	0.14	0.15	—	—	—
7000	0.12	0.13	0.15	0.16	—	—	—
7500	0.13	0.14	0.16	0.17	—	—	—
8000	0.14	0.15	0.17	—	—	—	—
8500	0.15	0.16	0.18	—	—	—	—
9000	0.16	0.17	0.19	—	—	—	—
9500	0.16	0.18	0.20	—	—	—	—

REDUCTION OF BAROMETER TO LATITUDE 45°

Metric Scale
For latitudes below 45°, subtract the correction; for latitudes greater than 45° it is to be added. Corrections in cm.
(From Smithsonian Meterological Tables.)

Latitude		Observed Height of Barometer in Centimeters					
		68	70	72	74	76	78
25°	65°	0.116	0.120	0.123	0.127	0.130	0.133
26	64	.111	.115	.118	.121	.125	.128
27	63	.106	.110	.113	.116	.119	.122
28	62	.101	.104	.107	.110	.113	.116
29	61	.096	.099	.102	.104	.107	.110
30	60	.091	.094	.096	.098	.101	.104
31	59	.085	.087	.090	.092	.095	.097
32	58	.079	.082	.084	.086	.089	.091
33	57	.074	.076	.078	.080	.082	.084
34	56	.068	.070	.072	.074	.076	.078
35	55	.062	.064	.066	.067	.069	.071
36	54	.056	.058	.059	.061	.063	.064
37	53	.050	.051	.053	.054	.056	.057
38	52	.044	.045	.046	.048	.049	.050
39	51	.038	.039	.040	.041	.042	.043
40	50	.031	.032	.033	.034	.035	.036
41	49	.025	.026	.027	.027	.028	.029
42	48	.019	.019	.020	.021	.021	.022
43	47	.013	.013	.013	14	.014	.014
44	46	.006	.007	.007	.007	.007	.007

English Scale (Corrections in inches)

Latitude		Observed Height in Inches					
		25	26	27	28	29	30
25°	65°	0.043	0.044	0.046	0.048	0.050	0.051
26	64	.041	.043	.044	.046	.048	.049
27	63	.039	.041	.042	.044	.045	.047
28	62	.037	.039	.040	.042	.043	.045
29	61	.035	.037	.038	.039	.041	.042
30	60	.033	.035	.036	.037	.039	.040
31	59	.031	.032	.034	.035	.036	.037
32	58	.029	.030	.032	.033	.034	.035
33	57	.027	.028	.029	.030	.031	.032
34	56	.025	.026	.027	.028	.029	.030
35	55	.023	.024	.025	.025	.026	.027
36	54	.021	.021	.022	.023	.024	.025
37	53	.018	.019	.020	.021	.021	.022
38	52	.016	.017	.017	.018	.019	.019
39	51	.014	.014	.015	.015	.016	.017
40	50	.012	.012	.012	.013	.013	.014
41	49	.009	.010	.010	.010	.011	.011
42	48	.007	.007	.008	.008	.008	.008
43	47	.005	.005	.005	.005	.005	.006
44	46	.002	.002	.003	.003	.003	.003

RELATIVE HUMIDITY — DEW-POINT

The table gives the relative humidity of the air for temperature t and dewpoint d.

(From Smithsonian Meterological Tables.)

Depression of dew-point t-d, °C	Dew-Point (d) −10	0	+10	+20	+30
0.0	100%	100%	100%	100%	100%
0.2	98	99	99	99	99
0.4	97	97	97	98	98
0.6	95	96	96	96	97
0.8	94	94	95	95	96
1.0	92	93	94	94	94
1.2	91	92	92	93	93
1.4	90	90	91	92	92
1.6	88	89	90	91	91
1.8	87	88	89	90	90
2.0	86	87	88	88	89
2.2	84	85	86	87	88
2.4	83	84	85	86	87
2.6	82	83	84	85	86
2.8	80	82	83	84	85
3.0	79	81	82	83	84
3.2	78	80	81	82	83
3.4	77	79	80	81	82
3.6	76	77	79	80	82
3.8	75	76	78	79	81
4.0	73	75	77	78	80
4.2	72	74	76	77	79
4.4	71	73	75	77	78
4.6	70	72	74	76	77
4.8	69	71	73	75	76
5.0	68	70	72	74	75
5.2	67	69	71	73	75
5.4	66	68	70	72	74
5.6	65	67	69	71	73
5.8	64	66	69	70	72
6.0	63	66	68	70	71
6.2	62	65	67	69	71
6.4	61	64	66	68	70
6.6	60	63	65	67	69
6.8	60	62	64	66	68
7.0	59	61	63	66	68
7.2	58	60	63	65	67
7.4	57	60	62	64	66
7.6	56	59	61	63	65
7.8	55	58	60	63	65
8.0	54	57	60	62	64
8.2	54	56	59	61	63
8.4	53	56	58	60	63
8.6	52	55	57	60	62
8.8	51	54	57	59	61
9.0	51	53	56	58	61
9.2	50	53	55	58	60
9.4	49	52	55	57	59
9.6	48	51	54	56	59
9.8	48	51	53	56	58
10.0	47	50	53	55	57
10.5	45	48	51	54	—
11.0	44	47	49	52	—
11.5	42	45	48	51	—
12.0	41	44	47	49	—
12.5	39	42	45	48	—
13.0	38	41	44	46	—
13.5	37	40	43	45	—
14.0	35	38	41	44	—
14.5	34	37	40	43	—
15.0	33	36	39	42	—
15.5	32	35	38	40	—
16.0	31	34	37	39	—
16.5	30	33	36	38	—
17.0	29	32	35	37	—
17.5	28	31	34	36	—
18.0	27	30	33	35	—
18.5	26	29	32	34	—
19.0	25	28	31	33	—
19.5	24	27	30	33	—
20.0	24	26	29	32	—
21.0	22	25	27	—	—
22.0	21	23	26	—	—
23.0	19	22	24	—	—
24.0	18	21	23	—	—
25.0	17	19	22	—	—
26.0	16	18	21	—	—
27.0	15	17	20	—	—
28.0	14	16	19	—	—
29.0	13	15	18	—	—
30.0	12	14	17	—	—

RELATIVE HUMIDITY FROM WET AND DRY BULB THERMOMETER (CENT. SCALE)

This table gives the approximate relative humidity directly from the reading of the air temperature (dry bulb) (t′°C) and the wet bulb (t°C). It is computed for a barometric pressure of 74.27 cm Hg. Errors resulting from the use of this table for air temperatures above −10°C and between 77.5 and 71 cm Hg will usually be within the errors of observation.

Condensed from Bulletin of the U.S. Weather Bureau No. 1071

t-t′ t	0.2	0.4	0.6	0.8	1.0	1.2	1.4	1.6	1.8	2.0	2.2	2.4	2.6	2.8	3.0	3.2	3.4	3.6	3.8	4.0	4.5	5.0	5.5	6.0	6.5	7.0	7.5	8.0	9.0	9.5	10.0	10.5	11.0
−10	93	87	80	74	67	61	54	48	41	35	28	22	16	9																			
−9	94	88	81	75	69	63	57	51	45	39	33	27	21	15	9																		
−8	94	88	83	77	71	65	60	54	48	43	37	32	26	20	15	10																	
−7	95	89	84	78	73	67	62	57	52	46	41	36	31	25	20	15	10	5															
−6	95	90	85	79	74	69	64	59	54	49	45	40	35	30	25	20	15	11	6														
−5	95	90	86	81	76	71	66	62	57	52	48	43	39	34	29	25	20	16	11	7													
−4	95	91	86	82	77	73	68	64	59	55	51	46	42	38	33	29	25	21	17	12													
−3	96	91	87	82	78	74	70	66	62	57	53	49	45	41	37	33	29	25	21	17	8												
−2	96	92	88	84	79	75	71	68	64	60	56	52	48	44	40	37	33	29	25	22	12												
−1	96	92	88	84	81	77	73	69	66	62	58	54	51	47	43	40	36	33	29	26	17	8											
0	96	93	89	85	81	78	74	71	67	64	60	57	53	50	46	43	40	36	33	29	22	13	5										
1	97	93	90	86	83	80	76	73	70	66	63	59	56	53	49	46	43	40	36	33	25	17	10										
2	97	93	90	87	84	81	78	74	71	68	65	62	59	56	53	49	46	43	40	37	29	22	14	7									
3	97	94	91	88	85	82	78	76	72	70	67	64	61	58	55	52	49	46	43	40	33	26	19	12	5								
4	97	94	91	88	85	82	79	77	74	71	68	65	62	60	57	54	51	48	46	43	36	29	22	16	9								
5	97	94	91	88	86	83	80	77	75	72	69	67	64	61	58	56	53	51	48	45	39	33	26	20	13	7							
6	97	94	92	89	86	84	81	78	76	73	70	68	65	63	60	58	55	53	50	48	41	35	29	24	17	11	5						
7	97	95	92	89	87	84	82	79	77	74	72	69	67	64	62	60	57	55	52	50	44	38	32	26	21	15	10						
8	97	95	92	90	87	85	82	80	77	75	73	70	68	65	63	61	58	56	54	51	46	40	35	29	24	19	14	8					
9	98	95	93	90	88	85	83	81	79	77	74	71	69	67	64	62	60	58	55	53	48	42	37	32	27	22	17	12	7				
10	98	95	93	90	88	86	83	81	79	77	75	72	70	68	66	63	61	59	57	55	50	44	39	34	29	24	20	15	10	6			
11	98	95	93	91	89	86	84	82	80	78	75	73	71	69	67	65	62	60	58	56	51	46	41	36	32	27	22	18	13	9	5		
12	98	96	93	91	89	87	85	83	81	78	76	74	72	70	68	66	64	62	60	58	53	48	43	39	34	29	25	21	16	12	8		
13	98	96	93	91	89	87	85	83	81	79	77	75	73	71	69	67	65	63	61	59	54	50	45	41	36	32	28	23	19	15	11	7	
14	98	96	94	92	90	88	86	84	82	80	78	76	74	72	70	68	66	64	62	60	57	53	48	44	40	36	32	27	22	18	14	10	6
15	98	96	94	92	90	88	86	84	82	80	78	76	74	73	71	69	67	65	63	61	57	53	51	48	44	40	36	32	27	24	20	16	13

Note: row 15 trailing values ... 9 6 (continuing beyond 11.0 column).

t-t′ t	0.5	1.0	1.5	2.0	2.5	3.0	3.5	4.0	4.5	5.0	5.5	6.0	6.5	7.0	7.5	8.0	8.5	9.0	9.5	10.0	10.5	11.0	11.5	12.0	12.5	13.0	13.5	14.0	14.5	15.0	16.0	17.0	18.0	19.0	20.0
16	95	90	85	81	76	71	67	63	58	54	50	46	42	38	34	30	26	23	19	15	12	8	5												
17	95	90	86	81	76	72	68	64	60	55	51	47	43	40	36	32	28	25	21	18	14	11	8												
18	95	91	86	82	77	73	69	65	61	57	53	49	45	41	38	34	30	27	23	20	17	14	10	7											
19	95	91	87	82	78	74	70	66	63	59	55	51	48	44	41	37	34	30	27	23	20	16	13	10	7										
20	96	91	87	83	78	74	70	66	63	59	55	51	48	44	41	37	34	31	27	24	21	18	15	12	9	6									
21	96	91	87	83	79	75	71	67	64	60	56	53	49	46	42	39	36	32	29	26	23	20	17	14	12	9	6								
22	96	92	87	83	80	76	72	68	64	61	57	54	50	47	44	40	37	34	31	28	25	22	19	16	14	11	8	6							
23	96	92	88	84	80	76	72	69	65	62	58	55	52	49	46	43	40	37	34	31	29	26	23	20	18	15	13	10	8	6					
24	96	92	88	84	80	77	73	69	66	62	59	56	53	49	46	43	40	37	34	31	29	26	23	20	18	15	13	10	8	5					
25	96	92	88	84	81	77	74	70	67	63	60	57	54	50	47	44	41	38	35	33	30	27	24	22	20	17	15	12	10	8					
26	96	92	88	85	81	78	74	71	67	64	61	58	55	52	50	47	44	42	39	37	34	31	29	26	24	21	19	17	14	12					
27	96	92	89	85	82	78	75	71	68	65	62	58	56	52	50	47	44	41	38	36	33	31	28	26	23	21	18	16	14	12					
28	96	93	89	85	82	79	75	72	69	66	62	59	56	53	51	48	45	42	40	37	35	32	29	27	25	22	20	18	16	13					
29	96	93	89	86	82	79	76	72	69	66	63	60	57	54	52	49	46	43	41	38	36	33	31	29	26	24	22	19	17	15					
30	96	93	89	86	83	79	76	73	70	67	64	61	58	55	52	50	47	44	42	39	37	35	32	30	28	25	23	21	19	17	13	9	5		
31	96	93	90	86	83	80	77	73	70	67	64	61	59	56	53	51	48	45	43	40	38	36	33	31	29	26	24	22	20	16	12	9	5		
32	96	93	90	86	83	80	77	74	71	68	65	62	60	57	54	51	49	46	44	41	39	37	35	32	30	28	26	24	22	18	12	9	5		
33	97	93	90	87	84	81	78	75	72	69	66	63	61	58	56	53	51	49	47	45	42	40	38	36	33	31	29	27	25	23	19	16	13	10	7
34	97	93	90	87	84	81	78	75	72	69	66	63	61	58	56	53	51	48	46	43	41	39	37	35	32	30	28	26	24	23	19	15	12	8	5
35	97	93	90	87	84	81	78	75	72	69	67	64	61	59	56	54	51	49	47	44	42	40	38	36	35	33	32	30	28	26	20	17	13	10	—
36	97	94	90	87	84	81	78	75	73	70	67	64	62	59	57	54	52	50	48	45	43	41	39	37	35	33	31	29	27	25	21	18	15	11	8
37	97	94	91	87	84	82	79	76	73	70	68	65	63	60	58	55	53	51	49	46	44	42	40	38	36	34	32	30	28	26	23	19	16	13	10
38	97	94	91	88	85	82	79	76	74	71	68	66	63	61	58	56	54	51	49	47	45	43	41	39	37	35	33	31	29	27	24	20	17	14	11
39	97	94	91	88	85	82	79	77	74	71	69	66	64	61	59	57	54	52	50	48	46	44	42	40	38	36	34	32	30	28	25	22	18	15	12
40	97	94	91	88	85	82	80	77	74	72	69	67	64	62	59	57	54	53	51	48	46	44	42	40	38	36	34	32	30	28	26	20	16	14	—

REDUCTION OF PSYCHROMETRIC OBSERVATION

For the reduction of observations with the wet and dry bulb thermometer. Assuming the relative velocity of the air to the thermometer bulbs is at least three meters per second; if t is the temperature of the air as indicated by the dry bulb, t_w, the temperature of the wet bulb, B, the barometric pressure, and E_w, the vapor tension of water corresponding to t_2, then the actual vapor tension is

$$E = E_w - 0.00066B(t - t_w) [1 + 0.00115(t - t_w)] \text{ millimeters}$$

The value of the term

$$0.00066B(t - t_w) [1 + 0.00115 (t - t_w)]$$

is given in the following table.

(From Miller's Laboratory Physics, Ginn & Co., Publishers, by permission.)

Barometric Pressure B in Millimeters

t−t_w °C	700 mm	710 mm	720 mm	730 mm	740 mm	750 mm	760 mm	770 mm
1	0.463	0.469	0.476	0.482	0.489	0.496	0.502	0.509
2	0.926	0.939	0.953	0.966	0.979	0.992	1.006	1.019
3	1.391	1.411	1.431	1.450	1.470	1.490	1.510	1.530
4	1.857	1.883	1.910	1.936	1.963	1.989	2.016	2.042
5	2.323	2.356	2.390	2.423	2.456	2.489	2.522	2.556
6	2.791	2.831	2.871	2.911	2.951	2.990	3.030	3.070
7	3.260	3.307	3.353	3.400	3.446	3.493	3.539	3.586
8	3.730	3.783	3.837	3.890	3.943	3.996	4.050	4.103
9	4.201	4.261	4.321	4.381	4.441	4.501	4.561	4.621
10	4.673	4.740	4.807	4.873	4.940	5.007	5.074	5.140
11	5.146	5.220	5.293	5.367	5.440	5.514	5.587	5.661
12	5.621	5.701	5.781	5.861	5.942	6.022	6.102	6.183
13	6.096	6.183	6.270	6.357	6.444	6.531	6.618	6.705
14	6.572	6.666	6.760	6.854	6.948	7.042	7.135	7.229
15	7.050	7.146	7.251	7.352	7.452	7.553	7.654	7.754
16	7.528	7.636	7.743	7.851	7.958	8.066	8.173	8.281
17	8.008	8.122	8.236	8.351	8.465	8.580	8.694	8.808
18	8.488	8.609	8.731	8.852	8.973	9.094	9.216	9.337
19	8.970	9.098	9.226	9.354	9.482	9.610	9.739	9.867
20	9.453	9.588	9.723	9.858	9.993	10.128	10.263	10.398

CONSTANT HUMIDITY

The following table shows the % humidity and the aqueous tension at the given temperature within a closed space when an excess of the substance indicated is in contact with a saturated aqueous solution of the given solid phase.

Solid phase	t°C.	% humidity	Aq. tension mm Hg	Solid phase	t°C.	% humidity	Aq. tension mm Hg
$H_3PO_4.\frac{1}{2}H_2O$	24	9	1.99	$NaClO_3$	20	75	13.0
$KC_2H_3O_2$	168	13	738	$(NH_4)_2SO_4$	108	75	754
$LiCl.H_2O$	20	15	2.60	$NaC_2H_3O_2.3H_2O$	20	76	13.2
$KC_2H_3O_2$	20	20	3.47	$H_2C_2O_4.2H_2O$	20	76	13.2
KF	100	22.9	174	$Na_2S_2O_3.5H_2O$	20	78	13.5
$NaBr$	100	22.9	174	NH_4Cl	20	79.5	13.8
$NaCl, KNO_3$ and $NaNO_3$	16.39	30.49	4.23	NH_4Cl	25	79.3	18.6
$CaCl_2.6H_2O$	24.5	31	7.08	NH_4Cl	30	77.5	24.4
$CaCl_2.6H_2O$	20	32.3	5.61	$(NH_4)_2SO_4$	20	81	14.1
$CaCl_2.6H_2O$	18.5	35	5.54	$(NH_4)_2SO_4$	25	81.1	19.1
CrO_3	20	35	6.08	$(NH_4)_2SO_4$	30	81.1	25.6
$CaCl_2.6H_2O$	10	38	3.47	KBr	20	84	14.6
$CaCl_2 6H_2O$	5	39.8	2.59	Tl_2SO_4	104.7	84.8	768
$Zn(NO_3)_2.6H_2O$	20	42	7.29	$KHSO_4$	20	86	14.9
$K_2CO_3.2H_2O$	24.5	43	9.82	$Na_2CO_3.10H_2O$	24.5	87	20.9
$K_2CO_3.2H_2O$	18.5	44	6.96	$BaCl_2.2H_2O$	24.5	88	20.1
KNO_2	20	45	7.81	K_2CrO_4	20	88	15.3
$KCNS$	20	47	8.16	$Pb(NO_3)_2$	103.5	88.4	760
NaI	100	50.4	383	$ZnSO_4.7H_2O$	20	90	15.6
$Ca(NO_3)_2.4H_2O$	24.5	51	11.6	$Na_2CO_3.10H_2O$	18.5	92	14.6
$NaHSO_4.H_2O$	20	52	9.03	$NaBrO_3$	20	92	16.0
$Na_2Cr_2O_7.2H_2O$	20	52	9.03	K_2HPO_4	20	92	16.0
$Mg(NO_3)_2.6H_2O$	24.5	52	11.9	$NH_4H_2PO_4$	30	92.9	29.3
$NaClO_3$	100	54	410	$NH_4H_2PO_4$	25	93	21.9
$Ca(NO_3)_2.4H_2O$	18.5	56	8.86	$Na_2SO_4.10H_2O$	20	93	16.1
$Mg(NO_3)_2.6H_2O$	18.5	56	8.86	$NH_4H_2PO_4$	20	93.1	16.2
KI	100	56.2	427	$ZnSO_4.7H_2O$	5	94.7	6.10
$NaBr.2H_2O$	20	58	10.1	$Na_2SO_3.7H_2O$	20	95	16.5
$Mg(C_2H_3O_2)_2.4H_2O$	20	65	11.3	$Na_2HPO_4.12H_2O$	20	95	16.5
$NaNO_2$	20	66	11.5	NaF	100	96.6	734
NH_4Cl and KNO_3	30	68.6	21.6	$Pb(NO_3)_2$	20	98	17.0
KBr	100	69.2	526	$CuSO_4.5H_2O$	20	98	17.0
NH_4Cl and KNO_3	25	71.2	16.7	$TlNO_3$	100.3	98.7	759
NH_4Cl and KNO_3	20	72.6	12.6	$TlCl$	100.1	99.7	761

CONSTANT HUMIDITY WITH SULFURIC ACID SOLUTIONS

The relative humidity and pressure of aqueous vapor of air in equilibrium conditions above aqueous solutions of sulfuric acid are given below.

Density of acid solution	Relative humidity	Vapor pressure at 20°C	Density of acid solution	Relative humidity	Vapor pressure at 20°C
1.00	100.0	17.4	1.30	58.3	10.1
1.05	97.5	17.0	1.35	47.2	8.3
1.10	93.9	16.3	1.40	37.1	6.5
1.15	88.8	15.4	1.50	18.8	3.3
1.20	80.5	14.0	1.60	8.5	1.5
1.25	70.4	12.2	1.70	3.2	0.6

For concentration of sulfuric acid solution refer to tables relating density to percent composition.

VELOCITY OF SOUND

Compiled by Gordon E. Becker, Bell Telephone Laboratories

The data for the Velocity of Sound is Various Materials were compiled from a variety of sources. For more extensive tables one is referred to the following books:
AIP Handbook, Smithsonian Tables.
Mason: Physical Acoustics and the Properties of Solids (1958).
Chalmers and Quarrell: Physical Examination of Metals (1960).
Mason: Piezoelectric Crystals and their Application to Ultrasonics (1950).
Bergmann: Der Ultraschall (Hirzel, 1954).

Definition of Terms:
V_l = Velocity of plane longitudinal wave in bulk material
V_s = Velocity of plane transverse (shear) wave
V_{ext} = Velocity of longitudinal wave (extensional wave) in thin rods.

SOLIDS

Substance	Density g/cc	V_l m/sec	V_s m/sec	V_{ext} m/sec
Metals				
Aluminum, rolled	2.7	6420	3040	5000
Beryllium	1.87	12890	8880	12870
Brass (70 Cu, 30 Zn)	8.6	4700	2110	3480
Copper, annealed	8.93	4760	2325	3810
Copper, rolled	8.93	5010	2270	3750
Duralumin 17S	2.79	6320	3130	5150
Gold, hard-drawn	19.7	3240	1200	2030
Iron, electrolytic	7.9	5950	3240	5120
Iron, Armco	7.85	5960	3240	5200
Lead, annealed	11.4	2160	700	1190
Lead, rolled	11.4	1960	690	1210
Magnesium, drawn, annealed	1.74	5770	3050	4940
Molybdenum	10.1	6250	3350	5400
Monel metal	8.90	5350	2720	4400
Nickel (unmagnetized)	8.85	5480	2990	4800
Nickel	8.9	6040	3000	4900
Platinum	21.4	3260	1730	2800
Silver	10.4	3650	1610	2680
Steel, mild	7.85	5960	3235	5200
Steel, 347 Stainless	7.9	5790	3100	5000
Steel (1%C)	7.84	5940	3220	5180
Steel (1%C, hardened)	7.84	5854	3150	5070
Tin, rolled	7.3	3320	1670	2730
Titanium	4.5	6070	3125	5080
Tungsten, annealed	19.3	5220	2890	4620
Tungsten, drawn	19.3	5410	2640	4320
Tungsten Carbide	13.8	6655	3980	6220
Zinc, rolled	7.1	4210	2440	3850
Various				
Fused silica	2.2	5968	3764	5760
Glass, pyrex	2.32	5640	3280	5170
Glass, heavy silicate flint	3.88	3980	2380	3720
Glass, light borate crown	2.24	5100	2840	4540
Lucite	1.18	2680	1100	1840
Nylon 6-6	1.11	2620	1070	1800
Polyethylene	0.90	1950	540	920
Polystyrene	1.06	2350	1120	2240
Rubber, butyl	1.07	1830		
Rubber, gum	0.95	1550		
Rubber neoprene	1.33	1600		
Brick	1.8			3650
Clay rock	2.2			3480
Cork	0.25			500
Marble	2.6			3810
Paraffin	0.9			1300
Tallow				390
Woods				
Ash, along the fiber				4670
Ash, across the rings				1390
Ash, along the rings				1260
Beech, along the fiber				3340
Elm, along the fiber				4120
Maple, along the fiber				4110
Oak, along the fiber				3850

LIQUIDS

Substance	Formula	Density g/cc	Velocity at 25°C m/sec	$-\delta v/\delta t$ m/sec °C
Acetone	C_3H_6O	0.79	1174	4.5
Benzene	C_6H_6	0.870	1295	4.65
Carbon disulphide	CS_2	1.26	1149	—
Carbon tetrachloride	CCl_4	1.595	926	2.7
Castor oil	$C_{11}H_{10}O_{10}$	0.969	1477	3.6
Chloroform	$CHCl_3$	1.49	987	3.4
Ethanol	C_2H_6O	0.79	1207	4.0
Ethanol amide	C_2H_7NO	1.018	1724	3.4
Ethyl ether	$C_4H_{10}O$	0.713	985	4.87
Ethylene glycol	$C_2H_6O_2$	1.113	1658	2.1
Glycerol	$C_3H_8O_3$	1.26	1904	2.2
Kerosene	—	0.81	1324	3.6
Mercury	Hg	13.5	1450	—
Methanol	CH_4O	0.791	1103	3.2
Nitrobenzene	$C_6H_4NO_2$	1.20	1463	3.6
Turpentine	—	0.88	1255	—
Water (distilled)	H_2O	0.998	1496.7 + ?	-2.4
Water (sea)	—	1.025	1531	-2.4
Xylene hexafluoride	$C_8H_4F_6$	1.37	879	

GASES AND VAPORS

Substance	Formula	Density g/l	Velocity m/sec	$\delta v/\delta t$ m/sec °C
Gases (0°C)				
Air, dry		1.293	331.45	0.59
Ammonia	NH_3	0.771	415	
Argon	Ar	1.783	319	0.56
Carbon monoxide	CO	1.25	338	0.6
Carbon dioxide	CO_2	1.977	259	0.4
Chlorine	Cl_2	3.214	206	
Deuterium	D_2		890	1.6
Ethane (10°C)	C_2H_6	1.356	308	
Ethylene	C_2H_4	1.260	317	
Helium	He	0.178	965	0.8
Hydrogen	H_2	0.0899	1284	2.2
Hydrogen bromide	HBr	3.50	200	
Hydrogen chloride	HCl	1.639	296	
Hydrogen iodide	HI	5.66	157	
Hydrogen sulfide	H_2S	1.539	289	
Illuminating (Coal gas)			453	
Methane	CH_4	0.7168	430	
Neon	Ne	0.900	435	0.8
Nitric oxide (10°C)	NO	1.34	324	
Nitrogen	N_2	1.251	334	0.6
Nitrous oxide	N_2O	1.977	263	0.5
Oxygen	O_2	1.429	316	0.56
Sulfur dioxide	SO_2	2.927	213	0.47
Vapors (97.1°C)				
Acetone	C_3H_6O		239	0.32
Benzene	C_6H_6		202	0.3
Carbon tetrachloride	CCl_4		145	
Chloroform	$CHCl_3$		171	0.24
Ethanol	C_2H_6O		269	0.4
Ethyl ether	$C_4H_{10}O$		206	0.3
Methanol	CH_4O		335	0.46
Water vapor (134°C)	H_2O		494	

SOUND VELOCITY IN WATER ABOVE 212°F

By permission from the Acoustical Society of America, Volume 31 (1959) and J. C. McDade, D. R. Pardue, A. L. Gedrich and F. Vrataric.

Temperature°F	Velocity m/sec	Velocity ft/sec	Temperature °F	Velocity m/sec	Velocity ft/sec
186.8	1552	5092	370	1368	4488
200	1548	5079	380	1353	4439
210	1544	5066	390	1337	4386
220	1538	5046	400	1320	4331
230	1532	5026	410	1302	4272
240	1524	5000	420	1283	4209
250	1516	4974	430	1264	4147
260	1507	4944	440	1244	4081
270	1497	4911	450	1220	4010
280	1487	4879	460	1200	3940
290	1476	4843	470	1180	3880
300	1465	4806	480	1160	3800
310	1453	4767	490	1140	3730
320	1440	4724	500	1110	3650
330	1426	4678	510	1090	3570
340	1412	4633	520	1070	3500
350	1398	4587	530	1040	3410
360	1383	4537	540	1010	3320
			550	980	3230

MUSICAL SCALES

EQUAL TEMPERED CHROMATIC SCALE
$A_4 = 440$

American Standard pitch. Adopted by the American Standards Association in 1936

Note	Frequency	Note	Frequency	Note	Frequency	Note	Frequency
C_0	16.35	C_2	65.41	C_4	261.63	C_6	1046.50
$C\#_0$	17.32	$C\#_2$	69.30	$C\#_4$	277.18	$C\#_6$	1108.73
D_0	18.35	D_2	73.42	D_4	293.66	D_6	1174.66
$D\#_0$	19.45	$D\#_2$	77.78	$D\#_4$	311.13	$D\#_6$	1244.51
E_0	20.60	E_2	82.41	E_4	329.63	E_6	1318.51
F_0	21.83	F_2	87.31	F_4	349.23	F_6	1396.91
$F\#_0$	23.12	$F\#_2$	92.50	$F\#_4$	369.99	$F\#_6$	1479.98
G_0	24.50	G_2	98.00	G_4	392.00	G_6	1567.98
$G\#_0$	25.96	$G\#_2$	103.83	$G\#_4$	415.30	$G\#_6$	1661.22
A_0	27.50	A_2	110.00	A_4	440.00	A_6	1760.00
$A\#_0$	29.14	$A\#_2$	116.54	$A\#_4$	466.16	$A\#_6$	1864.66
B_0	30.87	B_2	123.47	B_4	493.88	B_6	1975.53
C_1	32.70	C_3	130.81	C_5	523.25	C_7	2093.00
$C\#_1$	34.65	$C\#_3$	138.59	$C\#_5$	554.37	$C\#_7$	2217.46
D_1	36.71	D_3	146.83	D_5	587.33	D_7	2349.32
$D\#_1$	38.89	$D\#_3$	155.56	$D\#_5$	622.25	$D\#_7$	2489.02
E_1	41.20	E_3	164.81	E_5	659.26	E_7	2637.02
F_1	43.65	F_3	174.61	F_5	698.46	F_7	2793.83
$F\#1$	46.25	$F\#_3$	185.00	$F\#_5$	739.99	$F\#_7$	2959.96
G_1	49.00	G_3	196.00	G_5	783.99	G_7	3135.96
$G\#_1$	51.91	$G\#_3$	207.65	$G\#_5$	830.61	$G\#_7$	3322.44
A_1	55.00	A_3	220.00	A_5	880.00	A_7	3520.00
$A\#_1$	58.27	$A\#_3$	233.08	$A\#_5$	932.33	$A\#_7$	3729.31
B_1	61.74	B_3	246.94	B_5	987.77	B_7	3951.07
						C_8	4186.01

EQUAL TEMPERED CHROMATIC SCALE
$A_4 = 435$

International Pitch, adopted 1891

Note	Frequency	Note	Frequency	Note	Frequency	Note	Frequency
C_0	16.17	C_2	64.66	C_4	258.65	C_6	1034.61
$C\#_0$	17.13	$C\#_2$	68.51	$C\#_4$	274.03	$C\#_6$	1096.13
D_0	18.15	D_2	72.58	D_4	290.33	D_6	1161.31
$D\#_0$	19.22	$D\#_2$	76.90	$D\#_4$	307.59	$D\#_6$	1230.37
E_0	20.37	E_2	81.47	E_4	325.88	E_6	1303.53
F_0	21.58	F_2	86.31	F_4	345.26	F_6	1381.04
$F\#_0$	22.86	$F\#_2$	91.45	$F\#_4$	365.79	$F\#_6$	1463.16
G_0	24.22	G_2	96.89	G_4	387.54	G_6	1550.16
$G\#_0$	25.66	$G\#_2$	102.65	$G\#_4$	410.59	$G\#_6$	1642.34
A_0	27.19	A_2	108.75	A_4	435.00	A_6	1740.00
$A\#_0$	28.80	$A\#_2$	115.22	$A\#_4$	460.87	$A\#_6$	1843.47
B_0	30.52	B_2	122.07	B_4	488.27	B_6	1953.08
C_1	32.33	C_3	129.33	C_5	517.31	C_7	2069.22
$C\#_1$	34.25	$C\#_3$	137.02	$C\#_5$	548.07	$C\#_7$	2192.26
D_1	36.29	D_3	145.16	D_5	580.66	D_7	2322.62
$D\#_1$	38.45	$D\#_3$	153.80	$D\#_5$	615.18	$D\#_7$	2460.73
E_1	40.74	E_3	162.94	E_5	651.76	E_7	2607.05
F_1	43.16	F_3	172.63	F_5	690.52	F_7	2762.08
$F\#_1$	45.72	$F\#_3$	182.89	$F\#_5$	731.58	$F\#_7$	2926.32
G_1	48.44	G_3	193.77	G_5	775.08	G_7	3100.33
$G\#_1$	51.32	$G\#_3$	205.29	$G\#_5$	821.17	$G\#_7$	3284.68
A_1	54.38	A_3	217.50	A_5	870.00	A_7	3480.00
$A\#_1$	57.61	$A\#_3$	230.43	$A\#_5$	921.73	$A\#_7$	3686.93
B_1	61.03	B_3	244.14	B_5	976.54	B_7	3906.17
						C_8	4138.44

SCIENTIFIC OR JUST SCALE
$C_4 = 256$

Note	Frequency	Note	Frequency	Note	Frequency	Note	Frequency
C_0	16	C_2	64	C_4	256	C_6	1024
D_0	18	D_2	72	D_4	288	D_6	1152
E_0	20	E_2	80	E_4	320	E_6	1280
F_0	21.33	F_2	85.33	F_4	341.33	F_6	1365.33
G_0	24	G_2	96	G_4	384	G_6	1536
A_0	26.67	A_2	106.67	A_4	426.67	A_6	1706.67
B_0	30	B_2	120	B_4	480	B_6	1920
C_1	32	C_3	128	C_5	512	C_7	2048
D_1	36	D_3	144	D_5	576	D_7	2304
E_1	40	E_3	160	E_5	640	E_7	2560
F_1	42.67	F_3	170.67	F_5	682.67	F_7	2730.67
G_1	48	G_3	192	G_5	768	G_7	3072
A_1	53.33	A_3	213.33	A_5	853.33	A_7	3413.33
B_1	60	B_3	240	B_5	960	B_7	3840
						C_8	4096

ABSORPTION AND VELOCITY OF SOUND IN STILL AIR

The following data refer only to the temperature 20C (68F). They were abstracted from an extensive compilation prepared by L. B. Evans and H. E. Bass. The entire report, Tables of Absorption and Velocity of Sound in Still Air at 68F (20C), AD-738 576 is available from National Technical Information Service, U. S. Department of Commerce, 5285 Port Royal Road, Springfield, Va. 22151.

Frequency (Hz)	Absorption (dB/1000 ft)	Absorption (dB/Km)	Absorption (dB/sec)	Velocity (1000 ft/sec)
Relative Humidity = 0%				
20.	0.154	0.51	0.174	1.126892
40.	0.327	1.07	0.368	1.127013
50.	0.384	1.26	0.433	1.127050
63.	0.436	1.43	0.491	1.127085
100.	0.509	1.67	0.573	1.127131
200.	0.560	1.84	0.631	1.127161
400.	0.596	1.96	0.672	1.127169
630.	0.645	2.11	0.727	1.127171
800.	0.692	2.27	0.780	1.127172
1250.	0.861	2.82	0.970	1.127172
2000.	1.262	4.14	1.423	1.127173
4000.	2.696	8.84	3.039	1.127178
6300.	4.541	14.89	5.118	1.127182
10000.	8.013	26.28	9.032	1.127184
12500.	10.918	35.81	12.306	1.127186
16000.	15.901	52.15	17.923	1.127187
20000.	22.978	75.37	25.901	1.127187
40000.	81.405	267.01	91.759	1.127188
63000.	196.544	644.66	221.542	1.127188
80000.	314.677	1032.14	354.700	1.127188
Relative Humidity = 5%				
20.	0.031	0.10	0.034	1.126973
40.	0.074	0.24	0.083	1.126996
50.	0.092	0.30	0.101	1.127004
63.	0.114	0.37	0.129	1.127009
100.	0.179	0.59	0.202	1.127019
200.	0.449	1.47	0.506	1.127028
400.	1.451	4.76	1.635	1.127043
630.	3.211	10.53	3.619	1.127067
800.	4.774	15.66	5.380	1.127088
1250.	9.164	30.06	10.329	1.127147
2000.	15.175	49.77	17.106	1.127228
4000.	22.685	74.41	25.573	1.127321
6300.	26.245	86.08	29.587	1.127352
10000.	30.781	100.96	34.701	1.127365
12500.	34.306	112.52	38.676	1.127369
16000.	40.263	132.06	45.391	1.127372
20000.	48.653	159.58	54.850	1.127373
40000.	115.903	380.16	130.666	1.127378
63000.	242.070	793.99	272.905	1.127381
80000.	367.063	1203.97	413.821	1.127383
Relative Humidity = 10%				
20.	0.021	0.07	0.024	1.127167
40.	0.064	0.21	0.072	1.127183
50.	0.084	0.28	0.095	1.127191
63.	0.108	0.35	0.122	1.127199
100.	0.161	0.53	0.181	1.127213
200.	0.289	0.95	0.326	1.127225
400.	0.706	2.32	0.796	1.127230
630.	1.501	4.92	1.692	1.127234
800.	2.297	7.54	2.590	1.127238
1250.	5.155	16.91	5.811	1.127254
2000.	11.658	38.24	13.141	1.127287
4000.	31.023	101.76	34.975	1.127386
6300.	47.085	154.44	53.087	1.127462
10000.	61.578	201.98	69.431	1.127522
12500.	68.146	223.52	76.837	1.127540
16000.	76.231	250.04	85.955	1.127555
20000.	85.605	280.78	96.525	1.127563
40000.	151.938	498.36	171.321	1.127576
63000.	277.191	909.19	312.555	1.127581
80000.	403.662	1324.01	455.162	1.127583

ABSORPTION AND VELOCITY OF SOUND IN STILL AIR (*Continued*)

Frequency (Hz)	Absorption (dB/1000 ft)	Absorption (dB/Km)	Absorption (dB/sec)	Velocity (1000 ft/sec)
Relative Humidity = 20%				
20.	0.013	0.04	0.014	1.127568
40.	0.045	0.15	0.051	1.127577
50.	0.066	0.22	0.074	1.127582
63.	0.093	0.30	0.105	1.127587
100.	0.164	0.54	0.185	1.127603
200.	0.285	0.93	0.321	1.127624
400.	0.476	1.56	0.537	1.127633
630.	0.789	2.59	0.890	1.127636
800.	1.103	3.62	1.244	1.127637
1250.	2.277	7.47	2.568	1.127640
2000.	5.310	17.42	5.987	1.127645
4000.	18.991	62.29	21.416	1.127670
6300.	41.151	134.98	46.406	1.127710
10000.	79.657	261.28	89.836	1.127778
12500.	103.004	337.85	116.170	1.127817
16000.	130.465	427.93	147.146	1.127859
20000.	155.809	511.05	175.736	1.127893
40000.	251.952	826.40	284.192	1.127960
63000.	382.062	1253.16	430.957	1.127977
80000.	508.369	1667.45	573.431	1.127982
Relative Humidity = 30%				
20.	0.009	0.03	0.010	1.127976
40.	0.034	0.11	0.038	1.127980
50.	0.051	0.17	0.057	1.127984
63.	0.075	0.25	0.085	1.127987
100.	0.151	0.50	0.170	1.127999
200.	0.309	1.01	0.349	1.128023
400.	0.484	1.59	0.546	1.128037
630.	0.682	2.24	0.770	1.128041
800.	0.868	2.85	0.979	1.128044
1250.	1.552	5.09	1.751	1.128045
2000.	3.333	10.93	3.760	1.128047
4000.	11.856	38.89	13.374	1.128056
6300.	27.626	90.61	31.164	1.128072
10000.	62.493	204.98	70.498	1.128105
12500.	89.659	294.08	101.148	1.128131
16000.	128.814	422.51	145.324	1.128166
20000.	171.847	563.66	193.879	1.128204
40000.	338.710	1110.97	382.172	1.128316
63000.	499.838	1639.47	563.998	1.128361
80000.	635.085	2083.08	716.614	1.128375
Relative Humidity = 40%				
20.	0.007	0.02	0.008	1.128386
40.	0.027	0.09	0.030	1.128388
50.	0.041	0.13	0.046	1.128390
63.	0.062	0.20	0.070	1.128392
100.	0.134	0.44	0.151	1.128402
200.	0.318	1.04	0.359	1.128424
400.	0.524	1.72	0.592	1.128443
630.	0.692	2.27	0.781	1.128449
800.	0.829	2.72	0.935	1.128451
1250.	1.309	4.29	1.477	1.128453
2000.	2.544	8.34	2.870	1.128456
4000.	8.523	27.96	9.618	1.128460
6300.	19.995	65.58	22.564	1.128467
10000.	47.390	155.44	53.478	1.128484
12500.	70.857	232.41	79.962	1.128499
16000.	108.308	355.25	122.228	1.128521
20000.	154.838	507.87	174.742	1.128549
40000.	380.371	1247.62	429.310	1.128663
63000.	596.091	1955.18	672.827	1.128733
80000.	753.514	2471.53	850.536	1.128758
Relative Humidity = 50%				
20.	0.006	0.02	0.006	1.128795
40.	0.022	0.07	0.025	1.128797
50.	0.034	0.11	0.038	1.128798
63.	0.052	0.17	0.058	1.128800
100.	0.117	0.38	0.132	1.128807
200.	0.313	1.03	0.353	1.128826
400.	0.563	1.85	0.636	1.128849

Frequency (Hz)	Absorption (dB/1000 ft)	Absorption (dB/Km)	Absorption (dB/sec)	Velocity (1000 ft/sec)
630.	0.734	2.41	0.828	1.128858
800.	0.851	2.79	0.961	1.128860
1250.	1.231	4.04	1.390	1.128862
2000.	2.176	7.14	2.457	1.128865
4000.	6.752	22.15	7.622	1.128867
6300.	15.643	51.31	17.659	1.128871
10000.	37.521	123.07	42.357	1.128881
12500.	57.009	186.99	64.357	1.128890
16000.	89.594	293.87	101.143	1.128903
20000.	132.719	435.32	149.830	1.128922
40000.	382.722	1255.33	432.101	1.129020
63000.	654.606	2147.11	739.115	1.129099
80000.	844.024	2768.40	953.016	1.129133

Relative Humidity = 60%

20.	0.005	0.02	0.005	1.129207
40.	0.018	0.06	0.021	1.129208
50.	0.029	0.09	0.032	1.129209
63.	0.044	0.15	0.050	1.129210
100.	0.103	0.34	0.117	1.129215
200.	0.301	0.99	0.339	1.129232
400.	0.593	1.94	0.669	1.129254
630.	0.782	2.57	0.884	1.129266
800.	0.896	2.94	1.012	1.129269
1250.	1.223	4.01	1.381	1.129273
2000.	1.997	6.55	2.256	1.129275
4000.	5.711	18.73	6.449	1.129277
6300.	12.962	42.51	14.638	1.129279
10000.	31.050	101.84	35.064	1.129286
12500.	47.462	155.67	53.600	1.129291
16000.	75.544	247.78	85.312	1.129300
20000.	113.958	373.78	128.694	1.129313
40000.	364.443	1195.37	411.598	1.129389
63000.	677.022	2220.61	764.613	1.129461
80000.	899.912	2951.71	1016.458	1.129508

Relative Humidity = 70%

20.	0.004	0.01	0.005	1.129618
40.	0.016	0.05	0.018	1.129619
50.	0.025	0.08	0.028	1.129619
63.	0.039	0.13	0.044	1.129620
100.	0.092	0.30	0.104	1.129623
200.	0.284	0.93	0.321	1.129638
400.	0.611	2.01	0.691	1.129662
630.	0.829	2.72	0.937	1.129673
800.	0.947	3.11	1.070	1.129678
1250.	1.250	4.10	1.412	1.129683
2000.	1.915	6.28	2.163	1.129685
4000.	5.056	16.58	5.712	1.129687
6300.	11.197	36.73	12.649	1.129689
10000.	26.624	87.33	30.077	1.129693
12500.	40.758	133.69	46.044	1.129697
16000.	65.246	214.01	73.708	1.129704
20000.	99.365	325.92	112.254	1.129712
40000.	339.153	1112.42	383.165	1.129770
63000.	673.667	2209.63	761.137	1.129842
80000.	924.481	3032.30	1044.556	1.129884

Relative Humidity = 80%

20.	0.004	0.01	0.004	1.130030
40.	0.014	0.05	0.016	1.130030
50.	0.022	0.07	0.025	1.130031
63.	0.034	0.11	0.039	1.130032
100.	0.082	0.27	0.093	1.130034
200.	0.267	0.88	0.302	1.130047
400.	0.620	2.03	0.701	1.130069
630.	0.870	2.85	0.983	1.130082
800.	0.998	3.27	1.128	1.130088
1250.	1.293	4.24	1.461	1.130094
2000.	1.887	6.19	2.132	1.130096
4000.	4.626	15.17	5.228	1.130098
6300.	9.976	32.72	11.274	1.130100
10000.	23.466	76.97	26.519	1.130102
12500.	35.896	117.74	40.566	1.130106
16000.	57.588	188.89	65.081	1.130110
20000.	88.144	289.11	99.612	1.130116

Frequency (Hz)	Absorption (dB/1000 ft)	Absorption (dB/Km)	Absorption (dB/sec)	Velocity (1000 ft/sec)
40000.	313.525	1028.36	354.334	1.130161
63000.	655.282	2149.33	740.615	1.130223
80000.	925.566	3035.86	1046.134	1.130264

Relative Humidity = 90%

20.	0.003	0.01	0.004	1.130443
40.	0.013	0.04	0.014	1.130443
50.	0.019	0.06	0.022	1.130444
63.	0.031	0.10	0.035	1.130444
100.	0.074	0.24	0.084	1.130446
200.	0.250	0.82	0.283	1.130457
400.	0.621	2.04	0.702	1.130478
630.	0.903	2.96	1.021	1.130493
800.	1.045	3.43	1.182	1.130498
1250.	1.344	4.41	1.520	1.130505
2000.	1.892	6.21	2.139	1.130508
4000.	4.337	14.22	4.903	1.130511
6300.	9.099	29.84	10.286	1.130512
10000.	21.133	69.32	23.891	1.130514
12500.	32.258	105.81	36.468	1.130516
16000.	51.762	169.78	58.518	1.130520
20000.	79.424	260.51	89.791	1.130526
40000.	290.117	951.58	327.995	1.130561
63000.	629.701	2065.42	711.948	1.130612
80000.	911.286	2989.02	1030.346	1.130651

Relative Humidity = 100%

20.	0.003	0.01	0.003	1.130856
40.	0.011	0.04	0.013	1.130856
50.	0.018	0.06	0.020	1.130857
63.	0.028	0.09	0.031	1.130857
100.	0.068	0.22	0.076	1.130858
200.	0.234	0.77	0.265	1.130868
400.	0.616	2.02	0.696	1.130888
630.	0.928	3.05	1.050	1.130902
800.	1.087	3.57	1.230	1.130909
1250.	1.399	4.59	1.582	1.130917
2000.	1.917	6.29	2.168	1.130921
4000.	4.140	13.58	4.682	1.130924
6300.	8.451	27.72	9.557	1.130925
10000.	19.357	63.49	21.891	1.130926
12500.	29.461	96.63	33.318	1.130929
16000.	47.227	154.90	53.410	1.130931
20000.	72.538	237.93	82.036	1.130935
40000.	269.598	884.28	304.905	1.130963
63000.	601.714	1973.62	680.543	1.131007
80000.	888.113	2913.01	1004.492	1.131042

VELOCITY OF SOUND IN DRY AIR

Data in this table apply only to dry air. These data have been calculated with air being treated as a perfect gas.

Temp °C	0 m sec^{-1}	1 m sec^{-1}	2 m sec^{-1}	3 m sec^{-1}	4 m sec^{-1}	5 m sec^{-1}	6 m sec^{-1}	7 m sec^{-1}	8 m sec^{-1}	9 m sec^{-1}
60	366.05	366.60	367.14	367.69	368.24	368.78	369.33	369.87	370.42	370.96
50	360.51	361.07	361.62	362.18	362.74	363.29	363.84	364.39	364.95	365.50
40	354.89	355.46	356.02	356.58	357.15	357.71	358.27	358.83	359.39	359.95
30	349.18	349.75	350.33	350.90	351.47	352.04	352.62	353.19	353.75	354.32
20	343.37	343.95	344.54	345.12	345.70	346.29	346.87	347.44	348.02	348.60
10	337.46	338.06	338.65	339.25	339.84	340.43	341.02	341.61	342.20	342.78
0	331.45	332.06	332.66	333.27	333.87	334.47	335.07	335.67	336.27	336.87
−10	325.33	324.71	324.09	323.47	322.84	322.22	321.60	320.97	320.34	319.72
−20	319.09	318.45	317.82	317.19	316.55	315.92	315.28	314.64	314.00	313.36
−30	312.72	312.08	311.43	310.78	310.14	309.49	308.84	308.19	307.53	306.88
−40	306.22	305.56	304.91	304.25	303.58	302.92	302.26	301.59	300.92	300.25
−50	299.58	298.91	298.24	297.56	296.89	296.21	295.53	294.85	294.16	293.48
−60	292.79	292.11	291.42	290.73	290.03	289.34	288.64	287.95	287.25	286.55
−70	285.84	285.14	284.43	283.73	283.02	282.30	281.59	280.88	280.16	279.44
−80	278.72	278.00	277.27	276.55	275.82	275.09	274.36	273.62	272.89	272.15
−90	271.41	270.67	269.92	269.18	268.43	267.68	266.93	266.17	265.42	264.66

ELECTRICITY AND MAGNETISM

SPARK-GAP VOLTAGES
Based on results of the American Institute of Electric Engineers Air at 760 mm, 25°C.

Peak voltage, kilovolts	Diameter of spherical electrodes, cm				Needle points	Peak voltage, kilovolts	Diameter of spherical electrodes, cm				Needle points
	2.5	5	10	25			2.5	5	10	25	
	Length of spark gap cm						Length of spark gap cm				
5	0.13	0.15	0.15	0.16	0.42	110	—	5.79	4.25	3.88	17.7
10	0.27	0.29	0.30	0.32	0.85	120	—	7.07	4.78	4.28	19.8
15	0.42	0.44	0.46	0.48	1.30	130	—	—	5.35	4.69	22.0
20	0.58	0.60	0.62	0.64	1.75	140	—	—	5.97	5.10	24.1
25	0.76	0.77	0.78	0.81	2.20	150	—	—	6.64	5.52	26.1
30	0.95	0.94	0.95	0.98	2.69	160	—	—	7.37	5.95	28.1
35	1.17	1.12	1.12	1.15	2.69	170	—	—	8.16	6.39	30.1
40	1.41	1.30	1.29	1.32	3.81	180	—	—	9.03	6.84	32.0
45	1.68	1.50	1.47	1.49	4.49	190	—	—	10.0	7.30	33.9
50	2.00	1.71	1.65	1.66	5.20	200	—	—	11.1	7.76	35.7
60	2.82	2.17	2.02	2.01	6.81	210	—	—	12.3	8.24	37.6
70	4.05	2.68	2.42	2.37	8.81	220	—	—	13.7	8.73	39.5
80	—	3.26	2.84	2.74	11.1	230	—	—	15.3	9.24	41.4
90	—	3.94	3.28	3.11	13.3	240	—	—	—	9.76	43.4
100	—	4.77	3.75	3.49	15.5	250	—	—	—	10.3	45.2
						300	—	—	—	13.3	54.7

CORRECTIONS FOR TEMPERATURE AND PRESSURE

Values found in the above table may be corrected for temperature and pressure by multiplying the values given by the appropriate correction factor found below:

Temp. °C	Pressure mm			
	730	710	760	700
0	1.04	1.06	1.09	1.12
10	1.00	1.02	1.05	1.08
20	0.96	0.99	1.02	1.04
30	0.93	0.96	0.98	1.01

DIELECTRIC CONSTANTS

Dielectric Constants of Pure Liquids

The values listed in the following table were obtained from National Bureau of Standards Circular 514

The dielectric constants are intended to be the limiting values at low frequencies, the so-called static values. Temperature is the only variable considered explicitly. Usually pressure is atmospheric or insignificantly different with respect to its effect on the dielectric constant. Where data are listed above the normal boiling point, the pressure corresponds to the vapor pressure of the liquid unless otherwise noted in the footnote.

Symbols

ε = dielectric constant (ε vacuum -1)
t = temperature, Centigrade (°C)
T = temperature, Absolute (°K)
a = $-d\varepsilon/dt$
α = $-d\log_{10}\varepsilon \cdot dt$

t_1, t_2 = the limits of temperature between which a or α is considered applicable
mp = melting point
bp = boiling point
f = frequency of alternating current in cycles per second

Standard Liquids

		ε20°C	ε25°C	a(or α)*			ε20°C	ε25°C	a(or α)
C_6H_{12}	Cyclohexane	2.023	2.015	0.0016	CH_4O	Methanol	33.62	32.63	0.00260(α)
CCl_4	Carbon tetrachloride	2.238	2.228	0.0020	$C_6H_5NO_2$	Nitrobenzene	35.74	34.82	0.00225(α)
C_6H_6	Benzene	2.284	2.274	0.0020	H_2O	Water	80.37	78.54	0.00200(α)
C_6H_5Cl	Chlorobenzene	5.708	5.621	0.00133α	H_2	Hydrogen		1.228 at 20.4°K	0.0034
$C_2H_4Cl_2$	1,2-Dichloroethane	10.65	10.36	0.00240(α)	O_2	Oxygen		1.507 at 80.0°K	0.0024

* The values of a or α given in this table were derived from data in the vicinity of room temperature and are not necessarily identical with the values listed in the following sections of this table. They may be used to calculate values of dielectric constant between 15° and 30°C without introducing significant error.

DIELECTRIC CONSTANTS (Continued)
Dielectric Constants of Pure Liquids

Inorganic Liquids

Substance		ε	°C	a(or a)× 10^2	Range t_1,t_2
A	Argon	1.53$_8$	−191	0.34	−191,−184
AlBr$_3$	Aluminum bromide	3.38	100	0.33	100,240
AsH$_3$	Arsine	2.50	−100	0.43	−116,−72
BBr$_3$	Boron bromide	2.58	0	0.28	−70,80
Br$_2$	Bromine	3.09	20	0.7	0,50
CO$_2$	Carbon dioxide	1.60^c	20	—	—
Cl$_2$	Chlorine	2.10$_1$	−50	0.31	−65,−33
		1.91	14	0.32	−22,14
		1.7$_2$	77		
		1.5$_4$	142		
D$_2$	Deuterium	1.277	20°K	0.4	18.8,21.2°K
D$_2$O	Deuterium oxide	78.25	25	^{d}	0.4,98
F$_2$	Fluorine	1.54	−202	0.19	−216,−190
GeCl$_4$	Germanium tetrachloride	2.43^0	25	0.240	0.55
HBr	Hydrogen bromide	7.00	−85	0.26(a)	−85,−70
HCl	Hydrogen chloride	6.35	−15	0.288(a)	−85,−15
		12.	−113	—	—
		4.6	28		
HF	Hydrogen fluoride	17$_5$	−73	—	—
		13$_4$	−42		
		11$_1$	−27		
		84	0		
HI	Hydrogen iodide	3.39	−50	0.8	−51,−37
H$_2$	Hydrogen	1.228	20.4°K	0.34	14,21°K
H$_2$O	Water	78.54	25	^{e}	0,100
		34.5$_9$	200		100,370
H$_2$O$_2$	Hydrogen peroxide	84.2	0	^{f}	−30,20
H$_2$S	Hydrogen sulfide	9.26	−85.5	—	—
		9.05	−78.5		
He	Helium	1.055$_4$	2.06°K	—	—
		1.055$_4$	2.30^f		
		1.055$_3$	2.63		
		1.053$_9$	3.09		
		1.051$_4$	3.58		
		1.048	4.19		
I$_2$	Iodine	6.8	400	—	ε, = 13.3−0.0 0.016 (±0.002) T
		11.$_7$	140		
		13.$_0$	168		
NH$_3$	Ammonia	25.	−77.7	—	—
		22.4	−33.4	—	—
		18.9	5	—	—
		17.8	15		
		16.9	25		
		16.3	35		
NOBr	Nitrosyl bromide	13.$_4$	15		
NOCl	Nitrosyl chloride	18.$_2$	12	—	—
N$_2$	Nitrogen	1.454	−203	0.29	−210,−195
N$_2$H$_4$	Hydrazine	52.$_9$	20	0.21(a)	0,25
N$_2$O	Dinitrogen oxide	1.97	−90	—	—
		1.61	0	0.6	−6,14
N$_2$O$_4$	Dinitrogen tetroxide	2.5^b	15	—	—
O$_2$	Oxygen	1.507	−193	0.24	−218,−183
P	Phosphorus	4.10	34	—	—
		4.06	46		
		3.86	85		
PCl$_3$	Phosphorus trichloride	3.43	25	0.84	17,60
PCl$_4$	Phosphorus pentachloride	2.8$_5$	160	—	—
POCl$_3$	Phosphoryl chloride	13.$_3$	22	—	—
PSCl$_3$	Thophosphoryl chloride	5.8	22	—	—
PbCl$_4$	Lead tetrachloride	2.78	20	—	—
S	Sulfur	3.52	118	^{g}	—
		3.48	231		
SOBr$_2$	Thionyl bromide	9.06	20	3.0	at 20
SOCl$_2$	Thionyl chloride	9.25	20	3.9	at 20
SO$_2$	Sulfur dioxide	17.6	−20	0.287(a)	−65,−15
		15.0$_8$	0		
		14.$_1$	20	7.7	14,140
		2.1$_0$	154^h		
SO$_3$	Sulfur trioxide \	3.11	18	—	—
S$_2$Cl$_2$	Sulfur monochloride	4.79	15	0.146(a)	−41,15
SO$_2$Cl$_2$	Sulfuryl chloride	10.$_0$	22	—	—
SbCl$_5$	Antimony pentachloride	3.22	20	0.46	2,47
Se	Selenium	5.40	250	0.25	237,301
SiCl$_4$	Silicon tetrachloride	2.4^0	16	—	—
SnCl$_4$	Tin tetrachloride	2.87	20	0.30	−30,20
TiCl$_4$	Titanium tetrachloride	2.80	20	0.20	−20,20

Organic Liquids

Substance		ε	°C	a(or a)× 10^2	Range t_1,t_2
CCl$_4$	Carbon tetrachloride	2.238	20	0.200	−10,60
CN$_2$O$_8$	Tetranitromethane	2.52$_1$	25	—	—
CO$_2$	Carbon dioxide	1.60$_4$^{c}	0	—	—
CS$_2$	Carbon disulfide	2.641	20	0.268	−90,130
		3.001	−110		
		2.19	180		
CHBr$_3$	Bromoform	4.39	20	0.105(a)	10,70
CHCl$_3$	Chloroform	4.806	20	0.160(a)	0,50
		6.76	−60	—	—
		6.12	−40		
		5.61	−20		
		3.7$_1$	100	—	—
		3.3$_2$	140		
		2.9$_2$	180		
CHN	Hydrocyanic acid	158.$_1$	0	^{i}	−13,18
		114.$_9$	20	0.63(a)	18,26
CH$_2$Br$_2$	Dibromomethane	7.77	10		
		6.68	40		
CH$_2$Cl$_2$	Dichloromethane	9.08	20	^{j}	−80,25
CH$_2$I$_2$	Diiodomethane	5.32	25	—	—
CH$_2$O$_2$	Formic acid	58.$_5$^{a}	16	—	—
CH$_3$Br	Bromomethane	9.82	0	^{k}	−80,0
CH$_3$Cl	Chloromethane	12.6	−20	^{l}	−70,−20
CH$_3$I	Iodomethane	7.00	20	^{m}	−70,40
CH$_3$NO	Formamide	109.	20	72.	18,25
CH$_4$	Methane	1.70	−173	0.2	−181,−159
CH$_4$O	Methanol	32.63	25	0.264(a)	5,55
		64.	−113	—	—
		54.	−80		
		40.	−20		
CH$_5$N	Methylamine	11.4	−10	0.26(a)	−30,−10
		9.4	25	—	—
C$_2$					
C$_2$HCl$_3$	Trichloroethylene	3.4$_2$	ca 16	—	—
C$_2$HCl$_3$O	Chloral	4.9$_4$	20	0.17(a)	15,45
		7.6	−40		
		4.2	62		
C$_2$HCl$_3$O$_2$	Trichloroacetic acid	4.6	60	—	—
C$_2$H$_2$Br$_2$	cis-1,2Dibromoethylene	7.7$_2$	0	—	—
		7.0$_5$	25		
	trans-1,2-Dibromoethylene	2.9$_3$	0	—	—
		2.8$_5$	25		
C$_2$H$_2$Br$_4$	1,1,2,3-Tetrabromoethane	8.6	3	—	—
		7.0	22		
C$_2$H$_2$Cl$_2$	1,1-Dichloroethylene	4.6$_7$	16	—	—
	cis-1,2-Dichloroethylene	9.20	25	—	—
	trans-1,2-Dichloroethylene	2.14	25	—	—
C$_2$H$_2$Cl$_2$O$_2$	Dichloroacetic acid	8.2	22	—	—
		7.8	61		
C$_2$H$_3$ClO	Acetyl chloride	16.$_9$	2	—	—
		15.$_8$	22		
C$_2$H$_3$ClO$_2$	Chloroacetic acid	12.3	60	2.	60,80
C$_2$H$_4$O	Ethylene oxide	13.$_9$	−1	—	—
	Acetaldehyde	21.$_8$^{e}	10	—	—
C$_2$H$_4$O$_2$	Acetic acid	6.15	20	—	—
		6.29	40		
		6.62	70		
	Methyl formate	8.5	20	5.	0,20
C$_2$H$_4$ClO	2-Chloroethanol (ethylene chlorohydrin)	25.$_8$	25	—	—
		13.$_2$	132		
C$_2$H$_5$NO	Acetamide	50.^{e}	83	—	—
C$_2$H$_5$NO$_2$	Nitroethane	28.0$_6$	30	11.4	30,35
C$_2$H$_6$O	Ethanol	24.30	25	—	—
		24.3^5	25	0.270(a)	−5,70
		41.0^5	−60	0.297(a)	−110,−20
	Methyl ether	5.02	25	2.38	25,100
		2.97	110		
		2.64	120		
		2.37	125		
		2.26	126.1		
		1.90	127.6		
(C$_2$H$_6$OSi)$_n$ $n = 4$	Octamethylcyclotetrasiloxane	2.39	20	—	—

DIELECTRIC CONSTANTS (Continued)
Dielectric Constants of Pure Liquids

Organic Liquids

	Substance	ε	°C	a(or α)×10^2	Range t_1,t_2		Substance	ε	°C	a(or α)×10^2	Range t_1,t_2
n = 5	Decamethylcyclopentasi-loxane	2.50	20	—	—	C_5H_5N	Pyridine	12.3	25	—	—
								9.4	116	—	—
n = 6	Dodecamethylcyclohexasi-loxane	2.59	20	—	—	C_5H_8	1,3-Pentadiene'	2.32	25	—	—
n = 7	Tetradecamethylcyclohep-tasiloxane	2.68	20	—	—		2-Methyl-1,3-butadiene (Is-oprene)	2.10	25	0.24	−75,25
n = 8	Hexadecamethylcyclooctas-iloxane	2.74	20	—	—	$C_5H_8O_2$	2,4-Pentanedione (Acetyl-acetone)	$25._7$*	20	—	—
$C_2H_6O_2$	Glycol	$37._7$	25	0.224(α)	20,100	C_5H_{10}	1-Pentene	2.100	20	—	—
C_2H_7N	Dimethylamine	6.32	0	—	—		2-Methyl-1-butene	2.197	20	—	—
	C_3	5.26	25				Cyclopentane	1.965	20	—	—
C_3H_6	Propene	1.87_5	20	—	—		Ethylcyclopropane	1.933	20	—	—
		1.79^5	45			$C_5H_{10}O$	2-Pentanone	15.4_5	20	0.195(α)	−40,80
		1.69	65					22.0	−60		
		1.53_0	85				3-Pentanone	17.0_0	20	0.225(α)	0,80
		1.44_1	90					19.4	−20		
		1.33_1	91.9ʰ					19.8	−40		
C_3H_6O	2-Propen-1-ol(Allylalcohol)	$21._6$	15	—	—	$C_5H_{10}O_2$	Valeric acid	2.6_6	20	—	—
	Acetone	20.7_0	25	0.205(a)	−60,40		Isovaleric acid	2.6_4	20	—	—
		17.7	56				Methyl butyrate	5.6"	20	—	—
	Propionaldehyde	$18._5$*	17	—	—	$C_5H_{11}N$	Piperidine	$5.8^ʰ$	22	—	—
$C_3H_6O_2$	Propionic acid	3.30	10	—	—	C_5H_{12}	n-Pentane	1.844	20	0.160	−50,30
		3.44	40					2.011	−90		
	Ethyl formate	7.1_6	25	—	—			1.984	−70		
	Methyl acetate	6.68	25	2.2	25,40		2-Methylbutane	1.843	20	—	—
$C_3H_6O_3$	dl-Lactic acid	22.	17	—	—	$C_5H_{12}O$	1-Pentanol	13.9	25	0.23(α)	15,35
$C_3H_6NO_2$	Ethyl carbamate (Urethan)	14.2	50	5.2	50,70		3-Methyl-1-butanol	14.7	25	—	—
C_3H_8	Propane	1.61	0	0.20	−90,15			5.8^2	132		
C_3H_8O	1-Propanol	20.1	25	0.293(α)	20,90		2-Methyl-2-butanol	5.82	25		
		38.	−80				C_6				
		79	−90			C_6H_4Cl	o-Dichlorobenzene	9.93	25	0.194(α)	0,50
	2-Propanol	18.3	25	0.310(α)	20,70		m-Dichlorobenzene	5.04	25	0.120(α)	0,50
							p-Dichlorobenzene	2.41	50	0.18	50,80
$C_3H_8O_2$	1,2-Propanediol	$32._0$	20	0.27(α)	at 20	C_6H_5Br	Bromobenzene	5.40	25	0.115(α)	0,70
	1,3-Propanediol	$35._0$	20	0.23(α)	at 20	C_6H_5Cl	Chlorobenzene	5.708	20	—	—
$C_3H_8O_3$	Glycerol	42.5	25	0.208(α)	0,100			5.621	25		
C_3H_9N	Isopropylamine	$5.5^ʰ$	20	—	—			5.71	20	0.130(α)	0,80
	Trimethylamine	2.44	25	0.52	0,25			7.28	−50		
	C_4							6.30	−20		
$C_4H_2O_3$	Maleic anhydride	50°	60	—	—			4.21	130		
C_4H_4O	Furan	2.95	25	—	—	C_6H_5Cl	o-Chlorophenol	6.31	25	2.7	25,58
C_4H_4S	Thiophene	2.76	16	—	—		p-Chlorophenol	9.47	55	3.7	55,65
C_4H_5N	Pyrrole	7.48	18	—	—	C_6H_5I	Iodobenzene	4.63	20	—	—
C_4H_5NS	Allyl isothiocyanate	$17._2^ʰ$	18	—	—	$C_6H_5NO_2$	Nitrobenzene	34.82	25	0.225(α)	10,80
C_4H_6O	Vinyl ether	3.94	20	—	—			20.8	130	0.164(α)	130,211
$C_4H_6O_3$	Acetic anhydride	$22.^4$	1	—	—			24.9	90		
		$20._7$	19					22.7	110		
C_4H_8O	2-Butanone	18.5_1	20	0.207(α)	−60,60	$C_6H_5NO_3$	o-Nitrophenol	$17._3$	50	6.4	50,60
	Butyraldehyde	13.4	26	—	—	C_6H_6	Benzene	2.284	20	0.200	10,60
		10.8	77					2.073	129		
$C_4H_8O_2$	Butyric acid	2.97	20	−0.23	10,70			1.966	182		
	Isobutyric acid	2.71	10	—	—	C_6H_6BrN	m-Bromoaniline	$13._0$"	19	—	—
		2.73	40			C_6H_6ClN	m-Chloroaniline	$13.^{4n}$	19	—	—
	Propyl formate	7.7_2*	19	—	—	$C_6H_6N_2O_2$	o-Nitroaniline	$34._5$	90	3.	90,110
	Ethylacetate	6.02	25	1.5	at 25		p-Nitroaniline	$56._5$	160	6.	160,180
		5.3_0	77			C_6H_6O	Phenol	9.78	60	0.32(α)	40,70
	Methyl propionate	5.5*	19	—	—	C_6H_7N	Aniline	6.89	20	0.148(α)	0,50
	1,4-Dioxane	2.209	25	0.170	20,50			5.93	70		
C_4H_8NO	Morpho line	7.33	25	—	—			4.54	184.6	—	—
$C_4H_{10}O$	1-Butanol	17.8	20	0.300(α)	−40,20		2-Methylpyridine (α-Pico-line)	$9.8^ʰ$	20	—	—
		17.1	25	0.335(α)	25,70						
		8.2	118	—	—	$C_6H_8N_2$	Phenylhydrazine	7.2	23	—	—
	2-Methyl-1-propanol	17.7	25	0.377(α)	20,90	C_6H_{10}	Cyclohexene	2.220	25	—	—
		34.	−80	—	—			2.6_0	−105	—	—
		26.	−34	—	—	$C_6H_{10}O$	Cyclohexanone	18.3	20	—	—
	2-Butanol	15.8	25	—	—			$19._9$	−40		
	2-Methyl-2-propanol	10.9	30	—	—	$C_6H_{10}O_3$	Ethyl acetoacetate	$15._7$*	22	—	—
		8.49	50			C_6H_{12}	Cyclohexane	2.023	20	0.160	10,60
		6.89	70				Methylcyclopentane	1.985	20	—	—
	Ethyl ether	4.335	20	2.0	at 20		Ethylcyclobutane	1.965	20	—	—
		4.34^5	20	0.217(α)	−40,30	$C_6H_{12}O$	Cyclohexanol	15.0	25	0.437(α)	20,66
		10.4	−116	—	—			7.2_4	100		
		3.97	40	0.170(α)	40,140			4.8_1	150		
		2.1_2	180			$C_6H_{12}O_2$	Butyl acetate	5.01	20	1.4	20,40
		1.8^9	190					6.8_4	−73		
		1.5_3	193.3ʰ				Ethyl butyrate	5.10	18	1.0	at 20
$C_4H_{10}Zn$	Diethyl zinc	2.5_5	20	—	—	$C_6H_{12}O_3$	Paraldehyde	13.9	25	—	—
	C_5							6.29	128		
$C_5H_4O_2$	Furfural	$46.^9$	1	—	—	C_6H_{14}	n-Hexane	1.890	20	0.155	−10,50
		$41._9$	20					2.044	−90		
		$34._9$	50					1.990	−50		

Dielectric Constants of Pure Liquids

Organic Liquids

	Substance	ε	°C	a(or α) × 10^2	Range t_1,t_2
$C_6H_{14}O$	1-Hexanol	13.3	25	0.35(α)	15,35
		8.5_6	75	—	—
	Propyl ether	3.3_9	26	—	—
	Isopropyl ether	3.88	25	1.8	0,25
$C_6H_{15}Al$	Triethyl aluminum	2.9	20	—	—
$C_6H_{15}N$	Dipropylamine	2.9^b	21	—	—
	Triethylamine	2.42	25	—	—
$C_6H_{15}N$	Dipropylamine	2.9^b	21	—	—
	Triethylamine	2.42	25	—	—
$C_6H_{18}OSi_2$	$(CH_3)_3Si[OSi(CH_3)_2]\,n\,CH_3$				
$n = 1$	Hexamethyldisiloxane	2.17	20	—	—
$n = 2$	Octamethyltrisiloxane	2.30	20	—	—
$n = 3$	Decamethyltetrasiloxane	2.39	20	—	—
$n = 4$	Dodecamethylpentasiloxane	2.46	20	—	—
$n = 5$	Tetradecamethylhexasiloxane	2.50	20	—	—
$n = 66$"		2.72	20	—	—
	C_7				
C_7H_5ClO	Benzoyl chloride	29.	0	—	—
		23.	20	—	—
C_7H_5NO	Phenyl isocyanate	8.8^b	20	—	—
C_7H_5NS	Phenyl isothiocyanate	$10._4{}^a$	20	—	—
C_7H_6O	Benzaldehyde	$19._7$	0	—	—
		$17._8$	20	—	—
$C_7H_6O_2$	Salicylaldehyde	$17._1$	30	7.	30,40
C_7H_7Br	o-Bromotoluene	4.23	58	—	—
	m-Bromotoluene	5.36	58	—	—
	p-Bromotoluene	5.49	58	—	—
C_7H_7Cl	o-Chlorotoluene	4.45	20	—	—
		4.16	58	—	—
	m-Chlorotoluene	5.55	20	—	—
		5.04	58	—	—
	p-Chlorotoluene	6.08	20	—	—
		5.55	58	—	—
	α-Chlorotoluene	7.0	13	—	—
$C_7H_7NO_2$	o-Nitrotoluene	27.4	20	15.	at 20
		$21._6$	58	—	—
		11.8	222	—	—
	m-Nitrotoluene	$23._8$	20	—	—
		$21._9$	58	—	—
	p-Nitrotoluene	$22._2$	58	—	—
C_7H_8	Toluene	2.438	0	0.0455(α)	-90,0
		2.379	25	0.243	0,90
		2.15_7	127	—	—
		2.04_2	181	—	—
C_7H_8O	Benzyl alcohol	13.1	20	—	—
		9.47	70	—	—
		6.6	132	—	—
	o-Cresol	11.5	25	11	25,30
	m-Cresol	11.8	25	0.41(α)	15,50
	p-Cresol	9.9_1	58	—	—
C_7H_9N	o-Toluidine	6.34	18	—	—
		5.71	58	—	—
		4.00	200	—	—
	m-Toluidine	5.95	18	—	—
		5.45	58	—	—
	p-Toluidine	4.98	54	—	—
	N-Methylaniline	5.97	22	—	—
	1-Heptene	2.05	20	—	—
	C_8				
C_8H_8	Styrene (Phenylethylene)	2.43	25	—	—
		2.32	75	—	—
	Acetophenone	17.39	25	4.	at 25
		8.64	202	—	—
$C_8H_8O_3$	Phenyl acetate	5.23	20	0.7	at 20
	Methyl benzoate	6.59	20	0.14(α)	at 20
$C_8H_8O_3$	Methyl salicylate	9.41	30	3.1	30,40
C_8H_{10}	Ethylbenzene	2.412	20	—	—
	o-Xylene	2.568	20	0.266	-20,130
	m-Xylene	2.374	20	0.195	-40,180
	p-Xylene	2.270	20	0.160	20,130
C_8H_{18}	n-Octane	1.948	20	0.130	-50,50
		1.879	70	—	—
		1.817	110	—	—
	2,2,3-Trimethylpentane	1.96	20	—	—
	2,2,4-Trimethylpentane	1.940	20	0.142	-100,100
$C_8H_{18}O$	1-Octanol	10.3_4	20	0.410(α)	20,60
$C_8H_{20}O_4Si$	Tetraethyl silicate	4.1^b	ca 20	—	—
	C_9				
C_9H_7N	Quinoline	9.00	25	—	—
		5.05	238	—	—
	Isoquinoline	10.7	25	—	—
C_9H_8O	Cinnamaldehyde	16.9	24	—	—
$C_9H_{10}O_2$	Benzyl acetate	5.1^a	21	—	—
	Ethyl benzoate	6.02	20	2.1	20,40
$C_9H_{10}O_3$	Ethyl salicylate	7.99	30	2.	30,40
C_9H_{12}	Isopropylbenzene (Cumene)	2.38_0	20	—	—
	1,3,5-Trimethylbenzene (Mesitylene)	2.27_9	20	—	—
C_9H_{20}	n-Nonane	1.972	20	0.135	-10,90
		2.059	-50	—	—
		1.847	110	—	—
		1.787	150	—	—
	C_{10}				
$C_{10}H_8$	Naphthalene	2.54	85	—	—
$C_{10}H_{10}O_4$	Dimethyl phthalate	8.5	24	—	—
$C_{10}H_{16}$	d-Camphene	2.33	ca 40	—	—
	d-Pinene	2.64	25	—	—
	l-Pinene	2.76	20	—	—
	Terpinene	2.7^b	21	—	—
	d-Limonene	2.3_6	20	—	—
	d-Limonene (Dipentene)	2.3^0	20	—	—
$C_{10}H_{22}$	n-Decane	1.991	20	0.130	10,110
		2.050	-30	—	—
		1.844	130	—	—
		1.783	170	—	—
$C_{10}H_{22}$	1-Decanol	8.1	20	—	—
$C_{11}H_{24}$	n-Undecane	2.005	20	0.125	10,130
		2.039	-10	—	—
		1.838	150	—	—
		1.781	190	—	—
	C_{12}				
$C_{12}H_{10}$	Diphenyl	2.53	75	0.18	75,155
$C_{12}H_{10}O$	Phenyl ether	3.65	30	0.7	30,50
$C_{12}H_{11}N$	Diphenylamine	3.3	52	—	—
$C_{12}H_{26}$	n-Dodecane	2.015	20	0.120	10,150
		2.047	-10	—	—
		1.776	210	—	—
$C_{13}H_{10}$	Benzophenone	11.4	50	—	—
	C_{14}				
$C_{14}H_{15}N$	Dibenzylamine	3.6^b	20	—	—
	C_{16}				
$C_{16}H_{32}O_2$	Palmitic acid	2.30	71	—	—
	C_{18}				
$C_{18}H_{32}O_2$	Linoleic acid	2.61	0	—	—
		2.71	20	—	—
		2.70	70	—	—
		2.60	120	—	—
$C_{18}H_{34}O_2$	Oleic acid	2.46	20	—	—
		2.45	60	—	—
		2.41	100	—	—
$C_{18}H_{36}O_2$	Stearic acid	2.29	70	—	—
		2.26	100	—	—
	Ethyl palmitate	3.20	20	0.4	20,40
		2.71	104	—	—
		2.46	182	—	—
	C_{19}				
$C_{19}H_{16}$	Triphenylmethane	2.45	100	0.14	94,175
$C_{19}H_{38}O_4$	Monopalmitin	5.34	67	—	—
		5.09	80	—	—
	C_{20}				
$C_{20}H_{38}O_2$	Ethyl Oleate	3.17	28	0.48	28,122
$C_{20}H_{40}O_2$	Ethyl Stearate	2.98	40	0.6	32,50
		2.69	100	—	—
		2.48	167	—	—
	C_{21}				
$C_{21}H_{21}O_4P$	Tricresyl phosphate	6.9	40	—	—
	C_{22}				
$C_{22}H_{42}O_2$	Butyl oleate	4.0	25	—	—
$C_{22}H_{44}O_2$	Butyl stearate	3.11_1	30	0.53	30,35

DIELECTRIC CONSTANTS (Continued)
Dielectric Constants of Pure Liquids

Dielectric Constants of Solids
Compiled by Earle C. Gregg, Jr.

Solids[a] (17 to 22°C)

Material	Frequency	Dielectric constant	Material	Frequency	Dielectric constant
Acetamide	4×10^8	4.0	Phenanthrene	4×10^8	2.80
Acetanilide	—	2.9	Phenol (10°C)	4×10^8	4.3
Acetic acid (2°C)	4×10^8	4.1	Phosphorus, red	10^8	4.1
Aluminum oleate	4×10^8	2.40	Phosphorus, yellow	10^8	3.6
Ammonium bromide	10^8	7.1	Potassium aluminum		
Ammonium chloride	10^8	7.0	sulfate	10^6	3.8
Antimony trichloride	10^8	5.34	Potassium carbonate		
Apatite ⊥ optic axis	3×10^8	9.50	(15°C)	10^8	5.6
Apatite ∥ optic axis	3×10^8	7.41	Potassium chlorate	6×10^7	5.1
Asphalt	$<3 \times 10^6$	2.68	Potassium chloride	10^4	5.03
Barium chloride (anhyd.)	6×10^7	11.4	Potassium chromate	6×10^7	7.3
Barium chloride (2H$_2$O)	6×10^7	9.4	Potassium iodide	6×10^7	5.6
Barium nitrate	6×10^7	5.9	Potassium nitrate	6×10^7	5.0
Barium sulfate (15°C)	10^8	11.4	Potassium sulfate	6×10^7	5.9
Beryl ⊥ optic axis	10^4	7.02	Quartz ⊥ optic axis	3×10^7	4.34
Beryl ∥ optic axis	10^4	6.08	Quartz ∥ optic axis	3×10^7	4.27
Calcite ⊥ optic axis	10^4	8.3	Resorcinol	4×10^8	3.2
Calcite ∥ optic axis	10^4	8.0	Ruby ⊥ optic axis	10^4	13.27
Calcium carbonate	10^6	6.14	Ruby ∥ optic axis	10^4	11.28
Calcium fluoride	10^4	7.36	Rutile ⊥ optic axis	10^8	86
Calcium sulfate (2H$_2$O)	10^4	5.66	Rutile ∥ optic axis	10^8	170
Cassiterite ⊥ optic axis	10^{12}	23.4	Selenium	10^8	6.6
Cassiterite ∥ optic axis	10^{12}	24	Silver bromide	10^6	12.2
d-Cocaine	5×10^8	3.10	Silver chloride	10^6	11.2
Cupric oleate	4×10^8	2.80	Silver cyanide	10^6	5.6
Cupric oxide (15°C)	10^8	18.1	Smithsonite ⊥ optic	10^{12}	9.3
Cupric sulfate (anhyd.)	6×10^7	10.3	axis		
Cupric sulfate (5H$_2$O)	6×10^7	7.8	Smithsonite ∥ optic	10^{10}	9.4
Diamond	10^8	5.5	axis		
Diphenylmethane	4×10^8	2.7	Sodium carbonate (anhyd.)	6×10^7	8.4
Dolomite ⊥ optic axis	10^8	8.0			
Dolomite ∥	10^8	6.8	Sodium carbonate (10H$_2$O)	6×10^7	5.3
Ferrous oxide (15°C)	10^8	14.2			
Iodine	10^8	4	Sodium chloride	10^4	6.12
Lead acetate	10^6	2.6	Sodium nitrate	—	5.2
Lead carbonate (15°C)	10^6	18.6	Sodium oleate	4×10^8	2.75
Lead chloride	10^6	4.2	Sodium perchlorate	6×10^7	5.4
Lead monoxide (15°C)	10^8	25.9	Sucrose (mean)	3×10^8	3.32
Lead nitrate	6×10^7	37.7	Sulfur (mean)	—	4.0
Lead oleate	4×10^8	3.27	Thallium chloride	10^6	46.9
Lead sulfate	10^6	14.3	p-Toluidine	4×10^8	3.0
Lead sulfide (15°)	16^6	17.9	Tourmaline ⊥ optic	10^4	7.10
Malachite (mean)	10^{12}	7.2	axis		
Mercuric chloride	10^6	3.2	Tourmaline ∥ optic	10^4	6.3
Mercurous chloride	10^6	9.4	axis		
Naphthalene	4×10^8	2.52	Urea	4×10^8	3.5
			Zircon ⊥, ∥	10^8	12

a For plastics and other insulating materials, refer to table on Properties of Dielectrics.

DIELECTRIC CONSTANTS (Continued)

Table of Dielectric Constants of Reference Gases at 20°C and 1 Atmosphere

The listed values $(\epsilon - 1)$ refer to the gas at a temperature of 20°C and pressure of 1 atmosphere. The values can be adjusted to slightly different conditions without introducing more than 0.1% error by use of the following equation:

$$\frac{(\epsilon - 1)_{t,p}}{(\epsilon - 1)_{20°,1\,atm}} = \frac{p}{760[1 + 0.003411(t - 20)]}$$

where p = pressure in mm Hg
t = degrees C
t should be between 10 and 30°C and p between 700 and 800 mm Hg. From National Bureau of Standards Circular 537

Substance	$(\epsilon-1)\cdot10^6$	Ref.	Substance	$(\epsilon-1)\cdot10^6$	Ref.	Substance	$(\epsilon-1)\cdot10^6$	Ref.	Substance	$(\epsilon-1)\cdot10^6$	Ref.
Helium	Radio fre-		Oxygen	Radio fre-			Optical			536.1	Koster
	quency			quency			516.8	Cuthbertson		536.3	Perard
	67.8	Watson		494.₃	Watson		517.8	Quarder		535.8	Barrell
	63.7	Hector		496.2	Jelatis		517.0	Tausz	Nitrogen	Radio fre-	
	64.5	Jelatis		Micro-			516.7	Damkohler		quency	
	Microwave^a			wave^a		Air (dry,	Radio fre-			547.₂	Watson
	65.6	Birnbaum		494.9	Birnbaum	CO₂ Free)	quency			Microwave	
	65.2	Essen		495.0	Essen		537.₀	Watson		547.3	Birnbaum
	Optical			494.9	Essen		Micro-			548.0	Essen
	64.6	Koch		Optical			wave^a			548.0	Essen
	64.5	Cuthbertson		494.5	Cuthbertson		536.6	Birnbaum		Optical	
Hydrogen	Radio fre-			493.5	Lowery		536.6	Essen		548.9	Cuthbertson
	quency			494.7	Tausz		536.6	Essen		548.7	Koch
	254.0	Watson		494.4	Ladenberg		Optical			547.2	Tausz
	Microwave		Argon	Radio fre-			536.9	Koch	Carbon	Radio fre-	
	253.4	Essen		quency			535.8	Meggers	dioxide	quency	
	Optical			513.0	Watson		536.0	Traub		921.₅	Watson
	254.1	Cuthbertson		516.4	Jelatis		536.7	Quarder		Microwave	
	253.6	Koch		Microwave			536.4	Lowery		922.4	Birnbaum
	253.7	Kirn		517.7	Essen		536.5	Tausz		920.6	Essen
	254.3	Tausz										

^a These values were derived from measurements of the refractive index after making allowance for the magnetic permeability of oxygen.

DIELECTRIC CONSTANTS OF GASES AT 760 MM PRESSURE

Compiled by Earl C. Gregg, Jr.

Material	Temperature °C	Frequency cycles/sec	Dielectric Constant	Material	Temperature °C	Frequency cycles/sec	Dielectric Constant
Acetaldehyde	100	$<3\times10^6$	1.0213	n-Heptane	20	$<3\times10^6$	1.0035
	0	2×10^6		Hydrogen	100	$<3\times10^6$	1.000264
Acetone			1.0159	Hydrogen bromide	20	$<3\times10^6$	1.00313
Acetyl chloride	0	$<3\times10^6$	1.0217	Hydrogen chloride	0	$<3\times10^6$	1.0046
Acetylene	0	$<3\times10^6$	1.00134		0	$<3\times10^6$	
Air			1.000590	Hydrogen iodide	0	$<10^6$	1.00234
Ammonia	0	$<3\times10^6$	1.0072		100	$<3\times10^6$	
β-Amylene			1.0028	Hydrogen sulfide	0	$<10^6$	1.00030
Argon	23	10^{10}	1.000545		100	$<3\times10^6$	
Benzene	400	3×10^8	1.0028	Mercury	180	$<3\times10^6$	1.00074
Bromine	0	$<3\times10^6$	1.0128	Methane	0	$<3\times10^6$	1.000944
Butylene			1.00319	Methyl alcohol	0	$<10^6$	1.0057
Carbon dioxide	100	$<3\times10^6$	1.000985	Methyl bromide	0	$<3\times10^6$	1.00095
Carbon disulfide	23	10^{10}	1.0029	Methyl chloride	0	$<3\times10^6$	1.00094
Carbon monoxide	23	10^{10}	1.00070	Methyl iodide	110	$<3\times10^6$	1.0063
Carbon tetrachloride	100	$<3\times10^6$	1.0030	Methylamine	120	$<3\times10^6$	1.0038
Chloroform	100	$<3\times10^6$	1.0042	Methylene chloride	23	10^{10}	1.0065
Dichlorodifluoro-methane	100	$<3\times10^6$	1.00029	Neon			1.000127
	0	$<3\times10^6$		Nitrogen			1.000580
	0	$<3\times10^6$		Nitromethane	23	10^{10}	1.0247
Dichlorofluorome-thane	100	$<3\times10^6$	1.00049	Nitrous oxide (N₂O)	23	2.5×10^{10}	1.00113
Dimethylamine	0	2×10^6	1.00040		0	$<3\times10^6$	
Ethane	0	$<3\times10^6$	1.00150	Oxygen	100	$<3\times10^6$	1.000523
Ethylalcohol	100	$<3\times10^6$	1.0061	n-Pentane	23	10^{10}	1.0025
Ethylamine			1.00053	n-Propyl chloride	20	$<3\times10^6$	1.0143
Ethyl bromide	20	$<3\times10^6$	1.0139	iso-Propyl chloride	20	$<3\times10^6$	1.0152
Ethyl chloride	20	$<3\times10^6$	1.0132	Sulfur dioxide	100	$<3\times10^6$	1.00075
Ethyl ether	23	10^{10}	1.0049	Toluene	100	$<3\times10^6$	1.0043
Ethyl formate	126	$<3\times10^6$	1.0083	Vinyl bromide	20	$<3\times10^6$	1.0081
Ethyl iodide	20	$<3\times10^6$	1.0140		100	$<3\times10^6$	
			1.0089	Water (steam)	0	$<3\times10^6$	1.0126
Ethylene	110	$<3\times10^6$	1.00144		0	$<3\times10^6$	1.00785
Helium	140	$<3\times10^6$	1.0000684				

PROPERTIES OF DIELECTRICS

In most cases properties have been determined by A.S.T.M. (American Society for Testing Materials) test methods at room temperature under standard conditions. Values will in general change considerably with temperature.

DIELECTRIC CONSTANTS OF SOME PLASTICS AND RUBBERS

Name	°C	Frequency (hertz) 1 × 10³	1 × 10⁶	1 × 10⁸
Plastics				
Phenol-formaldehyde	25—27	5.15—8.61	4.45—5.05	4.1—4.5
	57	6.35	4.90	4.5
	88	8.5	5.2	4.7
Phenol-aniline-formaldehyde	25	4.50	4.31	4.11
	79	4.75	4.51	4.35
Melamine-formaldehyde	24—28	6.0—6.90	5.82—6.20	5.5—5.55
	57	6.95	5.40	4.90
	88	11.8	6.0	5.5
Urea-formaldehyde	24	6.7	6.0	5.2
	80	7.8	6.8	—
Polyamide resins				
Nylon 66	25	3.75	3.33	3.16
Nylon 610	25	3.50	3.14	3.0
	84	11.2	4.4	3.4
Cellulose acetate	26	3.50—4.48	3.28—3.90	3.05—3.40
Cellulose nitrate	27	8.4	6.6	5.2
	78	7.5	6.2	5.2
Methyl cellulose	22	6.8	5.7	4.3
Ethyl cellulose	25	3.09	3.01	2.90
Silicone resins	25	3.79—3.91	3.79—3.82	3.82
Polyethylene	−12	2.37	2.35	2.33
	23	2.26	2.26	2.26
Polyisobutylene	25	2.23	2.23	2.23
Vinylite QYNA	20	3.10	2.88	2.85
	76	3.83	3.0	2.8
	110	8.6		
Vinylite 5544	25	7.20	4.13	3.05
Vinylite 5901	25	5.5	3.4	3.0
Vinylite VU	21	5.65	3.10	3.00
	79	8.15	5.5	3.4
Vinylite VYHW	20	3.12	2.91	2.83
Vinylite VYNW	20	3.15	2.90	2.8
Polyvinyl chloride	25	4.55 (1 × 10⁴)	3.3	—

Name	°C	Frequency (hertz) 1 × 10³	1 × 10⁶	1 × 10⁸
Polyvinylidene and vinyl chloride	23	4.65	3.18	2.82
	84	4.94	4.40	3.2
Polychlorotrifluoroethylene	25	2.76	2.48	2.36
Polytetrafluoroethylene (Teflon)	22	2.1	2.1	2.1
	100	2.04	2.04	—
Polyvinylalcohol acetate	25	7.8	5.2	—
	85	100	10	—
Polyvinylacetals	26—27	3.02—3.12	2.86—2.92	2.67
	88	3.5	3.1	2.85
Polyacrylates				
Lucite	−12	2.9	2.63	2.50
	23	2.84	2.63	2.58
	81	3.45	2.72	2.59
Plexiglas	27	3.12	2.76	—
Polystyrene	25	2.54—2.56	2.54—2.56	2.55
	80	2.54	2.54	2.54
Styrene copolymers	25	2.55—2.95	2.55—2.80	2.55—2.77
Polyesters	25	3.22—4.3	3.12—4.0	2.94—2.98
Alkyd resins				
Alkyd isocyanate foam	25	1.223	1.218	1.20
Plaskon, clay filled	25	5.26	4.92	4.77
Plaskon, glass filled	25	5.04	4.73	4.50
Epoxy resins	25	3.63—3.67	3.52—3.62	3.32—3.35
Rubbers				
Hevea, vulcanized	27	2.94	2.74	2.42
Hevea compound	27	30	9	0.8
Gutta percha	25	2.60	2.53	2.47
Balata	25	2.50	2.50	2.42
Buna S	20	2.66	2.56	2.52
Butyl rubber compound	25	2.42	2.40	2.39
Neoprene	24	6.60	6.26	4.5
Silicon rubber	25	3.12—3.30	3.10—3.20	3.06—3.18

DIELECTRIC CONSTANTS OF CERAMICS

Material	Dielectric constant 10⁶ cycles	Dielectric strength volts mil	Volume resistivity Ohms-cm (23°C)	Loss factor[a]
Alumina	4.5—8.4	40—160	10^{11}—10^{14}	0.0002—0.01
Corderite	4.5—5.4	40—250	10^{12}—10^{14}	0.004—0.012
Forsterite	6.2	240	10^{14}	0.0004
Porcelain (dry process)	6.0—8.0	40—240	10^{12}—10^{14}	0.0003—0.02
Porcelain (wet process)	6.0—7.0	90—400	10^{12}—10^{14}	0.006—0.01
Porcelain, zircon	7.1—10.5	250—400	10^{13}—10^{15}	0.0002—0.008
Steatite	5.5—7.5	200—400	10^{13}—10^{15}	0.0002—0.004
Titanates (Ba, Sr, Ca, Mg, and Pb)	15—12.000	50—300	10^{8}—10^{15}	0.0001—0.02
Titanium dioxide	14—110	100—210	10^{13}—10^{18}	0.0002—0.005

DIELECTRIC CONSTANTS OF WAXES

Material	Dielectric constant 10⁶ Cycles	Dielectric strength volts mil	Volume resistivity Ohms-cm (23°C)	Loss factor[a]
Acrawax C	2.4	—	—	0.005
Beeswax, white	2.75—3.0	—	5×10^{14}	0.025
Beeswax, yellow	2.9	—	8×10^{14}	0.029
Candelilla	2.25—2.50	—		
Carnauba	2.75—3.0	—		
Cerese, brown G	2.27	—	—	0.0025
Ceresine	2.25—2.50	—	$>5 \times 10^{18}$	0.0011
Halowax 1001	∼4.10	—	2×10^{13}	0.014
Halowax 1013	∼4.75	—	—	0.036
Halowax 1014	∼4.40	—	—	0.035
Halowax 11-314	2.94	—	—	0.00094
Microcrystalline waxes	2.2—2.5			
Opalwax	3.1	—	—	0.34

[a] Power factor × dielectric constant equals loss factor.

Material	Dielectric constant 10⁶ Cycles	Dielectric strength volts mil	Volume resistivity Ohms-cm (23°C)	Loss factor[a]
Ozokerite wax	2.3	100—150	5×10^{14}	0.0018
Paraffin	2.0—2.5	250	10^{15}—10^{19}	0.003 (900 cps)
Parawax	2.25	—	10^{16}	0.00045
135 A.M.P. wax	2.25	—	—	0.00023

DIELECTRIC CONSTANTS OF GLASSES

Type	Dielectric constant at 100 mc (20°C)	Volume resistivity (350°C megohm-cm)	Loss factor[a]
Corning 0010	6.32	10	0.015
Corning 0080	6.75	0.13	0.058
Corning 0120	6.65	100	0.012
Pyrex 1710	6.00	2,500	0.025
Pyrex 3320	4.71	—	0.019
Pyrex 7040	4.65	80	0.013
Pyrex 7050	4.77	16	0.017
Pyrex 7052	5.07	25	0.019
Pyrex 7060	4.70	13	0.018
Pyrex 7070	4.00	1,300	0.0048
Vycor 7230	3.83	—	0.0061
Pyrex 7720	4.50	16	0.014
Pyrex 7740	5.00	4	0.040
Pyrex 7750	4.28	50	0.011
Pyrex 7760	4.50	50	0.0081
Vycor 7900	3.9	130	0.0023
Vycor 7910	3.8	1,600	0.00091
Vycor 7911	3.8	4,000	0.00072
Corning 8870	9.5	5,000	0.0085
G. E. Clear (silica glass)	3.81	4,000—30,000	0.00038
Quartz (fused)	3.75 4.1 (1 mc)	—	0.0002 (1 mc)

[a] Power factor × dielectric constant equals loss factor.

STATIC DIELECTRIC CONSTANT OF WATER SUBSTANCE[a]

The temperatures are in degrees kelvin and the pressures in megapascals. Some conversion factors for pressure which may be useful are:

Megapascals × 9.8692 = atmospheres (/60 mm Hg)
Megapascals × 14.504 = pounds per square inch
Megapascals × 10⁻⁶ = newtons per square meter (pascals)
Megapascals × 10.1972 = kilograms per square centimeter
Megapascals × 10 = bars
Megapascals × 7.501 × 10³ = mm Hg at 0°C
Megapascals × 4.014 × 10³ = inches of H_2O at 4°C

P/T	273.15	298.15	323.15	348.15	373.15	398.15	423.15	448.15	473.15	498.15	523.15	548.15	573.15	623.15	673.15	723.15	773.15	823.15
10	88.28	78.85	70.27	62.59	55.76	49.70	44.30	39.47	35.11	31.13	27.43	23.90	20.39	1.23	1.17	1.14	1.11	1.10
20	88.75	79.24	70.63	62.94	56.11	50.05	44.66	39.85	35.52	31.58	27.95	24.54	21.24	14.07	1.64	1.42	1.32	1.26
30	89.20	79.63	70.98	63.28	56.44	50.39	45.01	40.22	35.91	32.01	28.43	25.11	21.95	15.66	5.91	2.07	1.68	1.51
40	89.64	80.00	71.32	63.61	56.77	50.72	45.34	40.56	36.28	32.40	28.87	25.61	22.56	16.72	10.46	3.84	2.34	1.90
50	90.07	80.36	71.66	63.93	57.08	51.03	45.67	40.89	36.63	32.78	29.28	26.08	23.10	17.55	12.16	6.57	3.45	2.48
60	90.49	80.72	71.98	64.24	57.39	51.34	45.98	41.21	36.96	33.13	29.67	26.50	23.58	18.24	13.28	8.53	4.90	3.26
70	90.90	81.07	72.30	64.54	57.69	51.64	46.28	41.52	37.28	33.47	30.03	26.90	24.02	18.84	14.16	9.87	6.31	4.20
80	91.29	81.42	72.62	64.84	57.98	51.93	46.57	41.82	37.59	33.79	30.37	27.27	24.43	19.37	14.88	10.88	7.50	5.16
90	91.67	81.75	72.92	65.13	58.27	52.21	46.86	42.11	37.89	34.10	30.70	27.62	24.81	19.85	15.50	11.70	8.47	6.06
100	92.04	82.08	73.22	65.42	58.55	52.49	47.14	42.39	38.17	34.40	31.01	27.95	25.17	20.29	16.05	12.39	9.29	6.88
125	92.89	82.84	73.93	66.09	59.19	53.12	47.78	43.05	38.86	35.13	31.78	28.76	26.03	21.26	17.21	13.77	10.88	8.53
150	93.71	83.57	74.62	66.74	59.82	53.75	48.40	43.68	39.50	35.78	32.46	29.47	26.77	22.09	18.16	14.85	12.07	9.80
175	94.48	84.28	75.27	67.36	60.42	54.34	48.98	44.27	40.10	36.39	33.09	30.12	27.45	22.83	18.98	15.74	13.04	10.81
200	95.20	84.94	75.89	67.95	61.00	54.90	49.54	44.83	40.66	36.97	33.67	30.72	28.07	23.49	19.69	16.51	13.86	11.65
225	95.87	85.58	76.50	68.53	61.55	55.44	50.08	45.36	41.20	37.51	34.22	31.28	28.64	24.09	20.33	17.19	14.56	12.38
250	96.51	86.20	77.08	69.08	62.08	55.96	50.59	45.87	41.70	38.02	34.74	31.81	29.17	24.65	20.91	17.80	15.19	13.01
300	97.69	87.34	78.17	70.14	63.10	56.94	51.55	46.82	42.65	38.97	35.69	32.77	30.15	25.65	21.94	18.85	16.25	14.07
350	98.75	88.40	79.19	71.12	64.05	57.86	52.45	47.70	43.52	39.83	36.56	33.64	31.02	26.53	22.83	19.74	17.14	14.93
400	99.72	89.39	80.13	72.03	64.94	58.74	53.30	48.53	44.33	40.64	37.36	34.43	31.81	27.32	23.62	20.52	17.89	15.66
450	100.60	90.30	81.02	72.89	65.78	59.56	54.10	49.31	45.10	41.38	38.09	35.16	32.54	28.04	24.32	21.20	18.55	16.28
500	101.42	91.16	81.84	73.69	66.57	60.33	54.85	50.05	45.82	42.09	38.78	35.84	33.21	28.70	24.96	21.82	19.14	16.83

[a]   Prepared by International Association for the Properties of Steam.

DIELECTRIC CONSTANT OF DEUTERIUM OXIDE

t	c	$-\dfrac{dc}{dt}$	$-\dfrac{1}{c}\dfrac{dc}{dt}$
°C			
4	85.877	0.3974	4.627×10^{-3}
5	85.480	.3956	4.628
10	83.526	.3862	4.624
15	81.618	.3771	4.620
20	79.755	.3681	4.615
25	77.936	.3594	4.611
30	76.161	.3509	4.607
35	74.427	.3425	4.602
40	72.735	.3344	4.597
45	71.083	.3265	4.593
50	69.470	.3187	4.587
55	67.896	.3112	4.583
60	66.358	.3038	4.578
65	64.857	.2967	4.575
70	63.391	.2898	4.571
75	61.959	.2830	4.567
80	60.561	.2765	4.565
85	59.194	.2701	4.563
90	57.859	.2640	4.563
95	56.554	.2581	4.564
100	55.278	.2523	4.564

DIELECTRIC CONSTANTS (Continued)

Dielectric Constant of Liquid Parahydrogen vs. Temperature (°K) and Pressure (atm)

R. J. Corruccini

P atm \ T, °K	20	21	22	23	24	25	26	27	28	29	30	31	32
1	1.2297												
2	1.2302	1.2260	1.2216										
3	1.2306	1.2265	1.2221	1.2174	1.2122								
4	1.2311	1.2270	1.2227	1.2180	1.2129	1.2073	1.2010						
5	1.2315	1.2275	1.2233	1.2186	1.2136	1.2081	1.2020	1.1950					
6	1.2320	1.2280	1.2238	1.2192	1.2143	1.2089	1.2029	1.1962	1.1883				
7	1.2324	1.2285	1.2243	1.2198	1.2150	1.2097	1.2039	1.1973	1.1897	1.1805			
8	1.2329	1.2290	1.2249	1.2204	1.2157	1.2105	1.2048	1.1984	1.1911	1.1824			
9	1.2333	1.2295	1.2254	1.2210	1.2163	1.2112	1.2056	1.1994	1.1924	1.1842	1.1734		
10	1.2337	1.2300	1.2259	1.2216	1.2169	1.2119	1.2065	1.2004	1.1936	1.1857	1.1758	1.1621	
15	1.2358	1.2322	1.2284	1.2243	1.2200	1.2153	1.2103	1.2049	1.1990	1.1924	1.1847	1.1758	1.1645
20	1.2378	1.2343	1.2307	1.2268	1.2227	1.2184	1.2137	1.2088	1.2034	1.1976	1.1913	1.1839	1.1757
25	1.2396	1.2363	1.2328	1.2291	1.2253	1.2211	1.2168	1.2122	1.2073	1.2021	1.1964	1.1903	1.1832
30	1.2414	1.2382	1.2349	1.2313	1.2276	1.2237	1.2196	1.2153	1.2107	1.2059	1.2008	1.1952	1.1891
35	1.2431	1.2400	1.2368	1.2334	1.2298	1.2261	1.2222	1.2181	1.2138	1.2093	1.2046	1.1995	1.1942
40	1.2448	1.2418	1.2386	1.2354	1.2319	1.2284	1.2246	1.2208	1.2167	1.2124	1.2080	1.2033	1.1984
45	1.2464	1.2434	1.2404	1.2372	1.2339	1.2305	1.2269	1.2232	1.2193	1.2153	1.2111	1.2067	1.2021
50	1.2479	1.2450	1.2421	1.2390	1.2358	1.2325	1.2291	1.2255	1.2218	1.2179	1.2139	1.2098	1.2055
60	1.2508	1.2481	1.2453	1.2424	1.2394	1.2363	1.2331	1.2297	1.2263	1.2227	1.2191	1.2153	1.2114
70	1.2535	1.2510	1.2483	1.2455	1.2427	1.2397	1.2367	1.2336	1.2303	1.2270	1.2236	1.2201	1.2165
80	1.2561	1.2536	1.2511	1.2484	1.2457	1.2429	1.2400	1.2371	1.2340	1.2309	1.2277	1.2244	1.2211
90	1.2585	1.2561	1.2537	1.2512	1.2486	1.2459	1.2431	1.2403	1.2374	1.2345	1.2315	1.2284	1.2252
100	1.2608	1.2586	1.2562	1.2538	1.2513	1.2487	1.2461	1.2434	1.2406	1.2378	1.2349	1.2320	1.2290
120	1.2652	1.2631	1.2609	1.2586	1.2563	1.2539	1.2514	1.2489	1.2464	1.2438	1.2412	1.2385	1.2357
140	1.2693	1.2672	1.2651	1.2630	1.2608	1.2586	1.2563	1.2540	1.2516	1.2492	1.2467	1.2442	1.2417
160	1.2730	1.2711	1.2691	1.2671	1.2650	1.2629	1.2607	1.2585	1.2563	1.2540	1.2517	1.2494	1.2470
180	1.2766	1.2747	1.2728	1.2709	1.2689	1.2669	1.2649	1.2628	1.2606	1.2585	1.2563	1.2541	1.2518
200	1.2799	1.2781	1.2763	1.2745	1.2726	1.2707	1.2687	1.2667	1.2647	1.2626	1.2605	1.2584	1.2563
220	1.2831	1.2814	1.2796	1.2779	1.2760	1.2742	1.2723	1.2704	1.2685	1.2665	1.2645	1.2625	1.2605
240		1.2845	1.2828	1.2811	1.2793	1.2775	1.2757	1.2739	1.2720	1.2701	1.2682	1.2663	1.2643
260		1.2874	1.2858	1.2841	1.2824	1.2807	1.2790	1.2772	1.2754	1.2736	1.2717	1.2699	1.2680
280			1.2886	1.2870	1.2853	1.2837	1.2821	1.2803	1.2786	1.2768	1.2751	1.2733	1.2714
300			1.2914	1.2898	1.2882	1.2866	1.2850	1.2833	1.2817	1.2800	1.2782	1.2765	1.2747
320				1.2925	1.2910	1.2894	1.2878	1.2862	1.2846	1.2829	1.2813	1.2796	1.2779
340				1.2951	1.2936	1.2921	1.2905	1.2890	1.2874	1.2858	1.2842	1.2825	1.2809

Note: Values below the stepped line represent an extrapolation of p with density.

Selected Values of Electric Dipole Moments for Molecules in the Gas Phase

Ralph D. Nelson, Jr., David R. Lide, Jr., and Arthur A. Maryott

The following table was abstracted from the publication, "Selected Values of Electric Dipole Moments for Molecules in the Gas Phase" compiled by Nelson, Lide and Maryott and published as part of the National Reference Data Series—National Bureau of Standards (NSRDS—NBS 10). The publication is available from the Superintendent of Documents, U.S. Government Printing Office, Washington, D.C., 20402. Those desiring a complete listing of all compounds in the NSRDS—NBS 10, discussion of the bibliographic procedure and the principal methods of dipole moment measurement should obtain the publication.

Values of the dipole moment, μ, are expressed in the cgs system of units, since this is the system universally used by workers in the field. The numerical values are in debye units, D, ($1\ D = 10^{-18}$ electrostatic units of charge×centimeters). The conversion factor to the Système International is $1\ D = 3.33564 \times 10^{-30}$ coulomb-meter.

Code symbol	Estimated accuracy of value	Code symbol	Estimated accuracy of value
A	± 1% or, for $\mu < 1.0$ D, ± 0.01 D	i	The significance of these values may involve some ambiguity because of the possibility of different conformations or spatial isomers.
B	± 2% or, for $\mu < 1.0$ D, ± 0.02 D		
C	± 5% or, for $\mu < 1.0$ D, ± 0.05 D		
S	$\mu \cong 0$ on grounds of molecular symmetry		

Compounds not containing carbon

Formula	Compound name	Selected moment (debyes)	
AgCl	Silver chloride	5.73	C
AsCl₃	Arsenic trichloride	1.59	C
AsF₃	Arsenic trifluoride	2.59	B
AsH₃	Arsine	0.20	C
BCl₃	Boron trichloride	0	S
BF₃	Boron trifluoride	0	S
B₂H₆	Diborane	0	S
B₃H₆N₃	Triborotriazine (Borazine)	0	S
B₅H₉	Pentaborane	2.13	B
BaO	Barium oxide	7.95	A
BrH	Hydrogen bromide	0.82	B
BrH₃Si	Bromosilane	1.33	B
BrK	Potassium bromide	10.41	B
BrLi	Lithium bromide	7.27	A
Br₂Hg	Mercury dibromide	0	S
Br₄Sn	Tin tetrabromide	0	S
ClCs	Cesium chloride	10.42	A
ClF	Chlorine fluoride	0.88	C
ClFO₃	Perchloryl fluoride	0.023	A
ClGeH₃	Chlorogermane	2.13	A
ClH	Hydrogen chloride	1.08	B
ClH₃Si	Chlorosilane	1.31	A
ClK	Potassium chloride	10.27	A
ClLi	Lithium chloride	7.13	A
ClNa	Sodium chloride	9.00	A
ClNO₂	Nitryl chloride	0.53	A
ClTl	Thallium chloride	4.44	B
Cl₂F₃P	Dichlorotrifluorophosphorus	0.68	C
Cl₂H₂Si	Dichlorosilane	1.17	B
Cl₂Hg	Mercury dichloride	0	S
Cl₂OS	Thionyl chloride	1.45	B
Cl₂O₂S	Sulfuryl chloride	1.81	B
Cl₃F₂P	Trichlorodifluorophosphorus	0	S
Cl₃HSi	Trichlorosilane	0.86	B
Cl₃P	Phosphorus trichloride	0.78	C
Cl₄FP	Tetrachlorofluorophosphorus	0.21	B
Cl₄Ge	Germanium tetrachloride	0	S
Cl₄Si	Silicon tetrachloride	0	S
Cl₄Sn	Tin tetrachloride	0	S
Cl₄Ti	Titanium tetrachloride	0	S
CsF	Cesium fluoride	7.88	A
FH	Hydrogen fluoride	1.82	A
FH₃Si	Fluorosilane	1.27	B
FH₅Si₂	Fluorodisilane	1.26	A
FK	Potassium fluoride	8.60	A
FLi	Lithium fluoride	6.33	A
FNO	Nitrosyl fluoride	1.81	B
FNa	Sodium fluoride	8.16	A
FRb	Rubidium fluoride	8.55	A
FTl	Thallium fluoride	4.23	A
F₂HN	Difluoramine	1.92	A
F₂H₂Si	Difluorosilane	1.55	A
F₂N₂	cis-Difluorodiazine	0.16	A
F₂O	Oxygen difluoride	0.297	A
F₂OS	Thionyl fluoride	1.63	A
F₂O₂	Dioxygen difluoride	1.44	C
F₂O₂S	Sulfuryl fluoride	1.12	B
F₂S₂	Sulfur monofluoride (S = SF₂ isomer)	1.03	C
F₂S₂	Sulfur monofluoride (FSSF isomer)	1.45	B
F₂Si	Silicon difluoride	1.23	B
F₃HSi	Trifluorosilane	1.27	B
F₃N	Nitrogen trifluoride	0.235	A
F₃NS	Nitridotrifluorosulfur	1.91	B
F₃OP	Phosphoryl fluoride	1.76	A
F₃P	Phosphorus trifluoride	1.03	A
F₃PS	Thiophosphoryl fluoride	0.64	B

Compounds not containing carbon—Continued

Formula	Compound name	Selected moment (debyes)	
F₄N₂	Tetrafluorohydrazine, gauche conformation	0.26	B
F₄S	Sulfur tetrafluoride	0.632	A
F₄Si	Silicon tetrafluoride	0	S
F₅P	Phosphorus pentafluoride	0	S
F₅I	Iodine pentafluoride	2.18	C
F₆S	Sulfur hexafluoride	0	S
F₆Se	Selenium hexafluoride	0	S
F₆Te	Tellurium hexafluoride	0	S
F₆U	Uranium hexafluoride	0	S
HI	Hydrogen iodide	0.44	B
HLi	Lithium hydride	5.88	A
HN	Imidyl radical		
HNO₃	Nitric acid	2.17	A
HO	Hydroxyl radical	1.66	A
H₂O	Water	1.85	A
H₂O₂	Hydrogen peroxide	2.2	D
H₂S	Hydrogen sulfide	0.97	A
H₃N	Ammonia	1.47	A
H₃P	Phosphine	0.58	A
H₃Sb	Stibine	0.12	C
H₄N₂	Hydrazine	1.75	C
H₄Si	Silane	0	S
H₆OSi₂	Disilyl ether (disiloxane)	0.24	A
H₆Si₂	Disilane	0	S
HgI₂	Mercury diiodide	0	S
ILi	Lithium iodide	7.43	A
I₄Sn	Tin tetraiodide	0	S
NO	Nitrogen monoxide (nitric oxide)	0.153	A
NO₂	Nitrogen dioxide	0.316	A
N₂O	Dinitrogen oxide (nitrous oxide)	0.167	A
OS	Sulfur monoxide	1.55	A
OS₂	Disulfur monoxide	1.47	B
OSr	Strontium oxide	8.90	A
O₂S	Sulfur dioxide	1.63	A
O₃	Ozone	0.53	B
O₃S	Sulfur trioxide	0	S
O₄Os	Osmium tetroxide	0	S

Compounds containing carbon

Formula	Compound name	Selected moment (debyes)	
CBrF₃	Bromotrifluoromethane	0.65	C
CBr₂F₂	Dibromodifluoromethane	0.66	C
CClF₃	Chlorotrifluoromethane	0.50	A
CClN	Cyanogen chloride	2.82	B
CCl₂F₂	Dichlorodifluoromethane	0.51	C
CCl₂O	Carbonyl chloride (phosgene)	1.17	A
CCl₂S	Thiocarbonyl chloride	0.29	C
CCl₃F	Trichlorofluoromethane	0.45	C
CCl₃NO₂	Trichloronitromethane	1.89	C
CCl₄	Carbon tetrachloride	0	S
CFN	Cyanogen fluoride	2.17	C
CF₂	Carbon difluoride	0.46	B
CF₂O	Carbonyl fluoride	0.95	A
CF₃I	Iodotrifluoromethane	0.92	C
CF₃NO₂	Trifluoronitromethane	1.44	C
CF₄	Carbon tetrafluoride	0	S
CN₄O₈	Tetranitromethane	0	S
CO	Carbon monoxide	0.112	A
COS	Carbonyl sulfide	0.712	A
COSe	Carbonyl selenide	0.73	B
CO₂	Carbon dioxide	0	S
CS	Carbon monosulfide	1.98	A
CSTe	Thiocarbonyl telluride	0.17	A
CS₂	Carbon disulfide	0	S
CHBr₃	Tribromomethane	0.99	B
CHClF₂	Chlorodifluoromethane	1.42	B

Selected Values of Electric Dipole Moments for Molecules in the Gas Phase (Continued)

Compounds containing carbon—Continued

Formula	Compound name	Selected moment (debyes)	
CHCl2F	Dichlorofluoromethane	1.29	B
CHCl3	Trichloromethane (chloroform)	1.01	B
CHFO	Formyl fluoride	2.02	A
CHF3	Trifluoromethane	1.65	A
CHN	Hydrogen cyanide	2.98	A
CHP	Methylidyne phosphide (methinophosphide)	0.390	A
CH2Br2	Dibromomethane	1.43	B
CH2ClF	Chlorofluoromethane	1.82	B
CH2ClNO2	Chloronitromethane	2.91	B
CH2Cl2	Dichloromethane	1.60	A
CH2F2	Difluoromethane	1.97	A
CH2N2	Cyanogen amide (cyanamide)	4.27	C
CH2N2	Diazomethane	1.50	A
CH2N2	Diazirine	1.59	C
CH2O	Methanal (formaldehyde)	2.33	A
CH2O2	Methanoic acid (formic acid)	1.41	A
CH3BF2	Methyl difluoroborane	1.66	B
CH3BO	Carbonyl borane	1.80	B
CH3Br	Bromomethane	1.81	A
CH3Cl	Chloromethane	1.87	A
CH3F	Fluoromethane	1.85	A
CH3I	Iodomethane	1.62	B
CH3NO	Hydroxyliminomethane (formaldoxime)	0.44	A
CH3NO	Formyl amide (formamide)	3.73	B
CH3NOS	Methyl sulfinylamine	1.70	B
CH3NO2	Nitromethane	3.46	A
CH3NO3	Methyl nitrate	3.12	B
CH3N3	Methyl azide	2.17	B
CH4	Methane	0	S
CH4F2Si	Methyl difluorosilane	2.11	A
CH4O	Methanol	1.70	A
CH4S	Methanethiol (methyl mercaptan)	1.52	C
CH5FSi	Methyl monofluorosilane	1.71	A
CH5N	Methyl amine	1.31	B
CH5P	Methyl phosphine	1.10	A
CH6Ge	Methyl germane	0.643	A
CH6OSi	Methoxysilane	1.17	B
CH6Si	Methyl silane	0.735	A
CH6Sn	Methyl stannane	0.68	C
C2ClF5	Chloropentafluoroethane	0.52	C
C2F6	Hexafluoroethane	0	S
C2N2	Dicyanogen (cyanogen)	0	S
C2N2S	Dicyano sulfide	3.02	A
C2HCl	Chloroacetylene	0.44	C
C2HCl5	Pentachloroethane	0.92	C
C2HF	Fluoroacetylene	0.73	C
C2HF3	Trifluoroethylene	1.40	C
C2HF5	Pentafluoroethane	1.54	C
C2H2	Acetylene	0	S
C2H2Cl2	1,1-Dichloroethylene	1.34	A
C2H2Cl2	cis-1,2-Dichloroethylene	1.90	B
C2H2Cl2O	Chloroacetyl chloride	2.23	Ci
C2H2Cl4	1,1,2,2-Tetrachloroethane	1.32	Ci
C2H2FN	Fluorocyanomethane	3.43	C
C2H2F2	1,1-Difluoroethylene	1.38	A
C2H2F2	cis-1,2-Difluoroethylene	2.42	A
C2H2N2O	1,2,5-Oxadiazole	3.38	A
C2H2N2O	1,3,4-Oxadiazole	3.04	A
C2H2N2S	1,2,5-Thiadiazole	1.56	A
C2H2N2S	1,3,4-Thiadiazole	3.29	B
C2H2O	Methylene carbonyl (ketene)	1.42	B
C2H3Br	Bromoethylene	1.42	B
C2H3Cl	Chloroethylene	1.45	B
C2H3ClF2	1-Chloro-1,1-difluoroethane	2.14	B
C2H3ClO	Acetyl chloride	2.72	C
C2H3Cl3	1,1,1-Trichloroethane	1.78	B
C2H3F	Fluoroethylene	1.43	A
C2H3FO	Acetyl fluoride	2.96	A
C2H3F3	1,1,1-Trifluoroethane	2.32	B
C2H3F3	1,1,2-Trifluoroethane	1.58	B
C2H3N	Cyanomethane (acetonitrile)	3.92	A
C2H3N	Isocyanomethane	3.85	A
C2H4	Ethylene	0	S
C2H4ClF	1-Chloro-2-fluoroethane, gauche conformation	2.72	C
C2H4ClNO2	1-Chloro-1-nitroethane	3.27	B
C2H4Cl2	1,1-Dichloroethane	2.06	B
C2H4F2	1,1-Difluoroethane	2.27	B
C2H4Ge	Germyl acetylene	0.136	A
C2H4O	Oxirane (ethylene oxide)	1.89	A
C2H4O	Ethanal (acetaldehyde)	2.69	B
C2H4O2	Ethanoic acid (acetic acid)	1.74	C
C2H4O2	Methyl methanoate (methly formate)	1.77	B
C2H4S	Thiirane (ethylene sulfide)	1.85	A
C2H4Si	Silyl acetylene	0.316	A
C2H5Br	Bromoethane	2.03	A
C2H5BrO	Bromomethoxymethane	2.05	Ci
C2H5Cl	Chloroethane	2.05	A
C2H5ClO	2-Chloroethanol	1.78	Ci
C2H5F	Fluoroethane	1.94	B
C2H5I	Iodoethane	1.91	B
C2H5N	Iminoethane (ethyleneimine)	1.90	A
C2H5N	Methyliminomethane (CH3N = CH2)	1.53	B
C2H5NO	Acetyl amine (acetamide)	3.76	Bi
C2H5NO	Methylaminomethanal (N-methylformamide)	3.83	Bi
C2H5NO2	Nitritoethane (ethyl nitrite)	2.40	Ci
C2H5NO2	Nitroethane	3.65	B
C2H6	Ethane	0	S
C2H6BF	Dimethyl fluoroborane	1.32	C
C2H6O	Ethanol	1.69	Bi
C2H6O	Dimethyl ether	1.30	A
C2H6OS	Dimethylsulfoxide	3.96	A
C2H6O2	1,2-Ethanediol (ethylene glycol)	2.28	Ci
C2H6O2S	Dimethyl sulfoxylate (dimethyl sulfone)	4.49	B
C2H6S	Ethanethiol	1.58	Bi
C2H6S	Dimethyl sulfide	1.50	A
C2H6Si	Silyl ethylene	0.66	A
C2H7B5	2,4-Dicarbaheptaborane	1.32	B
C2H7N	Aminoethane (ethyl amine)	1.22	Ci
C2H7N	Dimethyl amine	1.03	B
C2H7P	Ethyl phosphine	1.17	A
C2H7P	Dimethyl phosphine	1.23	A
C2H8N2	1,2-Diaminoethane	1.99	Ci
C2H8Si	Dimethyl silane	0.75	A
C2H8Si	Ethyl silane	0.81	B
C3O2	Dicarbonyl carbon (carbon suboxide)	0	S
C3HF3	3,3,3-Trifluoropropyne	2.36	B
C3HN	Cyanoacetylene	3.72	A
C3HN2	Dicyanomethane	3.73	A
C3H2O	Propynal	2.47	A
C3H2O3	Vinylene carbonate	4.55	A
C3H3Br	3-Bromopropyne	1.54	C
C3H3Cl	3-Chloropropyne	1.68	C
C3H3F3	3,3,3-Trifluoropropene	2.45	B
C3H3N	Cyanoethylene	3.87	B
C3H3NO	Acetyl cyanide	3.45	B
C3H3NS	Thiazole	1.62	B
C3H4	Cyclopropene	0.45	A
C3H4	Propyne	0.781	A
C3H4	Propadiene (allene)	0	S
C3H4Cl2	1,1-Dichlorocyclopropane	1.58	B
C3H4O	Ethylidene carbonyl (methyl ketene)	1.79	B
C3H4O	Propenal, trans conformation (acrolein)	3.12	B
C3H4O2	2-Oxoöxetane (β-propiolactone)	4.18	A
C3H4O2	Vinyl formate	1.49	A
C3H5Cl	2-Chloropropene	1.66	B
C3H5Cl	cis-1-Chloropropene	1.67	C
C3H5Cl	trans-1-Chloropropene	1.97	C
C3H5Cl	3-Chloropropene	1.94	Ci
C3H5F	cis-1-Fluoropropene	1.46	B
C3H5F	2-Fluoropropene	1.61	B
C3H5F	3-Fluoropropene, cis conformation	1.70	A
C3H5F	3-Fluoropropene, gauche conformation	1.94	A
C3H5N	Cyanoethane (propionitrile)	4.02	A
C3H6	Cyclopropane	0	S
C3H6	Propene	0.366	A
C3H6ClNO2	1-Chloro-1-nitropropane	3.48	Bi
C3H6Cl2	1,2-Dichloropropane		i
C3H6Cl2	1,3-Dichloropropane	2.08	Bi
C3H6Cl2	2,2-Dichloropropane	2.27	C
C3H6O	Oxetane (trimethylene oxide)	1.94	A
C3H6O	Methyl oxirane (propylene oxide)	2.01	A
C3H6O	Propanone (acetone)	2.88	A
C3H6O	2-Propen-1-ol (allyl alcohol)	1.60	C
C3H6O	Propanal, cis conformation (propionaldehyde)	2.52	A
C3H6O2	Propanoic acid	1.75	Ci
C3H6O2	Methyl acetate	1.72	Ci
C3H6O2	Ethyl formate	1.93	Ci
C3H6O3	1,3,5-Trioxane	2.08	A
C3H6S	Thietane (trimethylene sulfide)	1.85	C
C3H6S	Methyl thiirane (propylene sulfide)	1.95	A
C3H7Br	1-Bromopropane	2.18	Ci
C3H7Br	2-Bromopropane	2.21	C
C3H7Cl	1-Chloropropane	2.05	Bi
C3H7Cl	2-Chloropropane	2.17	C
C3H7F	1-Fluoropropane, gauche conformation	1.90	C
C3H7F	1-Fluoropropane, trans conformation	2.05	B
C3H7I	1-Iodopropane	2.04	Ci
C3H7NO	N,N-Dimethylformamide	3.82	Bi
C3H7NO	Acetyl methylamine (N-Methylacetamide)	3.73	Bi
C3H7NO2	1-Nitropropane	3.66	Bi
C3H7NO2	2-Nitropropane	3.73	B
C3H8	Propane	0.084	A
C3H8O	1-Propanol	1.68	Bi
C3H8O	2-Propanol	1.66	Bi
C3H8O	Methoxyethane (methyl ethyl ether)	1.23	Ci
C3H9As	Trimethyl arsine	0.86	B
C3H9N	Trimethyl amine	0.612	A
C3H9N	1-Aminopropane (n-propylamine)	1.17	Ci
C3H9P	Trimethyl phosphine	1.19	A

Compounds containing carbon—Continued

Formula	Compound name	Selected moment (debyes)	
$C_3H_{10}Si$	Trimethyl silane	0.525	A
C_4F_8	Perfluorocyclobutane	0	S
$C_4H_2N_2$	trans-1,2-Dicyanoethylene	0	S
$C_4H_4Cl_2$	1,4-Dichloro-2-butyne	2.10	Bi
$C_4H_4F_2$	1,1-Difluoro-1,3-butadiene (trans conformation)	1.29	A
C_4H_4O	Furan	0.66	A
$C_4H_4O_2$	Diketene	3.53	B
C_4H_4S	Thiophene	0.55	C
C_4H_5Cl	4-Chloro-1,2-butadiene	2.02	Ci
C_4H_5Cl	1-Chloro-2-butyne	2.19	C
C_4H_5F	2-Fluoro-1,3-butadiene (trans conformation)	1.42	A
C_4H_5N	Pyrrole	1.84	C
C_4H_5N	cis-1-Cyanopropene	4.08	B
C_4H_5N	trans-1-Cyanopropene	4.50	B
C_4H_5N	2-Cyanopropene (methacrylonitrile)	3.69	C
C_4H_6	Cyclobutene	0.132	A
C_4H_6	1-Butyne	0.80	C
C_4H_6	1,2-Butadiene	0.403	A
C_4H_6	1,3-Butadiene	0	S
C_4H_6O	Cyclobutanone	2.99	B
C_4H_6O	trans-2-Butenal (crotonaldehyde)	3.67	Bi
C_4H_6O	2-Methylpropenal (methacrolein)	2.68	Ci
C_4H_6O	3-Butene-2-one	3.16	B
C_4H_7Cl	1-Chloro-2-methylpropene	1.95	Bi
$C_4H_7Cl_3$	1,1,2-Trichloro-2-methylpropane	1.86	Ci
C_4H_7F	Fluorocyclobutane	1.94	A
C_4H_7N	1-Cyanopropane	4.07	Bi
C_4H_8	1-Butene	0.34	Ci
C_4H_8	trans-2-Butene	0	S
C_4H_8	2-Methylpropene	0.50	A
$C_4H_8Cl_2$	1,4-Dichlorobutane	2.22	Ci
C_4H_8O	Tetrahydrofuran	1.63	C
C_4H_8O	cis-2,3-Dimethyloxirane	2.03	A
C_4H_8O	Butanal	2.72	Bi
$C_4H_8O_2$	1,4-Dioxane	0	S
$C_4H_8O_2$	Ethyl acetate	1.78	Ci
C_4H_9Br	1-Bromobutane	2.08	Ci
C_4H_9Br	2-Bromobutane	2.23	Ci
C_4H_9Cl	1-Chlorobutane	2.05	Bi
C_4H_9Cl	2-Chlorobutane	2.04	Ci
C_4H_9Cl	1-Chloro-2-methylpropane	2.00	Ci
C_4H_9Cl	2-Chloro-2-methylpropane	2.13	B
C_4H_9F	2-Fluoro-2-methylpropane	1.96	A
C_4H_9I	1-Iodobutane	2.12	Ci
C_4H_9NO	Propanoyl methylamine (N-methylpropionamide)	3.61	Bi
C_4H_9NO	Acetyl dimethylamine (N,N-dimethylacetamide)	3.81	Bi
$C_4H_9NO_2$	2-Nitrito-2-methylpropane (t-butyl nitrite)	2.74	Ci
$C_4H_9NO_2$	1-Nitrobutane	3.59	Bi
$C_4H_9NO_2$	2-Nitro-2-methylpropane	3.71	B
C_4H_{10}	Butane	≤0.05	Ci
C_4H_{10}	2-Methylpropane	0.132	A
$C_4H_{10}O$	1-Butanol	1.66	Bi
$C_4H_{10}O$	2-Methylpropan-1-ol (isobutanol)	1.64	C
$C_4H_{10}O$	Diethyl ether	1.15	Bi
$C_4H_{10}S$	Diethyl sulfide	1.54	Ci
$C_4H_{11}N$	Diethyl amine	0.92	Ci
C_5H_5N	Pyridine	2.19	B
C_5H_5N	1-Cyano-1,3-butadiene	3.90	Ci
C_5H_6	1,3-Cyclopentadiene	0.419	A
C_5H_8	Cyclopentene	0.20	B
C_5H_8	1-Pentyne	0.81	Ci
C_5H_8	2-Methyl-1,3-butadiene (trans conformation)	0.25	A
$C_5H_8O_2$	Acetylacetone		Ci
C_5H_9N	1-Cyanobutane	4.12	Bi
C_5H_9N	2-Cyano-2-methylpropane	3.95	A
$C_5H_{10}O_3$	Diethyl carbonate	1.10	Ci
$C_5H_{11}Br$	1-Bromopentane	2.20	Ci
$C_5H_{11}Cl$	1-Chloropentane	2.16	Ci
C_5H_{12}	2-Methylbutane	0.13	Ci
C_5H_{12}	2,2-Dimethylpropane	0	S
$C_6H_2Cl_2O_2$	2,5-Dichloro-1,4-cyclohexadienedione	0	S
$C_6H_4ClNO_2$	o-Chloronitrobenzene	4.64	B
$C_6H_4ClNO_2$	m-Chloronitrobenzene	3.73	B
$C_6H_4ClNO_2$	p-Chloronitrobenzene	2.83	B
$C_6H_4Cl_2$	o-Dichlorobenzene	2.50	B
$C_6H_4Cl_2$	m-Dichlorobenzene	1.72	C
$C_6H_4Cl_2$	p-Dichlorobenzene	0	S
$C_6H_4FNO_2$	p-Fluoronitrobenzene	2.87	B
$C_6H_4F_2$	m-Difluorobenzene	1.58	B
$C_6H_4N_2O_4$	p-Dinitrobenzene	0	S
$C_6H_4O_2$	1,4-Cyclohexadienedione (p-benzoquinone)	0	S
C_6H_5Br	Bromobenzene	1.70	B
C_6H_5Cl	Chlorobenzene	1.69	B
C_6H_5ClO	p-Chlorophenol	2.11	C
C_6H_5F	Fluorobenzene	1.60	C
C_6H_5I	Iodobenzene	1.70	C
$C_6H_5NO_2$	Nitrobenzene	4.22	B
C_6H_6	Benzene	0	S

Compounds containing carbon—Continued

Formula	Compound name	Selected moment (debyes)	
C_6H_6O	Phenol	1.45	C
C_6H_7N	Aminobenzene (aniline)	1.53	C
C_6H_8	1,3-Cyclohexadiene	0.44	B
C_6H_{10}	1-Hexyne	0.83	Ci
C_6H_{10}	3,3-Dimethyl-1-butyne	0.66	A
$C_6H_{10}Cl_2$	cis-1e,2a-Dichlorocyclohexane	3.11	C
$C_6H_{12}N_2$	Diisopropylidene hydrazine (dimethyl ketazine)	1.53	Bi
$C_6H_{12}O_2$	Pentyl formate (n-amyl formate)	1.90	Ci
$C_6H_{12}O_3$	2,4,6-Trimethyl-1,3,5-trioxane (paraldehyde)	1.43	C
$C_6H_{14}O$	Dipropyl ether	1.21	Ci
$C_6H_{14}O_2$	1,1-Diethoxyethane		i
$C_6H_{15}N$	Triethyl amine	0.66	Ci
$C_7H_4ClF_3$	o-Chloro(trifluoromethyl)benzene	3.46	B
$C_7H_4ClF_3$	p-Chloro(trifluoromethyl)benzene	1.58	B
$C_7H_5F_3$	(Trifluoromethyl)benzene	2.86	B
C_7H_5N	Cyanobenzene (benzonitrile)	4.18	B
C_7H_7Cl	o-Chlorotoluene	1.56	C
C_7H_7Cl	p-Chlorotoluene	2.21	B
C_7H_7F	o-Fluorotoluene	1.37	C
C_7H_7F	m-Fluorotoluene	1.86	C
C_7H_7F	p-Fluorotoluene	2.00	C
$C_7H_7NO_3$	o-Nitro(methoxy)benzene	4.83	Bi
$C_7H_7NO_3$	m-Nitro(methoxy)benzene	4.55	Bi
$C_7H_7NO_3$	p-Nitro(methoxy)benzene	5.26	B
C_7H_8	1,3,5-Cycloheptatriene	0.25	C
C_7H_8	Toluene	0.36	C
C_7H_8O	Phenylmethanol (benzyl alcohol)	1.71	C
C_7H_8O	Methoxybenzene (anisole)	1.38	C
C_7H_9NO	o-Amino(methoxy)benzene	1.61	Ci
$C_7H_{14}O_2$	Pentyl acetate (n-amyl acetate)	1.75	Ci
$C_7H_{15}Br$	1-Bromoheptane	2.16	Ci
$C_8H_4N_2$	p-Dicyanobenzene	0	S
C_8H_8O	Acetylbenzene (acetophenone)	3.02	B
$C_8H_8O_2$	2,5-Dimethyl-1,4-cyclohexadienedione	0	S
C_8H_{10}	Ethylbenzene	0.59	C
C_8H_{10}	o-Xylene	0.62	C
C_8H_{10}	p-Xylene	0	S
$C_8H_{12}O_2$	Tetramethylcyclobutane-1,3-dione	0	S
$C_8H_{18}O$	Dibutyl ether	1.17	Ci
C_9H_7N	Quinoline	2.29	C
C_9H_7N	Isoquinoline	2.73	C
$C_9H_{10}O_2$	Ethyl phenylformate (ethyl benzoate)	2.00	Ci
$C_{10}H_8$	Azulene	0.80	B
$C_{10}H_{14}BeO_4$	Bis(2,4-pentanedionato) beryllium	0	S
$C_{12}H_9BrO$	p-Bromophenoxybenzene	1.98	C
$C_{12}H_9NO_3$	p-Nitrophenoxybenzene	4.54	B
$C_{12}H_{10}$	Phenylbenzene (diphenyl)	0	S
$C_{13}H_{11}BrO$	p-Bromophenoxy-p-toluene	2.45	C
$C_{14}H_{14}O$	Bis(p-tolyl) ether	1.54	C
$C_{15}H_{21}AlO_6$	Tris(2,4-pentanedionato) aluminum	0	S
$C_{15}H_{21}CrO_6$	Tris(2,4-pentanedionato) chromium (III)	0	S
$C_{15}H_{21}FeO_6$	Tris(2,4-pentanedionato) iron (III)	0	S
$C_{20}H_{28}O_8Th$	Tetrakis(2,4-pentanedionato) thorium	0	S

DIPOLE MOMENTS

The method of measurement of the dipole moments is indicated in the following **two tables** by the symbols:

B   benzene solution
C   carbon tetrachloride solution
D   1,4-dioxane solution
H   n-heptane solution
St   measurement of Stark effect in microwave spectrum of gas

Dipole Moments for Some Inorganic Compounds

Compound	Dipole Moment $\times 10^{-18}$ e. s. u.	Method
AlBr₃	5.14	B
AlI₃	2.48	B
CsCl	10.42	St
CsF	7.875	St
HF	1.92 ± 0.02	..
HCl	$1.084 \pm 0.003-0.007$	..
NBr	0.78	..
HDSe	0.62	St
HI	0.38	..
DCl	$1.084 \pm 0.003-0.007$	..
HNO₃	2.16	St
HgBr₂	0.95	B
HgCl₂	1.23	B
H₂O	1.87	..
H₂O₂	2.13 ± 0.05	..
H₂S	1.10	..
SO₂	1.60	..
SO₃	0.00	..
SO₂F₂	1.110	St
NH₃	1.3	..
N₂H₄	1.84	..
NO	0.16	..
NO₂	0.29	..
N₂O₄	0.37	..
NOCl	1.83	..
NOBr	1.87	..
PCl₃	0.80 1.10	..
PCl₅	0.0	..
CO	0.10	..
CO₂	0.0	..
SiD₂F₂	1.53	St
SiH₂F₂	1.54	St
SnCl₄	0.95	B
SnI₄	0	B
TiCl₄	0	C

Dipole Moments for Some Organo-metallic Compounds

Compound	Dipole Moment $\times 10^{-18}$ e. s. u.	Method
Beryllium diethyl	1.0	H
Cadmium diethyl	0.3	H
Mercury diethyl	0.0	H
Magnesium diethyl	4.8	D
Zinc diethyl	0.0	H
Beryllium diphenyl	1.6	B
Cadmium diphenyl	0.6	B
Mercury diphenyl	0.2	B
Magnesium diphenyl	4.9	D
Zinc diphenyl	0.8	B
Chromium (0), diphenyl	0	B
Chromium, ditolyl	0	B
Cobalt, mononitrosyl tricarbonyl	0.72	B
Cyclopentadienyl, chromium dicarbonyl mono nitrosyl	3.23	B
Cyclopentadienyl, manganese tri-carbonyl	3.30	B
Cyclopentadienyl, cobalt ducarbonyl	2.87	B
Cyclopentadienyl, vanadium tetra-carbonyl	3.17	B
Penta cyclopentadienyl, dicobalt	0	B
Dicyclopentadienyl, iron (II)	0	B
Dicyclopentadienyl, lead (II)	1.63	B
Dicyclopentadienyl, tin (II)	1.02	B
Ethyl lithium	0.87	B
Iron, dinitrosyl dicarbonyl	0.05	B
Iron, tetracarbonyl-diiodide	3.68	B
Iron, tetracarbonyl mono-(methyl isonitrile)	5.07	B
Iron, pentacarbonyl	0.63	B
Iron, bis(p-chlorophenyl cyclo-pentadienyl	3.12	B
Ruthenium (II), di-indenyl	0	B

Dipole Moments

Dipole Moments of Amino Acid Esters

Substance	$\mu \cdot 10^{18}$ e. s. u.
Glycine ethyl ester	2.11
α-Alanine ethyl ester	2.09
α-Aminobutyric acid ethyl ester	2.13
α-Aminovaleric acid ethyl ester	2.13
Valine ethyl ester	2.11
α-Aminocaproic acid ethyl ester	2.13
β-Alanine ethyl ester	2.14
β-Aminobutyric acid ethyl ester	2.11

Accurate to $\pm 0.01 \cdot 10^{-18}$ e. s. u.
J. Wyman, Chem. Rev., 1936, **19**, 213.

Dipole Moments of Amides

Urea	4.56
Thiourea	4.89
Symm.-dimethylurea	4.8
Tetraethylurea	3.3
Propylurea	4.1
Acetamide	3.6
Sulfamide	3.9
Benzamide	3.6
Valeramide	3.7
Caproamide	3.9

For comprehensive list of dipole moments see Trans. Faraday Soc., 1934, **30**, General Discussion.

Dipole Moments of Some Hormones and Related Compounds in Dioxan

Cholestane-3(β) : 7(α)-diol	2.31
Cholestane-3(β) : 7(β)-diol	2.55
Cholestane	2.98
Δ⁵-Cholestane-3(β)ol-7 one	3.79
Androsterone	3.70
β-Androsterone	2.95
Δ⁵-Androstene-3(β) : 17(α)-diol	2.89
Δ⁵-Androstene-3(β) : 17(β)-diol	2.69
Δ⁵-Androstene-3(β)ol-17 one	2.46
Testosterone	4.32
cis-Testosterone	5.17
Δ⁴-Androstene-3 : 17 dione	3.32
Isophorone	3.96

Ethylenic $>C=C<$ in a six membered ring and conjugated with $>C=0$ increases the dipole moment approximately by 1 Debye. Non-conjugated $>C=C<$ in sterols decreases the dipole moment by approximately 0.49. Biological activity is not correlated with dipole moment. W. D. Kumler and G. M. Fohlen, J. Am. Chem. Soc., 1945, **67**, 437.

ELECTRON AFFINITIES

F. M. Page

The tabulated values are given in kJ mol^{-1}, eV and kcal mol^{-1} for the convenience of users. All values are experimental observations and refer to the gas phase at 0°K unless otherwise stated. Extended tables of atomic electron affinities are given by L. M. Branscombe in "Atomic and Molecular Processes", Ed. D. R. Bates, Academic Press, New York (1962), and for radical and molecular electron affinities in "Negative Ions and the Magnetron", Wiley —Interscience—London (1969). See also H. O. Pritchard, Chem. Rev. **52**, 529 (1953); F. H. Field and J. L. Franklin, "Electron Impact Phenomena", Academic Press, New York (1957); N. S. Buchel'nikova, Usp, fiz. nauk **65**, 357 (1958), A.E.C. Tr3657, Oak Ridge (1959), V. I. Vedeneyev et al, "Bond Energies, Ionisation Potentials and Electron Affinities", Arnold, London (1966) and C. J. Schexnayder, N.A.S.A. Technical Note D1791, Washington (1963).

Atoms

Ion Formed	Electron Affinity		
	kJ mol^{-1}	eV	kcal mol^{-1}
Br$^-$	324	3.363	77.4
C$^-$	108	1.12	25.8
Cl$^-$	348	3.613	83.2
F$^-$	333	3.448	79.6
H$^-$	77	0.80	18.4
*H$^-$ (1s^2)	72	0.747	17.2
I$^-$	296	3.063	70.8
O$^-$	142	1.466	33.8
S$^-$	200	2.07	47.7

* Quantum mechanical calculation.

Molecules

Ion Formed	Electron Affinity		
	kJ mol^{-1}	eV	kcal mol^{-1}
BF$_3^-$	255	2.65	61.0
(p-Benzoquinone)$^-$	129	1.34	30.9
(Chloranil)$^-$	231	2.40	55.3
(Fluoranil)$^-$	218	2.27	52.2
No$_2$	377	3.91	90.0
O$_2^-$	43	0.45	10.1
SF$_6^-$	138	1.43	33.0
(Tetracyanoethylene)$^-$	278	2.88	66.4
WF$_6^-$	264	2.74	63.0
UF$_6^-$	280	2.91	67.0

Radicals

Ion Formed	Electron Affinity		
	kJ mol^{-1}	eV	kcal mol^{-1}
CH$_3^-$	104	1.08	24.8
C$_2$H$_5^-$	86	0.89	20.5
C$_6$H$_5^-$	210	2.20	49.6
CF$_3^-$	179	1.85	42.7
CCl$_3^-$	117	1.22	28.0
SiF$_3^-$	323	3.35	77.2
NH$_2^-$	108	1.12	25.7
C$_6$H$_5$NH$^-$	149	1.55	35.7
(C$_6$H$_5$)$_2$N$^-$	114	1.19	23.3
PH$_2^-$	154	1.60	36.8
OH$^-$	176	1.83	42.1
CH$_3$O$^-$	36	0.38	8.7
CF$_3$O$^-$	131	1.35	31.2
SH$^-$	211	2.19	50.4
CH$_3$S$^-$	127	1.32	30.4
CN$^-$	305	3.17	7.30
SCN$^-$	209	2.17	49.9
SeCN$^-$	255	2.64	60.9

IONIZATION POTENTIALS*

Spectrum →

Z	Element	I	II	III	IV	V	VI	VII	VIII	IX	X	XI	XII	XIII	XIV	XV	XVI	XVII	XVIII	XIX	XX	XXI
1	H	13.598																				
2	He	24.587	54.416																			
3	Li	5.392	75.638	122.451																		
4	Be	9.322	18.211	153.893	217.713																	
5	B	8.298	25.154	37.930	259.368	340.217																
6	C	11.260	24.383	47.887	64.492	392.077	489.981															
7	N	14.534	29.601	47.448	77.472	97.888	552.057	667.029														
8	O	13.618	35.116	54.934	77.412	113.896	138.116	739.315	871.387													
9	F	17.422	34.970	62.707	87.138	114.240	157.161	185.182	953.886	1103.089												
10	Ne	21.564	40.962	63.45	97.11	126.21	157.93	207.27	239.09	1195.797	1362.164											
11	Na	5.139	47.286	71.64	98.91	138.39	172.15	208.47	264.18	299.87	1465.091	1648.659										
12	Mg	7.646	15.035	80.143	109.24	141.26	186.50	224.94	265.90	327.95	367.53	1761.802	1962.613									
13	Al	5.986	18.828	28.447	119.99	153.71	190.47	241.43	284.59	330.21	398.57	442.07	2085.983	2304.080								
14	Si	8.151	16.345	33.492	45.141	166.77	205.05	246.52	303.17	351.10	401.43	476.06	523.0	2437.676	2673.108							
15	P	10.486	19.725	30.18	51.37	65.023	220.43	263.22	309.41	371.73	424.50	479.57	560.8	611.85	2816.943	3069.762						
16	S	10.360	23.33	34.83	47.30	72.68	88.049	280.93	328.23	379.10	447.09	504.78	564.5	651.63	707.14	3223.836	3494.099					
17	Cl	12.967	23.81	39.61	53.46	67.8	97.03	114.193	348.28	400.05	455.62	529.26	591.7	656.69	749.74	809.39	3658.425	3946.193				
18	Ar	15.759	27.629	40.74	59.81	75.02	91.007	124.319	143.456	422.44	478.68	538.95	618.26	686.09	755.73	854.75	918	4120.778	4426.114			
19	K	4.341	31.625	45.72	60.91	82.66	100.0	117.56	154.86	175.814	503.44	564.13	629.09	714.02	787.13	861.77	968	1034	4610.955	4933.931		
20	Ca	6.113	11.871	50.908	67.10	84.41	108.78	127.7	147.24	188.54	211.270	591.25	656.39	726.03	816.61	895.12	974	1087	1157	5129.045	5469.738	
21	Sc	6.54	12.80	24.76	73.47	91.66	111.1	138.0	158.7	180.02	225.32	249.832		755.47	829.79	926.00		1266.1	1403.0	1547	1698	1856
22	Ti	6.82	13.58	27.491	43.266	99.22	119.36	140.8	168.5	193.2	215.91	265.23		787.33	861.33	940.36	1136.2					
23	V	6.74	14.65	29.310	46.707	65.23	128.12	150.17	173.7	205.8	230.5	265.04		336.267	384.30	974.02	489.5	546.8	607.2	671	738	
24	Cr	6.766	16.50	30.96	49.1	69.3	90.56	161.1	184.7	209.3	244.4	270.8		355	404	1010.64	512	571	633	698		
25	Mn	7.435	15.640	33.667	51.2	72.4	95	119.27	196.46	221.8	248.3	286.0		343.6	392.2	435.3	499	557	619			
26	Fe	7.870	16.18	30.651	54.8	75.0	99	125	151.06	235.04	262.1	290.4		361.0	411	457	520					
27	Co	7.86	17.06	33.50	51.3	79.5	102	129	157	186.13	276	305		379	430	444	542					
28	Ni	7.635	18.168	35.17	54.9	75.5	108	133	162	193	224.5	321		384	435	464						
29	Cu	7.726	20.292	36.83	55.2	79.9	103	139	166	199	232	265		401	435	484						
30	Zn	9.394	17.964	39.722	59.4	82.6	108	134	174	203	238	274		419.7	454	490						
31	Ga	5.999	20.51	30.71	64																	
32	Ge	7.899	15.934	34.22	45.71	93.5																
33	As	9.81	18.633	28.351	50.13	62.63	127.6															
34	Se	9.752	21.19	30.820	42.944	68.3	81.70	155.4														
35	Br	11.814	21.8	36	47.3	59.7	88.6	103.0	192.8													
36	Kr	13.999	24.359	36.95	52.5	64.7	78.5	111.0	126	230.39												
37	Rb	4.177	27.28	40	52.6	71.0	84.4	99.2	136	150	277.1	324										
38	Sr	5.695	11.030	43.6	57	71.6	90.8	106	122.3	162	177	206										
39	Y	6.38	12.24	20.52	61.8	77.0	93.0	116	129	146.52	191		37.0									
40	Zr	6.84	13.13	22.99	34.34	81.5																
41	Nb	6.88	14.32	25.04	38.3	50.55	102.6	125														
42	Mo	7.099	16.15	27.16	46.4	61.2	68	126.8	153													
43	Tc	7.28	15.26	29.54																		
44	Ru	7.37	16.76	28.47																		
45	Rh	7.46	18.08	31.06																		
46	Pd	8.34	19.43	32.93																		
47	Ag	7.576	21.49	34.83																		
48	Cd	8.993	16.908	37.48																		
49	In	5.786	18.869	28.03	54																	
50	Sn	7.344	14.632	30.502	40.734	72.28																
51	Sb	8.641	16.53	25.3	44.2	56	108															
52	Te	9.009	18.6	27.96	37.41	58.75	70.7	137														
53	I	10.451	19.131	33																		
54	Xe	12.130	21.21	32.1																		
55	Cs	3.894	25.1																			

IONIZATION POTENTIALS* (continued)

Spectrum

Z	Element	I	II	III	IV	V	VI	VII	VIII	IX	X	XI	XII	XIII	XIV	XV	XVI	XVII	XVIII	XIX	XX	XXI
56	Ba	5.212	10.004	19.175																		
57	La	5.577	11.06	20.20	36.72																	
58	Ce	5.47	10.85	21.62	38.95	57.45																
59	Pr	5.42	10.55																			
60	Nd	5.49	10.72																			
61	Pm	5.55	10.90																			
62	Sm	5.63	11.07																			
63	Eu	5.67	11.25																			
64	Gd	6.14	12.1																			
65	Tb	5.85	11.52																			
66	Dy	5.93	11.67																			
67	Ho	6.02	11.80																			
68	Er	6.10	11.93																			
69	Tm	6.18	12.05	23.71																		
70	Yb	6.254	12.17	25.2																		
71	Lu	5.426	13.9																			
72	Hf	7.0	14.9	23.3	33.3																	
73	Ta	7.89																				
74	W	7.98																				
75	Re	7.88																				
76	Os	8.7																				
77	Ir	9.1																				
78	Pt	9.0	18.563																			
79	Au	9.225	20.5																			
80	Hg	10.437	18.756	34.2																		
81	Tl	6.108	20.428	29.83																		
82	Pb	7.416	15.032	31.937	42.32	68.8																
83	Bi	7.289	16.69	25.56	45.3	56.0	88.3															
84	Po	8.42																				
85	At																					
86	Rn	10.748																				
87	Fr																					
88	Ra	5.279	10.147																			
89	Ac	6.9	12.1																			
90	Th		11.5	20.0	28.8																	
91	Pa																					
92	U																					
93	Np	5.8																				
94	Pu	6.0																				
95	Am																					

* Numerical values in this table are expressed in electron volts. The conversion factor used in converting the spectral data to electron volts was 1 eV = 8065.73 cm⁻¹.

From Moore, C. E., *Analyses of Optical Spectra*, NSRDS-NBS 34, Office of Standard Reference Data, National Bureau of Standards, Washington, D.C.

NUCLEAR SPINS, MOMENTS, AND MAGNETIC RESONANCE FREQUENCIES

Kenneth Lee and Weston A. Anderson

1967

This table contains the published values for the nuclear spins, magnetic moments, and quadrupole moments, and the calculated values for the nuclear magnetic resonance (NMR) frequency and for the relative sensitivities. Only those isotopes with both published spin and magnetic moment values are tabulated. The magnetic and quadrupole moment values were selected from results published during the period from January, 1955 to June, 1967. Earlier references were obtained from H. E. Walchli, A Table of Nuclear Moment Data, U.S. Atomic Energy Commission Report ORNL−1469, Supplement I (1953) and Supplement II (1955), and D. Strominger, J. M. Hollander, and G. T. Seaborg, Table of Isotopes, Rev. Mod. Phys. 30, 585 (1958). A table containing the known (1963) spin and electromagnetic moment values of nuclear ground and excited states has been compiled by I. Lindgren, Perturbed Angular Correlations; E. Karlsson, E. Mathias, and K. Siegbahn, editors; North-Holland Publishing Co. (1964). The magnetic moments given in this latter table are corrected for the diamagnetic effect. A more complete list of spin and moment results for nuclei in excited states are included in Lindgren's table.

In general, the results chosen for this table were selected with an inclination to NMR measurements and to the precision of the measurement. Only six significant figures are used in this table. Therefore, the number of figures may be less than those published. The experimental methods employed in determining the moments are indicated by the following symbols:

Ab = atomic beam magnetic resonance (hyperfine structure, double or triple resonance or other method)

E = electron spin resonance or electron-nuclear double resonance
M = microwave absorption in gases
Mb = molecular (or diamagnetic) beam magnetic resonance
Mc = miscellaneous
Mo = Mössbauer effect
N = nuclear magnetic resonance

No = nuclear orientation
O = optical spectroscopy (hyperfine structure, band structure, double resonance, or optical pumping)
Qr = quadrupole resonance

Other symbols used in the table are:

A = atomic weight (mass number)
El = element
I = nuclear spin in units of $h/2\pi$
μ = magnetic moment in units of the nuclear magneton $eh/4\pi Mc$
m = metastable excited state

Q = quadrupole moment in units of barns (10^{-24} cm²)
Z = atomic number
* = radioactive isotope
• = magnetic moment observed by NMR
( ) = assumed or estimated values

Assuming a nuclear magneton value of 5.0505×10^{-24} erg/gauss, the NMR frequency was calculated for a total field of 10 kilogauss. The sensitivities, relative to the proton, are calculated from the following expressions:

Sensitivity at constant field $= 7.652 \times 10^{-3} \mu^3 (I+1)/I^2$
Sensitivity at constant frequency $= 0.2387 \mu(I+1)$.

These expressions assume an equal number of nuclei, a constant temperature, and T_1 = (the longitudinal relaxation time) equals the transverse relaxation time). These sensitivities represent the ideal induced voltage in the receiver coil at saturation and with a constant noise source. The calculated values are therefore determined under complete optimum conditions and should be regarded as such.

Table (Z = 0 – 13)

Z	El	A	Spin I	NMR Freq (MHz, 10 kG)	Abund. %	Rel. Sens. const. field	Rel. Sens. const. freq.	Magnetic Moment μ	Ref	Method	Quadrupole Q (barns)	Ref	Method
0	n	1*	1/2	29.167		0.322	0.685	−1.91315	1	Ab			
1	•H	1	1/2	42.5759	99.985	1.00	1.00	2.79268	2	N			
1	•H	2	1	6.53566	1.5×10^{-2}	9.65×10^{-3}	0.409	0.857387	3	N	2.73×10^{-3}	4	Ab
1	•H	3*	1/2	45.4129		1.21	1.07	2.97877	5	N			
2	•He	3	1/2	32.433	1.3×10^{-4}	0.442	0.762	−2.1274	6	N			
3	•Li	6	1	6.2653	7.42	8.50×10^{-3}	0.392	0.82192	7	N	6.9×10^{-4}	9	N
3	•Li	7	3/2	16.546	92.58	0.293	1.94	3.2560	8	N	-3×10^{-2}	10	O
3	•Li	8*	2	6.300		2.59×10^{-2}	1.184	1.653	11	Ab			
4	•Be	9	3/2	5.9834	100	1.39×10^{-2}	0.703	−1.1774	12	N	5.2×10^{-2}	14	Ab
5	•B	10	3	4.5754	19.58	1.99×10^{-2}	1.72	1.8007	15	N	7.4×10^{-2}	16	Ab
5	•B	11	3/2	13.660	80.42	0.165	1.60	2.6880	15	N	3.55×10^{-2}	16	Ab
6	•C	13	1/2	10.7054	1.108	1.59×10^{-2}	0.251	0.702199	17	N			
7	•N	13*	1/2	4.91		1.53×10^{-3}	0.115	(−)0.322	18	Ab			
7	•N	14	1	3.0756	99.63	1.01×10^{-3}	0.193	0.40347	19	N	1.6×10^{-2}	20	Mc
7	•N	15	1/2	4.3142	0.37	1.04×10^{-4}	0.101	−0.28298	20	N			
8	•O	15*	1/2	11.0		1.70×10^{-2}	0.257	0.719	23	N			
8	•O	17	5/2	5.772	3.7×10^{-2}	2.91×10^{-2}	1.58	−1.8930	24	N	-2.6×10^{-2}	26	M
9	•F	17*	5/2	14.40		0.451	3.94	4.720	22				
9	•F	19	1/2	40.0541	100	0.833	0.941	2.62727	21	N			
10	•F	20*	2	7.977		5.26×10^{-2}	1.50	2.093	27	N			
10	•Ne	19*	1/2	28.75		0.308	0.675	−1.886	28	Ab			
10	•Ne	21	3/2	3.3611	0.257	2.50×10^{-3}	0.395	−0.66140	29	Ab	9×10^{-2}	30	Ab
11	•Na	21*	3/2	12.126		0.116	1.42	2.3861	30	Ab			
11	•Na	22*	3	4.436		1.81×10^{-2}	1.67	1.746	31				
11	•Na	23	3/2	11.262	100	9.25×10^{-2}	1.32	2.2161	17	N	0.14–0.15	32	O
11	•Na	24*	4	3.221		1.15×10^{-2}	2.02	1.690	34	Ab			
12	•Mg	25	5/2	2.6054	10.13	2.67×10^{-3}	0.714	−0.85449	13	N			
13	•Al	27	5/2	11.094	100	0.206	3.04	3.6385	35	N	0.149	36	Ab

Table (Z = 14 – 25)

Z	El	A	Spin I	NMR Freq (MHz, 10 kG)	Abund. %	Rel. Sens. const. field	Rel. Sens. const. freq.	Magnetic Moment μ	Ref	Method	Quadrupole Q (barns)	Ref	Method
14	•Si	29	1/2	8.4578	4.70	7.84×10^{-3}	0.199	−0.55477	37	N	—		
15	•P	31	1/2	17.235	100	6.63×10^{-4}	0.405	1.1305	38	N	—		
15	•P	32	1	1.923		2.46×10^{-3}	0.120	−0.2523	39	E			
16	•S	33	3/2	3.2654	0.76	2.26×10^{-3}	0.383	0.64257	37	M	-6.4×10^{-2}	40	M
17	•Cl	34	3	5.08		8.50×10^{-3}	0.597	1.00	41	M	4.5×10^{-2}	40	M
17	•Cl	35	3/2	4.1717	75.53	4.70×10^{-3}	0.490	0.82091	8	N	-7.89×10^{-2}	42	Ab
17	•Cl	36	2	4.8931		1.21×10^{-2}	0.919	1.2838	43	N	-1.72×10^{-2}	42	Ab
17	•Cl	37	3/2	3.472	24.47	2.71×10^{-3}	0.408	0.6833	45	N	-6.21×10^{-2}	44	Ab
18	•Ar	37	3/2	5.08		8.50×10^{-3}	0.597	1.0	240	O			
19	•K	38	3	3.491		8.82×10^{-3}	1.31	1.374	46	Ab			
19	•K	39	3/2	1.9868	93.10	5.08×10^{-4}	0.233	0.39097	47	N	0.11	49	O
19	•K	40	4	2.470	1.18×10^{-2}	5.21×10^{-3}	1.55	−1.296	48	N		50	
19	•K	41	3/2	1.0905	6.88	8.40×10^{-5}	0.128	0.21459	48	N			
20	•Ca	41	7/2	4.345		8.50×10^{-3}	0.816	−1.140	52	Ab			
20	•Ca	43	7/2	0.828	0.145	3.68×10^{-5}	9.73×10^{-2}	−0.163	53	Ab			
21	•Sc	43	7/2	2.8646		1.14×10^{-2}	1.71	−1.5924	54	N			
21	•Sc	44	2	10.04		6.40×10^{-2}	1.41	−1.3153	55	N			
21	•Sc	45	7/2	9.76		0.275	4.95	4.61	56	Ab	−0.26	56	Ab
21	•Sc	46	4	5.03		9.63×10^{-2}	1.83	2.56	57	Ab	0.14	57	Ab
21	•Sc	47	7/2	10.343		9.24×10^{-1}	6.62	3.96	57	Ab	0.37	57	Ab
21	•Sc	49	7/2	5.77		0.301	5.10	4.7492	60	Ab	−0.22	59	Ab
22	•Ti	47	5/2	11.6		6.65×10^{-1}	5.73	3.03	56	Ab	0.12	60	Ab
22	•Ti	49	7/2	0.207		2.40×10^{-6}	0.102	5.33	56	Ab	−0.22	56	Ab
23	•V	48	7/2	2.4000		2.09×10^{-3}	0.658	0.095	61	N	1.5×10^{-2}	61	Ab
23	•V	50	6	2.4005	0.24	3.76×10^{-3}	1.18	−0.78710	62	N			
23	•V	51	7/2	9.71	99.76	0.249	4.79	−1.1022	62	N			
24	•Cr	53	7/2	11.19	9.55	5.55×10^{-2}	5.58	4.46	63	E			
25	•Mn	53	6	12.4065		9.03×10^{-4}	5.52	3.3413	64	N			
25	•Mn	54	3/2	3.907		4.33×10^{-2}	5.14	5.139	66	N			
25	•Mn	52	2	0.030		2.9×10^{-3}	5.73×10^{-3}	−0.47354	67	Ab	-4×10^{-2}	65	O
25	•Mn	55	7/2	11.0	100	0.362	5.42	3.075	68	Ab			
								0.008	68	Ab			
								5.05	69	E			

NUCLEAR SPINS, MOMENTS, AND MAGNETIC RESONANCE FREQUENCIES (Continued)

First table

Z	El	A	Spin I	NMR Frequency in MHz for a 10 kilogauss field	Natural Abundance %	Rel. Sens. const. field	Rel. Sens. const. freq.	Magnetic Moment μ (eh/4πMc)	μ Reference	μ Method	Electric Quadrupole Moment Q (barns, 10⁻²⁴ cm²)	Q Reference	Q Method
25	Mn	54*	(2)	8.4	—	6.11x10⁻²	1.58	(2.2)	70	No			
25	•Mn	55	(3)	6.6	100	5.98x10⁻¹	2.48	(2.6)	71	E	0.55	72	M
25	Mn	56*	3	10.501		0.175	2.88	3.444	73	Ab			
26	•Fe	57	5/2	8.233	2.19	0.116	3.09	3.240	74	Ab			
27	Co	55*	1/2	1.3758		3.37x10⁻⁵	3.23x10⁻²	0.00024	75	No			
27	Co	56*	4	10.0		0.274	4.94	4.6	76	E			
27	Co	57*	5	7.34		0.136	4.60	3.85	76	E			
27	Co	57	7/2	10.1		0.283	4.99	4.65	77	E			
27	Co	58*	2	15.4		0.381	2.90	4.05		E			
27	•Co	59	7/2	10.054	100	0.277	4.96	4.6163	241	N	0.40	78	Ab
27	Co	60*	5	5.793		0.101	5.44	3.800	242	E			
28	•Ni	61	3/2	3.8047	1.19	3.57x10⁻³	0.447	-0.74868	79	E			
29	•Cu	61*	3/2	10.8		8.22x10⁻¹	1.27	2.13	80	Ab	-0.16	82	E
29	•Cu	63	3/2	11.285	69.09	9.31x10⁻²	1.33	2.2206	81	Ab			
29	Cu	64*	1	3.1		9.79x10⁻⁴	0.191	0.40	83	Ab			
29	•Cu	65	3/2	12.089	30.91	0.114	1.42	2.3789	8	Ab	-0.15	82	E
29	Cu	66*	1	1.65		1.54x10⁻⁴	0.103	-0.216	81	O			
30	Zn	65*	5/2	2.345		1.95x10⁻²	0.643	0.7692	83	O			
30	•Zn	67	5/2	2.663	4.11	2.83x10⁻³	0.730	0.8733	37	Ab			
31	Ga	68*	1	0.0892		2.45x10⁻⁵	5.59x10⁻³	0.0117	84	O	3.1x10⁻²	83	O
31	•Ga	69	3/2	10.22	60.4	6.91x10⁻²	1.20	2.011	84	Ab	0.178	84	Ab
31	Ga	71*	3/2	12.984	39.6	0.142	1.52	2.5549	86	Ab	0.112	85	Ab
31	•Ga	71	3	0.33591		7.80x10⁻⁶	0.126	-0.13220	86	O	0.72	85	Ab
31	Ga	72*	1/2	8.4		7.64x10⁻²	0.197	0.55	87	Ab			
32	•Ge	71*	9/2	1.4852		1.40x10⁻³	1.15	-0.87679	62	Ab	-0.2	88	M
32	•Ge	73	9/2	7.2919	7.76	2.51x10⁻³	0.856	1.4349	15	Ab	0.3	89	M
33	•As	75	2	3.45		4.27x10⁻³	0.649	-0.906	90	No			
33	As	76*	1/2	8.118	100	6.93x10⁻³	0.191	0.5325	91	M	0.9	40	M
34	Se	76*	2	2.22		2.98x10⁻³	1.10	-1.02	91	M	0.27	92	Ab
34	•Se	77	3/2	4.18	7.58	2.52x10⁻³	0.262	(-)0.548	8	M	0.33	93	Ab
34	Se	79*	3/2	10.667		7.86x10⁻²	1.25	2.0990	8	Ab	0.20	94	Ab
35	•Br	79	5	3.92	50.54	2.08x10⁻¹	0.245	0.514	94	O	0.76	94	Ab
35	Br	80*	5	2.008		4.20x10⁻¹	1.89	1.317	37	E	0.28	94	Ab
35	Br	80m*	3/2	11.498	49.46	9.88x10⁻²	1.35	2.2626	8	Ab	0.15	95	Ab
35	•Br	81	5/2	2.479		7.90x10⁻³	1.27	(+)1.626	95	O	0.25	97	O
35	Br	82*	5/2	1.638	11.55	1.88x10⁻³	1.31	-0.9671	96	N			
36	Kr	82*	9/2	1.6956		7.32x10⁻³	1.22	-1.001	99	Ab			
36	•Kr	83	5	10.4		6.20x10⁻³	2.15	2.05	99	Ab			
37	Rb	81*	3/2	2.29		1.23x10⁻¹	1.19	1.50	99	Ab			
37	Rb	83*	2	4.33		1.32x10⁻¹	0.945	1.42	100	Ab			
37	•Rb	85	5/2	5.03	72.15	1.05x10⁻²	1.13	-1.32	102	N	0.27	101	Ab
37	Rb	86*	1	6.44		2.77x10⁻¹	1.21	-1.69	103	Ab	0.13	101	Ab
37	•Rb	87	3/2	4.1108	27.85	2.69x10⁻²	1.64	2.7414	66	N	0.2	104	E
38	•Sr	87	9/2	13.931	7.02	0.175	1.43	-0.10893	47	Ab			
38	Sr	87m*	1/2	1.8452		1.18x10⁻⁴	4.90x10⁻⁴	-1.62	105	Ab	-0.16	105	Ab
39	Y	90*	2	2.0859		2.44x10⁻⁴	1.16	0.163	60	N			
39	•Y	91	1/2	6.17		1.99x10⁻⁴	8.07	-1.30284	106	Ab			
40	•Zr	91	5/2	2.49	11.23	9.48x10⁻⁴	1.09	0.9097	45	N	-0.2	108	O
41	•Nb	93	9/2	3.97249	100	0.482	0.776	6.1435	107	N	0.12	109	N
42	•Mo	95	5/2	10.407	15.72	3.23x10⁻³	0.776	-0.9289	45	N	1.1	109	N
42	•Mo	97	5/2	2.774	9.46	3.43x10⁻³	7.43	-0.284	110	Mo	0.3	111	O
43	Tc	99	9/2	2.833		1.96x10⁻²	0.169	5.6572	112	N			
44	•Ru	99	3/2	9.5830	12.72	1.41x10⁻⁴	0.576	-0.69	113	O			
44	•Ru	101	5/2	1.44	17.07								
45	•Rh	103	1/2	2.1	100	3.11x10⁻⁵	3.15x10⁻²	-0.08790	43	N			
46	•Pd	105	5/2	1.3401	22.23	1.12x10⁻³	0.534	-0.639	114	N			
47	Ag	104*	2	1.95		0.118	5.73	4.0	115	Ab			
47	Ag	104m*	11/2	6.1		0.291	2.65	3.7	116	Ab			
47	•Ag	105*	1/2	1.54		4.73x10⁻⁶	3.62x10⁻⁷	0.101	117	Ab			

Second table

Z	El	A	Spin I	NMR Frequency in MHz for a 10 kilogauss field	Natural Abundance %	Rel. Sens. const. field	Rel. Sens. const. freq.	Magnetic Moment μ (eh/4πMc)	μ Reference	μ Method	Electric Quadrupole Moment Q (barns, 10⁻²⁴ cm²)	Q Reference	Q Method
47	*Ag	107	1/2	1.7229	51.82	6.62x10⁻⁵	4.05x10⁻⁵	-0.11301	47	N	—		
47	Ag	108*	1	32.0	—	1.13	2.01	4.2	118	Ab			
47	*Ag	109	1/2	1.9807	48.18	1.01x10⁻⁴	4.65x10⁻⁵	-0.12992	47	N	—	118	O
47	Ag	110m*	6	4.557	—	6.87x10⁻¹	5.99	3.587	120	Ab			
47	Ag	111*	1/2	2.21	—	1.40x10⁻¹	5.19x10⁻²	-0.145	121	Ab			
47	Ag	112*	2	0.2077	—	9.00x10⁻⁷	3.90x10⁻¹	0.0545	122	Ab			
47	Ag	113*	1/2	1.879	—	1.81x10⁻⁴	5.66x10⁻³	0.158	122	Ab			
48	Cd	107*	5/2	2.529	—	1.00x10⁻³	0.515	-0.6162	123	O	0.8	123	O
48	•Cd	109*	5/2	9.028	—	2.44x10⁻³	0.693	-0.8293	124	O	0.8	124	O
48	•Cd	111	1/2	9.445	12.75	9.54x10⁻³	0.212	-0.5922	124	O			
48	•Cd	113	1/2	1.51	12.26	1.09x10⁻²	0.222	-0.6195	125	N			
48	Cd	113m*	11/2	9.862	—	2.13x10⁻³	1.69	-1.09	239	O	-0.79	239	O
48	Cd	115*	1/2	1.447	—	1.24x10⁻¹	0.232	-0.6469	127	N	-0.61	127	O
48	•In	113	9/2	9.3099	4.28	1.87x10⁻¹	1.62	-1.044	127	Ab	1.14	97	Ab
49	•In	113m*	1/2	3.209	—	4.28x10⁻⁴	7.22	5.4960	45	Ab			
49	•In	114*	5	7.2	—	0.191	7.54x10⁻²	-0.2105	129	N			
49	•In	115*	9/2	9.3301	95.72	0.347	6.73	4.7	128	Ab	1.16	97	Ab
49	In	115m*	1/2	3.715	—	0.137	7.23	5.5079	130	N			
49	In	116*	5	6.42	—	0.156	6.03	-0.2437	131	N			
49	In	116m*	1	6.7	—	3.50x10⁻²	6.30	4.21	129	N			
50	Sn	115	1/2	13.922	0.35	4.52x10⁻²	0.327	4.4	129	Ab			
50	•Sn	117	1/2	15.168	7.61	5.18x10⁻²	0.356	-0.91320	125	N			
50	•Sn	119	1/2	15.869	8.58	0.160	0.373	-0.99490	125	N			
50	Sb	121	5/2	10.189	57.25		2.79	-1.0409	125	N	-0.5	134	O
51	Sb	122*	2	7.24	—	3.94x10⁻²	1.36	3.3415	133	N			
51	•Sb	123	7/2	5.5176	42.75	4.57x10⁻²	2.72	2.5334	45	N	-0.7	134	O
52	•Te	119*	1/2	4.12	—	9.04x10⁻⁴	9.67x10⁻²	0.27	137	Ab			
52	•Te	123	1/2	11.16	0.87	1.8x10⁻²	0.262	-0.7319	37	N			
52	Te	125	1/2	13.45	6.99	3.15x10⁻²	0.316	-0.8824	37	N			
53	•I	127	5/2	9.0	100	0.116	2.51	3	138	M	-0.66	138	M
53	I	129*	7/2	8.5183	—	9.34x10⁻²	2.33	2.7937	139	N	-0.69	134	O
53	I	131*	7/2	5.963	—	4.96x10⁻²	2.80	2.6031	141	N	-0.48	140	O
54	•Xe	129	1/2	11.777	26.44	5.77x10⁻²	2.94	2.738	141	Ab	-0.41	141	Ab
54	•Xe	131	3/2	3.4911	21.18	2.12x10⁻³	0.277	-0.77247	142	N			
55	Cs	127*	1/2	21.8	—	2.76x10⁻³	0.410	0.68697	142	N	-0.12	143	O
55	Cs	129*	1/2	22.4	—	0.134	0.512	1.43	142	Ab			
55	Cs	130*	1	10.7	—	0.146	0.526	1.47	144	Ab			
55	Cs	131*	5/2	10.72	—	4.20x10⁻¹	0.668	1.4	144	Ab			
55	Cs	132*	2	8.46	—	0.186	1.59	2.22	144	Ab			
55	•Cs	133	7/2	5.58469	100	4.74x10⁻²	2.75	2.54422	8	N	-3x10⁻³	146	Ab
55	Cs	134*	4	5.666	—	6.28x10⁻¹	3.55	2.973	148	Ab	0.43	147	O
55	Cs	134m*	8	1.0447	—	1.42x10⁻¹	2.36	1.0964	148	Ab		149	O
55	Cs	135*	7/2	5.9096	—	5.62x10⁻²	2.91	2.7134	148	Ab			
55	Cs	137*	7/2	6.1459	—	6.32x10⁻²	3.03	2.8219	148	Ab			
56	•Ba	135	3/2	4.2296	6.59	4.90x10⁻³	0.497	0.83229	151	N	0.25	153	O
56	•Ba	137	3/2	4.7315	11.32	6.86x10⁻³	0.556	0.93107	152	N	0.2	153	O
57	•La	138	5*	5.6171	0.089	9.19x10⁻³	5.28	3.6844	154	N	2.7	154	N
57	•La	139	7/2	6.0144	99.911	5.92x10⁻²	2.97	2.7615	107	N	0.21	155	Ab
58	•Ce	137*	3/2	4.6	—	6.20x10⁻³	0.537	0.9	156	No			
58	•Ce	137m*	11/2	0.96	—	5.40x10⁻⁴	1.07	0.69	157	No			
58	•Ce	139*	3/2	5.1	—	8.50x10⁻³	0.597	1.0	156	No			
58	Ce	141*	7/2	2.1	—	2.57x10⁻³	1.04	0.97	158	E			

Left portion (Z = 58–75)

Z	El	A	Spin I	NMR Freq (MHz, 10 kG)	Nat. Abund. %	Rel. Sens. (const. field)	Rel. Sens. (const. freq.)	Magnetic Moment μ (nucl. magneton)	μ Ref.	μ Method	Q (barns, 10⁻²⁴ cm²)	Q Ref.	Q Method
58	Ce	143*	7/2	2.2	—	2.81x10⁻³	1.07	1.0	156	No			
59	*Pr	141	5/2	12.5	100	0.293	3.42	4.09	159	O	-5.9x10⁻²	158	Ab
59	Pr	142*	2	1.1	—		0.215	0.30	160	Ab	4x10⁻²	161	Ab
60	Nd	143	7/2	2.315	12.17	1.55x10⁻⁴	1.14	-1.063	161	Ab	-0.48	162	Ab
60	Nd	145	7/2	1.42	8.30	3.38x10⁻⁴	0.703	-0.654	162	Ab	-0.25	162	E
60	Nd	147*	5/2	1.77		7.86x10⁻⁴	0.484	0.579	162	E			
61	Pm	143*	(5/2)	11.6		8.32x10⁻³	3.17	(3.8)	163	No			
61	Pm	143*	(7/2)	8.5		0.235	4.19	(3.9)		No			
61	Pm	144*	(6)	2.6		0.167	2.43	(1.7)	164	No			
61	Pm	147*	7/2	2.3		9.02x10⁻³	2.77	(1.8)	165	Ab	0.7	165	O
61	Pm	148*	1	5.62		8.68x10⁻³	3.01	2.58	166	No	0.2	166	No
61	Pm	148m	7/2	16		4.83x10⁻³	1.00	2.1	163	No			
61	Pm	149*	7/2	2.3		0.142	3.01	1.8	167	Ab	1.9	167	Ab
61	Pm	151*	5/2	7.2		8.68x10⁻³	3.54	3.3	163	No			
62	Sm	147	7/2	5.5	14.97	2.50x10⁻²	1.50	1.8	167	Ab	-0.203	158	Ab
62	Sm	149	7/2	1.76	13.83	1.48x10⁻⁴	0.867	-0.807	168	Ab	6.0x10⁻²	158	Ab
62	Sm	151	5/2	1.40	47.82	7.47x10⁻⁴	0.691	-0.643	169	Ab	1.16	170	O
63	Eu	151	5/2	10.559	52.18	2.89	1.83	3.4630	169	Ab	2.9	170	O
63	*Eu	153	5/2	4.858		2.38x10⁻³	1.28	1.912	169	Ab			
63	Eu	153	5/2	4.6627		1.53x10⁻²		1.5292	172	Ab		173	O
63	Eu	154*	3	5.084		2.72x10⁻²	1.91	2.001	173	E			
64	Gd	155	3/2	1.6	14.73	2.79x10⁻⁴	0.191	-0.32	173	O	2	173	O
64	*Gd	157	3/2	2.0	15.68	5.44x10⁻⁴	0.239	-0.40	174	Ab			
65	*Tb	159	3/2	3.8	100	1.15x10⁻³	1.43	1.5	175	No	1.4	175	No
65	Tb	160*	3	9.66		5.83x10⁻²	1.13	1.90	158	E	1.3	176	E
65	Tb	160*	(3/2)	4.1		1.39x10⁻⁴	1.53	1.6	177	No	1.9	177	No
66	Dy	157*	5/2	1.1		7.87x10⁻⁵	0.125	0.21	178	No			
66	Dy	161	5/2	1.6	18.88	2.79x10⁻⁴	0.191	0.32	158	E	1.4	158	E
66	Dy	163	5/2	1.4	24.97	4.17x10⁻⁴	0.384	-0.46	158	E	1.6	158	Ab
67	*Ho	165	7/2	2.0	100	1.12x10⁻³	0.535	-0.64	158	Ab	2.82	158	Ab
68	Er	165*	7/2	8.73		0.181	4.31	4.01	166	Ab	2.2	166	Ab
68	Er	167	7/2	2.0	22.94	1.18x10⁻⁴	0.543	0.65	162	Ab	2.83	162	Ab
68	Er	169*	1/2	1.23		5.07x10⁻⁴	0.607	-0.565	179	Ab			
68	Er	171*	5/2	7.8		6.09x10⁻³	0.183	0.51	180	Ab	4.6	181	Ab
68	Tm	166*	2	2.1		1.47x10⁻⁴	0.585	0.70	180	Ab			
69	Tm	166*	2	0.19		7.17x10⁻⁷	3.58x10⁻²	0.05	181	Ab			
69	*Tm	169	1/2	3.52	100	5.66x10⁻⁴	8.27x10⁻²	-0.231	182	Ab			
69	Tm	170*	1	2.0		2.69x10⁻⁴	8.13x10⁻²	0.26	183	Ab	0.61	183	Ab
69	Tm	171*	1/2	3.46		5.37x10⁻⁴		0.227	180	Ab			
70	*Yb	171	1/2	7.4990	14.31	5.46x10⁻³	0.176	0.49188	184	O			
70	Yb	173	5/2	2.0659	16.13	1.33x10⁻³	0.566	-0.67755	185	O	2.8	186	O
70	Yb	175*	(7/2)	0.33		9.40x10⁻⁴	0.161	-0.15	187	No			
71	*Lu	175	7/2	4.86	97.41	3.12x10⁻²	2.40	2.23	188	N	5.68	189	Ab
71	Lu	176*	7	3.4	2.59	3.72x10⁻⁴	5.92	3.1	190	Ab	8.0	190	Ab
71	Lu	177*	7/2	4.84		3.08x10⁻²	2.38	2.22	191	Ab	5.51	191	Ab
72	Hf	177	7/2	1.3	18.50	6.38x10⁻⁴	0.655	0.61	192	Ab	3	192	Ab
72	Hf	179	9/2	0.80	13.75	2.16x10⁻⁴	0.617	-0.47	193	N	3	193	O
73	*Ta	181	7/2	5.096	99.988	3.60x10⁻²	2.51	2.340	194	N	3	195	N
74	*W	183	1/2	1.7716	14.40	7.20x10⁻⁵	4.16x10⁻²	0.116205	196	Ab			
75	*Re	185	5/2	9.5855	37.07	0.133	2.63	3.1437	13	N	2.8	97	O
75	Re	186*	1	13.17	—	7.90x10⁻²	0.825	1.728	197	Ab		241	
75	*Re	187	5/2	9.6837	62.93	0.137	2.65	3.1759	13	N	2.6	97	O

Right portion (Z = 75–95)

Z	El	A	Spin I	NMR Freq (MHz, 10 kG)	Nat. Abund. %	Rel. Sens. (const. field)	Rel. Sens. (const. freq.)	Magnetic Moment μ (nucl. magneton)	μ Ref.	μ Method	Q (barns, 10⁻²⁴ cm²)	Q Ref.	Q Method
75	Re	188*	1	13.55	—	8.59x10⁻³	0.848	1.777	197	Ab		195	O
76	*Os	187	1/2	0.98059	1.64	1.22x10⁻⁵	2.30x10⁻⁵	0.06432	198	N	0.8		
76	*Os	189	3/2	3.3034	16.1	2.34x10⁻³	0.388	0.65004	199	N	1.5	202	O
77	*Ir	191	3/2	0.7318	37.3	2.53x10⁻²	8.59x10⁻²	0.1440	200	N	1.5	202	O
77	*Ir	193	3/2	0.7968	62.7	3.27x10⁻⁵	9.36x10⁻²	0.1568	200	N			
78	*Pt	195	1/2	9.153	33.8	9.94x10⁻³	0.215	0.6004	45	N			
79	Au	190*	1	0.496		4.20x10⁻³	3.10x10⁻³	0.065	203	Ab			
79	Au	194*	1	0.56		5.95x10⁻³	3.49x10⁻³	0.073	204	Ab			
79	Au	195*	3/2	0.742		2.65x10⁻³	8.71x10⁻³	0.146	204	Ab	0.59	206	Ab
79	Au	196*	2	2.3		1.24x10⁻³	0.430	0.6	204	Ab			
79	*Au	197	3/2	0.729188	100	2.51x10⁻³	8.56x10⁻²	0.143489	205	Ab			
80	Au	198*	2	2.227		1.14x10⁻³	0.418	0.5842	201	O			
80	Au	199*	3/2	1.358		1.62x10⁻⁴	0.160	0.2673	207	O	1.37	209	O
80	Hg	199	3/2	3.1		1.93x10⁻³	0.364	-0.61	207	O			
80	Hg	201	13/2	1.23	16.84	1.57x10⁻³	1.88	-1.05	208	O	1.41	209	O
80	Hg	199	1/2	8.1		6.84x10⁻³	0.190	0.53	209	O			
81	Hg	201	13/2	1.22	13.22	1.53x10⁻³	1.86	-1.04	210	O	0.50	216	O
81	Hg	197*	1/2	7.9		6.46x10⁻³	0.186	0.52	211	N	0.5	217	O
81	*Hg	199	1/2	7.59012	16.84	5.67x10⁻³	0.178	0.497859	209	N			
81	Hg	201	3/2	2.5099	13.22	1.44x10⁻³	0.330	-0.55293	212	N			
82	Hg	203*	5/2	2.5		2.45x10⁻³	0.693	0.83	213	O			
83	Tl	203	1/2	23.6		0.171	0.555	1.55	45	N			
83	Tl	205	1/2	23.9		0.178	0.562	1.57	217	O			
83	Tl	201*	1/2	0.57		1.94x10⁻²	0.107	(0.15)	218	O			
83	Tl	202*	1/2	24.1		1.94x10⁻²	0.566	1.58	219	O			
83	*Tl	203	1/2	24.332	29.50	0.187	0.571	(0.15)	219	N			
84	Tl	204*	2	0.34	70.50	4.05x10⁻⁶	6.37x10⁻³	1.5960	219	N	-0.64	223	Ab
89	*Tl	205	1/2	24.570	22.6	0.192	0.577	0.089	221	Ab	-0.41	223	Ab
90	Pb	207	1/2	8.90771		9.16x10⁻³	0.209	1.6116	8	Ab	-0.19	223	Ab
91	Bi	209	9/2	7.78		0.201	6.03	0.584284	222	N	-0.4	97	Ab
92	Bi	203*	6	5.40		0.114	7.10	4.59	222	N	0.13	225	Ab
93	Bi	205*	6	9.3		0.346	7.22	4.25	223	Ab	0.17	226	Ab
94	Bi	206*	6	5.79		0.141	7.62	4.56	223	Ab	0.28	226	Ab
94	*Bi	209	9/2	6.84178	100	0.137	2.11x10⁻³	4.03896	223	Ab	-1.7	227	O
95	Bi	210*	1	0.337		1.32x10⁻⁴	0.217	0.0442	224	Ab	4.6	228	Ab
95	Po	207*	5/2	0.79		7.55x10⁻⁵	0.226	0.26	225	E	-3.0	230	Ab
	Ac	227*	5/2	0.82		8.43x10⁻⁵	0.656	0.27	226	E	3.5	231	E
	Th	229*	5/2	5.6		1.13x10⁻³	0.334	1.1	227	E	4.1	231	E
	Pa	231*	3/2	1.2		2.74x10⁻⁴	1.17	0.4	228	Ab			
	U	235*	7/2	9.96	0.72	6.40x10⁻³	2.03	1.96	229	E			
	Np	237*	5/2	17		6.75x10⁻⁴	0.451	3.4	230	E	4.9	236	Ab
	Pu	239	1/2	1.6		1.21x10⁻⁴	0.376	0.54	231	E	-2.8	237	O
	Pu	241*	5/2	0.76		0.926	7.16x10⁻³	0.35	232	E	4.9	236	O
	Am	241*	5/2	18		3.67x10⁻³	0.573	(6)	233	Ab			
	Am	242*	1	3.05		1.38x10⁻⁷	1.32	-0.200	234	Ab			
	Am	243*	5/2	2.09		1.69x10⁻²	0.182	-0.686	235	Ab			
				4.82		8.46x10⁻⁴	1.31	1.58	235	Ab			
				2.90		1.66x10⁻²		0.381	238	Ab			
				4.79				1.57					
	Free electron		1/2	2.80246x10⁴	—	2.84x10⁻⁸	657	-1836.09			—		
	γ th							g = 2.00232					

REFERENCES

(1) V. W. Cohen, et al., Phys. Rev. **104**, 283 (1956).
(2) H. Sommer, et al., Phys. Rev. **82**, 697 (1951).
(3) T. F. Wimett, Phys. Rev. **91**, 499A (1953).
(4) H. Kopfermann, et al., Z. Physik **144**, 9 (1956).
(5) F. Bloch, et al., Phys. Rev. **71**, 551 (1947).
(6) H. L. Anderson, Phys. Rev. **76**, 1460 (1949).
(7) M. P. Klein, et al., Phys. Rev. **106**, 837 (1957).
(8) H. E. Walchli, Thesis, M.S., U. of Tenn. (1954).   AEC Report ORNL-1775.
(9) N. A. Schuster, et al., Phys. Rev. **81**, 157 (1951).
(10) K. C. Brog, et al., Phys. Rev. **153**, 91 (1967).
(11) D. Connor, Phys. Rev. Letters **3**, 429 (1959).
(12) L. C. Brown, et al., J. Chem. Phys. **24**, 751 (1956).
(13) F. Alder, et al., Phys. Rev. **82**, 105 (1951).
(14) A. G. Blachman, et al., Bull Am. Phys. Soc. **11**, 343 (1966).
(15) Y. Ting, et al., Phys. Rev. **89**, 595 (1953).
(16) G. Wessel, Phys. Rev. **92**, 1581 (1953).
(17) G. Lindström, Arkiv Fysik **4**, 1 (1951).
(18) V. Royden, Phys. Rev. **96**, 543 (1954).
(19) A. M. Bernstein, et al., Phys. Rev. **136**, B27 (1964).
(20) L. W. Anderson, et al., Phys. Rev. **116**, 87 (1959).
(21) M. R. Baker, et al., Phys. Rev. **133**, A1533 (1964).
(22) A. Bassompiere, Compt. Rend. **240**, 285 (1955).
(23) E. D. Commins, et al., Phys. Rev. **131**, 700 (1963).
(24) F. Alder, et al., Phys. Rev. **81**, 1067 (1951).
(25) M. J. Stevenson, et al., Phys. Rev. **107**, 635 (1957).
(26) K. Sugimoto, et al., J. Phys. Soc. Japan **21**, 213 (1966).
(27) T. Tsang, et al., Phys. Rev. **132**, 1141 (1963).
(28) E. D. Commins, et al., Phys. Rev. Letters **10**, 347 (1963).
(29) J. T. LaTourette, et al., Phys. Rev. **107**, 1202 (1957).
(30) O. Ames, et al., Phys. Rev. **137**, B1157 (1965).
(31) L. Davis, et al., Phys. Rev. **76**, 1068 (1949).
(32) H. Ackermann, Z. Physik. **194**, 253 (1966).
(33) M. Baumann, et al., Z. Physik **194**, 270 (1966).
(34) Y. W. Chan, et al., Phys. Rev. **150**, 933 (1966).
(35) L. C. Brown, et al., J. Chem. Phys. **24**, 751 (1956).
(36) H. Lew, et al., Phys. Rev. **90**, 1 (1953).
(37) H. E. Weaver, Phys. Rev. **89**, 923 (1953).
(38) T. Kanda, et al., Phys. Rev. **85**, 938 (1952).
(39) G. Feher, et al., Phys. Rev. **107**, 1462 (1957).
(40) G. R. Bird, et al., Phys. Rev. **94**, 1203 (1954).
(41) B. F. Burke, et al., Phys. Rev. **93**, 193 (1954).
(42) V. Jaccarino, et al., Phys. Rev. **83**, 471 (1951).
(43) P. B. Sogo, et al., Phys. Rev. **98**, 1316 (1955).
(44) C. H. Townes, et al., Phys. Rev. **76**, 691 (1949).
(45) W. G. Proctor, et al., Phys. Rev. **81**, 20 (1951).
(46) E. A. Phillips, et al., Phys. Rev. **138**, B773 (1965).
(47) E. Brun, et al., Phys. Rev. **93**, 172 (1954).
(48) O. Lutz, et al., Phys. Letters **24A**, 122 (1967).
(49) G. W. Series, Phys. Rev. **105**, 1128 (1957).
(50) G. J. Ritter, et al., Proc. Roy. Soc. (London) **238**, 473 (1957).
(51) J. T. Eisinger, et al., Phys. Rev. **86**, 73 (1952).
(52) J. M. Kahn, et al., Phys. Rev. **134**, A45 (1964).
(53) F. R. Petersen, et al., Phys. Rev. **116**, 734 (1959).
(54) E. Brun, et al., Phys. Rev. Letters **9**, 166 (1962).
(55) C. D. Jeffries, Phys. Rev. **90**, 1130 (1953).
(56) R. G. Cornwall, et al., Phys. Rev. **141**, 1106 (1966).
(57) D. L. Harris, et al., Phys. Rev. **132**, 310 (1963).
(58) D. M. Hunten, Can. J. Phys. **29**, 463 (1951).
(59) G. Fricke, et al., Z. Physik, **156**, 416 (1959).
(60) F. R. Petersen, et al., Phys. Rev. **128**, 1740 (1962).
(61) R. G. Cornwall, et al., Phys. Rev. **148**, 1157 (1966).
(62) C. D. Jeffries, Phys. Rev. **92**, 1262 (1953).
(63) M. M. Weiss, et al., Bull. Am. Phys. Soc. **2**, 31 (1957).
(64) H. E. Walchli, et al., Phys. Rev. **87**, 541 (1952).
(65) K. Murakawa, et al., J. Phys. Soc. Japan **21**, 1466 (1966).
(66) C. D. Jeffries, et al., Phys. Rev. **91**, 1286 (1953).
(67) K. E. Adelroth, et al., Arkiv Fysik **31**, 549 (1966).
(68) E. A. Phillips, et al., Phys. Rev. **140**, B555 (1965).
(69) W. Dobrowolski, et al., Phys. Rev. **104**, 1378 (1956).
(70) R. W. Bauer, et al., Phys. Rev. **120**, 946 (1960).
(71) W. B. Mims, et al., Phys. Letters **24A**, 481 (1967).
(72) A. Javan, et al., Phys. Rev. **96**, 649 (1954).
(73) W. J. Childs, et al., Phys. Rev. **122**, 891 (1961).
(74) P. R. Locher, et al., Phys. Rev. **139**, A991 (1965).
(75) R. W. Bauer, et al., Nucl. Phys. **16**, 264 (1960).
(76) J. M. Baker, et al., Proc. Phys. Soc. (London), **A69**, 354 (1956).
(77) W. Dobrov, et al., Phys. Rev. **108**, 60 (1957).
(78) D. V. Ehrenstein, et al., Z. Physik **159**, 230 (1960).
(79) W. Dobrowolski, et al., Phys. Rev. **101**, 1001 (1956).
(80) L. E. Drain, Phys. Letters **11**, 114 (1964).
(81) B. M. Dodsworth, et al., Phys. Rev. **142**, 638 (1966).
(82) B. Bleaney, et al., Proc. Roy. Soc. (London) **228A**, 166 (1955).
(83) A Lemonick, et al., Phys. Rev. **95**, 1356 (1954).
(84) V. J. Ehlers, et al., Phys. Rev. **127**, 529 (1962).
(85) R. T. Daly, et al., Phys. Rev. **96**, 539 (1954).
(86) W. J. Childs, et al., Phys. Rev. **120**, 2138 (1960).
(87) W. J. Childs, et al., Phys. Rev. **141**, 15 (1966).
(88) J. M. Mays, et al., Phys. Rev. **81**, 940 (1951).
(89) B. P. Dailey, et al., Phys. Rev. **74**, 1245 (1948).
(90) F. M. Pipkin, et al., Phys. Rev. **106**, 1102 (1957).
(91) W. A. Hardy, et al., Phys. Rev. **92**, 1532 (1953).
(92) E. Lipworth, et al., Phys. Rev. **119**, 1053 (1960).
(93) J. G. King, et al., Phys. Rev. **94**, 1610 (1954).
(94) M. B. White, et al., Phys. Rev. **136**, B584 (1964).
(95) H. L. Garvin, et al., Phys. Rev. **116**, 393 (1959).
(96) E. Brun, et al., Helv. Phys. Acta **27**, 173A (1954).
(97) J. E. Mack, Rev. Mod. Phys. **22**, 64 (1950).
(98) E. Rasmussen, et al., Z. Physik **141**, 160 (1955).
(99) J. C. Hubbs, et al., Phys. Rev. **107**, 723 (1957).
(100) H. Walchli, et al., Phys. Rev. **85**, 922 (1952).

(101) B. Senitzky, et al., Phys. Rev. **103**, 315 (1956).
(102) E. H. Bellamy, et al., Phil. Mag. **44**, 33 (1953).
(103) E. Yasaitis, et al., Phys. Rev. **82**, 750 (1951).
(104) J. W. Culvahouse, et al., Phys. Rev. **140**, A1181 (1965).
(105) F. R. Petersen, et al., Phys. Rev. **125**, 284 (1962).
(106) E. Brun, et al., Phys. Rev. **105**, 1929 (1957).
(107) R. E. Sheriff, et al., Phys. Rev. **82**, 651 (1951).
(108) K. Murakawa, Phys. Rev. **98**, 1285 (1955).
(109) A. Narath, et al., Phys. Rev. **143**, 328 (1966).
(110) H. Walchli, et al., Phys. Rev. **85**, 479 (1952).
(111) K. G. Kessler, et al., Phys. Rev. **92**, 303 (1953).
(112) E. Matthias, et al., Phys. Rev. **139**, B532 (1965).
(113) K. Murakawa, J. Phys. Soc. Japan **10**, 919 (1955).
(114) J. A. Seitchik, et al., Phys. Rev. **138**, A148 (1965).
(115) J. A. Seitchik, et al., Phys. Rev. **136**, A1119 (1964).
(116) O. Ames, et al., Phys. Rev. **123**, 1793 (1960).
(117) W. B. Ewbank, et al., Phys. Rev. **129**, 1617 (1963).
(118) P. B. Sogo, et al., Phys. Rev. **93**, 174 (1954).
(119) G. K. Rochester, et al., Phys. Letters **8**, 266 (1964).
(120) S. G. Schmelling, et al., Phys. Rev. **154**, 1142 (1967).
(121) G. K. Woodgate, et al., Proc. Phys. Soc. (London) **69**, 581 (1956).
(122) Y. W. Chan, et al., Phys. Rev. **133**, B1138 (1964).
(123) F. W. Byron, et al., Phys. Rev. **132**, 1181 (1963).
(124) P. Thaddeus, et al., Phys. Rev. **132**, 1186 (1963).
(125) W. G. Proctor, Phys. Rev. **79**, 35 (1950).
(126) M. Leduc, et al., Compt. Rend. **B262**, 736 (1966).
(127) M. N. McDermott, et al., Phys. Rev. **134**, B25 (1964).
(128) W. J. Childs, et al., Phys. Rev. **118**, 1578 (1960).
(129) L. S. Goodman, et al., Phys. Rev. **108**, 1524 (1957).
(130) M. Rice, et al., Phys. Rev. **106**, 953 (1957).
(131) J. A. Cameron, et al., Can. J. Phys. **40**, 931 (1962).
(132) P. B. Nutter, Phil. Mag. **1**, 587 (1956).
(133) V. W. Cohen, et al., Phys. Rev. **79**, 191 (1950).
(134) K. Murakawa, Phys. Rev. **100**, 1369 (1955).
(135) F. M. Pipkin, Bull. Am. Phys. Soc. **3**, 8 (1958).
(136) P. C. B. Fernando, et al., Phil. Mag. **5**, 1309 (1960).
(137) K. E. Adelroth, et al., Arkiv Fysik **30**, 111 (1965).
(138) P. C. Fletcher, et al., Phys. Rev. **110**, 536 (1958).
(139) H. Walchli, et al., Phys. Rev. **82**, 97 (1951).
(140) R. Livingston, et al., Phys. Rev. **90**, 609 (1953).
(141) E. Lipworth, et al., Phys. Rev. **119**, 2022 (1960).
(142) D. Brinkmann, Helv. Phys. Acta **36**, 413 (1963).
(143) A. Bohr, et al., Arkiv Fysik **4**, 455 (1952).
(144) W. A. Nierenberg, et al., Phys. Rev. **112**, 186 (1958).
(145) R. D. Worley, et al., Phys. Rev. **140**, B1483 (1965).
(146) P. Buck, et al., Phys. Rev. **104**, 553 (1956).
(147) K. H. Althoff, et al., Naturwiss. **41**, 368 (1954).
(148) H. H. Stroke, et al., Phys. Rev. **105**, 590 (1957).
(149) G. Heinzelmann, et al., Phys. Letters **21**, 162 (1966).
(150) V. W. Cohen, et al., Phys. Rev. **127**, 517 (1962).
(151) H. E. Walchli, et al., Phys. Rev. **102**, 1334 (1956).
(152) L. Olschewski, et al., Z. Physik **196**, 77 (1966).
(153) N. I. Kaliteevskii, et al., Soviet Physics—JETP **12**, 661 (1961).
(154) P. B. Sogo, et al., Phys. Rev. **99**, 613 (1955).
(155) K. Murakawa, et al., J. Phys. Soc. Japan **16**, 2533 (1961).
(156) J. N. Haag, et al., Phys. Rev. **129**, 1601 (1963).
(157) J. Blok, et al., Phys. Rev. **143**, 78 (1966).
(158) B. Bleaney, Quantum Electronics Conf., Paris, 1963; P. Grivet and N. Bloembergen, editors; Columbia U. Press (1964).
(159) J. Reader, et al., Phys. Rev. **137**, B784 (1965).
(160) E. D. Jones, Phys. Rev. Letters **19**, 432 (1967).
(161) A. Y. Cabezas, et al., Phys. Rev. **126**, 1004 (1962).
(162) K. F. Smith, et al., Proc. Phys. Soc. (London) **86**, 1249 (1965).
(163) R. W. Grant, et al., Phys. Rev. **130**, 1100 (1963).
(164) D. A. Shirley, et al., Phys. Rev. **121**, 558 (1961).
(165) J. Reader, Phys. Rev. **141**, 1123 (1966).
(166) D. Ali, et al., Phys. Rev. **138**, B1356 (1965).
(167) B. Budick, et al., Phys. Rev. **132**, 723 (1963).
(168) G. K. Woodgate, Proc. Roy. Soc. (London) **A293**, 117 (1966).
(169) L. Evans, et al., Proc. Roy. Soc. (London) **A289**, 114 (1965).
(170) W. Müller, et al., Z. Physik **183**, 303 (1965).
(171) S. S. Alpert, Phys. Rev. **129**, 1344 (1963).
(172) E. L. Boyd, Phys. Rev. **145**, 174 (1966).
(173) N. I. Kaliteevskii, et al., Soviet Physics—JETP **37**, 629 (1960).
(174) E. L. Boyd, et al., Phys. Rev. Letters **12**, 20 (1964).
(175) C. A. Lovejoy, et al., Nucl. Phys. **30**, 452 (1962).
(176) C. Arnoult, et al., J. Opt. Soc. Am. **56**, 177 (1966).
(177) C. E. Johnson, et al., Phys. Rev. **120**, 2108 (1960).
(178) Q. O. Navarro, et al., Phys. Rev. **123**, 186 (1961).
(179) W. M. Doyle, et al., Phys. Rev. **131**, 1586 (1963).
(180) B. Budick, et al., Phys. Rev. **135**, B1281 (1964).
(181) J. C. Walker, Phys. Rev. **127**, 1739 (1962).
(182) D. Giglberger, et al., Z. Physik **199**, 244 (1967).
(183) A. Y. Cabezas, et al., Phys. Rev. **120**, 920 (1960).
(184) L. Olschewski, et al., Z. Physik **200**, 224 (1967).
(185) A. C. Gossard, et al., Phys. Rev. **133**, A881 (1964).
(186) J. S. Ross, et al., Phys. Rev. **128**, 1159 (1962).
(187) M. A. Grace, et al., Phil. Mag. **2**, 1079 (1957).
(188) A. H. Reddoch, et al., Phys. Rev. **126**, 1493 (1962).
(189) G. J. Ritter, Phys. Rev. **126**, 240 (1962).
(190) I. J. Spalding, et al., Proc. Phys. Soc. (London) **79**, 787 (1962).
(191) F. R. Petersen, et al., Phys. Rev. **126**, 252 (1962).
(192) D. R. Speck, Bull. Am. Phys. Soc. **1**, 282 (1956).
(193) D. R. Speck, et al., Phys. Rev. **101**, 1831 (1956).
(194) L. H. Bennett, et al., Phys. Rev. **120**, 1812 (1960).
(195) K. Murakawa, et al., Phys. Rev. **105**, 671 (1957).
(196) M. P. Klein, et al., Bull. Am. Phys. Soc. **6**, 104 (1961).
(197) L. Armstrong, et al., Phys. Rev. **138**, B310 (1965).
(198) J. Kaufmann, et al., Phys. Letters **24A**, 115 (1967).
(199) H. R. Loeliger, et al., Phys. Rev. **95**, 291 (1954).

NUCLEAR SPINS, MOMENTS, AND MAGNETIC RESONANCE FREQUENCIES (Continued)

REFERENCES (Continued)

(200) A. Narath, (to be published in Phys. Rev.).
(201) A. Narath, et al., Bull. Am. Phys. Soc. **12**, 314 (1967).
(202) W. von Siemens, Ann. Physik **13**, 136 (1953).
(203) Y. W. Chan, et al., Phys. Rev. **144**, 1020 (1966).
(204) Y. W. Chan, et al., Phys. Rev. **137**, B1129 (1965).
(205) H. Dahmen, et al., Z. Physik **200**, 456 (1967).
(206) W. J. Childs, et al., Phys. Rev. **141**, 176 (1966).
(207) P. A. Vanden Bout, et al., Phys. Rev. **158**, 1078 (1967).
(208) H. Kleiman, et al., Phys. Letters **13**, 212 (1964).
(209) W. J. Tomlinson, et al., Nucl. Phys. **60**, 614 (1964).
(210) H. Kleiman, et al., J. Opt. Soc. Am. **53**, 822 (1963).
(211) W. J. Tomlinson, et al., J. Opt. Soc. Am. **53**, 828 (1963).
(212) F. Bitter, et al., Phys. Rev. **96**, 1531 (1954).
(213) B. Cagnac, et al., Compt. Rend. **249**, 77 (1959).
(214) B. Cagnac, Ann. Phys. **6**, 467 (1961).
(215) J. C. Lehmann, et al., Compt. Rend. **257**, 3152 (1963).
(216) J. Blaise, et al., J. Phys. Radium **18**, 193 (1957).
(217) O. Redi, et al., Phys. Letters **8**, 257 (1964).
(218) S. P. Davis, et al., J. Opt. Soc. Am. **56**, 1604 (1966).
(219) R. J. Hull, et al., Phys. Rev. **122**, 1574 (1961).
(220) H. L. Poss, Phys. Rev. **75**, 600 (1949).
(221) G. O. Brink, et al., Phys. Rev. **107**, 189 (1957).
(222) E. B. Baker, J. Chem. Phys. **26**, 960 (1957).
(223) I. Lindgren, et al., Arkiv Fysik **15**, 445 (1959).
(224) Y. Ting, et al., Phys. Rev. **89**, 595 (1953).
(225) S. S. Alpert, et al., Phys. Rev. **125**, 256 (1962).
(226) C. M. Olsmats, et al., Arkiv Fysik **19**, 469 (1961).
(227) M. Fred, et al., Phys. Rev. **98**, 1514 (1955).
(228) V. N. Egorov, Opt. Spectry. (USSR) **16**, 301 (1964).
(229) J. D. Axe, et al., Phys. Rev. **121**, 1630 (1961).
(230) R. Marrus, et al., Nucl. Phys. **23**, 90 (1961).
(231) P. B. Dorain, et al., Phys. Rev. **105**, 1307 (1957).
(232) B. Bleaney, et al., Phil. Mag. **45**, 992 (1954).
(233) J. Faust, et al., Phys. Letters **16**, 71 (1965).
(234) R. J. Champeau, et al., Compt. Rend. **257**, 1238 (1963).
(235) L. Armstrong, et al., Phys. Rev. **144**, 994 (1966).
(236) T. E. Manning, et al., Phys. Rev. **102**, 1108 (1956).
(237) R. Marrus, et al., Phys. Rev. **124**, 1904 (1961).
(238) M. Fred, et al., J. Opt. Soc. Am. **47**, 1076 (1957).
(239) B. Perry, et al., Bull. Am. Phys. Soc. **8**, 345 (1963).
(240) M. M. Robertson, et al., Phys. Rev. **140**, B820 (1965).
(241) R. E. Walstedt, et al., Phys. Rev. **162**, 301 (1967).
(242) R. Freeman, et al., Proc. Roy Soc. (London) **242A**, 455 (1957).

IONIZATION POTENTIALS OF MOLECULES

From data published up to July, 1966
Condensed by J. L. Franklin and Pat Haug from a compilation entitled
"Ionization Potentials, Appearance Potentials and Heats of Formation of Positive Ions"
by
J. L. Franklin, J. G. Dillard, H. M. Rosenstock, J. T. Herron, K. Draxl and F. H. Field
Published by National Standard Reference Data System

The following symbols are employed for the principal important methods:

CTS = Charge Transfer Spectra, EI = Electron Impact, PE = Photoelectron Spectroscopy, PI = Photoionization, S = Optical Spectroscopy.

Molecule	Ionization Potential ev	Method	ΔH_f of Ion Kcal/mole	Reference
H_2	15.427±0.002	S	356	1
D_2	15.46±0.01	PI	356	2
BH	9.77±0.05	S	333	3
BH_3	11.4±0.2	EI	279	*
B_2H_6	12.0	EI	286	4, 5
B_5H_9	10.5	EI	262	*
B_6H_{10}	9.3±0.1	EI	237	6
C_2	12.0±0.6	EI	475	7
C_3	12.6	EI	480	7
CH	11.13±0.22	S	399	8
CH_2	10.396±0.003	EI, S, PI	333	9, 10, 11
CH_3	9.83	S, PI	259	12, 13
CD_3	9.832±0.002	S	259	12
CH_4	12.6	PI	274	*
CD_4	12.888	PI	280	11, 14
C_2H_2	11.4	PI, PE	317	*
C_2D_2	11.416±0.006	PI	317	15
C_2H_3	9.4	EI	269	*
C_2H_4	10.5	S, PE	253	16, 17
C_2H_5	8.4	PI	219	13
C_2H_6	11.5	PI, PE	245	14, 17
$HC≡C—CH_2$	8.25	PI, EI		18
$CH_2=C=CH_2$	10.16±0.02	EI	280	19
$CH_3C≡CH$	10.36±0.01	PI	283	20
cyclo-C_3H_4	9.95	EI	296	21
C_3H_5 (allyl)	8.15	EI	216	22, 8
cyclo-C_3H_5	8.05	EI	239	23
C_3H_6	9.73	S, PI	229	*
cyclo-C_3H_6	10.09±0.02	PI	245	24
n-C_3H_7	8.1	PI	209	13
iso-C_3H_7	7.5	PI	190	13
C_3H_8	11.1	PI, PE	231	*
C_4H_2	10.2±0.1	EI		25
C_4H_4	9.87	EI	294	25, 26
$CH_3CH=C=CH_2$	9.57±0.02	EI	259	19
$CH_2=CHCH=CH_2$	9.07	PI, PE	236	*
$C_2H_5C≡CH$	10.18±0.01	PI	274	20
$CH_3C≡CCH_3$	9.9±0.1	EI	263	25
cyclo-C_4H_7	7.88±0.05	EI	213	23
$CH_3CH=CHCH_2$	7.71±0.05	EI	203	27, 8
$CH_2=C(CH_3)CH_2$	8.03±0.05	EI	203	27, 8
1-C_4H_8	9.6	PI	221	*
cis-2-C_4H_8	9.13	PI	209	24, 28
trans-2-C_4H_8	9.13	PI	208	24, 28
iso-C_4H_8	9.23±0.02	PI	209	24, 28
cyclo-C_4H_8	10.58	EI	250	23
n-C_4H_9	8.64±0.05	EI	218	29
sec-C_4H_9	7.93±0.05	EI	192	29
iso-C_4H_9	8.35±0.05	EI	205	29
tert-C_4H_9	7.42±0.07	EI	176	29
n-C_4H_{10}	10.63±0.03	PI	215	30
iso-C_4H_{10}	10.57	PI	212	24
cyclo-C_5H_6	8.97	CTS	239	31, 32
$C_2H_5CH=C=CH_2$	9.42	EI	252	21
$CH_3CH=CHCH=CH_2$	8.68	EI	219	21
$CH_3CH=C=CHCH_3$	9.26	EI	247	21
$CH_2=CHCH_2CH=CH_2$	9.58	EI	246	21
$CH_2=CHC(CH_3)=CH_2$	8.845±0.005	PI	235	24
cyclo-C_5H_8	9.01±0.01	PI	216	24
cyclo-C_5H_9	7.79±0.02	EI	194	23
1-C_5H_{10}	9.50±0.02	PI	214	24
cis-2-C_5H_{10}	9.11	EI	203	21
trans-2-C_5H_{10}	9.06	EI	201	21
$(CH_3)_2CHCH=CH_2$	9.51±0.03	PI	212	24
$C_2H_5C(CH_3)=CH_2$	9.12±0.02	PI	202	24
$(CH_3)_2C=CHCH_3$	8.67±0.02	PI	189	24
cyclo-$C_3H_4(CH_3)_2$	9.77±0.02	EI	225	33
cyclo-C_5H_{10}	10.53±0.05	PI	224	24
tert-C_5H_{11}	7.12±0.1	EI	164	34, 35
neo-C_5H_{11}	8.33±0.1	EI	196	34
n-C_5H_{12}	10.35	PI	204	24
iso-C_5H_{12}	10.32	PI	201	24
neo-C_5H_{12}	10.35	PI	199	24
C_6H_4 (benzyne)	9.6	EI		*
cyclo-C_6H_5	9.2	PI, PE	284	36, 37
cyclo-C_6H_6	9.24	S, PI	233	*
$CH≡CCH=CHCH=CH_2$	9.50	EI	307	38
$C_2H_5C≡CC≡CH$	9.25	EI	307	38
$CH_3C≡CCH_2C≡CH$	9.75	EI	319	38
$CH_3C≡CC≡CCH_3$	9.20	EI	301	38
$CH≡CCH_2CH_2C≡CH$	10.35	EI	338	38
C_6H_8 (1-methylcyclopentadiene)	8.43±0.05	EI	218	31
C_6H_8 (2-methylcyclopentadiene)	8.46±0.05	EI	219	31
cyclo-C_6H_{10}	8.72	PE	199	17
cyclo-C_6H_{11}	7.7	EI	185	23
1-C_6H_{12}	9.45±0.02	PI	208	35, 24
$(CH_3)_2C=C(CH_3)_2$	8.30	PI	175	28
cyclo-C_6H_{12}	9.8	PI, PE	197	17, 24
n-C_6H_{14}	10.18	PI	195	24
iso-C_6H_{14}	10.12	PI	192	24
$(C_2H_5)_2CHCH_3$	10.08	PI	191	24
$C_2H_5C(CH_3)_3$	10.06	PI	188	24
$(CH_3)_2CHCH(CH_3)_2$	10.02	PI	189	24
cyclo-$C_6H_5CH_2$	7.76±0.08	EI	216	39
cyclo-C_6H_7	6.240±0.01	S	209	40
cyclo-$C_6H_5CH_3$	8.82±0.01	PI	215	*
cyclo-C_7H_8	8.5	EI	240	32, 41, 42
bicyclo-(2.2.1)C_7H_8	8.67	EI	267	42
bicyclo-(3.2.0)C_7H_8	9.37	EI	246	41
C_7H_{10} (1,2-dimethylcyclopentadiene)	8.1±0.1	EI	204	31
C_7H_{10} (5,5-dimethylcyclopentadiene)	8.22±0.05	EI	206	31
C_7H_{10} (1,3-cycloheptadiene)	8.55	EI	219	41
C_7H_{10} (norbornene)	8.95±0.15	EI	237	43
C_7H_{12} (4-methylcyclohexane)	8.91±0.01	PI	198	24
cyclo-$C_6H_{11}CH_3$	9.85±0.03	PI	190	24
n-C_7H_{16}	9.90±0.05	PI	183	37
cyclo-$C_6H_5C≡CH$	8.815±0.005	PI	279	24
C_8H_8 (styrene)	8.47±0.02	PI	232	24
C_8H_8 (cyclotatetraene)	8.0	PI, PE	255	17, 24
C_8H_8 (cubane)	8.74±0.15	EI	350	44
m-$C_6H_4CH_3CH_2$	7.65±0.03	EI	206	39
p-$C_6H_4CH_3CH_2$	7.46±0.03	EI	202	39
cyclo-$C_6H_5C_2H_5$	8.76±0.01	PI	209	24
o-$C_6H_4(CH_3)_2$	8.56	PI	202	*
m-$C_6H_4(CH_3)_2$	8.58	PI, PE	202	*
p-$C_6H_4(CH_3)_2$	8.44	PI	199	*
C_8H_{10} (7-methylcycloheptatriene)	8.39±0.1	EI	231	45
C_8H_{10} (1-methylspiroheptadiene)	8.02±0.1	EI	229	45
C_8H_{10} (2-methylspiroheptadiene)	8.07±0.1	EI	230	45
C_8H_{10} (6-methylspiroheptadiene)	8.40±0.1	EI	239	45
C_8H_{12} (1,2,3-trimethylcyclopentadiene)	7.96±0.05	EI	194	31
C_8H_{12} (1,5,5-trimethylcyclopentadiene)	8.00±0.1	EI	193	31

* Average of several values.

Molecule	Ionization Potential ev	Method	ΔH_f of Ion Kcal/mole	Reference	Molecule	Ionization Potential ev	Method	ΔH_f of Ion Kcal/mole	Reference
C_8H_{12} (4-vinylcyclo-hexene)	8.93 ± 0.02	PI	224	24	C_2N_2	13.6	EI	387	8, 64
cis-1,2-cyclo-$C_6H_{10}(CH_3)_2$	10.08 ± 0.02	EI	191	46	CH_5N	8.97	PI	201	67
trans-1,2-cyclo-$C_6H_{10}(CH_3)_2$	10.08 ± 0.03	EI	189	46	C_2H_2N	10.9	EI	298	68
C_8H_{18} (2,2,4-trimethyl-pentane)	9.86	PI	174	24	CH_3CN	12.2	PI	302	14, 24
C_8H_{18} (2,2,3,3-tetra-methylbutane)	9.79	EI	184	8	cyclo-C_2H_5N	9.94 ± 0.15	EI	255	69
C_9H_8 (indene)	8.81	EI	246	47	$C_2H_5NH_2$	8.86 ± 0.02	PI	193	24
$C_6H_5C(CH_3)=CH_2$	8.35 ± 0.01	PI	220	24	$(CH_3)_2NH$	8.24 ± 0.02	PI	186	24
cyclo-$C_6H_5(n-C_3H_7)$	8.72 ± 0.01	PI	203	24	$CH_2=CHCN$	10.91 ± 0.01	PI	296	24
cyclo-$C_6H_5(iso-C_3H_7)$	8.69 ± 0.01	PI	201	24	C_2H_5CN	11.84 ± 0.02	PI	289	24
1,2,3 cyclo-$C_6H_3(CH_3)_3$	8.48	PI	193	28	C_3H_7N	9.1 ± 0.15	EI	225	70
1,2,4 cyclo-$C_6H_3(CH_3)_3$	8.27	PI	187	28	$(CH_2)_3NH$	9.1 ± 0.15	EI	225	70
1,3,5 cyclo-$C_6H_3(CH_3)_3$	8.4	PI	190	*	$n-C_3H_7NH_2$	8.78 ± 0.02	PI	185	24
$C_{10}H_8$ (naphthalene)	8.12	PI	220	24, 48	$iso-C_3H_7NH_2$	8.72 ± 0.03	PI	183	24
$C_{10}H_8$ (azulene)	7.42	S	243	49, 50	$(CH_3)_3N$	7.82 ± 0.02	PI	175	24
$C_6H_4C_3H_7CH_2$	7.42		188	39	$CH_2=CHCH_2CN$	10.39 ± 0.01	PI	281	24
cyclo-C_6H_5 (n-C_4H_9)	8.69 ± 0.01	PI	197	24	C_4H_5N (pyrrole)	8.20 ± 0.01	PI	215	24
$C_{10}H_{14}$ (sec-butylbenzene)	8.68 ± 0.01	PI	196	24	$(CH_3)_2CCN$	9.15 ± 0.1	EI	239	68
$C_{10}H_{14}$ (tert-butylbenzene)	8.68 ± 0.01	PI	193	24	$n-C_3H_7CN$	11.67 ± 0.05	PI	280	24
$C_{10}H_{14}$ (1,2,3,5-tetramethylbenzene)	8.47 ± 0.05	EI	185	24	C_4H_9N (pyrrolidine)	8.41	PE	192	17
$C_{10}H_{14}$ (1,2,4,5-tetramethylbenzene)	8.03	PI, PE	174	*	$n-C_4H_9NH_2$	8.71 ± 0.03	PI	179	24
$C_{10}H_{18}$ (cis-decaline)	9.61 ± 0.02	EI	181	51	$sec-C_4H_9NH_2$	8.70	PI	177	24
$C_{10}H_{18}$ (trans-decaline)	9.61 ± 0.02	EI	178	51	$iso-C_4H_9NH_2$	8.70	PI	177	24
$C_{11}H_9$ (1-naphthyl methyl)	7.35	EI	208	52	$tert-C_4H_9NH_2$	8.64	PI	173	24
$C_{11}H_9$ (2-naphthyl methyl)	7.56 ± 0.05	EI	217	52	$(C_2H_5)_2NH$	8.01 ± 0.01	PI	163	24
$C_{11}H_{11}$ (methyl naphthalene)	7.96 ± 0.01	PI	209	24	C_5H_5N (pyridine)	9.3	S, PI	247	24, 71
$C_{11}H_{10}$ (2-methyl-naphthalene)	7.955 ± 0.01	PI	209	24	C_6H_7N (aniline)	7.7	PI	202	*
$C_{11}H(CH_3)_4$	7.92 ± 0.01	EI	165	53	C_6H_7N (2-picoline)	9.02 ± 0.03	PI	232	24
$C_{11}H_{18}$ (hexamethyl-cyclopentadiene)	7.74 ± 0.05	EI	165	31	C_6H_7N (3-picoline)	9.04 ± 0.03	PI	236	24
$C_{12}H_{10}$ (biphenyl)	8.27 ± 0.01	PI	230	24	C_6H_7N (4-picoline)	9.04 ± 0.03	PI	233	24
cyclo-$C_6(CH_3)_6$	7.85 ± 0.02	PI	152	28	$C_6H_{13}N$ (cyclo-hexylamine)	8.86	PE	181	17
$C_{13}H_{10}$ (fluorene)	8.63	EI	243	47	$(n-C_3H_7)_2NH$	7.84 ± 0.02	PI	153	24
$C_{14}H_{10}$ (diphenyl-acetylene)	8.85 ± 0.05	EI	303	54	$(iso-C_3H_7)_2NH$	7.73 ± 0.03	PI	148	24
$C_{14}H_{10}$ (anthracene)	7.55	EI	228	55	$(C_2H_5)_3N$	7.50 ± 0.02	PI	147	24
$C_{14}H_{10}$ (phenanthrene)	8.1	EI	233	54, 55	cyclo-C_6H_5CN	9.705 ± 0.01	PI	277	24
$C_{18}H_{12}$ (1,2-benzanthracene)	8.01	EI	251	56	C_7H_9N (n-methyl-aniline)	7.32	PI	192	72, 73
$C_{18}H_{30}$ (1-phenyl-dodecane)	9.05 ± 0.1	EI	165	57	C_7H_9N (m-toluidine)	7.50 ± 0.02	PI	189	48
$C_{18}H_{30}$ (3-phenyl-dodecane)	8.95 ± 0.1	EI	163	57	C_7H_9N (2,3-lutidine)	8.86 ± 0.02	PI	218	24
$C_{19}H_{32}$ (7-phenyl-tridecane)	8.91 ± 0.1	EI	157	57	C_7H_9N (2,4-lutidine)	8.85 ± 0.03	PI	218	24
$C_{26}H_{46}$ (1-phenyl-licosane)	9.34 ± 0.1	EI	132	57	C_7H_9N (2,6-lutidine)	8.85 ± 0.02	PI	218	24
$C_{26}H_{46}$ (2-phenylicosane)	9.22 ± 0.1	EI	129	57	$C_6H_5CH_2CN$	9.40 ± 0.5	EI	259	74
$C_{26}H_{46}$ (3-phenylicosane)	8.95 ± 0.1	EI	123	57	$m-C_6H_4CH_3CN$	9.66 ± 0.05	EI	271	74
$C_{26}H_{46}$ (4-phenylicosane)	9.01 ± 0.1	EI	125	57	$p-C_6H_4CH_3CN$	9.76	EI	273	75
$C_{26}H_{46}$ (5-phenylicosane)	9.04 ± 0.1	EI	125	57	cyclo-$C_6H_5NHC_2H_5$	7.56	CTS	193	76
$C_{26}H_{46}$ (7-phenylicosane)	8.97 ± 0.1	EI	124	57	cyclo-$C_6H_5N(CH_3)_2$	7.12	PI	185	72, 73
$C_{26}H_{46}$ (9-phenylicosane)	9.06 ± 0.1	EI	126	57	$(n-C_4H_9)_2NH$	7.69 ± 0.03	PI	140	24
$(CH_3)_3B$	8.8 ± 0.2	EI	173	58	$C_9H_{13}N$ (N-n-propyl-aniline)	7.54	CTS	188	76
$(C_2H_5)_3B$	9.0 ± 0.2	EI	170	58	$C_9H_{13}N$ (N-ethyl-N-methylaniline)	7.37	CTS	185	76
N_2	15.576	S	359	59	$C_9H_{13}N$ (N,N-dimethyl-o-toluidine)	7.37	CTS	184	76
NH	13.10 ± 0.05	EI	382	60	$C_9H_{13}N$ (N,N-dimethyl-m-toluidine)	7.35	CTS	181	76
NH_2	11.4 ± 0.1	EI	304	61	$C_9H_{13}N$ (N,N-dimethyl-p-toluidine)	7.33	CTS	181	76
NH_3	10.2	S, PI, PE	223	*	$(n-C_3H_7)_3N$	7.23	PI	207	24
N_2H_2	9.85 ± 0.1	EI		62	$C_{10}H_{15}N$ (N-n-butyl-aniline)	7.53	CTS	183	76
N_2H_4	8.74 ± 0.06	PI	224	63	$C_{10}H_{15}N$ (N,N-diethyl-aniline)	6.99	CTS	172	76
CN	14.3	EI	430	64, 65	$C_{10}H_{15}N$ (N,N-dimethyl-p-ethylaniline)	7.38	CTS	177	76
HCN	13.8	EI	351	26, 66	$C_{10}H_{15}N$ (N,N-2,4-tetra-methylaniline)	7.17	CTS	171	76
					$C_{10}H_{15}N$ (N,N,2,6-tetra-methylaniline)	7.22	CTS	173	76
					$C_{10}H_{15}N$ (N,N,3,5-tetra-methylaniline)	7.25	CTS	172	76
					$C_{11}H_{17}N$ (N,N-diethyl-p-toluidene)	6.93	CTS	164	76
					$C_{11}H_{17}N$ (N,N-dimethyl-p-isopropylaniline)	7.41	CTS	174	76
					$(C_6H_5)_2NH$	7.25 ± 0.03	PI	223	77
					$C_{12}H_{19}N$ (N,N-di-n-propylaniline)	6.96	CTS	163	76
					$C_{12}H_{19}N$ (N,N-dimethyl-p-tert-butylaniline)	7.43	CTS	165	76
					$C_{14}H_{23}N$ (N,N-di-n-butylaniline)	6.95	CTS	153	76

* Average of several values.

IONIZATION POTENTIALS OF MOLECULES (Continued)

Molecule	Ionization Potential ev	Method	ΔH_f of Ion Kcal/mole	Reference	Molecule	Ionization Potential ev	Method	ΔH_f of Ion Kcal/mole	Reference
$(C_6H_5)_3N$	6.86 ±0.03	PI	243	77	C_7H_6O (tropone)	9.68 ±0.02	EI	240	104
CH_2N_2 (diazirine)	10.18 ±0.05	EI	314	78	cyclo-$C_6H_5CH_2OH$	9.14 ±0.05	EI	186	74
CH_2N_2 (diazomethane)	8.999 ±0.001	S	257	79	cyclo-$C_6H_5OCH_3$	8.21 ±0.02	PI	173	24
CH_6N_2 (methyl-hydrazine)	8.00 ±0.06	PI	207	63	C_7H_8O (m-cresol)	8.52 ±0.05	EI	165	74
$CH_3N=NCH_3$	8.65 ±0.1	EI	243	80	$n-C_5H_{11}COCH_3$	9.33 ±0.03	PI	143	24
$C_2H_8N_2$ (1,1-dimethyl-hydrazine)	7.67 ±0.05	PI	197	81	C_8H_8O (acetophenone)	9.27 ±0.03	PI	191	24
$C_2H_8N_2$ (1,2-dimethyl-hydrazine)	7.75 ±0.1	EI	200	82	C_8H_8O (p-methyl-benzaldehyde)	9.33 ±0.05	EI	194	105
$(CH_3)_3N_2H$	7.93 ±0.1	EI	202	82	$C_8H_{10}O$ (benzyl methyl ether)	8.85 ±0.03	PI	184	30
$o-C_4H_4N_2$ (o-diazine)	9.9	EI	275	83	$C_8H_{10}O$ (phenyl ethyl ether)	8.13 ±0.02	PI	167	24
$m-C_4H_4N_2$ (m-diazine)	9.9	EI	277	83	$C_8H_{10}O$ (m-methylanisole)	8.31 ±0.05	EI	169	74
$p-C_4H_4N_2$ (p-diazine)	9.8	EI	274	83	$C_9H_{10}O$ (phenyl ethyl ketone)	9.27 ±0.05	EI	189	74
$1,1(C_2H_5)_2N_2H_2$	7.59 ±0.05	PI	184	63	$C_9H_{10}O$ (m-methyl-acetophenone)	9.15 ±0.05	EI	182	74
$(CH_3)_4N_2$	7.76 ±0.05	EI	196	82	$C_{12}H_{10}O$ (phenyl ether)	8.82 ±0.05	EI	220	106
$C_5H_4NNH_2$	8.97 ±0.05	EI	244	84	$C_{13}H_{10}O$ (benzophenone)	9.4	EI	229	106, 105
$C_5H_{14}N_2$ (1-methyl-1-n-butylhydrazine)	7.62 ±0.05	PI	180	63	$C_{14}H_{12}O$ (p-methyl-benzophenone)	9.13 ±0.05	EI	214	105
$(CH_3)_2NC_6H_4N(CH_3)_2$ [p-bis(dimethylamino)-benzene]	6.9	CTS	180	85	HCOOH	11.05 ±0.01	PI	164	24
CH_3N_3	9.5 ±0.1	EI	276	86	CH_3COOH	10.69±0.03	PI	135	24
O_2	12.063 ±0.001	PI	278	*	$HCOOCH_3$	10.815 ±0.005	PE	166	165
O_3	12.3 ±0.1	PE	318	87	C_2H_5COOH	10.24 ±0.03	PI	127	24
OH	13.17 ±0.1	EI	313	88	$HCOOC_2H_5$	10.61 ±0.01	PI	156	24
H_2O	12.6	PI	233	*	CH_3COOCH_3	10.27 ±0.02	PI	138	24
D_2O	12.6	PI	232	37	$(CH_3O)_2CH_2$	10.00 ±0.05	PI	145	24
HO_2	11.53 ±0.02	EI	271	89	$CH_3COOCH=CH_2$	9.19 ±0.05	PI	137	24
H_2O_2	11.0	EI	233	90, 91	$CH_3COCOCH_3$	9.24 ±0.03	PI	135	24
Li_2O	6.8	EI	120	92, 93	$n-C_3H_7COOH$	10.16 ±0.05	PI	121	24
CO	14.013 ±0.004	S	297	94	$iso-C_3H_7COOH$	10.02 ±0.05	PI	115	24
CO_2	13.769 ±0.03	S	223	96	$HCOOCH_2CH_2CH_3$	10.54 ±0.01	PI	149	24
NO	9.25	PI, S	235	*	$CH_3COOC_2H_5$	10.11 ±0.02	PI	126	24
N_2O	12.894	S	317	95	$C_2H_5COOCH_3$	10.15 ±0.03	PI	127	24
NO_2	9.79	PI	233	97, 98	$C_4H_8O_2$ (p-dioxane)	9.13 ±0.03	PI	126	24
CHO	9.8	EI	221	*	$(CH_3O)_2CHCH_3$	9.65 ±0.03	PI	129	24
CH_2O	10.88	PI	223	24, 48	$C_4H_4O_2$ (2-furfur-aldehyde)	9.21 ±0.01	PI	187	24
CH_3OH	10.84	PI, PE	202	17, 24	$CH_3COCH_2COCH_3$	8.87 ±0.03	PI	122	24
CH_3CO	10.3	PI	152	*	$HCOO(CH_2)_3CH_3$	10.50 ±0.02	PI	144	24
CH_3CHO	10.2	PI	196	*	$HCOOCH_2CH(CH_3)_2$	10.46 ±0.02	PI	139	24
C_2H_4O (ethylene oxide)	10.6	PI, S	231	*	$CH_3COOCH_2CH_2CH_3$	10.04 ±0.03	PI	121	24
C_2H_5OH	10.49	PI	185	24, 30	$CH_3COOCH(CH_3)_2$	9.99 ±0.03	PI	116	24
CH_3OCH_3	9.98	S, PI	186	24, 99	$C_2H_5COOC_2H_5$	10.00 ±0.02	PI	119	24
$CH_2=CHCHO$	10.10 ±0.01	PI	210	24	$n-C_3H_7COOCH_3$	10.07 ±0.03	PI	125	24
C_2H_5CHO	9.98	PI	181	24	$iso-C_3H_7COOCH_3$	9.98 ±0.02	PI	121	24
CH_3COCH_3	9.69	PI	171	*	$(C_2H_5O)_2CH_2$	9.70 ±0.05	PI	134	24
$CH_2=CHCH_2OH$	9.67 ±0.05	PI	191	24	$p-C_6H_4O_2$	9.67 ±0.02	PI	198	48
$CH_2=CHOCH_3$	8.93 ±0.02	PI	178	24	$CH_3COOC_4H_9$	9.56 ±0.03	PI	104	48
C_3H_6O (propyleneoxide)	10.22 ±0.02	PI	214	24	$CH_3COO(CH_2)_3CH_3$	10.01	PI	114	24
C_3H_6O (trimethylene-oxide)	9.667 ±0.005	S	199	100	$CH_3COOCH_2-CH(CH_3)_2$	9.97	PI	111	24
$n-C_3H_7OH$	10.1	PI	172	24, 101	$CH_3COOCH-(CH_3)C_2H_5$	9.91 ±0.03	PI	110	24
$iso-C_3H_7OH$	10.15	PI	169	24	C_6H_5COOH	9.73 ±0.09	EI	152	105
C_4H_4O (furan)	8.89	S, PI	197	24, 102	$p-HOC_6H_4CHO$	9.32 ±0.02	EI	157	105
$CH_3CH=CHCHO$	9.73 ±0.01	PI	194	24	$C_8H_8O_2$ (α-hydroxy-acetophenone)	9.33 ±0.05	EI	159	74
$n-C_3H_7CHO$	9.86 ±0.02	PI	174	24	$C_6H_5COOCH_3$	9.35 ±0.06	EI	144	105
$iso-C_3H_7CHO$	9.74 ±0.03	PI	169	24	$C_8H_8O_2$ (p-methoxy-benzaldehyde)	8.60 ±0.03	EI	150	105
$C_2H_5COCH_3$	9.5	PI	161	*	$C_8H_8O_2$ (m-hydroxy-acetophenone)	8.67 ±0.05	EI	134	74
cyclo-C_4H_8O	9.42	S	174	100	$C_8H_8O_2$ (p-hydroxy-acetophenone)	8.70 ±0.03	EI	135	105
$n-C_4H_9OH$	10.04	PI	165	24	$C_9H_{10}O_2$ (α-methoxy-acetophenone)	8.60 ±0.05	EI	142	74
$C_2H_5OC_2H_5$	9.6	PI	161	24	$C_9H_{10}O_2$ (m-methoxy-acetophenone)	8.53 ±0.05	EI	140	74
C_5H_8O (cyclo-pentanone)	9.26 ±0.01	PI	163	24	$C_9H_{10}O_2$ (p-methoxy-acetophenone)	8.62 ±0.05	EI	142	105
C_5H_8O (dihydropyran)	8.34 ±0.01	PI	164	24	$C_9H_{10}O_2$ (methyl p-toluate)	8.94 ±0.04	EI	130	105
$n-C_4H_9CHO$	9.82 ±0.05	PI	168	24	$C_{13}H_{10}O_2$ (p-hydroxy-benzophenone)	8.59 ±0.05	EI	165	105
$iso-C_4H_9CHO$	9.71 ±0.05	PI	164	24	$C_{13}H_{10}O_2$ (phenyl benzoate)	8.98 ±0.05	EI	177	106
$n-C_3H_7COCH_3$	9.37 ±0.02	PI	154	103	$C_{14}H_{10}O_2$ (benzil)	8.78 ±0.05	EI	181	106
$iso-C_3H_7COCH_3$	9.30 ±0.02	PI	151	24, 103					
$(C_2H_5)_2CO$	9.32 ±0.01	PI	153	24					
cyclo-$C_5H_{10}O$	9.25 ±0.01	S	161	100					
C_6H_5O	8.84	EI	226	36					
C_6H_6O (phenol)	8.51	PI	173	24, 48					
$(CH_3)_2C=CHCOCH_3$	9.08 ±0.03	PI	168	24					
$C_6H_{10}O$ (cyclo-hexanone)	9.14 ±0.01	PI	152	24					
$n-C_4H_9COCH_3$	9.35	PI	149	24, 103					
$iso-C_4H_9COCH_3$	9.30	PI	147	24, 103					
$tert-C_4H_9COCH_3$	9.17 ±0.03	PI	140	24					
$(n-C_3H_7)_2O$	9.27 ±0.05	PI	147	24					
$(iso-C_3H_7)_2O$	9.20 ±0.05	PI	142	24					
C_6H_5CHO	9.52	PI	209	24					

* Average of several values.

Molecule	Ionization Potential ev	Method	ΔH_f of Ion Kcal/mole	Reference	Molecule	Ionization Potential ev	Method	ΔH_f of Ion Kcal/mole	Reference
$CH_3OCH_2COOCH_3$	9.56 ±0.05	EI	88	74	$C_7H_5F_3$	9.68	S, PI	84	24, 125
$C_9H_{10}O_3$ (methyl p-methoxybenzoate)	8.43 ±0.04	EI	90	105	$C_7H_{11}F_3$ (trifluoro-methylcyclohexane)	10.46 ±0.02	PI	37	24
$(C_6H_5)_2CO_3$	9.01 ±0.05	EI	122	106	C_6H_5OF (o-fluoro-phenol)	8.66 ±0.01	PI	132	24
CH_3CONH_2	9.77 ±0.02	PI	171	24	Na_2	4.90 ±0.01	PI	147	126
$HCON(CH_3)_2$	9.12 ±0.02	PI	160	24	AlF	9.8	EI	166	127, 128
$CH_3CONHCH_3$	8.90 ±0.02	PI	150	24	$(CH_3)_4Si$	9.9	EI	171	129, 130
$CH_3CON(CH_3)_2$	8.81 ±0.03	PI	145	24	PH_3	9.98	PI	231	131
C_5H_4NOH	9.70 ±0.05	EI	209	84	PF_3	9.71	PI	4	131
$HCON(C_2H_5)_2$	8.89 ±0.02	PI	145	24	CH_3PH_2	9.72 ±0.15	EI	217	132
C_6H_5NO (2-pyridine-carboxaldehyde)	9.75 ±0.05	EI	227	84	$C_2H_5PH_2$	9.47 ±0.5	EI	206	132
C_6H_5NO (4-pyridine-carboxaldehyde)	10.12 ±0.05	EI	235	84	$(CH_3)_3P$	8.60 ±0.2	FI	175	132
$CH_3CON(C_2H_5)_2$	8.60 ±0.02	PI	130	24	$(C_6H_5)_3P$	7.36 ±0.05	PI	242	77
C_6H_5NCO	8.77 ±0.02	PI	222	24	S_6	9.7	EI	248	133
C_7H_7NO (benzamide)	9.4 ±0.2	EI	197	107, 108	S_7	9.2 ±0.3	EI		133
C_7H_7NO (p-amino-benzaldehyde)	8.25 ±0.02	EI	182	105	HS	10.50 ±0.1	EI	276	134
C_7H_9NO (p-methoxy-analine)	7.82	EI	169	75	H_2S	10.4	S	235	135
C_8H_9NO (acetanilide)	8.39 ±0.10	EI	171	108	CS_2	10.080	S	261	95
C_8H_9NO (m-amino-acetophenone)	8.09 ±0.05	EI	171	74	SO_2	12.34 ±0.02	PI	214	30
C_8H_9NO (p-amino-acetophenone)	8.17 ±0.02	EI	172	105	CH_3S	8.06 ±0.1	EI	218	134
C_9H_7NO (α-cyano-acetophenone)	9.56 ±0.05	EI	235	74	CH_3SH	9.440 ±0.005	PI	212	24
CH_3NO_2	11.1	PI	238	14, 24	C_2H_4S (ethylene sulfide)	8.87 ±0.15	EI	224	69
$C_2H_5NO_2$	10.88 ±0.05	PI	226	24	C_2H_5SH	9.285 ±0.005	PI	203	24
$n\text{-}C_3H_7NO_2$	10.81 ±0.03	PI	221	24	CH_3SCH_3	8.685 ±0.005	PI	191	24
$iso\text{-}C_3H_7NO_2$	10.71 ±0.05	PI	217	24	C_3H_6S (propylene sulfide)	8.6 ±0.2	EI	218	136
$C_6H_5NO_2$	9.92	PI	246	24	$(CH_2)_3S$	8.9 ±0.15	EI	220	70
$C_7H_6NO_2$ (m-nitro-benzyl radical)	8.56 ±0.1	EI	227	39	$n\text{-}C_3H_7SH$	9.195 ±0.005	PI	197	24
$C_7H_7NO_2$ (m-nitro-toluene)	9.65 ±0.05	EI	233	74	$C_2H_5SCH_3$	8.55 ±0.01	PI	183	24
$C_7H_7NO_2$ (p-nitro-toluene)	9.82	EI	237	75	C_4H_4S (thiophene)	8.860 ±0.005	PI	229	24
$C_8H_6NO_2$	8.08 ±0.01	EI	122	90	$CH_3SCH_2-CH=CH_2$	8.70 ±0.2	EI	211	137
$C_6H_6N_2O_2$ (o-nitro-aniline)	8.66	EI	215	75	$(CH_2)_4S$	8.57 ±0.75	EI	190	70
$C_6H_6N_2O_2$ (m-nitro-aniline)	8.7	EI	216	75	$n\text{-}C_4H_9SH$	9.14 ±0.02	PI	191	24
$C_6H_6N_2O_2$ (p-nitro-aniline)	8.85	EI	219	75	$C_2H_5SC_2H_5$	8.430 ±0.005	PI	175	24
$C_2H_5NO_3$	11.22	PI	222	24	$n\text{-}C_3H_7SCH_3$	8.8 ±0.15	EI	183	138
$n\text{-}C_3H_7ONO_2$	11.07 ±0.02	PI	213	24	$iso\text{-}C_3H_7SCH_3$	8.7 ±0.2	EI	179	137
$C_6H_5NO_3$ (p-nitro-phenol)	9.52	EI	187	75	C_6H_5S	8.63 ±0.1	EI	250	134
$C_7H_5NO_3$ (p-nitro-benzaldehyde)	10.27 ±0.01	EI	217	105	C_6H_5SH	8.32 ±0.01	PI	211	138
$C_8H_7NO_3$ (m-nitro-acetophenone)	9.89 ±0.05	EI	201	74	C_6H_8S (2-ethyl-thiophene)	8.8 ±0.2	EI	215	139
$C_8H_7NO_3$ (p-nitro-acetophenone)	10.07 ±0.02	EI	205	105	$(n\text{-}C_3H_7)_2S$	8.5	PI, EI	170	24, 143
$C_8H_7NO_4$ (methyl-p-nitrobenzoate)	10.20 ±0.03	EI	160	105	$C_6H_5S_2CH_3$	8.9	EI	229	137
F_2	15.7	S	362	109	$C_7H_{10}S$ (2-propyl-thiophene)	8.6 ±0.2	EI	205	139
HF	15.77 ±0.02	EI	299	110	$C_8H_{12}S$	8.8		221	137
BF	11.3	EI	233	111, 112	$C_8H_{12}S$ (2-butyl-thiophene)	8.5 ±0.2	EI	198	139
BF_3	15.5	EI	87	*	CH_3SSCH_3	8.46 ±0.03	PI	189	24
CF_2	11.8	EI	237	113, 114	$C_2H_5SSC_2H_5$	8.27 ±0.03	PI	173	24
C_2F_4	10.12	PI	78	28	CH_3SSSCH_3	8.80 ±0.15	EI	203	140
C_6F_6	9.97	PI		28	COS	11.17 ±0.01	PI	224	138
NF_2	11.9	EI	284	115, 116	SO_2F_2	13.3 ±0.1	EI	102	141
$trans\text{-}N_2F_2$	13.1 ±0.1	EI	322	115	CH_3NCS	9.25 ±0.03	PI	245	24
NF_3	13.2 ±0.2	EI	275	115, 123	CH_3SCN	10.065 ±0.01	PI	270	24
N_2F_4	12.04 ±0.10	EI	276	124	C_2H_5NCS	9.14 ±0.03	PI	237	24
OF_2	13.6	EI	309	117, 118	C_2H_5SCN	9.89 ±0.01	PI	261	24
XeF_2	11.5 ±0.2	S	239	119	C_7H_5NS (phenyl-isothiocyanate)	8.520 ±0.005	PI	260	24
CH_2F	9.35	EI	207	29	$C_6H_5CH_2SCN$	9.06 ±0.05	EI	274	74
CH_3F	12.85 ±0.01	EI	229	120	NH_2CSNH_2	8.50 ±0.05	EI	194	142
C_2H_3F	10.37	PI	211	28, 121	$NH_2CSNHCH_3$	8.29 ±0.05	EI	188	142
$cyclo\text{-}C_6H_5F$	9.2	S, PI	186	*	$NH_2CSNHCH=CH_2$	8.29 ±0.05	EI	213	142
C_7H_7F	8.9	PI	172	24	$NH_2CSN(CH_3)_2$	8.34 ±0.05	EI	186	142
CHF_2	9.45	EI	143	29, 122	$CH_3NHCSNHCH_3$	8.17 ±0.05	EI	184	142
$C_2H_2F_2$	10.30	PI	159	28, 121	$CH_3NHCSN(CH_3)_2$	7.93 ±0.05	EI	176	142
$o\text{-}C_6H_4F_2$	9.31	PI	147	121	$C_5H_{12}N_2S$	7.98 ±0.05	EI	170	142
$p\text{-}C_6H_4F_2$	9.15	PI	140	121	$(CH_3)_2NCSN(CH_3)_2$	7.95 ±0.05	EI	173	142
C_2HF_3	10.14	PI	122	28	CH_3COSH	10.00 ±0.02	PI	179	24
$CH_2{=}CHCF_3$	10.9	PI	93	28	Cl_2	11.48 ±0.01	PI	265	24
					HCl	12.74	PI	272	*
					LiCl	10.1	EI	186	144
					CCl_3	8.78 ±0.05	EI	214	*
					CCl_4	11.47 ±0.01	PI	240	24
					C_2Cl_4	9.32 ±0.01	PI	212	24, 28
					PCl_3	9.91	EI	160	131
					CH_2Cl	9.32	EI	244	29
					CH_3Cl	11.3	S, PI	239	14, 145
					C_2H_3Cl	9.996	S, PI	239	*
					C_2H_5Cl	10.97	PI	226	24
					$CH_3C{\equiv}CCl$	9.9 ±0.1	EI	267	25
					$n\text{-}C_3H_7Cl$	10.82 ±0.03	PI	219	24

* Average of several values.

Molecule	Ionization Potential ev	Method	ΔHf of Ion Kcal/mole	Reference	Molecule	Ionization Potential ev	Method	ΔHf of Ion Kcal/mole	Reference
iso-C3H7Cl	10.78 ±0.02	PI	211	24	Fe(CO)5	7.95 ±0.03	PI	8	155
n-C4H9Cl	10.67 ±0.03	PI	210	24	C5H5Co(CO)2	8.3 ±0.2	EI	136	154
sec-C4H9Cl	10.65 ±0.03	PI	210	24	Ni(CO)4	8.28 ±0.03	PI	47	155
iso-C4H9Cl	10.66 ±0.03	PI	208	24	(CH3)4Ge	9.2 ±0.2	EI	177	156
tert-C4H9Cl	10.61 ±0.03	PI	202	24	As4	9.07 ±0.07	EI	244	157
C6H5Cl	9.07	PI	222	24	AsH3	10.03	PI	247	131
C6H5CH2Cl	9.19 ±0.05	EI	219	74	AsCl3	11.7 ±0.1	EI	208	158
o-C6H4ClCH3	8.83 ±0.02	PI	208	24	(CH3)3As	8.3 ±0.1	EI	202	158
m-C6H4ClCH3	8.83 ±0.02	PI	207	24	(C6H5)3As	7.34 ±0.07	PI	250	77
p-C6H4ClCH3	8.69 ±0.02	PI	204	24	Br2	10.54 ±0.03	PI	250	24, 159
C7H9Cl (endo-5-chloro-2-norbornene)	9.1 ±0.15	EI	233	43	HBr	11.62 ±0.03	PI	259	24
C7H9Cl (exo-5-chloro-2-norbornene)	9.15 ±0.15	EI	234	43	MgBr2	10.65 ±0.15	EI	172	160
C7H9Cl (3-chloro-nortricyclene)	9.51 ±0.15	EI	234	43	BrCl	11.1 ±0.2	EI	259	161
CHCl2	9.30	EI	245	29	CH3Br	10.53	S, PI	234	*
CH2Cl2	11.35 ±0.02	PI	240	24	C2H3Br	9.80	PI	243	*
cis-C2H2Cl2	9.65	PI	223	*	C2H5Br	10.29	S	222	24, 162
trans-C2H2Cl2	9.64	PI	224	*	CH3C≡CBr	10.1 ±0.1	EI	283	25
CH2ClCH2Cl	11.12 ±0.05	PI	225	24	CH3CH=CHBr	9.30 ±0.05	PI	224	24
CH2=CClCH2Cl	9.82 ±0.03	PI	218	24	n-C3H7Br	10.18 ±0.01	PI	216	24
1,2-C3H6Cl2	10.87 ±0.05	PI	215	24	iso-C3H7Br	10.075 ±0.01	PI	208	24
1,3-C3H6Cl2	10.85 ±0.05	PI	215	24	n-C4H9Br	10.125 ±0.01	PI	208	24
o-C6H4Cl2	9.06	PI	217	24, 28	sec-C4H9Br	9.98 ±0.01	PI	206	24
m-C6H4Cl2	9.12 ±0.01	PI	217	24	iso-C4H9Br	10.09 ±0.02	PI	208	24
p-C6H4Cl2	8.95	PI	212	28	tert-C4H9Br	9.89 ±0.03	PI	201	24
CHCl3	11.42 ±0.03	PI	239	24	n-C5H11Br	10.10 ±0.02	PI	205	24
C2HCl3	9.45	PI	216	*	C6H5Br	8.98 ±0.02	PI	231	24
CHCl2CHCl2	11.10 ±0.05	EI	220	146	o-C6H4BrCH3	8.78 ±0.01	PI	218	24
CNCl	12.49 ±0.04	EI	321	147	m-C6H4BrCH3	8.81 ±0.02	PI	218	24
CF3Cl	12.91 ±0.03	PI	132	24	p-C6H4BrCH3	8.67 ±0.02	PI	215	24
C2F3Cl	10.4 ±0.2	EI	107	148	CH2Br2	10.49 ±0.02	PI	241	24
C6F5Cl	10.4 ±0.1	EI		149	cis-C2H2Br2	9.45	PI	241	*
CF2Cl2	12.31 ±0.05	PI	170	24	trans-C2H2Br2	9.46	PI	240	*
CF3CCl=CClCF3	10.36 ±0.1	PI		24	CH3CHBr2	10.19 ±0.03	PI	240	24
CFCl3	11.77 ±0.02	PI	205	24	1,3-C3H6Br2	10.07 ±0.02	PI	221	24
CF3CCl3	11.78 ±0.03	PI		24	CHBr3	10.51 ±0.02	PI	246	24
CFCl2CF2Cl	11.99 ±0.02	PI	95	24	C2HBr3	9.27	PI	240	24, 28
ClO3F	13.6 ±0.2	EI	308	150	CNBr	11.95 ±0.08	EI	320	147
C5H4NCl (2-chloro-pyridene)	9.91 ±0.05	EI	255	84	CF3Br	11.89	EI	121	*
C5H4NCl (4-chloro-pyridene)	10.15 ±0.05	EI	260	84	C5H4NBr (2-bromo-pyridine)	9.65 ±0.05	EI	261	84
CH3COCl	11.02 ±0.05	PI	196	24	C5H4NBr (4-bromo-pyridine)	9.94 ±0.05	EI	267	84
CH3COCH2Cl	9.99	EI	173	105, 151	CH3COBr	10.55 ±0.05	PI	197	24
o-C6H4(OH)Cl	9.28	EI	181	75	C6H4BrOH	9.04	EI	187	75
p-C6H4(OH)Cl	9.07	EI	175	75	CH2BrCOOCH3	10.37 ±0.05	EI	146	74
C6H5COCl	9.70 ±0.01	EI	195	105	C6H4FBr	8.99 ±0.03	PI	187	24
p-C6H4ClCHO	9.61 ±0.01	EI	201	105	CF2BrCH2Br	10.83 ±0.01	PI	160	24
C8H7OCl (α-chloro-acetophenone)	9.5	EI	195	74, 105	C4H3BrS	8.63 ±0.01	PI	228	24
C8H7OCl (p-chloro-acetophenone)	9.47 ±0.05	EI	190	105	CH2ClBr	10.77 ±0.01	PI	236	24
C8H7OCl (p-methyl-benzylchloride)	9.37 ±0.01	EI	187	105	CH2BrCH2Cl	10.63 ±0.03	PI	227	24
C6H4ClCOC6H5	9.68 ±0.01	EI	227	105	CHCl2Br	10.88 ±0.05	EI	237	146
CH2ClCOOCH3	10.53 ±0.05	EI	138	74	(CH3)3SiBr	10.24 ±0.02	EI	171	130
p-CH3OC6H4COCl	8.87 ±0.05	EI	149	105	Mo(CO)6	8.12 ±0.03	PI	-31	155
p-ClC6H4COCl	9.58 ±0.03	EI	192	105	RuO4	12.33 ±0.23	EI	240	163
cis-C2H2FCl	9.86	PI	191	24, 121	(CH3)4Sn	8.25 ±0.15	EI	186	129
trans-C2H2FCl	9.87	PI	191	24, 121	SbH3	9.58	PI	256	131
o-C6H4FCl	9.155 ±0.01	PI	180	24	(C6H5)3Sb	7.3 ±0.1	PI	255	77
m-C6H4FCl	9.21 ±0.01	PI	180	24	I2	9.28 ±0.02	PI	229	24
p-C6H4FCl	9.43 ±0.02	EI	185	152	HI	10.39	PI	246	24, 159
CHF2Cl	12.45 ±0.05	PI	174	24	ICl	10.31 ±0.02	EI	242	164
cis-C2HF2Cl	9.86 ±0.02	PI	147	121	IBr	9.98 ±0.03	EI	240	164
trans-C2HF2Cl	9.83 ±0.02	PI	147	121	CH3I	9.54	S, PI	223	24, 145
CH3CF2Cl	11.98 ±0.01	PI		24	C2H5I	9.33	S, PI	213	24, 162
n-C3F7CH2Cl	11.84 ±0.02	PI		24	n-C3H7I	9.26 ±0.01	PI	208	24
CHFCl2	12.39 ±0.20	EI	217	153	iso-C3H7I	9.17 ±0.02	PI	201	24
(CH3)3SiCl	10.58 ±0.04	EI	160	130	n-C4H9I	9.21 ±0.01	PI	202	24
CH3SiCl3	11.36 ±0.03	EI	136	153	sec-C4H9I	9.09 ±0.02	PI	198	24
CH2=CHSiCl3	10.79 ±0.02	PI	148	24	iso-C4H9I	9.18 ±0.02	PI	200	24
C2H5SiCl3	10.74 ±0.04	EI	117	153	tert-C4H9I	9.02 ±0.03	PI	193	24
iso-C3H7SiCl3	10.28 ±0.1	EI	100	153	n-C5H11I	9.19 ±0.01	PI	197	24
C4H3ClS	8.68 ±0.01	PI	217	24	C6H5I	8.73 ±0.03	PI	238	24
C6H4NO2COCl	10.66 ±0.01	EI	219	105	o-C6H4ICH3	8.62 ±0.01	PI	228	24
CaF	5.8	EI	75	*	m-C6H4ICH3	8.61 ±0.03	PI	226	24
C5H5V(CO)4	8.2 ±0.3	EI	9	154	p-C6H4ICH3	8.50 ±0.01	PI	224	24
Cr(CO)6	8.03 ±0.03	PI	-55	155	W(CO)6	8.18 ±0.03	PI	-20	155
C5H5Mn(CO)3	8.3 ±0.4	EI	83	154	OsO4	12.97 ±0.12	EI	219	163
					(CH3)2Hg	9.0	EI	233	143, 156
					(C2H5)2Hg	8.5 ±0.1	EI	221	143
					(iso-C3H7)2Hg	7.6 ±0.1	EI	188	143
					CH3HgCl	11.5 ±0.2	EI	253	143
					(CH3)4Pb	8.0 ±0.4	EI	217	129
					(C6H5)3Bi	7.3 ±0.1	PI	288	77

* Average of several values.

REFERENCES

1. Beutler, H. and Junger, H. O. *Zeit. f. Physik* **100**, 80 (1936).
2. Dibeler, V. H., Reese, R. M. and Krauss, M. *Adv. Mass Spectry.* **3**, 471 (1966).
3. Bauer, S. H., Herzberg, G. and Johns, J. W. C. *J. Mol. Spectr.* **13**, 256 (1964).
4. Margrave, J. L. *J. Phys. Chem.* **61**, 38 (1957).
5. Koski, W. S., Kaufman, J. J., Pachucki, C. F. and Shipko, F. J. *J. Am. Chem. Soc.* **80**, 3202 (1958).
6. Fehlner, T. P. and Koski, W. S. *J. Am. Chem. Soc.* **86**, 581 (1964).
7. Drowart, J., DeMaria, G. and Inghram, M. G. *J. Chem. Phys.* **29**, 1015 (1958).
8. Field, F. H. and Franklin, J. L. *Electron Impact Phenomena and the Properties of Gaseous Ions,* Academic Press, Inc., New York, N.Y. (1957).
9. Herzberg, G. *Can. J. Phys.* **39**, 1511 (1961).
10. Waldron, J. D., *Metropolitan Vickers Gazette* **27**, 66 (1956).
11. Dibeler, V.H., Krauss, M., Reese, R. M. and Harllee, F. N. *J. Chem. Phys.* **42**, 3791 (1965).
12. Herzberg, G. and Shoosmith, J., *Can J. Phys.* **34**, 523 (1956).
13. Elder, F. A., Giese, C., Steiner, B. and Inghram, M. *J. Chem. Phys.* **36**, 3292 (1962).
14. Nicholson, A. J. C., *J. Chem. Phys.* **43**, 1171 (1965).
15. Dibeler, V. H. and Reese, R. M. *J. Res. Natl. Bur. Std.* **A68**, 409 (1964).
16. Zelikoff, M. and Watanabe, K., *J. Opt. Soc. Am.* **43**, 756 (1953).
17. Al-Joboury, M. I. and Turner, D. W., *J. Chem. Soc. London* **4434** (1964).
18. Farmer, J. B. and Lossing, F. P., *Can. J. Chem.* **33**, 861 (1955).
19. Collin, J. and Lossing, F. P., *J. Am. Chem. Soc.* **79**, 5848 (1957).
20. Watanabe, K. and Namioka, T., *J. Chem. Phys.* **24**, 915 (1956).
21. Collin, J. and Lossing, F. P., *J. Am. Chem. Soc.* **81**, 2064 (1959).
22. Dorman, F. H. *J. Chem. Phys.* **43**, 3507 (1965).
23. Pottie, R. F., Harrison, A. G. and Lossing, F. P. *J. Am. Chem. Soc.* **83**, 3204 (1961).
24. Watanabe, K., Nakayama, T. and Mottl, J. *J. Quant. Spectrosc. Radiat. Transfer* **2**, 369 (1962).
25. Coats, F. H. and Anderson, R. C. *J. Am. Chem. Soc.* **79**, 1340 (1957).
26. Varsel, C. J., Morrell, F. A., Resnik, F. E. and Powell, W. A. *Anal. Chem.* **32**, 182 (1960).
27. McDowell, C. A., Lossing, F. P., Henderson, I. H. S. and Farmer, J. B. *Can. J. Chem.* **34**, 345 (1956).
28. Bralsford, R., Harris, P. V. and Price, W. C. *Proc. Roy. Soc.* **A258**, 459 (1960).
29. Lossing, F. P., Kebarle, P. and DeSousa, J. B. *Adv. Mass Spectrometry* **431** (1959).
30. Watanabe, K. *J. Chem. Phys.* **26**, 542 (1957).
31. Meyer, F. and Harrison, A. G. *Can. J. Chem.* **42**, 2256 (1964).
32. Harrison, A. G., Honnen, L. R., Dauben, H. J. and Lossing F. P. *J. Am. Chem. Soc.* **82**, 5593 (1960).
33. Natalis, P. and Laune, J. *Bull. Soc. Chim. Belg.* **73**, 944 (1964).
34. Taubert, R. and Lossing, F. P. *J. Am. Chem. Soc.* **84**, 1523 (1962).
35. Steiner, B., Giese, C. F. and Inghram, M. G. *J. Chem. Phys.* **34**, 189 (1961).
36. Fisher, I. P., Palmer, T. F. and Lossing, F. P. *J. Am. Chem. Soc.* **86**, 2741 (1964).
37. Brehm, B. *Z. Naturforschg.* **21a**, 196 (1966).
38. Momigny, J., Brakier, L. and D Or, L. *Bull. Classe Sci. Acad. Roy. Belg.* **48**, 1002 (1962).
39. Harrison, A. G., Kebarle, P. and Lossing, F. P. *J. Am. Chem. Soc.* **83**, 777 (1961).
40. Thrush, B. A. and Zwolenik, J. J. *Disc. Faraday Soc.* **35**, 196 (1963).
41. Lifshitz, C. and Bauer, S. H. *J. Phys. Chem.* **67**, 1629 (1963).
42. Meyerson, S., McCollum, J. D. and Rylander, P. N. *J. Am. Chem. Soc.* **83**, 1401 (1961).
43. Steele, W. C., Jennings, B. H., Botyos, G. L. and Dudek, G. O. *J. Org. Chem.* **30**, 2886 (1965).
44. Kybett, B. D., Carroll, S., Natalis, P., Bonnell, D. W., Margrave, J. L. and Franklin, J. L. *J. Am. Chem. Soc.* **88**, 626 (1966).
45. Meyer, F., Haynes, P., McLean, S. and Harrison, A. G. *Can. J. Chem.* **43**, 211 (1965).
46. Natalis, P. *Bull. Soc. Chim. Belg.* **73**, 961 (1964).
47. Pottie, R. F. and Lossing, F. P. *J. Am. Chem. Soc.* **85**, 269 (1963).
48. Vilesov, F. I. and Terenin, A. N. *Dokl. Phys. Chem., Proc. Acad. Sci.* **USSR 115**, 539 (1957).
49. Clark, L. B. *J. Chem. Phys.* **43**, 2566 (1965).
50. Kitagawa, T., Harada, Y., Inokuchi, H. and Kodera, K. *J. Mol. Spectr.* **19**, 1 (1966).
51. Natalis, P. *Bull. Soc. Roy. Sci. Liege* **31**, 803 (1962).
52. Harrison, A. G. and Lossing, F. P. *J. Am. Chem. Soc.* **82**, 1052 (1960).
53. Vilesov, F. I. *J. Phys. Chem.* **USSR 35**, 986 (1961).
54. Natalis, P. and Franklin, J. L. *J. Phys. Chem.* **69**, 2935 (1965).
55. Wacks, M. E. and Dibeler, V. H. *J. Chem. Phys.* **31**, 1557 (1959).
56. Wacks, M. E. *J. Chem. Phys.* **41**, 1661 (1964).
57. King, A. B. *J. Chem. Phys.* **42**, 3526 (1965).
58. Law, R. W. and Margrave, J. L. *J. Chem. Phys.* **25**, 1086 (1956).
59. Lofthus, A. *The Molecular Spectrum of Nitrogen* Department of Physics, University of Oslo, Blindern, **Norway, Spectroscopic Report No. 2**, 1 (1960).
60. Reed, R. I. and Snedden, W. *J. Chem. Soc.* **4132** (1959).
61. Foner, S. N. and Hudson, R. L. *J. Chem. Phys.* **29**, 442 (1958).
62. Foner, S. N. and Hudson, R. L. *J. Chem. Phys.* **28**, 719 (1958).
63. Akopyan, M. E. and Vilesov, F. I. *Kinetics and Catalysis*, **4**, 32 (1963).
64. Dibeler, V. H., Reese, R. M. and Franklin, J. L. *J. Am. Chem. Soc.* **83**, 1813 (1961).
65. Berkowitz, J. *J. Chem. Phys.* **36**, 2533 (1962).
66. Morrison, J. D. and Nicholson, A. J. C. *J. Chem. Phys.* **20**, 1021 (1952).
67. Watanabe, K. and Mottl, J. R. *J. Chem. Phys.* **26**, 1773 (1957).
68. Pottie, R. F. and Lossing, F. P. *J. Am. Chem. Soc.* **83**, 4737 (1961)
69. Gallegos, E. and Kiser, R. W. *J. Phys. Chem.* **65**, 1177 (1961).
70. Gallegos, E. J. and Kiser, R. W. *J. Phys. Chem.* **66**, 136 (1962).
71. Amr El-Sayed, M. F., Kasha, M. and Tanaka, Y. *J. Chem. Phys.* **34**, 334 (1961).
72. Akopyan, M. E. and Vilesov, F. I. *Dokl. Phys. Chem., Proc. Acad. Sci.* **USSR 158**, 965 (1964).
73. Kurbatov, B. L., Vilesov, F. I. and Terenin, A. N. *Soviet Physics-Doklady* **6**, 883 (1962).
74. Pignataro, S., Foffani, A., Innorta, G. and Distefano, G. *Z. Phys. Chem. Neue Folge* **49**, 291 (1966).
75. Crable, G. F. and Kearns, G. L. *J. Phys. Chem.* **66**, 436 (1962).
76. Farrell, P. G. and Newton, J. *J. Phys. Chem.* **69**, 3506 (1965).
77. Vilesov, F. I. and Zaitsev, V. M. *Dokl. Phys. Chem., Proc. Acad. Sci.* **USSR 154**, 117 (1964).
78. Paulett, G. S. and Ettinger, R. *J. Chem. Phys.* **39**, 825 (1963).
79. Merer, A. J., *Can. J. Phys.* **42**, 1242 (1964).
80. Gowenlock, B. G., Majer, J. R. and Snelling, D. R. *Trans. Faraday Soc.* **58**, 670 (1962).
81. Akopyan, M. E., Vilesov, F. I. and Terenin, A. N. *Izv. Akad. Nauk USSR, Ser. Fiz.* **27**, 1083 (1963).
82. Dibeler, V. H., Franklin, J. L. and Reese, R. M. *J. Am. Chem. Soc.* **81**, 68 (1959).
83. Momigny, J., Urbain, J. and Wankenne, H. *Bull. Soc. Roy. Sci. Liege* **34**, 337 (1965).
84. Basila, M. R. and Clancy, D. J. *J. Phys. Chem.* **67**, 1551 (1963).
85. Finch, A. C. M. *J. Chem. Soc.* **2272** (1964).
86. Franklin, J. L., Dibeler, V. H., Reese, R. M. and Krauss, M. *J. Am. Chem. Soc.* **80**, 298 (1958).
87. Radwan, T. N. and Turner, D. W. *J. Chem. Soc.* **Sect. A85** (1966).
88. Foner, S. N. and Hudson, R. L. *Advances in Chem.* **Ser. 34** (1962).
89. Foner, S. N. and Hudson, R. L. *J. Chem. Phys.* **23**, 1364 (1955).
90. Lindeman, L. P. and Guffy, J. C. *J. Chem. Phys.* **29**, 247 (1958).
91. Berkowitz, J. and Hudson, R. L. *J. Chem. Phys.* **36**, 2676 (1962).
92. Berkowitz, J., Chupka, W. A., Blue, G. D. and Margrave, J. L. *J. Phys. Chem.*, **63**, 644 (1959).
93. White, D., Seshadri, K. S., Dever, D. F., Mann, D. E. and Linevski M J *J. Chem. Phys.* **39**, 2463 (1963).
94. Krupenie, P. H. *The Band Spectrum of Carbon Monoxide* Institute for Basic Standards, National Bureau of Standards, Washington, D.C., **NSRDS-NBS 5**, 1 (1966).
95. Tanaka, Y., Jursa, A. S. and LeBlanc, F. J. *J. Chem. Phys.* **32**, 1205 (1960).
96. Tanaka, Y., Jursa, A. S. and LeBlanc, F. J, *Chem. Phys.* **32**, 1199 (1960).
97. Nakayama, T., Kitamura, M. Y. and Watanabe, K. *J. Chem. Phys.* **30**, 1180 (1959).
98. Frost, D. C., Mak, D. and McDowell, C. A. *Can. J. Chem.* **40**, 1064 (1962).
99. Hernandez, G. J. *J. Chem. Phys.* **38**, 1644 (1963).
100. Hernandez, G. J. *J. Chem. Phys.* **38**, 2233 (1963).
101. Chupka, W. A. *J. Chem. Phys.* **30**, 191 (1959).
102. Watanabe, K. and Nakayama, T. *J. Chem. Phys.* **29**, 48 (1958).
103. Murad, E. and Inghram, M. G. *J. Chem. Phys.* **40**, 3263 (1964).
104. Higasi, K., Nozoe, T. and Omura, I. *Bull. Chem. Soc. Japan* **30**, 408 (1957).
105. Foffani, A., Pignataro, S., Cantone, B. and Grasso, F. *Z. Phys. Chem. Neue Folge* **42**, 221 (1964).
106. Natalis, P. and Franklin, J. L. *J. Phys. Chem.* **69**, 2943 (1965).
107. Cotter, J. L. *J. Chem. Soc.* **5742** (1965).
108. Cotter, J. L. *J. Chem. Soc.* **5477** (1964).
109. Iczkowski, R. P. and Margrave, J. L. *J. Chem. Phys.* **30**, 403 (1959).
110. Frost, D. C. and McDowell, C. A. *Can. J. Chem.* **36**, 39 (1958).
111. Hildenbrand, D. L. and Muran, E. *J. Chem. Phys.* **43**, 1400 (1965).
112. Marriott, J. and Craggs, J. D. *J. Electronics and Control* **3**, 194 (1957).
113. Fisher, I. P., Homer, J. B. and Lossing, F. P. *J. Am. Chem. Soc.* **87**, 957 (1965).
114. Pottie, R. F. *J. Chem. Phys.* **42**, 2607 (1965).
115. Herron, J. T. and Dibeler, V. H. *J. Res. Natl. Bur. Std.* **65**, 405 (1961).
116. Loughran, E. D. and Mader, C. *J. Chem. Phys.* **32**, 1578 (1960).
117. Frost, D. C. and McDowell, C. A. *The Determination of Ionization and Dissociation Potentials of Molecules by Radiation with Electrons* Department of Chemistry, University of British Columbia, Vancouver 8, B. C., Canada, **AFCRL-TR-60-423** 1 (1960).
118. Dibeler, V. H., Reese, R. M. and Franklin, J. L. *J. Chem. Phys.* **27**, 1296 (1957).
119. Wilson, E. G. Jortner, J. and Rice, S. A. *J. Am. Chem. Soc.* **85**, 813 (1963).
120. Frost, D. C. and McDowell, C. A. *Proc. Roy. Soc.* **A241**, 194 (1957).
121. Momigny, J. *Nature* **199**, 1179 (1963).
122. Martin, R. H., Lampe, F. W. and Taft, R. W. *J. Am. Chem. Soc.* **88**, 1353 (1966).
123. Reese, R. M. and Dibeler, V. H. *J. Chem. Phys.* **24**, 1175 (1956).
124. Herron, J. T. and Dibeler, V. H. *J. Chem. Phys.* **33**, 1595 (1960).
125. Hammond, V. J., Price, W. C., Teegan, J. P. and Walsh, A. D. *Disc. Faraday Soc.* **9**, 53 (1950).
126. Hudson, R. D. *J. Chem. Phys.* **43**, 1790 (1965).
127. Margrave, J. L. *J. Chem. Phys.* **41**, 2250 (1964).

REFERENCES (Continued)

128. Ehlert, T. C. and Margrave, J. L. *J. Am. Chem. Soc.* **86**, 3901 (1964).
129. Hobrock, B. G. and Kiser, R. W. *J. Phys. Chem.* **65**, 2186 (1961).
130. Hess, G. G., Lampe, F. W. and Sommer, L. H. *J. Am. Chem. Soc.* **87**, 5327 (1965).
131. Price, W. C. and Passmore, T. R. *Disc. Faraday Soc.* **35**, 232 (1963).
132. Wada, Y. and Kiser, R. W. *J. Phys. Chem.* **68**, 2290 (1964).
133. Berkowitz, J. and Chupka, W. A. *J. Chem. Phys.* **40**, 287 (1964).
134. Palmer, T. F. and Lossing, F. P. *J. Am. Chem. Soc.* **84**, 4661 (1962).
135. Price, W. C. *Bull. Am. Phys. Soc.* **10**, 9 (1935).
136. Hobrock, B. G. and Kiser, R. W. *J. Phys. Chem.* **66**, 1551 (1962).
137. Hobrock, B. G. and Kiser, R. W. *J. Phys. Chem.* **67**, 648 (1963).
138. Hobrock, B. G. and Kiser, R. W. *J. Phys. Chem.* **66**, 1648 (1962).
139. Khvostenko, V. I. *Russ. J. Phys. Chem.* **36**, 197 (1962).
140. Hobrock, B. G. and Kiser, R. W. *J. Phys. Chem.* **67**, 1283 (1963).
141. Reese, R. M., Dibeler, V. H. and Franklin, J. L. *J. Chem. Phys.* **29**, 880 (1958).
142. Baldwin, M., Maccoll, A., Kirkien-Konasiewicz, A. and Saville, B. *Chem. Ind.* **286** (1966).
143. Gowenlock, B. G., Kay, J. and Majer, J. R. *Trans. Faraday Soc.* **59**, 2463 (1963).
144. Berkowitz, J., Tasman, H. A. and Chupka, W. A. *J. Chem. Phys.* **36**, 2170 (1962).
145. Price, W. C. *J. Chem. Phys.* **4**, 539 (1936).
146. Harrison, A. G. and Shannon, T. W. *Can. J. Chem.* **40**, 1730 (1962).

147. Herron, J. T. and Dibeler, V. H. *J. Am. Chem. Soc.* **82**, 1555 (1960).
148. Margrave, J. L. *J. Chem. Phys.* **31**, 1432 (1959).
149. Majer, J. R. and Patrick, C. R. *Trans. Faraday Soc.* **58**, 17 (1962).
150. Dibeler, V. H., Reese, R. M. and Mann, D. E. *J. Chem. Phys.* **27**, 176 (1957).
151. Foffani, A., Pignataro, S., Cantone, B. and Grasso, F. *Nuovo Cimento* **29**, 918 (1963).
152. Momigny, J. and Wirtz-Cordier, A. M. *Ann. Soc. Sci. Bruxelles* **76**, 164 (1962).
153. Hobrock, D. L. and Kiser, R. W. *J. Phys. Chem.* **68**, 575 (1964).
154. Winters, R. E. and Kiser, R. W. *J. Organometal. Chem.* **4**, 190 (1965).
155. Vilesov, F. I. and Kurbatov, B. L. *Dokl. Phys. Chem., Proc. Acad. Sci. USSR* **140**, 792 (1961).
156. Hobrock, B. G. and Kiser, R. W. *J. Phys. Chem.* **66**, 155 (1962).
157. Westmore, J. B., Mann, K. H. and Tickner, A. W. *J. Phys. Chem.* **68**, 606 (1964).
158. Cullen, W. R. and Frost, D. C. *Can. J. Chem.* **40**, 390 (1962).
159. Morrison, J. D., Hurzeler, H., Inghram, M. G. and Stanton, H. E. *J. Chem. Phys.* **33**, 821 (1960).
160. Berkowitz, J. and Marquart, J. R. *J. Chem. Phys.* **37**, 1853 (1962).
161. Irsa, A. P. and Friedman, L. *J. Inorg. Nucl. Chem.* **6**, 77 (1958).
162. Price, W. C. *J. Chem. Phys.* **4**, 547 (1936).
163. Dillard, J. G. and Kiser, R. W. *J. Phys. Chem.* **69**, 3893 (1965).
164. Frost, D. C. and McDowell, C. A. *Can. J. Chem.* **38**, 407 (1960).
165. Thomas, R. K., *Proc. R. Soc. London Ser. A,* **331**, 249, 1972.

ELECTRON WORK FUNCTIONS OF THE ELEMENTS

Compiled by Herbert B. Michaelson, 1977

The measured values cited for polycrystalline and single-crystal specimens are selected as being the best available data at this time. The selection is based on (1) The validity of the experimental technique (e.g., vacua of 10^{-9} or 10^{-10} Torr, clean surfaces, and identification of crystal-face distribution and other surface conditions), and (2) Best agreement with preferred values and theoretical values of the true work function (given variously by Fomenko,[1] Rivière,[2] Trasatti,[3] and Lang and Kohn[4]). Experimental data that are not well substantiated according to these criteria are listed in *italics*. Crystallographic directions for single-crystal data are indicated by parentheses.

Abbreviations apply to the experimental method: T, thermionic; P, photoelectric; CPD, contact potential difference; F, field emission. Important distinctions among such measurements are discussed in the Rivière[2] paper, pp. 180 to 198.

Element	Experimental value, ϕ (eV)	Experimental method	Ref.	Element	Experimental value, ϕ (eV)	Experimental method	Ref.
Ag	*4.26*	P	5	Hg	4.49	P	27
	4.64 (100)	P	5	In	4.12	P	28
	4.52 (110)	P	5	Ir	*5.27*	T	29
	4.74 (111)	P	6		5.42 (110)	F	30
Al	4.28	P	7		5.76 (111)	F	30
	4.41 (100)	P	8		5.67 (100)	F	31
	4.06 (110)	P	7		5.00 (210)	F	31
	4.24 (111)	P	8	K	2.30	P	32
As	*3.75*	P	9	La	3.5	P	10
Au	5.1	P	10	Li	*2.9*	F	33
	5.47 (100)	P	11	Lu	*3.3*	CPD	34
	5.37 (110)		11	Mg	*3.66*	P	35
	5.31 (111)		11	Mn	4.1	P	10
B	*4.45*	T	12	Mo	4.6	P	10
Ba	*2.7*	T	13		4.53 (100)	P	36
Be	4.98	P	14		4.95 (110)	P	36
Bi	*4.22*	P	15		4.55 (111)	P	36
C	*5.0*	CPD	16		4.36 (112)	P	36
Ca	2.87	P	17		4.50 (114)	P	36
Cd	*4.22*	CPD	18		4.55 (332)	P	36
Ce	2.9	P	10	Na	2.75	P	37
Co	5.0	P	10	Nb	4.3	P	10
Cr	4.5	P	10		4.02 (001)	T	38
Cs	2.14	P	19		4.87 (110)	T	38
Cu	4.65	P	10		4.36 (111)	T	38
	4.59 (100)	P	20		4.63 (112)	T	38
	4.48 (110)	P	20		4.29 (113)	T	38
	4.94 (111)	P	20		3.95 (116)	T	38
	4.53 (112)	P	20		4.18 (310)	T	38
Eu	2.5	P	10	Nd	3.2	P	10
Fe	4.5	P	10	Ni	5.15	P	10
	4.67 (100)	P	21		5.22 (100)	P	39
	4.81α (111)	P	22		5.04 (110)	P	39
	4.70α	P	23		5.35 (111)	P	39
	4.62β	P	23	Os	*4.83*	T	29
	4.68γ	P	23	Pb	4.25	P	40
Ga	4.2	CPD	24	Pd	5.12	P	31
Ge	5.0	CPD	25		5.6 (111)	P	41
	4.80 (111)	P	26	Pt	5.65	P	10
Gd	3.1	P	10		5.7 (111)	P	41
Hf	3.9	P	10	Rb	*2.16*	P	27

ELECTRON WORK FUNCTIONS OF THE ELEMENTS (continued)

Element	Experimental value, ϕ (eV)	Experimental method	Ref.	Element	Experimental value, ϕ (eV)	Experimental method	Ref.
Re	4.96	T	29	Te	4.95	P	44
	5.75 (1011)	F	33	Th	3.4	T	51
Rh	4.98	P	31	Ti	4.33	P	10
Ru	4.71	P	31	Tl	3.84	CPD	52
Sb	4.55 (amorph.)	–	42	U	3.63	P & CPD	53
	4.7 (100)	–	43		3.73 (100)	P & CPD	54
Sc	3.5	P	10		3.90 (110)	P & CPD	54
Se	5.9	P	44		3.67 (113)	P & CPD	54
Si	4.85n	CPD	40	V	4.3	P	10
	4.91p (100)	CPD	45	W	4.55	CPD	55
	4.60p (111)	P	46		4.63 (100)	F	30
Sm	2.7	P	10		5.25 (110)	F	30
Sn	4.42	CPD	47		4.47 (111)	F	30
Sr	2.59	T	48		4.18 (113)	CPD	56
Ta	4.25	T	29		4.30 (116)	T	57
	4.15 (100)	T	49	Y	3.1	P	10
	4.80 (110)	T	49	Zn	4.33	P	15
	4.00 (111)	T	49		4.9 (0001)	CPD	58
Tb	3.0	P	50	Zr	4.05	P	10

REFERENCES

1. Fomenko, V. S., *Emission Properties of Materials*, 3rd ed., Naukova Dumka, Kiev, 1970 (in Russian).
2. Rivière, J. C., *Solid State Surface Science*, Green, M., Ed., Vol. 1, Marcel Dekker, 1969, chap. 4.
3. Trasatti, S., *Chim. Ind.* (Milan), 53(6), 559, 1971.
4. Lang, N. D. and Kohn, W., *Phys. Rev. B*, 3(4), 1215, 1971.
5. Dweydari, A. W. and Mee, C. H. B., *Phys. Status Solidi A*, 27, 223, 1975.
6. Dweydari, A. W. and Mee, C. H. B., *Phys. Status Solidi A*, 17, 247, 1973.
7. Eastment, R. M. and Mee, C. H. B., *J. Phys. F*, 3, 1738, 1973.
8. Grepstad, J. K., Gartland, P. O., and Slagsvold, B. J., *Surf. Sci.*, 57, 348, 1976.
9. Raisin, C. and Pinchaux, R., *Solid State Commun.*, 16, 941, 1975.
10. Eastman, D. E., *Phys. Rev. Sect. B*, 2, 1, 1970.
11. Potter, H. C. and Blakeley, J. M., *J. Vac. Sci. Technol.*, 12, 635, 1975 and Potter, H. C., Ph.D. thesis Cornell University, Materials Science Center Rep. No. 1353, 1970.
12. Adlenvish, L. I. and Gol'dshtein, L. M., *Fiz. Tverdogo Tela* (Leningrad), 9, 1258, 1967.
13. Bondarenko, B. V. and Mukhov, V. I., *Sov. Phys. Solid State*, 12(7), 1522, 1971.
14. Gustafsson, Broden, and Nilsson, *J. Phys. F*, 4, 2351, 1974.
15. Suhrmann, R. and Wedler, G., *Z. Angew. Phys.*, 14, 70, 1962.
16. Koprieux, B., Faure, R., and Dussaulcy, J. P., *C. R. Acad. Sci. Ser. B*, 278(14), 659, 1974.
17. Gaudart, L. and Riviora, R., *Appl. Opt.*, 10, 2336, 1971.
18. Anderson, P. A., *Phys. Rev.*, 98, 1739, 1955.
19. Boutry, G. A. and Dormont, H., *Philips Tech. Rev.*, 30, 225, 1969.
20. Gartland, P. O., *Phys. Norv.*, 6(3,4), 201, 1972.
21. Ueda, K. and Shimizu, R., *Jp. J. Appl. Phys.*, 11(6), 916, 1972.
22. Kobayashi, H. and Kato, S., *Surf. Sci.*, 18(2), 341, 1969.
23. Cardwell, A., *Phys. Rev.*, 92, 554, 1953.
24. Osipova, E. V., Shurmovskaya, N. A., and Burshtein, R. Kh., *Elektrokhimiya*, 5(10), 1139, 1969 (in Russian).
25. Boiko, B. A., Gorodetskii, D. A., and Yas'ko, A. A., *Sov. Phys. Solid State*, 15(11), 2101, 1974.
26. Gobeli, G. W. and Allen, F. G., *Surf. Sci.*, 2, 402, 1964.
27. Lazarev, V. B. and Malov, Y. I., *Fiz. Met. Metalloved.*, 24(3), 565, 1967.
28. Peisner, J., Roboz, P., and Barna, P. B., *Phys. Stat. A*, 4, K187, 1971.
29. Wilson, R. G., *J. Appl. Phys.*, 37, 3170, 1966.
30. Strayer, R. W., Mackie, W., and Swanson, L. W., *Surf. Sci.*, 34, 225, 1973.
31. Nieuwenhuys, Bouwman, and Sachtler, *Thin Solid Films*, 21, 51, 1974.
32. Van Oirschot, Th. G. J., van den Brink, M., and Sachtler, W. H. M., *Surf. Sci.*, 29, 189, 1972.
33. Ovchinnikov, A. P. and Tsarev, B. M., *Sov. Phys. Solid State*, 9(12), 2766, 1968.
34. Bondarenko, B. V. and Makhov, V. I., *Sov. Phys. Solid State*, 12, 2986, 1971.
35. Garron, R., *C. R. Acad. Sci.*, 258, 1458, 1964.
36. Berge, Gartland, and Slagsvold, *Surf. Sci.*, 43, 275, 1974.
37. Whitefield, R. J. and Brady, J. J., *Phys. Rev. Lett.*, 26(7), 380, 1971.
38. Leblanc, R. P., Vanbrugghe, B. C., and Girouard, F. E., *Can. J. Phys.*, 52, 1589, 1974.
39. Baker, B. G., Johnson, E. B., and Maire, G. I. C., *Surf. Sci.*, 24, 572, 1971.
40. Thanailakis, A., *Inst. Phys. Conf. Ser.*, p. 59, 1974.
41. Demuth, J. E., *Chem. Phys. Lett.*, 45, 12, 1977.
42. Gorodetskii, D. A. and Yas'ko, A. A., *Sov. Phys. Solid State*, 13(11), 2928, 1972.
43. Gorodetskii, D. A. and Yas'ko, A. A., *Sov. Phys. Solid State*, 13(5), 1085, 1971.
44. Williams, R. H. and Polanco, J. I., *J. Phys. C*, 7, 2745, 1974.
45. Allen, F. G., *J. Phys. Chem. Solids*, 8, 119, 1959.
46. Allen, F. G. and Gobeli, G. W., *J. Appl. Phys.*, 35, 597, 1964.
47. Simmons, J. G., *Phys. Rev. Lett.*, 10, 10, 1963.
48. Alleau, T., *Surface Phenomena in Thermionic Emitters, Round Table Conf.*, Inst. Tech. Phys. Julich Nucl. Res. Establ., Julich, Germany, 1969, p. 54 (in English).
49. Protopopov, Mikheeva, Shreinberg, and Shuppe, *Fiz. Tverdogo Tela*, 8(4), 1140, 1966.
50. Nemchenok, R. L., Strakovskaya, S. E., and Titenskii, A. I., *Fiz. Tverdogo Tela* 11(9), 2692, 1969.
51. Estrup, P. J., Anderson, J. R., and Danforth, W. E., *Surf. Sci.*, 4, 286, 1966.
52. Klein, O. and Lange, E., *Z. Elektrochem.*, 44, 542, 1938.
53. Hopkins, B. J. and Sargood, A. J., *Nuovo Cimento*, 5, 459, 1967.
54. Lea, C. and Mee, C. H. B., *J. Appl. Phys.*, 39, 5890, 1968.
55. Hopkins, B. J. and Rivière, J. C., *Proc. Phys. Soc.* (London), 81, 590, 1963.
56. Love, H. M. and Dyer, G. L., *Can. J. Phys.*, 40, 1837, 1962.
57. Sultanov, V. M., *Radio Eng. Electron.*, 9, 252, 1964 (English translation).
58. Baker, J. M. and Blakeley, J. M., *Surf. Sci.*, 32, 45, 1972.

Metal.	Resistivity microhm-centimeters 20° C	Temp. coefficient 20° C.	Specific gravity.	Tensile strength, lbs./in.	Melting point ° C.
*Advance. See constantan					
Aluminum..........	2.824	0.0039	2.70	30,000	659
Antimony..........	41.7	.0036	6.6	......	630
Arsenic...........	33.3	.0042	5.73	......	......
Bismuth...........	120	.004	9.8	......	271
Brass.............	7	.002	8.6	70,000	900
Cadmium..........	7.6	.0038	8.6	......	321
*Calido. See nichrome					
Climax...........	87	.0007	8.1	150,000	1250
Cobalt...........	9.8	.0033	8.71	......	1480
Constantan.......	49	.00001	8.9	120,000	1190
Copper: annealed ..	1.7241	.00393	8.89	30,000	1083
hard-drawn.......	1.771	.00382	8.89	60,000	......
Eureka. See constantan					
Excello..........	92	.00016	8.9	95,000	1500
Gas Carbon........	5000	−.0005	......	......	3500
German silver, 18 % Ni...............	33	.0004	8.4	150,000	1100
Gold.............	2.44	.0034	19.3	20,000	1063
Ideal. See constantan					
Iron, 99.98 % pure...	10	.005	7.8	......	1530
Lead	22	.0039	11.4	3,000	327
Magnesium.........	4.6	.004	1.74	33,000	651
Manganin.........	44	.00001	8.4	150,000	910
Mercury..........	95.783	.00089	13.546	0	−38.9
Molybdenum, drawn	5.7	.004	9.0	......	2500
Monel metal.......	42	.0020	8.9	160,000	1300
*Nichrome.........	100	.0004	8.2	150,000	1500
Nickel............	7.8	.006	8.9	120,000	1452
Palladium.........	11	.0033	12.2	39,000	1550
Phosphor bronze....	7.8	.0018	8.9	25,000	750
Platinum..........	10	.003	21.4	50,000	1755
Silver............	1.59	.0038	10.5	42,000	960
Steel, E. B. B.......	10.4	.005	7.7	53,000	1510
Steel, B. B.........	11.9	.004	7.7	58,000	1510
Steel, Siemens-Martin	18	.003	7.7	100,000	1510
Steel, manganese....	70	.001	7.5	230,000	1260
Tantalum..........	15.5	.0031	16.6	......	2850
*Therlo............	47	.00001	8.2	......	......
Tin	11.5	.0042	7.3	4,000	232
Tungsten, drawn....	5.6	.0045	19	500,000	3400
Zinc..............	5.8	.0037	7.1	10,900	419

* Trade mark.

Superconductivity*

B.W. ROBERTS

General Electric Research Laboratory, Schenectady, New York

The following tables on superconductivity include superconductive properties of chemical elements, thin films, a selected list of compounds and alloys, and high-magnetic-field superconductors.

The historically first observed and most distinctive property of a superconductive body is the near total loss of resistance at a critical temperature (T_c) that is characteristic of each material. Figure 1(a) below illustrates schematically two types of possible transitions. The sharp vertical discontinuity in resistance is indicative of that found for a single crystal of a very pure element or one of a few well annealed alloy compositions. The broad transition, illustrated by broken lines, suggests the transition shape seen for materials that are not homogeneous and contain unusual strain distributions. Careful testing of the resistivity limits for superconductors shows that it is less than 4×10^{-23} ohm-cm, while the lowest resistivity observed in metals is of the order of 10^{-13} ohm-cm. If one compares the resistivity of a superconductive body to that of copper at room temperature, the superconductive body is at least 10^{17} times less resistive.

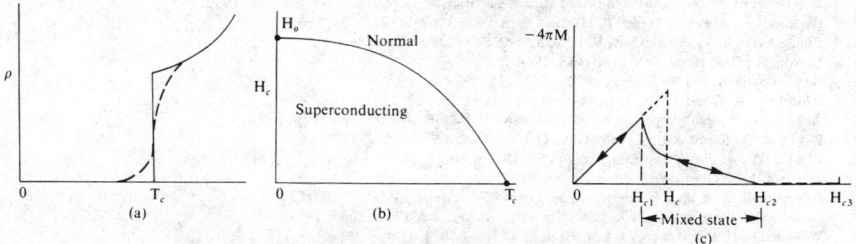

Figure 1. PHYSICAL PROPERTIES OF SUPERCONDUCTORS

(a) Resistivity versus temperature for a pure and perfect lattice (solid line). Impure and/or imperfect lattice (broken line).
(b) Magnetic-field temperature dependence for Type-I or "soft" superconductors.
(c) Schematic magnetization curve for "hard" or Type-II superconductors.

The temperature interval ΔT_c, over which the transition between the normal and superconductive states takes place, may be of the order of as little as 2×10^{-5} °K *or* several °K in width, depending on the material state. The narrow transition width was attained in 99.9999 percent pure gallium single crystals.

*Prepared for Office of Standard Reference Data, National Bureau of Standards, by Standard Reference Data Center on Superconductive Materials, Schenectady, N.Y.

A Type-I superconductor below T_c, as exemplified by a pure metal, exhibits perfect diamagnetism and excludes a magnetic field up to some critical field H_c, whereupon it reverts to the normal state as shown in the H-T diagram of Figure 1(b).

The difference in entropy near absolute zero between the superconductive and normal states relates directly to the electronic specific heat, γ: $(S_s - S_n)_{T \to 0} = -\gamma T$.

The magnetization of a typical high-field superconductor is shown in Figure 1(c). The discovery of the large current-carrying capability of Nb_3Sn and other similar alloys has led to an extensive study of the physical properties of these alloys. In brief, a high-field superconductor, or Type-II superconductor, passes from the perfect diamagnetic state at low magnetic fields to a mixed state and finally to a sheathed state before attaining the normal resistive state of the metal. The magnetic field values separating the four stages are given as H_{c1}, H_{c2}, and H_{c3}. The superconductive state below H_{c1} is perfectly diamagnetic, identical to the state of most pure metals of the "soft" or Type-I superconductor. Between H_{c1} and H_{c2} a "mixed superconductive state" is found in which fluxons (a minimal unit of magnetic flux) create lines of normal superconductor in a superconductive matrix. The volume of the normal state is proportional to $-4\pi M$ in the "mixed state" region. Thus at H_{c2} the fluxon density has become so great as to drive the interior volume of the superconductive body completely normal. Between H_{c2} and H_{c3} the superconductor has a sheath of current-carrying superconductive material at the body surface, and above H_{c3} the normal state exists. With several types of careful measurement, it is possible to determine H_{c1}, H_{c2}, and H_{c3}. Table 2-35 contains some of the available data on high-field superconductive materials.

High-field superconductive phenomena are also related to specimen dimension and configuration. For example, the Type-I superconductor, Hg, has entirely different magnetization behavior in high magnetic fields when contained in the very fine sets of filamentary tunnels found in an unprocessed Vycor glass. The great majority of superconductive materials are Type II. The elements in very pure form and a very few precisely stoichiometric and well annealed compounds are Type-I with the possible exceptions of vanadium and niobium.

Metallurgical Aspects. The sensitivity of superconductive properties to the material state is most pronounced and has been used in a reverse sense to study and specify the detailed state of alloys. The mechanical state, the homogeneity, and the presence of impurity atoms and other electron-scattering centers are all capable of controlling the critical temperature and the current-carrying capabilities in high-magnetic fields. Well annealed specimens tend to show sharper transitions than those that are strained or inhomogeneous. This sensitivity to mechanical state underlines a general problem in the tabulation of properties for superconductive materials. The occasional divergent values of the critical temperature and of the critical fields quoted for a Type-II superconductor may lie in the variation in sample preparation. Critical temperatures of materials studied early in the history of superconductivity must be evaluated in light of the probable metallurgical state of the material, as well as the availability of less pure starting elements. It has been noted that recent work has given extended consideration to the metallurgical aspects of sample preparation.

REFERENCES

References to the data presented in this section, to additional entries of superconductive materials, and to those materials specifically tested and found non-superconductive to some low temperature may be found in the following publications:

"Superconductive Materials and Some of Their Properties", *Progress in Cryogenics*, B.W. Roberts, Vol. IV, Heywood and Co., 1964, pp. 160–231.

"Superconductive Materials and Some of Their Properties", B.W. Roberts, National Bureau of Standards Technical Notes 408 and 482, U.S. Government Printing Office, 1966 and 1969.

SELECTIVE PROPERTIES OF SUPERCONDUCTIVE ELEMENTS

Element	T_c(K)	H_c(oersted)	θ_D(K)	γ(mJmol^{-1}K^{-1})	Element	T_c(K)	H_c(oersted)	θ_D(K)	γ(mJmol^{-1}K^{-1})
Al	1.175±0.002	104.9±0.3	420	1.35	Mo	0.915±0.005	96±3	460	1.83
Am* (α,?)	0.6				Nb	9.25±0.02	2060±50,HF	276	7.80
Am* (β,?)	1.0				Os	0.66±0.03	70	500	2.35
Be	0.026			0.21	Pa	1.4			
Cd	0.517±0.002	28±1	209	0.69	Pb	7.196±0.006	803±1	96	3.1
Ga	1.083±0.001	58.3±0.2	325	0.60	Re	1.697±0.006	200±5	4.5	2.35
Ga (β)	5.9,6.2	560			Ru	0.49±0.015	69±2	580	2.8
Ga (γ)	7	950,hf*			Sn	3.722±0.001	305±2	195	1.78
GA (Δ)	7.85	815,HF			Ta	4.47±0.04	829±6	258	6.15
Hf	0.128	12.7		2.21	Tc	7.8±0.1	1410,HF	411	6.28
Hg (α)	4.154±0.001	411±2	87,71.9	1.81	Th	1.38±0.02	160±3	165	4.32
Hg (β)	3.949	339	93	1.37	Ti	0.40±0.04	56	415	3.3
In	3.408±0.001	281.5±2	109	1.672	Tl	2.38±0.02	178±2	78.5	1.47
Ir	0.1125±0.001	16±0.05	425	3.19	V	5.40±0.05	1408	383	9.82
La (α)	4.88±0.02	800±10	151	9.8	W	0.0154±0.0005	1.15±0.03	383	0.90
La (β)	6.00±0.1	1096,1600	139	11.3	Zn	0.85±0.01	54±0.3	310	0.66
Lu	0.1±0.03	350±50			Zr	0.61±0.15	47	290	2.77
					Zr (ω)	0.65, 0.95			

* HF denotes high field superconductive properties.

From Roberts, B. W., Properties of Selected Superconductive Materials, 1978 Supplement, NBS Technical Note 983, U.S. Government Printing Office, Washington, D.C.

RANGE OF CRITICAL TEMPERATURES OBSERVED FOR SUPERCONDUCTIVE ELEMENTS IN THIN FILMS CONDENSED USUALLY AT LOW TEMPERATURES

Element	T_cRange (K)	H_c(oersted)	Element	T_cRange (K)	H_c(oersted)
			Mo	3.3—8.0	
Al	1.15—~5.7	HF*	Nb	2.0—10.1	
Be	5—9.75	HF	Pb	1.8—7.5	
Bi	6.17—2.6		Re	1.7—~7	
Cd			Sn	3.5—~6	
(Disor-dered)	0.79—0.91		Ta	<1.7—4.51	HF
(Ordered)	0.53—0.59		Tc	4.6—7.7	
Ga	?.5—8.5	HF	Ti	1.3 Max	
Hg	3.87—4.5		Tl	2.33—2.96	
In	2.2—5.6	HF	V	1.8—6.02	
La	3.55—6.74		W	<1.0—4.1	
			Zn	0.77—1.70,~1.9	

* HF denotes high magnetic field superconductive properties.

ELEMENTS EXHIBITING SUPERCONDUCTIVITY UNDER OR AFTER APPLICATION OF HIGH PRESSURE

Element	T_cRange (K)	Pressure (kbar)*
Al	1.98—0.075	0—62
As	0.31—0.5	220—140
	0.2—0.25	~140—100
Ba II	~1—1.8	~55—85
III	1.8—5	~85—144
IV	4.5—5.4	144—190
Bi II	3.9	25—27
III	6.55—7.25	28—38
IV	7.0,8.7—6.0	43,43—62
V	6.7,8.3	48—80
VI	8.55	90,92—101
VII(?)	8.2	30
Ce (α)	0.020—0.045	20—35
Ce (α')	1.9—1.3	45—125
Cs V	~1.5	>125
Ga II	6.38	≥35
II'	7.5	≥35 then P removed
Ge	5.35	115
La	~5.5—12.9	0—210
Lu	0.022—1.0	45—190
P	5.8	170
Pb II	3.55	160
Re II	2.3 Max.	"Plastic" compression
Sb (prepared 120 kbar, held below 77K)	2.6—2.7	
Sb III	3.55—3.40	85—~150
Se II	6.75,6.95	~130
Si	6.7—7.1	120—130
Sn II	5.2—4.85	125—160
III	5.30	113
Te II	2.4—5.1	38—55
III	4.1—4.2	~53—62
IV	4.72—4	63—80
( )	3.3—2.8	100—260
Tl (cubic form)	1.45	35
(Hexagonal form)	1.95	35
U	2.4—0.4	10—85
Y	2.3—1.7—2.5	110—125—160
Zr (omega form, metastable)	1—1.7	60—~130

* 1 kbar = 10^8 newton/meter^2 = 0.987 katm

From Roberts, B. W., Properties of Selected Superconductive Materials, 1978 Supplement, NBS Technical Note 983, U.S. Government Printing Office, Washington, D.C.

SELECTED SUPERCONDUCTIVE COMPOUNDS AND ALLOYS

All compositions are denoted on an atomic basis, i.e., AB, AB_2, or AB_3 for compounds, unless noted. Solid solutions or odd compositions may be denoted as A_zB_{1-z} or A_zB. A series of three or more alloys is indicated as A_xB_{1-x} or by actual indication of the atomic fraction range, such as $A_{0-0.6}B_{1-0.4}$. The critical temperature of such a series of alloys is denoted by a range of values or possibly the maximum value.

The selection of the critical temperature from a transition in the effective permeability, or the change in resistance, or possibly the incremental changes in frequency observed by certain techniques is not often obvious from the literature. Most authors choose the mid-point of such curves as the probable critical temperature of the idealized material, while others will choose the highest temperature at which a deviation from the normal state property is observed. In view of the previous discussion concerning the variability of the superconductive properties as a function of purity and other metallurgical aspects, it is recommended that appropriate literature be checked to determine the most probable critical temperature or critical field of a given alloy.

A very limited amount of data on critical fields, H_o, is available for these compounds and alloys; these values are given at the end of the table.

SYMBOLS: n = number of normal carriers per cubic centimeter for semiconductor superconductors.

Substance	T_c, °K	Crystal structure type††	Substance	T_c, °K	Crystal structure type††
$Ag_xAl_yZn_{1-x-y}$	0.5–0.845		$Al_{-0.8}Ge_{-0.7}Nb_3$	20.7	A15
$Ag_7BF_4O_8$	0.15	Cubic	$AlLa_3$	5.57	DO_{19}
$AgBi_2$	3.0–2.78		Al_2La	3.23	C15
$Ag_7F_{0.25}N_{0.75}O_{10.25}$	0.85–0.90		Al_3Mg_2	0.84	Cubic, f.c.
Ag_7FO_8	0.3	Cubic	$AlMo_3$	0.58	A15
Ag_2F	0.066		$AlMo_6Pd$	2.1	
$Ag_{0.8-0.3}Ga_{0.2-0.7}$	6.5–8		AlN	1.55	B4
Ag_4Ge	0.85	Hex., c.p.	Al_2NNb_3	1.3	A13
$Ag_{0.438}Hg_{0.562}$	0.64	$D8_2$	$AlNb_3$	18.0	A15
$AgIn_2$	~2.4	C16	Al_xNb_{1-x}	<4.2–13.5	$D8_b$
$Ag_{0.1}In_{0.9}Te$			Al_xNb_{1-x}	12–17.5	A15
$(n = 1.40 \times 10^{22})$	1.20–1.89	Bl	$Al_{0.27}Nb_{0.73-0.48}V_{0-0.25}$	14.5–17.5	A15
$Ag_{0.2}In_{0.8}Te$			$AlNb_xV_{1-x}$	<4.2–13.5	
$(n = 1.07 \times 10^{22})$	0.77–1.00	Bl	AlOs	0.39	B2
AgLa (9.5 kbar)	1.2	B2	Al_3Os	5.90	
Ag_7NO_{11}	1.04	Cubic	AlPb (films)	1.2–7	
Ag_xPb_{1-x}	7.2 max.		Al_2Pt	0.48–0.55	Cl
Ag_xSn_{1-x} (film)	2.0–3.8		Al_5Re_{24}	3.35	A12
Ag_xSn_{1-x}	1.5–3.7		Al_3Th	0.75	DO_{19}
$AgTe_3$	2.6	Cubic	$Al_xTi_yV_{1-x-y}$	2.05–3.62	Cubic
$AgTh_2$	2.26	C16	$Al_{0.108}V_{0.892}$	1.82	Cubic
$Ag_{0.03}Tl_{0.97}$	2.67		Al_xZn_{1-x}	0.5–0.845	
$Ag_{0.94}Tl_{0.06}$	2.32		$AlZr_3$	0.73	Ll_2
Ag_xZn_{1-x}	0.5–0.845		AsBiPb	9.0	
Al (film)	1.3–2.31		AsBiPbSb	9.0	
Al (1 to 21 katm)	1.170–0.687	Al	$As_{0.33}InTe_{0.67}$		
$AlAu_4$	0.4–0.7	Like Al3	$(n = 1.24 \times 10^{22})$	0.85–1.15	Bl
Al_2CMo_3	10.0	A13	$As_{0.5}InTe_{0.5}$		
Al_2CMo_3	9.8–10.2	A13 + trace 2nd phase	$(n = 0.97 \times 10^{22})$	0.44–0.62	Bl
			$As_{0.50}Ni_{0.06}Pd_{0.44}$	1.39	C2
Al_2CaSi	5.8		AsPb	8.4	
$Al_{0.131}Cr_{0.088}V_{0.781}$	1.46	Cubic	$AsPd_2$ (low-temperature phase)	0.60	Hexagonal
$AlGe_2$	1.75				
$Al_{0.5}Ge_{0.5}Nb$	12.6	A15	$AsPd_2$ (high-temp. phase)	1.70	C22

††See key at end of table.

SELECTED SUPERCONDUCTIVE COMPOUNDS AND ALLOYS (Continued)

Substance	T_c, °K	Crystal structure type††	Substance	T_c, °K	Crystal structure type††
$AsPd_5$	0.46	Complex	BW_2	3.1	Cl6
$AsRh$	0.58	B31	B_6Y	6.5–7.1	
$AsRh_{1.4-1.6}$	<0.03–0.56	Hexagonal	$B_{12}Y$	4.7	
$AsSn$	4.10		BZr	3.4	Cubic
$AsSn$			$B_{12}Zr$	5.82	
$(n = 2.14 \times 10^{22})$	3.41–3.65	Bl	$BaBi_3$	5.69	Tetragonal
$As_{\sim 2}Sn_{\sim 3}$	3.5–3.6,		$Ba_xO_3Sr_{1-x}Ti$		
	1.21–1.17		$(n = 4.2-11 \times 10^{19})$	<0.1–0.55	Tetragonal
As_3Sn_4			$Ba_{0.13}O_3W$	1.9	Tetragonal
$(n = 0.56 \times 10^{22})$	1.16–1.19	Rhombohedral	$Ba_{0.14}O_3W$	<1.25–2.2	Hexagonal
Au_5Ba	0.4–0.7	$D2_d$	$BaRh_2$	6.0	Cl5
$AuBe$	2.64	B20	$Be_{22}Mo$	2.51	Cubic,
Au_2Bi	1.80	Cl5			like $Be_{22}Re$
Au_5Ca	0.34–0.38	$Cl5_b$	$Be_8Nb_5Zr_2$	5.2	
$AuGa$	1.2	B31	$Be_{0.98-0.92}Re_{0.02-0.08}$		
$Au_{0.40-0.92}Ge_{0.60-0.08}$	<0.32–1.63	Complex	(quenched)	9.5–9.75	Cubic
$AuIn$	0.4–0.6	Complex	$Be_{0.957}Re_{0.043}$	9.62	Cubic,
$AuLu$	<0.35	B2			like $Be_{22}Re$
$AuNb_3$	11.5	A15	$BeTc$	5.21	Cubic
$AuNb_3$	1.2	A2	$Be_{22}W$	4.12	Cubic,
$Au_{0-0.3}Nb_{1-0.7}$	1.1–11.0				like $Be_{22}Re$
$Au_{0.02-0.98}Nb_3Rh_{0.98-0.02}$	2.53–10.9	A15	$Be_{13}W$	4.1	Tetragonal
$AuNb_{3(1-x)}V_{3x}$	1.5–11.0	A15	Bi_3Ca	2.0	
$AuPb_2$	3.15		$Bi_{0.5}Cd_{0.13}Pb_{0.25}Sn_{0.12}$		
$AuPb_2$ (film)	4.3		(weight fractions)	8.2	
$AuPb_3$	4.40		$BiCo$	0.42–0.49	
$AuPb_3$ (film)	4.25		Bi_2Cs	4.75	Cl5
Au_2Pb	1.18, 6–7	Cl5	Bi_xCu_{1-x}		
$AuSb_2$	0.58	C2	(electrodeposited)	2.2	
$AuSn$	1.25	$B8_1$	$BiCu$	1.33–1.40	
Au_xSn_{1-x} (film)	2.0–3.8		$Bi_{0.019}In_{0.981}$	3.86	
Au_5Sn	0.7–1.1	A3	$Bi_{0.05}In_{0.95}$	4.65	α-phase
Au_3Te_5	1.62	Cubic	$Bi_{0.10}In_{0.90}$	5.05	α-phase
$AuTh_2$	3.08	Cl6	$Bi_{0.15-0.30}In_{0.85-0.70}$	5.3–5.4	α- and β-phases
$AuTl$	1.92		$Bi_{0.34-0.48}In_{0.66-0.52}$	4.0–4.1	
AuV_3	0.74	A15	Bi_3In_5	4.1	
Au_xZn_{1-x}	0.50–0.845		$BiIn_2$	5.65	β-phase
$AuZn_3$	1.21	Cubic	Bi_2Ir	1.7–2.3	
Au_xZr_y	1.7–2.8	A3	Bi_2Ir (quenched)	3.0–3.96	
$AuZr_3$	0.92	A15	BiK	3.6	
$BCMo_2$	5.4	Orthorhombic	Bi_2K	3.58	Cl5
$B_{0.03}C_{0.51}Mo_{0.47}$	12.5		$BiLi$	2.47	Ll_0, α-phase
$BCMo_2$	5.3–7.0	Orthorhombic	$Bi_{4-9}Mg$	0.7–~1.0	
BHf	3.1	Cubic	Bi_3Mo	3–3.7	
B_6La	5.7		$BiNa$	2.25	Ll_0
$B_{12}Lu$	0.48		$BiNb_3$ (high pressure		
BMo	0.5 (extrap-		and temperature)	3.05	A15
	olated)		$BiNi$	4.25	$B8_1$
BMo_2	4.74	Cl6	Bi_3Ni	4.06	Orthorhombic
BNb	8.25	B_f	$Bi_{1-0}Pb_{0-1}$	7.26–9.14	
BRe_2	2.80, 4.6		$Bi_{1-0}Pb_{0-1}$ (film)	7.25–8.67	
$B_{0.3}Ru_{0.7}$	2.58	$D10_2$	$Bi_{0.05-0.40}Pb_{0.95-0.60}$	7.35–8.4	Hexagonal,
$B_{12}Sc$	0.39				c.p., to
BTa	4.0	B_f			ε-phase
B_6Th	0.74		$BiPbSb$	8.9	

††See key at end of table.

Substance	T_c, °K	Crystal structure type††	Substance	T_c, °K	Crystal structure type††
$Bi_{0.5}Pb_{0.31}Sn_{0.19}$ (weight fractions)	8.5		$C_{0.44}Mo_{0.56}$	1.3	Bl
$Bi_{0.5}Pb_{0.25}Sn_{0.25}$	8.5		$C_{0.5}Mo_xNb_{1-x}$	10.8–12.5	Bl
$BiPd_2$	4.0		$C_{0.6}Mo_{4.8}Si_3$	7.6	$D8_8$
$Bi_{0.4}Pd_{0.6}$	3.7–4	Hexagonal, ordered	$CMo_{0.2}Ta_{0.8}$	7.5	Bl
$BiPd$	3.7	Orthorhombic	$CMo_{0.5}Ta_{0.5}$	7.7	Bl
Bi_2Pd	1.70	Monoclinic, α-phase	$CMo_{0.75}Ta_{0.25}$	8.5	Bl
			$CMo_{0.8}Ta_{0.2}$	8.7	Bl
Bi_2Pd	4.25	Tetragonal, β-phase	$CMo_{0.85}Ta_{0.15}$	8.9	Bl
			CMo_xTi_{1-x}	10.2 max.	Bl
$BiPdSe$	1.0	C2	$CMo_{0.83}Ti_{0.17}$	10.2	Bl
$BiPdTe$	1.2	C2	CMo_xV_{1-x}	2.9–9.3	Bl
$BiPt$	1.21	$B8_1$	CMo_xZr_{1-x}	3.8–9.5	Bl
$BiPtSe$	1.45	C2	$C_{0.1-0.9}N_{0.9-0.1}Nb$	8.5–17.9	
$BiPtTe$	1.15	C2	$C_{0-0.38}N_{1-0.62}Ta$	10.0–11.3	
Bi_2Pt	0.155	Hexagonal	CNb (whiskers)	7.5–10.5	
Bi_2Rb	4.25	Cl5	$C_{0.984}Nb$	9.8	Bl
$BiRe_2$	1.9–2.2		CNb (extrapolated)	~14	
$BiRh$	2.06	$B8_1$	$C_{0.7-1.0}Nb_{0.3-0}$	6–11	Bl
Bi_3Rh	3.2	Orthorhombic, like NiB_3	CNb_2	9.1	
			CNb_xTa_{1-x}	8.2–13.9	
Bi_4Rh	2.7	Hexagonal	CNb_xTi_{1-x}	<4.2–8.8	Bl
Bi_3Sn	3.6–3.8		$CNb_{0.6-0.9}W_{0.4-0.1}$	12.5–11.6	Bl
$BiSn$	3.8		$CNb_{0.1-0.9}Zr_{0.9-0.1}$	4.2–8.4	Bl
Bi_xSn_y	3.85–4.18		CRe_x (gold)	0.023–0.131	Hexagonal
Bi_3Sr	5.62	Ll_2	$CRe_{0.01-0.08}W$	1.3 5.0	
Bi_3Te	0.75–1.0		$CRe_{0.06}W$	5.0	
Bi_5Tl_3	6.4		CTa	~11 (extrapolated)	
$Bi_{0.26}Tl_{0.74}$	4.4	Cubic, disordered	$C_{0.987}Ta$	9.7	
			$C_{0.848-0.987}Ta$	2.04–9.7	
$Bi_{0.26}Tl_{0.74}$	4.15	Ll_2, ordered?	CTa (film)	5.09	Bl
Bi_2Y_3	2.25		CTa_2	3.26	L'_3
Bi_3Zn	0.8–0.9		$CTa_{0.4}Ti_{0.6}$	4.8	Bl
$Bi_{0.3}Zr_{0.7}$	1.51		$CTa_{1-0.4}W_{0-0.6}$	8.5–10.5	Bl
$BiZr_3$	2.4–2.8		$CTa_{0.2-0.9}Zr_{0.8-0.1}$	4.6–8.3	Bl
CCs_x	0.020–0.135	Hexagonal	CTc (excess C)	3.85	Cubic
C_8K (gold)	0.55		$CTi_{0.5-0.7}W_{0.5-0.3}$	6.7–2.1	Bl
$CGaMo_2$	3.7–4.1	Hexagonal, H-phase	CW	1.0	
			CW_2	2.74	L'_3
$CHf_{0.5}Mo_{0.5}$	3.4	Bl	CW_2	5.2	Cubic, f.c.
$CHf_{0.3}Mo_{0.7}$	5.5	Bl	$CaIr_2$	6.15	Cl5
$CHf_{0.25}Mo_{0.75}$	6.6	Bl	$Ca_xO_3Sr_{1-x}Ti$ $(n = 3.7-11.0 \times 10^{19})$	<0.1–0.55	
$CHf_{0.7}Nb_{0.3}$	6.1	Bl	$Ca_{0.1}O_3W$	1.4–3.4	Hexagonal
$CHf_{0.6}Nb_{0.4}$	4.5	Bl	$CaPb$	7.0	
$CHf_{0.5}Nb_{0.5}$	4.8	Bl	$CaRh_2$	6.40	Cl5
$CHf_{0.4}Nb_{0.6}$	5.6	Bl	$Cd_{0.3-0.5}Hg_{0.7-0.5}$	1.70–1.92	
$CHf_{0.25}Nb_{0.75}$	7.0	Bl	$CdHg$	1.77, 2.15	Tetragonal
$CHf_{0.2}Nb_{0.8}$	7.8	Bl	$Cd_{0.0075-0.05}In_{1-x}$	3.24–3.36	Tetragonal
$CHf_{0.9-0.1}Ta_{0.1-0.9}$	5.0–9.0	Bl	$Cd_{0.97}Pb_{0.03}$	4.2	
Ck (excess K)	0.55	Hexagonal	$CdSn$	3.65	
C_8K	0.39	Hexagonal	$Cd_{0.17}Tl_{0.83}$	2.3	
$C_{0.40-0.44}Mo_{0.60-0.56}$	9–13		$Cd_{0.18}Tl_{0.82}$	2.54	
CMo	6.5, 9.26		$CeCo_2$	0.84	Cl5
CMo_2	12.2	Orthorhombic	$CeCo_{1.67}Ni_{0.33}$	0.46	Cl5

††See key at end of table.

Substance	T_c, °K	Crystal structure type††	Substance	T_c, °K	Crystal structure type††
$CeCo_{1.67}Rh_{0.33}$	0.47	C15	CuSSe	1.5–2.0	C18
$Ce_xGd_{1-x}Ru_2$	3.2–5.2	C15	$CuSe_2$	2.3–2.43	C18
$CeIr_3$	3.34		CuSeTe	1.6–2.0	C18
$CeIr_5$	1.82		Cu_xSn_{1-x} (film)	3.2–3.7	
$Ce_{0.005}La_{0.995}$	4.6		Cu_xSn_{1-x} (film)		
Ce_xLa_{1-x}	1.3–6.3		(made at 10°K)	3.6–7	
$Ce_xPr_{1-x}Ru_2$	1.4–5.3	C15	Cu_xSn_{1-x} (film)		
Ce_xPt_{1-x}	0.7–1.55		(made at 300°K)	2.8–3.7	
$CeRu_2$	6.0	C15	$CuTe_2$	<1.25–1.3	C18
$Co_xFe_{1-x}Si_2$	1.4 max.	C1	$CuTh_2$	3.49	C16
$CoHf_2$	0.56	$E9_3$	$Cu_{0-0.027}V$	3.9–5.3	A2
$CoLa_3$	4.28		Cu_xZn_{1-x}	0.5–0.845	
$CoLu_3$	~0.35		Er_xLa_{1-x}	1.4–6.3	
$Co_{0-0.01}Mo_{0.8}Re_{0.2}$	2–10		$Fe_{0-0.04}Mo_{0.8}Re_{0.2}$	1–10	
$Co_{0.02-0.10}Nb_3Rh_{0.98-0.90}$	2.28–1.90	A15	$Fe_{0.05}Ni_{0.05}Zr_{0.90}$	~3.9	
$Co_xNi_{1-x}Si_2$	1.4 max.	C1	Fe_3Th_7	1.86	D10
$Co_{0.5}Rh_{0.5}Si_2$	2.5		Fe_xTi_{1-x}	3.2 max.	Fe in α-Ti
$Co_xRh_{1-x}Si_2$	3.65 max.		Fe_xTi_{1-x}	3.7 max.	Fe in β-Ti
$Co_{~0.3}Sc_{~0.7}$	~0.35		$Fe_xTi_{0.6}V_{1-x}$	6.8 max.	
$CoSi_2$	1.40, 1.22	C1	FeU_6	3.86	$D2_c$
Co_3Th_7	1.83	$D10_2$	$Fe_{0.1}Zr_{0.9}$	1.0	A3
Co_xTi_{1-x}	2.8 max.	Co in α-Ti	$Ga_{0.5}Ge_{0.5}Nb_3$	7.3	A15
Co_xTi_{1-x}	3.8 max.	Co in β-Ti	$GaLa_3$	5.84	
$CoTi_2$	3.44	$E9_3$	Ga_2Mo	9.5	
CoTi	0.71	A2	$GaMo_3$	0.76	A15
CoU	1.7	B2, distorted	Ga_4Mo	9.8	
CoU_6	2.29	$D2_c$	GaN (black)	5.85	B4
$Co_{0.28}Y_{0.72}$	0.34		$GaNb_3$	14.5	A15
CoY_3	<0.34		$Ga_xNb_3Sn_{1-x}$	14–18.37	A15
$CoZr_2$	6.3	C16	$Ga_{0.7}Pt_{0.3}$	2.9	C1
$Co_{0.1}Zr_{0.9}$	3.9	A3	GaPt	1.74	B20
$Cr_{0.6}Ir_{0.4}$	0.4	Hexagonal, c.p.	GaSb (120 kbar, 77°K,		
$Cr_{0.65}Ir_{0.35}$	0.59	Hexagonal, c.p.	annealed)	4.24	A5
$Cr_{0.7}Ir_{0.3}$	0.76	Hexagonal, c.p.	GaSb (unannealed)	~5.9	
$Cr_{0.72}Ir_{0.28}$	0.83		$Ga_{0.1}Sn_{1-0}$ (quenched)	3.47–4.18	
Cr_3Ir	0.45	A15	$Ga_{0-1}Sn_{1-0}$ (annealed)	2.6–3.85	
$Cr_{0-0.1}Nb_{1-0.9}$	4.6–9.2	A2	Ga_5V_2	3.55	Tetragonal,
$Cr_{0.80}Os_{0.20}$	2.5	Cubic			Mn_2Hg_5 type
Cr_xRe_{1-x}	1.2–5.2		GaV_3	16.8	A15
$Cr_{0.40}Re_{0.60}$	2.15	$D8_b$	$GaV_{2.1-3.5}$	6.3–14.45	A15
$Cr_{0.8-0.6}Rh_{0.2-0.4}$	0.5–1.10	A3	$GaV_{4.5}$	9.15	
Cr_3Ru (annealed)	3.3	A15	Ga_3Zr	1.38	
Cr_2Ru	2.02	$D8_b$	Gd_xLa_{1-x}	<1.0–5.5	
$Cr_{0.1-0.5}Ru_{0.9-0.5}$	0.34–1.65	A3	$Gd_xOs_2Y_{1-x}$	1.4–4.7	
Cr_xTi_{1-x}	3.6 max.	Cr in α-Ti	$Gd_xRu_2Th_{1-x}$	3.6 max.	C15
Cr_xTi_{1-x}	4.2 max.	Cr in β-Ti	GeIr	4.7	B31
$Cr_{0.1}Ti_{0.3}V_{0.6}$	5.6		Ge_2La	1.49, 2.2	Orthorhombic,
$Cr_{0.0175}U_{0.9825}$	0.75	β-phase			distorted
$Cs_{0.32}O_3W$	1.12	Hexagonal			$ThSi_2$-type
$Cu_{0.15}In_{0.85}$ (film)	3.75		$GeMo_3$	1.43	A15
$Cu_{0.04-0.08}In_{1-x}$	4.4		$GeNb_2$	1.9	
CuLa	5.85		$GeNb_3$ (quenched)	6–17	A15
Cu_xPb_{1-x}	5.7–7.7		$Ge_{0.29}Nb_{0.71}$	6	A15
CuS	1.62	B18	$Ge_xNb_3Sn_{1-x}$	17.6–18.0	A15
CuS_2	1.48–1.53	C18	$Ge_{0.5}Nb_3Sn_{0.5}$	11.3	

††See key at end of table.

Substance	T_c, °K	Crystal structure type††	Substance	T_c, °K	Crystal structure type††
GePt	0.40	B3l	InSb	2.1	
Ge_3Rh_5	2.12	Orthorhombic, related to $InNi_2$	$(InSb)_{0.95-0.10}Sn_{0.05-0.90}$ (various heat treatments)	3.8–5.1	
Ge_2Sc	1.3		$(InSb)_{0-0.07}Sn_{1-0.93}$	3.67–3.74	
Ge_3Te_4 ($n \sim 1.06 \times 10^{22}$)	1.55–1.80	Rhombohedral	In_3Sn	~5.5	
Ge_xTe_{1-x} ($n = 8.5 - 64 \times 10^{20}$)	0.07–0.41	Bl	In_xSn_{1-x}	3.4–7.3	
GeV_3	6.01	A15	$In_{0.82-1}Te$ ($n = 0.83-1.71 \times 10^{22}$)	1.02–3.45	Bl
Ge_2Y	3.80	C_c	$In_{1.000}Te_{1.002}$	3.5–3.7	Bl
$Ge_{1.62}Y$	2.4		In_3Te_4 ($n = 0.47 \times 10^{22}$)	1.15–1.25	Rhombohedral
$H_{0.33}Nb_{0.67}$	7.28	Cubic, b.c.	In_xTl_{1-x}	2.7–3.374	
$H_{0.1}Nb_{0.9}$	7.38	Cubic, b.c.	$In_{0.8}Tl_{0.2}$	3.223	
$H_{0.05}Nb_{0.95}$	7.83	Cubic, b.c.	$In_{0.62}Tl_{0.38}$	2.760	
$H_{0.12}Ta_{0.88}$	2.81	Cubic, b.c.	$In_{0.78-0.69}Tl_{0.22-0.31}$	3.18–3.32	Tetragonal
$H_{0.08}Ta_{0.92}$	3.26	Cubic, b.c.	$In_{0.69-0.62}Tl_{0.31-0.38}$	2.98–3.3	Cubic, f.c.
$H_{0.04}Ta_{0.96}$	3.62	Cubic, b.c.	Ir_2La	0.48	Cl5
$HfN_{0.989}$	6.6	Bl	Ir_3La	2.32	$Dl0_2$
$Hf_{0-0.5}Nb_{1-0.5}$	8.3–9.5	A2	Ir_3La_7	2.24	$Dl0_2$
$Hf_{0.75}Nb_{0.25}$	>4.2		Ir_5La	2.13	
$HfOs_2$	2.69	Cl4	Ir_2Lu	2.47	Cl5
$HfRe_2$	4.80	Cl4	Ir_3Lu	2.89	Cl5
$Hf_{0.14}Re_{0.86}$	5.86	A12	IrMo	<1.0	A3
$Hf_{0.99-0.96}Rh_{0.01-0.04}$	0.85–1.51		$IrMo_3$	8.8	A15
$Hf_{0-0.55}Ta_{1-0.45}$	4.4–6.5	A2	$IrMo_3$	6.8	$D8_b$
HfV_2	8.9–9.6	Cl5	$IrNb_3$	1.9	A15
Hg_xIn_{1-x}	3.14–4.55		$Ir_{0.4}Nb_{0.6}$	9.8	$D8_b$
HgIn	3.81		$Ir_{0.37}Nb_{0.63}$	2.32	$D8_b$
Hg_2K	1.20	Orthorhombic	IrNb	7.9	$D8_b$
Hg_3K	3.18		$Ir_{0.02}Nb_3Rh_{0.98}$	2.43	A15
Hg_4K	3.27		$Ir_{0.05}Nb_3Rh_{0.95}$	2.38	A15
Hg_8K	3.42		$Ir_{0.287}O_{0.14}Ti_{0.573}$	5.5	$E9_3$
Hg_3Li	1.7	Hexagonal	$Ir_{0.265}O_{0.035}Ti_{0.65}$	2.30	$E9_3$
Hg_2Na	1.62	Hexagonal	Ir_xOs_{1-x}	0.3–0.98 (max.)–0.6	
Hg_4Na	3.05		IrOsY	2.6	Cl5
Hg_xPb_{1-x}	4.14–7.26		$Ir_{1.5}Os_{0.5}$	2.4	Cl4
HgSn	4.2		Ir_2Sc	2.07	Cl5
Hg_xTl_{1-x}	2.30–4.109		$Ir_{2.5}Sc$	2.46	Cl5
Hg_5Tl_2	3.86		$IrSn_2$	0.65–0.78	Cl
Ho_xLa_{1-x}	1.3–6.3		Ir_2Sr	5.70	Cl5
$InLa_3$	9.83, 10.4	Ll_2	$Ir_{0.5}Te_{0.5}$	~3	
$InLa_3$ (0–35, kbar)	9.75–10.55		$IrTe_3$	1.18	C2
$In_{1-0.86}Mg_{0-0.14}$	3.395–3.363		IrTh	<0.37	B_f
$InNb_3$ (high pressure and temp.)	4–8, 9.2	A15	Ir_2Th	6.50	Cl5
$In_{0-0.3}Nb_3Sn_{1-0.7}$	18.0–18.19	A15	Ir_3Th	4.71	
$In_{0.5}Nb_3Zr_{0.5}$	6.4		Ir_3Th_7	1.52	$Dl0_2$
$In_{0.11}O_3W$	<1.25–2.8	Hexagonal	Ir_5Th	3.93	$D2_d$
$In_{0.95-0.85}Pb_{0.05-0.15}$	3.6–5.05		$IrTi_3$	5.40	A15
$In_{0.98-0.91}Pb_{0.02-0.09}$	3.45–4.2		IrV_2	1.39	A15
InPb	6.65		IrW_3	3.82	
InPd	0.7	B2	$Ir_{0.28}W_{0.72}$	4.49	
InSb (quenched from 170 kbar into liquid N_2)	4.8	Like A5	Ir_2Y	2.18, 1.38	Cl5
			$Ir_{0.69}Y_{0.31}$	1.98, 1.44	Cl5
			$Ir_{0.70}Y_{0.30}$	2.16	Cl5

††See key at end of table.

Substance	T_c, °K	Crystal structure type[††]	Substance	T_c, °K	Crystal structure type[††]
Ir_2Y	1.09	Cl5	Mo_3Si	1.30	Al5
Ir_2Y_3	1.61		$MoSi_{0.7}$	1.34	
Ir_xY_{1-x}	0.3–3.7		Mo_xSiV_{3-x}	4.54–16.0	Al5
Ir_2Zr	4.10	Cl5	Mo_xTc_{1-x}	10.8–15.8	
$Ir_{0.1}Zr_{0.9}$	5.5	A3	$Mo_{0.16}Ti_{0.84}$	4.18, 4.25	
$K_{0.27-0.31}O_3W$	0.50	Hexagonal	$Mo_{0.913}Ti_{0.087}$	2.95	
$K_{0.40-0.57}O_3W$	1.5	Tetragonal	$Mo_{0.04}Ti_{0.96}$	2.0	Cubic
$La_{0.55}Lu_{0.45}$	2.2	Hexagonal, La type	$Mo_{0.025}Ti_{0.975}$	1.8	
$La_{0.8}Lu_{0.2}$	3.4	Hexagonal, La Type	Mo_xU_{1-x}	0.7–2.1	
			Mo_xV_{1-x}	0–~5.3	
$LaMg_2$	1.05	Cl5	Mo_2Zr	4.27–4.75	Cl5
LaN	1.35		NNb (whiskers)	10–14.5	
$LaOs_2$	6.5	Cl5	NNb (diffusion wires)	16.10	
$LaPt_2$	0.46	Cl5	NNb (film)	6–9	Bl
$La_{0.28}Pt_{0.72}$	0.54	Cl5	$N_{0.988}Nb$	14.9	Bl
$LaRh_3$	2.60		$N_{0.824-0.988}Nb$	14.4–15.3	Bl
$LaRh_5$	1.62		$N_{0.70-0.795}Nb$	11.3–12.9	Cubic and tetragonal
La_7Rh_3	2.58	Dl0$_2$	NNb_xO_y	13.5–17.0	Bl
$LaRu_2$	1.63	Cl5	NNb_xO_y	6.0–11	
La_3S_4	6.5	D7$_3$	$N_{100-42\ w/o}Nb_{0-58\ w/o}Ti$[†]	15–16.8	
La_3Se_4	8.6	D7$_3$	$N_{100-75\ w/o}Nb_{0-25\ w/o}Zr$[†]	12.5–16.35	
$LaSi_2$	2.3	C_c	NNb_xZr_{1-x}	9.8–13.8	Bl
La_xY_{1-x}	1.7–5.4		$N_{0.93}Nb_{0.85}Zr_{0.15}$	13.8	Bl
$LaZn$	1.04	B2	$N_xO_yTi_z$	2.9–5.6	Cubic
$LiPb$	7.2		$N_xO_yV_z$	5.8–8.2	Cubic
$LuOs_2$	3.49	Cl4	$N_{0.34}Re$	4–5	Cubic, f.c.
$Lu_{0.275}Rh_{0.725}$	1.27	Cl5	NTa	12–14	Bl
$LuRh_5$	0.49			(extrapolated)	
$LuRu_2$	0.86	Cl4			
$Mg_{~0.47}Tl_{~0.53}$	2.75	B2	NTa (film)	4.84	Bl
Mg_2Nb	5.6		$N_{0.6-0.987}Ti$	<1.17–5.8	Bl
Mn_xTi_{1-x}	2.3 max.	Mn in α-Ti	$N_{0.82-0.99}V$	2.9–7.9	Bl
Mn_xTi_{1-x}	1.1–3.0	Mn in β-Ti	NZr	9.8	Bl
MnU_6	2.32	D2$_c$	$N_{0.906-0.984}Zr$	3.0–9.5	Bl
MoN	12	Hexagonal	$Na_{0.28-0.35}O_3W$	0.56	Tetragonal
Mo_2N	5.0	Cubic, f.c.	$Na_{0.28}Pb_{0.72}$	7.2	
Mo_xNb_{1-x}	0.016–9.2		NbO	1.25	
Mo_3Os	7.2	Al5	$NbOs_2$	2.52	Al2
$Mo_{0.62}Os_{0.38}$	5.65	D8$_b$	Nb_3Os	1.05	Al5
Mo_3P	5.31	DO$_e$	$Nb_{0.6}Os_{0.4}$	1.89, 1.78	D8$_b$
$Mo_{0.5}Pd_{0.5}$	3.52	A3	$Nb_3Os_{0.02-0.10}Rh_{0.98-0.90}$	2.42–2.30	Al5
Mo_3Re	10.0		$Nb_{0.6}Pd_{0.4}$	1.60	D8$_f$ plus cubic
Mo_xRe_{1-x}	1.2–12.2		$Nb_3Pd_{0.02-0.10}Rh_{0.98-0.90}$	2.49–2.55	Al5
$MoRe_3$	9.25, 9.89	Al2	$Nb_{0.62}Pt_{0.38}$	4.21	D8$_b$
$Mo_{0.42}Re_{0.58}$	6.35	D8$_b$	Nb_3Pt	10.9	Al5
$Mo_{0.52}Re_{0.48}$	11.1		Nb_5Pt_3	3.73	D8$_b$
$Mo_{0.57}Re_{0.43}$	14.0		$Nb_3Pt_{0.02-0.98}Rh_{0.98-0.02}$	2.52–9.6	Al5
$Mo_{~0.60}Re_{0.395}$	10.6		$Nb_{0.38-0.18}Re_{0.62-0.82}$	2.43–9.70	Al2
$MoRh$	1.97	A3	Nb_3Rh	2.64	Al5
Mo_xRh_{1-x}	1.5–8.2	Cubic, b.c.	$Nb_{0.60}Rh_{0.40}$	4.21	D8$_b$ plus other
$MoRu$	9.5–10.5	A3	$Nb_3Rh_{0.98-0.90}Ru_{0.02-0.10}$	2.42–2.44	Al5
$Mo_{0.61}Ru_{0.39}$	7.18	D8$_b$	Nb_xRu_{1-x}	1.2–4.8	
$Mo_{0.2}Ru_{0.8}$	1.66	A3	NbS_2	6.1–6.3	Hexagonal, NbSe$_2$ type
Mo_3Sb_4	2.1				

[†]w/o denotes weight percent.   [††]See key at end of table.

Substance	T_c, °K	Crystal structure type††	Substance	T_c, °K	Crystal structure type††
NbS_2	5.0–5.5	Hexagonal, three-layer type	Os_2Zr	3.0	C14
			Os_xZr_{1-x}	1.50–5.6	
			PPb	7.8	
$Nb_3Sb_{0-0.7}Sn_{1-0.3}$	6.8–18	A15	$PPd_{3.0-3.2}$	<0.35–0.7	DO_{11}
$NbSe_2$	5.15–5.62	Hexagonal, NbS_2 type	P_3Pd_7 (high temperature)	1.0	Rhombohedral
$Nb_{1-1.05}Se_2$	2.2–7.0	Hexagonal, NbS_2 type	P_3Pd_7 (low temp.)	0.70	Complex
			PRh	1.22	
Nb_3Si	1.5	Ll$_2$	PRh_2	1.3	Cl
Nb_3SiSnV_3	4.0		PW_3	2.26	DO_e
Nb_3Sn	18.05	A15	Pb_2Pd	2.95	Cl6
$Nb_{0.8}Sn_{0.2}$	18.18, 18.5	A15	Pb_4Pt	2.80	Related to Cl6
Nb_xSn_{1-x} (film)	2.6–18.5		Pb_2Rh	2.66	Cl6
$NbSn_2$	2.60	Orthorhombic	PbSb	6.6	
Nb_3Sn_2	16.6	Tetragonal	PbTe (plus 0.1 w/o Pb)†	5.19	
$NbSnTa_2$	10.8	A15	PbTe (plus 0.1 w/o Tl)†	5.24–5.27	
Nb_2SnTa	16.4	A15	$PbTl_{0.27}$	6.43	
$Nb_{2.5}SnTa_{0.5}$	17.6	A15	$PbTl_{0.17}$	6.73	
$Nb_{2.75}SnTa_{0.25}$	17.8	A15	$PbTl_{0.12}$	6.88	
$Nb_{3x}SnTa_{3(1-x)}$	6.0–18.0		$PbTl_{0.075}$	6.98	
NbSnTaV	6.2	A15	$PbTl_{0.04}$	7.06	
$Nb_2SnTa_{0.5}V_{0.5}$	12.2	A15	$Pb_{1-0.26}Tl_{0-0.74}$	7.20–3.68	
$NbSnV_2$	5.5	A15	$PbTl_2$	3.75–4.1	
Nb_2SnV	9.8	A15	Pb_3Zr_5	4.60	$D8_8$
$Nb_{2.5}SnV_{0.5}$	14.2	A15	$PbZr_3$	0.76	A15
Nb_xTa_{1-x}	4.4–9.2	A2	$Pd_{0.9}Pt_{0.1}Te_2$	1.65	C6
$NbTa_3$	10.5	A15	$Pd_{0.05}Ru_{0.67}Zr_{0.9}$	~9	
Nb_xTi_{1-x}	0.6–9.8		$Pd_{2.2}S$ (quenched)	1.63	Cubic
$Nb_{0.6}Ti_{0.4}$	9.8		$PdSb_2$	1.25	C2
Nb_xU_{1-x}	1.95 max.		PdSb	1.50	$B8_1$
$Nb_{0.88}V_{0.12}$	5.7	A2	PdSbSe	1.0	C2
$Nb_{0.75}Zr_{0.25}$	10.8		PdSbTe	1.2	C2
$Nb_{0.66}Zr_{0.33}$	10.8		Pd_4Se	0.42	Tetragonal
$Ni_{0.3}Th_{0.7}$	1.98	$D10_2$	$Pd_{6-7}Se$	0.66	Like Pd_4Te
$NiZr_2$	1.52		$Pd_{2.8}Se$	2.3	
$Ni_{0.1}Zr_{0.9}$	1.5	A3	Pd_xSe_{1-x}	2.5 max.	
$O_3Rb_{0.27-0.29}W$	1.98	Hexagonal	PdSi	0.93	B3l
O_3SrTi			PdSn	0.41	B3l
(n = $1.7-12.0 \times 10^{19}$)	0.12–0.37		$PdSn_2$	3.34	
O_3SrTi			Pd_2Sn	0.41	C37
(n = $10^{18}-10^{21}$)	0.05–0.47		Pd_3Sn_2	0.47–0.64	$B8_2$
O_3SrTi			PdTe	2.3, 3.85	$B8_1$
(n = $\sim 10^{20}$)	0.47		$PdTe_{1.02-1.08}$	2.56–1.88	$B8_1$
OTi	0.58		$PdTe_2$	1.69	C6
$O_3Sr_{0.08}W$	2–4	Hexagonal	$PdTe_{2.1}$	1.89	C6
$O_3Tl_{0.30}W$	2.0–2.14	Hexagonal	$PdTe_{2.3}$	1.85	C6
OV_3Zr_3	7.5	$E9_3$	$Pd_{1.1}Te$	4.07	$B8_1$
OW_3 (film)	3.35, 1.1	A15	$PdTh_2$	0.85	Cl6
OsReY	2.0	C14	$Pd_{0.1}Zr_{0.9}$	7.5	A3
Os_2Sc	4.6	C14	PtSb	2.1	$B8_1$
OsTa	1.95	A12	PtSi	0.88	B3l
Os_3Th_7	1.51	$Dl0_2$	PtSn	0.37	$B8_1$
Os_xW_{1-x}	0.9–4.1		PtTe	0.59	Orthorhombic
OsW_3	~3		PtTh	0.44	B_f
Os_2Y	4.7	C14	Pt_3Th_7	0.98	$Dl0_2$
			Pt_5Th	3.13	

†w/o denotes weight percent.   ††See key at end of table.

Substance	T_c, °K	Crystal structure type††	Substance	T_c, °K	Crystal structure type††
$PtTi_3$	0.58	A15	Ru_2Y	1.52	C14
$Pt_{0.02}U_{0.98}$	0.87	β-phase	Ru_2Zr	1.84	C14
$PtV_{2.5}$	1.36	A15	$Ru_{0.1}Zr_{0.9}$	5.7	A3
PtV_3	2.87–3.20	A15	SbSn	1.30–1.42,	B1 or distorted
$PtV_{3.5}$	1.26	A15		1.42–2.37	B1
$Pt_{0.5}W_{0.5}$	1.45	A1	$SbTi_3$	5.8	A15
Pt_xW_{1-x}	0.4–2.7		Sb_2Tl_7	5.2	
Pt_2Y_3	0.90		$Sb_{0.01-0.03}V_{0.99-0.97}$	3.76–2.63	A2
Pt_2Y	1.57, 1.70	C15	SbV_3	0.80	A15
Pt_3Y_7	0.82	$D10_2$	Si_2Th	3.2	C_c, α-phase
PtZr	3.0	A3	Si_2Th	2.4	C32, β-phase
$Re_{0.64}Ta_{0.36}$	1.46	A12	SiV_3	17.1	A15
$Re_{24}Ti_5$	6.60	A12	$Si_{0.9}V_3Al_{0.1}$	14.05	A15
Re_xTi_{1-x}	6.6 max.		$Si_{0.9}V_3B_{0.1}$	15.8	A15
$Re_{0.76}V_{0.24}$	4.52	$D8_b$	$Si_{0.9}V_3C_{0.1}$	16.4	A15
$Re_{0.92}V_{0.08}$	6.8	A3	$SiV_{2.7}Cr_{0.3}$	11.3	A15
$Re_{0.6}W_{0.4}$	6.0		$Si_{0.9}V_3Ge_{0.1}$	14.0	A15
$Re_{0.5}W_{0.5}$	5.12	$D8_b$	$SiV_{2.7}Mo_{0.3}$	11.7	A15
Re_2Y	1.83	C14	$SiV_{2.7}Nb_{0.3}$	12.8	A15
Re_2Zr	5.9	C14	$SiV_{2.7}Ru_{0.3}$	2.9	A15
Re_6Zr	7.40	A12	$SiV_{2.7}Ti_{0.3}$	10.9	A15
$Rh_{17}S_{15}$	5.8	Cubic	$SiV_{2.7}Zr_{0.3}$	13.2	A15
$Rh_{\sim0.24}Sc_{\sim0.76}$	0.88, 0.92		Si_2W_3	2.8, 2.84	
Rh_xSe_{1-x}	6.0 max.		$Sn_{0.174-0.104}Ta_{0.826-0.896}$	6.5–<4.2	A15
Rh_2Sr	6.2	C15	$SnTa_3$	8.35	A15, highly ordered
$Rh_{0.4}Ta_{0.6}$	2.35	$D8_b$			
$RhTe_2$	1.51	C2	$SnTa_3$	6.2	A15, partially ordered
$Rh_{0.67}Te_{0.33}$	0.49				
Rh_xTe_{1-x}	1.51 max.		$SnTaV_2$	2.8	A15
RhTh	0.36	B_f	$SnTa_2V$	3.7	A15
Rh_3Th_7	2.15	$D10_2$	Sn_xTe_{1-x}		
Rh_5Th	1.07		$(n = 10.5-20 \times 10^{20})$	0.07–0.22	B1
Rh_xTi_{1-x}	2.25–3.95		Sn_xTl_{1-x}	2.37–5.2	
$Rh_{0.02}U_{0.98}$	0.96		SnV_3	3.8	A15
RhV_3	0.38	A15	$Sn_{0.02-0.057}V_{0.98-0.943}$	2.87–~1.6	A2
RhW	~3.4	A3	$Ta_{0.025}Ti_{0.975}$	1.3	Hexagonal
RhY_3	0.65		$Ta_{0.05}Ti_{0.95}$	2.9	Hexagonal
Rh_2Y_3	1.48		$Ta_{0.05-0.75}V_{0.095-0.25}$	4.30–2.65	A2
Rh_3Y	1.07	C15	$Ta_{0.8-1}W_{0.2-0}$	1.2–4.4	A2
Rh_5Y	0.56		$Tc_{0.1-0.4}W_{0.9-0.6}$	1.25–7.18	Cubic
$RhZr_2$	10.8	C16	$Tc_{0.50}W_{0.50}$	7.52	α plus σ
$Rh_{0.005}Zr$ (annealed)	5.8		$Tc_{0.60}W_{0.40}$	7.88	σ plus α
$Rh_{0-0.45}Zr_{1-0.55}$	2.1–10.8		Tc_6Zr	9.7	A12
$Rh_{0.1}Zr_{0.9}$	9.0	Hexagonal, c.p.	$Th_{0-0.55}Y_{1-0.45}$	1.2–1.8	
Ru_2Sc	1.67	C14	$Ti_{0.70}V_{0.30}$	6.14	Cubic
Ru_2Th	3.56	C15	Ti_xV_{1-x}	0.2–7.5	
RuTi	1.07	B2	$Ti_{0.5}Zr_{0.5}$ (annealed)	1.23	
$Ru_{0.05}Ti_{0.95}$	2.5		$Ti_{0.5}Zr_{0.5}$ (quenched)	2.0	
$Ru_{0.1}Ti_{0.9}$	3.5		V_2Zr	8.80	C15
$Ru_xTi_{0.6}V_y$	6.6 max.		$V_{0.26}Zr_{0.74}$	≈5.9	
$Ru_{0.45}V_{0.55}$	4.0	B2	W_2Zr	2.16	C15
RuW	7.5	A3			

††See key at end of table.

CRITICAL FIELD DATA

Substance	H_o, oersteds	Substance	H_o, oersteds
Ag_2F	2.5	InSb	1,100
Ag_7NO_{11}	57	In_xTl_{1-x}	252–284
Al_2CMo_3	1,700	$In_{0.8}Tl_{0.2}$	252
$BaBi_3$	740	$Mg_{\sim0.47}Tl_{\sim0.53}$	220
Bi_2Pt	10	$Mo_{0.16}Ti_{0.84}$	<985
Bi_3Sr	530	$NbSn_2$	620
Bi_5Tl_3	>400	$PbTl_{0.27}$	756
CdSn	>266	$PbTl_{0.17}$	796
$CoSi_2$	105	$PbTl_{0.12}$	849
$Cr_{0.1}Ti_{0.3}V_{0.6}$	1,360	$PbTl_{0.075}$	880
$In_{1-0.86}Mg_{0-0.14}$	272.4–259.2	$PbTl_{0.04}$	864

KEY TO CRYSTAL STRUCTURE TYPES

"Struck- turbericht" type*	Example	Class	"Struck- turbericht" type*	Example	Class
A1	Cu	Cubic, f.c.	$C15_b$	$AuBe_5$	Cubic
A2	W	Cubic, b.c.	C16	$CuAl_2$	Tetragonal, b.c.
A3	Mg	Hexagonal, close packed	C18	FeS_2	Orthorhombic
A4	Diamond	Cubic, f.c.	C22	Fe_2P	Trigonal
A5	White Sn	Tetragonal, b.c.	C23	$PbCl_2$	Orthorhombic
A6	In	Tetragonal, b.c. (f.c. cell usually used)	C32	AlB_2	Hexagonal
			C36	$MgNi_2$	Hexagonal
A7	As	Rhombohedral	C37	Co_2Si	Orthorhombic
A8	Se	Trigonal	C49	$ZrSi_2$	Orthorhombic
A10	Hg	Rhombohedral	C54	$TiSi_2$	Orthorhombic
A12	α-Mn	Cubic, b.c.	C_c	Si_2Th	Tetragonal, b.c.
A13	β-Mn	Cubic	DO_3	BiF_3	Cubic, f.c.
A15	"β-W" (W_3O)	Cubic	DO_{11}	Fe_3C	Orthorhombic
B1	NaCl	Cubic, f.c.	DO_{18}	Na_3As	Hexagonal
B2	CsCl	Cubic	DO_{19}	Ni_3Sn	Hexagonal
B3	ZnS	Cubic	DO_{20}	$NiAl_3$	Orthorhombic
B4	ZnS	Hexagonal	DO_{22}	$TiAl_3$	Tetragonal
$B8_1$	NiAs	Hexagonal	DO_e	Ni_3P	Tetragonal, b.c.
$B8_2$	Ni_2In	Hexagonal	$D1_3$	Al_4Ba	Tetragonal, b.c.
B10	PbO	Tetragonal	$D1_c$	$PtSn_4$	Orthorhombic
B11	γ-CuTi	Tetragonal	$D2_1$	CaB_6	Cubic
B17	PtS	Tetragonal	$D2_c$	MnU_6	Tetragonal, b.c.
B18	CuS	Hexagonal	$D2_d$	$CaZn_5$	Hexagonal
B20	FeSi	Cubic	$D5_2$	La_2O_3	Trigonal
B27	FeB	Orthorhombic	$D5_8$	Sb_2S_3	Orthorhombic
B31	MnP	Orthorhombic	$D7_3$	Th_3P_4	Cubic, b.c.
B32	NaTl	Cubic, f.c.	$D7_b$	Ta_3B_4	Orthorhombic
B34	PdS	Tetragonal	$D8_1$	Fe_3Zn_{10}	Cubic, b.c.
B_f	δ-CrB	Orthorhombic	$D8_2$	Cu_5Zn_8	Cubic, b.c.
B_g	MoB	Tetragonal, b.c.	$D8_3$	Cu_9Al_4	Cubic
B_h	WC	Hexagonal	$D8_8$	Mn_5Si_3	Hexagonal
B_i	γ-MoC	Hexagonal	$D8_b$	CrFe	Tetragonal
C1	CaF_2	Cubic, f.c.	$D8_i$	Mo_2B_5	Rhombohedral
$C1_b$	MgAgAs	Cubic, f.c.	$D10_2$	Fe_3Th_7	Hexagonal
C2	FeS_2	Cubic	$E2_1$	$CaTiO_3$	Cubic
C6	CdI_2	Trigonal	$E9_3$	Fe_3W_3C	Cubic, f.c.
C11b	$MoSi_2$	Tetragonal, b.c.	$L1_0$	CuAu	Tetragonal
C12	$CaSi_2$	Rhombohedral	$L1_2$	Cu_3Au	Cubic
C14	$MgZn_2$	Hexagonal	L'_{2b}	ThH_2	Tetragonal, b.c.
C15	Cu_2Mg	Cubic, f.c.	L'_3	Fe_2N	Hexagonal

*See "Handbook of Lattice Spacing and Structures of Metals", W.B. Pearson, Vol. I, Pergamon Press, 1958, p. 79. and Vol. II, Pergamon Press, 1967, p. 3.

HIGH CRITICAL MAGNETIC-FIELD SUPERCONDUCTIVE COMPOUNDS AND ALLOYS

With Critical Temperatures, H_{c1}, H_{c2}, H_{c3}, and the Temperature of Field Observations, T_{obs}

Substance	T_c, °K	H_{c1}, kg	H_{c2}, kg	H_{c3}, kg	T_{obs}, °K†
Al_2CMo_3	9.8–10.2	0.091	156		1.2
$AlNb_3$		0.375			
$Ba_xO_3Sr_{1-x}Ti$	<0.1–0.55	0.0039 max.			
$Bi_{0.5}Cd_{0.1}Pb_{0.27}Sn_{0.13}$			>24		3.06
Bi_xPb_{1-x}	7.35–8.4	0.122 max.	~30 max.		4.2
$Bi_{0.56}Pb_{0.44}$	8.8		15		4.2
$Bi_{7.5\ w/o}Pb_{92.5\ w/o}‡$			2.32		
$Bi_{0.099}Pb_{0.901}$		0.29	2.8		
$Bi_{0.02}Pb_{0.98}$		0.46	0.73		
$Bi_{0.53}Pb_{0.32}Sn_{0.16}$			>25		3.06
$Bi_{1-0.93}Sn_{0-0.07}$			0–0.032		3.7
Bi_5Tl_3	6.4		>5.56		3.35
C_8K (excess K)	0.55		0.160 (H⊥c)		0.32
			0.730 (H∥c)		0.32
C_8K	0.39		0.025 (H⊥c)		0.32
			0.250 (H∥c)		0.32
$C_{0.44}Mo_{0.56}$	12.5–13.5	0.087	98.5		1.2
CNb	8–10	0.12	16.9		4.2
$CNb_{0.4}Ta_{0.6}$	10–13.6	0.19	14.1		1.2
CTa	9–11.4	0.22	4.6		1.2
$Ca_xO_3Sr_{1-x}Ti$	<0.1–0.55	0.002–0.004			
$Cd_{0.1}Hg_{0.9}$ (by weight)		0.23	0.34		2.04
$Cd_{0.05}Hg_{0.95}$		0.28	0.31		2.16
$Cr_{0.10}Ti_{0.30}V_{0.60}$	5.6	0.071	84.4		0
GaN	5.85	0.725			4.2
Ga_xNb_{1-x}			>28		4.2
$GaSb$ (annealed)	4.24		2.64		3.5
$GaV_{1.95}$	5.3		73***		
$GaV_{2.1-3.5}$	6.3–14.45		230–300**		0
GaV_3		0.4	350***		0
			500**		
$GaV_{4.5}$	9 15		121*		0
Hf_xNb_y			>52->102		1.2
Hf_xTa_y			>28->86		1.2
$Hg_{0.05}Pb_{0.95}$		0.235	2.3		
$Hg_{0.101}Pb_{0.899}$		0.23	4.3		4.2
$Hg_{0.15}Pb_{0.85}$	~6.75		>13		2.93
$In_{0.98}Pb_{0.02}$	3.45	0.1		0.12	2.76
$In_{0.96}Pb_{0.04}$	3.68	0.1	0.12	0.25	2.94
$In_{0.94}Pb_{0.06}$	3.90	0.095	0.18	0.35	3.12
$In_{0.913}Pb_{0.087}$	4.2	~0.17	0.55	2.65	
$In_{0.316}Pb_{0.684}$		0.155	3.7		4.2
$In_{0.17}Pb_{0.83}$			2.8	5.5	4.2
$In_{1.000}Te_{1.002}$	3.5–3.7		1.2*		0
$In_{0.95}Tl_{0.05}$		0.263	0.263		3.3
$In_{0.90}Tl_{0.10}$		0.257	0.257		3.25
$In_{0.83}Tl_{0.17}$		0.242	0.39		3.21
$In_{0.75}Tl_{0.25}$		0.216	0.50		3.16
LaN	1.35	0.45			0.76
La_3S_4	6.5	≈0.15	>25		1.3
La_3Se_4	8.6	≈0.2	>25		1.25
$Mo_{0.52}Re_{0.48}$	11.1		14–21	22–33	4.2
			18–28	37–43	1.3
$Mo_{0.6\pm0.05}Re_{0.395}$	10.6		14–20	20–37	4.2
			19–26	26–37	1.3
$Mo_{~0.5}Tc_{~0.5}$			~75*		0
$Mo_{0.16}Ti_{0.84}$	4.18	0.028	98.7*		0
			36–38		3.0
$Mo_{0.913}Ti_{0.087}$	2.95	0.060	~15		4.2
$Mo_{0.1-0.3}U_{0.9-0.7}$	1.85–2.06		>25		
$Mo_{0.17}Zr_{0.83}$			~30		
$N_{(12.8\ w/o)}Nb$	15.2		>9.5		13.2
NNb (wires)	16.1		153*		0
			132		4.2
			95		8
			53		12
NNb_xO_{1-x}	13.5–17.0		~38		
NNb_xZr_{1-x}	9.8–13.8		4->130		4.2
$N_{0.93}Nb_{0.85}Zr_{0.15}$	13.8		>130		4.2
$Na_{0.086}Pb_{0.914}$		0.19	6.0		
$Na_{0.016}Pb_{0.984}$		0.28	2.05		

‡w/o denotes weight percent.

Substance	T_c, °K	H_{c1}, kg	H_{c2}, kg	H_{c3}, kg	T_{obs}, °K†
Nb	9.15		2.020		1.4
			1.710		4.2
Nb		0.4–1.1	3–5.5		4.2
Nb (unstrained)		1.1–1.8	3.40	6–9.1	4.2
Nb (strained)		1.25–1.92	3.44	6.0–8.7	4.2
Nb (cold-drawn wire)		2.48	4.10	≈10	4.2
Nb (film)			>25		4.2
NbSc			>30		
Nb_3Sn		0.170	221		4.2
			70		14.15
			54		15
			34		16
			17		17
$Nb_{0.1}Ta_{0.9}$		0.084	0.154		4.195
$Nb_{0.2}Ta_{0.8}$			10		4.2
$Nb_{0.65-0.73}Ta_{0.02-0.10}Zr_{0.25}$			>70->90		4.2
Nb_xTi_{1-x}			148 max.		1.2
			120 max.		4.2
$Nb_{0.222}U_{0.778}$		1.98	23		1.2
Nb_xZr_{1-x}			127 max.		1.2
			94 max.		4.2
O_3SrTi	0.43	.0049*	.504*		0
O_3SrTi	0.33	.00195*	.420*		0
$PbSb_{1\ w/o}$ (quenched)			>1.5		4.2
$PbSb_{1\ w/o}$ (annealed)			>0.7		4.2
$PbSb_{2.8\ w/o}$ (quenched)			>2.3		4.2
$PbSb_{2.8\ w/o}$ (annealed)			>0.7		4.2
$Pb_{0.871}Sn_{0.129}$		0.45	1.1		
$Pb_{0.965}Sn_{0.035}$		0.53	0.56		
$Pb_{1-0.26}Tl_{0-0.74}$	7.20–3.68		2–6.9*		0
$PbTl_{0.17}$	6.73		4.5*		0
$Re_{0.26}W_{0.74}$			>30		
$Sb_{0.93}Sn_{0.07}$			0.12		3.7
SiV_3	17.0	0.55	156***		
Sn_xTe_{1-x}		0.00043–0.00236	0.005–0.0775		0.012–0.079
Ta (99.95%)		0.425	1.850		1.3
		0.325	1.425		2.27
		0.275	1.175		2.66
		0.090	0.375		3.72
$Ta_{0.5}Nb_{0.5}$			3.55		4.2
$Ta_{0.65-0}Ti_{0.35-1}$	4.4–7.8		>14–138		1.2
$Ta_{0.5}Ti_{0.5}$			138		1.2
Te	~3.3	0.25*			0
Tc_xW_{1-x}	5.75–7.88		8–44		4.2
Ti				2.7	4.2
$Ti_{0.75}V_{0.25}$	5.3	0.029*	199*		0
$Ti_{0.775}V_{0.225}$	4.7	0.024*	172*		0
$Ti_{0.615}V_{0.385}$	7.07	0.050	~34		4.2
$Ti_{0.516}V_{0.484}$	7.20	0.062	~28		4.2
$Ti_{0.415}V_{0.585}$	7.49	0.078	~25		4.2
$Ti_{0.12}V_{0.88}$			17.3	28.1	4.2
$Ti_{0.09}V_{0.91}$			14.3	16.4	4.2
$Ti_{0.06}V_{0.94}$			8.2	12.7	4.2
$Ti_{0.03}V_{0.97}$			3.8	6.8	4.2
Ti_xV_{1-x}			108 max.		1.2
V	5.31	~0.8	~3.4		1.79
		~0.75	~3.15		2
		~0.45	~2.2		3
		~0.30	~1.2		4
$V_{0.26}Zr_{0.74}$	≈5.9	0.238			1.05
		0.227			1.78
		0.185			3.04
		0.165			3.5
W (film)	1.7–4.1		>34		1

†Temperature of critical field measurement.
*Extrapolated.
**Linear extrapolation.
***Parabolic extrapolation.

TABLES OF PROPERTIES OF SEMICONDUCTORS

Compiled by Dr. Brian Randall Pamplin

The term "semiconductor" is applied to a material in which electric current is carried by electrons or holes and whose electrical conductivity when extremely pure rises exponentially with temperature and may be increased from this low "intrinsic" value by many orders of magnitude by "doping" with electrically active impurities.

Semiconductors are characterised by an energy gap in the allowed energies of electrons in the material which separates the normally filled energy levels of the *valence band* (where "missing" electrons behave like positively charged current carriers "holes") and the *conduction band* (where electrons behave rather like a gas of free negatively charged carriers with an effective mass dependent on the material and the direction of the electrons' motion). This energy gap depends on the nature of the material and varies with direction in anisotropic crystals. It is slightly dependent on temperature and pressure, and this dependence is usually almost linear at normal temperatures and pressures.

The data is presented in three tables. Table I "General Properties of Semiconductors" lists the main crystallographic and semiconducting properties of a large number of semiconducting materials in three main categories; "Tetrahedral Semiconductors" in which every atom is tetrahedrally co-ordinated to four nearest neighbour atoms (or atomic sites) as for example in the diamond structure; "Octahedral Semiconductors" in which every atom is octahedrally co-ordinated to six nearest neighbour atoms—as for example in the halite structure; and "Other Semiconductors".

Table II gives more detailed information about some better known semiconductors, while Table III gives some information about the electronic energy band structure parameters of the best known materials.

TABLE 1

GENERAL PROPERTIES OF SEMICONDUCTORS
(listed by Crystal Structure)

Substance	Lattice Parameters (A° Room temperature)	Density (gm/cc)	Melting Point (°K)	Minimum Room Temperature Energy Gap (eV)	Thermal Conductivity	Heat of Formation k cal/mole	Mobility (Room Temperature) Electrons (cm²/V.s)	Holes	Remarks
PART A ADAMANTINE SEMICONDUCTORS									
§A1 Diamond Structure Elements (Strukturbericht symbol A4, Space Group Fd3m-O_h^7)									
C	3.5597	3.51	4300	5.4	2000	161	1800	1400	
Si	5.43072	2.3283	1685	1.107	1240	77.5	1900	500	
Ge	5.65754	5.3234	1231	0.67	640	69.5	3800	1820	
α-Sn	6.4912	5.765	503	0.08		64	2500	2400	
§A2 Sphalerite (Zinc Blende) Structure Compounds (Strukturbericht symbol B3 Space Group F $\bar{4}$ 3m-T_d^2)									
I VII Compounds									
CuF	4.255								
CuCl	5.4057	3.53	695						
CuBr	5.6905	4.72	770	2.94		115			
CuI	6.0427	5.63	878			105			
AgBr						102		4000	
AgI	6.473	5.67				93		30	
II VI Compounds									
BeS	4.865	2.36							
BeSe	5.139	4.315							
BeTe	5.626	5.090							
BePo	8.38	7.3							
ZnO	4.63								see § A3
ZnS	5.4093	4.079	2100	3.54	140	114	180	5(400°C)	see also § A3
ZnSe	5.6676	5.42	1790	2.58	140	101	540	28	
ZnTe	6.101	5.72	1568	2.26	140	90	340	100	
ZnPo									
CdS	5.5818		1750						see § A3
CdSe	6.05		1512						see § A3
CdTe	6.477	5.86	1365	1.44	55	81	1200	50	
CdPo									
HgS	5.8517	7.73	~2020						usually cinnabar
HgSe	6.084	8.25	1070	-0.10	10	59	20000		
HgTe	6.460	8.17	943	-0.15	20	58	25000	350	
III V Compounds									
BN	3.615	3.49	3000	~4	200	195			
BP(L.T.)	4.538	2.9		~6			500	70	
BAs	4.777								
AlP	5.451	2.85	1770	2.5					
AlAs	5.6622	3.81	1870	2.16		150	1200	420	
AlSb	6.1355	4.218	1330	1.60	600	140	200–400	550	
GaP	5.4505	4.13	1750	2.24	1100	152	300	100	
GaAs	5.65315	5.316	1510	1.35	370	128	8800	400	
GaSb	6.0954	5.619	980	0.67	270	118	4000	1400	
InP	5.86875	4.787	1330	1.27	800	134	4600	150	
InAs	6.05838	5.66	1215	0.36	290	114	33000	460	
InSb	6.47877	5.775	798	0.165	160	107	78000	750	
Other Sphalerite Structure Compounds									
MnS	5.60								see also § A3
MnSe	5.82								see also § A3
β-SiC	4.348	3.21	3070	2.3			4000		
Ga₂Te₃	5.899	5.75	1063	~1.0	~14	65			
In₂Te₃(H.T.)	6.150	5.8	940	~1.0	~8	47.4	~10		
MgGeP₂	5.652								
ZnSnP₂	5.65			2.1					
ZnSnAs₂(H.T.)	5.851	5.53	1050	~0.7	70				

TABLE 1

GENERAL PROPERTIES OF SEMICONDUCTORS (Continued)

§ A3 Wurtzite (Zincite) Structure Compounds (Strukturbericht symbol B4, Space Group P $6_3mc - C_{6v}^4$)

I VII Compounds

Substance	Lattice Parameters (A° Room temperature)	Density (gm/cc)	Melting Point (°K)	Minimum Room Temperature Energy Gap (eV)	Thermal Conductivity	Heat of Formation k cal/mole	Mobility Electrons (cm²/V·s)	Holes	Remarks
CuCl	3.91	6.42	T_c 680°K						
CuBr	4.06	6.66	T_c 658°K						
CuI	4.31	7.09							
AgI	4.580	7.494		2.63					

II VI Compounds

Substance	Lattice Parameters	Density	Melting Point	Energy Gap	Thermal Cond.	Heat of Form.	Electrons	Holes	Remarks
BeO	2.698 4.380		2800						
MgTe	4.54 7.39	3.85	~2800						
ZnO	3.24950 5.2069	5.66	2250	3.2	6	154	180		
ZnS	3.8140 6.2576	4.1	2100	3.67		110			
ZnSe	3.996 6.626		1793						
ZnTe	4.27 6.99		1568						
CdS	4.1348 6.7490	4.82	1748	2.42		96		400	
CdSe	4.299 7.010	5.66	1512	1.74		90		650	
CdTe	4.57 7.47			1.50					

III V Compounds

Substance	Lattice Parameters	Density	Melting Point	Energy Gap	Thermal Cond.	Heat of Form.	Electrons	Holes	Remarks
BP(H.T.)	3.562 5.900								
AlN	3.111 4.978	3.26	~2500	6.02		197			
GaN	3.190 5.189	6.10	1500	3.34		157			
InN	3.533 5.693	6.88	1200	2.0		133			

Other Wurtzite Structure Compounds

Substance	Lattice Parameters	Density	Melting Point	Energy Gap	Thermal Cond.	Heat of Form.	Electrons	Holes	Remarks
MnS	3.985 6.45								
MnSe	4.12 6.72								
SiC	3.076 5.048								
MnTe	4.078 6.701			~1.0					
Al₂S₃	3.579 5.829	2.55		4.1		426			
Al₂Se₃	3.890 6.30	3.91		3.1		367			

§ A4 Chalcopyrite Structure Compounds (Strukturbericht symbol E1₁, Space Group I $\bar{4}$ 2d $- D_{2d}^{12}$)

I III VI₂ Compounds

Substance	Lattice Parameters	Density	Melting Point	Energy Gap	Thermal Cond.	Heat of Form.	Electrons	Holes	Remarks
CuAlS₂	5.323 10.44	3.47		2.5					
CuAlSe₂	5.617 10.92	4.70	1270	1.1					
CuAlTe₂	5.976 11.80	5.50	1160	0.88					
CuGaS₂	5.360 10.49	4.35							
CuGaSe₂	5.618 11.01	5.56	1310	0.96, 1.63					
CuGaTe₂	6.012 11.00	5.99	1150	0.82, 1.0					
CuInS₂	5.528 11.08	4.75		1.2					
CuInSe₂	5.785 11.57	5.77	1250	0.86, 0.92		37			
CuInTe₂	6.179 12.365	6.10	970	0.95		49			
CuTlS₂	5.580 11.17	6.32							
CuTlSe₂(L.T.)	5.844 11.65	7.11	680	1.07					
CuFeS₂	5.25 10.32		1150	0.53					
CuFeSe₂			850	0.16					
CuLaS₂	5.65 10.86								
AgAlS₂	5.707 10.28	3.94							
AgAlSe₂	5.968 10.77	5.07	1220	0.7					
AgAlTe₂	6.309 11.85	6.18	1000	0.56					
AgGaS₂	5.755 10.28	4.72							
AgGaSe₂	5.985 10.90	5.84	1120	1.66					
AgGaTe₂	6.301 11.96	6.05	990	1.1		10			
AgInS₂(L.T.)	5.828 11.19	5.00		1.9					
AgInSe₂	6.102 11.69	5.81	1053	1.18		30			
AgInTe₂	6.42 12.59	6.12	965	0.96, 0.52					
AgFeS₂	5.66 10.30	4.53							

II IV V₂ Compounds

Substance	Lattice Parameters	Density	Melting Point	Energy Gap	Thermal Cond.	Heat of Form.	Electrons	Holes	Remarks
ZnSiP₂	5.400 10.441	3.39	1640	2.3			1000		
ZnGeP₂	5.465 10.771	4.17	1295	2.2					
CdSiP₂	5.678 10.431	4.00	~1470	2.2			1000		
CdGeP₂	5.741 10.775	4.48	~1060	1.8					
CdSnP₂	5.900 11.518			1.5					
ZnSiAs₂	5.61 10.88	4.70	~1350	1.7				50	
ZnGeAs₂	5.672 11.153	5.32	~1150	0.85		110			
ZnSnAs₂	5.8515 11.704	5.53	~ 910	0.65		150		300	disorders at 910°K
CdSiAs₂	5.884 10.882								
CdGeAs₂	5.9427 11.2172	5.60	~ 903	0.53		40	70	25	disorders at 903°K
CdSnAs₂	6.0944 11.9182	5.72	880	0.26		70	22000	250	

§ A5 "Defect Chalcopyrite" Structure Compounds (Strukturbericht symbol E3, Space Group I $\bar{4}$ $- S_4^2$)

Substance	Lattice Parameters	Density	Melting Point	Energy Gap	Thermal Cond.	Heat of Form.	Electrons	Holes	Remarks
ZnAl₂Se₄	5.503 10.90	4.37							
ZnAl₂Te₄(?)	5.904 12.05	4.95							
ZnGa₂S₄(?)	5.274 10.44	3.80							
ZnGa₂Se₄(?)	5.496 10.99	5.21							
ZnGa₂Te₄(?)	5.937 11.87	5.67		1.35					
ZnIn₂Se₄	5.711 11.42	5.44	1250	1.82					
ZnIn₂Te₄	6.122 12.24	5.83	1075	1.2					
CdAl₂S₄	5.564 10.32	3.06							
CdAl₂Se₄	5.747 10.68	4.54							
CdAl₂Te₄(?)	6.011 12.21	5.10							
CdGa₂S₄	5.577 10.08	4.03							
CdGa₂Se₄	5.743 10.73	5.32							
CdGa₂Te₄	6.093 11.81	5.77							
CdIn₂Te₄	6.205 12.41	5.9	1060	(1.26 or 0.9)					

TABLE I.

GENERAL PROPERTIES OF SEMICONDUCTORS (Continued)

Substance	Lattice Parameters (A° Room temperature)		Density (gm/cc)	Melting Point (°K)	Minimum Room Temperature Energy Gap (eV)	Thermal Conductivity	Heat of Formation k cal/mole	Mobility (Room Temperature) Electrons (cm²/V.s)	Holes	Remarks
$HgAl_2S_4$	5.488	10.26	4.11							
$HgAl_2Se_4$	5.708	10.74	5.05							
$HgAl_2Te_4(?)$	6.004	12.11	5.81							
$HgGa_2S_4$	5.507	10.23	5.00							
$HgGa_2Se_4$	5.715	10.78	6.18							
$HgIn_2Se_4$	5.764	11.80	6.3	1100	0.6					
$HgIn_2Te_4(?)$	6.186	12.37	6.3	980	0.86		200			

§ A6 Other Adamantine Compounds

Substance	Lattice Parameters		Density	Melting Point	Energy Gap	Thermal Cond.	Heat of Formation	Electrons	Holes	Remarks
αSiC	3.0817 15.1183		3.21	3070	2.86			400		6H structure
$Hg_5Ga_2Te_8$	6.235									B3 with super lattice
$Hg_5In_2Te_8$	6.328				0.7			2000		B3 with super lattice
$Cd\,In_2Se_4$	a=c=5.823				1.55					

PART B OCTAHEDRAL SEMICONDUCTORS

Halite Structure Semiconductors (Strukturbericht symbol B1, Space Group Fm3m – O_h^5)

Substance	Lattice Parameters	Density	Melting Point	Energy Gap	Thermal Cond.	Heat of Formation	Electrons	Holes	Remarks
SnSe	6.020		1133						
SnTe	6.313		1080(max)	0.5	91				
PbS	5.9362	7.61	1390	0.37	23	104	600	600	
PbSe	6.1243	8.15	1340	0.26	17	94	1000	900	
PbTe	6.454	8.16	1180	0.25	23	94	1600	600	

Selected Other Binary Chalcides

Substance	Lattice Parameters	Density	Melting Point	Energy Gap	Thermal Cond.	Heat of Formation	Electrons	Holes	Remarks
BiSe	5.99	7.98	880	0.4					
BiTe	6.47								
EuSe	6.191		2300	1.8	2.4				
GdSe	5.771		2400						
NiD	4.1684	6.6	2260	2.0 or 3.7			4	–	
CdO	4.6953		1700	2.5	7	127	100		
SrS	6.0199	3.643	3000	4.1					

Selected Multineny Compounds

Substance	Lattice Parameters	Density	Melting Point	Energy Gap	Thermal Cond.	Heat of Formation	Electrons	Holes	Remarks
$AgSbSe_2$	5.786	6.60	910	0.58	10.5				
$AgSbTe_2$ (or $Ag_{19}Sb_{29}Te_{52}$)	6.078	7.12	830	0.7, 0.27	8.6, 0.3				
$AgBiS_2$(H.T.)	5.648								
$AgBiSe_2$(H.T.)	5.82								
$AgBiTe_2$(H.T.)	6.155								
Cu_2CdSnS_4	5.586	10.83		1.16			<2		

TABLE II

SEMICONDUCTING PROPERTIES OF SELECTED MATERIALS

Substance	Minimum Energy Gap (eV) R.T.	0°K	$\frac{dE_g}{dT}$ x 10^4 eV/°C	$\frac{dE_g}{dP}$ x 10^6 eV.cm²/kg	Density of States Electron Effective Mass m_{d_n} (m_o)	Electron Mobility and Temperature Dependence μ_n cm²/V.s	$-x$	Density of States Hole Effective Mass m_{d_p} (m_o)	Hole Mobility and Temperature Dependence μ_p cm²/V.s	$-x$
Si	1.107	1.153	−2.3	−2.0	1.1	1,900	2.6	0.56	500	2.3
Ge	0.67	0.744	−3.7	+7.3	0.55	3,800	1.66	0.3	1,820	2.33
αSn	0.08	0.094	−0.5		0.02	2,500	1.65	0.3	2,400	2.0
Te	0.33				0.68	1,100		0.19	560	

III–V Compounds

Substance	R.T.	0°K	dE_g/dT	dE_g/dP	m_{d_n}	μ_n	$-x$	m_{d_p}	μ_p	$-x$
AlAs	2.2	2.3				1,200			420	
AlSb	1.6	1.7	−3.5	−1.6	0.09	200	1.5	0.4	500	1.8
GaP	2.24	2.40	−5.4	−1.7	0.35	300	1.5	0.5	150	1.5
GaAs	1.35	1.53	−5.0	+9.4	0.068	9,000	1.0	0.5	500	2.1
GaSb	0.67	0.78	−3.5	+12	0.050	5,000	2.0	0.23	1,400	0.9
InP	1.27	1.41	−4.6	+4.6	0.067	5,000	2.0		200	2.4
InAs	0.36	0.43	−2.8	+8	0.022	33,000	1.2	0.41	460	2.3
InSb	0.165	0.23	−2.8	+15	0.014	78,000	1.6	0.4	750	2.1

II–VI Compounds

Substance	R.T.	0°K	dE_g/dT	dE_g/dP	m_{d_n}	μ_n	$-x$	m_{d_p}	μ_p	$-x$
ZnO	3.2		−9.5	+0.6	0.38	180	1.5			
ZnS	3.54		−5.3	+5.7		180			5(400°C)	
ZnSe	2.58	2.80	−7.2	+6		540			28	
ZnTe	2.26			+6		340			100	
CdO	2.5 ± .1		−6		0.1	120				
CdS	2.42		−5	+3.3	0.165	400		0.8		
CdSe	1.74	1.85	−4.6		0.13	650	1.0	0.6		
CdTe	1.44	1.56	−4.1	+8	0.14	1,200		0.35	50	
HgSe	0.30				0.030	20,000	2.0			
HgTe	0.15		−1		0.017	25,000		0.5	350	

Halite Structure Compounds

Substance	R.T.	0°K	dE_g/dT	dE_g/dP	m_{d_n}	μ_n	$-x$	m_{d_p}	μ_p	$-x$
PbS	0.37	0.28	+4		0.16	800		0.1	1,000	2.2
PbSe	0.26	0.16	+4		0.3	1,500		0.34	1,500	2.2
PbTe	0.25	0.19	+4	−7	0.21	1,600		0.14	750	2.2

TABLE II
SEMICONDUCTING PROPERTIES OF SELECTED MATERIALS (Continued)

Substance	Minimum Energy Gap (eV) R.T.	0°K	dE_g/dT x 10⁴ eV/°C	dE_g/dP x 10⁶ eV.cm²/kg	Density of States Electron Effective Mass m_{d_n} (m_o)	Electron Mobility and Temperature Dependence μ_n cm²/V.s	$-x$	Density of States Hole Effective Mass m_{d_p} (m_o)	Hole Mobility and Temperature Dependence μ_p cm²/V.s	$-x$
Others										
ZnSb	0.50	0.56			0.15	10				1.5
CdSb	0.45	0.57	−5.4		0.15	300			2,000	1.5
Bi₂S₃	1.3					200			1,100	
Bi₂Se₃	0.27					600			675	
Bi₂Te₃	0.13			−0.95	0.58	1,200	1.68	1.07	510	1.95
Mg₂Si		0.77	−6.4		0.46	400	2.5		70	
Mg₂Ge		0.74	−9			280	2		110	
Mg₂Sn	0.21	0.33	−3.5		0.37	320			260	
Mg₃Sb₂		0.32				20			82	
Zn₃As₂	0.93					10	1.1		10	
Cd₃As₂	0.55				0.046	100,000	0.88			
GaSe	2.05		3.8						20	
GaTe	1.66	1.80	−3.6			14	−5			
InSe	1.8					900				
TlSe	0.57		−3.9		0.3	30		0.6	20	1.5
CdSnAs₂	0.23				0.05	25,000	1.7			
Ga₂Te₃	1.1	1.55	−4.8							
α-In₂Te₃	1.1	1.2			0.7				50	1.1
β-In₂Te₃	1.0								5	
Hg₅In₂Te₈	0.5								11,000	
SnO₂									78	

TABLE III

PART A. DATA ON VALENCE BANDS OF SEMICONDUCTORS (Room Temperature data)

Substance	Heavy Holes	Light Holes	"Split-off" Band Holes	Energy Separation of "Split-off" Band (eV)	Measured (Light) Hole Mobility cm²/V.s
	Band Curvature Effective Mass (Expressed as fraction of free electron mass)				

Semiconductors with Valence Band Maximum at Centre of Brillouin Zone ("Γ")

Substance	Heavy Holes	Light Holes	"Split-off" Band Holes	Energy Separation of "Split-off" Band (eV)	Measured (Light) Hole Mobility cm²/V.s
Si	0.50	0.16	0.25	0.011	500
Ge	0.31	0.043	0.08	0.3	1,820
Sn	0.3				2,400
AlAs					
AlSb	0.4			0.7	550
GaP				0.13	100
GaAs	0.8	0.12	0.20	0.34	400
GaSb	0.23	0.06		0.7	1,400
InP				0.21	150
InAs	0.41	0.025	0.083	0.43	460
InSb	0.4	0.015		0.85	750
CdTe	0.35				50
HgTe	0.5				350

Semiconductors with Multiple Valence Band Maxima

Substance	Number of Equivalent Valleys Direction	Curvature Effective Masses Longitudinal m_L	Transverse m_T	Anisotropy $\frac{m_L}{K = m_T}$	Measured (Light) Hole Mobility cm²/V.s
PbSe	4 "L" [111]	0.095	0.047	2.0	1,500
PbTe	4 "L" [111]	0.27	0.02	10	750
Bi₂Te₃	6	0.207	~0.045	4.5	515

PART B. DATA ON CONDUCTION BANDS OF SEMICONDUCTORS (Room Temperature Data)

Single Valley Semiconductors

Substance	Energy Gap (eV)	Effective Mass (m_o)	Mobility (cm²/V.s)	Comments
GaAs	1.35	0.067	8,500	3(or 6?) equivalent [100] valleys 0.36 eV above this maximum with a mobility of ~50
InP	1.27	0.067	5,000	3(or 6?) equivalent [100] valleys 0.4 eV above this minimum.
InAs	0.36	0.022	33,000	equivalent valleys ~1.0 eV above this minimum.
InSb	0.165	0.014	78,000	
CdTe	1.44	0.11	1,000	4(or 8?) equivalent [111] valleys 0.51 eV above this minimum.

Multivalley Semiconductors

Substance	Energy Gap	Number of Equivalent Valleys and Direction	Band Curvature Longitudinal m_L	Effective Mass Transverse m_T	Anisotropy $\frac{m_L}{K = m_T}$	Comments
Si	1.107	6 in [100] "Δ"	0.90	0.192	4.7	
Ge	0.67	4 in [111] at "L"	1.588	0.0815	19.5	
GaSb	0.67	as Ge (?)	~1.0	~0.2	~5	
PbSe	0.26	4 in [111] at "L"	0.085	0.05	1.7	
PbTe	0.25	4 in [111] at "L"	0.21	0.029	5.5	
Bi₂Te₃	0.13	6			~0.05	

STANDARD CALIBRATION TABLES FOR THERMOCOUPLES

The following tables which represent the Temperature-E.M.F. functions of various thermocouples should be used with appropriate correction curves if precise results are desired. These curves must be determined for each individual couple by plotting ΔE, the difference between the observed and the standard E.M.F., against the standard E.M.F. at three or more temperature points. The value ΔE as shown by such a correction curve is then subtracted algebraically from the observed E.M.F. to give the true E.M.F. reading.

In the following tables the fixed or "cold junction" is at 0°C.; when the cold junction is not maintained at 0°C. the readings of the E.M.F. must be corrected as follows: $Et = E_{(t-tc)} + Etc$ where $E_{(t-tc)}$ is the observed reading, Etc is the E.M.F. for the temperature corresponding to the cold junction temperature as read from the standard table and Et is the E.M.F. produced by the jot junction corrected to the value which would be obtained with the cold junction at 0°C. The temperature corresponding to Et is then obtained by reference to the standard table.

Since the E.M.F.-temperature function is not linear the cold junction should be maintained at a temperature very close to that at whch the thermocouple was calibrated. Otherwise considerable error will result despite the above correction.

PLATINUM VERSUS PLATINUM-10-PERCENT RHODIUM THERMOCOUPLES

(Electromotive Force in Absolute Millivolts. Temperatures in Degrees C (Int. 1948). Reference Junctions at 0°C.)

°C	0	10	20	30	40	50	60	70	80	90
					Millivolts					
0	0	0.06	0.11	0.17	0.24	0.30	0.36	0.43	0.50	0.57
100	0.64	0.72	0.79	0.87	0.95	1.03	1.11	1.19	1.27	1.35
200	1.44	1.52	1.61	1.69	1.78	1.87	1.96	2.05	2.14	2.23
300	2.32	2.41	2.50	2.59	2.69	2.78	2.87	2.97	3.06	3.16
400	3.25	3.35	3.44	3.54	3.64	3.73	3.83	3.93	4.02	4.12
500	4.22	4.32	4.42	4.52	4.62	4.72	4.82	4.92	5.02	5.12
600	5.22	5.33	5.43	5.53	5.64	5.74	5.84	5.95	6.05	6.16
700	6.26	6.37	6.47	6.58	6.68	6.79	6.90	7.01	7.11	7.22
800	7.33	7.44	7.55	7.66	7.77	7.88	7.99	8.10	8.21	8.32
900	8.43	8.55	8.66	8.77	8.88	9.00	9.11	9.23	9.34	9.46
1000	9.57	9.6	9.80	9.92	10.04	10.15	10.27	10.39	10.51	10.62
1100	10.74	10.86	10.98	11.10	11.22	11.34	11.46	11.58	11.70	11.82
1200	11.94	12.06	12.18	12.30	12.42	12.54	12.66	12.78	12.90	13.02
1300	13.14	13.26	13.38	13.50	13.62	13.74	13.86	13.98	14.10	14.22
1400	14.34	14.46	14.58	14.70	14.82	14.94	15.05	15.17	15.29	15.41
1500	15.53	15.65	15.77	15.89	16.01	16.12	16.24	16.36	16.48	16.60
1600	16.72	16.83	16.95	17.07	17.19	17.31	17.42	17.54	17.66	17.77
1700	17.89	18.01	18.12	18.24	18.36	18.47	18.59	—	—	—

(Electromotive Force in Absolute Millivolts. Temperatures in Degrees F.* Reference Junctions at 32°F.)

°F	0	10	20	30	40	50	60	70	80	90
					Millivolts					
0	—	—	—	—	0.02	0.06	0.09	0.12	0.15	0.19
100	0.22	0.26	0.29	0.33	0.36	0.40	0.44	0.48	0.52	0.56
200	0.60	0.64	0.68	0.72	0.76	0.80	0.84	0.89	0.93	0.97
300	1.02	1.06	1.11	1.15	1.20	1.24	1.29	1.33	1.38	1.43
400	1.47	1.52	1.57	1.62	1.66	1.71	1.76	1.81	1.86	1.91
500	1.96	2.01	2.06	2.11	2.16	2.21	2.26	2.31	2.36	2.41
600	2.46	2.51	2.56	2.61	2.66	2.72	2.77	2.82	2.87	2.92
700	2.98	3.03	3.08	3.14	3.19	3.24	3.29	3.35	3.40	3.45
800	3.51	3.56	3.61	3.67	3.72	3.78	3.83	3.88	3.94	3.99
900	4.05	4.10	4.16	4.21	4.26	4.32	4.37	4.43	4.49	4.54
1000	4.60	4.65	4.71	4.76	4.82	4.87	4.93	4.99	5.04	5.10
1100	5.16	5.21	5.27	5.33	5.38	5.44	5.50	5.56	5.61	5.67
1200	5.73	5.78	5.84	5.90	5.96	6.02	6.07	6.13	6.19	6.25
1300	6.31	6.37	6.42	6.48	6.54	6.60	6.66	6.72	6.78	6.84
1400	6.90	6.96	7.02	7.08	7.14	7.20	7.26	7.32	7.38	7.44
1500	7.50	7.56	7.62	7.68	7.74	7.80	7.86	7.93	7.99	8.05
1600	8.11	8.17	8.23	8.30	8.36	8.42	8.48	8.55	8.61	8.67
1700	8.73	8.80	8.86	8.92	8.98	9.05	9.11	9.17	9.24	9.30
1800	9.37	9.43	9.49	9.56	9.62	9.69	9.75	9.82	9.88	9.94
1900	10.01	10.07	10.14	1.20	10.27	10.33	10.40	10.47	10.53	10.60
2000	10.66	10.73	10.79	10.86	10.93	10.99	11.06	11.12	11.19	11.26
2100	11.32	11.39	11.46	11.52	11.59	11.66	11.72	11.79	11.86	11.92
2200	11.99	12.06	12.12	12.19	12.26	12.32	12.39	12.46	12.52	12.59
2300	12.66	12.72	12.79	12.86	12.92	12.99	13.06	13.12	13.19	13.26
2400	13.33	13.39	13.46	13.53	13.59	13.66	13.73	13.79	13.86	13.92
2500	13.99	14.06	14.12	14.19	14.26	14.32	14.39	14.46	14.52	14.59
2600	14.66	14.72	14.79	14.86	14.92	14.99	15.05	15.12	15.19	15.25
2700	15.32	15.39	15.45	15.52	15.58	15.65	15.72	15.78	15.85	15.91
2800	15.98	16.05	16.11	16.18	16.24	16.31	16.37	16.44	16.51	16.57
2900	16.64	16.70	16.77	16.83	16.90	16.97	17.03	17.10	17.16	17.23
3000	17.29	17.36	17.42	17.49	17.55	17.62	17.68	17.75	17.81	17.88
3100	17.94	18.01	18.07	18.14	18.20	18.27	18.33	18.40	18.46	18.53
3200	18.59	18.66	—	—	—	—	—	—	—	—

PLATINUM VERSUS PLATINUM-13-PERCENT RHODIUM THERMOCOUPLES

(Electromotive Force in Absolute Millivolts. Temperatures in Degrees C (Int. 1948) Reference Junctions at 0°C.)

°C	0	10	20	30	40	50	60	70	80	90
					Millivolts					
0	0.00	0.06	0.11	0.17	0.23	0.30	0.36	0.43	0.50	0.57
100	0.65	0.72	0.80	0.88	0.96	1.04	1.12	1.21	1.29	1.38
200	1.47	1.55	1.64	1.73	1.83	1.92	2.01	2.11	2.20	2.30
300	2.40	2.49	2.59	2.69	2.79	2.89	2.99	3.09	3.19	3.30
400	3.40	3.50	3.61	3.71	3.82	3.92	4.03	4.13	4.24	4.35
500	4.46	4.56	4.67	4.78	4.89	5.00	5.12	5.23	5.34	5.45
600	5.56	5.68	5.79	5.91	6.02	6.14	6.25	6.37	6.49	6.60
700	6.72	6.84	6.96	7.08	7.20	7.32	7.44	7.56	7.68	7.80
800	7.92	8.05	8.17	8.29	8.42	8.54	8.67	8.80	8.92	9.05
900	9.18	9.30	9.43	9.56	9.69	9.82	9.95	10.08	10.21	10.34
1000	10.47	10.60	10.74	10.87	11.00	11.14	11.27	11.41	11.54	11.68
1100	11.82	11.94	12.09	12.23	12.37	12.50	12.64	12.78	12.92	13.06
1200	13.19	13.33	13.47	13.61	13.75	13.89	14.03	14.17	14.30	14.44
1300	14.58	14.72	14.86	15.00	15.14	15.28	15.42	15.55	15.69	15.83
1400	15.97	16.11	16.25	16.39	16.52	16.66	16.80	16.94	17.08	17.22
1500	17.36	17.49	17.63	17.77	17.91	18.04	18.18	18.32	18.45	18.59
1600	18.73	18.86	19.00	19.14	19.27	19.41	19.55	19.68	19.82	19.95
1700	20.09	—	—	—	—	—	—	—	—	—

(Electromotive Force in Absolute Millivolts. Temperatures in Degrees F* Reference Junctions at 32°F.)

°F	0	10	20	30	40	50	60	70	80	90
					Millivolts					
0	—	—	—	—	0.02	0.06	0.09	0.12	0.15	0.19
100	0.22	0.26	0.29	0.33	0.36	0.40	0.44	0.48	0.52	0.56
200	0.60	0.64	0.68	0.72	0.76	0.81	0.85	0.89	0.94	0.98
300	1.03	1.08	1.12	1.17	1.21	1.26	1.31	1.36	1.41	1.46
400	1.50	1.55	1.60	1.65	1.70	1.75	1.81	1.86	1.91	1.96
500	2.01	2.07	2.12	2.17	2.22	2.28	2.33	2.38	2.44	2.49
600	2.55	2.60	2.66	2.71	2.77	2.82	2.88	2.94	2.99	3.05
700	3.10	3.16	3.22	3.27	3.33	3.39	3.45	3.50	3.56	3.62
800	3.68	3.74	3.79	3.85	3.91	3.97	4.03	4.09	4.14	4.21
900	4.26	4.32	4.38	4.44	4.50	4.56	4.62	4.69	4.75	4.81
1000	4.87	4.93	4.99	5.05	5.12	5.18	5.24	5.30	5.36	5.43
1100	5.49	5.55	5.61	5.68	5.74	5.81	5.87	5.93	6.00	6.06
1200	6.13	6.19	6.25	6.32	6.38	6.45	6.51	6.58	6.64	6.71
1300	6.77	6.84	6.90	6.97	7.04	7.10	7.17	7.24	7.30	7.37
1400	7.44	7.50	7.57	7.64	7.71	7.77	7.84	7.91	7.98	8.05
1500	8.12	8.18	8.25	8.32	8.39	8.46	8.53	8.60	8.67	8.74
1600	8.81	8.88	8.95	9.02	9.09	9.16	9.23	9.30	9.37	9.45
1700	9.52	9.59	9.66	9.73	9.80	9.87	9.95	10.02	10.09	10.16
1800	10.24	10.31	10.38	10.46	10.53	10.60	10.68	10.75	10.82	10.90
1900	10.97	11.05	11.12	11.20	11.27	11.35	11.42	11.50	11.58	11.65
2000	11.73	11.80	11.88	11.95	12.03	12.11	12.18	12.26	12.34	12.41
2100	12.49	12.56	12.64	12.72	12.80	12.87	12.95	13.03	13.10	13.18
2200	13.26	13.33	13.41	13.49	13.56	13.64	13.72	13.80	13.87	13.95
2300	14.03	14.10	14.18	14.26	14.34	14.41	14.49	14.57	14.64	14.72
2400	14.80	14.88	15.95	15.03	15.11	15.18	15.26	15.34	15.42	15.49
2500	15.57	15.65	15.72	15.80	15.88	15.95	16.03	16.11	16.19	16.26
2600	16.34	16.42	16.49	16.57	16.65	16.73	16.80	16.88	16.96	17.03
2700	17.11	17.19	17.26	17.34	17.42	17.49	17.57	17.65	17.72	17.80
2800	17.88	17.95	18.03	18.10	18.18	18.26	18.33	18.41	18.48	18.56
2900	18.64	18.71	18.79	18.86	18.94	19.02	19.09	19.17	19.24	19.32
3000	19.39	19.47	19.55	19.62	19.70	19.77	19.85	19.92	20.00	20.08
3100	20.15	—	—	—	—	—	—	—	—	—

* Based on the International Temperature Scale of 1948.

CHROMEL-ALUMEL THERMOCOUPLES

(Electromotive Force in Absolute Millivolts. Temperatures in Degrees C (Int. 1948) Reference Junctions at 0°C)

Millivolts

°C	0	1	2	3	4	5	6	7	8	9
−190	−5.60	−5.62	−5.63	−5.65	−5.67	−5.68	−5.70	−5.71	−5.73	−5.74
−180	−5.43	−5.45	−5.46	−5.48	−5.50	−5.52	−5.53	−5.55	−5.57	−5.58
−170	−5.24	−5.26	−5.28	−5.30	−5.32	−5.34	−5.35	−5.37	−5.39	−5.41
−160	−5.03	−5.05	−5.08	−5.10	−5.12	−5.14	−5.16	−5.18	−5.20	−5.22
−150	−4.81	−4.84	−4.86	−4.88	−4.90	−4.92	−4.95	−4.97	−4.99	−5.01
−140	−4.58	−4.60	−4.62	−4.65	−4.67	−4.70	−4.72	−4.74	−4.77	−4.79
−130	−4.32	−4.35	−4.37	−4.40	−4.42	−4.45	−4.48	−4.50	−4.52	−4.55
−120	−4.06	−4.08	−4.11	−4.14	−4.16	−4.19	−4.22	−4.24	−4.27	−4.30
−110	−3.78	−3.81	−3.84	−3.86	−3.89	−3.92	−3.95	−3.98	−4.00	−4.03
−100	−3.49	−3.52	−3.55	−3.58	−3.61	−3.64	−3.66	−3.69	−3.72	−3.75
−90	−3.19	−3.22	−3.25	−3.28	−3.31	−3.34	−3.37	−3.40	−3.43	−3.46
−80	−2.87	−2.90	−2.93	−2.96	−3.00	−3.03	−3.06	−3.09	−3.12	−3.16
−70	−2.54	−2.57	−2.61	−2.64	−2.67	−2.71	−2.74	−2.77	−2.80	−2.84
−60	−2.20	−2.24	−2.27	−2.30	−2.34	−2.37	−2.41	−2.44	−2.47	−2.51
−50	−1.86	−1.89	−1.93	−1.96	−2.00	−2.03	−2.07	−2.10	−2.13	−2.17
−40	−1.50	−1.54	−1.57	−1.61	−1.64	−1.68	−1.72	−1.75	−1.79	−1.82
−30	−1.14	−1.17	−1.21	−1.25	−1.28	−1.32	−1.36	−1.39	−1.43	−1.46
−20	−0.77	−0.80	−0.84	−0.88	−0.92	−0.95	−0.99	−1.03	−1.06	−1.10
−10	−0.39	−0.42	−0.46	−0.50	−0.54	−0.58	−0.62	−0.66	−0.69	−0.73
(−)0	−0.00	−0.04	−0.08	−0.12	−0.16	−0.19	−0.23	−0.27	−0.31	−0.35
(+)0	0.00	0.04	0.08	0.12	0.16	0.20	0.24	0.28	0.32	0.36
10	0.40	0.44	0.48	0.52	0.56	0.60	0.64	0.68	0.72	0.76
20	0.80	0.84	0.88	0.92	0.96	1.00	1.04	1.08	1.12	1.16
30	1.20	1.24	1.28	1.32	1.36	1.40	1.44	1.49	1.53	1.57
40	1.61	1.65	1.69	1.73	1.77	1.81	1.85	1.90	1.94	1.98
50	2.02	2.06	2.10	2.14	2.18	2.23	2.27	2.31	2.35	2.39
60	2.43	2.47	2.51	2.56	2.60	2.64	2.68	2.72	2.76	2.80
70	2.85	2.89	2.93	2.97	3.01	3.05	3.10	3.14	3.18	3.22
80	3.26	3.30	3.35	3.39	3.43	3.47	3.51	3.56	3.60	3.63
90	3.68	3.72	3.76	3.81	3.85	3.89	3.93	3.97	4.01	4.06
100	4.10	4.14	4.18	4.22	4.26	4.31	4.35	4.39	4.43	4.47
110	4.51	4.55	4.60	4.64	4.68	4.72	4.76	4.80	4.84	4.88
120	4.92	4.96	5.01	5.05	5.09	5.13	5.17	5.21	5.25	5.29
130	5.33	5.37	5.41	5.45	5.49	5.53	5.57	5.61	5.65	5.69
140	5.73	5.77	5.81	5.85	5.89	5.93	5.97	6.01	6.05	6.09
150	6.13	6.17	6.21	6.25	6.29	6.33	6.37	6.41	6.45	6.49
160	6.53	6.57	6.61	6.65	6.69	6.73	6.77	6.81	6.85	6.89
170	6.93	6.97	7.01	7.05	7.09	7.13	7.17	7.21	7.25	7.29
180	7.33	7.37	7.41	7.45	7.49	7.53	7.57	7.61	7.65	7.69
190	7.73	7.77	7.81	7.85	7.89	7.93	7.97	8.01	8.05	8.09
200	8.13	8.17	8.21	8.25	8.29	8.33	8.37	8.41	8.46	8.50
210	8.54	8.58	8.62	8.66	8.70	8.74	8.78	8.82	8.86	8.90
220	8.94	8.98	9.02	9.06	9.10	9.14	9.18	9.22	9.26	9.30
230	9.34	9.38	9.42	9.46	9.50	9.54	9.59	9.63	9.67	9.71
240	9.75	9.79	9.83	9.87	9.91	9.95	9.99	10.03	10.07	10.11
250	10.16	10.20	10.24	10.28	10.32	10.36	10.40	10.44	10.48	10.52
260	10.57	10.61	10.65	10.69	10.73	10.77	10.81	10.85	10.89	10.93
270	10.98	11.02	11.06	11.10	11.14	11.18	11.22	11.26	11.30	11.34
280	11.39	11.43	11.47	11.51	11.55	11.59	11.63	11.67	11.72	11.76
290	11.80	11.84	11.88	11.92	11.96	12.01	12.05	12.09	12.13	12.17
300	12.21	12.25	12.29	12.34	12.38	12.42	12.46	12.50	12.54	12.59
310	12.63	12.67	12.71	12.75	12.79	12.83	12.88	12.92	12.96	13.00
320	13.04	13.08	13.12	13.17	13.21	13.25	13.29	13.33	13.37	13.42
330	13.46	13.50	13.54	13.58	13.62	13.67	13.71	13.75	13.79	13.83
340	13.88	13.92	13.96	14.00	14.04	14.09	14.13	14.17	14.21	14.25
350	14.29	14.34	14.38	14.42	14.46	14.50	14.55	14.59	14.63	14.67
360	14.71	14.76	14.80	14.84	14.88	14.92	14.97	15.01	15.05	15.09
370	15.13	15.18	15.22	15.26	15.30	15.34	15.39	15.43	15.47	15.51
380	15.55	15.60	15.64	15.68	15.72	15.76	15.81	15.85	15.89	15.93
390	15.98	16.02	16.06	16.10	16.14	16.19	16.23	16.27	16.31	16.36
400	16.40	16.44	16.48	16.52	16.57	16.61	16.65	16.69	16.74	16.78
410	16.82	16.86	16.91	16.95	16.99	17.03	17.07	17.12	17.16	17.20
420	17.24	17.29	17.33	17.37	17.41	17.46	17.50	17.54	17.58	17.62
430	17.67	17.71	17.75	17.79	17.84	17.88	17.92	17.96	18.01	18.05
440	18.09	18.13	18.17	18.22	18.26	18.30	18.34	18.39	18.43	18.47
450	18.51	18.56	18.60	18.64	18.68	18.73	18.77	18.81	18.85	18.90
460	18.94	18.98	19.02	19.07	19.11	19.15	19.19	19.24	19.28	19.32
470	19.36	19.41	19.45	19.49	19.54	19.58	19.62	19.66	19.71	19.75
480	19.79	19.84	19.88	19.92	19.96	20.01	20.03	20.09	20.13	20.18
490	20.22	20.26	20.31	20.35	20.39	20.43	20.48	20.53	20.56	20.60
500	20.65	20.69	20.73	20.77	20.82	20.86	20.90	20.94	20.99	21.03
510	21.07	21.11	21.16	21.20	21.24	21.28	21.32	21.37	21.41	21.45
520	21.50	21.54	21.58	21.63	21.67	21.71	21.75	21.80	21.84	21.88
530	21.92	21.97	22.01	22.05	22.09	22.14	22.18	22.22	22.26	22.31
540	22.35	22.39	22.43	22.48	22.52	22.56	22.61	22.65	22.69	22.73
550	22.78	22.82	22.86	22.90	22.95	22.99	23.03	23.07	23.12	23.16
560	23.20	23.25	23.29	23.33	23.38	23.42	23.46	23.50	23.54	23.59
570	23.63	23.67	23.72	23.76	23.80	23.84	23.89	23.93	23.97	24.01

Millivolts

°C	0	1	2	3	4	5	6	7	8	9
580	24.06	24.10	24.14	24.18	24.23	24.27	24.31	24.36	24.40	24.44
590	24.49	24.53	24.57	24.61	24.65	24.70	24.74	24.78	24.83	24.87
600	24.91	24.95	25.00	25.04	25.08	25.12	25.17	25.21	25.25	25.29
610	25.34	25.38	25.42	25.47	25.51	25.55	25.59	25.64	25.68	25.72
620	25.76	25.81	25.85	25.89	25.93	25.98	26.02	26.06	26.10	26.15
630	26.19	26.23	26.27	26.32	26.36	26.40	26.44	26.48	26.53	26.57
640	26.61	26.65	26.70	26.74	26.78	26.82	26.86	26.91	26.95	26.99
650	27.03	27.07	27.12	27.16	27.20	27.24	27.28	27.33	27.37	27.41
660	27.45	27.49	27.54	27.58	27.62	27.66	27.71	27.75	27.79	27.83
670	27.87	27.92	27.96	28.00	28.04	28.08	28.13	28.17	28.21	28.25
680	28.29	28.34	28.38	28.42	28.46	28.50	28.55	28.59	28.63	28.67
690	28.72	28.76	28.80	28.84	28.88	28.93	28.97	29.01	29.05	29.10
700	29.14	29.18	29.22	29.26	29.30	29.35	29.39	29.43	29.47	29.52
710	29.56	29.60	29.64	29.68	29.72	29.77	29.81	29.85	29.89	29.93
720	29.97	30.02	30.06	30.10	30.14	30.18	30.23	30.27	30.31	30.35
730	30.39	30.44	30.48	30.52	30.56	30.60	30.65	30.69	30.73	30.77
740	30.81	30.85	30.90	30.94	30.98	31.02	31.06	31.10	31.15	31.19
750	31.23	31.27	31.31	31.35	31.40	31.44	31.48	31.52	31.56	31.60
760	31.65	31.69	31.73	31.77	31.81	31.85	31.90	31.94	31.98	32.02
770	32.06	32.10	32.15	32.19	32.23	32.27	32.31	32.35	32.39	32.43
780	32.48	32.52	32.56	32.60	32.64	32.68	32.72	32.76	32.81	32.85
790	32.89	32.93	32.97	33.01	33.05	33.09	33.13	33.18	33.22	33.26
800	33.30	33.34	33.38	33.42	33.46	33.50	33.54	33.59	33.63	33.67
810	33.71	33.75	33.79	33.83	33.87	33.91	33.95	33.99	34.04	34.08
820	34.12	34.16	34.20	34.24	34.28	34.32	34.36	34.40	34.44	34.48
830	34.53	34.57	34.61	34.65	34.69	34.73	34.77	34.81	34.85	34.89
840	34.93	34.97	35.02	35.06	35.10	35.14	35.18	35.22	35.26	35.30
850	35.34	35.38	35.42	35.46	35.50	35.54	35.58	35.63	35.67	35.71
860	35.75	35.79	35.83	35.87	35.91	35.95	35.99	36.03	36.07	36.11
870	36.15	36.19	36.23	36.27	36.31	36.35	36.39	36.43	36.47	36.51
880	36.55	36.59	36.63	36.67	36.72	36.76	36.80	36.84	36.88	36.92
890	36.96	37.00	37.04	37.08	37.12	37.16	37.20	37.24	37.28	37.32
900	37.36	37.40	37.44	37.48	37.52	37.56	37.60	37.64	37.68	37.72
910	37.76	37.80	37.84	37.88	37.92	37.96	38.00	38.04	38.08	38.12
920	38.16	38.20	38.24	38.28	38.32	38.36	38.40	38.44	38.48	38.52
930	38.56	38.60	38.64	38.68	38.72	38.76	38.80	38.84	38.88	38.92
940	38.95	38.99	39.03	39.07	39.11	39.15	39.19	39.23	39.27	39.31
950	39.35	39.39	39.43	39.47	39.51	39.55	39.59	39.63	39.67	39.71
960	39.75	39.79	39.83	39.86	39.90	39.94	39.98	40.02	40.06	40.10
970	40.14	40.18	40.22	40.26	40.30	40.34	40.38	40.41	40.45	40.49
980	40.53	40.57	40.61	40.65	40.69	40.73	40.77	40.81	40.85	40.89
990	40.92	40.96	41.00	41.04	41.08	41.12	41.16	41.20	41.24	41.28
1000	41.31	41.35	41.39	41.43	41.47	41.51	41.55	41.59	41.63	41.67
1010	41.70	41.74	41.78	41.82	41.86	41.90	41.94	41.98	42.02	42.05
1020	42.09	42.13	42.17	42.21	42.25	42.29	42.33	42.36	42.40	42.44
1030	42.48	42.52	42.56	42.60	42.63	42.67	42.71	42.75	42.79	42.83
1040	42.87	42.90	42.94	42.98	43.02	43.06	43.10	43.14	43.17	43.21
1050	43.25	43.29	43.33	43.37	43.41	43.44	43.48	43.52	43.56	43.60
1060	43.63	43.67	43.71	43.75	43.79	43.83	43.87	43.90	43.94	43.98
1070	44.02	44.06	44.10	44.13	44.17	44.21	44.25	44.29	44.33	44.36
1080	44.40	44.44	44.48	44.52	44.55	44.59	44.63	44.67	44.71	44.74
1090	44.78	44.82	44.86	44.90	44.93	44.97	45.01	45.05	45.09	45.12
1100	45.16	45.20	45.24	45.27	45.31	45.35	45.39	45.43	45.46	45.50
1110	45.54	45.58	45.62	45.65	45.69	45.73	45.77	45.80	45.84	45.88
1120	45.92	45.96	45.99	46.03	46.07	46.11	46.14	46.18	46.22	46.26
1130	46.29	46.33	46.37	46.41	46.44	46.48	46.52	46.56	46.59	46.63
1140	46.67	46.70	46.74	46.78	46.82	46.85	46.89	46.93	46.97	47.00
1150	47.04	47.08	47.12	47.15	47.19	47.23	47.26	47.30	47.34	47.38
1160	47.41	47.45	47.49	47.52	47.56	47.60	47.63	47.67	47.71	47.75
1170	47.78	47.82	47.86	47.89	47.93	47.97	48.00	48.04	48.08	48.12
1180	48.15	48.19	48.23	48.26	48.30	48.34	48.37	48.41	48.45	48.48
1190	48.52	48.56	48.59	48.63	48.67	48.70	48.74	48.78	48.81	48.85
1200	48.89	48.92	48.96	49.00	49.03	49.07	49.11	49.14	49.18	49.22
1210	49.25	49.29	49.32	49.36	49.40	49.43	49.47	49.51	49.54	49.58
1220	49.62	49.65	49.69	49.72	49.76	49.80	49.83	49.87	49.90	49.94
1230	49.98	50.01	50.05	50.08	50.12	50.16	50.19	50.23	50.26	50.30
1240	50.34	50.37	50.41	50.44	50.48	50.52	50.55	50.59	50.62	50.66
1250	50.69	50.73	50.77	50.80	50.84	50.87	50.91	50.94	50.98	51.02
1260	51.05	51.09	51.12	51.16	51.19	51.23	51.27	51.30	51.34	51.37
1270	51.41	51.44	51.48	51.51	51.55	51.58	51.62	51.66	51.69	51.73
1280	51.76	51.80	51.83	51.87	51.90	51.94	51.97	52.01	52.04	52.08
1290	52.11	52.15	52.18	52.22	52.25	52.29	52.32	52.36	52.39	52.43
1300	52.46	52.50	52.53	52.57	52.60	52.64	52.67	52.71	52.74	52.78
1310	52.81	52.85	52.88	52.92	52.95	52.99	53.02	53.06	53.09	53.13
1320	53.16	53.20	53.23	53.27	53.30	53.34	53.37	53.41	53.44	53.47
1330	53.51	53.54	53.58	53.61	53.65	53.68	53.72	53.75	53.79	53.82
1340	53.85	53.89	53.92	53.96	53.99	54.03	54.06	54.10	54.13	54.16
1350	54.20	54.23	54.27	54.30	54.34	54.37	54.40	54.44	54.47	54.51
1360	54.54	54.57	54.61	54.64	54.68	54.71	54.74	54.78	54.81	54.85
1370	54.88	54.91	—	—	—	—	—	—	—	—

* Based on the International Temperature Scale of 1948.

CHROMEL-ALUMEL THERMOCOUPLES (Continued)

(Electromotive Force in Absolute Millivolts. Temperatures in Degrees F* Reference Junctions at 32°F.)

Millivolts

°F	0	1	2	3	4	5	6	7	8	9
-300	-5.51	-5.52	-5.53	-5.54	-5.54	-5.55	-5.56	-5.57	-5.58	-5.59
-290	-5.41	-5.42	-5.43	-5.44	-5.45	-5.46	-5.47	-5.48	-5.49	-5.50
-280	-5.30	-5.31	-5.32	-5.34	-5.35	-5.36	-5.37	-5.38	-5.39	-5.40
-270	-5.20	-5.21	-5.22	-5.23	-5.24	-5.25	-5.26	-5.27	-5.28	-5.29
-260	-5.08	-5.09	-5.10	-5.12	-5.13	-5.14	-5.15	-5.16	-5.17	-5.18
-250	-4.96	-4.97	-4.99	-5.00	-5.01	-5.02	-5.03	-5.04	-5.06	-5.07
-240	-4.84	-4.85	-4.86	-4.88	-4.89	-4.90	-4.91	-4.92	-4.94	-4.95
-230	-4.71	-4.72	-4.74	-4.75	-4.76	-4.77	-4.79	-4.80	-4.81	-4.82
-220	-4.58	-4.59	-4.60	-4.62	-4.63	-4.64	-4.66	-4.67	-4.68	-4.70
-210	-4.44	-4.45	-4.46	-4.48	-4.49	-4.51	-4.52	-4.53	-4.55	-4.56
-200	-4.29	-4.31	-4.32	-4.34	-4.35	-4.36	-4.38	-4.39	-4.41	-4.42
-190	-4.15	-4.16	-4.18	-4.19	-4.21	-4.22	-4.24	-4.25	-4.26	-4.28
-180	-4.00	-4.01	-4.03	-4.04	-4.06	-4.07	-4.09	-4.10	-4.12	-4.13
-170	-3.84	-3.86	-3.88	-3.89	-3.91	-3.92	-3.94	-3.95	-3.97	-3.98
-160	-3.69	-3.70	-3.72	-3.73	-3.75	-3.76	-3.78	-3.80	-3.81	-3.83
-150	-3.52	-3.54	-3.56	-3.57	-3.59	-3.60	-3.62	-3.64	-3.65	-3.67
-140	-3.36	-3.38	-3.39	-3.41	-3.42	-3.44	-3.46	-3.47	-3.49	-3.51
-130	-3.19	-3.20	-3.22	-3.24	-3.25	-3.27	-3.29	-3.31	-3.32	-3.34
-120	-3.01	-3.03	-3.05	-3.06	-3.08	-3.10	-3.12	-3.13	-3.15	-3.17
-110	-2.83	-2.85	-2.87	-2.89	-2.90	-2.92	-2.94	-2.96	-2.98	-2.99
-100	-2.65	-2.67	-2.69	-2.71	-2.72	-2.74	-2.76	-2.78	-2.80	-2.82
-90	-2.47	-2.49	-2.50	-2.52	-2.54	-2.56	-2.58	-2.60	-2.62	-2.63
-80	-2.28	-2.30	-2.32	-2.34	-2.36	-2.37	-2.39	-2.41	-2.43	-2.45
-70	-2.09	-2.11	-2.13	-2.15	-2.17	-2.18	-2.20	-2.22	-2.24	-2.26
-60	-1.90	-1.92	-1.94	-1.96	-1.97	-1.99	-2.01	-2.03	-2.05	-2.07
-50	-1.70	-1.72	-1.74	-1.76	-1.78	-1.80	-1.82	-1.84	-1.86	-1.88
-40	-1.50	-1.52	-1.54	-1.56	-1.58	-1.60	-1.62	-1.64	-1.66	-1.68
-30	-1.30	-1.32	-1.34	-1.36	-1.38	-1.40	-1.42	-1.44	-1.46	-1.48
-20	-1.10	-1.12	-1.14	-1.16	-1.18	-1.20	-1.22	-1.24	-1.26	-1.28
-10	-0.89	-0.91	-0.93	-0.95	-0.97	-0.99	-1.01	-1.03	-1.06	-1.08
(-)0	-0.68	-0.70	-0.72	-0.75	-0.77	-0.79	-0.81	-0.83	-0.85	-0.87
(+)0	-0.68	-0.66	-0.64	-0.62	-0.60	-0.58	-0.56	-0.54	-0.52	-0.49
10	-0.47	-0.45	-0.43	-0.41	-0.39	-0.37	-0.34	-0.32	-0.30	-0.28
20	-0.26	-0.24	-0.22	-0.19	-0.17	-0.15	-0.13	-0.11	-0.09	-0.07
30	-0.04	-0.02	0.00	0.02	0.04	0.07	0.09	0.11	0.13	0.15
40	0.18	0.20	0.22	0.24	0.26	0.29	0.31	0.33	0.35	0.37
50	0.40	0.42	0.44	0.46	0.48	0.51	0.53	0.55	0.57	0.60
60	0.62	0.64	0.65	0.68	0.71	0.73	0.75	0.77	0.80	0.82
70	0.84	0.86	0.88	0.91	0.93	0.95	0.97	1.00	1.02	1.04
80	1.06	1.09	1.11	1.13	1.15	1.18	1.20	1.22	1.24	1.27
90	1.29	1.31	1.33	1.36	1.38	1.40	1.43	1.45	1.47	1.49
100	1.52	1.54	1.56	1.58	1.61	1.63	1.65	1.68	1.70	1.72
110	1.74	1.77	1.79	1.81	1.84	1.86	1.88	1.90	1.93	1.95
120	1.97	2.00	2.02	2.04	2.06	2.09	2.11	2.13	2.16	2.18
130	2.20	2.23	2.25	2.27	2.29	2.32	2.34	2.36	2.39	2.41
140	2.43	2.46	2.48	2.50	2.52	2.55	2.57	2.59	2.62	2.64
150	2.66	2.69	2.71	2.73	2.75	2.78	2.80	2.82	2.85	2.87
160	2.89	2.92	2.94	2.96	2.98	3.01	3.03	3.05	3.08	3.10
170	3.12	3.15	3.17	3.19	3.22	3.24	3.26	3.29	3.31	3.33
180	3.36	3.38	3.40	3.43	3.45	3.47	3.49	3.52	3.54	3.56
190	3.59	3.61	3.63	3.66	3.68	3.70	3.73	3.75	3.77	3.80
200	3.82	3.84	3.87	3.89	3.91	3.94	3.96	3.98	4.01	4.03
210	4.05	4.08	4.10	4.12	4.15	4.17	4.19	4.21	4.24	4.26
220	4.28	4.31	4.33	4.35	4.38	4.40	4.42	4.44	4.47	4.49
230	4.51	4.54	4.56	4.58	4.61	4.63	4.65	4.67	4.70	4.72
240	4.74	4.77	4.79	4.81	4.83	4.86	4.88	4.90	4.92	4.95
250	4.97	4.99	5.02	5.04	5.06	5.08	5.11	5.13	5.15	5.17
260	5.20	5.22	5.24	5.26	5.29	5.31	5.33	5.35	5.38	5.40
270	5.42	5.44	5.47	5.49	5.51	5.53	5.56	5.58	5.60	5.62
280	5.65	5.67	5.69	5.71	5.73	5.76	5.78	5.80	5.82	5.85
290	5.87	5.89	5.91	5.93	5.96	5.98	6.00	6.02	6.05	6.07
300	6.09	6.11	6.13	6.16	6.18	6.20	6.22	6.25	6.27	6.29
310	6.31	6.33	6.36	6.38	6.40	6.42	6.45	6.47	6.49	6.51
320	6.53	6.56	6.58	6.60	6.62	6.65	6.67	6.69	6.71	6.73
330	6.76	6.78	6.80	6.82	6.84	6.87	6.89	6.91	6.93	6.96
340	6.98	7.00	7.02	7.04	7.07	7.09	7.11	7.13	7.15	7.18
350	7.20	7.22	7.24	7.26	7.29	7.31	7.33	7.35	7.38	7.40
360	7.42	7.44	7.46	7.49	7.51	7.53	7.55	7.58	7.60	7.62
370	7.64	7.66	7.69	7.71	7.73	7.75	7.78	7.80	7.82	7.84
380	7.87	7.89	7.91	7.93	7.95	7.98	8.00	8.02	8.04	8.07
390	8.09	8.11	8.13	8.16	8.18	8.20	8.22	8.24	8.27	8.29
400	8.31	8.33	8.36	8.38	8.40	8.42	8.45	8.47	8.49	8.51
410	8.54	8.56	8.58	8.60	8.62	8.65	8.67	8.69	8.71	8.74
420	8.76	8.78	8.80	8.82	8.85	8.87	8.89	8.91	8.94	8.96
430	8.98	9.00	9.03	9.05	9.07	9.09	9.12	9.14	9.16	9.18
440	9.21	9.23	9.25	9.27	9.30	9.32	9.34	9.36	9.39	9.41
450	9.43	9.45	9.48	9.50	9.52	9.54	9.57	9.59	9.61	9.63
460	9.66	9.68	9.70	9.73	9.75	9.77	9.79	9.82	9.84	9.86
470	9.88	9.91	9.93	9.95	9.97	10.00	10.02	10.04	10.06	10.09
480	10.11	10.13	10.16	10.18	10.20	10.22	10.25	10.27	10.29	10.31
490	10.34	10.36	10.38	10.40	10.43	10.45	10.47	10.50	10.52	10.54
500	10.57	10.59	10.61	10.63	10.66	10.68	10.70	10.72	10.75	10.77

Millivolts

°F	0	1	2	3	4	5	6	7	8	9
510	10.79	10.82	10.84	10.86	10.88	10.91	10.93	10.95	10.98	11.00
520	11.02	11.04	11.07	11.09	11.11	11.13	11.16	11.18	11.20	11.23
530	11.25	11.27	11.29	11.32	11.34	11.36	11.39	11.41	11.43	11.45
540	11.48	11.50	11.52	11.55	11.57	11.59	11.61	11.64	11.66	11.68
550	11.71	11.73	11.75	11.78	11.80	11.82	11.84	11.87	11.89	11.91
560	11.94	11.98	11.98	12.01	12.03	12.05	12.07	12.10	12.12	12.14
570	12.17	12.19	12.21	12.24	12.26	12.28	12.30	12.33	12.35	12.37
580	12.40	12.42	12.44	12.47	12.49	12.51	12.53	12.56	12.58	12.60
590	12.63	12.65	12.67	12.70	12.72	12.74	12.76	12.79	12.81	12.83
600	12.86	12.88	12.90	12.93	12.95	12.97	13.00	13.02	13.04	13.06
610	13.09	13.11	13.13	13.16	13.18	13.20	13.23	13.25	13.27	13.30
620	13.32	13.34	13.36	13.39	13.41	13.44	13.46	13.48	13.50	13.53
630	13.55	13.57	13.60	13.62	13.63	13.67	13.69	13.71	13.74	13.76
640	13.78	13.81	13.83	13.85	13.88	13.90	13.92	13.95	13.97	13.99
650	14.02	14.04	14.06	14.09	14.11	14.13	14.15	14.18	14.20	14.22
660	14.25	14.27	14.29	14.32	14.34	14.36	14.39	14.41	14.43	14.46
670	14.48	14.50	14.53	14.55	14.57	14.60	14.62	14.64	14.67	14.69
680	14.71	14.74	14.76	14.78	14.81	14.83	14.85	14.88	14.90	14.92
690	14.95	14.97	14.99	15.02	15.04	15.06	15.09	15.11	15.13	15.16
700	15.18	15.20	15.23	15.25	15.27	15.30	15.32	15.34	15.37	15.39
710	15.41	15.44	15.46	15.48	15.51	15.53	15.55	15.58	15.60	15.62
720	15.65	15.67	15.69	15.72	15.74	15.76	15.79	15.81	15.83	15.86
730	15.88	15.90	15.93	15.95	15.98	16.00	16.02	16.05	16.07	16.09
740	16.12	16.14	16.16	16.19	16.21	16.23	16.26	16.28	16.30	16.33
750	16.35	16.37	16.40	16.42	16.45	16.47	16.49	16.52	16.54	16.56
760	16.59	16.61	16.63	16.66	16.68	16.70	16.73	16.75	16.77	16.80
770	16.82	16.84	16.87	16.89	16.92	16.94	16.96	16.99	17.01	17.03
780	17.06	17.08	17.10	17.13	17.15	17.17	17.20	17.22	17.24	17.27
790	17.29	17.31	17.34	17.36	17.39	17.41	17.43	17.46	17.48	17.50
800	17.53	17.55	17.57	17.60	17.62	17.64	17.67	17.69	17.71	17.74
810	17.76	17.78	17.81	17.83	17.86	17.88	17.90	17.93	17.95	17.97
820	18.00	18.02	18.04	18.07	18.09	18.11	18.14	18.16	18.18	18.21
830	18.23	18.25	18.28	18.30	18.33	18.35	18.37	18.40	18.42	18.44
840	18.47	18.49	18.51	18.54	18.56	18.58	18.61	18.63	18.65	18.68
850	18.70	18.73	18.75	18.77	18.80	18.82	18.84	18.87	18.89	18.91
860	18.94	18.96	18.90	19.01	19.03	19.06	19.08	19.10	19.13	19.15
870	19.18	19.20	19.22	19.25	19.27	19.29	19.32	19.34	19.36	19.39
880	19.41	19.44	19.46	19.48	19.51	19.53	19.55	19.58	19.60	19.63
890	19.65	19.67	19.70	19.72	19.75	19.77	19.79	19.82	19.84	19.86
900	19.89	19.91	19.94	19.96	19.98	20.01	20.03	20.05	20.08	20.10
910	20.13	20.15	20.17	20.20	20.22	20.24	20.27	20.29	20.32	20.34
920	20.36	20.39	20.41	20.43	20.46	20.48	20.50	20.53	20.55	20.58
930	20.60	20.62	20.65	20.67	20.69	20.72	20.74	20.76	20.79	20.81
940	20.84	20.86	20.88	20.91	20.93	20.95	20.98	21.00	21.03	21.05
950	21.07	21.10	21.12	21.14	21.17	21.19	21.21	21.24	21.26	21.28
960	21.31	21.33	21.36	21.38	21.40	21.43	21.45	21.47	21.50	21.52
970	21.54	21.57	21.59	21.62	21.64	21.66	21.69	21.71	21.73	21.76
980	21.78	21.81	21.83	21.85	21.88	21.90	21.92	21.95	21.97	21.99
990	22.02	22.04	22.07	22.09	22.11	22.14	22.16	22.18	22.21	22.23
1000	22.26	22.28	22.30	22.33	22.35	22.37	22.40	22.42	22.44	22.47
1010	22.49	22.52	22.54	22.56	22.59	22.61	22.63	22.66	22.68	22.71
1020	22.73	22.75	22.78	22.80	22.82	22.85	22.87	22.90	22.92	22.94
1030	22.97	22.99	23.01	23.04	23.06	23.08	23.11	23.13	23.16	23.18
1040	23.20	23.23	23.25	23.27	23.30	23.32	23.35	23.37	23.39	23.42
1050	23.44	23.46	23.49	23.51	23.54	23.56	23.58	23.61	23.63	23.65
1060	23.68	23.70	23.72	23.75	23.77	23.80	23.82	23.84	23.87	23.89
1070	23.91	23.94	23.96	23.99	24.01	24.03	24.06	24.08	24.10	24.13
1080	24.15	24.18	24.20	24.22	24.25	24.27	24.29	24.32	24.34	24.36
1090	24.39	24.41	24.44	24.46	24.49	24.51	24.53	24.55	24.58	24.60
1100	24.63	24.65	24.67	24.70	24.72	24.74	24.77	24.79	24.82	24.84
1110	24.86	24.89	24.91	24.93	24.96	24.98	25.01	25.03	25.05	25.08
1120	25.10	25.12	25.15	25.17	25.20	25.22	25.24	25.27	25.29	25.31
1130	25.34	25.36	25.38	25.41	25.43	25.46	25.48	25.50	25.53	25.55
1140	25.57	25.60	25.62	25.65	25.67	25.69	25.72	25.74	25.76	25.79
1150	25.81	25.83	25.86	25.88	25.91	25.93	25.95	25.98	26.00	26.02
1160	26.05	26.07	26.09	26.12	26.14	26.16	26.19	26.21	26.24	26.26
1170	26.28	26.31	26.33	26.35	26.38	26.40	26.42	26.45	26.47	26.49
1180	26.52	26.54	26.56	26.59	26.61	26.63	26.66	26.68	26.70	26.73
1190	26.75	26.77	26.80	26.82	26.85	26.87	26.89	26.91	26.94	26.96
1200	26.99	27.01	27.03	27.06	27.08	27.10	27.12	27.15	27.17	27.20
1210	27.22	27.24	27.27	27.29	27.31	27.34	27.36	27.38	27.40	27.43
1220	27.45	27.48	27.50	27.52	27.55	27.57	27.59	27.62	27.64	27.66
1230	27.69	27.71	27.73	27.76	27.78	27.80	27.83	27.85	27.87	27.90
1240	27.92	27.94	27.97	27.99	28.01	28.04	28.06	28.08	28.11	28.13
1250	28.15	28.18	28.20	28.22	28.25	28.27	28.29	28.32	28.34	28.37
1260	28.39	28.41	28.44	28.46	28.48	28.50	28.53	28.55	28.58	28.60
1270	28.62	28.65	28.67	28.69	28.72	28.74	28.76	28.79	28.81	28.83
1280	28.86	28.88	28.90	28.93	28.95	28.97	29.00	29.02	29.04	29.07
1290	29.09	29.11	29.14	29.16	29.18	29.21	29.23	29.25	29.28	29.30
1300	29.32	29.35	29.37	29.39	29.42	29.44	29.46	29.49	29.51	29.53
1310	29.56	29.58	29.60	29.63	29.65	29.67	29.70	29.72	29.74	29.77
1320	29.79	29.81	29.84	29.86	29.88	29.91	29.93	29.95	29.97	30.00

* Based on on the International Temperature Scale of 1948.

CHROMEL-ALUMEL THERMOCOUPLES (Continued)

Millivolts

°F	0	1	2	3	4	5	6	7	8	9
1330	30.02	30.05	30.07	30.09	30.11	30.14	30.16	30.18	30.21	30.23
1340	30.25	30.28	30.30	30.32	30.35	30.37	30.39	30.42	30.44	30.46
1350	30.49	30.51	30.53	30.56	30.58	30.60	30.63	30.65	30.67	30.70
1360	30.72	30.74	30.77	30.79	30.81	30.83	30.86	30.88	30.90	30.93
1370	30.95	30.97	31.00	31.02	31.04	31.07	31.09	31.11	31.14	31.16
1380	31.18	31.21	31.23	31.25	31.28	31.30	31.32	31.34	31.37	31.39
1390	31.42	31.44	31.46	31.48	31.51	31.53	31.55	31.58	31.60	31.62
1400	31.65	31.67	31.69	31.72	31.74	31.76	31.78	31.81	31.83	31.85
1410	31.88	31.90	31.92	31.95	31.97	31.99	32.02	32.04	32.06	32.08
1420	32.11	32.13	32.15	32.18	32.20	32.22	32.25	32.27	32.29	32.31
1430	32.34	32.36	32.38	32.41	32.43	32.45	32.48	32.50	32.52	32.54
1440	32.57	32.59	32.61	32.64	32.66	32.68	32.70	32.73	32.75	32.77
1450	32.80	32.82	32.84	32.86	32.89	32.91	32.93	32.96	32.98	33.00
1460	33.02	33.05	33.07	33.09	33.12	33.14	33.16	33.18	33.21	33.23
1470	33.25	33.28	33.30	33.32	33.34	33.37	33.39	33.41	33.43	33.46
1480	33.48	33.50	33.53	33.55	33.57	33.59	33.62	33.64	33.66	33.69
1490	33.71	33.73	33.75	33.78	33.80	33.82	33.84	33.87	33.89	33.91
1500	33.93	33.96	33.98	34.00	34.03	34.05	34.07	34.09	34.12	34.14
1510	34.16	34.18	34.21	34.23	34.25	34.28	34.30	34.32	34.34	34.37
1520	34.39	34.41	34.43	34.46	34.48	34.50	34.53	34.55	34.57	34.59
1530	34.62	34.64	34.66	34.68	34.71	34.73	34.75	34.77	34.80	34.82
1540	34.84	34.87	34.89	34.91	34.93	34.96	34.98	35.00	35.02	35.05
1550	35.07	35.09	35.11	35.14	35.16	35.18	35.21	35.23	35.25	35.27
1560	35.29	35.32	35.34	35.36	35.39	35.41	35.43	35.45	35.48	35.50
1570	35.52	35.54	35.57	35.59	35.61	35.63	35.66	35.68	35.70	35.72
1580	35.75	35.77	35.79	35.81	35.84	35.86	35.88	35.90	35.93	35.95
1590	35.97	35.99	36.02	36.04	36.06	36.08	36.11	36.13	36.15	36.17
1600	36.19	36.19	36.22	36.24	36.29	36.31	36.33	36.35	36.37	36.40
1610	36.42	36.44	36.46	36.49	36.51	36.53	36.55	36.58	36.60	36.62
1620	36.64	36.67	36.69	36.71	36.73	36.76	36.78	36.80	36.82	36.84
1630	36.87	36.89	36.91	36.93	36.96	36.98	37.00	37.02	37.05	37.07
1640	37.09	37.11	37.14	37.16	37.18	37.20	37.23	37.25	37.27	37.29
1650	37.31	37.34	37.36	37.38	37.40	37.43	37.45	37.47	37.49	37.52
1660	37.54	37.56	37.58	37.60	37.63	37.65	37.67	37.69	37.72	37.74
1670	37.76	37.79	37.81	37.83	37.85	37.87	37.89	37.92	37.94	37.96
1680	37.98	38.01	38.03	38.05	38.07	38.09	38.12	38.14	38.16	38.18
1690	38.20	38.23	38.25	38.27	38.29	38.32	38.34	38.36	38.38	38.40
1700	38.43	38.43	38.47	38.49	38.51	38.54	38.56	38.58	38.60	38.62
1710	38.65	38.67	38.69	38.71	38.73	38.76	38.78	38.80	38.82	38.84
1720	38.87	38.89	38.91	38.93	38.95	38.98	39.00	39.02	39.04	39.06
1730	39.09	39.11	39.13	39.15	39.17	39.20	39.22	39.24	39.26	39.28
1740	39.31	39.33	39.35	39.37	39.39	39.42	39.44	39.46	39.48	39.50
1750	39.53	39.55	39.57	39.59	39.61	39.64	39.66	39.68	39.70	39.72
1760	39.75	39.77	39.79	39.81	39.83	39.86	39.88	39.90	39.92	39.94
1770	39.96	39.99	40.01	40.03	40.05	40.07	40.10	40.12	40.14	40.16
1780	40.18	40.20	40.23	40.25	40.27	40.29	40.31	40.34	40.36	40.38
1790	40.40	40.42	40.44	40.47	40.49	40.51	40.53	40.55	40.58	40.60
1800	40.62	40.64	40.66	40.68	40.71	40.73	40.75	40.77	40.79	40.82
1810	40.84	40.86	40.88	40.90	40.92	40.95	40.97	40.99	41.01	41.03
1820	41.05	41.08	41.10	41.12	41.14	41.16	41.18	41.21	41.23	41.25
1830	41.27	41.29	41.31	41.34	41.36	41.38	41.40	41.42	41.45	41.47
1840	41.49	41.51	41.53	41.55	41.57	41.60	41.62	41.64	41.66	41.68
1850	41.70	41.73	41.75	41.77	41.79	41.81	41.83	41.85	41.88	41.90
1860	41.92	41.94	41.96	41.99	42.01	42.03	42.05	42.07	42.09	42.11
1870	42.14	42.16	42.18	42.20	42.22	42.24	42.26	42.29	42.31	42.33
1880	42.35	42.37	42.39	42.42	42.44	42.46	42.48	42.50	42.52	42.55
1890	42.57	42.59	42.61	42.63	42.65	42.67	42.69	42.72	42.74	42.76
1900	42.78	42.80	42.82	42.84	42.87	42.89	42.91	42.93	42.95	42.97
1910	42.99	43.01	43.04	43.06	43.08	43.10	43.12	43.14	43.17	43.19
1920	43.21	43.23	43.25	43.27	43.29	43.31	43.34	43.36	43.38	43.40
1930	43.42	43.44	43.47	43.49	43.51	43.53	43.55	43.57	43.59	43.61
1940	43.63	43.66	43.68	43.70	43.72	43.75	43.76	43.78	43.81	43.83
1950	43.85	43.87	43.89	43.91	43.93	43.95	43.98	44.00	44.02	44.04
1960	44.06	44.08	44.10	44.13	44.15	44.17	44.19	44.21	44.23	44.25
1970	44.27	44.30	44.32	44.34	44.36	44.38	44.40	44.42	44.44	44.47
1980	44.49	44.51	44.53	44.55	44.57	44.59	44.61	44.63	44.66	44.68
1990	44.70	44.72	44.74	44.76	44.78	44.80	44.82	44.85	44.87	44.89
2000	44.91	44.93	44.95	44.97	44.99	45.01	45.03	45.06	45.08	45.10
2010	45.12	45.14	45.16	45.18	45.20	45.22	45.24	45.27	45.29	45.31
2020	45.33	45.35	45.37	45.39	45.41	45.43	45.45	45.48	45.50	45.52
2030	45.54	45.56	45.58	45.60	45.62	45.64	45.66	45.69	45.71	45.73
2040	45.75	45.77	45.79	45.81	45.83	45.85	45.87	45.90	45.92	45.94
2050	45.96	45.98	46.00	46.02	46.04	46.06	46.08	46.11	46.13	46.15
2060	46.17	46.19	46.21	46.23	46.25	46.27	46.29	46.31	46.33	46.36
2070	46.38	46.40	46.42	46.44	46.46	46.48	46.50	46.52	46.54	46.56
2080	46.58	46.60	46.63	46.65	46.67	46.69	46.71	46.73	46.75	46.77
2090	46.79	46.81	46.83	46.85	46.87	46.90	46.92	46.94	46.96	46.98
2100	47.00	47.02	47.04	47.06	47.08	47.10	47.12	47.14	47.17	47.19
2110	47.21	47.23	47.25	47.27	47.29	47.31	47.33	47.35	47.37	47.39
2120	47.41	47.43	47.45	47.47	47.49	47.52	47.54	47.56	47.58	47.60
2130	47.62	47.64	47.66	47.68	47.70	47.72	47.74	47.76	47.78	47.80
2140	47.82	47.84	47.86	47.89	47.91	47.93	47.95	47.97	47.99	48.01
2150	48.03	48.05	48.07	48.09	48.11	48.13	48.15	48.17	48.19	48.21
2160	48.23	48.25	48.27	48.29	48.32	48.34	48.36	48.38	48.40	48.42
2170	48.44	48.46	48.48	48.50	48.52	48.54	48.56	48.58	48.60	48.62
2180	48.64	48.66	48.68	48.70	48.72	48.74	48.76	48.79	48.81	48.83
2190	48.85	48.87	48.89	48.91	48.93	48.95	48.97	48.99	49.01	49.03
2200	49.05	49.07	49.09	49.11	49.13	49.15	49.17	49.19	49.21	49.23
2210	49.25	49.27	49.29	49.31	49.33	49.35	49.37	49.39	49.41	49.43
2220	49.45	49.47	49.49	49.51	49.53	49.55	49.57	49.59	49.61	49.63
2230	49.65	49.67	49.69	49.71	49.73	49.76	49.78	49.80	49.82	49.84
2240	49.86	49.88	49.90	49.92	49.94	49.96	49.98	50.00	50.02	50.04
2250	50.06	50.08	50.10	50.12	50.14	50.16	50.18	50.20	50.22	50.24
2260	50.26	50.28	50.30	50.32	50.34	50.36	50.38	50.40	50.42	50.44
2270	50.46	50.48	50.50	50.52	50.54	50.56	50.57	50.59	50.61	50.63
2280	50.65	50.67	50.69	50.71	50.73	50.75	50.77	50.79	50.81	50.83
2290	50.85	50.87	50.89	50.91	50.93	50.95	50.97	50.99	51.01	51.03
2300	51.05	51.07	51.09	51.11	51.13	51.15	51.17	51.19	51.21	51.23
2310	51.25	51.27	51.29	51.31	51.33	51.35	51.37	51.39	51.41	51.43
2320	51.45	51.47	51.48	51.50	51.52	51.54	51.56	51.58	51.60	51.62
2330	51.64	51.66	51.68	51.70	51.72	51.74	51.76	51.78	51.80	51.82
2340	51.84	51.86	51.88	51.90	51.92	51.94	51.96	51.98	52.00	52.01
2350	52.03	52.05	52.07	52.09	52.11	52.13	52.15	52.17	52.19	52.21
2360	52.23	52.25	52.27	52.29	52.31	52.33	52.35	52.37	52.39	52.41
2370	52.42	52.44	52.46	52.48	52.50	52.52	52.54	52.56	52.58	52.60
2380	52.62	52.64	52.66	52.68	52.70	52.72	52.74	52.76	52.77	52.79
2390	52.81	52.83	52.85	52.87	52.89	52.91	52.93	52.95	52.97	52.99
2400	53.01	53.03	53.05	53.07	53.08	53.10	53.12	53.14	53.16	53.18
2410	53.20	53.22	53.24	53.26	53.28	53.30	53.32	53.34	53.35	53.37
2420	53.39	53.41	53.43	53.45	53.47	53.49	53.51	53.53	53.55	53.57
2430	53.59	53.60	53.62	53.64	53.66	53.68	53.70	53.72	53.74	53.76
2440	53.78	53.80	53.82	53.83	53.85	53.87	53.89	53.91	53.93	53.95
2450	53.97	53.99	54.01	54.03	54.04	54.06	54.08	54.10	54.12	54.14
2460	54.16	54.18	54.20	54.22	54.24	54.25	54.27	54.29	54.31	54.33
2470	54.35	54.37	54.39	54.41	54.43	54.44	54.46	54.48	54.50	54.52
2480	54.54	54.56	54.58	54.60	54.62	54.63	54.65	54.67	54.69	54.71
2490	54.73	54.75	54.77	54.79	54.81	54.82	54.84	54.86	54.88	54.90
2500	54.92	—	—	—	—	—	—	—	—	—

IRON-CONSTANTAN THERMOCOUPLES (MODIFIED 1913)

(Electromotive Force in Absolute Millivolts. Temperatures in Degrees C (Int. 1948). Reference Junctions at 0°C.)

Millivolts

°C	0	1	2	3	4	5	6	7	8	9
-190	-7.66	-7.69	-7.71	-7.73	-7.76	-7.78				
-180	-7.40	-7.43	-7.46	-7.49	-7.51	-7.54	-7.56	-7.59	-7.61	-7.64
-170	-7.12	-7.15	-7.18	-7.21	-7.24	-7.27	-7.30	-7.32	-7.35	-7.38
-160	-6.82	-6.85	-6.88	-6.91	-6.94	-6.97	-7.00	-7.03	-7.06	-7.09
-150	-6.50	-6.53	-6.56	-6.60	-6.63	-6.66	-6.69	-6.72	-6.76	-6.79
-140	-6.16	-6.19	-6.22	-6.26	-6.29	-6.33	-6.36	-6.40	-6.43	-6.46
-130	-5.80	-5.84	-5.87	-5.91	-5.94	-5.98	-6.01	-6.05	-6.08	-6.12
-120	-5.42	-5.46	-5.50	-5.54	-5.58	-5.61	-5.65	-5.69	-5.72	-5.76
-110	-5.03	-5.07	-5.11	-5.15	-5.19	-5.23	-5.27	-5.31	-5.35	-5.38
-100	-4.63	-4.67	-4.71	-4.75	-4.79	-4.83	-4.87	-4.91	-4.95	-4.99
-90	-4.21	-4.25	-4.30	-4.34	-4.38	-4.42	-4.46	-4.50	-4.55	-4.59
-80	-3.78	-3.82	-3.87	-3.91	-3.96	-4.00	-4.04	-4.08	-4.13	-4.17
-70	-3.34	-3.38	-3.43	-3.47	-3.52	-3.56	-3.60	-3.65	-3.69	-3.74
-60	-2.89	-2.94	-2.98	-3.03	-3.07	-3.12	-3.16	-3.21	-3.25	-3.30
-50	-2.43	-2.48	-2.52	-2.57	-2.62	-2.66	-2.71	-2.75	-2.80	-2.84
-40	-1.96	-2.01	-2.06	-2.10	-2.15	-2.20	-2.24	-2.29	-2.34	-2.38
-30	-1.48	-1.53	-1.58	-1.63	-1.67	-1.72	-1.77	-1.82	-1.87	-1.91
-20	-1.00	-1.04	-1.09	-1.14	-1.19	-1.24	-1.29	-1.34	-1.39	-1.43
-10	-0.50	-0.55	-0.60	-0.65	-0.70	-0.75	-0.80	-0.85	-0.90	-0.95
(-)0	0.00	-0.05	-0.10	-0.15	-0.20	-0.25	-0.30	-0.35	-0.40	-0.45
(+)0	0.00	0.05	0.10	0.15	0.20	0.25	0.30	0.35	0.40	0.45
10	0.50	0.56	0.61	0.66	0.71	0.76	0.81	0.86	0.91	0.97
20	1.02	1.07	1.12	1.17	1.22	1.28	1.33	1.38	1.43	1.48
30	1.54	1.59	1.64	1.69	1.74	1.80	1.85	1.90	1.95	2.00
40	2.06	2.11	2.16	2.22	2.27	2.32	2.37	2.42	2.48	2.53
50	2.58	2.64	2.69	2.74	2.80	2.85	2.90	2.96	3.01	3.06
60	3.11	3.17	3.22	3.27	3.33	3.38	3.43	3.49	3.54	3.60
70	3.65	3.70	3.76	3.81	3.86	3.92	3.97	4.02	4.08	4.13
80	4.19	4.24	4.29	4.35	4.40	4.46	4.51	4.56	4.62	4.67
90	4.73	4.78	4.83	4.89	4.94	5.00	5.05	5.10	5.16	5.21

IRON-CONSTANTAN THERMOCOUPLES (MODIFIED 1913) (Continued)

Millivolts

°C	0	1	2	3	4	5	6	7	8	9
100	5.27	5.32	5.38	5.43	5.48	5.54	5.59	5.65	5.70	5.76
110	5.81	5.86	5.92	5.97	6.03	6.08	6.14	6.19	6.25	6.30
120	6.36	6.41	6.47	6.52	6.58	6.63	6.68	6.74	6.79	6.85
130	6.90	6.96	7.01	7.07	7.12	7.18	7.23	7.29	7.34	7.40
140	7.45	7.51	7.56	7.62	7.67	7.73	7.78	7.84	7.89	7.95
150	8.00	8.06	8.12	8.17	8.23	8.28	8.34	8.39	8.45	8.50
160	8.56	8.61	8.67	8.72	8.78	8.84	8.89	8.95	9.00	9.06
170	9.11	9.17	9.22	9.28	9.33	9.39	9.44	9.50	9.56	9.61
180	9.67	9.72	9.78	9.83	9.89	9.95	10.00	10.06	10.11	10.17
190	10.22	10.28	10.34	10.39	10.45	10.50	10.56	10.61	10.67	10.72
200	10.78	10.84	10.89	10.95	11.00	11.06	11.12	11.17	11.23	11.28
210	11.34	11.39	11.45	11.50	11.56	11.62	11.67	11.73	11.78	11.84
220	11.89	11.95	12.00	12.06	12.12	12.17	12.23	12.28	12.34	12.39
230	12.45	12.50	12.56	12.62	12.67	12.73	12.78	12.84	12.89	12.95
240	13.01	13.06	13.12	13.17	13.23	13.28	13.34	13.40	13.45	13.51
250	13.56	13.62	13.67	13.73	13.78	13.84	13.89	13.95	14.00	14.06
260	14.12	14.17	14.23	14.28	14.34	14.39	14.45	14.50	14.56	14.61
270	14.67	14.72	14.78	14.83	14.89	14.94	15.00	15.06	15.11	15.17
280	15.22	15.28	15.33	15.39	15.44	15.50	15.55	15.61	15.66	15.72
290	15.77	15.83	15.88	15.94	16.00	16.05	16.11	16.16	16.22	16.27
300	16.33	16.38	16.44	16.49	16.55	16.60	16.66	16.71	16.77	16.82
310	16.88	16.93	16.99	17.04	17.10	17.15	17.21	17.26	17.32	17.37
320	17.43	17.48	17.54	17.60	17.65	17.71	17.76	17.82	17.87	17.93
330	17.98	18.04	18.09	18.15	18.20	18.26	18.32	18.37	18.43	18.48
340	18.54	18.59	18.65	18.70	18.76	18.81	18.87	18.92	18.98	19.03
350	19.09	19.14	19.20	19.26	19.31	19.37	19.42	19.48	19.53	19.59
360	19.64	19.70	19.75	19.81	19.86	19.92	19.97	20.03	20.08	20.14
370	20.20	20.25	20.31	20.36	20.42	20.47	20.53	20.58	20.64	20.69
380	20.75	20.80	20.86	20.91	20.97	21.02	21.08	21.13	21.19	21.24
390	21.30	21.35	21.41	21.46	21.52	21.57	21.63	21.68	21.74	21.79
400	21.85	21.90	21.96	22.02	22.07	22.13	22.18	22.24	22.29	22.35
410	22.40	22.46	22.51	22.57	22.62	22.68	22.73	22.79	22.84	22.90
420	22.95	23.01	23.06	23.12	23.17	23.23	23.28	23.34	23.39	23.45
430	23.50	23.56	23.61	23.67	23.72	23.78	23.83	23.89	23.94	24.00

Millivolts

°C	0	1	2	3	4	5	6	7	8	9
440	24.06	24.11	24.17	24.22	24.28	24.33	24.39	24.44	24.50	24.55
450	24.61	24.66	24.72	24.77	24.83	24.88	24.94	25.00	25.05	25.11
460	25.16	25.22	25.27	25.33	25.38	25.44	25.49	25.55	25.60	25.66
470	25.72	25.77	25.83	25.88	25.94	25.99	26.05	26.10	26.16	26.22
480	26.27	26.33	26.38	26.44	26.49	26.55	26.61	26.66	26.72	26.77
490	26.83	26.89	26.94	27.00	27.05	27.11	27.17	27.22	27.28	27.33
500	27.39	27.45	27.50	27.56	27.61	27.67	27.73	27.78	27.84	27.90
510	27.95	28.01	28.07	28.12	28.18	28.23	28.29	28.35	28.40	28.46
520	28.52	28.57	28.63	28.69	28.74	28.80	28.86	28.91	28.97	29.02
530	29.08	29.14	29.20	29.25	29.31	29.37	29.42	29.48	29.54	29.59
540	29.65	29.71	29.76	29.82	29.88	29.94	29.99	30.05	30.11	30.16
550	30.22	30.28	30.34	30.39	30.45	30.51	30.57	30.62	30.68	30.74
560	30.80	30.85	30.91	30.97	31.02	31.08	31.14	31.20	31.26	31.31
570	31.37	31.43	31.49	31.54	31.60	31.66	31.72	31.78	31.83	31.89
580	31.95	32.01	32.06	32.12	32.18	32.24	32.30	32.36	32.41	32.47
590	32.53	32.59	32.65	32.71	32.76	32.82	32.88	32.94	33.00	33.06
600	33.11	33.17	33.23	33.29	33.35	33.41	33.46	33.52	33.58	33.64
610	33.70	33.76	33.82	33.88	33.94	33.99	34.05	34.11	34.17	34.23
620	34.29	34.35	34.41	34.47	34.53	34.58	34.64	34.70	34.76	34.82
630	34.88	34.94	35.00	35.06	35.12	35.18	35.24	35.30	35.36	35.42
640	35.48	35.54	35.60	35.66	35.72	35.78	35.84	35.90	35.96	36.02
650	36.08	36.14	36.20	36.26	36.32	36.38	36.44	36.50	36.56	36.62
660	36.69	36.75	36.81	36.87	36.93	36.99	37.05	37.11	37.18	37.24
670	37.30	37.36	37.42	37.48	37.54	37.60	37.66	37.73	37.79	37.85
680	37.91	37.97	38.04	38.10	38.16	38.22	38.28	38.34	38.41	38.47
690	38.53	38.59	38.66	38.72	38.78	38.84	38.90	38.97	39.03	39.09
700	39.15	39.22	39.28	39.34	39.40	39.47	39.53	39.59	39.65	39.72
710	39.78	39.84	39.91	39.97	40.03	40.10	40.16	40.22	40.28	40.35
720	40.41	40.48	40.54	40.60	40.66	40.73	40.79	40.86	40.92	40.98
730	41.05	41.11	41.17	41.24	41.30	41.36	41.43	41.49	41.56	41.62
740	41.68	41.75	41.81	41.87	41.94	42.00	42.07	42.13	42.19	42.26
750	42.32	42.38	42.45	42.51	42.58	42.64	42.70	42.77	42.83	42.90
760	42.96	—	—	—	—	—	—	—	—	—

(Electromotive Force in Absolute Millivolts. Temperatures in Degrees F* Reference Junctions at 32°F.)

Millivolts

°F	0	1	2	3	4	5	6	7	8	9
-310	-7.66	-7.68	-7.69	-7.70	-7.71	-7.73	-7.74	-7.75	-7.76	-7.78
-300	-7.52	-7.54	-7.55	-7.57	-7.58	-7.59	-7.61	-7.62	-7.64	-7.65
-290	-7.38	-7.39	-7.40	-7.42	-7.44	-7.45	-7.46	-7.48	-7.49	-7.51
-280	-7.22	-7.24	-7.25	-7.27	-7.28	-7.30	-7.31	-7.33	-7.34	-7.36
-270	-7.06	-7.07	-7.09	-7.11	-7.12	-7.14	-7.15	-7.17	-7.19	-7.20
-260	-6.89	-6.90	-6.92	-6.94	-6.96	-6.97	-6.99	-7.01	-7.02	-7.04
-250	-6.71	-6.73	-6.75	-6.77	-6.78	-6.80	-6.82	-6.84	-6.85	-6.87
-240	-6.53	-6.55	-6.57	-6.59	-6.61	-6.62	-6.64	-6.66	-6.68	-6.70
-230	-6.35	-6.37	-6.38	-6.40	-6.42	-6.44	-6.46	-6.48	-6.50	-6.52
-220	-6.16	-6.18	-6.19	-6.21	-6.23	-6.25	-6.27	-6.29	-6.31	-6.33
-210	-5.96	-5.98	-6.00	-6.02	-6.04	-6.06	-6.08	-6.10	-6.12	-6.14
-200	-5.76	-5.78	-5.80	-5.82	-5.84	-5.86	-5.88	-5.90	-5.92	-5.94
-190	-5.55	-5.57	-5.59	-5.61	-5.63	-5.65	-5.67	-5.70	-5.72	-5.74
-180	-5.34	-5.36	-5.38	-5.40	-5.42	-5.44	-5.46	-5.49	-5.51	-5.53
-170	-5.12	-5.14	-5.16	-5.19	-5.21	-5.23	-5.25	-5.27	-5.30	-5.32
-160	-4.90	-4.92	-4.94	-4.97	-4.99	-5.01	-5.03	-5.06	-5.08	-5.10
-150	-4.68	-4.70	-4.72	-4.74	-4.76	-4.79	-4.81	-4.83	-4.86	-4.88
-140	-4.44	-4.47	-4.49	-4.51	-4.54	-4.56	-4.58	-4.61	-4.63	-4.65
-130	-4.21	-4.23	-4.26	-4.28	-4.30	-4.33	-4.35	-4.38	-4.40	-4.42
-120	-3.97	-4.00	-4.02	-4.04	-4.07	-4.09	-4.12	-4.14	-4.16	-4.19
-110	-3.73	3.76	-3.78	-3.81	-3.83	-3.85	-3.88	-3.90	-3.93	-3.95
-100	-3.49	-3.51	-3.54	-3.56	-3.59	-3.61	-3.64	-3.66	-3.68	-3.71
-90	-3.24	-3.27	-3.29	-3.32	-3.34	-3.36	-3.39	-3.41	-3.44	-3.46
-80	-2.99	-3.02	-3.04	-3.07	-3.09	-3.12	-3.14	-3.17	-3.19	-3.22
-70	-2.74	-2.76	-2.79	-2.81	-2.84	-2.86	-2.89	-2.92	-2.94	-2.97
-60	-2.48	-2.51	-2.53	-2.56	-2.58	-2.61	-2.64	-2.66	-2.69	-2.71
-50	-2.22	-2.25	-2.27	-2.30	-2.33	-2.35	-2.38	-2.40	-2.43	-2.46
-40	-1.96	-1.99	-2.01	-2.04	-2.06	-2.09	-2.12	-2.14	-2.17	-2.20
-30	-1.70	-1.72	-1.75	-1.78	-1.80	-1.83	-1.86	-1.88	-1.91	-1.94
-20	-1.43	-1.46	-1.48	-1.51	-1.54	-1.56	-1.59	-1.62	-1.64	-1.67
-10	-1.16	-1.19	-1.21	-1.24	-1.27	-1.29	-1.32	-1.35	-1.38	-1.40
(-)0	-0.89	-0.91	-0.94	-0.97	-1.00	-1.02	-1.05	-1.08	-1.10	-1.13
+(0)	-0.89	-0.86	-0.83	-0.80	-0.78	-0.75	-0.72	-0.70	-0.67	-0.64
10	-0.61	-0.58	-0.56	-0.53	-0.50	-0.48	-0.45	-0.42	-0.39	-0.36
20	-0.34	-0.31	-0.28	-0.25	-0.22	-0.20	-0.17	-0.14	-0.11	-0.09
30	-0.06	-0.03	0.00	0.03	0.05	0.08	0.11	0.14	0.17	0.19
40	0.22	0.25	0.28	0.31	0.34	0.36	0.39	0.42	0.45	0.48
50	0.50	0.53	0.56	0.59	0.62	0.65	0.67	0.70	0.73	0.76
60	0.79	0.82	0.84	0.87	0.90	0.93	0.96	0.99	1.02	1.04
70	1.07	1.10	1.13	1.16	1.19	1.22	1.25	1.28	1.30	1.33
80	1.36	1.39	1.42	1.45	1.48	1.51	1.54	1.56	1.59	1.62

Millivolts

°F	0	1	2	3	4	5	6	7	8	9
90	1.65	1.68	1.71	1.74	1.77	1.80	1.83	1.85	1.88	1.91
100	1.94	1.97	2.00	2.03	2.06	2.09	2.12	2.14	2.17	2.20
110	2.23	2.26	2.29	2.32	2.35	2.38	2.41	2.44	2.47	2.50
120	2.52	2.55	2.58	2.61	2.64	2.67	2.70	2.73	2.76	2.79
130	2.82	2.85	2.88	2.91	2.94	2.97	3.00	3.03	3.06	3.08
140	3.11	3.14	3.17	3.20	3.23	3.26	3.29	3.32	3.35	3.38
150	3.41	3.44	3.47	3.50	3.53	3.56	3.59	3.62	3.65	3.68
160	3.71	3.74	3.77	3.80	3.83	3.86	3.89	3.92	3.95	3.98
170	4.01	4.04	4.07	4.10	4.13	4.16	4.19	4.22	4.25	4.28
180	4.31	4.34	4.37	4.40	4.43	4.46	4.49	4.52	4.55	4.58
190	4.61	4.64	4.67	4.70	4.73	4.76	4.79	4.82	4.85	4.88
200	4.91	4.94	4.97	5.00	5.03	5.06	5.09	5.12	5.15	5.18
210	5.21	5.24	5.27	5.30	5.33	5.36	5.39	5.42	5.45	5.48
220	5.51	5.54	5.57	5.60	5.63	5.66	5.69	5.72	5.75	5.78
230	5.81	5.84	5.87	5.90	5.93	5.96	5.99	6.02	6.05	6.08
240	6.11	6.14	6.17	6.20	6.24	6.27	6.30	6.33	6.36	6.39
250	6.42	6.45	6.48	6.51	6.54	6.57	6.60	6.63	6.66	6.69
260	6.72	6.75	6.78	6.81	6.84	6.87	6.90	6.93	6.96	7.00
270	7.03	7.06	7.09	7.12	7.15	7.18	7.21	7.24	7.27	7.30
280	7.33	7.36	7.39	7.42	7.45	7.48	7.51	7.54	7.58	7.61
290	7.64	7.67	7.70	7.73	7.76	7.79	7.82	7.85	7.88	7.91
300	7.94	7.97	8.00	8.04	8.07	8.10	8.13	8.16	8.19	8.22
310	8.25	8.28	8.31	8.34	8.37	8.40	8.44	8.47	8.50	8.53
320	8.56	8.59	8.62	8.65	8.68	8.71	8.74	8.77	8.80	8.84
330	8.87	8.90	8.93	8.96	8.99	9.02	9.05	9.08	9.11	9.14
340	9.17	9.20	9.24	9.27	9.30	9.33	9.36	9.39	9.42	9.45
350	9.48	9.51	9.54	9.58	9.61	9.64	9.67	9.70	9.73	9.76
360	9.79	9.82	9.85	9.88	9.92	9.95	9.98	10.01	10.04	10.07
370	10.10	10.13	10.16	10.19	10.22	10.25	10.28	10.32	10.35	10.38
380	10.41	10.44	10.47	10.50	10.53	10.56	10.60	10.63	10.66	10.69
390	10.72	10.75	10.78	10.81	10.84	10.87	10.90	10.94	10.97	11.00
400	11.03	11.06	11.09	11.12	11.15	11.18	11.21	11.24	11.28	11.31
410	11.34	11.37	11.40	11.43	11.46	11.49	11.52	11.55	11.58	11.62
420	11.65	11.68	11.71	11.74	11.77	11.80	11.83	11.86	11.89	11.92
430	11.96	11.99	12.02	12.05	12.08	12.11	12.14	12.17	12.20	12.23
440	12.26	12.30	12.33	12.36	12.39	12.42	12.45	12.48	12.51	12.54
450	12.57	12.60	12.64	12.67	12.70	12.73	12.76	12.79	12.82	12.85
460	12.88	12.91	12.94	12.98	13.01	13.04	13.07	13.10	13.13	13.16
470	13.19	13.22	13.25	13.28	13.31	13.34	13.38	13.41	13.44	13.47
480	13.50	13.53	13.56	13.59	13.62	13.65	13.68	13.72	13.75	13.78
490	13.81	13.84	13.87	13.90	13.93	13.96	13.99	14.02	14.05	14.08

* Based on the International Temperature Scale of 1948.

IRON-CONSTANTAN THERMOCOUPLES (MODIFIED 1913) (Continued)

Millivolts

°F	0	1	2	3	4	5	6	7	8	9
500	14.12	14.15	14.18	14.21	14.24	14.27	14.30	14.33	14.36	14.39
510	14.42	14.45	14.48	14.52	14.55	14.58	14.61	14.64	14.67	14.70
520	14.73	14.76	14.79	14.82	14.85	14.88	14.91	14.94	14.98	15.01
530	15.04	15.07	15.10	15.13	15.16	15.19	15.22	15.25	15.28	15.31
540	15.34	15.37	15.40	15.44	15.47	15.50	15.53	15.56	15.59	15.62
550	15.65	15.68	15.71	15.74	15.77	15.80	15.84	15.87	15.90	15.93
560	15.96	15.99	16.02	16.05	16.08	16.11	16.14	16.17	16.20	16.23
570	16.26	16.30	16.33	16.36	16.39	16.42	16.45	16.48	16.51	16.54
580	16.57	16.60	16.63	16.66	16.69	16.72	16.75	16.78	16.82	16.85
590	16.88	16.91	16.94	16.97	17.00	17.03	17.06	17.09	17.12	17.15
600	17.18	17.21	17.24	17.28	17.31	17.34	17.37	17.40	17.43	17.46
610	17.49	17.52	17.55	17.58	17.61	17.64	17.68	17.71	17.74	17.77
620	17.80	17.83	17.86	17.89	17.92	17.95	17.98	18.01	18.04	18.08
630	18.11	18.14	18.17	18.20	18.23	18.26	18.29	18.32	18.35	18.38
640	18.41	18.44	18.47	18.50	18.54	18.57	18.60	18.63	18.66	18.69
650	18.72	18.75	18.78	18.81	18.84	18.87	18.90	18.94	18.97	19.00
660	19.03	19.06	19.09	19.12	19.15	19.18	19.21	19.24	19.27	19.30
670	19.34	19.37	19.40	19.43	19.46	19.49	19.52	19.55	19.58	19.61
680	19.64	19.67	19.70	19.74	19.77	19.80	19.83	19.86	19.89	19.92
690	19.95	19.98	20.01	20.04	20.07	20.10	20.13	20.16	20.20	20.23
700	20.26	20.29	20.32	20.35	20.38	20.41	20.44	20.47	20.50	20.53
710	20.56	20.59	20.62	20.66	20.69	20.72	20.75	20.78	20.81	20.84
720	20.87	20.90	20.93	20.96	20.99	21.02	21.05	21.08	21.11	21.14
730	21.18	21.21	21.24	21.27	21.30	21.33	21.36	21.39	21.42	21.45
740	21.48	21.51	21.54	21.57	21.60	21.64	21.67	21.70	21.73	21.76
750	21.79	21.82	21.85	21.88	21.91	21.94	21.97	22.00	22.03	22.06
760	22.10	22.13	22.16	22.19	22.22	22.25	22.28	22.31	22.34	22.37
770	22.40	22.43	22.46	22.49	22.52	22.55	22.58	22.62	22.65	22.68
780	22.71	22.74	22.77	22.80	22.83	22.86	22.89	22.92	22.95	22.98
790	23.01	23.04	23.08	23.11	23.14	23.17	23.20	23.23	23.26	23.29
800	23.32	23.35	23.38	23.41	23.44	23.47	23.50	23.53	23.56	23.60
810	23.63	23.66	23.69	23.72	23.75	23.78	23.81	23.84	23.87	23.90
820	23.93	23.96	23.99	24.02	24.06	24.09	24.12	24.15	24.18	24.21
830	24.24	24.27	24.30	24.33	24.36	24.39	24.42	24.45	24.48	24.52
840	24.55	24.58	24.61	24.64	24.67	24.70	24.73	24.76	24.79	24.82
850	24.85	24.88	24.91	24.94	24.98	25.01	25.04	25.07	25.10	25.13
860	25.16	25.19	25.22	25.25	25.28	25.32	25.35	25.38	25.41	25.44
870	25.47	25.50	25.53	25.56	25.59	25.62	25.65	25.68	25.72	25.75
880	25.78	25.81	25.84	25.87	25.90	25.93	25.96	25.99	26.02	26.06
890	26.09	26.12	26.15	26.18	26.21	26.24	26.27	26.30	26.33	26.36
900	26.40	26.43	26.46	26.49	26.52	26.55	26.58	26.61	26.64	26.67
910	26.70	26.74	26.77	26.80	26.83	26.86	26.89	26.92	26.95	26.98
920	27.02	27.05	27.08	27.11	27.14	27.17	27.20	27.23	27.26	27.30
930	27.33	27.36	27.39	27.42	27.45	27.48	27.51	27.54	27.58	27.61
940	27.64	27.67	27.70	27.73	27.76	27.80	27.83	27.86	27.89	27.92
950	27.95	27.98	28.02	28.05	28.08	28.11	28.14	28.17	28.20	28.23
960	28.26	28.30	28.33	28.36	28.39	28.42	28.45	28.48	28.52	28.55
970	28.58	28.61	28.64	28.67	28.70	28.74	28.77	28.80	28.83	28.86
980	28.89	28.92	28.96	28.99	29.02	29.05	29.08	29.11	29.14	29.18
990	29.21	29.24	29.27	29.30	29.33	29.37	29.40	29.43	29.46	29.49
1000	29.52	29.56	29.59	29.62	29.65	29.68	29.71	29.75	29.78	29.81
1010	29.84	29.87	29.90	29.94	29.97	30.00	30.03	30.06	30.10	30.13
1020	30.16	30.19	30.22	30.25	30.28	30.32	30.35	30.38	30.41	30.44
1030	30.48	30.51	30.54	30.57	30.60	30.64	30.67	30.70	30.73	30.76
1040	30.80	30.83	30.86	30.89	30.92	30.96	30.99	31.02	31.05	31.08
1050	31.12	31.15	31.18	31.21	31.24	31.28	31.31	31.34	31.37	31.40
1060	31.44	31.47	31.50	31.53	31.56	31.60	31.63	31.66	31.69	31.72
1070	31.76	31.79	31.82	31.85	31.88	31.92	31.95	31.98	32.01	32.05
1080	32.08	32.11	32.14	32.18	32.21	32.24	32.27	32.30	32.34	32.37
1090	32.40	32.43	32.47	32.50	32.53	32.56	32.60	32.63	32.66	32.69
1100	32.72	32.76	32.79	32.82	32.86	32.89	32.92	32.95	32.98	33.02
1110	33.05	33.08	33.11	33.15	33.18	33.21	33.24	33.28	33.31	33.34
1120	33.37	33.41	33.44	33.47	33.50	33.54	33.57	33.60	33.64	33.67
1130	33.70	33.73	33.76	33.80	33.83	33.86	33.90	33.93	33.96	33.99
1140	34.03	34.06	34.09	34.12	34.16	34.19	34.22	34.26	34.29	34.32
1150	34.36	34.39	34.42	34.45	34.49	34.52	34.55	34.58	34.62	34.65
1160	34.68	34.72	34.75	34.78	34.82	34.85	34.88	34.92	34.95	34.98
1170	35.01	35.05	35.08	35.11	35.15	35.18	35.21	35.25	35.28	35.31
1180	35.35	35.38	35.41	35.45	35.48	35.51	35.54	35.58	35.61	35.64
1190	35.68	35.71	35.74	35.78	35.81	35.84	35.88	35.91	35.94	35.98
1200	36.01	36.05	36.08	36.11	36.15	36.18	36.21	36.25	36.28	36.31
1210	36.35	36.38	36.42	36.45	36.48	36.52	36.55	36.58	36.62	36.65
1220	36.69	36.72	36.75	36.79	36.82	36.86	36.89	36.92	36.96	36.99
1230	37.02	37.06	37.09	37.13	37.16	37.20	37.23	37.26	37.30	37.33
1240	37.36	37.40	37.43	37.47	37.50	37.54	37.57	37.60	37.64	37.67
1250	37.71	37.74	37.78	37.81	37.84	37.88	37.91	37.95	37.98	38.02
1260	38.05	38.08	38.12	38.15	38.19	38.22	38.26	38.29	38.32	38.36
1270	38.39	38.43	38.46	38.50	38.53	38.57	38.60	38.64	38.67	38.70
1280	38.74	38.77	38.81	38.84	38.88	38.91	38.95	38.98	39.02	39.05
1290	39.08	39.12	39.15	39.19	39.22	39.26	39.29	39.33	39.36	39.40
1300	39.43	39.47	39.50	39.54	39.57	39.61	39.64	39.68	39.71	39.75
1310	39.78	39.82	39.85	39.89	39.92	39.96	39.99	40.03	40.06	40.10
1320	40.13	40.17	40.20	40.24	40.27	40.31	40.34	40.38	40.41	40.45
1330	40.48	40.52	40.55	40.59	40.62	40.66	40.69	40.73	40.76	40.80
1340	40.83	40.87	40.90	40.94	40.98	41.01	41.05	41.08	41.12	41.15
1350	41.19	41.22	41.26	41.29	41.33	41.36	41.40	41.43	41.47	41.50
1360	41.54	41.58	41.61	41.65	41.68	41.72	41.75	41.79	41.82	41.86
1370	41.90	41.93	41.97	42.00	42.04	42.07	42.11	42.14	42.18	42.22
1380	42.25	42.29	42.32	42.36	42.39	42.43	42.46	42.50	42.53	42.57
1390	42.61	42.64	42.68	42.71	42.75	42.78	42.82	42.85	42.89	42.92
1400	42.96	—	—	—	—	—	—	—	—	—

TEMPERATURE-E. M. F. VALUES FOR COPPER-CONSTANTAN

E. M. F. values are in millivolts; reference junctions at 0°C.; temperatures are in degrees C.
Roeser and Wensel, National Bureau of Standards

Millivolts

°C	0	10	20	30	40	50	60	70	80	90
-200	-5.54									
-100	-3.35	-3.62	-3.89	-4.14	-4.38	-4.60	-4.82	-5.02	-5.20	-5.38
0	0	-0.38	-0.75	-1.11	-1.47	-1.81	-2.14	-2.46	-2.77	-3.06
0	0	0.39	0.79	1.19	1.61	2.03	2.47	2.91	3.36	3.81
100	4.28	4.75	5.23	5.71	6.20	6.70	7.21	7.72	8.23	8.76
200	9.29	9.82	10.36	10.91	11.46	12.01	12.57	13.14	13.71	14.28
300	14.86	15.44	16.03	16.62	17.22	17.82	18.42	19.03	19.64	20.25
400	20.87									

E. M. F. values are in millivolts; reference junctions at 32°F.; temperatures are in degrees F.
Roeser and Wensel, National Bureau of Standards

Millivolts

°F	0	10	20	30	40	50	60	70	80	90
-300	-5.28									
-200	-4.11	-4.25	-4.38	-4.50	-4.63	-4.75	-4.86	-4.97	-5.08	-5.18
-100	-2.56	-2.73	-2.90	-3.06	-3.22	-3.38	-3.53	-3.68	-3.83	-3.97
0	-0.67	-0.87	-1.07	-1.27	-1.47	-1.66	-1.84	-2.03	-2.21	-2.39
0	-0.67	-0.46	-0.26	-0.04	+0.17	0.39	0.61	0.83	1.06	1.29
100	1.52	1.75	1.99	2.23	2.47	2.71	2.96	3.21	3.46	3.71
200	3.97	4.22	4.48	4.75	5.01	5.28	5.55	5.82	6.09	6.37
300	6.64	6.92	7.21	7.49	7.77	8.06	8.35	8.64	8.93	9.23
400	9.52	9.82	10.12	10.42	10.72	11.03	11.33	11.64	11.95	12.26
500	12.57	12.89	13.20	13.52	13.83	14.15	14.47	14.79	15.12	15.44
600	15.77	16.10	16.42	16.75	17.08	17.42	17.75	18.08	18.42	18.75
700	19.09	19.43	19.77	20.11	20.45	20.80				

REFERENCE TABLE FOR Pt TO Pt—10 PER CENT Rh THERMOCOUPLE

Emfs are expressed in microvolts and temperatures in °C. Cold junctions at 0°C. Roeser and Wensel, National Bureau of Standards

E(μv)	0	1,000	2,000	3,000	4,000	5,000	6,000	7,000	8,000	9,000	10,000	11,000	12,000	13,000	14,000	15,000	16,000	17,000
0	0	146.9	265.0	373.7	477.7	578.1	675.3	769.5	861.0	950.2	1,037.2	1,122.3	1,206.4	1,290.0	1,373.8	1,458.0	1,542.6	1,627.8
100	17.7	159.4	276.2	384.2	487.9	588.0	684.8	778.8	870.0	959.0	1,045.8	1,130.8	1,214.7	1,298.3	1,382.2	1,466.4	1,551.1	1,636.4
200	34.4	171.7	287.3	394.7	498.1	597.8	694.3	788.0	879.0	967.8	1,054.4	1,139.2	1,223.1	1,306.7	1,390.6	1,474.8	1,559.6	1,644.9
300	50.2	183.8	298.3	405.2	508.2	607.6	703.8	797.2	888.0	976.5	1,062.9	1,147.6	1,231.4	1,315.1	1,399.0	1,483.3	1,568.1	1,653.5
400	65.4	195.8	309.3	415.7	518.3	617.4	713.3	806.4	897.0	985.3	1,071.5	1,156.0	1,239.8	1,323.5	1,407.4	1,491.8	1,576.6	1,662.1
500	80.0	207.6	320.2	426.1	528.4	627.1	722.7	815.6	905.9	994.0	1,080.0	1,164.4	1,248.2	1,331.8	1,415.8	1,500.2	1,585.1	1,670.7
600	94.1	219.3	331.0	436.5	538.4	636.8	732.1	824.7	914.8	1,002.7	1,088.6	1,172.8	1,256.5	1,340.2	1,424.2	1,508.7	1,593.7	1,679.3
700	107.8	230.9	341.7	446.8	548.4	646.5	741.5	833.8	923.7	1,011.3	1,097.1	1,181.2	1,264.9	1,348.6	1,432.7	1,517.2	1,602.2	1,687.9
800	121.1	242.4	352.4	457.1	558.3	656.1	750.9	842.9	932.6	1,020.0	1,105.4	1,189.6	1,273.2	1,357.0	1,441.1	1,525.6	1,610.7	1,696.5
900	134.1	253.7	363.1	467.4	568.2	665.7	760.2	852.0	941.4	1,028.6	1,113.9	1,198.0	1,281.6	1,365.4	1,449.5	1,534.1	1,619.3	1,705.1
1,000	146.9	265.0	373.7	477.7	578.1	675.3	769.5	861.0	950.2	1,037.2	1,122.3	1,206.4	1,290.0	1,373.8	1,458.0	1,542.6	1,627.8	1,713.7

Emfs are expressed in microvolts and temperatures in °C. Cold junctions at 0°C. Roeser and Wensel, National Bureau of Standards

E(μv)	0	1,000	2,000	3,000	4,000	5,000	6,000	7,000	8,000	9,000	10,000	11,000	12,000	13,000	14,000	15,000	16,000	17,000
0	32.0	296.4	509.0	704.7	891.9	1,072.6	1,247.5	1,417.1	1,581.8	1,742.4	1,899.0	2,052.1	2,203.5	2,354.0	2,504.8	2,656.4	2,808.7	2,962.0
100	63.9	318.9	529.1	723.7	910.3	1,090.3	1,264.6	1,433.8	1,598.0	1,758.2	1,914.5	2,067.3	2,218.5	2,369.0	2,520.0	2,671.5	2,824.0	2,977.4
200	93.9	341.0	549.1	742.6	928.6	1,108.0	1,281.7	1,450.4	1,614.2	1,774.0	1,929.9	2,082.5	2,233.5	2,384.1	2,535.1	2,686.7	2,839.3	2,992.8
300	122.4	362.8	569.0	761.4	946.8	1,125.7	1,298.8	1,467.0	1,630.4	1,789.7	1,945.3	2,097.7	2,248.5	2,399.2	2,550.2	2,701.9	2,854.6	3,008.3
400	149.7	384.4	588.8	780.2	964.9	1,143.3	1,315.9	1,483.5	1,646.5	1,805.5	1,960.7	2,112.8	2,263.6	2,414.3	2,565.3	2,717.2	2,869.9	3,023.8
500	176.0	405.7	608.4	799.0	983.0	1,160.8	1,332.9	1,500.0	1,662.6	1,821.2	1,976.0	2,127.9	2,278.7	2,429.3	2,580.4	2,732.4	2,885.2	3,039.3
600	201.4	426.7	627.8	817.7	1,001.1	1,178.3	1,349.8	1,516.5	1,678.6	1,836.8	1,991.3	2,143.0	2,293.7	2,444.4	2,595.6	2,747.7	2,900.6	3,054.7
700	226.0	447.6	647.1	836.3	1,019.1	1,195.7	1,366.7	1,532.9	1,694.6	1,852.4	2,006.5	2,158.2	2,308.8	2,459.5	2,610.8	2,763.0	2,916.0	3,070.2
800	250.0	468.3	666.4	854.8	1,037.0	1,213.0	1,383.6	1,549.2	1,710.6	1,868.0	2,021.7	2,173.3	2,323.8	2,474.6	2,626.0	2,778.2	2,931.3	3,085.7
900	273.4	488.7	685.6	873.4	1,054.8	1,230.3	1,400.4	1,565.5	1,726.5	1,883.5	2,036.9	2,188.4	2,338.9	2,489.7	2,641.2	2,793.4	2,946.7	3,101.2
1,000	296.4	509.0	704.7	891.9	1,072.6	1,247.5	1,417.1	1,581.8	1,742.4	1,899.0	2,052.1	2,203.5	2,354.0	2,504.8	2,656.4	2,808.7	2,962.0	3,116.7

REFERENCE TABLE FOR Pt TO Pt—13 PER CENT Rh THERMOCOUPLE

Emfs are expressed in microvolts and temperatures in °C. Cold junctions at 0°C. Roeser and Wensel, National Bureau of Standards

E(μv)	0	1,000	2,000	3,000	4,000	5,000	6,000	7,000	8,000	9,000	10,000	11,000	12,000	13,000	14,000	15,000	16,000	17,000	18,000	19,000
0	0 (17.9)	145.3 (12.2)	258.8 (10.6)	361.0 (9.9)	457.4 (9.5)	549.8 (9.0)	638.3 (8.7)	723.5 (8.4)	806.0 (8.1)	886.1 (7.9)	964 (.6)	1,040.0 (7.4)	1,113.9 (7.3)	1,186.9 (7.3)	1,259.3 (7.3)	1,331.8 (7.2)	1,404.3 (7.3)	1,476.9 (7.3)	1,550.0 (7.4)	1,623.6 (7.4)
100	17.9 (16.7)	157.5 (12.0)	269.4 (10.5)	370.9 (9.8)	466.9 (9.4)	558.8 (9.0)	647.0 (8.6)	731.9 (8.4)	814.1 (8.1)	894.0 (7.8)	971.? (.7)	1,047.4 (7.5)	1,121.2 (7.4)	1,194.1 (7.3)	1,266.6 (7.2)	1,339.0 (7.2)	1,411.6 (7.3)	1,484.2 (7.3)	1,557.4 (7.4)	1,631.0 (7.4)
200	34.6 (15.8)	169.5 (11.7)	279.9 (10.4)	380.7 (9.8)	476.3 (9.4)	567.8 (9.0)	655.6 (8.6)	740.3 (8.3)	822.2 (8.1)	901.8 (7.9)	979.? (.6)	1,054.9 (7.4)	1,128.6 (7.3)	1,201.4 (7.3)	1,273.8 (7.3)	1,346.2 (7.3)	1,418.9 (7.2)	1,491.5 (7.3)	1,564.8 (7.3)	1,638.4 (7.4)
300	50.4 (15.1)	181.2 (11.5)	290.3 (10.3)	390.5 (9.7)	485.7 (9.3)	576.8 (8.9)	664.2 (8.4)	748.6 (8.3)	830.3 (8.0)	909.7 (7.8)	987.0 (.7)	1,062.3 (7.4)	1,135.9 (7.4)	1,208.7 (7.2)	1,281.1 (7.2)	1,353.5 (7.3)	1,426.1 (7.2)	1,498.8 (7.3)	1,572.1 (7.4)	1,645.8 (7.4)
400	65.5 (14.5)	192.7 (11.4)	300.6 (10.2)	400.2 (9.7)	495.0 (9.3)	585.7 (8.9)	672.6 (8.5)	756.9 (8.2)	838.3 (8.0)	917.5 (7.8)	994.? (.6)	1,069.7 (7.4)	1,143.2 (7.3)	1,215.9 (7.3)	1,288.3 (7.2)	1,360.8 (7.2)	1,433.3 (7.2)	1,506.1 (7.3)	1,579.5 (7.3)	1,653.2 (7.4)
500	80.0 (13.9)	204.1 (11.2)	310.8 (10.2)	409.9 (9.6)	504.3 (9.2)	594.6 (8.8)	681.3 (8.5)	765.1 (8.2)	846.3 (8.0)	925.3 (7.8)	1,002.3 (.6)	1,077.1 (7.3)	1,150.5 (7.3)	1,223.2 (7.3)	1,295.5 (7.2)	1,368.0 (7.3)	1,440.5 (7.3)	1,513.4 (7.3)	1,586.8 (7.4)	1,660.6 (7.4)
600	93.9 (13.4)	215.3 (11.1)	321.0 (10.1)	419.5 (9.5)	513.5 (9.1)	603.4 (8.8)	689.8 (8.5)	773.3 (8.2)	854.3 (8.0)	933.1 (7.8)	1,009.? (.6)	1,084.4 (7.4)	1,157.8 (7.3)	1,230.4 (7.3)	1,302.7 (7.2)	1,375.3 (7.3)	1447.8 (7.2)	1,520.7 (7.3)	1,594.2 (7.3)	1,668.0 (7.4)
700	107.3 (13.0)	226.4 (10.9)	331.1 (10.0)	429.0 (9.5)	522.6 (9.1)	612.2 (8.7)	698.3 (8.4)	781.5 (8.2)	862.3 (8.0)	940.9 (7.8)	1,017.? (.5)	1,091.8 (7.4)	1,165.1 (7.2)	1,237.6 (7.2)	1,309.9 (7.3)	1,382.6 (7.2)	1,455.0 (7.3)	1,528.0 (7.3)	1,601.5 (7.4)	1,675.4 (7.4)
800	120.3 (12.6)	237.3 (10.8)	341.1 (10.0)	438.5 (9.5)	531.7 (9.1)	620.9 (8.7)	706.7 (8.4)	789.7 (8.1)	870.3 (7.9)	948.7 (7.9)	1,025.0 (.5)	1,099.2 (7.3)	1,172.3 (7.3)	1,244.8 (7.3)	1,317.2 (7.3)	1,389.8 (7.3)	1,462.3 (7.3)	1,535.3 (7.3)	1,608.9 (7.3)	1,682.8 (7.4)
900	132.9 (12.4)	248.1 (10.7)	351.1 (9.9)	448.0 (9.4)	540.8 (9.0)	629.6 (8.7)	715.1 (8.4)	797.8 (8.2)	878.2 (7.9)	956.4 (7.7)	1,032.? (.5)	1,106.5 (7.4)	1,179.6 (7.3)	1,252.1 (7.2)	1,324.5 (7.3)	1,397.1 (7.2)	1,469.6 (7.3)	1,542.7 (7.3)	1,616.2 (7.4)	1,690.2 (7.4)
1,000	145.3	258.8	361.0	457.4	549.8	638.3	723.5	806.0	886.1	964.0	1,040.0	1,113.9	1,186.9	1,259.3	1,331.8	1,404.3	1,476.9	1,550.0	1,623.6	1,697.6

Emfs are expressed in microvolts and temperatures in °F. Cold junctions at 32°F. Roeser and Wensel, National Bureau of Standards

E(μv)	0	1,000	2,000	3,000	4,000	5,000	6,000	7,000	8,000	9,000	10,000	11,000	12,000	13,000	14,000	15,000	16,000	17,000	18,000	19,000
0	32.0 (32.2)	293.5 (22.0)	497.8 (19.1)	681.8 (17.8)	855.3 (17.0)	1,021.6 (16.2)	1,180.9 (15.6)	1,334.3 (15.1)	1,482.8 (14.6)	1,627.0 (14.2)	1,767.4 (13.8)	1,904.0 (13.4)	2,037.0 (13.2)	2,168.4 (13.0)	2,298.7 (13.0)	2,429.2 (13.0)	2,559.7 (13.1)	2,690.4 (13.2)	2,822.0 (13.3)	2,954.5 (13.3)
100	64.2 (30.1)	315.5 (21.6)	516.9 (18.9)	699.6 (17.7)	872.3 (17.0)	1,037.8 (16.2)	1,196.5 (15.6)	1,349.4 (15.1)	1,497.4 (14.6)	1,641.2 (14.1)	1,781.2 (13.7)	1,917.4 (13.4)	2,050.2 (13.2)	2,181.4 (13.1)	2,311.8 (13.0)	2,442.2 (13.0)	2,572.8 (13.0)	2,703.6 (13.1)	2,835.3 (13.3)	2,967.8 (13.3)
200	94.3 (28.4)	337.1 (21.1)	535.8 (18.7)	717.3 (17.6)	889.3 (16.9)	1,054.0 (16.2)	1,212.1 (15.5)	1,364.5 (15.0)	1,512.0 (14.5)	1,655.3 (14.1)	1,794.9 (13.7)	1,930.8 (13.3)	2,063.4 (13.2)	2,194.5 (13.1)	2,324.8 (13.1)	2,455.2 (13.1)	2,585.8 (13.1)	2,716.7 (13.2)	2,848.6 (13.2)	2,981.1 (13.3)
300	122.7 (27.2)	358.2 (20.7)	554.5 (18.6)	734.9 (17.5)	906.2 (16.8)	1,070.2 (16.1)	1,227.6 (15.4)	1,379.5 (14.9)	1,526.5 (14.4)	1,669.4 (14.1)	1,808.6 (13.8)	1,944.1 (13.3)	2,076.6 (13.2)	2,207.6 (13.0)	2,337.9 (13.0)	2,468.3 (13.1)	2,598.9 (13.0)	2,729.8 (13.2)	2,861.8 (13.3)	2,994.4 (13.4)
400	149.9 (26.1)	378.9 (20.4)	573.1 (18.4)	752.4 (17.4)	923.0 (16.7)	1,086.3 (16.0)	1,243.0 (15.3)	1,394.4 (14.8)	1,540.9 (14.4)	1,683.5 (14.0)	1,822.4 (13.7)	1,957.4 (13.3)	2,089.8 (13.1)	2,220.6 (13.1)	2,350.9 (13.0)	2,481.4 (13.0)	2,611.9 (13.0)	2,743.0 (13.1)	2,875.1 (13.2)	3,007.8 (13.3)
500	176.0 (25.0)	399.3 (20.2)	591.5 (18.3)	769.8 (17.3)	939.7 (16.6)	1,102.3 (16.0)	1,258.3 (15.3)	1,409.2 (14.8)	1,555.3 (14.4)	1,697.5 (14.1)	1,836.1 (13.7)	1,970.7 (13.3)	2,102.9 (13.1)	2,233.7 (13.0)	2,363.9 (13.0)	2,494.4 (13.1)	2,624.9 (13.1)	2,756.1 (13.1)	2,888.3 (13.3)	3,021.1 (13.3)
600	201.0 (24.1)	419.5 (20.0)	609.8 (18.2)	787.1 (17.2)	956.3 (16.4)	1,118.2 (15.9)	1,273.6 (15.3)	1,424.0 (14.8)	1,569.7 (14.4)	1,711.6 (14.0)	1,849.8 (13.6)	1,984.0 (13.3)	2,116.0 (13.1)	2,246.7 (13.0)	2,376.9 (13.0)	2,507.5 (13.0)	2,638.0 (13.0)	2,769.3 (13.1)	2,901.6 (13.2)	3,034.4 (13.3)
700	225.1 (23.4)	439.5 (19.7)	628.0 (18.0)	804.3 (17.1)	972.7 (16.4)	1,134.0 (15.8)	1,288.9 (15.2)	1,438.8 (14.7)	1,584.1 (14.4)	1,725.6 (14.0)	1,863.4 (13.6)	1,997.3 (13.3)	2,129.1 (13.1)	2,259.7 (13.0)	2,389.9 (13.0)	2,520.6 (13.0)	2,651.0 (13.1)	2,782.4 (13.1)	2,914.8 (13.2)	3,047.7 (13.4)
800	248.5 (22.7)	459.2 (19.4)	646.0 (18.0)	821.4 (17.0)	989.1 (16.3)	1,149.7 (15.7)	1,304.1 (15.1)	1,453.5 (14.6)	1,598.5 (14.3)	1,739.6 (13.9)	1,877.0 (13.5)	2,010.6 (13.2)	2,142.2 (13.1)	2,272.7 (13.0)	2,403.0 (13.0)	2,533.6 (13.0)	2,664.1 (13.0)	2,795.6 (13.2)	2,928.0 (13.2)	3,061.0 (13.4)
900	271.2 (22.3)	478.6 (19.2)	664.0 (17.8)	838.4 (16.9)	1,005.4 (16.2)	1,165.3 (15.6)	1,319.2 (15.1)	1,468.1 (14.7)	1,612.8 (14.2)	1,753.5 (13.9)	1,890.5 (13.5)	2,023.8 (13.2)	2,155.3 (13.1)	2,285.7 (13.0)	2,416.1 (13.1)	2,546.7 (13.0)	2,677.2 (13.2)	2,808.8 (13.2)	2,941.2 (13.3)	3,074.4 (13.3)
1,000	293.5	497.8	681.8	855.3	1,021.6	1,180.9	1,334.3	1,482.8	1,627.0	1,767.4	1,904.0	2,037.0	2,168.4	2,298.7	2,429.2	2,559.7	2,690.4	2,822.0	2,954.5	3,087.7

VALUES FOR THE LANGEVIN FUNCTION $\mathcal{L}(u)$

Compiled by Allen L. King

Because of random thermal rotations a dipole or dipole-like element ordinarily has an average dipole moment of zero. If it is placed in an orienting field F however it tends to align itself with the field so that the average component of the dipole moment parallel to the field equals p. Classically, if the system is in thermal equilibrium,

$$\overline{p} = p_o(\coth u - 1/u) = p_o\mathcal{L}(u)$$

Here p_o is the moment of the dipole itself and $\mathcal{L}(u)$ is the Langevin function of the argument u which equals Fp_o/kT.

$$\text{For } \mu \ll 1 \quad \mathcal{L}(\mu) = \mu/3$$

The Langevin function applies to permanent electric and magnetic dipoles.

(u)	0	1	2	3	4	5	6	7	8	9	10	Diff.
0.0	0000	0033	0066	0100	0133	0166	0200	0233	0266	0300	0333	33
0.1	0333	0366	0400	0433	0466	0499	0532	0565	0598	0631	0665	33
0.2	0665	0698	0731	0764	0797	0830	0862	0896	0929	0961	0994	33
0.3	0994	1027	1059	1092	1124	1157	1190	1222	1254	1287	1319	32
0.4	1319	1352	1384	1416	1448	1480	1512	1544	1576	1608	1640	32
0.5	1640	1671	1703	1734	1765	1797	1829	1860	1892	1923	1953	31
0.6	1953	1985	2016	2046	2077	2108	2138	2169	2200	2230	2260	31
0.7	2260	2290	2321	2351	2381	2411	2441	2471	2500	2530	2559	30
0.8	2559	2589	2618	2647	2676	2705	2734	2763	2792	2821	2850	29
0.9	2850	2878	2906	2934	2963	2991	3019	3047	3075	3103	3130	28
1.0	3130	3158	3185	3212	3240	3267	3294	3321	3348	3375	3401	27
1.1	3401	3428	3454	3480	3507	3533	3559	3585	3611	3637	3662	26
1.2	3662	3688	3713	3738	3763	3788	3813	3838	3863	3888	3913	25
1.3	3913	3937	3961	3985	4009	4033	4057	4081	4105	4129	4152	24
1.4	4152	4176	4199	4222	4245	4268	4291	4313	4336	4359	4381	23
1.5	4381	4403	4426	4448	4469	4491	4514	4536	4557	4579	4600	22
1.6	4600	4621	4642	4663	4684	4705	4726	4747	4768	4788	4808	21
1.7	4808	4828	4849	4869	4889	4909	4928	4948	4967	4987	5006	20
1.8	5006	5025	5044	5064	5083	5102	5121	5139	5158	5176	5195	19
1.9	5195	5213	5231	5249	5267	5285	5303	5321	5338	5356	5373	18
2.0	5373	5391	5408	5425	5442	5459	5476	5493	5509	5526	5542	17
2.1	5542	5559	5575	5591	5608	5624	5640	5656	5672	5688	5704	16
2.2	5704	5719	5734	5750	5765	5780	5795	5810	5825	5840	5855	15
2.3	5855	5870	5885	5899	5913	5928	5943	5957	5971	5985	6000	14
2.4	6000	6014	6027	6041	6055	6068	6082	6095	6109	6122	6136	13.5
2.5	6136	6149	6162	6175	6188	6201	6214	6227	6239	6252	6265	13
2.6	6265	6278	6290	6302	6314	6326	6339	6351	6363	6375	6387	12
2.7	6387	6399	6411	6422	6434	6446	6457	6469	6480	6492	6503	11.5
2.8	6503	6514	6525	6536	6548	6559	6570	6581	6591	6602	6613	11
2.9	6613	6624	6634	6644	6655	6665	6676	6686	6696	6707	6717	10
3.0	6717	6727	6737	6747	6757	6767	6777	6787	6796	6806	6815	10
3.1	6815	6825	6834	6844	6853	6862	6871	6880	6890	6899	6908	9
3.2	6908	6917	6926	6935	6944	6953	6962	6971	6979	6988	6997	9
3.3	6997	7006	7014	7023	7031	7040	7048	7056	7064	7073	7081	8.5
3.4	7081	7089	7097	7105	7113	7121	7129	7137	7145	7153	7161	8
3.5	7161	7169	7176	7184	7192	7200	7207	7215	7223	7230	7237	8
3.6	7237	7245	7252	7259	7267	7274	7281	7288	7295	7302	7309	7
3.7	7309	7317	7324	7330	7337	7344	7351	7358	7364	7371	7378	7
3.8	7378	7385	7392	7398	7405	7412	7418	7425	7431	7437	7444	6.5
3.9	7444	7450	7457	7463	7470	7476	7482	7488	7494	7501	7507	6

Note: Higher values of the Langevin function are nearly equal to $(u-1)/u$.

MAGNETIC PROPERTIES AND COMPOSITION OF SOME PERMANENT MAGNETIC ALLOYS

Name	Composition,* Weight percent					Remanence, B_r (Gauss)	Coercive force H_c (Oersteds)	Maximum energy product, $(BH)_{max}$ (Gauss-Oersteds $\times 10^{-6}$)
	Al	Ni	Co	Cu	Other			
U.S.A.								
Alnico I	12	20–22	5	.....	......	7,100	440	1.4
Alnico II	10	17	12.5	6	......	7,200	540	1.6
Alnico III	12	24–26	....	3	......	6,900	470	1.35
Alnico IV	12	27–28	5	....	......	5,500	700	1.3
Alnico V†	8	14	24	3	......	12,500	600	5.0
Alnico V DG†	8	14	24	3	......	13,100	640	6.0
Alnico VI†	8	15	24	3	1.25 Ti	10,500	750	3.75
Alnico VII†	8.5	18	24	3	5 Ti	7,200	1,050	2.75
Alnico XII	6	18	35	....	8 Ti	5,800	950	1.6
Carbon steel	.....	.....	....	....	1 Mn	10,000	50	0.2
Chromium steel	.....	.....	....	....	0.9 C 3.5 Cr	9,700	65	0.3
Cobalt steel	.....	.....	17		0.9 C 0.3 Mn 2.5 Cr 8 W 0.75 C	9,500	150	0.65

Name	Composition,* Weight percent					Remanence, B_r (Gauss)	Coercive force H_c (Oersteds)	Maximum energy product, $(BH)_{max}$ (Gauss-Oersteds $\times 10^{-6}$)
	Al	Ni	Co	Cu	Other			
Cunico	....	21	29	50	.....	3,400	660	0.80
Cunife	....	20		60	.....	5,400	550	1.5
Ferroxdur 1			$BaFe_{12}O_{19}$			2,200	1,800	1.0
Ferroxdur 2			$BaF_{12}O_{19}$ (oriented)			3,840	2,000	3.5
Platinum-Cobalt	....	....	23	....	77 Pt	6,000	4,300	7.5
Remalloy	....	....	12	....	17 Mo	10,500	250	1.1
Silmanol	4.4	....	....	....	86.6 Ag 8.8 Mn	550	6,000	0.075
Tungsten steel	....	....	....	....	5 W 0.3 Mn 0.7 C	10,300	70	0.32
Vicalloy I	....	....	52	....	10 V	8,800	300	1.0
Vicalloy II (wire)	....	....	52	....	14 V	10,000	510	3.5
Germany								
Alni 90	12	21	....	....	......	8,000	350	1.2
Alni 120	13	27	....	....	......	6,000	570	1.2
Alnico 130	12	23	5	....	......	6,300	620	1.4
Alnico 160	11	24	12	4	......	6,200	700	1.6
Alnico 190	12	21	15	4	......	7,000	700	1.8
Alnico 250	8	19	23	4	6 Ti	6,500	1,000	2.2
Alnico 400†	9	15	23	4	......	12,000	650	4.8
Alnico 580† (semicolumnar)	9	15	23	4	......	13,000	700	6.0
Oerstit 800	9	18	19	4	4 Ti	6,600	750	1.95
Great Britain								
Alcomax I	7.5	11	25	3	1.5 Ti	12,000	475	3.5
Alcomax II	8	11.5	24	4.5	......	12,400	575	4.7
Alcomax IISC (semicolumnar)	8	11	22	4.5	......	12,800	600	5.15
Alcomax III	8	13.5	24	3	0.8 Nb	12,500	670	5.10
Alcomax IIISC (semicolumnar)	8	13.5	24	3	0.8 Nb	13,000	700	5.80
Alcomax IV	8	13.5	24	3	2.5 Nb	11,200	750	4.30
Alcomax IVSC (semicolumnar)	8	13.5	24	3	2.5 Nb	11,700	780	5.10
Alni, high B_r	13	24	....	3.5	......	6,200	480	1.25
Alni, normal	....	....	....	....	......	5,600	580	1.25
Alni, high H_c	12	32	....	....	0–0.5 Ti	5,000	680	1.25
Alnico, high B_r	10	17	12	6	......	8,000	500	1.70
Alnico, normal	....	....	....	....	......	7,250	560	1.70
Alnico, high H_c	10	20	13.5	6	0.25 Ti	6,600	620	1.70
Columax (columnar)			similar to Alcomax III or IV			13,000–14,000	700–800	7.0–8.5
Hycomax	9	21	20	1.6	......	9,500	830	3.3

* Remainder of unlisted composition is either iron or iron plus trace impurities
† Cast anisotropic. Unmarked ones are cast isotropic.

HIGH PERMEABILITY MAGNETIC ALLOYS
(See separate table for magnetic properties)

Name	Composition,* Weight percent	Sp. gr., gm. per cc	Tensile strength		Remark	Use
			Kg/mm²**	Form		
Silicon iron AISI M 15	Si 4	7.68–7.64	44.3	....	Annealed 4 hrs 802–1093°C	Low core losses
Silicon iron AISI M 8	Si 3	7.68–7.64	44.2	Grain oriented	Annealed 4 hrs 802–1204°C	
45 Permalloy	Ni 45; Mn 0.3	8.17	....		....	Audio transformer, coils, relays
Monimax	Ni 47; Mo 3	8.27	....		....	High frequency coils
4-79 Permalloy	Ni 79; Mo 4; Mn 0.3	8.74	55.4	....	H₂ annealed 1121°C	Audio coils, transformers, magnetic shields
Sinimax	Ni 43; Si 3	7.70	....		....	High frequency coils
Nu-metal	Ni 75; Cr 2; Cu 5	8.58	44.8		H₂ annealed 1221°C	Audio coils, magnetic shields, transformers
Supermalloy	Ni 79; Mo 5; Mn 0.3	8.77	....		....	Pulse transformers, magnetic amplifiers, coils
2-V Permendur	Co 40; V 2	8.15	46.3	....	....	D-c electromagnets, pole tips

* Iron is additional alloying metal.
** kg/mm² $\times$ 1422.33 = lbs/in.²

Alloy name, country of manufacture*	Composition,** weight percent	Sp. gr., gm per cc	Thermal expansion		Tensile strength		Remark†	Use
			$\frac{Cm \times 10^{-6}}{cm \times °C}$	Between °C	*** Kg/mm²	Form		
Alnico I (USA)*	Al 12; Ni 20–22; Co 5	6.9	12.6	20–300	2.9	Cast	i.	Permanent magnets
Alnico II (USA)*	Al 10; Ni 17; Cu 6; Co 12.5	7.1	12.4	20–300	2.1 / 45.7	Cast / Sintered	i.	Temperature controls, magnetic toys and novelties
Alnico III (USA)*	Al 12; Ni 24–26; Cu 3	6.9	12	20–300	8.5	Cast	i.	Tractor magnetos
Alnico IV (USA)*	Al 12; Ni 27–28; Co 5	7.0	13.1	20–300	6.3 / 42.1	Cast / Sintered	i.	Application requiring high coercive force
Alnico V (USA)*	Al 8; Ni 14; Co 24; Cu 3	7.3	11.3	......	3.8 / 35	Cast / Sintered	a.	Application requiring high energy
Alnico V DG (USA)*	Al 8; Ni 14; Co 24; Cu 3	7.3	11.3	......	....	....	a., c.	
Alnico VI (USA)*	Al 8; Ni 15; Co 24; Cu 3; Ti 1.25	7.3	11.4	......	16.1	Cast	a.	Application requiring high energy
Alnico VII (USA)*	Al 8.5; Ni 18; Cu 3; Co 24; Ti 5	7.17	11.4	......	....	....	a.	
Alnico XII (USA)*	Al 6; Ni 18; Co 35; Ti 8	7.2	11	20–300	....	....	......	Permanent magnets
Comol (USA)*	Co 12; Mo 17	8.16	9.3	20–300	88.6	....	......	Permanent magnets
Cunife (USA)*	Cu 60; Ni 20	8.52	.....	......	70.3	....	......	Permanent magnets
Cunico (USA)*	Cu 50; Ni 21	8.31	.....	......	70.3	....	......	Permanent magnets
Barium ferrite Feroxdur (USA)*	Ba Fe$_{12}$O$_{19}$	4.7	10	......	70.3	....	....	Ceramics
Alcomax I (GB)*	Al 7.5; Ni 11; Co 25; Cu 3; Ti 1.5	....	....	......	....	....	a.	Permanent magnets
Alcomax II (GB)*	Al 8; Ni 11.5; Co 24; Cu 4.5	....	....	......	....	....	a.	Permanent magnets
Alcomax II SC(GB)*	Al 8; Ni 11; Co 22; Cu 4.5	7.3	....	......	....	....	a., sc.	
Alcomax III (GB)*	Al 8; Ni 13.5; Co 24; Nb 0.8	7.3	....	......	....	....	a.	Magnets for motors, loudspeakers
Alcomax IV (GB)*	Al 8; Ni 13.5; Cu 3; Co 24; Nb 2.5	....	....	......	....	....	......	Magnets for cycle-dynamos
Columax (GB)*	Similar to Alcomax III or IV	....	....	......	....	....	a., sc.	Permanent magnets, heat treatable
Hycomax (GB)*	Al 9; Ni 21; Co 20; Cu 1.6	....	....	......	....	....	a.	Permanent magnets
Alnico (high H$_c$) (GB)*	Al 10; Ni 20; Co 13.5; Cu 6; Ti 0.25	7.3	....	......	....	....	i.	
Alnico (high B$_r$) (GB)*	Al 10; Ni 17; Co 12; Cu 6	7.3	....	......	....	....	i.	
Alni (high H$_c$) (GB)*	Al 12; Ni 32; Ti 0–0.5	6.9	....	......	....	....	i.	
Alni (high B$_r$) (GB)*	Al 13; Ni 24; Cu 3.5	....	....	......	....	....	i.	
Alnico 580 (Ger)*	Al 9; Ni 15; Co 23; Cu 4	....	....	......	....	....	i.	
Alnico 400 (Ger)*	Al 9; Ni 15; Co 23; Cu 4	....	....	......	....	....	a.	
Oerstit 800 (Ger)*	Al 9; Ni 18; Co 19; Cu 4; Ti 4	....	....	......	....	....	i.	Permanent magnets
Alnico 250 (Ger)*	Al 8; Ni 19; Co 23; Cu 4; Ti 6	....	....	......	....	....	i.	
Alnico 190 (Ger)*	Al 12; Ni 21; Cu 4; Co 15	....	....	......	....	....	i.	
Alnico 160 (Austria)	Al 11; Ni 24; Co 12; Cu 4	....	....	......	....	....	i.	Permanent magnets, sintered
Alnico 130 (Ger)*	Al 12; Ni 23; Co 5	....	....	......	....	....	i.	
Alni 120 (Ger)*	Al 13; Ni 27	....	....	......	....	....	i.	
Alni 90 (Ger)*	Al 12; Ni 21	....	....	......	....	....	i.	

* USA—United States; GB—Great Britain; Ger—Germany.
** Iron is the additional alloying metal for each of the magnets listed.
*** kg/mm × 1422.33 = lbs/in.²
† i. = isotropic; a. = anisotropic; c. = columnar; sc. = semicolumnar.

MAGNETIC MATERIALS

Properties of Antiferromagnetic Compounds

Compound	Crystal Symmetry	θ_N(°K)	θ_P(°K)	$(P_A)_{eff}$ μ_B	P_A μ_B
CoCl₂.................	Rhombohedral	25	−38.1	5.18	3.1 ± 0.6
CoF₂.................	Tetragonal	38	50	5.15	3.0
CoO.................	Tetragonal	291	330	5.1	3.8
Cr...................	Cubic	475			
Cr₂O₃...............	Rhombohedral	307	485	3.73	3.0
CrSb................	Hexagonal	723	550	4.92	2.7
CuBr₂...............	Monoclinic	189	246	1.9	
CuCl₂·2H₂O...........	Orthorhombic	4.3	4–5	1.9	
CuCl₂...............	Monoclinic	~70	109	2.08	
FeCl₂...............	Hexagonal	24	−48	5.38	4.4 ± 0.7
FeF₂................	Tetragonal	79–90	117	5.56	4.64
FeO.................	Rhombohedral	198	507	7.06	3.32
α-Fe₂O₃.............	Rhombohedral	953	2940	6.4	5.0
α-Mn...............	Cubic	95			
MnBr₂·4H₂O...........	Monoclinic	2.1	$\begin{Bmatrix} 2.5 \\ 1.3 \end{Bmatrix}$	5.93	
MnCl₂·4H₂O...........	Monoclinic	1.66	1.8	5.94	
MnF₂................	Tetragonal	72–75	113.2	5.71	5
MnO................	Rhombohedral	122	610	5.95	5.0
β-MnS...............	Cubic	160	982	5.82	5.0
MnSe...............	Cubic	~173	361	5.67	
MnTe...............	Hexagonal	310–323	690	6.07	5.0
NiCl₂...............	Hexagonal	50	−68	3.32	
NiF₂................	Tetragonal	78.5–83	115.6	3.5	2.0
NiO.................	Rhombohedral	533–650	~2000	4.6	2.0
TiCl₃...............		100			
V₂O₃...............		170			

1. θ_N = Néel temperature, determined from susceptibility maxima or from the disappearance of magnetic scattering.
2. θ_P = a constant in the Curie-Weiss law written in the form $\chi_A = C_A/(T + \theta_P)$, which is valid for antiferromagnetic material for $T > \theta_N$.
3. $(P_A)_{eff}$ = effective moment per atom, derived from the atomic Curie constant $C_A = (P_A)_{eff}^2(N^2/3R)$ and expressed in units of the Bohr magneton, $\mu_B = 0.9273 \times 10^{-20}$ erg gauss⁻¹.
4. P_A = magnetic moment per atom, obtained from neutron diffraction measurements in the ordered state.

SATURATION CONSTANTS AND CURIE POINTS OF FERROMAGNETIC ELEMENTS

Element	σ_s (20°C)	M_s (20°C)	σ_s (0°K)	n_B	Curie Point (°C)
Fe	218.0	1,714	221.9	2.219	770
Co	161	1,422	162.5	1.715	1,131
Ni	54.39	484.1	57.50	0.604	358
Gd	0	0	253.5	7.12	16

σ_s = saturation magnetic moment/gram; M_s = saturation magnetic moment/cm³, in cgs units. n_B = magnetic moment per atom in Bohr magnetons.

From American Institute of Physics Handbook, McGraw-Hill Company (1963) by permission.

MAGNETIC PROPERTIES OF TRANSFORMER STEELS

Ordinary Transformer Steel

B (Gauss)	H (Oersted)	Permeability = B/H
2,000	0.60	3,340
4,000	0.87	4,600
6,000	1.10	5,450
8,000	1.48	5,400
10,000	2.28	4,380
12,000	3.85	3,120
14,000	10.9	1,280
16,000	43.0	372
18,000	149	121

HIGH SILICON TRANSFORMER STEEL

B	H	Permeability
2,000	0.50	4,000
4,000	0.70	5,720
6,000	0.90	6,670
8,000	1.28	6,250
10,000	1.99	5,020
12,000	3.60	3,340
14,000	9.80	1,430
16,000	47.4	338
18,000	165	109

SATURATION CONSTANTS FOR MAGNETIC SUBSTANCES

Substance	Field intensity	Induced magnetization	Substance	Field intensity	Induced magnetization
	(For saturation)			(For saturation)	
Cobalt	9000	1300	Nickel, hard	8000	400
Iron, wrought	2000	1700	annealed	7000	515
cast	4000	1200	Vicker's steel	15000	1600
Manganese steel	7000	200			

INITIAL PERMEABILITY OF HIGH PURITY IRON FOR VARIOUS TEMPERATURES

L. Alberts and B. J. Shepstone

Temperature °C	Permeability (gauss/oersted)
0	920
200	1040
400	1440
600	2550
700	3900
770	12580

MAGNETIC MATERIALS
High-permeability Materials

Material	Form	Fe	Ni	Co	Mo	Other	Typical heat treatment °C	Permeability at $B = 20$ gausses	Maximum permeability	Saturation flux density B gausses	Hysteresis ‡ loss, W_h ergs/cm³	Coercive ‡ force H_c oersteds	Resistivity microhm cm	Density, g/cm³
Cold rolled steel	Sheet	98.5	—	—	—	—	950 Anneal	180	2,000	21,000	—	1.8	10	7.88
Iron	Sheet	99.91	—	—	—	—	950 Anneal	200	5,000	21,500	5,000	1.0	10	7.88
Purified iron	Sheet	99.95	—	—	—	—	1480 H₂ + 880	5,000	180,000	21,500	300	.05	10	7.88
4% Silicon-iron	Sheet	96	—	—	—	4 Si	800 Anneal	500	7,000	19,700	3,500	.5	60	7.65
Grain oriented*	Sheet	97	—	—	—	3 Si	800 Anneal	1,500	30,000	20,000	—	.15	47	7.67
45 Permalloy	Sheet	54.7	45	—	—	.3 Mn	1050 Anneal	2,500	25,000	16,000	1,200	.3	45	8.17
45 Permalloy †	Sheet	54.7	45	—	—	.3 Mn	1200 H₂ Anneal	4,000	50,000	16,000	—	.07	45	8.17
Hipernik	Sheet	50	50	—	—	—	1200 H₂ Anneal	4,500	70,000	16,000	220	.05	50	8.25
Monimax	Sheet	—	—	—	—	—	1125 H₂ Anneal	2,000	35,000	15,000	—	.1	80	8.27
Sinimax	Sheet	—	—	—	—	—	1125 H₂ Anneal	3,000	35,000	11,000	—	—	90	—
78 Permalloy	Sheet	21.2	78.5	—	—	.3 Mn	1050 + 600 Q§	8,000	100,000	10,700	200	.05	16	8.60
4-79 Permalloy	Sheet	16.7	79	—	4	.3 Mn	1100 + Q	20,000	100,000	8,700	200	.05	55	8.72
Mu metal	Sheet	18	75	—	—	2 Cr, 5 Cu	1175 H₂	20,000	100,000	6,500	—	.05	62	8.58
Supermalloy	Sheet	15.7	79	—	5	.3 Mn	1300 H₂ + Q	100,000	800,000	8,000	—	.002	60	8.77
Permendur	Sheet	49.7	—	50	—	.3 Mn	800 Anneal	800	5,000	24,500	12,000	2.0	7	8.3
2V Permendur	Sheet	49	—	49	—	2 V	800 Anneal	800	4,500	24,000	6,000	2.0	26	8.2
Hiperco	Sheet	64	—	34	—	Cr	850 Anneal	650	10,000	24,200	—	1.0	25	8.0
2-81 Permalloy	Insulated powder	17	81	—	2	—	650 Anneal	125	130	8,000	—	<1.0	10⁶	7.8
Carbonyl iron	Insulated powder	99.9	—	—	—	—	—	55	132	—	—	—	—	7.86
Ferroxcube III	Sintered powder	MnFe₂O₄ + ZnFe₂O₄					—	1,000	1,500	2,500	—	.1	10⁸	5.0

*Properties in direction of rolling.
† Similar properties for Nicaloi, 4750 alloy, Carpenter 49, Armco 48.
‡ At saturation.
§ Q, quench or controlled cooling.

MAGNETIC MATERIALS (Continued)
Permanent Magnet Alloys

Material	Percent composition (remainder Fe)	Heat treatment* (temperature, °C)	Magnetizing force H_{max} oersteds	Coercive force H_c oersteds	Residual induction B_r gausses	Energy product BH_{max} $\times 10^{-6}$	Method of fabrication†	Mechanical properties‡	Weight lb/in.³
Carbon steel	1 Mn, 0.9 C	Q 800	300	50	10,000	.20	HR, M, P	H, S	.280
Tungsten steel	5 W, 0.3 Mn, 0.7 C	Q 850	300	70	10,300	.32	HR, M, P	H, S	.292
Chromium steel	3.5 Cr, 0.9 C, 0.3 Mn	Q 830	300	65	9,700	.30	HR, M, P	H, S	.280
17% Cobalt steel	17 Co, 0.75 C, 2.5 Cr, 8 W	—	1,000	150	9,500	.65	HR, M, P	H, S	—
36% Cobalt steel	36 Co, 0.7 C, 4 Cr, 5 W	Q 950	1,000	240	9,500	.97	HR, M, P	H, S	.296
Remalloy or Comol	17 Mo, 12 Co	Q 1200, B 700	1,000	250	10,500	1.1	HR, M, P	H	.295
Alnico I	12 Al, 20 Ni, 5 Co	A 1200, B 700	2,000	440	7,200	1.4	C, G	H, B	.249
Alnico II	10 Al, 17 Ni, 2.5 Co, 6 Cu	A 1200, B 600	2,000	550	7,200	1.6	C, G	H, B	.256
Alnico II (sintered)	10 Al, 17 Ni, 2.5 Co, 6 Cu	A 1300	2,000	520	6,900	1.4	Sn, G	H	.249
Alnico IV	12 Al, 28 Ni, 5 Co	Q 1200, B 650	3,000	700	5,500	1.3	Sn, C, G	H	.253
Alnico V	8 Al, 14 Ni, 24 Co, 3 Cu	AF 1300, B 600	2,000	550	12,500	4.5	C, G	H, B	.264
Alnico VI	8 Al, 15 Ni, 24 Co, 3 Cu, 1 Ti	—	3,000	750	10,000	3.5	C, G	H, B	.268
Alnico XII	6 Al, 18 Ni, 35 Co, 8 Ti	—	3,000	950	5,800	1.5	C, G	H, B	.26
Vicalloy I	52 Co, 10 V	B 600	1,000	300	8,800	1.0	C, CR, M, P	D	.295
Vicalloy II (wire)	52 Co, 14 V	CW + B 600	2,000	510	10,000	3.5	C, CR, M, P	D	.292
Cunife (wire)	60 Cu, 20 Ni	CW + B 600	2,400	550	5,400	1.5	C, CR, M, P	D, M	.311
Cunico	50 Cu, 21 Ni, 29 Co	—	3,200	660	3,400	.80	C, CR, M, P	D, M	.300
Vectolite	30 Fe$_2$O$_3$, 44 Fe$_3$O$_4$, 26 C$_2$O$_3$	—	3,000	1,000	1,600	.60	Sn, G	W	.113
Silmanal	86.8 Ag, 8.8 Mn, 4.4 Al	—	20,000	6,000ᵃ	550	.075	C, CR, M, P	D, M	.325
Platinum-cobalt	77 Pt, 23 Co	Q 1200, B 650	15,000	3,600	5,900	6.5	C, CR, M	D	—
Hyflux	Fine powder	—	2,000	390	6,600	.97	—	—	.176

ᵃ Value given is intrinsic H_c.
* Q—Quenched in oil or water. A—Air cooled. B—Baked. F—Cooled in magnetic field. CW—Cold worked.
† HR—Hot rolled or forged. CR—Cold rolled or drawn. M—Machined. G—Must be ground. P—Punched. C—Cast. Sn—Sintered.
‡ H—Hard. B—Brittle. S—Strong. D—Ductile. M—Malleable. W—Weak.

MAGNETIC SUSCEPTIBILITY OF THE ELEMENTS AND INORGANIC COMPOUNDS

The following table lists the magnetic susceptibilities of one gram formula weight of a number of paramagnetic and diamagnetic inorganic compounds as well as the magnetic susceptibilities of the elements.

In each instance the magnetic moment is expressed in cgs units.

A more extensive listing of the magnetic susceptibilities of inorganic compounds as well as those for organic compounds may be found in Constantes Selectionnes Diamagnetisme et Paramagnetisme Relaxation Paramagnetique, Volume 7. This table is abridged from the above publication by permission of the publishers.

Substance	Formula	Temp. °K.	Susceptibility 10^{-6} cgs	Substance	Formula	Temp. °K.	Susceptibility 10^{-6} cgs
Aluminum (s)	Al	ord.	+16.5	Chloride	BaCl$_2$	ord.	−72.6
(l)	Al		+12.0		BaCl$_2 \cdot$2H$_2$O	ord.	−100.0
Fluoride	AlF$_3$	302	−13.4	Fluoride	BaF$_2$	ord.	−51.0
Oxide	Al$_2$O$_3$	ord.	−37.0	Hydroxide	Ba(OH)$_2$	ord.	−53.2
Sulfate	Al$_2$(SO$_4$)$_3$	ord.	−93.0		Ba(OH)$_2 \cdot$8H$_2$O	ord.	−157.0
	Al$_2$(SO$_4$)$_3 \cdot$18H$_2$O	ord.	−323.0	Iodate	Ba(IO$_3$)$_2$	ord.	−122.5
Ammonia (g)	NH$_3$	ord.	−18.0	Iodide	BaI$_2$	ord.	−124.0
(aq)	NH$_3$	ord.	−17.0		BaI$_2 \cdot$2H$_2$O	ord.	−163.0
Ammonium				Nitrate	Ba(NO$_3$)$_2$	ord.	−66.5
Acetate	NH$_4$C$_2$H$_3$O$_2$	ord.	−41.1	Oxide	BaO	ord.	−29.1
Bromide	NH$_4$Br	ord.	−47.0		BaO$_2$	ord.	−40.6
Carbonate	(NH$_4$)$_2$CO$_3$	ord.	−42.50	Sulfate	BaSO$_4$	ord.	−71.3
Chlorate	NH$_4$ClO$_3$	ord.	−42.1	Beryllium (s)	Be	ord.	−9.0
Chloride	NH$_4$Cl	ord.	−36.7	Chloride	BeCl$_2$	ord.	−26.5
Fluoride	NH$_4$F	ord.	−23.0	Hydroxide	Be(OH)$_2$	ord.	−23.1
Hydroxide (aq)	NH$_4$OH	ord.	−31.5	Nitrate (aq)	Be(NO$_3$)$_2$	298	−41.0
Iodate	NH$_4$IO$_3$	ord.	−62.3	Oxide	BeO	ord.	−11.9
Iodide	NH$_4$I	ord.	−66.0	Sulfate	BeSO$_4$	ord.	−37.0
Nitrate	NH$_4$NO$_3$	ord.	−33.6	Bismuth (s)	Bi	ord.	−280.1
Sulfate	(NH$_4$)$_2$SO$_4$	ord.	−67.0	(l)	Bi		−10.5
Thiocyanate	NH$_4$SCN	ord.	−48.1	Bromide	BiBr$_3$	ord.	−147.0
Americium (s)	Am	300	+1000.0	Chloride	BiCl$_3$	ord.	−26.5
Antimony (s)	Sb	293	−99.0	Chromate	Bi$_2$(CrO$_4$)$_3$	ord.	+154.0
(l)	Sb		−2.5	Fluoride	BiF$_3$	303	−61.0
Bromide	SbBr$_3$	ord.	−115.0	Hydroxide	Bi(OH)$_3$	ord.	−65.8
Chloride, tri	SbCl$_3$	ord.	−86.7	Iodide	BiI$_3$	ord.	−200.5
Chloride, penta	SbCl$_5$	ord.	−120.0	Nitrate	Bi(NO$_3$)$_3$	ord.	−91.0
Fluoride	SbF$_3$	ord.	−46.0		Bi(NO$_3$)$_3 \cdot$5H$_2$O	ord.	−159.0
Iodide	SbI$_3$	ord.	−147.0	Oxide	BiO	ord.	−110.0
Oxide	Sb$_2$O$_3$	ord.	−69.4		Bi$_2$O$_3$	ord.	−83.0
Sulfide	Sb$_2$S$_3$	ord.	−86.0	Phosphate	BiPO$_4$	ord.	−77.0
Argon (g)	A	ord.	−19.6	Sulfate	Bi$_2$(SO$_4$)$_3$	ord.	−199.0
Arsenic (α)	As	293	−5.5	Sulfide	Bi$_2$S$_3$	ord.	−123.0
(β)	As	293	−23.7	Boric Acid	H$_3$BO$_3$	ord.	−34.1
(γ)	As	293	−23.0	Boron (s)	B	ord.	−6.7
Bromide	AsBr$_3$	ord.	−106.0	Chloride	BCl$_3$	ord.	−59.9
Chloride	AsCl$_3$	ord.	−79.9	Oxide	B$_2$O$_3$	ord.	−39.0
Iodide	AsI$_3$	ord.	−142.0	Bromine (l)	Br$_2$		−56.4
Sulfide	As$_2$S$_3$	ord.	−70.0	(g)	Br$_2$		−73.5
Arsenious Acid	H$_3$AsO$_3$	ord.	−51.2	Fluoride	BrF$_3$	ord.	−33.9
Barium	Ba	ord.	+20.6		BrF$_5$	ord.	−45.1
Acetate	Ba(C$_2$H$_3$O$_2$)$_2 \cdot$H$_2$O	ord.	−100.1	Cadmium (s)	Cd	ord.	−19.8
Bromate	Ba(BrO$_3$)$_2$	ord.	−105.8	(l)	Cd		−18.0
Bromide	BaBr$_2$	ord.	−92.0	Acetate	Cd(C$_2$H$_3$O$_2$)$_2$	ord.	−83.7
	BaBr$_2 \cdot$2H$_2$O	ord.	−119.0	Bromide	CdBr$_2$	ord.	−87.3
Carbonate	Ba(CO$_3$)	ord.	−58.9		CdBr$_2 \cdot$4H$_2$O	ord.	−140.0
Chlorate	Ba(ClO$_3$)$_2$	ord.	−87.5	Carbonate	CdCO$_3$	ord.	−46.7

MAGNETIC SUSCEPTIBILITY OF THE ELEMENTS AND INORGANIC COMPOUNDS
(Continued)

Substance	Formula	Temp. °K.	Susceptibility 10^{-6} cgs	Substance	Formula	Temp. °K.	Susceptibility 10^{-6} cgs
Chloride	$CdCl_2$	ord.	−68.7	Copper (s)	Cu	296	−5.46
	$CdCl_2 \cdot 2H_2O$	ord.	−99.0	(l)	Cu		−6.16
Chromate	$CdCrO_4$	ord.	−16.8	Bromide	CuBr	ord.	−49.0
Cyanide	$Cd(CN)_2$	ord.	−54.0		$CuBr_2$	341.6	+653.3
Fluoride	CdF_2	ord.	−40.6		$CuBr_2$	292.7	+685.5
Hydroxide	$Cd(OH)_2$	ord.	−41.0		$CuBr_2$	189	+736.9
Iodate	$Cd(IO_3)_2$	ord.	−108.4		$CuBr_2$	90	+658.7
Iodide	CdI_2	ord.	−117.2	Chloride	CuCl	ord.	−40.0
Nitrate	$Cd(NO_3)_2$	ord.	−55.1		$CuCl_2$	373.3	+1030.0
	$Cd(NO_3)_2 \cdot 4H_2O$	ord.	−140.0		$CuCl_2$	289	+1080.0
Oxide	CdO	ord.	−30.0		$CuCl_2$	170	+1815.0
Phosphate	$Cd_3(PO_4)_2$	ord.	−159.0		$CuCl_2$	69.25	+2370.0
Sulfate	$CdSO_4$	ord.	−59.2		$CuCl_2 \cdot 2H_2O$	293	+1420.0
Sulfide	CdS	ord.	−50.0	Cyanide	CuCN	ord.	−24.0
Calcium (s)	Ca		+40.0	Fluoride	CuF_2	293	+1050.0
Acetate	$Ca(C_2H_3O_2)_2$	ord.	−70.5		CuF_2	90	+1420.0
Bromate	$Ca(BrO_3)_2$	ord.	−84.0		$CuF_2 \cdot 2H_2O$	293	+1600.0
Bromide	$CaBr_2$	ord.	−73.8	Hydroxide	$Cu(OH)_2$	292	+1170.0
	$CaBr_2 \cdot 3H_2O$	ord.	−115.0	Iodide	CuI	ord.	−63.0
Carbonate	$CaCO_3$	ord.	−38.2	Nitrate	$Cu(NO_3)_2 \cdot 3H_2O$	293	+1570.0
Chloride	$CaCl_2$	ord.	−54.7		$Cu(NO_3)_2 \cdot 6H_2O$	293	+1625.0
Fluoride	CaF_2	ord.	−28.0	Oxide	Cu_2O	293	−20.0
Hydroxide	$Ca(OH)_2$	ord.	−22.0		CuO	780	+259.6
Iodate	$Ca(IO_3)_2$	ord.	−101.4		CuO	561	+267.3
Iodide	CaI_2	ord.	−109.0		CuO	397	+256.9
Nitrate (aq)	$Ca(NO_3)_2$	ord.	−45.9		CuO	289.6	+238.9
Oxide	CaO	ord.	−15.0		CuO	120	+156.2
	CaO_2	ord.	−23.8	Phosphide	Cu_3P	ord.	−33.0
Sulfate	$CaSO_4$	ord.	−49.7		CuP_2	ord.	−35.0
	$CaSO_4 \cdot 2H_2O$	ord.	−74.0	Sulfate	$CuSO_4$	293	+1330.0
Carbon (dia)	C	ord.	−5.9		$CuSO_4 \cdot H_2O$	293	+1520.0
(graph)	C	ord.	−6.0		$CuSO_4 \cdot 3H_2O$	ord.	+1480.0
Dioxide	CO_2	ord.	−21.0		$CuSO_4 \cdot 5H_2O$	293	+1460.0
Monoxide	CO	ord.	−9.8	Sulfide	CuS	293	−2.0
Cerium (α)	Ce	80.5	+5160.0	Thiocyanate	CuSCN	ord.	−48.0
(β)	Ce	293	+2450.0	Dysprosium	Dy	293.2	103,500.0
(β)	Ce	80.5	+6230.0	Oxide	Dy_2O_3	287.2	+89,600.0
(γ)	Ce	287.9	+2420.0	Sulfate	$Dy_2(SO_4)_3$	293	+91,400.0
(γ)	Ce	125.6	+4640.0		$Dy_2(SO_4)_3 \cdot 8H_2O$	291.2	+92,760.0
(γ)	Ce	80.5	+5200.0	Sulfide	Dy_2S_3	292	+95,200.0
Chloride	$CeCl_3$	287	+2490.0	Erbium	Er	291	+44,300.0
Fluoride	CeF_3	293	+2190.0	Oxide	Er_2O_3	286	+73,920.0
Nitrate	$Ce(NO_3)_3 \cdot 5H_2O$	292	+2310.0	Sulfate	$Er_2(SO_4)_3 \cdot 8H_2O$	293	+74,600.0
Oxide	CeO_2	293	+26.0	Sulfide	Er_2S_3	292	+77,200.0
Sulfate	$CeSO_4$	ord	+37.0	Europium	Eu	293	+34,000.0
	$Ce_2(SO_4)_3 \cdot 5H_2O$	293	+4540.0	Bromide	$EuBr_2$	292	+26,800.0
	$Ce(SO_4)_2 \cdot 4H_2O$	293	−97.0	Chloride	$EuCl_2$	292	+26,500.0
Sulfide	CeS	ord.	+2110.0	Fluoride	EuF_2	292	+23,750.0
	Ce_2S_3	292	+5080.0	Iodide	EuI_2	292	+26,000.0
Cesium (s)	Cs	ord.	+29.0	Oxide	Eu_2O_3	298	+10,100.0
(l)	Cs		+26.5	Sulfate	$EuSO_4$	293	+25,730.0
Bromate	$CsBrO_2$	ord.	−75.1		$Eu_2(SO_4)_3$	293	+10,400.0
Bromide	CsBr	ord.	−67.2		$Eu_2(SO_4)_3 \cdot 8H_2O$	293	+9,540.0
Carbonate	Cs_2CO_3	ord.	−103.6	Sulfide	EuS	293	+23,800.0
Chlorate	$CsClO_3$	ord.	−65.0		EuS	195	+35,400.0
Chloride	CsCl	ord.	−86.7	Gadolinium	Gd	300.6	+755,000.0
Fluoride	CsF	ord.	−44.5	Chloride	$GdCl_3$	293	+27,930.0
Iodate	$CsIO_3$	ord.	−83.1	Oxide	Gd_2O_3	293	+53,200.0
Iodide	CsI	ord.	−82.6	Sulfate	$Gd_2(SO_4)_3$	285.5	+54,200.0
Oxide	CsO_2	293	+1534.0		$Gd_2(SO_4)_3 \cdot 8H_2O$	293	+53,280.0
	CsO_2	90	+4504.0	Sulfide	Gd_2S_3	292	+55,500.0
Sulfate	Cs_2SO_4	ord.	−116.0	Gallium (s)	Ga	80	−24.4
Sulfide	Cs_2S	ord.	−104.0	(s)	Ga	290	−21.6
Chlorine (l)	Cl_2	ord.	−40.5	(l)	Ga	313	+2.5
Fluoride, tri	ClF_3	ord.	−26.5	Chloride	$GaCl_3$	ord.	−63.0
Chromium	Cr	273	+180.0	Iodide	GaI_3	ord.	−149.0
	Cr	1713	+224.0	Iodide	GaI_3	ord.	−149.0
Acetate	$Cr(C_2H_3O_2)_3$	293	+5104.0	Oxide	Ga_2O	ord.	−34.0
Chloride	$CrCl_2$	293	+7230.0	Sulfide	Ga_2S	ord.	−36.0
	$CrCl_3$	293	+6890.0		GaS	ord.	−23.0
Fluoride	CrF_3	293	+4370.0		Ga_2S_3	ord.	−80.0
Oxide	Cr_2O_3	300	+1960.0	Germanium	Ge	293	−76.84
	CrO_3	ord.	+40.0	Chloride	$GeCl_4$	ord.	−72.0
Sulfate	$Cr_2(SO_4)_3$	293	+11,800.0	Fluoride	GeF_4	ord.	−50.0
	$Cr_2(SO_4)_3 \cdot 8H_2O$	290	+12,700.0	Iodide	GeI_4	ord.	−174.0
	$Cr_2(SO_4)_3 \cdot 10H_2O$	290	+12,600.0	Oxide	GeO	ord.	−28.8
	$Cr_2(SO_4)_3 \cdot 14H_2O$	290	+12,160.0		GeO_2	ord.	−34.3
	$CrSO_4 \cdot 6H_2O$	293	+9690.0	Sulfide	GeS	189	−40.9
Sulfide	CrS	ord.	+2390.0		GeS_2	ord.	−53.3
Cobalt	Co		ferro	Gold (s)	Au	296	−28.0
Acetate	$Co(C_2H_3O_2)_2$	298	+11,000.0	(l)	Au		−34.0
Bromide	$CoBr_2$	293	+13,000.0	Bromide	AuBr	ord.	−61.0
Chloride	$CoCl_2$	293	+12,660.0	Chloride	AuCl	ord.	−67.0
	$CoCl_2 \cdot 6H_2O$	293	+9710.0		$AuCl_3$	ord.	−112.0
Cyanide	$Co(CN)_2$	303	+3825.0	Fluoride	AuF_3	ord.	+74.0
Fluoride	CoF_2	293	+9490.0	Iodide	AuI	ord.	−91.0
	CoF_3	293	+1900.0	Phosphide	AuP_3	ord.	−107.0
Iodide	CoI_2	293	+10,760.0	Hafnium (s)	Hf	298	+75.0
Oxide	CoO	260	+4900.0	(s)	Hf	1673	+104.0
	Co_2O_3	ord.	+4560.0	Oxide	HfO_2	ord.	−23.0
	Co_3O_4	ord.	+7380.0	Helium (g)	He	ord.	−1.88
Phosphate	$Co_3(PO_4)_2$	291	+28,110.0	Holmium	Ho		
Sulfate	$CoSO_4$	293	+10,000.0	Oxide	Ho_2O_3	293	+88,100.0
	$Co_2(SO_4)_3$	297	+1000.0	Sulfate	$Ho_2(SO_4)_3$	293	+91,700.0
Sulfide	CoS	688	+251.0		$Ho_2(SO_4)_3 \cdot 8H_2O$	293	+91,600.0
	CoS	293	+225.0	Hydrogen (g)	H_2		−3.98
Thiocyanate	$Co(SCN)_2$	303	+11,090.0				

MAGNETIC SUSCEPTIBILITY OF THE ELEMENTS AND INORGANIC COMPOUNDS
(Continued)

Substance	Formula	Temp. °K.	Susceptibility 10⁻⁶ cgs	Substance	Formula	Temp. °K.	Susceptibility 10⁻⁶ cgs
Bromide (*l*)	HBr	273		Acetate	Mg(C₂H₃O₂)₂·4H₂O	ord.	−116.0
(aq)	HBr	ord.		Bromide	MgBr₂	ord.	−72.0
Chloride (*l*)	HCl	273	−22.6	Carbonate	MgCO₃	ord.	−32.4
(aq)	HCl	300	−22.0		MgCO₃·3H₂O	ord.	−72.7
Fluoride (*l*)	HF	287	−8.6	Chloride	MgCl₂	ord.	−47.4
(aq)	HF	ord.	−9.3	Fluoride	MgF₂	ord.	−22.7
Iodide (*l*)	HI	281	−47.7	Hydroxide	Mg(OH)₂	288	−22.1
(*l*)	HI	233	−48.3	Iodide	MgI₂	ord.	−111.0
(s)	HI	195	−47.3	Oxide	MgO	ord.	−10.2
(aq)	HI	ord.	−50.2	Phosphate	Mg₃(PO₄)₂·4H₂O	ord.	−167.0
Oxide—See Water				Sulfate	MgSO₄	294	−50.0
Peroxide	H₂O₂	ord.	−17.7		MgSO₄·H₂O	ord.	−61.0
Sulfide	H₂S	ord.	−25.5		MgSO₄·5H₂O	ord.	−109.0
Indium	In	ord.	−64.0		MgSO₄·7H₂O	ord.	−133.7
Bromide	InBr₃	ord.	−107.0	Manganese (α)	Mn	293	+529.0
Chloride	InCl	ord.	−30.0	(β)	Mn	293	+483.0
	InCl₂	ord.	−56.0	Acetate	Mn(C₂H₃O₂)₂	293	+13,650.0
	InCl₃	ord.	−86.0	Bromide	MnBr₂	294	+13,900.0
Fluoride	InF₃	ord.	−61.0	Carbonate	MnCO₃	293	+11,400.0
Oxide	In₂O	ord.	−47.0	Chloride	MnCl₂	293	+14,350.0
	In₂O₃	ord.	−56.0		MnCl₂·4H₂O	293	+14,600.0
Sulfide	In₂S	ord.	−50.0	Fluoride	MnF₂	290	+10,700.0
	InS	ord.	−28.0		MnF₃	293	+10,500.0
	In₂S₃	ord.	−98.0	Hydroxide	Mn(OH)₂	293	+13,500.0
Iodic Acid	HIO₃	ord.	−48.0	Iodide	MnI₂	293	+14,400.0
metaper	HIO₄	ord.	−56.5	Oxide	MnO	293	+4850.0
orthoparaper	H₅IO₆	ord.	−71.4		Mn₂O₃	293	+14,100.0
Iodine (s)	I₂	ord.	−88.7		MnO₂	293	+2280.0
(atomic)	I	1303	+869.0		Mn₃O₄	298	+12,400.0
(atomic)	I	1400	+1120.0	Sulfate	MnSO₄	293	+13,660.0
Chloride	ICl	ord.	−54.6		MnSO₄·H₂O	293	+14,200.0
	ICl₃	ord.	−90.2		MnSO₄·4H₂O	293	+14,600.0
Fluoride	IF₅	ord.	−58.1		MnSO₄·5H₂O	293	+14,700.0
Oxide	I₂O₅	ord.	−79.4	Sulfide (α)	MnS	293	+5630.0
Iridium	Ir	298	+25.6	(β)	MnS	293	+3850.0
	Ir	698	+32.1	Mercury (s)	Hg		−24.1
Chloride	IrCl₃	ord.	−14.4	(*l*)	Hg	293	−33.44
Oxide	IrO₂	298	+224.0	(g)	Hg		−78.3
Iron	Fe		ferro	Acetate	HgC₂H₃O₂	ord.	−70.5
Bromide	FeBr₂	ord.	+13,600.0		Hg(C₂H₃O₂)₂	ord.	−100.0
Carbonate	FeCO₃	293	+11,300.0	Bromate	HgBrO₃	ord.	−57.7
Chloride	FeCl₂	293	+14,750.0	Bromide	HgBr	ord.	−57.2
	FeCl₂·4H₂O	293	+12,900.0		HgBr₂	ord.	−94.2
	FeCl₃	293	+13,450.0	Chloride	HgCl	ord.	−52.0
	FeCl₃	398	+9980.0		HgCl₂	ord.	−82.0
	FeCl₃·6H₂O	290	+15,250.0	Chromate	Hg₂CrO₄	ord.	−63.0
Fluoride	FeF₂	293	+9500.0		HgCrO₄	ord.	−12.5
	FeF₃	305	+13,760.0	Cyanide	Hg(CN)₂	ord.	−67.0
	FeF₃·3H₂O	293	+7870.0	Fluoride	HgF	ord.	−53.0
Iodide	FeI₂	293	+13,600.0		HgF₂	302	−62.0
Nitrate	Fe(NO₃)₃·9H₂O	293	+15,200.0	Hydroxide	Hg₂(OH)₂	ord.	−100.0
Oxide	FeO	293	+7200.0	Iodate	HgIO₃	ord.	−92.0
	Fe₂O₃	1033	+3586.0	Iodide	HgI	ord.	−83.0
Phosphate	FePO₄	ord.	+11,500.0		HgI₂	ord.	−128.6
Sulfate	FeSO₄	293	+10,200.0	Nitrate	HgNO₃	ord.	−55.9
	FeSO₄·H₂O	290	+10,500.0		Hg(NO₃)₂	ord.	−74.0
	FeSO₄·7H₂O	293	+11,200.0	Oxide	Hg₂O	ord.	−76.3
Sulfide	FeS	293	+1074.0		HgO	ord.	−44.0
Krypton	Kr		−28.8	Sulfate	Hg₂SO₄	ord.	−123.0
Lanthanum	La	ord.	<118.0		HgSO₄	ord.	−78.1
Oxide	La₂O₃	ord.	−78.0	Sulfide	HgS	ord.	−55.4
Sulfate	La₂(SO₄)₃·9H₂O	293	−262.0	Thiocyanate	Hg(SCN)₂	ord.	−96.5
Sulfide	La₂S₃	292	−37.0	Molybdenum	Mo	298	+89.0
	La₂S₄	293	−100.0		Mo	63.8	+108.0
Lead (s)	Pb	289	−23.0		Mo	20.4	+149.2
(*l*)	Pb	330	−15.5	Bromide	MoBr₃	293	+525.0
Acetate	Pb(C₂H₃O₂)₂	ord.	−89.1		MoBr₄	293	+520.0
Bromide	PbBr₂	ord.	−90.6		Mo₃Br₆	290.5	−46.0
Carbonate	PbCO₃	ord.	−61.2	Chloride	MoCl₃	290	+43.0
Chloride	PbCl₂	ord.	−73.8		MoCl₄	291	+1750.0
Chromate	PbCrO₄	ord.	−18.0		MoCl₅	289	+990.0
Fluoride	PbF₂	ord.	−58.1	Fluoride	MoF₆	ord.	−26.0
Iodate	Pb(IO₃)₂	ord.	−131.0	Oxide	Mo₂O₃	ord.	−42.0
Iodide	PbI₂	ord.	−126.5		MoO₂	289	+41.0
Nitrate	Pb(NO₃)₂	ord.	−74.0		MoO₃	292.5	+3.0
Oxide	PbO	ord.	−42.0		Mo₃O₈	ord.	+42.0
Phosphate	Pb₃(PO₄)₂	ord.	−182.0	Sulfide	MoS₃	289	−63.0
Sulfate	PbSO₄	ord.	−69.7	Neodymium	Nd	287.7	5628.0
Sulfide	PbS	ord.	−84.0	Fluoride	NdF₃	293	+4980.0
Thiocyanate	Lb(CNS)₂	ord.	−82.0	Nitrate	Nd(NO₃)₃	293	+5020.0
Lithium	Li	ord.	+14.2	Oxide	Nd₂O₃	292.0	+10,200.0
Acetate	LiC₂H₃O₂	ord.	−34.0	Sulfate	Nd₂(SO₄)₃	293	+9990.0
Bromate	LiBrO₃	ord.	−39.0	Sulfide	Nd₂S₃	292	+5550.0
Bromide	LiBr	ord.	−34.7	Neon	Ne	ord.	−6.74
Carbonate	Li₂CO₃	ord.	−27.0	Nickel	Ni		ferro
Chlorate (aq)	LiClO₃	ord.	−28.8	Acetate	Ni(C₂H₃O₂)₂	293	+4690.0
Chloride	LiCl	ord.	−24.3	Bromide	NiBr₂	293	+5600.0
Fluoride	LiF	ord.	−10.1	Chloride	NiCl₂	293	+6145.0
Hydride	LiH	ord.	−4.6		NiCl₂·6H₂O	293	4240.0
Hydroxide	LiOH	ord.	−12.3	Fluoride	NiF₂	293	+2410.0
Iodate	LiIO₃	ord.	−47.0	Hydroxide	Ni(OH)₂	ord.	+4500.0
Iodide	LiI	ord.	−50.0	Iodide	NiI₂	293	+3875.0
Nitrate	LiNO₃·3H₂O	ord.	−62.0	Nitrate	Ni(NO₃)₂·6H₂O	293.5	+4300.0
Sulfate	Li₂SO₄	ord.	−40.0	Oxide	NiO	293	+660.0
Lutetium	Lu	ord.	>0.0	Sulfate	NiSO₄	293	+4005.0
Magnesium	Mg	ord.	+13.1	Sulfide	NiS	293	+190.0

Substance	Formula	Temp. °K.	Susceptibility 10^{-6} cgs	Substance	Formula	Temp. °K.	Susceptibility 10^{-6} cgs
	Ni_3S_2	ord.	+ 1030.0	Oxide	Rh_2O_3	298	+ 104.0
Niobium	Nb	298	+ 195.0	Sulfate	$Rh_2(SO_4)_3 \cdot 6H_2O$	298	+ 104.0
Oxide	Nb_2O_5	ord.	−10.0		$Rh_2(SO_4)_3 \cdot 14H_2O$	298	+ 149.0
Nitric Acid	HNO_3		−19.9	Rubidium	Rb	303	+ 17.0
Nitrogen	N_2	ord.	−12.0	Bromide	RbBr	ord.	−56.4
Oxide	N_2O	285	−18.9	Carbonate	Rb_2CO_3	ord.	−75.4
(g)	NO	293	+ 1461.0	Chloride	RbCl	ord.	−46.0
(g)	NO	203.8	+ 1895.0	Fluoride	RbF	ord.	−31.9
(g)	NO	146.9	+ 2324.0	Iodide	RbI	ord.	−72.2
(l)	NO	117.64	+ 114.2	Nitrate	$RbNO_3$	ord.	−41.0
(s)	NO	90	+ 19.8	Oxide	RbO_2	293	+ 1527.0
	N_2O_3	291	−16.0	Sulfate	Rb_2SO_4	ord.	−88.4
	NO_2	408	+ 150.0	Sulfide	Rb_2S	ord.	−80.0
	N_2O_4	303.6	−22.1		Rb_2S_2	ord.	−90.0
	N_2O_4	295.1	−23.0	Ruthenium	Ru	298	+ 43.2
	N_2O_4	257	−25.4		Ru	723	+ 50.2
(aq)	N_2O_5	289	−35.6	Chloride	$RuCl_3$	290.9	+ 1998.0
Osmium	Os	298	+ 9.9	Oxide	RuO_2	298	+ 162.0
Chloride	$OsCl_2$	ord.	+ 41.3	Samarium	Sm	291	+ 1860.0
Oxygen (g)	O_2	293	+ 3449.0		Sm	195	+ 2230.0
(l)	O_2	90.1	+ 7699.0	Bromide	$SmBr_2$	293	+ 5337.0
(l)	O_2	70.8	+ 8685.0		$SmBr_3$	293	+ 972.0
(s, γ)	O_2	54.3	+ 10,200.0	Oxide	Sm_2O_3	292	+ 1988.0
(s, β)	O_2				Sm_2O_3	170	+ 1960.0
(s, α)	O_2	23.7	+ 1760.0		Sm_2O_3	85	+ 2282.0
Ozone (l)	O_3		+ 6.7	Sulfate	$Sm_2(SO_4)_3 \cdot 8H_2O$	293	+ 1710.0
Palladium	Pd	288	+ 567.4	Sulfide	Sm_2S_3	292	+ 3300.0
Chloride	$PdCl_2$	291.3	−38.0	Scandium	Sc	292	+ 315.0
Fluoride	PdF_3	293	+ 1760.0	Selenic Acid	H_2SeO_4	ord.	−51.2
Hydride	PdH	ord.	+ 1077.0	Selenious Acid	H_2SeO_3	ord.	−45.4
	Pd_4H	ord.	+ 2353.0	Selenium (s)	Se	ord.	−25.0
Phosphoric Acid (aq)	H_3PO_4	ord.	−43.8	(l)	Se	900	−24.0
Phosphorous Acid (aq)	H_3PO_3	ord.	−42.5	Bromide	Se_2Br_2	ord.	−113.0
Phosphorous (red)	P		−20.8	Chloride	Se_2Cl_2	ord.	−94.8
(black)	P		−26.6	Fluoride	SeF_6	ord.	−51.0
Chloride	PCl_3	ord.	−63.4	Oxide	SeO_2	ord.	−27.2
Platinum	Pt	290.3	+ 201.9	Silicon	Si	ord.	−3.9
Chloride	$PtCl_2$	298	−54.0	Bromide	$SiBr_4$	ord.	−128.6
	$PtCl_3$	ord.	−66.7	Carbide	SiC	ord.	−12.8
	$PtCl_4$	ord.	−93.0	Chloride	$SiCl_4$	ord.	−88.3
Fluoride	PtF_4	293	+ 455.0	Hydroxide	$Si(OH)_4$	ord.	−42.6
Oxide	Pt_2O_3	ord.	−37.70	Oxide	SiO_2	ord.	−29.6
Plutonium	Pu	293	+ 610.0	Silver (s)	Ag	296	−19.5
Fluoride	PuF_4	301	+ 1760.0	(l)	Ag		−24.0
	PuF_6	295	+ 173.0	Acetate	$AgC_2H_3O_2$	ord.	−60.4
Oxide	PuO_2	300	+ 730.0	Bromide	AgBr	283	−59.7
Potassium	K	ord.	+ 20.8	Carbonate	Ag_2CO_3	ord.	−80.90
Acetate (aq)	$KC_2H_3O_2$	28	−45.0	Chloride	AgCl	ord.	−49.0
Bromate	$KBrO_3$	ord.	−52.6	Chromate	Ag_2CrO_4	ord.	−40.0
Bromide	KBr	ord.	−49.1	Cyanide	AgCN	ord.	−43.2
Carbonate	K_2CO_3	ord.	−59.0	Fluoride	AgF	ord.	−36.5
Chlorate	$KClO_3$	ord.	−42.8	Iodide	AgI	ord.	−80.0
Chloride	KCl	ord.	−39.0	Nitrate	$AgNO_3$	ord.	−45.7
Chromate	K_2CrO_4	ord.	−3.9	Nitrite	$AgNO_2$	ord.	−42.0
	$K_2Cr_2O_7$	293	+ 29.4	Oxide	Ag_2O	ord.	−134.0
Cyanide	KCN	ord.	−37.0		AgO	287	−19.6
Ferricyanide	$K_3Fe(CN)_6$	297	+ 2290.0	Permanganate	$AgMnO_4$	300	−63.0
Ferrocyanide	$K_4Fe(CN)_6$	ord.	−130.0	Phosphate	Ag_3PO_4	ord.	−120.0
	$K_4Fe(CN)_6 \cdot 3H_2O$	ord.	−172.3	Sulfate	Ag_2SO_4	ord.	−92.90
Fluoride	KF	ord.	−23.6	Thiocyanate	AgSCN	ord.	−61.8
Hydroxide (aq)	KOH	ord.	−22.0	Sodium	Na	ord.	+ 16.0
Iodate	KIO_3	ord.	−63.1	Acetate	$NaC_2H_3O_2$	ord.	−37.6
Iodide	KI	ord.	−63.8	Borate, tetra	$Na_2B_4O_7$	ord.	−85.0
Nitrate	KNO_3	ord.	−33.7	Bromate	$NaBrO_3$	ord.	−44.2
Nitrite	KNO_2	ord.	−23.3	Bromide	NaBr	ord.	−41.0
Oxide	KO_2	293	+ 3230.0	Carbonate	Na_2CO_3	ord.	−41.0
	KO_3	ord.	+ 1185.0	Chlorate	$NaClO_3$	ord.	−34.7
Permanganate	K_4MnO_4	ord.	+ 20.0	Chloride	NaCl	ord.	−30.3
Sulfate	K_2SO_4	ord.	−67.0	Chromate, di	$Na_2Cr_2O_7$	ord.	+ 55.0
	$KHSO_4$	ord.	−49.8	Fluoride	NaF	ord.	−16.4
Sulfide	K_2S	ord.	−60.0	Hydroxide (aq)	NaOH	300	−16.0
	K_2S_2	ord.	−71.0	Iodate	$NaIO_3$	ord.	−53.0
	K_2S_3	ord.	−80.0	Iodide	NaI	ord.	−57.0
	K_2S_4	ord.	−89.0	Nitrate	$NaNO_3$	ord.	−25.6
	K_2S_5	ord.	−98.0	Nitrite	$NaNO_2$	ord.	−14.5
Sulfite	K_2SO_3	ord.	−64.0	Oxide	Na_2O	ord.	−19.8
Thiocyanate	KSCN	ord.	−48.0		Na_2O_2	ord.	−28.10
Praseodymium	Pr	293	+ 5010.0	Phosphate, meta	$NaPO_3$	ord.	−42.5
Chloride	$PrCl_3$	307.1	+ 44.5		Na_2HPO_4	ord.	−56.6
Oxide	PrO_2	293	+ 1930.0	Sulfate	Na_2SO_4	ord.	−52.0
	Pr_2O_3	827	+ 4000.0		$Na_2SO_4 \cdot 10H_2O$	ord.	−184.0
Oxide	Pr_2O_3	294.5	+ 8994.0	Sulfide	Na_2S	ord.	−39.0
Sulfate	$Pr_2(SO_4)_3$	291	+ 9660.0		Na_2S_2	ord.	−53.0
	$Pr_2(SO_4)_3 \cdot 8H_2O$	289	+ 9880.0		Na_2S_3	ord.	−68.0
Sulfide	Pr_2S_3	292	+ 10,770.0		Na_2S_4	ord.	−84.0
Rhenium	Re	293	+ 67.6		Na_2S_5	ord.	−99.0
Chloride	$ReCl_6$	293	+ 1225.0	Strontium	Sr	ord.	+ 92.0
Oxide	ReO_2	ord.	+ 44.0	Acetate	$Sr(C_2H_3O_2)_2$	ord.	−79.0
	$ReO_2 \cdot 2H_2O$	295	+ 74.0	Bromate	$Sr(BrO_3)_2$	ord.	−93.5
	ReO_3	ord.	+ 16.0	Bromide	$SrBr_2$	ord.	−86.6
	Re_2O_7	ord.	−16.0		$SrBr_2 \cdot 6H_2O$	ord.	−160.0
Sulfide	ReS_2	ord.	+ 38.0	Carbonate	$SrCO_3$	ord.	−47.0
Rhodium	Rh	298	+ 111.0	Chlorate	$Sr(ClO_3)_2$	ord.	−73.0
	Rh	723	+ 123.0	Chloride	$SrCl_2$	ord.	−63.0
Chloride	$RhCl_3$	298	−7.5		$SrCl_2 \cdot 6H_2O$	ord.	−145.0
Fluoride	RhF_4	293	+ 500.0	Chromate	$SrCrO_4$	ord.	−5.1

Substance	Formula	Temp. °K.	Susceptibility 10⁻⁶ cgs
Fluoride	SrF_2	ord.	−37.2
Hydroxide	$Sr(OH)_2$	ord.	−40.0
	$Sr(OH)_2 \cdot 8H_2O$	ord.	−136.0
Iodate	$Sr(IO_3)_2$	ord.	−108.0
Iodide	SrI_2	ord.	−112.0
Nitrate	$Sr(NO_3)_2$	ord.	−57.2
	$Sr(NO_3)_2 \cdot 4H_2O$	ord.	−106.0
Oxide	SrO	ord.	−35.0
	SrO_2	ord.	−32.3
Sulfate	$SrSO_4$	ord.	−57.9
Sulfur (α)	S	ord.	−15.5
(β)	S	ord.	−14.9
(l)	S		−15.4
(g)	S	828	1700.0
(g)	S	1023	+464.0
Chloride	S_2Cl_2	ord.	−62.2
	SCl_2	ord.	−49.4
	SCl_2	ord.	−49.4
Fluoride	SF_6	ord.	−44.0
Iodide	SI	ord.	−52.7
Oxide (l)	SO_2	ord.	−18.2
Sulfuric Acid	H_2SO_4	ord.	−39.8
Tantalum	Ta	293	+154.0
	Ta	2143	+124.0
Chloride	$TaCl_5$	304	+140.0
Fluoride	TaF_5	293	+795.0
Oxide	Ta_2O_5	ord.	−32.0
Technetium	Tc	402	250.0
	Tc	298	270.0
	Tc	78	290.0
Oxide	$TcO_2 \cdot 2H_2O$	300	+244.0
	Tc_2O_7	298	−40.0
Tellurium (s)	Te	ord.	−39.5
(l)	Te		−6.4
Bromide	$TeBr_2$	ord.	−106.0
Chloride	$TeCl_2$	ord.	−94.0
Fluoride	TeF_6	ord.	−66.0
Terbium	Tb	273	+146,000.0
Oxide	Tb_2O_3	288.1	+78,340.0
Sulfate	$Tb_2(SO_4)_3$	293	+78,200.0
	$Tb_2(SO_4)_3 \cdot 8H_2O$	293	+76,500.0
Thallium (α)	Tl	ord.	−50.9
(β)	Tl	500	−30.3
(l)	Tl	573	−26.8
Acetate	$TlC_2H_3O_2$	ord.	−69.0
Bromate	$TlBrO_3$	ord.	−75.9
Bromide	$TlBr$	ord.	−63.9
Carbonate	Tl_2CO_3	ord.	−101.6
Chlorate	$TlClO_3$	ord.	−65.5
Chloride	$TlCl$	ord.	−57.8
Chromate	Tl_2CrO_4	ord.	−39.3
Cyanide	$TlCN$	ord.	−49.0
Fluoride	TlF	ord.	−44.4
Iodate	$TlIO_3$	ord.	−86.8
Iodide	TlI	ord.	−82.2
Nitrate	$TlNO_3$	ord.	−56.5
Nitrite	$TlNO_2$	ord.	−50.8
Oxide	Tl_2O_3	ord.	+76.0
Phosphate	Tl_3PO_4	ord.	−145.2
Sulfate	Tl_2SO_4	ord.	−112.6
Sulfide	Tl_2S	ord.	−88.8
Thiocyanate	$TlCNS$	ord.	−66.7
Thorium	Th	293	+132.0
	Th	90	+153.0
Chloride	$ThCl_4 \cdot 8H_2O$	305.2	−180.0
Nitrate	$Th(NO_3)_4$	ord.	−108.0
Oxide	ThO_2	ord.	−16.0
Thulium	Tm	291	+25,500.0
Oxide	Tm_2O_3	296.5	+51,444.0
Tin (white)	Sn	ord.	+3.1
(gray)	Sn	280	−37.0
(gray)	Sn	100	−31.7
(l)	Sn		−4.5
Bromide	$SnBr_4$	ord.	−149.0
Chloride	$SnCl_2$	ord.	−69.0
	$SnCl_2 \cdot 2H_2O$	ord.	−91.4
	$SnCl_4$	ord.	−115.0
(l)	$SnCl_4$		
Hydroxide	$Sn(OH)_4$	ord.	−60.0
Oxide	SnO	ord.	−19.0
	SnO_2	ord.	−41.0
Titanium	Ti	293	+153.0
	Ti	90	+150.0
Bromide	$TiBr_2$	288	+640.0
	$TiBr_3$	441	+520.0
	$TiBr_3$	291	+660.0
	$TiBr_3$	195	+680.0
	$TiBr_3$	90	+220.0
Carbide	TiC	ord.	+8.0
Chloride	$TiCl_2$	288	+570.0
	$TiCl_3$	685	+705.0
	$TiCl_3$	373	+1030.0
	$TiCl_3$	292	+1110.0
	$TiCl_3$	212	+690.0
	$TiCl_3$	90	+220.0
	$TiCl_4$	ord.	−54.0
Fluoride	TiF_3	293	+1300.0

Substance	Formula	Temp. °K.	Susceptibility 10⁻⁶ cgs
Iodide	TiI_2	288	+1790.0
	TiI_3	434	+221.0
	TiI_3	292	+160.0
	TiI_3	195	+159.0
	TiI_3	90	+167.0
Oxide	Ti_2O_3	382	+152.0
	Ti_2O_3	298	+125.6
	TiO_3	248	+132.4
	TiO_2	ord.	+5.9
Sulfide	TiS	ord.	+432.0
Tungsten	W	298	+59.0
Bromide	WBr_5	293	+250.0
Carbide	WC	o-d.	+10.0
Chloride	WCl_2	293	−25.0
	WCl_5	293	+387.0
	WCl_6	ord.	−71.0
Fluoride	WF_6	ord.	−40.0
Oxide	WO_2	ord.	+57.0
	WO_3	ord.	−15.8
Sulfide	WS_2	303	+5850.0
Tungstic Acid, ortho	H_2WO_4	ord.	−28.0
Uranium (α)	U	78	+395.0
(α)	U	298	+409.0
(α)	U	623	+440.0
(β)	U		
(γ)	U	1393	+514.0
Bromide	UBr_3	294	+4740.0
	UBr_4	293	+3530.0
Chloride	UCl_3	300	+3460.0
	UCl_4	294	+3680.0
Fluoride	UF_4	300	+3530.0
	UF_6	ord.	+43.0
Hydrides	UH_3	462	+2821.0
	UH_3	391	+3568.0
	UH_3	295	+6244.0
	UH_3	255	+9306.0
Iodide	UI_3	293	+4460.0
Oxide	UO	293	+1600.0
	UO_2	293	+2360.0
	UO_3	ord.	+128.0
Sulfate	$U(SO_4)_2$	ord.	+31.0
Sulfide (α)	US_2	290.5	+3137.0
(β)	US_2	290.5	+3170.0
	U_2S_3	ord.	+5206.0
	U_2S_5	ord.	+11,220.0
Vanadium	V	298	+255.0
Bromide	VBr_2	293	+3230.0
	VBr_2	195	+3760.0
	VBr_2	90	+4470.0
	VBr_3	293	+2890.0
	VBr_3	195	+4110.0
	VBr_3	90	+8540.0
Chloride	VCl_2	293	+2410.0
	VCl_3	293	+3030.0
	VCl_4	293	+1130.0
	VCl_4	195	+1700.0
	VCl_4	90	+4360.0
Fluoride	VF_3	293	+2730.0
Oxide	VO_2	290	+270.0
	V_2O_3	293	+1976.0
	V_2O_5	ord.	+128.0
Sulfide	VS	ord.	+600.0
	V_2S_3	293	+1560.0
Water (g)	H_2O	>373	−13.1
(l)	H_2O	373	−13.09
(l)	H_2O	293	−12.97
(l)	H_2O	273	−12.93
(s)	H_2O	273	−12.65
(s)	H_2O	223	−12.31
(l)	DHO	302	−12.97
(l)	D_2O	293	−12.76
(l)	D_2O	276.8	−12.66
(s)	D_2O	276.8	−12.54
(s)	D_2O	213	−12.41
Xenon	Xe	ord.	−43.9
Ytterbium	Yb	292	+249.0
	Yb	90	+639.0
Sulfide	Yb_2S_3	292	+18,300.0
Yttrium	Y	292	+2.15
	Y	90	+2.43
Oxide	Y_2O_3	293	+44.4
Sulfide	Y_2S_3	ord.	+100.0
Zinc (s)	Zn	ord.	−11.4
(l)	Zn		−7.8
Acetate	$Zn(C_2H_3O_2)_2 \cdot 2H_2O$	ord.	−101.0
Carbonate	$ZnCO_3$	ord.	−34.0
Chloride	$ZnCl_2$	296	−65.0
Cyanide	$Zn(CN)_2$	ord.	−46.0
Fluoride	ZnF_2	299.6	−38.2
Hydroxide	$Zn(OH)_2$	ord.	−67.0
Iodide	ZnI_2	ord.	−98.0
Nitrate (aq)	$Zn(NO_3)_2$	ord.	−63.0
Oxide	ZnO	ord.	−46.0
Phosphate	$Zn_3(PO_4)_2$	ord.	−141.0
Sulfate	$ZnSO_4$	ord.	−45.0
	$ZnSO_4 \cdot H_2O$	ord.	−63.0

Substance	Formula	Temp. °K.	Susceptibility 10^{-6} cgs	Substance	Formula	Temp. °K.	Susceptibility 10^{-6} cgs
	$ZnSO_4 \cdot 7H_2O$	ord.	−143.0	Carbide	ZrC	ord.	−26.0
Sulfide	ZnS	ord.	−25.0	Nitrate	$Zr(NO_3)_4 \cdot 5H_2O$	ord.	−77.0
Zirconium	Zr	293	−122.0	Oxide	ZrO_2	ord.	−13.8
	Zr	90	+119.0				

DIAMAGNETIC SUSCEPTIBILITIES OF ORGANIC COMPOUNDS

Compiled by George W. Smith

This table and its supplement contain values for the molar susceptibility of χ_M, specific susceptibility χ, and volumetric susceptibility K. The cgs Gaussian system of units is employed. In the Gaussian units the relation between magnetic induction B and magnetic field strength H is

$$B = H + 4\pi I \tag{1}$$

where I is the magnetization or magnetic moment per unit volume. Actually, the quantities involved in equation (1) are all vectors, but one may assume that all three are collinear, a reasonable assumption for organic diamagnetic substances in the liquid state. For crystals I may vary with crystal orientation. Equation (I) may be written

$$B = H + 4\pi KH = (I + 4\pi K)H \tag{2}$$

Here, K is the magnetic susceptibility, often called the volumetric susceptibility and is a unitless quantity.
Other susceptibilities of use to chemists and physicists are the specific or mass susceptibility which is defined

$$\chi = K/\rho \tag{3}$$

where ρ is the density of the sample in grams per cc., and the molar susceptibility which is defined as

$$\chi_M = M_\chi = MK/\rho \tag{4}$$

where M is the molecular weight of the substance in grams.
Temperatures, when listed, are enclosed in parentheses and are listed in degrees C.
Literature references for values contained in this table may be found in General Motors Research Laboratories Bulletins GMR-317 and GMR-396.

Compound	$-\chi_M \times 10^6$	$-\chi \times 10^6$	$-K \times 10^6$
Acenaphthanthracene	184	.73	
Acenaphthene	109.3	(.709)	(.726)(99°)
Acetal	81.39	.688(32°)	(.568)(32°)
Acetaldehyde	22.70	$.515_3$	(.403)(18°)
Acetamide	34.1	.577	(.618)(20°)
Acetic acid	31.54	.525(32°)	(.551)(32°)
Acetic anhydride	(52.8)	.517	(.562)(15°)
Acetomainofluorene	(141)	.63	
Acetone	33.7_8	$.581_4$	(.460)(20°)
Acetonitrile	28.0	(.682)	(.534)(20°)
Acetonylacetone	62.51	$.547_6$	(.531)(20°)
Acetophenone	72.05	$.599_8$	(.615)(20°)
Acetophenone oxime	79.9_0	$.592_0$	
Acetophenone oxime-O-methyl ether	92.3_1	$.618_8$	
Acetoxime	44.4_2	$.607_6$	(.480)(20°)
Acetoxime-O-benzyl ether	104.8_2	$.642_7$	
Acetoxime-O-methyl-ether	54.8_7	$.629_6$	
Acetylacetone	54.88	$.548_1$	(.535)(20°)
Acetyl chloride	38.9	.496	(.548)(20°)
Acetylene	12.5	(.480)	
Acetylphenylacetylene	86.9	.508	
Acetylthiophene	71.7	(.568)	
Acridine	(123.3)	.688	(.757)(20°)
Adonitol	91.30	.600	
Alanine	50.5	(.567)	
Allyl acetate	(56.7)	.566	(.525)(20°)
Allyl alcohol	36.70	.632	.540(20°)
1-Allylpyrrole	73.80	.685(20°)	
Aminoazobenzene	(118.3)	.600	
Aminoazotoluene	(142.2)	.631	
o-Aminoazotoluene	(138)	.61 ± .02	
α-Aminobutyric acid	62.1	(.602)	
Aminomethyldietyldiazine	114.8	(.696)	
4-Aminostilbene	122.5	(.628)	
2-Aminothiazole	56.0	.564	
n-Amyl acetate	89.06	684_5	.5979(20.7°)
iso-Amyl acetate	89.40	.687	.599(20°)
n-Amyl alcohol	(67.5)	.766	(.624)(20°)
γ-Amyl alcohol	(71.0)	.8060	(.655)(25°)
Inactive Amyl alcohol	69.06	783_4(25°)	
iso-Amyl alcohol	68.96	$.782_3$(25°)	(.64)(15°)
sec-Amyl alcohol	69.1	.785	(.635)(20°)
tert-Amyl alcohol	(70.9)	.804	(.654)(15°)
n-Amylamine	69.4	(.796)	(.606)(20°)
iso-Amulamine	71.6	(.821)	(.616)(20°)
n-Amylbenzene	112.55	(.759)	(.652)(20°)
iso-Amyl bromide	(88.7)	.587	(.706)(20°)
iso-Amyl-n-butyrate	113.52	$.717_4$(25°)	(.616)(20°)
iso-Amyl chloride	(79.0)	.741	(.662)(20°)
iso-Amyl cyanide	73.4	(.755)	(.609)(20°)
iso-Amylene	53.7	.766	
Amylene bromide	(114.5)	.498	
Amylene chloride	(95.2)	.675	
iso-Amyl ether	(129)	.813	(.635)(15°)
iso-Amyl formate	78.38	$.674_8$	(.591)(25°)
Amylidene chloride	(93.4)	.662	
Amyl iodide	(118.7)	.5996(18°)	(.910)(20°)
n-Amyl methyl ketone	80.50	$.705_3$	(.580)(15°)
Amyl nitrate	(76.4)	.574	
iso-Amyl propionate	101.73	705_5(25°)	(.609)(25°)
n-Amyl valerate	124.55	$.723_9$	(.638)(0°)
Anethole	(96.0)	.648	(.644)(15°)
Aniline	62.95	(.676)	(.691)(20°)
Anisidine	(80.5)	.654	(.72)(20°)
Anisole	72.79	(.673)	(.672)(15°)
Anthanthrene	204.2	.739	
Anthanthrone	178.1	.581	
Anthracene	(130)	.731	(.914)(27°)
Anthracenedinitrile	154.6	(.678)	
Anthracenonitrile	142.1	(.700)	
Anthraquinone	(119.6)	.575	(.825)(20°)
Anthrazine	245.7	.646	
Arabinose	85.70	.571	(.905)(20°)
Arbutoside	158.0	(.589)	
Asarone	131.4	(.631)	(.735)(18°)
Asparagine	69.5	(.526)	(.812)(15°)
Aspartic acid	64.2±.4	(.482)	(.800)(12°)
Aurin	(161.4)	.556	
p-Azoanisole	(147.7)	.610	
Azobenzene	(106.8)	.586	.611(70.5°)
p-Azophenetole	(171.7)	.635	
m-Azotoluene	(127.8)	.608	.643(58°)
Azulene	98.5	.768	
Barbituric acid (Anh.)	53.8	(.420)	
Barbituric acid (.2H2O)	78.6	(4.79)	
Benzalazine	(123.7)	.594	
Benzaldehyde	60.78	(.573)	(.602)(15°)
Benzaldoxime	(69.8)	.576	(.639)(20°)
Benzamide	(72.3)	.597	(.801)(4°)
Benzanthrone	142.9	.620	
Benzene	54.84	.702(32°)	.611
Benzidine	110.9	.603	(.754)(20°)
Benzil	(118.6)	.564	.616(100°)
Benzoic acid	70.28	(.575)	(.728)(15°)
Benzoic anhydride	(124.9)	.552	(.662)(15°)
Benzonitile	65.19	(.632)	(.638)(15°)
Benzophenone	109.60	$.601_3$	(.66)(50°)
3,4-Benzopyrene	135.7	.538	
Benzopyrene	194.0		
Benzoyl acetone	(95.0)	.586	(.639)(60°)
Benzoyl chloride	(75.8)	.539(20°)	(.657)(15°)
Benzyl acetate	93.18	(.620)	(.655)(16°)
Benzyl alcohol	71.83	(.664)	(.697)(15°)
Benzylamine	75.26	(.702)	(.690)(19°)
Benzyl chloride	81.98	(.647)	(.713)(18°)
Benzyl formate	81.43	(.598)	(.646)(20°)
Benzylideneaniline	(100.4)	.554	
Benzylidene chloride	(97.9)	.608	(.763)(14°)
Benzylidenemethylamine	(73.1)	.613	
Benzylmethylketone	83.44	$.621_9$	(.624)(20°)
Bibenzyl	(126.8)	.696	.671 (54.5°)
3,4-Bis-(p-hydroxyphenyl)-2,4-hexadiene	(157)	.59	
m,m'-Bitolyl	(127.4)	.6993(27.4°)	(.699)(16°)
m,m'-Bitolyl sulfide	(140.0)	.6530(27.4°)	
Borneol	126.0	(.817)	(.826)(20°)
Bromobenzene	78.92	.5030(20°)	(.753)(20°)

DIAMAGNETIC SUSCEPTIBILITIES OF ORGANIC COMPOUNDS (Continued)

Compound	$-\chi_M \times 10^6$	$-\chi \times 10^6$	$-K \times 10^6$	Compound	$-\chi_M \times 10^6$	$-\chi \times 10^6$	$-K \times 10^6$
Bromobenzenediazocyan-ide	86.88	(.414)		o-Chlorophenol	77.4	(.602)	(.747) (18°)
Bromochloromethane	55.0 + .6	(.425)	(.846) (19°)	p-Chlorophenol	77.6	(.604)	(.789) (20°)
Bromodichloromethane	66.3 + .3	(.405)	(.812) (15°)	1-(o-Chlorophenylazo)-2-naphthol	161.0	(.570)	
Bromoform	82.60	.327	.948 (20°)	1-(p-Chlorphenylazo)-2-naphthol	161.4	(.571)	
Bromonaphthalene	(123.8)	.598		Chlorotrifluoroethylene	49.1	.422	
α-Bromonaphthalene	115.90	.560	.840 (20°)	Chlorotrifluoromethane	45.3 + 1.5	(.434)	
m-Bromotoluene	(93.4)	.546	(.770) (20°)	Cholesterol	(284.2)	.735	(.764) (20°)
Bromotrichloromethane	73.1 ± .7	(.369)	(.758) (0°)	Chrysene	166.67	.731	
Butane	57.4	(.998)		Chrysoidine	(126.3)	.595	
iso-Butane	51.7	(.890)		Cinnamic acid	78.36	.529	(.660) (4°)
1,4-Butanediol	61.5	(.682)	(.696) (20°)	Cinnamic acid (α-trans)	78.2	(.528)	
2-Butene (cis)	42.6	(.759)		Cinnamic acid (β-trans)	79.0	(.533)	
2-Butene (trans)	43.3	(.772)		Cinnamic acid (cis-MP 68°)	77.6	(.524)	
1-Butene-3,4-diacetate	95.5	(.555)		Cinnamic acid (cis-MP 58°)	77.9	(.526)	
2-Butene-1,4-diacetate (cis)	95.2	(.553)		Cinnamic acid (cis-MP 42°)	83.2	(.562)	
2-Butene-1,4-diacetate (trans)	95.1	(.552)		Cinnamic aldehyde	(74.8)	.566	(.629) (15°)
2-Butene-1,4-diol (cis)	54.3	(.616)		Cinnamyl alcohol	(87.2)	.650	(.679) (20°)
2-Butene-1,4-diol (trans)	53.5	(.607)		Cinnamylidencanileine	123.2	(.595)	
n-Butyl acetate	77.47	.666₉ (25°)	(.583) (25°)	Citral	(98.9)	.650	(.577) (20°)
iso-Butyl acetate	78.52	.676₇	(.584) (25°)	Coronene	(243.3)	.810	
n-Butyl alcohol	56.536 (20°)	(.7627)	(.6176) (20°)	Coumarin	82.5	(.565)	(.528) (20°)
iso-Butyl alcohol	57.704 (20°)	(.7785)	(.624) (20°)	o-Cresol	72.90	.675	(.706) (20°)
sec-Butyl alcohol	57.683 (20°)	(.7782)	(.629) (20°)	m-Cresol	72.02	.667 (26°)	(.690) (20°)
tert-Butyl alcohol	57.42	.774₇ (25°)	(.611) (20°)	p-Cresol	72.1	.667 (26°)	(.690) (20°)
n-Butylamine	58.9	(.805)	(.596) (20°)	o-Creaylmethyl ether	81.94	.671 (40°)	(.661) (15°)
iso-Butylamine	59.8	(.818)	(.599) (20°)	m-Creaylmethyl ether	77.91	.638 (40°)	(.623) (15°)
9-Butylanthracene	176.0	(.751)		p-Cresylmethyl ether	79.13	.648 (40°)	(.629) (19°)
n-Butylbenzene	100.79	(.751)	(.646) (20°)	Cumene	89.53	.744₆	(.642)
iso-Butylbenzene	101.81	(.759)	(.648) (20°)	Cyamelide	(56.1)	.435	(.490) (15°)
tert-Butylbenzene	102.5	(.764)	(.662) (20°)	Cyameluric acid	101.1 (10°)	(.457)	
n-Butyl benzoate	116.69	.654₈	(.656) (25°)	9-Cyanoanthracene	142.1	(.699)	
Butyl bromide	77.14	.563 (20°)	(.730) (20°)	Cyanogen	(21.6)	.415	(.359) (liq., 17°)
iso-Butyl bromide	79.88	.583 (20°)	(.737) (20°)	Cyanuric acid	61.5	.476	(.842) (20°)
1-n-Butyl chloride	67.10	.725	.642 (20°)	Cyclobutanecarboxylic acid	58.16	.5816 (30°)	(.613) (30°)
2-n-Butyl chloride	67.40	.728	.635 (20°)	1,3-Cyclohexadiene	48.6	(.607)	(.510) (20°)
n-Butyl cyanide	(62.8)	.7558 (27.4°)	(.606) (20°)	1,4-Cyclohexadiene	48.7	(.608)	(.515) (20°)
tert-Butyl cyclohexane	115.09	.8205	.6670 (20°)	Cyclohexane	68.13	.8100 (27.5°)	(.627) (20°)
1,4-Butyl diacetate	103.4	(.594)		Cyclohexanecarboxylic acid	83.24	.6499 (30°)	(.668) (30°)
n-Butyl ethyl ketone	80.73	.707₂	(.579) (20°)	Cyclohexanol	73.40	.732	.694 (20°)
n-Butyl formate	65.83	.644₆	(.571) (25°)	Cyclohexanone	62.0₁	.632₁	(.599) (20°)
iso-Butyl formate	66.77	.653 ?	(.571) (25°)	Cyclohexanone oxime	71.3₂	.632₁	
iso-Butylideneazine	(95.8)	.683		Cyclohexanoneoxime-O-methyl ether	82.9₆	.652₃	
Butyl iodide	(93.6)	.3086 (18°)	(.822) (20°)	Cyclohexene	57.5	(.700)	(.567) (20°)
iso-Butyl methyl ketone	70.05	.699₅	(.561) (20°)	Cyclohexenol	64.1	(.663)	
tert-Butyl methyl ketone	69.86	.697₈	(.558) (16°)	Cyclooctane	91.4	(.815)	(.684) (20°)
n-Butyl perfluor-n-butyrate	126.7	(.469)		Cyclooctene	84.6	(.769)	(.654) (20°)
p-tert-Butylphenol	108.0	(.719)	(.653) (114°)	Cyclooctatetraene	(53.9)	.518	
Butyl sulfide	(113.7)	.7774 (27.4°)	(.652) (16°)	Cyclopentane	59.18	.8439	.6290 (20°)
Butyl thiocyanate	(79.38)	.6891 (27.4°)	(.659) (25°)	Cyclopentanecarboxylic acid	73.48	.6446 (30°)	(.677) (30°)
2-Butyne-1,4-diacetate	95.9	(.564)		Cyclopentanone	51.63	.6141 (30°)	(.582) (30°)
2-Butyne-1,4-dibenzoate	169.0	(.561)		Cyclopropane	39.9	(.948)	(.683) (−79°)
2-Butyne-1,4-diol	50.3	(.584)		Cyclopropanecarboxylic acid	45.33	.5271 (30°)	(.569) (30°)
n-Butyraldehyde	46.08	.639₄	(.522) (20°)	p-Cymene	102.8	.766 (20°)	(.656) (20°)
iso-Butyraldehyde	46.38	.643₄	(.511) (20°)	Decalin	106,70	.7718	.6814 (20°)
iso-Butyraldoxine	56.1₂	.644₃	(.576) (20°)	cis-Decalin	(107.0)	.774	(.686) (35°)
n-Butyric acid	55.10	.625	.598 (20°)	trans-Decalin	(107.7)	.779	(.670) (35°)
iso-Butyric acid	56.06	.636₂ (25°)	(.601) (25°)	n-Decane	119.74	(.8416)	(.6143) (20°)
Butyronitrile	49.4	(.715)	(.569) (15°)	1-Deuterio pyrrole	48.75	.716 (20°)	
Butyrylphenylacetylene	(106.4)	.618		Deuteroindene	80.88	.690	(.692) (13°)
Cacodyl	(99.9)	.476	(.689) (15°)	Diacetal	(153.8)	.668	
Cacodylic acid	(79.9)	.579		Di-iso-amylamine	(133.1)	.846	(.649) (21°)
Camphor	(103)	.68	(.67) (25°)	Diazoacetic ester	57	(.50)	(.54) (24°)
Camphoric acid	129.0	(.644)	(.791) (20°)	Dibenzocoronene	289.4	.778	
Camphoric anhydride	(113)	.620	(.740) (20°)	1,2,5,6-Dibenzofluorene	184	.69 ± .03	
n-Caproic acid	78.55	.676₂	(.624) (25°)	3,4,5,6-Dibenzophenan-threne	(203)	.73 ± .03	
Caproylphenylacetylene	(130.4)	.651		Dibenzphenanthrone	200.5	.716	
n-Caprylic acid	101.60	.7053	(.642) (20°)	Dibenzpyrene	213.6	.706	
Carbanilide	134.05	(.6316)	(.783) (20°)	Dibenzpyrenequinone	183.1	.551	
Carbazole	117.4	(.702)		iso-Dibenzpyrenequinone	194.6	.586	
Carbon disulfide	42.2	.554	(.699) (22°)	Dibenzyl ketone	131.70	.626₂	
Carbon tetrabromide	93.73	.2826 (20°)	(.966)	p-Dibromobenzene	(101.4)	.430	.786 (100°)
Carbon tetrachloride	66.60	.433	.691 (20°)	2,3-Dibromo-2-butene-1,4-diol	94.2	(.383)	
Carbon tetraiodide	(136)	.261	(1.13) (20°)	Dibromodichloromethane	81.1 ± .4	(.334)	(.808) (25°)
Carvacrol	(109.1)	.726	(.709) (20°)	1,2-Dibromodiiodoethylene	(140.1)	.320	
Carvone	(92.2)	.614	(.590) (20°)	1,2-Dibromoethylene	(71.7)	.386	(.877) (17.5°)
Cetyl alcohol	(183.5)	.757 (17.5°)	(.619) (50°)	1,2-Dibromo-2-fluoroethane	(78.0)	.379	(.855) (17°)
Cetyl mercaptan	390.4	(1.510)		Dibromo-4-nitrophenol	(167.5)	.564	
Chloral	(67.7)	.459	(.694) (20°)	1,2-Dibromotetrachloroethane	(126.0)	.387	(1.049)
Chloranil	(112.6)	.458		Di-n-butylamine	103.7	(.802)	(.767) (20°)
Chloracetic acid	48.1	(.509)	(.804) (20°)	Di-iso-butylamine	105.7	(.817)	(.609) (20°)
Chloroacetone	(50.9)	.550	(.633) (20°)	Di-sec-butylamine	105.9	(.819)	(.641) (0°)
Chloroacetylchloride	53.7	(.475)	(.710) (0°)	Di-iso-butyl ketone	104.30	.733₃	(.591) (20°)
p-Chloranisole	89.1	(.625)		di-tert-butyl ketone	104.06	.732₄	
Chlorobenzene	69.97	(.6216)	(.688) (20°)	2,6-Di-tert-butyl-4-methyl phenol	165.3	(.750)	
Chlorobenzene diazocyan-ide	65.02	(.393)		2,4-Di-tert-butyl phenol	155.6	(.754)	
Chlorodibromomethane	75.1 + .4	(.361)	(.883) (15°)	Dibutyl phthalate	175.1	(.629)	(.657) (21°)
Chlorodifluoromethane	38.6	.446		Di-iso-butyralacetylene	(125.6)	.738	
1-Chloro-2,3-dihydroxy-propane	(77.9)	.604		Dicetyl sulfide	401.7	(.832)	
Chlorodiphenylmethane	131.9	(.651)		Dichloroacetic acid	58.2	(.451)	(.705) (20°)
Chloroethylene	35.9	.574	(.528) (liq., 15°)	Dichloroacetyl chloride	69.0	(.468)	
Chloroform	59.30	.497	.740 (20°)	o-Dichlorobenzene	84.26	.5734	(.748) (20°)
Chlorofumaric acid	67.02	(.445)					
p-Chloroiodi benzene	99.42	(.417)	(.756) (57°)				
Chloromaleic acid	67.36	(.448)					
Chloromethylstilbene	144.8	(.633)					
α-Chloromaphthalene	107.60	.661	.789 (20°)				
m-Chloronitrobenzene	(74.8)	.475	.638 (48°)				

Compound	$-\chi_M \times 10^6$	$-\chi \times 10^6$	$-K \times 10^6$
m-Dichlorobenzene	83.19	.5661	(.729) (20°)
p-Dichlorobenzene	82.93	.5644	(.823) (20.5°)
1,4-Dichlor-2-butyne	74.2	(.603)	
1,2-Dichloro-1,2-dibromoethane	(108.6)	.423	
1,1-Dichloro-difluoroethylene	60.0	.451	
Dichlorodifluoromethane	52.2	.432	(.642) (−30°)
1,1-Dichloroethylene	49.2	.508	(.635) (15°)
cis-1,2-Dichloroethylene	51.0	.526	(.679) (15°)
trans-1,2-Dichloroethylene	48.9	.504	(.638) (15°)
1,3-Dichloro-2-hydroxy-propane	(80.1)	.621	
Dicyandiamide	44.55	(.530)	(.742) (14°)
Dicyclohexanol acetylene	(151.6)	.682	
Dicyclohexyl	129.31	.7776	.6889 (20°)
1,1-Dicyclohexylnonane	231.98	.7930	.7001 (20°)
Diethanolacetate	(75.3)	.660	
Diethyl acetaldehyde	70.71	.705,	(.576) (20°)
Diethylallylacetophenone	146.2	(.676)	(.663) (16°)
Diethyl allylmalonate	118.8	(.593)	(.602) (14°)
Diethylamine	56.8	(.777)	(.552) (18°)
Diethylcyclohexylamine	(124.5)	.802	(.699) (0°)
Diethyl ethylmalonate	115.2	(.612)	(.614) (20°)
Diethyl ketone	58.1,	.675,	(.551) (19°)
Diethyl ketoxime	68.3,	.6754	
Diethyl malonate	(92.6)	.5782	(.611) (20°)
Diethyl-3-(1-methyl butane) ethyl-malonate	175	(.677)	
Diethyl oxalate	81.71	.5595	(.603) (15°)
Diethyl phthalate	127.5	(.574)	(.645) (25°)
Diethyl sebacate	(177.0)	.685	(.661) (20°)
Diethylstilbestrol	172.0	(.547)	
Diethylstilbestrol dipropionate	265.2	.720	
Diethyl succinate	105.07	.6035	(.628) (20°)
Diethyl sulfate	(86.8)	.563	(.667) (15°)
Diethyl sulfide	(67.9)	.753	(.630) (20°)
Diethyl tartrate	(113.4)	.550	(.662) (20°)
Difluoroacetamide	(41.2)	.433	
1-Difluoro-2-dibromoethane	(85.5)	.382	(.883) (20°)
1,1-Difluoro-2,2-dichloroethyl amyl ether	129.84	(.587)	(.694) (20°)
1,1-Difluoro-2,2-dichloroethyl butyl ether	119.48	(.577)	(.703) (20°)
1,1-Difluoro-2,2-dichloroethyl ethyl ether	96.13	(.537)	(.723) (20°)
1,1-Difluoro-2,2-dichloroethyl methyl ether	80.68	(.489)	(.696) (20°)
1,1-Difluoro-2,2-dichloroethyl propyl ether	107.19	(.555)	(.701) (20°)
Difluoroethanol	(41.3)	.503	
Di-n-heptylamine	171.5	(.805)	
Di-n-hexylamine	148.9	(.803)	
Dihydronaphthalene	(85.1)	.654	(.652) (12°)
o-Dimethoxybenzene	87.39	.6329	(.686) (25°)
m-Dimethoxybenzene	87.21	.6316	(.682) (0°)
p-Dimethoxybenzene	86.65	.6275	(.661) (55°)
o-(2,5-Dimethoxybenzoyl)-benzoic acid	161.0	(.562)	
Dimethoxymethane	(47.3)	.621	(.532)
Dimethylacetophenone	96.8	(.653)	(.645) (16°)
Dimethylallylacetophenone	122.4	(.650)	(.635) (16°)
Dimethylaniline	89.66	(.740)	
2,2-Dimethylbutane	76.24	.8848	.5744 (20°)
2,3-Dimethylbutane	76.22	.8845	.5853 (20°)
2,3-Dimethyl-2-butene	65.9	(.783)	(.557)
Dimethylcyclohexanone	(84.8)	.672	
1,2 and 1,3 Dimethylcyclopentanes	81.31	.8281	.6224 (20°)
2,5-Dimethyl-2,5-dibromo-3-hexine	(135.6)	.506	
Dimethyl diethylketo tetrahydrofurfurane	(116.2)	.753	
2,5-Dimethyl-1-ethylpyrrole	94.61	.768 (20°)	
2,5-Dimethyl-3-ethylpyrrole	93.87	.762 (20°)	
2,5-Dimethylfuran	66.37	.687 (20°)	(.620) (18°)
Dimethyl furazan	57.27	.584	
2,5-Dimethyl-4-heptene	100.6	(.797)	
2,4-Dimethyl-2,4-hexadiene	(78.7)	.714	
2,3-Dimethylhexane	98.77	.8648	.6164 (20°)
2,5-Dimethylhexane	98.15	.8593	.5969 (20°)
3,4-Dimethylhexane	99.06	.8673	.6420 (20°)
2,6-Dimethyl-4 hexanol	116.9	.812	
2,5-Dimethyl-3-hexine-2,5-diol	(103.0)	.724	
Dimethyl isoxazole	59.7	(.615)	
Dimethylketo tetrahydrofurfurane	(68.5)	.600	
Dimethyl malonate	69.69	.5277	(.609) (20°)
1,6-Dimethylnaphthalene	113.3	(.725)	
2,4-Dimethylnonane	134.68	(.862)	(.636) (20°)
3,4-Dimethylnonane	134.70	(.862)	(.647) (20°)
4,5-Dimethylnonane	134.52	(.861)	(.647) (20°)
2,6-Dimethyl-2,6,8-nonatriene	(108.8)	.724	
Dimethyl-2,4-nonatriene	(148.8)	.990	
2,6-Dimethyloctane	122.54	(.861)	(.627) (20°)
3,4-Dimethyloxadiazole	57.17	.583	
Dimethyl oxalate	(55.7)	.472	(.542) (54°)
Dimethyl oxamide	(63.2)	.544	
2,2-Dimethylpentane	86.97	.8680	.5849 (20°)
2,3-Dimethylpentane	87.51	.8733	.6070 (20°)
2,4-Dimethylpentane	87.48	.8732	.5876 (20°)
2,2-Dimethylpropane	63.1	(.875)	(.536) (0°)
2,5-Dimethyl-3-propyl-pyrrole	106.07	.773 (20°)	
2,4-Dimethylpyrrole	69.64	.732 (20°)	(.679) 14°)
2,5-Dimethylpyrrole	71.92	.756 (20°)	(.707) (20°)
α,ω-Dimethyl styrene	(90.7)	.686	
Dimethyl succinate	81.50	.5581	(.625) (18°)
Dimethyl sulfate	(62.2)	.493	(.657) (20°)
Dimethyl sulfide	(44.9)	.723	(.612) (21°)
Dimethyltrichloromethylcarbinol	(105)	.59	
N,N-Dimethyl urea	55.1	(.625)	(.784)
N,N′-Dimethyl urea	56.3	(.639)	(.730)
o-Dinitrobenzene	65.98	.3921	(.614) (17°)
m-Dinitrobenzene	70.53	.4197	(.659) (0°)
p-Dinitrobenzene	68.30	.4064	(.660) (30°)
2,4-Dinitrophenol	(73.1)	.397	(.668) (24°)
Dinitroresorciniol	(62.4)	.312	
1,4-Dioxane	52.16	.592 (32°)	(.606) (32°)
Diphenyl	103.25	.6695	(.664) (73°)
1,1-Diphenylallyl-3-chloride	146.1	(.639)	
1,3-Diphenylallyl-3-chloride	140.7	(.615)	
Diphenylamine	(109.7)	.648	.686 (55.5°)
Diphenyl-bis-diazocyanide	85.03	(.327)	
Diphenylbutadiene	129.6	(.629)	
Diphenylchloroarsine	(145.5)	.550	(.871) (40°)
Diphenyldecapentaene	180.5	(.635)	
Diphenyldiacetylene	(134.6)	.640	
Diphenyldiazomethane	115	(.592)	
Diphenyldihydrotetrazine	129.9	(.545)	
1,1-Diphenylethylene	(118.0)	.655	(.680) (14°)
1,6-Diphenylhexane	171.81	.7208	.6877 (20°)
Diphenylhexatriene	146.9	(.632)	
Diphenylmethane	(115.7)	.688	.684 (35.5°)
Diphenylmethanol	119.1	.647	
1,1-Diphenylnonane	206.32	.7357	.6935 (20°)
Diphenyloctatetraene	164.3	(.636)	
Diphenylphenoxyarsine	(225.2)	.567	
N,N-Diphenyl urea	126.3	(.595)	(.759)
N,N′-Diphenyl urea	127.5	(.600)	(.743) (20°)
Di-n-propyl ketone	80.45	.705,	(.576) (20°)
Di-iso-propyl ketone	81.14	.711,	(.573) (20°)
Dipropyl oxalate	105.27	.6046	(.628) (o°)
Di-iso-propyl oxalate	106.02	.6089	
Dodecyl alcohol	147.70	.7849 (20.7°)	(.652) (24°)
Dulcitol	112.40	.617	(.905) (15°)
Elaidic acid	204.8	(.725)	(.619) (79°)
Erythritol	73.80	.604	(.876) (20°)
Ethane	37.3,	(.910)	(.511) (−100°)
4-Ethoxy-3-methoxybenzyl acetate	138.5	.619	
4-Ethoxy-3-methoxybenzyl benzoate	177.3	.620	
1-Ethoxynaphthalene	119.9	(.696)	(.738) (20°)
2-Ethoxynaphthalene	119.2	(.692)	(.734) (25°)
Ethyl acetate	54.10	.614	.554 (20°)
Ethyl acetoacetate	71.67	.550,	(.565) (20°)
Ethylacetophenone	95.5	(.644)	(.639) (16°)
Ethyl alcohol	33.60	.728	.575 (20°)
Ethylally acetophenone	122.5	(.651)	(.634) (16°)
Ethyl amylpropiolate	(112.7)	.670	
Ethylaniline	89.30	(.737)	(.709)
9-Ethyl anthracene	153.0	(.741)	(.771) (99°)
Ethylbenzene	77.20	.7272	.6341 (20°)
Ethyl benzoate	93.32	.621,	(.648) (25°)
Ethyl benzoylacetate	(115.3)	.600	(.673) (20°)
Ethyl benzylidenecyanoacetate	(116.3)	.578	
Ethyl benzylmalonate	(154.5)	.6172	(.663) (20°)
Ethyl bromide	54.70	.502	.719 (20°)
Ethyl bromoacetate	(82.8)	.496	(.747) (20°)
Ethyl-1-isobutylacetoacetate	(121.4)	.652	
Ethyl butylmalonate	139.3	.644,	(.629) (20°)
Ethyl-n-butyrate	(77.7)	.6693	(.585) (25°)
Ethyl-iso-butyrate	78.32	.674;	(.583) (25°)
Ethyl chloroacetate	(72.3)	.590	(.684) (20°)
Ethyl cinnamate	(107.5)	.610	(.640) (20°)
Ethyl-iso-cyanate	(45.6)	.642	(.582) (16°)
Ethyl cyanoacetate	(67.3)	.595	(.632) (20°)
Ethylcyclohexane	91.09	.8118	.6324 (20°)
Ethyldiallylacetophenone	147.4	(.646)	(.636) (16°)
Ethyl dibromocinnamate	174.5	.519	
Ethyl dicloracetate	85.2	(.543)	(.698) (20°)
Ethyl diethylacetoacetate	(117.9)	.6328	(.615) (20°)
Ethyl diethylmalonate	(140.4)	.6492	(.641) (20°)
Ethyl dithiolacetate	(71.0)	.5904 (27.4°)	
Ethylene	12.0	(.428)	(.242) (−102°)
Ethylene	15.30	.546 (32°)	(.309) (−102°)
Ethylene bromide	78.80	.419	.915 (20°)
Ethylene chloride	59.62	.602 (32°)	(.757) (20°)
Ethylenediamine	46.26	.771 (32°)	(.686) (20°)
Ethylene iodide	104.7	.371 (32°)	(.791) (10°)

Compound	$-\chi_M \times 10^6$	$-\chi \times 10^6$	$-K \times 10^6$	Compound	$-\chi_M \times 10^6$	$-\chi \times 10^6$	$-K \times 10^6$
Ethylene oxide	30.7	(.697)	(.618) (7°)	n-Hexyl benzene	124.23 (20°)	(.767)	(.658) (20°)
Ethyl ether	55.10	.743	.534 (20°)	n-Hexyl methyl ketone	91.4₂	.713₁	(.583)
Ethyl ethylacetoacetate	93.9	.5937	(.582) (20°)	n-Hexyl methyl ketoxime	102.5₈	.716₂	(.634) (20°)
Ethyl ethylbutylmalonate	(163.3)	.6683	(.650) (20°)	Hexylpropiolanide	(103.7)	.677	
Ethyl ethylpropylmalonate	(152.4)	.6619	(.648) (20°)	Hydrinidene	(78.5)	.664	(.639) (16°)
Ethyl formate	43.00	.580	.531 (20°)	Hydroquinone	64.63	.587	(.797) (20°)
Ethyl hexylpropiolate	129.9	.713		Hydroxyazobenzene	(99.7)	.503	
Ethyl hydroxylamine	(4370)	.704		p-Hydroxybenzaldehyde	66.8	(.547)	(.618) (130°)
Ethylidene chloride	(57.4)	.580	(.681) (20°)	4-Hydroxy-2-butanone	48.5	.55	(.573) (14°)
Ethyl iodide	(69.7)	.4470 (17.5)	(.884) (20°)	Indene (natural)	84.79	(.730)	(.723) (25°)
Ethyl iodoacetate	(97.6)	.456	(.829) (13°)	Indene (synthetic)	80.89	(.696)	(.690) (25°)
Ethyl lactate	(72.6)	.615	(.633) (25°)	Indole	85.0	(.726)	
Ethyl methylacetoacetate	(81.9)	.5684	(.569) (20°)	Iodobenzene	92.00	.451	.826 (20°)
Ethyl methyl ketoxime	57.3₂	.658₀	(.530) (20°)	Iodoform (in sol'n)	117.1	.2974 (20°)	(1.192) (17°)
Ethyl-1-methyl-2-oxocy-clohexane-carboxylate	112.1	(.608)		1-Iodo-2-phenylacetylene	(110.1)	.483	
Ethyl methylphenylmalon-ate	(153.2)	.6121	(.658) (20°)	o-Iodotoluene	(112.3)	.5145 (30°)	(.874) (20°)
				m-Iodotoluene	(112.3)	.5152 (30°)	(.875) (20°)
Ethyl nitrophenylpropiolate	114.6	.523		p-Iodotoluene	101.31	(.465)	(.780) (40°)
Ethyl oxamate	62.0	(.529)	(.427) (19°)	Leucine	84.9	(.647)	
Ethyl perfluor-n-butyrate	103.5	(.427)		iso-Leucine	84.9	(.647)	
Ethyl phenylacetate	104.27	(.635)	(.656) (20°)	Maleic acid	49.71	(.428)	(.681) (20°)
Ethyl phenylmalonate	(142.2)	.6017	(.659) (20°)	Maleic anhydride	(35.8)	.365	(.341) (20°)
Ethyl phenylpropiolate	(104.2)	.598	(.636) (13°)	Malonic acid	(46.3)	.4453	(.726) (15°)
Ethyl phosphate	(98.2)	.539	(.576) (20°)	Mannitol	111.20	.610	(.908) (20°)
Ethyl propionate	(66.5)	.6514	(.584) (15°)	Mannose	102.90	.571	(.879)
Ethyl propylacetoacetate	(105.7)	.6135	(.593) (20°)	Mesitylene	92.32	.7682 (20°)	(.665) (20°)
Ethyl-n-propyl ketone	69.03	.689₁	(.560) (22°)	Methane	13.2₇	(.765)	
Ethyl succinimide	(72.0)	.566		Methione	91.0	(.610)	
Ethyl sulfide	(67.0)	.631		p-Methoxyazobenzene	(118.9)	.560	
Ethyl sulfite	(75.4)	.546	(.604) (0°)	o-Methoxybenzaldehyde	76.0	(.558)	(.632) (20°)
Ethylsulfone ethyl ether	(81.8)	.592		p-Methoxybenzaldehyde	78.0	(.572)	(.642) (20°)
Ethyl thiocyanate	(55.7)	.6392 (27.4°)	(.637) (25°)	o-Methoxybenzyl alcohol	87.9	.637	(.664) (25°)
Ethyl isothiocyanate	(59.0)	.6772 (27.4°)	(.680) (15°)	I-Methoxynaphthalene	107.0	(.676)	(.741) (14°)
Ethyl thiolacetate	(62.7)	.6019 (27.4°)	(.586) (25°)	2-Methoxynaphthalene	107.6	(.680)	
Ethyl thionacetate	(63.5)	.6098 (27.4°)		1-(o-Methoxyphenylazo)-2-naphthol	163.6	(.588)	
Ethyl tribromoacetate	(119.5)	.368	(.821) (20°)	Methoxysaligenin acetate	110.3	.613	
Ethyl trichloracetate	(99.6)	.520	(.719) (20°)	Methyl acetate	42.60	.575	.537 (20°)
N-Ethyl urea	55.5	(.630)	(.764) (18°)	Methyl acetoacetate	59.60	.513₂	(.553) (20°)
Ethyl iso-valerate	(91.1)	.700	(.607) (20°)	Methylacetylacetone	(65.0)	.569	
Eucalyptol	(116.3)	.754	(.699) (20°)	Methyl alcohol	21.40	.668	.530 (20°)
Eugenol and iso-eugenol	(102.1)	.622	(.663) (20°)	Methylallylaketone	111.9	(.1.330)	
Flavanthrone	241.0	.590		Methylamine	(27.0)	.870	(.608) (−11°)
Fluorene	110.3	.655		N-Methylaniline	82.7₄	(.773)	(.762) (20°)
Fluorenone	99.4	.552	(.623) (100°)	9-Methylanthracene	146.5	.762	(.812) (99°)
Fluorobenzene	(58.4)	.608	(.623) (20°)	Methyl benzoate	81.59	.599₃	(.651) (25°)
Fluorobromacetic acid	(59.5)	.379		Methyl-o-benzoylbenzoate	139.4	(.580)	(.690) (19°)
Fluorodichloromethane	48.8	.171	(.676) (0°)	Methylbenzylaniline	(132.2)	.670	
p-Fluorophenetole	(88.0)	.628		2-Methylbutane	64.40	.8925	.5531 (20°)
Fluoro trichloroethylene	72.5	.485	(.742) (25°)	2-Methyl-2-butene	54.14	(.772)	(.516) (13°)
Fluorotrichloromethane	58.7	.427	(.638) (17°)	Methyl butyl ketone	69.1	.690	(.563) (15°)
Formaldehyde	(18.6)	.62	(.51) (−20°)	Methyl iso-butyl ketone	(69.3)	.692	(.554) (20°)
Formamide	(21.9)	.486	(.551) (20°)	Methyl tert-butyl ketone	(70.4)	.703	(.562) (16°)
Formic acid	19.90	.432	.527 (20°)	p-Methyl-o-tert-butyl-phenol	120.3	(.732)	
β-Formylpropionic acid	55.3	(.542)		Methyl butyrate	(66.4)	.6498	(.588) (16°)
Fructose	102.60	.570		o-Methylcarbanilde	154.0	(.681)	
Fulvene (Bensene χM meas-ured to be 49)	42.9	(.549)	(.452) (20°)	Methyl chloride	(32.0)	.633	
Fulvene χ 54.8/49	48.0	(.614)	(.505) (20°)	Methyl chloroacetate	58.1	(.535)	(.661) (20°)
Fumaric acid	49.11	(.423)	(.692) (20°)	Methylcholanthrene	(182)	.68 ± .04	
Furan	43.09	.633 (20°)	(.598) (15°)	3-Methylcholanthrene	194.0	(.723)	
Furfural	47.1	(.490)	(.568) (20°)	Methylcyclohexane	78.91	.8038	.6181 (20°)
Galactose	103.00	.572		2-Methylcyclohexanone	(74.0)	.660	(.610) (18°)
Gallic acid	90.0	(.529)	(.896) (4°)	3-Methylcyclohexanone	(74.8)	.667	(.610) (20°)
Geraniol formate	(119.9)	.658	(.610) (20°)	4-Methylcyclohexanone	(63.5)	.566	(.516) (24°)
Glucose	102.60	.570		Methylcyclopentane	70.17	.8338	.6245 (20°)
D-Glucose	101.5	(.563)	(.869) (25°)	4-Methyl-2,6-di-tert-butyl-phenol	167.6	(.761)	
Glutamic acid	78.5	(.533)					
Glycerol	57.06	.619	.779 (20°)	Methyl dichloracetate	73.1	(.511)	
Glycine	40.3	(.537)	(.846) (50°)	Methyldiphenoxyphosphine oxide	(152.9)	.616	
Glycol	38.80	.624	.698 (20°)				
Gluaiscol	(79.2)	.638	(.720) (21°)	Methyldiphenyltriazine	(155.1)	.627	
Helianthrone	189.9	.497		Methylene bromide	65.10	.375	.935 (20°)
1,2-Heptadiene	73.5	(.764)		Methylene chloride	(46.6)	.549	(.733) (20°)
2,3-Heptadiene	72.1	(.749)		Methylene iodide	93.10	.348	1.156 (20°)
Heptaldehyde	81.02	.709₆	(.603) (20°)	Methylene succinic acid	57.57	(.443)	(.723)
n-Heptane	85.24	.8507	.5817 (20°)	Methyl ether	26.3	.571	
4-Heptanol	91.5	.789	(.647) (20°)	Methylethylallyacetophen-one	133.3	(.659)	(.643) (16°)
n-Heptanoic acid	88.60	.680	.626 (20°)				
n-Heptyl amine	93.1	(.808)	(.628) (20°)	Methyl ethyl ketone	45.5₈	.632₁	(.509) (20°)
n-Heptyl benzene	134.41	.7625	.6528 (20°)	Methyl formate	(32.0)	.5327	(.519) (20°)
Heptyl cyclohexane	147.40	.8084	.6559 (20°)	Methylfurmaic acid	56.98	(.438)	(.642)
n-Heptylic acid	89.74	.6900	(.630) (25°)	3-Methylheptane	97.99	.8580	.6056 (20°)
I-Heptyne	77.0	(.801)	(.584) (20°)	2-Methyl-4-heptene	88.0	(.784)	
2-Heptyne	79.5	(.826)	(.615) (25°)	5-Methyl-1,2-hexadiene	73.6	(.765)	(.553) (19°)
Hexabromoethane	(148.0)	.294	(1.124) (20°)	2-Methylhexane	86.24	.8607	.5841 (20°)
Hexachlorobenzene	(147.5)	.518	(1.059) (24°)	Methyl hexyl ketone	(93.3)	.728	(.596) (20°)
Hexachloroethane	(112.7)	.476	(.995) (20°)	Methyl-m-hydroxybenzoate	88.4	(.581)	
Hexachlorohexatrione	(145.0)	.433		Methyl-p-hydroxybenzoate	88.7	(.583)	
n-Hexadecane	187.63	.8286	.6421 (20°)	Methyl iodide	(57.2)	.403	(.918) (20°)
1,5-Hexadiene	(55.1)	.671	(.462) (20°)	Methylmaleic acid	57.84	(.446)	(.721)
2,3-Hexadiene	60.9	(.741)		9-Methyl-10-methoxyan-thracene	158.1	(.711)	
n-Hexaldehyde	69.40	.693₀					
2,2,4,7,9,9-Hexamethylde-cane	191.52	.8458	.6596 (20°)	Methyl-o-methoxybenzoate	95.6	(.575)	(.665) (19°)
Hexamethyl dialoxane	118.9	.7324		Methyl-p-methoxybenzoate	98.6	(.593)	
Hexamethylene glycol	84.30	.713		Methyl-α-methoxy-isobu-tyrate	(81.9)	.620	
n-Hexane	(74.6)	.8654 (27.4°)	(.565)				
Hexene	65.7	(.781)		1-Methylnaphthalene	102.8	(.723)	(.741) (14°)
Hexestrol	(165)	.61 ± .02		2-Methylnaphthalene	102.6	(.722)	(.743) (20°)
n-Hexyl alcohol	79.20	.774	.637 (20°)	4-Methylnonane	121.39	(.853)	(.625) (20°)

Compound	$-\chi_M \times 10^6$	$-\chi \times 10^6$	$-K \times 10^6$
5-Methyl-5-nonene	111.6	(.796)	
4-Methyloctane	109.63	(.855)	(.618) (20°)
Methylol urea	48.3	(.493)	
2-Methylpentane	75.26	.8734	.5707 (20°)
3-Methylpentane	75.52	.8764	.5823 (20°)
4-Methyl-2-pentanol	80.4	.788	(.641) (20°)
Methyl perfluor-n-butyrate	92.5	(.406)	
Methyl phenylacetate	92.73	(.618)	(.645) (16°)
Methyl phenylpropriolate	95.6	.597	
2-Methylpropene	44.4	(.791)	
Methyl propionate	(55.0)	.6240	(.571) (20°)
Methyl-n-propyl ketone	57.41	.666,	(.541) (15°)
Methyl-iso-propyl ketone	58.45	.679,	(.545) (20°)
1-Methylpyrrole	58.56	.722 (20°)	(.664) (10°)
2-Methylpyrrole	60.10	.741 (20°)	(.700)
Methyl salicylate	86.30	.567	.668 (20°)
Methyl silicone	(172.7)	.730	
α-Methyl styrene	(80.1)	.678	(.620) (20°)
2-Methylthiazole	59.56	.601 (20°)	(.689) (20°)
2-Methylthiophene	66.35	.676 (20°)	(.689) (20°)
Methyl trichloroacetate	84.2	(.475)	(.707) (19°)
N-Methyl urea	44.6	(.602)	(.725)
Morpholine	55.0	(.631)	(.631)
Myleran	169.7	.69	(.834) (4°)
Myristic acid	176.0	(.7741)	(.661) (60°)
Naphthalaldehydic acid	117.6	(.588)	
Naphthalene	(91.9)	.717	(.821) (20°)
Naphthalene picrate	185.9	(.523)	
2-Naphthalenesulfonylam-ine	127.6	(.616)	
2-Naphthalenesulfonyl chloride	121.91	(.538)	
meso-Naphthodianthrene	214.6	.612	
meso-Naphthodianthrone	221.8	.583	
1-Naphthol	98.2	.681	(.834) (4°)
	97.0	.673	(.819) (4°)
2-Naphthol	98.25	(.682)	(.829) (4°)
α-Naphthonitrile	103.3	(.674)	(.753) (5°)
β-Naphthonitrile	101.0	(.659)	(.721) (60°)
α-Naphthoquinone	73.5	(.465)	(.661)
β-Naphthoquinone	67.9	(.429)	
N-1-Naphthylacetamide	117.8	(.636)	
N-2-Naphthylacetamide	117.8	(.636)	
1-Naphthylamine	98.8	.690	.757 (54°)
2-Naphthylamine	98.00	(.684)	(.726) (98°)
1-Naphthylamine hydro-chloride	(127.6)	.710	
Nicotine	113.328	(.699)	
o-Nitroaniline	66.47	(.481)	(.694) (15°)
m-Nitroaniline	70.09	(.507)	(.725) (20°)
p-Nitroaniline	66.43	(.481)	(.691) (14°)
o-Nitrobenzaldehyde	68.23	.4517	
m-Nitrobenzaldehyde	68.55	.4538	
p-Nitrobenzaldehyde	66.57	.4407	(.507) (0°)
Nitrobenzene	61.80	.502	.604 (20°)
Nitrobenzene diazo cyanide	59.22	(.336)	
o-Nitrobenzoic acid	76.11	.4556	(.718) (4°)
m-Nitrobenzoic acid	80.22	.4802	(.717) (4°)
p-Nitrobenzoic acid	78.81	.4718	(.731) (32°)
o-Nitrobromobenzene	87.3	(.432)	(.700) (80°)
m-Nitrobromobenzene	89.5	(.443)	(.755) (20°)
p-Nitrobromobenzene	89.6	(.444)	(.859) (22°)
m-Nitro carbanilide	148.1	(.576)	
Nitroethane	(35.4)	.472	(.497) (20°)
Nitromethane	21.1	.3457	(.391) (25°)
1-Nitronaphthalene	98.47	(.569)	(.696) (62°)
o-Nitrophenol	73.3	.527 (24°)	(.873) (20°)
m-Nitrophenol	70.8	.509 (25°)	(.756)
p-Nitrophenol	69.5	.500 (22°)	(.740)
1-(m-Nitrophenylazo)-2-naphthol	142.0	(.484)	
1-(p-Nirophenylazo)-2-naphthol	141.7	(.483)	
Nitrophenylfluoroform	(84.1)	.440	
2-Nitropropane	45.73	.5135	(.509) (20°)
Nitrosobenzene	59.1	(.552)	
N-Nitrosodiethylamine	59.3	(.580)	(.546) (20°)
p-Nitrosodiethylaniline	92.6	(.520)	(.644) (15°)
p-Nitrosodimethylaniline	73.3	(.488)	
N-Nitrosodiphenylamine	110.7	(.558)	
1-Nitroso-2-naphthol	83.9	(.485)	
2-Nitroso-1-naphthol	82.7	(.478)	
4-Nitroso-1-naphthol	91.8	(.530)	
m-Nitrosonitrobenzene	66.0	(.433)	
p-Nitrosonitrobenzene	65.8	(.433)	
p-Nitrosophenol	50.7	(.412)	
Nitrosopiperidine	(63.4)	.555	(.590) (20°)
p-Nitrosotoluene	70.4	(.581)	
o-Nitrotoluene	72.28	.5272	(.613) (20°)
m-Nitrotoluene	72.71	.5304	(.614) (20°)
p-Nitrotoluene (in sol'n)	72.06	.5257	(.676) (20°)
n-Nonane	108.13	.8431	.6057 (20°)
1,2-Octadiene	83.6	(.759)	
n-Octane	96.63	.8460	.5949 (20°)
Octanoxime	(102.7)	.717	
Oxtyl alcohol	102.65	.7766 (20°)	(.640) (20°)
Oxtyl chloride	(114.9)	.773	(.676) (20°)
Oxtycyclohexane	158.09	.8051	.6578 (20°)
Oxtylene	(89.5)	.798	(.576) (17°)
Oxtylene bromide	(150.4)	.553	
n-Octyl mercaptan	(115.1)	.7866 (27.4°)	
Oenanthylidene chloride	(116.5)	.689	
Oleic acid	208.5	(.738)	(.661) (18°)
Opianic acid	111.5	(.530)	
Ovalene	353.8	.888	
Oxalic acid (anh.)	33.8	(.375)	
Oxalic acid	60.05	(.4763)	(.787)
Oxamide	(39.0)	.443	(.738)
Palmitic acid	198.6	(.775)	(.661) (62°)
Paraldehyde	(86.2)	.652	(.648) (20°)
Pentabromophenol	(194.0)	.397	
Pentacene	(205.4)	.738	
Pentachloroethane	(99.1)	.490	(.819) (25°)
Pentachlorohexadione	(129.5)	.452	
2,3-Pentadiene	49.1	(.721)	(.501) (20°)
n-Pentane	63.05	.8739	.5472 (20°)
2,4-Pentanediol	70.4	.677	
Perfluoroacetic acid	43.3	(.380)	
Perfluoro-n-butyric acid	81.0	(.378)	
Perfluorobutyric anhydride	149.4	(.387)	
Perfluorocyclooctane oxide	157.6	(.379)	
Perfluoropropionic acid	61.0	(.372)	
Perhydroanthracene	146.01	.7592	.7178 (20°)
Perylene	166.8	.662	
Phenanthrene	(127.9)	.718	(.763) (100°)
Phenanthrenequinone	104.5	.502	(.698)
Phenanthrenonitrile	139.0	(.685)	
o-Phenetidine	(101.7)	.741 (25°)	
p-Phenetidine	(96.8)	.706 (25°)	(.749) (15°)
Phenetole	(84.5)	.692	(.689) (20°)
Phenol	60.21	(.640)	(.675) (45°)
Phenothiazine	114.8	(.576)	
Phenylacetaldehyde	72.01	.599,	(.614) (20°)
Phenyl acetate	82.04	(.603)	(.647) (25°)
Phenylacetic acid	82.72	(.608)	(.657) (80°)
Phenylacetylene	72.01	(.705)	(.655) (20°)
1-Phenylazo-2-naphthol	137.6	(.554)	
2-Phenylbensour	130.5	(.672)	
1	(140.3)	.599	
Phenylbutadiene	(85.7)	.658	
4-Phenyl-1-butene	93.49	(.7077)	(.6239) (20°)
Phenylbutyl acetate	134.5	.653	
Phenyl n-butyrate	105.46	(.643)	
Phenyl iso-cyanate	(72.7)	.610	(.699) (20°)
o-Phenylenediamine	71.98	.6662	
m-Phenylenediamine	70.53	.6529	(.723) (58°)
p-Phenylenedianine	70.28	.6503	
Phenyl ether	(108.1)	.635	(.681) (20°)
Phenylethyl sulfide	(94.4)	.6826 (27.4°)	
Phenylfluoroform	(77.3)	.529	
Phenylhydrazine	67.82	(.627)	(.688) (23°)
Phenylhydroxylamine	(68.2)	.625	
Phenyl mercaptan	(70.8)	.6425 (27.4°)	(.693) (20°)
1-Phenyl-2-Methylbutane	113.53	(.766)	(.660) (20°)
Phenylmethyl sulfide	(83.2)	.6695 (27.4°)	
Phenylpropiolamide	(83.3)	.574	
Phenyl propionate	93.79	(.625)	(.654) (25°)
Phenylsulfone	(129.0)	.591	(.740) (20°)
Phenyl thiocyanate	(81.5)	.6027 (27.4°)	(.677) (24°)
Phenyl isothiocyanate	(86.0)	.6365 (27.4°)	(719) (24°)
1-Phenyl-4,6,6-trimethyl-heptane	173.90	(1796)	(.682) (20°)
N-Phenyl urea	82.1	(.603)	(.785)
Phloroglucinol	(73.4)	.582	
Phthalamide	(91.3)	.556	
Phthalic acid	83.61	.5035	(.802) (20°)
iso-Phthalic acid	84.64	.5097	
tere-Phthalic acid	83.51	.5029	(.759)
Phthalic anhydride	67.31	(.454)	(.694) (4°)
Phthalimide	(78.4)	.533	
Picric acid	84.38	(.368)	(.649)
Piperazine	56.8	(.659)	
Piperidine	64.2	(.754)	(.650) (20°)
Propane	40.5	(.919)	(.538) (−45°)
Propene	31.5	(.749)	(.456) (−47°)
Propionaldehyde	34.32	.591,	(.477) (20°)
Propionic acid	43.50	.586	.582 (20°)
Propionitrile	38.5	(.699)	(.547) (21°)
Propionylphenylacetylene	(95.1)	.601	
Propiophenone	83.73	.624,	(.631) (20°)
n-Propyl acetate	65.91	.645, (25°)	(.569) (25°)
iso-Propyl acetate	67.04	.656,	(.566) (25°)
n-Propyl alcohol	45.176 (20°)	(.7518)	(.6047) (20°)
iso-Propyl alcohol	45.794 (20°)	(.7621)	(.5985) (20°)
9-Propylanthracene	164.0	(.744)	
n-Propylbenzene	89.24	(.742)	(.640) (20°)
n-Propylbenzoate	105.00	(.640)	(.646) (25°)
n-Propl bromide	(65.6)	.533	(.721) (20°)
iso-Propyl bromide	(65.1)	.529	(.693) (20°)
Propyl butyrate	89.4	.6867	(.640) (15°)
prim-Propyl chloride	56.10	.715	.633 (20°)
iso-Propylcyclohexane	102.65	.8131	.6528 (20°)
Propylenediamine	(58.1)	.784	(.688) (15°)
Propylenediamine	(58.1)	.784	(.688) (15°)
Propylene oxide	42.5	(.732)	(.629) (0°)
Propyl formate	(55.0)	.6248	(.563) (20°)
Propyl hexylpropiolate	(136.8)	.697	
Propyl iodide	(84.3)	.4958 (30°)	(.864) (20°)
Propyl propionate (extrap.)	(77.95)	.6711	(.593) (20°)
Propyl sulfide	(92.1)	.7787 (27.4°)	(.634) (17°)

Compound	$-\chi_M \times 10^6$	$-\chi \times 10^6$	$-K \times 10^6$	Compound	$-\chi_M \times 10^6$	$-\chi \times 10^6$	$-K \times 10^6$
N-Propyl urea	67.4	(.660)		Triallylacetophenone	152.5	(.634)	
Pseudocumene	(101.6)	.845 (20°)	(.740) (20°)	Tri-iso-amylamine	192	.845	(.647) (25°)
Pyramidone	149.0	(.645)		Trianilinophosphine oxide	(201.7)	.624	
Pyranthrene	266.9	.709		1,2,3-Tribromopropane	(117.9)	.420	(1.023) (23°)
Pyranthrone	250.3	.616		Tri-iso-butylamine	(156.8)	.846	(.646) (25°)
Pyrazine	37.6	(.469)	(.484) (61°)	Trichloroacetic acid (in sol'n)	73.0	(.44)	(.723) (46°)
Pyrene	147.9	.731	(.933) (0°)	Trichlorobenzene	(106.5)	.587	
Pyridine	49.21	(.622)	(.611) (20°)	Trichlor-tert-butyl alcohol (in sol'n)	98.01	.552	
Pyrocatechol	68.76	.6248	(.857) (15°)	Trichloroethylene	65.8	.501	(.734) (20°)
Pyrrole	47.6	(.709)	(.688) (20°)	Trichloronitromethane	(75.3)	.458	(.756) (20°)
Pyrrolidine	54.8	(.771)	(.657) (23°)	Triethylamine	81.4	(.804)	(.586) (20°)
Quinoline	86.0	(.666)	(.729) (20°)	Triethyl citrate	(161.9)	.586	(.666) (20°)
Quinone	38.4	(.355)	(.468) (20°)	Triethyl phosphate	(125.3)	.688	(.735) (20°)
Quinonoxime	(50.4)	.409		Triethylphosphine	(90.0)	.762	(.610) (15°)
Resorcinol	67.26	.6112	(.785) (15°)	Triethylphosphine oxide	(91.6)	.683	
Rhamnose	99.20	.605	(.890) (20°)	Triethyl phosphite	(104.8)	.631	(.611) (20°)
Safrol and iso-Safro	(97.5)	.601	(.66) (20°)	Triethyl triazinetricarbon-ate	(164.1)	.552	
Salicylaldehyde	64.4	(.527)	(.615) (20°)	Trifluorocresol	(83.8)	.517	
Salicylic acid	72.23	.523	(.755) (20°)	Tri-n-heptylmaine	251.3	(.806)	
Saligenin	76.9	.620	(.720) (25°)	Tri-n-hexylamine	221.7	(.823)	
Salol	(123.2)	.575	.678 (45°)	Trimethylacetophenone	108.2	(.667)	(.648) (16°)
Salvarsan dihydrochloride	(246.1)	.518		2,2,3-Trimethylbutane	88.36	.8818	.6086 (20°)
Selenophene	66.82	.510		2,2,3-Trimethylpentane	99.86	.8743	.6261 (20°)
iso-Selenophene	110-111	.84-.85		2,2,4-Trimethylpentane	98.34	.8610	.5958 (20°)
trans-Selenophene	70-77	.53-.59		2,3,5-Trimethylpyrrole	82.31	.754 (20°)	
Sorbitol	107.80	.592		1,3,5-Trinitrobenzene	74.55	(.350)	(.591) (20°)
Stearic acid	220.8	(.776)	(.657) (69°)	Triperfluorobutylamine	253.0	(.377)	
Stilbene	(120.0)	.666	(.646) (125°)	Triphenoxyarsine	(195.2)	.551	
Stilbestrol	(130)	.62, .63		Triphenylarsine	(177.0)	.578	
Styrene	(68.2)	.655	(.594) (20°)	Triphenylarsine dihydrox-ide	(270.5)	.795	
Succinic acid	(57.9)	.4902	(.767) (15°)	Triphenylarsine oxide	(199.1)	.618	(.708) (20°)
Succinic anhydride	(47.5)	.475	(.524)	Triphenylbismuthine	(196.8)	.447	
Succinimide	(47.3)	.477	(.674) (16°)	Triphenylbismuthine dini-trate	(254.5)	.451	
Sulfamide	44.4	(.462)	(.832)	Triphenylcarbinol	(175.7)	.675	(.802) (20°)
p-Sulfanilamide	80.15	(.465)		Triphenylmethane	(165.6)	.678	.686 (100°)
Terpineol	111.9	(.725)	(.678) (room temp.)	Triphenylphosphine	(166.8)	.636	(.759)
Tetrabenzylmonosilane	266.2	(.678)		Triphenyl phosphite	(183.7)	.592	(.701) (18°)
1,1,2,2-Tetrabromoethane	(123.4)	.357	(1.058) (20°)	Triphenylstilbine	(182.2)	.516	
Tetrabromoethylene	(114.8)	.334		Triphenylstibine dihydrox-ide	(238.5)	.616	
Tetracene	(168.0)	.736					
1,1,2,2-Tetrachloroethane	(89.8)	.535	(.856) (20°)	N,N′,N-Triphenyl urea	176.5	(.613)	
Tetrachloroethylene	81.6	.492	(.803) (15°)	Triquinoyl	(133.0)	.426	
Tetrahydroquinoline	(89.0)	.668	(.715) (4°)	Tropolone	81	.38	
Tetraiodoethylene	(164.3)	.309	(.922) (20°)	Tryptophan	132.0	(.646)	
Tetraiodopyrrole	(188.9)	.331		Tyrosine	105.3	(.581)	
Tetramethylketotetrahy-drothiopene	(104.7)	.736		Undecane	131.84	(.8435)	(.6247) (20°)
Tetranitromethane	43.02	.2195	(.360) (20°)	Urea	33.4	(.556)	(.742) (20°)
Tetraphenylbutadiene	228.0	(.636)		Urethan	(57)	.64	(.63) (21°)
Tetraphenyldecapentaene	280.8	(.643)		iso-Valeraldehyde	(57.5)	.668	(.536) (17°)
Tetraphenylhexatriene	246.4	(.641)		n-Valeric acid	66.85	.6548	(.617) (20°)
Tetraphenyloctatetraene	264.1	(.643)		iso-Valeric acid	(67.7)	.663	(.631) (15°)
Tetraphenyltubene	344.0	(.646)		Valerylphenylacetylene	(119.0)	639	
Tetra-p-tolymonosilane	276.4	(.704)		Valine	74.3	(.634)	
Tetrolic acetal	(97.8)	.688		Violanthrene	273.5	.641	
Tetronic acid	(52.5)	.525		Violanthrone	204.8	.449	
Thiacoumerin	93.6	.577		iso-Violanthrone	215.9	.473	
Thiazole	50.55	.595 (20°)	(.714) (17°)	Water	(13.00)	.7218 (20°)	(.7205) (20°)
Thiobarbituric acid	72.9	(.506)		Water (value usually used as standard)	(12.97)	.720 (20°)	(.719) (20°)
Thiophene	57.38	.682 (20°)	(.726) (20°)	Xanthone	(108.1)	.551	
Tolane	(118.9)	.667	(.644) (100°)	o-Xylene	77.78	.7327	.6440 (20°)
Toluene	66.11	.7176	.6179 (20°)	m-Xylene	76.56	.7212	.6235 (20°)
o-Toluidine	76.0	.710 (24°)	(.709) (20°)	p-Xylene	76.78	.7232	.6226 (20°)
m-Toluidine	74.6	.697 (25°)	(.689) (20°)	Xylose	84.80	.565	(.862) (20°)
p-Toluidine	72.1	.673 (25°)	(.704) (20°)				
α-Tolunitrile	76.87	(.656)	(.666) (18°)				
1-(o-Tolylazo)-2-naphthol	148.7	(.567)					
1-(p-Tolylazo)-2-naphthol	157.6	(.601)					

Supplementary Table

Formula	Compound	$-\chi_M \times 10^6$	$-\chi \times 10^6$	$-K \times 10^6$
$H_6N_3B_3$	Borazine	49.6	.616	
CO	Carbon monoxide	9.8	.350	
COS	Carbon oxysulfide	32.4	.539	
$COCl_2$	Phosgene	48	.485	(.675) (19°)
CO_2	Carbon dioxide	20	.45	
CNCl	Cyanogen chloride	32.4	.527	(.642) (4°)
CCl_2S	Thiophosgene	50.6	.440	(.664) (15°)
CH_2N_2	Cyaniamide	24.8	.590	(.639)
CH_3SiCl_3	Methyl trichlorsilane	87.45	$.584_8$	
CH_4N_2S	Thiourea	42.4	.557	(.782) (20°)
C_2HF_3	Trifluoroethylene	32.2	.393	
$C_2HF_3Cl_2$	1,2-Dichloro-1,1,2-trifluoroethane	66.2	.433	
$C_2HF_3Br_2$	1,2-Dibromo-1,1,2-trifluoroethane	90.9	.376	
$C_2H_2O_2N_3$	Urasole	46.2	.456	
C_2H_4OS	Thioacetic acid	38.4	.505	(.542) (10°)
$C_2H_4O_2S$	Mercaptoacetic acid	50.0	.543	(.720) (20°)
C_2H_4NS	Thioacetamide	42.45	.565	
$C_2H_5SiCl_3$	Ethyl trichlorosilane	98.84	$.604_6$	
$C_2H_6O_2S_2$	Methyl thiosulfite	62.3	.494	
$C_2H_6O_3S$	Methyl sulfite	54.2	.492	
$C_2H_6N_2S$	N-Methyl thiourea	53.6	.595	
C_2H_6S	Ethyl mercaptan	47.0	.756	(.635)

Formula	Compound	$-\chi_M \times 10^6$	$-\chi \times 10^6$	$-K \times 10^6$
$C_2H_6SiCl_2$	Dimethyl dichlorasilane	82.45	$.639_2$	
$C_2H_8N_2 \cdot HCl$	1,2-Diaminoethane hydrochloride	76.2	.789	
C_3H_3ON	Pyruvic nitrile	33.2	.481	
$C_3H_5OF_3$	1-Methoxy-1,1,2-trifluoroethane	55.9	.490	
C_3H_5OCl	Propionyl chloride	51	.55	(.59)
$C_3H_5O_2Br$	Methylbromoacetate	71.1	.465	
C_3H_5Cl	1-Chloro-2-propene	47.8	.625	
C_3H_5Br	1-Bromo-2-propene	58.6	.484	
C_3H_5I	1-Iodo-2-propene	72.8	.433	
$C_3H_6N_6$	Melamine	61.8	.490	(.771) (250°)
C_3H_7N	1-Amino-2-propene	40.1	.702	
$C_3H_7SiCl_3$	n-Propyl trichlorosilane	110.2	$.620_0$	
$C_3H_8O_2N_2$	N-Hydroxymethyl-N-methylurea	68.0	.653	
$C_3H_8N_2S$	N,N-Dimethyl thiourea	64.2	.616	
$C_3H_8N_2S$	N,N'-Dimethyl thiourea	64.7	.621	
$C_3H_8N_2S$	N-Ethyl thiourea	62.8	.603	
C_3H_8S	Propyl mercaptan	58.5	.768	(.642) (25°)
C_3H_9SiCl	Trimethyl chlorosilane	77.36	$.712_2$	
$C_3H_{12}N_3B_3$	N-Trimethyl borazine	78.6	.641	
$C_4H_5ON_3$	Cytosine	55.8	.502	
$C_4H_6O_2$	Vinylacetate	46.4	.539	(.502)
$C_4H_6O_2NCl_3$	Nitro tri(chloromethyl) methane	107.8	.522	
$C_4H_6O_6$	Tartaric acid	67.5	.402	
C_4H_7OCl	Butyryl chloride	62.1	.582	(.598) (20°)
C_4H_7OCl	iso-Butryl chloride	63.9	.599	
$C_4H_8N_2S$	N-Allyl thiourea	69.0	.595	(.725) (20°)
$C_4H_9ON_3$	Acetone semicarbazone	66.29	$.575_8$	(.716) (20°)
$C_4H_{10}O_2S_2$	Ethyl thiosulfite	86.2	.559	
$C_4H_{10}S_2$	Ethyl disulfide	83.6	.684	(.679) (20°)
$C_4H_{10}SiCl_2$	Diethyl dichlorosilane	105.80	$.673_5$	
$C_5H_4O_3N_4$	Uric acid	66.2	.394	(.746)
C_5H_5	Cyclopentadiene	44.5	.673	
$C_5H_6O_2N_2$	Thymine	57.1	.453	
$C_5H_8O_2$	Methyl methacrylate	57.3	.572	(.535) (20°)
$C_5H_8N_2$	3,5-Dimethyl pyrazole	56.2	.585	
$C_5H_8Cl_4$	Tetra (chloromethyl) methane	129	.652	
$C_5H_9S_3Na$	Sodium butyl thiocarbonate	104.5	.555	
$C_5H_9Cl_3$	1,1,1-Tri(chloromethyl) ethane	114	.650	
$C_5H_{10}O$	Cyclopentanol	64.0	.743	(.705) (20°)
$C_5H_{10}O_2$	Ethyl carbonate	75.4	.738	
$C_5H_{10}O_6$	Di-Ribose	84.6	.564	
$C_5H_{10}NS_2Na$	Sodium diethyldithiocarbonate	99.2	.579	
$C_5H_{10}Cl_2$	2,2-Di-(Chloromethyl) propane	96.9	.687	
$C_5H_{11}ON$	Methyl n-propyl ketoxime	68.82	$.679_6$	(.618) (20°)
$C_5H_{11}ON_3$	Ethyl methyl ketone semicarbazone	77.93	$.603_4$	(.710) (20°)
$C_5H_{12}ON_2$	N,N'-Diethyl urea	74.1	.638	
$C_5H_{12}ON_2$	N,N,N',N'-Tetramethyl urea	75.7	.652	(.634) (15°)
$C_5H_{14}O_2Si$	Methyl diethoxysilane	92.99	$.692_4$	
$C_6H_3O_4N_2Cl$	1-Chlor-2,4-dinitrobenzene	84.4	.417	(.708) (22°)
$C_6H_3Cl_2I$	3,4-Dichloro-1-iodobenzene	118	.432	
$C_6H_5SiCl_3$	Phenyl trichlorosilane	120.4	$.569_4$	
C_6H_6NBr	o-Bromoaniline	87.32	.508	
C_6H_6NBr	m-Bromoaniline	84.89	.494	(.780) (20.4°)
C_6H_6NBr	p-Bromoaniline	84.06	.489	(.880)
$C_6H_7O_3NS \cdot H_2O$	Sulfanlic acid	90.1	.471	
C_6H_7N	2-Methyl pyridine	60.3	.648	(.616) (15°)
C_6H_7N	3-Methyl pyridine	59.8	.642	(.617) (15°)
C_6H_7N	4-Methyl pyridine	59.8	.642	(.614) (15°)
$C_6H_7N \cdot HI$	Aniline hydroiodide	113.6	.514	
C_6H_{10}	2,3-Dimethyl-1,3-butadiene	57.2	.696	(.539) (0°)
C_6HuOCl	2-Ethyl butyryl chloride	87.2	.648	
$C_6HuSiCl_3$	Cyclohexytrichlorosilane	138.1	$.634_3$	
$C_6H_{12}N_2S_3$	Tetramethyl thiuran monosulfide	118.4	.568	(.795)
$C_6H_{12}N_2S_4$	Tetramethl thiuran disulfide	140.6	.585	(.755) (20°)
$C_6H_{13}ON$	Methyl-n-butyl ketoxime	79.97	$.694_2$	(.623) (20°)
$C_6H_{13}ON_3$	Methyl-n-propyl ketone semicarbazone	89.47	$.624_9$	(.669) (20°)
$C_6H_{13}ON_3$	Diethyl ketone semicarbazone	90.68	$.633_3$	(.730) (20°)
C_6H_{14}	3-Ethyl pentane	86.21	$.861_0$	(.601) (20°)
$C_6H_{14}O$	iso-Propyl ether	79.4	.777	(.564) (20°)
$C_6H_{14}S_2$	Propyl disulfide	106.2	.707	
$C_6H_{16}O_2Si$	Dimethyl diethoxysilane	104.6	$.705_2$	
$C_6H_{18}O_3Si_3$	Hexamethylcyclotrisolaxane	140.7	$.632_4$	
$C_6H_{18}N_3B_3$	Hexamethylborazine	119	.723	
$C_7H_5O_2Cl$	o-Chlorobenzoic acid	83.56	.534	(.824) (20°)
$C_7H_5O_4N$	Quinoleic acid	72.3	.433	
$C_7H_5NS_2$	2-Mercaptobenzothiazole	99.4	.495	(.843) (20°)
$C_7H_5F_2Cl$	Chlorodifluoromethylbenzene	87.2	.536	
$C_7H_7O_2N$	o-Aminobenzoic acid	77.18	.563	
C_7H_7Cl	o-Chlorotoluene	81.98	.648	(.701) (20°)
C_7H_7Cl	m-Chlorotoluene	80.07	.633	(.679) (20°)
C_7H_7Cl	p-Chlorotoluene	80.07	.633	(.677) (20°)
$C_7H_8N_2S$	N-phenylthiourea	87.4	.574	(.75)
C_7H_9ON	o-Anisidine	80.44	.654	(.714) (20°)
C_7H_9ON	m-Anisidine	79.95	.650	(.712) (20°)
C_7H_9ON	p-Anisidine	80.56	.655	(.702) (55°)
C_7H_9N	2,4-Dimethyl pyridine	71.50	.667	(.633) (0°)

Formula	Compound	$-\chi_M \times 10^6$	$-\chi \times 10^6$	$-K \times 10^6$
C_7H_9N	2,6-Dimethyl pyridine	71.72	.669	(.630) (0°)
$C_7H_{15}ON$	Methyl amyl ketoxime	91.24	.706$_2$	(.630) (20°)
$C_7H_{15}ON_3$	Methyl-n-butyl ketone semicarbazone	100.40	.638$_6$	(.643) (20°)
$C_7H_{16}O$	1-Heptanol	91.7	.790	(.649) (20°)
$C_7H_{18}O_3Si$	Methyl triethoxysilane	120.6	.676$_2$	
C_8H_7OCl	Phenylacetyl chloride	88.5	.572	(.668) (20°)
$C_8H_8O_2$	o-Toluic acid	80.83	.594	(.626) (112°)
$C_8H_{10}ON_2$	N-Methyl, N′Phenyl urea	93.35	.622	
$C_8H_{10}O_4$	Butyne diacetate	95.9	.563	
$C_8H_{11}ON$	m-Phenetidine	90.28	.659	
$C_8H_{11}N$	2,4,6-Trimethyl pyridine	83.22	.687	(.630) (20°)
$C_8H_{16}O_2$	Hexylacetate	100.9	.700	(.623) (0°)
$C_8H_{17}ON_3$	Methyl amyl ketone semicarbazone	112.35	.655$_5$	(.687) (20°)
C_8H_{18}	3-Methyl-3-ethyl pentane	99.9	.875	(.623)
C_8H_{18}	4-Methyl heptane	97.30	.851$_7$	(.614)
C_8H_{18}	3-Ethyl hexane	97.76	.855$_8$	(.614) (20°)
C_8H_{18}	2,3,4-Trimethyl pentane	99.75	.872$_3$	
$C_8H_{18}S_2$	Butyl disulfide	129.6	.727	
$C_8H_{18}S_3$	Butyl trisulfide	144.5	.687	
$C_8H_{18}S_4$	Butyl tetrasulfide	158.5	.653	
$C_8H_{20}O_4Si$	Tetraethoxysilane	137.1	.657$_9$	
$C_8H_{24}O_4Si_4$	Octamethyl cyclotetrasiloxane	187.4	.632$_2$	
C_9H_5N	Phenyl propionitrile	78.2	.615	
$C_9H_6O_2$	Phenyl propiolic acid	81.0	.554	
$C_9H_6O_3$	7-Hydroxycoumarin	88.22	.544$_9$	
C_9H_7N	iso-Quinoline	83.9	.650	(.714) (20°)
$C_9H_{10}O_4N_4$	Acetone-2,4-dinitrophenylhydrazone	110.62	.464$_4$	(.653) (20°)
$C_9H_{11}ON$	Benzylidenaemino-1-hydroxyethane	91.0	.610	
$C_9H_{11}N$	Benzylideneaminoethane	85.5	.642	
$C_9H_{12}O_3S$	Ethyl-p-toluene sulfonate	115	.574	
$C_9H_{12}O_6N_2$	Unidine	106.9	.438	
$C_9H_{13}O_5N_3$	Cytidine	123.7	.509	
$C_9H_{14}O_3$	1-Acetyl-1 ethoxycarbonyl cyclobutane	103.4	.608	
$C_9H_{14}O_4$	1,1-Di(ethoxycarbonyl) cyclopropane	110.4	.593	
$C_9H_{14}Si$	Trimethylphenylsilane	109.1	.726	
$C_9H_{19}ON_3$	Methyl-n-hexyl ketone semicarbazone	123.60	.662$_2$	(.716) (20°)
$C_9H_{20}O$	Di-isobutyl carbinol	116.9	.810	(.667) (0°)
$C_9H_{20}ON_2$	N,N,N′N′ Tetraethylurea	122.4	.710	
$C_9H_{20}N_2S$	N,N,N′,N′-Tetraethylthiourea	132.6	.704	
$C_9H_{24}N_3B_3$	B-Triethyl-N-Trimethyl borzine	146	.706	
$C_{10}H_6O_5N_2$	6,7-Dinitro-7-hydroxy-4-methylcoumarin	111.9	.420$_6$	
$C_{10}H_7O_3Br$	7-Hydroxy-4-methyl-3-bromocoumarin	127.3	.498$_7$	
$C_{10}H_7O_5N$	5-Nitro-6-hydroxy-4-methylcoumarin	105.6	.477$_9$	
$C_{10}H_7O_5N$	6-Nitro-7-hydroxy-4-met-ylcoumarin	105.5	.477$_5$	
$C_{10}H_7O_6N$	8-Nitro-7-hydroxy-4-methylcoumarin	106.0	.479$_8$	
$C_{10}H_8O_3$	6-Hydroxy-4-methylcoumarin	98.69	.560$_7$	
$C_{10}H_8O_3$	7-Hydroxy-4-methylcoumarin	99.96	.563$_7$	
$C_{10}H_8O_4$	Benzylidene malonic acid	97.5	.507	
$C_{10}H_8O_4$	5,7-Dihydroxy-4-methylcoumarin	106.9	.556$_7$	
$C_{10}H_8O_4$	7,8-Dihydroxy-4-methylcoumarin	105.1	.550$_4$	
$C_{10}H_8S$	1-Mercaptonaphthalene	109.5	.684	
$C_{10}H_8S_2$	1,8-Dimercaptonaphthalene	118.0	.614	
$C_{10}H_{10}O_4$	Methyl terephthalate	101.6	.523	
$C_{10}H_{10}O_8$	Bis cyclopentadienyl osmium	193 (20°)	.602	
$C_{10}H_{12}$	1,2,3,4-T etrahydronaphthalene	93.3	.706	(.685)
$C_{10}H_{12}O_4N_4$	Ethyl methyl ketone-2,4-dinitrophenylhydrazone	124.11	.492$_1$	(.631) (20°)
$C_{10}H_{13}O_4N_5$	Adenosine	137.5	.514	
$C_{10}H_{13}O_5N_5$	Guanosine	149.1	.526	
$C_{10}H_{14}$	Durene	101.2	.754	(.632) (81°)
$C_{10}H_{14}$	see-Butyl benzene	101.31	.754$_9$	(.651) (20°)
$C_{10}H_{16}O_4$	1,1-Di(ethoxycarbonyl) cyclobutane	118	.589	
$C_{10}H_{19}N$	Camphylamine	122.0	.796	
$C_{11}H_9BrS$	1-Bromo-4-Methylthionaphthalene	147.0	.581	
$C_{11}H_{10}O_2$	4,6-Dimethylcoumarin	106.2	.610$_3$	
$C_{11}H_{10}O_2$	4,7-Dimethylcoumarin	107.6	.619$_0$	
$C_{11}H_{10}O_3$	5-Hydroxy-4,7-dimethytcoumarin	113.5	.597$_3$	
$C_{11}H_{10}O_3$	6-Methoxy-4-methylcoumarin	109.5	.576$_3$	
$C_{11}H_{10}O_3$	7-Methoxy-4-methylcoumarin	110.5	.581$_4$	
$C_{11}H_{10}S$	Methyl-α-naphthyl sulfide	120.2	.690	
$C_{11}H_{14}O_2$	Benzyl butyrate	116.3	.653	(.663) (17.5°)
$C_{11}H_{14}O_4N_4$	Diethyl ketone-2,4-dinitrophenylhydrazone	135.16	.507$_6$	(.670) (20°)
$C_{11}H_{14}O_4N_4$	Methyl-n-propyl ketone-2,4-dinitrophenylhydrazone	135.70	.509$_7$	(.623) (20°)
$C_{11}H_{16}ON_2$	N,N-Diethyl-N′-phenyl urea	126.4	.657$_3$	
$C_{12}H_6O_4N_2Cl_2$	4,4′-Diechloro-2,2′-dinitro-1,1′-biphenyl	150	.479	
$C_{12}H_8$	Acenaphthylene	111.6	.733	(.659) (16°)
$C_{12}H_8O_4N_2$	2,4′-Dinitro-1,1′-biphenyl	116.5	.477	(.703) (20°)
$C_{12}H_8Cl_2$	4,4′-Dichloro-1,1′-biphenyl	133.1	.497	(.859) (20°)
$C_{12}H_8Br_2$	4,4′-Dibromo-1,1′-biphenyl	151.3	.485	(.920) (20°)
$C_{12}H_9O_2N$	2-Nitro-1,1′-biphenyl	109	.547	(.788) (20°)
$C_{12}H_9O_2N$	5-Nitroacenaphthene	116.0	.582	
$C_{12}H_{10}O_4$	Cinnamylidene malonic acid	105.4	.483	
$C_{12}H_{10}O_4$	Quinhydrone	105	.481	(.674) (20°)
$C_{12}H_{10}S$	Phenyl sulfide	119.2	.640	(.716) (20°)
$C_{12}H_{10}S_2$	Phenyl disulfide	122.5	.561	
$C_{12}H_{10}CbSi$	Diphenyl dichlorosilane	153.5	.606$_3$	

Formula	Compound	$-\chi_M \times 10^6$	$-\chi \times 10^6$	$-K \times 10^6$
$C_{12}H_{12}O_2Si$	Diphenyl silanediol	131.6	.608$_2$	
$C_{12}H_{16}O_2$	Amyl benzoate	128.5	.668	
$C_{12}H_{16}O_4N_4$	Methyl-n-butyl ketone-2,4-dinitrophenylhydra-zone	147.65	.526$_8$	(.643) (20°)
$C_{12}H_{18}$	Hexamethyl benzene	122.5	.755	
$C_{12}H_{18}$	Disopropyl benzene	124.77	.767$_1$	
$C_{12}H_{22}Ou$	Sucrose (saccharose)	189.1	.552	(.877) (15°)
$C_{12}H_{25}Cl_3Si$	Dodecyl trichlorosilane	209.2	.688$_9$	
$C_{12}H_{30}N_3B_3$	Hexaethylborazine	189	.759	
$C_{13}H_8OS$	Thioxanthone	130	.612	
$C_{13}H_{10}O_2$	Phenyl benzoate	117.3	.592	
$C_{13}H_{10}S$	Thiobenzophenone	118.1	.596	
$C_{13}H_{11}NS$	N-Thiobenzoyl aniline	123.0	.577	
$C_{13}H_{12}N_2S$	N,N'-Diphenyl thiourea	136.5	.598	
$C_{13}H_{12}N_2S$	N,N-Diphenyl thiourea	134.9	.591	
$C_{13}H_{18}O_4N_4$	Methyl amyl ketone-2,4-dinitrophenylhydra-zone	156.84	.532$_9$	(.651) (20°)
$C_{14}H_8N_2S_4$	Di-2-benzothiazolyl disulfide	189.0	.568	
$C_{14}H_{10}O$	Anthrone	118	.608	
$C_{14}H_{12}O_2$	Benzyl benzoate	132.2	.622	(.693) (18°)
$C_{14}H_{12}O_2$	Diphenyl acetic acid	124.5	.587	
$C_{14}H_{14}O_2S_2$	p-Toyl thiosulfonate	157	.564	
$C_{14}H_{14}N_2$	p-Azotoluene	135.1	.643	
$C_{14}H_{16}Si$	Dimethyl diphenyl silane	146.6	.690	
$C_{14}H_{20}O_4N_4$	Methyl n-hexyl keton-2,4-dinitrophenylhydra-zone	171.73	.557$_0$	(.655) (20°)
$C_{15}H_{10}O_2$	Flavone	120	.540	
$C_{15}H_{12}$	9-Ethylidene fluorene	124.8	.649	
$C_{15}H_{12}O$	Chalcone	125.7	.604	(.646) (62°)
$C_{15}H_{14}O_2$	Benzyl phenylacetate	143.7	.635	
$C_{15}H_{12}ON_2$	N,N'-Dimethyl-N,N'-Diphenyl urea	148.9	.620	
$C_{16}H_{10}$	Fluoranthene	138.0	.682	
$C_{16}H_{12}$	1-Benzylidene indene	130.5	.716	
$C_{16}H_8$	1,2-ibenzylamine ethane	164	.787	
$C_{16}H_{33}SiCl_3$	Hexadecyl trichlorosilane	252.4	.701$_3$	
$C_{16}H_{48}O_6Si_7$	Hexadecamethyl heptasiloxane	351.1	.658$_8$	
$C_{17}H_{16}$	9-Cutydiene fluorene	147.5	.670	
$C_{18}H_8S_4$	Tetrathioetracene	202	.573	
$C_{18}H_{10}Cl_2$	5,11-Dichloro tetracene	202	.680	
$C_{18}H_{10}Br_2$	,11-Dibromo tetracene	216	.559	
$C_{18}H_{12}$	Triphenylene	156.6	.686	
$C_{18}H_{14}$	o-Diphenyl benzene	150.4	.653	
$C_{18}H_{16}$	p-Diphenyl benzene (Terphenyl)	152	.660	
$C_{18}H_{15}SiCl$	Triphenyl chlorosilane	186.7	.633$_3$	(.814) (0°)
$C_{18}H_{16}OSi$	Triphenyl silanol	176.6	.634$_5$	
$C_{18}H_{16}Si$	Triphenylsilane	174	.668	
$C_{18}H_{29}$	5-Cyclohexyl acenaphthene	165.2	.699	
$C_{18}H_{22}$	Dimesitylene	171.8	.721	
$C_{18}H_{24}O_2$	Estradiol	186.6	.717	
$C_{18}H_{37}SiCl_3$	Octadecyl trichlorosilane	273.7	.705$_7$	
$C_{19}H_{14}O$	4-Benzoyl-1,1'-Biphenyl	158	.612	
$C_{19}H_{18}Si$	Methyl triphenylsilane	186	.678	
$C_{20}H_{14}O_2$	1,4-Dibenzoyl benzene	153	.534	
$C_{21}H_{24}N_3B_3$	B-Trimethyl-N-triphenylborazine	234	.667	
$C_{24}H_{18}$	4,4'-Diphenyl-1,1'-Biphenyl	201.3	.657	
$C_{24}H_{25}i$	Tetraphenyl silane	224	.666	
$C_{24}H_{30}N_3B_3$	B-Triethyl-N-triphenylborazine	264	.693	
$C_{24}H_{40}O_4$	Desoxychlolic acid	272.0	.715	
$C_{24}H_{40}O_5$	Cholic acid	282.3	.691	
$C_{25}H_{20}N_2S$	N,N,N',N'-Tetraphenylthiourea	229.7	.604	
$C_{26}H_{20}$	Tetraphenyl ethylene	217.4	.654	
$C_{28}H_{16}O_2$	Bianthrone (Dianthraquinone)	220	.572	
$C_{28}H_8O_2$	Dianthrone (Diant-ronol)	229	.593	
$C_{28}H_{20}$	1,1,4,4-Tetraphenyl-1,2,3-butatriene	227	.659	
$C_{28}H_{44}O$	Calciferol	273.3	.689	
$C_{28}H_{44}O$	Ergosterol	279.6	.704	
$C_{36}H_{30}O_3Si_3$	Hexaphenyl cyclotrisiloxane	364.1	.612$_9$	
$C_{48}H_{40}O_4Si_4$	Octaphenyl cyclotetrasiloxane	485.3	.613$_2$	
$C_{60}H_{36}$	Hexabenzocoronene	346.0	.457	

DIAMAGNETIC SUSCEPTIBILITY DATA OF ORGANOSILICON COMPOUNDS

R. R. Gupta

The molecular susceptibility is represented by χ_M and expressed in c.g.s. units. For some of the compounds volume or specific susceptibility is available in the literature and in such cases χ_M has been calculated by the relation[1]

$$\chi_M = M \times \chi = M\kappa/\zeta$$

where M is the molecular weight of the substance in grams, χ is the specific or mass susceptibility, κ is the volume susceptibility and ζ is the density of the substance in grams per cubic centimeter.

The compounds are arranged in the increasing order of C, H, Si, O, Cl, Br, I, S, and N.

Compound	Molecular formula	Structural formula	$\chi_M \times 10^{-6}$	Ref.
Methyl trichlorosilane	CH_3SiCl_3	$CH_3-Si-Cl_3$	85.40, 87.45	2, 8
Methyl tribromosilane	CH_3SiBr_3	$CH_3-Si-Br_3$	126.00, 128.00, 155.50	10, 11, 4

Compound	Molecular formula	Structural formula	$\chi Mx-10^{-6}$	Ref.
Methyl dichlorosilane	CH_4SiCl_2	$CH_3-SiH-Cl_2$	66.90	4
Ethyl trichlorosilane	$C_2H_5SiCl_3$	$C_2H_5-Si-Cl_3$	98.00, 98.84, 96.40	2, 9, 7
Ethyl tribromosilane	$C_2H_5SiBr_3$	$C_2H_5-Si-Br_3$	147.50	4
Ethyl triiodosilane	$C_2H_5SiI_3$	$C_2H_5-Si-I_3$	171.50	4
Ethyl dichlorosilane	$C_2H_6SiCl_2$	$C_2H_5-SiH-Cl_2$	78.90	4
Dimethyl dichlorosilane	$C_2H_6SiCl_2$	$(CH_3)_2-Si-Cl_2$	81.30, 82.45	4, 8
Dimethyl diiodosilane	$C_2H_6SiI_2$	$(CH_3)_2-Si-I_2$	131.80	4
Dimethyl silanediol	$C_2H_8SiO_2$	$(CH_3)_2-Si-(OH)_2$	58.40	17, 18
n-Propyl trichlorosilane	$C_3H_7SiCl_3$	$CH_3-(CH_2)_2-Si-Cl_3$	108.00, 110.20	4, 8
Trimethyl chlorosilane	C_3H_9SiCl	$(CH_3)_3-SiCl$	79.00, 73.76	4, 8
Trimethyl bromosilane	C_3H_9SiBr	$(CH_3)_3-Si-Br$	91.30	4
Trimethyl iodosilane	C_3H_9SiI	$(CH_3)_3-Si-I$	104.10	4
1,4,6,9-Tetraoxa-5-silaspiro[4,4]nonane	$C_4H_8SiO_4$	(spiro structure)	80.94	5
Diethyl dichlorosilane	$C_4H_{10}SiCl_2$	$(C_2H_5)_2-SiCl_2$	105.80	8
Tetramethylsilane	$C_4H_{12}Si$	$Si-(CH_3)_4$	74.80	1
Trimethyl methoxysilane	$C_4H_{12}SiO$	$(CH_3)_3-Si-OCH_3$	78.80	1
Dimethyl dimethoxysilane	$C_4H_{12}SiO_2$	$(CH_3)_2-Si-(OCH_3)_2$	81.70, 81.60, 81.95	1, 17, 5
Diethyl silanediol	$C_4H_{12}SiO_2$	$(C_2H_5)_2-Si-(OH)_2$	81.00	17, 18
Methyl trimethoxysilane	$C_4H_{12}SiO_3$	$CH_3-Si-(OCH_3)_3$	85.6	17
Tetramethoxysilane	$C_4H_{12}SiO_4$	$Si-(OCH_3)_4$	83.68	5
Trimethylacetoxysilane	$C_5H_{12}SiO_2$	$(CH_3)_3-Si-OCOCH_3$	85.00, 86.09	16, 5
Trimethylsilylchloromethylamidoxime	$C_5H_{13}SiOClN_2$	$NH_2-C-(CH_2Cl)=N-O-Si-(CH_3)_3$	109.60	15
Trimethylethylsilane	$C_5H_{14}Si$	$C_2H_5-Si-(CH_3)_3$	86.00	2
Trimethylethoxysilane	$C_5H_{14}SiO$	$(CH_3)_3-Si-OC_2H_5$	89.50	1, 17
Phenyl trichlorosilane	$C_6H_5SiCl_3$	$C_6H_5-Si-Cl_3$	120.4	8
Dimethyl diacetoxysilane	$C_6H_{12}SiO_4$	$(CH_3)_2-Si-(OCOCH_3)_2$	98.3, 101.18	16, 5
1,4,6,9-Tetraoxa-2,7-dimethyl-5-silaspiro[4,4]nonane	$C_6H_{12}SiO_4$	(spiro structure)	104.34	5
Trimethylpropionoxysilane	$C_6H_{14}Si-O_2$	$(CH_3)_3-Si-O-COC_2H_5$	99.09	5
Dimethyldiethylsilane	$C_6H_{16}Si$	$(C_2H_5)_2-Si-(CH_3)_2$	97.90	1
Triethylsilane	$C_6H_{16}Si$	$(C_2H_5)_3-Si-H$	95.00	3
Trimethylpropoxysilane	$C_6H_{16}SiO$	$(CH_3)_3-Si-O-(CH_2)_2-CH_3$	101.1	1
Trimethylsilylethylamidoxime	$C_6H_{16}SiON_2$	$(NH_2)-C(C_2H_5)=N-O-Si(CH_3)_3$	101.00	15
Dimethyldiethoxysilane	$C_6H_{16}SiO_2$	$(CH_3)_2-Si-(OC_2H_5)_2$	103.60, 104.70, 104.1, 104.6	1, 18, 5, 8
Trimethylthio-n-propylsilane	$C_6H_{16}SiS$	$(CH_3)_3-Si-S-(CH_2)_2-CH_3$	114.09	12
Trimethylthio-isopropylsilane	$C_6H_{16}SiS$	$(CH_3)_3-Si-S-(CH_3)_2$	115.11	12
Hexamethyldisiloxane	$C_6H_{18}Si_2O$	$(CH_3)_3-Si-O-Si-(CH_3)_3$	118.9	8
Hexamethylcyclotrisiloxane	$C_6H_{18}Si_3O_3$	(cyclic structure)	140.5	8
Methyltriacetoxysilane	$C_7H_{12}SiO_6$	$(CH_3)_3-Si-(OCOCH_3)_3$	112.50	5
Triethylmethylsilane	$C_7H_{18}Si$	$(C_2H_5)_3-Si-CH_3$	109.30	2
Trimethylbutylsilane	$C_7H_{18}Si$	$(CH_3)_3-Si-(CH_2)_3-CH_3$	109.00	2
Trimethylbutoxysilane	$C_7H_{18}SiO$	$(CH_3)_3-Si-O-(CH_2)_3-CH_3$	114.2	5
Trimethylsilyl-n-propylamidoxime	$C_7H_{18}SiON_2$	$NH_2-C-(CH_2-CH_2-CH_3)=N-O-Si-(CH_3)_3$	123.8	15
Methyltriethoxysilane	$C_7H_{18}SiO_3$	$CH_3-Si-(OC_2H_5)_3$	120.3	17
Trimethylthio-n-butylsilane	$C_7H_{18}SiS$	$(CH_3)_3-Si-S-(CH_2)_3.CH_3$	125.73	12
Trimethylthio-tert-butylsilane	$C_7H_{18}SiS$	$(CH_3)_3-Si-S-C-(CH_3)_3$	126.55	12
Trimethyl N-dimethylsilane	$C_7H_{19}SiN$	$(CH_3)_3-Si-N-(CH_3)_2$	91.3	1
2,2-Dimethyl-2-sila-2,3-dihydrobenzothiazole	$C_8H_{11}SiN$	(ring structure)	144.44	12
Tetraacetoxysilane	$C_8H_{12}SiO$	$Si-(OCOCH_3)_4$	129.26	5
Dimethyldiproprionoxysilane	$C_8H_{16}SiO$	$(CH_3)_2-Si-(OCOC_2H_5)_2$	122.0, 123.97	16, 5
1,4,6,9-Tetraoxo-2,3,7,8-tetramethyl-5-silaspiro[4,4]nonane	$C_8H_{16}SiO_4$	(spiro structure)	127.12	5
1,5,7,11-Tetraoxa-2,8-dimethyl-6-silaspiro[5,5]undecane	$C_8H_{16}SiO_4$	(spiro structure)	126.34	5

Compound	Molecular formula	Structural formula	$-\chi_M \times 10^{-6}$	Ref.
Dimethyldi-n-propylsilane	$C_8H_{20}Si$	$(CH_3)_2-Si-[(CH_2)_2-CH_3]_2$	121.70	1
Tetraethylsilane	$C_8H_{20}Si$	$Si-(C_2H_5)_4$	120.30, 120.40, 117.00	1, 4, 5
Dimethyl-di-n-propoxysilane	$C_8H_{20}OSiO_2$	$(CH_3)_2-Si-[O(CH_2)_2-CH_3]_2$	125.70, 126.90	1, 5
Tetraethoxysilane	$C_8H_{20}OSiO_4$	$Si-(OC_2H_5)_4$	83.68, 134.50	5, 17
Bis-(trimethylsilyl)-methylamidoxime	$C_8H_{22}SiO_2N_2$	$(CH_3)_3-Si-NH-C(CH_3)=N-O-Si-(CH_3)_3$	156.60	15
Octamethyltrisiloxane	$C_8H_{24}Si_3O_2$	$(CH_3)_3-Si-O-Si-O-Si(CH_3)_3(CH_3)_2$	165.00	16
Octamethyltetracyclosiloxane	$C_8H_{24}Si_4O_4$		188.00	5

Compound	Molecular formula	Structural formula	$-\chi_M \times 10^{-6}$	Ref.
Tri-n-propylsilane	$C_9H_{22}Si$	$CH_3-(CH_2)_2{}_3-Si-H$	130.00	3
Trimethyl N-(n-propyl)silane	$C_9H_{23}SiN$	$(CH_3)_3-Si-N-[(CH_2)_2-CH_3]_2$	113.4	1
Bis(trimethyl)-ethylamidoxime	$C_9H_{24}Si_2ON_2$	$(CH_3)_3-Si-NH-C(C_2H_5)=N-O-Si(CH_3)_3$	168.00	15
Bis(trimethyl)-phenylamidoxime	$C_{10}H_{16}SiON_2$	$NH_2-C-(C_6H_5)=N-O-Si(CH_3)_3$	141.7	15
2,2-Diacetoxy-4,4,6,6,tetramethyl-1,3-dioxa-2-sila-cyclohexane	$C_{10}H_{20}SiO_4$		150.65	5

Compound	Molecular formula	Structural formula	$-\chi_M \times 10^{-6}$	Ref.
Dimethyl di-n-butylsilane	$C_{10}H_{24}Si$	$(CH_3)_2-Si-[(CH_2)_3-CH_3]_2$	145.80	1
Methyl tri-n-butylsilane	$C_{10}H_{24}Si$	$CH_3-Si-[(CH_2)_2-CH_3]_3$	143.90	1
Dimethyl di-n-butoxysilane	$C_{10}H_{24}SiO_2$	$(CH_3)_2-Si-[O(CH_2)_3CH_3]_2$	147.40, 149.40	1, 5
Dimethyl di-iso-butoxysilane	$C_{10}H_{24}SiO_2$	$(CH_3)_2-Si-[O-CH_2-CH-(CH_3)_2]_2$	150.73	5
Dimethyl di-(1-methylpropoxy)silane	$C_{10}H_{24}SiO_2$	$(CH_3)_2-Si-[O-CH(CH_3)-C_2H_5]_2$	150.83	5
Di-n-propyldiethoxysilane	$C_{10}H_{24}SiO_2$	$[CH_3-(CH_2)_2]_2-Si-(OC_2H_5)_2$	150.80	17, 18
Bis(trimethylsilyl) n-propyl amidoxime	$C_{10}H_{26}SiON_2$	$(CH_3)_3-Si-NH-C(CH_2)_2-CH_3=N-O-Si-(CH_3)_3$	179.00	15
Trimethyl-n-octylsilane	$C_{11}H_{26}Si$	$(CH_3)_3-Si-(CH_2)_7-CH_3$	156.00	2
Trimethyl-n-octoxysilane	$C_{11}H_{26}SiO$	$(CH_3)_3-Si-O(CH_2)_7-CH_3$	149.00	1
Trimethyl N-(di-N-butyl)silane	$C_{11}H_{27}SiN$	$(CH_3)_3-Si-N-[(CH_2)_3-CH_3]_2$	158.5	1
2,2′Spirobis(6-chloro-2,3-dihydrobenzothiazole)silane	$C_{12}H_8SiCl_2S_2N_2$		183.29	12

Compound	Molecular formula	Structural formula	$-\chi_M \times 10^{-6}$	Ref.
2,2′Spirobis(6-bromo-2,3-dihydrobenzothiazole)silane	$C_{12}H_8SiBr_2S_2N_2$		199.37	12

Compound	Molecular formula	Structural formula	$-\chi_M \times 10^{-6}$	Ref.
2,2′Spirobis(2,3-dihydrobenzothiazole)silane	$C_{12}H_{10}SiS_2N_2$		153.71	12

Compound	Molecular formula	Structural formula	$-\chi_M \times 10^{-6}$	Ref.
Diphenylsilanediol	$C_{12}H_{12}SiO_2$	$(C_6H_5)_2-Si-(OH)_2$	131.60	8
Tetrapropionoxysilane	$C_{12}H_{20}SiO_8$	$Si-(OCOC_2H_5)_4$	129.26	5
1,4,6,9-Tetraoxo-2,3,7,8-octamethyl-1-5-silaspiro[4,4]nonane	$C_{12}H_{24}SiO_4$		127.45	5

Compound	Molecular formula	Structural formula	$-\chi_M \times 10^{-6}$	Ref.
1,5,7,11-Tetraoxa-2,4,8,10-hexamethyl-1-6-silaspiro[5,5]undecane	$C_{12}H_{24}SiO_4$		172.48	5

Compound	Molecular formula	Structural formula	$-\chi_M \times 10^{-6}$	Ref.
Dodeconyltrichlorosilane	$C_{12}H_{25}SiCl_3$	$CH_3-(CH_2)_{11}-Si-Cl_3$	210.40, 209.20	2, 8
Dimethyl di-n-pentylsilane	$C_{12}H_{28}Si$	$(CH_3)_2-Si-[(CH_2)_4CH_3]_2$	167.60	1
Tetra-n-propylsilane	$C_{12}H_{28}Si$	$Si-[(CH_2)_2-CH_3]_4$	165.80	1
Tri-n-butylsilane	$C_{12}H_{28}Si$	$H-Si-[(CH_2)_3-CH_3]_3$	172.00	3

DIAMAGNETIC SUSCEPTIBILITY DATA OF ORGANO-SILCON COMPOUNDS (Continued)

Compound	Molecular formula	Structural formula	$-Mx \cdot 10^{-6}$	Ref.
Dimethyl di-n-pentoxysilane	$C_{12}H_{28}SiO_2$	$(CH_3)_2-Si[O-(CH_2)_4-CH_3]_2$	172.37	5
Hexaethoxydisiloxane	$C_{12}H_{30}Si_2O_7$	$(H_5C_2O)_3-Si-O-Si-(OC_2H_5)_3$	217.00	13
Bis-(trimethylsilyl)-phenylamidoxime	$C_{13}H_{24}Si_2ON_2$	$(CH_3)_3-Si-NH-C(C_6H_5)=N-O-Si-(CH_3)_3$	199.50	15
Methyltri-n-butylsilane	$C_{13}H_{30}Si$	$CH_3-Si-[(CH_2)_3-CH_3]_3$	178.50	1
2,2′-Spirobis(6-methyl-2,3-dihydrobenzothiazole)silane	$C_{14}H_{14}S_2N_2$		170.06	12
Diphenyldimethoxysilane	$C_{14}H_{16}SiO_2$	$(C_6H_5)_2-Si-(OCH_3)_2$	156.80	19
Trimethyldodeconylsilane	$C_{15}H_{34}Si$	$(CH_3)_3-Si-(CH_2)_{11}-CH_3$	192.00	2
Trimethyl-n-dodeconoxysilane	$C_{15}H_{34}SiO$	$(CH_3)_3-Si-O-(CH_2)_{11}-CH_3$	205.6	1
Diphenyl diethoxysilane	$C_{16}H_{20}SiO_2$	$(C_6H_5)_2-Si-(OC_2H_5)_2$	179.25	19
Hexadodeconyltrichlorosilane	$C_{16}H_{23}SiCl_3$	$C_{16}H_{33}-Si-Cl_3$	259.00, 252.40	2, 8
Tetra-n-butylsilane	$C_{16}H_{36}Si$	$Si-[(CH_2)_3-CH_3]_4$	202.50	3
Hexa-dodecamethylheptasiloxane	$C_{16}H_{44}O_6$	$(CH_3)_3-Si-(O-Si)_5-O-Si-(CH_3)_3$ $(CH_3)_2$	351.3	8
Triphenylhydroxysilane	$C_{18}H_{16}SiO$	$(C_6H_5)_3-Si-OH$	176.0	8
Dephenyl di-n-propoxysilane	$C_{18}H_{24}SiO_2$	$(C_6H_5)_2-Si-[O(CH_2)_2-CH_3]_2$	201.58	19
Diphenyl di-isopropoxysilane	$C_{18}H_{24}SiO_2$	$(C_6H_5)_2-Si-[OCH-(CH_3)]_2$	202.46	19
n-Octadeconyltrichlorosilane	$C_{18}H_{37}SiCl_3$	$CH_3-(CH_2)_{17}-Si-Cl_3$	278.00, 273.7	2, 8
Diphenyl di-n-butoxysilane	$C_{20}H_{28}SiO_2$	$(C_6H_5)_2-Si-[O-(CH_2)_3-CH_3]_2$	223.62	19
Diphenyl di-iso-butoxy silane	$C_{20}H_{28}SiO_2$	$(C_6H_5)_2-Si-[O-CH_2-CH(CH_3)]_2$	224.46	19
Trimethyl n-octadeconyl silane	$C_{21}H_{46}Si$	$CH_3-(CH_2)_{17}-Si-(CH_3)_3$	278.00	2
Tetraphenylsilane	$C_{24}H_{20}Si$	$Si-(C_6H_5)_4$	212.20	7
Tetrabenzylsilane	$C_{28}H_{28}Si$	$Si-(CH_2-C_6H_5)_4$	266.30	6
Tetra-p-tolylsilane	$C_{28}H_{28}Si$	$Si-(C_6H_4-CH_3)_4$	276.40	6
Other silicon compounds				
Carborundum (silicon carbide)	CSi	$C-Si$	12.80	20
Silane	SiH_4	$Si-H_4$	20.4	13
Silicon dioxide	SiO_2	$O=Si=O$	29.6	11
Tetrachlorosilane	$SiCl_4$	$Si-Cl_4$	87.40, 84.60, 88.30	4, 11, 10
Tetrabromosilane	$SiBr_4$	$Si-Br_4$	123.30	4
Tetraiodosilane	SiI_4	$Si-I_4$	186.20	4
Trichlorosilane	$HSiCl_3$	$H-Si-Cl_3$	70.50, 71.30, 70.60	2, 7, 4
Silicic acid	H_2SiO_3	$O=S=(OH)_2$	33.30	11
Ortho silicic acid	H_4SiO_4	$Si-(OH)_4$	42.60	21
Disilane	Si_2H_6	$H_3-Si-Si-H_3$	37.3	4
Hexachlorodisilane	Si_2Cl_6	$Cl_3-Si-Si-Cl_3$	138.00	7

REFERENCES

1. Abel, E. W., Bush, R. P., Jenkins, C. R., and Zobel, T. *Trans. Faraday Soc.*, 60, 1214, 1964.
2. Abel, E. W. and Bush, R. P. *Trans. Faraday Soc.*, 59, 630, 1963.
3. Dorfman, Y. G. and Lependina, O. L., *Zh. Strukt. Khim.*, 5, 632, 1964.
4. Lister, M. W. and Marson, R., *Can. J. Chem.*, 42(9), 2101, 1964.
5. Mital, R. L. and Gupta, R. R., *J. Am. Chem. Soc.*, 91, 4664, 1969.
6. Asai, K., *Sci. Rep. Res. Inst. Tohoku Univ.*, 2, 205, 1950.
7. Pascal, P., *Compt. Rend.*, 218, 57, 1944.
8. Mathur, R. M., *Trans. Faraday Soc.*, 54, 1577, 1958.
9. Mathur, R. M., *J. Vikram Univ.*, 2, 55, 1958.
10. Nevgi, M. B., *J. Univ. Bombay*, 7(3), 82, 1958.
11. Fox, G., Gorter, C. J., and Smits, L. J., *Tables Constantes Selectioness Diamagnetisme et Para-magnetisme*, Masson et cie, Paris, 1957, 122.
12. Mital, R. L., Goyal, R. D., and Gupta, R. R., *Inorg. Chem.*, 11, 1924, 1972.
13. Barter, C., Meisenheimer, R. G., and Stevenson, D. P., *J. Phys. Chem.*, 64, 1312, 1960.
14. Foex, G., Gorter, C. J., and Smits, L. J., *Compt. Rend.*, 175, 125, 1922.
15. Goel, A. B., and Gupta, V. D., *Indian J. Chem.*, 13(2), 181, 1975.
16. Pacault, A., *Compt. Rend.*, 232, 1352, 1951.
17. Mital, R. L., *Bull. Chem. Soc. Jpn.*, 37(10), 1440, 1964.
18. Mital, R. L., *J. Phy. Chem.*, 68, 1613, 1964.
19. Goyal, R. D., Gupta, R. R., and Mital, R. L., *J. Phys. Chem.*, 76, 1579, 1972.
20. Sigamony, A., *Proc. Indian Acad. Sci.*, 19A, 377, 1944.
21. Pascal, P., *Compt. Rend.*, 175, 814, 1922.

MASS ATTENUATION COEFFICIENTS

Wavelength	Formvar ($C_5H_7O_2$)$_x$	Collodion ($C_{12}H_{11}O_{22}N_6$)$_x$	Polypropylene (CH_2)$_x$	Cellulose acetate ($C_{10}H_{21}O_{15}$)$_x$	Mylar ($C_{10}H_8O_4$)$_x$	Teflon (CF_2)$_x$	Energy (eV)
2.0	14	20	8	19	14	28	6199.0
4.0	113	156	69	150	116	220	3099.5
6.0	372	510	234	489	384	700	2066.3
8.0	850	1140	550	1100	870	1540	1549.8
10.0	1580	2110	1040	2020	1630	2800	1239.8
12.0	2600	3450	1740	3310	2680	4520	1033.2
14.0	3920	5200	2660	4950	4040	6700	885.6
16.0	5600	7300	3830	7000	5800	9400	774.9
18.0	7500	9800	5200	9400	7800	12600	688.8
20.0	9900	12800	6900	12300	10200	2780	619.9
22.0	12500	16200	8800	15500	12900	3540	563.5
24.0	8200	6400	11000	4850	8500	4430	516.6
26.0	10100	7900	13500	5900	10400	5400	476.8
28.0	12000	9400	16200	7100	12400	6500	442.8
30.0	14300	11100	19200	8400	14700	7800	413.3
32.0	16700	8200	22400	9900	17200	9100	387.4
34.0	19300	9500	25900	11400	19900	10600	364.6
36.0	22100	10800	29600	13100	22800	12100	344.4
38.0	25000	12300	33600	14900	25800	13800	326.3
40.0	28200	13900	37800	16800	29100	15600	309.9
42.0	31500	15500	42200	18700	32500	17500	295.2
44.0	3250	4540	1940	4370	3350	6900	281.8
46.0	3640	5100	2180	4900	3760	7800	269.5
48.0	4050	5600	2430	5400	4170	8600	258.3
50.0	4450	6200	2690	6000	4590	9500	248.0
52.0	4910	6800	2960	6600	5100	10500	238.4
54.0	5400	7500	3240	7200	5600	11500	229.6
56.0	5900	8200	3540	7900	6100	12500	221.4
58.0	6400	8900	3860	8600	6600	13600	213.8
60.0	7000	9700	4190	9300	7200	14800	206.6
62.0	7500	10500	4540	10100	7800	16000	200.0
64.0	8100	11300	4880	10900	8400	17300	193.7
66.0	8700	12100	5200	11700	9000	18600	187.8
68.0	9400	13100	5700	12600	9700	19900	182.3
70.0	10000	14000	6000	13500	10300	21300	177.1
72.0	10700	14900	6400	14400	11100	22700	172.2
74.0	11400	15900	6800	15400	11800	24200	167.5
76.0	12200	17000	7300	16300	12500	25700	163.1
78.0	12900	18000	7700	17300	13300	27200	158.9
80.0	13600	19100	8100	18400	14100	28800	155.0
82.0	14500	20200	8600	19500	14900	30500	151.2
84.0	15300	21400	9100	20600	15800	32100	147.6
86.0	16100	22600	9600	21700	16600	33800	144.2
88.0	17000	23700	10100	22900	17500	35500	140.9
90.0	17800	25000	10600	24100	18400	37300	137.8
92.0	18800	26300	11100	25300	19400	39100	134.8
94.0	19700	27600	11700	26500	20300	40900	131.9
96.0	20600	28900	12200	27800	21300	43000	129.1
98.0	21600	30300	12800	29200	22300	44600	126.5
100.0	22600	31700	13400	30500	23300	46300	124.0
105.0	25200	35300	15000	34000	26000	51000	118.1
110.0	27900	39100	16600	37600	28800	56000	112.7
115.0	30700	43000	18200	41300	31600	61000	107.8
120.0	33600	47000	20000	45100	34600	67000	103.3
125.0	36800	52000	21900	49600	38000	72000	99.2
130.0	39800	56000	23900	53000	41100	78000	95.4
135.0	43200	60000	25900	58000	44600	83000	91.8
140.0	46700	65000	28100	63000	48200	89000	88.6
145.0	50000	70000	30300	67000	52000	95000	85.5
150.0	54000	75000	32600	72000	55000	101000	82.7
155.0	57000	80000	35000	76000	59000	106000	80.0
160.0	61000	85000	37600	82000	63000	112000	77.5
165.0	65000	90000	40200	87000	67000	118000	75.1
170.0	69000	96000	42800	92000	72000	124000	72.9
175.0	73000	101000	45400	97000	75000	130000	70.8
180.0	77000	107000	47900	102000	80000	136000	68.9
185.0	82000	113000	51000	108000	84000	141000	67.0
190.0	86000	118000	54000	112000	88000	147000	65.3
195.0	90000	124000	57000	118000	93000	153000	63.6
200.0	95000	130000	61000	124000	98000	159000	62.0

Wavelength	Polystyrene (CH)$_x$	Nylon (C$_{12}$H$_{22}$O$_3$N$_2$)$_x$	Vyns (C$_{22}$H$_{33}$O$_2$Cl$_9$)$_x$	Saran (C$_2$H$_2$Cl$_2$)$_x$	Aluminum oxide Al$_2$O$_3$	Quartz (SiO$_2$)$_x$	Energy (eV)
2.0	9	12	114	164	68	77	6199.0
4.0	74	96	750	1070	480	530	3099.5
6.0	252	318	350	375	1440	1600	2066.3
8.0	590	730	780	820	860	980	1549.8
10.0	1120	1370	1440	1520	1580	1810	1239.8
12.0	1870	2270	2370	2490	2570	2950	1033.2
14.0	2870	3440	3590	3760	3840	4400	885.6
16.0	4120	4910	5100	5400	5500	6200	774.9
18.0	5600	6700	6900	7200	7300	8400	688.8
20.0	7500	8800	9000	9300	9600	11000	619.9
22.0	9500	11100	11400	11800	12100	13800	563.5
24.0	11900	10600	12800	14400	4290	4920	516.6
26.0	14600	13000	15500	17300	5200	6000	476.8
28.0	17400	15500	18400	20400	6300	7200	442.8
30.0	20700	18400	21400	23700	7500	8500	413.3
32.0	24200	17300	24700	27100	8700	9900	387.4
34.0	28000	20000	28100	30800	10100	11300	364.6
36.0	31900	22800	31600	34400	11500	12900	344.4
38.0	36200	25800	35400	38200	13100	14500	326.3
40.0	40700	29100	39100	42100	14700	16200	309.9
42.0	45500	32500	43100	46000	16300	18000	295.2
44.0	2090	2730	25700	37000	18100	19800	281.8
46.0	2350	3060	23600	33600	19900	21700	269.5
48.0	2620	3400	25300	36100	21800	23600	258.3
50.0	2900	3750	27300	38800	23700	25500	248.0
52.0	3190	4130	28900	41100	25700	27500	238.4
54.0	3500	4540	30600	43400	27700	29400	229.6
56.0	3820	4950	32200	45700	29800	31800	221.4
58.0	4160	5400	33900	47900	31800	33700	213.8
60.0	4510	5800	35100	49500	33900	35600	206.6
62.0	4890	6300			35900	37500	200.0
64.0	5300	6800			38000	39900	193.7
66.0	5600	7300			40600	41900	187.8
68.0	6100	7900			42800	44000	182.3
70.0	6500	8400			44900	46000	177.1
72.0	6900	9000			47100	48000	172.2
74.0	7400	9600			49300	50000	167.5
76.0	7800	10200			51000	52000	163.1
78.0	8300	10800			54000	54000	158.9
80.0	8800	11400			56000	56000	155.0
82.0	9300	12100			58000	52000	151.2
84.0	9800	12800			61000	53000	147.6
86.0	10300	13500			62000	56000	144.2
88.0	10900	14200			65000	57000	140.9
90.0	11400	15000			67000	59000	137.8
92.0	12000	15700			69000	61000	134.8
94.0	12600	16500			71000	63000	131.9
96.0	13200	17300			73000	65000	129.1
98.0	13800	18100			75000	67000	126.5
100.0	14500	19000			77000	69000	124.0
105.0	16100	21100			83000	73000	118.1
110.0	17900	23400			79000	78000	112.7
115.0	19600	25800			84000	82000	107.8
120.0	21600	28300			88000	86000	103.3
125.0	23600	31000			93000		99.2
130.0	25700	33600			97000		95.4
135.0	28000	36500			101000		91.8
140.0	30300	39400			106000		88.6
145.0	32700	42400			110000		85.5
150.0	35100	45600			114000		82.7
155.0	37700	48900			118000		80.0
160.0	40500	52000			122000		77.5
165.0	43300	56000			125000		75.1
170.0	46100	59000					72.9
175.0	48900	63000					70.8
180.0	52000	66000					68.9
185.0	55000	70000					67.0
190.0	58000	74000					65.3
195.0	62000	78000					63.6
200.0	65000	82000					62.0

MASS ATTENUATION COEFFICIENTS (Continued)

Wavelength	Stearate CH$_3$(CH$_2$)$_{16}$COO$^-$	Animal proteins C = 52.5 % H = 7 % S = 1.5 % O = 22.5 % N = 16.5 %	Air O = 21 % N = 78 % Ar = 1 %	P 10 CH$_4$ = 10 % Ar = 90 %	Methane CH$_4$	Q Gas C$_4$H$_{10}$ = 1.3 % He = 98.7 %	Energy (eV)
2.0	10	16	21	230	7	1	6199.0
4.0	84	126	148	162	60	12	3099.5
6.0	281	361	481	467	205	41	2066.3
8.0	650	820	1090	1020	479	97	1549.8
10.0	1220	1530	2020	1850	910	185	1239.8
12.0	2030	2530	3310	3010	1520	313	1033.2
14.0	3080	3830	4980	4500	2330	484	885.6
16.0	4410	5500	7100	6400	3350	700	774.9
18.0	6000	7400	9500	8400	4570	970	688.8
20.0	7900	9700	12400	10900	6100	1300	619.9
22.0	10100	12300	15700	13500	7700	1670	563.5
24.0	10000	10300	14100	16400	9700	2110	516.6
26.0	12200	12600	17100	19600	11800	2620	476.8
28.0	14600	15100	20400	22800	14100	3160	442.8
30.0	17300	17800	24000	26300	16800	3780	413.3
32.0	20300	15500	2290	29700	19600	4460	387.4
34.0	23400	17900	2650	33300	22700	5200	364.6
36.0	26800	20400	3040	36900	25900	6000	344.4
38.0	30300	23100	3460	40500	29300	6900	326.3
40.0	34100	26000	3810	37600	33000	7800	309.9
42.0	38200	29100	4270	40900	36900	8800	295.2
44.0	2390	3750	4780	42600	1700	2960	281.8
46.0	2680	4180	5300	45600	1910	3380	269.5
48.0	2980	4610	5900	48900	2130	3840	258.3
50.0	3290	5000	6400	52000	2350	4340	248.0
52.0	3620	5500	6300		2590	4910	238.4
54.0	3970	5900	7000		2840	5500	229.6
56.0	4340	6400	7600		3100	6100	221.4
58.0	4730	6900	8200		3380	6700	213.8
60.0	5100	7500	8900		3660	7400	206.6
62.0	5600	8100	9700		3970	8200	200.0
64.0	6000	8700	10400		4270	8900	193.7
66.0	6400	9300	11200		4570	9700	187.8
68.0	6900	10000	12100		4940	10600	182.3
70.0	7400	10600	12900		5200	11500	177.1
72.0	7900	11300	13800		5600	12500	172.2
74.0	8400	12000	14700		6000	13500	167.5
76.0	8900	11600	15600		6400	14600	163.1
78.0	9500	12300	16600		6700	15600	158.9
80.0	10000	13000	17600		7100	16800	155.0
82.0	10600	13800	18600		7600	18000	151.2
84.0	11200	14600	19700		7900	19300	147.6
86.0	11800	15400	20800		8400	20500	144.2
88.0	12400	16200	21900		8800	21900	140.9
90.0	13100	17000	23100		9300	23300	137.8
92.0	13700	17900	24200		9700	24700	134.8
94.0	14400	18800	25400		10300	26200	131.9
96.0	15100	19700	26700		10700	27800	129.1
98.0	15800	20700	28000		11200	29400	126.5
100.0	16600	21600	29200		11800	31000	124.0
105.0	18400	24100	32700		13100	35400	118.1
110.0	20400	26700	36200		14500	40100	112.7
115.0	22500	29300	39900		15900	44800	107.8
120.0	24600	32200	43800		17500	50000	103.3
125.0	27000	35300	48000		19200	56000	99.2
130.0	29300	38200	52000		20900	62000	95.4
135.0	31800	41500	57000		22700	68000	91.8
140.0	34400	44800	61000		24600	75000	88.6
145.0	37000	48200	65000		26500	82000	85.5
150.0	39800	52000	71000		28500	89000	82.7
155.0	42600	55000	76000		30600	96000	80.0
160.0	45700	59000	80000		32900	104000	77.5
165.0	48600	63000	86000		35100	112000	75.1
170.0	52000	67000	91000		37400	121000	72.9
175.0	55000	71000	96000		39700	130000	70.8
180.0	58000	75000	102000		41900	138000	68.9
185.0	62000	79000	108000		44900	147000	67.0
190.0	65000	83000	114000		47200	157000	65.3
195.0	68000	88000	119000		50000	167000	63.6
200.0	72000	92000	125000		53000	177000	62.0

INTRODUCTION TO X-RAY CROSS SECTIONS

Alex F. Burr

These tables are part of an extensive report published by W. H. McMaster, et al. as UCRL 50174 and available from the National Technical Information Service, Springfield, Va. 22151. Section I of UCRL 50174 describes the data base and the treatment given it. Section II contains the total cross sections between 1 and 1000 keV for all the elements. Section III contains results used in producing Section II, and Section IV contains total cross sections for selected energies and is reproduced in part here. To obtain these values existing experimental x-ray total cross section data and theoretical cross section calculations were surveyed. The coherent (Rayleigh) scattering cross sections and the incoherent (Compton) scattering cross sections were computed. The photo-electric cross sections were obtained by least square fitting of experimental data, theory, and interpolation of experiment and theory. The following table contains cross sections interpolated from the basic compilation at those wavelengths of most use to x-ray crystallographers. The wavelengths chosen were selected to correspond to those given in the International Tables for X-Ray Crystallography. The energy-to-wavelength conversion is given below.

Table I. Energy-to-wavelength conversion

Target radiation	Å	keV		Target radiation	Å	keV
Ag $K\bar{\alpha}$	0.5608	22.105		Ni $K\bar{\alpha}$	1.6591	7.472
$K\beta_1$	0.4970	24.942		$K\beta_1$	1.5001	8.265
Pd $K\bar{\alpha}$	0.5869	21.125		Co $K\bar{\alpha}$	1.7902	6.925
$K\beta_1$	0.5205	23.819		$K\beta_1$	1.6208	7.649
Rh $K\bar{\alpha}$	0.6147	20.169		Fe $K\bar{\alpha}$	1.9373	6.400
$K\beta_1$	0.5456	22.724		$K\beta_1$	1.7565	7.058
Mo $K\bar{\alpha}$	0.7107	17.444		Mn $K\bar{\alpha}$	2.1031	5.895
$K\beta_1$	0.6323	19.608		$K\beta_1$	1.9102	6.490
Zn $K\bar{\alpha}$	1.4364	8.631		Cr $K\bar{\alpha}$	2.2909	5.412
$K\beta_1$	1.2952	9.572		$K\beta_1$	2.0848	5.947
Cu $K\bar{\alpha}$	1.5418	8.041		Ti $K\bar{\alpha}$	2.7496	4.509
$K\beta_1$	1.3922	8.905		$K\beta_1$	2.5138	4.932

Table II. Total Cross Section in cm²/g

Z KEV	1 H	2 He	3 Li	4 Be	5 B	6 C	7 N	8 O	9 F	10 Ne	11 Na	12 Mg	13 Al	14 Si	15 P	16 S	17 Cl	18 Ar	19 K	20 Ca
4.51	432	661	2.10	5.63	12.9	25.6	43.0	64.8	91.8	125	168	220	264	336	386	464	518	577	679	805
4.93	421	550	1.62	4.25	9.69	19.4	32.6	49.4	70.3	95.9	129	171	206	263	304	364	411	456	542	638
5.41	412	465	1.24	3.18	7.23	14.5	24.4	37.2	53.1	72.7	98.5	131	158	203	236	282	322	356	427	500
5.90	405	405	.986	2.45	5.53	11.1	18.7	28.6	41.1	56.4	76.6	102	124	160	186	223	256	282	342	398
5.95	405	400	.964	2.39	5.39	10.8	18.2	27.9	40.0	54.9	74.7	99.6	121	156	182	217	250	276	334	389
6.40	400	362	.798	1.92	4.28	8.55	14.5	22.2	32.0	44.0	59.9	80.2	97.5	126	148	177	205	225	275	319
6.49	400	355	.770	1.84	4.10	8.18	13.9	21.3	30.6	42.2	57.5	77	93.7	121	142	170	197	217	264	307
6.93	397	329	.659	1.52	3.36	6.68	11.3	17.4	25.1	34.7	47.3	63.5	77.5	100	118	141	165	181	222	257
7.06	396	322	.631	1.44	3.17	6.30	10.7	16.5	23.7	32.8	44.7	60.1	73.4	95	112	134	156	172	211	244
7.47	394	303	.555	1.23	2.67	5.28	8.96	13.8	19.9	27.6	37.7	50.8	62.2	80.7	95.2	114	134	147	181	209
7.65	393	297	.528	1.15	2.49	4.92	8.33	12.9	18.6	25.7	35.2	47.4	58.1	75.4	89.1	107	125	137	170	196
8.04	391	284	.477	1.01	2.14	4.22	7.14	11.0	16.0	22.1	30.3	40.9	50.2	65.3	77.3	92.5	109	120	148	171
8.27	390	277	.452	.936	1.98	3.88	6.56	10.1	14.7	20.4	27.9	37.6	46.3	60.2	71.3	85.5	101	111	138	159
8.63	389	268	.417	.837	1.74	3.40	5.74	8.87	12.8	17.9	24.4	33.0	40.7	53.0	62.9	75.4	89.4	97.7	122	141
8.91	388	262	.394	.774	1.59	3.09	5.22	8.06	11.7	16.2	22.2	30.1	37.1	48.4	57.4	68.9	81.8	89.3	112	129
9.57	386	250	.349	.651	1.30	2.49	4.19	6.47	9.36	13.1	17.9	24.2	30.0	39.1	46.6	55.9	66.7	72.7	91.4	106
17.44	373	202	.197	.245	.345	.535	.755	1.07	1.58	2.20	2.94	3.98	5.04	6.5	7.87	9.63	11.6	12.6	16.2	19
19.61	370	197	.187	.222	.293	.429	.605	.855	1.15	1.60	2.10	2.83	3.59	4.62	5.57	6.84	8.26	8.95	11.5	13.6
20.17	369	196	.185	.217	.283	.408	.570	.799	1.07	1.48	1.94	2.60	3.30	4.26	5.13	6.30	7.61	8.24	10.6	12.5
21.13	368	195	.182	.210	.268	.379	.519	.717	.952	1.31	1.70	2.28	2.89	3.71	4.47	5.50	6.63	7.18	9.24	10.9
22.11	366	193	.179	.205	.256	.354	.476	.648	.851	1.16	1.50	2.00	2.54	3.25	3.91	4.82	5.80	6.28	8.08	9.57
22.72	366	192	.177	.201	.249	.340	.452	.610	.795	1.08	1.39	1.86	2.35	3.00	3.61	4.45	5.35	5.79	7.45	8.84
23.82	364	191	.175	.196	.239	.319	.416	.553	.711	.963	1.23	1.63	2.06	2.62	3.15	3.88	4.66	5.04	6.49	7.71
24.94	363	189	.173	.192	.229	.301	.385	.504	.640	.861	1.09	1.44	1.81	2.30	2.76	3.40	4.08	4.41	5.67	6.75

Z KEV	21 Sc	22 Ti	23 V	24 Cr	25 Mn	26 Fe	27 Co	28 Ni	29 Cu	30 Zn	31 Ga	32 Ge	33 As	34 Se	35 Br	36 Kr	37 Rb	38 Sr	39 Y	40 Zr
4.51	819	111	125	143	160	188	206	240	257	280	309	329	368	403	435	464	508	552	599	648
4.93	658	86.8	97.3	111	125	147	161	188	201	220	242	258	288	317	342	366	400	436	473	511
5.41	521	571	75.1	85.7	96.1	113	125	146	155	172	187	200	224	246	266	285	312	339	369	399
5.90	420	459	513	67.4	75.6	88.9	98.4	115	123	137	148	158	178	195	211	226	248	270	294	317
5.95	411	449	501	65.8	73.8	86.8	96.1	113	120	134	144	155	173	190	206	221	242	263	287	310
6.40	339	370	411	462	59.9	70.4	78.3	91.8	97.4	110	117	126	142	156	169	181	198	215	235	254
6.49	327	357	396	445	57.6	67.7	75.3	88.3	93.6	106	113	122	136	150	162	174	191	207	227	244
6.93	276	301	333	375	405	56.3	62.9	73.8	78.1	88.7	94.2	102	114	125	136	146	160	174	190	205
7.06	262	286	316	357	385	53.3	59.6	70.0	74.1	84.3	89.3	96.8	108	119	129	138	152	165	181	195
7.47	226	246	271	307	331	367	50.9	59.8	63.2	72.4	76.2	82.9	92.5	101	110	119	130	141	155	167
7.65	212	231	255	288	311	346	47.7	56.1	59.2	68.0	72.1	77.8	86.8	95.1	104	111	122	132	145	157
8.04	186	202	223	252	273	304	339	48.8	51.5	59.5	62.1	67.9	75.7	82.9	90.3	97	106	115	127	137
8.27	173	188	206	234	253	283	315	45.2	47.7	55.3	57.5	63.0	70.1	77.8	83.7	89.9	98.5	107	118	127
8.63	153	167	183	208	225	253	281	306	42.3	49.2	50.9	56.0	62.2	68.0	74.2	79.8	87.4	94.7	105	113
8.91	141	153	168	191	207	234	259	283	38.7	45.3	46.7	51.4	57.0	62.3	68.1	73.2	80.2	86.8	96.2	103
9.57	116	126	138	157	170	194	214	236	245	37.4	38.1	42.3	46.7	51.0	55.8	60.0	65.7	71.0	79.0	84.8
17.44	21.0	23.3	25.2	29.3	31.9	37.7	41.0	47.2	49.3	55.5	56.9	60.5	66.0	68.8	74.7	79.1	83.0	88.0	97.6	16.1
19.61	15.0	16.7	18.1	21.0	22.9	27.2	29.5	34.2	35.8	40.3	41.7	44.3	48.6	51.2	55.6	58.6	62.1	65.6	72.6	75.2
20.17	13.8	15.4	16.7	19.4	21.1	25.1	27.2	31.6	33.1	37.2	38.6	41.0	45.1	47.6	51.7	54.5	57.8	61.0	67.5	70.1
21.13	12.1	13.4	14.6	17.0	18.5	22.0	23.8	27.7	29.0	32.7	34.0	36.1	39.7	42.1	45.6	48.1	51.1	54.0	59.7	62.1
22.11	10.5	11.8	12.8	14.9	16.2	19.3	20.9	24.3	25.5	28.7	29.9	31.8	35.0	37.2	40.4	42.5	45.3	47.9	53.0	55.2
22.72	9.72	10.9	11.8	13.7	15.0	17.8	19.3	22.5	23.6	26.6	27.7	29.4	32.4	34.6	37.5	39.5	42.1	44.5	49.2	51.4
23.82	8.47	9.48	10.3	12.0	13.1	15.6	16.9	19.7	20.7	23.3	24.3	25.8	28.5	30.5	33.1	34.8	37.2	39.3	43.5	45.5
24.94	7.40	8.30	9.02	10.5	11.5	13.7	14.8	17.3	18.2	20.4	21.4	22.7	25.1	26.9	29.2	30.7	32.9	34.8	38.5	40.4

Z KEV	41 Nb	42 Mo	43 Tc	44 Ru	45 Rh	46 Pd	47 Ag	48 Cd	49 In	50 Sn	51 Sb	52 Te	53 I	54 Xe	55 Cs	56 Ba	57 La	58 Ce	59 Pr	60 Nd
4.51	697	738	786	832	892	928	987	1064	1151	1128	997	753	293	300	324	334	355	383	414	433
4.93	552	585	621	660	708	739	785	842	906	899	926	843	921	683	259	266	282	304	330	344
5.41	432	457	486	518	555	581	617	659	706	709	733	769	835	755	803	587	223	240	261	271
5.90	344	365	387	414	444	466	495	526	561	569	592	617	666	701	742	660	677	521	210	218
5.95	336	357	378	404	434	455	484	514	548	557	579	603	651	685	725	645	662	509	205	213
6.40	276	293	310	333	357	375	398	422	449	460	479	497	535	565	597	615	636	592	450	464
6.49	266	282	299	320	344	361	384	407	433	443	462	479	516	545	575	593	613	571	610	447
6.93	223	237	251	269	289	304	324	342	363	374	391	404	434	459	484	499	519	559	596	532
7.06	212	225	238	256	275	289	308	325	345	356	373	385	413	437	460	475	494	532	567	506
7.47	182	193	204	220	236	249	265	279	295	307	322	331	355	376	395	408	427	459	488	476
7.65	170	181	192	207	222	234	249	262	277	289	303	312	333	353	372	384	402	432	459	418
8.04	149	158	168	181	194	205	218	229	242	253	267	273	292	310	325	336	354	379	402	389
8.27	138	147	156	168	180	190	203	213	225	236	248	254	271	288	302	312	329	352	374	347
8.63	122	130	138	149	160	169	180	189	200	210	221	226	241	256	269	278	294	314	333	320
8.91	112	120	127	137	147	156	166	174	183	193	204	208	222	236	248	256	271	290	307	290
9.57	92.0	98.2	104	113	121	128	137	143	151	159	168	172	183	195	204	211	224	240	253	265
17.44	17.0	18.4	19.8	21.3	23.1	24.4	26.4	27.7	29.1	31.2	33.0	33.9	36.3	37.9	40.4	42.4	45.3	48.6	50.8	53.3
19.61	81.2	13.3	14.3	15.4	16.7	17.6	19.1	20.1	21.2	22.6	23.9	24.7	26.5	27.9	29.5	31.0	33.1	35.5	37.1	38.9
20.17	75.6	79.3	13.2	14.2	15.4	16.3	17.7	18.6	19.6	20.9	22.1	22.8	24.6	25.8	27.3	28.7	30.7	33.0	34.4	36.0
21.13	67.1	70.3	73.0	12.5	13.5	14.3	15.5	16.4	17.3	18.4	19.4	20.1	21.7	22.7	24.1	25.4	27.0	29.1	30.3	31.7
22.11	59.6	62.5	65.0	11.0	11.9	12.6	13.7	14.5	15.3	16.2	17.1	17.7	19.2	20.0	21.3	22.5	23.9	25.7	26.8	28.0
22.72	55.5	58.2	60.6	63.8	11.0	11.6	12.7	13.4	14.1	15.0	15.8	16.4	17.7	18.6	19.8	20.9	22.1	23.9	24.9	26.0
23.82	49.1	51.5	53.7	56.6	58.3	10.2	11.1	11.8	12.4	13.2	13.8	14.4	15.7	16.3	17.4	18.4	19.5	21.0	21.9	22.9
24.94	43.6	45.7	47.7	50.3	52.0	57.0	97.8	10.4	11.0	11.6	12.1	12.7	13.9	14.4	15.4	16.3	17.2	18.6	19.3	20.2

X-RAY CROSS SECTIONS (Continued)

Table II. Total Cross Section in cm²/g (Continued)

Z / KEV	61 Pm	62 Sm	63 Eu	64 Gd	65 Tb	66 Dy	67 Ho	68 Er	69 Tm	70 Yb	71 Lu	72 Hf	73 Ta	74 W	75 Re	76 Os	77 Ir	78 Pt	79 Au	80 Hg
4.51	455	473	503	510	546	568	589	615	644	664	696	720	736	753	796	824	871	934	906	958
4.93	361	375	398	405	432	449	465	485	508	524	545	568	581	598	631	653	688	734	720	760
5.41	285	295	313	319	339	352	363	380	397	410	430	445	455	470	496	512	540	572	568	598
5.90	229	237	251	256	271	281	290	303	317	327	342	355	363	378	397	410	431	455	457	480
5.95	224	232	245	250	265	275	283	296	310	319	335	347	355	369	388	401	422	444	447	469
6.40	186	192	203	207	219	227	234	244	255	263	276	286	293	306	321	332	348	365	371	388
6.49	476	185	195	200	211	219	225	235	246	254	266	276	282	295	310	320	336	351	358	375
6.93	401	412	165	170	179	185	190	198	207	214	225	233	238	250	262	270	283	295	303	317
7.06	536	392	420	161	170	176	181	189	197	204	214	222	227	238	249	257	270	281	289	302
7.47	535	475	361	369	147	152	156	163	170	175	184	191	195	206	215	222	233	241	250	261
7.65	503	454	477	347	367	143	146	153	160	165	173	180	184	194	203	209	219	226	236	246
8.04	441	422	418	427	322	337	128	134	140	145	152	158	162	171	178	184	192	198	208	216
8.27	410	376	450	397	420	313	333	125	131	135	142	147	150	159	166	171	179	184	194	202
8.63	366	347	401	410	375	392	296	318	117	120	125	131	134	142	149	153	160	164	174	180
8.91	336		369	377	400	360	272	292	289	111	117	121	124	132	137	141	148	151	161	167
9.57	278	287	305	312	330	344	364	336	239	232	257	236	103	110	114	118	123	125	134	139
17.44	55.5	58.0	61.2	62.8	66.8	68.9	72.1	75.6	79.0	80.2	84.2	86.3	89.5	95.8	98.7	100	103	109	111	115
19.61	40.5	42.4	44.7	46.0	48.9	50.4	52.8	55.1	57.9	59.2	62.0	64.2	66.1	70.6	72.5	74.1	77.2	80.2	82.3	85.3
20.17	37.6	39.3	41.5	42.6	45.3	46.7	48.9	51.0	53.8	55.0	57.6	59.7	61.4	65.5	67.3	68.9	71.9	74.6	76.5	79.4
21.13	33.1	34.7	36.6	37.6	40.0	41.2	43.2	45.0	47.5	48.7	51.0	52.9	54.4	58.0	59.5	61.1	63.8	66.0	67.8	70.4
22.11	29.3	30.7	32.4	33.3	35.4	36.5	38.3	39.8	42.1	43.2	45.3	46.9	48.3	51.4	52.8	54.2	56.8	58.6	60.2	62.6
22.72	27.1	28.5	30.1	30.9	32.9	33.9	35.6	37.0	39.1	40.2	42.1	43.7	44.9	47.8	49.1	50.5	52.9	54.5	56.0	58.3
23.82	23.9	25.1	26.5	27.3	29.0	29.9	31.4	32.6	34.5	35.5	37.2	38.6	39.7	42.2	43.3	44.6	46.8	48.2	49.5	51.7
24.94	21.1	22.2	23.4	24.1	25.6	26.4	27.8	28.8	30.6	31.5	33.0	34.2	35.2	37.4	38.4	39.6	41.6	42.7	43.9	45.9

Z / KEV	81 Tl	82 Pb	83 Bi	86 Rn	90 Th	92 U	95 Pu
4.51	991	1035	1066	1174	1098	1084	960
4.93	785	820	847	930	993	862	1023
5.41	617	645	667	731	844	774	803
5.90	494	517	536	586	678	672	731
5.95	483	505	524	573	663	657	772
6.40	400	418	435	474	549	545	638
6.49	386	403	419	458	530	526	615
6.93	326	341	355	387	449	446	520
7.06	311	325	338	369	428	425	495
7.47	268	280	293	318	370	368	427
7.65	253	264	276	300	348	347	402
8.04	222	232	243	264	307	306	353
8.27	207	216	227	246	286	285	329
8.63	185	194	203	220	256	256	294
8.91	171	179	188	203	237	236	271
9.57	142	149	156	169	197	197	225
17.44	119	123	126		99.5	96.7	48.8
19.61	88.3	90.6	93.5	101	73.3	72.6	79.0
20.17	82.0	84.1	87.0	93.8	95.0	67.8	73.7
21.13	72.7	74.5	77.2	83.3	97.2	84.2	65.7
22.11	64.6	66.1	68.7	74.1	86.3	86.9	58.8
22.72	60.1	61.5	64.0	69.1	80.3	81.1	76.4
23.82	53.2	54.4	56.7	61.2	71.0	72.1	78.3
24.94	47.2	48.2	50.4	54.4	62.9	64.2	69.8

X-RAY WAVELENGTHS

J. A. Bearden

These tables were originally published as the final report to the U.S. Atomic Energy Commission as Report NYO-10586 in partial fulfillment of Contract AT(30-1)-2543. The tables were later reproduced in *Review of Modern Physics*. The data may also be obtained from the Superintendent of Documents, U.S. Government Printing Office, Washington, D. C. 20402 in the publication NSRDS-NBS 14. Persons seeking discussion of the experimental work, conventions, secondary standards, etc. will find these in *Review of Modern Physics*, Vol. 39, No. 1, 78-124, January 1967.

THE W $K\alpha_1$ WAVELENGTH STANDARD

A wavelength standard should possess characteristics which permit its ready redetermination in other laboratories by different techniques. Considering all of the factors involved in the selection of a wavelength standard, the W $K\alpha_1$ line is superior to any other x-ray or γ-ray wavelength. Its advantages as the x-ray wavelength standard are discussed in *Review of Modern Physics* Vol. 39, page 82 (1967).

$$\lambda W\ K\alpha_1 = 0.2090100\ \text{Å} \pm 5\ \text{ppm}.$$

This numerical value of the wavelength of the W $K\alpha_1$ line is used to define the *x-ray wavelength standard* by the relation

$$\lambda(W\ K\alpha_1) = 0.2090100\ \text{Å}^*.$$

This is a new unit of length which may differ from the angstrom by ± 5 ppm (probable error), *but as a wavelength standard it has no error*. In order to clearly indicate that this unit is not exactly an angstrom, it has been designated Å*.

Wavelengths tabulated normally refer to the pure element in its solid form. However, there are many instances in which such data are not available. For example, rare gases are of necessity almost always used in the gaseous form, while the rare-earth elements were customarily used in the form of salts.

In high precision work there is some ambiguity as to exactly what feature of a line profile should be taken to be the "true wavelength." In double-crystal work the line peak is usually employed. In crystallography the centroid is widely used; in photographic work with visual observation of the plates, there is involved some subjective criterion of the observer which is difficult to define precisely. In this survey the peak of the line profile has been adopted as the standard criterion.

In the study of the X-ray literature, the wavelengths of a number of lines were noted which appeared inconsistent with the remaining data. A Moseley-type diagram was constructed, and if the value was clearly outside estimated probable error, it was assumed that an experimental or typographical error had occurred, and the interpolated value was listed in the table. Such cases are marked with a dagger (†) as a superscript to the wavelength. For elements of atomic number 85 through 89 and 91, there are no measured lines of the K series and very few of other series except for 88 radium and 91 protactinium. Likewise there are very few measurements for 43 technetium and 54 xenon. In these cases, interpolated values are listed for the more prominent lines and marked with a dagger (†).

X-ray wavelengths in Å* units and in keV. The probable error (p.e.) is the error in the last digit of wavelength. Designation indicates both conventional Siegbahn notation (if applicable) and transition, e.g., $\beta_1\ L_{II}M_{IV}$ denotes a transition between the L_{II} and M_{IV} levels, which is the $L\beta_1$ line in Siegbahn notation.

3 Lithium / 4 Beryllium

Designation	Å*	p.e.	keV	Å*	p.e.	keV
$\alpha\ KL$	228.	1	0.0543	114.	1	0.1085

5 Boron / 6 Carbon

Designation	Å*	p.e.	keV	Å*	p.e.	keV
$\alpha\ KL$	67.6	3	0.1833	44.7	3	0.277

7 Nitrogen / 8 Oxygen

Designation	Å*	p.e.	keV	Å*	p.e.	keV
$\alpha\ KL$	31.6	4	0.3924	23.62	3	0.5249

9 Fluorine / 10 Neon

Designation	Å*	p.e.	keV	Å*	p.e.	keV
$\alpha_{1,2}\ KL_{II,III}$	18.32	2	0.6768	14.610	3	0.8486
$\beta\ KM$				14.452	5	0.8579

11 Sodium / 12 Magnesium

Designation	Å*	p.e.	keV	Å*	p.e.	keV
$\alpha_{1,2}\ KL_{II,III}$	11.9101	9	1.0410	9.8900	2	1.25360
$\beta\ KM$	11.575	2	1.0711	9.521	2	1.3022
$L_{II,III}M$	407.1	5	0.03045	251.5	5	0.0493
$L_I L_{II,III}$	376	1	0.0330	317	1	0.0392

13 Aluminum / 14 Silicon

Designation	Å*	p.e.	keV	Å*	p.e.	keV
$\alpha_2\ KL_{II}$	8.34173	9	1.48627	7.12791	9	1.73938
$\alpha_1\ KL_{III}$	8.33934	9	1.48670	7.12542	9	1.73998
$\beta\ KM$	7.960	2	1.5574	6.753	1	1.8359
$L_{II,III}$	171.4	5	0.0724	135.5	4	0.0915
$L_I L_{II,III}$	290.	1	0.0428			

15 Phosphorus / 16 Sulfur

Designation	Å*	p.e.	keV	Å*	p.e.	keV
$\alpha_2\ KL_{II}$	6.160†	1	2.0127	5.37496	8	2.30664
$\alpha_1\ KL_{III}$	6.157†	1	2.0137	5.37216	7	2.30784
$\beta\ KM$	5.796	2	2.1390			
$\beta_1\ KM$				5.0316	2	2.4640
$\beta_2\ KM$				5.0233	3	2.4681
$L_{II,III}M$	103.8	4	0.1194			
$l,\eta\ L_{II,III}M_I$				83.4	3	0.1487

17 Chlorine / 18 Argon

Designation	Å*	p.e.	keV	Å*	p.e.	keV
$\alpha_2\ KL_{II}$	4.7307	1	2.62078	4.19474	5	2.95563
$\alpha_1\ KL_{III}$	4.7278	1	2.62239	4.19180	5	2.95770
$\beta\ KM$	4.4034	3	2.8156			
$\beta_{1,3}\ KM_{II,III}$				3.8860	2	3.1905
$\eta\ L_{II}M_I$	67.33	9	0.1841	55.9†	1	0.2217
$l\ L_{III}M_I$	67.90	9	0.1826	56.3†	1	0.2201

19 Potassium / 20 Calcium

Designation	Å*	p.e.	keV	Å*	p.e.	keV
$\alpha_2\ KL_{II}$	3.7445	2	3.3111	3.36166	3	3.68809
$\alpha_1\ KL_{III}$	3.7414	2	3.3138	3.35839	3	3.69168
$\beta_{1,3}\ KM_{II,III}$	3.4539	2	3.5896	3.0897	2	4.0127
$\beta_5\ KM_{IV,V}$	3.4413	4	3.6027	3.0746	3	4.0325

19 Potassium (Cont.) / 20 Calcium (Cont.)

Designation	Å*	p.e.	keV	Å*	p.e.	keV
$\eta\ L_{II}M_I$	47.24	2	0.2625	40.46	2	0.3064
β_1				35.94	2	0.3449
$l\ L_{II}M_I$	47.74	1	0.25971	40.96	2	0.3027
$\alpha_{1,2}\ L_{III}M_{IV,V}$				36.33	2	0.3413
$M_{II,III}N_I$	692	9	0.0179	525.	9	0.0236

21 Scandium / 22 Titanium

Designation	Å*	p.e.	keV	Å*	p.e.	keV
$\alpha_2\ KL_{II}$	3.0342	1	4.0861	2.75216	2	4.50486
$\alpha_1\ KL_{III}$	3.0309†	1	4.0906	2.74851	2	4.51084
$\beta_{1,3}\ KM_{II,III}$	2.7796	2	4.4605	2.51391	2	4.93181
$\beta_5\ KM_{IV,V}$	2.7634	3	4.4865	2.4985	2	4.9623
$\eta\ L_{II}M_I$	35.13	2	0.3529	30.89	3	0.4013
$\beta_1\ L_{II}M_{IV}$	31.02	2	0.3996	27.05	2	0.4584
$l\ L_{III}M_I$	35.59	3	0.3483	31.36	2	0.3953
$\alpha_{1,2}\ L_{III}M_{IV,V}$	31.35	3	0.3954	27.42	2	0.4522

23 Vanadium / 24 Chromium

Designation	Å*	p.e.	keV	Å*	p.e.	keV
$\alpha_2\ KL_{II}$	2.50738	2	4.94464	2.293606	3	5.40551
$\alpha_1\ KL_{III}$	2.50356	2	4.95220	2.28970	2	5.41472
$\beta_{1,3}\ KM_{II,III}$	2.28440	2	5.42729	2.08487	2	5.94671
$\beta_5\ KM_{IV,V}$	2.26951	6	5.4629	2.07087	6	5.9869
$\beta_{2,4}\ L_I M_{II,III}$	21.19†	9	0.585	18.96	2	0.654
$\eta\ L_{II}M_I$	27.34	3	0.4535	24.30	3	0.5102
$\beta_1\ L_{II}M_{IV}$	23.88	4	0.5192	21.27	1	0.5828
$l\ L_{III}M_I$	27.77	1	0.4465	24.78	1	0.5003
$\alpha_{1,2}\ L_{III}M_{IV,V}$	24.25	3	0.5113	21.64	3	0.5728
$M_{II,III}M_{IV,V}$	337.	9	0.037	309.	9	0.040

25 Manganese / 26 Iron

Designation	Å*	p.e.	keV	Å*	p.e.	keV
$\alpha_2\ KL_{II}$	2.10578	2	5.88765	1.939980	9	6.39084
$\alpha_1\ KL_{III}$	2.101820	9	5.89875	1.936042	9	6.40384
$\beta_{1,3}\ KM_{II,III}$	1.91021	2	6.49045	1.75661	2	7.05798
$\beta_5\ KM_{IV,V}$	1.8971	1	6.5352	1.7442	1	7.1081
$\beta_{2,4}\ L_I M_{II,III}$	17.19	2	0.721	15.65	2	0.792
$\eta\ L_{II}M_I$	21.85	2	0.5675	19.75	4	0.628
$\beta_1\ L_{II}M_{IV}$	19.11	2	0.6488	17.26	1	0.7185
$l\ L_{III}M_I$	22.29	2	0.5563	20.15	1	0.6152
$\alpha_{1,2}\ L_{III}M_{IV,V}$	19.45	1	0.6374	17.59	2	0.7050
$M_{II,III}M_{IV,V}$	273.	6	0.045	243.	5	0.051

27 Cobalt / 28 Nickel

Designation	Å*	p.e.	keV	Å*	p.e.	keV
$\alpha_2\ KL_{II}$	1.792850	9	6.91530	1.661747	8	7.46089
$\alpha_1\ KL_{III}$	1.788965	9	6.93032	1.657910	8	7.47815
$\beta_{1,3}\ KM_{II,III}$	1.62079	2	7.64943	1.500135	8	8.26466
$\beta_5\ KM_{IV,V}$	1.60891	3	7.7059	1.48862	4	8.3286
$\beta_{2,4}\ L_I M_{II,III}$	14.31	3	0.870	13.18	1	0.941
$\eta\ L_{II}M_I$	17.87	3	0.694	16.27	3	0.762
$\beta_1\ L_{II}M_{IV}$	15.666	8	0.7914	14.271	6	0.8688
$l\ L_{III}M_I$	18.292	8	0.6778	16.693	9	0.7427
$\alpha_{1,2}\ L_{III}M_{IV,V}$	15.972	6	0.7762	14.561	3	0.8515
$M_{II,III}M_{IV,V}$	214.	6	0.058	190.	2	0.0651

29 Copper / 30 Zinc

Designation	Å*	p.e.	keV	Å*	p.e.	keV
$\alpha_2\,KL_{II}$	1.544390	2	8.02783	1.439000	8	8.61578
$\alpha_1\,KL_{III}$	1.540562	2	8.04778	1.435155	7	8.63886
$\beta_3\,KM_{II}$	1.3926	1	8.9029			
$\beta_{1,3}\,KM_{II,III}$	1.392218	9	8.90529	1.29525	2	9.5720
$\beta_2\,KN_{II,III}$				1.28372	2	9.6580
$\beta_5\,KM_{IV,V}$	1.38109	3	8.9770	1.2848	1	9.6501
$\beta_{3,4}\,L_IM_{II,III}$	12.122	8	1.0228	11.200	7	1.1070
$\eta\,L_{II}M_I$	14.90	2	0.832	13.68	2	0.906
$\beta_1\,L_{II}M_{IV}$	13.053	3	0.9498	11.983	3	1.0347
$l\,L_{III}M_I$	15.286	9	0.8111	14.02	2	0.884
$\alpha_{1,2}\,L_{III}M_{IV,V}$	13.336	3	0.9297	12.254	3	1.0117
$M_{II,III}M_{V,V}$	173.	3	0.072	157.	3	0.079

31 Gallium / 32 Germanium

Designation	Å*	p.e.	keV	Å*	p.e.	keV
$\alpha_2\,KL_{II}$	1.34399	1	9.22482	1.258011	9	9.85532
$\alpha_1\,KL_{III}$	1.340083	9	9.25174	1.254054	9	9.88642
$\beta_3\,KM_{II}$	1.20835	5	10.2603	1.12936	9	10.9780
$\beta_1\,KM_{III}$	1.20789	2	10.2642	1.12894	2	10.9821
$\beta_2\,KN_{II,III}$	1.19600	2	10.3663	1.11686	2	11.1008
$\beta_5\,KM_{IV,V}$	1.1981	2	10.348	1.1195	1	11.0745
$\beta_4\,L_IM_{II}$				9.640	2	1.2861
$\beta_3\,L_IM_{III}$				9.581	2	1.2941
$\beta_{3,4}\,L_IM_{II,III}$	10.359†	8	1.197			
$\eta\,L_{II}M_I$	12.597	2	0.9842	11.609	2	1.0680
$\beta_1\,L_{II}M_{IV}$	11.023	2	1.1248	10.175	1	1.2185
$l\,L_{III}M_I$	12.953	2	0.9572	11.965	4	1.0362
$\alpha_{1,2}\,L_{III}M_{IV,V}$	11.292	1	1.09792	10.4361	8	1.18800

33 Arsenic / 34 Selenium

Designation	Å*	p.e.	keV	Å*	p.e.	keV
$\alpha_2\,KL_{II}$	1.17987	1	10.50799	1.10882	2	11.1814
$\alpha_1\,KL_{III}$	1.17588	1	10.54372	1.10477	2	11.2224
$\beta_3\,KM_{II}$	1.05783	5	11.7203	0.99268	5	12.4896
$\beta_1\,KM_{III}$	1.05730	2	11.7262	0.99218	3	12.4959
$\beta_2\,KN_{II,III}$	1.04500	3	11.8642	0.97992	5	12.6522
$\beta_5\,KM_{IV,V}$	1.0488	1	11.822	0.9843	1	12.595
$\beta_{3,4}\,L_IM_{II,III}$	8.929	1	1.3884	8.321†	9	1.490
$\eta\,L_{II}M_I$	10.734	1	1.1550	9.962	1	1.2446
$\beta_1\,L_{II}M_{IV}$	9.4141	8	1.3170	8.7358	5	1.41923
$l\,L_{III}M_I$	11.072	1	1.1198	10.294	1	1.2044
$\alpha_{1,2}\,L_{III}M_{IV,V}$	9.6709	8	1.2820	8.9900	5	1.37910
M_VN_{III}				230.	2	0.0538

35 Bromine / 36 Krypton

Designation	Å*	p.e.	keV	Å*	p.e.	keV
$\alpha_2\,KL_{II}$	1.04382	2	11.8776	0.9841	1	12.598
$\alpha_1\,KL_{III}$	1.03974	2	11.9242	0.9801	1	12.649
$\beta_3\,KM_{II}$	0.93327	5	13.2845	0.8790	1	14.104
$\beta_1\,KM_{III}$	0.93279	2	13.2914	0.8785	1	14.112
$\beta_2\,KN_{II,III}$	0.92046	2	13.4695	0.8661	1	14.315
$\beta_5\,KM_{IV,V}$	0.9255	1	13.396	0.8708	2	14.238
$\beta_4\,KN_{IV,V}$				0.8653	2	14.328
$\beta_4\,L_IM_{II}$				7.304	5	1.697
$\beta_3\,L_IM_{III}$				7.264	5	1.707

35 Bromine (*Cont.*) / 36 Krypton (*Cont.*)

Designation	Å*	p.e.	keV	Å*	p.e.	keV
$\beta_{3,4}\,L_IM_{II,III}$	7.767†	9	1.596			
$\eta\,L_{II}M_I$	9.255	1	1.3396			
$\beta_1\,L_{II}M_{IV}$	8.1251	5	1.52590	7.576†	3	1.6366
γ_5				7.279	5	1.703
$l\,L_{III}M_I$	9.585	1	1.2935			
$\alpha_{1,2}\,L_{III}M_{IV,V}$	8.3746	5	1.48043	7.817†	3	1.5860
β_6				7.510	4	1.6510
$L_{II}N_{III}$				7.250	5	1.710
M_IM_{II}	184.6	3	0.0672			
M_IM_{III}	164.7	3	0.0753			
$M_{II}M_{IV}$	109.4	3	0.1133			
$M_{II}N_I$	76.9	2	0.1613			
$M_{III}M_{IV,V}$	113.8	3	0.1089			
	79.8	3	0.1554			
$\zeta_2\,M_{IV}N_{II}$	191.1	2	0.06488			
$M_{IV}N_{III}$	189.5	3	0.0654			
$\zeta_1\,M_VN_{III}$	192.6	2	0.06437			

37 Rubidium / 38 Strontium

Designation	Å*	p.e.	keV	Å*	p.e.	keV
$\alpha_2\,KL_{II}$	0.92969	1	13.3358	0.87943	1	14.0979
$\alpha_1\,KL_{III}$	0.925553	9	13.3953	0.87526	1	14.1650
$\beta_3\,KM_{II}$	0.82921	3	14.9517	0.78345	3	15.8249
$\beta_1\,KM_{III}$	0.82868	2	14.9613	0.78292	2	15.8357
$\beta_2\,KN_{II,III}$	0.81645	3	15.1854	0.77081	3	16.0846
$\beta_5\,KM_{IV,V}$	0.8219	1	15.085	0.7764	1	15.969
$\beta_4\,KN_{IV,V}$	0.8154	2	15.205	0.76989	5	16.104
$\beta_4\,L_IM_{II}$	6.8207	3	1.81771	6.4026	3	1.93643
$\beta_3\,L_IM_{III}$	6.7876	3	1.82659	6.3672	3	1.94719
$\gamma_{2,3}\,L_IN_{II,III}$	6.0458	3	2.0507	5.6445	3	2.1965
$\eta\,L_{II}M_I$	8.0415	4	1.54177	7.5171	3	1.64933
$\beta_1\,L_{II}M_{IV}$	7.0759	3	1.75217	6.6239	3	1.87172
$\gamma_5\,L_{II}N_{IV}$	6.7553	3	1.83532	6.2961	3	1.96916
$l\,L_{III}M_I$	8.3636	4	1.48238	7.8362	3	1.58215
$\alpha_2\,L_{III}M_{IV}$	7.3251	3	1.69256	6.8697	3	1.80474
$\alpha_1\,L_{III}M_V$	7.3183	2	1.69413	6.8628	2	1.80656
$\beta_6\,L_{III}N_I$	6.9842	3	1.77517	6.5191	3	1.90181
M_IM_{III}	144.4	3	0.0859			
$M_{II}M_{IV}$	91.5	2	0.1355	85.7	2	0.1447
$M_{II}N_I$	57.0	2	0.2174	51.3	1	0.2416
$M_{III}M_{IV,V}$	96.7	2	0.1282	91.4	2	0.1357
$M_{III}N_I$	59.5	2	0.2083	53.6	1	0.2313
$\zeta_2\,M_{IV}N_{II}$	127.8	2	0.0970			
$M_{IV}N_{III}$	126.8	2	0.0978			
$\zeta_2\,M_{IV}N_{II,III}$				108.0	2	0.1148
$\zeta_1\,M_VN_{III}$	128.7	2	0.0964	108.7	1	0.1140

39 Yttrium / 40 Zirconium

Designation	Å*	p.e.	keV	Å*	p.e.	keV
$\alpha_2\,KL_{II}$	0.83305	1	14.8829	0.79015	1	15.6909
$\alpha_1\,KL_{III}$	0.82884	1	14.9584	0.78593	1	15.7751
$\beta_3\,KM_{II}$	0.74126	3	16.7258	0.70228	4	17.654
$\beta_1\,KM_{III}$	0.74072	2	16.7378	0.70173	3	17.6678
$\beta_2\,KN_{II,III}$	0.72864	4	17.0154	0.68993	4	17.970
$\beta_5\,KM_{IV,V}$	0.7345	1	16.879	0.6959	1	17.815

39 Yttrium (Cont.) / 40 Zirconium (Cont.)

Designation	Å*	p.e.	keV	Å*	p.e.	keV
$\beta_4 KN_{IV,V}$	0.72776	5	17.036	0.68901	5	17.994
$\beta_4 L_I M_{II}$	6.0186	3	2.0600	5.6681	3	2.1873
$\beta_3 L_I M_{III}$	5.9832	3	2.0722	5.6330	3	2.2010
$\gamma_{2,3} L_I N_{II,III}$	5.2830	3	2.3468	4.9536	3	2.5029
$\eta\ L_I M_I$	7.0406	3	1.76095	6.6069	3	1.87654
$\beta_1 L_{III} M_{IV}$	6.2120	3	1.99584	5.8360	3	2.1244
$\gamma_5 L_{II} N_I$	5.8754	3	2.1102	5.4977	3	2.2551
$\gamma_1 L_{II} N_{IV}$				5.3843	3	2.3027
$l\ L_{III} M_I$	7.3563	3	1.68536	6.9185	3	1.79201
$\alpha_2 L_{III} M_{IV}$	6.4558	3	1.92047	6.0778	3	2.0399
$\alpha_1 L_{III} M_V$	6.4488	2	1.92256	6.0705	2	2.04236
$\beta_6 L_{III} N_I$	6.0942	3	2.0344	5.7101	3	2.1712
$\beta_{2,15}$				5.5863	3	2.2194
$M_{II} M_{IV}$	81.5	2	0.1522	76.7	2	0.1617
$M_{II} N_I$	46.48	9	0.267			
$M_{III} M_V$				80.9	3	0.1533
$M_{III} N_I$	48.5	2	0.256			
$M_{III} M_{IV,V}$	86.5	2	0.1434			
$\zeta\ M_{IV,V} N_{II,III}$	93.4	2	0.1328	82.1	2	0.1511
$M_{IV,V} O_{II,III}$				70.0	4	0.177

41 Niobium / 42 Molybdenum

Designation	Å*	p.e.	keV	Å*	p.e.	keV
$\alpha_2 KL_{II}$	0.75044	1	16.5210	0.713590	6	17.3743
$\alpha_1 KL_{III}$	0.74620	1	16.6151	0.709300	1	17.47934
$\beta_3 KM_{II}$	0.66634	3	18.6063	0.632872	9	19.5903
$\beta_1 KM_{III}$	0.66576	2	18.6225	0.632288	9	19.6083
β_5^{II}				0.62107	5	19.963
$\beta_2 KN_{II,III}$	0.65416	4	18.953	0.62099	2	19.9652
$\beta_4 KN_{IV,V}$	0.65318	5	18.981			
$\beta_5^{II} KM_{IV}$				0.62708	5	19.771
$\beta_5^{I} KM_V$				0.62692	5	19.776
$\beta_4 KN_{IV,V}$				0.62001	9	19.996
$\beta_4 L_I M_{II}$	5.3455	3	2.3194	5.0488	3	2.4557
$\beta_3 L_I M_{III}$	5.3102	3	2.3348	5.0133	3	2.4730
$\gamma_{2,3} L_I N_{II,III}$	4.6542	2	2.6638	4.3800	2	2.8306
$\eta\ L_{II} M_I$	6.2109	3	1.99620	5.8475	3	2.1202
$\beta_1 L_{II} M_{IV}$	5.4923	3	2.2574	5.17708	8	2.39481
$\gamma_5 L_{II} N_I$	5.1517	3	2.4066	4.8369	2	2.5632
$\gamma_1 L_{II} N_{IV}$	5.0361	3	2.4618	4.7258	2	2.6235
$l\ L_{III} M_I$	6.5176	3	1.90225	6.1508	3	2.01568
$\alpha_2 L_{III} M_{IV}$	5.7319	3	2.1630	5.41437	8	2.28985
$\alpha_1 L_{III} M_V$	5.7243	2	2.16589	5.40655	8	2.29316
$\beta_6 L_{III} N_I$	5.3613	3	2.3125	5.0488	5	2.4557
$\beta_{2,15} L_{III} N_{IV,V}$	5.2379	3	2.3670	4.9232	2	2.5183
$M_{II} M_{IV}$	72.1	3	0.1718	68.9	2	0.1798
$M_{II} N_I$	38.4	3	0.323	35.3	3	0.351
$M_{II} N_{IV}$	33.1	2	0.375			
$M_{III} M_V$	78.4	2	0.1582	74.9	1	0.1656
$M_{III} N_I$	40.7	2	0.305	37.5	2	0.331
$\gamma\ M_{III} N_{IV,V}$	34.9	2	0.356			
$\zeta\ M_{IV,V} N_{II,III}$	72.19	9	0.1717	64.38	7	0.1926
$M_{IV,V} O_{II,III}$	61.9	2	0.2002	54.8	2	0.2262

43 Technetium / 44 Ruthenium

Designation	Å*	p.e.	keV	Å*	p.e.	keV
$\alpha_2 KL_{II}$	0.67932†	3	18.2508	0.647408	5	19.1504
$\alpha_1 KL_{III}$	0.67502†	3	18.3671	0.643083	4	19.2792
$\beta_3 KM_{II}$	0.60188†	4	20.599	0.573067	4	21.6346
$\beta_1 KM_{III}$	0.60130†	4	20.619	0.572482	4	21.6568
$\beta_2 KN_{II,III}$	0.59024†	5	21.005	0.56166	3	22.074
$\beta_5^{II} KM_{IV}$				0.5680	2	21.829
$\beta_5^{I} KM_V$				0.56785	9	21.834
β_4				0.56089	9	22.104
$\beta_4 L_{II} M_{IV}$				4.5230	2	2.7411
$\beta_3 L_I M_{III}$				4.4866	3	2.7634
$\gamma_{2,3} L_I N_{II,III}$				3.8977	2	3.1809
$\eta\ L_{II} M_I$				5.2050	2	2.38197
$\beta_1 L_{II} M_{IV}$	4.8873†	8	2.5368	4.62058	3	2.68323
$\gamma_5 L_{II} N_I$				4.2873	2	2.8918
$\gamma_1 L_{II} N_{IV}$				4.1822	2	2.9645
$l\ L_{III} M_I$				5.5035	3	2.2528
$\alpha_2 L_{III} M_{IV}$				4.85381	7	2.55431
$\alpha_1 L_{III} M_V$	5.1148†	3	2.4240	4.84575	5	2.55855
$\beta_6 L_{III} N_I$				4.4866	3	2.7634
$\beta_{2,15} L_{III} N_{IV,V}$				4.3718	2	2.8360
$M_{II} M_{IV}$				62.2	1	0.1992
$M_{II} N_I$				32.3	2	0.384
$M_{II} N_{IV}$				25.50	9	0.486
$M_{III} M_V$				68.3	1	0.1814
$\gamma\ M_{III} N_{IV,V}$				26.9	1	0.462
$\zeta\ M_{IV,V} N_{II,III}$				52.34	7	0.2369
$M_{IV,V} O_{II,III}$				44.8	1	0.2768

45 Rhodium / 46 Palladium

Designation	Å*	p.e.	keV	Å*	p.e.	keV
$\alpha_2 KL_{II}$	0.617630	4	20.0737	0.589821	3	21.0201
$\alpha_1 KL_{III}$	0.613279	4	20.2161	0.585448	3	21.1771
$\beta_3 KM_{II}$	0.546200	4	22.6989	0.521123	4	23.7911
$\beta_1 KM_{III}$	0.545605	4	22.7236	0.520520	4	23.8187
$\beta_2^{II} KN_{II}$	0.53513	5	23.168			
$\beta_2 KN_{II,III}$	0.53503	2	23.1728	0.510228	4	24.2991
$\beta_5^{II} KM_{IV}$	0.54118	9	22.909			
$\beta_5^{I} KM_V$	0.54101	9	22.917			
$\beta_4 KN_{IV,V}$	0.53401	9	23.217	0.5093	2	24.346
$\beta_5 KM_{IV,V}$				0.51670	9	23.995
$\beta_4 L_I M_{II}$	4.2888	2	2.8908	4.0711	2	3.0454
$\beta_3 L_I M_{III}$	4.2522	2	2.9157	4.0346	2	3.0730
$\gamma_{2,3} L_I N_{II,III}$	3.6855	2	3.3640	3.4892	2	3.5533
$\eta\ L_{II} M_I$	4.9217	2	2.5191	4.6605	2	2.6603
$\beta_1 L_{II} M_{IV}$	4.37414	4	2.83441	4.14622	5	2.99022
$\gamma_5 L_{II} N_I$	4.0451	2	3.0650	3.8222	2	3.2437
$\gamma_1 L_{II} N_{IV}$	3.9437	2	3.1438	3.7246	2	3.3287
$l\ L_{III} M_I$	5.2169	3	2.3765	4.9525	3	2.5034
$\alpha_2 L_{III} M_{IV}$	4.60545	9	2.69205	4.37588	7	2.83329
$\alpha_1 L_{III} M_V$	4.59743	9	2.69674	4.36767	5	2.83861
$\beta_6 L_{III} N_I$	4.2417	2	2.9229	4.0162	2	3.0870
$\beta_{2,15} L_{III} N_{IV,V}$	4.1310	2	3.0013	3.90887	4	3.17179
$\beta_{10} L_I M_{IV}$				3.7988	2	3.2637

45 Rhodium (*Cont.*) / 46 Palladium (*Cont.*)

Designation	Å*	p.e.	keV	Å*	p.e.	keV
$\beta_9\ L_I M_V$				3.7920	2	3.2696
$M_I N_{II,III}$				20.1	2	0.616
$M_{II} M_{IV}$	59.3	1	0.2090	56.5	1	0.2194
$M_{II} N_I$	28.1	2	0.442	26.2	2	0.474
$M_{II} N_{IV}$				22.1	1	0.560
$M_{II}\ M_V$	65.5	1	0.1892	62.9	1	0.1970
$M_{II}'N_I$	29.8	1	0.417	27.9	1	0.445
$\gamma\ M_{III} N_{IV,V}$	25.01	9	0.496	23.3†	1	0.531
$\zeta\ M_{IV,V} N_{II,III}$	47.67	9	0.2601	43.6	1	0.2844
$M_{IV,V} O_{II,III}$	40.9	2	0.303	37.4	2	0.332

47 Silver / 48 Cadmium

Designation	Å*	p.e.	keV	Å*	p.e.	keV
$\alpha_2\ K L_{II}$	0.563798	4	21.9903	0.539422	3	22.9841
$\alpha_1\ K L_{III}$	0.5594075	6	22.16292	0.535010	3	23.1736
$\beta_3\ K M_{II}$	0.497685	4	24.9115	0.475730	5	26.0612
$\beta_1\ K M_{III}$	0.497069	4	24.9424	0.475105	6	26.0955
$\beta_2\ K N_{II,III}$	0.487032	4	25.4564	0.465328	7	26.6438
$\beta_5\ K M_{IV,V}$	0.49306	2	25.145			
$\beta_4\ K N_{IV,V}$	0.48598	3	25.512			
$\beta_3\ L_I M_{II}$	3.87023	5	3.20346	3.68203	9	3.36719
$\beta_3\ L_I M_{III}$	3.83313	9	3.23446	3.64495	9	3.40145
$\gamma_2\ L_I N_{II}$	3.31216	9	3.7432	3.1377	2	3.9513
$\gamma_3\ L_I N_{III}$	3.30635	9	3.7498			
$\eta\ L_{II} M_I$	4.4183	2	2.8061	4.19315	9	2.95675
$\beta_1\ L_{II} M_{IV}$	3.93473	3	3.15094	3.73823	4	3.31657
$\gamma_5\ L_{II} N_I$	3.61638	9	3.42832	3.42551	9	3.61935
$\gamma_1\ L_{III} N_{IV}$	3.52260	4	3.51959	3.33564	6	3.71686
$l\ L_{III} M_I$	4.7076	2	2.6337	4.48014	9	2.76735
$\alpha_2\ L_{III} M_{IV}$	4.16294	5	2.97821	3.96496	6	3.12691
$\alpha_1\ L_{III} M_V$	4.15443	3	2.98431	3.95635	4	3.13373
$\beta_6\ L_{III} N_I$	3.80774	9	3.25603	3.61467	9	3.42994
$\beta_{2,15}\ L_{III} N_{IV,V}$	3.70335	9	3.34781	3.51408	4	3.52812
$\beta_{10}\ L_I M_{IV}$	3.61158	9	3.43287	3.4367	2	3.6075
$\beta_9\ L_I M_V$	3.60497	9	3.43917	3.43015	9	3.61445
$M_I N_{II,III}$	18.8	2	0.658			
$M_{II} M_{IV}$	54.0	1	0.2295	52.0	2	0.2384
$M_{II} N_I$				22.9	2	0.540
$M_{II} N_{IV}$	20.66	7	0.600	19.40	7	0.639
$M_{III} M_V$	60.5	1	0.2048	58.7	2	0.2111
$M_{III} N_I$	26.0	1	0.478	24.5	1	0.507
$\gamma\ M_{III} N_{IV,V}$	21.82	7	0.568	20.47	7	0.606
$M_{IV} O_{II,III}$				30.4	1	0.408
$\zeta\ M_{IV,V} N_{II,III}$	39.77	7	0.3117	36.8	1	0.3371
$M_V N_I$	24.4	2	0.509			
$M_V O_{III}$				30.8	1	0.403
$M_{IV,V} O_{II,III}$	33.5	3	0.370			

49 Indium / 50 Tin

Designation	Å*	p.e.	keV	Å*	p.e.	keV
$\alpha_2\ K L_{II}$	0.516544	3	24.0020	0.495053	3	25.0440
$\alpha_1\ K L_{III}$	0.512113	3	24.2097	0.490599	3	25.2713
$\beta_3\ K M_{II}$	0.455181	4	27.2377	0.435877	5	28.4440

49 Indium (*Cont.*) / 50 Tin (*Cont.*)

Designation	Å*	p.e.	keV	Å*	p.e.	keV
$\beta_1\ K M_{III}$	0.454545	4	27.2759	0.435236	5	28.4860
$\beta_2\ K N_{II,III}$	0.44500	1	27.8608	0.425915	8	29.1093
$K O_{II,III}$	0.44374	3	27.940	0.42467	3	29.195
$\beta_5\ II\ K M_{IV}$	0.45098	2	27.491	0.43184	3	28.710
$\beta_5\ I\ K M_V$	0.45086	2	27.499	0.43175	3	28.716
$\beta_4\ K N_{IV,V}$	0.44393	4	27.928	0.42495	3	29.175
$\beta_3\ L_I M_{II}$	3.50697	9	3.5353	3.34335	9	3.7083
$\beta_3\ L_I M_{III}$	3.46984	9	3.5731	3.30585	3	3.7500
$\gamma_{2,3}\ L_I N_{II,III}$	2.9800	2	4.1605	2.8327	2	4.3768
$\gamma_4\ L_I O_{II,III}$	2.9264	2	4.2367	2.7775	2	4.4638
$\eta\ L_{II} M_I$	3.98327	9	3.11254	3.78876	9	3.27234
$\beta_1\ L_{II} M_{IV}$	3.55531	4	3.48721	3.38487	3	3.66280
$\gamma_5\ L_{II} N_I$	3.24907	9	3.8159	3.08475	9	4.0192
$\gamma_1\ L_{II} N_{IV}$	3.16213	4	3.92081	3.00115	3	4.13112
$l\ L_{III} M_I$	4.26873	9	2.90440	4.07165	9	3.04499
$\alpha_2\ L_{III} M_{IV}$	3.78073	6	3.27929	3.60891	4	3.43542
$\alpha_1\ L_{III} M_V$	3.77192	4	3.28694	3.59994	3	3.44398
$\beta_6\ L_{III} N_I$	3.43606	9	3.60823	3.26901	9	3.7926
$\beta_{2,15}\ L_{III} N_{IV,V}$	3.33838	3	3.71381	3.17505	3	3.90486
$\beta_7\ L_{III} O_I$	3.324	4	3.730	3.1564	3	3.9279
$\beta_{10}\ L_I M_{IV}$	3.27404	9	3.7868	3.12170	9	3.9716
$\beta_9\ L_I M_V$	3.26763	9	3.7942	3.11513	9	3.9800
$M_{II} M_{IV}$				47.3	1	0.2621
$M_{II} N_I$				20.0	1	0.619
$M_{II} N_{IV}$				16.93	5	0.733
$M_{III} M_V$				54.2	1	0.2287
$M_{III} N_I$				21.5	1	0.575
$\gamma\ M_{III} N_{IV,V}$				17.94	5	0.691
$M_{IV} O_{II,III}$				25.3	1	0.491
$\zeta\ M_{IV,V} N_{II,III}$				31.24	9	0.397
$M_V O_{III}$				25.7	1	0.483

51 Antimony / 52 Tellurium

Designation	Å*	p.e.	keV	Å*	p.e.	keV
$\alpha_2\ K L_{II}$	0.474827	3	26.1108	0.455784	3	27.2017
$\alpha_1\ K L_{III}$	0.470354	3	26.3591	0.451295	3	27.4723
$\beta_3\ K M_{II}$	0.417737	4	29.6792	0.400659	4	30.9443
$\beta_1\ K M_{III}$	0.417085	3	29.7256	0.399995	5	30.9957
$\beta_2\ K N_{II,III}$	0.407973	5	30.3895	0.391102	6	31.7004
$K O_{II,III}$	0.40666	1	30.4875	0.38974	1	31.8114
$\beta_5\ II\ K M_{IV}$	0.41388	1	29.9560			
$\beta_5\ I\ K M_V$	0.41378	1	29.9632			
$\beta_4\ K N_{IV,V}$	0.40702	1	30.4604			
$\beta_3\ L_I M_{II}$	3.19014	9	3.8864	3.04661	9	4.0695
$\beta_3\ L_I M_{III}$	3.15258	9	3.9327	3.00893	9	4.1204
$\gamma_{2,3}\ L_I N_{II,III}$	2.6953	2	4.5999	2.5674	2	4.8290
$\gamma_4\ L_I O_{II,III}$	2.6398	2	4.6967	2.5113	2	4.9369
$\eta\ L_{II} M_I$	3.60765	9	3.43661	3.43832	9	3.60586
$\beta_1\ L_{II} M_{IV}$	3.22567	4	3.84357	3.07677	6	4.02958
$\gamma_5\ L_{II} N_I$	2.93187	9	4.2287	2.79007	9	4.4437
$\gamma_1\ L_{II} N_{IV}$	2.85159	3	4.34779	2.71241	6	4.5709
$l\ L_{III} M_I$	3.88826	9	3.18860	3.71696	9	3.33555
$\alpha_2\ L_{III} M_{IV}$	3.44840	6	3.59532	3.29846	9	3.7588

Designation	Å*	p.e.	keV	Å*	p.e.	keV
51 Antimony (Cont.)				**52 Tellurium** (Cont.)		
$\alpha_1\ L_{III}M_V$	3.43941	4	3.60472	3.28920	6	3.76933
$\beta_6\ L_{III}N_I$	3.11513	9	3.9800	2.97088	9	4.1732
$\beta_{2,15}\ L_{III}N_{IV,V}$	3.02335	3	4.10078	2.88217	8	4.3017
$\beta_7\ L_{III}O_I$	3.0052	3	4.1255	2.8634	3	4.3298
$\beta_{10}\ L_I M_{IV}$	2.97917	9	4.1616	2.84679	9	4.3551
$\beta_9\ L_I M_V$	2.97261	9	4.1708	2.83897	9	4.3671
$M_{II}M_{IV}$	45.2	1	0.2743			
$M_{II}N_I$	18.8	1	0.658	17.6	1	0.703
$M_{II}N_{IV}$	15.98	5	0.776			
$M_{III}M_V$	52.2	1	0.2375	50.3	1	0.2465
$M_{III}N_I$	20.2	1	0.612	19.1	1	0.648
$\gamma\ M_{III}N_{IV,V}$	16.92	4	0.733	15.93	4	0.778
$M_{IV}O_{II,III}$				21.34	5	0.581
$\zeta\ M_{IV,V}N_{II,III}$	28.88	8	0.429	26.72	9	0.464
$M_V O_{III}$				21.78	5	0.569
53 Iodine				**54 Xenon**		
$\alpha_2\ K L_{II}$	0.437829	7	28.3172	0.42087†	2	29.458
$\alpha_1\ K L_{III}$	0.433318	5	28.6120	0.41634†	2	29.779
$\beta_3\ K M_{II}$	0.384561	1	32.2391	0.36941†	2	33.562
$\beta_1\ K M_{III}$	0.383905	4	32.2947	0.36872†	2	33.624
$\beta_2\ K N_{II,III}$	0.37523†	2	33.042	0.36026†	3	34.415
$\beta_4\ L_I M_{II}$	2.91207	9	4.2573			
$\beta_3\ L_I M_{III}$	2.87429	9	4.3134			
$\gamma_{2,3}\ L_I N_{II,III}$	2.4475	2	5.0657			
$\gamma_4\ L_I O_{II,III}$	2.3913	2	5.1848			
$\eta\ L_{II}M_I$	3.27979	9	3.7801			
$\beta_1\ L_{II}M_{IV}$	2.93744	6	4.22072			
$\gamma_5\ L_{II}N_I$	2.65710	9	4.6660			
$\gamma_1\ L_{II}N_{IV}$	2.58244	8	4.8009			
$l\ L_{III}M_I$	3.55754	9	3.48502			
$\alpha_2\ L_{III}M_{IV}$	3.15791	6	3.92604			
$\alpha_1\ L_{III}M_V$	3.14860	6	3.93765	3.0166†	2	4.1099
$\beta_6\ L_{III}N_I$	2.83672	9	4.3706			
$\beta_{2,15}\ L_{III}N_{IV,V}$	2.75053	8	4.5075			
$\beta_7\ L_{III}O_I$	2.7288	3	4.5435			
$\beta_{10}\ L_I M_{IV}$	2.72104	9	4.5564			
$\beta_9\ L_I M_V$	2.71352	9	4.5690			
55 Cesium				**56 Barium**		
$\alpha_2\ K L_{II}$	0.404835	4	30.6251	0.389668	5	31.8171
$\alpha_1\ K L_{III}$	0.400290	4	30.9728	0.385111	4	32.1936
$\beta_3\ K M_{II}$	0.355050	4	34.9194	0.341507	4	36.3040
$\beta_1\ K M_{III}$	0.354364	7	34.9869	0.340811	3	36.3782
$\beta_2\ K N_{II,III}$	0.34611	2	35.822	0.33277	1	37.257
$K O_{II,III}$				0.33127	2	37.426
$\beta_5^{II}\ K M_{IV}$				0.33835	2	36.643
$\beta_5^{I}\ K M_V$				0.33814	2	36.666
$\beta_4\ K N_{IV,V}$				0.33229	2	37.311
$\beta_4\ L_I M_{II}$	2.6666	2	4.6494	2.5553	2	4.8519
$\beta_3\ L_I M_{III}$	2.6285	2	4.7167	2.5164	2	4.9269
$\gamma_2\ L_I N_{II}$	2.2371	2	5.5420	2.1387	2	5.7969
$\gamma_3\ L_I N_{III}$	2.2328	2	5.5527	2.1342	2	5.8092

Designation	Å*	p.e.	keV	Å*	p.e.	keV
55 Cesium (Cont.)				**56 Barium** (Cont.)		
$\gamma_4\ L_I O_{II,III}$	2.1741	2	5.7026	2.0756	3	5.9733
$\eta\ L_{II}M_I$	2.9932	2	4.1421	2.8627	3	4.3309
$\beta_1\ L_{II}M_{IV}$	2.6837	2	4.6198	2.56821	5	4.82753
$\gamma_5\ L_{II}N_I$	2.4174	2	5.1287	2.3085	3	5.3707
$\gamma_1\ L_{II}N_{IV}$	2.3480	2	5.2804	2.2415	2	5.5311
$l\ L_{III}M_I$	3.2670	2	3.7950	3.1355	2	3.9541
$\alpha_2\ L_{III}M_{IV}$	2.9020	2	4.2722	2.78553	5	4.45090
$\alpha_1\ L_{III}M_V$	2.8924	2	4.2865	2.77595	5	4.46626
$\beta_6\ L_{III}N_I$	2.5932	2	4.7811	2.4826	2	4.9939
$\beta_{2,15}\ L_{III}N_{IV,V}$	2.5118	2	4.9359	2.40435	6	5.1565
$\beta_7\ L_{III}O_I$	2.4849	2	4.9893	2.3806	2	5.2079
$\beta_{10}\ L_I M_{IV}$	2.4920	2	4.9752	2.3869	2	5.1941
$\beta_9\ L_I M_V$	2.4783	2	5.0026	2.3764	2	5.2171
$\gamma\ M_{III}N_{IV,V}$				12.75	3	0.973
$M_{IV}O_{II}$				15.91	5	0.779
$M_{IV}O_{III}$				15.72	9	0.789
$\zeta\ M_V N_{III}$				20.64	4	0.601
$M_V O_{III}$				16.20	5	0.765
$N_{IV}O_{II}$	188.6	1	0.06574	163.3	2	0.07590
$N_{IV}O_{III}$	183.8	1	0.06746	159.0	2	0.07796
$N_V O_{III}$	190.3	1	0.06515	164.6	2	0.07530
57 Lanthanum				**58 Cerium**		
$\alpha_2\ K L_{II}$	0.375313	2	33.0341	0.361683	2	34.2789
$\alpha_1\ K L_{III}$	0.370737	2	33.4418	0.357092	2	34.7197
$\beta_3\ K M_{II}$	0.328686	4	37.7202	0.316520	4	39.1701
$\beta_1\ K M_{III}$	0.327983	3	37.8010	0.315816	2	39.2573
$\beta_2\ K N_{II,III}$	0.320117	7	38.7299	0.30816	1	40.233
$K O_{II,III}$	0.31864	2	38.909	0.30668	2	40.427
$\beta_5^{II}\ K M_{IV}$	0.32563	2	38.074	0.31357	2	39.539
$\beta_5^{I}\ K M_V$	0.32546	2	38.094	0.31342	2	39.558
$\beta_4\ K N_{IV,V}$	0.31931	2	38.828	0.30737	2	40.337
$\beta_4\ L_I M_{II}$	2.4493	3	5.0620	2.3497	4	5.2765
$\beta_3\ L_I M_{III}$	2.4105	3	5.1434	2.3109	3	5.3651
$\gamma_2\ L_I N_{II}$	2.0460	4	6.060	1.9602	3	6.3250
$\gamma_3\ L_I N_{III}$	2.0410	4	6.074	1.9553	3	6.3409
$\gamma_4\ L_I O_{II,III}$	1.9830	4	6.252	1.8991	4	6.528
$\eta\ L_{II}M_I$	2.740	3	4.525	2.6203	4	4.7315
$\beta_1\ L_{II}M_{IV}$	2.45891	5	5.0421	2.3561	3	5.2622
$\gamma_5\ L_{II}N_I$	2.2056	4	5.621	2.1103	3	5.8751
$\gamma_1\ L_{II}N_{IV}$	2.1418	3	5.7885	2.0487	4	6.052
$\gamma_8\ L_{II}O_I$				2.0237	4	6.126
$l\ L_{III}M_I$	3.006	3	4.124	2.8917	4	4.2875
$\alpha_2\ L_{III}M_{IV}$	2.67533	5	4.63423	2.5706	3	4.8230
$\alpha_1\ L_{III}M_V$	2.66570	5	4.65097	2.5615	2	4.8402
$\beta_6\ L_{III}N_I$	2.3790	4	5.2114	2.2818	3	5.4334
$\beta_{2,15}\ L_{III}N_{IV,V}$	2.3030	3	5.3835	2.2087	3	5.6134
$\beta_7\ L_{III}O_I$	2.275	2	5.450	2.1701	2	5.7132
$\beta_{10}\ L_I M_{IV}$	2.290	3	5.415	2.1958	5	5.646
$\beta_9\ L_I M_V$	2.282	3	5.434	2.1885	3	5.6650
$\gamma\ M_{III}N_{IV,V}$	12.08	4	1.027	11.53	1	1.0749
$\beta\ M_{IV}N_{VI}$	14.51	5	0.854	13.75	4	0.902
$\zeta\ M_V N_{III}$	19.44	5	0.638	18.35	4	0.676
$\alpha\ M_V N_{VI,VII}$	14.88	5	0.833	14.04	2	0.883

57 Lanthanum (Cont.) | 58 Cerium (Cont.)

Designation	Å*	p.e.	keV	Å*	p.e.	keV
$M_V O_{II,III}$				14.39	5	0.862
$N_{IV,V} O_{II,III}$	152.6	6	0.0812	144.4	6	0.0859

59 Praseodymium | 60 Neodymium

Designation	Å*	p.e.	keV	Å*	p.e.	keV
$\alpha_2 K L_{II}$	0.348749	2	35.5502	0.336472	2	36.8474
$\alpha_1 K L_{III}$	0.344140	2	36.0263	0.331846	2	37.3610
$\beta_3 K M_{II}$	0.304975	5	40.6529	0.294027	3	42.1665
$\beta_1 K M_{III}$	0.304261	4	40.7482	0.293299	2	42.2713
$\beta_2 K N_{II,III}$	0.29679	2	41.773	0.2861†	1	43.33
$\beta_4 L_I M_{II}$	2.2550	4	5.4981	2.1669	3	5.7216
$\beta_3 L_I M_{III}$	2.2172	3	5.5918	2.1268	2	5.8294
$\gamma_2 L_I N_{II}$	1.8791	4	6.598	1.8013	4	6.883
$\gamma_3 L_I N_{III}$	1.8740	4	6.616	1.7964	4	6.902
$\gamma_4 L_I O_{II,III}$	1.8193	4	6.815	1.7445	4	7.107
$\eta L_{II} M_I$	2.512	3	4.935	2.4094	4	5.1457
$\beta_1 L_{II} M_{IV}$	2.2588	3	5.4889	2.1669	2	5.7216
$\gamma_5 L_{II} N_I$	2.0205	4	6.136	1.9355	4	6.406
$\gamma_1 L_{II} N_{IV}$	1.9611	3	6.3221	1.8779	2	6.6021
$\gamma_8 L_{II} O_I$	1.9362	4	6.403	1.8552	6	6.683
$l L_{III} M_I$	2.7841	4	4.4532	2.6760	4	4.6330
$\alpha_2 L_{III} M_{IV}$	2.4729	3	5.0135	2.3807	3	5.2077
$\alpha_1 L_{III} M_V$	2.4630	2	5.0337	2.3704	2	5.2304
$\beta_6 L_{III} N_I$	2.1906	4	5.660	2.1039	3	5.8930
$\beta_{2,15} L_{III} N_{IV,V}$	2.1194	4	5.850	2.0360	3	6.0894
$\beta_7 L_{III} O_I$	2.0919	4	5.927	2.0092	3	6.1708
$\beta_{10} L_I M_{IV}$	2.1071	4	5.884	2.0237	3	6.1265
$\beta_9 L_I M_V$	2.1004	4	5.903	2.0165	3	6.1484
$\gamma M_{III} N_{IV,V}$	10.998	9	1.1273	10.505	9	1.180
$\beta M_{IV} N_{VI}$	13.06	2	0.950	12.44	2	0.997
$\zeta M_V N_{III}$	17.38	4	0.714	16.46	4	0.753
$\alpha M_V N_{VI,VII}$	13.343	5	0.9292	12.68	2	0.978
$N_{IV,V} N_{VI,VII}$	113.	1	0.1095	107.	1	0.116
$N_{IV,V} O_{II,III}$	136.5	4	0.0908	128.9	7	0.0962

61 Promethium | 62 Samarium

Designation	Å*	p.e.	keV	Å*	p.e.	keV
$\alpha_2 K L_{II}$	0.324803	4	38.1712	0.313698	2	39.5224
$\alpha_1 K L_{III}$	0.320160	4	38.7247	0.309040	2	40.1181
$\beta_3 K M_{II}$	0.28363†	4	43.713	0.27376	2	45.289
$\beta_1 K M_{III}$	0.28290†	3	43.826	0.27301	2	45.413
$\beta_2 K N_{II,III}$	0.2759†	1	44.94	0.2662	1	46.58
$K O_{II,III}$				0.26491	3	46.801
$\beta_5 K M_{IV,V}$				0.27111	3	45.731
$\beta_4 L_I M_{II}$				2.00095	6	6.1963
$\beta_3 L_I M_{III}$	2.0421	4	6.071	1.96241	3	6.3180
$\gamma_2 L_I N_{II}$				1.66044	6	7.4668
$\gamma_3 L_I N_{III}$				1.65601	3	7.4867
$\gamma_4 L_I O_{II,III}$				1.60728	3	7.7137
$\eta L_{II} M_I$				2.21824	3	5.5892
$\beta_1 L_{II} M_{IV}$	2.0797	4	5.961	1.99806	3	6.2051
$\gamma_5 L_{II} N_I$				1.77934	3	6.9678
$\gamma_1 L_{II} N_{IV}$	1.7989	9	6.892	1.72724	3	7.1780
$\gamma_8 L_{II} O_{IV}$				1.6966	9	7.3076
$l L_{III} M_I$				2.4823	4	4.9945

61 Promethium (Cont.) | 62 Samarium (Cont.)

Designation	Å*	p.e.	keV	Å*	p.e.	keV
$\alpha_2 L_{III} M_{IV}$	2.2926	4	5.4078	2.21062	3	5.6084
$\alpha_1 L_{III} M_V$	2.2822	3	5.4325	2.1998	2	5.6361
$\beta_6 L_{III} N_I$				1.94643	3	6.3697
$\beta_{2,15} L_{III} N_{IV,V}$	1.9559	6	6.339	1.88221	3	6.5870
$\beta_7 L_{III} O_I$				1.85626	3	6.6791
$\beta_5 L_{III} O_{IV,V}$				1.84700	9	6.7126
$\beta_{10} L_I M_{IV}$				1.86990	3	6.6304
$\beta_9 L_I M_V$				1.86166	3	6.6597
$\gamma M_{III} N_{IV,V}$				9.600	9	1.291
$\beta M_{IV} N_{VI}$				11.27	1	1.0998
$\zeta M_V N_{III}$				14.91	4	0.831
$\alpha M_V N_{VI,VII}$				11.47	3	1.081
$N_{IV,V} N_{VI,VII}$				98.	1	0.126
$N_{IV,V} O_{II,III}$				117.4	4	0.1056

63 Europium | 64 Gadolinium

Designation	Å*	p.e.	keV	Å*	p.e.	keV
$\alpha_2 K L_{II}$	0.303118	2	40.9019	0.293038	2	42.3089
$\alpha_1 K L_{III}$	0.298446	2	41.5422	0.288353	2	42.9962
$\beta_3 K M_{II}$	0.264332	5	46.9036	0.25534	2	48.555
$\beta_1 K M_{III}$	0.263577	5	47.0379	0.25460	2	48.697
$\beta_2 K N_{II,III}$	0.256923	8	48.256	0.24816	3	49.959
$K O_{II,III}$	0.255645	7	48.497	0.24687	3	50.221
$\beta_5 K M_{IV,V}$				0.25275	3	49.052
$\beta_4 L_I M_{II}$	1.9255	2	6.4389	1.8540	2	6.6871
$\beta_3 L_I M_{III}$	1.8867	2	6.5713	1.8150	2	6.8311
$\gamma_2 L_I N_{II}$	1.5961	2	7.7677	1.5331	2	8.087
$\gamma_3 L_I N_{III}$	1.5903	2	7.7961	1.5297	2	8.105
$\gamma_4 L_I O_{II,III}$	1.5439	1	8.0304	1.4839	2	8.355
$\eta L_{II} M_I$	2.1315	2	5.8166	2.0494	1	6.0495
$\beta_1 L_{II} M_{IV}$	1.9203	2	6.4564	1.8468	2	6.7132
$\gamma_5 L_{II} N_I$	1.7085	2	7.2566	1.6412	2	7.5543
$\gamma_1 L_{II} N_{IV}$	1.6574	2	7.4803	1.5924	2	7.7858
$\gamma_8 L_{II} O_I$	1.6346	2	7.5849	1.5707	2	7.894
$\gamma_6 L_{II} O_{IV}$	1.6282	2	7.6147	1.5644	2	7.925
$l L_{III} M_I$	2.3948	2	5.1772	2.3122	2	5.3621
$\alpha_2 L_{III} M_{IV}$	2.1315	2	5.8166	2.0578	2	6.0250
$\alpha_1 L_{III} M_V$	2.1209	2	5.8457	2.0468	2	6.0572
$\beta_6 L_{III} N_I$	1.8737	2	6.6170	1.8054	2	6.8671
$\beta_{2,15} L_{III} N_{IV,V}$	1.8118	2	6.8432	1.7455	2	7.1028
$\beta_7 L_{III} O_I$	1.7851	2	6.9453	1.7203	2	7.2071
$\beta_5 L_{III} O_{IV,V}$	1.7772	2	6.9763	1.7130	2	7.2374
$\beta_{10} L_I M_{IV}$	1.7993	3	6.890	1.7315	3	7.160
$\beta_9 L_I M_V$	1.7916	3	6.920	1.7240	3	7.192
$L_I O_{IV,V}$				1.4807	3	8.373
$\gamma M_{III} N_{IV,V}$	9.211	9	1.346	8.844	9	1.402
$\beta M_{IV} N_{VI}$	10.750	7	1.1533	10.254	6	1.2091
$\zeta M_V N_{III}$	14.22	2	0.872	13.57	2	0.914
$\alpha M_V N_{VI,VII}$	10.96	3	1.131	10.46	3	1.185
$N_{IV,V} O_{II,III}$	112.0	6	0.1107			

65 Terbium | 66 Dysprosium

Designation	Å*	p.e.	keV	Å*	p.e.	keV
$\alpha_2 K L_{II}$	0.283423	2	43.7441	0.274247	2	45.2078
$\alpha_1 K L_{III}$	0.278724	2	44.4816	0.269533	2	45.9984
$\beta K M_{II}$	0.24683	2	50.229	0.23862	2	51.957

X-Ray Wavelengths
TABLE A (Continued)

Designation	Å*	p.e.	keV	Å*	p.e.	keV
	65 Terbium (*Cont.*)			**66 Dysprosium** (*Cont.*)		
$\beta_1\,KM_{III}$	0.24608	2	50.382	0.23788	2	52.119
$\beta_2\,KN_{II,III}$	0.2397†	2	51.72	0.2317†	2	53.51
$KO_{II,III}$	0.23858	3	51.965	0.23056	3	53.774
$\beta_5\,KM_{IV,V}$				0.23618	3	52.494
$\beta_4\,L_IM_{II}$	1.7864	2	6.9403	1.72103	2	7.2039
$\beta_3\,L_IM_{III}$	1.7472	2	7.0959	1.6822	2	7.3702
$\gamma_2\,L_IN_{II}$	1.4764	2	8.398	1.42278	7	8.7140
$\gamma_3\,L_IN_{III}$	1.4718	2	8.423	1.41640	7	8.7532
$\gamma_4\,L_IO_{II,III}$	1.4276	2	8.685	1.37459	7	9.0195
$\eta\,L_{II}M_I$	1.9730	2	6.2839	1.89743	7	6.5342
$\beta_1\,L_{II}M_{IV}$	1.7768	3	6.978	1.71062	7	7.2477
$\gamma_5\,L_{II}N_I$	1.5787	2	7.8535	1.51824	7	8.1661
$\gamma_1\,L_{II}N_{IV}$	1.5303	2	8.102	1.47266	7	8.4188
$\gamma_8\,L_{II}O_I$	1.5097	2	8.212			
$\gamma_6\,L_{II}O_{IV}$	1.5035	2	8.246	1.44579	7	8.5753
$l\,L_{III}M_I$	2.2352	2	5.5467	2.15877	7	5.7431
$\alpha_2\,L_{III}M_{IV}$	1.9875	2	6.2380	1.91991	3	6.4577
$\alpha_1\,L_{II}M_V$	1.9765	2	6.2728	1.90881	3	6.4952
$\beta_6\,L_{III}N_I$	1.7422	2	7.1163	1.68213	7	7.3705
$\beta_{2,15}\,L_{III}N_{IV,V}$	1.6830	2	7.3667	1.62369	7	7.6357
$\beta_7\,L_{III}O_I$	1.6585	2	7.4753	1.60447	7	7.7272
$\beta_5\,L_{III}O_{IV,V}$	1.6510	2	7.5094	1.58837	7	7.8055
$\beta_{10}\,L_IM_{IV}$	1.6673	3	7.436	1.60743	9	7.7130
$\beta_9\,L_IM_V$				1.59970	9	7.7501
$L_IO_{IV,V}$	1.4228	3	8.714			
$\gamma\,M_{III}N_{IV,V}$	8.486	9	1.461	8.144	9	1.522
$\beta\,M_{IV}N_{VI}$	9.792	6	1.2661	9.357	6	1.3250
$\zeta\,M_VN_{III}$	12.98	2	0.955	12.43	2	0.998
$\alpha\,M_VN_{VI,VII}$	10.00	2	1.240	9.59	2	1.293
$N_{IV,V}N_{VI,VII}$	86.	1	0.144	83.	1	0.149
$N_{IV,V}O_{II,III}$	102.2	4	0.1213	97.2	8	0.128
	67 Holmium			**68 Erbium**		
$\alpha_2\,KL_{II}$	0.265486	2	46.6997	0.257110	2	48.2211
$\alpha_1\,KL_{III}$	0.260756	2	47.5467	0.252365	2	49.1277
$\beta_3\,KM_{II}$	0.23083	2	53.711	0.22341	2	55.494
$\beta_1\,KM_{III}$	0.23012	2	53.877	0.22266	2	55.681
$\beta_2\,KN_{II,III}$	0.2241†	2	55.32	0.2167†	2	57.21
$KO_{II,III}$	0.22305	3	55.584	0.21581	3	57.450
$\beta_5\,KM_{IV,V}$	0.22855	3	54.246	0.22124	3	56.040
$\beta_4\,L_IM_{II}$	1.6595	2	7.4708	1.6007	1	7.7453
$\beta_3\,L_IM_{III}$	1.6203	2	7.6519	1.5616	1	7.9392
$\gamma_2\,L_IN_{II}$	1.3698	2	9.051	1.3210	2	9.385
$\gamma_3\,L_IN_{III}$	1.3643	2	9.087	1.3146	2	9.4309
$\gamma_4\,L_IO_{II,III}$	1.3225	2	9.374	1.2752	2	9.722
$\eta\,L_{II}M_I$	1.8264	2	6.7883	1.7566	1	7.0579
$\beta_1\,L_{II}M_{IV}$	1.6475	2	7.5253	1.5873	1	7.8109
$\gamma_5\,L_{II}N_I$	1.4618	2	8.481	1.4067	3	8.814
$\gamma_1\,L_{II}N_{IV}$	1.4174	2	8.747	1.3641	2	9.089
$\gamma_8\,L_{II}O_I$	1.3983	2	8.867			
$\gamma_6\,L_{II}O_{IV}$	1.3923	2	8.905	1.3397	3	9.255
$l\,L_{III}M_I$	2.0860	2	5.9434	2.015	1	6.152
$\alpha_2\,L_{III}M_{IV}$	1.8561	2	6.6795	1.7955	2	6.9050
$\alpha_1\,L_{III}M_V$	1.8450	2	6.7198	1.78425	9	6.9487
$\beta_6\,L_{III}N_I$	1.6237	2	7.6359	1.5675	2	7.909
$\beta_{2,15}\,L_{III}N_{IV,V}$	1.5671	2	7.911	1.51399	9	8.1890
$\beta_7\,L_{III}O_I$				1.4941	3	8.298
$\beta_5\,L_{III}O_{IV,V}$	1.5378	2	8.062	1.4848	3	8.350

Right table

Designation	Å*	p.e.	keV	Å*	p.e.	keV
	67 Holmium (*Cont.*)			**68 Erbium** (*Cont.*)		
$\beta_{10}\,L_IM_{IV}$	1.5486	3	8.006	1.4941	3	8.298
$L_IO_{IV,V}$	1.3208	3	9.387			
$\beta_9\,L_IM_V$				1.4855	5	8.346
$M\,N_{IV}$				7.60	1	1.632
$\gamma\,M_{III}N_{IV,V}$	7.865	9	1.576			
$\gamma\,M_{III}\;v$				7.546	8	1.643
$\beta\,M_{IV}N_{VI}$	8.965	4	1.3830	8.592	3	1.4430
$\zeta\,M_VN_{III}$	11.86	1	1.0450	11.37	1	1.0901
$\alpha\,M_VN_{VI,VII}$	9.20	2	1.348	8.82	1	1.406
$N_{IV}N_{VI}$				72.7	9	0.171
$N_VN_{VI,VII}$				76.3	7	0.163
	69 Thulium			**70 Ytterbium**		
$\alpha_2\,KL_{II}$	0.249095	2	49.7726	0.241424	2	51.3540
$\alpha_1\,KL_{III}$	0.244338	2	50.7416	0.236655	2	52.3889
$\beta_3\,KM_{II}$	0.21636	2	57.304	0.2096†	1	59.14
$\beta_1\,KM_{III}$	0.21556	2	57.517	0.20884	8	59.37
$\beta_2\,KN_{II,III}$	0.2098†	2	59.09	0.2033†	2	60.98
$KO_{II,III}$	0.20891	2	59.346	0.20226	2	61.298
$\beta_5\,KM_{IV,V}$	0.21404	2	57.923	0.20739	2	59.782
$\beta_4\,L_IM_{II}$	1.5448	2	8.026	1.49138	3	8.3132
$\beta_3\,L_IM_{III}$	1.5063	2	8.231	1.45233	5	8.5367
$\gamma_2\,L_IN_{II}$	1.2742	2	9.730	1.22879	7	10.0897
$\gamma_3\,L_IN_{III}$	1.2670	2	9.779	1.22232	5	10.1431
$\gamma_4\,L_IO_{II,III}$	1.2294	2	10.084	1.1853	1	10.4603
$\eta\,L_{II}M_I$	1.6963	2	7.3088	1.63560	5	7.5802
$\beta_1\,L_{II}M_{IV}$	1.5304	2	8.101	1.47565	5	8.4018
$\gamma_5\,L_{II}N_I$	1.3558	2	9.144	1.3063	1	9.4910
$\gamma_1\,L_{II}N_{IV}$	1.3153	2	9.426	1.26769	5	9.8701
$\gamma_8\,L_{II}O_I$				1.24923	5	9.9246
$\gamma_6\,L_{II}O_{IV}$	1.2905	2	9.607	1.24271	3	9.9766
$l\,L_{III}M_I$	1.9550	2	6.3419	1.89415	5	6.5455
$\alpha_2\,L_{III}M_{IV}$	1.7381	2	7.1331	1.68285	5	7.3673
$\alpha_1\,L_{III}M_V$	1.7268†	2	7.1799	1.67189	4	7.4156
$\beta_6\,L_{III}N_I$	1.5162	2	8.177	1.4661	1	8.4563
$\beta_{2,15}\,L_{III}N_{IV,V}$	1.4640	2	8.468	1.41550	5	8.7588
$\beta_7\,L_{III}O_I$				1.3948	1	8.8889
$\beta_5\,L_{III}O_{IV,V}$	1.4349	2	8.641	1.38696	7	8.9390
$\beta_{10}\,L_IM_{IV}$	1.4410	3	8.604	1.3915	1	8.9100
$\beta_9\,L_IM_V$	1.4336	3	8.648	1.3838	1	8.9597
L_IO_I				1.1886	1	10.4312
$L_IO_{IV,V}$	1.2263	3	10.110	1.1827	1	10.4833
$L_{II}M_{II}$				1.58844	9	7.8052
$L_{II}O_{II,III}$				1.2453	1	9.9561
$t\,L_{III}M_{II}$				1.83091	9	6.7715
$L_{III}O_{II,III}$				1.3898	1	8.9209
$M_{III}N_I$				8.470	9	1.464
$\gamma\,M_{III}N_V$				7.024	8	1.765
$\beta\,M_{IV}N_{VI}$	8.249	7	1.503	7.909	2	1.5675
$\zeta\,M_VN_{III}$				10.48	1	1.183
$\alpha\,M_VN_{VI,VII}$	8.48	1	1.462	8.149	5	1.5214
$N_{IV}N_{VI}$				65.1	7	0.190
$N_VN_{VI,VII}$				69.3	5	0.179
	71 Lutetium			**72 Hafnium**		
$\alpha_2\,KL_{II}$	0.234081	2	52.9650	0.227024	3	54.6114
$\alpha_1\,KL_{III}$	0.229298	2	54.0698	0.222227	3	55.7902
$\beta_3\,KM_{II}$	0.20309†	4	61.05	0.19686†	4	62.98

Designation	Å*	p.e.	keV	Å*	p.e.	keV
	71 Lutetium (*Cont.*)			**72 Hafnium** (*Cont.*)		
β₁ KM_III	0.20231†	3	61.283	0.19607†	3	63.234
β₂ KN_II,III	0.1969†	2	62.97	0.1908†	2	64.98
KO_II,III	0.19589	2	63.293			
β₅ KM_IV,V	0.20084	2	61.732			
β₄ L_IM_II	1.44056	5	8.6064	1.39220	5	8.9054
β₃ L_IM_III	1.40140	5	8.8469	1.35300	5	9.1634
γ₂ L_IN_II	1.1853	2	10.460	1.14442	5	10.8335
γ₃ L_IN_III	1.17953	4	10.5110	1.13841	5	10.8907
γ'₄ L_IO_II				1.10376	5	11.2326
γ₄ L_IO_II,III	1.1435	1	10.8425	1.10303	5	11.2401
η L_IIM_I	1.5779	1	7.8575	1.52325	5	8.1393
β₁ L_IIM_IV	1.42359	3	8.7090	1.37410	5	9.0227
γ₅ L_IIN_I	1.2596	1	9.8428	1.21537	5	10.2011
γ₁ L_IIN_IV	1.22228	4	10.1434	1.17900	5	10.5158
γ₈ L_IIO_I	1.2047	1	10.2915	1.16138	5	10.6754
γ₆ L_IIO_IV	1.1987	1	10.3431	1.15519	5	10.7325
l L_IIIM_I	1.8360	1	6.7528	1.78145	5	6.9596
α₂ L_IIIM_IV	1.63029	5	7.6049	1.58046	5	7.8446
α₁ L_IIIM_IV	1.61951	3	7.6555	1.56958	5	7.8990
β₆ L_IIIN_I	1.4189	1	8.7376	1.37410	5	9.0227
β₁₅ L_IIIN_IV	1.3715	1	9.0395	1.32783	5	9.3371
β₂ L_IIIN_V	1.37012	3	9.0489	1.32639	5	9.3473
β₇ L_IIIO_I	1.34949	5	9.1873	1.30564	5	9.4958
β₅ L_IIIO_IV,V	1.34183	7	9.2397	1.29761	5	9.5546
L_IM_I				1.43025	9	8.6685
β₁₀ L_IM_IV	1.3430	2	9.232	1.29819	9	9.5503
β₉ L_IM_V	1.3358	1	9.2816	1.29025	9	9.6090
L_IN_IV	1.16227	9	10.6672	1.12250	9	11.0451
γ₁₁ L_IN_V	1.16107	9	10.6782	1.12146	9	11.0553
L_IO_I				1.10664	9	11.2034
L_IO_IV				1.10086	9	11.2622
L_IIM_II	1.53333	9	8.0858	1.48064	9	8.3735
β₁₇ L_IIM_III				1.43643	9	8.6312
L_IIN_V				1.17788	9	10.5258
v L_IIN_VI				1.15830	9	10.7037
L_IIO_II,III	1.2014	1	10.3198			
t L_IIIM_II	1.7760	1	6.9810	1.72305	9	7.1954
s L_IIIM_III				1.66346	9	7.4532
L_IIIN_II				1.35887	9	9.1239
L_IIIN_III				1.35053	9	9.1802
u L_IIIN_VI,VII				1.30165	9	9.5249
L_IIIO_II,III	1.34524	9	9.2163			
M_IIIN_I				7.887	9	1.572
γ M_IIIN_V	6.768	6	1.832	6.544	4	1.895
ζ₂				9.686	7	1.2800
β M_IVN_VI	7.601	2	1.6312	7.303	1	1.6976
ζ₁				9.686	7	1.2800
α M_VN_VI,VII	7.840	2	1.5813	7.539	1	1.6446
N_IVN_VI	63.0	5	0.197			
N_VN_VI,VII	65.7	2	0.1886			
	73 Tantalum			**74 Tungsten**		
α₂ KL_II	0.220305	8	56.277	0.213828	2	57.9817
α₁ KL_III	0.215497	4	57.532	0.2090100	Std	59.31824
β₃ KM_II	0.190890	2	64.9488	0.185181	2	66.9514
β₁ KM_III	0.190089	4	65.223	0.184374	2	67.2443
β₂^II KN_II	0.185188	9	66.949	0.17960	1	69.031
β₄^I KN_III	0.185011	8	67.013	0.179421	7	69.101

Designation	Å*	p.e.	keV	Å*	p.e.	keV
	73 Tantalum (*Cont.*)			**74 Tungsten** (*Cont.*)		
KO_II,III	0.184031	7	67.370	0.178444	5	69.479
KL_I				0.21592	4	57.42
β₅^II KM_IV	0.188920	6	65.626	0.183264	5	67.652
β₅^I KM_V	0.188757	6	65.683	0.183092	7	67.715
β₄ KN_IV,V	0.18451	1	67.194	0.17892	2	69.294
β₄ L_IM_II	1.34581	3	9.2124	1.30162	5	9.5252
β₃ L_IM_III	1.30678	3	9.4875	1.26269	5	9.8188
γ₂ L_IN_II	1.1053	1	11.217	1.06806	3	11.6080
γ₃ L_IN_III	1.09936	4	11.2776	1.06200	6	11.6743
γ'₄ L_IO_II	1.06544	3	11.6366	1.02863	3	12.0530
γ₄ L_IO_III	1.06467	3	11.6451	1.02775	3	12.0634
η L_IIM_I	1.47106	5	8.4280	1.42110	3	8.7243
β₁ L_IIM_IV	1.32698	3	9.3431	1.281809	9	9.67235
γ₅ L_IIN_I	1.1729	1	10.5702	1.13235	3	10.9490
γ₁ L_IIN_IV	1.13794	3	10.8952	1.09855	3	11.2859
γ₈ L_IIO_I	1.1205	1	11.0646	1.08113	4	11.4677
γ₆ L_IIO_IV	1.11388	3	11.1306	1.07448	5	11.5387
l L_IIIM_I	1.72841	5	7.1731	1.6782	1	7.3878
α₂ L_IIIM_IV	1.53293	2	8.0879	1.48743	2	8.3352
α₁ L_IIIM_V	1.52197	2	8.1461	1.47639	2	8.3976
β₆ L_IIIN_I	1.33094	8	9.3153	1.28989	7	9.6117
β₁₅ L_IIIN_IV	1.28619	5	9.6394	1.24631	3	9.9478
β₂ L_IIIN_V	1.28454	2	9.6518	1.24460	3	9.9615
β₇ L_IIIO_I	1.26385	5	9.8098	1.22400	4	10.1292
β₅ L_IIIO_IV,V	1.2555	1	9.8750	1.21545	3	10.2004
L_IM_I				1.3365	3	9.277
β₁₀ L_IM_IV	1.2537	2	9.889	1.21218	3	10.2279
β₉ L_IM_V	1.2466	2	9.946	1.20479	7	10.2907
L_IN_I	1.11521	9	11.1173			
L_IN_IV	1.08377	7	11.4398	1.0468	2	11.844
γ₁₁ L_IN_V	1.08205	7	11.4580	1.0458	1	11.856
L_IN_VI,VII	1.06357	9	11.6570			
L_IO_I	1.06771	9	11.6118	1.0317	3	12.017
L_IO_IV,V	1.06192	9	11.6752	1.0250	2	12.095
L_IIM_II	1.43048	9	8.6671			
β₁₇ L_IIM_III	1.3864	1	8.9428	1.3387	2	9.261
L_IIM_V	1.31897	9	9.3998	1.2728	2	9.741
L_IIN_II	1.1600	2	10.688	1.1218	3	11.052
L_IIN_III	1.1553	1	10.7316	1.1149	2	11.120
L_IIN_V	1.13687	9	10.9055			
v L_IIN_VI	1.1158	1	11.1113	1.0771	1	11.510
L_IIO_II	1.11789	9	11.0907			
L_IIO_III	1.11693	9	11.1001	1.0792	2	11.488
t L_IIIM_II	1.67265	9	7.4123	1.6244	3	7.632
s L_IIIM_III	1.61264	9	7.6881	1.5642	3	7.926
L_IIIN_II	1.3167	1	9.4158	1.2765	2	9.712
L_IIIN_III	1.3086	1	9.4742	1.2672	2	9.784
u L_IIIN_VI,VII	1.25778	4	9.8572	1.21868	5	10.1733
L_IIIO_II,III	1.2601	3	9.839	1.2211	2	10.153
M_IN_III	5.40	2	2.295	5.172	9	2.397
M_IO_II,III				4.44	2	2.79
M_IIN_I				6.28	2	1.973
M_IIN_IV	5.570	4	2.226	5.357	4	2.314
M_IIIN_I	7.612	9	1.629	7.360	8	1.684
M_IIIN_IV	6.353	5	1.951	6.134	4	2.021
γ M_IIIN_V	6.312	4	1.964	6.092	3	2.035
M_IIIO_I	5.83	2	2.126	5.628	8	2.203
M_IIIO_IV,V	5.67	3	2.19			

73 Tantalum (*Cont.*) | 74 Tungsten (*Cont.*)

Designation	Å*	p.e.	keV	Å*	p.e.	keV
$\zeta_2\,M_{IV}N_{II}$	9.330	5	1.3288	8.993	5	1.3787
$M_{IV}N_{III}$	8.90	2	1.393	8.573	8	1.446
$\beta\,M_{IV}N_{VI}$	7.023	1	1.7655	6.757	1	1.8349
$M_{IV}O_{II}$	7.09	2	1.748	6.806	9	1.822
$\zeta_1\,M_VN_{III}$	9.316	4	1.3308	8.962	4	1.3835
$\alpha\,M_VN_{VI,VII}$	7.252	1	1.7096			
$\alpha_2\,M_VN_{VI}$				6.992	2	1.7731
$\alpha_1\,M_VN_{VII}$				6.983	1	1.7754
M_VO_{III}	7.30	2	1.700	7.005	9	1.770
$N_{II}N_{IV}$				54.0	2	0.2295
$N_{IV}N_{VI}$	58.2	1	0.2130	55.8	1	0.2221
$N_VN_{VI,VII}$	61.1	2	0.2028			
N_VN_{VI}				59.5	3	0.208
N_VN_{VII}				58.4	1	0.2122

75 Rhenium | 76 Osmium

Designation	Å*	p.e.	keV	Å*	p.e.	keV
$\alpha_2\,KL_{II}$	0.207611	1	59.7179	0.201639	2	61.4867
$\alpha_1\,KL_{III}$	0.202781	2	61.1403	0.196794	2	63.0005
$\beta_3\,KM_{II}$	0.179697	3	68.994	0.174431	3	71.077
$\beta_1\,KM_{III}$	0.178880	3	69.310	0.173611	3	71.413
$\beta_2^{II}\,KN_{II}$	0.17425	1	71.151	0.16910	1	73.318
$\beta_2^{I}\,KN_{III}$	0.174054	6	71.232	0.168906	6	73.402
$KO_{II,III}$	0.17308	1	71.633	0.16798	1	73.808
$\beta_5^{II}\,KM_{IV}$	0.17783	1	69.719	0.17262	1	71.824
$\beta_5^{I}\,KM_V$	0.17766	1	69.786	0.17245	1	71.895
$\beta_4\,KN_{IV,V}$	0.17362	2	71.410	0.16842	2	73.615
$\beta_4\,L_IM_{II}$	1.25917	5	9.8463	1.21844	5	10.1754
$\beta_3\,L_IM_{III}$	1.22031	5	10.1598	1.17955	7	10.5108
$\gamma_2\,L_IN_{II}$	1.03233	5	12.0098	0.99805	5	12.4224
$\gamma_3\,L_IN_{III}$	1.02613	7	12.0824	0.99186	5	12.4998
$\gamma'_4\,L_IO_{II}$	0.99334	5	12.4813	0.96033	8	12.910
$\gamma_4\,L_IO_{III}$	0.99249	5	12.4920	0.95938	8	12.923
$\eta\,L_{II}M_I$	1.37342	5	9.0272	1.32785	7	9.3370
$\beta_1\,L_{II}M_{IV}$	1.23858	2	10.0100	1.19727	7	10.3553
$\gamma_5\,L_{II}N_I$	1.09388	5	11.3341	1.05693	5	11.7303
$\gamma_1\,L_{II}N_{IV}$	1.06099	5	11.6854	1.02503	5	12.0953
$\gamma_8\,L_{II}O_I$	1.04398	5	11.8758	1.00788	5	12.3012
$\gamma_6\,L_{II}O_{IV}$	1.03699	9	11.956	1.00107	5	12.3848
$l\,L_{III}M_I$	1.63056	5	7.6036	1.58498	7	7.8222
$\alpha_2\,L_{III}M_{IV}$	1.44396	5	8.5862	1.40234	5	8.8410
$\alpha_1\,L_{III}M_V$	1.43290	4	8.6525	1.39121	5	8.9117
$\beta_6\,L_{III}N_I$	1.25100	5	9.9105	1.21349	5	10.2169
$\beta_{15}\,L_{III}N_{IV}$	1.20819	5	10.2617	1.17167	5	10.5816
$\beta_2\,L_{III}N_V$	1.20660	4	10.2752	1.16979	8	10.5985
$\beta_7\,L_{III}O_I$	1.18610	5	10.4529	1.14933	8	10.7872
$\beta_5\,L_{III}O_{IV,V}$	1.17721	5	10.5318	1.1405	1	10.8711
$\beta_{10}\,L_IM_{IV}$	1.17218	5	10.5770	1.13353	5	10.9376
$\beta_9\,L_IM_V$	1.16487	4	10.6433	1.12637	6	11.0071
L_IN_I	1.0420	1	11.899			
L_IN_{IV}	1.0119	1	12.252	0.9772	3	12.687
$\gamma_{11}\,L_IN_V$	1.0108	1	12.266	0.9765	3	12.696
L_IO_I	0.9965	1	12.442	0.96318	7	12.8721
$L_IO_{IV,V}$	0.9900	1	12.524	0.95603	5	12.9683
$L_{II}M_{II}$	1.3366	1	9.2761	1.2934	2	9.586
$\beta_{17}\,L_{II}M_{III}$	1.2927	1	9.5910	1.2480	2	9.934

75 Rhenium (*Cont.*) | 76 Osmium (*Cont.*)

Designation	Å*	p.e.	keV	Å*	p.e.	keV
$L_{II}M_V$	1.2305	1	10.0753	1.18977	7	10.4205
$L_{II}N_{II}$	1.0839	1	11.438			
$L_{II}N_{III}$	1.0767	1	11.515	1.03973	5	11.9243
$v\,L_{II}N_{VI}$	1.0404	1	11.917	1.0050	2	12.337
$L_{II}O_{III}$	1.0397	1	11.925	1.0047	2	12.340
$t\,L_{III}M_{II}$	1.5789	1	7.8525	1.5347	2	8.079
$s\,L_{III}M_{III}$	1.5178	1	8.1682	1.4735	2	8.414
$L_{III}N_I$				1.20086	7	10.3244
$L_{III}N_{III}$	1.2283	1	10.0933			
$u\,L_{III}N_{VI,VII}$	1.1815	1	10.4931	1.14537	7	10.8245
M_IN_{III}				4.79	2	2.59
$M_{II}N_I$				5.81	2	2.133
$M_{II}N_{IV}$				4.955	4	2.502
$M_{III}N_I$				6.89	2	1.798
$M_{III}N_{IV}$	5.931	5	2.090	5.724	5	2.166
$\gamma\,M_{III}N_V$	5.885	2	2.1067	5.682	4	2.182
$\zeta_2\,M_{IV}N_{II}$	8.664	5	1.4310	8.359	5	1.4831
$M_{IV}N_{III}$	8.239	8	1.505			
$\beta\,M_{IV}N_{VI}$	6.504	1	1.9061	6.267	1	1.9783
$\zeta_1\,M_VN_{III}$	8.629	4	1.4368	8.310	4	1.4919
$\alpha\,M_VN_{VI,VII}$	6.729	1	1.8425	6.490	1	1.9102
$N_{IV}N_{VI}$				51.9	1	0.2388
$N_VN_{VI,VII}$				54.7	2	0.2266

77 Iridium | 78 Platinum

Designation	Å*	p.e.	keV	Å*	p.e.	keV
$\alpha_2\,KL_{II}$	0.195904	2	63.2867	0.190381	4	65.122
$\alpha_1\,KL_{III}$	0.191047	2	64.8956	0.185511	4	66.832
$\beta_3\,KM_{II}$	0.169367	2	73.2027	0.164501	3	75.368
$\beta_1\,KM_{III}$	0.168542	2	73.5608	0.163675	3	75.748
$\beta_2^{II}\,KN_{II}$	0.16415	1	75.529	0.15939	1	77.785
$\beta_2^{I}\,KN_{III}$	0.163956	7	75.619	0.15920	1	77.878
$KO_{II,III}$	0.163019	5	76.053	0.15826	1	78.341
$\beta_5^{II}\,KM_{IV}$	0.16759	2	73.980	0.16271	2	76.199
$\beta_5^{I}\,KM_V$	0.167373	9	74.075	0.16255	3	76.27
$\beta_4\,KN_{IV,V}$	0.16352	2	75.821	0.15881	2	78.069
$\beta_4\,L_IM_{II}$	1.17958	3	10.5106	1.14223	5	10.8543
$\beta_3\,L_IM_{III}$	1.14085	3	10.8674	1.10394	5	11.2308
$\gamma_2\,L_IN_{II}$	0.96545	3	12.8418	0.93427	5	13.2704
$\gamma_3\,L_IN_{III}$	0.95931	5	12.9240	0.92791	5	13.3613
$\gamma'_4\,L_IO_{II}$	0.92831	3	13.3555	0.89747	4	13.8145
$\gamma_4\,L_IO_{III}$	0.92744	3	13.3681	0.89659	4	13.8281
$\eta\,L_{II}M_I$	1.28448	3	9.6522	1.2429	2	9.975
$\beta_1\,L_{II}M_{IV}$	1.15781	3	10.7083	1.11990	2	11.0707
$\gamma_5\,L_{II}N_I$	1.02175	5	12.1342	0.9877	2	12.552
$\gamma_1\,L_{II}N_{IV}$	0.99085	3	12.5126	0.95797	3	12.9420
$\gamma_8\,L_{II}O_I$	0.97409	3	12.7279	0.9411	1	13.173
$\gamma_6\,L_{II}O_{IV}$	0.96708	4	12.8201	0.9342	2	13.271
$l\,L_{III}M_I$	1.54094	3	8.0458	1.4995	2	8.268
$\alpha_2\,L_{III}M_{IV}$	1.36250	5	9.0995	1.32432	2	9.3618
$\alpha_1\,L_{III}M_V$	1.35128	3	9.1751	1.31304	3	9.4423
$\beta_6\,L_{III}N_I$	1.17796	3	10.5251	1.14355	5	10.8418
$\beta_{15}\,L_{III}N_{IV}$	1.13707	3	10.9036			
$\beta_2\,L_{III}N_V$	1.13532	3	10.9203	1.10200	3	11.2505
$\beta_7\,L_{III}O_I$	1.11489	3	11.1205	1.08168	3	11.4619

Designation	Å*	p.e.	keV	Å*	p.e.	keV	Designation	Å*	p.e.	keV	Å*	p.e.	keV
	77 Iridium (*Cont.*)			**78 Platinum** (*Cont.*)				**79 Gold** (*Cont.*)			**80 Mercury** (*Cont.*)		
β_5 $L_{III}O_{IV,V}$	1.10585	3	11.2114	1.0724	2	11.561	γ_2 L_IN_{II}	0.90434	3	13.7095	0.87544	7	14.162
L_IM_I	1.2102	2	10.245	1.16962	9	10.6001	γ_3 L_IN_{III}	0.89783	5	13.8090	0.86915	7	14.265
β_{10} L_IM_{IV}	1.09702	4	11.3016	1.06183	7	11.6762	γ'_4 L_IO_{II}	0.86816	4	14.2809	0.84013	7	14.757
β_9 L_IM_V	1.08975	5	11.3770	1.05446	5	11.7577	γ_4 L_IO_{III}	0.86703	4	14.2996	0.83894	7	14.778
L_IN_I	0.9766	2	12.695	0.9455	2	13.113	η $L_{II}M_I$	1.20273	3	10.3083	1.1640	1	10.6512
L_IN_{IV}	0.9459	2	13.108				β_1 $L_{II}M_{IV}$	1.08353	3	11.4423	1.04868	5	11.8226
γ_{11} L_IN_V	0.9446	2	13.126	0.9143	2	13.560	γ_5 $L_{II}N_I$	0.95559	3	12.9743	0.92453	7	13.410
$L_IO_{IV,V}$	0.9243	3	13.413				γ_1 $L_{II}N_{IV}$	0.92650	3	13.3817	0.89646	5	13.8301
L_IO_I				0.8995	2	13.784	γ_8 $L_{II}O_I$	0.90989	5	13.6260	0.87995	7	14.090
L_IO_{IV}				0.8943	1	13.864	γ_6 $L_{II}O_{IV}$	0.90297	3	13.7304	0.87319	7	14.199
L_IO_V				0.8934	1	13.878	l $L_{III}M_I$	1.45964	9	8.4939	1.4216	1	8.7210
$L_{II}M_{II}$	1.2502	3	9.917	1.213	1	10.225	α_2 $L_{III}M_{IV}$	1.28772	3	9.6280	1.25264	5	9.8976
β_{17} $L_{II}M_{III}$	1.2069	2	10.273	1.1667	1	10.6265	α_1 $L_{III}M_V$	1.27640	3	9.7133	1.24120	5	9.9888
$L_{II}M_V$	1.1489	2	10.791	1.1129	2	11.140	β_6 $L_{III}N_I$	1.11092	3	11.1602	1.07975	7	11.4824
$L_{II}N_{II}$	1.0120	2	12.251	0.9792	2	12.661	β_{15} $L_{III}N_{IV}$	1.07188	5	11.5667	1.04151	7	11.9040
$L_{II}N_{III}$	1.0054	3	12.332	0.97173	4	12.7588	β_2 $L_{III}N_V$	1.07022	3	11.5847	1.03975	7	11.9241
v $L_{II}N_{VI}$	0.97161	6	12.7603	0.93931	5	13.1992	β_7 $L_{III}O_I$	1.04974	8	11.8106	1.01937	7	12.1625
$L_{II}O_{II}$	0.96979	5	12.7843				β_5 $L_{III}O_{IV,V}$	1.04044	3	11.9163	1.00987	7	12.2769
t $L_{III}M_{II}$	1.4930	3	8.304	1.4530	2	8.533	L_IM_I	1.13525	5	10.9210	1.0999	2	11.272
s $L_{III}M_{III}$	1.4318	2	8.659	1.3895	2	8.923	β_{10} L_IM_{IV}	1.02789	7	12.0617	0.9962	2	12.446
$L_{III}N_{II}$	1.16545	5	10.6380	1.1310	2	10.962	β_9 L_IM_V	1.02063	7	12.1474	0.9871	2	12.560
$L_{III}N_{III}$	1.1560	3	10.725	1.1226	2	11.044	L_IN_I	0.9131	1	13.578	0.8827	2	14.045
u $L_{III}N_{VI,VII}$	1.11145	4	11.1549	1.07896	5	11.4908	L_IN_{IV}	0.88563	7	13.999			
$L_{III}O_{II,III}$	1.10923	6	11.1772	1.0761	3	11.521	γ_{11} L_IN_V	0.88433	7	14.020	0.85657	7	14.474
M_IN_{III}	4.631†	9	2.677	4.460	9	2.780	L_IO_I	0.87074	5	14.2385	0.8452	2	14.670
$M_{II}N_{IV}$	4.780	4	2.594	4.601	4	2.695	$L_IO_{IV,V}$	0.86400	5	14.3497	0.8350	2	14.847
$M_{III}N_I$	6.669	9	1.859	6.455	9	1.921	$L_{II}M_{II}$	1.1708	1	10.5892	1.1387	5	10.888
$M_{III}N_{IV}$	5.540	5	2.238	5.357	5	2.314	β_{17} $L_{II}M_{III}$	1.12798	5	10.9915	1.0916	5	11.358
γ $M_{II}N_V$	5.500	4	2.254	5.319	4	2.331	$L_{II}M_V$	1.0756	2	11.526			
$M_{III}O_I$				4.876	9	2.543	$L_{II}N_{III}$	0.9402	2	13.186	0.90894	7	13.640
$M_{III}O_{IV}$ v	4.869	9	2.546	4.694	8	2.641	v $L_{II}N_{VI}$	0.90837	5	13.6487	0.87885	7	14.107
ζ_2 $M_{IV}N_{II}$	8.065	5	1.5373	7.790	5	1.592	$L_{II}O_{II}$	0.90746	7	13.662	0.8784	1	14.114
$M_{IV}N_{III}$	7.645	8	1.622	7.371	8	1.682	$L_{II}O_{III}$	0.90638	7	13.679	0.8758	1	14.156
β $M_{IV}N_{VI}$	6.038	1	2.0535	5.828	1	2.1273	t $L_{III}M_{II}$	1.41366	7	8.7702	1.3746	2	9.019
ζ_1 M_VN_{III}	8.021	4	1.5458	7.738	4	1.6022	s $L_{III}M_{III}$	1.35131	7	9.1749	1.3112	2	9.455
α_2 M_VN_{VI}	6.275	3	1.9758	6.058	3	2.047	$L_{III}N_{II}$	1.09968	7	11.2743	1.0649	2	11.642
α_1 M_VN_{VII}	6.262	1	1.9799	6.047	1	2.0505	$L_{III}N_{III}$	1.09026	7	11.3717	1.0585	1	11.713
M_VO_{III}				5.987	9	2.071	u $L_{III}N_{VI,VII}$	1.04752	5	11.8357			
$N_{IV}N_{VI}$	50.2	1	0.2470	48.1	2	0.258	u' $L_{III}N_{VI}$				1.01769	7	12.1826
$N_VN_{VI,VII}$	52.8	1	0.2348	50.9	1	0.2436	u $L_{III}N_{VII}$				1.01674	7	12.1940
	79 Gold			**80 Mercury**			$L_{III}O_{II,III}$	1.0450	2	11.865			
							$L_{III}O_{II}$				1.01558	7	12.2079
α_2 KL_{II}	0.185075	2	66.9895	0.179958	3	68.895	$L_{III}O_{III}$				1.01404	7	12.2264
α_1 KL_{III}	0.180195	2	68.8037	0.175068	3	70.819	$L_{III}P_{II,III}$	1.03876	7	11.9355			
β_1 KM_{II}	0.159810	2	77.580	0.155321	3	79.822	M_IN_{III}	4.300	9	2.883			
β_1 KM_{III}	0.158982	3	77.984	0.154487	3	80.253	$M_{II}N_{IV}$	4.432	4	2.797			
β_2^{II} KN_{II}	0.15483	2	80.08	0.15040	2	82.43	$M_{III}N_I$	6.259	9	1.981	6.09	2	2.036
β_2^I KN_{III}	0.154618	9	80.185	0.15020	2	82.54	$M_{III}N_{IV}$	5.186	5	2.391			
$KO_{II,III}$	0.153694	7	80.667	0.14931	2	83.04	γ $M_{III}N_V$	5.145	4	2.410	4.984†	2	2.4875
KL_I	0.18672	4	66.40				$M_{III}O_I$	4.703	9	2.636			
β_5^{II} KM_{IV}	0.158062	7	78.438				$M_{III}O_{IV,V}$	4.522	6	2.742			
β_5^I KM_V	0.157880	5	78.529				ζ_2 $M_{IV}N_{II}$	7.523	5	1.648			
β_5 $KM_{IV,V}$				0.15353	2	80.75	$M_{IV}N_{III}$	7.101	8	1.746	6.87	2	1.805
β_4 $KN_{IV,V}$	0.154224	5	80.391	0.14978	2	82.78	β $M_{IV}N_{VI}$	5.624	1	2.2046	5.4318†	9	2.2825
β_4 L_IM_{II}	1.10651	3	11.2047	1.07222	7	11.5630	ζ_1 M_VN_{III}	7.466	4	1.6605			
β_3 L_IM_{III}	1.06785	9	11.6103	1.03358	7	11.9953	α_2 M_VN_{VI}	5.854	3	2.118			

Designation	Å*	p.e.	keV	Å*	p.e.	keV
79 Gold (Cont.)				**80 Mercury (Cont.)**		
$\alpha_1\ M_V N_{VII}$	5.840	1	2.1229	5.6476†	9	2.1953
$M_V O_{III}$	5.767	9	2.150			
$N_{IV} N_{VI}$	46.8	2	0.265	45.2†	3	0.274
$N_V N_{VI,VII}$	49.4	1	0.2510	47.9†	3	0.259
81 Thallium				**82 Lead**		
$\alpha_2\ K L_{II}$	0.175036	2	70.8319	0.170294	2	72.8042
$\alpha_1\ K L_{III}$	0.170136	2	72.8715	0.165376	2	74.9694
$\beta_3\ K M_{II}$	0.150980	6	82.118	0.146810	4	84.450
$\beta_1\ K M_{III}$	0.150142	5	82.576	0.145970	6	84.936
$\beta_2^{II}\ K N_{II}$	0.14614	1	84.836	0.14212	2	87.23
$\beta_2^{I}\ K N_{III}$	0.14595	1	84.946	0.14191	1	87.364
$K O_{II,III}$	0.14509	1	85.451	0.141012	8	87.922
$K P$				0.1408	1	88.06
$\beta_5\ K M_{IV,V}$	0.14917	1	83.114			
$\beta_5^{II}\ K M_{IV}$				0.14512	2	85.43
$\beta_5^{I}\ K M_V$				0.14495	3	85.53
$\beta_4\ K N_{IV,V}$	0.14553	2	85.19	0.14155	3	87.59
$\beta_4\ L_I M_{II}$	1.03918	3	11.9306	1.0075	1	12.306
$\beta_3\ L_I M_{III}$	1.00062	3	12.3904	0.96011	7	12.7930
$\gamma_2\ L_I N_{II}$	0.84773	5	14.6251	0.8210	2	15.101
$\gamma_3\ L_I N_{III}$	0.84130	4	14.7368	0.8147	1	15.218
$\gamma'_4\ L_I O_{II}$	0.81308	5	15.2482	0.78706	7	15.752
$\gamma_4\ L_I O_{III}$	0.81184	5	15.2716	0.7858	1	15.777
$\eta\ L_{II} M_I$	1.12769	3	10.9943	1.09241	7	11.3493
$\beta_1\ L_{II} M_{IV}$	1.01513	4	12.2133	0.98291	3	12.6137
$\gamma_5\ L_{II} N_I$	0.89500	4	13.8526	0.86655	5	14.3075
$\gamma_1\ L_{II} N_{IV}$	0.86752	3	14.2915	0.83973	3	14.7644
$\gamma_8\ L_{II} O_I$	0.8513	2	14.564	0.82365	5	15.0527
$\gamma_6\ L_{II} O_{IV}$	0.8442	2	14.685	0.81683	5	15.1783
$L_{II} P_I$				0.81583	5	15.1969
$l\ L_{III} M_I$	1.38477	3	8.9532	1.34990	7	9.1845
$\alpha_2\ L_{III} M_{IV}$	1.21875	3	10.1728	1.18648	5	10.4495
$\alpha_1\ L_{III} M_V$	1.20739	4	10.2685	1.17501	2	10.5515
$\beta_6\ L_{III} N_I$	1.04963	5	11.8118	1.0210	1	12.143
$\beta_{15}\ L_{III} N_{IV}$	1.01201	3	12.2510	0.98389	7	12.6011
$\beta_2\ L_{III} N_V$	1.01031	3	12.2715	0.98221	7	12.6226
$\beta_7\ L_{III} O_I$	0.99017	5	12.5212	0.9620	1	12.888
$\beta_5\ L_{III} O_{IV,V}$	0.98058	3	12.6436	0.9526	1	13.015
$L_I M_I$	1.0644	2	11.648	1.0323	2	12.010
$\beta_{10}\ L_I M_{IV}$	0.96389	7	12.8626	0.9339	2	13.275
$\beta_9\ L_I M_V$	0.95675	7	12.9585	0.9268	1	13.377
$L_I N_I$	0.8519	1	14.503	0.82859	7	14.963
$L_I N_{IV}$	0.83001	7	14.937	0.80364	7	15.427
$\gamma_{11}\ L_I N_V$	0.82879	5	14.9593	0.80233	9	15.453
$L_I N_{VI,VII}$				0.7884	1	15.725
$L_I O_I$	0.8158	1	15.198	0.7897	1	15.699
$L_I O_{IV,V}$	0.80861	5	15.3327	0.78257	7	15.843
$L_{II} M_{II}$	1.0997	1	11.274	1.0644	2	11.648
$\beta_{17}\ L_{II} M_{III}$	1.05609	7	11.7397	1.0223	1	12.127
$L_{II} M_V$	1.00722	5	12.3093	0.9747	1	12.720
$L_{II} N_{II}$	0.882	2	14.057	0.8585	3	14.442

Designation	Å*	p.e.	keV	Å*	p.e.	keV
81 Thallium (Cont.)				**82 Lead (Cont.)**		
$L_{II} N_{III}$	0.87996	5	14.0893	0.85192	7	14.553
$L_{II} N_V$				0.8382	2	14.791
$v\ L_{II} N_{VI}$	0.85048	5	14.5777	0.82327	7	15.060
$L_{II} O_{II}$	0.8490	1	14.604			
$L_{II} O_{III}$				0.8200	1	15.120
$t\ L_{III} M_{II}$	1.34154	5	9.2417	1.30767	7	9.4811
$s\ L_{III} M_{III}$	1.27807	5	9.7007	1.24385	7	9.9675
$L_{III} N_{II}$				1.01040	7	12.2705
$L_{III} N_{III}$	1.0286	1	12.053	1.0005	1	12.392
$u\ L_{III} N_{VI,VII}$	0.9888	1	12.538	0.96133	7	12.8968
$L_{III} O_{II}$	0.98738	5	12.5566	0.9586	1	12.934
$L_{III} O_{III}$	0.98538	5	12.5820	0.9578	1	12.945
$L_{III} P_{II,III}$	0.97926	5	12.6607	0.95118	7	13.0344
$M_I N_{III}$	4.013	9	3.089	3.872	9	3.202
$M_{II} N_I$				4.655	8	2.664
$M_{II} N_{IV}$	4.116	4	3.013	3.968	5	3.124
$M_{III} N_I$	5.884	8	2.107	5.704	5	2.174
$M_{III} N_{IV}$	4.865	5	2.548	4.715	3	2.630
$\gamma\ M_{III} N_V$	4.823	4	2.571	4.674	1	2.6527
$M_{III} O_I$				4.244	9	2.921
$M_{II} O_{IV,V}$	4.216	6	2.941	4.069	6	3.047
$\xi_2\ M_{IV} N_{II}$	7.032	5	1.763	6.802	5	1.823
$M_{IV} N_{III}$	5.584	7	1.342	5.404	7	1.342
$\beta\ M_{IV} N_{VI}$	5.249	1	2.3621	5.076	1	2.4427
$M_{IV} O_{II}$	5.196	9	2.386	5.004	9	2.477
$\eta\ M_V N_{III}$	6.974	4	1.778	6.740	3	1.8393
$\alpha_2\ M_V N_{VI}$	5.472	2	2.2656	5.299	2	2.3397
$\alpha_1\ M_V N_{VII}$	5.460	1	2.2706	5.286	1	2.3455
$M_V O_{III}$				5.168	9	2.399
$N_{IV} N_{VI}$				42.3	2	0.293
$N_V N_{VI,VII}$	46.5	2	0.267	45.0	1	0.2756
$N_{VI} O_{IV}$	115.3	2	0.1075	102.4	1	0.1211
$N_{VI} O_V$	113.0	1	0.10968	100.2	2	0.1237
$N_{VII} O_V$	117.7	1	0.10530	104.3	1	0.1189
83 Bismuth				**84 Polonium**		
$\alpha_2\ K L_{II}$	0.165717	2	74.8148	0.16130†	1	76.862
$\alpha_1\ K L_{III}$	0.160789	2	77.1079	0.15636†	1	79.290
$\beta_3\ K M_{II}$	0.142779	7	86.834	0.13892†	2	89.25
$\beta_1\ K M_{III}$	0.141948	3	87.343	0.13807†	2	89.80
$\beta_2^{II}\ K N_{II}$	0.13817	1	89.733	0.13438†	2	92.26
$\beta_2^{I}\ K N_{III}$	0.13797	1	89.864	0.13418†	2	92.40
$K O_{II,III}$	0.13709	1	90.435			
$\beta_5\ K M_{IV,V}$	0.14111	1	87.860			
$\beta_4\ K N_{IV,V}$	0.13759	2	90.11			
$\beta_4\ L_I M_{II}$	0.97690	4	12.6912	0.9475	3	13.086
$\beta_3\ L_I M_{III}$	0.93855	3	13.2098	0.9091	3	13.638
$\gamma_2\ L_I N_{II}$	0.79565	3	15.5824	0.772	1	16.07
$\gamma_3\ L_I N_{III}$	0.78917	5	15.7102			
$\gamma'_4\ L_I O_{II}$	0.76198	3	16.2709			
$\gamma_4\ L_I O_{III}$	0.76087	3	16.2947			
$\gamma_{18}\ L_I P_{II,III}$	0.75690	3	16.3802			
$\eta\ L_{II} M_I$	1.05856	3	11.7122			
$\beta_1\ L_{II} M_{IV}$	0.951978	9	13.0235	0.9220	2	13.447
$\gamma_5\ L_{II} N_I$	0.83923	5	14.7732			

83 Bismuth (*Cont.*) / 84 Polonium (*Cont.*)

Designation	Å*	p.e.	keV	Å*	p.e.	keV
$\gamma_1\ L_{II}N_{IV}$	0.81311	2	15.2477	0.78748	9	15.744
$\gamma_8\ L_{II}O_{I}$	0.7973	1	15.551			
$\gamma_6\ L_{II}O_{IV}$	0.79043	3	15.6853	0.7645	2	16.218
$l\ L_{III}M_{I}$	1.31610	7	9.4204	1.2829	5	9.664
$\alpha_2\ L_{III}M_{IV}$	1.15536	1	10.73091	1.12548†	5	11.0158
$\alpha_1\ L_{III}M_{V}$	1.14386	2	10.8388	1.11386	4	11.1308
$\beta_6\ L_{III}N_{I}$	0.99331	3	12.4816	0.9672	2	12.819
$\beta_{15}\ L_{III}N_{IV}$	0.95702	5	12.9549	0.9312	2	13.314
$\beta_2\ L_{III}N_{V}$	0.95518	4	12.9799	0.92937	5	13.3404
$\beta_7\ L_{III}O_{I}$	0.93505	5	13.2593			
$\beta_5\ L_{III}O_{IV,V}$	0.92556	3	13.3953	0.8996	2	13.782
L_1M_1	1.0005	9	12.39			
$\beta_{10}\ L_1M_{IV}$	0.90495	4	13.7002			
$\beta_9\ L_1M_{V}$	0.89791	3	13.8077			
L_1N_1	0.8022	1	15.456			
L_1N_{IV}	0.7795	5	15.904			
$\gamma_{11}\ L_1N_{V}$	0.77728	5	15.951			
$L_1N_{VI,VII}$	0.7641	5	16.23			
$L_1O_{IV,V}$	0.75791	5	16.358			
$L_{II}M_{II}$	1.0346	9	11.98			
$\beta_{17}\ L_{II}M_{III}$	0.98913	5	12.5344			
$L_{II}M_{V}$	0.94419	5	13.1310			
$L_{II}N_{II}$	0.8344	9	14.86			
$L_{II}N_{III}$	0.8248	1	15.031			
$v\ L_{II}N_{VI}$	0.79721	9	15.552			
$L_{II}O_{III}$	0.79384	5	15.6178			
$t\ L_{III}M_{II}$	1.2748	1	9.7252			
$s\ L_{III}M_{III}$	1.2105	1	10.2421			
$L_{III}N_{II}$	0.98280	5	12.6151			
$L_{III}N_{III}$	0.97321	5	12.7394			
$u\ L_{III}N_{VI,VII}$	0.93505	5	13.2593			
$L_{III}O_{II}$	0.9323	2	13.298			
$L_{III}O_{III}$	0.9302	2	13.328			
$L_{III}P_{II,III}$	0.92413	4	13.4159			
M_1N_{II}	3.892	9	3.185			
M_1N_{III}	3.740	9	3.315			
$M_{II}N_{IV}$	3.834	4	3.234			
$M_{III}N_{I}$	5.537	8	2.239			
$M_{III}N_{IV}$	4.571	5	2.712			
$\gamma\ M_{III}N_{V}$	4.532	2	2.735			
$M_{III}O_{I}$	4.105	9	3.021			
$M_{III}O_{IV,V}$	3.932	6	3.153			
$\zeta_2\ M_{IV}N_{II}$	6.585	5	1.883			
$M_{IV}N_{III}$	6.162	8	2.012			
$\beta\ M_{IV}N_{VI}$	4.909	1	2.5255			
$M_{IV}O_{II}$	4.823	3	2.571			
$M_{IV}P_{II,III}$	4.59	2	2.70			
$\zeta_1\ M_{V}N_{III}$	6.521	4	1.901			
$\alpha_2\ M_{V}N_{VI}$	5.130	2	2.4170			
$\alpha_1\ M_{V}N_{VII}$	5.118	1	2.4226			
$N_1P_{II,III}$	13.30	6	0.932			
$N_{VI}O_{IV}$	91.6	1	0.1354			
$N_{VII}O_{V}$	93.2	1	0.1330			

85 Astatine / 86 Radon

Designation	Å*	p.e.	keV	Å*	p.e.	keV
$\alpha_2\ KL_{II}$	0.15705†	2	78.95	0.15294†	3	81.07
$\alpha_1\ KL_{III}$	0.15210†	2	81.52	0.14798†	3	83.78
$\beta_3\ KM_{II}$	0.13517†	4	91.72	0.13155†	5	94.24
$\beta_1\ KM_{III}$	0.13432†	4	92.30	0.13069†	5	94.87
$\beta_2^{II}\ KN_{II}$	0.13072†	4	94.84	0.12719†	5	97.47
$\beta_2^{I}\ KN_{III}$	0.13052†	4	94.99	0.12698†	5	97.64
$\beta_3\ L_1M_{III}$	0.88135†	9	14.067	0.85436†	9	14.512
$\beta_1\ L_{II}M_{IV}$	0.89349†	9	13.876	0.86605†	9	14.316
$\gamma_1\ L_{II}N_{IV}$	0.76289†	9	16.251	0.73928†	9	16.770
$\alpha_2\ L_{III}M_{IV}$	1.09671†	5	11.3048	1.06899†	5	11.5979
$\alpha_1\ L_{III}M_{V}$	1.08500†	5	11.4268	1.05723†	5	11.7270

87 Francium / 88 Radium

Designation	Å*	p.e.	keV	Å*	p.e.	keV
$\alpha_2\ KL_{II}$	0.14896†	3	83.23	0.14512†	2	85.43
$\alpha_1\ KL_{III}$	0.14399†	3	86.10	0.14014†	2	88.47
$\beta_3\ KM_{II}$	0.12807†	5	96.81	0.12469†	3	99.43
$\beta_1\ KM_{III}$	0.12719†	5	97.47	0.12382†	3	100.13
$\beta_2^{II}\ KN_{II}$	0.12379†	5	100.16	0.12050†	3	102.89
$\beta_2^{I}\ KN_{III}$	0.12358†	5	100.33	0.12029†	3	103.07
$\beta_4\ L_1M_{II}$				0.84071	3	14.7472
$\beta_3\ L_1M_{III}$	0.82789†	9	14.976	0.80273	5	15.4449
$\gamma_2\ L_1N_{II}$				0.68199	5	18.179
$\gamma_3\ L_1N_{III}$				0.67538	5	18.357
$\gamma'_4\ L_1O_{II}$				0.65131	5	19.036
$\gamma_4\ L_1O_{III}$				0.64965	5	19.084
$\gamma_{13}\ L_1P_{II,III}$				0.64513	5	19.218
$\eta\ L_{II}M_{I}$				0.90742	5	13.6630
$\beta_1\ L_{II}M_{IV}$	0.83940†	9	14.770	0.81375	5	15.2358
$\gamma_5\ L_{II}N_{I}$				0.71774	5	17.274
$\gamma_1\ L_{II}N_{IV}$	0.71652†	9	17.303	0.69463	5	17.849
γ_8				0.6801	1	18.230
$\gamma_6\ L_{II}O_{IV}$				0.67328	5	18.414
$L_{II}P_{I}$				0.6724	1	18.439
$l\ L_{III}M_{I}$				1.16719	5	10.6222
$\alpha_2\ L_{III}M_{IV}$	1.04230	5	11.8950	1.01656	5	12.1962
$\alpha_1\ L_{III}M_{V}$	1.03049	5	12.0313	1.00473	5	12.3397
$\beta_6\ L_{III}N_{I}$				0.87088	5	14.2362
$\beta_{15}\ L_{III}N_{IV}$				0.83722	5	14.8086
$\beta_2\ L_{III}N_{V}$	0.858	2	14.45	0.83537	5	14.8414
$\beta_7\ L_{III}O_{I}$				0.8162	1	15.190
$\beta_5\ L_{III}O_{IV,V}$				0.80627	5	15.3771
$L_{III}P_{I}$				0.8050	1	15.402
$\beta_{10}\ L_1M_{IV}$				0.77546	5	15.988
$\beta_9\ L_1M_{V}$				0.76857	5	16.131
L_1N_1				0.6874	1	18.036
L_1N_{IV}				0.6666	1	18.600
$\gamma_{11}\ L_1N_{V}$				0.6654	1	18.633
$L_1O_{IV,V}$				0.6468	1	19.167
$\beta_{17}\ L_{II}M_{III}$				0.8438	1	14.692
$L_{II}N_{III}$				0.7043	1	17.604
$L_{II}N_{V}$				0.6932	1	17.884
$L_{II}O_{II}$				0.6780	1	18.286

Designation	Å*	p.e.	keV	Å*	p.e.	keV	Designation	Å*	p.e.	keV	Å*	p.e.	keV
87 Francium (Cont.)				**88 Radium (Cont.)**			**89 Actinium (Cont.)**				**90 Thorium (Cont.)**		
$L_{II}O_{III}$				0.6764	1	18.330	$v\ L_{II}N_{VI}$				0.64064	9	19.353
$L_{II}P_{II,III}$				0.6714	1	18.466	$L_{II}O_{II}$				0.6369	1	19.466
$L_{III}N_{II}$				0.8618	1	14.387	$L_{II}O_{III}$				0.6356	1	19.506
$L_{III}N_{III}$				0.8512	1	14.566	$L_{II}P_{II,III}$				0.6312	1	19.642
$u\ L_{III}N_{VI,VII}$				0.8186	1	15.146	$t\ L_{III}M_1$				1.08009	9	11.4788
$L_{III}P_{II,III}$				0.8038	1	15.425	$s\ L_{III}M_{II}$				1.0112	1	12.261
							$L_{III}N_{II}$				0.8190	2	15.138
89 Actinium				**90 Thorium**			$L_{III}N_{III}$				0.8082	1	15.341
$\alpha_2 KL_{II}$	0.14141†	2	87.67	0.137829	2	89.953	$u\ L_{III}N_{VI,VII}$				0.77661	5	15.964
$\alpha_1 KL_{III}$	0.136417†	8	90.884	0.132813	2	93.350	$L_{III}O_{II}$				0.7713	1	16.074
$\beta_3 KM_{II}$	0.12143†	2	102.10	0.118268	3	104.831	$L_{III}O_{III}$				0.7690	1	16.123
$\beta_1 KM_{III}$	0.12055†	2	102.85	0.117396	9	105.609	$L_{III}P_{II,III}$				0.7625	2	16.260
$\beta_2^{II} KN_{II}$	0.11732†	2	105.67	0.11426	1	108.511	M_1N_{III}				2.934	8	4.23
$\beta_2^{I} KN_{III}$	0.11711†	2	105.86	0.114040	9	108.717	M_1O_{III}				2.442	9	5.08
$KO_{II,III}$				0.11322	1	109.500	$M_{II}N_1$				3.537	9	3.505
$\beta_5 KM_{IV,V}$				0.116667	9	106.269	$M_{II}N_{IV}$				3.011	2	4.117
$\beta_4 KN_{IV,V}$				0.11366	2	109.08	$M_{II}O_{IV}$				2.618	5	4.735
$\beta_4 L_1M_{II}$				0.79257	4	15.6429	$M_{III}N_1$				4.568	5	2.714
$\beta_3 L_1M_{III}$	0.77822†	9	15.931	0.75479	3	16.4258	$M_{III}N_{IV}$				3.718	3	3.335
$\gamma_2 L_1N_{II}$				0.64221	4	19.305	$\gamma M_{II}N_V$				3.679	2	3.370
$\gamma_3 L_1N_{III}$				0.63559	4	19.507	$M_{III}O_1$				3.283	9	3.78
$\gamma'_4 L_1O_{II}$				0.61251	4	20.241	$M_{III}O_{IV,V}$				3.131	3	3.959
$\gamma_4 L_1O_{III}$				0.61098	4	20.292	$\zeta_2 M_VN_{II}$				5.340	5	2.322
$\gamma_{13} L_1P_{II,III}$				0.60705	8	20.424	$M_{IV}N_{III}$				4.911	5	2.524
$\eta L_{II}M_1$				0.85446	4	14.5099	$\beta M_{IV}N_{VI}$				3.941	1	3.1458
$\beta_1 L_{II}M_{IV}$	0.78903†	9	15.713	0.765210	9	16.2022	$M_{IV}O_{II}$				3.808	4	3.256
$\gamma_5 L_{II}N_1$				0.67491	4	18.370	$\zeta_1 M_VN_{III}$				5.245	5	2.364
$\gamma_1 L_{II}N_{IV}$	0.67351†	9	18.408	0.65313	3	18.9825	$\alpha_2 M_VN_{VI}$				4.151	2	2.987
$\gamma_8 L_{II}O_1$				0.63898	5	19.403	$\alpha_1 M_VN_{VII}$				4.1381	9	2.9961
$\gamma_6 L_{II}O_{IV}$				0.63258	4	19.599	M_VP_{III}				3.760	9	3.298
$L_{II}P_1$				0.6316	1	19.629	$N\ P_{II}$				9.44	7	1.313
$L_{II}P_{IV}$				0.62991	9	19.682	$N\ P_{III}$				9.40	7	1.1319
$l\ L_{III}M_1$				1.11508	4	11.1186	$N_{II}O_{IV}$				11.56	5	1.072
$\alpha_2 L_{III}M_{IV}$	0.99178†	5	12.5008	0.96788	2	12.8096	N_1P_1				11.07	7	1.120
$\alpha_1 L_{III}M_V$	0.97993†	5	12.6520	0.95600	3	12.9687	$N_{III}O_V$				13.8	1	0.897
$\beta_6 L_{III}N_1$				0.82790	8	14.975	$N_{IV}N_V$				33.57	9	0.3693
$\beta_{15} L_{III}N_{IV}$				0.79539	5	15.5875	$N_VN_{VI,VII}$				36.32	9	0.3414
$\beta_2 L_{III}N_V$				0.79354	3	15.6237	$N_{VI}O_{IV}$				49.5	1	0.2505
$\beta_7 L_{III}O_1$				0.77437	4	16.0105	$N_{VI}O_V$				48.2	1	0.2572
$\beta_5 L_{III}O_{IV,V}$				0.76468	5	16.213	$N_{VII}O_V$				50.0	1	0.2479
$L_{III}P_1$				0.76338	5	16.241	$O_{III}P_{IV,V}$				68.2	3	0.1817
$L_{III}P_{IV,V}$				0.76087	9	16.295	$O_{IV,V}Q_{II,III}$				181.	5	0.068
$\beta_{10} L_1M_{IV}$				0.7301	1	16.981							
$\beta_9 L_1M_V$				0.7234	1	17.139							
L_1N_1				0.64755	5	19.146	**91 Protactinium**				**92 Uranium**		
L_1N_{IV}				0.6276	1	19.755	$\alpha_2 KL_{II}$	0.134343†	9	92.287	0.130968	4	94.665
$\gamma_{11} L_1N_V$				0.62636	9	19.794	$\alpha_1 KL_{III}$	0.129325†	3	95.868	0.125947	3	98.439
$L_1N_{VI,VII}$				0.6160	1	20.128	$\beta_3 KM_{II}$	0.11523†	2	107.60	0.112296	4	110.406
L_1O_1				0.6146	1	20.174	$\beta_1 KM_{III}$	0.114345†	8	108.427	0.111394	5	111.300
$L_1O_{IV,V}$				0.6083	1	20.383	$\beta_2^{II} KN_{II}$	0.11129†	2	111.40	0.10837	1	114.40
$L_{II}M_{II}$				0.8338	1	14.869	$\beta_2^{I} KN_{III}$	0.11107†	2	111.62	0.10818	1	114.60
$\beta_{17} L_{II}M_{III}$				0.79257	4	15.6429	$KO_{II,III}$				0.10744	1	115.39
$L_{II}M_V$				0.7579	1	16.359	$\beta_5 KM_{IV,V}$				0.11069	1	112.01
$L_{II}N_{III}$				0.6620	1	18.729	$\beta_4 KN_{IV,V}$				0.10780	2	115.01
$L_{II}N_V$				0.6521	1	19.014	$\beta_4 L_1M_1$	0.7699	1	16.104	0.747985	9	16.5753
							$\beta_3 L_1M_{III}$	0.73230	5	16.930	0.71029	2	17.4550

Left section

Designation	Å*	p.e.	keV	Å*	p.e.	keV
	91 Protactinium (*Cont.*)			**92 Uranium** (*Cont.*)		
$\gamma_2\ L_I N_{II}$	0.6239	1	19.872	0.605237	9	20.4847
$\gamma_3\ L_I N_{III}$	0.6169	1	20.098	0.598574	9	20.7127
$\gamma'_4\ L_I O_{II}$				0.576700	9	21.4984
$\gamma_4\ L_I O_{II,III}$	0.5937	1	20.882	0.57499	9	21.562
γ_8				0.5706	1	21.729
$\eta\ L_{II} M_I$	0.8295	1	14.946	0.80509	2	15.3997
$\beta_1\ L_{II} M_{IV}$	0.74232	5	16.702	0.719984	8	17.2200
$\gamma_5\ L_{II} N_I$	0.6550	1	18.930	0.63557	2	19.5072
$\gamma_1\ L_{II} N_{IV}$	0.63358†	9	19.568	0.614770	9	20.1671
$\gamma_8\ L_{II} O_I$				0.60125	5	20.621
$\gamma_6\ L_{II} O_{IV}$	0.6133	1	20.216	0.594845	9	20.8426
$L_{II} P_{IV}$				0.59203	5	20.942
$l\ L_{III} M_I$	1.0908	1	11.366	1.06712	2	11.6183
$\alpha_2\ L_{III} M_{IV}$	0.94482†	5	13.1222	0.922558	9	13.4388
$\alpha_1\ L_{III} M_V$	0.93284	5	13.2907	0.910639	9	13.6147
$\beta_6\ L_{III} N_I$	0.8079	1	15.347	0.78838	2	15.7260
$\beta_{15}\ L_{III} N_{IV}$				0.756642	9	16.3857
$\beta_2\ L_{III} N_V$	0.7737	1	16.024	0.754681	9	16.4283
$\beta_7\ L_{III} O_I$	0.7546	2	16.431	0.73602	6	16.845
$\beta_5\ L_{III} O_{IV,V}$	0.7452	2	16.636	0.726305	9	17.0701
$L_{III} P_I$				0.72521	5	17.096
$L_{III} P_{IV,V}$				0.72240	5	17.162
$\beta_{10}\ L_I M_V$	0.7088	2	17.492	0.68760	5	18.031
$\beta_9\ L_I M_V$	0.7018	1	17.667	0.681014	8	18.2054
$L_I N_{IV}$				0.59096	5	20.979
$\gamma_{11}\ L_I N_V$				0.58986	5	21.019
$L_I O_{IV,V}$				0.5725	1	21.657
$\beta_{17}\ L_{II} M_{III}$				0.74503	5	16.641
$v\ L_{II} N_{VI}$				0.6228	1	19.907
$L_{III} O_{III}$				0.6031	1	20.556
$L_{II} P_{II,III}$				0.59728	5	20.758
$t\ L_{III} M_{II}$				0.5930	2	20.906
$s\ L_{III} M_{III}$				1.0347	1	11.982
$L_{III} N_{II}$				0.9636	1	12.866
$L_{III} N_{III}$				0.78017	9	15.892
$u\ L_{III} N_{VI,VII}$				0.7691	1	16.120
$L_{III} O_{II}$				0.738603	9	16.7859
$L_{III} O_{III}$				0.7333	1	16.907
$L_{III} P_{II,III}$				0.7309	1	16.962
$M_I N_{II}$				0.72426	5	17.118
$M_I N_{III}$				2.92	2	4.25
$M_I O_{III}$				2.753	8	4.50
$M_I P_{III}$				2.304	7	5.38
$M_{II} N_I$				2.253	6	5.50
$M_{II} N_I$	3.441	5	3.603	3.329	4	3.724
$M_{II} N_{IV}$	2.910	2	4.260	2.817	2	4.401
$M_{II} O_{IV}$	2.527	4	4.906	2.443	4	5.075
$M_{III} N_I$	4.450	4	2.786	4.330	2	2.863
$M_{III} N_{IV}$	3.614		3.430	3.521	2	3.521
$\gamma\ M_{III} N_V$	3.577	1	3.4657	3.479	1	3.563
$M_{III} O_I$	3.245	9	3.82	3.115	7	3.980
$M_{III} O_{IV,V}$	3.038	2	4.081	2.948	2	4.205
$\zeta_2\ M_{IV} N_{II}$	5.193	2	2.3876	5.050	2	2.4548
$M_{IV} N_{III}$				4.625	5	2.681
$\beta\ M_{IV} N_{VI}$	3.827	1	3.2397	3.716	1	3.3367

Designation	Å*	p.e.	keV	Å*	p.e.	keV
	91 Protactinium (*Cont.*)			**92 Uranium** (*Cont.*)		
$M_{IV} O_{II}$	3.691	2	3.359	3.576	1	3.4666
$\zeta_1\ M_V N_{III}$	5.092	2	2.4350	4.946	2	2.507
$\alpha_2\ M_V N_{VI}$	4.035	3	3.072	3.924	1	3.1595
$\alpha_1\ M_V N_{VII}$	4.022	1	3.0823	3.910	1	3.1708
$N_I O_{III}$				10.09	7	1.229
$N_I P_{II}$				8.81	7	1.41
$N_I P_{III}$				8.76	7	1.42
$N_{II} P_I$				10.40	7	1.192
$N_{III} O_V$				12.90	9	0.961
$N_{IV} N_{VI}$				31.8	1	0.390
$N_V N_{VI,VII}$				34.8	1	0.357
$N_{IV} O_{IV}$				43.3	2	0.286
$N_{VI} O_V$				42.1	2	0.295
$N_I P_{IV,V}$				8.60	7	1.44
	93 Neptunium			**94 Plutonium**		
$\beta_4\ L_I M_{II}$	0.72671	2	17.0607	0.70620	2	17.5560
$\beta_3\ L_I M_{III}$	0.68920†	9	17.989	0.66871	2	18.5405
$\gamma_2\ L_I N_{II}$	0.5873	5	21.11	0.57068	2	21.7251
$\gamma_3\ L_I N_{III}$	0.5810	5	21.34	0.564001	9	21.9824
$\gamma'_4\ L_I O_{II}$				0.5432	1	22.823
$\gamma_4\ L_I O_{II,III}$	0.5585	5	22.20	0.5416	1	22.891
$\eta\ L_{II} M_I$	0.7809	2	15.876	0.7591	1	16.333
$\beta_1\ L_{II} M_{IV}$	0.698478	9	17.7502	0.67772	2	18.2937
$\gamma_5\ L_{II} N_I$	0.616	1	20.12	0.5988	1	20.704
$\gamma_1\ L_{II} N_{IV}$	0.596498	9	20.7848	0.578882	9	21.4173
γ_8				0.5658	1	21.914
$\gamma_6\ L_{II} O_{IV}$	0.57699	5	21.488	0.55973	2	22.1502
$l\ L_{III} M_I$	1.0428	6	11.890	1.0226	1	12.124
$\alpha_2\ L_{III} M_{IV}$	0.901045	9	13.7597	0.88028	2	14.0842
$\alpha_1\ L_{III} M_V$	0.889128	9	13.9441	0.86830	2	14.2786
$\beta_6\ L_{III} N_I$	0.769	1	16.13	0.75148	2	16.4983
$\beta_{15}\ L_{III} N_{IV}$				0.7205	1	17.208
$\beta_2\ L_{III} N_V$	0.736230	9	16.8400	0.71851	2	17.2553
$\beta_7\ L_{III} O_I$				0.7003	1	17.705
$\beta_5\ L_{III} O_{IV,V}$	0.70814	2	17.5081	0.69068	2	17.9506
$\beta_{10}\ L_I M_{IV}$				0.6482	1	19.126
$\beta_9\ L_I M_V$				0.6416	1	19.323
$u\ L_{III} N_{VI,VII}$				0.7031	1	17.635
	95 Americium					
$\beta_4\ L_I M_{II}$	0.68639	2	18.0627			
$\beta_3\ L_I M_{III}$	0.64891	2	19.1059			
$\gamma_2\ L_I N_{II}$	0.5544	2	22.361			
$\beta_1\ L_{II} M_{IV}$	0.657655	9	18.8520			
$\gamma_1\ L_{II} N_{IV}$	0.561886	9	22.0652			
$\gamma_6\ L_{II} O_{IV}$	0.54311	2	22.8282			
$l\ L_{III} M_I$	1.0012	6	12.384			
$\alpha_2\ L_{III} M_{IV}$	0.860266	9	14.4119			
$\alpha_1\ L_{III} M_V$	0.848187	9	14.6172			
$\beta_6\ L_{III} N_I$	0.73418	2	16.8870			
$\beta_{15}\ L_{III} N_{IV}$	0.70341	2	17.6258			
$\beta_2\ L_{III} N_V$	0.701390	9	17.6765			
$\beta_5\ L_{III} O_{IV,V}$	0.67383	2	18.3996			

Wavelength Å*	p.e.	Element		Designation	keV	Wavelength Å*	p.e.	Element		Designation	keV
0.10723	1	92 U	K	Abs. Edge	115.62	0.1408	1	82 Pb		KP	88.06
0.10744	1	92 U		$KO_{II,III}$	115.39	0.140880	5	82 Pb	K	Abs. Edge	88.005
0.10780	2	92 U	$K\beta_4$	$KN_{IV,V}$	115.01	0.141012	8	82 Pb		$KO_{II,III}$	87.922
0.10818	1	92 U	$K\beta_2{}^I$	KN_{III}	114.60	0.14111	1	83 Bi	$K\beta_5$	$KM_{IV,V}$	87.860
0.10837	1	92 U	$K\beta_2{}^{II}$	KN_{II}	114.40	0.14141	2	89 Ac	$K\alpha_2$	KL_{II}	87.67
0.11069	1	92 U	$K\beta_5$	$KM_{IV,V}$	112.01	0.14155	3	82 Pb	$K\beta_4$	$KN_{IV,V}$	87.59
0.11107	2	91 Pa	$K\beta_2{}^I$	KN_{III}	111.62	0.14191	1	82 Pb	$K\beta_2{}^I$	KN_{III}	87.364
0.11129	2	91 Pa	$K\beta_2{}^{II}$	KN_{II}	111.40	0.141948	3	83 Bi	$K\beta_1$	KM_{III}	87.343
0.111394	5	92 U	$K\beta_1$	KM_{III}	111.300	0.14212	2	82 Pb	$K\beta_2{}^{II}$	KN_{II}	87.23
0.112296	4	92 U	$K\beta_3$	KM_{II}	110.406	0.142779	7	83 Bi	$K\beta_3$	KM_{II}	86.834
0.11307	1	90 Th	K	Abs. Edge	109.646	0.14399	3	87 Fr	$K\alpha_1$	KL_{III}	86.10
0.11322	1	90 Th		$KO_{II,III}$	109.500	0.14495	1	81 Tl	K	Abs. Edge	85.533
0.11366	2	90 Th	$K\beta_4$	$KN_{IV,V}$	109.08	0.14495	3	82 Pb	$K\beta_5{}^I$	KM_V	85.53
0.114040	9	90 Th	$K\beta_2{}^I$	KN_{III}	108.717	0.14509	1	81 Tl		$KO_{II,III}$	85.451
0.11426	1	90 Th	$K\beta_2{}^{II}$	KN_{II}	108.511	0.14512	2	82 Pb	$K\beta_5{}^{II}$	KM_{IV}	85.43
0.114345	8	91 Pa	$K\beta_1$	KM_{III}	108.427	0.14512	2	88 Ra	$K\alpha_2$	KL_{II}	85.43
0.11523	2	91 Pa	$K\beta_3$	KM_{II}	107.60	0.14553	2	81 Tl	$K\beta_4$	$KN_{IV,V}$	85.19
0.116667	9	90 Th	$K\beta_5$	$KM_{IV,V}$	106.269	0.14595	1	81 Tl	$K\beta_2{}^I$	KN_{III}	84.946
0.11711	2	89 Ac	$K\beta_2{}^I$	KN_{III}	105.86	0.145970	6	82 Pb	$K\beta_1$	KM_{III}	84.936
0.11732	2	89 Ac	$K\beta_2{}^{II}$	KN_{II}	105.67	0.14614	1	81 Tl	$K\beta_2{}^{II}$	KN_{II}	84.836
0.117396	9	90 Th	$K\beta_1$	KM_{III}	105.609	0.146810	4	82 Pb	$K\beta_3$	KM_{II}	84.450
0.118268	3	90 Th	$K\beta_3$	KM_{II}	104.831	0.14798	3	86 Rn	$K\alpha_1$	KL_{III}	83.78
0.12029	3	88 Ra	$K\beta_2{}^I$	KN_{III}	103.07	0.14896	3	87 Fr	$K\alpha_2$	KL_{II}	83.23
0.12050	3	88 Ra	$K\beta_2{}^{II}$	KN_{II}	102.89	0.14917	1	81 Tl	$K\beta_5$	$KM_{IV,V}$	83.114
0.12055	3	89 Ac	$K\beta_1$	KM_{III}	102.85	0.14918	1	80 Hg	K	Abs. Edge	83.109
0.12143	2	89 Ac	$K\beta_3$	KM_{II}	102.10	0.14931	2	80 Hg		$KO_{II,III}$	83.04
0.12358	5	87 Fr	$K\beta_2{}^I$	KN_{III}	100.33	0.14978	2	80 Hg	$K\beta_4$	$KN_{IV,V}$	82.78
0.12370	5	87 Fr	$K\beta_2{}^{II}$	KN_{II}	100.16	0.150112	5	81 Tl	$K\beta_1$	KM_{III}	82.576
0.12382	3	88 Ra	$K\beta_1$	KM_{III}	100.13	0.15020	2	80 Hg	$K\beta_2{}^I$	KN_{III}	82.54
0.12469	3	88 Ra	$K\beta_3$	KM_{II}	99.43	0.15040	2	80 Hg	$K\beta_2{}^{II}$	KN_{II}	82.43
0.125947	3	92 U	$K\alpha_1$	KL_{III}	98.439	0.150980	6	81 Tl	$K\beta_3$	KM_{II}	82.118
0.12698	5	86 Rn	$K\beta_2{}^I$	KN_{III}	97.64	0.15210	2	85 At	$K\alpha_1$	KL_{III}	81.52
0.12719	5	86 Rn	$K\beta_2{}^{II}$	KN_{II}	97.47	0.15294	3	86 Rn	$K\alpha_2$	KL_{II}	81.07
0.12719	5	87 Fr	$K\beta_1$	KM_{III}	97.47	0.15353	2	80 Hg	$K\beta_5$	$KM_{IV,V}$	80.75
0.12807	5	87 Fr	$K\beta_3$	KM_{II}	96.81	0.153593	5	79 Au	K	Abs. Edge	80.720
0.129325	3	91 Pa	$K\alpha_1$	KL_{III}	95.868	0.153694	7	79 Au		$KO_{II,III}$	80.667
0.13052	4	85 At	$K\beta_2{}^I$	KN_{III}	94.99	0.154224	5	79 Au	$K\beta_4$	$KN_{IV,V}$	80.391
0.13069	5	86 Rn	$K\beta_1$	KM_{III}	94.87	0.154487	3	80 Hg	$K\beta_1$	KM_{III}	80.253
0.13072	4	85 At	$K\beta_2{}^{II}$	KN_{II}	94.84	0.154618	9	79 Au	$K\beta_2{}^I$	KN_{III}	80.185
0.130968	4	92 U	$K\alpha_2$	KL_{II}	94.665	0.15483	2	79 Au	$K\beta_2{}^{II}$	KN_{II}	80.08
0.13155	5	86 Rn	$K\beta_3$	KM_{II}	94.24	0.155321	3	80 Hg	$K\beta_3$	KM_{II}	79.822
0.132813	2	90 Th	$K\alpha_1$	KL_{III}	93.350	0.15636	1	84 Po	$K\alpha_1$	KL_{III}	79.290
0.13418	2	84 Po	$K\beta_2{}^I$	KN_{III}	92.40	0.15705	2	85 At	$K\alpha_2$	KL_{II}	78.95
0.13432	4	85 At	$K\beta_1$	KM_{III}	92.30	0.157880	5	79 Au	$K\beta_5{}^I$	KM_V	78.529
0.134343	9	91 Pa	$K\alpha_2$	KL_{II}	92.287	0.158062	7	79 Au	$K\beta_5{}^{II}$	KM_{IV}	78.438
0.13438	2	84 Po	$K\beta_2{}^{II}$	KN_{II}	92.26	0.15818	1	78 Pt	K	Abs. Edge	78.381
0.13517	4	85 At	$K\beta_3$	KM_{II}	91.72	0.15826	1	78 Pt		$KO_{II,III}$	78.341
0.136417	8	89 Ac	$K\alpha_1$	KL_{III}	90.884	0.15881	2	78 Pt	$K\beta_4$	$KN_{IV,V}$	78.069
0.13694	1	83 Bi	K	Abs. Edge	90.534	0.158982	3	79 Au	$K\beta_1$	KM_{III}	77.984
0.13709	1	83 Bi		$KO_{II,III}$	90.435	0.15920	1	78 Pt	$K\beta_2{}^I$	KN_{III}	77.878
0.13759	2	83 Bi	$K\beta_4$	$KN_{IV,V}$	90.11	0.15939	1	78 Pt	$K\beta_2{}^{II}$	KN_{II}	77.785
0.137829	2	90 Th	$K\alpha_2$	KL_{II}	89.953	0.159810	2	79 Au	$K\beta_3$	KM_{II}	77.580
0.13797	1	83 Bi	$K\beta_2{}^I$	KN_{III}	89.864	0.160789	2	83 Bi	$K\alpha_1$	KL_{III}	77.1079
0.13807	2	84 Po	$K\beta_1$	KM_{III}	89.80	0.16130	1	84 Po	$K\alpha_2$	KL_{II}	76.862
0.13817	1	83 Bi	$K\beta_2{}^{II}$	KN_{II}	89.733	0.16255	3	78 Pt	$K\beta_5{}^I$	KM_V	76.27
0.13892	2	84 Po	$K\beta_3$	KM_{II}	89.25	0.16271	2	78 Pt	$K\beta_5{}^{II}$	KM_{IV}	76.199
0.14014	2	88 Ra	$K\alpha_1$	KL_{III}	88.47	0.16292	1	77 Ir	K	Abs. Edge	76.101

Wavelength Å*	p.e.	Element	Designation		keV	Wavelength Å*	p.e.	Element	Designation		keV
0.163019	5	77 Ir		$KO_{II,III}$	76.053	0.190381	4	78 Pt	$K\alpha_2$	KL_{II}	65.122
0.16352	2	77 Ir	$K\beta_4$	$KN_{IV,V}$	75.821	0.1908	2	72 Hf	$K\beta_2$	$KN_{II,III}$	64.98
0.163675	3	78 Pt	$K\beta_1$	KM_{III}	75.748	0.190890	2	73 Ta	$K\beta_3$	KM_{II}	64.9488
0.163956	7	77 Ir	$K\beta_2^I$	KN_{III}	75.619	0.191047	2	77 Ir	$K\alpha_1$	KL_{III}	64.8956
0.16415	1	77 Ir	$K\beta_2^{II}$	KN_{II}	75.529	0.19585	5	71 Lu	K	Abs. Edge	63.31
0.164501	3	78 Pt	$K\beta_3$	KM_{II}	75.368	0.19589	2	71 Lu		$KO_{II,III}$	63.293
0.165376	2	82 Pb	$K\alpha_1$	KL_{III}	74.9694	0.195904	2	77 Ir	$K\alpha_2$	KL_{II}	63.2867
0.165717	2	83 Bi	$K\alpha_2$	KL_{II}	74.8148	0.19607	3	72 Hf	$K\beta_1$	KM_{III}	63.234
0.167373	9	77 Ir	$K\beta_5^I$	KM_V	74.075	0.196794	2	76 Os	$K\alpha_1$	KL_{III}	63.0005
0.16759	2	77 Ir	$K\beta_5^{II}$	KM_{IV}	73.980	0.19686	4	72 Hf	$K\beta_3$	KM_{II}	62.98
0.16787	1	76 Os	K	Abs. Edge	73.856	0.1969	2	71 Lu	$K\beta_2$	$KN_{II,III}$	62.97
0.16798	1	76 Os		$KO_{II,III}$	73.808	0.20084	2	71 Lu	$K\beta_5$	$KM_{IV,V}$	61.732
0.16842	2	76 Os	$K\beta_4$	$KN_{IV,V}$	73.615	0.201639	2	76 Os	$K\alpha_2$	KL_{II}	61.4867
0.168542	2	77 Ir	$K\beta_1$	KM_{III}	73.5608	0.20224	5	70 Yb	K	Abs. Edge	61.30
0.168906	6	76 Os	$K\beta_2^I$	KN_{III}	73.402	0.20226	2	70 Yb		$KO_{II,III}$	61.298
0.16910	1	76 Os	$K\beta_2^{II}$	KN_{II}	73.318	0.20231	3	71 Lu	$K\beta_1$	KM_{III}	61.283
0.169367	2	77 Ir	$K\beta_3$	KM_{II}	73.2027	0.202781	2	75 Re	$K\alpha_1$	KL_{III}	61.1403
0.170136	2	81 Tl	$K\alpha_1$	KL_{III}	72.8715	0.20309	4	71 Lu	$K\beta_3$	KM_{II}	61.05
0.170294	2	82 Pb	$K\alpha_2$	KL_{II}	72.8042	0.2033	2	70 Yb	$K\beta_2$	$KN_{II,III}$	60.89
0.17245	1	76 Os	$K\beta_5^I$	KM_V	71.895	0.20739	2	70 Yb	$K\beta_5$	$KM_{IV,V}$	59.782
0.17262	1	76 Os	$K\beta_5^{II}$	KM_{IV}	71.824	0.207611	1	75 Re	$K\alpha_2$	KL_{II}	59.7179
0.17302	1	75 Re	K	Abs. Edge	71.658	0.20880	5	69 Tm	K	Abs. Edge	59.38
0.17308	1	75 Re		$KO_{II,III}$	71.633	0.20884	8	70 Yb	$K\beta_1$	KM_{III}	59.37
0.173611	3	76 Os	$K\beta_1$	KM_{III}	71.413	0.20891	2	69 Tm		$KO_{II,III}$	59.346
0.17362	2	75 Re	$K\beta_4$	$KN_{IV,V}$	71.410	0.2090100	Std.	74 W	$K\alpha_1$	KL_{III}	59.31824
0.174054	6	75 Re	$K\beta_2^I$	KN_{III}	71.232	0.2096	1	70 Yb	$K\beta_3$	KM_{II}	59.14
0.17425	1	75 Re	$K\beta_2^{II}$	KN_{II}	71.151	0.2098	2	69 Tm	$K\beta_2$	$KN_{II,III}$	59.09
0.174431	3	76 Os	$K\beta_3$	KM_{II}	71.077	0.213828	2	74 W	$K\alpha_2$	KL_{II}	57.9817
0.175036	2	81 Tl	$K\alpha_2$	KL_{II}	70.8319	0.21404	2	69 Tm	$K\beta_5$	$KM_{IV,V}$	57.923
0.175068	3	80 Hg	$K\alpha_1$	KL_{III}	70.819	0.215497	4	73 Ta	$K\alpha_1$	KL_{III}	57.532
0.17766	1	75 Re	$K\beta_5^I$	KM_V	69.786	0.21556	2	69 Tm	$K\beta_1$	KM_{III}	57.517
0.17783	1	75 Re	$K\beta_5^{II}$	KM_{IV}	69.719	0.21567	1	68 Er	K	Abs. Edge	57.487
0.17837	1	74 W	K	Abs. Edge	69.508	0.21581	3	68 Er		$KO_{II,III}$	57.450
0.178444	5	74 W		$KO_{II,III}$	69.479	0.21592	4	74 W		KL_I	57.42
0.178880	3	75 Re	$K\beta_1$	KM_{III}	69.310	0.21636	2	69 Tm	$K\beta_3$	KM_{II}	57.304
0.17892	2	74 W	$K\beta_4$	$KN_{IV,V}$	69.294	0.2167	2	68 Er	$K\beta_2$	$KN_{II,III}$	57.21
0.179421	7	74 W	$K\beta_2^I$	KN_{III}	69.101	0.220305	8	73 Ta	$K\alpha_2$	KL_{II}	56.277
0.17960	1	74 W	$K\beta_2^{II}$	KN_{II}	69.031	0.22124	3	68 Er	$K\beta_5$	$KM_{IV,V}$	56.040
0.179697	3	75 Re	$K\beta_3$	KM_{II}	68.994	0.222227	3	72 Hf	$K\alpha_1$	KL_{III}	55.7902
0.179958	3	80 Hg	$K\alpha_2$	KL_{II}	68.895	0.22266	2	68 Er	$K\beta_1$	KM_{III}	55.681
0.180195	2	79 Au	$K\alpha_1$	KL_{III}	68.8037	0.22291	1	67 Ho	K	Abs. Edge	55.619
0.183092	7	74 W	$K\beta_5^I$	KM_V	67.715	0.22305	3	67 Ho		$KO_{II,III}$	55.584
0.183264	5	74 W	$K\beta_5^{II}$	KM_{IV}	67.652	0.22341	2	68 Er	$K\beta_3$	KM_{II}	55.494
0.18394	1	73 Ta	K	Abs. Edge	67.403	0.2241	2	67 Ho	$K\beta_2$	$KN_{II,III}$	55.32
0.184031	7	73 Ta		$KO_{II,III}$	67.370	0.227024	3	72 Hf	$K\alpha_2$	KL_{II}	54.6114
0.184374	2	74 W	$K\beta_1$	KM_{III}	67.2443	0.22855	3	67 Ho	$K\beta_5$	$KM_{IV,V}$	54.246
0.18451	1	73 Ta	$K\beta_4$	$KN_{IV,V}$	67.194	0.229298	2	71 Lu	$K\alpha_1$	KL_{III}	54.0698
0.185011	8	73 Ta	$K\beta_2^I$	KN_{III}	67.013	0.23012	2	67 Ho	$K\beta_1$	KM_{III}	53.877
0.185075	2	79 Au	$K\alpha_2$	KL_{II}	66.9895	0.23048	1	66 Dy	K	Abs. Edge	53.793
0.185181	2	74 W	$K\beta_3$	KM_{II}	66.9514	0.23056	3	66 Dy		$KO_{II,III}$	53.774
0.185188	9	73 Ta	$K\beta_2^{II}$	KN_{II}	66.949	0.23083	2	67 Ho	$K\beta_3$	KM_{II}	53.711
0.185511	4	78 Pt	$K\alpha_1$	KL_{III}	66.832	0.2317	2	66 Dy	$K\beta_2$	$KN_{II,III}$	53.47
0.18672	4	79 Au		KL_I	66.40	0.234081	2	71 Lu	$K\alpha_2$	KL_{II}	52.9650
0.188757	6	73 Ta	$K\beta_5^I$	KM_V	65.683	0.23618	3	66 Dy	$K\beta_5$	$KM_{IV,V}$	52.494
0.188920	6	73 Ta	$K\beta_5^{II}$	KM_{IV}	65.626	0.236655	2	70 Yb	$K\alpha_1$	KL_{III}	52.3889
0.18982	5	72 Hf	K	Abs. Edge	65.31	0.23788	2	66 Dy	$K\beta_1$	KM_{III}	52.119
0.190089	4	73 Ta	$K\beta_1$	KM_{III}	65.223	0.23841	1	65 Tb	K	Abs. Edge	52.002

Wavelength Å*	p.e.	Element	Designation		keV
0.23858	3	65 Tb		$KO_{II,III}$	51.965
0.23862	2	66 Dy	$K\beta_3$	KM_{II}	51.957
0.2397	2	65 Tb	$K\beta_2$	$KN_{II,III}$	51.68
0.241424	2	70 Yb	$K\alpha_2$	KL_{II}	51.3540
0.244338	2	69 Tm	$K\alpha_1$	KL_{III}	50.7416
0.24608	2	65 Tb	$K\beta_1$	KM_{III}	50.382
0.24681	1	64 Gd	K	Abs. Edge	50.233
0.24683	2	65 Tb	$K\beta_3$	KM_{II}	50.229
0.24687	3	64 Gd		$KO_{II,III}$	50.221
0.24816	3	64 Gd	$K\beta_2$	$KN_{II,III}$	49.959
0.249095	2	69 Tm	$K\alpha_2$	KL_{II}	49.7726
0.252365	2	68 Er	$K\alpha_1$	KL_{III}	49.1277
0.25275	3	64 Gd	$K\beta_5$	$KM_{IV,V}$	49.052
0.25460	2	64 Gd	$K\beta_1$	KM_{III}	48.697
0.25534	2	64 Gd	$K\beta_3$	KM_{II}	48.555
0.25553	1	63 Eu	K	Abs. Edge	48.519
0.255645	7	63 Eu		$KO_{II,III}$	48.497
0.256923	8	63 Eu	$K\beta_2{}^{I}$	$KN_{II,III}$	48.256
0.257110	2	68 Er	$K\alpha_2$	KL_{II}	48.2211
0.260756	2	67 Ho	$K\alpha_1$	KL_{III}	47.5467
0.263577	5	63 Eu	$K\beta_1$	KM_{III}	47.0379
0.264332	5	63 Eu	$K\beta_3$	KM_{II}	46.9036
0.26464	5	62 Sm	K	Abs. Edge	46.849
0.26491	3	62 Sm		$KO_{II,III}$	46.801
0.265486	2	67 Ho	$K\alpha_2$	KL_{II}	46.6997
0.2662	1	62 Sm	$K\beta_2$	$KN_{II,III}$	46.57
0.269533	2	66 Dy	$K\alpha_1$	KL_{III}	45.9984
0.27111	3	62 Sm	$K\beta_5$	$KM_{IV,V}$	45.731
0.27301	2	62 Sm	$K\beta_1$	KM_{III}	45.413
0.27376	2	62 Sm	$K\beta_3$	KM_{II}	45.289
0.274247	2	66 Dy	$K\alpha_2$	KL_{II}	45.2078
0.27431	5	61 Pm	K	Abs. Edge	45.198
0.2759	1	61 Pm	$K\beta_2$	$KN_{II,III}$	44.93
0.278724	2	65 Tb	$K\alpha_1$	KL_{III}	44.4816
0.28290	3	61 Pm	$K\beta_1$	KM_{III}	43.826
0.283423	2	65 Tb	$K\alpha_2$	KL_{II}	43.7441
0.28363	4	61 Pm	$K\beta_3$	KM_{II}	43.713
0.28453	5	60 Nd	K	Abs. Edge	43.574
0.2861	1	60 Nd		$KN_{II,III}$	43.32
0.288353	2	64 Gd	$K\alpha_1$	KL_{III}	42.9962
0.293038	2	64 Gd	$K\alpha_2$	KL_{II}	42.3089
0.293299	2	60 Nd	$K\beta_1$	KM_{III}	42.2713
0.294027	3	60 Nd	$K\beta_3$	KM_{II}	42.1665
0.29518	5	59 Pr	K	Abs. Edge	42.002
0.29679	2	59 Pr	$K\beta_2$	$KN_{II,III}$	41.773
0.298446	2	63 Eu	$K\alpha_1$	KL_{III}	41.5422
0.303118	2	63 Eu	$K\alpha_2$	KL_{II}	40.9019
0.304261	4	59 Pr	$K\beta_1$	KM_{III}	40.7482
0.304975	5	59 Pr	$K\beta_3$	KM_{II}	40.6529
0.30648	5	58 Ce	K	Abs. Edge	40.453
0.30668	2	58 Ce		$KO_{II,III}$	40.427
0.30737	2	58 Ce	$K\beta_4{}^{I}$	$KN_{IV,V}$	40.337
0.30816	1	58 Ce	$K\beta_2$	$KN_{II,III}$	40.233
0.309040	2	62 Sm	$K\alpha_1$	KL_{III}	40.1181
0.31342	2	58 Ce	$K\beta_5{}^{I}$	KM_{V}	39.558
0.31357	2	58 Ce	$K\beta_5{}^{II}$	KM_{IV}	39.539
0.313698	2	62 Sm	$K\alpha_2$	KL_{II}	39.5224

Wavelength Å*	p.e.	Element	Designation		keV
0.315816	2	58 Ce	$K\beta_1$	KM_{III}	39.2573
0.316520	4	58 Ce	$K\beta_3$	KM_{II}	39.1701
0.31844	5	57 La	K	Abs. Edge	38.934
0.31864	2	57 La		$KO_{II,III}$	38.909
0.31931	2	57 La	$K\beta_4{}^{I}$	$KN_{IV,V}$	38.828
0.320117	7	57 La	$K\beta_2$	$KN_{II,III}$	38.7299
0.320160	4	61 Pm	$K\alpha_1$	KL_{III}	38.7247
0.324803	4	61 Pm	$K\alpha_2$	KL_{II}	38.1712
0.32546	2	57 La	$K\beta_5{}^{I}$	KM_{V}	38.094
0.32563	2	57 La	$K\beta_5{}^{II}$	KM_{IV}	38.074
0.327983	3	57 La	$K\beta_1$	KM_{III}	37.8010
0.328686	4	57 La	$K\beta_3$	KM_{II}	37.7202
0.33104	1	56 Ba	K	Abs. Edge	37.452
0.33127	2	56 Ba		$KO_{II,III}$	37.426
0.331846	2	60 Nd	$K\alpha_1$	KL_{III}	37.3610
0.33229	2	56 Ba	$K\beta_4{}^{II}$	KN_{IV}	37.311
0.33277	1	56 Ba	$K\beta_2$	$KN_{II,III}$	37.257
0.336472	2	60 Nd	$K\alpha_2$	KL_{II}	36.8474
0.33814	2	56 Ba	$K\beta_5{}^{I}$	KM_{V}	36.666
0.33835	2	56 Ba	$K\beta_5{}^{II}$	KM_{IV}	36.643
0.340811	3	56 Ba	$K\beta_1$	KM_{III}	36.3782
0.341507	4	56 Ba	$K\beta_3$	KM_{II}	36.3040
0.344140	2	59 Pr	$K\alpha_1$	KL_{III}	36.0263
0.34451	1	55 Cs	K	Abs. Edge	35.987
0.34611	2	55 Cs	$K\beta_2$	$KN_{II,III}$	35.822
0.348749	2	59 Pr	$K\alpha_2$	KL_{II}	35.5502
0.354364	7	55 Cs	$K\beta_1$	KM_{III}	34.9869
0.355050	4	55 Cs	$K\beta_3$	KM_{II}	34.9194
0.357092	2	58 Ce	$K\alpha_1$	KL_{III}	34.7197
0.3584	5	54 Xe	K	Abs. Edge	34.59
0.36026	3	54 Xe	$K\beta_2$	$KN_{II,III}$	34.415
0.361683	2	58 Ce	$K\alpha_2$	KL_{II}	34.2789
0.36872	2	54 Xe	$K\beta_1$	KM_{III}	33.624
0.36941	2	54 Xe	$K\beta_3$	KM_{II}	33.562
0.370737	2	57 La	$K\alpha_1$	KL_{III}	33.4418
0.37381	1	53 I	K	Abs. Edge	33.1665
0.37523	2	53 I	$K\beta_2$	$KN_{II,III}$	33.042
0.375313	2	57 La	$K\alpha_2$	KL_{II}	33.0341
0.383905	4	53 I	$K\beta_1$	KM_{III}	32.2947
0.384564	4	53 I	$K\beta_3$	KM_{II}	32.2394
0.385111	4	56 Ba	$K\alpha_1$	KL_{III}	32.1936
0.389668	5	56 Ba	$K\alpha_2$	KL_{II}	31.8171
0.38974	1	52 Te		$KO_{II,III}$	31.8114
0.38974	1	52 Te	K	Abs. Edge	31.8114
0.391102	6	52 Te	$K\beta_2$	$KN_{II,III}$	31.7004
0.399995	5	52 Te	$K\beta_1$	KM_{III}	30.9957
0.400290	4	55 Cs	$K\alpha_1$	KL_{III}	30.9728
0.400659	4	52 Te	$K\beta_3$	KM_{II}	30.9443
0.404835	4	55 Cs	$K\alpha_2$	KL_{II}	30.6251
0.40666	1	51 Sb		$KO_{II,III}$	30.4875
0.40668	1	51 Sb	K	Abs. Edge	30.4860
0.40702	1	51 Sb	$K\beta_4{}^{I}$	$KN_{IV,V}$	30.4604
0.407973	5	51 Sb	$K\beta_2$	$KN_{II,III}$	30.3895
0.41378	1	51 Sb	$K\beta_5{}^{I}$	KM_{V}	29.9632
0.41388	1	51 Sb	$K\beta_5{}^{II}$	KM_{IV}	29.9560
0.41634	2	54 Xe	$K\alpha_1$	KL_{III}	29.779
0.417085	3	51 Sb	$K\beta_1$	KM_{III}	29.7256

Wavelength Å*	p.e.	Element		Designation	keV
0.417737	4	51 Sb	$K\beta_3$	KM_{II}	29.6792
0.42087	2	54 Xe	$K\alpha_2$	KL_{II}	29.458
0.42467	3	50 Sn		$KO_{II,III}$	29.195
0.42467	1	50 Sn	K	Abs. Edge	29.1947
0.42495	3	50 Sn	$K\beta_4{}^{I}$	$KN_{IV,V}$	29.175
0.425915	8	50 Sn	$K\beta_2$	$KN_{II,III}$	29.1093
0.43175	3	50 Sn	$K\beta_5{}^{I}$	KM_V	28.716
0.43184	3	50 Sn	$K\beta_5{}^{II}$	KM_{IV}	28.710
0.433318	5	53 I	$K\alpha_1$	KL_{III}	28.6120
0.435236	5	50 Sn	$K\beta_1$	KM_{III}	28.4860
0.435877	5	50 Sn	$K\beta_3$	KM_{II}	28.4440
0.437829	7	53 I	$K\alpha_2$	KL_{II}	28.3172
0.44371	1	49 In	K	Abs. Edge	27.9420
0.44374	3	49 In		$KO_{II,III}$	27.940
0.44393	4	49 In	$K\beta_4{}^{I}$	$KN_{IV,V}$	27.928
0.44500	1	49 In	$K\beta_2$	$KN_{II,III}$	27.8608
0.45086	2	49 In	$K\beta_5{}^{I}$	KM_V	27.499
0.45098	2	49 In	$K\beta_5{}^{II}$	KM_{IV}	27.491
0.451295	3	52 Te	$K\alpha_1$	KL_{III}	27.4723
0.454545	4	49 In	$K\beta_1$	KM_{III}	27.2759
0.455181	4	49 In	$K\beta_3$	KM_{II}	27.2377
0.455784	3	52 Te	$K\alpha_2$	KL_{II}	27.2017
0.46407	1	48 Cd	K	Abs. Edge	26.7159
0.465328	7	48 Cd	$K\beta_2$	$KN_{II,III}$	26.6438
0.470354	3	51 Sb	$K\alpha_1$	KL_{III}	26.3591
0.474827	3	51 Sb	$K\alpha_2$	KL_{II}	26.1108
0.475105	6	48 Cd	$K\beta_1$	KM_{III}	26.0955
0.475730	5	48 Cd	$K\beta_3$	KM_{II}	26.0612
0.48589	1	47 Ag	K	Abs. Edge	25.5165
0.4859	9	47 Ag	$K\beta_4$	$KN_{IV,V}$	25.512
0.487032	4	47 Ag	$K\beta_2$	$KN_{II,III}$	25.4564
0.490599	3	50 Sn	$K\alpha_1$	KL_{III}	25.2713
0.49306	2	47 Ag	$K\beta_5$	$KM_{IV,V}$	25.145
0.495053	3	50 Sn	$K\alpha_2$	KL_{II}	25.0440
0.497069	4	47 Ag	$K\beta_1$	KM_{III}	24.9424
0.497685	4	47 Ag	$K\beta_3$	KM_{II}	24.9115
0.5092	1	46 Pd	K	Abs. Edge	24.348
0.5093	2	46 Pd	$K\beta_4$	$KN_{IV,V}$	24.346
0.510228	4	46 Pd	$K\beta_2$	$KN_{II,III}$	24.2991
0.512113	3	49 In	$K\alpha_1$	KL_{III}	24.2097
0.516544	3	49 In	$K\alpha_2$	KL_{II}	24.0020
0.51670	9	46 Pd	$K\beta_5$	$KM_{IV,V}$	23.995
0.520520	4	46 Pd	$K\beta_1$	KM_{III}	23.8187
0.521123	4	46 Pd	$K\beta_3$	KM_{II}	23.7911
0.53395	1	45 Rh	K	Abs. Edge	23.2198
0.53401	9	45 Rh	$K\beta_4{}^{I}$	$KN_{IV,V}$	23.217
0.535010	3	48 Cd	$K\alpha_1$	KL_{III}	23.1736
0.53503	2	45 Rh	$K\beta_2$	$KN_{II,III}$	23.1728
0.53513	5	45 Rh	$K\beta_2{}^{II}$	KN_{II}	23.168
0.5365	1	94 Pu	L_I	Abs. Edge	23.109
0.539422	3	48 Cd	$K\alpha_2$	KL_{II}	22.9841
0.54101	9	45 Rh	$K\beta_5{}^{I}$	KM_V	22.917
0.54118	9	45 Rh	$K\beta_5{}^{II}$	KM_{IV}	22.909
0.5416	1	94 Pu	$L\gamma_4$	L_IO_{III}	22.891
0.54311	2	95 Am	$L\gamma_6$	$L_{II}O_{IV}$	22.8282
0.5432	1	94 Pu	$L\gamma_4'$	L_IO_{II}	22.823
0.545605	4	45 Rh	$K\beta_1$	KM_{III}	22.7236

Wavelength Å*	p.e.	Element		Designation	keV
0.546200	4	45 Rh	$K\beta_3$	KM_{II}	22.6989
0.5544	2	95 Am	$L\gamma_2$	L_IN_{II}	22.361
0.5572	1	94 Pu	L_{II}	Abs. Edge	22.253
0.5585	5	93 Np	$L\gamma_4$	$L_IO_{II,III}$	22.20
0.5594075	6	47 Ag	$K\alpha_1$	KL_{III}	22.16292
0.55973	2	94 Pu	$L\gamma_6$	$L_{II}O_{IV}$	22.1502
0.56051	1	44 Ru	K	Abs. Edge	22.1193
0.56089	9	44 Ru	$K\beta_4$	$KN_{IV,V}$	22.104
0.56166	3	44 Ru	$K\beta_2$	$KN_{II,III}$	22.074
0.561886	9	95 Am	$L\gamma_1$	$L_{II}N_{IV}$	22.0652
0.563798	4	47 Ag	$K\alpha_2$	KL_{II}	21.9903
0.564001	9	94 Pu	$L\gamma_3$	L_IN_{III}	21.9824
0.5658	1	94 Pu	$L\gamma_8$	$L_{II}O_I$	21.914
0.56785	9	44 Ru	$K\beta_5{}^{I}$	KM_V	21.834
0.5680	2	44 Ru	$K\beta_5{}^{II}$	KM_{IV}	21.829
0.5695	1	92 U	L_I	Abs. Edge	21.771
0.5706	1	92 U	$L\gamma_{13}$	$L_IP_{II,III}$	21.729
0.57068	2	94 Pu	$L\gamma_2$	L_IN_{II}	21.1251
0.572482	4	44 Ru	$K\beta_1$	KM_{III}	21.6568
0.5725	1	92 U		$L_IO_{IV,V}$	21.657
0.573067	4	44 Ru	$K\beta_3$	KM_{II}	21.6346
0.57499	9	92 U	$L\gamma_4$	L_IO_{III}	21.562
0.576700	9	92 U	$L\gamma_4'$	L_IO_{II}	21.4984
0.57699	5	93 Np	$L\gamma_6$	$L_{II}O_{IV}$	21.488
0.578882	9	94 Pu	$L\gamma_1$	$L_{II}N_{IV}$	21.4173
0.5810	5	93 Np	$L\gamma_3$	L_IN_{III}	21.34
0.585448	3	46 Pd	$K\alpha_1$	KL_{III}	21.1771
0.5873	5	93 Np	$L\gamma_2$	L_IN_{II}	21.11
0.58906	1	43 Te	K	Abs. Edge	21.0473
0.589821	3	46 Pd	$K\alpha_2$	KL_{II}	21.0201
0.58986	5	92 U	$L\gamma_{11}$	L_IN_V	21.019
0.59024	5	43 Tc	$K\beta_2$	$KN_{II,III}$	21.005
0.59096	5	92 U		L_IN_{IV}	20.979
0.5919	1	92 U	L_{II}	Abs. Edge	20.945
0.59203	5	92 U		$L_{II}P_{IV}$	20.942
0.5930	2	92 U		$L_{II}P_{II,III}$	20.906
0.5937	1	91 Pa	$L\gamma_4$	$L_IO_{II,III}$	20.882
0.594845	9	92 U	$L\gamma_6$	$L_{II}O_{IV}$	20.8426
0.596498	9	93 Np	$L\gamma_1$	$L_{II}N_{IV}$	20.7848
0.59728	5	92 U		$L_{II}O_{III}$	20.758
0.598574	9	92 U	$L\gamma_3$	L_IN_{III}	20.7127
0.5988	1	94 Pu	$L\gamma_5$	$L_{II}N_I$	20.704
0.60125	5	92 U	$L\gamma_8$	$L_{II}O_I$	20.621
0.60130	4	43 Tc	$K\beta_1$	KM_{III}	20.619
0.60188	4	43 Tc	$K\beta_3$	KM_{II}	20.599
0.6031	1	92 U	Lv	$L_{II}N_{VI}$	20.556
0.605237	9	92 U	$L\gamma_2$	L_IN_{II}	20.4847
0.6059	1	90 Th	L_I	Abs. Edge	20.464
0.60705	8	90 Th	$L\gamma_{13}$	$L_IP_{II,III}$	20.424
0.6083	1	90 Th		$L_IO_{IV,V}$	20.383
0.61098	4	90 Th	$L\gamma_4$	L_IO_{III}	20.292
0.61251	4	90 Th	$L\gamma_4'$	L_IO_{II}	20.242
0.6133	1	91 Pa	$L\gamma_6$	$L_{II}O_{IV}$	20.216
0.613279	4	45 Rh	$K\alpha_1$	KL_{III}	20.2161
0.6146	1	90 Th		L_IO_I	20.174
0.614770	9	92 U	$L\gamma_1$	$L_{II}N_{IV}$	20.1671
0.6160	1	90 Th		$L_IN_{VI,VII}$	20.128

Wavelength Å*	p.e.	Element		Designation	keV	Wavelength Å*	p.e.	Element		Designation	keV
0.616	1	93 Np	$L\gamma_5$	$L_{II}N_I$	20.12	0.67383	2	95 Am	$L\beta_5$	$L_{III}O_{IV,V}$	18.3996
0.6169	1	91 Pa	$L\gamma_3$	$L_I N_{III}$	20.098	0.67491	4	90 Th	$L\gamma_5$	$L_{II}N_I$	18.370
0.617630	4	45 Rh	$K\alpha_2$	KL_{II}	20.0737	0.67502	3	43 Tc	$K\alpha_1$	KL_{III}	18.3671
0.61978	1	42 Mo	K	Abs. Edge	20.0039	0.67538	5	88 Ra	$L\gamma_3$	$L_I N_{III}$	18.357
0.62001	9	42 Mo	$K\beta_4{}^I$	$KN_{IV,V}$	19.996	0.6764	1	88 Ra		$L_{II}O_{III}$	18.330
0.62099	2	42 Mo	$K\beta_2$	$KN_{II,III}$	19.9652	0.67772	2	94 Pu	$L\beta_1$	$L_{II}M_{IV}$	18.2937
0.62107	5	42 Mo	$K\beta_2{}^{II}$	KN_{II}	19.963	0.6780	1	88 Ra		$L_{II}O_{II}$	18.286
0.6228	1	92 U		$L_{II}N_{III}$	19.907	0.67932	3	43 Tc	$K\alpha_2$	KL_{II}	18.2508
0.6239	1	91 Pa	$L\gamma_6$	$L_I N_{II}$	19.872	0.6801	1	88 Ra	$L\gamma_8$	$L_{II}O_I$	18.230
0.62636	9	90 Th	$L\gamma_{11}$	$L_I N_V$	19.794	0.681014	8	92 U	$L\beta_9$	$L_I M_V$	18.2054
0.62692	5	42 Mo	$K\beta_5{}^I$	KM_V	19.776	0.68199	5	88 Ra	$L\gamma_2$	$L_I N_{II}$	18.179
0.62708	5	42 Mo	$K\beta_5{}^{II}$	KM_{IV}	19.771	0.68639	2	95 Am	$L\beta_4$	$L_I M_{II}$	18.0627
0.6276	1	90 Th		$L_I N_{IV}$	19.755	0.6867	1	94 Pu	L_{III}	Abs. Edge	18.054
0.6299	1	90 Th	L_{II}	Abs. Edge	19.683	0.6874	1	88 Ra		$L_I N_I$	18.036
0.62991	9	90 Th		$L_{II}P_{IV}$	19.682	0.68760	5	92 U	$L\beta_{10}$	$L_I M_{IV}$	18.031
0.6312	1	90 Th		$L_{II}P_{II,III}$	19.642	0.68883	1	40 Zr	K	Abs. Edge	17.9989
0.6316	1	90 Th		$L_{II}P_I$	19.629	0.68901	5	40 Zr	$K\beta_4$	$KN_{IV,V}$	17.994
0.632288	9	42 Mo	$K\beta_1$	KM_{III}	19.6083	0.68920	9	93 Np	$L\beta_3$	$L_I M_{III}$	17.989
0.63258	4	90 Th	$L\gamma_6$	$L_{II}O_{IV}$	19.599	0.68993	4	40 Zr	$K\beta_2$	$KN_{II,III}$	17.970
0.632872	2	42 Mo	$K\beta_3$	KM_{II}	19.5903	0.69068	2	94 Pu	$L\beta_5$	$L_{III}O_{IV,V}$	17.9506
0.63358	9	91 Pa	$L\gamma_1$	$L_{II}N_{IV}$	19.568	0.6932	1	88 Ra		$L_{II}N_V$	17.884
0.63557	2	92 U	$L\gamma_5$	$L_{II}N_I$	19.5072	0.69463	5	88 Ra	$L\gamma_1$	$L_{II}N_{IV}$	17.849
0.63559	4	90 Th	$L\gamma_3$	$L_I N_{III}$	19.507	0.6959	1	40 Zr	$K\beta_5$	$KM_{IV,V}$	17.815
0.6356	1	90 Th		$L_{II}O_{III}$	19.506	0.698478	9	93 Np	$L\beta_1$	$L_{II}M_{IV}$	17.7502
0.6369	1	90 Th		$L_{II}O_{II}$	19.466	0.7003	1	94 Pu	$L\beta_7$	$L_{III}O_I$	17.705
0.63898	3	90 Th	$L\gamma_8$	$L_{II}O_I$	19.403	0.701338	3	95 Am	$L\beta_2$	$L_{III}N_V$	17.6766
0.64064	9	90 Th	Lv	$L_{II}N_{VI}$	19.353	0.70173	3	40 Zr	$K\beta_1$	KM_{III}	17.6678
0.6416	1	94 Pu	$L\beta_9$	$L_I M_V$	19.323	0.7018	1	91 Pa	$L\beta_9$	$L_I M_V$	17.667
0.64221	4	90 Th	$L\gamma_2$	$L_I N_{II}$	19.305	0.70228	4	40 Zr	$K\beta_3$	KM_{II}	17.654
0.643083	4	44 Ru	$K\alpha_1$	KL_{III}	19.2792	0.7031	1	94 Pu	Lu	$L_{III}N_{VI,VII}$	17.635
0.6445	1	88 Ra	L_I	Abs. Edge	19.236	0.70341	2	95 Am	$L\beta_{15}$	$L_{III}N_{IV}$	17.6258
0.64513	5	88 Ra	$L\gamma_{13}$	$L_I P_{II,III}$	19.218	0.7043	1	88 Ra		$L_{II}N_{III}$	17.604
0.6468	1	88 Ra		$L_I O_{IV,V}$	19.167	0.70620	2	94 Pu	$L\beta_4$	$L_I M_{II}$	17.5560
0.647408	5	44 Ru	$K\alpha_2$	KL_{II}	19.1504	0.70814	2	93 Np	$L\beta_5$	$L_{III}O_{IV,V}$	17.5081
0.64755	5	90 Th		$L_I N_I$	19.146	0.7088	2	91 Pa	$L\beta_{10}$	$L_I M_{IV}$	17.492
0.6482	1	94 Pu	$L\beta_{10}$	$L_I M_{IV}$	19.126	0.709300	1	42 Mo	$K\alpha_1$	KL_{III}	17.47934
0.64891	2	95 Am	$L\beta_3$	$L_I M_{III}$	19.1059	0.71029	2	92 U	$L\beta_3$	$L_I M_{III}$	17.4550
0.64965	5	88 Ra	$L\gamma_4$	$L_I O_{III}$	19.084	0.713590	6	42 Mo	$K\alpha_2$	KL_{II}	17.3743
0.65131	5	88 Ra	$L\gamma_4'$	$L_I O_{II}$	19.036	0.71652	9	87 Fr	$L\gamma_1$	$L_{II}N_{IV}$	17.303
0.6521	1	90 Th		$L_{II}N_V$	19.014	0.71774	5	88 Ra	$L\gamma_5$	$L_{II}N_I$	17.274
0.65298	1	41 Nb	K	Abs. Edge	18.9869	0.71851	2	94 Pu	$L\beta_2$	$L_{III}N_V$	17.2553
0.65313	3	90 Th	$L\gamma_1$	$L_{II}N_{IV}$	18.9825	0.719984	8	92 U	$L\beta_1$	$L_{II}M_{IV}$	17.2200
0.65318	5	41 Nb	$K\beta_4$	$KN_{IV,V}$	18.981	0.7205	1	94 Pu	$L\beta_{15}$	$L_{III}N_{IV}$	17.208
0.65416	4	41 Nb	$K\beta_2$	$KN_{II,III}$	18.953	0.7223	1	92 U	L_{III}	Abs. Edge	17.165
0.6550	1	91 Pa	$L\gamma_5$	$L_{II}N_I$	18.930	0.72240	5	92 U		$L_{III}P_{IV,V}$	17.162
0.657655	9	95 Am	$L\beta_1$	$L_{II}M_{IV}$	18.8520	0.7234	1	90 Th	$L\beta_9$	$L_I M_V$	17.139
0.6620	1	90 Th		$L_{II}N_{III}$	18.729	0.72426	5	92 U		$L_{III}P_{II,III}$	17.118
0.6654	1	88 Ra	$L\gamma_{11}$	$L_I N_V$	18.633	0.72521	5	92 U		$L_{III}P_I$	17.096
0.66576	2	41 Nb	$K\beta_1$	KM_{III}	18.6225	0.726305	9	92 U	$L\beta_5$	$L_{III}O_{IV,V}$	17.0701
0.66634	3	41 Nb	$K\beta_3$	KM_{II}	18.6063	0.72671	2	93 Np	$L\beta_4$	$L_I M_{II}$	17.0607
0.6666	1	88 Ra		$L_I N_{IV}$	18.600	0.72766	5	39 Y	K	Abs. Edge	17.038
0.66871	2	94 Pu	$L\beta_3$	$L_I M_{III}$	18.5405	0.72776	5	39 Y	$K\beta_4$	$KN_{IV,V}$	17.036
0.6707	1	88 Ra	L_{II}	Abs. Edge	18.486	0.72864	4	39 Y	$K\beta_2$	$KN_{II,III}$	17.0154
0.6714	1	88 Ra		$L_{II}P_{II,III}$	18.466	0.7301	1	90 Th	$L\beta_{10}$	$L_I M_{IV}$	16.981
0.6724	1	88 Ra		$L_{II}P_I$	18.439	0.7309	1	92 U		$L_{III}O_{III}$	16.962
0.67328	5	88 Ra	$L\gamma_6$	$L_{II}O_{IV}$	18.414	0.73230	5	91 Pa	$L\beta_3$	$L_I M_{III}$	16.930
0.67351	9	89 Ac	$L\gamma_1$	$L_{II}N_{IV}$	18.408	0.7333	1	92 U		$L_{III}O_{II}$	16.907

Wavelength Å*	p.e.	Element	Designation		keV	Wavelength Å*	p.e.	Element	Designation		keV
0.73418	2	95 Am	$L\beta_6$	$L_{III}N_I$	16.8870	0.78292	2	38 Sr	$K\beta_1$	KM_{III}	15.8357
0.7345	1	39 Y	$K\beta_5$	$KM_{IV,V}$	16.879	0.78345	3	38 Sr	$K\beta_3$	KM_{II}	15.8249
0.73602	6	92 U	$L\beta_7$	$L_{III}O_I$	16.845	0.7858	1	82 Pb	$L\gamma_4$	L_IO_{III}	15.777
0.736230	9	93 Np	$L\beta_2$	$L_{III}N_V$	16.8400	0.78593	1	40 Zr	$K\alpha_1$	KL_{III}	15.7751
0.738603	9	92 U	Lu	$L_{III}N_{VI,VII}$	16.7859	0.78706	7	82 Pb	$L\gamma_4'$	L_IO_{II}	15.752
0.73928	9	86 Rn	$L\gamma_1$	$L_{II}N_{IV}$	16.770	0.78748	9	84 Po	$L\gamma_1$	$L_{II}N_{IV}$	15.744
0.74072	2	39 Y	$K\beta_1$	KM_{III}	16.7378	0.78838	2	92 U	$L\beta_6$	$L_{III}N_I$	15.7260
0.74126	3	39 Y	$K\beta_3$	KM_{II}	16.7258	0.7884	1	82 Pb		$L_IN_{VI,VII}$	15.725
0.74232	5	91 Pa	$L\beta_1$	$L_{II}M_{IV}$	16.702	0.7887	1	83 Bi	L_{II}	Abs. Edge	15.719
0.74503	5	92 U	$L\beta_{17}$	$L_{II}M_{III}$	16.641	0.78903	9	89 Ac	$L\beta_1$	$L_{II}M_{IV}$	15.713
0.7452	2	91 Pa	$L\beta_5$	$L_{III}O_{IV,V}$	16.636	0.78917	5	83 Bi	$L\gamma_3$	L_IN_{III}	15.7102
0.74620	1	41 Nb	$K\alpha_1$	KL_{III}	16.6151	0.7897	1	82 Pb		L_IO_I	15.699
0.747985	9	92 U	$L\beta_4$	L_IM_{II}	16.5753	0.79015	1	40 Zr	$K\alpha_2$	KL_{II}	15.6909
0.75044	1	41 Nb	$K\alpha_2$	KL_{II}	16.5210	0.79043	3	83 Bi	$L\gamma_6$	$L_{II}O_{IV}$	15.6853
0.75148	2	94 Pu	$L\beta_6$	$L_{III}N_I$	16.4983	0.79257	4	90 Th	$L\beta_4$	L_IM_{II}	15.6429
0.7546	2	91 Pa	$L\beta_7$	$L_{III}O_I$	16.431	0.79257	4	90 Th	$L\beta_{17}$	$L_{II}M_{III}$	15.6429
0.754681	9	92 U	$L\beta_2$	$L_{III}N_V$	16.4283	0.79354	3	90 Th	$L\beta_2$	$L_{III}M_V$	15.6237
0.75479	3	90 Th	$L\beta_3$	L_IM_{III}	16.4258	0.79384	5	83 Bi		$L_{II}O_{III}$	15.6178
0.756642	9	92 U	$L\beta_{15}$	$L_{III}N_{IV}$	16.3857	0.79539	5	90 Th	$L\beta_{15}$	$L_{III}N_{IV}$	15.5875
0.75690	3	83 Bi	$L\gamma_{13}$	$L_IP_{II,III}$	16.3802	0.79565	3	83 Bi	$L\gamma_2$	L_IN_{II}	15.5824
0.7571	1	83 Bi	L_I	Abs. Edge	16.376	0.79721	9	83 Bi	Lv	$L_{II}N_{VI}$	15.552
0.7579	1	90 Th		$L_{II}M_V$	16.359	0.7973	1	83 Bi	$L\gamma_8$	$L_{II}O_I$	15.551
0.75791	5	83 Bi		$L_IO_{IV,V}$	16.358	0.8022	1	83 Bi		L_IN_I	15.456
0.7591	1	94 Pu	$L\eta$	$L_{II}M_I$	16.333	0.80233	9	82 Pb	$L\gamma_{11}$	L_IN_V	15.453
0.7607	1	90 Th	L_{III}	Abs. Edge	16.299	0.80273	5	88 Ra	$L\beta_3$	L_IM_{III}	15.4449
0.76087	9	90 Th		$L_{III}P_{IV,V}$	16.295	0.8028	1	88 Ra	L_{III}	Abs. Edge	15.444
0.76087	3	83 Bi	$L\gamma_4$	L_IO_{III}	16.2947	0.80364	7	82 Pb		L_IN_{IV}	15.427
0.76198	3	83 Bi	$L\gamma_4'$	L_IO_{II}	16.2709	0.8038	1	88 Ra		$L_{III}P_{II,III}$	15.425
0.7625	2	90 Th		$L_{III}P_{II,III}$	16.260	0.8050	1	88 Ra		$L_{III}P_I$	15.402
0.76289	9	85 At	$L\gamma_1$	$L_{II}N_{IV}$	16.251	0.80509	2	92 U	$L\eta$	$L_{II}M_I$	15.3997
0.76338	5	90 Th		$L_{III}P_I$	16.241	0.80627	5	88 Ra	$L\beta_5$	$L_{III}O_{IV,V}$	15.3771
0.7641	5	83 Bi		$L_IN_{VI,VII}$	16.23	0.8079	1	91 Pa	$L\beta_6$	$L_{III}N_I$	15.347
0.7645	2	84 Po	$L\gamma_6$	$L_{II}O_{IV}$	16.218	0.8081	1	81 Tl	L_I	Abs. Edge	15.343
0.76468	5	90 Th	$L\beta_5$	$L_{III}O_{IV,V}$	16.213	0.8082	1	90 Th		$L_{III}N_{III}$	15.341
0.765210	9	90 Th	$L\beta_1$	$L_{II}M_{IV}$	16.2022	0.80861	5	81 Tl		$L_IO_{IV,V}$	15.3327
0.76857	5	88 Ra	$L\beta_9$	L_IM_V	16.131	0.81163	9	90 Th		L_IM_I	15.276
0.769	1	93 Np	$L\beta_6$	$L_{III}N_I$	16.13	0.81184	5	81 Tl	$L\gamma_4$	L_IO_{III}	15.2716
0.7690	1	90 Th		$L_{III}O_{III}$	16.123	0.81308	5	81 Tl	$L\gamma_4'$	L_IO_{II}	15.2482
0.7691	1	92 U		$L_{III}N_{III}$	16.120	0.81311	2	83 Bi	$L\gamma_1$	$L_{II}N_{IV}$	15.2477
0.76973	5	38 Sr	K	Abs. Edge	16.107	0.81375	5	88 Ra	$L\beta_1$	$L_{II}M_{IV}$	15.2358
0.7699	1	91 Pa	$L\beta_4$	L_IM_{II}	16.104	0.8147	1	82 Pb	$L\gamma_3$	L_IN_{III}	15.218
0.76989	5	38 Sr	$K\beta_4$	$KN_{IV,V}$	16.104	0.81538	5	82 Pb	L_{II}	Abs. Edge	15.2053
0.77081	3	38 Sr	$K\beta_2$	$KN_{II,III}$	16.0846	0.8154	2	37 Rb	$K\beta_4$	$KN_{IV,V}$	15.205
0.7713	1	90 Th		$L_{III}O_{II}$	16.074	0.81554	5	37 Rb	K	Abs. Edge	15.2023
0.772	1	84 Po	$L\gamma_2$	L_IN_{II}	16.07	0.8158	1	81 Tl		L_IO_I	15.198
0.7737	1	91 Pa	$L\beta_2$	$L_{III}N_V$	16.024	0.81583	5	82 Pb		$L_{II}P_I$	15.1969
0.77437	4	90 Th	$L\beta_7$	$L_{III}O_I$	16.0105	0.8162	1	88 Ra	$L\beta_7$	$L_{III}O_I$	15.190
0.77546	5	88 Ra	$L\beta_{10}$	L_IM_{IV}	15.988	0.81645	3	37 Rb	$K\beta_2$	$KN_{II,III}$	15.1854
0.7764	1	38 Sr	$K\beta_5$	$KM_{IV,V}$	15.969	0.81683	5	82 Pb	$L\gamma_6$	$L_{II}O_{IV}$	15.1783
0.77661	5	90 Th	Lu	$L_{III}N_{VI,VII}$	15.964	0.8186	1	88 Ra	Lu	$L_{III}N_{VI,VII}$	15.146
0.77728	5	83 Bi	$L\gamma_{11}$	L_IN_V	15.951	0.8190	2	90 Th		$L_{III}N_{II}$	15.138
0.77822	9	89 Ac	$L\beta_3$	L_IM_{III}	15.931	0.8200	1	82 Pb		$L_{II}O_{III}$	15.120
0.77954	5	83 Bi		L_IN_{IV}	15.904	0.8210	2	82 Pb	$L\gamma_2$	L_IN_{II}	15.101
0.78017	9	92 U		$L_{III}N_{II}$	15.892	0.8219	1	37 Rb	$K\beta_5$	$KM_{IV,V}$	15.085
0.7809	2	93 Np	$L\eta$	$L_{II}M_I$	15.876	0.82327	7	82 Pb	Lv	$L_{II}N_{VI}$	15.060
0.78196	5	82 Pb	L_I	Abs. Edge	15.855	0.82365	5	82 Pb	$L\gamma_8$	$L_{II}O_I$	15.0527
0.78257	7	82 Pb		$L_IO_{IV,V}$	15.843	0.8248	1	83 Bi		$L_{II}N_{III}$	15.031

Wavelength Å*	p.e.	Element	Designation		keV
0.82789	9	87 Fr	$L\beta_2$	L_IM_{III}	14.976
0.82790	8	90 Th	$L\beta_6$	$L_{III}N_I$	14.975
0.82859	7	82 Pb		L_IN_I	14.963
0.82868	2	37 Rb	$K\beta_1$	KM_{III}	14.9613
0.82879	5	81 Tl	$L\gamma_{11}$	L_IN_V	14.9593
0.82884	1	39 Y	$K\alpha_1$	KL_{III}	14.9584
0.82921	3	37 Rb	$K\beta_3$	KM_{II}	14.9517
0.8295	1	91 Pa	$L\eta$	$L_{II}M_I$	14.946
0.83001	7	81 Tl		L_IN_{IV}	14.937
0.83305	1	39 Y	$K\alpha_2$	KL_{II}	14.8829
0.8338	1	90 Th		$L_{II}M_{II}$	14.869
0.8344	9	83 Bi		$L_{III}N_I$	14.86
0.8350	2	80 Hg		$L_IO_{IV,V}$	14.847
0.8353	1	80 Hg	L_I	Abs. Edge	14.842
0.83537	5	88 Ra	$L\beta_2$	$L_{III}N_V$	14.8414
0.83722	5	88 Ra	$L\beta_{15}$	$L_{III}N_{IV}$	14.8086
0.8382	2	82 Pb		$L_{II}N_V$	14.791
0.83894	7	80 Hg	$L\gamma_4$	L_IO_{III}	14.778
0.83923	5	83 Bi	$L\gamma_5$	$L_{II}N_I$	14.7732
0.83940	9	87 Fr	$L\beta_1$	$L_{II}M_{IV}$	14.770
0.83973	3	82 Pb	$L\gamma_1$	$L_{II}N_{IV}$	14.7644
0.84013	7	80 Hg	$L\gamma_4'$	L_IO_{II}	14.757
0.84071	5	88 Ra	$L\beta_1$	L_IM_{II}	14.7472
0.84130	4	81 Tl	$L\gamma_3$	L_IN_{III}	14.7368
0.8434	1	81 Tl	L_{II}	Abs. Edge	14.699
0.8438	1	88 Ra	$L\beta_{17}$	$L_{II}M_{III}$	14.692
0.8442	2	81 Tl	$L\gamma_6$	$L_{II}O_{IV}$	14.685
0.8452	2	80 Hg		L_IO_I	14.670
0.84773	5	81 Tl	$L\gamma_2$	L_IN_{II}	14.6251
0.848187	9	95 Am	$L\alpha_1$	$L_{III}M_V$	14.6172
0.8490	1	81 Tl		$L_{II}O_{II}$	14.604
0.85048	5	81 Tl	Lv	$L_{II}N_{VI}$	14.5777
0.8512	1	88 Ra		$L_{III}N_{III}$	14.566
0.8513	2	81 Tl	$L\gamma_8$	$L_{II}O_I$	14.564
0.85192	7	82 Pb		$L_{II}N_{III}$	14.553
0.85436	9	86 Rn	$L\beta_3$	L_IM_{III}	14.512
0.85446	4	90 Th	$L\eta$	$L_{III}M_I$	14.5099
0.8549	1	81 Tl		L_IN_I	14.503
0.85657	7	80 Hg	$L\gamma_{11}$	L_IN_V	14.474
0.858	2	87 Fr	$L\beta_2$	$L_{III}N_V$	14.45
0.8585	3	82 Pb		$L_{II}N_{II}$	14.442
0.860266	9	95 Am	$L\alpha_2$	$L_{III}M_{IV}$	14.4119
0.8618	1	88 Ra		$L_{III}N_{II}$	14.387
0.86376	5	79 Au	L_I	Abs. Edge	14.3537
0.86400	5	79 Au		$L_IO_{IV,V}$	14.3497
0.8653	2	36 Kr	$K\beta_4$	$KN_{IV,V}$	14.328
0.86552	1	36 Kr	K	Abs. Edge	14.3244
0.86605	9	86 Rn	$L\beta_1$	$L_{II}M_{IV}$	14.316
0.8661	1	36 Kr	$K\beta_2$	$KN_{II,III}$	14.315
0.86655	5	82 Pb	$L\gamma_5$	$L_{II}N_I$	14.3075
0.86703	4	79 Au	$L\gamma_4$	L_IO_{III}	14.2996
0.86752	3	81 Tl	$L\gamma_1$	$L_{II}N_{IV}$	14.2915
0.86816	4	79 Au	$L\gamma_4'$	L_IO_{II}	14.2809
0.86830	2	94 Pu	$L\alpha_1$	$L_{III}M_V$	14.2786
0.86915	7	80 Hg	$L\gamma_3$	L_IN_{III}	14.265
0.87074	5	79 Au		L_IO_I	14.2385
0.8708	2	36 Kr	$K\beta_5$	$KM_{IV,V}$	14.238
0.87088	5	88 Ra	$L\beta_6$	$L_{III}N_I$	14.2362
0.8722	1	80 Hg	L_{II}	Abs. Edge	14.215
0.87319	7	80 Hg	$L\gamma_6$	$L_{II}O_{IV}$	14.199
0.87526	1	38 Sr	$K\alpha_1$	KL_{III}	14.1650
0.87544	7	80 Hg	$L\gamma_2$	L_IN_{II}	14.162
0.8758	1	80 Hg		$L_{II}O_{III}$	14.156
0.8784	1	80 Hg		$L_{III}O_{II}$	14.114
0.8785	1	36 Kr	$K\beta_1$	KM_{III}	14.112
0.87885	7	80 Hg	Lv	$L_{II}N_{VI}$	14.107
0.8790	1	36 Kr	$K\beta_3$	KM_{II}	14.104
0.87943	1	38 Sr	$K\alpha_2$	KL_{II}	14.0979
0.87995	7	80 Hg	$L\gamma_8$	$L_{II}O_I$	14.090
0.87996	5	81 Tl		$L_{II}N_{III}$	14.0893
0.88028	2	94 Pu	$L\alpha_2$	$L_{III}M_{IV}$	14.0842
0.88135	9	85 At	$L\beta_3$	L_IM_{III}	14.067
0.8827	2	80 Hg		L_IN_I	14.045
0.88433	7	79 Au	$L\gamma_{11}$	L_IN_V	14.020
0.88563	7	79 Au		L_IN_{IV}	13.999
0.8882	2	81 Tl		$L_{II}M_{II}$	13.959
0.889128	9	93 Np	$L\alpha_1$	$L_{III}M_V$	13.9441
0.8931	1	78 Pt	L_I	Abs. Edge	13.883
0.8934	1	78 Pt		L_IO_V	13.878
0.89349	9	85 At	$L\beta_1$	$L_{II}M_{IV}$	13.876
0.8943	1	78 Pt		L_IO_{IV}	13.861
0.89500	4	81 Tl	$L\gamma_5$	$L_{II}N_I$	13.8526
0.89646	5	80 Hg	$L\gamma_1$	$L_{II}N_{IV}$	13.8301
0.89659	4	78 Pt	$L\gamma_4$	L_IO_{III}	13.8281
0.89747	4	78 Pt	$L\gamma_4'$	L_IO_{II}	13.8145
0.89783	5	79 Au	$L\gamma_8$	L_IN_{III}	13.8090
0.89791	3	83 Bi	$L\beta_9$	L_IM_V	13.8077
0.8995	2	78 Pt		L_IO_I	13.784
0.8996	2	84 Po	$L\beta_6$	$L_{III}O_{IV,V}$	13.782
0.901045	9	93 Np	$L\alpha_2$	$L_{III}M_{IV}$	13.7597
0.90259	5	79 Au	L_{II}	Abs. Edge	13.7361
0.90297	3	79 Au	$L\gamma_6$	$L_{II}O_{IV}$	13.7304
0.90434	3	79 Au	$L\gamma_2$	L_IN_{II}	13.7095
0.90495	4	83 Bi	$L\beta_{10}$	L_IM_{IV}	13.7002
0.90638	7	79 Au		$L_{II}O_{III}$	13.679
0.90742	5	88 Ra	$L\eta$	$L_{II}M_I$	13.6630
0.90746	7	79 Au		$L_{II}O_{II}$	13.662
0.90837	5	79 Au	Lv	$L_{II}N_{VI}$	13.6487
0.90894	7	80 Hg		$L_{II}N_{III}$	13.640
0.9091	3	84 Po	$L\beta_3$	L_IM_{III}	13.638
0.90989	5	79 Au	$L\gamma_8$	$L_{II}O_I$	13.6260
0.910639	9	92 U	$L\alpha_1$	$L_{III}M_V$	13.6147
0.9131	1	79 Au		L_IN_I	13.578
0.9143	2	78 Pt	$L\gamma_{11}$	L_IN_V	13.560
0.9204	1	35 Br	K	Abs. Edge	13.470
0.92046	2	35 Br	$K\beta_2$	$KN_{II,III}$	13.4695
0.9220	2	84 Po	$L\beta_1$	$L_{II}M_{IV}$	13.447
0.922558	9	92 U	$L\alpha_2$	$L_{III}M_{IV}$	13.4388
0.9234	1	83 Bi	L_{III}	Abs. Edge	13.426
0.9236	1	77 Ir	L_I	Abs. Edge	13.423
0.92413	4	83 Bi		$L_{III}P_{II,III}$	13.4159
0.9243	3	77 Ir		$L_IO_{IV,V}$	13.413
0.92453	7	80 Hg	$L\gamma_5$	$L_{II}N_I$	13.410
0.9255	1	35 Br	$K\beta_5$	$KM_{IV,V}$	13.396

Wavelength Å*	p.e.	Element	Designation		keV
0.925553	9	37 Rb	$K\alpha_1$	KL_{III}	13.3953
0.92556	3	83 Bi	$L\beta_5$	$L_{III}O_{IV,V}$	13.3953
0.92650	3	79 Au	$L\gamma_1$	$L_{II}N_{IV}$	13.3817
0.9268	1	82 Pb	$L\beta_9$	L_IM_V	13.377
0.92744	2	77 Ir	$L\gamma_4$	L_IO_{III}	13.3681
0.92791	5	78 Pt	$L\gamma_3$	L_IN_{III}	13.3613
0.92831	3	77 Ir	$L\gamma_4'$	L_IO_{II}	13.3555
0.92937	5	84 Po	$L\beta_2$	$L_{III}N_V$	13.3404
0.92969	1	37 Rb	$K\alpha_2$	KL_{II}	13.3358
0.9302	2	83 Bi		$L_{III}O_{III}$	13.328
0.9312	2	84 Po	$L\beta_{15}$	$L_{III}N_{IV}$	13.314
0.9323	2	83 Bi		$L_{III}O_{II}$	13.298
0.93279	2	35 Br	$K\beta_1$	KM_{III}	13.2914
0.93284	5	91 Pa	$L\alpha_1$	$L_{III}M_V$	13.2907
0.93327	5	35 Br	$K\beta_3$	KM_{II}	13.2845
0.9339	2	82 Pb	$L\beta_{10}$	L_IM_{IV}	13.275
0.93414	5	78 Pt	L_{II}	Abs. Edge	13.2723
0.9342	2	78 Pt	$L\gamma_6$	$L_{II}O_{IV}$	13.271
0.93427	5	78 Pt	$L\gamma_2$	L_IN_{II}	13.2704
0.93505	5	83 Bi	$L\beta_7$	$L_{II}O_I$	13.2593
0.93505	5	83 Bi	Lu	$L_{III}N_{VI,VII}$	13.2593
0.93855	3	83 Bi	$L\beta_3$	L_IM_{III}	13.2098
0.93931	5	78 Pt	Lv	$L_{II}N_{VI}$	13.1992
0.9402	2	79 Au		$L_{II}N_{III}$	13.186
0.9411	1	78 Pt	$L\gamma_8$	$L_{II}O_I$	13.173
0.9446	2	77 Ir	$L\gamma_{11}$	L_IN_V	13.126
0.94482	5	91 Pa	$L\alpha_2$	$L_{III}M_{IV}$	13.1222
0.9455	2	78 Pt		L_IN_I	13.113
0.9459	2	77 Ir		L_IN_{IV}	13.108
0.9475	3	84 Po	$L\beta_4$	L_IM_{II}	13.086
0.95073	5	82 Pb	L_{III}	Abs. Edge	13.0406
0.95118	7	82 Pb		$L_{III}P_{II,III}$	13.0344
0.951978	9	83 Bi	$L\beta_1$	$L_{II}M_{IV}$	13.0235
0.9526	1	82 Pb	$L\beta_5$	$L_{III}O_{IV,V}$	13.015
0.95518	4	83 Bi	$L\beta_2$	$L_{III}N_V$	12.9799
0.95559	3	79 Au	$L\gamma_5$	$L_{II}N_I$	12.9743
0.9558	1	76 Os	L_I	Abs. Edge	12.972
0.95600	3	90 Th	$L\alpha_1$	$L_{III}M_V$	12.9687
0.95603	5	76 Os		$L_IO_{IV,V}$	12.9683
0.95675	7	81 Tl	$L\beta_9$	L_IM_V	12.9585
0.95702	5	83 Bi	$L\beta_{15}$	$L_{III}N_{IV}$	12.9549
0.9578	1	82 Pb		$L_{III}O_{III}$	12.945
0.95797	3	78 Pt	$L\gamma_1$	$L_{II}N_{IV}$	12.9420
0.9586	1	82 Pb		$L_{III}O_{II}$	12.934
0.95931	5	77 Ir	$L\gamma_3$	L_IN_{III}	12.9240
0.95938	8	76 Os	$L\gamma_4$	L_IO_{III}	12.923
0.96033	8	76 Os	$L\gamma_4'$	L_IO_{II}	12.910
0.96133	7	82 Pb	Lu	$L_{III}N_{VI,VII}$	12.8968
0.9620	1	82 Pb	$L\beta_7$	$L_{III}O_I$	12.888
0.96318	7	76 Os		L_IO_I	12.8721
0.9636	3	92 U	Ls	$L_{III}M_{III}$	12.866
0.96389	7	81 Tl	$L\beta_{10}$	L_IM_{IV}	12.8626
0.96545	3	77 Ir	$L\gamma_2$	L_IN_{II}	12.8418
0.96708	4	77 Ir	$L\gamma_6$	$L_{II}O_{IV}$	12.8201
0.9671	1	77 Ir	L_{II}	Abs. Edge	12.820
0.9672	2	84 Po	$L\beta_6$	$L_{III}N_I$	12.819
0.96788	2	90 Th	$L\alpha_2$	$L_{III}M_{IV}$	12.8096
0.96911	7	82 Pb	$L\beta_3$	L_IM_{III}	12.7933
0.96979	5	77 Ir		$L_{II}O_{III}$	12.7843
0.97161	6	77 Ir	Lv	$L_{II}N_{VI}$	12.7603
0.97173	4	78 Pt		$L_{II}N_{III}$	12.7588
0.97321	5	83 Bi		$L_{III}N_{III}$	12.7394
0.97409	3	77 Ir	$L\gamma_8$	$L_{II}O_I$	12.7279
0.9747	1	82 Pb		$L_{II}M_V$	12.720
0.9765	3	76 Os	$L\gamma_{11}$	L_IN_V	12.696
0.9766	2	77 Ir		L_IN_I	12.695
0.97690	4	83 Bi	$L\beta_4$	L_IM_{II}	12.6912
0.9772	3	76 Os		L_IN_{IV}	12.687
0.9792	2	78 Pt		$L_{II}N_{II}$	12.661
0.97926	5	81 Tl		$L_{III}P_{II,III}$	12.6607
0.9793	1	81 Tl	L_{III}	Abs. Edge	12.660
0.97974	1	34 Se	K	Abs. Edge	12.6545
0.97992	5	34 Se	$K\beta_2$	$KN_{II,III}$	12.6522
0.97993	5	89 Ac	$L\alpha_1$	$L_{III}M_V$	12.6520
0.9801	1	36 Kr	$K\alpha_1$	KL_{III}	12.649
0.98058	3	81 Tl	$L\beta_5$	$L_{III}O_{IV,V}$	12.6436
0.98221	7	82 Pb	$L\beta_2$	$L_{III}N_V$	12.6226
0.98280	5	83 Bi		$L_{III}N_{II}$	12.6151
0.98291	3	82 Pb	$L\beta_1$	$L_{II}M_{IV}$	12.6137
0.98389	7	82 Pb	$L\beta_{15}$	$L_{III}N_{IV}$	12.6011
0.9841	1	36 Kr	$K\alpha_2$	KL_{II}	12.598
0.9843	1	34 Se	$K\beta_5$	$KM_{IV,V}$	12.595
0.98538	5	81 Tl		$L_{III}O_{III}$	12.5820
0.9871	2	80 Hg	$L\beta_9$	L_IM_V	12.560
0.98738	5	81 Tl		$L_{III}O_{II}$	12.5566
0.9877	2	78 Pt	$L\gamma_5$	$L_{II}N_I$	12.552
0.9888	1	81 Tl	Lu	$L_{III}N_{VI,VII}$	12.538
0.98913	5	83 Bi	$L\beta_{17}$	$L_{II}M_{III}$	12.5344
0.9894	1	75 Re	L_I	Abs. Edge	12.530
0.9900	1	75 Re		$L_IO_{IV,V}$	12.524
0.99017	5	81 Tl	$L\beta_7$	$L_{III}O_I$	12.5212
0.99085	3	77 Ir	$L\gamma_1$	$L_{II}N_{IV}$	12.5126
0.99178	5	89 Ac	$L\alpha_2$	$L_{III}M_{IV}$	12.5008
0.99186	5	76 Os	$L\gamma_3$	L_IN_{III}	12.4998
0.99218	3	34 Se	$K\beta_1$	KM_{III}	12.4959
0.99249	5	75 Re	$L\gamma_4$	L_IO_{III}	12.4920
0.99268	5	34 Se	$K\beta_3$	KM_{II}	12.4896
0.99331	3	83 Bi	$L\beta_6$	$L_{III}N_I$	12.4816
0.99334	5	75 Re	$L\gamma_4'$	L_IO_{II}	12.4813
0.9962	2	80 Hg	$L\beta_{10}$	L_IM_{IV}	12.446
0.9965	1	75 Re		L_IO_I	12.442
0.99805	5	76 Os	$L\gamma_2$	L_IN_{II}	12.4224
1.0005	1	82 Pb		$L_{III}N_{III}$	12.392
1.0005	9	83 Bi		L_IM_I	12.39
1.00062	3	81 Tl	$L\beta_3$	L_IM_{III}	12.3904
1.00107	5	76 Os	$L\gamma_6$	$L_{II}O_{IV}$	12.3848
1.0012	6	95 Am	Ll	$L_{III}M_I$	12.384
1.0014	1	76 Os	L_{II}	Abs. Edge	12.381
1.0047	2	76 Os		$L_{III}O_{III}$	12.340
1.00473	5	88 Ra	$L\alpha_1$	$L_{III}M_V$	12.3397
1.0050	2	76 Os	Lv	$L_{II}N_{VI}$	12.337
1.0054	3	77 Ir		$L_{II}N_{III}$	12.332
1.00722	5	81 Tl		$L_{III}M_V$	12.3093

Wavelength Å*	p.e.	Element	Designation		keV	Wavelength Å*	p.e.	Element	Designation		keV
1.0075	1	82 Pb	$L\beta_4$	$L_I M_{II}$	12.306	1.04500	3	33 As	$K\beta_2$	$KN_{II,III}$	11.8642
1.00788	5	76 Os	$L\gamma_8$	$L_{II}O_I$	12.3012	1.0458	1	74 W	$L\gamma_{11}$	$L_I N_V$	11.856
1.0091	1	80 Hg	L_{III}	Abs. Edge	12.286	1.0468	2	74 W		$L_I N_{IV}$	11.844
1.00987	7	80 Hg	$L\beta_5$	$L_{III}O_{IV,V}$	12.2769	1.04752	5	79 Au	Lu	$L_{III}N_{VI,VII}$	11.8357
1.01031	3	81 Tl	$L\beta_2$	$L_{III}N_V$	12.2715	1.04868	5	80 Hg	$L\beta_1$	$L_{II}M_{IV}$	11.8226
1.01040	7	82 Pb		$L_{III}N_{II}$	12.2705	1.0488	1	33 As	$K\beta_5$	$KM_{IV,V}$	11.822
1.0108	1	75 Re	$L\gamma_{11}$	$L_I N_V$	12.266	1.04963	5	81 Tl	$L\beta_6$	$L_{III}N_I$	11.8118
1.0112	1	90 Th	Ls	$L_{III}M_{III}$	12.261	1.04974	8	79 Au	$L\beta_7$	$L_{III}O_I$	11.8100
1.0119	1	75 Re		$L_{II}N_{IV}$	12.252	1.05446	5	78 Pt	$L\beta_9$	$L_I M_V$	11.7577
1.0120	2	77 Ir		$L_{II}N_{II}$	12.251	1.05609	7	81 Tl	$L\beta_{17}$	$L_{II}M_{III}$	11.7397
1.01201	3	81 Tl	$L\beta_{15}$	$L_{III}N_{IV}$	12.2510	1.05693	5	76 Os	$L\gamma_5$	$L_{II}N_I$	11.7303
1.01404	7	80 Hg		$L_{III}O_{III}$	12.2264	1.05723	5	86 Rn	$L\alpha_1$	$L_{III}M_V$	11.7270
1.01513	4	81 Tl	$L\beta_1$	$L_{II}M_{IV}$	12.2133	1.05730	2	33 As	$K\beta_1$	KM_{III}	11.7262
1.01558	7	80 Hg		$L_{III}O_{II}$	12.2079	1.05783	5	33 As	$K\beta_3$	KM_{II}	11.7203
1.01656	5	88 Ra	$L\alpha_2$	$L_{III}M_{IV}$	12.1962	1.0585	1	80 Hg		$L_{III}N_{III}$	11.713
1.01674	7	80 Hg	Lu	$L_{III}N_{VII}$	12.1940	1.05856	3	83 Bi	$L\eta$	$L_{II}M_I$	11.7122
1.01769	7	80 Hg	Lu'	$L_{III}N_{VI}$	12.1826	1.06099	5	75 Re	$L\gamma_1$	$L_{II}N_{IV}$	11.6854
1.01937	7	80 Hg	$L\beta_7$	$L_{III}O_I$	12.1625	1.0613	1	73 Ta	L_I	Abs. Edge	11.682
1.02063	7	79 Au	$L\beta_9$	$L_I M_V$	12.1474	1.06183	7	78 Pt	$L\beta_{10}$	$L_I M_{IV}$	11.6762
1.0210	1	82 Pb	$L\beta_6$	$L_{III}N_I$	12.143	1.06192	9	73 Ta		$L_I O_{IV,V}$	11.6752
1.02175	5	77 Ir	$L\gamma_5$	$L_{II}N_I$	12.1342	1.06200	6	74 W	$L\gamma_3$	$L_I N_{III}$	11.6743
1.0223	1	82 Pb	$L\beta_{17}$	$L_{II}M_{III}$	12.127	1.06357	9	73 Ta		$L_I N_{VI,VII}$	11.6570
1.0226	1	94 Pu	Ll	$L_{III}M_I$	12.101	1.0644	?	82 Pb		$L_{II}M_{II}$	11.648
1.02467	5	74 W	L_I	Abs. Edge	12.0996	1.0644	2	81 Tl		$L_I M_I$	11.648
1.0250	2	74 W		$L_I O_{IV,V}$	12.095	1.06467	3	73 Ta	$L\gamma_4$	$L_I O_{III}$	11.6451
1.02503	5	76 Os	$L\gamma_1$	$L_{II}N_{IV}$	12.0953	1.0649	?	80 Hg		$L_{III}N_{II}$	11.642
1.02613	7	75 Re	$L\gamma_3$	$L_I N_{III}$	12.0824	1.06544	3	73 Ta	$L\gamma_4'$	$L_I O_{II}$	11.6366
1.02775	3	74 W	$L\gamma_4$	$L_I O_{III}$	12.0634	1.06712	2	92 U	Ll	$L_{III}M_I$	11.6183
1.02789	7	79 Au	$L\beta_{10}$	$L_I M_{IV}$	12.0617	1.06771	9	73 Ta		$L_I O_I$	11.6118
1.0286	1	81 Tl		$L_{III}N_{III}$	12.053	1.06785	9	79 Au	$L\beta_3$	$L_I M_{III}$	11.6103
1.02863	3	74 W	$L\gamma_4'$	$L_I O_{II}$	12.0530	1.06806	3	74 W	$L\gamma_2$	$L_I N_{II}$	11.6080
1.03049	5	87 Fr	$L\alpha_1$	$L_{III}M_V$	12.0313	1.06899	5	86 Rn	$L\alpha_2$	$L_{III}M_{IV}$	11.5979
1.0317	3	74 W		$L_I O_I$	12.017	1.07022	3	79 Au	$L\beta_2$	$L_{III}N_V$	11.5847
1.03233	5	75 Re	$L\gamma_2$	$L_I N_{II}$	12.0098	1.07188	5	79 Au	$L\beta_{15}$	$L_{III}N_{IV}$	11.5667
1.0323	2	82 Pb		$L_I M_I$	12.010	1.07222	7	80 Hg	$L\beta_4$	$L_I M_{II}$	11.5630
1.03358	7	80 Hg	$L\beta_3$	$L_I M_{III}$	11.9953	1.0723	1	78 Pt	L_{III}	Abs. Edge	11.562
1.0346	9	83 Bi		$L_I M_{II}$	11.98	1.0724	2	78 Pt	$L\beta_5$	$L_{III}O_{IV,V}$	11.561
1.0347	1	92 U	Lt	$L_{III}M_{II}$	11.982	1.07448	5	74 W	$L\gamma_6$	$L_{II}O_{IV}$	11.5387
1.03699	9	75 Re	$L\gamma_6$	$L_{II}O_{IV}$	11.956	1.0745	1	74 W	L_{II}	Abs. Edge	11.538
1.0371	1	75 Re	L_{II}	Abs. Edge	11.954	1.0756	2	79 Au		$L_{III}M_V$	11.526
1.03876	7	79 Au		$L_{III}P_{II,III}$	11.9355	1.0761	3	78 Pt		$L_{III}O_{II,III}$	11.521
1.03918	3	81 Tl	$L\beta_4$	$L_{II}M_{II}$	11.9306	1.0767	1	75 Re		$L_{II}N_{III}$	11.515
1.0397	1	75 Re		$L_{II}O_{III}$	11.925	1.0771	1	74 W	Lv	$L_{II}N_{VI}$	11.510
1.03973	5	76 Os		$L_{III}N_{III}$	11.9243	1.07896	5	78 Pt	Lu	$L_{III}N_{VI,VII}$	11.4908
1.03974	2	35 Br	$K\alpha_1$	KL_{III}	11.9242	1.0792	2	74 W		$L_{II}O_{III}$	11.488
1.03975	7	80 Hg	$L\beta_2$	$L_{III}N_V$	11.9241	1.07975	7	80 Hg	$L\beta_6$	$L_{III}N_I$	11.4824
1.04000	5	79 Au	L_{III}	Abs. Edge	11.9212	1.08009	9	90 Th	Lt	$L_{III}M_{II}$	11.4788
1.0404	1	75 Re	Lv	$L_{II}N_{VI}$	11.917	1.08113	4	74 W	$L\gamma_8$	$L_{II}O_I$	11.4677
1.04044	3	79 Au	$L\beta_5$	$L_{III}O_{IV,V}$	11.9163	1.08168	3	78 Pt	$L\beta_7$	$L_{III}O_I$	11.4619
1.04151	7	80 Hg	$L\beta_{15}$	$L_{III}N_{IV}$	11.9040	1.08205	7	73 Ta	$L\gamma_{11}$	$L_I N_V$	11.4580
1.0420	1	75 Re		$L_I N_I$	11.899	1.08353	3	79 Au	$L\beta_1$	$L_{II}M_{IV}$	11.4423
1.04230	5	87 Fr	$L\alpha_2$	$L_{III}M_{IV}$	11.8950	1.08377	7	73 Ta		$L_I N_{IV}$	11.4398
1.0428	6	93 Np	Ll	$L_{III}M_I$	11.890	1.0839	1	75 Re		$L_{II}N_{II}$	11.438
1.04382	2	35 Br	$K\alpha_2$	KL_{II}	11.8776	1.08500	5	85 At	$L\alpha_1$	$L_{III}M_V$	11.4268
1.04398	5	75 Re	$L\gamma_8$	$L_{II}O_I$	11.8758	1.08975	5	77 Ir	$L\beta_9$	$L_I M_V$	11.3770
1.0450	2	79 Au		$L_{III}O_{II,III}$	11.865	1.09026	7	79 Au		$L_{III}N_{III}$	11.3717
1.0450	1	33 As	K	Abs. Edge	11.865	1.0908	1	91 Pa	Ll	$L_{III}M_I$	11.366

Wavelength Å*	p.e.	Element		Designation	keV	Wavelength Å*	p.e.	Element		Designation	keV
1.0916	5	80 Hg	$L\beta_{17}$	$L_{II}M_{III}$	11.358	1.13687	9	73 Ta		$L_{II}N_V$	10.9055
1.09241	7	82 Pb	$L\eta$	$L_{II}M_I$	11.3493	1.13707	3	77 Ir	$L\beta_{15}$	$L_{III}N_{IV}$	10.9036
1.09388	5	75 Re	$L\gamma_5$	$L_{II}N_I$	11.3341	1.13794	3	73 Ta	$L\gamma_1$	$L_{II}N_{IV}$	10.8952
1.09671	5	85 At	$L\alpha_2$	$L_{III}M_{IV}$	11.3048	1.13841	5	72 Hf	$L\gamma_3$	L_IN_{III}	10.8907
1.09702	4	77 Ir	$L\beta_{10}$	L_IM_{IV}	11.3016	1.1387	5	80 Hg		$L_{II}M_{II}$	10.888
1.09855	3	74 W	$L\gamma_1$	$L_{II}N_{IV}$	11.2859	1.1402	1	71 Lu	L_I	Abs. Edge	10.8740
1.09936	4	73 Ta	$L\gamma_3$	L_IN_{III}	11.2776	1.1405	1	76 Os	$L\beta_5$	$L_{III}O_{IV,V}$	10.8711
1.0997	1	81 Tl		$L_{II}M_{II}$	11.274	1.1408	1	76 Os	L_{III}	Abs. Edge	10.8683
1.0997	1	72 Hf	L_I	Abs. Edge	11.274	1.14085	3	77 Ir	$L\beta_3$	L_IM_{III}	10.8674
1.09968	7	79 Au		$L_{III}N_{II}$	11.2743	1.14223	5	78 Pt	$L\beta_4$	L_IM_{II}	10.8543
1.0999	2	80 Hg		L_IM_I	11.272	1.1435	1	71 Lu	$L\gamma_4$	$L_IO_{II,III}$	10.8425
1.10086	9	72 Hf		L_IO_{IV}	11.2622	1.14355	5	78 Pt	$L\beta_6$	$L_{III}N_I$	10.8418
1.10200	3	78 Pt	$L\beta_2$	$L_{III}N_V$	11.2505	1.14386	2	83 Bi	$L\alpha_1$	$L_{III}M_V$	10.8388
1.10303	5	72 Hf	$L\gamma_4$	L_IO_{III}	11.2401	1.14442	5	72 Hf	$L\gamma_2$	L_IN_{II}	10.8335
1.10376	5	72 Hf	$L\gamma_4'$	L_IO_{II}	11.2326	1.14537	7	76 Os	Lu	$L_{III}N_{VI,VII}$	10.8245
1.10394	5	78 Pt	$L\beta_3$	L_IM_{III}	11.2308	1.1489	2	77 Ir		$L_{II}M_V$	10.791
1.10477	2	34 Se	$K\alpha_1$	KL_{III}	11.2224	1.14933	8	76 Os	$L\beta_7$	$L_{III}O_I$	10.7872
1.1053	1	73 Ta	$L\gamma_2$	L_IN_{II}	11.217	1.1548	1	72 Hf	L_{II}	Abs. Edge	10.7362
1.1058	1	77 Ir	L_{III}	Abs. Edge	11.212	1.15519	5	72 Hf	$L\gamma_6$	$L_{II}O_{IV}$	10.7325
1.10585	3	77 Ir	$L\beta_5$	$L_{III}O_{IV,V}$	11.2114	1.1553	1	73 Ta		$L_{II}N_{III}$	10.7316
1.10651	3	79 Au	$L\beta_4$	L_IM_{II}	11.2047	1.15536	1	83 Bi	$L\alpha_2$	$L_{III}M_{IV}$	10.73091
1.10664	9	72 Hf		L_IO_I	11.2034	1.1560	3	77 Ir		$L_{III}N_{III}$	10.725
1.10882	2	34 Se	$K\alpha_2$	KL_{II}	11.1814	1.15781	3	77 Ir	$L\beta_1$	$L_{II}M_{IV}$	10.7083
1.10923	6	77 Ir		$L_{III}O_{II,III}$	11.1772	1.15830	9	72 Hf	Lv	$L_{II}N_{VI}$	10.7037
1.11092	3	79 Au	$L\beta_6$	$L_{III}N_I$	11.1602	1.1600	2	73 Ta		$L_{II}N_{II}$	10.688
1.11145	4	77 Ir	Lu	$L_{III}N_{VI,VII}$	11.1549	1.16107	9	71 Lu	$L\gamma_{11}$	L_IN_V	10.6782
1.1129	2	78 Pt		$L_{II}M_V$	11.140	1.16138	5	72 Hf	$L\gamma_8$	$L_{II}O_I$	10.6754
1.1137	1	73 Ta	L_{II}	Abs. Edge	11.132	1.16227	9	71 Lu		L_IN_{IV}	10.6672
1.11386	4	84 Po	$L\alpha_1$	$L_{III}M_V$	11.1308	1.1640	1	80 Hg	$L\eta$	$L_{II}M_I$	10.6512
1.11388	3	73 Ta	$L\gamma_6$	$L_{II}O_{IV}$	11.1306	1.16487	4	75 Re	$L\beta_9$	L_IM_V	10.6433
1.11489	3	77 Ir	$L\beta_7$	$L_{III}O_I$	11.1205	1.16545	5	77 Ir		$L_{III}N_{II}$	10.6380
1.1149	2	74 W		$L_{II}N_{III}$	11.120	1.1667	1	78 Pt	$L\beta_{17}$	$L_{II}M_{III}$	10.6265
1.11508	4	90 Th	Ll	$L_{III}M_I$	11.1186	1.16719	5	88 Ra	Ll	$L_{III}M_I$	10.6222
1.11521	9	73 Ta		L_IN_I	11.1173	1.16962	9	78 Pt		L_IM_I	10.6001
1.1158	1	73 Ta	Lv	$L_{II}N_{VI}$	11.1113	1.16979	8	76 Os	$L\beta_2$	$L_{III}N_V$	10.5985
1.11658	5	32 Ge	K	Abs. Edge	11.1036	1.1708	1	79 Au		$L_{II}M_{II}$	10.5892
1.11686	2	32 Ge	$K\beta_2$	$KN_{II,III}$	11.1008	1.17167	5	76 Os	$L\beta_{15}$	$L_{III}N_{IV}$	10.5816
1.11693	9	73 Ta		$L_{II}O_{III}$	11.1001	1.17218	5	75 Re	$L\beta_{10}$	L_IM_{IV}	10.5770
1.11789	9	73 Ta		$L_{II}O_{II}$	11.0907	1.1729	1	73 Ta	$L\gamma_5$	$L_{II}N_I$	10.5702
1.1195	1	32 Ge	$K\beta_5$	$KM_{IV,V}$	11.0745	1.17501	2	82 Pb	$L\alpha_1$	$L_{III}M_V$	10.5515
1.11990	2	78 Pt	$L\beta_1$	$L_{II}M_{IV}$	11.0707	1.17588	1	33 As	$K\alpha_1$	KL_{III}	10.54372
1.1205	1	73 Ta	$L\gamma_8$	$L_{II}O_I$	11.0646	1.17721	5	75 Re	$L\beta_5$	$L_{III}O_{IV,V}$	10.5318
1.12146	9	72 Hf	$L\gamma_{11}$	L_IN_V	11.0553	1.1773	1	75 Re	L_{III}	Abs. Edge	10.5306
1.1218	3	74 W		$L_{II}N_{II}$	11.052	1.17788	9	72 Hf		$L_{II}N_V$	10.5258
1.12250	9	72 Hf		L_IN_{IV}	11.0451	1.17796	3	77 Ir	$L\beta_6$	$L_{III}N_I$	10.5251
1.1226	2	78 Pt		$L_{III}N_{III}$	11.044	1.17900	5	72 Hf	$L\gamma_1$	$L_{II}N_{IV}$	10.5158
1.12548	5	84 Po	$L\alpha_2$	$L_{III}M_{IV}$	11.0158	1.17953	4	71 Lu	$L\gamma_3$	L_IN_{III}	10.5110
1.12637	6	76 Os	$L\beta_9$	L_IM_V	11.0071	1.17955	7	76 Os	$L\beta_3$	L_IM_{III}	10.5108
1.12769	3	81 Tl	$L\eta$	$L_{II}M_I$	10.9943	1.17958	3	77 Ir	$L\beta_4$	L_IM_{II}	10.5106
1.12798	5	79 Au	$L\beta_{17}$	$L_{II}M_{III}$	10.9915	1.17987	1	33 As	$K\alpha_2$	KL_{II}	10.50799
1.12894	2	32 Ge	$K\beta_1$	KM_{III}	10.9821	1.1815	1	75 Re	Lu	$L_{III}N_{VI,VII}$	10.4931
1.12936	9	32 Ge	$K\beta_3$	KM_{II}	10.9780	1.1818	1	70 Yb	L_I	Abs. Edge	10.4904
1.1310	2	78 Pt		$L_{III}N_{II}$	10.962	1.1827	1	70 Yb		$L_IO_{IV,V}$	10.4833
1.13235	3	74 W	$L\gamma_5$	$L_{II}N_I$	10.9490	1.1853	1	70 Yb	$L\gamma_4$	$L_IO_{II,III}$	10.4603
1.13353	5	76 Os	$L\beta_{10}$	L_IM_{IV}	10.9376	1.1853	2	71 Lu	$L\gamma_2$	L_IN_{II}	10.460
1.13525	5	79 Au		L_IM_I	10.9210	1.18610	5	75 Re	$L\beta_7$	$L_{III}O_I$	10.4529
1.13532	3	77 Ir	$L\beta_2$	$L_{III}N_V$	10.9203	1.18648	5	82 Pb	$L\alpha_2$	$L_{III}M_{IV}$	10.4495

Wavelength Å*	p.e.	Element	Designation		keV
1.1886	1	70 Yb		L_IO_I	10.4312
1.18977	7	76 Os		$L_{II}M_V$	10.4205
1.1958	1	31 Ga	K	Abs. Edge	10.3682
1.19600	2	31 Ga	$K\beta_2$	$KN_{II,III}$	10.3663
1.19727	7	76 Os	$L\beta_1$	$L_{II}M_{IV}$	10.3553
1.1981	2	31 Ga	$K\beta_5$	$KM_{IV,V}$	10.348
1.1985	1	71 Lu	L_{II}	Abs. Edge	10.3448
1.1987	1	71 Lu	$L\gamma_6$	$L_{III}O_{IV}$	10.3131
1.20086	7	76 Os		$L_{III}N_{II}$	10.3244
1.2014	1	71 Lu		$L_{II}O_{II,III}$	10.3198
1.20273	3	79 Au	$L\eta$	$L_{II}M_I$	10.3083
1.2047	1	71 Lu	$L\gamma_8$	L_IO_I	10.2915
1.20479	7	74 W	$L\beta_9$	L_IM_V	10.2907
1.20660	4	75 Re	$L\beta_2$	$L_{III}N_V$	10.2752
1.2069	2	77 Ir	$L\beta_{17}$	$L_{II}M_{III}$	10.273
1.20739	4	81 Tl	$L\alpha_1$	$L_{III}M_V$	10.2685
1.20789	2	31 Ga	$K\beta_1$	KM_{III}	10.2642
1.20819	5	75 Re	$L\beta_{15}$	$L_{III}N_{IV}$	10.2617
1.20835	5	31 Ga	$K\beta_3$	KM_{II}	10.2603
1.2102	2	77 Ir		L_IM_I	10.245
1.2105	1	83 Bi	Ls	$L_{III}M_{III}$	10.2421
1.21218	3	74 W	$L\beta_{10}$	L_IM_{IV}	10.2279
1.213	1	78 Pt		$L_{II}M_{II}$	10.225
1.21349	5	76 Os	$L\beta_6$	$L_{III}N_I$	10.2109
1.21537	5	72 Hf	$L\gamma_6$	$L_{II}N_I$	10.2011
1.21545	3	74 W	$L\beta_5$	$L_{III}O_{IV,V}$	10.2004
1.2155	1	74 W	L_{III}	Abs. Edge	10.1999
1.21844	5	76 Os	$L\beta_4$	L_IM_{II}	10.1754
1.21868	5	74 W	Lu	$L_{III}N_{VI,VII}$	10.1733
1.21875	3	81 Tl	$L\alpha_2$	$L_{III}M_{IV}$	10.1728
1.22031	5	75 Re	$L\beta_3$	L_IM_{III}	10.1598
1.2211	2	74 W		$L_{III}O_{II,III}$	10.153
1.22228	4	71 Lu	$L\gamma_1$	$L_{II}N_{IV}$	10.1434
1.22232	5	70 Yb	$L\gamma_3$	L_IN_{III}	10.1431
1.22400	4	74 W	$L\beta_7$	$L_{III}O_I$	10.1292
1.2250	1	69 Tm	L_I	Abs. Edge	10.1206
1.2263	3	69 Tm		$L_IO_{IV,V}$	10.110
1.2283	1	75 Re		$L_{III}N_{III}$	10.0933
1.22879	7	70 Yb	$L\gamma_2$	L_IN_{II}	10.0897
1.2294	2	69 Tm	$L\gamma_4$	$L_IO_{II,III}$	10.084
1.2305	1	75 Re		$L_{II}M_V$	10.0753
1.23858	2	75 Re	$L\beta_1$	$L_{II}M_{IV}$	10.0100
1.24120	5	80 Hg	$L\alpha_1$	$L_{III}M_V$	9.9888
1.24271	3	70 Yb	$L\gamma_6$	$L_{II}O_{IV}$	9.9766
1.2428	1	70 Yb	L_{II}	Abs. Edge	9.9761
1.2429	2	78 Pt	$L\eta$	$L_{II}M_I$	9.975
1.24385	7	82 Pb	Ls	$L_{III}M_{III}$	9.9675
1.24460	3	74 W	$L\beta_2$	$L_{III}N_V$	9.9615
1.2453	1	70 Yb		$L_{II}O_{II,III}$	9.9561
1.24631	3	74 W	$L\beta_{15}$	$L_{III}N_{IV}$	9.9478
1.2466	2	73 Ta	$L\beta_9$	L_IM_V	9.946
1.2480	2	76 Os	$L\beta_{17}$	$L_{II}M_{III}$	9.934
1.24923	5	70 Yb	$L\gamma_8$	$L_{II}O_I$	9.9246
1.2502	3	77 Ir		$L_{II}M_{II}$	9.917
1.25100	5	75 Re	$L\beta_6$	$L_{III}N_I$	9.9105
1.25264	7	80 Hg	$L\alpha_2$	$L_{III}M_{IV}$	9.8976
1.2537	2	73 Ta	$L\beta_{10}$	L_IM_{IV}	9.889
1.254054	9	32 Ge	$K\alpha_1$	KL_{III}	9.88642
1.2553	1	73 Ta	L_{III}	Abs. Edge	9.8766
1.2555	1	73 Ta	$L\beta_5$	$L_{III}O_{IV,V}$	9.8750
1.25778	4	73 Ta	Lu	$L_{II}N_{VI,VII}$	9.8572
1.258011	9	32 Ge	$K\alpha_2$	KL_{II}	9.85532
1.25917	5	75 Re	$L\beta_4$	L_IM_{II}	9.8463
1.2596	1	71 Lu	$L\gamma_5$	$L_{II}N_I$	9.8428
1.2601	3	73 Ta		$L_{III}O_{II,III}$	9.839
1.26269	5	74 W	$L\beta_3$	L_IM_{III}	9.8188
1.26385	5	73 Ta	$L\beta_7$	$L_{III}O_I$	9.8098
1.2672	2	74 W		$L_{III}N_{III}$	9.784
1.26769	5	70 Yb	$L\gamma_1$	$L_{II}N_{IV}$	9.7801
1.2678	2	69 Tm	$L\gamma_3$	L_IN_{III}	9.779
1.2706	1	68 Er	L_I	Abs. Edge	9.7574
1.2728	2	74 W		$L_{II}M_V$	9.741
1.2742	2	69 Tm	$L\gamma_2$	L_IN_{II}	9.730
1.2748	1	83 Bi	Lt	$L_{III}M_{II}$	9.7252
1.2752	2	68 Er	$L\gamma_4$	$L_IO_{II,III}$	9.722
1.27640	3	79 Au	$L\alpha_1$	$L_{III}M_V$	9.7133
1.2765	2	74 W		$L_{III}N_{II}$	9.712
1.27807	5	81 Tl	Ls	$L_{III}M_{III}$	9.7007
1.281809	9	74 W	$L\beta_1$	$L_{II}M_{IV}$	9.67235
1.2829	5	84 Po	Ll	$L_{III}M_I$	9.664
1.2831	1	30 Zn	K	Abs. Edge	9.6607
1.28372	2	30 Zn	$K\beta_2$	$KN_{II,III}$	9.6580
1.28448	3	77 Ir	$L\eta$	$L_{II}M_I$	9.6522
1.28454	2	73 Ta	$L\beta_4$	$L_{III}N_V$	9.6518
1.2848	1	30 Zn	$K\beta_5$	$KM_{IV,V}$	9.6501
1.28619	5	73 Ta	$L\beta_{15}$	$L_{III}N_{IV}$	9.6394
1.28772	3	79 Au	$L\alpha_2$	$L_{III}M_{IV}$	9.6280
1.2892	1	69 Tm	L_{II}	Abs. Edge	9.6171
1.28989	7	74 W	$L\beta_6$	$L_{III}N_I$	9.6117
1.29025	9	72 Hf	$L\beta_9$	L_IM_V	9.6090
1.2905	2	69 Tm	$L\gamma_6$	$L_{II}O_{IV}$	9.607
1.2927	1	75 Re	$L\beta_{17}$	$L_{II}M_{III}$	9.5910
1.2934	2	76 Os		$L_{II}M_{II}$	9.586
1.29525	2	30 Zn	$K\beta_{1,3}$	$KM_{II,III}$	9.5720
1.2972	1	72 Hf	L_{III}	Abs. Edge	9.5577
1.29761	5	72 Hf	$L\beta_5$	$L_{III}O_{IV,V}$	9.5546
1.29819	9	72 Hf	$L\beta_{10}$	L_IM_{IV}	9.5503
1.30162	5	74 W	$L\beta_4$	L_IM_{II}	9.5252
1.30165	9	72 Hf	Lu	$L_{III}N_{VI,VII}$	9.5249
1.30564	5	72 Hf	$L\beta_7$	$L_{III}O_I$	9.4958
1.3063	1	70 Yb	$L\gamma_5$	$L_{II}N_I$	9.4910
1.30678	3	73 Ta	$L\beta_3$	L_IM_{III}	9.4875
1.30767	7	82 Pb	Lt	$L_{III}M_{II}$	9.4811
1.3086	1	73 Ta		$L_{III}N_{III}$	9.4742
1.3112	2	80 Hg	Ls	$L_{III}M_{III}$	9.455
1.31304	3	78 Pt	$L\alpha_1$	$L_{III}M_V$	9.4423
1.3146	1	68 Er	$L\gamma_3$	L_IN_{III}	9.4309
1.3153	2	69 Tm	$L\gamma_1$	$L_{II}N_{IV}$	9.426
1.31610	7	83 Bi	Ll	$L_{III}M_I$	9.4204
1.3167	1	73 Ta		$L_{III}N_{II}$	9.4158
1.31897	9	73 Ta		$L_{II}M_V$	9.3998
1.3190	1	67 Ho	L_I	Abs. Edge	9.3994
1.3208	3	67 Ho		$L_IO_{IV,V}$	9.387
1.3210	2	68 Er	$L\gamma_2$	L_IN_{II}	9.385

Wavelength Å*	p.e.	Element	Designation		keV
1.3225	2	67 Ho	$L\gamma_4$	$L_IO_{II,III}$	9.374
1.32432	2	78 Pt	$L\alpha_2$	$L_{III}M_{IV}$	9.3618
1.32639	5	72 Hf	$L\beta_2$	$L_{III}N_V$	9.3473
1.32698	3	73 Ta	$L\beta_1$	$L_{II}M_{IV}$	9.3431
1.32783	5	72 Hf	$L\beta_{15}$	$L_{III}N_{IV}$	9.3371
1.32785	7	76 Os	$L\eta$	$L_{II}M_I$	9.3370
1.33094	8	73 Ta	$L\beta_6$	$L_{III}N_I$	9.3153
1.3358	1	71 Lu	$L\beta_9$	L_IM_V	9.2816
1.3365	3	74 W		L_IM_I	9.277
1.3366	1	75 Re		$L_{II}M_{II}$	9.2761
1.3386	1	68 Er	L_{II}	Abs. Edge	9.2622
1.3387	2	74 W	$L\beta_{17}$	$L_{II}M_{III}$	9.261
1.3397	3	68 Er	$L\gamma_6$	$L_{II}O_{IV}$	9.255
1.340083	9	31 Ga	$K\alpha_1$	KL_{III}	9.25174
1.3405	1	71 Lu	L_{III}	Abs. Edge	9.2490
1.34154	5	81 Tl	Lt	$L_{III}M_{II}$	9.2417
1.34183	7	71 Lu	$L\beta_5$	$L_{III}O_{IV,V}$	9.2397
1.3430	2	71 Lu	$L\beta_{10}$	L_IM_{IV}	9.232
1.34399	1	31 Ga	$K\alpha_2$	KL_{II}	9.22482
1.34524	9	71 Lu		$L_{III}O_{II,III}$	9.2163
1.34581	3	73 Ta	$L\beta_4$	L_IM_{II}	9.2124
1.34949	5	71 Lu	$L\beta_7$	$L_{III}O_I$	9.1873
1.34990	7	82 Pb	Ll	$L_{III}M_I$	9.1845
1.35053	9	72 Hf		$L_{III}N_{III}$	9.1802
1.35128	3	77 Ir	$L\alpha_1$	$L_{III}M_V$	9.1751
1.35131	7	79 Au	Ls	$L_{III}M_{III}$	9.1749
1.35300	5	72 Hf	$L\beta_3$	L_IM_{III}	9.1634
1.3558	2	69 Tm	$L\gamma_5$	$L_{II}N_I$	9.144
1.35887	9	72 Hf		$L_{III}N_{II}$	9.1239
1.36250	5	77 Ir	$L\alpha_2$	$L_{III}M_{IV}$	9.0995
1.3641	2	68 Er	$L\gamma_1$	$L_{II}N_{IV}$	9.089
1.3643	2	67 Ho	$L\gamma_3$	L_IN_{III}	9.087
1.3692	1	66 Dy	L_I	Abs. Edge	9.0548
1.3698	2	67 Ho	$L\gamma_2$	L_IN_{II}	9.051
1.37012	3	71 Lu	$L\beta_2$	$L_{III}N_V$	9.0489
1.3715	1	71 Lu	$L\beta_{15}$	$L_{III}N_{IV}$	9.0395
1.37342	5	75 Re	$L\eta$	$L_{II}M_I$	9.0272
1.37410	5	72 Hf	$L\beta_1$	$L_{II}M_{IV}$	9.0227
1.37410	5	72 Hf	$L\beta_6$	$L_{III}N_I$	9.0227
1.37459	7	66 Dy	$L\gamma_4$	$L_IO_{II,III}$	9.0195
1.3746	2	80 Hg	Lt	$L_{III}M_{II}$	9.019
1.38059	5	29 Cu	K	Abs. Edge	8.9803
1.38109	3	29 Cu	$K\beta_2$	$KM_{IV,V}$	8.9770
1.3838	1	70 Yb	$L\beta_9$	L_IM_V	8.9597
1.38477	3	81 Tl	Ll	$L_{III}M_I$	8.9532
1.3862	1	70 Yb	L_{III}	Abs. Edge	8.9441
1.3864	1	73 Ta	$L\beta_{17}$	$L_{II}M_{III}$	8.9428
1.38696	7	70 Yb	$L\beta_5$	$L_{III}O_{IV,V}$	8.9390
1.3895	2	78 Pt	Ls	$L_{III}M_{III}$	8.923
1.3898	1	70 Yb		$L_{III}O_{II,III}$	8.9209
1.3905	1	67 Ho	L_{II}	Abs. Edge	8.9164
1.39121	5	76 Os	$L\alpha_1$	$L_{III}M_V$	8.9117
1.3915	1	70 Yb	$L\beta_{10}$	L_IM_{IV}	8.9100
1.39220	5	72 Hf	$L\beta_4$	L_IM_{II}	8.9054
1.392218	9	29 Cu	$K\beta_{1,3}$	$KM_{II,III}$	8.90529
1.3923	2	67 Ho	$L\gamma_6$	$L_{II}O_{IV}$	8.905
1.3926	1	29 Cu	$K\beta_3$	KM_{II}	8.9029
1.3948	1	70 Yb	$L\beta_7$	$L_{III}O_I$	8.8889
1.3983	2	67 Ho	$L\gamma_8$	$L_{II}O_I$	8.867
1.40140	5	71 Lu	$L\beta_3$	L_IM_{III}	8.8469
1.40234	5	76 Os	$L\alpha_2$	$L_{III}M_{IV}$	8.8410
1.4067	3	68 Er	$L\gamma_5$	$L_{II}N_I$	8.814
1.41366	7	79 Au	Lt	$L_{III}M_{II}$	8.7702
1.41550	5	70 Yb	$L\beta_{2,15}$	$L_{III}N_{IV,V}$	8.7588
1.41640	7	66 Dy	$L\gamma_3$	L_IN_{III}	8.7532
1.4174	2	67 Ho	$L\gamma_1$	$L_{II}N_{IV}$	8.747
1.4189	1	71 Lu	$L\beta_6$	$L_{III}N_I$	8.7376
1.42110	3	74 W	$L\eta$	$L_{II}M_I$	8.7243
1.4216	1	80 Hg	Ll	$L_{III}M_I$	8.7210
1.4223	1	65 Tb	L_I	Abs. Edge	8.7167
1.42278	7	66 Dy	$L\gamma_2$	L_IN_{II}	8.7140
1.4228	3	65 Tb		$L_IO_{IV,V}$	8.714
1.42359	3	71 Lu	$L\beta_1$	$L_{II}M_{IV}$	8.7090
1.4276	2	65 Tb	$L\gamma_4$	$L_IO_{II,III}$	8.685
1.43025	9	72 Hf		L_IM_I	8.6685
1.43048	9	73 Ta		$L_{II}M_{II}$	8.6671
1.4318	2	77 Ir	Ls	$L_{III}M_{III}$	8.659
1.43290	4	75 Re	$L\alpha_1$	$L_{III}M_V$	8.6525
1.4334	1	69 Tm	L_{III}	Abs. Edge	8.6496
1.4336	3	69 Tm	$L\beta_9$	L_IM_V	8.648
1.4349	2	69 Tm	$L\beta_5$	$L_{III}O_{IV,V}$	8.641
1.435155	7	30 Zn	$K\alpha_1$	KL_{III}	8.63886
1.43643	9	72 Hf	$L\beta_{17}$	$L_{II}M_{III}$	8.6312
1.439000	8	30 Zn	$K\alpha_2$	KL_{II}	8.61578
1.44056	5	71 Lu	$L\beta_4$	L_IM_{II}	8.6064
1.4410	3	69 Tm	$L\beta_{10}$	L_IM_{IV}	8.604
1.44396	5	75 Re	$L\alpha_2$	$L_{III}M_{IV}$	8.5862
1.4445	1	66 Dy	L_{II}	Abs. Edge	8.5830
1.44579	7	66 Dy	$L\gamma_6$	$L_{II}O_{IV}$	8.5753
1.45233	5	70 Yb	$L\beta_3$	L_IM_{III}	8.5367
1.4530	2	78 Pt	Lt	$L_{III}M_{II}$	8.533
1.45964	9	79 Au	Ll	$L_{III}M_I$	8.4939
1.4618	2	67 Ho	$L\gamma_5$	$L_{II}N_I$	8.481
1.4640	2	69 Tm	$L\beta_{2,15}$	$L_{III}N_{IV,V}$	8.468
1.4661	1	70 Yb	$L\beta_6$	$L_{III}N_I$	8.4563
1.47106	5	73 Ta	$L\eta$	$L_{II}M_I$	8.4280
1.4718	2	65 Tb	$L\gamma_3$	L_IN_{III}	8.423
1.47266	7	66 Dy	$L\gamma_1$	$L_{II}N_{IV}$	8.4188
1.4735	2	76 Os	Ls	$L_{III}M_{III}$	8.414
1.47565	5	70 Yb	$L\beta_1$	$L_{II}M_{IV}$	8.4018
1.4764	2	65 Tb	$L\gamma_2$	L_IN_{II}	8.398
1.47639	2	74 W	$L\alpha_1$	$L_{III}M_V$	8.3976
1.4784	1	64 Gd	L_I	Abs. Edge	8.3864
1.48064	9	72 Hf		$L_{II}M_I$	8.3735
1.4807	3	64 Gd		$L_IO_{IV,V}$	8.373
1.4835	1	68 Er	L_{III}	Abs. Edge	8.3575
1.4839	2	64 Gd	$L\gamma_4$	$L_IO_{II,III}$	8.355
1.4848	3	68 Er	$L\beta_5$	$L_{III}O_{IV,V}$	8.350
1.4855	5	68 Er	$L\beta_9$	L_IM_V	8.346
1.48743	2	74 W	$L\alpha_2$	$L_{III}M_{IV}$	8.3352
1.48807	1	28 Ni	K	Abs. Edge	8.33165
1.48862	4	28 Ni	$K\beta_5$	$KM_{IV,V}$	8.3286
1.49138	3	70 Yb	$L\beta_4$	L_IM_{II}	8.3132
1.4930	3	77 Ir	Lt	$L_{III}M_{II}$	8.304

Wavelength Å*	p.e.	Element		Designation	keV	Wavelength Å*	p.e.	Element		Designation	keV
1.4941	3	68 Er	$L\beta_7$	$L_{III}O_I$	8.298	1.60891	3	27 Co	$K\beta_5$	$KM_{IV,V}$	7.7059
1.4941	3	68 Er	$L\beta_{10}$	L_IM_{IV}	8.298	1.61264	9	73 Ta	Ls	$L_{III}M_{III}$	7.6881
1.4995	2	78 Pt	Ll	$L_{III}M_I$	8.268	1.61951	3	71 Lu	$L\alpha_1$	$L_{III}M_V$	7.6555
1.500135	8	28 Ni	$K\beta_{1,3}$	$KM_{II,III}$	8.26466	1.6203	2	67 Ho	$L\beta_3$	L_IM_{III}	7.6519
1.5023	1	65 Tb	L_{II}	Abs. Edge	8.2527	1.62079	2	27 Co	$K\beta_{1,3}$	$KM_{II,III}$	7.64943
1.5035	2	65 Tb	$L\gamma_6$	$L_{II}O_{IV}$	8.246	1.6237	2	67 Ho	$L\beta_6$	$L_{III}N_I$	7.6359
1.5063	2	69 Tm	$L\beta_3$	L_IM_{III}	8.231	1.62369	7	66 Dy	$L\beta_{2,15}$	$L_{III}N_{IV,V}$	7.6357
1.5097	2	65 Tb	$L\gamma_8$	$L_{II}O_I$	8.212	1.6244	3	74 W	Ll	$L_{III}M_{II}$	7.6324
1.51399	9	68 Er	$L\beta_{2,15}$	$L_{III}N_{IV,V}$	8.1890	1.6271	1	63 Eu	L_{II}	Abs. Edge	7.6199
1.5162	2	69 Tm	$L\beta_6$	$L_{III}N_I$	8.177	1.6282	2	63 Eu	$L\gamma_6$	$L_{II}O_{IV}$	7.6147
1.5178	1	75 Re	Ls	$L_{III}M_{III}$	8.1682	1.63029	5	71 Lu	$L\alpha_2$	$L_{III}M_{IV}$	7.6049
1.51824	7	66 Dy	$L\gamma_5$	$L_{II}N_I$	8.1661	1.63056	5	75 Re	Ll	$L_{III}M_I$	7.6036
1.52197	2	73 Ta	$L\alpha_1$	$L_{III}M_V$	8.1461	1.6346	2	63 Eu	$L\gamma_8$	$L_{II}O_I$	7.5849
1.52325	5	72 Hf	$L\eta$	$L_{II}M_I$	8.1393	1.63560	5	70 Yb	$L\eta$	$L_{II}M_I$	7.5802
1.5297	2	64 Gd	$L\gamma_3$	L_IN_{III}	8.105	1.6412	2	64 Gd	$L\gamma_5$	$L_{II}N_I$	7.5543
1.5303	2	65 Tb	$L\gamma_1$	$L_{II}N_{IV}$	8.102	1.6475	2	67 Ho	$L\beta_1$	$L_{II}M_{IV}$	7.5253
1.5304	2	69 Tm	$L\beta_1$	$L_{II}M_{IV}$	8.101	1.6497	1	65 Tb	L_{III}	Abs. Edge	7.5153
1.53293	2	73 Ta	$L\alpha_2$	$L_{III}M_{IV}$	8.0879	1.6510	2	65 Tb	$L\beta_5$	$L_{III}O_{IV,V}$	7.5094
1.5331	2	64 Gd	$L\gamma_2$	L_IN_{II}	8.087	1.65601	3	62 Sm	$L\gamma_3$	L_IN_{III}	7.487
1.53333	9	71 Lu		$L_{II}M_{II}$	8.0858	1.6574	2	63 Eu	$L\gamma_1$	$L_{II}N_{IV}$	7.4803
1.5347	2	76 Os	Ll	$L_{III}M_{II}$	8.079	1.657910	8	28 Ni	$K\alpha_1$	KL_{III}	7.47815
1.5368	1	67 Ho	L_{III}	Abs. Edge	8.0676	1.6585	2	65 Tb	$L\beta_7$	$L_{III}O_I$	7.4753
1.5378	2	67 Ho	$L\beta_5$	$L_{III}O_{IV,V}$	8.062	1.6595	8	67 Ho	$L\beta_9$	L_IM_{II}	7.4708
1.5381	1	63 Eu	L_I	Abs. Edge	8.0607	1.66044	6	62 Sm	$L\gamma_2$	L_IN_{II}	7.467
1.540562	2	29 Cu	$K\alpha_1$	KL_{III}	8.04778	1.661747	8	28 Ni	$K\alpha_2$	KL_{II}	7.46089
1.54094	3	77 Ir	Ll	$L_{III}M_I$	8.0458	1.66346	9	72 Hf	Ls	$L_{III}M_{III}$	7.4532
1.5439	1	63 Eu	$L\gamma_4$	$L_IO_{II,III}$	8.0304	1.6673	3	65 Tb	$L\beta_{10}$	L_IM_{IV}	7.436
1.544390	9	29 Cu	$K\alpha_2$	KL_{II}	8.02783	1.6674	5	61 Pm	L_I	Abs. Edge	7.436
1.5448	2	69 Tm	$L\beta_4$	L_IM_{II}	8.026	1.67189	4	70 Yb	$L\alpha_1$	$L_{III}M_V$	7.4156
1.5486	3	67 Ho	$L\beta_{10}$	L_IM_{IV}	8.006	1.67265	9	73 Ta	Ll	$L_{III}M_{II}$	7.4123
1.5616	1	68 Er	$L\beta_3$	L_IM_{III}	7.9392	1.6782	1	74 W	Ll	$L_{III}M_I$	7.3878
1.5632	1	64 Gd	L_{II}	Abs. Edge	7.9310	1.68213	7	66 Dy	$L\beta_6$	$L_{III}N_I$	7.3705
1.5642	3	74 W	Ls	$L_{III}M_{III}$	7.926	1.6822	2	66 Dy	$L\beta_3$	L_IM_{III}	7.3702
1.5644	2	64 Gd	$L\gamma_6$	$L_{II}O_{IV}$	7.925	1.68285	5	70 Yb	$L\alpha_2$	$L_{III}M_{IV}$	7.3673
1.5671	2	67 Ho	$L\beta_{2,15}$	$L_{III}N_{IV,V}$	7.911	1.6830	2	65 Tb	$L\beta_{2,15}$	$L_{III}N_{1v,V}$	7.3667
1.5675	2	68 Er	$L\beta_6$	$L_{III}N_I$	7.909	1.6953	1	62 Sm	L_{II}	Abs. Edge	7.3132
1.56958	5	72 Hf	$L\alpha_1$	$L_{III}M_V$	7.8990	1.6963	2	69 Tm	$L\eta$	$L_{II}M_I$	7.3088
1.5707	2	64 Gd	$L\gamma_8$	$L_{II}O_I$	7.894	1.6966	9	62 Sm	$L\gamma_6$	$L_{II}O_{IV}$	7.308
1.5779	1	71 Lu	$L\eta$	$L_{II}M_I$	7.8575	1.7085	2	63 Eu	$L\gamma_5$	$L_{II}N_I$	7.2566
1.5787	2	65 Tb	$L\gamma_5$	$L_{II}N_I$	7.8535	1.71062	7	66 Dy	$L\beta_1$	$L_{II}M_{IV}$	7.2477
1.5789	1	75 Re	Ll	$L_{III}M_{II}$	7.8525	1.7117	1	64 Gd	L_{III}	Abs. Edge	7.2430
1.58046	5	72 Hf	$L\alpha_2$	$L_{III}M_{IV}$	7.8446	1.7130	2	64 Gd	$L\beta_5$	$L_{III}O_{IV,V}$	7.2374
1.58498	7	76 Os	Ll	$L_{III}M_I$	7.8222	1.7203	2	64 Gd	$L\beta_7$	$L_{III}O_I$	7.2071
1.5873	1	68 Er	$L\beta_1$	$L_{II}M_{IV}$	7.8109	1.72103	7	66 Dy	$L\beta_4$	L_IM_{II}	7.2039
1.58837	7	66 Dy	$L\beta_5$	$L_{III}O_{IV,V}$	7.8055	1.72305	9	72 Hf	Ll	$L_{III}M_{II}$	7.1954
1.58844	9	70 Yb		$L_{II}M_{II}$	7.8052	1.7240	3	64 Gd	$L\beta_9$	L_IM_V	7.192
1.5903	2	63 Eu	$L\gamma_3$	L_IN_{III}	7.7961	1.72724	3	62 Sm	$L\gamma_1$	$L_{II}N_{IV}$	7.178
1.5916	1	66 Dy	L_{III}	Abs. Edge	7.7897	1.7268	2	69 Tm	Ll	$L_{III}M_V$	7.1799
1.5924	2	64 Gd	$L\gamma_1$	$L_{II}N_{IV}$	7.7858	1.72841	5	73 Ta	Ll	$L_{III}M_I$	7.1731
1.5961	2	63 Eu	$L\gamma_2$	L_IN_{II}	7.7677	1.7315	3	64 Gd	$L\beta_{10}$	L_IM_{IV}	7.160
1.59973	9	66 Dy	$L\beta_9$	L_IM_V	7.7501	1.7381	2	69 Tm	$L\alpha_2$	$L_{III}M_{IV}$	7.1331
1.6002	1	62 Sm	L_I	Abs. Edge	7.7478	1.7390	1	60 Nd	L_I	Abs. Edge	7.1294
1.6007	1	68 Er	$L\beta_4$	L_IM_{II}	7.7453	1.7422	2	65 Tb	$L\beta_6$	$L_{III}N_I$	7.1163
1.60447	7	66 Dy	$L\beta_7$	$L_{III}O_I$	7.7272	1.74346	1	26 Fe	K	Abs. Edge	7.11120
1.60728	3	62 Sm	$L\gamma_4$	$L_IO_{II,III}$	7.714	1.7442	1	26 Fe	$K\beta_5$	$KM_{IV,V}$	7.1081
1.60743	9	66 Dy	$L\beta_{10}$	L_IM_{IV}	7.7130	1.7445	4	60 Nd	$L\gamma_4$	$L_IO_{II,III}$	7.107
1.60815	1	27 Co	K	Abs. Edge	7.70954	1.7455	2	64 Gd	$L\beta_{2,15}$	$L_{III}N_{IV,V}$	7.1028

Wavelength Å*	p.e.	Element	Designation		keV	Wavelength Å*	p.e.	Element	Designation		keV
1.7472	2	65 Tb	$L\beta_2$	L_IM_{III}	7.0959	1.9255	2	63 Eu	$L\beta_4$	L_IM_{II}	6.4389
1.75661	2	26 Fe	$K\beta_{1,3}$	$KM_{II,III}$	7.05798	1.9255	5	59 Pr	L_{II}	Abs. Edge	6.439
1.7566	1	68 Er	$L\eta$	$L_{II}M_I$	7.0579	1.9355	4	60 Nd	$L\gamma_5$	$L_{II}N_I$	6.406
1.7676	5	61 Pm	L_{II}	Abs. Edge	7.014	1.936042	9	26 Fe	$K\alpha_1$	KL_{III}	6.40384
1.7760	1	71 Lu	Ll	$L_{III}M_{II}$	6.9810	1.9362	4	59 Pr	$L\gamma_8$	$L_{II}O_I$	6.403
1.7761	1	63 Eu	L_{III}	Abs. Edge	6.9806	1.939980	9	26 Fe	$K\alpha_2$	KL_{II}	6.39084
1.7768	3	65 Tb	$L\beta_1$	$L_{II}M_{IV}$	6.978	1.94643	3	62 Sm	$L\beta_6$	$L_{III}N_I$	6.3693
1.7772	2	63 Eu	$L\beta_5$	$L_{III}O_{IV,V}$	6.9763	1.9550	2	69 Tm	Ll	$L_{III}M_I$	6.3419
1.77934	3	62 Sm	$L\gamma_5$	$L_{II}N_I$	6.968	1.9553	3	58 Ce	$L\gamma_3$	L_IN_{III}	6.3409
1.78145	5	72 Hf	Ll	$L_{III}M_I$	6.9596	1.9559	6	61 Pm	$L\beta_{2,15}$	$L_{III}N_{IV,V}$	6.339
1.78425	9	68 Er	$L\alpha_1$	$L_{III}M_V$	6.9487	1.9602	3	58 Ce	$L\gamma_2$	L_IN_{II}	6.3250
1.7851	2	63 Eu	$L\beta_7$	$L_{III}O_I$	6.9453	1.9611	3	59 Pr	$L\gamma_1$	$L_{II}N_{IV}$	6.3221
1.7864	2	65 Tb	$L\beta_4$	L_IM_{II}	6.9403	1.96241	3	62 Sm	$L\beta_3$	L_IM_{III}	6.318
1.788965	9	27 Co	$K\alpha_1$	KL_{III}	6.93032	1.9730	2	65 Tb	$L\eta$	$L_{II}M_I$	6.2839
1.7916	3	63 Eu	$L\beta_9$	L_IM_V	6.920	1.9765	2	65 Tb	$L\alpha_1$	$L_{III}M_V$	6.2728
1.792850	9	27 Co	$K\alpha_2$	KL_{II}	6.91530	1.9780	5	57 La	L_I	Abs. Edge	6.268
1.7955	2	68 Er	$L\alpha_2$	$L_{III}M_{IV}$	6.9050	1.9830	4	57 La	$L\gamma_4$	$L_IO_{II,III}$	6.252
1.7964	4	60 Nd	$L\gamma_3$	L_IN_{III}	6.902	1.9875	2	65 Tb	$L\alpha_2$	$L_{III}M_{IV}$	6.2380
1.7989	9	61 Pm	$L\gamma_1$	$L_{II}N_{IV}$	6.892	1.9967	1	60 Nd	L_{III}	Abs. Edge	6.2092
1.7993	3	63 Eu	$L\beta_{10}$	L_IM_{IV}	6.890	1.99806	3	62 Sm	$L\beta_1$	$L_{II}M_{IV}$	6.2051
1.8013	4	60 Nd	$L\gamma_2$	L_IN_{II}	6.883	2.00095	6	62 Sm	$L\beta_4$	L_IM_{II}	6.196
1.8054	2	64 Gd	$L\beta_6$	$L_{III}N_I$	6.8671	2.0092	3	60 Nd	$L\beta_7$	$L_{III}O_I$	6.1708
1.8118	2	63 Eu	$L\beta_{2,15}$	$L_{III}N_{IV,V}$	6.8432	2.0124	5	58 Ce	L_{II}	Abs. Edge	6.161
1.8141	5	59 Pr	L_I	Abs. Edge	6.834	2.015	1	68 Er	Ll	$L_{III}M_I$	6.152
1.8150	2	64 Gd	$L\beta_3$	L_IM_{III}	6.8311	2.0165	3	60 Nd	$L\beta_9$	L_IM_V	6.1484
1.8193	4	59 Pr	$L\gamma_4$	$L_IO_{II,III}$	6.815	2.0205	4	59 Pr	$L\gamma_5$	$L_{II}N_I$	6.136
1.8264	2	67 Ho	$L\eta$	$L_{II}M_I$	6.7883	2.0237	4	58 Ce	$L\gamma_8$	$L_{II}O_I$	6.126
1.83091	9	70 Yb	Ll	$L_{III}M_{II}$	6.7715	2.0237	3	60 Nd	$L\beta_{10}$	L_IM_{IV}	6.1265
1.8360	1	71 Lu	Ll	$L_{III}M_I$	6.7528	2.0360	3	60 Nd	$L\beta_{2,15}$	$L_{III}N_{IV,V}$	6.0894
1.8440	1	60 Nd	L_{II}	Abs. Edge	6.7234	2.0410	4	57 La	$L\gamma_3$	L_IN_{III}	6.074
1.8450	2	67 Ho	$L\alpha_1$	$L_{III}M_V$	6.7198	2.0421	4	61 Pm	$L\beta_3$	L_IM_{III}	6.071
1.8457	1	62 Sm	L_{III}	Abs. Edge	6.7172	2.0460	4	57 La	$L\gamma_2$	L_IN_{II}	6.060
1.8468	2	64 Gd	$L\beta_1$	$L_{II}M_{IV}$	6.7132	2.0468	2	64 Gd	$L\alpha_1$	$L_{III}M_V$	6.0572
1.84700	9	62 Sm	$L\beta_5$	$L_{III}O_{IV,V}$	6.7126	2.0487	2	58 Ce	$L\gamma_1$	$L_{II}N_{IV}$	6.052
1.8540	2	64 Gd	$L\beta_4$	L_IM_{II}	6.6871	2.0494	1	64 Gd	$L\eta$	$L_{II}M_I$	6.0495
1.8552	5	60 Nd	$L\gamma_8$	$L_{II}O_I$	6.683	2.0578	2	64 Gd	$L\alpha_2$	$L_{III}M_{IV}$	6.0250
1.8561	2	67 Ho	$L\alpha_2$	$L_{III}M_{IV}$	6.6795	2.0678	5	56 Ba	L_I	Abs. Edge	5.996
1.85626	3	62 Sm	$L\beta_7$	$L_{III}O_I$	6.679	2.07020	5	24 Cr	K	Abs. Edge	5.9888
1.86166	3	62 Sm	$L\beta_9$	L_IM_V	6.660	2.07087	6	24 Cr	$K\beta_5$	$KM_{IV,V}$	5.9869
1.86990	3	62 Sm	$L\beta_{10}$	L_IM_{IV}	6.634	2.0756	3	56 Ba	$L\gamma_4$	$L_IO_{II,III}$	5.9733
1.8737	2	63 Eu	$L\beta_6$	$L_{III}N_I$	6.6170	2.0791	5	59 Pr	L_{III}	Abs. Edge	5.963
1.8740	4	59 Pr	$L\gamma_3$	L_IN_{III}	6.616	2.0797	4	61 Pm	$L\beta_1$	$L_{II}M_{IV}$	5.961
1.8779	2	60 Nd	$L\gamma_1$	$L_{II}N_{IV}$	6.6021	2.08487	2	24 Cr	$K\beta_{1,3}$	$KM_{II,III}$	5.94671
1.8791	4	59 Pr	$L\gamma_2$	L_IN_{II}	6.598	2.0860	2	67 Ho	Ll	$L_{III}M_I$	5.9434
1.8821	3	62 Sm	$L\beta_{2,15}$	$L_{III}N_{IV,V}$	6.586	2.0919	4	59 Pr	$L\beta_7$	$L_{III}O_I$	5.927
1.8867	2	63 Eu	$L\beta_3$	L_IM_{III}	6.5713	2.1004	4	59 Pr	$L\beta_9$	L_IM_V	5.903
1.8934	5	58 Ce	L_I	Abs. Edge	6.548	2.101820	9	25 Mn	$K\alpha_1$	KL_{III}	5.89875
1.89415	5	70 Yb	Ll	$L_{III}M_I$	6.5455	2.1039	3	60 Nd	$L\beta_6$	$L_{III}N_I$	5.8930
1.89643	5	25 Mn	K	Abs. Edge	6.5376	2.1053	5	57 La	L_{II}	Abs. Edge	5.889
1.8971	1	25 Mn	$K\beta_5$	$KM_{IV,V}$	6.5352	2.10578	2	25 Mn	$K\alpha_2$	KL_{II}	5.88765
1.89743	7	66 Dy	$L\eta$	$L_{II}M_I$	6.5342	2.1071	4	59 Pr	$L\beta_{10}$	L_IM_{IV}	5.884
1.8991	4	58 Ce	$L\gamma_4$	$L_IO_{II,III}$	6.528	2.1103	3	58 Ce	$L\gamma_5$	$L_{II}N_I$	5.8751
1.90881	3	66 Dy	$L\alpha_1$	$L_{III}M_V$	6.4952	2.1194	4	59 Pr	$L\beta_{2,15}$	$L_{III}N_{IV,V}$	5.850
1.91021	2	25 Mn	$K\beta_{1,3}$	$KM_{II,III}$	6.49045	2.1209	2	63 Eu	$L\alpha_1$	$L_{III}M_V$	5.8457
1.9191	1	61 Pm	L_{III}	Abs. Edge	6.4605	2.1268	2	60 Nd	$L\beta_3$	L_IM_{III}	5.8294
1.91991	3	66 Dy	$L\alpha_2$	$L_{III}M_{IV}$	6.4577	2.1315	2	63 Eu	$L\eta$	$L_{II}M_I$	5.8166
1.9203	2	63 Eu	$L\beta_1$	$L_{II}M_{IV}$	6.4564	2.1315	2	63 Eu	$L\alpha_2$	$L_{III}M_{IV}$	5.8166

Wavelength Å*	p.e.	Element	Designation		keV
2.1342	2	56 Ba	$L\gamma_3$	$L_I N_{III}$	5.8092
2.1387	2	56 Ba	$L\gamma_2$	$L_I N_{II}$	5.7969
2.1418	3	57 La	$L\gamma_1$	$L_{II} N_{IV}$	5.7885
2.15877	7	66 Dy	Ll	$L_{III} M_I$	5.7431
2.166	1	58 Ce	L_{III}	Abs. Edge	5.723
2.1669	3	60 Nd	$L\beta_4$	$L_I M_{II}$	5.7216
2.1669	2	60 Nd	$L\beta_1$	$L_{II} M_{IV}$	5.7216
2.1673	5	55 Cs	L_I	Abs. Edge	5.721
2.1701	2	58 Ce	$L\beta_7$	$L_{III} O_I$	5.7132
2.1741	2	55 Cs	$L\gamma_4$	$L_I O_{II,III}$	5.7026
2.1885	3	58 Ce	$L\beta_9$	$L_I M_V$	5.6650
2.1906	4	59 Pr	$L\beta_6$	$L_{III} N_I$	5.660
2.1958	5	58 Ce	$L\beta_{10}$	$L_I M_{IV}$	5.646
2.1998	2	62 Sm	$L\alpha_1$	$L_{III} M_V$	5.6361
2.2048	1	56 Ba	L_{II}	Abs. Edge	5.6233
2.2056	4	57 La	$L\gamma_5$	$L_{II} N_I$	5.621
2.2087	2	58 Ce	$L\beta_{2,15}$	$L_{III} N_{IV,V}$	5.6134
2.21062	3	62 Sm	$L\alpha_2$	$L_{III} M_{IV}$	5.6090
2.2172	3	59 Pr	$L\beta_3$	$L_I M_{III}$	5.5918
2.21824	3	62 Sm	$L\eta$	$L_{II} M_I$	5.589
2.2328	2	55 Cs	$L\gamma_3$	$L_I N_{III}$	5.5527
2.2352	2	65 Tb	Ll	$L_{III} M_I$	5.5467
2.2371	2	55 Cs	$L\gamma_2$	$L_I N_{II}$	5.5420
2.2415	2	56 Ba	$L\gamma_1$	$L_{III} N_{IV}$	5.5311
2.253	6	92 U		$M_I P_{III}$	5.50
2.2550	4	59 Pr	$L\beta_4$	$L_I M_{II}$	5.4981
2.2588	3	59 Pr	$L\beta_1$	$L_{II} M_{IV}$	5.4889
2.261	1	57 La	L_{III}	Abs. Edge	5.484
2.2691	1	23 V	K	Abs. Edge	5.4639
2.26951	6	23 V	$K\beta_5$	$K M_{IV,V}$	5.4629
2.2737	1	54 Xe	L_I	Abs. Edge	5.4528
2.275	3	57 La	$L\beta_7$	$L_{III} O_I$	5.450
2.282	3	57 La	$L\beta_9$	$L_I M_V$	5.434
2.2818	3	58 Ce	$L\beta_6$	$L_{III} N_I$	5.4334
2.2822	3	61 Pm	$L\alpha_1$	$L_{III} M_V$	5.4325
2.28440	2	23 V	$K\beta_{1,3}$	$K M_{II,III}$	5.42729
2.28970	2	24 Cr	$K\alpha_1$	$K L_{III}$	5.41472
2.290	3	57 La	$L\beta_{10}$	$L_I M_{IV}$	5.415
2.2926	4	61 Pm	$L\alpha_2$	$L_{III} M_{IV}$	5.4078
2.293606	3	24 Cr	$K\alpha_2$	$K L_{II}$	5.405509
2.3030	3	57 La	$L\beta_{2,15}$	$L_{III} N_{IV,V}$	5.3835
2.304	7	92 U		$M_I O_{III}$	5.38
2.3085	3	56 Ba	$L\gamma_5$	$L_{II} N_I$	5.3707
2.3109	3	58 Ce	$L\beta_3$	$L_I M_{III}$	5.3651
2.3122	2	64 Gd	Ll	$L_{III} M_I$	5.3621
2.3139	1	55 Cs	L_{II}	Abs. Edge	5.3581
2.3480	2	55 Cs	$L\gamma_1$	$L_{II} N_{IV}$	5.2804
2.3497	4	58 Ce	$L\beta_4$	$L_I M_{II}$	5.2765
2.3561	3	58 Ce	$L\beta_1$	$L_{II} M_{IV}$	5.2622
2.3629	1	56 Ba	L_{III}	Abs. Edge	5.2470
2.3704	2	60 Nd	$L\alpha_1$	$L_{III} M_V$	5.2304
2.3764	2	56 Ba	$L\beta_9$	$L_I M_V$	5.2171
2.3790	4	57 La	$L\beta_6$	$L_{III} N_I$	5.2114
2.3806	2	56 Ba	$L\beta_7$	$L_{III} O_I$	5.2079
2.3807	3	60 Nd	$L\alpha_2$	$L_{III} M_{IV}$	5.2077
2.3869	2	56 Ba	$L\beta_{10}$	$L_I M_{IV}$	5.1941
2.3880	5	53 I	L_I	Abs. Edge	5.192

Wavelength Å*	p.e.	Element	Designation		keV
2.3913	2	53 I	$L\gamma_4$	$L_I O_{II,III}$	5.1848
2.3948	2	63 Eu	Ll	$L_{III} M_I$	5.1772
2.40435	6	56 Ba	$L\beta_{2,15}$	$L_{III} N_{IV,V}$	5.1565
2.4094	4	60 Nd	$L\eta$	$L_{II} M_I$	5.1457
2.4105	3	57 La	$L\beta_3$	$L_I M_{III}$	5.1434
2.4174	2	55 Cs	$L\gamma_5$	$L_{II} N_I$	5.1287
2.4292	1	54 Xe	L_{II}	Abs. Edge	5.1037
2.442	9	90 Th		$M_I O_{III}$	5.08
2.443	4	92 U		$M_{II} O_{IV}$	5.075
2.4475	2	53 I	$L\gamma_{2,3}$	$L_I N_{II,III}$	5.0657
2.4493	3	57 La	$L\beta_4$	$L_I M_{II}$	5.0620
2.45891	5	57 La	$L\beta_1$	$L_{II} M_{IV}$	5.0421
2.4630	2	59 Pr	$L\alpha_1$	$L_{III} M_V$	5.0337
2.4729	3	59 Pr	$L\alpha_2$	$L_{III} M_{IV}$	5.0135
2.4740	1	55 Cs	L_{III}	Abs. Edge	5.0113
2.4783	2	55 Cs	$L\beta_9$	$L_I M_V$	5.0026
2.4823	4	62 Sm	Ll	$L_{III} M_I$	4.9945
2.4826	2	56 Ba	$L\beta_6$	$L_{III} N_I$	4.9939
2.4849	2	55 Cs	$L\beta_7$	$L_{III} O_I$	4.9893
2.4920	2	55 Cs	$L\beta_{10}$	$L_I M_{IV}$	4.9752
2.49734	5	22 Ti	K	Abs. Edge	4.96452
2.4985	2	22 Ti	$K\beta_5$	$K M_{IV,V}$	4.9623
2.50356	2	23 V	$K\alpha_1$	$K L_{III}$	4.95220
2.50738	2	23 V	$K\alpha_2$	$K L_{II}$	4.94464
2.5099	1	52 Te	L_I	Abs. Edge	4.9397
2.5113	2	52 Te	$L\gamma_4$	$L_I O_{II,III}$	4.9369
2.5118	2	55 Cs	$L\beta_{2,15}$	$L_{III} N_{IV,V}$	4.9339
2.512	3	59 Pr	$L\eta$	$L_{II} M_I$	4.935
2.51391	2	22 Ti	$K\beta_{1,3}$	$K M_{II,III}$	4.93181
2.5164	2	56 Ba	$L\beta_3$	$L_I M_{III}$	4.9269
2.527	4	91 Pa		$M_{II} O_{IV}$	4.906
2.5542	5	53 I	L_{II}	Abs. Edge	4.8540
2.5553	2	56 Ba	$L\beta_4$	$L_I M_{II}$	4.8519
2.5615	2	58 Ce	$L\alpha_1$	$L_{III} M_V$	4.8402
2.5674	2	52 Te	$L\gamma_{2,3}$	$L_I N_{II,III}$	4.8290
2.56821	5	56 Ba	$L\beta_1$	$L_{II} M_{IV}$	4.82753
2.5706	3	58 Ce	$L\alpha_2$	$L_{III} M_{IV}$	4.8230
2.58244	8	53 I	$L\gamma_1$	$L_{II} N_{IV}$	4.8009
2.5926	1	54 Xe	L_{III}	Abs. Edge	4.7822
2.5932	2	55 Cs	$L\beta_6$	$L_{III} N_I$	4.7811
2.618	5	90 Th		$M_{II} O_{IV}$	4.735
2.6203	4	58 Ce	$L\eta$	$L_{II} M_I$	4.7315
2.6285	2	55 Cs	$L\beta_3$	$L_I M_{III}$	4.7167
2.6388	1	51 Sb	L_I	Abs. Edge	4.6984
2.6398	2	51 Sb	$L\gamma_4$	$L_I O_{II,III}$	4.6967
2.65710	9	53 I	$L\gamma_5$	$L_{II} N_I$	4.6660
2.66570	5	57 La	$L\alpha_1$	$L_{III} M_V$	4.65097
2.6666	2	55 Cs	$L\beta_4$	$L_I M_{II}$	4.6494
2.67533	5	57 La	$L\alpha_2$	$L_{III} M_{IV}$	4.63423
2.6760	4	60 Nd	Ll	$L_{III} M_I$	4.6330
2.6837	2	55 Cs	$L\beta_1$	$L_{II} M_{IV}$	4.6198
2.6879	1	52 Te	L_{II}	Abs. Edge	4.6126
2.6953	2	51 Sb	$L\gamma_{2,3}$	$L_I N_{II,III}$	4.5999
2.71241	6	52 Te	$L\gamma_1$	$L_{II} N_{IV}$	4.5709
2.71352	9	53 I	$L\beta_9$	$L_I M_V$	4.5690
2.7196	5	53 I	L_{III}	Abs. Edge	4.5587
2.72104	9	53 I	$L\beta_{10}$	$L_I M_{IV}$	4.5564

Wavelength Å*	p.e.	Element		Designation	keV
2.7288	3	53 I	$L\beta_7$	$L_{III}O_I$	4.5435
2.740	3	57 La	$L\eta$	$L_{II}M_I$	4.525
2.74851	2	22 Ti	$K\alpha_1$	KL_{III}	4.51084
2.75053	8	53 I	$L\beta_{2,15}$	$L_{III}N_{IV,V}$	4.5075
2.75216	2	22 Ti	$K\alpha_2$	KL_{II}	4.50486
2.753	8	92 U		M_IN_{III}	4.50
2.762	1	21 Sc	K	Abs. Edge	4.489
2.7634	3	21 Sc	$K\beta_5$	$KM_{IV,V}$	4.4865
2.77595	5	56 Ba	$L\alpha_1$	$L_{III}M_V$	4.46626
2.7769	1	50 Sn	L_I	Abs. Edge	4.4648
2.7775	2	50 Sn	$L\gamma_4$	$L_IO_{II,III}$	4.4638
2.7796	2	21 Sc	$K\beta_{1,3}$	$KM_{II,III}$	4.4605
2.7841	4	59 Pr	Ll	$L_{III}M_I$	4.4532
2.78553	5	56 Ba	$L\alpha_2$	$L_{III}M_{IV}$	4.45090
2.79007	9	52 Te	$L\gamma_5$	$L_{II}N_I$	4.4437
2.817	2	92 U		$M_{II}N_{IV}$	4.401
2.8294	5	51 Sb	L_{II}	Abs. Edge	4.3819
2.8327	2	50 Sn	$L\gamma_{2,3}$	$L_IN_{II,III}$	4.3768
2.83672	9	53 I	$L\beta_6$	$L_{III}N_I$	4.3706
2.83897	9	52 Te	$L\beta_9$	L_IM_V	4.3671
2.84679	9	52 Te	$L\beta_{10}$	L_IM_{IV}	4.3551
2.85159	3	51 Sb	$L\gamma_1$	$L_{II}N_{IV}$	4.34779
2.8555	1	52 Te	L_{III}	Abs. Edge	4.3418
2.8627	3	56 Ba	$L\eta$	$L_{II}M_I$	4.3309
2.8634	3	52 Te	$L\beta_7$	$L_{III}O_I$	4.3298
2.87429	9	53 I	$L\beta_2$	$L_{III}N_V$	4.3134
2.88217	8	52 Te	$L\beta_{2,15}$	$L_{III}N_{IV,V}$	4.3017
2.884	5	92 U	M_{III}	Abs. Edge	4.299
2.8917	4	58 Ce	Ll	$L_{III}M_I$	4.2875
2.8924	2	55 Cs	$L\alpha_1$	$L_{III}M_V$	4.2865
2.9020	2	55 Cs	$L\alpha_2$	$L_{III}M_{IV}$	4.2722
2.910	2	91 Pa		$M_{II}N_{IV}$	4.260
2.91207	9	53 I	$L\beta_4$	L_IM_{II}	4.2575
2.92	2	92 U		M_IN_{II}	4.25
2.9260	1	49 In	L_I	Abs. Edge	4.2373
2.9264	2	49 In	$L\gamma_4$	$L_IO_{II,III}$	4.2367
2.93187	9	51 Sb	$L\gamma_5$	$L_{II}N_I$	4.2287
2.934	8	90 Th		M_IN_{III}	4.23
2.93744	6	53 I	$L\beta_1$	$L_{II}M_{IV}$	4.22072
2.948	2	92 U		$M_{III}O_{IV,V}$	4.205
2.97088	9	52 Te	$L\beta_6$	$L_{III}N_I$	4.1732
2.97261	9	51 Sb	$L\beta_9$	L_IM_V	4.1708
2.97917	9	51 Sb	$L\beta_{10}$	L_IM_{IV}	4.1616
2.9800	2	49 In	$L\gamma_{2,3}$	$L_IN_{II,III}$	4.1605
2.9823	1	50 Sn	L_{II}	Abs. Edge	4.1573
2.9932	2	55 Cs	$L\eta$	$L_{II}M_I$	4.1421
3.0003	1	51 Sb	L_{III}	Abs. Edge	4.1323
3.00115	3	50 Sn	$L\gamma_1$	$L_{II}N_{IV}$	4.13112
3.0052	3	51 Sb	$L\beta_7$	$L_{III}O_I$	4.1255
3.006	3	57 La	Ll	$L_{III}M_I$	4.124
3.00893	9	52 Te	$L\beta_3$	L_IM_{III}	4.1204
3.011	2	90 Th		$M_{II}N_{IV}$	4.117
3.0166	2	54 Xe	$L\alpha_1$	$L_{III}M_V$	4.1099
3.02335	3	51 Sb	$L\beta_{2,15}$	$L_{III}N_{IV,V}$	4.10078
3.0309	1	21 Sc	$K\alpha_1$	KL_{III}	4.0906
3.0342	1	21 Sc	$K\alpha_2$	KL_{II}	4.0861
3.038	2	91 Pa		$M_{III}O_{IV,V}$	4.081
3.04661	9	52 Te	$L\beta_4$	L_IM_{II}	4.0695
3.068	5	90 Th	M_{III}	Abs. Edge	4.041
3.0703	1	20 Ca	K	Abs. Edge	4.0381
3.0746	3	20 Ca	$K\beta_5$	$KM_{IV,V}$	4.0325
3.07677	6	52 Te	$L\beta_1$	$L_{II}M_{IV}$	4.02958
3.08475	9	50 Sn	$L\gamma_5$	$L_{II}N_I$	4.0192
3.0849	1	48 Cd	L_I	Abs. Edge	4.0190
3.0897	2	20 Ca	$K\beta_{1,3}$	$KM_{II,III}$	4.0127
3.094	5	83 Bi	M_I	Abs. Edge	4.007
3.11513	9	50 Sn	$L\beta_9$	L_IM_V	3.9800
3.11513	9	51 Sb	$L\beta_6$	$L_{III}N_I$	3.9800
3.115	7	92 U		$M_{III}O_I$	3.980
3.12170	9	50 Sn	$L\beta_{10}$	L_IM_{IV}	3.9716
3.131	3	90 Th		$M_{III}O_{IV,V}$	3.959
3.1355	2	56 Ba	Ll	$L_{III}M_I$	3.9541
3.1377	2	48 Cd	$L\gamma_2$	L_IN_{II}	3.9513
3.1473	1	49 In	L_{II}	Abs. Edge	3.9393
3.14860	6	53 I	$L\alpha_1$	$L_{III}M_V$	3.93765
3.15258	9	51 Sb	$L\beta_3$	L_IM_{III}	3.9327
3.1557	1	50 Sn	L_{III}	Abs. Edge	3.9288
3.1564	3	50 Sn	$L\beta_7$	$L_{III}O_I$	3.9279
3.15791	6	53 I	$L\alpha_2$	$L_{III}M_{IV}$	3.92604
3.16213	4	49 In	$L\gamma_1$	$L_{II}N_{IV}$	3.92081
3.17505	3	50 Sn	$L\beta_{2,15}$	$L_{III}N_{IV,V}$	3.90486
3.19014	9	51 Sb	$L\beta_4$	L_IM_{II}	3.8364
3.217	5	82 Pb	M_I	Abs. Edge	3.854
3.22567	4	51 Sb	$L\beta_1$	$L_{II}M_{IV}$	3.84357
3.245	9	91 Pa		$M_{III}O_I$	3.82
3.24907	9	49 In	$L\gamma_5$	$L_{II}N_I$	3.8159
3.2564	1	47 Ag	L_I	Abs. Edge	3.8072
3.2670	2	55 Cs	Ll	$L_{III}M_I$	3.7950
3.26763	9	49 In	$L\beta_9$	L_IM_V	3.7942
3.26901	9	50 Sn	$L\beta_6$	$L_{III}N_I$	3.7926
3.27404	9	49 In	$L\beta_{10}$	L_IM_{IV}	3.7868
3.27979	9	53 I	$L\eta$	$L_{II}M_I$	3.7801
3.283	9	90 Th		$M_{III}O_I$	3.78
3.28920	6	52 Te	$L\alpha_1$	$L_{III}M_V$	3.76933
3.29846	9	52 Te	$L\alpha_2$	$L_{III}M_{IV}$	3.7588
3.30585	3	50 Sn	$L\beta_3$	L_IM_{III}	3.7500
3.30635	9	47 Ag	$L\gamma_3$	L_IN_{III}	3.7498
3.31216	9	47 Ag	$L\gamma_2$	L_IN_{II}	3.7432
3.3237	1	49 In	L_{III}	Abs. Edge	3.7302
3.324	4	49 In	$L\beta_7$	$L_{III}O_I$	3.730
3.3257	1	48 Cd	L_{II}	Abs. Edge	3.7280
3.329	4	92 U		$M_{II}N_I$	3.724
3.333	5	92 U	M_{IV}	Abs. Edge	3.720
3.33564	6	48 Cd	$L\gamma_1$	$L_{II}N_{IV}$	3.71686
3.33838	3	49 In	$L\beta_{2,15}$	$L_{III}N_{IV,V}$	3.71381
3.34335	9	50 Sn	$L\beta_4$	L_IM_{II}	3.7083
3.346	5	81 Tl	M_I	Abs. Edge	3.705
3.35839	3	20 Ca	$K\alpha_1$	KL_{III}	3.69168
3.359	5	83 Bi	M_{II}	Abs. Edge	3.691
3.36166	3	20 Ca	$K\alpha_2$	KL_{II}	3.68809
3.38487	3	50 Sn	$L\beta_1$	$L_{II}M_{IV}$	3.66280
3.42551	9	48 Cd	$L\gamma_5$	$L_{II}N_I$	3.61935
3.43015	9	48 Cd	$L\beta_9$	L_IM_V	3.61445
3.43606	9	49 In	$L\beta_6$	$L_{III}N_I$	3.60823

Wavelength Å*	p.e.	Element		Designation	keV	Wavelength Å*	p.e.	Element		Designation	keV
3.4365	1	19 K	K	Abs. Edge	3.6078	3.77192	4	49 In	$L\alpha_1$	$L_{III}M_V$	3.28694
3.4367	2	48 Cd	$L\beta_{10}$	L_IM_{IV}	3.6075	3.78073	6	49 In	$L\alpha_2$	$L_{III}M_{IV}$	3.27929
3.437	1	46 Pd	L_I	Abs. Edge	3.607	3.783	5	80 Hg	M_{II}	Abs. Edge	3.277
3.43832	9	52 Te	$L\eta$	$L_{II}M_I$	3.60586	3.78876	9	50 Sn	$L\eta$	$L_{II}M_I$	3.27234
3.43941	4	51 Sb	$L\alpha_1$	$L_{III}M_V$	3.60472	3.7920	2	46 Pd	$L\beta_9$	L_IM_V	3.2696
3.441	5	91 Pa		$M_{II}N_I$	3.603	3.7988	2	46 Pd	$L\beta_{10}$	L_IM_{IV}	3.2637
3.4413	4	19 K	$K\beta_5$	$KM_{IV,V}$	3.6027	3.80774	9	47 Ag	$L\beta_6$	$L_{III}N_I$	3.25603
3.44840	6	51 Sb	$L\alpha_2$	$L_{III}M_{IV}$	3.59532	3.808	4	90 Th		$M_{IV}O_{II}$	3.256
3.4539	2	19 K	$K\beta_{1,3}$	$KM_{II,III}$	3.5896	3.8222	2	46 Pd	$L\gamma_5$	$L_{II}N_I$	3.2437
3.46984	9	49 In	$L\beta_3$	L_IM_{III}	3.57311	3.827	1	91 Pa	$M\beta$	$M_{IV}N_{VI}$	3.2397
3.478	5	80 Hg	M_I	Abs. Edge	3.565	3.83313	9	47 Ag	$L\beta_3$	L_IM_{III}	3.23446
3.479	1	92 U	$M\gamma$	$M_{III}N_V$	3.563	3.834	4	83 Bi		$M_{II}N_{IV}$	3.234
3.4892	2	46 Pd	$L\gamma_{2,3}$	$L_IN_{II,III}$	3.5533	3.835	5	44 Ru	L_I	Abs. Edge	3.233
3.492	5	82 Pb	M_{II}	Abs. Edge	3.550	3.87023	5	47 Ag	$L\beta_4$	L_IM_{II}	3.20346
3.497	5	92 U	M_V	Abs. Edge	3.545	3.87090	5	18 A	K	Abs. Edge	3.20290
3.5047	1	48 Cd	L_{III}	Abs. Edge	3.5376	3.872	9	82 Pb		M_IN_{III}	3.202
3.50697	9	49 In	$L\beta_4$	L_IM_{II}	3.53528	3.8860	2	18 A	$K\beta_{1,3}$	$KM_{II,III}$	3.1905
3.51408	4	48 Cd	$L\beta_{2,15}$	$L_{III}N_{IV,V}$	3.52812	3.88826	9	51 Sb	Ll	$L_{III}M_I$	3.18860
3.5164	1	47 Ag	L_{II}	Abs. Edge	3.5258	3.892	9	83 Bi		M_IN_{II}	3.185
3.521	2	92 U		$M_{III}N_{IV}$	3.521	3.8977	2	44 Ru	$L\gamma_{2,3}$	$L_IN_{II,III}$	3.1809
3.52260	4	47 Ag	$L\gamma_1$	$L_{II}N_{IV}$	3.51959	3.904	5	83 Bi	M_{III}	Abs. Edge	3.176
3.537	9	90 Th		$M_{II}N_I$	3.505	3.9074	1	46 Pd	L_{III}	Abs. Edge	3.17298
3.55531	4	49 In	$L\beta_1$	$L_{II}M_{III}$	3.48721	3.90887	4	46 Pd	$L\beta_{2,15}$	$L_{III}N_{IV,V}$	3.17170
3.557	5	90 Th	M_{IV}	Abs. Edge	3.485	3.910	1	92 U	$M\alpha_1$	$M_{III}N_{VII}$	3.1708
3.55754	9	53 I	Ll	$L_{III}M_I$	3.48502	3.915	5	77 Ir	M_I	Abs. Edge	3.167
3.576	1	92 U		$M_{IV}O_{II}$	3.4666	3.924	1	92 U	$M\alpha_2$	M_VN_{VI}	3.1595
3.577	1	91 Pa	$M\gamma$	$M_{III}N_V$	3.4657	3.932	6	83 Bi		$M_{III}O_{IV,V}$	3.153
3.59994	3	50 Sn	$L\alpha_1$	$L_{III}M_V$	3.44398	3.93473	3	47 Ag	$L\beta_1$	$L_{II}M_{IV}$	3.15094
3.60497	9	47 Ag	$L\beta_9$	L_IM_V	3.43917	3.936	5	79 Au	M_{II}	Abs. Edge	3.150
3.60765	9	51 Sb	$L\eta$	$L_{II}M_I$	3.43661	3.941	1	90 Th	$M\beta$	$M_{IV}N_{VI}$	3.1458
3.60891	4	50 Sn	$L\alpha_2$	$L_{III}M_{IV}$	3.43542	3.9425	5	45 Rh	L_{II}	Abs. Edge	3.1448
3.61158	9	47 Ag	$L\beta_{10}$	L_IM_{IV}	3.43287	3.9437	2	45 Rh	$L\gamma_1$	$L_{II}N_{IV}$	3.1438
3.614	2	91 Pa		$M_{III}N_{IV}$	3.430	3.95635	4	48 Cd	$L\alpha_1$	$L_{III}M_V$	3.13373
3.61467	9	48 Cd	$L\beta_6$	$L_{III}N_I$	3.42994	3.96496	6	48 Cd	$L\alpha_2$	$L_{III}M_{IV}$	3.12691
3.61638	9	47 Ag	$L\gamma_5$	$L_{II}N_I$	3.42832	3.968	5	82 Pb		$M_{III}N_{IV}$	3.124
3.616	5	79 Au	M_I	Abs. Edge	3.428	3.98327	9	49 In	$L\eta$	$L_{II}M_I$	3.11254
3.629	5	45 Rh	L_I	Abs. Edge	3.417	4.013	9	81 Tl		M_IN_{III}	3.089
3.634	5	81 Tl	M_{II}	Abs. Edge	3.412	4.0162	2	46 Pd	$L\beta_6$	$L_{III}N_I$	3.0870
3.64495	9	48 Cd	$L\beta_3$	L_IM_{III}	3.40145	4.022	1	91 Pa	$M\alpha_1$	M_VN_{VII}	3.0823
3.679	5	90 Th	$M\gamma$	$M_{III}N_V$	3.370	4.0346	2	46 Pd	$L\beta_3$	L_IM_{III}	3.0730
3.68203	9	48 Cd	$L\beta_4$	L_IM_{II}	3.36719	4.035	3	91 Pa	$M\alpha_2$	M_VN_{VI}	3.072
3.6855	2	45 Rh	$L\gamma_{2,3}$	$L_IN_{II,III}$	3.3640	4.0451	2	45 Rh	$L\gamma_5$	$L_{II}N_I$	3.0650
3.691	2	91 Pa		$M_{IV}O_{II}$	3.359	4.047	1	82 Pb	M_{III}	Abs. Edge	3.0632
3.6999	1	47 Ag	L_{III}	Abs. Edge	3.35096	4.058	5	43 Te	L_I	Abs. Edge	3.055
3.70335	3	47 Ag	$L\beta_{2,15}$	$L_{III}N_{IV,V}$	3.34781	4.069	6	82 Pb		$M_{III}O_{IV,V}$	3.047
3.716	1	92 U	$M\beta$	$M_{IV}N_{VI}$	3.3367	4.0711	2	46 Pd	$L\beta_4$	L_IM_{II}	3.0454
3.71696	9	52 Te	Ll	$L_{III}M_I$	3.33555	4.071	5	76 Os	M_I	Abs. Edge	3.045
3.718	3	90 Th		$M_{III}N_{IV}$	3.335	4.07165	9	50 Sn	Ll	$L_{III}M_I$	3.04499
3.7228	1	46 Pd	L_{II}	Abs. Edge	3.33031	4.093	5	78 Pt	M_{II}	Abs. Edge	3.029
3.7246	2	46 Pd	$L\gamma_1$	$L_{II}N_{IV}$	3.3287	4.105	9	83 Bi		$M_{III}O_I$	3.021
3.729	5	90 Th	M_V	Abs. Edge	3.325	4.116	4	81 Tl		$M_{II}N_{IV}$	3.013
3.73823	4	48 Cd	$L\beta_1$	$L_{II}M_{IV}$	3.31657	4.1299	5	45 Rh	L_{III}	Abs. Edge	3.0021
3.740	9	83 Bi		M_IN_{III}	3.315	4.1310	2	45 Rh	$L\beta_{2,15}$	$L_{III}N_{IV,V}$	3.0013
3.7414	2	19 K	$K\alpha_1$	KL_{III}	3.3138	4.1381	9	90 Th	$M\alpha_1$	M_VN_{VII}	2.9961
3.7445	2	19 K	$K\alpha_2$	KL_{II}	3.3111	4.14622	5	46 Pd	$L\beta_1$	$L_{II}M_{IV}$	2.99022
3.760	9	90 Th		M_VP_{III}	3.298	4.151	2	90 Th	$M\alpha_2$	M_VN_{VI}	2.987
3.762	5	78 Pt	M_I	Abs. Edge	3.296	4.15443	3	47 Ag	$L\alpha_1$	$L_{III}M_V$	2.98431

Wavelength Å*	p.e.	Element		Designation	keV	Wavelength Å*	p.e.	Element		Designation	keV
4.16294	5	47 Ag	$L\alpha_2$	$L_{III}M_{IV}$	2.97821	4.6542	2	41 Nb	$L\gamma_{2,3}$	$L_IN_{II,III}$	2.6638
4.180	1	44 Ru	L_{II}	Abs. Edge	2.9663	4.655	8	82 Pb		$M_{II}N_I$	2.664
4.1822	2	44 Ru	$L\gamma_1$	$L_{II}N_{IV}$	2.9645	4.6605	2	46 Pd	$L\eta$	$L_{II}M_I$	2.6603
4.19180	5	18 A	$K\alpha_1$	KL_{III}	2.95770	4.674	1	82 Pb	$M\gamma$	$M_{III}N_V$	2.6527
4.19315	9	48 Cd	$L\eta$	$L_{II}M_I$	2.95675	4.686	1	78 Pt	M_{III}	Abs. Edge	2.6459
4.19474	5	18 A	$K\alpha_2$	KL_{II}	2.95563	4.694	8	78 Pt		$M_{III}O_{IV,V}$	2.641
4.198	1	81 Tl	M_{III}	Abs. Edge	2.9535	4.703	9	79 Au		$M_{III}O_I$	2.636
4.216	6	81 Tl		$M_{III}O_{IV,V}$	2.941	4.7076	2	47 Ag	Ll	$L_{III}M_I$	2.6337
4.236	5	75 Re	M_I	Abs. Edge	2.927	4.715	3	82 Pb		$M_{III}N_{IV}$	2.630
4.2417	2	45 Rh	$L\beta_6$	$L_{III}N_I$	2.9229	4.719	1	42 Mo	L_{II}	Abs. Edge	2.6274
4.244	9	82 Pb		$M_{III}O_I$	2.921	4.7258	2	42 Mo	$L\gamma_1$	$L_{II}N_{IV}$	2.6235
4.2522	2	45 Rh	$L\beta_3$	L_IM_{III}	2.9157	4.7278	1	17 Cl	$K\alpha_1$	KL_{III}	2.62239
4.260	5	77 Ir	M_{II}	Abs. Edge	2.910	4.7307	1	17 Cl	$K\alpha_2$	KL_{II}	2.62078
4.26873	9	49 In	Ll	$L_{III}M_I$	2.90440	4.757	5	82 Pb	M_{IV}	Abs. Edge	2.606
4.2873	2	44 Ru	$L\gamma_5$	$L_{II}N_I$	2.8918	4.764	5	83 Bi	M_V	Abs. Edge	2.603
4.2888	2	45 Rh	$L\beta_4$	L_IM_{II}	2.8908	4.780	4	77 Ir		$M_{II}N_{IV}$	2.594
4.300	9	79 Au		M_IN_{III}	2.883	4.79	2	76 Os		M_IN_{III}	2.59
4.304	5	42 Mo	L_I	Abs. Edge	2.881	4.815	5	74 W	M_{II}	Abs. Edge	2.575
4.330	2	92 U		$M_{III}N_I$	2.863	4.823	3	83 Bi		$M_{IV}O_{II}$	2.571
4.355	1	80 Hg	M_{III}	Abs. Edge	2.8469	4.823	4	81 Tl	$M\gamma$	$M_{III}N_V$	2.571
4.36767	5	46 Pd	$L\alpha_1$	$L_{III}M_V$	2.83861	4.8369	2	42 Mo	$L\gamma_5$	$L_{II}N_I$	2.5632
4.369	1	44 Ru	L_{III}	Abs. Edge	2.8377	4.84575	5	44 Ru	$L\alpha_1$	$L_{III}M_V$	2.55855
4.3718	2	44 Ru	$L\beta_{2,15}$	$L_{III}N_{IV,V}$	2.8360	4.85381	7	44 Ru	$L\alpha_2$	$L_{III}M_{IV}$	2.55431
4.37414	4	45 Rh	$L\beta_1$	$L_{II}M_{IV}$	2.83441	4.861	1	77 Ir	M_{III}	Abs. Edge	2.5505
4.37588	7	46 Pd	$L\alpha_2$	$L_{III}M_{IV}$	2.83329	4.865	5	81 Tl		$M_{III}N_{IV}$	2.548
4.3800	2	42 Mo	$L\gamma_{2,3}$	$L_IN_{II,III}$	2.8306	4.869	9	77 Ir		$M_{III}O_{IV,V}$	2.546
4.3971	1	17 Cl	K	Abs. Edge	2.81960	4.876	9	78 Pt		$M_{III}O_I$	2.543
4.4034	3	17 Cl	$K\beta$	KM	2.8156	4.879	5	40 Zr	L_I	Abs. Edge	2.541
4.407	5	74 W	M_I	Abs. Edge	2.813	4.8873	8	43 Tc	$L\beta_1$	$L_{II}M_{IV}$	2.5368
4.4183	2	47 Ag	$L\eta$	$L_{II}M_I$	2.8061	4.909	1	83 Bi	$M\beta$	$M_{IV}N_{VI}$	2.5255
4.432	4	79 Au		$M_{III}N_{IV}$	2.797	4.911	5	90 Th		$M_{IV}N_{III}$	2.524
4.433	5	76 Os	M_{II}	Abs. Edge	2.797	4.913	1	42 Mo	L_{III}	Abs. Edge	2.5234
4.436	1	43 Te	L_{II}	Abs. Edge	2.7948	4.9217	2	45 Rh	$L\eta$	$L_{II}M_I$	2.5191
4.44	2	74 W		$M_IO_{II,III}$	2.79	4.9232	2	42 Mo	$L\beta_{2,15}$	$L_{III}N_{IV,V}$	2.5183
4.450	4	91 Pa		$M_{III}N_I$	2.786	4.946	2	92 U	$M\zeta_1$	M_VN_{III}	2.507
4.460	9	78 Pt		M_IN_{III}	2.780	4.952	5	81 Tl	M_{IV}	Abs. Edge	2.504
4.48014	9	48 Cd	Ll	$L_{III}M_I$	2.76735	4.9525	3	46 Pd	Ll	$L_{III}M_I$	2.5034
4.4866	3	44 Ru	$L\beta_3$	L_IM_{III}	2.7634	4.9536	3	40 Zr	$L\gamma_{2,3}$	$L_IN_{II,III}$	2.5029
4.4866	3	44 Ru	$L\beta_6$	$L_{III}N_I$	2.7634	4.955	4	76 Os		$M_{II}N_{IV}$	2.502
4.518	1	79 Au	M_{III}	Abs. Edge	2.7439	4.955	5	82 Pb	M_V	Abs. Edge	2.502
4.522	6	79 Au		$M_{III}O_{IV,V}$	2.742	4.984	2	80 Hg	$M\gamma$	$M_{III}N_V$	2.4875
4.5230	2	44 Ru	$L\beta_4$	L_IM_{II}	2.7411	5.004	9	82 Pb		$M_{IV}O_{II}$	2.477
4.532	2	83 Bi	$M\gamma$	$M_{III}N_V$	2.735	5.0133	3	42 Mo	$L\beta_3$	L_IM_{III}	2.4730
4.568	5	90 Th		$M_{III}N_I$	2.714	5.0185	1	16 S	K	Abs. Edge	2.47048
4.571	5	83 Bi		$M_{III}N_{IV}$	2.712	5.020	5	73 Ta	M_{II}	Abs. Edge	2.470
4.572	5	83 Bi	M_{IV}	Abs. Edge	2.711	5.0233	3	16 S	$K\beta_x$	KM	2.4681
4.575	5	41 Nb	L_I	Abs. Edge	2.710	5.031	1	41 Nb	L_{II}	Abs. Edge	2.4641
4.585	5	73 Ta	M_I	Abs. Edge	2.704	5.0316	2	16 S	$K\beta_1$	KM	2.46404
4.59	2	83 Bi		$M_{IV}P_{II,III}$	2.70	5.0361	3	41 Nb	$L\gamma_1$	$L_{II}N_{IV}$	2.4618
4.59743	9	45 Rh	$L\alpha_1$	$L_{III}M_V$	2.69674	5.043	5	76 Os	M_{III}	Abs. Edge	2.458
4.601	4	78 Pt		$M_{II}N_{IV}$	2.695	5.0488	3	42 Mo	$L\beta_4$	L_IM_{II}	2.4557
4.60545	9	45 Rh	$L\alpha_2$	$L_{III}M_{IV}$	2.69205	5.0488	5	42 Mo	$L\beta_6$	$L_{III}N_I$	2.4557
4.620	5	75 Re	M_{II}	Abs. Edge	2.684	5.050	2	92 U	$M\zeta_2$	$M_{IV}N_{II}$	2.4548
4.62058	9	44 Ru	$L\beta_1$	$L_{II}M_{IV}$	2.68323	5.076	1	82 Pb	$M\beta$	$M_{IV}N_{VI}$	2.4427
4.625	5	92 U		$M_{IV}N_{III}$	2.681	5.092	2	91 Pa	$M\zeta_1$	M_VN_{III}	2.4350
4.630	1	43 Tc	L_{III}	Abs. Edge	2.6780	5.1148	3	43 Tc	$L\alpha_1$	$L_{III}M_V$	2.4240
4.631	9	77 Ir		M_IN_{III}	2.677	5.118	1	83 Bi	$M\alpha_1$	M_VN_{VII}	2.4226

Wavelength Å*	p.e.	Element	Designation		keV
5.130	2	83 Bi	$M\alpha_2$	$M_V N_{VI}$	2.4170
5.145	4	79 Au	$M\gamma$	$M_{III} N_V$	2.410
5.1517	3	41 Nb	$L\gamma_5$	$L_{II} N_I$	2.4066
5.153	5	81 Tl	M_V	Abs. Edge	2.406
5.157	5	80 Hg	M_{IV}	Abs. Edge	2.404
5.168	9	82 Pb		$M_V O_{III}$	2.399
5.172	9	74 W		$M_I N_{III}$	2.397
5.17708	8	42 Mo	$L\beta_1$	$L_{II} M_{IV}$	2.39481
5.186	5	79 Au		$M_{III} N_{IV}$	2.391
5.193	2	91 Pa	$M\zeta_2$	$M_{IV} N_{II}$	2.3876
5.196	9	81 Tl		$M_{IV} O_{II}$	2.386
5.2050	2	44 Ru	$L\eta$	$L_{II} M_I$	2.38197
5.217	5	39 Y	L_I	Abs. Edge	2.377
5.2169	3	45 Rh	Ll	$L_{III} M_I$	2.3765
5.230	1	41 Nb	L_{III}	Abs. Edge	2.3706
5.234	5	75 Re	M_{III}		2.369
5.2379	3	41 Nb	$L\beta_{2,15}$	$L_{III} N_{IV,V}$	2.3670
5.245	5	90 Th	$M\zeta_1$	$M_V N_{III}$	2.364
5.249	1	81 Tl	$M\beta$	$M_{IV} N_{VI}$	2.3621
5.2830	3	39 Y	$L\gamma_{2,3}$	$L_I N_{II,III}$	2.3468
5.286	1	82 Pb	$M\alpha_1$	$M_V N_{VII}$	2.3455
5.299	2	82 Pb	$M\alpha_2$	$M_V N_{VI}$	2.3397
5.3102	3	41 Nb	$L\beta_3$	$L_I M_{III}$	2.3348
5.319	4	78 Pt	$M\gamma$	$M_{III} N_V$	2.331
5.340	5	90 Th	$M\zeta_2$	$M_{IV} N_{II}$	2.322
5.3455	3	41 Nb	$L\beta_4$	$L_I M_{II}$	2.3194
5.351	4	74 W		$M_{II} N_{IV}$	2.314
5.357	5	78 Pt		$M_{III} N_{IV}$	2.314
5.36	1	80 Hg	M_V	Abs. Edge	2.313
5.3613	3	41 Nb	$L\beta_6$	$L_{III} N_I$	2.3125
5.37216	7	16 S	$K\alpha_1$	$K L_{III}$	2.30784
5.374	5	79 Au	M_{IV}	Abs. Edge	2.307
5.37496	8	16 S	$K\alpha_2$	$K L_{II}$	2.30664
5.378	1	40 Zr	L_{II}	Abs. Edge	2.3053
5.3843	3	40 Zr	$L\gamma_1$	$L_{II} N_{IV}$	2.3027
5.40	2	73 Ta		$M_I N_{III}$	2.295
5.40655	8	42 Mo	$L\alpha_1$	$L_{III} M_V$	2.29316
5.41437	8	42 Mo	$L\alpha_2$	$L_{III} M_{IV}$	2.28985
5.4318	9	80 Hg	$M\beta$	$M_{IV} N_{VI}$	2.2825
5.435	1	74 W	M_{III}	Abs. Edge	2.2811
5.460	1	81 Tl	$M\alpha_1$	$M_V N_{VII}$	2.2706
5.472	2	81 Tl	$M\alpha_2$	$M_V N_{VI}$	2.2656
5.4923	3	41 Nb	$L\beta_1$	$L_{II} M_{IV}$	2.2574
5.4977	3	40 Zr	$L\gamma_5$	$L_{II} N_I$	2.2551
5.500	4	77 Ir	$M\gamma$	$M_{III} N_V$	2.254
5.5035	3	44 Ru	Ll	$L_{III} M_I$	2.2528
5.537	8	83 Bi		$M_{III} N_I$	2.239
5.540	5	77 Ir		$M_{III} N_{IV}$	2.238
5.570	4	73 Ta		$M_{II} N_{IV}$	2.226
5.579	1	40 Zr	L_{III}	Abs. Edge	2.2225
5.584	5	79 Au	M_V	Abs. Edge	2.220
5.5863	3	40 Zr	$L\beta_{2,15}$	$L_{III} N_{IV,V}$	2.2194
5.59	1	78 Pt	M_{IV}	Abs. Edge	2.217
5.592	5	38 Sr	L_I	Abs. Edge	2.217
5.624	1	79 Au	$M\beta$	$M_{IV} N_{VI}$	2.2046
5.628	8	74 W		$M_{III} O_I$	2.203
5.6330	3	40 Zr	$L\beta_3$	$L_I M_{III}$	2.2010
5.6445	3	38 Sr	$L\gamma_{2,3}$	$L_I N_{II,III}$	2.1965
5.6476	9	80 Hg	$M\alpha_1$	$M_V N_{VII}$	2.1953
5.650	5	73 Ta	M_{III}	Abs. Edge	2.194
5.6681	3	40 Zr	$L\beta_4$	$L_I M_{II}$	2.1873
5.67	3	73 Ta		$M_{III} O_{IV,V}$	2.19
5.682	4	76 Os	$M\gamma$	$M_{III} N_V$	2.182
5.704	8	82 Pb		$M_{III} N_I$	2.174
5.7101	3	40 Zr	$L\beta_6$	$L_{III} N_I$	2.1712
5.724	5	76 Os		$M_{III} N_{IV}$	2.166
5.7243	2	41 Nb	$L\alpha_1$	$L_{III} M_V$	2.16589
5.7319	3	41 Nb	$L\alpha_2$	$L_{III} M_{IV}$	2.1630
5.756	1	39 Y	L_{II}	Abs. Edge	2.1540
5.767	9	79 Au		$M_V O_{III}$	2.150
5.784	1	15 P	K	Abs. Edge	2.1435
5.796	2	15 P	$K\beta$	$K M$	2.1391
5.81	2	76 Os		$M_{II} N_I$	2.133
5.81	1	78 Pt	M_V	Abs. Edge	2.133
5.828	1	78 Pt	$M\beta$	$M_{IV} N_{VI}$	2.1273
5.83	2	73 Ta		$M_{III} O_I$	2.126
5.83	1	77 Ir	M_{IV}	Abs. Edge	2.126
5.8360	3	40 Zr	$L\beta_1$	$L_{II} M_{IV}$	2.1244
5.840	1	79 Au	$M\alpha_1$	$M_V N_{VII}$	2.1229
5.8475	3	42 Mo	$L\eta$	$L_{II} M_I$	2.1202
5.854	3	79 Au	$M\alpha_2$	$M_V N_{VI}$	2.118
5.8754	3	39 Y	$L\gamma_5$	$L_{II} N_I$	2.1102
5.884	8	81 Tl		$M_{III} N_I$	2.107
5.885	2	75 Re	$M\gamma$	$M_{III} N_V$	2.1067
5.931	5	75 Re		$M_{III} N_{IV}$	2.090
5.962	1	39 Y	L_{III}	Abs. Edge	2.0794
5.9832	3	39 Y	$L\beta_3$	$L_I M_{III}$	2.0722
5.987	9	78 Pt		$M_V O_{III}$	2.071
6.008	5	37 Rb	L_I	Abs. Edge	2.063
6.0186	3	39 Y	$L\beta_4$	$L_I M_{II}$	2.0600
6.038	1	77 Ir	$M\beta$	$M_{IV} N_{VI}$	2.0535
6.0458	3	37 Rb	$L\gamma_{2,3}$	$L_I N_{II,III}$	2.0507
6.047	1	78 Pt	$M\alpha_1$	$M_V N_{VII}$	2.0505
6.05	1	77 Ir	M_V	Abs. Edge	2.048
6.058	3	78 Pt	$M\alpha_2$	$M_V N_{VI}$	2.047
6.0705	2	40 Zr	$L\alpha_1$	$L_{III} M_V$	2.04236
6.073	5	76 Os	M_{IV}	Abs. Edge	2.042
6.0778	3	40 Zr	$L\alpha_2$	$L_{III} M_{IV}$	2.0399
6.09	2	80 Hg		$M_{III} N_I$	2.036
6.092	3	74 W	$M\gamma$	$M_{III} N_V$	2.035
6.0942	3	39 Y	$L\beta_6$	$L_{III} N_I$	2.0344
6.134	4	74 W		$M_{III} N_{IV}$	2.021
6.1508	3	42 Mo	Ll	$L_{III} M_I$	2.01568
6.157	1	15 P	$K\alpha_1$	$K L_{III}$	2.0137
6.160	1	15 P	$K\alpha_2$	$K L_{II}$	2.0127
6.162	8	83 Bi		$M_{IV} N_{III}$	2.012
6.173	1	38 Sr	L_{II}	Abs. Edge	2.0085
6.2109	3	41 Nb	$L\eta$	$L_{II} M_I$	1.99620
6.2120	3	39 Y	$L\beta_1$	$L_{II} M_{IV}$	1.99584
6.259	9	79 Au		$M_{III} N_I$	1.981
6.262	1	77 Ir	$M\alpha_1$	$M_V N_{VII}$	1.9799
6.267	1	76 Os	$M\beta$	$M_{IV} N_{VI}$	1.9783
6.275	3	77 Ir	$M\alpha_2$	$M_V N_{VI}$	1.9758
6.28	2	74 W		$M_{II} N_I$	1.973

Wavelength Å*	p.e.	Element	Designation		keV
6.2961	3	38 Sr	$L\gamma_5$	$L_{II}N_I$	1.96916
6.30	1	76 Os	M_V	Abs. Edge	1.967
6.312	4	73 Ta	$M\gamma$	$M_{III}N_V$	1.964
6.33	1	75 Re	M_{IV}	Abs. Edge	1.958
6.353	5	73 Ta		$M_{III}N_{IV}$	1.951
6.3672	3	38 Sr	$L\beta_3$	$L_I M_{III}$	1.94719
6.384	7	82 Pb		$M_{IV}N_{III}$	1.942
6.387	1	38 Sr	L_{III}	Abs. Edge	1.9411
6.4026	3	38 Sr	$L\beta_4$	$L_I M_{II}$	1.93643
6.4488	2	39 Y	$L\alpha_1$	$L_{III}M_V$	1.92256
6.455	9	78 Pt		$M_{III}N_I$	1.921
6.4558	3	39 Y	$L\alpha_2$	$L_{III}M_{IV}$	1.92047
6.47	1	36 Kr	L_I	Abs. Edge	1.915
6.490	1	76 Os	$M\alpha$	$M_V N_{VI,VII}$	1.9102
6.504	1	75 Re	$M\beta$	$M_{IV}N_{VI}$	1.9061
6.5176	3	41 Nb	Ll	$L_{III}M_I$	1.90225
6.5191	3	38 Sr	$L\beta_6$	$L_{III}N_I$	1.90181
6.521	4	83 Bi	$M\zeta_1$	$M_V N_{III}$	1.901
6.544	4	72 Hf	$M\gamma$	$M_{III}N_V$	1.895
6.560	5	75 Re	M_V	Abs. Edge	1.890
6.585	5	83 Bi	$M\zeta_2$	$M_{IV}N_{II}$	1.883
6.59	1	74 W	M_{IV}	Abs. Edge	1.880
6.6069	3	40 Zr	$L\eta$	$L_{II}M_I$	1.87654
6.6239	3	38 Sr	$L\beta_1$	$L_{II}M_{IV}$	1.87172
6.644	1	37 Rb	L_{II}	Abs. Edge	1.8661
6.669	9	77 Ir		$M_{III}N_I$	1.859
6.729	1	75 Re	$M\alpha$	$M_V N_{VI,VII}$	1.8425
6.738	1	14 Si	K	Abs. Edge	1.8400
6.740	3	82 Pb	$M\zeta_1$	$M_V N_{III}$	1.8395
6.7530	1	14 Si	$K\beta$	KM	1.83594
6.755	3	37 Rb	$L\gamma_5$	$L_{II}N_{IV}$	1.83532
6.757	1	74 W	$M\beta$	$M_{IV}N_{VI}$	1.8349
6.768	6	71 Lu	$M\gamma$	$M_{III}N_V$	1.832
6.7876	3	37 Rb	$L\beta_3$	$L_I M_{III}$	1.82659
6.802	5	82 Pb	$M\zeta_2$	$M_{IV}N_{II}$	1.823
6.806	9	74 W		$M_{IV}O_{II}$	1.822
6.8207	3	37 Rb	$L\beta_4$	$L_I M_{II}$	1.81771
6.83	1	74 W	M_V	Abs. Edge	1.814
6.862	1	37 Rb	L_{III}	Abs. Edge	1.8067
6.8628	2	38 Sr	$L\alpha_1$	$L_{III}M_V$	1.80656
6.8697	3	38 Sr	$L\alpha_2$	$L_{III}M_{IV}$	1.80474
6.87	1	73 Ta	M_{IV}	Abs. Edge	1.804
6.87	2	80 Hg	δ	$M_{IV}N_{III}$	1.805
6.89	2	76 Os		$M_{III}N_I$	1.798
6.9185	3	40 Zr	Ll	$L_{III}M_I$	1.79201
6.959	5	35 Br		Abs. Edge	1.781
6.974	4	81 Tl	$M\zeta_1$	$M_V N_{III}$	1.778
6.983	1	74 W	$M\alpha_1$	$M_V N_{VII}$	1.7754
6.9842	3	37 Rb	$L\beta_6$	$L_{III}N_I$	1.77517
6.992	2	74 W	$M\alpha_2$	$M_V N_{VI}$	1.7731
7.005	9	74 W		$M_V O_{III}$	1.770
7.023	1	73 Ta	$M\beta$	$M_{IV}N_{VI}$	1.7655
7.024	8	70 Yb	$M\gamma$	$M_{III}N_V$	1.765
7.032	5	81 Tl	$M\zeta_2$	$M_{IV}N_{II}$	1.763
7.0406	3	39 Y	$L\eta$	$L_{II}M_I$	1.76095
7.0759	3	37 Rb	$L\beta_1$	$L_{II}M_{IV}$	1.75217
7.09	2	73 Ta		$M_{IV}O_{II,III}$	1.748
7.101	8	79 Au		$M_{IV}N_{III}$	1.746
7.11	1	73 Ta	M_V	Abs. Edge	1.743
7.12542	9	14 Si	$K\alpha_1$	KL_{III}	1.73998
7.12791	9	14 Si	$K\alpha_2$	KL_{II}	1.73938
7.168	1	36 Kr	L_{II}	Abs. Edge	1.7297
7.250	5	36 Kr		$L_{II}N_{III}$	1.710
7.252	1	73 Ta	$M\alpha$	$M_V N_{VI,VII}$	1.7096
7.264	5	36 Kr	$L\beta_3$	$L_I M_{III}$	1.707
7.279	5	36 Kr	$L\gamma_5$	$L_{II}N_I$	1.703
7.30	2	73 Ta		$M_V O_{III}$	1.700
7.303	1	72 Hf	$M\beta$	$M_{IV}N_{VI}$	1.6976
7.304	5	36 Kr	$L\beta_4$	$L_I M_{II}$	1.697
7.3183	2	37 Rb	$L\alpha_1$	$L_{III}M_V$	1.69413
7.3251	3	37 Rb	$L\alpha_2$	$L_{III}M_{IV}$	1.69256
7.3563	3	39 Y	Ll	$L_{III}M_I$	1.68536
7.360	8	74 W		$M_{III}N_I$	1.684
7.371	8	78 Pt		$M_{IV}N_{III}$	1.682
7.392	1	36 Kr	L_{III}	Abs. Edge	1.6772
7.466	4	79 Au	$M\zeta_1$	$M_V N_{III}$	1.6605
7.503	1	34 Se	L_I	Abs. Edge	1.6525
7.510	4	36 Kr	$L\beta_6$	$L_{III}N_I$	1.6510
7.5171	3	38 Sr	$L\eta$	$L_{II}M_I$	1.64933
7.523	5	79 Au	$M\zeta_2$	$M_{IV}N_{II}$	1.648
7.539	1	72 Hf	$M\alpha$	$M_V N_{VI,VII}$	1.6446
7.546	8	68 Er	$M\gamma$	$M_{III}N_V$	1.643
7.576	3	36 Kr	$L\beta_1$	$L_{II}M_{IV}$	1.6366
7.60	1	68 Er		$M_{III}N_{IV}$	1.632
7.601	2	71 Lu	$M\beta$	$M_{IV}N_{VI}$	1.6312
7.612	9	73 Ta		$M_{III}N_I$	1.629
7.645	8	77 Ir		$M_{IV}N_{III}$	1.622
7.738	4	78 Pt	$M\zeta_1$	$M_V N_{III}$	1.6022
7.753	5	35 Br	L_{II}	Abs. Edge	1.599
7.767	9	35 Br	$L\beta_{3,4}$	$L_I M_{II,III}$	1.596
7.790	5	78 Pt	$M\zeta_2$	$M_{IV}N_{II}$	1.592
7.817	3	36 Kr	$L\alpha_{1,2}$	$L_{III}M_{IV,V}$	1.5860
7.8362	3	38 Sr	Ll	$L_{III}M_I$	1.58215
7.840	2	71 Lu	$M\alpha$	$M_V N_{VI,VII}$	1.5813
7.865	9	67 Ho	$M\gamma$	$M_{III}N_{IV,V}$	1.576
7.887	2	72 Hf		$M_{III}N_I$	1.572
7.909	2	70 Yb	$M\beta$	$M_{IV}N_{VI}$	1.5675
7.94813	5	13 Al	K	Abs. Edge	1.55988
7.960	2	13 Al	$K\beta$	KM	1.55745
7.984	5	35 Br	L_{III}	Abs. Edge	1.5530
8.021	4	77 Ir	$M\zeta_1$	$M_V N_{III}$	1.5458
8.0415	4	37 Rb	$L\eta$	$L_{II}M_I$	1.54177
8.065	5	77 Ir	$M\zeta_2$	$M_{IV}N_{II}$	1.5373
8.107	1	33 As	L_I	Abs. Edge	1.5293
8.1251	5	35 Br	$L\beta_1$	$L_{II}M_{IV}$	1.52590
8.144	9	66 Dy	$M\gamma$	$M_{III}N_{IV,V}$	1.522
8.149	5	70 Yb	$M\alpha$	$M_V N_{VI,VII}$	1.5214
8.239	8	75 Re		$M_{IV}N_{III}$	1.505
8.249	7	69 Tm	$M\beta$	$M_{IV}N_{VI}$	1.503
8.310	4	76 Os	$M\zeta_1$	$M_V N_{III}$	1.4919
8.321	9	34 Se	$L\beta_{3,4}$	$L_I M_{II,III}$	1.490
8.33934	9	13 Al	$K\alpha_1$	KL_{III}	1.48670
8.34173	9	13 Al	$K\alpha_2$	KL_{II}	1.48627
8.359	5	76 Os	$M\zeta_2$	$M_{IV}N_{II}$	1.4831

Wavelength Å*	p.e.	Element	Designation		keV
8.3636	4	37 Rb	Ll	$L_{III}M_I$	1.48238
8.3746	5	35 Br	$L\alpha_{1,2}$	$L_{III}M_{IV,V}$	1.48043
8.407	1	34 Se	L_{II}	Abs. Edge	1.4747
8.470	9	70 Yb		$M_{III}N_I$	1.464
8.48	1	69 Tm	$M\alpha$	$M_V N_{VI,VII}$	1.462
8.486	9	65 Tb	$M\gamma$	$M_{III}N_{IV,V}$	1.461
8.487	5	69 Tm	M_V	Abs. Edge	1.4609
8.573	8	74 W		$M_{IV}N_{III}$	1.446
8.592	3	68 Er	$M\beta$	$M_{IV}N_{VI}$	1.4430
8.60	7	92 U		$N_I P_{IV,V}$	1.44
8.601	5	68 Er	M_{IV}	Abs. Edge	1.4415
8.629	4	75 Re	$M\zeta_1$	$M_V N_{III}$	1.4368
8.646	1	34 Se	L_{III}	Abs. Edge	1.4340
8.664	5	75 Re	$M\zeta_2$	$M_{IV}N_{II}$	1.4310
8.7358	5	34 Se	$L\beta_1$	$L_{II}M_{IV}$	1.41923
8.76	7	92 U		$N_I P_{III}$	1.42
8.773	1	32 Ge	L_I	Abs. Edge	1.4132
8.81	7	92 U		$N_I P_{II}$	1.41
8.82	1	68 Er	$M\alpha$	$M_V N_{VI,VII}$	1.406
8.844	9	64 Gd	$M\gamma$	$M_{III}N_{IV,V}$	1.402
8.847	5	68 Er	M_V	Abs. Edge	1.4013
8.90	2	73 Ta		$M_{IV}N_{III}$	1.393
8.929	1	33 As	$L\beta_{3,4}$	$L_I M_{II,III}$	1.3884
8.962	4	74 W	$M\zeta_1$	$M_V N_{III}$	1.3835
8.965	4	67 Ho	$M\beta$	$M_{IV}N_{VI}$	1.3830
8.9900	5	34 Se	$L\alpha_{1,2}$	$L_{III}M_{IV,V}$	1.37910
8.993	5	74 W	$M\zeta_2$	$M_{IV}N_{II}$	1.3787
9.125	1	33 As	L_{II}	Abs. Edge	1.3587
9.20	2	67 Ho	$M\alpha$	$M_V N_{VI,VII}$	1.348
9.211	9	63 Eu	$M\gamma$	$M_{III}N_{IV,V}$	1.346
9.255	1	35 Br	$L\eta$	$L_{II}M_I$	1.3396
9.316	4	73 Ta	$M\zeta_1$	$M_V N_{III}$	1.3308
9.330	5	73 Ta	$M\zeta_2$	$M_{IV}N_{II}$	1.3288
9.357	6	66 Dy	$M\beta$	$M_{IV}N_{VI}$	1.3250
9.367	1	33 As	L_{III}	Abs. Edge	1.3235
9.40	7	90 Th		$N_I P_{III}$	1.319
9.4141	8	33 As	$L\beta_1$	$L_{II}M_{IV}$	1.3170
9.44	7	90 Th		$N_I P_{II}$	1.313
9.5122	1	12 Mg	K	Abs. Edge	1.30339
9.517	5	31 Ga	L_I	Abs. Edge	1.3028
9.521	2	12 Mg	$K\beta$	KM	1.3022
9.581	2	32 Ge	$L\beta_3$	$L_I M_{III}$	1.2941
9.585	1	35 Br	Ll	$L_{III}M_I$	1.2935
9.59	2	66 Dy	$M\alpha$	$M_V N_{VI,VII}$	1.293
9.600	9	62 Sm	$M\gamma$	$M_{III}N_{IV,V}$	1.291
9.640	2	32 Ge	$L\beta_4$	$L_I M_{II}$	1.2861
9.6709	8	33 As	$L\alpha_{1,2}$	$L_{III}M_{IV,V}$	1.2820
9.686	7	72 Hf	$M\zeta_2$	$M_{IV}N_{II}$	1.2800
9.686	7	72 Hf	$M\zeta_1$	$M_V N_{III}$	1.2800
9.792	6	65 Tb	$M\beta$	$M_{IV}N_{VI}$	1.2661
9.8900	2	12 Mg	$K\alpha_{1,2}$	$KL_{II,III}$	1.25360
9.924	1	32 Ge	L_{II}	Abs. Edge	1.2494
9.962	1	34 Se	$L\eta$	$L_{II}M_I$	1.2446
10.00	2	65 Tb	$M\alpha$	$M_V N_{VI,VII}$	1.240
10.09	7	92 U		$N_I O_{III}$	1.229
10.175	1	32 Ge	$L\beta_1$	$L_{II}M_{IV}$	1.2185
10.187	1	32 Ge	L_{III}	Abs. Edge	1.2170
10.254	6	64 Gd	$M\beta$	$M_{IV}N_{VI}$	1.2091
10.294	1	34 Se	Ll	$L_{III}M_I$	1.2044
10.359	9	31 Ga	$L\beta_{3,4}$	$L_I M_{II,III}$	1.197
10.40	7	92 U		$N_{II}P_I$	1.192
10.4361	8	32 Ge	$L\alpha_{1,2}$	$L_{III}M_{IV,V}$	1.18800
10.46	3	64 Gd	$M\alpha$	$M_V N_{VI,VII}$	1.185
10.48	1	70 Yb	$M\zeta$	$M_V N_{III}$	1.183
10.505	9	60 Nd	$M\gamma$	$M_{III}N_{IV,V}$	1.180
10.711	5	63 Eu	M_{IV}	Abs. Edge	1.1575
10.734	1	33 As	$L\eta$	$L_{II}M_I$	1.1550
10.750	7	63 Eu	$M\beta$	$M_{IV}N_{VI}$	1.1533
10.828	5	31 Ga	L_{II}	Abs. Edge	1.1450
10.96	3	63 Eu	$M\alpha$	$M_V N_{VI,VII}$	1.131
10.998	9	59 Pr	$M\gamma$	$M_{III}N_{IV,V}$	1.1273
11.013	5	63 Eu	M_V	Abs. Edge	1.1258
11.023	2	31 Ga	$L\beta_1$	$L_{II}M_{IV}$	1.1248
11.072	1	33 As	Ll	$L_{III}M_I$	1.1198
11.07	7	90 Th		$N_{II}P_I$	1.120
11.100	1	31 Ga	L_{III}	Abs. Edge	1.1169
11.200	7	30 Zn	$L\beta_{3,4}$	$L_I M_{II,III}$	1.1070
11.27	1	62 Sm	$M\beta$	$M_{IV}N_{VI}$	1.0998
11.288	5	62 Sm	M_{IV}	Abs. Edge	1.0983
11.292	1	31 Ga	$L\alpha_{1,2}$	$L_{III}M_{IV,V}$	1.09792
11.37	1	68 Er	$M\zeta$	$M_V N_{III}$	1.0901
11.47	3	62 Sm	$M\alpha$	$M_V N_{VI,VII}$	1.081
11.53	1	58 Ce	$M\gamma$	$M_{III}N_{IV,V}$	1.0749
11.552	5	62 Sm	M_V	Abs. Edge	1.0732
11.56	5	90 Th		$N_{II}O_{IV}$	1.072
11.569	1	11 Na	K	Abs. Edge	1.07167
11.575	2	11 Na	$K\beta$	KM	1.0711
11.609	2	32 Ge	$L\eta$	$L_{II}M_I$	1.0680
11.862	1	30 Zn	L_{II}	Abs. Edge	1.04523
11.86	1	67 Ho	$M\zeta$	$M_V N_{III}$	1.0450
11.9101	9	11 Na	$K\alpha_{1,2}$	$KL_{II,III}$	1.04098
11.965	2	32 Ge	Ll	$L_{III}M_I$	1.0362
11.983	3	30 Zn	$L\beta_1$	$L_{II}M_{IV}$	1.0347
12.08	4	57 La	$M\gamma$	$M_{III}N_{IV,V}$	1.027
12.122	3	29 Cu	$L\beta_{3,4}$	$L_I M_{II,III}$	1.0228
12.131	1	30 Zn	L_{III}	Abs. Edge	1.02201
12.254	3	30 Zn	$L\alpha_{1,2}$	$L_{III}M_{IV,V}$	1.0117
12.43	2	66 Dy	$M\zeta$	$M_V N_{III}$	0.998
12.44	2	60 Nd	$M\beta$	$M_{IV}N_{VI}$	0.997
12.459	5	60 Nd	M_{IV}	Abs. Edge	0.9951
12.597	2	31 Ga	$L\eta$	$L_{II}M_I$	0.9842
12.68	2	60 Nd	$M\alpha$	$M_V N_{VI,VII}$	0.978
12.737	5	60 Nd	M_V	Abs. Edge	0.9734
12.75	3	56 Ba	$M\gamma$	$M_{III}N_{IV,V}$	0.973
12.90	9	92 U		$N_{III}O_V$	0.961
12.953	2	31 Ga	Ll	$L_{III}M_I$	0.9572
12.98	2	65 Tb	$M\zeta$	$M_V N_{III}$	0.955
13.014	1	29 Cu	L_{II}	Abs. Edge	0.95268
13.053	3	29 Cu	$L\beta_1$	$L_{II}M_{IV}$	0.9498
13.06	2	59 Pr	$M\beta$	$M_{IV}N_{VI}$	0.950
13.06	1	30 Zn	L_I	Abs. Edge	0.9495
13.122	5	59 Pr	M_{IV}	Abs. Edge	0.9448
13.18	2	28 Ni	$L\beta_{3,4}$	$L_I M_{II,III}$	0.941
13.288	1	29 Cu	L_{III}	Abs. Edge	0.93306

Wavelength Å*	p.e.	Element	Designation		keV
13.30	6	83 Bi		$N_I P_{II,III}$	0.932
13.336	3	29 Cu	$L\alpha_{1,2}$	$L_{III}M_{IV,V}$	0.9297
13.343	5	59 Pr	$M\alpha$	$M_V N_{VI,VII}$	0.9292
13.394	5	59 Pr	M_V	Abs. Edge	0.9257
13.57	2	64 Gd	$M\zeta$	$M_V N_{III}$	0.914
13.68	2	30 Zn	$L\eta$	$L_{II}M_I$	0.906
13.75	4	58 Ce	$M\beta$	$M_{IV}N_{VI}$	0.902
13.8	1	90 Th		$N_{III}O_V$	0.897
14.02	2	30 Zn	Ll	$L_{III}M_I$	0.884
14.04	2	58 Ce	$M\alpha$	$M_V N_{VI,VII}$	0.883
14.22	2	63 Eu	$M\zeta$	$M_V N_{III}$	0.872
14.242	5	28 Ni	L_{II}	Abs. Edge	0.8706
14.271	6	28 Ni	$L\beta_1$	$L_{II}M_{IV}$	0.8688
14.3018	1	10 Ne	K	Abs. Edge	0.866889
14.31	3	27 Co	$L\beta_{3,4}$	$L_I M_{II,III}$	0.870
14.39	5	58 Ce		$M_V O_{II,III}$	0.862
14.452	5	10 Ne	$K\beta$	KM	0.8579
14.51	5	57 La	$M\beta$	$M_{IV}N_{VI}$	0.854
14.525	5	28 Ni	L_{III}	Abs. Edge	0.8536
14.561	3	28 Ni	$L\alpha_{1,2}$	$L_{III}M_{IV,V}$	0.8515
14.610	3	10 Ne	$K\alpha_{1,2}$	$KL_{II,III}$	0.8486
14.88	5	57 La	$M\alpha$	$M_V N_{VI,VII}$	0.833
14.90	2	29 Cu	$L\eta$	$L_{II}M_I$	0.832
14.91	4	62 Sm	$M\zeta$	$M_V N_{III}$	0.831
15.286	9	29 Cu	Ll	$L_{III}M_I$	0.8111
15.56	1	56 Ba	M_{IV}	Abs. Edge	0.7967
15.618	5	27 Co	L_{II}	Abs. Edge	0.7938
15.65	4	26 Fe	$L\beta_{3,4}$	$L_I M_{II,III}$	0.792
15.666	8	27 Co	$L\beta_1$	$L_{II}M_{IV}$	0.7914
15.72	9	56 Ba		$M_{IV}O_{III}$	0.789
15.89	1	56 Ba	M_V	Abs. Edge	0.7801
15.91	5	56 Ba		$M_{IV}O_{II}$	0.779
15.915	5	27 Co	L_{III}	Abs. Edge	0.7790
15.93	4	52 Te	$M\gamma$	$M_{III}N_{IV,V}$	0.778
15.972	6	27 Co	$L\alpha_{1,2}$	$L_{III}M_{IV,V}$	0.7762
15.98	5	51 Sb		$M_{II}N_{IV}$	0.776
16.20	5	56 Ba		$M_V O_{III}$	0.765
16.27	3	28 Ni	$L\eta$	$L_{II}M_I$	0.762
16.46	4	60 Nd	$M\zeta$	$M_V N_{III}$	0.753
16.693	9	28 Ni	Ll	$L_{III}M_I$	0.7427
16.7	1	24 Cr	L_I	Abs. Edge	0.741
16.92	4	51 Sb	$M\gamma$	$M_{III}N_{IV,V}$	0.733
16.93	5	50 Sn		$M_{II}N_{IV}$	0.733
17.19	4	25 Mn	$L\beta_{3,4}$	$L_I M_{II,III}$	0.721
17.202	5	26 Fe	L_{II}	Abs. Edge	0.7208
17.26	1	26 Fe	$L\beta_1$	$L_{II}M_{IV}$	0.7185
17.38	4	59 Pr	$M\zeta$	$M_V N_{III}$	0.714
17.525	5	26 Fe	L_{III}	Abs. Edge	0.7074
17.59	2	26 Fe	$L\alpha_{1,2}$	$L_{III}M_{IV,V}$	0.7050
17.6	1	52 Te		$M_{II}N_I$	0.703
17.87	3	27 Co	$L\eta$	$L_{II}M_I$	0.694
17.94	5	50 Sn	$M\gamma$	$M_{III}N_{IV,V}$	0.691
17.9	1	24 Cr	L_{II}	Abs. Edge	0.691
18.292	8	27 Co	Ll	$L_{III}M_I$	0.6778
18.32	2	9 F	$K\alpha$	KL	0.6768
18.35	4	58 Ce	$M\zeta$	$M_V N_{III}$	0.676
18.8	1	51 Sb		$M_{II}N_I$	0.658

Wavelength Å*	p.e.	Element	Designation		keV
18.8	2	47 Ag		$M_I N_{II,III}$	0.658
18.96	4	24 Cr	$L\beta_{3,4}$	$L_I M_{II,III}$	0.654
19.11	2	25 Mn	$L\beta_1$	$L_{II}M_{IV}$	0.6488
19.1	1	52 Te		$M_{III}N_I$	0.648
19.40	7	48 Cd		$M_{II}N_{IV}$	0.639
19.44	5	57 La	$M\zeta$	$M_V N_{III}$	0.638
19.45	1	25 Mn	$L\alpha_{1,2}$	$L_{III}M_{IV,V}$	0.6374
19.66	5	53 I	$M_{IV,V}$	Abs. Edge	0.631
19.75	4	26 Fe	$L\eta$	$L_{II}M_I$	0.628
20.0	1	50 Sn		$M_{III}N_I$	0.619
20.1	2	46 Pd		$M_I N_{II,III}$	0.616
20.15	1	26 Fe	Ll	$L_{III}M_I$	0.6152
20.2	1	51 Sb		$M_{III}N_I$	0.612
20.47	7	48 Cd	$M\gamma$	$M_{III}N_{IV,V}$	0.606
20.64	4	56 Ba	$M\zeta$	$M_V N_{III}$	0.601
20.66	7	47 Ag		$M_{III}N_{IV}$	0.600
20.7	1	24 Cr	L_{III}	Abs. Edge	0.598
21.19	5	23 Va	$L\beta_{3,4}$	$L_I M_{II,III}$	0.585
21.27	1	24 Cr	$L\beta_1$	$L_{II}M_{IV}$	0.5828
21.34	5	52 Te		$M_{IV}O_{II,III}$	0.581
21.5	1	50 Sn		$M_{III}N_I$	0.575
21.64	3	24 Cr	$L\alpha_{1,2}$	$L_{III}M_{IV,V}$	0.5728
21.78	5	52 Te		$M_V O_{III}$	0.569
21.82	7	47 Ag	$M\gamma$	$M_{III}N_{IV,V}$	0.568
21.85	2	25 Mn	$L\eta$	$L_{II}M_I$	0.5675
22.1	1	46 Pd		$M_{II}N_{IV}$	0.560
22.29	1	25 Mn	Ll	$L_{III}M_I$	0.5563
22.9	2	48 Cd		$M_{II}N_I$	0.540
23.32	1	8 O	K	Abs. Edge	0.5317
23.3	1	46 Pd	$M\gamma$	$M_{III}N_{IV,V}$	0.531
23.62	3	8 O	$K\alpha$	KL	0.5249
23.88	4	23 Va	$L\beta_1$	$L_{II}M_{IV}$	0.5192
24.25	3	23 Va	$L\alpha_{1,2}$	$L_{III}M_{IV,V}$	0.5113
24.28	5	50 Sn	$M_{IV,V}$	Abs. Edge	0.511
24.30	3	24 Cr	$L\eta$	$L_{II}M_I$	0.5102
24.4	2	47 Ag		$M_V N_I$	0.509
24.5	1	48 Cd		$M_{III}N_I$	0.507
24.78	1	24 Cr	Ll	$L_{III}M_I$	0.5003
25.01	9	45 Rh	$M\gamma$	$M_{III}N_{IV,V}$	0.496
25.3	1	50 Sn		$M_{IV}O_{II,III}$	0.491
25.50	9	44 Ru		$M_{II}N_{IV}$	0.486
25.7	1	50 Sn		$M_V O_{III}$	0.483
26.0	1	47 Ag		$M_{III}N_I$	0.478
26.2	2	46 Pd		$M_{II}N_I$	0.474
26.72	9	52 Te	$M\zeta$	$M_{IV,V}N_{II,III}$	0.464
26.9	1	44 Ru	$M\gamma$	$M_{III}N_{IV,V}$	0.462
27.05	2	22 Ti	$L\beta_1$	$L_{II}M_{IV}$	0.4584
27.29	1	22 Ti	$L_{II,III}$	Abs. Edge	0.4544
27.34	3	23 Va	$L\eta$	$L_{II}M_I$	0.4535
27.42	2	22 Ti	$L\alpha_{1,2}$	$L_{III}M_{IV,V}$	0.4522
27.77	1	23 Va	Ll	$L_{III}M_I$	0.4465
27.9	1	46 Pd		$M_{III}N_I$	0.445
28.1	2	45 Rh		$M_{II}N_I$	0.442
28.13	5	48 Cd	$M_{IV,V}$	Abs. Edge	0.4408
28.88	8	51 Sb	$M\zeta$	$M_{IV,V}N_{II,III}$	0.429
29.8	1	45 Rh		$M_{III}N_I$	0.417
30.4	1	48 Cd		$M_{IV}O_{II,III}$	0.408

Wavelength Å*	p.e.	Element		Designation	keV
30.8	1	48 Cd		$M_V O_{III}$	0.403
30.82	5	47 Ag	M_{IV}	Abs. Edge	0.4022
30.89	3	22 Ti	$L\eta$	$L_{II} M_I$	0.4013
30.99	1	7 N	K	Abs. Edge	0.4000
31.02	2	21 Sc	$L\beta_1$	$L_{II} M_{IV}$	0.3996
31.14	5	47 Ag	M_V	Abs. Edge	0.3981
31.24	9	50 Sn	$M\zeta$	$M_{IV,V} N_{II,III}$	0.397
31.35	3	21 Sc	$L\alpha_{1,2}$	$L_{III} M_{IV,V}$	0.3954
31.36	2	22 Ti	Ll	$L_{III} M_I$	0.3953
31.60	4	7 N	$K\alpha$	KL	0.3924
31.8	1	92 U		$N_{IV} N_{VI}$	0.390
32.3	2	44 Ru		$M_{II} N_I$	0.384
33.1	2	41 Nb		$M_{II} N_{IV}$	0.375
33.5	3	47 Ag		$M_{IV,V} O_{II,III}$	0.370
33.57	9	90 Th		$N_{IV} N_{VI}$	0.3693
34.8	1	92 U		$N_V N_{VI,VII}$	0.357
34.9	2	41 Nb	$M\gamma$	$M_{III} N_{IV,V}$	0.356
35.13	2	21 Sc	$L\eta$	$L_{II} M_I$	0.3529
35.13	1	20 Ca	L_{II}	Abs. Edge	0.3529
35.3	3	42 Mo		$M_{II} N_I$	0.351
35.49	1	20 Ca	L_{III}	Abs. Edge	0.34931
35.59	3	21 Sc	Ll	$L_{III} M_I$	0.3483
35.63	1	20 Ca	$L_{II,III}$	Abs. Edge	0.34793
35.94	2	20 Ca	$L\beta_1$	$L_{II} M_{IV}$	0.3449
36.32	9	90 Th		$N_V N_{VI,VII}$	0.3414
36.33	2	20 Ca	$L\alpha_{1,2}$	$L_{III} M_{IV,V}$	0.3413
36.8	1	48 Cd	$M\zeta$	$M_{IV,V} N_{II,III}$	0.3371
37.4	2	46 Pd		$M_{IV,V} O_{II,III}$	0.332
37.5	2	42 Mo		$M_{III} N_I$	0.331
38.4	3	41 Nb		$M_{II} N_I$	0.323
39.77	7	47 Ag	$M\zeta$	$M_{IV,V} N_{II,III}$	0.3117
40.46	2	20 Ca	$L\eta$	$L_{II} M_I$	0.3064
40.7	2	41 Nb		$M_{III} N_I$	0.305
40.9	2	45 Rh		$M_{IV,V} O_{II,III}$	0.303
40.96	2	20 Ca	Ll	$L_{III} M_I$	0.3027
42.1	2	92 U		$N_{VI} O_V$	0.295
42.1	1	19 K	$L_{II,III}$	Abs. Edge	0.2946
42.3	2	82 Pb		$N_{IV} N_{VI}$	0.293
43.3	2	92 U		$N_{VI} O_{IV}$	0.286
43.6	1	46 Pd	$M\zeta$	$M_{IV,V} N_{II,III}$	0.2844
43.68	1	6 C	K	Abs. Edge	0.28384
44.7	3	6 C	$K\alpha$	KL	0.277
44.8	1	44 Ru		$M_{IV,V} O_{II,III}$	0.2768
45.0	1	82 Pb		$N_V N_{VI,VII}$	0.2756
45.2	3	80 Hg		$N_{IV} N_{VI}$	0.274
45.2	1	51 Sb		$M_{II} M_{IV}$	0.2743
46.48	9	39 Y		$M_{II} N_I$	0.267
46.5	2	81 Tl		$N_V N_{VI,VII}$	0.267
46.8	2	79 Au		$N_{IV} N_{VI}$	0.265
47.24	2	19 K	Ll	$L_{II} M_I$	0.2625
47.3	1	50 Sn		$M_{II} M_{IV}$	0.2621
47.67	9	45 Rh	$M\zeta$	$M_{IV,V} N_{II,III}$	0.2601
47.74	1	19 K	Ll	$L_{III} M_I$	0.25971
47.9	3	80 Hg		$N_V N_{VI,VII}$	0.259
48.1	2	78 Pt		$N_{IV} N_{VI}$	0.258
48.2	1	90 Th		$N_{VI} O_V$	0.2572
48.5	2	39 Y		$M_{III} N_I$	0.256
49.4	1	79 Au		$N_V N_{VI,VII}$	0.2510
49.5	1	90 Th		$N_{VI} O_{IV}$	0.2505
50.0	1	90 Th		$N_{VII} O_V$	0.2479
50.2	1	77 Ir		$N_{IV} N_{VI}$	0.2470
50.3	1	52 Te		$M_{III} M_V$	0.2465
50.9	1	78 Pt		$N_V N_{VI,VII}$	0.2436
51.3	1	38 Sr		$M_{II} N_I$	0.2416
51.9	1	76 Os		$N_{IV} N_{VI}$	0.2388
52.0	2	48 Cd		$M_{II} M_{IV}$	0.2384
52.2	1	51 Sb		$M_{III} M_V$	0.2375
52.34	7	44 Ru	$M\zeta$	$M_{IV,V} N_{II,III}$	0.2369
52.8	1	77 Ir		$N_V N_{VI,VII}$	0.2348
53.6	1	38 Sr		$M_{III} N_I$	0.2313
54.0	2	74 W		$N_{II} N_{IV}$	0.2295
54.0	1	47 Ag		$M_{II} M_{IV}$	0.2295
54.2	1	50 Sn		$M_{III} M_V$	0.2287
54.7	2	76 Os		$N_V N_{VI,VII}$	0.2266
54.8	2	42 Mo		$M_{IV,V} O_{II,III}$	0.2262
55.8	1	74 W		$N_{IV} N_{VI}$	0.2221
55.9	1	18 A	$L\eta$	$L_{II} M_I$	0.2217
56.3	1	18 A	Ll	$L_{III} M_I$	0.2201
56.5	1	46 Pd		$M_{II} M_{IV}$	0.2194
57.0	2	37 Rb		$M_{II} N_I$	0.2174
58.2	1	73 Ta		$N_{IV} N_{VI}$	0.2130
58.4	1	74 W		$N_V N_{VII}$	0.2122
58.7	2	48 Cd		$M_{III} M_V$	0.2111
59.3	1	45 Rh		$M_{II} M_{IV}$	0.2090
59.5	3	74 W		$N_V N_{VI}$	0.208
59.5	2	37 Rb		$M_{III} N_I$	0.2083
60.5	1	47 Ag		$M_{III} M_V$	0.2048
61.1	2	73 Ta		$N_V N_{VI,VII}$	0.2028
61.9	2	41 Nb		$M_{IV,V} O_{II,III}$	0.2002
62.2	1	44 Ru		$M_{II} M_{IV}$	0.1992
62.9	1	46 Pd		$M_{III} M_V$	0.1970
63.0	5	71 Lu		$N_{IV} N_{VI}$	0.197
64.38	7	42 Mo	$M\zeta$	$M_{IV,V} N_{II,III}$	0.1926
65.1	7	70 Yb		$N_{IV} N_{VI}$	0.190
65.5	1	45 Rh		$M_{III} M_V$	0.1892
65.7	2	71 Lu		$N_V N_{VI,VII}$	0.1886
67.33	9	17 Cl	$L\eta$	$L_{II} M_I$	0.1841
67.6	3	5 B	$K\alpha$	KL	0.1833
67.90	9	17 Cl	Ll	$L_{III} M_I$	0.1826
68.2	3	90 Th		$O_{III} P_{IV,V}$	0.1817
68.3	1	44 Ru		$M_{III} M_V$	0.1814
68.9	2	42 Mo		$M_{II} M_{IV}$	0.1798
69.3	5	70 Yb		$N_V N_{VI,VII}$	0.179
70.0	4	40 Zr		$M_{IV,V} O_{II,III}$	0.177
72.1	3	41 Nb		$M_{II} M_{IV}$	0.1718
72.19	9	41 Nb	$M\zeta$	$M_{IV,V} N_{II,III}$	0.1717
72.7	9	68 Er		$N_{IV} N_{VI}$	0.171
74.9	1	42 Mo		$M_{III} M_V$	0.1656
76.3	7	68 Er		$N_V N_{VI,VII}$	0.163
76.7	2	40 Zr		$M_{II} M_{IV}$	0.1617
76.9	2	35 Br		$M_{II} N_I$	0.1613
78.4	2	41 Nb		$M_{III} N_I$	0.1582
79.8	3	35 Br		$M_{III} N_I$	0.1554
80.9	3	40 Zr		$M_{III} M_V$	0.1533

Wavelength Å*	p.e.	Element		Designation	keV
81.5	2	39 Y		$M_{II}M_{IV}$	0.1522
82.1	2	40 Zr	$M\zeta$	$M_{IV,V}N_{II,III}$	0.1511
83.	1	66 Dy		$N_{IV,V}N_{VI,VII}$	0.149
83.4	3	16 S	Ll,η	$L_{II,III}M_I$	0.1487
85.7	2	38 Sr		$M_{II}M_{IV}$	0.1447
86.	1	65 Tb		$N_{IV,V}N_{VI,VII}$	0.144
86.5	2	39 Y		$M_{III}M_{IV,V}$	0.1434
91.4	2	38 Sr		$M_{III}M_{IV,V}$	0.1357
91.5	2	37 Rb		$M_{II}M_{IV}$	0.1355
91.6	1	83 Bi		$N_{VI}O_{IV}$	0.1354
93.2	1	83 Bi		$N_{VII}O_V$	0.1330
93.4	2	39 Y	$M\zeta$	$M_{IV,V}N_{II,III}$	0.1328
94.	1	15 P	$L_{II,III}$	Abs. Edge	0.132
96.7	2	37 Rb		$M_{III}M_{IV,V}$	0.1282
97.2	8	66 Dy		$N_{IV,V}O_{II,III}$	0.128
98.	1	62 Sm		$N_{IV,V}N_{VI,VII}$	0.126
100.2	2	82 Pb		$N_{VI}O_V$	0.1237
102.2	4	65 Tb		$N_{IV,V}O_{II,III}$	0.1213
102.4	1	82 Pb		$N_{VI}O_{IV}$	0.1211
103.8	4	15 P		$L_{II,III}M$	0.1194
104.3	1	82 Pb		$N_{VII}O_V$	0.1189
107.	1	60 Nd		$N_{IV,V}N_{VI,VII}$	0.116
108.0	2	38 Sr	$M\zeta_2$	$M_{IV}N_{II,III}$	0.1148
108.7	1	38 Sr	$M\zeta_1$	$M_V N_{III}$	0.1140
109.4	3	35 Br		$M_{II}M_{IV}$	0.1133
110.6	5	29 Cu	M_I	Abs. Edge	0.1121
111.	1	4 Be	K	Abs. Edge	0.111
112.0	6	63 Eu		$N_{IV,V}O_{II,III}$	0.1107
113.0	1	81 Tl		$N_{VI}O_V$	0.10968
113.	1	59 Pr		$N_{IV,V}N_{VI,VII}$	0.1095
113.8	3	35 Br		$M_{III}M_{IV,V}$	0.1089
114.	1	4 Be	$K\alpha$	KL	0.1085
115.3	2	81 Tl		$N_{VI}O_{IV}$	0.1075
117.4	4	62 Sm		$N_{IV,V}O_{II,III}$	0.1056
117.7	1	81 Tl		$N_{VII}O_V$	0.10530
123.	1	14 Si	$L_{II,III}$	Abs. Edge	0.1006
126.8	2	37 Rb		$M_{IV}N_{III}$	0.0978
127.8	2	37 Rb	$M\zeta_2$	$M_{IV}N_{II}$	0.0970
128.7	2	37 Rb	$M\zeta_1$	$M_V N_{III}$	0.0964
128.9	7	60 Nd		$N_{IV,V}O_{II,III}$	0.0962
135.5	4	14 Si		$L_{II,III}M$	0.0915
136.5	4	59 Pr		$N_{IV,V}O_{II,III}$	0.0908
137.0	5	30 Zn	M_{II}	Abs. Edge	0.0905
142.5	1	13 Al	L_I	Abs. Edge	0.08701
143.9	5	30 Zn	M_{III}	Abs. Edge	0.0862
144.4	6	58 Ce		$N_{IV,V}O_{II,III}$	0.0859
144.4	3	37 Rb		$M_I M_{III}$	0.0859
152.6	6	57 La		$N_{IV,V}O_{II,III}$	0.0812
157.	3	30 Zn		$M_{II,III}M_{IV,V}$	0.079
159.0	2	56 Ba		$N_{IV}O_{III}$	0.07796
159.5	5	29 Cu	M_{II}	Abs. Edge	0.0777
163.3	2	56 Ba		$N_{IV}O_{II}$	0.07590
164.6	2	56 Ba		$N_V O_{III}$	0.07530
164.7	3	35 Br		$M_I M_{III}$	0.0753
166.0	5	29 Cu	M_{III}	Abs. Edge	0.0747
170.4	1	13 Al	$L_{II,III}$	Abs. Edge	0.07278
171.4	5	13 Al		$L_{II,III}M$	0.0724
173.	3	29 Cu		$M_{II,III}M_{IV,V}$	0.072
181.	5	90 Th		$O_{IV,V}Q_{II,III}$	0.068
183.8	1	55 Cs		$N_{IV}O_{III}$	0.06746
184.6	3	35 Br		$M_I M_{II}$	0.0672
188.4	1	28 Ni	M_{III}	Abs. Edge	0.06581
188.6	1	55 Cs		$N_{IV}O_{II}$	0.06574
189.5	3	35 Br		$M_{IV}N_{III}$	0.0654
190.3	1	55 Cs		$N_V O_{III}$	0.06515
190.	2	28 Ni		$M_{II,III}M_{IV,V}$	0.0651
191.1	2	35 Br	$M\zeta_2$	$M_{IV}N_{III}$	0.06488
192.6	2	35 Br	$M\zeta_1$	$M_V N_{III}$	0.06437
197.3	1	12 Mg	L_I	Abs. Edge	0.06284
202.	5	27 Co	$M_{II,III}$	Abs. Edge	0.061
203.	1	16 S		$L_I L_{II,III}$	0.061
214.	6	27 Co		$M_{II,III}M_{IV,V}$	0.058
224.	1	53 I	$N_{IV,V}$	Abs. Edge	0.0552
226.5	1	3 Li	K	Abs. Edge	0.05475
227.8	1	34 Se	M_V	Abs. Edge	0.05443
228.	1	3 Li	$K\alpha$	KL	0.0543
230.	2	34 Se		$M_V N_{III}$	0.0538
230.	1	26 Fe	$M_{II,III}$	Abs. Edge	0.0538
243.	5	26 Fe		$M_{II,III}M_{IV,V}$	0.051
249.3	1	12 Mg	L_{II}	Abs. Edge	0.04973
250.7	1	12 Mg	L_{III}	Abs. Edge	0.04945
251.5	5	12 Mg		$L_{II,III}M$	0.04929
273.	6	25 Mn		$M_{II,III}M_{IV,V}$	0.045
290.	1	13 Al		$L_I L_{II,III}$	0.0428
309.	9	24 Cr		$M_{II,III}M_{IV,V}$	0.040
317.	1	12 Mg		$L_I L_{II,III}$	0.0392
337.	9	23 V		$M_{II,III}M_{IV,V}$	0.0368
376.	1	11 Na		$L_I L_{II,III}$	0.03299
399.	5	35 Br	N_I	Abs. Edge	0.0311
405.	5	11 Na	$L_{II,III}$	Abs. Edge	0.0306
407.1	5	11 Na		$L_{II,III}M$	0.03045
417.	5	17 Cl	M_I	Abs. Edge	0.0297
444.	5	53 I	O_I	Abs. Edge	0.0279
525.	9	20 Ca		$M_{II,III}N_I$	0.0236
692.	9	19 K		$M_{II,III}N_I$	0.0179

X-RAY ATOMIC ENERGY LEVELS

J. A. Bearden and A. F. Burr

These tables were originally published as the final report to the U.S. Atomic Energy Commission as Report NYO-2543-1 in partial fulfillment of Contract AT(30-1)-2543. The tables were later reproduced in *Review of Modern Physics*. The data may also be obtained from the Superintendent of Documents, U.S. Government Printing Office, Washington, D. C. 20402 in the publication NSRDS-NBS 14. Persons seeking discussion of the details of calculations, sources of energy level information and the problem of properly interpreting the experimental measurements should refer to the original publication or to *Review of Modern Physics*, Vol. 39, 125–142, January 1967.

All of the x-ray emission wavelengths have recently been reevaluated and placed on a consistent Å* scale. For most elements these data give a highly overdetermined set of equations for energy level differences, which have been solved by least squares adjustment for each case. This procedure makes "best" use of all x-ray wavelength data, and also permits calculation of the probable error for each energy difference. Photoelectron measurements of absolute energy levels are more precise than x-ray absorption edge data. These have been used to establish the absolute scale for eighty-one elements and, in many cases, to provide additional energy level difference data. The x-ray absorption wavelengths were used for eight elements and ionization measurements for two; the remaining five were interpolated by a Moseley diagram involving the output values of energy levels from adjacent elements. Probable errors are listed on an absolute energy basis. In the original source of the present data, a table of energy levels in Rydberg units is given. Difference tables in volts, Rydbergs, and milli-Å* wavelength units, with the respective probable errors, are also included there.

Recommended values of the atomic energy levels, and probable errors in eV. Where available, photoelectron direct measurements are listed in brackets [ ] immediately under the recommended values. The measured values of the x-ray absorption energies are shown in parentheses ( ). Interpolated values are enclosed in angle brackets ⟨ ⟩.

Level	1 H	2 He	3 Li	4 Be	5 B	6 C	7 N	8 O
K	13.59811[a]	24.58678[b]	54.75±0.02 (54.75)	111.0±1.0 (111.0)	188.0±0.4 [188.0][e]	283.8±0.4 [283.8][e]	401.6±0.4 [401.6][e]	532.0±0.4 [532.0][e]
L_I								23.7±0.4 [23.7][d]
$L_{II,III}$					4.7±0.9	6.4±1.9	9.2±0.6	7.1±0.9

Level	9 F	10 Ne	11 Na	12 Mg	13 Al	14 Si	15 P	16 S
K	685.4±0.4 [685.4][e]	866.9±0.3 (866.9)	1072.1±0.4 [1072.1][e] (1072.)	1305.0±0.4 [1305.0][e] (1303.)	1559.6±0.4 [1559.6][e] (1559.8)	1838.9±0.4 [1838.9][e]	2145.5±0.4 [2145.5][d]	2472.0±0.4 [2472.0][d] (2470.)
L_I	⟨31.⟩	⟨45.⟩	63.3±0.4 [63.3][d]	89.4±0.4 [89.4][d] (63.)	117.7±0.4 [117.7][d] (87.)	148.7±0.4 [148.7][d]	189.3±0.4 [189.3][d]	229.2±0.4 [229.2][d]
$L_{II,III}$	8.6±0.8	18.3±0.4	31.1±0.4 (31.)	51.4±0.5 (50.)	73.1±0.5 (72.8)	99.2±0.5 (100.6)	132.2±0.5 (132.)	164.8±0.7

Level	17 Cl	18 Ar	19 K	20 Ca	21 Sc	22 Ti	23 V	24 Cr
K	2822.4±0.3 [2822.4][e] (2020.)	3202.9±0.3 (3202.9)	3607.4±0.4 [3607.4][e] (3607.8)	4038.1±0.4 [4038.1][e] (4038.1)	4492.8±0.4 [4492.8][e]	4966.4±0.4 [4966.4][d] (4964.5)	5465.1±0.3 [5465.1][e] (5464.)	5989.2±0.3 [5989.2][e] (5989.)
L_I	270.2±0.4 [270.2][d]	320. (320.)[d]	377.1±0.4 [377.1][d]	437.8±0.4 [437.8][d]	500.4±0.4 [500.4][d]	563.7±0.4 [563.7][d]	628.2±0.4 [628.2][d]	694.6±0.4 [694.6][d]
L_{II}	201.6±0.3	247.3±0.3	296.3±0.4	350.0±0.4	406.7±0.4	461.5±0.4	520.5±0.3	583.7±0.3
L_{III}	200.0±0.3	245.2±0.3	293.6±0.4	346.4±0.4	402.2±0.4	455.5±0.4	512.9±0.3	574.5±0.3
M_I	17.5±0.4	25.3±0.4	33.9±0.4	43.7±0.4	53.8±0.4	60.3±0.4	66.5±0.4	74.1±0.4
$M_{II,III}$	6.8±0.4	12.4±0.3	17.8±0.4	25.4±0.4	32.3±0.5	34.6±0.4	37.8±0.3	42.5±0.3
$M_{IV,V}$					6.6±0.5	3.7	2.2±0.3	2.3±0.4

Level	25 Mn	26 Fe	27 Co	28 Ni	29 Cu	30 Zn	31 Ga	32 Ge
K	6539.0±0.4 [6539.0][e] (6538.)	7112.0±0.9 [7111.3][e,f] (7111.2)	7708.9±0.3 [7708.9][e] (7709.5)	8332.8±0.4 [8332.8][e] (8331.6)	8978.9±0.4 [8978.9][e,g] (8980.3)	9658.6±0.6 [9658.6][e] (9660.7)	10367.1±0.5 [10367.1][e] (10368.2)	11103.1±0.7 [11103.8][e] (11103.6)
L_I	769.0±0.4 [769.0][d]	846.1±0.4 [846.1][d]	925.6±0.4 [925.6][d]	1008.1±0.4 [1008.1][d]	1096.1±0.4 [1096.0][d]	1193.6±0.9	1297.7±1.1	1414.3±0.7 [1413.6][e]
L_{II}	651.4±0.4	721.1±0.9 (720.8)	793.6±0.3 (793.8)	871.9±0.4 (870.6)	951.0±0.4 [950.0][h] (953.)	1042.8±0.6 (1045.)	1142.3±0.5	1247.8±0.7 (1249.)
L_{III}	640.3±0.4	708.1±0.9 (707.4)	778.6±0.3 (779.0)	854.7±0.4 (853.6)	931.1±0.4 [931.4][h] (933.)	1019.7±0.6 (1022.)	1115.4±0.5 (1117.)	1216.7±0.7 (1217.0)

	25 Mn	26 Fe	27 Co	28 Ni	29 Cu	30 Zn	31 Ga	32 Ge
M_I	83.9±0.5	92.9±0.9	100.7±0.4	111.8±0.6	119.8±0.6	135.9±1.1	158.1±0.5	180.0±0.8
M_{II}	48.6±0.4	54.0±0.9	59.5±0.3	68.1±0.4	73.6±0.4	86.6±0.6	106.8±0.7	127.9±0.9
M_{III}		(54.)	(61.)	(66.)	(75.)	(86.)	102.9±0.5	120.8±0.7
$M_{IV,V}$	3.3±0.5	3.6±0.9	2.9±0.3	3.6±0.4	1.6±0.4	8.1±0.6	17.4±0.5	28.7±0.7

	33 As	34 Se	35 Br	36 Kr	37 Rb	38 Sr	39 Y	40 Zr
K	11866.7±0.7 [11866.7]i (11865.)	12657.8±0.7 [12657.8]= (12654.5)	13473.7±0.4 (13470.)	14325.6±0.8 (14324.4)	15199.7±0.3 (15202.)	16104.6±0.3 (16107.)	17038.4±0.3 (17038.)	17997.6±0.4 (17999.)
L_I	1526.5±0.8 (1529.)	1653.9±3.5 (1652.5)	1782.0±0.4 [1782.0]j	1921.0±0.6 [1921.2]k	2065.1±0.3 [2065.4]j	2216.3±0.3 [2216.2]l	2372.5±0.3 [2372.7]l	2531.6±0.3 [2531.6]l
L_{II}	1358.6±0.7 (1358.7)	1476.2±0.7 (1474.7)	1596.0±0.4 [1596.2]j	1727.2±0.5 [1727.2]k (1730.)	1863.9±0.3 [1863.4]j	2006.8±0.3 [2006.6]l (2008.5)	2155.5±0.3 [2155.0]l (2154.0)	2306.7±0.3 [2306.5]l (2305.3)
L_{III}	1323.1±0.7 (1323.5)	1435.8±0.7 (1434.0)	1549.9±0.5 [1549.7]j	1674.9±0.5 [1674.8]k (1677.)	1804.4±0.3 [1804.6]j	1939.6±0.3 [1939.9]l (1941.)	2080.0±0.3 [2080.2]l (2079.4)	2222.3±0.3 [2222.5]l (2222.5)
M_I	203.5±0.7	231.5±0.7	256.5±0.4		322.1±0.3	357.5±0.3	393.6±0.3	430.3±0.3
M_{II}	146.4±1.2	168.2±1.3	189.3±0.4	222.7±1.1	247.4±0.3	279.8±0.3	312.4±0.4	344.2±0.4
M_{III}	140.5±0.8	161.9±1.0	181.5±0.4	213.8±1.1	238.5±0.3	269.1±0.3	300.3±0.4	330.5±0.4
M_{IV}	41.2±0.7	56.7±0.8	70.1±0.4	88.9±0.8	111.8±0.3	135.0±0.3	159.6±0.3	182.4±0.3
M_V			69.0±0.4		110.3±0.3	133.1±0.3	157.4±0.3	180.0±0.3
N_I			27.3±0.5	24.0±0.8	29.3±0.3	37.7±0.3	45.4±0.3	51.3±0.3
N_{II}	2.5±1.0	5.6±1.3	5.2±0.4	10.6±1.9	14.8±0.4	19.9±0.3	25.6±0.4	28.7±0.4
N_{III}			4.6±0.4		14.0±0.3			

	41 Nb	42 Mo	43 Tc	44 Ru	45 Rh	46 Pd	47 Ag	48 Cd
K	18985.6±0.4 (18987.)	19999.5±0.3 (20004.)	21044.0±0.7	22117.2±0.3 (22119.)	23219.9±0.3 (23219.8)	24350.3±0.3 (24348.)	25514.0±0.3 (25516.)	26711.2±0.3 (26716.)
L_I	2697.7±0.3 [2697.7]l	2865.5±0.3 [2866.0]l	3042.5±0.4 [3042.5]l	3224.0±0.3 [3224.3]l	3411.9±0.3 [3412.0]l (3417.)	3604.3±0.3 [3604.6]l (3607.)	3805.8±0.3 [3806.2]m (3807.)	4018.0±0.3 [4018.1]m (4019.)
L_{II}	2464.7±0.3 [2464.7]l	2625.1±0.3 [2624.5]l (2627.)	2793.2±0.4 [2973.2]l	2966.9±0.3 [2966.8]l (2966.3)	3146.1±0.3 [3146.3]l (3145.)	3330.3±0.3 [3330.3]l (3330.3)	3523.7±0.3 [3523.6]=,m (3526.)	3727.0±0.3 [3727.1]m (3728.)
L_{III}	2370.5±0.3 [2370.6]l	2520.2±0.3 [2520.2]l (2523.2)	2676.9±0.3 [2676.9]l	2837.9±0.3 [2837.7]l (2837.7)	3003.8±0.3 [3003.5]=,l (3002.)	3173.3±0.3 [3173.0]=,l (3173.0)	3351.1±0.3 [3350.8]= (3351.0)	3537.5±0.3 [3537.3]= (3537.6)
M_I	468.4±0.3	504.6±0.3		585.0±0.3	627.1±0.3	669.9±0.3	717.5±0.3	770.2±0.3
M_{II}	378.4±0.4	409.7±0.4	444.9±1.5	482.8±0.3	521.0±0.3	559.1±0.3	602.4±0.3	650.7±0.3
M_{III}	363.0±0.4	392.3±0.3	425.0±1.5	460.6±0.3	496.2±0.3	531.5±0.3	571.4±0.3	616.5±0.3
M_{IV}	207.4±0.3	230.3±0.3	256.4±0.5	283.6±0.3	311.7±0.3	340.0±0.3	372.8±0.3	410.5±0.3
M_V	204.6±0.3	227.0±0.3	252.9±0.4	279.4±0.3	307.0±0.3	334.7±0.3	366.7±0.3	403.7±0.3
N_I	58.1±0.3	61.8±0.3		74.9±0.3	81.0±0.3	86.4±0.3	95.2±0.3	107.6±0.3
N_{II}	33.9±0.4	34.8±0.4	38.9±1.9	43.1±0.4	47.9±0.4	51.1±0.4	62.6±0.3	66.9±0.4
N_{III}							55.9±0.3	
$N_{IV,V}$	3.2±0.3	1.8±0.3		2.0±0.3	2.5±0.4	1.5±0.3	3.3±0.3	9.3±0.3

	49 In	50 Sn	51 Sb	52 Te	53 I	54 Xe	55 Cs	56 Ba
K	27939.9±0.3	29200.1±0.4 (29195.)	30491.2±0.3 (30486.)	31813.8±0.3 (31811.)	33169.4±0.4 (33167.)	34561.4±1.1 (34590.)	35984.6±0.4 (35987.)	37440.6±0.4 (37452.)
L_I	4237.5±0.3 [4237.7]m (4237.3)	4464.7±0.3 [4464.5]= (4464.8)	4698.3±0.3 [4698.3]m (4698.4)	4939.2±0.3 [4939.3]m (4939.7)	5188.1±0.3 [5188.1]j	5452.8±0.4 (5452.8)	5714.3±0.4 [5712.7]j (5721.)	5988.8±0.4 [5986.8]j (5996.)
L_{II}	3938.0±0.3 [3937.8]m (3939.3)	4156.1±0.3 [4156.2]= (4157.)	4380.4±0.3 [4380.6]m (4382.)	4612.0±0.3 [4612.0]m (4612.6)	4852.1±0.3 [4852.0]j	5103.7±0.4 (5103.7)	5359.4±0.3 [5359.5]j (5358.)	5623.6±0.3 [5623.6]j (5623.3)
L_{III}	3730.1±0.3 [3730.0]= (3730.2)	3928.8±0.3 [3928.8]= (3928.8)	4132.2±0.3 [4132.2]= (4132.3)	4341.4±0.3 [4341.2]= (4341.8)	4557.1±0.3 [4557.1]j	4782.2±0.4 (4782.2)	5011.9±0.3 [5012.0]j (5011.3)	5247.0±0.3 [5247.3]j (5247.0)
M_I	825.6±0.3	883.8±0.3	943.7±0.3	1006.0±0.3	1072.1±0.3		1217.1±0.4	1292.8±0.4
M_{II}	702.2±0.3	756.4±0.4	811.9±0.3	869.7±0.3	930.5±0.3	999.0±2.1	1065.0±0.5	1136.7±0.5

	49 In	50 Sn	51 Sb	52 Te	53 I	54 Xe	55 Cs	56 Ba
M_{III}	664.3±0.3	714.4±0.3	765.6±0.3	818.7±0.3	874.6±0.3	937.0±2.1	997.6±0.5	1062.2±0.5
M_{IV}	450.8±0.3	493.3±0.3	536.9±0.3	582.5±0.3	631.3±0.3		739.5±0.4	796.1±0.3
M_{V}	443.1±0.3	484.8±0.3	527.5±0.3	572.1±0.3	619.4±0.3	672.3±0.5	725.5±0.5	780.7±0.3
N_{I}	121.9±0.3	136.5±0.4	152.0±0.3	168.3±0.3	186.4±0.3		230.8±0.4	253.0±0.5
N_{II}	77.4±0.4	88.6±0.4	98.4±0.5	110.2±0.5	122.7±0.5	146.7±3.1	172.3±0.6	191.8±0.7
N_{III}							161.6±0.6	179.7±0.6
N_{IV}	16.2±0.3	23.9±0.3	31.4±0.3	39.8±0.3	49.6±0.3		78.8±0.5	92.5±0.5
N_{V}							76.5±0.5	89.9±0.5
O_{I}	0.1±4.5	0.9±0.5	6.7±0.5	11.6±0.6	13.6±0.6		22.7±0.5	39.1±0.6
O_{II}	0.8±0.4	1.1±0.5	2.1±0.4	2.3±0.5	3.3±0.5		13.1±0.5	16.6±0.5
O_{III}							11.4±0.5	14.6±0.5

	57 La	58 Ce	59 Pr	60 Nd	61 Pm	62 Sm	63 Eu	64 Gd
K	38924.6±0.4 (38934.)	40443.0±0.4 (40453.)	41990.6±0.5 (42002.)	43568.9±0.4 (43574.)	45184.0±0.7 (45198.)	46834.2±0.5 (46849.)	48519.0±0.4 (48519.)	50239.1±0.5 (50233.)
L_{I}	6266.3±0.5 [6266.3][a]	6548.8±0.5 [6548.5][a]	6834.8±0.5 [6834.9][a]	7126.0±0.4 [7125.8][a] (7129.)	7427.9±0.8 [7427.9][b] (7748.)	7736.8±0.5 [7736.2][a] (7748.)	8052.0±0.4 [8051.7][a] (8061.)	8375.6±0.5 [8375.4][a] (8386.)
L_{II}	5890.6±0.4 [5890.7][a]	6164.2±0.4 [6164.3][a]	6440.4±0.5 [6440.2][a]	6721.5±0.4 [6721.8][a] (6723.)	7012.8±0.6 [7012.8][b] (7313.)	7311.8±0.4 [7312.0][a] (7313.)	7617.1±0.4 [7617.6][a] (7620.)	7930.3±0.4 [7930.5][a] (7931.)
L_{III}	5482.7±0.4 [5482.6][a]	5723.4±0.4 [5723.6][a]	5964.3±0.4 [5964.3][a]	6207.9±0.4 [6208.][a] (6209.)	6459.3±0.6 [6459.4][b] (6717.)	6716.2±0.5 [6716.8][a] (6717.)	6976.9±0.4 [6976.7][a] (6981.)	7242.8±0.4 [7242.8][a] (7243.)
M_{I}	1361.3±0.3	1434.6±0.6	1511.0±0.8	1575.3±0.7		1722.8±0.8	1800.0±0.5	1880.8±0.5
M_{II}	1204.4±0.6	1272.8±0.6	1337.4±0.7	1402.8±0.6	1471.4±6.?	1540.7±1.3	1613.0±0.8	1688.0±0.?
M_{III}	1123.4±0.5	1185.4±0.5	1242.2±0.6	1297.4±0.5	1356.9±1.4	1419.8±1.1	1480.6±0.6	1544.0±0.8
M_{IV}	848.5±0.4	901.3±0.6	951.1±0.6	999.9±0.6	1051.5±0.9	1106.0±0.8	1160.6±0.6	1217.2±0.6
M_{V}	831.7±0.4	883.3±0.5	931.0±0.6	977.7±0.6	1026.9±1.0	1080.2±0.6	1130.9±0.6	1185.2±0.6
N_{I}	270.4±0.8	289.6±0.7	304.5±0.9	315.2±0.8		345.7±0.9	360.2±0.7	375.8±0.7
N_{II}	205.8±1.2	223.3±1.1	236.3±1.5	243.3±1.6	242.±16.	265.6±1.9	283.9±1.0	288.5±1.2
N_{III}	191.4±0.9	207.2±0.9	217.6±1.1	224.6±1.3		247.4±1.5	256.6±0.8	270.9±0.9
$N_{IV,V}$	98.9±0.8	110.0±0.6	113.2±0.7	117.5±0.7	120.4±2.0	129.0±1.2	133.2±0.6	140.5±0.8
$N_{VI,VII}$		0.1±1.2	2.0±0.6	1.5±0.9		5.5±1.1	0.0±3.2	0.1±3.5
O_{I}	32.3±7.2	37.8±1.3	37.4±1.0	37.5±0.8		37.4±1.5	31.8±0.7	36.1±0.8
$O_{II,III}$	14.4±1.2	19.8±1.2	22.3±0.7	21.1±0.8		21.3±1.5	22.0±0.6	20.3±1.2

	65 Tb	66 Dy	67 Ho	68 Er	69 Tm	70 Yb	71 Lu	72 Hf
K	51995.7±0.5 (52002.)	53788.5±0.5 (53793.)	55617.7±0.5 (55619.)	57485.5±0.5 (57487.)	59389.6±0.5 (61300.)	61332.3±0.5 (61300.)	63313.8±0.5 (63310.)	65350.8±0.6 (65310.)
L_{I}	8708.0±0.5 [8707.6][a] (8717.)	9045.8±0.5 [9046.5][a]	9394.2±0.4 [9394.3][a] (9399.)	9751.3±0.4 [9751.5][a] (9757.)	10115.7±0.4 [10115.6][a] (10121.)	10486.4±0.4 [10487.3][a] (10490.)	10870.4±0.4 [10870.1][a] (10874.)	11270.7±0.4 [11271.6][b] (11274.)
L_{II}	8251.6±0.4 [8251.8][a] (8253.)	8580.6±0.4 [8580.4][a] (8583.)	8917.8±0.4 [8918.2][a] (8916.)	9264.3±0.4 [9264.3][a] (9262.)	9616.9±0.4 [9617.1][a] (9617.1)	9978.2±0.4 [9977.9][a] (9976.)	10348.6±0.4 [10349.0][a] (10345.)	10739.4±0.4 [10738.9][b] (10736.)
L_{III}	7514.0±0.4 [7514.2][a] (7515.)	7790.1±0.4 [7789.6][a] (7789.7)	8071.1±0.4 [8070.6][a] (8068.)	8357.9±0.4 [8357.6][a] (8357.5)	8648.0±0.4 [8647.8][a] (8649.6)	8943.6±0.4 [8942.6][a] (8944.1)	9244.1±0.4 [9243.8][a]	9560.7±0.4 [9560.4][b] (9558.)
M_{I}	1967.5±0.6	2046.8±0.4	2128.3±0.6	2206.5±0.6	2306.8±0.7	2398.1±0.4	2491.2±0.5	2600.9±0.4
M_{II}	1767.7±0.9	1841.8±0.5	1922.8±1.0	2005.8±0.6	2089.8±1.1	2173.0±0.4	2263.5±0.4	2365.4±0.4
M_{III}	1611.3±0.8	1675.6±0.9	1741.2±0.9	1811.8±0.6	1884.5±1.1	1949.8±0.5	2023.6±0.5	2107.6±0.4
M_{IV}	1275.0±0.6	1332.5±0.4	1391.5±0.7	1453.3±0.5	1514.6±0.7	1576.3±0.4	1639.4±0.4	1716.4±0.4
M_{V}	1241.2±0.7	1294.9±0.4	1351.4±0.8	1409.3±0.6	1467.7±0.9	1527.8±0.4	1588.5±0.4	1661.7±0.4
N_{I}	397.9±0.8	416.3±0.5	435.7±0.8	449.1±1.0	471.7±0.9	487.2±0.6	506.2±0.6	538.1±0.4
N_{II}	310.2±1.2	331.8±0.6	343.5±1.4	366.2±1.5	385.9±1.6	396.7±0.7	410.1±1.8	437.0±0.5
N_{III}	385.0±1.0	292.9±0.6	306.6±0.9	320.0±0.7	336.6±1.6	343.5±0.5	359.3±0.5	380.4±0.5
N_{IV}	147.0±0.8	154.2±0.5	161.0±1.0	176.7±1.2	179.6±1.2	198.1±0.5	204.8±0.5	223.8±0.4
N_{V}				167.6±1.5		184.9±1.3	195.0±0.4	213.7±0.5

	65 Tb	66 Dy	67 Ho	68 Er	69 Tm	70 Yb	71 Lu	72 Hf
$N_{VI,VII}$	2.6±1.5	4.2±1.6	3.7±3.0	4.3±1.4	5.3±1.9	6.3±1.0	6.9±0.5	17.1±0.5
O_I	39.0±0.8	62.9±0.5	51.2±1.3	59.8±1.7	53.2±3.0	54.1±0.5	56.8±0.5	64.9±0.4
O_{II}	25.4±0.8	26.3±0.6	20.3±1.5	29.4±1.6	32.3±1.6	23.4±0.6	28.0±0.6	38.1±0.6
O_{III}								30.6±0.6

	73 Ta	74 W	75 Re	76 Os	77 Ir	78 Pt	79 Au	80 Hg
K	67416.4±0.6 (67403.)	69525.0±0.3 (69508.)	71676.4±0.4 (71658.)	73870.8±0.5	76111.0±0.5	78394.8±0.7 (78381.)	80724.9±0.5 (80720.)	83102.3±0.8
L_I	11681.5±0.3 [11680.2]p (11682.)	12099.8±0.3 [12098.2]p (12099.6)	12526.7±0.4 (12530.)	12968.0±0.4 (12972.)	13418.5±0.3 (13423.)	13879.9±0.4 (13883.)	14352.8±0.4 (14353.7)	14839.3±1.0 (14842.)
L_{II}	11136.1±0.3 [11136.1]p (11132.)	11544.0±0.3 [11541.4]p (11538.)	11958.7±0.3 [11956.9]p (11954.)	12385.0±0.4 (12381.)	12824.1±0.3 [12824.0]e,p (12820.)	13272.6±0.3 [13272.6]e,p (13272.3)	13733.6±0.3 [13733.5]e,p (13736.)	14208.7±0.7 (14215.)
L_{III}	9881.1±0.3 [9880.3]p (9877.7)	10206.8±0.3 [10204.2]p (10200.)	10535.3±0.3 [10534.2]p (10531.)	10870.9±0.3 [10870.7]p (10868.)	11215.2±0.3 [11215.1]e,p (11212.)	11563.7±0.3 [11563.7]e,p (11562.)	11918.7±0.3 [11918.2]e,p (11921.)	12283.9±0.4 [12284.0]e,p (12286.)
M_I	2708.0±0.4	2819.6±0.4	2931.7±0.4	3048.5±0.4	3173.7±1.7	3296.0±0.9	3424.9±0.3 [3424.8]p	3561.6±1.1
M_{II}	2468.7±0.3 [2468.6]p	2574.9±0.3 [2575.0]p	2681.6±0.4	2792.2±0.3 [2791.9]p	2908.7±0.3 [2909.1]p	3026.5±0.4 [3026.5]p (3029.)	3147.8±0.4 [3149.5]p	3278.5±1.3
M_{III}	2194.0±0.3 [2194.1]p	2281.0±0.3 [2281.0]p	2367.3±0.3 [2367.3]p	2457.2±0.4 [2457.4]p	2550.7±0.3 [2550.5]p (2550.5)	2645.4±0.4 [2645.5]p (2645.9)	2743.0±0.3 [2743.1]p (2744.0)	2847.1±0.4 [2847.1]p
M_{IV}	1793.2±0.3 [1793.1]p	1871.6±0.3 [1871.4]p	1948.9±0.3 [1948.9]p	2030.8±0.3 [2031.0]p	2116.1±0.3 [2116.1]p	2201.9±0.3 [2201.9]p	2291.1±0.3 [2291.2]p (2307.)	2384.9±0.3 [2384.9]p
M_V	1735.1±0.3 [1735.2]p	1809.2±0.3 [1809.3]p	1882.9±0.3 [1882.9]p	1960.1±0.3 [1960.2]p	2040.4±0.3 [2040.5]p	2121.6±0.3 [2121.6]p	2205.7±0.3 [2206.1]p (2220.)	2294.9±0.3 [2294.9]p
N_I	565.5±0.5	595.0±0.4	625.0±0.4	654.3±0.5	690.1±0.4	722.0±0.6	758.8±0.4	800.3±1.0
N_{II}	464.8±0.5	491.6±0.4	517.9±0.5	546.5±0.5	577.1±0.4	609.2±0.6	643.7±0.5	676.9±2.4
N_{III}	404.5±0.4	425.3±0.5	444.4±0.5	468.2±0.6	494.3±0.6	519.0±0.6	545.4±0.5	571.0±1.4
N_{IV}	241.3±0.4	258.8±0.4	273.7±0.5	289.4±0.5	311.4±0.4	330.8±0.5	352.0±0.4	378.3±1.0
N_V	229.3±0.3	245.4±0.4	260.2±0.4	272.8±0.6	294.9±0.4	313.3±0.4	333.9±0.4	359.8±1.2
N_{VI}	25.0±0.4	36.5±0.4	40.6±0.4	46.3±0.6	63.4±0.4	74.3±0.4	86.4±0.4	102.2±0.5
N_{VII}		33.6±0.4			60.5±0.4	71.1±0.5	82.8±0.5	98.5±0.5
O_I	71.1±0.5	77.1±0.4	82.8±0.5	83.7±0.6	95.2±0.4	101.7±0.4	107.8±0.7	120.3±1.3
O_{II}	44.9±0.4	46.8±0.5	45.6±0.7	58.0±1.1	63.0±0.6	65.3±0.7	71.7±0.7	80.5±1.3
O_{III}	36.4±0.4	35.6±0.5	34.6±0.6	45.4±1.0	50.5±0.6	51.7±0.7	53.7±0.7	57.6±1.3
$O_{IV,V}$	5.7±0.4	6.1±0.4	3.5±0.5		3.8±0.4	2.2±1.3	2.5±0.5	6.4±1.4

	81 Tl	82 Pb	83 Bi	84 Po	85 At	86 Rn	87 Fr	88 Ra
K	85530.4±0.6	88004.5±0.7 (88005.)	90525.9±0.7 (90534.)	93105.0±3.8	95729.9±7.7	98404.±12.	101137.±13.	103921.9±7.2
L_I	15346.7±0.4 (15343.)	15860.8±0.5 (15855.)	16387.5±0.4 (16376.)	16939.3±9.8	17493.±29.	18049.±38.	18639.±40.	19236.7±1.5 (19236.0)
L_{II}	14697.9±0.3 [14697.3]p (14699.)	15200.0±0.4 (15205.)	15711.1±0.3 [15708.4]p (15719.)	16244.3±2.4	16784.7±2.5	17337.1±3.4	17906.5±3.5	18484.3±1.5 (18486.0)
L_{III}	12657.5±0.3 [12656.3]e,p (12660.)	13035.2±0.3 [13034.9]e,p (13041.)	13418.6±0.3 [13418.3]e,p (13426.)	13813.8±1.0 ⟨13813.8⟩	14213.5±2.0 (14213.5)	14619.4±3.0 (14619.4)	15031.2±3.0 (15031.2)	15444.4±1.5 (15444.0)
M_I	3704.1±0.4	3850.7±0.5	3999.1±0.3 [3999.1]p	4149.4±3.9	⟨4317.⟩	⟨4482.⟩	⟨4652.⟩	4822.0±1.5
M_{II}	3415.7±0.3 [3415.7]p	3554.2±0.3 [3554.2]p	3696.3±0.3 [3696.4]p	3854.1±9.8	4008.±28.	4159.±38.	4327.±40.	4489.5±1.8
M_{III}	2956.6±0.3 [2956.5]p	3066.4±0.4 [3066.3]p	3176.9±0.3 [3176.8]p	3301.9±9.9	3426.±29.	3538.±38.	3663.±40.	3791.8±1.7
M_{IV}	2485.1±0.3 [2485.2]p	2585.6±0.3 [2585.5]p (2606.)	2687.6±0.3 [2687.4]p	2798.0±1.2	2908.7±2.1	3021.5±3.1	3136.2±3.1	3248.4±1.6

	81 Tl	82 Pb	83 Bi	84 Po	85 At	86 Rn	87 Fr	88 Ra
M_V	2389.3±0.3 [2389.4]^p	2484.0±0.3 [2484.2]^p (2502.)	2579.6±0.3 [2579.5]^p	2683.0±1.1	2786.7±2.1	2892.4±3.1	2999.9±3.1	3104.9±1.6
N_I	845.5±0.5	893.6±0.7	938.2±0.3 [938.7]^p	995.3±2.9	(1042.)	(1097.)	⟨1153.⟩	1208.4±1.6
N_{II}	721.3±0.8	763.9±0.8	805.3±0.3 [805.3]^p	851.±12.	886.±30.	929.±40.	980±42.	1057.6±1.8
N_{III}	609.0±0.5	644.5±0.6	678.9±0.3 [678.9]^p	705.±14.	740.±30.	768.±40.	810±43.	879.1±1.8
N_{IV}	406.6±0.4	435.2±0.5	463.6±0.3 [463.6]^p	500.?±2.4	533.2±3.2	566.6±4.0	603.3±4.1	635.9±1.6
N_V	386.2±0.5	412.9±0.6	440.0±0.3 [440.1]^p	473.4±1.3			577.±34.	602.7±1.7
N_{VI}	122.8±0.4	142.9±0.4	161.9±0.5 }					298.9±2.4
N_{VII}	118.5±0.4	138.1±0.4	157.4±0.6 }					
O_I	136.3±0.7	147.3±0.8	159.3±0.7					254.4±2.1
O_{II}	99.6±0.6	104.8±1.0	116.8±0.7					200.4±2.0
O_{III}	75.4±0.6	86.0±1.0	92.8±0.6					152.8±2.0
O_{IV}	15.3±0.4	21.8±0.4	26.5±0.5 }	31.4±3.2				67.2±1.7
O_V	13.1±0.4	19.2±0.4	24.4±0.6 }					
P_I		3.1±1.0						43.5±2.2
$P_{II,III}$		0.7±1.0	2.7±0.7					18.8±1.8

	89 Ac	90 Th	91 Pa	92 U	93 Np	94 Pu	95 Am	96 Cm
K	106755.3±5.3	109650.9±0.9	112601.4±3.4	115605.1±1.0	118678.±33.	121818.±44.	125027.±55.	128220
L_I	19840.±18.	20472.1±0.5 (20464.)	21104.6±1.8 (21128.)	21757.4±0.3 (21771.)	22426.8±0.9	23097.2±1.6 (23100.)	23772.9±2.0 (23772.9)	24460
L_{II}	19083.2±2.8	19693.2±0.4 (19683.)	20313.7±1.5 (20319.)	20947.6±0.3 (20945.)	21600.5±0.4	22266.2±0.7 (22253.)	22944.0±1.0	23779
L_{III}	15871.0±2.0 ⟨15871.0⟩	16300.3±0.3 [16299.6]^a (16299.)	16733.1±1.4 (16733.)	17166.3±0.3 [17168.5]^r (17165.)	17610.0±0.4 (17606.2)	18056.8±0.6 (18053.1)	18504.1±0.9 (18504.1)	18930
M_I	⟨5002.⟩	5182.3±0.3 [5182.3]^a	5366.9±1.6	5548.0±0.4	5723.2±3.6	5932.9±1.4	6120.5±7.5	6288
M_{II}	4656.±18.	4830.4±0.4 [4830.6]^a	5000.9±2.3	5182.2±0.4 [5180.9]^r	5366.2±0.7 [5366.4]^e	5541.2±1.7	5710.2±2.1	5895
M_{III}	3909.±18.	4046.1±0.4 [4046.1]^a (4041.)	4173.8±1.8	4303.4±0.3 [4303.6]^r (4299.)	4434.7±0.5 [4434.6]^e	4556.6±1.5	4667.0±2.1	4797
M_{IV}	3370.2±2.1	3490.8±0.3 [3490.7]^a (3485.)	3611.2±1.4 (3608.)	3727.6±0.3 [3728.1]^r (3720.)	3850.3±0.4 [3849.8]^e	3972.6±0.6 [3972.7]^t	4092.1±1.0	4227
M_V	3219.0±2.1	3332.0±0.3 [3332.1]^a (3325.)	3441.8±1.4 (3436.)	3551.7±0.3 [3551.7]^r (3545.)	3665.8±0.4 [3664.2]^e	3778.1±0.6 [3778.0]^t	3886.9±1.0	3971
N_I	⟨1269.⟩	1329.5±0.4 [1329.8]^a	1387.1±1.9	1440.8±0.4 [1441.3]^r	1500.7±0.8 [1500.7]^e	1558.6±0.8	1617.1±1.1	1643
N_{II}	1080.±19.	1168.2±0.4 [1168.3]^a	1224.3±1.6	1272.6±0.4 [1272.5]^r	1327.7±0.8 [1327.7]^e	1372.1±1.8	1411.8±8.3	1440
N_{III}	890.±19.	967.3±0.4 [967.6]^a	1006.7±1.7	1044.9±0.3 [1044.9]^r	1086.8±0.7 [1086.8]^e	1114.8±1.6	⟨1135.7⟩	1154
N_{IV}	674.9±3.7	714.1±0.4 [714.4]^a	743.4±2.1	780.4±0.3 [779.7]^r	815.9±0.5 [817.1]^e	848.9±0.6 [848.9]^t	878.7±1.0	
N_V		676.4±0.4 [676.6]^a	708.2±1.8	737.7±0.3 [737.6]^r	770.3±0.4 [773.2]^e	801.4±0.6 [801.4]^t	827.6±1.0	
N_{VI}		344.4±0.3 [344.2]^a	371.2±1.6	391.3±0.6	415.0±0.8 [415.0]^e	445.8±1.7		
N_{VII}		335.2±0.4 [335.0]^a	359.5±1.6	380.9±0.9	404.4±0.5 [404.4]^e	432.4±2.1		
O_I		290.2±0.8	309.6±4.3	323.7±1.1		351.9±2.4		385
O_{II}		229.4±1.1 }	222.9±3.9 }	259.3±0.5 }	283.4±0.8 [283.4]^e	274.1±4.7		
O_{III}		181.8±0.4 [181.8]^a }		195.1±1.3 }	206.1±0.7 [206.1]^e	206.5±4.7		

	89 Ac	90 Th	91 Pa	92 U	93 Np	94 Pu	95 Am	96 Cm
O_{IV}		94.3 ± 0.4 [94.4]^q		105.0 ± 0.5	109.3 ± 0.7 [108.8]^s	116.0 ± 1.2	115.8 ± 1.3	
			94.1 ± 2.8					
O_V		87.9 ± 0.3 [88.1]^q		96.3 ± 1.4	101.3 ± 0.5 [101.4]^s	105.4 ± 1.0	103.3 ± 1.1	
P_I		59.5 ± 1.1		70.7 ± 1.2				
P_{II}		49.0 ± 2.5		42.3 ± 9.0				
P_{III}		43.0 ± 2.5		32.3 ± 9.0				

	97 Bk	98 Cf	99 Es	100 Fm	101 Md	102 No	103 Lw
K	[131590$\pm$40]^u	135960	139490	143090	146780	150540	154380
L_I	[25275$\pm$17]^u	26110	26900	27700	28530	29380	30240
L_{II}	[24385$\pm$17]^u	25250	26020	26810	27610	28440	29280
L_{III}	[19452$\pm$20]^u	19930	20410	20900	21390	21880	22360
M_I	[6556$\pm$21]^u	6754	6977	7205	7441	7675	7900
M_{II}	[6147$\pm$31]^u	6359	6574	6793	7019	7245	7460
M_{III}	[4977$\pm$31]^u	5109	5252	5397	5546	5688	5710
M_{IV}	4366	4497	4630	4766	4903	5037	5150
M_V	4132	4253	4374	4498	4622	4741	4860
N_I	[1755$\pm$22]^u	1799	1868	1937	2010	2078	2140
N_{II}	1554	1616	1680	1747	1814	1876	1930
N_{III}	1235	1279	1321	1366	1410	1448	1480
O_I	[398$\pm$22]^u	419	435	454	472	484	490

[a] J. E. Mack, 1949, as given in C. E. Moore, *Atomic Energy Levels* (U. S. National Bureau of Standards, Washington, D. C., 1949), Vol. 1, p. 1.

[b] G. Herzberg, 1957, as given in C. E. Moore, *Atomic Energy Levels* (U. S. National Bureau of Standards, Washington, D. C., 1958), Vol. 3, p. 238.

[c] See Ref. 18.

[d] A. Fahlman, D. Hamrin, R. Nordberg, C. Nordling, and K. Siegbahn, Phys. Rev. Letters **14**, 127 (1965). See also Ref. 26.

[e] See Ref. 15.

[f] See Ref. 11.

[g] C. Nordling, Arkiv Fysik **15**, 397 (1959).

[h] E. Sokolowski, C. Nordling, and K. Siegbahn, Arkiv Fysik **12**, 301 (1957).

[i] C. Nordling and S. Hagström, Arkiv Fysik **16**, 515 (1960).

[j] I. Andersson and S. Hagström, Arkiv Fysik **27**, 161 (1964).

[k] M. O. Krause, Phys. Rev. **140**, A1845 (1965).

[l] A. Fahlman, O. Hörnfeldt, and C. Nordling, Arkiv Fysik **23**, 75 (1962).

[m] P. Bergvall, O. Hörnfeldt, and C. Nordling, Arkiv Fysik **17**, 113 (1960).

[n] P. Bergvall and S. Hagström, Arkiv Fysik **17**, 61 (1960).

[o] S. Hagström, Z. Physik **178**, 82 (1964).

[p] A. Fahlman and S. Hagström, Arkiv Fysik **27**, 69 (1964).

[q] C. Nordling and S. Hagström, Z. Physik **178**, 418 (1964).

[r] C. Nordling and S. Hagström, Arkiv Fysik **15**, 431 (1959).

[s] S. Hagström, Bull. Am. Phys. Soc. **11**, 389 (1966).

[t] A. Fahlman, K. Hamrin, R. Nordberg, C. Nordling, K. Siegbahn, and L. W. Holm, Phys. Letters **19**, 643 (1966).

[u] J. M. Hollander, M. D. Holtz, T. Novakov, and R. L. Graham, Arkiv Fysik **28**, 375 (1965).

LATTICE SPACING OF COMMON ANALYSING CRYSTALS

Crystal	Reflection plane	*d* Spacing	Crystal	Reflection plane	*d* Spacing
ADP^a	101	5.31	Lead stearate		51
ADP^a	110	5.325	LiF	200	2.014
ADP^a	200	3.75	LiF	220	1.424
Beryl	10$\bar{1}$0	7.98	Mica	002	9.96
Calcite	100	3.036	NaCl	200	2.820
EDDT^b	020	4.404	Oxalic acid	001	5.85
Germanium	111	3.265	PET^d	002	4.371
Graphite	001	6.69	Quartz	10$\bar{1}$0	4.255
Gypsum	010	7.600	Quartz	10$\bar{1}$1	3.343
KAP^c	001	13.32	Quartz	11$\bar{2}$0	2.456
KBr	200	3.29	Silicon	111	3.13
KCl	200	3.14	Topaz	303	1.356

While several of the above spacings have been measured to more than four significant figures, no more than four figures are given here because complications introduced by the index of refraction, anomalous dispersion, temperature coefficient of expansion, and crystal impurities must be considered before the additional figures are useful.

[a] Ammonium dihydrogen phosphate.

[b] Ethylenediamine d-tartrate.

[c] Potassium acid phthalate.

[d] Pentaerythritol.

RADIATIVE TRANSITION PROBABILITIES FOR K X-RAY LINES

$$K\beta_1' = KM_2 + KM_3 + KM_{4,5} \qquad K\beta_2' = KN_{2,3} + KO_{2,3}$$
$$K\alpha = K\alpha_1 + K\alpha_2 \qquad K\beta = K\beta_1' + K\beta_2'$$

Element	$K\alpha_2/K\alpha_1$	$K\alpha_3/K\alpha_1$	$K\beta_1/K\alpha_1$	$K\beta_2'/K\alpha_1$	$K\beta_4/K\alpha_1$	$K\beta_5/K\alpha_1$	$K\beta_3/K\beta_1$	$K\beta/K\alpha$
12^{Mg}								0.013
14^{Si}								0.027
16^{S}								0.059
18^{Ar}								0.105
20^{Ca}	0.502							0.128
22^{Ti}	0.503							0.134
24^{Cr}	0.504							0.135
26^{Fe}	0.506							0.135
28^{Ni}	0.508							0.135
30^{Zn}	0.510							0.138
32^{Ge}	0.513							0.147
34^{Se}	0.515							0.157
36^{Kr}	0.517			0.019				0.172
38^{Sr}	0.520			0.030				0.180
40^{Br}	0.523			0.037				0.190
42^{Mo}	0.525			0.041				0.197
44^{Ru}	0.527			0.045				0.204
46^{Pd}	0.529			0.048				0.210
48^{Cd}	0.532			0.053			0.519	0.213
50^{Sn}	0.534			0.055			0.519	0.220
52^{Te}	0.537			0.058			0.519	0.225
54^{Xe}	0.539			0.064			0.518	0.232
56^{Ba}	0.543			0.070			0.518	0.237
58^{Ce}	0.546			0.076			0.518	0.242
60^{Nd}	0.549	0.11×10^{-3}		0.083			0.518	0.247
62^{Sm}	0.552	0.14×10^{-3}		0.086			0.517	0.250
64^{Gd}	0.556	0.17×10^{-3}	0.192	0.089	0.85×10^{-3}	3.02×10^{-3}	0.517	0.255
66^{Dy}	0.560	0.21×10^{-3}	0.198	0.089	0.92×10^{-3}	3.43×10^{-3}	0.517	0.257
68^{Er}	0.564	0.26×10^{-3}	0.202	0.088	0.96×10^{-3}	3.85×10^{-3}	0.518	0.260
70^{Yb}	0.567	0.30×10^{-3}	0.207	0.087	1.04×10^{-3}	4.23×10^{-3}	0.518	0.264
72^{Hf}	0.572	0.36×10^{-3}	0.213	0.005	1.16×10^{-3}	4.62×10^{-3}	0.518	0.267
74^{W}	0.576	0.43×10^{-3}	0.216	0.086	1.28×10^{-3}	5.04×10^{-3}	0.518	0.269
76^{Os}	0.580	0.51×10^{-3}	0.222	0.087	1.43×10^{-3}	5.44×10^{-3}	0.519	0.273
78^{Pt}	0.583	0.63×10^{-3}	0.226	0.091	1.61×10^{-3}	5.84×10^{-3}	0.520	0.275
80^{Hg}	0.588	0.76×10^{-3}	0.228	0.096	1.80×10^{-3}	6.24×10^{-3}	0.520	0.278
82^{Pb}	0.593	0.91×10^{-3}	0.228	0.102	2.02×10^{-3}	6.64×10^{-3}	0.521	0.280
84^{Po}	0.597	1.12×10^{-3}	0.228	0.108	2.26×10^{-3}	7.05×10^{-3}	0.522	0.283
86^{Rn}	0.602	1.32×10^{-3}	0.228	0.113	2.52×10^{-3}	7.48×10^{-3}	0.523	0.286
88^{Ra}	0.608	1.58×10^{-3}	0.230	0.117	2.80×10^{-3}	7.80×10^{-3}	0.524	0.287
90^{Th}	0.613	1.85×10^{-3}	0.232	0.120	3.13×10^{-3}	8.25×10^{-3}	0.525	0.288
92^{U}	0.619	2.15×10^{-3}	0.234	0.123	3.47×10^{-3}	8.65×10^{-3}	0.527	0.289
94^{Pu}	0.625		0.234	0.125			0.528	0.291
96^{Cm}	0.632		0.234	0.128			0.529	0.293
98^{Cf}	0.642		0.238	0.132			0.531	0.295
100^{Fm}	0.648		0.240	0.135			0.533	0.297

From Salem, S. I., Panossian, S. L., and Krause, R. A., *At. Data Nucl. Data Tables*, 14, 91, 1974. Reproduced by permission of the copyright owner, Academic Press.

RADIATIVE TRANSITION PROBABILITIES FOR
L X-RAY LINES

The following three tables present data for the radiative transition probabilities for the L_1, L_2, and L_3 X-ray lines. The data are normalized respectively to $L\beta_3 = 100$, $L\beta_1 = 100$ and $L\alpha_1 = 100$.

L_1 X-RAY LINES NORMALIZED TO $L\beta_3 = 100$

Element	$L\beta_3$	$L\beta_{4,4}$	$L\gamma_2$	$L\gamma_3$	Element	$L\beta_3$	$L\beta_4$	$L\gamma_2$	$L\gamma_3$
32^{Ge}	100			17.3	66^{Dy}	100	61.8	19.5	28.0
34^{Se}	100			18.0	68^{Er}	100	63.5	19.8	29.0
36^{Kr}	100			18.2	70^{Yb}	100	65.5	20.7	29.8
38^{Sr}	100			18.8	72^{Hf}	100	67.8	21.2	30.7
40^{Zr}	100			19.0	74^{W}	100	70.5	21.8	31.8
42^{Mo}	100	70.6		19.6	76^{Os}	100	73.2	23.0	32.8
44^{Ru}	100	67.8		20.2	78^{Pt}	100	76.5	24.5	33.8
46^{Pd}	100	65.5		20.6	80^{Hg}	100	80.3	26.3	35.0
48^{Cd}	100	63.5		21.3	82^{Pb}	100	84.2	28.6	36.0
50^{Sn}	100	62.1		22.0	84^{Po}	100	88.5	31.3	37.2
52^{Te}	100	60.7		22.6	86^{Rn}	100	93.4	34.2	38.2
54^{Xe}	100	59.8		23.3	88^{Ra}	100	98.9	37.5	39.6
56^{Ba}	100	59.5		24.0	90^{Th}	100	104.5	41.2	41.0
58^{Ce}	100	59.2		24.6	92^{U}	100	110.2	45.0	42.6
60^{Nd}	100	59.4		25.4	94^{Pu}	100	116.2	49.5	44.0
62^{Sm}	100	60.0		26.3	96^{Cm}	100	123.0	55.7	45.7
64^{Gd}	100	60.8	19.2	27.0					

RADIATIVE TRANSITION PROBABILITIES FOR
L X-RAY LINES (Continued)

L_2 X-RAY LINES NORMALIZED TO $L\beta_1$ = 100

Element	$L\beta_1$	L_η	$L\gamma_1$	$L\gamma_6$	Element	$L\beta_1$	L_η	$L\gamma_1$	$L\gamma_6$
28 Ni	100	7.60			64 Gd	100	2.35	17.00	
30 Zn	100	6.80			66 Dy	100	2.25	17.40	
32 Ge	100	6.28			68 Er	100	2.16	17.80	
34 Se	100	5.80			70 Yb	100	2.10	18.17	
36 Kr	100	5.35			72 Hf	100	2.08	18.43	
38 Sr	100	4.93			74 W	100	2.10	18.80	0.72
40 Zr	100	4.60	3.30		76 Os	100	2.12	19.34	1.65
42 Mo	100	4.30	5.50		78 Pt	100	2.18	19.73	2.40
44 Ru	100	4.00	7.33		80 Hg	100	2.25	20.35	3.10
46 Pd	100	3.75	10.67		82 Pb	100	2.30	20.93	3.65
48 Cd	100	3.55	10.60		84 Po	100	2.40	21.54	4.15
50 Sn	100	3.35	11.80		86 Rn	100	2.46	22.20	4.55
52 Te	100	3.20	12.70		88 Ra	100	2.50	22.87	4.87
54 Xe	100	3.00	14.00		90 Th	100	2.60	23.43	5.02
56 Ba	100	2.85	14.50		92 U	100	2.65	24.10	5.12
58 Ce	100	2.70	15.30		94 Pu	100	2.70	24.40	5.16
60 Nd	100	2.60	16.00		96 Cm	100	2.75	25.07	5.20
62 Sn	100	2.45	16.50						

L_3 X-RAY LINES NORMALIZED TO $L\alpha_1$ = 100

Element	$L\alpha_1$	$L\beta_{2,15}$	$L\alpha_2$	$L\beta_5$	$L\beta_6$	$L\ell$
26 Fe	100					12.22
28 Ni	100					8.95
30 Zn	100					7.34
32 Ge	100					6.45
34 Se	100					7.76
36 Kr	100					5.28
38 Sr	100					4.92
40 Zr	100	0.70	11.10			4.67
42 Mo	100	5.17	11.10			4.45
44 Ru	100	9.30	11.12			4.28
46 Pd	100	11.80	11.12			4.11
48 Cd	100	14.33	11.12			4.07
50 Sn	100	16.00	11.13			4.00
52 Te	100	18.00	11.13			4.00
54 Xe	100	19.40	11.13			4.00
56 Ba	100	20.67	11.13			4.02
58 Ce	100	21.00	11.14			4.09
60 Nd	100	21.33	11.14		0.875	4.13
62 Sm	100	21.07	11.14		0.925	4.16
64 Gd	100	20.83	11.14		0.99	4.20
66 Dy	100	20.50	11.14		1.05	4.26
68 Er	100	20.04	11.15		1.12	4.33
70 Yb	100	19.40	11.15		1.17	4.47
72 Hf	100	21.33	11.15	0.30	1.21	4.59
74 W	100	22.74	11.16	0.50	1.25	4.76
76 Os	100	23.40	11.16	1.32	1.37	4.95
78 Pt	100	24.00	11.16	1.98	1.43	5.14
80 Hg	100	24.50	11.17	2.62	1.50	5.37
82 Pb	100	24.83	11.17	3.21	1.56	5.58
84 Po	100	25.13	11.17	3.73	1.62	5.80
86 Rn	100	25.60	11.18	4.25	1.68	6.00
88 Ra	100	25.92	11.18	4.73	1.76	6.26
90 Th	100	26.17	11.18	5.18	1.82	6.54
92 U	100	26.40	11.18	5.58	1.89	6.79
94 Pu	100	26.67	11.18	5.92	1.95	7.02
96 Cm	100	26.93	11.18	6.26	2.01	7.34

From Salem, S. I., Panossian, S. L., and Krause, R. A., *At. Data Nucl. Data Tables*, 14, 91, 1974. Reproduced by permission of the copyright owner. Academic Press.

K AND L X-RAY ABSORPTION EDGES

The wavelengths in Angstroms for the K absorption edge of elements having atomic numbers 3 to 11 and 83 to 100 were calculated from critical X-ray absorption energies listed in the Review of Scientific Instruments, 23, 523-528. (1952)

All wavelengths for the L absorption edges were calculated from absorption energies given in the above reference.

Values for the M, N, and O absorption edges are also listed in the above reference. The wavelengths in Angstroms for the K absorption edge of the elements having atomic numbers 12 to 82 are reproduced by permission from Structure of Metals, Barrett, 2nd edition, McGraw-Hill Book Company, Inc. (1952).

ABSORPTION EDGES—ANGSTROMS

Atomic Number	Element	K	L_I	L_{II}	L_{III}	Atomic Number	Element	K	L_I	L_{II}	L_{III}
3	Li	226.62				52	Te	0.38972	2.50272	2.68253	2.85162
4	Be	110.68				53	I	0.373_{79}	2.38982	2.55324	2.71901
5	B	66.289				54	Xe	0.35849	2.27449	2.43058	2.59330
6	C	43.648				55	Cs	0.34473	2.17245	2.31268	2.47227
7	N	30.990				56	Ba	0.33137	2.08336	2.20216	2.36114
8	O	23.301				57	La	0.31842	1.97892	2.10030	2.25792
9	F	17.913				58	Ce	0.30647	1.89078	2.00940	2.16410
10	Ne	14.183				59	Pr	0.29517	1.81307	1.92305	2.07707
11	Na	11.478				60	Nd	0.28451	1.73759	1.84244	1.99452
12	Mg	9.5117	197.39	247.92		61	Pm	0.27425	1.66837	1.76581	1.91888
13	Al	7.9511	142.48	172.16		62	Sm	0.26462	1.60113	1.69436	1.84464
14	Si	6.7446	105.05	126.48		63	Eu	0.25552	1.53815	1.62591	1.77491
15	P	5.7866	81.02	96.843		64	Gd	0.24680	1.47870	1.56081	1.70955
16	S	5.0182	64.228	76.049	76.519	65	Tb	0.23840	1.42270	1.50108	1.64840
17	Cl	4.3969	52.084	61.366	61.672	66	Dy	0.23046	1.36926	1.44357	1.59025
18	A	3.87068	43.192	50.390	50.803	67	Ho	0.22290	1.31942	1.38999	1.53529
19	K	3.43645	36.352	42.020	42.452	68	Er	0.21565	1.27086	1.33721	1.48242
20	Ca	3.07016	31.068	35.417	35.827	69	Tm	0.2089	1.22708	1.28856	1.43140
21	Sc	2.7572	26.831	30.161	30.457	70	Yb	0.20223	1.18136	1.24146	1.38518
22	Ti	2.497_{30}	23.389	26.831	27.184	71	Lu	0.19584	1.14007	1.19745	1.34039
23	Va	2.269_{02}	20.523	23.702	24.0699	72	Hf	0.18981	1.09864	1.15311	1.29570
24	Cr	2.070_{12}	18.256	21.226	21.5958	73	Ta	0.18393	1.06084	1.11264	1.25427
25	Mn	1.896_{36}	16.268	18.896	19.2484	74	W	0.17837	1.02497	1.07436	1.21529
26	Fe	1.743_{14}	14.601	17.169	17.4938	75	Re	0.17311	0.99009	1.03645	1.17720
27	Co	1.608_{11}	13.343	15.534	15.8314	76	Os	0.16780	0.95574	1.00129	1.14143
28	Ni	1.488_{11}	12.267	14.135	14.4476	77	Ir	0.16286	0.92425	0.96700	1.10599
29	Cu	1.380_{43}	11.269	12.994	13.2578	78	Pt	0.15817	0.89405	0.93484	1.07306
30	Zn	1.283_{7}	10.330	11.9205	13.1055	79	Au	0.15344	0.86378	0.90277	1.04028
31	Ga	1.195_{67}	9.535	10.6130	10.8546	80	Hg	0.14923	0.83531	0.87790	1.00944
32	Ge	1.116_{52}	8.729	9.9646	10.2277	81	Tl	0.14470	0.80787	0.84355	0.97968
33	As	1.044_{97}	8.107	9.1281	9.3767	82	Pb	0.14077	0.78153	0.81552	0.95112
34	Se	0.979_{78}	7.467	8.4212	8.6624	83	Bi	0.13691	0.75649	0.78910	0.92459
35	Br	0.91994	6.925	7.7523	7.9871	84	Po	0.13306	0.73219	0.76377	0.89761
36	Kr	0.86546	6.456	7.1653	7.4227	85	At	0.12949	0.70915	0.73873	0.87234
37	Rb	0.81549	6.006	6.6538	6.8752	86	Rn	0.12591	0.68675	0.71529	0.84845
38	Sr	0.76969	5.604	6.1856	6.3996	87	Fr	0.12261	0.66537	0.69290	0.82529
39	Y	0.72762	5.19312	5.70981	5.91412	88	Ra	0.11931	0.64461	0.67114	0.80284
40	Zr	0.68877	4.89380	5.37088	5.57374	89	Ac	0.11618	0.62479	0.65002	0.78158
41	Nb	0.65291	4.59111	5.02472	5.22596	90	Th	0.11290	0.60610	0.63010	0.76151
42	Mo	0.61977	4.32066	4.71330	4.90930	91	Pa	0.11028	0.58748	0.61064	0.74138
43	Tc	0.58888	4.06426	4.42714	4.62537	92	U	0.10775	0.56974	0.59186	0.72225
44	Ru	0.560_{47}	3.84133	4.17654	4.36632	93	Np	0.10487	0.55314	0.57415	0.70391
45	Rh	0.533_{78}	3.64159	3.94902	4.13889	94	Pu	0.10228	0.53662	0.55712	0.68637
46	Pd	0.509_{15}	3.42999	3.71360	3.89688	95	Am	0.09972	0.52084	0.54036	0.66933
47	Ag	0.4858_{2}	3.23824	3.49478	3.67288	96	Cm	0.09674	0.50595	0.52458	0.65276
48	Cd	0.46408	3.08434	3.32243	3.50070	97	Bk	0.09506	0.49131	0.50928	0.63667
49	In	0.44397	2.93327	3.15500	3.32154	98	Cf	0.09278	0.47713	0.49445	0.62135
50	Sn	0.42468	2.78875	2.99492	3.16952	99		0.09068	0.46357	0.48009	0.60645
51	Sb	0.40663	2.63296	2.82304	2.99637	100		0.08861	0.45059	0.46654	0.59225

NATURAL WIDTH, IN eV OF THE INDICATED K X-RAY LINES

Element	$K\alpha_1$	$K\alpha_2$	$K\beta_1$	$K\beta_3$	Element	$K\alpha_1$	$K\alpha_2$	$K\beta_1$	$K\beta_3$
20Ca	1.00	0.98			60Nd	21.50	21.50	23.25	21.33
22Ti	1.45	2.13			62Sm	26.00	24.70	25.65	24.65
24Cr	2.05	2.64			64Gd	29.50	28.00	29.37	28.00
26Fe	2.45	3.20			66Dy	33.90	32.20	32.73	32.00
28Ni	3.00	3.70			68Er	37.40	35.50	36.20	35.70
30Zn	3.40	3.96			70Yb	42.00	40.60	41.43	41.15
32Ge	3.75	4.18			72Hf	45.30	44.30	46.00	46.10
34Se	4.10	4.43			74W	47.75	48.00	51.83	51.50
36Kr	4.23	4.62			76Os	53.00	49.40	55.90	55.95
38Sr	5.17	4.97			78Pt	60.30	54.30	59.98	62.13
40Zn	5.70	5.25			80Hg	64.75	68.20	65.75	68.95
42Mo	6.82	6.80			82Pb	68.30	79.00	72.20	74.90
44Ru	7.41	7.96			84Po	73.20	86.30	78.60	82.85
46Pd	8.80	9.20			86Rn	80.00	89.50	85.50	91.20
48Cd	9.80	10.40			88Ra	87.00	91.20	94.20	98.95
50Sn	11.20	12.40	11.80	11.00	90Th	94.70	97.00	99.70	105.00
52Te	12.80	14.20	13.30	12.30	92U	103.00	106.00	115.00	120.00
54Xe	14.20	15.10	15.30	13.43					
56Ba	16.10	16.80	18.15	16.00					
58Ce	18.60	19.50	20.60	17.95					

From Salem, S. I. and Lee, P. L., *At. Data Nucl. Data Tables,* 18, 233, 1976.

NATURAL WIDTH, IN eV OF THE INDICATED L X-RAY LINES

Element	$L\alpha_1$	$L\alpha_2$	$L\beta_1$	$L\beta_2$	$L\beta_3$	$L\beta_4$	$L\gamma_1$
40Zn	1.68	1.52	1.87	5.13	5.50	5.60	3.34
42Mo	1.86	1.80	2.03	5.30	5.90	5.78	3.76
44Ru	2.03	1.98	2.18	5.45	6.35	5.96	4.15
46Pd	2.21	2.16	2.36	5.63	6.80	6.18	4.50
48Cd	2.43	2.40	2.54	5.82	7.23	6.28	4.83
50Sn	2.62	2.62	2.75	6.10	7.70	6.60	5.23
52Tc	2.88	2.88	2.96	6.25	8.22	6.82	5.60
54Xe	3.15	3.15	3.20	6.43	8.70	7.15	5.95
56Ba	3.39	3.45	3.45	6.70	9.20	7.42	6.35
58Ce	3.70	3.78	3.73	6.86	9.70	7.82	6.75
60Nd	3.93	4.08	4.00	7.18	10.30	8.15	7.16
62Sm	4.13	4.50	4.33	7.42	10.80	8.60	7.50
64Gd	4.46	4.90	4.63	7.70	11.20	9.08	7.83
66Dy	4.81	5.35	5.03	7.90	11.50	9.60	8.30
68Er	5.17	5.73	5.45	8.28	11.85	10.03	8.75
70Yb	5.40	6.22	5.90	8.58	12.20	11.00	9.20
72Hf	5.83	6.70	6.36	8.92	12.40	12.80	9.63
74W	6.50	7.20	6.90	9.06	13.10	14.60	10.20
76Os	7.04	7.70	7.42	9.60	14.60	16.50	10.65
78Pt	7.60	8.28	8.00	9.95	16.10	18.00	11.20
80Hg	8.10	8.80	8.70	10.40	17.40	19.70	11.80
82Pb	8.82	9.35	9.35	10.75	18.65	21.30	12.30
84Po	9.50	9.95	10.10	11.25	19.90	22.70	13.05
86Rn	10.03	10.50	10.65	11.65	21.00	24.00	13.55
88Ra	11.00	11.20	11.60	12.20	22.00	25.20	14.30
90Th	11.90	11.80	12.40	12.80	22.85	26.35	15.00
92U	12.40	12.40	13.50	13.30	23.70	27.50	15.70
94Pu	13.20	13.00	14.10	13.90	24.10	28.30	16.40
96Cm	14.80	13.60	15.70	14.60	25.00	29.40	17.10

DIFFRACTION DATA FOR CUBIC ISOMORPHS

From Volume 14, pages 689, 690, and 691 of the Analytical Edition of *Industrial and Engineering Chemistry*, with permission.

X Units	Substance	X Units	Substance
	A 4	4.44	ScN
3.56	C (diamond)	4.446	TaC
5.42	Si	4.458	HfC
5.62	Ge	4.615	NaF
6.46	α-Sn	4.62	ZrN
	A 1	4.69	CdO
3.517	Ni	4.69	ZrC
3.554	α-Co	4.80	CaO
3.60	Taenite (57.7% Fe, 40.8% Ni, 0.5% P)	4.82	(Na₂CeO₃)
3.608	Cu	4.84	(Na₂PrO₃)
3.63	γ-Fe (1370°K.)	4.88	NaH
3.797	Rh	4.92	AgF
3.831	Ir	5.006	CaNH
3.880	Pd	5.13	SrO
3.88—4.04	Pd-H	5.14	LiCl
3.912	Pt	5.14	NdN
4.041	Al	5.19	MgS
4.070	Au	5.192	MnS (130°K.)
4.077	Ag	5.210	MnS (299°K.)
4.30	Co-N	5.33	KF
4.40—4.46	Ti-H	5.45	MgSe
4.52	Ne (4°K.)	5.45	MnSe
4.66	Zr-H	5.45	SrNH
4.84	β-Tl	5.49	LiBr
4.939	Pb	5.52	BaO
5.08	Th	5.545	AgCl
5.14	α-Ce	5.55—5.76	AgCl-AgBr
5.296	β-La	5.627	NaCl
5.43	A (4°K.)	5.63	RbF
5.56	Ca	5.68	CaS
5.59	Kr (20°K.)	5.60	SmAs
5.70	Kr (92°K.)	5.70	KH
6.05	Sr	5.755	AgBr
6.20	X (88°K.)	5.76—5.92	AgBr-AgI
	A 2	5.83	NdP
2.861	α-Fe	5.83	NaCN
2.875	α-Cr	5.84	BaNH
2.90	β-Fe (1070°K.)	5.87	SrS
2.93	δ-Fe (1700°K.)	5.91	CaSe
3.03	V	5.94	PbS
3.03—3.41	V-C	5.95	NaBr
3.140	Mo	5.957	EuS
3.157	W	5.96	NdAs
3.295	Cb	6.00	PrAs
3.30	Ta	6.00	LiI
3.32	β-Ti (1200°K.)	6.01	CsF
3.46	Li (~80°K.)	6.04	RbH
3.50	Li	6.05	β-NaSH (>360°K.)
3.61	β-Zr (1120°K.)	6.06	CeAs
4.24	Na (~80°K.)	6.13	LaAs
4.29	Na	6.14	PbSe
5.02	Ba	6.23	SrSe
5.20	K (120°K.)	6.278	KCl
5.33	K	6.285	SnTe
5.62	Rb (~80°K.)	6.31	NdSb
6.05	Cs (~80°K.)	6.345	CaTe
	B 1	6.35	PrSb
4.018	LiF	6.36	BaS
4.065	LiD	6.38	CsH
4.08	VO	6.40	CeSb
4.09	LiH	6.44	PbTe
4.12	(Li₂TiO₃)	6.462	NaI
4.12—4.20	(Li₂TiO₃-MgO)	6.45	PrBi
4.13	VN	6.48	LaSb
4.14	CrN	6.49	CeBi
4.14	VC	6.53	KCN
4.142	(63Li₂Fe₂O₄·37Li₂TiO₃)	6.53	NH₄Cl (>457°K.)
4.173	NiO	6.56	RbCl
4.207	MgO	6.57	LaBi
4.282	MgO (1570°K.)	6.58	KBr
4.225	TiN	6.59	BaSe
4.235	TiO	6.60	β-KSH (>440°K.)
4.24	80 TiN-20 TiC	6.65	SrTe
4.27	CoO	6.82	RbCN
4.28	V-N	6.86	RbBr
4.283	FeO (160°K.)	6.90	NH₄Br (>411°K.)
4.290	FeO (299°K.)	6.93	β-RbSH (470°K.)
4.30	VC (ε-phase)	6.99	BaTe
4.315	TiC	7.052	KI
4.40	CbC	7.10	β-CsCl (>730°K.)
4.41	CbN	7.24	NH₄I (>255°K.)
4.426	MnO (117°K.)	7.325	RbI
4.436	MnO (299°K.)		H 0s
		6.96 ± 0.04	AgClO₄ (453 ± 20°K.)

DIFFRACTION DATA FOR CUBIC ISOMORPHS (Continued)

X Units	Substance	X Units	Substance
7.16 ± 0.10	NaClO$_4$ (618 $\pm$ 35°K.)	3.86	CeCd
7.49 ± 0.02	KClO$_4$ (598 $\pm$ 15°K.)	3.88	MgPr
7.65 ± 0.05	TlClO$_4$(553°K.)	3.90	LaCd
7.65 ± 0.02	NH$_4$ClO$_4$(528 $\pm$ 15°K.)	3.97	TlBr
7.68 ± 0.03	RbClO$_4$(583 $\pm$ 10°K.)	3.98	TlBi
7.97 ± 0.01	CsClO$_4$ (513 $\pm$ 10°K.)	4.024	SrTl
	B 3	4.05	NH$_4$Br (<411°K.)
4.255	CuF	4.112	CsCl
4.36	CSi IV	4.20	TlI
4.855	BeS	4.20	CsCl (<720°K.)
5.10	BeSe	4.25	CsCN
5.304	(Cu, Fe, Mo, Sn)$_4$(S, As, Te)$_2$, cousite	4.287	CsBr
5.41	CuCl	4.29	CsSH
5.425	β-ZnS	4.37	NH$_4$I (290°K.)
5.43	AlP	4.56	CsI
5.44	GaP		**D 2$_1$**
5.58	BeTe	4.07	YB$_6$
5.60	MnS (red)	4.07	ErB$_6$
5.63	AlAs	4.10	NdB$_6$
5.635	GaAs	4.12	GdB$_6$
5.655	ZnSe	4.12	PrB$_6$
5.68	CuBr	4.13	CeB$_6$
5.82	β-CdS	4.13	YbB$_6$
5.84	HgS	4.14	CaB$_6$
5.86	InP	4.15	LaB$_6$
6.04	CdSe	4.15	ThB$_6$
6.04	InAs	4.19	SrB$_6$
6.05	CuI	4.33	BaB$_6$
6.07	HgSe		**C 1**
6.08	ZnTe	4.33	Be$_2$C
6.103	α-Cu$_2$HgI$_4$	4.619	Li$_2$O
6.12	AlSb	5.06	(3ZrO$_2$ · MgO)
6.12	GaSb	5.07	ZrO$_2$
6.13	SnSb	5.08	(95ZrO$_2$ · 5CeO$_2$)
6.383	α-Ag$_2$HgI$_4$	5.13	(95HiO$_2$ · 5CeO$_2$)
6.40	HgTe	5.38	PrO$_2$
6.43	CdTe	5.40	CeO$_2$
6.45	InSb	5.40	CdF$_2$
6.48	AgI	5.406	CuF$_2$
	B 32	5.45	CaF$_2$
6.195	LiGa	5.47	UO$_2$
6.209	LiZn	5.526	(66CaF$_2$ · 33YF$_3$)
6.36	LiAl	5.53	(91CaF$_2$ · 9ThF$_4$)
6.687	LiCd	5.54	HgF$_2$
6.786	LiIn	5.55	Na$_2$O
7.297	NaIn	5.58	ThO$_2$
7.373	(CeMg$_3$)	5.59	Cu$_2$S
7.373	(PrMg$_3$)	5.704	Li$_2$S
7.473	NaTl	5.749	Cu$_2$Se
	B 20	5.782	SrF$_2$
4.437	NiSi	5.796	EuF$_2$
4.438	FeSi	5.838	(66SrF$_2$ · 33LaF$_3$
4.438	CoSi	5.91	PtAl$_2$
4.548	MnSi	5.91	PtGa$_2$
4.620	CrSi	5.935	β-PbF$_2$(520°K.)
	B 2	5.99	Al$_2$Au
2.603	NiBe	6.005	Li$_2$Se
2.606	CoBe	6.06	AuGa$_2$
2.69	CuBe	6.19	BaF$_2$
2.813	PdBe	6.34	Mg$_2$Si
2.82	AlNi	6.35	PtIn$_2$
2.945	CuZn	6.368	RaF$_2$
2.989	CuPd	6.379	Mg$_2$Ge
3.146	AuZn	6.436	K$_2$O
3.156	AgZn	6.50	Li$_2$Te
3.168	AgLi	6.50	AuIn$_2$
3.259	AuMg	6.526	Na$_2$S
3.275	AgMg	6.763	Mg$_2$Sn
3.287	HgLi	6.809	Na$_2$Se
3.325	AgCd	6.81	Mg$_2$Pb
3.34	AuCd (670°K.)	6.98	SrCl$_2$
3.424	LiTl	7.314	Na$_2$Te
3.442	HgMg	7.38	K$_2$S
3.628	MgTl	7.65	RbS$_2$
3.67	PrZn	7.676	K$_2$Se
3.70	CeZn	8.152	K$_2$Te
3.73	AlNd		**C 15**
3.74	α-RbCl (83°K.)	5.94	Be$_2$Cu
3.75	LaZn	6.287	Be$_2$Ag
3.82	TlCn	6.435	Be$_2$Ti
3.82	PrCd	6.96	MgNiZn
3.84	TlSb	7.03	Cu$_2$Mg
3.835	TlCl	7.61	W$_2$Zr
3.847	CaTl	7.79	Au$_2$Na
3.86	NH$_4$Cl (<457°K.)		

DIFFRACTION DATA FOR CUBIC ISOMORPHS (Continued)

X Units	Substance	X Units	Substance
7.91	Au_2Pb	3.83	$NaWO_3$
7.94	Au_2Bi	3.85	(Na, Ce, Ca)(Ti, Cb)O_3 Loparite
8.02	Al_2Ca	3.88	$NaTaO_3$
8.04	Al_2Ce	3.89	$LaGaO_3$
8.16	Al_2La	3.89	$NaCbO_3$
9.50	Bi_2K	3.91	$SrTiO_3$
C 2		3.92	$CaSnO_3$
5.41	FeS_2	3.97	$BaTiO_3$
5.42	(Fe, Ni)S_2 (6.5% Ni)	3.98	$KTaO_3$
5.57	RbS_2	3.99	$CaZrO_3$
5.57	RuS_2	4.00	$KMgF_3$
5.57	Bravoite (53.8% NiS_2, 39.1% FeS_2, 7.1% CoS_2)	4.005	$KNiF_3$
		4.01	$KCbO_3$
5.62	OsS_2	4.03	$SrSnO_3$
5.64	CoS_2	4.05	$KZnF_3$
5.65	(Cu, Ni, Co, Fe)(S, Se)$_2$	4.07	$KCoF_3$
5.68	PtP_2	4.07	$SrHfO_3$
5.74	NiS_2	4.09	$SrZrO_3$
5.85	$CoSe_2$	4.18	$BaZrO_3$
5.92	$RuSe_2$	4.35	$BaPrO_3$
5.93	$OsSe_2$	4.38	$BaCeO_3$
5.94	$PtAs_2$	4.46	KIO_3
5.97	$PdAs_2$	4.48	$BaThO_3$
6.02	$NiSe_2$	4.5	NH_4IO_3
6.096	MnS_2	4.52	$RbIO_3$
6.36	$RuTe_2$	4.66	$CsIO_3$
6.37	$OsTe_2$	5.12	$MgZrO_3$
6.43	$PtSb_2$	5.20	$CsCdCl_3$
6.44	$PdSb_2$	5.33	$CsCdBr_3$
6.64	$AuSb_2$	5.44	$CsHgCl_3$
6.94	$MnTe_2$	5.77	$CsHgBr_3$
C 3		**G 0₁**	
4.26	Cu_2O	6.57	$NaClO_3$
4.73	Ag_2O	6.71	$NaBrO_3$
F 1		**G 2₁**	
5.55	CoAsS	7.60	$Cu(NO_3)_2$
5.60	NiAsS	7.81	$Sr(NO_3)_2$
5.90	NiSbS	7.84	$Pb(NO_3)_2$
	(Ni, Fe)AsS, plessite	8.11	$Ba(NO_3)_2$
	Ni(As, Sb)S, corynite	**H 1₁**	
	Ni(Sb, Bi)S, kallilite	8.045	$NiAl_2O_4$
	(Co, Ni)SbS, willyamite	8.07	$CuAl_2O_4$
D 5₃		8.07	$CoCo_2O_4$
8.13	Be_3N_2	8.07	$MgAl_2O_4$
9.37	(Mn, Fe)$_2O_3$	8.08	$CoAl_2O_4$
9.42	Mn_2O_3	8.08	$ZnAl_2O_4$
9.74	Zn_3N_2	8.10	$FeAl_2O_4$
9.79	Sc_2O_3	8.11	(Ni, Co)(Co, Ni)$_2O_4$
9.94	Mg_3N_2	8.11	(Zn, Co)Co_2O_4
10.12	In_2O_3	8.11	$MgCo_2O_4$
10.15	Be_3P_2	8.27	$MnAl_2O_4$
10.37	Lu_2O_3	8.27	(Mn, Co)(Co, Mn)$_2O_4$
10.39	Yb_2O_3		
10.52	Tm_2O_3	**H 1₁**	
10.54	Er_2O_3	8.28	$MgGa_2O_4$
10.57	Tl_2O_3	8.30	$NiCr_2O_4$
10.58	Ho_2O_3	8.30	$MgCr_2O_4$
10.60	Y_2O_3	8.31	$ZnCr_2O_4$
10.63	Dy_2O_3	8.32	$CoCr_2O_4$
10.70	Tb_2O_3	8.32	$ZnGa_2O_4$
10.79	Gd_2O_3	8.35	$NiFe_2O_4$
10.79	Cd_2N_2	8.35	$Cu_2Cr_2O_4$
10.84	Eu_2O_3	8.35	$FeCr_2O_4$
10.85	Sm_2O_3	8.36	$MgFe_2O_4$
11.05	Nd_2O_3	8.38	$CoFe_2O_4$
11.40	α-Ca_2N_2	8.38	$NiMn_2O_4$
12.02	Mg_3P_2	8.40	$ZnFe_2O_4$
12.33	Mg_3As_2	8.40	$FeFe_2O_4$
D 6₁		8.42	(Mn, Mg)Fe_2O_4
11.05	As_4O_6	8.42	$TiCo_2O_4$
11.14	Sb_4O_6	8.43	$MnCr_2O_4$
D 1₁		8.43	$TiMg_2O_4$
10.32	$ZrCl_4$	8.43	$TiZn_2O_4$
11.25	$TiBr_4$	8.44	$CuFe_2O_4$
(11.34)	(CBr_4) (>320°K.)	8.47	FeV_2O_4
(11.62)	(CI_4)	8.49	$MnCr_2O_4$
11.89	GeI_4	8.50	$TiFe_2O_4$
11.99	SiI_4	8.54	$MnFe_2O_4$
12.00	TiI_4	8.58	$CdCr_2O_4$
12.23	SnI_4	8.58	$SnMg_2O_4$
E 2₁		8.61	$SnCo_2O_4$
3.67	$YAlO_3$	8.63	$SnZn_2O_4$
3.75	$CdTiO_3$	8.67	$CdFe_2O_4$
3.78	$LaAlO_3$	8.67	$TiMn_2O_4$
3.80	$CaTiO_3$	8.81	$MgIn_2O_4$
		9.26	Ag_2MoO_4

X Units	Substance	X Units	Substance
9.4	$CoCo_2S_4$	10.47	$Mg(NH_3)_6Br_2$
9.45	$(Co, Ni)_3S_4$	10.48	K_2SnBr_6
9.46	$CuCo_2O_4$	10.51	$Co(NH_3)_6SO_4Br$
9.5	NiN_2S_4	10.52	$Mn(NH_3)_6Br_2$
9.92	$ZnCr_2S_4$	10.54	$Sr_2Ni(NO_2)_6$
10.05	$MnCr_2S_4$	10.55	$Pb_2Ni(NO_2)_6$
10.19	$CdCr_2S_4$	10.57	$(NH_4)SnBr_6$
12.54	$K_2Zn(Cn)_4$	10.58	Rb_2SnBr_6
12.76	$K_2Hg(CN)_4$	10.62	$Co(NH_3)_5H_2OSO_4I$
12.84	$K_2Cd(CN)_4$	10.63	$Co(NH_3)_5SeO_4Br$
H 5₈		10.67	$Ba_2Ni(NO_2)_6$
10.08	$2Na_2SO_4 \cdot NaCl \cdot NaF$	10.71	$Ca(NH_3)_6Br_2$
H 4₁₃		10.71	$Co(NH_3)_5SO_4I$
12.11	$KCr(SO_4)_2 \cdot 12H_2O$	10.77	Cs_2SnBr_6
12.12	$KAl(SO_4)_2 \cdot 12H_2O$	10.79	$Co(NH_3)_5SeO_4I$
12.15	$NH_4Al(SO_4)_2 \cdot 12H_2O$	10.9	$Ni(NH_3)_6I_2$
12.20	$RbAl(SO_4)_2 \cdot 12H_2O$	10.91	$Co(NH_3)_6I_2$
12.21	$TlAl(SO_4)_2 \cdot 12H_2O$	10.96	$Zn(NH_3)_6I_2$
12.31	$CsAl(SO_4)_2 \cdot 12H_2O$	10.97	$Fe(NH_3)_6I_2$
12.44	$NH_3 \cdot CH_3Al(SO_4)_2 \cdot 12H_2O$ (β-alum)	10.98	$Mg(NH_2)_6I_2$
Langbeinite		11.04	$Mn(NH_3)_6I_2$
9.93	$K_2Mg(SO_4)_3$	11.04	$Cd(NH_3I_2$
10.2	$K_2(Ca_2Mg)SO_3)_3$	11.24	$Ca(NH_3)_6I_2$
H 2₁		11.27	$Ni(NH_3)_6(BF_4)_2$
6.00	Ag_3PO_4	11.3	$Co(NH_3)_6(BF_4)_2$
6.120	$Ag_3AsO_4(90°K.)$	(11.3)	$Zn(NH_3)_6(ClO_4)_2$
6.130	$Ag_3AsO_4 (380°K.)$	11.34	$Mg(NH_3)_6(BF_4)_2$
H 2₄		11.34	$Fe(NH_3)_6(BF_4)_2$
5.37	Cu_3VS_4	11.37	$Mn(NH_3)_6(BF_4)_2$
J 1₁		11.38	$Cd(NH_3)_6(BF_4)_2$
8.17	K_2SiF_6	11.41	$Ni(NH_3)_6(ClO_4)_2$
8.35	$(NH_4)_2SiF_6$	11.43	$Co(NH_3)_6(ClO_4)_2$
8.38	$Rb_2CrF_5H_2O$	11.46	$Ni(NH_3)_6(SO_3F)_2$
8.41	$Tl_2CrF_5H_2O$	11.49	$Co(NH_2)_6(SO_3F)_2$
8.42	$(NH_4)_2VF_5H_2O$	11.52	$Fe(NH_3)_6(ClO_4)_2$
8.42	$Rb_2VF_5H_2O$	11.53	$Mg(NH_3)_6(ClO_4)_2$
8.45	$Tl_2VF_5H_2O$	11.54	$Cd(NH_3)_6Br_2$
8.45	Rb_2SiF_6	11.54	$Fe(NH_3)_6(SO_2F)_2$
8.58	Tl_2SiF_6	11.58	$Mn(NH_3)_6(ClO_4)_2$
8.87	Cs_2SiF_6	11.59	$Cd(NH_3)_6(ClO_4)_2$
8.99	Cs_2GeF_6	11.59	$Mn(NH_2)_6(SO_2F)_2$
9.73	K_2PtCl_6	11.62	$Cd(NH_2)_6(SO_3F)_2$
9.73	K_2OsCl_6	11.91	$Ni(NH_3)_6(PF_6)_2$
9.76	Tl_2PtCl_6	11.94	$Co(NH_2)_6(PF_6)_2$
9.84	$(NH_4)_2PtCl_6$	12.03	$Ni(NH_2 \cdot CH_3)_6I_2$
9.86	K_2ReCl_6	12.05	$Co(NH_2 \cdot CH_3)_6I_2$
9.88	Rb_2PtCl_6	12.19	$\{NH(CH_3)_3\}_2SnCl_6$
9.92	Rb_2TiCl_6	12.41	$\{S(CH_3)_2\}_2SnCl_6$
9.94	$(NH_4)_2SeCl_6$	12.65	$\{N(CH_3)_4\}_2PtCl_6$
9.97	K_2SnCl_6	12.80	$\{S(CH_3)_2C_2H_5\}_2SnCl_6$
9.97	Tl_2SnCl_6	12.87	$\{N(CH_3)_4\}_2SnCl_6$
9.98	Rb_2SeCl_6	13.17	$\{N(CH_3)_3CH_5\}_2SnCl_6$
10.02	Rb_2PdBr_6	13.51	$\{N(CH_3)(C_2H_5)_3\}_2SnCl_6$
10.04	$(NH_4)SnCl_6$	13.93	$\{P(CH_3)(C_2H_5)_3\}_2SnCl_6$
10.08	$Ni(NH_3)_6Cl_2$		
10.10	Rb_2SnCl_6	**J 2₁ and Related Structures**	
10.10	$Co(NH_3)_6Cl_2$	8.88	Li_3FeF_6
10.11	Tl_2TeCl_6	8.90	$(NH_4)_3AlF_6$
10.14	$(NH_4)_2PBCl_6$	9.01	$(NH_4)_3CrF_6$
10.14	K_2TeCl_6	9.04	$(NH_4)_3VF_6$
10.15	$Fe(NH_3)_6Cl_2$	9.10	$(NH_4)_3FeF_6$
10.16	$Mg(NH_3)_6Cl_2$	9.10	$(NH_4)MoO_3F_2$
10.17	Cs_2PtCl_6	9.26	Na_2FeF_6
J 1₁		9.93	K_2FeF_6
10.18	Rb_2ZrCl_6	9.96	$CuLi_2Fe(CN)_6$
10.18	$(NH_4)_2TeCl_6$	10.0	$CuR_2Fe(CN)_6$ R = Na, K, Rb, NH₄, Tl
10.20	$Mn(NH_3)_6Cl_2$	10.15	$K_2CdFe(NO_2)_6$
10.20	Rb_2PbCl_6	10.17	$K_2CaCo(NO_2)_6$
10.22	Cs_2TiCl_6	10.19	$K_2CaFe(NO_2)_6$
10.23	Rb_2TeCl_6	10.2	$Fe^{III}RFe^{II}(CN)_6$ R = Na, K, Rb, NH₄
10.25	$Zn(NH_3)_6(ClI_4)_3$	10.22	$K_2HgFe(NO_2)_6$
10.26	Cs_2SeCl_6	10.23	$Ba_2SrCo(NO_2)_6$
10.30	K_2OSBr_6	10.25	$(NH_4)_2CaFe(NO_2)_6$
10.35	Cs_2SnCl_6	10.25	$NaTl_2Co(NO_2)_6$
10.36	K_2SeBr_6	10.28	$K_2CdNi(NO_2)_6$
10.36	K_2PtBr_6	10.28	$(NH_4)_2CdFe(NO_2)_6$
10.39	$Co(NH_3)_6Br_2$	10.29	$K_2HgNi(NO_2)_6$
10.4	$Ni(NH_3)_6Br_2$	10.30	$K_2SrFe(NO_2)_6$
10.41	Cs_2ZrCl_6	10.30	$Tl_2CaFe(NO_2)_6$
10.42	Cs_2PbCl_6	10.31	$K_2PbFe(NO_2)_6$
10.45	Cs_2TeCl_6	10.32	$K_2CaNi(NO_2)_6$
10.45	$Co(NH_3)H_2OSO_4Br$	10.34	$(NH_4)_2SrFe(NO_2)_6$
10.46	$(NH_4)SeBr_6$	10.37	$(NH_4)_2PbFe(NO_2)_6$
10.46	$Zn(NH_3)_6Br_2$	10.37	$Tl_2CdNi(NO_2)_6$
10.47	$Fe(NH_3)_6Br_2$	10.39	$Tl_2PbFe(NO_2)_6$

X Units	Substance
10.39	$NaRb_2Co(NO_2)_6$
10.40	$Tl_2SrFe(NO_2)_6$
10.4	$K_2PbCo(NO_2)_6$
10.41	$(NH_4)_2CdNi)NO_2)_6$
10.42	$Tl_2HgNi(NO_2)_6$
10.43	$K_2BaFe(NO_2)_6$
10.45	$K_2BaCo(NO_2)_6$
10.45	$K_2Co(NO_2)_6$
10.46	$(NH_4)_2HgNi(NO_2)_6$
10.47	$Rh_2HgNi(NO_2)_6$
10.49	$K_2SrNi(NO_2)_6$
10.49	$K_2Ni(NO_2)_6$
10.50	$(NH_4)_2BaFe(NO_2)_6$
10.54	$K_2LiBi(NO_2)_6$
10.55	$K_2PbNi(NO_2)_6$
10.55	$Tl_2BaFe(NO_2)_6$
10.58	$Rb_2CdNi, Cd(NO_2)_6$
10.58	$K_3Ir(NO_2)_6$
10.59	$Rb_2LiBi(NO_2)_6$
10.6	$K_2PbCu(NO_2)_6$
10.63	$K_3Rh(NO_2)_6$
10.63	$(NH_4)_2LiBi(NO_2)_6$
10.64	$Tl_2LiBi(NO_2)_6$
10.67	$K_2BaNi(NO_2)_6$
10.70	$NaCs_2Co(NO_2)_6$
10.70	$Ba_3\{Rh(NO_2)_6\}_2$
10.72	$Tl_2Co(NO_2)_6$
10.73	$Rb_3Co(NO_2)_6$
10.73	$(NH_4)_3Ir(NO_2)_6$
10.73	$Tl_3Ir(NO_2)_6$
10.77	$Rb_2Ir(NO_2)_6$
10.8	$(NH_4)_3Co(NO_2)_6$
10.81	$Cs_2Cd\{Ni, Cd(NO_2)_6\}$
10.82	$Co(NH_3)_6I_2I_3O_3$
10.83	$Rb_3Rh(NO_2)_6$
10.88	$K_2NaBi(NO_2)_6$
10.89	$Co(NH_3)_6I_3$
10.91	$Tl_3Rh(NO_2)_6$
10.91	$(NH_4)_3Rh (NO_2)_6$
10.94	$Cs_2LiBi(NO_2)_6$
10.95	$K_2AgBi(NO_2)_6$
10.98	$Rb_2NaBi(NO_2)_6$
10.99	$(NH_4)_2NaBi(NO_2)_6$
11.01	$Tl_2NaBi(NO_2)_6$
11.05	$Rb_2AgBi(NO_2)_6$
11.06	$Tl_2AgBi(NO_2)_6$
11.10	$(NH_4)_2AgBi(NO_2)_6$
11.15	$Cs_2NaBi(NO_2)_6$
11.15	$Cs_3Co(NO_2)_6$
11.17	$Cs_3Ir(NO_2)_6$
11.19	$Cs_3Bi(NO_2)_6$
11.19	$Cs_2AgBi(NO_2)_6$
11.21	$Co(NH_3)_6(BF_4)_3$
11.30	$Cs_3Rh(NO_2)_6$
11.32	$\{Co(NH_3)_6 \cdot H_2O\}(ClO_4)_3$
11.39	$Co(NH_3)_6(ClO_4)_3$
11.67	$Co(NH_3)_6(PF_6)$

K 6₁	
7.46	SiP_2O_7
7.80	TiP_2O_7
7.98	SnP_2O_7
8.18	HfP_2O_7
8.20	ZrP_2O_7
8.61	UP_2O_7

S 1₄	
11.51	$Al_2(Mg, Fe)_3(SiO_4)_3$, pyrope
11.51	$Al_2Fe_2(SiO_4)_3$, almandite
11.60	$Al_2Mn_3(SiO_4)_3$, spessartite
11.87	$Al_2Ca_2(SiO_4)_3$, grossularite
11.89	$(Al, Fe)_2Ca_3(SiO_4)_3$, hessonite
11.95	$Cr_2Ca_3(SiO_4)_3$, uvarovite
12.03	$Fe_2Ca_3(SiO_4)_3$, andradite
12.10	$(Na, Li)_3AlF_6$, cryolithionite
12.35—12.46	$(Mg, Mn)_2(Ca, Na)_3AsO_4)_3$, berzelite

S 6₁	
13.68	$NaAlSi_2O_6H_2O$

S 0₆	
13.82	$Al_{13}Si_5O_{20}(OH, F)_{18}Cl$, zunyite

S 6₂	
8.87	$Na_4(AlSiO_4)_3Cl$, sodalite

Tetrahedrite	
10.19	$(Cu, Fe)_{12}As_4S_{13}$, binnite
10.2—10.6	$(Cu, Ag)_{10} (Zn, Fe)_2(Sb, As)_4S_{13}$

PHOTOMETRIC QUANTITIES, UNITS AND STANDARDS

Photometric quantities and units are also given in the section Quantities and Units under the sub-division Light.

Candela-cd (formerly candle): The International System (SI) unit of luminous intensity. One candela is defined as the luminous intensity of one sixtieth of one square centimeter of projected area of a black body radiator operating at the temperature of freezing platinum (2042K).

Footcandle-fc: The unit of illumination when a foot is the unit of length. Footcandle is the illumination on a surface one square foot in area on which there is a uniformly distributed flux of one lumen, or the illumination on a surface all points of which are at a distance of one foot from a directionally uniform point source of one candela. (1 footcandle = 10.76 lux.)

Lumen-lm: The unit of luminous flux. It is equal to the flux through a unit solid angle (steradian), from a uniform point source of one candela, or to the flux on a unit surface all points of which are at unit distance from a uniform source of one candela.

Lux-lx: The International System (SI) unit of illumination in which the meter is the unit of length. (See Footcandle above.) 1 Lux = 0.0929 footcandles.

Luminous Efficacy of a Source of Light: The quotient of the total luminous flux emitted by the total lamp power input. It is expressed in lumens per watt.

Luminance-L: The luminous intensity (photometric brightness) of any surface in a given direction per unit of projected area of the surface as viewed from that direction. In International System (SI) units the luminance of light sources is commonly expressed in candelas per square centimeter.

EFFICACIES OF ILLUMINANTS

The theoretical maximum efficacy of 680 lm/W is that which would be obtained if all the power input were emitted as green light at a wavelength of 555 nm (at which the eye is most sensitive). If all power could be emitted uniformly over the visible spectrum as white light, the efficacy would be of the order of 220 lm/W.

Fluorescent life-ratings are based upon 3 hr of operation per start. Life increases as lamps are burned for longer periods. Mercury, metal-halide, and high-pressure sodium life-ratings are based upon 10 hr per start.

Lamp	Nominal lamp watts	Approximate hours life	Approximate initial lumens	Approximate initial efficacy, lm/lamp W
A. Tungsten filament (120 V)				
Vacuum	25	2,500	235	9.4
Gas filled	40	1,500	455	11.4
Gas filled	60	1,000	870	14.5
Gas filled	75	750	1,190	15.9
Gas filled	100	750	1,750	17.5
Gas filled	200	750	4,010	20.1
Gas filled	500	1,000	10,850	21.7
Gas filled	1,000	1,000	23,740	23.7
Gas filled (3200 K)	5,000	150	145,000	29.0
Gas filled (3200 K)	10,000	75	335,000	33.5
B. Fluorescent (*cool* white)				
Preheat (T12)	20	9,000	1,300	65.0
Rapid start (T12)	40	20,000	3,150	78.8
Rapid start (T12, shielded cathode)	40	15,000	3,250	81.3
Rapid start (T12, U shaped)	40	12,000	2,900	72.5
Slimline (96T12, 425 mA)	75	12,000	6,300	84.0
Preheat (T17)	90	9,000	6,400	71.1
High output (96T12, 800 mA)	110	12,000	9,200	83.6
Grooved lamp (96PG17, 1500 mA)	215	12,000	16,000	74.4
C. Fluorescent (Watt-miser II™ 'Lite-white')*				
Rapid start (T12; 48″)	35	20,000ᵇ	3,050	87.1
Slimline (T12; 96″)	60	12,000ᵇ	6,000	100.0
High output (T12; 96″)	95	12,000ᵇ	9,100	95.8
D. Fluorescent (other)				
Rapid start (T12, deluxe cool white)	40	20,000 +	2,200	55.0
Rapid start (T12, white)	40	20,000 +	3,200	80.0
Rapid start (T12, warm white)	40	20,000 +	3,150	78.7
Rapid start (T12, deluxe warm-white)	40	20,000 +	2,150	53.7
Rapid start (T12, daylight)	40	20,000 +	2,600	65.0
Rapid start (T12, natural)	40	20,000 +	2,100	52.5
Rapid start (T12, 5,000 K)	40	20,000 +	2,200	55.0
Rapid start (T12, 7,500 K)	40	20,000 +	2,000	50.0
Rapid start (T12, blue)	40	20,000 +	1,200	30.0
Rapid start (T12, gold)	40	20,000 +	2,400	60.0
Rapid start (T12, green)	40	20,000 +	4,350	108.8
Rapid start (T12, pink)	40	20,000 +	1,160	29.0
Rapid start (T12, red)	40	20,000 +	200	5.0

EFFICACIES OF ILLUMINANTS (continued)

Lamp	Nominal lamp watts	Approximate hours life	Approximate initial lumens	Approximate initial efficacy, lm/lamp W
E. High-intensity discharge (HID)				
1. Mercury-vapor (E-37)				
Clear	400	24,000	21,000	52.5
Deluxe white	400	24,000	20,500	51.3
Warm deluxe white	400	24,000	19,500	48.8
2. Mercury-vapor (BT-36)				
Clear	1,000	24,000	57,000	57.0
Deluxe white	1,000	24,000	63,000	63.0
Warm deluxe white	1,000	24,000	58,000	58.0
3. Metal halide (Multi-vapor® (clear lamps)				
E-28 (vertical)	175	7,500	14,000	80.0
E-28 (vertical)	250	7,500	20,500	82.0
E-28 (horizontal)	250	7,500	19,500	78.0
E-37 (vertical)	400	15,000	34,000	85.0
E-37 (horizontal)	400	15,000	32,000	80.0
BT-56 (vertical)	1,000	10,000	110,000	110.0
BT-56 (horizontal)	1,000	10,000	107,800	107.8
BT-56 (vertical)	1,500	1,500	155,000	103.3
BT-56 (horizontal)	1,500	1,500	145,000	96.6
4. High-pressure sodium (Lucalox®) (clear lamps)				
E23 1/2	50	24,000	3,300	66.0
E23 1/2	70	24,000	5,800	82.9
E23 1/2	100	24,000	9,500	95.0
E23 1/2	150	24,000	16,000	106.7
E-28 (retrofit)	150	12,000	12,000	80.0
E-18	200	24,000	22,000	110.0
E-28 (retrofit)	215	12,000	19,000	88.4
E-18	250	24,000	27,500	110.0
E-18 ('S')	250	24,000	30,000	120.0
E-18	310	24,000	37,000	119.4
E-18	400	24,000	50,000	125.0
E-25	1,000	24,000	140,000	140.0

Note: ® is the registered trademark of the General Electric Company.
™ is the trademark of the General Electric Company.

"Color improved" deleted from mercury-vapor.

*   'Lite-white' phosphor — closely matches 'cool-white' phosphor.
b   Life ratings estimated, based on engineering design.

APPROXIMATE LUMINANCE
OF VARIOUS LIGHT SOURCES

Compiled by C.J. Allen and G.D. Rowe

Luminance of source is given in candelas per square centimeter

Source		Approx. avg. luminance* (Cd cm^{-2})
Natural sources		
Clear sky	Average luminance	0.8
Sun (as observed from earth's surface)	At meridian	160000
Sun (as observed from earth's surface)	Near horizon	600
Moon (as observed from earth's surface)	Bright spot	0.25
Combustion sources		
Candle flame (sperm)	Bright spot	1.0
Welsbach mantle	Bright spot	6.2
Acetylene flame	Mees burner	10.5
Photoflash lamps		16000-40000 peak
Incandescent electric lamps		
Carbon filament	3 Lumens per watt	52
Tungsten filament	Vacuum lamp— 10 lumens per watt	200
Tungsten filament	Gas-filled lamp— 20 lumens per watt	1200
Tungsten filament	750-watt projection lamp—26 lumens per watt	7500
Fluorescent lamps (cool white)		
Rapid start (40wT12)	430 mA	0.82
High-output (96T12)	800 mA	1.13
Grooved bulb (96T17)	1500 mA	1.50
Fluorescent lamps (other than cool white)		
Daylight (40wT12)	430 mA	0.62
Blue (40wT12)	430 mA	0.30
Green (40wT12)	430 mA	1.17
Red (40wT12)	430 mA	0.05
High-intensity discharge (HID)		
Mercury-vapor (E-37)		
Clear	400 watt	970
Color-improved	400 watt	11.0
Deluxe-white	400 watt	12.1
Mercury-vapor (BT-56)		
Clear	1000 watt	980
Color-improved	1000 watt	15.0
Deluxe-white	1000 watt	17.2
Metal halide (E-37)		
Clear	400 watt	810
Color-improved	1000 watt	930
Deluxe white	1500 watt	1620
High pressure sodium (Lucalox®)		
	250 watt	520
	400 watt	780
	1000 watt	810

®Registered trademark of the General Electric Company
*Luminance values perpendicular to lamp axis

FLAME STANDARDS

Value of Various Former Standards in International Candles

Standard Pentane lamp, burning pentane	10.0 candles
Standard Hefner lamp, burning amyl acetate	0.9 candles
Standard Carcel lamp, burning colza oil	9.6 candles

The *Carcel unit* is the horizontal intensity of the carcel lamp, burning 42 grams of colza oil per hour. For a consumption between 38 and 46 grams per hour the intensity may be considered proportional to the consumption.

The *Hefner unit* is the horizontal intensity of the Hefner lamp burning amyl acetate, with a flame 4 cm. high. If the flame is l mm. high, the intensity $I = 1 + 0.027(l - 40)$.

WAVE LENGTHS OF VARIOUS RADIATIONS

	Ångstroms
Cosmic rays	0.0005
Gamma rays	0.005–1.40
X-rays	0.1–100
Ultra violet, below	4000
Limit of sun's U.V. at earth's surface	2920
Visible spectrum	4000–7000
Violet, representative, 4100, limits	4000–4240
Blue, representative, 4700, limits	4240–4912
Green, representative, 5200, limits	4912–5750
Maximum visibility	5560
Yellow, representative, 5800, limits	5750–5850
Orange, representative, 6000, limits	5850–6470
Red, representative, 6500, limits	6470–7000
Infra red, greater than	7000
Hertzian waves, beyond	2.20×10^6

WAVE LENGTHS OF THE FRAUNHOFER LINES

Sun's Spectrum

At 15°C and 76 cm pressure. Wave length in Ångström units (Fabry and Buisson system).

Line	Due to	Wave length	Line	Due to	Wave length
U	Fe	2947.9	h	H	4101.750
t	Fe	2994.4	g	Ca	4226.742
T	Fe	3021.067	G {	Fe	4307.914
s	Fe	3047.623	{	Ca	4307.749
S_1 {	Fe	3100.683	G'	H	4340.477
S_2 {	Fe	3100.326	F	H	4861.344
{	Fe	3099.943	b_4 {	Fe	5167.510
R {	Ca	3181.277	{	Mg	5167.330
{	Ca	3179.343	b_2	Mg	5172.700
Q	Fe	3286.773	b_1	Mg	5183.621
P	Ti	3361.194	E_2	Fe	5269.557
O	Fe	3441.020	D_2	Na	5889.977
N	Fe	3581.210	D_1	Na	5895.944
M	Fe	3727.636	C	H	6562.816
L	Fe	3820.438	B	O	6869.955
K	Ca	3933.684	A {	O	7621
H	Ca	3968.494	{	O	7594
			Z	....	8228.5
			Y	....	8990.0

BRIGHTNESS OF TUNGSTEN

Characteristics of Straight Tungsten Wire in a Vacuum (Forsythe and Worthing, 1924).

Temperature °K			Brightness candles/cm²	$\dfrac{B}{dB}\dfrac{dt}{T}$
Absolute	Brightness	Color		
1000	966	1006	0.00012	22.0
1200	1149	1210	0.006	20.0
1400	1330	1414	0.11	17.2
1600	1509	1619	0.92	15.2
1800	1684	1825	5.05	13.7
2000	1857	2033	20.0	12.3
2200	2026	2242	61.3	11.2
2400	2192	2452	157.0	10.3
2600	2356	2663	347.0	9.6
2800	2516	2878	694.0	8.9
3000	2673	3094	1257.0	8.3
3200	2827	3311	2110.0	7.8
3400	2978	3533	3370.0	7.6
3655*	3165	3817	5740.0	7.3

* Melting-point of tungsten.

WAVE LENGTHS FOR SPECTROSCOPE CALIBRATION

Source	Wave length	Source	Wave length
Potassium flame	0.7699μ	E, solar	0.5270μ
Potassium flame	0.7665	b_1, solar or magnesium flame	0.5184
Mercury I arc	0.6907		
B, solar	0.6869	b_2, solar or magnesium flame	0.5173
Lithium flame	0.6708		
C, solar or hydrogen tube	0.6563	Mercury I arc	0.4960
Mercury I arc	0.6234	Mercury I arc	0.4916
D_1, solar or sodium flame	0.5896	F, solar or hydrogen tube	0.4861
D_2, solar or sodium flame	0.5890	Strontium flame	0.4608
Mercury I arc	0.5791	Mercury I arc	0.4358
Mercury I arc	0.5770	G, solar or hydrogen tube	0.4340
Mercury I arc	0.5461	Mercury I arc	0.4047
Thallium flame	0.5351	H_1, solar	0.3969
		K, solar	0.3934

STANDARD UNITS, SYMBOLS, AND DEFINING EQUATIONS FOR FUNDAMENTAL PHOTOMETRIC AND RADIOMETRIC QUANTITIES

Submitted by Abraham Abramowitz
from Z-7.1-1967

Radiometric Quantities
(See Note at bottom of page)

Quantity*	Symbol*	Defining Equation**	Commonly Used Units	Symbol
Radiant energy	$Q, (Q_e)$		erg	
			†joule	J
			kilowatt-hour	kWh
Radiant density	$w, (w_e)$	$w = dQ/dV$	†joule per cubic meter	J/m^3
			erg per cubic centimeter	erg/cm^3
Radiant flux	$\Phi, (\Phi_e)$	$\Phi = dQ/dt$	erg per second	erg/s
			†watt	W
Radiant flux density at a surface				
Radiant exitance (Radiant emittance)†	$M, (M_e)$	$M = d\Phi/dA$	watt per square centimeter	W/cm^2
Irradiance	$E, (E_e)$	$E = d\Phi/dA$	‡watt per square meter, etc.	W/m^2
Radiant intensity	$I, (I_e)$	$I = d\Phi/d\omega$ (ω = solid angle through which flux from point source is radiated)	‡watt per steradian	W/sr
Radiance	$L, (L_e)$	$L = d^2\Phi/d\omega\,(dA\cos\theta)$ $= dI/(dA\cos\theta)$ (θ = angle between line of sight and normal to surface considered)	watt per steradian and square centimeter †watt per steradian and square meter	$W \cdot sr^{-1}\,cm^{-2}$ $W \cdot sr^{-1}\,m^{-2}$
Emissivity	ε	$\varepsilon = M/M_{blackbody}$ (M and $M_{blackbody}$ are respectively the radiant exitance of the measured specimen and that of a blackbody at the same temperature as the specimen)	one (numeric)	—

Note: The symbols for photometric quantities (see following table) are the same as those for the corresponding radiometric quantities (see above). When it is necessary to differentiate them the subscripts v and e respectively should be used, e.g., Q_v and Q_e.

*Quantities may be restricted to a narrow wavelength band by adding the word spectral and indicating the wavelength. The corresponding symbols are changed by adding a subscript λ, e.g., Q_λ for a spectral concentration or a λ in parentheses, e.g., $K(\lambda)$, for a function of wavelength.

**The equations in this column are given merely for identification.

***Φ_i = incident flux
   Φ_a = absorbed flux
   Φ_r = reflected flux
   Φ_t = transmitted flux
†to be deprecated.
‡International System (SI) unit.

PHOTOMETRIC QUANTITIES

Quantity*	Symbol*	Defining Equation**	Commonly Used Units	Symbol
Absorptance	$\alpha, (\alpha_v, \alpha_e)$	$\alpha = \Phi_a/\Phi_i$***	one (numeric)	—
Reflectance	$\rho, (\rho_v, \rho_e)$	$\rho = \Phi_r/\Phi_i$***	one (numeric)	—
Transmittance	$\tau, (\tau_v, \tau_e)$	$\tau = \Phi_t/\Phi_i$***	one (numeric)	—
Luminous energy (quantity of light)	$Q, (Q_v)$	$Q_v = \int_{380}^{760} K(\lambda) Q_e \lambda\, d\lambda$	lumen-hour ‡lumen-second (talbot)	lm·h lm·s
Luminous density	$w, (w_v)$	$w = dQ/dV$	‡lumen-second per cubic meter	lm·s·m⁻³
Luminous flux	$\Phi, (\Phi_v)$	$\Phi = dQ/dt$	‡lumen	lm
Luminous flux density at a surface				
Luminous exitance (Luminous emittance)†	$M, (M_v)$	$M = d\Phi/dA$	lumen per square foot	lm/ft²
Illumination (illuminance)	$E, (E_v)$	$E = d\Phi/dA$	footcandle (lumen per square foot) ‡lux (lm/m²) phot (lm/cm²)	fc lx ph
Luminous intensity (candlepower)	$I, (I_v)$	$I = d\Phi/d\omega$ (ω = solid angle through which flux from point source is radiated)	‡candela (lumen per steradian)	cd
Luminance (photometric brightness)	$L, (L_v)$	$L = d^2\Phi/d\omega\,(dA\cos\theta)$ $= dI/(dA\cos\theta)$ (θ = angle between line of sight and normal to surface considered)	candela per unit area stilb (cd/cm²) nit (†cd/m²) footlambert (cd/πft²) lambert (cd/πcm²) apostilb (cd/πm²)	cd/in², etc. sb nt fL L asb
Luminous efficacy	K	$K = \Phi_n/\Phi_n$	†lumen per watt	lm/W
Luminous efficiency	V	$V = K/K_{maximum}$ ($K_{maximum}$ = maximum value of $K(\lambda)$ function)	one (numeric)	—

ILLUMINATION CONVERSION FACTORS

1 lumen = 1/680 lightwatt (sec 3.7.2)
1 lumen-hour = 60 lumen-minutes
1 footcandle = 1 lumen/ft²

1 watt-second = 1 joule = 10^7 ergs
1 phot = 1 lumen/cm²
1 lux = 1 lumen/m²

Number of →
Multiplied by↘

Equals Number of ↓	Footcandles	*Lux	Phots	Milliphots
Footcandles	1	0.0929	929	0.929
*Lux	10.76	1	10,000	10
Phot	0.00108	0.0001	1	0.001
Milliphot	1.076	0.1	1,000	1

*The International Standard (SI) unit.

LINE SPECTRA OF THE ELEMENTS

Edited by Joseph Reader and Charles H. Corliss*
National Bureau of Standards

These tables were prepared under the auspices of the Committee on Line Spectra of the Elements of the National Academy of Sciences—National Research Council. They contain the outstanding spectral lines of neutral (I), singly ionized (II), doubly ionized (III), triply ionized (IV), and quadruply ionized (V) atoms. Listed are lines that appear in emission from the vacuum ultraviolet to the far infrared. For most atoms these lines were selected from much larger lists in such a way as to include the stronger observed lines in each spectral region. In a few cases prominent monoxide band heads are also given.

The data were compiled by the following contributors, whose initials are given in the headings of the tables that they prepared:

- J. G. Conway—Lawrence Berkeley Laboratory
- C. H. Corliss—National Bureau of Standards
- R. D. Cowan—Los Alamos Scientific Laboratory
- C. R. Cowley—University of Michigan
- Henry M. and Hannah Crosswhite—Argonne National Laboratory
- S. P. Davis—University of California, Berkeley
- V. Kaufman—National Bureau of Standards
- R. L. Kelly—Naval Postgraduate School
- J. F. Kielkopf—University of Louisville
- W. C. Martin—National Bureau of Standards
- T. K. McCubbin—Pennsylvania State University
- L. J. Radziemski—Los Alamos Scientific Laboratory
- J. Reader—National Bureau of Standards
- C. J. Sansonetti—National Bureau of Standards
- G. V. Shalimoff—Lawrence Berkeley Laboratory
- R. W. Stanley—Purdue University
- J. O. Stoner, Jr.—University of Arizona
- H. H. Stroke—New York University
- D. R. Wood—Wright State University
- E. F. Worden—Lawrence Livermore Laboratory
- J. J. Wynne—International Business Machines Corporation
- R. Zalubas—National Bureau of Standards

The literature references are collected at the end of the entire set of tables.

All wavelengths are given in Angstroms. Below 2000 Å the wavelengths are in vacuum; above 2000 Å the wavelengths are in air. Wavelengths given to three decimal places have an uncertainty of less than 0.001 Å and are therefore suitable for the calibration of most spectrographs. In the air region, the elements used most commonly for calibration purposes are Ne, Ar, Kr, Fe, Th, and Hg; in the vacuum region, the most common are C, N, O, Si, Cu.

A large number of the lines for neutral and singly ionized atoms were extracted from the National Bureau of Standards (NBS) Tables of Spectral-Line Intensities.[1] The intensities of these lines represent quantitative estimates of relative line strengths that take account of varying detection sensitivity at different wavelengths. They are on a linear scale. For nearly all of the other lines the intensities represent qualitative estimates of the relative strengths of lines not greatly separated in wavelength. Because different observers frequently use different scales for their intensity estimates, these intensities are useful only as a rough indication of the appearance of a spectrum. In some cases the intensity scale is not intended to be linear. In the tables of first and second spectra the intensities of the lines of the singly ionized atom relative to those of the neutral atom should be used with caution, inasmuch as the concentration of ions in a light source depends greatly on the excitation conditions.

Descriptive symbols used in the tables have the following meanings:

- c—complex
- d—line consists of two unresolved lines
- h—hazy
- l—shaded to longer wavelengths
- s—shaded to shorter wavelengths
- p—perturbed by a close line
- b—band head
- r—easily reversed
- w—wide

ACTINIUM (Ac)
Z = 89

Ac I and II
Ref. 193 — J.G.C.

Intensity		Wavelength (Air)	Spectrum
8	h	2100.00	II
20		2712.50	II
10		2726.23	II
10	h	2760.18	II
10	h	2781.56	II
20		2797.59	II
20		2806.76	II
8		2833.47	II
150	h	2847.16	II
8		2895.20	II
30		2896.82	II
30		2923.02	II
200		2994.17	II
500		3043.30	II
200		3069.36	II
100		3078.07	II
100		3086.04	II
100		3087.37	II
200		3112.83	II
100		3120.16	II
500	s	3153.09	II
600	s	3154.41	II
200	s	3164.81	II
300	s	3230.59	II
150	s	3237.70	II
500		3260.91	II
100	s	3318.01	II
200	s	3383.53	II
200		3413.84	II
500		3417.77	II
500	s	3481.16	II
200		3489.53	II
100		3529.24	II
100		3534.63	II
200	s	3554.99	II
1000	s	3565.59	II
100		3694.88	II
300	s	3756.67	II
200		3799.82	II
2000	s	3863.12	II
100		3914.47	II
400	s	4061.60	II
3000		4088.44	II
3000	s	4168.40	II
100		4179.76	I
20		4183.12	I
20		4194.40	I
300	s	4209.69	II
300		4359.13	II
20	1	4384.53	I
1000	1	4386.41	II
20		4396.71	I
20		4462.73	I
1000	1	4507.20	II
500		4605.45	II
10		4716.58	I
400	s	4720.16	II
300		4812.22	II
100		4945.18	II
100		4958.23	II
100		4960.87	II
150		5446.38	II
300	1	5732.05	II
400		5758.97	II
1000		5910.85	II
600	1	6164.75	II
200	1	6167.83	II
400		6242.83	II
20		6359.86	I
20	1	6691.27	I
6		7290.40	I
6	1	7866.10	I

Ac III
Ref. 193 — J.G.C.

Intensity		Wavelength (Air)	Spectrum
1000	h	2626.44	III
50	h	2682.90	III
2000	h	2952.55	III
2000	h	3392.78	III
3000		3487.59	III
2000	h	4413.09	III
3000	h	4569.87	III
8	h	5193.21	III

Ac IV
Ref. 193 — J.G.C.

Intensity		Wavelength (Air)	Spectrum
20	h	2062.00	IV
30	h	2502.12	IV
100	h	2558.08	IV
5	h	2790.83	IV
50	h	2793.90	IV
20	1	3224.7	IV

ALUMINUM (Al)
Z = 13

Al I and II
Ref. 81, 89, 144, 227, 228, 282 — E.F.W.

Intensity	Wavelength (Vacuum)	Spectrum
40	1177.43	II
50	1191.812	II
150	1350.18	II
800	1539.830	II
100	1569.385	II
125	1596.059	II
150	1625.627	II
100	1644.235	II
100	1644.809	II
1000	1670.787	II
100	1686.250	II
800	1719.440	II
500	1721.244	II
900	1721.271	II
500	1724.952	II
900	1724.984	II
350	1760.104	II
300	1761.975	II
290	1763.00	I
500	1763.869	II
700	1763.952	II
450	1765.64	I

* Charles H. Corliss is now retired.

Aluminum (Cont.)

Intensity	Wavelength	
300	1765.815	II
450	1766.38	I
400	1767.731	II
450	1769.14	I
600	1828.588	II
400	1832.837	II
250	1834.808	II
300	1855.929	II
700	1858.026	II
120	1859.980	II
1000	1862.311	II
200	1929.978	II
150	1931.048	II
200	1932.377	II
400	1934.503	II
150	1934.713	II
150	1936.907	II
220	1939.261	II
700	1990.531	II

Air

Intensity	Wavelength	
150	2016.052	II
150	2016.234	II
100	2016.368	II
200	2074.008	II
700	2094.264	II
150	2094.744	II
300	2094.791	II
100	2095.104	II
200	2095.141	II
60	2150.70	I
60	2181.00	I
400	2269.10	I
120	2269.22	I
60	2312.49	I
70	2313.53	I
90	2317.48	I
60	2319.06	I
140	2321.56	I
460	2367.05	I
110	2367.61	I
110	2368.11	I
100	2369.30	I
140	2370.22	I
70	2370.73	I
160	2372.07	I
850	2373.13	I
170	2373.35	I
110	2373.57	I
60	2378.40	I
60	2513.30	I
240	2567.98	I
480	2575.10	I
60	2575.40	I
80	2631.55	II
110	2637.70	II
150	2652.48	I
200	2660.39	I
160	2669.17	II
650	2816.19	II
90	2837.96	I
90	2840.10	I
150	3041.28	II
360	3050.07	I
60	3054.68	I
450	3057.14	I
90	3064.29	I
60	3066.14	I
150	3074.64	II
4500 rS	3082.153	I
7200 r	3092.710	I
1800 r	3092.839	I
150	3428.92	II
70	3439.35	I
150	3443.64	I
70	3444.86	I
70	3458.22	I
60	3479.81	I
60	3482.63	I
450	3586.56	II
360	3587.07	II
290	3587.45	II
220	3651.06	II
110	3651.10	II
150	3654.98	II
290	3655.00	II
450	3900.68	II
450	3932.00	II
4500 r	3944.006	I
9000 r	3961.520	I
110	3995.86	II
290	4226.81	II
150	4585.82	II
110	4588.19	II
550	4666.80	II
110	4898.76	II
110	4902.77	II
150	5280.21	II
70	5107.52	I
290	5283.77	II
150	5285.85	II
110	5312.32	II
220	5316.07	II
150	5371.84	II
180	5557.06	I
110	5557.95	I
450	5593.23	II
110	5853.62	II
220	5971.94	II
290	6001.76	II
220	6001.88	II
450	6006.42	II
150	6061.11	II
290	6068.43	II
110	6068.53	II
450	6073.23	II
110	6181.57	II
150	6181.68	II
290	6182.28	II
220	6182.45	II
450 h	6183.42	II
450	6201.52	II
360	6201.70	II
290	6226.18	II
360	6231.78	II
450	6243.36	II
450	6335.74	II
360	6696.02	I
230	6698.67	I
60	7083.97	I
70	7084.64	I
110	7361.57	I
140	7362.30	I
60	7606.16	I
90	7614.82	I
230	7835.31	I
290	7836.13	I
60	7993.05	I
90	8003.19	I
70	8065.97	I
110	8075.35	I
290	8640.70	II
360	8772.87	I
450	8773.90	I
110	8900.31	I
180	8941.28	I
90	8912.90	I
140	8923.56	I
60	9089.91	I
70	9139.95	I
150	9290.65	II
110	9290.75	II
150	10076.29	II
110	10768.36	I
140	10782.04	I
110	10872.98	I
230	10891.73	I
450	11253.19	I
570	11254.88	I
570	13123.41	I
450	13150.76	I
230	16718.96	I
300	16750.56	I
140	16763.36	I
300	21093.04	I
360	21163.75	I

Al III
Ref. 127 — E.F.W.

Intensity	Wavelength	
	Vacuum	
70	486.884	III
30	486.912	III
250	511.138	III
150	511.191	III
500	560.317	III
200	560.433	III
100	670.068	III
200	671.118	III
500	695.829	III
400	696.217	III
200	725.683	III
300	726.915	III
400	855.034	III
500	856.746	III
400	892.024	III
50	893.887	III
450	893.897	III
10	1162.59	III
5	1162.62	III
100	1352.81	III
5	1352.82	III
70	1352.86	III
600	1379.67	III
800	1384.13	III
700	1605.766	III
100	1611.814	III
800	1611.874	III
1000	1854.716	III
600	1862.790	III
300	1935.840	III
15	1935.863	III
200	1935.949	III

Air

Intensity	Wavelength	
110	2399.00	III
285	2762.77	III
220	2762.87	III
450	2906.93	III
360	3348.52	III
290	3350.88	III
870	3601.63	III
550	3601.93	III
750	3612.36	III
450	3702.11	III
550	3713.12	III
110 h	3980.14	III
110	4082.45	III
150	4088.61	III
110 h	4142.37	III
650	4149.92	III
650	4150.17	III
110 h	4364.64	III
650	4479.89	III
650	4479.97	III
760	4512.56	III
550	4528.94	III
870	4529.19	III
110	4701.15	III
150	4701.41	III
110 h	4904.10	III
110 h	5151.01	III
110 h	5163.89	III
1200	5696.60	III
1000	5722.73	III
110	6055.21	III
220 h	7635.37	III
150	7660.26	III
220	7681.97	III
360	7881.79	III
150	7882.52	III
290	7905.51	III
290 h	8243.59	III
360 h	8275.11	III
290	9571.52	III
360	9605.99	III

Al IV
Ref. 8, 146 — E.F.W.

Intensity	Wavelength	
	Vacuum	
400	124.03	IV
700	129.73	IV
800	160.07	IV
700	161.69	IV
500	1027.34	IV
800	1042.17	IV
700	1048.52	IV
500	1058.90	IV
500	1061.43	IV
600	1064.89	IV
500	1066.57	IV
600	1069.44	IV
400	1105.74	IV
600	1118.82	IV
500	1125.61	IV
400	1136.82	IV
400	1198.50	IV
400	1220.55	IV
900	1237.19	IV
600	1240.21	IV
700	1240.86	IV
700	1248.79	IV
900	1257.62	IV
800	1264.18	IV
1000	1272.76	IV
400	1337.90	IV
500	1376.62	IV
500	1388.79	IV
600	1431.94	IV
700	1441.82	IV
800	1447.51	IV
600	1457.96	IV
700	1486.89	IV
800	1494.79	IV
400	1519.07	IV
800	1537.54	IV
500	1550.19	IV
1000	1557.25	IV
500	1559.03	IV
700	1564.16	IV
900	1582.04	IV
800	1584.46	IV
400	1589.29	IV
400	1606.65	IV
400	1617.81	IV
600	1627.54	IV
500	1636.82	IV
800	1639.06	IV
1000	1818.56	IV
700	1881.16	IV

Air

Intensity	Wavelength	
400	2515.87	IV
500	3208.20	IV
600	3285.13	IV
400	3344.46	IV
500	3473.54	IV
900	3492.23	IV
800	3508.46	IV
500	3511.28	IV
700	3517.56	IV
400	3527.03	IV
500	3541.08	IV

Al V
Ref. 6 — E.F.W.

Intensity	Wavelength	
	Vacuum	
300	103.80	V
400	103.88	V
250	104.07	V
250	104.18	V
600	107.95	V
300	108.06	V
300	108.11	V
250	118.50	V
900	125.53	V
800	126.07	V
800	130.41	V
1000	130.85	V
900	131.00	V
900	131.44	V
500	132.63	V
1000	278.69	V
900	281.39	V
250	1068.26	V
500	1088.67	V
300	1090.14	V
300	1150.30	V
350	1165.42	V
250	1168.48	V
300	1287.70	V
400	1330.06	V
400	1350.52	V
400	1363.35	V
600	1369.20	V
300	1373.70	V
400	1445.87	V
300	1455.26	V
600	1475.64	V
300	1486.05	V
700	1508.37	V
1000	1526.14	V
500	1539.12	V
300	1577.90	V
350	1589.87	V

AMERICIUM (Am)
Z = 95

Am I and II
Ref. 92 — J.G.C.

Intensity	Wavelength	
	Air	
100 s	2706.35	II
100 s	2728.69	II
200 s	2756.55	II
100 s	2812.10	II
200 s	2812.92	II
1000 l	2815.28	II
100 s	2815.98	II
100 l	2831.24	II
5000 l	2832.26	II
100 l	2833.95	II
100 l	2861.92	II
100 s	2866.20	II
1000 l	2888.51	II
100 l	2893.29	II
200 l	2899.56	II
100 l	2909.86	II
200 l	2911.13	II
1000 s	2920.59	II
200 l	2927.53	II
200 l	2936.99	II
100 l	2939.08	II
500 l	2950.39	II
100 s	2957.05	II
100 l	2958.39	II
100 l	2963.02	II
1000 l	2966.71	II
1000 l	2969.29	II
1000 s	2987.24	II
500 l	2993.51	II
1000 s	3004.25	II

Americium (Cont.)

Intensity		Wavelength	Spectrum
500	s	3027.99	II
100	l	3028.86	II
500	l	3038.36	II
200	s	3053.69	II
2000		3120.49	II
200	s	3161.83	II
100		3167.86	II
100	l	3203.26	II
500	l	3282.32	II
200	l	3286.67	II
100		2343.87	I
500	s	3362.55	II
200	l	3395.01	I
200	s	3419.66	II
200		3446.19	I
1000	l	3452.10	II
5000	l	3483.31	II
5000		3510.13	I
1000	l	3530.95	I
200	s	3562.68	II
5000		3569.16	I
100	s	3596.07	II
500		3603.41	I
5000		3673.12	I
100	l	3684.57	II
1000	s	3696.42	II
100	l	3707.86	II
5000	l	3777.50	II
5000	l	3926.25	II
1000	s	3952.58	II
100	s	4020.25	I
100	s	4035.81	I
500	s	4036.37	II
5000	s	4089.29	II
100	l	4089.32	II
100		4140.96	I
1000	s	4188.12	II
1000		4265.55	I
5000		4289.26	I
200	s	4309.65	II
2000	s	4324.57	II
2000		4441.36	II
5000	l	4509.45	II
5000	l	4575.59	II
1000	s	4593.31	II
100	l	4649.12	I
100	l	4653.45	I
5000	l	4662.79	I
2000	l	4681.65	I
2000	l	4699.70	II
1000	l	4706.80	I
2000	l	4872.22	II
200	l	4990.79	I
100	s	5000.21	I
1000	l	5020.96	II
200	l	5215.99	II
1000	s	5402.62	I
1000	l	5424.70	I
1000	l	5584.21	II
1000	l	5598.13	I
10000	l	6054.64	I
1000	l	6405.11	I
500	l	6544.16	I
500	s	6955.58	I

Intensity		Wavelength	Spectrum
6		1565.51	II
8		1576.11	II
7		1581.36	II
80	r	1599.96	I
10		1606.98	II
200	w	1612.8	I
100		1623.3	I
50	h	1651.20	I
20		1657.04	II
100	w	1662.6	I
50		1698.85	I
80	r	1716.93	I
150	r	1717.45	I
150	r	1723.43	I
100	r	1736.19	I
8		1736.43	II
50		1757.79	I
100	h	1765.76	I
100		1780.87	I
100	r	1788.24	I
150		1800.18	I
50	r	1810.50	I
80	r	1814.20	I
100		1829.50	I
60		1858.89	I
50	r	1868.17	I
300	r	1871.15	I
150	r	1882.56	I
70		1891.28	I
70		1899.39	I
100		1927.08	I
200	r	1950.39	I
80	h	1964.3	I
60		1986.05	I
6		1990.60	II

Air

Intensity		Wavelength	Spectrum
50		2024.00	I
60	r	2029.49	I
70	r	2039.77	I
150	r	2049.57	I
50		2063.43	I
1000	r	2068.33	I
100		2079.56	I
50	r	2098.41	I
80	r	2118.48	I
100	r	2127.39	I
50	r	2137.05	I
100	r	2139.69	I
10		2141.80	II
50	r	2141.83	I
100	r	2144.86	I
50		2158.91	I
1500	r	2175.81	I
250	r	2179.19	I
6		2179.25	II
200	r	2201.32	I
300	r	2208.45	I
6		2208.50	II
150	r	2220.73	I
100		2221.98	I
120	r	2224.93	I
6		2225.15	II
300	r	2262.51	I
120		2288.98	I
150	r	2293.44	I
300	r	2306.46	I
2500	r	2311.47	I
150		2315.89	I
400	h	2373.67	I
300	h	2383.64	I
100		2395.22	I
150		2422.13	I
250		2426.35	I
400	r	2445.51	I
400		2478.32	I
150		2480.44	I
8		2480.46	II
100		2510.54	I
2000	r	2528.52	I
15		2528.54	II
10		2567.75	II
150		2574.06	I
1500	r	2598.05	I
500	r	2598.09	I
300	r	2612.31	I
200	r	2652.60	I
12		2656.55	II
300	r	2670.64	I
200	r	2682.76	I
120		2692.25	I
150	r	2718.90	I
400	r	2769.95	I
12		2851.09	II
100		2851.11	I
1000	r	2877.92	I
12		2966.10	II
15		2980.96	II
500	r	3029.83	I
12		3034.01	II
12		3040.67	II

Intensity		Wavelength	Spectrum
600	r	3232.52	I
20		3241.28	II
700	r	3267.51	I
12		3383.09	II
100		3383.15	I
15		3498.46	II
12		3520.47	II
25		3637.80	II
250		3637.83	II
20		3722.78	II
200	r	3722.79	I
20		3850.22	II
200		4033.55	I
20		4033.56	II
20		4133.63	II
15		4140.54	II
15		4195.17	II
20		4219.07	II
20		4314.32	II
12		4344.83	II
12		4411.42	II
12		4446.48	II
12		4506.92	II
15		4514.50	II
30		4596.90	II
20		4599.09	II
15		4604.77	II
30		4647.32	II
20		4675.74	II
40		4711.26	II
12		4735.44	II
20		4757.81	II
20		4765.36	II
12		4766.91	II
30		4784.03	II
20		4802.01	II
20		4832.82	II
20		4877.24	II
15		4947.40	II
15		5044.56	II
12		5166.32	II
15		5176.55	II
20		5238.94	II
20		5354.24	II
15		5464.08	II
30	h	5490.32	I
40	h	5556.10	I
15		5568.13	II
30	l	5602.19	I
100	l	5632.02	I
30		5639.75	II
60	h	5830.34	I
15		5895.09	II
100		6005.21	II
20		6053.41	II
30		6079.80	II
50		6130.04	II
20		6154.94	II
12		6302.76	II
20		6611.49	I
30		6647.44	II
15		6688.01	II
6		6806.67	II
30	h	7648.28	I
80		7844.44	I
200		7924.65	I
40	h	7969.55	I
60		8411.69	I
150		8572.64	I
100		8619.55	I
30	h	8682.7	I
30		9132.21	I
400		9518.68	I
30		9866.78	I
400		9949.14	I
200		10078.49	I
300		10261.01	I
50		10364.33	I
50	h	10488.3	I
200		10585.60	I
1000		10677.41	I
800		10741.94	I
80		10794.11	I
600		10839.17	I
200		10868.58	I
400		10879.55	I
300		11012.79	I
40	h	11079.95	I
30	h	11084.98	I
50	h	11104.84	I
50	h	11108.52	I
30		11189.61	I
150		11266.23	I
30		11863.37	I
1		11957.7	I
5		12116.06	I
2		12276.6	I
2		12466.75	I

Sb III
Ref. 164 — L.J.R. and J.R.

Intensity   Wavelength

Intensity	Wavelength	Spectrum
	Vacuum	
10	691.18	III
10	698.69	III
15	722.86	III
8	724.81	III
15	732.33	III
15	999.62	III
40	1011.94	III
10	1056.58	III
40	1065.90	III
20	1069.93	III
20	1070.43	III
5	1073.76	III
30	1075.82	III
5	1078.10	III
20	1084.06	III
10	1098.34	III
10	1135.43	III
30	1151.49	III
40	1157.74	III
12	1166.96	III
50	1205.20	III
50	1210.64	III
20	1306.69	III
8	1379.58	III
20	1404.18	III
10	1429.57	III
15	1673.89	III
3	1710.23	III
15	1711.84	III
15	1725.33	III
12	1762.30	III
12	1839.32	III
10	1946.13	III
	Air	
3	2054.10	III
2	2091.85	III
5	2127.00	III
5	2507.71	III
15	2590.13	III
1	2614.20	III
12	2617.17	III
1	2617.63	III
20	2669.39	III
5	2785.87	III
20	2790.27	III
20	3336.61	III
50	3504.07	III
15	3519.06	III
15	3533.45	III
40	3559.18	III
40	3566.25	III
30	3738.90	III
40	4265.09	III
50	4352.16	III
30	4591.89	III
30	4692.91	III
1	5247.71	III
1	5690.8	III
1	5717.3	III
3	5845.5	III
5	6246.7	III
3	6287.6	III

Sb IV
Ref. 386 — L.J.R.

Intensity	Wavelength		
	Vacuum		
861.5	SB	IV	1
873.5	SB	IV	1
888.3	SB	IV	1
891.2	SB	IV	1
1087.6	SB	IV	1
1099.3	SB	IV	1
1115.1	SB	IV	1
1120.4	SB	IV	1
1145.9	SB	IV	1
1151.5	SB	IV	1
1171.4	SB	IV	1
1192.9	SB	IV	1
1199.1	SB	IV	1
1499.2	SB	IV	1

Sb V
Ref. 406 — L.J.R.

Intensity	Wavelength	
	Vacuum	
3	699.22	V
1	746.06	V
6	831.00	V
1	898.02	V
8	1104.32	V
12	1226.00	V
12	1505.70	V
12	1524.47	V

ANTIMONY (Sb)
Z = 51

Sb I and II
Ref. 167, 194 — L.J.R. and J.R.

Intensity		Wavelength	Spectrum
		Vacuum	
1		691.20	II
1		764.43	II
1		814.85	II
1		849.39	II
2		855.08	II
4		876.84	II
4		921.07	II
6		983.57	II
6		1001.13	II
6		1009.43	II
6		1052.21	II
8		1056.27	II
8		1057.32	II
6		1073.81	II
6		1230.30	II
8		1274.98	II
8		1327.40	II
8		1358.04	II
8		1384.70	II
6		1407.83	II
10		1430.76	II
8		1436.49	II
10	h	1464.19	I
20	r	1486.57	I
40	h	1491.36	I
50	r	1512.57	I
120	r	1532.74	I
80	r	1535.06	I

ARGON (Ar)
Z = 18

Ar I and II
Ref. 190, 203, 204, 219, — E.F.W.

Intensity		Wavelength	

Vacuum

Intensity		Wavelength	
30		487.227	II
50		490.650	II
30		490.701	II
30		519.327	II
30		542.912	II
200		543.203	II
70		547.461	II
70		556.817	II
70		573.362	II
30		576.736	II
70		580.263	II
30		583.437	II
70		597.700	II
30		602.858	II
30		612.372	II
500		661.867	II
30		664.562	II
200		666.011	II
1000		670.946	II
3000		671.851	II
70		676.242	II
30		677.952	II
30		679.218	II
200		679.401	II
200		718.090	II
3000		723.361	II
500		725.548	II
70		730.930	II
200		740.269	II
200		744.925	II
70		745.322	II
20		803.059	I
100		806.471	I
60		806.869	I
30		807.218	I
40		807.653	I
50		809.927	I
120		816.232	I
70		816.464	I
80		820.124	I
120		825.346	I
120		826.365	I
150		834.392	I
100		835.002	I
100		842.805	I
180		866.800	I
150		869.754	I
180	r	876.058	I
180	r	879.947	I
150		894.310	I
1000		919.781	II
1000		932.054	II
1000	r	1048.220	I
500	r	1066.660	I

Air

Intensity		Wavelength	
5		2420.456	II
10		2516.789	II
10		2534.709	II
15		2562.087	II
25		2891.612	II
200		2942.893	II
100		2979.050	II
50		3033.508	II
50		3093.402	II
8		3200.37	I
20		3243.689	II
25		3293.640	II
20		3307.228	II
7		3319.34	I
25		3350.924	II
7		3373.47	I
25		3376.436	II
25		3388.531	II
7		3393.73	I
7		3461.07	I
70		3476.747	II
20		3478.232	II
50		3491.244	II
100		3491.536	II
70		3509.778	II
70		3514.388	II
70		3545.596	II
70		3545.845	II
7		3554.306	I
100		3559.508	II
100		3561.030	II
70		3576.616	II
25		3581.608	II
50		3582.355	II
70		3588.441	II
7		3606.522	I
25		3622.138	II
20		3639.833	II
35		3718.206	II
70		3729.309	II
50		3737.889	II
150		3765.270	II
50		3766.119	II
20		3770.369	I
20		3770.520	II
25		3780.840	II
25		3803.172	II
50		3809.456	II
7		3834.679	I
70		3850.581	II
35		3868.528	II
35		3925.719	II
50		3928.623	II
25		3932.547	II
70		3946.097	II
7		3947.505	I
35		3948.979	I
20		3979.356	II
35		3994.792	II
50		4013.857	II
50		4033.809	II
20		4035.460	II
150		4042.894	II
50		4044.418	I
100		4052.921	II
200		4072.005	II
70		4072.385	II
25		4076.628	II
35		4079.574	II
25		4082.387	II
150		4103.912	II
300		4131.724	II
35		4156.086	II
400		4158.590	I
50		4164.180	I
25		4179.297	II
50		4181.884	I
100		4190.713	I
50		4191.029	I
200		4198.317	I
400		4200.674	I
25		4218.665	II
25		4222.637	II
25		4226.988	II
100		4228.158	II
100		4237.220	II
25		4251.185	I
200		4259.362	I
100		4266.286	I
70		4266.527	II
150		4272.169	I
550		4277.528	II
20		4282.898	II
100		4300.101	I
25		4300.650	II
70		4309.239	II
200		4331.200	II
50		4332.030	II
100		4333.561	I
50		4335.338	I
25		4345.168	I
800		4348.064	II
50		4352.205	II
25		4362.066	II
50		4367.832	II
200		4370.753	II
70		4371.329	II
50		4375.954	II
150		4379.667	II
50		4385.057	II
70		4400.097	II
200		4400.986	II
400		4426.001	II
150		4430.189	II
50		4430.996	II
50		4433.838	II
20		4439.461	II
35		4448.879	II
100		4474.759	II
200		4481.811	II
100		4510.733	I
20		4522.323	I
20		4530.552	II
400		4545.052	II
20		4564.405	II
400		4579.350	II
400		4589.898	II
15		4596.097	I
550		4609.567	II
7		4628.441	I
35		4637.233	II
400		4657.901	II
15		4702.316	I
20		4721.591	II
550		4726.868	II
50		4732.053	II
300		4735.906	II
800		4764.865	II
550		4806.020	II
150		4847.810	II
50		4865.910	II
800		4879.864	II
70		4889.042	II
20		4904.752	II
35		4933.209	II
200		4965.080	II
50		5009.334	II
70		5017.163	II
70		5062.037	II
20		5090.495	II
100		5141.783	II
70		5145.308	II
5		5151.391	I
15		5162.285	I
25		5165.773	II
20		5187.746	I
20		5216.814	I
7		5221.271	I
5		5421.352	I
10		5451.652	I
25		5495.874	I
5		5506.113	I
25		5558.702	I
5		5572.541	I
35		5606.733	I
20		5650.704	I
10		5739.520	I
5		5834.263	I
10		5860.310	I
15		5882.624	I
25		5888.584	I
50		5912.085	I
15		5928.813	I
5		5942.669	I
7		5987.302	I
5		5998.999	I
5		6025.150	I
70		6032.127	I
35		6043.223	I
10		6052.723	I
20		6059.372	I
7		6098.803	I
10		6105.635	I
100		6114.923	II
10		6145.441	I
7		6170.174	I
150		6172.278	II
10		6173.096	I
10		6212.503	I
5		6215.938	I
25		6243.120	II
7		6296.872	I
15		6307.657	I
7		6369.575	I
20		6384.717	I
70		6416.307	I
25		6483.082	II
15		6538.112	I
15		6604.853	I
25		6638.221	II
20		6639.740	II
50		6643.698	II
5		6660.676	I
5		6664.051	I
25		6666.359	II
100		6677.282	I
35		6684.293	II
150		6752.834	I
5		6756.163	I
15		6766.612	I
20		6861.269	II
150		6871.289	I
5		6879.582	I
10		6888.174	I
50		6937.664	I
7		6951.478	I
7		6960.250	I
10000		6965.431	I
150		7030.251	I
10000		7067.218	I
100		7068.736	I
25		7107.478	I
25		7125.820	I
1000		7147.042	I
15		7158.839	I
70		7206.980	I
15		7265.172	I
7		7270.664	I
2000		7272.936	I
35		7311.716	I
25		7316.005	I
5		7350.814	I
70		7353.293	I
200		7372.118	I
20		7380.426	II
10000		7383.980	I
20		7392.980	I
15		7412.337	I
10		7425.294	I
25		7435.368	I
10		7436.297	I
20000		7503.869	I
15000		7514.652	I
25000		7635.106	I
15000		7723.761	I
10000		7724.207	I
10		7891.075	I
20000		7948.176	I
20000		8006.157	I
25000		8014.786	I
7		8053.308	I
20000		8103.693	I
35000		8115.311	I
10000		8264.522	I
20		8392.21	I
15000		8408.210	I
20000		8424.648	I
15000		8521.442	I
7		8605.776	I
4500		8667.944	I
20		8771.860	II
180		8849.91	I
20		9075.394	I
35000		9122.967	I
550		9194.638	I
15000		9224.499	I
400		9291.531	I
1600		9354.220	I
25000		9657.786	I
4500		9784.503	I
180		10052.06	I
30		10332.12	I
100		10467.177	II
1600		10470.054	I
13		10478.034	I
180		10506.50	I
200		10673.565	I
11		10681.773	I
7		10683.034	II
30		10722.07	I
30		10759.16	I
7		10812.896	II
11		11078.869	I
30		11106.46	I
12		11441.832	I
400		11488.109	I
200		11668.710	I
12		11719.488	I
200		12112.326	I
50		12139.738	I
50		12343.393	I
200		12402.827	I
200		12439.321	I
100		12456.12	I
200		12487.663	I
150		12702.281	I
30		12733.418	I
12		12746.232	I
200		12802.739	I
50		12933.195	I
500		12956.659	I
200		13008.264	I
200		13213.99	I
200		13228.107	I
100		13230.90	I
500		13272.64	I
1000		13313.210	I
1000		13367.111	I
30		13499.41	I
1000		13504.191	I
11		13573.617	I
30		13599.333	I
400		13622.659	I
200		13678.550	I
1000		13718.577	I
10		13825.715	I
10		13907.478	I
200		14093.640	I
100		15046.50	I
25		15172.69	I
10		15329.34	I
30		15989.49	I
30		16519.86	I
500		16940.58	I
12		18427.76	I
50		20616.23	I
30		20986.11	I
20		23133.20	I
20		23966.52	I

Ar III
Ref. 367, 372, 373, 375 — E.F.W.

Intensity	Wavelength	

Vacuum

Intensity	Wavelength	
12	769.15	III

Argon (Cont.)

Intensity	Wavelength	
10	871.10	III
9	875.53	III
12	878.73	III
8	879.62	III
9	883.18	III
10	887.40	III
7	1669.67	III
7	1673.42	III
7	1675.48	III
9	1914.40	III
7	1915.56	III

Air

Intensity	Wavelength	
10	2125.16	III
15	2133.87	III
10	2138.59	III
10	2148.73	III
15	2166.19	III
10	2168.26	III
20	2170.23	III
25	2177.22	III
8	2184.06	III
10	2188.22	III
15	2192.06	III
7	2248.73	III
10	2279.10	III
7	2281.22	III
7	2282.21	III
12	2293.03	III
10	2300.85	III
15	2302.17	III
9	2317.00	III
15	2317.47	III
12	2318.04	III
10	2319.13	III
10	2319.37	III
9	2345.17	III
7	2351.67	III
9	2360.26	III
10	2395.63	III
12	2399.15	III
10	2413.20	III
7	2415.61	III
10	2418.82	III
12	2423.52	III
12	2423.93	III
7	2443.69	III
8	2472.95	III
7	2476.10	III
12	2488.86	III
7	2631.90	III
10	2654.63	III
8	2674.02	III
9	2678.38	III
10	2724.84	III
7	2762.23	III
7	2842.88	III
8	2855.29	III
9	2884.12	III
10	3010.02	III
12	3024.05	III
12	3054.82	III
10	3064.77	III
10	3078.15	III
7	3110.41	III
7	3127.90	III
25	3285.85	III
20	3301.88	III
15	3311.25	III
7	3323.59	III
25	3336.13	III
20	3344.72	III
15	3358.49	III
7	3361.28	III
15	3391.85	III
7	3417.49	III
9	3424.25	III
8	3438.04	III
9	3471.32	III
20	3480.55	III
12	3499.67	III
15	3503.58	III
8	3511.12	III
20	3795.37	III
10	3858.32	III
7	3907.84	III
8	3960.53	III
6	4023.60	III
5	4146.70	III

Ar IV
Ref. 367, 368, 374 — E.F.W.

Intensity	Wavelength	
	Vacuum	
4	396.87	IV
4	398.55	IV
6	623.77	IV
10	683.28	IV
7	688.39	IV
12 p	689.01	IV
6	699.41	IV
8	700.28	IV
4	754.20	IV
5	761.47	IV
5	800.57	IV
10	801.09	IV
10	801.41	IV
5	801.91	IV
15	840.03	IV
20	843.77	IV
25	850.60	IV
5	900.36	IV
9	901.17	IV

Air

Intensity	Wavelength	
4	2299.72	IV
8	2447.71	IV
12	2513.28	IV
6	2518.40	IV
9	2525.69	IV
12	2562.17	IV
10	2568.07	IV
7	2569.53	IV
12	2599.47	IV
10	2608.06	IV
7	2608.44	IV
12	2615.68	IV
6	2619.98	IV
12	2621.36	IV
12	2624.92	IV
15	2640.34	IV
9	2682.63	IV
14	2757.92	IV
10	2776.26	IV
12	2784.47	IV
14	2788.96	IV
7	2797.11	IV
16	2809.44	IV
10	2830.25	IV
6	2874.40	IV
12	2913.00	IV
11	2926.33	IV
6	3037.98	IV
8	3077.40	IV

Ar V
Ref. 414, 421 — E.F.W.

Intensity	Wavelength	
	Vacuum	
3	336.56	V
3	337.56	V
6	338.00	V
2	338.43	V
2	339.01	V
3	339.89	V
3	350.88	V
2	436.67	V
5	446.00	V
8	446.95	V
4	447.53	V
18	449.06	V
4	449.49	V
3	458.12	V
2	458.98	V
6 p	461.23	V
3	462.42	V
7	463.94	V
3	522.09	V
5	524.19	V
6	527.69	V
2	554.50	V
5	558.48	V
3	635.12	V
3	705.35	V
5	709.20	V
4	715.60	V
3	715.65	V
2	725.11	V
4	822.16	V
5	827.05	V
3	827.35	V
4 p	834.88	V
2	836.13	V

ARSENIC (As)
Z = 33
As I and II
Ref. 168, 197 — R.L.K.

Intensity	Wavelength	
	Vacuum	
165	761.24	II
165	802.83	II
340	1021.96	II
340	1082.35	II
500	1139.40	II
615	1149.31	II
555	1181.51	II
555	1189.87	II
615	1196.38	II
615	1196.56	II
340	1207.44	II
800	1211.17	II
800	1218.10	II
340	1223.15	II
760	1241.31	II
965	1243.08	II
870	1245.67	II
800	1258.58	II
965	1263.77	II
800	1266.34	II
800	1267.59	II
715	1280.99	II
715	1287.54	II
715	1305.70	II
340	1307.74	II
760	1333.15	II
965	1341.55	II
760	1355.93	II
965	1369.77	II
800	1373.65	II
1000	1375.07	II
760	1375.78	II
800	1394.64	II
800	1400.31	II
500	1448.59	II
500	1558.88	II
500	1570.99	II
100 r	1593.60	I
500	1660.55	II
100	1758.60	I
170	1806.15	I
340	1860.34	II
1000 r	1890.42	I
500	1912.94	II
800 r	1937.59	I
585 r	1972.62	I
170	1990.35	I
100 r	1991.13	I
100 r	1995.43	I

Air

Intensity	Wavelength	
230 r	2003.34	I
100 r	2009.19	I
100	2013.32	I
100	2112.99	I
100	2144.08	I
135	2165.52	I
350 r	2288.12	I
350 r	2349.84	I
100 r	2370.77	I
135 r	2381.18	I
170 r	2456.53	I
340	2602.00	II
170 r	2780.22	I
300	2830.359	II
300	2831.164	II
100 r	2860.44	I
300	2884.406	II
615	2959.572	II
300	3003.819	II
300	3116.516	II
340	3842.60	II
715	4190.082	II
615	4197.40	II
615	4242.982	II
500	4315.657	II
500	4323.867	II
500	4336.64	II
500	4352.145	II
425	4352.864	II
375	4371.17	II
615	4427.106	II
615	4431.562	II
715	4458.469	II
340	4461.075	II
715	4466.348	II
500	4474.46	II
800	4494.230	II
850	4507.659	II
615	4539.74	II
715	4543.483	II
615	4602.427	II
340	4629.787	II
340	4707.586	II
340	4730.67	II
340	4888.557	II
100	5068.98	I
340	5105.58	II
500	5107.55	II
100	5121.34	I
100	5141.63	I
425	5231.38	II
500	5331.23	II
100	5408.13	I
135	5451.32	I
340	5497.727	II
425	5558.09	II
425	5651.32	II
425	6110.07	II
500	6170.27	II
300	6511.74	II
300	7092.27	II
300	7102.72	II
340	7990.53	II
300	8174.51	II
100	8428.91	I
100	8564.71	I
100	8654.14	I
135	8821.73	I
100	8869.66	I
135	9267.28	I
200	9300.61	I
230	9597.95	I
290	9626.70	I
230	9833.76	I
100	9886.05	I
140	9900.55	I
170	9915.71	I
290	9923.05	I
100	10010.63	I
290	10024.04	I
100	10453.09	I
100	10575.02	I
170	10614.07	I

As III
Ref. 163 — R.L.K.

Intensity	Wavelength	
	Vacuum	
185	849.9	III
185	866.3	III
510	871.7	III
325	889.0	III
325	927.5	III
325	953.6	III
325	937.2	III
325	963.8	III
120	1172.2	III
185	1209.3	III

Air

Intensity	Wavelength	
80	2926.3	III
185	2982.0	III
325	3922.6	III
185	4037.2	III

As IV
Ref. 244 — R.L.K.

Intensity	Wavelength	
	Air	
150	2253.1	IV
200	2263.2	IV
200	2301.0	IV
250	2417.5	IV
150	2446.1	IV
250	2454.0	IV
200	2461.4	IV
150	3108.8	IV

As V
Ref. 280 — R.L.K.

Intensity	Wavelength	
	Vacuum	
25	600.7	V
40	616.0	V
120	715.5	V
150	734.8	V
60	737.2	V
250	987.7	V
250	1029.5	V
40	1051.6	V
60	1056.6	V

ASTATINE (At)
Z = 85
At I
Ref. 188 — E.F.W.

Astatine (Cont.)

Intensity		Wavelength	
		Air	
8		2162.25	I
10		2244.01	I

BARIUM (Ba)
Z = 56

Ba I and II
Ref. 1, 252, 277, 279 — J.J.W.

Intensity		Wavelength	
		Vacuum	
200		1486.72	II
400		1504.01	II
300		1554.38	II
200		1572.73	II
		1573.92	II
		1630.40	II
100		1674.51	II
400		1694.37	II
		1697.16	II
		1761.75	II
		1771.03	II
		1786.93	II
100		1904.15	II
500		1924.70	II
		1985.60	II
300		1999.54	II
		Air	
		2009.20	II
400		2023.95	II
		2052.68	II
		2054.57	II
500		2214.7	II
800		2245.61	II
1000		2254.73	II
1400		2304.24	II
2000		2335.27	II
190		2347.58	II
60		2500.51	II
8	h	2596.64	I
100		2634.78	II
8		2702.63	I
18		2771.36	II
15		2785.28	I
100	r	3071.58	I
10	h	3108.21	I
8		3132.60	I
8	h	3135.72	I
10		3137.70	I
10		3155.34	I
10		3155.67	I
12		3158.05	I
12	h	3158.54	I
25		3165.60	I
15	h	3173.69	I
30		3183.16	I
15		3183.96	I
10		3193.91	I
25	h	3203.70	I
30		3221.63	I
40		3222.19	I
50		3261.96	I
60	r	3262.34	I
40		3281.50	I
15		3281.77	I
50		3322.80	I
80	h	3356.80	I
60	r	3377.08	I
20		3377.39	I
70	r	3420.32	I
25		3421.01	I
30	h	3421.48	I
40		3463.74	I
200	r	3501.11	I
80	h	3524.97	I
30	h	3531.35	I
80	h	3544.66	I
20	h	3547.68	I
100		3552.45	II
200		3567.73	II
100		3576.28	II
30		3577.62	I
80	h	3579.67	I
200		3596.57	II
40		3630.64	I
40	h	3636.83	I
20	h	3688.47	I
400		3735.75	II
200		3816.69	II
200		3842.80	II
100		3854.76	II
20		3889.33	I
1400	l	3891.78	II
20		3892.65	I
40		3909.91	I

Intensity		Wavelength	
500		3914.73	II
50		3935.72	I
20		3937.87	I
200		3939.67	II
500		3949.51	II
80		3993.40	I
30		3995.66	I
300		4036.26	II
200		4083.77	II
30	h	4084.86	I
1500	h	4130.66	II
20		4132.43	I
200		4166.00	II
500		4216.04	II
800		4267.95	II
100		4283.10	II
300		4287.80	II
200		4297.60	II
800		4309.32	II
20	h	4323.00	I
600		4325.73	II
200		4326.74	II
300		4329.62	II
80		4350.33	I
60		4402.54	I
400		4405.23	II
60	h	4488.98	I
50	h	4493.64	I
40		4505.92	I
200		4509.63	II
60	h	4523.17	I
130		4524.93	II
65000		4554.03	II
40		4573.85	I
80		4579.64	I
30		4599.75	I
20	h	4619.92	I
25	h	4628.33	I
300		4644.10	II
30		4673.62	I
35		4691.62	I
20		4700.43	I
800		4708.94	II
40		4726.44	I
800		4843.46	II
300		4847.14	II
200		4850.84	II
30	h	4877.65	I
400		4899.97	II
15		4902.90	I
20000		4934.09	II
8		4947.35	I
1000		4957.15	II
300		4997.81	II
1000		5013.00	II
20	h	5159.94	I
20		5267.03	I
800		5361.35	II
1000		5391.60	II
200		5421.05	II
100		5424.55	I
200		5428.79	II
300		5480.30	II
200		5519.05	II
1000	r	5535.48	II
20	h	5620.40	I
10		5680.18	I
400		5777.62	I
800		5784.18	II
100		5800.23	I
20		5805.69	I
150		5826.28	I
2800		5853.68	II
15		5907.64	I
100		5971.70	I
800		5981.25	II
100		5997.09	I
300		5999.85	II
100		6019.47	I
200		6063.12	I
300		6110.78	I
400		6135.83	II
20000		6141.72	II
150		6341.68	I
500		6378.91	II
90		6450.85	I
150		6482.91	I
12000		6496.90	II
300		6498.76	I
150		6527.31	I
3000		6595.33	I
150		6654.10	I
1500		6675.27	I
1800		6693.84	I
1000		6769.62	II
600		6865.69	I
300	h	6867.85	I
1000		6874.09	II
6000		7059.94	I
2400	hS	7120.33	I
600		7195.24	I
600	hL	7228.84	I
3000		7280.30	I

Intensity		Wavelength	
1200		7392.41	I
300		7417.53	I
900	hL	7459.78	I
600		7488.08	I
450	hL	7636.90	I
600	hL	7642.91	I
1800		7672.09	I
1200		7780.48	I
180	h	7839.57	I
1500		7905.75	I
600		7911.34	I
900	h	8210.24	I
1800	h	8559.97	I
100		8710.74	II
100		8737.71	II
300	h	8799.76	I
300		8860.98	I
450		8914.99	I
300		9219.69	I
300		9308.08	I
300	h	9324.58	I
1500		9370.06	I
300		9455.92	I
450		9589.37	I
900		9608.88	I
300	h	9645.72	I
1500	hL	9830.37	I
900		10001.08	I
600		10032.10	I
1200	h	10233.23	I
300		10471.26	I
120	hL	10791.25	I
180	hL	11012.69	I
150	h	11114.42	I
240		11303.04	I
120	h	11697.45	I
120		13207.30	I
120		13810.50	I
120		14077.90	I
120		15000.40	I
120		20712.00	I
150		25515.70	I
150		28222.00	II

Ba III
Ref. III — J.J.W.

Intensity		Wavelength	
		Vacuum	
5		403.82	III
2		407.12	III
7		420.12	III
4		423.84	III
9		448.95	III
8		456.96	III
14		555.48	III
14		587.57	III
18		647.27	III
9		653.36	III
15		743.12	III
12		1097.41	III
15		1113.67	III
11		1116.01	III
14		1133.05	III
12		1151.76	III
12		1170.62	III
13		1207.29	III
11		1218.92	III
12		1224.55	III
12		1288.53	III
11		1299.18	III
12		1307.40	III
12		1308.87	III
12		1315.72	III
12		1334.01	III
11		1354.71	III
12		1369.53	III
11		1416.61	III
12		1478.85	III
12		1510.68	III
12		1514.22	III
12		1565.61	III
12		1566.12	III
12		1574.55	III
12		1596.80	III
12		1610.95	III
12		1615.78	III
12		1711.53	III
12		1861.74	III
12		1883.92	III
11		1974.76	III
		Air	
10		2001.30	III
15		2008.40	III
13		2022.45	III
10		2038.84	III
12		2070.43	III
12		2071.68	III
10		2076.00	III
12		2081.35	III
10		2134.87	III
16		2156.37	III
10		2160.76	III

Intensity		Wavelength	
20		2230.33	III
30		2280.68	III
35		2323.51	III
60		2331.10	III
25		2476.73	III
25		2505.07	III
40		2512.28	III
40		2523.83	III
50		2530.92	III
25		2559.54	III
40		2570.48	III
40		2681.89	III
30		2745.78	III
25		2938.95	III
25		2960.05	III
30		2962.48	III
20		3014.22	III
30		3043.42	III
40		3079.14	III
30		3103.92	III
30		3119.22	III
30		3152.70	III
25		3195.17	III
25		3235.04	III
25		3281.65	III
20		3286.79	III
50		3368.18	III
30		3369.68	III
25		3649.18	III
25		3926.85	III
25		3993.06	III
18	p	4053.71	III
15		4697.44	III
10		5049.55	III
10		5097.54	III
12	p	5102.25	III
10		5134.54	III
10		5998.00	III
13		6101.99	III
10		6377.11	III
10		6383.76	III
8		6526.17	III
8		7095.49	III
8		8308.69	III
8		9521.76	III

Ba IV
Ref. 78 — J.J.W.

Intensity		Wavelength	
		Vacuum	
40000		794.89	IV
50000		923.74	IV

Ba V
Ref. 259—J.R.

Intensity		Wavelength	
		Vacuum	
15		612.55	V
5		658.11	V
5		681.09	V
300		719.86	V
150		721.85	V
100		760.45	V
1000		766.87	V
100		783.61	V
100		816.41	V
40		875.69	V
300		877.41	V
15		892.28	V
200		946.26	V

BERKELIUM (Bk)
Z = 97

Bk I and II
Ref.53, 339—J.G.C.

Intensity		Wavelength	
		Air	
10000	s	2748.02	II
10000	s	2827.57	II
10000	l	2872.11	II
10000	s	2878.57	II
10000	l	2884.77	II
10000	s	2889.80	II
10000	s	2893.66	II
10000	l	2910.65	II
10000	s	2926.49	
10000	s	2927.91	
10000	l	2941.71	II
10000	l	2951.76	II
10000	s	2969.13	II
10000	l	2987.76	II
10000	l	3178.47	II
10000		3239.72	
10000	s	3247.26	II
10000		3252.19	I
10000	s	3263.47	II
10000	l	3288.75	I

Berkelium (Cont.)

Intensity		Wavelength	Species
10000		3289.35	I
10000	s	3302.35	II
10000		3335.26	
10000	s	3387.45	II
10000		3408.28	I
10000	l	3412.13	II
10000		3426.95	I
10000	l	3432.62	I
10000	l	3437.47	I
10000		3442.66	I
10000	s	3461.24	II
10000	l	3453.90	I
10000		3464.13	II
10000	s	3472.02	II
10000	s	3477.62	II
10000		3528.72	I
10000	l	3531.40	I
10000	l	3535.73	I
10000	s	3542.19	II
10000		3553.60	I
10000	s	3555.88	I
10000		3556.52	I
10000		3565.41	I
10000	l	3567.25	II
10000		3590.32	I
10000		3595.88	I
10000		3601.12	I
10000	s	3603.20	II
10000		3604.78	
10000	l	3608.49	I
10000		3609.61	I
10000		3611.03	
10000	l	3611.93	I
10000		3613.91	I
10000		3616.62	I
10000		3619.37	I
10000	s	3621.81	I
10000		3627.61	I
10000		3633.28	I
10000	l	3637.05	I
10000		3640.26	I
10000	s	3640.93	II
10000	l	3675.59	I
10000	s	3681.22	II
10000	l	3684.43	I
10000	s	3685.21	I
10000		3686.74	I
10000	l	3692.73	I
10000		3695.37	I
10000		3703.28	I
10000	s	3704.02	I
10000		3705.26	I
10000	s	3711.14	II
10000		3712.93	I
10000		3725.39	I
10000		3739.92	I
10000		3743.05	I
10000		3745.40	I
10000		3750.08	I
10000		3751.91	I
10000		3757.35	I
10000		3757.85	I
10000	s	3771.06	II
10000		3780.72	I
10000		3781.17	I
10000		3785.38	I
10000		3788.21	I
10000		3791.42	I
10000		3796.21	I
10000		3797.12	
10000		3798.63	I
10000		3801.08	II
10000	s	3802.35	II
10000		3802.47	I
10000		3815.29	I
10000	s	3823.10	II
10000	s	3824.08	II
10000		3825.19	I
10000	s	3825.84	II
10000		3827.41	I
10000		3830.55	I
10000	l	3831.57	II
10000	s	3833.48	I
10000	s	3835.97	II
10000		3842.19	I
10000	l	3846.62	I
10000		3847.63	I
10000		3855.03	I
10000		3859.89	II
10000	l	3877.94	II
10000		3880.11	I
10000	l	3882.60	I
10000		3894.55	II
10000	s	3906.09	II
10000	l	3912.16	II
10000	l	3916.37	II
10000		3921.42	I
10000		3928.05	II
10000	l	4147.13	II
10000		4189.69	II
10000	s	4197.44	II
10000	l	4329.58	I
10000		4351.50	I
10000		4363.64	I
10000		4423.01	I
10000		4466.46	I
10000		4685.70	
10000	l	4765.40	I
10000	s	5056.73	I
10000	l	5118.24	I
10000	s	5135.53	II
10000		5170.61	I
10000		5197.55	I
10000		5212.53	I
10000		5271.95	I
10000		5392.03	I
10000		5394.24	I
10000	l	5404.62	I
10000		5449.63	I
10000	l	5467.47	I
10000	s	5484.58	I
10000		5512.22	II
10000	l	5537.93	I
10000	l	5556.80	I
10000	s	5557.09	I
10000		5581.21	I
10000	l	5656.54	I
10000		5659.03	I
10000		5702.24	I
10000		5910.71	I
10000		7040.85	I
10000		7107.85	I
10000	l	7176.22	I
10000		7249.26	I
10000		7252.50	I
10000	l	7257.21	I
10000		7306.94	I
10000		7394.26	
10000		7511.26	I
10000		7551.12	I
10000		7579.77	I
10000	l	7729.93	I
10000	s	7903.90	I
10000		9319.30	I
10000	s	9429.13	I
10000	l	9801.18	I
10000	l	9862.39	II
10000	l	9879.29	I
10000	l	9892.38	I
10000		10126.20	I
10000		10186.58	I
10000		10292.44	I
10000	l	10527.71	I
10000	l	10570.53	I
10000	l	11293.14	I
10000	l	11500.30	I
10000	s	11575.34	I
10000	s	11793.09	I
10000	s	12159.05	I
10000		13061.13	I
10000		13498.36	I
10000	s	14196.93	I
10000		15136.10	I
10000		18352.31	
10000	l	19273.87	I
10000	s	19653.22	I
10000	s	23902.85	I
10000	s	24192.62	I

BERYLLIUM (Be)
Z = 4

Be I and II
Ref. 15, 44, 115, 134, 135, 198, 335 — J.O.S.

Intensity	Wavelength	Species
	Vacuum	
	82.58	II
	83.66	II
	89.16	II
	89.80	II
	90.04	II
	90.21	I
	90.67	I
	91.06	II
	91.36	II
	91.74	II
	92.19	I
	92.61	II
	93.14	II
	93.42	II
	93.93	II
	94.78	II
	95.76	II
	96.29	I
	97.24	I
	97.44	I
	97.86	I
	97.97	I
	98.12	I
	98.37	I
	98.66	I
	98.94	I
	99.19	I
	100.86	I
	101.20	I
	102.13	I
	102.49	II
	104.40	II
	104.67	II
	105.80	I
	107.26	I
	107.38	I
	714.0	II
	725.71	II
	743.58	II
8	775.37	II
20	842.06	II
	865.3	II
2	925.25	II
10	943.56	II
10	973.27	II
	981.4	II
	1020.1	II
8	1026.93	II
5	1036.32	II
15	1048.23	II
20	1143.03	II
	1155.9	II
60	1197.19	II
	1426.12	I
	1491.76	I
20	1512.30	II
60	1512.43	II
100	1661.49	II
15	1776.12	II
20	1776.34	II
	1907.	I
	1909.0	II
	1912.	I
	1919.	I
5	1929.67	I
10	1943.68	I
	1956.	I
50	1964.59	I
5	1985.13	I
	1997.95	I
	1997.98	I
60	1998.01	I
	Air	
	2033.25	I
	2033.28	I
	2033.38	I
50	2055.90	I
100	2056.01	I
10	2125.57	I
20	2125.68	I
25	2145.	I
55	2174.99	I
55	2175.10	I
	2273.5	II
	2324.6	II
	2337.0	I
950	2348.61	I
20	2350.66	I
60	2350.71	I
200	2350.83	I
2	2413.34	II
16	2413.46	II
20	2453.84	II
	2480.6	I
35	2494.54	I
35	2494.58	I
100	2494.73	I
16	2507.43	II
5	2617.99	II
20	2618.13	II
100	2650.45	I
60	2650.55	I
200	2650.62	I
60	2650.69	I
100	2650.76	I
5	2697.46	II
20	2697.58	II
20	2728.88	II
30	2738.05	I
	2764.2	II
20	2898.13	I
10	2898.19	I
20	2898.25	I
30	2986.06	I
10	2986.42	I
60	3019.33	I
30	3019.49	I
30	3019.53	I
20	3019.60	I
10	3046.56	II
30	3046.69	II
	3090.3	I
10	3110.81	I
10	3110.92	I
20	3110.99	I
	3120.	I
480	3130.42	II
320	3131.07	II
	3136.	I
	3150.	I
	3160.6	I
	3163.	I
	3168.	I
	3180.7	II
	3187.	I
20	3193.81	I
20	3197.10	II
30	3197.15	II
20	3208.60	I
	3220.	I
60	3229.63	I
2	3233.52	II
10	3241.62	II
30	3241.83	II
15	3269.02	I
100	3274.58	II
30	3274.67	II
30	3282.91	I
30	3321.01	I
30	3321.09	I
220	3321.34	I
20	3345.43	I
60	3367.63	I
	3405.6	II
5	3451.37	I
300	3455.18	I
20	3476.56	I
300	3515.54	I
-10	3555.	I
100	3736.30	I
700	3813.45	I
40	3865.13	I
80	3865.42	I
1	3865.51	I
6	3865.72	I
100	3866.03	I
100	4253.05	I
60	4253.76	I
300	4360.66	II
500	4360.99	II
400	4407.94	I
	4526.6	I
	4548.	I
12	4572.66	I
700	4673.33	II
1000	4673.42	II
6	4709.37	I
200	4828.16	II
40	4849.16	I
2 h	4858.22	II
80	5087.75	I
8	5218.12	II
20	5218.33	II
3	5255.86	II
64	5270.28	II
500	5270.81	II
20	5403.04	II
20	5410.21	II
	5558.	I
10	6229.11	I
16	6279.43	II
30	6279.73	II
30	6473.54	I
60	6547.89	II
60	6558.36	II
30	6564.52	I
2 h	6636.44	II
1 h	6756.72	II
2 h	6757.13	II
30	6786.56	I
1 h	6884.22	I
6 h	6884.44	I
100	6982.75	I
6 h	7154.40	I
40 h	7154.65	I
100	7209.13	I
3	7401.20	II
2	7401.43	II
10	7551.90	I
10 h	7618.68	I
20 h	7618.88	I
60	8090.06	I
5 h	8158.99	I
10 h	8159.24	I
4	8254.07	I
10 h	8287.07	I
30	8547.36	I
60	8547.67	I
300	8801.37	I
6	8882.18	I
40	9190.45	I
20 h	9243.92	I
1 h	9343.89	II
40	9392.74	I
2	9476.43	I
16	9477.03	II
20	9847.32	I
10 h	9895.63	I
20 h	9895.96	I
80	9939.78	I
16	10095.52	II

Beryllium (Cont.)

Intensity		Wavelength	
20		10095.73	II
60		10119.92	II
80		10331.03	I
30		11066.46	I
		11173.	II
1		11173.73	II
120		11496.39	I
2	h	11625.16	II
		11659.	II
2		11660.25	II
100		12095.36	II
30		12098.18	II
100		14643.92	I
60		14644.75	I
200		16157.72	I
80		17855.38	I
120		17856.63	I
100		18143.54	I
160		31775.05	I
200		31778.70	I

Be III
Ref. 73, 102, 175 — J.O.S.

Intensity		Wavelength	
		Vacuum	
1	h	76.10	III
2		76.48	III
3		78.53	III
4		78.66	III
1	h	78.92	III
5		81.89	III
10		82.38	III
20		83.20	III
30		84.76	III
50		88.31	III
100		100.25	III
3		509.99	III
2		549.31	III
6		582.08	III
4		661.32	III
8		675.59	III
4		725.59	III
7		746.23	III
2		767.75	III
1		1114.69	III
?		1312.18	III
1		1214.32	III
2		1362.25	III
1		1401.52	III
10		1421.26	III
5		1422.86	III
1		1435.17	III
2		1440.77	III
2	h	1754.69	III
3		1917.03	III
60	h	1954.97	III
		Air	
75	h	2076.94	III
60	h	2080.38	III
25		2118.56	III
15	h	2122.27	III
15	h	2127.20	III
5		2137.25	III
5		2191.57	III
100		3720.36	III
		3720.92	III
		3722.98	III
90	h	4249.14	III
2		4485.52	III
100		4487.30	III
1		4495.09	III
140	h	4497.8	III
140	h	6142.01	III

Be IV
Ref. 171 — J.O.S.

Intensity	Wavelength	
	Vacuum	
	58.13	IV
	58.57	IV
	59.32	IV
	60.74	IV
	64.06	IV
	75.93	IV

BISMUTH (Bi)
Z = 83

Bi I and II
Ref. 1, 357-359 — C.H.C.

Intensity		Wavelength	
		Vacuum	
15		1058.88	II
20		1085.47	II
10		1099.20	II
8		1163.19	II
8		1167.06	II
10		1225.43	II
15		1232.78	II
10		1241.05	II
10		1265.35	II
15		1283.73	II
10		1306.18	II
20		1325.46	II
20		1329.47	II
20		1350.07	II
25		1372.61	II
15		1376.02	II
20		1393.92	II
45		1436.83	II
25		1447.94	II
50		1455.11	II
25		1462.14	II
35		1486.93	II
20		1502.50	II
40		1520.57	II
40		1533.17	II
30		1536.77	II
35		1538.06	II
20		1563.67	II
40		1573.70	II
60		1591.79	II
25		1601.58	II
40		1609.70	II
40		1611.38	II
20		1652.81	II
20		1749.29	II
80		1777.11	II
60		1787.47	II
70		1791.93	II
70		1822.00	II
100		1902.41	II
9000		1954.53	I
7000		1960.13	I
25		1989.35	II
		Air	
7000		2021.21	I
9000		2061.70	I
45	h	2068.9	II
4600		2110.26	I
2500		2133.63	I
15		2143.40	II
15		2143.46	II
60		2186.9	II
40	h	2214.0	II
360		2228.25	I
1700		2230.61	I
340		2276.58	I
16		2368.12	II
12		2368.25	II
190		2400.88	I
10		2501.0	II
25		2515.69	I
70		2524.49	I
20	h	2544.5	II
700		2627.91	I
12		2693.0	II
280	c	2696.76	I
20		2713.3	II
140	d	2730.50	I
360		2780.52	I
15		2803.42	II
11		2803.70	II
12		2805.3	II
140	c	2809.62	I
4000		2897.98	I
15		2936.7	II
3200		2938.30	I
20		2950.4	II
12		2963.4	II
2800		2989.03	I
700		2993.34	I
2400		3024.64	I
60		3034.87	I
9000	c	3067.72	I
140		3076.66	I
550	c	3397.21	I
10		3430.83	II
12		3431.23	II
500	c	3510.85	I
380	c	3596.11	I
12		3654.2	II
70	h	3792.5	II
12		3811.1	II
20		3815.8	II
10		3845.8	II
30		3863.9	II
40	h	4079.1	II
10		4097.2	II
140		4121.53	I
140		4121.86	I
75	h	4259.4	II
25		4272.0	II
70	h	4301.7	II
12	h	4339.8	II
25		4340.5	II
12	h	4379.4	II
25	h	4476.8	II
60	h	4705.3	II
600	c	4722.52	I
30		4730.3	II
20		4749.7	II
12		4908.2	II
10		4916.6	II
12		4969.7	II
20		4993.6	II
10		5091.6	II
50	h	5124.3	II
60	h	5144.3	II
20		5201.5	II
75	h	5209.2	II
40	h	5270.3	II
10		5397.8	II
10	c	5552.35	I
3		5599.41	I
20		5655.2	II
40	h	5719.2	II
6		5742.55	I
12		5818.3	II
20		5860.2	II
20		5973.0	II
15		6059.1	II
15		6128.0	II
6		6134.82	I
3		6475.73	I
3		6476.24	I
15		6497.7	II
10		6577.2	II
40	h	6600.2	II
50	h	6808.6	II
4	h	6991.12	I
12		7033.	II
2		7036.15	I
10	h	7381.	II
2		7502.33	I
10	h	7637.	II
10		7750.	II
3		7838.70	I
2		7840.33	I
20		7965.	II
12	h	8050.	II
15		8328.	II
15		8388.	II
30		8532.	II
2		8544.54	I
1		8579.74	I
25		8653.	II
2		8754.88	I
3		8761.54	I
25		8863.	II
2		8907.81	I
2000	d	9657.04	I
40		9827.78	I
20		10104.5	I
15		10138.8	I
20		10300.6	I
20		10536.19	I
50		11072.44	I
15		11551.6	I
1500	d	11710.37	I
40		11999.49	I
200		12165.08	I
10		12374.64	I
200		12690.04	I
100		12817.8	I
200		14330.5	I
50		16001.5	I
60		22551.6	I

Bi III
Ref. 359 — C.H.C.

Intensity		Wavelength	
		Vacuum	
1		590.73	III
5		670.76	III
4		775.16	III
6		803.65	III
7		920.93	III
6		925.48	III
25		1039.99	III
50	h	1045.76	III
30		1051.81	III
20		1139.01	III
15		1145.91	III
50		1224.64	III
40		1326.84	III
60		1346.12	III
35		1423.33	III
35		1423.52	III
60	h	1461.00	III
60	h	1606.40	III
20	h	1691.5	III
20		1834.32	III
10		1863.9	III
10		1912.12	III
10		1988.26	III
		Air	
20		2020.75	III
20		2021.15	III
10		2073.22	III
14		2073.37	III
15		2103.42	III
30		2213.55	III
75	h	2414.6	III
10		2437.6	III
30	h	2847.4	III
80	h	2855.6	III
35		3115.0	III
40	h	3451.0	III
40		3473.8	III
35		3485.5	III
15		3540.8	III
45		3613.4	III
50		3695.32	III
50		3695.68	III
12		4224.6	III
25		4327.8	III
30		4560.84	III
30		4561.54	III
40	h	4797.4	III
45	h	5079.3	III
12		6623.4	III
10	h	7381.	III
12		7551.	III
25		7598.	III
10	h	7637.	III
10		8008.	III
50		8070.	III
20		8100.	III
15		8671.	III
20		8934.	III

Bi IV
Ref. 360 — C.H.C.

Intensity	Wavelength	
	Vacuum	
6	420.7	IV
6	431.2	IV
6	790.5	IV
6	790.6	IV
8	792.5	IV
10	820.3	IV
9	822.9	IV
12	824.9	IV
15	872.6	IV
8	876.8	IV
9	916.7	IV
12	923.9	IV
15	943.3	IV
9	967.6	IV
8	968.8	IV
8	989.8	IV
24	1103.4	IV
7	1128.8	IV
6	1138.6	IV
6	1139.8	IV
7	1149.7	IV
60	1317.	IV
30	1910.0	IV
	Air	
30	2093.	IV
100	2311.	IV
100	2326.	IV
100	2376.	IV
100	2629.	IV
100	2677.	IV
100	2767.	IV
100	2772.	IV
100	2786.	IV
100	2842.	IV
100	2924.	IV
100	2933.	IV
100	2936.	IV
100	3012.	IV
100	3239.	IV
100	3643.	IV
100	3682.	IV
100	3734.	IV
100	3868.	IV
30	4342.	IV

Bismuth (Cont.)

Intensity	Wavelength	
30	5347.	IV

Bi V
Ref. 361 — C.H.C.

Intensity	Wavelength	
	Vacuum	
1	355.77	V
1	369.52	V
1	429.78	V
1	435.63	V
2	488.39	V
1	492.72	V
3	563.62	V
2	678.87	V
6	686.88	V
1	706.54	V
5	730.71	V
10	738.17	V
6	849.86	V
5	855.68	V
15 d	864.45	V
6	880.17	V
6	929.81	V
15 d	1139.46	V

BORON (B)
Z = 5

B I and II
Ref. 66, 104, 171, 222 — R.L.K.

Intensity	Wavelength	
	Vacuum	
70	693.95	II
40	731.36	II
40	731.44	II
110	882.54	II
110	882.68	II
40	984.67	II
110	1081.88	II
110	1082.07	II
110	1230.16	II
220	1362.46	II
70	1600.46	I
120	1600.73	I
160	1623.58	II
110	1623.77	II
220	1624.02	II
70	1624.16	II
160	1624.34	II
100	1663.04	I
150	1666.87	I
200	1667.29	I
150	1817.86	I
200	1818.37	I
300	1825.91	I
300	1826.41	I
110	1842.81	II
	Air	
250	2066.38	I
250	2066.65	I
100	2066.93	I
300	2067.19	I
500	2088.91	I
500	2089.57	I
70	2220.30	II
40	2323.03	II
40	2328.67	II
40	2393.20	II
220	2395.05	II
40	2459.69	II
40	2459.90	II
1000	2496.77	I
1000	2497.73	I
160	2918.08	II
110	3032.26	II
70	3179.33	II
110	3323.18	II
110	3323.60	II
450	3451.29	II
285	4121.93	II
110	4194.79	II
110	4472.10	II
110	4472.85	II
70	4784.21	II
110	4940.38	II
110	6080.44	II
70	6285.47	II
70	7030.20	II
40	7031.90	II
70	8668.57	I
20	8667.22	I
800	11660.04	I
570	11662.47	I
125	15629.08	I
200	16240.38	I
250	16244.67	I
235	18994.33	I

B III
Ref. 69, 221 — R.L.K.

Intensity	Wavelength	
	Vacuum	
150	518.24	III
75	518.27	III
40	411.80	III
20	510.77	III
40	510.85	III
110	677.00	III
160	677.14	III
40	758.48	III
70	758.67	III
20	1953.83	III
	Air	
550	2065.78	III
450	2067.23	III
160	2077.09	III
40	2234.09	III
70	2234.59	III
40	4242.98	III
70	4243.61	III
220	4487.05	III
360	4497.73	III
110	7835.25	III
70	7841.41	III

B IV
Ref. 74 — R.L.K.

Intensity	Wavelength	
	Vacuum	
10	52.68	IV
30	60.31	IV
160	344.0	IV
450	385.0	IV
285	418.7	IV
70	1112.2	IV
450	1168.9	IV
70	1170.9	IV
	Air	
70	2524.4	IV
160	2530.3	IV
450	2821.68	IV
70	2824.57	IV
285	2825.85	IV

B V
Ref. 94 — R.L.K.

Intensity	Wavelength	
	Vacuum	
30	41.00	V
	48.59	V
	194.37	V
	262.37	V
	512.53	V
	749.74	V

BROMINE (Br)
Z = 35

Br I and II
Ref. 122, 124, 240, 248, 316
G.V.S.

Intensity	Wavelength	
	Vacuum	
300	711.68	II
250	815.48	II
350	856.19	II
1000	889.23	II
500	896.64	II
500	905.99	II
300	922.56	II
1000	948.97	II
500	984.93	II
500	1012.10	II
1000	1015.54	II
500	1037.02	II
1000	1049.01	II
450	1064.76	II
500	1071.87	II
250	1101.50	I
300	1134.59	I
250	1136.29	I
250	1177.23	I
400	1178.90	I
1000	1189.28	I
250	1189.38	I
1000	1189.50	I
500	1198.37	I
800	1209.76	I
1000	1210.73	I
750	1216.01	I
1000	1221.13	I
900	1221.87	I
1000	1223.24	I
1200	1224.41	I
1200	1226.90	I
750	1228.05	I
7500	1232.43	I
1200	1243.90	I
800	1249.59	I
1500	1251.66	I
1000	1255.80	I
1500	1259.20	I
1200	1261.66	I
1200	1266.20	I
1000	1279.48	I
1000	1286.26	I
3000	1309.91	I
3000	1316.14	I
1000	1317.37	I
2000	1317.70	I
12000	1384.60	I
3000	1449.90	I
50000	1488.45	I
30000	1531.74	I
25000	1540.65	I
30000	1574.84	I
20000	1576.39	I
25000	1582.31	I
75000	1633.40	I
	Air	
350	2285.17	II
350	2287.60	II
500 h	2317.30	II
400	2336.93	II
350	2386.45	II
500	2386.70	II
300	2388.69	II
450	2388.96	II
500	2389.69	II
350	2392.21	II
400	2392.42	II
300	2488.50	II
300 h	2495.22	II
450	2521.70	II
400	2541.48	II
400	2556.92	II
350	2690.17	II
400	2713.77	II
350 h	2746.52	II
300 h	2807.55	II
400 h	2893.40	II
400 h	2917.18	II
400	2967.21	II
500 h	2972.26	II
300	2981.86	II
300 h	2985.87	II
300 h	2986.53	II
300 h	3016.48	II
350	3423.82	II
300 h	3606.80	II
350	3714.30	II
1200	3815.65	I
350	3834.69	II
300	3871.21	II
400	3891.63	II
300	3901.24	II
300	3914.20	II
500	3914.38	II
350	3919.51	II
400	3924.09	II
300	3929.55	II
350	3939.69	II
350	3950.61	II
500	3980.38	II
1500	3992.36	I
300	4024.04	II
300	4135.16	II
300	4140.20	II
400	4179.63	II
300	4193.45	II
1000	4223.89	II
300	4236.86	II
300	4291.39	II
2000	4365.14	I
1000	4365.60	II
1500	4425.14	I
10000	4441.74	I
10000	4472.61	I
20000	4477.72	I
1000	4490.42	I
3000	4513.44	I
15000	4525.59	I
300	4529.60	II
500	4542.92	II
3000	4575.74	I
300	4601.36	II
2500	4614.58	I
350	4622.70	II
300	4642.02	II
300	4651.98	II
500	4678.70	II
400	4693.17	II
500	4704.85	II
350	4719.76	II
400	4720.36	II
300	4728.20	II
300	4735.41	II
400	4742.64	II
2500	4752.28	I
350	4766.00	II
400	4779.40	II
4000	4780.31	I
1600	4785.19	I
500	4785.50	II
300	4802.33	II
500	4816.70	II
300	4818.46	II
350	4844.81	II
350	4848.75	II
400	4921.12	II
400	4928.79	II
450	4930.66	II
300	4945.51	II
4000	4979.76	I
300	5038.74	II
300	5054.64	II
400	5164.38	II
300	5180.01	II
500	5182.35	II
300	5193.90	II
500	5238.23	II
300	5272.68	II
350	5304.10	II
400	5330.57	II
500	5332.05	II
1200	5395.48	I
400	5422.78	II
350	5424.99	II
300	5435.07	II
1200	5466.22	I
350	5478.47	II
300	5488.79	II
300	5495.06	II
500	5506.69	II
350	5589.94	II
300	5718.71	II
300	5830.78	II
1800	5852.08	I
1600	5940.48	I
2400	6122.14	I
40000	6148.60	I
300	6161.74	II
2000	6177.39	I
1500	6335.48	I
60000	6350.73	I
400	6352.94	II
2500	6410.32	I
1800	6483.56	I
1000	6514.62	I
20000	6544.57	I
1500	6548.09	I
50000 c	6559.80	I
1000	6571.31	I
1800	6579.14	I
20000	6582.17	I
1500	6620.47	I
50000 c	6631.62	I
20000	6682.28	I
10000	6692.13	I
8000	6728.28	I
2000	6760.06	I
2000	6779.48	I
2200	6786.74	I
6500	6790.04	I
1600 c	6791.48	I
1800	6861.15	I
10000	7005.19	I
2000	7260.45	I
10000	7348.51	I
40000	7512.96	I
1600	7591.61	I
1800	7595.07	I
2000	7616.41	I
30000	7803.02	I
1200	7827.23	I
2500 s	7881.45	I
2500	7881.57	I
2500	7925.81	I
30000 c	7938.68	I
3000	7947.94	I
3000	7950.18	I

Bromine (Cont.)

Intensity		Wavelength	
8000		7978.44	I
10000		7978.57	I
30000		7989.94	I
2000		8026.35	I
2500		8026.54	I
30000		8131.52	I
1000	c	8152.65	I
10000		8153.75	I
25000		8154.00	I
5000		8246.86	I
15000		8264.96	I
75000	c	8272.44	I
20000		8334.70	I
10000		8343.70	I
1200		8384.04	I
40000		8446.55	I
4000		8477.45	I
1500		8513.38	I
1000		8557.73	I
1000		8566.28	I
20000		8638.66	I
4000		8698.53	I
10000	c	8793.47	I
15000		8819.96	I
25000		8825.22	I
4000		8888.98	I
30000		8897.62	I
6000		8932.40	I
1800		8949.39	I
9000		8964.00	I
350		9024.42	II
30000		9166.06	I
15000		9173.63	I
20000		9178.16	I
40000		9265.42	I
15000		9320.86	I
300		9434.04	II
6000		9793.48	I
10000		9896.40	I
3000		10140.08	I
6000		10237.74	I
1000		10299.62	I
1500		10377.65	I
30000		10457.96	I
1000		10742.14	I
3000		10755.92	I
1700		13217.17	I
1800		14354.57	I
1250		14888.70	I
1800		16731.19	I
1200		18568.31	I
3500		19733.62	I
1000		20281.73	I
1000		20624.67	I
1200		21787.24	I
4000		22865.65	I
1000		23513.15	I
500		28346.50	I
500		30380.85	I
600		31630.13	I
120		32693.90	I
150		34181.87	I
150		38345.75	I
120		39964.36	I

Br III
Ref. 246, 250 — G.V.S.

Intensity		Wavelength	
		Vacuum	
450		611.1	III
300		620.4	III
500		665.54	III
500		677.19	III
300		677.8	III
450		687.68	III
400		690.2	III
350		696.99	III
300		727.0	III
300		736.4	III
250		769.63	III
250		817.79	III
350		949.0	III
400		960.4	III
450		984.9	III
250		1313.5	III
250		1402.9	III
		Air	
400	h	2293.44	III
300		2313.29	III
300		2462.39	III
300		2482.60	III
350		2499.25	III
350		2529.49	III
350	h	2551.09	III
350	h	2570.83	III
300	h	2573.17	III
400	h	2584.99	III
500	h	2589.14	III
300	h	2594.48	III
400	h	2595.98	III
450	h	2606.20	III
350	h	2608.15	III
500	h	2613.13	III
350	h	2616.26	III
500	h	2626.52	III
350	h	2629.23	III
350	h	2639.60	III
350	h	2671.53	III
350	h	2735.83	III
300		2770.50	III
300	h	2785.28	III
300		2804.16	III
400		2926.96	III
300		2936.22	III
350		2969.00	III
400		2994.04	III
500		3020.76	III
300		3033.63	III
350		3036.45	III
500		3074.42	III
350		3091.94	III
350		3117.29	III
300		3147.81	III
400		3174.08	III
300		3321.08	III
450		3333.07	III
500		3349.64	III
300		3385.25	III
450		3447.36	III
400		3487.58	III
300		3506.47	III
450		3517.36	III
500		3540.16	III
300		3551.08	III
500		3562.43	III
450		3600.71	III
250		3693.53	III
450		3820.26	III
200		3903.95	III
250		4506.55	III
200		4519.74	III
150		5175.87	III
100		5446.80	III
100		7192.8	III
100		7673.1	III

Br IV
Ref. 139, 142, 243, 249
G.V.S.

Intensity		Wavelength	
		Vacuum	
700		379.73	IV
700		400.37	IV
1000		545.43	IV
1000		559.76	IV
1000		569.19	IV
1000		576.59	IV
1000		585.10	IV
1000		586.71	IV
1000		597.51	IV
1000		600.09	IV
1000		601.27	IV
1000		607.03	IV
1000		617.85	IV
1000		619.87	IV
1000		630.14	IV
1000		642.23	IV
1000		661.53	IV
1000		683.51	IV
1000		697.72	IV
1000		715.39	IV
1000		731.00	IV
1000		800.12	IV
1000		813.66	IV
900		1274.82	IV
1000		1703.51	IV
		Air	
1000		2133.79	IV
1000		2145.02	IV
1000		2257.21	IV
1000		2272.73	IV
1000		2307.40	IV
1000		2408.16	IV
1000		2411.58	IV
700		2491.14	IV
1000		2581.19	IV
600		2661.40	IV
700		2820.87	IV
1000		2842.88	IV
1100	h	2907.71	IV
500		3041.18	IV
500		3380.56	IV

Br V
Ref. 42 — G.V.S.

Intensity	Wavelength	
	Vacuum	
600	468.37	V
800	482.11	V
900	531.97	V
1000	547.90	V
700	549.77	V
800	621.03	V
800	632.22	V
700	645.44	V
400	652.64	V
800	657.54	V
800	679.62	V
700	812.95	V
1000	850.81	V
150	855.27	V
600	1041.60	V
1000	1069.15	V
500	1080.54	V
900	1112.13	V
1000	1143.56	V
150	1429.75	V
400	1442.60	V
150	1470.35	V

CADMIUM (Cd)
Z = 48
Cd I and II
Ref. 44, 285, 296 — R.D.C.

Intensity		Wavelength	
		Vacuum	
100		1256.00	II
150		1296.63	II
100		1326.50	II
150		1370.91	II
200		1514.26	II
200		1571.58	II
100		1668.60	II
50		1702.47	II
50		1724.41	II
100		1785.84	II
100		1827.70	II
300		1922.23	II
100		1943.54	II
40		1965.54	II
30		1986.89	II
200		1995.43	II
		Air	
100		2007.49	II
50		2032.45	II
75		2036.23	II
150		2096.00	II
1000	r	2144.41	II
50		2155.06	II
100		2187.79	II
1000		2194.56	II
1000		2265.02	II
1500	r	2288.022	I
1000		2312.77	II
200		2321.07	II
40		2376.82	II
50		2418.69	II
50		2469.73	II
40		2487.93	II
3		2491.00	I
40		2495.58	II
10		2508.91	I
50		2509.11	II
30		2516.22	II
15	h	2518.59	I
25	h	2525.196	I
50		2544.613	I
50		2551.98	II
25		2553.465	I
3		2565.789	I
500		2572.93	II
50		2580.106	I
3		2584.87	I
30		2592.026	I
25	h	2602.048	I
50		2628.979	I
40		2632.190	I
75		2639.420	I
40		2659.23	II
50	h	2660.325	I
25		2668.20	II
50		2672.62	II
100		2677.540	I
25		2677.748	I
50		2707.00	II
75		2712.505	I
50		2733.820	I
1000		2748.54	II
100	h	2763.894	I
50	h	2764.230	I
50		2774.958	I
30		2823.19	II
200		2836.900	I
25		2856.46	II
100		2868.180	I
200	r	2880.767	I
50	r	2881.224	I
200		2914.67	II
50		2927.87	II
200		2929.27	II
1000	r	2980.620	I
200	r	2981.362	I
50		2981.845	I
50		3030.60	II
150		3080.822	I
25		3081.48	II
30		3082.593	I
100		3092.34	II
200		3133.167	I
50		3146.79	II
150		3250.33	I
300		3252.524	I
300		3261.055	I
50		3343.21	II
50		3385.49	II
30		3388.88	II
800		3403.652	I
50		3417.49	II
50		3442.42	II
100		3464.43	II
1000		3466.200	I
800		3467.655	I
25		3483.08	II
150		3495.44	II
25		3499.952	I
100		3524.11	II
100		3535.69	II
1000		3610.508	I
800		3612.873	I
60		3614.453	I
20		3649.558	I
10		3981.926	I
100		4029.12	II
200		4134.77	II
50		4141.49	II
100		4285.08	II
8		4306.672	I
100		4412.41	II
3		4412.989	I
1000		4415.63	II
30		4440.45	II
8		4662.352	I
200		4678.149	I
30		4744.69	II
300		4799.912	I
50		4881.27	II
50		5025.50	II
1000	h	5085.822	I
6		5154.660	I
100		5268.01	II
100		5271.60	II
1000		5337.48	II
1000		5378.13	II
200		5381.89	II
40		5843.30	II
50		5880.22	II
300		6099.142	I
100		6111.49	II
100		6325.166	I
30		6330.013	I
400		6354.72	II
500		6359.98	II
2000		6438.470	I
400		6464.94	II
25		6567.65	II
500		6725.78	II
100		6759.19	II
30		6778.116	I
50		7237.01	II
100		7284.38	II
1000		7345.670	I
50		8066.99	II
5		8200.309	I
20		9292	I
15		11655	I
35		14491	I
80		15712	I
55	d	19125	I
25		24378	I
35		25455	I

Cd III
Ref. 296 — R.D.C.

Intensity	Wavelength	
	Vacuum	
8	677.39	III

Cadmium (Cont.)

Intensity	Wavelength	
15	684.58	III
10	720.70	III
5	1383.60	III
15	1392.10	III
10	1396.78	III
25	1416.28	III
5	1420.29	III
8	1420.54	III
5	1432.86	III
20	1446.08	III
25	1447.55	III
30	1455.74	III
8	1466.14	III
5	1471.97	III
15	1491.81	III
5	1511.01	III
10	1511.65	III
5	1513.13	III
10	1523.55	III
15	1528.40	III
30	1529.30	III
5	1532.10	III
50	1545.17	III
25	1547.57	III
10	1550.07	III
20	1550.45	III
15	1550.89	III
5	1552.18	III
5	1556.48	III
15	1560.66	III
5	1566.03	III
15	1568.98	III
10	1582.39	III
40	1601.59	III
20	1604.87	III
20	1606.64	III
10	1607.28	III
15	1608.91	III
10	1609.61	III
15	1612.51	III
15	1625.27	III
25	1628.54	III
20	1651.87	III
25	1655.63	III
30	1678.15	III
10	1699.70	III
40	1707.16	III
10	1721.93	III
40	1722.95	III
30	1725.66	III
25	1739.00	III
5	1745.69	III
40	1747.67	III
15	1748.15	III
30	1768.82	III
40	1773.06	III
30	1789.19	III
75	1793.40	III
15	1796.10	III
5	1800.57	III
40	1823.41	III
50	1844.66	III
40	1851.13	III
20	1851.37	III
40	1855.85	III
200	1856.67	III
150	1874.08	III
15	1886.49	III
15	1903.48	III
25	1909.98	III
15	1910.57	III
10	1939.59	III
15	1988.81	III

Air

Intensity	Wavelength	
20	2000.60	III
15	2004.07	III
15	2016.12	III
40	2039.83	III
50	2045.61	III
10	2061.25	III
75	2087.91	III
10	2097.45	III
5	2100.47	III
50	2111.60	III
5	2188.13	III
5	2218.43	III
5	2224.43	III
7	2418.24	III
10	2426.36	III
25	2499.81	III
15	2618.81	III
5	2630.56	III
20	2766.99	III
30	2805.59	III
10	3035.72	III

Cd IV
Ref. 353, 399 — R.D.C.

Vacuum

Intensity	Wavelength	
50	427.01	IV
20	437.88	IV
50	447.85	IV
60	480.90	IV
10	489.49	IV
70	493.00	IV
70	495.13	IV
70	498.14	IV
70	498.53	IV
80	504.09	IV
70	504.20	IV
70	504.50	IV
80	506.31	IV
60	508.01	IV
50	508.95	IV
70	509.55	IV
25	509.81	IV
70	511.40	IV
80	513.00	IV
70	514.50	IV
60	519.42	IV
15	520.97	IV
80	524.41	IV
70	524.47	IV
50	524.77	IV
70	525.10	IV
60	525.19	IV
70	527.07	IV
50	530.79	IV
80	531.09	IV
80	531.51	IV
70	534.29	IV
70	536.77	IV
50	537.24	IV
60	540.90	IV
70	541.74	IV
80	542.60	IV
80	546.55	IV
40	548.01	IV
20	548.33	IV
15	548.90	IV
25	551.27	IV
20	552.90	IV
60	553.06	IV
80	554.05	IV
25	560.26	IV
10	564.16	IV
60	567.01	IV
20	1062.23	IV
150	1118.16	IV
30	1126.00	IV
20	1134.08	IV
15	1139.04	IV
20	1154.64	IV
10	1155.73	IV
100	1164.65	IV
20	1165.78	IV
40	1167.30	IV
20	1179.73	IV
15	1183.07	IV
100	1183.40	IV
40	1194.13	IV
20	1195.63	IV
30	1196.47	IV
20	1198.93	IV
15	1215.38	IV
20	1223.52	IV
20	1246.06	IV
15	1246.56	IV
15	1249.94	IV
30	1266.47	IV
15	1274.41	IV
20	1285.63	IV
20	1287.58	IV
20	1299.46	IV
40	1304.36	IV
30	1306.07	IV
15	1316.89	IV
15	1321.85	IV
20	1325.55	IV
30	1340.97	IV
15	1346.15	IV
30	1354.78	IV
20	1358.11	IV
30	1362.55	IV
60	1370.48	IV
30	1380.98	IV
20	1397.65	IV
30	1403.68	IV
15	1406.58	IV
60	1418.89	IV
15	1429.83	IV
20	1447.54	IV
20	1452.63	IV
15	1465.97	IV
15	1466.67	IV
15	1482.95	IV
20	1491.79	IV
20	1570.20	IV
20	1598.73	IV
20	1600.42	IV
15	1622.87	IV

CALCIUM (Ca)
Z = 20

Ca I and II
Ref. 70, 150, 270 — J.J.W. and H.H.S.

Vacuum

Intensity	Wavelength	
24	1341.89	II
12	1342.54	II
20	1433.75	II
12	1432.50	II
20	1553.18	II
32	1554.64	II
4	1642.80	II
20	1643.77	II
36	1644.44	II
60	1649.86	II
32	1651.99	II
12	1673.86	II
20	1680.05	II
2	1680.13	II
8	1691.78	II
16	1698.18	II
20	1807.34	II
40	1814.50	II
4	1814.65	II
60	1840.06	II
40	1838.01	II
20	1843.09	II
40	1850.69	II

Air

Intensity	Wavelength	
	2103.24	II
	2112.76	II
	2113.15	II
	2128.75	II
	2131.51	II
	2132.30	II
2	2150.80	I
	2197.79	II
5	2200.73	I
	2208.61	II
6	2275.46	I
8	2398.56	I
7	2721.65	I
9	2994.96	I
8	2997.31	I
8	2999.64	I
9	3000.86	I
10	3006.86	I
9	3009.21	I
2	3024.94	I
2	3034.54	I
2	3045.74	I
3	3055.32	I
2	3071.57	I
2	3076.95	I
2	3080.79	I
2	3099.30	I
10	3125.18	II
5	3136.02	I
6	3140.79	I
7	3150.75	I
170	3158.87	II
180	3179.33	II
5	3180.52	I
150	3181.28	II
7	3209.96	I
8	3215.17	I
6	3215.34	I
9	3225.90	I
6	3226.15	I
5	3274.67	I
6	3286.07	I
10	3308.02	II
20	3316.51	II
10	3344.51	I
10	3347.04	II
11	3350.21	I
9	3350.36	I
12	3361.92	I
9	3362.14	I
10	3452.66	II
20	3461.87	II
9	3468.48	I
11	3474.76	I
10	3485.61	II
13	3487.60	I
10	3495.16	II
15	3624.11	I
17	3630.75	I
14	3630.97	I
20	3644.41	I
14	3644.77	I
8	3644.99	I
5	3675.29	I
6	3678.21	I
30	3683.70	II
40	3694.11	II
10	3694.36	II
170	3706.03	II
180	3736.90	II
10	3739.38	II
6	3748.35	I
8	3750.29	I
9	3753.34	I
20	3755.67	II
30	3758.39	II
9	3870.48	I
11	3872.54	I
11	3872.56	I
12	3875.78	I
12	3875.80	I
6	3889.10	I
6	3923.48	I
230	3933.66	II
9	3935.29	I
6	3946.04	I
15	3948.90	I
17	3957.05	I
220	3968.47	II
8	3972.57	I
18	3973.71	I
50	4097.10	II
15	4098.53	I
15	4098.57	I
60	4109.82	II
30	4110.28	II
40	4206.18	II
50	4220.07	II
50	4226.73	I
15	4240.46	I
24	4283.01	I
22	4289.36	I
22	4298.99	I
25	4302.53	I
23	4307.74	I
22	4318.65	I
20	4355.08	I
25	4425.44	I
26	4434.96	I
25	4435.69	I
30	4454.78	I
28	4455.89	I
20	4456.61	I
20	4472.04	II
10	4479.23	II
20	4489.18	II
23	4526.94	I
22	4578.55	I
23	4581.40	I
23	4581.47	I
24	4585.87	I
24	4585.96	I
20	4685.27	I
30	4716.74	II
40	4721.03	II
40	4799.97	II
25	4878.13	I
70	5001.48	II
80	5019.97	II
40	5021.14	II
23	5041.62	I
25	5188.85	I
22	5261.71	I
23	5262.24	I
22	5264.24	I
24	5265.56	I
25	5270.27	I
60	5285.27	II
70	5307.22	II
50	5339.19	II
27	5349.47	I
23	5512.98	I
25	5581.97	I
27	5588.76	I
24	5590.12	I
26	5594.47	I
25	5598.49	I
24	5601.29	I
24	5602.85	I
30	5857.45	I
10	5922.72	II
10	5923.69	II
27	6102.72	I
29	6122.22	I
22	6161.29	I
30	6162.17	I
22	6163.76	I
24	6166.44	I
26	6169.06	I
28	6169.56	I
35	6439.07	I
30	6449.81	I
22	6455.60	I
80	6456.87	II
34	6462.57	I
29	6471.66	I

Calcium (Cont.)

Intensity	Wavelength	
32	6493.78	I
28	6499.65	I
23	6572.78	I
30	6717.69	I
33	7148.15	I
31	7202.19	I
	7291.47	II
	7323.89	II
33	7326.15	I
30	7575.81	II
60	7581.11	II
80	7601.30	II
20	7602.32	II
40	7820.78	II
60	7843.38	II
20	8017.50	II
20	8020.50	II
70	8133.05	II
100	8201.72	II
110	8248.80	II
70	8254.73	II
14	8256.67	I
10	8338.04	I
12	8339.12	I
10	8352.39	I
11	8357.17	I
130	8498.02	II
170	8542.09	II
10	8633.95	I
160	8662.14	II
12	8842.61	I
15	8909.18	I
100	8912.07	II
110	8927.36	II
12	8967.47	I
16	9099.10	I
13	9105.62	I
12	9108.82	I
10	9171.14	I
110	9213.90	II
90	9312.00	II
100	9319.56	II
110	9320.65	II
25	9416.97	I
10	9456.80	I
10	9534.88	I
11	9548.38	I
100	9567.97	II
110	9599.26	II
80	9601.82	II
10	9604.28	I
12	9663.65	I
10	9664.41	I
14	9676.30	I
14	9688.67	I
13	9701.94	I
80	9854.74	II
110	9890.63	II
90	9931.39	II
100	10223.04	II
20	10343.81	I
13	10838.97	I
13	10861.58	I
13	10863.87	I
14	10869.50	I
14	10879.87	I
20	11838.99	II
10	11949.72	II
25	12216.04	I
24	12823.86	I
25	12909.10	I
30	13033.57	I
21	13086.44	I
24	13134.95	I
20	16150.77	I
22	16157.36	I
21	16197.04	I
20	18925.47	I
24	18970.14	I
30	19046.14	I
48	19309.20	I
49	19452.99	I
47	19505.72	I
50	19776.79	I
35	19853.10	I
34	19862.22	I
23	19917.19	I
24	19933.70	I
	21389.00	II
	21428.90	II
20	22607.93	I
25	22624.93	I
30	22651.23	I

Ca III
Ref. 25, 16 — J.J.W. and H.H.S.

Intensity	Wavelength	
	Vacuum	
6	296.96	III
9	403.72	III
7	409.95	III
5	439.69	III
5	633.59	III
5	685.41	III
5	697.55	III
5	699.09	III
5	699.89	III
6	701.39	III
5	727.66	III
8	740.55	III
6	746.25	III
5	747.98	III
5	779.61	III
5	800.30	III
5	809.93	III
5	817.06	III
6	821.57	III
6	840.56	III
6	1020.07	III
5	1034.65	III
5	1187.30	III
8	1188.61	III
8	1188.61	III
5	1190.86	III
10	1262.65	III
11	1278.39	III
10	1281.55	III
12	1286.52	III
11	1298.04	III
11	1317.70	III
10	1328.95	III
11	1335.13	III
10	1360.01	III
11	1385.43	III
11	1397.69	III
13	1453.16	III
12	1459.79	III
11	1461.88	III
15	1463.34	III
16	1484.87	III
12	1496.88	III
11	1506.88	III
20	1535.23	III
15	1555.53	III
10	1562.47	III
13	1571.27	III
13	1586.13	III
10	1762.26	III
10	1783.93	III
10	1794.22	III
12	1800.21	III
13	1807.89	III
14	1812.15	III
11	1813.59	III
12	1830.06	III
10	1860.43	III
14	1870.26	III
14	1872.37	III
10	1894.12	III
11	1910.10	III
12	1935.72	III
10	1939.68	III
11	1943.01	III
12	1948.26	III
10	1953.55	III
10	1958.97	III
13	1964.61	III
13	1967.94	III
12	1972.82	III
10	1977.01	III
10	1978.55	III
11	1981.19	III
	Air	
12	2033.36	III
12	2041.53	III
13	2078.92	III
13	2098.49	III
15	2114.41	III
17	2123.03	III
14	2129.19	III
14	2133.96	III
13	2140.36	III
16	2152.43	III
12	2171.57	III
12	2276.52	III
15	2312.08	III
14	2497.74	III
15	2541.50	III
13	2587.15	III
12	2590.41	III
15	2620.82	III
15	2634.14	III
12	2686.72	III
16	2687.76	III
15	2704.86	III
14	2771.28	III
15	2791.59	III
16	2813.98	III
17	2866.54	III
18	2869.95	III
19	2881.78	III
21	2899.79	III
19	2924.33	III
20	2988.63	III
18	2989.27	III
15	3028.59	III
19	3119.67	III
15	3367.79	III
19	3372.67	III
18	3537.77	III
15	4081.77	III
15	4153.57	III
15	4164.31	III
15	4184.20	III
18	4207.24	III
17	4233.74	III
16	4240.74	III
15	4284.39	III
20	4302.81	III
15	4329.19	III
16	4333.57	III
15	4358.38	III
19	4399.59	III
17	4406.29	III
17	4431.30	III
19	4499.88	III
18	4516.59	III
18	4572.12	III
11	4708.83	III
11	4716.27	III
10	4859.17	III
10	5008.95	III
10	5050.07	III
10	5231.82	III
11	5247.37	III
13	5271.98	III
10	5301.32	III
11	5321.29	III
10	5328.06	III
11	5570.58	III
10	5579.06	III
13	6069.98	III
10	6173.22	III
12	6213.98	III
11	6294.89	III
11	6370.11	III
10	6397.33	III
12	6424.51	III
12	6485.35	III
10	6538.78	III
10	6542.24	III
10	7308.69	III
10	7843.06	III
12	7898.46	III
10 1	8217.20	III

Ca IV
Ref. 150 — J.J.W. and H.H.S.

Intensity	Wavelength	
	Vacuum	
150	249.41	IV
150	250.15	IV
150	251.35	IV
250	296.55	IV
200	299.32	IV
200	318.09	IV
50	318.39	IV
120	321.59	IV
250	329.12	IV
150	329.39	IV
200	331.44	IV
250	331.99	IV
235	332.53	IV
150	332.81	IV
200	338.83	IV
150	339.79	IV
200	340.29	IV
200	341.29	IV
200	341.46	IV
250 c	342.45	IV
100	343.19	IV
200	343.44	IV
250	343.93	IV
200	344.96	IV
215	345.13	IV
250	374.74	IV
600	434.57	IV
100	437.27	IV
250	437.77	IV
200	438.93	IV
750	443.82	IV
50	445.02	IV
500	450.57	IV
50	454.55	IV
250	456.98	IV
250	461.09	IV
150	565.46	IV
750	656.00	IV
500	669.70	IV

Ca V
Ref. 150 — J.J.W. and H.H.S.

Intensity	Wavelength	
	Vacuum	
200	190.36	V
250	190.46	V
250	196.97	V
300	199.55	V
250	200.51	V
265	257.98	V
165	260.45	V
400	267.77	V
300	270.31	V
200	271.16	V
250	272.27	V
200	272.98	V
400	280.99	V
300	284.98	V
450 c	286.96	V
500	322.17	V
250	322.76	V
300	323.22	V
250	324.48	V
250	325.28	V
300	330.94	V
200	333.44	V
200	333.57	V
300	334.55	V
250 c	335.34	V
200	336.55	V
200	337.54	V
250	338.06	V
200	343.64	V
450	352.92	V
250	356.25	V
250	377.18	V
200	387.08	V
750	425.00	V
500	558.60	V
400	637.93	V
300	643.12	V
400	646.57	V
250	647.88	V
250	651.55	V
300	656.76	V

CALIFORNIUM (Cf)
Z = 98

Cf I and II
Ref. 52, 331 — J.G.C.

Intensity		Wavelength	
		Air	
10000		2739.31	
10000	s	2759.10	
10000		2774.52	
10000	1	2852.03	
10000	s	2855.24	
10000		3298.14	
10000		3352.71	
10000	1	3367.79	
10000		3392.22	I
10000		3481.07	
10000	1	3513.47	I
10000		3531.49	I
10000		3540.98	I
10000		3598.77	I
10000		3605.32	I
10000		3612.11	II
10000		3617.49	I
10000	s	3626.76	II
10000		3659.46	
10000		3662.70	I
10000	s	3699.49	
10000	1	3722.11	II
10000		3739.35	I
10000		3785.61	I
10000	1	3789.04	II
10000	s	3893.23	II
10000	1	3993.57	II
10000		4035.45	
10000		4099.12	I
10000		4242.38	I
10000		4329.03	I
10000		4335.22	I
10000		5173.96	I
10000		5179.08	I
10000		5219.24	I
10000	s	5279.01	
10000	s	5320.09	
10000	s	5339.13	
10000		5408.88	I
10000		5726.05	I
10000		6622.83	I
10000		6631.26	I

Intensity		Wavelength	
10000		6677.90	I
10000		6894.59	I
10000		6927.10	II
10000	1	7074.52	I
10000	s	7307.90	I
10000		8141.29	I
10000		8241.77	I
10000		8333.85	II
10000		8423.49	II
10000		8568.83	II
10000	1	9228.52	I
10000		9337.70	I
10000		9649.51	I
10000	s	10308.41	I
10000	s	10568.83	I
10000	s	10614.84	I
10000		11300.19	I
10000		11681.85	I
10000		11941.33	I
10000	1	12183.05	I
10000	s	12352.72	I
10000		12437.48	I
10000	1	12789.41	I
10000	1	13329.98	I
10000	s	13362.98	I
10000	1	13376.89	I
10000	1	13474.44	I
10000	1	14772.49	I
10000	1	15281.32	I
10000		15587.12	I
10000		15675.92	I
10000		16759.06	I
10000	s	17626.25	I
10000		18718.69	I
10000	h	19068.71	I
10000		19336.96	I
10000	1	19576.84	I
10000	1	20393.38	I
10000	s	20869.98	I

CARBON (C)
Z = 6

C I and II
Ref. 211 — R.L.K.

Intensity	Wavelength	
	Vacuum	
9	595.022	II
30	687.053	II
50	687.345	II
10	858.092	II
20	858.559	II
30	903.624	II
60	903.962	II
150	904.142	II
30	904.480	II
9	1009.86	II
10	1010.08	II
10	1010.37	II
80	1036.337	II
150	1037.018	II
150	1157.910	I
150	1158.019	I
150	1158.035	I
150	1188.992	I
150	1189.447	I
200	1189.631	I
300	1193.009	I
300	1193.031	I
300	1193.240	I
300	1193.264	I
100	1193.393	I
150	1193.649	I
150	1193.679	I
100	1194.064	I
100	1194.488	I
100	1261.552	I
250	1277.245	I
250	1277.282	I
300	1277.513	I
300	1277.550	I
200	1280.333	I
100	1311.363	I
9	1323.951	II
120	1329.578	I
120	1329.600	I
150	1334.532	II
300	1335.708	II
100	1354.288	I
150	1355.84	I
120	1364.164	I
100	1459.032	I
200	1463.336	I
120	1467.402	I
150	1481.764	I
150	1560.310	I
400	1560.683	I
400	1560.708	I
100	1561.341	I

Intensity		Wavelength	
400		1561.438	I
150		1656.266	I
120		1656.928	I
300		1657.008	I
120		1657.380	I
120		1657.907	I
150		1658.122	I
500		1751.823	I
1000		1930.905	I
		Air	
800		2478.56	I
250		2509.12	II
350		2512.06	II
250	h	2574.83	II
350	1	2741.28	II
250		2746.49	II
1000		2836.71	II
800		2837.60	II
800	h	2992.62	II
350		3876.19	II
350		3876.41	II
350		3876.66	II
570		3918.98	II
800		3920.69	II
250		4074.52	II
350	1	4075.85	II
800		4267.00	II
1000		4267.26	II
200		4771.75	I
200		4932.05	I
200		5052.17	I
350		5132.94	II
350		5133.28	II
350		5143.49	II
570		5145.16	II
400		5151.09	II
300		5380.34	I
250		5648.07	II
350		5662.47	II
570		5889.77	II
350		5891.59	II
200		6001.13	I
250		6006.03	I
110		6007.18	I
150		6010.68	I
300		6013.22	I
250		6014.84	I
800		6578.05	II
570		6582.88	II
200		6587.61	I
250		6783.90	II
250		7113.18	I
250		7115.19	I
250		7115.63	II
200		7116.99	I
350		7119.90	II
800		7231.32	II
1000		7236.42	II
200		7860.89	I
200		8058.62	I
520		8335.15	I
250		9061.43	I
200		9062.47	I
200		9078.28	I
250		9088.51	I
450		9094.83	I
300		9111.80	I
800		9405.73	I
150		9603.03	I
250		9620.80	I
300		9658.44	I
200		10683.08	I
300		10691.25	I
12		11619.29	I
23		11628.83	I
13		11658.85	I
47		11659.68	I
24		11669.63	I
85		11748.22	I
142		11753.32	I
114		11754.76	I
11		11777.54	I
17		11892.91	I
30		11895.75	I
26		12614.10	I
20		13502.27	I
38		14399.65	I
16		14403.25	I
61		14420.12	I
12		14429.03	I
13		14442.24	I
12		16559.66	I
50		16890.38	I
10		17338.56	I
11		17448.60	I
13		18139.80	I
23		19721.99	I

C III
Ref. 22, 211 — R.L.K.

Intensity		Wavelength	
		Vacuum	
250		371.69	III
250		371.75	III
150		371.78	III
500		386.203	III
200		450.734	III
400		459.46	III
500		459.52	III
570		459.63	III
250		511.522	III
250		535.288	III
300		538.080	III
350		538.149	III
400		538.312	III
350		574.281	III
800		977.03	III
370		1174.93	III
350		1175.26	III
330		1175.59	III
500		1175.71	III
350		1175.99	III
370		1176.37	III
		Air	
250		2162.94	III
800		2296.87	III
150		2697.75	III
110	1	2724.85	III
150	1	2725.30	III
150	1	2725.90	III
200		2982.11	III
150		4056.06	III
200		4067.94	III
250		4068.91	III
250		4070.26	III
150		4162.86	III
250	h	4186.90	III
200		4325.56	III
600		4647.42	III
520		4650.25	III
375		4651.47	III
200		4665.86	III
450		5695.92	III
150		5826.42	III
150		6744.38	III
150	h	7037.25	III
150		7612.65	III
300	h	8196.48	III
150		8332.99	III
300		8500.32	III

C IV
Ref. 66, 211 — R.L.K.

Intensity		Wavelength	
		Vacuum	
250		244.91	IV
200		289.14	IV
250		289.23	IV
570		312.42	IV
500		312.46	IV
650		384.03	IV
700		384.18	IV
400		419.52	IV
500		419.71	IV
1000		1548.202	IV
900		1550.774	IV
		Air	
200	1	2524.41	IV
300	s	2529.98	IV
200	w	4658.30	IV
250		5801.33	IV
200		5811.98	IV
90	w	7726.2	IV

C V
Ref. 211 — R.L.K.

Intensity	Wavelength	
	Vacuum	
110	34.973	V
450	40.268	V
110	227.19	V
160	248.66	V
160	248.74	V
	Air	
40	2270.91	V

5	2277.25	V
20	2277.92	V
5	4943.88	V
5	4944.56	V

CERIUM (Ce)
Z = 58

Ce I and II
Ref. 1 — C.H.C.

Intensity	Wavelength	
	Air	
130	2462.97	II
110	2518.51	II
200	2548.68	II
340	2651.01	II
120	2696.07	II
120	2706.88	II
110	2723.38	II
100	2741.96	II
150	2750.89	II
120	2761.42	II
100	2784.27	II
100	2785.35	II
270	2790.53	II
140	2791.42	II
100	2830.90	II
250	2833.31	II
110	2874.14	II
100	2908.42	II
120	2918.67	II
110	2955.94	II
100	2964.80	II
400	2972.58	II
150	2976.91	II
120	2977.46	II
250	2980.41	II
110	2990.87	II
320	2994.42	II
400	2995.64	II
370	3008.79	II
210	3017.20	II
200	3037.73	II
350	3051.98	II
320	3055.24	II
680	3056.78	II
320	3063.01	II
250	3083.67	II
200	3084.44	II
370	3090.37	II
200	3103.38	II
320	3107.47	II
300	3110.28	II
220	3111.17	II
200	3127.53	II
240	3130.33	II
200	3130.87	II
290	3144.60	II
290	3145.28	II
290	3146.41	II
290	3164.15	II
290	3169.18	II
290	3171.61	II
480	3183.52	II
240	3186.13	II
200	3190.34	II
710	3194.83	II
200	3199.28	II
990	3201.71	II
200	3218.38	II
710	3218.94	II
880	3221.17	II
330	3225.67	II
710	3227.11	II
240	3229.36	II
480	3231.24	II
710	3234.16	II
330	3234.89	II
390	3236.74	II
390	3243.37	II
200	3246.67	II
200	3260.98	II
990	3263.88	II
330	3272.25	II
200	3274.86	II
330	3279.84	II
240	3285.22	II
200	3295.28	II
220	3296.88	II
240	3300.15	II
240	3304.84	II
200	3312.22	II
240	3314.72	II
200	3317.80	II
330	3334.46	II
240	3341.87	II
330	3343.86	II
440	3344.76	II
200	3355.02	II

Cerium (Cont.)

Int.	λ	Sp.	Int.	λ	Sp.	Int.	λ	Sp.	Int.	λ	Sp.
240	3357.22	II	370	3857.64	II	670	4081.22	II	700	4349.79	II
200	3360.54	II	200	3862.46	II	910	4083.23	II	560	4352.71	II
240	3366.55	II	200	3868.13	II	450	4085.23	II	910	4364.66	II
200	3371.18	II	270	3874.68	II	250	4087.36	II	350	4373.82	II
200	3373.46	II	620	3876.97	II	230	4088.85	II	530	4375.92	II
480	3373.73	II	1100	3878.36	II	450	4101.77	II	910	4382.17	II
200	3377.13	II	1500	3882.45	II	250	4105.00	II	700	4386.84	II
200	3383.68	II	1000	3889.98	II	510	4107.42	II	310	4388.01	II
200	3404.91	II	210	3890.75	II	200	4110.38	II	1700	4391.66	II
240	3405.98	II	210	3890.98	II	250	4111.39	II	200	4398.79	II
290	3417.45	II	620	3895.11	II	420	4115.37	II	510	4399.20	II
600	3422.71	II	590	3896.80	II	250	4117.01	II	350	4410.64	II
390	3426.21	II	490	3898.27	II	200	4117.29	II	350	4410.76	II
290	3441.21	II	270	3898.94	II	200	4117.59	II	310	4416.90	II
480	3476.84	II	200	3903.34	II	770	4118.14	II	980	4418.78	II
240	3482.35	II	250	3904.34	II	250	4119.02	II	200	4423.68	II
710	3485.05	II	200	3906.92	II	310	4119.79	II	310	4427.07	II
210	3507.94	II	770	3907.29	II	310	4119.08	II	480	4427.92	II
600	3517.38	II	560	3908.41	II	450	4120.83	II	310	4428.44	II
210	3520.52	II	390	3908.54	II	510	4123.24	II	650	4429.27	II
330	3521.88	II	270	3909.31	II	510	4123.49	II	480	4444.39	II
210	3526.68	II	230	3912.19	II	980	4123.87	II	450	4444.70	II
600	3534.05	II	980	3912.44	II	510	4124.79	II	770	4449.34	II
770	3539.08	II	390	3915.52	II	980	4127.37	II	620	4450.73	II
210	3545.60	II	390	3916.14	II	250	4127.74	II	2400	4460.21	II
290	3546.19	II	230	3917.64	II	200	4128.07	II	450	4461.14	II
240	3552.73	II	770	3918.28	II	530	4130.71	II	420	4463.41	II
420	3555.00	II	480	3919.81	II	480	4131.10	II	280	4467.54	II
1200	3560.80	II	590	3921.73	II	2700	4133.80	II	1400	4471.24	II
210	3576.23	II	560	3923.11	II	270	4135.44	II	450	4472.72	II
1000	3577.45	II	450	3924.64	II	270	4137.47	II	700	4479.36	II
330	3590.60	II	770	3931.09	II	2000	4137.65	II	700	4483.90	II
390	3607.63	II	310	3931.37	II	270	4138.10	II	840	4486.91	II
550	3609.69	II	230	3931.83	II	210	4138.35	II	250	4497.85	II
420	3613.70	II	310	3933.73	II	770	4142.40	II	100	4506.41	I
440	3622.15	II	560	3938.09	II	390	4144.49	II	110	4515.86	II
380	3623.74	II	770	3940.34	II	670	4145.00	II	100	4519.59	II
440	3623.84	II	310	3940.97	II	480	4146.23	II	770	4523.08	II
200	3631.19	II	2000	3942.15	II	280	4148.90	II	840	4527.35	II
350	3646.97	II	2700	3942.75	II	420	4149.79	II	840	4528.47	II
260	3647.75	II	770	3943.89	II	980	4149.94	II	110	4532.49	II
260	3647.95	II	310	3947.97	II	420	4150.91	II	111	4533.10	II
470	3652.11	II	1111	3951.54	II	1400	4151.97	II	840	4539.75	II
660	3653.67	II	340	3953.66	II	230	4153.13	II	210	4544.96	II
310	3654.97	II	310	3953.36	II	450	4159.03	II	250	4551.30	II
1800	3655.85	II	230	3956.06	II	310	4163.52	II	650	4560.28	II
440	3659.23	II	980	3956.28	II	1300	4165.61	II	310	4560.96	II
330	3659.97	II	230	3958.27	II	620	4166.88	II	2100	4562.36	II
880	3660.64	II	230	3958.87	II	250	4167.80	II	420	4565.84	II
880	3667.98	II	770	3960.91	II	320	4169.77	II	1100	4572.28	II
220	3672.18	I	390	3964.50	II	320	4169.88	II	420	4582.50	II
350	3672.79	II	770	3967.05	II	340	4176.70	II	130	4591.12	II
220	3679.42	II	450	3971.68	II	340	4181.08	II	840	4593.93	II
300	3694.91	II	270	3972.07	II	340	4185.33	II	420	4606.40	II
220	3704.98	II	270	3975.07	II	3500	4186.60	II	420	4624.90	II
1000	3709.29	II	770	3978.65	II	530	4187.32	II	1700	4628.16	II
1000	3709.93	II	560	3980.88	II	560	4193.09	II	170	4632.32	I
1400	3716.37	II	560	3982.89	II	370	4193.28	II	110	4650.51	I
420	3718.19	II	310	3983.29	II	370	4193.87	II	130	4654.29	II
420	3718.38	II	770	3984.68	II	630	4196.34	II	110	4669.50	II
210	3719.80	II	370	3989.44	II	280	4198.00	II	150	4680.13	II
420	3725.68	II	700	3992.39	II	280	4198.67	II	270	4684.61	II
490	3728.02	II	370	3992.91	II	840	4198.72	II	200	4714.00	II
800	3728.42	II	910	3993.82	II	240	4201.24	II	100	4714.81	II
320	3748.06	II	2800	3999.24	II	910	4202.94	II	110	4725.09	II
250	3751.45	II	230	4001.56	II	270	4209.41	II	100	4733.52	II
200	3755.43	II	910	4003.77	II	370	4214.04	II	310	4737.28	II
300	3762.98	II	370	4005.64	II	310	4217.59	II	100	4739.53	II
680	3764.12	II	210	4007.59	II	1500	4222.60	II	160	4747.17	II
200	3765.04	II	2700	4012.39	II	770	4227.75	II	110	4757.84	II
300	3768.76	II	910	4014.90	II	390	4231.74	II	100	4768.77	II
210	3770.76	II	250	4015.88	II	240	4234.21	II	230	4773.94	II
300	3771.60	II	200	4019.04	II	200	4236.02	II	110	4822.55	I
250	3776.61	II	240	4022.27	II	980	4239.92	II	140	4847.77	I
620	3781.62	II	840	4024.49	II	390	4242.72	II	180	4882.46	II
440	3782.52	II	240	4025.15	II	310	4245.89	II	110	4943.44	I
200	3783.58	II	840	4028.41	II	310	4245.98	II	130	4971.50	II
860	3786.63	II	250	4030.34	II	390	4246.72	II	130	4994.63	I
520	3788.75	II	840	4031.34	II	1100	4248.68	II	210	5009.10	I
300	3792.32	II	340	4037.67	II	390	4253.37	II	100	5011.77	II
2500	3801.52	II	2100	4040.76	II	620	4255.79	II	120	5022.87	II
800	3803.09	II	910	4042.58	II	200	4263.43	II	120	5037.78	II
1000	3808.11	II	230	4045.21	II	620	4270.19	II	120	5040.85	I
490	3809.21	II	620	4046.34	II	390	4270.72	II	180	5044.02	II
250	3812.20	II	210	4051.43	II	200	4278.86	II	120	5071.78	I
490	3815.85	II	210	4051.99	II	280	4285.37	II	240	5075.35	II
470	3817.46	II	700	4053.51	II	200	4288.66	II	470	5079.68	II
300	3819.02	II	450	4054.99	II	200	4289.44	II	130	5112.70	I
470	3823.90	II	280	4062.22	II	2000	4289.94	II	160	5117.17	II
470	3830.55	II	230	4062.94	II	200	4296.07	II	170	5129.57	I
490	3831.08	II	280	4067.28	II	1500	4296.67	II	110	5147.57	II
490	3834.55	II	420	4068.84	II	420	4296.78	II	100	5149.99	I
270	3836.10	II	1100	4071.81	II	590	4299.36	II	280	5159.69	I
1100	3838.54	II	270	4072.92	II	770	4300.33	II	280	5161.48	I
200	3843.76	II	1800	4073.48	II	420	4305.14	II	190	5174.55	I
220	3846.52	II	210	4073.74	II	770	4306.72	II	370	5187.46	II
250	3848.10	II	1500	4075.71	II	390	4309.74	II	210	5191.66	II
860	3848.59	II	1500	4075.85	II	560	4320.72	II	190	5211.92	I
860	3853.15	II	210	4076.24	II	310	4330.45	II	260	5223.46	I
1200	3854.18	II	420	4077.47	II	310	4332.71	II	180	5229.75	I
1200	3854.31	II	530	4078.32	II	240	4336.23	II	140	5232.92	II
620	3855.29	II	270	4078.52	II	980	4337.77	II	260	5245.92	I
390	3857.02	II	270	4080.44	II	340	4339.31	II	130	5265.71	II

Cerium (Cont.)

Intensity	Wavelength	
340	5274.23	II
130	5296.56	I
130	5328.08	I
190	5330.54	II
450	5353.53	II
300	5393.40	II
150	5397.64	I
280	5409.23	II
110	5420.38	I
140	5449.24	I
140	5468.37	II
140	5472.29	II
260	5512.08	II
110	5556.25	I
170	5564.97	I
130	5565.97	I
100	5595.88	I
240	5601.28	I
190	5655.14	I
240	5669.96	I
120	5677.75	I
120	5692.94	I
300	5696.99	I
370	5699.23	I
240	5719.03	I
140	5773.12	I
120	5788.15	I
120	5812.92	I
230	5940.86	I
11	6000.18	I
55	6001.90	I
55	6005.86	I
15	6006.20	I
55	6006.82	I
19	6007.37	I
75	6013.42	I
23	6016.59	I
110	6024.20	I
15	6027.16	I
11	6031.26	I
23	6033.58	II
35	6034.20	II
23	6034.41	I
35	6035.49	II
110	6043.39	II
28	6045.42	I
55	6047.40	I
19	6051.80	II
23	6057.50	I
35	6058.00	I
23	6066.75	I
19	6069.46	I
35	6069.48	I
35	6072.00	I
35	6076.61	I
17	6077.16	I
17	6080.37	I
17	6081.28	I
19	6088.86	I
19	6088.96	I
35	6093.19	I
45	6098.34	II
11	6099.80	I
28	6108.74	II
15	6118.56	I
17	6118.90	I
45	6123.67	I
19	6132.00	II
19	6132.18	I
11	6135.45	I
23	6139.03	I
15	6142.92	I
35	6143.36	II
23	6146.43	I
19	6147.84	I
23	6151.72	I
19	6159.82	I
19	6162.14	I
19	6165.45	I
19	6175.28	I
35	6186.17	I
15	6187.97	I
15	6195.23	I
19	6195.53	I
19	6198.05	I
35	6208.98	I
11	6216.82	I
35	6228.94	I
19	6229.13	I
23	6232.45	II
28	6237.45	I
13	6238.71	I
11	6241.87	I
13	6242.91	I
15	6253.65	I
13	6257.99	I
15	6264.27	I
45	6272.05	II
15	6276.47	I
35	6295.58	I
28	6299.51	II
23	6300.21	I
13	6306.64	I

Intensity	Wavelength		
35	6310.01		I
15	6335.40		I
11	6337.21		I
13	6340.70		I
35	6343.95		II
35	6371.11		II
28	6386.84		I
23	6393.02		II
11	6395.16		I
11	6425.29		II
35	6430.07		I
19	6434.39		I
23	6436.40		I
19	6446.12		I
35	6458.03		I
19	6466.88		II
28	6467.39		I
35	6473.72		I
17	6490.97		I
11	6503.27		II
11	6507.16		II
23	6513.59		II
19	6517.31		I
19	6551.70		I
45	6555.65		I
23	6579.10		I
15	6606.35		I
15	6606.86		II
22	6612.06		I
10	6623.00		I
30	6628.93		I
13	6650.89		I
22	6652.72		II
10	6661.41		I
13	6665.59		I
10	6675.54		II
15	6686.60		I
26	6700.66		I
35	6704.27		I
13	6704.52		II
10	6706.04		II
15	6728.71		I
15	6729.57		I
15	6744.70		II
10	6746.90		I
30	6774.25		II
35	6775.59		I
10	6778.28		I
18	6807.81		I
10	6808.82		I
15	6818.23		I
10	6829.73		II
13	6847.25		I
12	6856.55		I
10	6893.66		I
10	6898.45		II
30	6924.81		I
12	6939.45		I
19	6973.50		II
10	6983.82		II
30	6986.02		I
12	7054.51		I
11	7058.68		II
11	7060.00		I
35	7061.75		II
11	7064.49		I
35	7086.35		I
11	7105.04		II
11	7115.08		II
10	7124.73		I
16	7141.42		I
19	7150.23		II
10	7151.67		I
16	7155.25		I
16	7156.99		II
16	7189.40		II
10	7191.72		I
11	7201.56		II
16	7201.89		I
10	7203.55		I
12	7210.67		I
19	7217.36		I
16	7235.71		II
22	7238.36		II
12	7241.73		I
25	7252.75		I
12	7262.64		I
11	7277.90	h	I
11	7296.07		I
19	7301.42		II
19	7313.45		II
25	7329.91		I
16	7334.68		II
12	7343.44		I
25	7397.77		I
11	7401.27		I
12	7417.94		II
11	7424.70		C
12	7433.08		I
11	7438.56		I
12	7444.44		I
10	7472.41		I
16	7486.57		II

Intensity	Wavelength		
11	7527.46		I
11	7527.68		I
10	7533.73		I
10	7551.25		I
12	7562.44		I
10	7562.86		I
10	7563.60	h	I
10	7603.10		I
25	7616.11		II
12	7646.08		I
10	7647.88		I
12	7682.47		I
25	7689.17		II
10	7732.33		I
16	7748.35		I
10	7797.70		I
12	7842.59		I
22	7844.94		II
16	7850.02		II
16	7851.18		II
22	7857.54		II
12	7864.49		I
10	7866.04		I
16	7898.96		II
11	7913.52		I
10	7927.30		C
10	7927.72		I
10	7934.50		II
30	8025.56		II
16	8070.71		I
10	8094.43		I
16	8120.36		I
10	8241.55		II
12	8261.09		I
16	8418.23		I
11	8495.82		I
12	8539.08		II
10	8612.64		I
10	8647.66	h	I
11	8702.38	h	II
25	8772.14		II
12	8810.84		I
30	8891.20		II

Ce III
Ref. 136, 305 — J.R.

Intensity	Wavelength	
	Vacuum	
100	840.24	III
20	844.11	III
40	845.02	III
20	847.88	III
200	851.18	III
200	852.63	III
200	853.47	III
60	853.78	III
200	855.16	III
400	860.15	III
200	862.25	III
40	868.74	III
20	869.51	III
40	869.84	III
20	871.15	III
40	871.27	III
20	880.68	III
60	881.75	III
60	884.04	III
20	885.22	III
30	888.39	III
80	892.75	III
20	899.32	III
100	912.77	III
40	937.04	III
40	999.26	III
20	1025.25	III
20	1025.29	III
20	1026.28	III
40	1029.37	III
40	1034.55	III
20	1041.14	III
20	1042.74	III
100	1051.61	III
50	1057.40	III
70	1057.66	III
100	1058.46	III
50	1062.99	III
30	1063.26	III
50	1063.51	III
100	1067.76	III
100	1068.69	III
20	1070.54	III
200	1072.79	III
40	1073.69	III
30	1079.35	III
20	1080.82	III
30	1088.70	III
30	1090.03	III
20	1092.48	III

Intensity	Wavelength	
20	1099.25	III
40	1100.71	III
20	1107.09	III
20	1111.19	III
20	1116.30	III
20	1125.58	III
20	1129.73	III
20	1132.74	III
100	1142.55	III
20	1192.41	III
50	1201.87	III
20	1204.05	III
20	1719.43	III
20	1796.89	III
20	1836.66	III
20	1836.99	III
30	1862.32	III
30	1950.36	III
20	1990.54	III
	Air	
100	2033.34	III
100	2057.65	III
100	2077.87	III
200	2083.32	III
100	2089.96	III
400	2109.07	III
100	2122.55	III
500	2136.95	III
1000	2151.44	III
3000	2166.88	III
2000	2169.48	III
5000	2180.64	III
1000	2183.71	III
2000	2203.15	III
3000	2218.11	III
5000	2222.01	III
5000	2225.08	III
3000	2227.84	III
3000	2228.05	III
5000	2242.29	III
2000	2249.25	III
2000	2264.85	III
2000	2268.20	III
2000	2287.82	III
2000	2298.70	III
3000	2300.65	III
4000	2302.09	III
5000	2317.34	III
10000	2318.64	III
5000	2324.31	III
2000	2337.66	III
5000	2350.10	III
2000	2362.54	III
2000	2367.77	III
10000	2372.34	III
5000	2377.07	III
5000	2377.48	III
10000	2380.12	III
3000	2382.28	III
3000	2385.06	III
5000	2395.04	III
3000	2406.15	III
4000	2408.08	III
2000	2410.26	III
5000	2415.60	III
2000	2417.01	III
2000	2423.02	III
3000	2428.64	III
5000	2430.24	III
10000	2431.45	III
15000	2439.80	III
3000	2441.55	III
2000	2444.78	III
10000	2454.32	III
10000	2469.95	III
3000	2471.66	III
5000	2477.25	III
8000	2479.44	III
3000	2479.51	III
10000	2483.82	III
10000	2497.50	III
3000	2503.56	III
2000	2504.43	III
20000	2531.99	III
3000	2539.27	III
3000	2557.49	III
4000	2577.67	III
2000	2578.30	III
2000	2584.71	III
10000	2603.59	III
2000	2607.96	III
2000	2615.79	III
2000	2649.38	III
2000	2662.81	III
3000	2719.30	III
2000	2730.04	III
3000	2743.71	III
4000	2748.90	III
4000	2754.87	III
4000	2768.28	III
3000	2849.40	III

Cerium (Cont.)

Intensity	Wavelength			Intensity	Wavelength		
2000	2861.39	III		42	18579.82	III	
4000	2907.05	III		38	19141.29	III	
10000	2923.81	III		27	19377.15	III	
5000	2925.26	III		26	19466.14	III	
10000	2931.54	III		20	19498.14	III	
2000	2948.53	III		55	19524.18	III	
5000	2973.72	III		30	20685.63	III	
10000	3022.75	III		12	21380.23	III	
50000	3031.58	III					
95000	3055.59	III					

Ce IV
Ref. 166 — J.R.

Intensity	Wavelength	
	Vacuum	
2	447.58	IV
1	443.11	IV
8	558.92	IV
8	571.59	IV
40	741.79	IV
30	754.60	IV
12	755.75	IV
6	975.20	IV
5	1009.31	IV
2	1022.12	IV
9	1057.67	IV
1	1059.64	IV
50	1289.41	IV
75	1332.16	IV
75	1372.72	IV
2	1577.60	IV
1	1572.62	IV
15	1641.58	IV
20	1775.30	IV
20	1779.03	IV
35	1914.75	IV
10	1937.21	IV
	Air	
100	2000.42	IV
35	2003.11	IV
100	2009.94	IV
3	2433.50	IV
5	2445.50	IV

Ce V
Ref. 261 — J.R.

Intensity	Wavelength	
	Vacuum	
100	365.66	V
300	399.36	V
150	404.21	V
200	482.96	V
100	552.13	V

Continuing Cerium III (first column):

Intensity	Wavelength	
20000	3056.56	III
40000	3057.23	III
20000	3057.58	III
40000	3085.10	III
20000	3106.98	III
30000	3110.53	III
30000	3121.56	III
20000	3141.29	III
20000	3143.96	III
20000	3147.06	III
20000	3228.57	III
3000	3234.20	III
4000	3267.76	III
3000	3267.94	III
20000	3353.29	III
10000	3395.77	III
4000	3398.91	III
30000	3427.36	III
40000	3443.63	III
30000	3454.39	III
40000	3459.39	III
60000	3470.92	III
50000	3497.81	III
60000	3504.64	III
500	3514.41	III
50000	3544.07	III
3000	3784.29	III
800	3936.80	III
300	3957.10	III
500	4169.42	III
300	4191.70	III
500	4194.83	III
300	4213.26	III
300	4217.13	III
400	4284.77	III
300	4304.71	III
600	4346.35	III
400	4389.97	III
600	4448.32	III
500	4485.77	III
1000	4521.92	III
1000	4535.73	III
300	4576.90	III
500	4627.60	III
300	4766.07	III
500	4976.45	III
500	5650.97	III
1000	5664.20	III
500	5691.08	III
300	5710.59	III
500	5749.47	III
500	5949.83	III
2000	5962.22	III
500	5962.71	III
400	5979.56	III
1000	5983.40	III
3000	6002.63	III
10000	6032.54	III
10000	6060.91	III
500	6061.79	III
500	6097.35	III
500	6098.87	III
500	6135.10	III
300	6287.79	III
500	6308.16	III
300	6341.75	III
1000	6944.94	III
700	7739.04	III
300	7758.27	III
500	7826.80	III
300	7948.64	III
500	7960.31	III
500	7991.01	III
400	8030.80	III
300	8084.12	III
400	8177.33	III
300	8186.03	III
300	8222.16	III
300	9056.53	III
400	9079.58	III
400	9328.20	III
300	9367.03	III
300	9567.37	III
400	10458.37	III
400	10494.42	III
300	10534.36	III
400	10684.46	III
15	12756.96	III
12	12821.62	III
80	15847.58	III
80	15956.79	III
12	15960.59	III
87	16128.75	III

CESIUM (Cs)
Z = 55
Cs I and II
Ref. 82, 154, 155, 200, 263, 325—
C.J.S.

Intensity		Wavelength			Intensity		Wavelength			Intensity		Wavelength	
		Vacuum											
250		591.04	II		330		2315.69	II		8800		6646.57	II
2000		639.36	II		1400		2332.46	II		3300	c	6723.28	I
500		668.39	II		240		2375.86	II		9600		6724.47	II
15000		718.14	II		220		2379.27	II		200		6824.65	I
15000		808.76	II		1000		2392.86	II		300		6870.45	I
15000		813.84	II		370		2425.17	II		37000		6955.50	II
35000		901.27	II		220		2543.93	II		4800		6973.30	I
40000		926.66	II		630		2596.99	II		16000		6979.67	II
3		1656.15	II		450		2609.43	II		980		6983.49	I
7		1689.46	II		170		2628.86	II		2300		7130.54	II
2		1691.83	II		300		2699.18	II		13000	w	7149.54	II
7		1717.64	II		180		2789.78	II		1600		7160.90	II
7		1718.97	II		150		2793.31	II		630		7188.37	II
3		1727.79	II		680	w	2816.92	II		1100		7206.04	II
3		1736.77	II		150		2829.04	II		790		7228.53	I
8		1807.83	II		140	c	2866.36	II		960		7248.88	II
8		1813.75	II		1600		2931.08	II		130		7279.90	I
7		1815.16	II		290		3151.19	II		1100		7279.96	I
18		1840.50	II		520	c	3265.91	II		780		7369.36	II
13		1859.16	II		650	w	3267.12	II		550		7437.78	II
7		1864.83	II		850	c	3271.63	II		440	w	7523.39	II
7		1876.72	II		570		3368.57	II		2600		7608.90	I
18		1883.93	II		150		3514.05	II		910	w	7651.95	II
5		1914.61	II		110	w	3615.01	II		310	h	7746.98	II
29		1935.19	II		170	w	3680.11	II		2400		7852.52	II
		Air			130		3732.56	II		3300		7943.88	II
72		2024.97	II		1100	w	3785.44	II		22000		7997.44	II
46		2028.32	II		2700		3805.12	II		2100		8012.98	II
180		2080.05	II		520	c	3861.50	II		3500		8015.73	I
160		2087.20	II		2100	c	3876.15	I		8200	c	8047.13	II
100		2102.22	II		600	c	3888.61	I		1300		8078.50	II
160		2205.52	II		3400		3896.99	II		510		8078.94	I
190		2220.53	II		4200		3959.51	II		4500		8079.04	I
120		2245.81	II		2500		3965.20	II		59000	cr	8521.13	I
1600		2267.65	II		2100		3974.25	II		680		8521.62	II
750		2273.84	II		8000		4039.85	II		1700		8608.31	II
220		2285.41	II		2000		4068.78	II		420	w	8695.60	II
250		2286.15	II		320	w	4151.27	II		15000	c	8761.41	I
					2000		4213.14	II		840		8775.42	II
					1900		4232.20	II		340	w	8857.39	II
					980		4234.41	II		61000	cr	8943.47	I
					14000		4264.70	II		18000		9172.32	I
					18000	w	4277.13	II		5200		9208.53	I
					5100		4288.38	II		540	c	9212.36	II
					1900		4300.65	II		3300	c	9220.75	II
					7600		4363.30	II		310	c	9718.11	II
					3700		4373.04	II		630	c	9932.91	II
					960		4384.44	II		1400		9994.79	II
					7000	w	4405.10	II		19000		10024.36	I
					12000		4501.55	II		4800		10123.41	I
					20000		4526.74	II		26000		10123.60	I
					4100		4538.97	II		3000	c	10176.02	II
					1000	c	4555.28	I		2700		10379.66	II
					460	c	4593.17	I		1300		10480.93	II
					100000		4603.79	II		4700		10504.51	II
					4200	c	4616.17	II		1900	w	10807.88	II
					2800		4646.52	II		610	w	11324.34	II
					990		4670.29	II		530	c	11496.56	II
					1500		4732.99	II		410	c	11704.18	II
					7000		4763.64	II		330	c	11797.93	II
					1900		4786.38	II		710	c	11840.84	II
					25000		4830.19	II		1100		12604.29	II
					19000		4870.04	II		2000	c	12735.52	II
					4000	c	4880.05	II		400	c	12746.34	II
					37000		4952.85	II		850		13406.83	II
					8200		4972.60	II		2900		13424.31	I
					27000		5043.80	II		38000	c	13588.29	I
					2800	c	5059.87	II		8400		13602.56	I
					2900		5096.60	II		4200	c	13692.91	II
					6500		5209.58	II		5700		13758.81	I
					75000		5227.04	II		1400	c	13868.82	II
					29000		5249.38	II		880		14482.21	II
					11000		5274.05	II		50000	c	14694.91	I
					1300		5306.60	II		350		14906.91	II
					10000	c	5349.13	II		1900	c	15293.80	II
					22000		5370.99	II		1600	c	15356.61	II
					1200		5402.78	II		620		15445.47	II
					2900		5419.67	II		940	c	15735.48	II
					60	c	5465.94	I		1100		16426.14	II
					37		5502.88	I		820		16535.63	I
					39000		5563.02	II		1500		17012.32	I
					100		5635.21	I		340		18160.84	II
					210	c	5664.02	I		430		18179.35	II
					27		5745.72	I		710		18221.36	II
					4500		5814.16	II		390		18222.40	II
					24000		5831.14	II		510		18404.77	II
					59	c	5838.83	I		310		18407.23	II
					300		5845.14	I		590		18509.52	II
					51000		5925.63	II		610		18509.85	II
					1400	c	5984.39	II		100	c	18742.18	II
					640		6010.49	I		530	c	18921.76	II
					86		6034.09	I		150	c	18986.48	II
					760		6076.72	II		180		19924.85	II
					9800		6128.61	II		390		20110.77	II
					1000		6213.10	I		760		20138.47	I
					170		6217.60	I		200		20301.54	II
					320	c	6354.55	I		220	c	20443.87	II
					2000		6419.52	II		92	w	21103.42	II
					8300		6495.53	II		79	c	22344.98	II
					10000	w	6536.44	II		78		22448.98	II
					490		6586.51	I		880		22811.86	I
					530	c	6628.01	II		1100		23037.98	I
					97		6628.66	I		3900		23344.47	I

Cesium (Cont.)

Intensity		Wavelength	
320	c	23408.41	II
180		24132.52	II
4400		24251.21	I
850		24374.96	I
120	c	24528.78	II
140		24810.89	II
170		25189.83	II
900		25220.37	II
240		25733.29	II
890	d	25763.51	II
500		25764.73	I
120		26448.83	II
340		26503.26	II
190		26727.33	II
17		26954.56	II
22		26978.68	II
23		27187.98	II
9		28649.38	II
680	c	29310.06	I
7		29542.08	II
2800		30103.27	I
9		30487.16	II
610	c	30953.06	I
1100		34900.13	I
190		36131.00	I
2	c	39177.28	I
2	d	39421.25	I
1		39424.11	I

Cs III
Ref. 78, 200, 201—C.J.S.

Intensity		Wavelength	
		Vacuum	
10000		614.01	III
2000		638.17	III
2500		666.25	III
5000		691.60	III
3500		703.89	III
20000		721.79	III
20000		722.20	III
5000		731.56	III
12000		740.29	III
7500		830.39	III
15000		920.35	III
25000	c	1054.79	III
17	c	1673.99	III
12	c	1705.25	III
10		1801.83	III
20	c	1822.40	III
11		1823.93	III
12		1824.70	III
12		1841.80	III
25		1915.50	III
25	c	1923.29	III
12		1961.33	III
17		1996.56	III
		Air	
710		2035.11	III
120		2056.43	III
330		2076.43	III
540		2077.30	III
410		2088.68	III
210		2101.63	III
200		2141.47	III
1000		2316.88	III
230		2325.95	III
390		2340.49	III
1600		2455.81	III
1600		2477.57	III
890		2485.45	III
410		2495.07	III
1400		2525.67	III
430		2573.05	III
16000		2596.86	III
390		2610.12	III
6200		2630.51	III
370		2700.32	III
710		2701.20	III
390		2776.44	III
270		2810.87	III
630		2845.70	III
3100		2859.32	III
200		2893.85	III
180		2921.13	III
3200		2976.86	III
210		3001.28	III
1700		3066.59	III
1100	c	3149.36	III
1400		3152.36	III
8400		3268.32	III
1300		3315.51	III
550		3340.60	III
430		3344.02	III
1200		3349.46	III
400		3463.45	III
580		3476.83	III
480		3559.82	III
7200		3597.45	III
1300		3608.31	III
2300		3618.19	III
300	c	3641.34	III
520		3651.08	III
4800		3661.40	III
640		3699.50	III
430		3837.46	III
2900		3888.37	III
2700		3925.60	III
680	c	4001.70	III
3100		4006.55	III
420		4006.78	III
520		4043.42	III
370		4403.86	III
1200		4410.22	III
940		4425.68	III
530		4471.48	III
1200		4506.72	III
590		4522.86	III
420	h	4620.61	III
210		4665.52	III
140		4851.59	III
370		5035.72	III
230		5380.79	III
140		5950.14	III
110		5979.97	III
150		6043.99	III
870		6079.86	III
330		6150.42	III
450		6242.96	III
510		6456.33	III
400		6753.12	III
1900	c	7219.60	III

Cs IV
Ref. 259 — J.R.

Intensity	Wavelength	
	Vacuum	
35	707.20	IV
60	759.57	IV
5	778.21	IV
500	824.80	IV
350	828.86	IV
400	868.18	IV
1000	874.84	IV
400	896.92	IV
400	923.02	IV
600	986.14	IV
300	995.14	IV
60	1019.13	IV
550	1068.91	IV
200	1282.66	IV

CHLORINE (Cl)
Z = 17
Cl I and II
Ref 238, 239 — L.J.R.

Intensity		Wavelength	
		Vacuum	
350		559.305	II
400		571.904	II
800		574.406	II
500		586.24	II
700		618.057	II
600		619.982	II
800		620.298	II
700		626.735	II
800		635.881	II
1000		636.626	II
1000		650.894	II
1000		659.811	II
1300		661.841	II
2000		663.074	II
1500		682.053	II
1500		687.656	II
1500		693.594	II
2000		725.271	II
2500		728.951	II
2000		777.562	II
5000		787.580	II
5000		788.740	II
5000		793.342	II
6000		839.297	II
8000		839.599	II
7000	p	841.41	II
5000		851.691	II
2000		888.026	II
2000		893.549	II
2000		961.499	II
30		969.92	I
40		978.284	I
25		998.372	I
25		998.432	I
75		1002.346	I
150		1013.664	I
90		1025.553	I
6000		1063.831	II
3000		1067.945	II
9000		1071.036	II
6000		1071.767	II
5000		1075.230	II
5000		1079.080	II
200		1084.667	I
200		1085.171	I
250		1085.304	I
400		1088.06	I
350		1090.271	I
250		1090.982	I
250		1092.437	I
400		1094.769	I
350		1095.148	I
350		1095.662	I
400		1095.797	I
250		1096.810	I
300		1097.369	I
200		1098.068	I
200		1099.523	I
500		1107.528	I
800		1139.214	II
800		1167.148	I
3000		1179.293	I
1200		1188.774	I
900		1201.353	I
3000		1335.726	I
10000		1347.240	I
5000		1351.657	I
12000		1363.447	I
2500		1373.116	I
20000		1379.528	I
25000		1389.693	I
20000		1389.957	I
12000		1396.527	I
500		1441.470	II
500		1528.569	II
500		1542.942	II
500		1558.144	II
500		1565.050	II
500		1857.488	II
450	h	1997.370	II
		Air	
450		2032.116	II
350	h	2088.583	II
350	h	2091.458	II
170		2427.79	II
360		2434.07	II
340		2498.53	II
470		2502.74	II
260		2546.96	II
500		2549.88	II
460		2564.84	II
320		2603.31	II
410		2912.05	II
950		2996.65	II
500		3006.06	II
950		3057.96	II
1300		3071.32	II
1400		3092.19	II
1200		3123.72	II
1900		3315.43	II
1200		3329.10	II
2500		3353.35	II
20		3726.54	I
1200		3749.96	II
1000		3781.17	II
1500		3798.76	II
1900		3805.18	II
1300		3809.46	II
1700		3820.20	II
2800		3827.59	II
4500		3833.35	II
2500		3843.20	II
3100		3845.37	II
3900		3845.65	II
1500		3845.80	II
10000		3850.99	II
7900		3851.37	II
1200		3851.65	II
25000		3860.83	II
4400		3860.99	II
1000		3861.37	II
1500		3913.87	II
1100		3916.63	II
20		3944.82	I
20		4104.79	I
10000	h	4132.50	II
65		4209.67	I
50		4226.42	I
60		4264.58	I
100		4363.27	I
100		4369.50	I
5000		4372.93	II
100		4379.90	I
100		4389.75	I
90		4390.40	I
90		4403.03	I
100		4438.49	I
90		4475.30	I
1500		4489.91	II
100		4526.19	I
80		4600.98	I
40		4623.938	I
50		4654.040	I
80		4661.208	I
45		4691.523	I
40		4721.255	I
45		4740.729	I
4300		4768.65	II
13000		4781.32	II
99000		4794.55	II
29000		4810.06	II
16000		4819.47	II
81000		4896.77	II
47000		4904.78	II
26000		4917.73	II
10000		4995.48	II
26000		5078.26	II
30		5099.789	I
56000		5217.94	II
23000		5221.36	II
15000		5392.12	II
99000		5423.23	II
10000		5423.51	II
19000		5443.37	II
10000		5444.21	II
5600		5457.02	II
40		5532.162	I
50	d	5796.305	I
45		5799.914	I
30		5856.742	I
100	d	5948.58	I
50		6019.812	I
35		6082.61	I
1900		6094.69	II
160		6114.43	I
200		6140.245	I
160		6194.757	I
160		6398.66	I
150		6434.833	I
150		6531.43	I
1400		6661.67	II
150		6678.43	I
1300		6686.02	II
1200		6713.41	II
150		6840.29	I
300		6932.903	I
300		6981.886	I
600		7086.814	I
7500		7256.62	I
5000		7414.11	I
550		7462.370	I
550		7489.47	I
700		7492.118	I
11000		7547.072	I
2300		7672.42	I
450		7702.828	I
7000		7717.581	I
10000		7744.97	I
2200		7769.16	I
650		7771.09	I
2200		7821.36	I
1700		7830.75	I
3000		7878.22	I
220		7893.34	I
2300		7899.31	I
1800		7915.08	I
3000		7924.645	I
2100		7933.89	I
1700		7935.012	I
650		7952.52	I
1500		7974.72	I
1300		7976.97	I
600		7980.60	I
2900		7997.85	I
2200		8015.61	I
1100		8023.33	I
400		8051.07	I
1700		8084.51	I
2200		8085.56	I
3000		8086.67	I
1300		8087.73	I
250		8094.67	I
2500		8194.42	I
2200		8199.13	I
2200		8200.21	I
800		8203.78	I
18000		8212.04	I
3000		8220.45	I
20000		8221.74	I
18000		8333.31	I
1000		8360.71	II
560		8361.84	II
99900		8375.94	I
180		8382.67	II
100		8392.02	II
400		8406.199	I
15000		8428.25	I
2200		8467.34	I
2200		8550.44	I
20000		8575.24	I
750		8578.02	I
75000		8585.97	I
450		8628.54	I
300		8641.71	I
3500		8686.26	I
2200		8912.92	I
3000		8948.06	I
2000		9038.982	I
2500		9045.43	I
1000		9069.656	I
2000		9073.17	I
7500		9121.15	I

Chlorine (Cont.)

Intensity	Wavelength	
3000	9191.731	I
500	9197.596	I
4000	9288.86	I
1500	9393.862	I
3500	9452.10	I
500	9486.964	I
1000	9584.801	I
3500	9592.22	I
250	9632.509	I
1000	9702.439	I
250	9744.426	I
200	9807.057	I
400	9875.970	I
331	10392.549	I
38	10432.83	II
10	10506.62	II
14	10509.12	II
19	10512.46	II
25	10514.17	II
9	10801.47	II
5	10885.42	II
1	10955.71	II
300	11123.05	I
231	11392.62	I
269	11409.69	I
1000	11436.33	I
180	11720.56	I
195	11866.76	I
172	12021.7	I
350	13243.8	I
310	13296.0	I
550	13346.8	I
525	13821.7	I
148	14369.7	I
294	14931.7	I
269	15108.0	I
381	15465.1	I
169	15467.6	I
1094	15520.3	I
1487	15730.1	I
193	15818.4	I
2780	15869.7	I
277	15883.3	I
342	15928.9	I
735	15960.0	I
283	16030.5	I
129	16077.6	I
259	16190.5	I
227	19370.3	I
717	19755.3	I
185	19766.8	I
227	20199.4	I
85	20370.1	I
100	24470.0	I
	39603.7	I
	39615.3	I
	39716.0	I
	39744.0	I
	39750.9	I
	39875.3	I
	39881.0	I
	39985.7	I
	40085.5	I
	40089.5	I
	40171.0	I
	40310.3	I
	40335.4	I
	40532.2	I

Cl III
Ref. 28, 30 — L.J.R.

Intensity	Wavelength	
	Vacuum	
100	406.27	III
400	411.37	III
400	411.81	III
600	556.23	III
700	556.61	III
700	557.12	III
600	2965.56	III
600	3104.46	III
800	3139.34	III
900	3191.45	III
700	3289.80	III
700	3320.57	III
800	3329.06	III
900	3340.42	III
800	3392.89	III
800	3393.45	III
900	3530.03	III
800	3560.68	III
900	3602.10	III
800	3612.85	III
700	3622.69	III
700	3656.95	III
700	3670.28	III
700	3682.05	III
600	3705.45	III
600	3707.34	III
800	3720.45	III
800	3748.81	III
500	3779.35	III
500	3925.87	III
700	3991.50	III
600	4018.50	III
600	4059.07	III
500	4104.23	III
500	4106.83	III
400	4370.91	III
500	4608.22	III
300	4703.14	III
100	4863.75	III
10	4971.64	III
700	561.53	III
700	561.68	III
700	561.74	III
500	606.35	III
400	621.28	III
300	670.38	III
300	673.13	III
100	936.28	III
500	1005.28	III
600	1009.70	III
700	1015.02	III
600	1822.50	III
500	1828.40	III
500	1901.61	III
500	1983.61	III
	Air	
400	2006.84	III
700	2253.07	III
500	2268.95	III
500	2278.34	III
700	2283.93	III
600	2323.50	III
500	2336.45	III
600	2340.64	III
600	2359.67	III
600	2370.37	III
700	2416.42	III
600	2447.14	III
600	2448.58	III
500	2486.91	III
500	2532.48	III
600	2580.67	III
500	2603.59	III
500	2632.67	III
500	2633.18	III
600	2665.54	III
700	2710.37	III

Cl IV
Ref. 11, 28, 30, 31 — L.J.R.

Intensity	Wavelength	
	Vacuum	
300	319.62	IV
200	331.84	IV
400	437.83	IV
400	464.86	IV
800	486.17	IV
800	534.73	IV
700	535.67	IV
600	536.15	IV
900	537.61	IV
600	538.12	IV
500	549.22	IV
400	550.02	IV
700	552.02	IV
600	553.30	IV
700	554.62	IV
500	601.50	IV
500	604.59	IV
400	608.90	IV
400	612.07	IV
400	653.70	IV
400	745.21	IV
400	831.43	IV
500	834.84	IV
500	834.97	IV
400	840.81	IV
600	840.93	IV
	865.3	IV
500	973.21	IV
600	977.56	IV
400	977.90	IV
700	984.95	IV
400	985.75	IV
200	1537.21	IV
200	1539.30	IV
200	1545.19	IV
200	1549.15	IV
200	1622.86	IV
	Air	
400	2701.36	IV
500	2724.03	IV
500	2751.23	IV
400	2770.64	IV
700	2782.47	IV
400	2835.4	IV
500	3063.13	IV
600	3076.68	IV
200	3167.87	IV

Cl V
Ref. 11, 28, 30, 85, 233 — L.J.R.

Intensity	Wavelength	
	Vacuum	
300	287.33	V
300	373.78	V
400	390.15	V
500	392.43	V
300	536.53	V
400	537.01	V
300	537.46	V
500	538.03	V
400	538.68	V
800	542.23	V
600	542.30	V
400	542.87	V
1000	545.11	V
600	546.33	V
1000	547.63	V
400	633.19	V
400	635.32	V
400	681.92	V
400	683.17	V
400	688.93	V
	715.55	V
	716.19	V
400	883.13	V
400	894.34	V
100	894.91	V
	914.5	V

CHROMIUM (Cr)
Z = 24

Cr I and II
Ref. 1 — C.H.C.

Intensity	Wavelength	
	Vacuum	
19000	2055.32	II
14000	2061.49	II
8900	2065.42	II
80 h	2364.71	I
130	2383.33	I
140	2408.62	I
170	2496.31	I
110	2502.53	I
190	2504.31	I
50	2508.11	I
60	2508.98	I
40	2513.62	I
110	2516.92	I
80	2518.71	I
390	2519.52	I
190	2527.12	I
40	2530.45	I
70	2534.34	II
50	2545.64	I
160	2549.54	I
40	2553.06	I
80	2557.15	I
130	2560.69	I
150	2571.74	I
100	2577.65	I
50	2588.20	I
380	2591.85	I
35	2603.57	I
35	2622.86	I
22	2625.32	I
18	2626.60	I
18	2629.82	I
35	2642.12	I
250	2653.59	II
250	2658.59	II
70	2661.73	II
320	2663.42	II
70	2663.68	II
440	2666.02	II
280	2668.71	II
350	2671.81	II
280	2672.83	II
1800	2677.16	II
35	2678.16	I
320	2678.79	II
18	2680.34	II
230	2687.09	II
60	2688.04	I
55	2688.29	II
26	2690.26	I
280	2691.04	II
35	2693.52	II
35	2697.91	II
180	2698.41	II
180	2698.69	II
18	2700.60	I
110	2701.99	I
18	2702.53	I
70	2703.48	I
35	2703.55	II
18	2703.86	II
60	2705.43	I
35	2708.79	II
140	2709.31	II
45	2712.31	II
55	2716.18	I
45	2717.51	II
170	2718.43	II
18	2722.75	II
420 h	2724.04	II
45	2726.51	I
280 h	2727.26	II
170 h	2731.91	I
70	2736.47	I
70	2739.38	II
95	2740.10	I
95	2741.07	II
250	2742.03	II
35	2742.17	I
110 h	2743.64	II
330	2746.21	II
390	2748.29	I
45	2748.98	II
280	2750.73	II
110 h	2751.60	I
35	2751.87	II
110 h	2752.88	I
35	2754.28	II
22	2754.90	I
22	2755.27	I
22	2756.75	I
150	2757.10	I
350	2757.72	II
60	2758.98	II
80	2759.39	II
45	2759.73	II
90 h	2761.76	I
750	2762.59	II
22	2763.06	I
80 h	2764.35	I
750	2766.54	II
22	2767.54	I
250 h	2769.82	II
18	2771.45	I
45	2778.06	II
22	2779.14	I
80	2780.30	II
810	2780.70	I
70	2785.70	II
35	2787.63	II
35	2787.84	II
90	2792.16	II
55	2798.67	II
70	2800.77	II
80	2812.01	II
60	2818.36	II
45	2822.01	II
180	2822.37	II
22	2826.75	I
180	2830.47	II
70	2834.26	II
2500	2835.63	II
45	2836.48	II
55	2838.79	II
110	2840.02	II
1700	2843.25	II
22	2846.02	I
45	2849.29	II
1200	2849.84	II
120	2851.36	II
55	2853.22	II
55	2855.07	II
880	2855.68	II
90	2856.77	II
70	2857.40	II
610	2858.91	II
440	2860.93	II
790	2862.57	II
750	2865.11	II
55	2865.33	II
610	2866.74	II
90	2867.10	II
480	2867.65	II
210	2870.44	II
110	2871.63	I
160	2873.48	II
90	2873.82	II
320	2875.99	II
230	2876.24	II
180	2877.98	II
70	2878.45	II
120	2879.27	I
95	2880.87	II
30	2881.14	I
170	2887.00	II
55	2888.74	I
700	2889.29	II
55	2889.82	II
55	2891.42	I
370	2893.25	I
190	2894.17	I
55	2896.46	II

Chromium (Cont.)

Int	λ		Int	λ		Int	λ		Int	λ	
210	2896.75	I	30	3208.59	II	220	3641.83	I	40	3993.97	I
55 d	2897.67	II	170	3209.18	II	45	3646.16	I	160	4001.44	I
	2897.73	II	140	3217.40	II	85	3648.53	I	120	4012.47	II
90	2898.54	II	30	3229.20	I	220	3649.00	I	30	4014.67	I
80	2899.21	I	28	3234.06	II	170	3653.91	I	85	4022.26	I
55	2899.48	II	65	3237.73	II	220	3656.26	I	70	4025.01	I
26	2903.97	II	120	3245.54	I	45	3662.84	I	120	4026.17	I
55	2904.68	I	130	3251.84	I	130	3663.21	I	85	4027.10	I
180	2905.49	I	130	3257.82	I	45	3665.98	I	85	4030.68	I
260	2909.05	I	95	3259.98	I	95	3666.64	I	190	4039.10	I
260	2910.90	I	30	3295.43	II	55	3668.03	I	160	4048.78	I
250	2911.14	I	24	3307.02	II	65	3676.32	I	120	4058.77	I
45	2911.68	II	55	3324.06	II	40	3677.68	II	40	4065.72	I
60	2913.73	I	28	3326.59	I	55	3677.89	II	85	4066.94	I
22	2915.23	II	30	3328.35	II	40	3679.82	I	35	4074.86	I
22	2915.46	II	30	3329.05	I	19	3681.69	I	40	4076.06	I
90	2921.24	II	95	3336.33	II	120	3685.55	I	40	4077.09	I
60	2921.82	II	130	3339.80	II	130	3686.80	I	40	4077.68	I
60	2927.08	II	110	3342.59	II	130	3687.25	I	40	4104.87	I
80	2928.15	II	30	3343.34	I	75	3687.54	I	40	4109.58	I
95	2928.30	II	95	3346.02	I	19	3688.46	I	40	4120.61	I
26	2929.44	II	95	3346.74	I	75	3712.95	II	40	4121.82	I
35	2930.85	II	95	3347.84	II	40	3716.53	I	35	4122.16	I
26	2932.70	II	65	3349.07	I	130	3730.81	I	40	4123.39	I
55	2933.97	II	30	3349.32	I	150	3732.03	I	140	4126.52	I
90	2935.14	II	55	3351.60	I	95	3742.97	I	35	4127.30	I
45	2940.22	II	55 h	3351.97	I	480	3743.58	I	40	4127.64	I
60	2946.84	II	55 h	3353.03	I	570	3743.88	I	30	4131.36	I
55	2953.36	II		3353.13	II	85	3744.49	I	30	4152.78	I
45	2953.71	II	170	3358.50	II	55	3748.61	I	120	4153.82	I
55	2961.73	II	160	3360.30	II	340	3749.00	I	85	4161.42	I
45	2966.05	II	65	3361.77	II	50	3757.17	I	140	4163.62	I
480	2967.64	I	55	3362.21	I	230	3757.66	I	70	4165.52	I
480	2971.11	I	430	3368.05	II	60	3758.04	I	40	4169.84	I
210	2971.91	II	30	3376.40	I	24	3767.43	I	35	4170.20	I
480	2975.48	I	55	3378.34	II	260	3768.24	I	40	4172.77	I
30	2976.72	II	30	3379.17	I	95	3768.73	I	170	4174.80	I
190	2979.74	I	30	3379.37	II	95	3788.86	I	30	4175.94	I
350	2980.79	I	95	3379.83	II	95	3790.45	I	170	4179.26	I
110	2985.32	II	140	3382.68	II	130	3791.38	I	35	4184.90	I
480	2985.85	I	95	3391.43	II	130	3792.14	I	30	4186.36	I
1500	2986.00	I	55	3392.99	II	120	3793.29	I	35	4190.13	I
2100	2986.47	I	70	3393.84	II	130	3793.88	I	85	4191.27	I
660	2988.65	I	55	3394.30	II	85	3794.61	I	35	4192.10	I
160	2989.19	II	30	3402.40	II	140	3797.13	I	85	4193.66	I
480	2991.89	I	170	3403.32	II	200	3797.72	I	70	4194.95	I
230	2994.07	I	360	3408.76	II	530	3804.80	I	40	4197.23	I
300	2995.10	I	210	3421.21	II	110	3806.83	I	85	4198.52	I
700	2996.58	I	270	3422.74	II	110	3807.93	I	60	4203.59	I
210	2998.79	I	140	3433.31	II	180	3815.43	I	40	4204.47	I
1100	3000.89	I	270	3433.60	I	70	3818.48	I	35	4208.36	I
750	3005.06	I	55	3434.11	I	180	3819.56	I	110	4209.37	I
140	3013.03	I	160	3436.19	I	70	3823.52	I	40	4209.76	I
710	3013.71	I	70	3441.12	I	130	3826.42	I	40	4211.35	I
710	3014.76	I	140	3441.44	I	130	3830.03	I	40	4216.36	I
1400	3014.92	I	30	3443.79	I	380	3841.28	I	85	4217.63	I
710	3015.19	I	170	3445.62	I	190	3848.98	I	40	4221.57	I
2800	3017.57	I	30	3447.02	I	140	3849.36	I	40	4222.73	I
430	3018.50	I	170	3447.43	I	290	3850.04	I	40	4238.96	I
240	3018.82	I	70	3447.76	I	140	3852.22	I	60	4240.70	I
430	3020.67	I	190	3453.33	I	190	3854.22	I	20000	4254.35	I
2800	3021.56	I	40	3453.74	I	110	3855.29	I	70	4255.50	I
1100	3024.35	I	130	3455.60	I	140	3855.57	I	60	4261.35	I
85	3026.65	II	100	3460.43	I	260	3857.63	I	110	4263.14	I
170	3029.16	I	65	3465.25	I	70	3874.53	I	30	4271.06	I
710	3030.24	I	40	3467.02	I	660	3883.29	I	40	4272.91	I
140	3031.35	I	70	3467.72	I	50	3883.66	I	16000	4274.80	I
28	3032.93	II	45	3469.59	I	570	3885.22	I	85	4280.40	I
390	3034.19	I	16	3472.76	I	380	3886.79	I	10000	4289.72	I
550	3037.04	I	24	3472.91	I	60	3891.93	I	40	4291.96	I
80	3039.78	I	40	3473.61	I	260	3894.04	I	85	4295.76	I
550	3040.85	I	70	3481.30	I	40	3897.65	I	70	4297.74	I
	3040.91	II	55	3481.54	I	35	3902.11	I	35	4300.51	I
55	3041.74	II	55	3494.97	I	360	3902.92	I	50	4301.18	I
110	3050.14	II	40	3495.38	II	60	3903.16	I	30	4305.45	I
710	3053.88	I	80	3510.54	I	960	3908.76	I	35	4319.64	I
24	3059.52	II	40	3511.84	II	120 Hd	3911.82	I	60	4325.08	I
85	3065.07	I	120	3550.64	I		3912.00	I	780	4337.57	I
28	3067.16	II	80	3558.52	I	120	3915.84	I	1100	4339.45	I
85	3073.68	I	130	3566.16	I	190	3916.24	I	380	4339.72	I
55	3077.83	I	130	3573.64	I	35	3917.60	I	60	4340.13	I
28	3095.86	I	80	3574.04	I	1900	3919.16	I	1900	4344.51	I
28	3109.34	I	330 h	3574.80	I	600	3921.02	I	70	4346.83	I
28	3110.86	I		3574.94	I	30	3926.65	I	380	4351.05	I
240	3118.65	II	19000	3578.69	I	600	3928.64	I	2300	4351.77	I
45	3119.25	I	160 h	3584.33	I	410	3941.49	I	570	4359.63	I
40	3119.71	II	130	3585.30	II	30	3951.10	I	70	4363.13	I
430	3120.37	II	17000	3593.49	I	40	3952.40	I	530	4371.28	I
28	3122.60	II	350	3601.67	I	35	3953.16	I	70	4373.25	I
470	3124.94	II	40	3602.57	I	1900	3963.69	I	110	4374.16	I
	3125.02	II	85	3603.74	I	120	3969.06	I	70	4375.33	I
120	3128.70	II		3603.78	II	1600	3969.75	I	50	4381.11	I
590	3132.06	II	13000	3605.33	I	85	3971.26	I	530	4384.98	I
140	3136.68	II	40	3608.40	I	1600	3976.66	I	60	4387.50	I
140	3147.23	II	40	3609.48	I	85	3978.68	I	70	4391.75	I
85	3148.44	II	40	3610.05	I	40	3979.80	I	60	4403.50	I
100	3155.15	I	70	3612.61	I	85	3981.23	I	24	4410.30	I
100	3163.76	I	85	3615.64	I	960	3983.91	I	60	4411.09	I
240	3180.70	II	130	3632.84	I	190	3984.34	I	35	4412.25	I
30	3181.43	II	350	3636.59	I	160	3989.99	I	50	4413.87	I
65 h	3188.01	I	630	3639.80	I	960	3991.12	I	60	4424.28	I
220	3197.08	II	85	3640.39	I	160	3991.67	I	24	4428.50	I
24	3198.11	I	70	3641.47	I	190	3992.84	I	50	4430.49	I

Chromium (Cont.)

Intensity	Wavelength		Intensity	Wavelength		Intensity	Wavelength	
50	4432.18	I	120	4801.03	I	22	6362.87	I
110	4458.54	I	110	4829.38	I	19	6661.08	I
30	4459.74	I	14	4836.86	I	11	6669.26	I
30	4465.36	I	17	4861.20	I	5 h	6881.62	I
30	4482.88	I	70	4861.84	I	10 h	6882.38	I
40	4488.05	I	140	4870.80	I	21 h	6883.03	I
50	4489.47	I	35	4885.78	I	27 h	6924.13	I
60	4492.31	I	19	4885.96	I	17 h	6925.20	I
660	4496.86	I	130	4887.01	I	30 h	6978.48	I
50	4498.73	I	19	4888.53	I	11 h	6979.82	I
70	4500.30	I	35	4903.24	I	7	7185.52	I
50	4501.11	I	260	4922.27	I	6 h	7236.20	I
22	4501.79	I	110	4936.33	I	85	7355.90	I
24	4506.85	I	70	4942.50	I	130	7400.21	I
95	4511.90	I	110	4954.81	I	150	7462.31	I
12	4514.37	I	35	4964.93	I	11 h	7942.04	I
35	4514.53	I	60	5013.32	I	5 h	8163.18	I
24	4521.14	I	17	5051.90	I	9	8348.28	I
24	4526.11	I	17	5065.91	I	6	8450.26	I
380	4526.47	I	40	5067.71	I	3	8433.24	I
70 d	4527.34	I	40	5072.92	I	6	8548.86	I
	4527.47	I	30	5110.75	I	40	8947.15	I
24	4529.85	I	17	5113.13	I	19	8976.83	I
380	4530.74	I	17	5123.46	I			
50	4535.15	I	50	5139.65	I			
240	4535.72	I	14	5144.67	I			
40	4539.79	I	70	5166.23	I			
240	4540.50	I	35	5177.43	I			
240	4540.72	I	70	5184.59	I			
35	4541.07	I	70	5192.00	I			
19	4541.51	I	12	5193.49	I			
24	4542.62	I	85	5196.44	I			
140	4544.62	I	35	5200.19	I			
24	4545.34	I	5300	5204.52	I			
600	4545.96	I	8400	5206.04	I			
50	4556.17	I	11000	5208.44	I			
22	4558.66	II	19	5214.13	I			
19	4564.17	I	30	5221.75	I			
120	4565.51	I	85	5224.94	I			
95	4569.64	I	12	5226.89	I			
120	4571.68	I	19	5238.97	I			
22	4575.12	I	30	5243.40	I			
360	4580.06	I	290	5247.56	I			
24	4586.14	I	60	5254.92	I			
360	4591.39	I	60	5255.13	I			
70	4595.59	I	15	5261.73	I			
50	4600.10	I	330	5264.15	I			
480	4600.75	I	30	5265.16	I			
50	4601.02	I	180	5265.72	I			
240	4613.37	I	35	5272.01	I			
800	4616.16	I	30	5273.44	I			
70	4619.55	I	95 h	5275.17	I			
85	4621.96	I	35 h	5275.69	I			
70	4622.49	I	70 h	5276.03	I			
24	4622.76	I	19	5280.29	I			
550	4626.19	I	10	5287.19	I			
24	4632.18	I	340	5296.69	I			
40	4637.18	I	70 h	5297.36	I			
50	4637.77	I	660	5298.27	I			
50 d	4639.52	I	85	5300.75	I			
	4639.70	I	17	5304.21	I			
1600	4646.17	I	24	5312.88	I			
24	4646.81	I	24	5318.78	I			
24	4648.13	I	340 h	5328.34	I			
24	4648.87	I	70 h	5329.17	I			
35	4649.46	I	17 h	5329.72	I			
570	4651.28	I	14	5340.44	I			
840	4652.16	I	10	5344.76	I			
35	4654.74	I	780	5345.81	I			
19	4656.19	I	380	5348.32	I			
40	4663.33	I	30	5386.98	I			
70	4663.83	I	22	5387.57	I			
95	4664.80	I	10	5390.39	I			
35	4665.90	I	40	5400.61	I			
22	4666.22	I	22	5405.00	I			
70	4666.51	I	1400	5409.79	I			
50	4669.34	I	12	5442.41	I			
40	4680.54	I	19	5463.97	I			
19	4680.87	I	19	5480.50	I			
70	4689.37	I	24	5628.64	I			
60	4693.95	I	7	5642.36	I			
24	4695.15	I	12 h	5649.37	I			
60	4697.06	I	24	5664.04	I			
240 d	4698.46	I	7 h	5681.20	I			
	4698.62	CR I	7 h	5682.48	I			
35	4700.61	I	24	5694.73	I			
190	4708.04	I	40	5698.33	I			
240	4718.43	I	24	5702.31	I			
50	4723.10	I	12	5712.64	I			
50	4724.42	I	24	5712.78	I			
50	4727.15	I	7	5719.82	I			
24	4729.72	I	7	5746.43	I			
120	4730.71	I	7	5753.69	I			
140	4737.35	I	12 h	5781.20	I			
19	4745.31	I	6 h	5781.81	I			
70	4752.08	I	24 h	5783.11	I			
340	4756.11	I	30 h	5783.93	I			
50	4764.29	I	24 h	5785.00	I			
22	4766.63	I	19 h	5785.82	I			
30	4767.86	I	60 h	5787.99	I			
190	4789.32	I	180 h	5791.00	I			
95	4792.51	I	35	6330.10	I			

Cr III
Ref. 412 — C.H.C.

Intensity	Wavelength		Intensity	Wavelength	
	Vacuum				
20	969.26	III	50	2039.63	III
40	1000.86	III	80	2047.23	III
40	1001.04	III	100	2113.73	III
30	1002.96	III	100	2113.83	III
50	1017.14	III	50	2114.26	III
50	1017.31	III	50	2114.53	III
50	1017.57	III	100	2114.87	III
30	1028.33	III	100	2117.53	III
60	1030.47	III	80	2123.53	III
30	1030.89	III	80	2139.11	III
50	1033.23	III	100	2141.15	III
50	1033.45	III	80	2144.15	III
100	1033.69	III	50	2147.16	III
50	1035.83	III	50	2147.56	III
100	1036.03	III	50	2148.65	III
30	1040.17	III	50	2149.48	III
40	1040.53	III	50	2152.76	III
40	1045.06	III	100	2157.17	III
40	1045.14	III	50	2163.86	III
60	1059.13	III	60	2166.25	III
60	1060.15	III	100	2170.70	III
60	1061.04	III	50	2183.71	III
50	1062.68	III	100	2185.01	III
30	1064.32	III	50	2190.09	III
50	1064.43	III	100	2190.76	III
50	1066.23	III	100	2191.58	III
80	1068.41	III	100	2197.89	III
30	1100.61	III	100	2198.62	III
30	1101.43	III	100	2203.22	III
30	1102.88	III	60	2208.70	III
30	1117.19	III	200	2226.72	III
30	1132.75	III	100	2231.81	III
50	1136.67	III	100	2233.81	III
50	1161.43	III	200	2235.91	III
30	1187.65	III	150	2237.59	III
60	1206.38	III	150	2244.10	III
80	1209.13	III	80	2251.45	III
80	1211.12	III	50	2257.92	III
40	1221.07	III	100	2273.30	III
40	1221.90	III	80	2275.43	III
30	1225.65	III	100	2276.38	III
30	1228.65	III	80	2277.47	III
30	1231.88	III	150	2284.44	III
50	1232.96	III	50	2289.23	III
40	1236.20	III	80	2290.66	III
40	1238.51	III	60	2295.55	III
50	1252.61	III	50	2309.99	III
40	1259.02	III	80	2314.63	III
40	1261.86	III	100	2319.17	III
30	1262.34	III	150	2324.88	III
35	1263.61	III	60	2340.51	III
35	1264.21	III	50	2456.83	III
40	1287.05	III	100	2472.88	III
30	1455.27	III	100	2479.77	III
40	1584.60	III	100	2483.06	III
30	1603.19	III	60	2488.26	III
30	1679.25	III	80	2506.41	III
30	1690.28	III	80	2530.99	III
60	1692.89	III	80	2537.73	III
60	1696.64	III	80	2544.37	III
60	1701.48	III	50	2545.17	III
80	1707.43	III	80	2564.76	III
40	1707.78	III	80	2616.50	III
45	1762.81	III	100	2626.08	III
30	1766.92	III	100	2640.73	III
30	1769.17	III	50	2647.50	III
30	1827.26	III	40	2655.28	III
			40	2916.57	III
	AIR				
60	2036.39	III			

Cr IV
Ref. 379, 412 — C.H.C.

Intensity	Wavelength	
	Vacuum	
50	575.05	IV
30	576.24	IV
30	576.62	IV
30	595.09	IV
50	612.64	IV
40	613.75	IV
40	614.03	IV
40	614.90	IV
30	615.34	IV
30	615.60	IV
50	616.82	IV
40	618.23	IV
40	619.13	IV
100	620.66	IV
60	621.36	IV
40	622.09	IV
30	623.54	IV
40	625.04	IV
40	625.99	IV
100	629.26	IV
50	629.74	IV
80	630.30	IV
30	632.62	IV
30	637.34	IV
50	637.55	IV
50	638.13	IV

Chromium (Cont.)

Intensity		Wavelength	
30		638.54	IV
100		666.55	IV
75		667.30	IV
40		677.55	IV
40		687.12	IV
50		688.46	IV
100		693.92	IV
50		695.21	IV
50		705.98	IV
30		712.90	IV
80		1055.89	IV
60		1057.85	IV
30		1367.39	IV
40		1375.05	IV
70		1401.82	IV
100		1417.42	IV
30		1485.05	IV
80		1595.04	IV
90	d	1595.59	IV
100		1658.08	IV
120		1672.66	IV
90		1686.07	IV
100		1690.88	IV
80		1725.26	IV
90		1727.07	IV
100		1732.04	IV
40		1733.98	IV
80		1734.16	IV
50		1739.19	IV
70		1746.88	IV
80		1747.13	IV
110		1755.64	IV
120		1758.51	IV
100		1769.64	IV
100		1777.82	IV
40		1791.09	IV
140		1802.72	IV
130		1812.41	IV
60		1819.23	IV
30		1826.21	IV
30		1826.86	IV
100		1840.14	IV
50		1851.89	IV
100		1863.11	IV
140		1873.89	IV
35		1883.16	IV
40		1937.63	IV
30		1946.59	IV
140		1967.18	IV
120		1972.07	IV
40		1990.25	IV

Air

Intensity		Wavelength	
50	d	2042.91	IV
40		2055.73	IV
70		2299.21	IV
90		2299.59	IV
100		2316.85	IV
40		2324.06	IV
50		2360.40	IV
70		2405.15	IV
60		2423.32	IV

Cr V
Ref. 380 — C.H.C.

Intensity	Wavelength	

Vacuum

Intensity	Wavelength	
100	438.62	V
100	464.02	V
50	469.64	V
50	825.60	V
50	968.70	V
50	1045.04	V
60	1060.65	V
50	1112.45	V
60	1114.35	V
100	1116.48	V
80	1117.56	V
50	1118.16	V
150	1121.07	V
150	1127.63	V
100	1193.95	V
80	1196.04	V
50	1210.50	V
50	1259.99	V
100	1263.50	V
150	1465.86	V
50	1481.65	V
50	1482.76	V
50	1484.67	V
100	1489.71	V
150	1497.97	V
170	1519.03	V
220	1579.70	V
170	1591.72	V
150	1603.19	V
60	1638.50	V
50	1639.40	V
200	1837.44	V

COBALT (Co)
Z = 27

Co I and II
Ref. 1, 125, 276 — C.R.C.

Intensity		Wavelength	

Vacuum

Intensity		Wavelength	
20		1265.93	II
40		1271.94	II
20		1276.90	II
30		1293.97	II
25		1295.53	II
30		1295.86	II
20		1297.10	II
80		1299.58	II
30		1302.39	II
20		1306.76	II
80		1306.95	II
40		1311.12	II
40		1311.86	II
30		1315.42	II
30		1316.09	II
20		1318.19	II
30		1318.60	II
20		1319.84	II
20		1409.33	II
20		1466.21	II
20		1471.87	II
30		1472.90	II
30		1475.81	II
20		1484.26	II
30		1486.50	II
20		1509.23	II
20		1590.54	II
20		1595.77	II
20		1599.30	II
20		1693.34	II
8		1706.05	II
10		1723.01	II
15	h	1740.55	II
15		1743.39	II
8		1754.21	II
20	d	1808.01	II
10		1837.56	II
15		1839.37	II
1500		1842.34	I
1800		1847.89	I
1800		1852.71	I
2400		1855.05	I
1500		1878.28	I
10		1917.62	II
1800		1936.58	I
1500		1946.79	I
30		1950.09	II
1500		1951.90	I
1800		1954.22	I
1800		1955.17	I
30		1957.42	II
1500		1958.55	I
1500		1961.59	I
1500	h	1968.69	I
1500	h	1968.93	I
3000		1970.71	I
1800	h	1971.16	I
1800	h	1972.52	I
1500		1973.85	I
1800		1976.97	I
2400	h	1980.89	I
1500		1989.80	I
1800		1990.34	I
1500	l	1998.49	I

Air

Intensity		Wavelength	
20		2000.79	II
1500		2002.32	I
900		2008.04	I
50		2011.51	II
1200	h	2014.58	I
900		2016.17	I
50		2022.35	II
40		2025.76	II
50		2027.04	II
900		2031.96	I
30		2036.58	II
1500		2039.95	I
1200		2041.11	I
20		2049.17	II
40		2058.82	II
40		2063.78	II
50		2065.54	II
1500	h	2077.76	I
900		2085.67	I
900		2087.55	I
900		2089.35	I
900		2093.40	I
900		2094.86	I
900		2095.77	I
1200		2097.51	I
1500		2104.73	I
1500		2106.80	I
900		2108.98	I
30		2111.44	II
900	s	2117.68	I
20		2128.79	II
900		2137.78	I
900		2138.97	I
900		2163.03	I
20		2164.44	II
30		2173.33	II
1100		2174.60	II
20		2181.99	II
30		2187.01	II
20		2190.68	II
20		2192.50	II
200		2193.60	II
20		2200.40	II
40	p	2202.95	II
200		2256.73	II
150		2260.00	II
200		2283.52	II
1000		2286.15	II
200		2291.98	II
300	d	2293.38	II
300		2301.40	II
800	d	2307.85	II
2600		2309.02	I
500		2311.60	II
500		2314.05	II
300		2314.96	II
200	p	2317.06	II
2400		2323.14	I
300	p	2324.31	II
200	d	2326.11	II
500		2326.47	II
1400		2335.99	I
1600		2338.67	I
200		2347.39	II
1600		2352.85	I
200	d	2353.41	II
2000		2353.42	I
500		2363.80	II
400		2378.62	II
1400		2380.48	I
200		2381.76	II
300	p	2383.45	II
1400		2384.86	I
200		2386.36	II
500		2388.92	II
200		2397.38	II
1100	d	2402.06	I
200	p	2404.16	II
5300		2407.25	I
5300		2411.62	I
1600		2412.76	I
4800		2414.46	I
4800		2415.30	I
300		2417.65	II
4100		2424.93	I
3300		2432.21	I
2900		2436.66	I
2400		2439.05	I
200		2442.63	II
200	d	2446.03	II
200	p	2447.69	II
200		2450.00	II
200		2464.20	II
200		2486.44	II
200		2498.82	II
570		2504.52	I
500		2506.46	II
360		2506.88	I
200		2511.16	II
860		2517.87	I
500		2519.82	II
4300		2521.36	I
200	h	2524.65	II
300		2524.97	II
500		2528.62	II
2900		2528.97	I
200	p	2530.09	II
720		2530.13	I
860		2532.18	II
200	d	2533.82	II
2900		2535.96	I
860		2536.49	I
300		2541.94	II
1700		2544.25	I
200		2546.74	II
340		2548.34	I
310		2553.37	I
310		2555.07	I
300		2559.41	II
200		2560.03	II
960		2562.15	I
500		2564.04	II
1100		2567.35	I
960		2574.35	I
800		2580.32	II
300	d	2582.22	II
500		2587.22	II
500		2587.52	II
200		2588.91	II
100	p	2605.71	II
100		2612.50	II
100		2614.36	II
100	p	2628.77	II
100		2632.26	II
100		2636.07	II
310		2646.42	I
770		2648.64	I
100		2653.72	II
100		2663.53	II
200		2666.73	II
100		2675.85	II
100		2684.42	II
200		2702.02	II
200		2706.62	II
200		2707.35	II
190		2715.99	I
100		2727.78	II
80		2734.54	II
190		2745.10	I
100		2753.22	II
190		2764.19	I
100		2766.70	II
100		2774.97	II
100		2791.00	II
100		2793.73	II
150		2815.56	I
80		2835.63	II
80		2847.35	II
80		2871.22	II
190		2886.44	I
100		2918.38	II
100		2930.24	II
100		2954.73	II
690		2987.16	I
690		2989.59	I
20		3008.86	II
60		3022.59	II
30		3035.13	II
3100		3044.00	I
1700		3061.82	I
20		3352.79	II
80		3387.70	II
1100		3388.17	I
2200		3395.38	I
11000		3405.12	I
4500		3409.18	I
6700		3412.34	I
2200		3412.63	I
30		3415.77	II
2700		3417.16	I
50		3423.84	II
2500		3431.58	I
4500		3433.04	I
1600		3442.93	I
8800		3443.64	I
50		3446.39	II
4100		3449.17	I
2100		3449.44	I
21000		3453.50	I
1000		3455.23	I
5100		3462.80	I
5100		3465.80	I
8000		3474.02	I
1900		3483.41	I
4800		3489.40	I
2400		3495.69	I
40		3497.33	II
50		3501.72	II
9600		3502.28	I
7000		3506.32	I
50		3507.77	II
2900		3509.84	I
1400		3510.43	I
4800		3512.64	I
3800		3513.48	I
10		3514.23	II
30		3517.50	II
4800		3518.35	I
1300		3520.08	I
2700		3521.57	I
3800		3523.43	I
60		3523.51	II
6400		3526.85	I
2700		3529.03	I
7300		3529.81	I
1900		3533.36	I
40		3535.92	II
50		3545.03	II
20		3555.93	II
1100		3560.89	I
80		3561.07	II
20		3566.98	II
8800		3569.38	I
50		3574.95	II

Cobalt (Cont.)

Intensity		Wavelength	
1600		3574.96	I
60		3575.32	II
2500		3575.36	I
60		3577.96	II
1000		3585.16	I
6700		3587.19	I
1900		3594.87	I
1600		3602.08	I
100		3621.21	II
1000		3627.81	I
80		3643.61	II
10		3656.75	II
60		3681.35	II
5		3695.32	II
5		3714.73	II
1100		3745.50	I
20		3754.69	II
1400		3842.05	I
6900		3845.47	I
5500		3873.12	I
2800		3873.96	I
7900		3894.08	I
20	h	3911.40	II
1500		3935.97	I
80	h	3963.10	II
40	h	3976.74	II
10		3983.02	II
6000		3995.31	I
970		3997.91	I
350		4020.90	I
10	h	4036.14	II
20	h	4037.37	II
4		4040.02	II
370		4045.39	I
5	h	4050.23	II
10	h	4052.40	II
20	h	4062.73	II
2	h	4064.50	II
350		4066.37	I
5	h	4074.34	II
830		4092.39	I
1		4096.57	II
550		4110.54	I
2800		4118.77	I
4400		4121.32	I
3		4130.88	II
3		4145.13	II
3	d	4160.67	II
1		4181.13	II
90		4190.71	I
1		4208.61	II
30	s	4244.25	II
8		4272.33	II
20	h	4288.25	II
3		4328.86	II
2		4384.26	II
2	h	4396.94	II
3		4413.91	II
90		4469.56	I
10		4482.50	II
2	h	4489.12	II
4	h	4497.44	II
10	d	4500.54	II
0		4516.65	II
690		4530.96	I
2	h	4533.22	II
0	h	4537.95	II
90		4549.66	I
1	h	4559.29	II
140		4565.59	I
1		4569.26	II
190		4581.60	I
5		4616.30	II
120		4629.38	I
25	h	4660.66	II
85		4663.41	I
110		4792.86	I
10	d	4831.16	II
100		4840.27	I
150		4867.88	I
80	h	4964.18	II
10	h	4970.05	II
10	h	4990.47	II
20	h	4995.98	II
35		5146.74	I
50		5212.71	I
50		5230.22	I
45		5235.21	I
50		5247.93	I
26		5266.30	I
45		5266.49	I
26		5268.52	I
45	h	5280.65	I
26		5301.06	I
50		5342.71	I
26		5343.39	I
50		5352.05	I
26		5353.48	I
35		5369.58	I
45		5483.34	I
17		5530.77	I
17		5647.22	I
17		5991.88	I

Intensity	Wavelength	
17	6082.44	I
17	6282.63	I
45	6450.24	I
21	6455.00	I
15	6563.42	I
15	6632.44	I
14	6814.94	I
14	6872.40	I
21	7052.89	I
45	7084.99	I
8	7417.38	I
8	7712.68	I
7	7908.71	I
9	7987.38	I
13	8007.27	I
9	8093.96	I
9	8372.84	I
4	8575.35	I
3	8819.15	I

Co III
Ref. 291 — C.R.C.

Intensity	Wavelength	
	Vacuum	
1000	1696.01	III
800	1697.99	III
500	1702.79	III
1000	1707.35	III
500	1707.95	III
500	1723.97	III
400	1745.67	III
500	1755.98	III
5000	1760.35	III
500	1769.96	III
500	1773.22	III
5000	1773.57	III
500	1774.42	III
1000	1777.14	III
2000	1780.05	III
3000	1782.97	III
500	1784.06	III
1000	1787.08	III
1000	1789.07	III
500	1790.26	III
500	1791.28	III
500	1798.06	III
500	1805.56	III
400	1811.47	III
400	1821.26	III
400	1821.69	III
400	1821.77	III
1000	1823.08	III
400	1825.36	III
750	1825.95	III
2000	1830.09	III
2000	1831.44	III
750	1831.92	III
400	1832.20	III
5000	1835.00	III
1000	1837.63	III
500	1846.16	III
500	1852.92	III
400	1854.39	III
400	1854.76	III
1000	1861.78	III
2000	1863.83	III
400	1864.19	III
500	1871.87	III
1000	1881.70	III
500	1895.37	III
500	1919.12	III
500	1928.57	III
500	1940.15	III
500	1953.94	III
500	1959.41	III
400	1989.60	III
	Air	
100	2001.09	III
200	2011.62	III
200	2013.88	III
100	2031.81	III
200	2053.11	III
100	2056.21	III
10	2062.17	III
10	2079.74	III
15	2088.58	III
10	2090.51	III
10	2097.63	III
10	2134.15	III
10	2452.16	III
20	2811.75	III
10	2888.31	III
10	2933.27	III
10	2978.01	III
20	2991.89	III
25	3010.92	III
10	3116.68	III
2	3151.40	III
2	3180.64	III
20	3232.11	III
2	3249.24	III
20	3259.68	III
2	3269.23	III
10	3287.68	III
15	3305.38	III
10	3451.25	III
2	3526.24	III
1	3634.21	III
1	3636.31	III
2	3667.52	III
2	3677.23	III
3	3680.74	III
1	3762.50	III
15	3782.27	III

Co IV
Ref. 236 — C.R.C.

Intensity	Wavelength	
	Vacuum	
81	606.79	IV
74	607.59	IV
55	608.24	IV
66	609.16	IV
70	609.21	IV
64	609.28	IV
43	610.04	IV
37	610.25	IV
24	610.79	IV

Co V
Ref. 100, 159 — C.R.C.

Intensity	Wavelength	
	Vacuum	
20	355.52	V
18	355.88	V
12	356.06	V
4	1006.86	V
10	1007.31	V
15	1009.02	V
10	1010.94	V
10	1013.80	V
1	1017.43	V
10	1018.36	V
10	1021.14	V
1	1028.08	V
3	1226.31	V
8	1228.19	V
15	1231.73	V
2	1234.55	V
20	1236.95	V
2	1239.85	V
8	1246.91	V
6	1258.61	V
5	1263.28	V
20	1270.70	V
20	1272.23	V
2	1275.52	V
50	1277.01	V
30	1281.63	V
15	1284.00	V
28	1286.95	V
25	1295.55	V
40	1295.87	V
35	1301.12	V
50	1345.67	V
6	1351.22	V
15	1353.42	V
40	1355.20	V
30	1357.67	V
20	1361.32	V
30	1362.46	V
30	1364.17	V
30	1368.24	V
4	1369.30	V
10	1371.01	V
30	1373.09	V
30	1375.20	V
25	1378.12	V
10	1379.05	V
10	1380.21	V
32	1389.11	V
15	1459.77	V
35	1468.98	V
30	1476.65	V
25	1482.62	V
30	1482.91	V
25	1486.02	V
20	1488.73	V

COPPER (Cu)
Z = 29
Cu I and II

Ref. 273, 290—V.K.

Intensity	Wavelength	
	Vacuum	
80	685.141	II
100	709.313	II
100	718.179	II
150	724.489	II
200	735.520	II
250	736.032	II
80	779.295	II
100	797.455	II
150	810.998	II
200	813.883	II
300	826.996	II
150	848.808	II
250	851.303	II
250	858.487	II
400	861.994	II
400	865.390	II
250	869.336	II
150	873.263	II
200	876.723	II
250	877.012	II
200	877.555	II
500	878.699	II
100	884.133	II
250	885.847	II
600	886.943	II
600	890.567	II
500	892.414	II
800	893.678	II
400	894.227	II
600	896.759	II
400	896.976	II
600	901.073	II
400	906.113	II
800	914.213	II
600	922.019	II
500	924.239	II
400	935.232	II
600	935.898	II
600	943.333	II
600	945.525	II
500	945.965	II
200	954.383	II
250	956.290	II
400	958.154	II
200	960.414	II
250	968.042	II
200	974.759	II
250	977.567	II
100	987.657	II
250	992.953	II
300	1004.055	II
300	1008.569	II
300	1008.728	II
300	1010.269	II
250	1012.597	II
500	1018.707	II
500	1027.831	II
250	1028.328	II
200	1030.263	II
600	1036.470	II
600	1039.348	II
600	1039.582	II
800	1044.519	II
800	1044.744	II
500	1049.755	II
600	1054.690	II
400	1055.797	II
600	1056.955	II
400	1058.799	II
600	1059.096	II
600	1060.634	II
600	1063.005	II
200	1065.782	II
200	1066.134	II
500	1069.195	II
300	1073.745	II
200	1088.395	II
300	1094.402	II
250	1097.053	II
150	1119.947	II
200	1142.640	II
300	1144.856	II
100	1250.048	II
150	1265.506	II
300	1275.572	II
150	1282.455	II
150	1287.468	II
150	1298.395	II
300	1308.297	II
300	1314.337	II
100	1320.686	II
100	1326.395	II
150	1350.594	II
250	1351.837	II
150	1355.305	II
300	1358.773	II
200	1359.009	II

Copper (Cont.)

Intensity		Wavelength	Sp
200		1362.600	II
250		1367.951	II
200		1371.840	II
100		1393.128	II
100		1398.642	II
150		1402.777	II
150		1407.169	II
100		1414.898	II
250		1418.426	II
250		1421.759	II
200		1427.829	II
400		1430.243	II
250		1434.904	II
150		1436.236	II
150		1442.139	II
200		1445.984	II
200		1449.058	II
250		1450.304	II
200		1452.294	II
300		1458.002	II
250		1459.412	II
200		1463.752	II
400		1463.838	II
200		1466.070	II
400		1470.697	II
200		1472.395	II
250		1473.978	II
200		1474.935	II
150		1476.059	II
200		1481.544	II
200		1485.328	II
750		1488.831	II
300		1492.834	II
250		1493.366	II
250		1495.430	II
350		1496.687	II
150		1503.368	II
250		1504.757	II
200		1505.388	II
300		1508.632	II
350		1510.506	II
200		1512.465	II
200		1513.366	II
500		1514.492	II
200		1517.631	II
500		1519.492	II
600		1519.837	II
200		1520.540	II
200		1524.860	II
150		1525.764	II
500		1531.856	II
300		1532.131	II
250		1533.986	II
250		1535.002	II
500		1537.559	II
200		1540.239	II
300		1540.389	II
300		1540.588	II
750		1541.703	II
400		1544.677	II
100		1547.958	II
300		1550.653	II
300		1551.389	II
500		1552.646	II
250		1553.896	II
400		1555.134	II
500		1555.703	II
300		1558.345	II
400		1565.924	II
400		1566.415	II
100		1569.416	II
300		1579.492	II
300		1580.626	II
400		1581.995	II
500		1583.682	II
400		1590.165	II
600		1593.556	II
400		1598.402	II
400		1602.388	II
200		1604.848	II
300		1605.281	II
400		1606.834	II
250		1608.639	II
150		1610.296	II
200		1617.915	II
600		1621.426	II
400		1622.428	II
250		1630.268	II
100		1636.605	II
250		1649.458	II
30	r	1655.32	II
200		1656.322	II
200		1660.001	II
300		1663.002	II
100		1672.776	II
30		1688.09	I
30		1691.08	I
30	r	1703.84	I
50	r	1713.36	I
150		1717.721	II
50	r	1725.66	I
100		1736.551	II
50	r	1741.57	I
150		1753.281	II

Intensity		Wavelength	Sp
200	r	1774.82	I
100	r	1825.35	I
250		1929.751	II
250		1944.597	II
100		1946.493	II
200		1957.518	II
150		1970.495	II
150		1977.027	II
500		1979.956	II
300		1989.855	II

Air

Intensity		Wavelength	Sp
250		1999.698	II
270		2035.854	II
250		2037.127	II
350		2043.802	II
300		2054.980	II
100		2078.663	II
110		2098.398	II
320		2104.797	II
300		2112.100	II
320		2117.310	II
350		2122.980	II
350		2126.044	II
420		2134.341	II
900		2135.981	II
400		2148.984	II
150		2161.320	II
1300	r	2165.09	I
250		2174.982	II
1600	r	2178.94	I
700		2179.410	II
1700	r	2181.72	I
700		2189.630	II
900		2192.268	II
400		2195.683	II
1700	r	2199.58	I
1300	r	2199.75	I
100		2200.509	II
200		2209.806	II
750		2210.268	II
1600	r	2214.58	I
250		2215.106	II
1000	r	2215.65	I
750		2218.108	II
2100	r	2225.70	I
150		2226.780	II
1600	r	2227.78	I
350		2228.868	II
2500	r	2230.08	I
1100		2238.45	I
900		2242.618	II
2300	r	2244.26	I
1000		2247.002	II
1300	r	2260.53	I
2200	r	2263.08	I
150		2263.786	II
200		2276.258	II
100		2286.645	II
2500	r	2293.84	I
170		2294.368	II
1000		2303.12	I
150		2369.890	II
2500	r	2392.63	I
120		2403.337	II
1500		2406.66	I
1000	r	2441.64	I
100		2485.792	II
2000	r	2492.15	I
150		2506.273	II
120		2526.593	II
300		2544.805	II
100		2571.756	II
150		2590.529	II
200		2600.270	II
2500	r	2618.37	I
200		2666.291	II
750		2689.300	II
700		2700.962	II
650		2703.184	II
700		2713.508	II
650		2718.778	II
300		2721.677	II
120		2737.342	II
270		2745.271	II
2500	r	2766.37	I
800		2769.669	II
200		2791.795	II
170		2799.528	II
100		2810.804	II
1250	r	2824.37	I
350		2837.368	II
100		2857.748	II
600		2877.100	II
270		2884.946	II
2500	r	2961.16	I
100		2986.335	II
2000		2997.36	I
2000		3010.84	I
2500		3036.10	I
2500		3063.41	I
1400		3073.80	I
1500		3093.99	I

Intensity		Wavelength	Sp
1250		3099.93	I
2000		3108.60	I
1400	h	3126.11	I
1500		3194.10	I
1400		3208.23	I
1500	h	3243.16	I
10000	r	3247.54	I
10000	r	3273.96	I
1400		3282.72	I
400		3290.418	II
1500	h	3290.54	I
110		3300.881	II
250		3301.229	II
2500	h	3307.95	I
200		3316.276	II
1500		3337.84	I
150		3338.648	II
200		3365.648	II
450		3370.454	II
300		3374.952	II
200		3380.712	II
100		3384.945	II
1250	h	3483.76	I
1250		3524.23	I
2000		3530.38	I
1400		3599.13	I
1400		3602.03	I
1000		3686.555	II
150		3786.270	II
170		3797.849	II
100		3818.879	II
140		3826.921	II
160		3864.137	II
280		3884.131	II
150		3892.924	II
170		3903.177	II
140		3920.654	II
120		3933.268	II
120		3987.024	II
150		3993.302	II
140		4003.476	II
1250		4022.63	I
100		4032.647	II
600		4043.484	II
500		4043.751	II
2000		4062.64	I
120		4068.106	II
500		4131.363	II
200		4143.017	II
300		4153.623	II
500		4161.140	II
370		4164.284	II
400		4171.851	II
500		4179.512	II
500		4211.866	II
320		4230.449	II
200		4255.635	II
950		4275.11	I
300		4279.962	II
500		4292.470	II
400		4365.370	II
100		4444.831	II
400		4506.002	II
150		4516.049	II
150		4541.032	II
500		4555.920	II
100		4596.906	II
120		4649.271	II
2000		4651.12	I
120		4661.363	II
320		4671.702	II
300		4673.577	II
450		4681.994	II
100		4758.433	II
400		4812.948	II
120		4851.262	II
300		4854.988	II
100		4873.304	II
150		4901.427	II
1000		4909.734	II
500		4918.376	II
200		4926.424	II
900		4931.698	II
120		4943.026	II
700		4953.724	II
500		4985.506	II
400		5006.801	II
350		5009.851	II
400		5012.620	II
350		5021.279	II
200		5039.016	II
300		5047.348	II
900		5051.793	II
400		5058.910	II
500		5065.459	II
450		5067.046	II
350		5072.302	II
450		5088.277	II
420		5093.816	II
350		5100.067	II
1500		5105.54	I
250		5124.476	II
2000		5153.24	I
100		5158.093	II

Intensity	Wavelength	Sp
100	5183.367	II
2500	5218.20	I
100	5269.991	II
100	5276.525	II
1650	5292.52	I
100	5368.383	II
1500	5700.24	I
1500	5782.13	I
150	5805.989	II
100	5833.515	II
200	5897.971	II
120	5937.577	II
400	5941.196	II
100	5993.260	II
650	6000.120	II
100	6023.264	II
250	6072.218	II
150	6080.343	II
150	6099.990	II
160	6107.412	II
300	6114.493	II
600	6150.384	II
750	6154.222	II
500	6172.037	II
550	6186.884	II
400	6188.676	II
300	6198.092	II
470	6204.261	II
450	6208.457	II
750	6216.939	II
700	6219.844	II
500	6261.848	II
1000	6273.349	II
350	6288.696	II
900	6301.009	II
550	6305.972	II
400	6312.492	II
120	6326.466	II
400	6373.268	II
750	6377.840	II
400	6403.384	II
850	6423.884	II
200	6442.965	II
750	6448.559	II
170	6466.246	II
950	6470.168	II
750	6481.437	II
400	6484.421	II
220	6517.317	II
400	6530.083	II
120	6551.286	II
200	6577.080	II
750	6624.292	II
800	6641.396	II
450	6660.962	II
100	6770.362	II
300	6806.216	II
400	6809.647	II
320	6823.202	II
250	6844.157	II
320	6868.791	II
270	6872.231	II
270	6879.404	II
220	6937.553	II
150	6952.871	II
150	6977.572	II
200	7022.860	II
300	7194.896	II
400	7326.008	II
300	7331.694	II
250	7382.277	II
1000	7404.354	II
270	7434.156	II
500	7562.015	II
700	7652.333	II
1000	7664.648	II
150	7681.788	II
450	7744.097	II
800	7778.738	II
750	7805.184	II
1500	7807.659	II
1000	7825.654	II
350	7860.577	II
300	7890.567	II
700	7902.553	II
1500	7933.13	I
400	7944.438	II
400	7972.033	II
1200	7988.163	II
2000	8092.63	I
500	8277.560	II
800	8283.160	II
250	8503.396	II
750	8511.061	II
200	8609.134	II
500	9813.213	II
250	9827.978	II
200	9830.798	II
600	9861.280	II
600	9864.137	II
200	9883.969	II
550	9916.419	II
500	9917.954	II
550	9925.594	II

Copper (Cont.)

Intensity	Wavelength	
450	9938.998	II
500	9960.354	II
450	10006.588	II
550	10022.969	II
550	10038.093	II
650	10054.938	II
450	10080.354	II

Cu III
Ref. 295—V.K.

Intensity	Wavelength	
	Vacuum	
75	542.90	III
200	615.67	III
150	616.03	III
150	687.98	III
150	713.53	III
125	730.38	III
250	788.07	III
250	788.46	III
250	791.36	III
150	801.14	III
100	829.34	III
40	1048.88	III
50	1186.80	III
50	1200.96	III
300	1219.30	III
200	1244.38	III
100	1279.14	III
200	1312.39	III
300	1332.97	III
200	1339.48	III
150	1363.08	III
300 r	1376.79	III
200 r	1377.49	III
150	1423.48	III
300 r	1481.23	III
200	1543.46	III
500 r	1593.75	III
1000 r	1642.21	III
300	1679.14	III
600	1700.10	III
500	1722.37	III
600	1741.37	III
200	1768.86	III
200	1840.91	III
100	1971.95	III
	Air	
200	2013.22	III
150	2157.28	III
100	2299.47	III
500	2368.17	III
400	2391.74	III
800	2405.50	III
700	2412.34	III
2000	2444.44	III
500	2468.41	III
1000	2482.36	III
700	2486.46	III
500	2508.49	III
500	2522.38	III
500	2538.66	III
400	2566.37	III
400	2573.33	III
500	2609.32	III
200	2643.92	III
200	2696.38	III
20	2751.33	III
100	2812.94	III
100	2978.87	III
75	3548.87	III
100	3639.42	III
500	3702.92	III
800	3744.70	III
400	3748.27	III
600	3752.06	III
1000	3776.97	III
800	3790.80	III
600	3804.13	III
600	3809.18	III
300	3881.68	III
150	3953.81	III
100	4090.49	III
200	4283.40	III
500	4351.97	III
1000	4352.80	III
500	4355.24	III
500	4370.84	III
500	4371.40	III
500	4373.43	III
1000	4377.11	III
200	4386.42	III
150	4927.41	III
400	5094.28	III
200	5168.97	III
400	5208.34	III
600	5219.21	III
200	5268.59	III
400	5317.78	III
300	5369.79	III
350	5418.48	III
250 d	5494.94	III
50	5573.94	III
100	5609.00	III
75	5702.12	III
100	5768.56	III
100	5850.72	III
200	5965.25	III
30	6100.87	III
50	6369.27	III
20	6512.54	III
20	6644.13	III
50	6793.20	III

Cu IV
Ref. 199—V.K.

Intensity	Wavelength	
	Vacuum	
30	360.86	IV
20	374.40	IV
30	405.24	IV
80	406.45	IV
40	413.45	IV
70	443.68	IV
80	451.16	IV
80	463.72	IV
80	484.53	IV
90	497.00	IV
90	504.60	IV
70	509.38	IV
40	519.51	IV
40	540.65	IV
60	550.92	IV
20	584.85	IV
60	1056.13	IV
30	1074.72	IV
50	1091.65	IV
30	1105.50	IV
25	1119.63	IV
40 p	1152.18	IV
60	1227.44	IV
70	1228.87	IV
70	1258.69	IV
90	1274.84	IV
90	1293.46	IV
90	1309.41	IV
70	1321.17	IV
100 d	1340.08	IV
100	1350.42	IV
100	1362.05	IV
100	1372.14	IV
100	1377.82	IV
100	1388.80	IV
90	1405.49	IV
90	1415.27	IV
90	1434.34	IV
90	1449.69	IV
80	1466.18	IV
70	1482.77	IV
80	1499.81	IV
90	1515.28	IV
90	1535.12	IV
80	1551.12	IV
90	1567.35	IV
80	1583.47	IV
70	1595.12	IV
80 p	1608.14	IV
90	1639.75	IV
20	1650.16	IV
70	1704.37	IV
30	1797.99	IV
70	1817.56	IV
70	1819.23	IV
60	1837.04	IV
80	1849.62	IV
30	1867.24	IV
30	1918.71	IV
40	1966.31	IV

Cu V
Ref. 324—V.K.

Intensity	Wavelength	
	Vacuum	
9 h	258.95	V
49	271.33	V
49	283.97	V
22	293.41	V
56	299.64	V
65	305.83	V
51	312.51	V
66	321.05	V
74	326.57	V
82	333.56	V
81	339.88	V
81	346.00	V
86	355.41	V
77	363.96	V
65	370.63	V
74	377.76	V
70	387.40	V
51	396.06	V
25	406.94	V
13	1097.10	V
42	1106.24	V
77	1113.22	V
67	1121.20	V
76	1128.80	V
59	1133.86	V
63	1142.38	V
54	1149.06	V
72	1157.54	V
64	1167.35	V
77	1176.53	V
84	1183.63	V
76	1192.54	V
83	1201.22	V
77	1204.90	V
71	1214.36	V
70	1221.34	V
76	1230.11	V
80	1239.73	V
79	1246.99	V
65	1253.07	V
70	1260.24	V
77	1269.35	V
78	1278.20	V
68	1286.13	V
67	1292.08	V
73	1299.22	V
65	1309.72	V
55	1318.89	V
65	1323.28	V
64	1329.22	V

CURIUM (Cm)
Z = 96
Cm I and II
Ref. 51, 332—J.G.C.

Intensity	Wavelength	
	Air	
10000	2462.76	II
10000	2617.17	II
10000	2636.28	II
10000	2651.17	II
10000	2653.80	II
10000	2725.68	II
10000	2736.89	II
10000	2748.04	II
10000	2784.83	II
10000	2811.62	II
10000	2824.20	II
10000	2833.58	II
10000	2899.90	II
10000	2912.97	II
10000	2928.92	II
10000	2996.18	II
10000	2999.39	I
10000 b	3014.87	II
10000	3044.85	II
10000	3109.69	I
10000	3116.41	I
10000	3137.16	II
10000	3147.33	I
10000	3155.10	I
10000	3158.60	I
10000	3169.98	II
10000	3177.55	I
10000	3179.10	I
10000	3186.41	I
10000	3188.11	I
10000	3207.12	II
10000	3207.71	I
10000	3209.89	II
10000	3209.94	I
10000	3210.05	II
10000	3220.76	II
10000	3224.23	I
10000	3225.11	I
10000	3226.41	II
10000	3230.28	I
10000	3230.35	II
10000	3236.74	I
10000	3238.55	II
10000	3242.66	II
10000	3246.25	I
10000	3252.68	I
10000	3265.81	I
10000	3280.45	I
10000	3296.71	II
10000	3304.85	I
10000	3317.14	I
10000	3374.70	I
10000	3452.92	I
10000	3458.34	I
10000	3510.28	I
10000	3522.36	I
10000	3524.94	I
10000	3542.06	I
10000	3547.02	I
10000	3547.92	I
10000	3561.44	I
10000	3572.95	II
10000	3600.62	I
10000	3639.94	I
10000	3664.34	I
10000	3709.43	I
10000	3729.00	I
10000	3732.35	I
10000	3747.86	I
10000	3763.05	I
10000	3775.75	I
10000	3816.30	I
10000	3825.14	I
10000	3833.32	I
10000	3837.59	I
10000	3842.00	I
10000	3849.92	I
10000	3854.11	I
10000	3900.25	I
10000	3904.06	II
10000	3908.24	II
10000	3936.67	I
10000	3942.03	I
10000	3944.15	I
10000	3948.68	I
10000	3953.36	I
10000	3964.83	I
10000	3995.10	I
10000	4016.17	I
10000	4031.76	I
10000	4048.29	I
10000	4049.65	I
10000	4113.29	I
10000	4129.71	I
10000	4207.66	II
10000	4211.62	I
10000	4266.45	I
10000	4293.00	I
10000	4330.82	I
10000	4345.69	I
10000	4447.77	I
10000	4659.16	I
10000	4608.40	I
10000	5846.07	I
10000	5952.41	I
10000	6058.90	I
10000	6243.35	I
10000	6376.71	I
10000	6510.16	I
10000	6554.41	I
10000	6640.17	I
10000	6663.25	I
10000	6686.87	I
10000	6706.85	I
10000	6726.68	I
10000	6793.15	I
10000	7162.69	I
10000	7577.80	I
10000	7673.79	I
10000	7720.47	I
10000	8392.37	I
10000	9293.25	I
10000	9567.08	I
10000	9657.12	I
10000	10310.83	I
10000	10351.73	I
10000	10424.49	I
10000	10508.11	I
10000	10542.98	I
10000	10792.25	I
10000	10897.45	I
10000	11507.45	I
10000	11707.73	I
10000	11780.95	I
10000	11834.68	I
10000	12017.85	I
10000	12394.16	I
10000	12454.98	I
10000	12464.99	I
10000	13004.56	I
10000	13258.18	I
10000	13289.84	I
10000	13344.62	I
10000	13480.54	I
10000	13590.01	I
10000	13644.77	I
10000	13789.52	I
10000	13840.18	I
10000	13908.46	I
10000	13964.14	I
10000	14235.27	I
10000	14334.52	I
10000	14563.41	I
10000	14580.23	I
10000	15018.13	I
10000	15222.27	I
10000	15642.59	I

Intensity	Wavelength	
10000	15757.23	I
10000	15793.31	I
10000	16008.41	I
10000	17148.22	I
10000	17453.18	I
10000	17619.28	I
10000	18069.02	I
10000	19572.64	I
10000	19975.98	I
10000	20526.32	I
10000	20853.49	I
10000	20911.52	I
10000	20968.11	I
10000	21241.06	I
10000	21393.23	I

DYSPROSIUM (Dy)
Z = 66

Dy I and II
Ref. 1—C.H.C.

Intensity	Wavelength (Air)	
260	2356.91	II
65	2381.95	
130	2387.36	II
150	2392.15	
180	2402.29	II
240	2410.01	II
150	2422.75	II
260	2439.84	II
90	2455.15	II
110	2459.99	II
90	2471.40	II
110	2480.93	II
170	2490.61	II
90	2510.31	II
170	2513.55	II
170	2517.61	II
130	2543.81	II
90	2545.12	II
150	2552.29	II
180	2557.94	II
90	2560.21	II
90	2566.25	II
220	2585.30	I
90	2591.56	II
75	2592.54	II
130	2600.16	II
130	2600.76	II
75	2608.69	II
370	2623.69	II
440	2634.80	II
110	2642.15	I
110	2645.35	II
110	2667.94	I
55	2676.84	II
50	2677.34	II
85	2689.31	II
85	2692.83	II
55	2709.01	II
55	2727.17	II
85	2729.50	II
55	2735.79	I
40	2739.30	II
85	2740.70	II
220	2755.75	II
55	2757.08	II
70	2766.50	II
70	2772.42	II
110	2772.61	II
40	2779.58	II
55	2791.44	II
120	2800.33	II
110	2800.53	II
110	2801.41	II
300	2816.39	II
140	2825.42	II
140	2862.70	I
190	2877.88	II
110	2884.28	II
120	2885.53	I
120	2890.74	II
120	2900.82	II
110	2904.62	II
190	2906.39	II
390	2913.95	II
110	2934.31	II
250	2934.52	II
110	2941.05	II
140	2944.56	II
150	2947.06	II
150	2947.21	II
250	2948.31	II
170	2950.33	II
110	2952.12	II
140	2953.70	II
220	2964.60	I
110	2977.42	II

Intensity	Wavelength	
110	2985.97	II
220	3015.68	II
390	3026.16	II
210	3029.81	II
610	3038.28	II
280	3043.13	II
210	3047.56	II
280	3060.64	II
390	3062.62	II
220	3066.99	II
330	3071.91	II
280	3073.54	II
220	3078.68	II
280	3101.93	II
220	3103.24	II
410	3109.76	II
330	3128.41	II
830	3135.38	II
360	3140.64	II
500	3141.14	II
220	3143.83	II
250	3146.16	II
1200	3156.52	II
670	3162.83	II
1000	3169.99	II
400	3177.89	II
220	3178.37	II
200	3184.79	II
330	3186.38	II
240	3187.68	II
330	3193.30	II
240	3206.40	II
220	3207.12	II
290	3208.85	II
470	3215.19	II
830	3216.63	II
240	3221.49	II
290	3223.28	II
240	3225.08	II
330	3225.95	II
490	3235.89	II
290	3236.69	II
490	3245.12	II
200	3248.36	II
1200	3251.27	II
200	3252.51	II
290	3256.26	II
200	3266.21	II
240	3269.11	II
200	3272.73	II
890	3280.09	II
490	3282.77	II
200	3287.94	II
200	3293.88	II
200	3296.30	II
200	3305.40	II
200	3305.51	II
240	3306.19	II
440	3308.79	II
1100	3308.88	II
510	3312.72	II
780	3316.32	II
240	3317.12	II
1000	3319.88	II
270	3326.19	II
780	3341.00	II
270	3341.88	II
200	3347.83	II
270	3352.69	II
510	3353.58	II
240	3359.46	II
510	3368.11	II
5300	3385.02	II
210	3386.57	II
610	3388.85	II
210	3391.96	II
3800	3393.57	II
1300	3396.16	II
380	3407.16	II
5300	3407.80	II
420	3408.14	II
1300	3413.78	II
530	3414.82	II
780	3419.63	II
530	3425.06	II
420	3429.44	II
1900	3434.37	II
330	3438.94	II
560	3440.93	II
1300	3441.45	II
3800	3445.57	II
830	3446.99	II
440	3449.89	II
2700	3454.32	II
440	3454.51	II
1300	3456.56	II
4400	3460.97	II
720	3468.43	II
560	3471.14	II
560 d	3471.53	II
380	3473.70	II
1300	3477.07	II
4400	3494.49	II
560	3496.34	II

Intensity	Wavelength	
400	3497.81	II
830	3498.71	II
400	3501.50	II
830	3504.53	II
830	3505.45	II
1300	3506.81	II
560	3517.26	II
4400	3523.98	II
22000	3531.70	II
4400	3534.96	II
5500	3536.02	II
4400	3538.52	II
400	3539.37	II
1700	3542.33	II
400	3544.20	II
400	3544.35	II
1400	3546.83	II
330	3548.19	II
4400	3550.22	II
2200	3551.62	II
440 h	3558.23	II
440	3559.30	II
2200	3563.15	II
560	3563.69	II
780	3573.83	II
1400	3574.15	II
4400	3576.24	II
1700	3576.87	II
830	3577.98	II
440	3580.04	II
400	3584.42	II
3300	3585.06	II
1400	3585.78	II
560	3586.11	II
360	3590.07	II
1100	3591.41	II
560	3591.81	II
560	3592.11	II
1800	3595.08	II
400	3596.06	II
560	3600.38	II
360	3602.82	II
1800	3606.12	II
440	3618.51	II
560	3620.16	II
470	3624.27	II
1100	3629.42	II
4000	3630.24	II
440	3632.78	II
400	3635.27	II
360	3637.28	II
1100	3640.25	II
400	3643.92	II
11000	3645.40	II
360	3645.86	II
1000	3648.78	II
700	3664.62	II
400	3666.84	I
990	3672.30	II
420	3672.70	II
400	3673.14	II
1400	3674.08	II
2200	3676.59	II
640	3678.51	I
820	3684.85	I
1300	3685.78	I
4700	3694.81	II
370	3697.31	II
990	3698.21	II
540	3701.63	II
330	3707.40	II
440	3707.57	II
440	3708.22	II
420	3710.07	II
330	3711.66	II
1600	3724.45	II
300	3728.00	I
930	3739.34	I
1200	3747.82	II
1400	3753.51	II
1400	3753.75	II
1200	3757.05	I
4700	3757.37	II
640	3767.63	II
330	3771.11	I
640	3773.05	I
370	3774.71	I
420	3781.47	I
330	3785.41	II
3300	3786.18	II
1600	3788.44	II
700	3791.87	II
510	3804.14	II
580	3806.27	II
470	3812.27	I
470	3813.67	II
1400	3816.76	II
700	3825.68	II
2300	3836.50	II
370	3840.89	I
1400	3841.31	II
330	3842.00	II
330	3844.36	I

Intensity	Wavelength	
420	3846.34	II
420	3847.02	I
330	3849.39	II
1200	3853.03	II
420	3858.40	I
370	3866.58	II
560	3868.45	II
1600	3868.81	I
300	3869.42	II
820	3869.86	II
7000	3872.11	II
1200	3873.99	II
470	3879.11	II
300	3881.99	II
5800	3898.53	II
540	3914.87	II
540	3915.59	II
540 d	3917.29	I
320	3923.38	II
420	3927.86	I
540	3930.14	I
2100	3931.52	II
320	3932.22	II
370	3933.00	II
320	3934.21	II
420	3936.70	I
540	3942.53	II
10000	3944.68	II
420	3946.93	II
540	3950.39	II
420	3954.55	II
800	3957.79	II
370	3962.59	I
320	3967.51	I
14000	3968.39	II
2700	3978.57	II
1400	3981.92	II
1600	3983.65	II
800	3984.21	II
540	3991.32	II
1600	3996.69	II
8000	4000.45	II
420	4005.84	I
320	4006.07	I
540	4011.29	II
540	4013.82	I
540	4014.70	II
370	4023.71	I
420	4027.78	II
520 d	4028.32	II
520	4032.47	II
420	4033.65	II
420	4036.32	II
320	4041.98	II
12000	4045.97	I
1600	4050.56	II
520	4055.14	II
2500	4073.12	II
7400	4077.96	II
370	4085.34	I
390	4096.10	I
3900	4103.30	II
860	4103.87	I
1500	4111.34	II
490	4124.63	II
390	4128.24	II
350	4129.12	I
990	4129.42	II
350	4130.35	I
390	4133.85	I
470	4141.50	II
1200	4143.10	II
990	4146.06	I
5700	4167.97	I
370	4171.93	I
930	4183.72	I
12000	4186.82	I
320	4190.94	I
2200	4191.64	I
6800	4194.84	I
320	4195.19	II
800	4198.02	I
680	4201.30	I
680	4202.24	I
230	4205.06	I
370	4206.54	II
440	4211.24	I
16000	4211.72	I
1800	4213.18	I
3700	4215.16	I
4400	4218.09	I
4400	4221.11	I
540	4222.21	I
2700	4225.16	I
680	4232.02	I
680	4239.85	I
440	4245.91	I
440	4256.33	II
250	4276.69	I
370 d	4294.93	II
	4295.04	I
1000	4308.63	II
320	4325.86	I
200	4358.44	II

Dysprosium (Cont.)

Intensity		Wavelength	Spectrum
320		4374.24	II
320		4374.76	II
540		4409.38	II
150		4444.58	I
740		4449.70	II
110		4455.60	II
250		4468.14	II
100	d	4527.58	I
		4527.76	II
100		4541.66	II
140		4565.09	I
420		4577.78	I
2100		4589.36	I
990		4612.26	I
50		4613.83	I
50		4614.82	I
60		4617.26	II
140		4620.03	II
50		4662.72	I
110		4664.66	II
85		4673.60	II
50		4682.03	II
50		4689.75	II
95		4698.68	II
85		4721.22	I
70		4727.13	II
170		4731.84	II
40		4745.73	II
60		4754.99	II
50		4760.04	II
60		4771.94	I
50		4774.80	I
120	h	4775.79	I
75		4786.92	II
95		4791.29	I
29		4800.64	I
50		4807.94	I
40		4810.28	I
50		4812.80	I
75		4819.04	I
85		4824.96	I
75		4828.88	I
50		4829.68	II
70		4832.39	I
35		4833.75	II
75		4841.75	I
40		4856.24	II
40		4868.05	II
40		4873.93	I
85		4880.16	I
40		4884.55	I
95		4888.08	I
40		4889.33	II
75		4890.10	II
50		4893.68	I
24		4899.24	I
55		4916.41	I
50		4922.22	II
65		4923.16	II
480		4957.34	II
24		4959.59	I
28		4973.57	I
40		4985.52	I
50		5003.87	I
55		5004.28	II
24		5010.60	I
24		5017.98	II
70		5022.12	I
30		5024.03	I
24		5024.54	I
40		5027.87	I
50		5033.00	I
160		5042.63	I
24		5047.25	I
50	h	5050.21	I
30		5053.35	I
24		5055.46	I
95		5070.68	I
120		5077.67	I
80		5090.38	II
80		5110.32	I
130	h	5120.04	I
30		5135.02	I
190		5139.60	II
40		5161.03	II
40		5164.12	II
50		5165.34	I
110		5169.69	II
20		5172.90	II
80		5185.30	I
40		5188.45	II
290		5192.86	II
95		5197.66	II
50		5246.94	II
70		5259.88	I
130		5260.56	I
55	B1	5263.3	D
65		5267.11	I
50		5272.25	II
50		5275.29	II
50		5279.70	II
55		5282.07	I
28		5284.99	II
40		5297.82	II

Intensity		Wavelength	Spectrum
160		5301.58	I
40		5309.02	II
50		5324.69	I
24		5337.43	II
65		5340.30	I
30		5352.11	I
30		5368.20	II
20		5385.63	II
85		5389.58	II
40		5395.57	I
20	h	5398.26	D
24		5399.93	II
50		5404.19	I
80		5419.13	I
70		5423.32	I
30		5424.27	I
40		5426.70	II
30		5443.34	I
95		5431.11	I
13		5453.47	II
30		5469.10	II
24		5496.83	I
28		5502.79	I
24		5506.52	I
28		5515.41	II
30		5528.01	I
65		5547.27	I
40	d	5600.65	II
24		5605.53	I
30		5613.23	I
20		5627.49	I
100		5639.50	I
55	h	5645.99	I
80		5652.01	I
24		5685.58	I
28	h	5693.67	D
24	h	5694.10	I
28	Cw	5694.54	I
28		5698.72	II
24		5702.91	I
70	h	5718.46	I
28	h	5725.84	h
55	h	5728.64	D
24	h	5738.33	D
50		5740.20	I
55		5745.53	I
24		5750.48	I
24		5758.79	I
80	h	5832.01	D
55	h	5833.85	D
40	h	5834.86	D
28	h	5844.41	D
24		5845.65	D
40	h	5848.05	D
40		5855.56	D
55	h	5868.11	II
40		5915.16	II
20		5924.56	II
70		5945.80	I
50	l	5964.46	I
120		5974.49	I
24		5984.86	I
140		5988.56	I
24	h	6005.75	D
24	h	6006.54	I
24	h	6006.97	D
30		6008.94	I
65		6010.82	I
24		6017.26	I
24		6030.98	I
24	B1	6042.49	D
24		6058.18	I
30		6085.06	I
140		6088.26	I
24		6127.15	I
24		6133.64	I
24		6158.28	I
100		6168.43	I
20		6196.23	II
270		6259.09	I
30		6260.36	I
14		6343.32	I
40		6386.80	I
24		6396.60	II
50		6421.92	I
13	h	6436.55	I
8		6460.83	I
10		6468.58	II
11		6474.91	I
20		6483.59	II
28		6486.59	I
20		6558.02	I
160		6579.37	I
14		6594.14	II
15		6643.37	I
22		6658.36	I
29		6661.64	I
75		6667.86	I
10		6700.64	II
29		6747.93	I
10		6757.62	I
45		6765.89	I
12		6818.20	I
180		6835.42	I

Intensity		Wavelength	Spectrum
80		6852.96	I
22		6856.46	I
22		6888.83	I
15		6897.97	II
65		6899.32	II
22		6906.53	II
15		6929.55	I
29		6950.28	II
11		6951.42	I
40		6958.08	I
13	h	6982.44	I
13		6991.30	I
45		6998.10	I
20		7017.42	I
35		7055.95	II
24		7075.14	II
17		7109.26	II
11		7120.81	II
13		7175.11	II
11		7213.27	I
17	h	7230.04	I
13		7250.01	I
17		7345.13	II
11		7370.23	II
20		7376.04	I
11		7407.59	I
24		7412.37	I
55		7426.86	II
20		7457.05	II
17		7516.61	II
55		7543.73	I
17	h	7553.00	I
27		7559.78	I
40		7562.96	II
20	h	7577.46	II
27	h	7591.30	I
13	h	7611.55	I
11	h	7617.70	I
35	h	7641.09	I
17		7645.86	I
13		7646.64	I
80		7662.36	I
11		7666.78	II
11	h	7715.00	II
45		7729.76	II
20		7751.62	II
35		7812.06	I
27		7909.38	I
11		7968.63	I
12		7982.05	II
13		8147.29	I
27		8198.77	II
100		8201.57	II
11		8218.62	I
20		8265.53	I
35		8326.10	I
35		8392.01	II
12		8405.85	II
20		8416.64	II
24		8438.58	II
11		8630.12	I
27		8655.94	II
17		8657.68	II
17		8678.49	II
11		8696.83	II
11	h	8715.95	II
20		8750.40	II
12		8780.83	I
45		8791.39	II
13		8833.08	II
24		8850.37	II

EINSTEINIUM (Es)
Z = 99

Es I and II
Ref. 333—J.G.C.

Intensity		Wavelength	Spectrum
		Air	
300	1	2694.32	II
10000	s	2708.66	II
100	s	2716.02	II
1000	s	2724.57	II
1000	1	2765.76	II
3000	1	2787.10	II
1000	1	2796.11	II
3000	1	2815.15	II
100	s	2885.84	II
100	s	2886.44	
3000		2907.03	
100		3003.28	
1000	s	3065.40	
3000		3135.25	I
1000	1	3154.27	II
10000		3413.17	
300	1	3423.12	
100	1	3424.28	
10000	s	3428.48	I
300		3437.31	

Dysprosium (Cont.)

Intensity		Wavelength	Spectrum
100		3437.34	
3000	s	3445.25	
300	1	3446.93	
3000		3452.36	
100		3453.16	
300	s	3470.77	
3000	s	3484.59	I
300		3494.30	
10000	s	3498.11	I
10000	1	3514.33	I
10000	s	3521.38	
10000	s	3523.49	I
300	1	3528.58	
10000	s	3536.01	
10000	s	3547.75	II
300		3549.97	
10000	1	3555.34	I
100		3555.53	
3000	s	3558.65	
3000		3560.92	
1000	s	3575.68	
100	s	3578.56	
300	s	3579.38	
1000	s	3582.95	
3000		3590.28	
1000		3595.47	
10000	s	3602.43	II
300		3605.58	
1000	s	3606.75	
100	1	3624.52	
3000	1	3631.09	
10000	1	3632.87	
100	1	3634.41	
1000	1	3651.94	
10000	1	3670.01	II
100	1	3672.32	
100	s	3713.56	
1000	1	3720.56	
300		3722.32	
10000	1	3728.55	II
100	h	3737.47	
300	s	3776.27	
10000	1	3792.99	
10000	1	3815.00	
300	s	3929.10	
3000		3930.77	
100		3957.19	
100		3995.35	
300		4077.71	
10000	s	4082.24	
3000	1	4107.59	
1000	1	4176.94	
100		4496.25	
100	1	4631.66	
300	1	4650.86	
1000		4789.93	
300	h	4802.17	
1000	h	4802.21	
3000	h	4958.29	
10000	s	5052.08	
1000	s	5102.93	
100	s	5155.82	
10000	s	5161.74	I
10000	1	5204.40	I
3000	L	5615.51	I
100	S	6539.71	ES

ERBIUM (Er)
Z = 68

Er I and II
Ref. 1—C.H.C.

Wavelength		Intensity	Spectrum
		Air	
110		2358.51	II
100		2386.58	II
120		2387.17	II
110		2396.38	III
140		2446.39	II
100		2537.02	II
110		2547.28	II
290		2586.73	II
110		2587.04	II
130		2592.57	II
120		2595.03	II
140		2624.18	II
490		2670.26	II
330		2672.25	II
100		2675.35	II
270		2739.31	II
310		2750.19	II
230		2755.01	II
610		2755.63	II
510		2770.02	II
230		2778.97	II
230		2802.53	II
310		2804.35	II
410		2820.19	II
270	d	2833.91	II

Erbium (Cont.)

Intensity	Wavelength	Note	Ion
390	2838.71		II
270	2848.37		II
250	2855.41		II
310	2859.84		II
310	2896.96		II
390	2897.52		II
1000	2904.47		II
210	2909.58		II
1500	2910.36		II
270	2915.62		II
350	2929.27		II
270	2945.28		II
230	2946.62		II
1500	2964.52		II
410	2968.76		II
210	2974.47		II
230	2975.68		II
270	2983.80		II
1200	3002.41		II
310	3002.65		II
230	3012.47		II
230	3016.84		II
290	3025.95		II
270	3028.27		II
370	3031.31		II
310	3036.22		II
210	3054.42		II
230	3066.22		II
450	3070.74		II
560	3072.53		II
610	3073.34		II
210	3078.87		II
720	3082.08		II
610	3084.02		II
370	3099.19		II
230	3106.78		II
310	3113.43	d	II
	3113.54		II
770	3122.72		II
290	3132.52		II
470	3132.77		II
410	3141.10		II
	3141.15		II
250	3144.33		II
410	3154.29		II
870	3181.92		II
410	3183.42		II
250	3185.25		II
310	3200.58		II
230	3205.15		II
270	3214.44		II
870	3220.73		II
610	3223.31		II
210	3227.16		II
2300	3230.58		II
250	3232.03		II
330	3237.98		II
330	3249.34		II
560	3259.05		II
	3259.11		II
2700	3264.78		II
430	3267.10		II
	3267.18		II
330	3269.41		II
250	3278.22		II
720	3279.33		II
720	3280.22		II
470	3286.77		II
330	3303.88	d	II
	3303.95		II
370	3305.56		II
2300	3312.42		II
560	3316.39		II
770	3323.19		II
290	3329.66		II
770	3332.70		II
370	3337.25		II
290	3337.79		II
250	3340.03		II
290	3341.84		II
1300	3346.04		II
470	3350.06		II
350	3350.26		II
1400	3364.08		II
1400	3368.02	d	II
	3368.13		I
450	3370.55		II
7700	3372.71		II
970	3374.17		II
290	3381.32		II
230	3382.06		I
1700	3385.08		II
450	3389.74		II
2300	3392.00		II
350	3396.07		II
290	3396.84		II
390	3401.83		II
350	3417.63		II
490	3428.39		II
770	3441.13		II
390	3442.68		I
490	3469.51		I
970	3471.71		II
610	3479.41		II
970	3485.85		II
350	3486.82		II
350	3496.86		II
6700	3499.10		II
610	3502.78		I
390	3508.38		II
490	3514.89		II
390	3518.18		II
610	3524.91		II
410	3539.59		I
310	3548.26		II
820	3549.84		II
310	3553.20		II
1500	3558.02		I
510	3558.71		II
1000	3559.90		II
310	3565.17		II
920	3570.75		II
310	3578.24		I
1000	3580.52		II
370	3586.60		I
610	3590.76		II
410	3595.84		I
610	3599.50		II
1000	3599.83		II
510	3604.90		II
410	3607.42		I
3100	3616.56		II
510	3617.85		II
510	3618.92		II
720	3628.04		I
310	3629.37		II
1000	3633.54		II
510	3634.67		I
1600	3638.68		I
900	3645.94		II
520	3650.41		II
360	3652.58		II
500	3652.87		II
360	3664.45		I
470	3669.02		II
500	3682.70		II
320	3684.01		I
380	3684.28		II
7900	3692.65		II
450	3696.25		II
380	3697.58		I
540	3700.72		II
520	3707.64		II
520	3712.39		II
320	3719.35		I
1300	3729.52		II
450	3731.26		II
540	3738.16		II
340	3741.10		II
900	3742.64		II
900	3747.43		I
540	3756.05		I
410	3781.01		II
1800	3786.84		II
560	3787.86		II
560	3791.83		II
500	3792.79		I
560	3797.06		II
1600	3810.33		I
3600	3830.48		II
540	3849.91		I
320	3851.60		II
680	3855.90		I
540	3858.39		II
7500	3862.85		I
1500	3880.61		II
1200	3882.89		I
400	3890.61		II
4200	3892.68		I
5200	3896.23		II
810	3902.76		II
250	3903.98		I
250	3904.56		II
1200	3905.40		I
11000	3906.31		II
280	3918.05		I
210	3918.35		II
280	3921.88		II
810	3932.25		I
3200	3937.01		I
2100	3938.63		II
3200	3944.42		I
550	3948.06		I
250	3951.48		I
320	3956.42		I
280	3966.35		I
2700	3973.04		I
3200	3973.58		I
1400	3974.72		II
280	3976.73		I
810	3977.02		I
1100	3982.33		I
280	3987.53		I
810	3987.66		I
230	3991.15		I
230	4004.05		I
14000	4007.96		I
230	4008.18		II
280	4009.16		II
1100	4012.58		I
350	4015.57		II
3000	4020.51		I
450	4021.55		I
230	4043.01		II
1000	4046.96		I
280	4048.34		II
200	4049.49		II
940	4055.47		II
550	4059.51		I
690	4059.78		II
420	4077.88		II
550	4081.24		II
3500	4087.63		I
210	4092.90		I
1100	4098.10		I
350	4100.56		II
320	4116.36		I
320	4118.55		I
600	4131.50		I
550	4142.91		II
6900	4151.11		I
280	4189.98		II
1000	4190.70		I
130	4205.32		I
1400	4218.43		I
200	4220.99		I
320	4230.20		II
140	4234.78		II
200	4251.94		II
140	4276.48		II
690	4286.56		I
320	4298.91		I
320	4301.60		II
140	4303.81		II
110	4319.94		II
130	4328.81		I
110	4331.36		I
140	4340.92		I
190	4348.34		I
110	4369.39		I
160	4382.17		I
300	4384.70		II
300	4386.40		I
100	4403.17		II
810	4409.34		I
180	4418.70		I
570	4419.61		II
110	4422.51		II
320	4424.57		I
370	4426.77		I
110	4437.66		II
100	4459.24		II
100	4473.50		II
130	4496.39		I
200	4500.75		II
130	4522.74		I
160	4563.26		II
1000	4606.61		I
160	4630.88		II
110	4640.60		II
110	4665.44		II
310	4673.16		I
570	4675.62		II
150	4679.06		II
230	4722.69		I
150	4729.05		I
130	4751.52		II
170	4759.65		II
190	4820.35		II
140	4857.44		I
150	4872.09		II
210	4900.08		II
210	4934.11		II
130	4944.36		I
180	4951.74		II
130	4976.42		I
250	5007.25		I
140	5028.33		I
120	5028.91		II
200	5035.94		II
210	5042.05		II
130	5043.86		I
130	5044.89		I
130	5077.59		II
120	5124.56		I
130	5127.41		II
120	5131.53		I
130	5133.83		II
170	5164.77		II
130	5172.78		I
160	5188.90		II
150	5206.52		II
60	5212.91		II
30	5215.13		II
30	5218.26		II
45	5229.34		II
140	5255.93		II
22	5256.47		II
27	5257.02		II
35	5264.77		II
80	5272.91		I
55	5277.71		I
27	5279.34		II
45	5302.30		II
55	5333.06		I
27	5333.33		II
22	5343.94		II
30	5344.50		II
90	5348.06		I
45	5350.47		II
35	5368.85		I
35	5395.87		II
60	5414.63		II
18	5422.81		II
18	5451.30	h	I
35	5454.27		II
180	5456.62		I
35	5462.43		II
90	5468.32		I
18	5477.47		II
80	5485.97		II
27	5497.44		II
27	5516.02		I
80	5593.46		I
45	5601.14	d	I
	5601.32		I
45	5609.94	h	I
60	5611.82		I
70	5622.01		I
80	5626.53		II
30	5636.20		I
90	5640.36		I
22	5641.42		I
22	5658.63	h	II
70	5664.95		I
45	5665.44		I
55	5675.48		II
14	5695.53		II
27	5710.87		II
55	5717.48		I
70	5719.55		I
55	5726.97		I
22	5733.43		II
22	5736.56		I
22	5736.94		I
100	5739.19		I
35	5740.61		I
60	5748.65		I
55	5752.53		I
70	5757.63		II
290	5762.80		I
70	5769.92		I
45	5782.82		I
70	5784.66		I
22	5791.15		II
70	5800.79		I
22	5806.10		I
430	5826.79		I
45	5835.84		I
100	5850.07		I
120	5855.31		I
140	5872.35		I
120	5881.14		I
27	5886.30		II
27	5902.08		II
55	5906.06		I
45	5909.24		I
35	5933.50		I
22	5946.37		I
55	5968.68		I
27	5975.49		I
35	6006.79		II
22	6008.75		I
55	6014.83		I
35	6015.74		II
70	6022.56		I
22	6032.12		II
22	6045.63		II
22	6048.14		II
45	6054.85		I
70	6061.25		I
60	6076.45		II
35	6116.01		I
35	6125.32		I
30	6170.06		II
27	6183.21		II
360	6221.02		I
35	6230.90		I
55	6262.56		I
45	6267.93		I
60	6268.87		I
35	6274.94		I
30	6286.86		I
45	6299.42		I
130	6308.77		I
55	6326.13		I
22	6347.16		II
45	6388.19		I
22	6432.53		I
27	6485.87		I
55	6492.35		I

Erbium (Cont.)

Intensity		Wavelength	
22		6541.57	I
60		6583.48	I
70		6601.11	I
27		6721.91	I
70		6759.87	I
22		6762.92	I
27		6773.37	I
35		6790.92	I
22		6825.44	I
22		6825.98	I
70		6848.10	I
55		6865.13	I
27		6879.98	I
22		7001.40	I
12		7058.55	I
12	h	7065.04	I
11		7070.99	II
18		7101.27	I
8		7109.67	I
11		7155.40	II
5		7161.91	I
14	h	7197.00	I
7	h	7264.82	II
7	h	7283.95	I
14		7329.73	II
18		7355.37	I
11		7356.34	I
18		7428.67	I
55		7459.55	I
9		7460.42	I
120		7469.51	I
22		7532.34	I
6	h	7539.18	I
27		7556.26	I
6	h	7574.21	I
5	h	7590.51	I
11		7597.33	I
6		7607.23	I
11		7613.52	I
6		7623.48	I
16	h	7645.67	I
8		7650.63	I
22		7654.45	II
12	h	7658.05	I
22		7659.25	I
4		7665.64	I
35		7680.01	I
9		7722.14	I
8		7726.19	II
11	h	7747.44	I
22		7754.63	I
4		7762.16	I
9		7796.69	I
35		7797.47	I
9		7838.80	I
11		7844.00	I
16		7847.55	I
5		7875.36	I
5		7879.36	I
18		7899.55	I
8		7913.08	I
35		7921.85	I
30		7937.84	I
8		7952.93	I
12		7964.51	I
8		7979.03	II
8		7980.87	I
5		8023.03	I
12		8035.91	I
12		8181.85	I
35		8312.82	I
18		8328.57	II
5		8367.58	II
55		8409.90	I
11		8466.18	II
35		8472.42	I
14		8517.71	II
18		8521.37	II
22		8768.64	I
11	h	8776.63	II
9		8866.84	II

Er III
Ref. 301—J.R.

Intensity	Wavelength (Air)	
2	2165.26	III
3	2190.77	III
10	2198.15	III
1	2223.98	III
4	2232.35	III
60	2235.28	III
8	2245.60	III
2	2255.95	III
80	2269.36	III
600	2277.65	III
100	2309.19	III
40	2358.69	III
10	2358.79	III
50	2359.33	III
200	2367.64	III
20	2375.50	III
10	2377.07	III
80	2381.25	III
20	2381.40	III
40	2381.75	III
6	2391.96	III
60	2393.08	III
5	2393.60	III
250	2396.40	III
10	2398.91	III
2	2402.75	III
80	2404.58	III
100	2410.47	III
200	2419.81	III
200	2422.47	III
40	2431.51	III
60	2464.60	III
2	2492.04	III
100	2508.59	III
2	2531.03	III
100	2532.36	III
8	2536.76	III
80	2540.91	III
3	2543.31	III
10	2545.95	III
50	2557.22	III
80	2570.74	III
40	2580.02	III
2	2589.55	III
80	2590.72	III
20	2591.56	III
200	2591.83	III
20	2598.39	III
3	2599.18	III
100	2603.62	III
40	2604.91	III
25	2614.53	III
30	2617.64	III
2	2618.40	III
2	2618.94	III
8	2625.19	III
20	2626.37	III
4	2637.52	III
300	2647.77	III
5	2651.49	III
25	2683.10	III
400	2723.29	III
100	2738.53	III
500	2739.27	III
8	2741.41	III
80	2746.03	III
6	2752.20	III
80	2756.20	III
400	2759.23	III
150	2761.92	III
60	2762.66	III
15	2767.11	III
100	2768.72	III
60	2772.07	III
60	2774.80	III
20	2775.55	III
2	2780.60	III
80	2783.11	III
500	2792.54	III
10	2804.10	III
100	2805.87	III
6	2808.44	III
50	2824.75	III
150	2830.34	III
1	2831.95	III
8	2833.03	III
8	2845.29	III
60	2846.08	III
6	2849.63	III
1	2869.52	III
8	2878.24	III
1	2955.93	III
1	2958.63	III
1000	3055.10	III
1000	3070.40	III
500	3100.40	III
1500	3166.25	III
3	3172.47	III
1	3173.45	III
50	3175.74	III
400	3214.95	III
2000	3301.23	III
8	3341.00	III
200	3480.54	III
8	3592.96	III
600	3715.67	III
200	3739.43	III
4000	3816.78	III
600	3962.87	III
40	4009.70	III
2	4088.58	III
1000	4288.18	III
40000	4290.06	III
300	4338.24	III
20000	4386.86	III
30	4612.93	III
15000	4735.56	III
2000	4783.12	III
8	4876.07	III
8000	5903.30	III

EUROPIUM (Eu)
Z = 63

Eu I and II
Ref. 1—C.H.C.

Intensity		Wavelength (Air)	
21		2499.39	II
26		2554.78	II
26		2559.18	II
160		2564.17	II
110		2568.17	II
26		2574.76	II
230		2577.14	II
26		2581.86	II
26		2604.61	I
30		2635.50	II
1000		2638.77	II
380		2641.27	II
40		2653.61	II
640		2668.34	II
110		2673.42	II
250		2678.29	II
250		2685.66	II
550		2692.03	II
700		2701.14	II
800		2701.90	II
240		2705.28	II
180		2709.99	I
700		2716.98	II
70		2723.96	I
4200		2727.78	II
190		2729.33	II
380		2729.44	II
50		2731.37	I
40		2732.61	I
80		2735.25	I
160		2740.02	II
70		2743.28	I
120		2744.26	II
40		2745.61	I
70		2747.29	II
80		2747.83	I
90		2752.17	II
480		2781.89	II
1900		2802.84	II
220		2811.75	II
30		2813.08	II
3400		2813.94	II
550		2816.18	II
2000		2820.78	II
400	Cw	2828.72	II
120		2829.30	II
140		2833.26	II
80		2843.96	II
60		2852.05	II
260		2859.67	II
280		2862.57	II
25		2864.42	II
60		2876.06	II
100		2878.87	I
80		2887.85	II
200		2892.54	I
140		2893.03	II
360		2893.83	II
3200		2906.68	II
160		2908.99	I
30		2917.44	II
850		2925.04	II
60		2947.29	II
200	Cw	2952.68	II
30		2958.91	I
35		2959.47	II
260		2960.21	II
300		2991.33	II
35		2995.22	II
40		3006.26	II
35		3022.15	I
30		3040.77	II
320	Cw	3054.94	II
120		3058.98	I
35		3069.11	II
35		3076.07	II
220		3077.36	II
35		3089.35	II
120		3097.45	II
320		3106.18	I
950		3111.43	II
120		3130.73	I
40		3132.16	I
45		3149.88	II
85		3173.61	II
40		3185.54	I
420		3210.57	I
1000		3212.81	I
420		3213.75	I
45		3235.13	I
95		3241.40	I
45		3246.03	I
45		3247.32	II
100		3247.55	I
100		3266.39	II
150		3272.77	II
210		3277.78	II
150		3301.95	II
45		3304.50	II
140		3308.02	II
140		3313.33	II
65		3319.89	II
95		3321.86	II
85		3322.26	I
950		3334.33	II
45		3338.75	II
110		3350.40	I
40		3351.56	II
40		3354.38	II
45		3367.64	II
140		3369.06	II
65		3380.25	II
75		3390.78	II
190		3391.99	II
280		3396.58	II
45		3419.84	II
65		3423.09	II
150		3425.02	II
45		3426.44	II
45		3435.05	II
65		3435.20	II
40		3435.72	II
45		3440.82	II
150		3441.00	II
45		3445.18	II
85		3457.05	I
45		3457.56	II
130		3461.38	II
85		3467.88	I
75		3477.07	I
75	h	3505.30	II
470	Cw	3521.09	II
75		3531.15	II
45		3533.22	II
65		3538.08	II
150		3542.15	II
85		3543.85	II
180		3552.52	II
75		3589.27	I
45		3591.31	II
150		3603.20	II
75		3611.57	II
45		3616.15	II
95		3622.54	II
95		3632.18	II
45		3673.19	II
45		3674.63	II
45		3678.26	II
6400		3688.42	II
60		3710.87	II
95		3713.45	II
95		3714.90	II
35		3716.94	II
35		3717.69	II
40		3719.16	I
20000	Cw	3724.94	II
45		3729.68	II
45		3729.74	II
21		3732.20	I
45		3738.08	II
350		3741.31	II
100		3743.56	II
260		3761.12	II
95		3765.93	II
40		3774.10	I
60		3781.40	II
40		3788.76	II
45		3791.50	II
130		3799.01	II
70		3801.36	II
95		3807.54	II
120		3811.33	I
120		3815.50	II
39000	Cw	3819.67	II
120		3826.68	II
140		3844.23	II
190		3865.57	I
45		3872.72	II
70		3877.27	II
150		3884.75	I
23		3896.78	I
23		3900.18	I
70		3900.51	II
28000	Cw	3907.10	II
45		3915.24	II
45		3916.00	I
230		3917.29	I
23		3917.70	II
40		3918.52	I
100		3919.09	II
40		3928.87	II
32000	Cw	3930.48	II
55		3941.56	II

Europium (Cont.)

Intensity	Note	Wavelength	Species
30	h	3942.21	II
60		3942.94	II
120		3943.08	II
30		3944.59	II
30		3945.67	II
30		3949.13	II
60		3949.60	I
45		3950.76	II
55		3951.33	II
60		3955.75	I
40		3957.92	II
30		3963.61	I
120		3964.90	II
150		3966.59	II
45		3967.18	I
30000	Cw	3971.96	II
60		3978.42	I
30		3979.63	II
55	h	3986.60	I
40		3988.24	II
30		3993.93	II
55		3995.98	II
60		4003.71	II
180		4011.69	II
150		4017.58	II
120		4039.19	II
45	h	4078.24	I
120		4085.38	II
75		4096.80	II
60		4106.88	I
90	h	4112.04	II
45		4119.30	II
75		4127.28	I
33000	Cw	4129.70	II
30		4136.59	II
40		4137.07	I
30		4141.02	II
60		4141.72	II
30		4151.52	II
45		4151.64	II
30		4157.72	I
110		4172.80	II
30		4175.16	II
110		4182.22	I
40		4196.18	II
60000	Cw	4205.05	II
45		4221.08	II
40		4223.88	II
90	h	4227.40	II
75		4229.33	II
75		4232.45	II
90		4237.51	II
45		4238.69	II
45		4244.74	I
45		4247.06	II
45		4253.80	II
30		4270.24	II
150		4298.73	I
90		4329.36	I
75		4329.97	I
60		4330.61	II
40		4331.18	I
90		4337.68	I
240		4355.09	II
27		4361.57	II
55		4369.47	II
45	h	4372.20	II
75		4383.17	II
90		4387.88	I
21	h	4405.27	II
55		4407.07	II
18		4419.66	II
120		4434.81	II
14000	Cw	4435.56	II
75	h	4464.97	II
24		4485.15	II
3000		4522.57	II
45	h	4535.59	I
11000		4594.03	I
21		4602.63	I
9800		4627.22	I
8300		4661.88	I
30		4713.59	I
27		4740.50	I
45		4792.59	I
40	h	4829.30	I
60		4830.33	I
40	h	4840.47	I
60	h	4849.64	I
110		4867.62	I
40	h	4884.05	I
90		4894.68	I
60		4900.86	I
150		4907.18	I
180		4911.40	I
55		4953.52	I
55		4960.21	I
55		4962.55	I
45		4975.76	I
180		5013.17	I
170		5022.91	I
110		5029.54	I
90		5033.55	I
75		5067.95	I

Intensity	Note	Wavelength	Species
75	h	5092.69	I
90	h	5096.44	I
170		5114.37	I
90		5124.77	I
170		5129.10	I
90		5130.08	I
210		5133.52	I
270		5160.07	I
210		5166.07	I
60		5193.74	I
200		5199.85	I
110		5200.96	I
120		5206.44	I
750		5215.10	I
300		5223.49	I
120		5239.24	I
200		5266.40	I
390		5271.96	I
110		5272.48	I
150		5282.82	I
55		5287.25	I
60		5289.25	I
120		5291.26	I
60		5293.68	I
120		5294.64	I
90		5303.85	I
30	h	5350.41	I
75	h	5351.69	I
40		5352.84	I
90		5355.10	I
540		5357.61	I
60		5360.83	I
120		5361.61	I
110		5376.94	I
120		5392.94	I
450		5402.77	I
45		5405.33	I
45		5411.86	I
55		5421.07	I
90		5426.94	I
40		5443.56	I
380		5451.51	I
260		5452.94	I
40		5457.62	I
90		5472.32	I
120		5488.65	I
45		5495.20	I
15		5500.83	I
120		5510.52	I
30		5526.63	I
30		5533.25	I
30		5542.54	I
200		5547.44	I
150		5570.33	I
200		5577.14	I
75		5579.63	I
120		5580.03	I
90		5586.24	I
75		5586.83	I
18		5592.25	I
18		5599.80	I
18		5605.86	I
40		5618.81	I
60		5622.44	I
75		5632.54	I
210		5645.80	I
15		5651.11	I
60		5673.85	I
27		5681.10	I
27		5684.24	I
60		5730.87	I
60		5739.00	I
330		5765.20	I
180		5783.69	I
15		5792.72	I
60		5800.27	I
170		5818.74	II
600	Cw	5830.98	I
27		5845.77	I
27		5860.97	I
15		5864.77	I
90		5872.98	II
15		5895.31	I
27		5902.97	I
12		5909.94	I
75		5915.74	I
12		5925.30	I
27		5926.52	I
45		5942.72	I
27		5953.49	I
27		5953.84	II
30		5954.28	I
90		5963.76	I
330		5966.07	II
480	Cw	5967.10	I
15	h	5968.43	I
30		5971.69	I
170		5972.95	I
15		5980.47	I
27		5983.14	I
27		5983.78	I
240		5992.83	I
60		6004.36	I
15	h	6005.61	I

Intensity	Note	Wavelength	Species
60	h	6012.20	I
110		6012.56	I
60		6015.58	I
420		6018.15	I
60		6023.15	I
170		6029.00	I
60		6044.66	I
420		6049.51	II
140		6057.36	?
90		6075.58	I
30		6077.38	I
240		6083.84	I
240		6099.35	I
60		6108.15	I
120		6118.78	I
60		6124.67	I
330		6173.05	II
110		6178.76	I
260	Cw	6188.13	I
140		6195.07	I
15	h	6207.60	I
15		6230.51	I
90	h	6233.73	I
55		6250.47	I
240		6262.25	I
55		6266.95	I
15	h	6285.95	I
60		6291.34	I
170		6299.77	I
230		6303.41	II
24	h	6313.78	I
15		6318.58	I
75		6335.62	I
120	Cw	6350.04	I
60		6355.89	I
60		6369.25	I
55		6382.73	I
75		6383.86	I
120	Cw	6400.93	I
40		6406.11	I
180		6410.04	I
140		6411.32	I
55		6428.29	I
830		6437.64	II
18		6439.93	I
120		6457.96	I
12		6470.70	I
18		6483.02	I
45		6501.55	I
60		6519.59	I
15		6522.72	I
8	h	6549.12	I
75		6567.87	I
45		6593.79	I
18	h	6603.55	I
1400		6645.11	II
26		6685.21	I
95		6693.96	I
7	h	6701.06	I
12	h	6710.45	I
30		6744.88	I
30	h	6782.54	I
14	h	6787.48	I
140		6802.72	I
35		6816.06	I
11	h	6834.30	I
17		6840.93	I
17	h	6844.83	I
14	h	6847.04	I
360		6864.54	I
21		6898.21	I
60	h	6903.67	I
14	h	6910.17	I
30	h	6914.82	I
120		7040.20	I
12		7074.54	I
330		7077.10	II
100		7106.48	I
6		7164.66	I
30		7175.55	I
570		7194.81	II
570		7217.55	II
11	h	7224.68	I
15		7258.72	I
30		7262.77	I
11	h	7281.53	I
6	h	7297.56	I
540		7301.17	II
11		7310.46	I
12		7313.63	I
55	Cw	7336.18	I
4		7346.25	I
4		7356.65	I
11		7362.25	I
55	Cw	7369.60	I
720		7370.22	II
4		7387.36	I
12		7389.16	I
11	h	7404.41	I
300		7426.57	II
21	h	7436.59	I
8		7470.53	I
5	h	7491.00	I
50	Cw	7528.70	I

Intensity	Note	Wavelength	Species
5	h	7533.02	I
6		7547.32	I
160		7583.91	I
60	Cw	7742.57	I
70		7746.19	I
8	h	7803.32	I
8		7818.21	I
35		7887.99	I
7		8015.47	I
24	Cw	8209.80	I
15	Cw	8226.81	I
6	h	8464.71	I
21	Cw	8642.67	I
7		8727.77	I
6		8782.46	I
12	Cw	8790.88	I
18		8870.30	I

Eu III
Ref. 312—J.R.

Intensity	Wavelength Air	
10	2073.40	III
10	2093.50	III
30	2124.69	III
10	2167.12	III
10	2173.59	III
10	2184.68	III
10	2190.59	III
10	2194.81	III
20	2211.85	III
20	2212.63	III
20	2214.66	III
10	2215.34	III
30	2217.23	III
30	2219.33	III
20	2219.42	III
10	2223.13	III
10	2235.17	III
10	2240.14	III
10	2261.88	III
20	2265.74	III
10	2269.39	III
20	2276.85	III
40	2291.62	III
20	2304.37	III
10	2311.92	III
10	2327.69	III
10	2334.56	III
10	2336.96	III
10	2339.84	III
10	2343.10	III
10	2346.83	III
10	2347.64	III
10	2350.38	III
200	2350.51	III
10	2352.28	III
10	2357.87	III
10	2359.08	III
10	2360.65	III
20	2363.76	III
20	2368.04	III
20	2374.08	III
10	2375.20	III
4000	2375.46	III
10	2376.42	III
20	2377.23	III
10	2381.81	III
10	2383.62	III
20	2387.29	III
20	2389.11	III
10	2389.98	III
10	2391.11	III
20	2391.90	III
20	2392.59	III
10	2394.66	III
20	2395.62	III
20	2398.79	III
20	2401.00	III
20	2402.34	III
20	2404.08	III
10	2406.14	III
20	2407.30	III
20	2408.32	III
10	2409.63	III
10	2410.08	III
40	2412.02	III
20	2412.96	III
20	2413.26	III
10	2413.41	III
10	2419.11	III
10	2419.25	III
10	2419.58	III
30	2422.00	III
10	2422.90	III
10	2425.33	III
50	2425.68	III
40	2427.67	III
40	2429.32	III
10 d	2429.66	III
10	2430.04	III

Europium (Cont.)

Intensity		Wavelength	
10		2431.49	III
10		2431.76	III
10		2432.55	III
10		2433.65	III
10		2434.19	III
100	d	2435.14	III
20		2436.39	III
10		2436.77	III
10		2438.83	III
10		2440.26	III
50		2440.67	III
1000		2444.38	III
4000		2445.99	III
30	h	2446.43	III
20		2448.57	III
20		2451.24	III
10		2451.73	III
30		2455.22	III
10		2461.79	III
30		2463.30	III
30		2466.67	III
40		2470.51	III
10		2474.94	III
10		2476.24	III
20		2476.45	III
10		2477.78	III
20		2480.02	III
20		2483.29	III
10		2486.92	III
10		2488.91	III
10		2490.50	III
10		2491.08	III
10		2492.48	III
10		2496.92	III
10		2499.17	III
2000		2513.76	III
20		2517.94	III
200		2522.14	III
10		2539.14	III
10		2548.30	III
20		2548.59	III
10		2554.50	III
10		2558.07	III
10		2560.36	III
10		2590.71	III
20		2594.76	III
10		2596.34	III
10		2604.44	III
30		2608.34	III
30		2610.09	III
50	c	2616.11	III
20		2616.26	III
10		2616.33	III
10		2616.35	III
20		2620.79	III
20		2623.33	III
10		2626.98	III
20	c	2628.46	III
10		2628.82	III
10		2631.98	III
30	c	2642.27	III
20		2645.22	III
20	c	2650.93	III
10		2653.19	III
10		2655.09	III
20		2662.24	III
20	c	2666.86	III
20	c	2668.21	III
40	c	2676.09	III
20	c	2683.21	III
90		2686.13	III
20	c	2687.74	III
40	c	2693.51	III
10		2694.80	III
20		2699.87	III
20	c	2700.78	III
10	c	2708.25	III
20	c	2708.84	III
10		2712.08	III
50	c	2720.67	III
20	c	2725.54	III
10		2743.94	III
10		2752.68	III
20		2755.12	III
20		2757.75	III
20	c	2760.21	III
10		2761.72	III
10	c	2766.26	III
20	c	2768.38	III
10	c	2768.54	III
10		2769.71	III
20	c	2780.48	III
20		2792.51	III
10		2808.09	III
10		2817.58	III
20	c	2839.56	III
20		2844.99	III
10	c	2848.44	III
10		2850.39	III
10	c	2892.60	III
40		2912.23	III
40	c	2912.64	III
10		2913.04	III

Intensity		Wavelength	
10		2928.91	III
20	c	2931.00	III
10	c	2950.20	III
20		2956.74	III
10		2956.90	III
10	c	2972.30	III
30	c	2982.29	III
20	c	3000.11	III
20		3006.37	III
20	c	3013.28	III
10	c	3018.43	III
20	c	3022.08	III
50	c	3022.69	III
20	c	3023.40	III
100	c	3023.93	III
10		3025.32	III
10	c	3026.09	III
200	c	3026.79	III
50	c	3029.92	III
20	c	3031.24	III
10	c	3034.04	III
20	c	3036.98	III
10		3038.64	III
10		3039.05	III
20	c	3039.98	III
10		3054.07	III
10	c	3054.97	III
20		3076.43	III
10	c	3089.09	III
10		3105.25	III
10		3109.67	III
10		3129.31	III
10		3142.54	III
50	c	3171.00	III
10		3178.08	III
20	h	3178.87	III
50	c	3183.78	III
10	h	3191.46	III
20	c	3194.34	III
10		3206.30	III
10		3208.95	III
10	h	3213.84	III
10		4837.98	III
50		6666.35	III
30		7331.86	III
20		7690.44	III
10		8079.07	III

FLUORINE (F)
Z = 9

F I and II
Ref. 169, 224—G.V.S.

Intensity	Wavelength (Vacuum)	
30	375.30	II
30	380.90	II
40	407.04	II
50	430.91	II
40	431.55	II
40	435.64	II
70	457.18	II
40	471.95	II
60	472.00	II
50	472.71	II
40	473.02	II
90	484.60	II
50	513.64	II
70	514.94	II
70	546.85	II
60	547.87	II
50	548.32	II
40	548.52	II
90	605.67	II
80	606.29	II
100	606.80	II
70	606.92	II
80	607.47	II
90	608.06	II
15	780.39	I
10	780.52	I
10	782.38	I
12	791.88	I
10	792.54	I
10	794.42	I
150	806.96	I
125	809.60	I
500	951.87	I
1000	954.83	I
750	955.55	I
500	958.52	I
20	972.40	I
350	973.90	I
100	976.22	I
40	976.51	I
100	977.75	I
40	1129.76	II
40	1327.06	II
50	1328.11	II

Intensity		Wavelength	
40		1333.59	II
50		1343.60	II
40		1344.04	II
50		1400.61	II
40		1407.14	II
60		1493.09	II
50		1493.24	II
40		1493.31	II
40		1702.13	II
40		1744.75	II
50		1745.55	II
60		1747.39	II

Air

Intensity		Wavelength	
100		2556.11	II
100		2871.40	II
120		3059.99	II
140		3153.49	II
170		3202.76	II
140		3264.08	II
140		3414.65	II
150		3416.45	II
140		3416.80	II
160		3417.00	II
160		3472.96	II
150		3473.31	II
170		3474.78	II
190		3501.39	II
200		3501.45	II
200		3501.57	II
180		3502.84	II
200		3502.96	II
210		3503.11	II
170		3505.37	II
200		3505.52	II
220		3505.63	II
160		3522.89	II
150		3536.87	II
160		3541.77	II
160		3590.52	II
6		3594.10	I
170		3598.69	II
180		3601.39	II
111		3601.74	II
12		3668.17	I
180		3704.53	II
160		3710.35	II
160		3720.67	II
140		3805.83	II
270		3847.09	II
260		3849.99	II
250		3851.67	II
5		3898.48	I
190		3898.83	II
180		3901.93	II
170		3903.82	II
8		3930.69	I
5		3934.26	I
5		3948.56	I
150		3972.04	II
160		3972.67	II
170		3974.78	II
240		4024.73	II
220		4025.01	II
230		4025.49	II
160		4083.91	II
190		4103.07	II
170		4103.22	II
200		4103.51	II
180		4103.71	II
170		4103.87	II
170		4109.16	II
160		4116.54	II
150		4119.21	II
140		4207.15	II
170	h	4225.16	II
150	h	4244.12	II
200		4246.23	II
190		4246.39	II
180		4246.59	II
170		4246.77	II
160		4246.84	II
170	h	4275.36	II
160	h	4277.53	II
160	h	4278.93	II
200		4299.17	II
160		4446.53	II
170		4446.72	II
180		4447.19	II
140		4734.38	II
170		4859.39	II
160		4933.26	II
6		4960.65	I
140		5002.00	II
150		5173.25	II
15		5230.41	I
12		5279.01	I
18		5540.52	I
12		5552.43	I
10		5577.33	I
160		5589.27	II
20		5624.06	I

Intensity		Wavelength	
12		5626.93	I
15		5659.15	I
40		5667.53	I
90		5671.67	I
18		5689.14	I
25		5700.82	I
25		5707.31	I
12		5950.15	I
25		5959.19	I
70		5965.28	I
50		5994.43	I
150		6015.83	I
80		6038.04	I
900		6047.54	I
100		6080.11	I
800		6149.76	I
400		6210.87	I
13000		6239.65	I
140		6247.90	II
10000		6348.51	I
8000		6413.65	I
450		6569.69	I
300		6580.39	I
400		6650.41	I
1800		6690.48	I
400		6708.28	I
7000		6773.98	I
1500		6795.53	I
9000		6834.26	I
50000		6856.03	I
8000		6870.22	I
15000		6902.48	I
6000		6909.82	I
4000		6966.35	I
45000		7037.47	I
30000		7127.89	I
130		7179.90	II
15000		7202.36	I
130	h	7211.79	II
1000		7309.03	I
15000		7311.02	I
700		7314.30	I
5000		7331.96	I
10000		7398.69	I
4000		7425.65	I
2200		7482.72	I
2500		7489.16	I
900		7514.92	I
5000		7552.24	I
5000		7573.38	I
7000		7607.17	I
18000		7754.70	I
15000		7800.21	I
300		7879.18	I
500		7898.59	I
350		7936.31	I
300		7956.32	I
80		8016.01	II
1000		8040.93	I
900		8075.52	I
350		8077.52	I
350		8126.56	I
600		8129.26	I
300		8159.51	I
600		8179.34	I
300		8191.24	I
350		8208.63	I
2500		8214.73	I
3000		8230.77	I
500		8232.19	I
1500		8274.62	I
2000		8298.58	I
600		8302.40	I
900		8807.58	I
1000		8900.92	I
300		8912.78	I
350		9025.49	I
400		9042.10	I
350		9178.68	I
200		9433.67	I
25		9505.30	I
12		9662.04	I
25		9734.34	I
15		9822.11	I
12		9902.65	I
80	h	10047.98	II
15		10285.45	I
20		10862.31	I

F III
Ref. 225—G.V.S.

Intensity	Wavelength (Vacuum)	
50 h	230.12	III
50	255.72	III
60	255.77	III
70	255.86	III
70	261.71	III
60	261.75	III
80	263.81	III

Fluorine (Cont.)

<table>
<tr><td>

Int.	Wavelength	
70	279.69	III
80	315.22	III
70	315.54	III
60	315.75	III
100	429.51	III
110	430.15	III
80	430.22	III
90	464.29	III
100	465.11	III
120	508.39	III
120	567.69	III
110	567.75	III
80	630.14	III
90	630.20	III
120	656.12	III
130	656.87	III
140	658.33	III
80	1219.03	III
80	1266.87	III
90	1267.71	III
70	1297.54	III
70	1359.92	III
110	1498.93	III
120	1502.01	III
110	1504.18	III
140	1504.79	III
130	1506.30	III
110	1506.77	III
100	1553.02	III
110	1557.59	III
100	1563.73	III
100	1565.54	III
100	1623.40	III
100	1650.76	III
130	1670.39	III
140	1677.40	III
100	1716.99	III
120	1770.09	III
150	1770.67	III
110	1772.93	III
140	1773.36	III
160	1791.65	III
110	1803.03	III
100	1804.70	III
170	1805.90	III
110	1839.30	III
120	1839.97	III
110	1840.14	III
80	1900.76	III

Air

100	2027.44	III
120	2030.32	III
120	2217.17	III
120	2452.07	III
130	2464.85	III
130	2470.29	III
120	2478.73	III
150	2484.37	III
120	2542.77	III
120	2580.04	III
130	2583.81	III
120	2593.23	III
130	2595.53	III
140	2599.28	III
130	2625.01	III
140	2629.70	III
120	2656.44	III
130	2755.55	III
160	2759.63	III
120	2788.15	III
160	2811.45	III
140	2833.99	III
150	2835.63	III
150	2860.33	III
120	2862.86	III
140	2887.58	III
150	2889.45	III
120	2905.30	III
140	2913.29	III
160	2916.34	III
140	2932.49	III
140	2994.28	III
120 h	2997.21	III
130	2997.53	III
120	2999.47	III
130	3039.25	III
120	3039.75	III
160	3042.80	III
150	3049.14	III
140	3113.62	III
160	3115.70	III
180	3121.54	III
140	3124.79	III
140	3134.23	III
140	3146.99	III
180	3174.17	III
170	3174.76	III
120	3214.00	III
140 h	4420.30	III
120 h	4427.35	III
120 h	4432.32	III

</td><td>

140 h	4479.99	III
150	5012.54	III
160	5110.99	III
140	5753.17	III
120	5761.20	III
150	6091.82	III
140	6125.50	III
130	6233.57	III
140	6363.05	III
120	7336.77	III
130	7354.94	III

F IV
Ref. 68, 226—G.V.S.

Intensity	Wavelength	

Vacuum

30	169.79	IV
30	169.84	IV
30	171.07	IV
40	176.37	IV
40	181.52	IV
40	181.57	IV
30	187.24	IV
50	196.39	IV
60	196.45	IV
50	199.76	IV
50	199.80	IV
50	199.85	IV
50	199.93	IV
50	200.00	IV
70	200.09	IV
60	201.01	IV
70	201.06	IV
60	201.10	IV
80	201.16	IV
60	201.22	IV
90	208.25	IV
70	213.85	IV
70	214.06	IV
70	220.77	IV
60	226.94	IV
50	227.10	IV
60	233.22	IV
50	233.39	IV
70	239.86	IV
70	240.02	IV
90	240.08	IV
70	240.15	IV
70	240.28	IV
70	240.37	IV
100	251.03	IV
60	270.23	IV
140	419.65	IV
150	420.05	IV
160	420.73	IV
150	430.76	IV
130	490.57	IV
160	491.00	IV
50	497.38	IV
60	497.83	IV
70	498.80	IV
140	570.64	IV
140	571.30	IV
150	571.39	IV
160	572.66	IV
140	676.12	IV
130	677.15	IV
150	677.22	IV
130	678.99	IV
160	679.21	IV

Air

40	2171.44	IV
50	2298.29	IV
40	2451.58	IV
50	2456.92	IV
40	2820.74	IV
50	2826.13	IV

F V
Ref. 68, 226—G.V.S.

Intensity	Wavelength	

Vacuum

40	134.54	V
40	147.95	V
50	148.00	V
40	152.51	V
40	158.54	V
40	162.27	V
40	163.50	V
50	163.56	V
90	165.98	V
100	166.18	V
40	174.70	V
50	178.43	V

</td><td>

40	178.59	V
40	182.98	V
40	186.72	V
40	186.79	V
50	186.84	V
40	186.97	V
40	187.01	V
60	190.57	V
70	190.84	V
40	191.97	V
40	205.55	V
100	464.37	V
110	465.37	V
120	465.98	V
100	466.99	V
90	506.16	V
100	508.08	V
70	513.97	V
60	514.08	V
80	524.59	V
90	525.29	V
100	526.30	V
70	647.67	V
100	647.77	V
110	647.87	V
70	647.97	V
130	654.03	V
110	657.23	V
140	657.33	V
60	757.04	V
60	1082.31	V
70	1088.39	V

AIR

10	2229.18	V
20	2252.72	V
20	2450.63	V
10	2461.33	V
10	2693.98	V
10	2702.30	V
10	2703.96	V
20	2707.17	V

FRANCIUM (Fr)
Z = 87

Fr I
Ref. 408 — C.H.C.

Intensity	Wavelength	

Air

| | 7177. | I |

GADOLINIUM (Gd)
Z = 64

Gd I and II
Ref. 1—C.H.C.

Intensity	Wavelength	

Air

100	2468.22	II
55	2471.58	II
35	2485.67	II
70	2487.46	II
110	2488.72	II
55	2493.29	II
35	2496.35	II
45	2499.04	II
28	2543.68	II
28	2586.13	II
28	2661.50	II
70	2720.50	II
430	2750.22	II
460	2764.08	II
40	2768.51	II
320	2769.81	II
230	2770.17	II
21	2770.98	II
45	2778.76	I
45	2779.14	II
440	2781.40	II
70	2787.68	I
390	2791.96	II
100	2794.66	II
930	2796.93	II
60	2808.38	II
750	2809.72	II
160	2810.93	II
45	2814.01	II
300	2833.75	II
35	2836.69	II
70	2837.00	II
560	2840.23	II
140	2841.33	II
40	2853.91	II

</td><td>

60	2856.52	II
19	2859.78	II
120	2862.48	II
60	2865.06	II
40	2866.33	II
40	2871.75	II
460	2881.33	II
40	2882.13	II
130	2885.60	II
35	2907.44	II
170	2910.53	II
60	2913.08	II
45	2918.52	II
95	2923.32	II
35	2924.25	II
35	2928.34	II
35	2947.80	II
70	2948.01	II
35	2952.43	I
35	2955.60	II
70	2960.93	II
130	2963.60	II
80	2965.43	II
29	2972.74	II
560	2980.15	II
35	2983.74	II
40	2991.52	II
95	2993.04	II
1200	2999.04	II
370	3002.86	II
100	3005.09	II
2100	3010.13	II
130	3012.19	II
1900	3027.60	II
120	3028.98	II
2100	3032.84	II
1600	3034.05	II
130	3043.01	I
160	3046.48	II
280	3053.57	II
100	3059.92	I
1000	3068.64	II
560	3072.56	II
640	3076.92	II
150	3077.08	II
2100	3081.99	II
140	3084.01	II
280	3089.95	II
140	3092.06	II
460	3098.64	II
190	3098.90	II
3500	3100.50	II
120	3101.18	II
230	3101.91	II
580	3102.55	II
130	3108.36	II
170	3111.19	I
160	3113.17	II
120	3118.60	II
120	3119.01	I
510	3119.94	II
100	3120.18	II
370	3123.99	II
120	3124.25	II
130	3128.56	II
130	3130.81	II
100	3133.09	II
460	3133.85	II
210	3135.03	II
190	3136.93	I
190	3137.30	II
120	3138.71	I
230	3143.13	II
930	3145.00	II
370	3145.52	II
230	3146.88	II
980	3156.53	II
200	3158.63	I
140	3160.69	II
980	3161.37	II
220	3190.28	I
220	3199.30	I
160	3199.58	I
110	3203.41	II
690	3223.74	II
	3223.78	I
110	3225.46	II
160	3226.32	II
220	3232.78	I
100	3250.19	II
110	3259.25	II
540	3266.73	I
250	3267.64	I
140	3268.34	II
110	3274.18	II
110	3279.53	II
100	3281.61	II
250	3282.25	I
	3282.30	II
430	3291.48	I
370	3292.21	II
430	3294.08	I
330	3313.73	II
200	3315.59	II

</td></tr>
</table>

Gadolinium (Cont.)

Int.	λ		Int.	λ		Int.	λ		Int.	λ	
430	3330.34	II	6100	3646.19	II	1200	3957.67	II	1700	4251.73	II
1400	3331.38	II	310	3649.44	II	750	3959.44	II	860	4253.37	II
830	3332.13	II	450	3650.95	II		3959.52	II	650	4253.61	II
1100	3336.18	II	620	3652.54	II	220	3963.66	II	810	4260.12	I
590	3345.98	II	3900	3654.62	II	590	3966.28	I	1600	4262.09	I
200	3350.10	II	3100	3656.15	II	590	3968.26	II	650	4266.60	I
5400	3350.47	II	210	3658.19	I	750	3969.00	I	470	4267.00	I
220	3357.61	I	1400	3662.26	II	270	3969.29	II	300	4274.17	I
270	3358.43	II	2700	3664.60	II	450	3971.75	II	910	4280.49	II
4300	3358.62	II	2000	3671.20	II	390	3972.71	I	430	4285.82	I
780	3360.71	II	1000	3674.05	I	590	3973.98	II	300	4286.12	I
5400	3362.23	II	350	3679.21	II	300	3974.81	I	540	4296.08	II
270	3364.24	II	2000	3684.13	I	750	3979.33	I	220	4297.17	II
200	3365.59	II	720	3686.33	II	450	3987.21	II	430	4299.29	I
220	3374.69	II	3100	3687.74	II	470	3987.84	II	1100	4306.34	I
220	3379.76	II	210	3694.03	II	320	3992.69	I	260	4309.29	I
220	3380.52	II	2000	3697.73	II	220	3993.21	II	1800	4313.84	I
1100	3392.53	II	1300	3699.73	II	650	3994.16	II	520	4314.40	I
540	3395.12	II	2700	3712.70	II	700	3996.37	II	520	4316.05	II
220 d	3397.11	I	2000	3713.57	I	320	3997.76	II	370	4320.32	I
	3397.32	I	1400	3716.36	II	470	4001.26	II	750	4321.11	II
200	3399.41	II	2000	3717.48	I	260	4004.94	II		4321.20	I
540	3399.99	II	1800 d	3719.45	II	320	4008.33	I	2600 d	4325.57	II
540	3402.07	II		3719.53	II	300	4008.91	II		4325.69	I
200	3406.92	I	250	3722.07	II	300	4013.80	II	1900	4327.12	I
1100 d	3407.56	II	430	3725.47	II	200	4015.58	I	370	4329.58	I
	3407.61	II	1500	3730.84	II	300	4017.25	II	340	4330.61	II
250	3409.30	II	270	3732.32	I	430	4017.71	I	240	4331.38	I
220	3411.02	I	230	3732.45	II	300	4019.73	I	450	4341.28	I
220	3413.27	II	230	3732.67	I	300	4022.33	II	910	4342.18	II
1400	3416.95	II	510	3733.08	II	1100	4023.14	I	1000	4344.30	II
1400	3418.73	II	490	3739.76	I	810	4023.35	I	2200	4346.46	I
6900	3422.47	II	330	3740.02	II	220	4027.61	I	910	4346.62	I
390	3422.75	II	4500	3743.47	II	1100	4028.15	I	220	4347.31	II
1100	3423.90	I	620	3744.83	I	860	4030.88	I	300	4369.77	II
	3423.92	II	230	3757.74	II	700	4033.49	I	970	4373.83	I
830	3424.59	II	1000	3757.94	I	340	4035.40	I	280	4392.06	I
390	3425.93	II	1400	3758.31	II	260	4036.84	I	1400	4401.86	I
220	3428.47	II	820	3759.00	II	1400	4037.33	II	520	4403.14	I
690	3432.99	II	620	3760.71	II	700	4037.90	II	260	4406.67	II
1700	3439.21	II	290	3760.92	II	410	4043.71	I	260	4408.25	II
830	3439.78	II	870	3762.20	I	1600	4045.01	I	220	4409.25	I
2700	3439.99	II	210	3763.33	II	270	4046.84	II	520	4411.16	I
390	3449.62	II	370	3764.20	II	270	4047.09	I	860	4414.16	I
1600	3450.39	II	870	3767.04	II	270	4049.20	I	700	4414.73	I
1100	3451.23	II	8700	3768.39	II	1300	4049.43	II	340	4419.03	II
540	3454.14	II	620	3769.45	II	2200	4049.86	II	1400	4422.41	I
880	3454.90	II	1400	3770.69	II	270	4050.37	I	1100	4430.63	I
200	3455.27	I	250	3771.26	T	910	4053.23	II	240 d	4436.10	I
200	3457.05	II	210	3773.45	I	2600	4053.64	I		4436.22	II
220	3461.95	II	210	3776.83	I	810	4054.72	I	300	4464.74	I
220	3463.00	II	1000	3782.34	II	2600	4058.22	I	300	4466.55	II
2700	3463.98	II	2900	3783.05	I	650	4059.88	I		4466.60	I
330	3466.95	II	1100	3787.56	II	270	4061.30	II	520	4467.08	I
1700	3467.27	II	200	3790.63	I	650	4062.59	II	700	4474.13	I
1700	3468.99	II	770	3791.17	II	1900	4063.39	II	860	4476.12	I
1400	3473.22	II	490	3792.39	II	540	4063.59	II	220	4478.80	II
2200	3481.28	II	5100	3796.37	II	260	4066.04	I	280	4481.06	II
1700	3481.80	II	720	3801.29	II	520	4068.35	I	220	4483.33	II
490	3482.60	II	210	3804.39	I	260	4068.74	I	220	4484.70	I
220	3486.20	I	210	3805.09	II	750	4070.20	II	280	4486.90	I
980	3491.95	II	560	3805.52	II		4070.39	II	500	4497.13	I
1700	3494.40	II	3700	3813.97	II	650	4073.20	II	220	4497.32	I
1400	3505.51	II	430	3814.74	II	300	4073.76	II	430	4506.21	I
780	3512.22	II	770	3816.64	II	1300	4078.44	II	140	4506.33	II
1100	3512.50	II	430	3818.75	II	2800	4078.70	II	140	4514.50	II
830	3513.65	I	350	3826.05	II	520	4083.70	I	1100	4519.66	I
980	3524.20	II	230	3827.33	II	1500	4085.56	II	300	4522.82	II
430	3528.54	II	230	3829.46	II	260	4087.69	II	150	4524.12	I
540	3542.77	II	370	3831.80	II	650	4090.41	I	910	4537.81	I
4300	3545.80	II	210	3832.97	I	1100	4092.71	I	220	4540.02	II
3900	3549.36	II	330	3834.99	II	260	4093.72	I	300	4542.03	I
1400	3557.05	II	970	3836.91	II	260	4094.48	II	240	4548.00	I
540	3558.19	II	1000	3839.64	II	2600	4098.61	II	120	4558.08	II
430	3558.47	II	1200	3842.20	II	520	4098.90	II	130	4573.81	I
200	3564.05	II	1400	3843.28	I	650	4100.26	I	260	4575.91	I
690	3571.93	II	1400	3844.58	II	390	4111.44	II	280	4579.59	I
330	3574.74	II	3300	3850.69	II	2200	4130.37	II	410	4581.29	I
390	3578.36	II	5100	3850.97	II	270	4131.48	II	130	4582.53	II
980	3581.91	II	4300	3852.45	II	1100	4132.28	II	410	4583.07	I
5400	3584.96	II	470	3855.56	II	750	4134.16	I	160	4586.99	I
540	3590.47	II	250	3863.05	II	410	4137.10	II	220	4596.98	II
1100	3592.71	II	1600	3866.99	I	280	4148.86	I	320	4597.91	II
200	3593.44	II	250	3873.57	I	540	4162.73	II	410	4598.90	I
540	3600.96	II	220	3875.46	II	280	4163.09	II	340	4601.05	II
1100	3604.87	I	1500	3894.70	II	280	4167.16	II	240	4602.93	I
270	3605.26	II	450	3895.79	II		4167.27	I	520	4614.50	I
250	3605.66	II	750	3902.40	II	2400	4175.54	I	140	4624.42	I
830	3608.75	II	300	3902.71	I	2400	4184.25	II	430	4636.64	I
830	3610.76	II	240	3904.29	I	2200	4190.78	I	110	4639.00	II
220	3610.91	II	450	3905.65	I	750	4191.07	II	170	4640.04	I
540	3613.39	II	2200	3916.51	II	750	4191.63	I	170	4646.00	I
270 d	3614.21	II	450	3923.25	II	450	4197.68	II	170	4647.64	I
	3614.42	I	1200	3934.79	II	590	4204.86	II	170 d	4648.59	I
430	3617.16	II		3934.82	II	1300	4212.00	II		4648.70	I
390	3620.46	II	220	3935.38	I	970	4215.02	II	430	4653.54	I
270	3624.89	II	450	3941.80	I	650	4217.20	II	140 h	4670.87	I
250	3629.51	II	590	3942.63	I	320	4225.03	I	170	4679.18	I
330	3634.76	II	270	3943.24	I	4800	4225.85	I	260	4680.04	I
220	3639.05	II	220	3943.62	I	220	4227.14	II	430	4683.33	I
250	3640.18	II	1400	3945.54	I	220	4229.80	II	140	4688.12	I
330	3641.39	II	300	3952.00	II	650	4238.78	II	700	4694.33	I
870	3645.62	II	590	3953.37	I	200	4246.57	II	170	4695.49	I

Gadolinium (Cont.)

Intensity		Wavelength	Spectrum
430		4697.42	I
170		4703.13	I
200		4709.78	I
110		4721.46	I
150		4728.47	II
220		4732.60	II
260		4735.75	I
410		4743.65	I
110		4745.82	I
320		4758.70	I
110		4760.74	I
130		4763.82	I
470		4767.24	I
180		4781.92	I
300		4784.62	I
110		4786.75	I
140		4801.05	II
220		4807.45	I
320		4821.69	I
130		4835.26	I
110		4848.10	I
110		4862.59	I
170		4865.02	II
120		4871.50	I
280		4934.12	I
220		4938.61	I
110		4952.47	I
130		4958.79	I
65		5010.82	II
55		5011.74	I
750		5015.04	I
55		5023.13	II
65		5031.29	II
75		5039.09	I
65		5050.88	II
55		5073.74	I
55		5082.80	I
95		5092.25	I
65		5096.06	II
130		5098.38	II
55		5100.94	II
910		5103.45	I
180		5108.91	II
120		5125.56	II
65		5130.28	II
65		5135.59	I
75		5136.04	I
85		5140.84	II
75		5141.50	I
75		5142.68	I
860		5155.84	I
55		5156.76	II
75		5158.48	I
75		5163.70	I
55		5164.54	II
190		5176.28	II
55		5187.24	II
55		5187.88	I
55		5191.08	II
410		5197.77	I
55		5210.49	II
85		5217.48	I
280		5219.40	I
75		5220.30	II
130		5233.93	I
65		5246.87	I
320		5251.18	I
120		5252.14	II
85		5254.75	I
140		5255.80	I
65		5268.78	I
55		5272.91	I
55		5282.48	I
280		5283.08	I
280		5301.67	I
220		5302.76	I
55		5306.70	I
280		5307.30	I
130		5321.50	I
280		5321.78	I
110		5327.32	I
65		5328.30	I
170		5333.30	I
55		5337.53	I
300		5343.00	I
85		5345.13	I
75		5345.68	I
200		5348.67	I
300		5350.38	I
240		5353.26	I
55		5361.66	I
95		5365.38	I
95		5369.92	I
150		5370.63	I
85		5389.50	I
85		5413.20	I
85		5415.69	I
65		5453.46	I
55		5583.68	II
55	d	5591.85	I
190		5617.91	I
65		5629.55	I
110		5632.25	I

Intensity		Wavelength	Spectrum
260		5643.24	I
55	B1	5680.89	G
390		5696.22	I
95		5701.35	I
65		5709.42	I
120		5733.86	II
85		5746.36	I
85	d	5754.17	I
75		5776.02	I
240		5791.38	I
65	h	5796.80	I
55		5802.92	I
55	Hs	5807.72	I
55	h	5809.22	I
55		5815.85	II
65	Hs	5819.51	G
55		5840.47	II
220		5851.63	I
55		5855.24	II
280		5856.22	I
55		5860.73	I
65		5877.26	II
55		5886.46	I
55		5904.07	II
110		5904.56	I
170		5911.45	II
65		5913.55	II
55		5916.77	I
85		5930.29	I
85		5936.84	I
65		5937.71	I
55	h	5940.95	G
55	h	5942.78	G
55		5951.60	II
55		5956.48	I
85		5977.25	I
110	h	5988.02	I
85		5999.08	I
65		6000.96	G
75	h	6001.87	G
55		6004.57	II
55		6008.71	I
55		6021.13	I
55		6080.65	II
430		6114.07	I
55		6180.42	II
110	B1	6182.68	G
110	B1	6200.86	G
110	B1	6211.71	G
110	B1	6220.93	G
55	B1	6231.62	G
75	B1	6241.66	G
55	b	6252.12	G
55	B1	6262.64	G
45	b	6273.00	G
85		6289.73	II
30		6292.87	I
75		6305.15	II
30		6309.11	II
27		6317.19	I
40		6331.35	I
17		6333.75	I
17		6336.34	I
27		6346.65	II
27	h	6351.96	I
17		6363.23	I
40		6380.95	II
17		6382.19	I
22		6408.55	I
22		6422.42	II
17		6424.52	I
19	h	6470.29	I
15		6480.11	II
40	h	6538.15	I
22		6549.25	I
55		6564.78	I
10		6568.00	II
10		6573.80	I
30		6591.60	I
15		6593.42	I
10		6610.04	II
50		6634.36	I
35		6640.08	I
10		6642.76	I
30		6643.98	I
10		6646.85	I
10		6653.55	I
10		6679.56	II
35		6681.23	I
10		6692.86	I
10		6704.18	II
14		6718.14	II
17		6727.83	I
85		6730.73	I
50		6752.67	II
14		6753.91	I
14		6783.39	I
26		6786.33	I
10		6787.18	I
12		6814.56	I
26		6816.49	I
17		6820.90	I
100		6828.25	I

Intensity		Wavelength	Spectrum
35		6846.60	II
30		6857.13	II
15		6864.25	I
21		6887.63	II
14		6900.73	II
100		6916.57	I
21		6920.62	II
15		6924.99	II
21		6926.49	II
17		6945.98	II
15		6957.74	II
15		6959.24	II
14		6964.33	I
15		6971.66	II
12		6976.35	II
10		6978.27	II
26	h	6980.86	I
50		6985.89	II
10		6988.75	II
75		6991.92	I
21		6993.18	II
60		6996.76	II
17		7000.75	II
45		7006.16	II
10		7016.56	I
21		7037.26	II
14		7051.00	II
13		7054.62	II
10		7058.02	II
10		7068.09	II
18		7071.00	I
18		7073.63	I
14		7098.11	I
14		7098.73	I
10	h	7116.77	II
21		7118.86	II
35		7122.57	I
13		7135.73	II
18		7147.31	II
13		7158.28	I
170		7168.37	I
21		7172.26	II
28		7189.57	II
13		7197.08	II
13		7201.41	II
10		7228.02	I
25		7233.45	I
14		7252.70	II
28		7262.66	I
14		7291.35	I
21		7313.28	I
18		7324.89	II
14		7373.81	I
14		7376.41	I
13		7377.27	II
13		7380.28	I
13		7394.90	II
13		7430.19	I
35		7441.85	I
40		7464.36	I
55		7562.97	I
10		7563.19	II
10		7588.20	I
10		7611.78	I
21		7621.96	I
21		7650.32	I
25		7672.56	I
10		7676.06	I
13		7694.45	I
80		7733.50	I
35		7749.30	I
10		7755.97	I
10		7766.48	II
11	h	7844.87	I
10		7845.80	I
35		7846.35	II
35		7856.93	I
14		7869.72	I
25		7930.21	II
13		8077.59	I
18		8146.15	I
11		8218.08	I
10		8275.42	I
10		8349.73	I
11		8398.30	I
10		8445.47	I
13	h	8527.88	I
21		8668.63	I
11		8770.36	I
13		8784.85	I
10		8795.76	I
21	h	8832.06	II
14	h	8849.14	I
18	h	8867.31	I

Gd III
Ref. 46, 137, 151—J.F.K.

Intensity	Wavelength	Spectrum
	Vacuum	
600	1813.47	III

Intensity	Wavelength	Spectrum
900	1946.26	III
1100	1974.34	III
2200	1975.24	III
	Air	
900	2008.79	III
3400	2018.07	III
1800	2027.82	III
800	2046.02	III
500	2057.79	III
1500	2080.08	III
1800	2098.20	III
1300	2125.68	III
1400	2148.03	III
1700	2176.84	III
1700	2223.95	III
1700	2236.73	III
1200	2239.84	III
1500	2243.75	III
1700	2250.18	III
1300	2257.05	III
1300	2292.51	III
1000	2300.38	III
1200	2303.72	III
1500	2307.03	III
1100	2313.50	III
1700	2313.56	III
1000	2315.09	III
1400	2323.12	III
1700	2323.18	III
2200	2323.78	III
1900	2329.35	III
2100	2335.01	III
1600	2336.02	III
1900	2338.97	III
1600	2339.88	III
2100	2342.74	III
2500	2346.52	III
2800	2359.31	III
1900	2360.87	III
2300	2361.91	III
1400	2362.38	III
2100	2363.26	III
1600	2365.22	III
2000	2373.38	III
1300	2374.29	III
1300	2381.38	III
1600	2387.82	III
2000	2388.77	III
1200	2393.86	III
1200	2397.34	III
1200	2405.03	III
1300	2408.41	III
1200	2409.35	III
1500	2466.84	III
1100	2469.14	III
1300	2499.53	III
1600	2520.38	III
1600	2534.11	III
1300	2536.10	III
1600	2551.56	III
2200	2553.90	III
2100	2554.04	III
2500	2563.33	III
2100	2564.46	III
1000	2565.04	III
2400	2565.95	III
1800	2569.27	III
1800	2573.57	III
2000	2576.06	III
1300	2576.15	III
1400	2578.13	III
1600	2578.76	III
1700	2583.62	III
2000	2588.21	III
2000	2588.46	III
1300	2595.81	III
1800	2609.77	III
1200	2619.40	III
1200	2621.52	III
1400	2623.52	III
1400	2625.48	III
2000	2628.10	III
1300	2628.99	III
2400	2629.83	III
2100	2632.30	III
1800	2633.32	III
1400	2635.71	III
1600	2636.44	III
1700	2637.15	III
2100	2637.97	III
2100	2638.06	III
2200	2640.53	III
1600	2641.65	III
1300	2643.71	III
1600	2644.52	III
1800	2646.04	III
1800	2646.84	III
1600	2651.48	III
2000	2655.59	III
1900	2656.55	III
2200	2660.83	III

Intensity	Wavelength	
1800	2675.75	III
1800	2679.44	III
1700	2680.63	III
1800	2682.52	III
1500	2683.91	III
1600	2692.78	III
1900	2692.86	III
1500	2694.43	III
2800	2697.39	III
1500	2702.91	III
2800	2703.28	III
1600	2704.53	III
1800	2717.35	III
2700	2727.89	III
450	2751.24	III
1800	2833.83	III
9000	2904.73	III
1800	2918.40	III
9500	2955.53	III
1000	2975.42	III
1000	2984.10	III
1000	3116.59	III
2500	3118.04	III
4000	3176.66	III
400	3253.53	III
400	3330.34	III
400	3371.05	III
400	3402.97	III
450	3624.90	III
250	3700.47	III
300	3831.73	III
300	3910.24	III
300	4016.91	III
600	4177.26	III
400	4279.96	III
300	4314.28	III
300	4445.91	III
600	4684.25	III
600	4715.06	III
600	4782.79	III
250	4976.72	III
5000	5091.70	III
300	5124.06	III
1800	5347.95	III
3000	5365.96	III
1100	5412.62	III
4000	5553.30	III
3000	5587.88	III
3000	5658.98	III
1000	5706.96	III
1500	5862.09	III
1500	5987.85	III
5000	14332.88	III
2000	17474.78	III
800	19996.34	III
800	21259.44	III
600	22493.33	III

Gd IV
Ref. 152—J.F.K.

Intensity	Wavelength	
	Vacuum	
1000	967.92	IV
1000	983.42	IV
1000	987.10	IV
1000	987.91	IV
1000	995.04	IV
1000	995.80	IV
1000	996.49	IV
1000	999.24	IV
1000	1000.36	IV
1000	1002.73	IV
1000	1004.46	IV
1000	1005.66	IV
1000	1006.55	IV
1200	1007.24	IV
1200	1063.84	IV
500	1228.37	IV
500	1307.23	IV
500	1313.29	IV
500	1316.71	IV
600	1321.42	IV
500	1330.79	IV
1100	1393.24	IV
1600	1476.98	IV
1500	1705.03	IV
1600	1706.01	IV
2000	1736.24	IV
1500	1815.32	IV
400	1997.89	IV
	Air	
800	2049.28	IV
800	2061.30	IV
800	2070.40	IV
800	2076.66	IV
1000	2094.29	IV
1000	2296.89	IV
1000	2352.66	IV
800	2379.17	IV
900	2385.65	IV
500	2390.07	IV
700	2392.30	IV
700	2393.29	IV
700	2395.76	IV
500	2396.22	IV
600	2396.27	IV
1400	2397.87	IV
700	2402.70	IV
500	2412.21	IV
900	2419.26	IV
500	2439.84	IV
600	2440.38	IV
600	2468.60	IV

GALLIUM (Ga)
Z = 31

Ga I and II
Ref. 19, 132, 195, 281—L.J.R.

Intensity		Wavelength	
		Vacuum	
2		829.60	II
2		958.67	II
1		960.57	II
2		969.19	II
2		998.52	II
3		1002.95	II
5		1012.38	II
3		1019.10	II
5		1023.80	II
8		1033.69	II
1		1113.87	II
3		1119.25	II
5		1130.81	II
1		1167.62	II
2		1173.78	II
3		1186.81	II
1		1277.13	II
5		1286.38	II
5		1327.81	II
20		1414.44	II
5		1449.49	II
2		1463.85	II
3		1473.73	II
3		1483.52	II
3		1485.95	II
3		1495.21	II
3		1504.41	II
3		1505.01	II
5		1514.57	II
3		1515.19	II
8		1535.40	II
5		1536.37	II
1		1536.91	II
3		1669.83	II
5		1695.85	II
5		1799.42	II
10		1813.98	II
15		1845.30	II
		Air	
20		2091.34	II
1		2218.04	I
1		2255.03	I
1		2259.23	I
2		2294.19	I
1		2297.87	I
3		2338.24	I
1		2338.60	I
3		2371.29	I
2		2377.53	II
4		2418.69	I
6		2438.88	II
7		2450.08	I
3		2500.19	I
3		2500.71	I
5		2513.55	II
3		2514.15	II
2		2551.26	II
3		2552.87	II
4		2555.28	II
5		2607.47	I
8		2624.82	I
10		2632.66	I
3		2659.87	I
10		2665.05	I
8		2691.29	I
20		2700.47	II
3		2719.66	II
15		2780.15	II
6		2874.24	I
1		2886.45	II
		2893.65	II
2		2910.77	II
6		2943.64	I
6		2944.17	I
3		2969.41	II
1		2971.01	II
3		2971.60	II
5		2974.77	II
1		2992.84	II
2		3011.90	II
1		3158.18	II
4		3374.94	II
1		3375.95	II
2		3436.66	II
3		3446.46	II
2		3447.26	II
5		3470.34	II
1		3471.46	II
1		3472.52	II
2		3583.60	II
1		3693.93	II
2		3705.85	II
4		3734.85	II
9		3924.39	II
10		4032.99	I
10		4172.04	I
4		4251.11	II
15		4251.16	II
10	h	4254.04	II
4	h	4255.64	II
5		4255.70	II
10		4255.77	II
40		4262.00	II
3		5218.21	II
1		5338.3	II
2		5353.49	I
2		5360.6	II
1		5363.5	II
3		5416.8	II
1		5421.6	II
1		5425.6	II
10		6334.2	II
2000		6396.56	I
1000		6413.44	I
5		6419.4	II
3		6456.3	II
1		7000.0	II
3	h	7051.24	I
5	h	7106.87	I
5	h	7116.3	I
2	h	7172.9	I
5	h	7193.6	I
7		7198.7	II
10	h	7231.4	I
3	h	7289.6	I
5	h	7349.3	I
20	h	7403.0	I
30	h	7464.0	I
6	h	7556.6	I
10	h	7620.5	I
50	h	7734.77	I
2		7793.0	II
100	h	7800.01	I
4	h	7801.6	I
15	h	8002.55	I
20	h	8074.25	I
3	h	8167.5	I
5	h	8171.6	I
100	h	8311.86	I
200	h	8386.49	I
10	h	8389.30	I
7		8415.51	I
10	h	8419.91	I
20	h	8808.75	I
30	h	8813.56	I
20	h	8856.37	I
30	h	8944.33	I
200	h	9492.92	I
200	h	9493.12	I
300	h	9589.36	I
20		9594.25	I
60	h	10898.10	I
100	h	10905.95	I
10		10968.27	I
20		11103.51	I
400		11949.12	I
200		12109.78	I
40		12885.05	I
50		13057.50	I
50		14982.75	I
60		14996.64	I
20		17757.91	I
10		17868.96	I
60		22016.81	I
70		22568.71	I

Ga III
Ref. 141—L.J.R.

Intensity	Wavelength	
	Vacuum	
50	620.00	III
40	622.01	III
90	806.51	III
90	817.30	III
50	828.70	III
80	1085.00	III
60	1105.61	III
90	1150.27	III
90	1267.16	III
80	1293.46	III
60	1295.36	III
60	1323.15	III
70	1353.92	III
90	1495.07	III
50	1534.46	III
	Air	
90	2417.70	III
90	2423.98	III
15	2424.36	III
50	3521.77	III
80	3581.19	III
100	3589.34	III
10	3731.10	III
10	3806.60	III
100	4380.69	III
150	4381.76	III
100	4863.00	III
150	4993.78	III
10	5808.28	III
20	5848.25	III
15	5993.51	III

Ga IV
Ref. 141, 143—L.J.R.

Intensity	Wavelength	
	Vacuum	
14	294.53	IV
61	295.67	IV
41	304.99	IV
4	422.12	IV
25	423.18	IV
16	439.92	IV
67	1137.06	IV
70	1156.10	IV
70	1163.60	IV
75	1170.58	IV
48	1171.71	IV
68	1185.23	IV
40	1186.06	IV
73	1190.89	IV
73	1193.02	IV
75	1195.02	IV
69	1201.54	IV
72	1206.89	IV
63	1216.15	IV
50	1228.03	IV
60	1236.38	IV
60	1238.59	IV
45	1241.81	IV
75	1245.53	IV
83	1258.77	IV
81	1264.66	IV
82	1267.15	IV
81	1279.24	IV
80	1285.33	IV
82	1295.86	IV
83	1299.46	IV
82	1303.53	IV
80	1309.68	IV
80	1314.82	IV
85	1338.09	IV
77	1347.03	IV
76	1351.06	IV
74	1364.63	IV
60	1395.54	IV
77	1402.55	IV
70	1405.32	IV
73	1465.12	IV

Ga V
Ref. 2, 62, 140—L.J.R.

Intensity	Wavelength	
	Vacuum	
5	290.53	V
1	296.13	V
5	296.82	V
30	298.44	V
20	299.47	V
30	300.01	V
25	300.57	V
10	300.78	V
20	301.19	V
30	302.86	V
20	303.84	V
30	307.03	V
30	308.26	V
15	309.64	V
30	311.79	V
25	312.41	V
30	313.68	V

Gallium (Cont.)

Intensity	Wavelength	
15	315.95	V
20	316.48	V
40	319.41	V
12	320.53	V
40	322.31	V
50	322.99	V
30	323.10	V
40	324.25	V
40	324.95	V
40	326.14	V
30	326.77	V
30	328.65	V
5	336.61	V
20	878.17	V
40	973.21	V
10	977.89	V
15	979.60	V
20	984.95	V
40	989.75	V
90	1014.47	V
90	1019.71	V
20	1033.55	V
30	1038.76	V
30	1047.50	V
120	1050.48	V
80	1054.56	V
90	1058.12	V
80	1066.69	V
35	1068.59	V
30	1069.45	V
60	1069.60	V
55	1071.19	V
45	1071.41	V
80	1073.77	V
90	1078.83	V
110	1079.60	V
60	1080.99	V
250	1085.01	V
80	1087.37	V
40	1090.53	V
90	1091.71	V
100	1094.36	V
80	1095.10	V
70	1101.62	V
160	1102.83	V
140	1103.03	V
60	1104.93	V
75	1105.62	V
70	1106.17	V
40	1115.55	V
80	1118.34	V
55	1123.18	V
80	1123.66	V
120	1126.40	V
80	1127.75	V
130	1128.10	V
120	1128.53	V
100	1129.94	V
80	1131.43	V
40	1133.91	V
130	1136.07	V
65	1138.20	V
60	1144.30	V
50	1145.70	V
30	1148.42	V
45	1150.09	V
130	1150.23	V
120	1156.51	V
35	1157.74	V
25	1169.40	V
40	1178.95	V
80	1213.17	V
30	1265.45	V
30	1276.85	V
15	1283.64	V
10	1311.35	V

Intensity		Wavelength	
500		1181.65	II
200		1188.73	II
100		1189.62	II
300		1191.26	II
50		1191.72	II
500		1237.059	II
500		1261.905	II
100		1264.710	II
100		1380.42	II
50		1392.26	II
200		1401.24	II
200		1538.091	II
500		1576.855	II
75		1581.070	II
100		1602.486	II
3	r	1615.57	I
2	r	1624.130	I
2	r	1630.173	I
3	r	1636.31	I
2		1638.96	I
4	r	1639.730	I
2		1647.531	I
200		1649.194	II
2		1651.528	I
4	r	1651.955	I
3		1661.345	I
4	r	1663.539	I
10	h	1665.275	I
4		1667.802	I
3	r	1670.608	I
100	r	1691.090	I
200	r	1716.784	I
100	h	1739.102	I
100		1742.195	I
50		1746.065	I
200		1750.043	I
100		1758.279	I
100	h	1764.185	I
100	h	1765.284	I
50	h	1766.433	I
200		1774.176	I
200		1785.046	I
100	h	1793.071	I
75	h	1801.432	I
200	h	1841.328	I
200	h	1842.410	I
100	h	1844.410	I
100	h	1845.872	I
100	h	1846.958	I
200		1853.134	I
500	r	1860.086	I
100		1865.052	I
300	r	1874.256	I
100		1895.197	I
500	r	1904.702	I
50	h	1908.434	I
30		1912.409	I
300	r	1917.592	I
100	h	1923.467	I
500	r	1929.826	I
10	h	1934.048	I
100	r	1937.483	I
500		1938.008	II
100	r	1938.300	I
500		1938.891	II
30	s	1944.116	I
200		1944.731	I
200		1955.115	I
500		1962.013	I
30	h	1963.373	I
30		1965.383	I
200		1970.880	I
200		1979.274	II
300	h	1987.849	I
300		1988.267	I
500	r	1998.887	I

Intensity		Wavelength	
15		2397.885	I
130		2417.367	I
30		2436.412	I
100		2478.66	II
90		2497.962	I
500		2500.54	II
70		2533.230	I
3		2556.298	I
28		2589.188	I
500		2592.534	I
8		2644.184	I
1200		2651.172	I
500		2651.568	I
500		2691.341	I
200		2704.03	II
850		2709.624	I
400		2729.78	II
40		2740.426	I
650		2754.588	I
50		2770.59	II
75		2772.35	II
70		2793.925	I
80		2829.008	I
1000		2831.843	II
50		2834.28	II
75		2839.68	II
1000		2845.527	II
75		2853.97	II
750		3039.067	I
600		3067.021	I
20		3124.816	I
50		3186.72	II
100		3221.64	II
110		3269.489	II
50		3312.56	II
75		3323.64	II
100		3455.72	II
300		3499.21	II
30		3845.11	II
70		4226.562	I
10		4685.829	I
75	h	4689.87	II
50	h	4690.02	II
1000		4741.806	II
1000		4814.608	II
50		4824.097	II
100		5131.752	II
200		5178.648	II
3		5194.583	I
6		5265.892	I
6		5513.263	I
8		5564.741	I
8		5607.010	I
6		5616.135	I
7		5621.426	I
8		5655.96	I
6		5664.226	I
5		5664.842	I
9		5691.954	I
6		5701.776	I
5		5717.877	I
6		5801.029	I
9		5802.093	I
1000		5893.389	II
500		6021.041	II
150		6078.39	II
50		6267.14	II
150		6268.07	II
100		6268.34	II
75		6283.452	II
100		6336.377	II
100		6484.181	II
6		6557.488	I
50		6780.51	II
50		7049.369	II
6		7130.12	I
30		7145.390	II
7		7330.38	I
5		7353.334	I
7		7384.208	I
6		7402.64	I
7		7511.57	I
5		7776.20	I
10		7833.575	I
7		7837.63	I
6		7853.77	I
7		7878.12	I
5		7962.26	I
5		7983.33	I
10		8031.039	I
6		8044.165	I
5		8095.29	I
5		8225.22	I
7		8226.09	I
10		8256.013	I
5		8264.15	I
5		8280.09	I
6		8281.04	I
8		8367.81	I
7		8391.70	I
5		8396.36	I
5		8429.42	I
10		8482.21	I

Intensity	Wavelength	
8	8506.70	I
5	8507.66	I
8	8564.89	I
6	8599.27	I
6	8652.42	I
5	8669.60	I
9	8700.60	I
5	8712.90	I
6	8734.78	I
6	8789.88	I
5	9068.785	I
5	9095.957	I
6	9398.868	I
20	9474.993	II
20	9475.645	II
4	9492.559	I
7	9625.664	I
5	10039.436	I
4	10200.952	I
10	10382.427	I
10	10404.913	I
8	10734.068	I
8	10947.416	I
10	11125.130	I
230	11252.83	I
24	11293.40	I
33	11318.13	I
55	11459.05	I
150	11483.77	I
175	11614.81	I
600	11714.76	I
10	11839.77	I
55	11917.01	I
10	12025.64	I
10	12055.49	I
30	12061.41	I
45	12065.76	I
1300	12069.20	I
30	12198.88	I
20	12207.73	I
60	12286.75	I
55	12338.76	I
1050	12391.58	I
48	12540.41	I
15	12636.80	I
150	12676.58	I
40	12681.28	I
115	12800.66	I
175	12836.38	I
12	12847.92	I
120	12955.73	I
15	13028.64	I
235	13107.61	I
20	13492.28	I
42	13534.85	I
28	13724.48	I
42	14116.70	I
42	14297.15	I
40	14569.84	I
12	14667.52	I
470	14822.38	I
16	14921.97	I
15	15001.75	I
13	15041.21	I
20	15504.34	I
14	16424.77	I
12	16626.64	I
70	16699.29	I
150	16759.79	I
135	17214.34	I
16	18428.30	I
35	18495.54	I
10	18764.11	I
70	18811.86	I
62	19279.24	I
28	20673.66	I
4	21518.30	I
9	22091.84	I
5	23921.92	I

GERMANIUM (Ge)
Z = 32

Ge I and II
Ref. 5, 119, 293, 340—C.H.C.

Intensity	Wavelength	
	Vacuum	
1	822.97	II
3	835.08	II
10	850.50	II
10	862.234	II
15	875.493	II
15	905.977	II
20	920.554	II
50	999.101	II
100	1016.638	II
100	1075.072	II
300	1085.51	II
200	1098.71	II
500	1106.74	II
500	1120.46	II
200	1164.27	II
500	1181.19	II

Intensity		Wavelength	
		Air	
50		2007.04	II
200		2011.29	I
1700		2019.068	I
2400	r	2041.712	I
1600	r	2043.770	I
420		2054.461	I
220	h	2057.238	I
750		2065.215	I
2600	r	2068.656	I
420		2086.021	I
2000		2094.258	I
240		2105.824	I
95	h	2124.744	I
50	h	2186.451	I
100		2197.62	II
340	r	2198.714	I
100		2205.85	II
15		2220.375	I
18		2256.001	I
18		2314.201	I
24		2327.918	I
15		2359.233	I
20		2379.144	I
10		2389.472	I

Ge III
Ref. 341—C.H.C.

Intensity	Wavelength	
	Vacuum	
2	542.90	III
2	663.77	III
3	670.88	III
2	680.28	III
2	952.76	III
12	988.96	III
15	995.72	III
10	996.50	III
15	1011.21	III
10	1012.31	III
8	1032.62	III
12	1040.99	III
12	1058.91	III
40	1088.45	III
10	1137.92	III
12	1150.55	III

Germanium (Cont.)

Intensity	Wavelength	
8	1159.15	III
8	1159.62	III
8	1160.79	III
10	1173.78	III
8	1212.47	III
4	1323.24	III
10	1525.32	III
2	1527.15	III
9	1600.09	III
6	1883.26	III
2	1978.22	III

Air

Intensity	Wavelength	
2	2019.22	III
4	2022.25	III
3	2062.14	III
15	2100.05	III
15	2102.42	III
25	2104.45	III
3	2922.86	III
25	3197.56	III
35	3211.86	III
25	3214.95	III
40	3255.05	III
20	3259.90	III
5	3369.57	III
20	3414.27	III
40	3434.03	III
8	3464.59	III
40	3489.08	III
2	3724.51	III
15	3884.78	III
200	4178.96	III
12	4245.41	III
200	4260.85	III
150	4291.71	III
10	4674.36	III
10	5016.88	III
18	5134.75	III
5	5229.37	III
3	5256.61	III

Ge IV
Ref. 341—C.H.C.

Intensity	Wavelength	
	Vacuum	
1	440.11	IV
1	441.95	IV
3	847.80	IV
3	868.30	IV
8	915.00	IV
8	936.70	IV
4	938.90	IV
1	1073.44	IV
20	1188.99	IV
20	1229.81	IV
2	1494.89	IV
6	1500.61	IV
3	1648.14	IV

Air

Intensity	Wavelength	
2	2293.0	IV
2	2343.37	IV
15	2445.38	IV
15	2445.71	IV
30	2488.25	IV
20	2542.44	IV
5	2631.78	IV
3	2698.08	IV
15	2717.44	IV
30	2736.09	IV
30	2788.61	IV
5	3071.84	IV
60	3554.19	IV
50	3676.65	IV

Ge V
Ref. 342—C.H.C.

Intensity	Wavelength	
	Vacuum	
700	294.51	V
1000	295.64	V
200	304.98	V
20	621.52	V
35	716.26	V
50	724.21	V
35	733.54	V
35	735.35	V
35	741.52	V
60	746.88	V
40	750.26	V
35	755.84	V
60	760.05	V
60	958.51	V
300	971.35	V
150	984.92	V
200	988.13	V
300	990.66	V
300	1004.38	V
300	1016.66	V
250	1038.40	V
900	1045.71	V
400	1050.05	V
300	1054.59	V
300	1068.43	V
400	1069.13	V
700	1072.66	V
600	1086.65	V
500	1087.85	V
800	1089.49	V
300	1092.09	V
1000	1116.94	V
300	1122.01	V
700	1163.39	V
300	1165.26	V
200	1176.69	V
700	1222.30	V

GOLD (Au)
Z = 79
Au I and II
Ref. 38, 72, 234—C.H.C.

Intensity		Wavelength	
		Vacuum	
		925.72	II
		946.03	II
		950.39	II
20		957.78	II
3		967.94	II
3		974.47	II
2		982.24	II
		1062.67	II
8		1066.96	II
		1085.00	II
8		1090.78	II
6		1094.92	II
20		1102.31	II
3		1166.76	II
5		1210.86	II
20		1224.57	II
40	h	1305.34	I
25		1310.47	I
100	h	1328.37	I
3		1336.26	II
10	h	1338.37	I
10	h	1342.80	I
20		1350.09	I
20		1350.84	I
22		1351.74	I
25		1352.82	I
25		1354.14	I
30		1355.79	I
35		1357.86	I
40		1360.51	I
6		1362.33	II
20		1362.47	I
8		1363.15	II
50		1363.98	I
25		1364.15	I
10		1364.74	II
60		1368.62	I
35		1368.98	I
70		1374.82	I
50		1375.76	I
30		1378.87	I
8		1380.53	II
80		1382.75	I
50		1385.33	I
20		1389.14	I
60		1392.27	I
6		1393.80	II
50		1402.12	I
30		1405.12	II
70		1407.38	I
100		1408.45	I
10		1410.69	II
10		1415.22	II
80		1429.19	I
50		1435.79	I
20		1436.61	II
10		1468.85	II
10		1469.17	II
10		1469.28	II
100		1481.76	I
25		1486.55	II
20		1532.82	I
20		1532.86	I
10		1562.04	II
200		1587.16	I
12		1593.41	II
70		1598.24	I
12		1611.11	II
		1616.65	II
2		1622.83	II
100		1624.34	I
2		1632.53	II
50		1639.90	I
150		1646.67	I
10		1656.99	II
100		1665.76	I
25		1673.59	II
7		1694.38	II
2		1698.65	II
200		1699.34	I
30		1700.69	II
10		1720.04	II
25		1725.75	II
45		1740.52	II
10		1749.80	II
35		1756.15	II
60		1783.22	II
35		1793.31	II
35		1800.58	II
25		1823.24	II
100		1879.83	I
20		1919.64	I
20		1921.64	II
45		1942.31	I
25		1951.93	I
30		1978.19	I

Air

Intensity		Wavelength	
25		2000.81	II
11000		2012.00	I
2600		2021.38	I
50		2044.54	II
150		2082.09	II
35		2095.13	II
20		2098.14	II
60		2110.68	II
30		2125.29	II
15		2126.63	I
20		2170.75	I
35		2188.81	II
25		2201.32	II
35		2215.63	II
45		2228.88	II
30		2231.18	II
25		2240.16	II
70		2248.56	II
80		2263.62	II
18		2263.88	II
25		2277.32	II
25		2283.30	II
25		2291.40	II
45		2304.69	II
25		2314.55	II
20		2315.75	II
25		2340.06	II
180		2352.65	I
20		2376.28	I
120		2387.75	I
2600		2427.95	I
60		2533.52	II
16		2544.19	I
45		2552.67	II
20		2589.25	I
30		2590.04	I
50		2616.40	II
20		2627.02	II
250		2641.48	I
3400		2675.95	I
20		2687.63	II
20		2688.16	II
30		2688.71	I
80		2700.89	I
1100		2748.25	I
20		2748.71	II
100		2780.82	I
30		2800.93	I
1000		2802.04	II
300		2819.79	II
100		2822.55	II
30		2823.13	II
100	h	2825.44	II
30		2833.03	II
300		2837.85	II
100		2846.92	II
100		2856.74	II
3		2872.36	I
300		2883.45	I
3		2886.96	I
10		2888.40	I
300		2891.96	I
100		2893.25	II
3		2905.74	I
30		2905.90	I
100		2907.04	II
300		2913.52	II
3		2914.82	I
300		2918.24	II
16		2932.19	I
30		2940.67	I
100		2954.22	II
10		2973.33	II
100	h	2990.27	II
300		2994.80	II
10		3002.65	I
3		3005.85	I
10		3024.67	I
320		3029.20	I
30		3033.25	I
300		3065.42	I
10		3102.63	I
10	h	3117.01	I
100		3122.50	II
1600		3122.78	I
30		3126.86	II
30		3127.03	I
10		3164.88	I
10		3172.35	II
30		3191.76	I
100		3194.72	I
30		3200.37	I
30		3204.74	I
1		3221.86	I
30		3225.25	I
300		3230.63	I
10		3253.94	I
10	h	3265.10	I
30	h	3267.07	I
10		3271.63	I
10		3273.47	I
300		3308.30	I
300		3309.64	I
100		3320.12	I
100		3355.15	I
30	h	3368.44	I
10		3381.90	I
100	h	3391.31	I
100		3395.40	I
30	h	3440.36	I
100		3467.21	I
30		3471.61	I
30	h	3509.04	I
10		3510.82	I
3		3523.34	II
3		3545.61	I
30		3553.57	I
300		3557.36	I
10		3558.22	I
30		3566.07	I
30	h	3584.37	I
300		3586.73	I
30	h	3588.79	I
30		3598.06	I
100		3611.57	I
30	h	3614.00	I
30		3622.74	I
100	h	3631.31	I
10		3633.22	II
30		3634.53	I
10		3635.12	I
300		3637.90	I
3		3639.87	I
100	h	3645.02	I
10		3649.09	I
100		3650.74	I
10		3653.53	I
3		3654.69	I
10		3655.30	I
10		3656.90	I
30		3706.55	II
100		3709.62	I
10		3766.61	I
10		3770.76	I
100		3796.01	I
30		3801.92	I
30		3804.01	II
10		3821.85	I
10		3825.70	I
100		3874.73	I
30		3880.25	I
30		3889.48	I
100	h	3892.26	I
400		3897.86	I
30		3901.09	I
300		3909.38	I
100		3927.69	I
10	h	3959.10	I
30		3966.23	I
30		3976.65	I
30		3979.68	I
30		3991.37	I
3		4012.57	I
10		4016.07	II
400		4040.93	I
30		4052.79	II
700		4065.07	I
10		4076.35	II
3		4083.28	II
100		4084.10	I
30		4101.70	I
30		4128.59	I
30		4201.13	I
30	h	4227.88	I
100		4241.80	I
200		4315.11	I
30		4361.04	II
10		4420.61	II
120	h	4437.27	I

Gold (Cont.)

Intensity		Wavelength	
250		4488.25	I
900	h	4607.51	I
100	h	4620.56	I
1		4663.92	I
3		4663.97	I
10		4694.69	I
3		4760.17	II
500		4792.58	I
100		4811.60	I
10		4822.96	I
30	h	4950.82	I
30		5064.59	I
30	h	5108.84	I
100		5147.44	I
300		5230.26	I
100	h	5261.76	I
100		5655.77	I
100	h	5721.36	I
300		5837.37	I
100	h	5862.93	I
300	h	5956.96	I
30	h	5962.68	I
600		6278.17	I
100		6562.68	I
30		6652.89	I
600		7510.73	I
10		8145.06	I
10		9254.28	I

Au III
Ref. 72, 393, 395 — R.D.C.

Intensity		Wavelength	

Vacuum

30		779.73	III
30		788.78	III
50		799.93	III
40		811.83	III
50		817.95	III
40		820.06	III
80		833.16	III
100		843.44	III
100		845.14	III
80		855.49	III
80		859.90	III
80		863.42	III
80		901.03	III
80		910.45	III
80		924.02	III
200		945.10	III
100		1040.63	III
80		1044.49	III
80		1046.81	III
100	h	1239.96	III
100		1278.51	III
100		1314.84	III
200		1336.72	III
180		1341.68	III
100		1348.89	III
150		1350.32	III
150		1355.61	III
150		1356.13	III
80		1362.06	III
500		1365.40	III
200		1367.17	III
180		1377.73	III
150		1378.69	III
150		1379.98	III
125		1380.53	III
200		1381.36	III
300		1385.79	III
100		1389.41	III
180		1391.46	III
180		1396.00	III
100		1402.91	III
225		1409.50	III
250		1413.80	III
100		1414.27	III
80		1415.54	III
100		1417.09	III
125		1417.39	III
150		1427.42	III
300		1428.93	III
250		1430.06	III
275		1433.37	III
250		1435.81	III
80		1436.12	III
300		1439.12	III
200		1441.21	III
150		1446.37	III
80		1446.69	III
250		1448.42	III
250		1454.95	III
100		1464.72	III
150		1471.28	III
80		1473.32	III
100		1474.73	III
150		1481.10	III
300		1487.15	III
250		1487.91	III
200		1489.47	III
250		1500.37	III
200		1502.47	III
200		1503.74	III
80		1540.26	III
100		1542.00	III
80		1542.25	III
100		1548.50	III
80		1554.61	III
80		1562.33	III
80		1562.41	III
200		1567.54	III
80		1571.94	III
200		1574.85	III
200		1579.44	III
150		1584.10	III
200		1589.56	III
80		1589.68	III
150		1593.41	III
200		1600.51	III
250		1617.16	III
100		1617.78	III
500		1621.93	III
300	d	1629.13	III
250		1638.88	III
100		1644.17	III
250		1652.74	III
250		1664.77	III
100		1668.11	III
125		1673.93	III
1000		1693.94	III
150		1697.09	III
200		1698.98	III
200		1700.00	III
200		1702.25	III
100		1707.53	III
250		1710.16	III
200		1715.69	III
100		1716.71	III
300		1717.83	III
500		1727.31	III
100	d	1733.17	III
300		1738.48	III
150		1744.39	III
500		1746.10	III
500		1756.92	III
500		1761.95	III
300		1767.42	III
100		1774.42	III
800		1775.17	III
200		1776.40	III
100		1780.57	III
300		1786.11	III
150		1792.65	III
500		1793.76	III
200		1801.98	III
400		1805.24	III
100		1809.81	III
400		1821.17	III
400		1844.89	III
150		1848.83	III
80		1850.15	III
500		1861.80	III
150		1871.92	III
150		1918.28	III
100		1932.04	III
100		1935.42	III
200		1948.79	III
100		1958.47	III
400		1989.63	III
150		1996.85	III

Air

300	2083.09	III
80	2085.45	III
100	2159.08	III
80	2167.33	III
200	2172.20	III
100	2184.11	III
500	2188.97	III
300	2322.27	III
100	2382.40	III
150	2402.71	III
150	2405.12	III
100	3227.99	III
100	3309.86	III

HAFNIUM (Hf)
Z = 72

Hf I and II
Ref. 1—C.H.C.

Intensity	Wavelength	

Air

6200	2012.78	II
8500	2028.18	II
1200	2096.18	II
540	2210.82	II
320	2254.01	II
160	2255.15	II
250	2266.83	II
620	2277.16	II
230	2321.14	II
580	2322.47	II
300	2323.25	II
120	2324.50	II
300	2324.89	II
200	2332.97	II
200	2337.33	II
230	2343.32	II
320	2347.44	II
540	2351.22	II
110	2353.02	I
90	2365.98	II
250	2380.30	II
100	2381.00	II
170	2393.18	II
450	2393.36	II
670	2393.83	II
130	2400.78	II
70	2404.56	II
540	2405.42	II
130	2406.44	II
370	2410.14	II
90	2413.33	II
55	2415.96	II
320	2417.69	II
120	2425.98	II
45	2428.75	I
120	2428.99	II
130	2433.57	II
45	2434.74	II
35	2444.99	I
390	2447.25	II
140	2449.44	II
35	2452.30	II
110	2453.34	II
450	2460.49	II
70	2463.97	II
430	2464.19	II
90	2465.06	II
35	2465.67	I
140	2467.97	II
210	2469.18	II
100	2473.92	II
55	2481.44	II
55	2482.65	I
55	2487.16	I
290	2496.99	II
580	2512.69	II
580	2513.03	II
130	2515.48	II
890	2516.88	II
340	2531.19	II
200	2537.33	II
110	2548.20	II
320	2551.40	II
130	2559.19	II
250	2563.61	II
890	2571.67	II
320	2573.90	II
320	2576.82	II
300	2578.14	II
320	2582.54	II
130	2591.33	II
390	2606.37	II
450	2607.03	II
120	2608.45	I
230	2613.60	II
450	2622.74	II
160	2637.00	I
1100	2638.71	II
1100	2641.41	II
160	2642.75	I
670	2647.29	II
100	2651.16	II
160	2657.84	II
210	2661.88	II
290	2683.35	II
670	2705.61	I
110	2706.73	II
210	2712.42	II
140	2713.84	I
250	2718.59	II
120	2730.85	I
710	2738.76	II
200	2743.64	I
360	2751.81	II
450	2761.63	I
160	2766.96	I
170	2773.02	I
980	2773.36	II
180	2774.02	II
390	2779.37	I
100	2789.50	II
140	2789.73	II
230	2808.00	II
230	2813.86	II
170	2814.48	II
230	2817.68	I
140	2818.94	I
200	2819.74	I
1200	2820.22	II
490	2822.68	II
180	2833.28	I
110	2834.13	I
410	2845.83	I
270	2849.21	II
270	2850.96	I
180	2851.21	I
180	2860.56	I
760	2861.01	II
760	2861.70	II
2100	2866.37	I
130	2869.82	II
150	2876.33	II
210	2887.14	I
100	2887.54	I
800	2889.62	I
1800	2898.26	I
130	2898.71	II
1200	2904.41	I
890	2904.75	I
140	2909.91	II
2000	2916.48	I
580	2918.58	I
320	2919.59	II
180	2924.62	I
490	2929.63	II
450	2929.90	I
710	2937.80	II
2000	2940.77	I
160	2944.71	I
1200	2950.68	I
1100	2954.20	I
540	2958.02	I
120	2961.80	II
1400	2964.88	I
620	2966.93	I
140	2967.23	II
710	2968.81	II
110	2973.37	I
890	2975.88	II
150	2979.28	I
1100	2980.81	I
210	2982.72	I
170	3000.10	II
800	3005.56	I
1100	3012.90	II
540	3016.78	I
1100	3016.94	II
980	3018.31	I
1200	3020.53	I
140	3025.29	II
410	3031.16	II
110	3046.08	II
710	3050.76	I
1100	3057.02	I
130	3063.78	II
130	3064.68	II
850	3067.41	I
2100	3072.88	I
170	3074.10	I
250	3074.79	I
150	3080.66	II
430	3080.84	I
200	3096.76	I
340	3101.40	II
710	3109.12	II
130	3110.87	II
130	3119.98	I
710	3131.81	I
850	3134.72	II
130	3137.51	I
170	3139.65	II
120	3140.76	II
220	3145.32	I
220	3148.41	I
120	3151.63	I
450	3156.63	I
270	3159.82	I
710	3162.61	II
450	3168.39	I
890	3172.94	I
450	3176.86	I
220	3181.01	I
120	3181.15	I
130	3189.62	I
360	3193.53	II
670	3194.19	II
200	3196.93	I
130	3199.99	II
310	3206.11	I
180	3210.98	I
180	3217.30	II
180	3220.61	II
130	3230.06	I
130	3239.44	I
130	3243.35	I
360	3247.66	I
220	3249.53	I
890	3253.70	II
270	3255.28	II
120	3262.47	I
180	3273.66	II
270	3279.98	II

Intensity	Note	Wavelength	Spectrum
160		3291.05	I
210		3306.12	I
120		3309.19	I
340		3310.27	I
670		3312.86	I
180		3317.99	II
130		3328.21	II
890		3332.73	I
370		3352.06	II
130		3356.78	I
230		3358.91	I
180		3360.06	I
140		3366.68	I
180		3378.93	I
140		3384.14	II
230		3384.70	II
170		3386.21	I
800		3389.83	II
230		3392.81	I
230		3394.59	II
140		3394.98	II
230		3397.26	I
230		3397.60	I
2300		3399.80	II
170		3400.21	I
180		3402.51	I
140		3407.76	II
230		3410.17	II
230		3417.34	I
410		3419.18	I
140		3427.44	I
200		3428.37	II
250		3438.24	II
140		3438.43	I
100		3441.84	I
100		3452.31	I
140		3462.64	II
140		3467.60	I
710		3472.40	I
200		3478.99	II
480		3479.28	II
250		3495.75	II
250		3497.16	I
980		3497.49	I
100		3498.98	I
1200		3505.23	II
130		3513.28	I
130		3518.75	II
980		3523.02	I
100		3530.87	I
100		3531.23	I
980		3535.54	II
760		3536.62	I
180		3548.81	I
540		3552.70	II
150		3554.00	I
1300		3561.66	II
150		3564.31	I
270		3567.36	I
1100		3569.04	II
150		3579.90	I
110		3583.28	I
210		3597.42	II
540		3599.87	I
110		3615.04	I
800		3616.89	I
110		3617.68	I
110		3624.00	II
320		3630.87	II
100		3635.43	I
800		3644.36	II
320		3649.10	I
200		3651.84	I
140		3661.05	II
220		3665.35	II
100		3668.21	I
200		3672.27	I
480		3675.74	I
2200		3682.24	I
280		3696.51	I
100		3698.40	II
240		3699.72	II
340		3701.15	II
100		3704.92	I
120		3705.40	II
1000		3717.80	I
650		3719.28	II
140		3726.49	I
160		3729.10	I
460		3733.79	I
160		3737.88	II
120		3739.04	I
100		3744.98	II
400		3746.80	I
140		3753.22	I
100		3765.05	I
100		3765.56	I
170		3766.92	II
200		3768.25	I
1400		3777.64	I
1400		3785.46	I
650		3793.37	II
100		3798.66	I
850	d	3800.38	I
140		3806.07	II
320		3811.78	I
100		3817.20	II
100		3819.38	I
1300		3820.73	I
140		3829.67	I
280		3830.02	I
800		3849.18	I
140		3849.52	II
600		3858.31	I
230		3860.91	I
200		3872.55	II
160		3877.10	II
380		3880.82	II
200		3882.52	I
150		3883.77	II
200		3889.29	I
200		3889.33	I
620		3899.94	I
620		3918.09	II
200		3923.90	II
120		3926.42	I
150		3927.57	I
110		3929.54	II
320		3931.38	I
120		3935.65	II
120		3939.04	I
410		3951.83	I
160		3968.01	I
150	B1	3970.05	H
200		3973.48	I
180		4032.27	I
100		4047.96	II
230		4062.84	I
140		4066.21	I
180		4083.35	I
540		4093.16	II
110		4104.23	I
140		4106.58	I
110		4113.53	II
110		4118.60	I
150		4127.80	II
140		4145.21	I
150		4162.36	II
110		4163.60	I
1100		4174.34	I
120		4190.95	I
160		4206.58	II
190		4209.70	I
170		4228.08	I
170		4232.44	II
120	B1	4252.08	H
170		4260.98	I
200		4263.39	I
170		4272.85	II
320		4294.79	I
120		4318.14	I
160		4330.27	I
180		4336.66	II
150		4350.51	II
250		4356.23	I
110		4367.90	I
180		4370.97	II
120		4417.35	II
160		4417.91	I
200		4438.04	I
140		4457.34	I
140		4461.18	I
140		4540.93	I
250		4565.94	I
500	d	4598.80	I
230		4620.86	I
210		4655.19	I
120		4699.01	I
160		4782.74	I
310		4800.50	I
130		4859.24	I
120		4975.25	I
95		5018.20	I
15		5021.75	I
55		5040.82	II
95		5047.45	I
55	b	5074.74	H
30		5079.65	II
55	b	5093.88	H
15		5112.13	I
19		5128.53	II
30		5157.96	I
55		5167.42	I
75		5170.18	I
230		5181.86	I
30		5186.84	I
30		5187.75	II
110		5243.99	I
55		5247.10	II
25		5260.44	II
30		5264.95	II
55		5275.04	I
22		5286.09	I
120		5294.87	I
45		5298.06	II
30		5307.82	I
45		5309.68	I
55		5311.60	II
12		5324.26	II
9		5346.30	II
110		5354.73	I
110		5373.86	I
40		5389.34	I
19		5391.36	II
19		5404.47	I
28		5424.02	I
12		5435.78	I
40		5438.74	I
14		5444.07	II
75		5452.92	I
30		5463.38	II
15		5497.30	I
15		5510.12	I
15		5510.45	I
19		5524.35	II
45		5538.02	I
28		5538.26	I
230		5550.60	I
230		5552.12	I
55		5575.86	I
14		5600.77	I
95		5613.27	I
25		5614.01	I
8		5628.27	I
19		5650.83	I
40	B1	5698.03	H
25		5713.28	I
160		5719.18	I
25	B1	5720.16	H
12		5748.72	I
14		5767.48	II
12		5809.50	II
19		5817.47	I
25		5842.23	II
25		5845.87	I
19		5847.77	I
22		5883.66	I
15		5926.47	I
60		5933.69	I
75		5974.28	I
25		5974.72	I
60		5978.66	I
50		5991.90	I
45	e	6010.79	I
28	b	6021.12	H
25	b	6043.19	H
25		6054.17	I
95		6098.67	I
95		6185.13	I
55		6210.70	I
28		6216.82	I
45		6238.58	I
60		6248.95	II
22	h	6299.54	I
25	h	6311.85	I
19		6318.33	I
30		6338.10	I
19	h	6380.19	I
60		6386.23	I
19	h	6409.52	I
15	h	6556.50	I
28		6587.23	I
45		6644.60	II
19		6647.06	II
11		6659.40	I
30		6713.48	I
17		6754.61	II
11		6769.95	I
85		6789.27	I
160		6818.94	I
15		6826.56	I
13		6850.07	I
35		6858.70	I
45		6911.40	I
10		6926.19	I
19		6979.59	I
21		6980.91	II
7		7019.25	I
7		7030.33	II
7		7035.13	I
11		7061.90	I
15		7062.87	I
160		7063.83	I
11	h	7094.40	I
15		7100.54	I
55		7119.52	I
570		7131.81	I
650		7237.10	I
410		7240.87	I
6		7262.62	I
75		7320.05	I
16		7321.76	I
6		7356.10	I
6		7365.28	I
20		7390.70	I
6		7423.69	I
25		7437.56	I
13		7463.86	I
7		7484.56	I
15		7556.37	I
75		7562.93	I
15		7564.22	I
11		7576.95	I
11		7592.96	I
13		7608.59	I
360		7624.40	I
20		7645.64	I
110		7740.17	I
8		7743.57	I
5		7757.89	II
40		7790.90	I
7		7796.81	I
35		7814.55	I
310		7845.35	I
7		7846.56	I
130		7920.71	I
29		7938.06	I
250		7994.73	I
7		8010.58	I
25		8056.52	I
25		8080.32	I
16		8173.89	I
130		8204.58	I
7		8248.81	I
55		8276.95	I
13		8305.91	II
25		8344.25	I
5		8380.06	I
5		8382.98	I
35		8460.01	I
150		8546.48	I
160		8640.06	I
40		8711.24	I
65		9004.73	I

Hf III
Ref. 404 — R.L.K.

Intensity	Note	Wavelength	Spectrum
		Vacuum	
20		1449.83	III
30		1507.82	III
50		1683.95	III
50		1756.91	III
60		1843.64	III
60		1870.58	III
50		1874.81	III
150		1885.15	III
100		1991.44	III
		Air	
100		2037.76	III
300		2070.94	III
150	h	2085.33	III
200	h	2099.30	III
200	h	2110.31	III
100	h	2119.69	III
200		2155.66	III
200		2183.50	III
200		2195.44	III
100		2213.54	III
200		2234.59	III
200	h	2313.44	III
300		2336.47	III
150	h	2355.48	III
100		2373.30	III
120		2377.57	III
250		2383.540	III
400		2461.74	III
2000		2495.16	III
1000		2515.16	III
100		2534.33	III
400	h	2560.74	III
300	h	2567.46	III
200		2687.22	III
500		2753.60	III
100	h	3060.08	III
200	h	3279.67	III
100	h	3741.94	III

Hf IV
Ref. 369, 425 — R.L.K.

Intensity	Wavelength	Spectrum
	Vacuum	
40	520.04	IV
50	569.19	IV
100	596.56	IV
100	600.90	IV
100	603.16	IV
200	618.27	IV
100	620.19	IV
100	633.58	IV
100	643.05	IV
200	644.54	IV
400	647.39	IV
600	665.65	IV
100	671.36	IV
200	673.49	IV

Hafnium (Cont.)

Intensity	Wavelength	
15	1305.24	IV
12	1357.40	IV
40	1390.39	IV
50	1491.67	IV
35	1528.82	IV
15	1560.18	IV
25	1572.03	IV
100	1717.21	IV
20	1718.57	IV

Air

100	2054.46	IV
7	2014.06	IV
20	7751.29	IV
10	7267.58	IV

Hf V
Ref. 410 — R.L.K.

Intensity	Wavelength	

Vacuum

220	545.41	V
180	600.00	V
100	816.81	V
100	830.69	V
100	836.74	V
100	846.87	V
100	856.32	V
100	861.80	V
135	865.16	V
270	867.25	V
180	875.88	V
135	877.87	V
135	880.37	V
100	880.85	V
180	885.58	V
135	885.80	V
135	894.24	V
100	894.41	V
180	896.14	V
100	896.47	V
135	899.70	V
180	901.54	V
135	901.92	V
135	904.95	V
135	909.70	V
135	913.68	V
135	918.48	V
180	919.10	V
270	921.67	V
135	928.01	V
135	931.50	V
135	947.12	V
245	951.62	V
180	960.12	V
180	964.74	V
160	971.51	V
135	974.62	V
120	984.64	V
135	991.50	V
100	1078.42	V
100	1079.92	V
160	1092.76	V
135	1097.28	V
135	1137.49	V
160	1201.76	V
135	1208.88	V
135	1224.62	V
135	1227.98	V
135	1230.21	V
270	1232.03	V
200	1233.59	V
160	1237.42	V
100	1238.85	V
160	1239.53	V
160	1244.46	V
100	1259.25	V
440	1396.66	V
270	1400.09	V
160	1401.70	V
135	1405.77	V
370	1407.17	V
370	1408.38	V
270	1412.28	V
270	1413.51	V
160	1421.96	V
220	1422.53	V
370	1433.43	V
370	1437.27	V
500	1437.73	V
370	1445.40	V
270	1457.91	V
270	1719.32	V
550	1729.08	V
750	1731.83	V
750	1733.96	V
440	1741.74	V
1000	1749.11	V
1000	1750.19	V
500	1760.89	V
370	1765.62	V
270	1774.02	V
135	1792.39	V

HELIUM (He)
Z = 2
He I and II
Ref. 16, 94, 173, 183, 317
W.C.M.

Intensity	Wavelength	

Vacuum

15	231.454	II
20	232.584	II
30	234.347	II
50	237.331	II
100	243.027	II
300	256.317	II
1000	303.780	II
500	303.786	II
10	320.293	I
2	505.500	I
3	505.684	I
4	505.912	I
5	506.200	I
7	506.570	I
10	507.058	I
15	507.718	I
20	508.643	I
25	509.998	I
35	512.098	I
50	515.616	I
100	522.213	I
400	537.030	I
1000	584.334	I
50	591.412	I
5	958.70	II
6	972.11	II
8	992.36	II
15	1025.27	II
30	1084.94	II
35	1215.09	II
50	1215.17	II
120	1640.34	II
180	1640.47	II

Air

7	2385.40	II
9	2511.20	II
50	2577.6	I
1	2723.19	I
12	2733.30	II
2	2763.80	I
10	2818.2	I
4	2829.08	I
10	2945.11	I
40	3013.7	I
20	3187.74	I
3	3202.96	II
15	3203.10	II
1	3354.55	I
2	3447.59	I
1	3587.27	I
3	3613.64	I
2	3634.23	I
3	3705.00	I
1	3732.86	I
10	3819.607	I
1	3819.76	I
500	3888.65	I
20	3964.729	I
1	4009.27	I
50	4026.191	I
5	4026.36	I
12	4120.82	I
2	4120.99	I
3	4143.76	I
10	4387.929	I
3	4437.55	I
200	4471.479	I
25	4471.68	I
6	4685.4	II
30	4685.7	II
30	4713.146	I
4	4713.38	I
20	4921.931	I
100	5015.678	I
10	5047.74	I
5	5411.52	II
500	5875.65	I
100	5875.97	I
8	6560.10	II
100	6678.15	I
3	6867.48	I
200	7065.19	I
30	7065.71	I
50	7281.35	I
1	7816.15	I
2	8361.69	I
2	9063.27	I
2	9210.34	I
10	9463.61	I
4	9516.60	I
3	9526.17	I
1	9529.27	I
1	9603.42	I
3	9702.60	I
6	10027.73	I
2	10031.16	I
15	10123.6	II
1	10138.50	I
10	10311.23	I
2	10311.54	I
3	10667.65	I
300	10829.09	I
1000	10830.25	I
2000	10830.34	I
9	10913.05	I
3	10917.10	I
4	11626.4	II
30	11969.12	I
20	12527.52	I
50	12784.99	I
20	12790.57	I
7	12845.96	I
10	12968.45	I
2	12984.89	I
12	15083.64	I
200	17002.47	I
1	18555.55	I
6	18636.8	II
500	18685.34	I
200	18697.23	I
100	19089.38	I
20	19543.08	I
1000	20581.30	I
80	21120.07	I
10	21121.43	I
20	21132.03	I
3	30908.5	II
4	40478.90	I

HOLMIUM (Ho)
Z = 67
Ho I and II
Ref. 1 — C.H.C.

Intensity		Wavelength	

Air

170		2502.91	II
80		2508.53	II
110		2513.55	II
95		2518.73	II
170		2533.16	I
130		2536.86	II
80		2556.84	I
80		2567.73	II
80		2586.52	I
60		2591.05	II
95		2592.99	I
190		2605.86	II
110		2610.51	II
95		2613.99	II
60		2625.20	II
80		2640.09	I
80		2640.30	II
60		2649.68	II
80		2666.24	II
70		2689.03	II
210		2713.65	II
230		2733.95	II
270		2750.35	II
110	c	2759.35	II
110		2766.85	II
270		2769.89	II
110		2772.83	II
140		2777.10	II
140		2794.41	II
100		2799.99	II
100		2806.72	II
160	c	2809.99	II
220		2811.36	II
180		2812.00	II
190		2814.74	II
300		2824.20	II
140		2826.64	II
270	c	2831.69	II
210		2834.99	II
110		2835.85	II
110		2844.18	II
100		2844.68	II
270		2849.10	II
100		2861.23	II
250		2861.69	II
150		2862.72	II
210		2871.99	II
230		2874.06	II
160		2874.43	II
360		2880.26	II
460		2880.98	II
340		2894.99	II
160		2895.62	II
170		2900.84	II
570	c	2909.41	II
170		2915.82	II
300		2919.62	II
110		2925.35	II
160		2926.09	II
300	c	2928.30	II
220		2942.05	II
300		2944.49	II
250	c	2953.11	II
390		2973.00	II
410	c	2979.63	II
180		2981.46	II
140		2985.48	II
410		2987.64	II
250		2990.27	II
110		2995.86	II
320	c	3008.10	II
220		3014.60	II
270		3038.69	II
480	c	3049.38	II
410	c	3054.00	II
500	c	3057.45	II
230		3074.30	II
500	c	3082.34	II
910		3084.36	II
430	c	3086.54	II
200		3108.31	II
200	c	3109.91	II
760		3118.50	II
300		3130.99	II
200	c	3134.39	II
300	c	3144.36	II
200		3156.18	II
270		3156.97	II
200	c	3159.67	II
580	c	3166.62	II
390	D1	3171.72	II
810		3173.78	II
390		3174.84	II
270	c	3176.97	II
810	c	3181.50	II
390		3183.84	II
270	Cw	3184.48	II
200		3186.37	I
390	c	3197.83	II
390		3201.76	II
200		3206.86	II
270	c	3210.41	II
200	c	3221.42	II
320		3233.34	II
200		3236.90	II
200		3237.40	II
200	c	3257.45	II
390	c	3278.15	II
270		3279.25	II
980	c	3281.97	II
390		3288.46	II
270	c	3290.96	II
200	c	3305.16	II
200		3319.87	II
230		3320.25	II
200		3331.93	II
630	c	3337.23	II
390	c	3338.86	II
980	c	3343.58	II
200		3344.47	II
360		3350.49	II
320		3352.10	II
320	Cw	3353.55	II
320		3354.58	II
320		3357.91	II
320		3364.27	II
290	c	3370.87	II
230		3374.16	II
290	c	3390.75	II
320	c	3394.60	II
8100	c	3398.98	II
810	c	3410.26	II
1200		3421.63	II
3200		3453.14	II
390	c	3410.65	II
1400	c	3414.90	II
5400		3416.46	II
2000	c	3425.34	II
2000	c	3428.13	II
630	c	3429.18	II
320	c	3432.10	II
390		3449.35	I
810	c	3455.70	II
16000	c	3456.00	II
1600		3461.97	II
360	c	3467.07	II
810	c	3473.91	II
5400	c	3474.26	II
6300		3484.84	II
490		3489.58	II
580	c	3493.09	II

Holmium (Cont.)

Intensity	Code	λ (Å)	Spectrum
2500	c	3494.76	II
810	c	3498.88	II
410	c	3506.95	II
320		3509.37	II
810		3510.73	I
4100	c	3515.59	II
410	c	3519.94	II
630		3540.76	II
1600		3546.05	II
1100	c	3556.78	II
410		3560.15	II
410	c	3573.24	II
630	c	3574.80	II
810		3579.12	I
410		3580.75	II
410		3581.83	II
630	c	3592.23	II
1100	Cw	3598.77	II
340		3599.48	I
540	c	3600.95	II
340		3613.31	II
410		3618.43	I
430	c	3626.69	II
490		3627.25	II
430	c	3631.76	II
430	c	3638.30	II
1600	c	3662.69	I
430		3662.99	I
720		3666.65	I
1400		3667.97	I
320		3669.05	II
450		3669.52	I
450	c	3674.77	II
720		3679.19	I
670		3679.70	I
720		3682.65	I
430		3685.16	II
580		3690.65	I
340		3691.95	I
410		3700.04	I
490	c	3702.35	II
320		3709.76	I
430		3712.88	I
450		3720.72	I
1100		3731.40	I
360		3732.09	I
810		3736.35	I
3200	Cw	3748.17	II
320		3753.73	II
340		3769.09	I
370		3780.00	II
8900	c	3796.75	II
8900	c	3810.73	II
490		3811.86	I
900	c	3813.25	II
300		3821.73	II
390		3829.27	I
320	Cw	3831.9	II
410	c	3835.35	II
1300	c	3837.51	II
410	c	3842.05	II
1100		3843.86	II
490	c	3846.73	II
300		3849.88	I
320		3852.40	II
1800	c	3854.07	II
390	Cw	3856.94	II
720		3857.72	II
2700	c	3861.68	II
540		3862.62	I
360		3872.05	II
320	c	3874.09	II
630		3874.68	II
540		3881.61	II
3000	c	3888.96	II
490		3890.42	I
13000	c	3891.02	II
540		3896.76	II
290		3902.23	II
320		3904.44	I
1300	Cw	3905.68	II
320		3911.80	I
320		3919.45	I
320	c	3936.44	II
220		3938.85	I
320	Cw	3940.53	II
220		3950.56	I
580		3955.73	I
230	c	3959.51	II
490		3959.68	I
220		3975.88	I
390	c	3976.93	I
220	Cw	3985.71	II
220		3993.73	II
380		3999.58	I
160	Cw	4002.59	II
220		4003.39	I
110		4013.50	I
320		4014.20	II
160	c	4018.09	II
160	c	4022.76	II
160	c	4023.94	II
110		4025.39	I
320		4027.21	I
270		4028.86	I
180	c	4031.80	I
220		4037.62	I
220	c	4038.87	II
2700		4040.81	I
5400		4045.44	II
220	c	4047.52	I
8100		4053.93	I
540		4054.48	II
270		4057.57	I
220		4060.31	I
1700		4065.09	II
170		4067.57	I
720		4068.05	I
270		4071.83	I
270		4073.31	I
290		4073.51	I
120	c	4080.23	II
230		4083.67	I
140		4085.09	I
170		4087.33	I
200		4087.59	I
140		4091.64	I
120		4094.78	I
230		4100.22	I
8900		4103.84	I
120		4105.04	I
270		4106.50	I
100		4107.36	I
2900		4108.62	I
300		4112.00	I
100		4112.72	I
270		4116.73	I
1500		4120.20	I
1300		4125.65	I
4300		4127.16	I
300		4134.54	I
1500		4136.22	I
130		4139.34	I
230		4142.19	I
290		4148.97	I
980	Cw	4152.61	II
8100		4163.03	I
160		4172.23	I
2500		4173.23	I
540		4194.35	I
100		4198.08	I
130		4203.21	I
100		4211.30	II
290		4222.29	I
290		4223.47	I
2000		4227.04	I
390		4229.52	II
130	h	4231.24	I
290		4243.78	I
1300	Cw	4254.43	I
130	c	4258.61	II
490		4264.05	I
300		4266.04	I
100		4273.63	II
200		4311.04	I
250		4330.64	II
300		4337.13	II
100	Cw	4346.84	I
1300		4350.73	I
290		4356.73	II
140		4363.93	II
170		4379.14	II
180	c	4384.83	II
150		4400.59	II
120		4401.24	II
180		4403.27	I
200		4420.56	II
130		4444.63	I
100		4473.59	II
300		4477.64	II
120		4484.57	II
140		4510.82	I
100	c	4526.14	II
170		4530.08	II
170	c	4531.28	I
130	c	4531.65	II
170		4534.58	I
200		4562.52	I
120	Cw	4609.32	II
130		4613.37	I
100		4618.84	I
100	c	4628.22	I
290		4629.10	II
200		4649.77	II
130	c	4661.33	II
140	c	4674.62	II
70		4701.17	II
80		4701.69	I
130		4709.84	II
65		4711.39	I
130	c	4717.52	I
35	c	4728.72	I
35		4738.00	II
290		4742.04	II
35	c	4749.09	I
35		4751.40	I
100	c	4757.01	I
35		4762.39	II
35		4763.57	II
55		4777.48	II
30		4779.14	I
70	c	4781.19	I
65		4782.92	I
55	c	4786.29	I
35		4791.48	II
35		4795.92	II
45	h	4798.87	I
27		4812.92	II
55		4832.31	II
30		4833.32	I
30		4855.54	II
45		4860.39	I
27	c	4889.67	II
30		4892.35	I
35		4896.44	II
55		4906.99	II
45		4922.73	I
55	c	4934.89	I
290		4939.01	I
27	c	4946.80	I
45		4948.18	II
65	c	4959.42	II
35		4961.03	II
55	Cw	4966.73	I
250	c	4967.21	II
220		4979.97	I
35	c	4988.96	I
90		4995.05	I
35	c	5012.42	I
55		5013.28	II
65		5026.53	I
30		5028.17	I
55		5032.95	II
65	c	5037.60	I
130		5042.37	I
35		5044.73	I
30		5051.44	II
30		5054.92	II
35	c	5060.75	I
65		5074.34	I
80		5093.07	I
140		5127.81	I
55		5135.27	II
130		5142.59	II
110		5143.22	II
160		5149.59	II
90	c	5167.88	I
130		5182.11	I
55		5187.85	I
90		5190.11	II
18		5195.23	I
45		5221.54	I
35	c	5244.47	I
65		5251.82	I
55		5275.48	I
90		5301.25	I
35		5319.24	I
35		5319.65	I
80		5330.11	I
90		5359.99	I
55		5381.40	I
30		5384.56	I
30		5384.97	I
18	h	5393.85	I
70		5403.17	I
100		5407.08	I
14		5413.62	II
16		5434.39	II
18		5435.87	I
30		5445.39	I
18		5449.8	II
30	h	5451.90	I
14		5454.0	I
30	c	5498.57	I
30		5504.51	I
27		5515.56	II
18		5516.45	II
30		5534.33	I
27		5553.14	I
35	c	5560.94	I
35	b	5563.6	H
70		5566.52	I
18		5573.96	II
35	B1	5584.7	H
55	b	5591.1	H
55	B1	5592.3	H
30	b	5607.1	H
27		5613.64	I
45	b	5626.4	H
65		5627.60	I
30		5628.24	I
55		5640.62	I
70	Bs	5655.9	H
65	b	5658.9	H
140		5659.58	I
70	c	5671.84	I
65		5674.70	
140	c	5691.47	I
70	Bs	5696.3	H
140	c	5696.57	I
27		5734.02	I
45		5736.4	H
55		5739.24	I
22		5749.58	I
30		5766.64	
27	b	5803.8	H
45	b	5819.2	H
27	h	5821.90	I
22		5839.47	I
45	b	5849.4	H
140	c	5860.28	I
27	h	5864.42	I
45		5870.85	I
27	b	5879.6	H
70	c	5882.99	I
35	c	5892.56	I
22		5904.29	I
70		5921.76	I
30	c	5933.71	I
70	Cw	5948.03	I
45		5955.98	I
70		5977.76	I
90		5973.52	I
22		5981.43	I
230	c	5982.90	I
55		6002.04	I
27		6005.33	I
35		6021.43	I
16		6038.97	I
27		6050.71	I
45		6060.31	I
120		6081.79	I
70	Cw	6133.60	I
35		6156.38	I
27		6156.58	I
55		6191.68	I
70		6208.65	I
18		6234.17	I
45	c	6255.75	I
70	c	6305.36	I
22		6306.68	I
30		6321.94	I
30	c	6354.35	I
30	c	6372.59	I
14	h	6373.86	I
22	h	6413.41	I
27	c	6471.77	I
13		6479.17	I
11		6515.30	I
11	h	6538.99	I
70		6550.37	I
15		6560.08	I
35	d	6600.58	I
260		6604.94	I
55		6607.47	I
13		6628.35	I
120		6628.99	I
15		6632.24	I
9	h	6652.98	I
15		6662.52	I
19	c	6680.46	I
24	c	6681.62	I
15	h	6682.02	I
55	Cw	6694.32	I
15	Cw	6722.34	I
40		6745.05	I
13		6766.74	I
28	c	6774.68	I
55	c	6785.43	I
13		6793.7	I
13	Cw	6811.04	I
15	Cw	6820.38	I
24		6821.64	I
17	c	6825.72	I
8	h	6826.62	I
8	h	6852.97	I
17	Cw	6865.85	I
9		6883.36	I
13		6888.50	I
15	c	6892.96	I
17		6897.95	I
15	h	6903.80	I
15	Cw	6913.47	I
9		6916.70	I
40	Cw	6939.49	I
45	Cw	6950.39	I
13	H1	6955.3	I
19		6976.7	II
10		6985.11	I
9		6994.38	I
14	h	7000.71	I
10		7079.07	I
12		7098.58	I
9		7242.08	I
9		7250.60	I
14		7308.55	I
25		7341.43	I
18		7389.40	I
5	h	7496.20	I
10	h	7510.74	I
140		7555.09	I
18		7589.20	I
25		7591.87	I
9	h	7593.64	I
7	h	7594.35	I

Holmium (Cont.)

Intensity		Wavelength	
12		7602.31	II
16		7605.35	I
12		7617.05	I
14		7627.98	I
40	c	7628.42	I
9	c	7641.14	I
4		7648.16	I
14		7653.80	I
12	c	7667.30	I
20		7690.43	I
50	c	7693.15	I
40	Cw	7715.06	I
16		7719.05	I
16	h	7738.98	I
8	c	7752.01	I
60	Cw	7815.48	I
40	Cw	7823.63	I
8	h	7879.22	I
60		7894.64	I
10	h	8464.66	I
10	h	8482.67	I
50		8512.94	I
20		8545.61	II
18		8601.84	II
40		8670.19	I
8	h	8697.32	I
16	h	8805.48	II
20	c	8834.49	I
90		8915.98	II

HYDROGEN (H)
Z = 1

H 1
Ref. 214 — W.C.M.

Intensity	Wavelength	
	Vacuum	
15	926.226	I
20	930.748	I
30	937.803	I
50	949.743	I
100	972.537	I
300	1025.722	I
1000	1215.668	I
500	1215.674	I
	Air	
5	3835.384	I
6	3889.049	I
8	3970.072	I
15	4101.74	I
30	4340.47	I
80	4861.33	I
120	6562.72	I
180	6562.852	I
5	9545.97	I
7	10049.4	I
12	10938.1	I
20	12818.1	I
40	18751.0	I
5	21655.3	I
8	26251.5	I
15	40511.6	I
4	46525.1	I
6	74578	I
3	123685	I

INDIUM (In)
Z = 49

In I and II
Ref. 1, 132, 348—350 — C.H.C.

Intensity		Wavelength	
		Vacuum	
2		1648.00	I
1	h	1676.16	I
5	h	1711.54	I
2	h	1741.23	I
1	h	1758.49	I
		Air	
10		2103.89	II
10		2166.88	II
2		2179.90	I
2		2182.40	I
2		2187.40	I
2		2190.84	I
15		2195.67	II
2		2197.41	I
2		2202.24	I
50		2205.28	II
3		2211.14	I
5		2230.70	I
3		2241.66	I
30		2255.79	II
10		2259.99	I
5		2278.20	I
40		2281.64	II
2		2283.75	I
2		2298.33	I
2		2298.70	I
2		2302.40	I
100	c	2306.05	II
25		2306.86	I
3		2309.42	I
2		2309.75	I
90	d	2313.21	II
2		2315.09	I
30		2323.40	II
5		2324.41	II
3		2324.92	I
70	d	2327.95	II
3		2332.76	I
80	h	2334.57	II
10		2340.19	I
8		2345.90	I
5		2346.56	I
50	d	2350.75	II
5		2358.70	I
15		2378.14	II
10		2379.00	I
110	d	2382.63	II
40		2389.54	I
40		2393.18	I
10		2399.18	I
50	h	2406.47	II
50		2408.76	II
50		2419.06	II
50		2419.20	II
70	h	2427.07	II
20		2429.86	II
10		2430.99	I
50		2432.73	II
60		2442.63	II
100		2447.90	II
60		2453.23	II
60		2460.08	I
30	h	2468.02	II
70		2486.15	II
110	d	2488.62	II
90		2488.95	II
80		2498.59	II
100		2499.60	II
90	d	2500.99	II
60		2508.16	II
110	d	2512.31	II
100		2521.37	I
10		2522.98	I
70		2553.56	II
160	d	2554.44	II
1100		2560.19	I
70		2565.13	II
70	d	2598.75	II
200		2601.76	I
50		2604.04	II
90	d	2654.70	II
100		2662.63	II
140	d	2668.65	II
140		2674.56	II
80		2683.12	II
1600		2710.26	I
300		2713.94	II
130	d	2749.75	II
700		2753.88	I
40		2775.37	I
60		2798.76	II
90	d	2818.97	II
180	c	2836.92	I
30		2858.14	II
80		2865.68	II
120	d	2890.18	II
1100		2932.63	I
100		2941.05	II
20		2957.01	I
60	d	2966.17	II
110	c	2999.40	II
8000		3039.36	I
8		3051.15	I
110		3099.80	II
180	c	3101.8	II
130		3138.60	II
80		3142.75	II
130	d	3146.70	II
150		3155.77	II
100	c	3158.40	II
90	c	3176.30	II
90	d	3198.11	II
13000		3256.09	I
3000		3258.56	I
90	c	3338.50	II
75	c	3376.59	II
100	c	3404.28	II
110	d	3438.40	II
180	c	3693.91	II
95	c	3708.13	II
380	w	3716.14	II
120	c	3718.30	II
160	c	3718.72	II
160	c	3723.40	II
170	w	3795.21	II
230	c	3799.21	II
250	c	3834.65	II
200	c	3842.18	II
100		3889.78	II
100	c	3902.07	II
60	d	3922.12	II
65	c	3934.40	II
250	w	3962.35	II
120		4004.66	II
140	d	4013.92	II
410		4056.94	II
17000		4101.76	I
140	c	4205.14	II
100	d	4213.04	II
110	d	4219.66	II
150		4372.87	II
150	c	4500.78	II
18000		4511.31	I
110	c	4549.01	II
140		4570.85	II
180	w	4578.02	II
180	w	4578.40	II
140		4616.08	II
170	c	4617.17	II
250	c	4620.14	II
150	w	4620.70	II
170		4627.30	II
140	c	4637.04	II
380	c	4638.16	II
220	c	4644.58	II
360	c	4655.62	II
320	w	4656.74	II
190	c	4681.11	II
450	w	4684.8	II
3		4878.37	I
90	d	4907.06	II
70	d	4924.93	II
150	c	4973.77	II
80	h	5109.36	II
100	w	5115.14	II
140	c	5117.40	II
270	c	5120.80	II
200	w	5121.75	II
80	c	5129.85	II
240	c	5175.42	II
140	c	5184.44	II
30		5254.32	I
12		5262.74	I
150	c	5309.45	II
80		5411.41	II
140	c	5418.45	II
220	w	5436.70	II
130	c	5497.50	II
140	c	5507.08	II
320	c	5513.00	II
250	w	5523.28	II
130	c	5536.50	II
190	w	5555.45	II
240	c	5576.90	II
200	w	5636.70	II
160	c	5708.50	II
50		5709.91	I
100	c	5721.80	II
50		5727.68	I
210	c	5853.15	II
490	w	5903.4	II
260	w	5915.4	II
120	c	5918.78	II
130	c	6062.9	II
250	c	6095.95	II
210	c	6108.66	II
180	w	6115.9	II
230	w	6128.7	II
240	w	6129.4	II
320	w	6132.1	II
150	c	6140.0	II
90		6143.23	II
140	c	6148.10	II
190	w	6149.5	II
80		6161.15	II
180	w	6162.45	II
100	c	6224.28	II
280	w	6228.3	II
140	c	6231.1	II
270	w	6304.8	II
290	w	6362.3	II
300	w	6469.0	II
210	c	6541.20	II
190	c	6751.88	II
180	c	6765.9	II
100	c	6783.72	II
8	h	6847.44	I
320	w	6891.5	II
4	h	6900.13	I
380	w	7182.9	II
180	c	7255.0	II
210	c	7276.5	II
180	c	7303.4	II
320	c	7350.6	II
100	c	7632.7	II
100	c	7682.9	II
210	c	7740.7	II
100	c	7776.96	II
180	c	7789.0	II
70	c	7806.8	II
70	c	7814.5	II
90	c	7840.9	II
20	h	8050.78	I
240	c	8227.0	II
30	h	8238.66	I
15	h	8314.92	I
50	c	8434.55	II
30		8678.95	I
20		8682.63	I
50		8700.25	I
100	w	8813.5	II
80	c	8832.6	II
40		8894.47	I
10		9170.08	I
120	c	9197.7	II
120	c	9202.0	II
220	w	9213.0	II
160	d	9241.1	II
40	c	9349.83	I
60	h	9370.27	I
20		9427.99	I
100		9977.86	I
200		10257.03	I
60	h	10717.42	I
100	h	10744.31	I
20		11334.72	I
20		11731.48	I
10		12912.59	I
9		13429.96	I
5		13824.48	I
6		14316.25	I
3		14419.20	I
6		14668.66	I
7		14719.08	I
2		16504.31	I
6		22291.06	I
7		23879.13	I

In III
Ref. 351 — C.H.C.

Intensity	Wavelength	
	Vacuum	
7	685.31	III
5	691.62	III
1	782.17	III
10	882.24	III
10	890.84	III
10	915.87	III
2	917.45	III
5	926.83	III
30	1403.08	III
30	1434.85	III
20	1487.70	III
20	1494.14	III
10	1524.78	III
20	1530.21	III
30	1532.95	III
100	1625.42	III
20	1642.28	III
20	1702.53	III
100	1748.83	III
2	1767.88	III
1	1810.71	III
30	1842.41	III
40	1850.30	III
15	1862.98	III
	Air	
30	2154.08	III
2	2154.42	III
10	2199.52	III
5	2232.18	III
20	2261.26	III
5	2266.26	III
5	2272.41	III
5	2272.84	III
10	2300.90	III
100	2527.41	III
50	2725.52	III
80	2726.15	III
100	2982.80	III
100	3008.08	III
30	3008.82	III
30	3293.55	III
8	3350.91	III
5	3562.32	III
100	3852.82	III
100	4023.77	III
150	4032.32	III
50	4062.30	III
100	4071.57	III
100	4072.93	III
100	4252.68	III
40	4509.58	III
200	5248.77	III

Indium (Cont.)

Intensity	Wavelength	
100	5645.15	III
40	5723.17	III
100	5819.50	III
200	6197.72	III

In IV
Ref. 352 — C.H.C.

Intensity	Wavelength	
	Vacuum	
8	677.33	IV
15	684.53	IV
8	720.64	IV
4	1470.34	IV
5	1512.62	IV
5	1523.50	IV
7	1532.03	IV
8	1568.92	IV
3	1569.33	IV
12	1601.50	IV
7	1606.59	IV
10	1628.48	IV
8	1651.80	IV
10	1655.62	IV
1	1667.38	IV
5	1670.64	IV
12	1678.07	IV
4	1702.02	IV
15	1707.11	IV
7	1721.89	IV
12	1747.65	IV
10	1748.10	IV
10	1768.77	IV
12	1773.01	IV
8	1789.15	IV
12	1793.38	IV
12	1844.67	IV
9	1851.10	IV
9	1855.84	IV
15	1856.64	IV
15	1874.08	IV
5	1903.52	IV
5	1939.58	IV
3	1943.31	IV
	Air	
6	2004.73	IV
1	2101.17	IV
0	2112.31	IV
4	2225.13	IV
1	1605.08	IV

In V
Ref. 353 — C.H.C.

Intensity	Wavelength	
	Vacuum	
6	368.67	V
6	370.10	V
10	372.82	V
10	372.94	V
2	374.95	V
6	375.84	V
6	376.07	V
10	376.79	V
17	378.61	V
3	379.24	V
9	380.27	V
11	381.56	V
9	382.14	V
11	382.76	V
10	383.05	V
17	386.21	V
10	386.70	V
3	388.66	V
14	388.91	V
11	390.03	V
11	390.92	V
9	392.29	V
9	392.46	V
1	393.60	V
25	393.89	V
11	395.74	V
3	397.73	V
10	399.79	V
9	400.05	V
25	400.57	V
25	402.39	V
3	405.33	V
9	407.28	V
3	407.36	V
9	407.95	V
9	417.43	V
2	418.45	V
2	423.16	V

IODINE (I)

Z = 53

I I and II
Ref. 124, 153, 176, 184
L.J.R.

Intensity	Wavelength	
	Vacuum	
2	655.80	II
6	659.00	II
8	663.98	II
8	664.52	II
8	665.06	II
150	665.70	II
1000	719.55	II
1000	722.98	II
1000	798.16	II
1200	834.10	II
600	847.80	II
1000	873.49	II
1000	875.94	II
2000	879.84	II
1500	881.88	II
1000	891.00	II
1000	893.17	II
1200	1000.57	II
1000	1003.35	II
4000	1018.58	II
10000	1034.66	II
1500	1054.74	II
2000	1066.34	II
3000	1075.21	II
5000	1105.00	II
2500	1111.16	II
1500	1117.22	II
3500	1125.25	II
2000	1131.50	II
1200	1139.75	II
10000	1139.80	II
1500	1154.67	II
1000	1159.87	II
10000	1160.56	II
20000	1166.48	II
1500	1167.05	II
5000	1175.84	II
10000	1170.65	II
15000	1187.34	II
10000	1190.85	II
18	1193.29	I
5000	1198.88	II
7000	1200.22	II
200	1218.41	I
20000	1220.89	II
600	1224.05	I
600	1224.08	I
500	1228.89	I
20000	1234.06	II
600	1251.34	I
2500	1259.15	I
3000	1259.51	I
800	1261.27	I
600	1267.57	I
600	1267.60	I
1500	1275.26	I
3000	1289.40	I
10000	1300.34	I
3000	1302.98	I
3000	1313.95	I
3000	1317.54	I
2000	1330.19	I
20000	1336.52	II
5000	1355.10	I
3000	1357.97	I
5000	1360.97	I
3000	1361.11	I
2500	1367.71	I
2500	1368.22	I
4000	1383.23	I
3000	1390.75	I
2000	1392.90	I
2000	1400.01	I
8000	1425.49	I
5000	1446.26	I
5000	1453.18	I
5000	1457.39	I
5000	1457.47	I
10000	1457.98	I
2500	1458.79	I
4000	1459.15	I
2500	1465.83	I
1000	1485.92	I
5000	1492.89	I
5000	1507.04	I
5000	1514.68	I
15000	1518.05	I
2500	1526.45	I
5000	1593.58	I
5000	1617.60	I
2500	1640.78	I
15000	1702.07	I
12000	1782.76	I
5000	1799.09	I
75000	1830.38	I
15000	1844.45	I
	Air	
2000	2061.63	I
100	2408.01	II
100	2419.18	II
100	2494.74	II
100	2533.60	II
200	2534.27	II
1000	2566.24	II
2000	2582.79	II
300	2593.46	II
200 c	2688.98	II
500	2730.12	II
20	2765.15	II
200	2808.59	II
1500	2878.63	II
1000	2993.87	II
5000	3078.73	II
200	3161.03	II
1000	3175.07	II
300	3355.53	II
250 c	3424.99	II
300 c	3497.41	II
500	3526.90	II
200	3742.14	II
200	4102.23	I
200	4129.21	I
100 d	4134.15	I
500 d	4321.84	I
300	4452.86	II
200	4599.77	II
300 c	4632.45	II
500 d	4666.48	II
1000	4675.53	II
250	4763.31	I
1000	4862.32	I
200	4916.94	I
1000	4986.92	II
400 c	5065.37	II
10000	5119.29	I
200	5149.73	II
3000 c	5161.90	II
300	5176.19	II
600	5216.27	II
500 d	5228.97	II
1000	5234.57	I
3000 p	5265.71	II
500	5269.36	II
400	5299.78	II
400	5322.80	II
10000	5338.22	II
5000 c	5345.15	II
1000 c	5369.86	II
800 c	5405.42	II
800 c	5407.36	II
600	5427.06	I
3000	5435.83	II
1000	5438.00	II
2000 c	5464.62	II
800	5491.50	II
1000 c	5496.94	II
1000	5504.72	II
600 c	5522.06	II
600 c	5598.52	II
1000	5600.32	II
1500	5612.89	II
10000	5625.69	II
1000	5678.08	II
2000 c	5690.91	II
500	5702.05	II
4000 c	5710.53	II
1000	5738.27	II
1000	5760.72	II
1000 d	5764.33	I
500	5774.83	II
500 c	5787.02	II
2000	5894.86	I
5000	5950.25	II
300	5984.86	I
2000 d	6024.08	I
500	6068.93	II
2000 c	6074.98	II
1000	6082.43	I
2000 c	6127.49	II
800	6191.88	I
1000	6204.86	II
500	6213.10	I
800	6244.48	I
900 c	6257.49	II
1000	6293.98	I
500	6313.13	I
800	6330.37	I
400	6333.50	I
2000	6337.85	I
1000	6339.44	I
500	6359.16	I
1000	6566.49	I
2000	6583.75	I
1000	6585.27	I
5000	6619.66	I
500	6661.11	I
600	6665.96	II
500 c	6697.29	I
300	6718.83	II
400	6732.03	I
4000	6812.57	II
1000	6958.78	II
500	6989.78	I
200 c	7085.21	II
500	7120.05	I
1200	7122.05	I
2000	7142.06	I
1000	7164.79	I
400 d	7191.66	I
700	7227.30	I
1000	7236.78	I
500	7237.84	I
500	7351.35	II
5000	7402.06	I
1000	7410.50	I
500	7616.69	I
5000	7468.99	I
500 c	7490.52	I
2000	7554.18	I
500 d	7556.65	I
2000	7700.20	I
500	7798.98	II
600	7897.98	I
500	7969.48	I
1000	8003.63	I
99000	8043.74	I
300 d	8065.70	I
1000	8090.76	I
800 c	8169.38	I
500	8222.57	I
4000	8240.05	I
10000 c	8393.30	I
150	8414.60	II
1000	8486.11	I
1500 c	8664.95	I
500 c	8700.80	I
250 c	8748.22	I
1000	8853.24	I
2000	8853.80	I
3000	8857.50	I
1000 d	8898.50	I
400	8964.69	I
400	8993.13	I
5000	9022.40	I
15000	9116.33	I
1000	9098.86	I
12000	9113.91	I
600	9128.03	I
30	9195.30	II
600	9227.74	I
1000	9335.05	I
4000	9426.71	I
3000	9427.15	I
10 c	9480.33	II
2000	9598.22	I
2000	9649.61	I
3000 d	9653.06	I
5000	9731.73	I
500	10003.05	I
750	10131.16	I
1000	10238.82	I
400	10375.20	I
400	10391.74	I
6	10405.49	II
5000	10466.54	I
1	11084.68	II
400	11236.56	I
350	11558.46	I
320	11778.34	I
450	11996.86	I
300	12033.69	I
150	12304.58	I
60	13149.16	I
140	13958.27	I
200	14287.02	I
100	14460.00	I
225	15032.57	I
105	15528.65	I
150	16037.33	I
15	18275.71	I
20	18348.52	I
15	18982.41	I
35	19070.17	I
110	19105.12	I
50	19370.02	I
10	20648.69	I
220	22183.03	I
150	22226.53	I
30	22309.21	I
32	24420.82	I
12	27365.42	I
9	27573.05	I
10	30361.93	I
8	30383.88	I
10	34295.73	I
9	34513.11	I
3	40228.54	I
2	41633.80	I

Iodine (Cont.)

I III
Ref. 20, 21, 161 — L.J.R.

Intensity	Wavelength	
	Vacuum	
6	666.81	III
8	705.11	III
7	784.64	III
7	784.80	III
8	795.52	III
5	865.97	III
5	920.38	III
6	961.17	III
6	1078.58	III
8	1094.20	III
4	1244.66	III
8	1252.35	III
5	1306.93	III
	Air	
1	2224.43	III
1	2238.12	III
3	2249.31	III
2	2309.38	III
3	2340.85	III
3	2350.43	III
2	2353.46	III
4	2367.74	III
2	2371.45	III
3	2372.45	III
4	2376.47	III
4	2387.12	III
3	2392.01	III
2	2403.06	III
2	2403.63	III
2	2414.85	III
2	2418.49	III
2	2418.85	III
2	2423.91	III
5	2426.12	III
3	2434.88	III
2	2462.50	III
3	2466.69	III
3	2466.99	III
6	2475.36	III
4	2489.27	III
2	2493.21	III
2	2494.27	III
3	2495.16	III
2	2496.07	III
3	2501.41	III
2	2516.82	III
6	2519.75	III
2	2521.72	III
3	2531.99	III
2	2537.56	III
7	2545.71	III
4	2640.77	III
4	2642.11	III
6	2652.25	III
2	2818.48	III
2	2839.44	III
4	2864.67	III
4	2885.15	III
3	2910.98	III
3	2917.35	III
2	2931.11	III
2	3005.68	III
3	3069.23	III
3	3153.88	III
3	3170.14	III
3	3181.66	III
3	3210.14	III
4	3213.49	III
4	3224.93	III
2	3300.47	III
2	3479.53	III
3	3546.92	III
3	3613.81	III
2	3754.40	III
2	3754.55	III
3	3963.16	III
3	4077.14	III

I IV
Ref. 21, 58 — L.J.R.

Intensity	Wavelength	
	Vacuum	
5	601.86	IV
6	612.46	IV
4	615.17	IV
4	654.22	IV
4	654.56	IV
7	919.28	IV
	Air	
5	2249.30	IV

4	2340.84	IV
7	2361.13	IV
5	2367.75	IV
6	2372.45	IV
7	2376.46	IV
4	2385.28	IV
8	2387.11	IV
6	2392.00	IV
4	2403.05	IV
2	2418.45	IV
3	2423.89	IV
9	2426.10	IV
6	2434.85	IV
3	2466.68	IV
3	2466.96	IV
8	2475.35	IV
4	2485.51	IV
5	2489.24	IV
4	2493.20	IV
2	2501.38	IV
3	2513.74	IV
8	2519.74	IV
6	2521.72	IV
4	2531.98	IV
5	2537.54	IV
8	2545.67	IV
4	2640.77	IV
5	2642.11	IV
8	2652.23	IV
3	2818.45	IV
6	2864.68	IV
4	2910.97	IV
5	2917.33	IV
4	3069.17	IV
4	3170.11	IV
4	3181.64	IV
4	3210.12	IV
6	3213.48	IV
6	3224.90	IV
4	3546.90	IV

IV
Ref. 84 — L.J.R.

Intensity	Wavelength	
	Vacuum	
30	363.78	V
36	380.74	V
45	565.53	V
50	607.57	V

IRIDIUM (Ir)
Z = 77

Ir I and II
Ref. 1 — C.H.C.

Intensity	Wavelength	
	Air	
9900	2010.65	I
8700	2022.35	I
15000	2033.57	I
6200	2052.22	I
5000	2060.64	I
3700	2083.22	I
3100	2085.74	I
17000	2088.82	I
14000	2092.63	I
2700	2112.68	I
1800	2119.54	I
2000	2125.44	I
4500	2126.81	II
2000	2127.52	I
4500	2127.94	I
3700	2148.22	I
2500	2150.54	I
3500	2152.68	II
2900	2155.81	I
7900	2158.05	I
2100	2162.88	I
5800	2169.42	II
4500	2175.24	I
2700	2178.17	I
1600	2187.43	II
1100	2190.38	II
740	2191.64	II
910	2208.09	II
1300	2220.37	I
790	2221.07	II
2500	2242.68	II
620	2245.76	II
2100	2253.38	I
	2253.49	I
2100	2255.10	I
1400	2255.81	I
350	2258.51	I
1400	2258.86	I
830	2264.61	I
1100	2266.33	I
1000	2268.90	I
660	2280.00	I
950	2281.02	II
660	2281.91	I
330	2284.60	I
330	2295.08	I
790	2298.05	I
	2298.16	I
460	2299.53	I
910	2300.50	I
2700	2304.22	I
410	2305.47	I
210	2307.27	I
910	2308.93	I
460	2315.38	I
410	2321.45	I
410	2321.58	I
21`	2327.98	I
54υ	2333.30	I
740	2333.84	I
580	2334.50	I
1600	2343.18	I
740	2343.61	I
100	2352.62	I
580	2355.00	I
230	2357.53	II
410	2358.16	I
500	2360.73	I
2500	2363.04	I
370	2368.04	II
3500	2372.77	I
290	2375.09	II
250	2377.28	I
250	2377.98	I
500	2379.38	I
540	2381.62	I
210	2383.17	I
120	2386.58	II
1300	2386.89	I
2500	2390.62	I
2700	2391.18	I
230	2407.59	I
290	2409.37	I
290	2410.17	I
290	2410.73	I
540	2413.31	I
370	2415.86	I
620	2418.11	I
120	2424.32	I
120	2424.66	I
210	2424.89	I
370	2424.99	I
290	2425.66	I
170	2426.53	II
540	2427.61	I
540	2431.24	I
1300	2431.94	I
170	2432.36	I
100	2432.58	I
270	2435.14	I
250	2445.34	I
250	2447.76	I
190	2448.23	I
910	2452.81	I
1300	2455.61	I
230	2455.87	I
210	2457.03	I
210	2457.23	I
120	2465.09	I
870	2467.30	I
3300	2475.12	I
210	2478.11	I
2100	2481.18	I
100	2485.38	II
620	2493.08	I
210	2496.27	I
250	2502.63	I
4100	2502.98	I
170	2504.37	I
120	2505.74	I
120	2507.63	I
170	2509.71	I
170	2511.94	I
170	2512.58	II
210	2513.71	I
120	2515.36	I
40	2524.88	II
170	2525.05	I
120	2532.52	I
990	2533.13	I
1100	2534.46	I
580	2537.22	I
170	2537.68	I
100	2541.48	I
580	2542.02	I
40	2542.80	II
7900	2543.97	I
150	2545.54	I
790	2546.03	I
120	2547.20	I
120	2547.69	I
210	2551.40	I
190	2554.40	I
210	2555.35	I
170	2555.88	I
150	2563.28	I
910	2564.18	I
210	2569.88	I
100	2570.62	I
230	2572.70	I
740	2577.26	I
100	2578.71	I
35	2579.49	II
740	2592.06	I
740	2599.04	I
150	2602.04	I
190	2604.55	I
190	2607.52	I
700	2608.25	I
1800	2611.30	I
210	2614.98	I
330	2617.78	I
210	2619.88	I
70	2623.64	II
250	2625.32	I
100	2626.76	I
700	2634.17	I
170	2635.27	I
250	2639.42	I
3500	2639.71	I
210	2644.19	I
170	2653.76	I
100	2656.81	I
1800	2661.98	I
350	2662.63	I
2700	2664.79	I
140	2668.99	I
520	2669.91	I
520	2671.84	I
330	2673.61	I
120	2676.83	I
110	2684.04	I
270	2692.34	I
3000	2694.23	I
110	2704.03	I
160	2712.74	I
140	2744.00	I
330	2772.46	I
250	2775.55	I
520	2781.29	I
330	2785.22	I
540	2797.35	I
1600	2797.70	I
380	2798.18	I
410	2800.82	I
680	2823.18	I
1200	2824.45	I
110	2833.24	II
110	2835.66	I
820	2836.40	I
160	2837.33	I
1100	2839.16	I
820	2840.22	I
160	2842.28	I
3800	2849.72	I
110	2863.84	I
380	2875.60	I
380	2875.98	I
270	2877.68	I
140	2879.41	I
820	2882.64	I
650	2897.15	I
260	2901.95	I
260	2904.80	I
200	2907.24	I
440	2916.36	I
230	2918.57	I
4400	2924.79	I
1200	2934.64	I
880	2936.68	I
250	2938.47	I
190	2939.27	I
140	2940.54	I
2700	2943.15	I
230	2946.97	I
200	2949.76	I
1200	2951.22	I
150	2962.99	I
200	2974.95	I
440	2980.65	I
150	2985.80	I
190	2990.62	I
300	2996.08	I
180	2997.41	I
220	3002.25	I
600	3003.63	I
160	3011.69	I
120	3016.43	I
270	3017.31	I
140	3019.23	I
110	3025.82	I
380	3029.36	I
330	3039.26	I
35	3042.65	II
300	3047.16	I
300	3049.44	I
300	3057.28	I

Iridium (Cont.)

Intensity	Wavelength	
1600	3068.89	I
190	3069.09	I
190	3069.71	I
170	3076.69	I
320	3083.22	I
240	3086.44	I
390	3088.04	I
510	3100.29	I
510	3100.45	I
340	3120.76	I
200	3121.78	I
3400	3133.32	I
190	3150.61	I
190	3154.74	I
190	3159.15	I
140	3168.18	I
490	3168.88	I
370	3177.58	I
170	3180.35	I
370	3198.92	I
610	3212.12	I
370	3219.51	I
5100	3220.78	I
100	3221.28	I
300	3229.28	I
100	3230.76	I
470	3241.52	I
200	3262.01	I
390	3266.44	I
160	3277.28	I
100	3287.59	I
160	3310.52	I
200	3322.60	I
130	3334.16	I
560	3368.48	I
660	3437.02	I
100	3437.50	I
410	3448.97	I
3200	3513.64	I
220	3515.95	I
410	3522.03	I
160	3557.17	I
320	3558.99	I
1200	3573.72	I
320	3594.39	I
111	3603.77	I
190	3617.21	I
160	3626.29	I
660	3628.67	I
220	3636.20	I
300	3661.71	I
300	3664.62	I
320	3674.98	I
200	3687.08	I
140	3725.38	I
200	3731.36	II
130	3738.53	I
530	3747.20	I
120	3793.79	I
3100	3800.12	I
230	3817.24	I
170	3865.64	I
480	3902.51	I
480	3915.38	I
400	3934.84	I
120	3946.27	I
590	3976.31	I
460	3992.12	I
180	4020.03	I
350	4033.76	I
130	4040.08	I
370	4069.92	I
150	4070.68	I
100	4092.61	I
140	4115.78	I
23	4127.92	I
27	4155.70	I
15	4166.04	I
90	4172.56	I
35	4182.47	I
15 h	4183.21	I
18	4185.66	I
23	4197.54	I
27	4217.76	I
13	4220.80	I
75	4259.11	I
27	4265.30	I
260	4268.10	I
23	4286.62	I
75	4301.60	I
55	4310.59	I
220	4311.50	I
18	4351.30	I
18	4352.56	I
18	4392.59	I
160	4399.47	I
65	4403.78	I
110	4426.27	I
15	4450.18	I
55	4478.48	I
16	4495.35	I
11 h	4496.03	I
55	4545.68	I
30	4548.48	I
13	4550.78	I
35	4568.09	I
18	4570.02	I
18	4604.48	I
75	4616.39	I
26	4656.18	I
17 h	4668.99	I
21	4708.88	I
50	4728.86	I
21	4731.86	I
26	4756.41	I
13	4757.96	I
65	4778.16	I
30	4795.67	I
10	4807.14	I
21	4809.47	I
10	4840.77	I
17	4845.38	I
50	4938.09	I
26	4970.48	I
25	4999.74	I
25	5002.74	I
17	5009.17	I
30	5014.98	I
17	5046.06	I
30	5123.66	I
20	5177.95	I
22	5238.92	I
12	5340.74	I
35	5364.32	I
75	5449.50	I
30	5454.50	I
7	5469.40	I
10	5620.04	I
45	5625.55	I
10	5828.55	I
10	5882.50	I
7	5887.36	I
35	5894.06	I
7	6026.10	I
12	6067.83	I
20	6110.67	I
12	6288.28	I
7	6376.66	I
5	6624.73	I
10	6686.08	I
5	6830.01	I
5	6929.88	I
4	7183.71	I
6	7834.32	I

IRON (Fe)
Z = 26

Fe I and II
Ref. 56, 63, 105, 138, 174, 278
— H.M.C. and H.C.

Intensity	Wavelength	

Vacuum

Intensity	Wavelength	
12	1055.27	II
15	1068.36	II
15	1071.60	II
15	1096.89	II
12	1099.12	II
18	1112.09	II
12	1121.99	II
12	1122.86	II
12	1128.07	II
12	1130.43	II
15	1133.41	II
15	1133.68	II
12	1138.64	II
12	1142.33	II
12	1143.23	II
18	1144.95	II
12	1147.41	II
15	1148.29	II
12	1151.16	II
12	1267.44	II
12	1272.00	II
12	1371.02	II
12	1563.79	II
12	1580.62	II
18	1608.46	II
12	1618.47	II
15	1621.68	II
15	1629.15	II
15	1631.12	II
18	1635.40	II
15	1636.32	II
15	1639.40	II
12	1641.76	II
12	1647.16	II
12	1670.74	II
12	1702.04	II
12	1761.38	II
20	1785.26	II
20	1786.74	II
18	1788.07	II
30	1934.538	I
25	1937.269	I
50	1946.988	I
25	1951.571	I
30	1952.59	I
30	1953.005	I
60	1957.823	I
60	1960.144	I
30	1961.25	I
50	1962.111	I
12	1963.11	II

Air

Intensity	Wavelength	
100	2084.122	I
50	2157.794	I
15	2162.02	II
40	2166.773	I
300	2178.118	I
150	2186.488	I
60	2186.892	I
120	2187.195	I
250	2191.839	I
150	2196.043	I
80	2200.390	I
80	2200.724	I
15	2208.41	II
20	2213.65	II
12	2218.26	II
20	2220.38	II
25	2245.58	II
50	2250.790	I
60	2251.874	I
25	2255.77	II
300	2259.511	I
60	2264.389	I
80	2267.085	I
80	2267.469	I
50	2270.862	I
150	2272.070	I
150	2276.026	I
80	2279.937	I
150	2284.086	I
150	2288.xxx	I
300	2292.524	I
80	2294.41	I
200	2297.787	I
600	2298.169	I
80	2299.220	I
300	2300.142	I
50	2301.684	I
100	2303.424	I
150	2303.581	I
120	2308.999	I
150	2313.104	I
200	2320.358	I
100	2327.40	II
15	2327.88	II
100	2331.31	II
15	2331.97	II
300	2332.80	II
200	2338.01	II
600	2343.49	II
80	2343.96	II
150	2344.28	II
25	2344.98	II
50	2345.34	II
200	2348.11	II
250	2348.30	II
50	2351.20	II
15	2351.67	II
25	2352.31	II
30	2353.47	II
15	2353.68	II
50	2354.48	II
40	2354.89	II
200	2359.12	II
15	2359.59	II
150	2360.00	II
120	2360.29	II
30	2360.51	II
40	2362.02	II
60	2363.86	II
200	2364.83	II
80	2365.76	II
25	2366.59	II
80	2368.59	II
80	2369.04	I
80	2369.95	II
25	2370.50	II
120	2371.430	I
300	2373.624	I
150	2373.74	II
120	2374.518	I
60	2375.19	II
120	2376.43	II
20	2378.13	II
80	2379.27	II
20	2379.41	II
40	2380.20	II
120	2380.76	II
150	2381.835	I
1000	2382.04	II
20	2382.90	II
20	2383.06	II
60	2383.25	II
50	2384.39	II
40	2388.37	II
300	2388.63	II
200	2389.973	I
30	2390.10	II
20	2390.77	II
15	2391.48	II
20	2392.58	II
40	2395.42	II
1000	2395.62	II
15	2396.72	II
300	2399.24	II
20	2400.05	II
15	2401.29	II
50	2404.43	II
800	2404.88	II
250	2406.66	II
80	2406.97	II
300	2410.52	II
200	2411.07	II
50	2411.81	II
150	2413.31	II
20	2416.45	II
80	2417.87	II
15	2418.44	II
60	2420.396	I
60	2422.69	II
60	2423.089	I
40	2423.21	II
150	2424.14	II
15	2424.39	II
30	2424.59	II
30	2428.29	II
120	2428.36	II
25	2428.80	II
25	2429.03	II
20	2429.39	II
30	2429.86	II
120	2430.08	II
25	2431.02	II
80	2432.26	II
60	2432.87	II
25	2434.06	II
20	2434.24	II
20	2434.65	II
50	2434.73	II
50	2434.95	II
25	2436.62	II
60	2438.182	I
150	2439.30	II
150	2439.74	I
80	2440.11	I
40	2440.42	II
30	2442.37	II
100	2442.57	I
60	2443.71	II
250	2443.872	I
100	2444.51	II
50	2445.11	II
50	2445.212	I
100	2445.57	II
40	2445.80	II
50	2446.11	II
30	2446.47	II
40	2447.20	II
25	2447.33	II
60	2447.709	I
30	2447.75	II
25	2449.96	II
25	2450.20	II
100	2453.476	I
20	2453.98	II
30	2454.58	II
15	2455.71	II
15	2455.90	II
15	2457.09	II
1500	2457.598	I
150	2458.78	II
40	2458.97	II
60	2460.44	II
80	2461.28	II
100	2461.86	II
100	2462.181	I
1500	2462.647	I
50	2463.29	II
50	2463.730	I
40	2464.01	II
40	2464.90	II
800	2465.149	I
50	2465.91	II
15	2466.50	II
60	2466.67	II
60	2466.82	II
60	2467.732	I
15	2468.29	II
600	2468.879	I
60	2469.51	II
25	2470.41	II
80	2470.67	II

Iron (Cont.)

Int	λ		Int	λ		Int	λ		Int	λ	
80	2470.965	I	50	2537.14	II	250	2718.436	I	250	2957.364	I
800	2472.336	I	50	2538.20	II	4000	2719.027	I	80	2959.99	I
40	2472.43	II	40	2538.50	II	100	2719.420	I	150	2965.254	I
40	2472.60	II	100	2538.80	II	50	2720.197	I	1500	2966.898	I
1000	2472.895	I	100	2538.91	II	1500	2720.903	I	120	2969.36	I
200	2473.16	I	150	2538.99	II	400	2723.578	I	800	2970.099	I
50	2473.32	II	50	2539.357	I	30	2724.88	II	15	2970.52	II
30	2474.05	II	200	2540.66	II	150	2724.953	I	1200	2973.132	I
600	2474.814	I	600	2540.972	II	80	2726.05	I	500	2973.235	I
50	2475.12	II	80	2541.10	II	50	2726.235	I	600	2981.445	I
40	2475.54	II	60	2541.84	II	25	2727.38	II	1000	2983.570	I
15	2476.26	II	300	2542.10	I	80	2727.54	II	60	2984.77	I
60	2476.657	I	25	2542.78	II	200	2728.020	I	50	2984.82	II
25	2477.34	II	60	2543.38	II	50	2728.820	I	13	2985.54	II
60	2478.57	II	250	2543.92	I	80	2728.90	II	1000	2994.427	I
120	2479.480	I	150	2544.70	II	40	2730.73	II	250	2994.502	I
1200	2479.776	I	40	2544.97	II	1000	2733.581	I	500	2999.512	I
100	2480.16	II	40	2545.22	II	60	2734.005	I	120	3000.451	I
15	2481.05	II	800	2545.978	I	50	2734.268	I	800	3000.948	I
80	2482.12	II	40	2546.44	II	500	2735.475	I	60	3001.655	I
25	2482.32	II	80	2546.67	II	50	2735.612	I	15	3002.64	II
100	2482.66	II	80	2546.87	I	500	2737.310	I	200	3007.282	I
10000	2483.271	I	100	2548.74	II	120	2737.83	I	500	3008.14	I
300	2483.533	I	80	2549.08	II	400	2739.55	II	120	3009.569	I
1000	2484.185	I	80	2549.39	II	250	2742.254	I	60	3017.627	I
60	2484.24	II	60	2549.46	II	800	2742.405	I	60	3018.983	I
30	2484.44	II	600	2549.613	I	200	2743.20	II	60	3020.01	II
50	2485.990	I	40	2549.77	II	150	2743.565	I	500	3020.491	I
800	2486.373	I	60	2550.03	II	200	2744.068	I	1500	3020.639	I
100	2486.691	I	25	2550.15	II	80	2744.527	I	600	3021.073	I
100	2487.066	I	50	2550.68	II	300	2746.48	II	500	3024.032	I
120	2487.370	I	40	2560.28	II	100	2749.32	II	150	3025.638	I
4000	2488.143	I	25	2562.09	II	500	2749.48	II	500	3025.842	I
100	2488.945	I	400	2562.53	II	1200	2750.140	I	80	3030.148	I
80	2489.48	II	200	2563.48	II	20	2751.13	II	60	3031.214	I
1000	2489.750	I	60	2566.91	II	20	2752.15	II	60	3034.484	I
50	2489.83	II	25	2570.52	II	80	2753.29	II	40	3036.96	II
50	2489.913	I	30	2570.85	II	50	2753.69	I	800	3037.389	I
3000	2490.644	I	150	2574.36	II	150	2754.032	I	80	3041.637	I
100	2490.71	II	50	2575.74	I	100	2754.426	I	800	3047.604	I
60	2490.86	II	300	2576.691	I	30	2754.89	II	600	3057.446	I
2000	2491.155	I	25	2576.86	II	800	2755.73	II	1000	3059.086	I
100	2491.40	II	60	2577.92	II	250	2756.328	I	250	3067.244	I
25	2492.34	II	50	2582.30	I	100	2757.316	I	120	3075.719	I
100	2493.18	II	100	2582.58	II	50	2759.81	I	120	3091.577	I
500	2493.26	II	1500	2584.54	I	120	2761.780	I	80	3098.189	I
60	2494.000	I	650	2585.88	II	150	2761.81	II	100	3099.895	I
50	2494.251	I	90	2588.00	I	150	2762.026	I	100	3099.968	I
100	2495.87	I	90	2591.54	II	120	2762.772	I	60	3100.303	I
600	2496.533	I	30	2592.78	II	120	2763.109	I	100	3100.665	I
50	2497.82	II	60	2593.51	I	20	2763.66	II	12	3154.20	II
150	2498.90	I	90	2593.73	II	25	2765.13	II	80	3175.445	I
40	2500.92	II	650	2598.37	II	80	2766.910	I	150	3184.895	I
1000	2501.132	I	2000	2599.40	II	250	2767.522	I	250	3191.659	I
40	2501.31	II	300	2599.57	II	50	2769.30	I	500	3193.226	I
50	2501.693	I	20	2605.34	II	25	2769.35	II	800	3193.299	I
60	2502.39	II	20	2605.42	II	300	2772.07	I	12	3196.08	II
40	2503.33	II	60	2605.657	I	50	2773.23	I	200	3196.928	I
60	2503.87	II	300	2606.51	II	20	2774.69	II	80	3199.500	I
80	2506.09	II	800	2606.827	I	15	2776.91	II	60	3200.47	I
40	2506.80	II	650	2607.09	II	60	2778.07	I	50	3205.398	I
500	2507.900	I	20	2611.07	II	600	2778.220	I	50	3211.67	I
30	2508.34	II	600	2611.87	II	40	2779.30	II	100	3211.88	I
50	2508.753	I	320	2613.82	II	50	2783.69	II	13	3213.31	II
1000	2510.835	I	320	2617.62	II	30	2785.19	II	200	3214.011	I
120	2511.76	II	250	2618.018	I	3000	2788.10	I	200	3214.396	I
80	2512.275	I	20	2619.07	II	20	2793.89	II	60	3215.938	I
400	2512.365	I	90	2620.41	II	200	2797.78	I	50	3217.377	I
50	2514.38	II	20	2620.69	II	30	2799.29	II	80	3219.583	I
80	2516.570	I	40	2621.67	II	400	2804.521	I	60	3219.766	I
50	2517.13	II	400	2623.53	I	1500	2806.98	I	300	3222.045	I
300	2517.661	I	50	2625.49	II	2500	2813.287	I	600	3225.78	I
800	2518.102	I	200	2625.67	II	300	2823.276	I	13	3227.73	II
60	2519.05	II	150	2628.29	II	600	2825.56	I	80	3227.796	I
150	2519.629	I	20	2630.07	II	50	2825.687	I	20	3230.42	II
40	2521.09	II	250	2631.05	II	120	2828.808	I	80	3233.05	I
30	2521.82	II	250	2631.32	II	25	2831.56	II	50	3233.967	I
50	2522.480	I	50	2631.61	II	1500	2832.436	I	120	3234.613	I
4000	2522.849	I	100	2632.237	I	120	2835.950	I	300	3236.222	I
200	2523.66	I	300	2635.809	I	200	2838.119	I	100	3239.433	I
500	2524.293	I	50	2641.646	I	30	2839.51	II	80	3244.187	I
100	2525.02	I	200	2643.998	I	20	2839.80	II	80	3246.005	I
200	2525.39	II	60	2664.66	II	15	2840.65	II	60	3254.36	I
25	2526.07	II	30	2666.64	II	200	2843.631	I	80	3265.046	I
300	2526.29	II	300	2666.812	I	1000	2843.977	I	50	3265.617	I
2000	2527.435	I	60	2666.965	I	100	2845.594	I	50	3271.000	I
30	2527.70	II	600	2679.062	I	15	2848.11	II	50	3280.26	I
800	2529.135	I	500	2684.75	II	15	2848.32	II	150	3286.75	I
25	2529.23	II	400	2689.212	I	800	2851.797	I	120	3305.97	I
80	2529.31	I	60	2692.60	II	30	2856.91	II	200	3306.343	I
250	2529.55	II	50	2696.28	I	25	2858.34	II	400	3355.227	I
150	2529.836	I	200	2699.106	I	50	2869.307	I	80	3355.517	I
40	2530.11	II	60	2703.99	II	50	2872.334	I	60	3369.546	I
200	2530.687	I	80	2706.012	I	80	2874.172	I	120	3370.783	I
120	2533.63	II	400	2706.582	I	50	2894.504	I	50	3378.678	I
60	2533.80	I	60	2708.571	I	120	2912.157	I	50	3380.110	I
100	2534.42	II	20	2709.05	II	120	2929.007	I	60	3383.978	I
120	2535.49	II	200	2711.655	I	1200	2936.903	I	12	3388.13	II
400	2535.607	I	80	2714.41	II	60	2941.343	I	50	3392.304	I
60	2536.67	II	50	2716.22	II	12	2944.40	II	150	3392.651	I
200	2536.792	I	50	2716.257	I	1000	2947.876	I	150	3399.333	I
200	2536.80	II	50	2717.786	I	60	2950.24	I	80	3404.353	I
50	2536.84	II	50	2717.87	II	600	2953.940	I	500	3407.458	I

Iron (Cont.)

Int.	λ	Sp.	Int.	λ	Sp.	Int.	λ	Sp.	Int.	λ	Sp.
250	3413.131	I	3000	3748.262	I	80	4191.430	I	100	5266.555	I
60	3424.284	I	80	3748.964	I	40	4195.329	I	1200	5269.537	I
500	3427.119	I	3000	3749.485	I	150	4198.304	I	800	5270.357	I
60	3428.748	I	1500	3758.232	I	40	4199.095	I	30	5281.789	I
6000	3440.606	I	400	3760.05	I	300	4202.029	I	60	5283.621	I
2500	3440.989	I	1500	3763.788	I	40	4203.984	I	25	5302.299	I
1000	3443.876	I	400	3765.54	I	80	4206.696	I	11	5306.18	II
200	3445.149	I	600	3767.191	I	80	4210.343	I	13	5316.23	II
15	3453.61	II	60	3776.452	I	400	4216.183	I	150	5324.178	I
1200	3465.860	I	250	3785.95	I	100	4219.360	I	800	5328.038	I
2000	3475.450	I	100	3786.68	I	50	4222.212	I	300	5328.531	I
500	3476.702	I	250	3787.880	I	50	4225.956	I	100	5332.899	I
2500	3490.574	I	250	3790.092	I	200	4227.423	I	14	5339.59	II
500	3497.840	I	150	3794.34	I	11	4233.17	II	80	5339.928	I
250	3513.817	I	400	3795.002	I	100	4233.602	I	500	5341.023	I
300	3521.261	I	120	3797.518	I	250	4235.936	I	25	5364.87	I
400	3526.040	I	250	3798.511	I	50	4238.809	I	40	5367.47	I
100	3526.166	I	400	3799.547	I	50	4247.425	I	50	5369.96	I
60	3526.237	I	200	3805.345	I	200	4250.118	I	400	5371.489	I
60	3526.381	I	80	3806.696	I	300	4250.787	I	60	5383.37	I
60	3526.467	I	600	3812.964	I	40	4258.315	I	14	5387.06	II
100	3533.199	I	60	3813.059	I	800	4260.473	I	40	5393.167	I
200	3536.556	I	1500	3815.840	I	250	4271.153	I	12	5395.86	II
300	3541.083	I	2500	3820.425	I	1200	4271.759	I	300	5397.127	I
250	3542.075	I	150	3821.179	I	1200	4282.402	I	15	5402.06	II
80	3553.739	I	80	3824.306	I	80	4291.462	I	60	5404.12	I
400	3554.925	I	2500	3824.444	I	250	4299.234	I	250	5405.774	I
200	3556.878	I	1500	3825.880	I	1200	4307.901	I	30	5410.91	I
400	3558.515	I	1200	3827.823	I	150	4315.084	I	60	5415.20	I
1000	3565.379	I	1000	3834.222	I	1500	4325.761	I	60	5424.07	I
1200	3570.097	I	120	3839.257	I	80	4352.734	I	30	5427.83	II
800	3570.25	I	500	3840.437	I	80	4369.771	I	250	5429.695	I
120	3571.996	I	800	3841.047	I	800	4375.929	I	13	5429.99	II
100	3573.393	I	120	3843.256	I	3000	4383.544	I	100	5434.523	I
60	3573.829	I	80	3846.800	I	1200	4404.750	I	200	5446.871	I
60	3573.888	I	200	3849.96	I	300	4415.122	I	25	5455.45	I
4000	3581.19	I	120	3850.817	I	600	4427.299	I	120	5455.609	I
150	3582.199	I	2500	3856.372	I	400	4461.652	I	16	5465.93	II
150	3584.660	I	150	3859.212	I	120	4466.551	I	20	5466.94	II
120	3584.929	I	10000	3859.911	I	80	4476.017	I	16	5482.31	II
300	3585.319	I	150	3865.523	I	80	4482.169	I	14	5493.83	II
150	3585.705	I	60	3867.215	I	200	4482.252	I	25	5497.516	I
200	3586.103	I	250	3872.501	I	50	4489.739	I	20	5501.464	I
400	3586.984	I	150	3873.761	I	50	4528.613	I	18	5506.20	II
100	3594.633	I	250	3878.018	I	11	4583.83	II	30	5506.778	I
—	3603.204	I	250	3878.573	I	30	4647.433	I	12	5510.78	II
200	3605.454	I	2000	———	I	30	4736.771	I	12	5529.06	II
500	3606.680	I	4000	3886.282	I	50	4859.741	I	13	5544.76	II
1500	3608.859	I	200	3887.048	I	120	4871.317	I	30	5569.610	I
250	3610.16	I	300	3888.513	I	60	4872.138	I	60	5572.841	I
60	3612.068	I	800	3893.038	I	30	4878.208	I	120	5586.755	I
150	3617.788	I	1200	3899.707	I	100	4890.754	I	200	5615.644	I
1500	3618.768	I	400	3902.945	I	250	4891.492	I	20	5624.541	I
200	3621.462	I	250	3906.479	I	30	4903.309	I	12	5645.40	II
150	3622.004	I	80	3916.731	I	150	4918.992	I	50	5662.515	I
150	3623.19	I	600	3920.258	I	500	4920.502	I	20	5762.990	I
100	3631.096	I	1200	3922.911	I	12	4923.92	II	11	5783.63	II
1200	3631.463	I	1200	3927.920	I	1500	4957.597	I	30	5862.353	I
60	3632.041	I	2000	3930.296	I	11	4990.50	II	13	5885.02	II
100	3638.298	I	60	3948.774	I	80	5001.862	I	16	5902.82	II
200	3640.389	I	60	3949.953	I	18	5001.71	II	30	5914.114	I
80	3643.717	I	50	3951.164	I	11	5004.20	II	14	5955.70	II
1500	3647.842	I	50	3952.601	I	30	5005.711	I	30	5956.956	I
250	3649.506	I	60	3956.454	I	100	5006.117	I	18	5961.71	II
80	3650.279	I	250	3956.68	I	60	5012.067	I	30	5962.4	II
200	3651.467	I	60	3966.614	I	30	5014.941	I	13	5965.63	II
120	3670.024	I	100	3969.257	I	12	5018.43	II	40	6065.482	I
150	3670.089	I	80	3977.741	I	11	5030.64	II	30	6102.159	I
100	3676.311	I	40	3981.771	I	25	5030.77	I	40	6136.614	I
150	3677.629	I	50	3983.956	I	12	5035.71	II	40	6137.694	I
1500	3679.913	I	60	3994.114	I	150	5041.755	I	30	6147.73	II
200	3682.242	I	200	3997.392	I	30	5049.819	I	20	6149.24	II
120	3683.054	I	40	3998.053	I	30	5051.634	I	15	6175.16	II
150	3684.107	I	400	4005.241	I	25	5074.748	I	40	6191.558	I
120	3685.998	I	60	4009.713	I	18	5100.73	II	30	6213.429	I
500	3687.456	I	80	4014.53	I	15	5100.95	II	30	6219.279	I
120	3689.477	I	100	4021.867	I	150	5110.357	I	40	6230.726	I
150	3694.008	I	50	4040.638	I	40	5133.69	I	20	6238.37	II
120	3695.051	I	4000	4045.813	I	40	5139.251	I	20	6246.317	I
150	3701.086	I	1500	4063.594	I	100	5139.462	I	80	6247.56	I
80	3704.462	I	50	4066.975	I	11	5144.36	II	30	6252.554	I
1200	3705.566	I	50	4067.977	I	12	5149.46	II	15	6305.32	II
60	3707.041	I	1200	4071.737	I	25	5151.910	I	12	6331.97	II
150	3707.821	I	40	4076.629	I	30	5162.27	I	15	6383.75	II
300	3707.919	I	40	4100.737	I	80	5166.281	I	20	6393.602	I
600	3709.246	I	40	4107.489	I	2500	5167.489	I	30	6399.999	I
120	3716.442	I	150	4118.544	I	80	5168.897	I	20	6411.647	I
8000	3719.935	I	40	4127.608	I	12	5169.03	II	20	6416.90	II
1500	3722.563	I	400	4132.058	I	500	5171.595	I	20	6421.349	I
120	3724.377	I	80	4134.676	I	50	5191.454	I	30	6430.844	I
60	3725.491	I	40	4136.997	I	80	5192.343	I	20	6446.43	II
60	3727.093	I	200	4143.415	I	200	5194.941	I	200	6456.38	II
500	3727.619	I	800	4143.869	I	30	5204.582	I	60	6494.981	I
150	3732.396	I	40	4153.898	I	25	5215.179	I	20	6516.05	II
1200	3733.317	I	50	4154.500	I	150	5216.274	I	20	6546.239	I
5000	3734.864	I	60	4156.799	I	18	5216.85	II	20	6592.913	I
120	3735.324	I	50	4172.744	I	60	5226.862	I	40	6677.989	I
6000	3737.131	I	60	4174.912	I	1000	5227.150	I	15	6855.18	I
100	3738.306	I	50	4175.635	I	13	5227.49	II	15	6945.21	I
400	3743.362	I	50	4177.593	I	250	5232.939	I	20	7067.44	II
80	3743.47	I	120	4181.754	I	13	5247.95	II	15	7130.94	I
6000	3745.561	I	50	4184.891	I	13	5251.23	II	25	7164.443	I
1200	3745.899	I	120	4187.038	I	18	5260.26	II	80	7187.313	I
			120	4187.795	I	11	5264.18	II			

Iron (Cont.)

Intensity		Wavelength	
30		7207.381	I
12		7224.51	II
50		7307.97	II
40		7320.70	II
20		7376.46	II
30		7445.746	I
20		7462.38	II
40		7495.059	I
60		7511.045	I
15		7586.04	I
15		7711.71	II
30		7780.59	I
40		7832.22	I
80		7937.131	I
60		7945.984	I
80		7998.939	I
60		8046.047	I
50		8085.176	I
150		8220.41	I
120		8327.053	I
20		8331.908	I
120		8387.770	I
30		8468.404	I
15		8514.069	I
60		8661.898	I
150		8688.621	I
12		8793.38	I
12		8824.23	I
20		8866.96	I
15		8999.56	I
15		10216.32	I
13		10469.65	I
21		11119.80	I
14		11374.08	I
52		11422.32	I
87		11439.12	I
91		11593.59	I
255		11607.57	I
160		11638.26	I
230		11689.98	I
160		11783.26	I
580		11882.84	I
225		11884.08	I
1030		11973.05	I
15		12638.71	I
14		12879.76	I
17		13565.04	I
30		14236.25	I
24		14285.11	I
14		14292.38	I
16		14308.69	I
96		14400.56	I
20		14442.28	I
72		14512.23	I
50		14555.06	I
14		14565.95	I
40		14826.43	I
37		15051.77	I
28		15207.55	I
94		15294.58	I
16		15335.40	I
30		15621.67	I
25		15631.97	I
14		15723.59	I
41		15769.42	I
28		15813.13	I
13		16444.82	I
20		16486.69	I
105		18856.65	I
47		18987.01	I
25		19113.68	I
22		19791.88	I
14		22380.82	I
21		22619.85	I
38		26222.04	I
17		26659.22	I

Fe III
Ref. 71, 101 — J.R.

Intensity		Wavelength	
		Vacuum	
6		728.81	III
5		730.00	III
5		737.71	III
5		739.26	III
9		807.55	III
8		807.86	III
8		808.84	III
8	p	811.28	III
10		813.38	III
8		838.05	III
10		844.28	III
9		845.41	III
8	w	847.42	III
8		859.72	III
8	p	861.76	III
10	p	861.83	III
8		873.46	III
9		890.76	III
10		891.17	III
8		891.44	III
8		899.42	III
10		950.33	III
10		981.37	III
10	w	983.88	III
8		985.82	III
9		991.23	III
8		1017.25	III
8		1017.74	III
8		1018.29	III
8		1032.12	III
8		1063.87	III
9		1122.53	III
9		1124.88	III
8		1128.02	III
10	h	1505.17	III
10	h	1538.63	III
12	h	1550.20	III
10	h	1601.21	III
10		1869.83	III
12		1877.99	III
10		1882.05	III
12		1886.76	III
13		1890.67	III
11		1893.98	III
20		1895.46	III
10	s	1907.58	III
19		1914.06	III
15		1915.08	III
15		1922.79	III
10	p	1926.01	III
18		1926.30	III
15		1930.39	III
14		1931.51	III
14		1937.34	III
10	l	1938.90	III
14	s	1943.48	III
12		1945.34	III
10		1950.33	III
12		1951.01	III
11		1952.65	III
13		1953.32	III
10		1953.49	III
10	w	1954.22	III
11		1958.58	III
13		1960.32	III
15		1987.50	III
14		1991.61	III
13		1994.07	III
12		1995.56	III
12		1996.42	III
		Air	
10		2061.55	III
12		2068.24	III
14		2078.99	III
10		2084.35	III
12		2090.14	III
15		2097.48	III
12		2097.69	III
12		2103.80	III
10		2107.32	III
15		2151.78	III
12		2157.71	III
12		2158.47	III
10		2161.27	III
12		2166.95	III
12		2171.04	III
15		2174.66	III
12		2180.41	III
10	p	2208.85	III
10		2221.83	III
10		2229.27	III
10		2232.43	III
10		2232.69	III
10		2235.91	III
10		2238.16	III
12	p	2241.54	III
12		2261.59	III
10		2267.42	III
10		2293.06	III
15		2295.86	III
10	p	2317.70	III
10		2319.22	III
10	p	2321.71	III
10		2326.95	III
10	p	2336.77	III
10		2338.96	III
8		2389.53	III
8	p	2582.37	III
8		2595.62	III
8		2617.15	III
9	p	2645.39	III
10	h	2695.13	III
9	h	2695.34	III
8	h	2700.02	III
8	h	2701.13	III
8		2773.31	III
10	p	2813.24	III
8	p	2895.08	III
9	p	2902.47	III
12		2904.43	III
8	p	2905.80	III
10		2907.50	III
12		2907.70	III
8		2923.90	III
8		2948.39	III
8		2963.23	III
12		3001.62	III
12	h	3007.28	III
15		3013.17	III
10	p	3136.43	III
10		3174.09	III
10		3175.99	III
10		3178.01	III
13		3266.88	III
11		3276.08	III
10		3288.81	III
9		3305.22	III
9		3339.39	III
9		3499.59	III
9		3500.28	III
10		3501.76	III
10		3586.04	III
11		3600.94	III
11		3603.88	III
16		3954.33	III
11		3968.72	III
9		3969.49	III
10	w	3979.42	III
10		4035.42	III
11		4053.11	III
12		4081.00	III
10		4120.90	III
11		4122.02	III
11		4122.78	III
15		4137.76	III
13		4139.35	III
9		4140.48	III
9		4154.96	III
18		4164.73	III
9		4164.92	III
13		4166.84	III
13		4174.26	III
9		4210.67	III
11		4222.27	III
13		4235.56	III
9		4238.62	III
12		4243.75	III
12	h	4273.40	III
12		4279.72	III
14	h	4286.16	III
16	h	4296.85	III
18	h	4304.78	III
20	h	4310.36	III
9		4323.68	III
9	h	4372.04	III
9	h	4372.14	III
11	h	4372.31	III
14	h	4372.53	III
18	h	4372.81	III
9		4395.76	III
12		4419.60	III
9		4431.02	III
9		5111.07	III
9		5127.35	III
12		5156.12	III
10		5199.08	III
10		5235.66	III
18		5243.31	III
13	l	5260.34	III
9		5272.37	III
14		5272.98	III
15		5276.48	III
16		5282.30	III
12		5284.83	III
11		5298.12	III
12		5299.93	III
14	w	5302.60	III
10		5306.76	III
9		5310.88	III
10		5322.74	III
11		5346.68	III
12		5353.77	III
12		5363.76	III
10		5368.06	III
11	l	5375.47	III
11		5719.88	III
9		5744.19	III
10		5756.38	III
18		5833.93	III
9		5848.76	III
10		5854.62	III
9		5876.26	III
15		5891.91	III
9		5898.68	III
9		5918.96	III
10	p	5920.13	III
18	p	5929.69	III
10		5952.31	III
14		5953.62	III
9		5968.48	III
12		5979.32	III
9	h	5981.01	III
12	h	5989.08	III
18		5999.54	III
9		6031.02	III
16		6032.59	III
13		6036.56	III
11		6048.72	III
9		6054.18	III
9		6056.36	III
9		6149.99	III
9		6169.74	III
9		6185.26	III
7		6186.56	III
7		6194.79	III
6		6195.43	III
6		6201.37	III
5	s	6203.04	III
5		6259.81	III
6	p	6294.50	III
5		6357.81	III
5	h	7317.63	III
6	h	7320.14	III
5	w	7921.17	III
5	w	8230.88	III
5	w	8231.79	III
9	w	8235.45	III
8	w	8236.75	III
6	w	8238.98	III
5		8563.49	III

Fe IV
Ref. 382 — J.R.

Intensity	Wavelength	
	Vacuum	
10	502.42	IV
11	506.69	IV
11	505.35	IV
17	525.69	IV
15	526.29	IV
10	526.57	IV
13	526.63	IV
10	530.91	IV
10	531.78	IV
10	535.55	IV
14	536.61	IV
10	536.74	IV
15	537.10	IV
13	537.26	IV
14	537.79	IV
13	537.94	IV
10	538.44	IV
10	544.20	IV
10	546.22	IV
10	548.80	IV
11	550.32	IV
10	551.77	IV
13	552.14	IV
11	552.74	IV
10	554.26	IV
10	555.66	IV
10	572.88	IV
10	576.76	IV
10	579.76	IV
14	607.53	IV
13	608.80	IV
10	609.65	IV
12	1425.73	IV
13	1431.43	IV
12	1473.20	IV
13	1489.53	IV
12	1495.18	IV
13	1526.60	IV
13	1530.26	IV
14	1532.63	IV
13	1532.91	IV
15	1533.86	IV
13	1533.95	IV
14	1536.58	IV
12	1538.29	IV
13	1542.16	IV
14	1542.70	IV
12	1546.40	IV
12	1552.35	IV
12	1552.71	IV
12	1562.46	IV
13	1566.26	IV
14	1568.27	IV
12	1570.18	IV
12	1570.42	IV
12	1571.24	IV
12	1577.20	IV
12	1577.76	IV
12	1590.62	IV
13	1591.51	IV
13	1592.05	IV
12	1596.67	IV
13	1598.01	IV
12	1600.50	IV
13	1600.58	IV
13	1601.67	IV

Iron (Cont.)

Intensity	Wavelength	
12	1602.08	IV
13	1603.18	IV
13	1603.73	IV
13	1604.88	IV
13	1605.68	IV
15	1605.97	IV
13	1606.98	IV
17	1609.10	IV
14	1609.83	IV
13	1610.47	IV
13	1611.20	IV
13	1613.64	IV
15	1614.02	IV
13	1614.64	IV
13	1615.00	IV
12	1615.61	IV
16	1616.68	IV
14	1617.68	IV
14	1619.02	IV
12	1620.91	IV
13	1621.16	IV
14	1621.57	IV
13	1623.38	IV
13	1623.53	IV
15	1626.47	IV
14	1626.90	IV
13	1628.54	IV
13	1630.18	IV
17	1631.08	IV
12	1632.08	IV
14	1632.40	IV
13	1634.01	IV
12	1638.07	IV
12	1638.30	IV
14	1639.40	IV
16	1640.04	IV
14	1640.16	IV
15	1641.87	IV
12	1642.88	IV
15	1647.09	IV
15	1651.58	IV
15	1652.90	IV
13	1653.41	IV
13	1656.11	IV
15	1656.65	IV
12	1657.82	IV
12	1658.43	IV
14	1660.10	IV
12	1661.57	IV
13	1662.32	IV
13	1662.52	IV
13	1663.54	IV
13	1668.09	IV
12	1669.61	IV
14	1671.04	IV
12	1672.86	IV
13	1673.68	IV
14	1675.66	IV
12	1676.78	IV
12	1677.12	IV
13	1681.36	IV
12	1681.95	IV
15	1687.69	IV
15	1698.88	IV
12	1700.40	IV
12	1704.93	IV
13	1709.81	IV
15	1711.41	IV
14	1712.76	IV
12	1717.11	IV
14	1717.90	IV
14	1718.16	IV
12	1718.42	IV
13	1719.46	IV
14	1722.71	IV
14	1724.06	IV
12	1724.26	IV
16	1725.63	IV
13	1761.08	IV
12	1764.92	IV
12	1767.36	IV
13	1792.10	IV
13	1796.93	IV
12	1805.32	IV
12	1820.42	IV
13	1827.98	IV
12	1840.24	IV
12	1860.42	IV
12	1869.64	IV
12	1874.23	IV

Fe V
Ref. 381 — J.R.

Intensity	Wavelength	
	Vacuum	
300	361.28	V
300	365.43	V
300	365.86	V
300	374.24	V
300	374.87	V

Intensity	Wavelength	
300	375.98	V
300	379.59	V
300	380.31	V
300	381.27	V
300	384.96	V
300	384.97	V
300	385.03	V
300	385.11	V
300	385.25	V
300	385.26	V
300	385.75	V
300	385.88	V
350	386.16	V
300	386.74	V
300	386.78	V
300	386.85	V
300	386.88	V
350	386.88	V
400	387.20	V
400	387.50	V
300	387.62	V
400	387.76	V
400	387.78	V
300	387.98	V
300	388.61	V
300	388.82	V
300	390.11	V
300	390.19	V
300	390.78	V
300	391.94	V
300	392.38	V
300	392.50	V
300	392.51	V
300	392.70	V
300	392.91	V
300	393.27	V
300	393.72	V
300	393.73	V
300	393.91	V
300	393.97	V
300	394.04	V
300	394.11	V
300	395.15	V
300	395.79	V
400	395.90	V
300	399.84	V
300	400.11	V
300	400.51	V
300	400.52	V
300	400.63	V
300	401.04	V
300	401.64	V
300	401.86	V
300	402.87	V
300	403.06	V
400	404.62	V
400	405.50	V
800	407.42	V
600	407.44	V
400	407.49	V
500	407.75	V
400	409.71	V
400	410.20	V
600	411.55	V
300	415.01	V
300	416.66	V
300	416.84	V
700	417.39	V
700	418.04	V
500	418.47	V
300	420.56	V
700	421.06	V
500	421.78	V
300	422.28	V
500	422.31	V
300	423.23	V
500	426.06	V
500	426.11	V
300	426.83	V
350	426.97	V
300	434.42	V
300	439.22	V
300	444.70	V
300	445.44	V
300	446.04	V
300	458.16	V
300	486.17	V
300	1317.86	V
300	1318.35	V
300	1320.41	V
300	1321.34	V
300	1321.49	V
400	1323.27	V
400	1330.40	V
300	1345.61	V
400	1359.01	V
300	1361.28	V
300	1361.45	V
600	1361.82	V
300	1363.08	V
300	1363.64	V

Intensity	Wavelength	
300	1365.57	V
700	1373.59	V
600	1373.67	V
300	1374.12	V
500	1376.34	V
300	1376.46	V
500	1378.56	V
300	1385.68	V
800	1387.94	V
400	1397.97	V
600	1400.24	V
800	1402.39	V
400	1406.67	V
500	1406.82	V
400	1407.25	V
300	1409.03	V
300	1409.22	V
600	1409.45	V
400	1415.20	V
300	1418.12	V
600	1420.46	V
800	1430.57	V
800	1440.53	V
300	1440.79	V
400	1442.22	V
800	1446.62	V
700	1448.85	V
400	1449.93	V
300	1455.56	V
700	1456.16	V
500	1459.83	V
400	1460.73	V
500	1462.63	V
700	1464.68	V
500	1465.38	V
400	1466.65	V
500	1469.00	V
300	1475.60	V
500	1479.47	V
300	1554.22	V

KRYTPON (Kr)
7 — 36

Kr I and II
Ref. 61, 121, 123, 147, 208, 232
— E.F.W.

Intensity		Wavelength	
		Vacuum	
60		729.40	II
200		761.18	II
100		763.98	II
60		766.20	II
200		771.03	II
60	p	773.69	II
200		782.10	II
100		783.72	II
60		818.15	II
60		830.38	II
100		844.06	II
60		864.82	II
60		868.87	II
200		884.14	II
1000		886.30	II
400		891.01	II
200		911.39	II
2000		917.43	II
50		945.44	I
50		946.54	I
20		951.06	I
50		953.40	I
50		963.37	I
2000		964.97	II
100		1001.06	I
100		1003.55	I
100		1030.02	I
200		1164.87	I
650		1235.84	I
		Air	
100	h	2464.77	II
60		2492.48	II
80	h	2712.40	II
100		2833.00	II
100	h	3607.88	II
200		3631.889	II
250		3653.928	II
80		3665.324	I
150		3669.01	II
100		3679.559	I
80		3686.182	II
300	h	3718.02	II
200		3718.595	II
150		3721.350	II
200		3741.638	II
150		3744.80	II
80		3754.245	II
500		3778.089	II
500		3783.095	II

Intensity		Wavelength	
150	h	3875.44	II
150		3906.177	II
200		3920.081	II
100		3994.840	II
100	h	3997.793	II
300		4057.037	II
300		4065.128	II
500		4088.337	II
250		4098.729	II
100		4109.248	II
250		4145.122	II
150		4250.580	II
1000		4273.969	I
100		4282.967	I
600		4292.923	II
200		4300.49	II
500	h	4317.81	II
400		4318.551	I
1000		4319.579	I
150	h	4322.98	II
100		4351.359	I
3000		4355.477	II
500		4362.641	I
200		4369.69	II
800		4376.121	I
300	h	4386.54	II
200		4399.965	I
100		4425.189	I
500		4431.685	II
600		4436.812	II
600		4453.917	I
800		4463.689	I
800		4475.014	II
400	h	4489.88	II
600		4502.353	I
400	h	4523.14	II
200	h	4556.61	II
800		4577.209	II
300		4582.978	II
150	h	4592.80	II
500		4615.292	II
1000		4619.166	II
800		4633.885	II
2000		4658.876	II
500		4680.406	II
100		4691.301	II
200		4694.360	II
3000		4739.002	II
300		4762.435	II
1000		4765.744	II
300		4811.76	II
300		4825.18	II
800		4832.077	II
700		4846.612	II
150		4857.20	II
300		4945.59	II
200		5022.40	II
250		5086.52	II
400	h	5125.73	II
500		5208.32	II
200		5308.66	II
500		5333.41	II
200		5468.17	II
500		5562.224	I
2000		5570.288	I
80		5580.386	I
100		5649.561	I
400		5681.89	II
200	h	5690.35	II
100		5832.855	I
3000		5870.914	I
200		5992.22	II
60		5993.849	I
60		6056.125	I
300		6420.18	II
100		6421.026	I
200		6456.288	I
150		6570.07	II
60		6699.228	I
100		6904.678	I
250		7213.13	II
100		7224.104	I
80		7287.258	I
400		7289.78	II
400		7407.02	II
60		7425.541	I
200		7435.78	II
100		7486.862	I
300		7524.46	II
1000		7587.411	I
2000		7601.544	I
150		7641.16	II
1000		7685.244	I
1200		7694.538	I
250		7735.69	II
150		7746.827	I
800		7854.821	I
200		7913.423	I
180		7928.597	I
200		7933.22	II
120		7973.62	II
100		7982.401	I
1500		8059.503	I
4000		8104.364	I

Krypton (Cont.)

Intensity		Wavelength		Intensity		Wavelength		Intensity	Wavelength		Intensity		Wavelength	
6000		8112.899	I	1400		39588.4	I	30	3285.89	III			Ref. 1 — C.H.C.	
60		8132.967	I	1100		39589.6	I	30	3304.75	III				
3000		8190.054	I	500		39954.8	I	50	3311.47	III	Intensity		Wavelength	
200		8202.72	II	300		39966.6	I	200	3325.75	III				
80		8218.365	I	1300		40306.1	I	60	3330.76	III			Air	
3000		8263.240	I	250		40685.16	I	50	3342.48	III	240	2187.87	II		
100		8272.353	I					100	3351.93	III	770	2256.76	II		
5000		8298.107	I			**Kr III**		40	3374.96	III	200	2319.44	II		
1500		8281.050	I	Ref. 208, 366, 390, 421 — E.F.W.				100	3439.46	III	400	2610.34	II		
100		8412.430	I					70	3474.65	III	420	2808.39	II		
3000		8508.870	I	Intensity		Wavelength		100	3488.59	III	130	2885.14	II		
150		8764.110	I					200	3507.42	III	160	2893.07	II		
6000		8776.748	I			Vacuum		100	3564.23	III	110	2950.05	II		
2000		8928.692	I	30		467.35	III	30	3641.34	III	180	3104.59	II		
500		9238.48	II	30		540.86	III	30	3690.65	III	130	3142.76	II		
500	hL	9293.82	II	30		565.64	III	40	h	3868.70	III	510	3245.13	II	
200	h	9320.99	II	30		569.16	III	50	4067.37	III	260	3249.35	II		
300		9361.95	II	30		571.98	III	40	4131.33	III	550	3265.67	II		
100		9362.082	I	30		579.83	III	40	4154.46	III	800	3303.11	II		
200	h	9402.82	II	30		585.14	III	20	h	5016.45	III	1500	3337.49	II	
200	h	9470.93	II	30		585.96	III	10	5501.43	III	870	3344.56	II		
500		9577.52	II	30		593.70	III	10	h	6037.17	III	200	3376.33	II	
500	h	9605.80	II	30		594.10	III	10	h	6078.38	III	1500	3380.91	II	
400	h	9619.61	II	30		596.41	III	10	6310.22	III	130	3452.18	II		
200		9663.34	II	40		600.17	III					180	3453.17	II	
200	h	9711.60	II	30		603.67	III		**Kr IV**			200	3574.43	I	
2000		9751.758	I	50		605.86	III	Ref. 366, 409, 417 — E.F.W.				320	3628.83	II	
500		9803.14	II	35		606.47	III					120	3637.15	II	
500		9856.314	II	50		611.12	III	Intensity		Wavelength		170	d	3641.53	I
1000		10221.46	II	35		616.72	III							3641.66	II
100		11187.108	I	40		621.45	III			Vacuum		1000	3645.42	II	
200		11257.711	I	45		622.80	III			793.44	IV	390	3650.18	II	
150		11259.126	I	50		625.02	III			794.11	IV	170	3662.08	II	
500		11457.481	I	30		625.76	III	7		805.76	IV	120	3704.54	I	
150		11792.425	I	45		628.59	III	18		816.82	IV	320	3705.82	II	
1500		11819.377	I	50		630.04	III	22		842.04	IV	550	3713.54	II	
600		11997.105	I	35		633.09	III			Air		140	3714.87	II	
160		12077.224	I	50		639.98	III	3		2237.34	IV	270	3715.53	II	
100		12861.892	I	60		646.41	III	6		2291.26	IV	2400	3759.08	II	
1100		13177.412	I	50		651.20	III	3		2329.3	IV	120	3780.67	II	
1000		13622.415	I	50		659.72	III	4		2336.5	IV	3700	3790.83	II	
2400		13634.220	I	30		664.86	III	4		2348.27	IV	3900	3794.78	II	
800		13658.394	I	40		672.34	III	3		2358.5	IV	190	3835.08	II	
200		13711.036	I	35		672.85	III	3		2388.05	IV	600	3840.72	II	
600		13738.851	I	35		676.57	III	4		2416.9	IV	120	3846.00	II	
150		13974.027	I	35		680.13	III	3		2428.04	IV	1600	3849.00	II	
550		14045.657	I	35		683.68	III	5		2442.68	IV	130	3854.91	II	
140		14104.298	I	45		686.25	III	4		2451.7	IV	3400	3871.64	II	
180		14402.22	I	45		687.98	III	6		2459.74	IV	1700	3886.37	II	
2000		14426.793	I	45		691.93	III	5		2474.06	IV	1300	3916.05	II	
100		14517.84	I	50		695.61	III	4		2517.0	IV	1100	3921.54	II	
1600		14734.436	I	30		698.05	III	5		2518.02	IV	160	3927.56	I	
550		14762.672	I	50		708.36	III	6		2519.38	IV	2200	3929.22	II	
450		14765.472	I	50		714.00	III	5		2524.5	IV	180	3936.22	II	
400		14961.894	I	100	p	722.04	III	5		2546.0	IV	9000	3949.10	II
120		15005.307	I	30		746.70	III	6		2547.0	IV	4400	3988.52	II	
140		15209.526	I	60		785.97	III	4		2558.08	IV	3600	3995.75	II	
1700		15239.615	I	50		837.66	III	3		2586.9	IV	180	4015.39	I	
130		15326.480	I	50		854.73	III	5		2606.17	IV	250	4025.88	II	
1500		15334.958	I	60		862.58	III	10		2609.5	IV	2800	4031.69	II	
700		15372.037	I	40		870.84	III	8		2615.3	IV	140	4037.21	I	
200		15474.026	I	50		876.08	III	7		2621.11	IV	3000	4042.91	II	
180		15681.02	I	75		897.81	III	3		2730.55	IV	320	4050.08	II	
120		15820.09	I	50		987.29	III	8		2748.18	IV	220	4060.33	I	
200		16726.513	I	30		1158.74	III	6		2774.70	IV	160	4064.79	I	
2000		16785.128	I	6		1638.82	III	3		2829.60	IV	850	4067.39	II	
1000		16853.488	I	6		1914.09	III	3		2836.08	IV	110	4076.71	II	
2400		16890.441	I					5		2853.0	IV	2800	4077.35	II	
1600		16896.753	I			Air		3		2859.3	IV	120	4079.18	I	
1800		16935.806	I	40		2393.94	III	3		3142.01	IV	5500	4086.72	II	
600		17098.771	I	40		2494.01	III	6		3224.99	IV	180	4089.61	I	
700		17367.606	I	30		2563.25	III	3		3261.70	IV	280	4099.54	II	
120		17404.443	I	60		2639.76	III	3		3809.30	IV	110	4104.87	I	
150		17616.854	I	30		2680.32	III	5		3860.58	IV	4400	4123.23	II	
650		17842.737	I	40		2681.19	III	5		3934.29	IV	110	4137.04	I	
700		18002.229	I	30		2841.00	III					550	4141.74	II	
2600		18167.315	I	30		2851.16	III		**Kr V**			1100	4151.97	II	
100		18399.786	I	50		2870.61	III	Ref. 409, 421 — E.F.W.				220	4152.78	II	
150		18580.896	I	100		2892.18	III					100	4160.26	I	
300		18696.294	I	30		2909.17	III	Intensity		Wavelength		280	4187.32	I	
170		18785.460	I	50		2952.56	III					280	4192.36	II	
200		18797.703	I	60		2992.22	III			Vacuum		1500	4196.55	II	
140		20209.878	I	50		3022.30	III	150		472.16	V	240	4204.04	II	
300		20423.964	I	80		3024.45	III	100		484.39	V	300	4217.56	II	
140		20446.971	I	50		3046.93	III	250		496.25	V	200	4230.95	II	
600		21165.471	I	30		3056.72	III	120		500.77	V	1600	4238.38	II	
1800		21902.513	I	60		3063.13	III	200		507.20	V	140	4249.99	II	
120		22485.775	I	40		3097.16	III	60		548.04	V	320	4263.59	II	
180		23340.416	I	60		3112.25	III	120		637.87	V	480	4269.50	II	
120		24260.506	I	30		3120.61	III			690.86	V	240	4275.64	II	
180		24292.221	I	100		3124.39	III			691.75	V	300	4280.27	I	
600		25233.820	I	60		3141.35	III	600		708.85	V	600	4286.97	II	
180		28610.55	I	100		3189.11	III			810.70	V	600	4296.05	II	
1000		28655.72	I	80		3191.21	III					120	4300.44	II	
150		28769.71	I	40		3239.52	III		**LANTHANUM (La)**			440	4322.51	II	
140		28822.49	I	40		3240.44	III		**Z = 57**			4600	4333.74	II	
300		29236.69	I	300		3245.69	III					550	4354.40	II	
300		30663.54	I	150		3264.81	III		La I and II			110	4364.67	II	
300		30979.16	I	100		3268.48	III					110	B1	4371.97	L
500		39300.6	I	30		3271.65	III					110	B1	4375.84	L
1100		39486.52	I									110	4378.10	II	
220		39557.25	I									280	4383.44	II	
100		39572.60	I									100	4385.20	II	

Lanthanum (Cont.)

Intensity		Wavelength		Intensity		Wavelength	
220	B1	4418.24	L	250		6455.99	I
160	B1	4423.17	L	110		6526.99	II
160		4423.90	I	130		6543.16	I
260		4427.55	II	140		6578.51	I
100	B1	4428.10	L	180		6709.50	I
2000		4429.90	II	120		6774.26	II
160	B1	4432.98	L	13	b	7011.22	L
100	B1	4438.01	L	75		7023.67	I
100		4452.15	I	26		7032.05	I
100		4455.80	II	26	b	7040.84	L
850		4522.37	II	110		7045.96	I
170		4525.31	II	13	b	7054.80	L
420		4526.12	II	160		7066.23	II
400		4558.46	II	65		7068.37	I
110		4559.29	II	21	B1	7070.79	L
160		4567.91	I	13		7076.38	I
200		4570.02	I	21	B1	7085.40	I
400		4574.88	II	26	B1	7101.02	L
200		4580.06	II	10	h	7116.8	II
160		4605.78	II	19	B1	7131.58	L
410		4613.39	II	10		7149.77	I
410		4619.88	II	40	h	7158.08	I
110		4645.28	II	50		7161.25	I
540		4655.50	II	10		7162.60	L
360		4662.51	II	21		7219.91	I
230		4663.76	II	10	b	7257.16	L
200		4668.91	II	26		7270.09	I
160		4671.83	II	10		7270.30	I
230		4692.50	II	110	Cw	7282.34	II
140		4703.28	II	10		7320.91	I
170		4716.44	II	110	Cw	7334.18	I
140		4719.94	II	65		7345.34	I
230		4728.42	II	50	B1	7379.71	L
500		4740.28	II	85	B1	7380.61	L
390		4743.09	II	35		7382.73	I
320		4748.73	II	110	B1	7403.52	L
160		4766.89	I	210	B1	7403.75	L
160		4804.04	II	50	b	7411.34	L
160		4809.01	II	65	B1	7434.28	L
200		4824.06	II	110	B1	7434.36	L
320		4860.91	II	30	b	7442.92	L
850		4899.92	II	50	h	7463.08	I
1000		4920.98	II	50	B1	7465.25	L
1000		4931.70	II	0F	B1	7466.40	L
140		4934.83	II	75	Cw	7483.50	II
110		4946.47	II	40	B1	7496.40	L
370		4949.77	I	95	B1	7496.78	L
340		4970.39	II	50		7498.83	I
370		4986.83	II	30	b	7506.79	L
140		4991.28	II	19	B1	7528.21	L
720		4999.47	II	50	B1	7528.39	L
140		5046.88	I	30		7533.59	I
210		5050.57	I	85		7539.23	I
170		5056.46	I	35	B1	7560.09	L
200		5106.23	I	35	B1	7592.26	L
470		5114.56	II	19	h	7612.94	II
470		5122.99	II	19	b	7624.99	L
450		5145.42	I	21		7664.34	I
180		5156.74	II	15	h	7841.80	I
180		5157.43	II	21	h	7876.87	I
290		5158.69	I	75	B1	7877.22	L
120		5163.62	II	75	B1	7910.19	L
580		5177.31	I	150	B1	7910.54	L
850		5183.42	II	50	b	7944.61	L
260		5188.22	II	110	B1	7944.95	L
170		5204.15	II	40		7964.83	I
720		5211.86	I	35	b	7979.34	L
520		5234.27	I	75	B1	7979.70	L
340		5253.46	I	35	h	8001.89	I
110		5259.39	II	21	b	8014.43	L
370		5271.19	I	65	B1	8014.79	L
140		5290.84	II	30	b	8019.48	L
370		5301.98	II	35	Hc	8051.39	I
140		5302.62	II	75		8086.05	I
180		5303.55	II	15	b	8122.20	L
110		5340.67	II	15	b	8159.02	L
110		5357.86	I	7	h	8203.38	I
130		5377.09	II	50		8247.44	I
140		5380.99	II	13	h	8316.04	I
500		5455.15	I	85		8324.69	I
470		5501.34	I	95		8346.53	I
110	B1	5602.50	L	8	h	8379.80	L
160		5631.22	I	8	b	8453.55	L
240		5648.25	I	8	h	8467.62	I
130		5657.72	I	26		8476.48	I
180		5740.66	I	13	h	8507.37	I
160		5744.41	I	13	h	8513.57	I
160		5761.84	I	8	h	8514.65	II
160		5769.07	II	17	b	8526.59	L
370		5769.34	I	17	c	8543.46	I
320		5789.24	I	65		8545.44	I
450		5791.34	I	15	b	8563.54	L
220		5797.58	II	9	h	8590.94	I
160		5805.78	II	9	b	8600.81	L
140		5821.99	I	7	h	8624.22	I
320		5930.62	I	15		8638.47	I
720		6249.93	I	19	Hw	8672.11	I
260	d	6262.30	II	40		8674.43	I
180		6296.09	II	13	h	8720.41	I
160		6320.39	II	35		8748.38	I
110		6325.91	I	19		8818.93	I
170		6390.48	II	35		8825.82	I
450		6394.23	I	21		8839.63	I
210		6410.99	I				

La III
Ref. 220, 309 — J.R.

Intensity	Wavelength	
	Vacuum	
3	744.19	III
10	753.03	III
1	786.64	III
200	787.14	III
1	796.03	III
400	796.99	III
1	797.20	III
10	835.03	III
30	845.62	III
1	850.73	III
1	860.39	III
2	860.88	III
5	865.04	III
2000	870.40	III
30	872.43	III
1000	882.34	III
20	882.72	III
200	929.71	III
400	942.86	III
30	967.69	III
10	974.33	III
50	979.99	III
10	980.29	III
200	1058.63	III
1000	1072.59	III
5000	1076.91	III
50000	1081.61	III
95000	1099.73	III
5000	1100.70	III
30	1208.80	III
30	1212.29	III
200	1236.54	III
100	1253.99	III
2000	1255.63	III
100	1259.55	III
100	1322.42	III
5000	1330.04	III
10000	1349.18	III
5000	1459.49	III
2000	1466.44	III
10000	1523.79	III
500	1518.33	III
5000	1536.17	III
200	1923.34	III
500	1938.57	III
	Air	
60	2216.07	III
20	2238.36	III
25	2258.61	III
5	2260.30	III
250	2297.74	III
400	2379.37	III
10	2387.99	III
20	2392.49	III
100	2476.60	III
50	2478.65	III
2	2513.43	III
4	2588.87	III
2	2604.83	III
400	2651.50	III
100	2682.34	III
150	2684.76	III
110	2897.88	III
160	2904.58	III
7	2950.84	III
10	2953.77	III
40	2992.10	III
4	3006.19	III
15	3009.22	III
100	3075.17	III
4	3085.38	III
25	3093.03	III
15	3096.26	III
200	3111.97	III
50	3116.74	III
1000	3171.63	III
1500	3171.74	III
50	3172.69	III
20	3196.84	III
70	3289.11	III
15	3301.48	III
35	3327.66	III
500	3517.09	III
600	3517.22	III
3	4129.24	III
5	4137.43	III
200	4482.97	III
300	4499.05	III
5	5145.73	III
8	5158.41	III
6	5467.81	III
55	5491.90	III
2	5511.72	III
1	5518.19	III
45	5529.54	III
1	5744.09	III
200	5778.14	III
2	5813.45	III
3	5875.63	III
55	5888.62	III
3	5932.71	III
2	6017.11	III
20	6055.84	III
35	6119.25	III
120	6141.99	III
55	6220.00	III
60	6348.21	III
3	8114.42	III
2	8135.96	III
250	8252.60	III
100	8275.39	III
200	8287.75	III
250	8321.11	III
300	8583.45	III
120	9184.38	III
100	9212.63	III
80	9923.99	III
140	10284.79	III
20	10370.34	III
12	10937.90	III
13	894.47	III
14	096.18	III
17	898.09	III

La IV
Ref. 79 — J.R.

Intensity		Wavelength	
		Vacuum	
100		344.12	IV
7000		453.50	IV
10000		463.14	IV
15000		499.54	IV
40000		552.02	IV
30000		631.26	IV
25		724.92	IV
15		733.29	IV
10		797.03	IV
10	c	980.03	IV
50		1039.30	IV
60	p	1062.09	IV
75		1158.35	IV
50		1164.29	IV
400		1230.90	IV
75	p	1260.79	IV
300		1261.12	IV
150		1283.19	IV
2000		1302.31	IV
1200		1333.53	IV
3500		1334.96	IV
1000		1352.76	IV
25000		1368.04	IV
8000		1377.49	IV
3000		1394.32	IV
5000		1414.58	IV
7000		1432.55	IV
7000		1441.63	IV
7500		1462.15	IV
20000		1463.47	IV
7500		1467.54	IV
15000		1507.87	IV
5000		1527.19	IV
2500		1575.92	IV
1500		1583.61	IV
750		1585.11	IV
750		1637.42	IV
750	d	1645.21	IV
1000		1664.84	IV
750		1684.17	IV
2000	p	1767.65	IV
4000		1808.66	IV
1000		1851.81	IV
1500		1852.77	IV
750		1879.79	IV
1000		1881.57	IV
800		1889.22	IV
1000		1891.47	IV
5000		1902.97	IV
800		1907.44	IV
1500	c	1950.80	IV
1200	c	1957.57	IV
		Air	
3000		2012.42	IV
750		2037.43	IV
2000	c	2066.50	IV
3000	c	2073.18	IV
1500		2143.23	IV
4000	c	2197.45	IV
1000	w	2221.12	IV
900		2227.34	IV
3000		2244.95	IV
7500	w	2265.91	IV
2000		2315.89	IV
750		2348.36	IV
750		2355.31	IV
2000	c	2407.10	IV

Lanthanum (Cont.)

Intensity		Wavelength	
25000	w	2417.58	IV
1200	p	2443.92	IV
18000	c	2502.81	IV
15000		2515.02	IV
50000		2532.75	IV
900	d	2535.76	IV
45000		2582.05	IV
18000	c	2591.30	IV
95000	w	2597.50	IV
5000		2608.01	IV
70000	w	2662.75	IV
50000		2848.30	IV
12000	c	2863.30	IV
30000	c	2962.58	IV
70000	w	3009.51	IV
90000	c	3056.68	IV
3500		3522.28	IV
2000		3650.40	IV
2000	p	4270.76	IV
1500	w	4549.80	IV
500		4836.89	IV

La V
Ref. 78 — J.R.

Intensity	Wavelength	
	Vacuum	
2	389.03	V
400	390.72	V
1	398.53	V
30	399.34	V
350	405.10	V
50	416.13	V
3	421.55	V
50	423.07	V
400	424.78	V
1000	432.11	V
2500	435.28	V
700	436.14	V
700	436.84	V
300	437.11	V
700	437.55	V
20	444.01	V
10	444.07	V
1250	450.40	V
600	457.30	V
1000	463.85	V
150	476.67	V
5000	482.16	V
200	482.43	V
2000	483.30	V
7000	498.08	V
4000	499.03	V
10000	503.58	V
40	508.15	V
1500	525.71	V
12000	526.76	V
10000	531.07	V
15000	533.23	V
4000	540.20	V
6000	544.80	V
8000	547.44	V
3000	570.90	V
2500	593.18	V
750	597.70	V
2000	600.01	V
5000	600.24	V
700	611.70	V
500	617.60	V

LEAD (Pb)
Z = 82

Pb I and II
Ref. 64, 274, 283, 329, 330 — D.R.W.

Intensity		Wavelength	
		Vacuum	
2		846.04	II
2	h	849.88	II
3		855.57	II
3		863.00	II
6		873.71	II
2		877.96	II
8		889.68	II
3		896.30	II
5		926.44	II
2		958.76	II
2		960.21	II
3		965.36	II
10		967.23	II
9		972.56	II
8		982.17	II
10		986.71	II
10		995.89	II
6		1001.81	II
10		1016.61	II
10		1049.82	II
10		1050.77	II
10		1060.66	II
9		1065.58	II
10		1103.94	II
10		1108.43	II
10		1109.84	II
10		1119.57	II
10		1121.36	II
10		1133.14	II
4		1145.91	II
10		1203.63	II
10		1231.20	II
10		1331.65	II
10		1335.20	II
10		1348.37	II
10		1433.96	II
3		1449.35	II
10		1512.42	II
10		1671.53	II
10		1682.15	II
20		1726.75	II
2		1740.00	I
2		1766.64	I
2		1794.67	I
10		1796.670	II
5		1812.97	I
10		1822.050	II
4		1868.76	I
10		1904.77	II
7		1921.471	II
4		1972.44	I
2		1977.88	I
2		1991.60	I
2		1992.31	I
		Air	
5	r	2022.02	I
5		2050.88	I
8	r	2053.28	I
6		2111.758	I
10		2115.066	I
500	r	2170.00	I
7		2175.580	I
7		2187.888	I
8		2189.603	I
10		2203.534	II
20		2237.425	I
20		2246.86	I
25		2246.89	I
150		2332.418	I
180		2388.797	I
550	r	2393.792	I
140		2399.597	I
320	r	2401.940	I
320	r	2411.734	I
150	r	2443.829	I
130	r	2476.378	I
8	c	2526.69	II
8	c	2576.60	II
80	r	2577.260	I
2	c	2608.38	II
500		2613.655	I
900	r	2614.175	I
160		2628.262	I
4		2634.256	II
10		2657.094	I
700		2663.154	I
10		2697.541	1
25000	r	2801.995	I
100		2822.58	I
14000	r	2823.189	I
35000	r	2833.053	I
6		2840.557	II
14000	r	2873.311	I
3	c	2887.30	II
3		2914.442	II
2	c	2947.43	II
3	c	2948.52	II
15		2966.460	I
15		2972.991	I
15		2980.157	I
4		2986.876	II
10	c	3016.39	II
150		3118.894	I
600		3220.528	I
100		3229.613	I
400		3240.186	I
200		3262.355	I
35000		3572.729	I
50000	r	3639.568	I
20000		3671.491	I
70000	r	3683.462	I
10		3713.982	II
25000		3739.935	I
15000		4019.632	I
95000		4057.807	I
14000		4062.136	I
5		4110.76	II
4		4113.35	II
10		4152.82	II
10		4157.814	I
10000		4168.033	I
9	c	4242.14	II
20	c	4244.92	II
7		4293.82	II
6		4296.65	II
200		4340.413	I
10		4352.74	II
20	c	4386.46	II
10		4579.051	II
10		4582.27	II
1000		5005.416	I
100		5006.572	I
50		5042.58	II
10		5070.58	II
10		5074.53	II
10		5076.35	I
50		5089.484	I
20		5090.01	I
10		5107.242	II
10		5111.64	II
2000		5201.437	I
10		5367.64	II
10		5372.099	II
10	c	5544.25	II
20	c	5608.85	II
40		5692.346	I
200		5895.624	I
2000		6001.862	I
9	c	6009.58	II
500		6011.667	I
8	c	6041.17	II
500		6059.356	I
40		6075.74	II
40		6081.409	II
50		6110.520	I
10		6159.89	II
100		6235.266	I
50	c	6660.20	II
10		6892.11	I
5		7128.94	I
20		7193.60	II
20000		7228.965	I
5		7304.68	I
8		7330.15	I
10		7346.676	I
10		7558.97	II
10		7632.56	II
4		7732.96	II
20		7809.259	I
5		7817.97	I
6		7829.01	I
5		7896.737	I
2	d	8156.91	II
10		8168.001	I
6		8191.886	I
5		8217.711	I
8		8255.61	I
40		8272.690	I
6		8335.54	II
10		8395.68	II
20		8409.384	I
10		8478.492	I
8		8532.17	I
7	c	8544.95	II
7		8709.90	II
5		8719.39	II
5		8722.810	I
10		8857.457	I
10		9050.82	II
10		9063.43	II
2	d	9245.28	II
8		9293.476	I
5		9384.189	I
5		9385.89	I
15		9438.05	I
15		9604.297	I
6		9608.73	I
15		9674.351	I
200		10290.458	I
5		10434.32	I
100		10498.965	I
50		10649.249	I
5		10759.41	I
7		10759.74	I
15		10886.688	I
40		10969.53	I
6		11059.22	I
3		11333.08	I
2	d	11479.49	II
2		11488.76	I
5		11627.91	II
1		12561.37	I
		13495.3	I
		13498.2	I
		13512.6	I
		15314.8	I
		15327.6	I
		15331.0	I
		15349.6	I
		38831.1	I
		38950.1	I
		38958.6	I
		39039.4	I

Pb III
Ref. 54, 256, 297 — D.R.W.

Intensity	Wavelength	
	Vacuum	
1	961.01	III
3	1030.5	III
12	1048.9	III
4	1069.2	III
3	1074.7	III
4	1118.67	III
4	1167.0	III
4	1250.6	III
1	1266.9	III
20	1553.1	III
1	1610.1	III
4	1711.23	III
	Air	
10	3043.85	III
4	3089.08	III
4	3102.74	III
10	3137.81	III
10	3176.50	III
5	3242.84	III
1	3530.17	III
7	3589.87	III
7	3689.31	III
3	3706.02	III
5	3728.69	III
12	3854.08	III
8	3951.92	III
3	4031.16	III
3	4094.54	III
2	4128.11	III
8	4272.66	III
6	4499.34	III
7	4571.21	III
1	4596.45	III
6	4761.12	III
4	4798.59	III
1	4826.86	III
2	4855.06	III
3	5065.12	III
4	5191.56	III
5	5523.97	III
3	5779.41	III
6	5857.96	III

Pb IV
Ref. 106 — D.R.W.

Intensity	Wavelength	
	Vacuum	
8	475.36	IV
7	478.35	IV
10	496.38	IV
12	499.94	IV
9	515.07	IV
14	529.78	IV
20	570.16	IV
8	573.90	IV
8	584.52	IV
10	648.50	IV
9	656.10	IV
10	761.09	IV
18	802.07	IV
12	802.82	IV
10	812.59	IV
8	822.07	IV
10	827.41	IV
12	832.60	IV
8	840.99	IV
8	842.88	IV
12	845.94	IV
18	857.64	IV
8	859.02	IV
16	862.33	IV
14	870.44	IV
12	879.96	IV
7	880.35	IV
14	884.96	IV
14	884.99	IV
16	890.72	IV
12	908.51	IV
12	917.90	IV
10	922.12	IV
12	922.49	IV
7	924.52	IV
10	927.64	IV
14	932.20	IV
8	937.00	IV
7	952.85	IV

Lead (Cont.)

Intensity	Wavelength	
8	1012.44	IV
14	1028.61	IV
20	1032.05	IV
16	1041.24	IV
18	1044.14	IV
15	1056.53	IV
12	1072.09	IV
7	1079.88	IV
18	1080.81	IV
20	1084.17	IV
6	1089.94	IV
7	1099.47	IV
6	1115.30	IV
20	1116.08	IV
18	1137.84	IV
8	1142.77	IV
14	1144.93	IV
20	1189.95	IV
8	1267.55	IV
8	1290.82	IV
10	1291.10	IV
20	1313.05	IV
8	1323.92	IV
12	1343.06	IV
16	1388.94	IV
6	1397.02	IV
18	1400.26	IV
10	1404.34	IV
7	1510.76	IV
14	1535.71	IV
8	1798.39	IV
8	1893.19	IV
12	1959.34	IV
16	1973.16	IV

Air

Intensity	Wavelength	
10	2042.58	IV
12	2049.34	IV
12	2079.22	IV
8	2151.96	IV
15	2154.01	IV
12	2177.46	IV
16	2359.53	IV
4	2864.24	IV
4	2864.50	IV
16	2417.61	IV
4	2978.14	IV
4	3052.56	IV
4	3221.17	IV
4	3962.48	IV
4	4049.80	IV
10	4496.15	IV
16	4534.60	IV
8	4605.40	IV
2	5914.54	IV

Pb V
Ref. 106 — D.R.W.

Intensity	Wavelength	
	Vacuum	
2	367.40	V
2	372.53	V
2	387.87	V
2	394.38	V
5	424.64	V
3	431.03	V
3	436.60	V
2	438.47	V
6	438.91	V
4	453.45	V
3	461.70	V
3	496.20	V
4	694.42	V
8	696.20	V
20	703.73	V
4	706.29	V
6	707.66	V
5	730.85	V
12	749.46	V
10	752.52	V
10	755.80	V
6	762.76	V
10	765.87	V
18	767.45	V
18	769.49	V
14	771.42	V
14	782.79	V
10	787.05	V
15	797.02	V
5	799.80	V
18	809.63	V
5	812.32	V
8	814.10	V
8	820.09	V
5	825.52	V
8	829.32	V
5	851.98	V
20	863.97	V
10	867.10	V

Intensity	Wavelength	
6	880.50	V
18	883.90	V
14	888.37	V
14	894.40	V
12	896.08	V
8	915.09	V
14	915.71	V
12	918.09	V
12	920.28	V
12	920.66	V
6	940.74	V
8	946.20	V
6	950.93	V
12	954.35	V
4	954.95	V
10	955.28	V
4	964.38	V
6	989.14	V
8	1005.42	V
10	1051.26	V
4	1059.26	V
10	1088.86	V
9	1096.52	V
6	1104.79	V
4	1121.33	V
10	1137.50	V
4	1152.36	V
12	1157.88	V
14	1185.43	V
11	1233.50	V
10	1248.47	V
8	1635.75	V
2	1802.87	V
2	1843.00	V
2	1888.67	V
2	1897.02	V
2	1914.33	V
4	1919.74	V
5	1957.96	V
2	1998.58	V
10	1998.83	V

Air

Intensity	Wavelength	
8	2078.43	V
10	2142.55	V
10	2167.97	V
20	2259.01	V
10	2276.66	V
8	2301.49	V
15	2424.81	V
4	4809.36	V
5	6650.99	V
4	6753.20	V

LITHIUM (Li)
Z = 3

Li I and II
Ref. 3, 15, 17, 18, 37, 44, 112, 284, 321, 335 — J.D.S.

Intensity	Wavelength	
	Vacuum	
	125.5	II
	136.5	II
	140.5	II
	167.21	II
	168.74	II
	171.58	II
	178.02	II
	199.28	II
	207.5	II
	456.	II
	483.	II
	540.	II
	729.	II
	800.	II
	820.	II
	861.	II
	905.5	II
	917.5	II
	936.	II
	945.	II
	965.	II
	972.	II
	988.	II
	1018.	II
	1032.	II
	1036.	II
	1093.	II
	1103.	II
	1109.	II
	1116.	II
	1132.1	II
	1141.	II
	1166.4	II
	1198.09	II
	1215.	II
	1238.	II
	1253.8	II
	1420.89	II

Intensity	Wavelength	
	1424.	II
3	1492.93	II
5	1492.97	II
1	1493.04	II
	1555.	II
3	1653.08	II
5	1653.13	II
1	1653.21	II
	1681.66	II
	1755.33	II

Air

Intensity	Wavelength	
	2009.	II
	2039.	I
	2068.	II
	2131.	II
	2164.	II
	2173.4	I
	2183.	II
	2214.	II
	2222.	II
	2237.	II
H	2249.21	II
	2286.82	II
	2302.57	II
	2303.33	II
	2304.59	I
	2304.92	I
	2305.36	I
	2305.83	I
	2306.29	I
	2306.82	I
	2307.44	I
	2308.97	I
	2309.88	I
	2310.94	I
	2312.11	I
	2313.49	I
	2315.08	I
	2316.95	I
	2319.18	I
	2321.88	I
	2325.11	I
4	2328.02	II
4	2329.84	II
1	2333.94	I
3	2336.88	II
5	2336.91	II
2	2337.00	II
	2340.15	I
	2348.22	I
	2358.93	I
	2373.54	I
	2381.54	II
	2383.20	II
1	2394.39	I
	2402.33	II
	2410.84	II
3	2425.43	I
	2429.81	II
	2460.2	I
10	2475.06	I
	2506.94	II
	2508.78	II
	2518.	I
	2539.49	II
24	2551.7	II
	2559	II
15	2562.31	I
	2605.08	II
	2640.	I
2	2657.29	II
3	2657.30	II
	2674.46	II
0	2728.24	II
5	2728.29	II
2	2728.32	II
3	2730.47	II
1	2730.55	II
5	2741.20	I
	2766.99	II
	2790.31	II
	2801.	I
	2846.	I
	2868.	I
	2895.	I
2	2934.02	II
2	2934.07	II
5	2934.12	II
1	2934.25	II
	2968.	I
3	3029.12	II
3	3029.14	II
	3144.	I
3	3155.31	II
4	3155.33	II
1	3196.26	II
9	3196.33	II
4	3196.36	II
5	3199.33	II
2	3199.43	II
17	3232.66	I
	3249.87	II

Intensity	Wavelength		
	3306.28		II
	3393.		
	3488.		I
	3579.8		I
	3618.		I
	3662.		I
	3684.32		II
1	3714.00		II
5	3714.16		II
6	3714.27	d	II
8	3714.29		II
7	3714.40	d	II
10	3714.41		II
1	3714.51		II
0	3714.58		II
3	3718.7		I
6	3794.72		I
20	3915.30		I
20	3915.35		I
10	3985.48		I
10	3985.54		I
40	4132.56		I
40	4132.62		I
	4196.		I
20	4273.07		I
20	4273.13		I
5	4325.42		II
5	4325.47		II
1	4325.54		II
	4516.45		II
	4590.		
13	4602.83		I
13	4602.89		I
	4607.34		II
0	4671.51		II
6	4671.65		II
2	4671.70		II
3	4678.06		II
1	4678.29		II
	4760.		I
	4763.		I
	4788.36		II
	4843.0		II
4	4881.32		II
4	4881.39		II
1	4881.49		II
8	4971.66		I
8	4971.75		I
	5037.92		II
	5095.		
	5114.		
	5190.		
	5271.		I
	5315.		I
	5395.		I
	5440.		I
600	5483.55	c	II
600	5485.65	c	II
320	6103.54		I
320	6103.65		I
3600	6707.76		I
3600	6707.91		I
48	8126.23		I
48	8126.45		I
	8517.37		II
	9581.42		II
	10120.		II
	12232.		I
	12782.		I
	13566.		I
	17552.		I
	18697.		I
	19290.		I
	24467.		I
	40475.		I

Li III
Ref. 335 — J.O.S.

Intensity	Wavelength	
	Vacuum	
	102.9	III
	103.4	III
	104.1	III
	105.5	III
	108.0	III
	113.9	III
	135.0	III
	540.0	III
	729.1	III

LUTETIUM (Lu)
Z = 71

Lu I and II
Ref. 1 — C.H.C.

Intensity	Wavelength
	Air

Lutetium (Cont.)

Intensity		Wavelength	Spectrum
1700	h	2195.54	II
95		2276.94	II
190		2297.41	II
1300		2392.19	II
120		2399.14	II
80		2419.21	II
55		2430.26	II
130		2459.64	II
80		2469.27	II
21	h	2481.72	II
370		2536.95	II
40		2546.87	II
20		2549.44	I
20		2549.72	I
35		2561.80	II
930		2571.23	II
1700		2578.79	II
80	h	2582.13	II
1800		2613.40	II
18000		2615.42	II
1800		2619.26	II
90		2657.05	
2700		2657.80	II
90	h	2677.25	I
570		2685.08	I
90	h	2685.54	I
4200		2701.71	II
90	h	2715.91	I
180	d	2719.09	I
480	h	2728.95	I
75	c	2738.17	II
3600		2754.17	II
750	h	2765.74	I
2700		2796.63	II
35		2821.23	II
270	c	2834.35	II
330		2845.13	I
3000		2847.51	II
570	h	2885.14	I
6300		2894.84	II
4500		2900.30	II
300		2903.05	I
9000		2911.39	II
270	h	2949.73	I
1200		2951.69	II
60		2955.78	II
4200		2963.32	II
2400		2969.82	II
1800		2989.27	I
3000		3020.54	II
120		3027.29	II
2100		3056.72	II
7500		3077.60	II
390		3080.11	I
5100	h	3081.47	I
3000		3118.43	I
2400		3171.36	I
100		3183.73	II
260		3191.80	II
1400		3198.12	II
4800		3254.31	II
3800		3278.97	I
7600		3281.74	I
6200		3312.11	I
7600		3359.56	I
6200		3376.50	I
950		3385.50	I
160	h	3391.55	I
1400		3396.82	I
4100		3397.07	II
4800		3472.48	II
8300	c	3507.39	II
1600		3508.42	I
4800		3554.43	II
4800		3567.84	I
340		3596.34	I
800		3623.99	II
680		3636.25	I
2600		3647.77	I
60		3684.32	I
60		3710.95	I
110		3756.70	I
110		3756.79	I
30	h	3786.18	I
150		3800.67	I
75	h	3829.07	I
2700		3841.18	I
75		3843.61	I
95		3853.29	I
40		3874.61	I
530		3876.65	II
29		3911.77	I
50		3918.86	I
35	h	3926.62	I
480		3968.46	I
50		3991.38	I
670		4054.45	I
75	B1	4096.13	L
35	h	4107.44	I
95	h	4112.67	I
310		4122.49	I
3100		4124.73	I
150	c	4131.79	I

Intensity		Wavelength	Spectrum
460		4154.08	I
24		4158.98	I
1600		4184.25	II
150		4277.50	I
250		4281.03	I
330	d	4295.97	I
		4296.09	I
150		4309.57	I
75		4332.72	I
29		4341.98	II
65	h	4420.96	I
190	c	4430.48	I
35		4438.79	I
190		4450.81	I
50	h	4471.55	I
60	h	4498.85	I
3300		4518.57	I
24	b	4560.95	L
24	b	4575.31	L
85	c	4605.39	I
95	h	4645.47	I
100	h	4648.21	I
95	h	4648.85	I
65	b	4654.03	L
1000		4658.02	I
85	h	4659.03	I
630	B1	4661.75	L
310	B1	4672.31	L
420	B1	4684.16	L
270	B1	4695.46	L
190	B1	4708.00	L
30		4716.70	I
65	b	4720.86	L
65	h	4726.20	I
100	B1	4735.00	L
75	B1	4749.11	L
40	B1	4764.22	L
150		4785.42	II
85		4815.05	I
50	c	4839.62	II
18		4865.36	II
460		4904.88	I
180		4942.34	I
800		4994.13	II
800		5001.14	I
55	h	5057.60	I
140		5134.05	I
2700		5135.09	I
130	B1	5170.11	L
170		5196.61	I
90		5206.47	I
40		5304.40	I
80		5349.12	I
500		5402.57	I
140	c	5421.90	I
100		5437.88	I
35		5453.57	I
2100		5476.69	II
9		5664.89	II
14		5713.49	II
550		5736.55	I
55		5775.40	I
80		5800.59	I
160		5860.79	I
9		5866.30	I
690	Cw	5983.9	II
140		5997.13	I
1400		6004.52	I
35	h	6041.66	I
440		6055.03	I
11		6140.71	I
150		6159.94	II
160		6199.66	I
2100		6221.87	II
35		6228.14	I
80		6235.36	I
160		6242.34	II
16	h	6248.80	I
70	h	6345.35	I
18	h	6354.85	I
9		6365.79	I
16		6366.00	I
22		6441.14	I
11		6444.89	II
1100		6463.12	II
29		6477.67	I
55	c	6523.18	I
35	Cw	6611.28	II
		6611.58	LU II
		6611.81	LU II
		6611.95	LU II
		6612.04	LU II
11		6619.15	II
23	c	6677.14	I
9	h	6735.76	I
30	c	6793.77	I
11		6826.59	II
45		6917.31	I
8		6943.96	II
23		7031.24	I
14	c	7096.34	I
45		7125.84	II
9		7142.79	I

Intensity		Wavelength	Spectrum
7		7143.10	I
8		7165.94	II
14	Ch	7237.98	I
5		7409.70	II
11	c	7441.52	I
8		7456.96	II
7	Ch	7640.08	I
7	Cw	7758.30	I
7	h	7815.9	I
9	c	8178.16	I
17		8382.08	I
35		8459.19	II
10	d	8478.50	I
29		8508.08	I
35	c	8610.98	I

Lu III
Ref. 148 — J.R.

Intensity	Wavelength	
	Vacuum	
1	677.34	III
7	691.05	III
30	700.25	III
50	714.89	III
100	738.76	III
200	755.03	III
3	755.16	III
500	810.73	III
100	830.53	III
2000	832.28	III
10	972.66	III
2	991.26	III
100	996.44	III
400	1001.18	III
1	1022.40	III
100	1029.83	III
200	1030.33	III
100	1031.54	III
50	1056.53	III
20	1061.99	III
3	1092.84	III
200	1187.34	III
50	1228.7	III
200	1277.53	III
30	1283.41	III
100	1331.93	III
1000	1854.57	III
	Air	
40	2050.72	III
1500	2065.35	III
1500 c	2070.56	III
100	2083.34	III
200	2099.44	III
1000	2236.14	III
2000	2236.22	III
500 c	2381.59	III
300	2563.51	III
4500 c	2603.35	III
200	2721.65	III
2000	2772.55	III
20	2781.16	III
10	2788.37	III
500	2800.90	III
20 p	2993.21	III
1000	3057.86	III
200	4251.44	III
300	4271.91	III
200	4490.00	III
400	4956.43	III
10	5046.12	III
150 c	5145.86	III
70	5419.42	III
60	5519.88	III
5	5526.80	III
5	5748.71	III
70	5786.46	III
80	5869.71	III
60	5889.76	III
300	6197.96	III
600	6198.13	III
100	7309.95	III
200	7310.25	III
50	7534.27	III
70 1	7936.45	III
3	8008.59	III

Lu IV
Ref. 310 — J.R.

Intensity	Wavelength	
	Vacuum	
400	876.80	IV
100	902.06	IV
300	1015.18	IV
20	1136.17	IV

Intensity		Wavelength	
50		1189.27	IV
20	p	1194.59	IV
60		1213.08	IV
15		1220.74	IV
20		1223.75	IV
20		1240.07	IV
20		1248.10	IV
40		1266.27	IV
100		1272.42	IV
20		1273.02	IV
40		1274.77	IV
20		1276.54	IV
50		1289.38	IV
60		1310.08	IV
50		1323.02	IV
15		1331.04	IV
800		1333.79	IV
300		1334.94	IV
50		1338.20	IV
100		1339.49	IV
300		1342.58	IV
200		1351.68	IV
300		1353.74	IV
200		1355.85	IV
20		1359.67	IV
100		1363.24	IV
50		1363.37	IV
20		1367.34	IV
20		1373.54	IV
15		1375.36	IV
100		1376.02	IV
30		1379.56	IV
15		1383.18	IV
15		1389.85	IV
60		1390.07	IV
200		1390.30	IV
50		1390.69	IV
40		1392.38	IV
40		1397.18	IV
30		1401.32	IV
100		1401.46	IV
200		1406.64	IV
100		1407.00	IV
250		1407.04	IV
20		1420.32	IV
100		1421.59	IV
200		1429.08	IV
400		1429.38	IV
40		1430.80	IV
100		1440.62	IV
100		1448.14	IV
200		1452.33	IV
30		1462.65	IV
100		1483.79	IV
200		1493.24	IV
400		1511.26	IV
200		1521.06	IV
100		1522.21	IV
100		1537.77	IV
30		1549.35	IV
20		1551.59	IV
20		1562.06	IV
30		1592.55	IV
15		1594.92	IV
100	c	1607.72	IV
50	c	1631.65	IV
20		1684.50	IV
60	c	1693.67	IV
400		1721.42	IV
100		1725.14	IV
100	c	1735.79	IV
100		1736.78	IV
100		1741.74	IV
50	c	1743.84	IV
40	c	1752.60	IV
200		1759.61	IV
300		1772.08	IV
600		1772.57	IV
200		1782.45	IV
100		1797.52	IV
20	c	1901.63	IV
100		1983.92	IV
20		1990.52	IV
40		1996.18	IV
		Air	
100		2003.18	IV
100		2020.94	IV
20		2071.10	IV
100		2081.09	IV
400	c	2085.70	IV
600	c	2086.47	IV
400	c	2092.16	IV
100		2103.63	IV
1000	c	2104.41	IV
200	c	2107.85	IV
1000	c	2108.31	IV
100	c	2127.43	IV

Lu V
Ref. 401 — J.R.

Lutetium (Cont.)

Intensity		Wavelength	
		Vacuum	
40		555.44	V
100		563.72	V
50		601.54	V
60		614.23	V
40		628.79	V
50	p	637.44	V
40		637.53	V
50		663.29	V
60		850.06	V
100		861.92	V
70		866.93	V
40		870.84	V
50		875.89	V
50		876.45	V
100		880.32	V
40		884.21	V
70		886.16	V
50		886.32	V
50		886.44	V
100		891.81	V
60		895.01	V
40		895.15	V
40		898.42	V
100		914.72	V
40		918.26	V
50		920.92	V
40		921.32	V
60		921.90	V
40		922.73	V
80		925.79	V
50		927.22	V
40		947.80	V
50	w	1420.02	V
50		1432.50	V
100		1432.77	V
200		1441.76	V
40		1443.64	V
100		1448.14	V
40		1449.32	V
100		1450.36	V
100		1450.85	V
100		1452.64	V
200		1453.35	V
100		1454.38	V
100		1455.21	V
100		1460.11	V
40	c	1467.81	V
200		1468.99	V
50		1469.45	V
100		1471.20	V
400		1472.12	V
200		1473.71	V
40		1475.77	V
200		1485.58	V
50		1709.02	V
40		1728.90	V
50	c	1775.92	V
40	c	1777.68	V
40	c	1784.71	V
100	c	1786.25	V
60	c	1787.58	V
60	c	1793.85	V
60		1809.73	V
60		1814.24	V

MAGNESIUM (Mg)
Z = 12

Mg I and II
Ref. 49, 83, 103, 217, 269, 315, 335
— J.O.S.

Intensity	Wavelength	
	Vacuum	
	184.05	II
	184.31	II
	184.68	II
	184.81	II
	185.26	II
	185.59	II
	185.98	II
	186.47	II
	186.84	II
	187.19	II
	187.38	II
	188.54	II
	188.91	II
	189.01	II
	189.23	II
	189.37	II
	191.30	II
	191.56	II
	191.65	II
	192.40	II
	192.55	II
	192.84	II
	193.09	II
	193.31	II

Intensity	Wavelength	
	193.40	II
	193.64	II
	197.76	II
	199.31	II
	200.29	I
	202.00	II
	202.27	II
	202.51	II
	202.94	II
	203.15	I
	203.42	II
	203.53	II
	204.22	I
	209.09	II
	209.43	II
	209.84	I
	213.53	I
	215.12	I
	215.31	I
	215.45	I
	215.66	I
	215.79	I
	216.22	I
	216.36	I
	216.68	I
	217.21	I
	217.37	I
	218.19	I
	218.34	I
	218.42	I
	218.74	I
	219.04	I
	219.28	I
	220.03	I
	220.33	I
	222.03	I
	222.67	I
	223.45	I
	223.74	I
	225.18	I
	225.54	I
	226.26	I
	247.14	II
	248.47	II
8	884.70	II
8	884.72	II
	907.38	II
12	907.41	II
8	946.70	II
9	946.77	II
14	1025.96	II
12	1026.11	II
25	1239.94	II
20	1240.40	II
6	1248.51	II
8	1249.93	II
8	1271.24	II
9	1271.94	II
8	1272.72	II
11	1273.43	II
11	1306.71	II
12	1307.88	II
12	1308.28	II
14	1309.44	II
14	1365.45	II
	1365.74	II
15	1367.26	II
15	1367.70	II
18	1369.42	II
20	1476.00	II
25	1478.01	II
20	1480.89	II
30	1482.90	II
16	1625.22	I
2	1625.50	I
4	1625.81	I
3	1626.16	I
	1626.36	I
2	1626.56	I
3	1626.79	I
10	1627.02	I
12	1627.27	I
13	1627.53	I
1	1627.82	I
1	1628.12	I
20	1628.46	I
22	1628.80	I
14	1629.21	I
9	1629.59	I
8	1630.52	I
6	1631.62	I
6	1632.93	I
7	1634.52	I
2	1636.48	I
17	1638.90	I
6	1641.97	I
9	1645.93	I
7	1651.16	I
8	1658.31	I
7	1668.43	I
8	1683.41	I
140	1707.06	I
300	1734.84	II
500	1737.62	II

Intensity	Wavelength	
20	1747.80	I
40	1750.65	II
50	1753.46	II
30	1827.93	I
	Air	
9	2025.82	I
3	2329.58	II
6	2449.57	II
1	2557.23	I
1	2560.94	I
1	2562.26	I
1	2564.94	I
1	2570.91	I
1	2572.25	I
2	2574.94	I
1	2577.89	I
1	2580.59	I
1	2584.22	I
2	2585.56	I
3	2588.28	I
1	2591.89	I
1	2593.23	I
2	2595.97	I
2	2602.50	I
4	2603.85	I
5	2606.62	I
1	2613.36	I
2	2614.73	I
3	2617.51	I
3	2628.66	I
6	2630.05	I
8	2632.87	I
2	2644.80	I
3	2646.21	I
4	2649.06	I
8	2660.76	II
8	2660.82	II
6	2668.12	I
8	2669.55	I
10	2672.46	I
3	2693.72	II
5	2695.18	I
8	2731.99	I
10	2733.49	I
12	2736.53	I
5	2765.22	I
7	2768.34	I
4	2776.69	I
7	2778.27	I
6	2779.83	I
10	2781.29	I
9	2781.42	I
7	2782.97	I
10	2790.79	II
1000	2795.53	II
16	2798.06	II
600	2802.70	II
12	2846.75	I
14	2848.42	I
16	2851.65	I
6000	2852.13	I
3	2915.45	I
10	2936.74	I
12	2938.47	I
2	2942.00	I
3	2809.76	I
2	2811.11	I
1	2811.78	I
12	2846.72	I
14	2848.34	I
16	2851.66	I
2	2902.92	I
4	2906.36	I
3	2915.45	I
2	2928.75	II
3	2936.54	II
10	2936.74	I
12	2938.47	I
13	2942.00	I
1	2967.87	II
1	2971.70	II
20	3091.08	I
22	3092.99	I
14	3096.90	I
9	3104.71	II
8	3104.81	II
6	3168.98	II
6	3172.71	II
7	3175.78	II
2	3197.62	I
17	3329.93	I
6	3332.15	I
9	3336.68	I
7	3535.04	II
8	3538.86	II
7	3549.52	II
8	3553.37	II
140	3829.30	I
300	3832.30	I
500	3838.29	I

Intensity	Wavelength	
8	3848.24	II
1	3848.91	I
7	3850.40	II
2	3853.96	I
1	3854.96	I
2	3858.86	I
3	3878.31	I
2	3891.91	I
2	3893.30	I
3	3895.57	I
4	3903.86	I
6	3938.40	I
1	3984.21	I
8	3986.75	I
2	4054.69	I
10	4057.50	I
3	4075.06	I
2	4081.83	I
4	4165.10	I
15	4167.27	I
20	4351.91	I
6	4354.53	I
6	4380.38	I
9	4384.64	II
10	4390.59	II
8	4428.00	II
9	4433.99	II
5	4436.49	II
4	4436.60	II
14	4481.16	II
13	4481.33	II
6	4534.29	II
28	4571.10	I
3	4621.30	I
7	4702.99	I
10	4730.03	I
6	4739.59	II
5	4739.71	II
7	4851.10	II
75	5167.33	I
220	5172.68	I
400	5183.61	I
8	5264.21	II
7	5264.37	II
1	5345.98	I
9	5401.54	II
2	5509.60	I
6	5528.41	I
30	5711.09	I
5	5785.31	I
1	5785.56	I
7	5916.43	II
6	5918.16	II
10	6318.72	I
9	6319.24	I
7	6319.49	I
10	6346.74	II
9	6346.96	II
11	6545.97	II
5	6620.44	II
6	6620.57	II
2	6630.83	I
7	6781.45	II
8	6787.85	II
7	6812.86	II
8	6819.27	II
4	6894.90	I
6	6965.40	I
8	7060.41	I
10	7193.17	I
10	7291.06	I
5	7387.00	I
12	7387.69	I
4	7580.76	II
20	7657.60	I
19	7659.15	I
17	7659.90	I
8	7690.16	I
15	7691.55	I
1	7722.61	I
1	7746.34	I
1	7759.30	I
5	7786.50	II
4	7790.98	II
3	7811.14	II
12	7877.05	I
2	7881.67	I
13	7896.37	II
7	7930.81	I
3	8047.73	I
5	8049.85	I
7	8054.23	I
10	8098.72	I
9	8115.22	II
8	8120.43	II
1	8154.64	I
2	8159.13	I
10	8209.84	I
20	8213.03	I
10	8213.99	II
7	8222.92	II
7	8233.19	II
11	8234.64	II

Magnesium (Cont.)

Intensity		Wavelength	
7		8303.31	I
9		8305.60	I
10		8310.26	I
15		8346.12	I
2		8466.48	I
5		8468.84	I
7		8473.69	I
10		8710.18	I
12		8712.69	I
13		8717.83	I
10		8734.99	II
17		8736.02	I
11		8745.66	II
14		8806.76	I
10		8824.32	II
11		8835.08	II
20		8923.57	I
7		8989.03	I
9		8991.69	I
10		8997.16	I
14		9218.25	II
13		9244.27	II
12		9246.50	I
30		9255.78	I
10		9327.54	II
10		9340.54	I
25		9414.96	I
17		9429.81	I
19		9432.76	I
20		9438.78	I
8		9502.45	I
7		9503.11	I
5		9503.43	I
12		9631.89	II
11		9632.43	II
15		9953.20	I
15		9983.20	I
17		9986.47	I
18		9993.21	I
14		10092.16	II
5		10391.76	II
6		10392.23	II
35		10811.08	I
11		10914.23	II
7		10915.27	II
10		10951.78	II
25		10953.32	I
27		10957.30	I
28		10965.45	I
15		11032.10	I
14		11033.66	I
5		11255.93	II
4		11256.35	II
45		11828.18	I
30		12083.66	I
28		14877.62	I
35		15024.99	I
30		15040.24	I
25		15047.70	I
6		15740.71	I
8		15748.99	I
10		15765.84	I
30		17108.66	I
5		26392.90	I

Mg III
Ref. 4, 83, 177 — J.O.S.

Intensity		Wavelength	
		Vacuum	
		106.30	III
		106.92	III
		108.08	III
		110.16	III
		114.32	III
		126.50	III
15		170.80	III
15		171.39	III
15		182.24	III
12		182.97	III
20		186.51	III
20		187.20	III
10		188.53	III
100		231.73	III
80		234.26	III
10		1274.83	III
11	h	1280.70	III
12		1391.27	III
15		1393.39	III
10		1431.14	III
16		1572.71	III
12		1586.24	III
13		1687.09	III
13		1697.28	III
10		1722.04	III
22		1738.84	III
12		1747.56	III
18		1748.93	III
15		1772.98	III
20		1783.25	III
14		1794.58	III
15		1800.66	III
13		1858.19	III
12		1879.49	III
10		1908.50	III
12		1923.90	III
13		1930.67	III
11		1937.84	III
		Air	
15		2039.55	III
15		2055.49	III
25		2064.90	III
15		2085.90	III
20		2091.96	III
13		2097.93	III
15		2112.77	III
16		2134.06	III
20		2177.70	III
20		2395.15	III
15		2467.75	III
10		2490.54	III
10		2529.19	III
12		3299.05	III
13		3306.39	III
12		3335.90	III
11		3342.58	III
12		3361.41	III
10		3381.24	III
11		3382.90	III
11		3387.37	III
10		3706.74	III
10		4916.00	III
10		5839.82	III
15		6256.75	III

Mg IV
Ref. 7, 128, 129 — J.O.S.

Intensity		Wavelength	
		Vacuum	
40		118.16	IV
80	p	118.81	IV
70		123.59	IV
240		124.65	IV
300		129.86	IV
300		132.81	IV
400		146.95	IV
300		147.41	IV
300		147.54	IV
350		180.07	IV
400		180.62	IV
400		180.80	IV
350		181.34	IV
4000		320.99	IV
3000		323.31	IV
40		800.41	IV
150		857.29	IV
30		866.74	IV
50		919.03	IV
30		929.78	IV
40		1008.76	IV
30		1026.41	IV
250		1037.41	IV
80		1044.37	IV
60		1055.76	IV
300		1210.99	IV
300		1342.19	IV
800		1346.57	IV
300		1346.68	IV
600		1352.05	IV
900		1384.46	IV
500		1385.77	IV
800		1387.53	IV
300		1404.68	IV
1000		1409.36	IV
500		1437.53	IV
1000		1437.64	IV
300		1447.42	IV
300		1459.54	IV
400		1459.62	IV
400		1481.51	IV
350		1490.45	IV
300		1495.50	IV
300		1607.11	IV
500		1683.02	IV
400		1698.81	IV
300		1844.17	IV
		Air	
12	p	2518.40	IV
4		2534.79	IV

Mg V
Ref. 128 — J.O.S.

Intensity		Wavelength	
		Vacuum	
5		251.58	V
35		276.58	V
10		312.30	V
20		351.09	V
18		352.20	V
30		353.09	V
15		353.30	V
18		354.22	V
20		355.33	V

MANGANESE (Mn)
Z = 25

Mn I and II
Ref. 1, 126 — C.H.C.

Intensity		Wavelength	
		Vacuum	
20		1726.47	II
30		1732.70	II
50		1733.55	II
40		1734.49	II
30		1737.93	II
20		1740.16	II
20		1742.00	II
30		1853.27	II
20		1857.92	II
50		1902.95	II
20		1907.84	II
30		1911.41	II
20	d	1914.68	II
100		1915.10	II
20		1918.64	II
30		1919.64	II
80		1921.25	II
20		1923.07	II
20		1923.34	II
30		1925.52	II
50		1926.59	II
30		1931.40	II
20		1945.15	II
20		1947.93	II
20		1950.14	II
30		1953.23	II
20	d	1954.81	II
30		1959.25	II
20		1969.24	II
30		1994.23	II
9700		1996.06	I
14000		1999.51	I
		Air	
18000		2003.85	I
50		2037.31	II
40		2037.64	II
40		2039.97	II
30		2076.21	II
1500		2092.16	I
20		2097.46	II
20		2102.50	II
1700		2109.58	I
30		2113.96	II
290		2208.81	I
540		2213.85	I
770		2221.84	I
20		2373.36	II
20		2427.38	II
50		2427.72	II
30		2427.94	II
30		2437.37	II
20		2437.74	II
30		2452.49	II
50		2499.00	II
30		2507.60	II
20		2516.60	II
30		2516.74	II
20		2521.66	II
20		2530.72	II
20		2531.80	II
50		2532.78	II
75		2533.06	I
50		2533.33	II
30		2534.10	II
80		2534.22	II
100		2535.66	II
30		2535.98	II
100		2537.92	II
50		2541.11	II
80		2542.92	II
50		2543.45	II
100		2548.75	II
50		2551.85	II
30		2553.27	II
75		2556.56	II
3u		2556.89	II
50		2557.54	II
95		2558.59	II
30		2559.41	II
150		2563.65	II
30		2565.22	II
580		2572.76	I
480		2575.51	I
12000		2576.10	II
550		2584.31	I
30		2588.97	II
45		2589.71	II
250		2592.94	I
6200		2593.73	II
250		2595.76	I
95		2598.90	II
40		2602.14	I
30		2602.72	II
45		2603.72	II
4300		2605.69	II
190		2610.20	II
500		2618.14	II
140		2622.90	I
150		2624.04	I
40		2624.80	II
200		2625.58	II
95		2626.64	I
30		2630.26	I
60		2630.57	I
190		2632.35	II
130		2638.17	II
80		2639.84	II
27		2650.99	II
60		2655.91	II
30		2666.77	II
45		2667.00	I
30		2667.03	II
110		2672.59	II
55		2673.37	II
55		2674.43	II
30		2676.33	I
45		2680.34	II
30		2680.68	II
30		2681.25	II
40		2681.72	I
45		2683.02	I
23		2683.75	I
55		2684.55	II
55		2685.94	II
110		2688.25	II
85		2692.66	I
27		2693.19	II
55		2695.36	II
27		2698.97	II
85		2701.00	II
50		2701.17	II
160		2701.70	II
100		2703.98	II
130		2705.74	II
80		2707.53	II
110		2708.45	II
45		2709.96	II
80		2710.33	II
110		2711.58	II
30		2716.80	II
30		2717.53	II
30		2719.01	II
50		2719.74	II
30		2722.10	II
30		2724.46	II
55		2728.61	II
30		2738.86	I
45	h	2760.93	I
30		2771.44	I
30	h	2776.23	I
30		2780.00	I
55		2789.20	I
60		2790.36	I
60		2791.08	I
6200		2794.82	I
5100		2798.27	I
220		2799.84	I
3700		2801.06	I
70		2804.10	I
60		2806.14	I
55		2808.02	I
110		2809.11	I
60		2812.84	I
70		2813.47	I
60		2815.02	II
30		2816.33	II
85		2817.97	I
40		2818.77	I
55		2821.45	I
55		2822.55	I
80		2830.79	I
27		2836.31	I
60		2870.08	II
30		2872.94	II
80		2879.49	II
40		2882.90	I
70		2886.68	II
160		2889.58	II
55		2892.39	II
50		2898.70	II
80		2900.16	II
40		2907.22	I
140	h	2914.60	I
190	h	2925.57	I

Manganese (Cont.)

Intensity		Wavelength	
27		2928.68	I
1100		2933.06	II
27		2934.02	I
1500		2939.30	II
250	h	2940.39	I
		2940.48	I
60		2941.04	I
1900		2949.20	II
40		3007.66	I
40		3011.16	I
40		3011.38	I
40		3014.67	I
60		3016.45	I
30		3019.92	II
70		3022.75	I
55		3031.06	II
30		3035.35	II
95		3040.60	I
27		3042.73	I
85		3043.36	I
330		3044.57	I
120		3045.59	I
200		3047.04	I
40		3048.86	I
30		3050.65	II
250		3054.36	I
140		3062.12	I
170		3066.02	I
170		3070.27	I
160		3073.13	I
90		3079.63	I
50		3081.33	I
23		3082.05	I
40		3097.06	I
40		3110.68	I
60	h	3148.18	I
90	h	3161.04	I
140	h	3178.50	I
220		3212.88	I
65		3216.95	I
1000		3228.09	I
300		3230.72	I
850		3236.78	I
330		3243.78	I
650		3248.52	I
100		3251.16	I
310		3252.95	I
65		3254.04	I
310		3256.14	I
220		3258.41	I
180		3260.23	I
180		3264.71	I
65		3296.88	I
65		3298.22	I
65		3320.69	I
70		3330.67	I
200		3330.78	II
100		3336.39	II
30		3365.02	II
30		3400.12	II
50		3438.97	II
720		3441.99	II
50		3460.03	II
360		3460.33	II
360	h	3474.04	II
		3474.13	II
290		3482.91	II
180		3488.68	II
140		3495.84	II
50		3496.81	II
100		3497.54	II
360		3531.85	I
		3532.00	I
1100		3532.12	I
1300		3547.80	I
1100		3548.03	I
390		3548.20	I
2200		3569.49	I
720		3569.80	I
		3570.04	I
1400		3577.88	I
720		3586.54	I
290		3595.12	I
420		3607.54	I
420		3608.49	I
360		3610.30	I
290		3619.28	I
220		3623.79	I
140		3629.74	I
100		3660.40	I
70		3670.52	I
70		3676.96	I
50		3682.09	I
280		3693.67	I
180		3696.57	I
70		3701.73	I
210		3706.08	I
130		3718.93	I
55		3728.89	I
130		3731.93	I
260		3790.22	I
55		3799.26	I
110		3800.55	I
55		3801.91	I
3200		3806.72	I
700		3809.59	I
55		3810.69	I
90		3816.75	I
2100		3823.51	I
390		3823.68	I
200		3829.68	II
480		3833.86	I
1300		3834.36	I
350		3839.78	I
670		3841.08	I
350		3843.98	I
65		3918.32	I
120		3926.47	I
65		3952.84	I
55		3975.89	I
65		3977.08	I
130		3982.58	I
150		3985.24	I
190		3986.83	I
150		3987.10	I
1500		4018.10	I
150		4026.44	I
27000		4030.76	I
19000		4033.07	I
11000		4034.49	I
1500		4035.73	I
55		4038.73	I
5600		4041.36	I
210	d	4045.13	I
		4045.21	I
1100		4048.76	I
80		4049.00	I
55		4051.73	I
65		4052.47	I
150		4055.21	I
1900		4055.54	I
210		4057.95	I
1100		4058.93	I
150		4059.39	I
730		4061.74	I
730		4063.53	I
80		4065.08	I
80		4068.00	I
290		4070.28	I
730		4079.24	I
730		4079.42	I
1100		4082.94	I
1100		4082.60	I
65		4089.94	I
55		4105.36	I
200		4110.90	I
150		4131.12	I
120		4135.04	I
80		4141.06	I
55		4147.53	I
80		4148.80	I
150		4176.60	I
120		4189.99	I
65		4201.76	I
65		4211.75	I
370		4235.14	I
510		4235.29	I
190		4239.72	I
290		4257.66	I
290		4265.92	I
270		4281.10	I
65		4284.08	I
65		4312.55	I
50		4323.63	II
45		4374.95	I
45		4381.70	I
55		4411.88	I
350		4414.88	I
55		4419.78	I
210		4436.35	I
800		4451.59	I
160		4453.00	I
130		4455.01	I
160		4455.32	I
110		4455.82	I
55		4457.04	I
210		4457.55	I
270		4458.26	I
55		4460.38	I
150		4461.08	I
510		4462.02	I
290		4464.68	I
200		4470.14	I
130		4472.79	I
40		4479.64	I
170		4490.08	I
240		4498.90	I
240		4502.22	I
80		4605.36	I
80		4626.54	I
35		4671.69	I
50		4701.16	I
160		4709.72	I
180		4727.48	I
130		4739.11	I
1000		4754.04	I
180		4761.53	I
750		4762.38	I
300		4765.86	I
500		4766.43	I
940		4783.42	I
1000		4823.52	I
25		4844.32	I
35		4965.88	I
19		5004.91	I
30		5074.79	I
60		5117.94	I
50		5150.89	I
50		5196.59	I
85		5255.32	I
160		5341.06	I
19		5349.88	I
95		5377.40	I
95		5394.67	I
50		5399.49	I
93		5407.42	I
35		5413.69	I
85		5420.36	I
35		5432.55	I
12		5457.47	I
60		5470.64	I
40		5481.40	I
30		5505.87	I
50		5516.77	I
40		5537.76	I
21		5551.98	I
8		5567.76	I
7		5573.01	I
8		5573.68	I
7		5738.29	I
7		5780.19	I
7		5816.84	I
140		6013.50	I
200		6016.64	I
290		6021.80	I
7		6384.67	I
17		6440.97	I
24		6491.71	I
14	h	6942.52	I
12		6989.96	I
16		7065.04	I
12		7184.25	I
10		7247.02	I
24	h	7283.82	I
35	h	7302.89	I
50		7326.51	I
12		7680.20	I
10		7712.42	I
10	h	7764.72	I
10	h	8670.92	I
12	h	8672.06	I
10	h	8673.97	I
12	h	8701.05	I
17	h	8703.76	I
30	h	8740.93	I

Mn III
Ref. 385 — C.H.C.

Intensity		Wavelength	
		Vacuum	
20		892.39	III
20		1108.16	III
30		1183.30	III
25	w	1183.86	III
30		1198.49	III
30		1219.80	III
100		1228.97	III
500		1283.58	III
400		1287.59	III
300		1291.62	III
1000		1360.72	III
800		1365.20	III
400		1369.43	III
300		1371.65	III
300		1596.95	III
500	h	1609.17	III
1000		1614.14	III
2000		1620.60	III
300		1623.91	III
400		1629.12	III
500		1633.80	III
250		1647.46	III
400		1653.57	III
400		1804.06	III
300		1806.47	III
300		1811.02	III
400		1877.62	III
300		1885.21	III
500		1941.28	III
250	w	1942.89	III
800		1943.21	III
500		1952.36	III
1000		1952.52	III
300		1956.61	III
250		1962.04	III
500		1978.95	III
400		1982.76	III
400		1989.59	III

Air

Intensity		Wavelength	
300		2022.19	III
1000	w	2027.83	III
500	w	2028.14	III
300		2044.57	III
400		2048.93	III
500		2049.68	III
300		2056.80	III
500		2066.38	III
1000		2069.02	III
900		2077.38	III
300		2078.13	III
800		2084.23	III
600		2090.06	III
300		2090.25	III
300		2094.14	III
500		2094.78	III
500		2097.93	III
500		2099.97	III
300		2123.25	III
1000		2169.78	III
700		2174.15	III
900		2176.87	III
800		2181.86	III
800	w	2184.87	III
600		2185.13	III
400		2211.95	III
600		2212.42	III
800		2215.21	III
900		2220.55	III
1000		2227.42	III
100		3287.49	III
100		3540.52	III
150		3601.72	III
100		3616.00	III
100		4246.17	III
200		5079.20	III
150		5100.03	III
100		5117.03	III
100		5252.23	III
100		5365.59	III
150		5454.07	III
200		5474.68	III
100		5671.10	III
200		5946.65	III
100	s	6213.11	III
200		6231.21	III
100		6238.64	III
100		6273.71	III

Mn IV
Ref. 433 — C.H.C.

Intensity		Wavelength	
		Vacuum	
60		579.79	IV
60		581.44	IV
60		581.65	IV
60		585.21	IV
90		1242.25	IV
90		1244.50	IV
85		1247.73	IV
95		1251.93	IV
95		1257.28	IV
90		1264.41	IV
70		1603.60	IV
70		1611.10	IV
75		1653.83	IV
70		1656.39	IV
70		1659.25	IV
75		1664.73	IV
80	b	1667.00	IV
70		1670.08	IV
75		1691.68	IV
75		1693.15	IV
80		1698.30	IV
75		1698.70	IV
70		1699.06	IV
75		1707.43	IV
65		1718.67	IV
75	b	1720.52	IV
75		1720.74	IV
75		1721.41	IV
65		1722.94	IV
75		1724.83	IV
85	b	1742.10	IV
85		1751.59	IV
75		1759.82	IV
70		1762.17	IV
75		1762.94	IV
85	d	1766.27	IV
75		1767.09	IV
65		1772.11	IV
75		1773.51	IV
75		1782.21	IV
75		1786.02	IV

Manganese (Cont.)

Intensity	Wavelength	
75	1787.04	IV
75	1787.38	IV
75	1788.64	IV
75	1790.44	IV
80	1795.65	IV
80	1795.79	IV
60	1907.03	IV
75	1910.25	IV
65	1997.54	IV

Mn V
Ref. 405 — C.H.C.

Intensity	Wavelength	
	Vacuum	
300	404.36	V
380	406.02	V
300	406.40	V
600	410.30	V
600	410.60	V
480	410.98	V
400	411.32	V
460	412.74	V
460	413.75	V
600	415.62	V
650	415.98	V
350	419.80	V
600	428.59	V
500	429.05	V
400	433.54	V
600	435.67	V
350	436.16	V
500	436.18	V
450	438.74	V
350	439.35	V
1000	441.72	V
850	442.49	V
400	467.32	V
300	474.82	V

MERCURY (198) (Hg)
Z = 80

Hg I and II (198)
Ref. 43, 50, 69, 145, 229, 242 — R.W.S.

Intensity	Wavelength	
	Vacuum	
80	1250.564	I
8	1259.242	I
100	1268.825	I
5	1307.751	I
20	1402.619	I
10	1435.503	I
1000	1849.492	I
	Air	
60	2262.210	II
20	2302.065	I
20	2345.440	I
100	2378.325	I
20	2380.004	I
40	2399.349	I
20	2399.729	I
20	2446.900	I
15	2464.064	I
40	2481.999	I
30	2482.713	I
40	2483.821	I
90	2534.769	I
15000	2536.506	I
25	2563.861	I
25	2576.290	I
250	2652.043	I
400	2653.683	I
100	2655.130	I
50	2698.831	I
80	2752.783	I
20	2759.710	I
40	2803.471	I
30	2804.438	I
750	2847.675	II
50	2856.939	I
150	2893.598	I
150	2916.227	II
60	2925.413	I
1200	2967.283	I
300	3021.500	I
120	3023.476	I
30	3025.608	I
50	3027.490	I
400	3125.670	I
320	3131.551	I
320	3131.842	I
80	3341.481	I
2800	3650.157	I
300	3654.839	I
80	3662.883	I
240	3663.281	I
30	3701.432	I
35	3704.170	I
30	3801.660	I
20	3901.867	I
60	3906.372	I
200	3983.839	II
1800	4046.572	I
150	4077.838	I
40	4108.057	I
250	4339.224	I
400	4347.496	I
4000	4358.337	I
80	4916.068	I
1100	5460.753	I
160	5675.922	I
240	5769.598	I
280	5790.663	I
20	6072.713	I
30	6234.402	I
160	6716.429	I
250	6907.461	I
240	11287.407	I

MERCURY (NATURAL) (Hg)
Z = 80

Hg I and II (nat.)
Ref. 34, 45, 90, 117, 133, 189, 235, 304, 327, 328 — R.W.S.

Intensity	Wavelength	
	Vacuum	
400	893.08	II
300	915.83	II
150	923.39	II
200	940.80	II
100	962.74	II
50	969.13	II
800	1099.26	II
80	1250.58	I
8	1259.24	I
100	1268.82	I
5	1307.75	I
300	1307.93	II
400	1321.71	II
400	1331.74	II
80	1350.07	II
200	1361.27	II
20	1402.62	I
200	1414.43	II
10	1435.51	I
15	1619.46	II
120	1623.95	II
20	1628.25	II
150	1649.94	II
50	1653.64	II
200	1672.41	II
100	1702.73	II
100	1707.40	II
120	1727.18	II
250	1732.14	II
20	1775.68	I
40	1783.70	II
30	1796.22	II
200	1796.90	II
60	1798.74	II
30	1803.89	II
40	1808.29	II
400	1820.34	II
5	1832.74	I
1000	1849.50	I
160	1869.23	II
300	1870.55	II
200	1875.54	II
20	1900.28	II
30	1927.60	II
300	1942.27	II
100	1972.94	II
200	1973.89	II
150	1987.98	II
	Air	
90	2026.97	II
90	2052.93	II
70	2148.00	II
5	2247.55	I
60	2262.23	II
20	2302.06	I
15	2323.20	I
5	2340.57	I
20	2345.43	I
20	2352.48	I
100	2378.32	I
20	2380.00	I
40	2399.38	I
20	2399.73	I
10	2400.49	I
60	2407.35	II
50	2414.13	II
5	2441.06	I
20	2446.90	I
15	2464.06	I
40	2482.00	I
30	2482.72	I
40	2483.82	I
90	2534.77	I
15000	2536.52	I
25	2563.86	I
25	2576.29	I
5	2578.91	I
15	2625.19	I
5	2639.78	I
250	2652.04	I
400	2653.69	I
100	2655.13	I
5	2674.91	I
50	2698.83	I
50	2699.38	I
80	2705.36	II
80	2752.78	I
20	2759.71	I
40	2803.46	I
30	2804.43	I
2	2805.34	I
2	2806.77	I
150	2814.93	II
750	2847.68	I
50	2856.94	I
150	2893.60	I
150	2916.27	II
60	2925.41	I
150	2935.94	I
400	2947.08	II
1200	2967.28	I
300	3021.50	I
120	3023.47	I
30	3025.61	I
50	3027.49	I
400	3125.67	I
320	3131.55	I
320	3131.84	I
400	3208.20	II
400	3264.06	II
80	3341.48	I
100	3385.25	II
400	3451.69	II
200	3549.42	II
2800	3650.15	I
300	3654.84	I
80	3662.88	I
240	3663.28	I
30	3701.44	I
35	3704.17	I
30	3801.66	I
100	3806.38	II
20	3901.87	I
60	3906.37	I
100	3918.92	II
200	3983.96	II
1800	4046.56	I
150	4077.83	I
40	4108.05	I
250	4339.22	I
400	4347.49	I
4000	4358.33	I
100	4398.62	II
90	4660.28	II
80	4855.72	II
5	4883.00	I
5	4889.91	I
80	4916.07	I
5	4970.37	I
5	4980.64	I
20	5102.70	I
40	5120.64	I
100	5128.45	II
20	5137.94	I
20	5290.74	I
5	5316.78	I
60	5354.05	I
30	5384.63	I
1100	5460.74	I
30	5549.63	I
160	5675.86	I
240	5769.60	I
100	5789.66	I
280	5790.66	I
140	5803.78	I
60	5859.25	I
60	5871.73	II
20	5871.98	I
20	6072.72	I
1000	6149.50	II
30	6234.40	I
80	6521.13	II
160	6716.43	I
250	6907.52	I
250	7081.90	I
200	7091.86	I
40	7346.37	II
100	7485.87	II
20	7728.82	I
100	7944.66	II
2000	10139.75	I
240	11287.40	I
120	13209.95	I
140	13426.57	I
60	13468.38	I
80	13505.58	I
500	13570.21	I
450	13673.51	I
200	13950.55	I
500	15295.82	I
100	16881.48	I
400	16920.16	I
300	16942.00	I
500	17072.79	I
400	17109.93	I
20	17116.75	I
20	17198.67	I
20	17213.20	I
70	17329.41	I
30	17436.18	I
50	18130.38	I
40	19700.17	I
250	23253.07	I
	32148.06	I
	36303.03	I

Hg III
Ref. 343 — C.H.C.

Intensity	Wavelength	
	Vacuum	
3	621.44	III
2	679.68	III
2	878.59	III
1	886.48	III
1	988.89	III
2	1009.29	III
5	1068.03	III
2	1161.95	III
9	1681.40	III
15	1759.75	III
1	1894.77	III
	Air	
7	2314.15	III
4	2380.55	III
8	2431.65	III
5	2480.56	III
7	2484.50	III
2	2612.92	III
4	2617.97	III
3	2670.49	III
70	2724.43	III
6	2769.22	III
3	2844.76	III
15	3090.05	III
5	3283.02	III
12	3312.28	III
8	3389.01	III
5	3450.77	III
3	3500.35	III
4	3538.88	III
5	3557.24	III
15	3803.51	III
70	4122.07	III
10	4140.34	III
100	4216.74	III
15	4470.58	III
12	4552.84	III
50	4797.01	III
10	4869.85	III
80	4973.57	III
30	5210.82	III
6	5695.71	III
25	6220.35	III
35	6418.98	III
40	6501.38	III
10	6584.26	III
6	6610.12	III
30	6709.29	III
12	7517.46	III
7	7808.10	III
25	7946.75	III
50	7984.51	III
5	8151.64	III

MOLYBDENUM (Mo)
Z = 42

Mo I and II
Ref. 1 — C.H.C.

Intensity   Wavelength

AIR

Intensity	Wavelength		Intensity	Wavelength		Intensity	Wavelength		Intensity	Wavelength	
19000	2015.11	II	250	2607.37	I	1700	2871.51	II	290	3195.96	I
40000	2020.30	II	190	2611.20	I	85	2872.88	II	120	3198.85	I
21000	2038.44	II	290	2613.08	I	220	2879.05	II	40	3201.50	II
17000	2045.98	II	130	2615.39	I	65	2888.15	II	330	3205.22	I
4800	2081.68	II	400	2616.78	I	95	2891.28	II	880	3205.88	I
2400	2089.52	II	70	2619.34	II	1300	2890.99	II	3000	3208.83	I
2200	2092.50	II	140	2621.07	I	190	2892.81	II	240	3210.97	I
4000	2093.11	II	320	2627.55	I	950	2894.45	II	560	3215.07	I
2700	2100.84	II	160	2628.74	I	140	2897.63	II	350	3221.74	I
1500	2104.29	II	440	2629.85	I	70	2900.80	II	880	3228.22	I
1400	2108.02	II	330	2636.67	II	290	2903.07	II	600	3229.79	I
400	2269.69	II	250	2638.30	I	160	2905.27	I	1100	3233.14	I
160	2304.25	II	720	2638.76	II	80	2907.12	II	950	3237.08	I
160	2306.97	II	410	2640.99	I	600	2909.12	II	65	3240.71	II
130	2325.94	I	600	2644.35	II	1100	2911.92	II	950	3256.01	I
240	2330.46	II	370	2646.49	II	55	2913.81	II	300	3262.63	I
110	2332.12	II	640	2649.46	I	120	2918.83	II	480	3264.40	I
190	2340.47	I	480	2653.35	II	1300	2923.39	II	800	3270.90	I
190	2341.59	II	560 h	2655.03	I	140	2924.32	II	240	3285.02	I
80	2352.61	I	290	2658.11	I	65	2927.34	II	320	3285.36	I
80	2355.22	I	640	2660.58	II	50	2930.06	II	1100	3289.02	I
80	2355.42	II	110	2665.10	I	1100	2930.50	II	950	3290.82	I
70	2364.37	I	55	2671.83	II	55	2930.77	II	190	3292.31	II
50	2366.09	II	720	2672.84	II	800	2934.30	II	320	3305.56	I
140	2372.27	I	250	2673.27	II	65	2935.20	II	320	3307.12	I
100	2380.41	II	1000	2679.85	I	120	2937.66	I	100	3313.62	II
150	2383.52	II	95	2681.36	II	40	2938.30	II	190	3320.90	II
110	2389.20	II	640	2683.23	II	95	2940.10	II	640	3323.95	I
140	2403.61	II	880	2684.14	II	110	2941.22	II	360	3325.67	I
80	2404.66	II	560	2687.99	II	140	2944.21	I	360	3327.30	I
140	2405.86	I	30	2692.61	II	150	2944.82	II	240	3340.17	I
10	2100.93	I	55	2695.22	II	140	2945.66	I	1300	3344.75	I
40	2412.04	II	30	2696.83	II		2945.95	II	95	3346.40	II
120	2413.01	II	55	2699.41	II	190	2946.01	I	320	3347.02	I
70	2415.33	I	140	2701.03	I	140	2946.42	I	1600	3358.12	I
80	2417.96	II	480	2701.42	II	140	2946.69	II	250	3361.37	I
65	2419.01	II	30	2701.87	II	95	2947.28	II	950	3363.78	I
80	2420.18	II	30	2704.93	II	95	2955.84	II	950	3379.97	I
70	2424.00	II	40	2710.19	II	240	2956.06	II	320	3382.48	I
65	2430.43	I	30	2711.49	II	70	2956.90	II	1900	3384.62	I
65	2435.96	II	50	2712.35	II	95	2960.24	II	130	3395.36	II
65	2440.28	II	190	2713.51	II	140	2962.89	I	640	3404.34	I
40	2461.81	II	290	2717.35	II	250	2963.79	II	1300	3405.94	I
50	2466.68	II	110	2724.41	I	50	2964.06	II	240	3418.52	I
50	2466.97	II	180	2725.15	I	210	2965.27	II	250	3420.04	I
50	2468.78	II	85	2726.97	II	70	2971.91	II	250	3422.31	I
30	2470.04	II	140	2729.68	II	250	2972.61	II	380	3434.79	I
150 h	2471.97	I	80	2730.20	II	80	2975.60	II	320	3435.65	I
70	2477.57	II	930	2732.88	II	180	2978.28	II	640	3437.22	I
70 h	2481.81	I	250	2733.39	I	120	2981.52	I	250	3438.87	I
65	2482.57	II	160	2736.96	II	110	2987.92	I	250	3441.44	I
40	2484.75	II	80 h	2737.88	II	160	2988.68	I	250	3443.26	I
40 h	2485.31	I	50	2738.60	II	190	2989.80	I	130	3446.08	II
24	2496.24	II	40	2741.32	II	95	2992.84	II	3200	3447.12	I
85	2498.28	II	55	2741.62	II	50	2993.52	II	640	3449.07	I
40	2500.44	II	240	2743.07	I	190	3002.21	I	300	3451.75	I
65	2502.84	II	290	2746.30	II	40	3004.46	II	250	3452.60	I
50	2511.80	II	320	2751.47	I	130	3013.39	I	950	3456.39	I
65	2515.08	II	110	2756.07	II	140 h	3013.76	I	640	3460.78	I
70	2527.14	II	65 d	2758.63	II	250	3025.00	I	320	3466.83	I
50	2530.34	II	20	2760.53	II	95	3027.77	II	250	3467.85	I
70	2532.31	II	190	2761.53	I	100	3036.31	I	320	3469.22	I
440	2538.46	II	220	2763.62	II	300	3041.70	I	240	3485.93	I
50	2539.44	II	110	2766.26	I	150	3046.80	I	800	3504.41	I
110 h	2540.45	I	240	2769.76	II	210	3047.31	I	240	3505.32	I
330	2542.67	II	160	2773.78	II	210	3055.32	I	560	3508.12	I
40	2543.61	II	190	2774.39	II	100	3060.78	II	480	3521.41	I
330	2548.22	I	1700	2775.40	II	160	3061.59	I	240	3524.98	I
110 h	2550.85	I	130	2777.74	II	800	3064.28	I	640	3537.28	I
65	2555.42	II	65	2777.86	II	250	3065.04	II	320	3542.17	I
40	2556.75	II	880	2780.04	II	100	3068.00	I	520	3558.10	I
80	2558.88	II	400	2784.99	II	250	3070.90	I	400	3563.14	I
65	2562.08	II	180	2787.83	II	800	3074.37	I	300	3566.05	I
85	2564.34	II	40	2791.54	II	85	3077.66	II	240	3570.65	I
40	2566.26	II	240 d	2797.93	I	150	3079.88	I	320	3573.88	I
250	2567.05	I	220	2801.47	I	210	3080.41	I	1400	3581.89	I
20	2571.45	II	400	2807.76	II	800	3085.62	I	200	3590.74	I
320	2572.34	I	28	2812.58	II	270	3087.62	II	210	3598.88	I
50	2574.42	II	24	2814.67	II	100	3089.12	I	270	3602.94	I
40	2576.56	II	1700	2816.15	II	100	3089.71	I	210	3608.37	I
40	2578.36	II	220	2817.44	II	190	3092.07	II	200	3623.23	I
250	2582.16	I	50	2822.03	II	560	3094.66	I	1400	3624.46	I
30	2585.95	II	240	2826.54	II	110	3099.93	I	330	3626.18	I
65	2588.78	II	80	2827.74	II	110	3100.88	I	28	3635.14	II
40	2591.77	II	40	2831.44	II	560	3101.34	I	1000	3635.43	I
250	2593.70	II	30	2832.07	II	1400	3112.12	I	400	3657.35	I
100	2595.40	I	80	2834.39	II	290	3122.00	II	540	3664.81	I
40	2597.38	II	80	2835.33	II	14000	3132.59	I	290	3666.72	I
250	2602.80	II	160	2842.15	II	110	3138.72	II	590	3672.82	I
40	2605.08	II	24	2843.73	II	220	3147.35	I	1300	3680.60	I
40	2605.93	II	220	2844.39	I	220	3152.82	II	45	3684.22	II
			1700	2848.23	II	55	3155.64	II	65	3688.31	II
			160	2849.38	I	6000	3158.16	I	240	3690.59	I
			370	2853.23	II	120	3164.53	I	180	3692.64	II
			50	2856.00	II	8700	3170.35	I	1400	3694.94	I
			24	2863.20	II	95	3172.03	II	220	3702.03	I
			370	2863.81	II	160	3172.74	II	220	3715.65	I
			160	2864.31	I	370	3183.03	I	500	3727.69	I
			140	2864.66	I	120	3184.57	I	330 d	3732.71	I
			40	2865.62	II	370	3185.10	I	240	3742.28	I
			220	2866.69	II	180	3185.71	I	80	3744.37	II
			40	2868.11	II	120 d	3187.59	II	360	3770.45	I
			40	2868.32	II	7600	3193.97	I	220	3779.77	I

Molybdenum (Cont.)

Intensity	Wavelength		Spectrum
360	3781.59		
250	3797.30		I
29000	3798.25		I
290	3801.84		I
520	3826.70		I
940	3828.87		I
1700	3833.75		I
380	3847.25		I
29000	3864.11		I
580	3869.08		I
580	3886.82		I
380	3901.77		I
19000	3902.96		I
65	3941.48		II
230	3943.04		I
270	4056.01		I
1400	4062.08		I
2300	4069.88		I
1300	4081.44		I
940	4084.38		I
250	4102.15		I
730	4107.47		I
630	4120.10		I
2900	4143.55		I
230	4148.94		I
250	4155.28		I
200	4157.40		I
200	4178.27		I
480	4185.82		I
2500	4188.32		I
250	4194.56		I
1500	4232.59		I
270	4269.28		I
890	4276.91		I
1200	4277.24		I
1400	4288.64		I
680	4292.13		I
890	4293.21		I
360	4293.88		I
840	4326.14		I
250	4326.74		I
230	4350.34		I
230	4369.04		I
1900	4381.64		I
2500	4411.57		I
210	4423.62		I
990	4434.95		I
200	4442.20		I
340	4449.74		I
480	4457.36		I
630	4474.56		I
230	4491.28		I
120	4504.90		I
140	4512.15		I
230	4517.13		I
230	4524.34		I
120	4529.40		I
400	4536.80		I
110	4558.11		I
210	4576.50		I
170	4595.16		I
360	4609.88		I
100	4621.38		I
460	4626.47		I
100	4627.48		I
220	4662.76		I
130	4671.90		I
130	4688.22		I
640	4707.26		I
150	4708.22		I
220	4717.92		I
100	4729.14		I
700	4731.44		I
100	4750.39		I
770	4760.19		I
150	4776.34		I
100	4796.52		I
410	4819.25		I
410	4830.51		I
360	4868.00		I
110	4950.62		I
150	4957.54		I
210	4979.12		I
110	4999.91		I
20	5010.81		I
180	5014.60		I
26	5016.78		I
20	5019.85		I
80	5029.00		I
65	5030.78		I
23	5038.91		I
26	5046.52		I
100	5047.71		I
50	5055.00		I
35	5058.07		I
200	5059.88		I
35	5062.52		I
29	5064.64		I
35	5079.87		I
100	5080.02		I
35	5081.26		I
40	5090.97		I
35	5091.34		I

Intensity	Wavelength		Spectrum
35	5092.16		I
40	5095.89		I
100	5096.65		I
130	5097.52		I
35	5098.03		I
130	5109.71		I
80	5114.97		I
35	5116.97		I
29	5123.83		I
150	5145.38		I
110	5147.39		I
80	5163.19		I
100	5167.76		I
160	5171.08	d	I
	5171.25		I
230	5172.94	h	I
160	5174.18	h	I
40	5191.44		I
110	5200.17		I
50	5200.74		I
26	5210.44		I
50	5211.86		I
80	5219.40		I
65	5231.06		I
26	5232.36		I
100	5234.26		I
460	5238.20	h	I
230	5240.88	h	I
110	5242.81	h	I
100	5245.51		I
150	5259.04		I
16	5260.17		I
65	5261.14		I
20	5268.95		I
35	5271.80		I
35	5276.28		I
65	5279.65		I
210	5280.86		I
20	5283.84		I
55	5292.08		I
35	5293.46		I
55	5295.47		I
16	5306.26		I
55	5313.89		I
35	5315.04		I
20	5319.89		I
20	5324.47		I
35	5327.06		I
20	5352.35		I
80	5354.88		I
35	5355.51		I
65	5356.48		I
560	5360.56	H1	I
110	5364.28	H1	I
35	5367.11	H1	I
35	5372.40		I
26	5388.69		I
65	5394.52		I
35	5397.38		I
50	5400.47		I
35	5405.79		I
35	5406.39		I
40	5417.38		I
23	5426.89		I
55	5435.68		I
65	5437.75		I
40	5450.51		I
35	5456.46		I
26	5460.53		I
23	5465.57		I
35	5475.90		I
35	5490.28		I
20	5492.17		I
26	5493.80	h	I
26	5498.49		I
50	5501.54		I
23	5501.87		I
26	5503.54	h	I
7800	5506.49		I
23	5520.04		I
26	5520.64		I
40	5526.52		I
40	5526.97		I
5200	5533.05		I
40	5539.41		I
50	5543.12		I
40	5544.49		I
55	5556.28		I
26	5556.72		I
20	5564.05		I
40	5568.62		I
26	5569.48		I
2500	5570.45		I
35	5575.19		I
20	5591.68		I
40	5602.76		I
23	5608.62		I
23	5609.23		I
100	5610.93		I
23	5613.07		I
20	5618.45		I
23	5619.38		I
330	5632.47		I

Intensity	Wavelength		Spectrum
50	5634.86		I
230	5650.13		I
23	5673.63		I
55	5674.47		I
40	5677.89		I
35	5682.89		I
460	5689.14		I
23	5699.28		I
80	5705.72		I
23	5711.80		I
210	5722.74		I
23	5728.77		I
26	5729.45	d	I
	5729.59		I
620	5751.40		I
23	5774.55		I
40	5779.36		I
23	5783.33	h	I
520	5791.85		I
23	5795.77	h	I
26	5800.46		I
35	5802.67		I
23	5825.20		I
23	5835.59		I
20	5839.99		I
20	5848.86	h	I
55	5849.73		I
50	5851.52	h	I
520	5858.27		I
20	5861.38		I
50	5869.33		I
26	5876.59		I
820	5888.33		I
23	5892.29		I
50	5893.38	h	I
20	5898.78		I
	5898.82	MO	I
40	5901.47		I
40	5926.36	h	I
160	5928.88	h	I
40	5988.17		I
35	6025.49		I
16	6027.27		I
1300	6030.66		I
20	6047.83		I
20	6054.81		I
20	6079.58		I
10	6081.27		I
40	6101.87		I
10	6130.63		I
10	6197.66		I
20	6217.89		I
10	6264.27		I
16	6265.88		I
15	6290.74		I
13	6301.75		I
11	6323.54		I
40	6357.22		I
16	6389.11		I
11	6391.12		I
35	6401.07		I
26	6409.11		I
10	6412.39		I
100	6424.37		I
20	6446.34		I
20	6471.20		I
20	6473.99		I
10	6493.13		I
23	6519.84		I
15	6611.20	h	I
230	6619.13		I
10	6624.57		I
50	6650.38		I
13	6659.68		I
18	6690.47		I
110	6733.98		I
21	6746.08		I
50	6746.27		I
35	6753.97		I
13	6763.50		I
10	6799.88	h	I
10	6802.62		I
10	6812.03		I
13	6825.63		I
18	6828.87	d	I
	6829.05		I
40	6838.88		I
16	6848.92		I
21	6886.28		I
16	6892.36		I
10	6898.01		I
10	6898.98		I
13	6908.20		I
35	6914.01		I
13	6934.10		I
10	6947.39		I
10	6960.64		I
16	6978.71		I
26	6988.94		I
12	6999.13		I
16	7001.60		I
22	7037.98		I
22	7060.21		I

Intensity	Wavelength		Spectrum
13	7063.34		I
13	7081.22		I
110	7109.87		I
27	7134.08		I
150	7242.50		I
40	7245.85		I
22	7267.62		I
17	7300.19		I
13	7348.49		I
13	7361.65		I
10	7364.41	h	I
40	7391.36		I
10	7434.10		I
13	7447.34		I
13	7452.85	h	I
140	7485.74		I
13	7504.47		I
11	7572.64		I
11	7595.16	h	I
11	7601.84		I
17	7656.76	h	I
13	7679.49		I
27	7720.77		I
17	7829.65		I
15	7854.45		I
11	7923.15		I
15	7986.60		I
22	8245.06	h	I
40	8328.44	h	I
45	8389.32	h	I
45	8483.39	h	I

Mo III
Ref. 420 — C.H.C.

Intensity	Wavelength	Spectrum
	Vacuum	
50	1166.07	III
100	1169.33	III
50	1173.67	III
50	1209.60	III
50	1225.46	III
50	1230.34	III
50	1234.63	III
30	1236.10	III
100	1254.93	III
100	1262.21	III
100	1263.74	III
100	1274.37	III
50	1274.94	III
100	1276.40	III
200	1277.40	III
200	1277.58	III
100	1278.06	III
200	1278.40	III
150	1281.90	III
150	1283.60	III
100	1258.52	III
100	1286.42	III
50	1288.07	III
50	1288.25	III
100	1290.49	III
100	1299.82	III
50	1305.58	III
50	1437.37	III
50	1452.38	III
50	1534.86	III
50	1751.22	III
50	1760.57	III
50	1807.60	III
100	1854.73	III
	Air	
75	2165.19	III
75	2170.57	III
50	2172.46	III
75	2179.37	III
100	2184.37	III
100	2211.02	III
50	2223.19	III
100	2253.18	III
150	2269.71	III
50	2275.47	III
50	2275.64	III
200	2294.97	III
80	2304.26	III
50	2326.75	III
150	2330.93	III
100	2359.76	III
80	2386.96	III
50	2403.61	III
70	2412.71	III
50	2422.18	III
200	2506.19	III
90	2597.13	III
50	2756.06	III
100	2807.74	III
125	2947.32	III

Molybdenum (Cont.)

Intensity	Wavelength	
75	2983.94	III
80	3254.70	III
200	3271.69	III

Mo IV
Ref. 383 -- C.H.C.

Intensity	Wavelength	
	Vacuum	
10	857.75	IV
10	859.72	IV
25	865.24	IV
20	865.53	IV
20	863.63	IV
50	867.92	IV
10	878.43	IV
100	884.19	IV
10	884.82	IV
60	886.05	IV
50	891.74	IV
40	894.80	IV
15	895.41	IV
20	1819.50	IV
30	1821.59	IV
30	1850.69	IV
20	1877.88	IV
80	1926.26	IV
100	1929.24	IV
20	1949.44	IV
80	1971.06	IV
20	1991.41	IV
	Air	
70	2010.92	IV
20	2023.78	IV
25	2055.64	IV
50	2060.38	IV
15	2091.89	IV
10	2113.78	IV
5	2140.33	IV

NEODYMIUM (Nd)
Z = 60

Nd I and II
Ref. 1 -- C.H.C.

Intensity		Wavelength	
		Air	
75		2702.46	
75		2704.54	
75		2764.98	I
60		2785.79	I
50		2863.95	
50		2921.26	
55		2962.88	II
65		2963.58	II
80		2993.20	II
40		2994.73	
95		3007.97	II
95		3014.19	II
95		3018.35	II
80		3026.47	II
50		3038.98	II
50		3043.29	
50		3051.11	III
80		3052.15	III
140		3056.71	III
130		3069.73	II
65	d	3071.43	II
		3071.50	II
160		3075.38	II
95		3079.38	II
95		3080.94	II
95		3092.73	
240		3092.92	II
140		3098.48	II
55		3099.52	II
130		3105.43	II
95		3106.18	II
65		3108.01	II
260		3115.18	II
190		3116.15	II
50		3119.75	II
160		3123.06	II
190		3124.58	II
290		3133.60	II
220		3134.90	II
100		3137.24	II
170		3141.46	II
170		3142.44	II
100		3144.55	II
100		3144.82	II
100		3148.51	II
100		3149.29	I
100		3149.51	II

Intensity		Wavelength	
100		3162.62	II
100		3175.99	II
50		3181.54	II
50		3188.73	II
50		3200.62	II
150		3203.47	II
85		3211.00	II
100		3217.12	II
50		3222.62	I
50		3228.04	II
60		3234.62	II
40		3237.91	II
100		3254.08	II
50		3256.91	II
220		3259.24	II
100		3260.66	II
220		3265.12	II
50		3265.38	II
170		3267.25	II
100		3273.18	II
320		3275.22	II
50		3281.49	II
50		3282.78	II
290		3285.10	II
100		3286.62	II
50		3289.52	
100		3290.65	
70		3293.84	II
70		3294.68	II
70		3298.61	II
300		3300.16	II
200		3312.75	II
200		3325.90	II
410		3328.28	II
250		3331.57	II
290		3334.48	II
290		3339.07	II
320		3353.59	II
200		3355.93	II
270		3364.96	II
290		3393.63	II
120	h	3484.88	I
200		3527.53	II
290		3543.35	II
200		3555.77	II
410		3560.75	II
340		3568.87	II
470		3587.51	II
300		3592.59	II
360		3598.01	II
300		3600.91	II
320		3609.79	II
370		3615.82	II
300		3618.96	II
300		3631.02	II
340		3634.30	II
240		3637.00	II
240		3637.23	II
240		3640.24	II
240		3645.78	II
340		3648.20	II
240		3649.46	II
240		3650.42	II
410		3653.15	II
240		3654.16	II
470		3662.26	II
540		3665.18	II
540		3672.36	II
580		3673.54	II
240		3678.18	II
1200		3685.80	II
440		3687.30	II
410		3689.69	II
300		3694.81	II
410		3697.56	II
240		3702.84	II
240		3704.95	II
200		3712.81	II
470		3713.70	II
370		3714.20	II
640	d	3714.73	II
250		3715.04	II
200		3715.39	II
470		3715.68	II
410		3718.54	II
410		3721.35	II
220		3722.42	II
780		3723.50	II
410		3724.87	II
250		3726.90	II
710		3728.13	II
470		3730.58	II
270		3732.78	II
1000	d	3735.54	II
		3735.60	II
440		3737.10	II
1000		3738.06	II
270		3741.42	II
200		3749.85	II
320		3750.31	II
580		3752.49	II
370		3752.67	II
250		3754.83	II

Intensity		Wavelength	
370		3755.60	II
510		3757.82	II
930		3758.95	II
300		3759.79	II
930		3763.47	II
300		3766.59	II
510		3769.65	II
1400		3775.50	II
250		3776.34	II
710		3779.47	II
580		3780.40	II
510		3781.32	II
300		3783.78	II
2400		3784.25	II
270		3784.73	II
340		3791.50	II
340		3795.45	II
240		3799.55	II
370		3801.12	II
200		3801.38	II
340		3802.30	II
1200		3803.47	II
200		3804.10	II
2500		3805.36	II
340		3805.55	II
470		3807.23	II
540		3808.77	II
440		3809.06	II
580		3810.49	II
240		3811.06	II
270		3811.77	II
200		3812.53	II
710		3814.73	II
240		3819.70	II
410		3822.47	II
1200		3826.42	II
240		3828.00	II
540		3828.85	II
440		3829.16	II
510		3830.47	II
740		3836.54	II
340		3837.91	II
1700		3838.98	II
340		3839.51	II
110	d	3841.82	II
		3841.08	II
1700	d	3848.24	II
		3848.31	II
1500		3848.52	II
470		3850.22	II
2400	d	3851.66	II
		3851.74	II
340		3858.55	II
270		3860.94	II
300		3862.52	II
3700	d	3863.33	II
		3863.40	II
240		3866.52	II
220		3866.81	II
850		3869.07	II
240		3875.74	II
470		3875.87	II
1100		3878.58	II
1000		3879.55	II
780		3880.38	II
1200		3880.78	II
200		3881.59	
540		3887.87	II
370	h	3889.66	II
1300		3889.93	II
1300		3890.58	II
1300		3890.94	II
580		3891.51	II
470		3892.06	II
810		3894.63	II
270		3896.13	II
440		3897.63	II
2000		3900.21	II
1300		3901.84	II
1700		3905.89	II
200		3907.70	II
510		3907.84	II
2000		3911.16	II
850		3912.23	II
340		3913.69	II
440		3915.13	II
610		3915.95	II
340		3917.65	II
220		3919.92	II
1100		3920.96	II
510		3927.10	II
200		3929.26	II
610		3934.82	II
410		3936.11	II
510		3938.86	II
2000		3941.51	II
2000		3951.16	II
810		3952.20	II
320		3952.87	II
320		3953.52	II
240		3957.45	II
590		3958.00	II
510		3962.21	II

Intensity		Wavelength	
1400		3963.12	II
270		3963.90	II
1100		3973.30	II
740		3973.69	II
740		3976.85	II
740		3979.49	II
320		3982.36	II
470		3986.25	II
1400		3990.10	II
1000		3991.74	II
1100		3994.68	II
410		4000.50	II
540		4004.02	II
410		4007.43	II
3700		4012.25	II
540		4012.70	II
370		4018.81	II
1000		4020.87	II
1000		4021.34	II
1000		4021.78	II
1200		4023.00	II
340		4024.78	II
410		4030.47	II
1200		4031.82	II
270		4038.12	II
3000		4040.80	II
200		4041.06	II
410		4043.59	II
410		4048.81	II
850		4051.15	II
850		4059.96	II
4700		4061.09	II
1100		4069.28	II
710		4075.12	II
470		4075.28	II
240		4077.62	II
470		4080.23	II
240		4085.82	II
270		4096.13	II
220		4098.18	II
200		4106.59	II
1400		4109.08	II
2500		4109.46	II
510		4110.48	II
300	h	4113.83	II
410		4123.88	II
470		4133.36	II
510		4135.33	II
3000		4156.09	II
510		4156.26	II
340		4160.57	II
410		4168.00	II
810		4175.61	II
2400		4177.32	II
200		4178.64	II
640		4179.59	II
250		4184.98	II
470		4205.60	II
470		4211.29	II
290		4220.25	II
440		4227.73	II
1300		4232.38	II
250		4234.19	II
290	h	4235.24	II
290		4239.84	II
2000		4247.38	II
850		4252.44	II
290		4254.29	
410		4261.84	II
340		4266.71	II
240		4270.56	II
340		4272.79	II
340		4275.09	II
470		4282.44	II
240		4282.57	II
710		4284.52	II
270		4297.80	II
5400		4303.58	II
340		4304.45	II
200		4307.78	II
470		4314.52	II
1100		4325.76	II
510		4327.93	II
540		4338.70	II
680		4351.29	II
850		4358.17	II
240		4366.38	II
340		4368.64	II
470	d	4374.93	II
		4375.04	II
710		4385.66	II
250		4390.66	II
540		4400.83	II
510		4411.06	II
580		4446.39	II
1400		4451.57	II
200		4451.99	II
300		4456.40	II
740		4462.99	II
410		4501.82	II
200		4506.59	II
170		4513.34	II
250		4516.36	II

Neodymium (Cont.)

Intensity	Wavelength	Spectrum
120	4527.25	I
340	4541.27	II
340	4542.61	II
100	4556.14	II
170	4559.67	I
340	4563.22	II
200	4578.89	II
200	4579.32	II
100	4586.62	I
200	4597.02	II
100	4603.82	I
100	4609.87	I
300	4621.94	I
100	4627.98	I
510	4634.24	I
340	4641.10	I
250	4645.77	II
200	4646.40	I
300	4649.67	I
200	4654.73	I
130	4670.56	II
170	4680.74	II
310	4683.45	I
110	4684.04	I
110	4690.35	I
190	4696.44	I
130	4703.57	II
470	4706.54	II
140	4706.96	I
190	4709.71	II
190	4715.59	II
240	4719.02	I
190	4724.35	II
140	4731.77	I
120	4779.46	I
170	4789.41	II
120	4797.15	II
240	4811.34	II
140	4820.34	II
350	4825.48	II
130	4832.28	II
110	4849.06	II
280	4859.02	II
190	4866.74	I
350	4883.81	I
140	4889.10	II
220	4890.70	II
240	4891.07	I
280	4896.93	I
120	4901.53	I
210	4901.84	I
110	4902.03	II
190	4913.41	I
170	4914.37	II
330	4920.68	II
470	4924.53	I
260	4944.83	I
290	4954.78	I
290	4959.13	II
150	4961.39	II
250	4989.94	II
150	5033.52	II
110	5063.73	II
360	5076.59	II
150 h	5089.84	II
360	5092.80	II
180	5102.39	II
150 d	5105.21	II
	5105.35	I
360	5107.59	II
340	5123.79	II
680	5130.60	II
170	5132.33	II
170	5165.14	II
130	5181.17	II
120	5182.60	II
500	5191.45	II
630	5192.62	II
330	5200.12	II
310	5212.37	II
150	5213.23	I
130	5225.05	II
130	5228.43	II
450	5234.20	II
250	5239.79	II
720	5249.59	II
200	5250.82	II
360	5255.51	II
120	5269.48	II
590	5273.43	II
150	5276.88	II
110	5291.67	II
680	5293.17	II
160	5302.28	II
110	5306.47	II
220	5311.46	II
500	5319.82	II
180	5356.98	II
290	5361.47	II
150	5371.94	II
110	5385.90	II
160	5431.53	II
110	5451.12	II

Intensity	Wavelength	Spectrum
170	5485.70	II
35	5501.47	I
45	5525.72	I
90	5533.82	I
55	5535.27	II
55	5543.24	I
55	5548.47	II
55	5561.17	I
27	5575.50	I
27	5576.70	I
27	5577.70	I
27	5587.61	I
240	5594.43	II
55	5601.43	I
45	5601.92	I
220	5620.54	I
65	5635.76	I
45	5639.54	I
35	5653.57	I
70	5668.87	II
65	5669.77	I
140 d	5675.97	I
55	5676.33	I
220	5688.53	II
23	5689.51	I
30	5701.57	I
130	5702.24	II
80	5706.21	II
160	5708.28	II
80	5718.12	II
65	5726.83	II
100	5729.29	I
23	5734.45	I
70	5740.86	II
55	5749.19	I
27	5749.66	I
23	5767.33	I
45	5776.12	I
45	5784.96	I
45	5788.22	I
45	5800.09	I
160	5804.02	II
80	5811.57	II
45	5813.89	I
27	5820.37	I
70	5825.87	II
30	5826.74	I
80	5842.39	II
30	5844.66	I
23	5845.95	I
55	5858.91	I
35	5867.08	II
30	5868.90	I
27	5871.04	I
30	5883.29	I
23	5886.24	I
30	5887.91	I
27	5921.22	I
27	5955.87	I
30	5994.76	I
27	5996.47	I
45	6007.67	I
35	6031.27	II
27	6033.29	I
45	6034.24	II
55	6066.03	I
27	6071.70	I
30	6073.97	I
23 d	6133.47	II
27	6149.28	I
27	6155.06	I
35	6157.83	II
23	6166.67	I
35	6170.49	II
45	6178.59	I
27	6183.91	II
27	6208.24	I
45	6223.39	I
27	6226.50	I
23	6238.50	II
35	6244.08	I
23	6257.49	I
27	6258.73	II
23	6277.29	II
27	6285.79	I
23	6292.84	I
23	6297.07	I
55	6310.49	I
27	6341.51	II
23	6382.07	II
65	6385.20	II
35	6485.69	I
45	6630.14	I
35	6637.96	II
45	6650.57	II
30	6655.67	I
25	6737.79	II
40	6740.11	II
25	6742.54	II
30	6790.37	II
30	6804.00	II
25	6846.72	II
40	6900.43	II

Intensity	Wavelength	Spectrum
24	6941.39	II
17 h	7010.80	II
8	7018.85	II
17	7020.92	II
17	7024.58	II
10	7033.21	II
35	7037.30	II
7	7052.14	II
7	7054.74	II
7	7061.47	II
40	7066.89	II
8	7082.93	II
12 h	7089.71	II
12 h	7092.09	II
12 h	7092.74	II
12 h	7092.94	II
17 h	7093.98	I
20 h	7095.42	I
29	7129.35	II
12 h	7142.04	II
10	7143.72	II
8	7151.03	II
6	7153.09	I
6 h	7185.01	II
10	7189.09	II
24	7189.42	II
20	7192.01	II
10	7199.00	II
8 h	7227.01	I
15	7236.54	II
7 h	7261.64	II
9	7285.29	II
9	7288.56	II
6	7291.38	II
7	7298.72	II
12	7316.81	II
7 h	7321.43	I
7	7323.12	II
6	7334.54	I
6	7357.10	II
6	7374.04	II
7	7381.79	II
9	7401.31	II
10	7406.62	II
6	7411.20	II
10	7418.18	II
9	7427.41	II
9	7448.71	II
5	7481.28	II
12	7511.16	II
17	7513.73	II
7 h	7514.44	II
7	7516.02	II
7	7526.45	II
12	7528.99	II
10	7538.26	II
5	7540.97	II
7	7547.00	II
5	7577.54	II
7	7587.65	II
6	7590.75	II
6	7603.73	II
5	7605.92	II
5	7614.72	I
9	7639.79	II
8	7646.00	II
6	7663.52	II
12	7696.56	II
6	7718.20	II
4	7743.90	II
4	7748.92	II
10	7750.95	II
6	7773.06	II
7	7792.22	II
6	7796.40	II
8	7797.32	II
5	7798.32	II
10	7808.47	II
7	7818.83	II
5	7825.20	II
12	7863.04	II
5 h	7872.03	I
7	7886.60	II
4 h	7896.50	II
9	7900.40	II
5 h	7906.03	I
12	7917.01	II
10	7925.03	II
5	7947.93	II
10	7949.68	II
5 h	7955.38	II
12	7958.95	I
12	7965.73	II
15	7982.09	II
12	7982.68	II
12	8000.76	II
9	8007.70	I
4 h	8020.07	II
8	8026.35	II
10	8043.24	I
8	8051.33	II
5	8064.00	II
10	8099.17	I

Intensity	Wavelength	Spectrum
10	8120.93	II
12	8122.07	II
12	8141.75	II
12	8143.27	II
7 h	8164.97	I
8	8172.56	II
9	8179.83	II
9	8182.41	II
4	8185.58	II
7 h	8205.38	II
10	8231.52	II
4	8248.76	II
5 h	8249.68	II
4 h	8262.80	II
7 h	8266.72	II
4	8272.79	II
4 h	8302.74	II
10	8307.72	II
6	8324.50	II
4	8332.01	II
12	8346.36	II
4	8375.16	II
4	8375.33	II
4	8394.71	II
7	8400.85	II
5 h	8456.87	II
4	8530.53	II
5	8582.03	II
5	8591.53	II
7	8594.87	II
8 c	8643.43	II
5	8667.07	II
5	8677.48	II
6	8691.29	II
6	8695.07	II
6	8712.82	II
6	8715.03	I
17	8839.10	II

NEON (Ne)
Z = 10

Ne I and II
Ref. 56, 58, 118, 150, 230 — S.P.D.

Intensity	Wavelength	Spectrum
	Vacuum	
90	352.956	I
60	354.962	I
90	361.433	II
60	362.455	II
150	405.854	II
120	407.138	II
200	445.040	II
300	446.256	II
250	446.590	II
180	447.815	II
150	454.654	II
200	455.274	II
10	456.275	II
120	456.348	II
90	456.896	II
1000	460.728	II
500	462.391	II
35	587.213	I
35	589.179	I
35	589.911	I
70	591.830	I
100	595.920	I
75	598.706	I
35	598.891	I
70	600.036	I
170	602.726	I
170	615.628	I
170	618.672	I
120	619.102	I
200	626.823	I
200	629.739	I
1000	735.896	I
400	743.720	I
60	993.88	I
70	1068.65	I
90	1131.72	I
100	1131.85	II
90	1229.83	I
90	1418.38	I
90	1428.58	I
90	1436.09	I
120	1681.68	II
180	1688.36	II
100	1888.11	II
100	1889.71	II
200	1907.49	II
500	1916.08	II
300	1930.03	II
200	1938.83	II
100 c	1945.46	II
	AIR	

Neon (Cont.)

Intensity		Wavelength	Spectrum
80		2007.01	II
80		2025.56	II
150		2085.47	II
180		2096.11	II
120		2096.25	II
80	p	2562.12	II
90	w	2567.12	II
80		2623.11	II
80		2629.89	II
90	w	2636.07	II
80		2638.29	II
80		2644.10	II
80		2762.92	II
90		2792.02	II
80		2794.22	II
100		2809.48	II
80		2906.59	II
80		2906.82	II
90		2910.06	II
90		2910.41	II
80		2911.14	II
80		2915.12	II
80		2925.62	II
80	w	2932.10	II
80		2940.65	II
90		2946.04	II
150		2955.72	II
150		2963.24	II
150		2967.18	II
100		2973.10	II
15		2974.72	I
100		2979.46	II
12		2982.67	I
150		3001.67	II
120	P	3017.31	II
300		3027.02	II
300		3028.36	II
100		3030.79	II
120		3034.46	II
100		3035.92	II
100		3037.72	II
100		3039.59	II
100		3044.09	II
100		3045.56	II
120		3047.56	II
100		3054.34	II
100		3054.68	II
100		3059.11	II
100		3062.49	II
100		3063.30	II
100		3070.89	II
100		3071.53	II
100		3075.73	II
120		3088.17	II
100		3092.09	II
120		3092.90	II
100		3094.01	II
100		3095.10	II
100		3097.13	II
100		3117.98	II
120		3118.16	II
10		3126.199	I
300		3141.33	II
100		3143.72	II
100	p	3148.68	II
100		3164.43	II
100		3165.65	II
100		3188.74	II
120		3194.58	II
500		3198.59	II
60		3208.96	II
120		3209.36	II
120		3213.74	II
150		3214.33	II
150		3218.19	II
120		3224.82	II
120		3229.57	II
200		3230.07	II
120		3230.42	II
120		3232.02	II
150		3232.37	II
100		3243.40	II
100		3244.10	II
100		3248.34	II
100		3250.36	II
150		3297.73	II
150		3309.74	II
300		3319.72	II
1000		3323.74	II
150		3327.15	II
100		3329.16	II
200		3334.84	II
150		3344.40	II
300		3345.45	II
150		3345.83	II
200		3355.02	II
120		3357.82	II
200		3360.60	II
120		3362.16	II
100		3362.71	II
120		3367.22	II
12		3369.808	I
40		3369.908	I
100		3371.80	II
500		3378.22	II
150		3388.42	II
120		3388.94	II
300		3392.80	II
100		3404.82	II
120		3406.95	II
100		3413.15	II
120		3416.91	II
120		3417.69	II
50		3417.904	I
15		3418.006	I
120		3428.69	II
60		3447.703	I
50		3454.195	I
100		3456.61	II
100		3459.32	II
25		3460.524	I
30		3464.339	I
30		3466.579	I
60		3472.571	I
150		3479.52	II
200		3480.72	II
200		3481.93	II
25		3498.064	I
30		3501.216	I
25		3515.191	I
150		3520.472	I
120		3542.85	II
120		3557.80	II
100		3561.20	II
250		3568.50	II
100		3574.18	II
200		3574.61	II
50		3593.526	I
30		3593.640	I
15		3600.169	I
20		3633.665	I
150		3643.93	II
200		3664.07	II
20		3682.243	I
12		3685.736	I
200		3694.21	II
10		3701.225	I
150		3709.62	II
250		3713.08	II
250		3727.11	II
800		3766.26	II
1000		3777.13	II
100		3818.43	II
120		3829.75	II
150		4219.74	II
100		4233.85	II
120		4250.65	II
120		4369.86	II
70		4379.40	II
150		4379.55	II
100		4385.06	II
200		4391.99	II
150		4397.99	II
150		4409.30	II
100		4413.22	II
100		4421.39	II
100	P	4428.52	II
100	P	4428.63	II
150	P	4430.90	II
150	P	4430.94	II
120		4457.05	II
100		4522.72	II
10		4537.754	I
10		4540.380	I
100		4569.06	II
15		4704.395	I
12		4708.862	I
10		4710.067	I
10		4712.066	I
15		4715.347	I
10		4752.732	I
12		4788.927	I
10		4790.22	I
10		4827.344	I
10		4884.917	I
4		5005.159	I
10		5037.751	I
10		5144.938	I
25		5330.778	I
20		5341.094	I
8		5343.283	I
60		5400.562	I
5		5562.766	I
10		5656.659	I
5		5719.225	I
12		5748.298	I
80		5764.419	I
12		5804.450	I
40		5820.156	I
500		5852.488	I
100		5872.828	I
100		5881.895	I
60		5902.462	I
60		5906.429	I
100		5944.834	I
100		5965.471	I
100		5974.627	I
120		5975.534	I
80		5987.907	I
100		6029.997	I
100		6074.338	I
80		6096.163	I
60		6128.450	I
100		6143.063	I
120		6163.594	I
250		6182.146	I
150		6217.281	I
150		6266.495	I
60		6304.789	I
100		6334.428	I
120		6382.992	I
200		6402.246	I
150		6506.528	I
60		6532.882	I
150		6598.953	I
70		6652.093	I
90		6678.276	I
20		6717.043	I
100		6929.467	I
90		7024.050	I
100		7032.413	I
50		7051.292	I
80		7059.107	I
100		7173.938	I
150		7213.20	II
150		7235.19	II
100		7245.167	I
150		7343.94	II
40		7472.439	I
90		7488.871	I
100		7492.10	II
150		7522.82	II
80		7535.774	I
60		7544.044	I
100		7724.628	I
120		7740.74	II
300		7839.055	I
120		7926.20	II
400		7927.118	I
700		7936.996	I
2000		7943.181	I
2000		8082.458	I
100		8084.34	II
1000		8118.549	I
600		8128.911	I
3000		8136.406	I
2500		8259.379	I
100		8264.81	II
2500		8266.077	I
800		8267.117	I
6000		8300.326	I
100		8315.00	II
1500		8365.749	I
100		8372.11	II
8000		8377.606	I
1000		8417.159	I
4000		8418.427	I
1500		8463.358	I
800		8484.444	I
5000		8495.360	I
600		8544.696	I
1000		8571.352	I
4000		8591.259	I
6000		8634.647	I
3000		8647.041	I
15000		8654.383	I
4000		8655.522	I
100		8668.26	II
5000		8679.492	I
5000		8681.921	I
2000		8704.112	I
4000		8771.656	I
12000		8780.621	I
10000		8783.753	I
500		8830.907	I
7000		8853.867	I
1000		8865.306	I
1000		8865.755	I
3000		8919.501	I
2000		8988.57	I
100		9079.46	II
6000		9148.67	I
6000		9201.76	I
4000		9220.06	I
2000		9221.58	I
2000		9226.69	I
1000		9275.52	I
200		9287.56	II
6000		9300.85	I
1500		9310.58	I
3000		9313.97	I
6000		9326.51	I
2000		9373.31	I
5000		9425.38	I
3000		9459.21	I
5000		9486.68	I
5000		9534.16	I
3000		9547.40	I
120		9577.01	II
1000		9665.42	I
100		9808.86	II
800		10295.42	I
2000		10562.41	I
1500		10798.07	I
2000		10844.48	I
3000		11143.020	I
3500		11177.528	I
1600		11390.434	I
1100		11409.134	I
3000		11522.746	I
1500		11525.020	I
950		11536.344	I
500		11601.537	I
1200		11614.081	I
300		11688.002	I
2000		11766.792	I
1500		11789.044	I
500		11789.889	I
1000		11984.912	I
3000		12066.334	I
800		12459.389	I
1000		12689.201	I
1100		12912.014	I
700		13219.241	I
800		15230.714	I
400		17161.930	I
400		18035.80	I
1000		18083.21	I
350		18221.11	I
250		18227.02	I
2500		18276.68	I
2000		18282.62	I
1200		18303.97	I
250		18359.12	I
1200		18384.85	I
2000		18389.95	I
1000		18402.84	I
1200		18422.39	I
300		18458.65	I
400		18475.79	I
900		18591.55	I
1600		18597.70	I
350		18618.96	I
550		18625.16	I
1200		21041.295	I
750		21708.145	I
300		22247.35	I
350		22428.13	I
2230		22530.40	I
400		22661.81	I
600		23100.51	I
1000		23260.30	I
1050		23373.00	I
850		23565.36	I
3500		23636.52	I
300		23701.64	I
1100		23709.2	I
1800		23951.42	I
600		23956.46	I
1000		23978.12	I
200		24098.54	I
500		24161.42	I
600		24249.64	I
1500		24365.05	I
800		24371.60	I
400		24447.85	I
700		24459.4	I
300		24776.46	I
550		24928.88	I
250		25161.69	I
650		25524.37	I
125		28386.21	I
150		30200.	I
250		33173.09	I
450		33352.35	I
1300		33901.	I
2200		33912.10	I
600		34131.31	I
100		34471.44	I
120		35834.78	I

Ne III
Ref. 365, 371, 402 — R.L.K.

Intensity	Wavelength	
	Vacuum	
20	251.14	III
20	251.56	III
20	251.73	III
40	267.06	III
40	267.52	III
20	267.71	III
40	283.18	III
160	283.21	III
110	283.69	III
40	283.89	III
220	301.12	III
220	313.05	III
220	313.68	III
40	313.95	III

Neon (Cont.)

Intensity	Wavelength	
220	379.31	III
285	488.10	III
220	488.87	III
450	489.50	III
70	489.64	III
220	490.31	III
360	491.05	III
20	1255.03	III
110	1255.68	III
160	1257.19	III

Air

Intensity	Wavelength	
200	2086.96	III
300	2089.43	III
240	2092.44	III
400	2095.54	III
200	2161.22	III
300	2163.77	III
200	2180.89	III
200	2209.35	III
200	2211.85	III
240	2213.76	III
300	2216.07	III
240	2263.21	III
200	2264.91	III
300	2412.73	III
240	2412.94	III
200	2413.78	III
200	2473.40	III
800	2590.04	III
600	2593.60	III
400	2595.68	III
300	2610.03	III
240	2613.41	III
200	2615.87	III
200	2638.70	III
200	2641.07	III
600	2677.90	III
500	2678.64	III

Ne IV
Ref. 69, 364, 388, 400, 413, 430, — R.L.K.

Intensity	Wavelength	

Vacuum

Intensity	Wavelength	
15	151.82	IV
15	152.23	IV
15	158.65	IV
15	158.82	IV
80	172.62	IV
80	177.16	IV
150	186.58	IV
100	194.28	IV
100	208.48	IV
100	208.73	IV
80	208.90	IV
150	212.56	IV
140	223.24	IV
120	223.60	IV
140	234.32	IV
120	234.70	IV
50	357.83	IV
200	358.72	IV
125	387.14	IV
100	388.22	IV
150	421.61	IV
140	469.77	IV
200	469.82	IV
180	469.87	IV
140	469.92	IV
120	521.74	IV
140	521.82	IV
80	541.13	IV
100	542.07	IV
150	543.89	IV

Air

Intensity	Wavelength	
65	2018.44	IV
110	2022.19	IV
30	2203.88	IV
10	2220.81	IV
250	2258.02	IV
175	2262.08	IV
110	2264.54	IV
550	2285.79	IV
30	2293.14	IV
250	2293.49	IV
250	2363.28	IV
110	2365.49	IV
250	2350.84	IV
450	2352.52	IV
700	2357.96	IV
250	2362.68	IV
350	2372.16	IV
65	2384.20	IV
350	2384.95	IV

Ne V
Ref. 69, 388, 389, 400, 413 — R.L.K.

Vacuum

Intensity	Wavelength	
66	119.01	V
200	122.52	V
66	125.12	V
45	131.99	V
50	132.04	V
150	140.76	V
150	140.79	V
100	142.44	V
100	142.50	V
150	142.72	V
100	143.27	V
150	143.34	V
150	147.13	V
66	151.23	V
120	151.42	V
45	154.50	V
100	164.02	V
100	164.14	V
500	173.93	V
400	357.96	V
500	358.47	V
500	359.38	V
1000	365.59	V
800	416.20	V
250	480.41	V
150	481.28	V
250	481.36	V
500	482.99	V
400	568.42	V
250	569.76	V
500	569.83	V
250	572.11	V
800	572.34	V

Air

Intensity	Wavelength	
75	2227.42	V
110	2232.41	V
65	2245.48	V
65	2259.57	V
65	2263.39	V
250	2265.71	V

NEPTUNIUM (Np)
Z = 93
Np I and II
Ref. 93 = J.G.C.

Air

Intensity		Wavelength	
300		3481.93	I
300	h	3501.50	I
300	l	3986.89	I
300	l	5044.66	I
300	l	5601.70	I
300	l	5652.75	I
300	l	5784.39	I
300	l	5878.04	I
300	s	6011.22	I
300		6056.09	I
300	s	6073.90	I
300	s	6080.05	I
300	l	6120.49	I
300		6188.59	I
300	l	6200.00	I
300	s	6215.90	I
300	s	6317.84	I
300	l	6341.38	I
300	l	6566.11	I
300	l	6720.68	I
300	l	6751.32	I
300	s	6795.21	I
300	l	6802.62	I
300	l	6805.81	I
300	l	6816.44	I
300	l	6865.45	I
300	s	6907.13	I
300	h	6912.91	I
1000	s	6930.31	I
300	l	6963.63	I
3000	s	6972.09	I
300		7014.21	I
300	l	7018.91	I
300	s	7039.11	I
300	s	7080.01	I
300	l	7174.83	I
300	l	7184.93	I
300	l	7284.28	I
300	l	7292.29	I
300	l	7332.52	I
300	s	7370.60	I
300	l	7381.03	I
300	l	7381.65	I
300	l	7402.70	I
300	s	7512.22	I
300	l	7515.15	I
300	l	7546.05	I
300	l	7624.83	I
300		7626.85	I
300	s	7681.01	I
300	s	7685.25	I
1000	l	7735.14	I
300	l	7761.61	I
1000	l	7765.75	I
300	s	7776.07	I
300		7787.46	I
1000	l	7791.38	I
300	l	7851.44	I
300	l	7887.88	I
300	l	7901.71	I
300	l	7975.98	I
300	h	8080.32	I
300	s	8124.59	I
300		8155.11	I
300	l	8167.42	I
300	l	8183.06	I
300	l	8188.61	I
300	l	8247.82	I
300	l	8287.11	I
300	s	8287.65	I
300	l	8306.22	I
300	s	8313.66	I
1000	l	8339.12	I
300		8356.79	I
300	l	8367.11	I
3000		8372.88	I
3000		8529.96	I
1000	s	8696.23	I
1000	s	8906.02	I
1000		8942.70	I
1000	s	9004.75	I
1000	l	9006.31	I
10000	l	9016.18	I
3000	l	9141.30	I
3000	s	9379.33	I
3000	l	9468.66	I
3000	s	9679.13	I
3000	l	9930.55	I
10000	s	10091.99	I
10000	s	10817.45	I
10000	l	11695.15	I
10000	l	11776.64	I
10000	s	12148.18	I
10000	s	12377.42	I
10000	l	12407.99	I
10000	l	13834.33	I

NICKEL (Ni)
Z = 28
Ni I and II
Ref. 1,294 = C.H.C.

Vacuum

Intensity	Wavelength	
500	1317.22	II
400	1335.20	II
500	1370.14	II
1000	1741.55	II
500	1748.28	II

Air

Intensity	Wavelength	
1000	2165.55	II
2000	2169.10	II
2000	2174.67	II
1500	2175.15	II
500	2177.09	II
400	2177.35	II
400	2179.35	II
800	2180.47	II
800	2184.60	II
2500	2185.50	II
3000	2192.09	II
600	2201.41	II
5000	2205.55	II
4000	2206.72	II
6000	2216.48	II
800	2220.40	II
500	2221.06	II
900	2222.96	II
500	2242.68	II
500	2253.85	II
1000	2264.46	II
2000	2270.21	II
800	2277.28	II
400	2278.32	II
800	2278.77	II
500	2287.65	II
1600	2289.98	I
400	2296.55	II
400	2297.14	II
630	2300.78	I
1000	2303.00	II
2000	2310.96	I
1700	2312.34	I
1400	2313.66	I
1400	2313.98	I
1000	2316.04	II
1400	2317.16	I
500	2319.75	II
2600	2320.03	I
1900	2321.38	I
240	2322.68	I
1400	2325.79	I
940	2329.96	I
500	2334.58	II
460	2337.49	I
160	2337.82	I
500	2341.20	II
1200	2345.54	I
190	2346.63	I
400	2347.52	I
160	2360.63	I
200	2362.06	I
1000	2375.42	II
240	2386.58	I
1000	2394.52	II
2000	2416.13	II
240	2419.31	I
85	2421.23	I
70	2423.33	I
70	2423.66	I
70	2424.03	I
500	2437.89	II
85	2453.99	I
160	2472.06	I
85	2476.87	I
500	2510.87	II
500	2565.92	II
500	2606.26	II
500	2609.94	II
500	2615.06	II
45	2696.49	I
150	2798.65	I
250	2821.29	I
500	2864.02	II
50	2865.50	I
60	2907.46	I
25	2914.01	I
500	2943.91	I
570	2981.65	I
250	2984.13	I
500	2992.60	I
1000	2994.46	I
4000	3002.49	I
2200	3003.63	I
3700	3012.00	I
350	3019.14	I
120	3031.87	I
1700	3037.94	I
150	3045.01	I
3500	3050.82	I
1500	3054.32	I
1900	3057.64	I
500	3064.62	I
420	3080.76	I
260	3097.12	I
210	3099.12	I
2600	3101.55	I
1300	3101.88	I
220	3105.47	I
270	3114.12	I
2900	3134.11	I
55	3145.72	I
55	3181.74	I
100	3184.37	I
55	3195.57	I
150	3197.11	I
55	3202.14	I
180	3214.06	I
180	3217.83	I
100	3221.27	I
150	3221.65	I
210	3225.02	I
1100	3232.96	I
290	3234.65	I
600	3243.06	I
100	3248.46	I
120	3250.74	I
100	3271.12	I
120	3282.70	I
400	3292.87	II
500	3297.60	II
400	3305.71	II
660	3315.66	I
330	3320.26	I
310	3322.31	I
2000	3331.88	II
400	3335.64	II
500	3338.09	II
500	3348.84	II
500	3349.24	II

Nickel (Cont.)

Intensity	Wavelength	Spectrum
600	3358.68	II
330	3361.56	I
500	3363.45	II
330	3365.77	I
330	3366.17	I
65	3366.81	I
65	3367.89	I
2900	3369.57	I
400	3371.99	I
260	3374.22	I
130	3374.64	I
500	3378.97	II
3300	3380.57	I
240	3380.85	I
1300	3391.05	I
3300	3392.99	I
500	3401.05	II
130	3409.58	I
330	3413.48	I
330	3413.94	I
8200	3414.76	I
1800	3423.71	I
2600	3433.56	I
990	3437.28	I
4800	3446.26	I
1300	3452.89	I
5000	3458.47	I
5000	3461.65	I
200	3467.50	I
240	3469.49	I
1600	3472.54	I
550	3483.77	I
130	3485.89	I
5500	3492.96	I
660	3500.85	I
65	3502.60	I
55	3507.69	I
2600	3510.34	I
260	3513.93	I
6600	3515.05	I
660	3519.77	I
8200	3524.54	I
110	3527.98	I
330	3548.18	I
55	3551.53	I
11	3561.33	I
5000	3566.37	I
990	3571.87	I
130	3587.93	I
1300	3597.70	I
1300	3610.46	I
530	3612.74	I
6600	3619.39	I
130	3624.73	I
200	3664.10	I
130	3669.24	I
180	3670.43	I
260	3674.15	I
160	3688.42	I
80	3693.93	I
120	3722.48	I
150	3736.81	I
60	3739.23	I
600	3775.57	I
700	3783.53	I
700	3807.14	I
110	3831.69	I
1200	3858.30	I
30	3889.67	I
35	3972.17	I
110	3973.56	I
110	4401.55	I
85	4459.04	I
18	4462.46	I
55	4470.48	I
35	4592.53	I
18	4600.37	I
65	4605.00	I
18	4606.23	I
75	4648.66	I
23	4686.22	I
110	4714.42	I
22	4715.78	I
30	4756.52	I
15	4763.95	I
45	4786.54	I
22	4807.00	I
22 h	4829.03	I
19	4831.18	I
45	4855.41	I
30	4866.27	I
17	4873.44	I
40	4904.41	I
22	4918.36	I
16	4935.83	I
45	4980.16	I
45	4984.13	I
500	4992.02	II
16 h	5000.34	I
18	5012.46	I
50	5017.59	I
100	5035.37	I
16	5048.85	I
100	5080.52	I
65	5081.11	I
26 h	5084.08	I
18	5099.32	I
26 h	5099.95	I
21	5115.40	I
18 h	5129.38	I
23	5137.08	I
23 h	5142.77	I
40 h	5146.48	I
40 h	5155.76	I
16	5168.66	I
13	5176.56	I
8	5435.87	I
180	5476.91	I
6	5510.00	I
6	5578.71	I
9	5587.86	I
13	5592.28	I
9	5614.79	I
5 h	5625.33	I
4	5649.70	I
5	5664.02	I
12	5682.20	I
8	5695.00	I
23	5709.56	I
10	5711.90	I
10	5715.09	I
16	5754.68	I
8	5760.85	I
10	5857.76	I
10	5892.88	I
10	6108.12	I
10	6176.81	I
10	6191.18	I
13	6256.36	I
10	6314.66	I
16	6643.64	I
22	6767.77	I
9	6772.32	I
10	6914.56	I
5	7110.90	I
26	7122.20	I
6	7182.00	I
5	7197.07	I
5	7261.93	I
5	7291.45	I
4	7385.24	I
16	7393.60	I
10	7409.35	I
5	7414.51	I
23	7422.28	I
13	7522.76	I
9	7525.12	I
19	7555.60	I
8	7574.05	I
23	7617.00	I
9	7619.21	I
16	7714.32	I
5 h	7715.58	I
19	7727.61	I
19	7748.89	I
10	7788.94	I
13	7797.59	I
2	7917.44	I
1000	8096.75	II
500	8114.21	II
700	8121.48	II
2	8809.42	I
9	8862.55	I
500 w	9900.92	II

Ni III
Ref. 422 — C.H.C.

Intensity	Wavelength (Vacuum)	Spectrum
100	625.68	III
500	630.71	III
200	637.54	III
200	662.37	III
150	663.57	III
500	676.94	III
200	700.17	III
300	713.33	III
300	713.38	III
500	718.48	III
200	721.26	III
300	722.09	III
250	725.20	III
500	729.82	III
250	730.11	III
400	731.70	III
300	732.16	III
200	738.26	III
300	747.99	III
200	749.68	III
300	750.05	III
200	752.02	III
300	757.80	III
250	758.73	III
250	758.27	III
400	770.22	III
200	772.04	III
500	778.81	III
200 d	785.02	III
300	788.04	III
200	788.30	III
500	811.57	III
500	826.14	III
200	826.50	III
500	842.14	III
400	845.24	III
300	847.43	III
200	857.09	III
300	860.64	III
300	862.88	III
300	863.22	III
300	867.51	III
800	869.70	III
200	870.84	III
300	973.79	III
400	979.59	III
200	1428.87	III
200	1434.31	III
200	1451.50	III
300	1604.54	III
300	1652.87	III
200	1653.12	III
250	1656.13	III
200	1661.79	III
400	1687.90	III
1000	1692.51	III
200	1707.35	III
200	1707.43	III
800	1709.90	III
650	1715.30	III
500	1719.46	III
200	1721.26	III
400	1722.28	III
250	1733.13	III
500	1738.25	III
300	1739.78	III
300	1741.30	III
550	1747.01	III
300	1752.43	III
400	1753.01	III
800	1764.69	III
500	1767.94	III
2000	1769.64	III
400	1776.07	III
200	1788.30	III
250	1790.40	III
200	1790.93	III
200	1791.64	III
200	1794.90	III
300	1807.24	III
200	1811.69	III
300	1819.28	III
800	1823.06	III
400	1830.01	III
200	1830.08	III
650	1847.28	III
800	1854.15	III
300	1858.75	III
200	1930.43	III
200	1952.54	III

Ni IV
Ref. 415 — C.H.C.

Intensity	Wavelength (Vacuum)	Spectrum
33	392.68	IV
32	393.24	IV
49	424.40	IV
57	444.21	IV
67	469.67	IV
65	471.24	IV
65	485.42	IV
66	536.28	IV
67	537.96	IV
58	1345.72	IV
69	1357.07	IV
76	1398.19	IV
74	1411.45	IV
69	1419.58	IV
74	1421.22	IV
70	1427.45	IV
67	1428.93	IV
67	1435.24	IV
70	1438.82	IV
73	1449.01	IV
76	1452.22	IV
70	1455.42	IV
69	1472.63	IV
68	1476.82	IV
73	1482.25	IV
67	1489.53	IV
72	1489.83	IV
69	1493.01	IV
74	1493.67	IV
68	1498.71	IV
71	1498.77	IV
72	1498.90	IV
67	1499.97	IV
70	1512.74	IV
70	1516.66	IV
73	1520.63	IV
75	1525.31	IV
74	1527.68	IV
74	1527.80	IV
76	1534.71	IV
73	1537.25	IV
69	1538.93	IV
75	1543.41	IV
74	1546.23	IV
68	1548.04	IV
69	1557.28	IV
67	1560.18	IV

Ni V
Ref. 416 — C.H.C.

Intensity	Wavelength (Vacuum)	Spectrum
29	304.02	V
55	315.24	V
56	315.71	V
63	336.79	V
68	343.93	V
78	347.34	V
70	347.46	V
67	347.72	V
71	348.10	V
69	350.77	V
69	353.59	V
72	354.18	V
76	354.42	V
68	356.60	V
68	355.61	V
70	355.78	V
65	357.37	V
69	358.57	V
60	358.58	V
66	359.47	V
69	365.62	V
70	370.62	V
67	371.31	V
68	371.76	V
67	373.60	V
72	377.68	V
70	393.91	V
66	394.31	V
66	395.24	V
41	400.59	V

NIOBIUM (Nb)
Z = 41

Nb I and II
Ref. I = C.H.C.

Intensity	Wavelength (Air)	Spectrum
3300	2029.32	II
3000	2032.99	II
2000	2109.42	II
1700	2125.21	II
1100	2126.54	II
1500	2131.18	II
370	2295.68	II
280	2302.08	II
170	2376.40	II
110	2387.09	II
140	2387.52	II
45	2388.27	II
160	2398.48	II
55	2405.34	II
55	2405.85	II
140	2412.46	II
160	2416.99	II
140	2418.69	II
75	2433.80	II
40	2435.95	II
35	2436.33	I
45	2437.42	II
40	2442.14	II
28	2442.68	II
65	2451.87	II
65	2453.95	II
55	2458.09	II
65	2462.89	I
35	2466.73	I
55	2469.08	I
110	2477.38	II
65	2478.29	II

Niobium (Cont.)

Int		λ		Int		λ		Int		λ		Int		λ		Int		λ	
65		2479.94	II	140		3099.19	II	250		3544.65	I	870		4192.07	I	870		4195.09	I
35		2483.88	II	150		3111.45	I	300		3550.45	I	1300		4195.66	I	310		4198.51	I
110		2504.65	I	270		3127.53	II	250		3554.52	I	350		4201.52	I	870		4205.31	I
110		2511.00	II	1500		3130.79	II	1000		3554.66	I	350		4214.73	I	420		4217.94	I
110		2521.40	II	390		3145.40	II	630		3563.50	I	420		4229.15	I	250		4255.44	I
390		2544.80	II	140		3151.87	I	630		3563.62	I	770		4262.05	I	420		4266.02	I
110		2551.38	II	1200		3163.40	II	1500		3575.85	I	290		4270.69	I	400		4286.99	I
130		2556.94	II	150		3175.78	II	200		3577.72	I	580		4299.60	I	580		4300.99	I
130		2562.41	II	390		3180.29	II	5000		3580.27	I	120		4309.56	I	390		4311.27	I
130		2565.41	II	200		3187.49	I	500		3584.97	I	120		4312.45	I	350		4326.33	I
100		2569.03	I	300		3191.10	II	750		3589.11	I	120		4327.38	I	390		4331.37	I
110		2571.33	II	150		3191.43	II	500		3589.36	I	140		4342.82	I	140		4348.65	I
200		2578.74	I	1000		3194.98	II	500		3593.97	I	110		4349.03	I	290		4351.57	I
390		2583.99	II	120		3203.35	II	500		3602.56	I	210		4368.43	I	140		4377.96	I
390		2590.94	II	300		3206.34	II	300		3619.51	II	130		4388.36	I	160		4392.69	I
270		2592.20	I	390		3215.60	II	200		3621.03	I	330		4410.21	I	190		4419.44	I
130		2616.48	I	800		3225.48	II	200		3639.33	I	230	c	4437.22	I	290		4447.18	I
130		2623.51	I	140		3229.56	II	420		3649.85	I	140		4456.80	I	140		4457.42	I
130		2627.44	I	400		3236.40	II	250		3650.81	I	140		4469.71	I	140		4471.29	I
130		2628.49	I	200		3247.47	II	400		3651.19	II	140		4472.53	I	150		4503.04	I
200		2642.24	I	120		3248.94	II	200		3659.61	II	530		4523.41	I	480		4546.82	I
320		2646.26	II	160		3249.52	I	630		3660.37	I	370		4564.53	I	720		4573.08	I
330		2647.50	I	320		3254.07	II	900		3664.70	I	480		4581.62	I	1200		4606.77	I
240		2649.52	I	230		3260.56	II	220		3669.01	I	170		4616.17	I	450		4630.11	I
330		2654.45	I	160		3263.37	II	270		3674.78	I	450		4648.95	I	110		4649.27	I
310		2656.08	II	160		3264.59	I	1500		3697.85	I	450		4663.83	I	340		4666.24	I
160		2657.62	I	120		3270.47	I	330		3711.34	I	240		4667.22	I	580		4672.09	I
110		2665.25	II	100		3270.76	I	3300		3713.01	I	530		4675.37	I	110		4678.48	I
110		2666.59	II	200		3272.07	I	480		3716.99	I	320		4685.14	I	130	c	4706.14	I
110		2667.30	II	160		3277.67	I	2700		3726.24	I	260		4708.29	I	150		4713.50	I
130		2668.29	I	200		3283.46	II	270		3738.42	I	110	c	4733.89	I	220	c	4749.70	I
400		2671.93	II	230		3285.66	I	2700		3739.80	I	110		4816.38	I	110	c	4848.37	I
200		2673.57	II	200		3287.59	I	670		3740.73	II	130	c	4967.78	I	110		4973.14	I
200		2675.94	II	160		3287.92	I	270		3741.78	I	190		4988.97	I	85		5000.95	I
130		2687.15	I	160		3292.02	II	1700		3742.39	I	65		5002.25	I	40		5013.27	I
160		2691.77	II	320		3296.01	I	250		3753.18	I	230		5017.75	I	40		5019.51	I
1000		2697.06	II	160		3299.61	I	210		3755.77	I	150		5026.36	I	40		5030.13	I
320		2698.86	II	120		3304.83	I	530		3763.49	I	210		5039.04	I	40		5047.96	I
320		2702.20	II	120		3308.05	I	350		3765.08	I	170		5058.01	I	65		5059.35	I
150		2702.52	II	120		3310.47	I	250		3766.13	I	130		5065.25	I	40		5077.40	I
470		2716.62	II	400		3312.60	I	530		3771.85	I	750		5078.96	I	40	c	5094.41	I
470		2721.98	II	200		3315.22	I	870		3781.01	I	420		5095.30	I	170		5100.16	I
310		2733.26	II	200		3318.98	I	1700		3787.06	I	170		5120.30	I	85		5121.80	I
110		2737.09	II	120		3319.26	I	1300		3790.15	I	85		5127.66	I	40		5133.34	I
200		2746.91	I	120		3319.58	II	3500		3791.21	I	210		5134.75	I	75		5140.58	I
200		2748.85	I	240		3326.62	I	2700		3798.12	I	75		5147.54	I	40		5150.64	I
190		2753.01	I	170		3329.36	I	270		3801.30	I	75		5152.63	I	250		5160.33	I
280		2758.61	I	110		3332.16	I	2700		3802.92	I	250		5164.38	I	230		5180.31	I
240		2768.13	II	130		3341.60	II	670		3803.88	I	110		5186.98	I	190		5189.20	I
310		2773.20	I	1300		3341.97	I	530		3804.74	I	170		5193.08	I	150		5195.84	I
270		2780.24	II	1300		3343.71	I	670		3810.49	I	65		5203.22	I	35		5205.13	I
130		2782.36	I	130		3346.93	I	530		3811.03	I	85	c	5219.10	I				
110		2793.05	II	1700		3349.06	I	530		3815.51	I								
190		2827.08	II	420		3349.52	I	210		3818.86	II								
150		2836.24	I	340		3354.74	I	210		3819.15	I								
110		2840.94	I	130		3357.04	I	670		3824.88	I								
250		2841.15	II	1700		3358.42	I	350		3835.18	I								
280		2842.65	II	130		3365.58	II	250		3836.45	I								
160		2846.28	II	340		3366.96	I	210		3845.90	I								
110		2851.45	I	130		3369.16	II	290		3858.95	I								
240		2861.09	II	170		3371.33	I	350		3863.38	I								
100		2864.32	I	350		3374.92	I	270		3867.92	I								
100		2865.61	II	270		3380.41	I	530		3877.56	I								
500		2868.52	II	130		3380.86	I	870		3878.82	I								
800		2875.39	II	170		3386.24	II	670		3883.14	I								
270		2876.95	II	350		3392.44	I	1100		3885.44	I								
530		2877.03	II	170		3395.93	I	670		3885.68	I								
100		2880.72	II	120		3399.40	I	210		3886.07	I								
570		2883.18	II	230		3405.41	I	580		3891.30	I								
280		2888.83	II	130		3406.13	I	210		3908.97	I								
470		2897.81	II	270		3408.38	I	670		3914.70	I								
400		2899.24	II	230		3408.68	II	530		3920.20	I								
470		2908.24	II	180		3409.19	II	670		3937.44	I								
670		2910.59	II	230		3412.94	II	520		3943.67	I								
470		2911.74	II	180		3415.97	I	250		3965.69	I								
1100		2927.81	II	180		3423.76	I	910	d	3966.09	I								
110		2931.47	II	230		3425.42	II	210		3971.85	I								
870		2941.54	II	130		3425.85	I	1100		4032.52	I								
110	h	2945.88	II	230		3426.57	II	250		4039.53	I								
110		2946.12	II	230		3427.45	II	16000	c	4058.94	I								
110		2946.90	II	130		3429.04	I	210		4059.51	I								
1100		2950.88	II	180		3432.70	II	350		4060.79	I								
400		2972.57	II	180		3440.59	II	210		4068.26	I								
320		2974.10	II	170		3463.81	I	12000		4079.73	I								
210		2977.68	II	180		3465.86	I	270		4084.86	I								
200		2982.11	II	130		3469.44	I	440		4100.40	I								
330		2990.26	II	100		3471.19	I	6700		4100.92	I								
470		2994.73	II	140		3473.02	I	310		4116.90	I								
140		3024.74	II	290		3478.69	I	5300		4123.81	I								
350		3028.44	II	200		3479.56	II	670		4129.43	I								
300		3032.77	II	100		3484.05	II	770		4129.93	I								
100		3044.76	II	230		3491.03	I	2300		4137.10	I								
150		3048.10	I	200		3497.81	I	440		4139.44	I								
110		3053.09	I	500		3498.63	I	2700		4139.71	I								
100		3055.52	II	460		3507.96	I	350		4143.21	I								
220		3064.53	II	200		3510.26	II	870		4150.12	I								
110		3069.68	II	200		3515.42	II	4400		4152.58	I								
100		3070.90	II	200		3517.67	II	870		4163.47	I								
110		3071.56	II	200		3520.06	I	4400		4163.66	I								
100		3073.24	II	2000		3535.30	I	4000		4164.66	I								
400		3076.87	II	1300		3537.48	I	3500		4168.13	I								
110		3080.35	II	250		3540.96	II	310		4184.44	I								
1800		3094.18	II	500		3544.02	I	1200		4190.88	I								

Niobium (Cont.)

Intensity		Wavelength	
65		5225.16	I
150		5232.81	I
85	c	5237.43	I
29		5240.39	I
150	d	5251.62	I
		5251.81	I
75		5253.03	I
85		5253.93	I
50		5269.92	I
270		5271.53	I
25		5272.48	I
130	c	5276.20	I
29	c	5279.43	I
50		5285.26	I
35		5296.34	I
50		5315.55	I
17		5317.01	I
250		5318.60	I
50		5319.49	I
75		5334.87	I
25		5336.81	I
50		5340.80	I
25		5343.58	I
460		5344.17	I
340		5350.74	I
40		5353.28	I
25		5355.31	I
40		5355.70	I
29		5359.19	I
17		5362.01	I
40		5375.27	I
40		5381.34	I
17		5388.30	I
21		5395.86	I
29		5396.33	I
29		5411.24	I
21		5416.30	I
65		5422.44	I
21		5431.26	I
110		5437.27	I
19		5448.31	I
19		5456.19	I
40		5458.04	I
19	h	5468.10	I
40		5481.00	I
13		5483.09	I
19		5483.49	I
13		5491.06	I
17		5499.53	I
40		5504.30	I
17		5509.12	I
35	c	5512.82	I
17		5517.39	I
50		5523.57	I
25		5541.47	I
85		5551.35	I
29		5563.00	I
17	c	5571.44	I
35	c	5576.16	I
35		5578.29	I
50		5586.97	I
17	c	5590.95	I
13		5594.89	I
17	c	5599.59	I
40		5603.52	I
13		5603.93	I
25		5628.26	I
65		5629.17	I
35	c	5635.42	I
170		5642.11	I
35		5645.30	I
17		5654.14	I
130		5664.71	I
170		5665.63	I
17		5666.86	I
65	Cw	5671.02	I
85		5671.91	I
25		5677.47	I
25		5693.09	I
35	d	5697.90	I
		5698.03	I
40		5706.16	I
85		5706.48	I
29		5709.33	I
17		5715.59	I
65		5716.35	I
25		5725.66	I
130		5729.19	I
21		5737.36	I
13		5738.20	I
85		5751.44	I
110		5760.34	I
65		5764.99	I
29		5771.08	I
50	c	5776.07	I
17		5780.34	I
85		5787.54	I
17		5789.79	I
50		5794.24	I
50		5804.03	I
29	h	5815.33	I
110		5819.43	I
35		5820.62	I
75		5834.90	I
25		5838.15	I
130	d	5838.64	I
50		5842.47	I
17		5846.09	I
65		5866.47	I
35		5874.70	I
17		5877.79	I
40		5893.44	I
190	Cw	5900.62	I
40	c	5903.80	I
29		5927.41	I
40	c	5934.16	I
40		5957.70	I
150		5983.22	I
65		5986.08	I
85	Cw	5997.93	I
50		6029.75	I
50		6031.84	I
50		6045.50	I
25		6048.72	I
29		6056.65	I
29		6107.71	I
40		6142.51	I
50		6148.13	I
50		6164.42	I
29		6213.06	I
75		6221.96	I
40	c	6251.76	I
21		6260.77	I
85	c	6430.46	I
50	c	6433.22	I
17		6497.84	I
65		6544.61	I
15		6574.73	I
19	Cw	6591.00	I
19		6606.16	I
19		6607.28	I
35		6614.15	I
19		6626.98	I
210	Cw	6660.84	I
150	Cw	6677.33	I
65		6701.20	I
130	c	6723.62	I
75		6770.88	I
25		6795.31	I
85		6828.11	I
25	c	6849.35	I
19		6870.92	I
40		6876.36	I
25	c	6902.89	I
35		6908.07	I
40		6918.32	I
17		6946.07	I
17		6972.49	I
25		6986.09	I
85		6990.32	I
17	c	6996.11	I
21		7023.48	I
17		7038.04	I
190	c	7046.81	I
8		7066.41	I
8		7075.23	I
40	c	7098.94	I
17	Cw	7102.01	I
19		7119.31	I
15		7122.95	I
35		7126.17	I
17		7130.06	I
130		7159.43	I
17		7191.37	I
19	c	7208.94	I
50		7252.35	I
15		7274.81	I
13		7317.03	I
17	c	7323.92	I
29	Cw	7328.38	I
65	c	7353.16	I
190	Cw	7372.50	I
13		7419.83	I
15		7436.02	I
19		7478.20	I
65		7515.93	I
29	c	7519.77	I
170	c	7574.58	I
17	c	7583.21	I
13		7639.81	I
13		7647.71	I
25		7703.33	I
75	c	7726.68	I
25		7757.31	I
6		7787.11	I
13	Cw	7873.41	I
35		7885.31	I
25		7938.89	I
8		7954.76	I
40		8135.20	I
13	Cw	8240.00	I
29	Cw	8320.93	I
29		8346.08	I
10		8350.04	I
17		8439.77	I
17	Cw	8475.98	I
25		8526.99	I
13	c	8547.25	I
17	c	8560.54	I
17		8575.87	I
21	c	8697.55	I
21		8740.96	I
21		8767.97	I
29	Cw	8815.56	I
35		8905.78	I

Nb III
Ref. 392 — C.H.C.

Intensity	Wavelength	
	Vacuum	
60	1314.56	III
30	1319.15	III
60	1431.92	III
60	1433.39	III
50	1435.26	III
80	1445.43	III
80	1445.98	III
80	1447.09	III
50	1448.50	III
60	1451.63	III
100	1456.68	III
80	1484.73	III
50	1486.79	III
100	1495.94	III
80	1498.02	III
80	1499.45	III
50	1501.53	III
100	1501.99	III
60	1505.03	III
50	1509.71	III
50	1512.34	III
50	1513.25	III
80	1513.81	III
50	1517.38	III
100	1524.91	III
60	1532.98	III
80	1557.50	III
50	1566.92	III
50	1570.19	III
50	1586.82	III
100	1590.21	III
80	1598.86	III
80	1604.72	III
80	1639.51	III
100	1682.77	III
60	1684.40	III
100	1705.44	III
100	1707.14	III
50	1739.30	III
50	1758.63	III
60	1763.72	III
50	1808.70	III
50	1863.13	III
100	1892.92	III
100	1938.84	III
60 h	1979.07	III
50	1985.15	III
50	1997.11	III
	Air	
50 h	2007.28	III
50	2032.47	III
50	2060.29	III
80 h	2130.24	III
60	2206.01	III
60	2240.31	III
60	2244.19	III
60	2265.63	III
80	2273.92	III
100	2275.23	III
80	2279.36	III
100	2281.51	III
80	2284.40	III
100	2290.36	III
60	2304.78	III
50	2309.92	III
100	2313.30	III
50	2330.22	III
100	2338.09	III
80	2344.12	III
90	2349.21	III
80	2355.54	III
100	2362.06	III
80	2362.50	III
80	2365.70	III
100	2372.73	III
100	2387.41	III
80	2388.23	III
50	2404.23	III
80	2404.89	III
100	2413.94	III
60	2414.50	III
100	2421.91	III
50	2437.74	III
60	2446.45	III
100	2456.99	III
50	2460.34	III
50	2460.45	III
50	2463.72	III
80	2468.72	III
60	2469.39	III
80	2475.87	III
50	2486.02	III
60	2488.74	III
60	2493.02	III
100	2499.73	III
60	2508.53	III
50	2511.95	III
100	2545.64	III
80	2557.94	III
50	2567.44	III
60	2593.75	III
80	2598.86	III
50	2628.67	III
80	2633.17	III
80	2657.99	III
50	2937.71	III
80	3001.84	III
80	3142.26	III
60	3266.11	III

Nb IV
Ref. 407 — C.H.C.

Intensity	Wavelength	
	Vacuum	
12	542.38	IV
12	543.09	IV
12	545.21	IV
10	559.94	IV
10	566.22	IV
18	981.27	IV
60	993.54	IV
18	996.16	IV
50	1002.76	IV
400	1005.72	IV
500	1007.05	IV
500	1010.19	IV
45	1030.27	IV
100	1116.00	IV
150	1120.02	IV
50	1447.48	IV
40	1473.43	IV
40	1487.23	IV
60	1502.30	IV
60	1524.36	IV
50	1534.06	IV
40	1635.68	IV
40	1910.70	IV
60	1922.41	IV
60	1978.22	IV
	Air	
40	2027.50	IV
65	2032.53	IV
40	2034.67	IV
40	2068.62	IV
55	2084.07	IV
35	2093.12	IV
50	2093.30	IV
20	2122.68	IV
25	2130.23	IV
45	2146.36	IV
18	2249.98	IV

Nb V
Ref. 431 — C.H.C.

Intensity	Wavelength	
	Vacuum	
80	464.55	V
80	468.32	V
60	753.01	V
80	763.77	V
80	774.02	V
40	1007.02	V
50	1044.90	V
70	1212.21	V
100	1258.87	V
40	1267.60	V
100	1758.33	V
100	1877.34	V

NITROGEN (N)
Z = 7

N I and II
Ref. 213 = R.L.K.

Intensity	Wavelength

Nitrogen (Cont.)

Vacuum

Intensity	Wavelength		Ion
285	644.634		II
360	644.837		II
450	645.178		II
140	647.50		I
360	660.286		II
170	671.016		II
285	671.386		II
150	671.630		II
160	671.773		II
170	672.001		II
350	692.70		I
285	746.984		II
650	775.965		II
90	885.67		I
90	909.697		I
80	910.278		I
40	910.645		I
450	915.612		II
450	915.962		II
550	916.012		II
650	916.701		II
90	953.415		I
100	953.655		I
130	953.970		I
130	963.990		I
115	964.626		I
70	965.041		I
90	1067.614		I
60	1068.612		I
450	1083.990		II
600	1084.580		II
430	1085.546		II
650	1085.701		II
175	1097.237		I
115	1098.095		I
115	1098.260		I
105	1100.360		I
40	1100.465		I
90	1101.291		I
360	1134.165		I
385	1134.415		I
410	1134.980		I
105	1143.65		I
130	1163.884		I
60	1164.206		I
105	1164.325		I
270	1167.448		I
105	1168.334		I
60	1168.417		I
195	1168.536		I
230	1176.510		I
105	1176.630		I
195	1177.695		I
410	1199.550		I
385	1200.223		I
360	1200.710		I
175	1225.026		I
160	1225.37		I
130	1228.41		I
160	1228.79		I
360	1243.179		I
315	1243.306		I
290	1310.540		I
250	1310.95		I
230	1319.00		I
315	1319.68		I
115	1326.57		I
115	1327.92		I
360	1411.94		I
700	1492.625		I
490	1492.820		I
640	1494.675		I
775	1742.729		I
700	1745.252		I

Air

Intensity	Wavelength		Ion
160	2095.53		II
70	2096.20		II
110	2096.86		II
110	2130.18		II
160	2142.78		II
160	2206.09		II
160	2286.69		II
110	2288.44		II
220	2316.49		II
160	2316.69		II
285	2317.05		II
160	2461.27		II
160	2496.83		II
70	2496.97		II
110	2520.22		II
160	2520.79		II
220	2522.23		II
110	2590.94		II
160	2709.84		II
110	2799.22		II
110	2823.64		II
160	2885.27		II
220	3006.83		II
360	3437.15		II
285	3838.37		II
360	3919.00		II
450	3955.85		II
1000	3995.00		II
360	4035.08		II
550	4041.31		II
360	4043.53		II
140	4099.94		I
185	4109.95		I
285	4176.16		II
285	4227.74		II
285	4236.91		II
220	4237.05		II
450	4241.78		II
285	4432.74		II
650	4447.03		II
360	4530.41		II
550	4601.48		II
450	4607.16		II
360	4613.87		II
450	4621.39		II
870	4630.54		II
550	4643.08		II
285	4788.13		II
450	4803.29		II
180	4847.38		I
285	4895.11		II
160	4914.94		I
210	4935.12		I
160	4950.23		I
350	4963.98		I
285	4987.37		II
450	4994.36		II
650	5001.48		II
360	5002.70		II
870	5005.15		II
550	5007.32		II
450	5010.62		II
360	5016.39		II
360	5025.66		II
550	5045.10		II
185	5281.20		I
140	5292.68		I
450	5495.67		II
285	5535.36		II
650	5666.63		II
550	5676.02		II
870	5679.56		II
450	5686.21		II
450	5710.77		II
285	5747.30		II
700	5752.50		I
240	5764.75		I
265	5829.54		I
235	5854.04		I
360	5927.81		II
550	5931.78		II
285	5940.24		II
650	5941.65		II
285	5952.39		II
160	5999.43		I
210	6008.47		I
285	6167.76		II
360	6379.62		II
185	6411.65		I
210	6420.64		I
210	6423.02		I
210	6428.32		I
185	6437.68		I
235	6440.94		I
185	6457.90		I
300	6468.44		I
750	6482.05		II
360	6482.70		I
300	6483.75		I
265	6481.71		I
325	6484.80		I
160	6491.22		I
210	6499.54		I
185	6506.31		I
750	6610.56		II
185	6622.54		I
185	6636.94		I
235	6644.96		I
185	6646.50		I
235	6653.46		I
210	6656.51		I
185	6722.62		I
210	7398.64		I
160	7406.12		I
265	7406.24		I
685	7423.64		I
785	7442.29		I
900	7468.31		I
185	7608.80		I
450	7762.24		II
400	8184.87		I
400	8188.02		I
250	8200.36		I
300	8210.72		I
570	8216.34		I
400	8223.14		I
400	8242.39		I
550	8438.74		II
500	8567.74		I
570	8594.00		I
650	8629.24		I
500	8655.89		I
220	8676.08		II
700	8680.28		I
650	8683.40		I
500	8686.15		I
110	8687.43		II
110	8699.00	h	II
500	8703.25		I
160	8710.54	h	II
570	8711.70		I
500	8718.83		I
250	8728.89		I
200	8747.36		I
500	9386.80		I
570	9392.79		I
250	9460.68		I
200	9863.33		I
160	9865.41	h	II
110	9868.21	h	II
160	9887.39	h	II
220	9891.09	h	II
160	9961.86	h	II
220	9969.34	h	II
160	10070.12	h	II
285	10023.27	h	II
220	10035.45	h	II
220	10065.15	h	II
250	10105.13		I
300	10108.89		I
350	10112.48		I
400	10114.64		I
110	10126.27	h	II
250	10539.57		I
200	12074.51		I
380	12186.82		I
225	12288.97		I
290	12328.76		I
310	12381.65		I
180	12438.40		I
510	12461.25		I
920	12469.62		I
500	13429.61		I
840	13581.33		I
180	13587.73		I
180	13602.27		I
290	13624.18		I
250	14757.07		I
100	14868.87		I
160	14966.60		I
180	15582.27		I
120	17516.58	s	I
100	17584.86	l	I
100	17878.26		I

N III
Ref. 66, 213 = R.L.K.

Intensity	Wavelength		Ion

Vacuum

500	257.95		III
650	258.50		III
700	259.19		III
800	260.09		III
800	261.28		III
500	262.91		III
500	265.23		III
500	265.27		III
500	268.70		III
150	314.715		III
200	314.850		III
90	314.877		III
600	323.26		III
500	338.35		III
500	340.20		III
120	362.833		III
150	362.881		III
150	362.946		III
90	362.985		III
300	374.204		III
350	374.441		III
500	387.48		III
250	451.869		III
300	452.226		III
500	684.996		III
570	685.513		III
650	685.816		III
500	686.335		III
500	763.336		III
570	764.359		III
250	771.544		III
300	771.901		III
350	772.385		III
200	772.891		III
150	772.975		III
650	979.842		III
700	979.919		III
900	989.790		III
700	991.514		III
1000	991.579		III
500	1183.031		III
570	1184.550		III
150	1387.371		III
250	1729.945		III
570	1747.848		III
350	1751.218		III
650	1751.657		III
150	1804.486		III
200	1805.669		III
150	1846.42		III
350	1885.06		III
400	1885.22		III
200	1907.99		III
150	1919.55		III
150	1919.77		III
300	1920.65		III
150	1920.84		III
200	1921.30		III

Air

200	2064.01		III
250	2064.42		III
120	2068.68		III
90	2071.09		III
90	2117.59		III
90	2121.50		III
90	2147.31		III
200	2188.20		III
150	2188.38		III
250	2682.18	w	III
90	2689.20		III
120	3367.34		III
90	3754.67		III
120	3771.05		III
90	3938.52		III
150	3998.63		III
200	4003.58		III
250	4097.33		III
200	4103.43		III
120	4195.76		III
150	4200.10		III
90	4332.91		III
120	4345.68		III
300	4379.11		III
90	4510.91		III
120	4514.86		III
90	4634.14		III
120	4640.64		III
90	4858.82		III
150	4867.15		III
90	5314.35		III
200	5320.82		III
150	5327.18		III
90	6454.11		III
120	6467.02		III

N IV
Ref. 108, 212 = R.L.K.

Intensity	Wavelength		Ion

Vacuum

400	181.75		IV
400	191.7		IV
400	192.9		IV
500	196.87		IV
500	197.23		IV
500	202.60		IV
500	205.94		IV
500	205.97		IV
500	206.03		IV
500	217.20		IV
500	217.90	d	IV
500	223.4	d	IV
800	225.12	w	IV
800	225.21		IV
600	234.12	w	IV
600	234.20	w	IV
600	234.25	w	IV
550	236.07		IV
500	237.99		IV
500	238.7	w	IV
600	238.80		IV
500	239.62	w	IV
900	247.20		IV
500	248.43	w	IV
500	248.46	w	IV
500	248.48	w	IV
600	260.45		IV
650	270.99		IV
250	283.42		IV
300	283.48		IV
350	283.58		IV
600	285.56		IV
600	297.7	w	IV
700	297.82		IV
650	300.32		IV
90	303.123		IV

Nitrogen (Cont.)

Intensity		Wavelength	Species
500		303.28	IV
150		315.053	IV
120		322.503	IV
150		322.570	IV
200		322.724	IV
120		323.175	IV
300		335.050	IV
500	w	351.93	IV
700		353.06	IV
500		420.77	IV
650		463.74	IV
570		765.148	IV
520		921.992	IV
500		922.519	IV
480		923.057	IV
520		921.992	IV
500		922.519	IV
520		924.283	IV
1000		955.335	IV
150	w	1036.16	IV
90		1078.71	IV
90		1188.01	IV
1000		1718.55	IV

Air

Intensity		Wavelength	Species
90		2080.34	IV
90	w	2318.09	IV
150		2477.69	IV
250		2645.65	IV
300		2646.18	IV
350		2646.96	IV
90		3078.25	IV
90		3463.37	IV
570		3478.71	IV
500		3482.99	IV
400		3484.96	IV
90		3747.54	IV
150		4057.76	IV
90		4606.33	IV
150		6380.77	IV

N V
Ref. 66, 107, 318 = R.L.K.

Intensity		Wavelength	Species
		Vacuum	
52		166.947	V
32		186.069	V
62		186.153	V
90		209.303	V
90		247.561	V
120		247.706	V
150		266.196	V
200		266.379	V
90		713.518	V
150		713.860	V
150		748.195	V
200		748.291	V
1000		1238.821	V
900		1242.804	V
90		1549.336	V
200	l	1616.33	V
350	l	1619.69	V
90	w	1860.37	V

Air

Intensity		Wavelength	Species
60	l	2859.16	V
90	l	2974.52	V
150	w	2980.78	V
250	w	2981.31	V
60	w	2998.43	V
350		4603.73	V
250		4619.98	V
200	w	4944.56	V
60	w	7618.46	V

OSMIUM (Os)
Z = 76

Os I and II
Ref. 1 = C.H.C.

Intensity	Wavelength	Species
	Air	
9600	2001.45	I
13000	2003.73	I
9000	2004.78	I
17000	2010.15	I
29000	2018.14	I
29000	2020.26	I
14000	2022.76	I
14000	2028.23	I
18000	2034.44	I
26000	2045.36	I
8600	2058.69	I
	2058.78	I

Intensity	Wavelength	Species
13000	2061.69	I
7800	2067.21	II
4200	2070.67	II
7200	2076.95	I
7200	2078.09	
14000	2079.97	I
2900	2082.54	I
2900	2089.03	I
2900	2089.21	I
6000	2097.60	I
5300	2100.63	I
2100	2117.66	I
4800	2117.96	I
6600	2119.79	
1900	2123.84	I
5300	2137.11	I
2400	2149.97	I
2600	2154.59	I
1300	2157.84	I
1200	2158.53	I
2400	2161.00	
3100	2166.90	I
1100	2167.75	I
2100	2171.65	I
960	2184.68	I
840	2194.39	II
760	2202.49	I
600	2227.98	I
1100	2234.61	I
1300	2252.15	I
2000	2255.85	II
1400	2264.60	I
360	2268.28	I
960	2270.17	I
1400	2282.26	II
840	2283.67	I
570	2289.32	I
380	2297.31	I
660	2308.31	I
190	2313.75	II
550	2320.18	I
310	2323.98	I
660	2324.24	I
330	2326.99	I
310	2334.56	I
710	2336.62	II
430	2338.63	I
290	2340.69	I
430	2343.74	I
260	2345.75	I
430	2347.38	I
230	2350.23	II
360	2352.99	I
120	2355.28	II
240	2356.92	I
240	2357.25	I
310	2362.41	I
900	2362.77	I
500	2367.35	II
290	2369.24	I
500	2370.70	I
480	2371.18	I
95	2375.06	II
2600	2377.03	I
260	2377.61	I
900	2379.39	I
240	2384.62	I
1700	2387.29	I
330	2394.29	I
290	2395.39	I
1100	2395.88	I
220	2396.78	I
960	2401.13	I
260	2402.23	I
200	2403.54	I
330	2403.85	I
95	2405.08	II
290	2405.45	I
200	2405.96	I
360	2408.67	I
240	2410.98	I
290	2414.52	I
530	2417.99	I
530	2418.53	I
95	2420.02	II
200	2423.07	II
70	2424.02	II
500	2424.56	I
1400	2424.97	I
240	2426.81	I
70	2427.90	II
380	2431.19	I
380	2431.61	I
360	2446.02	I
900	2450.74	I
530	2451.73	I
530	2453.90	I
110	2454.91	II
530	2456.46	I
1800	2461.42	I
110	2468.90	II
290	2472.28	I
290	2474.78	I

Intensity	Wavelength	Species
900	2476.84	I
360	2482.43	I
530	2486.24	II
4500	2488.55	I
290	2491.02	I
290	2491.69	I
360	2492.42	I
2600	2498.41	I
330	2499.92	I
330	2502.29	I
500	2504.39	I
260	2504.51	I
35	2507.18	II
70	2509.71	II
660	2512.87	I
2400	2513.25	I
660	2515.04	I
500	2517.92	I
660	2518.44	I
200	2519.29	I
330	2519.79	I
200	2532.44	I
780	2538.00	II
240	2538.10	I
1000	2542.51	I
30	2548.83	II
310	2554.46	I
190	2563.16	II
600	2566.49	I
290	2566.88	I
480	2568.83	I
340	2571.78	I
150	2578.32	II
130	2580.03	II
360	2581.05	I
740	2581.96	I
1000	2590.76	I
200	2591.98	I
170	2596.00	II
210	2609.20	I
380	2609.56	I
400	2610.78	I
470	2612.63	I
1800	2613.06	I
800	2619.94	I
230	2620.62	II
530	2621.82	I
380	2628.48	I
27	2631.22	II
310	2637.13	I
1900	2644.11	I
340	2646.89	I
380	2647.73	I
380	2649.34	I
490	2656.68	I
1900	2658.60	I
640	2659.83	I
380	2661.18	I
40	2664.29	II
580	2674.57	I
400	2674.88	I
2100	2689.82	I
510	2699.59	I
580	2706.70	I
3000	2714.64	I
580	2715.36	I
1300	2720.04	I
850	2721.86	I
580	2730.61	I
40	2731.36	II
580	2732.80	I
690	2761.42	I
470	2763.27	I
340	2765.04	I
960	2770.71	I
300	2776.91	I
740	2782.55	I
40	2783.88	II
640	2786.31	I
230	2793.99	I
230	2794.19	I
530	2796.73	I
320	2804.07	I
2800	2806.91	I
470	2808.94	I
420	2813.84	I
740	2814.20	I
300	2815.78	I
420	2829.27	I
230	2837.42	I
470	2838.17	I
5100	2838.63	I
740	2841.60	I
2300	2844.40	I
420	2846.39	I
420	2848.25	I
1500	2850.76	I
1500	2860.96	I
35	2863.37	II
360	2874.96	I
300	2878.40	I
35	2879.39	II
30	2880.20	II

Intensity		Wavelength	Species
260		2896.06	I
9600		2909.06	I
2100		2912.33	I
530		2917.26	I
2100		2919.79	I
300		2925.57	I
360		2929.51	I
510		2931.28	I
260		2934.64	I
200		2942.85	I
1100	h	2948.23	I
1400		2949.53	I
210	d	2949.81	I
300		2961.01	I
530		2962.15	I
450		2964.06	I
740		2970.97	I
450		2977.64	I
510		2982.90	I
340		2983.49	I
260		2997.65	I
330		3013.07	I
570		3017.25	I
4400		3018.04	I
480		3019.38	I
1100		3030.70	I
2900		3040.90	I
210		3043.50	I
120		3042.74	II
230		3049.46	I
210		3050.39	I
8600		3058.66	I
290		3060.30	I
570		3062.19	I
210		3069.94	I
360		3074.08	I
290		3074.96	I
290		3077.44	I
1100		3077.72	I
360		3078.11	I
230		3078.38	I
230		3090.08	I
270		3093.59	I
310		3101.53	I
810		3103.99	I
310		3108.98	I
620		3109.38	I
250		3111.09	I
310		3118.22	I
480		3131.12	I
250		3152.67	I
290		3153.61	I
3100		3156.25	I
250		3156.78	I
310		3166.51	I
180		3173.93	II
420		3178.06	I
230		3181.88	I
230		3185.33	I
310		3186.98	I
310		3189.46	I
310		3194.23	I
150		3213.31	II
1900		3232.06	I
290		3238.63	I
190		3241.04	I
120		3248.00	I
190		3254.91	I
190		3256.92	I
190		3260.30	I
3100		3262.29	I
380		3262.75	I
3100		3267.94	I
620		3269.21	I
190		3272.16	I
530		3275.20	I
330		3277.97	I
190		3288.84	I
1200		3290.26	I
7600		3301.56	I
250		3306.23	I
620		3310.91	I
120		3315.42	I
250		3324.33	I
310		3327.42	I
960		3336.15	I
110		3351.74	I
120		3353.91	I
230		3357.97	I
250		3361.15	I
190		3364.12	I
120		3370.20	I
960		3370.59	I
160		3372.08	I
120		3378.68	I
310		3384.00	I
190		3385.94	I
620		3387.84	I
120		3401.17	I
620		3401.86	I
250		3402.51	I
120		3408.76	I
120		3412.74	I

Osmium (Cont.)

Intensity	Wavelength		Intensity	Wavelength		Intensity	Wavelength	
120	3421.69	I	120	4184.13	I	170	5780.82	I
150	3427.67	I	320	4189.91	I	40	5800.60	I
250	3440.60	I	180	4201.45	I	8	5842.49	I
120	3444.46	I	250	4202.06	I	110	5857.76	I
160	3445.55	I	1200	4211.86	I	28	5860.64	I
310	3449.20	I	120	4213.86	I	11	5882.92	I
120	3458.38	I	100	4215.16	I	11	5903.98	I
120	3465.44	I	170	4233.46	I	11	5906.84	I
120	3478.53	I	4900	4260.85	I	7	5908.95	I
120	3482.11	I	100	4264.75	I	7	5981.36	I
120	3487.46	I	120	4269.61	I	11	5983.22	I
120	3490.33	I	100	4285.90	I	65	5996.00	I
160	3498.54	I	560	4293.95	I	20	6015.79	I
250	3501.16	I	560	4311.40	I	7	6054.63	I
620	3504.66	I	110	4326.25	I	20	6144.53	I
440	3512.99	I	340	4328.68	I	11	6158.03	I
310	3518.72	I	100	4338.75	I	35	6227.70	I
120	3520.00	I	100	4351.53	I	7	6241.70	I
480	3523.64	I	210	4365.67	I	22	6269.41	I
120	3526.04	I	110	4370.66	I	11	6274.94	I
1200	3528.60	I	520	4394.86	I	11	6286.83	I
230	3530.06	I	160	4397.26	I	9	6398.86	I
230	3532.80	I	160	4402.74	I	22	6403.15	I
120	3533.41	I	4900	4420.47	I	9	6448.13	I
230	3542.71	I	100	4432.41	I	6	6520.85	I
960	3559.79	I	290	4436.32	I	7	6528.87	I
1200	3560.86	I	100	4439.64	I	7	6533.14	I
120	3562.34	I	230	4447.35	I	11	6538.30	I
310	3569.78	I	120	4484.76	I	11	6576.83	I
120	3574.08	I	110	4548.66	I	8	6614.56	I
120	3587.32	I	540	4550.41	I	4	6615.43	I
620	3598.11	I	140	4551.30	I	7	6661.81	I
190	3601.83	I	170	4616.78	I	27	6729.56	I
95	3604.48	II	170	4631.83	I	18	6791.53	I
250	3616.57	I	140	4663.82	I	14	6806.61	I
120	3619.43	I	670	4793.99	I	5	6878.70	I
450	3640.33	I	110	4865.60	I	4	6901.58	I
230	3654.49	I	55	5031.83	I	11	6956.02	I
330	3656.90	I	45	5039.12	I	6	6984.95	I
120	3666.31	I	35	5072.88	I	15	7060.67	I
480	3670.89	I	35	5074.77	I	22	7145.54	I
120	3675.45	I	35	5079.09	I	10	7149.89	I
250	3689.06	I	90	5103.50	I	4	7184.10	I
190	3703.25	I	55	5110.81	I	10	7206.33	I
120	3706.56	I	22	5122.23	I	5	7209.96	I
120	3709.14	I	22	5145.54	I	9	7251.16	I
230	3713.73	I	140	5149.74	I	6	7253.49	I
210	3719.52	I	28	5152.01	I	6	7375.07	I
230	3720.13	I	28	5168.98	I	9	7407.95	I
180	3746.47	I	40	5193.52	I	26	7602.95	I
3700	3752.52	I	270	5202.63	I	4	7701.46	I
100	3757.12	I	35	5203.23	I	7	7789.96	I
130	3766.30	I	20	5250.46	I	7	7852.17	I
120	3768.14	I	45	5255.82	I	6	7981.20	I
120	3774.40	I	55	5265.15	I	7	8041.29	I
110	3774.62	I	20	5283.89	I			
120	3776.25	I	20	5295.65	I			
290	3776.99	I	40	5298.78	I			
2100	3782.20	I	13	5302.58	I			
620	3790.14	I	18	5336.23	I			
180	3790.73	I	11	5346.03	I			
370	3793.91	I	13	5352.25	I			
250	3836.06	I	110	5376.79	I			
150	3840.30	I	16	5403.43	I			
150	3841.29	I	13	5412.14	I			
190	3849.94	I	120	5416.34	I			
230	3857.09	I	45	5416.69	I			
230	3865.47	I	28	5417.51	I			
730	3876.77	I	16	5441.82	I			
250	3881.86	I	55	5443.31	I			
140	3900.39	I	22	5446.93	I			
190	3901.71	I	11	5447.76	I			
100	3930.00	I	20	5449.37	I			
250	3938.59	I	20	5453.40	I			
100	3949.78	I	22	5457.30	I			
200	3961.02	I	28	5470.00	I			
1000	3963.63	I	13	5474.58	I			
100	3964.96	I	13	5475.13	I			
150	3969.67	I	9	5477.27	I			
110 h	3975.44	I	16	5481.85	I			
730	3977.23	I	22	5509.33	I			
100	3988.18	I	9	5516.01	I			
150	4003.48	I	270	5523.53	I			
100	4004.02	I	22	5546.82	I			
150	4005.16	I	9	5549.79	I			
160	4018.26	I	13	5552.88	I			
100	4037.84	I	11	5560.62	I			
280	4041.92	I	16	5580.66	I			
160	4048.05	I	80	5584.44	I			
960	4066.69	I	8	5600.50	I			
250	4070.86	I	35	5620.08	I			
190	4071.56	I	9	5637.41	I			
230	4074.68	I	22	5642.56	I			
490	4091.82	I	28	5645.25	I			
120	4100.30	I	7	5648.98	I			
1200	4112.02	I	9	5660.21	I			
180	4124.60	I	7	5674.38	I			
180	4128.96	I	28	5680.88	I			
2500	4135.78	I	11	5709.37	I			
150	4137.84	I	170	5721.93	I			
180	4172.57	I	8	5737.89	I			
1200	4173.23	I	8	5739.72	I			
620	4175.63	I	22	5765.05	I			

OXYGEN (O)
Z = 8
O I and II
Ref. 66, 69, 209, 210, 215 — R.L.K.

Intensity	Wavelength		Intensity	Wavelength	
	Vacuum				
250	537.83	II	300	1306.029	I
300	538.26	II		Air	
220	539.09	II	30 d	2283.42	II
200	539.55	II	30 d	2284.89	II
150	539.85	II	110	2293.32	II
150	644.148	II	200	2300.35	II
200	672.95	II	30 d	2313.05	II
150	673.77	II	30 d	2316.12	II
70	685.544	I	30 d	2316.79	II
900	718.484	II	50 d	2319.68	II
600	718.562	II	30 d	2322.15	II
70	744.794	I	30 d	2339.31	II
70	770.793	I	110	2411.60	II
90	771.056	I	80	2425.55	II
70	775.321	I	250	2433.56	II
70	791.973	I	80 d	2436.06	II
300	796.66	II	80	2444.26	II
90	804.267	I	300	2445.55	II
70	804.848	I	300	2733.34	II
70	805.295	I	110	2747.46	II
80	805.810	I	265	2972.29	I
240	832.762	II	160	3122.62	II
450	833.332	II	220	3129.44	II
600	834.467	II	450	3134.82	II
40	877.879	I	285	3138.44	II
80	922.008	I	220	3270.98	II
90	935.193	I	220	3273.52	II
40	948.686	I	220	3277.69	II
90	971.738	I	360	3287.59	II
40	976.448	I	160	3305.15	II
160	988.773	I	160	3306.60	II
40	990.204	I	220	3377.20	II
250	1025.762	I	285	3390.25	II
90	1027.431	I	220	3407.38	II
160	1039.230	I	160	3409.84	II
60	1040.942	t	285	3470.81	II
40	1152.152	I	220	3712.75	II
900	1302.168	I	285	3727.33	II
600	1304.858	t	160	3739.92	II
			360	3749.49	II
			160	3803.14	II
			120	3823.41	I
			450	3911.96	II
			160	3919.29	II
			185	3947.29	I
			160	3947.48	I
			140	3947.59	I
			220	3954.37	II
			100	3954.61	I
			450	3973.26	II
			220	3982.20	II
			160	4069.90	II
			285	4072.16	II
			450	4075.87	II
			80 d	4083.91	II
			50 d	4087.14	II
			150 d	4089.27	II
			110	4097.24	II
			220	4105.00	II
			285	4119.22	II
			160	4132.81	II
			50	4146.06	II
			220	4153.30	II
			285	4185.46	II
			450	4189.79	II
			80	4233.27	I
			50 d	4253.74	II
			50 d	4253.98	II
			50 d	4275.47	II
			50 d	4303.78	II
			285	4317.14	II
			160	4336.86	II
			220	4345.56	II
			285	4349.43	II
			220	4366.90	II
			100	4368.25	I
			220	4395.95	II
			450	4414.91	II
			285	4416.98	II
			160	4448.21	II
			160	4452.38	II
			50	4465.45	II
			50 d	4466.28	II
			50	4467.83	II
			50	4469.41	II
			360	4590.97	II
			285	4596.17	II
			80 d	4609.39	II
			160	4638.85	II
			360	4641.81	II
			450	4649.14	II
			160	4650.84	II
			360	4661.64	II
			285	4676.23	II
			220	4699.21	II
			285	4705.36	II
			160	4924.60	II
			220	4943.06	II
			135	5329.10	I
			160	5329.68	I
			190	5330.74	I
			90	5435.18	I
			110	5435.78	I

Oxygen (Cont.)

Intensity		Wavelength	
135		5436.86	I
120		5577.34	I
160		5958.39	I
190		5958.58	I
80		5995.28	I
160		6046.23	I
190		6046.44	I
110		6046.49	I
100		6106.27	I
400		6155.98	I
450		6156.77	I
490		6158.18	I
80		6256.83	I
100		6261.55	I
100		6366.34	I
100		6374.32	I
320		6453.60	I
360		6454.44	I
400		6455.98	I
80		6604.91	I
100		6653.83	I
360		7001.92	I
450		7002.23	I
210		7156.70	I
400		7254.15	I
450		7254.45	I
320		7254.53	I
210		7476.44	I
100		7477.24	I
120		7479.08	I
120		7480.67	I
100		7706.75	I
870		7771.94	I
810		7774.17	I
750		7775.39	I
80		7886.27	I
100		7943.15	I
100		7947.17	I
235		7947.55	I
210		7950.80	I
185		7952.16	I
110		7981.94	I
135		7982.40	I
190		7986.98	I
135		7987.33	I
250		7995.07	I
400		8221.82	I
265		8227.65	I
265		8230.02	I
325		8233.00	I
120		8235.35	I
120		8426.16	I
810		8446.25	I
1000		8446.36	I
935		8446.76	I
325		8820.43	I
160	d	9057.01	I
120		9118.29	I
80		9134.71	I
80		9150.14	I
80		9151.48	I
235		9156.01	I
450		9260.81	I
490		9260.84	I
450		9260.94	I
400		9262.58	I
540		9262.67	I
590		9262.77	I
490		9265.94	I
640		9266.01	I
185		9399.19	I
120		9481.16	I
120	d	9482.88	I
235		9487.43	I
140		9492.71	I
265		9497.97	I
160		9499.30	I
235		9505.59	I
210		9521.96	I
120		9523.36	I
120		9523.96	I
100		9528.28	I
100		9622.13	I
120		9625.29	I
160		9677.38	I
80		9694.66	I
65		9694.91	I
235		9741.50	I
235		9760.65	I
120		9909.05	I
140		9936.98	I
120		9940.41	I
160		9995.31	I
120	d	10421.18	I
590		11286.34	I
640		11286.91	I
490		11287.02	I
490		11287.32	I
490		11295.10	I
540		11297.68	I
590		11302.38	I
265		11358.69	I
490		12464.02	I
450		12570.04	I
120		12990.77	I
160		13076.91	I
700		13163.89	I
750		13164.85	I
640		13165.11	I
160		16212.06	I
120		17966.70	I
590		18021.21	I
120		18041.48	I
120		18042.19	I
120		18046.23	I
140		18229.23	I
540		18243.63	I
140		26173.56	I

O III
Ref. 23, 66, 210 — R.L.K.

Intensity		Wavelength	
		Vacuum	
80	d	264.34	III
110		264.48	III
110		266.97	III
150		266.98	III
150		267.03	III
150		277.38	III
80		295.62	III
110		295.66	III
120		295.72	III
150		303.41	III
150		303.46	III
140		303.52	III
160		303.62	III
160		303.69	III
250		303.80	III
200		305.60	III
250		305.66	III
190		305.70	III
300		305.77	III
190		305.84	III
450		320.979	III
300		328.45	III
250		328.74	III
110		345.31	III
110		355.14	III
90		355.33	III
80		355.47	III
200		359.02	III
190		359.22	III
150		369.30	III
210		373.80	III
200		374.00	III
300		374.08	III
190		374.16	III
200		374.33	III
210		374.44	III
450		395.558	III
300		434.98	III
800		507.391	III
900		507.683	III
1000		508.182	III
1000		525.795	III
700		597.818	III
1000		599.598	III
110		609.70	III
160		610.04	III
200		610.75	III
100		610.85	III
800		702.332	III
800		702.822	III
900		702.899	III
1000		703.850	III
600		832.927	III
780		833.742	III
600		835.096	III
800		835.292	III
160		1476.89	III
285		1590.01	III
160		1591.33	III
220		1760.12	III
110		1760.42	III
220		1763.22	III
220		1764.48	III
750		1767.78	III
550		1768.24	III
360		1771.67	III
110		1773.00	III
110		1773.85	III
220		1779.16	III
160		1781.03	III
160		1784.85	III
220		1789.66	III
110		1848.26	III
110		1856.62	III
285		1872.78	III
285		1872.87	III
285		1874.94	III
160		1920.04	III
110		1920.75	III
110		1921.52	III
220		1923.49	III
110		1923.82	III
110		1926.94	III

Intensity		Wavelength	
		Air	
360		2013.21	III
160		2026.96	III
220		2045.67	III
160		2052.74	III
200	d	2390.44	III
80		2394.33	III
80		2422.84	III
80	d	2438.83	III
200		2454.99	III
200		2558.06	III
80		2687.53	III
110		2695.49	III
80		2959.68	III
250		2983.78	III
80		3017.63	III
80		3023.45	III
80		3043.02	III
200		3047.13	III
110		3059.30	III
80		3121.71	III
110		3132.86	III
80		3238.57	III
200		3260.98	III
300		3265.46	III
80		3267.31	III
80		3312.30	III
110		3340.74	III
80		3444.10	III
80		3455.12	III
80		3698.70	III
80		3702.75	III
80		3703.37	III
110		3707.24	III
110		3715.08	III
110		3744.00	III
150		3754.65	III
80		3757.21	III
250		3759.87	III
110		3791.26	III
200		3961.59	III
110		5592.37	III

O IV
Ref. 36, 66 = R.L.K.

Intensity	Wavelength	
	Vacuum	
150	195.86	IV
200	196.01	IV
110	207.18	IV
150	207.24	IV
140	233.46	IV
150	233.50	IV
110	233.52	IV
200	233.56	IV
110	233.60	IV
90	238.36	IV
180	238.57	IV
110	252.56	IV
110	252.95	IV
150	253.08	IV
300	260.39	IV
250	260.56	IV
300	279.63	IV
375	279.94	IV
110	285.71	IV
150	285.84	IV
200	306.62	IV
150	306.88	IV
700	553.330	IV
775	554.075	IV
850	554.514	IV
700	555.261	IV
580	608.398	IV
640	609.829	IV
270	616.952	IV
150	617.005	IV
200	617.036	IV
520	624.617	IV
580	625.130	IV
640	625.852	IV
200	779.734	IV
315	779.821	IV
360	779.912	IV
200	779.997	IV
640	787.711	IV
520	790.109	IV
700	790.199	IV
200	802.200	IV
160	802.255	IV
130	921.296	IV
160	921.366	IV
200	923.367	IV
130	923.433	IV
200	1338.612	IV
130	1342.992	IV
230	1343.512	IV
	Air	
200	2449.372	IV
200	2450.040	IV
200	2493.44	IV
200	2493.77	IV
200	2507.73	IV
230	2509.19	IV
200	2517.2	IV
160	2836.26	IV
160	2921.45	IV
460	3063.42	IV
410	3071.61	IV
160	3209.66	IV
230	3348.08	IV
270	3349.11	IV
160	3354.27	IV
200	3375.40	IV
130	3378.06	IV
360	3381.20	IV
360	3385.52	IV
270	3396.79	IV
360	3403.52	IV
230	3409.66	IV
410	3411.69	IV
230	3413.64	IV
200	3489.83	IV
160	3492.24	IV
230	3560.39	IV
270	3563.33	IV

Intensity		Wavelength	
315	w	3725.93	IV
360		3729.03	IV
410		3736.85	IV
230		3744.89	IV

O V
Ref. 24, 66 = R.L.K.

Intensity		Wavelength	
		Vacuum	
80		124.616	V
110		135.523	V
80		138.109	V
110		139.029	V
80		151.447	V
110		151.477	V
150		151.546	V
80		164.574	V
110		164.657	V
80		164.709	V
80		166.735	V
150		167.99	V
110		170.219	V
450		172.169	V
250		185.745	V
575		192.751	V
450		192.799	V
520		192.906	V
80		193.003	V
200		194.593	V
80		202.161	V
80		202.224	V
80		202.283	V
80		202.334	V
150		202.393	V
110		203.78	V
150		203.82	V
100		203.85	V
200		203.89	V
100		203.94	V
300		207.794	V
150		215.040	V
200		215.103	V
250		215.245	V
250		216.018	V
520		220.352	V
80		227.372	V
80		227.469	V
150		227.511	V
80		227.549	V
80		227.634	V
80		227.689	V
150		231.823	V
110		248.459	V
110		286.448	V
1000		629.730	V
230		681.272	V
700		758.678	V
640		759.441	V
580		760.228	V
775		760.445	V
640		761.128	V
700		762.003	V
520		774.518	V
640		1371.292	V
160	w	1506.72	V
315	w	1643.68	V
160		1707.996	V
		Air	
1000		2781.01	V
920		2786.99	V
775		2789.85	V
200		2941.33	V
210		2941.65	V
160		3144.66	V
100		4123.99	V
230	w	4930.27	V
130		5597.91	V
130		6500.24	V

PALLADIUM (Pd)
Z = 46

Pd I and II
Ref. 1, 287 — C.H.C.

Intensity		Wavelength	
		Air	
50		2162.27	II
50		2182.35	II
50		2212.15	II
100	r	2231.59	II
200	r	2296.53	II
50		2351.32	II
50		2362.31	II
75		2367.92	II
60		2372.16	II
50		2388.29	II
60		2414.73	II
75		2418.72	II
75		2424.49	II
100		2426.87	II
100		2430.94	II
100		2433.11	II
100		2435.32	II
150		2446.17	II
75		2446.72	II
1100		2447.91	I
80		2448.15	II
100		2457.29	II
60		2457.76	II
150		2469.29	II
80		2470.06	II
100		2471.18	II
50		2472.55	II
1700		2476.42	I
250		2486.52	II
300		2488.92	II
75		2489.61	II
200		2498.81	II
150		2505.73	II
50		2514.47	II
80		2534.57	II
50	h	2539.44	II
150		2551.84	II
150		2565.51	II
100		2569.56	II
60		2593.24	II
50		2628.24	II
70		2635.92	II
150		2658.75	II
1900		2763.09	I
150	h	2776.85	II
100	h	2787.92	II
50	h	2800.64	II
50	h	2807.59	II
200		2854.59	II
100	h	2871.37	II
100	h	2878.01	II
520		2922.49	I
50		2980.63	II
650		3002.65	I
45		3009.78	I
1500		3027.91	I
1100		3065.31	I
2600		3114.04	I
270		3142.81	I
11000		3242.70	I
2700		3251.64	I
3500		3258.78	I
460		3287.25	I
3600		3302.13	I
5000		3373.00	I
24000		3404.58	I
13000		3421.24	I
5000		3433.45	I
6400		3441.40	I
7700		3460.77	I
10000		3481.15	I
2000		3489.77	I
12000		3516.94	I
12000		3553.08	I
4500		3571.16	I
20000		3609.55	I
20000		3634.70	I
5500		3690.34	I
1400		3718.91	I
1500		3799.19	I
1500		3832.29	I
2200		3894.20	I
1500		3958.64	I
290		4087.34	I
90		4169.84	I
2500		4212.95	I
180		4473.59	I
55	h	4788.18	I
45	h	4817.51	I
35		4875.43	I
55		5110.81	I
75		5117.02	I
160		5163.84	I
55		5234.86	I
120		5295.63	I
18		5312.57	I
15		5345.10	I
35		5395.24	I
55		5542.80	I
35		5547.02	I
27		5619.44	I
15		5642.69	I
14		5655.42	I
75		5670.07	I
11		5690.14	I
55	h	5695.09	I
18		5736.61	I
23		6774.54	I
65		6784.52	I
4	h	6833.42	I
11		7016.44	I
13	h	7310.06	I
75		7368.12	I
27		7391.92	I
16		7486.90	I
120		7764.03	I
27		7786.67	I
45		7915.80	I
18		7961.08	I
55		8132.82	I
45		8300.83	I
9	h	8353.58	I
18	h	8532.74	I
16	h	8599.10	I
65		8761.35	I

Pd III
Ref. 424 — L.J.R.

Intensity		Wavelength	
		Vacuum	
10		688.74	III
20		689.46	III
50		689.54	III
50		695.91	III
200		705.49	III
150		707.80	III
100		709.89	III
100		717.90	III
100		719.47	III
200		727.72	III
150		738.79	III
100		756.85	III
100		757.41	III
500		763.06	III
500		766.42	III
200		772.11	III
200		776.51	III
2000		781.02	III
200		784.99	III
200		787.31	III
200		787.95	III
200		789.58	III
500		794.08	III
500		797.52	III
500		800.03	III
500		800.10	III
500		803.67	III
500		825.35	III
500		840.58	III
500		856.47	III
500		864.04	III
500		880.59	III
500		888.84	III
1000		889.29	III
300		947.78	III
300		965.52	III
100		1505.40	III
200		1517.18	III
200	h	1526.88	III
100		1542.63	III
200	h	1545.95	III
300		1596.89	III
200	h	1606.10	III
150		1630.84	III
50		1679.73	III
100		1704.33	III
50		1706.40	III
200		1719.86	III
500		1741.62	III
400		1758.19	III
4000		1782.55	III
400		1804.91	III
400		1843.49	III
1500		1851.59	III
2000		1852.27	III
1000		1859.21	III
1500		1874.63	III
2000		1885.83	III
1000		1887.40	III
1500		1891.34	III
4000		1914.62	III
1000		1930.33	III
2000		1941.64	III
400		1951.56	III
300		1972.29	III
300		1977.53	III
		Air	
800		2002.16	III
1000		2004.47	III
500		2055.11	III
500		2149.82	III
500		2177.55	III
500		2177.63	III
100		2291.45	III
100		2452.42	III
100		2633.22	III

PHOSPHORUS (P)
Z = 15

P I and II
Ref. 182 = R.L.K.

Intensity	Wavelength	
	Vacuum	
10	810.24	II
10	865.44	II
20	1249.82	II
20	1301.87	II
20	1304.47	II
15	1304.68	II
35	1305.48	II
25	1309.87	II
60	1310.70	II
15	1372.033	I
15	1373.500	I
10	1374.732	I
15	1377.080	I
15	1377.937	I
25	1379.429	I
25	1381.469	I
15	1381.637	I
30	1452.89	II
80	1532.51	II
120	1535.90	II
80	1536.39	II
120	1542.29	II
140	1671.070	I
100	1671.510	I
180	1671.680	I
140	1672.035	I
140	1672.474	I
600	1674.591	I
600	1679.695	I
140	1685.976	I
100	1694.028	I
100	1694.486	I
100	1706.376	I
100	1707.553	I
600	1774.951	I
500	1782.838	I
400	1787.656	I
140	1834.801	I
140	1847.165	I
100	1849.820	I
140	1851.194	I
100	1852.069	I
500	1858.886	I
400	1859.393	I
140	1864.348	I
180	1905.481	I
140	1906.403	I
280	1907.665	I
	Air	
280	2023.489	I
180	2024.516	I
400	2032.432	I
400	2033.477	I
400	2135.465	I
400	2136.182	I
400	2149.145	I
280	2152.940	I
500	2154.080	I
180	2235.732	I
100	2484.19	II
750	2533.976	I
950	2535.603	I
750	2553.262	I
500	2554.915	I
150	2606.06	II
100	2626.18	II
90	2636.76	II
150	3308.92	II
125	3419.34	II
100	3425.00	II
100	4178.48	II
200	4288.60	II
200	4385.35	II
400	4420.71	II
100	4452.46	II
150	4463.00	II
120	4467.98	II
200	4475.26	II
200	4499.24	II
120	4530.81	II
120	4554.83	II
120	4558.07	II
120	4581.71	II
500	4588.04	II
500	4589.86	II
600	4602.08	II
300	4626.70	II
300	4658.31	II
200	4864.42	II
150	4927.20	II
500	4943.53	II
300	4954.39	II
300	4969.71	II
100	5079.381	I
100	5098.221	I
100	5100.974	I
140	5109.628	I
140	5154.844	I
180	5162.290	I
150	5191.41	II
300	5253.52	II
140	5293.539	I
400	5296.13	II
250	5316.07	II
300	5344.75	II
180	5345.851	I
100	5364.631	I
250	5378.20	II
300	5386.88	II
200	5409.72	II
400	5425.91	II
100	5428.094	I
400	5450.74	II
140	5458.305	I
125	5461.20	I
180	5477.672	I
140	5477.860	I
140	5478.267	I
200	5483.55	II
200	5499.73	II
200	5507.19	II
100	5514.774	I
100	5516.997	I
200	5541.14	II
200	5583.27	II
250	5588.34	II
100	5727.71	II
500	6024.18	II
400	6034.04	II
500	6043.12	II
250	6055.50	II
100	6057.86	II
350	6087.82	II
180	6097.690	I
350	6165.59	II
500	6199.024	I
180	6210.499	I
100	6232.29	II
200	6367.27	II
140	6375.681	I
100	6388.579	I
250	6435.32	II
130	6436.31	II
600	6459.99	II
600	6503.46	II
600	6507.97	II
150	6713.28	II
100	6717.411	I
100	7102.200	I
100	7158.367	I
180	7165.465	I
180	7175.102	I
180	7176.660	I
250	7845.63	II
100	8046.801	I
140	8278.058	I
100	8367.856	I
140	8531.475	I
140	8613.835	I
180	8637.578	I
400	8741.529	I
100	8872.174	I
140	9153.34	I
180	9175.819	I
950	9193.85	I
600	9278.88	I
1250	9304.94	I
500	9323.50	I
100	9327.13	I
140	9372.09	I
950	9435.069	I
950	9441.86	I
600	9452.83	I
100	9481.84	I
100	9492.12	I
1250	9493.56	I
180	9521.78	I
1700	9525.73	I
1500	9545.18	I
280	9556.81	I
1700	9563.439	I
280	9593.50	I
750	9609.04	I
180	9625.80	I
180	9628.42	I
400	9638.939	I
140	9675.41	I
500	9676.24	I
180	9706.533	I
1500	9734.750	I
280	9736.680	I
1500	9750.77	I

Phosphorus (Cont.)

Intensity	Wavelength	
100	9760.77	I
100	9776.85	I
100	9779.11	I
600	9790.21	I
1700	9796.85	I
280	9834.80	I
400	9903.68	I
280	9976.67	I
229	10084.27	I
174	10432.66	I
132	10455.87	I
458	10511.58	I
962	10529.52	I
1235	10581.57	I
415	10596.90	I
435	10681.40	I
265	10813.13	I
134	10932.72	I
103	10967.37	I
180	11160.05	I
764	11183.23	I
402	11186.75	I
76	13438.43	I
86	13485.19	I
479	14241.64	I
150	14272.75	I
256	14307.83	I
173	14430.50	I
135	14470.62	I
98	14646.42	I
714	15711.52	I
228	15962.53	I
296	16254.77	I
203	16292.97	I
1627	16482.92	I
588	16590.07	I
225	16613.05	I
221	16738.68	I
419	16803.39	I
471	17112.48	I
104	17223.28	I
289	17286.91	I
145	17359.00	I
299	17423.67	I
95	17665.68	I
186	18007.63	I
106	18518.90	I
92	18881.16	I
125	20841.62	I
124	23038.83	I
287	23844.97	I
98	26134.44	I
118	27959.52	I
188	28049.42	I
127	28154.52	I
132	28284.16	I
98	28288.69	I
311	29097.16	I
92	31483.21	I
91	32270.90	I
146	35551.93	I
90	35582.27	I
192	35802.53	I
146	36417.43	I

P III
Ref. 180 = R.L.K.

Intensity	Wavelength	
	Vacuum	
90	471.146	III
90	484.278	III
120	498.180	III
200	569.853	III
200	581.831	III
200	844.646	III
150	845.038	III
250	845.664	III
300	847.669	III
200	848.016	III
120	848.465	III
150	848.639	III
250	852.686	III
350	855.624	III
200	859.406	III
500	859.652	III
250	859.729	III
300	913.971	III
300	917.120	III
350	918.665	III
250	921.849	III
200	997.999	III
250	1003.598	III
500	1334.808	III
650	1344.327	III
300	1344.845	III
250	1380.463	III
150	1381.089	III
350	1502.228	III
250	1504.663	III
150	1618.632	III
200	1618.907	III
	Air	
200	2611.147	III

Intensity	Wavelength	
300	2632.713	III
200	2680.133	III
250	2895.241	III
250	3186.186	III
300	3219.307	III
150	3233.536	III
400	3233.602	III
200	3556.546	III
200	3577.526	III
200	3904.812	III
250	3914.314	III
300	3957.641	III
350	3978.307	III
200	4057.440	III
400	4059.312	III
300	4080.084	III
500	4222.195	III
350	4246.720	III
200	4428.171	III
200	4463.668	III
250	4479.776	III
150	6083.409	III
150	6409.204	III
150	6484.440	III
150	6486.381	III
150	6992.690	III
150	8113.528	III

P IV
Ref. 336 = R.L.K.

Intensity	Wavelength	
	Vacuum	
90	282.301	IV
90	304.996	IV
120	359.293	IV
150	359.899	IV
120	361.514	IV
150	361.629	IV
120	371.299	IV
150	371.504	IV
200	372.001	IV
500	388.318	IV
120	414.604	IV
200	414.999	IV
250	415.805	IV
830	444.245	IV
300	445.158	IV
250	568.038	IV
350	629.008	IV
400	629.914	IV
500	631.779	IV
350	648.482	IV
300	756.510	IV
300	776.353	IV
650	823.179	IV
700	824.730	IV
800	827.932	IV
250	847.019	IV
350	849.799	IV
200	850.392	IV
700	877.476	IV
1000	950.655	IV
570	1025.563	IV
500	1028.096	IV
570	1030.517	IV
500	1033.111	IV
500	1035.517	IV
570	1118.551	IV
200	1206.422	IV
200	1335.705	IV
500	1366.695	IV
400	1372.674	IV
350	1377.282	IV
500	1484.507	IV
400	1487.788	IV
300	1489.098	IV
250	1862.762	IV
120	1862.893	IV
200	1863.580	IV
650	1888.523	IV
200	1910.183	IV
120	1985.682	IV
150	1985.951	IV
200	1986.114	IV
150	1987.022	IV
	Air	
200	2477.823	IV
150	2478.070	IV
250	2478.256	IV
250	2605.506	IV
400	2644.295	IV
300	2724.764	IV
400	2728.770	IV
200	2729.120	IV
500	2739.309	IV
250	2739.872	IV
200	2740.223	IV
200	2961.242	IV
650	3347.736	IV
570	3364.467	IV
400	3371.122	IV
200	3413.543	IV
200	3733.393	IV

Intensity	Wavelength	
300	4249.656	IV
250	4540.288	IV
250	4541.112	IV
150	4548.056	IV
200	4548.449	IV
150	5235.499	IV
150	5989.774	IV
150	6142.605	IV
150	6713.939	IV
120	6715.906	IV
200	7443.657	IV

P V
Ref. 179 = R.L.K.

Intensity	Wavelength	
	Vacuum	
80	255.59	V
50	255.67	V
110	310.58	V
150	311.34	V
300	328.47	V
250	328.78	V
150	347.23	V
200	348.20	V
110	378.56	V
250	389.50	V
300	390.70	V
150	410.03	V
375	475.60	V
110	534.63	V
80	534.99	V
520	542.57	V
600	544.92	V
450	673.90	V
450	865.45	V
600	871.39	V
250	997.62	V
150	1000.38	V
900	1117.98	V
700	1128.01	V
150	1379.62	V
250	1385.45	V
375	1447.83	V
450	1610.50	V
	•	
200	2180.29	V
150	2186.42	V
375	2424.40	V
450	2440.93	V
200	2441.24	V
300	2961.00	V
450	2978.55	V
700	3175.09	V
520	3204.04	V
150	4083.18	V
110	4094.95	V
110	5156.72	V

PLATINUM (Pt)
Z = 78
Pt I and II
Ref. 1, 288 — C.H.C.

Intensity		Wavelength	
		Vacuum	
30		1621.66	II
30		1723.13	II
30		1751.70	II
50	r	1777.09	II
30		1781.86	II
30		1879.09	II
40		1883.05	II
50		1889.52	II
50		1911.70	II
30		1929.25	II
30		1929.68	II
30		1939.80	II
30		1949.90	II
30		1983.74	II
		Air	
40		2014.93	II
3200		2030.63	I
4400		2032.41	I
100		2036.46	II
40		2041.57	II
5500		2049.37	I
1500		2067.50	I
3000		2084.59	I
1000		2103.33	I
30		2115.57	II
950		2128.61	I
30		2130.69	II
1900		2144.23	I
100		2144.24	II
600		2165.17	I
1500		2174.67	I
30		2190.32	II
400		2202.22	I
50	h	2202.58	II
320		2222.61	I
50	h	2233.11	II
150		2249.30	I

Intensity		Wavelength	
30	h	2240.99	II
100		2245.52	II
30		2251.52	II
30	h	2251.92	II
190		2268.84	I
30	h	2271.72	II
280		2274.38	I
50	h	2287.50	II
30		2288.20	II
150		2289.27	I
150		2292.40	I
240		2308.04	I
50		2310.96	II
90		2315.50	I
220		2318.29	I
100		2326.10	I
170		2340.18	I
280		2357.10	I
180		2368.28	I
50		2377.28	II
130		2383.64	I
40		2386.81	I
120		2389.53	I
35		2396.17	I
70		2401.87	I
200		2403.09	I
100		2418.06	I
50		2424.87	II
80		2428.04	I
50		2428.20	I
25		2429.10	I
180		2436.69	I
650		2440.06	I
60		2450.97	I
440		2467.44	I
35		2471.01	I
1000		2487.17	I
25		2488.74	II
200		2490.12	I
160		2495.82	I
240		2498.50	I
50		2505.93	I
120		2508.50	I
50		2514.07	I
60		2514.11	I
240		2515.58	I
140		2524.30	I
40		2529.41	I
50		2536.49	I
160		2539.20	I
18		2549.46	I
50		2552.25	I
50		2596.00	I
70		2603.14	I
30		2616.76	II
50		2619.57	I
30		2625.34	II
1100		2628.03	I
130		2639.35	I
1000		2646.89	I
500		2650.86	I
20		2658.17	I
2800		2659.45	I
40		2674.57	I
200		2698.43	I
2000		2702.40	I
1600		2705.89	I
60		2713.13	I
1300		2719.04	I
130		2729.92	I
1800		2733.96	I
70		2738.48	I
70		2747.61	I
80		2753.86	I
200		2754.92	I
30		2769.84	I
500		2771.67	I
40		2773.24	I
20		2774.00	I
50		2774.77	II
50		2793.27	I
100		2794.21	II
40	h	2799.98	II
140		2803.24	I
10		2808.51	I
50		2818.25	I
30	h	2822.27	II
1400		2830.30	I
70		2834.71	I
16		2853.11	I
80	h	2860.68	II
40	h	2865.05	II
40	h	2875.85	II
100	h	2877.52	II
25		2888.20	I
25		2893.22	I
600		2893.86	I
300		2897.87	I
60		2905.90	I
120		2912.26	I
120		2913.54	I
70		2919.34	I
30		2921.38	I

Platinum (Cont.)

Intensity		Wavelength	Spec.
1700		2929.79	I
30		2942.76	I
30		2944.75	I
25		2959.10	I
60		2960.75	I
1800		2997.97	I
35		3001.17	II
220		3002.27	I
30		3017.88	I
30	h	3031.22	II
130		3036.45	I
800		3042.64	I
3200		3064.71	I
30		3071.94	I
130		3100.04	I
320		3139.39	I
140		3156.56	I
120		3200.71	I
320		3204.04	I
30		3230.29	I
20		3233.42	I
20		3250.36	I
40		3251.98	I
160		3255.92	I
25		3268.42	I
25		3281.97	I
120		3290.22	I
500		3301.86	I
60		3315.05	I
35		3323.80	I
340		3408.13	I
35		3427.93	I
60		3483.43	I
160		3485.27	I
120		3628.11	I
70		3638.79	I
70		3643.17	I
50		3663.10	I
80		3671.99	I
80		3674.04	I
35		3699.91	I
18		3706.53	I
20		3801.05	
80		3818.69	I
40		3900.73	I
110		3922.96	I
35		3948.40	I
100		3966.36	I
20		3996.57	I
110		4118.69	I
80		4164.56	I
40		4192.43	I
18		4327.06	I
18		4391.83	I
80		4442.55	I
14		4445.55	I
25		4498.76	I
12		4520.90	I
35		4552.42	I
12		4879.53	I
14		5044.04	I
30		5059.48	I
35		5227.66	I
40		5301.02	I
12		5368.99	I
12		5390.79	I
14		5475.77	I
14		5478.50	I
6		5763.57	I
20		5840.12	I
8		5844.84	I
6		6026.04	I
7		6318.37	I
8		6326.58	I
9		6523.45	I
10		6710.42	I
20		6760.02	I
60		6842.60	I
20		7113.73	I
10		8224.74	I

PLUTONIUM (Pu)
Z = 94
Pu I and II
Ref. 91 — J.G.C.

Intensity	Wavelength Air	Spec.
10000	2781.40	II
10000	2784.48	II
10000	2806.11	II
10000	2815.77	II
10000	2897.97	II
10000	2898.94	II
10000	2904.25	II
10000	2904.94	II
10000	2910.40	II
10000	2918.00	II
10000	2918.80	II
10000	2926.08	II
10000	2928.25	II
10000	2929.71	II
10000	2930.98	II
10000	2932.32	II
10000	2933.30	II
10000	2938.54	II
10000	2938.95	II
10000	2941.39	II
10000	2945.26	II
10000	2946.00	II
10000	2950.06	II
10000	2951.62	II
10000	2954.46	II
10000	2963.47	II
10000	2966.84	II
10000	2967.54	II
10000	2972.50	II
10000	2977.81	II
10000	2978.37	II
10000	2980.23	II
10000	2981.23	II
10000	2986.95	II
10000	2988.21	II
10000	2991.31	II
10000	2996.40	II
10000	3000.31	II
10000	3009.57	II
10000	3028.85	II
10000	3042.61	II
10000	3043.12	II
10000	3060.32	II
10000	3069.32	II
10000	3091.33	II
10000	3091.94	II
10000	3092.59	II
10000	3104.12	II
10000	3105.04	II
10000	3106.03	
10000	3123.87	II
10000	3159.21	II
10000	3161.73	II
10000	3163.18	II
10000	3174.49	II
10000	3179.41	II
10000	3185.12	II
10000	3187.60	II
10000	3189.23	II
10000	3193.54	
10000	3193.55	II
10000	3194.56	II
10000	3198.47	II
10000	3200.23	II
10000	3201.00	II
10000	3201.66	II
10000	3204.48	II
10000	3206.80	II
10000	3207.97	II
10000	3215.08	I
10000	3216.15	II
10000	3220.94	II
10000	3224.87	II
10000	3231.86	II
10000	3232.24	
10000	3232.63	II
10000	3241.39	II
10000	3242.96	II
10000	3243.40	II
10000	3244.16	I
10000	3245.25	II
10000	3245.71	
10000	3246.35	II
10000	3247.50	
10000	3247.56	II
10000	3252.08	I
10000	3260.54	II
10000	3265.17	
10000	3273.11	II
10000	3274.71	II
10000	3275.24	I
10000	3292.56	I
10000	3293.61	I
10000	3296.91	I
10000	3297.87	I
10000	3298.47	II
10000	3301.76	I
10000	3306.59	I
10000	3306.66	I
10000	3307.66	II
10000	3308.75	I
10000	3312.65	II
10000	3315.34	II
10000	3316.96	II
10000	3320.61	I
10000	3320.84	I
10000	3323.48	
10000	3327.19	I
10000	3330.11	
10000	3331.52	II
10000	3332.34	I
10000	3333.03	
10000	3337.71	II
10000	3338.40	II
10000	3338.94	I
10000	3347.87	II
10000	3349.63	I
10000	3351.82	II
10000	3356.61	II
10000	3358.41	II
10000	3358.84	II
10000	3362.26	II
10000	3365.20	I
10000	3365.66	
10000	3368.86	I
10000	3370.64	II
10000	3371.19	I
10000	3375.80	I
10000	3376.76	II
10000	3376.94	II
10000	3377.37	II
10000	3379.51	I
10000	3381.82	I
10000	3381.97	II
10000	3382.70	I
10000	3390.33	II
10000	3391.41	II
10000	3393.67	I
10000	3394.32	I
10000	3418.88	II
10000	3465.10	II
10000	3473.64	II
10000	3483.20	I
10000	3585.87	II
10000	3632.21	II
10000	3699.19	II
10000	3720.59	I
10000	3725.98	I
10000	3726.11	II
10000	3726.79	II
10000	3732.03	II
10000	3744.78	I
10000	3753.63	I
10000	3755.94	I
10000	3757.82	I
10000	3758.34	I
10000	3774.38	I
10000	3776.71	I
10000	3792.22	I
10000	3799.37	I
10000	3805.93	I
10000	3811.40	I
10000	3812.30	II
10000	3827.57	I
10000	3835.52	I
10000	3836.96	I
10000	3838.92	I
10000	3842.10	I
10000	3851.01	I
10000	3851.85	I
10000	3878.54	I
10000	3895.89	I
10000	3928.53	I
10000	3975.43	II
10000	4097.12	I
10000	4101.96	I
10000	4105.95	II
10000	4111.07	I
10000	4114.91	I
10000	4128.12	I
10000	4129.93	II
10000	4133.01	I
10000	4135.97	I
10000	4140.04	I
10000	4141.20	II
10000	4151.09	I
10000	4151.45	I
10000	4155.46	I
10000	4159.39	II
10000	4167.77	I
10000	4170.95	I
10000	4178.28	II
10000	4189.90	II
10000	4190.06	II
10000	4196.20	II
10000	4206.48	I
10000	4208.23	I
10000	4221.87	I
10000	4224.20	I
10000	4229.77	II
10000	4254.76	II
10000	4261.88	I
10000	4269.77	I
10000	4273.34	II
10000	4281.17	I
10000	4289.08	I
10000	4337.18	II
10000	4352.71	II
10000	4367.41	I
10000	4379.91	II
10000	4385.35	II
10000	4393.93	II
10000	4404.90	I
10000	4441.65	II
10000	4468.54	II
10000	4472.79	II
10000	4493.78	II
10000	4504.91	II
10000	4536.15	II
10000	4735.40	I
10000	4989.34	I
10000	5269.86	I
10000	5381.02	I
10000	5498.50	I
10000	5510.72	I
10000	5537.59	I
10000	5549.62	I
10000	5590.54	I
10000	5592.33	I
10000	5712.39	I
10000	5770.26	I
10000	5839.05	I
10000	5983.35	I
10000	6012.78	I
10000	6192.80	I
10000	6304.66	I
10000	6449.75	I
10000	6486.71	I
10000	6488.86	I
10000	6488.89	
10000	6535.27	I
10000	6544.21	I
10000	6608.95	I
10000	6672.72	I
10000	6784.66	I
10000	6880.16	I
10000	6891.38	I
10000	7059.23	I
10000	7068.90	I
10000	7092.46	I
10000	7116.88	I
10000	7141.66	I
10000	7177.14	I
10000	7231.09	I
10000	7258.06	I
10000	7322.23	I
10000	7325.97	I
10000	7331.81	I
10000	7431.18	I
10000	7447.99	I
10000	7507.80	I
10000	7526.93	I
10000	7547.45	I
10000	7564.50	I
10000	7571.87	I
10000	7572.93	I
10000	7609.77	I
10000	7689.40	I
10000	7758.20	I
10000	7798.54	I
10000	7953.17	I
10000	8102.54	I
10000	8130.86	I
10000	8309.61	I
10000	8435.47	I
10000	8476.13	I
10000	8495.75	I
10000	8597.26	I
10000	8665.02	I
10000	8691.94	I
3000	8729.82	I
3000	8836.16	I
3000	9533.07	I
3000	10046.75	I
3000	11114.82	I
3000	12144.46	I
3000	12231.22	I
3000	15377.31	I
3000	16397.38	I

POLONIUM (Po)
Z = 84
Po I and II
Ref. 47, 48 — E.F.W.

Intensity		Wavelength Air	Spec.
250	w	2139.02	I
300	h	2203.80	
300		2220.67	I
200		2222.13	I
200		2284.22	
250		2344.61	I
250		2421.72	I
300		2426.09	I
1500	w	2450.08	I
700		2483.94	I
700		2490.53	I
200	h	2502.18	
300		2534.95	I
300		2557.33	I
1500	w	2558.01	I
400		2562.31	
300		2578.80	
400		2587.64	I
200		2637.01	
300		2645.36	I
700	h	2663.33	
200		2671.67	I
600		2761.92	I
400		2800.26	I
250		2824.11	I
300		2866.01	I
400		2919.31	I
600		2958.92	I
2500	w	3003.21	I
450		3069.31	I
200		3115.95	

Potassium (Cont.)

Intensity		Wavelength	
400		3189.02	I
600		3240.24	I
250		3286.38	I
600		3328.60	I
300		3489.79	I
200		3493.65	
400		3588.33	
200		3671.36	
500		3861.93	I
200		4051.98	
1200		4170.52	I
250		4236.13	
200	h	4415.58	
800		4493.21	I
350		4611.44	I
200		4867.12	
400		4876.24	I
450		4946.81	
350		5323.23	I
300		5744.85	I
600		7962.62	I
300		8433.87	I
500		8618.26	I
250		9227.87	

POTASSIUM (K)
Z = 19
K I and II
Ref. 59, 76, 172, 268 — L.J.R.

Intensity	Wavelength	
	Vacuum	
5	261.20	II
25	441.81	II
5	465.08	II
	469.50	II
10	476.03	II
30	495.14	II
30	600.77	II
25	607.93	II
30	612.62	II
3	1725.0	II
	Air	
6	2190.00	II
4	2210.53	II
5	2265.04	II
h	2550.02	II
4	2743.55	II
	2992.12	I
	2992.42	I
	3034.76	
	3034.92	
5	3062.18	II
4	3101.79	I
3	3102.04	I
6	3105.00	II
5	3190.07	II
7	3217.16	I
6	3217.62	I
4	3220.60	II
5	3290.65	II
6	3345.32	II
6	3373.60	II
6	3380.62	II
6	3384.86	II
6	3404.24	II
7	3440.05	II
11	3446.37	I
10	3447.38	I
6	3481.11	II
7	3530.75	II
5	3608.88	II
6	3618.49	II
4	3626.42	II
3	3648.84	I
4	3648.98	I
6	3681.54	II
5	3716.60	II
5	3721.34	II
5	3739.13	II
5	3744.42	II
6	3767.36	II
6	3783.19	II
6	3800.14	II
6	3816.56	II
7	3817.50	II
5	3873.74	II
4	3878.62	II
8	3897.92	II
5	3923.00	II
5	3926.36	II
6	3942.53	II
6	3955.21	II
6	3966.72	II
6	3972.58	II
6	3995.10	II
7	4001.24	II
5	4012.10	II
6	4042.59	II
18	4044.14	I
17	4047.21	I
5	4093.69	II
6	4114.99	II
7	4134.72	II
7	4149.19	II
8	4186.24	II
7	4222.97	II
7	4225.67	II
7	4263.40	II
7	4305.00	II
7	4309.10	II
5	4340.03	II
7	4388.16	II
5	4466.65	II
6	4505.33	II
5	4595.65	II
8	4608.45	II
10	4641.88	I
11	4642.37	I
5	4659.38	II
4	4740.91	I
6	4744.35	I
5	4753.93	I
7	4757.39	I
5	4786.49	I
7	4791.05	I
6	4799.75	I
8	4804.35	I
9	4829.23	II
7	4849.86	I
8	4856.09	I
8	4863.48	I
9	4869.76	I
8	4942.02	I
6	4943.29	II
9	4950.82	I
9	4956.15	I
10	4965.03	I
8	5005.60	II
7	5056.21	II
10	5084.23	I
11	5097.17	I
11	5099.20	I
12	5112.25	I
5	5310.24	II
12	5323.28	I
13	5339.69	I
12	5342.97	I
5	5463.57	I
6	5470.10	II
5	5642.75	II
4	5772.32	II
16	5782.38	I
17	5801.75	I
15	5812.15	I
17	5831.89	I
2	5969.64	II
8	6120.27	II
6	6246.59	II
7	6307.29	II
5	6427.96	II
2	6595.00	II
19	6911.08	I
12	6936.28	I
20	6938.77	I
7	6964.18	I
12	6964.67	I
25	7664.90	I
24	7698.96	I
5	7955.37	I
4	7956.83	I
7	8078.11	I
6	8079.62	I
9	8250.18	I
8	8251.74	I
3	8390.22	I
	8391.44	I
2	8417.54	I
1	8420.00	I
11	8503.45	I
10	8505.11	I
4	8763.96	I
3	8767.05	I
13	8902.19	I
12	8904.02	I
5	8923.31	I
4	8925.46	I
7	9347.24	I
3	9349.25	I
6	9351.59	I
15	9595.70	I
14	9597.83	I
6	9949.67	I
5	9954.14	I
9	10479.63	I
5	10482.15	I
8	10487.11	I
17	11019.48	I
16	11022.67	I
17	11690.01	I
16	11769.62	I
17	11772.83	I
	12432.24	I
	12522.11	I
	13377.86	I
	13397.09	I
	15163.08	I
	15168.40	I
	40158.37	I

K III
Ref. 60, 76 — L.J.R.

Intensity	Wavelength	
	Vacuum	
2	325.28	III
5	327.60	III
25	330.68	III
30	341.92	III
15	348.00	III
30	379.12	III
25	380.48	III
30	382.23	III
15	398.63	III
20	402.10	III
30	406.48	III
40	408.96	III
50	413.79	III
30	414.87	III
30	416.00	III
30	417.54	III
30	418.62	III
75	434.72	III
50	435.68	III
75	444.34	III
75	448.60	III
75	466.79	III
100	470.09	III
75	471.57	III
45	474.92	III
40	479.18	III
10	482.11	III
10	482.41	III
75	497.10	III
10	514.94	III
50	520.61	III
25	523.79	III
40	529.80	III
15	539.71	III
15	546.12	III
20	708.84	III
30	765.31	III
30	765.64	III
35	778.53	III
20	872.31	III
10	873.86	III
15	874.04	III
	Air	
6	2550.02	III
5	2635.11	III
1	2736.96	III
5	2689.90	III
1	2898.90	III
5	2938.45	III
1	2948.94	III
5	2986.20	III
6	2992.24	III
6	3052.07	III
6	3056.84	III
5	3201.95	III
6	3209.34	III
6	3278.79	III
6	3289.06	III
6	3322.40	III
6	3364.22	III
6	3420.82	III
4	3421.83	III
6	3468.32	III
6	3481.11	III
5	3513.88	III
1	3885.50	III

K IV
Ref. 32, 76, 86, 150, 160, 314, 322 — L.J.R.

Intensity	Wavelength	
	Vacuum	
150	271.82	IV
100	273.06	IV
	279.88	IV
300	340.46	IV
150	340.74	IV
300	354.93	IV
150	356.26	IV
300	359.73	IV
200	359.91	IV
250	362.08	IV
150	362.15	IV
150	363.02	IV
300	375.96	IV
300	379.88	IV
250	380.48	IV
200	381.70	IV
300	382.23	IV
150	382.49	IV
200	382.65	IV
300	382.91	IV
300	384.10	IV
200	386.61	IV
250	388.92	IV
250	389.07	IV
250	390.42	IV
300	390.57	IV
200	391.46	IV
200	392.47	IV
500	393.14	IV
400	400.21	IV
300	402.91	IV
250	403.97	IV
150	404.41	IV
250	408.08	IV
150	417.28	IV
200	442.30	IV
300	443.57	IV
200	445.61	IV
250	446.83	IV
750	448.60	IV
400	456.33	IV
250	523.00	IV
200	526.45	IV
150	527.62	IV
750	646.19	IV
500	737.14	IV
500	741.95	IV
500	745.26	IV
400	746.35	IV
300	749.99	IV
150	754.19	IV
400	754.67	IV

K V
Ref. 32, 75, 76, 150, 322 — L.J.R.

Intensity	Wavelength	
	Vacuum	
100	214.35	V
150	282.35	V
150	293.33	V
300	294.84	V
200	296.17	V
200	297.06	V
200	300.25	V
200	300.50	V
200	311.24	V
250	312.77	V
200	315.18	V
250	327.38	V
250	389.07	V
200	349.50	V
500	372.15	V
200	372.46	V
200	372.77	V
300	375.96	V
250	377.76	V
300	379.12	V
300	387.80	V
250	390.11	V
250	395.40	V
200	398.36	V
200	398.88	V
200	399.75	V
250	415.05	V
200	415.79	V
400	422.18	V
300	425.16	V
500	425.59	V
250	438.02	V
200	449.71	V
200	452.90	V
250	455.67	V
400	456.33	V
200	482.71	V
200	483.75	V
750	580.32	V
250	585.51	V
500	586.32	V
250	602.27	V
400	603.43	V
250	638.67	V
300	687.50	V
300	720.43	V
400	724.42	V
600	731.86	V
150	770.29	V
150	771.46	V
	1035.60	V

PRASEODYMIUM (Pr)
Z = 59
Pr I and II
Ref. 1 — C.H.C.

Intensity		Wavelength	
		Air	
25		2558.58	II
25		2578.27	I
30		2579.31	I
40	h	2598.04	II
25		2608.92	II
25		2615.75	II
25		2648.48	II
30		2654.75	II
25		2666.70	II
25		2672.52	II
20		2685.19	II
45		2685.70	II

Praseodymium (Cont.)

Intensity	Code	Wavelength	Spectrum
50		2698.92	II
60		2700.38	II
30		2702.25	II
100	h	2707.37	II
20		2714.16	II
60		2720.17	II
30		2721.90	II
50		2726.50	II
12		2731.78	II
25		2733.12	II
50		2734.30	II
25		2737.90	II
40		2742.12	II
25		2744.66	II
20		2746.28	II
60		2760.35	II
50		2769.60	II
50	d	2775.94	II
		2776.03	II
40		2778.80	II
50		2783.31	II
30		2789.05	II
35		2792.51	II
50		2802.05	II
20		2823.17	II
20		2824.14	II
20		2828.29	II
20		2842.98	I
20		2844.01	II
20		2850.62	I
25		2853.99	II
30		2865.64	II
50		2881.60	I
30		2882.31	II
30		2884.89	II
30		2943.97	II
30		2967.58	II
30		2971.13	II
40	d	2971.40	II
		2971.46	II
50		2984.98	II
30		2986.18	II
30		2990.22	II
110		3082.11	II
100		3111.34	II
140		3121.58	II
140		3163.73	II
270		3168.24	II
160		3172.31	II
110		3191.42	II
200	d	3195.99	II
110		3199.04	II
100		3207.89	II
190		3219.48	II
100		3234.27	II
100		3245.48	II
140		3355.67	II
110		3394.62	II
110		3465.74	II
200		3584.21	II
130		3611.94	II
170		3630.96	II
100		3645.55	II
250		3645.66	II
250		3646.30	II
100		3648.30	II
150	c	3660.36	II
100		3661.62	II
370		3668.83	II
250		3687.03	II
150		3687.19	II
100		3689.71	II
150		3698.06	II
230		3706.75	II
170	c	3711.10	II
290		3714.05	II
120	c	3733.03	II
210	c	3734.41	II
250		3735.76	II
190		3736.49	II
410		3739.18	II
150		3740.99	II
120		3743.98	II
190		3750.98	II
140		3759.60	II
120		3760.08	II
680		3761.87	II
230		3764.77	II
230		3768.94	II
170	c	3772.82	II
170		3774.06	II
140		3777.62	II
170		3780.66	II
150		3785.46	II
150		3786.86	II
210		3792.51	II
190		3794.93	II
680		3800.30	II
290		3804.84	II
140		3809.18	II
390		3811.84	II
1300	h	3816.02	II
120		3817.66	II
680		3818.28	II
120		3819.14	II
310		3821.80	II
150	c	3823.18	II
120		3826.67	II
960		3830.72	II
140		3834.93	II
480		3840.99	II
270		3842.34	II
150	c	3844.54	II
580		3846.59	II
1200		3850.79	II
720	c	3851.55	II
960		3852.80	II
120		3858.25	II
110		3859.14	II
480	c	3865.45	II
210		3867.52	II
210		3870.72	II
480		3876.19	II
1700	c	3877.18	II
270		3879.20	II
680		3880.47	II
440	c	3885.19	II
440	c	3889.34	II
120	c	3891.71	II
190		3897.25	II
210		3898.84	II
250		3902.45	II
770	c	3908.05	II
630		3912.90	II
310		3913.55	II
210		3914.76	II
1300	c	3918.85	II
420		3919.63	II
250		3920.53	II
960		3925.47	II
480		3927.46	II
370		3929.29	II
370		3935.82	II
250		3938.30	II
730	c	3947.63	II
900		3949.43	II
900		3953.51	II
380		3956.75	II
190		3959.44	I
470		3962.45	II
560		3964.26	II
1600	c	3964.81	II
560		3966.57	II
500		3971.16	II
320		3971.67	II
620	c	3972.14	II
320		3974.85	II
1300	c	3989.68	II
230		3991.91	II
340		3992.16	II
1600		3994.79	II
270		3995.83	II
560	c	3997.04	II
230		3997.96	II
320		3999.12	II
620	c	4000.17	II
730		4004.70	II
1900		4008.69	II
620		4010.60	II
730		4015.39	II
620		4020.96	II
470		4022.71	II
360		4025.54	II
230		4026.83	II
230		4029.00	II
360	c	4029.72	II
730	c	4031.75	II
230		4032.47	II
960		4033.83	II
230		4034.33	II
230		4038.22	II
730		4038.45	II
470		4039.34	II
1300		4044.81	II
230		4045.70	II
230		4046.63	II
340		4047.08	II
450		4051.13	II
2200		4054.88	II
2200		4056.54	II
450		4058.80	II
230		4062.22	II
3400		4062.81	II
210		4068.80	II
500	c	4079.77	II
500	c	4080.98	II
790		4081.85	II
500		4083.34	II
200	c	4087.21	II
560		4096.82	II
380		4098.40	II
2900	c	4100.72	II
270	c	4113.89	II
1700	c	4118.46	II
250		4129.15	II
340		4130.77	II
200		4133.61	II
1500	c	4141.22	II
2700		4143.11	II
270	c	4146.50	II
270		4148.44	II
200		4156.50	II
1700	c	4164.16	II
270		4168.04	II
230		4169.45	II
620		4171.82	II
730		4172.25	II
250		4175.32	II
250		4175.62	II
200		4178.63	II
5200		4179.39	II
2500		4189.48	II
560	c	4191.60	II
290		4201.17	II
2500	c	4206.72	II
500		4208.32	II
320		4211.86	II
320		4217.81	II
3800		4222.93	II
3800		4225.35	II
320		4233.11	II
320	c	4236.15	II
270		4240.02	II
960		4241.01	II
340		4243.51	II
840	c	4247.63	II
500		4254.40	II
270	c	4263.78	II
320		4269.09	II
790	c	4272.27	II
470		4280.07	II
790	c	4282.42	II
450	c	4298.98	II
290		4303.61	II
1500		4305.76	II
210		4323.55	II
270		4329.41	II
1300		4333.97	II
200		4335.74	II
360		4338.70	II
620	Cw	4344.30	II
470	c	4347.49	II
340		4350.40	II
450		4354.91	II
410	c	4359.79	II
1200		4368.33	II
320		4371.62	II
270		4396.08	II
170		4403.60	II
100		4405.12	II
430		4405.83	II
1700		4408.82	II
410		4413.77	II
160		4419.04	II
190		4419.65	II
160	c	4421.22	II
160		4424.58	II
1200	c	4429.13	II
110		4432.28	II
730		4449.83	II
140		4451.90	II
140		4454.68	II
100		4465.97	II
960		4468.66	II
140	c	4477.26	II
1100		4496.46	II
790		4510.15	II
200	c	4517.58	II
340	c	4534.15	II
340		4535.92	II
200		4563.12	II
140		4612.08	II
270	c	4628.74	II
140		4632.28	I
140		4635.68	I
200		4639.55	I
110	c	4643.49	II
140		4646.05	II
200	c	4651.50	II
140		4664.65	II
270	c	4672.09	II
180		4687.80	I
290		4695.77	I
140	Cw	4708.07	II
140		4709.52	I
180		4730.67	I
250		4736.69	I
100		4744.16	I
150		4746.92	II
100		4762.72	II
110		4783.35	II
110		4906.99	I
140		4914.02	I
200		4924.60	I
140		4936.00	I
320		4939.74	I
160		4940.30	I
380		4951.37	I
110		4975.75	I
120		5018.59	I
200		5019.76	I
200		5026.96	I
100		5033.38	I
270		5034.41	II
110		5043.83	I
320		5045.52	I
160		5053.40	I
180		5087.12	I
360		5110.38	II
560		5110.76	II
410		5129.52	II
270		5133.44	I
270		5135.14	II
100		5139.81	I
100	c	5152.30	II
200		5161.74	II
620		5173.90	II
200		5191.32	II
120		5194.43	I
150		5195.11	II
200		5195.31	II
360		5206.55	II
150		5207.90	II
360		5219.05	II
560		5220.11	II
110		5227.97	I
680		5259.73	II
180		5263.88	II
340	c	5292.02	II
340		5292.62	II
230		5298.09	II
430		5322.76	II
200		5352.40	II
16		5501.50	I
40		5508.79	II
65		5509.15	II
16	c	5511.63	II
55		5513.58	II
28		5515.12	II
13		5519.38	II
20	c	5520.31	II
45		5522.79	II
28	c	5524.15	I
28	c	5525.91	II
16	c	5527.93	I
13		5530.21	I
45		5531.16	I
150		5535.17	II
28		5538.37	I
20		5538.78	II
55		5545.01	II
20		5548.33	II
11		5553.42	II
22		5561.46	II
45	c	5562.06	I
13		5565.52	II
45		5571.83	II
11		5574.61	II
11		5578.81	I
13		5582.35	II
11		5584.02	II
22		5594.92	I
22		5597.29	II
13		5601.30	II
90		5605.65	II
13		5606.68	I
28		5608.93	II
55		5610.22	II
11		5620.06	II
20		5620.26	I
45	c	5621.89	II
110		5623.05	II
90		5624.45	II
11	h	5633.03	I
22		5636.46	II
55	c	5638.79	II
16		5640.37	II
16	Cw	5643.16	I
22		5645.41	II
35		5654.23	II
55		5659.84	II
35	h	5661.57	I
16		5662.19	II
65	c	5668.46	I
45		5669.55	II
35		5669.99	II
16		5674.14	II
16		5677.03	II
55		5681.89	II
13		5685.60	II
16		5686.52	I
22	h	5687.17	II
65		5688.44	II
22		5689.21	II
55	h	5690.97	II
22		5695.90	II
22		5704.38	I
65		5707.61	II
40		5711.63	II
22		5713.83	II
16		5716.08	II
45		5719.08	II
45	d	5719.63	II
		5719.80	II
11		5728.38	I

Praseodymium (Cont.)

Intensity		Wavelength	
40		5731.88	II
20		5747.13	II
11		5747.74	I
11		5747.95	II
22		5753.02	II
90		5756.17	II
16		5759.40	II
22		5760.20	I
22		5769.16	II
16		5769.79	II
45		5773.16	II
11		5775.91	II
16		5777.29	II
90		5779.28	I
65	c	5785.28	II
65		5786.17	II
16	h	5788.29	II
16		5788.92	II
16		5790.86	II
45		5791.36	II
22		5797.95	I
40		5810.58	II
16		5813.55	II
160	d	5815.17	II
		5815.33	II
55		5818.57	II
40		5820.62	II
16	h	5821.36	I
55		5822.59	II
90		5823.72	II
45		5830.94	II
40		5835.13	I
35	c	5844.65	II
40		5844.98	II
65		5847.13	II
65	c	5850.64	II
45		5852.63	II
11	c	5854.44	I
45		5856.07	II
55		5856.90	II
90		5859.68	II
80		5868.83	II
22		5873.83	II
35		5874.72	I
35		5878.10	II
35		5879.04	I
80		5879.25	II
35	c	5884.72	I
55		5892.23	II
22		5894.22	II
40		5903.11	II
45		5904.45	II
40		5908.67	II
11		5915.31	I
11		5915.97	I
40		5920.76	I
40		5930.66	II
16		5936.33	II
160		5939.90	II
65		5940.72	II
22		5941.65	I
35		5947.16	II
22	c	5949.76	I
55		5951.27	II
20		5951.76	II
90		5956.60	II
		5956.70	I
13		5959.25	I
20		5962.18	II
28		5963.00	I
110		5967.82	II
13	c	5976.95	I
13		5978.88	I
65		5981.19	II
40		5986.14	I
45	c	5987.14	I
		5987.29	II
13		5991.27	I
13	c	5994.89	I
11		5996.06	I
29		6002.44	II
90		6006.33	II
13		6008.54	I
55		6016.48	II
150		6017.80	II
28	c	6019.85	I
150		6025.72	II
35		6042.87	II
55		6046.66	II
35		6049.26	I
28		6050.04	II
11		6050.88	I
140		6055.13	I
13		6067.27	II
13		6085.81	I
28		6086.16	II
65		6087.52	II
20		6090.38	II
28		6093.09	II
18		6096.28	I
22		6106.72	II
18		6109.08	I
65		6114.38	II
22	c	6118.02	I

Intensity		Wavelength	
22	c	6122.15	I
35		6141.51	II
65		6148.23	I
		6148.24	II
22		6157.82	II
13		6159.10	II
190		6161.18	II
18		6165.38	I
270		6165.94	II
55		6182.34	II
13		6187.96	I
35		6197.45	II
35		6200.81	II
13		6205.63	II
13		6210.59	I
22		6212.73	I
18		6218.06	I
20	h	6236.80	I
20	h	6241.05	I
45		6244.35	II
35		6255.10	II
40		6262.55	II
18		6264.54	II
22	c	6274.66	II
		6274.81	II
40		6278.68	II
110		6281.28	II
18	c	6289.02	I
11	c	6298.01	I
11		6302.05	I
35		6302.35	II
16		6304.05	I
35		6305.23	II
11	h	6318.10	II
45	c	6322.36	I
22	h	6343.88	I
28		6347.11	II
18	c	6350.98	I
22	c	6357.20	I
55	c	6359.03	I
11	h	6363.62	II
16		6377.61	I
16		6378.59	I
11		6389.57	I
16	c	6391.03	II
40		6392.18	I
45		6397.96	II
10	h	6410.69	I
55		6411.23	I
40		6413.68	II
10		6415.43	I
45		6429.63	II
45		6431.84	II
7	h	6442.78	II
9	h	6443.91	II
16	c	6453.44	I
9		6454.84	II
9		6456.18	I
9	h	6460.19	I
18		6467.72	II
9	h	6475.26	II
35	Cw	6478.02	II
45		6486.55	I
9	h	6486.97	II
40	h	6491.75	I
9		6493.49	I
11		6494.89	I
22	c	6497.11	I
18		6498.94	I
22		6500.72	I
9		6504.09	I
8		6517.14	I
16		6518.79	II
8		6534.52	I
16		6540.47	I
7	h	6553.30	I
22		6564.62	II
45		6566.77	II
7		6571.03	I
6		6578.00	I
6		6584.56	II
9	h	6593.74	II
11		6595.48	I
15		6609.86	I
55		6616.67	I
11		6618.34	II
7	h	6631.00	I
13	h	6632.06	I
14		6647.12	I
75		6656.83	II
55		6673.41	II
75		6673.78	II
5	h	6687.51	II
4	h	6699.25	I
13		6736.79	I
35	c	6747.09	I
19	c	6749.19	II
7	c	6784.99	II
55	Cw	6798.60	II
11		6811.76	II
17	Cw	6812.87	II
13		6814.04	II
9		6817.61	I
35	Cw	6827.60	II

Intensity		Wavelength	
19		6830.50	II
9		6844.39	I
9	h	6845.47	II
9		6846.59	II
17	c	6850.46	II
11		6852.77	I
11	c	6870.44	I
7		6884.66	I
8		6892.71	I
8	h	6970.38	I
8	c	6980.12	I
40		7021.51	II
10		7024.53	I
13		7042.40	I
8		7044.45	II
7		7051.07	I
10		7079.99	I
11	c	7095.18	I
20		7114.55	I
10	h	7116.90	I
11		7118.24	II
7		7137.33	II
10	h	7159.88	I
7		7167.77	II
7	h	7189.95	I
10	c	7208.83	II
24		7227.70	II
13		7231.53	I
7	c	7243.26	I
7	c	7259.21	I
7	c	7287.61	I
7	h	7289.19	I
7	Cw	7324.42	I
7		7328.47	I
7		7344.86	
16		7407.56	II
20	c	7451.74	II
11	h	7495.59	I
6	h	7499.42	I
14		7541.02	II
6		7574.86	I
20		7645.66	II
7		7704.98	I
16		7721.84	I
11	h	7731.33	I
6	Cw	7841.27	I
14		7871.67	I
6		7881.09	I
6	Cw	7888.56	II
6		7915.19	II
6		8031.92	I
6		8055.13	I
14		8067.44	I
10	Cw	8122.78	II
11		8141.10	I
5		8181.34	II
5	c	8211.93	I
6		8289.93	I
6		8379.84	I
6	h	8427.82	I
6	h	8605.27	II
6		8714.59	II

Pr III
Ref. 306, 308 — J.R.

Intensity		Wavelength	
		Vacuum	
25		1008.61	III
50		1021.35	III
25		1026.18	III
100		1029.03	III
50		1038.29	III
50		1042.96	III
25		1043.80	III
25		1044.03	III
25		1046.20	III
150		1047.24	III
100		1049.09	III
50		1052.63	III
25		1061.60	III
25		1066.03	III
25	p	1068.85	III
25		1069.88	III
25		1084.42	III
25		1088.66	III
100		1104.84	III
25		1108.82	III
30		1352.70	III
25		1881.22	III
		Air	
50		2031.46	III
100		2033.30	III
25		2043.12	III
100		2052.30	III
50		2052.87	III
50		2053.85	III
200		2058.59	III
50		2064.08	III
25		2075.08	III
200		2090.75	III
200		2093.49	III
25		2096.85	III
50		2096.94	III

Intensity		Wavelength	
10	w	2148.14	III
10	w	2194.24	III
10	w	2197.25	III
10	w	2205.48	III
10	w	2206.26	III
10	w	2214.45	III
10	w	2215.25	III
10	w	2217.12	III
10	w	2223.23	III
10	w	2230.35	III
10	w	2237.26	III
10		2239.06	III
10		2239.42	III
10	w	2242.15	III
10	w	2284.62	III
10		2307.59	III
10		2307.77	III
10		2308.41	III
10		2311.29	III
10		2311.44	III
10		2314.18	III
10	w	2315.46	III
10		2318.15	III
10		2318.36	III
10		2318.64	III
10		2318.82	III
10		2318.97	III
10	w	2319.40	III
10		2320.41	III
10	w	2328.56	III
10	w	2336.13	III
10		2365.52	III
10		2368.78	III
10		2369.08	III
10	w	2378.06	III
10	w	2378.97	III
10		2395.44	III
10		2399.70	III
10	w	2405.56	III
10		2408.19	III
10		2409.80	III
10		2412.40	III
10		2417.69	III
10		2418.95	III
10		2426.05	III
10	w	2430.32	III
10	w	2434.18	III
10		2436.39	III
10		2435.91	III
10		2436.89	III
10		2438.63	III
10	w	2444.93	III
10		2445.49	III
10		2446.77	III
10		2448.16	III
10		2452.02	III
10	w	2452.81	III
10	w	2452.85	III
10		2454.60	III
10		2454.82	III
10	w	2459.77	III
10		2460.72	III
10		2462.18	III
10		2462.90	III
10		2468.20	III
10		2468.97	III
10	w	2473.42	III
10	w	2478.32	III
10		2479.98	III
10		2481.02	III
10		2483.30	III
10		2483.99	III
10	w	2484.60	III
10	w	2485.16	III
10	w	2488.72	III
10		2491.97	III
10	w	2494.20	III
10	w	2495.37	III
10		2495.51	III
10		2499.97	III
20	w	2587.71	III
40	w	2624.91	III
20	w	2644.62	III
20	w	2656.88	III
20	w	2667.51	III
70	w	2679.47	III
40	w	2710.30	III
20	w	2718.65	III
100	w	2724.03	III
20	l	2841.94	III
70	s	2910.61	III
50	s	2911.77	III
100	l	2914.49	III
50	s	2930.19	III
70	s	2942.43	III
70	s	2953.58	III
90	w	2954.40	III
90	s	2964.85	III
150	s	2968.83	III
80	s	2969.41	III
150	l	2976.86	III
150	s	2977.06	III
500	s	2980.54	III

Praseodymium (Cont.)

Intensity		Wavelength	
100	l	2981.65	III
150	c	2982.42	III
500	s	2985.82	III
150	s	2997.12	III
70	l	2998.79	III
150	l	3000.46	III
150	s	3003.20	III
60	l	3006.47	III
150	s	3008.04	III
150	l	3010.61	III
90	s	3014.60	III
100	l	3015.13	III
90	s	3016.26	III
70	s	3021.77	III
70	s	3025.26	III
100	l	3029.38	III
100	l	3033.31	III
90	l	3034.25	III
70		3040.02	III
60	c	3040.94	III
60	l	3041.78	III
100	s	3042.35	III
100	l	3045.81	III
70	l	3046.98	III
120	l	3050.30	III
70	s	3055.30	III
150	l	3058.90	III
150	l	3066.71	III
70	s	3078.68	III
150	l	3080.20	III
50	w	3248.39	III
90	s	3280.92	III
90	l	3292.58	III
90	s	3296.10	III
100	l	3306.14	III
60	s	3333.26	III
90	s	3340.58	III
500	s	3341.43	III
70	s	3341.68	III
70	s	3345.38	III
70	l	3345.44	III
50	w	3351.07	III
50	s	3353.87	III
250	s	3354.91	III
500	l	3357.56	III
500	l	3359.41	III
100	l	3364.52	III
50		3364.88	III
50	s	3365.80	III
200	s	3367.35	III
500	s	3367.58	III
100	l	3371.92	III
100	s	3377.14	III
50	s	3379.13	III
150	s	3380.21	III
150	s	3381.26	III
300	s	3381.84	III
100	s	3391.08	III
1000	w	3394.22	III
600	d	3396.07	III
300	s	3396.62	III
300	l	3397.46	III
50	l	3402.97	III
500	c	3413.21	III
300	s	3415.15	III
150	s	3420.07	III
300	l	3422.22	III
50	c	3426.27	III
500	l	3427.02	III
300	l	3436.36	III
150	l	3440.62	III
50	l	3445.29	III
70	s	3454.05	III
180		3653.58	III
60		3817.25	III
60	l	3861.80	III
150		3980.51	III
200		4000.20	III
90		4018.36	III
180		4029.60	III
90		4142.46	III
120		4144.48	III
90		4147.85	III
90		4172.15	III
150		4179.77	III
180		4184.18	III
240		4197.01	III
120		4219.45	III
180		4231.45	III
180		4275.07	III
120		4286.32	III
90		4298.27	III
90		4301.73	III
90		4316.34	III
60		4354.28	III
90		4379.82	III
90		4381.47	III
120		4404.71	III
120		4421.10	III
120		4431.85	III
150	w	4447.93	III
180	w	4450.14	III
120	w	4451.00	III
120		4461.02	III
200		4461.81	III
300	w	4500.31	III
450	w	4612.02	III
600	w	4625.18	III
120		4654.16	III
600	w	4713.70	III
300	w	4725.55	III
270		4728.21	III
300		4747.11	III
300	w	4771.83	III
450	w	4775.30	III
600		4857.39	III
150		5208.51	III
150		5261.68	III
1000		5264.44	III
1500		5284.70	III
1500		5299.99	III
1500		5340.02	III
100		5427.70	III
100		5581.74	III
150		5646.80	III
600		5765.27	III
1500		5844.41	III
200		5947.98	III
7000	w	5956.05	III
900		5998.94	III
1500	w	6053.01	III
900		6071.09	III
9000	w	6090.02	III
5000		6160.24	III
1500	w	6161.22	III
100		6195.05	III
2000		6195.63	III
200		6310.36	III
100		6361.65	III
300		6429.26	III
300		6444.74	III
600		6500.04	III
300		6501.49	III
200		6578.90	III
100		6616.46	III
600		6706.70	III
100		6727.63	III
200		6827.96	III
100		6854.63	III
200		6857.30	III
1000		6866.80	III
1000		6899.06	III
500		6903.52	III
7000		6910.14	III
150	w	6934.55	III
500		6970.96	III
100		6979.83	III
5000		7030.39	III
100		7075.21	III
4500		7076.62	III
100		7083.99	III
500		7112.53	III
100	w	7165.64	III
250		7231.62	III
100	w	7238.26	III
250		7240.21	III
100	w	7262.32	III
150	w	7340.69	III
350		7343.70	III
200		7349.75	III
300	w	7350.61	III
100	w	7355.52	III
2000		7426.48	III
4000		7429.05	III
100		7463.96	III
250		7487.40	III
200		7493.20	III
100		7511.17	III
500		7529.11	III
100		7549.20	III
150	w	7588.64	III
300		7596.41	III
100	w	7625.63	III
100	w	7648.34	III
100	w	7670.65	III
200	w	7674.65	III
500		7742.34	III
250	w	7745.54	III
500		7754.31	III
100	w	7755.48	III
3000		7781.98	III
200	w	7814.74	III
1500	w	7866.14	III
1000		7888.12	III
400		7897.09	III
1000		7914.00	III
100		7923.16	III
1000	w	7972.75	III
250		8001.14	III
3000		8102.90	III
250		8119.54	III
400		8132.23	III
100		8138.34	III
150	w	8235.33	III
250	w	8244.89	III
100		8409.10	III
100	w	8494.99	III
200	w	8567.63	III
5000	w	8602.74	III
500		8691.58	III
500		8771.38	III
1000		8854.05	III
125		8886.17	III
100		8908.70	III
250		9099.98	III
250		9131.90	III
200		9222.32	III
250	w	9265.56	III
125		9320.54	III
250		9334.33	III
175		9377.44	III
175		9388.56	III
175		9549.77	III
100		9579.74	III
150		9802.98	III
175		9806.37	III
500	w	9991.16	III
500		10031.10	III
500		10160.33	III
500		10238.63	III
500		10301.58	III
500		10324.59	III
500		10716.58	III

Pr IV
Ref. 337, 338 — J.R.

Intensity	Wavelength	
	Vacuum	
20	718.23	IV
30	721.34	IV
60	722.41	IV
30	722.58	IV
20	726.04	IV
50	730.37	IV
30	731.77	IV
30	734.86	IV
20	735.04	IV
20	736.19	IV
50	736.32	IV
100	737.17	IV
100	741.45	IV
20	743.15	IV
20	743.89	IV
20	746.14	IV
40	763.16	IV
20	764.00	IV
300	1226.40	IV
2000	1228.59	IV
500	1230.69	IV
400	1238.19	IV
200	1249.35	IV
200	1255.64	IV
200	1261.27	IV
400	1268.32	IV
300	1270.58	IV
1000	1275.10	IV
200	1275.40	IV
1000	1278.65	IV
200	1279.34	IV
1000	1287.44	IV
300	1290.93	IV
1000	1292.30	IV
5000	1293.22	IV
5000	1295.28	IV
400	1296.50	IV
200	1298.26	IV
300	1298.54	IV
200	1304.71	IV
300	1306.86	IV
200	1308.08	IV
200	1310.71	IV
500	1314.96	IV
300	1315.28	IV
300	1316.96	IV
500	1320.10	IV
1000	1320.70	IV
5000	1321.36	IV
500	1322.51	IV
500	1326.38	IV
5000	1333.57	IV
300	1335.96	IV
500	1339.29	IV
1000	1340.74	IV
200	1341.32	IV
300	1344.23	IV
1000	1347.07	IV
1000	1352.81	IV
500	1354.35	IV
5000	1354.66	IV
2000	1360.64	IV
1000	1364.81	IV
2000	1365.77	IV
400	1368.90	IV
5000	1374.41	IV
1000	1382.62	IV
200	1384.23	IV
500	1385.91	IV
300	1394.11	IV
500	1397.11	IV
1000	1399.31	IV
1000	1400.96	IV
400	1410.90	IV
1000	1424.36	IV
500	1426.59	IV
5000	1435.56	IV
200	1459.95	IV
200	1461.76	IV
1000	1474.91	IV
200	1477.32	IV
500	1485.88	IV
500	1503.35	IV
400	1516.86	IV
200	1520.71	IV
2000	1520.98	IV
400	1523.46	IV
500	1553.62	IV
500	1559.49	IV
500	1570.13	IV
200	1572.80	IV
5000	1574.55	IV
5000	1575.10	IV
3000	1578.38	IV
500	1585.10	IV
300	1613.00	IV
400	1613.65	IV
1000	1618.03	IV
2000	1622.30	IV
300	1634.77	IV
400	1676.08	IV
200	1688.49	IV
200	1713.53	IV
500	1732.86	IV
300	1762.86	IV
1000	1766.88	IV
1000	1771.14	IV
500	1841.08	IV
10000	1884.87	IV
400	1951.23	IV
200	1954.61	IV
	Air	
200	2025.06	IV
1000	2039.15	IV
200	2047.05	IV
200	2050.73	IV
200	2058.48	IV
2000	2083.23	IV
500	2100.42	IV
1000	2154.31	IV
300	2193.37	IV
1000	2205.13	IV
300	2265.70	IV
200 c	2334.46	IV
200 c	2339.08	IV
500 c	2376.09	IV
2000 c	2378.98	IV
1000 c	2379.66	IV
500 c	2427.07	IV
500 c	2428.13	IV
500 c	2438.57	IV
500 c	2455.64	IV
500 c	2705.19	IV
200 c	2708.01	IV
200 c	2753.47	IV
200 c	2767.60	IV

Pr V
Ref. 149 — J.R.

Intensity	Wavelength	
	Vacuum	
200	843.78	V
7000	865.90	V
5000	869.17	V
80	869.66	V
1000	896.65	V
750	922.29	V
250	1234.07	V
250	1342.78	V
200	1958.09	V
400	1958.20	V
	AIR	
300	2246.06	V
300	2246.20	V

PROMETHIUM (Pm)
Z = 61
Pm I and II
Ref. 196, 260 — C.H.C.

Intensity		Wavelength	
		Air	
40	w	2502.12	II
40		2608.24	II
150		2632.00	II
70		2638.46	II
100		2671.05	II
50	w	2787.72	II
40		2808.05	II
100	h	2820.10	II
100	w	2840.82	II
150	w	2841.86	II
200	c	2857.46	II
100		3004.59	II
100		3008.85	II
300		3072.41	II
150		3086.02	II

Promethium (Cont.)

Intensity		Wavelength	Spectrum
120		3090.19	II
150		3091.86	II
150		3108.11	II
100		3115.36	II
100		3117.22	II
100		3118.76	II
35		3162.23	I
35		3168.82	I
100		3172.77	II
35		3222.04	I
35		3238.55	I
60		3239.62	II
75		3296.63	I
60		3311.76	II
50		3313.38	I
50		3329.22	I
75		3331.57	I
100		3354.45	I
100		3358.14	II
80		3360.21	II
80		3364.44	II
300		3366.03	I
90		3377.68	II
100		3391.28	II
100		3408.06	II
500		3427.40	II
120		3441.15	II
400		3449.80	II
250		3460.25	II
200		3462.91	II
200		3480.61	II
150		3497.13	II
100		3514.85	II
200		3546.81	I
100		3559.43	II
200		3565.31	II
150		3580.10	II
200		3610.76	II
200		3629.84	II
300		3634.20	II
300		3659.39	II
300	r	3669.22	I
200		3674.85	I
200		3678.51	II
300	r	3679.85	I
200		3687.65	II
400		3689.79	II
300		3692.50	II
300		3697.50	II
300	r	3697.63	I
400		3702.63	II
800		3711.72	II
200		3715.75	II
200		3721.72	II
500		3726.01	I
200		3738.43	I
300		3740.68	I
300		3742.52	II
300	r	3742.97	I
500		3745.86	II
300		3747.09	II
500		3750.09	II
200		3761.68	I
300		3765.75	I
300		3775.42	I
300		3780.77	I
200		3781.43	I
400		3795.66	II
250		3806.06	II
300	r	3809.20	I
400		3810.93	I
200		3819.26	II
300		3820.53	II
300		3839.52	I
200		3842.88	II
300		3842.98	II
250		3845.38	II
300		3874.03	I
800		3877.62	II
300	r	3885.79	I
250		3890.97	I
1000		3892.15	II
300		3898.73	I
400		3899.78	II
250		3909.50	II
1000		3910.26	II
1000		3919.10	II
800		3936.48	II
300		3944.21	II
300	r	3954.76	I
1000		3957.74	II
500		3980.74	II
300		3995.05	II
1000	r	3998.96	II
500		4009.96	II
200		4012.72	II
250		4014.20	II
200		4019.34	II
250		4028.20	II
200		4045.36	II
300		4051.54	II
600	r	4055.20	II
200	r	4056.56	I
600		4075.84	II
200	r	4085.31	I
500		4086.10	II
250		4140.46	II
200		4185.74	II
300		4192.92	II
200		4194.70	II
200		4222.15	II
300	r	4264.32	I
300		4284.37	I
600		4297.78	II
200		4303.89	II
200	r	4305.64	I
400		4318.80	I
250		4325.92	II
200		4332.05	II
300		4336.54	II
200		4337.48	II
300		4342.12	II
200		4347.72	I
350	r	4363.92	I
300	r	4369.64	II
200		4381.88	II
400		4388.49	I
200		4388.76	II
400	r	4409.42	I
500	r	4412.47	I
1000		4417.96	II
400		4432.51	II
250	r	4435.86	I
300	r	4436.55	I
300	r	4438.68	I
500		4445.41	II
600		4446.90	II
800		4453.95	II
200		4459.97	II
250	r	4468.16	I
200		4471.48	II
300		4473.23	II
200		4477.46	II
350	r	4478.58	I
300	r	4481.60	I
300	r	4485.05	I
300	r	4490.50	I
250		4493.05	II
600		4500.15	II
350	r	4500.33	I
250		4506.84	I
100		4509.38	II
100		4513.56	II
200		4517.31	I
200		4523.32	I
600		4525.20	II
250		4526.12	I
250	r	4526.76	I
400	r	4527.70	I
800		4529.21	II
300		4540.06	I
300	r	4541.42	I
450	r	4541.75	I
500	r	4544.08	I
200		4545.17	I
400	r	4549.78	I
300	r	4554.03	I
200		4554.63	I
500	r	4555.34	I
200		4556.06	I
300		4557.03	I
300		4559.21	I
100		4564.83	II
300		4568.14	I
200		4570.37	I
300	r	4572.15	I
400		4575.27	I
300		4578.28	I
200		4578.41	I
300		4579.48	I
300		4581.14	I
300		4585.49	I
200		4593.82	I
400		4595.82	I
800		4597.55	I
500		4600.25	I
400		4602.96	I
400		4604.59	I
600	r	4605.66	I
500	r	4609.85	I
100		4615.87	II
600		4617.02	I
200		4618.40	I
400		4618.49	I
500		4619.75	I
500		4621.57	I
500		4623.31	I
700		4623.68	I
900		4624.41	I
500		4625.29	I
400		4627.60	I
200		4630.93	I
600	r	4633.45	I
400	r	4640.96	I
700	r	4643.36	I
700	r	4643.76	I
400		4645.94	I
600	r	4647.03	I
600		4650.42	I
500		4650.52	I
400	r	4653.41	I
400		4654.50	I
500	r	4655.05	I
300		4659.38	I
500		4660.79	I
300		4663.26	I
600		4663.46	I
400		4665.19	I
500	r	4671.23	I
400		4671.76	I
500	r	4674.42	I
200		4677.46	I
500	r	4677.92	I
400	r	4678.09	I
700	r	4682.92	I
500	r	4696.80	I
200		4699.51	I
250		4722.06	II
300		4727.06	I
900	r	4728.36	I
400		4728.76	I
800	r	4734.27	I
200	r	4737.99	I
100		4739.08	II
200		4739.78	I
350	r	4745.13	I
500		4757.73	I
800	r	4759.00	I
700		4762.57	I
700		4773.46	I
900	r	4781.29	I
250		4794.59	I
200		4795.43	I
700	r	4798.98	I
900	r	4801.36	I
700	r	4809.54	I
900	r	4811.96	I
400	r	4817.12	I
400	r	4827.72	I
800		4837.66	I
100		4838.31	I
300		4839.62	I
200		4844.01	I
350	r	4852.73	I
400	r	4860.62	I
700	r	4860.74	I
300	r	4865.30	I
500	r	4865.72	I
400	r	4869.80	I
700	r	4872.42	I
500	r	4887.02	I
700	r	4892.52	I
400	r	4900.30	I
300	r	4904.28	I
400	r	4918.28	I
600	r	4932.99	I
700	r	4959.46	I
100		4971.40	II
500	r	4997.10	I
200		5030.80	I
300		5058.31	I
100		5067.35	II
150		5080.52	II
150		5089.35	II
200		5092.42	I
400	r	5094.83	I
200		5096.18	I
150		5097.30	II
400	r	5100.77	I
250		5121.47	II
400	r	5127.34	I
200		5129.75	I
400	r	5145.13	I
500	r	5146.30	I
400		5153.86	II
300		5169.71	II
500		5171.58	II
300		5194.05	II
500		5208.09	II
150		5215.56	II
250		5225.12	II
500		5236.26	II
300		5236.66	II
400		5246.33	II
150		5262.42	II
500		5270.64	II
200		5293.92	II
100		5308.86	II
150		5318.58	II
200		5410.45	II
200		5424.54	II
180		5424.79	II
150		5429.04	II
150		5467.64	II
150		5495.45	II
100		5516.42	II
180		5534.96	II
200		5537.38	I
800		5546.08	II
120		5556.88	II
150		5558.39	II
200		5561.73	II
800		5576.02	II
200		5641.29	II
200		5730.81	I
200		5768.16	II
200		5776.99	I
500		5823.93	II
300	c	5868.79	II
200	c	5875.31	II
100	c	5878.76	II
150		5899.76	II
250		5904.71	I
100		5905.90	I
125		5914.96	I
250	c	5927.17	II
150		5939.66	I
400	c	5946.49	II
800		5956.42	I
200		5956.69	I
100	c	5960.08	II
150		5963.00	II
400		5967.89	I
200		5979.73	I
200		5984.82	I
100	c	5987.13	II
400		5997.12	I
200		6027.11	II
300		6030.06	I
400		6031.32	I
500		6043.39	I
150	c	6052.57	II
100	c	6067.00	II
500		6069.06	I
100		6076.40	II
200		6085.41	II
900		6100.21	I
400		6106.40	I
100		6114.90	II
400	h	6151.76	I
100		6159.53	II
400		6163.16	I
100		6184.52	II
200		6208.91	II
500		6229.64	I
600		6237.70	I
100		6263.25	II
400		6272.69	I
400		6286.06	I
500		6308.29	I
100		6314.20	II
700		6323.84	I
500		6390.31	I
100		6429.64	II
500	h	6431.93	I
100		6436.57	II
400		6487.61	I
400		6510.34	I
500		6517.25	I
200		6519.43	II
1000	d	6520.45	I
500		6542.20	I
100	h	6558.48	II
100		6586.39	II
100		6592.29	II
900		6598.15	I
800		6598.66	I
700		6606.37	I
800	w	6625.23	I
100	h	6625.54	II
700		6649.81	I
400		6659.05	II
100		6661.25	II
500		6661.68	I
400		6663.76	I
800	c	6667.51	I
700	h	6677.47	I
200		6680.89	II
500		6685.55	I
500		6685.68	I
150		6690.09	II
600		6700.33	I
100		6706.27	II
700		6714.67	I
500		6717.26	I
500		6720.71	I
700		6727.50	I
600		6743.71	I
900		6749.91	I
900		6750.48	I
200		6756.45	II
300		6772.29	II
400		6778.78	I
100		6783.09	II
100		6796.87	II
200		6811.68	II
800		6833.30	I
400		6848.37	I
50		6858.58	II

PROTACTINIUM (Pa)
Z = 91

Pa I and II
Ref. 96 — J.G.C.

Intensity		Wavelength
		Air
3000	h	2466.85	
3000	h	2492.85	
3000		2599.16	II
3000		2699.22	II
3000		2822.79	II
3000	l	2832.14	
3000		2870.01	
3000	h	2871.42	II
3000	h	2891.14	II
3000	h	2906.93	
3000	l	3011.10	II
3000	s	3033.59	II
3000	l	3071.24	II
3000	h	3083.19	
3000	l	3093.23	II
3000	l	3126.23	II
3000	l	3146.28	II
3000	l	3170.89	II
3000	l	3171.54	II
3000		3204.16	
3000	l	3240.58	II
3000		3274.46	II
3000	l	3332.69	II
3000	s	3346.66	II
3000	l	3394.49	
3000	l	3452.82	II
3000		3504.97	I
3000	s	3530.65	II
3000		3570.56	I
3000		3571.82	I
3000		3618.07	
10000		3636.52	I
3000		3702.74	I
3000		3752.67	I
3000		3873.35	I
3000		3931.83	I
3000	s	3952.62	II
10000	l	3957.85	II
3000	s	3970.07	II
3000		3981.82	I
10000		3982.23	I
3000	l	4012.96	II
3000	s	4018.21	II
3000		4030.16	II
3000	s	4046.93	II
10000		4056.20	II
10000	s	4070.40	II
3000		4117.62	
3000	l	4176.18	II
10000	l	4217.23	II
10000	s	4248.08	II
3000	s	4291.34	II
3000		4400.77	
3000		4436.13	
3000	s	4601.43	II
3000		4628.19	
3000		4820.34	
3000	s	4861.49	
3000	l	6035.78	I
3000		6162.56	I
3000		6216.35	
3000	l	6358.61	I
3000		6379.25	I
3000	l	6438.97	I
3000	h	6792.75	I
10000		6945.72	I
3000		6960.09	I
3000	h	6961.78	I
3000	s	6992.73	I
3000		7076.27	I
3000	h	7100.94	I
10000		7114.89	I
3000	h	7171.55	I
3000		7227.13	I
3000		7318.79	I
10000	l	7368.25	I
3000		7471.89	I
10000	h	7493.15	I
3000		7558.26	I
10000	h	7608.20	I
10000		7626.79	I
10000	s	7635.18	I
10000		7669.34	I
3000		7679.20	I
10000	h	7749.19	I
3000		7872.95	I
3000	l	7945.56	I
10000		8039.34	I
10000	h	8099.84	I
10000		8199.04	I
10000		8271.87	I
3000	s	8358.98	I
3000	s	8369.60	I
3000	h	8441.04	I
10000	h	8532.66	I
10000	s	8572.96	I
3000	h	8639.91	I
3000	h	8653.51	I
10000		8735.27	I
3000		10594.38	I
3000		10923.32	I
3000		11646.78	
10000		11791.73	I
3000		12279.01	
3000		13234.09	
10000		13522.40	
10000		14344.76	I
3000		18478.61	I

RADIUM (Ra)
Z = 88

Ra I and II
Ref. 253, 254 — E.F.W.

Intensity		Wavelength
		Air
8		2369.73	II
8		2460.55	II
10		2475.50	II
8		2586.61	II
10		2643.73	II
20		2708.96	II
10		2795.21	II
30		2813.76	II
10		3033.44	II
5		3101.80	I
100		3649.55	II
200		3814.42	II
8		4194.09	II
8		4244.72	II
100		4340.64	II
20		4436.27	II
30		4533.11	II
8		4641.29	I
100		4682.28	II
8		4699.28	I
100		4825.91	I
10		4856.07	I
10		4859.41	II
10		4927.53	II
10		5097.56	I
10		5205.93	I
10		5283.28	I
10		5320.29	I
10		5399.80	I
20		5400.23	I
20		5406.81	I
8		5482.13	I
10		5501.98	I
10		5553.57	I
20		5555.85	I
10		5616.66	I
50		5660.81	I
20		5813.63	II
30		6200.30	I
10	p	6336.90	I
20		6446.20	I
20		6487.32	I
10		6593.34	II
10		6719.32	II
20		6980.22	I
20		7118.50	I
50		7141.21	I
20		7225.16	I
10		7310.27	I
20		7838.12	I
50		8019.70	II
6		8177.31	I
5		8335.07	I
5		9932.21	I

RADON (Rn)
Z = 86

Rn I
Ref. 251 — E.F.W.

Intensity	Wavelength
	Air
5	3514.60	I
10	3739.46	I
20	3753.65	I
10	3917.20	I
10	3941.72	I
10	3952.36	I
10	4226.06	I
80	4307.76	I
7	4335.78	I
100	4349.60	I
40	4435.05	I
50	4459.25	I
50	4508.48	I
50	4577.72	I
50	4609.38	I
30	4721.76	I
6	5722.58	I
10	6061.92	I
6	6200.75	I
6	6380.45	I
10	6557.49	I
10	6606.43	I
15	6627.23	I
6	6669.60	I
8	6704.28	I
20	6751.81	I
6	6806.79	I
8	6836.95	I
8	6837.57	I
10	6891.16	I
10	6998.90	I
200	7055.42	I
100	7268.11	I
20	7291.00	I
6	7320.98	I
10	7419.04	I
300	7450.00	I
8	7470.89	I
8	7483.13	I
8	7514.13	I
8	7516.92	I
6	7523.93	I
6	7597.55	I
8	7601.28	I
10	7657.48	I
10	7738.43	I
20	7746.64	I
100	7809.82	I
20	8049.00	I
100	8099.51	I
6	8173.84	I
100	8270.96	I
8	8314.51	I
6	8349.74	I
10	8381.05	I
10	8487.48	I
10	8494.89	I
20	8520.95	I
100	8600.07	I
10	8639.76	I
15	8675.83	I
10	8807.75	I
50	9327.02	I
6	9948.57	I
5	10106.13	I

RHENIUM (Re)
Z = 75

Re I and II
Ref. 1 — C.H.C.

Intensity		Wavelength
		Air
25000		2003.53	I
16000		2017.87	I
27000		2049.08	I
4200		2074.70	I
3700		2083.92	I
10000		2085.59	I
4700		2092.41	II
9800		2097.12	I
2700		2109.22	I
3400		2139.04	II
1600		2142.74	II
		2142.97	I
3700		2156.67	I
4900		2167.94	I
3400		2176.21	I
4200	c	2214.26	II
2200		2214.58	I
1700		2226.42	I
920		2235.44	I
440		2255.73	I
860		2256.19	I
2000		2264.39	I
2100		2274.62	I
5200	c	2275.25	II
1600		2281.62	I
2900		2287.51	I
2700		2294.49	I
390		2298.09	II
390		2299.77	I
610		2302.99	I
680		2306.54	I
230		2312.97	I
220		2313.34	I
220		2319.19	I
370		2320.16	I
800		2322.49	I
300		2328.66	I
270		2334.33	I
270		2335.73	I
220		2336.10	I
270		2337.95	I
860		2344.78	I
230		2349.39	I
220	d	2350.46	I
680		2352.07	I
210	d	2353.95	I
250		2356.50	I
200		2365.32	I
1200		2365.90	I
570		2367.68	I
180		2368.53	II
520		2369.27	I
220		2370.76	II
210		2371.52	I
150		2373.48	II
320		2375.07	I
75		2378.53	II
370		2379.77	I
180		2386.90	II
340		2388.57	I
230		2393.65	I
320		2394.37	I
320		2396.79	I
200		2397.31	I
210	d	2400.72	I
		2400.89	I
210		2401.68	I
75		2403.04	II
1500		2405.06	I
740		2405.60	I
320		2406.70	I
270		2410.37	I
60		2418.20	II
1200		2419.81	I
300		2421.73	I
300		2421.88	I
60		2423.84	II
2500		2428.58	I
490		2431.54	I
420		2432.18	I
340	c	2441.47	I
230		2442.51	I
250		2444.94	I
610		2446.98	I
85		2449.03	II
85		2449.52	II
610		2449.71	I
200		2455.83	II
390		2461.20	I
800	c	2461.84	II
200		2467.57	II
120		2467.85	I
150	c	2469.36	II
120		2470.61	II
75		2471.05	II
150		2473.72	II
160		2475.17	II
75		2477.43	II
200		2479.02	I
1200		2483.92	I
390		2485.81	I
980		2487.33	I
75		2490.16	II
200		2492.84	I
370		2496.04	I
200		2498.22	I
370		2501.72	I
570		2502.35	II
230		2504.60	II
270		2505.94	I
1800	c	2508.99	I
570		2520.01	I
540		2521.50	I
150		2534.10	II
370		2534.80	I
570		2540.51	I
740	d	2544.74	I
370		2545.48	I
160		2550.09	II
300		2552.02	I
150	c	2553.59	II
370		2554.63	II
1000		2556.51	I
250		2559.08	I
340		2564.19	I
540		2568.64	II
370		2571.81	II
380		2586.79	I
290		2599.86	I
290		2603.89	I
660		2608.50	II
610	d	2611.54	I
160	c	2616.72	II
200		2622.76	I
310		2635.83	I
550		2636.64	I
190		2637.01	II
90		2641.02	II
270		2642.75	I
65		2648.46	II
270		2649.05	I
660		2651.90	I

Intensity	Flag	Wavelength	Species
400		2654.12	I
220		2663.63	I
940		2674.34	I
220		2688.53	I
1300		2715.47	I
200		2731.56	II
220		2732.21	I
610		2733.04	II
110	h	2753.64	II
220		2758.00	I
210		2763.79	I
200		2766.39	I
310		2767.74	I
220		2768.85	I
220		2769.32	I
350		2770.42	I
550		2783.57	I
220		2791.29	I
120		2803.28	II
220		2814.68	I
75		2819.78	II
880		2819.95	I
310		2834.08	I
200		2837.55	I
200		2840.35	I
220		2843.00	I
270		2850.98	I
240		2867.19	I
200		2875.28	I
200		2883.44	I
2900		2887.68	I
130	c	2888.06	II
490		2896.01	I
830	c	2902.48	I
210		2905.58	I
550		2909.82	I
65	h	2916.73	II
830	c	2927.42	I
270		2930.61	I
440		2943.14	I
130	h	2957.91	II
270		2962.27	I
720		2965.11	I
1500		2965.76	I
90		2968.98	II
310		2976.29	
210		2978.15	I
220		2980.82	I
220		2982.19	I
220		2988.47	I
1800		2992.36	I
5500		2999.60	I
350		3001.14	I
220		3004.14	I
200		3006.42	I
500		3016.02	I
300		3016.49	I
380		3030.45	I
240		3047.25	I
200		3058.78	I
1600		3067.40	I
320		3069.94	I
260		3071.16	I
200		3072.96	I
550		3082.43	I
340		3088.76	I
200		3093.64	I
200		3095.06	I
700		3100.67	I
140		3103.06	II
700		3108.81	I
340		3110.86	I
340	c	3118.19	I
340		3121.36	I
420		3128.94	I
260		3134.02	I
250		3141.38	I
440		3151.64	I
330		3153.79	I
360	c	3158.31	I
220		3164.52	I
700		3168.37	I
220		3174.61	I
440		3177.71	I
260		3178.61	I
600		3182.87	I
1100		3184.76	I
1100		3185.57	I
260		3190.78	I
260		3192.36	I
200		3194.50	I
220		3198.58	I
1100	c	3204.25	I
380		3235.94	I
600		3258.85	I
600		3259.55	I
200		3261.56	I
300		3268.89	I
200		3294.83	I
280		3296.70	I
280		3296.99	I
280		3301.60	I
240		3302.23	I
320		3303.21	II

Intensity	Flag	Wavelength	Species
280		3303.75	I
240		3313.95	I
600		3322.48	I
200		3331.52	I
2000		3338.18	I
1600		3342.24	I
810		3344.32	I
320		3346.20	I
240	d	3356.33	I
200		3358.02	I
200		3362.74	I
240		3377.74	I
320		3379.06	II
320		3379.70	I
200		3385.76	I
240		3389.43	I
200		3390.25	I
4000		3399.30	I
650		3404.72	I
650		3405.89	I
240		3408.67	I
320		3409.83	I
320		3417.77	I
810		3419.41	I
8000		3424.62	I
400		3426.19	I
300		3427.61	I
320		3437.71	I
400		3449.37	I
16000	c	3451.88	I
240		3453.50	I
55000	c	3460.46	I
40000	c	3464.73	I
400		3467.96	I
240		3476.44	I
400		3480.38	I
320		3480.85	I
240		3482.23	I
560		3503.06	I
100	c	3512.28	I
320		3516.65	I
320		3517.33	I
120		3534.82	I
320		3537.46	I
160		3546.88	I
240		3549.09	I
160		3551.29	I
160		3553.65	I
160		3558.94	I
160		3568.23	I
240		3570.26	I
360		3579.12	I
810	c	3580.15	II
650		3580.97	I
810		3583.02	I
160		3596.39	I
160		3610.49	I
320		3617.08	I
160		3621.46	I
160		3625.91	I
140		3637.06	I
810		3637.84	I
440		3651.97	I
120		3669.78	I
320		3670.53	I
860	c	3689.50	I
1500	c	3691.48	I
100		3697.71	I
520		3703.24	I
100		3705.02	I
240		3709.93	I
360	c	3717.28	I
4000		3725.76	I
140		3731.87	I
140		3732.28	I
240	c	3735.01	I
810		3735.31	I
910		3740.10	I
140		3740.41	I
130		3742.26	II
300	Cw	3745.44	I
140		3766.48	I
120		3768.26	I
140		3777.66	I
700		3787.64	I
160		3796.59	I
160		3797.59	I
190		3807.74	I
120		3815.66	I
120		3836.30	I
240		3869.94	I
240		3875.26	I
240		3876.86	I
100		3908.21	I
130		3913.92	I
380	c	3917.27	I
550		3929.85	I
140		3936.90	I
110		3944.72	I
180		3945.91	I
280		3961.04	I
350	c	3962.48	I
100		4004.93	I
140		4022.96	I

Intensity	Flag	Wavelength	Species
100		4023.31	I
110	c	4029.63	I
220		4033.31	I
110		4037.49	I
200		4048.99	I
240		4081.43	I
140		4104.42	I
240	c	4110.89	I
190		4121.64	I
240	Cw	4133.42	I
1800		4136.45	I
700		4144.36	I
140		4149.96	I
160		4170.40	I
220		4182.90	I
220		4183.06	I
650		4221.08	I
3600	c	4227.46	I
150		4241.39	T
260	c	4257.60	I
120	c	4291.17	I
200		4304.40	I
200		4332.25	I
40		4357.98	II
380		4358.69	I
190		4367.58	I
140		4391.34	I
360	Cw	4394.38	I
110	Cw	4406.40	I
180		4415.82	I
150		4475.08	I
120		4478.39	I
120	c	4507.04	I
2600		4513.31	I
260		4516.64	I
500		4522.73	I
120		4523.88	I
120		4529.95	I
100		4545.17	I
120		4580.68	I
120		4605.73	I
100		4621.38	I
190	c	4791.42	I
2200	Cw	4889.14	I
110		4923.90	I
40		5058.56	I
70		5096.50	I
20		5120.32	I
25		5161.65	I
40	c	5178.89	I
20		5181.74	I
35		5234.31	I
50		5248.86	I
1300		5270.95	I
1600	Cw	5275.56	I
100		5278.24	I
30		5305.56	I
20		5317.28	I
35		5321.28	I
50		5327.46	I
20		5331.90	I
20		5332.76	I
20		5333.85	I
35		5369.48	I
50	c	5369.80	I
100	c	5377.10	I
25		5431.90	I
14		5437.03	I
14		5447.92	I
25		5460.64	I
14	h	5520.05	I
25		5521.10	I
50	c	5532.68	I
50		5563.24	I
25		5573.47	I
25		5584.72	I
10	h	5607.21	I
12	h	5612.27	I
100		5667.88	I
25		5711.43	I
18		5716.95	I
110	c	5752.93	I
110	Cw	5776.83	I
18		5791.60	I
10	c	5815.92	I
550		5834.31	I
10		5919.86	I
60		5943.24	I
10		5950.21	I
18	h	5969.77	I
10		5989.99	I
18	h	5995.73	I
30		6114.22	I
35	c	6145.81	I
50		6146.82	I
18		6203.24	I
25		6217.97	I
30	Cw	6229.42	I
35	Cw	6243.24	I
35	d	6260.02	I
		6260.24	I
18	c	6271.37	I
18		6278.76	I
10		6286.41	I

Intensity	Flag	Wavelength	Species
10		6303.42	I
200		6307.70	I
200		6321.90	I
80	d	6350.75	I
16	h	6382.94	I
14		6411.47	I
50		6511.47	I
14		6515.25	I
12		6544.91	I
35	c	6577.11	I
40	Cw	6592.52	I
100	Cw	6605.19	I
30	c	6623.91	I
10	c	6637.25	I
27	Cw	6652.39	I
15	h	6683.28	I
9	c	6711.30	I
30		6751.22	I
5	c	6761.19	I
180	c	6813.41	I
260		6829.90	I
85		6971.53	I
35	Cw	7006.63	I
65	Cw	7024.15	I
65	Cw	7246.67	I
13	Cw	7292.72	I
40	Cw	7578.73	I
13		7611.89	I
7	Cw	7620.25	I
50	Cw	7640.94	I
65	Cw	7912.94	I
35	Cw	7980.77	I
40		8417.13	I
29	Cw	8527.73	I

RHODIUM (Rh)
Z = 45

Rh I and II
Ref. 1 — C.H.C.

Intensity	Flag	Wavelength (Air)	Species
150		2276.21	II
140		2288.57	I
110		2309.82	I
55		2318.36	I
95		2319.10	I
95		2321.73	I
350		2322.58	I
140		2326.47	I
80		2328.64	I
190		2334.49	II
55		2345.41	I
55		2352.47	I
55		2359.18	I
300		2361.92	I
110		2368.34	I
270		2382.89	I
230		2383.40	I
40		2384.65	I
270		2386.14	II
80		2407.88	I
27		2408.19	I
27		2410.25	I
80		2415.84	II
55		2418.64	I
45		2419.75	I
45		2420.18	II
65		2420.98	II
75		2423.94	I
65		2427.11	II
130		2427.68	I
230		2429.52	I
40		2431.85	II
40		2432.66	I
18		2437.08	I
110		2437.90	I
330		2440.34	I
50	h	2444.27	I
65		2448.84	I
50		2449.04	I
75		2450.56	I
30		2455.70	II
65		2458.90	II
90		2461.04	II
30		2463.61	I
75		2470.39	I
90		2471.47	I
30		2472.51	I
130		2473.09	I
15		2475.64	II
15		2477.54	II
25		2482.04	I
50		2483.33	I
150		2487.47	I
100		2490.77	II
30		2492.30	I
75	h	2494.51	I
15		2499.02	I
40		2500.58	I
130		2502.46	I

Rhodium (Cont.)

Intensity	Note	Wavelength	Spectrum
15		2503.84	II
300		2504.29	II
40		2505.10	II
150		2505.67	I
350		2509.70	I
50		2510.66	II
300		2511.03	II
75		2513.36	II
200		2515.75	I
130		2520.53	II
13		2525.99	I
13		2531.74	I
50		2532.66	
13		2533.59	
50		2534.07	
110		2536.71	
110		2537.04	II
30		2539.72	
40		2544.22	
350		2545.70	I
13		2548.60	
550		2555.36	I
25		2558.62	I
50		2565.79	I
45		2566.04	I
25		2566.92	II
50		2567.28	
25		2574.66	
25		2575.75	I
13		2576.23	
40		2587.29	II
30		2598.07	
30		2603.32	II
75		2606.44	II
75		2613.60	
150		2622.58	I
230		2625.88	I
100		2630.42	I
40		2634.99	I
30		2638.74	II
75		2643.00	I
110		2647.28	I
400		2652.66	I
30		2659.01	I
30		2671.06	I
65		2676.11	I
25		2680.28	I
100		2680.63	I
30		2681.78	I
30	h	2686.50	
30	h	2686.91	
50		2694.31	I
400		2703.73	I
40		2705.63	II
40		2707.23	I
75		2714.41	I
100		2715.31	II
75		2717.51	I
180		2718.54	I
65		2720.14	I
30		2720.52	I
160		2728.94	I
40		2736.76	I
75		2741.75	I
50		2767.73	I
100		2771.51	I
50		2778.06	I
75		2779.54	I
130		2783.03	I
25		2791.16	I
75		2796.63	I
150		2826.43	I
180		2826.68	I
30		2827.31	I
75		2834.12	I
45		2835.44	I
75		2836.69	I
50		2856.16	I
50	d	2860.68	I
		2860.76	I
280		2862.94	I
65		2864.40	I
50		2871.35	I
30		2873.62	I
110		2878.66	I
75		2880.76	I
140		2882.37	I
75		2885.97	I
75		2889.11	I
75		2889.84	I
65		2899.96	I
25		2904.81	I
160		2907.21	I
65		2910.17	II
75		2912.62	I
90		2915.42	I
30		2923.10	I
180		2924.02	I
130		2929.11	I
130		2931.94	I
30		2955.41	I
230		2968.66	I
25		2974.03	I
160		2977.68	I
450		2986.20	I
90		2986.99	I
50		2987.45	I
110		3004.46	I
50		3019.54	I
130		3023.91	I
50		3028.43	I
30		3045.77	I
30		3046.76	I
25		3057.89	I
65		3067.30	I
180		3083.96	I
29		3087.42	I
70		3114.91	I
140		3121.76	I
240		3123.70	I
35		3130.79	I
95		3137.71	I
45		3151.36	I
45		3152.60	I
130		3155.78	I
70		3179.73	I
80		3185.59	I
140		3189.05	I
470		3191.19	I
190		3197.13	I
70		3214.32	I
80		3237.66	I
520		3263.14	I
520		3271.61	I
2300		3280.55	I
110		3281.70	I
2300		3283.57	I
280		3289.14	I
45		3289.64	I
210		3294.28	I
45		3296.72	I
260		3300.46	I
4200		3323.09	I
60		3331.09	I
45		3331.24	I
330		3338.54	I
70		3342.90	I
80		3344.20	I
60		3359.90	I
280		3360.80	I
60		3362.18	I
420		3368.38	I
45		3369.68	I
1100		3372.25	I
110		3377.14	I
80		3377.71	I
110		3385.78	I
5600		3396.82	I
820		3399.70	I
160		3406.55	I
820		3412.27	I
60		3420.16	I
330		3421.22	I
120	d	3424.38	I
8200		3434.89	I
1400		3440.53	I
35		3442.63	I
120		3447.74	I
60		3448.58	I
120		3450.29	I
60		3451.15	I
400		3455.22	I
60		3455.42	I
180		3457.07	I
220		3457.93	I
5900		3462.04	I
180		3469.62	I
4700		3470.66	I
120		3472.25	I
4700		3474.78	I
2100		3478.91	I
95		3484.04	I
80		3491.07	I
110		3494.44	I
1200		3498.73	I
5900		3502.52	I
60		3505.41	I
2800		3507.32	I
60		3511.78	I
60		3513.10	I
60		3519.54	I
8800		3528.02	I
880	d	3538.14	I
		3538.26	I
280		3541.91	I
1200		3543.95	I
1800		3549.54	I
240		3564.13	I
1200		3570.18	I
4700		3583.10	I
120		3583.53	I
4700		3596.19	I
5900		3597.15	I
310		3605.86	I
3100		3612.47	I
240		3614.78	I
200		3620.46	I
1800		3626.59	I
95		3627.80	I
310		3639.51	I
350		3654.87	I
8200		3657.99	I
280		3661.86	I
1300		3666.22	I
180		3666.91	I
140		3674.76	I
560		3681.04	I
1900		3690.70	I
9400		3692.36	I
60		3694.95	I
940		3695.52	I
280		3698.26	I
380		3698.60	I
7600		3700.91	I
940		3713.02	I
60		3713.43	I
45		3714.83	I
16		3724.94	I
650		3735.28	I
420		3737.27	I
420		3744.17	I
1200		3748.22	I
240		3754.12	I
380		3754.27	I
490		3755.58	I
1000		3760.40	I
2300		3765.08	I
490		3769.97	I
70		3775.72	I
380		3778.13	I
1000		3788.47	I
1300		3792.18	I
3800		3793.22	I
4900		3799.31	I
760		3805.92	I
1300		3806.76	I
45		3809.50	I
95		3812.45	I
470		3815.01	I
760		3816.47	I
1300		3818.19	I
3800		3822.26	I
2300		3828.48	I
2000		3833.89	I
45		3834.75	I
5900		3856.52	I
490		3870.01	I
70		3872.39	I
380		3877.34	I
70		3888.34	I
29		3904.22	I
23		3912.83	I
120		3913.51	I
240		3922.19	I
2000		3934.23	I
45		3934.98	I
50		3935.84	I
590		3942.72	I
95		3958.24	I
3800		3958.86	I
45		3964.54	II
380		3975.31	I
240		3984.40	I
240		3995.61	I
380		3996.15	I
120		4023.14	I
60		4048.41	I
23		4049.04	I
40		4053.44	I
23		4056.34	I
70		4077.57	I
560		4082.78	I
19		4084.28	I
45		4087.79	I
60		4088.50	I
140		4097.52	I
45		4107.49	I
70		4116.33	I
120		4119.68	I
1100		4121.68	I
1500		4128.87	I
2100		4135.27	I
240		4154.37	I
330		4196.50	I
70		4206.62	I
3300		4211.14	I
29		4230.20	I
40		4244.44	I
60		4273.43	I
60		4278.60	I
820		4288.71	I
70		4296.77	I
23		4342.44	I
45		4373.04	I
4200		4374.80	I
95		4379.92	I
23		4433.32	I
35		4492.47	I
29		4503.78	I
23		4528.72	I
16		4544.27	I
35		4548.73	I
40		4551.64	I
19		4560.89	I
16		4565.19	I
130		4569.00	I
14		4571.31	I
29		4608.12	I
14		4619.91	I
23		4643.18	I
150		4675.03	I
19		4721.00	I
70		4745.11	I
12		4755.58	I
23		4810.49	I
21		4842.43	I
45		4843.99	I
60		4851.63	I
60		4963.71	I
60		4977.75	I
40		4979.18	I
14		5085.52	I
70		5090.63	I
23		5120.69	I
19		5130.76	I
60		5155.54	I
14		5157.09	I
40		5158.69	I
60		5175.97	I
12		5177.27	I
35		5184.19	I
95		5193.14	I
16		5206.95	I
16		5211.52	I
19		5212.73	I
16		5214.79	I
19		5222.66	I
19		5230.62	I
45		5237.16	I
9		5237.80	I
14		5269.27	I
11	h	5280.12	I
14		5292.14	I
14		5314.79	I
40	h	5329.74	I
14	h	5331.08	I
9		5349.31	I
130		5354.40	I
23		5356.47	I
45		5379.10	I
95		5390.44	I
23	h	5404.73	I
60	h	5424.07	I
19		5424.72	I
19	h	5425.45	I
12		5439.58	I
12	h	5441.36	I
9	h	5444.32	I
35	h	5445.23	I
23	h	5468.11	I
35	h	5470.85	I
12		5476.12	I
12		5481.42	I
16		5484.23	I
9		5504.65	I
29		5535.04	I
21	l	5544.58	I
160		5599.42	I
7		5607.71	I
16		5608.35	I
5		5632.77	I
9		5659.62	I
40		5686.38	I
9	h	5702.47	I
6		5727.30	I
29		5792.66	I
9		5795.79	I
9		5803.34	I
40		5806.91	I
6		5821.86	I
35		5831.58	I
7		5907.31	I
9		5918.54	I
7		5941.46	I
130		5983.60	I
9		5991.19	I
35		6102.72	I
6		6116.15	I
8		6128.06	I
8		6186.89	I
14		6199.99	I
16		6253.72	I
5		6276.66	I
8		6277.46	I
6		6293.38	I
29		6319.53	I
12		6414.72	I
16		6510.41	I
19		6519.70	I
9		6627.80	I
19		6630.16	I
40		6752.35	I
9		6796.65	I
13		6827.33	I
11		6857.68	I
20		6879.94	I

Rhodium (Cont.)

Intensity		Wavelength	Spectrum
65		6965.67	I
8		6972.91	I
16		6979.15	I
16		7001.58	I
11		7038.76	I
18		7101.64	I
15		7104.45	I
6		7142.55	I
9		7219.06	I
18		7268.18	I
35		7270.82	I
12		7271.94	I
5		7273.03	I
9		7375.57	I
5	h	7386.64	I
9		7430.80	I
18	h	7442.39	I
7		7446.77	I
12		7475.74	I
12		7495.24	I
8		7542.02	I
11		7557.67	I
8		7577.22	I
11		7690.05	I
18		7772.90	I
29		7791.61	I
55		7824.91	I
15		7830.05	I
15		7846.50	I
21		8029.91	I
11	h	8036.09	I
29		8045.36	I
7		8063.50	I
15		8136.20	I
7	h	8193.67	I
5	h	8369.67	I
8		8425.59	I

Rh III
Ref. 396 — L.J.R.

Intensity		Wavelength	Spectrum
		Vacuum	
10		746.78	III
30		759.54	III
50		813.44	III
30		826.01	III
30		843.63	III
30		849.08	III
40		852.70	III
40		854.77	III
40		859.89	III
40		861.34	III
50		863.78	III
50		865.22	III
50		870.40	III
80		882.51	III
100		925.75	III
150		937.28	III
100		976.12	III
500		991.62	III
400		992.48	III
500	d	1009.60	III
200		1012.22	III
200		1015.17	III
100		1050.00	III
100		1058.97	III
200		1073.87	III
100		1100.58	III
100		1113.79	III
100		1768.43	III
150		1784.24	III
200		1784.94	III
150		1796.50	III
200		1816.03	III
1000		1832.05	III
500		1859.85	III
100		1874.70	III
800		1880.66	III
500		1884.91	III
500		1887.36	III
700		1888.62	III
800		1901.32	III
500		1910.16	III
600		1919.37	III
500		1927.07	III
700		1931.79	III
500		1954.25	III
400		1965.16	III
500		1994.26	III
		Air	
400		2005.14	III
800		2013.71	III
500		2017.47	III
500		2028.53	III
800		2036.72	III
600		2037.61	III
1000		2040.18	III
3000		2048.67	III
2000		2064.11	III
800		2076.84	III
1000		2118.53	III
1000		2118.63	III
1000		2139.44	III
1000		2152.23	III
3000		2158.17	III
3000		2163.19	III
3000		2167.33	III
100		2207.00	III
100		2230.66	III
50		2250.84	III
30		2374.84	III
20		2470.65	III
50		3006.43	III
50		3052.44	III
50		3310.69	III
1		3852.98	III

RUBIDIUM (Rb)
Z = 37
Rb I and II
Ref. 12, 130, 241, 257, 264 — J.R.

Intensity		Wavelength	Spectrum
		Vacuum	
10		474.88	II
40		481.118	II
90		497.430	II
20		508.434	II
150		513.266	II
300		530.173	II
75		533.801	II
40		542.887	II
200		555.036	II
2500		589.419	II
1500		643.878	II
3000		697.049	II
6000		711.187	II
10000		741.456	II
1000		1604.12	II
200		1644.96	II
200		1707.52	II
600		1716.85	II
5000		1760.50	II
200		1803.47	II
500		1809.60	II
500		1865.33	II
500		1889.42	II
500		1954.24	II
300		1956.54	II
200		1971.42	II
500		1983.19	II
		Air	
300		2042.23	II
300		2052.21	II
500		2052.80	II
2000		2068.92	II
1000		2071.50	II
10000		2075.95	II
1000		2090.29	II
200		2108.06	II
300		2116.50	II
1000		2125.25	II
400		2129.82	II
200		2143.10	II
30000		2143.83	II
200		2190.36	II
600		2197.99	II
600		2198.26	II
300		2207.86	II
10000		2217.08	II
200		2223.79	II
400		2237.72	II
500		2250.65	II
200		2251.43	II
800		2254.19	II
200		2254.55	II
200		2263.54	II
500		2263.94	II
500		2286.82	II
5000		2291.71	II
300		2298.80	II
250		2333.01	II
2000		2333.39	II
350		2353.11	II
300		2353.96	II
400		2356.97	II
300		2358.04	II
300		2364.27	II
200		2364.32	II
200		2365.15	II
300		2367.51	II
200		2373.21	II
2000		2385.34	II
250		2405.94	II
400		2434.17	II
800		2459.14	II
50000		2472.20	II
300		2484.56	II
700		2484.70	II
2000		2496.38	II
200		2502.67	II
250		2514.18	II
1000		2524.24	II
200		2594.56	II
400		2623.76	II
400		2645.58	II
1000		2684.10	II
1000		2711.76	II
250		2741.01	II
500		2812.15	II
350		2838.51	II
750		2873.88	II
1000		3051.36	II
2		3082.02	I
250		3088.58	II
10		3112.57	I
3		3113.06	I
5000	c	3148.90	II
25		3157.54	I
5		3158.26	I
1200		3161.00	II
50		3227.98	I
6		3229.16	I
2000		3270.99	II
1500		3321.49	II
1200		3340.55	II
60		3348.72	I
75		3350.82	I
750		3353.89	II
1200		3393.03	II
750		3415.58	II
1000		3434.18	II
1500		3461.50	II
3000		3521.39	II
3000	1	3531.55	II
1000		3541.15	II
100		3587.05	I
40		3591.57	I
5000		3600.60	II
10000		3600.64	II
600	c	3639.80	II
100	c	3640.20	II
350	c	3647.56	II
1000	c	3662.74	II
900	c	3663.81	II
350		3666.72	II
300		3675.66	II
2500	c	3699.58	II
350		3746.33	II
3500		3796.81	II
2500		3801.90	II
1000		3826.66	II
450		3860.74	II
250		3907.29	II
500		3922.20	II
2500	1	3926.44	II
25000		3940.51	II
1000	c	3978.15	II
1700		4029.49	II
2500	c	4083.88	II
2000		4104.28	II
1700	c	4136.11	II
3500		4193.08	II
1000		4201.80	I
500		4215.53	II
90000		4244.40	II
500		4266.58	II
250	c	4270.25	II
15000		4273.14	II
2500	c	4287.97	II
1500		4293.97	II
500	c	4306.26	II
1000		4346.96	II
2500		4377.12	II
300		4440.10	II
1000		4469.47	II
400	c	4493.92	II
700		4519.04	II
3000		4530.34	II
500	1	4533.79	II
400		4540.74	II
20000		4571.77	II
3000	c	4622.42	II
350	c	4631.89	II
10000		4648.57	II
500		4659.28	II
1000		4730.45	II
1000		4755.30	II
400	c	4757.82	II
30000		4775.95	II
5000	c	4782.83	II
300	c	4855.34	II
1500	c	4885.59	II
2		5087.987	I
2		5132.471	I
10		5150.134	I
10000		5152.08	II
300		5164.58	II
1		5165.023	I
2		5165.142	I
1		5169.65	I
15		5195.278	I
2		5233.968	I
20		5260.034	I
1		5260.228	I
200		5270.51	II
3		5322.380	I
40		5362.601	I
4		5390.568	I
75		5431.532	I
3		5431.830	I
500		5512.55	II
5000		5522.78	II
6		5578.788	I
5000	c	5635.99	II
40		5647.774	I
20		5653.750	I
3000	d	5699.15	II
60		5724.121	I
3		5724.614	I
200		5739.64	II
75		6070.755	I
200		6135.27	II
30	c	6159.626	I
1000	c	6199.08	II
75	c	6206.309	I
300		6269.40	II
120	c	6298.325	I
5		6299.224	I
10000		6458.33	II
1000		6555.62	II
5000		6560.81	II
3000	1	6775.07	II
100	1	7279.997	I
300	c	7316.52	II
150		7408.173	I
200	1	7618.933	I
300		7757.651	I
60		7759.436	I
90000	c	7800.27	I
5	1	7925.26	I
4		7925.54	I
45000	c	7947.60	I
40	1	8271.41	I
30		8271.71	I
2000		8603.96	II
40	1	8868.512	I
30		8868.852	I
200		9021.77	II
3		9224.64	I
2		9234.25	I
500	c	9246.41	II
300		9338.87	II
200	w	9373.50	II
300		9391.36	II
1000		9479.32	II
700	1	9493.72	II
30	1	9522.65	I
5		9523.05	I
20	1	9540.18	I
300		9612.99	II
300		9671.54	II
2000	c	9689.05	II
200		9776.06	II
200		9934.76	II
35	1	10075.282	I
30	1	10075.708	I
100		13235.17	I
20		13442.81	I
30		13443.57	I
75		13665.01	I
1000		14752.41	I
800		15288.43	I
150		15289.48	I
20		22529.65	I
10		22932.47	I
4		27314.31	I
2		27905.37	I

Rb III
Ref. 258, 262 — J.R.

Intensity		Wavelength	Spectrum
		Vacuum	
30		465.85	III
35	p	482.43	III
30	p	482.47	III
500		482.83	III
300		484.84	III
500		489.66	III
100		489.96	III
600		493.48	III
50		497.82	III
100		500.28	III
30		508.33	III
400		516.79	III
800		533.64	III
1200		535.86	III
1200		556.19	III
500		558.36	III

Rubidium (Cont.)

Intensity		Wavelength	Spectrum
700		564.77	III
1500		566.71	III
1000		572.82	III
1500		576.65	III
2500		579.63	III
1500		581.26	III
500		582.34	III
800		586.77	III
100		591.42	III
900		593.65	III
1000		594.94	III
1300		595.88	III
1200		598.49	III
450		602.09	III
50		605.51	III
500		607.28	III
400		613.31	III
500		619.67	III
20		620.83	III
100		622.24	III
250		630.06	III
500		645.67	III
20		674.81	III
5000		769.04	III
2500		815.28	III

Air

Intensity		Wavelength	Spectrum
100		2153.21	III
250		2164.59	III
100		2268.00	III
150		2300.12	III
500		2304.14	III
150		2304.45	III
250		2312.46	III
200		2337.07	III
100		2341.90	III
200		2345.37	III
100		2349.81	III
150		2380.44	III
100		2381.29	III
150		2418.46	III
300		2561.86	III
100		2573.71	III
100		2577.07	III
200		2586.83	III
1000		2631.75	III
350		2636.83	III
100		2656.68	III
100		2713.86	III
500		2798.86	III
150		2800.27	III
500		2807.58	III
100		2845.44	III
150		2869.77	III
500		2903.69	III
150		2949.62	III
100		2951.01	III
2000		2956.07	III
500	l	2967.45	III
150		2968.13	III
500		2970.74	III
250		2987.40	III
350		3023.61	III
200		3039.62	III
200		3041.48	III
250		3070.70	III
500		3086.84	III
100		3098.49	III
500		3111.36	III
250	s	3114.82	III
120		3118.92	III
100		3169.34	III
200		3222.60	III
500		3286.41	III
100		3330.16	III
200		3346.92	III
250		3439.26	III
100		3492.68	III

Rb IV
Ref. 109 — J.R.

Intensity	Wavelength	Spectrum
	Vacuum	
10	595.18	IV
25	663.76	IV
25	716.24	IV
20	733.41	IV
50	740.85	IV
20	749.86	IV
20	753.75	IV
10	771.54	IV
25	776.89	IV
9	817.92	IV
15	850.18	IV
10	988.00	IV

RUTHENIUM (Ru)
Z = 44

Ru I and II
Ref. 1 — C.H.C.

Intensity		Wavelength	Spectrum
		Air	
2400		2076.43	I
2600		2083.77	I
2400		2090.89	I
690		2255.52	I
290		2259.53	I
780		2272.09	I
240		2278.19	I
780		2279.57	I
170		2285.38	I
290		2302.54	I
480		2317.80	I
150		2322.01	I
120		2334.96	II
240		2340.69	I
190	h	2342.85	II
190		2349.34	I
310		2351.33	I
170		2357.91	II
140		2360.56	I
170		2370.17	I
240		2375.27	I
80		2375.63	II
160		2392.42	I
95		2396.71	II
780		2402.72	II
150		2407.92	II
55		2410.89	I
55		2414.82	II
130		2420.82	I
55		2422.92	I
45		2429.60	I
65		2432.93	I
30		2447.45	I
30		2450.58	I
65		2454.92	I
180		2455.53	II
150		2456.44	II
370		2456.57	II
65	h	2458.62	I
55		2462.94	I
85		2464.70	I
30		2474.04	I
110		2475.41	I
100		2476.88	I
280		2478.93	II
28		2481.11	II
30		2489.91	I
18		2491.78	I
65		2493.69	II
85		2494.02	I
45		2494.48	II
85		2495.69	II
65		2496.56	I
140		2498.42	II
140		2498.57	II
85		2499.78	I
260		2507.01	II
130		2508.27	I
110		2509.07	I
110		2512.81	I
110		2513.32	II
110		2517.32	II
150		2535.59	II
65		2543.25	II
280		2544.22	I
120		2546.67	I
280		2549.48	I
550		2549.58	I
130		2560.26	I
120		2560.83	I
110		2563.15	I
160		2568.77	I
100		2570.97	I
100		2578.57	I
100		2579.53	I
100		2589.57	I
170		2591.12	I
120		2592.02	I
100		2593.70	I
110		2594.85	I
370		2609.06	I
830		2612.07	I
100		2615.09	I
220		2631.30	I
220		2635.86	I
170		2636.67	I
110		2640.33	I
460		2642.96	I
110		2647.32	I
110		2651.29	I
330		2651.84	I
28		2656.25	II
400		2659.62	I
23		2661.17	II
330		2661.61	II
200		2664.76	I
30		2667.40	II
690		2678.76	II
220		2686.29	I
28		2687.50	II
		2688.16	II
330		2692.06	II
110		2701.34	I
110		2702.83	I
170		2709.20	I
200		2712.41	II
690		2719.52	I
130		2722.65	I
140		2725.47	II
310		2734.35	II
1800		2735.72	I
170		2739.22	I
130		2744.45	I
35		2747.97	II
75		2752.45	II
75		2752.77	II
260		2763.42	I
35		2765.44	II
90		2768.93	II
100		2778.38	II
110		2787.83	II
140		2802.81	I
35		2806.74	II
350		2810.03	I
4900		2810.55	I
350		2818.36	I
110		2822.03	I
200		2827.87	I
400		2829.16	I
130		2834.00	I
150		2840.54	I
35		2841.68	II
640		2854.07	I
180		2860.02	I
420		2861.41	I
550		2866.64	I
110		2868.31	I
1800		2874.98	I
220		2879.76	I
55		2882.12	II
130		2883.60	I
740		2886.54	I
180		2892.56	I
110		2901.94	I
140		2905.65	I
370		2908.88	I
1100		2916.26	I
150		2919.61	I
35		2927.54	II
180		2945.67	II
180		2946.99	I
370		2949.50	I
150		2954.49	I
18		2963.40	II
550		2965.16	I
170		2965.55	II
140		2976.59	II
550		2976.92	I
45		2977.23	II
75		2979.96	II
1400		2988.95	I
35		2991.62	II
110		2993.27	I
460		2994.96	I
440		3006.59	I
330		3017.24	I
310		3020.88	I
240		3033.45	I
200		3040.31	I
220		3042.48	I
110		3045.71	I
110		3048.78	I
150		3054.94	I
390		3064.84	I
170		3089.14	I
120		3089.80	I
330		3096.57	I
120		3097.60	I
830		3099.28	I
740		3100.84	I
120		3125.96	I
120		3153.82	I
290		3159.92	I
200		3168.52	I
60		3177.05	II
180		3186.04	I
240		3188.34	I
240		3189.98	I
180		3196.59	I
180		3223.27	I
110		3226.37	I
100		3227.88	I
220		3228.53	I
220		3238.53	I
120		3241.24	I
120		3243.50	I
280		3260.35	I
120	d	3264.55	I
120		3266.44	I
200		3268.21	I
200		3273.08	I
200		3274.71	I
100		3277.57	I
490		3294.11	I
370		3301.59	I
220		3306.17	I
290		3315.23	I
290		3316.39	I
100		3325.00	I
120		3335.69	I
930		3339.55	I
240		3341.66	I
200		3361.15	I
370		3368.45	I
100		3371.86	I
130		3374.65	I
120		3378.02	I
100		3379.60	I
130		3380.18	I
130		3385.14	I
130		3388.71	I
100		3389.50	I
370		3392.54	I
310		3401.74	I
310		3409.28	I
3100		3417.35	I
4900		3428.31	I
490		3430.77	I
310		3432.74	I
6400		3436.74	I
260		3438.37	I
220		3440.20	I
260		3473.75	I
240		3481.30	I
8300		3498.94	I
640		3514.49	I
330		3519.64	I
200		3528.68	I
240		3532.81	I
390		3537.95	I
790		3539.37	I
200		3541.63	I
690		3570.59	I
200		3574.58	I
390		3587.20	I
6400		3589.22	I
6900		3593.02	I
6400		3596.18	I
1300		3599.76	I
350		3625.20	I
370		3626.74	I
3100		3634.93	I
210		3637.47	I
200		3640.64	I
290		3650.32	I
310		3654.40	I
6200		3661.35	I
830		3663.37	I
650		3669.49	I
240		3678.32	I
260		3696.59	I
410		3717.00	I
260		3719.33	I
550		3726.10	I
8700		3726.93	I
11000		3728.03	I
7100		3730.43	I
280		3737.40	I
410		3739.46	I
3500		3742.28	I
870		3742.78	I
280		3744.22	I
410		3744.40	I
2800		3745.59	I
760		3753.54	I
310		3755.09	I
870		3755.93	I
1200		3759.84	I
370		3760.03	I
600		3761.51	I
600		3767.35	I
1500		3777.59	I
460		3781.18	I
600		3782.74	I
3900		3786.06	I
6000		3790.51	I
240		3794.92	I
760		3798.05	I
7600		3798.90	I
7600		3799.35	I
310		3800.26	I
310		3808.68	I
600		3812.72	I
760		3817.27	I
760		3819.03	I
650		3822.09	I
550		3824.93	I
760		3831.80	I
220		3835.05	I

Ruthenium (Cont.)

Intensity	Wavelength		Intensity	Wavelength		Intensity	Wavelength		Intensity	Wavelength	
310	3838.07	I	120	4733.52	I	65	5814.98	I	50	921.78	III
930	3839.70	I	500	4757.84	I	8	5828.06	I	250	928.08	III
480	3846.68	I	260	4815.52	I	16	5833.21 h	I	150	937.16	III
760	3850.43	I	120	4844.56	I	55	5919.34	I	500	940.09	III
480	3856.46	I	550	4869.61	I	80	5921.45	I	250	940.68	III
1300	3857.55	I	160	4895.60	I	21	5926.87	I	50	941.85	III
220	3860.72	I	470	4903.05	I	26	5932.38	I	50	942.63	III
650	3862.69	I	120	4907.89	I	8	5936.65	I	50	943.06	III
1300	3867.84	I	260	4921.07	I	8	5951.15	I	150	945.68	III
260	3873.52	I	180	4938.43	I	21	5973.38 h	I	100	946.05	III
650	3892.21	I	160	4968.90	I	8	5974.17	I	100	947.14	III
760	3909.08	I	160	4980.35	I	16	5988.67	I	100	949.83	III
260	3920.92	I	120	4992.74	I	35	5993.65	I	50	950.35	III
1500	3923.47	I	160	5011.23	I	18	6116.77	I	100	950.45	III
3300	3925.92	I	90	5014.95	I	26	6199.42	I	50	952.59	III
600	3931.76	I	90	5026.18	I	26	6225.20	I	50	957.06	III
310	3933.55	I	65	5028.16	I	9	6284.49	I	50	957.18	III
760	3945.57	I	35	5040.35	I	18	6295.22	I	50	961.58	III
460	3950.21	I	35	5040.74	I	13	6330.62	I	250	961.68	III
310	3952.68	I	65	5047.31	I	9	6336.12	I	100	962.56	III
460	3964.90	I	430	5057.33	I	9	6363.41 h	I	500	966.54	III
600	3978.44	I	21	5062.64	I	9	6376.45	I	250	967.09	III
600	3979.42	I	90	5072.97	I	16	6390.23	I	150	967.85	III
870	3984.86	I	120	5076.32	I	8	6417.57	I	150	967.92	III
280	3995.98	I	200	5093.83	I	26	6444.84 h	I	150	971.83	III
1500	4022.16	I	80	5107.07	I	8	6496.44	I	250	972.40	III
600	4023.83	I	24	5123.73	I	11	6528.74	I	100	973.54	III
310	4039.21	I	55	5127.26	I	4	6560.45	I	150	973.78	III
1400	4051.40	I	65	5133.89	I	4	6593.74	I	750	974.14	III
710	4054.05	I	530	5136.55	I	9	6618.20	I	250	974.46	III
370	4064.46	I	170	5142.76	I	21	6663.14	I	250	977.51	III
200	4067.61	I	250	5147.24	I	55	6690.00	I	100	978.18	III
760	4068.37	I	110	5151.07	I	11	6707.52	I	900	979.43	III
200	4073.00	I	55	5153.20	I	15	6718.30	I	500	981.35	III
980	4076.73	I	500	5155.14	I	15	6730.45	I	250	983.81	III
6000	4080.60	I	55	5160.00	I	7	6756.54	I	250	983.91	III
310	4085.43	I	920	5171.03	I	21	6766.95	I	250	985.55	III
930	4097.79	I	180	5195.02	I	30	6775.02	I	900	986.84	III
350	4101.74	I	80	5199.87	I	13	6787.23	I	200	987.87	III
1900	4112.74	I	45	5202.12	I	8	6813.51	I	250	991.67	III
2000	4144.16	I	45	5213.43	I	15	6823.88	I	250	992.75	III
650	4145.74	I	65	5223.55	I	21	6824.17	I	900	994.56	III
260	4146.77	I	40	5242.38	I	7	6831.52	I	250	995.30	III
870	4167.51	I	55	5251.61	I	26	6911.48	I	200	1000.78	III
550	4197.58	I	40	5257.07	I	110	6923.23	I	300	1001.66	III
550	4198.88	I	10	5260.47	I	26	6982.01	I	250	1004.29	III
7600	4199.90	I	40	5266.83	I	26	7027.98	I	500	1009.13	III
1500	4206.02	I	40	5280.82	I	9	7086.06	I	900	1009.87	III
5400	4212.06	I	130	5284.08	I	12	7087.35	I	500	1014.68	III
760	4214.44	I	40	5291.16	I	4	7141.72	I	100	1018.72	III
930	4217.27	I	80	5304.86	I	6	7219.26	I	100	1019.33	III
370	4220.68	I	260	5309.27	I	35	7238.92	I	100	1020.77	III
550	4230.31	I	13	5315.33	I	7	7266.96	I	200	1080.00	III
760	4241.05	I	40	5332.93	I	8	7323.56	I	100	1184.37	III
760	4243.06	I	45	5334.70 h	I	16	7393.93	I	800	1190.51	III
370	4246.73	I	110	5335.93	I	18	7468.91	I	500	1200.07	III
310	4258.99	I	130	5361.77	I	12	7475.40	I	100	1204.57	III
760	4284.33	I	65	5377.84	I	26	7485.79	I	200	1204.88	III
220	4293.28	I	65	5385.88	I	70	7499.75	I	500	1207.17	III
260	4294.79	I	110	5401.04 h	I	7	7532.07	I	500	1209.77	III
550	4295.93	I	40	5401.39	I	26	7559.61	I	300	1211.31	III
3700	4297.71	I	40	5418.86	I	5	7612.94	I	200	1232.57	III
930	4307.60	I	55	5427.59	I	18	7621.50	I	100	1653.77 h	III
370	4318.43	I	26	5439.21 l	I	18	7722.87	I	200	1699.84	III
550	4319.87	I	13	5452.71	I	5	7729.91	I	100	1715.97 h	III
550	4342.07	I	80	5454.82 h	I	22	7791.86	I	200	1759.49	III
350	4349.70	I	90	5456.13	I	4	7797.89	I	200	1880.95 h	III
710	4354.13	I	13	5475.18 h	I	4	7806.82	I	100	1883.56	III
870	4361.21	I	55	5479.40	I	3	7813.43	I	200	1899.04 h	III
2400	4372.21	I	26	5480.30	I	4	7829.81	I	100	1899.42 h	III
870	4385.39	I	80	5484.32	I	5	7833.39 h	I	100	1908.31	III
1300	4385.65	I	18	5484.64	I	6	7841.90 h	I	500	1941.35	III
1700	4390.44	I	26	5496.69	I	30	7847.80	I	100	1981.82	III
1600	4410.03	I	13	5501.02	I	80	7881.49	I	100	1982.10	III
160	4421.46	I	130	5510.71	I	16	7890.37	I	200	1989.22	III
330	4428.46	I	20	5512.37	I	16	7924.43	I	200	1993.32	III
460	4439.76	I	8	5517.86	I	5	7948.15	I	100	1997.55	III
440	4449.34	I	12	5521.78	I	9	7967.84	I			
1100	4460.04	I	12	5530.99	I	9	8112.47	I		Air	
190	4473.93	I	24	5540.66	I	18	8264.96	I			
150	4480.45	I	12	5556.52	I	11	8348.98	I	200	2005.71	III
350	4498.14	I	90	5559.75	I	6	8352.94	I	100	2006.46	III
120	4510.10	I	11	5569.03	I	4	8435.77	I	500	2009.28	III
220	4516.89	I	21	5578.40	I	11	8473.64	I	100	2011.17	III
220	4517.82	I	21	5603.14	I	11	8483.56	I	50	2011.56	III
110	4520.95	I	8	5603.55	I	22	8710.84	I	50	2011.66	III
170	4547.33	I	13	5606.73	I	14	8724.98	I	100	2015.20	III
110	4547.85	I	11	5629.79	I	9	8777.36	I	50	2018.58	III
5400	4554.51	I	290	5636.24	I				100	2044.59	III
110	4559.98	I	11	5641.66	I						
1700	4584.44	I	7	5649.56	I		Ru III				
110	4591.10	I	7	5653.30	I		Ref. 423 — C.H.C.				
150	4592.52	I	11	5665.20	I						
330	4599.08	I	16	5679.63	I	Intensity	Wavelength				
170	4635.69	I	180	5699.05	I						
200	4645.09	I	13	5724.82	I		Vacuum				
720	4647.61	I	13	5725.73	I						
290	4654.32	I	16	5745.99	I	250	850.09	III			
290	4681.79	I	16	5747.47	I	200	850.30	III			
190	4684.02	I	11	5752.49	I	50	851.22	III			
290	4690.11	I	11	5756.83	I	50	852.49	III			
1400	4709.48	I	11	5767.92	I	150	856.32	III			
140	4731.33	I	16	5804.39	I	50	867.48	III			
						250	919.74	III			

SAMARIUM (Sm)
Z = 62

Sm I and II
Ref. 1 — C.H.C.

Intensity	Wavelength
	Air
45	2610.07
90	2640.27

Samarium (Cont.)

Int	λ	c	Sp	Int	λ	c	Sp	Int	λ	c	Sp	Int	λ	c	Sp
35	2649.17			430	3264.94		II	3400	3609.49		II	450	3881.38		II
45	2657.68			180	3270.49		II	240	3620.58		II	450	3881.79		II
70	2662.42			180	3270.68		II	1700	3621.23		II	320	3882.50		II
120	2675.15			430	3272.48	d	II	240	3623.32		II	3700	3885.29		II
100	2688.60				3272.60		II	850	3627.01		II	660	3889.16		II
45	2690.90		II	430	3272.81		II	850	3631.13		II		3889.22		II
130	2693.34			430	3273.48		II	3400	3634.29		II	610	3890.08		II
45	2693.74			430	3276.75		II	240	3634.93		II	320	3391.21		II
60	2696.08			270	3280.84		II	410	3638.77		II	400	3894.05		II
85	2707.96			180	3285.66		II	360	3645.29		II	1600	3896.98		II
50	2732.42			430	3286.23		II	300	3645.39		II	1300	3903.42		II
35	2739.87			720	3290.28	d	II	660	3649.53		II	620	3917.44		II
29	2762.28				3290.39		II	340	3650.19		II	2500	3922.40		II
35	2764.18			180	3290.65		II	340	3656.22		II	1900	3928.28		II
85	2767.85		II	240	3293.37		II	2200	3661.36		II	470	3935.76		II
60	2774.77			360	3295.44		II	220	3662.69		II	1300	3941.87		II
85	2776.11			430	3295.81		II	340	3667.93		II	620	3943.24		II
85	2779.23		II	720	3298.10		II	340	3670.66		II	500	3946.51		II
85	2786.64			170	3300.98		II	2200	3670.84		II	740	3948.11		II
150	2789.38		II	340	3301.68		II	340	3677.79		II	470	3951.89		I
130	2796.70	h		340	3304.52		II	270	3681.73		II	370	3959.53		II
85	2807.36			340	3305.18		II	270	3688.42		II	1500	3963.00		II
150	2809.50			1700	3306.39		II	270	3692.22		II	620	3966.04		II
120	2810.86		II	170	3306.61		II	1100	3693.99		II	470	3967.68		II
85	2817.20		II	850	3307.02		II	480	3706.75		II	740	3970.53		II
29	2820.96		II	340	3309.52		II	480	3706.98		II	1500	3971.40		II
220	2830.94			850	3310.66		II	480	3708.41		II	620	3974.66		I
60	2840.30			600	3312.42		II	930	3708.65		II	960	3976.27		II
60	2847.49		II	410	3316.58		II	480	3711.54		II	1000	3976.43		II
60	2851.35			430	3320.16		II	350	3712.76		II	960	3979.20		II
120	2866.09		II	110	3320.59		II	930	3718.88		II	740	3983.14		II
70	2868.40		II	1200	3321.18		II	930	3721.85		II	740	3986.68		II
70	2881.34			340	3323.77		II	420	3724.90		II	370	3987.43		II
85	2881.68			340	3325.26		II	1600	3728.47		II	1500	3990.00		II
60	2883.09			170	3325.48		II	2100	3731.26		II		3990.02		I
45	2889.06			340	3327.88		II	1600	3735.98		II	740	3993.31		II
60	2891.34			170	3333.64		II	800	3737.14		II	280	4003.46		II
85	2907.88	d		170	3336.12		II	320	3737.48		II	470	4007.48		II
	2907.99		II	850	3340.58		II	2900	3739.12		II	280	4019.98		II
130	2910.28		II	240	3343.49		II		3739.20		II	880	4023.23		II
85	2937.48		II	110	3343.64		II	800	3741.29		II	740	4035.11		II
70	2943.49		II	240	3344.35		II	1200	3743.87		II	590	4041.68		II
150	2953.19		II	170	3347.30		II	930	3745.46		I	740	4042.72		II
85	2962.74		II	240	3348.68		II		3745.60		II	880	4042.90		II
160	2969.02		II	220	3350.88		II	480	3747.62		II	240	4044.11		II
100	2983.43		II	410	3354.18	d	II	800	3755.28		II	560	4045.05		II
60	2991.57		II		3354.30		II	800	3756.41		I	440	4046.16		II
100	3021.01			170	3354.72		II	1200	3757.53		II	740	4047.16		II
150	3034.84		II	1200	3365.86		II	450	3758.45		II	210	4048.62		II
100	3039.13		II	150	3367.27		II	660	3758.97		II	590	4049.81		II
120	3046.93		II	340	3368.57		II	350	3760.04		II	440	4058.87		II
150	3067.54		II	340	3369.46		II	1900	3760.69		II	560	4063.54		II
120	3071.29		II	170	3370.59		II	660	3762.59		II	280	4064.32		II
100	3086.45		II	340	3371.21		II	1100	3764.37		II	1400	4064.58		II
120	3096.88		II	150	3376.48		II	480	3767.36		II	810	4066.74		II
100	3102.30	h	II	1200	3382.40		II	480	3767.76		II	710	4068.33		II
250	3106.52		II	510	3384.66		II	370	3773.33	d	I	810	4075.84		II
220	3110.20		II	150	3384.86				3773.42		II	280	4076.65		II
200	3117.72		II	150	3387.66		II	1100	3778.14		II	240	4080.56		II
270	3136.30		II	410	3389.32		II	660	3780.76		II	410	4082.60		II
150	3139.97		II	150	3391.11		II	420	3780.93		II	280	4083.58		II
150	3147.19		II	410	3396.19		II	320	3787.20		II	220	4084.40		II
180	3152.10		II	150	3397.76		II	1500	3788.12		II	1000	4092.27		II
410	3152.52		II	150	3399.84		II	1600	3793.97		II	290	4094.05		II
150	3162.30		II	600	3402.46		II	420	3797.28		II	240	4104.13		II
360	3169.88		II	210	3403.09		II	1600	3797.73		II	810	4107.28		II
180	3178.12		II	850	3408.68		II	500	3799.54		II		4107.39		II
720	3183.92		II	270	3418.15		II	800	3800.89		II	410	4109.40		II
310	3187.01		II	430	3418.51		II	320	3805.63		II	280	4110.19		II
430	3187.22		II	170	3419.77		II	420	3808.46		II	410	4113.90		II
360	3187.79		II	120	3424.78		II	320	3809.75		II	1900	4118.55		II
360	3193.01		II	170	3426.20		II	320	3809.88		II	410	4121.36		II
360	3196.18		II	170	3433.68		II	420	3810.43		II	280	4122.51		II
150	3201.80		II	150	3437.10		II	500	3812.07		II	710	4123.96		II
150	3204.90		II	170	3438.06		II	480	3813.63		II	280	4129.23		II
360	3207.18		II	240	3440.50		II	420	3814.63		II	250	4135.14		II
180	3208.17		II	170	3453.56		II	930	3820.82	d		320	4147.71		II
600	3211.73		II	170	3459.20		II	530	3824.18		II	810	4149.83		II
150	3214.12		II	120	3459.42		II	1600	3826.20		II	1200	4152.21		II
270	3215.26		II	240	3461.13		II	530	3830.29		II	530	4153.33		II
530	3216.85		II	120	3464.07		II	1100	3831.50		II	560	4155.22		II
600	3218.61		II	170	3467.87		II	530	3833.83		II	810	4169.48		II
150	3219.43		II	130	3473.96		II	560	3834.48		I	410	4171.57		II
270	3226.84		II	130	3479.53		II	560	3834.60		II	440	4178.02		II
180	3228.50		II	130	3480.26		II	370	3835.72		II	530	4181.10		II
270	3228.78		II	170	3480.56		II	500	3838.94		II	210	4183.33		I
720	3230.56		II	170	3487.41		II	400	3840.45		II	530	4183.76		II
360	3231.53		II	170	3493.61		II	1600	3843.50		II	1000	4188.13		II
150	3231.95		II	220	3499.84		II	530	3847.51		II	410	4191.93		II
430	3233.68		II	340	3511.23		II	640	3848.78		II	270	4199.45		II
720	3236.64		II	310	3530.60		II	420	3851.88		II	650	4202.92		II
150	3237.89		II	220	3532.57		II	530	3853.30		I	1100	4203.05		II
720	3239.66		II	270	3535.65		II	2700	3854.21		II	660	4206.13		II
530	3241.16		II	240	3554.15		II	480	3854.56		I	270	4206.62		II
180	3241.59		II	510	3559.10		II	800	3855.90		II	660	4210.35		II
180	3242.04		II	220	3566.64		II	480	3857.91		II	740	4220.66		II
150	3244.69		II	4200	3568.27		II	400	3858.74		I	1000	4225.33		II
240	3249.75		II	270	3577.79		II	660	3862.05		II	740	4229.70		II
720	3250.37		II	390	3580.94		II	350	3862.23		II	620	4234.57		II
360	3253.40		II	310	3583.39		II	320	3865.24		II	1200	4236.74		II
270	3253.94		II	4200	3592.60		II	800	3871.78		II	500	4237.66		II
850	3254.38		II	340	3601.69		II	400	3875.19		II	620	4244.70		II
110	3255.63		II	1700	3604.28		II	560	3875.54		II	210	4249.55		II
360	3262.28		II					800	3880.77		II	250	4251.78		II

Samarium (Cont.)

Intensity	Wavelength		Spectrum
2100	4256.39		II
210	4258.58		II
1300	4262.68		II
500	4265.08		II
1200	4279.68		II
	4279.75		II
240	4279.94		II
2200	4280.79		II
710	4282.21		I
470	4282.83		II
240	4283.50		I
350	4286.64		II
350	4292.18		II
1600	4296.74		I
320	4304.94		II
880	4309.01		II
240	4312.85		I
1900	4318.94		II
470	4319.53		I
590	4323.28		II
240	4324.46		I
1800	4329.02		II
440	4330.02		I
1300	4334.15		II
880	4336.14		I
560	4345.86		II
1100	4347.80		II
560	4350.46		II
560	4352.10		II
560	4360.72		II
220	4361.07		II
810	4362.04		II
440	4362.91		I
220	4363.45		II
500	4368.03		II
210	4369.92		II
440	4373.46		II
320	4374.98		II
880	4378.24		II
530	4380.42		I
290	4384.29		II
1600	4390.86		II
210	4393.35		I
290	4397.34		I
410	4401.17		I
810	4403.06	d	II
	4403.13		I
410	4403.36		II
520	4409.33		II
290	4411.58		I
380	4417.58		II
470	4419.33		I
1500	4420.53		II
960	4421.14		II
2900	4424.34		II
470	4429.66		I
1600	4433.88		II
1800	4434.32		II
530	4441.81		I
440	4442.28		I
710	4444.26		II
710	4445.15		I
1300	4452.73		II
250	4452.95		I
1200	4454.63		II
1000	4458.52		II
250	4459.29		I
2200	4467.34		II
810	4470.89		I
470	4472.43		II
620	4473.02		II
740	4478.66		II
370	4499.11		I
370	4499.48		II
240	4503.38		I
180	4505.05		II
120	4511.33		I
560	4511.83		II
440	4515.09		II
880	4519.63		II
440	4523.04		II
	4523.18		I
650	4523.91		II
290	4533.80		I
270	4536.51		II
710	4537.95		II
150	4538.53		II
290	4540.19		II
380	4542.06		II
810	4543.95		II
100	4544.83		II
410	4552.66		II
270	4554.45		II
240	4560.43		II
470	4566.21		II
590	4577.69		II
290	4581.58		I
440	4581.73		I
560	4584.83		II
290	4591.82		II
380	4593.54		II
560	4595.29		II
240	4596.74		I
220	4604.18		II
290	4606.51		II
290	4615.44		II
470	4615.69		II
150	4630.21		II
880	4642.24		II
290	4645.40		I
290	4646.68		II
240	4648.16		II
380	4649.49		I
150	4655.13		II
290	4663.56		I
740	4669.40		II
620	4669.65		II
470	4670.75	d	II
	4670.83		I
1100	4674.60		II
680	4676.91		II
210	4681.55		I
370	4687.18		II
370	4688.73		I
130	4693.02	h	II
120	4699.34		II
530	4704.00		II
270	4713.06		II
130	4715.26		II
730	4716.10		I
270	4717.07		I
210	4717.72		II
190	4718.33	h	II
270	4719.84		II
130	4726.02		II
770	4728.42		I
470	4745.68		II
150	4750.72		I
730	4760.27		I
110	4770.20		I
110	4774.15		II
190	4777.85		II
580	4783.10		II
350	4785.86		I
160	4789.96		I
230	4791.58		II
430	4815.81		II
130	4829.57		II
970	4841.78		I
310	4844.21		II
140	4847.76		II
270	4848.32		I
120	4854.36		II
210	4883.77		I
730	4883.97		I
170	4904.97		I
630	4910.40		I
350	4913.25		II
430	4918.99		I
110	4924.04		I
120	4938.10		II
170	4948.63		II
120	4952.37		II
170	4961.94		II
170	4975.98		II
140	5028.44		II
400	5044.28		I
200	5052.76		II
170	5069.46		II
540	5071.20		I
170	5100.22		II
	5100.39		I
260	5103.09		II
140	5104.48		II
140	5116.70		II
510	5117.16		I
350	5122.14		I
360	5155.03		II
250	5172.74		I
470	5175.42		II
250	5200.59		I
260	5251.92		I
400	5271.40		I
250	5282.91		I
190	5320.60		I
110	5341.29		I
140	5368.36		I
130	5405.23		I
220	5453.00		I
140	5466.72		I
230	5493.72		I
80	5512.10		I
230	5516.09		I
50	5548.95		I
140	5550.40		I
45	5573.42		I
35	5588.20		I
50	5600.86		II
50	5621.79		I
70	5626.01		I
85	5644.10		I
140	5659.86		I
120	5696.73		I
85	5706.20		I
35	5710.93		I
50	5732.95		I
50	5743.35		II
45	5759.52		I
70	5773.77		I
60	5778.33		I
45	5779.24		I
45	5781.93		II
70	5786.98	d	II
60	5788.38		I
60	5800.52		I
65	5802.84		I
45	5814.89		I
45	5831.02		II
45	5836.37		II
35	5860.78		I
65	5867.79		I
45	5868.61		I
35	5871.06		I
50	5874.60		I
45	5897.39		II
50	5898.96		I
35	5938.90		II
65	5965.71		II
35	5968.02	h	II
35	5984.29		I
50	6045.00		I
45	6045.39		I
50	6070.06		I
45	6084.12		I
35	6091.40	h	I
45	6110.66		II
45	6159.56	h	I
45	6246.76		II
45	6256.54		I
45	6256.66		II
100	6267.28		II
50	6291.82		II
35	6307.06		II
70	6327.47		II
45	6426.64		II
45	6472.34		II
35	6484.52		II
35	6498.67		II
50	6542.76		II
140	6569.51		II
35	6570.67	h	II
40	6585.21	h	II
110	6600.70		II
40	6601.83		II
95	6604.56		II
40	6632.28	h	II
50	6671.51		I
70	6679.21		II
70	6693.55		II
40	6723.07	d	I
120	6731.84	d	II
70	6734.06	d	II
40	6734.81	d	II
55	6741.47		II
40	6778.61	h	II
60	6790.00		II
95	6794.20		II
55	6844.71		II
75	6856.07		II
120	6860.93		I
40	6862.82		II
30	6950.51		II
120	6955.29		II
90	7020.00		II
13	7036.73		II
90	7039.22		II
90	7042.24		II
90	7049.15		II
90	7051.52		II
16	7054.97		II
19	7074.67		I
90	7082.37		II
40	7085.52	d	II
26	7088.30		I
16	7091.16		I
30	7095.50		I
16	7096.33		I
30	7104.54		I
19	7106.23		I
26	7115.96		I
23	7117.51		II
26	7119.81	h	II
12	7122.40		II
23	7125.11	h	II
13	7131.80		I
10	7136.01		I
12	7139.39		I
40	7143.98	d	II
85	7149.60	d	II
10	7172.67		I
10	7189.57		II
9	7210.95		I
23	7213.82		I
26	7218.09	d	II
13	7220.07		I
13	7237.02		II
60	7240.90		II
9	7257.11		II
9	7261.52	d	II
13	7279.25		I
26	7281.47		II
8	7282.21		I
19	7283.33		II
16	7288.92		II
13	7290.23		I
26	7300.72	h	II
13	7327.08		II
13	7332.65		I
8	7338.04		I
26	7347.30		I
26	7376.69		II
13	7393.98		II
30	7444.56		I
26	7445.41		I
26	7453.03	d	II
13	7470.76		I
26	7481.99		II
23	7502.39	h	II
10	7517.00	h	II
23	7541.42	h	II
9	7544.74		I
10	7546.57		I
12	7560.03	h	II
19	7562.94		II
23	7570.95		II
23	7572.29		II
19	7578.09		II
30	7585.85		II
23	7588.31		II
10	7598.01		I
23	7607.48	d	II
	7607.74		I
12	7613.94		II
10	7631.77	h	II
23	7637.94		II
45	7645.09		II
12	7645.82		I
19	7648.02		II
10	7655.78		II
19	7667.20		II
8	7672.49		II
10	7678.79	h	II
10	7695.78	h	I
23	7712.04		II
30	7728.56		II
30	7736.76		II
30	7749.30		II
23	7755.20		II
10	7794.50		I
10	7801.54		I
8	7812.75	h	II
16	7820.15		II
10	7831.40		II
40	7835.08	w	II
26	7837.27		II
10	7844.82		II
6	7859.53		I
19	7863.65		II
10	7880.01	h	II
16	7895.96		I
26	7914.96		II
90	7928.14		II
9	7931.92		I
19	7937.09		II
16	7948.12		II
19	8001.61	w	II
19	8014.92	w	II
23	8025.12		II
23	8026.32	w	II
16	8032.03		II
40	8048.70		II
16	8065.16		I
45	8068.46		II
9	8117.16	w	II
9	8125.12		II
26	8161.82		II
19	8195.50	w	II
6	8206.30		II
26	8218.76	w	II
9	8230.33		I
16	8240.98		II
19	8289.26	w	II
10	8300.88		II
40	8305.79	w	II
10	8315.45		I
19	8348.68	w	II
19	8383.71		I
19	8387.77		II
30	8432.64	w	II
19	8473.54	w	II
45	8485.99	w	II
30	8510.90	w	II
23	8543.22		II
23	8617.03	w	II
23	8632.82	w	II
12	8677.81	w	II
13	8706.32		II
45	8708.44	w	II
30	8717.89	w	II
30	8758.28	w	II
16	8780.59	w	II
23	8788.83	w	II
26	8859.76		II
95	8913.66		II

SCANDIUM (Sc)
Z = 21

Sc I and II
Ref. 1, 88 — C.H.C.

Intensity	Notes	Wavelength (Air)	Spectrum
65		2429.16	I
110		2438.62	I
560		2545.22	II
2900		2552.37	II
560		2555.82	II
2300		2560.25	II
1100		2563.21	II
40		2611.22	II
19		2684.23	II
120		2692.78	I
360		2706.77	I
210		2707.95	I
580		2711.35	I
30		2819.54	II
35		2822.15	II
60		2826.68	II
340		2965.86	I
1200		2974.01	I
1400		2980.75	I
340		2988.95	I
2200		3015.36	I
2700		3019.34	I
360		3030.76	I
30		3039.93	II
70		3045.72	II
85		3052.93	II
120	h	3056.31	I
130		3065.11	II
45		3139.75	II
990		3251.32	II
1500		3255.69	I
4400		3269.91	I
5500		3273.63	I
110	d	3343.28	II
270		3352.05	II
9900		3353.73	II
65	d	3357.30	II
2000		3359.68	II
1700		3361.27	II
1700		3361.94	II
4000		3368.95	II
6600		3372.15	II
90		3416.68	I
130		3418.51	I
65		3419.36	I
200		3429.21	I
200		3429.48	I
270		3431.36	I
530		3435.56	I
90		3439.41	I
65		3440.18	I
65		3448.49	I
270		3457.45	I
180		3462.19	I
130	d	3469.65	I
110		3471.13	I
200		3498.91	I
2700		3535.73	II
6600		3558.55	II
6100		3567.70	II
13000		3572.53	II
9900		3576.35	II
7700		3580.94	II
4000		3589.64	II
4000		3590.48	II
28000		3613.84	II
110		3617.43	I
20000		3630.75	II
13000		3642.79	II
6600		3645.31	II
110		3646.90	I
5300		3651.80	II
110		3664.25	II
290		3666.54	II
55		3675.26	II
40		3678.35	II
75	h	3717.10	I
270		3833.07	II
610		3843.03	II
90		3894.97	I
20000		3907.49	I
23000		3911.81	I
45		3923.51	II
4400		3933.38	I
45		3952.27	II
45		3989.06	II
5500		3996.61	I
530		4014.49	II
20000		4020.40	I
20000		4023.69	I
220		4030.67	I
140		4031.39	I
100		4034.23	I
220		4043.80	I
200		4046.48	I
2700		4047.79	I
120		4049.95	I
5500		4054.55	I
220		4056.59	I
160	h	4074.97	I
160		4078.57	I
6100		4082.40	I
200		4086.67	I
400		4087.16	I
40	h	4093.13	I
65		4094.85	I
55	h	4098.35	I
65		4100.33	I
440	h	4133.00	I
530	h	4140.30	I
65	h	4147.40	I
720		4152.36	I
55	h	4154.72	I
90	Hd	4161.88	I
1100		4165.19	I
65	h	4171.56	I
45	h	4186.45	I
65	h	4187.62	I
75		4205.20	I
65		4212.34	I
45		4212.49	I
75	h	4216.10	I
110	h	4218.26	I
110	h	4219.73	I
40		4221.88	I
90	d	4225.59	I
180		4231.93	I
200		4233.61	I
100		4237.82	I
400		4238.05	I
90		4239.57	I
100		4246.12	I
15000		4246.83	II
55		4283.56	I
290		4294.77	II
350		4305.71	II
4200		4314.09	II
3300		4320.74	II
2400		4325.01	II
28		4348.53	I
180		4354.61	II
110		4358.64	I
55		4359.08	I
28		4364.92	I
2000		4374.46	II
130		4384.81	II
45	h	4389.60	I
1100		4400.37	II
880		4415.56	II
28		4420.66	II
45		4431.36	II
65		4542.55	I
90		4544.68	I
120	h	4557.24	I
160	h	4573.99	I
65	h	4592.94	I
65	h	4598.45	I
55	h	4604.72	I
45		4609.53	I
45		4609.95	I
350		4670.40	II
40	h	4680.49	I
50		4698.29	II
120		4706.97	I
120		4709.34	I
200		4728.77	I
490		4729.23	I
40	h	4732.30	I
590		4734.10	I
60		4735.08	
690		4737.65	
790		4741.02	I
1200		4743.81	I
200		4753.16	I
220		4779.35	I
90		4791.50	I
100		4827.28	I
100		4833.67	I
170		4839.44	I
40		4840.47	I
80		4847.68	I
80		4852.68	I
140	BLd	4857.79	S
		4858.09	SCO
80		4906.67	I
90		4909.76	I
90		4922.84	I
90		4934.25	I
45		4935.74	I
70		4941.33	I
170		4954.06	I
120		4973.66	I
150		4980.37	I
80		4983.45	I
140		4991.92	I
80		5018.39	I
70		5020.14	I
80		5021.51	I
530		5031.02	II
55		5032.74	I
250		5064.32	I
80		5068.86	I
530		5070.23	I
250		5075.81	I
2100		5081.56	I
1200		5083.72	I
1100		5085.55	I
750		5086.95	I
390		5087.14	I
270		5089.89	I
45		5092.46	I
390		5096.73	I
620		5099.23	I
370		5101.12	I
180		5109.06	I
150		5112.86	I
320		5116.69	I
70	b	5133.68	S
45	b	5171.06	S
390		5210.52	I
45		5211.28	I
280		5219.67	I
350		5239.82	II
280		5258.33	I
35		5284.97	I
210		5285.76	I
35		5301.94	I
22		5318.35	II
70		5331.77	I
14		5334.23	II
95		5339.41	I
120		5341.05	I
95		5342.96	I
350		5349.30	I
120		5349.71	I
60		5350.30	I
210		5355.75	I
530		5356.10	I
14		5357.19	II
270		5375.35	I
370		5392.08	I
45		5416.12	I
45		5425.57	I
45		5429.41	I
35		5432.94	I
55		5433.23	I
45		5438.22	I
55		5439.03	I
55	h	5442.60	I
270		5446.20	I
18		5447.39	I
120		5451.34	I
30		5455.21	I
18		5465.20	I
55		5468.40	I
60		5472.19	I
18		5474.64	I
750		5481.99	I
530		5484.62	I
570		5514.22	I
16		5515.39	I
660		5520.50	I
45		5526.06	I
660		5526.82	II
55		5541.04	I
30		5546.40	I
18		5550.40	I
5		5552.25	II
35		5553.59	I
16		5561.10	I
70		5564.86	I
18		5571.24	I
14		5579.76	I
110		5591.33	I
35	h	5593.38	I
22		5604.19	I
22		5631.02	I
80		5640.98	II
45		5646.36	I
16		5647.60	I
55		5649.56	I
250		5657.88	II
60		5658.34	II
55		5667.16	II
70		5669.04	II
1500		5671.81	I
95		5684.20	II
1200		5686.84	I
1100		5700.21	I
190		5708.61	I
880		5711.75	I
230		5717.28	I
180		5724.08	I
55	B1	5736.85	S
55	B1	5764.45	S
95	B1	5772.74	S
55	B1	5775.32	S
70	B1	5809.84	S
70	B1	5811.60	S
95	B1	5847.73	S
70	B1	5849.07	S
70	b	5887.38	S
35	B1	5918.04	S
30		5919.11	I
60	B1	5928.10	S
35		5961.49	I
60	B1	5968.25	S
35		5969.19	I
90		5988.42	I
160	B1	6017.07	S
60		6026.18	I
620	B1	6036.17	S
490	B1	6064.31	S
440	B1	6072.65	S
620	B1	6079.30	S
320	B1	6101.87	S
370	B1	6109.93	S
370	B1	6115.97	S
180	b	6148.70	S
150	b	6153.93	S
150	b	6188.09	S
150	b	6192.90	S
620		6210.68	I
90		6239.41	I
320		6239.78	S
120		6245.63	II
110		6249.96	I
250		6258.96	I
60		6262.25	I
55		6276.31	I
45		6279.76	II
18		6300.70	II
750		6305.67	I
26		6309.90	II
16		6320.85	II
26		6344.83	I
60		6378.82	I
55	B1	6408.41	S
90		6413.35	I
26	b	6437.08	S
55	B1	6446.24	S
26	b	6457.78	S
35	b	6485.40	S
26	b	6495.90	S
55	b	6525.62	S
22	b	6535.30	S
45	b	6557.84	S
35	b	6566.88	S
18	b	6575.85	S
60		6604.60	II
26	B1	6609.99	S
18	B1	6617.94	S
18	B1	6645.08	S
22	B1	6654.42	S
26	B1	6661.01	S
18	b	6700.48	S
18	b	6705.93	S
65		6737.87	I
35		6739.40	I
35		6817.08	I
50		6819.52	I
29		6829.54	I
50		6835.03	I
5		6963.12	S
5	B1	6990.68	S
5	B1	7025.72	S
8	b	7035.77	S
5	b	7072.37	S
5	b	7094.38	S
12	h	7138.14	I
14		7169.13	I
12		7257.57	I
8		7275.57	I
3	h	7300.62	I
12	h	7524.13	I
14	h	7553.96	I
15	h	7574.44	I
11		7617.45	I
14	h	7665.72	I
30		7697.73	I
18		7729.72	I
55	h	7741.17	I
5	h	7750.37	I
5		7752.72	I
6	h	7771.06	I
15		7785.17	I
8		7794.68	I
30		7800.44	I
11		7821.64	I
11	h	8196.98	I
15		8241.13	I
19	h	8761.40	I
11	h	8774.8	I
15	h	8794.72	I
15	h	8823.8	I
30	h	8834.35	I
70		20616.32	I
30		20985.81	I
400		22051.86	I
150		22065.05	I

Scandium (Cont.)

Sc III
Ref. 323 — C.H.C.

Intensity	Wavelength	
	Vacuum	
10	730.60	III
15	731.65	III
15	1148.24	III
20	1154.52	III
20	1162.44	III
25	1168.61	III
10	1168.88	III
80	1598.00	III
180	1603.06	III
150	1610.19	III
40	1895.44	III
60	1912.62	III
90	1993.89	III
	Air	
160	2010.42	III
50	2012.26	III
350	2699.07	III
230	2734.05	III
10	2831.75	III
80	4061.21	III
100	4068.66	III
40	4309.47	III
10	4740.95	III
15	4780.87	III
50	4992.89	III
60	5032.09	III
80	6256.01	III
60	6307.60	III
90	7449.16	III
70	7548.15	III
70	7868.65	III
35	8814.29	III
50	8829.78	III
30	8865.89	III
15	8881.58	III

Sc IV
Ref. 298 — C.H.C.

Intensity		Wavelength	
		Vacuum	
8		220.28	IV
15		289.85	IV
15		296.31	IV
15		299.04	IV
10		371.16	IV
9		438.80	IV
8		557.50	IV
8		584.83	IV
9		617.08	IV
8		761.43	IV
8		769.70	IV
10		785.12	IV
8		789.00	IV
8		791.71	IV
8		861.24	IV
8		861.30	IV
8		890.87	IV
8		1219.40	IV
9		1228.20	IV
9		1424.66	IV
8		1444.10	IV
8		1489.64	IV
8		1514.96	IV
8		1535.76	IV
9		1543.86	IV
9		1549.55	IV
15		1550.80	IV
8		1555.72	IV
9		1563.81	IV
10		1574.92	IV
9		1583.41	IV
8		1584.64	IV
8		1592.23	IV
8		1660.71	IV
10		1665.92	IV
8		1746.23	IV
		Air	
10		2056.06	IV
8		2078.93	IV
12		2118.97	IV
9		2164.43	IV
11		2185.43	IV
11		2205.46	IV
14		2222.22	IV
11		2271.33	IV
9		2464.45	IV
8		2520.93	IV
11		2586.93	IV
9		2595.17	IV
8		2678.01	IV
8		2723.52	IV
8		2773.04	IV
8	d	4594.42	IV
8	d	4639.96	IV
8		5501.74	IV
9		5620.72	IV
10		5706.82	IV
14		5771.63	IV
9		6548.03	IV

Sc V
Ref. 150 — C.H.C.

Intensity		Wavelength	
		Vacuum	
150		179.42	V
350		180.14	V
200		180.82	V
200		180.96	V
50		181.55	V
200		182.39	V
300		228.56	V
100		230.85	V
40		243.82	V
500		243.87	V
400		246.42	V
400		250.98	V
500		252.85	V
500		253.73	V
50		255.38	V
300		255.64	V
200		257.16	V
150		258.24	V
40		258.81	V
50		260.05	V
400		281.00	V
900		283.91	V
800		284.45	V
600		288.29	V
900		289.59	V
1000	d	291.93	V
800		293.25	V
400		296.17	V
700		300.00	V
400		375.05	V
100		378.60	V
200		388.68	V
400		395.32	V
200		399.10	V
1000		573.36	V
600		587.94	V

SELENIUM (Se)
Z = 34

Se I and II
Ref. 80, 181, 216, 275 — R.L.K.

Intensity	Wavelength	
	Vacuum	
285	828.5	II
360	832.7	II
285	906.6	II
360	912.9	II
360	1013.4	II
360	1014.0	II
450	1033.6	II
450	1049.6	II
360	1057.4	II
285	1097.8	II
360	1141.9	II
220	1156.0	II
285	1156.9	II
285	1168.5	II
450	1192.0	II
220	1205.7	II
220	1234.9	II
285	1291.0	II
285	1308.9	II
100	1405.4	I
100	1406.4	I
100	1406.6	I
120	1435.3	I
120	1435.8	I
100	1444.8	I
100	1446.8	I
100	1447.0	I
150	1449.2	I
120	1456.3	I
150	1500.9	I
250	1530.4	I
150	1531.3	I
200	1531.8	I
120	1547.1	I
150	1560.3	I
150	1575.3	I
150	1577.6	I
150	1577.9	I
150	1579.5	I
200	1580.0	I
150	1587.5	I
150	1593.2	I
250	1606.5	I
100	1610.7	I
100	1611.3	I
200	1617.4	I
150	1621.2	I
100	1622.7	I
120	1626.2	I
150	1643.4	I
250	1671.2	I
250	1675.3	I
250	1690.7	I
250	1793.3	I
300	1795.3	I
300	1855.2	I
250	1858.8	I
400	1898.6	I
350	1913.8	I
300	1919.2	I
500	1960.9	I
150	1995.1	I
	Air	
500	2039.8	I
500	2074.8	I
500	2164.2	I
150	2332.8	I
600	2413.5	I
300	2548.0	I
220	3038.7	II
220	3041.3	II
285	4070.2	II
360	4175.3	II
450	4180.9	II
120	4328.7	I
100	4330.3	I
285	4382.9	II
285	4446.0	II
220	4449.2	II
285	4467.6	II
500	4730.8	I
400	4739.0	I
300	4742.2	I
285	4804.0	II
360	4841.0	II
450	5227.5	II
360	5305.4	II
100	5365.5	I
120	5369.9	I
110	5374.1	I
285	5522.4	II
285	5566.9	II
285	5866.3	II
450	6056.0	II
200	6325.6	I
360	6444.2	II
285	6490.5	II
285	6535.0	II
150	6831.3	I
120	6990.690	I
100	6991.792	I
200	7010.809	I
150	7013.875	I
300	7062.065	I
200	7575.1	I
250	7583.4	I
150	7592.2	I
120	7606.8	I
300	8001.0	I
200	8036.4	I
120	8060.9	I
120	8065.3	I
120	8081.1	I
150	8093.2	I
150	8094.7	I
180	8149.3	I
150	8152.0	I
200	8157.7	I
180	8163.1	I
150	8182.9	I
100	8185.0	I
120	8194.6	I
150	8440.47	I
150	8450.38	I
150	8742.33	I
300	8918.86	I
100	8969.69	I
200	9001.97	I
200	9038.61	I
80	9083.14	I
120	9088.79	I
80	9140.83	I
60	9181.88	I
60	9271.12	I
100	9432.50	I
60	9825.58	I
200	10217.25	I
377	10307.45	I
900	10327.26	I
640	10386.36	I
124	10650.30	I
125	11934.56	I
275	11946.87	I
100	11947.92	I
105	11952.27	I
170	11952.64	I
100	11966.04	I
205	11972.93	I
115	11973.07	I
315	14817.93	I
410	14917.47	I
500	15151.44	I
115	15469.06	I
320	15471.00	I
265	15520.97	I
395	15618.40	I
115	15620.38	I
360	16659.44	I
505	16813.78	I
165	16817.76	I
205	16866.54	I
115	16972.71	I
235	21374.24	I
680	21442.56	I
415	21473.48	I
270	21716.36	I
240	21730.60	I
105	23133.66	I
150	23388.85	I
110	23628.17	I
265	24148.18	I
170	24159.23	I
185	24204.44	I
375	24385.99	I
160	24413.67	I
225	24471.17	I
255	25017.51	I
510	25127.43	I

Se III
Ref. 9, 247 — R.L.K.

Intensity	Wavelength	
	Vacuum	
220	700.2	III
220	709.4	III
220	720.6	III
360	724.3	III
285	726.4	III
110	737.2	III
220	741.9	III
285	777.3	III
220	790.8	III
360	843.0	III
220	879.2	III
285	953.7	III
220	954.4	III
220	954.7	III
160	974.1	III
360	974.8	III
285	1079.8	III
360	1099.1	III
450	1119.2	III
	Air	
285	2057.5	III
285	2767.2	III
220	2773.8	III
285	3379.8	III
450	3387.2	III
450	3413.9	III
285	3428.4	III
450	3457.8	III
360	3543.6	III
285	3570.2	III
450	3637.6	III
360	3711.7	III
450	3738.7	III
285	3743.0	III
450	3800.9	III
360	4046.7	III
220	4083.2	III
450	4169.1	III
220	4637.9	III
285	6303.8	III

Se IV
Ref. 245 — R.L.K.

Intensity	Wavelength	
	Vacuum	
285	636.0	IV
285	654.2	IV
360	652.7	IV
450	670.1	IV
285	671.9	IV
220	722.8	IV
285	734.6	IV
450	746.4	IV
285	759.0	IV
285	776.5	IV
285	803.8	IV

Selenium (Cont.)

Intensity		Wavelength	
360		959.6	IV
450		996.7	IV
220		1307.2	IV
285		1314.4	IV

Air

220		2090.0	IV
285		2136.6	IV
160		2165.2	IV
160		2166.6	IV
360		2665.5	IV
285		2724.3	IV
160		2951.6	IV

Se V
Ref. 245 — R.L.K.

Intensity   Wavelength

Vacuum

Intensity		Wavelength	
285		596.0	V
285		601.0	V
220		608.7	V
360		613.0	V
285		614.3	V
450		759.1	V
285		785.8	V
285		804.3	V
360		808.7	V
220		814.8	V
220		820.7	V
360		830.3	V
450		839.5	V
360		845.8	V
360		1094.7	V
220		1151.0	V
450		1227.6	V

SILICON (Si)
Z = 14

Si I and II
Ref. 170, 237, 292 — L.J.R.

Intensity   Wavelength

Vacuum

Intensity		Wavelength	
10	h	805.10	II
20	h	820.52	II
20	h	843.72	II
40	h	845.77	II
10		850.14	II
100		889.72	II
200		892.00	II
10		899.41	II
20		901.74	II
10		913.01	II
20		913.85	II
20		929.81	II
100		989.87	II
200		992.68	II
25		1020.70	II
50		1023.69	II
30		1057.05	II
15		1057.50	II
20	h	1127.44	II
40	h	1127.91	II
100		1190.42	II
200		1193.28	II
250		1194.50	II
100		1197.39	II
10	h	1216.12	II
20		1223.91	II
20		1224.25	II
10		1224.97	II
50		1226.81	II
20		1226.89	II
40		1226.99	II
100		1227.60	II
10		1228.44	II
25		1228.62	II
150		1228.75	II
200		1229.39	II
10		1235.92	II
100		1246.74	II
150		1248.43	II
100		1250.09	II
150		1250.43	II
200		1251.16	II
10		1255.28	I
40		1256.49	I
50		1258.80	I
1000		1260.42	II
2000		1264.73	II
200		1265.02	II
100		1304.37	II
50	h	1305.59	II
200		1309.27	II
20	h	1309.46	II
100		1346.87	II
100		1348.54	II
150		1350.06	II
20		1350.52	II
20		1350.66	II
100		1352.64	II
100		1353.72	II
10	h	1409.07	II
20	h	1410.22	II
10	h	1416.97	II
15	h	1474.65	II
15		1484.87	II
90	h	1485.02	II
30		1485.22	II
100	h	1485.51	II
100	h	1509.10	II
50	h	1512.07	II
30	p	1513.57	II
60	p	1516.91	II
500		1526.72	II
1000		1533.45	II
10		1562.45	II
15		1562.85	II
10		1563.77	II
50		1573.87	I
50		1574.82	I
50		1592.41	I
150		1594.55	I
50		1594.93	I
30		1597.95	I
100		1622.87	I
30		1625.71	I
300		1629.43	I
200		1629.92	I
75		1631.13	I
30		1633.98	I
30	h	1653.35	I
30		1664.52	I
50		1666.37	I
100		1667.62	I
100		1668.52	I
100		1672.59	I
200		1675.20	I
30		1682.68	I
30		1686.82	I
50	h	1689.29	I
30	h	1690.79	I
50		1693.29	I
50		1695.51	I
200		1696.20	I
200		1697.94	I
50		1700.42	I
30		1700.63	I
30		1702.86	I
50		1704.43	I
10	h	1710.83	II
20	h	1711.30	II
30	h	1743.88	I
50		1747.40	I
30	h	1753.11	I
50		1763.66	I
40		1765.03	I
30	h	1765.60	I
30		1766.06	I
30		1770.63	I
100	h	1770.92	I
100	h	1776.83	I
50	h	1783.23	I
100	h	1799.21	I
150		1808.00	II
50	h	1809.09	I
500	h	1814.07	I
200		1816.92	II
10		1817.45	II
50		1822.45	I
200		1836.51	I
30	h	1838.01	I
100	h	1841.15	I
200		1841.44	I
200		1843.77	I
300		1845.51	I
100		1846.10	I
400		1847.47	I
200		1848.14	I
100		1848.74	I
500		1850.67	I
30	h	1851.79	I
200		1852.46	I
50		1853.15	I
20		1869.32	II
15		1870.23	II
100		1873.10	I
500	h	1874.84	I
100		1875.81	I
200		1881.85	I
200		1887.70	I
200		1893.15	I
1000	h	1901.33	I
100	h	1902.46	II
50	h	1904.66	I
50	h	1910.62	II
50		1941.67	II
15		1944.59	II
10		1949.33	II
100		1949.56	II
100		1954.97	I
30		1984.43	I
50		1991.85	I

Air

Intensity		Wavelength	
30		2010.97	I
50		2054.83	I
50		2058.65	II
50		2059.01	II
40		2061.19	I
30		2065.52	I
200		2072.02	II
200		2072.70	II
30	h	2103.21	I
30		2114.63	I
100		2124.12	I
10	h	2133.99	II
30	h	2136.40	II
50	h	2136.56	II
50	h	2147.91	I
110		2207.98	I
115		2210.89	I
110		2211.74	I
120		2216.67	I
120		2218.06	I
50		2218.91	I
35		2291.03	I
55		2303.06	I
30		2334.40	II
30		2334.61	II
10		2344.20	II
10	h	2349.54	II
20		2350.17	II
20	h	2353.09	II
100	h	2356.30	II
30	h	2357.18	II
50	h	2357.97	II
10	h	2360.20	II
30		2366.97	II
20		2374.26	II
10	h	2428.45	II
300		2435.15	I
65		2438.77	I
65		2443.36	I
70		2452.12	I
425		2506.90	I
375		2514.32	I
500		2516.113	I
350		2519.202	I
425		2524.108	I
450		2528.509	I
110		2532.381	I
30		2563.679	I
85		2568.641	I
45		2577.151	I
190		2631.282	I
10	h	2682.21	II
1000		2881.579	I
10	h	2887.51	II
300		2904.28	II
500		2905.69	II
55		2970.355	I
150		2987.645	I
50		3006.739	I
100	h	3030.00	II
75		3020.004	I
20	h	3021.55	II
20	h	3041.57	II
30	h	3042.19	II
100	h	3043.69	II
10	h	3043.85	II
10	h	3045.77	II
50	h	3048.30	II
150	h	3053.18	II
150		3188.97	II
50		3192.25	II
150		3193.09	II
50		3194.21	II
50		3194.69	II
100		3195.41	II
200		3199.51	II
20		3202.49	II
100	h	3203.87	II
200	h	3210.03	II
75		3214.66	II
15	h	3217.99	II
10		3220.44	II
20		3223.01	II
300		3333.14	II
500		3339.82	II
100	h	3853.66	II
500	h	3856.02	II
200	h	3862.60	II
300		3905.523	I
10	h	3955.74	II
10	h	3977.46	II
15	h	3991.77	II
10		3998.01	II
20	h	4075.45	II
15	h	4076.78	II
70		4102.936	I
300	h	4128.07	II
500	h	4130.89	II
10	h	4183.35	II
100	h	4190.72	II
50		4198.13	II
100		4621.42	II
150		4621.72	II
50		4782.991	I
35		4792.212	I
80		4792.324	I
15	h	4883.20	II
20	h	4906.99	II
20	h	4932.80	II
30		4947.607	I
40		5006.061	I
1000		5041.03	II
1000		5055.98	II
100		5181.90	II
100	h	5185.25	II
200	h	5192.86	II
500	h	5202.41	II
30		5295.19	II
100		5405.34	II
15		5417.24	II
15	h	5428.92	II
15	h	5432.89	II
100		5438.62	II
20	h	5447.26	II
15		5454.49	II
100		5456.45	II
500		5466.43	II
500	h	5466.87	II
100	h	5469.21	II
40		5493.23	I
200	h	5496.45	II
35		5517.535	I
100	h	5540.74	II
150	h	5576.66	II
30		5622.221	I
100	h	5632.97	II
200	h	5639.48	II
90		5645.611	I
150	h	5660.66	II
80		5665.554	I
1000		5669.56	II
30	h	5681.44	II
120		5684.484	I
300	h	5688.81	II
100		5690.425	I
90		5701.105	I
200	h	5701.37	II
100	h	5706.37	II
160		5708.397	I
45		5747.667	I
45		5753.625	I
45		5754.220	I
45		5762.977	I
70		5772.145	I
70		5780.384	I
30	h	5785.73	II
90		5793.071	I
30	h	5794.90	II
100		5797.859	I
150	h	5800.47	II
200		5806.74	II
30		5827.80	II
50		5846.13	II
10		5867.48	II
300	h	5868.40	II
40		5873.764	I
150		5915.22	II
200		5948.545	I
500		5957.56	II
500		5978.93	II
10	h	6067.45	II
20	h	6080.06	II
10	h	6086.67	II
90		6125.021	I
85		6131.574	I
90		6131.850	I
100		6142.487	I
100		6145.015	I
160		6155.134	I
160		6237.320	I
40		6238.287	I
125		6243.813	I
125		6244.468	I
180		6254.188	I
45		6331.954	I
1000		6347.10	II
1000		6371.36	II
45		6526.609	I
45		6527.199	I
45		6555.462	I
50	h	6660.52	II
15		6665.00	II
100		6671.88	II
20		6699.38	II
50	h	6717.04	II
100		6721.853	I
30		6741.64	II
20	h	6750.28	II
30		6818.45	II
50		6829.82	II
30		6848.568	I
80		6976.523	I

Silicon (Cont.)

Intensity	Wavelength		Spectrum
180	7003.567		I
180	7005.883		I
30	7017.28		I
90	7017.646		I
250	7034.903		I
70	7164.69		I
200	7165.545		I
70	7184.89		I
65	7193.58		I
30	7193.90		I
100	7226.206		I
100	7235.326		I
60	7235.82		I
180	7250.625		I
160	7275.294		I
40	7282.81		I
400	7289.173		I
55	7290.26		I
35	7373.00		I
375	7405.774		I
200	7409.082		I
40	7415.35		I
275	7415.946		I
425	7423.497		I
85	7424.60		I
100	7680.267		I
40	7742.71		I
30	7800.008		I
400	7848.80		II
500	7849.72		II
30	7849.967		I
90	7918.386		I
120	7932.349		I
140	7944.001		I
35	7970.306		I
35	8035.619		I
70	8093.241		I
35	8230.642		I
40	8443.982		I
40	8501.547		I
60	8502.221		I
40	8536.165		I
120	8556.780		I
50	8648.462		I
40	8728.011		I
75	8742.451		I
100	8752.009		I
35	8790.389		I
100	9412.72		II
100	9413.506		I
30	10371.269		I
120	10585.141		I
120	10603.431		I
120	10660.975		I
30	10694.251		I
30	10727.408		I
60	10749.384		I
30	10784.550		I
80	10786.856		I
140	10827.091		I
60	10843.854		I
30	10868.79		I
130	10869.541		I
30	10882.802		I
30	10885.336		I
80	10979.308		I
30	10982.061		I
80	11017.965		I
13	11187.60		I
12	11289.84		I
12	11611.09		I
370	11984.19		I
220	11991.57		I
440	12031.51		I
150	12103.53		I
120	12270.68		I
11	13176.90		I
190	15888.39		I
40	15960.04		I
95	16060.03		I
20	16094.80		I
60	16163.71		I
11	16215.68		I
16	16381.55		I
29	16680.77		I
28	17327.29		I
26	18722.90		I
15	19385.94		I
48	19432.97		I
13	19493.38		I
110	19722.50		I
31	19928.88		I
12	20917.13		I
21	21354.24		I
12	22062.71		I

Si III
Ref. 320 — L.J.R.

Intensity	Wavelength		Spectrum
	Vacuum		
8	566.61		III
6	652.22		III
8	653.33		III
5	673.48		III
5	800.07		III
9	823.41		III
5	883.40		III
9	939.09		III
9	967.95		III
10	993.52		III
13	994.79		III
16	997.39		III
7	1005.37		III
7	1031.16		III
8	1033.92		III
7	1037.05		III
6	1083.22		III
14	1108.37		III
16	1109.97		III
18	1113.23		III
6	1140.55		III
7	1141.58		III
6	1142.28		III
8	1144.31		III
6	1144.96		III
8	1145.11		III
7	1145.18		III
6	1155.00		III
6	1155.96		III
7	1158.10		III
6	1160.26		III
8	1161.58		III
5	1174.37		III
6	1174.43		III
8	1178.00		III
30	1206.51		III
30	1206.53		III
9	1207.52		III
10	1210.46		III
7	1235.43		III
6	1280.35		III
17	1294.54		III
14	1296.73		III
15	1298.89		III
18	1298.96		III
14	1301.15		III
16	1303.32		III
13	1312.59		III
8	1341.47		III
7	1342.39		III
6	1343.39		III
8	1361.60		III
5	1362.37		III
7	1363.47		III
8	1365.26		III
7	1367.05		III
5	1369.44		III
5	1373.03		III
5	1387.99		III
13	1417.24		III
6	1433.69		III
8	1435.77		III
7	1436.17		III
5	1441.73		III
6	1447.20		III
5	1457.25		III
12	1500.24		III
10	1501.19		III
9	1501.87		III
6	1506.06		III
7	1673.32		III
9	1842.55		III
	Air		
5	2176.89		III
6	2295.48		III
10	2296.87		III
8	2300.93		III
10	2308.19		III
11	2449.48		III
6	2483.20		III
25	2541.82		III
10	2546.09		III
14	2559.21		III
11	2640.79		III
14	2655.51		III
9	2817.11		III
7	2831.49		III
5	2839.62		III
5	2959.15		III
5	2980.52		III
5	3013.09		III
6	3034.73		III
8	3037.29		III
9	3040.93		III
7	3043.93		III
5	3045.08		III
7	3068.24		III
25	3086.24		III
6	3086.46		III
20	3093.42		III
5	3093.65		III
16	3096.83		III
6	3126.27		III
7	3147.37		III
8	3161.61		III
16	3185.13		III
13	3186.02		III
14	3196.50		III
15	3210.55		III
7	3216.25		III
12	3230.50		III
14	3233.95		III
15	3241.62		III
7	3253.40		III
5	3253.74		III
7	3254.80		III
12	3258.66		III
6	3270.46		III
10	3276.26		III
7	3279.26		III
15	3486.91		III
9	3525.94		III
8	3569.67		III
20	3590.47		III
8	h	3622.54	III
5	h	3639.45	III
6	h	3645.12	III
7	h	3681.40	III
5	h	3682.15	III
20	c	3791.41	III
25		3796.11	III
30		3806.54	III
7		3842.46	III
20		3924.47	III
6	h	3947.49	III
6		3963.84	III
5		3981.24	III
5		4101.86	III
8		4102.42	III
5	h	4115.50	III
9		4338.50	III
8		4341.40	III
8	h	4377.63	III
6	h	4405.90	III
8	h	4406.72	III
8		4494.05	III
30		4552.62	III
8		4554.00	III
25		4567.82	III
20		4574.76	III
7		4619.66	III
7		4638.28	III
8		4665.87	III
9		4683.02	III
7		4683.80	III
16		4716.65	III
7		4730.52	III
8		4800.43	III
15		4813.33	III
16		4819.72	III
18		4828.97	III
10	h	5091.42	III
7	h	5113.76	III
8	h	5114.12	III
5		5197.26	III
6		5451.46	III
7		5473.05	III
7		5704.60	III
8		5716.29	III
20		5739.73	III
10	h	5898.79	III
7		6314.46	III
6	h	6524.36	III
6	h	6831.56	III
7	h	6851.65	III
5	h	7461.89	III
8	h	7462.62	III
9	h	7466.32	III
12	h	7612.36	III
9	h	8102.86	III
11	h	8103.45	III
7	h	8190.43	III
6	h	8191.16	III
8	h	8191.68	III
9	h	8262.57	III
5	h	8265.64	III
8	h	8269.32	III
5	h	8271.38	III
6	h	8271.94	III

Si IV
Ref. 319 — L.J.R.

Intensity	Wavelength		Spectrum
	Vacuum		
4	457.82		IV
3	458.16		IV
2	515.12		IV
3	516.35		IV
2	645.76		IV
5	749.94		IV
7	815.05		IV
8	818.13		IV
8	1066.63		IV
8	1122.49		IV
10	1128.34		IV
15	1393.76		IV
12	1402.77		IV
1	1634.61		IV
6	1722.53		IV
5	1727.38		IV
	Air		
3	2120.18		IV
4	2127.47		IV
5	h	2287.04	IV
2	h	2328.56	IV
2		2366.76	IV
3		2370.99	IV
2		2482.82	IV
1		2485.38	IV
7		2517.51	IV
1		2672.19	IV
4		2675.12	IV
4		2675.25	IV
1		2677.57	IV
3	h	2723.81	IV
3		2895.13	IV
2	h	2904.47	IV
1	h	2971.52	IV
7		3149.56	IV
9		3165.71	IV
1	h	3244.19	IV
8		3762.44	IV
6		3773.15	IV
1	h	4031.39	IV
2	h	4038.06	IV
10		4088.85	IV
9		4116.10	IV
7	h	4212.41	IV
3		4314.10	IV
5		4328.18	IV
2	h	4403.73	IV
1	h	4411.65	IV
1	h	4611.27	IV
3	h	4628.62	IV
9	h	4631.24	IV
10	h	4654.32	IV
3		4656.92	IV
1	h	4667.14	IV
2	h	4673.30	IV
1	h	4947.45	IV
3		4950.11	IV
2	h	5304.97	IV
1	h	5309.49	IV
5		6667.56	IV
7		6701.21	IV
3	h	6998.36	IV
6	h	7047.94	IV
4	h	7068.41	IV
2		7630.50	IV
4		7654.56	IV
4		7678.75	IV
5	h	7718.79	IV
6	h	7723.82	IV
2	h	7725.64	IV
1	h	7730.47	IV
1		7752.91	IV
1	h	8240.61	IV
2	h	8957.25	IV
1	h	9018.16	IV

Si V
Ref. 87 — L.J.R.

Intensity	Wavelength		Spectrum
	Vacuum		
1	78.61		V
1	78.90		V
2	80.81		V
2	81.11		V
10	85.18		V
6	85.58		V
4	90.45		V
4	90.85		V
15	96.44		V
10	97.14		V
2	98.21		V
20	117.86		V
20	118.97		V

SILVER (Ag)
Z = 47

Ag I and II
Ref. 13, 99, 255, 286, 289 — C.H.C.

Intensity	Wavelength		Spectrum
	Vacuum		
25	730.83		II

Silver (Cont.)

Intensity		Wavelength	
30		752.80	II
15		1005.32	II
10		1065.49	II
12		1072.23	II
250		1074.22	II
150		1107.03	II
150		1112.46	II
60		1195.83	II
50		1223.33	II
50		1240.80	II
50		1246.87	II
55		1256.81	II
55		1257.55	II
50		1266.63	II
70		1273.67	II
65		1297.51	II
85		1311.20	II
55		1313.81	II
50		1314.61	II
60		1323.84	II
60		1342.09	II
50		1342.57	II
70		1346.62	II
50		1353.54	II
150		1364.50	II
100		1396.00	II
100		1410.93	II
90		1419.72	II
95		1432.60	II
100		1464.72	II
50		1466.23	II
50	r	1507.37	I
100	r	1515.63	I
50	r	1548.58	I
100		1555.16	II
100		1644.50	II
60		1651.52	I
50		1652.10	I
120		1682.82	II
10		1708.11	I
50		1709.27	I
125		1736.44	II
10	h	1766.14	I
75		1790.37	II
20		1847.71	I
100		1967.38	II

Air

Intensity		Wavelength	
150		2015.96	II
150		2033.98	II
200		2061.17	I
100		2069.85	I
80	r	2113.82	II
60		2145.60	II
15		2170.00	I
50		2186.76	II
60		2229.53	II
100	r	2246.43	II
75	r	2248.74	II
75		2280.03	II
30	h	2309.56	I
10	h	2312.60	I
70	r	2317.05	II
80	r	2320.29	II
70	r	2324.68	II
80	r	2331.40	II
70		2357.92	II
50	h	2375.02	I
75		2411.41	II
90	r	2413.23	II
100	r	2437.81	II
80		2447.93	II
80		2473.84	II
60		2506.63	II
50	h	2575.63	I
60		2660.49	II
60		2721.77	I
75		2767.54	II
100	h	2824.39	I
10	h	2926.77	I
20	h	2938.42	I
20		3099.10	I
30	h	3130.02	I
10	h	3170.58	I
90		3180.70	II
15	h	3215.67	I
10		3225.15	I
15		3233.18	I
100		3267.35	II
55000	r	3280.68	I
10	h	3305.67	I
28000	r	3382.89	I
10	h	3403.78	I
30		3469.16	I
70		3475.82	II
80		3495.28	II
20	h	3501.92	I
20		3508.03	I
15	h	3513.38	I
10		3521.12	I
50		3542.61	I
10	h	3547.16	I
10	h	3557.01	I

Intensity		Wavelength	
20	h	3586.67	I
10	h	3623.49	I
50	h	3624.68	I
75		3682.46	II
30		3682.50	I
80		3683.34	II
50		3709.20	I
10	h	3727.42	I
20	h	3753.14	I
200		3810.94	I
50		3811.78	I
100	h	3840.74	I
15		3847.85	I
50		3907.41	I
50		3909.31	II
50		3914.40	I
70		3920.10	II
10		3928.01	I
10		3940.43	I
10		3942.97	I
60		3949.43	II
100		3981.58	I
70		3985.19	II
10		3992.15	I
100	h	4055.48	I
10	h	4083.43	I
80		4085.91	II
100		4185.48	II
90	h	4210.96	I
100		4212.82	II
50		4311.07	I
20		4396.23	I
50		4476.04	II
20	h	4556.0	I
30	h	4615.69	I
80		4620.04	II
50		4620.46	II
60	h	4668.48	I
30	h	4677.60	I
100		4788.40	II
20	h	4796.2	I
30	h	4847.82	I
100		4874.10	I
20		4888.21	I
10	h	4917.5	I
10		4935.75	I
20	h	4992.89	I
80		5027.35	II
15	h	5123.50	I
1000		5209.08	I
10	h	5333.62	I
1000		5465.50	I
100		5471.55	I
20		5475.38	I
20	h	5545.67	I
10	h	5559.58	I
100		5667.34	I
10		6083.78	I
10	h	6268.50	I
20		6621.08	I
20		7359.96	I
320		7687.78	I
25		8005.4	II
15		8254.7	II
500		8273.52	I
20		8324.4	II
15		8379.5	I
25		8403.8	II
15		8492.5	I
30	h	8645.70	I
10	h	8704.85	I
12		8747.6	II
15		9000.9	II
10		12551.0	I
60		16819.5	I
20		17416.7	I
15		18307.9	I
15		18382.3	I

Ag III
Ref. 363, 387, 398 — R.D.C.

Intensity   Wavelength

Vacuum

Intensity	Wavelength	
200	709.80	III
200	713.85	III
100	717.73	III
200	718.53	III
300	726.96	III
350	730.04	III
150	730.28	III
150	730.94	III
200	736.57	III
100	738.13	III
200	740.98	III
200	742.29	III
200	748.30	III
150	755.73	III
100	758.27	III
200	767.19	III
250	768.33	III
150	769.61	III
350	776.38	III
200	782.91	III
150	785.76	III
200	789.08	III
250	792.35	III
200	796.54	III
250	797.91	III
400	799.41	III
300	808.88	III
150	816.12	III
180	822.39	III
200	838.11	III
120	1373.22	III
120	1374.76	III
110	1404.93	III
120	1413.90	III
120	1414.29	III
120	1428.61	III
120	1452.74	III
200	1456.41	III
100	1471.44	III
300	1489.01	III
150	1515.08	III
100	1524.23	III
120	1527.04	III
150	1541.14	III
150	1550.89	III
130	1553.04	III
130	1587.41	III
120	1589.28	III
100	1613.79	III
130	1619.14	III
100	1634.46	III
100	1652.24	III
130	1653.60	III
300	1654.43	III
700	1656.18	III
150	1657.10	III
100	1661.54	III
130	1670.75	III
150	1676.14	III
100	1681.07	III
500	1693.51	III
200	1705.06	III
150	1708.86	III
130	1717.68	III
200	1722.27	III
150	1726.76	III
250	1728.14	III
200	1747.34	III
120	1749.64	III
150	1750.89	III
750	1751.03	III
100	1760.57	III
150	1762.62	III
150	1768.70	III
100	1771.81	III
100	1783.85	III
100	1791.70	III
100	1792.69	III
150	1793.90	III
150	1802.24	III
150	1802.26	III
100	1802.77	III
300	1808.23	III
250	1816.83	III
150	1822.45	III
350	1828.83	III
250	1832.33	III
120	1832.50	III
100	1834.31	III
250	1836.10	III
150	1838.64	III
400	1840.14	III
100	1846.96	III
120	1849.93	III
150	1856.33	III
120	1858.91	III
100	1860.39	III
100	1860.64	III
350	1867.12	III
150	1868.10	III
100	1872.55	III
400	1873.45	III
250	1880.36	III
300	1889.57	III
400	1916.92	III
600	1917.08	III
200	1925.30	III
150	1946.32	III
100	1948.44	III
700	1957.62	III
120	1959.27	III
100	1960.86	III
400	1966.89	III
600	1975.92	III
500	1977.03	III
150	1981.87	III
200	1987.02	III
130	1995.16	III
600	2000.24	III

Air

Intensity	Wavelength	
300	2007.30	III
200	2011.49	III
150	2013.65	III
150	2041.33	III
150	2053.17	III
150	2053.83	III
200	2056.99	III
200	2081.04	III
300	2146.47	III
150	2149.19	III
600	2161.89	III
150	2166.21	III
150	2211.23	III
100	2238.40	III
500	2246.51	III
100	2286.50	III
700	2310.04	III
100	2386.85	III
300	2395.69	III
100	2469.62	III
100	2562.87	III

SODIUM (Na)
Z = 11

Na I and II
Ref. 268, 334 — T.K.M.

Intensity	Wavelength	

Vacuum

160	300.15	II
160	300.20	II
90	301.32	II
100	301.44	II
60	302.45	II
300	372.08	II
350	376.38	II
60	1293.97	II
50	1327.74	II
45	1347.54	II
90	1374.69	II
90	1404.68	II
45	1495.21	II
45	1497.73	II
80	1506.41	II
60	1506.91	II
70	1513.10	II
60	1519.63	II
60	1657.92	II
90	1776.57	II
60	1783.04	II
80	1787.19	II
45	1788.85	II
80	1798.41	II
45	1801.26	II
90	1807.09	II
80	1808.38	II
50	1821.70	II
45	1833.87	II
80	1835.22	II
45	1837.89	II
60	1841.82	II
70	1845.02	II
45	1850.15	II
70	1851.19	II
80	1853.17	II
45	1866.45	II
45	1873.37	II
60	1875.08	II
160	1881.91	II
50	1885.09	II
45	1885.74	II

Air

80	2228.53	II
80	2303.58	II
300	2315.65	II
130	2393.28	II
100	2401.01	II
300	2420.99	II
300	2424.73	II
200	2439.14	II
250	2441.50	II
200	2448.72	II
200	2452.18	II
1000	2493.15	II
300	2502.84	II
450	2506.30	II
600	2515.46	II
600	2531.54	II
20	2543.84	I
10	2543.87	I
550	2586.31	II
70	2593.87	I
35	2593.92	I
600	2594.96	II
850	2611.81	II
300	2627.41	II

Sodium (Cont.)

Intensity	Wavelength	
850	2661.00	II
350	2666.46	II
1000	2671.83	II
850	2678.09	II
200	2680.34	I
100	2680.43	I
650	2808.71	II
850	2809.52	II
600	2829.87	II
800	2839.56	II
1000	2841.72	II
400	2852.81	I
200	2853.01	I
650	2856.51	II
800	2859.49	II
750	2871.28	II
650	2872.95	II
900	2881.15	II
850	2886.26	II
2	2893.62	I
700	2893.95	II
900	2901.14	II
800	2904.72	II
1100	2904.92	II
1100	2917.52	II
1100	2919.05	II
1200	2919.85	II
1300	2920.95	II
1000	2923.49	II
750	2930.88	II
850	2934.08	II
950	2937.74	II
800	2945.70	II
950	2947.50	II
1200	2951.24	II
1100	2952.40	II
850	2960.12	II
500	2970.73	II
600	2974.24	II
750	2974.99	II
1000	2977.13	II
1100	2979.66	II
1100	2980.63	II
1300	2984.19	II
550	3004.15	II
750	3007.66	II
750	3009.14	II
600	3015.40	II
550	3053.67	II
550	3055.35	II
550	3056.16	II
550	3057.38	II
550	3057.95	II
550	3058.72	II
700	3060.25	II
800	3061.35	II
500	3064.38	II
500	3066.22	II
500	3066.54	II
550	3074.33	II
550	3078.32	II
550	3080.25	II
550	3087.06	II
550	3092.04	II
550	3092.73	II
650	3094.45	II
650	3095.55	II
500	3103.58	II
500	3104.40	II
500	3113.69	II
1700	3124.42	II
600	3125.21	II
600	3129.38	II
2500	3135.48	II
1700	3137.86	II
950	3145.71	II
2000	3149.28	II
2000	3163.74	II
700	3175.09	II
1000	3179.06	II
1700	3189.79	II
1600	3212.19	II
700	3234.93	II
1500	3257.96	II
650	3260.21	II
950	3274.22	II
1700	3285.60	II
1700	3301.35	II
1200	3302.37	I
600	3302.98	II
1500	3304.96	II
1000	3318.04	II
950	3327.69	II
50	3426.86	I
1500	3533.05	II
1200	3631.27	II
850	3711.07	II
300	4113.70	II
250	4123.08	II
250	4233.26	II
6	4238.99	I
250	4240.90	II
10	4242.08	I

Intensity	Wavelength	
1	4249.41	I
2	4252.52	I
15	4273.64	I
20	4276.79	I
2	4287.84	I
3	4291.01	I
250	4292.48	II
250	4292.86	II
250	4308.81	II
250	4309.04	II
250	4320.91	II
30	4321.40	I
40	4324.62	I
250	4337.29	II
3	4341.49	I
250	4344.11	II
5	4344.74	I
200	4368.60	II
200	4375.22	II
200	4387.49	II
40	4390.03	I
250	4392.81	II
60	4393.34	I
200	4405.12	II
5	4419.88	I
8	4423.25	I
200	4446.70	II
200	4447.41	II
200	4454.74	II
200	4455.23	II
200	4457.21	II
200	4474.63	II
200	4478.80	II
200	4481.67	II
200	4490.15	II
200	4490.87	II
60	4494.18	I
100	4497.66	I
200	4499.62	II
200	4506.97	II
200	4519.21	II
200	4524.98	II
200	4533.32	II
10	4541.63	I
15	4545.19	I
200	4551.53	II
160	4590.92	II
120	4664.811	I
200	4668.560	I
160	4722.23	II
160	4731.10	II
160	4741.67	II
20	4747.941	I
30	4751.822	I
160	4768.79	II
100	4788.79	II
200	4978.541	I
400	4982.813	I
40	5148.838	I
80	5153.402	I
100	5191.65	II
80	5208.55	II
70	5400.46	II
90	5414.55	II
280	5682.633	I
70	5688.193	I
560	5688.205	I
80000	5889.950	I
40000	5895.924	I
120	6154.225	I
240	6160.747	I
60	6175.25	II
70	6199.26	II
70	6234.68	II
80	6260.01	II
80	6274.74	II
70	6361.15	II
70	6366.41	II
90	6514.21	II
80	6524.68	II
130	6530.70	II
130	6544.04	II
130	6545.75	II
80	6552.43	II
20	7373.23	I
10	7373.49	I
50	7809.78	I
25	7810.24	I
4400	8183.256	I
800	8194.790	I
8800	8194.824	I
100	8649.92	I
60	8650.89	I
25	8942.96	I
40	9153.88	I
60	9465.94	I
80	9961.28	I
20	10566.00	I
60	10572.28	I
200	10746.44	I
80	10749.29	I
120	10834.87	I
35	11190.19	I

Intensity	Wavelength	
50	11197.21	I
400	11381.45	I
1000	11403.78	I
400	12679.17	I
60	14767.48	I
100	14779.73	I
60	16373.85	I
100	16388.85	I
400	18465.25	I
50	22056.44	I
25	22083.67	I
60	23348.41	I
100	23379.13	I

Na III
Ref. 178, 205, 207 — T.K.M.

Intensity		Wavelength	
		Vacuum	
5		183.95	III
5	h	189.35	III
5		193.80	III
5	h	194.04	III
5	h	194.17	III
5		194.29	III
6		194.68	III
6		195.53	III
6		202.15	III
6		202.19	III
8		202.49	III
5	d	202.71	III
7	d	202.72	III
8		202.76	III
8	p	203.06	III
8		203.28	III
8		203.33	III
10		207.30	III
10	c	215.34	III
12		215.86	III
12		216.12	III
15		229.87	III
12		230.59	III
50	c	250.52	III
30		251.27	III
25		266.90	III
70		267.65	III
50		267.87	III
50		268.63	III
20	p	272.08	III
20		272.45	III
100		378.14	III
70		380.10	III
7		1336.76	III
7		1337.36	III
8		1340.67	III
9	d	1342.39	III
10		1342.73	III
11		1355.28	III
12		1361.90	III
11		1372.34	III
10		1420.89	III
10		1444.19	III
12		1449.31	III
11		1562.87	III
10		1565.29	III
10		1598.18	III
11		1688.94	III
10		1699.29	III
10		1711.12	III
11		1728.27	III
10		1731.11	III
10		1755.48	III
15		1807.07	III
10		1810.77	III
11		1811.67	III
10		1816.81	III
10	d	1835.22	III
10		1838.94	III
11		1844.36	III
12	d	1847.53	III
10	d	1847.59	III
15		1849.56	III
12		1850.38	III
10		1855.92	III
10		1856.71	III
10		1861.21	III
10		1880.66	III
10	d	1887.39	III
20	d	1887.47	III
15	d	1890.75	III
15		1900.16	III
10		1918.45	III
11		1923.96	III
14		1926.26	III
12		1927.24	III
12		1932.74	III
13		1933.89	III
10		1943.52	III
12		1946.43	III
12		1950.91	III
14		1951.24	III
10		1977.16	III
13		1985.57	III

Intensity		Wavelength	
10		1995.68	III
		Air	
10		2004.21	III
11		2005.22	III
11		2008.47	III
15		2011.87	III
11		2014.17	III
12		2017.03	III
12		2028.56	III
12		2031.13	III
11		2035.90	III
12		2041.66	III
12		2043.29	III
10		2044.82	III
10		2045.44	III
11		2051.48	III
10		2060.36	III
15		2066.60	III
13		2082.91	III
15		2140.72	III
14		2144.54	III
15		2202.83	III
15		2225.93	III
30		2230.33	III
16		2232.19	III
20	h	2246.70	III
14		2251.47	III
15		2278.42	III
13		2285.66	III
15		2309.99	III
18		2386.99	III
17		2394.03	III
15		2406.59	III
25		2459.31	III
18		2468.85	III
20		2474.73	III
25		2497.03	III
17		2510.26	III
15		2530.25	III
14		2542.80	III

Na IV
Ref. 206 — T K M

Intensity		Wavelength	
		Vacuum	
4		136.551	IV
4		136.854	IV
4		139.961	IV
7		142.232	IV
6		142.359	IV
8		146.064	IV
7		146.302	IV
9		150.298	IV
7		150.543	IV
7	c	150.64	IV
8		150.687	IV
7		151.299	IV
7		155.083	IV
7		155.240	IV
7		155.448	IV
8		155.510	IV
8		156.537	IV
12		162.448	IV
10		163.190	IV
12		168.411	IV
10		168.546	IV
10		190.445	IV
10		199.772	IV
10	c	205.49	IV
10		319.644	IV
10		360.76	IV
12		408.684	IV
10		409.614	IV
15		410.372	IV
10		411.334	IV
13		412.242	IV
10		1580.50	IV
11		1582.18	IV
10		1582.33	IV
11	d	1583.98	IV
12		1584.14	IV
10	d	1586.99	IV
12	d	1587.05	IV
10		1613.95	IV
11		1615.92	IV
12		1618.57	IV
11		1655.47	IV
15	c	1701.97	IV
10		1702.41	IV
12		1960.76	IV
11		1965.08	IV
10		1967.60	IV
		Air	
10		2018.39	IV
12	d	2106.33	IV
10		2114.53	IV

Sodium (Cont.)

Intensity	Wavelength	
10	2155.76	IV

Na V
Ref. 299 — T.K.M.

Intensity		Wavelength	
		Vacuum	
100		106.28	V
100		106.30	V
100		106.40	V
100		106.49	V
200	c	107.93	V
200	c	108.02	V
200	c	110.82	V
200		110.88	V
100		111.51	V
300	c	112.01	V
100	h	114.70	V
100		114.74	V
400		117.99	V
100		120.04	V
400		125.18	V
400		125.22	V
500		125.29	V
300		125.43	V
300		125.46	V
200		125.90	V
100		126.21	V
200		126.56	V
100		126.61	V
400		127.44	V
400		127.47	V
400		128.03	V
400		128.05	V
200		130.68	V
300		131.35	V
200		131.41	V
300	h	131.64	V
500		133.16	V
400		133.39	V
200		134.27	V
300		135.79	V
300		135.85	V
200		138.81	V
300		138.92	V
400		148.64	V
300		148.86	V
400		151.13	V
300		157.21	V
300		163.62	V
800		307.15	V
1000		308.26	V
800		332.55	V
900		333.91	V
800		360.32	V
800		360.37	V
1000		400.72	V
500		445.05	V
600		445.19	V
600		459.90	V
850		461.05	V
1000		463.26	V

STRONTIUM (Sr)
Z = 38

Sr I and II
Ref. 1, 218, 279, 313 — J.J.W.

Intensity		Wavelength	
		Air	
1400		2152.84	II
1400		2165.96	II
160		2428.10	I
120		2569.47	I
200		2931.83	I
300		3301.73	I
300		3329.99	I
400		3351.25	I
300		3366.33	I
650		3380.71	II
950		3464.46	II
120		3474.89	II
300	h	3940.80	I
600		3969.26	I
300		3970.04	I
1300		4030.38	I
300		4032.38	I
46000		4077.71	II
200		4161.80	II
32000		4215.52	II
340		4305.45	II
350	h	4438.04	II
		4526.10	II
		4585.91	II
65000		4607.33	I
3200		4722.28	I
2200		4741.92	I
1400		4784.32	I
4800		4811.88	I
3600		4832.08	I
500		4855.04	I
600		4868.70	I
3000		4872.49	I
600		4876.06	I
2000		4876.32	I
1000		4891.98	I
8000		4962.26	t
1300		4967.94	I
800	h	5156.07	I
1400		5222.20	I
2000		5225.11	I
2000		5229.27	I
2800		5238.55	I
4800		5256.90	I
		5303.13	II
350	h	5329.82	I
		5379.13	II
		5385.45	II
1500		5450.84	I
7000		5480.84	I
1100		5486.12	I
3500		5504.17	I
2600		5521.83	I
2000		5534.81	I
2000		5540.05	I
250	h	5543.36	I
		5622.94	II
		5650.54	II
		5723.70	II
		5819.00	II
200	h	5970.10	I
250	h	6345.75	I
250	h	6363.94	I
350	h	6369.96	I
1000		6380.75	I
900	h	6386.50	I
600	h	6388.24	I
9000		6408.47	I
250		6446.68	I
250	h	6465.79	I
		6483.17	II
5500		6504.00	I
		6509.20	II
1000		6546.79	I
1700		6550.26	I
3000		6617.26	I
800		6643.54	I
1800		6791.05	I
4800		6878.38	I
1200		6892.59	I
5500		7070.10	I
60		7153.09	I
250	h	7167.24	I
200		7232.27	I
2500		7309.41	I
500		7621.50	I
400	h	7673.06	I
50	hL	7850.00	I
30	h	7866.90	I
20	hL	7874.00	I
200	h	8422.80	I
120		8505.69	II
200		8688.91	II
30		8719.56	II
40		9170.00	I
30		9204.50	I
20		9283.90	I
100		9294.10	I
15		9306.60	I
30		9319.20	I
60		9380.45	I
40	h	9411.25	I
400	h	9448.95	I
600		9596.00	I
300		9624.70	I
100		9638.10	I
100	h	9647.70	II
300		10036.66	II
1000		10327.31	II
7		10872.70	II
200		10914.88	II
10		10984.00	I
13		11224.57	II
700		11241.25	I
100		12014.76	II
20		12236.20	I
60		12445.90	II
20		12479.60	I
40		12495.00	I
15		12652.20	I
75		12974.70	II
100		13123.80	II
15		13522.80	I
15		17140.90	I
30		17170.50	I
50		17447.40	I
4		17626.00	II
30		17743.00	I
15		19759.60	I
230		20261.40	I
120		20700.70	I
40		20764.50	I
15		20778.70	I
30		26023.60	I

Sr III
Ref. 231, 265 — J.J.W.

Intensity		Wavelength	
		Vacuum	
25		307.18	III
50		316.11	III
50		321.61	III
125		330.67	III
500		351.62	III
75		358.80	III
250		363.49	III
150		371.21	III
1000		437.24	III
1875		491.79	III
1250		507.04	III
3750		514.38	III
2500		562.75	III
20		968.37	III
20		975.78	III
25		992.98	III
50		1025.23	III
35		1044.91	III
20		1057.74	III
25		1060.20	III
20		1098.77	III
35		1125.49	III
20		1140.24	III
20		1168.27	III
20		1182.09	III
50		1236.23	III
20		1940.58	III
30	p	1958.44	III
30		1966.92	III
		Air	
25		2068.63	III
50		2099.59	III
25		2114.31	III
30		2118.48	III
50		2119.52	III
50		2133.12	III
30		2142.80	III
20		2145.74	III
30		2178.91	III
30		2180.14	III
50		2190.88	III
50		2203.86	III
50		2219.50	III
50		2220.05	III
50		2267.03	III
100		2273.71	III
50		2277.87	III
30		2310.33	III
50		2314.95	III
50		2334.79	III
100		2340.13	III
50		2404.17	III
30		2410.52	III
50		2454.03	III
100		2486.52	III
50		2503.59	III
30		2599.10	III
35		2622.69	III
30		2642.96	III
30		2648.51	III
35		2654.66	III
40		2722.47	III
50		2786.00	III
50		2821.42	III
30		2874.86	III
30		2929.34	III
30		2983.00	III
100		3002.61	III
200		3012.32	III
100		3021.73	III
30		3059.83	III
50		3061.43	III
30		3104.25	III
50		3182.61	III
100		3235.39	III
30		3302.72	III
50		3430.76	III
30		3874.26	III
30		3936.40	III
30		3936.72	III
30		3958.75	III
30		4094.03	III
30		4097.02	III
30		4105.63	III
35		4335.80	III
30		5071.09	III
30		5130.34	III
35		5158.26	III
40		5257.71	III
30		5262.21	III
30		5288.32	III
30		5391.03	III
40		5443.48	III
30		5463.90	III
30		5664.66	III
30		5689.72	III

Sr IV
Ref. 110 — J.J.W.

Intensity		Wavelength	
		Vacuum	
12		284.31	IV
12		291.09	IV
12		291.19	IV
12		293.22	IV
15		298.12	IV
15		300.12	IV
12		300.27	IV
12		301.67	IV
20		378.53	IV
75		392.44	IV
50		393.00	IV
45		394.90	IV
50		396.22	IV
40		399.92	IV
35		403.85	IV
35		406.94	IV
30		412.93	IV
40		413.07	IV
40		415.32	IV
30		419.78	IV
25		430.21	IV
30		430.65	IV
25		442.73	IV
25		471.76	IV
25		484.20	IV
25	p	508.14	IV
25		534.19	IV
200		664.43	IV
100		710.35	IV
30		1189.21	IV
30		1244.14	IV
20	p	1244.75	IV
20	p	1244.87	IV
20	p	1257.78	IV
20		1268.62	IV
20		1331.13	IV
30		1347.90	IV
20		1361.15	IV
25		1408.67	IV
20		1592.74	IV
25		1677.03	IV
20		1705.16	IV
20		1724.23	IV
25		1729.53	IV
20		1732.12	IV
20		1777.25	IV
20		1994.61	IV
		Air	
20		2104.38	IV
20		2117.90	IV
20		2217.99	IV
20		2230.41	IV
20		2240.49	IV
20		2253.38	IV
50		2346.97	IV
20		2357.34	IV
20		2438.93	IV
25		2441.41	IV
30		2482.79	IV
25		2483.57	IV
18		2500.57	IV
20		2508.02	IV
20		2534.03	IV
18		2548.02	IV
40		2555.60	IV
40		2571.04	IV
25		2571.58	IV
15		2589.34	IV
25		2620.35	IV
20		2621.16	IV
20		2642.16	IV
15		2830.53	IV
9		2934.60	IV
10		3019.29	IV
9		3266.52	IV
9		3566.43	IV
9		3741.05	IV
9		4298.57	IV
9		4685.08	IV

Sr V
Ref. 109 — J.J.W.

Intensity	Wavelength
	Vacuum

Strontium (Cont.)

Intensity	Wavelength	
10	517.28	V
6	540.51	V
25	578.01	V
30	624.93	V
25	642.23	V
50	649.21	V
20	659.15	V
25	660.94	V
9	669.93	V
35	686.23	V
6	715.79	V
12	747.82	V
9	862.32	V

SULFUR (S)
Z = 16
S I and II
Ref. 144, 209, 210, 266 — R.L.K.

Intensity	Wavelength	
	Vacuum	
40	906.9	II
40	910.5	II
40	912.7	II
40	937.4	II
40	937.7	II
20	996.0	II
20	1000.5	II
20	1014.4	II
20	1019.5	II
20	1096.6	II
40	1102.3	II
20	1131.0	II
20	1131.6	II
40	1234.1	II
40	1250.5	II
110	1253.8	II
110	1259.5	II
275	1270.782	I
250	1277.216	I
280	1295.653	I
275	1302.337	I
235	1302.862	I
235	1303.110	I
245	1303.430	I
260	1305.883	I
265	1310.194	I
355	1316.542	I
290	1316.618	I
375	1323.515	I
355	1326.643	I
775	1381.552	I
710	1385.510	I
960	1388.435	I
640	1389.154	I
775	1392.588	I
1000	1396.112	I
300	1409.337	I
510	1425.030	I
425	1433.280	I
300	1436.968	I
300	1448.229	I
425	1472.972	I
550	1473.995	I
300	1474.380	I
355	1481.665	I
485	1483.039	I
300	1483.233	I
330	1485.622	I
390	1487.150	I
680	1666.688	I
640	1687.530	I
710	1807.311	I
680	1820.343	I
640	1826.245	I
710	1900.286	I
550	1914.698	I
	Air	
20	2629.1	II
40	2670.0	II
40	2847.7	II
285	3867.6	I
285	3902.0	I
360	3933.3	II
450	4120.8	I
280	4142.3	II
360	4145.1	II
450	4153.1	II
450	4162.7	II
450	4694.1	I
285	4695.4	I
160	4696.2	I
280	4716.2	II
450	4815.5	II
360	4924.1	II
450	4925.3	II
285	4993.5	I
360	5428.6	II
650	5432.8	II
1000	5453.8	II
1000	5473.6	II
1000	5509.7	II
280	5564.9	II
1000	5606.1	II
450	5640.0	II
450	5640.3	II
280	5647.0	II
650	5659.9	II
450	5664.7	II
160	5706.1	I
450	5819.2	II
450	6052.7	I
280	6286.4	II
450	6287.1	II
450	6305.5	II
450	6312.7	II
280	6384.9	II
280	6397.3	II
280	6398.0	II
360	6413.7	II
160	6743.6	I
285	6748.8	I
450	6757.2	I
450	7579.0	I
450	7629.8	I
285	7686.1	I
450	7696.7	I
1000	7924.0	I
160	7928.8	I
285	7930.3	I
450	7931.7	I
450	7967.4	I
450	7967.4	II
450	8314.7	I
450	8314.7	II
450	8585.6	I
285	8680.5	I
450	8694.7	I
360	8874.5	I
110	8882.5	I
220	8884.2	I
160	9035.9	I
650	9111.1	I
450	9228.1	I
450	9237.5	I
285	9413.5	I
285	9421.9	I
285	9437.1	I
650	9649.9	I
450	9672.3	I
450	9680.8	I
450	9693.7	I
285	9697.3	I
285	9739.7	I
110	9741.9	I
285	9932.3	I
285	9949.8	I
285	9958.9	I
285	10455.5	I
70	10456.8	I
285	10459.5	I

S III
Ref. 209, 210 — R.L.K.

Intensity	Wavelength	
	Vacuum	
70	729.5	III
110	732.42	III
70	735.2	III
70	738.5	III
70	789.0	III
70	796.7	III
70	824.9	III
70	836.3	III
285	1077.1	III
70	1194.0	III
70	1201.0	III
	Air	
110	2460.5	III
110	2489.6	III
160	2496.2	III
160	2499.1	III
220	2508.2	III
70	2636.9	III
220	2665.4	III
70	2680.5	III
110	2691.8	III
220	2702.8	III
110	2718.9	III
220	2721.4	III
220	2726.8	III
220	2731.1	III
110	2741.0	III
285	2756.9	III
110	2775.2	III
160	2785.5	III
70	2797.4	III
70	2856.0	III
110	2863.5	III
160	2904.3	III
70	2964.8	III
160	2986.0	III
70	3234.2	III
70	3324.9	III
110	3497.3	III
160	3632.0	III
70	3662.0	III
110	3709.4	III
160	3717.8	III
160	3838.3	III
160	3928.6	III
360	4253.6	III
110	4285.0	III
70	4332.7	III

S IV
Ref. 29, 202, 209 — R.L.K.

Intensity	Wavelength	
	Vacuum	
20	519.3	IV
20	520.1	IV
40	520.8	IV
20	522.0	IV
20	522.5	IV
20	551.2	IV
40	652.5	IV
40	653.0	IV
70	653.6	IV
40	654.0	IV
70	655.6	IV
20	655.9	IV
110	657.3	IV
40	660.9	IV
160	661.4	IV
40	663.7	IV
40	664.8	IV
70	666.1	IV
110	744.9	IV
110	748.4	IV
110	750.2	IV
110	753.8	IV
40	798.3	IV
70	800.5	IV
70	804.0	IV
70	809.7	IV
110	816.0	IV
160	1062.7	IV
160	1073.0	IV
70	1073.5	IV
20	1108.4	IV
20	1110.9	IV
20	1624.0	IV
20	1629.2	IV
	Air	
20	2387.0	IV
40	2398.9	IV
110	3097.5	IV
40	3117.7	IV

S V
Ref. 29 — R.L.K.

Intensity	Wavelength	
	Vacuum	
5	437.4	V
5	438.2	V
5	439.6	V
40	658.3	V
70	659.8	V
110	663.2	V
5	676.2	V
5	677.3	V
20	678.1	V
40	680.3	V
110	680.9	V
40	681.6	V
5	686.2	V
5	686.9	V
5	689.8	V
5	691.7	V
20	693.5	V
285	786.5	V
160	849.2	V
110	852.2	V
110	854.8	V
110	857.9	V
110	860.5	V
20	883.6	V
20	884.5	V
5	885.8	V
20	900.9	V
5	902.8	V
20	905.9	V

TANTALUM (Ta)
Z = 73
Ta I and II
Ref. 1 — C.H.C.

Intensity		Wavelength	
		Air	
1100		2140.13	II
1500		2146.87	II
740		2150.62	II
600		2165.01	II
740		2178.03	II
1200		2182.71	II
540		2193.20	II
1100		2193.88	II
1500		2196.03	II
1500		2199.67	II
500		2207.14	II
1400	d	2210.03	II
		2210.19	II
420		2215.60	II
1400		2239.48	II
240		2248.48	II
480		2249.79	II
1200		2250.76	II
260		2254.86	II
440		2255.77	II
360		2256.51	II
500		2258.71	II
840		2261.42	II
260		2261.62	II
990		2262.30	II
220		2269.56	II
740		2271.85	II
990		2272.59	II
200		2279.85	I
320		2282.10	II
130		2285.02	II
790		2285.25	II
600		2286.59	II
240		2287.27	II
990		2289.16	II
180		2292.54	II
160		2295.18	II
160		2301.47	II
440		2302.24	II
440		2302.93	II
300		2303.49	II
100		2308.46	II
440		2312.60	II
420		2315.46	II
260		2319.16	II
100		2331.29	II
690		2331.98	II
550		2332.19	II
110		2334.13	II
180		2334.88	II
140		2335.75	II
300		2338.28	II
200		2340.94	II
200		2341.61	II
130		2343.64	II
100		2346.42	II
90		2351.99	II
170		2353.86	II
120		2355.22	II
170		2356.05	II
140		2356.90	II
250		2357.30	I
170		2359.16	II
260		2361.09	I
160		2362.78	II
130		2363.32	II
600		2364.24	II
50		2367.24	II
150		2369.32	II
300		2370.76	II
320		2371.58	I
100		2373.94	II
70		2375.91	I
150		2378.31	II
440		2381.13	II
240		2381.52	II
170		2383.72	II
240		2384.28	II
130		2385.73	I
1400		2387.06	II
80		2388.37	II
160		2389.11	II
70		2396.30	I
110		2399.15	II
50		2399.92	II
2400		2400.63	II
140		2402.13	II
100		2403.68	II
130		2406.55	I

Tantalum (Cont.)

Int.		λ	Sp.
130		2408.26	II
120		2414.32	I
240		2415.21	II
320		2416.89	II
220		2417.86	II
150		2418.77	II
140		2421.03	I
150		2421.85	II
170		2423.48	II
130		2425.91	II
360		2427.64	I
360		2429.71	II
170		2431.06	II
480		2432.70	II
130		2433.59	II
130		2436.51	II
110		2437.07	I
110		2438.64	II
200		2439.91	I
130		2442.39	I
100		2444.13	II
100		2447.17	I
100		2454.48	I
100		2458.68	I
100		2460.55	I
160		2463.82	II
130		2466.99	II
130		2467.37	II
380		2470.90	II
120		2471.38	I
120		2472.13	I
150		2473.13	I
120		2473.31	II
600		2474.62	I
120		2475.33	I
200		2476.67	II
150		2478.22	I
120		2481.86	II
100		2482.10	I
100		2482.58	II
100		2484.04	II
500		2484.95	I
120		2486.70	I
600		2488.70	II
500		2490.46	I
600		2504.45	I
600		2507.45	I
240		2512.65	I
1200	d	2526.35	I
600		2532.12	II
240		2545.49	II
240		2546.80	I
460	d	2551.07	I
460		2554.62	II
240		2555.05	I
1200		2559.43	I
460		2562.10	I
340		2571.51	II
430		2573.54	I
390		2573.79	I
600		2577.37	II
340		2577.78	I
210		2580.16	I
340		2584.03	II
430		2593.08	I
410		2593.66	II
310		2594.25	II
560		2595.26	I
310	l	2596.45	II
220		2600.14	I
600		2603.49	II
1400		2608.63	I
210		2609.00	I
310	d	2611.34	I
340		2615.46	I
310		2615.66	I
1200		2635.58	II
470		2636.67	I
860		2636.90	I
510		2646.22	I
600		2646.37	I
2400		2647.47	I
270		2651.22	II
2600		2653.27	I
1900		2656.61	I
1500		2661.34	I
220		2665.60	II
220		2668.07	I
600		2668.62	I
770		2675.90	II
270		2680.06	II
220		2680.66	II
600		2684.28	I
1500		2685.17	II
340		2691.31	I
260		2692.40	I
470		2694.52	II
240		2696.81	I
1000		2698.30	I
470		2706.69	I
310		2709.27	II
1200		2710.13	I
2600		2714.67	I
240		2717.18	I

Int.		λ	Sp.
470		2720.76	I
470		2727.44	II
410		2727.78	I
310		2736.25	II
210		2739.26	II
210		2743.59	I
510		2746.68	I
1200		2748.78	I
860		2749.83	I
410		2752.49	II
1000		2758.31	I
430		2761.68	II
770		2775.88	I
390		2787.69	I
680		2796.34	I
680		2797.76	II
380		2802.07	I
430		2806.30	I
510		2806.58	I
260		2817.10	II
260		2842.82	I
640		2844.25	I
290		2844.46	II
290	c	2845.35	I
560		2848.52	I
1500		2850.49	I
1900		2850.98	I
220		2858.44	II
360		2861.98	I
310		2868.65	I
470		2871.42	I
270		2873.36	I
260		2873.56	I
210		2874.17	I
380		2880.02	I
770		2891.84	I
260		2899.04	I
560		2902.05	I
210		2914.12	I
310		2915.49	I
410		2925.19	I
310		2932.70	I
1700		2933.55	I
470		2940.06	I
1200		2940.22	I
240		2942.14	I
510		2951.92	I
340		2953.56	I
1500		2963.32	I
770		2965.13	II
770		2965.54	I
340		2969.47	I
430		2975.56	I
210		3011.88	I
1800		3012.54	II
290	d	3027.48	I
290		3042.06	II
530		3049.56	I
530		3069.24	I
360		3077.24	I
560		3103.25	I
380		3124.97	I
380		3130.58	I
270		3132.64	I
320		3170.29	I
270		3173.59	I
200		3176.29	I
600		3180.95	I
240		3184.55	I
200		3198.67	I
200		3213.91	II
300		3223.83	I
230		3229.24	I
200		3242.05	I
200		3242.83	I
210		3274.95	II
1100		3311.16	I
210		3317.93	I
680		3318.84	I
330	d	3330.99	II
230		3358.47	I
640		3371.54	I
360		3385.05	I
230		3398.33	I
450		3406.94	I
230		3463.77	I
490		3480.52	I
380		3497.85	I
240		3503.87	I
490		3511.04	I
200		3513.61	I
750		3607.41	I
980		3626.62	I
500		3642.06	I
100		3686.18	I
100		3689.73	I
130		3731.02	I
140		3736.76	I
130		3746.36	I
110		3754.52	I
110		3777.10	I
110		3792.02	I
210		3833.74	II
100		3848.05	I

Int.		λ	Sp.
100		3885.20	I
210		3918.51	I
140		3922.78	I
140		3922.92	I
210		3970.10	I
210		3996.17	I
100		3999.28	I
190		4006.84	I
190		4026.94	I
140		4029.94	I
120		4040.87	I
410		4061.40	I
210		4064.63	I
100		4067.24	I
310		4067.91	I
120		4105.02	I
210		4129.38	I
230		4136.20	I
230		4147.89	I
210		4175.21	I
100		4177.92	I
130		4181.15	I
300		4205.88	I
120		4206.40	I
130		4245.35	I
130		4268.26	I
160	c	4302.98	I
110		4355.14	I
100		4378.82	I
150		4386.07	I
110		4398.45	I
180		4402.50	I
130		4415.74	I
360	c	4510.98	I
190		4530.85	I
130		4551.95	I
170		4565.85	I
340		4574.31	I
260		4619.51	I
130		4669.14	I
450		4681.88	I
130		4691.90	I
150		4740.16	I
220		4756.51	I
120		4768.98	I
220		4812.75	I
110		4920.11	I
100		4921.27	I
110		4926.00	I
150		4936.42	I
200		5037.37	I
100		5067.87	I
110		5115.84	I
100		5141.62	I
100		5143.69	I
330		5156.56	I
110		5212.74	I
110	d	5218.45	I
		5218.66	I
140		5341.05	I
200		5402.51	I
130		5419.19	I
18		5500.68	I
20		5505.66	I
15	c	5516.27	I
90		5518.91	I
9		5521.15	I
10		5523.98	I
13		5528.36	I
10	l	5545.20	I
20		5548.32	I
30		5584.02	I
15		5598.75	I
30		5599.52	I
9	c	5605.50	I
9	c	5617.71	I
40		5620.68	I
13		5628.20	I
20		5635.71	I
40		5640.18	I
150		5645.91	I
130		5664.90	I
30		5688.25	I
40		5699.24	I
15		5704.31	I
25		5706.28	I
30		5715.24	I
8		5716.53	I
23		5746.71	I
30		5755.81	I
15	h	5761.61	I
25		5766.56	I
30	c	5767.91	I
10		5771.93	I
130		5776.77	I
25		5780.02	I
90		5780.71	I
130		5811.10	I
25		5816.51	I
45	c	5843.94	I
13		5849.68	I
15		5866.61	I
240		5877.36	I
130		5882.30	I

Int.		λ	Sp.
90		5901.91	I
30		5916.51	I
90		5918.95	I
15		5925.90	I
15		5930.62	I
23		5931.05	I
20		5931.68	I
18		5935.54	I
130		5939.76	I
240		5944.02	I
25		5951.78	I
18		5960.13	I
190	c	5997.23	I
25	h	6009.89	I
25		6015.90	I
100		6020.72	I
250		6045.39	I
100		6047.25	I
25		6053.70	I
30		6090.82	I
18		6092.06	I
100		6101.58	I
25		6140.07	I
65		6144.56	I
30		6152.54	I
130		6154.50	I
40		6158.84	I
15		6170.46	I
15		6189.66	I
15		6193.11	I
25		6208.37	I
40		6249.79	I
150		6256.68	I
150		6268.70	I
50		6278.34	I
65		6281.33	I
15		6287.36	I
40		6287.91	I
40		6289.34	I
50		6309.06	I
150		6309.58	I
25	c	6312.22	I
75		6325.08	I
50		6332.91	I
65		6341.17	I
30		6346.02	I
75		6356.16	I
65		6360.84	I
40		6373.06	I
15		6379.07	I
90		6389.45	I
23	h	6392.21	I
65		6428.60	I
250		6430.79	I
13		6437.36	I
40		6444.61	I
30		6445.87	I
200		6450.36	I
20		6455.83	I
30		6459.92	I
380		6485.37	I
18	h	6502.43	I
65		6505.52	I
100		6514.39	I
100		6516.10	I
25	h	6561.60	I
25	Cw	6564.26	I
100		6574.84	I
10		6585.13	I
15	c	6587.16	I
110		6611.95	I
75		6621.30	I
15	Cw	6662.24	I
100		6673.73	I
180		6675.53	I
30		6684.00	I
15		6693.61	I
15		6706.46	I
25		6709.39	I
10		6714.44	I
15	h	6723.61	I
75	c	6740.73	I
40		6754.91	I
13	c	6755.85	I
13		6770.37	I
75		6771.74	I
40	Cw	6774.25	I
40	Cw	6788.99	I
13	c	6790.06	I
13		6799.27	I
40	c	6810.46	I
160	c	6813.25	I
20		6819.36	I
18	c	6824.96	I
13		6832.00	I
15		6850.83	I
15		6865.13	I
210		6866.23	I
180		6875.27	I
40		6877.49	I
15		6896.77	I
40		6900.55	I
150		6902.10	I
140		6927.38	I

Tantalum (Cont.)

Intensity		Wavelength	
140		6928.54	I
8	h	6939.33	I
20	c	6946.87	I
65		6951.26	I
45		6953.88	I
180		6966.13	I
8		6969.49	I
8	c	6971.31	I
9		6971.53	I
23		6983.52	I
110	d	6995.22	I
		6995.49	I
20		7000.21	I
40		7005.07	I
75		7006.96	I
50		7025.03	I
13		7031.51	I
40		7039.07	I
15	h	7081.30	I
20		7085.40	I
23		7093.02	I
8		7108.05	I
15	c	7117.52	I
20		7121.27	I
40		7125.72	I
150		7148.63	I
110		7172.90	I
13	c	7174.91	I
13		7191.35	I
8		7233.45	I
30	h	7250.27	I
11	h	7264.82	I
6		7272.29	I
30		7276.96	I
5		7277.54	I
9		7286.36	I
13	c	7296.32	I
140		7301.74	I
20		7319.84	I
11	c	7322.72	I
13		7325.95	I
11	Cw	7340.19	I
160		7346.41	I
140	c	7352.86	I
100		7356.96	I
90	Cw	7369.09	I
160		7407.09	I
11	c	7435.19	I
25		7440.17	I
30		7467.75	I
23		7486.01	I
30		7520.56	I
6		7569.23	I
9		7590.22	I
6		7649.62	I
11	h	7722.02	I
11		7763.11	I
9	c	7779.67	I
20	c	7842.76	I
100		7882.37	I
30		7950.19	I
5		7952.07	I
6		7998.75	I
6		8022.09	I
75		8026.50	I
5		8029.04	I
15		8039.08	I
8		8053.93	I
15		8068.98	I
5		8100.11	I
13		8128.76	I
9		8158.54	I
5	c	8180.74	I
13	c	8248.95	I
20	d	8264.85	I
75		8281.62	I
11		8389.06	I
5		8415.73	I
25	Cw	8447.62	I
11	h	8550.49	I
15		8575.92	I
10	Cw	8595.84	I

Ta IV
Ref. 411 — R.L.K.

Intensity	Wavelength	
	Vacuum	
10	763.14	IV
32	934.41	IV
67	999.34	IV
65	1063.53	IV
68	1067.17	IV
71	1074.47	IV
71	1086.39	IV
72	1094.60	IV
68	1100.13	IV
79	1116.10	IV
71	1118.83	IV
78	1136.17	IV
66	1138.26	IV
75	1149.72	IV
75	1150.42	IV
75	1151.92	IV
76	1172.51	IV
85	1175.51	IV
80	1189.28	IV
78	1192.52	IV
80	1192.67	IV
78	1211.94	IV
80	1212.68	IV
85	1213.09	IV
70	1214.66	IV
85	1215.53	IV
80	1220.73	IV
80	1220.96	IV
90	1223.73	IV
88	1238.12	IV
95	1240.06	IV
87	1258.34	IV
94	1264.91	IV
98	1272.42	IV
94	1275.48	IV
86	1275.94	IV
92	1308.51	IV
85	1311.35	IV
87	1315.58	IV
81	1325.19	IV
92	1332.38	IV
86	1343.30	IV
75	1350.46	IV
92	1365.88	IV
79	1376.62	IV
78	1388.80	IV
91	1398.78	IV
93	1413.40	IV
79	1430.11	IV
83	1441.54	IV
91	1454.32	IV
92	1464.41	IV
93	1469.82	IV
90	1495.25	IV
95	1514.19	IV
70	1525.69	IV
82	1565.97	IV
82	1584.64	IV
84	1594.91	IV
85	1607.70	IV
84	1631.65	IV
79	1639.82	IV
84	1668.76	IV
84	1678.43	IV
85	1712.16	IV
85	1716.13	IV
82	1753.90	IV
82	1759.04	IV
79	1763.03	IV
83	1865.92	IV
84	1901.63	IV
84	1907.66	IV
84	1924.75	IV
77	1940.25	IV
79	1985.68	IV
82	1989.44	IV
	Air	
85	2055.75	IV
83	2079.01	IV
75	2111.53	IV
90	2199.58	IV
90	2207.64	IV
68	2697.42	IV
10	3076.06	IV

Ta V
Ref. 426 — R.L.K.

Intensity	Wavelength	
	Vacuum	
20	478.29	V
60	493.07	V
200	841.31	V
1000	890.87	V
500	947.30	V
100	990.29	V
200	1066.64	V
200	1140.49	V
500	1213.42	V
100	1242.98	V
5000	1392.56	V
7000	1709.10	V

TECHNETIUM (Tc)
Z = 43

Tc I and II
Ref. 35 — C.H.C.

Intensity		Wavelength	
		Air	
15		2106.23	II
20		2116.44	II
15		2119.41	I
30		2156.27	I
30		2185.39	I
30		2189.06	I
40		2193.35	I
10		2266.22	II
10		2282.12	I
10		2282.71	I
50		2285.45	I
100		2298.08	I
30		2416.22	I
50		2423.23	I
20		2424.54	I
20		2435.83	I
10		2436.99	I
80	w	2463.69	I
20		2465.09	I
30		2466.87	I
20		2475.11	I
50		2480.70	I
50		2483.22	I
20		2486.50	I
25		2492.72	I
20		2493.43	I
100		2496.77	II
30		2510.17	I
80		2529.34	II
500		2543.23	II
60		2544.81	II
50		2547.92	II
50		2558.61	II
50		2567.01	II
30		2575.06	II
80		2576.28	II
40		2577.86	II
300	H1	2578.79	I
200		2589.86	I
20	w	2590.19	I
100		2592.82	I
20		2597.19	II
500		2600.06	I
1000	c	2609.99	II
1500		2614.23	I
1000		2615.87	I
30		2618.28	I
200		2634.91	II
80		2636.36	I
30		2641.26	I
100		2642.37	I
40		2644.50	II
1000	c	2647.01	II
300	c	2649.21	I
100		2652.35	II
30		2653.57	I
100		2654.31	I
120		2660.88	I
100		2662.30	I
80		2681.19	II
60		2683.14	I
80		2683.89	I
80		2693.11	I
50		2696.64	I
70		2702.27	I
40		2702.96	II
100		2707.90	II
1000		2708.78	I
30		2715.20	I
30		2723.55	I
1000		2726.69	I
30	c	2728.47	I
500		2730.53	I
300		2732.87	I
150		2736.23	I
60	c	2736.83	II
100		2737.97	I
20	c	2738.83	II
100		2755.76	I
100		2762.13	I
200		2762.34	I
60		2765.95	I
500		2766.89	I
20		2777.31	II
150		2778.91	I
25		2781.22	I
1000		2782.05	I
500		2785.59	I
40		2788.89	I
500		2789.25	I
100		2794.53	I
80		2795.65	I
200		2795.78	II
1000		2802.81	I
150		2803.02	I
500		2808.36	I
50	c	2809.65	II
500		2811.61	II
30		2814.86	I
40		2819.46	I
100		2821.35	II
200		2828.04	I
60		2831.18	II
50		2840.38	II
60		2845.04	I
10		2846.39	II
60		2849.20	I
150		2850.96	I
500	h	2857.13	I
2000	c	2859.11	I
500		2864.49	I
100		2868.09	I
1000		2887.73	I
100		2888.46	I
30		2889.20	II
200		2893.16	I
150		2893.45	I
200		2894.32	I
1000		2896.34	I
40		2903.81	I
1000		2913.15	I
500		2921.91	I
20	c	2923.34	II
1000		2928.20	I
80		2933.89	I
200		2955.93	I
200		2973.65	I
100		2979.34	I
150		2985.36	I
100		3010.83	I
300		3017.23	I
150		3021.56	I
100		3022.66	I
200		3023.68	I
80		3025.26	I
300	w	3026.89	I
50		3033.16	I
80		3034.57	I
40		3036.88	I
20		3037.90	II
100		3038.23	I
40		3042.64	I
100	h	3051.55	I
40		3052.47	I
80		3062.11	I
200		3062.36	I
300		3064.67	I
100	c	3066.60	I
120	c	3068.34	I
80		3076.24	I
150		3089.34	I
1000		3099.10	I
200		3099.52	I
60		3108.25	I
40		3109.15	I
60		3115.98	I
80		3119.17	I
40		3119.66	I
700		3122.64	I
1500		3131.23	I
40	c	3150.26	I
300		3161.67	I
3000		3173.30	I
200		3180.30	I
2000		3182.37	I
2000		3183.11	I
800	c	3195.20	II
40	w	3197.53	I
300	c	3202.83	I
1000		3212.02	II
40		3220.74	I
60		3230.02	I
1000		3237.02	II
100		3241.84	I
500		3244.19	I
300		3252.05	I
40		3256.10	I
40		3261.94	I
100		3287.14	I
30		3298.84	II
100		3300.77	I
80		3305.89	I
200		3310.65	I
150		3313.65	I
200		3325.55	I
150	c	3327.10	I
100		3330.77	I
50		3332.47	I
60		3350.56	I
50	c	3350.83	I
400		3366.75	I
40		3386.67	I
50		3392.23	I
300		3394.18	I
60		3396.90	I
40		3397.83	I
300		3398.33	I
200		3402.10	I
200	c	3403.93	I
80		3405.33	I
80		3407.28	I
50		3408.33	I
40		3411.80	I
40	c	3418.20	I
100		3419.10	I

Technetium (Cont.)

Int.		λ	Sp.
60		3427.85	I
40		3431.75	I
200		3434.70	I
40		3435.68	I
150		3437.44	I
80		3438.73	I
200	c	3443.47	I
200		3451.05	I
200		3456.85	I
400		3457.24	I
40		3457.60	I
5000	c	3466.28	I
150		3470.51	I
80		3475.18	I
1000		3475.59	I
60		3484.62	I
1000	c	3486.23	I
100		3490.30	I
400		3493.39	I
500		3494.62	I
40		3499.14	I
1000		3500.70	I
200		3501.24	I
800	c	3502.70	I
100		3507.19	I
100		3508.27	I
100		3510.91	I
800		3525.83	I
300		3526.18	I
100		3529.83	I
150		3534.88	I
500		3535.51	I
300		3538.12	I
800		3538.68	I
2000	c	3541.77	I
6000	c	3549.72	I
4000	c	3550.64	I
300		3559.75	I
800		3560.32	I
100		3565.22	I
800		3568.85	I
100		3570.65	I
100		3575.42	I
1000		3580.06	I
600		3581.26	I
800		3582.08	I
2000		3582.63	I
4000		3587.94	I
200		3593.47	I
300		3594.57	I
1000	c	3595.66	I
1000	c	3607.32	I
200		3607.62	I
2000	c	3608.27	I
200		3618.94	I
1000	c	3627.36	I
200		3630.39	I
3000	c	3635.15	I
10000	c	3636.07	I
1000		3638.22	I
200		3638.85	I
900		3639.38	I
400		3640.23	I
1000	c	3648.04	I
600		3651.47	I
1000	c	3658.59	I
400	c	3661.45	I
200		3664.92	I
1000		3679.15	I
300		3680.32	I
5000		3684.74	I
300		3692.76	I
800		3703.83	I
300		3704.80	I
200		3706.70	I
200		3707.63	I
200		3708.26	I
1000		3712.26	I
300		3712.82	I
500		3715.94	I
10000		3718.86	I
1500		3723.67	I
2000		3724.40	I
5000		3726.35	I
200		3727.36	I
400	c	3729.18	I
500		3731.74	I
300		3737.42	I
400		3745.01	I
1000		3746.15	I
5000		3746.84	I
1000		3752.13	I
4000		3754.37	I
1000		3758.54	I
2000		3761.81	I
5000		3768.77	I
3000		3771.03	I
500		3777.27	I
2000		3779.37	I
3000	c	3780.68	I
500		3784.06	I
200		3786.06	I
500		3791.28	I
300		3791.73	I
200		3797.44	I
1000		3797.77	I
200		3814.67	I
300		3816.89	I
300		3824.47	I
500		3828.54	I
200		3830.35	I
200		3832.45	I
600		3832.82	I
1500		3837.56	I
800		3841.31	I
800		3845.97	I
500		3847.60	I
300		3851.22	I
500	c	3856.73	I
200		3863.07	I
400		3864.11	I
1000		3868.24	I
200		3875.66	I
500	c	3879.16	I
600	c	3880.72	I
300	w	3892.12	I
200		3893.22	I
600		3899.83	I
300		3919.38	I
300	c	3923.66	I
200		3927.57	I
200		3933.70	I
4000	c	3946.57	I
2000		3947.09	I
200		3955.73	I
300		3979.64	I
500		3980.35	I
10000	c	3984.97	I
400		3987.78	I
300		3994.04	I
2000		3994.51	I
200		3996.97	I
300		4004.69	I
500		4007.14	I
1000		4012.00	I
400		4016.68	I
600		4017.22	I
2000		4020.76	I
20000	c	4031.63	I
1000		4039.25	I
200		4041.78	I
10000	c	4049.11	I
500		4051.95	I
200		4053.18	I
200	c	4056.08	I
400		4083.54	I
10000		4088.71	I
2000		4093.69	I
15000		4095.67	I
1000		4110.22	I
10000		4115.08	I
600		4119.27	I
8000		4124.22	I
1000		4128.27	I
300		4134.81	I
300		4139.12	I
800		4139.85	I
400		4141.27	I
6000		4144.95	I
3000		4145.08	I
200		4147.62	I
10000		4165.61	I
500		4167.42	I
1000		4169.68	I
4000		4170.27	I
5000		4172.53	I
1000		4176.28	I
800		4186.51	I
300		4218.61	I
10000	c	4238.19	I
20000		4262.27	I
1000		4262.69	I
800		4274.97	II
800		4278.90	I
30000		4297.06	I
400	c	4336.86	I
400		4358.49	I
200		4359.26	I
1000		4429.59	I
1000		4481.53	I
3000		4487.06	I
400		4495.03	I
1000		4515.98	I
10000		4522.84	I
2000		4539.53	I
400		4542.09	I
400		4552.20	I
800		4552.85	I
1000		4557.05	I
2000		4564.54	I
1000		4578.45	I
1000		4593.35	I
300	c	4609.16	I
1000		4616.86	I
200	c	4622.69	I
300		4624.96	I
1000		4630.57	I
200		4633.15	I
3000		4637.50	I
500		4643.28	I
2000		4648.33	I
2000	c	4660.21	I
2000		4669.30	I
400		4672.17	I
200		4678.90	I
400		4689.36	I
300		4694.28	I
1000		4706.92	I
200		4714.22	I
2000		4717.77	I
500	c	4719.02	I
4000	c	4719.28	I
200	c	4736.51	I
10000		4740.61	I
500		4749.61	I
1000		4752.72	I
200		4762.36	I
4000		4771.54	I
200		4773.89	I
200		4783.94	I
500		4785.60	I
200	c	4790.48	I
250		4791.62	I
300		4799.98	I
100		4805.69	I
100		4809.42	I
500		4816.79	I
10000		4820.74	I
300		4831.35	I
1000		4834.37	I
1000		4835.39	I
100		4841.36	I
20000		4853.59	I
100		4857.21	I
100	c	4862.19	I
10000		4866.73	I
200		4870.77	I
100		4888.70	I
150	c	4890.88	I
8000		4891.92	I
150		4892.49	I
1000		4908.51	I
2000		4909.57	I
500		4913.02	I
150	c	4914.70	I
200		4920.67	I
300		4923.60	I
400		4948.06	I
5000		4976.34	I
400		4995.00	I
200		5002.67	I
100		5005.74	I
200	c	5014.52	I
500		5026.24	I
300		5026.79	I
150		5027.89	I
80		5032.45	I
300		5055.27	I
60		5058.33	I
500		5060.69	I
80		5090.74	I
5000		5096.28	I
200	c	5103.24	I
500		5104.32	I
200		5109.81	I
100		5120.60	I
500		5139.26	I
500		5150.63	I
2000		5161.81	I
2000		5174.81	I
100		5206.56	I
200		5225.55	I
200	c	5260.22	I
200	c	5261.44	I
1000		5275.51	I
800		5285.07	I
100		5305.31	I
400		5314.96	I
600		5320.20	I
200		5334.79	I
500	c	5353.48	I
200		5356.63	I
300		5358.65	I
200		5360.14	I
500	h	5375.20	I
150		5423.05	I
200		5447.40	I
500	c	5451.90	I
100		5455.95	I
300		5471.96	I
70		5483.01	I
60		5485.37	I
80		5506.89	I
150		5524.11	I
100		5528.23	I
200		5541.94	I
80		5543.63	I
100		5550.53	I
3000	c	5589.02	I
200		5602.23	I
2000	c	5620.45	I
300		5629.94	I
1500		5642.13	I
800		5644.94	I
100		5656.00	I
60		5672.15	I
200		5687.30	I
200		5689.05	I
700		5725.31	I
500	c	5771.47	I
100		5794.65	I
80	c	5799.85	I
100		5814.24	I
200		5831.48	I
150	c	5836.33	I
150		5923.36	I
1000	c	5924.47	I
200		5926.29	I
600	c	5931.93	I
60		6032.36	I
60		6047.99	I
200		6065.09	I
800		6085.23	I
300		6099.39	I
500	c	6102.96	I
1000		6120.68	I
1000		6130.80	I
150	c	6132.23	I
100		6184.70	I
800		6192.66	I
600	c	6244.18	I
100		6312.18	I
100		6354.86	I
100		6356.73	I
80		6389.87	I
100		6408.83	I
1000		6455.90	I
600	c	6461.93	I
100		6470.27	I
200	c	6491.68	I
200		6526.82	I
150		6579.24	I
500	c	6625.57	I
300	c	6673.66	I
100		6687.10	I
80		6786.00	I
70		6798.63	I
60		6856.90	I
150		7002.37	I
100		7016.57	I
500	c	7086.18	I
60		7093.12	I
200		7141.28	I
200	c	7157.62	I
70		7256.08	I
100		7322.38	I
80		7329.14	I
100		7396.80	I
100	c	7402.61	I
200		7405.36	I
60		7427.15	I
150		7434.12	I
600		7452.49	I
60		7461.59	I
80		7534.95	I
800		7540.26	I
80		7543.39	I
200		7574.02	I
500		7579.26	I
90		7624.53	I
100		7684.45	I
500		7697.37	I
80	c	7698.19	I
800		7793.04	I
60		7798.28	I
60		7816.74	I
800		7817.72	I
100		7856.38	I
200		7861.44	I
400	d	7871.25	I
60	c	7874.76	I
70		7965.45	I
500		7999.73	I
200		8126.55	I
200		8170.55	I
150		8205.27	I
100		8206.49	I
150		8211.31	I
500	c	8237.08	I
200		8308.15	I
200		8309.16	I
60		8315.50	I
100		8531.06	I
100		8543.61	I
100	c	8707.21	I
100	c	8737.93	I
200	c	8829.82	I

TELLURIUM (Te)
Z = 52

Te I and II
Ref. 1, 344—347—C.H.C.

Tellurium (Cont.)

TERBIUM (Tb)
Z = 65

Tb I and II
Ref. 1 — C.H.C.

Vacuum

Intensity	Wavelength	
6	799.60	II
8	802.28	II
6	942.62	II
6	1003.73	II
6	1007.80	II
5	1014.27	II
5	1022.79	II
6	1057.00	II
8	1059.51	II
6	1068.86	II
8	1077.66	II
6	1090.11	II
6	1144.04	II
5	1153.10	II
10	1161.42	II
10	1174.34	II
12	1175.79	II
9	1208.54	II
5	1213.00	II
9	1220.98	II
9	1253.62	II
9	1270.52	II
7	1274.76	II
8	1306.53	II
10	1324.92	II
7	1336.42	II
7	1345.20	II
9	1363.24	II
8	1366.73	II
10	1374.80	II
6	1395.22	II
6	1439.52	II
6	1465.25	II
7	1489.56	II
8	1607.99	II
10	1608.41	II
10	1613.15	II
6	1638.91	II
5	1655.4	I
5	1688.5	I
6	1700.0	I
6	1701.58	II
3	1708.0	I
6	1751.0	I
5	1759.4	I
5	1775.0	I
6	1795.7	I
6	1796.3	I
10	1822.4	I
6	1825.5	I
6	1850.6	I
6	1852.1	I
6	1853.8	I
8	1857.2	I
6	1860.4	I
3	1962.88	II
7	1994.83	I

Air

Intensity	Wavelength	
6	2000.2	I
26000	2002.02	I
8	2070.9	I
6500	2081.16	I
18000	2142.81	I
3200	2147.25	I
360	2159.85	I
9	2208.74	I
10	2255.49	I
500	2259.02	I
10	2265.52	I
20	2373.06	II
1200	2383.26	I
1500	2385.78	I
20	2387.82	II
10	2401.63	II
10	2436.47	II
50	2438.69	II
120	2530.72	I
20	2567.82	II
10	2574.96	II
5	2576.10	II
7	2579.24	II
10	2591.12	II
10	2592.85	II
10	2605.72	II
5	2621.92	II
10	2624.86	II
20	2627.96	II
20	2641.89	II
20	2648.48	II
100	2649.66	II
40	2657.70	II
80	2661.10	II
110	2677.13	I
20	2711.58	II
6	2769.65	I
10	2841.17	II
10	2846.15	II

Intensity	Wavelength	
100	2858.29	II
20	2861.00	II
40	2868.82	II
150	2895.41	II
30	2919.89	II
50	2942.11	II
50	2946.68	II
70	2967.29	II
20	2973.67	II
50	2975.90	II
15	2997.04	II
15	3012.02	II
50	3017.58	II
20	3023.31	II
70	3047.00	II
20	3052.46	II
10	3063.16	II
15	3073.56	II
8	3104.44	II
10	3132.58	II
20	3160.66	II
100	3175.14	I
10	3189.83	II
5	3211.21	II
60	3256.80	II
30	3268.77	II
30	3282.67	II
40	3321.92	II
40	3323.11	II
60	3329.22	II
60	3352.10	II
60	3362.79	II
25	3374.10	II
150	3406.79	II
20	3419.63	II
50	3442.25	II
40	3455.12	II
20	3456.88	II
20	3480.32	II
40	3483.67	II
20	3486.11	II
50	3521.11	II
50	3552.19	II
100	3611.78	II
50	3617.57	II
40	3644.46	II
20	3679.13	II
30	3725.66	II
40	3797.22	II
20	3800.92	II
20	3905.67	II
20	3918.54	II
30	3931.49	II
20	3947.98	II
40	3969.22	II
25	3975.94	II
20	3981.77	II
50	4006.52	II
20	4011.69	II
30	4029.73	II
40	4047.17	II
30	4048.88	II
15	4073.48	II
30	4101.04	II
70	4127.32	II
30	4163.55	II
100	4169.77	II
30	4179.29	II
25	4211.31	II
80	4225.73	II
30	4246.47	II
20	4251.15	II
100	4261.11	II
30	4264.36	II
60	4273.43	II
80	4285.85	II
40	4320.90	II
30	4361.28	II
150	4364.00	II
30	4377.12	II
75	4385.10	II
60	4396.00	II
170	4478.63	II
80	4537.07	II
100	4557.78	II
70	4630.62	II
100	4641.12	II
180	4654.37	II
200	4686.91	II
100	4696.38	II
100	4706.53	II
100	4766.05	II
70	4771.56	II
100	4784.87	II
100	4827.14	II
150	4831.28	II
150	4842.90	II
130	4865.12	II
200	4866.24	II
80	4885.22	II
80	4904.44	II
60	4961.88	II
60	5000.82	II
8	5083.0	I

Intensity	Wavelength		
7	5148.7		I
50	5449.84		II
50	5487.95		II
150	5576.35		II
150	5649.26		II
100	5666.20		II
200	5708.12		II
7	5733.5		I
150	5755.85		II
8	5789.1		I
50	5936.15		II
100	5974.68		II
8	6273.5	d	I
8	6349.7	h	I
50	6367.13		II
8	6405.9		I
7	6456.7	h	I
8	6613.4	h	I
10	6648.58		II
8	6660.2		I
8	6690.0	h	I
10	6790.0	h	I
20	6837.6	h	I
20	6854.7	h	I
10	7016.06		II
10	7039.13		II
15	7191.1	h	I
10	7236.62		II
20	7263.5	h	I
8	7280.9		I
10	7289.26		II
10	7445.39		II
12	7460.98		II
15	7468.75		II
10	7481.26		II
10	7556.8		I
6	7688.61		II
15	7759.1		I
8	7818.79		II
8	7861.61		II
15	7921.69		II
15	7943.14		II
10	7950.34		II
20	7972.9		I
6	8056.15		II
30	8061.46	h	I
10	8082.5	h	I
10	8122.44		II
8	8130.39		II
8	8154.47		II
20	8166.44		II
6	8190.94		II
10	8251.5		I
15	8273.53		II
10	8276.6		I
10	8291.1		I
15	8355.8		I
10	8372.12		II
7	8469.8		I
8	8492.2		I
8	8500.8		I
12	8521.4		I
10	8535.68		II
12	8575.78		II
10	8604.63		II
8	8621.68		II
7	8632.1	h	I
15	8672.95		II
12	8701.09		I
10	8733.81		II
205	8758.18		I
12	8831.52		I
18	8851.15		I
6	8897.92		II
81	9004.37		I
18	9043.39		I
12	9196.80		I
15	9206.78		I
17	9207.64		I
12	9469.00		I
5660	9722.74		I
185	9785.54		I
109	9842.30		I
532	9868.92		I
118	9902.61		I
689	9956.30		I
37	9959.93		I
325	9977.13		I
136	9979.31		I
45	9985.85		I
5950	10051.41		I
4097	10091.01		I
104	10099.57		I
279	10106.05		I
381	10118.08		I
296	10151.06		I
397	10300.56		I
205	10323.05		I
745	10493.57		I
197	10509.86		I
1880	10918.34		I
298	11007.80		I
10200	11089.56		I
508	11163.74		I

Intensity	Wavelength	
6620	11487.23	I
280	11978.96	I
188	12566.24	I
389	12589.19	I
161	12805.50	I
400	13104.18	I
1580	13247.75	I
483	13316.63	I
217	14037.09	I
144	14072.53	I
434	14335.74	I
220	14417.46	I
1050	14513.51	I
129	14554.68	I
1480	15452.45	I
2430	15546.23	I
3760	16403.90	I
1960	17303.54	I
2780	18291.59	I
394	18777.30	I
269	19623.52	I
239	20147.54	I
1020	21043.73	I
464	21602.50	I
37	21799.64	I
74	22555.29	I
48	22755.66	I
27	23294.94	I
17	23978.70	I
25	24059.04	I
13	26428.62	I
38	26539.17	I
15	26553.74	I
7	27179.26	I

TERBIUM (Tb)
Z = 65

Tb I and II
Ref. 1 — C.H.C.

Air

Intensity	Wavelength	
29	2577.73	II
110	2584.61	II
29	2590.31	II
29	2591.42	II
24	2592.64	II
55	2597.71	II
40	2602.93	II
110	2608.57	II
40	2616.90	
130	2628.69	II
55	2655.96	II
50	2661.40	II
24	2661.64	II
55	2667.64	II
50	2668.86	II
140	2669.29	II
40	2674.13	II
40	2674.69	II
29	2678.15	II
40	2683.97	II
35	2687.82	II
50	2691.90	II
35	2693.05	II
55	2693.41	II
35	2695.46	II
50	2696.83	II
190	2704.07	II
130	2736.24	II
160	2759.47	II
270	2769.53	II
130	2784.49	II
180	2800.51	II
250	2802.75	II
250	2809.30	II
180	2812.64	II
190	2852.14	II
110	2857.68	II
230	2886.29	II
160	2894.45	II
320	2897.44	II
160	2898.86	II
110	2901.54	II
110	2910.30	II
160	2914.75	II
160	2915.30	II
190	2915.60	II
120	2916.24	II
120	2918.89	II
120	2924.16	II
120	2924.53	II
160	2932.89	II
150	2940.05	II
250	2956.21	II
170	2968.87	II
170	2977.78	II
110	2987.03	II
110	2988.57	II

Terbium (Cont.)

Int.		Wavelength	Sp.
130		2996.00	II
110		2999.03	II
130		3005.52	II
170		3009.30	II
230		3010.59	II
230		3016.18	II
130		3019.17	II
170		3020.29	II
110		3023.43	II
170		3027.33	II
230		3031.60	II
230		3044.96	II
190		3051.13	II
130		3053.24	II
460		3053.55	II
130		3062.78	II
230		3064.09	II
110		3065.69	II
230		3067.20	II
270		3069.03	II
460		3070.05	II
270		3072.60	II
670		3078.86	II
480		3082.36	II
120		3086.78	II
250		3088.43	II
480		3089.58	II
230		3102.54	II
480		3102.96	II
290		3117.89	II
290		3119.62	II
230		3121.94	II
230		3123.05	II
160		3124.54	II
110		3131.35	II
250		3134.26	II
440		3139.64	II
190		3140.06	II
230		3145.22	II
150		3146.67	II
310		3147.04	II
310		3147.15	II
310		3148.71	II
120		3155.62	II
130		3162.42	II
290		3162.93	II
190		3165.74	II
380		3167.52	II
140		3168.32	II
230		3169.84	II
190		3173.76	II
380		3174.66	II
380		3180.54	II
140		3183.88	II
480		3187.26	II
290		3188.03	II
190		3194.69	II
380		3195.60	II
480		3199.56	II
1100		3218.93	II
1200		3219.98	II
250		3230.03	II
250		3231.06	II
210	d	3239.60	II
250		3240.00	II
480		3252.32	II
250		3262.97	II
230	d	3263.87	II
230		3264.90	II
400		3266.40	II
250		3274.14	II
250		3274.33	II
210		3277.32	II
760		3280.31	II
760		3281.40	II
520		3283.10	II
1000		3285.04	II
310		3287.55	II
310		3291.56	II
1500		3293.07	II
210		3295.33	II
310		3298.66	II
210		3304.95	II
420	d	3307.44	II
210		3308.51	II
210		3314.38	II
340	d	3321.15	II
420		3322.28	II
210		3323.38	II
210		3323.89	II
3800		3324.40	II
520		3329.08	II
210		3334.48	II
250		3336.70	II
310		3338.03	II
250		3339.00	II
210		3347.27	II
210		3348.07	II
760		3349.42	II
320		3362.25	II
760		3364.93	II
230	d	3370.61	II
320		3371.50	II
520		3372.36	II

Int.		Wavelength	Sp.
460	d	3372.72	II
520		3375.03	II
320		3378.73	II
520		3378.86	II
320		3382.80	II
210		3390.60	II
380		3391.28	II
270		3398.35	II
320		3399.10	II
270		3400.53	II
210	d	3400.86	II
420		3402.33	II
210		3410.40	II
210		3410.68	II
520		3413.76	II
270		3416.24	II
400	d	3420.34	II
210		3430.61	II
320		3433.26	II
270		3439.72	II
520		3440.37	II
320		3444.58	II
210		3446.40	II
270		3449.46	II
810		3454.06	II
380		3460.38	II
230	d	3462.97	II
620		3468.03	II
270		3471.73	II
270		3472.37	II
810	d	3472.79	II
210		3473.00	II
380		3480.17	II
230		3483.04	II
230		3483.69	II
290	d	3489.51	II
210	d	3492.00	II
270		3494.21	II
270		3495.36	II
810		3500.84	II
570		3507.45	II
5700		3509.17	II
380		3510.10	II
320		3513.10	II
570		3519.76	II
1300		3523.66	II
380		3525.14	II
440		3525.61	II
440		3536.32	II
570		3537.94	II
1100		3540.24	II
810		3543.89	II
310	d	3551.03	
320		3551.96	II
460	d	3558.77	
3200		3561.74	II
480		3562.90	II
570		3565.74	II
810		3567.35	II
4200		3568.52	II
1600		3568.98	II
320		3572.07	II
1100		3579.20	II
710		3585.03	II
570		3587.44	II
810		3596.38	II
440		3598.06	II
1600		3600.44	II
320		3604.90	II
320		3611.33	II
320		3614.63	II
320	d	3615.66	II
320		3616.58	II
380		3617.86	II
380		3619.73	II
810		3625.54	II
570		3626.50	II
380		3629.44	II
670		3633.29	II
670		3638.46	II
670		3641.66	II
440		3647.06	II
570		3647.75	II
2300		3650.40	II
810		3654.88	II
2000		3658.88	II
450		3663.12	II
3800		3676.35	II
300		3677.89	II
810		3682.26	II
320		3688.15	II
610		3691.15	II
300		3692.95	II
450		3693.58	I
320		3696.85	II
450		3700.12	I
4700		3702.86	II
300		3703.12	I
2400		3703.72	II
370		3709.30	II
1000	d	3711.76	II
300		3719.05	II
650		3729.91	II
430		3732.39	II

Int.		Wavelength	Sp.
430		3743.09	II
650		3745.04	I
870		3747.17	II
870		3747.34	II
1100		3755.24	II
430		3757.44	II
430	d	3757.90	II
650		3759.35	I
350		3761.14	I
1700		3765.14	I
2100		3776.49	II
330		3779.22	II
600		3783.53	I
410	d	3787.22	II
410		3789.92	I
390		3792.20	I
600		3793.55	II
330		3801.80	II
760	d	3806.85	II
1500		3830.26	I
540		3833.42	II
920	d	3842.50	II
370	d	3845.61	II
3700		3848.73	II
450	d	3869.75	II
3500	w	3874.17	II
330		3883.34	I
480		3888.22	I
490		3894.64	I
330		3895.99	I
330		3896.58	I
330		3897.89	I
2400		3899.20	II
1600		3901.33	I
480		3908.06	I
380		3909.14	I
330		3909.55	I
650		3915.43	I
480		3919.52	II
300		3922.10	II
480		3922.74	II
760		3925.45	II
650		3935.24	II
810	d	3939.52	II
650		3946.89	II
350	d	3958.36	II
2200	d	3976.84	II
1800		3981.87	II
300		3983.85	II
350		3999.40	II
350	d	4002.19	II
970		4002.59	II
1900		4005.47	II
300		4010.04	I
760		4012.75	II
330		4013.26	I
370		4019.14	II
540		4020.47	II
220		4022.88	I
370		4024.77	I
520		4031.66	II
870		4032.28	I
2100		4033.03	II
350		4036.22	I
210		4038.86	I
300		4051.86	II
300		4052.87	II
430		4054.12	I
410		4060.37	I
220		4060.87	II
1300		4061.58	II
220		4063.89	II
390		4066.22	II
260		4075.24	I
390		4081.24	I
210		4086.60	I
210		4092.19	I
260		4094.37	II
260		4094.49	I
260		4103.90	II
650		4105.37	I
300		4112.50	I
260		4119.92	I
280		4143.51	I
1100		4144.41	II
350		4158.53	I
240		4169.09	I
240		4169.32	I
240		4171.05	I
240		4172.60	I
240		4172.82	I
260		4173.47	I
240		4186.21	I
300		4187.16	I
390		4196.74	I
450		4201.00	II
650		4203.74	I
600		4206.49	I
300	Cw	4213.50	I
300		4214.42	II
480		4215.09	I
300		4217.56	I
260		4219.16	I
260		4224.28	I

Int.		Wavelength	Sp.
480	Cw	4226.45	II
260		4231.89	I
480		4232.82	I
300		4235.35	I
370		4255.24	I
480		4258.23	II
260		4263.66	I
650		4266.34	I
330		4269.69	I
220		4275.21	I
760	Cw	4278.52	II
300		4285.13	II
300		4289.70	I
370		4298.36	I
300		4299.90	I
240		4302.95	I
240		4307.18	II
450		4310.42	I
300		4311.56	I
370		4313.25	I
2200		4318.83	I
600		4322.23	I
600		4325.83	II
3000		4326.43	I
240		4328.90	I
600		4332.12	I
870		4336.43	I
600		4337.64	I
1700		4338.41	I
700		4340.62	I
430	Cw	4342.53	I
430	d	4353.20	II
280		4356.09	I
870		4356.81	I
280		4360.16	I
220		4367.30	II
220		4372.02	I
330		4382.45	I
300		4388.23	I
260		4390.91	I
200		4416.27	II
140		4420.19	I
350		4423.10	I
110		4432.72	I
240		4436.12	I
110		4439.38	I
240		4448.04	I
110		4467.69	I
430		4493.07	I
45	d	4509.04	II
150	h	4511.52	I
45		4512.96	II
75		4514.31	II
45		4519.72	II
45		4525.01	II
45		4529.76	
45	h	4531.83	II
45		4534.13	I
45		4537.14	I
45		4537.23	I
110		4549.07	I
45		4549.72	II
110		4550.45	I
110		4556.46	I
55		4562.24	II
110		4563.69	II
30		4564.85	II
55		4573.19	II
210		4578.69	II
65		4584.84	II
65		4591.56	II
45		4592.38	I
45	h	4604.10	II
30		4611.96	I
45	h	4615.92	II
27		4617.49	I
30		4619.36	II
75	d	4626.32	II
95		4626.94	II
65		4632.07	I
65	h	4636.59	I
30		4636.99	II
85		4641.00	II
210		4641.98	II
260	Cw	4645.31	II
80		4647.23	I
60		4658.38	I
20		4658.73	
80		4662.79	I
50	c	4665.45	I
40		4669.40	I
80		4676.90	I
70	c	4681.87	I
50		4682.52	I
25	c	4682.79	II
80		4688.63	II
80		4693.11	II
30	h	4693.39	II
200		4702.41	II
110		4707.94	II
40	w	4716.07	II
40		4728.16	II
60	Cw	4734.20	I
80		4739.93	I

Terbium (Cont.)

Intensity		Wavelength	Spectrum
70		4747.80	I
410	Cw	4752.53	II
40		4758.44	II
40		4760.19	II
30		4762.37	II
25		4764.47	II
35		4778.36	II
35		4778.80	II
180		4786.78	I
40	Cw	4789.91	II
30		4801.87	II
100		4813.77	I
60		4837.59	II
25		4840.39	I
30	c	4842.69	II
30		4844.89	II
30		4854.81	I
20		4856.54	II
30		4858.87	II
80		4875.57	II
25		4876.12	II
80		4881.15	II
29		4894.33	
95		4915.90	I
35		4924.09	I
35		4926.83	I
50		4928.93	I
65		4931.79	I
29		4970.99	II
29		4971.42	I
29		4973.04	I
29		4980.16	II
29		4980.56	I
85		4993.82	II
50		4995.84	II
55		4997.95	I
29		5006.10	II
50		5022.16	I
29		5024.24	II
29		5024.65	I
50		5033.12	I
50	w	5042.06	II
55		5054.30	I
55		5065.79	I
110		5078.25	I
24		5080.05	II
24		5081.11	I
75		5089.12	II
24		5089.66	I
24		5101.09	I
24		5108.56	I
35		5118.39	I
24		5120.18	I
50	w	5131.69	I
50	w	5141.08	II
50		5147.58	I
24		5164.27	I
29		5170.13	I
24		5170.61	I
50		5176.51	I
50		5179.97	I
50		5184.59	I
85		5186.13	I
50		5188.48	I
50		5198.86	I
35	w	5202.77	I
40		5204.55	I
40		5207.97	I
40		5214.28	I
40		5221.99	I
120		5228.12	I
40		5235.11	I
75		5248.71	I
75	w	5262.11	II
24		5275.03	I
75		5281.05	I
65		5304.72	I
29		5308.19	I
29		5309.46	I
110		5319.23	I
35		5331.04	I
65	w	5337.90	I
35	d	5338.59	I
24		5347.83	II
160		5354.88	I
75		5369.72	I
75		5375.98	I
29	d	5402.06	II
29		5413.65	I
29		5416.20	I
50		5424.10	II
29		5426.43	I
35		5443.38	I
29		5457.00	I
55		5459.81	I
29	w	5470.34	II
24		5481.45	I
55		5509.61	I
50		5514.54	I
65		5524.12	I
24	c	5525.62	I
35		5565.93	I
29	c	5638.80	I
29	c	5685.74	II

Intensity		Wavelength	Spectrum
40	c	5686.48	I
85	c	5747.58	I
24		5762.66	I
24		5785.18	II
75		5795.64	I
75		5803.13	II
65		5815.36	I
29		5842.97	I
65		5851.07	I
65		5870.62	I
35		5898.84	I
24		5902.40	I
35		5904.71	I
65	c	5920.78	I
50	c	5939.38	I
35		5940.17	I
24		5951.17	I
75		5967.34	II
29		6038.97	I
29		6039.38	I
24		6104.79	II
24	c	6292.43	I
35		6331.68	II
24		6334.91	II
24		6446.87	II
35	Cw	6518.68	I
24	c	6574.04	II
35		6581.82	I
30		6607.11	II
90		6677.94	II
40	Cw	6702.61	I
20	c	6706.79	II
30		6785.12	I
130		6794.58	II
40		6874.18	II
55		6896.37	I
45	h	6899.95	I
40		6901.98	I
9		7005.99	II
17		7082.85	II
11		7089.22	II
11		7112.69	I
10		7187.48	I
10	h	7195.89	II
65		7204.28	I
19	h	7234.98	I
40		7257.73	I
17		7311.94	I
45		7348.88	II
10		7398.27	II
15	h	7424.24	II
10	h	7429.62	II
9		7472.15	I
22		7484.54	I
9		7495.45	I
45		7496.12	I
17		7499.69	II
27		7511.40	I
9		7519.77	II
6		7557.59	II
27	h	7582.03	II
27		7587.49	II
45		7590.24	I
65		7596.44	I
17	h	7601.18	II
17		7616.01	II
22	h	7624.05	I
30		7627.81	I
9	h	7639.05	II
8		7672.72	II
8	h	7694.74	II
22	h	7706.16	II
22	h	7726.97	II
30		7737.63	I
22		7793.20	I
8		7807.33	II
16		7832.91	II
30		7855.79	II
15		7864.99	I
6	h	7885.70	I
6		7913.11	I
27		7927.90	II
13		7955.31	I
11		7998.03	I
17		8001.04	I
13	h	8010.16	II
30		8025.42	II
6		8053.80	I
19		8067.35	II
30		8085.06	II
27		8164.17	I
13		8171.70	I
65		8194.82	II
95		8212.57	I
11		8214.33	I
8		8259.08	I
40		8450.06	II
8	h	8465.80	II
13		8502.70	I
30	h	8511.80	I
45		8583.45	II
30		8603.40	I
9		8678.25	I
65		8765.74	II

Tb IV
Ref. 302 — J.R.

Intensity	Wavelength	
	Vacuum	
30	1176.58	IV
30	1192.01	IV
70	1200.58	IV
50	1213.94	IV
80	1221.22	IV
500	1235.04	IV
1000	1259.40	IV
300	1301.48	IV
300	1308.30	IV
600	1311.70	IV
700	1315.12	IV
500	1325.56	IV
1000	1327.67	IV
100	1367.56	IV
400	1367.71	IV
700	1369.64	IV
1000	1373.86	IV
400	1376.46	IV
200	1378.23	IV
300	1381.00	IV
100	1382.83	IV
20	1389.92	IV
200	1516.17	IV
50	1530.10	IV
5000	1595.39	IV
2000	1633.19	IV
300	1649.38	IV
400	1654.75	IV
400	1667.58	IV
200	1672.55	IV
400	1681.98	IV
5	1684.46	IV
100	1685.37	IV
400	1691.95	IV
10	1695.25	IV
300	1698.36	IV
30	1701.60	IV
50	1704.79	IV
20	1705.05	IV
3	1943.94	IV
50	1970.90	IV
	Air	
2000	2027.79	IV
200	2029.22	IV
400	2048.88	IV
200	2078.83	IV
1000	2089.98	IV
1000	2332.54	IV
100	2436.01	IV

THALLIUM (Tl)
Z = 81

Tl I and II
Ref. 1, 195, 348, 354 — C.H.C.

Intensity		Wavelength	
		Vacuum	
3		650.90	II
5	r	670.87	II
4		674.10	II
15	r	696.30	II
5	r	709.23	II
10	r	817.18	II
5	r	836.34	II
8	r	1018.85	II
10	r	1049.73	II
8	r	1050.30	II
5	r	1074.97	II
10	r	1130.17	II
15	r	1162.55	II
10	r	1167.43	II
10	r	1183.41	II
12	r	1194.84	II
8		1231.81	II
5	r	1246.00	II
15	r	1307.50	II
8	r	1310.20	II
25	r	1321.71	II
8	r	1330.40	II
10	r	1373.52	II
1		1423.2	I
8	r	1489.65	II
5		1490.50	II
10	r	1499.30	II
10	r	1507.82	II
15	r	1561.58	II
10	r	1568.57	II
7	r	1593.26	II
5	h	1616.	I
1		1650.2	I

Intensity		Wavelength	
5		1685.40	I
1		1728.	I
10	r	1792.76	II
12	r	1814.85	II
3	h	1847.	I
8		1892.72	II
25	r	1908.64	II
		Air	
100	r	2007.56	I
2		2209.75	I
100	r	2210.71	I
3		2287.6	I
30		2298.04	II
140		2315.98	I
900	h	2379.69	I
8		2451.83	II
6		2469.03	II
1		2508.2	I
20		2530.86	II
700		2580.14	I
60		2608.99	I
80		2665.57	I
420		2709.23	I
50	h	2710.67	I
4400	d	2767.87	I
280		2826.16	I
10		2849.80	II
2800		2918.32	I
440		2921.52	I
5		3029.01	II
20		3091.56	II
15		3185.51	II
15		3186.56	II
15		3187.74	II
1200		3229.75	I
15		3291.01	II
12		3319.91	II
12		3321.04	II
8		3322.25	II
15		3369.15	II
8		3381.00	II
6		3381.80	II
6	d	3460.48	II
20000		3519.24	I
5000		3529.43	I
8		3540.08	II
9		3560.68	II
5		3567.67	II
12000	Cw	3775.72	II
8		3793.95	II
10		3832.30	II
6		3869.15	II
10		3887.15	II
8		4223.05	II
20		4274.98	II
40		4306.80	II
2		4359.9	I
8		4490.77	II
20		4737.05	II
15		4981.35	II
25		5078.54	II
25		5152.14	II
6		5181.95	II
6		5183.10	II
18000		5350.46	I
15	d	5384.85	II
7		5409.92	II
10		5410.97	II
25		5949.48	II
10		6179.98	II
8	d	6239.03	II
10		6378.32	II
16	h	6549.84	I
6	h	6713.80	I
10		6966.5	I
3		7493.6	I
2		7678.93	I
10		7815.80	I
8		8130.0	I
20		8373.6	I
8		8445.8	II
10		8474.27	I
8		8632.9	II
10		8664.1	II
4		8850.4	I
5		8976.75	I
3		9038.4	I
20		9130.	II
20		9130.5	I
2	h	9183.1	I
4		9225.	II
2	h	9252.6	I
3		9254.	II
40		9509.4	I
10		9863.4	I
20		9930.4	I
2		9937.4	I
30		10011.9	I
40		10488.80	I
5		11101.61	I
4		11483.7	I
1000		11512.82	I

Thallium (Cont.)

Intensity	Wavelength	
5	11592.9	I
15	12491.8	I
150	12736.4	I
700	13013.2	I

Tl III
Ref 355 — C.H.C.

Intensity	Wavelength	
	Vacuum	
7	1231.57	III
10	1266.33	III
4	1332.36	III
10	1477.14	III
4	1506.37	III
8	1558.67	III
8	1660.05	III
	Air	
6	3163.53	III
3	3300.80	III
9	3456.34	III
4	3507.41	III
6	3933.05	III
2	3946.02	III
7	4109.85	III
4	4155.75	III
6	4269.81	III
2	4380.57	III
4	5086.99	III
4	5362.40	III
2	5499.4	III
5	5927.8	III
4	8001.	III

Tl IV
Ref. 356 — C.H.C.

Intensity	Wavelength	
	Vacuum	
7	531.26	IV
10	570.49	IV
4	597.01	IV
1	868.99	IV
3	912.74	IV
8	917.31	IV
30	1028.69	IV
20	1034.73	IV
20	1036.61	IV
10	1049.48	IV
10	1057.56	IV
20	1068.04	IV
20	1070.47	IV
30	1079.68	IV
5	1079.70	IV
2	1092.90	IV
4	1094.41	IV
6	1099.60	IV
4	1125.52	IV
5	1139.30	IV
3	1144.07	IV
3	1225.45	IV
6	1273.03	IV
3	1304.55	IV
6	1323.66	IV
7	1337.10	IV
7	1358.56	IV
5	1374.62	IV
7	1377.75	IV
8	1404.60	IV
5	1412.93	IV
6	1434.72	IV
5	1449.37	IV
5	1883.2	IV
3	1974.6	IV

THORIUM (Th)
Z = 90
Th I and II
Ref. 1, 97, 98, 434 — J.G.C. and R.Z.

Intensity	Wavelength	
	Air	
100	2326.926	II
190	2377.84	II
90	2404.504	II
100	2413.409	II
30	2439.433	I
5	2532.894	II
150	2547.901	II
500	2565.593	II
270	2566.588	II
200	2576.688	II
230	2589.059	II
230	2597.047	II
230	2600.882	II
100	2609.855	II
230	2618.91	II
270	2623.448	II
270	2625.737	II
55	2628.812	II
270	2641.488	II
170	2650.583	II
150	2658.663	II
50	2680.692	I
360	2684.288	II
480	2692.415	II
100	2695.553	II
270	2703.958	II
170	2708.176	II
230	2721.691	II
170	2722.380	II
250	2729.327	II
250	2732.808	II
28	2735.834	II
520	2747.156	II
100	2749.530	II
410	2752.166	II
130	2760.391	II
100	2765.123	II
270	2768.841	II
200	2770.816	II
70	2773.951	II
50	2774.066	II
70	2778.706	II
35	2791.496	I
90	2794.255	II
70	2797.737	II
55	2799.114	II
110	2807.827	II
180	2808.998	II
70	2814.319	II
45	2816.071	II
100	2819.322	II
100	2820.336	II
100	2822.025	II
170	2826.855	II
70	2830.442	II
800	2832.315	II
1200	2837.295	II
320	2842.812	II
100	2848.084	I
270	2851.260	II
50	2854.342	II
30	2860.490	I
220	2861.42	II
35	2868.461	I
70	2869.916	II
550	2870.406	II
30	2878.657	I
40	2882.511	II
320	2884.289	II
360	2885.049	II
360	2887.817	II
40	2892.172	II
250	2899.720	II
45	2903.167	II
50	2908.506	I
200	2910.594	II
90	2911.320	II
90	2912.009	II
140	2919.840	II
250	2925.050	II
250	2928.254	II
50	2931.281	I
55	2934.135	II
100	2936.086	I
35	2940.589	II
340	2942.860	II
100	2943.729	I
150	2949.068	II
80	2950.438	II
35	2955.849	II
170	2957.580	II
28	2959.853	I
28	2963.607	I
270	2968.686	II
110	2971.481	II
220	2974.011	II
50	2976.104	I
55	2980.334	II
160	2985.243	II
360	2988.232	II
150	2991.062	II
110	2996.986	II
50	3002.686	I
30	3004.248	I
180	3008.497	II
50	3010.736	I
20	3018.644	I
50	3021.056	II
150	3026.575	II
40	3030.487	I
370	3034.065	II
170	3035.110	II
85	3038.598	II
130	3045.564	II
420	3049.092	II
30	3056.692	I
220	3061.699	II
450	3067.729	II
370	3072.114	II
670	3078.828	II
480	3080.217	II
240	3088.470	II
130	3090.093	II
50	3093.711	I
140	3097.266	II
200	3102.664	II
510	3108.296	II
50	3115.538	I
100	3116.263	I
510	3119.526	II
510	3122.963	II
370	3124.387	II
480	3125.507	II
150	3131.070	II
100	3136.216	I
420	3139.306	II
420	3142.835	II
310	3146.044	II
150	3150.455	II
310	3154.300	II
50	3157.221	I
30	3161.364	I
140	3166.099	II
110	3169.328	II
420	3175.726	II
270	3179.048	II
1100	3180.193	II
310	3184.948	II
770	3188.233	II
85	3191.221	II
55	3192.585	I
55	3195.689	I
30	3202.520	I
170	3210.308	II
55	3214.380	I
560	3221.292	II
30	3223.168	I
560	3229.009	II
110	3230.868	II
480	3235.84	II
590	3238.116	II
240	3241.108	II
110	3244.448	I
280	3251.915	II
910	3256.274	II
180	3257.366	I
910	3262.668	II
180	3267.003	II
110	3272.027	II
30	3278.733	I
50	3281.048	I
130	3285.752	I
620	3287.789	II
910	3291.739	II
620	3292.520	II
240	3297.832	II
240	3301.650	I
480	3304.238	I
130	3309.365	I
50	3314.790	I
30	3318.390	I
510	3321.450	II
390	3324.752	II
840	3325.120	II
55	3328.255	II
250	3330.476	I
620	3334.604	II
620	3337.870	II
30	3342.073	II
180	3346.557	II
310	3348.768	I
980	3351.228	II
310	3354.179	II
620	3358.602	II
75	3361.738	II
390	3367.819	II
250	3374.974	I
390	3378.573	II
130	3380.859	I
310	3385.531	II
310	3386.501	II
110	3387.920	II
1300	3392.035	II
200	3396.727	I
250	3398.544	I
200	3405.558	I
250	3413.012	I
50	3417.497	I
390	3421.210	II
270	3423.989	I
50	3428.622	I
980	3433.998	II
770	3435.976	II
340	3438.949	II
110	3442.578	I
50	3446.547	I
130	3451.702	I
50	3457.068	I
340	3462.850	II
130	3465.924	II
390	3468.219	II
1300	3469.920	II
170	3471.218	I
250	3479.173	II
70	3480.052	II
200	3486.552	I
100	3489.184	I
270	3493.518	II
70	3496.810	I
130	3498.621	I
70	3503.786	I
50	3506.645	I
110	3511.157	I
140	3518.404	I
70	3521.059	I
70	3526.633	I
140	3531.450	I
670	3539.587	II
180	3544.018	I
170	3549.595	I
140	3551.401	I
200	3555.013	I
530	3559.451	II
110	3563.375	I
70	3569.820	I
200	3576.557	I
100	3583.101	I
170	3589.750	I
270	3592.780	I
270	3598.120	I
390	3601.034	II
170	3608.377	I
980	3609.445	II
200	3612.427	I
480	3615.133	II
400	3617.118	II
270	3623.970	II
390	3625.627	II
140	3632.831	I
270	3635.943	II
70	3638.644	I
210	3642.248	I
170	3649.735	I
50	3658.808	I
100	3659.629	I
220	3663.202	I
140	3668.140	I
280	3669.968	I
700	3675.567	II
150	3682.486	I
50	3688.658	I
100	3690.624	I
170	3692.566	I
180	3698.105	I
50	3703.229	I
340	3706.767	I
280	3711.305	II
590	3719.435	I
50	3719.836	I
770	3721.825	II
110	3727.902	I
50	3733.672	I
1300	3741.183	I
310	3747.539	I
650	3752.569	II
140	3757.694	I
110	3765.240	I
180	3770.056	I
50	3772.649	I
85	3776.271	I
50	3780.966	I
340	3785.600	II
100	3789.167	I
85	3795.386	I
50	3800.197	I
590	3803.075	I
50	3807.273	I
340	3813.068	II
50	3818.685	I
75	3825.133	I
450	3828.384	I
70	3836.584	I
840	3839.746	II
280	3841.960	II
100	3846.887	I
85	3852.135	I
390	3854.511	II
140	3859.840	II
450	3863.405	II
100	3869.663	I
210	3875.374	I
140	3879.644	I
100	3886.915	I
340	3895.419	I
50	3901.661	I
110	3903.102	I
170	3905.186	II
50	3908.750	I
85	3911.909	I
50	3916.417	I

Int	λ	Sp	Int	λ	Sp	Int	λ	Sp	Int	λ	Sp
110	3919.023	I	25	4452.565	I	70	5277.501	II	8	6317.185	I
140	3925.093	I	85	4458.002	I	15	5281.069	I	21	6327.278	I
590	3929.669	II	220	4465.341	II	10	5294.397	I	35	6342.860	I
200	3932.911	I	30	4475.221	I	30	5297.743	I	50	6355.911	II
140	3937.040	II	75	4482.169	I	30	5307.466	II	14	6369.140	I
50	3942.072	I	50	4489.664	I	35	5312.002	I	40	6376.931	I
50	3948.030	I	110	4498.940	I	20	5317.494	I	30	6411.899	I
200	3948.964	II	55	4505.216	I	60	5325.145	II	24	6413.615	I
50	3952.760	I	280	4510.527	II	50	5326.976	I	15	6437.762	I
110	3959.300	I	70	4521.194	I	20	5330.080	I	15	6450.005	I
390	3967.392	I	22	4530.319	I	60	5343.581	I	60	6457.283	I
200	3972.155	I	40	4535.255	I	14	5351.126	I	50	6462.614	I
150	3980.089	I	30	4545.915	II	30	5358.707	I	5	6466.717	I
110	3991.730	I	70	4555.812	I	20	5369.281	I	14	6490.738	I
530	3994.549	II	40	4563.660	I	30	5378.836	I	5	6501.992	I
50	3998.061	I	65	4570.972	I	70	5390.466	II	20	6512.364	I
240	4003.309	II	50	4588.426	I	50	5392.572	I	5	6522.044	I
250	4007.021	II	75	4595.421	I	20	5399.175	I	50 h	6531.342	I
220	4008.210	I	26	4612.554	II	24	5417.486	I	6	6554.160	I
220	4009.056	I	30	4621.163	I	60	5425.678	II	5	6558.876	I
280	4012.495	I	140	4631.761	II	50	5435.893	II	3	6565.070	I
4200	4019.129	I	140	4631.761	II	40	5449.479	II	5	6577.215	I
210	4025.656	II	30	4641.254	I	30	5462.615	II	24	6583.907	I
140	4027.009	I	140	4651.558	II	15	5470.759	I	24	6588.540	I
250	4030.842	I	23	4663.202	I	24	5484.147	II	24	6593.940	I
250	4036.047	I	50	4669.984	I	10	5496.137	I	24	6605.416	II
240	4036.565	II	65	4676.056	I	19	5504.302	I	24	6619.946	II
240	4041.204	II	50	4686.195	I	35	5509.994	I	24	6619.947	I
55	4048.287	I	140	4694.091	II	12	5524.584	I	21	6644.650	II
110	4050.887	I	50	4703.990	I	50	5539.262	I	6	6658.678	II
140	4059.253	I	20	4712.841	I	70	5539.911	II	30	6662.269	I
250	4063.407	I	90	4723.438	I	35	5548.176	I	6	6674.697	I
300	4069.201	II	30	4729.128	I	50	5558.342	I	3	6683.367	I
100	4069.461	I	190	4740.529	II	60	5564.203	II	8	6692.724	II
55	4075.503	I	140	4752.414	II	40	5573.354	I	5	6697.712	I
110	4081.368	I	13	4766.600	I	60	5587.046	I	16	6727.459	I
85	4085.434	I	50	4778.294	I	24	5595.064	I	3	6735.126	I
700	4086.520	II	20	4786.531	I	50	5604.515	II	5	6742.884	I
70	4088.726	I	40	4789.387	I	35	5615.320	I	20	6756.453	I
700	4094.747	II	45	4808.134	I	7	5630.297	I	6	6765.677	I
150	4100.341	I	20	4819.193	I	70	5639.746	II	15	6778.313	I
270	4105.330	II	26	4822.855	I	7	5648.991	I	15	6780.413	I
840	4108.421	II	40	4826.700	I	12	5657.925	I	6	6791.236	I
240	4112.754	I	45	4831.121	I	15	5667.128	I	3	6798.747	I
280	4115.758	I	50	4840.843	I	20	5677.053	I	3	6809.511	I
1100	4116.713	II	30	4848.362	I	10	5683.198	I	3	6823.509	I
30	4123.600	I	15	4852.868	I	65	5700.918	II	11	6829.036	I
200	4127.411	I	40	4858.333	II	95	5707.103	II	14	6834.925	I
110	4131.002	I	280	4863.163	II	50	5720.183	I	4	6862.873	I
340	4132.753	II	40	4872.917	I	30	5732.975	II	5	6866.767	I
200	4134.067	I	76	4878.733	I	24	5742.084	I	8	6874.754	I
220	4140.235	II	45	4894.955	I	30	5749.388	II	20	6889.303	II
250	4142.701	II	20	4907.209	I	70	5760.551	I	3	6908.988	I
200	4148.182	II	240	4919.816	II	15	5773.946	I	24	6911.227	I
450	4149.986	II	18	4927.780	I	20	5789.645	I	5	6916.129	I
110	4158.535	I	40	4939.642	I	35	5804.141	I	5	6936.652	I
140	4165.766	I	60	4947.575	II	19	5815.422	II	35	6943.611	I
620	4178.060	II	50	4954.659	II	10	5832.370	I	5	6954.657	I
250	4179.714	II	30	4965.731	I	10	5845.919	I	15	6965.947	I
30	4184.138	I	35	4975.950	II	15	5854.121	I	3	6981.086	I
130	4193.017	I	24	4985.372	I	15	5868.373	I	55	6989.656	I
620	4208.890	II	50	5002.097	I	10	5878.933	I	24	6993.038	II
130	4210.923	I	50	5002.097	I	8	5891.451	I	18	7000.806	I
28	4214.828	I	50	5015.889	II	15	5899.844	I	18	7000.806	I
55	4220.065	I	260	5017.255	II	20	5914.387	II	3	7015.319	I
55	4227.387	I	20	5029.892	I	19	5925.893	II	10	7018.569	I
30	4230.824	I	24	5039.230	I	10	5937.162	I	3	7026.462	I
85	4235.463	I	50	5044.719	I	10	5944.648	I	7	7036.281	I
20	4241.094	I	240	5049.796	II	8	5957.587	I	30	7045.795	II
30	4247.989	II	85	5055.347	II	30	5973.665	I	15	7053.619	I
110	4253.538	I	70	5058.562	II	85	5989.044	II	6	7060.654	I
70	4256.254	I	110	5067.974	I	24	5994.129	I	24	7075.333	II
110	4260.333	I	30	5081.446	I	21	6007.072	I	30	7084.171	I
28	4269.942	I	50	5090.051	I	30	6015.426	II	24	7089.339	II
280	4273.357	II	50	5098.043	II	17	6021.036	I	10	7100.512	II
480	4277.313	II	40	5101.130	I	17	6037.698	I	3	7109.861	I
700	4282.042	II	50	5110.867	II	24	6044.431	II	11	7124.562	I
28	4288.669	I	30	5115.044	I	10	6053.381	I	3	7132.613	I
55	4297.306	I	10	5125.950	I	5	6061.536	I	5	7148.560	I
85	4299.839	I	20	5134.746	I	30	6077.106	I	10	7154.954	I
100	4307.176	I	95	5143.267	II	30	6087.296	II	30	7168.896	I
200	4309.991	II	120	5148.211	II	24	6099.083	I	15	7173.373	I
55	4315.254	I	50	5151.612	I	30	6104.580	I	40	7191.132	II
110	4318.416	I	50	5154.243	I	40	6112.837	I	7	7200.046	I
30	4325.274	I	85	5158.604	I	30	6120.557	II	35	7208.006	I
28	4330.844	I	70	5160.730	I	10	6124.480	I	11	7212.69	I
130	4337.277	I	20	5168.922	I	14	6151.993	I	10	7217.755	II
85	4342.256	II	50	5176.961	I	10	6161.354	I	11	7218.054	I
130	4344.326	II	35	5183.990	II	60	6169.822	I	3	7242.355	I
55	4349.072	I	50	5190.871	II	50	6182.622	I	5	7255.354	I
55	4354.484	I	50 h	5195.814	I	12	6191.906	I	3	7270.558	I
85	4359.372	I	50	5198.800	I	24	6193.858	II	7	7284.904	I
85	4365.930	I	95	5199.164	I	12	6203.493	I	5	7298.143	I
85	4374.123	I	50	5211.230	I	12	6224.528	I	11	7305.405	II
1300	4381.860	II	95	5216.596	II	24 h	6234.856	I	5	7315.067	I
1100	4391.110	II	50	5218.528	II	8	6240.954	I	7	7324.808	I
55	4392.974	I	35	5219.110	I	10	6257.424	I	5	7339.606	I
55	4401.580	I	110	5231.160	I	21	6261.063	I	8	7341.152	I
85	4408.882	I	85	5233.225	II	21	6261.418	I	5	7361.349	I
210	4412.741	II	85	5233.229	I	50	6274.116	II	5	7384.175	I
28	4416.845	I	95	5247.654	II	50	6274.117	II	18	7385.501	I
50	4422.048	I	10	5255.573	I	30	6279.172	II	3	7393.431	II
250	4432.963	II	35	5258.360	I	8	6291.192	I	5	7402.252	I
140	4440.866	II	12	5266.710	I	10	6303.251	II	3	7418.550	I

Thorium (Cont.)

Intensity	Wavelength		Intensity	Wavelength	
21	7428.940	I	10	9203.963	I
10	7430.254	I	10	9266.208	I
3	7444.749	I	10	9276.276	I
2	7462.993	I	10	9289.563	I
10	7481.355	I	4	9317.722	I
2	7493.427	I	7	9340.706	I
50	7525.508	II	15	9399.085	I
7	7549.314	I	10	9431.603	I
18	7567.740	I	15	9461.030	I
12	7585.69	I	15	9474.882	I
12	7585.792	I	15	9495.501	I
4	7598.204	I	10	9505.392	I
2	7607.824	I	10	9561.24	I
5	7627.176	I	8	9613.689	II
3	7636.176	I	12	9632.647	I
30	7647.380	I	10	9664.700	I
4	7658.324	I	4	9676.106	I
7	7676.219	I	15	9700.564	I
21	7685.305	I	15	9746.46	I
4	7710.269	I	10	9812.70	I
4	7728.951	I	10	9826.45	I
10	7731.72	II	20	9833.42	I
4	7771.948	I	15	10039.364	I
15	7787.79	II	15	10089.138	I
15	7788.937	I	15	10133.56	II
5	7798.360	I	15	10419.57	II
4	7810.625	I	15	10556.45	I
21	7817.771	I	15	10723.92	II
8	7834.459	II	20	10726.93	I
15	7847.540	I	20	10942.24	II
4	7864.023	I	15	11051.90	I
12	7865.95	I	30	11230.259	I
6	7886.284	I	20	11354.719	I
11	7900.31	I	15	11703.46	I
4	7937.732	I	15	11864.25	I
11	7941.72	I	15	11940.64	I
5	7954.594	I	20	11984.67	II
7	7972.598	I	15	12018.72	I
24	7978.974	I	20	12127.30	I
11	7987.97	I	20	12194.16	II
5	7993.680	I	15	12206.89	I
2	8014.502	I	20	12231.94	I
5	8024.253	II	15	12338.00	I
11	8032.433	I	15	12477.30	I
11	8062.64	I	20	12646.54	I
5	8085.220	I	15	12866.64	I
5	8093.626	I	15	12940.65	II
5	8129.407	I	15	12959.82	I
11	8138.477	I	15	13145.90	II
18	8143.139	I	15	13565.67	I
12	8159.729	I	15	14090.25	I
10	8163.125	II	15	14168.67	I
7	8169.788	I	15	14424.54	I
15	8186.914	I	15	14618.98	I
12	8203.199	II	15	14654.91	I
2	8231.408	I	15	14940.49	II
5	8252.395	I	15	15240.24	II
3	8259.512	I	15	15429.78	I
18	8275.629	I	15	15831.75	I
3	8292.529	I	20	17208.22	II
15	8320.857	I	15	17307.66	I
30	8330.451	I	15	17381.91	I
4	8358.726	I	15	17481.04	I
6	8387.104	II	15	17584.52	I
18	8403.767	II	15	17936.43	II
15	8416.729	I	15	18811.88	I
12	8421.227	I	15	19145.60	II
21	8446.509	I	15	19338.98	II
5	8464.230	I	10	19774.30	I
18	8478.360	I	10	20634.36	I
5	8510.621	I	10	20692.06	I
5	8516.557	I	10	22264.35	II
5	8539.795	I			
6	8554.946	I			
11	8573.122	I			
12	8591.838	I			
3	8621.325	I			
3	8638.363	I			
10	8665.487	I			
4	8668.116	I			
2	8701.127	I			
5	8709.236	I			
8	8732.401	II			
18	8748.033	I			
15	8758.244	I			
8	8775.573	I			
4	8792.058	I			
3	8804.590	I			
5	8841.185	I			
18	8842.073	II			
15	8868.834	I			
4	8875.233	I			
2	8907.038	I			
5	8955.848	I			
15	8957.97	II			
40	8967.641	I			
3	8987.408	I			
5	9031.819	I			
25	9048.252	I			
5	9063.953	I			
6	9094.831	I			
3	9107.225	I			
2	9118.140	I			
3	9170.825	I			

Th III
Ref. 157 — J.G.C.

Intensity	Wavelength		Intensity	Wavelength	
	Vacuum				
100	1888.12	III			
	Air				
50	2149.18	III	200	2427.94	III
50	2162.82	III	200	2431.68	III
50	2199.74	III	200	2441.24	III
50	2206.62	III	100	2463.66	III
50	2291.59	III	50	2473.93	III
100	2301.18	III	100	2501.08	III
100	2319.52	III	60	2512.69	III
80	2324.68	III	50	2514.31	III
150	2335.50	III	100	4555.73	III
100	2340.58	III	50	4589.28	III
100	2363.06	III	100	5376.13	III
50	2368.91	III	50	5447.18	III
100	2371.42	III	50	6242.95	III
80	2381.47	III	50	6599.39	III
100	2391.48	III	50	7461.59	III
200	2413.50	III	50	8105.14	III
50	2424.54	III			

Th IV
Ref. 156, 165 — J.G.C.

Intensity	Wavelength	
	Vacuum	
4	797.53	IV
1	835.55	IV
30	846.91	IV
1	854.02	IV
30	882.39	IV
12	886.66	IV
100	1565.85	IV
70	1682.22	IV
30	1684.01	IV
150	1707.37	IV
200	1959.02	IV
	Air	
200	2002.34	IV
100	2066.70	IV
20	2143.91	IV
30	2146.81	IV
1	2242.11	IV
2	2261.26	IV
5	2296.81	IV
100	2693.99	IV
2	4937.09	IV
4	4952.52	IV
3	5420.38	IV
2	6711.87	IV
3	6740.37	IV
50	6901.16	IV
	9839.25	IV
	10875.05	IV

THULIUM (Tm)
Z = 69

Tm I and II
Ref. 1 — C.H.C.

Intensity		Wavelength		Intensity	Wavelength	
		Air				
360		2284.79	II	210	2640.76	II
120		2329.77	II	130	2646.45	II
70		2340.92	II	160	2650.27	II
120		2363.91	II	190	2658.48	II
45		2365.96	II	250	2660.09	II
160		2367.11	II	140	2668.20	II
150		2383.68	II	310	2679.57	II
110		2388.95	II	170	2697.50	II
450		2409.02	II	540	2721.19	II
110		2412.44	II	200	2744.08	II
120		2421.65	II	270	2779.55	II
450		2426.17	II	350	2785.07	II
140		2445.47	II	680	2794.60	II
770		2480.13	II	730	2797.27	II
150		2481.15	II	250	2818.47	II
130		2487.52	II	250	2827.02	II
250		2491.60	II	580	2827.92	II
100		2499.54	II	200	2831.55	II
130		2507.15	II	310	2844.67	II
1300		2509.08	II	200	2854.17	I
200		2520.87	II	200	2860.12	II
250		2522.17	II	200	2861.74	II
180		2524.11	II	1600	2869.23	II
130		2527.02	I	630	2890.94	II
110		2527.42	II	210	2918.27	II
120		2542.66	II	270	2925.65	II
360		2552.76	I	680	2926.74	II
540		2561.65	II	630	2935.99	II
150		2563.86	II	350	2951.26	II
430		2582.27	II	430	2965.86	II
170	h	2596.49	I	490	2973.22	I
110		2601.09	I	540	2981.48	II
220		2606.02	II	350	2986.52	II
810		2607.06	II	630	2990.54	II
730		2624.33	II	200	2993.26	II
				230	3013.71	II
				430	3014.65	II
				1500	3015.30	II
				270	3017.09	II
				330	3026.07	II
				280	3042.35	II
				340 d	3046.76	II
				320	3050.73	II
				340	3056.07	II
				580	3073.08	II
				360	3081.12	I
				740	3098.60	II
				7400	3131.26	II
				2300	3133.89	II
				230	3144.90	II
				230	3146.16	II
				1900	3151.04	II
				1500	3157.34	II
				450	3172.65	I
				2300	3172.83	II
				380	3173.58	II
				230	3195.33	II
				320	3210.56	II
				320	3210.82	II
				320	3212.01	II
				230	3231.51	II
				470	3235.44	II
				1200	3236.81	II
				1600	3240.23	II
				2300	3241.54	II
				320	3246.96	I
				420	3247.46	II
				1900	3258.05	II
				400	3261.65	II
				320	3264.10	II
				1600	3266.64	II
				1200	3267.40	II
				790	3268.99	II
				1100	3276.81	II
				1200	3283.40	II
				1200	3285.61	II
				2300	3291.00	II
				2000	3302.46	II
				210	3306.01	II
				210	3306.91	II
				210	3308.01	II
				1200	3309.80	II
				640	3310.59	II
				400	3316.88	II
				210	3318.65	II
				230	3349.99	I
				230	3354.86	II
				4000	3362.61	II
				490	3374.50	II
				420 d	3384.99	II
				1700	3397.50	II
				420	3399.95	II
				850	3410.05	I
				340	3412.59	I
				340	3416.59	I
				6400	3425.08	II
				950	3425.63	II
				340	3429.33	I
				850	3429.96	II
				420	3431.19	II
				4900	3441.50	II
				4900	3453.66	II
				8500	3462.20	II
				210	3467.51	I

Thulium (Cont.)

Intensity	Wavelength	Spectrum
340	3476.69	I
340	3480.98	I
340	3481.75	II
420	3487.38	I
210	3492.58	II
340	3499.95	I
250	3513.02	II
250	3517.60	I
250	3534.85	II
1700	3535.52	II
490	3536.21	II
850	3536.58	II
420	3537.91	I
210	3555.82	I
420	3557.79	II
340	3560.92	I
420	3563.88	I
490	3565.91	II
1300	3566.47	II
420	3567.36	I
280	3574.06	II
280	3586.07	I
2100	3608.77	II
250	3609.53	II
380	3638.41	I
950	3643.65	II
240	3647.72	II
600	3653.61	II
500	3665.81	II
1100	3668.09	II
410	3677.98	II
450 d	3678.85	II
410	3694.74	II
4800	3700.26	II
3800	3701.36	II
330	3704.85	II
7700	3717.91	I
890	3725.06	II
2400	3734.12	II
5000	3744.06	I
1700	3751.81	I
310	3756.86	II
6000	3761.33	II
4800	3761.91	II
260	3783.55	II
380	3795.16	II
7100	3795.75	II
770	3798.54	I
240	3798.75	II
600	3807.72	I
380	3810.72	II
550	3817.39	II
290	3826.39	I
1300	3838.20	II
290	3840.87	I
8900	3848.02	II
140	3857.84	II
6800	3883.13	I
1800	3883.44	II
5400	3887.35	I
440	3890.53	II
440	3896.62	I
680	3900.79	II
3500	3916.48	I
120	3928.66	II
570	3929.58	II
1500	3949.27	I
1500	3958.10	II
440	3995.58	II
1800	3996.52	II
220	4024.23	I
380	4044.47	I
10000	4094.19	I
9500	4105.84	I
120	4132.69	II
1100	4138.33	I
120	4149.14	I
120	4158.60	I
8800	4187.62	I
520	4199.92	II
6000	4203.73	I
220	4206.00	II
380	4222.67	I
3000	4242.15	II
270	4271.71	I
150	4298.36	I
2700	4359.93	I
1400	4386.43	I
200	4394.42	I
120	4395.96	I
140	4396.50	I
55	4437.40	II
80	4442.74	I
50	4447.58	I
120	4454.03	I
80	4459.99	I
50	4467.98	I
540	4481.26	II
80	4489.70	II
150	4519.60	I
260	4522.57	II
180	4529.38	II
80	4532.15	I
110	4548.60	I

Intensity	Wavelength	Spectrum
40	4556.68	II
40	4561.86	I
80	4564.68	I
40	4567.11	II
95	4596.63	I
270	4599.02	I
35	4601.29	II
55	4603.43	II
40	4604.85	I
50	4613.97	I
40	4614.47	II
300	4615.94	II
35	4619.06	II
40	4621.72	I
80	4626.33	II
95	4626.56	II
40	4626.97	I
110	4634.26	II
40	4642.96	II
95	4643.12	I
35	4644.58	I
120	4655.09	I
35	4666.70	II
35	4671.99	II
35	4675.10	I
80	4675.31	II
40	4677.86	II
160	4681.92	I
70	4685.11	I
120	4691.11	I
110	4724.26	I
680	4733.34	I
35	4750.75	II
70	4759.90	I
27	4789.92	II
27	4807.48	I
35	4808.68	I
35	4813.50	I
27	4826.99	II
27	4828.97	I
80	4831.20	II
35	4835.75	I
27 d	4851.76	I
27	4851.90	II
19	4872.28	II
27	4879.19	I
27	4901.61	I
24	4909.74	I
55	4923.83	I
140	4957.18	I
40	4970.87	II
27	4971.26	I
40	4975.12	II
50	4978.90	I
40	4980.68	II
55	4989.32	II
27	4993.79	II
19	4994.72	II
35	5001.02	I
27	5001.59	I
160	5009.77	II
35	5014.56	II
27	5017.87	II
160	5034.22	II
27 h	5041.00	II
22	5043.50	I
35	5045.41	I
27	5060.42	II
150	5060.90	I
27	5062.25	I
27	5065.88	I
80	5066.67	I
27	5072.42	I
27	5076.36	I
27	5077.18	I
35	5085.09	I
40	5107.53	I
95	5113.97	I
50	5114.55	II
22	5120.67	I
40	5140.28	II
40	5149.40	II
19	5182.68	I
40	5185.25	I
14	5204.51	II
80	5213.38	I
22	5228.23	II
14	5260.93	II
24	5267.34	II
40	5291.14	I
40	5294.32	I
35	5300.21	I
35	5302.69	I
55	5305.87	II
650	5307.12	I
16	5322.99	II
35	5338.90	I
540	5339.03	I
80	5346.49	II
27	5372.98	II
14	5391.96	II
27	5400.46	II
27	5402.23	I
14	5405.98	II

Intensity	Wavelength	Spectrum
14	5461.95	II
14	5464.14	I
14	5465.54	II
16	5500.30	II
14	5526.82	II
24	5528.34	I
14	5539.03	II
27	5566.00	I
22	5581.37	I
14	5586.65	II
14	5589.94	II
14	5606.64	I
270	5631.41	I
40	5642.60	I
27	5645.40	I
70	5658.30	I
520	5675.84	I
14	5683.59	I
40	5684.76	II
14	5696.42	II
35	5709.97	II
22	5715.79	I
14	5733.81	II
11 d	5737.20	II
	5737.25	II
14 h	5738.92	II
27	5758.02	I
55	5760.20	I
190	5764.29	I
5	5778.82	II
19	5782.36	II
22	5784.46	II
11	5799.97	II
14	5811.19	II
14 h	5816.46	I
35	5838.76	II
240	5895.63	I
35	5899.47	I
24	5901.57	I
8	5912.58	I
11	5931.70	I
27	5935.90	I
140	5971.26	I
27	5975.02	I
11	5984.87	I
19	6025.44	I
11	6067.78	II
16	6131.53	I
14	6175.29	II
14	6181.41	II
14	6299.46	II
27	6352.66	I
22	6401.44	II
8	6430.94	II
14	6440.54	I
200	6460.26	I
14	6490.70	I
14	6519.78	I
8	6575.54	I
95	6604.96	I
8	6627.25	I
35	6657.72	I
11	6658.64	I
11	6692.93	I
30	6721.36	I
9	6726.34	I
9	6727.94	II
18	6739.22	I
9 h	6767.48	I
9	6777.93	I
110	6779.77	I
14 h	6782.00	I
18	6788.52	I
13 h	6820.27	I
14	6826.95	I
14	6829.12	II
23	6831.09	I
120	6844.26	I
80	6845.76	I
18	6854.12	I
6	6898.56	I
6	6915.86	I
10	6937.37	I
5	6949.54	I
5 h	6976.69	II
5	7010.79	I
6 h	7014.31	II
10	7017.90	I
6 h	7029.40	I
12	7034.34	I
10	7056.43	II
5	7060.97	I
6	7079.78	II
10	7106.14	I
5 h	7231.33	I
5	7233.74	II
4	7257.72	I
17	7272.62	I
8	7284.30	I
11 h	7286.16	I
14	7310.51	I
11	7336.63	II
14	7432.18	I
5	7434.51	II

Intensity	Wavelength	Spectrum
5	7439.95	II
75	7481.08	I
75	7490.20	I
10 h	7507.28	I
14	7545.78	I
140	7558.33	I
17 h	7580.61	I
20 h	7593.74	I
17	7595.07	II
5	7629.85	I
5 h	7648.76	II
17	7655.00	I
4	7660.32	I
7 h	7666.24	I
8	7676.04	II
8 h	7701.46	I
80	7731.51	I
4 h	7778.27	I
12 h	7782.35	I
8 h	7785.51	I
	7785.90	I
17	7803.93	I
4	7829.22	I
40	7856.08	I
3	7861.67	I
5	7918.10	I
55	7927.51	I
110	7930.84	I
6	7971.56	I
11 h	7985.93	I
14 h	8014.77	I
95	8017.90	I
3 h	8021.33	I
14	8194.19	I
5	8294.52	I
7	8365.75	I
7	8460.79	II
27	8472.01	II
7 h	8546.07	II
11	8565.73	II

Tm III
Ref. 307 — J.R.

Intensity	Wavelength	
	Air	
500	2099.11	III
500	2107.10	III
200	2136.67	III
200	2156.29	III
800	2182.98	III
300	2183.91	III
5000	2185.94	III
100	2212.25	III
300	2230.86	III
400	2231.25	III
200	2243.34	III
400	2243.98	III
200	2246.68	III
200	2269.39	III
1000	2276.91	III
100	2280.08	III
100	2281.27	III
100	2282.86	III
200	2282.98	III
200	2286.57	III
400	2287.21	III
500	2294.73	III
20000	2296.21	III
200	2297.43	III
100	2304.64	III
400	2304.82	III
5000	2305.03	III
20000	2311.16	III
5000	2312.72	III
200	2314.88	III
400	2317.35	III
500	2320.96	III
200	2322.83	III
100	2323.71	III
100	2323.77	III
100	2324.43	III
500	2324.62	III
5000	2326.19	III
100	2327.02	III
300	2327.25	III
6000	2328.50	III
6000	2329.29	III
200	2330.87	III
3000	2331.80	III
400	2335.01	III
1000	2338.36	III
500	2341.74	III
300	2342.04	III
100	2344.59	III
500	2345.61	III
300	2347.43	III
400	2353.10	III
100	2355.65	III
3000	2357.05	III

Thulium (Cont.)

Intensity	Wavelength		Intensity	Wavelength	
1000	2361.23	III	100	2974.85	III
500	2363.97	III	1000	2998.28	III
1000	2375.32	III	100	3048.11	III
700	2375.83	III	700	3078.87	III
400	2389.52	III	200	3120.15	III
4000	2406.63	III	200	3277.26	III
500	2435.31	III	200	3407.73	III
500	2457.86	III	100	3415.40	III
500	2471.23	III	100	3415.96	III
30000	2489.44	III	100	3436.93	III
200	2496.25	III	400	3467.93	III
2000	2504.71	III	200	3529.29	III
3000	2519.78	III	200	3533.28	III
10000	2552.46	III	100	3537.47	III
500	2557.90	III	300	3562.41	III
1000	2574.52	III	100	3563.42	III
500	2574.98	III	200	3587.74	III
100	2581.84	III	600	3617.96	III
100	2585.48	III	1000	3629.09	III
300	2589.20	III	100	3706.11	III
500	2608.96	III	100	3799.41	III
300	2609.66	III	300	3998.84	III
500	2617.22	III	200	4021.92	III
500	2618.78	III	200	4026.03	III
1000	2621.12	III	700	4032.13	III
400	2621.35	III	100	4076.15	III
400	2622.31	III	200	4335.47	III
100	2627.09	III	500	4385.41	III
100	2628.83	III			
300	2634.66	III			
200	2636.68	III			
200	2637.30	III			
100	2640.32	III			
500	2643.58	III			
100	2645.05	III			
200	2649.27	III			
100	2650.82	III			
100	2654.05	III			
100	2656.30	III			
500	2661.51	III			
1000	2663.00	III			
500	2664.76	III			
500	2664.88	III			
200	2665.05	III			
1000	2666.93	III			
200	2668.59	III			
100	2668.66	III			
200	2669.18	III			
100	2671.42	III			
100	2675.30	III			
1000	2676.64	III			
500	2676.91	III			
100	2678.28	III			
100	2680.49	III			
5000	2682.32	III			
300	2682.64	III			
300	2687.14	III			
300	2695.69	III			
400	2698.21	III			
1000	2699.49	III			
1000	2699.80	III			
100	2703.63	III			
200	2703.68	III			
100	2704.93	III			
2000	2707.03	III			
300	2707.19	III			
200	2707.44	III			
500	2707.60	III			
1000	2709.74	III			
200	2710.79	III			
1000	2713.38	III			
200	2715.81	III			
300	2717.56	III			
100	2718.02	III			
3000	2719.47	III			
3000	2724.44	III			
4000	2727.56	III			
200	2728.13	III			
1000	2731.38	III			
300	2732.11	III			
400	2737.98	III			
800	2744.74	III			
400	2745.99	III			
500	2752.46	III			
400	2753.20	III			
400	2756.15	III			
800	2765.98	III			
700	2769.92	III			
200	2772.64	III			
100	2777.43	III			
400	2781.12	III			
2000	2806.77	III			
300	2821.12	III			
200	2849.52	III			
700	2882.02	III			
100	2899.29	III			
100	2912.33	III			
400	2921.08	III			
200	2947.02	III			
1000	2947.72	III			
500	2953.18	III			
100	2966.15	III			
500	2966.85	III			
400	2972.61	III			

TIN (Sn)
Z = 50

Sn I and II
Ref. 187, 191 — C.H.C.

Intensity		Wavelength	
		Vacuum	
1		899.92	II
2		917.40	II
1		935.63	II
3		945.83	II
4		954.50	II
7		985.13	II
4		997.21	II
2		1016.26	II
4		1040.78	II
1		1041.32	II
3		1062.10	II
8		1108.19	II
4		1159.05	II
10		1161.43	II
3		1162.94	II
4		1180.51	II
9		1219.07	II
13		1223.70	II
11		1243.00	II
20		1290.86	II
20		1316.59	II
25		1400.52	II
20		1475.15	II
9		1489.22	II
7		1609.47	II
10	r	1737.21	I
15	r	1751.46	I
10	h	1753.3	I
7		1758.00	I
20	r	1764.98	I
20		1773.40	I
30	r	1790.75	I
80	r	1804.60	I
15		1811.34	II
30		1813.04	I
40	r	1815.74	I
25		1819.31	I
120	r	1823.00	I
9		1831.89	II
50	r	1848.75	I
30		1852.00	I
200	r	1860.32	I
20		1861.42	I
20		1865.52	I
30		1865.96	I
15		1873.29	I
30		1882.64	I
80		1886.05	I
100		1891.40	I
20		1897.29	I
12		1899.91	II
50		1909.30	I
40		1911.61	I
20		1913.52	I
80		1925.31	I
20		1926.77	I
15		1927.95	I
40	h	1928.9	I
25		1933.17	I
20		1942.69	I
150		1952.15	I
15		1960.21	I
30		1971.46	I

Intensity		Wavelength	
50	h	1977.6	I
80		1984.20	I
15		1991.88	I
20		1994.98	I
		Air	
25		2008.05	I
30		2015.76	I
30		2026.98	I
50		2040.66	I
20		2040.90	I
50		2054.03	I
70		2058.31	I
20		2064.00	I
80		2068.58	I
100		2072.89	I
100		2073.08	I
25		2080.62	I
30		2091.58	I
40		2094.35	I
200		2096.39	I
100		2100.93	I
100	r	2113.93	I
50		2121.26	I
25		2140.73	I
20		2141.43	I
15		2148.46	I
1		2148.63	II
40	r	2148.73	I
20	r	2151.43	I
30		2151.54	II
80		2171.32	I
150	r	2194.49	I
300	r	2199.34	I
400	r	2209.65	I
4		2209.67	II
40		2211.05	I
80	r	2231.72	I
400	r	2246.05	I
6		2246.07	II
60		2251.17	I
30		2267.19	I
400	r	2268.91	I
20		2282.26	I
200	r	2286.68	I
600	r	2317.23	I
300	r	2334.80	I
1000	r	2354.84	I
20		2357.79	I
3		2360.34	II
22		2368.33	II
60		2380.72	I
4		2384.54	II
100		2408.15	I
800	r	2421.70	I
1000	r	2429.49	I
1		2433.52	II
15		2448.98	II
60		2455.24	I
20		2476.40	I
300		2483.39	I
13		2483.48	II
10		2486.99	II
200		2495.70	I
5		2522.60	II
90		2523.92	I
80	h	2531.17	I
400		2546.55	I
40	h	2558.01	I
500	r	2571.58	I
200		2594.42	I
50	h	2636.94	I
200	r	2661.24	I
2		2664.93	II
700	r	2706.51	I
2		2727.82	II
20		2761.78	I
150		2779.81	I
80		2785.03	I
60		2787.96	I
60		2812.59	I
80		2813.58	I
2		2825.52	II
1400	r	2839.99	I
1		2846.42	II
200		2850.62	I
1000	r	2863.32	I
1		2912.80	II
200		2913.54	I
6		2919.82	II
3		2991.00	II
7		2994.44	II
700	r	3009.14	I
1		3012.18	II
8		3023.94	II
200		3032.80	I
850	r	3034.12	I
12		3047.50	II
6		3094.69	II
60		3141.84	I
550	r	3175.05	I
40		3218.71	I

Intensity		Wavelength	
550	r	3262.34	I
50		3283.21	II
110		3330.62	I
60		3351.97	II
2		3407.48	II
10		3472.46	II
7		3537.57	II
11		3575.45	II
3		3582.39	II
2		3620.08	II
6		3620.54	II
40		3655.78	I
6		3715.23	II
280	r	3801.02	I
4		3841.44	II
1		4294.65	II
40		4524.74	I
1		4579.13	II
1		4580.29	II
2		4877.22	II
3		4944.31	II
20		4979.73	I
2		5071.14	II
2		5072.67	II
20		5174.54	I
10		5332.36	II
20		5561.95	II
25		5588.92	II
2		5596.20	II
500		5631.71	I
15		5753.59	I
1		5797.20	II
15		5799.18	II
50		5925.44	I
100		5970.30	I
150		6037.70	I
200		6054.86	+
250		6069.00	I
100		6073.46	I
6		6077.48	II
5		6079.70	II
400		6149.71	I
200		6154.60	I
150		6171.50	I
100		6310.78	I
40		6354.35	I
70		6453.50	II
8		6761.45	II
25		6844.05	II
20		7191.40	II
10		7387.79	II
20	h	7398.6	I
1		7408.62	II
30		7685.30	I
13		7741.80	II
100		7754.97	I
3		7904.00	II
100	h	8030.5	I
30	h	8039.3	I
200		8114.09	I
30	h	8121.0	I
30		8349.35	I
80		8357.04	I
300		8422.72	I
400		8552.60	I
50	h	8681.7	I
30	h	9018.95	I
50	h	9410.86	I
80	h	9415.37	I
150		9616.40	I
50		9741.1	I
100	h	9742.8	I
300	h	9805.38	I
500		9850.52	I
25		10456.47	I
11		10807.58	I
54		10894.00	I
70		11191.85	I
56		11277.66	I
17		11336.97	I
200		11454.59	I
200		11616.26	I
76		11670.77	I
25		11694.45	I
258		11739.78	I
96		11825.18	I
106		11835.82	I
254		11932.99	I
48		12009.50	I
111		12313.24	I
33		12335.6	I
42		12530.87	I
42		12536.5	I
37		12788.2	I
89		12888.5	I
187		12981.7	I
20		13000.3	I
187		13018.5	I
68		13081.5	I
378		13460.2	I
144		13608.2	I
13		15018.2	I
30		15464.2	I
20		17000.5	I

Intensity	Wavelength	
10	17807.5	I
20	20622.2	I
40	20861.7	I
8	21686.2	I
4	22131.7	I
3	22997.2	I
4	24327.2	I
4	24738.2	I

Sn III
Ref. 423 — C.H.C.

Intensity	Wavelength	
	Vacuum	
100	753.01	III
50	760.62	III
75	775.79	III
50	784.68	III
200	910.92	III
50	1010.92	III
50	1048.84	III
1000	1139.29	III
1000	1158.33	III
200	1161.09	III
100	1161.58	III
100	1180.62	III
1000	1184.25	III
200	1189.99	III
200	1204.06	III
2000	1210.52	III
100	1215.10	III
100	1218.14	III
100	1230.17	III
100	1231.38	III
500	1243.63	III
1000	1259.92	III
40	1276.31	III
1000	1305.97	III
1000	1327.34	III
200	1334.70	III
200	1346.05	III
1000	1347.65	III
200	1365.71	III
1000	1386.74	III
500	1410.61	III
200	1449.77	III
1000	1570.36	III
50	1674.29	III
500	1811.71	III
500	1941.86	III
50	1955.52	III

Sn IV
Ref. 423 — C.H.C.

Intensity	Wavelength	
	Vacuum	
50	605.23	IV
50	619.04	IV
50	628.73	IV
50	908.22	IV
500	956.25	IV
500	1019.72	IV
1000	1044.49	IV
100	1058.37	IV
50	1058.59	IV
1000	1073.41	IV
200	1087.50	IV
300	1096.92	IV
50	1103.24	IV
000	1119.34	IV
200	1120.68	IV
1000	1314.55	IV
1000	1437.52	IV
100	1532.90	IV

Sn V
Ref. 399, 423 — C.H.C.

Intensity	Wavelength	
	Vacuum	
120	355.14	V
150	361.01	V
100	372.55	V
200	1089.35	V
100	1132.79	V
200	1160.74	V
100	1176.26	V
100	1189.92	V
100	1205.72	V
2000	1251.38	V
100	1283.81	V
200	1294.36	V
100	1302.20	V

TITANIUM (Ti)
Z = 22
Ti I and II
Ref. 1 — C.H.C.

Intensity	Wavelength	
	Air	
140	2272.61	I
180	2273.28	I
130	2276.70	I
190	2279.96	I
150	2299.85	I
140	2302.73	I
190	2305.67	I
65	2380.81	I
35	2384.52	I
55	2418.36	I
75	2421.30	I
95	2424.24	I
40	2428.23	I
35	2433.22	I
19	2434.40	I
35	2440.21	II
65	2440.98	I
24	2450.44	II
24	2504.54	I
75	2517.43	II
40	2519.04	I
140	2520.54	I
75	2524.64	II
360	2525.60	II
29	2527.98	I
210	2529.85	I
190	2531.25	II
190	2534.62	II
130	2535.87	II
190	2541.92	I
65	2555.99	II
110	2571.03	II
50	2572.65	II
50	2580.82	I
35	2590.26	I
100	2593.64	I
65	2596.58	I
270	2599.92	I
340	2605.15	I
510	2611.28	I
75	2611.48	I
300	2619.94	I
170	2631.54	I
170	2632.42	I
640	2641.10	I
800	2644.26	I
950	2646.64	I
30	2649.30	I
15	2654.93	I
35	2657.19	I
85	2661.97	I
95	2669.60	I
130	2679.93	I
26	2684.80	I
30	2685.14	I
65	2688.82	I
26	2716.25	II
85	2725.07	I
75	2727.42	I
21	2731.13	I
40	2731.58	I
170	2733.26	I
55	2735.29	I
40	2735.61	I
85	2739.81	I
250	2742.32	I
40	2749.06	I
65	2757.40	I
95	2758.08	I
15	2761.29	II
250	2802.50	I
55	2805.70	I
30	2806.50	II
40	2809.17	II
75	2810.30	II
30	2812.98	I
30	2817.40	I
65	2817.84	I
	2817.87	II
65	2828.07	I
	2828.15	II
130	2832.16	II
190	2841.94	II
110	2851.10	II
40	2853.93	II
95	2862.32	II
55	2868.74	II
180	2877.44	II
280	2884.11	II
65	2888.93	II
55	2891.07	II
55	2905.66	I
30	2909.92	II
450	2912.08	I
340	2928.34	I
15	2931.03	II
180	2933.55	I
26	2935.96	II
150	2937.32	I
1100	2942.00	I
1300	2948.26	I
30	2954.58	II
1600	2956.13	II
170	2956.80	I
30	2958.77	II
26	2959.71	I
35	2959.99	I
170	2965.71	I
190	2967.22	I
26	2968.23	I
75	2970.38	I
30	2974.93	I
170	2983.31	I
35	3000.87	I
120	3017.19	II
140	3029.73	II
110	3046.68	II
130	3056.74	II
130	3057.00	II
170	3058.09	II
85	3059.74	II
1300 d	3066.22	II
	3066.35	II
70	3071.24	II
600	3072.11	II
1100	3072.97	II
1600	3075.22	II
2300	3078.64	II
3600	3088.02	II
180	3089.40	II
180	3097.19	II
180	3100.67	I
230	3103.80	II
230	3105.08	II
360	3106.23	II
70	3106.81	I
50	3110.67	II
50	3112.48	II
140	3117.67	II
720	3119.72	I
	3119.80	II
190	3123.07	I
240	3130.80	II
140	3141.54	I
95	3141.67	I
220	3143.76	II
240	3148.04	II
240	3152.25	II
240	3154.20	II
240	3155.67	II
500	3161.20	II
780	3161.77	II
1000	3162.57	II
1600	3168.52	II
2400	3186.45	I
1000	3190.87	II
3100	3191.99	I
50	3197.52	II
3800	3199.92	I
780	3202.54	II
50	3203.44	II
240	3203.83	I
50	3204.87	I
110	3213.14	II
260	3214.24	I
190	3214.75	II
1100	3217.06	II
110	3217.94	I
260	3218.27	II
110	3219.21	I
110	3221.38	I
1300	3222.84	II
220	3223.52	I
240	3224.24	II
140	3226.13	I
530	3228.60	II
780	3229.19	II
530	3229.42	II
110	3231.32	II
240	3232.28	II
6600	3234.52	II
220	3236.12	II
5200	3236.57	II
4100	3239.04	II
220	3239.66	II
2600	3241.99	II
1200	3248.60	II
950	3251.91	II
1200	3252.91	II
1200	3254.25	II
1200	3261.60	II
310	3271.65	II
310	3272.08	II
200	3278.29	II
260	3278.92	II
220	3282.33	II
530	3287.66	II
290	3292.08	I
170	3299.41	I
170	3306.88	I
220	3308.39	I
220	3308.81	II
260	3309.50	I
60	3309.73	I
110	3312.69	II
840	3314.42	I
	3314.52	I
290	3315.32	II
330	3318.02	II
550	3321.70	II
2900	3322.94	II
380	3326.76	II
2100	3329.46	II
550	3332.11	II
1800	3335.20	II
1100	3340.34	II
5700	3341.88	I
120	3342.15	I
260	3343.77	II
330	3346.73	II
4300	3349.04	II
12000	3349.41	II
120	3352.94	I
4100	3354.64	II
290	3358.28	I
290	3360.99	I
7200	3361.21	II
	3361.26	I
120	3361.84	I
1100	3370.44	I
4300	3371.45	I
140	3372.21	II
5700	3372.80	II
60	3374.35	II
2900 d	3377.48	I
	3377.58	I
290	3379.22	I
1400	3380.28	II
170	3382.31	I
5700	3383.76	II
170	3385.66	I
1400	3385.95	I
1400	3387.84	II
60	3388.76	II
140	3390.68	I
140	3392.71	I
1100	3394.58	II
60	3398.63	I
60	3402.42	II
60	3407.20	II
95	3409.81	II
60	3439.30	I
890	3444.31	II
60	3452.47	II
180	3456.39	I
600	3461.50	II
95	3467.26	I
600	3477.18	II
60	3478.92	I
240	3480.53	I
60	3485.69	I
60	3489.74	II
480	3491.05	II
60	3495.75	I
95	3499.10	I
890	3504.89	II
120	3506.64	I
600	3510.84	II
60	3520.25	I
310	3535.41	II
190	3547.03	I
120	3573.74	II
60	3574.24	I
240	3587.13	II
190	3596.05	II
600	3598.72	I
600	3610.16	I
190	3624.82	II
95	3635.20	I
4800	3635.46	I
120	3637.97	I
190	3641.33	II
6600	3642.68	I
180	3646.20	I
7200	3653.50	I
290	3654.59	I
660	3658.10	I
120	3659.76	II
380	3660.63	I
190	3662.24	II
380	3668.97	I
600	3671.67	I
3100	3685.20	I
120	3685.96	I
95	3687.35	I
600	3689.91	I

Titanium (Cont.)

Intensity	Note	Wavelength	Species
140		3694.45	I
30		3698.18	I
60		3698.43	I
60		3700.08	I
120		3702.29	I
190		3704.30	I
140		3706.23	II
50		3707.53	I
290		3709.96	I
30		3715.40	I
450		3717.40	I
140		3721.64	II
330		3722.57	I
600		3724.57	I
380		3725.16	I
2900		3729.82	I
50		3735.67	I
60		3738.90	I
3300		3741.06	I
330		3741.64	II
160		3748.10	I
5200		3752.86	I
600		3753.64	I
140		3757.69	II
3300		3759.30	II
2900		3761.32	II
50		3761.89	II
60		3766.45	I
600		3771.66	I
30		3776.06	II
840		3786.04	I
120		3789.30	I
70		3795.90	I
60		3798.31	I
70		3818.22	I
60		3822.03	I
240		3828.19	I
95		3833.68	I
95		3836.78	I
60		3846.45	I
130		3853.05	I
130		3853.73	I
170		3858.14	I
240		3866.44	I
170		3868.40	I
120		3873.21	I
260		3875.26	I
170		3882.15	I
170		3882.33	I
500		3882.89	I
60	h	3888.02	I
70		3889.95	I
200	h	3895.25	I
85		3898.49	I
530		3900.54	II
180		3900.96	I
2600		3904.78	I
110	h	3911.19	I
500		3913.46	II
500		3914.34	I
24		3914.74	I
35		3919.82	I
290		3921.42	I
1100		3924.53	I
110		3926.32	I
890		3929.88	I
35		3932.02	II
70		3934.24	I
1100		3947.78	I
4500		3948.67	I
4500		3956.34	I
5200		3958.21	I
950		3962.85	I
950		3964.27	I
4800		3981.76	I
570		3982.48	I
60		3984.33	I
35		3985.25	I
60		3985.59	I
5700		3989.76	I
35		3994.70	I
7800		3998.64	I
70		3999.36	I
70		4002.49	I
70		4003.81	I
35		4005.97	I
70		4008.06	I
950		4008.93	I
190		4009.66	I
70		4012.39	II
180		4013.58	I
70		4015.38	I
35		4016.28	I
120	h	4017.77	I
140		4021.83	I
1200		4024.57	I
40		4025.14	II
190	h	4026.54	I
40		4027.48	I
40		4028.34	II
190	h	4030.51	I
40		4033.91	I
30		4034.91	I
110		4035.83	I
35	h	4040.32	I
290		4055.02	I
85		4057.62	I
85		4058.14	I
410		4060.26	I
200		4064.22	I
200		4065.10	I
840		4078.47	I
40		4079.72	I
290		4082.46	I
85		4099.17	I
220		4112.71	I
85		4122.17	I
40		4123.31	I
85		4123.57	I
130		4127.54	I
40		4129.17	I
40		4131.25	I
140		4137.29	I
85		4143.05	I
170		4150.96	I
85		4159.64	I
70		4163.65	II
35		4164.14	I
40		4166.32	I
85		4169.35	I
120		4171.03	I
40		4171.90	II
35		4183.30	I
360		4186.12	I
40		4188.69	I
70		4200.75	I
85		4203.46	I
35		4211.73	I
40		4224.79	I
40		4227.65	I
130		4237.89	I
85		4249.12	I
130		4256.04	I
70		4258.54	I
70		4261.60	I
330		4263.13	I
35		4265.71	I
40		4266.22	I
70		4270.14	I
85		4272.43	I
240		4274.58	I
120		4276.43	I
120		4278.23	I
30		4278.81	I
110		4281.38	I
220		4282.71	I
160		4284.99	I
890		4286.01	I
840		4287.40	I
30		4288.16	I
950		4289.07	I
120		4290.23	II
840		4290.94	I
120		4291.14	I
140		4294.12	II
840		4295.76	I
2000		4298.66	I
200		4299.23	I
200		4299.64	I
200		4300.05	II
2900		4300.56	I
4100		4301.09	I
85		4301.93	II
6000		4305.92	I
180		4307.90	II
35		4308.50	I
40		4311.65	I
85		4312.87	II
85		4314.35	I
1200		4314.80	I
360		4318.64	I
180		4321.66	I
190		4325.13	I
160		4326.36	I
30		4334.84	I
160		4337.92	II
24		4344.29	II
70		4346.11	I
35		4354.06	I
95		4360.49	I
24		4368.94	I
95		4369.68	I
60		4372.38	I
30		4388.08	I
170		4393.92	I
330		4395.04	II
60		4399.77	II
240		4404.28	I
60		4404.90	I
30		4405.68	I
60		4416.54	I
220		4417.28	I
60		4417.72	II
120		4421.76	I
120		4422.82	I
24		4424.39	I
30		4425.83	I
120		4426.06	I
890		4427.10	I
21		4430.02	I
85		4430.37	I
50		4431.28	I
30		4432.60	I
24		4433.58	I
170		4434.00	I
70		4436.59	I
30		4438.23	I
130		4440.35	I
50		4441.27	I
230		4443.80	II
24		4444.27	I
840		4449.15	I
30		4450.49	II
550		4450.90	I
840		4453.32	I
290		4453.71	I
950		4455.33	I
1100		4457.43	I
21		4462.09	I
70		4463.38	I
95		4463.54	II
290		4465.81	I
240		4468.50	II
240		4471.24	I
95		4474.78	I
95		4479.70	I
50		4480.59	I
530		4481.26	I
95		4482.69	I
19		4488.32	II
260		4489.09	I
24		4492.55	I
40		4495.01	I
240		4496.15	I
24		4497.73	I
200		4501.27	II
40		4503.78	I
21		4506.36	I
50		4511.17	I
780		4512.74	I
19		4515.62	I
1000		4518.03	I
95		4518.70	I
1000		4522.80	I
780		4527.31	I
6000		4533.24	I
240		4533.97	II
3600		4534.78	I
2400		4535.58	I
1200		4535.92	I
1200		4536.05	I
24		4537.23	I
24		4539.10	I
720		4544.69	I
950		4548.77	I
240		4549.63	II
950		4552.46	I
24		4555.08	I
720		4555.49	I
19		4557.86	I
19		4558.11	I
60		4559.92	I
50		4562.63	I
35		4563.43	I
110		4563.77	II
35		4570.91	I
240		4571.98	II
19		4585.84	I
24		4589.95	II
60		4599.23	I
21		4609.37	I
950		4617.27	I
24		4619.52	I
480		4623.09	I
190		4629.34	I
50	d	4634.87	
60		4637.88	I
240		4639.37	I
220		4639.67	I
190		4639.95	I
140		4645.19	I
120		4650.02	I
24		4656.04	I
720		4656.47	I
840		4667.59	I
70		4675.12	I
950		4681.92	I
21		4686.92	I
24		4690.80	I
190		4691.34	I
40		4693.68	I
24		4696.94	I
190		4698.76	I
120		4710.19	I
24		4715.30	I
65		4722.62	I
65		4723.17	I
55		4731.17	I
45		4733.43	I
18		4734.68	I
22		4742.11	I
170		4742.79	I
22		4747.68	I
310		4758.12	I
310		4759.28	I
45		4766.33	I
28		4769.77	I
65		4778.26	I
45		4781.72	I
110		4792.49	I
45		4796.22	I
35		4797.98	I
110		4799.80	I
28		4805.10	II
110		4805.43	I
45		4808.53	I
22		4811.08	I
40		4812.25	I
200		4820.42	I
22		4825.46	I
40		4836.13	I
470		4840.87	I
65		4848.47	I
290		4856.01	I
35		4864.18	I
200		4868.26	I
250		4870.14	I
28		4880.91	I
45		4882.35	I
400		4885.08	I
380		4899.91	I
320		4913.62	I
55		4915.24	I
130		4919.87	I
180		4921.77	I
55		4925.41	I
30		4926.16	I
150		4928.34	I
30		4937.74	I
95		4938.29	I
30		4941.58	I
21		4948.19	I
21		4958.25	I
55		4964.75	I
21		4966.04	I
65		4968.58	I
75		4973.05	I
120		4975.35	I
65		4977.74	I
120		4978.20	I
5800		4981.73	I
150		4989.15	I
4600		4991.07	I
30		4995.08	I
140		4997.10	I
4000		4999.51	I
230		5001.01	I
3600		5007.21	I
120		5009.65	I
230		5013.30	I
3200	d	5014.19	I
		5014.24	I
580		5016.17	I
840		5020.03	I
840		5022.87	I
580		5024.84	I
300		5025.58	I
1200		5035.91	I
840		5036.47	I
740		5038.40	I
1200		5039.95	I
75		5040.62	I
85		5043.59	I
35		5044.27	I
55		5045.41	I
26		5048.21	I
110		5052.87	I
21		5054.08	I
110		5062.11	I
35		5064.07	I
1400		5064.66	I
95		5065.99	I
35	h	5068.33	I
65		5069.35	I
130		5071.48	I
40		5085.34	I
130		5087.07	I
21		5103.15	I
55		5109.44	I
190		5113.44	I
270		5120.42	I
30		5129.15	II
270		5145.47	I
230		5147.48	I
210		5152.20	I
21	B1	5166.86	T
1100		5173.75	I
40		5186.34	I
85		5188.70	II
30		5189.58	I
1300		5192.98	I
85	h	5194.04	I
65		5201.10	I
120		5206.08	I
75		5207.87	I
65		5208.42	

Titanium (Cont.)

Intensity	Note	Wavelength	Species
1400		5210.39	I
65		5212.29	I
150		5219.71	I
95		5222.69	I
85		5223.64	I
250		5224.32	I
95		5224.56	I
190		5224.95	I
65		5226.56	II
120		5238.58	I
21		5246.15	I
55		5246.57	I
75		5247.31	I
21		5250.95	I
110		5252.11	I
75		5255.83	I
55		5259.99	I
55		5263.50	I
150		5265.98	I
40		5282.39	I
140		5283.45	I
35		5284.39	I
26		5288.81	
65		5295.79	I
120		5297.26	I
65		5298.44	I
26		5336.81	II
17		5341.50	I
75		5351.08	I
26		5366.65	I
55		5369.64	I
40		5389.18	I
55		5389.99	I
17		5396.60	I
85		5397.09	I
35		5404.02	I
110		5409.61	I
40		5426.26	I
75		5429.15	I
26		5436.73	I
17		5438.32	I
40		5446.64	I
11	B1	5448.34	T
30		5448.90	I
21		5449.16	I
35		5453.65	I
55		5460.51	I
75		5471.21	I
35		5472.70	I
40	h	5473.55	I
85		5474.23	I
30		5474.66	I
120	h	5477.71	I
110		5481.43	I
75		5481.87	I
85	h	5488.20	I
150		5490.15	I
26		5490.84	I
110		5503.90	I
40		5511.78	I
340		5512.53	I
270		5514.35	I
320		5514.54	I
26		5530.49	I
110		5565.49	I
13		5579.16	
21	h	5582.98	I
30	h	5585.68	I
65	B1	5597.85	T
55	B1	5629.28	T
17		5635.84	
250		5644.14	I
75		5648.58	I
26	B1	5661.55	T
190		5662.16	I
75		5662.91	I
21		5673.42	I
130		5675.44	I
30	h	5679.94	I
95		5689.47	I
75		5702.68	I
35		5708.23	I
65		5711.88	I
40	h	5713.92	I
95		5715.13	I
55		5716.48	I
35		5720.48	I
85		5739.51	I
40		5740.02	I
19		5741.22	I
21		5752.84	I
19		5756.86	I
40	h	5762.27	I
55	h	5766.35	I
75	h	5774.05	I
30		5780.78	I
75	h	5785.98	I
65	H1	5804.26	I
21	B1	5814.96	T
40		5823.71	I
21	h	5841.18	I
21		5852.34	I
400		5866.46	I
65		5880.31	I
21	h	5888.68	
230		5899.32	I
55		5903.33	I
120		5918.55	I
150		5922.12	I
75		5937.82	I
120		5941.76	I
300		5953.17	I
200		5965.84	I
270		5978.56	I
340		5999.04	I
65		5999.68	I
21		6012.73	
110		6064.63	I
120		6085.23	I
120		6091.17	I
40		6092.81	I
40	h	6098.67	I
35	h	6121.01	I
120		6126.22	I
19		6138.38	I
30		6146.22	I
21		6149.74	I
30	B1	6162.23	T
35		6186.15	I
95	h	6215.28	I
75	h	6220.49	I
65	h	6221.41	I
380		6258.10	I
380		6258.70	I
300		6261.10	I
65		6303.75	I
55		6312.24	I
26		6318.03	I
30		6336.10	I
35		6366.35	I
11		6419.10	I
17		6497.69	I
19		6508.14	I
55		6546.28	I
65		6554.23	I
11	h	6554.83	I
75		6556.07	I
19	h	6565.62	I
14	h	6575.18	I
35		6599.11	I
18	B1	6651.46	T
18	h	6666.55	I
22	h	6667.74	
9		6668.39	
19		6677.18	I
22	b	6691.21	T
26		6716.68	I
16	B1	6723.95	T
80		6743.12	I
22		6745.52	I
18		6844.64	
18		6860.39	
35		6861.47	I
9		6873.92	I
12		6913.19	I
14	h	6933.15	I
14	h	6943.70	I
23		6996.63	I
15		7004.66	I
14		7008.35	I
14		7010.94	I
14	h	7035.86	I
40		7038.80	I
14		7050.65	I
40	B1	7054.51	T
23		7069.11	I
23		7072.05	I
45	B1	7087.89	T
30	b	7124.9	T
40	B1	7125.61	T
26		7138.91	I
26		7167.13	
23		7171.53	
55		7189.89	I
26	b	7203.64	T
260		7209.44	I
60		7216.20	I
130		7244.86	I
130		7251.72	I
19		7263.40	
19		7266.29	I
19	b	7269.05	T
15		7315.56	I
26		7318.39	I
120		7344.72	I
11		7352.16	I
90		7357.74	I
60		7364.11	I
26		7440.60	I
9		7474.94	I
26		7489.61	I
19		7496.12	I
12		7580.55	I
9	B1	7589.62	T
15		7614.50	I
23		7654.44	I
11	B1	7705.21	T
30		7949.17	I
26	h	7961.58	I
60		7978.88	I
9		7979.07	I
30		7996.53	I
7	h	8003.55	
55		8024.84	I
30		8068.24	I
8		8267.62	
14	h	8306.31	I
9	h	8307.41	I
9	h	8311.76	I
8	h	8312.85	I
12		8334.37	I
14		8353.15	I
75		8364.24	I
100		8377.85	I
100		8382.54	I
55		8382.82	I
75		8396.87	I
120		8412.36	I
19		8416.98	I
15		8424.41	I
170		8426.52	I
490		8434.94	I
240		8435.70	I
40		8438.93	I
40		8450.89	I
9	h	8457.10	I
19	h	8467.15	I
45		8468.50	I
15		8496.04	I
19	h	8518.05	I
40		8518.32	I
14		8539.38	I
40		8548.12	I
9		8569.77	I
9	h	8598.18	I
90		8675.39	I
45		8682.99	I
23		8692.33	I
19		8734.69	I
23		8766.64	I
15	h	8778.71	I

Ti III
Ref. 378 — C.H.C.

Intensity	Note	Wavelength	Species
		Vacuum	
6		1282.48	III
6		1286.23	III
15		1286.36	III
10		1289.30	III
10		1291.62	III
10		1293.23	III
15		1294.70	III
10		1295.88	III
20		1298.66	III
20		1298.97	III
12		1327.59	III
10		1420.04	III
10		1420.44	III
10		1421.63	III
10		1421.77	III
12		1422.40	III
10		1424.14	III
23		1455.19	III
10		1498.70	III
		Air	
10		2199.22	III
12		2237.77	III
10		2327.02	III
15		2331.35	III
15		2331.66	III
15		2334.34	III
17		2339.00	III
18		2346.79	III
18		2374.99	III
22		2413.99	III
25		2516.05	III
24		2527.84	III
23		2540.06	III
24		2563.44	III
23		2565.42	III
22		2567.56	III
15		2576.47	III
15		2580.46	III
10		2692.16	III
12		2701.96	III
22		2984.75	III
12	d	3354.71	III
12		3872.50	III
12		3881.21	III
12		3893.63	III
10		3896.33	III
15		3915.47	III
12		3921.38	III
10		3921.61	III
12		3922.95	III
10		3924.86	III
10		4060.21	III
10		4119.14	III
11		4215.52	III
11		4269.84	III
11		4296.70	III
10		4348.04	III
11		4433.91	III
10		4540.22	III
15		4549.84	III
10	d	4555.46	III
15	d	4572.20	III
10		4649.45	III
12		4652.86	III
10		4874.00	III
10		4950.10	III
10		4971.19	III
10		5083.80	III
14		5147.31	III
12	d	5226.28	III
11		5247.49	III
17		5278.12	III
10		5278.70	III
12		5298.43	III
16		5301.20	III
15		5306.88	III
10		5395.69	III
12		5533.01	III
10		5817.44	III
12		6611.38	III
18		6621.58	III
10		6629.37	III
14		6647.47	III
18		6667.99	III
15		6674.19	III
14		6707.76	III
12		6724.80	III
16		6734.10	III
15		6862.26	III
12		6874.35	III
10		6896.12	III
12		7015.38	III
10		7071.93	III
20		7072.64	III
18		7084.57	III
15		7124.13	III
11		7171.79	III
10		7175.92	III
10		7217.50	III
9		7225.55	III
12		7270.67	III
14		7316.30	III
10		7316.68	III
12		7379.96	III
10		7408.13	III
10		7457.85	III
15		7506.87	III
17		7507.68	III
10		7523.85	III
12		7544.29	III
9		7566.25	III
10	h	8172.21	III
9	h	8173.37	III
9	h	8178.00	III
10		8182.42	III
9	h	8192.68	III
9	h	8194.75	III
9	h	8263.67	III
15	h	8267.32	III
10		8338.54	III
12		8394.20	III
20		8466.87	III
5		8699.85	III
3		9017.10	III

Ti IV
Ref. 428 — C.H.C.

Intensity	Wavelength	Species
	Vacuum	
10	776.76	IV
18	779.07	IV
16	781.73	IV
8	1183.64	IV
10	1195.21	IV
18	1451.74	IV
20	1467.34	IV
12	1469.19	IV
	Air	
20	2067.56	IV
18	2103.16	IV
10	2359.14	IV
10	2359.50	IV

Titanium (Cont.)

Intensity		Wavelength	
8		2541.79	IV
10		2546.88	IV
5		2862.60	IV
6		2929.96	IV
14		2937.33	IV
12		2957.31	IV
15		3541.36	IV
17		3576.44	IV
10		3581.39	IV
13		4131.22	IV
14		4133.78	IV
10		4397.33	IV
9		4403.45	IV
15		4618.11	IV
20		5398.93	IV
8		5470.98	IV
18		5492.51	IV
10		5517.72	IV
14		5877.79	IV
15		5885.96	IV
7		5891.15	IV
6		6231.62	IV
17		6246.65	IV
11		6247.74	IV
15		6292.41	IV
12		6913.85	IV
15		6978.51	IV
9		7491.37	IV
8		7494.77	IV
5	h	7652.12	IV
8	h	7706.85	IV

Ti V
Ref. 427 — C.H.C.

Intensity		Wavelength	
		Vacuum	
12		225.35	V
10		228.91	V
17		252.96	V
7		323.36	V
7		461.41	V
8		474.69	V
8		483.99	V
10	d	488.58	V
15		498.26	V
14		502.08	V
7		502.71	V
12		504.66	V
7		506.47	V
8		513.37	V
7		523.05	V
12		524.58	V
13		526.57	V
8		529.32	V
10		535.84	V
10		535.89	V
8	d	540.14	V
8		541.46	V
9		541.71	V
7		543.10	V
7		543.34	V
7		1128.55	V
8		1192.35	V
9		1198.66	V
9		1222.36	V
10		1230.36	V
11		1239.96	V
10		1241.67	V
7		1246.13	V
8		1268.49	V
8		1306.11	V
8		1411.31	V
9		1675.15	V
8		1687.16	V
11		1717.40	V
8		1759.76	V
7		1771.45	V
10		1841.49	V
7		1864.45	V
7		1881.89	V
7		1920.16	V
7		1988.75	V

TUNGSTEN (W)
Z = 74

W I and II
Ref. 1 — C.H.C.

Intensity		Wavelength	
		Air	
5800		2001.71	II
13000		2008.07	II
5100		2009.98	II
4100		2010.23	II
4100		2014.23	II
7300		2026.08	II
15000		2029.98	II
2700		2035.03	II
5300		2049.63	II
2300		2065.57	II
3400		2071.21	II
2200		2075.59	II
9700		2079.11	II
3600		2088.19	II
2200		2089.14	II
1700		2090.48	I
6100		2094.75	II
2400		2098.60	II
2200		2100.67	II
1500		2101.54	I
1500		2106.18	II
1300		2110.34	II
2100		2118.87	II
2400		2121.59	II
850		2153.56	II
850		2157.80	II
1500		2166.32	II
480		2182.90	I
440		2194.52	II
1300		2204.48	II
460		2248.75	II
460		2249.80	I
180		2270.24	II
95		2271.37	I
510		2277.58	I
160		2284.91	I
320		2285.17	I
530	d	2294.49	I
		2294.54	II
270		2298.33	I
240		2303.83	II
240		2306.59	I
340		2309.02	I
440		2313.17	I
220		2314.17	I
190		2315.02	II
460		2321.63	I
290		2326.09	II
390	d	2326.56	I
		2326.70	I
75		2328.31	II
130		2333.77	II
210		2341.37	I
75		2349.26	II
120	d	2350.37	II
320		2354.61	I
60		2358.81	II
580		2360.44	I
850		2363.07	I
60		2364.22	II
510		2374.47	I
210		2382.99	I
670		2384.82	I
240		2389.08	I
120		2390.37	II
120		2392.93	II
730		2397.09	II
560		2397.73	II
560		2397.98	I
75		2404.24	II
1700	d	2405.58	I
		2405.69	I
75		2411.54	II
320		2414.04	I
610		2415.68	II
50		2419.34	II
50		2421.01	II
870		2424.21	I
190		2427.49	II
170		2429.39	II
580		2431.08	I
630		2433.98	I
60		2435.01	II
1800		2435.96	I
250		2436.62	I
580		2444.06	I
160		2446.39	II
270		2448.39	I
270		2451.35	I
780		2451.48	II
870		2452.00	I
430		2454.72	I
630		2454.98	I
780		2455.51	I
780		2456.53	I
1100		2459.30	I
270		2460.16	I
480		2462.79	I
270		2464.30	I
230		2466.52	II
1400		2466.85	I
75		2470.80	II
480		2472.51	I
1200		2474.15	I
290		2477.80	II
870		2480.13	I
390		2480.96	I
1500		2481.44	I
480	d	2482.10	I
		2482.21	I
29		2484.40	II
580		2484.74	I
390		2487.50	I
270		2488.77	II
390		2489.23	II
75		2492.93	II
630		2495.26	I
230		2496.64	II
95		2497.48	II
140		2499.69	II
40		2500.11	II
		2501.90	II
680		2504.70	I
270		2506.06	II
24		2508.00	II
250		2510.17	I
75		2510.47	II
60		2518.14	II
310		2520.46	I
780		2521.32	II
270		2522.04	II
430		2527.76	I
780		2533.64	I
50		2534.82	II
580		2545.34	I
1200		2547.14	I
50		2549.09	I
40		2550.10	II
780		2550.38	I
2700		2551.35	I
450		2553.82	I
410		2554.86	II
580		2555.09	I
		2555.21	I
310		2556.75	I
290		2560.12	I
730		2561.97	I
230		2563.16	II
110		2563.91	II
530		2571.44	II
170	d	2572.24	II
		2572.35	II
75		2573.95	II
190		2579.26	II
290		2580.34	I
870		2580.49	I
40		2581.20	II
390		2584.39	I
390		2589.17	II
170		2591.49	II
110		2598.74	II
370		2601.96	I
75		2602.51	II
75		2603.02	II
270		2603.54	I
680		2606.39	I
320		2607.38	I
370		2608.32	I
970		2613.08	I
480		2613.82	I
230		2615.12	I
70		2615.44	II
210		2619.18	I
400		2620.25	I
400		2622.21	I
400		2625.22	I
210		2628.26	I
400		2632.48	I
400		2632.70	I
810		2633.13	I
290		2636.54	I
400	d	2638.62	I
		2638.75	I
210		2645.69	I
650		2646.18	I
400		2646.73	I
75		2647.74	II
40		2653.42	II
80		2653.57	II
1600		2656.54	I
400		2657.38	I
400	d	2658.04	II
		2658.18	I
810		2662.84	I
260		2664.97	I
75		2666.49	II
210		2669.30	I
810		2671.47	I
80		2673.59	II
650		2677.28	I
160	d	2677.79	II
		2677.91	I
400		2678.88	I
2100		2681.42	I
290		2683.35	I
210		2691.09	I
650		2695.67	I
210		2697.71	II
650		2699.59	I
400		2700.01	I
40		2701.48	II
160		2702.11	II
210		2702.52	I
400		2706.58	I
400		2708.59	I
400	d	2708.80	I
		2708.93	I
80		2709.58	II
40		2710.78	II
400		2715.50	I
80		2716.32	I
80		2718.04	II
2100		2718.91	I
320		2719.33	I
210		2719.86	I
2600		2724.35	I
210		2724.62	I
400		2725.03	I
		2725.06	I
80		2729.62	II
75		2740.79	II
650		2748.84	I
40	d	2760.74	I
80		2761.59	II
400		2762.34	I
400		2764.27	II
210		2768.98	I
400		2769.74	I
810		2770.88	I
210		2773.70	I
810		2774.00	I
810		2774.48	I
160		2776.50	II
40		2778.69	II
210		2787.98	I
340		2791.96	I
810		2792.70	I
80		2799.03	II
400		2799.93	I
160	d	2801.05	II
		2801.17	I
130		2805.92	II
40		2812.25	II
810		2818.06	I
160		2822.57	II
260		2829.82	I
1600		2831.38	I
810		2833.63	I
210		2835.64	I
400		2841.57	I
810		2848.02	I
650		2856.03	I
650		2866.06	I
230		2878.72	I
610		2879.11	I
610		2879.40	I
440		2896.01	I
1500		2896.44	I
230		2910.48	I
270		2911.00	I
360		2918.25	I
50		2918.63	II
360		2923.10	I
230		2923.54	I
230		2925.13	I
690		2935.00	I
2400		2944.40	I
2400		2946.99	I
480		2947.39	I
210		2952.29	II
440		2964.52	I
480		2977.11	I
		2977.21	I
730	d	2979.71	I
		2979.86	I
400		2993.61	I
240		2995.26	I
190		3009.09	I
360		3013.79	I
520		3016.47	I
770		3017.44	I
110		3024.50	II
210		3024.93	I
310	d	3026.67	I
		3026.79	I
160		3033.56	I
160		3034.19	I
160		3039.31	I
440	d	3041.73	I
		3041.86	I
270		3043.80	I
440		3046.44	I
110		3048.66	I
810		3049.69	I
110		3064.93	I
180		3073.28	I
110		3077.52	II
180	d	3084.83	I
		3084.91	I
370		3093.50	I
240		3107.23	I
240		3108.02	I

Tungsten (Cont.)

Intensity	Wavelength	sp	Intensity	Wavelength	sp	Intensity	Wavelength	sp	Intensity	Wavelength	sp
230	3117.57	I	140 h	3897.91	I	640	4659.87	I	4	6994.06	I
260	3120.18	I	150	3935.03	I	640	4680.51	I	8	7017.88	I
160	3133.88	I	120	3936.97	I	100	4693.72	I	3	7028.68	I
130	3141.42	I	120	3947.98	I	140	4757.54	I	3	7098.22	I
65	3149.85	II	120	3952.52	I	790	4843.81	I	4 h	7111.18	I
290	3163.42	I	120	3952.90	I	380	4886.90	I	15	7140.52	I
130	3164.44	I	160	3953.15	I	220	4982.59	I	9	7162.64	I
130	3165.38	I	200	3955.30	I	330	5006.15	I	5 h	7191.33	I
320	3176.60	I	160	3965.14	I	220	5015.30	I	5	7198.62	I
130	3179.06	I	130	3968.59	I	820	5053.28	I	11	7200.16	I
190	3181.82	I	150 h	3970.80	I	210	5054.60	I	5	7216.35	I
130	3184.05	I	130	3979.29	I	210	5069.12	I	4	7226.06	I
130	3184.42	I	130	3980.64	I	120	5071.74	I	8	7237.12	I
65	3189.24	II	250	3983.29	I	770	5224.66	I	5	7274.47	I
390	3191.57	I	8600	4008.75	I	27	5500.49	I	10	7278.24	I
390	3198.84	I	540	4015.22	I	27	5503.44	I	15	7285.81	I
520	3207.25	I	170 h	4016.52	I	10	5508.61	I	15	7296.55	I
140	3208.28	I	220	4019.23	I	220	5514.68	I	3	7298.25	I
1000	3215.56	I	130	4022.12	I	15	5531.38	I	7	7385.08	I
140	3221.21	I	180	4028.79	I	15	5537.72	I	4	7451.39	I
140	3221.91	I	180	4036.86	I	13	5568.09	I	3	7456.37	I
190	3232.49	I	140	4039.85	I	13	5604.31	I	8	7483.35	I
140	3237.09	I	140 h	4044.28	I	11	5631.27	I	7	7504.13	I
140	3242.03	I	910	4045.59	I	27	5631.94	I	10	7509.00	I
140	3252.29	I	180	4064.79	I	65	5648.37	I	3	7520.66	I
210	3254.36	I	150	4069.79	I	35	5660.72	I	9	7537.45	I
140	3259.43	I	730	4069.93	I	27	5674.39	I	9	7550.48	I
210	3259.66	I	340	4070.61	I	13	5676.60	I	17	7569.92	I
210 d	3266.62	I	100	4071.93	I	15	5676.90	I	5	7582.88	I
	3266.77	I	5000	4074.36	I	15	5697.79	I	3	7612.18	I
150	3281.94	I	150	4082.96	I	55	5735.09	I	17	7614.15	I
150	3293.71	I	130	4088.33	I	13	5749.24	I	3	7631.29	I
730	3300.82	I	100	4095.69	I	11	5756.10	I	3	7654.81	I
440	3311.38	I	1000	4102.70	I	13	5793.06	I	13	7688.97	I
440	3326.20	I	150	4109.75	I	13	5796.49	I	4	7701.01	I
440	3331.69	I	100	4111.82	I	45	5804.85	I	5	7761.16	I
150	3354.45	I	150	4118.05	I	13 d	5806.05	I	3	7776.73	I
150	3371.04	I	100	4118.19	I		5806.24	I	11	7784.15	I
390	3373.75	I	100	4120.85	I	13	5833.61	I	7	7808.96	I
150	3412.96	I	100	4125.16	I	13	5838.97	I	2	7823.82	I
150	3413.53	I	150	4126.80	I	17	5845.27	I	4	7863.47	I
150	3422.42	I	100	4133.48	I	28	5851.58	I	2	7867.04	I
150	3427.71	I	540	4137.46	I	11	5856.61	I	4	7880.40	I
230	3429.59	I	150	4138.02	I	22	5864.63	I	5	7886.48	I
240	3443.00	I	110	4142.25	I	11	5874.22	I	9	7940.92	I
160	3477.94	I	140	4145.16	I	13	5880.31	I	13	7937.06	I
600	3503.14	I	110	4145.95	I	13	5891.61	I	22	8017.19	I
160	3508.73	I	100	4154.66	I	13	5901.20	I	7	8054.89	I
160	3510.02	I	160	4170.53	I	40	5902.64	I	22	8055.64	I
160	3526.85	I	450	4171.17	I	13	5928.58	I	5	8060.38	I
160	3535.54	I	160	4204.40	I	35	5947.57	I	13	8123.82	I
160	3537.45	I	220	4207.05	I	13	5953.96	I	5	8143.19	I
650	3545.22	I	110	4215.38	I	13	5956.19	I	3	8165.72	I
160	3568.04	I	250	4219.37	I	27	5960.83	I	5	8210.22	I
240	3570.65	I	110	4222.04	I	55	5965.86	I	4	8322.05	I
80	3572.48	II	150	4234.34	I	27	5972.51	I	10	8338.08	I
160	3575.22	I	290	4241.44	I	20	5978.86	I	4	8348.81	I
80	3592.42	II	540	4244.36	I	20	5983.82	I	7	8358.72	I
240	3606.06	I	290	4259.35	I	13	6009.01	I	3	8382.94	I
80	3613.79	II	200	4260.29	I	55	6012.78	I	4	8402.60	I
1900	3617.52	I	200	4263.30	I	40	6021.52	I	4	8475.14	I
160	3622.34	I	1400	4269.38	I	20	6028.32	I	27	8585.11	I
130	3627.24	I	110	4269.77	I	20	6043.31	I	10	8594.42	I
320	3631.94	I	220	4274.55	I	13	6049.92	I	8	8613.27	I
240	3641.41	II	160	4275.49	I	13	6065.08	I	3	8614.50	I
80	3646.52	II	160	4276.74	I	22	6081.44	I	13	8865.53	I
80	3657.59	II	110	4282.34	I	13	6111.66	I			
160	3675.55	I	110	4286.01	I	13	6115.52	I			
650	3682.08	I	110	4294.10	I	22	6128.25	I			
400	3683.30	I	4100	4294.61	I	13	6143.94	I			
	3683.39	I	2200	4302.11	I	20	6153.72	I			
160	3683.93	I	160	4306.87	I	20	6154.87	I			
570	3688.06	I	110	4307.64	I	20	6203.51	I			
810	3707.92	I	110	4332.13	I	20	6254.28	I			
60	3716.08	II	100	4347.00	I	27	6285.88	I			
100	3719.39	I	150	4355.17	I	45	6292.02	I			
50	3736.22	II	100	4361.81	I	20	6303.21	I			
120	3741.71	I	150	4364.78	I	13	6386.47	I			
510	3757.92	I	100 d	4365.95	I	35	6404.21	I			
680	3760.13	I		4366.07	I	40	6445.12	I			
1000	3768.45	I	150	4372.52	I	11	6508.05	I			
120	3769.21	I	200	4378.48	I	15	6532.39	I			
120	3769.86	I	180	4384.85	I	13	6538.11	I			
340	3773.71	I	100	4403.95	I	13	6563.20	I			
1000	3780.77	I	200	4408.28	I	20	6573.93	I			
170	3792.76	I	130	4412.19	I	11	6607.13	I			
290	3809.22	I	160	4436.90	I	11	6609.05	I			
190	3810.38	I	140	4460.49	I	17	6611.62	I			
260	3810.79	I	140	4466.34	I	11	6621.74	I			
1400	3817.48	I	140	4466.74	I	13	6678.42	I			
110	3829.13	I	640	4484.19	I	15	6693.08	I			
1100	3835.06	I	160	4504.84	I	5	6746.56	I			
290	3838.51	I	130	4512.88	I	5	6764.45	I			
730	3846.22	I	120	4513.25	I	7	6805.31	I			
250	3847.49	I	150	4543.54	I	9	6814.92	I			
27	3851.57	II	150	4546.47	I	9	6820.27	I			
150	3855.55	I	150	4551.82	I	8	6828.43	I			
150	3859.30	I	140	4570.64	I	4	6853.74	I			
180	3864.34	I	170	4588.73	I	4	6876.01	I			
1800	3867.99	I	140	4599.94	I	5	6908.29	I			
250	3872.84	I	140	4609.89	I	9	6934.23	I			
110	3874.41	I	160	4613.30	I	8	6964.12	I			
730	3881.41	I	100	4642.53	I	13	6984.27	I			
110	3892.72	I	130	4657.42	I	8	6993.27	I			

URANIUM (U)
Z = 92

U I and II
Ref. 1, 303 — J.G.C.

Intensity	Wavelength (Air)	sp
440	2565.41	II
340	2569.71	II
340	2591.25	II
610	2635.53	II
470	2645.47	II
340	2669.17	II
470	2683.28	II
320	2691.04	II
370	2706.95	II
370	2733.97	II
470	2754.16	II
340	2762.85	II
390	2770.04	II
410	2784.45	II
830	2793.94	II
870	2802.56	II
630	2807.05	II
440	2808.98	II
630	2817.96	II
870	2821.12	II
390	2824.37	II
680	2828.90	II
920	2832.06	II
360	2837.19	II
460	2839.89	II

Intensity		Wavelength	Spectrum
360		2842.09	II
360		2849.48	II
390		2860.47	II
970		2865.68	II
340		2870.97	II
490		2882.74	II
460		2887.25	II
410		2888.26	II
1200		2889.62	II
320		2894.14	II
410		2894.51	II
780		2906.80	II
780		2908.28	II
320		2914.25	II
360		2914.63	II
440	p	2921.68	II
320		2927.38	II
490		2928.60	II
580		2931.41	II
440		2932.61	II
340		2933.86	II
530	p	2940.37	II
1300		2941.92	II
830		2943.90	II
340		2948.09	II
390		2954.77	II
580		2956.06	II
460		2965.03	II
580		2967.94	II
580		2971.06	II
410		2976.35	II
320		2982.74	II
530		2984.61	II
410		2992.72	II
360		3007.91	II
320		3021.22	II
630		3022.21	II
320		3024.51	II
320		3028.19	II
630		3031.99	II
490		3033.18	II
490		3044.16	II
580		3050.20	II
630		3057.91	II
460		3061.62	II
630		3062.54	II
580		3072.78	II
580		3093.01	II
320		3095.75	II
320		3098.01	II
580		3102.39	II
460		3104.15	II
970		3111.62	II
530		3119.35	II
680		3124.95	II
530		3139.61	II
410		3144.97	II
490		3145.56	II
680		3149.24	II
530		3153.11	II
340		3176.21	II
340		3177.33	II
340		3206.05	II
730		3229.50	II
680		3232.16	II
440		3244.22	II
340		3265.79	II
440		3270.12	II
440		3288.21	II
730		3291.33	II
1100		3305.89	II
390		3337.79	II
440		3341.66	II
390		3357.84	I
730		3390.38	I
340		3394.77	II
580		3424.56	II
580		3435.49	I
360		3453.55	II
320		3454.23	II
320		3457.05	II
320		3457.71	II
360		3459.92	I
320		3462.22	I
460		3463.55	I
630		3466.30	I
390		3472.52	II
320		3473.43	I
360		3480.36	I
680		3482.49	II
1600		3489.37	I
390		3493.33	II
340		3494.00	I
320		3494.84	II
530		3496.41	II
630		3500.08	I
320		3504.01	II
320		3505.07	II
780		3507.34	I
320		3508.84	II
1600		3514.61	I
390		3509.66	II
320		3513.67	I
390		3519.96	II
390		3531.11	II
630		3533.57	II
320		3534.33	I
530		3540.47	II
320		3542.57	I
390		3547.19	II
320		3549.20	I
1200		3550.82	II
320		3552.17	II
680		3555.32	I
320		3561.41	I
1200		3561.80	I
390		3563.66	I
2300		3566.59	I
530		3569.08	I
320		3574.76	I
360		3577.92	I
630		3578.72	II
360		3581.84	II
3200		3584.88	I
320		3590.50	II
390		3591.74	I
460		3593.52	II
460		3605.27	I
360		3606.32	II
320		3616.33	I
320		3616.76	II
320		3620.08	I
320		3622.70	I
390		3623.06	II
460		3630.73	II
840		3638.20	I
310		3640.76	II
420		3644.24	I
310		3645.03	II
660		3651.54	I
490		3652.06	I
960		3659.15	I
2800		3670.07	II
380		3678.75	II
540		3691.92	II
330		3693.70	II
540		3700.57	II
1100		3701.52	II
350		3713.55	I
300		3717.42	II
350		3718.11	II
350		3729.82	II
350		3732.62	II
350		3733.07	II
600		3738.04	II
300		3744.25	II
680		3746.42	II
350		3747.14	II
950		3748.68	II
600		3751.17	I
350		3752.66	II
350		3755.48	II
490		3758.35	I
350		3759.24	II
330		3763.26	I
490		3764.57	II
430		3766.89	I
330		3769.53	II
540		3773.43	I
300		3776.48	I
380		3780.71	II
1900		3782.84	II
430		3783.84	II
570		3793.10	II
380		3793.26	I
380		3793.57	II
380		3808.92	I
380		3809.22	II
1900		3811.99	I
380		3813.79	II
380		3814.06	II
750		3826.51	II
2000		3831.46	II
1200		3839.63	I
490	p	3848.60	II
620		3854.22	I
2400		3854.64	II
4900		3859.57	II
490		3861.17	II
1900		3865.92	II
380		3866.80	II
1500		3871.03	I
620		3874.04	II
620		3878.08	II
1000		3881.45	II
490		3882.36	II
380		3883.28	II
2200		3890.36	II
620		3892.68	II
490		3894.12	II
490		3896.77	II
620		3899.78	II
410		3902.55	II
460		3904.30	II
380		3906.45	I
330		3911.67	II
380		3915.88	II
330		3926.21	I
330		3926.72	I
430		3930.98	II
2000		3932.02	II
490		3935.38	II
330		3940.48	II
1200		3943.82	I
300		3948.44	I
300		3953.58	II
360		3954.67	II
350		3964.21	I
600		3966.52	II
1200		3985.79	II
460		3990.42	II
380		3992.53	II
350		3998.24	II
350		4004.06	II
430		4005.21	I
570		4017.72	II
300		4018.99	II
1000		4042.75	I
520		4044.41	II
410		4047.61	I
1600		4050.04	II
540		4051.91	II
300		4054.30	II
430		4058.19	II
880		4062.54	II
520		4067.75	II
410		4071.12	II
300		4074.48	II
330		4076.69	II
330		4080.60	II
2200		4090.13	II
460		4093.03	II
380		4106.38	II
810		4116.10	II
410		4124.73	II
410		4128.34	II
460		4141.22	II
880		4153.97	I
380		4156.65	I
350		4163.68	II
1400		4171.59	II
300		4189.27	II
350		4222.37	I
1000		4241.67	II
520		4244.37	II
680		4341.69	II
430	h	4355.74	I
430		4362.05	I
330		4393.59	I
600		4472.33	II
240		4515.28	II
620		4543.63	II
300		4620.21	I
240		4627.07	II
210		4631.62	I
220		4646.60	II
140		4666.85	II
100		4671.40	II
170		4689.07	II
100		4702.51	II
160		4722.72	II
120		4731.59	II
100		4755.74	II
150		4756.81	I
100		4772.70	II
100		4860.99	II
110		5008.21	II
170		5027.38	I
70		5117.24	II
80		5160.32	II
55		5164.14	I
55		5184.57	II
45		5204.31	II
45		5247.75	II
45		5257.04	II
70		5280.38	I
55		5386.19	II
80		5475.70	II
70		5480.26	II
70		5481.20	II
45		5482.53	II
160		5492.95	II
70		5527.82	II
70		5564.17	I
45		5581.59	II
55		5620.78	I
70		5780.59	I
70		5798.53	II
45		5836.02	I
55		5837.68	II
230		5915.39	I
55		5971.50	I
100		5976.32	I
45		5997.31	I
28		6017.38	II
55		6051.74	II
45		6067.22	II
90		6077.29	II
28		6087.34	II
40		6171.86	I
35		6175.39	II
28		6280.18	II
28		6359.29	I
55		6372.46	I
28		6378.52	II
28		6392.77	I
90		6395.42	I
110		6449.16	I
35		6464.98	I
90		6826.92	I
35		6876.74	II
23		7074.79	I
27		7101.61	I
30		7128.90	I
16		7147.89	I
16		7254.45	I
23		7425.50	I
45		7533.93	I
16		7619.35	I
50		7881.94	I
18		7970.46	I
16		8174.66	I
16		8262.06	I
16		8318.35	I
16		8337.50	II
18		8381.87	I
16		8441.21	I
35		8445.39	I
18		8450.03	I
16		8570.52	I
75		8607.95	I
23		8691.28	I
18		8710.76	I
18		8753.69	I
30		8757.76	I
16		8951.96	I
16		8989.92	I
10		9093.67	I
10		9139.56	I
10		9201.51	I
10		9265.34	I
10		9276.44	I
10		9385.90	I
10		9653.26	I
10		9819.00	I
10		9819.05	I
10		9868.36	I
10		9932.76	I
10		9964.11	I
50		10157.91	I
50		10259.55	I
100		10554.93	I
50		10799.78	I
25		10823.93	I
25		11095.77	I
75		11167.84	I
50		11294.13	I
100		11384.13	I
25		11410.43	I
50		11503.38	I
25		11568.81	I
20		11784.72	II
100		11859.42	I
100		11908.83	I
100		12250.46	I
25		13088.28	I
100		13185.16	I
75		13306.23	I
100		13961.58	I
50		16906.00	I
50		17451.11	I
50		18136.65	I
50		18366.96	I
75		18634.43	I
25		19029.39	I
10		20201.13	I
10		20271.41	I
10		20374.13	I
10		20517.29	I
10		20690.64	I
10		20772.19	I
10		21008.38	I
10		21099.98	I
10		21112.14	I
20		21144.90	II
10		21674.51	I
10		21693.38	I
75		21910.22	I
10		22110.73	I
10		23156.76	I
10		23948.19	I
10		29557.07	I

VANADIUM (V)
Z = 23

V I and II
Ref. 1 — C.H.C.

Intensity	Wavelength
	Air
2100	2092.44	I

Vanadium (Cont.)

Int.	λ (Å)	Sp.	Int.	λ (Å)	Sp.	Int.	λ (Å)	Sp.	Int.	λ (Å)	Sp.
40	2384.00	II	120	2917.37	II	560	3592.53	I	150	4035.63	II
40	2384.28	I	210	2919.99	II	270	3593.33	II	120	4042.64	I
60	2386.96	I	380	2920.38	II	110	3606.69	I	360	4050.96	I
60	2388.92	I	710	2923.62	I	110	3639.02	I	360	4051.35	I
75	2390.87	I	2400	2924.02	II	110	3644.71	I	280	4057.07	I
75	2391.26	I	1700	2924.64	II	250	3663.59	I	130	4057.82	I
85	2392.90	I	710	2930.81	II	250	3667.74	I	230	4063.93	I
70	2397.78	I	210	2934.40	II	110	3669.41	II	230	4071.54	I
70	2398.27	I	110	2935.87	I	170	3671.20	I	1100	4090.58	I
70	2399.96	I	900	2941.37	II	280	3673.40	I	180	4092.41	I
120	2406.75	I	450	2941.49	II	280	3675.70	I	1800	4092.69	I
110	2407.90	I	230 d	2942.33	I	170	3676.68	I	120	4093.50	I
120	2415.33	I	230	2943.20	I	300	3680.11	I	890	4095.49	I
120	2416.75	I	1100	2944.57	II	570	3683.13	I	2800	4099.80	I
100	2420.12	I	110	2946.53	I	190	3686.26	I	590	4102.16	I
100	2421.06	I	230	2949.63	I	470	3687.47	I	230	4104.40	I
100	2421.98	I	300	2950.35	II	1300	3688.07	I	260	4104.78	I
110	2428.28	I	640	2952.08	II	1000	3690.28	I	2800	4105.17	I
110	2435.52	I	120	2954.33	I	1500	3692.22	I	120	4108.22	I
140	2501.61	I	260	2957.52	II	450	3695.34	I	2300	4109.79	I
150	2506.90	I	410	2962.77	I	1000	3695.86	I	8900	4111.78	I
240	2507.78	I	600	2968.38	I	3800	3703.58	I	120	4112.33	I
180	2511.65	I	120	2972.25	II	1800	3704.70	I	230	4113.52	I
180	2511.95	I	120	2976.20	II	570	3705.04	I	4300	4115.18	I
180	2517.14	I	380	2976.52	II	130	3708.72	I	1800	4116.47	I
240	2519.62	I	240	2977.54	I	320	3715.54	II	180	4118.18	I
410	2526.22	I	260	3001.20	II	250	3727.34	II	180	4118.64	I
210	2527.90	II	140	3014.82	II	280	3732.76	II	230	4119.46	I
120	2528.47	II	180	3016.78	II	150	3734.43	I	180	4120.54	I
150	2528.84	II	270	3033.45	II	230	3745.80	II	180	4123.19	I
240	2530.18	I	290	3033.82	II	210	3750.87	II	2000	4123.57	I
110	2549.28	II	230	3043.12	I	210	3770.97	II	120	4124.07	I
120	2552.65	I	230	3043.56	I	270	3778.68	I	3100	4128.07	I
210	2562.13	I	230	3044.94	I	520	3790.32	I	120	4128.86	I
110	2564.82	I	230	3048.22	II	1100	3794.96	I	3100	4132.02	I
230	2574.02	I	170	3050.89	I	570	3799.91	I	2300	4134.49	I
140	2630.67	II	180	3053.39	II	570	3803.47	I	150	4159.69	I
130	2642.21	II	450	3053.65	I	190	3806.80	I	100	4174.01	I
150	2645.26	I	1200	3056.33	I	300	3807.50	I	230	4179.42	I
140	2651.90	I	1400	3060.46	I	520	3808.52	I	150	4182.59	I
150	2656.22	I	140	3063.25	II	230	3809.60	I	180	4189.84	I
180	2661.42	II	2400	3066.38	I	1000	3813.49	I	180	4191.56	I
290	2672.00	II	200	3067.12	II	140 d	3817.84	I	230	4209.86	I
380	2677.80	II	140	3069.64	II	1300	3818.24	I	120	4226.62	I
270	2678.57	II	170	3073.82	I	230	3819.96	I	360	4222.66	I
380	2679.32	II	100	3075.27	I	230	3821.49	I	180	4232.95	I
180	2682.87	II	150	3082.11	I	570	3822.01	I	180	4234.00	I
180	2683.09	II	3000	3093.11	II	450	3822.89	I	120	4235.76	I
1100	2687.96	II	200	3094.20	II	300	3823.21	I	100	4257.37	I
170	2688.72	II	180	3100.94	II	1700	3830.58	I	120	4259.31	I
130	2689.88	II	3000	3102.30	II	280	3834.22	I	120	4262.16	I
230	2690.24	II	2600	3110.71	II	160	3839.00	I	560	4268.64	I
240	2690.79	II	2000	3118.38	II	110	3839.38	I	460	4271.55	I
120	2696.99	I	380	3121.14	II	570	3840.44	I	460	4276.96	I
120	2697.74	I	150	3122.90	II	2600	3840.75	I	430	4284.06	I
680	2700.94	II	1500	3125.28	II	110	3841.89	I	330	4291.82	I
380	2702.19	II	260	3126.22	II	380	3844.44	I	220	4296.11	I
530	2706.17	II	530	3130.27	II	320	3847.33	I	170	4297.68	I
150	2706.70	II	410	3133.33	II	110	3849.32	I	170	4298.03	I
110	2707.86	II	210	3134.93	II	1200	3855.37	I	170	4306.21	I
170	2711.74	II	150	3136.51	II	3000	3855.84	I	140	4307.18	I
120	2714.20	II	150	3139.74	II	150	3862.22	I	170	4309.80	I
640	2715.69	II	200	3142.48	II	130	3863.87	I	460	4330.02	I
150	2722.56	I	150	3145.34	II	1300	3864.86	I	510	4332.82	I
240	2728.64	II	3200	3183.41	I	230	3867.60	I	760	4341.01	I
180	2731.35	I	5300	3183.98	I	170	3871.08	I	1000	4352.87	I
100	2739.71	II	3800	3185.40	I	1500	3875.08	I	130	4354.98	I
140	2753.40	II	410	3187.71	II	420	3875.90	I	150	4355.94	I
140	2765.67	II	530	3188.51	II	570	3876.09	I	150	4368.04	I
140	2777.73	II	750	3190.68	II	130	3878.71	II	140 d	4373.23	I
120	2803.47	II	530	3198.01	I	700	3890.18	I	100	4375.30	I
120	2846.57	I	750	3202.38	I	460	3892.86	I	12000	4379.24	I
110	2847.57	II	450	3205.58	I	280 h	3898.02	I	100	4380.55	I
140	2852.87	I	450	3207.41	I	140	3899.13	II	7000	4384.72	I
140	2854.34	II	410	3212.43	I	140 h	3900.18	I	4800	4389.97	I
200	2855.22	I	210	3217.11	II	140 h	3901.15	I	3600	4395.23	I
180	2859.97	I	150	3237.87	II	2400	3902.25	I	1400	4400.58	I
240	2864.36	I	140	3254.77	II	100	3906.75	I	2300	4406.64	I
170	2866.59	I	140	3263.24	I	700	3909.89	I	2800	4407.64	I
210	2868.10	I	1100	3267.70	II	100	3910.79	I	3600	4408.20	I
140	2869.13	II	900	3271.12	II	220	3912.21	I	4600	4408.51	I
210	2870.55	I	750	3276.12	II	140	3914.33	II	140	4412.14	I
110	2877.69	II	110	3279.84	II	100	3916.41	II	640	4416.47	I
110	2879.16	II	140	3298.14	II	100	3920.49	I	120	4419.94	I
350	2880.03	II	110	3329.86	I	230	3921.90	I	640	4421.57	I
380	2882.50	II	110	3365.55	I	240	3924.66	I	460	4426.00	I
380	2884.78	II	110	3377.62	I	150	3925.24	I	120	4427.31	I
140	2888.25	II	170	3400.40	I	200	3927.93	I	310	4428.52	I
380	2889.62	II	110	3425.07	I	260	3930.02	I	230	4429.80	I
900	2891.64	II	110	3485.92	II	150	3931.34	I	430	4436.14	I
530	2892.44	II	210	3504.44	II	260	3934.01	I	640	4437.84	I
900	2892.66	II	560	3517.30	II	150	3935.14	I	830	4441.68	I
1400	2893.32	II	150	3520.02	II	100	3936.28	I	640	4444.21	I
360	2896.21	II	110	3524.72	II	150	3943.66	I	610	4452.01	I
110	2899.60	I	230	3529.74	I	100	3950.23	I	410	4457.48	I
360	2903.08	II	230	3530.77	II	140	3951.97	II	120	4457.76	I
150	2906.13	I	560	3533.68	I	100	3973.64	II	1000	4459.76	I
900	2906.46	II	110	3543.50	II	540	3990.57	I	2000	4460.29	I
490	2907.47	II	560	3545.20	II	260	3992.80	I	610	4462.36	I
2400	2908.82	II	110	3553.27	II	430	3998.73	I	120	4468.01	I
710	2910.02	II	560	3556.80	II	170	4005.71	II	380	4469.71	I
530	2910.39	II	110	3566.18	I	120	4023.39	II	120	4474.04	I
560	2911.06	II	560	3589.76	II	120	4031.83	I	200	4474.71	I
380	2914.93	I	490	3592.02	II				380	4488.89	I

Vanadium (Cont.)

Intensity		Wavelength	
100		4496.06	I
120		4501.95	I
140		4524.22	I
360		4545.39	I
100		4549.65	I
280		4560.71	I
200		4571.78	I
510		4577.17	I
140		4578.73	I
640		4580.40	I
830		4586.36	I
170		4591.22	I
1300		4594.11	I
100		4606.15	I
30		4609.65	I
25		4611.74	I
230		4619.77	I
65		4624.41	I
50		4626.48	I
100		4635.18	I
65		4640.07	I
65		4640.74	I
130		4646.40	I
30		4648.89	I
30		4666.14	I
160		4670.49	I
24		4684.45	I
35		4686.92	I
55		4706.16	I
80		4706.57	I
80		4710.56	I
65		4714.12	I
35		4715.89	I
55		4717.69	I
40		4721.51	I
40		4722.86	I
40		4729.53	I
27		4730.38	I
27		4742.63	I
24		4746.63	I
40		4748.52	I
45		4750.98	I
35		4751.56	I
40		4753.93	I
65		4757.48	I
55		4766.63	I
130		4776.36	I
		4776.52	I
110		4786.51	I
130		4796.92	I
19		4799.77	I
130		4807.53	I
130		4827.45	I
150		4831.64	I
120		4832.43	I
19		4833.02	I
19		4848.81	I
320		4851.48	I
35		4862.61	I
480		4864.74	I
21		4871.26	I
620		4875.48	I
55		4880.56	I
740		4881.56	I
27		4891.60	I
21		4894.21	I
55		4900.62	I
95	d	4904.29	I
		4904.34	I
85		4925.65	I
35		4932.03	I
23		4966.12	I
70		5002.33	I
85		5014.62	I
28		5051.63	I
35		5064.12	I
35		5105.14	I
110		5128.53	I
110		5138.42	I
25		5139.53	I
70		5148.72	I
40		5159.35	I
23		5169.94	I
70		5176.77	I
20		5192.01	I
110		5192.99	I
23		5193.62	I
110		5194.83	I
55		5195.36	I
20		5206.61	I
40		5216.59	I
35		5225.77	I
35		5233.75	I
110		5234.07	I
20		5240.20	I
110		5240.87	I
17		5260.98	I
40		5353.41	I
35		5383.43	I
40		5385.14	I
14		5388.30	I
11		5397.87	I
100		5401.93	I
140		5415.26	I

Intensity		Wavelength	
28		5418.09	I
50		5424.08	I
40		5434.18	I
11		5437.66	I
17		5458.12	I
13		5471.33	
25		5487.22	I
85		5487.92	I
25		5489.94	I
28		5504.87	I
70		5507.75	I
14		5511.18	I
23		5545.93	I
70		5547.07	I
35		5558.75	I
28		5561.66	I
140		5584.50	I
23		5586.00	I
100		5592.42	I
28		5601.38	I
70		5604.94	I
13		5624.20	I
200		5624.60	I
70		5624.89	I
55		5626.01	I
400		5627.64	I
13		5632.46	I
10		5633.90	I
13		5635.51	I
85		5646.11	I
110		5657.44	I
110		5668.36	I
310		5670.85	I
20		5683.22	I
1200		5698.52	I
920		5703.56	I
570		5706.98	I
11		5708.95	I
11	h	5716.21	I
70		5725.64	I
850		5727.03	I
170		5727.66	I
230		5731.25	I
40		5734.01	I
230		5737.06	I
110		5743.45	I
17		5747.70	I
40		5748.87	I
17		5752.74	I
17		5761.41	I
70		5772.42	I
35		5776.64	I
11		5782.61	I
11		5783.50	I
40	h	5784.38	I
55	h	5786.16	I
23		5788.56	I
35	h	5807.14	I
23		5817.06	I
35	h	5817.53	I
55	h	5830.72	I
85	h	5846.30	I
11		5850.32	I
40		5924.57	I
28		5978.91	I
20		5980.78	I
28		6002.31	I
55		6002.63	I
28		6016.12	I
20		6025.41	I
450		6039.73	I
100		6058.14	I
20		6067.26	I
480		6081.44	I
1300		6090.22	I
28		6106.98	I
280		6111.67	I
600		6119.52	I
20		6128.34	I
280		6135.38	I
180		6150.15	I
85		6170.36	I
23		6189.35	I
450		6199.19	I
130		6213.87	I
450		6216.37	I
28	h	6218.31	I
130		6224.50	I
430		6230.74	I
100		6233.20	I
55		6240.13	I
170		6242.81	I
710		6243.10	I
280		6251.82	I
85		6256.90	I
85		6258.57	I
55		6261.22	I
85		6266.32	I
130		6268.82	I
170		6274.65	I
17	h	6282.33	I
200		6285.16	I
200		6292.83	I
170		6296.49	I

Intensity		Wavelength	
28	h	6311.50	I
14		6324.66	I
70		6326.84	I
55		6339.09	I
50		6349.48	I
14		6355.58	I
50		6357.30	I
25		6358.82	I
35		6361.27	I
23		6379.36	I
14		6393.28	I
35		6430.47	I
23		6431.63	I
14		6433.18	I
11		6435.16	I
70		6452.34	I
11		6488.05	I
55		6504.17	I
110		6531.43	I
28		6543.51	I
17		6558.02	I
11		6565.88	I
50		6605.97	I
15		6607.83	I
10		6623.54	I
50		6624.85	I
13		6633.26	I
13		6643.79	I
8		6693.66	I
8		6708.07	I
65	c	6753.00	I
10		6760.12	I
50	c	6766.49	I
40		6784.98	I
15		6786.32	I
26		6812.40	I
9	c	6829.94	I
15		6832.44	I
12		6839.58	I
12		6841.90	I
10	c	6870.88	I
8		6871.56	I
7		6894.00	I
12		6974.50	I
21		7026.07	I
7		7063.69	I
11	h	7092.05	I
6		7102.58	I
24		7148.15	I
7		7151.36	I
7		7182.08	I
14		7264.29	I
8		7321.44	I
40		7338.92	I
35		7356.54	I
11		7358.66	I
24		7361.39	I
12		7362.49	I
24		7363.16	I
9		7385.95	I
6	h	7393.49	
12	h	7485.90	I
12	h	7488.08	I
12	h	7492.44	I
12	h	7578.75	I
9	h	7591.24	I
14	h	7596.92	I
12	h	7598.28	I
24		7624.81	I
5		7701.37	I
8		7704.81	I
8	h	7851.18	I
14	B1	7865.51	VO
12		7896.40	I
14	h	7898.81	
24		7937.92	I
29	c	8027.39	I
14		8028.13	I
14	h	8035.38	I
14	h	8045.71	I
12		8051.89	I
14		8093.48	I
8		8102.44	I
12		8108.59	I
9	h	8109.07	I
120	Cw	8116.80	I
11	h	8136.79	I
29		8144.59	I
9		8154.55	I
70	c	8161.07	I
14		8171.35	I
7		8180.21	I
35		8186.71	I
24		8187.33	I
29		8198.87	I
35		8203.07	I
24		8241.61	I
29	c	8253.51	I
29		8255.88	I
5	h	8280.39	I
19		8282.37	I
8		8324.42	I
14	h	8331.23	I
14		8342.03	I

Intensity		Wavelength	
7		8402.81	I
12		8499.52	I
6		8534.49	I
6	B1	8624.86	VO
60	c	8919.85	I
29	c	8932.93	I
12		8971.62	I

V III
Ref. 394 — C.H.C.

Intensity		Wavelength	
		Vacuum	
25		616.09	III
50		633.94	III
40		635.41	III
100		864.27	III
75		948.84	III
500		1006.46	III
500		1149.94	III
400		1157.18	III
300		1160.77	III
500		1252.11	III
400		1254.01	III
500		1287.87	III
400		1289.42	III
300		1290.77	III
400		1313.35	III
300		1313.27	III
500		1331.99	III
500		1335.12	III
1000		1643.03	III
1000		1650.14	III
300		1668.03	III
300		1670.66	III
300		1679.19	III
1000		1694.78	III
400		1721.98	III
300		1724.63	III
500	d	1751.68	III
500		1757.73	III
1000		1760.07	III
300		1773.43	III
400		1778.02	III
500		1779.72	III
400		1784.44	III
1000		1788.26	III
500		1793.82	III
1000		1794.60	III
300		1796.77	III
500		1798.15	III
300		1802.55	III
500		1804.13	III
1000		1812.19	III
400		1831.15	III
400		1831.64	III
300		1845.07	III
300		1850.69	III
400		1852.01	III
500		1854.42	III
300		1855.06	III
500		1856.64	III
300		1864.51	III
300		1878.68	III
400		1880.41	III
300		1895.01	III
400		1899.81	III
500		1902.23	III
300		1934.00	III
		Air	
500		2232.91	III
400		2241.53	III
1000		2292.86	III
400		2314.18	III
500		2318.06	III
400		2319.00	III
500		2323.82	III
2500		2330.42	III
500		2331.75	III
400		2334.20	III
500		2343.10	III
500		2358.73	III
500		2366.31	III
2500		2371.06	III
1000		2382.46	III
500		2393.58	III
500		2404.18	III
500		2516.14	III
250		2521.16	III
250		2521.55	III
250		2548.21	III
150		2554.22	III
150		2563.32	III
250		2593.05	III
250		2595.10	III
100		3679.86	III
50	h	3705.35	III

Vanadium (Cont.)

Intensity		Wavelength	
40		4714.89	III
50		6597.20	III

V IV
Ref. 397 — C.H.C.

Intensity		Wavelength	
		Vacuum	
200		677.34	IV
60		678.74	IV
50		679.65	IV
500		684.37	IV
100		684.45	IV
100		691.53	IV
50		693.13	IV
400		737.85	IV
150		750.11	IV
60		1226.52	IV
50		1308.06	IV
80		1355.13	IV
60		1395.00	IV
50		1414.41	IV
80		1419.58	IV
100		1426.65	IV
60		1520.14	IV
80		1601.92	IV
80		1611.88	IV
80		1806.18	IV
60		1809.85	IV
100		1817.68	IV
200		1825.84	IV
300		1861.56	IV
500		1939.06	IV
400		1951.43	IV
300		1963.10	IV
500		1997.72	IV
200		1999.32	IV
		Air	
100		2002.48	IV
50	h	2088.74	IV
50	h	2146.83	IV
100	h	2155.34	IV
500		2268.30	IV
50	h	2401.02	IV
50	h	2433.53	IV
50	h	2446.80	IV
50	h	2450.87	IV
50	h	2556.92	IV
80	h	2570.72	IV
50	h	2624.21	IV
80	h	2645.54	IV
50	h	2655.41	IV
50	h	2656.87	IV
50		3284.56	IV
60		3334.79	IV
50		3448.41	IV
50		3496.42	IV
80	h	3514.25	IV
50	h	4985.65	IV
50	h	5130.78	IV
50	h	5262.16	IV
60	h	5352.32	IV
40	h	5940.12	IV

V V
Ref. 432 — C.H.C.

Intensity		Wavelength	
		Vacuum	
18		224.91	V
20		225.46	V
17		227.88	V
16		239.41	V
19		239.48	V
20		251.66	V
18		252.44	V
18		285.98	V
20		286.84	V
17		312.39	V
35		483.01	V
25		484.51	V
20		820.86	V
30		829.48	V
15		962.03	V
15		1142.74	V
25		1157.58	V
20		1490.11	V
100		1680.20	V
50		1716.72	V
15		1724.99	V
25		1792.99	V
30		1811.42	V
		Air	
20	d	2319.66	V
20		2577.90	V
15		2775.82	V
15		3617.97	V
12	d	3746.36	V
20		4200.32	V
15	w	4930.53	V
8		5356.07	V
7		6628.80	V
3		7595.51	V

XENON (Xe)
Z = 54

Xe I and II
Ref. 33, 116, 118, 120, 232 — S.P.D.

Intensity		Wavelength	
		Vacuum	
350		740.41	II
350		803.07	II
600		880.80	II
350		885.54	II
600		925.87	II
250		935.40	II
800		972.77	II
700		976.68	II
500		1032.64	II
700		1037.68	II
1100		1041.31	II
1000		1048.27	II
1200		1051.92	II
2000		1074.48	II
600		1083.86	II
1200		1100.43	II
600		1158.47	II
250		1169.63	II
800	p	1183.05	II
950		1131.04	I
600		1244.76	II
250		1250.20	I
1000		1295.59	I
600		1469.61	I
		Air	
200		2864.73	II
150	h	2895.22	II
400		2979.32	II
100	h	3017.43	II
300		3128.87	II
200	h	3366.72	II
2		3400.07	I
2		3418.37	I
2		3420.00	I
3		3442.66	I
100	h	3461.26	II
4		3469.81	I
4		3472.36	I
5		3506.74	I
10		3549.86	I
10		3554.04	I
15		3610.32	I
8		3613.06	I
6		3633.06	I
10		3669.91	I
40		3685.90	I
40		3693.49	I
100	l	3907.91	II
100		4037.59	II
200	l	4057.46	II
100	h	4098.89	II
200	l	4158.04	II
1000	h	4180.10	II
500	h	4193.15	II
300	h	4208.48	II
100	h	4209.47	II
300	h	4213.72	II
100		4215.60	II
300	h	4223.00	II
400	h	4238.25	II
500	h	4245.38	II
100	l	4251.57	II
500	h	4296.40	II
500	h	4310.51	II
1000	l	4330.52	II
200	h	4369.20	II
100	l	4373.78	II
500	h	4393.20	II
500	l	4395.77	II
200	l	4406.88	II
150	l	4416.07	II
500	h	4448.13	II
1000	h	4462.19	II
500	l	4480.86	II
100	l	4521.86	II
600		4734.152	I
150		4792.619	I
500		4807.02	I
400		4829.71	I
300		4843.29	I
500		4916.51	I
500		4923.152	I
200	l	4971.71	II
400		4972.71	I
300		4988.77	II
100	l	4991.17	II
200		5028.280	I
200		5044.92	II
1000		5080.62	II
300		5122.42	II
100		5125.70	II
100		5178.82	II
300		5188.04	II
400		5191.37	II
100		5192.10	II
500		5260.44	II
500		5261.95	II
2000		5292.22	II
300		5309.27	II
1000		5313.87	II
2000		5339.33	II
200		5363.20	II
200		5368.07	II
500		5372.39	II
100		5392.80	I
3000		5419.15	II
800		5438.96	II
300		5445.45	II
200		5450.45	II
400		5460.39	II
1000		5472.61	II
100	l	5494.86	II
200		5525.53	II
600		5531.07	II
100		5566.62	I
300		5616.67	II
300		5659.38	II
600		5667.56	II
150		5670.91	II
100		5695.75	I
200		5699.61	II
200		5716.10	II
500		5726.91	II
500		5751.03	II
300		5758.65	II
300		5776.39	II
100		5815.96	II
300		5823.89	I
150		5824.80	I
100		5875.02	I
300		5893.29	II
100		5894.99	I
200		5905.13	II
100		5934.17	II
500		5945.53	II
300		5971.13	II
2000		5976.46	II
200		6008.92	II
1000		6036.20	II
2000		6051.15	II
600		6093.50	II
1500		6097.59	II
400		6101.43	II
100		6115.08	II
100		6146.45	II
150		6178.30	I
120		6179.66	I
300		6182.42	I
500		6194.07	II
100		6198.26	I
200		6220.02	II
500		6270.82	II
400		6277.54	II
100		6284.41	II
100		6286.01	II
250		6300.86	II
500		6318.06	I
400		6343.96	II
600		6356.35	II
200		6375.28	II
100		6397.99	II
300		6469.70	I
150		6472.84	II
120		6487.76	I
100		6498.72	I
200	h	6504.18	I
300		6512.83	II
200		6528.65	II
100		6533.16	I
1000		6595.01	II
100		6595.56	I
400		6597.25	II
100		6598.84	II
150		6668.92	I
300		6694.32	II
200		6728.01	II
150		6788.71	II
100		6790.37	II
1000		6805.74	II
200		6827.32	I
100		6872.11	I
300		6882.16	I
80		6910.22	II
100		6925.53	I
800	h	6942.11	II
100		6976.18	I
2000		6990.88	II
150		7082.15	II
500		7119.60	I
50	s	7147.50	II
200		7149.03	II
500		7164.83	II
100		7284.34	II
200		7301.80	II
200		7339.30	II
100		7386.00	I
150		7393.79	I
300		7548.45	II
200		7584.68	I
80		7618.37	II
500		7642.02	I
100		7643.91	I
200		7670.66	II
60		7787.04	II
100		7802.65	I
300		7881.32	I
300		7887.40	I
500		7967.34	I
100		8029.67	I
200		8057.26	I
150		8061.34	I
100		8101.98	I
150	h	8151.80	II
100		8171.02	I
700		8206.34	I
10000		8231.635	I
500		8266.52	I
7000		8280.116	I
2000		8346.82	I
100		8347.24	II
2000		8409.19	I
50	h	8515.19	II
200		8576.01	I
50	h	8604.23	II
250		8648.54	I
100		8692.20	I
200		8696.86	I
50	h	8716.19	II
300		8739.39	I
100		8758.20	I
5000		8819.41	I
300		8862.32	I
200		8908.73	I
200		8930.83	I
1000		8952.25	I
100		8981.05	I
200		8987.57	I
400		9045.45	I
500		9162.65	I
100		9167.52	I
100		9374.76	I
200		9513.38	I
50	h	9591.35	II
150		9685.32	I
50	l	9698.68	II
100		9718.16	I
2000		9799.70	I
3000		9923.19	I
100		10838.37	I
90		11742.01	I
375		12235.24	I
100		12257.76	I
300		12590.20	I
2500		12623.391	I
250		13544.15	I
2000		13657.055	I
1250		14142.444	I
800		14240.96	I
375		14364.99	I
140		14660.81	I
3000		14732.806	I
100		15099.72	I
2500		15418.394	I
150		15557.13	I
250		15979.54	I
100		16039.90	I
1000		16053.28	I
125		16554.49	I
1500		16728.15	I
1500		17325.77	I
350		18788.13	I
150		20187.19	I
3000		20262.242	I
250		21470.09	I
1250		23193.33	I
110		23279.54	I
1800		24824.71	I
175		25145.84	I
2000		26269.08	I
2500		26510.86	I
250		28381.54	I
750		28582.25	I
300		29384.41	I
150		29448.06	I

Xenon (Cont.)

Intensity	Wavelength	Species
100	29649.58	I
100	29813.62	I
600	30253.14	I
1500	30475.46	I
100	30504.12	I
500	30794.18	I
6000	31069.23	I
125	31336.01	I
550	31607.91	I
100	32293.08	I
1800	32739.26	I
3500	33666.69	I
150	34014.67	I
450	34335.27	I
170	34744.00	I
5000	35070.25	I
110	35246.92	I
250	36209.21	I
150	36231.74	I
450	36508.36	I
850	36788.83	I
140	38685.98	I
175	38737.82	I
270	38939.60	I
120	39955.14	I

Xe III
Ref. 33, 384, 391, 429 — R.L.K.

Intensity		Wavelength	Species

Vacuum

Intensity		Wavelength	Species
8		657.8	III
8		660.1	III
9		673.8	III
9		674.0	III
9		676.6	III
10		694.0	III
20		698.5	III
12		705.1	III
10		721.2	III
15		731.0	III
10		733.3	III
15		742.6	III
10		756.0	III
10		761.5	III
10		769.1	III
25		779.1	III
15		792.9	III
12		796.1	III
15		802.0	III
25		823.2	III
30		824.9	III
25		853.0	III
15		889.3	III
20		894.0	III
20		896.0	III
10		965.5	III
35		1003.4	III
35		1017.7	III
10		1047.8	III
12		1066.4	III
30		1130.3	III
25		1232.1	III

Air

Intensity		Wavelength	Species
80		2668.98	III
100		2717.33	III
30		2814.45	III
40		2815.91	III
30		2827.45	III
40		2847.65	III
30		2862.40	III
80	w	2871.10	III
60	w	2871.24	III
30		2871.7	III
30		2896.62	III
50		2906.6	III
40		2911.89	III
80	w	2912.36	III
40		2940.2	III
60		2945.2	III
40		2947.5	III
40		2948.1	III
80	w	2970.47	III
40		2992.87	III
30		3004.25	III
100		3023.81	III
40		3083.5	III
50		3091.1	III
30		3106.46	III
100	w	3138.3	III
80	c	3150.82	III
40		3185.2	III
100		3242.86	III
80		3268.98	III
30		3287.82	III
80	w	3301.55	III
40		3331.6	III
30		3358.0	III
80		3384.12	III
60		3444.2	III
70		3454.2	III
100	w	3458.7	III
40		3468.22	III
80		3522.83	III
50		3542.3	III
50		3552.1	III
40		3561.4	III
100		3579.7	III
80		3583.6	III
100	w	3595.4	III
100		3606.06	III
40		3607.0	III
100	w	3615.9	III
40		3623.1	III
600		3624.08	III
50		3676.67	III
40		3776.3	III
300		3781.02	III
100		3841.5	III
200		3877.8	III
60		3880.5	III
500		3922.55	III
300		3950.59	III
200		4050.07	III
60		4060.4	III
100		4109.1	III
100		4145.7	III
30		4285.9	III
50		4434.2	III
100	w	4462.1	III
100	w	4569.1	III
100	w	4570.1	III
100	w	4641.4	III
30		4673.7	III
60		4683.57	III
30		4723.60	III
100	w	4757.3	III
40		4869.5	III
60		5239.0	III
30		5367.1	III
50		5401.00	III
40		5524.4	III
60		6205.97	III
25		6221.7	III
60		6238.2	III
60		6259.05	III

YTTERBIUM (Yb)
Z = 70

Yb I and II
Ref. 1 = C.H.C.

Air

Intensity		Wavelength	Species
2500		2116.67	II
3000		2126.74	II
370		2161.60	II
850		2185.71	II
640		2224.46	II
140		2320.81	II
50		2362.89	II
170		2390.74	II
18		2398.02	II
28		2421.35	II
25		2447.26	II
28		2460.25	II
460		2464.50	I
14		2484.89	II
70		2502.02	II
28		2505.48	II
11		2508.07	II
140		2512.06	II
18		2516.35	II
50		2522.44	II
65		2537.65	II
270		2538.67	II
14		2550.06	II
70		2552.15	II
55		2552.70	II
21		2565.57	II
28		2571.36	II
13		2573.15	II
18		2596.16	II
28		2596.32	II
21		2615.26	II
100		2617.01	II
55		2634.31	II
45		2639.45	II
85		2641.89	II
110		2644.31	II
28		2646.44	II
28		2647.46	II
28		2648.80	II
50		2649.79	II
28		2650.73	II
990		2653.75	II
35		2656.12	II
21		2659.27	II
200		2665.04	II
55		2668.75	II
390		2671.96	I
390		2672.66	II
21		2680.40	II
14		2683.42	II
70		2684.75	II
25		2687.98	II
28		2695.43	II
14		2696.62	II
18		2700.80	II
21		2708.84	II
65		2710.54	II
25		2711.78	II
55		2712.66	II
170		2718.35	II
21		2722.20	II
110		2732.74	II
21		2734.09	II
55		2741.71	II
55		2747.58	II
18		2748.04	II
230		2748.66	II
1300		2750.48	II
85		2751.45	II
21		2759.00	II
65		2760.78	II
55		2761.37	II
35		2764.41	II
85		2771.32	II
170		2776.28	II
100		2784.66	II
18		2787.96	II
45		2793.28	II
25		2794.44	II
21		2795.01	II
18		2795.29	II
35		2797.80	II
100		2798.21	II
45		2799.38	II
50		2800.00	II
35		2800.06	II
14		2810.72	II
65		2814.53	II
28		2816.32	II
140		2821.15	II
100		2824.97	II
190		2830.99	II
18		2832.20	II
28		2834.97	II
14		2842.59	II
230	h	2847.18	II
100		2848.44	II
21		2849.34	II
360		2851.13	II
55		2851.86	II
21		2853.41	II
18		2853.68	II
55		2854.14	II
45		2854.49	II
45		2858.33	II
45		2858.46	II
100		2859.39	II
430		2859.80	II
55		2860.39	II
140		2861.21	II
100		2861.34	II
200		2867.06	II
25		2870.06	II
45		2873.49	I
28		2885.97	II
70		2886.26	II
200		2888.04	II
3600		2891.38	II
45		2893.62	II
28		2896.90	II
85		2899.70	II
18		2902.41	II
21		2902.92	II
21		2906.88	II
28		2908.93	II
35		2909.19	II
55		2909.48	II
85		2911.52	II
18		2912.86	II
170		2914.21	II
140		2915.28	II
18		2916.43	II
280		2919.35	II
55		2921.12	II
45		2924.24	II
25		2927.85	II
35		2934.36	I
55		2935.11	II
21		2937.19	II
45		2939.53	II
45		2940.52	II
28		2942.04	II
140		2945.91	II
45		2946.30	II
18		2946.76	II
28		2950.33	II
45		2955.32	II
18		2957.63	II
65		2962.52	II
21		2963.26	II
45		2963.46	II
130		2964.76	II
2000		2970.56	II
45		2982.49	II
21		2982.66	II
28		2983.70	II
200		2983.99	II
90		2985.08	II
35		2985.88	II
45		2990.37	II
65		2991.87	II
28		2993.94	II
170		2994.80	II
28		2995.86	II
70		3000.46	II
25		3002.61	II
310		3005.77	II
100		3009.39	II
65		3010.62	II
55		3014.43	II
160		3017.56	II
160		3026.67	II
920		3031.11	II
55		3034.64	II
25		3037.99	II
55		3039.67	II
80		3042.65	II
21		3044.00	II
45		3046.48	II
35		3047.05	II
45		3063.12	II
21		3063.67	II
110		3065.04	II
18		3076.01	II
100		3089.10	II
70		3093.87	II
28		3100.74	I
45		3101.36	II
28		3102.07	II
55		3107.76	II
170		3107.90	II
85		3115.34	II
55		3116.70	II
190		3117.81	II
50		3136.76	II
230		3140.94	II
80		3141.73	II
80		3145.06	II
28		3145.54	II
28		3153.18	II
90		3153.88	II
50		3155.18	II
28		3162.29	I
70		3163.80	II
50		3165.21	II
120		3169.06	II
120		3180.92	II
390		3192.88	II
70		3198.65	II
240		3201.16	II
80		3217.18	II
50		3218.32	II
50		3225.88	II
45		3239.20	II
35		3239.58	I
35		3246.06	II
130		3261.51	II
18000		3289.37	II
130		3305.25	I
140		3305.73	II
50		3315.10	II
80		3319.41	I
50		3333.06	II
240		3337.17	II
280	d	3342.93	II
		3343.07	II
80		3346.50	II
50		3347.54	II
35		3351.09	II
50		3351.26	II
100		3352.49	II
100		3362.44	II
50		3363.64	II
240		3375.48	II
50		3376.62	II
28		3382.54	II
140		3387.50	I
50		3390.25	II
28		3390.42	II
50		3391.10	II
50	h	3394.44	II
50		3401.01	II
35		3404.10	II
50		3412.45	I
140		3418.39	I
360		3426.04	II
80		3428.46	II
240		3431.11	I
45		3434.61	II

Ytterbium (Cont.)

Intensity	Note	Wavelength	Spectrum
50		3438.71	II
100		3438.85	II
35		3443.59	
35		3446.89	II
85		3452.40	I
500		3454.08	II
190	d	3458.29	II
		3458.39	I
360		3460.27	I
35		3462.34	II
2400		3464.37	I
500		3476.30	II
500		3478.84	II
50		3482.56	II
85		3485.76	II
85		3488.43	II
100	Hw	3495.90	II
85		3507.83	II
50		3517.00	I
230		3520.29	II
50		3545.72	II
100		3549.82	II
35		3559.03	I
200		3560.33	II
170		3560.70	II
50	h	3563.94	II
85		3570.57	II
50		3572.50	II
50		3574.58	II
360		3585.47	II
130		3606.48	II
50		3610.23	II
70		3611.30	II
200		3619.80	II
110		3634.52	
240		3637.76	II
70		3648.15	I
90		3655.73	II
240		3669.69	II
50		3670.69	II
140		3675.08	II
50		3690.56	II
32000		3694.19	II
70		3698.60	II
70		3700.58	I
50		3710.34	II
60		3724.21	II
180		3734.69	I
550		3770.10	I
80		3774.32	I
60	h	3791.74	I
170		3839.91	I
340		3872.85	I
340		3900.85	I
50		3904.81	II
140		3911.27	I
32000		3987.99	II
930		3990.88	I
50		4007.36	I
70		4052.28	I
85		4077.28	II
440		4089.68	I
120	h	4119.25	II
70		4135.09	II
470		4149.07	I
120		4174.56	I
340		4180.81	II
150	d	4218.56	II
		4218.69	I
120		4231.97	I
70		4277.74	II
120		4305.97	I
70		4316.95	II
60	h	4393.69	I
60	h	4430.21	I
440		4439.19	I
85	h	4482.42	I
85		4515.16	II
35		4553.58	II
85	h	4563.95	I
640		4576.21	I
200		4582.36	II
70		4589.21	I
140		4590.83	II
40		4598.36	II
35		4683.81	II
40		4684.27	I
190		4726.08	II
170	h	4781.87	I
170		4786.61	II
35		4816.43	I
40		4820.24	II
35		4836.96	I
40		4837.46	I
17		4851.15	II
40	h	4894.60	I
27		4912.36	I
710		4935.50	I
24		4937.22	II
140		4966.90	I
24		5009.52	I
17		5067.30	I
30		5067.80	I
70		5069.14	I
220		5074.34	I
50		5076.74	I
20		5135.98	II
14		5147.02	II
20		5184.15	II
60		5196.08	I
85		5211.60	I
35		5240.51	II
100		5244.11	I
40		5257.49	II
150	h	5277.04	I
35		5279.53	I
17		5300.94	II
170		5335.15	II
30	d	5345.66	II
		5345.83	II
60		5347.22	II
30	h	5351.29	I
150		5352.95	II
30		5358.64	II
30		5363.66	I
17		5389.84	II
14		5432.71	II
40		5449.27	II
14		5478.50	II
60		5481.92	I
40		5505.49	I
17		5524.54	I
85	h	5539.05	I
2400		5556.47	I
35		5562.09	I
20		5568.11	I
20		5586.36	I
40		5588.45	II
60		5651.98	II
7		5686.53	II
220		5719.99	I
10		5749.91	II
10	h	5755.89	I
27		5771.66	II
10		5803.44	I
10		5819.41	II
35		5833.99	II
35		5837.14	II
27		5854.51	I
8		5887.11	II
20		5908.36	II
17		5989.33	I
40		5991.51	II
10		6052.88	II
10		6054.57	I
60		6152.57	II
30		6246.97	II
60		6274.78	II
14		6308.15	II
35	h	6400.35	I
35	h	6417.91	I
20		6432.73	II
17	h	6463.15	II
340		6489.06	I
20		6643.55	II
180		6667.82	I
15		6678.17	I
25		6727.61	II
25		6768.70	I
690		6799.60	I
18		6934.05	II
20		6999.88	II
10		7043.78	II
9	h	7244.41	I
8	h	7305.22	I
10	h	7313.05	I
16	h	7350.04	I
25		7448.28	I
30	h	7527.46	I
750		7699.48	I
7		7895.08	I
70	h	8922.56	II

Yb III
Ref. 40, 192 — J.R.

Vacuum

Intensity	Note	Wavelength	Spectrum
5		968.46	III
20		973.16	III
10		994.56	III
10		1560.66	III
80		1561.42	III
30		1669.60	III
50		1670.78	III
50		1719.82	III
60		1739.18	III
70		1762.80	III
80	h	1765.21	III
65		1775.29	III
70		1779.74	III
70		1781.31	III
20		1793.70	III
60		1798.85	III
65		1810.88	III
60		1826.41	III
20		1826.77	III
30		1838.01	III
30		1847.30	III
30		1849.24	III
10		1849.42	III
75		1852.36	III
75		1852.94	III
90		1854.80	III
80		1857.16	III
100		1863.32	III
5		1864.85	III
10		1867.23	III
10		1867.63	III
10		1868.19	III
5		1868.92	III
10		1870.07	III
15		1870.83	III
10		1871.15	III
200		1872.03	III
800		1873.91	III
100		1875.41	III
75		1875.92	III
70		1880.30	III
80		1884.22	III
70		1885.07	III
70		1887.23	III
10		1890.34	III
10		1890.87	III
15		1892.42	III
5		1895.50	III
100		1896.18	III
10		1897.57	III
500		1898.25	III
7		1906.74	III
10		1908.50	III
100		1909.66	III
70		1910.86	III
20		1920.53	III
10		1926.76	III
70		1928.09	III
15		1930.63	III
55		1942.59	III
40		1950.34	III
13		1962.80	III
80		1967.13	III
20		1969.47	III
30		1969.73	III
10		1973.96	III
10		1974.18	III
25		1976.46	III
2		1981.74	III
5		1983.88	III
25		1984.62	III
7		1985.74	III
80		1986.43	III
80	h	1989.82	III
50		1991.14	III
45		1995.05	III
55		1997.28	III
55		1997.66	III
500		1998.82	III

Air

Intensity	Note	Wavelength	Spectrum
20		2054.80	III
10		2066.49	III
10		2073.64	III
30		2078.05	III
10		2087.37	III
50		2087.98	III
20		2091.23	III
20		2092.26	III
10		2094.77	III
80		2095.31	III
15		2096.79	III
30		2098.36	III
10		2106.71	III
50		2109.54	III
20		2119.18	III
20		2198.14	III
80		2202.27	III
300		2240.11	III
100		2244.28	III
200		2257.03	III
100		2262.26	III
200		2265.67	III
150		2282.99	III
100		2283.99	III
300		2305.32	III
100		2309.27	III
200		2314.49	III
200		2337.97	III
40		2361.08	III
200		2365.43	III
50		2367.46	III
30		2369.99	III
20		2377.22	III
50		2403.95	III
20		2410.04	III
60		2412.33	III
10		2429.18	III
20		2433.43	III
100		2438.27	III
20		2439.31	III
20		2440.43	III
10		2458.64	III
10		2464.59	III
200		2490.42	III
20		2491.69	III
40		2506.25	III
300		2516.82	III
15		2522.07	III
20		2529.14	III
40		2550.39	III
300		2555.29	III
100		2560.56	III
10		2561.66	III
100		2566.78	III
2000		2567.61	III
1000		2579.57	III
100		2588.62	III
70		2592.69	III
500		2597.23	III
800		2599.14	III
30		2609.14	III
600		2621.11	III
300		2627.07	III
30		2635.37	III
500		2638.06	III
300		2640.48	III
1000		2642.56	III
100		2643.62	III
1000		2651.74	III
700		2652.25	III
100		2659.98	III
70		2664.89	III
2000		2666.13	III
2000		2666.99	III
30		2673.33	III
500		2677.39	III
500		2691.01	III
30		2708.04	III
400		2712.32	III
500		2749.91	III
200		2755.94	III
200		2756.76	III
100		2765.50	III
300		2788.24	III
600		2795.60	III
400		2803.32	III
1000		2803.43	III
10		2807.11	III
50		2808.51	III
600		2816.92	III
1000		2818.72	III
15		2826.01	III
300		2842.96	III
400		2875.86	III
600		2898.30	III
1000		2906.31	III
300		2928.97	III
50		2977.84	III
800		2998.00	III
2000		3029.49	III
100		3031.65	III
30		3040.65	III
3000		3092.50	III
20		3102.18	III
4000		3126.01	III
1000		3138.58	III
100		3151.44	III
70		3179.34	III
800		3191.35	III
50		3216.27	III
2000		3228.58	III
2000		3325.51	III
50		3358.25	III
20		3364.30	III
2000		3384.01	III
150		3392.56	III
100		3397.66	III
80		3432.94	III
40		3456.18	III
150		3463.51	III
20		3469.98	III
300		3550.87	III
200		3613.89	III
30		3659.84	III
30		3663.74	III
200		3664.74	III
20		3675.78	III
400		3711.91	III
20		3879.98	III
10		3882.58	III
20		3887.17	III
150		3896.55	III
15		3912.75	III
20		3913.23	III
500		3931.23	III
100		3985.56	III
10		3991.74	III
10		3997.67	III
2000		4028.14	III
10		4033.03	III
20		4074.53	III
20		4090.67	III

Yb III (continued)

Intensity	Wavelength	
20	4098.23	III
15	4121.06	III
10	4150.04	III
15	4153.11	III
100	4162.72	III
60	4172.95	III
30	4194.34	III
100	4194.95	III
10	4198.74	III
300	4213.64	III
10	4220.83	III
15	4231.07	III
20	4289.64	III
40	4301.14	III
20	4304.01	III
15	4350.80	III
10	4380.07	III
100	4517.58	III
40	4639.14	III
10	4834.93	III
15	5054.94	III
20	5256.85	III
20	5331.54	III
15	5740.83	III
10 d	5949.02	III
20	5973.05	III
40	6055.85	III
100	6214.22	III
200	6328.52	III
10	6365.88	III
150	6378.33	III
25	6466.33	III
20	6985.15	III
10	7037.04	III
15	7157.72	III
10	7311.02	III
10	7399.98	III
80	7410.01	III
15	7456.86	III
70	7664.41	III
80	7892.39	III
20	7893.10	III
100	7971.46	III
20	8056.02	III
10	8117.44	III
10	8326.86	III
30	8327.88	III
20	8400.01	III
30	8489.90	III
200	10110.60	III
100	10830.36	III

Yb IV
Ref. 40, 311 — J.R.

Intensity	Wavelength	
	Vacuum	
200	828.96	IV
200	870.35	IV
300	902.46	IV
300	927.01	IV
300	936.22	IV
400	943.04	IV
400	946.20	IV
200	975.21	IV
1000	1050.24	IV
1000	1054.46	IV
400	1092.51	IV
200	1109.96	IV
200	1110.55	IV
5000	1134.43	IV
300	1136.24	IV
500	1166.01	IV
600	1185.58	IV
200	1290.24	IV
600	1305.58	IV
900	1316.04	IV
200	1326.32	IV
800	1326.36	IV
200	1340.06	IV
300	1345.36	IV
900	1350.26	IV
200	1353.43	IV
400	1356.15	IV
200	1361.75	IV
300	1365.88	IV
300	1369.72	IV
400	1375.42	IV
300	1376.66	IV
200	1384.41	IV
250	1393.93	IV
350	1398.77	IV
400	1407.05	IV
300	1413.14	IV
400	1416.15	IV
400	1417.72	IV
200	1423.99	IV
300	1430.29	IV
200	1440.61	IV
400	1477.92	IV
300	1491.57	IV
400	1765.03	IV
200	1776.18	IV
200	1778.20	IV
200	1779.34	IV
300	1789.71	IV
800	1791.06	IV
200	1801.67	IV
250	1809.63	IV
600	1813.84	IV
400	1816.07	IV
250	1817.58	IV
300	1819.02	IV
200	1824.22	IV

Intensity	Wavelength	
	Air	
300	2106.48	IV
900	2116.65	IV
500	2121.29	IV
250	2122.84	IV
800	2123.32	IV
600	2125.72	IV
200	2129.65	IV
500	2135.21	IV
300	2137.58	IV
500	2138.35	IV
200	2138.53	IV
800	2139.99	IV
400	2141.04	IV
200	2142.20	IV
300	2143.42	IV
300	2143.89	IV
20000	2144.77	IV
400	2148.10	IV
300	2148.52	IV
15000	2154.18	IV
250	2165.55	IV
300	2169.12	IV
200	2172.16	IV
300	2177.53	IV
400	2183.32	IV
270 h	2186.13	IV
270	2187.17	IV
90 h	2189.90	IV
120	2193.34	IV
150	2198.27	IV
90	2224.64	IV
150	2231.28	IV
90	2233.30	IV
90	2244.20	IV
140	2331.36	IV

YTTRIUM (Y)
Z = 39

Y I and II
Ref. 1 — C.H.C.

Intensity	Wavelength	
	Air	
350	2243.06	II
50	2354.20	I
30	2373.83	
50	2385.24	
25	2413.93	II
560	2422.20	II
60	2460.61	II
25	2490.42	I
12	2540.28	
14	2547.57	
10	2550.17	
20	2681.65	I
60	2694.21	I
26	2695.39	I
95	2723.00	I
22	2730.08	I
22	2734.85	II
70	2742.53	I
140	2760.10	I
30	2785.21	II
12	2785.59	II
12	2791.20	I
30	2800.11	II
26	2813.64	I
18	2818.86	I
45	2822.56	I
22	2825.37	II
45	2826.38	II
70	2854.43	II
26	2856.30	II
11	2857.87	II
95	2886.48	I
18	2897.69	II
14	2898.82	II
160	2919.05	I
18 h	2930.03	II
390	2948.40	I
350	2964.96	I
18	2973.91	II
480	2974.59	I
30	2980.55	II
750	2984.26	I
70	2995.26	I
140	2996.94	I
70	3005.26	I
55	3018.95	I
130	3021.73	I
90	3022.28	I
26	3026.49	II
30	3036.59	II
45	3044.84	I
190	3045.37	I
22	3047.11	I
60	3055.22	II
60	3086.85	II
55 h	3091.70	I
22	3093.76	II
95	3095.88	II
45	3111.81	I
55	3112.04	II
22	3114.28	I
60	3128.77	II
80	3129.93	II
95	3135.17	II
110	3173.06	II
220	3179.41	II
70	3191.31	I
2300	3195.62	II
2200	3200.27	II
2200	3203.32	II
3900	3216.69	II
6200	3242.28	II
310	3280.91	II
19	3308.47	II
4700	3327.89	II
55	3340.38	I
160	3362.00	II
85	3388.59	I
45	3397.04	I
85	3412.47	I
200	3448.82	II
70	3450.95	I
110	3467.88	II
170	3485.73	I
1700	3496.09	II
80	3521.53	I
45	3546.01	II
3900	3549.01	II
130	3551.80	I
540	3552.69	II
170	3558.76	I
190	3571.43	I
260	3576.05	I
3300	3584.52	II
300	3587.75	I
100	3589.69	I
2800	3592.92	I
10000	3600.73	II
6200	3601.92	II
7800	3611.05	II
4300	3620.94	I
1900	3628.71	II
7800	3633.12	II
3000	3664.61	II
45	3668.49	II
170	3692.53	I
13000	3710.30	II
60	3718.12	I
60	3738.61	I
1200	3747.55	II
50	3749.89	I
10000	3774.33	II
1400	3776.56	II
50	3782.30	II
7400	3788.70	II
1300	3818.35	II
4000	3832.88	II
70	3847.87	I
80	3876.82	I
480	3878.28	II
30	3887.77	I
60 h	3904.59	I
50 h	3918.25	I
60 h	3930.11	I
240	3930.66	II
4400	3950.36	II
150	3951.60	II
60 h	3955.09	I
3600	3982.60	II
40	3987.50	I
940	4039.83	I
2400	4047.64	I
9400	4077.38	I
90 h	4081.22	I
2000	4083.71	I
9900	4102.38	I
60 h	4106.39	I
80	4110.81	I
320	4124.92	II
8900	4128.31	I
7500	4142.85	I
100 h	4157.63	I
2400	4167.52	I
2000	4174.14	I
8000	4177.54	II
120	4199.28	II
380	4204.70	II
80	4213.02	I
40	4213.54	I
160	4217.80	I
280 h	4220.63	I
80	4224.25	I
600	4235.73	II
2200	4235.94	I
300	4251.20	I
360 h	4302.30	I
2800	4309.63	II
50	4316.30	I
110	4330.78	I
30	4337.29	I
60	4344.65	I
440 h	4348.79	I
60	4352.33	I
60	4352.70	I
120	4357.73	I
800	4358.73	II
12000	4374.94	II
150 h	4375.61	I
80	4379.33	I
30	4385.48	I
100	4387.74	I
30	4394.01	I
30	4394.67	I
1800	4398.02	II
890	4422.59	II
80	4437.34	I
100	4443.66	I
130	4446.63	I
20	4465.27	I
40	4473.89	I
170	4475.72	I
180	4476.96	I
160	4477.45	I
110	4487.28	I
300	4487.47	I
30	4491.75	I
25	4492.42	I
500	4505.95	I
50	4513.58	I
80	4514.01	I
40 h	4522.05	I
890	4527.25	I
440	4527.80	I
100	4544.32	I
100	4559.37	I
30	4564.39	I
60 h	4573.56	I
35	4581.32	I
30	4581.77	I
130	4596.55	I
95	4604.80	I
40	4613.00	I
2000	4643.70	I
200 h	4658.32	I
70	4658.89	I
85	4667.47	I
60	4670.82	I
2000	4674.84	I
60	4678.35	I
260	4682.32	II
85	4689.77	I
180	4696.81	I
35	4708.85	I
60	4725.85	I
170	4728.53	I
60 h	4732.37	I
85	4741.40	I
160	4752.79	I
410	4760.98	I
17	4780.18	I
120	4781.04	I
160	4786.58	II
170	4786.89	I
180	4799.30	I
50	4804.31	I
70	4804.81	I
85 B1	4817.38	YO
140 B1	4818.20	YO
140	4819.64	I
120	4822.13	I
190	4823.31	II
60	4839.15	I
770	4839.87	I
550	4845.68	I
410	4852.69	I
120	4854.25	I
890	4854.87	II
50	4856.70	I
330	4859.84	I
50	4879.65	I
1900	4883.69	II
50	4886.28	I
40	4886.65	I
95	4893.44	I
1100	4900.12	II
100	4906.11	I

Yttrium (Cont.)

Intensity		Wavelength	Species
45		4909.00	I
150		4921.87	I
35		4930.93	I
45		4950.66	I
120		4974.30	I
120		4982.13	II
100		5006.97	I
75		5070.21	I
75		5072.19	I
1100		5087.42	II
30		5088.18	I
210		5119.11	II
450		5123.21	II
180		5135.20	I
120		5196.43	II
960		5200.41	II
1500		5205.72	II
180		5240.81	I
60		5289.82	II
45		5320.78	II
75		5380.62	I
220		5400.70	II
24		5417.03	I
90		5424.37	I
190		5438.24	I
710		5466.46	I
100		5468.47	I
90		5473.39	II
90		5480.74	II
60		5493.17	I
35		5495.59	I
240		5497.41	II
300		5503.45	I
250		5509.90	II
60		5513.64	I
120		5521.63	I
		5521.70	II
24		5526.76	I
740		5527.54	I
35		5541.63	I
120		5544.50	I
		5544.61	II
90		5546.02	II
75		5556.43	I
60		5567.75	I
180		5577.42	I
24		5581.08	I
620		5581.87	I
21		5590.96	I
21		5594.12	I
100		5608.33	I
15		5623.91	I
560		5630.13	I
24		5632.25	I
21		5632.89	I
120		5644.69	I
120		5648.47	I
740		5662.94	II
90		5675.27	I
18		5693.63	I
160		5706.73	I
24		5720.61	I
75		5728.89	I
150	B1	5730.12	YO
21		5732.09	I
90		5743.85	I
18	B1	5746.93	YO
24	B1	5764.22	YO
75		5765.64	I
35		5773.95	I
100		5781.69	II
15	B1	5800.00	YO
15	B1	5818.58	YO
30		5821.87	I
21		5832.27	I
9	b	5838.07	YO
15		5858.83	YO
15		5871.83	I
24		5876.14	YO
24		5879.96	I
24	b	5893.94	YO
35		5902.96	I
24	b	5912.19	YO
24	b	5931.10	YO
90	B1	5939.08	YO
45		5945.72	I
24		5950.02	I
75	b	5956.41	YO
1300	B1	5972.04	YO
50		5981.86	I
1000	B1	5987.64	YO
740	B1	6003.60	YO
120		6004.65	
120		6009.19	I
620	B1	6019.87	YO
120		6023.41	I
500	B1	6036.60	YO
420	B1	6053.81	YO
130	B1	6072.78	YO
50		6088.00	I
210	B1	6089.35	YO
160	B1	6096.78	YO
130	B1	6107.82	YO
130	B1	6114.73	YO
75	B1	6127.38	YO
1400	B1	6132.06	YO
120		6135.04	I
150		6138.43	I
1100	B1	6148.36	YO
120		6151.72	YO
820	B1	6165.08	YO
560	B1	6182.23	YO
1200		6191.73	I
590	B1	6199.82	YO
450	B1	6217.96	YO
300		6222.59	I
270	B1	6236.72	YO
45		6251.05	I
120	B1	6275.01	YO
60	b	6295.46	YO
24	b	6316.20	YO
24	b	6338.10	YO
15	b	6359.48	YO
15		6369.87	YO
75		6402.01	I
1000		6435.00	
24		6437.18	I
18	h	6501.23	YO
18	h	6518.33	YO
18	h	6535.84	YO
90		6538.60	I
12	h	6553.84	YO
70		6557.39	I
12	h	6572.58	
35		6576.85	I
23		6584.87	I
95		6613.75	II
14		6622.49	I
19	h	6636.49	I
40		6650.61	I
21		6664.40	I
150		6687.58	I
14	h	6691.83	I
7		6694.75	I
16	h	6699.26	I
70		6700.71	I
35		6713.20	I
40		6735.99	I
190		6793.71	I
70	h	6795.41	II
12	h	6803.13	I
21		6815.16	I
14		6832.49	II
45		6865.26	I
14		6858.24	II
29		6887.22	I
21		6896.00	II
9		6908.26	I
14		6933.52	I
24	h	6950.31	I
10		6951.68	II
10		6958.04	I
24		6979.88	I
13	h	7008.97	I
10		7009.93	I
19	h	7035.18	I
29		7052.94	I
13	h	7054.28	I
9		7075.13	I
11		7127.92	I
35		7191.66	I
10	h	7195.93	I
35		7264.17	II
9	h	7293.08	I
9	h	7330.62	I
5		7332.96	II
50		7346.46	I
11	h	7398.77	I
29		7450.30	II
17		7494.88	I
7	h	7536.71	I
35		7563.13	I
8	h	7617.72	I
19	h	7622.94	I
7		7652.89	I
5		7689.49	I
8	h	7698.00	I
19		7719.89	I
19		7724.08	I
13		7788.42	I
13		7796.32	I
6		7802.52	I
17		7812.16	I
29		7855.52	I
110		7881.90	II
10	h	7999.33	I
9		8329.61	I
24		8344.43	I
8	h	8365.64	I
17		8450.36	I
8	h	8528.94	I
95		8800.62	I
19	h	8835.85	II

Y III
Ref. 77 — J.R.

Intensity		Wavelength	Species

Vacuum

1		643.68	III
4		646.69	III
6		653.87	III
10		656.98	III
25		668.74	III
40		671.98	III
100		691.72	III
4		693.85	III
200		695.20	III
9		727.91	III
4		728.47	III
2		728.83	III
20		729.73	III
600		730.49	III
15		732.70	III
800		734.38	III
15		770.78	III
10		771.79	III
20		804.26	III
5000		805.20	III
75		806.18	III
150		808.97	III
7000		809.92	III
100		855.64	III
60		857.82	III
25		984.23	III
15		987.96	III
15000		989.21	III
25000		996.37	III
20		999.19	III
150		1000.56	III
25		1003.35	III
1000		1006.58	III
1200		1007.86	III
120		1077.52	III
500		1081.35	III
75	p	1084.63	III
350	p	1088.39	III
250		1095.25	III
25		1098.07	III
150		1102.21	III
3000		1289.74	III
2500		1306.96	III
5000		1314.51	III
1500		1316.10	III
4000		1334.04	III
8		1549.08	III
15	p	1553.81	III
30		1635.14	III
75		1640.43	III
200		1779.80	III
600		1786.05	III

Air

10		2041.93	III
5		2042.07	III
1500		2060.58	III
4000		2068.98	III
10000		2127.98	III
16000		2191.16	III
8000		2200.76	III
8000		2206.03	III
150		2261.41	III
80		2261.57	III
10000		2284.34	III
3		2319.92	III
10000		2327.31	III
50000		2367.23	III
40000		2414.64	III
100	p	2710.30	III
90	h	2710.54	III
5		2780.11	III
70		2791.44	III
20		2803.27	III
100		2807.00	III
90000		2817.04	III
6000		2867.67	III
6000		2913.41	III
1500		2917.74	III
1600		2918.56	III
15		2940.53	III
99000		2946.01	III
20	p	2948.48	III
6000		2970.42	III
1400		3013.93	III
1500		3018.85	III
3		3267.10	III
25	l	3276.80	III
500		3866.69	III
3000		3900.74	III
4000		3914.58	III
3800		4039.60	III
3000		4040.11	III
120	h	4121.61	III
2000	c	4737.62	III
7500		5102.88	III
1300		5120.40	III
10000		5238.10	III
3000		5263.58	III
4000		5383.64	III
6000		5562.81	III
600		5567.27	III
4000		5572.24	III
400		5595.48	III
3000		5602.08	III
2000	h	7254.58	III
9000		7558.71	III
6000		7864.53	III
8000		7916.71	III
400		7989.41	III
10000		7991.43	III
8000		8171.41	III
4000		8645.09	III
10000		8796.21	III
8000		9116.59	III

Y IV
Ref. 265 — J.R.

Intensity	Wavelength	Species

Vacuum

3	211.80	IV
3	214.51	IV
6	215.97	IV
6	217.39	IV
12	221.71	IV
3	222.18	IV
6	222.98	IV
2	228.84	IV
20	228.94	IV
3	229.78	IV
3	235.17	IV
25	235.77	IV
1	242.12	IV
30	242.30	IV
3	244.14	IV
3	263.72	IV
150	264.04	IV
30	270.40	IV
150	273.03	IV
10	278.60	IV
900	355.86	IV
300	370.42	IV
500	386.82	IV
600	425.03	IV
300	473.10	IV

Y V
Ref. 419 — J.R.

Intensity	Wavelength	Species

Vacuum

5	289.18	V
50	299.99	V
3	312.89	V
200	313.35	V
40	320.47	V
40	321.69	V
150	325.58	V
200	326.57	V
175	328.34	V
50	330.40	V
900	333.09	V
500	333.80	V
100	335.12	V
400	335.14	V
500	336.62	V
500	339.02	V
200	340.02	V
5	340.42	V
500	344.59	V
100	349.65	V
2	349.75	V
10	351.36	V
100	353.98	V
100	355.56	V
300	372.05	V
400	379.96	V
200	397.77	V
300	403.45	V
1	408.81	V
200	409.31	V
200	415.03	V
100	418.18	V
150	418.59	V
300	419.79	V
300	420.74	V
1	427.87	V
50	430.75	V
100	437.66	V
3	441.62	V
30	442.96	V

Yttrium (Cont.)

Intensity		Wavelength	
2		451.97	V
15		455.84	V
85		457.84	V
4000		584.98	V
2000		630.97	V

ZINC (Zn)
Z = 30

Zn I and II
Ref. 39, 55, 113, 131, 185, 186 — R.D.C.

Intensity		Wavelength	
		Vacuum	
60		1193.23	II
60		1277.31	II
60	d	1366.68	II
		1404.12	I
60		1410.44	II
60		1439.09	II
60		1445.04	II
50		1456.91	II
		1457.57	I
60		1477.02	II
50		1514.76	II
90		1572.99	II
		1589.57	I
60		1617.68	II
60		1658.25	II
50		1713.25	II
60		1715.76	II
80	d	1735.61	II
60		1736.89	II
50		1737.90	II
75		1747.12	II
80	c	1762.19	II
75		1774.04	II
80		1790.76	II
100		1797.64	II
100	d	1811.05	II
80		1816.48	II
80		1831.38	II
100	d	1833.57	II
70		1836.01	II
75		1836.65	II
75		1847.56	II
100		1864.12	II
100		1866.08	II
100		1872.13	II
75		1894.26	II
60		1901.52	II
60		1914.81	II
100	d	1918.96	II
70		1920.27	II
100	d	1929.67	II
60		1945.58	II
60		1951.91	II
80		1953.00	II
75		1954.87	II
80		1964.54	II
100		1969.40	II
100		1982.11	II
70		1985.61	II
100		1986.99	II
50		1993.37	II
50		1996.92	II
		Air	
100		2011.94	II
500		2025.48	II
60		2039.31	II
500		2062.00	II
200		2064.23	II
120		2079.08	I
50		2079.93	II
60		2087.33	I
80		2096.93	I
300		2099.94	II
200		2102.18	II
150		2104.42	I
75		2122.74	II
800	r	2138.56	I
75		2147.42	II
60		2210.18	II
50		2273.15	II
1000		2501.99	II
150		2515.81	I
50		2527.96	II
1000		2557.95	II
50		2567.80	II
50		2567.98	II
100	h	2569.87	I
100		2582.44	I
300		2582.49	I
200		2608.56	I
300		2608.64	I
200		2670.53	I
300		2684.16	I
300		2712.49	I
200		2756.45	I
300		2770.86	I
300		2770.98	I
400		2800.87	I
100		2801.06	I
5		2801.17	I
100		2801.96	II
100		2902.30	II
125		3018.36	I
200		3035.78	I
200		3072.06	I
150		3075.90	I
100		3171.45	II
100		3172.23	II
300		3196.31	II
100		3197.10	II
500	r	3282.33	I
50		3299.42	II
800		3302.58	I
700	r	3302.94	I
75		3306.01	II
800		3345.02	I
500		3345.57	I
150		3345.94	I
5		3799.00	I
50		3806.34	II
100		3840.29	II
50		3883.34	I
15		3965.43	I
10		4113.21	I
25		4292.88	I
25		4298.33	I
35		4629.81	I
300		4680.14	I
400		4722.15	I
400		4810.53	I
800		4911.62	II
500		4924.03	II
7		5068.66	I
15		5069.58	I
200		5181.98	I
8		5308.65	I
7		5310.24	I
7		5311.02	I
4		5772.10	I
4		5775.50	I
10		5777.11	I
500		5894.33	II
500		6021.18	II
500		6102.49	II
100		6111.53	II
500		6214.61	II
8		6237.90	I
8		6239.17	I
1000	h	6362.34	I
10		6479.18	I
15		6928.32	I
8		6938.47	I
3		6943.20	I
200		7478.8	II
300		7588.5	II
100		7612.9	II
300		7732.5	II
200		7757.9	II
10		7799.36	I
100		11054.25	I
100		13053.63	I
100		13150.59	I
20		13196.61	I
100		14038.70	I
20		15680.29	I
20		16483.45	I
20		16491.98	I
20		16505.23	I
5		24044.16	I
10		24375.02	I

Zn III
Ref. 376, 377 — R.D.C.

Intensity	Wavelength	
	Vacuum	
1000	677.63	III
750	677.96	III
200	713.90	III
100	1432.15	III
200	1456.72	III
100	1464.20	III
100	1465.75	III
300	1473.41	III
100	1489.26	III
100	1490.96	III
200	1498.79	III
300	1499.42	III
300	1500.42	III
300	1505.92	III
300	1515.85	III
30	1533.09	III
300	1552.30	III
200	1552.94	III
80	1553.11	III
200	1560.79	III
150	1562.55	III
200	1581.53	III
100	1582.06	III
100	1598.52	III
100	1600.87	III
100	1619.61	III
150	1622.51	III
200	1629.19	III
200	1639.33	III
150	1644.82	III
100	1651.74	III
200	1673.05	III
100	1688.59	III
80	1695.40	III
80	1706.65	III
80	1749.63	III
50	1753.84	III
100	1767.69	III
80	1839.32	III

Zn IV
Ref. 370, 377 — R.D.C.

Intensity	Wavelength	
	Vacuum	
30	412.67	IV
30	423.42	IV
50	423.54	IV
200	425.90	IV
200	428.54	IV
80	428.79	IV
150	429.30	IV
200	430.59	IV
150	431.54	IV
120	431.62	IV
10	434.41	IV
150	435.02	IV
150	435.76	IV
80	436.25	IV
30	436.38	IV
50	436.82	IV
50	441.15	IV
20	441.52	IV
20	441.56	IV
100	441.70	IV
200	442.39	IV
150	444.39	IV
100	444.46	IV
100	446.58	IV
30	447.85	IV
10	449.13	IV
200	449.98	IV
200	450.99	IV
30	451.62	IV
20	452.80	IV
80	456.67	IV
150	457.32	IV
200	466.93	IV
150	468.43	IV
200	472.09	IV
200	472.66	IV
200	473.02	IV
200	473.51	IV
200	474.56	IV
50	475.78	IV
200	476.42	IV
200	478.65	IV
200	478.90	IV
200	482.10	IV
10	482.68	IV
10	485.48	IV
10	489.19	IV
10	490.96	IV
10	493.37	IV
10	496.72	IV
100	497.70	IV
15	1193.29	IV
15	1203.44	IV
8	1212.71	IV
5	1214.14	IV
25	1224.35	IV
20	1227.62	IV
40	1228.65	IV
30	1231.46	IV
30	1237.26	IV
50	1239.12	IV
3	1246.26	IV
15	1247.01	IV
50	1249.69	IV
30	1253.46	IV
30	1257.31	IV
8	1259.68	IV
500	1265.74	IV
5	1269.15	IV
100	1272.21	IV
200	1272.98	IV
30	1275.78	IV
25	1278.51	IV
100	1280.47	IV
100	1291.83	IV
100	1292.49	IV
10	1294.32	IV
100	1296.62	IV
100	1296.73	IV
20	1301.88	IV
500	1306.66	IV
200	1318.00	IV
200	1320.74	IV
200	1321.22	IV
200	1322.33	IV
200	1322.43	IV
200	1326.74	IV
200	1329.11	IV
100	1333.32	IV
150	1344.08	IV
200	1347.98	IV
200	1349.90	IV
50	1352.27	IV
100	1356.20	IV
200	1357.82	IV
100	1363.43	IV
200	1363.95	IV
40	1368.17	IV
200	1369.53	IV
50	1370.42	IV
100	1375.33	IV
50	1375.98	IV
200	1377.65	IV
50	1391.24	IV
80	1393.07	IV
50	1394.54	IV
30	1400.14	IV
15	1403.98	IV
30	1409.40	IV
5	1410.33	IV
30	1419.60	IV
50	1427.79	IV
20	1438.58	IV
80	1455.65	IV
200	1459.98	IV
100	1476.43	IV
100	1481.25	IV
100	1529.84	IV
50	1533.68	IV

ZIRCONIUM (Zr)
Z = 40

Zr I and II
Ref. 1 — C.H.C.

Intensity	Wavelength	
	Air	
60	2374.42	I
60	2384.17	I
50	2388.01	I
50	2389.21	I
45	2405.52	I
60	2419.41	II
150	2449.85	II
21	2457.44	II
75	2487.29	II
45	2496.48	II
180	2532.46	II
90	2539.65	I
220	2542.10	II
45	2550.51	I
220	2550.74	II
45	2556.43	I
60	2567.45	I
570	2567.64	II
1600	2568.87	II
2100	2571.39	II
75	2583.40	II
130	2589.07	II
22	2589.65	I
45	2609.43	I
150	2630.91	II
80	2635.42	I
210	2639.09	II
70	2643.40	II
55	2647.78	I
110	2650.38	II
70	2658.69	I
180	2667.80	II
55	2669.49	II
120	2670.96	II
1800	2678.63	II
35	2681.76	II
90	2687.75	I
90	2692.60	II
22	2692.92	I
160	2693.53	II
180	2694.06	II
70	2695.43	II

Zirconium (Cont.)

Int.	λ		Int.	λ		Int.	λ		Int.	λ	
95	2699.60	II	880	3182.86	II	960	3698.17	II	200	4496.97	II
750	2700.13	II	540	3191.21	I	720	3709.26	II	550	4507.12	I
280	2711.51	II	210	3191.90	II	270	3731.26	II	610	4535.75	I
140	2712.42	II	540	3212.01	I	560	3745.98	II	490	4542.22	I
140	2714.26	II	760	3214.19	II	880	3751.60	II	200	4553.01	I
1300	2722.61	II	110	3222.47	II	480	3764.39	I	200	4555.13	I
140	2725.47	I	200	3228.81	I	480	3766.72	I	140	4555.52	I
800	2726.49	II	630	3231.69	II	340	3766.82	II	490	4575.52	I
490	2732.72	II	630	3234.12	I	720	3780.54	I	100	4582.29	I
1400	2734.86	II	110	3236.58	II	560	3791.40	I	140	4590.55	I
110	2740.51	II	760	3241.05	II	210	3817.58	II	350	4602.57	I
140	2741.55	II	320	3250.39	I	560	3822.41	I	140	4604.42	I
1100	2742.56	II	200	3254.28	I	2200	3835.96	I	210	4626.41	I
660	2745.86	II	200	3260.11	I	1300	3836.76	II	700	4633.98	I
660	2752.21	II	190	3269.66	I	550	3843.02	II	210	4644.83	I
530	2758.81	II	150	3271.13	II	550	3847.01	I	260	4683.42	I
200 d	2768.73	II	540	3272.22	II	550	3849.25	I	2300	4687.80	I
	2768.85	II	1000	3273.05	II	2900	3863.87	I	510	4688.45	I
170 d	2774.04	I	320 d	3282.73	I	770	3864.34	I	110	4707.79	I
	2774.16	II	880	3284.71	II	990	3877.60	I	1900	4710.08	I
200	2790.14	I	140	3285.88	II	200	3879.05	I	160	4711.97	I
170	2792.04	I	150	3288.80	II	1500	3885.42	I	120	4717.62	I
160	2796.90	II	540	3305.15	II	2900	3890.32	I	210	4719.12	I
110	2799.15	II	880	3306.28	II	2000	3891.38	I	300	4732.33	I
180	2810.91	II	150	3313.70	II	400	3900.52	I	1400	4739.48	I
620	2814.90	I	210	3314.50	II	310	3915.94	II	190	4762.78	I
390	2818.74	II	150	3319.02	II	610	3921.79	I	870	4772.31	I
530	2825.56	II	380	3322.99	II	1200	3929.53	I	210	4784.92	I
110	2833.91	II	380	3326.80	II	200	3934.12	II	160	4788.67	I
710	2837.23	I	380	3334.25	II	200	3934.79	II	260	4805.87	I
120	2839.34	II	210	3334.62	II	940	3958.22	II	140	4809.47	I
130	2843.52	II	190	3338.41	II	490	3966.66	I	190	4815.04	I
660	2844.58	II	760	3340.56	II	990	3968.26	I	700	4815.63	I
210	2848.19	II	380	3344.79	II	660	3973.50	I	280	4824.29	I
350	2848.52	I	130	3353.66	I	200	3975.29	I	190	4828.04	I
350	2851.97	II	180	3354.39	II	200 h	3981.60	I	110	4838.78	I
340	2869.81	II	760	3356.09	II	770	3991.13	II	210	4851.36	I
490	2875.98	I	540	3357.26	II	770	3998.97	II	160	4866.06	I
120	2892.26	I	180	3359.96	II	200	4007.60	I	110	4881.24	I
160	2905.23	II	150	3360.46	I	200	4012.25	I	110	4883.60	I
300	2915.99	II	150	3363.82	II	400	4023.98	I	100	4994.76	I
110	2916.64	II	150	3367.82	II	770	4024.92	I	30	5011.46	I
270	2918.24	II	150	3370.59	I	990	4027.20	I	250	5046.58	I
320	2926.99	II	180	3373.42	II	240	4028.95	I	85	5060.39	I
160	2934.61	II	380	3374.73	II	400	4029.68	II	360	5064.91	I
160	2936.31	II	110	3376.27	II	490	4030.04	I	110	5065.22	I
320	2948.94	II	150	3377.46	II	400	4035.89	I	100	5070.26	I
210	2951.48	II	570	3387.87	II	240	4042.22	I	75	5073.98	I
320	2955.78	II	760	3388.30	II	610	4043.58	I	470	5078.25	I
320	2960.87	I	5700	3391.98	II	490	4044.56	I	85	5091.06	I
510	2962.68	II	570	3393.12	II	400	4045.61	II	50	5112.27	II
320	2968.96	II	160	3396.33	II	610	4048.67	II	140	5115.24	I
120	2969.19	I	380	3399.35	II	200	4050.33	II	50	5120.42	I
230	2969.63	II	570	3404.83	II	200	4050.48	I	85	5133.40	I
130	2976.61	II	760	3410.25	II	770	4055.03	I	300	5155.45	I
320	2978.05	II	380	3414.66	I	600	4055.71	I	200	5158.00	I
230	2979.18	II	1000	3430.53	II	330	4061.53	I	35	5158.67	I
160	2981.02	II	380	3437.14	II	1500	4064.16	I	75	5160.99	I
820	2985.39	I	4700	3438.23	II	2000	4072.70	I	85	5165.96	I
320	3003.74	II	600	3447.36	I	310	4074.93	I	17	5178.99	I
100	3005.37	I	200	3455.91	I	200	4076.53	I	100	5183.70	I
160	3005.50	I	410	3457.56	II	240	4078.31	I	30	5187.03	I
820	3011.75	II	200	3458.93	II	2000	4081.22	I	100	5191.60	II
100	3013.32	II	820	3463.02	II	200	4108.40	I	100	5201.15	I
160	3019.84	II	600	3471.19	I	400	4121.46	I	85	5209.30	I
350	3020.47	II	200	3478.79	I	1200	4149.20	II	85	5224.93	I
500	3028.04	II	1200	3479.39	II	200	4152.64	I	30	5243.47	I
880	3029.52	I	1300	3481.15	II	290	4156.24	II	120	5277.41	I
180	3030.92	II	760	3483.54	II	400	4161.21	II	75	5280.05	I
350 d	3036.39	II	4100	3496.21	II	400	4166.36	I	60	5294.82	I
100	3045.83	I	350	3505.48	II	200	4183.32	I	120	5296.79	I
690	3054.84	II	820	3505.67	II	660	4187.56	I	60	5301.97	I
100	3060.11	II	1000	3509.32	I	400	4194.76	I	110	5311.40	I
100	3064.63	II	200	3510.46	II	610	4199.09	I	25	5321.26	I
110	3085.34	I	2000	3519.60	I	610	4201.46	I	22	5330.84	I
110	3094.80	I	440	3525.81	II	610	4208.98	II	12	5338.43	I
250	3095.07	II	440	3533.22	I	200	4211.88	II	30	5350.09	II
110	3095.82	I	210	3535.16	I	400	4213.86	I	30	5350.35	II
280	3099.23	II	630	3542.62	II	2000	4227.76	I	25	5350.90	I
690	3106.58	II	1800	3547.68	I	200	4236.06	I	25	5351.92	I
110	3108.37	I	210	3549.74	II	2000	4239.31	I	75	5362.56	I
210	3110.88	II	630	3550.46	I	770	4240.34	I	12	5363.35	I
350	3120.74	I	1800	3551.95	II	770	4241.20	I	17	5369.39	I
320	3125.92	II	2100	3556.60	II	1200	4241.69	I	20	5382.37	I
500	3129.18	II	1100	3566.10	I	310	4268.02	I	270	5385.14	I
500	3129.76	II	210	3568.88	I	550	4282.20	I	30	5386.65	I
140	3131.11	I	2100	3572.47	II	550	4294.79	I	17	5391.18	I
350	3132.07	I	210	3573.08	II	310	4302.89	I	17	5395.88	I
110	3133.23	I	1100	3575.79	I	550	4341.13	I	25	5405.13	I
350	3133.48	II	1300	3576.85	II	1000	4347.89	I	85	5407.62	I
180	3136.96	I	880	3586.29	II	290	4359.74	II	17	5413.93	I
690	3138.68	II	440	3587.98	II	310	4360.81	I	20	5421.86	I
140	3139.80	I	3500	3601.19	I	350	4366.45	I	15	5426.36	I
180	3148.82	I	690	3611.89	II	240	4379.78	II	25	5428.42	I
290	3155.67	II	1100	3613.10	II	190	4413.04	I	25	5437.76	I
150	3157.00	II	1100	3614.77	II	240	4420.46	I	15	5440.41	I
320	3157.82	I	1100	3623.86	II	120	4427.24	I	35	5448.57	I
540	3164.31	II	320	3634.15	I	160	4431.49	I	10	5474.92	I
150	3165.45	II	260	3661.00	I	140	4443.00	II	10	5477.40	I
880	3165.97	II	1100	3663.65	I	110	4457.43	I	35	5478.33	I
150	3166.26	II	390	3671.27	II	110	4466.91	I	35	5480.83	I
190	3178.09	II	800	3674.72	II	110	4470.31	I	10	5481.16	I
190	3181.58	II	390	3697.46	II	190	4470.56	I	30	5486.09	I
150	3181.92	II							140	5502.12	I

Zirconium (Cont.)

Intensity		Wavelength		Intensity	Wavelength		Intensity	Wavelength		Intensity	Wavelength	
25		5507.87	I	85	6769.16	I	12	8188.77	I	50	1989.83	III
30		5517.11	I	27	6772.89	I	40	8194.77	I	30	1990.95	III
10		5518.05	I	15	6787.15	II	60	8201.73	I	30	1994.46	III
75		5528.41	I	35	6790.85	I	280	8212.53	I			
20		5532.30	I	45	6828.78	I	20	8240.37	I		Air	
45		5537.46	I	45	6832.89	I	40	8283.81	I			
50		5545.32	I	13	6845.33	I	140	8305.90	I	45	2000.23	III
22	B1	5551.75	Z	17	6846.34	I	14	8332.44	I	100	2006.82	III
25	B1	5553.17	Z	100	6846.97	I	50	8370.23	l	30	2013.30	III
12		5612.11	I	27	6849.26	I	120	8389.41	I	35	2016.63	III
120		5620.14	I	13	6852.56	I	70	8414.00	I	30	2021.52	III
35		5623.53	I	120	6888.29	I	50	8453.17	t	60	2026.78	III
25	B1	5629.02	Z	29	6900.69	I	50	8464.65	I	100	2035.42	III
25	B1	5629.58	Z	20	6904.36	I	40	8498.44	I	50	2036.92	III
160		5664.51	I	29	6907.37	I	18	8584.21	I	75	2056.13	III
20		5666.28	I	20	6916.87	I	10	8734.86	I	40	2058.73	III
120		5680.90	I	16	6932.38	I	12	8749.48	I	75	2060.83	III
15		5685.42	I	29	6948.46	I	10	8786.23	I	50	2061.47	III
30		5708.89	I	150	6953.84	I	16	8804.98	I	125	2070.43	III
75	B1	5718.21	Z	60	6966.44	I	70	8836.09	I	50	2074.12	III
120		5735.70	I	10	6975.91	I	60	8899.52	I	100	2077.92	III
35	B1	5748.17	Z	150	6990.84	I				100	2080.99	III
17	B1	5778.57	Z	80	6994.32	I				30	2081.81	III
160		5797.74	I	10	7005.46	I				25	2085.35	III
30		5847.32	I	100	7027.40	I		Zr III		200	2086.78	III
50		5868.27	I	25	7057.36	I		Ref. 403 — J.R.		40	2089.50	III
110		5869.50	I	14	7057.96	I				40	2097.03	III
340		5879.80	I	140	7087.30	I	Intensity	Wavelength		40	2102.30	III
85		5885.62	I	25	7089.43	I				50	2103.16	III
50		5901.09	I	35	7094.46	I		Vacuum		25	2104.23	III
30	B1	5908.61	Z	50	7095.59	I				40	2113.98	III
140		5925.13	I	540	7097.70	I	25	687.64	III	35	2114.10	III
100		5935.20	I	280	7102.91	I	50	690.39	III	40	2125.06	III
110		5955.35	I	170	7103.72	I	25	819.59	III	35	2137.90	III
30	B1	5977.80	Z	140	7111.68	I	30	820.21	III	35	2138.45	III
100		5984.23	I	40	7112.82	I	35	823.69	III	25	2139.85	III
17		5995.37	I	18	7113.52	I	25	829.50	III	40	2159.24	III
50		6001.05	I	12	7132.95	I	30	850.61	III	40	2162.20	III
30		6025.36	I	16	7140.74	I	25	859.56	III	100	2175.80	III
85		6032.61	I	12	7144.47	I	30	864.86	III	100	2191.15	III
170		6045.85	I	590	7169.09	I	25	868.64	III	35	2192.05	III
100		6049.24	I	50	7201.62	I	25	868.99	III	60	2206.33	III
140		6062.84	I	12	7258.17	I	25	919.59	III	40	2206.97	III
50		6120.83	I	35	7264.76	I	30	1320.81	III	30	2231.00	III
170		6121.91	I	20	7306.21	I	25	1375.13	III	50	2245.36	III
85		6124.84	I	25	7311.62	I	30	1378.93	III	25	2251.14	III
680		6127.44	I	35	7313.72	I	40	1403.48	III	40	2257.83	III
340		6134.55	I	90	7318.08	I	35	1420.12	III	30	2281.43	III
100		6140.46	I	10	7327.82	I	25	1420.87	III	100	2301.60	III
440		6143.20	I	50	7335.97	I	25	1465.44	III	75	2308.12	III
30		6155.61	I	50	7343.96	I	25	1563.24	III	35	2405.81	III
75		6157.71	I	20	7373.50	I	40	1593.59	III	40	2406.21	III
25		6160.20	I	25	7383.63	I	100	1612.38	III	75	2420.65	III
35		6189.40	I	14	7400.90	I	50	1620.62	III	25	2438.70	III
60		6192.96	I	10	7411.39	I	75	1631.31	III	50	2444.58	III
85		6213.05	I	10	7422.75	I	50	1638.33	III	100	2448.86	III
100		6214.69	I	10	7433.10	I	35	1675.06	III	100	2593.64	III
170	B1	6226.51	Z	110	7439.86	I	35	1675.75	III	250	2620.56	III
100		6257.26	I	18	7467.57	I	40	1703.36	III	50	2621.28	III
50	b	6261.05	Z	16	7479.58	I	40	1725.03	III	60	2628.26	III
35		6267.06	I	14 h	7515.70	I	40	1754.38	III	200	2643.79	III
45	B1	6292.84	Z	12	7517.95	I	35	1759.12	III	100	2656.46	III
120		6299.66	I	20 h	7540.62	I	30	1764.75	III	150	2664.26	III
15		6304.34	I	20	7544.59	I	30	1771.96	III	100	2682.16	III
300		6313.02	I	29	7551.46	I	40	1773.90	III	75	2686.28	III
30		6314.71	I	40	7554.70	I	100	1779.51	III	70	2690.49	III
50		6321.35	I	25	7558.45	I	40	1783.35	III	60	2698.31	III
22		6340.36	I	12	7560.09	I	200	1790.19	III	50	2709.05	III
50	B1	6345.10	Z	12	7562.12	I	150	1793.56	III	45	2715.76	III
75		6345.22	I	80	7607.15	I	125	1798.13	III	40	2720.07	III
75	B1	6378.56	Z	14	7612.08	I	75	1800.03	III	75	2735.76	III
35		6407.00	I	20	7621.17	I	25	1801.67	III	25	2775.23	III
50	b	6412.39	Z	29	7658.60	I	100	1805.26	III	40	2836.18	III
12		6426.17	I	18	7690.83	I	35	1831.89	III	40	3278.86	III
35		6434.33	I	14	7704.27	I	40	1850.06	III			
60		6445.74	I	10	7708.42	I	40	1853.38	III			
20		6451.62	I	10	7766.55	I	30	1859.12	III		Zr IV	
20		6457.63	I	12	7816.32	I	25	1860.47	III		Ref. 362 — J.R.	
110		6470.21	I	110	7819.35	I	25	1861.77	III			
60	B1	6473.79	Z	35	7822.94	I	75	1864.06	III	Intensity	Wavelength	
11		6484.35	I	40	7826.72	I	30	1865.45	III			
110		6489.64	I	90	7849.35	I	35	1877.00	III		Vacuum	
22		6493.10	I	35	7869.99	I	50	1892.07	III			
50		6503.26	I	14	7876.25	I	65	1914.25	III	15	478.97	IV
50		6506.36	I	16	7882.18	I	75	1921.96	III	60	480.66	IV
50	B1	6508.15	Z	10	7897.98	I	75	1932.54	III	60	497.23	IV
30	B1	6542.90	Z	16	7908.46	I	50	1934.32	III	60	500.22	IV
35		6550.54	I	20	7940.47	I	40	1935.20	III	4	500.34	IV
30		6569.43	I	160	7944.61	I	75	1936.48	III	4	584.65	IV
20		6576.56	I	80	7956.66	I	75	1936.67	III	15	585.42	IV
30	b	6578.06	Z	80	7959.98	I	80	1937.27	III	2	586.42	IV
50		6591.99	I	20	7963.63	I	200	1940.25	III	60	588.89	IV
10		6596.71	I	160	8005.27	I	100	1941.08	III	100	589.74	IV
10		6598.84	I	25	8046.05	I	25	1946.12	III	600	628.66	IV
50		6603.27	I	16	8053.06	I	80	1946.61	III	500	633.56	IV
15		6620.56	I	20	8055.29	I	50	1946.99	III	200	633.63	IV
11		6678.01	II	20	8055.76	I	100	1953.95	III	8	712.49	IV
22		6688.18	I	60	8058.08	I	50	1961.32	III	90	754.39	IV
11		6702.12	I	150	8063.09	I	100	1962.01	III	90	760.15	IV
17		6709.61	I	790	8070.08	I	40	1962.92	III	150	846.40	IV
27		6717.88	I	10	8114.28	I	85	1966.22	III	300	863.65	IV
40		6752.73	I	20	8120.17	I	25	1967.81	III	500	864.59	IV
75		6762.38	I	390	8132.99	I	60	1974.99	III	200	881.30	IV
				20	8152.58	I	40	1983.14	III			

Zirconium (Cont.)

Intensity	Wavelength	
200	882.59	IV
100	1099.76	IV
150	1100.00	IV
9000	1183.97	IV
9000	1201.77	IV
200	1212.71	IV
100	1213.01	IV
10000	1219.86	IV
500	1285.89	IV
500	1290.56	IV
70	1291.70	IV
500	1417.70	IV
500	1440.65	IV
500	1441.06	IV
1000	1469.47	IV
10000	1546.17	IV
10	1596.29	IV
10000	1598.95	IV
150	1605.26	IV
5000	1607.95	IV
90	1609.50	IV
35	1818.06	IV
500	1836.15	IV
500	1846.37	IV
200	1848.03	IV
70	1851.91	IV

Air

20	2045.12	IV
40	2047.15	IV
10000	2091.49	IV
10000	2092.36	IV
10000	2163.68	IV
10000	2286.67	IV
200	2473.75	IV
300	2476.71	IV
5	2572.32	IV
400	2573.66	IV
400	2583.32	IV

Zr V
Ref. 418 — J.R.

Intensity		Wavelength	
		Vacuum	
300		292.19	V
500		304.01	V
300		305.24	V
400	p	368.18	V
200		519.25	V
200		536.50	V
200	p	674.13	V
200		675.53	V
300		679.39	V
200		688.27	V
200	p	703.03	V
300		717.24	V
200		740.33	V
2000		740.61	V
500		742.74	V
300		752.48	V
300		764.58	V
500		766.20	V
200		773.80	V
200		775.58	V
200		779.21	V
400		784.50	V
400		797.23	V
10000		800.00	V
10000		806.89	V
200		809.75	V
10000		812.05	V
500		822.06	V
300	p	823.46	V
200		823.78	V
200		825.13	V
200		830.59	V

400		836.57	V
3000		841.40	V
300	p	852.87	V
700		853.68	V
200		873.11	V
300		885.68	V
300		900.48	V
500		906.66	V
500		915.30	V
400		923.10	V
400		940.41	V
200		949.70	V
200		978.06	V
200		980.70	V
200		984.18	V
300		995.59	V
400		1002.48	V
400		1038.69	V
200		1044.41	V
300		1047.77	V
400		1068.55	V
300		1072.25	V
300		1083.45	V
200		1087.05	V
200		1093.54	V
200		1108.79	V
300		1194.24	V
300		1200.76	V
200		1233.91	V
300		1238.93	V
300		1253.61	V
200		1259.70	V
400		1260.91	V
500		1265.38	V
300		1295.81	V
400		1302.80	V
500		1303.93	V
400		1306.76	V
200		1315.14	V
400		1320.74	V

500	p	1323.81	V
300		1332.06	V
300		1337.34	V
300		1355.21	V
500		1355.98	V
300		1361.39	V
500		1376.54	V
200		1396.79	V
300		1410.03	V
200		1413.40	V
300		1460.05	V
300		1486.90	V
300		1491.33	V
200		1520.47	V
200		1550.12	V
500		1633.03	V
500		1654.46	V
700		1725.02	V
500		1786.20	V
200	p	1790.81	V
500		1806.09	V
500		1860.48	V
600		1860.86	V
500		1878.33	V
400		1926.24	V
500		1927.43	V
400		1934.88	V

Air

500	2009.29	V
600	2028.54	V
600	2132.42	V
200	2150.18	V
200	2336.94	V

REFERENCES

1. Meggers, W. F., Corliss, C. H., and Scribner, B. F., *Natl. Bur. Stand. (U.S.) Monogr.*, 145, Washington, D.C., 1975.
2. Aksenov, V. P. and Ryabtsev, A. N., *Opt. Spectrosc.*, 37, 860, 1970.
3. Andersen, N., Bickel, W. S., Carriveau, G. W., Jensen, K., and Veje, E., *Phys. Scr.*, 4, 113, 1971.
4. Andersson, E. and Johannesson, G. A., *Phys. Scr.*, 3, 203, 1971.
5. Andrew, K. L. and Meissner, K. W., *J. Opt. Soc. Am.*, 49, 146, 1959.
6. Artru, M. C. and Brillet, W. U. L., *J. Opt. Soc. Am.*, 64, 1063, 1974.
7. Artru, M. C. and Kaufman, V., *J. Opt. Soc. Am.*, 62, 949, 1972.
8. Artru, M. C. and Kaufman, V., *J. Opt. Soc. Am.*, 65, 594, 1975.
9. Badami, J. S. and Rao, K. R., *Proc. R. Soc. London*, 140(A), 387, 1933.
10. Baird, K. M. and Smith, D. S., *J. Opt. Soc. Am.*, 48, 300, 1958.
11. Bashkin, S. and Martinson, I., *J. Opt. Soc. Am.*, 61, 1686, 1971.
12. Beacham, J. R., Ph.D. thesis, Purdue University, 1970.
13. Benschop, H., Joshi, Y. N., and van Kleef, T. A. M., *Can. J. Phys.*, 53, 700, 1975.
14. Berry, H. G., Bromander, J., and Buchta, R., *Phys. Scr.*, 1, 181, 1970.
15. Berry, H. G., Bromander, J., Martinson, I., and Buchta, R., *Phys. Scr.*, 3, 63, 1971.
16. Berry, H. G., Desesquelles, J., and Dufay, M., *Phys. Rev. Sect. A.*, 6, 600, 1972.
17. Berry, H. G., Desesquelles, J., and Dufay, M., *Nucl. Instrum. Methods*, 110, 43, 1973.
18. Berry, H. G., Pinnington, E. H., and Subtil, J. L., *J. Opt. Soc. Am.*, 62, 767, 1972.
19. Bidelman, W. P. and Corliss, C. H., *Astrophys. J.*, 135, 968, 1962.
20. Bloch, L. and Bloch, E., *Ann. Phys. (Paris)*, 10(11), 141, 1929.
21. Bloch, L., Bloch, E., and Felici, N., *J. Phys. Radium*, 8, 355, 1937.
22. Bockasten, K., *Ark. Fys.*, 9, 457, 1955.
23. Bockasten, K., Hallin, R., Johansson, K. B., and Tsui, P., *Phys. Lett.* (Netherlands), 8, 181, 1964.
24. Bockasten, K. and Johansson, K. B., *Ark. Fys.*, 38, 563, 1969.
25. Borgstrom, A., *Ark. Fys.*, 38, 243, 1968.
26. Borgstrom, A., *Phys. Scr.*, 3, 157, 1971.
27. Bowen, I. S., *Phys. Rev.*, 29, 231, 1927.
28. Bowen, I. S., *Phys. Rev.*, 31, 34, 1928.
29. Bowen, I. S., *Phys. Rev.*, 39, 8, 1932.
30. Bowen, I. S., *Phys. Rev.*, 45, 401, 1934.
31. Bowen, I. S., *Phys. Rev.*, 46, 377, 1934.
32. Bowen, I. S., *Phys. Rev.*, 46, 791, 1934.
33. Boyce, J. C., *Phys. Rev.*, 49, 730, 1936.
34. Boyce, J. C. and Robinson, H. A., *J. Opt. Soc. Am.*, 26, 133, 1936.
35. Bozman, W. R., Meggers, W. F., and Corliss, C. H., *J. Res. Natl. Bur. Stand. Sect. A*, 71, 547, 1967.
36. Bromander, J., *Ark. Fys.*, 40, 257, 1969.
37. Bromander, J. and Buchta, R., *Phys. Scr.*, 1, 184, 1970.
38. Brown, C. M. and Ginter, M. L., *J. Opt. Soc. Am.*, 68, 243, 1978.
39. Brown, C. M., Tilford, S. G., and Ginter, M. L., *J. Opt. Soc. Am.*, 65, 1404, 1975.
40. Bryant, B. W., *Johns Hopkins Spectroscopic Report* No. 21, 1961.
41. Buchet, J. P., Buchet-Poulizac, M. C., Berry, H. G., and Drake, G. W. F., *Phys. Rev. Sect. A*, 7, 922, 1973.
42. Budhiraja, C. J. and Joshi, Y. N., *Can. J. Phys.*, 49, 391, 1971.

43. Burns, K. and Adams, K. B., *J. Opt. Soc. Am.*, 42, 56, 1952.
44. Burns, K. and Adams, K. B., *J. Opt. Soc. Am.*, 46, 94, 1956.
45. Burns, K., Adams, K. B., and Longwell, J., *J. Opt. Soc. Am.*, 40, 339, 1950.
46. Callahan, W. R., Ph.D. thesis, Johns Hopkins University, 1962.
47. Charles, G. W., *J. Opt. Soc. Am.*, 56, 1292, 1966.
48. Charles, G. W., Hunt, D. J., Pish, G., and Timma, D. L., *J. Opt. Soc. Am.*, 45, 869, 1955.
49. Codling, K., *Proc. Phys. Soc.*, 77, 797, 1961.
50. Comite Consulatif Pour La Definition du Metre, *J. Phys. Chem. Ref. Data*, 3, 852, 1974.
51. Conway, J. G., Blaise, J., and Verges, J., *Spectrochim. Acta Part B*, 31, 31, 1976.
52. Conway, J. G., Worden, E. F., Blaise, J., and Verges, J., *Spectrochim. Acta Part B*, 32, 97, 1977.
53. Conway, J. G., Worden, E. F., Blaise, J., Camus, P., and Verges, J., *Spectrochim. Acta Part B*, 32, 101, 1977.
54. Crooker, A. M., *Can. J. Res. Sect. A*, 14, 115, 1936.
55. Crooker, A. M. and Dick, K. A., *Can. J. Phys.*, 46, 1241, 1968.
56. Crosswhite, H. M., *J. Res. Natl. Bur. Stand. Sect. A*, 79, 17, 1975.
58. Crosswhite, H. M. and Dieke, G. H., *American Institute of Physics Handbook*, Section 7, 1972.
59. de Bruin, T. L., *Z. Phys.*, 38, 94, 1926.
60. de Bruin, T. L., *Z. Phys.*, 53, 658, 1929.
61. de Bruin, T. L., Humphreys, C. J., and Meggers, W. F., *J. Res. Natl. Bur. Stand.*, 11, 409, 1933.
62. Dick, K. A., *J. Opt. Soc. Am.*, 64, 702, 1973.
63. Dobbie, J. C., *Ann. Solar Phys. Observ.* (Cambridge), 5, 1, 1938.
64. Earls, L. T. and Sawyer, R. A., *Phys. Rev.*, 47, 115, 1935.
65. Edlen, B., *Z. Phys.*, 85, 85, 1933.
66. Edlen, B., *Nova Acta Reglae Soc. Sci. Ups.*, (IV) 9, No. 6, 1934.
67. Edlen, B., *Z. Phys.*, 93, 726, 1935.
68. Edlen, B., *Z. Phys.*, 94, 47, 1935.
69. Edlen, B., *Rep. Prog. Phys.*, 26, 181, 1963.
70. Edlen, B. and Risberg, P., *Ark. Fys.*, 10, 553, 1956.
71. Edlen, B. and Swings, P., *Astrophys. J.*, 95, 532, 1942.
72. Ehrhardt, J. C. and Davis, S. P., *J. Opt. Soc. Am.*, 61, 1342, 1971.
73. Eidelsberg, M., *J. Phys. B*, 5, 1031, 1972.
74. Eidelsberg, M., *J. Phys. B*, 7, 1476, 1974.
75. Ekberg, J. O. and Svensson, L. A., *Phys. Scr.*, 2, 283, 1970.
76. Ekefors, E., *Z. Phys.*, 71, 53, 1931.
77. Epstein, G. L. and Reader, J., *J. Opt. Soc. Am.*, 65, 310, 1975.
78. Epstein, G. L. and Reader, J., *J. Opt. Soc. Am.*, 66, 590, 1976.
79. Epstein, G. L. and Reader, J., unpublished.
80. Eriksson, K. B. S., *Phys. Lett. A.*, 41, 97, 1972.
81. Eriksson, K. B. S. and Isberg, H. B. S., *Ark. Fys.*, 23, 527, 1963.
82. Eriksson, K. B. S. and Wenaker, I., *Phys. Scr.*, 1, 21, 1970.
83. Esteva, J. M. and Mehlman, G., *Astrophys. J.*, 193, 747, 1974.
84. Even-Zohar, M. and Fraenkel, B. S., *J. Phys. B*, 5, 1596, 1972.
85. Fawcett, B. C., *J. Phys. B*, 3, 1732, 1970.
86. Fawcett, B. C., Culham Laboratory Report ARU-R4, 1971.
87. Ferner, E., *Ark. Mat. Astron. Fys.*, 28(A), 4, 1941.
88. Fischer, R. A., Knopf, W. C., and Kinney, F. E., *Astrophys. J.*, 130, 683, 1959.
89. Fowler, A., *Report on Series in Line Spectra*, Fleetway Press, London, 1922.
90. Fowles, G. R., *J. Opt. Soc. Am.*, 44, 760, 1954.
91. Fred, M., *Argonne Natl. Lab.*, unpublished, 1977.
92. Fred, M. and Tomkins, F. S., *J. Opt. Soc. Am.*, 47, 1076, 1957.
93. Fred, M., Tomkins, F. S., Blaise, J. E., Camus, P., and Verges, J., Argonne National Laboratory Report No. 76-68, 1976.
94. Garcia, J. D. and Mack, J. E., *J. Opt. Soc. Am.*, 55, 654, 1965.
96. Giacchetti, A., *Argonne Natl. Lab.*, unpublished, 1975.
97. Giacchetti, A., Blaise, J., Corliss, C. H., and Zalubas, R., *J. Res. Natl. Bur. Stand. Sect. A*, 78, 247, 1974.
98. Giacchetti, A., Stanley, R. W., and Zalubas, R., *J. Opt. Soc. Am.*, 69, 474, 1970.
99. Gilbert, W. P., *Phys. Rev.*, 47, 847, 1935.
100. Gilroy, H. T., *Phys. Rev.*, 38, 2217, 1931.
101. Glad, S., *Ark. Fys.*, 10, 291, 1956.
102. Goldsmith, S., *J. Phys. B*, 2, 1075, 1969.
103. Goorvitch, D., Mehlman-Balloffet, G., and Valero, F. P. J., *J. Opt. Soc. Am.*, 60, 1458, 1970.
104. Goorvitch, D. and Valero, F. P. J., *Astrophys. J.*, 171, 643, 1972.
105. Green, L. C., *Phys. Rev.*, 55, 1209, 1939.
106. Gutman, F., *Diss. Abstr. Int. B*, 31, 363, 1970.
107. Hallin, R., *Ark. Fys.*, 31, 511, 1966.
108. Hallin, R., *Ark. Fys.*, 32, 201, 1966.
109. Hansen, J. E. and Persson, W., *J. Opt. Soc. Am.*, 64, 696, 1974.
110. Hansen, J. E. and Persson, W., *Phys. Scr.*, 13, 166, 1976.
111. Hellintin, P., *Phys. Scr.*, 13, 155, 1976.
112. Herzberg, G. and Moore, H. R., *Can. J. Phys.*, 37, 1293, 1959.
113. Hetzler, C. W., Boreman, R. W., and Burns, K., *Phys. Rev.*, 48, 656, 1935.
114. Holmstrom, J. E. and Johansson, L., *Ark. Fys.*, 40, 133, 1969.
115. Hontzeas, S., Martinson, I., Erman, P., and Buchta, R., *Nucl. Instrum. Methods*, 110, 51, 1973.
116. Humphreys, C. J., *J. Res. Natl. Bur. Stand.*, 22, 19, 1939.
117. Humphreys, C. J., *J. Opt. Soc. Am.*, 43, 1027, 1953.
118. Humphreys, C. J., *J. Phys. Chem. Ref. Data*, 2, 519, 1973.
119. Humphreys, C. J. and Andrew, K. L., *J. Opt. Soc. Am.*, 54, 1134, 1964.
120. Humphreys, C. J. and Meggers, W. F., *J. Res. Natl. Bur. Stand.*, 10, 139, 1933.
121. Humphreys, C. J. and Paul, E., Jr., *J. Opt. Soc. Am.*, 60, 200, 1970.
122. Humphreys, C. J. and Paul, E., Jr., *J. Opt. Soc. Am.*, 62, 432, 1972.
123. Humphreys, C. J., Paul, E., Jr., Cowan, R. D., and Andrew, K. L., *J. Opt. Soc. Am.*, 57, 855, 1967.
124. Humphreys, C. J., Paul, E., Jr., and Minnhagen, L., *J. Opt. Soc. Am.*, 61, 110, 1971.
125. Iglesias, L., Inst. of Optics, Madrid, unpublished, 1977.

126. Iglesias, L. and Velasco, R., *Publ. Inst. Opt. Madrid,* No. 23, 1964.
127. Isberg, B., *Ark. Fys.,* 35, 551, 1967.
128. Johannesson, G. A., Lundstrom, T., and Minnhagen, L., *Phys. Scr.,* 6, 129, 1972.
129. Johannesson, G. A. and Lundstrom, T., *Phys. Scr.,* 8, 53, 1973.
130. Johansson, I., *Ark. Fys.,* 20, 135, 1961.
131. Johansson, I. and Contreras, R., *Ark. Fys.,* 37, 513, 1968.
132. Johansson, I. and Litzen, U., *Ark. Fys.,* 34, 573, 1967.
133. Johansson, I. and Svensson, K. F., *Ark. Fys.,* 16, 353, 1960.
134. Johansson, L., *Ark. Fys.,* 20, 489, 1961.
135. Johansson, L., *Ark. Fys.,* 23, 119, 1963.
136. Johansson, S. and Litzen, U., *Phys. Scr.,* 6, 139, 1972.
137. Johansson, S. and Litzen, U., *Phys. Scr.,* 8, 43, 1973.
138. Johansson, S. and Litzen, U., *Phys. Scr.,* 10, 121, 1974.
139. Joshi, Y. N., St. Francis Xavier Univ., Nova Scotia, unpublished.
140. Joshi, Y. N., Bhatia, K. S., and Jones, W. E., *Sci. Light Tokyo,* 21, 113, 1972.
141. Joshi, Y. N., Bhatia, K. S., and Jones, W. E., *Spectrochim. Acta Part B,* 28, 149, 1973.
142. Joshi, Y. N. and Budhiraja, C. J., *Can. J. Phys.,* 49, 670, 1971.
143. Joshi, Y. N. and van Kleef, T. A. M., *Can. J. Phys.,* 52, 1891, 1974.
144. Kaufman, V., *Natl. Bur. Stand.,* unpublished.
145. Kaufman, V., *J. Opt. Soc. Am.,* 52, 866, 1962.
146. Kaufman, V., Artru, M. C., and Brillet, W. U. L., *J. Opt. Soc. Am.,* 64, 197, 1974.
147. Kaufman, V. and Humphreys, C. J., *J. Opt. Soc. Am.,* 59, 1614, 1969.
148. Kaufman, V. and Sugar, J., *J. Opt. Soc. Am.,* 61, 1693, 1971.
149. Kaufman, V. and Sugar, J., *J. Res. Natl. Bur. Stand. Sect. A,* 71, 583, 1967.
150. Kelly, R. L. and Palumbo, L. J., *Naval Research Laboratory Report 7599, Washington, D. C.,* 1973.
151. Kielkopf, J. F., *Univ. of Louisville,* unpublished, 1975.
152. Kielkopf, J. F., *Univ. of Louisville,* unpublished, 1976.
153. Kiess, C. C. and Corliss, C. H., *J. Res. Natl. Bur. Stand. Sect. A,* 63, 1, 1959.
154. Kleiman, H., *J. Opt. Soc. Am.,* 52, 441, 1962.
155. Eriksson, K. B., Johansson, I., and Norlen, G., *Ark. Fys.,* 28, 233, 1964.
156. Klinkenberg, P. F. A., *Physica,* 15, 774, 1949.
157. Klinkenberg, P. F. A., *Physica,* 16, 618, 1950.
158. Krishnamurty, S. G., *Proc. Phys. Soc. London,* 48, 277, 1936.
159. Kruger, P. G. and Gilroy, H. T., *Phys. Rev.,* 48, 720, 1935.
160. Kruger, P. G. and Pattin, H. S., *Phys. Rev.,* 52, 621, 1937.
161. Lacroute, P., *Ann. Phys.* (Paris), 3, 5, 1935.
162. Lang, R. J., *Phys. Rev.,* 30, 762, 1927.
163. Lang, R. J., *Phys. Rev.,* 32, 737, 1928.
164. Lang, R. J., *Phys. Rev.,* 35, 445, 1930.
165. Lang, R. J., *Can. J. Res. Sect. A,* 14, 43, 1936.
166. Lang, R. J., *Can. J. Res. Sect. A,* 14, 127, 1936.
167. Lang, R. J. and Vastine, D. H., *Phys. Rev.,* 42, 233, 1932.
168. Li, H. and Andrew, K. L., *J. Opt. Soc. Am.,* 61, 96, 1971.
169. Liden, K., *Ark. Fys.,* 1, 229, 1949.
170. Litzen, U., *Ark. Fys.,* 28, 239, 1965.
171. Litzen, U., *Phys. Scr.,* 1, 251, 1970.
172. Litzen, U., *Phys. Scr.,* 1, 253, 1970.
173. Litzen, U., *Phys. Scr.,* 2, 103, 1970.
174. Litzen, U. and Verges, J., *Phys. Scr.,* 13, 240, 1976.
175. Lofstrand, B., *Phys. Scr.,* 8, 57, 1973.
176. Luc-Koenig, E., Morillon, C., and Verges, J., *Phys. Scr.,* 12, 199, 1975.
177. Lundstrom, T., *Phys. Scr.,* 7, 62, 1973.
178. Lundstrom, T. and Minnhagen, L., *Phys. Scr.,* 5, 243, 1972.
179. Magnusson, C. E. and Zetterberg, P. O., *Phys. Scr.,* 10, 177, 1974.
180. Magnusson, C. E., and Zetterberg, P. O., *Phys. Scr.,* 15, 237, 1977.
181. Martin, D. C., *Phys. Rev.,* 48, 938, 1935.
182. Svendenius, N., *Phys. Scr.,* 22, 240, 1980.
183. Martin, W. C., *J. Res. Natl. Bur. Stand. Sect. A,* 64, 19, 1960.
184. Martin, W. C. and Corliss, C. H., *J. Res. Natl. Bur. Stand. Sect. A,* 64, 443, 1960.
185. Martin, W. C. and Kaufman, V., *J. Res. Natl. Bur. Stand. Sect. A,* 74, 11, 1970.
186. Martin, W. C. and Kaufman, V., *J. Opt. Soc. Am.,* 60, 1096, 1970.
187. McCormick, W. W. and Sawyer, R. A., *Phys. Rev.,* 54, 71, 1938.
188. McLaughlin, R., *J. Opt. Soc. Am.,* 54, 965, 1964.
189. McLennan, J. C., McLay, A. B., and Crawford, M. F., *Proc. R. Soc. London Ser. A,* 134, 41, 1931.
190. Meissner, K. W., *Z. Phys.,* 39, 172, 1926.
191. Meggers, W. F., *J. Res. Natl. Bur. Stand.,* 24, 153, 1940.
192. Meggers, W. F. and Corliss, C. H., *J. Res. Natl. Bur. Stand. Sect. A,* 70, 63, 1966.
193. Meggers, W. F., Fred, M., and Tomkins, F. S., *J. Res. Natl. Bur. Stand.,* 58, 297, 1957.
194. Meggers, W. F. and Humphreys, C. J., *J. Res. Natl. Bur. Stand.,* 28, 463, 1942.
195. Meggers, W. F. and Murphy, R. J., *J. Res. Natl. Bur. Stand.,* 48, 334, 1952.
196. Meggers, W. F., Scribner, B. F., and Bozman, W. R., *J. Res. Natl. Bur. Stand.,* 46, 85, 1951.
197. Meggers, W. F., Shenstone, A. G., and Moore, C. E., *J. Res. Natl. Bur. Stand.,* 45, 346, 1950.
198. Mehlman, G. and Esteva, J. M., *Astrophys. J.,* 188, 191, 1974.
199. Meinders, E., *Physica,* 84(C), 117, 1976.
200. Sansonetti, C. J., Dissertation, Purdue University, 1981.
201. Sansonetti, C. J., *Natl. Bur. Stand. (U.S.),* unpublished.
202. Millikan, R. A. and Bowen, I. S. *Phys. Rev.,* 25, 600, 1925.
203. Minnhagen, L., *J. Opt. Soc. Am.,* 61, 1257, 1925.
204. Minnhagen, L., *J. Opt. Soc. Am.,* 63, 1185, 1973.
205. Minnhagen, L., *Phys. Scr.,* 11, 38, 1975.
206. Minnhagen, L., *J. Opt. Soc. Am.,* 66, 659, 1976.
207. Minnhagen, L. and Nietsche, H., *Phys. Scr.,* 5, 237, 1972.
208. Minnhagen, L., Strihed, H., and Petersson, B., *Ark. Fys.,* 39, 471, 1969.
209. Moore, C. E., *Natl. Bur. Stand. (U.S.) Circ.,* 488, 1950.
210. Moore, C. E., *Revised Multiplet Table,* Princeton University Observatory No. 20, 1945.

211. **Moore, C. E.,** National Standard Reference Data Series — National Bureau of Standards 3, Sect. 3, 1970.
212. **Moore, C. E.,** National Standard Reference Data Series — National Bureau of Standards 3, Sect. 4, 1971.
213. **Moore, C. E.,** National Standard Reference Data Series — National Bureau of Standards 3, Sect. 5, 1975.
214. **Moore, C. E.,** National Standard Reference Data Series — National Bureau of Standards 3, Sect. 6, 1972.
215. **Moore, C. E.,** *National Standard Reference Data Series — National Bureau of Standards 3, Sect. 7,* 1975.
216. **Morillon, C. and Verges, J.,** *Phys. Scr.,* 10, 227, 1974.
217. **Newsom, G. H.,** *Astrophys. J.,* 166, 243, 1971.
218. **Newsom, G. H., O'Connor, S., and Learner, R. C. M.,** *J. Phys. B,* 6, 2162, 1973.
219. **Norlen, G.,** *Phys. Scr.,* 8, 249, 1973.
220. **Odabasi, H.,** *J. Opt. Soc. Am.,* 57, 1459, 1967.
221. **Olme, A.,** *Ark. Fys.,* 40, 35, 1969.
222. **Olme, A.,** *Phys. Scr.,* 1, 256, 1970.
223. **Johansson, S., and Litzen, U.,** *J. Opt. Soc. Am.,* 61, 1427, 1971.
224. **Palenius, H. P.,** *Ark. Fys.,* 39, 15, 1969.
225. **Palenius, H. P.,** *Phys. Scr.,* 1, 113, 1970.
226. **Palenius, H. P.,** *Univ. of Lund, Sweden,* unpublished.
227. **Paschen, F.,** *Ann. Phys.,* Series 5, 12, 509, 1932.
228. **Paschen, F. and Ritschl, R.,** *Ann. Phys.,* Series 5, 18, 867, 1933.
229. **Peck, E. R., Khanna, B. N., and Anderholm, N. C.** *J. Opt. Soc. Am.,* 52, 53, 1962.
230. **Persson, W.,** *Phys. Scr.,* 3, 133, 1971.
231. **Persson, W. and Valind S.,** *Phys. Scr.,* 5, 187, 1972.
232. **Petersson, B.,** *Ark. Fys.,* 27, 317, 1964.
233. **Phillips, L. W. and Parker, W. L.,** *Phys. Rev.,* 60, 301, 1941.
234. **Platt, J. R. and Sawyer, R. A.,** *Phys. Rev.,* 60, 866, 1941.
235. **Plyer, E. K., Blaine, L. R., and Tidwell, E.,** *J. Res. Natl. Bur. Stand.,* 55, 279, 1955.
236. **Poppe, R., van Kleef, T. A. M., and Raassen, A. J. J.,** *Physica,* 77, 165, 1974.
237. **Radziemski, L. J., Jr. and Andrew, K. L.,** *J. Opt. Soc. Am.,* 55, 474, 1965.
238. **Radziemski, L. J., Jr. and Kaufman, V.,** *J. Opt. Soc. Am.,* 59, 424, 1969.
239. **Radziemski, L. J., Jr. and Kaufman, V.,** *J. Opt. Soc. Am.,* 64, 366, 1974.
240. **Ramanadham, R. and Rao, K. R.,** *Indian J. Phys.,* 18, 317, 1944.
241. **Ramb, R.,** *Ann. Phys.,* 10, 311, 1931.
242. **Rank, D. H., Bennett, J. M., and Bennett, H. E.,** *J. Opt. Soc. Am.,* 40, 477, 1950.
243. **Rao, A. S. and Krishnamurty, S. G.,** *Proc. Phys. Soc. London,* 46, 531, 1943.
244. **Rao, K. R.,** *Proc. R. Soc. London, Ser. A,* 134, 604, 1932.
245. **Rao, K. R. and Badami, J. S.,** *Proc. R. Soc. London Ser. A,* 131, 154, 1931.
246. **Rao, K. R. and Krishnamurty, S. G.,** *Proc. R. Soc. London Ser. A,* 161, 38, 1937.
247. **Rao, K. R. and Murti, S. G. K.,** *Proc. R. Soc. London Ser. A,* 145, 681, 1934.
248. **Rao, Y. B.,** *Indian J. Phys.,* 32, 497, 1958.
249. **Rao, Y. B.,** *Indian J. Phys.,* 33, 546, 1959.
250. **Rao, Y. B.,** *Indian J. Phys.,* 35, 386, 1961.
251. **Rasmussen, E.,** *Z. Phys.,* 80, 726, 1933.
252. **Rasmussen, E.,** *Z. Phys.,* 83, 404, 1933.
253. **Rasmussen, E.,** *Z. Phys.,* 86, 24, 1934.
254. **Rasmussen, E.,** *Z. Phys.,* 87, 607, 1934.
255. **Rasmussen, E.,** *Phys. Rev.,* 57, 840, 1940.
256. **Rau, A. S. and Narayan, A. L.,** *Z. Phys.,* 59, 687, 1930.
257. **Reader, J.,** *J. Opt. Soc. Am.,* 65, 286, 1975.
258. **Reader, J.,** *J. Opt. Soc. Am.,* 65, 988, 1975.
259. **Reader, J.,** *J. Opt. Soc. Am.,* 73, 1983, (in press).
260. **Reader, J. and Davis, S.,** *J. Res. Natl. Bur. Stand. Sect. A,* 71, 587, 1967, and unpublished.
261. **Reader, J. and Ekberg, J. O.,** *J. Opt. Soc. Am.,* 62, 464, 1972.
262. **Reader, J. and Epstein, G. L.,** *J. Opt. Soc. Am.,* 62, 1467, 1972.
263. **Reader, J. and Epstein, G. L.,** *J. Opt. Soc. Am.,* 65, 638, 1975.
264. **Reader, J. and Epstein, G. L.,** *Natl. Bur. Stand.,* unpublished.
265. **Reader, J., Epstein, G. L., and Ekberg, J. O.,** *J. Opt. Soc. Am.,* 62, 273, 1972.
266. **Kaufman, V.,** *Phys. Scr.,* 26, 439, 1982.
267. **Ricard, R., Givord, M., and George, F.,** *C. R. Acad. Sci. Paris,* 205, 1229, 1937.
268. **Risberg, P.,** *Ark. Fys.,* 10, 583, 1956.
269. **Risberg, G.,** *Ark. Fys.,* 28, 381, 1965.
270. **Risberg, G.,** *Ark. Fys.,* 37, 231, 1968.
271. **Robinson, H. A.,** *Phys. Rev.,* 49, 297, 1936.
272. **Robinson, H. A.,** *Phys. Rev.,* 50, 99, 1936.
273. **Ross, C. B., Jr.,** Doctoral dissertation, Purdue University, 1969.
274. **Ross, C. B., Wood, D. R., and Scholl, P. S.,** *J. Opt. Soc. Am.,* 66, 36, 1976.
275. **Ruedy, J. E. and Gibbs, R. C.,** *Phys. Rev.,* 46, 880, 1934.
276. **Russell, H. N., King, R. B., and Moore, C. E.,** *Phys. Rev.,* 58, 407, 1940.
277. **Russell, H. N. and Moore, C. E.,** *J. Res. Natl. Bur. Stand.,* 55, 299, 1955.
278. **Russell, H. N., Moore, C. E., and Weeks, D. W.,** *Trans. Am. Philos. Soc.,* 34(2), 111, 1944.
279. **Saunders, F., Schneider, E., and Buckingham, E.,** *Proc. Natl. Acad. Sci.,* 20, 291, 1934.
280. **Sawyer, R. A. and Humphreys, C. J.,** *Phys. Rev.,* 32, 583, 1928.
281. **Sawyer, R. A. and Lang, R. J.,** *Phys. Rev.,* 34, 712, 1929.
282. **Sawyer, R. A. and Paschen, F.** *Ann. Phys.,* 84(4), 1, 1927.
283. **Scholl, P. S.,** M.S. thesis, Wright State Univ., 1975.
284. **Schurmann, D.,** *Z. Phys.,* 17, 4, 1975.
285. **Seguier, J.,** *C. R. Acad. Sci. Paris,* 256, 1703, 1963.
286. **Shenstone, A. G.,** *Phys. Rev.,* 31, 317, 1928.
287. **Shenstone, A. G.,** *Phys. Rev.,* 32, 30, 1928.
288. **Shenstone, A. G.,** *Trans. R. Soc. London,* 237(A), 57, 1938.
289. **Shenstone, A. G.,** *Phys. Rev.,* 57, 894, 1940.
290. **Shenstone, A. G.** *Philos. Trans. R. Soc. London Ser. A,* 241, 297, 1948.
291. **Shenstone, A. G.,** *Can. J. Phys.,* 38, 677, 1960.
292. **Shenstone, A. G.,** *Proc. R. Soc. London,* 261(A), 153, 1961.

293. Shenstone, A. G., *Proc. R. Soc. London*, 276(A), 293, 1963.
294. Shenstone, A. G., *J. Res. Natl. Bur. Stand. Sect. A*, 74, 801, 1970.
295. Shenstone, A. G., *J. Res. Natl. Bur. Stand. Sect. A*, 79, 497, 1975.
296. Shenstone, A. G. and Pittenger, J. T., *J. Opt. Soc. Am.*, 39, 219, 1949.
297. Smith, S., *Phys. Rev.*, 36, 1, 1930.
298. Smitt, R., *Phys. Scr.*, 8, 292, 1973.
299. Soderqvist, J., *Ark. Mat. Astronom. Fys.*, 32(A), 1, 1946.
300. Sommer, L. A., *Ann. Phys.*, 75, 163, 1924.
301. Spector, N., *J. Opt. Soc. Am.*, 63, 358, 1973.
302. Spector, N. and Sugar, J., *J. Opt. Soc. Am.*, 66, 436, 1976.
303. Steinhaus, D. W., Radziemski, L. J., Jr., and Blaise, J., *Los Alamos Sci. Lab.*, unpublished, 1975.
304. Subbaraya, T. S., *Z. Phys.*, 78, 541, 1932.
305. Sugar, J., *J. Opt. Soc. Am.*, 55, 33, 1965.
306. Sugar, J., *J. Res. Natl. Bur. Stand. Sect. A*, 73, 333, 1969.
307. Sugar, J., *J. Opt. Soc. Am.*, 60, 454, 1970.
308. Sugar, J., *J. Res. Natl. Bur. Stand. Sect. A*, 78, 555, 1974.
309. Sugar, J. and Kaufman, V., *J. Opt. Soc. Am.*, 55, 1283, 1965.
310. Sugar, J. and Kaufman, V., *J. Opt. Soc. Am.*, 62, 562, 1972.
311. Sugar, J., Kaufman, V., and Spector, N., *J. Res. Natl. Bur. Stand.*, Sect. A, 83, 233, 1978.
312. Sugar, J. and Spector, N., *J. Opt. Soc. Am.*, 64, 1484, 1974.
313. Sullivan, F. J. *Univ. Pittsburgh Bull.*, 35, 1, 1938.
314. Svensson, L. A. and Ekberg, J. O., *Ark. Fys.*, 37, 65, 1968.
315. Swensson, J. W. and Risberg, G., *Ark. Fys.*, 31, 237, 1966.
316. Tech. J. L., *J. Res. Natl. Bur. Stand. Sect. A*, 67, 505, 1963.
317. Tech, J. L. and Ward, J. F., *Phys. Rev. Lett.*, 27, 367, 1971.
318. Tilford, S. G., *J. Opt. Soc. Am.*, 53, 1051, 1963.
319. Toresson, Y. G., *Ark. Fys.*, 17, 179, 1960.
320. Toresson, Y. G., *Ark. Fys.*, 18, 389, 1960.
321. Toresson, Y. G. and Edlen, B., *Ark. Fys.*, 23, 117, 1963.
322. Tsien, W. Z., *Chin. J. Phys.*, Peiping, 3, 117, 1939.
323. van Deurzen, C. H. H., Conway, J., and Davis, S. P., *J. Opt. Soc. Am.*, 63, 158, 1973.
324. van Kleef, T. A. M., Raassen, A. J. J., and Joshi, Y. N., *Physica*, 84(C), 401, 1976.
325. Sansonetti, C. J., Andrew, K. L., and Verges, J., *J. Opt. Soc. Am.*, 71, 423, 1981.
326. Wheatley, M. A. and Sawyer, R. A., *Phys. Rev.*, 61, 591, 1942.
327. Wilkinson, P. G., *J. Opt. Soc. Am.*, 45, 862, 1955.
328. Wilkinson, P. G. and Andrew, K. L., *J. Opt. Soc. Am.*, 53, 710, 1963.
329. Wood, D. and Andrew, K. L., *J. Opt. Soc. Am.*, 58, 818, 1968.
330. Wood, D. R., Ross, C. B., Scholl, P. S., and Hoke, M., *J. Opt. Soc. Am.*, 64, 1159, 1974.
331. Worden, E. F. and Conway, J. G., *Lawrence Livermore Lab.*, unpublished, 1977.
332. Worden, E. F., Hulet, E. K., Gutmacher, R. G., Conway, J. G., *At. Data Nucl. Data Tables*, 18, 459, 1976.
333. Worden, E. F., Lougheed, R. W., Gutmacher, R. G., and Conway, J. G., *J. Opt. Soc. Am.*, 64, 77, 1974.
334. Wu, C. M., Ph.D. thesis, University of British Columbia, 1971.
335. Zaidel, A. N., Prokofev, V. K., Raiskii, S. M., Slavnyi, V. A., and Schreider, E. Y., *Tables of Spectral Lines*, 3rd ed., Plenum, New York, 1970.
336. Zetterberg, P. O. and Magnusson, C. E., *Phys. Scr.*, 15, 189, 1977.
337. Sugar, J., *J. Opt. Soc. Am.*, 55, 1058, 1965.
338. Sugar, J., *J. Opt. Soc. Am.*, 61, 727, 1971.
339. Worden, E. F., and Conway, J. G., *At. Data Nucl. Data Tables*, 22, 329, 1978.
340. Kaufman, V. and Edlen, B., *J. Phys. Chem. Ref. Data*, 3, 825, 1974.
341. Lang, R. J., *Phys. Rev.*, 34, 697, 1929.
342. Ryabtsev, A. N., *Opt. Spectros.*, 39, 455, 1975.
343. Foster, E. W., *Proc. R. Soc. London*, 200(A), 429, 1950.
344. Morillon, C. and Verges, J., *Phys. Scr.*, 12, 129, 1975.
345. Ruedy, J. E., *Phys. Rev.*, 41, 588, 1932.
346. McLennan, J. C., McLay, A. B., and McLeod, J. H., *Philos. Mag.*, 4, 486, 1927.
347. Handrup, M. B. and Mack, J. E., *Physica*, 30, 1245, 1964.
348. Clearman, H. E., *J. Opt. Soc. Am.*, 42, 373, 1952.
349. Paschen, F., *Ann. Physik*, 424, 148, 1938.
350. Paschen, F. and Campbell, J. S., *Ann. Phys.*, 31(5), 29, 1938.
351. Nodwell, R., *Univ. of British Columbia, Vancouver*, unpublished, 1955.
352. Gibbs, R. C. and White, H. E., *Phys. Rev.*, 31, 776, 1928.
353. Green, M., *Phys. Rev.*, 60, 117, 1941.
354. Ellis, C. B. and Sawyer, R. A., *Phys. Rev.*, 49, 145, 1936.
355. McLennan, J. C., McLay, A. B., and Crawford, M. F., *Proc. R. Soc. London Ser. A*, 125, 50, 1929.
356. Mack, J. E. and Fromer, M., *Phys. Rev.*, 48, 346, 1935.
357. Humphreys, C. J. and Paul, E., *U.S. Nav. Ord. Lab.*, Navord Rep. 4589, 25, 1956.
358. Walters, F. M., *Sci. Pap. Bur. Stand.*, 17, 161, 1921.
359. Crawford, M. F. and McLay, A. B., *Proc. R. Soc. London Ser. A*, 143, 540, 1934.
360. McLay, A. B. and Crawford, M. F., *Phys. Rev.*, 44, 986, 1933.
361. Schoepfle, G. K., *Phys. Rev.*, 47, 232, 1935.
362. Acquista, N., and Reader, J., *J. Opt. Soc. Am.*, 70, 789, 1980.
363. Benschop, H., Joshi, Y. N., and Van Kleef, T. A. M., *Can. J. Phys.*, 53, 498, 1975.
364. Bockasten, K., Hallin, R., and Hughes, T. P., *Proc. Phys. Soc.*, 81, 522, 1963.
365. Boyce, J. C., *Phys. Rev.*, 46, 378, 1934.
366. Boyce, J. C., *Phys. Rev.*, 47, 718, 1935.
367. Boyce, J. C., *Phys. Rev.*, 48, 396, 1935.
368. Boyce, J. C., *Phys. Rev.*, 49, 351, 1936.
369. Corliss, C. H. and Meggers, W. F., *J. Res. Natl. Bur. Stand.*, 61, 269, 1958.
370. Crooker, A. M. and Dick, K. A., *Can. J. Phys.*, 42, 766, 1964.
371. De Bruin, T. L., *Z. Physik*, 77, 505, 1932.
372. De Bruin, T. L., *Proc. Roy. Acad. Amsterdam*, 36, 727, 1933.
373. De Bruin, T. L., *Zeeman Verhandelingen*, (The Hague), 1935, p. 415.
374. De Bruin, T. L., *Physica*, 3, 809, 1936.
375. De Bruin, T. L., *Proc. Roy. Acad. Amsterdam*, 40, 339, 1937.

376. Dick, K. A., *Can. J. Phys.*, 46, 1291, 1968.
377. Dick, K. A., unpublished, 1978.
378. Edlen, B. and Swensson, J. W., *Phys. Scr.*, 12, 21, 1975.
379. Ekberg, J. O., *Phys. Scr.*, 7, 55, 1973.
380. Ekberg, J. O., *Phys. Scr.*, 7, 59, 1973.
381. Ekberg, J. O., *Phys. Scr.*, 12, 42, 1975.
382. Ekberg, J. O. and Edlen, B., *Phys. Scr.*, 18, 107, 1978.
383. Eliason, A. Y., *Phys. Rev.*, 43, 745, 1933.
384. Gallardo, M., Massone, C. A., Tagliaferri, A. A., Garavaglia, M., and Persson, W., *Phys. Scr.*, to be published, 1979.
385. Garcia-Riquelme, O., *Optica Pura Y Aplicada*, 1, 53, 1968.
386. Gibbs, R. C., Vieweg, A. M., and Gartlein, C. W., *Phys. Rev.*, 34, 406, 1929.
387. Gilbert, W. P., *Phys. Rev.*, 48, 338, 1935.
388. Goldsmith, S. and Kaufman, A. S., *Proc. Phys. Soc.*, 81, 544, 1963.
389. Hermansdorfer, H., *J. Opt. Soc. Am.*, 62, 1149, 1972.
390. Humphreys, C. J., *Phys. Rev.*, 47, 712, 1935.
391. Humphreys, C. J., *J. Res. Natl. Bur. Stand.*, 16, 639, 1936.
392. Iglesias, L., *J. Opt. Soc. Am.*, 45, 856, 1955.
393. Iglesias, L., *J. Res. Natl. Bur. Stand.*, 64A, 481, 1960.
394. Iglesias, L., *Anales Fisica Y Quimica*, 58A, 191, 1962.
395. Iglesias, L., *J. Res. Natl. Bur. Stand.*, 70A, 465, 1966.
396. Iglesias, L., *Can. J. Phys.*, 44, 895, 1966.
397. Iglesias, L., *J. Res. Natl. Bur. Stand.*, 72A, 295, 1968.
398. Joshi, Y. N., *Can. Spectrosc.*, 15, 96, 1970.
399. Joshi, Y. N. and Van Kleef, T. A. M., *Can. J. Phys.*, 55, 714, 1977.
400. Kaufman, A. S., Hughes, T. P., and Williams, R. V., *Proc. Phys. Soc.*, 76, 17, 1960.
401. Kaufman, V. and Sugar, J., *J. Opt. Soc. Am.*, 68, 1529, 1978.
402. Keussler, V., *Z. Physik*, 85, 1, 1933.
403. Kiess, C. C., *J. Res. Natl. Bur. Stand.*, 56, 167, 1956.
404. Klinkenberg, P. F. A., Van Kleef, T. A. M., and Noorman, P. E., *Physica*, 27, 1177, 1961.
405. Kovalev, V. I., Romanos, A. A., and Ryabtsev, A. N., *Opt. Spectrosc.*, 43, 10, 1977.
406. Lang, R. J., *Proc. Natl. Acad. Sci.*, 13, 341, 1927.
407. Lang, R. J., *Zeeman Verhandelingen*, (The Hague), 44, 1935.
408. Liberman, S., et al., *C. R. Acad. Sci. (Paris)*, 286, 253, 1978.
409. Livingston, A. E., *J. Phys.*, B9, L215, 1976.
410. Meijer, F. G., *Physica*, 72, 431, 1974.
411. Meijer, F. G. and Metsch, B. C., *Physica*, 94C, 259, 1978.
412. Moore, F. L., thesis, Princeton, 1949.
413. Paul, F. W. and Polster, H. D., *Phys. Rev.*, 59, 424, 1941.
414. Phillips, L. W. and Parker, W. L., *Phys. Rev.*, 60, 301, 1941.
415. Poppe, R., *Physica*, 81C, 351, 1976.
416. Raassen, A. J. J., Van Kleef, T. A. M., and Metsch, B. C., *Physica*, 84C, 133, 1976.
417. Rao, A. B. and Krishnamurty, S. G., *Proc. Phys. Soc. (London)*, 51, 772, 1939.
418. Reader, J. and Acquista, N., *J. Opt. Soc. Am.*, 69, 239, 1979.
419. Reader, J. and Epstein, G. L., *J. Opt. Soc. Am.*, 62, 619, 1972.
420. Rico, F. R., *Anales, Real Soc. Esp. Fis. Quim.*, 61, 103, 1965.
421. Schonheit, E., *Optik*, 23, 409, 1966.
422. Shenstone, A. G., *J. Opt. Soc. Am.*, 44, 749, 1954.
423. Shenstone, A. G., unpublished, 1958.
424. Shenstone, A. G., *J. Res. Natl. Bur. Stand.*, 67A, 87, 1963.
425. Sugar, J. and Kaufman, V., *J. Opt. Soc. Am.*, 64, 1656, 1974.
426. Sugar, J. and Kaufman, V., *Phys. Rev.*, C12, 1336, 1975.
427. Svensson, L. A., *Phys. Scr.*, 13, 235, 1976.
428. Swensson, J. W. and Edlen, B., *Phys. Scr.*, 9, 335, 1974.
429. Tagliaferri, A. A., Gallego Lluesma, E., Garavaglia, M., Gallardo, M., and Massone, C. A., *Optica Pura Y Aplica*, 7, 89.
430. Tilford, S. G. and Giddings, L. E., *Astrophys. J.*, 141, 1222, 1965.
431. Trawick, M. W., *Phys. Rev.*, 46, 63, 1934.
432. Van Deurzen, C. H. H., *J. Opt. Soc. Am.*, 67, 476, 1977.
433. Yarosewick, S. L. and Moore, F. L., *J. Opt. Soc. Am.*, 57, 1381, 1967.
434. Zalubas, R., unpublished, 1979.

ATOMIC TRANSITION PROBABILITIES

Compiled by W. L. Wiese and G. A. Martin

These tables were prepared under the auspices of the Committee on Line Spectra of the Elements of the National Academy of Sciences — National Research Council. They contain critically evaluated atomic transition probabilities for about 5000 selected lines of all elements for which reliable data are available on an absolute scale. The material is largely for neutral and singly ionized spectra, but includes a number of prominent lines of more highly charged ions.

Many of the data are obtained from comprehensive compilations of the National Bureau of Standards Data Center on Atomic Transition Probabilities. Specifically, data have been taken from critical compilations on V,[1] Cr,[1] Mn,[1] Fe,[2] Co,[2] and Ni[2] without changes. Material from earlier compilations for the elements H through Ne,[3] Na through Ca,[4] and Sc and Ti[5] was supplemented by more recent material taken directly from the original literature. For the higher ions, many data were derived from studies of the systematic behavior of transition probabilities.[6-8] The original literature is cited in a recent bibliography;[9] for lack of space, individual literature references are not cited here.

The wavelength range for the neutral species has normally been restricted to the visible or shorter wavelengths; only the very prominent near infrared lines are included. For the higher ions, most of the strong lines are located in the far UV. The tabulation is limited to electric dipole — including intercombination — lines and comprises essentially the fairly strong transitions with estimated uncertainties of 50% or less. With the exception of hydrogen, helium, and the alkalis, most transitions are between states with low principal quantum numbers.

The transition probability, A, is given in units of 10^8 s^{-1} and is listed to as many digits as is consistent with the indicated accuracy. A number in parentheses following the tabulated value of the transition probability indicates the power of 10 by which this value has to be multiplied. The estimated uncertainties of the A-values are indicated by code letters as follows: AA — for uncertainties within 1%; A — within 3%; B — within 10%; C — within 25%; D — within 50%.

Each transition is identified by the wavelength, λ, in angstroms; the energy level of the upper atomic state, E_k, in cm^{-1}; and the statistical weights, g_i and g_k, of the lower (i) and upper (k) states [the product g_kA (or g_if) is needed in many applications]. Whenever the wavelengths of individual lines within a multiplet are extremely close, an average wavelength for the multiplet is given, and is indicated by an asterisk (*) to the right of the g_k value. This has also been done when the transition probability for an entire multiplet has been taken from the literature and values for individual lines cannot be determined because of insufficient knowledge of the coupling of electrons. Wavelength and energy level data have been taken either from recent compilations or from the original literature cited in bibliographies published by the National Bureau of Standards Atomic Energy Levels Data Center.[10,11] Wavelength values are consistent with those given in Section E of this Handbook.

In the table for hydrogen, the energy level and uncertainty columns have been eliminated since the transition probabilities and energy levels are known very precisely for this element. Because of the hydrogen degeneracy, a "transition" is actually the sum of all transitions between the principal quantum numbers listed in the transition column, and the tabulation represents the properly weighted A values.

In addition to the transition probability A, the atomic oscillator-strength f and the line-strength S are often used in the literature. The conversion factors between these quantities are:

$$g_i f = 1.499 \times 10^{-8} \lambda^2 g_k A = 303.8 \lambda^{-1} S$$

where λ is in angstroms, A is in 10^8 s^{-1}, and S is in atomic units, which are $a_o^2 e^2 = 7.188 \times 10^{-59}$ m^{2}C^2 for electric dipole transitions.

We acknowledge the valuable preparatory work by D. Trahan, W. Croom, and F. Farley in arranging and compiling the numerical data.

Table 1a
TRANSITION PROBABILITIES FOR ALLOWED LINES OF HYDROGEN

Wavelength λ [Å]	Transition	Statistical weights g_i	g_k	Average transition probability A [10^8 s^{-1}]	Wavelength λ[Å]	Transition	Statistical weights g_i	g_k	Average transition probability A [10^8 s^{-1}]
914.039	1—20	2	800	3.928(−5)	8413.32	3—19	18	722	1.964(−5)
914.286	1—19	2	722	5.077(−5)	8437.96	3—18	18	648	2.580(−5)
914.576	1—18	2	648	6.654(−5)	8467.26	3—17	18	578	3.444(−5)
914.919	1—17	2	578	8.858(−5)	8502.49	3—16	18	512	4.680(−5)
915.329	1—16	2	512	1.200(−4)	8545.39	3—15	18	450	6.490(−5)
915.824	1—15	2	450	1.657(−4)	8598.40	3—14	18	392	9.211(−5)
916.429	1—14	2	392	2.341(−4)	8665.02	3—13	18	338	1.343(−4)
917.181	1—13	2	338	3.393(−4)	8750.48	3—12	18	288	2.021(−4)
918.129	1—12	2	288	5.066(−4)	8862.79	3—11	18	242	3.156(−4)
919.351	1—11	2	242	7.834(−4)	9014.91	3—10	18	200	5.156(−4)
920.963	1—10	2	200	1.263(−3)	9229.02	3— 9	18	162	8.905(−4)
923.150	1— 9	2	162	2.143(−3)	9545.97	3— 8(Pε)	18	128	1.651(−3)
926.226	1— 8	2	128	3.869(−3)	10049.4	3— 7(Pδ)	18	98	3.358(−3)
930.748	1— 7	2	98	7.568(−3)	10938.1	3— 6(Pγ)	18	72	7.783(−3)
937.803	1— 6(Lε)	2	72	1.644(−2)	12818.1	3— 5(Pβ)	18	50	2.201(−2)
949.743	1—5 (Lδ)	2	50	4.125(−2)	16407.2	4—12	32	288	1.620(−4)
972.537	1— 4(Lγ)	2	32	1.278(−1)	16806.5	4—11	32	242	2.556(−4)
1025.72	1— 3(Lβ)	2	18	5.575(−1)	17362.1	4—10	32	200	4.235(−4)
1215.67	1— 2(Lα)	2	8	4.699	18174.1	4— 9	32	162	7.459(−4)
3682.81	2—20	8	800	2.172(−5)	18751.0	3— 4(Pα)	18	32	8.986(−2)
3686.83	2—19	8	722	2.809(−5)	19445.6	4— 8	32	128	1.424(−3)
3691.55	2—18	8	648	3.685(−5)	21655.3	4— 7	32	98	3.041(−3)
3697.15	2—17	8	578	4.910(−5)	26251.5	4— 6	32	72	7.711(−3)
3703.85	2—16	8	512	6.658(−5)	27575	5—12	50	288	1.402(−4)
3711.97	2—15	8	450	9.210(−5)	28722	5—11	50	242	2.246(−4)
3721.94	2—14	8	392	1.303(−4)	30384	5—10	50	200	3.800(−4)
3734.37	2—13	8	338	1.893(−4)	32961	5— 9	50	162	6.908(−4)
3750.15	2—12	8	288	2.834(−4)	37395	5— 8	50	128	1.388(−3)
3770.63	2—11	8	242	4.397(−4)	40511.5	4— 5	32	50	2.699(−2)
3797.90	2—10	8	200	7.122(−4)	43753	6—12	72	288	1.288(−4)
3835.38	2— 9	8	162	1.216(−3)	46525	5— 7	50	98	3.253(−3)
3889.05	2— 8	8	128	2.215(−3)	46712	6—11	72	242	2.110(−4)
3970.07	2— 7(Hε)	8	98	4.389(−3)	51273	6—10	72	200	3.688(−4)
4101.73	2— 6(Hδ)	8	72	9.732(−3)	59066	6— 9	72	162	7.065(−4)
4340.46	2— 5(Hγ)	8	50	2.530(−2)	74578	5— 6	50	72	1.025(−2)
4861.32	2— 4(Hβ)	8	32	8.419(−2)	75004	6— 8	72	128	1.561(−3)
6562.80	2— 3(Hα)	8	18	4.410(−1)	123680	6— 7	72	98	4.561(−3)
8392.40	3—20	18	800	1.517(−5)					

For hydrogen-like ions of nuclear charge Z, the following scaling laws hold:

$$A_z = Z^4 A_{Hydrogen}; \quad f_z = f_H; \quad S_z = Z^{-2} S_H$$
(For wavelengths $\lambda_z = Z^{-2} \lambda_H$).

For very highly charged ions, relativistic effects need to be taken into account.[12]

Table 1b
TRANSITION PROBABILITIES FOR SELECTED ATOMIC AND IONIC SPECIES

Wavelength (λ[Å])	Upper energy level (Ek[cm⁻¹])	Statistical weights gi	gk	Transition Probability (A[10⁸ s⁻¹])	Uncertainty
Aluminum					
Al I					
2263.5	44166	2	4	.66	C
2269.1	44169	4	6	.79	C
2269.2	44166	4	4	.13	C
2367.1	42234	2	4	.72	C
2373.1	42238	4	6	.86	C
2373.4	42234	4	4	.14	C
2568.0	38929	2	4	.23	C
2575.1	38934	4	6	.28	C
2575.4	38929	4	4	.044	C
2652.5	37689	2	2	.133	C
2660.4	37689	4	2	.264	C
3082.2	32435	2	4	.63	C
3092.7	32437	4	6	.74	C
3092.8	32435	4	4	.12	C
3944.0	25348	2	2	.493	C
3961.5	25348	4	2	.98	C
6696.0	40278	2	4	.0169	C
6698.7	40272	2	2	.0169	C
7835.3	45195	4	6	.057	D
7836.1	45195	6	8	.062	D
Al II					
1047.9	132823	1	3	.36	D
1048.6	132823	3	5	.48	D
1539.8	124794	3	5	8.8	D
1670.8	59852	1	3	14.6	B
1719.4	95551	1	3	6.79	B
1764.0	94269	5	5	9.8	C
1772.8	150493	1	3	9.5	D
1777.0	150544	5	7	17.	D
1819.0	150525	15	15*	5.6	D
1855.9	91275	1	3	.772	B
1858.0	91275	3	3	2.48	B
1862.3	91275	5	3	4.12	B
1931.0	111637	3	1	10.8	C
1990.5	110090	3	5	14.7	C
2816.2	95351	3	1	3.83	C
4663.1	106921	5	3	.53	C
6226.2	121484	1	3	.62	D
6231.8	121484	3	5	.84	D
6243.4	121484	5	7	1.1	D
6335.7	125869	5	3	.14	D
6823.4	120093	3	3	.34	C
6837.1	120093	5	3	.57	D
6920.3	121367	3	1	.96	C
7042.1	105471	3	5	.59	C
7056.7	105442	3	3	.58	C
7471.4	123471	5	7	.94	D
Al III					
560.36	178458	2	6*	.40	D
695.83	143714	2	4	.74	C
696.22	143633	2	2	.72	C
1352.8	189876	10	14*	4.40	C
1379.7	126164	2	2	4.59	C
1384.1	126164	4	2	9.1	C
1605.8	115959	2	4	12.2	B
1611.8	115959	4	4	2.42	B
1611.9	115956	4	6	14.5	B
1854.7	53917	2	4	5.40	B
1862.8	53683	2	2	5.33	B
1935.9	167613	10	14*	12.2	C
3601.6	143714	6	4	1.34	C
3601.9	143714	4	4	.149	C
3612.4	143633	4	2	1.5	C
Al X					
39.925	2504700	1	3	2220.	C
51.979	1923850	1	3	4800.	C
55.227	1965860	1	3	5200.	D
55.272	1966030	3	5	7200.	D
55.376	1966270	5	7	9500.	D
59.107	1992340	3	5	4600.	C
332.78	300490	1	3	56.	C
394.83	553783	3	1	83.	C
395.36	409690	3	5	12.	C
397.76	406517	1	3	17.	C
400.43	406517	3	3	13.	C
401.12	409690	5	5	36.	C
403.55	404574	3	1	49.	C

Wavelength (λ[Å])	Upper energy level (Ek[cm⁻¹])	Statistical weights gi	gk	Transition Probability (A[10⁸ s⁻¹])	Uncertainty
Al X					
406.31	406517	5	3	19.	C
670.06	449732	3	5	9.8	C
2535.	1923850	1	3	.38	D
Al XI					
36.675	2726700	2	6*	1500.	C
39.091	2734100	2	4	2600.	C
39.180	2734500	4	6	3100.	C
39.530	2705700	2	2	180.	D
39.623	2705700	4	2	370.	D
48.298	2070520	2	4	3090.	B
48.338	2068770	2	2	3080.	B
52.299	2088100	2	4	8100.	C
52.446	2088530	4	6	9600.	C
52.458	2088100	4	4	1600.	C
54.217	2020450	2	2	480.	C
54.388	2020450	4	2	960.	C
99.083	3029700	2	6*	220.	C
103.6	3033700	2	4	420.	C
103.8	3033700	4	6	500.	C
141.6	2726700	2	6*	407.	C
150.31	2734100	2	4	850.	C
150.61	2734500	4	6	990.	C
157.0	2705700	2	2	130.	C
157.4	2705700	4	2	260.	C
205.0	3193600	2	6*	63.	C
308.6	3029700	2	6*	99.	C
341.3	3019700	6	2*	130.	C
550.05	181802	2	4	0.55	D
568.12	176019	2	2	7.73	B
1997.	2070520	2	4	1.07	C
2069.	2068770	2	2	.97	C
4761.	2726700	2	6*	.233	C
5172.	2088100	2	4	.0395	C
5551.	2088530	4	6	.0385	C
5687.	2088100	4	4	.0060	D
Argon					
Ar I					
3554.3	121271	5	5	.0029	D
3567.7	121165	5	7	.0012	D
3606.5	121470	3	1	.0081	D
3649.8	122791	3	1	.0085	D
3834.7	121470	3	1	.0080	D
3949.0	118460	3	5	.00467	C
4044.4	118469	3	5	.00346	C
4158.6	117184	5	5	.0145	C
4164.2	117151	5	3	.00295	C
4181.9	118460	1	3	.0058	C
4190.7	116999	5	5	.00254	C
4191.0	118407	1	3	.0056	C
4198.3	117563	3	1	.0276	C
4200.7	116943	5	7	.0103	C
4259.4	118871	3	1	.0415	C
4266.3	117184	3	5	.00333	C
4272.2	117151	3	3	.0084	C
4300.1	116999	3	5	.00394	C
4333.6	118469	3	5	.0060	C
4335.3	118460	3	3	.00387	C
4345.2	118407	3	3	.00313	C
4510.7	117563	3	1	.0123	C
4768.7	125066	3	5	.0090	D
4836.7	124772	3	5	.00106	C
4876.3	124604	3	5	.0081	D
4887.9	124555	3	3	.014	D
4894.7	124527	3	1	.019	D
5048.8	123903	3	5	.0048	D
5054.2	123882	3	3	.0047	D
5056.5	123873	3	1	.0059	D
5118.2	125150	5	7	.0028	D
5151.4	123509	3	1	.0249	C
5152.3	123505	3	5	.0011	D
5162.3	123468	3	3	.0198	C
5177.5	124772	7	5	.0025	D
5187.7	123373	3	5	.0138	C
5194.1	125335	3	1	.0081	D
5210.5	124650	7	7	.0011	D
5214.8	124783	5	3	.0022	D
5216.3	124783	3	5	.0014	D
5221.3	124610	7	9	.0092	D
5241.1	124692	5	5	.0014	D

TRANSITION PROBABILITIES FOR SELECTED ATOMIC AND IONIC SPECIES

Wavelength (λ[Å])	Upper energy level (E_k[cm⁻¹])	g_i	g_k	Transition Probability (A[10⁸ s⁻¹])	Uncertainty
			Ar I		
5252.8	124650	5	7	.0056	D
5373.5	124692	3	5	.0028	D
5394.0	124772	5	5	.0010	D
5410.5	124715	5	7	.0021	D
5421.4	123903	7	5	.0062	D
5440.0	122479	3	3	.0020	D
5442.2	123832	7	7	9.7 (–4)	D
5451.7	122440	3	5	.0049	D
5457.4	123936	5	3	.0037	D
5467.2	123903	5	5	7.9 (–4)	D
5473.5	123882	5	3	.0021	D
5490.1	123827	5	5	8.9 (–4)	D
5495.9	123653	7	9	.0176	C
5506.1	123774	5	7	.0037	D
5525.0	123557	7	7	.0018	D
5534.5	125353	5	3	.0028	D
5558.7	122087	3	5	.0148	C
5559.7	125113	3	5	.0023	D
5572.5	123557	5	7	.0069	C
5588.7	123505	5	5	.0016	D
5606.7	121933	3	3	.0229	C
5618.0	123882	3	3	.0022	D
5620.9	123873	3	1	.0038	D
5623.8	125066	5	5	.0015	D
5635.6	123827	3	5	.0010	D
5639.1	124783	1	3	.0022	D
5641.4	123809	3	5	9.1 (–4)	D
5648.7	123936	5	3	.0013	D
5650.7	121794	3	1	.0333	C
5659.1	123903	5	5	.0027	D
5681.9	123832	5	7	.0021	D
5683.7	123827	5	5	.0021	D
5700.9	123774	5	7	.0061	D
5739.5	123505	3	5	.0091	D
5772.1	123557	5	7	.0021	D
5774.0	124604	5	5	.0011	D
5783.5	123373	3	5	8.4 (–4)	D
5802.1	123468	5	3	.0044	D
5834.3	123373	5	5	.0052	C
5860.3	121161	3	3	.00285	C
5882.6	121097	3	1	.0128	C
5888.6	122440	7	5	.0134	C
5912.1	121012	3	3	.0105	C
5928.8	122479	5	3	.011	D
5940.9	123882	1	3	.0012	D
5942.7	122440	5	5	.0019	D
5949.3	123936	3	3	.0016	D
5968.3	123882	3	3	.0019	D
5971.6	123873	3	1	.011	D
5987.3	122160	7	7	.0013	D
5988.1	123827	3	5	6.4 (–4)	D
5999.0	122282	5	5	.0015	D
6005.7	123936	5	3	.0015	D
6013.7	122087	7	5	.0015	D
6025.2	123882	5	5	.0094	D
6032.1	122036	7	9	.0246	C
6043.2	122160	5	7	.0153	C
6052.7	120619	3	5	.0020	D
6059.4	120601	3	5	.00423	C
6064.8	123774	5	7	6.0 (–4)	D
6090.8	123468	1	3	.0031	D
6098.8	122479	3	3	.0054	D
6101.2	123882	3	3	.0034	D
6104.6	123873	3	1	.0035	D
6105.6	123505	3	5	.0126	C
6127.4	121933	5	3	.0011	D
6128.7	123809	3	5	9.0 (–4)	D
6145.4	123557	5	7	.0079	C
6155.2	122479	5	3	.0053	D
6165.1	123505	5	5	.00103	C
6170.2	122440	5	5	.0052	C
6173.1	122282	3	5	.0070	C
6212.5	122330	5	7	.0041	D
6215.9	123373	5	5	.0059	D
6248.4	122087	3	5	7.1 (–4)	D
6296.9	123373	3	5	.0094	D
6307.7	122087	5	5	.0063	D
6364.9	121794	3	1	.0058	D
6369.6	121933	5	3	.0044	D
6384.7	119760	3	3	.00439	C
6416.3	119683	3	5	.0121	C
			Ar I		
6466.6	122514	1	3	.0016	D
6481.1	122479	1	3	9.8 (–4)	D
6538.1	120753	7	7	.0011	D
6604.0	120601	7	5	.0029	D
6660.7	121097	3	1	.0081	D
6664.1	120619	5	5	.0016	D
6677.3	108723	3	1	.00241	C
6698.9	121161	5	3	.0017	D
6719.2	121933	1	3	.0025	D
6752.8	118907	3	5	.0201	D
6754.4	121933	3	3	.0022	D
6756.2	122087	5	5	.0038	D
6766.6	121012	5	3	.0042	D
6779.9	123468	1	3	.00126	C
6827.2	121933	5	3	.0025	D
6851.9	122087	3	5	7.0 (–4)	D
6871.3	118651	3	3	.0290	C
6879.6	120619	3	5	.0019	D
6887.1	120753	5	7	.0014	D
6888.2	120601	3	5	.0026	D
6925.0	121933	3	3	.0012	D
6937.7	118512	3	1	.0321	C
6951.5	120619	5	5	.0023	D
6960.3	120601	5	5	.0025	D
6965.4	107496	5	3	.067	C
6992.2	121794	3	1	.0078	D
7030.3	119683	7	5	.0278	C
7067.2	107290	5	5	.0395	C
7068.7	119760	5	3	.021	D
7086.7	121161	1	3	.0016	D
7107.5	119683	5	5	.0047	D
7125.8	121161	3	3	.0063	D
7147.0	107132	5	3	.0065	C
7158.8	121097	3	1	.022	D
7207.0	121161	5	3	.0258	C
7229.9	119445	5	5	6.9 (–4)	D
7265.2	119848	3	3	.0018	D
7270.7	119213	7	7	.0011	D
7272.9	107496	3	3	.0200	C
7285.4	121012	5	3	.0013	D
7311.7	119760	3	3	.018	D
7316.0	121161	3	3	.010	D
7350.8	121097	3	1	.012	D
7353.3	119213	5	7	.010	D
7372.1	119024	7	9	.020	D
7384.0	107290	5	5	.087	C
7393.0	119760	5	3	.0075	D
7412.3	120619	3	5	.0041	D
7422.3	120601	3	5	6.9 (–4)	D
7425.3	120753	5	7	.0032	D
7435.4	119683	5	5	.0094	D
7436.3	118907	7	5	.0028	D
7484.3	119445	3	5	.0035	D
7503.9	108723	3	1	.472	C
7510.4	120601	5	5	.0047	D
7514.7	107054	3	1	.430	C
7618.3	120619	3	5	.0030	D
7628.9	120601	3	5	.0030	D
7635.1	106238	5	5	.274	C
7670.0	118651	5	3	.0029	D
7704.8	119213	5	7	6.6 (–4)	D
7723.8	106087	5	3	.057	C
7724.2	107496	1	3	.127	C
7798.6	118907	3	5	9.1 (–4)	D
7868.2	119760	1	3	.00365	C
7891.1	118907	5	5	.0099	D
7916.4	119760	3	3	.0013	D
7948.2	107132	1	3	.196	C
			Ar II		
3000.4	192712	4	4	1.5	D
3028.9	192712	2	4	2.3	D
3093.4	192557	4	6	4.4	D
3139.0	186891	6	6	1.0	D
3161.4	192712	2	4	1.8	D
3169.7	186891	4	6	.82	D
3181.0	186470	6	4	.63	D
3236.8	190592	2	4	.52	D
3243.7	186171	4	2	2.0	D
3249.8	186470	2	4	1.0	D
3293.6	190592	4	4	1.7	D

TRANSITION PROBABILITIES FOR SELECTED ATOMIC AND IONIC SPECIES

Wavelength (λ[Å])	Upper energy level (E_k[cm⁻¹])	Statistical weights g_i	g_k	Transition Probability (A[10⁸ s⁻¹])	Uncertainty
		Ar II			
3307.2	189935	2	2	3.4	D
3350.9	200235	6	6	1.5	D
3376.4	200139	8	8	1.5	D
3388.5	190592	2	4	1.9	D
3454.1	183986	6	4	.45	D
3464.1	187589	6	6	.37	D
3476.7	183797	6	6	1.34	C
3491.2	183986	4	4	2.2	D
3509.8	184192	2	2	2.5	D
3514.4	183797	4	6	1.23	C
3520.0	186074	6	6	.80	D
3521.3	185625	8	8	.23	D
3535.3	183986	2	4	.82	D
3545.6	187589	4	6	3.4	D
3545.8	198595	6	8	3.9	D
3548.5	186341	4	4	1.1	D
3559.5	186816	6	8	3.9	D
3565.0	186470	2	4	1.1	D
3576.6	185625	6	8	2.77	C
3581.6	186341	2	4	1.8	D
3582.4	186074	4	6	3.72	C
3588.4	185093	8	10	3.39	C
3600.2	199525	4	4	2.2	D
3622.1	182951	4	2	.64	D
3639.8	199680	4	6	1.4	D
3655.3	187589	4	6	.23	D
3671.0	199447	4	2	.71	D
3680.1	199982	2	4	1.2	D
3718.2	200235	4	6	2.0	D
3724.5	200235	6	6	.34	D
3729.3	161049	6	4	.60	D
3737.9	200139	6	8	2.3	D
3765.3	161594	6	6	.90	D
3770.5	182222	2	4	.41	D
3780.8	183676	8	8	.94	D
3796.6	199680	4	6	.25	D
3803.2	199680	6	6	1.5	D
3809.5	181594	4	6	.44	D
3825.7	199525	6	4	.76	D
3850.6	161049	4	4	.47	D
3868.5	186891	4	6	1.9	D
3925.7	195867	6	4	1.4	D
3928.6	161049	2	4	.30	D
3932.5	186470	4	4	1.1	D
3946.1	195865	8	6	1.4	D
3952.7	186341	4	4	.35	D
3979.4	186171	4	2	1.3	D
4033.8	182951	4	2	.98	D
4042.9	173348	4	4	1.4	D
4072.0	173393	6	6	.57	C
4076.6	182951	2	2	.80	D
4076.9	183915	4	2	.99	D
4079.6	173348	6	4	.26	D
4131.7	172816	4	2	1.4	D
4156.1	182222	4	4	.39	D
4179.3	181594	6	6	.13	D
4218.7	183091	4	4	.36	D
4222.6	183915	4	2	.69	D
4227.0	195865	4	6	.41	D
4266.5	157673	6	6	.156	C
4275.2	183091	2	4	.26	D
4277.5	172214	6	4	1.0	D
4331.2	158168	4	4	.56	C
4337.1	195867	2	4	.34	D
4348.1	157234	6	8	1.24	C
4370.8	173348	4	4	.65	C
4371.3	155351	6	4	.233	C
4379.7	158428	2	2	1.04	C
4401.0	155043	8	6	.322	C
4426.0	157673	4	6	.83	C
4430.2	158168	2	4	.53	D
4448.9	195865	6	6	.65	D
4481.8	173393	6	6	.494	C
4545.1	160239	4	4	.413	B
4547.8	182222	4	4	.077	D
4589.9	170401	4	6	.82	C
4609.6	170530	6	8	.91	C
4637.2	170401	6	6	.090	D
4657.9	159707	4	2	.81	C
4726.9	159393	4	4	.50	C
4735.9	155351	6	4	.58	C
		Ar II			
4764.9	160239	2	4	.575	B
4806.0	155043	6	6	.79	C
4847.8	155708	4	2	.85	C
4865.9	181594	4	6	.15	D
4879.9	158730	4	6	.78	C
4965.1	159393	2	4	.347	C
5009.3	155043	4	6	.147	C
5017.2	170401	4	6	.231	C
5062.0	155351	2	4	.221	C
5141.8	170530	6	8	.095	C
6638.2	158168	6	4	.129	C
6643.7	157234	10	8	.167	C
6684.3	157673	8	6	.113	C
		Ar III			
769.15	144023	5	3	6.0	D
871.10	114798	5	3	1.59	C
875.53	115328	3	1	3.74	C
878.73	113801	5	5	2.79	C
879.62	114798	3	3	.92	C
883.18	114798	1	3	1.22	C
887.40	113801	3	5	.90	C
3024.1	240292	5	7	2.6	D
3027.2	240258	5	5	.64	D
3054.8	240258	3	5	1.9	D
3064.8	240151	3	3	1.0	D
3078.2	240151	1	3	1.4	D
3285.9	204797	5	7	2.0	D
3301.9	204649	5	5	2.0	D
3311.3	204564	5	3	2.0	D
3336.1	226646	7	9	2.0	D
3344.7	226503	5	7	1.8	D
3352.1	226503	7	7	.22	D
3358.5	226356	3	5	1.6	D
3361.3	226356	5	5	.30	D
3472.6	225403	5	7	.20	D
3480.6	225403	7	7	1.6	D
3499.7	225155	3	3	1.3	D
3500.6	225148	3	5	.26	D
3502.7	225155	5	5	.43	D
3503.6	225148	5	5	1.2	D
3511.7	225148	7	5	.26	D
		Ar IV			
840.03	119044	4	2	2.73	C
843.77	118515	4	4	2.70	C
850.60	117564	4	6	2.63	C
		Ar VI			
292.15	342286	2	2	69.	C
294.05	342286	4	2	136.	C
		Ar VII			
250.41	514083	9	3*	278.	C
477.54	324151	9	15*	99.2	B
585.75	170720	1	3	78.3	B
637.30	271657	9	9*	67.	C
		Ar VIII			
158.92	629237	2	4	110.	C
159.18	628240	2	2	111.	C
229.44	575910	2	2	112.	C
230.88	575910	4	2	221.	C
337.09	629237	4	4	12.	D
337.26	629237	6	4	100.	C
338.22	628240	4	2	110.	C
519.43	332576	2	4	63.	C
526.46	332727	4	6	72.	C
526.87	332576	4	4	12.	D
700.24	142776	2	4	25.5	C
713.81	140058	2	2	24.	C
		Ar IX			
48.739	2051750	1	3	1690.	B
		Ar XIII			
162.96	698650	5	3	340.	C
163.08	628610	9	3*	530.	C
184.90	625840	5	5	166.	C
186.38	698650	1	3	88.	C
207.89	496450	9	9*	95.	C
245.10	423420	9	15*	37.	C
		Ar XIV			
180.29	554660	2	4	45.	C

TRANSITION PROBABILITIES FOR SELECTED ATOMIC AND IONIC SPECIES

Wavelength (λ[Å])	Upper energy level (E_k[cm⁻¹])	g_i	g_k	Transition Probability (A[10⁸ s⁻¹])	Uncertainty	Wavelength (λ[Å])	Upper energy level (E_k[cm⁻¹])	g_i	g_k	Transition Probability (A[10⁸ s⁻¹])	Uncertainty

Wavelength (λ[Å])	Upper energy level (E_k[cm⁻¹])	g_i	g_k	Transition Probability (A[10^8 s⁻¹])	Uncertainty	Wavelength (λ[Å])	Upper energy level (E_k[cm⁻¹])	g_i	g_k	Transition Probability (A[10^8 s⁻¹])	Uncertainty
Ar XIV						**Ba I**					
183.41	545230	2	2	169.	C	4579.6	35344	5	5	1.8	D
187.95	554660	4	4	197.	C	4591.8	30987	5	5	.016	D
191.35	545230	4	2	75.	C	4599.8	34371	3	1	1.0	D
194.39	514430	2	2	46.	C	4605.0	30744	3	1	.077	D
203.35	514430	4	2	78.	C	4619.9	33905	1	3	.093	D
Ar XV						4628.3	30816	5	3	.060	D
25.05	3992000	1	3	1.7 (+4)	B	4673.6	30987	7	5	.065	D
221.10	452280	1	3	95.5	B	4691.6	34823	5	3	1.6	D
265.3	621500	9	9*	81.	C	4700.4	33905	3	3	.24	D
Ar XVI						4726.4	32547	5	3	.46	D
23.52	4251000	2	6*	1.43(+4)	B	5519.1	30751	3	5	.50	D
24.96	4281000	6	10*	4.4 (+4)	B	5535.5	18060	1	3	1.15	B
353.88	282580	2	4	15.	B	5777.6	30818	5	7	.64	D
389.11	257000	2	2	11.	B	5800.2	30751	5	5	.099	D
1268.	4254000	2	4	1.9	B	5805.7	26816	7	7	.011	D
1401.	4246000	2	2	1.4	B	5826.3	28554	5	3	.56	D
2975.	4280000	2	4	.090	C	5907.6	25957	3	5	.036	D
3514.	4281000	4	6	.065	C	5971.7	25957	5	5	.29	D
Arsenic						5997.1	25704	3	3	.27	D
As I						6019.5	25642	3	1	1.4	D
1890.4	52898	4	6	2.0	D	6063.1	25704	5	3	.57	D
1937.6	51610	4	4	2.0	D	6110.8	25957	7	5	1.0	D
1972.6	50694	4	2	2.0	D	6341.7	24980	5	7	.19	D
2288.1	54605	6	4	2.8	D	6450.9	24532	3	5	.11	D
2344.0	60835	2	4	.35	D	6482.9	26816	5	7	.44	D
2349.8	53136	4	2	3.1	D	6498.8	24980	7	7	.86	D
2369.7	60835	4	4	.60	D	6527.3	24532	5	5	.59	D
2370.8	60815	4	6	.42	D	6595.3	24192	3	3	.39	D
2456.5	51610	6	4	.072	D	6675.3	24192	5	3	.19	D
2492.9	50694	4	2	.12	D	6693.8	24532	7	5	.28	D
2745.0	54605	2	4	.26	D	6865.7	25957	5	5	.078	D
2780.2	54605	4	4	.78	D	7059.9	23757	7	9	.71	D
2860.4	53136	2	2	.55	D	7120.3	23074	3	5	.21	D
2898.7	53136	4	2	.099	D	7195.2	26160	1	3	.24	D
Barium						7280.3	22947	5	7	.53	D
Ba I						7392.4	26160	3	3	.50	D
2409.2	41494	1	3	8.6 (–4)	C	7417.5	23074	7	5	.025	D
2414.1	41411	1	3	.0015	C	7488.1	22947	7	7	.10	D
2420.1	41308	1	3	.0023	C	7672.1	22065	3	5	.31	D
2427.4	41184	1	3	.0056	C	7780.5	22065	5	5	.13	D
2432.5	41097	1	3	.0072	C	7905.8	26160	5	3	.63	D
2438.8	40991	1	3	.0014	C	7911.3	12637	1	3	.00298	C
2444.6	40893	1	3	.0045	C						
2452.4	40764	1	3	8.1 (–4)	C	**Ba II**					
2473.2	40421	1	3	.0046	C	1413.4	76429	6	8	.017	D
2500.2	39985	1	3	.015	D	1417.1	75438	4	6	.038	D
2543.2	39309	1	3	.041	D	1444.9	74091	4	6	.081	D
2596.6	38500	1	3	.12	D	1461.5	74109	6	8	.087	D
2646.5	37775	1	3	.011	D	1487.0	72143	4	6	.14	D
2702.6	36990	1	3	.025	D	1503.9	72170	6	8	.15	D
2739.2	36496	1	3	.0091	D	1554.4	69212	4	6	.26	D
2785.3	35893	1	3	.028	D	1572.7	69260	6	8	.24	D
3071.6	32547	1	3	.41	C	1573.9	69212	6	6	.016	D
3501.1	28554	1	3	.19	D	1630.4	61336	2	2	.017	D
3889.3	25704	1	3	.0088	D	1674.5	64596	4	6	.22	D
3909.9	34603	3	5	.49	D	1694.4	64697	6	8	.21	D
3935.7	34617	5	7	.47	D	1697.2	64596	6	6	.017	D
3937.9	34603	5	5	.11	D	1761.8	61636	4	4	.0039	D
3993.4	34631	7	9	.55	D	1771.0	61336	4	2	.034	D
3995.7	34617	7	7	.088	D	1786.9	61636	6	4	.044	D
4132.4	24192	1	3	.0071	D	1892.7	73102	2	4	.090	D
4239.6	37095	5	3	.24	D	1904.2	57391	4	6	.011	D
4242.6	36200	3	5	.056	D	1906.8	72705	2	2	.051	D
4264.4	35709	1	3	.15	D	1924.7	57632	6	8	.031	D
4283.1	34736	5	7	.64	D	1954.2	73122	4	6	.13	D
4323.0	35762	3	5	.15	D	1955.1	73102	4	4	.018	D
4325.2	36629	5	7	.071	D	1970.2	72705	4	2	.067	D
4332.9	35709	3	3	.15	D	1985.6	70620	2	4	.25	D
4350.3	35617	3	5	.60	D	1999.5	50011	2	4	.10	D
4402.5	35344	3	5	.70	D	2009.2	70015	2	2	.086	D
4406.8	36200	5	5	.10	D	2052.7	70652	4	6	.20	D
4431.9	34823	1	3	1.2	D	2054.6	70620	4	4	.029	D
4467.1	35894	5	7	.066	D	2080.0	70015	4	2	.10	D
4489.0	35785	5	7	.42	D	2153.9	66674	2	4	.53	D
4493.6	35762	5	5	.36	D	2200.9	65683	2	2	.20	D
4505.9	34823	3	3	1.1	D	2232.8	66725	4	6	.29	D
4523.2	35617	5	5	.96	D	2235.4	66674	4	4	.044	D
4573.9	34494	3	1	2.9	D	2286.0	65683	4	2	.13	D
						2528.5	59800	2	4	.71	D
						2634.8	59895	4	6	.76	D
						2641.4	59800	4	4	.12	D

Wavelength (λ[Å])	Upper energy level (E_k[cm⁻¹])	Statistical weights g_i	g_k	Transition Probability (A[10⁸ s⁻¹])	Uncertainty	Wavelength (λ[Å])	Upper energy level (E_k[cm⁻¹])	Statistical weights g_i	g_k	Transition Probability (A[10⁸ s⁻¹])	Uncertainty
		Ba II						Be II			
2647.3	58025	2	2	.20	D	6757.1	133556	4	2	.102	C
2771.4	58025	4	2	.40	D	7401.2	128972	2	4	.030	C
3816.7	72143	4	6	.0023	D	7401.4	128972	2	2	.030	C
3842.8	72170	6	8	.0022	D			Bismuth			
3891.8	45949	2	4	1.67	B			Bi I			
4024.1	73102	6	4	.0053	D	1954.5	51159	4	6	1.2	D
4057.5	73122	8	6	.012	D	2021.2	49461	4	4	.060	D
4130.7	46155	4	6	1.80	B	2061.7	48490	4	6	.99	D
4166.0	45949	4	4	.37	D	2110.3	47373	4	2	.91	D
4216.0	73102	2	4	.058	D	2177.3	45916	4	2	.026	D
4287.8	72705	2	2	.024	D	2228.3	44865	4	4	.89	D
4325.7	73122	4	6	.059	D	2230.6	44817	4	6	2.6	D
4329.6	73102	4	4	.0088	D	2276.6	43913	4	4	.25	D
4405.2	72705	2	4	.039	D	2515.7	51159	4	6	.043	D
4470.7	70620	6	4	.014	D	2627.9	49461	4	4	.47	D
4509.6	70652	8	6	.012	D	2696.8	48490	4	6	.064	D
4524.9	42355	2	2	.72	D	2780.5	47373	4	2	.309	C
4554.0	21952	2	4	1.17	A	2798.7	51159	6	6	.036	D
4708.9	70620	2	4	.097	D	2898.0	45916	4	2	1.53	C
4843.5	70652	4	6	.093	D	2938.3	49461	6	4	1.23	C
4847.1	70015	2	2	.041	D	2989.0	44865	4	4	.55	C
4850.8	70620	4	4	.014	D	2993.3	44817	4	6	.16	C
4900.0	42355	4	2	.775	B	3024.6	48490	6	6	.88	C
4934.1	20262	2	2	.955	B	3067.7	32588	4	2	2.07	C
4997.8	70015	4	2	.061	D	3076.7	43913	4	4	.035	D
5185.0	61636	2	4	.018	D	3397.2	44865	6	4	.181	C
5361.4	64596	4	6	.048	D	3402.9	44817	6	6	.016	D
5391.6	64697	6	8	.052	D	3510.9	43913	6	4	.068	D
5413.6	66725	6	6	8.4 (–4)	D	3596.1	49461	2	4	.198	C
5421.1	64596	8	6	.0019	D	3888.2	47373	2	2	.009	D
5428.8	66674	6	4	.023	D	4121.5	45916	2	2	.164	D
5480.3	66725	8	6	.018	D	4308.5	44865	2	4	.016	D
5784.2	66674	2	4	.20	D	4493.0	43913	2	4	.015	D
5853.7	21952	4	4	.048	B	4722.5	32588	4	2	.117	C
5981.3	66725	4	6	.16	D	6134.8	49461	4	4	.018	D
5999.9	66674	4	4	.026	D			Boron			
6135.8	65683	2	2	.085	D			B I			
6141.7	21952	6	4	.37	B	1378.6	72535	2	4	3.50	C
6363.2	73102	6	4	.0029	D	1378.9	72523	2	2	14.0	C
6372.9	61636	4	4	6.7 (–4)	D	1378.9	72535	4	4	17.5	C
6378.9	65683	4	2	.099	D	1379.2	72523	4	2	7.0	C
6457.7	61636	6	4	.0030	D	1465.5	97000	2	4	3.34	C
6496.9	20262	4	2	.332	C	1465.7	97000	4	4	6.7	C
7556.8	70620	6	4	.0016	D	1465.8	97000	6	4	10.0	C
7678.2	70652	8	6	6.6 (–4)	D	1825.9	54767	2	4	2.0	C
8710.7	57632	6	8	.80	D	1826.4	54767	4	6	2.4	C
8737.7	57391	4	6	.93	D	2088.9	47857	2	4	.28	D
		Beryllium				2089.5	47857	4	6	.33	D
		Be I				2496.8	40040	2	2	.85	C
1491.8	67035	1	3	.013	D	2497.7	40040	4	2	1.69	C
1661.5	60187	1	3	.20	D			Bromine			
2348.6	42565	1	3	5.56	B			Br I			
2494.7	62054	9	15*	1.6	C	1488.5	67184	4	4	1.2	D
2650.6	59696	9	9*	4.31	C	1540.7	64907	4	4	1.4	D
4572.7	64428	3	5	.79	C	1574.8	67184	2	4	.20	D
						1576.4	63436	4	6	.021	D
		Be II				1633.4	64907	2	4	.081	D
1197.1	115464	2	2	.47	D	4365.1	89786	2	4	.0075	D
1197.2	115464	4	2	.94	D	4425.1	87499	4	2	.0042	D
1512.3	98055	2	4	9.2	C	4441.7	85944	6	4	.0075	D
1512.4	98055	4	6	11.	C	4472.6	87259	4	4	.0093	D
1776.1	88232	2	2	1.4	C	4477.7	85763	6	8	.013	D
1776.3	88232	4	2	2.9	C	4513.4	85586	6	4	.0028	D
2453.8	128972	2	6*	.142	C	4525.6	85527	6	6	.0072	D
3046.5	129310	2	4	.48	C	4575.7	89032	4	4	.016	D
3046.7	129310	4	6	.59	C	4614.6	88848	4	6	.0054	D
3130.4	31935	2	4	1.14	B	4979.8	87259	4	4	.0026	D
3131.1	31929	2	2	1.15	B	5245.1	85944	2	4	.0031	D
3241.6	127335	2	2	.141	C	5345.4	85586	2	4	.0076	D
3241.8	127335	4	2	.28	C	7348.5	78512	4	6	.12	D
3274.6	118761	2	4	.19	C	7513.0	76743	6	4	.12	D
3274.7	118761	2	2	.19	C	7803.0	79696	2	4	.053	D
4360.7	119421	2	4	.92	C	7938.7	88483	6	6	.19	D
4361.0	119421	4	6	1.1	C	8131.5	79178	2	4	.038	D
5255.9	134485	2	6*	.0256	C	8343.7	78866	2	2	.22	D
5270.3	115464	2	2	.330	C	8446.6	76743	4	4	.12	D
5270.8	115464	4	2	.66	C	8638.7	75009	6	4	.097	D
6279.4	134681	2	4	.12	C						
6279.7	134681	4	6	.143	C						
6756.7	133556	2	2	.051	C						

Table 1b (Continued)
TRANSITION PROBABILITIES FOR SELECTED ATOMIC AND IONIC SPECIES

Wavelength (λ[Å])	Upper energy level (E_k[cm⁻¹])	g_i	g_k	Transition Probability (A[10⁸ s⁻¹])	Uncertainty
			Br II		
4704.9	115176	5	7	1.1	D
4785.5	114818	5	5	.94	D
4816.7	114683	5	3	1.1	D
			Cadmium		
			Cd I		
2288.0	43692	1	3	5.3	C
2836.9	65354	1	3	.28	D
2880.8	65359	3	5	.42	D
2881.2	65353	3	3	.24	D
2980.6	65367	5	7	.59	D
2981.4	65359	5	5	.15	D
3261.1	30656	1	3	.00406	C
3403.7	59486	1	3	.77	C
3466.2	59498	3	5	1.2	D
3467.7	59486	3	3	.67	D
3610.5	59516	5	7	1.3	D
3612.9	59498	5	5	.35	D
4140.5	67838	3	5	.047	D
4662.4	65135	3	5	.055	C
4678.1	51484	1	3	.13	C
4799.9	51484	3	3	.41	C
5085.8	51484	5	3	.56	C
6438.5	59220	3	5	.59	C
			Cd II		
2144.4	46619	2	4	2.8	C
2265.0	44136	2	2	3.0	C
2572.9	82991	2	2	1.7	C
2748.5	82991	4	2	2.8	C
4415.6	69259	4	6	.014	B
			Calcium		
			Ca I		
2275.5	43933	1	3	.301	C
2995.0	48538	1	3	.367	C
2997.3	48564	3	5	.241	C
2999.6	48538	3	3	.279	C
3000.9	48524	3	1	1.58	C
3006.9	48564	5	5	.75	C
3009.2	48538	5	3	.430	C
3344.5	45049	1	3	.151	C
3350.2	45050	3	5	.178	C
3361.9	45052	5	7	.223	C
3624.1	42743	1	3	.212	C
3630.8	42745	3	5	.297	C
3631.0	42743	3	3	.153	C
3644.4	42747	5	7	.355	C
3644.8	42745	5	5	.094	C
3870.5	46165	3	5	.072	D
3957.1	40474	3	3	.098	C
3973.7	40474	5	3	.175	C
4092.6	44763	3	5	.11	D
4094.9	44763	5	7	.12	D
4098.5	44763	7	9	.13	D
4108.5	46182	5	7	.90	D
4226.7	23652	1	3	2.18	B
4283.0	38552	3	5	.434	C
4289.4	38465	1	3	.60	C
4299.0	38465	3	3	.466	C
4302.5	38552	5	5	1.36	C
4307.7	38418	3	1	1.99	C
4318.7	38465	5	3	.74	C
4355.1	44805	5	7	.19	D
4425.4	37748	1	3	.498	C
4435.0	37752	3	5	.67	C
4435.7	37748	3	3	.342	C
4454.8	37757	5	7	.87	C
4455.9	37752	5	5	.20	C
4526.9	43933	5	3	.41	D
4578.6	42170	3	5	.176	C
4581.4	42171	5	7	.209	C
4585.9	42171	7	9	.229	C
4685.3	44990	3	5	.080	D
4878.1	42344	5	7	.188	C
5041.6	41679	5	3	.33	D
5188.9	42919	3	5	.40	D
5261.7	39335	3	3	.15	D
5262.2	39333	3	1	.60	D
5264.2	39340	5	5	.091	D
5265.6	39335	5	3	.44	D
			Ca I		
5270.3	39340	7	5	.50	D
5582.0	38259	5	7	.060	D
5588.8	38259	7	7	.49	D
5590.1	38219	3	5	.083	D
5594.5	38219	5	5	.38	D
5598.5	38192	3	3	.43	D
5601.3	38219	7	5	.086	D
5602.9	38192	5	3	.14	D
5857.5	40720	3	5	.66	D
6102.7	31539	1	3	.096	C
6122.2	31539	3	3	.287	C
6161.3	36575	5	5	.033	D
6162.2	31539	5	3	.477	C
6163.8	36555	3	3	.056	D
6166.4	36548	3	1	.22	D
6169.1	36555	5	3	.17	D
6169.6	36575	7	5	.19	D
6439.1	35897	7	9	.53	D
6449.8	35835	3	5	.090	D
6462.6	35819	5	7	.47	D
6471.7	35819	7	7	.059	D
6493.8	35730	3	5	.44	D
6499.7	35730	5	5	.081	D
			Ca II		
1341.9	74522	2	4	.015	D
1342.5	74485	2	2	.015	D
1649.9	60611	2	4	.0032	D
1652.0	60533	2	2	.0031	D
1673.9	84934	2	4	.224	C
1680.1	84936	4	6	.265	C
1680.1	84934	4	4	.0441	C
1807.3	80522	2	4	.354	C
1814.5	80526	4	6	.42	C
1814.7	80522	4	4	.070	C
1843.1	79448	2	2	.16	C
1850.7	79448	4	2	.308	C
2103.2	72722	2	4	.82	C
2112.8	72731	4	6	.97	C
2113.2	72722	4	4	.16	C
2197.8	70678	2	2	.31	C
2208.6	70678	4	2	.62	C
3158.9	56839	2	4	3.1	C
3179.3	56858	4	6	3.6	C
3181.3	56839	4	4	.58	C
3706.0	52167	2	2	.88	C
3736.9	52167	4	2	1.7	C
3933.7	25414	2	4	1.47	C
3968.5	25192	2	2	1.4	C
			Ca III		
357.97	279354	1	3	880.	D
439.69	227432	1	3	.19	D
490.55	203852	1	3	.016	D
			Ca V		
558.60	197845	5	3	22.	D
637.93	156760	5	3	3.9	D
643.12	157901	3	1	9.1	D
646.57	154671	5	5	6.9	D
647.88	156760	3	3	2.3	D
651.55	156760	1	3	2.9	D
656.76	154671	3	5	2.1	D
			Ca VII		
550.20	203616	5	5	18.	D
624.39	160158	1	3	3.3	D
630.54	160220	3	5	4.5	D
630.79	160158	3	3	2.2	D
639.15	160529	5	7	5.7	D
640.41	160220	5	5	1.3	D
			Ca VIII		
182.71	547322	2	2	160.	C
184.16	547322	4	2	320.	C
			Ca IX		
163.23	758974	5	3	376.	C
371.89	410514	1	3	88.	C
373.81	410627	3	5	116.	C
378.08	410841	5	7	150.	C
395.03	467631	3	5	220.	D
466.24	214482	1	3	112.	B
498.01	343908	3	5	24.9	C

Table 1b (Continued)
TRANSITION PROBABILITIES FOR SELECTED ATOMIC AND IONIC SPECIES

Wavelength (λ[Å])	Upper energy level (E_k[cm⁻¹])	Statistical weights g_i	g_k	Transition Probability (A[10⁸ s⁻¹])	Uncertainty
		Ca IX			
506.18	343908	5	5	72.	C
515.57	340308	5	3	37.5	C
		Ca X			
110.96	901200	2	4	290.	C
111.20	899290	2	2	292.	C
151.84	832790	2	2	230.	C
153.02	832790	4	2	450.	C
206.57	901200	4	4	29.	D
206.75	901200	6	4	260.	C
207.39	899290	4	2	280.	C
411.70	417112	2	4	83.	C
419.75	417522	4	6	95.	C
420.47	417112	4	4	16.	D
557.76	179287	2	4	35.0	C
574.01	174213	2	2	32.	C
		Ca XI			
30.448	3284300	1	3	6200.	D
30.867	3239700	1	3	4.9 (+4)	D
35.212	2839940	1	3	2000.	D
		Ca XII			
140.05	709000	4	2	370.	C
147.27	709000	2	2	160.	C
		Ca XV			
141.69	814370	5	3	408.	C
142.23	728910	9	3*	630.	C
161.00	729720	5	5	190.	C
		Ca XVII			
19.558	5113000	1	3	3.8 (+4)	C
21.198	5236000	3	5	4.9 (+4)	C
192.82	518620	1	3	121.	C
218.82	726450	3	5	27.6	C
223.02	706680	1	3	34.4	C
228.72	706680	3	3	23.7	C
232.83	726450	5	5	65.	C
244.06	706680	5	3	32.8	C
		Ca XVIII			
18.71	5346000	2	6*	2.31(+4)	B
19.74	5383000	6	10*	7.0 (+4)	B
302.19	330920	2	4	20.	B
344.76	290060	2	2	13.	B
		Carbon			
		C I			
945.19	105799	1	3	6.2	D
945.34	105799	3	3	18.	D
945.58	105799	5	3	31.	D
1260.7	79319	1	3	.40	D
1260.9	79323	3	1	1.2	D
1261.0	79319	3	3	.31	D
1261.1	79311	3	5	.30	D
1261.4	79319	5	3	.50	D
1261.6	79311	5	5	.93	D
1274.1	78530	5	7	.0068	D
1277.2	78293	1	3	.88	D
1277.3	78308	3	5	1.2	D
1277.5	78293	3	3	.65	D
1277.6	78318	5	7	1.5	D
1277.7	78308	5	5	.39	D
1278.0	78293	5	3	.042	D
1279.2	78216	5	7	.11	D
1279.9	78148	3	5	.21	D
1280.1	78117	1	3	.27	D
1280.3	78148	5	5	.62	D
1280.4	78117	3	3	.20	D
1280.6	78105	3	1	.81	D
1280.8	78117	5	3	.35	D
1328.8	75254	1	3	.49	D
1364.2	83498	5	5	.047	D
1431.6	103587	5	7	1.5	D
1432.1	103563	5	5	1.4	D
1432.5	103542	5	3	1.3	D
1459.0	78731	5	3	.37	D
1463.3	78530	5	7	2.1	D
1467.4	78340	5	3	.46	D
1468.4	78293	5	3	.019	D
1470.1	78216	5	7	.0088	D
1472.2	78117	5	3	.0051	D
1481.8	77680	5	5	.33	D

Wavelength (λ[Å])	Upper energy level (E_k[cm⁻¹])	Statistical weights g_i	g_k	Transition Probability (A[10⁸ s⁻¹])	Uncertainty
		C I			
1560.3	64090	1	3	.82	D
1561.3	64091	5	5	.36	D
1561.4	64087	5	7	1.4	D
1656.3	60393	3	5	.80	D
1656.9	60353	1	3	1.1	D
1657.0	60393	5	5	2.4	D
1657.4	60353	3	3	.80	D
1657.9	60333	3	1	3.2	D
1658.1	60353	5	3	1.3	D
1751.8	78731	1	3	.57	D
1763.9	78340	1	3	.022	D
1765.4	78293	1	3	.0071	D
1930.9	61982	5	3	3.7	D
2478.6	61982	1	3	.18	D
2902.3	105799	1	3	.0066	D
2903.3	105799	3	3	.017	D
2905.0	105799	5	3	.022	D
4269.0	85400	3	5	.0032	D
4371.4	84852	3	3	.0097	D
4762.3	81326	1	3	.0052	D
4762.5	81344	3	5	.0038	D
4766.7	81326	3	3	.0039	D
4770.0	81311	3	1	.015	D
4771.8	81344	5	5	.012	D
4775.9	81326	5	3	.0062	D
4812.9	81105	1	3	9.7 (−4)	D
4817.4	81105	3	3	.0028	D
4826.8	81105	5	3	.0047	D
4932.1	82252	3	1	.046	D
5050.0	81170	3	5	.017	D
5380.3	80563	3	3	.016	D
5793.1	81344	7	5	.0033	D
5794.5	81344	5	5	5.8 (−4)	D
5800.2	81326	3	3	9.7 (−4)	D
5800.6	81326	5	3	.0029	D
5805.2	81311	3	1	.0039	D
6587.6	84032	3	3	.024	D
		C II			
687.35	145551	4	6	27.0	C
858.09	116538	2	2	.369	C
858.56	116538	4	2	1.11	C
903.62	110666	2	4	6.6	C
903.96	110624	2	2	26.3	C
904.14	110666	4	4	33.0	C
904.48	110624	4	2	13.3	C
1009.9	142027	2	4	5.8	C
1010.1	142027	4	4	11.5	C
1010.4	142027	6	4	17.3	C
1036.3	96494	2	2	8.0	C
1037.0	96494	4	2	15.9	C
1323.9	150467	4	4	4.53	C
1324.0	150462	6	6	4.71	C
1334.5	74933	2	4	2.41	C
1335.7	74930	4	6	2.89	C
2509.1	150467	2	4	.54	C
2511.7	150467	4	4	.106	C
2512.1	150462	4	6	.64	C
6578.1	131736	2	4	.36	C
6582.9	131724	2	2	.36	C
7231.3	145549	2	4	.36	C
7236.4	145551	4	6	.44	C
7237.2	145549	4	4	.072	C
		C III			
310.17	322404	1	3	18.	C
386.20	258931	1	3	32.2	B
459.46	270011	1	3	55.	C
459.52	270012	3	5	75.	C
459.63	270015	5	7	98.	C
574.28	276483	3	5	63.	C
977.03	102352	1	3	17.5	B
1174.9	137502	3	5	3.42	C
1175.3	137454	1	3	4.55	C
1175.6	137454	3	3	3.41	C
1175.7	137502	5	5	10.2	C
1176.0	137426	3	1	13.6	C
1176.4	137454	5	3	5.7	C
1247.4	182503	3	1	18.6	C
2296.9	145876	3	5	1.46	C
4647.4	259724	3	5	.73	C
4650.3	259711	3	3	.74	C

TRANSITION PROBABILITIES FOR SELECTED ATOMIC AND IONIC SPECIES

Wavelength (λ[Å])	Upper energy level (E_k[cm⁻¹])	g_i	g_k	Transition Probability (A[10^8 s⁻¹])	Uncertainty
			C III		
4651.5	259706	3	1	.74	C
			C IV		
312.43	320071	2	6*	44.9	B
384.13	324886	6	10*	180.	B
1548.2	64592	2	4	2.66	B
1550.8	64484	2	2	2.64	B
5801.3	320082	2	4	.319	B
5812.0	320050	2	2	.316	B
			C V		
34.973	2859375	1	3	2554.	AA
40.268	2483371	1	3	8873.	AA
227.19	2851418	3	9*	136.3	AA
247.31	2859375	1	3	127.9	AA
248.70	2857310	9	15*	425.	A
260.19	2839562	9	3*	66.83	AA
267.27	2857529	3	5	396.	A
2273.9	2455225	3	9*	.5650	AA
3526.7	2483371	1	3	.1663	AA
8432.2	2851418	3	9*	.06870	AA
			Cesium — Cs I		
3203.5	31207	2	4	7.6 (–6)	C
3205.3	31189	2	4	7.9 (–6)	C
3207.5	31168	2	4	8.5 (–6)	C
3210.0	31144	2	4	9.4 (–6)	C
3212.8	31116	2	4	1.19(–5)	C
3216.2	31084	2	4	1.49(–5)	C
3220.1	31046	2	4	1.7 (–5)	C
3220.2	31045	2	2	1.07(–7)	C
3224.8	31001	2	4	2.0 (–5)	C
3225.0	30999	2	2	1.43(–7)	C
3230.5	30946	2	4	2.5 (–5)	C
3230.7	30944	2	2	1.97(–7)	C
3237.4	30880	2	4	2.8 (–5)	C
3237.6	30878	2	2	2.63(–7)	C
3245.9	30799	2	4	3.45(–5)	C
3246.2	30796	2	2	3.7 (–7)	C
3256.7	30698	2	4	4.25(–5)	C
3257.1	30694	2	2	7.0 (–7)	C
3270.5	30568	2	4	5.6 (–5)	C
3271.0	30563	2	2	9.8 (–7)	C
3288.6	30399	2	4	1.0 (–4)	C
3289.3	30393	2	2	2.7 (–6)	C
3313.1	30175	2	4	1.6 (–4)	C
3314.0	30166	2	2	5.2 (–6)	C
3347.5	29865	2	4	2.2 (–4)	C
3348.8	29853	2	2	1.1 (–5)	C
3397.9	29421	2	4	4.0 (–4)	C
3400.0	29404	2	2	2.4 (–5)	C
3476.8	28754	2	4	6.6 (–4)	C
3480.0	28727	2	2	6.6 (–5)	C
3611.4	27682	2	4	.0015	C
3617.3	27637	2	2	2.5 (–4)	C
3876.1	25792	2	4	.0038	C
3888.6	25709	2	2	9.7 (–4)	C
4555.3	21946	2	4	.0188	C
4593.2	21765	2	2	.0080	C
			Chlorine — Cl I		
1188.8	84120	4	6	2.33	C
1188.8	84122	4	4	.271	C
1201.4	84122	2	4	2.39	C
1335.7	74866	4	2	1.74	C
1347.2	74226	4	4	4.19	C
1351.7	74866	2	2	3.23	C
1363.4	74226	2	4	.75	C
4323.3	95612	4	4	.011	D
4363.3	95401	4	6	.0068	D
4379.9	95313	4	4	.014	D
4389.8	94732	6	8	.014	D
4526.2	96313	4	4	.051	C
4601.0	96594	2	2	.042	D
4661.2	96313	2	4	.012	D
7256.6	85735	6	4	.15	D
7414.1	85442	6	4	.047	D
7547.1	85735	4	4	.12	D
7717.6	85442	4	4	.030	D
7745.0	85735	2	4	.063	D

Wavelength (λ[Å])	Upper energy level (E_k[cm⁻¹])	g_i	g_k	Transition Probability (A[10^8 s⁻¹])	Uncertainty
			Cl I		
7769.2	95787	6	6	.060	D
7821.4	95701	6	8	.098	D
7830.8	95898	4	4	.097	D
7878.2	84648	6	6	.018	D
7899.3	95787	4	6	.051	D
7924.6	85442	2	4	.021	D
7935.0	96731	6	8	.039	D
7997.9	84988	4	4	.021	D
			Cl II		
3329.1	161798	5	7	1.5	D
3522.1	174855	7	7	1.4	D
3798.8	172652	5	7	1.6	D
3805.2	172743	7	9	1.8	D
3809.5	172575	3	5	1.5	D
3851.0	154624	5	7	1.8	D
3851.4	154621	5	5	1.6	D
3854.7	184660	3	5	2.2	D
3861.9	184657	5	7	2.4	D
3868.6	184630	7	9	2.7	D
3913.9	172743	9	9	.82	D
3990.2	174855	5	7	.84	D
4132.5	153259	5	5	1.6	D
4276.5	170577	9	7	.76	D
4768.7	158771	3	5	.77	D
4781.3	158788	5	7	1.0	D
4794.6	128731	5	7	1.04	C
4810.1	128644	5	5	.99	C
4819.5	128623	5	3	1.00	C
4904.8	147128	5	7	.81	D
4917.7	147056	3	5	.75	D
5078.3	146471	7	7	.77	D
5219.1	131765	3	9*	.86	C
5392.1	147607	5	7	1.0	D
			Cl III		
2298.5	248528	4	4	4.2	D
2340.6	248658	6	6	4.2	D
2370.4	258886	8	6	2.8	D
2531.8	248528	2	4	4.4	D
2532.5	248658	4	6	5.3	D
2577.1	243828	4	6	4.3	D
2580.7	244685	6	8	4.7	D
2601.2	239506	2	4	4.6	D
2603.6	239730	4	6	5.0	D
2609.5	240075	6	8	5.7	D
2617.0	240568	8	10	6.6	D
2661.6	241685	4	6	3.4	D
2665.5	242046	6	8	4.8	D
2691.5	243081	4	4	3.5	D
2710.4	242823	4	6	3.5	D
3340.4	204541	6	6	1.5	D
3392.9	217913	4	4	1.9	D
3393.5	217850	6	6	1.9	D
3530.0	216710	6	8	1.8	D
3560.7	216525	4	6	1.7	D
3602.1	202368	6	8	1.7	D
3612.9	201765	4	6	1.2	D
3720.5	205947	4	6	1.7	D
			Chromium — Cr I		
2000.0	58293	9	9	1.4	D
2383.3	50253	9	11	.41	D
2385.7	50211	9	9	.17	D
2389.2	49653	3	5	.23	D
2408.6	49812	9	7	.67	D
2408.7	49598	7	5	.29	D
2492.6	47918	3	5	.45	C
2496.3	47975	5	7	.56	C
2502.5	48043	7	9	.22	D
2504.3	48014	7	9	.45	D
2549.5	47022	3	3	.48	D
2560.7	46968	5	5	.43	D
2577.7	46879	7	7	.26	D
2591.9	46879	9	7	.65	C
2622.9	46422	9	9	.13	D
2702.0	45306	9	11	.21	C
2726.5	44259	5	7	.75	C
2731.9	44187	5	5	.78	C
2736.5	44126	5	3	.75	D
2752.9	44126	3	3	.87	D

TRANSITION PROBABILITIES FOR SELECTED ATOMIC AND IONIC SPECIES

Wavelength (λ[Å])	Upper energy level (E_k[cm^{-1}])	Statistical weights g_i	g_k	Transition Probability (A[10^8 s^{-1}])	Uncertainty
		Cr I			
2757.1	44187	5	5	.68	C
2761.8	44126	5	3	.68	D
2764.4	44259	7	7	.37	D
2769.9	44187	7	5	1.1	C
2780.7	44259	9	7	1.4	C
2871.6	42909	7	9	.12	D
2879.3	42648	5	7	.21	D
2887.0	42439	3	5	.27	D
2889.3	42909	9	9	.66	C
2899.2	42293	3	3	.15	D
2909.1	42293	5	3	.68	C
2967.6	41782	7	9	.39	D
2971.1	41575	5	7	.71	C
2975.5	41409	3	5	.89	C
2980.8	41289	1	3	.85	D
2988.7	41043	5	7	.52	C
2991.9	41225	3	1	3.0	D
2995.1	40971	5	5	.43	D
2996.6	41289	5	3	2.0	C
2998.8	40930	5	3	.59	D
3000.9	41409	7	5	1.6	C
3005.1	41575	9	7	.92	C
3013.7	40983	3	5	.83	C
3020.7	40906	3	3	1.5	D
3021.6	41393	9	11	3.2	C
3024.4	40983	5	5	2.3	C
3030.2	41086	7	7	1.1	C
3034.2	41043	7	7	.35	C
3037.0	41225	9	9	.54	C
3040.9	40971	7	5	.74	D
3053.9	41043	9	7	1.2	C
3148.4	55686	9	11	.59	C
3155.2	55741	11	13	.54	C
3163.8	55799	13	15	.32	C
3237.7	54811	9	9	1.3	D
3238.1	54930	11	11	.20	D
3578.7	27935	7	9	1.48	B
3593.5	27820	7	7	1.50	B
3605.3	27729	7	5	1.62	B
3639.8	47986	13	11	1.8	D
3768.7	47047	7	5	.22	D
3804.8	50558	9	9	.69	D
3879.2	50058	3	5	.56	D
3963.7	45741	13	15	1.3	D
3969.8	45707	11	13	1.2	D
3981.2	46968	3	5	.11	D
3983.9	45615	7	9	1.05	C
4001.4	56362	9	11	.65	D
4030.7	56155	3	5	.79	D
4039.1	55799	15	15	.68	C
4048.8	55741	13	13	.65	D
4058.8	55686	11	11	.69	D
4161.4	59957	13	15	.80	D
4165.5	59884	11	13	.75	D
4211.4	48043	7	9	.085	D
4224.5	48562	9	9	.067	D
4239.0	47866	9	9	.071	D
4254.4	23499	7	9	.315	B
4274.8	23386	7	7	.306	B
4280.4	54405	13	15	.47	D
4289.7	23305	7	5	.313	B
4297.7	54317	11	13	.49	D
4344.5	31106	7	9	.11	C
4351.8	31280	9	11	.12	C
4374.2	47055	13	11	.10	C
4375.3	46905	11	9	.072	D
4381.1	44667	5	3	.10	D
4387.5	46986	13	11	.066	D
4413.9	51287	7	5	.41	D
4432.2	45719	1	3	.17	D
4458.5	46705	9	11	.13	D
4482.9	49477	3	3	.30	D
4488.1	56210	7	7	.63	D
4506.9	55945	13	11	.27	D
4511.9	47055	9	9	.13	D
4526.5	42606	13	13	.20	D
4530.7	42589	11	11	.20	D
4544.6	42515	5	5	.26	D
4595.6	55517	13	13	.47	D
4632.2	46688	7	7	.071	D
4639.5	46637	5	7	.095	D
4646.2	29825	9	7	.087	C
4689.4	46525	7	5	.23	D
4708.0	46783	11	9	.37	D
4718.4	46959	13	11	.42	D
4723.1	46000	7	7	.093	D
4724.4	46058	9	9	.063	D
4727.2	45349	13	13	.051	D
4729.7	46077	5	3	.17	D
4730.7	43900	7	5	.28	D
4737.4	46000	9	7	.24	D
4789.3	41393	13	11	.076	D
4792.5	45966	7	5	.26	D
4801.0	46000	9	7	.23	D
4870.8	45359	7	9	.35	D
4887.0	45354	9	11	.32	D
4922.3	45349	11	13	.40	D
4936.3	45359	7	9	.14	D
4954.8	45354	9	11	.12	D
5139.7	47048	7	7	.13	D
5177.4	46959	9	11	.061	D
5184.6	46783	7	9	.11	D
5192.0	46637	5	7	.14	D
5196.6	47055	11	9	.12	D
5200.2	46525	3	5	.16	D
5204.5	26802	5	3	.55	D
5206.0	26796	5	5	.53	D
5208.4	26788	5	7	.51	D
5243.4	46449	5	3	.20	D
5261.8	48825	7	9	.13	D
5272.0	46783	7	9	.11	D
5287.2	46637	5	7	.078	D
5304.2	46783	9	9	.066	D
5312.9	46637	7	7	.11	D
5318.8	46525	5	5	.13	D
5328.3	42261	9	11	.60	D
5340.4	46449	5	3	.16	D
5400.6	45734	5	5	.16	D
5409.8	26788	9	7	.062	D
		Cr II			
2653.6	49706	4	6	.35	D
2658.6	49565	2	4	.58	D
2666.0	49646	6	8	.59	D
2668.7	49493	4	2	1.4	D
2671.8	49565	6	4	1.0	D
2672.8	49706	8	6	.55	D
2693.5	67334	10	8	1.4	D
2727.3	67876	10	8	1.7	D
2740.1	48632	6	8	.11	D
2745.0	66727	4	6	.85	D
2768.6	74424	6	8	2.8	D
2774.4	75717	8	8	1.7	D
2778.1	75810	10	10	3.2	D
2782.4	69348	6	4	1.6	D
2785.7	69506	10	8	2.1	D
2787.6	66727	6	6	1.5	D
2792.2	69498	12	10	2.3	D
2800.8	69388	12	14	2.2	D
2818.4	68993	8	10	2.2	D
2822.0	68844	6	8	2.3	D
2832.5	70108	12	10	1.3	D
2838.8	73486	8	8	2.7	D
2840.0	65420	10	12	2.7	D
2843.3	47465	8	10	.64	D
2849.8	47228	6	8	.92	D
2851.4	65218	8	10	2.2	D
2856.8	54626	4	6	.43	D
2857.4	54785	6	8	.28	D
2860.9	46906	2	4	.69	D
2862.6	47228	8	8	.63	D
2866.7	46906	4	4	1.2	D
2867.1	54500	4	4	1.1	D
2867.7	46824	2	2	1.1	D
2870.4	54626	6	6	1.3	D
2873.8	54418	4	2	.88	D
2878.5	47228	10	8	.074	D
2880.9	54500	6	4	.79	D
2888.7	70880	10	12	.88	D
2898.5	65710	10	12	1.2	D
2921.8	65384	8	10	.90	D
2927.1	72717	10	10	2.8	D

Wavelength (λ[Å])	Upper energy level (E_k[cm^{-1}])	g_i	g_k	Transition Probability (A[10^8 s^{-1}])	Uncertainty
Cr II					
2930.9	64062	2	4	1.1	D
2935.1	64924	6	8	1.8	D
2953.4	63802	2	2	1.8	D
2953.7	68477	10	10	.92	D
2966.1	64924	10	8	.54	D
2971.9	64031	14	14	2.0	D
2979.7	63849	12	12	1.8	D
2985.3	63707	10	10	2.2	D
2989.2	63601	8	8	2.2	D
3040.9	67506	10	12	4.8	D
3041.7	68477	10	10	3.1	D
3050.1	67589	12	14	1.8	D
3093.5	70880	10	12	.67	D
3096.1	70852	10	8	.75	D
3107.6	70679	8	10	.62	D
3118.7	51584	2	4	1.7	D
3120.4	51670	4	6	1.5	D
3122.6	65710	12	12	.44	D
3128.7	51584	4	4	.81	D
3136.7	51670	6	6	.64	D
3152.2	67070	4	4	1.8	D
3180.7	51943	12	10	.70	D
3183.3	67012	8	6	.87	D
3209.2	51670	8	6	.68	D
3217.4	51584	6	4	.77	D
3234.1	65543	10	8	.92	D
3238.8	65680	12	10	.54	D
3295.4	64031	12	14	.32	D
3336.3	49493	2	2	.42	D
3339.8	49565	4	4	.49	D
3342.6	49706	6	6	.39	D
3347.8	49493	4	2	.52	D
3358.5	49565	6	4	1.1	D
3360.3	54785	8	8	1.3	D
3368.1	49706	8	6	1.4	D
3378.3	54626	8	6	.41	D
3379.4	54626	4	6	.48	D
3382.7	49352	6	6	.45	D
3391.4	49006	2	4	.19	D
3393.0	54500	2	4	.46	D
3393.8	54500	4	4	.66	D
3394.3	54500	6	4	.75	D
3402.4	54418	2	2	.80	D
3408.8	49352	8	6	.95	D
3421.2	48750	2	2	1.7	D
3422.7	49006	6	4	1.4	D
3433.3	48750	4	2	1.3	D
3511.8	48491	8	6	.079	D
4242.4	54785	10	8	.12	D
Cr XXI					
149.90	667110	1	3	160.	C
Cr XXII					
8.51	11800000	2	6*	1.2 (+4)	C
9.493	10534000	2	6*	2.5 (+4)	C
9.809	10553000	2	4	4.1 (+4)	B
9.865	10590000	4	6	4.9 (+4)	B
12.620	7924000	2	4	5.13(+4)	B
12.662	7898000	2	2	5.28(+4)	B
13.147	7964000	2	4	1.3 (+5)	B
13.294	7966000	4	6	1.6 (+5)	B
13.306	7964000	4	4	2.6 (+4)	C
25.2	11800000	2	6*	3750.	C
36.93	10534000	2	6*	7000.	C
37.52	10580000	6	10*	1.7 (+4)	B
223.00	448430	2	4	33.	B
279.69	357540	2	2	17.	B
Cr XXIII					
2.182	45830000	1	3	3.3 (+6)	B
Cobalt — Co I					
2987.2	33467	10	8	.050	C
2989.6	33440	10	10	.037	C
3013.6	33173	10	8	.016	C
3017.5	33946	8	6	.072	C
3042.5	33674	8	6	.020	C
3044.0	32842	10	10	.19	C
3048.9	34196	6	4	.078	C
3061.8	33467	8	8	.15	C
Co I					
3064.4	33440	8	10	.0068	C
3082.6	32431	10	12	.026	C
3086.8	34196	4	4	.19	C
3089.6	33173	8	8	.024	C
3098.2	33674	6	6	.027	C
3139.9	32655	8	6	.028	C
3147.1	33173	6	8	.045	C
3149.3	33151	6	4	.031	C
3395.4	34134	6	8	.26	C
3405.1	32842	10	10	.98	C
3409.2	33467	8	8	.42	C
3412.3	33440	8	10	.64	C
3412.6	29295	10	8	.12	C
3417.2	33946	6	6	.32	C
3431.6	29949	8	6	.11	C
3433.0	34196	4	4	1.1	C
3442.9	30444	6	4	.12	C
3443.6	33173	8	8	.63	C
3449.2	33674	6	6	.73	C
3449.4	32465	10	10	.16	C
3455.2	30743	4	2	.18	C
3462.8	33946	4	6	.87	C
3465.8	28845	10	12	.097	C
3483.4	32842	8	10	.062	C
3489.4	36092	8	6	1.6	C
3491.3	30444	4	4	.053	C
3495.7	33674	6	6	.45	C
3496.7	32733	8	6	.036	C
3518.4	36875	6	4	1.7	C
3520.1	29216	8	6	.034	C
3521.6	31871	10	8	.12	C
3523.4	33449	4	2	1.2	C
3526.9	28346	10	10	.12	C
3529.0	29735	6	8	.090	C
3529.8	32465	8	10	.48	C
3533.4	30103	4	6	.091	C
3550.6	29563	6	6	.042	C
3560.9	33151	4	4	.24	C
3564.9	32733	6	8	.086	C
3569.4	35451	8	6	1.6	C
3575.0	32655	6	6	.18	C
3575.4	28777	8	8	.094	C
3585.2	32028	8	8	.076	C
3587.2	36330	6	6	1.9	C
3594.9	29216	6	6	.086	C
3602.1	29563	4	4	.10	C
3605.4	31871	8	8	.039	C
3627.8	31700	8	10	.052	C
3631.4	28346	8	10	.0065	C
3647.7	29216	4	6	.012	C
3652.5	28777	6	8	.0095	C
3704.1	35451	6	4	.18	C
3745.5	34134	8	8	.077	C
3842.1	33463	8	8	.31	C
3845.5	33440	8	10	.49	C
3873.1	29295	10	8	.12	C
3874.0	29949	8	6	.12	C
3881.9	30444	6	4	.11	C
3894.1	34134	6	8	.81	C
3895.0	30743	4	2	.11	C
3909.9	25569	10	12	.0019	C
3936.0	32842	8	10	.15	C
3995.3	32465	8	10	.36	C
3997.9	33467	6	8	.079	C
4020.9	28346	10	10	.0092	D
4092.4	31871	8	8	.14	D
4118.8	32733	6	8	.34	C
4121.3	31700	8	10	.24	C
Copper — Cu I					
2024.3	49383	2	6*	.098	C
2165.1	46173	2	4	.51	B
2178.9	45879	2	4	.913	B
2181.7	45821	2	2	1.0	C
2225.7	44916	2	2	.46	C
2244.3	44544	2	4	.0119	B
2441.6	40944	2	2	.020	C
2492.2	40114	2	4	.0311	B
2618.4	49383	6	4	.307	C
2766.4	49383	4	4	.096	C

TRANSITION PROBABILITIES FOR SELECTED ATOMIC AND IONIC SPECIES

Wavelength ($\lambda[\text{Å}]$)	Upper energy level ($E_k[\text{cm}^{-1}]$)	Statistical weights g_i	g_k	Transition Probability ($A[10^8\text{ s}^{-1}]$)	Uncertainty
Cu I					
2824.4	46598	6	6	.078	C
2961.2	44963	6	8	.0376	C
3063.4	45879	4	4	.0155	C
3194.1	44544	4	4	.0155	C
3247.5	30784	2	4	1.39	B
3274.0	30535	2	2	1.37	B
3337.8	41153	6	8	.0038	C
4022.6	55388	2	4	.190	C
4062.6	55391	4	6	.210	C
4249.0	64472	2	2	.195	C
4275.1	62403	6	8	.345	C
4480.4	52849	2	2	.030	C
4509.4	64472	4	2	.275	C
4530.8	52849	4	2	.084	C
4539.7	63585	6	4	.212	C
4587.0	62948	8	6	.320	C
4651.1	62403	10	8	.380	C
4704.6	62403	8	8	.055	C
5105.5	30784	6	4	.020	C
5153.2	49935	2	4	.60	C
5218.2	49942	4	6	.75	C
5220.1	49935	4	4	.150	C
5292.5	62403	8	8	.109	C
5700.2	30784	4	4	.0024	C
5782.1	30535	4	2	.0165	C
Cu II					
2489.7	66419	5	5	.015	D
2544.8	108015	9	7	1.1	D
2689.3	108015	7	7	.41	D
2701.0	110380	3	5	.67	D
2703.2	110084	3	3	1.2	D
2713.5	108336	5	5	.68	D
2769.7	108015	7	7	.61	D
Dysprosium Dy I					
2862.7	34922	17	15	.065	D
2964.6	33722	17	17	.065	D
3147.7	35894	15	17	.11	D
3263.2	34770	15	13	.14	D
3511.0	32608	15	13	.31	D
3571.4	32126	15	13	.20	D
3757.1	34175	17	19	3.0	D
3868.8	33406	17	17	3.1	D
3967.5	32763	17	19	.87	D
4046.0	24709	17	15	1.5	D
4103.9	31411	13	11	1.7	D
4186.8	23878	17	17	1.32	C
4194.8	23832	17	17	.72	D
4211.7	23737	17	19	2.08	C
4218.1	27835	15	15	1.85	C
4221.1	27818	15	17	1.52	C
4225.2	30712	13	15	4.5	D
4268.3	27556	15	15	.036	D
4276.7	30427	13	13	.73	D
4292.0	27427	15	15	.058	D
4577.8	21839	17	19	.022	D
4589.4	21783	17	15	.13	D
4612.3	21675	17	15	.082	D
5077.7	19689	17	17	.0057	D
5301.6	18857	17	15	.011	D
5547.3	18022	17	17	.0027	C
5639.5	17727	17	19	.0047	C
5974.5	16733	17	17	.0040	C
5988.6	16694	17	15	.0053	C
6010.8	20766	15	15	.026	D
6088.3	20555	15	13	.035	D
6168.4	20341	15	17	.025	D
6259.1	15972	17	19	.0085	C
6579.4	15195	17	15	.0075	D
Erbium Er I					
3862.9	25880	13	13	2.5	D
4008.0	24943	13	15	2.6	D
4151.1	24083	13	11	1.8	D
Europium Eu I					
2372.9	42131	8	6	.19	D
Eu I					
2375.3	42087	8	8	.20	D
2379.7	42010	8	10	.20	D
2619.3	38167	8	10	.0070	D
2643.8	37813	8	8	.0066	D
2659.4	37591	8	10	.012	D
2682.6	37266	8	6	.012	D
2710.0	36890	8	10	.14	D
2724.0	36700	8	8	.12	D
2731.4	36601	8	8	.031	D
2732.6	36584	8	6	.037	D
2735.3	36549	8	10	.047	D
2738.6	36505	8	10	.013	D
2743.3	36442	8	6	.11	D
2745.6	36411	8	6	.050	D
2747.8	36382	8	6	.052	D
2772.9	36053	8	6	.010	D
2878.9	34726	8	10	.028	D
2892.5	34562	8	8	.10	D
2893.0	34556	8	6	.10	D
2909.0	34366	8	10	.069	D
2958.9	33787	8	6	.016	D
3059.0	32681	8	8	.038	D
3067.0	32596	8	10	.0091	D
3106.2	32185	8	10	.055	D
3111.4	32130	8	10	.30	D
3168.3	31554	8	10	.069	D
3185.5	31383	8	10	.0058	D
3210.6	31138	8	8	.11	D
3212.8	31116	8	8	.29	D
3219.0	31107	8	6	.18	D
3235.1	30902	8	10	.010	D
3241.4	30842	8	8	.023	D
3246.0	30798	8	6	.014	D
3247.6	30784	8	8	.023	D
3322.3	30091	8	6	.035	D
3334.3	29983	8	6	.34	D
3350.4	29839	8	10	.015	D
3353.7	29809	8	8	.0058	D
3457.1	28918	8	8	.0084	D
3467.9	28828	8	8	.010	D
3589.3	27853	8	6	.0069	D
4594.0	21761	8	10	1.4	D
4627.2	21605	8	8	1.3	D
4661.9	21445	8	6	1.3	D
5645.8	17707	8	6	.0054	D
5765.2	17341	8	8	.011	D
6018.2	16612	8	10	.0085	D
6291.3	15891	8	6	.0018	D
6864.5	14564	8	10	.0058	D
7106.5	14068	8	8	.0026	D
Fluorine F I					
806.96	123921	4	6	3.3	C
809.60	123922	2	4	2.8	C
951.87	105056	4	2	2.6	C
954.83	104731	4	4	6.4	C
955.55	105056	2	2	5.1	C
958.52	104731	2	4	1.3	C
6239.7	118428	6	4	.25	D
6348.5	118428	4	4	.18	D
6413.7	118428	2	4	.11	D
6708.3	117309	6	4	.014	D
6774.0	117164	6	6	.10	D
6795.5	117392	4	2	.052	D
6834.3	117309	4	4	.21	D
6856.0	116987	6	8	.42	D
6870.2	117392	2	2	.38	D
6902.5	117164	4	6	.32	D
6909.8	117309	2	4	.22	D
6966.4	119082	4	2	.11	D
7037.5	118937	4	4	.30	D
7127.9	119082	2	2	.38	D
7309.0	137599	6	8	.47	D
7311.0	118405	4	2	.39	D
7314.3	137590	4	6	.48	D
7332.0	116041	6	4	.31	D
7398.7	115918	6	6	.31	D
7425.7	116144	4	2	.34	D
7482.7	116041	4	4	.056	D
7489.2	118405	2	2	.11	D

Table 1b (Continued)
TRANSITION PROBABILITIES FOR SELECTED ATOMIC AND IONIC SPECIES

Wavelength (λ[Å])	Upper energy level (E_k[cm⁻¹])	g_i	g_k	Transition Probability (A[10⁸ s⁻¹])	Uncertainty
F I					
7514.9	116144	2	2	.052	D
7552.2	115918	4	6	.078	D
7573.4	116041	2	4	.10	D
7607.2	117873	4	4	.070	D
7754.7	117623	4	6	.30	D
7800.2	117873	2	4	.21	D
Gallium					
Ga I					
2195.4	45537	2	2	.019	C
2199.7	46274	4	2	.033	C
2214.4	45972	4	6	.012	C
2235.9	45537	4	2	.043	C
2255.0	44332	2	2	.031	C
2259.2	45076	4	6	.031	C
2294.2	43575	2	4	.070	C
2297.9	44332	4	2	.058	C
2338.2	43581	4	6	.098	C
2371.3	42159	2	2	.057	C
2418.7	42159	4	2	.10	C
2450.1	40803	2	4	.28	C
2500.2	40811	4	6	.34	C
2659.9	37585	2	2	.12	C
2719.7	37585	4	2	.23	C
2874.2	34782	2	4	1.2	C
2943.6	34788	4	6	1.4	C
2944.2	34782	4	4	.27	C
4033.0	24789	2	2	.49	C
4172.0	24789	4	2	.92	C
Ga II					
829.60	120540	1	3	.22	D
1414.4	70700	1	3	18.8	C
Germanium					
Ge I					
1944.7	51978	3	1	.70	C
1955.1	51705	3	3	.28	C
1988.3	51705	5	3	.25	C
1998.9	51438	5	5	.55	C
2041.7	48963	1	3	1.1	C
2065.2	48963	3	3	.85	C
2068.7	48882	3	5	1.2	C
2086.0	48480	3	5	.40	C
2094.3	49144	5	7	.97	C
2105.8	48882	5	5	.17	C
2256.0	51438	3	5	.032	C
2417.4	48480	5	5	.96	C
2498.0	40021	1	3	.13	C
2533.2	40021	3	3	.10	C
2589.2	40021	5	3	.051	C
2592.5	39118	3	5	.71	C
2651.2	39118	5	5	2.0	C
2651.6	37702	1	3	.85	C
2691.3	37702	3	3	.61	C
2709.6	37452	3	1	2.8	C
2754.6	37702	5	3	1.1	C
3039.1	40021	5	3	2.8	C
3124.8	39118	5	5	.031	C
3269.5	37702	5	3	.29	C
4226.6	40021	1	3	.21	C
4685.8	37702	1	3	.095	C
Ge II					
999.10	100090	2	4	1.9	D
1016.6	100131	4	6	2.1	D
1017.1	100090	4	4	.35	D
1055.0	94784	2	2	.69	D
1075.1	94784	4	2	1.3	D
1237.1	80837	2	4	19.	D
1261.9	81013	4	6	22.	D
1264.7	80837	4	4	3.5	D
1602.5	62403	2	2	3.4	D
1649.2	62403	4	2	6.5	D
4741.8	100090	2	4	.46	D
4814.6	100131	4	6	.51	D
4824.1	100090	4	4	.086	D
5131.8	100318	4	6	1.9	D
5178.5	100318	6	6	.13	D
5178.6	100317	6	8	2.0	D
5893.4	79367	2	4	.92	D
6021.0	79007	2	2	.84	D

Wavelength (λ[Å])	Upper energy level (E_k[cm⁻¹])	g_i	g_k	Transition Probability (A[10⁸ s⁻¹])	Uncertainty
Ge II					
6336.4	94784	2	2	.44	D
6484.2	94784	4	2	.85	D
Gold					
Au I					
2428.0	41174	2	4	1.5	D
2676.0	37359	2	2	1.1	D
Helium					
He I					
510.00	196079	1	3	.462	B
512.10	195275	1	3	.717	B
515.62	193943	1	3	1.3	B
522.21	191493	1	3	2.46	A
537.03	186209	1	3	5.66	A
584.33	171135	1	3	17.99	AA
2696.1	196935	3	9*	.00550	B
2723.2	196567	3	9*	.00780	B
2763.8	196027	3	9*	.0111	B
2829.1	195193	3	9*	.017	B
2945.1	193801	3	9*	.0320	A
3187.7	191217	3	9*	.05639	AA
3354.6	196079	1	3	.0130	B
3447.6	195275	1	3	.0232	A
3554.4	197213	9	15*	.0131	A
3587.3	196955	9	15*	.0205	C
3613.6	193943	1	3	.0390	A
3634.2	196595	9	15*	.0261	A
3705.0	196070	9	15*	.0444	C
3819.6	195260	9	15*	.0636	A
3833.6	197213	3	5	.00971	B
3867.5	194936	9	3*	.025	B
3871.8	196956	3	5	.0126	C
3888.7	185565	3	9*	.09478	AA
3926.5	196596	3	5	.0195	A
3964.7	191493	1	3	.0719	A
4009.3	196070	3	5	.0279	C
4026.2	193917	9	15*	.116	A
4120.8	193347	9	3*	.0444	A
4143.8	195261	3	5	.0485	A
4387.9	193918	3	5	.0894	A
4437.6	193664	3	1	.033	B
4471.5	191445	9	15*	.246	A
4713.2	190298	9	3*	.0955	A
4921.9	191447	3	5	.198	A
5015.7	186210	1	3	.1338	AA
5047.7	190940	3	1	.0675	A
5875.7	186102	9	15*	.7053	AA
6678.2	186105	3	5	.6339	AA
7065.2	183237	9	3*	.2786	AA
7281.4	184865	3	1	.1829	AA
8361.7	195193	3	9*	.00334	A
9463.6	193801	3	9*	.00501	A
9603.4	195275	1	3	.00610	A
9702.6	195868	3	9*	.00858	B
10311.	195260	9	15*	.0201	A
10668.	194936	9	3*	.0152	A
10830.	169087	3	9*	.1022	AA
10913.	195262	15	21*	.0212	B
10917.	195262	5	7	.0212	B
10997.	195193	15	9*	.0013	A
11013.	193943	1	3	.0100	A
11045.	195261	3	5	.0185	A
11226.	195115	3	1	.0108	A
11969.	193917	9	15*	.0358	A
12528.	191217	3	9*	.00710	A
12756.	193943	5	3	.0012	B
12785.	193921	15	21*	.0462	B
12791.	193921	5	7	.0461	B
12846.	193347	9	3*	.0289	A
12968.	193918	3	5	.0343	A
12985.	193801	15	9*	.0025	B
Indium					
In I					
2710.3	39098	4	6	.4	D
2560.2	39048	2	4	.4	D
3256.1	32915	4	6	1.3	D
3039.4	32892	2	4	1.3	D
4101.8	24373	2	2	.56	C
4511.3	24373	4	2	1.02	C

Table 1b (Continued)
TRANSITION PROBABILITIES FOR SELECTED ATOMIC AND IONIC SPECIES

Wavelength (λ[Å])	Upper energy level (E_k[cm^{-1}])	g_i	g_k	Transition Probability (A[10^8 s^{-1}])	Uncertainty
In II					
2941.1	97025	3	1	1.4	D
Iodine					
I I					
1782.8	56093	4	4	2.71	C
1830.4	54633	4	6	.16	D
Iron					
Fe I					
1934.5	51692	9	7	.25	C
1937.3	51619	9	7	.22	C
1940.7	51945	7	5	.26	C
2084.1	47967	9	7	.37	C
2102.4	47967	7	7	.088	C
2113.0	48290	1	3	.19	C
2132.0	46889	9	9	.076	C
2166.8	46137	9	7	2.7	C
2191.8	46314	5	5	1.2	C
2196.0	46410	3	3	1.2	C
2200.7	46314	3	5	.28	C
2259.5	44244	9	11	.070	C
2276.0	43923	9	7	.17	C
2277.1	51630	7	5	37.	C
2287.3	44411	5	3	.34	C
2294.4	44459	3	1	.61	C
2300.1	44166	5	7	.080	C
2309.0	44184	3	5	.15	C
2313.1	43923	5	7	.14	C
2320.4	43499	7	9	.12	C
2373.6	42533	7	7	.067	C
2374.3	43079	1	3	.29	C
2462.2	41018	7	5	.15	C
2462.6	40594	9	9	.58	C
2479.8	41018	5	5	1.8	C
2483.3	40257	9	11	4.9	C
2488.1	40594	7	9	4.7	C
2490.6	40842	5	7	3.8	C
2491.2	41018	3	5	3.0	C
2501.1	39970	9	7	.68	C
2510.8	40231	7	5	1.3	C
2518.1	40405	5	3	1.9	C
2522.8	39626	9	9	2.9	C
2524.3	40491	3	1	3.4	C
2527.4	39970	7	7	1.9	C
2529.1	40231	5	5	.98	C
2535.6	40405	1	3	.97	C
2541.0	40231	3	5	.92	C
2546.0	39970	5	7	.67	C
2549.6	39626	7	9	.36	C
2584.5	45608	11	13	.46	C
2719.0	36767	9	7	1.4	C
2720.9	37158	7	5	1.1	C
2723.6	37410	5	3	.64	C
2737.3	37410	3	3	.85	C
2742.4	37158	5	5	.63	C
2744.1	37410	1	3	.35	C
2750.1	36767	7	7	.39	C
2756.3	37158	3	5	.20	C
2788.1	42784	11	13	.63	C
2894.5	52916	5	5	.63	C
2899.4	52858	5	3	.61	C
2923.8	60549	11	11	1.7	C
2925.4	56423	7	9	.19	C
2936.9	34040	9	9	.14	C
2954.7	52213	5	7	.12	C
2966.9	33695	9	11	.272	B
2980.5	55791	7	7	.22	C
2983.6	33507	9	7	.280	B
2990.4	55430	9	11	.40	C
2996.4	52916	3	5	.19	C
2999.5	40257	11	11	.23	C
3000.9	34017	5	3	.642	B
3008.1	34122	3	1	1.07	B
3009.1	52613	13	11	.079	C
3009.6	40594	9	9	.18	C
3011.5	55446	7	9	.48	C
3019.0	40842	7	7	.15	C
3037.4	33802	3	5	.32	C
3042.7	40842	5	7	.066	C
Fe I					
3047.6	33507	5	7	.284	B
3053.1	52297	3	5	.18	C
3057.4	39626	11	9	.45	C
3059.1	33096	7	9	.18	C
3067.2	39970	9	7	.35	C
3075.7	40231	7	5	.30	C
3083.7	40405	5	3	.35	C
3098.2	53983	11	11	.11	C
3100.7	39970	7	7	.16	C
3119.5	51668	11	9	.096	C
3120.4	51826	9	7	.10	C
3160.7	51192	9	9	.19	C
3161.9	50968	11	13	.12	C
3166.4	52213	9	7	.14	C
3175.4	50833	11	11	.13	C
3199.5	50808	9	9	.27	C
3205.4	51208	3	3	1.2	C
3215.9	50999	5	5	.81	C
3217.4	50423	11	9	.23	C
3225.8	50342	11	13	1.0	C
3231.0	50699	7	5	.39	C
3233.1	57028	13	15	.55	C
3234.0	50475	9	9	.20	C
3248.2	50534	7	7	.22	C
3253.6	56951	7	9	.18	C
3254.4	57070	11	13	.51	C
3265.6	48163	7	5	.39	C
3271.0	48290	5	3	.67	C
3280.3	57104	9	11	.55	C
3282.9	56859	3	5	.31	C
3292.0	56593	7	9	.62	C
3292.6	48290	3	3	.31	C
3306.0	47067	5	7	.48	C
3307.2	56334	13	13	.20	C
3314.7	56783	5	7	.70	C
3323.7	52916	5	5	.31	C
3328.9	56383	11	11	.27	C
3337.7	51668	11	9	.067	C
3355.2	56423	9	9	.33	C
3369.5	51668	9	9	.25	C
3370.8	51374	11	11	.34	C
3380.1	51826	7	7	.24	C
3384.0	47093	7	7	.11	C
3392.7	47017	7	7	.26	C
3399.3	47136	5	5	.39	C
3402.3	55490	13	13	.29	C
3406.4	55754	3	5	.30	C
3407.5	46889	7	9	.60	C
3410.2	56859	3	5	.48	C
3411.4	51305	9	9	.065	C
3413.1	47017	5	7	.37	C
3417.8	47177	3	3	.52	C
3418.5	47171	3	1	1.3	C
3424.3	46745	7	7	.21	C
3425.0	53763	9	7	.29	C
3427.1	46721	7	9	.56	C
3428.2	46889	5	5	.22	C
3445.1	46745	5	7	.28	C
3447.3	46727	5	5	.11	C
3450.3	46902	3	3	.24	C
3495.3	49243	9	7	.13	C
3497.1	46317	7	7	.15	C
3521.8	46314	3	5	.11	C
3524.1	49243	7	5	.091	C
3527.8	51335	9	9	.20	C
3529.8	51567	3	3	.78	C
3536.6	51461	5	7	.80	C
3540.1	51350	7	9	.12	C
3541.1	51229	9	11	.64	C
3542.1	51335	7	9	.76	C
3552.8	51331	5	5	.17	C
3553.7	56951	11	9	.83	C
3556.9	51103	9	11	.45	C
3559.5	52858	3	3	.22	C
3560.7	54301	7	9	.077	C
3565.4	35768	7	9	.39	C
3570.1	35379	9	11	.677	B
3572.0	50833	11	11	.25	C
3581.2	34844	11	13	1.02	B
3582.2	54014	13	11	.25	C
3587.0	35856	5	5	.17	C

Table 1b (Continued)
TRANSITION PROBABILITIES FOR SELECTED ATOMIC AND IONIC SPECIES

Wavelength (λ[Å])	Upper energy level (E_k[cm^{-1}])	Statistical weights g_i	g_k	Transition Probability (A[10^8 s^{-1}])	Uncertainty	Wavelength (λ[Å])	Upper energy level (E_k[cm^{-1}])	Statistical weights g_i	g_k	Transition Probability (A[10^8 s^{-1}])	Uncertainty
				Fe I						Fe I	
3594.6	50808	9	9	.28	C	3840.4	34017	5	3	.470	B
3599.6	56593	11	9	.19	C	3841.0	38996	5	3	1.4	C
3603.2	49461	11	11	.27	C	3843.3	50587	9	7	.48	C
3605.5	49727	9	9	.65	C	3846.8	52213	7	7	.67	C
3606.7	49434	11	13	.84	C	3850.0	34122	3	1	.606	B
3608.9	35856	3	5	.814	B	3859.2	45295	13	11	.087	C
3610.2	50342	13	13	.50	C	3859.9	25900	9	9	.0970	B
3612.1	50523	11	13	.077	C	3865.5	34017	3	3	.155	B
3617.8	51969	5	7	.66	C	3867.2	50187	5	5	.35	C
3618.8	35612	5	7	.73	C	3871.8	49604	11	11	.070	C
3621.5	49604	9	11	.52	C	3872.5	33802	5	5	.105	B
3622.0	49851	7	7	.53	C	3873.8	45428	11	9	.082	C
3623.2	46982	13	13	.076	C	3878.0	33507	7	7	.0772	B
3631.5	35257	7	9	.52	C	3883.3	51969	7	7	.17	C
3632.0	52297	3	5	.50	C	3884.4	47453	11	9	.042	C
3638.3	49727	7	9	.27	C	3888.5	38678	5	5	.27	C
3640.4	49461	9	11	.39	C	3891.9	53230	3	3	.40	C
3645.8	52512	1	3	.58	C	3893.4	49461	11	11	.14	C
3647.8	34782	9	11	.292	B	3900.5	51771	7	7	.086	C
3649.5	49109	11	9	.43	C	3902.9	38175	7	7	.24	C
3651.5	49628	7	9	.64	C	3903.9	49727	9	9	.097	C
3655.5	50187	5	5	.12	C	3916.7	51630	13	11	.12	C
3659.5	47106	9	9	.068	C	3919.1	49628	9	9	.045	C
3669.5	49243	9	7	.30	C	3942.4	48305	3	5	.11	C
3670.1	51023	11	13	.078	C	3951.2	51708	3	5	.36	C
3676.3	47835	9	11	.061	C	3952.6	47008	11	11	.052	C
3677.6	49433	7	5	.82	C	3953.2	49628	7	9	.043	C
3682.2	55754	5	5	1.7	C	3963.1	51705	3	5	.17	C
3684.1	49135	9	7	.34	C	3967.4	51826	9	7	.24	C
3686.0	50833	9	11	.26	C	3969.3	37163	9	7	.24	C
3687.5	34040	11	9	.0801	B	3971.3	46889	11	9	.068	C
3690.7	55907	11	11	.28	C	3973.7	53763	5	7	.080	C
3694.0	51570	5	7	.70	C	3977.7	42860	5	5	.082	C
3697.4	51219	7	7	.21	C	3981.8	47106	9	9	.046	C
3701.1	51192	7	9	.49	C	3984.0	47093	9	7	.089	C
3704.5	48703	11	9	.14	C	3985.4	51708	5	5	.082	C
3709.2	34329	9	7	.156	B	3997.0	54811	9	9	.074	C
3719.9	26875	9	11	.163	B	3997.4	47008	11	9	.16	C
3724.4	45221	5	7	.13	C	3998.1	46721	11	9	.075	C
3727.6	34547	7	5	.225	B	4005.2	37521	7	5	.22	C
3730.4	51374	9	11	.13	C	4014.5	53722	11	9	.24	C
3732.4	44512	5	5	.28	C	4017.2	49461	9	11	.053	C
3734.9	33695	11	11	.902	B	4021.9	47106	7	9	.10	C
3737.1	27167	7	9	.142	C	4032.0	51201	3	5	.086	C
3738.3	53094	11	13	.38	C	4045.8	36686	9	9	.75	C
3740.2	52954	7	7	.19	C	4062.4	47556	3	3	.23	C
3742.6	50423	9	9	.11	C	4063.6	37163	7	7	.69	C
3744.1	51208	5	3	.38	C	4068.0	50475	9	9	.17	C
3745.6	27395	5	7	.115	B	4070.8	50699	7	5	.14	C
3749.5	34040	9	9	.764	B	4071.7	37521	5	5	.80	C
3753.6	44184	7	5	.11	C	4073.8	50880	5	3	.19	C
3756.9	55430	11	11	.25	C	4074.8	49109	9	9	.056	C
3758.2	34329	7	7	.634	B	4076.6	50423	9	9	.20	C
3760.1	45978	13	15	.057	C	4084.5	51350	11	9	.12	C
3763.8	34547	5	5	.544	B	4085.3	50611	7	7	.12	C
3765.5	52655	13	15	.99	C	4098.2	50534	7	7	.082	C
3767.2	34692	3	3	.640	B	4107.5	47177	5	3	.25	C
3787.2	55754	5	5	.12	C	4109.8	47272	3	3	.19	C
3787.9	34547	3	5	.129	B	4113.0	58002	11	13	.15	C
3794.3	46136	9	11	.046	C	4127.6	47272	1	3	.16	C
3795.0	34329	5	7	.115	B	4132.9	47136	3	5	.11	C
3798.5	33695	9	11	.0323	B	4134.7	47017	5	7	.18	C
3799.5	34040	7	9	.0732	B	4137.0	51708	3	5	.23	C
3804.0	53155	11	9	.052	C	4143.9	36686	7	9	.16	C
3805.3	52899	9	11	1.0	C	4149.4	50968	11	13	.043	C
3806.2	53808	3	3	.25	C	4153.9	51462	7	9	.24	C
3806.7	52613	11	11	.55	C	4154.8	51229	9	11	.15	C
3807.5	44184	3	5	.097	C	4156.8	46889	5	5	.19	C
3810.8	52858	5	3	.24	C	4170.9	48305	5	5	.072	C
3813.0	33947	7	5	.0792	B	4172.1	50187	7	5	.12	C
3813.9	55526	13	11	.091	C	4175.6	46889	3	5	.17	C
3815.8	38175	9	7	1.3	C	4184.9	46727	5	5	.12	C
3820.4	33096	11	9	.668	B	4187.0	43634	7	5	.23	C
3821.2	52514	11	13	.70	C	4187.8	43435	9	7	.16	C
3821.8	47197	5	5	.089	C	4196.2	51219	7	7	.11	C
3825.9	33507	9	7	.598	B	4210.3	43764	3	3	.20	C
3827.8	38678	7	5	1.1	C	4217.6	51370	3	5	.24	C
3833.3	46721	9	9	.059	C	4219.4	52514	11	13	.38	C
3834.2	33802	7	5	.453	B	4222.2	43435	7	7	.063	C
3836.3	52683	5	5	.39	C	4224.2	50833	9	11	.14	C
3839.3	50614	9	9	.29	C	4225.5	51219	5	7	.17	C

Wavelength ($\lambda[\text{Å}]$)	Upper energy level ($E_k[\text{cm}^{-1}]$)	g_i	g_k	Transition Probability ($A[10^8\ \text{s}^{-1}]$)	Uncertainty
Fe I					
4233.6	43634	3	5	.20	C
4238.8	50980	7	9	.22	C
4246.1	52916	7	5	.069	C
4250.1	43435	5	7	.23	C
4282.4	40895	7	5	.13	C
4307.9	35768	7	9	.35	C
4315.1	40895	5	5	.090	C
4325.8	36079	5	7	.51	C
4327.1	51708	5	5	.094	C
4369.8	47453	9	9	.074	C
4383.5	34782	9	11	.46	C
4388.4	51837	7	7	.13	C
4401.3	51771	7	7	.069	C
4404.8	35257	7	9	.25	C
4415.1	35612	5	7	.13	C
4443.2	45552	1	3	.13	C
4466.6	45221	5	7	.13	C
4469.4	51837	5	7	.27	C
4528.6	39626	7	9	.063	C
4547.9	50587	5	7	.078	C
4736.8	47006	9	11	.050	C
4789.7	49477	5	5	.084	C
4859.7	43764	5	3	.15	C
4871.3	43634	7	5	.22	C
4872.1	43764	3	3	.24	C
4878.2	43764	1	3	.11	C
4890.8	43634	5	5	.21	C
4891.5	43435	9	7	.30	C
4903.3	43634	3	5	.054	C
4919.0	43435	7	7	.17	C
4920.5	43163	11	9	.36	C
4966.1	47006	11	11	.037	C
4973.1	52040	3	3	.12	C
4989.0	53546	7	7	.050	C
5001.9	51294	9	7	.40	C
5014.9	51740	7	5	.31	C
5022.2	52040	5	3	.27	C
5074.7	53739	9	11	.15	C
5090.8	53967	7	5	.21	C
5121.6	54067	5	5	.086	C
5137.4	53155	11	9	.11	C
5208.6	45334	7	5	.060	C
5232.9	42816	9	11	.15	C
5242.5	48383	13	11	.032	C
5263.3	45334	5	5	.061	C
5266.6	43163	7	9	.088	C
5281.8	43435	5	7	.038	C
5302.3	45334	3	5	.073	C
5324.2	44677	9	9	.15	C
5339.9	45061	5	7	.071	C
5367.5	54237	7	9	.59	C
5370.0	53874	9	11	.48	C
5383.4	53353	11	13	.59	C
5393.2	44677	7	9	.037	C
5410.9	54555	7	9	.49	C
5415.2	53841	11	13	.68	C
5463.3	54067	9	9	.33	C
5473.9	51771	7	7	.057	C
5569.6	45509	5	3	.21	C
5572.8	45334	7	5	.22	C
5586.8	45061	9	7	.19	C
5615.6	44677	11	9	.17	C
5624.5	45334	5	5	.062	C
5658.8	45061	7	7	.042	C
5753.1	51740	3	5	.072	C
5763.0	51294	5	7	.10	C
Fe II					
2029.2	65110	10	8	.076	D
2040.7	64832	10	10	.46	D
2051.0	65110	8	8	.42	D
2296.7	65110	10	8	.037	D
2303.3	64832	12	10	.054	D
2369.2	64832	10	10	.026	D
2379.0	64832	8	10	.064	D
2388.4	62662	10	12	.14	D
2433.5	62662	10	12	.091	D
2555.0	65110	8	8	.019	D
2559.8	65110	6	8	.22	D
2561.6	64832	10	10	.0081	D
2573.2	64832	8	10	.11	D

Wavelength ($\lambda[\text{Å}]$)	Upper energy level ($E_k[\text{cm}^{-1}]$)	g_i	g_k	Transition Probability ($A[10^8\ \text{s}^{-1}]$)	Uncertainty
Fe II					
2591.5	46967	6	6	.52	C
2592.8	71433	14	16	2.25	C
2598.0	64832	10	10	.020	D
2598.4	38859	8	6	1.3	C
2599.4	38459	10	10	2.22	C
2623.1	70987	14	14	.092	D
2625.5	70987	12	14	2.04	C
2625.7	38459	8	10	.34	C
2664.7	64832	8	10	1.50	C
2666.6	65110	6	8	1.62	C
2684.9	62662	12	12	.0043	D
2712.4	62662	10	12	.11	D
2753.3	62662	10	12	1.71	C
2879.2	65110	10	8	.029	D
2902.5	64832	10	10	.038	D
2910.8	65110	8	8	.0055	D
3002.3	65110	6	8	.018	D
3044.8	64832	8	10	.011	D
3131.7	64832	12	10	.012	D
3162.8	65110	8	8	.042	D
3186.7	45044	4	4	.039	C
3187.3	64832	10	10	.028	D
3213.3	44785	4	6	.065	C
3277.3	38459	8	10	.0023	D
3360.1	62662	12	12	.0084	D
4515.3	45080	6	6	.0018	D
4520.2	44754	10	8	.0010	D
4583.8	44447	10	8	.0063	C
4629.3	44233	10	10	.0013	D
4920.9	45021	6	4	.030	C
4954.0	65110	6	8	.0016	D
5018.4	43239	6	6	.026	C
5019.5	64832	8	10	.0015	D
Fe XVI					
50.350	1986100	2	4	2120.	B
50.555	1978040	2	2	2100.	B
62.879	1867530	2	2	1110.	B
63.719	1867530	4	2	2140.	B
335.41	298140	2	4	77.5	B
360.80	277160	2	2	62.3	B
Fe XVIII					
93.931	1064610	4	2	690.	D
103.95	1064610	2	2	260.	D
Fe XXI					
98.37	1265800	5	3	700.	D
99.43	1095500	9	3*	1000.	D
113.34	1132280	5	5	470.	D
Fe XXIII					
132.83	752840	1	3	195.	B
Fe XXIV					
10.619	9417100	2	4	7.28(+4)	B
10.663	9378200	2	2	7.51(+4)	B
11.124	9467100	6	10*	2.18(+5)	B
192.04	520720	2	4	43.	B
255.10	392000	2	2	18.	B
Krypton					
Kr I					
4274.0	103363	5	5	.026	D
4351.4	108822	3	1	.032	D
4362.6	102887	5	3	.0084	D
4376.1	103762	3	1	.056	D
4400.0	108568	3	5	.020	D
4410.4	108514	3	3	.0044	D
4425.2	108438	3	3	.0097	D
4453.9	103363	3	5	.0078	D
4463.7	103314	3	3	.023	D
4502.4	103121	3	5	.0092	D
5562.2	97945	5	5	.0028	D
5570.3	97919	5	3	.021	D
5649.6	102887	1	3	.0037	D
5870.9	97945	3	5	.018	D
6904.7	105648	3	5	.013	D
7224.1	105007	3	5	.014	D
7587.4	94093	3	1	.51	D
7601.5	93123	5	5	.31	D
7685.2	98855	3	1	.49	D
7694.5	92964	5	3	.056	D

Table 1b (Continued)
TRANSITION PROBABILITIES FOR SELECTED ATOMIC AND IONIC SPECIES

Wavelength (λ[Å])	Upper energy level (E_k[cm^{-1}])	Statistical weights g_i	g_k	Transition Probability (A[10^8 s^{-1}])	Uncertainty
Kr I					
7854.8	97919	1	3	.23	D
8059.5	97596	1	3	.19	D
8104.4	92307	5	5	.13	D
8112.9	92294	5	7	.36	D
8190.1	93123	3	5	.11	D
8263.2	97945	3	5	.35	D
8281.1	97919	3	3	.19	D
8298.1	92964	3	3	.32	D
8508.9	97596	3	3	.24	D
8776.7	92307	3	5	.27	D
8928.7	91169	5	3	.37	D
Kr II					
4250.6	141996	4	4	.12	D
4292.9	138381	4	4	.96	D
4355.5	135783	6	8	1.0	D
4431.7	140163	2	2	1.8	D
4436.8	140137	2	2	.66	D
4577.2	149705	6	8	.96	D
4583.0	157885	6	4	.76	D
4615.3	140137	4	4	.54	D
4619.2	140119	4	6	.81	D
4633.9	149173	4	4	.71	D
4658.9	134288	6	4	.65	D
4739.0	133926	6	6	.76	D
4762.4	141996	4	2	.42	D
4765.7	136071	4	4	.67	D
4811.8	138381	2	4	.17	D
4825.2	141723	2	4	.19	D
4832.1	135783	4	2	.73	D
5208.3	134288	4	4	.14	D
5308.7	133926	4	6	.024	D
7407.0	133926	6	6	.070	D
Lead — Pb I					
2022.0	49440	1	3	.052	D
2053.3	48687	1	3	.12	D
2170.0	46068	1	3	1.5	D
2401.9	49440	3	3	.19	D
2446.2	48687	3	3	.25	D
2476.4	48189	3	5	.28	D
2577.3	49440	5	3	.50	D
2613.7	46068	3	3	.27	D
2614.2	46061	3	5	1.9	D
2628.3	48687	3	5	.031	D
2657.1	45443	3	5	9.8 (−4)	D
2663.2	48189	5	5	.71	D
2802.0	46329	5	7	1.6	D
2823.2	46061	5	5	.26	D
2833.1	35287	1	3	.58	D
2873.3	45443	5	5	.37	D
3572.7	49440	5	3	.99	D
3639.6	35287	3	3	.34	D
3671.5	48687	5	3	.44	D
3683.5	34960	3	1	1.5	D
3739.9	48189	5	5	.73	D
4019.6	46329	5	7	.035	D
4057.8	35287	5	3	.89	D
4062.1	46068	5	3	.92	D
4168.0	45443	5	5	.012	D
5005.4	49440	1	3	.27	D
5201.4	48687	1	3	.19	D
7229.0	35287	5	3	.0089	D
Lithium — Li I					
2741.2	36470	2	6*	.013	D
3232.7	30925	2	6*	.012	B
4602.8	36623	2	4	.197	C
4602.9	36623	4	6	.24	C
6103.5	31283	2	4	.60	C
6103.7	31283	4	6	.71	C
6103.7	31283	4	4	.12	C
6707.8	14904	2	4	.372	B
6707.9	14904	2	2	.372	B
Lutetium — Lu I					
3376.5	29608	4	4	2.23	B
3567.8	28020	4	6	.59	C

Wavelength (λ[Å])	Upper energy level (E_k[cm^{-1}])	Statistical weights g_i	g_k	Transition Probability (A[10^8 s^{-1}])	Uncertainty
Lu I					
3620.3	29608	6	4	.011	D
3841.2	28020	6	6	.25	C
4518.6	22125	4	4	.21	B
Magnesium — Mg I					
2025.8	49347	1	3	.84	D
2779.8	57854	9	9*	5.2	C
2850.0	56968	9	15*	.23	C
2852.1	35051	1	3	**4.95**	**B**
3094.9	54192	9	15*	.52	C
3329.9	51873	1	3	.033	C
3332.2	51873	3	3	.097	C
3336.7	51873	5	3	.16	C
3835.3	47957	9	15*	1.68	B
4703.0	56308	3	5	.255	C
5167.3	41197	1	3	.116	B
5172.7	41197	3	3	.346	B
5183.6	41197	5	3	.575	B
5528.4	53135	3	5	.199	C
Mg II					
1239.9	80650	2	4	.014	C
1240.4	80620	2	2	.014	C
2660.8	109062	10	14*	.38	D
2790.8	71491	2	4	4.0	C
2795.5	35671	2	4	2.6	C
2797.9	71491	4	4	.79	D
2798.1	71490	4	6	4.8	C
2802.7	35669	2	2	2.6	C
2928.8	69805	2	2	1.2	C
2936.5	69805	4	2	2.3	C
3104.8	103690	10	14*	.81	C
3848.2	97469	6	4	.028	C
3848.3	97469	4	4	.0030	D
3850.4	97455	4	2	.030	D
4481.2	93800	10	14*	2.23	C
9218.3	80650	2	4	.36	C
9244.3	80620	2	2	.36	C
Mg IV					
320.99	311532	4	2	120.	D
323.31	311532	2	2	59.	D
1219.0	679098	6	6	5.9	D
1375.5	677361	4	4	4.5	D
1459.6	612231	6	4	4.6	D
1495.5	679098	4	4	6.4	D
1510.7	678426	4	4	6.7	D
1683.0	603137	6	8	5.8	D
1698.8	604002	4	4	3.9	D
1893.9	596521	6	6	2.8	D
Mg VI					
269.92	424600	10	6*	310.	D
292.53	424600	6	6*	90.	D
314.64	400600	6	2*	180.	D
349.15	340600	10	10*	61.	C
387.94	340600	6	10*	13.	D
399.29	250445	4	2	28.	C
400.68	249578	4	4	28.	C
403.32	247945	4	6	27.	C
Mg VII					
277.01	362128	3	3	95.	C
278.41	362128	5	3	150.	C
280.74	397700	5	3	200.	D
319.02	354900	5	5	89.	C
366.42	274922	9	9*	44.	C
433.04	232934	9	15*	16.	C
1334.3	1125850	5	5	5.3	C
1410.0	1196770	5	5	2.57	C
1487.0	1192185	3	5	3.02	C
1487.9	1193061	5	7	3.66	C
Mg VIII					
74.976	1335965	6	10*	4300.	D
315.02	320742	4	4	120.	C
342.29	524437	10	6*	63.	D
353.86	414380	4	4	38.9	C
356.00	414380	6	4	57.	C
428.52	465654	10	10*	32.4	C
434.62	232290	6	10*	16.	C
489.33	524437	6	6*	39.	D

Wavelength (λ[Å])	Upper energy level (E_k[cm^-1])	g_i	g_k	Transition Probability (A[10^8 s^{-1}])	Uncertainty
Mg VIII					
686.92	465654	6	10*	9.4	D
Mg IX					
62.751	1593600	1	3	2870.	B
67.189	1631500	9	15*	6200.	C
71.965	1532700	9	3*	1220.	C
72.312	1654583	3	5	4430.	C
77.737	1558076	3	1	392.	C
368.07	271687	1	3	52.7	B
438.69	499640	3	1	70.	C
443.74	368500	9	9*	41.9	C
749.55	405100	3	5	8.2	C
1639.8	1654583	3	5	2.1	D
2814.2	1593600	1	3	.335	C
Mg X					
57.876	1727832	2	4	2090.	B
57.920	1726519	2	2	2090.	B
63.152	1743410	2	4	5600.	B
63.295	1743880	4	6	6700.	B
609.79	163976	2	4	7.53	B
624.94	159929	2	2	7.01	B
2212.5	1727832	2	4	.964	B
2278.7	1726519	2	2	.882	B
5918.7	1743410	2	4	.0320	C
6229.6	1743880	4	6	.0330	C
Mg XI					
7.310	13680600	1	3	1.15(+4)	B
7.473	13381100	1	3	2.27(+4)	B
7.850	12738400	1	3	5.50(+4)	B
9.169	10007300	1	3	1.97(+5)	B
Manganese					
Mn I					
2794.8	35770	6	8	3.7	C
2798.3	35726	6	6	3.6	C
2801.1	35690	6	4	3.7	C
3007.7	58520	6	8	.18	D
3011.4	58486	8	10	.31	D
3016.5	58427	10	12	.29	D
3043.4	58137	8	8	.59	D
3044.6	49888	10	8	.57	D
3045.6	58110	10	10	.67	D
3045.8	58110	8	10	.17	D
3046.6	67753	10	12	.13	D
3047.0	58075	12	12	.61	D
3054.4	50013	8	6	.46	D
3066.0	49888	8	8	.16	D
3073.1	50099	4	4	.37	D
3082.7	66569	14	14	.29	D
3110.7	59340	6	8	.27	D
3113.8	66356	12	10	.26	D
3122.9	66356	10	10	.19	D
3126.9	66395	8	6	.23	D
3132.3	66855	10	10	.21	D
3132.8	66334	8	8	.27	D
3160.2	66574	10	12	.14	D
3175.6	66523	8	10	.18	D
3175.7	66419	10	12	.12	D
3190.0	66454	6	8	.16	D
3201.1	66395	4	6	.22	D
3212.9	48168	10	10	.16	D
3228.1	48021	10	12	.64	D
3230.2	65887	10	12	.19	D
3230.7	48226	8	8	.35	D
3238.7	65909	8	10	.12	D
3243.8	48271	6	6	.53	D
3256.1	48271	4	6	.50	D
3258.4	48318	2	2	.97	D
3260.2	48301	2	4	.38	D
3264.7	47904	8	10	.14	D
3267.8	64732	14	14	.35	D
3268.7	64410	6	8	.33	D
3270.4	64820	12	12	.26	D
3273.0	64888	10	10	.27	D
3278.6	47775	8	8	.0091	D
3298.2	57512	6	4	.28	D
3420.8	63364	14	14	.12	D
3463.7	66600	8	8	.32	C
3470.0	66600	6	8	.24	C
3511.8	65887	12	12	.27	C
3535.3	65909	10	10	.17	C
3559.8	65873	6	6	.21	C
3577.9	44994	10	8	.94	C
3601.3	65769	12	10	.23	C
3607.5	44994	8	8	.23	C
3608.5	45156	6	6	.36	C
3610.3	45259	4	4	.42	C
3635.7	65617	10	8	.21	C
3660.4	61730	10	14	.91	C
3677.0	64820	10	12	.73	C
3680.2	64585	12	10	.19	C
3682.1	64888	8	10	.76	D
3684.9	64920	6	8	.26	C
3706.1	61226	12	14	1.4	C
3718.9	61226	10	12	.96	C
3729.5	61744	10	12	.066	D
3731.9	61211	8	10	1.0	C
3746.6	60934	12	12	.16	C
3756.6	60956	10	10	.14	C
3767.7	60957	8	8	.14	C
3768.2	61469	10	12	.071	C
3771.4	69561	14	14	.19	C
3773.9	69630	12	12	.25	C
3800.6	57306	6	8	.27	C
3801.9	51561	12	12	.064	C
3806.7	43314	10	12	.38	C
3809.6	43524	8	8	.20	C
3823.5	43429	8	10	.44	C
3823.9	43596	6	6	.36	C
3833.9	43044	4	4	.52	C
3834.4	43524	6	8	.52	D
3839.8	43673	2	2	.58	C
3844.0	43644	2	4	.29	C
3872.1	63449	10	12	.077	C
3873.2	63548	8	10	.11	C
3889.5	68843	12	14	.31	C
3898.4	59470	6	8	.17	C
3899.3	60102	4	6	.24	C
3911.1	56562	6	6	.13	D
3919.3	66738	8	8	.088	D
3923.3	59732	12	10	.13	C
3924.1	56602	2	4	.94	D
3926.5	56462	6	8	.54	C
3929.7	59784	10	8	.092	C
3931.5	63548	10	10	.082	C
3936.8	59818	8	6	.12	C
3952.8	59117	6	6	.41	D
3975.9	59990	2	4	.18	D
3977.1	59600	4	6	.16	D
3980.1	66149	10	8	.13	D
3982.2	59568	4	2	.35	D
3982.6	50383	6	4	.23	D
3982.9	66504	6	4	.55	D
3986.8	50341	12	10	.11	D
3987.1	50359	10	8	.10	D
4003.3	62393	12	10	.11	D
4011.9	66149	8	8	.23	D
4018.1	41933	10	8	.33	C
4026.4	50095	12	14	.089	D
4030.8	24802	6	8	.19	C
4031.8	50081	10	12	.073	D
4033.1	24788	6	6	.18	C
4034.5	24779	6	4	.18	C
4041.4	41789	10	10	1.0	C
4048.8	42144	6	4	.75	C
4052.5	59784	6	8	.38	D
4055.5	41933	8	8	.61	C
4058.9	42199	4	2	1.0	C
4061.7	49415	8	6	.19	D
4063.5	42054	6	6	.22	C
4065.1	58843	12	14	.25	D
4066.2	65617	10	8	.22	D
4079.4	42144	2	4	.38	C
4082.9	42054	4	6	.37	C
4083.6	41933	6	8	.28	C
4089.9	58867	8	10	.17	D
4092.4	59470	8	8	.14	D
4099.4	65617	8	8	.11	D
4105.4	59290	10	8	.17	D
4107.9	61485	12	10	.097	D
4114.4	59340	8	8	.15	D

Wavelength (λ[Å])	Upper energy level (E_k[cm⁻¹])	Statistical weights g_i	g_k	Transition Probability (A[10⁸ s⁻¹])	Uncertainty	Wavelength (λ[Å])	Upper energy level (E_k[cm⁻¹])	Statistical weights g_i	g_k	Transition Probability (A[10⁸ s⁻¹])	Uncertainty
Mn I						**Hg I**					
4116.6	62075	6	6	.12	D	4916.1	74405	3	1	.13	D
4125.8	65262	10	10	.070	D	5460.7	62350	5	3	.56	D
4132.3	68716	8	10	.15	D	5769.6	71396	3	5	.61	D
4135.0	58427	12	12	.30	D	6234.4	79964	1	3	.0053	D
4141.1	58486	10	10	.26	D	6716.4	78813	1	3	.0043	D
4147.5	51305	6	6	.066	D	6907.5	76824	3	5	.028	D
4148.8	58520	8	8	.23	D	7728.8	76863	1	3	.0097	D
4176.6	58075	14	12	.21	D	**Neodymium**					
4182.3	61913	12	14	.092	D	**Nd II**					
4190.0	58110	12	10	.20	D	3780.4	30247	16	18	.14	D
4201.8	58137	10	8	.23	D	3805.4	28857	14	16	.69	D
4220.6	57512	6	4	.16	D	3807.2	26772	10	12	.049	D
4239.7	47299	4	2	.39	D	3863.3	25877	8	10	.15	D
4257.7	47299	2	2	.37	D	3941.5	25877	10	10	.61	D
4261.3	61469	12	12	.081	D	3951.2	26772	12	12	.60	D
4265.9	47155	4	4	.35	D	3973.3	30247	18	18	.63	D
4278.7	61485	10	10	.068	D	3979.5	26772	10	12	.27	D
4281.1	46901	6	6	.23	D	3990.1	28857	16	16	.52	D
4300.2	60668	12	10	.087	D	4012.3	30002	18	20	.55	D
4381.7	61485	8	10	.14	D	4061.1	28419	16	18	.44	D
4411.9	60668	12	10	.26	D	4106.6	28857	14	16	.068	D
4414.9	45941	8	6	.18	D	4109.5	26913	14	16	.37	D
4419.8	60739	10	8	.21	D	4133.4	26772	14	12	.15	D
4451.6	45754	8	8	.71	D	4156.1	25524	12	14	.34	D
4452.5	59617	14	14	.059	D	4205.6	28857	18	16	.18	D
4455.8	47216	4	6	.17	D	4284.5	28419	18	18	.085	D
4457.0	47218	6	4	.20	D	4303.6	23230	8	10	.47	D
4457.6	47216	6	6	.38	D	4325.8	26913	16	16	.16	D
4458.3	47212	6	8	.28	D	4358.2	25524	14	14	.15	D
4461.1	47212	8	8	.17	D	4382.7	25877	12	10	.040	D
4462.0	47207	8	10	.43	D	4400.8	23230	10	10	.068	D
4464.7	45941	6	6	.26	D	4451.6	25524	12	14	.25	D
4479.4	63548	8	10	.34	D	4456.4	28419	16	18	.064	D
4498.9	45941	4	6	.11	D	4463.0	26913	14	16	.18	D
4502.2	45754	6	8	.078	D	4958.1	23230	12	10	.012	D
4503.9	59617	12	14	.083	D	5130.6	30002	22	20	.16	D
4605.4	59828	10	12	.36	D	5192.6	28419	20	18	.17	D
4626.5	59617	12	14	.36	D	5249.6	26913	18	16	.18	D
4709.7	44523	8	8	.077	D	5276.9	25877	12	10	.12	D
4727.5	44696	6	6	.084	D	5293.2	25524	16	14	.12	D
4754.0	39431	6	8	.38	D	5302.3	30247	20	18	.11	D
4761.5	44815	2	4	.28	D	5311.5	26772	14	12	.11	D
4762.4	44289	8	10	.57	D	5319.8	23230	12	10	.16	D
4765.9	44696	4	6	.28	D	5357.0	28857	18	16	.18	D
4766.4	44523	6	8	.45	D	5371.9	30002	20	20	.051	D
4783.4	39431	8	8	.39	D	5485.7	28419	18	18	.057	D
4823.5	39431	10	8	.45	D	5594.4	26913	16	16	.070	D
Mn II						5620.6	30247	18	18	.13	D
2933.1	43557	5	3	1.7	C	5688.5	25524	14	14	.059	D
2939.3	43485	5	5	1.8	C	5718.1	28857	16	16	.087	D
2949.2	43370	5	7	1.7	C	5726.8	25877	10	10	.056	D
3439.0	38543	5	7	.0041	D	5740.9	26772	12	12	.072	D
3442.0	43370	9	7	.43	C	5804.0	23230	10	10	.046	D
3460.3	43485	7	5	.32	C	5865.1	28419	16	18	.013	D
3474.0	43370	7	7	.079	C	6051.9	25877	12	10	.011	D
3474.1	43557	5	3	.15	C	**Neon**					
3482.9	43485	5	5	.20	C	**Ne I**					
3488.7	43557	3	3	.25	C	615.63	162436	1	3	.38	C
3495.8	43557	1	3	.11	C	618.67	161637	1	3	.93	C
3496.8	43370	5	7	.016	C	619.10	161524	1	3	.33	C
3497.5	43485	3	5	.051	C	626.82	159535	1	3	.74	C
Mn XXIII						629.74	158796	1	3	.48	C
12.03	8687000	2	4	1.5 (+5)	B	735.90	135889	1	3	6.11	B
12.158	8708000	4	6	1.8 (+5)	B	743.72	134459	1	3	.486	B
Mercury						3369.8	163709	5	5	.0010	D
Hg I						3369.9	163708	5	3	.0076	D
2536.5	39412	1	3	.13	D	3375.6	163657	5	3	.0022	D
2752.8	73961	1	3	.057	D	3417.9	163709	3	5	.0092	D
2856.9	74405	3	1	.012	D	3418.0	163708	3	3	.0022	D
2893.6	73961	3	3	.16	D	3423.9	163657	3	3	.0010	D
2925.4	78216	5	3	.077	D	3447.7	163038	5	5	.021	D
2967.3	71336	1	3	.45	D	3450.8	163013	5	3	.0049	D
3125.7	71396	3	5	.51	D	3454.2	163401	3	1	.037	D
3341.5	73961	5	3	.27	D	3460.5	163708	1	3	.0070	D
4046.6	62350	1	3	.18	D	3464.3	162899	5	5	.0067	D
4077.8	63928	3	1	.041	D	3466.6	163657	1	3	.013	D
4108.1	78404	3	1	.030	D	3472.6	162831	5	7	.017	D
4339.2	77108	3	5	.080	D	3498.1	163038	3	5	.0051	D
4358.3	62350	3	3	.40	D	3501.2	163013	3	3	.012	D

Table 1b (Continued)
TRANSITION PROBABILITIES FOR SELECTED ATOMIC AND IONIC SPECIES

Wavelength (λ[Å])	Upper energy level (E_k[cm^-1])	g_i	g_k	Transition Probability (A[10^8 s^-1])	Uncertainty	Wavelength (λ[Å])	Upper energy level (E_k[cm^-1])	g_i	g_k	Transition Probability (A[10^8 s^-1])	Uncertainty
Ne I						Ne I					
3510.7	162518	5	3	.0022	D	8647.0	162420	5	5	.0391	C
3515.2	162899	3	5	.0069	D	8681.9	161637	3	3	.21	D
3520.5	164286	3	1	.093	D	8767.5	161524	3	3	.0011	D
3593.5	163709	3	5	.0099	D	8771.7	162436	3	3	.16	D
3593.6	163708	3	3	.0066	D	8783.8	162420	3	5	.313	C
3600.2	163657	3	3	.0043	D	8865.3	159535	3	3	.0094	D
3633.7	163401	3	1	.011	D	9201.8	161637	3	3	.091	D
3682.2	163038	3	5	.0016	D	9433.0	161637	3	3	.0011	D
3685.7	163013	3	3	.0039	D	9486.7	158796	3	3	.025	D
3701.2	162899	3	5	.0022	D	9534.2	161524	3	3	.063	D
4536.3	170296	3	3	.0050	D	10621.	159535	3	3	.0024	D
4702.5	169517	3	3	.0021	D	11409.	159535	3	3	.042	D
4708.9	169488	3	3	.042	D	11525.	158796	3	3	.084	D
4955.4	170296	3	3	.0033	D	11767.	159535	3	3	.069	D
5113.7	167808	3	3	.010	D	12459.	158796	3	3	.015	D
5120.5	170296	3	3	.0056	D						
5154.4	169517	3	3	.019	D	Ne II					
5191.3	170296	3	3	.013	D	357.03	280351	6	10*	38.	C
5326.4	167027	3	3	.0068	D	361.77	276677	6	2*	16.	C
5333.3	169517	3	3	.0053	D	406.28	246395	6	10*	18.	C
5341.1	166975	3	3	.11	D	446.37	224291	6	6*	40.7	C
5400.6	152971	3	1	.0090	B	460.73	217048	4	2	47.	B
5418.6	169488	3	3	.0052	D	462.39	217048	2	2	23.	B
5433.7	166657	3	3	.00283	B	1907.5	276512	4	2	.28	D
5652.6	167808	3	3	.0089	D	1916.1	276277	4	4	.69	D
5662.5	165913	3	3	.0069	D	1930.0	276512	2	2	.57	D
5852.5	152971	3	1	.682	B	1938.8	276277	2	4	.13	D
5868.4	167808	3	3	.014	B	2858.0	281171	6	6	.79	D
5881.9	151038	5	3	.115	B	2870.0	281026	6	6	.17	D
5913.6	167027	3	3	.040	D	2873.0	280990	6	4	.39	D
5939.3	166657	3	3	.00200	B	2876.3	281171	4	6	.78	D
5944.8	150859	5	5	.113	B	2876.5	280947	6	4	.33	D
5961.6	167808	3	3	.033	D	2878.1	281333	2	2	.069	D
5975.5	150772	5	3	.0351	B	2888.4	281026	4	6	.070	D
6030.0	151038	3	3	.0561	B	2891.5	280990	4	4	.061	D
6046.1	166657	3	3	.00226	B	2897.0	280701	6	8	.052	D
6074.3	150917	3	1	.603	B	2906.8	280990	2	4	.55	D
6096.2	150859	3	5	.181	B	2910.1	280769	4	2	1.7	D
6118.0	166657	5	3	.00609	B	2910.4	280947	2	4	.59	D
6128.5	150772	3	3	.0067	B	2916.2	280474	6	4	.096	D
6143.1	150316	5	5	.282	B	2925.6	280769	2	2	.56	D
6150.3	167027	3	3	.015	D	2933.7	280269	6	6	.069	D
6163.6	151038	1	3	.146	B	2955.7	252954	6	4	1.2	D
6217.3	150122	5	3	.0637	B	3001.7	252954	4	4	.87	D
6266.5	150772	1	3	.249	B	3017.3	279325	6	4	.35	D
6273.0	166975	3	3	.0097	D	3027.0	279219	6	6	1.4	D
6293.7	166657	3	3	.00639	B	3028.7	279423	4	2	.85	D
6304.8	150316	3	5	.0416	B	3028.9	252954	2	4	.47	D
6328.2	166657	3	3	.0339	B	3034.5	279138	6	8	3.1	D
6330.9	165913	3	3	.023	D	3037.7	279325	4	4	2.1	D
6334.4	149824	5	5	.161	B	3045.6	279423	2	2	2.5	D
6351.9	166657	1	3	.00345	B	3047.6	279219	4	6	1.8	D
6383.0	150122	3	3	.321	B	3054.7	279325	2	4	.94	D
6401.1	166657	3	3	.0139	B	3092.9	306687	6	6	1.3	D
6402.2	149657	3	7	.514	B	3097.1	306688	8	8	1.3	D
6506.5	149824	3	5	.300	B	3118.0	281171	8	6	.042	D
6532.9	150122	1	3	.108	B	3134.1	306263	6	4	.26	D
6599.0	151038	3	3	.232	B	3140.4	306243	8	6	.24	D
6602.9	165913	3	3	.0059	D	3151.1	281171	6	6	.048	D
6652.1	150917	3	1	.0029	B	3154.8	280797	8	6	.018	D
6678.3	150859	3	5	.233	B	3164.4	280701	8	8	.16	D
6717.0	150772	3	3	.217	B	3165.7	281026	6	6	.12	D
6721.1	165913	3	3	4.9 (-4)	D	3173.6	280947	6	4	.045	D
6929.5	150316	3	5	.174	B	3176.1	281171	4	6	.060	D
7024.1	150122	3	3	.0189	B	3187.6	251011	4	6	.014	D
7032.4	148258	5	3	.253	B	3188.7	280797	6	6	.39	D
7051.3	162436	3	3	.030	B	3190.9	281026	4	6	.15	D
7059.1	162420	3	5	.068	C	3194.6	280990	4	4	.52	D
7173.9	149824	3	5	.0287	B	3198.6	280701	6	8	1.7	D
7245.2	148258	3	3	.0935	B	3198.9	280947	4	4	.23	D
7304.8	166657	1	3	.00255	B	3209.0	280262	8	8	.16	D
7438.9	148258	1	3	.0231	B	3209.4	280990	2	4	.60	D
7472.4	161637	3	3	.040	D	3213.7	280947	2	4	1.7	D
7535.8	161524	3	3	.43	D	3214.3	280797	4	6	2.2	D
7937.0	162420	5	5	.0078	C	3218.2	280173	8	10	3.6	D
8082.5	148258	3	3	.0012	B	3224.8	305365	6	8	3.5	D
8118.5	162436	3	3	.049	D	3229.5	305365	8	8	.13	D
8128.9	162420	3	5	.0072	C	3229.6	305364	8	10	3.6	D
8259.4	162420	5	5	.0203	D	3230.1	277344	6	6	1.8	D
8571.4	162436	3	3	.055	D	3230.4	277344	4	6	.14	D
8582.9	162420	3	5	.0100	C	3232.0	277326	6	4	.27	D

TRANSITION PROBABILITIES FOR SELECTED ATOMIC AND IONIC SPECIES

Wavelength (λ[Å])	Upper energy level (E_k[cm⁻¹])	Statistical weights g_i	g_k	Transition Probability (A[10⁸ s⁻¹])	Uncertainty	Wavelength (λ[Å])	Upper energy level (E_k[cm⁻¹])	Statistical weights g_i	g_k	Transition Probability (A[10⁸ s⁻¹])	Uncertainty
			Ne II						Ne II		
3232.4	277326	4	4	1.6	D	3694.2	246192	6	6	1.0	D
3243.4	280269	6	6	.23	D	3697.1	281333	2	2	.28	D
3244.1	280262	6	8	1.5	D	3701.8	281171	4	6	.27	D
3248.1	280474	4	4	.24	D	3709.6	246598	4	2	1.1	D
3255.4	281720	6	4	.038	D	3713.1	251011	4	6	1.3	D
3263.4	280474	2	4	.39	D	3721.8	281026	4	6	.20	D
3269.9	280269	4	6	.51	D	3726.9	280990	4	4	.12	D
3270.8	249696	6	4	.057	D	3727.1	251522	2	4	.98	D
3297.7	249446	6	6	.43	D	3734.9	246415	4	4	.19	D
3309.7	254292	4	2	.31	D	3744.6	280990	2	4	.26	D
3310.5	281720	4	4	.069	D	3751.2	246598	2	2	.18	D
3311.3	249840	4	2	.26	D	3753.8	280797	4	6	.45	D
3314.7	281171	6	6	.044	D	3766.3	246192	4	6	.29	D
3319.7	276512	4	2	1.6	D	3777.1	246415	2	4	.42	D
3320.2	279219	8	6	.21	D	3800.0	280474	4	4	.37	D
3323.7	254165	4	4	1.6	D	3818.4	280474	2	4	.61	D
3327.2	249696	4	4	.91	D	3829.8	280269	4	6	.84	D
3329.2	279138	8	8	.88	D	3942.3	249446	4	6	.010	D
3330.7	281026	6	6	.039	D				Ne V		
3334.8	249109	6	8	1.8	D	142.61	701945	9	9*	670.	C
3336.1	306243	4	6	1.1	D	143.32	698517	9	15*	1200.	C
3344.4	249840	2	2	1.5	D	147.13	709956	5	7	1500.	C
3345.5	276277	6	4	1.4	D	151.23	691540	5	5	338.	C
3345.8	276277	4	4	.22	D	154.50	711210	1	3	700.	C
3353.6	281333	4	2	.12	D	167.69	597083	9	9*	150.	C
3355.0	249446	4	6	1.3	D	358.93	279365	9	3*	210.	C
3356.3	280797	6	6	.20	D	365.59	303812	5	3	135.	C
3357.8	279219	6	6	.50	D	482.15	208161	9	9*	30.1	C
3360.3	306263	2	4	.86	D	571.04	175876	9	15*	10.	C
3360.6	249696	2	4	.82	D	2259.6	640868	3	5	1.9	D
3362.9	279423	4	2	.35	D	2265.7	641646	5	7	2.4	D
3371.8	281171	4	6	.22	D				Ne VII		
3374.1	279325	4	4	.30	D	97.502	1025600	1	3	1070.	B
3378.2	254292	2	2	1.7	D	115.46	978320	9	3*	480.	C
3379.3	279423	2	2	.30	D	116.69	1071920	3	5	1600.	C
3386.2	279219	4	6	.055	D	127.66	998280	3	1	190.	C
3388.4	281026	4	6	2.2	D	465.22	214952	1	3	40.9	B
3390.6	279325	2	4	.077	D	558.61	290740	3	5	8.11	B
3392.8	254165	4	4	.44	D	559.95	289850	1	3	10.7	B
3404.8	306687	4	6	1.9	D	561.38	289850	3	3	7.99	B
3407.0	306688	6	8	2.3	D	561.73	290740	5	5	23.9	B
3411.4	305582	4	2	.61	D	562.99	289340	3	1	31.7	B
3413.2	305567	4	4	1.8	D	564.53	289850	5	3	13.1	B
3414.9	280797	4	6	.018	D				Ne VIII		
3416.9	280269	6	6	.64	D	88.09	1135000	2	6*	840.	B
3417.7	280262	6	8	1.6	D	98.208	1147500	6	10*	2770.	B
3438.9	305582	2	2	1.4	D	770.41	129800	2	4	5.90	B
3440.7	305567	2	4	.35	D	780.32	128150	2	2	5.69	B
3453.1	280474	4	4	.46	D	2820.7	1135000	2	4	.720	B
3454.8	306263	4	4	1.6	D	2860.1	1135000	2	2	.688	B
3456.6	281720	2	4	.96	D				Nickel		
3457.1	306243	4	6	.099	D				Ni I		
3459.3	306243	6	6	1.6	D	1976.9	50790	7	9	1.1	C
3475.2	281720	4	4	.012	D	1990.3	51125	5	7	.83	C
3477.6	280269	4	6	.43	D	2007.0	50689	5	5	.17	C
3481.9	252798	4	2	1.4	D	2014.3	51344	3	5	.93	C
3503.6	281333	2	2	2.0	D	2026.6	49328	9	7	.24	C
3522.7	281333	4	2	.023	D	2047.4	49033	7	7	.13	C
3538.0	305582	4	2	.76	D	2055.5	50851	5	3	.33	C
3539.9	305567	4	4	.036	D	2124.8	50458	5	3	.38	C
3542.2	305567	6	4	.60	D	2147.8	47425	5	3	.47	C
3542.9	281171	4	6	1.2	D	2158.3	46523	7	5	.69	C
3546.2	280990	2	4	.063	D	2190.2	46523	5	5	.30	C
3551.6	280947	2	4	.037	D	2197.4	47208	3	3	.78	C
3557.8	252798	2	2	.19	D	2201.6	48818	5	3	.73	C
3561.2	281026	4	6	.21	D	2244.5	45419	5	5	.38	C
3565.8	280990	4	4	.62	D	2253.6	44565	7	7	.19	C
3568.5	274409	6	8	1.4	D	2258.2	44475	7	5	.17	C
3571.2	280947	4	4	.63	D	2290.0	43655	9	7	2.1	C
3574.2	274365	6	6	.10	D	2300.8	43655	7	7	.75	C
3574.6	274365	4	4	1.3	D	2303.0	45122	3	3	.45	C
3590.4	280797	4	6	.036	D	2312.3	44565	7	7	5.5	C
3594.2	280769	4	2	1.3	D	2317.2	44475	7	5	3.8	C
3612.3	280474	2	4	.26	D	2320.0	43090	9	11	6.9	C
3628.0	281720	4	4	.60	D	2325.8	44315	7	9	3.5	C
3632.7	280474	4	4	.13	D	2330.0	45122	5	5	5.3	C
3643.9	251522	4	4	.32	D	2345.5	42621	9	7	2.2	C
3644.9	281720	2	4	.99	D	2943.9	34163	7	5	.11	C
3659.9	280269	4	6	.067	D						
3664.1	246415	6	4	.70	D						
3679.8	281333	4	2	.32	D						

TRANSITION PROBABILITIES FOR SELECTED ATOMIC AND IONIC SPECIES

Ni I

Wavelength (λ[Å])	Upper energy level (E_k[cm^-1])	g_i	g_k	Transition Probability (A[10^8 s^-1])	Uncertainty
3012.0	36601	5	5	1.5	C
3037.9	33112	7	7	.32	C
3064.6	33501	5	7	.075	C
3080.8	34163	3	5	.093	C
3101.6	33112	5	7	.72	C
3101.9	35639	5	7	.49	C
3134.1	33611	3	5	.71	C
3225.0	34409	5	3	.11	C
3233.0	30923	9	11	.053	C
3271.1	31442	5	5	.0072	C
3315.7	31031	5	7	.053	C
3320.3	31442	7	5	.048	C
3361.6	30619	5	5	.045	C
3369.6	29669	9	7	.17	C
3380.6	32982	5	3	1.2	C
3393.0	29669	7	7	.24	C
3413.5	30619	7	5	.038	C
3414.8	29481	7	9	.55	C
3423.7	30913	3	3	.35	C
3433.6	29321	7	7	.17	C
3446.3	29889	5	5	.44	C
3452.9	29833	5	5	.098	C
3458.5	30619	3	5	.61	C
3461.7	29084	7	9	.27	C
3472.5	29669	5	7	.12	C
3483.8	30913	5	3	.14	C
3493.0	29501	5	3	.98	C
3510.3	30192	3	1	1.2	C
3515.1	29321	5	7	.44	C
3519.0	30619	5	5	.041	C
3524.5	28569	7	5	1.0	C
3566.4	31442	5	5	.56	C
3571.9	29321	7	7	.052	C
3597.7	29501	3	3	.14	C
3610.5	28569	5	5	.072	C
3619.4	31031	5	7	.73	C
3664.1	29501	5	3	.019	D
3807.1	29669	5	7	.043	C
3831.7	29501	5	3	.015	C
3858.3	29321	5	7	.069	C
4295.9	54251	9	7	.17	D
4401.6	48467	9	11	.38	D
4714.4	48467	13	11	.46	D
4786.5	48467	11	11	.18	D
4817.9	54251	7	7	.070	D
4838.7	54251	9	7	.22	D
4855.4	49159	5	5	.57	D
4980.2	49159	9	11	.19	D
5017.6	48467	11	11	.20	D
5080.5	49159	9	11	.32	D
5129.4	49159	7	5	.12	D
5371.3	54251	7	7	.16	D
5664.0	54251	5	7	.11	D
6176.8	49159	9	11	.047	D
6314.7	31442	5	5	.0057	D
6767.8	29501	1	3	.0033	D
7714.3	28569	5	5	.0014	D

Ni II

Wavelength (λ[Å])	Upper energy level (E_k[cm^-1])	g_i	g_k	Transition Probability (A[10^8 s^-1])	Uncertainty
2053.3	57081	10	8	.041	D
2080.8	58706	4	4	.13	D
2090.1	58493	4	6	.11	D
2093.6	57081	8	8	.11	D
2125.1	56371	8	8	.10	D
2125.9	55418	10	8	.073	D
2128.6	57081	8	8	.40	D
2138.6	56075	8	6	.28	D
2158.7	56424	6	4	.57	D
2161.2	56371	6	8	.32	D
2165.6	54557	10	10	3.8	D
2169.1	55418	8	8	2.3	D
2174.7	55300	8	10	2.4	D
2175.2	56075	6	6	2.8	D
2184.6	56424	4	4	4.7	D
2188.1	55019	8	6	.090	D
2201.4	56075	4	6	2.1	D
2206.7	55418	6	8	2.5	D
2210.4	54557	8	10	.64	D
2216.5	53496	10	12	5.5	D
2220.4	68131	6	8	3.7	D
2223.0	53365	10	10	1.6	D
2224.4	58493	8	6	.51	D
2224.9	54263	8	8	2.5	D
2226.3	55019	6	6	2.0	D
2253.9	55019	4	6	3.2	D
2264.5	54263	6	8	2.4	D
2270.2	53365	8	10	2.5	D
2278.8	57420	8	6	4.5	D
2287.1	58706	6	4	4.5	D
2296.6	57081	8	8	3.2	D
2297.1	53635	6	4	4.6	D
2297.5	54176	4	2	5.3	D
2298.3	58493	6	6	4.5	D
2303.0	52738	8	6	4.7	D
2316.0	51558	10	8	4.9	D
2326.4	53635	4	4	1.0	D
2334.6	56371	8	8	1.3	D
2356.4	57420	6	6	.45	D
2367.4	51558	8	8	.13	D
2375.4	57081	6	8	1.1	D
2387.8	55418	8	8	.23	D
2394.5	55300	8	10	2.9	D
2413.0	56424	6	4	.13	D
2416.1	56371	6	8	3.3	D
2433.6	56075	6	6	.12	D
2437.9	54557	8	10	.87	D
2510.9	53365	8	10	.94	D
2545.9	54263	6	8	.26	D

Nitrogen — N I

Wavelength (λ[Å])	Upper energy level (E_k[cm^-1])	g_i	g_k	Transition Probability (A[10^8 s^-1])	Uncertainty
1163.9	103144	6	8	.43	D
1164.0	105144	4	6	.032	D
1164.2	105120	6	4	.048	D
1164.3	105120	4	4	.43	D
1167.4	104881	6	8	1.1	D
1168.4	104810	6	6	.095	D
1168.5	104810	4	6	1.3	D
1169.7	104717	6	8	.030	D
1176.5	104222	6	4	.95	D
1176.6	104222	4	4	.11	D
1177.7	104145	4	2	1.3	D
1199.6	83365	4	6	5.5	D
1200.2	83318	4	4	5.3	D
1200.7	83284	4	2	5.5	D
1310.5	105144	4	6	1.3	D
1316.3	104810	4	6	.025	D
1492.6	86221	6	4	5.3	D
1492.8	86221	4	4	.58	D
1494.7	86137	4	2	5.0	D
4099.9	110521	2	4	.034	D
4110.0	110545	4	6	.040	D
4114.0	110521	4	4	.0068	D
4137.6	107446	2	4	.0039	D
4143.4	107446	4	4	.0078	D
4151.5	107446	6	4	.013	D
4214.8	107037	4	6	.022	D
4216.1	106996	2	4	.031	D
4218.9	106980	2	2	.012	D
4222.1	106996	4	4	.0098	D
4223.1	107037	6	6	.051	D
4224.9	106980	4	2	.061	D
4230.5	106996	6	4	.033	D
4385.5	120566	2	2	.0052	C
4392.4	120566	4	2	.0102	C
4914.9	106478	2	2	.00759	B
4935.1	106478	4	2	.0158	B
5169.6	107446	6	4	.00209	C
5181.4	107446	4	4	.00144	C
5186.6	107446	2	4	7.3 (–4)	C
5199.8	112808	2	2	.023	D
5201.6	112801	2	4	.023	D
5281.2	107037	6	6	.00282	C
5292.7	106996	6	4	.00167	C
5293.5	107037	4	6	.00113	C
5309.4	106980	4	2	.00273	C
5310.5	106996	2	4	.00137	C
5344.0	106814	6	6	6.2 (–4)	C
5356.6	106814	4	6	.00189	C
5367.0	106778	4	4	.00118	C
5372.6	106778	2	4	.00107	C
5378.3	106759	2	2	.00210	C

Table 1b (Continued)
TRANSITION PROBABILITIES FOR SELECTED ATOMIC AND IONIC SPECIES

Wavelength (λ[Å])	Upper energy level (E_k[cm⁻¹])	g_i	g_k	Transition Probability (A[10^8 s⁻¹])	Uncertainty
				N I	
5816.5	112681	4	6	.00278	C
5829.5	112681	6	6	.0064	C
5834.6	112610	2	4	.00383	C
5840.9	112610	4	4	.00122	C
5849.7	112565	2	2	.00152	C
5854.0	112610	6	4	.00409	C
5856.0	112565	4	2	.0076	C
6606.2	109927	4	6	7.9 (−4)	C
6622.5	109927	6	6	.0071	C
6627.0	109857	2	4	.00197	C
6636.9	109857	4	4	.0125	C
6645.0	109927	8	6	.0311	C
6646.5	109812	2	2	.0194	C
6653.5	109857	6	4	.0244	C
6656.5	109812	4	2	.0193	C
6926.7	109927	4	6	.0064	C
6945.2	109927	6	6	.0149	C
6951.6	109857	2	4	.0088	C
6960.5	109857	4	4	.00281	C
6973.1	109812	2	2	.00350	C
6979.2	109857	6	4	.0094	C
6982.0	109812	4	2	.0174	C
7423.6	96751	2	4	.052	C
7442.3	96751	4	4	.106	C
7468.3	96751	6	4	.161	C
				N II	
474.89	210705	5	5	4.5	D
475.65	210240	1	3	14.	D
475.70	210266	3	5	20.	D
475.76	210240	3	3	11.	D
475.80	210302	5	7	26.	D
475.88	210266	5	5	6.8	D
508.70	196712	5	5	2.8	D
510.76	211104	5	7	31.	D
513.85	209926	5	5	6.8	D
529.36	188909	1	3	6.5	D
529.41	188937	3	1	20.	D
529.49	188909	3	3	4.9	D
529.64	188857	3	5	4.9	D
529.72	188909	5	3	8.1	D
529.87	188857	5	5	15.	D
533.51	187438	1	3	20.	D
533.58	187462	3	5	27.	D
533.65	187438	3	3	15.	D
533.73	187492	5	7	36.	D
533.82	187462	5	5	9.1	D
547.82	197859	5	3	5.2	D
559.76	211336	1	3	12.	D
574.65	189335	5	7	35.	D
582.16	187091	5	5	13.	D
635.20	190120	1	3	18.	D
644.63	155127	1	3	12.	C
644.84	155127	3	3	35.	C
645.18	155127	5	5	58.	C
660.29	166766	5	3	40.	C
671.02	149077	3	5	2.9	D
671.39	149077	5	5	8.9	D
671.41	148940	1	3	3.6	D
671.63	148940	3	3	2.7	D
671.77	148909	3	1	12.	D
672.00	148940	5	3	4.4	D
745.84	166766	1	3	10.	C
746.98	149188	5	3	40.	C
748.37	148940	5	3	2.0	D
775.97	144188	5	5	35.	D
915.61	109217	1	3	3.6	C
915.96	109224	3	1	11.	C
1084.0	92252	1	3	2.0	C
1085.5	92250	5	5	.90	C
1085.7	92237	5	7	3.6	C
2139.5	211336	3	3	.29	D
3593.6	196712	3	5	.24	D
3609.1	196592	3	3	.24	D
3615.9	196540	3	1	.24	D
3829.8	196712	3	5	.15	D
3838.4	196712	5	5	.45	D
3842.2	196592	1	3	.20	D
3847.4	196592	3	3	.15	D
3855.1	196540	3	1	.60	D
3856.1	196592	5	3	.25	D

Wavelength (λ[Å])	Upper energy level (E_k[cm⁻¹])	g_i	g_k	Transition Probability (A[10^8 s⁻¹])	Uncertainty
				N II	
3919.0	190120	3	3	1.00	C
3995.0	174212	3	5	1.3	D
4114.4	188909	3	3	.0019	D
4447.0	187091	3	5	1.30	C
4477.7	188909	5	3	.035	D
4507.6	188857	7	5	.038	D
4601.5	170666	3	5	.270	C
4607.2	170608	1	3	.340	C
4613.9	170608	3	3	.196	C
4621.4	170573	3	1	.90	C
4630.5	170666	5	5	.84	C
4643.1	170608	5	3	.466	C
4774.2	187462	3	5	.054	C
4779.7	187438	3	3	.269	C
4781.2	187492	5	7	.040	C
4788.1	187462	5	5	.248	C
4793.7	187438	5	3	.089	C
4803.3	187492	7	7	.313	C
4810.3	187462	7	5	.055	C
4987.4	188937	3	1	.63	C
4994.4	188909	3	3	.74	C
5001.1	186512	3	5	1.02	C
5001.5	186571	5	7	1.08	C
5002.7	168892	1	3	.085	C
5005.2	186652	7	9	1.22	C
5007.3	188857	3	5	.77	C
5010.6	168892	3	3	.268	C
5025.7	186571	7	7	.134	C
5040.7	186512	7	5	.0053	C
5045.1	168892	5	3	.410	C
5452.1	188909	1	3	.14	D
5454.2	188937	3	1	.41	D
5462.6	188909	3	3	.10	D
5478.1	188857	3	5	.10	D
5480.1	188909	5	3	.17	D
5495.7	188857	5	5	.30	D
5666.6	166582	3	5	.423	C
5676.0	166522	1	3	.310	C
5679.6	166579	5	7	.56	C
5686.2	166522	3	3	.231	C
5710.8	166582	5	5	.137	C
5927.8	187438	1	3	.315	C
5931.8	187462	3	5	.425	C
5940.2	187438	3	3	.235	C
5941.7	187492	5	7	.56	C
5952.4	187462	5	5	.140	C
6482.1	164611	3	3	.37	D
6610.6	189335	5	7	.59	D
				N III	
374.20	267238	2	4	101.	C
451.87	221302	2	2	8.9	C
452.23	221302	4	2	17.8	C
685.00	145986	2	4	9.3	C
685.51	145876	2	2	37.1	C
685.82	145986	4	4	46.8	C
686.34	145876	4	2	19.0	C
763.34	131004	2	2	9.6	C
764.36	131004	4	2	18.7	C
771.54	186797	2	4	8.2	C
771.90	186797	4	4	16.5	C
772.39	186797	6	4	24.7	C
772.89	230409	6	4	20.3	C
772.98	230404	4	2	22.7	C
979.84	203089	4	4	8.9	C
979.92	203075	6	6	9.3	C
989.79	101031	2	4	4.15	C
991.51	101031	4	4	.82	C
991.58	101024	4	6	4.96	C
1747.8	203089	2	4	1.31	C
1751.2	203089	4	4	.258	C
1751.7	203075	4	6	1.57	C
4097.3	245701	2	4	.82	C
4103.4	245665	2	2	.82	C
4634.1	267238	2	4	.65	C
4640.6	267244	4	6	.78	C
4641.9	267238	4	4	.130	C
				N IV	
247.20	404522	1	3	114.	C
283.52	420053	9	15*	290.	C
322.64	377285	9	3*	84.	C

Wavelength ($\lambda[\text{Å}]$)	Upper energy level ($E_k[\text{cm}^{-1}]$)	g_i	g_k	Transition Probability ($A[10^8\ \text{s}^{-1}]$)	Uncertainty
			N IV		
335.05	429160	3	5	185.	C
387.35	388855	3	1	28.	D
765.15	130694	1	3	24.0	B
923.16	175669	9	9*	18.	B
955.34	235369	3	1	30.	C
1718.6	188883	3	5	2.37	C
3480.8	406005	3	9*	1.1	C
4057.8	429160	3	5	.68	C
6380.8	404522	1	3	.14	B
7116.7	420053	9	15*	.12	C
			N V		
209.29	477817	2	6*	118.	B
247.66	484418	6	10*	430.	C
1238.8	80722	2	4	3.41	B
1242.8	80463	2	2	3.38	B
4603.7	477842	2	4	.412	B
4620.0	477766	2	2	.408	B
			N VI		
24.898	4016390	1	3	5158.	AA
28.787	3473840	1	3	180.9(+2)	AA
161.22	4006180	3	9*	285.9	AA
173.29	4016390	1	3	269.7	AA
173.92	4013460	9	15*	876.	A
185.19	4013820	3	5	824.	A
1901.5	3438490	3	9*	.6777	AA
2896.4	3473840	1	3	.2080	AA
		Oxygen			
			O I		
1028.2	97488	1	3	.09	D
1152.2	102662	5	5	5.5	D
1217.6	115918	1	3	1.8	C
1302.2	76795	5	3	3.3	C
1304.9	76795	3	3	2.0	C
1306.0	76795	1	3	.66	C
5435.2	105019	3	5	.0061	C
5435.8	105019	5	5	.0102	C
5436.9	105019	7	5	.0142	C
6453.6	102117	3	5	.0142	B
6454.4	102117	5	5	.0237	B
6456.0	102117	7	5	.0331	B
6653.8	130943	3	1	.600	B
7156.7	116631	5	5	.473	B
7471.4	127292	5	3	.0114	B
7473.2	127288	5	5	.102	B
7476.4	127283	5	7	.408	B
7477.2	127292	3	3	.170	B
7479.1	127288	3	5	.306	B
7480.7	127292	1	3	.226	B
7771.9	86631	5	7	.340	B
7774.2	86628	5	5	.340	B
7775.4	86626	5	3	.340	B
7886.3	128595	3	5	.370	B
7939.5	113727	7	5	.00165	C
7943.2	113721	7	7	.0417	C
7947.2	113727	5	5	.058	C
7947.6	113714	7	9	.373	C
7950.8	113721	5	7	.331	C
7952.2	113727	3	5	.313	C
7981.9	101155	3	3	.12	D
7982.4	101155	1	3	.16	D
7987.0	101148	3	5	.21	D
7987.3	101148	5	5	.072	D
7995.1	101135	5	7	.29	D
			O II		
429.92	232603	4	2	39.	D
430.04	232536	4	4	39.	D
430.18	232463	4	6	39.	D
483.75	233544	4	2	.84	D
483.98	233430	6	4	.76	D
484.03	233430	4	4	.084	D
485.09	232959	6	8	25.	D
485.47	232796	6	6	1.6	D
485.52	232796	4	6	23.	D
3007.1	265999	8	10	.84	C
3007.7	265985	6	8	.72	C
3013.4	265639	6	8	.74	C
3032.1	265930	8	10	.85	C
3032.5	265763	6	8	.82	C
			O II		
3134.8	238893	8	6	1.23	C
3273.5	259286	8	6	1.14	C
3377.2	233544	2	2	1.88	C
3390.3	233430	2	4	1.86	C
3407.4	259286	6	6	.75	C
3749.5	212162	6	4	.90	C
3882.2	232754	8	8	.493	C
3912.0	232527	6	4	1.27	C
3919.3	232480	4	2	1.40	C
3973.3	214229	4	4	1.27	C
4069.6	231296	2	4	1.39	C
4069.9	231350	4	6	1.49	C
4072.2	231428	6	8	1.70	C
4075.9	231530	8	10	1.98	C
4085.1	231530	6	6	.478	C
4087.2	255756	4	6	2.24	C
4089.3	255978	10	12	2.62	C
4095.6	255759	6	8	2.23	C
4097.2	255828	8	10	2.37	C
4104.7	232748	4	6	1.04	C
4105.0	232746	4	4	.80	C
4108.8	255759	8	8	.349	C
4119.2	232754	6	8	1.48	C
4120.3	232748	6	6	.443	C
4132.8	232536	2	4	.84	C
4153.3	232463	4	6	.77	C
4276.7	256123	6	8	1.82	C
4277.4	256084	2	4	1.49	C
4277.9	256123	8	8	.302	C
4281.4	255913	0	0	.00	C
4282.8	255913	4	4	1.06	C
4283.0	256088	4	6	1.58	C
4283.1	256088	6	6	.51	C
4283.8	256084	4	4	.59	C
4294.8	255813	4	6	1.39	C
4303.8	255911	6	8	1.97	C
4328.6	255622	4	2	1.21	C
4340.4	255829	6	8	2.23	C
4347.4	229968	4	4	.94	C
4349.4	208484	6	6	.74	C
4351.3	229947	6	6	.97	C
4396.0	234454	6	6	.398	C
4414.9	211713	4	6	1.15	C
4417.0	211522	2	4	.95	C
4443.1	251466	6	6	.57	C
4448.2	251221	8	8	.57	C
4489.5	255812	2	4	1.51	C
4491.3	255690	4	6	1.81	C
4596.2	228723	4	6	1.03	C
4602.1	256126	4	6	1.70	C
4609.4	256143	6	8	1.82	C
4641.8	206878	4	6	.79	C
4661.6	206786	4	4	.52	C
4701.2	253792	4	4	.87	C
4703.2	251224	4	6	.82	C
4705.4	232959	6	8	1.38	C
4871.6	253048	4	6	.435	C
4906.9	232536	4	4	.68	C
4924.6	232463	4	6	.67	C
4941.1	234402	2	4	.83	C
4943.1	234454	4	6	1.06	C
5206.7	233430	4	4	.391	C
6627.6	248514	4	4	.089	C
6666.9	248425	4	2	.0349	C
6678.2	248514	2	4	.0173	C
6718.1	248425	2	2	.068	C
6721.4	203942	4	2	.189	C
6810.6	246029	6	8	.00180	C
6844.1	245903	4	6	.00325	C
6847.0	246029	8	8	.0347	C
6869.7	245903	6	6	.059	C
6885.1	245816	4	4	.067	C
6895.3	246029	10	8	.298	C
6906.5	245903	8	6	.272	C
6908.1	245768	4	2	.332	C
6910.8	245816	6	4	.267	C
			O III		
262.88	380706	5	5	40.	D
263.69	379232	1	3	52.	D
263.73	379293	3	5	73.	D

TRANSITION PROBABILITIES FOR SELECTED ATOMIC AND IONIC SPECIES

Wavelength (λ[Å])	Upper energy level (E_k[cm^{-1}])	g_i	g_k	Transition Probability (A[10^8 s^{-1}])	Uncertainty
			O III		
263.77	379232	3	3	40.	D
263.82	379356	5	7	96.	D
263.86	379293	5	5	24.	D
277.38	380782	5	7	110.	D
279.79	377687	5	5	26.	D
295.94	381086	1	3	48.	D
303.41	329582	1	3	34.	D
303.46	329643	3	1	100.	D
303.52	329582	3	3	26.	D
303.62	329468	3	5	25.	D
303.69	329582	5	3	42.	D
303.80	329468	5	5	76.	D
305.60	327228	1	3	100.	D
305.66	327277	3	5	140.	D
305.70	327228	3	3	76.	D
305.77	327351	5	7	180.	D
305.84	327277	5	5	46.	D
320.98	331820	5	7	190.	D
328.45	324734	5	5	61.	D
345.31	332777	1	3	99.	D
374.08	267633	5	5	26.	D
395.56	273080	5	3	49.	D
507.39	197087	1	3	16.	C
507.68	197087	3	3	47.	C
508.18	197087	5	3	79.	C
525.80	210459	5	3	88.	C
597.82	210459	1	3	18.	C
599.60	187049	5	5	55.	C
702.33	142383	1	3	5.7	C
702.82	142397	3	1	17.	C
832.93	120059	1	3	3.2	C
835.10	120053	5	5	1.4	C
835.29	120025	5	7	5.7	C
1109.5	381086	3	3	2.8	D
1679.1	357111	3	5	.94	D
1686.9	356838	3	3	.94	D
1760.4	357111	3	3	.60	D
1764.5	357111	5	5	1.8	D
1766.3	356838	1	3	.78	D
1772.3	356732	3	1	2.3	D
1773.0	356838	5	3	.98	D
2390.4	332777	3	3	2.2	D
2959.7	324734	3	3	2.1	D
2996.5	327228	3	3	.51	D
3004.4	327277	5	5	.47	D
3017.6	327351	7	7	.59	D
3115.7	329643	3	1	1.5	D
3121.7	329582	3	3	1.5	D
3132.9	329468	3	5	1.4	D
3261.0	324658	5	7	1.8	D
3265.5	324836	7	9	2.1	D
3267.3	324462	3	5	1.7	D
3281.9	324462	5	5	.32	D
3284.6	324658	7	7	.23	D
3405.7	329582	1	3	.27	D
3408.1	329643	3	1	.81	D
3415.3	329582	3	3	.20	D
3428.7	329468	3	5	.20	D
3430.6	329582	5	3	.33	D
3444.1	329468	5	5	.59	D
3702.8	327228	1	3	.62	D
3707.2	327277	3	5	.83	D
3714.0	327228	3	3	.46	D
3715.1	327351	5	7	1.1	D
3725.3	327277	5	5	.27	D
3961.6	331820	5	7	1.3	D
5268.1	332777	1	3	.31	D
5508.1	324734	5	5	.11	D
5592.4	290957	3	3	.36	D
			O IV		
238.36	419534	2	4	288.	C
238.57	419551	4	6	346.	C
238.58	419534	4	4	58.	C
279.63	357614	2	2	23.7	C
279.94	357614	4	2	47.7	C
553.33	180724	2	4	12.0	C
554.08	180481	2	2	47.3	C
554.51	180724	4	4	60.	C
555.26	180481	4	2	25.0	C
608.40	164366	2	2	12.5	C
609.83	164366	4	2	23.6	C
616.95	289024	6	4	25.5	C
617.01	289024	4	4	2.93	C
617.04	289015	4	2	28.6	C
624.62	231538	2	4	10.7	C
625.13	231538	4	4	21.5	C
625.85	231538	6	4	32.2	C
779.73	255185	6	4	1.53	C
779.82	255185	4	4	13.1	C
779.91	255156	6	6	13.7	C
780.00	255156	4	6	1.0	C
787.71	126950	2	4	5.9	C
790.11	126950	4	4	1.15	C
790.20	126936	4	6	7.1	C
921.30	289024	2	4	2.11	C
921.37	289015	2	2	9.2	C
923.37	289024	4	4	11.4	C
923.43	289015	4	2	4.48	C
1338.6	255185	2	4	2.22	C
1343.0	255185	4	4	.428	C
1343.5	255156	4	6	2.64	C
			O V		
172.17	580825	1	3	296.	B
192.85	600766	9	15*	690.	B
215.17	546973	9	3*	170.	C
220.35	612616	3	5	440.	C
248.46	561276	3	1	65.	D
629.73	158798	1	3	28.0	B
758.68	213887	3	5	5.68	B
759.44	213618	1	3	7.55	B
760.23	213618	3	5	5.64	B
760.45	213887	5	5	16.9	B
761.13	213463	3	1	22.5	B
762.00	213618	5	3	9.34	B
774.52	287910	3	1	35.4	C
1371.3	231721	3	5	3.29	C
2784.0	582882	3	9*	1.6	D
3144.7	612616	3	5	.93	B
5114.1	580825	1	3	.17	B
5589.9	600766	9	15*	.15	C
			O VI		
150.10	666218	2	6*	254.	B
173.03	674656	6	10*	885.	B
1031.9	96908	2	4	4.15	B
1037.6	96375	2	2	4.08	B
3811.4	666270	2	4	.513	B
3834.2	666113	2	2	.505	B
			O VII		
18.627	5368550	1	3	9362.	AA
21.602	4629200	1	3	330.9(+2)	AA
120.33	5355670	3	9*	533.5	AA
128.20	5368550	1	3	505.3	AA
128.46	5364420	9	15*	1620.	A
135.82	5365470	3	5	1530.	A
1630.2	4585980	3	9*	.7935	AA
2450.0	4629200	1	3	.2514	AA
			Phosphorus		
			P I		
1671.7	59820	4	2	.39	D
1674.6	59716	4	4	.40	D
1679.7	59535	4	6	.39	D
1775.0	56340	4	6	2.17	C
1782.9	56090	4	4	2.14	C
1787.7	55939	4	2	2.13	C
2135.5	58174	4	4	.211	C
2136.2	58174	6	4	2.83	C
2149.1	57877	4	2	3.18	C
2152.9	65156	2	4	.485	C
2154.1	65156	4	4	.173	C
2154.1	65157	4	6	.58	C
2534.0	58174	2	4	.200	C
2535.6	58174	4	4	.95	C
2553.3	57877	2	2	.71	C
2554.9	57877	4	2	.300	C
			P II		
1301.9	76813	1	3	.50	C
1304.5	76824	3	1	1.5	C
1304.7	76813	3	3	.37	C

Table 1b (Continued)
TRANSITION PROBABILITIES FOR SELECTED ATOMIC AND IONIC SPECIES

Wavelength (λ[Å])	Upper energy level (E_k[cm^{-1}])	g_i	g_k	Transition Probability (A[10^8 s^{-1}])	Uncertainty
P II					
1305.5	76765	3	5	.38	C
1309.9	76813	5	3	.62	C
1310.7	76765	5	5	1.1	C
4475.3	127889	5	7	1.3	D
4499.2	130143	5	7	1.4	D
4530.8	127368	3	5	1.0	D
4554.8	127951	3	5	.96	D
4588.0	125130	5	7	1.7	D
4589.9	124948	3	5	1.6	D
4602.1	125392	7	9	1.9	D
4943.5	123892	7	5	.63	D
5253.5	107924	3	5	1.0	D
5425.9	105550	5	5	.69	D
6024.2	103340	3	5	.51	D
6043.1	103669	5	7	.68	D
P III					
1334.8	74917	2	4	.55	D
1344.3	74946	4	6	.64	D
1344.8	74917	4	4	.11	D
4057.4	141514	4	4	.10	D
4059.3	141514	6	4	.90	D
4080.1	141377	4	2	.99	D
Potassium					
K I					
4044.1	24720	2	4	.0124	C
4047.2	24701	2	2	.0124	C
5084.2	32648	2	2	.00350	C
5099.2	32648	4	2	.0070	C
5323.3	31765	2	2	.0063	C
5339.7	31765	4	2	.0126	C
5343.0	31696	2	4	.0040	D
5359.6	31696	4	6	.0046	D
5782.4	30274	2	2	.0123	C
5801.8	30274	4	2	.0246	C
5812.2	30186	2	2	.0028	D
5831.9	30185	4	6	.0032	D
6911.1	27451	2	2	.0272	C
6938.8	27451	4	2	.054	C
7664.9	13043	2	4	.387	B
7699.0	12985	2	2	.382	B
K II					
607.93	164496	1	3	.013	D
K III					
2550.0	246626	6	4	2.0	D
2635.1	246626	4	4	1.2	D
2992.4	240830	6	8	2.5	D
3052.1	241444	4	6	1.7	D
3202.0	243947	4	4	1.8	D
3289.1	243121	4	6	2.0	D
3322.4	237512	6	6	1.3	D
3421.8	243448	2	4	1.5	D
K XVI					
206.27	484800	1	3	94.	C
K XVII					
22.020	4814800	2	4	4.7 (+4)	C
22.163	4818000	4	6	5.6 (+4)	C
22.18	4814800	4	4	9300.	C
22.60	4699000	2	2	2500.	D
22.76	4699000	4	2	4700.	C
Praseodymium					
Pr II					
3997.0	28010	15	15	.187	C
4062.8	28010	13	15	1.00	C
4100.7	28816	17	19	.84	C
4143.1	27128	15	17	.58	C
4179.4	25569	13	15	.52	C
4222.9	24116	11	13	.391	C
4241.0	28010	17	15	.230	C
4359.8	28010	15	15	.11	D
4405.8	27128	17	17	.090	D
4429.3	25569	15	15	.228	C
4449.8	24116	13	13	.124	C
4468.7	24116	11	13	.154	C
4510.2	25569	13	15	.116	C
4534.2	27128	15	17	.049	D
4734.2	24116	15	13	.025	D

Wavelength (λ[Å])	Upper energy level (E_k[cm^{-1}])	g_i	g_k	Transition Probability (A[10^8 s^{-1}])	Uncertainty
Pr II					
4879.1	25569	15	15	.018	D
4886.0	25569	15	15	.013	D
4912.6	28010	17	15	.057	D
5034.4	28816	19	19	.11	D
5110.8	28816	21	19	.278	C
5135.1	27128	17	17	.125	C
5173.9	27128	19	17	.318	C
5219.1	25569	15	15	.095	D
5220.1	25569	17	15	.235	C
5251.7	24116	15	13	.011	D
5259.7	24116	15	13	.224	C
5292.6	24116	13	13	.093	D
5810.6	28816	17	19	.023	D
5879.3	28010	15	15	.076	D
6200.8	27128	15	17	.018	D
6278.7	25569	13	15	.026	D
6398.0	24116	11	13	.019	D
Rubidium					
Rb I					
3022.5	33076	2	4	4.13(−5)	C
3032.0	32972	2	4	4.93(−5)	C
3044.2	32840	2	4	8.2 (−5)	C
3060.2	32668	2	4	1.05(−4)	C
3082.0	32437	2	4	1.49(−4)	C
3112.6	32119	2	4	2.5 (−4)	C
3113.1	32114	2	2	1.3 (−4)	C
3157.5	31661	2	4	3.38(−4)	C
3158.3	31654	2	2	2.0 (−4)	C
3228.0	30970	2	4	6.4 (−4)	C
3229.2	30939	2	2	3.8 (−4)	C
3348.7	29854	2	4	.00137	C
3350.8	29835	2	2	8.9 (−4)	C
3587.1	27870	2	4	.00397	C
3591.6	27835	2	2	.0029	C
4201.8	23793	2	4	.018	C
4215.5	23715	2	2	.015	C
7800.3	12817	2	4	.370	B
7947.6	12579	2	2	.340	B
Scandium					
Sc I					
2113.5	47315	4	6	.032	C
2263.0	44189	4	4	.058	C
2267.3	44105	4	2	.48	C
2271.6	44189	6	4	.46	C
2335.4	42819	4	2	.17	C
2346.8	42780	6	4	.13	C
2439.4	41163	6	6	.21	C
2439.9	41153	6	4	.022	C
2711.4	37040	6	6	.29	C
2739.8	36666	6	6	.0056	C
2974.0	33615	6	6	.45	C
2980.8	33707	6	6	.44	C
3015.4	33154	4	6	.66	C
3019.3	33278	6	8	.81	C
3269.9	30573	4	2	3.1	C
3273.6	30707	6	4	2.7	C
3907.5	25585	4	6	1.28	C
3911.8	25725	6	8	1.37	C
4020.4	24866	4	4	1.65	C
4023.7	25014	6	6	1.44	C
4753.2	21033	4	6	.010	C
4779.4	21086	6	8	.0084	C
4791.5	21033	6	6	.0023	C
5301.9	18856	4	4	.0013	C
5343.0	18711	4	2	.0051	C
5349.7	18856	6	4	.0040	C
6210.7	16097	4	4	.012	C
6239.8	16022	4	4	.0065	C
6305.7	16023	6	6	.015	C
6378.8	15673	4	4	.0016	C
6448.1	15673	6	4	2.6 (−4)	C
Sc II					
2273.1	55716	1	3	7.7	D
3353.7	32350	5	7	2.0	D
3372.2	29824	7	5	1.2	D
3535.7	30816	5	3	.83	D
3572.5	28161	7	7	1.8	D
3576.4	28021	5	5	1.4	D

Table 1b (Continued)
TRANSITION PROBABILITIES FOR SELECTED ATOMIC AND IONIC SPECIES

Wavelength (λ[Å])	Upper energy level (E_k[cm⁻¹])	Statistical weights g_i	g_k	Transition Probability (A[10⁸ s⁻¹])	Uncertainty
		Sc II			
3613.8	27841	7	9	1.9	D
3630.8	27602	5	7	1.6	D
3642.8	27444	3	5	1.5	D
4246.8	26081	5	5	1.5	D
4314.1	28161	9	7	.41	D
4374.5	27841	9	9	.14	D
4670.4	32350	5	7	.18	D
5031.0	30816	5	3	.49	D
5239.8	30816	1	3	.14	D
5526.8	32350	9	7	.42	D
5657.9	29824	5	5	.13	D
		Silicon			
		Si I			
1977.6	50566	1	3	.18	D
1979.2	50602	3	1	.51	D
1980.6	50566	3	3	.13	D
1983.2	50500	3	5	.14	D
1986.4	50566	5	3	.21	D
1989.0	50500	5	5	.41	D
2208.0	45276	1	3	.311	C
2210.9	45294	3	5	.416	C
2211.7	45276	3	3	.232	C
2216.7	45322	5	7	.55	C
2218.1	45294	5	5	.138	C
2506.9	39955	3	5	.466	C
2514.3	39760	1	3	.61	C
2516.1	39955	5	5	1.21	C
2519.2	39760	3	3	.456	C
2524.1	39683	3	1	1.81	C
2528.5	39760	3	5	.77	C
2532.4	54871	1	3	.26	D
2631.3	53387	1	3	.97	D
2881.6	40992	5	3	1.89	C
3905.5	40992	1	3	.118	C
4738.8	60857	3	3	.010	D
4783.0	60857	5	3	.017	D
4792.3	60816	5	5	.017	D
4818.1	60705	5	7	.011	D
4821.2	60496	3	5	.0080	D
4947.6	61198	3	1	.042	D
5006.1	60962	3	5	.028	D
5622.2	57542	3	3	.016	D
5690.4	57329	3	3	.012	D
5708.4	57468	5	5	.014	D
5754.2	57329	5	5	.015	D
5772.1	58312	3	1	.036	D
5948.5	57798	3	5	.022	D
7226.2	59111	3	5	.0079	D
7405.8	58775	3	5	.037	D
7409.1	58787	5	7	.023	D
7680.3	60301	3	5	.046	D
7918.4	60645	3	5	.052	D
7932.3	60705	5	7	.051	D
7944.0	60849	7	9	.058	D
7970.3	60645	5	5	.0071	D
		Si II			
989.87	101023	2	4	6.7	D
992.68	101025	4	6	8.0	D
1020.7	97972	2	2	1.3	D
1190.4	84005	2	4	6.9	D
1193.3	83802	2	2	28.	D
1194.5	84005	4	4	36.	D
1197.4	83802	4	2	14.	D
1248.4	123034	4	4	13.	D
1251.2	123034	6	4	19.	D
1260.4	79339	2	4	20.	D
1264.7	79355	4	6	23.	D
1304.4	76666	2	2	3.6	D
1309.3	76666	4	2	7.0	D
1526.7	65501	2	2	3.73	C
1533.5	65501	4	2	7.4	C
1808.0	55310	2	4	.037	D
2904.3	113761	4	6	.67	D
2905.7	113760	6	8	.71	D
3210.0	112395	4	6	.46	D
4128.1	103556	4	6	1.32	C
4130.9	103556	6	8	1.42	C
5041.0	101023	2	4	.98	D
5056.0	101025	4	6	1.2	D
5957.6	97972	2	2	.42	D
5978.9	97972	4	2	.81	D
6347.1	81252	2	4	.70	C
6371.4	81192	2	2	.69	C
7848.8	113761	4	6	.39	D
7849.7	113760	6	8	.42	D
		Si III			
883.40	235414	5	7	63.	D
994.79	153377	3	3	7.89	B
997.39	153377	5	3	13.1	B
1141.6	217440	3	5	30.	D
1144.3	217489	5	7	39.	D
1161.6	216190	5	5	16.	D
1206.5	82884	1	3	25.9	B
1206.5	165765	3	5	48.9	B
1207.5	205029	5	5	19.	D
1294.5	130101	3	5	5.42	B
1296.7	129842	1	3	7.19	B
1298.9	129842	3	3	5.36	B
1299.0	130101	5	5	16.1	B
1301.2	129842	3	1	21.3	B
1303.3	129842	5	3	8.85	B
1328.8	228700	1	3	27.	D
1417.2	153444	3	1	26.0	C
1435.8	235414	5	7	21.	D
1589.0	228700	5	3	11.	D
1778.7	199164	7	9	4.4	D
1783.1	199026	5	7	3.8	D
3241.6	206176	5	3	2.3	D
3486.9	230270	15	21*	1.8	D
3590.5	204331	3	5	3.9	D
4552.6	175336	3	5	1.26	C
4554.0	248773	5	3	.76	D
4567.8	175263	3	3	1.25	C
4683.0	248168	5	5	.95	D
4716.7	225526	5	7	2.8	D
5451.5	244866	3	5	.60	D
5473.1	245087	5	7	.79	D
5716.3	227089	9	7	.19	D
5739.7	176487	1	3	.47	D
7462.6	214995	5	3	.49	D
7466.3	214989	7	5	.54	D
7612.4	227665	3	5	1.1	D
		Si IV			
457.82	218429	2	4	3.6	D
458.16	218267	2	2	3.6	D
515.12	265418	2	2	4.1	D
516.35	265418	4	2	8.2	D
560.50	250008	6	10*	1.0	D
749.94	293719	10	14*	14.5	C
815.05	193979	2	2	12.3	C
818.13	193979	4	2	24.4	C
860.74	276554	10	6*	1.8	D
1066.6	254128	10	14*	39.	C
1122.5	160376	2	4	20.5	C
1128.3	160376	4	4	4.03	C
1128.3	160374	4	6	24.2	C
1393.8	71749	2	4	7.73	B
1402.8	71288	2	2	7.58	B
1724.1	218375	10	6*	5.5	C
		Si V			
96.439	1036915	1	3	480.	D
97.143	1029407	1	3	2000.	D
117.86	848511	1	3	300.	D
		Si VI			
246.00	406497	4	2	170.	C
249.12	406497	2	2	85.	C
		Si VII			
217.83	506080	5	3	430.	D
272.64	366780	5	3	51.	C
274.18	368760	3	1	120.	C
275.35	363170	5	5	89.	C
275.67	366780	3	3	30.	C
276.84	366780	1	3	39.	C
278.45	363170	3	5	29.	C
		Si VIII			
214.76	534810	4	2	410.	D
216.92	530420	6	4	360.	D

Wavelength (λ[Å])	Upper energy level (E_k[cm⁻¹])	g_i	g_k	Transition Probability (A[10⁸ s⁻¹])	Uncertainty
			Si VIII		
232.86	534810	2	2	80.	D
235.56	530420	4	4	97.	D
250.45	504630	2	2	77.	D
250.79	504630	4	2	160.	D
314.31	318160	4	2	52.	D
316.20	316250	4	4	50.	D
319.83	312670	4	6	49.	D
			Si IX		
223.73	446967	1	3	42.	C
225.03	446967	3	3	120.	C
227.01	446967	5	3	200.	C
227.30	492890	5	3	230.	D
258.10	440390	5	5	104.	C
294.37	344100	9	9*	59.	C
347.36	292301	9	15*	22.	C
			Si X		
253.77	394040	2	4	29.	C
256.57	390050	2	2	110.	C
258.35	394040	4	4	140.	C
261.05	390050	4	2	54.	C
272.00	367670	2	2	30.	C
277.26	367670	4	2	57.	C
287.08	510300	2	4	26.	C
289.19	510300	4	4	50.	C
292.22	510300	6	4	73.	C
347.73	575450	10	10*	43.	D
353.09	287870	6	10*	21.	C
			Si XI		
43.763	2205040	1	3	6110.	B
49.116	2210220	9	3*	2450.	C
49.222	2361010	3	5	8900.	C
52.296	2241590	3	1	760.	C
303.30	329090	1	3	64.2	B
358.29	608790	3	1	103.	C
358.63	451000	3	5	13.8	C
361.41	446510	1	3	18.0	C
364.50	446510	3	3	13.2	C
365.42	451000	5	5	39.0	C
368.28	443690	3	1	51.	C
371.48	446510	5	3	20.7	C
604.14	495210	3	5	11.2	C
2300.8	2285040	1	3	.434	C
			Si XII		
40.924	2443500	2	6*	4420.	B
44.118	2464130	6	10*	1.4 (+4)	B
499.43	200290	2	4	9.56	B
520.72	191900	2	2	8.47	B
1862.	2444300	2	4	1.15	B
1949.	2441900	2	2	1.0	B
4620.	2463400	2	4	.046	C
4942.	2464530	4	6	.045	C
			Silver Ag I		
2061.2	48501	2	4	.031	D
2069.9	48297	2	2	.015	D
3280.7	30473	2	4	1.4	B
3382.9	29552	2	2	1.3	B
5209.1	48744	2	4	.75	D
5465.5	48764	4	6	.86	D
5471.6	48744	4	4	.14	D
			Sodium Na I		
3302.4	30273	2	4	.0281	C
3303.0	30267	2	2	.0281	C
4390.0	39729	2	4	.0077	D
4393.3	39729	4	4	.0016	D
4393.3	39729	4	6	.0092	D
4494.2	39201	2	4	.012	C
4497.7	39201	4	6	.014	C
4497.7	39201	4	4	.0024	D
4664.8	38387	2	4	.0233	C
4668.6	38387	4	4	.0041	D
4668.6	38387	4	6	.025	D
4747.9	38012	2	2	.0063	D
4751.8	38012	4	2	.0127	C
4978.5	37037	2	4	.041	C
4982.8	37037	4	4	.0082	D
			Na I		
4982.8	37037	4	6	.0489	C
5148.8	36373	2	2	.0117	C
5153.4	36373	4	2	.0233	C
5682.6	34549	2	4	.103	C
5688.2	34549	4	6	.12	C
5688.2	34549	4	4	.021	D
5890.0	16973	2	4	.622	A
5895.9	16956	2	2	.618	A
6154.2	33201	2	2	.026	C
6160.8	33201	4	2	.052	C
8183.3	29173	2	4	.453	C
8194.8	29173	4	6	.54	C
8194.8	29173	4	4	.090	D
11381.	25740	2	2	.089	C
11404.	25740	4	2	.176	C
			Na II		
300.15	333163	1	3	30.	D
301.44	331745	1	3	49.	D
372.08	268763	1	3	34.	D
			Na III		
378.14	264455	4	2	77.	C
380.10	264455	2	2	37.	C
1991.0	465399	4	6	8.3	D
2004.2	466788	2	4	4.6	D
2011.9	463971	6	8	8.4	D
2151.5	465018	2	4	4.4	D
2174.5	464390	4	6	5.3	D
2230.3	410977	6	8	3.7	D
2292.2	418418	4	4	3.3	D
2246.7	411536	4	6	2.4	D
2459.3	414282	4	6	3.0	D
2468.9	415172	2	4	2.4	D
2497.0	406190	6	6	1.7	D
			Na V		
307.89	372400	10	6*	200.	C
333.46	372400	6	6*	56.	C
369.01	568100	10	6*	120.	D
400.72	297100	10	10*	50.	C
445.14	297100	6	10*	7.1	C
459.90	217440	4	2	23.	D
461.05	216896	4	4	23.	D
463.26	215860	4	6	22.	D
510.10	569200	2	2	56.	D
511.19	567600	4	4	68.	D
			Na VI		
313.75	320589	5	3	130.	C
361.25	312175	5	5	77.	C
416.53	241341	9	9*	37.	C
492.80	204187	9	15*	13.	C
1550.6	873287	5	5	4.35	C
1567.8	872577	5	3	2.68	C
1608.5	934745	3	1	2.6	C
1649.4	933915	5	5	2.05	C
1741.5	929999	3	5	2.59	C
1747.5	930510	5	7	3.1	C
			Na VII		
94.409	1060651	6	10*	2700.	C
105.27	951347	6	2*	450.	C
353.29	285189	4	4	100.	C
381.30	264400	4	2	40.	C
397.49	367500	4	4	35.	C
399.18	367500	6	4	52.	C
483.28	412345	10	10*	29.	C
486.74	205448	2	4	11.	C
491.95	205412	4	6	13.	C
555.80	465111	4	4	23.	D
777.83	412311	4	6	6.8	D
			Na VIII		
83.34	1327500	9	15*	3940.	C
89.88	1240300	9	3*	809.	C
90.536	1347756	9	5	2860.	C
411.15	243223	1	3	44.2	C
1239.4	1432991	3	3	3.02	C
1802.7	1481521	3	1	2.70	C
1867.7	1347756	3	5	2.01	C
2059.1	1474598	3	5	1.80	C
2558.2	1513677	5	3	.0226	C
2772.0	1469055	3	5	.419	C

Wavelength ($\lambda[\text{Å}]$)	Upper energy level ($E_k[\text{cm}^{-1}]$)	g_i	g_k	Transition Probability ($A[10^8\ s^{-1}]$)	Uncertainty
Na VIII					
3021.0	1507690	5	7	.490	C
3108.9	1513677	1	3	.258	C
3182.3	1294214	1	3	.292	C
Na IX					
70.615	1416130	2	4	1350.	B
70.653	1415368	2	2	1350.	B
77.764	1429980	2	4	3600.	B
77.911	1430204	4	6	4300.	B
681.72	146688	2	4	6.63	B
694.17	144038	2	2	6.30	B
2487.7	1416130	2	4	.832	B
2535.8	1415368	2	2	.789	B
6841.8	1429980	2	4	.0259	C
7103.4	1430204	4	6	.0278	C
Strontium					
Sr I					
2206.2	45312	1	3	.0066	C
2211.3	45208	1	3	.0085	C
2217.8	45075	1	3	.012	C
2226.3	44904	1	3	.016	C
2237.7	44676	1	3	.023	C
2253.3	44366	1	3	.037	C
2275.5	43937	1	3	.067	C
2307.3	43328	1	3	.12	C
2354.3	42462	1	3	.18	C
2428.1	41172	1	3	.17	C
2569.5	38907	1	3	.053	C
2931.8	34098	1	3	.019	C
4607.3	21698	1	3	2.01	B
Sr II					
2018.7	73237	2	2	.12	D
2051.9	73237	4	2	.24	D
2282.0	67523	4	2	.83	D
2322.4	67563	4	6	.91	D
2324.5	67523	4	4	.15	D
2423.5	64964	2	2	.24	D
2471.6	64964	4	2	.48	D
3464.5	53373	4	6	3.1	D
3474.9	53286	4	4	.51	D
4077.7	24517	2	4	1.42	C
4161.8	47737	2	2	.65	D
4215.5	23715	2	2	1.27	C
4305.5	47737	4	2	1.4	D
4414.8	78702	4	6	.11	D
4417.5	78689	4	4	.018	D
4585.9	77858	4	2	.070	D
5303.1	74621	2	4	.19	D
5379.1	74643	4	6	.22	D
5385.5	74621	4	4	.037	D
5723.7	73237	2	2	.071	D
5819.0	73237	4	2	.14	D
8688.9	67563	4	6	.55	D
8719.6	67523	4	4	.097	D
Sulfur					
S I					
1295.7	77181	5	5	4.9	D
1296.2	77150	5	3	2.7	D
1302.3	77181	3	5	1.8	D
1302.9	77150	3	3	1.6	D
1303.1	77136	3	1	6.6	D
1303.4	76721	5	3	1.9	D
1305.9	77150	1	3	2.4	D
1401.5	71351	5	3	.91	D
1409.3	71351	3	3	.50	D
1412.9	71351	1	3	.16	D
1425.0	70174	5	7	4.5	D
1425.2	70166	5	5	1.2	D
1433.3	70166	3	5	3.3	D
1433.3	70165	3	3	1.9	D
1437.0	70165	1	3	2.4	D
1448.2	78288	5	3	7.3	D
1473.0	67890	5	7	.42	D
1474.0	67843	5	7	1.6	D
1474.4	67825	5	5	.50	D
1474.6	67816	5	3	.062	D
1481.7	67888	3	5	.17	D
1483.0	67825	3	5	1.2	D
1483.2	67816	3	3	.75	D
S I					
1487.2	67816	1	3	.87	D
1666.7	69238	5	5	6.3	C
1687.5	81438	1	3	.94	D
1782.3	78288	1	3	1.9	D
1807.3	55331	5	3	3.8	C
1820.3	55331	3	3	2.2	C
1826.2	55331	1	3	.72	C
4694.1	73921	5	7	.0067	D
4695.4	73915	5	5	.0067	D
4696.2	73911	5	3	.0065	D
6403.6	79058	3	5	.0057	D
6408.1	79058	5	5	.0095	D
6415.5	79058	7	5	.013	D
6751.2	78271	15	25*	.079	D
7679.6	76464	3	5	.012	D
7686.1	76464	5	5	.020	D
7696.7	76464	7	5	.028	D
S II					
1124.4	113461	2	4	1.0	D
1125.0	113461	4	4	4.6	D
1131.0	112937	2	2	3.5	D
1131.6	112937	4	2	1.4	D
1250.5	79968	4	2	.46	C
1253.8	79758	4	4	.42	C
1259.5	79395	4	6	.34	C
4463.6	150996	8	6	.53	D
4483.4	150531	6	4	.31	D
4486.7	150258	4	2	.66	D
4524.7	143623	4	4	.093	D
4525.0	143623	6	4	1.2	D
4552.4	143489	4	2	1.2	D
4656.7	131029	2	4	.09	D
4716.2	131029	4	4	.29	D
4815.5	131029	6	4	.88	D
4885.6	133400	2	4	.17	D
4917.2	133269	2	2	.66	D
4924.1	130134	4	6	.22	D
4925.3	129858	2	4	.24	D
4942.5	129788	2	2	.15	D
4991.9	129858	4	4	.15	D
5009.5	129788	4	2	.70	D
5014.0	133400	4	4	.84	D
5027.2	125485	4	2	.26	D
5032.4	130134	6	6	.81	D
5047.3	133269	4	2	.36	D
5103.3	129858	6	4	.50	D
5142.3	125485	2	2	.19	D
5201.0	140750	4	4	.75	D
5201.3	140750	6	4	.065	D
5212.6	140709	4	6	.098	D
5212.6	140709	6	6	.85	D
5320.7	140319	6	8	.92	D
5345.7	140230	4	6	.88	D
5345.7	140230	6	6	.11	D
5428.6	127976	2	4	.42	D
5432.8	128233	4	6	.68	D
5453.8	128599	6	8	.85	D
5473.6	127825	2	2	.73	D
5509.7	127976	4	4	.40	D
5526.2	128599	8	8	.081	D
5536.8	128233	4	6	.066	D
5556.0	127825	4	2	.11	D
5564.9	128233	6	6	.17	D
5578.8	128233	6	6	.11	D
5606.1	128599	10	8	.54	D
5616.6	127976	4	4	.12	D
5640.0	131187	4	6	.66	D
5645.6	127976	6	4	.018	D
5647.0	130641	2	4	.57	D
5659.9	127976	6	4	.46	D
5664.7	127825	4	2	.58	D
5819.2	130641	4	4	.085	D
6305.5	130134	8	6	.18	D
6312.7	130641	6	4	.30	D
S III					
2496.2	210698	7	5	2.5	D
2508.2	209926	5	3	2.3	D
2636.9	210698	3	5	.45	D
2665.4	210698	5	5	1.4	D
2680.5	209926	1	3	.62	D

Wavelength (λ[Å])	Upper energy level (E_k[cm⁻¹])	Statistical weights g_i	g_k	Transition Probability (A[10⁸ s⁻¹])	Uncertainty
S III					
2691.8	209926	3	3	.46	D
2702.8	209773	3	1	1.9	D
2718.9	206539	3	3	1.2	D
2721.4	209926	5	3	.77	D
2726.8	210698	5	5	.60	D
2731.1	206672	5	5	1.1	D
2756.9	206911	7	7	1.4	D
2785.5	209926	3	3	.61	D
2856.0	205071	5	7	5.1	D
2863.5	205561	7	9	5.7	D
2872.0	204579	3	5	4.7	D
2950.2	206672	3	5	3.0	D
2964.8	206911	5	7	4.0	D
3662.0	174036	3	3	.64	D
3717.8	174036	5	5	1.0	D
3778.9	173192	3	5	.44	D
3831.8	172786	1	3	.56	D
3837.8	172786	3	3	.42	D
3838.3	173192	5	5	1.3	D
3860.6	172631	3	1	1.6	D
3899.1	172786	5	3	.67	D
4253.6	170649	5	7	1.2	D
4285.0	170067	3	5	.90	D
S IV					
551.17	181432	2	2	20.6	C
554.07	181432	4	2	40.8	C
3097.5	213717	2	4	2.6	D
3117.7	213507	2	2	2.5	D
S V					
437.37	311700	1	3	11.2	C
438.19	311700	3	3	33.3	C
439.65	311700	5	3	55	C
661.52	235000	9	15*	64.4	B
679.01	348100	9	15*	86.	D
690.75	345600	9	9*	50.	D
786.48	127149	1	3	52.5	B
854.85	200800	9	9*	41.8	C
S VI					
248.99	401621	2	4	31.	C
249.27	401164	2	2	31.	C
388.94	362983	2	2	45.	C
390.86	362983	4	2	88.	C
706.48	247420	2	4	41.7	C
712.68	247452	4	6	48.5	C
712.84	247420	4	4	8.1	D
933.38	107137	2	4	17.	C
944.52	105874	2	2	16.	C
S VII					
60.161	1662210	1	3	9460.	B
60.804	1644630	1	3	510.	B
72.029	1388330	1	3	861.	B
S VIII					
198.55	503590	4	2	250.	C
202.61	503590	2	2	120.	C
S XI					
189.90	535250	9	3*	430.	C
190.37	592500	5	3	280.	C
215.95	530180	5	5	140.	C
217.63	592500	1	3	72.	C
239.81	417040	1	3	26.	D
242.57	417420	3	5	19.	D
242.82	417040	3	3	19.	D
246.90	417420	5	5	54.	D
247.12	417040	5	3	30.	D
288.49	355260	9	15*	29.	C
S XII					
212.14	471480	2	4	37.	C
215.18	464750	2	2	140.	C
218.20	471480	4	4	170.	C
221.44	464750	4	2	64.	C
227.50	439540	2	2	37.	C
234.48	439540	4	2	68.	C
S XIII					
32.236	3102150	1	3	1.09(+4)	B
37.600	3049260	3	1	1300.	C
256.66	389660	1	3	87.	C
299.89	537520	3	5	17.8	C

Wavelength (λ[Å])	Upper energy level (E_k[cm⁻¹])	Statistical weights g_i	g_k	Transition Probability (A[10⁸ s⁻¹])	Uncertainty
S XIII					
303.37	529420	1	3	22.8	C
307.36	529420	3	3	16.4	C
308.91	537520	5	5	48.2	C
312.68	523880	3	1	63.	C
316.84	529420	5	3	25.0	C
500.42	3012200	3	5	14.3	C
S XIV					
30.434	3285500	2	6*	8280.	B
32.517	3309780	6	10*	2.6 (+4)	B
417.67	239460	2	4	12.	B
445.71	224330	2	2	10.	B
1550.	3287000	2	4	1.4	B
1663.	3282640	2	2	1.2	B
3967.	3307840	2	4	.054	C
4153.	3314070	4	6	.057	C
Thallium					
Tl I					
2104.6	47500	2	4	.040	D
2118.9	47179	2	2	.020	D
2129.3	46950	2	4	.058	D
2151.9	46457	2	2	.031	D
2168.6	46099	2	4	.098	D
2237.8	44673	2	4	.19	D
2316.0	43166	2	2	.078	D
2379.7	42011	2	4	.44	C
2507.9	47655	4	2	.011	C
2538.2	47179	4	2	.016	C
2580.1	38746	2	2	.18	D
2609.0	46110	4	6	.10	C
2609.8	46099	4	4	.019	C
2665.6	45297	4	2	.057	C
2709.2	44693	4	6	.17	C
2710.7	44673	4	4	.037	C
2767.9	36118	2	4	1.26	C
2826.2	43166	4	2	.080	C
2918.3	42049	4	6	.42	C
2921.5	42011	4	4	.076	C
3229.8	38746	4	2	.173	C
3519.2	36200	4	6	1.24	C
3529.4	36118	4	4	.220	C
3775.7	26478	2	2	.625	B
5350.5	26478	4	2	.705	B
Thulium					
Tm I					
2513.8	39769	8	10	.069	D
2527.0	39560	8	8	.17	D
2596.5	38502	8	10	.16	D
2601.1	38434	8	6	.17	D
2622.5	38121	8	10	.061	D
2841.1	43958	6	6	.20	D
2854.2	35026	8	6	.27	D
2914.8	34297	8	8	.077	D
2933.0	34085	8	6	.10	D
2973.2	33624	8	8	.23	D
3046.9	32811	8	8	.18	D
3081.1	32446	8	8	.19	D
3122.5	40787	6	6	.52	D
3142.4	40585	6	6	.088	D
3172.7	31510	8	8	.18	D
3233.7	30915	8	10	.051	D
3247.0	39560	6	8	.30	D
3251.8	39515	6	4	.52	D
3380.7	38343	6	8	.20	D
3406.0	38123	6	8	.15	D
3410.1	29317	8	10	.10	D
3416.6	29261	8	8	.057	D
3418.6	38014	6	6	.11	D
3563.9	28051	8	6	.098	D
3567.4	28024	8	10	.042	D
3744.1	26701	8	8	.95	D
3751.8	26646	8	10	.19	D
3798.5	35090	6	4	1.2	D
3807.7	35026	6	6	.39	D
3883.1	25745	8	6	1.0	D
3887.4	25717	8	8	.38	D
3916.5	34297	6	8	1.5	D
3949.3	34085	6	6	1.0	D
4022.6	33624	6	8	.040	D

Wavelength (λ[Å])	Upper energy level (E_k[cm^{-1}])	Statistical weights g_i	g_k	Transition Probability (A[10^8 s^{-1}])	Uncertainty
		Tm I			
4044.5	33489	6	4	.29	D
4094.2	24418	8	6	.90	C
4105.8	24349	8	10	.60	C
4138.3	32929	6	4	.70	D
4158.6	32811	6	8	.055	D
4187.6	23873	8	8	.61	C
4203.7	23782	8	10	.25	C
4222.7	32446	6	8	.15	D
4271.7	32174	6	6	.11	D
4359.9	22930	8	6	.13	D
4386.4	22791	8	8	.042	D
4394.4	31521	6	4	.11	D
4643.1	30302	6	6	.034	D
4681.9	30124	6	8	.039	D
4691.1	30082	6	6	.039	D
5307.1	18837	8	10	.023	D
5658.3	26439	6	8	.010	D
5675.8	17614	8	10	.013	D
5760.2	26127	6	6	.013	D
		Sn I (Tin)			
2073.1	48222	1	3	.036	D
2199.3	47146	3	5	.29	D
2209.7	48670	5	5	.56	D
2246.1	44509	1	3	1.6	D
2268.9	47488	5	7	1.2	D
2286.7	47146	5	5	.31	D
2317.2	51755	5	7	2.0	D
2334.8	44509	3	5	.66	D
2354.8	44145	3	5	1.7	D
2380.7	43683	3	5	.031	D
2408.2	50126	5	3	.18	D
2421.7	49894	5	7	2.5	D
2429.5	44576	5	7	1.5	C
2433.5	44509	5	3	.0080	D
2455.2	44145	5	5	.011	D
2476.4	48982	5	3	.011	D
2483.4	43683	5	5	.21	D
2491.8	57282	1	3	.17	D
2495.7	48670	5	5	.62	D
2523.9	48222	5	3	.074	D
2546.6	39257	1	3	.21	D
2558.0	56244	1	3	.34	D
2571.6	47488	5	7	.45	D
2594.4	47146	5	5	.30	D
2636.9	55074	1	3	.11	D
2661.2	39257	3	5	.11	D
2706.5	38629	3	5	.66	D
2761.8	39626	5	5	.0037	D
2779.8	44576	5	7	.18	D
2785.0	44509	5	3	.14	D
2788.0	53021	1	3	.14	D
2812.6	52707	1	3	.23	D
2813.6	44145	5	5	.12	D
2840.0	38629	5	5	1.7	D
2850.6	43683	5	5	.33	D
2863.3	34914	1	3	.54	D
2913.5	51475	1	3	.83	D
3009.1	34914	3	3	.38	D
3032.8	50126	1	3	.62	D
3034.1	34641	3	1	2.0	D
3141.8	48982	1	3	.19	D
3175.1	34914	5	3	1.0	D
3218.7	48222	1	3	.047	D
3223.6	39626	5	5	.0012	D
3262.3	39257	5	3	2.7	D
3330.6	38629	5	5	.20	D
3655.8	44509	1	3	.041	D
3801.0	34914	5	3	.28	D
4524.7	39257	1	3	.26	D
5631.7	34914	1	3	.024	D
5970.3	55374	5	3	.096	D
6037.7	55187	5	5	.050	D
6069.0	51113	1	3	.046	D
6073.5	51375	3	1	.063	D
6171.5	51113	3	3	.049	D
		Sn II			
2368.3	46464	4	2	.0044	D
2449.0	99663	4	6	.37	D

Wavelength (λ[Å])	Upper energy level (E_k[cm^{-1}])	Statistical weights g_i	g_k	Transition Probability (A[10^8 s^{-1}])	Uncertainty
		Sn II			
2487.0	99659	6	8	.55	D
3283.2	89292	4	6	1.0	D
3352.0	89286	6	8	1.0	D
3472.5	100285	2	4	.16	D
3575.5	100339	4	6	.13	D
5332.4	90242	2	4	.86	D
5562.0	90354	4	6	1.2	D
5588.9	89292	4	6	.85	D
5596.2	90242	4	4	.15	D
5797.2	89292	6	6	.28	D
5799.2	89286	6	8	.81	D
6453.5	72377	2	4	1.2	D
6761.5	86280	2	2	.32	D
6844.1	74494	2	2	.66	D
		Ti I (Titanium)			
2276.7	44080	7	5	1.3	C
2299.9	43468	5	5	.69	C
2384.5	42311	9	7	.090	C
2424.2	41624	9	9	.17	C
2441.0	41342	9	11	.072	C
2520.5	39662	5	3	.38	C
2529.9	39686	7	5	.38	C
2541.9	39715	9	7	.43	C
2599.9	38451	5	5	.67	C
2605.2	38544	7	7	.64	C
2611.3	38671	9	9	.64	C
2611.5	38451	7	5	.33	C
2632.4	37977	5	5	.27	C
2641.1	37852	5	3	1.8	C
2644.3	37977	7	5	1.4	C
2646.6	38160	9	7	1.5	C
2662.0	37555	5	7	.089	C
2669.6	37618	7	9	.10	C
2679.9	37690	9	11	.13	C
2735.3	45041	3	1	4.1	D
2912.1	41585	5	7	1.3	D
2933.6	34079	5	7	.096	C
2942.0	33981	5	5	1.0	C
2948.3	34079	7	7	.93	C
2956.1	34205	9	9	.97	C
2967.2	34079	9	7	.11	C
2983.3	33680	7	7	.11	C
3000.9	33701	9	9	.12	C
3186.5	31374	5	7	.80	C
3192.0	31489	7	9	.85	C
3199.9	31629	9	11	.94	C
3203.8	31374	7	7	.072	C
3214.2	31489	9	9	.065	C
3341.9	29915	5	7	.65	C
3354.6	29971	7	9	.64	C
3370.4	29661	5	3	.76	C
3371.5	30039	9	11	.67	C
3377.6	29769	7	5	.69	C
3385.7	29915	9	7	.052	C
3635.5	27499	5	7	.72	C
3642.7	27615	7	9	.67	C
3653.5	27750	9	11	.66	C
3654.6	27355	5	3	.087	C
3671.7	27615	9	9	.048	C
3689.9	27480	9	7	.045	C
3717.4	26893	5	7	.043	C
3724.6	38959	9	9	.91	D
3729.8	26803	5	5	.40	C
3741.1	26893	7	7	.38	C
3752.9	27026	9	9	.47	C
3771.7	26893	9	7	.066	C
3786.0	33661	5	3	1.4	D
3924.5	25644	7	7	.073	C
3929.9	25439	5	5	.075	C
3948.7	25318	5	3	.53	C
3958.2	25644	9	7	.43	C
3981.8	25107	5	5	.38	C
3989.8	25227	7	7	.36	C
3998.6	25388	9	9	.39	C
4008.9	25107	7	5	.071	C
4024.6	25227	9	7	.063	C
4065.1	33085	3	1	.70	D
4186.1	36000	9	9	.28	D
4285.0	37359	5	5	.32	D

Wavelength (λ[Å])	Upper energy level (Ek[cm⁻¹])	Statistical weights gi	gk	Transition Probability (A[10⁸ s⁻¹])	Uncertainty
				Ti I	
4295.8	29829	3	1	1.3	D
4393.9	41040	9	11	.33	D
4417.3	37852	11	9	.36	D
4449.2	37690	11	11	.97	D
4455.3	34079	7	7	.48	D
4457.4	34205	9	9	.56	D
4481.3	36415	7	7	.57	D
4527.3	28639	3	5	.22	D
4544.7	28596	5	3	.33	D
4656.5	21409	5	7	.022	C
4667.6	21588	7	9	.023	C
4681.9	21740	9	11	.025	C
4981.7	26911	11	13	.59	C
4991.1	26773	9	11	.50	C
4999.5	26657	7	9	.50	C
5007.2	26564	5	7	.48	C
5014.2	26494	3	5	.68	C
5022.9	26564	7	7	.15	C
5024.8	26494	5	5	.15	C
5040.0	20006	7	5	.036	C
5064.7	20126	9	7	.043	C
5173.8	19323	5	5	.037	C
5193.0	19422	7	7	.032	C
5210.4	19574	9	9	.034	C
5866.5	25644	5	7	.046	C
5899.3	25439	3	5	.031	C
6258.1	27615	7	9	.091	C
7251.7	25318	5	3	.072	C
8024.8	27615	9	9	.0083	C
8068.2	27499	7	7	.0077	C
				Ti II	
2635.6	68768	4	4	1.9	D
2638.7	68845	6	6	1.7	D
2641.0	68930	8	8	1.8	D
2646.1	69081	10	10	2.7	D
2752.8	69081	8	10	1.1	D
2800.6	66997	10	8	1.8	D
2805.0	65185	6	8	4.6	D
2810.3	65307	8	10	5.1	D
2817.9	65446	10	12	3.8	D
2827.2	65094	8	10	1.0	D
2828.2	65589	12	14	4.4	D
2828.9	65307	10	10	.92	D
2834.1	65242	10	12	.79	D
2836.6	64978	8	8	1.2	D
2839.7	65446	12	12	.83	D
2846.1	65094	10	10	1.2	D
2856.2	65242	12	12	1.5	D
2931.3	65313	6	6	3.2	D
2936.2	64885	4	6	2.7	D
2938.7	64978	6	8	2.4	D
2942.0	65094	8	10	1.8	D
2943.1	65459	8	8	1.1	D
2945.5	65242	10	12	2.7	D
2954.8	68582	10	12	4.0	D
2959.0	68329	10	10	4.0	D
3081.6	62410	10	8	1.1	D
3088.0	32767	10	8	1.3	C
3089.4	47625	8	6	1.3	C
3103.8	47467	10	8	1.1	C
3127.9	63168	6	6	1.6	D
3128.6	63445	8	8	1.2	D
3190.9	40075	6	8	1.3	C
3202.5	39927	4	6	1.1	C
3224.2	43781	12	10	.70	C
3234.5	31301	10	10	1.3	C
3236.6	31114	8	8	1.1	D
3287.7	45674	8	10	1.4	C
3341.9	34543	6	8	.96	D
3349.0	34749	8	10	1.0	D
3349.4	30241	10	12	1.3	D
3361.2	29968	8	10	1.2	D
3372.8	29734	6	8	1.1	C
3383.8	29544	4	6	1.1	C
3483.8	63445	10	8	.97	D
3492.4	63168	8	6	.98	D
3504.9	43781	10	10	.82	D
3510.8	43741	8	8	.93	D
3759.3	31491	8	8	.96	D
3761.3	31207	6	6	1.0	D
				Ti III	
2375.0	83797	5	3	4.0	D
2414.0	83117	5	7	3.8	D
2516.1	78159	7	9	3.4	D
2527.8	77746	5	7	2.2	D
2540.1	77422	3	5	2.0	D
2563.4	77424	7	7	2.1	D
2565.4	77167	5	5	1.6	D
2567.6	77000	3	3	2.3	D
2576.5	77000	5	3	.92	D
2984.8	75198	5	5	1.9	D
				Ti XIX	
169.33	590580	1	3	129.	C
				Ti XX	
11.452	8732000	2	6*	1.7 (+4)	C
11.872	8749000	2	4	2.8 (+4)	C
11.958	8751000	4	6	3.4 (+4)	C
15.211	6574000	2	4	3.50(+4)	B
15.253	6556000	2	2	3.58(+4)	B
15.907	6612000	2	4	8.8 (+4)	C
16.049	6619000	4	6	1.1 (+5)	C
16.067	6612000	4	4	1.7 (+4)	C
				Uranium	
				U I	
3553.0	32413	13	13	.020	C
3553.0	32591	9	7	.014	C
3553.4	31935	15	13	.022	C
3554.5	28746	11	9	.0084	C
3554.9	31923	15	17	.0079	C
3555.3	30119	13	13	.027	C
3555.8	28115	13	11	.0041	C
3556.9	32282	13	11	.0075	C
3557.8	28099	13	13	.029	C
3558.0	33860	11	13	.016	C
3558.6	32546	9	7	.039	C
3559.4	31955	7	9	.015	C
3560.3	34071	9	7	.064	C
3561.4	31872	15	13	.055	C
3561.5	34062	9	9	.025	C
3561.8	28068	13	11	.057	C
3563.7	28053	13	13	.029	C
3563.8	31920	7	7	.011	C
3565.0	32318	13	11	.029	C
3566.0	32310	13	15	.017	C
3566.6	28650	11	11	.24	B
3568.8	32289	13	13	.038	C
3569.1	35656	17	15	.11	C
3569.4	32462	9	9	.015	C
3570.1	32278	13	11	.013	C
3570.2	28622	11	9	.0053	C
3570.6	35004	13	15	.027	C
3570.7	31798	15	15	.012	C
3571.2	28614	11	11	.0063	C
3571.6	38338	17	15	.13	C
3572.9	32256	13	15	.015	C
3573.9	27973	13	11	.040	C
3574.1	34977	13	15	.035	C
3574.8	27966	13	15	.019	C
3577.1	35594	17	15	.043	C
3577.5	31745	15	13	.0078	C
3577.8	28563	11	11	.0083	C
3577.9	27941	13	13	.023	C
3578.3	27938	13	11	.020	C
3580.0	32379	9	9	.012	C
3580.2	28543	11	9	.029	C
3580.4	28542	11	13	.0075	C
3580.9	32194	13	13	.021	C
3582.6	32180	13	13	.029	C
3584.6	31757	7	5	.024	C
3584.9	27887	13	15	.18	B
3585.4	28503	11	11	.019	C
3585.8	28500	11	9	.028	C
3587.8	32318	9	11	.013	C
3588.3	31729	7	9	.018	C
3589.7	28470	11	13	.021	C
3589.8	31650	15	13	.059	C
3590.7	32295	9	7	.022	C
3591.7	28454	11	9	.053	C

TRANSITION PROBABILITIES FOR SELECTED ATOMIC AND IONIC SPECIES

Wavelength (λ[Å])	Upper energy level (E_k[cm⁻¹])	Statistical weights g_i	g_k	Transition Probability (A[10⁸ s⁻¹])	Uncertainty
U I					
3593.0	28445	11	11	.014	C
3593.2	32098	13	15	.042	C
3593.7	33581	11	11	.072	C
Vanadium					
V I					
3050.4	43874	10	8	.47	D
3053.7	32738	4	4	1.1	D
3056.3	32847	6	6	1.0	D
3060.5	32989	8	8	1.1	D
3066.4	33155	10	10	1.6	D
3088.1	42010	4	6	.43	D
3089.1	41999	4	4	.45	D
3093.8	42138	6	6	.36	D
3112.9	41752	4	2	.43	D
3183.4	31541	6	8	1.3	D
3185.4	31937	10	12	1.4	D
3198.0	31398	6	6	.31	D
3202.4	31541	8	8	.29	D
3205.6	42079	8	10	1.1	D
3212.4	42221	10	12	1.2	D
3218.9	41950	8	6	.31	D
3233.2	42021	10	8	.28	D
3273.0	41437	8	8	.24	D
3284.4	41539	10	10	.24	D
3309.2	39847	4	4	.28	D
3329.9	39847	6	4	.69	D
3356.4	39423	4	6	.27	D
3365.6	39249	2	4	.41	D
3376.1	39249	4	4	.28	D
3377.4	39237	4	2	.80	D
3377.6	39423	6	6	.53	D
3400.4	38116	8	8	.22	D
3529.7	37960	4	6	.36	D
3533.7	38116	6	8	.44	D
3533.8	37835	2	4	.32	D
3543.5	37757	2	2	.58	D
3545.3	37835	4	4	.32	D
3663.6	43649	4	6	2.7	D
3667.7	43707	6	8	2.4	D
3671.2	38124	8	10	.18	D
3672.4	44140	12	12	.79	D
3673.4	43788	8	10	2.4	D
3676.7	44327	14	14	1.1	D
3680.1	43894	10	12	1.9	D
3686.3	38221	10	12	.20	D
3687.5	44028	12	14	2.6	D
3688.1	29418	8	8	.28	D
3690.3	29203	2	4	.37	D
3692.2	29296	6	6	.46	D
3695.3	44190	14	16	2.5	D
3695.9	29203	4	4	.54	D
3703.6	29418	10	8	.79	D
3704.7	29296	8	6	.58	D
3705.0	29203	6	4	.31	D
3706.0	42079	10	10	.46	D
3708.7	42221	12	12	.39	D
3790.5	37475	10	8	.20	D
3795.0	28768	10	10	.21	D
3806.8	37362	10	10	.22	D
3818.2	26183	4	2	.56	C
3828.6	26249	6	4	.431	C
3840.1	36926	8	8	.18	D
3840.8	26353	8	6	.46	D
3855.4	25931	4	4	.28	D
3855.8	26480	10	8	.451	C
3863.9	36766	8	6	.27	D
3864.3	36763	8	6	.17	D
3864.9	26004	6	6	.208	C
3871.1	36926	10	8	.24	D
3875.1	26122	8	8	.17	D
3886.6	36823	10	8	.14	D
3902.3	26172	10	10	.217	C
3921.9	33967	4	2	.23	D
3922.4	34066	6	6	.23	D
3930.0	36539	10	10	.29	D
3934.0	34128	8	8	.56	D
3992.8	40039	12	10	1.1	D
3998.7	40064	14	12	.92	D
4051.0	41861	10	10	1.2	D
4051.4	41918	12	12	1.2	D
V I					
4090.6	33155	8	10	.77	D
4092.7	26738	8	10	.21	D
4095.5	32989	6	8	.54	D
4099.8	26605	6	8	.39	D
4102.2	32847	4	6	.50	D
4104.8	40126	10	8	1.9	D
4105.2	26506	4	6	.42	D
4109.8	26438	2	4	.47	D
4111.8	26738	10	10	.91	D
4113.5	34128	6	8	.15	D
4115.2	26605	8	8	.59	D
4116.5	26506	6	6	.24	D
4123.6	26397	4	2	.94	D
4128.1	26438	6	4	.70	D
4132.0	26506	8	6	.52	D
4134.5	26605	10	8	.27	D
4232.5	39391	10	10	.86	D
4233.0	39342	8	8	.67	D
4268.6	38483	14	14	1.0	D
4271.6	38405	12	12	.84	D
4277.0	38324	10	10	.84	D
4284.1	38246	8	8	1.0	D
4291.8	40536	12	14	.78	D
4296.1	40452	10	12	.69	D
4297.7	40379	8	10	.61	D
4298.0	40315	6	8	.70	D
4342.8	38124	10	10	.13	D
4355.0	38221	12	12	.11	D
4379.2	25254	10	12	1.2	D
4384.7	25112	8	10	.97	D
4390.0	24993	6	8	.70	D
4395.2	24899	4	6	.48	D
4400.6	24830	2	4	.33	D
4406.6	25112	10	10	.19	D
4407.6	24993	8	8	.38	D
4408.2	24899	6	6	.51	D
4416.5	24789	4	2	.21	D
4452.0	37518	14	16	.80	D
4457.8	37530	10	12	.24	D
4460.3	24839	10	8	.26	D
4462.4	37404	12	14	.66	D
4468.0	37285	8	10	.20	D
4469.7	37316	10	12	.53	D
4474.0	38116	10	8	.41	D
4490.8	37211	10	12	.10	D
4496.1	37960	8	6	.35	D
4514.2	37835	6	4	.29	D
4524.2	37362	12	10	.26	D
4525.2	37757	4	2	.36	D
4529.6	37175	10	8	.21	D
4545.4	37765	10	12	.67	D
4560.7	37644	10	8	.62	D
4571.8	37556	6	8	.53	D
4578.7	37499	4	6	.59	D
4579.2	37556	8	8	.13	D
4706.2	36815	6	4	.21	D
4706.6	38483	12	14	.15	D
4710.6	38405	10	12	.17	D
4746.6	37423	4	4	.17	D
4751.0	37615	8	8	.15	D
4754.0	37758	10	10	.13	D
4757.5	37375	4	2	.65	D
4766.6	37423	6	4	.48	D
4776.4	37503	8	6	.43	D
4786.5	37615	10	8	.40	D
4796.9	37758	12	10	.42	D
4807.5	37931	14	12	.51	D
5193.0	37931	12	12	.35	D
5195.4	37615	8	8	.21	D
5234.1	38124	10	10	.41	D
5240.9	38221	12	12	.39	D
5415.3	37606	12	14	.27	D
5487.9	37362	12	10	.25	D
5507.8	37175	10	8	.30	D
5559.9	31786	6	2*	.14	D
5698.5	26122	6	8	.28	D
5703.6	26004	4	6	.19	D
5707.0	25931	2	4	.19	D
5725.6	36539	8	10	.18	D
5727.0	26172	8	10	.18	D
6090.2	25131	8	6	.13	D

TRANSITION PROBABILITIES FOR SELECTED ATOMIC AND IONIC SPECIES

Wavelength (λ[Å])	Upper energy level (E_k[cm^{-1}])	Statistical weights g_i	g_k	Transition Probability (A[10^8 s^{-1}])	Uncertainty
		V II			
2503.0	48580	5	7	.23	C
2506.2	48731	7	9	.23	C
2514.6	48853	9	11	.24	C
2672.0	37521	5	7	.18	D
2677.8	37369	3	5	.29	D
2679.3	37521	7	7	.26	D
2688.0	37531	9	9	.60	D
2690.8	37259	5	3	.43	D
2700.9	37352	9	11	.29	C
2702.2	37205	7	7	.20	D
2706.2	37151	7	9	.29	C
2728.6	36673	3	5	.20	C
2768.6	48731	11	9	.82	C
2774.3	48580	9	7	.88	C
2799.5	49202	5	5	.60	C
2802.8	49211	7	7	.46	C
2803.5	49269	9	9	.58	C
2836.5	48853	9	11	.25	C
2841.0	48731	7	9	.27	C
2877.7	49202	7	5	.67	C
2880.0	37521	7	7	.15	D
2882.5	37369	5	5	.27	D
2888.3	49269	11	9	.54	C
2889.6	37201	3	1	1.4	D
2891.6	37259	5	3	.92	D
2893.3	37521	9	7	.70	D
2903.1	37041	3	5	.21	D
2906.5	37205	7	7	.53	D
2907.5	37352	9	11	.10	C
2908.8	37531	11	9	1.1	D
2910.0	37041	5	5	.72	D
2910.4	36955	3	3	.87	D
2911.1	37151	7	9	.30	C
2924.6	37151	9	9	.91	C
2930.1	48580	7	7	.27	C
2944.6	36919	9	7	.65	C
2950.4	36489	3	3	.34	C
2952.1	36673	7	5	.58	C
2957.5	36489	5	3	.44	C
3048.9	49211	9	7	.54	D
3093.1	35483	11	13	1.8	D
3100.9	48580	7	7	1.0	C
3102.3	35193	9	11	1.6	D
3110.7	34947	7	9	1.5	D
3118.4	34746	5	7	1.5	D
3121.1	35193	11	11	.22	D
3125.3	34593	3	5	1.6	D
3126.2	34947	9	9	.41	D
3130.3	34746	7	7	.50	D
3133.3	34593	5	5	.48	D
3187.7	40002	5	5	.85	D
3188.5	40196	7	7	.80	D
3190.7	40430	9	9	.88	D
3208.4	40002	7	5	.17	D
3232.0	49202	3	5	.23	C
3267.7	39234	5	7	1.2	C
3271.1	39404	7	9	1.1	C
3276.1	39613	9	11	1.1	C
3321.5	49211	9	7	.15	C
3517.3	37521	9	7	.14	D
3530.8	36955	5	3	.19	D
3545.2	37041	7	5	.18	D
3556.8	37205	9	7	.20	D
3589.6	36489	5	3	.37	D
3592.0	36673	7	5	.23	D
3715.5	39613	13	11	.16	D
3727.2	40430	9	9	.22	C
3732.8	39404	11	9	.16	C
3745.8	39234	9	7	.17	C
3750.9	40196	7	7	.20	C
3771.0	40002	5	5	.22	C
		V XX			
14.360	6964000	1	3	6.9 (+4)	C
160.0	625000	1	3	150.	C
		V XXI			
8.843	11308000	2	6*	6000.	C
8.882	11675000	4	6	6400.	C
9.111	11316000	2	4	8900.	C

Wavelength (λ[Å])	Upper energy level (E_k[cm^{-1}])	Statistical weights g_i	g_k	Transition Probability (A[10^8 s^{-1}])	Uncertainty
		V XXI			
9.175	11316000	4	6	1.04(+4)	C
9.352	10693000	2	6*	1.0 (+4)	C
9.633	10721000	2	4	1.63(+4)	C
9.704	10722000	4	6	1.91(+4)	C
10.413	9603400	2	6*	2.0 (+4)	C
10.768	9627300	2	4	3.5 (+4)	B
10.853	9630600	4	6	4.2 (+4)	B
13.828	7231700	2	4	4.26(+4)	B
13.870	7209800	2	2	4.37(+4)	B
14.435	7268900	2	4	1.1 (+5)	B
14.578	7276300	4	6	1.3 (+5)	B
14.592	7268900	4	4	2.1 (+4)	C
27.95	10693000	2	6*	3000.	C
28.59	10722000	6	10*	6700.	C
40.20	9603400	2	6*	5800.	C
41.58	9629300	6	10*	1.4 (+4)	B
48.32	11673000	6	10*	1060.	C
58.39	11316000	6	10*	1660.	C
89.40	10722000	6	10*	2900.	C
240.0	416600	2	4	29.	B
293.7	340500	2	2	16.	B
		V XXII			
2.382	41986020	1	3	2.9 (+6)	B
		Xenon			
		Xe I			
1043.8	95801	1	3	.59	D
1047.1	95499	1	3	1.3	D
1050.1	95229	1	3	.085	D
1056.1	94686	1	3	2.45	C
1061.2	94229	1	3	.19	D
1068.2	93610	1	3	3.99	C
1085.4	92129	1	3	.410	C
1099.7	90933	1	3	.434	C
1110.7	90033	1	3	1.5	C
1129.3	88550	1	3	.044	C
1170.4	85441	1	3	1.6	C
1192.0	83890	1	3	6.2	C
1250.2	79987	1	3	.14	D
1295.6	77186	1	3	2.5	C
1469.6	68046	1	3	2.8	B
4501.0	89279	5	3	.0062	D
4524.7	89163	5	5	.0021	D
4624.3	88687	5	5	.0072	D
4671.2	88470	5	7	.010	D
4807.0	88843	3	1	.024	D
4844.3	89279	5	7	.066	D
7119.6	92445	7	9	.066	D
7967.3	88745	1	3	.0030	D
8409.2	78957	5	3	.010	D
		Xe II			
4180.1	135708	4	4	2.2	D
4330.5	136598	6	8	1.4	D
4414.8	132208	6	6	1.0	D
4603.0	116783	4	4	.82	D
4844.3	113705	6	8	1.1	D
4876.5	130064	6	8	.63	D
5260.4	123255	2	4	.22	D
5262.0	131924	4	4	.85	D
5292.2	111959	6	6	.89	D
5372.4	113673	4	2	.71	D
5419.2	113512	4	6	.62	D
5439.0	121180	4	2	.74	D
5472.6	113705	8	8	.099	D
5531.1	113512	4	6	.088	D
5719.6	113512	4	6	.061	D
5976.5	111792	4	4	.28	D
6036.2	111959	6	6	.075	D
6051.2	111959	8	6	.17	D
6097.6	111792	6	4	.26	D
6270.8	128867	4	6	.18	D
6277.5	111959	4	6	.036	D
6805.7	113512	6	6	.061	D
6990.9	113705	10	8	.27	D
		Ytterbium			
		Yb I			
2464.5	40564	1	3	.91	C
2672.0	37415	1	3	.118	C
3464.4	28857	1	3	.62	C
3988.0	25068	1	3	1.76	C

TRANSITION PROBABILITIES FOR SELECTED ATOMIC AND IONIC SPECIES

Wavelength ($\lambda[\text{Å}]$)	Upper energy level ($E_k[\text{cm}^{-1}]$)	Statistical weights g_i	g_k	Transition Probability ($A[10^8\ \text{s}^{-1}]$)	Uncertainty
		Yb I			
5556.5	17992	1	3	.0114	C
		Yb II			
3289.4	30392	2	4	1.8	C
3694.2	27062	2	2	1.4	C
		Zinc			
		Zn I			
748.29	133638	1	3	.060	C
765.60	130617	1	3	.076	C
792.05	126255	1	3	.057	C
793.85	125968	1	3	.18	C
809.92	123470	1	3	.26	C
1109.1	90158	1	3	.305	C
2138.6	46745	1	3	7.09	B
3075.9	32501	1	3	3.29(−4)	B
3282.3	62769	1	3	.90	B
3302.6	62769	3	5	1.2	B
3302.9	62769	3	3	.67	B
3345.0	62777	5	7	1.7	B
3345.6	62772	5	5	.40	B
3345.9	62769	5	3	.045	B
6362.3	62459	3	5	.474	C
11054.	55789	3	1	.243	C
		Zn II			
2025.5	49355	2	4	3.3	C
2064.2	96910	2	4	4.6	D
2099.9	96960	4	6	5.6	D
2102.2	96910	4	4	.93	D
4911.6	117264	4	6	1.6	D

REFERENCES

1. Younger, S. M., Fuhr, J. R., Martin, G. A., and Wiese, W. L., *J. Phys. Chem. Ref. Data*, 7, 495, 1978.
2. Fuhr, J. R., Martin, G. A., Wiese, W. L., and Younger, S. M., *J. Phys. Chem. Ref. Data,* to be published, 1980.
3. Wiese, W. L., Smith, M. W., and Glennon, B. M., Atomic Transition Probabilities (H through Ne — A Critical Data Compilation), National Standard Reference Data Series, National Bureau of Standards 4, Vol. I, U.S. Government Printing Office, Washington, D.C., 1966.
4. Wiese, W. L., Smith, M. W., and Miles, B. M., Atomic Transition Probabilities (Na through Ca — A Critical Data Compilation), National Standard Reference Data Series, National Bureau of Standards 22, Vol. II, U.S. Government Printing Office, Washington, D.C., 1969.
5. Wiese, W. L. and Fuhr, J. R., *J. Phys. Chem. Ref. Data,* 4, 263, 1975.
6. Wiese, W. L. and Weiss, A. W., *Phys. Rev.,* 175, 50, 1968.
7. Smith, M. W. and Wiese, W. L., *Astrophys. J. Suppl. Ser.,* 23, No. 196, 103, 1971.
8. Martin, G. A. and Wiese, W. L., *J. Phys. Chem. Ref. Data,* 5, 537, 1976.
9. Fuhr, J. R., Miller, B. J., and Martin, G. A., Bibliography on Atomic Transition Probabilities (1914 through October 1977), National Bureau of Standards Special Publication 505, 1978; Miller, B. J., Fuhr, J. R., and Martin, G. A., Bibliography on Atomic Transition Probabilities (November 1977 through February 1980), National Bureau of Standards Special Publication 505, Supplement 1, 1980.
10. Moore, C. E., Bibliography on the Analyses of Optical Atomic Spectra, National Bureau of Standards Special Publication 306 — Section 1, 1968; Sections 2—4, 1969.
11. Hagan, L. and Martin, W. C., Bibliography on Atomic Energy Levels and Spectra (July 1968 through June 1971), National Bureau of Standards Special Publication 363, 1972; Hagan, L., Bibliography on Atomic Energy Levels and Spectra (July 1971 through June 1975), National Bureau of Standards Special Publication 363, Supplement 1, 1977; Zalubas, R. and Albright, A., Bibliography on Atomic Energy Levels and Spectra (July 1975 through June 1979), National Bureau of Standards Special Publication 363, Supplement 2, 1980.
12. Younger, S. M. and Weiss, A. W., *J. Res. Natl. Bur. Stand.,* 79A, 629, 1975.

NOMOGRAPH AND TABLE FOR DOPPLER LINEWIDTHS
Sidney O. Kastner

The Doppler width of a spectral line is given by the well-known relation $\Delta\lambda = (7.162 \times 10^{-7})\, \lambda T^{1/2} M^{-1/2}$, where wavelength units are in Angstroms, temperature is in degrees Kelvin and M is the atomic mass. This relation between four variables is amenable to representation by a nomograph, which does not appear to have been constructed but which would seem to be of practical value. Therefore such a nomograph is presented here. Its construction is briefly described so that the reader who has not made a plot of this type may follow the steps.

The Doppler relation is first rewritten as

$$\ln \Delta\lambda = \ln k\lambda + \tfrac{1}{2}\ln T - \tfrac{1}{2}\ln M$$

or equivalently $u_4 = u_1 + u_2 + u_3$, with $k = 7.162 \times 10^{-7}$. Putting $\xi - u_1 + u_2$, first, one has a linear relation $\xi - u_1 - u_2 = 0$ between three variables which can be represented (Menzel*) by the determinantal equation

$$\begin{vmatrix} \xi/2 & \tfrac{1}{2} & 1 \\ u_1 & 1 & 1 \\ u_2 & 0 & 1 \end{vmatrix} = 0$$

so that in the Cartesian (x,y) plane the function u_2 lies along the x axis, the function u_1 along the line $y = 1$ and the function $\xi/2$ along the line $y = \tfrac{1}{2}$. These lines form three parallel scales with which to obtain a value of ξ, given the pair (u_1, u_2).

The original equation $u_4 = \xi + u_3$ provides a second linear relation $u_4 - \xi - u_3 = 0$, so that a second set of scales similarly results with u_3 along the x axis, ξ along the line $y = 1$ and $u_4/2$ along the line $y = \tfrac{1}{2}$. The function $u_4 = \log \Delta\lambda$ can then be obtained from the known pair (ξ, u_3).

In practice, for the nomograph scales to represent useful ranges of the physical variables, some shifting and magnification of the individual scales will be found to be necessary, after completion of the first diagram. The nomograph arrived at here is shown in Figure 1. The x axis runs horizontally through the values $x = 0$, $T = 10^5$, and the ordinates of the six scales from left to right are given by:

$$y_1 = 5\ln(10^{5/2}\, k\lambda) \qquad y_2 = 10\ln(k\lambda T^{1/2})$$

$$y_3 = \frac{5}{2}\ln(k\lambda T^{1/2}) \qquad y_4 = 5\ln\Delta\lambda + \frac{5}{2}\ln 10$$

$$y_5 = 5\ln(T^{1/2}/10^{5/2}) \qquad y_6 = 5\ln(10/M)$$

The approximate range of temperatures covered by the figure is a slowly varying function of wavelength, being $1000°K$-$10^6°K$ for $\lambda = 10,000$Å; $4000°K$-$5\times10^6°K$ for $\lambda = 5000$Å; $80,000°K$-$10^7°K$ for $\lambda = 1000$Å; and $300,000°K$-$10^7°K$ for $\lambda = 500$Å.

In use, the left-hand sides of the three lines are used to find the value of the intermediate variable x corresponding to given λ and T. The right-hand sides then give the Doppler width $\Delta\lambda$ in Å, for this value of x and the given M.

For example, suppose one wishes to find the Doppler width of the solar coronal forbidden line of Fe XIV (atomic mass M = 56) at 5303Å, emitted at a temperature of about $T = 2\times10^6°K$. A straight line drawn between $\lambda = 5300$Å and $T = 2\times10^6$ intersects the x scale at about 4.2. A second line then drawn between $x = 4.2$ and $M = 56$ intersects the middle line at $\Delta\lambda \approx 0.73$Å.

To use the nomograph, one must know the appropriate atomic weight M for a given element of interest. An alternative and more accurate procedure, if a calculator is available, for obtaining any required Doppler width is to use Table I, which gives the value of the constant $k_z \equiv 7.162\, M_z^{-1/2}$ for any given atomic number Z, in the equivalent Doppler relation $\lg D\lambda = k_z(\lambda T^{1/2}) \times 10^{-7}$. This table thereby avoids the necessity of looking up the atomic weight M. For the forbidden line example above, where $Z = 26$, the Doppler width obtained from the table in $\Delta\lambda = (0.9584)(5303)(2,000,000)^{1/2}(10^{-7}) = 0.719$Å.

---

* D. H. Menzel, *Fundamental Formulas of Physics*, Dover Publications, New York, 1960, ch. 3.

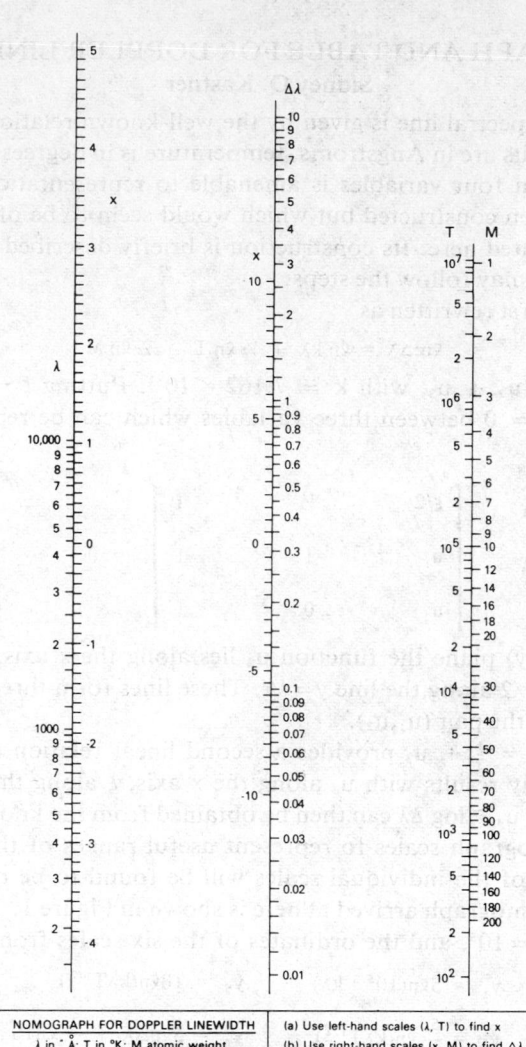

NOMOGRAPH FOR DOPPLER LINEWIDTH
λ in Å; T in °K; M atomic weight
(a) Use left-hand scales (λ, T) to find x
(b) Use right-hand scales (x, M) to find Δλ

Table 1
VALUES OF Kz (Z ATOMIC NUMBER)
FOR USE IN DOPPLER LINEWIDTH
FORMULA Δλ = Kz(λT^{1/2}) × 10^{-7}

Z	kz	Z	kz	Z	kz	Z	kz
1	7.1335	21	1.0682	41	0.7430	61	0.5927
2	3.5798	22	1.0348	42	0.7312	62	0.5840
3	2.7185	23	1.0035	43	0.7202	63	0.5810
4	2.3857	24	0.9932	44	0.7124	64	0.5711
5	2.1782	25	0.9663	45	0.7060	65	0.5681
6	2.0665	26	0.9584	46	0.6943	66	0.5618
7	1.9137	27	0.9329	47	0.6896	67	0.5577
8	1.7905	28	0.9347	48	0.6755	68	0.5538
9	1.6431	29	0.8984	49	0.6684	69	0.5510
10	1.5944	30	0.8858	50	0.6574	70	0.5492
11	1.4937	31	0.8577	51	0.6491	71	0.5414
12	1.4527	32	0.8406	52	0.6340	72	0.5361
13	1.3788	33	0.8274	53	0.6358	73	0.5324
14	1.3514	34	0.8060	54	0.6250	74	0.5282
15	1.2869	35	0.8012	55	0.6212	75	0.5249
16	1.2649	36	0.7824	56	0.6111	76	0.5193
17	1.2028	37	0.7747	57	0.6077	77	0.5166
18	1.1331	38	0.7651	58	0.6050	78	0.5128
19	1.1453	39	0.7596	59	0.6033	79	0.5103
20	1.1313	40	0.7499	60	0.5963	80	0.5057

SECONDARY ELECTRON EMISSION

N. R. Whetten

General Electric Research Laboratory, Schenectady, New York

By permission from "Methods of Experimental Physics" Vol. IV (1962)

Academic Press

The secondary emission yield, or secondary emission ratio, δ, is the average number of secondary electrons emitted from a bombarded material for every incident primary electron. The secondary emission yield is a function of the primary electron energy. δ_{max} is the maximum yield corresponding to a primary electron energy $E_p max$ (see figure). The two primary electron energies corresponding to a yield of unity are denoted the first and second crossovers (E_I and E_{II}). An insulating target, or a

PRIMARY ELECTRON ENERGY (Ep)

conducting target that is electrically floating, will charge positively or negatively depending on the primary electron energy. For $E_I < E_p < E_{II}$, $\delta > 1$ and the surface charges positively provided there is a collector present that is positive with respect to the target. For $E_p < E_I$, $\delta < 1$, and the surface charges negatively towards the potential of the source of primary electrons. For $E_p > E_{II}$, $\delta < 1$, and the surface charges negatively to the second crossover.

The secondary emission yield is very sensitive to surface contamination, such as oxide films and carbon deposits. Whenever possible, yields believed to be most typical of clean surfaces have been selected. The yields are for measurements at room temperature and normal incidence of the primary electrons.

Table I

Secondary Electron Emission Properties of Elements and Compounds. δ_{max} is the maximum secondary emission yield. $\varepsilon_{p\ max}$ the primary electron energy for maximum yield, and E_I and E_{II} are the first and second crossovers.

Elements	δ	$E_{p\ max}$ (ev)	E_I (ev)	E_{II} (ev)	Ref.	Compounds	δ_{max}	$E_{p\ max}$ (ev)	Ref.
Ag	1.5	800	200	>2000	a, b, c	Alkali Halides			
Al	1.0	300	300	300	b	CsCl	6.5		h'
Au	1.4	800	150	>2000	a, c, d	KBr (crystal)	14	1800	c',e'
B	1.2	150	50	600	f	KCl (crystal)	12	1600	p',d'
Ba	0.8	400	None	None	b	(layer)	7.5	1200	b',f'
Bi	1.2	550			g, s	KI (crystal)	10	1600	c',d',e'
Be	0.5	200	None	None	b, h, i, d	(layer)	5.6		b'
C (diamond)	2.8	750		>5000	j	LiF (crystal)	8.5		g'
(graphite)	1.0	300	300	300	k	(layer)	5.6	700	b'
(soot)	0.45	500	None	None	k	NaBr (crystal)	24	1800	g',h',d'
Cd	1.1	450	300	700	l, d	(layer)	6.3		b'
Co	1.2	600	200		m, n	NaCl (crystal)	14	1200	i',e',c',d',g'
Cs	0.7	400	None	None	b, o	(layer)	6.8	600	b',j'
Cu	1.3	600	200	1500	a, l, b	NaF (crystal)	14	1200	g'
Fe	1.3	400	120	1400	n, c, p	(layer)	5.7		b'
Ga	1.55	500	75		g	NaI (crystal)	19	1300	g'
Ge	1.15	500	150	900	f, r, s	(layer)	5.5		b'
Hg	1.3	600	350	>1200	q	RbCl (layer)	5.8		b'
K	0.7	200	None	None	t, u	Oxides			
Li	0.5	85	None	None	b	Ag_2O	1.0		l'
Mg	0.95	300	None	None	o, b	Al_2O_3 (layer)	2 to 9		k',m',i',b
Mo	1.25	375	150	1200	a, w, c, e, p	BaO (layer)	2.3 to 4.8	400	m',b
						BeO	3.4	2000	m'
Na	0.82	300	None	None	x	CaO	2.2	500	m'
Nb	1.2	375	150	1050	a, c	Cu_2O	1.2	400	b',n'
Ni	1.3	550	150	>1500	a, n, m, w, p	MgO (crystal)	20 to 25	1500	t',u',v'
						(layer)	3 to 15	400 to 1500	w,m',r',s',t'
Pb	1.1	500	250	1000	g, q	MoO_2	1.2		l'
Pd	>1.3	>250	120		v	SiO_2 (quartz)	2.1 to 4	400	k',g'
Pt	1.8	700	350	3000	c	SnO_2	3.2	640	o'
Rb	0.9	350	None	None	t	Sulfides			
Sb	1.3	600	250	2000	y	MoS_2	1.1		b'
Si	1.1	250	125	500	f	PbS	1.2	500	p'
Sn	1.35	500			g, x	WS_2	1.0		b'
Ta	1.3	600	250	>2000	a	ZnS	1.8	350	q'
Th	1.1	800			b	Others			
Ti	0.9	280	None	None	k	BaF_2 (layer)	4.5		b'
Tl	1.7	650	70	>1500	s	CaF_2 (layer)	3.2		b'
W	1.4	650	250	>1500	z, a, a', c	$BiCs_3$	6	1000	p'
						BiCs	1.9	1000	p'
Zr	1.1	350			k	GeCs	7	700	p'
						Rb_3Sb	7.1	450	p'
						$SbCs_3$	6	700	p',x'
						Mica	2.4	350	k'
						Glasses	2 to 3	300 to 450	k',y'

References

a   R. Warnecke J. Phys. Radium, 7, 270 (1936).
b   H. Bruining and J. H. deBoer, Physica, 5, 17 (1938).
c   R. Kollath, Physik. Z., 38, 202 (1937).
d   R. Suhrmann and W. Kundt, Z. Physik, 121, 118 (1943).
e   P. L. Copeland, J. Franklin Inst., 215, 593 (1933).

f   L. R. Koller and J. S. Burgess, Phys. Rev., 70, 571 (1946).
g   P. M. Morozov, J. Exptl. Theoret. Phys. USSR, 11, 410, (1941).
h   R. Kollath, Ann. Physik, 33, 285 (1938).
i   E. G. Schneider, Phys. Rev., 54, 185 (1938).
j   J. B. Johnson, Phys. Rev., 92, 843 (1953).

Table I (Continued)
References

k H. Bruining, Philips Tech. Rev., 3, 80 (1938).
l R. Suhrmann and W. Kundt, Z. Physik, 120, 363 (1943).
m D. E. Woolridge, Phys. Rev., 56, 1062 (1939).
n L. R. G. Treloar and D. H. Landon, Proc. Soc. (London), B50, 625 (1938).
o N. S. Klebnikov, Tech. Phys. USSR, 5, 593 (1938).
p R. L. Petry, Phys. Rev., 26, 346 (1925).
q J. J. Brophy, Phys. Rev., 83, 534 (1951).
r J. B. Johnson and K. G. McKay, Phys. Rev., 93, 668 (1954).
s H. Gobrecht and F. Spear, Z. Physik, 135, 602 (1953).
t A. Afanasjewa and P. W. Timofeew, Tech. Phys. USSR, 4, 953 (1937).
u M. S. Joffe and I. V. Nechlaev, J. Exptl. Theoret. Phys. USSR, 11, 93 (1941).
v H. E. Farnsworth, Phys. Rev., 25, 41 (1925).
w G. Blankenfeld, Ann. Physik. 9, 48 (1951).
x J. Woods, Proc. Phys. Soc. (London), B67, 843 (1954).
y R. Kollath, Handbuch der Physik, vol. 21 (1956), p. 232.
z R. L. Petry, Phys. Rev. 28, 362 (1926).
a' E. A. Coomes, Phys. Rev., 55, 519 (1939).
b' H. Bruining and J. H. deBoer, Physica, 6, 834 (1939).
c' D. N. Dobretzov and A. S. Titkow, Doklady Acad. Nauk USSR, 100, 33 (1955).
d' N. R. Whetten, Bull. Am. Phys. Soc. Ser. II, 5, 347 (1960).
e' A. R. Shulman and B. P. Dementyev, J. Tech. Phys. USSR, 25, 2256 (1955).
f' M. Knoll, O. Hachenberg, and J. Randmer, Z. Physik. 122, 137 (1944).

g' D. N. Dobretzov and T. L. Matskevich, J. Tech. Phys. USSR, 27, 734 (1957).
h' T. L. Matskevich, J. Tech. Phys. USSR, 26, 2399 (1956).
i' A. R. Shulman, W. L. Makedonsky and J. D. Yaroshetsky, J. Tech. Phys. USSR 23, 1152 (1953).
j' M. M. Vudinsky, J. Tech. Phys. USSR, 9, 271 (1939).
k' H. Salow, Z. Tech. Phys., 21, 8 (1940).
l' A. Afanasjewa, P. Timofeew, and A. Ignaton, Phys. Z. Sowjet, 10, 831 (1936).
m' K. H. Geyer, Ann Phys., 42, 241 (1942).
n' N. B. Gornij, J. Exptl. Theoret. Phys. USSR, 26, 79 (1954).
o' H. E. Mendenhall, Phys. Rev., 72, 532 (1947).
p' O. Hachenberg and W. Braner in Advances in Electronics and Electron Physics, Vol. XI, Academic Press, New York (1959), p. 438.
q' N. B. Gornij, J. Exptl. Theoreto. Phys. USSR, 26, 88 (1954).
r' P. Wargo, B. V. Haxby, and W. G. Shepherd, J. Appl. Phys. 27, 1311 (1956).
s' P. Rappaport, J. Appl. Phys., 25, 288 (1954).
t' N. R. Whetten and A. B. Laponsky, J. Appl. Phys., 30, 432 (1959).
n' J. R. Johnson and K. G. McKay, Phys. Rev., 91, 582 (1953).
v' R. G. Lye, Phys. Rev., 99, 1647 (1955).
w' N. R. Whetten and A. B. Laponsky, J. Appl. Phys., 28, 515 (1957).
x' N. D. Morgulis and B. I. Djatlowitskaja, J. Tech. Phys. USSR, 10, 657 (1940).
y' C. W. Mueller, J. Appl. Phys., 16, 453 (1945).

EMISSIVITY OF TUNGSTEN

Wavelengths in μ

Temperature °K	0.25	0.30	0.35	0.40	0.50	0.60	0.70
1600	.448*	.482	.478	.481	.469	.455	.444
1800	.442*	.478*	.476	.477	.465	.452	.44
2000	.436*	.474	.473	.474	.462	.448	.436
2200	.429*	.470	.470	.471	.458	.445	.431
2400	.422	.465	.466	.468	.455	.441	.427
2600	.418	.461	.464	.464	.451	.437	.423
2800	.411	.456	.461	.461	.448	.434	.419

Temperature °K	0.80	0.90	1.0	1.1	1.2	1.3	1.4
1600	.431	.413	.39	.366	.345	.322*	.300*
1800	.425	.407	.385	.364	.344	.323*	.302*
2000	.419	.401	.381	.361	.343	.323	.305
2200	.415	.396	.378	.359	.342	.324	.306

Temperature °K	0.80	0.90	1.0	1.1	1.2	1.3	1.4
2400	.408	.391	.372	.355	.340	.324	.309
2600	.404	.386	.369	.352	.338	.325	.310
2800	.400	.383	.367	.352	.337	.325	.313

Temperature °K	1.5	1.6	1.8	2.0	2.2	2.4	2.6
1600	.279*	.263*	.234*	.210*	.19*	.175*	.164*
1800	.282	.267*	.241*	.218*	.20*	.182*	.174*
2000	.288	.273	.247	.227	.209*	.197	.175
2200	.291	.278	.254	.235	.218	.205	.194
2400	.296	.283	.262	.244	.228	.215	.205
2600	.299	.288	.269	.251	.236	.224	.214
2800	.302	.292	.274	.259	.245	.233	.224

* Values by extrapolation.

LIQUIDS FOR INDEX BY IMMERSION METHOD

Liquid	N_D 24°C	Liquid	N_D 24°C
Trimethylene chloride	1.446	Iodobenzene + Bromobenzene	1.603
Cineole	1.456	Iodobenzene + Bromobenzene	1.613
Hexahydrophenol	1.466	Quinoline	1.622
Decahydronaphthalene	1.477	α-Chloronaphthalene	1.633
Isoamylphthalate	1.486	α-Bromophthalene + α-Chloronaphthalene	1.650-1.650
Tetrachloroethane	1.492	α-Bromophthalene + α-Iodonaphthalene	1.660-1.690
Pentachloroethane	1.501	Methylne iodide + Iodobenzene	1.700-1.730
Trimethylene bromide	1.513	Methylene iodide	1.738
Chlorobenzene	1.523	Methylene iodide saturated with sulfur	1.78
Ethylene bromide + Chlorobenzene	1.533	Yellow phosphorus, sulfur and methylene iodide (8:1:1 by weight)	2.06
α-Nitrotoluene	1.544	Can be diluted with methylene iodide to cover range 1.74-2.06. For precautions in use, cf. West, Am. Mineral, 21, p. 245-9 (1936).	
Xylidine	1.557		
α-Toluidine	1.570		
Aniline	1.584		
Bromoform	1.595		

HEAVY LIQUIDS FOR MINERAL SEPARATION

Liquid	Density
Tetrabromoethane (sym.)	2.964, 20°/4°
Can be diluted with carbon tetrachloride (1.595) or benzene (0.894)	
Methylene iodide	3.325, 20°/4°
Can be diluted with carbon tetrachloride or benzene.	
Thallium formate, sq.	3.5
Can be diluted with water.	
Thallium malonate-thallium formate, aq.	4.9
Can be diluted with water.	

For preparation and recovery of these liquids, cf. U. S. Bureau Mines, Rept. Inv. #2897 (1928).

INDEX OF REFRACTION

Indices of refraction for elements, inorganic, metal-organic and organic compounds and minerals will be found in the tables of physical constants for the various classes of substances in the section Properties and Physical Constants.

Values for compounds not there listed and data subsequently collected are given below.

Indices not otherwise indicated are for sodium light, $\lambda = 589.3$ mμ. Other wave lengths are indicated by the value in millimicrons or symbol in parentheses which follows the index. Wave lengths are indicated as follows: He, $\lambda = 587.6$ mμ; Li, $\lambda = 670.8$ mμ; Hg, $\lambda = 579.1$ mμ; A, $\lambda = 759.4$ mμ; C, $\lambda = 656.3$ mμ; D, $\lambda = 589.3$ mμ; F, $\lambda = 486.1$ mμ.

Temperatures are understood to be 20°C for liquids, or ordinary room temperatures in the case of solids. Other temperatures appear as superior figures with the index.

Indices for the elements and inorganic compounds will be understood to be for the solid form except as indicated by the abbreviation liq.

See also under Physical Constants of Inorganic Compounds and index of Refraction of Gases.

Elements

Name	Formula	Index	Name	Formula	Index
Bromine (liq.)	Br_2	1.661_{15}	Oxygen (liq.)	O_2	1.221^{-131}
Cadmium (liq.)	Cd	0.82 (579 mμ)	Phosphorous (yel.) (sol.)		2.1442^{25}
(sol.)		1.13	Selenium	Ses	3.00, 4.04
Chlorine (liq.)	Cl_2	1.385	(amor.) (sol.)		2.92
(gas)		1.00768	Sodium (liq.)	Na	0.0045
Hydrogen (liq.)	H_2	$1.10974^{-252.83}$(579 mμ)	(sol.)		4.22
			Sulfur (liq.)	S_6	1.929^{110}
Iodine (sol.)	I_2	3.34	(amor.) (sol.)		.1998
(gas)		1.001920	(rhombic, α)		1.957, 2.0377, 2,2454
Lead	Pb	2.6 (579 mμ)			
Mercury (liq.)	Hg	1.6-1.9	Tin (liq.)	Sn	2.1
Nitrogen (liq.)	N_2	1.2053^{-190}			

Inorganic Compounds
See also under Physical Constants of Inorganic Compounds

Name	Formula	Index	Name	Formula	Index
Aluminum carbide	AlC_3	2.7, 2.75 (700 mμ)	potassium selenate	$CoSeO_4 \cdot K_2SeO_4 \cdot 6H_2O$	1.5135, 1.5195, 1.5358
chloride	$AlCl_3 \cdot 6H_2O$	1.560, 1.507	rubidium sulfate	$CoSO_4 \cdot Rb_2SO_4 \cdot 6H_2O$	1.4859, 1.4916, 1.5014
oxide	Al_2O_3	1.665-1.680, 1.63-1.65	selenate	$CoSeO_4 \cdot 6H_2O$	α1.5225, γ1.5227
Alums. See under appropriate element.			Copper ammonium selenate	$CuSeO_4 \cdot (NH_4)_2SeO_4 \cdot 6H_2O$	1.5213, 1.5355, 1.5395
Ammonium antimony tartrate	$2(NH_4 \cdot SbO \cdot C_4H_4O_6)H_2O$	β1.6229 (C)	ammonium sulfate	$CuSO_4 \cdot (NH_4)_2SO_4 \cdot 6H_2O$	1.4910, 1.5007, 1.5054
orthoarsenate, di-H	$NH_4^2H_2AsO_4$	1.5766, 1.5217	cesium sulfate	$CuSO_4 \cdot Cs_2SO_4 \cdot 6H_2O$	1.5048, 1.5061, 1.5153
bromide	NH_4Br	1.7108	chloride (ic)	$CuCl_2 \cdot 2H_2O$	1.644, 1.684, 1.742
perchlorate	NH_4ClO_4	1.4818, 1.4833, 1.4881	formate	$Cu(CHO_2)_2 4H_2O$	1.4133, 1.5423, 1.5571
chloroplatinate	$(NH_4)_2PtCl_6$	1.8	Copper oxide (ous) (cuprite)	Cu_2O	2.705
fluoride	NH_4F	ω<1.328	potassium chloride	$CuCl_2 \cdot 2KCl \cdot 2H_2O$	1.6365, 1.6148
acid	NH_4HF_2	1.385, 1.390, 1.394	potassium cyanide (ous)	$CuK_3(CN)_4$	1.5215
hydrogen malate (d)	$NH_4C_4H_5O_5$	β1.503	potassium selenate	$CuSeO_4 \cdot K_2SeO_4 \cdot 6H_2O$	1.5096, 1.5235, 1.5387
nitrate	NH_4NO_3	1.413, 1.611(IIc), 1.63	potassium sulfate	$CuSO_4 K_2SO_4 \cdot 6H_2O$	1.4936, 1.4964, 1.5028
Ammonium sulfate, acid	NH_4HSO_4	1.463, 1.473, 1.510	strontium formate	$Cu(HCO_2)_2 \cdot 2[SrHCO_2)_2] 8H_2O$	1.4995, 1.5199, 1.5801
tartrate (dl)	$(NH_4)_2C_4H_4O_6 \cdot 2H_2O$	β1.564	sulfate (ic)	$CuSO_4$	1.724, 1.733, 1.739
thiocyanate	NH_4CNS	1.546, 1.685, 1.692	Cyanogen	C_2N_2	1.327^{18} (liq.)
uranyl acetate	$NH_4C_2H_3O_2 \cdot UO_2(C_2H_3O_2)_2$	1.4808, 1.4933	Germanium bromide, tetra-	$GeBr_4$	1.6269
Antimony bromide	$SbBr_3$	>1.74+	Gold sodium chloride	$AuNaCl_4 \cdot 2H_2O$	α1.545, γ1.75+
iodide, tri-	SbI_3	2.78 (Li), 2.36	Hafnium oxychloride	$HfOCl_2 \cdot 8H_2O$	1.557, 1.543
Barium cadmium bromide	$BaCdBr_4 \cdot 4H_2O$	β1.702	Ice	H_2O	1.3049, 1.3062 (A), 1.3001, 1.3104 (D), 1.3133, 1.3147 (F)
cadmium chloride	$BaCdCl_4 \cdot 4H_2O$	β1.651			
calcium propionate	$BaCa_2(C_3H_5O_2)_6$	1.4442			
fluochloride	$BaCl_2 \cdot BaF_2$	1.640, 1.633	Iron ammonium chloride	$Fe(NH_4)_2Cl_4$	1.6439
fluoride	BaF_2	1.475 also 1.4741	ammonium selenate	$FeSeO_4 \cdot (NH_4)_2SeO_4 \cdot 6H_2O$	1.5201, 1.5260, 1.5356
Barium oxide	BaO	1.980	cesium sulfate (ic)	$FeCs_2(SO_4)_2 \cdot 12H_2O$	1.4839
orthophosphate, di-	$BaHPO_4$	1.617, 1.63±, 1.635	cesium sulfate (ous)	$FeSO_4 \cdot Cs_2SO_4 \cdot 6H_2O$	1.5003, 1.5035, 1.5094
propionate	$Ba(C_3H_5CO_2)_2 \cdot H_2O$	β1.5175	rubidium sulfate	$FeRb_2(SO_4)_2 \cdot 12H_2O$	1.48234
sulfide, mono-	BaS	2.155	sulfate (ic)	$Fe_2(SO_4)_3$	1.802, 1.814, 1.818
Cadmium ammonium chloride	$CdCl_2 \cdot 4NH_4Cl$	1.6038, 1.6042	thallium sulfate	$FeTl(SO_4)_2 \cdot 12H_2O$	1.52365
cesium sulfate	$CdSO_4 \cdot Cs_2SO_4 \cdot 6H_2O$	1.498, 1.500, 1.506	Lanthanum sulfate	$La_2(SO_4)_3 \cdot 9H_2O$	1.564, 1.569
fluoride	CdF_2	1.56	Lead orthoarsenate, di-	$PbHAsO_4$	1.8903, 1.9097, 1.9765
magnesium chloride	$(CdCl_2)_2 \cdot MgCl_2 \cdot 12H_2O$	1.49, 1.5331, 1.5769	nitrate	$Pb(NO_3)_2$	1.782
oxide	CdO	2.49 (Li)	Lithium ammonium sulfate	$LiNH_4SO_4$	β1.437 (Li)
potassium chloride	$CdCl_2 \cdot 4KCl$	1.5906, 1.5907	ammonium tartrate (d)	$LiNH_4(C_4H_4O_6) \cdot H_2O$	β1.567, γ1.5673
cyanide	$Cd(CN)_2 \cdot 2KCN$	1.4213	ammonium tartrate (dl)	$LiNH_4(C_4H_4O_6) \cdot H_2O$	β1.5287
rubidium sulfate	$CdSO_4 \cdot Rb_2SO_4 \cdot 6H_2O$	1.4798, 1.4848, 1.4948	bromide	LiBr	1.784
Calcium aluminate	$Ca_3Al_2O_6$	1.710	chloride	LiCl	1.662
borate	$Cao \cdot B_2O_3$	1.540, 1.656, 1.682	dithionate	$Li_2S_2O_6 \cdot H_2O$	1.5487, 1.5602, 1.5788
carbide	CaC_2	<1.75	oxide	Li_2O	1.644
copper acetate	$CaCu(C_2H_3O_2)_4 \cdot 6H_2O$	1.436, 1.478	potassium sulfate	$LiKSO_4$	1.4723, 1.4717
cyanamide	$CaCN_2$	1.60, <1.95	potassium tartrate	$LiK(C_4H_4O_6) \cdot H_2O$	β1.5226 (red)
dithionate	$CaS_2O_6 \cdot 4H_2O$	1.5516, 1.5414	rubidium tartrate (a)	$LiRb(C_4H_4O_6) \cdot H_2O$	β1.552
pyrophosphate	$Ca_2P_2O_7$	1.585, 1.60±, 1.605	sodium tartrate (dl)	$LiNa(C_4H_4O_6) \cdot 2H_2O$	β1.4904
platinocyanide	$CaPt(CN)_4 \cdot 5H_2O$	1.623, 1.644, 1.767	Magnesium ammonium selenate	$MgSeO_4 \cdot (NH_4)_2SeO_4 \cdot 6H_2O$	1.5070, 1.5093, 1.5169
stromtium propionate	$Ca_2Sr(C_3H_5O_2)_6$	1.4871, 1.4956	ammonium sulfate	$Mg(NH_4)_2 \cdot (SO_4)_2 \cdot 6H_2O$	1.4716, 1.4730, 1.4786
sulfide (oldhamite)	CaS	2.137	ortho borate	$3MgO \cdot B_2O_3$	1.6527, 1.6537, 1.6748
sulfite	$CaSO_3 \cdot 2H_2O$	1.590, 1.595, 1.628	cesium sulfate	$MgCs_2(SO_4)_2 \cdot 6H_2O$	1.4857, 1.4858, 1.4916
thiosulfate	$CaS_2O_3 \cdot 6H_2O$	1.545, 1.560, 1.605	chlorostannate	$MgSnCl_6 \cdot 6H_2O$	1.5885, 1.5970
Carbon dioxide (liq.)	CO_2	1.195^{15}	fluosilicate	$MgSiF_6 \cdot 6H_2O$	1.3439, 1.3602
Cerium dithionate	$Ce_2(S_2O_6)_3 \cdot 15H_2O$	β1.507	platinocyanide	$MgPt(CN)_4 \cdot 7H_2O$	1.5608, 1.91
Cesium perchlorate	$CsClO_4$	1.4752, 1.4788, 1.4804	Magnesium potassium selenate	$MgK_2(SeO_4)_2 \cdot 6H_2O$	1.4969, 1.4991, 1.5139
nitrate	$CsNO_3$	1.55, 1.56	potassium sulfate	$MgK_2(SO_4)_2 \cdot 6H_2O$	1.407, 1.4629, 1.4755
selenate	Cs_2SeO_4	1.5989, 1.5999, 1.6003	rubidium sulfate	$MgRb_2(SO_4)_2 \cdot 6Hi2O$	1.4672, 1.4689, 1.4779
thallium chloride	$Cs_2Tl_2Cl_9$	1.784, 1.774	silicate	$MgSiO_3$	1.651, 1.654 (calc.), 1.660
Chromium cesium sulfate	$CrCs(SO_4)_2 \cdot 12H_2O$	1.4810	sulfide	MgS	2.271 also 2.268
oxide (ic)	Cr_2O_3	2.5	Manganese borate	$Mn_3B_4O_9$	1.617, 1.738, 1.776
potassium cyanide (ic)	$CrK_3(CN)_6$	4.5221, 1.5244, 1.5373	cesium sulfate	$MnCs_2(SO_4)_2 \cdot 6H_2O$	1.4946, 1.4966, 1.5025
sulfate (ic)	$Cr_2(SO_4)_3 \cdot 18H_2O$	1.564	chloride	$MnCl_2 \cdot 4H_2O$	1.555, 1.575, 1.607
thallium sulfate	$CrTl(SO_4)_2 \cdot 12H_2O$	1.5228	rubidium sulfate	$MnRb_2(SO_4)_2 \cdot 6H_2O$	1.4767, 1.4807, 1.4907
Cobalt acetate	$Co(C_2H_3O_2)_2 \cdot 4H_2O$	β1.542	sulfate (ous)	$MnSO_4 \cdot 4H_2O$	1.508, 1.518, 1.522
aluminate (Thenard's Blue)	$Co(AlO_2)_2$	<1.78 (red), 1.74 (blue)		$MnSO_4 \cdot 5H_2O$	1.495, 1.508, 1.514
ammonium selenate	$CoSeO_4 \cdot (NH_4)_2SeO_4 \cdot 6H_2O$	1.5246, 1.5311, 1.5396	Mercury chloride (ic)	$HgCl_2$	1.725, 1.859, 1.965
cesium sulfate	$CoCs_2(SO_4)_2 \cdot 6H_2O$	1.5057, 5.5085, 1.5132	cyanide (ic)	$Hg(CN)_2$	1.645, 1.492
chloride (ous)	$CoCl_2 \cdot 2H_2O$	<1.624, <1.671, >1.67	iodide (ic) (red)	HgI_2	2.748, 2.455

Name	Formula	Index	Name	Formula	Index
Nickel ammonium selenate	Ni(NH₄)₂·(SeO₄)₂·6H₂O	1.5291, 1.5372, 1.5466		NaH₂AsO₄·2H₂O	1.4794, 1.5021, 1.5265
cesium sulfate	NiCs₂(SO₄)₂·6H₂O	1.5087, 1.5129, 1.5162	bromide	NaBr	1.6412
Nickel chloride	NiCl₂·6H₂O	α1.53, γ1.61	carbonate	Na₂CO₃	1.415, 1.535, 1.546
fluoride, acid	NiF₂·5HF·6H₂O	1.392, 1.408	Sodium carbonate, acid	NaHCO₃	1.376, 1.500, 1.582
potassium selenate	NiK₂(SeO₄)₂·6H₂O	1.5199, 1.5248, 1.5339	cyanide	NaCN	1.452
rubidium sulfate	NiRb₂(SO₄)₂·6H₂O	1.4895, 1.4961, 1.505	iodide	NaI	1.7745
selenate	NiSeO₄·6H₂O	1.593, 1.5125	molybdate	3Na₂O.7MoO₃·22H₂O	β1.627
Platinum potassium dibromo-			nitrate	NaNO₃	1.5874, 1.3361
nitrite	PtK₂(NO₂)₂)Br₂·H₂O	1.626, 1.6684, 1.757	phosphate	NaH₂PO₄·2H₂O	1.4401, 1.4629, 1.4815
Potassium carbonate	K₂CO₃	1.426, 1.531, 1.541		Na₂HPO₄·7H₂O	1.4412, 1.4424, 1.4526
carbonate, acid	KHCO₃	1.380, 1.482, 1.598	hypophosphate	Na₄HP₂O₆·9H₂O	1.4653, 1.4738, 1.4804
perchlorate	KClO₄	1.4731, 1.4737, 1.4769	silicate	Na₂SiO₃	1.513, 1.520, 1.528
chloroplatinate	K₂PtCl₆	1.827 (577 mμ)	sulfate, acid	NaHSO₄·H₂O	1.43, 1.46, 1.47
chloroplatinite	K₂PtCl₄	1.64, 1.67	sulfite	Na₂SO₃	1.565, 1.515
dichromate	K₂Cr₂O₇	1.7202, 1.7280, 1.8197	acid	Nahso₃	1.474, 1.526, 1.685
cyanide	KCN	1.410	tartrate, acid (d)	NaH(C₄H₄O₆)·H₂O	β1.533
fluoborate	KBF₄	1.3239, 1.3245, 1.3247	thiocyanate	NaCNS	1.545, 1.625, 1.695
fluoride	KF	1.352 (1.361)	Sodium tungstate	Na₂WO₄·2H₂O	1.5526, 1.5533, 1.5695
	KF·2H₂O	1.345, 1.352, 1.363	vanadate	Na₃VO₄·10H₂O	1.5305, ω1.5398, ε1.5475
fluosilicate	K₂SiF₆	1.3391		Na₃VO₄·12H₂O	1.5095, 1.5232
periodate	KClO₄	1.6205, 1.6479	Strontium dichromate	SrCr₂O₇·3H₂O	1.7146, 1.7174, 1.812
lithium ferrocyanide	K₂Li₂Fe(CN)₆·3H₂O	1.5883, 1.6007, 1.6316	fluoride	SrF₂	1.442 (1.438)
hypophosphate	K₂H₂P₂O₆·2H₂O	1.4893, 1.5314, 1.5363	oxide	SrO	1.870
	K₂H₂P₂O₆·3H₂O	1.4768, 1.4843, 1.4870	orthophosphate, acid	SrHPO₄	1.608, 1.62±, 1.625
ruthenium cyanide	K₄Ru(CN)₆·3H₂O	β1.5837	sulfide, mono-	SrS	2.107
silicate	K₂SiO₃	1.520, 1.521, 1.528	Sulfur nitride	S₄N₄	α1.908, β2.046
thiocyanate	KCNS	1.532, 1.660, 1.730	Thallium chloride, mono-	TlCl	2.247
thionate, tetra-	K₂S₄O₆	1.5896, 1.6057, 1.6435	iodide, mono-	TlI	2.78
penta-	2K₂S₅O₆·3H₂O	1.565, 1.63, 1.655	Tin iodide (ic)	SnI₄	2.106
Rhodium cesium sulfate	RhCs₂(SO₄)₂·12H₂O	1.5077	Uranyl potassium sulfate	UO₂·SO₄·k₂SO₄·2H₂O	1.5144, 1.5266, 1.5705
Rubidium perchlorate	RbClO₄	1.4692, 1.4701, 1.4731			(580 mμ)
chromate	Rb₂CrO₄	β1.71, γ1.72	Vanadium ammonium sulfate	VNH₄(SO₄)₂·12H₂O	1.475
dithionate	Rb₂S₂O₆	1.4574, 1.5078	Zinc ammonium selenate	Zn(SeO₄)·(NH₄)₂SeO₄·6H₂O	1.5240, 1.5300, 1.5385
fluoride	RbF	1.396	bromate	Zn(BrO₃)₂·6H₂O	1.5452
selenate	RbSeO₄	1.5515, 1.5537, 1.5582	cesium sulfate	ZnCs₂(SO₄)₂·6H₂O	1.5022, 1.5048, 1.5093
Ruthenium sodium nitrate	RuNa₂(NO₂)₅·2H₂O	1.5889, 1.5943, 1.7163	chloride	ZnCl₂	1.687, 1.713
Selenium oxide	SeO₂	>1.76	fluosilicate	ZnSiF₆·6H₂O	1.3824, 1.3956
Silver cyanide	AgCN	1.685, 1.94	potassium cyanide	ZnK₂(CN)₄	1.4115
nitrate	AgNO₃	1.729, 1.744, 1.788	selenate	ZnK₂(SeO₄)₂·6H₂O	1.5121, 1.5181, 1.5335
phosphate	Ag₂HPO₄	1.8036, 1.7983	sulfate	ZnK₂(SO₄)₂·6H₂O	1.4775, 1.4833, 1.4969
potassium cyanide	AgK(CN)₂	1.625, 1.63	rubidium sulfate	ZnRb₂(SO₄)₂·6H₂O	1.4833, 1.4884, 1.4975
Sodium ammonium tartrate (d)	NaNH₄(C₄H₄O₆)·4H₂O	1.495, 1.498, 1.499	silicate	ZnSiO₃	1.616, 1.62±, 1.623
ammonium tartrate (dl)	NaNH₄(C₄H₄O₆)·H₂O	β1.473 (red)	Zirconium ammonium fluoride	Zr(NH₄)₄F₇	1.433
orthoarsenate	NaH₂AsO₄·H.O	1.5382, 1.5535, 1.5607			

ORGANIC COMPOUNDS

See also under Physical Constants of Organic Compounds.

Name	Index
Allontoin, solid	γ1.579, λ1.660
Dimethyl thiophene (α, α'), liq	1.51693[13.4] (He)
(β, β'), liq	1.52217[15] (He)
Ethyl carbylamine, liq	1.3659[24]
Ethylidene cyanhydrin, liq	1.40582[18.4]
Hexyl acetylene (n), liq	1.4208[12.5]

Miscellaneous

Albite glass	1.4890	Magdala red	1.90
Amber	1.546	Obsidian	1.482-1.496
Anorthite glass	1.5755	Paraffin	1.43295[38.3] (C)
Asphalt	1.635	Quartz, fused	
Bell metal	1.0052		1.45640 (656 mμ)
Borax, amorphous, fused	1.4630		1.45843 (589 mμ)
Canada balsam	1.530		1.46190 (509 mμ)
Ebonite	1.66 (red)		1.47503 (361 mμ)
Fuchsin	2.70		1.49634 (275 mμ)
Gelatin, Nelson's No. 1	1.530		1.53386 (214 mμ)
Gelatin, various	1.516-1.534		1.57464 (185 mμ)
Gum Arabic	1.480 (1.5,4)	Resin, aloes	1.619 (red)
	(red)	colophony	1.548 (red)
Hoffman's violet	2.20	copal	1.528 (red)
Ivory	1.539, 1.541	mastic	1.535 (red)
		Peru balsam	1.593

INDEX OF REFRACTION OF ORGANIC COMPOUNDS

(See also that section of this book which contains data of Physical Constants of Organic Compounds)

The following table contains a list of organic compounds arranged in order of increasing refractive index. Measurements were made at 25°C.

Compound	n_D	Compound	n_D
Trifluoroacetic acid	1.283	Propionaldehyde	1.371
2,2,2-Trifluoroethanol	1.290	n-Hexane	1.372
Octofluoropentanol-1	1.316	2,3-Dimethylbutane	1.372
Dodecafluoroheptanol-1	1.316	3-Methylpentane	1.374
Methanol	1.326	2-Propanol	1.375
Acetonitrile	1.342	Isopropyl acetate	1.375
Ethyl ether	1.352	Propyl formate	1.375
Acetone	1.357	2-Chloropropane	1.376
Ethyl formate	1.358	2-Butanone	1.377
Ethanol	1.359	2-Chloropropane	1.377
Methyl acetate	1.360	Methylethyl ketone	1.377
Propionitrile	1.363	Butyraldehyde	1.378
2,2-Dimethylbutane	1.366	2,4-Dimethylpentane	1.379
Isopropyl ether	1.367	Propyl ether	1.379
2-methylpentane	1.369	Acetaldehyde-diethylacetal	1.379
Ethyl acetate	1.370	Butylethyl ether	1.380
Acetic acid	1.370	Nitromethane	1.380

Compound	n_D	Compound	n_D
Trifluoropropanol	1.381	Ethylcyanoacetate	1.415
2-Methylhexane	1.382	Dibutylamine	1.416
Butyronitrile	1.382	2-Pentanol	1.416
Propyl acetate	1.382	1,1-Dichloroethane	1.416
Ethyl propionate	1.382	Heptachlorodiethyl ether	1.416
2-Methyl-2-propanol	1.383	1-Hexanol	1.416
1-Propanol	1.383	1-Amino-3-methoxy propane	1.417
Isobutyl formate	1.383	Octyl nitrile	1.418
Diethyl carbonate	1.383	2-Heptanol	1.418
Heptane	1.385	2-Propenyl amine	1.419
2-Methyl-2-propanol	1.385	1,2-Propyleneglycol carbonate	1.419
Propionic acid	1.385	Methylpentyl carbinol	1.420
3-Methylhexane	1.386	2-Ethyl-1-butanol	1.420
n-Propyl amine	1.386	1-Chloro-2-mehyl-1-propene	1.420
1,1-Dimethyl-2-propanone	1.386	p-Dioxane	1.420
1-Chloropropane	1.386	Methylcyclohexane	1.421
2,2,3-Trimethylbutane	1.387	4-Hydroxy-4-methyl-2-pentanone	1.421
Methylpropyl ketone	1.387	1-Heptanol	1.422
sec-Butyl acetate	1.387	3-Isorpopyl-2-heptanone	1.423
Butyl formate	1.387	Cyclohexane	1.424
β-Methylpropyl ethanoate	1.388	2-Bromopropane	1.424
2,2,4-Trimethyl pentane	1.389	3-Chloro-2-methylprop-1-ene	1.425
2,3-Dimethyl pentane	1.389	Caproic acid	1.426
Acetic anhydride	1.389	Glycol carbonate	1.426
Diisopropyl amine	1.390	1-Octanol	1.427
2-Aminobutane	1.390	1,1-Dimethylhexanol	1.427
2-Pentanone	1.390	N,N-Dimethylformamide	1.427
3-Pentanone	1.390	Sulfuric acid	1.427
Nitroethane	1.390	1-Chlorooctane	1.428
Methyl-b-butyrate	1.391	Triisobutylene	1.429
Butyl acetate	1.392	N-Methylaniline nitrile	1.429
2-Nitropropane	1.392	Etbylene glycol	1.429
4-Methyl-2-pentnone	1.394	1-Chloro-2-ethylhexane	1.430
2-Methyl-1-propanol	1.394	Ethylcyclohexane	1.431
Octane	1.395	1.2-Propanediol	1.431
1-Amino-2-methylpropane	1.395	1-Bromopropane	1.431
Valeronitrile	1.395	2-Methyl-7-ethyl-4-nonanone	1.433
2-Butanol	1.395	Ethyleneglycol-mono-allyl ether	1.434
5-Hexanone	1.395	Butyral lactone	1.434
5-Methyl-3-hexanone	1.395	2-Methyl-7-ethyl-4-undecanone	1.435
2-Chlorobutane	1.395	4-n-Propyl-5-ethyldioxane	1.435
Butyric acid	1.396	1,2-Dichloro-2-methylpropane	1.435
2,2,2-Trimethylhexane	1.397	1,2-Propyleneglycol sulfite	1.435
n-Dibutyl ether	1.397	N-Methylmorpholine	1.436
1-Butanol	1.397	1-Chloro-3-methyl-2-propanol	1.436
Acrolein	1.397	Epichlorohydrin	1.436
1-Chloro-2-methylpropane	1.397	Triethyleneglycol-mono-butyl ether	1.437
Methacrylonitrile	1.398	4-Ethyl-7,7,7-trimethyl-1-heptanol	1.438
3-Methyl-2-pentanone	1.398	1-Methyl-3-ethyloctan-1-ol	1.438
Triethyl amine	1.399	1-Ethyl-3-ethylhexan-1-ol	1.438
n-Butyl amine	1.399	Diethyl maleate	1.438
1,1,3,3-Tetramethyl-2-propanone	1.399	1-Butanethiol	1.440
Isobutyl-n-butyrate	1.399	2-Chloroethanal	1.440
1-Nitropropane	1.399	Dibutyl sebacate	1.440
n-Dodecane	1.400	1-Ethyl-3-ethyloctan-1-ol	1.441
Amyl acetate	1.400	Dimethylmaleate	1.441
1-Chlorobutane	1.400	3-Methylpentane-2,4-diol	1.441
2-Methoxy ethanol	1.400	Ethyl sulfide	1.442
Propionic acid anhydride	1.400	Mesityl oxide	1.442
2,2,3-Trimethylpentane	1.401	Butyl stearate	1.442
1-Chlorobutane	1.401	1,2-Dichloroethane	1.444
β-Methoxypropionitrile	1.401	Chloroform	1.444
3-Methyl butanoic acid	1.402	trans-1,2-Dichloroethylene	1.444
n-Nonane	1.403	Diethyleneglycol	1.445
Dipropylamine	1.403	cis-1,2-Dichloroethylene	1.445
Isoamylacetate	1.403	3-(α-Butyloctyl)-oxypropyl-1-amine	1.446
Cyclopentane	1.404	2-Methylmorpholine	1.446
2-Methyl-2-butanol	1.404	Dipropyleneglycol-monoethyl ether	1.446
3-Methyl-1-butanol	1.404	Formamide	1.446
Tetrahydrofuran	1.404	3-Lauryloxypropyl-1-amine	1.447
Capronitrile	1.405	Cyclohexanone	1.448
2-Pentanone	1.405	1-Aminopropan-1-ol	1.448
2-Ethoxyethanol	1.405	Diethyleneglycol-mono-β-oxypropyl ether	1.448
2-Heptanone	1.406	1-Amino-2-methylpentan-1-ol	1.449
Valeric acid	1.406	Tetrahydrofurfural alcohol	1.450
Diisobutylene	1.407	2-Propylcyclohexa-1-one	1.452
Methylcyclopentane	1.407	2-Aminoethanol	1.452
Isoamyl ether	1.407	2-Butylcyclohexan-1-one	1.453
Methylpropyl carbinol	1.407	Ethylenediamine	1.454
Tributyl borate	1.407	2-(β-Methyl)-propylcyclohexan-1-one	1.454
1-Pentanol	1.408	4-Methylcyclohexanol	1.454
3-Methyl-2-butanol	1.408	3-Methylcyclohexanol	1.455
Diethyl oxalate	1.408	bis-2-Chloroethyl ether	1.455
n-Decane	1.409	Cyclohexylamine	1.456
4-Methyl-2-pentanol	1.409	1,8-Cineol	1.456
3-Isopropyl-2-pentanone	1.409	2,2′-Dimethyl-2,2′-dipropyldieththanol amine	1.456
2-Methyl-1-butanol	1.409	1,1′,2,2′-Tetramethyldiethanol amine	1.459
Butyric acid anhydride	1.410	1-Aminopropan-3-ol	1.459
Amyl ether	1.410	Carbon tetrachloride	1.459
Isoamyl isovalerate	1.410	3-Methyl-5-ethylheptan-2,4-diol	1.459
1-Chloropentane	1.410	2-(β-Ethyl)-butylcyclohexan-1-one	1.461
2-Propene-1-ol	1.411	2-Methylcyclohexanol	1.461
2,4-Dimethyl dioxane	1.412	N-(n-Butyl)-diethanol amine	1.461
Ethyl lactate	1.412	4,5-Chloro-1,3-dioxolane-2	1.461
Diethyl malonate	1.412	2-Butylcyclohexan-1-ol	1.462
3-Chloropropene	1.413		
Ethyleneglycol diacetate	1.413		
2-Octanone	1.414		
3-Octanone	1.414		
3-Methyl-2-heptanone	1.415		
Caproic acid	1.415		
4-Methyldioxane	1.415		
1,2-Propyleneglycol-1-monobutyl ether	1.415		

Compound	n_D	Compound	n_D
N-β-Oxypropyl morpholine	1.462	Propyl benzoate	1.498
2-(β-Ethyl)-hexylcyclohexanone	1.463	α-Picoline	1.499
2-Ethylcyclohexan-1-ol	1.463	1,2-Diethylbenzene	1.501
Fluorobenzene	1.463	Pentachloroethane	1.501
d-α-Pinene	1.464	1-Iodopropane	1.502
1-α-Pinene	1.465	1,2-Dimethylbenzene	1.503
Cyclohexanol	1.465	Ethyl benzoate	1.503
m-Fluorotoluene	1.465	β-Picoline	1.504
p-Fluorotoluene	1.467	Tetrachloroethylene	1.504
trans-Decahydronaphthalene	1.468	Phenetole	1.505
o-Fluorotoluene	1.468	Pyridine	1.507
3-Alloxy-2-oxypropylamine-1	1.469	Iodoethane	1.512
Ethanol-1-methylisopropanol amine	1.470	Phenylmethallyl ether	1.514
d-Limonene	1.471	Anisole	1.515
1,2,3-Trichloroisobutane	1.473	Methyl benzoate	1.515
Decahydronaphthalene	1.474	Diallylphthalate	1.517
1,2,3-Propanetriol	1.474	Benzylacetae	1.518
Trichloroethylene	1.475	2-Methyl-4-tertiarybtylphenetol	1.521
N-β-Oxyethylmorpholine	1.476	Phenylacetonitrile	1.521
Dimethylsulfoxide	1.476	Methyl salicylate	1.522
cis-Decahydronaphthalene	1.479	Chlorobenzene	1.523
N-β-Chlorallylmorpholine	1.481	Fufural	1.524
n-Dodecyl-4-tertiarybutylphenyl ether	1.482	Benzonitrile	1.526
n-Dodecylphenyl ether	1.482	Thiophene	1.526
n-Dodecyl-4-methylphenyl ether	1.483	Nonachlorodiethyl ether	1.529
2-Ethylidene cyclohexanone	1.486	Iodomethane	1.530
n-Butylbenzene	1.487	4-Phenyldioxane	1.530
p-Cymene	1.488	3-Phenylpropan-1-ol	1.532
iso-Propylbenzene	1.489	Acetophenone	1.532
Furfuralcohol	1.489	Benzyl alcohol	1.538
tert-Butylcumene	1.490	1,2-Dibromoethane	1.538
n-Propylbenzene	1.490	1,2,3,4-Tetrahydronaphthalene	1.539
sec-Butylbenzene	1.490	m-Cresol	1.542
tert-Butylbenzene	1.490	1,3-Dichlorobenzene	1.543
Dibutylphthalate	1.490	Benzaldehyde	1.544
tert-Butyltoluene	1.491	Styrene	1.545
1-Penyl-1-oxyphenylethane	1.491	Nitrobenzene	1.550
n-Hexylcumene	1.492	o-Dichlorobenzene	1.551
n-Octyltoluene	1.492	Bromobenzene	1.557
n-Octylcumene	1.492	o-Nitroanisole	1.560
p-Xylene	1.493	m-Toluidine	1.566
1,31-Diethylbenzene	1.493	Benzyl benzoate	1.568
Ethylbenzene	1.493	o-Toluidine	1.570
1,3-Dimorpholylpropan-2-ol	1.493	1-Methoxyphenyl-1-phenyl-ethane	1.571
1,12,2-Tetrachloroethane	1.493	Aniline	1.583
Toluene	1.494	o-Chloroaniline	1.586
Benzylethyl ether	1.494	Bromoform	1.587
m-Xylene	1.495	Benzenethiol	1.588
1,4-Diethylbenzene	1.496	2,4-Bis(β-phenylethyl)-phenylmethyl ether	1.590
2,3-Dichlorodioxane	1.496	Carbondisulfide	1.628
Mesitylene	1.497	1,12,2-Tetrabromomethane	1.633
2-Iodopropane	1.497	Diiodomethane	1.749
Benzene	1.498		

MOLAR REFRACTION OF ORGANIC COMPOUNDS

The molar refraction, R, is defined as:

$$R = \left(\frac{n^2 - 1}{n^2 + 2} \right) \left(\frac{M}{d} \right)$$

where n = refractive index; M = molecular weight; d = density in grams per cm³; and (M/d) is the volume occupied by 1 gram molecular weight of the compound. The units of R will then be CM³.

R is, to a first approximation, independent of temperature or physical state, and it provides an approximate measure of the actual total volume (without free space) of the molecules in one gram mole.

For a very large number of compounds R is approximately additive for the bonds present in the molecule. Using R_d based on n_D (sodium light), the following atomic, group and structural contributions to R_D are based on Vogel's extensive modern measurements published in the Journal of the Chemical Society, 1948.

INDEX OF REFRACTION OF WATER

Alcohol and Carbon Bisulfide
For sodium light, $\lambda = .5893$

Temp. °C	Water, pure relative to air	Ethyl Alcohol 99.8 relative to air	Carbon Bisulfide relative to air	Temp. °C	Water, pure relative to air	Ethyl Alcohol 99.8 relative to air	Carbon Bisulfide relative to air
14	1.33348			34	1.33136	1.35474	1.61413
15	1.33341		1.62935	36	1.33107	1.35390	1.61247
16	1.33333	1.36210	1.62858	38	1.33079	1.35306	1.61080
18	1.33317	1.36129	1.62704	40	1.33051	1.35222	1.60914
20	1.33299	1.36048	1.62546	42	1.33023	1.35138	1.60748
22	1.33281	1.35967	1.62387	44	1.32992	1.35054	1.60582
24	1.33262	1.35885	1.62226	46	1.32959	1.34969	
26	1.33241	1.35803	1.62064	48	1.32927	1.34885	
28	1.33219	1.35721	1.61902	50	1.32894	1.34800	
30	1.33192	1.35639	1.61740	52	1.32860	1.34715	
32	1.33164	1.35557	1.61577	54	1.32827	1.34629	

INDEX OF REFRACTION OF WATER (continued)

Temp. °C	Water, pure relative to air	Ethyl Alcohol 99.8 relative to air	Carbon Bisulfide relative to air	Temp. °C	Water, pure relative to air	Ethyl Alcohol 99.8 relative to air	Carbon Bisulfide relative to air
56	1.32792	1.34543		78	1.32332		
58	1.32755	1.34456		80	1.32287		
60	1.32718	1.34368		82	1.32241		
62	1.32678	1.34279		84	1.32195		
64	1.32636	1.34189		86	1.32148		
66	1.32596	1.34096		88	1.32100		
68	1.32555	1.34004		90	1.32050		
70	1.32511	1.33912		92	1.32000		
72	1.32466	1.33820		94	1.31949		
74	1.32421	1.33728		96	1.31897		
76	1.32376	1.33626		98	1.31842		
				100	1.31783		

C	2.591	Cl	5.844	=S	7.921
H	1.028	Br	8.741	C≡N	5.459
=O	2.122	I	13.954	N (primary aliphatic)	2.376
		C_6H_5	25.463		
O	1.643	$C_{10}H_7$	43.00	N (secondary aliphatic)	2.582
OH	2.553	S	7.729		
F	0.81				

Ethylenic bond	1.575	Four membered ring	0.317
Acetylenic bond	1.977	Three membered ring	0.614
N (aromatic)	3.550		

Example: −For C_2H_5COOH: $R_{calc.}$ = 3(2.591) + 5(1.028)
+ 2.122 + 2.553 = 17.588
$R_{obs.}$ for this compound is 17.51
For $C_6H_5NHCH_3$: $R_{calc.}$ = 25.463 + 3.550
+ 2.591 + 4(1.028) = 35.716
$R_{obs.}$ for this compound is 35.67

ABSOLUTE INDEX FOR PURE WATER FOR SODIUM LIGHT

Temperature	Index	Temperature	Index
15°C.	1.33377	60°C.	1.32754
20	1.33335	65	1.32652
25	1.33287	70	1.32547
30	1.33228	75	1.32434
35	1.33157	80	1.32323
40	1.33087	85	1.32208
45	1.33011	90	1.32086
50	1.32930	95	1.31959
55	1.32846	100	1.31819

INDEX OF REFRACTION OF ROCK SALT, SYLVINE, CALCITE, FLUORITE AND QUARTZ

(Compiled from data of Martens, Paschen, and others.)

Wave length	Rock salt	Silvine, KCl	Fluorite	Calcspar, ordinary ray	Calcspar, extraordinary ray	Quartz, ordinary ray	Quartz, extraordinary ray
0.185	1.893	1.827				1.676	1690
0.198			1.496		1.578	1.651	1.664
0.340				1.701	1.506	1.567	1.577
0.589	1.544	1.490	1.434	1.658	1.486	1.544	1.553
0.760			1.431	1.650	1.483	1.539	1.548
0.884	1.534	1.481	1.430				
1.179	1.530	1.478	1.428				
1.229				1.639	1.479		
2.324					1.474	1.516	
2.357	1.526	1.475	1.421				
3.536	1.523	1.473	1.414				
5.893	1.516	1.469	1.387				
8.840	1.502	1.461	1.331				

INDEX OF REFRACTION OF GLASS

Relative to Air

Variety	.361	.434	.486	.589 (Na)	.656	.768	1.20	2.00
Zinc crown	1.539	1.528	1.523	1.517	1.514	1.511	1.505	1.497
Higher dispersion crown	1.546	1.533	1.527	1.520	1.517	1.514	1.507	1.497
Light flint	1.614	1.594	1.585	1.575	1.571	1.567	1.559	1.549
Heavy flint	1.705	1.675	1.664	1.650	1.644	1.638	1.628	1.617
Heaviest flint		1.945	1.919	1.890	1.879	1.867	1.848	1.832

INDEX OF REFRACTION, AQUEOUS SOLUTIONS

Substance	Density	Temp. °C	Index for λ = .5893 (Na)	Observer	Substance	Density	Temp. °C	Index for λ = .5893 (Na)	Observer
Ammonium chloride	1.067	27.05	1.379	Willigen	Soda (caustic)	1.376	21.6	1.413	Willigen
Ammonium chloride	1.025	29.75	1.351	Willigen	Sodium chloride	1.189	18.07	1.378	Schutt
Calcium chloride	1.398	25.65	1.443	Willigen	Sodium chloride	1.109	18.07	1.360	Schutt
Calcium chloride	1.215	22.9	1.397	Willigen	Sodium chloride	1.035	18.07	1.342	Schutt
Calcium chloride	1.143	25.8	1.374	Willigen	Sodium nitrate	1.358	22.8	1.385	Willigen
Hydrochloric acid	1.166	20.75	1.411	Willigen	Sulfuric acid	1.811	18.3	1.437	Willigen
Nitric acid	1.359	18.75	1.402	Willigen	Sulfuric acid	1.632	18.3	1.425	Willigen
Potash (caustic)	1.416	11.0	1.403	Frauenhofer	Sulfuric acid	1.221	18.3	1.370	Willigen
Potassium chloride	Nonal solution		1.343	Bender	Sulfuric acid	1.028	18.3	1.339	Willigen
Potassium chloride	Double normal		1.352	Bender	Zinc chloride	1.359	26.6	1.402	Willigen
Potassium chloride	Triple normal		1.360	Bender	Zinc chloride	1.209	26.4	1.375	Willigen

INDEX OF REFRACTION OF FUSED QUARTZ

λ mμ, 15°C	n, 18°C	λ mμ, 15°C	n, 18°C	λ mμ, 15°C	n, 18°C	λ mμ, 15°C	n, 18°C
185.467	1.57436	434.047	1.46690	250.329	1.50745	546.072	1.46013
193.583	1.55999	435.834	1.46675	257.304	1.50379	589.29	1.45845
202.55	1.54727	467.815	1.46435	274.867	1.49617	643.847	1.45674
214.439	1.53386	479.991	1.46355	303.412	1.48594	656.278	1.45640
219.462	1.52907	486.133	1.46318	340.365	1.47867	706.520	1.45517
226.503	1.52308	508.582	1.46191	396.848	1.47061	794.763	1.45340
231.288	1.51941	533.85	1.46067	404.656	1.46968		

INDEX OF REFRACTION OF AIR (15°C, 76 cm Hg)

Corrections for reducing wavelengths and frequencies in air (15°C, 76 cm Hg) to vacuo

The indices were computed from the Cauchy formula $(n - 1)10^7 = 2726.43 + 12.288/(\lambda_2 \times 10^{-8}) + 0.3555/(\lambda^4 \times 10^{-16})$. For 0°C and 76 cm Hg the constants of the equation become 2875.66, 13.412 and 0.3777 respectively, and for 30°C and 76 cm Hg 2589.72, 12.259 and 0.2576. Sellmeier's formula for but one absorption band closely fits the observations: $n_2 = 1 + 0.00057378\lambda^2/(\lambda^2 - 595260)$. If $n - 1$ were strictly proportional to the density, then $(n - 1)_v/(n - 1)t$ would equal $1 + \alpha t$ where α should be 0.00367. The following values of α were found to hold:

λ	0.85μ	0.75μ	0.65μ	0.55μ	0.45μ	0.35μ	0.25μ
α	0.003672	0.003674	0.003678	0.003685	0.003700	0.003738	0.003872

The indices are for dry air (0.05 ± % CO_2). Corrections to reduce to dry air the indices for moist air may be made for any wavelength by Lorenz's formula, $+0.000041(m/760)$, where m is the vapor pressure in mm. The corresponding frequencies in waves per cm and the corrections to reduce wavelengths and frequencies in air at 15°C and 76 cm Hg pressure to vacuo are given. E.g., a light wave of 5000 angstroms in dry air at 15°C, 76 cm Hg becomes 5001.391 Å in vacuo; a frequency of 20,000 waves per cm correspondingly becomes 19994.44.

Wavelength, λ angstroms	Dry air (n — 1) × 10⁷ 15°C 76 cm Hg	Vacuo correction for λ in air (nλ — λ) add	Frequency waves per cm 1/λ in air	Vacuo correction or 1/λ in air (1/nλ — 1/λ) subtract	Wavelength, λ angstroms	Dry air (n — 1) × 10⁷ 15°C 76 cm Hg	Vacuo correction for λ in air (nλ — λ) add	Frequency waves per cm 1/λ in air	Vacuo correction for 1/λ in air (1/nλ — 1/λ) subtract
2000	3256	.651	50,000	16.27	5500	2771	1.524	18,181	5.04
2100	3188	.670	47,619	15.18	5600	2769	1.551	17,857	4.94
2200	3132	.689	45,454	14.23	5700	2768	1.578	17,543	4.85
2300	3086	.710	43,478	13.41	5800	2766	1.604	17,241	4.77
2400	3047	.731	41,666	12.69	5900	2765	1.631	16,949	4.68
2500	3014	.754	40,000	12.05	6000	2763	1.658	16,666	4.60
2600	2986	.776	38,461	11.48	6100	2762	1.685	16,393	4.53
2700	2962	.800	37,037	10.97	6200	2761	1.712	16,129	4.45
2800	2941	.824	35,714	10.50	6300	2760	1.739	15,873	4.38
2900	2923	.848	34,482	10.08	6400	2759	1.766	15,625	4.31
3000	2907	.872	33,333	9.69	6500	2758	1.792	15,384	4.24
3100	2893	.897	32,258	9.33	6600	2757	1.819	15,151	4.18
3200	2880	.922	31,250	9.00	6700	2756	1.846	14,925	4.11
3300	2869	.947	30,303	8.69	6800	2755	1.873	14,705	4.05
3400	2859	.972	29,411	8.41	6900	2754	1.900	14,492	3.99
3500	2850	.998	28,571	8.14	7000	2753	1.927	14,285	3.93
3600	2842	1.023	27,777	7.89	7100	2752	1.954	14,084	3.88
3700	2835	1.049	27,027	7.66	7200	2751	1.981	13,888	3.82
3800	2829	1.075	26,315	7.44	7300	2751	2,008	13,698	3.77
3900	2823	1.101	25,641	7.24	7400	2750	2,035	13,513	3.72
4000	2817	1.127	25,000	7.04	7500	2749	2,062	13,333	3.66
4100	2812	1.153	24,390	6.86	7600	2749	2,089	13,157	3.62
4200	2808	1.179	23,809	6.68	7700	2748	2,116	12,987	3.57
4300	2803	1.205	23,255	6.52	7800	2748	2,143	12,820	3.52
4400	2799	1.232	22,727	6.36	7900	2747	2,170	12,658	3.48
4500	2796	1.258	22,222	6.21	8000	2746	2,197	12,500	3.43
4600	2792	1.284	21,739	6.07	8100	2746	2,224	12,345	3.39
4700	2789	1.311	21,276	5.93	8250	2745	2,265	12,121	3.33
4800	2786	1.338	20,833	5.80	8500	2744	2,332	11,764	3.23
4900	2784	1.364	20,406	5.68	8750	2743	2,400	11,428	3.13
5000	2781	1.391	20,000	5.56	9000	2742	2,468	11,111	3.05
5100	2779	1.417	19,607	5.45	9250	2741	2,536	10,810	2.96
5200	2777	1.444	19,230	5.34	9500	2740	2,604	10,526	2.88
5300	2775	1.471	18,867	5.23	9750	2740	2,671	10,256	2.81
5400	2773	1.497	18,518	5.13	10000	2739	2,739	10,000	2.74

INDEX OF REFRACTION, GASES

Values are relative to a vacuum and for a Temp. of 0°C, and 760 mm pressure.
(From Smithsonian Tables)

Substance	Kind of light	Indices of refraction	Observer	Substance	Kind of light	Indices of refraction	Observer
etone	D	1.001079-1.001100		Hydrochloric acid	D	1.000447	Mascart
	D	1.0002926	Perreau	Hydrogen	white	1.000138-1.000143	
mmonia	white	1.000381-1.000385		Hydrogen	D	1.000132	Burton
mmonia	D	1.000373-1.000379		sulfide	D	1.000644	Dulong
gon	D	1.000281	Rayleigh	sulfide	D	1.000623	Mascart
nzene	D	1.001700-1.001823		Methane	white	1.000443	Dulong
omine	D	1.001132	Mascart	Methane	D	1.000444	Mascart
rbon dioxide	white	1.000449-1.000450		Methyl alcohol	D	1.000549-1.000623	
oxide	D	1.000448-1.000454		Methyl ether	D	1.000891	Marcast
ulfide	white	1.001500	Dulong	Nitric oxide	white	1.000303	Dulong
sulfide	D	1.001478-1.001485		Nitric oxide	D	1.000297	Mascart
onoxide	White	1.000340	Dulong	Nitrogen	white	1.000295 1.000300	
onoxide	white	1.000335	Mascart	Nitrogen	D	1.000296-1.000298	
lorine	white	1.000772	Dulong	Nitrous oxide	white	1.000503-1.000507	
lorine	D	1.000773	Mascart	Nitrous oxide	D	1.000516	Mascart
loroform	D	1.001436-1.001464		Oxygen	white	1.000272-1.000280	
anogen	white	1.000834	Dulong	Oxygen	D	1.000271-1.000272	
anogen	D	1.000784-1.000825		Pentane	D	1.001711	Mascart
yl alcohol	D	1.000871-1.000885		Sulfur dioxide	white	1.000665	Dulong
her	D	1.001521-1.001544		Sulfur dioxide	D	1.000686	Ketteler
lium	D	1.000036	Ramsay	Water	white	1.000261	Jamin
drochloric acid	white	1.000449	Mascart	Water	D	1.000249-1.000259	

COEFFICIENT OF TRANSPARENCY OF UVIOL
GLASS FOR THE ULTRA-VIOLET

For a thickness of 1 mm.

Wave length microns	0.280	0.309	0.325	0.346	0.361	0.383	0.397
Uviol crown	0.56	0.95	0.990	0.996	0.999	1.000	1.000

INDEX OF REFRACTION OF AQUEOUS SOLUTIONS OF SUCROSE
(CANE SUGAR)

The table gives the index of refraction for λ = 0.5893 of aqueous sugar solutions at 20°C from 0—85% sugar.
Corrections for temperatures other than 20 are given at the end of the table

Percent sugar		.0	.1	.2	.3	.4	.5	.6	.7	.8	.9
00	1.3	330	331	333	334	336	337	338	340	341	342
1		344	345	347	348	350	351	353	355	356	357
2		359	361	362	363	365	367	368	369	371	373
3		374	375	377	378	380	381	382	384	385	387
4		388	389	391	393	394	395	397	399	400	401
5		403	405	406	407	409	411	412	413	415	417
6		418	419	421	423	424	425	427	429	430	431
7		433	435	436	437	439	441	442	443	445	447
8		448	450	451	453	454	456	458	459	461	462
9		464	465	467	469	470	471	473	475	476	477
10		479	481	482	483	485	487	488	489	491	493
11		494	496	497	499	500	502	504	505	507	508
12		510	512	513	515	516	518	520	521	523	524
13		526	527	529	531	532	533	535	537	538	539
14		541	543	544	546	547	549	551	552	554	555
15		557	559	560	562	563	565	567	568	570	571
16		573	575	576	578	580	582	583	585	587	588
17		590	592	593	595	596	598	600	601	603	604
18		606	608	609	611	612	614	616	617	619	620
19		622	624	625	627	629	631	632	634	636	637
20		639	641	652	644	645	647	649	650	652	653
21		655	657	658	660	662	663	665	667	669	670
22		672	674	675	677	679	681	682	684	686	687
23		689	691	692	694	696	698	699	701	703	704
24		706	708	709	711	713	715	716	718	720	721
25		723	725	726	728	730	731	733	735	737	738
26		740	742	744	745	747	749	751	753	754	756
27		758	760	761	763	765	767	768	770	772	773
28		775	777	779	780	782	784	786	788	789	791
29		793	795	797	798	800	802	804	806	807	809
30		811	813	815	816	818	820	822	824	825	827
31		829	831	833	834	836	838	840	842	843	845
32		847	849	851	852	854	856	858	860	861	863
33		865	867	869	870	872	874	876	878	879	881
34		883	885	887	889	891	893	894	896	898	900
35		902	904	906	907	909	911	913	915	916	918
36		920	922	924	926	928	929	931	933	935	937
37		939	941	943	945	947	949	950	952	954	956
38		958	960	962	964	966	968	970	972	974	976
39		978	980	982	984	986	987	989	991	993	995

INDEX OF REFRACTION OF AQUEOUS SOLUTIONS OF SUCROSE
(CANE SUGAR) (continued)

Percent sugar		.0	.1	.2	.3	.4	.5	.6	.7	.8	.9
40		997	999	*001	*003	*005	*007	*008	*010	*012	*014
41	1.4	016	018	020	022	024	026	028	030	032	034
42		036	038	040	042	044	046	048	050	052	054
43		056	058	060	062	064	066	068	070	072	074
44		076	078	080	082	084	086	088	090	092	094
45		096	098	100	102	104	107	109	111	113	115
46		117	119	121	123	125	127	129	131	133	135
47		137	139	141	143	145	147	150	152	154	156
48		158	160	162	164	166	169	171	173	175	177
49		179	181	183	185	187	189	192	194	196	198
50		200	202	204	206	208	211	213	215	217	219
51		221	223	225	227	229	231	234	236	238	240
52		242	244	246	249	251	253	255	257	260	262
53		264	266	268	270	272	275	277	279	281	283
54		285	287	289	292	294	296	298	300	303	305
55		307	309	311	313	316	318	320	322	325	327
56		329	331	333	336	338	340	342	344	347	349
57		351	353	355	358	360	362	364	366	369	371
58		373	375	378	380	382	385	387	389	391	394
59		396	398	400	403	405	407	409	411	414	416
60		418	420	423	425	427	429	432	434	436	439
61		441	443	446	448	450	453	455	457	459	462
62		464	466	468	471	473	475	477	479	482	484
63		486	488	491	493	495	497	500	502	504	507
64		509	511	514	516	518	521	523	525	527	530
65		532	534	537	539	541	544	546	548	550	553
66		558	561	563	565	567	570	572	574	577	579
67		581	584	586	588	591	593	595	598	600	602
68		605	607	609	612	614	616	619	621	623	625
69		628	630	632	635	637	639	642	644	646	649
70		651	653	656	658	661	663	666	668	671	673
71		676	678	681	683	685	688	690	693	695	698
72		700	703	705	708	710	713	715	717	720	722
73		725	727	730	732	735	737	740	742	744	747
74		749	752	754	757	759	762	764	767	769	772
75		774	777	779	782	784	787	789	792	794	797
76		799	802	804	807	810	812	815	817	820	822
77		825	827	830	832	835	838	840	843	845	848
78		850	853	855	858	860	863	865	868	871	873
79		876	878	881	883	886	888	891	893	896	898
80		901	904	906	909	912	914	917	919	922	925
81		927	930	933	935	938	941	943	946	949	951
82		954	956	959	962	964	967	970	972	975	978
83		980	983	985	988	991	993	996	999	*001	*004
84	1.5	007	009	012	015	017	020	022	025	028	030
85		033									

Correction table for determining the percentage of sucrose by means of the refractometer when the readings are made at temperatures other than 20°C[1]

Percentage of sucrose

Tem-perature	0	5	10	15	20	25	30	35	40	45	50	55	60	65	70

Subtract from the percentage of sucrose

°C.															
10	0.50	0.54	0.58	0.61	0.64	0.66	0.68	0.70	0.72	0.73	0.74	0.75	0.76	0.78	0.79
11	.46	.49	.53	.55	.58	.60	.62	.64	.65	.66	.67	.68	.69	.70	.71
12	.42	.45	.48	.50	.52	.54	.56	.57	.58	.59	.60	.61	.61	.63	.63
13	.37	.40	.42	.44	.46	.48	.49	.50	.51	.52	.53	.54	.54	.55	.55
14	.33	.35	.37	.39	.40	.41	.42	.43	.44	.45	.45	.46	.46	.47	.48
15	.27	.29	.31	.33	.34	.34	.35	.36	.37	.37	.38	.39	.39	.40	.40
16	.22	.24	.25	.26	.27	.28	.29	.30	.30	.30	.31	.31	.32	.32
17	.17	.18	.19	.20	.21	.21	.21	.22	.22	.23	.23	.23	.23	.24	.24
18	.12	.13	.13	.14	.14	.14	.14	.15	.15	.15	.15	.16	.16	.16	.16
19	.06	.06	.06	.07	.07	.07	.07	.08	.08	.08	.08	.08	.08	.08	.08

Add to the percentage of sucrose

21	0.06	0.07	0.07	0.07	0.07	0.08	0.08	0.08	0.08	0.08	0.08	0.08	0.08	0.08	0.08
22	.13	.13	.14	.14	.15	.15	.15	.15	.15	.16.	16	.16	.16	.16	.16
23	.19	.20	.21	.22	.22	.23	.23	.23	.23	.24	.24	.24	.24	.24	.24
24	.26	.27	.28	.29	.30	.30	.31	.31	.31	.31	.31	.32	.32	.32	.32
25	.33	.35	.36	.37	.38	.38	.39	.40	.40	.40	.40	.40	.40	.40	.40
26	.40	.42	.43	.44	.45	.46	.47	.48	.48	.48	.48	.48	.48	.48	.48
27	.48	.50	.52	.53	.54	.55	.55	.56	.56	.56	.56	.56	.56	.56	.56
28	.56	.57	.60	.61	.62	.63	.63	.64	.64	.64	.64	.64	.64	.64	.64
29	.64	.66	.68	.69	.71	.72	.72	.73	.73	.73	.73	.73	.73	.73	.73
30	.72	.74	.77	.78	.79	.80	.80	.81	.81	.81	.81	.81	.81	.81	.81

International Temperature Correction Table, 1936, adopted by the International Commission for Uniform Methods of Sugar Analysis (Int. Sugar J. **39**, 24s, 1937).

INDEX OF REFRACTION OF MALTOSE HYDRATE SOLUTIONS

McDonald: Journal of Research of the National Bureau of Standards, Vol. **46**, 165, 1951.
The following table gives the refractive index of maltose solutions for $\lambda = 5893$ A at 20 and 25 C.

Percent	n_D^{20}	n_D^{25}	Δn	$\Delta n/\Delta t$	Percent	n_D^{20}	n_D^{25}	Δn	$\Delta n/\Delta t$
1	1.33438	1.33389	0.00049	0.00010	33	1.38470	1.38388	.00082	.00016
2	1.33579	1.33528	.00051	.00010	34	1.38647	1.38565	.00081	.00016
3	1.33720	1.33668	.00052	.00010	35	1.38825	1.38744	.00081	.00016
4	1.33862	1.33810	.00052	.00010	36	1.39004	1.38924	.00080	.00016
5	1.34006	1.33952	.00054	.00011	37	1.39185	1.39105	.00080	.00016
6	1.34150	1.34095	.00055	.00011	38	1.39367	1.39287	.00080	.00016
7	1.34295	1.34239	.00056	.00011	39	1.39551	1.39471	.00080	.00016
8	1.34442	1.34384	.00058	.00012	40	1.39735	1.39656	.00079	.00016
9	1.34589	1.34530	.00059	.00012	41	1.39922	1.39843	.00079	.00016
10	1.34738	1.34677	.00061	.00012	42	1.40109	1.40031	.00078	.00016
11	1.34887	1.34825	.00062	.00012	43	1.40298	1.40221	.00077	.00016
12	1.35039	1.34974	.00065	.00013	44	1.40488	1.40411	.00077	.00015
13	1.35190	1.35124	.00066	.00013	45	1.40680	1.40603	.00077	.00015
14	1.35343	1.35276	.00067	.00013	46	1.40873	1.40797	.00076	.00015
15	1.35497	1.35428	.00069	.00014	47	1.41067	1.40992	.00075	.00015
16	1.35652	1.35582	.00070	.00014	48	1.41263	1.41188	.00075	.00015
17	1.35808	1.35737	.00071	.00014	49	1.41460	1.41385	.00075	.00015
18	1.35965	1.35893	.00072	.00014	50	1.41658	1.41584	.00074	.00015
19	1.36124	1.36051	.00073	.00015	51	1.41858	1.41784	.00074	.00015
20	1.36283	1.36209	.00074	.00015	52	1.42059	1.41986	.00073	.00015
21	1.36444	1.36369	.00075	.00015	53	1.42262	1.42189	.00073	.00015
22	1.36606	1.36530	.00076	.00015	54	1.42466	1.42392	.00074	.00015
23	1.36770	1.36692	.00078	.00016	55	1.42672	1.42598	.00074	.00015
24	1.36934	1.36856	.00078	.00016	56	1.42878	1.42804	.00074	.00015
25	1.37100	1.37021	.00079	.00016	57	1.43087	1.43012	.00075	.00015
26	1.37267	1.37187	.00080	.00016	58	1.43296	1.43221	.00075	.00015
27	1.37435	1.37355	.00080	.00016	59	1.43508	1.43431	.00077	.00015
28	1.37604	1.37524	.00080	.00016	60	1.43720	1.43643	.00077	.00015
29	1.37775	1.37694	.00081	.00016	61	1.43934	1.43855	.00079	.00016
30	1.37946	1.37865	.00081	.00016	62	1.44150	1.44069	.00081	.00016
31	1.38120	1.38038	.00082	.00016	63	1.44367	1.44283	.00084	.00017
32	1.38294	1.38213	.00082	.00016	64	1.44585	1.44499	.00086	.00017
					65	1.44805	1.44716	.00089	.00018

RADIATION FROM AN IDEAL BLACK BODY

From NASA TT-F-783

Temperature dependence of the specific power radiated, Q_T, and of λ_{max} for an ideal black body according to Kirchhoff's law ($\sigma_o = 5.68 \cdot 10^{-8}$ W/m$^2 \cdot$deg^4 = $4.88 \cdot 10^{-8}$ kcal/m$^2 \cdot$deg^4)

T, °K	t, °C	Q_T, W/cm^2	Q_T kcal/m$^2 \cdot$hr	λ_{max} μ	T, °K	t, °C	Q_T W/cm^2	Q_T kcal/m$^2 \cdot$hr	λ_{max} μ
100	−173	$5.680 \cdot 10^{-4}$	$4.880 \cdot 10^0$	28.96	630	357	8.948	7.687	4.597
200	−73	$9.088 \cdot 10^{-3}$	$7.808 \cdot 10^1$	14.48	640	367	9.529	8.187	4.525
273	0	$3.155 \cdot 10^{-2}$	$2.711 \cdot 10^2$	10.608	650	377	$1.014 \cdot 10^0$	8.711	4.455
300	27	4.601	3.953	9.655	660	387	1.078	9.260	4.388
310	37	5.246	4.507	9.342	670	397	1.145	9.831	4.322
320	47	5.956	5.117	9.050	680	407	1.214	$1.013 \cdot 10^4$	4.259
330	57	6.736	5.787	8.766	690	417	1.287	1.106	4.197
340	67	7.590	6.521	8.518	700	427	1.364	1.172	4.137
350	77	8.524	7.323	8.274	710	437	1.443	1.240	4.069
360	87	9.540	8.196	8.044	720	447	1.526	1.311	4.022
370	97	1.065	9.146	7.827	730	457	$1.613 \cdot 10^0$	$1.386 \cdot 10^1$	3.967
380	107	$1.184 \cdot 10^{-1}$	$1.018 \cdot 10^3$	7.621	740	467	1.703	1.463	3.914
390	117	1.314	1.128	7.426	750	477	1.797	1.544	3.861
400	127	1.454	1.249	7.270	760	487	1.895	1.628	3.811
410	137	1.605	1.379	7.053	770	497	1.997	1.715	3.761
420	147	1.76	1.519	6.865	780	507	2.102	1.806	3.713
430	157	1.942	1.668	6.735	790	517	2.212	1.901	3.666
440	167	2.129	1.829	6.562	800	527	2.327	1.999	3.620
450	177	2.329	2.001	6.436	810	537	2.445	2.101	3.565
460	187	2.543	2.185	6.266	820	547	2.568	2.206	3.532
470	197	2.772	2.381	6.162	830	557	2.696	2.316	3.489
480	207	3.015	2.591	6.033	840	567	2.828	2.430	3.448
490	217	3.274	2.813	5.910	850	577	2.965	2.547	3.407
500	227	3.550	3.05	5.792	860	587	3.107	2.670	3.367
510	237	3.843	3.301	5.668	870	597	3.254	2.796	3.329
520	247	4.163	3.568	5.559	880	607	3.406	2.927	3.291
530	257	4.482	3.851	5.454	890	617	3.564	3.062	3.254
540	267	4.830	4.150	5.363	900	627	3.727	3.202	3.218
550	277	5.198	4.465	5.255	910	637	3.895	3.346	3.162
560	287	5.586	4.799	5.161	920	647	4.069	3.496	3.148
570	297	5.996	5.151	5.061	930	657	4.249	3.650	3.114
580	307	6.428	5.522	4.963	940	667	4.435	3.810	3.081
590	317	6.883	5.913	4.908	950	677	4.626	3.975	3.048
600	327	7.361	6.324	4.827	960	687	4.824	4.145	3.017
610	337	7.864	6.757	4.748	970	697	5.028	4.320	2.986
620	347	8.393	7.211	4.671	980	707	5.239	4.501	2.955

E-367

RADIATION FROM AN IDEAL BLACK BODY *(Continued)*

T, °K	t, °C	Q_T, W/cm²	Q_T kcal/m²·hr	λmax μ	T, °K	t, °C	Q_T W/cm²	Q_T kcal/m²·hr	λmax μ
990	717	5.456	4.688	2.925	1540	1267	3.195	2.745	1.881
1000	727	5.680	4.880	2.896	1550	1277	3.278	2.817	1.866
1010	737	5.909	5.07	2.866	1560	1287	3.363	2.890	1.856
1020	747	6.143	5.278	2.836	1570	1297	3.451	2.965	1.845
1030	757	6.394	5.494	2.812	1580	1307	3.540	3.041	1.833
1040	767	6.645	5.709	2.785	1590	1317	3.630	3.198	1.821
1050	777	6.904	5.932	2.758	1600	1327	3.722		1.810
1060	787	7.171	6.161	2.732	1610	1337	3.816	3.279	1.797
1070	797	7.445	6.397	2.707	1620	1347	3.912	3.361	1.787
1080	807	7.728	6.640	2.681	1630	1357	4.010	3.445	1.777
1090	817	8.018	6.888	2.657	1640	1367	$4.109 \cdot 10^1$	$3.530 \cdot 10^5$	1.766
1100	827	8.316	7.145	2.633	1650	1377	4.210	3.617	1.755
1110	837	8.623	7.408	2.609	1660	1387	4.313	3.706	1.745
1120	847	8.937	7.679	2.586	1670	1397	4.418	3.796	1.734
1130	857	9.261	7.956	2.563	1680	1407	4.525	3.887	1.724
1140	867	9.593	8.242	2.540	1690	1417	4.633	3.981	1.714
1150	877	9.934	8.535	2.516	1700	1427	4.744	4.076	1.704
1160	887	$1.028 \cdot 10^1$	8.836	2.497	1710	1437	4.858	4.183	1.694
1170	897	1.064	9.145	2.475	1720	1447	4.971	4.271	1.684
1180	907	1.101	9.461	2.454	1730	1457	5.088	4.371	1.674
1190	917	$1.139 \cdot 10^1$	$9.786 \cdot 10^4$	2.434	1740	1467	5.206	4.473	1.664
1200	927	1.178	$1.042 \cdot 10^5$	2.413	1750	1477	5.327	4.577	1.655
1210	937	1.218	1.046	2.393	1760	1487	5.450	4.682	1.645
1220	947	1.258	1.081	2.374	1770	1497	5.575	4.790	1.636
1230	957	1.300	1.117	2.354	1780	1507	5.703	4.900	1.627
1240	967	1.343	1.154	2.335	1790	1517	5.831	5.010	1.617
1250	977	1.387	1.191	2.317	1800	1527	5.963	5.123	1.607
1260	987	1.431	1.230	2.298	1810	1537	6.096	5.238	1.600
1270	997	1.478	1.270	2.280	1820	1547	6.232	5.354	1.591
1280	1007	1.525	1.310	2.263	1830	1557	6.370	5.473	1.583
1290	1017	1.573	1.351	2.245	1840	1567	6.511	5.594	1.574
1300	1027	1.622	1.394	2.227	1850	1577	6.653	5.717	1.565
1310	1037	1.673	1.437	2.211	1860	1587	6.798	5.841	1.557
1320	1047	1.724	1.482	2.194	1870	1597	6.946	5.967	1.549
1330	1057	1.777	1.527	2.177	1880	1607	7.095	6.096	1.540
1340	1067	1.831	1.573	2.161	1890	1617	7.248	6.227	1.532
1350	1077	1.887	1.621	2.145	1900	1627	7.402	6.360	1.524
1360	1087	1.943	1.669	2.129	1910	1637	7.559	6.495	1.516
1370	1097	2.001	1.719	2.114	1920	1647	7.720	6.632	1.509
1380	1107	2.058	1.768	2.099	1930	1657	7.881	6.771	1.501
1390	1117	2.120	1.822	2.083	1940	1667	8.046	6.912	1.493
1400	1127	2.182	1.875	2.067	1950	1677	8.213	7.056	1.485
1410	1137	2.245	1.929	2.054	1960	1687	8.382	7.202	1.478
1420	1147	2.309	1.984	2.036	1970	1697	8.555	7.350	1.470
1430	1157	2.375	2.041	2.025	1980	1707	8.730	7.500	1.463
1440	1167	2.442	2.098	2.011	1990	1717	8.907	7.653	1.455
1450	1177	2.511	2.157	1.997	2000	1727	9.088	7.808	1.448
1460	1187	2.581	2.217	1.984	2500	2227	$2.219 \cdot 10^2$	$1.906 \cdot 10^6$	1.156
1470	1197	2.652	2.279	1.970	3000	2727	4.601	3.953	0.965
1480	1207	2.725	2.341	1.957	3500	3227	8.524	7.323	0.826
1490	1217	2.800	2.405	1.944	4000	3727	$1.454 \cdot 10^3$	1.249	0.722
1500	1227	2.876	2.471	1.931	4500	4227	2.329	2.001	0.644
1510	1237	2.953	2.537	1.917	5000	4727	3.550	3.050	0.579
1520	1247	3.032	2.605	1.905	5500	5227	5.198	4.465	0.527
1530	1257	3.113	2.674	1.893	6000	5727	7.361	6.324	0.483

OPTICAL PROPERTIES OF METALS

J. H. Weaver

These tables list the optical properties of metals index of refraction (n), the extinction coefficient (k), and the normal-incidence reflection (R) as a function of photon energy which is expressed in electron volts (eV). To convert the energy (in eV) to wavelength (in Å) use $\lambda = 12398/E$. To compute the dielectric function, $\bar{\varepsilon} = \varepsilon_1 + i\varepsilon_2$, from the index of refraction, $N = n + ik$, use $\varepsilon_1 = 2nk$ and $\varepsilon_2 = n^2 + k^2$.

The optical constants in these tables are abridged from a more extensive tabulation (J. H. Weaver, C. Krafka, D. W. Lynch, and E. E. Koch, "Optical Properties of Metals", Volumes I and II, *Physics Data*, Nr. 18-1 and 18-2, Fachinformationzentrum, 7514 Eggenstein-Leopoldshafen 2, Karlsruhe, West Germany). The extensive tabulation provides detailed comparison of all optical data available in the literature at the time of the compilation of the tables for *Physics Data*. In the extensive tables, metals in addition to those listed below are examined, other optical properties are given (dielectric function, loss function) and other photon energies are included. For critical applications the reader should refer to the original work. References for metals listed in this abridged table are listed at the end of the tables. Generally, tabulated values for the optical properties are accurate to 10%.

Aluminum[1]

Energy (eV)	n	k	R($\phi = 0$)
0.040	98.595	203.701	0.9923
0.050	74.997	172.199	0.9915
0.060	62.852	150.799	0.9906
0.070	53.790	135.500	0.9899
0.080	45.784	123.734	0.9895
0.090	39.651	114.102	0.9892
0.100	34.464	105.600	0.9889
0.125	24.965	89.250	0.9884
0.150	18.572	76.960	0.9882
0.175	14.274	66.930	0.9879
0.200	11.733	59.370	0.9873
0.250	8.586	48.235	0.9858
0.300	6.759	40.960	0.9844
0.350	5.438	35.599	0.9834
0.400	4.454	31.485	0.9826
0.500	3.072	25.581	0.9817
0.600	2.273	21.403	0.9806
0.700	1.770	18.328	0.9794
0.800	1.444	15.955	0.9778
0.900	1.264	14.021	0.9749
1.000	1.212	12.464	0.9697
1.100	1.201	11.181	0.9630
1.200	1.260	10.010	0.9521
1.300	1.468	8.949	0.9318
1.400	2.237	8.212	0.8852
1.500	2.745	8.309	0.8678
1.600	2.625	8.597	0.8794
1.700	2.143	8.573	0.8972
1.800	1.741	8.205	0.9069
1.900	1.488	7.821	0.9116
2.000	1.304	7.479	0.9148
2.200	1.018	6.846	0.9200
2.400	0.826	6.283	0.9228
2.600	0.695	5.800	0.9238
2.800	0.598	5.385	0.9242
3.000	0.523	5.024	0.9241
3.200	0.460	4.708	0.9243
3.400	0.407	4.426	0.9245
3.600	0.363	4.174	0.9246
3.800	0.326	3.946	0.9247
4.000	0.294	3.740	0.9248
4.200	0.267	3.552	0.9248
4.400	0.244	3.380	0.9249
4.600	0.223	3.222	0.9249
4.800	0.205	3.076	0.9249
5.000	0.190	2.942	0.9244
6.000	0.130	2.391	0.9257
6.500	0.110	2.173	0.9260
7.000	0.095	1.983	0.9262
7.500	0.082	1.814	0.9265
8.000	0.072	1.663	0.9269
8.500	0.063	1.527	0.9272
9.000	0.056	1.402	0.9277
9.500	0.049	1.286	0.9282
10.000	0.044	1.178	0.9286
10.500	0.040	1.076	0.9293
11.000	0.036	0.979	0.9298
11.500	0.033	0.883	0.9283
12.000	0.033	0.791	0.9224
12.500	0.034	0.700	0.9118
13.000	0.038	0.609	0.8960
13.500	0.041	0.517	0.8789
14.000	0.048	0.417	0.8486
14.200	0.053	0.373	0.8312
14.400	0.058	0.327	0.8102
14.600	0.067	0.273	0.7802
14.800	0.086	0.211	0.7202
15.000	0.125	0.153	0.6119
15.200	0.178	0.108	0.4903
15.400	0.234	0.184	0.3881
15.600	0.280	0.073	0.3182
15.800	0.318	0.065	0.2694
16.000	0.351	0.060	0.2326
16.200	0.380	0.055	0.2031
16.400	0.407	0.050	0.1789
16.750	0.448	0.045	0.1460
17.000	0.474	0.042	0.1278
17.250	0.498	0.040	0.1129
17.500	0.520	0.038	0.1005
17.750	0.540	0.036	0.0899
18.000	0.558	0.035	0.0809
18.500	0.591	0.032	0.0664
19.000	0.620	0.030	0.0554
19.500	0.646	0.028	0.0467
20.000	0.668	0.027	0.0398
20.500	0.689	0.025	0.0342
21.000	0.707	0.024	0.0296
21.500	0.724	0.023	0.0258
22.000	0.739	0.022	0.0226
22.500	0.753	0.021	0.0199
23.000	0.766	0.021	0.0177
23.500	0.778	0.020	0.0157
24.000	0.789	0.019	0.0140
24.500	0.799	0.018	0.0126
25.000	0.809	0.018	0.0113
25.500	0.817	0.017	0.0102
26.000	0.826	0.016	0.0092
27.000	0.840	0.015	0.0076
28.000	0.854	0.014	0.0063
29.000	0.865	0.014	0.0053
30.000	0.876	0.013	0.0044
35.000	0.915	0.010	0.0020
40.000	0.940	0.008	0.0010
45.000	0.957	0.007	0.0005
50.000	0.969	0.006	0.0003
55.000	0.979	0.005	0.0001
60.000	0.987	0.004	0.0000
65.000	0.995	0.004	0.0000
70.000	1.006	0.004	0.0000
72.500	1.025	0.004	0.0002
75.000	1.011	0.024	0.0002
77.500	1.008	0.025	0.0002
80.000	1.007	0.024	0.0002
85.000	1.007	0.028	0.0002
90.000	1.003	0.031	0.0002
95.000	0.999	0.036	0.0003
100.000	0.991	0.030	0.0002
110.000	0.994	0.025	0.0002
120.000	0.991	0.024	0.0002
130.000	0.987	0.021	0.0001
140.000	0.989	0.016	0.0001
150.000	0.990	0.015	0.0001
160.000	0.989	0.014	0.0001
170.000	0.989	0.011	0.0001
180.000	0.990	0.010	0.0000
190.000	0.990	0.009	0.000
200.000	0.991	0.007	0.0000
220.000	0.992	0.006	0.0000
240.000	0.993	0.005	0.0000
260.000	0.993	0.004	0.0000
280.000	0.994	0.003	0.0000
300.000	0.995	0.002	0.0000

Chromium[2]

Energy (eV)	n	k	R($\phi = 0$)
0.06	21.19	42.00	0.962
0.10	11.81	29.76	0.955
0.14	15.31	26.36	0.936
0.18	8.73	25.37	0.953
0.22	5.30	20.62	0.954
0.26	3.91	17.12	0.951
0.30	3.15	14.28	0.943
0.42	3.47	8.97	0.862
0.54	3.92	7.06	0.788
0.66	3.96	5.95	0.736
0.78	4.13	5.03	0.680
0.90	4.43	4.60	0.650
1.00	4.47	4.43	0.639
1.12	4.53	4.31	0.631
1.24	4.50	4.28	0.629
1.36	4.42	4.30	0.631
1.46	4.31	4.32	0.632
1.77	3.84	4.37	0.639
2.00	3.48	4.36	0.644
2.20	3.18	4.41	0.656
2.40	2.75	4.46	0.677
2.60	2.22	4.36	0.698
2.80	1.80	4.06	0.703
3.00	1.54	3.71	0.695
3.20	1.44	3.40	0.670
3.40	1.39	3.24	0.657
3.60	1.26	3.12	0.661
3.80	1.12	2.95	0.660
4.00	1.02	2.76	0.651
4.20	0.94	2.58	0.639
4.40	0.90	2.42	0.620
4.50	0.89	2.35	0.607
4.60	0.88	2.28	0.598
4.70	0.86	2.21	0.586
4.80	0.86	2.13	0.572
4.90	0.86	2.07	0.557
5.00	0.85	2.01	0.542
5.10	0.86	1.94	0.523
5.20	0.87	1.87	0.503
5.40	0.93	1.80	0.466
5.60	0.95	1.74	0.443
5.80	0.97	1.74	0.437
6.00	0.94	1.73	0.444
6.20	0.89	1.69	0.446
6.40	0.85	1.66	0.447
6.60	0.80	1.59	0.444
6.80	0.75	1.51	0.439
7.00	0.74	1.45	0.425
7.20	0.71	1.39	0.414
7.40	0.69	1.33	0.404
7.60	0.66	1.23	0.378
7.80	0.67	1.15	0.347
8.00	0.68	1.07	0.315
8.20	0.71	1.00	0.278
8.50	0.74	0.92	0.235
9.0	0.83	0.81	0.170
9.50	0.92	0.74	0.132
10.00	0.98	0.73	0.120
10.50	1.01	0.72	0.112
11.00	1.05	0.69	0.103
11.50	1.09	0.69	0.100
12.00	1.13	0.70	0.101
12.50	1.15	0.73	0.108
13.00	1.15	0.77	0.119
13.50	1.12	0.80	0.128
14.00	1.09	0.82	0.135
14.50	1.03	0.82	0.142
15.00	1.00	0.82	0.143
15.50	0.96	0.80	0.141
16.00	0.92	0.77	0.139
16.50	0.91	0.75	0.134
17.00	0.90	0.73	0.132
17.50	0.88	0.72	0.130
18.00	0.87	0.70	0.129
18.50	0.84	0.69	0.130
19.00	0.82	0.68	0.131
20.00	0.77	0.64	0.130
20.5	0.76	0.63	0.129
21.0	0.74	0.58	0.121
21.5	0.72	0.55	0.116
22.0	0.71	0.52	0.112
22.5	0.70	0.50	0.109
23.0	0.69	0.48	0.105
23.5	0.68	0.45	0.101
24.0	0.68	0.43	0.096
24.5	0.67	0.39	0.089
25.0	0.68	0.36	0.080
25.5	0.68	0.33	0.072
26.0	0.70	0.31	0.063
26.5	0.71	0.28	0.055
27.0	0.72	0.26	0.048
27.5	0.73	0.25	0.043
28.0	0.75	0.23	0.037
29.0	0.77	0.22	0.032
30.0	0.78	0.21	0.030

Cobalt (Single crystal, $\vec{E} \parallel \hat{c}$)[3]

Energy (eV)	n	k	R($\phi = 0$)
0.10	6.71	37.87	0.982

Energy (eV)	n	k	R(φ = 0)
0.15	4.66	25.47	0.973
0.20	3.55	18.78	0.962
0.25	3.98	14.59	0.933
0.30	4.04	12.16	0.907
0.40	4.24	9.13	0.847
0.50	4.41	7.19	0.782
0.60	4.91	6.13	0.729
0.70	5.24	5.85	0.713
0.80	5.17	5.89	0.716
0.90	4.94	5.95	0.720
1.00	4.46	5.86	0.722
1.10	4.07	5.61	0.715
1.20	3.81	5.36	0.706
1.30	3.60	5.20	0.701
1.40	3.37	5.09	0.701
1.50	3.10	4.96	0.701
1.60	2.84	4.77	0.697
1.70	2.66	4.57	0.690
1.80	2.45	4.41	0.687
1.90	2.31	4.18	0.675
2.00	2.21	4.00	0.664
2.10	2.13	3.85	0.654
2.20	2.07	3.70	0.642
2.30	2.01	3.59	0.634
2.40	1.95	3.49	0.627
2.50	1.88	3.40	0.622
2.60	1.81	3.32	0.618
2.70	1.73	3.24	0.615
2.80	1.66	3.13	0.607
2.90	1.61	3.05	0.600
3.00	1.55	2.96	0.594
3.20	1.46	2.80	0.579
3.40	1.38	2.64	0.563
3.60	1.31	2.48	0.544
3.80	1.28	2.33	0.519
4.00	1.26	2.20	0.495
4.20	1.25	2.10	0.471
4.40	1.24	2.01	0.452
4.60	1.24	1.94	0.435
4.80	1.23	1.88	0.423
5.00	1.22	1.83	0.411
5.20	1.21	1.79	0.403
5.40	1.19	1.77	0.399
5.60	1.16	1.75	0.400
5.80	1.10	1.73	0.406
6.00	1.03	1.68	0.407
6.20	0.97	1.62	0.401
6.40	0.94	1.53	0.386
6.60	0.91	1.46	0.368
6.80	0.91	1.38	0.345
7.00	0.91	1.32	0.326
7.00	0.91	1.26	0.305
7.40	0.92	1.21	0.286
7.60	0.93	1.17	0.269
7.80	0.94	1.13	0.253
8.00	0.95	1.09	0.239

Cobalt (Single crystal, $\vec{E} \perp \hat{c}$)[3]

Energy (eV)	n	k	R(φ = 0)
0.10	5.83	32.36	0.979
0.15	4.24	21.37	0.965
0.20	3.87	15.53	0.042
0.30	4.34	10.01	0.865
0.40	4.66	7.39	0.785
0.50	5.17	5.75	0.709
0.60	5.77	5.17	0.682
0.70	6.15	5.20	0.685
0.80	6.08	5.61	0.702
0.90	5.57	5.93	0.715
1.00	4.83	5.94	0.721
1.10	4.31	5.60	0.711
1.20	4.02	5.34	0.701
1.30	3.78	5.16	0.694
1.40	3.55	5.05	0.692
1.50	3.26	4.93	0.692
1.60	3.03	4.74	0.687
1.70	2.83	4.60	0.684
1.80	2.61	4.45	0.683
1.90	2.41	4.27	0.677
2.00	2.25	4.09	0.670
2.10	2.13	3.89	0.659
2.20	2.04	3.72	0.646
2.30	1.99	3.56	0.632
2.40	1.95	3.44	0.620
2.50	1.90	3.34	0.611
2.60	1.86	3.26	0.605
2.70	1.79	3.19	0.602
2.80	1.72	3.11	0.596
2.90	1.66	3.03	0.591
3.00	1.60	2.94	0.586
3.20	1.50	2.78	0.571
3.40	1.42	2.62	0.553
3.60	1.36	2.47	0.533
3.80	1.33	2.33	0.511
4.00	1.31	2.21	0.488
4.20	1.28	2.12	0.471
4.40	1.27	2.03	0.452
4.60	1.26	1.95	0.435
4.80	1.25	1.90	0.423
5.00	1.24	1.84	0.411
5.20	1.22	1.80	0.403
5.40	1.21	1.78	0.399
5.60	1.17	1.76	0.400
5.80	1.11	1.74	0.406
6.00	1.04	1.69	0.407
6.20	0.98	1.62	0.401
6.40	0.94	1.54	0.386
6.60	0.92	1.46	0.368
6.80	0.91	1.38	0.345
7.00	0.91	1.32	0.326
7.20	0.91	1.26	0.305
7.40	0.92	1.21	0.285
7.60	0.93	1.17	0.269
7.80	0.94	1.13	0.253

Copper[4]

Energy (eV)	n	k	R(φ = 0)
0.10	29.69	71.57	0.980
0.50	1.71	17.63	0.979
1.00	0.44	8.48	0.976
1.50	0.26	5.26	0.965
1.70	0.22	4.43	0.958
1.75	0.21	4.25	0.956
1.80	0.21	4.04	0.952
1.85	0.22	3.85	0.947
1.90	0.21	3.67	0.943
2.00	0.27	3.24	0.910
2.10	0.47	2.81	0.814
2.20	0.83	2.60	0.673
2.30	1.04	2.59	0.618
2.40	1.12	2.60	0.602
2.60	1.15	2.50	0.577
2.80	1.17	2.36	0.545
3.00	1.18	2.21	0.509
3.20	1.23	2.07	0.468
3.40	1.27	1.95	0.434
3.60	1.31	1.87	0.407
3.80	1.34	1.81	0.387
4.00	1.34	1.72	0.364
4.20	1.42	1.64	0.336
4.40	1.49	1.64	0.329
4.60	1.52	1.67	0.334
4.80	1.53	1.71	0.345
5.00	1.47	1.78	0.366
5.20	1.38	1.80	0.380
5.40	1.28	1.78	0.389
5.60	1.18	1.74	0.391
5.80	1.10	1.67	0.389
6.00	1.04	1.59	0.380
6.50	0.96	1.37	0.329
7.00	0.97	1.20	0.271
7.50	1.00	1.09	0.230
8.00	1.03	1.03	0.206
8.50	1.03	0.98	0.189
9.00	1.03	0.92	0.171
9.50	1.03	0.87	0.154
10.00	1.04	0.82	0.1
11.00	1.07	0.75	0.1
12.00	1.09	0.73	0.1
13.00	1.08	0.72	0.10
14.00	1.06	0.72	0.1
14.50	1.03	0.72	0.1
15.00	1.01	0.71	0.1
15.50	0.98	0.69	0.10
16.00	0.95	0.67	0.10
17.00	0.91	0.62	0.00
18.00	0.89	0.56	0.00
19.00	0.88	0.51	0.0
20.00	0.88	0.45	0.03
21.00	0.90	0.41	0.0
22.00	0.92	0.38	0.0
23.00	0.94	0.37	0.03
24.00	0.96	0.37	0.0
25.00	0.96	0.40	0.04
26.00	0.92	0.40	0.0
27.00	0.88	0.38	0.04
28.00	0.86	0.35	0.03
29.00	0.85	0.30	0.0
30.00	0.86	0.26	0.0
31.00	0.88	0.24	0.02
32.00	0.89	0.22	0.01
33.00	0.90	0.21	0.0
34.00	0.91	0.20	0.01
35.00	0.92	0.20	0.01
36.00	0.92	0.19	0.01
37.00	0.92	0.19	0.01
38.00	0.93	0.18	0.01
39.00	0.93	0.17	0.00
40.00	0.93	0.17	0.00
41.00	0.94	0.16	0.00
42.00	0.94	0.16	0.00
43.00	0.94	0.15	0.00
44.00	0.95	0.15	0.00
45.00	0.95	0.15	0.00
46.00	0.95	0.15	0.00
47.00	0.95	0.14	0.00
48.00	0.95	0.14	0.00
49.00	0.95	0.14	0.00
50.00	0.95	0.13	0.00
51.00	0.95	0.13	0.00
52.00	0.95	0.13	0.00
53.00	0.96	0.12	0.00
54.00	0.96	0.12	0.00
55.00	0.96	0.12	0.00
56.00	0.96	0.11	0.00
57.00	0.96	0.11	0.00
58.00	0.96	0.11	0.00
59.00	0.97	0.11	0.00
60.00	0.97	0.11	0.00
61.00	0.97	0.11	0.00
62.00	0.97	0.11	0.00
63.00	0.96	0.10	0.00
64.00	0.96	0.10	0.00
65.00	0.97	0.10	0.00
66.00	0.97	0.10	0.00
67.00	0.97	0.09	0.00
68.00	0.97	0.09	0.00
69.00	0.97	0.09	0.00
70.00	0.97	0.09	0.00
75.00	0.98	0.09	0.00
80.00	0.98	0.09	0.00
85.00	0.97	0.09	0.00
90.00	0.96	0.08	0.00

Gold (Electropolished, Au [110])[5]

Energy (eV)	n	k	R(φ = 0
0.10	8.17	82.83	0.99
0.20	2.13	41.73	0.99
0.30	0.99	27.82	0.99
0.40	0.59	20.83	0.99
0.50	0.39	16.61	0.99
0.60	0.28	13.78	0.99

Energy (eV)	n	k	R
0.70	0.22	11.75	0.994
0.80	0.18	10.21	0.993
0.90	0.15	9.01	0.993
1.00	0.13	8.03	0.992
1.20	0.10	6.54	0.991
1.40	0.08	5.44	0.989
1.60	0.08	4.56	0.986
1.80	0.09	3.82	0.979
2.00	0.13	3.16	0.953
2.10	0.18	2.84	0.925
2.20	0.24	2.54	0.880
2.30	0.32	2.22	0.807
2.40	0.50	1.86	0.647
2.50	0.82	1.59	0.438
2.60	1.24	1.54	0.331
2.70	1.43	1.72	0.356
2.80	1.46	1.77	0.368
2.90	1.50	1.79	0.368
3.00	1.54	1.80	0.369
3.10	1.54	1.81	0.371
3.20	1.54	1.80	0.368
3.30	1.55	1.78	0.362
3.40	1.56	1.76	0.356
3.50	1.58	1.73	0.349
3.60	1.62	1.73	0.346
3.70	1.64	1.75	0.351
3.80	1.63	1.79	0.360
3.90	1.59	1.81	0.366
4.00	1.55	1.81	0.369
4.10	1.51	1.79	0.368
4.20	1.48	1.78	0.367
4.30	1.45	1.77	0.368
4.40	1.41	1.76	0.370
4.50	1.35	1.74	0.370
4.60	1.30	1.69	0.364
4.70	1.27	1.64	0.354
4.80	1.25	1.59	0.344
4.90	1.23	1.54	0.332
5.00	1.22	1.49	0.319
5.20	1.21	1.40	0.295
5.40	1.21	1.33	0.275
5.60	1.21	1.27	0.256
5.80	1.21	1.20	0.236
6.00	1.22	1.14	0.218
6.20	1.24	1.09	0.203
6.40	1.25	1.05	0.190
6.60	1.27	1.01	0.177
6.80	1.30	0.97	0.167
7.00	1.34	0.95	0.162
7.20	1.36	0.95	0.161
7.40	1.38	0.96	0.164
7.60	1.38	0.98	0.169
7.80	1.35	0.99	0.171
8.00	1.31	0.96	0.165
8.20	1.30	0.92	0.155
8.40	1.30	0.89	0.147
8.60	1.31	0.88	0.144
8.80	1.31	0.86	0.140
9.00	1.30	0.83	0.133
9.20	1.31	0.81	0.126
9.40	1.33	0.78	0.122
9.60	1.36	0.78	0.121
9.80	1.37	0.79	0.124
10.00	1.37	0.80	0.126
10.20	1.36	0.80	0.127
10.40	1.35	0.80	0.125
10.60	1.34	0.79	0.123
10.80	1.34	0.77	0.120
11.00	1.34	0.76	0.116
11.20	1.34	0.74	0.113
11.40	1.35	0.73	0.111
11.60	1.36	0.72	0.109
11.80	1.38	0.71	0.108
12.00	1.39	0.71	0.109
12.40	1.44	0.73	0.115
12.80	1.45	0.79	0.127
13.20	1.42	0.84	0.137
13.60	1.37	0.86	0.140
14.00	1.33	0.86	0.140
14.40	1.29	0.86	0.139
14.80	1.26	0.84	0.135
15.20	1.24	0.83	0.132
15.60	1.22	0.81	0.127
16.00	1.21	0.79	0.123
16.40	1.20	0.78	0.119
16.80	1.19	0.76	0.116
17.20	1.19	0.75	0.114
17.60	1.19	0.74	0.111
18.00	1.19	0.74	0.109
18.40	1.19	0.73	0.109
18.80	1.20	0.74	0.110
19.20	1.21	0.76	0.116
19.60	1.21	0.80	0.125
20.00	1.18	0.83	0.133
20.40	1.14	0.85	0.141
20.80	1.10	0.87	0.149
21.20	1.05	0.88	0.156
21.60	1.00	0.88	0.162
22.00	0.94	0.86	0.164
22.40	0.89	0.83	0.163
22.80	0.85	0.79	0.157
23.20	0.82	0.75	0.149
23.60	0.80	0.70	0.138
24.00	0.80	0.66	0.125
24.40	0.80	0.62	0.113
24.80	0.80	0.58	0.101
25.20	0.82	0.56	0.090
25.60	0.83	0.54	0.084
26.00	0.84	0.52	0.079
26.40	0.85	0.51	0.074
26.80	0.85	0.50	0.071
27.20	0.86	0.49	0.068
27.60	0.86	0.49	0.065
28.00	0.87	0.48	0.063
28.40	0.88	0.48	0.062
28.80	0.88	0.48	0.062
29.20	0.88	0.48	0.062
29.60	0.87	0.48	0.061
30.00	0.86	0.48	0.064

Hafnium (Single crystal, $\vec{E} \parallel \hat{c}$)[6]

Energy (eV)	n	k	R($\phi = 0$)
0.52	1.48	4.11	0.747
0.56	1.84	3.29	0.615
0.60	2.34	2.62	0.486
0.66	3.21	2.13	0.428
0.70	3.70	2.03	0.441
0.76	4.31	2.10	0.476
0.80	4.61	2.31	0.504
0.86	4.71	2.70	0.533
0.90	4.64	2.85	0.541
0.95	4.54	2.96	0.545
1.00	4.45	3.00	0.545
1.10	4.28	3.08	0.547
1.20	4.08	3.10	0.544
1.30	3.87	3.04	0.536
1.40	3.72	2.95	0.525
1.50	3.60	2.85	0.514
1.60	3.52	2.73	0.500
1.70	3.52	2.61	0.488
1.80	3.57	2.56	0.485
1.90	3.63	2.59	0.489
2.00	3.65	2.67	0.498
2.10	3.64	2.81	0.511
2.20	3.53	2.99	0.526
2.30	3.34	3.09	0.534
2.40	3.15	3.11	0.537
2.50	2.99	3.13	0.540
2.60	2.83	3.12	0.542
2.70	2.68	3.10	0.542
2.80	2.54	3.08	0.543
2.90	2.40	3.04	0.544
3.00	2.27	3.00	0.544
3.10	2.14	2.95	0.544
3.20	2.00	2.89	0.544
3.30	1.87	2.79	0.538
3.40	1.78	2.68	0.528
3.50	1.71	2.58	0.517
3.60	1.66	2.48	0.503
3.70	1.63	2.40	0.491
3.80	1.60	2.33	0.481
3.90	1.56	2.27	0.473
4.00	1.52	2.21	0.466
4.10	1.48	2.14	0.455
4.20	1.45	2.07	0.442
4.30	1.43	2.01	0.431
4.40	1.41	1.95	0.420
4.50	1.39	1.89	0.407
4.60	1.39	1.83	0.394
4.70	1.39	1.79	0.382
4.80	1.38	1.75	0.373
4.90	1.38	1.71	0.364
5.00	1.37	1.68	0.356
5.20	1.36	1.61	0.341
5.40	1.35	1.55	0.324
5.60	1.35	1.51	0.314
5.80	1.32	1.48	0.308
6.00	1.28	1.41	0.295
6.20	1.26	1.35	0.278
6.40	1.26	1.28	0.258
6.60	1.27	1.22	0.240
6.80	1.28	1.16	0.224
7.00	1.31	1.13	0.212
7.20	1.33	1.10	0.204
7.40	1.34	1.07	0.197
7.60	1.36	1.05	0.191
7.80	1.37	1.02	0.183
8.00	1.40	1.01	0.179
8.20	1.43	1.01	0.178
8.40	1.45	1.01	0.180
8.60	1.47	1.02	0.183
8.80	1.48	1.04	0.186
9.00	1.49	1.07	0.193
9.20	1.50	1.10	0.201
9.40	1.48	1.14	0.211
9.60	1.46	1.18	0.222
9.80	1.41	1.21	0.230
10.00	1.36	1.22	0.235
10.20	1.32	1.22	0.238
10.40	1.28	1.22	0.240
10.60	1.24	1.21	0.241
10.80	1.20	1.20	0.242
11.00	1.16	1.19	0.242
11.20	1.13	1.17	0.241
11.40	1.10	1.16	0.241
11.60	1.07	1.14	0.239
11.80	1.04	1.12	0.238
12.00	1.02	1.10	0.236
12.40	0.96	1.06	0.232
12.80	0.92	1.01	0.225
13.20	0.88	0.96	0.218
13.60	0.84	0.90	0.205
14.00	0.83	0.83	0.186
14.40	0.83	0.80	0.172
14.80	0.81	0.76	0.167
15.20	0.79	0.70	0.153
15.60	0.79	0.64	0.132
16.00	0.83	0.60	0.111
16.40	0.81	0.60	0.114
16.80	0.79	0.55	0.105
17.20	0.79	0.50	0.089
17.60	0.80	0.46	0.077
18.00	0.81	0.42	0.064
18.40	0.84	0.38	0.051
18.80	0.87	0.34	0.040
19.00	0.89	0.33	0.036
19.60	0.93	0.32	0.030
20.00	0.94	0.31	0.027
20.60	0.97	0.30	0.023
21.00	0.99	0.29	0.022
21.60	1.01	0.28	0.020
22.00	1.03	0.28	0.020
22.60	1.06	0.28	0.020
23.00	1.07	0.28	0.021
23.60	1.09	0.29	0.022
24.00	1.09	0.30	0.023
24.60	1.10	0.31	0.024

Hafnium (Single crystal, $\vec{E} \perp \hat{c}$)[6]

Energy (eV)	n	k	R(φ = 0)
0.52	2.25	4.65	0.723
0.56	2.34	3.66	0.623
0.60	2.84	2.89	0.512
0.66	3.71	2.35	0.469
0.70	4.26	2.21	0.482
0.76	4.97	2.33	0.521
0.80	5.41	2.62	0.554
0.86	5.46	3.36	0.593
0.90	5.22	3.62	0.601
0.95	4.95	3.72	0.602
1.00	4.76	3.76	0.602
1.10	4.43	3.80	0.601
1.20	4.07	3.74	0.594
1.30	3.79	3.55	0.578
1.40	3.61	3.36	0.561
1.50	3.55	3.13	0.540
1.60	3.58	3.01	0.529
1.70	3.63	2.98	0.526
1.80	3.66	3.02	0.530
1.90	3.63	3.14	0.541
2.00	3.51	3.26	0.551
2.10	3.35	3.33	0.558
2.20	3.18	3.36	0.563
2.30	2.99	3.39	0.568
2.40	2.78	3.35	0.569
2.50	2.65	3.26	0.562
2.60	2.54	3.22	0.560
2.70	2.42	3.17	0.559
2.80	2.31	3.13	0.558
2.90	2.20	3.08	0.558
3.00	2.08	3.05	0.561
3.10	1.94	2.98	0.560
3.20	1.83	2.88	0.555
3.30	1.74	2.78	0.547
3.40	1.68	2.69	0.538
3.50	1.62	2.61	0.529
3.60	1.57	2.52	0.519
3.70	1.53	2.45	0.510
3.80	1.49	2.38	0.501
3.90	1.45	2.32	0.493
4.00	1.41	2.25	0.484
4.10	1.38	2.18	0.474
4.20	1.35	2.11	0.462
4.30	1.33	2.05	0.451
4.40	1.31	1.99	0.438
4.50	1.30	1.93	0.427
4.60	1.29	1.88	0.415
4.70	1.28	1.82	0.402
4.80	1.28	1.77	0.389
4.90	1.27	1.73	0.379
5.00	1.27	1.69	0.367
5.20	1.27	1.62	0.349
5.40	1.27	1.57	0.335
5.60	1.26	1.52	0.322
5.80	1.24	1.48	0.313
6.00	1.21	1.42	0.302
6.20	1.19	1.36	0.285
6.40	1.18	1.29	0.265
6.60	1.19	1.22	0.244
6.80	1.21	1.18	0.230
7.00	1.22	1.14	0.217
7.20	1.23	1.10	0.206
7.40	1.26	1.06	0.194
7.60	1.28	1.04	0.187
7.80	1.30	1.02	0.180
8.00	1.33	1.00	0.174
8.20	1.35	0.99	0.173
8.40	1.38	0.99	0.173
8.60	1.40	1.00	0.174
8.80	1.42	1.02	0.178
9.00	1.43	1.04	0.184
9.20	1.45	1.08	0.193
9.40	1.43	1.12	0.204
9.60	1.40	1.16	0.214
9.80	1.37	1.19	0.223
10.00	1.32	1.21	0.230
10.20	1.27	1.21	0.234
10.40	1.23	1.20	0.235
10.60	1.19	1.20	0.237
10.80	1.15	1.19	0.237
11.00	1.12	1.17	0.237
11.20	1.08	1.16	0.237
11.40	1.05	1.14	0.236
11.60	1.03	1.12	0.235
11.80	1.00	1.10	0.233
12.00	0.97	1.08	0.231
12.40	0.92	1.04	0.226
12.80	0.88	0.99	0.219
13.20	0.83	0.94	0.211
13.60	0.80	0.88	0.196
14.00	0.79	0.81	0.177
14.40	0.80	0.77	0.160
14.80	0.77	0.73	0.154
15.20	0.76	0.68	0.140
15.60	0.76	0.61	0.119
16.00	0.81	0.58	0.099
16.40	0.78	0.57	0.102
16.80	0.77	0.53	0.092
17.20	0.77	0.48	0.077
17.60	0.79	0.44	0.065
18.00	0.80	0.39	0.053
18.40	0.82	0.36	0.041
18.80	0.86	0.33	0.032
19.00	0.88	0.32	0.030
19.60	0.91	0.31	0.025
20.00	0.93	0.30	0.023
20.60	0.96	0.29	0.021
21.00	0.97	0.29	0.020
21.60	1.00	0.28	0.019
22.00	1.01	0.28	0.019
22.60	1.03	0.27	0.018
23.00	1.05	0.28	0.019
23.60	1.06	0.28	0.020
24.00	1.07	0.29	0.021
24.60	1.09	0.30	0.022

Iridium[7]

Energy (eV)	n	k	R(φ = 0)
0.10	28.49	60.62	0.975
0.15	15.32	45.15	0.973
0.20	9.69	35.34	0.972
0.25	6.86	28.84	0.969
0.30	5.16	24.25	0.967
0.35	4.11	20.79	0.064
0.40	3.42	18.06	0.960
0.45	3.05	15.82	0.954
0.50	2.98	14.06	0.944
0.60	2.79	11.58	0.925
0.70	2.93	9.78	0.895
0.80	3.14	8.61	0.862
0.90	3.19	7.88	0.840
1.00	3.15	7.31	0.822
1.10	3.04	6.84	0.808
1.20	2.96	6.41	0.791
1.30	2.85	6.07	0.779
1.40	2.72	5.74	0.767
1.50	2.65	5.39	0.750
1.60	2.68	5.08	0.728
1.70	2.69	4.92	0.716
1.80	2.64	4.81	0.710
1.90	2.57	4.68	0.704
2.00	2.50	4.57	0.699
2.10	2.40	4.48	0.697
2.20	2.29	4.38	0.695
2.30	2.18	4.26	0.692
2.40	2.07	4.14	0.689
2.50	1.98	4.00	0.682
2.60	1.91	3.86	0.673
2.70	1.85	3.73	0.665
2.80	1.81	3.61	0.655
2.90	1.77	3.51	0.646
3.00	1.73	3.43	0.640
3.20	1.62	3.26	0.629
3.40	1.53	3.05	0.610
3.60	1.52	2.81	0.573
3.80	1.61	2.69	0.541
4.00	1.64	2.68	0.535
4.20	1.58	2.71	0.549
4.40	1.45	2.68	0.561
4.60	1.31	2.60	0.567
4.80	1.18	2.49	0.570
5.00	1.10	2.35	0.559
5.20	1.04	2.22	0.543
5.40	1.00	2.09	0.522
5.60	0.98	1.98	0.499
5.80	0.96	1.86	0.474
6.00	0.95	1.78	0.454
6.20	0.94	1.68	0.427
6.40	0.94	1.59	0.401
6.60	0.94	1.50	0.375
6.80	0.95	1.42	0.345
7.00	0.97	1.34	0.318
7.20	0.99	1.27	0.290
7.40	1.02	1.20	0.262
7.60	1.03	1.14	0.241
7.80	1.08	1.06	0.208
8.00	1.13	1.03	0.191
8.20	1.18	1.00	0.179
8.40	1.22	0.98	0.171
8.60	1.26	0.96	0.164
8.80	1.29	0.95	0.160
9.00	1.33	0.94	0.157
9.20	1.36	0.95	0.159
9.40	1.39	0.95	0.161
9.60	1.42	0.97	0.163
9.80	1.44	0.99	0.169
10.00	1.45	1.01	0.175
10.20	1.45	1.04	0.182
10.40	1.44	1.07	0.187
10.60	1.43	1.09	0.193
10.80	1.41	1.12	0.200
11.00	1.38	1.13	0.206
11.20	1.34	1.14	0.208
11.40	1.31	1.13	0.208
11.60	1.28	1.12	0.206
11.80	1.25	1.10	0.203
12.00	1.24	1.08	0.199
12.40	1.21	1.05	0.191
12.80	1.19	1.01	0.181
13.20	1.18	0.98	0.173
13.60	1.17	0.95	0.165
14.00	1.16	0.91	0.155
14.40	1.17	0.88	0.147
14.80	1.18	0.87	0.142
15.20	1.19	0.84	0.136
15.60	1.20	0.83	0.133
16.00	1.21	0.83	0.131
16.40	1.23	0.82	0.129
16.80	1.25	0.82	0.127
17.20	1.28	0.83	0.131
17.60	1.30	0.87	0.140
18.00	1.30	0.93	0.154
18.40	1.27	0.97	0.166
18.80	1.24	1.00	0.176
19.20	1.20	1.03	0.187
19.60	1.15	1.05	0.197
20.00	1.10	1.06	0.205
20.50	1.04	1.05	0.210
21.00	0.99	1.04	0.215
21.50	0.94	1.02	0.220
22.00	0.89	1.00	0.222
22.50	0.84	0.99	0.228
23.00	0.79	0.96	0.232
23.50	0.76	0.92	0.228
24.00	0.73	0.87	0.223
24.50	0.70	0.83	0.218
25.00	0.69	0.79	0.209
25.50	0.68	0.76	0.200
26.00	0.67	0.72	0.192
26.50	0.67	0.69	0.181
27.00	0.66	0.66	0.174
27.50	0.66	0.63	0.166
28.00	0.66	0.61	0.158
28.50	0.66	0.59	0.151
29.00	0.65	0.57	0.148

Energy (eV)	n	k	R(φ = 0)
29.50	0.64	0.55	0.145
30.00	0.64	0.53	0.140
32.00	0.62	0.44	0.119
34.00	0.64	0.35	0.091
36.00	0.69	0.27	0.059
38.00	0.73	0.24	0.044
40.00	0.76	0.22	0.034

Iron[3]

Energy (eV)	n	k	R(φ = 0)
0.10	6.41	33.07	0.978
0.15	6.26	22.82	0.956
0.20	3.68	18.23	0.958
0.26	4.98	13.68	0.911
0.30	4.87	12.05	0.892
0.36	4.68	10.44	0.867
0.40	4.42	9.75	0.858
0.50	4.14	8.02	0.817
0.60	3.93	6.95	0.783
0.70	3.78	6.17	0.752
0.80	3.65	5.60	0.725
0.90	3.52	5.16	0.700
1.00	3.43	4.79	0.678
1.10	3.33	4.52	0.660
1.20	3.24	4.26	0.641
1.30	3.16	4.07	0.626
1.40	3.12	3.87	0.609
1.50	3.05	3.77	0.601
1.60	3.00	3.60	0.585
1.70	2.98	3.52	0.577
1.80	2.92	3.46	0.573
1.90	2.89	3.37	0.563
2.00	2.85	3.36	0.563
2.10	2.80	3.34	0.562
2.20	2.74	3.33	0.563
2.30	2.65	3.34	0.567
2.40	2.56	3.31	0.567
2.50	2.46	3.31	0.570
2.60	2.34	3.30	0.576
2.70	2.23	3.25	0.575
2.80	2.12	3.23	0.580
2.90	2.01	3.17	0.580
3.00	1.88	3.12	0.583
3.10	1.78	3.04	0.580
3.20	1.70	2.96	0.576
3.30	1.62	2.87	0.572
3.40	1.55	2.79	0.565
3.50	1.50	2.70	0.556
3.60	1.47	2.63	0.548
3.70	1.43	2.56	0.542
3.83	1.38	2.49	0.534
4.00	1.30	2.39	0.527
4.17	1.26	2.27	0.510
4.33	1.23	2.18	0.494
4.50	1.20	2.10	0.482
4.67	1.16	2.02	0.470
4.83	1.14	1.93	0.451
5.00	1.14	1.87	0.435
5.17	1.12	1.81	0.425
5.33	1.11	1.75	0.408
5.50	1.09	1.17	0.401
5.67	1.09	1.65	0.383
5.83	1.10	1.61	0.373
6.00	1.09	1.59	0.366
6.17	1.08	1.57	0.365
6.33	1.04	1.55	0.365
6.50	1.02	1.51	0.358
6.67	1.00	1.47	0.351
6.83	0.97	1.43	0.346
7.00	0.96	1.39	0.333
7.17	0.94	1.35	0.327
7.33	0.94	1.30	0.311
7.50	0.94	1.26	0.298
7.67	0.94	1.23	0.288
7.83	0.94	1.21	0.279
8.00	0.94	1.18	0.272
8.17	0.94	1.16	0.265
8.33	0.94	1.14	0.258
8.50	0.94	1.12	0.251
8.67	0.94	1.10	0.246
8.83	0.93	1.08	0.240
9.00	0.93	1.07	0.236
9.17	0.92	1.06	0.233
9.33	0.91	1.04	0.231
9.50	0.90	1.02	0.226
9.67	0.90	1.00	0.221
9.83	0.89	0.99	0.218
10.00	0.88	0.97	0.213
10.17	0.87	0.94	0.203
10.33	0.87	0.91	0.196
10.50	0.87	0.89	0.189
10.67	0.88	0.87	0.179
10.83	0.89	0.85	0.170
11.00	0.91	0.83	0.162
11.17	0.92	0.83	0.159
11.33	0.93	0.84	0.159
11.50	0.93	0.84	0.160
11.67	0.93	0.84	0.162
11.83	0.92	0.84	0.163
12.00	0.91	0.84	0.163
12.17	0.90	0.84	0.165
12.33	0.89	0.83	0.164
12.50	0.98	0.83	0.165
12.67	0.87	0.82	0.166
12.83	0.86	0.81	0.166
13.00	0.85	0.80	0.162
13.17	0.84	0.79	0.161
13.33	0.84	0.78	0.160
13.50	0.83	0.77	0.159
13.67	0.82	0.76	0.157
13.83	0.81	0.75	0.154
14.00	0.81	0.73	0.151
14.17	0.80	0.72	0.149
14.33	0.80	0.71	0.146
14.50	0.79	0.70	0.144
14.67	0.79	0.69	0.141
14.83	0.78	0.67	0.139
15.00	0.78	0.66	0.135
15.17	0.78	0.65	0.131
15.33	0.78	0.64	0.128
15.50	0.77	0.63	0.126
15.67	0.77	0.62	0.123
15.83	0.77	0.61	0.119
16.00	0.77	0.60	0.116
16.17	0.78	0.58	0.112
16.33	0.78	0.58	0.110
16.50	0.78	0.57	0.107
16.67	0.77	0.56	0.106
16.83	0.78	0.55	0.103
17.00	0.78	0.55	0.102
17.17	0.78	0.54	0.100
17.33	0.78	0.54	0.098
17.50	0.77	0.53	0.097
17.67	0.77	0.52	0.095
17.83	0.78	0.51	0.092
18.00	0.78	0.51	0.091
18.17	0.78	0.51	0.090
18.33	0.78	0.50	0.089
18.50	0.77	0.50	0.089
18.67	0.77	0.50	0.088
18.83	0.77	0.49	0.087
19.00	0.77	0.49	0.087
19.17	0.76	0.49	0.088
19.33	0.76	0.48	0.087
19.50	0.75	0.47	0.086
19.67	0.75	0.47	0.085
19.83	0.75	0.46	0.084
20.00	0.74	0.45	0.083
20.17	0.74	0.44	0.081
20.33	0.74	0.44	0.081
20.50	0.74	0.43	0.080
20.67	0.73	0.43	0.079
20.83	0.73	0.42	0.078
21.00	0.73	0.41	0.077
21.17	0.72	0.40	0.076
21.33	0.72	0.39	0.074
21.50	0.72	0.38	0.073
21.67	0.72	0.38	0.071
21.83	0.72	0.37	0.070
22.00	0.72	0.36	0.068
22.17	0.71	0.35	0.067
22.33	0.72	0.34	0.064
22.50	0.72	0.34	0.063
22.67	0.72	0.33	0.062
22.83	0.72	0.32	0.059
23.00	0.72	0.31	0.058
23.17	0.72	0.30	0.056
23.33	0.72	0.29	0.054
23.50	0.73	0.28	0.050
23.67	0.73	0.28	0.049
23.83	0.74	0.27	0.047
24.00	0.74	0.27	0.045
24.17	0.74	0.26	0.044
24.33	0.74	0.26	0.043
24.50	0.74	0.25	0.042
24.67	0.75	0.25	0.040
24.83	0.75	0.24	0.039
25.00	0.75	0.24	0.038
26.00	0.76	0.21	0.031
27.00	0.78	0.18	0.026
28.00	0.79	0.16	0.021
29.00	0.81	0.14	0.017
30.00	0.82	0.13	0.014

Manganese[8]

Energy (eV)	n	k	R(φ = 0)
0.64	3.89	5.95	0.738
0.77	3.78	5.41	0.710
0.89	3.65	5.02	0.688
1.02	3.48	4.74	0.673
1.14	3.30	4.53	0.662
1.26	3.10	4.35	0.653
1.39	2.97	4.18	0.643
1.51	2.83	4.03	0.634
1.64	2.70	3.91	0.627
1.76	2.62	3.78	0.617
1.88	2.56	3.65	0.606
2.01	2.51	3.54	0.596
2.13	2.47	3.43	0.585
2.26	2.39	3.33	0.577
2.38	2.32	3.23	0.567
2.50	2.25	3.14	0.559
2.63	2.19	3.06	0.552
2.75	2.11	2.98	0.545
2.88	2.06	2.90	0.536
3.00	2.00	2.82	0.528
3.12	1.96	2.74	0.518
3.25	1.92	2.67	0.509
3.37	1.89	2.59	0.498
3.50	1.89	2.51	0.484
3.62	1.87	2.45	0.475
3.74	1.86	2.38	0.463
3.87	1.86	2.32	0.451
3.99	1.86	2.25	0.438
4.12	1.86	2.19	0.427
4.24	1.85	2.14	0.417
4.36	1.85	2.08	0.406
4.49	1.86	2.03	0.395
4.61	1.85	1.99	0.388
4.74	1.84	1.94	0.378
4.86	1.83	1.91	0.372
4.98	1.82	1.86	0.362
5.11	1.82	1.82	0.354
5.23	1.81	1.79	0.348
5.36	1.78	1.76	0.342
5.48	1.74	1.73	0.337
5.60	1.73	1.70	0.331
5.73	1.72	1.67	0.325
5.85	1.70	1.64	0.319
5.98	1.67	1.61	0.313
6.10	1.63	1.58	0.307
6.22	1.62	1.55	0.301
6.35	1.59	1.52	0.295
6.47	1.55	1.50	0.292
6.60	1.48	1.47	0.288

Molybdenum[9]

Energy (eV)	n	k	R(φ = 0)
0.10	18.53	68.51	0.985
0.15	8.78	47.54	0.985
0.20	5.10	35.99	0.985
0.25	3.36	28.75	0.984
0.30	2.44	23.80	0.983
0.34	2.00	20.84	0.982
0.38	1.70	18.44	0.980
0.42	1.57	16.50	0.978
0.46	1.46	14.91	0.975
0.50	1.37	13.55	0.971
0.54	1.35	12.36	0.966
0.58	1.34	11.34	0.960
0.62	1.38	10.44	0.952
0.66	1.43	9.67	0.942
0.70	1.48	8.99	0.932
0.74	1.51	8.38	0.921
0.78	1.60	7.83	0.906
0.82	1.64	7.35	0.892
0.86	1.70	6.89	0.876
0.90	1.74	6.48	0.859
1.00	1.94	5.58	0.805
1.10	2.15	4.85	0.743
1.20	2.44	4.22	0.671
1.30	2.77	3.74	0.608
1.40	3.15	3.40	0.562
1.50	3.53	3.30	0.550
1.60	3.77	3.41	0.562
1.70	3.84	3.51	0.570
1.80	3.81	3.58	0.576
1.90	3.74	3.58	0.576
2.00	3.68	3.52	0.571
2.10	3.68	3.45	0.565
2.20	3.76	3.41	0.562
2.30	3.79	3.61	0.578
2.40	3.59	3.78	0.594
2.50	3.36	3.73	0.591
2.60	3.22	3.61	0.582
2.70	3.13	3.51	0.573
2.80	3.08	3.42	0.565
2.90	3.05	3.33	0.566
3.00	3.04	3.27	0.550
3.10	3.03	3.21	0.544
3.20	3.05	3.18	0.540
3.30	3.06	3.18	0.540
3.40	3.06	3.19	0.541
3.50	3.06	3.21	0.543
3.60	3.05	3.23	0.546
3.70	3.04	3.27	0.550
3.80	3.04	3.31	0.554
3.90	3.04	3.40	0.564
4.00	3.01	3.51	0.576
4.20	2.77	3.77	0.610
4.40	2.39	3.88	0.640
4.60	2.06	3.84	0.658
4.80	1.75	3.76	0.678
5.00	1.46	3.62	0.695
5.20	1.22	3.42	0.706
5.40	1.07	3.20	0.706
5.60	0.96	2.99	0.700
5.80	0.89	2.80	0.688
6.00	0.85	2.64	0.674
6.20	0.81	2.50	0.660
6.40	0.79	2.36	0.641
6.60	0.78	2.24	0.619
6.80	0.78	2.13	0.592
7.00	0.80	2.04	0.568
7.20	0.81	1.98	0.548
7.40	0.81	1.95	0.542
7.60	0.75	1.90	0.552
7.80	0.71	1.81	0.542
8.00	0.69	1.73	0.530
8.20	0.67	1.65	0.512
8.40	0.66	1.57	0.495
8.60	0.65	1.49	0.475
8.80	0.65	1.41	0.450
9.00	0.65	1.33	0.420
9.20	0.67	1.25	0.385
9.40	0.69	1.19	0.355
9.60	0.71	1.12	0.320
9.80	0.74	1.05	0.285
10.00	0.77	0.99	0.250
10.20	0.81	0.93	0.217
10.40	0.86	0.88	0.188
10.60	0.91	0.83	0.162
10.80	0.98	0.79	0.138
11.00	1.05	0.77	0.125
11.20	1.12	0.78	0.123
11.40	1.18	0.80	0.125
11.60	1.23	0.85	0.135
11.80	1.25	0.89	0.145
12.00	1.26	0.92	0.154
12.40	1.25	0.98	0.168
12.80	1.23	1.00	0.178
13.20	1.20	1.02	0.185
13.60	1.17	1.02	0.187
14.00	1.15	1.01	0.185
14.40	1.13	1.00	0.182
14.80	1.13	0.99	0.179
15.00	1.14	0.99	0.179
15.60	1.15	1.01	0.184
16.00	1.14	1.04	0.194
16.60	1.10	1.10	0.216
17.00	1.04	1.12	0.233
17.60	0.94	1.14	0.257
18.00	0.87	1.12	0.270
18.60	0.77	1.08	0.283
19.00	0.71	1.02	0.284
19.60	0.66	0.94	0.275
20.00	0.64	0.89	0.264
20.60	0.62	0.81	0.245
21.00	0.61	0.77	0.234
21.60	0.61	0.71	0.215
22.00	0.60	0.69	0.207
22.60	0.59	0.63	0.195
23.00	0.58	0.60	0.185
23.60	0.58	0.53	0.166
24.00	0.58	0.49	0.151
24.60	0.60	0.43	0.124
25.00	0.62	0.39	0.106
25.60	0.66	0.35	0.085
26.00	0.68	0.33	0.072
26.50	0.71	0.31	0.060
27.00	0.73	0.29	0.050
27.50	0.76	0.28	0.041
28.00	0.79	0.27	0.036
28.50	0.81	0.26	0.031
29.00	0.83	0.26	0.028
29.50	0.86	0.26	0.025
30.00	0.88	0.26	0.023
31.00	0.92	0.29	0.024
32.00	0.92	0.32	0.030
33.00	0.90	0.33	0.032
34.00	0.91	0.34	0.034
35.00	0.87	0.37	0.043
36.00	0.82	0.34	0.043
37.00	0.81	0.30	0.038
38.00	0.81	0.27	0.033
39.00	0.82	0.25	0.029
40.00	0.83	0.23	0.025

Nickel[10]

Energy (eV)	n	k	(Rφ = 0)
0.10	9.54	45.82	0.983
0.15	5.45	30.56	0.978
0.20	4.12	22.48	0.969
0.25	4.25	17.68	0.950
0.30	4.19	15.05	0.934
0.35	4.03	13.05	0.918
0.40	3.84	11.43	0.900
0.50	4.03	9.64	0.864
0.60	3.84	8.35	0.835
0.70	3.59	7.48	0.813
0.80	3.38	6.82	0.794
0.90	3.18	6.23	0.774
1.00	3.06	5.74	0.753
1.10	2.97	5.38	0.734
1.20	2.85	5.10	0.721
1.30	2.74	4.85	0.708
1.40	2.65	4.63	0.695
1.50	2.53	4.47	0.688
1.60	2.43	4.31	0.679
1.70	2.28	4.18	0.677
1.80	2.14	4.01	0.670
1.90	2.02	3.82	0.659
2.00	1.92	3.65	0.649
2.10	1.85	3.48	0.634
2.20	1.80	3.33	0.620
2.30	1.75	3.19	0.605
2.40	1.71	3.06	0.590
2.50	1.67	2.93	0.575
2.60	1.65	2.81	0.557
2.70	1.64	2.71	0.542
2.80	1.63	2.61	0.525
2.90	1.62	2.52	0.509
3.00	1.61	2.44	0.495
3.10	1.61	2.36	0.480
3.20	1.61	2.30	0.467
3.30	1.61	2.23	0.454
3.40	1.62	2.17	0.441
3.50	1.63	2.11	0.428
3.60	1.64	2.07	0.416
3.70	1.66	2.02	0.405
3.80	1.69	1.99	0.397
3.90	1.72	1.98	0.393
4.00	1.73	1.98	0.392
4.20	1.74	2.01	0.396
4.40	1.71	2.06	0.409
4.60	1.63	2.09	0.421
4.80	1.53	2.11	0.435
5.00	1.40	2.10	0.449
5.20	1.27	2.04	0.454
5.40	1.16	1.94	0.449
5.60	1.09	1.83	0.435
5.80	1.04	1.73	0.417
6.20	1.00	1.54	0.371
6.40	1.01	1.46	0.345
6.60	1.01	1.40	0.325
6.80	1.02	1.35	0.308
7.00	1.03	1.30	0.291
7.20	1.03	1.27	0.282
7.40	1.03	1.24	0.273
7.60	1.02	1.22	0.265
7.80	1.01	1.18	0.256
8.00	1.01	1.15	0.248
8.20	1.00	1.13	0.242
8.40	0.99	1.11	0.235
8.60	0.98	1.08	0.228
8.80	0.97	1.05	0.220
9.00	0.97	1.01	0.211
9.20	0.96	0.99	0.203
9.40	0.95	0.96	0.194
9.60	0.95	0.93	0.185
9.80	0.95	0.89	0.175
10.00	0.95	0.87	0.166
10.20	0.95	0.83	0.155
10.40	0.95	0.80	0.145
10.60	0.97	0.76	0.129
10.80	0.99	0.75	0.123
11.00	1.01	0.73	0.115
11.25	1.04	0.72	0.111
11.50	1.05	0.71	0.109
11.75	1.07	0.71	0.108
12.00	1.07	0.71	0.108
12.25	1.07	0.71	0.107
12.50	1.08	0.71	0.106
12.75	1.08	0.71	0.106
13.00	1.08	0.71	0.105
13.25	1.08	0.71	0.105
13.50	1.07	0.70	0.105
13.75	1.07	0.70	0.105
14.00	1.07	0.71	0.106
14.25	1.06	0.70	0.106
14.50	1.05	0.70	0.106
14.75	1.04	0.70	0.107
15.00	1.03	0.70	0.107
15.25	1.02	0.69	0.106
15.50	1.01	0.69	0.105

Energy (eV)	n	k	R(φ = 0)
15.75	1.00	0.68	0.104
16.00	0.99	0.67	0.103
16.50	0.98	0.66	0.101
17.00	0.96	0.64	0.098
17.50	0.94	0.63	0.096
18.00	0.92	0.61	0.092
18.50	0.91	0.58	0.087
19.00	0.90	0.56	0.082
19.50	0.90	0.54	0.077
20.00	0.89	0.51	0.071
20.50	0.89	0.49	0.066
21.00	0.90	0.47	0.061
21.50	0.91	0.46	0.057
22.00	0.91	0.45	0.055
22.50	0.91	0.44	0.053
23.00	0.92	0.44	0.051
23.50	0.91	0.44	0.052
24.00	0.90	0.43	0.051
24.50	0.90	0.43	0.051
25.00	0.89	0.42	0.050
26.00	0.88	0.39	0.046
27.00	0.87	0.37	0.042
28.00	0.87	0.35	0.040
29.00	0.86	0.34	0.037
30.00	0.86	0.32	0.034
35.00	0.86	0.24	0.022
40.00	0.87	0.18	0.014
45.00	0.88	0.13	0.008
50.00	0.92	0.10	0.004
60.00	0.96	0.08	0.002
65.00	0.98	0.09	0.002
68.00	0.96	0.12	0.004
70.00	0.94	0.11	0.004
75.00	0.94	0.09	0.003
80.00	0.94	0.07	0.002
90.00	0.94	0.06	0.002

Niobium[11]

Energy (eV)	n	k	R(φ = 0)
0.12	15.99	53.20	0.979
0.20	7.25	34.14	0.976
0.24	5.47	28.88	0.975
0.28	4.26	24.95	0.974
0.35	3.11	20.03	0.970
0.45	2.28	15.58	0.964
0.55	1.83	12.67	0.956
0.65	1.57	10.59	0.947
0.75	1.41	9.00	0.935
0.85	1.35	7.74	0.918
0.95	1.35	6.70	0.893
1.05	1.44	5.86	0.857
1.15	1.55	5.18	0.814
1.25	1.65	4.63	0.768
1.35	1.76	4.13	0.715
1.45	1.95	3.68	0.650
1.55	2.15	3.37	0.595
1.65	2.36	3.13	0.552
1.75	2.54	2.99	0.527
1.85	2.69	2.89	0.510
1.95	2.82	2.86	0.505
2.05	2.89	2.87	0.505
2.15	2.92	2.87	0.505
2.25	2.93	2.87	0.505
2.35	2.92	2.88	0.506
2.45	2.89	2.90	0.509
2.55	2.83	2.92	0.512
2.65	2.74	2.90	0.511
2.75	2.66	2.86	0.507
2.85	2.58	2.80	0.500
3.00	2.51	2.68	0.485
3.10	2.48	2.60	0.475
3.20	2.45	2.53	0.465
3.30	2.44	2.45	0.453
3.40	2.46	2.38	0.442
3.50	2.48	2.33	0.435
3.60	2.52	2.29	0.428
3.70	2.56	2.27	0.426
3.80	2.59	2.28	0.427
3.90	2.62	2.29	0.429
4.00	2.64	2.33	0.434
4.20	2.64	2.42	0.447
4.40	2.53	2.56	0.467
4.60	2.39	2.56	0.470
4.80	2.32	2.52	0.465
5.00	2.26	2.57	0.475
5.20	2.16	2.62	0.487
5.40	2.00	2.68	0.505
5.60	1.81	2.67	0.518
5.80	1.63	2.60	0.522
6.00	1.49	2.49	0.520
6.20	1.38	2.38	0.512
6.40	1.31	2.25	0.496
6.60	1.26	2.14	0.480
6.80	1.24	2.04	0.460
7.00	1.23	1.96	0.441
7.20	1.22	1.91	0.430
7.40	1.20	1.88	0.427
7.60	1.14	1.85	0.430
7.80	1.07	1.78	0.428
8.00	1.02	1.69	0.412
8.20	1.00	1.60	0.390
8.40	0.99	1.51	0.365
8.60	0.99	1.43	0.340
8.70	0.99	1.39	0.328
8.80	1.00	1.36	0.315
9.00	1.01	1.29	0.290
9.20	1.04	1.22	0.265
9.40	1.07	1.18	0.245
9.60	1.10	1.13	0.227
9.80	1.13	1.09	0.209
10.00	1.18	1.05	0.194
10.20	1.23	1.04	0.187
10.40	1.27	1.04	0.185
10.60	1.30	1.06	0.190
10.80	1.32	1.08	0.195
11.00	1.32	1.10	0.200
11.20	1.31	1.12	0.204
11.40	1.30	1.13	0.207
11.60	1.28	1.13	0.209
11.80	1.27	1.13	0.210
12.00	1.25	1.12	0.209
12.40	1.24	1.10	0.204
12.80	1.24	1.09	0.200
13.20	1.24	1.09	0.201
13.60	1.23	1.12	0.208
14.00	1.20	1.13	0.216
14.40	1.16	1.15	0.225
14.80	1.11	1.16	0.234
15.00	1.08	1.16	0.238
15.60	0.99	1.14	0.247
16.00	0.92	1.11	0.250
16.60	0.85	1.04	0.245
17.00	0.80	0.99	0.240
17.20	0.79	0.96	0.236
17.40	0.77	0.93	0.230
17.80	0.75	0.87	0.217
18.00	0.74	0.85	0.209
18.60	0.73	0.77	0.185
19.00	0.72	0.72	0.170
19.60	0.72	0.66	0.150
20.00	0.72	0.62	0.137
20.60	0.71	0.55	0.119
21.00	0.72	0.50	0.100
21.60	0.75	0.43	0.075
22.00	0.78	0.40	0.063
22.60	0.82	0.35	0.045
23.00	0.85	0.33	0.038
23.60	0.88	0.30	0.029
24.00	0.91	0.29	0.025
24.60	0.94	0.28	0.022
25.00	0.96	0.27	0.020
25.60	0.99	0.26	0.018
26.00	1.00	0.26	0.017
26.60	1.03	0.25	0.016
27.00	1.04	0.25	0.015
27.60	1.06	0.25	0.015
28.00	1.08	0.24	0.015
28.60	1.11	0.24	0.016
29.00	1.13	0.25	0.017
29.60	1.16	0.26	0.020
30.00	1.18	0.28	0.023
31.00	1.18	0.31	0.026
32.00	1.20	0.34	0.031
33.00	1.21	0.38	0.038
34.00	1.20	0.42	0.044
35.20	1.17	0.47	0.051
36.00	1.15	0.50	0.056
37.50	1.07	0.53	0.064
39.50	0.95	0.50	0.063
40.50	0.92	0.47	0.059

Osmium (Polycrystalline)[5]

Energy (eV)	n	k	R(φ = 0)
0.10	4.08	50.23	0.994
0.15	2.90	33.60	0.990
0.20	2.44	25.11	0.985
0.25	2.35	19.99	0.977
0.30	2.23	16.54	0.969
0.35	2.33	14.06	0.955
0.40	2.45	12.32	0.940
0.45	2.43	11.02	0.927
0.50	2.41	9.97	0.913
0.55	2.33	9.12	0.901
0.60	2.21	8.37	0.890
0.65	2.11	7.68	0.877
0.70	2.02	7.04	0.862
0.75	2.00	6.46	0.842
0.80	2.00	5.95	0.820
0.85	2.01	5.51	0.796
0.90	2.03	5.10	0.769
0.95	2.05	4.74	0.742
1.00	2.09	4.41	0.712
1.10	2.15	3.84	0.651
1.20	2.16	3.35	0.592
1.30	2.25	2.77	0.506
1.40	2.49	2.23	0.419
1.50	2.84	1.80	0.369
1.60	3.36	1.62	0.379
1.70	3.70	1.75	0.411
1.80	3.78	1.83	0.423
1.90	3.81	1.75	0.418
2.00	3.98	1.60	0.418
2.10	4.26	1.54	0.432
2.20	4.58	1.62	0.457
2.30	4.84	1.76	0.479
2.40	5.10	2.01	0.506
2.50	5.28	2.38	0.532
2.60	5.36	2.82	0.557
2.70	5.30	3.29	0.580
2.80	5.07	3.78	0.603
2.90	4.65	4.18	0.624
3.00	4.05	4.40	0.639
3.20	3.29	3.96	0.614
3.40	2.93	3.79	0.607
3.60	2.75	3.45	0.577
3.80	2.73	3.32	0.562
4.00	2.71	3.34	0.565
4.20	2.53	3.44	0.584
4.40	2.24	3.44	0.599
4.60	2.01	3.31	0.598
4.80	1.88	3.19	0.592
5.00	1.74	3.12	0.596
5.20	1.58	3.00	0.597
5.40	1.46	2.88	0.593
5.60	1.36	2.77	0.589
5.80	1.27	2.65	0.582
6.00	1.20	2.54	0.575
6.20	1.13	2.44	0.571
6.40	1.06	2.33	0.562
6.60	1.01	2.21	0.548
6.80	0.97	2.11	0.532
7.00	0.95	2.00	0.514
7.20	0.92	1.91	0.497
7.40	0.91	1.81	0.476
7.60	0.90	1.72	0.451
7.80	0.90	1.63	0.426
8.00	0.91	1.55	0.400

Energy (eV)	n	k	R(φ = 0)
8.20	0.91	1.48	0.375
8.40	0.94	1.40	0.344
8.60	0.96	1.34	0.319
8.80	0.98	1.29	0.296
9.00	1.01	1.24	0.274
9.20	1.04	1.19	0.255
9.40	1.08	1.16	0.238
9.60	1.10	1.14	0.229
9.80	1.13	1.11	0.217
10.00	1.16	1.10	0.209
10.20	1.19	1.08	0.203
10.30	1.20	1.08	0.201
10.40	1.22	1.08	0.200
10.50	1.23	1.09	0.201
10.60	1.24	1.10	0.203
10.80	1.25	1.11	0.206
11.00	1.24	1.13	0.213
11.20	1.23	1.14	0.217
11.40	1.19	1.15	0.223
11.60	1.17	1.12	0.216
11.80	1.16	1.10	0.211
12.00	1.15	1.08	0.205
12.40	1.14	1.03	0.191
12.80	1.15	1.01	0.183
13.20	1.16	0.98	0.174
13.60	1.17	0.97	0.170
14.00	1.17	0.96	0.169
14.40	1.16	0.94	0.165
14.80	1.16	0.91	0.156
15.20	1.17	0.89	0.148
15.60	1.20	0.86	0.140
16.00	1.25	0.87	0.140
16.40	1.28	0.90	0.147
16.80	1.28	0.94	0.157
17.20	1.27	0.97	0.167
17.60	1.26	1.01	0.178
18.00	1.23	1.04	0.189
18.40	1.19	1.08	0.200
18.80	1.14	1.10	0.210
19.20	1.10	1.10	0.219
19.60	1.05	1.11	0.227
20.00	0.96	1.10	0.239
20.40	0.93	1.09	0.240
20.80	0.89	1.05	0.240
21.20	0.86	1.02	0.237
21.60	0.83	0.99	0.235
22.00	0.80	0.96	0.230
22.40	0.78	0.93	0.226
22.80	0.77	0.90	0.220
23.20	0.75	0.88	0.217
23.60	0.75	0.86	0.211
24.00	0.73	0.84	0.209
24.40	0.72	0.82	0.207
24.80	0.70	0.80	0.205
25.20	0.69	0.77	0.202
25.60	0.67	0.75	0.199
26.00	0.66	0.72	0.195
26.40	0.65	0.69	0.189
26.80	0.63	0.66	0.183
27.20	0.65	0.62	0.165
28.00	0.64	0.59	0.156
28.40	0.64	0.57	0.148
28.80	0.65	0.55	0.140
29.20	0.65	0.53	0.134
29.60	0.65	0.51	0.128
30.00	0.65	0.49	0.121
31.00	0.65	0.45	0.111
32.00	0.66	0.41	0.095
33.00	0.68	0.37	0.079
34.00	0.70	0.34	0.068
35.00	0.72	0.31	0.057
36.00	0.74	0.29	0.048
37.00	0.77	0.27	0.040
38.00	0.79	0.26	0.035
39.00	0.81	0.26	0.031
40.00	0.84	0.26	0.026

Palladium[12]

Energy (eV)	n	k	R(φ = 0)
0.10	4.13	54.15	0.994
0.15	3.13	35.82	0.990
0.20	3.07	26.59	0.983
0.26	3.11	20.15	0.971
0.30	3.56	17.27	0.955
0.36	3.98	14.41	0.932
0.40	4.27	13.27	0.916
0.46	4.27	12.11	0.902
0.50	4.10	11.44	0.896
0.56	3.92	10.49	0.883
0.60	3.80	9.96	0.876
0.72	3.51	8.70	0.854
0.80	3.35	8.06	0.840
1.00	2.99	6.89	0.811
1.10	2.81	6.46	0.800
1.20	2.65	6.10	0.790
1.30	2.50	5.78	0.781
1.40	2.34	5.50	0.774
1.50	2.17	5.22	0.767
1.60	2.08	4.95	0.755
1.70	2.00	4.72	0.745
1.80	1.92	4.54	0.737
1.90	1.82	4.35	0.729
2.00	1.75	4.18	0.721
2.10	1.67	4.03	0.714
2.20	1.60	3.88	0.707
2.30	1.53	3.75	0.700
2.40	1.47	3.61	0.693
2.50	1.41	3.48	0.685
2.60	1.37	3.36	0.676
2.70	1.32	3.25	0.668
2.80	1.29	3.13	0.658
2.90	1.26	3.03	0.648
3.00	1.23	2.94	0.639
3.10	1.20	2.85	0.630
3.20	1.17	2.77	0.622
3.30	1.14	2.68	0.613
3.40	1.12	2.60	0.602
3.50	1.10	2.52	0.591
3.60	1.08	2.45	0.581
3.70	1.07	2.38	0.570
3.80	1.06	2.31	0.558
3.90	1.05	2.25	0.547
4.00	1.03	2.19	0.537
4.20	1.04	2.09	0.510
4.40	1.03	2.01	0.493
4.60	1.03	1.94	0.476
4.80	1.01	1.90	0.470
5.00	0.96	1.86	0.472
5.20	0.90	1.79	0.474
5.40	0.85	1.70	0.463
5.60	0.81	1.62	0.449
5.80	0.78	1.54	0.437
6.00	0.76	1.45	0.418
6.20	0.74	1.37	0.397
6.40	0.73	1.29	0.375
6.60	0.72	1.21	0.350
6.80	0.73	1.13	0.316
7.00	0.73	1.05	0.287
7.20	0.75	0.98	0.255
7.40	0.77	0.91	0.223
7.60	0.79	0.85	0.195
7.80	0.83	0.78	0.163
8.00	0.88	0.73	0.133
8.20	0.94	0.70	0.117
8.40	0.96	0.70	0.114
8.60	1.00	0.65	0.097
8.80	1.04	0.65	0.094
9.00	1.07	0.64	0.090
9.50	1.12	0.65	0.089
10.00	1.14	0.65	0.088
10.50	1.16	0.65	0.087
11.00	1.18	0.64	0.086
11.50	1.19	0.65	0.087
12.00	1.20	0.66	0.089
12.50	1.19	0.67	0.091
13.00	1.18	0.67	0.091
13.50	1.18	0.67	0.092
14.00	1.17	0.67	0.093
14.50	1.15	0.68	0.095
15.00	1.13	0.69	0.098
15.50	1.10	0.68	0.096
16.00	1.08	0.66	0.092
16.50	1.06	0.63	0.086
17.00	1.07	0.61	0.081
17.50	1.06	0.61	0.080
18.00	1.07	0.59	0.077
18.50	1.07	0.59	0.077
19.00	1.08	0.59	0.077
19.50	1.08	0.61	0.080
20.00	1.07	0.67	0.090
20.50	1.03	0.67	0.098
21.00	0.99	0.67	0.103
21.50	0.95	0.66	0.103
22.00	0.91	0.64	0.103
22.50	0.88	0.62	0.101
23.00	0.86	0.59	0.097
23.50	0.85	0.56	0.091
24.00	0.84	0.54	0.086
25.00	0.81	0.51	0.084
26.40	0.80	0.43	0.066
27.80	0.81	0.38	0.052
29.20	0.82	0.35	0.046

Platinum[13]

Energy (eV)	n	k	R(φ = 0)
0.10	13.21	44.72	0.976
0.15	8.18	31.16	0.969
0.20	5.90	23.95	0.962
0.25	4.70	19.40	0.954
0.30	3.92	16.16	0.945
0.35	3.28	13.66	0.936
0.40	2.81	11.38	0.922
0.45	3.03	9.31	0.882
0.50	3.91	7.71	0.813
0.55	4.58	7.14	0.777
0.60	5.13	6.75	0.753
0.65	5.52	6.66	0.746
0.70	5.71	6.83	0.751
0.75	5.57	7.02	0.759
0.80	5.31	7.04	0.762
0.85	5.05	6.98	0.763
0.90	4.77	6.91	0.765
0.95	4.50	6.77	0.763
1.00	4.25	6.62	0.762
1.10	3.86	6.24	0.753
1.20	3.55	5.92	0.746
1.30	3.29	5.61	0.736
1.40	3.10	5.32	0.725
1.50	2.92	5.07	0.716
1.60	2.76	4.84	0.706
1.70	2.63	4.64	0.697
1.80	2.51	4.43	0.686
1.90	2.38	4.26	0.678
2.00	2.30	4.07	0.664
2.10	2.23	3.92	0.654
2.20	2.17	3.77	0.642
2.30	2.10	3.67	0.636
2.40	2.03	3.54	0.626
2.50	1.96	3.42	0.616
2.60	1.91	3.30	0.605
2.70	1.87	3.20	0.595
2.80	1.83	3.10	0.585
2.90	1.79	3.01	0.575
3.00	1.75	2.92	0.565
3.20	1.68	2.76	0.546
3.40	1.63	2.62	0.527
3.60	1.58	2.48	0.507
3.80	1.53	2.37	0.491
4.00	1.49	2.25	0.472
4.20	1.45	2.14	0.452
4.40	1.43	2.04	0.432
4.60	1.39	1.95	0.415
4.80	1.38	1.85	0.392
5.00	1.36	1.76	0.372
5.20	1.36	1.67	0.350
5.40	1.36	1.61	0.332
5.60	1.36	1.54	0.315
5.80	1.36	1.47	0.295
6.00	1.38	1.40	0.276
6.20	1.39	1.35	0.261
6.40	1.42	1.29	0.246
6.60	1.45	1.26	0.236

Energy (eV)	n	k	R(φ = 0)
6.80	1.48	1.24	0.231
7.00	1.50	1.24	0.230
7.20	1.50	1.25	0.231
7.40	1.49	1.23	0.228
7.60	1.48	1.22	0.225
7.80	1.48	1.20	0.221
8.00	1.47	1.18	0.216
8.20	1.47	1.17	0.212
8.40	1.47	1.15	0.209
8.60	1.47	1.14	0.205
8.80	1.47	1.13	0.202
9.00	1.48	1.12	0.200
9.20	1.49	1.11	0.198
9.40	1.49	1.12	0.200
9.60	1.49	1.13	0.203
9.80	1.48	1.15	0.207
10.00	1.46	1.15	0.209
10.20	1.43	1.16	0.211
10.40	1.40	1.15	0.210
10.60	1.37	1.14	0.207
10.80	1.35	1.12	0.203
11.00	1.33	1.10	0.199
11.20	1.31	1.08	0.194
11.40	1.30	1.06	0.188
11.60	1.29	1.04	0.183
11.80	1.29	1.01	0.177
12.00	1.29	1.00	0.173
12.40	1.29	0.97	0.165
12.80	1.29	0.94	0.158
13.20	1.31	0.93	0.155
13.60	1.31	0.93	0.155
14.00	1.31	0.93	0.155
14.40	1.30	0.93	0.156
14.80	1.27	0.93	0.157
15.20	1.27	0.93	0.153
15.60	1.25	0.92	0.151
16.00	1.24	0.89	0.146
16.50	1.24	0.87	0.142
17.00	1.25	0.86	0.138
17.50	1.27	0.85	0.135
18.00	1.31	0.88	0.142
18.50	1.30	0.94	0.157
19.00	1.28	0.99	0.171
19.50	1.23	1.03	0.184
20.00	1.18	1.06	0.197
20.50	1.11	1.09	0.212
21.00	1.03	1.10	0.226
21.50	0.94	1.08	0.238
22.00	0.87	1.04	0.240
22.50	0.81	0.98	0.235
23.00	0.77	0.92	0.226
23.50	0.75	0.87	0.213
24.00	0.74	0.82	0.201
24.50	0.73	0.77	0.187
25.00	0.73	0.73	0.174
25.50	0.73	0.70	0.162
26.00	0.74	0.67	0.150
26.50	0.74	0.65	0.142
27.00	0.74	0.63	0.136
27.50	0.74	0.62	0.130
28.00	0.75	0.60	0.125
28.50	0.75	0.59	0.121
29.00	0.75	0.58	0.118
29.50	0.74	0.58	0.120
30.00	0.73	0.58	0.124

Rhenium (Single crystal with $\vec{E} \parallel \hat{c}$)[5]

Energy (eV)	n	k	R(φ = 0)
0.10	6.06	51.03	0.991
0.15	4.66	33.96	0.984
0.20	4.16	25.36	0.975
0.25	4.03	20.10	0.962
0.30	4.37	16.69	0.943
0.35	4.50	14.53	0.925
0.40	4.53	12.96	0.909
0.45	4.53	11.78	0.893
0.50	4.53	10.88	0.878
0.55	4.50	10.26	0.867
0.60	4.29	9.75	0.861
0.65	4.07	9.35	0.856
0.70	3.80	8.94	0.853
0.75	3.48	8.55	0.850
0.80	3.21	8.10	0.846
0.85	2.96	7.68	0.841
0.90	2.73	7.24	0.835
0.95	2.56	6.79	0.826
1.00	2.45	6.36	0.813
1.10	2.38	5.61	0.778
1.20	2.35	5.02	0.742
1.30	2.39	4.54	0.702
1.40	2.44	4.13	0.662
1.50	2.50	3.79	0.624
1.60	2.59	3.49	0.587
1.70	2.70	3.27	0.557
1.80	2.82	3.10	0.535
1.90	2.90	3.00	0.520
2.00	2.97	2.91	0.510
2.10	3.03	2.86	0.504
2.20	3.06	2.84	0.501
2.30	3.07	2.82	0.499
2.40	3.06	2.81	0.498
2.50	3.02	2.80	0.497
2.60	2.96	2.77	0.493
2.70	2.89	2.68	0.482
2.80	2.89	2.57	0.468
2.90	2.99	2.47	0.457
3.00	3.11	2.57	0.470
3.20	2.90	2.68	0.482
3.40	2.83	2.50	0.459
3.60	2.93	2.48	0.457
3.80	2.86	2.56	0.467
4.00	2.81	2.51	0.460
4.20	2.86	2.55	0.466
4.40	2.81	2.74	0.489
4.60	2.56	2.83	0.504
4.80	2.41	2.71	0.493
5.00	2.39	2.68	0.488
5.20	2.34	2.75	0.500
5.40	2.20	2.81	0.515
5.60	2.02	2.84	0.530
5.80	1.83	2.80	0.538
6.00	1.65	2.71	0.541
6.20	1.54	2.59	0.532
6.40	1.45	2.50	0.526
6.80	1.32	2.31	0.508
7.00	1.26	2.23	0.500
7.20	1.20	2.15	0.493
7.40	1.16	2.06	0.480
7.60	1.12	1.99	0.470
7.80	1.08	1.89	0.454
8.00	1.05	1.80	0.435
8.20	1.05	1.71	0.411
8.40	1.05	1.62	0.386
8.60	1.06	1.55	0.360
8.80	1.09	1.48	0.336
9.00	1.11	1.43	0.317
9.20	1.13	1.39	0.301
9.40	1.16	1.34	0.281
9.60	1.18	1.32	0.274
9.80	1.20	1.29	0.264
10.00	1.23	1.26	0.252
10.20	1.25	1.25	0.246
10.40	1.28	1.25	0.242
10.60	1.29	1.25	0.242
10.80	1.30	1.26	0.244
11.00	1.30	1.27	0.247
11.20	1.29	1.28	0.249
11.40	1.28	1.28	0.252
11.60	1.26	1.28	0.252
11.80	1.24	1.26	0.249
12.00	1.23	1.24	0.244
12.40	1.22	1.21	0.237
12.80	1.21	1.18	0.230
13.20	1.22	1.16	0.222
13.60	1.22	1.13	0.215
14.00	1.24	1.12	0.209
14.40	1.27	1.11	0.204
14.80	1.29	1.15	0.213
15.20	1.29	1.19	0.225
15.60	1.26	1.22	0.236
16.00	1.23	1.25	0.248
16.40	1.19	1.27	0.259
16.80	1.14	1.29	0.269
17.00	1.12	1.30	0.275
17.40	1.07	1.30	0.286
18.00	0.99	1.30	0.300
18.40	0.93	1.29	0.311
18.80	0.87	1.28	0.321
19.20	0.81	1.25	0.330
19.60	0.77	1.21	0.332
20.00	0.73	1.18	0.333
20.40	0.70	1.14	0.332
20.80	0.67	1.11	0.332
21.20	0.64	1.08	0.334
21.60	0.61	1.04	0.335
22.00	0.58	1.01	0.340
22.40	0.55	0.97	0.341
22.80	0.53	0.93	0.338
23.20	0.51	0.89	0.334
23.60	0.50	0.85	0.329
24.00	0.48	0.80	0.319
24.40	0.48	0.76	0.307
24.80	0.47	0.72	0.296
25.20	0.47	0.68	0.282
25.60	0.47	0.65	0.270
26.00	0.47	0.61	0.255
26.40	0.48	0.57	0.240
26.80	0.48	0.54	0.225
27.20	0.49	0.51	0.208
27.60	0.50	0.48	0.193
28.00	0.51	0.45	0.176
29.00	0.54	0.39	0.145
30.00	0.57	0.33	0.114
31.00	0.62	0.29	0.086
32.00	0.66	0.26	0.065
33.00	0.68	0.24	0.054
34.00	0.72	0.21	0.041
35.00	0.76	0.20	0.031
36.00	0.79	0.20	0.025
37.00	0.82	0.19	0.021
38.00	0.85	0.20	0.018
39.00	0.89	0.21	0.016
40.00	0.88	0.26	0.022
42.00	0.88	0.26	0.022
44.00	0.89	0.29	0.026
46.00	0.85	0.32	0.035
48.00	0.82	0.30	0.036
50.00	0.80	0.30	0.038
52.00	0.78	0.30	0.044
54.00	0.72	0.30	0.055
56.00	0.66	0.24	0.061
58.00	0.65	0.16	0.055

Rhenium (Single crystal, $\vec{E} \perp \hat{c}$)[5]

Energy (eV)	n	k	R(φ = 0)
0.10	4.25	42.83	0.991
0.15	3.28	28.08	0.984
0.20	3.28	20.66	0.971
0.25	3.47	16.27	0.951
0.30	3.73	13.44	0.926
0.35	3.93	11.54	0.900
0.40	3.99	10.15	0.875
0.45	4.17	9.03	0.846
0.50	4.34	8.26	0.821
0.55	4.45	7.73	0.801
0.60	4.53	7.40	0.788
0.65	4.44	7.26	0.784
0.70	4.13	7.09	0.784
0.75	3.77	6.75	0.779
0.80	3.55	6.32	0.766
0.85	3.39	5.95	0.752
0.90	3.26	5.61	0.737
0.95	3.17	5.27	0.719
1.00	3.09	4.96	0.701
1.10	3.05	4.39	0.658
1.20	3.08	3.89	0.613
1.30	3.20	3.56	0.578
1.40	3.23	3.38	0.559
1.50	3.23	3.12	0.532

Energy	n	k	R
1.60	3.29	2.88	0.507
1.70	3.38	2.72	0.491
1.80	3.47	2.59	0.480
1.90	3.54	2.50	0.473
2.00	3.63	2.43	0.469
2.10	3.74	2.40	0.470
2.20	3.83	2.38	0.472
2.30	3.93	2.44	0.481
2.40	4.00	2.55	0.492
2.50	4.01	2.70	0.505
2.60	3.90	2.84	0.514
2.70	3.74	2.92	0.517
2.80	3.57	2.88	0.511
2.90	3.49	2.75	0.497
3.00	3.53	2.71	0.493
3.20	3.55	2.84	0.506
3.40	3.34	2.88	0.508
3.60	3.25	2.83	0.501
3.80	3.24	2.84	0.502
4.00	3.19	2.94	0.513
4.20	3.05	3.06	0.526
4.40	2.88	3.15	0.539
4.60	2.67	3.18	0.548
4.80	2.44	3.17	0.554
5.00	2.25	3.12	0.556
5.20	2.10	3.04	0.555
5.40	1.96	2.96	0.553
5.60	1.84	2.88	0.551
5.80	1.73	2.81	0.549
6.00	1.61	2.74	0.549
6.20	1.51	2.64	0.545
6.40	1.42	2.56	0.541
6.80	1.28	2.37	0.526
7.00	1.22	2.28	0.517
7.20	1.16	2.19	0.508
7.40	1.12	2.08	0.493
7.60	1.12	1.98	0.468
7.80	1.08	1.93	0.463
8.00	1.05	1.83	0.443
8.20	1.05	1.74	0.418
8.40	1.05	1.66	0.397
8.60	1.06	1.58	0.372
8.80	1.07	1.52	0.351
9.00	1.09	1.46	0.327
9.20	1.11	1.41	0.309
9.40	1.14	1.36	0.290
9.60	1.17	1.31	0.273
9.80	1.20	1.27	0.258
10.00	1.24	1.24	0.244
10.20	1.29	1.22	0.234
10.40	1.33	1.23	0.233
10.60	1.36	1.25	0.238
10.80	1.38	1.28	0.245
11.00	1.37	1.31	0.253
11.20	1.36	1.33	0.259
11.40	1.33	1.34	0.264
11.60	1.31	1.34	0.266
11.80	1.28	1.33	0.266
12.00	1.26	1.32	0.264
12.40	1.23	1.29	0.257
12.80	1.22	1.26	0.251
13.20	1.20	1.23	0.245
13.60	1.19	1.20	0.236
14.00	1.20	1.16	0.225
14.40	1.22	1.13	0.214
14.80	1.27	1.12	0.207
15.20	1.31	1.17	0.218
15.60	1.31	1.23	0.234
16.00	1.28	1.28	0.251
16.40	1.24	1.33	0.270
16.80	1.17	1.37	0.288
17.00	1.14	1.38	0.297
17.40	1.06	1.39	0.314
18.00	0.95	1.38	0.334
18.40	0.88	1.36	0.346
18.80	0.82	1.33	0.355
19.20	0.76	1.29	0.360
19.60	0.72	1.25	0.363
20.00	0.67	1.21	0.369
20.40	0.64	1.15	0.364
20.80	0.61	1.10	0.357
21.20	0.60	1.06	0.349
21.60	0.58	1.02	0.342
22.00	0.57	0.98	0.336
22.40	0.56	0.95	0.328
22.80	0.55	0.92	0.325
23.20	0.53	0.89	0.322
23.60	0.52	0.85	0.317
24.00	0.50	0.82	0.314
24.40	0.49	0.79	0.309
24.80	0.48	0.75	0.303
25.20	0.47	0.72	0.295
25.60	0.47	0.68	0.286
26.00	0.46	0.64	0.276
26.40	0.46	0.61	0.263
26.80	0.46	0.57	0.249
27.20	0.47	0.53	0.231
27.60	0.48	0.50	0.216
28.00	0.49	0.47	0.198
29.00	0.51	0.41	0.164
30.00	0.55	0.34	0.129
31.00	0.59	0.29	0.097
32.00	0.64	0.26	0.072
33.00	0.67	0.24	0.060
34.00	0.70	0.22	0.047
35.00	0.74	0.20	0.036
36.00	0.77	0.19	0.029
37.00	0.80	0.19	0.023
38.00	0.84	0.19	0.018
39.00	0.88	0.21	0.016
40.00	0.87	0.25	0.023
42.00	0.87	0.25	0.023
44.00	0.88	0.28	0.026
46.00	0.84	0.31	0.035
48.00	0.82	0.30	0.036
50.00	0.80	0.30	0.039
52.00	0.77	0.30	0.044
54.00	0.71	0.29	0.055
56.00	0.66	0.23	0.061
58.00	0.64	0.16	0.055

Rhodium[7]

Energy (eV)	n	k	R(φ = 0)
0.10	18.48	69.43	0.986
0.20	8.66	37.46	0.977
0.30	5.85	25.94	0.967
0.40	4.74	19.80	0.955
0.50	4.20	16.07	0.941
0.60	3.87	13.51	0.925
0.70	3.67	11.72	0.908
0.80	3.63	10.34	0.887
0.90	3.62	9.36	0.867
1.00	3.71	8.67	0.848
1.10	3.67	8.26	0.837
1.20	3.51	7.94	0.832
1.30	3.26	7.63	0.829
1.40	3.01	7.31	0.827
1.50	2.78	6.97	0.823
1.60	2.60	6.64	0.818
1.70	2.42	6.33	0.813
1.80	2.30	6.02	0.805
1.90	2.20	5.76	0.798
2.00	2.12	5.51	0.789
2.10	2.05	5.30	0.780
2.20	2.00	5.11	0.772
2.30	1.94	4.94	0.765
2.40	1.90	4.78	0.756
2.50	1.88	4.65	0.748
2.60	1.85	4.55	0.743
2.70	1.80	4.49	0.742
2.90	1.63	4.36	0.748
3.00	1.53	4.29	0.753
3.10	1.41	4.20	0.760
3.20	1.30	4.09	0.764
3.30	1.20	3.97	0.767
3.40	1.11	3.84	0.769
3.50	1.04	3.71	0.768
3.60	0.99	3.58	0.764
3.70	0.95	3.45	0.759
3.80	0.91	3.34	0.753
3.90	0.88	3.23	0.747
4.00	0.86	3.12	0.739
4.20	0.83	2.94	0.722
4.40	0.80	2.76	0.706
4.60	0.78	2.60	0.684
4.80	0.79	2.46	0.659
5.00	0.79	2.34	0.635
5.20	0.79	2.23	0.613
5.40	0.80	2.14	0.591
5.60	0.80	2.06	0.573
5.80	0.79	2.00	0.561
6.00	0.76	1.93	0.556
6.20	0.73	1.85	0.544
6.40	0.70	1.77	0.534
6.60	0.68	1.69	0.518
6.80	0.67	1.60	0.498
7.00	0.66	1.52	0.476
7.20	0.66	1.43	0.452
7.40	0.66	1.35	0.423
7.60	0.67	1.27	0.394
7.80	0.68	1.20	0.363
8.00	0.69	1.12	0.329
8.20	0.71	1.04	0.288
8.40	0.74	0.97	0.252
8.60	0.78	0.89	0.212
8.80	0.83	0.83	0.179
9.00	0.88	0.77	0.148
9.20	0.95	0.73	0.125
9.40	1.01	0.71	0.110
9.60	1.07	0.69	0.102
9.80	1.12	0.69	0.098
10.00	1.17	0.69	0.098
10.60	1.26	0.73	0.106
11.00	1.29	0.76	0.113
11.60	1.32	0.80	0.124
12.00	1.32	0.82	0.127
12.60	1.32	0.82	0.129
13.00	1.32	0.83	0.131
13.60	1.32	0.85	0.134
14.00	1.32	0.86	0.138
14.60	1.30	0.89	0.144
15.00	1.28	0.90	0.147
15.60	1.25	0.90	0.147
16.00	1.24	0.89	0.147
16.50	1.23	0.88	0.145
17.00	1.22	0.88	0.144
17.50	1.22	0.87	0.143
18.00	1.23	0.88	0.145
18.50	1.25	0.92	0.155
19.00	1.24	0.98	0.172
19.50	1.18	1.05	0.193
20.00	1.10	1.09	0.213
20.50	1.00	1.09	0.230
21.00	0.91	1.05	0.234
21.50	0.86	1.00	0.228
22.00	0.83	0.95	0.219
22.50	0.81	0.92	0.214
23.00	0.79	0.90	0.213
23.50	0.75	0.87	0.214
24.00	0.73	0.84	0.210
24.50	0.70	0.81	0.208
25.00	0.69	0.77	0.202
25.50	0.67	0.74	0.195
26.00	0.66	0.70	0.188
26.50	0.65	0.66	0.176
27.00	0.65	0.64	0.168
27.50	0.65	0.61	0.159
28.00	0.65	0.59	0.152
29.00	0.65	0.54	0.137
30.00	0.66	0.51	0.127
31.00	0.64	0.49	0.127
32.00	0.61	0.44	0.126
33.00	0.60	0.37	0.110
34.00	0.65	0.30	0.074
35.00	0.69	0.28	0.058
36.00	0.73	0.27	0.049
37.00	0.74	0.28	0.047
38.00	0.74	0.27	0.045
39.00	0.75	0.25	0.041

Ruthenium (Single crystal, $\vec{E} \parallel \hat{c}$)[5]

Energy (eV)	n	k	R($\phi = 0$)
0.10	11.50	51.38	0.984
0.20	5.93	27.14	0.970
0.30	4.33	18.50	0.953
0.40	3.60	13.97	0.933
0.50	3.18	11.04	0.909
0.60	3.28	8.89	0.865
0.70	3.62	7.73	0.822
0.80	3.42	7.02	0.801
0.90	3.25	6.12	0.766
1.00	3.39	5.33	0.715
1.10	3.66	4.83	0.675
1.20	3.84	4.57	0.654
1.30	3.94	4.38	0.638
1.40	4.02	4.19	0.624
1.50	4.16	4.07	0.614
1.60	4.33	4.08	0.615
1.70	4.42	4.21	0.624
1.80	4.40	4.38	0.636
1.90	4.29	4.61	0.651
2.00	4.04	4.81	0.667
2.10	3.69	4.90	0.679
2.20	3.35	4.82	0.683
2.30	3.09	4.70	0.681
2.40	2.89	4.55	0.677
2.50	2.74	4.40	0.671
2.60	2.64	4.25	0.663
2.70	2.58	4.14	0.656
2.80	2.54	4.05	0.650
2.90	2.48	4.03	0.650
3.00	2.38	4.03	0.656
3.10	2.26	4.00	0.661
3.20	2.13	3.96	0.666
3.30	2.00	3.91	0.671
3.40	1.87	3.83	0.673
3.50	1.76	3.74	0.674
3.60	1.66	3.65	0.675
3.70	1.57	3.55	0.673
3.80	1.49	3.45	0.672
3.90	1.42	3.35	0.668
4.00	1.37	3.24	0.661
4.20	1.29	3.08	0.649
4.40	1.22	2.93	0.639
4.60	1.16	2.79	0.628
4.80	1.11	2.67	0.617
5.00	1.06	2.56	0.607
5.20	1.01	2.46	0.600
5.40	0.95	2.35	0.593
5.60	0.92	2.23	0.576
5.80	0.90	2.14	0.559
6.00	0.88	2.05	0.545
6.20	0.87	1.98	0.531
6.40	0.84	1.91	0.521
6.60	0.82	1.84	0.510
6.80	0.79	1.77	0.500
7.00	0.76	1.69	0.489
7.20	0.75	1.61	0.472
7.40	0.73	1.54	0.455
7.60	0.73	1.46	0.433
7.80	0.73	1.39	0.411
8.00	0.72	1.33	0.391
8.20	0.72	1.26	0.366
8.40	0.73	1.20	0.342
8.60	0.74	1.14	0.318
8.80	0.74	1.08	0.295
9.00	0.75	1.02	0.267
9.20	0.77	0.97	0.243
9.40	0.79	0.91	0.217
9.60	0.82	0.86	0.190
9.80	0.85	0.81	0.167
10.00	0.88	0.76	0.144
10.20	0.92	0.72	0.125
10.40	0.96	0.69	0.110
10.60	1.01	0.67	0.100
10.80	1.05	0.66	0.094
11.00	1.09	0.65	0.090
11.20	1.12	0.65	0.088
11.40	1.15	0.65	0.087
11.60	1.18	0.65	0.088
11.80	1.21	0.66	0.090
12.00	1.23	0.67	0.092
12.40	1.26	0.69	0.098
12.80	1.27	0.72	0.104
13.20	1.28	0.74	0.108
13.60	1.28	0.75	0.111
14.00	1.28	0.76	0.114
14.40	1.27	0.76	0.114
14.80	1.27	0.76	0.114
15.00	1.27	0.76	0.114
15.60	1.28	0.77	0.115
16.00	1.30	0.78	0.118
16.50	1.32	0.80	0.123
17.00	1.34	0.85	0.136
17.50	1.32	0.93	0.155
18.00	1.26	0.99	0.173
18.50	1.18	1.02	0.185
19.00	1.11	1.02	0.192
19.50	1.05	1.02	0.199
20.00	0.99	1.02	0.208
20.50	0.92	0.99	0.212
21.00	0.86	0.94	0.209
21.50	0.83	0.90	0.203
22.00	0.81	0.86	0.193
23.00	0.77	0.79	0.182
24.00	0.74	0.74	0.171
25.00	0.71	0.69	0.163
26.00	0.68	0.63	0.154
27.00	0.67	0.57	0.140
28.00	0.66	0.51	0.124
29.00	0.67	0.46	0.107
30.00	0.67	0.43	0.097
31.00	0.67	0.37	0.084
32.00	0.69	0.33	0.070
33.00	0.71	0.30	0.058
34.00	0.73	0.27	0.048
35.00	0.75	0.25	0.039
36.00	0.77	0.24	0.035
37.00	0.79	0.23	0.030
38.00	0.80	0.22	0.027
39.00	0.82	0.22	0.024
40.00	0.83	0.22	0.022

Ruthenium (Single crystal, $\vec{E} \perp \hat{c}$)[5]

Energy (eV)	n	k	R($\phi = 0$)
0.10	11.85	50.81	0.983
0.20	6.68	27.18	0.966
0.30	4.94	18.92	0.950
0.40	3.90	14.51	0.933
0.50	3.27	11.63	0.915
0.60	2.98	9.54	0.888
0.70	2.82	7.99	0.856
0.80	2.73	6.71	0.815
0.90	2.82	5.54	0.751
1.00	3.17	4.59	0.670
1.10	3.69	3.91	0.604
1.20	4.28	3.66	0.585
1.30	4.66	3.72	0.593
1.40	4.86	3.79	0.601
1.50	4.99	3.89	0.609
1.60	5.08	4.03	0.618
1.70	5.12	4.22	0.629
1.80	5.10	4.45	0.642
1.90	4.96	4.78	0.660
2.00	4.61	5.06	0.677
2.10	4.21	5.09	0.682
2.20	3.94	5.00	0.681
2.30	3.69	4.97	0.684
2.40	3.44	4.88	0.684
2.50	3.27	4.77	0.681
2.60	3.14	4.66	0.677
2.70	3.06	4.59	0.674
2.80	2.99	4.59	0.676
2.90	2.87	4.64	0.686
3.00	2.64	4.69	0.701
3.10	2.40	4.64	0.710
3.20	2.18	4.55	0.717
3.30	2.00	4.43	0.721
3.40	1.84	4.30	0.723
3.50	1.71	4.16	0.723
3.60	1.60	4.03	0.722
3.70	1.50	3.90	0.721
3.80	1.41	3.77	0.718
3.90	1.35	3.64	0.713
4.00	1.29	3.53	0.707
4.20	1.21	3.31	0.694
4.40	1.16	3.13	0.679
4.60	1.13	2.97	0.662
4.80	1.09	2.86	0.652
5.00	1.03	2.75	0.648
5.20	0.97	2.64	0.643
5.40	0.91	2.52	0.635
5.60	0.88	2.40	0.622
5.80	0.86	2.29	0.605
6.00	0.84	2.20	0.591
6.20	0.82	2.11	0.576
6.40	0.81	2.04	0.564
6.60	0.78	1.97	0.556
6.80	0.76	1.89	0.545
7.00	0.73	1.82	0.538
7.20	0.70	1.75	0.527
7.40	0.68	1.67	0.513
7.60	0.67	1.59	0.496
7.80	0.66	1.51	0.476
8.00	0.66	1.44	0.454
8.20	0.65	1.36	0.430
8.40	0.66	1.29	0.403
8.60	0.66	1.22	0.378
8.80	0.68	1.15	0.346
9.00	0.69	1.09	0.317
9.20	0.70	1.02	0.286
9.40	0.73	0.95	0.251
9.60	0.77	0.89	0.216
9.80	0.82	0.84	0.185
10.00	0.86	0.81	0.163
10.20	0.90	0.77	0.143
10.40	0.94	0.74	0.127
10.60	0.99	0.72	0.115
10.80	1.04	0.71	0.108
11.00	1.08	0.70	0.104
11.20	1.11	0.70	0.102
11.40	1.14	0.70	0.101
11.60	1.17	0.71	0.102
11.80	1.20	0.72	0.104
12.00	1.22	0.73	0.107
12.40	1.25	0.76	0.113
12.80	1.26	0.78	0.118
13.20	1.27	0.81	0.124
13.60	1.27	0.83	0.129
14.00	1.26	0.84	0.132
14.40	1.25	0.84	0.132
14.80	1.25	0.84	0.133
15.00	1.25	0.84	0.133
15.60	1.25	0.85	0.134
16.00	1.27	0.85	0.134
16.50	1.28	0.89	0.145
17.00	1.28	0.94	0.158
17.50	1.25	1.00	0.175
18.00	1.19	1.04	0.190
18.50	1.12	1.05	0.200
19.00	1.07	1.05	0.205
19.50	1.02	1.04	0.212
20.00	0.97	1.04	0.219
20.50	0.91	1.03	0.228
21.00	0.85	1.01	0.234
21.50	0.80	0.97	0.234
22.00	0.77	0.94	0.233
23.00	0.71	0.87	0.229
24.00	0.67	0.79	0.218
25.00	0.64	0.73	0.205
26.00	0.61	0.66	0.194
27.00	0.60	0.59	0.177
28.00	0.60	0.53	0.155
29.00	0.61	0.48	0.134
30.00	0.62	0.45	0.123
31.00	0.61	0.40	0.114
32.00	0.63	0.34	0.093
33.00	0.65	0.31	0.077
34.00	0.67	0.28	0.065

35.00	0.70	0.26	0.054	44.00	0.90	0.33	0.033	7.40	1.13	1.82	0.425
36.00	0.72	0.25	0.047	46.00	0.90	0.32	0.031	7.60	1.12	1.75	0.406
37.00	0.73	0.23	0.041	48.00	0.89	0.31	0.030	7.80	1.11	1.68	0.390
38.00	0.75	0.22	0.035	50.00	0.88	0.29	0.027	8.00	1.11	1.61	0.370
39.00	0.77	0.22	0.031	52.00	0.89	0.28	0.024	8.20	1.12	1.55	0.350
40.00	0.79	0.22	0.028	54.00	0.88	0.27	0.024	8.40	1.13	1.50	0.332

Silver[4]

Energy (eV)	n	k	R(φ = 0)
0.10	9.91	90.27	0.995
0.20	2.84	45.70	0.995
0.30	1.41	30.51	0.994
0.40	0.91	22.89	0.993
0.50	0.67	18.32	0.992
1.00	0.28	9.03	0.987
1.50	0.27	5.79	0.969
2.00	0.27	4.18	0.944
2.50	0.24	3.09	0.914
3.00	0.23	2.27	0.864
3.25	0.23	1.86	0.816
3.50	0.21	1.42	0.756
3.60	0.23	1.13	0.671
3.70	0.30	0.77	0.475
3.77	0.53	0.40	0.154
3.80	0.73	0.30	0.053
3.90	1.30	0.36	0.040
4.00	1.61	0.60	0.103
4.10	1.73	0.85	0.153
4.20	1.75	1.06	0.194
4.30	1.73	1.13	0.208
4.50	1.69	1.28	0.238
4.75	1.61	1.34	0.252
5.00	1.55	1.36	0.257
5.50	1.45	1.34	0.257
6.00	1.34	1.28	0.246
6.50	1.25	1.18	0.225
7.00	1.18	1.06	0.196
7.50	1.14	0.91	0.157
8.00	1.16	0.75	0.114
9.00	1.33	0.56	0.074
10.00	1.46	0.56	0.082
11.00	1.52	0.56	0.088
12.00	1.61	0.59	0.100
13.00	1.66	0.64	0.112
14.00	1.72	0.78	0.141
14.50	1.64	0.88	0.152
15.00	1.56	0.92	0.156
16.00	1.42	0.91	0.151
17.00	1.33	0.86	0.139
18.00	1.28	0.80	0.124
19.00	1.27	0.75	0.111
20.00	1.29	0.71	0.103
21.00	1.35	0.75	0.112
21.50	1.37	0.80	0.124
22.00	1.34	0.87	0.141
22.50	1.26	0.93	0.157
23.00	1.17	0.94	0.163
23.50	1.10	0.93	0.165
24.00	1.04	0.90	0.165
24.50	0.99	0.87	0.160
25.00	0.95	0.83	0.154
25.50	0.91	0.78	0.144
26.00	0.90	0.74	0.133
26.50	0.89	0.69	0.121
27.00	0.89	0.65	0.109
27.50	0.89	0.62	0.099
28.00	0.90	0.59	0.090
28.50	0.91	0.57	0.084
29.00	0.92	0.56	0.079
30.00	0.93	0.54	0.074
31.00	0.93	0.53	0.072
32.00	0.92	0.53	0.072
33.00	0.90	0.51	0.071
34.00	0.88	0.49	0.067
35.00	0.86	0.45	0.061
36.00	0.89	0.44	0.055
38.00	0.89	0.39	0.043
40.00	0.90	0.37	0.039
42.00	0.90	0.35	0.036

Energy (eV)	n	k	R (φ = 0)
44.00	0.90	0.33	0.033
46.00	0.90	0.32	0.031
48.00	0.89	0.31	0.030
50.00	0.88	0.29	0.027
52.00	0.89	0.28	0.024
54.00	0.88	0.27	0.024
56.00	0.87	0.26	0.024
58.00	0.87	0.24	0.021
60.00	0.87	0.22	0.018
62.00	0.88	0.21	0.016
64.00	0.88	0.21	0.016
66.00	0.88	0.21	0.016
68.00	0.87	0.21	0.017
70.00	0.83	0.20	0.021
72.00	0.85	0.18	0.016
74.00	0.85	0.17	0.014
76.00	0.85	0.16	0.013
78.00	0.85	0.15	0.013
80.00	0.85	0.14	0.012
85.00	0.85	0.11	0.011
90.00	0.85	0.08	0.009
95.00	0.86	0.06	0.007
100.00	0.87	0.04	0.005

Tantalum[9]

Energy (eV)	n	k	R (φ = 0)
0.10	10.14	66.39	0.984
0.15	9.45	46.41	0.983
0.20	5.77	35.46	0.982
0.26	3.67	27.53	0.981
0.30	2.87	23.90	0.980
0.38	2.03	18.87	0.978
0.50	1.37	14.26	0.974
0.58	1.15	12.19	0.970
0.70	0.96	9.92	0.962
0.78	0.89	8.77	0.956
0.90	0.84	7.38	0.942
1.00	0.89	6.47	0.992
1.10	0.93	5.75	0.899
1.20	0.98	5.14	0.872
1.30	1.00	4.62	0.842
1.40	1.04	4.15	0.805
1.50	1.09	3.73	0.762
1.60	1.15	3.33	0.707
1.70	1.24	2.95	0.640
1.80	1.35	2.60	0.560
1.90	1.57	2.24	0.460
2.00	1.83	1.99	0.388
2.10	2.10	1.84	0.354
2.20	2.36	1.81	0.351
2.30	2.56	1.86	0.365
2.40	2.68	1.92	0.378
2.50	2.75	1.98	0.388
2.60	2.80	2.02	0.395
2.70	2.84	2.08	0.405
2.80	2.85	2.14	0.412
2.90	2.84	2.20	0.420
3.00	2.81	2.24	0.425
3.20	2.73	2.31	0.432
3.40	2.61	2.33	0.435
3.60	2.49	2.30	0.430
3.80	2.40	2.22	0.418
4.00	2.36	2.14	0.406
4.20	2.35	2.06	0.392
4.40	2.39	2.01	0.384
4.60	2.45	2.00	0.384
4.80	2.53	2.06	0.394
5.00	2.58	2.20	0.416
5.20	2.52	2.44	0.450
5.40	2.31	2.61	0.480
5.60	2.06	2.67	0.501
5.80	1.83	2.63	0.510
6.00	1.63	2.56	0.515
6.20	1.48	2.45	0.512
6.40	1.37	2.33	0.504
6.60	1.29	2.22	0.492
6.80	1.23	2.11	0.478
7.00	1.18	2.01	0.462
7.20	1.15	1.91	0.445

Energy (eV)	n	k	R (φ = 0)
7.40	1.13	1.82	0.425
7.60	1.12	1.75	0.406
7.80	1.11	1.68	0.390
8.00	1.11	1.61	0.370
8.20	1.12	1.55	0.350
8.40	1.13	1.50	0.332
8.60	1.14	1.45	0.317
8.80	1.17	1.41	0.301
9.00	1.19	1.40	0.294
9.20	1.21	1.38	0.289
9.40	1.21	1.38	0.287
9.60	1.21	1.38	0.285
9.80	1.21	1.37	0.285
10.00	1.20	1.37	0.286
10.20	1.19	1.37	0.286
10.40	1.18	1.37	0.287
10.60	1.16	1.36	0.288
10.80	1.15	1.36	0.289
11.00	1.13	1.35	0.290
11.20	1.11	1.35	0.292
11.40	1.09	1.34	0.293
11.60	1.07	1.33	0.294
11.80	1.05	1.32	0.295
12.00	1.02	1.31	0.296
12.20	1.00	1.29	0.295
12.40	0.98	1.28	0.294
12.60	0.96	1.26	0.292
12.80	0.94	1.24	0.289
13.00	0.93	1.22	0.286
13.60	0.91	1.16	0.272
14.00	0.90	1.15	0.272
14.60	0.85	1.15	0.285
15.00	0.80	1.13	0.293
15.60	0.72	1.08	0.301
16.00	0.68	1.04	0.304
16.60	0.63	0.97	0.301
17.00	0.60	0.92	0.296
17.60	0.60	0.92	0.296
18.00	0.55	0.79	0.274
18.60	0.53	0.71	0.254
19.00	0.53	0.65	0.236
19.60	0.53	0.57	0.207
20.00	0.54	0.52	0.185
20.60	0.55	0.44	0.153
21.00	0.57	0.39	0.127
21.60	0.64	0.34	0.089
22.00	0.64	0.32	0.081
22.60	0.69	0.27	0.058
23.00	0.73	0.24	0.043
23.60	0.80	0.26	0.033
24.00	0.80	0.26	0.034
24.60	0.82	0.25	0.029
25.00	0.83	0.25	0.026
25.60	0.86	0.24	0.022
26.00	0.88	0.25	0.022
26.60	0.87	0.26	0.023
27.00	0.87	0.25	0.022
27.60	0.89	0.23	0.019
28.00	0.90	0.23	0.017
28.60	0.91	0.22	0.015
29.00	0.92	0.22	0.014
29.60	0.94	0.22	0.014
30.00	0.95	0.22	0.014
31.00	0.97	0.23	0.014
32.00	0.98	0.24	0.015
33.00	0.98	0.25	0.015
34.00	0.99	0.25	0.016
35.00	0.99	0.26	0.017
36.00	0.99	0.27	0.018
37.00	0.99	0.28	0.019
38.00	0.98	0.28	0.021
39.00	0.97	0.29	0.022
40.00	0.95	0.29	0.023

Titanium (Polycrystalline)[14]

Energy (eV)	n	k	R (φ = 0)
0.10	5.03	23.38	0.965
0.15	3.00	15.72	0.954
0.20	2.12	11.34	0.939

0.25	2.05	8.10	0.890
0.30	6.39	9.94	0.833
0.35	2.74	6.21	0.792
0.40	2.49	4.68	0.708
0.45	3.35	3.25	0.545
0.50	4.43	3.22	0.555
0.60	4.71	3.77	0.597
0.70	4.38	3.89	0.603
0.80	4.04	3.82	0.596
0.90	3.80	3.65	0.582
1.00	3.62	3.52	0.570
1.10	3.47	3.40	0.560
1.20	3.35	3.30	0.550
1.30	3.28	3.25	0.546
1.40	3.17	3.28	0.549
1.50	2.98	3.32	0.557
1.60	2.74	3.30	0.559
1.70	2.54	3.23	0.557
1.80	2.36	3.11	0.550
1.90	2.22	2.99	0.540
2.00	2.11	2.88	0.530
2.10	2.01	2.77	0.520
2.20	1.92	2.67	0.509
2.30	1.86	2.56	0.495
2.40	1.81	2.47	0.483
2.50	1.78	2.39	0.471
2.60	1.75	2.34	0.462
2.70	1.71	2.29	0.456
2.80	1.68	2.25	0.451
2.90	1.63	2.21	0.447
3.00	1.59	2.17	0.444
3.10	1.55	2.15	0.442
3.20	1.50	2.12	0.442
3.30	1.44	2.09	0.442
3.40	1.37	2.06	0.443
3.50	1.30	2.01	0.443
3.60	1.24	1.96	0.441
3.70	1.17	1.90	0.436
3.80	1.11	1.83	0.430
3.85	1.08	1.78	0.423
3.90	1.06	1.73	0.413
4.00	1.04	1.62	0.389
4.20	1.05	1.45	0.333
4.40	1.13	1.33	0.284
4.60	1.17	1.29	0.265
4.80	1.21	1.23	0.244
5.00	1.24	1.21	0.236
5.20	1.27	1.20	0.228
5.40	1.17	1.16	0.228
5.60	1.24	1.21	0.234
5.80	1.21	1.22	0.241
6.00	1.15	1.21	0.244
6.20	1.11	1.18	0.240
6.40	1.08	1.14	0.232
6.60	1.04	1.06	0.212
6.80	1.05	1.02	0.198
7.00	1.06	0.97	0.182
7.20	1.07	0.95	0.175
7.40	1.11	0.94	0.167
7.60	1.09	0.92	0.165
7.80	1.11	0.93	0.165
8.00	1.10	0.94	0.169
8.20	1.10	0.95	0.171
8.40	1.08	0.95	0.175
8.60	1.04	0.96	0.181
8.80	1.02	0.95	0.181
9.00	1.00	0.94	0.182
9.20	0.97	0.93	0.182
9.40	0.95	0.91	0.181
9.60	0.94	0.90	0.179
9.80	0.91	0.88	0.179
10.00	0.89	0.88	0.180
10.20	0.86	0.85	0.178
10.40	0.85	0.83	0.175
10.60	0.81	0.79	0.167
10.80	0.80	0.76	0.162
11.00	0.79	0.72	0.152
11.20	0.81	0.69	0.139
11.40	0.81	0.69	0.139
11.60	0.79	0.68	0.139
11.80	0.78	0.67	0.137
12.00	0.77	0.65	0.132
12.80	0.76	0.55	0.106
13.20	0.76	0.52	0.097
13.60	0.76	0.48	0.087
14.00	0.77	0.45	0.077
14.40	0.77	0.42	0.069
14.80	0.79	0.38	0.058
15.20	0.79	0.36	0.052
15.60	0.79	0.32	0.045
16.00	0.83	0.31	0.037
16.40	0.84	0.28	0.030
16.80	0.87	0.27	0.025
17.20	0.90	0.25	0.020
17.60	0.93	0.25	0.017
18.00	0.94	0.24	0.165
18.40	0.94	0.23	0.017
18.80	0.95	0.24	0.016
19.20	0.96	0.25	0.016
19.60	0.97	0.25	0.017
20.00	0.98	0.27	0.018
20.40	0.98	0.27	0.019
20.60	1.00	0.29	0.020
21.20	0.99	0.31	0.023
21.60	0.99	0.31	0.024
22.00	0.98	0.32	0.025
22.40	0.98	0.33	0.027
22.80	0.97	0.33	0.028
23.20	0.96	0.34	0.030
23.60	0.95	0.35	0.031
24.00	0.92	0.35	0.033
24.5	0.91	0.34	0.032
25.0	0.91	0.33	0.032
25.5	0.89	0.33	0.032
26.0	0.89	0.33	0.032
26.5	0.88	0.32	0.032
27.0	0.86	0.31	0.032
27.5	0.85	0.30	0.033
28.0	0.84	0.29	0.027
28.5	0.82	0.26	0.029
29.0	0.83	0.25	0.027
30.0	0.84	0.22	0.022

Tungsten[15]

Energy (eV)	n	k	R (φ = 0)
0.10	14.06	54.71	0.983
0.20	3.87	28.30	0.981
0.25	2.56	22.44	0.980
0.30	1.83	18.32	0.979
0.34	1.71	15.71	0.973
0.38	1.86	13.88	0.963
0.42	1.92	12.63	0.954
0.46	1.69	11.59	0.952
0.50	1.40	10.52	0.952
0.54	1.23	9.45	0.948
0.58	1.17	8.44	0.938
0.62	1.28	7.52	0.917
0.66	1.45	6.78	0.888
0.70	0.59	6.13	0.856
0.74	1.83	5.52	0.810
0.78	2.12	—	0.759
0.82	2.36	4.61	0.710
0.86	2.92	4.37	0.661
0.90	3.11	4.44	0.660
0.94	3.15	4.43	0.658
0.98	3.15	4.36	0.653
1.00	3.14	4.32	0.649
1.10	3.05	4.04	0.627
1.20	3.00	3.64	0.590
1.30	3.12	3.24	0.545
1.40	3.29	2.96	0.515
1.50	3.48	2.79	0.500
1.60	3.67	2.68	0.494
1.70	3.84	2.79	0.507
1.80	3.82	2.91	0.518
1.90	3.70	2.94	0.518
2.00	3.60	2.89	0.512
2.10	3.54	2.84	0.506
2.20	3.49	2.76	0.497
2.30	3.49	2.72	0.494
2.40	3.45	2.72	0.493
2.50	3.38	2.68	0.487
2.60	3.34	2.62	0.480
2.70	3.31	2.55	0.472
2.80	3.31	2.49	0.466
2.90	3.32	2.45	0.461
3.00	3.35	2.42	0.459
3.10	3.39	2.41	0.460
3.20	3.43	2.45	0.465
3.30	3.45	2.55	0.476
3.40	3.39	2.66	0.485
3.50	3.24	2.70	0.488
3.60	3.13	2.67	0.482
3.70	3.05	2.62	0.476
3.80	2.99	2.56	0.468
3.90	2.96	2.50	0.460
4.00	2.95	2.43	0.451
4.20	3.02	2.33	0.440
4.40	3.13	2.32	0.442
4.60	3.24	2.41	0.455
4.80	3.33	2.57	0.475
5.00	3.40	2.85	0.505
5.20	3.27	3.27	0.548
5.40	2.92	3.58	0.586
5.60	2.43	3.70	0.618
5.80	2.00	3.61	0.637
6.00	1.70	3.42	0.643
6.20	1.47	3.24	0.646
6.40	1.32	3.04	0.640
6.60	1.21	2.87	0.631
6.80	1.12	2.70	0.619
7.00	1.06	2.56	0.607
7.20	1.01	2.43	0.593
7.40	0.98	2.30	0.573
7.60	0.95	2.18	0.556
7.80	0.93	2.06	0.533
8.00	0.94	1.95	0.505
8.20	0.94	1.86	0.481
8.40	0.96	1.76	0.449
8.60	0.99	1.70	0.422
8.80	1.01	1.65	0.401
9.00	1.01	1.60	0.388
9.20	1.02	1.55	0.369
9.40	1.03	1.50	0.352
9.60	1.05	1.44	0.329
9.80	1.09	1.38	0.307
10.00	1.13	1.34	0.287
10.20	1.19	1.33	0.274
10.40	1.24	1.34	0.270
10.60	1.27	1.36	0.274
10.80	1.29	1.39	0.282
11.00	1.28	1.42	0.290
11.20	1.27	1.44	0.297
11.40	1.25	1.46	0.305
11.60	1.22	1.48	0.313
11.80	1.20	1.48	0.318
12.00	1.16	1.48	0.323
12.40	1.10	1.47	0.329
12.80	1.04	1.44	0.333
13.20	0.98	1.40	0.332
13.60	0.94	1.35	0.325
14.00	0.91	1.28	0.312
14.40	0.90	1.23	0.296
14.80	0.90	1.17	0.276
15.20	0.93	1.13	0.255
15.60	0.97	1.12	0.246
16.00	0.98	1.14	0.249
16.40	0.97	1.17	0.260
16.80	0.94	1.19	0.273
17.20	0.90	1.21	0.289
17.60	0.85	1.21	0.304
18.00	0.80	1.20	0.317
18.40	0.74	1.18	0.330
18.80	0.69	1.15	0.340
19.20	0.64	1.11	0.347
19.60	0.60	1.07	0.353
20.00	0.56	1.02	0.354
20.40	0.54	0.97	0.350
20.80	0.52	0.92	0.342
21.20	0.50	0.87	0.331
21.60	0.50	0.82	0.318

Energy (eV)	n	k	R (φ = 0)
22.00	0.49	0.77	0.303
22.40	0.49	0.73	0.287
22.80	0.49	0.69	0.272
23.20	0.49	0.66	0.263
23.60	0.48	0.62	0.252
24.00	0.49	0.57	0.234
24.40	0.50	0.53	0.213
24.80	0.51	0.49	0.191
25.20	0.53	0.46	0.171
25.60	0.55	0.43	0.150
26.00	0.57	0.40	0.132
26.40	0.59	0.38	0.117
26.80	0.61	0.37	0.105
27.00	0.62	0.36	0.099
27.50	0.64	0.34	0.085
28.00	0.67	0.32	0.073
28.50	0.69	0.31	0.065
29.00	0.71	0.30	0.057
29.50	0.73	0.30	0.052
30.00	0.75	0.29	0.047
31.00	0.78	0.29	0.042
32.00	0.79	0.29	0.040
33.00	0.82	0.28	0.033
34.00	0.84	0.29	0.032
35.00	0.85	0.31	0.033
36.00	0.85	0.32	0.036
37.00	0.84	0.33	0.039
38.00	0.83	0.33	0.040
39.00	0.81	0.33	0.042
40.00	0.80	0.33	0.045

Vanadium[9]

Energy (eV)	n	k	R (φ = 0)
0.10	12.83	45.89	0.978
0.20	3.90	24.30	0.975
0.28	2.13	17.35	0.973
0.36	1.54	13.32	0.966
0.44	1.28	10.74	0.957
0.52	1.16	8.93	0.945
0.60	1.10	7.59	0.929
0.68	1.07	6.54	0.909
0.76	1.08	5.67	0.882
0.80	1.10	5.30	0.864
0.90	1.18	4.50	0.811
1.00	1.34	3.80	0.730
1.10	1.60	3.26	0.632
1.20	1.93	2.88	0.543
1.30	2.25	2.71	0.498
1.40	2.48	2.72	0.491
1.50	2.57	2.79	0.499
1.60	2.57	2.84	0.507
1.70	2.52	2.88	0.512
1.80	2.45	2.88	0.515
1.90	2.36	2.85	0.514
2.00	2.34	2.81	0.509
2.10	2.31	2.78	0.506
2.20	2.28	2.80	0.510
2.30	2.23	2.83	0.516
2.40	2.15	2.88	0.528
2.50	2.02	2.91	0.540
2.60	1.89	2.92	0.552
2.70	1.74	2.89	0.561
2.80	1.61	2.85	0.569
2.90	1.48	2.80	0.577
3.00	1.36	2.73	0.582
3.20	1.16	2.55	0.585
3.40	0.99	2.37	0.586
3.60	0.87	2.17	0.575
3.80	0.80	1.96	0.547
4.00	0.78	1.76	0.503
4.20	0.80	1.60	0.449
4.40	0.83	1.47	0.400
4.60	0.87	1.38	0.355
4.80	0.90	1.31	0.326
5.00	0.91	1.26	0.304
5.25	0.93	1.18	0.271
5.50	0.94	1.14	0.258
5.75	0.96	1.09	0.235
6.00	0.98	1.06	0.223
6.25	0.97	1.02	0.212
6.50	0.97	0.98	0.199
6.75	0.97	0.94	0.185
7.00	0.98	0.91	0.175
7.33	0.97	0.89	0.170
7.66	0.98	0.87	0.162
8.00	0.98	0.85	0.155
8.33	0.98	0.81	0.146
8.66	0.98	0.81	0.145
9.00	0.96	0.79	0.142
9.50	0.94	0.77	0.136
10.00	0.91	0.74	0.133
10.50	0.89	0.71	0.126
11.00	0.87	0.65	0.112
11.50	0.88	0.58	0.091
12.00	0.90	0.58	0.089
12.50	0.89	0.57	0.086
13.00	0.88	0.55	0.082
13.50	0.87	0.53	0.079
14.00	0.86	0.51	0.075
14.50	0.86	0.49	0.070
15.00	0.86	0.47	0.065
15.50	0.86	0.46	0.062
16.00	0.85	0.45	0.061
16.50	0.84	0.43	0.059
17.00	0.84	0.41	0.056
17.50	0.83	0.40	0.054
18.00	0.82	0.38	0.051
18.50	0.82	0.37	0.048
19.00	0.82	0.35	0.045
19.50	0.82	0.34	0.043
20.00	0.81	0.32	0.041
20.50	0.81	0.31	0.038
21.00	0.81	0.29	0.036
21.50	0.81	0.28	0.033
22.00	0.81	0.27	0.032
22.50	0.81	0.25	0.029
23.00	0.82	0.24	0.027
23.50	0.82	0.23	0.025
24.00	0.82	0.22	0.024
24.50	0.83	0.21	0.022
25.00	0.83	0.20	0.020
25.50	0.83	0.19	0.019
26.00	0.83	0.18	0.018
26.50	0.84	0.17	0.016
27.00	0.84	0.16	0.015
27.50	0.85	0.16	0.014
28.00	0.85	0.15	0.013
28.50	0.86	0.14	0.012
29.00	0.86	0.14	0.011
29.50	0.86	0.13	0.010
30.00	0.87	0.13	0.009
31.00	0.88	0.12	0.008
32.00	0.90	0.11	0.007
33.00	0.90	0.10	0.005
34.00	0.91	0.10	0.005
35.00	0.92	0.09	0.004
36.00	0.94	0.10	0.004
37.00	0.94	0.10	0.004
38.00	0.95	0.11	0.004
39.00	0.95	0.12	0.004
40.00	0.95	0.13	0.005

Zirconium (Polycrystalline)[a][14]

Energy (eV)	n	k	R (φ = 0)
0.10	6.18	1.76	0.300
0.15	3.37	1.30	0.123
0.20	2.34	1.08	0.058
0.26	2.24	1.06	0.052
0.30	2.59	1.14	0.073
0.36	3.17	1.26	0.110
0.40	3.09	1.24	0.105
0.46	3.36	1.30	0.123
0.50	4.13	1.44	0.175
0.56	5.01	1.58	0.231
0.60	5.18	1.61	0.242
0.70	4.54	1.51	0.202
0.80	4.03	1.42	0.168
0.90	3.74	1.37	0.149
0.96	3.69	1.36	0.145
1.00	3.66	1.35	0.143
1.10	3.65	1.35	0.142
1.20	3.53	1.33	0.134
1.30	3.25	1.27	0.116
1.40	3.10	1.25	0.106
1.50	3.02	1.23	0.100
1.60	2.88	1.20	0.091
1.70	2.68	1.16	0.078
1.80	2.49	1.12	0.067
2.00	2.14	1.03	0.047
2.10	1.99	1.00	0.040
2.20	1.87	0.97	0.034
2.30	1.78	0.94	0.030
2.40	1.71	0.92	0.027
2.50	1.62	0.90	0.024
2.60	1.54	0.88	0.022
2.70	1.46	0.86	0.019
2.80	1.40	0.84	0.018
2.90	1.34	0.82	0.016
3.00	0.30	0.81	0.016
3.10	1.26	0.80	0.015
3.30	1.19	0.77	0.014
3.40	1.16	0.76	0.013
3.50	1.13	0.75	0.013
3.60	1.10	0.74	0.013
3.70	1.07	0.73	0.013
3.80	1.04	0.72	0.012
3.90	1.01	0.71	0.012
4.00	0.98	0.70	0.012
4.20	0.94	0.68	0.013
4.40	0.89	0.67	0.013
4.60	0.85	0.65	0.014
4.80	0.81	0.64	0.014
5.00	0.78	0.63	0.015
5.20	0.77	0.62	0.016
5.40	0.77	0.62	0.016
5.60	0.80	0.63	0.014
5.80	0.87	0.66	0.013
6.00	1.00	0.71	0.012
6.20	1.11	0.75	0.013
6.40	1.23	0.78	0.014
6.60	1.33	0.81	0.016
6.80	1.42	0.84	0.018
7.00	1.49	0.86	0.020
7.20	1.54	0.88	0.022
7.40	1.58	0.89	0.023
7.60	1.61	0.90	0.024
7.80	1.63	0.90	0.025
8.00	1.66	0.91	0.026
8.20	0.67	0.91	0.026
8.40	1.68	0.92	0.026
8.60	1.68	0.92	0.026
8.80	1.66	0.91	0.026
9.00	1.65	0.91	0.025
9.20	1.63	0.90	0.025
9.40	1.60	0.89	0.024
9.60	1.57	0.89	0.023
9.80	1.52	0.87	0.021
10.00	1.47	0.86	0.020
10.20	1.42	0.84	0.018
10.40	1.35	0.82	0.016
10.50	1.32	0.81	0.016
10.60	1.28	0.80	0.015
10.80	1.23	0.78	0.014
11.00	1.19	0.77	0.014
11.20	1.16	0.76	0.013
11.40	1.13	0.75	0.013
11.60	1.11	0.74	0.013
11.80	1.09	0.74	0.013
12.00	1.08	0.73	0.013
12.40	1.05	0.72	0.012
12.80	1.01	0.71	0.012
13.20	0.98	0.70	0.012
13.60	0.95	0.69	0.013
14.00	0.92	0.68	0.013
14.40	0.89	0.67	0.013
14.80	0.90	0.67	0.013

[a] Data for single crystal is in original publication.

15.20	0.92	0.68	0.013	22.00	1.20	0.77	0.014
15.60	0.95	0.69	0.013	22.60	1.15	0.76	0.013
16.00	0.98	0.70	0.012	23.00	1.12	0.75	0.013
16.40	1.01	0.71	0.012	23.60	1.08	0.73	0.013
16.80	1.04	0.72	0.012	24.00	1.05	0.73	0.013
17.20	1.09	0.74	0.013	24.60	1.02	0.71	0.012
17.60	1.13	0.75	0.013	25.00	1.00	0.71	0.012
18.00	1.17	0.76	0.014	25.60	0.97	0.69	0.012
18.40	1.21	0.78	0.014	26.00	0.95	0.69	0.013
18.80	1.24	0.79	0.014	26.60	0.91	0.67	0.013
19.20	1.27	0.80	0.015	27.00	0.88	0.66	0.013
19.60	1.29	0.80	0.015	27.60	0.84	0.65	0.014
20.00	1.30	0.81	0.015	28.00	0.83	0.64	0.014
20.60	1.29	0.80	0.015	28.60	0.82	0.64	0.014
21.00	1.27	0.80	0.013	29.00	0.81	0.64	0.014
21.60	1.23	0.78	0.014	29.60	0.82	0.64	0.014
				30.00	0.82	0.64	0.014

REFERENCES

1. Shiles, E., Sasaki, T., Inokuti, M., and Smith, D. Y., *Phys. Rev. Sect. B*, 22, 1612, 1980.
2. Bos, L. W. and Lynch, D. W., *Phys. Rev. Sect. B*, 2, 4567, 1970; Olson, C. G. and Lynch, D. W., unpublished.
3. Weaver, J. H., Colavita, E., Lynch, D. W., and Rosei, R., *Phys. Rev. Sect. B*, 19, 3850, 1979.
4. Hagemann, H. J., Gudat, W., and Kunz, C., *J. Opt. Soc. Am.*, 65, 742, 1975.
5. Olson, C. G., Lynch, D. W., and Weaver, J. H., unpublished.
6. Lynch, D. W., Olson, C. G., and Weaver, J. H., unpublished. Supersedes those data by the same authors in *Phys. Rev. Sect. B*, 11, 3617, 1975.
7. Weaver, J. H., Olson, C. G., and Lynch, D. W., *Phys. Rev. Sect. B*, 15, 4115, 1977.
8. Johnson, P. B. and Christy, R. W., *Phys. Rev. Sect. B*, 9, 5056, 1974.
9. Weaver, J. H., Lynch, D. W., and Olson, C. G., *Phys. Rev. Sect. B*, 10, 501, 1973.
10. Lynch, D. W., Rosei, R., and Weaver, J. H., *Solid State Commun.*, 9, 2195, 1971.
11. Weaver, J. H., Lynch, D. W., and Olson, C. G., *Phys. Rev. Sect. B*, 7, 4311, 1973.
12. Weaver, J. H. and Benbow, R. L., *Phys. Rev. Sect. B*, 12, 3509, 1975.
13. Weaver, J. H., *Phys. Rev. Sect. B*, 11, 1416, 1975.
14. Lynch, D. W., Olson, C. G., and Weaver, J. H., *Phys. Rev. Sect. B*, 11, 3671, 1975.
15. Weaver, J. H., Lynch, D. W., and Olson, C. G., *Phys. Rev. Sect. B*, 12, 1293, 1975.

REFLECTION COEFFICIENTS

Coefficients of Reflection of Miscellaneous Surfaces for
Monochromatic Radiation in the Visible Spectrum
(J. L. Michaelson)

	Wave lengths (μ)			
Material	0.400	0.500	0.600	0.700
Carbon black in oil	0.003	0.003	0.003	0.003
Clay				
Kaolin (treated)	0.82	0.81	0.82	0.82
Kaolin (untreated)	0.75	0.79	0.85	0.86
White georgia	0.94	0.92	0.93	0.94
Magnesium oxide	0.97	0.98	0.98	0.98
Paint				
Lithopone	0.95	0.98	0.98	0.98
$MgCO_3$-Vynal acetate lacquer	0.90	0.88	0.88	0.88
ZnO-Milk	0.74	0.84	0.85	0.86
Paper				
Blotting	0.64	0.72	0.79	0.79
Calendered	0.64	0.69	0.73	0.76
Crepe, green	0.23	0.49	0.19	0.48
Crepe, red	0.03	0.02	0.21	0.69
Crepe, yellow	0.17	0.44	0.75	0.79
News print stock	0.38	0.61	0.63	0.78
Peach				
Green	0.18	0.17	0.62	0.63
Ripe	0.10	0.10	0.41	0.42
Pear				
Green	0.04	0.12	0.29	0.41
Ripe	0.08	0.19	0.46	0.53
Pigment				
Chrome yellow	0.05	0.13	0.70	0.77
French ochre	0.06	0.14	0.50	0.56
Porcelain enamel				
Blue	0.44	0.10	0.05	0.23
Orange	0.09	0.09	0.59	0.69
Red	0.05	0.03	0.08	0.62
White	0.77	0.73	0.72	0.70
Yellow	0.11	0.46	0.62	0.62
Talcum, Italian	0.94	0.89	0.88	0.88
Wheat flour	0.75	0.87	0.94	0.97

REFLECTION COEFFICIENTS OF SURFACES FOR "INCANDESCENT" LIGHT

Material	Nature of Surface	Coefficient	Authority
Aluminum, "Alzak"	Diffusing	0.77—0.81	3
"Alzak"	Specular	0.79—0.83	3
On glass	First surface	0.82—0.86	4
Polished	Specular	0.69	3
Black paper	Diffusing	0.05—0.06	4
Chromium	Specular	0.62	4
Copper	Specular	0.63	4
Gold	Specular	0.75	1
Magnesium oxide	Diffusing	0.98	5
Nickel	Specular	0.62—0.64	1,3
Platinum	Specular	0.62	1
Porcelain enamel	Glossy	0.76—0.79	3
Porcelain enamel	Ground	0.81	3
Porcelain enamel	Matt.	0.72—0.76	3
Silver	Polished	0.93	1
Silvered glass	Second surface	0.88—0.93	3
Snow	Diffusing	0.93	2
Steel	Specular	0.55	1
Stellite	Specular	0.58—0.65	4

(1) Hagen and Rubena; (2) Nutting, Jones, and Elliot; (3) J. E. Bock; (4) Frank Benford; (5) J. L. Michaelson.

EMISSIVITY AND ABSORPTION

These data are the result of investigations made by the Bureau of Standards, the British National Physical Laboratory, General Electric Research Laboratories, and several eastern universities, and were collected by W. J. King of the General Electric Company.

Low Temperature Total Emissivities

Silver, highly polished	0.02	Brass, polished	0.60
Platinum, highly polished	0.05	Oxidized copper	0.60
Zinc, highly polished	0.05	Oxidized steel	0.70
Aluminum, highly polished	0.08	Bronze paint	0.80
Monel metal, polished	0.09	Black gloss paint	0.90
Nickel, polished	0.12	White lacquer	0.95
Copper, polished	0.15	White vitreous enamel	0.95
Stellite, polished	0.18	Asbestos paper	0.95
Cast iron, polished	0.25	Green paint	0.95
Monel metal, oxidized	0.43	Gray paint	0.95
Aluminum paint	0.55	Lamp black	0.95

Coefficient of Absorption of Solar Radiation

Silver, highly polished	0.07	Stellite, polished	0.30
Platinum, highly polished	0.10	Light cream paint	0.35
Nickel, highly polished	0.15	Monel metal, polished	0.40
Aluminum*	0.15	Light yellow paint	0.45
Magnesium carbonate	0.15	Light green paint	0.50
Zinc oxide	0.15	Aluminum paint	0.55
Steel*	0.20	Zinc, polished metal	0.55
Copper	0.25	Gray paint	0.75
White lead paint	0.25	Black matte	0.97
Zinc oxide paint	0.30		

* Questionable because of scant or inconsistent data.

EMISSIVITY OF TOTAL RADIATION, ε_{tot}, FOR VARIOUS MATERIALS

Material	Temperature (°C)	ε_{tot}	Material	Temperature (°C)	ε_{tot}
Alloys			**Lamp black**	25	0.94
Nickel-Chromium				20—400	0.96
20 Ni—25 Cr—55 Fe, oxidized	200	0.90	**Lead**		
	500	0.97	Oxidized	200	0.05
60 Ni—12 Cr—28 Fe, oxidized	270	0.89	Unoxidized	200	0.63
	560	0.82	**Mercury**		
80 Ni—20 Cr	100	0.87	Unoxidized	25	0.10
	600	0.87		100	0.12
	1300	0.89	**Molybdenum**	600—1000	0.08—0.13
Aluminum				1500—2200	0.19—0.26
Polished	50—500	0.04—0.06	**Monel metal**		
Rough surface	20—50	0.06—0.07	Oxidized		
Strongly oxidized	55—500	0.2—0.3		200	0.43
	25	0.022		600	0.43
	100	0.028	**Nichrome**		
	500	0.060	Wire		
Oxidized	200	0.11	Clean	50	0.65
	600	0.19		500—1000	0.71—0.79
Asbestos board	20	0.96	Oxidized	50—500	0.95—0.98
Bismuth			**Nickel**		
Unoxidized	25	0.048	Industrial, polished	200—400	0.07—0.09
Brass			Oxidized	200	0.37
Dull tarnished	200	0.61	Oxidized at 600°C	200—600	0.37—0.48
Oxidized at 600° C	200	0.61	Unoxidized	25	0.045
	600	0.59		100	0.06
Unoxidized	25	0.035		500	0.12
	100	0.035		1000	0.19
Polished	200	0.03	**Platinum**		
Rolled sheet	20	0.06	Clean, polished	200—600	0.05—0.1
Bronze			Unoxidized	25	0.037
Polished	50	0.1		100	0.047
Carbon				500	0.096
Filament	1000—1400	0.53		1000	0.152
Graphite	0—3600	0.7—0.8		1500	0.191
Lamp black	20—400	0.96	Wire	50—200	0.06—0.07
Soot applied to solid	50—1000	0.96		500—1000	0.1—0.16
Soot with water glass	20—200	0.96		1100	0.18
Unoxidized	100	0.81	**Porcelain**		
Chromium			Glazed	20	0.92
Polished	50	0.1	**Rubber**		
	500—1000	0.28—0.38	Hard	20	0.95
Colbalt			Soft, gray, rough	20	0.86
Unoxidized	500	0.13	**Silica brick**	1000	0.80
	1000	0.23		1100	0.85
Columbium			**Silver**		
Unoxidized	1500	0.19	Clean, polished	200—600	0.02—0.03
Copper			Unoxidized	100	0.02
Calorized	100	0.26		500	0.035
Calorized, oxidized	200	0.18	**Soot applied to a solid surface**	50—1000	0.94—0.91
	600	0.19	**Soot with water glass**	20—200	0.96
Commercial, scoured to a shine	20	0.07	**Steel**		
Oxidized	50	0.6—0.7	Alloyed (8% Ni, 18% Cr)	500	0.35
	500	0.88	Aluminized	50—500	0.79
Polished	50—100	0.02	Dull nickel plated	20	0.11
Unoxidized	100	0.02	Flat, rough surface	50	0.95—0.98
Unoxidized, liquid	—	0.15	Cast, polished	750—1050	0.52—0.56
Fire brick	1000	0.75	Sheet, ground	50	0.56
Glass	20—100	0.94—0.91		950—1100	0.55—0.61
	250—1000	0.87—0.72	Oxidized	200—600	0.8
	1100—1500	0.7—0.67	Calorized, oxidized	200	0.52
Gold				600	0.57
Carefully polished	200—600	0.02—0.03	Sheet with shiny layer of oxide	20	0.82
Unoxidized	100	0.02	Strongly oxidized	50	0.88
Enamel	100	0.37		500	0.98
Graphite	0—3600	0.7—0.8	Unoxidized	100	0.08
Gypsum	20	0.93	Unoxidized, liquid	—	0.28
Iron			**Tantalum**		
Cast			Unoxidized	1500	0.21
Oxidized	200	0.64		2000	0.26
	600	0.78	**Tungsten**		
Strongly oxidized	40	0.95	Unoxidized	25	0.024
	250	0.95		100	0.032
Unoxidized	100	0.21		500	0.071
Unoxidized, liquid	—	0.29		1000	0.15
Oxidized	100	0.74		1500	0.23
	500	0.84		2000	0.28
	1200	0.89	**Varnish**	40—100	0.8—0.95
Rusted	25	0.65	Dull black	40—100	0.96—0.98
Wrought, dull	100	0.05	Glossy black sprayed on iron	20	0.87
			Zinc		
			Polished	200—300	0.04—0.05
			Unoxidized	300	0.05

SPECTRAL EMISSIVITY

Prepared by Roeser and Wensel, National Bureau of Standards
Spectral Emissivity of Materials, Surface Unoxidized for 0.65μ

Element	Solid	Liquid	Element	Solid	Liquid
Beryllium	0.61	0.61	Thorium	0.36	0.40
Carbon	0.80—0.93	—	Titanium	0.63	0.65
Chromium	0.34	0.39	Tungsten	0.43	—
Cobalt	0.36	0.37	Uranium	0.54	0.34
Columbium	0.37	0.40	Vanadium	0.35	0.32
Copper	0.10	0.15	Yttrium	0.35	0.35
Erbium	0.55	0.38	Zirconium	0.32	0.30
Gold	0.14	0.22	Steel	0.35	0.37
Iridium	0.30	—	Cast Iron	0.37	0.40
Iron	0.35	0.37	Constantan	0.35	—
Manganese	0.59	0.59	Monel	0.37	—
Molybdenum	0.37	0.40	Chromel P (90Ni-10Cr)	0.35	—
Nickel	0.36	0.37	80Ni-20Cr	0.35	—
Palladium	0.33	0.37	60Ni-24Fe-16Cr	0.36	—
Platinum	0.30	0.38	Alumel (95Ni; Bal. Al,		
Rhodium	0.24	0.30	Mn, Si)	0.37	—
Silver	0.07	0.07	90Pt-10Rh	0.27	
Tantalum	0.49				

SPECTRAL EMISSIVITY OF OXIDES

The emissivity of oxides and oxidized metals depends to a large extent upon the roughness of the surface. In general, higher values of emissivity are obtained on the rougher surfaces.

Material	Range of observed values	Probable value for oxide formed on smooth metal
Oxide		
Aluminum	0.22—0.40	0.30
Beryllim	0.07—0.37	0.35
Cerium	0.58—0.80	—
Chromium	0.60—0.80	0.70
Cobalt	—	0.75
Columbium	0.55—0.71	0.70
Copper	0.60—0.80	0.70
Iron	0.63—0.98	0.70
Magnesium	0.10—0.43	0.20
Nickel	0.85—0.96	0.90
Thorium	0.20—0.57	0.50
Tin	0.32—0.60	—
Titanium	—	0.50
Uranium	—	0.30
Vanadium	—	0.70
Yttrium	—	0.60
Zirconium	0.18—0.43	0.40
Oxidized		
Alumel	—	0.87
Cast Iron	—	0.70
Chromel P (90Ni-10Cr)	—	0.87
80Ni-20Cr	—	0.90
60Ni-24Fe-16Cr	—	0.83
55Fe-37.5Cr-7.5Al	—	0.78
70Fe-23Cr-5Al-2Co	—	0.75
Constantan (55Cu-45Ni)	—	0.84
Carbon Steel	—	0.80
Stainless Steel (18-8)	—	0.85
Porcelain	0.25—0.50	—

PROPERTIES OF TUNGSTEN

Jones and Langmuir, General Electric Review

Temp. °K	Resistivity microhm cm	Electron emission amp./cm²	Evaporation g/cm² sec	Vapor pressure dynes/cm²	Thermal expansion per cent l_0 at 293°	Atomic heat cal./g. atom./°C.
300	5.65	—	—	—	.003	6.0
400	8.06	—	—	—	.044	6.0
500	10.56	—	—	—	.086	6.1
600	13.23	—	—	—	.130	6.1
700	16.09	—	—	—	.175	6.2
800	19.00	—	—	—	.222	6.2
900	21.94	—	—	—	.270	6.3
1000	24.93	1.07×10^{-15}	5.32×10^{-34}	1.98×10^{-29}	.320	6.4
1100	27.94	1.52×10^{-13}	2.17×10^{-30}	1.22×10^{-25}	.371	6.4
1200	30.98	9.73×10^{-12}	3.21×10^{-27}	1.87×10^{-22}	.424	6.5
1300	34.08	3.21×10^{-10}	1.35×10^{-24}	8.18×10^{-20}	.479	6.7
1400	37.19	6.62×10^{-9}	2.51×10^{-22}	1.62×10^{-17}	.535	6.8
1500	40.36	1.14×10^{-7}	2.37×10^{-20}	1.54×10^{-15}	.593	7.0
1600	43.55	9.27×10^{-7}	1.25×10^{-18}	8.43×10^{-14}	.652	7.1
1700	46.78	7.08×10^{-6}	4.17×10^{-17}	2.82×10^{-12}	.713	7.2
1800	50.05	4.47×10^{-5}	8.81×10^{-16}	6.31×10^{-11}	.775	7.4
1900	53.35	2.28×10^{-4}	1.41×10^{-14}	1.01×10^{-9}	.839	7.6
2000	56.67	1.00×10^{-3}	1.76×10^{-13}	1.33×10^{-8}	.904	7.7
2100	60.06	3.93×10^{-3}	1.66×10^{-12}	1.28×10^{-7}	.971	7.8
2200	63.48	1.33×10^{-2}	1.25×10^{-11}	9.88×10^{-7}	1.039	8.0
2300	66.91	4.07×10^{-2}	8.00×10^{-11}	6.47×10^{-6}	1.109	8.2
2400	70.39	1.16×10^{-1}	4.26×10^{-10}	3.52×10^{-5}	1.180	8.3
2500	73.91	2.98×10^{-1}	2.03×10^{-9}	1.71×10^{-4}	1.253	8.4
2600	77.49	7.16×10^{-1}	8.41×10^{-9}	7.24×10^{-4}	1.328	8.6
2700	81.04	1.63	3.19×10^{-8}	2.86×10^{-3}	1.404	8.7
2800	84.70	3.54	1.10×10^{-7}	9.84×10^{-3}	1.479	8.9
2900	88.33	7.31	3.30×10^{-7}	3.00×10^{-2}	1.561	9.0
3000	92.04	1.42×10	9.95×10^{-7}	9.20×10^{-2}	1.642	9.2
3100	95.76	2.64×10	2.60×10^{-6}	2.50×10^{-1}	1.724	9.4
3200	99.54	4.78×10	6.38×10^{-6}	6.13×10^{-1}	1.808	9.5
3300	103.3	8.44×10	1.56×10^{-5}	1.51	1.893	9.6
3400	107.2	1.42×10^2	3.47×10^{-5}	3.41	1.980	9.8
3500	111.1	2.33×10^2	7.54×10^{-5}	7.52	2.068	9.9
3600	115.0	3.73×10^2	1.51×10^{-4}	1.53×10	2.158	10.1
3655	117.1	4.79×10^2	2.28×10^{-4}	2.33×10	2.209	10.2

Roeser and Wensel, National Bureau of Standards

Temp. °K	Normal brightness new candles per cm²	Spectral emissivity 0.65μ	0.467μ	Color emissivity	Total emissivity	Brightness temp. 0.65μ	Color temp.
300	—	0.472	0.505	—	0.032	—	—
400	—	—	—	—	.042	—	—
500	—	—	—	—	.053	—	—
600	—	—	—	—	.064	—	—
700	—	—	—	—	.076	—	—
800	—	—	—	—	.088	—	—
900	—	—	—	—	.101	—	—
1000	0.0001	.458	.486	.395	.114	966	1007
1100	0.001	.456	.484	.392	.128	1059	1108
1200	0.006	.454	.482	.390	.143	1151	1210
1300	0.029	.452	.480	.387	.158	1242	1312
1400	0.11	.450	.478	.385	.175	1332	1414
1500	0.33	.448	.476	.382	.192	1422	1516
1600	0.92	.446	.475	.380	.207	1511	1619
1700	2.3	.444	.473	.377	.222	1599	1722
1800	5.1	.442	.472	.374	.236	1687	1825
1900	10.4	.440	.470	.371	.249	1774	1928
2000	20.0	.438	.469	.368	.260	1861	2032
2100	36	.436	.467	.365	.270	1946	2136
2200	61	.434	.466	.362	.279	2031	2241
2300	101	.432	.464	.359	.288	2115	2345
2400	157	.430	.463	.356	.296	2198	2451
2500	240	.428	.462	.353	.303	2280	2556
2600	350	.426	.460	.349	.311	2362	2662
2700	500	.424	.459	.346	.318	2443	2769
2800	690	.422	.458	.343	.323	2523	2876
2900	950	.420	.456	.340	.329	2602	2984
3000	1260	.418	.455	.336	.334	2681	3092
3100	1650	.416	.454	.333	.337	2759	3200
3200	2100	.414	.452	.330	.341	2837	3310
3300	2700	.412	.451	.326	.344	2913	3420
3400	3400	.410	.450	.323	.348	2989	3530
3500	4200	.408	.449	.320	.351	3063	3642
3600	5200	.406	.447	.317	.354	3137	3754

TRANSMISSION OF CORNING COLORED FILTERS

Supplied by R. G. Saxton

If I_o is the intensity of radiation entering a layer of some medium and I the intensity reaching the opposite surface, the ratio I/I_o is called the transmittance. In practice the ratio of intensity of radiation passing through a glass sample to that incident on its surface is often measured and plotted as transmission. The transmission is the result of two factors, the transmittance of the glass and the losses by reflection. These losses amount to about 4 % for each glass-air surface; the transmission of a sample is about 92 % of its transmittance. Since the reflection losses differ slightly with different samples, the correction is often determined and applied when the transmission is measured. Values in this table have been corrected for reflection losses.

The identifying glass number, CS number, color and properties, and nominal thickness for the Corning glasses in this table are:

Glass No.	CS	Color and properties	Nominal thickness
0160	0-54	Clear; Ultraviolet transmitting	2.0
2030	2-64	Red; Sharp cut	3.0
2403	2-58	Red; Sharp cut	3.0
2404	2-59	Red; Sharp cut	3.0
2408	2-60	Red; Sharp cut	3.0
2412	2-61	Red; Sharp cut	3.0
2418	2-62	Red; Sharp cut	3.0
2424	2-63	Red; Sharp cut	3.0
2434	2-73	Red; Sharp cut	3.0
2540	7-56	Black; IR transmitting; Visible absorbing	2.5
2550	7-57	Black; IR transmitting; Visible absorbing	2.0
2600	7-69	Black; IR transmitting; Visible absorbing	3.0
3060	3-75	Straw	2.0
3304	3-76	Dark amber	3.0
3307	3-77	Dark amber	3.0
3384	3-70	Yellow	3.0
3385	3-71	Yellow	3.0
3387	3-72	Straw	3.0
3389	3-73	Straw	3.0
3391	3-74	Straw	3.0
3480	3-66	Yellow; Sharp cut	3.0
3482	3-67	Yellow; Sharp cut	3.0
3484	3-68	Yellow; Sharp cut	3.0
3486	3-69	Yellow; Sharp cut	3.0
3718	3-94	Yellow	3.0
3750	3-79	Yellow; Yellow green fluorescing	3.0
3780	3-00	Yellow	2.0
3850	0-51	Clear; UV transmitting	4.0
3961	1-56	Bluish; IR absorbing; Visible transmitting	2.5
3962	1-57	Bluish; IR absorbing; Visible transmitting	2.5
3965	1-58	Bluish; IR absorbing; Visible transmitting	2.5
3966	1-59	Bluish; IR absorbing; Visible transmitting	2.5
4010	4-64	Green	4.0
4015	4-65	Yellow green	3.0
4060	4-67	Green	4.0
4084	4-68	Green	4.5
4303	4-72	Blue green	4.0
4305	4-71	Blue green	4.0
4308	4-70	Blue green	4.0
4309	4-69	Blue green	4.0
4445	4-74	Green	2.5
4602	1-75	Bluish; IR absorbing; Visible transmitting	3.0
4784	4-94	Blue green	5.0
5030	5-57	Blue	5.0
5031	5-56	Blue	4.5
5070	7-62	Amethyst	3.9
5071	7-63	Amethyst	3.9
5073	7-64	Amethyst	3.9
5113	5-58	Blue	4.0
5120	1-60	Smoky violet; Absorbs yellow	5.2
5300	4-106	Green	3.9
5330	1-64	Blue	4.5
5433	5-59	Blue	5.0
5543	5-60	Blue	5.0
5562	5-61	Blue	5.0
5572	1-61	Blue	5.0
5840	7-60	Black; UV transmitting; Visible absorbing	4.5
5850	7-59	Purple; UV transmitting; Visible absorbing	4.0
5860	7-37	Black; UV transmitting; Visible absorbing	5.0
5874	7-39	Black; UV transmitting; Visible absorbing	5.0
5900	1-62	Blue	5.5
5970	7-51	Black; UV transmitting; Visible absorbing	5.0
7380	0-52	Clear; UV transmitting	2.0
7740	0-53	Clear; UV transmitting	2.0
7905	9-30	Clear; UV transmitting; Long Range IR transmitting	2.0
7910	9-54	Clear; UV transmitting	2.0
8364	7-98	Gray	2.0
9780	4-76	Blue green	5.0
9782	4-96	Blue green	5.0
9788	4-97	Blue green	5.0
9830	4-77	Green	3.4
9863	7-54	Black; UV transmitting; Visible absorbing	3.0

TRANSMISSION OF CORNING COLORED FILTERS (Continued)

Transmittance

$\lambda(\mu)$	0160	2030	2403	2404	2408	2412	2418	2424	2434	2540	2550	2600
						Corning Glass Number						
.22	.000	.000	.000	.000	.000	.000	.000	.000	.000	.000	.000	.000
.24	.000	.000	.000	.000	.000	.000	.000	.000	.000	.000	.000	.000
.26	.000	.000	.000	.000	.000	.000	.000	.000	.000	.000	.000	.000
.28	.000	.000	.000	.000	.000	.000	.000	.000	.000	.000	.000	.000
.30	.005	.000	.000	.000	.000	.000	.000	.000	.000	.000	.000	.000
.32	.642	.000	.000	.000	.000	.000	.000	.000	.000	.000	.000	.000
.34	.850	.000	.000	.000	.000	.000	.000	.000	.000	.000	.000	.000
.36	.882	.000	.000	.000	.000	.000	.000	.000	.000	.000	.000	.000
.38	.890	.000	.000	.000	.000	.000	.000	.000	.000	.000	.000	.000
.40	.892	.000	.000	.000	.000	.000	.000	.000	.000	.000	.000	.000
.41	.893	.000	.000	.000	.000	.000	.000	.000	.000	.000	.000	.000
.42	.896	.000	.000	.000	.000	.000	.000	.000	.000	.000	.000	.000
.43	.896	.000	.000	.000	.000	.000	.000	.000	.000	.000	.000	.000
.44	.898	.000	.000	.000	.000	.000	.000	.000	.000	.000	.000	.000
.45	.899	.000	.000	.000	.000	.000	.000	.000	.000	.000	.000	.000
.46	.900	.000	.000	.000	.000	.000	.000	.000	.000	.000	.000	.000
.47	.900	.000	.000	.000	.000	.000	.000	.000	.000	.000	.000	.000
.48	.900	.000	.000	.000	.000	.000	.000	.000	.000	.000	.000	.000
.49	.900	.000	.000	.000	.000	.000	.000	.000	.000	.000	.000	.000
.50	.900	.000	.000	.000	.000	.000	.000	.000	.000	.000	.000	.000
.51	.900	.000	.000	.000	.000	.000	.000	.000	.000	.000	.000	.000
.52	.900	.000	.000	.000	.000	.000	.000	.000	.000	.000	.000	.000
.53	.900	.000	.000	.000	.000	.000	.000	.000	.000	.000	.000	.000
.54	.900	.000	.000	.000	.000	.000	.000	.000	.000	.000	.000	.000
.55	.900	.000	.000	.000	.000	.000	.000	.000	.000	.000	.000	.000
.56	.901	.000	.000	.000	.000	.000	.000	.000	.000	.000	.000	.000
.57	.904	.000	.000	.000	.000	.000	.000	.000	.005	.000	.000	.000
.58	.904	.000	.000	.000	.000	.000	.000	.005	.200	.000	.000	.000
.59	.908	.000	.000	.000	.000	.000	.008	.170	.615	.000	.000	.000
.60	.910	.000	.000	.000	.000	.006	.250	.575	.808	.000	.000	.000
.61	.910	.000	.000	.000	.018	.190	.660	.790	.856	.000	.000	.000
.62	.910	.000	.000	.015	.265	.625	.822	.848	.872	.000	.000	.000
.63	.910	.000	.018	.295	.670	.828	.870	.881	.000	.000	.000
.64	.910	.006	.260	.660	.828	.868	.874	.880	.887	.000	.001	.000
.65	.910	.028	.675	.796	.866	.881	.881	.887	.892	.000	.003	.000
.66	.910	.110	.838	.828	.877	.885	.885	.893	.895	.000	.005	.000
.67	.910	.305	.871	.842	.883	.887	.887	.897	.897	.000	.006	.000
.68	.910	.550	.880	.847	.886	.889	.889	.900	.899	.000	.009	.000
.69	.910	.735	.885	.851	.888	.900	.900	.901	.900	.000	.012	.000
.70	.910	.820	.886	.852	.888	.900	.900	.903	.900	.000	.017	.000
.71	.910	.853	.888	.854	.888	.889	.889	.903	.900	.000	.023	.000
.72	.910	.864	.889	.853	.888	.888	.888	.903	.899	.000	.031	.040
.73	.910	.867	.900	.851	.887	.887	.887	.903	.897	.000	.041	.175
.74	.910	.867	.900	.850	.885	.886	.886	.903	.896	.000	.055	.372
.75	.910	.866	.900	.849	.884	.885	.885	.902	.895	.000	.069	.547
.80	.910	.839	.875	.827	.870	.858	.857	.881	.866	.005	.225	.770
1.00	.912	.801	.840	.772	.840	.828	.822	.857	.842	.562	.780	.350
1.20	.908	.799	.845	.786	.845	.837	.827	.859	.849	.790	.870	.000
1.40	.909	.811	.854	.809	.854	.848	.840	.862	.857	.850	.895	.000
1.60	.913	.839	.873	.837	.873	.869	.858	.880	.879	.872	.904	.000
1.80	.909	.844	.870	.829	.870	.864	.854	.877	.872	.880	.900	.000
2.00	.904	.841	.868	.827	.868	.861	.851	.874	.871	.880	.897	.000
2.20	.888	.833	.820	.773	.825	.820	.810	.837	.835	.860	.875	.005
2.40	.875	.832	.803	.757	.809	.803	.792	.818	.812	.868	.870	.049
2.60	.868	.822	.750	.695	.754	.750	.723	.772	.767	.858	.850	.058
2.80	.690	.600	.100	.100	.100	.100	.050	.050	.260	.450	.480	.030
3.00	.630	.470	.070	.020	.070	.070	.072	.122	.400	.465	.383	.022
3.20	.500	.340	.140	.074	.150	.140	.142	.180	.470	.390	.330	.020
3.40	.379	.260	.140	.078	.140	.140	.120	.150	.350	.310	.245	.018
3.60	.320	.247	.000	.000	.000	.000	.000	.006	.015	.280	.220	.021
3.80	.310	.257	.000	.000	.000	.000	.000	.000	.020	.285	.250	.035
4.00	.311	.274	.000	.000	.000	.000	.000	.000	.015	.285	.275	.068
4.20	.251	.200	.000	.000	.000	.000	.000	.000	.017	.190	.190	.065
4.40	.110	.060	.000	.000	.000	.000	.000	.000	.002	.050	.100	.020
4.60	.012	.000	.000	.000	.000	.000	.000	.000	.000	.000	.000	.000
4.80	.004	.000	.000	.000	.000	.000	.000	.000	.000	.000	.000	.000
5.00	.000	.000	.000	.000	.000	.000	.000	.000	.000	.000	.000	.000

Transmittance

λ(μ)	Corning Glass Number											
	3060	3304	3307	3384	3385	3387	3389	3391	3480	3482	3484	3486
.22	.000	.000	.000	.000	.000	.000	.000	.000	.000	.000	.000	.000
.24	.000	.000	.000	.000	.000	.000	.000	.000	.000	.000	.000	.000
.26	.000	.000	.000	.000	.000	.000	.000	.000	.000	.000	.000	.000
.28	.000	.000	.000	.000	.000	.000	.000	.000	.000	.000	.000	.000
.30	.000	.000	.000	.000	.000	.000	.000	.000	.000	.000	.000	.000
.32	.000	.000	.000	.000	.000	.000	.005	.000	.000	.000	.000	.000
.34	.000	.000	.038	.000	.000	.000	.010	.000	.000	.000	.000	.000
.36	.000	.000	.050	.000	.005	.000	.015	.000	.000	.000	.000	.000
.38	.060	.000	.027	.005	.010	.010	.020	.000	.000	.000	.000	.000
.40	.410	.000	.016	.011	.016	.020	.026	.075	.000	.000	.000	.005
.41	.517	.000	.014	.011	.016	.020	.025	.425	.000	.000	.000	.005
.42	.604	.000	.014	.010	.015	.019	.105	.655	.000	.000	.000	.005
.43	.665	.000	.016	.009	.013	.017	.437	.747	.000	.000	.000	.005
.44	.710	.000	.022	.005	.011	.050	.620	.801	.000	.000	.000	.005
.45	.748	.000	.033	.003	.010	.325	.714	.838	.000	.000	.000	.005
.46	.778	.000	.049	.002	.008	.565	.780	.860	.000	.000	.000	.004
.47	.800	.000	.070	.001	.060	.690	.820	.874	.000	.000	.000	.003
.48	.819	.000	.101	.005	.410	.763	.848	.884	.000	.000	.000	.003
.49	.836	.003	.143	.088	.640	.803	.866	.890	.000	.000	.000	.002
.50	.850	.009	.193	.350	.727	.834	.878	.895	.000	.000	.000	.001
.51	.860	.019	.250	.595	.780	.854	.886	.898	.000	.000	.000	.045
.52	.870	.037	.315	.725	.817	.868	.890	.900	.000	.000	.003	.425
.53	.875	.063	.379	.789	.840	.876	.892	.901	.000	.000	.175	.710
.54	.881	.102	.447	.825	.856	.883	.894	.902	.000	.015	.600	.792
.55	.884	.146	.504	.846	.866	.887	.895	.903	.000	.230	.774	.823
.56	.886	.200	.560	.860	.873	.889	.894	.902	.020	.675	.818	.844
.57	.885	.255	.607	.869	.876	.890	.893	.901	.325	.850	.839	.859
.58	.883	.310	.648	.873	.878	.889	.892	.900	.710	.885	.854	.868
.59	.882	.360	.680	.876	.877	.887	.890	.898	.829	.894	.865	.876
.60	.882	.404	.705	.877	.877	.884	.886	.896	.858	.900	.873	.882
.61	.882	.438	.722	.877	.877	.881	.884	.893	.869	.903	.880	.886
.62	.882	.466	.735	.875	.876	.875	.880	.890	.876	.905	.885	.888
.63	.882	.100	.744	.871	.874	.871	.876	.886	.881	.906	.888	.889
.64	.883	.505	.748	.865	.872	.866	.872	.884	.884	.907	.890	.890
.65	.885	.519	.750	.860	.867	.860	.868	.881	.885	.908	.892	.890
.66	.886	.531	.750	.856	.863	.856	.865	.876	.886	.908	.893	.890
.67	.888	.543	.749	.850	.858	.851	.860	.873	.886	.908	.894	.890
.68	.890	.552	.745	.844	.853	.846	.856	.869	.886	.908	.893	.889
.69	.891	.561	.740	.837	.847	.839	.852	.865	.885	.907	.892	.887
.70	.892	.569	.734	.831	.842	.834	.847	.860	.884	.907	.891	.885
.71	.893	.574	.727	.825	.837	.827	.842	.856	.882	.906	.890	.883
.72	.893	.575	.720	.819	.831	.822	.837	.852	.880	.905	.888	.880
.73	.892	.576	.712	.813	.825	.816	.831	.848	.877	.905	.886	.877
.74	.891	.574	.702	.807	.820	.810	.826	.844	.874	.904	.885	.875
.75	.890	.570	.694	.800	.814	.805	.820	.840	.870	.903	.882	.873
.80	.871	.526	.642	.770	.865	.780	.775	.815	.837	.878	.846	.829
1.00	.830	.435	.516	.715	.830	.725	.716	.772	.801	.857	.811	.781
1.20	.860	.429	.500	.718	.858	.735	.730	.782	.807	.859	.819	.793
1.40	.901	.475	.540	.750	.900	.768	.768	.810	.828	.870	.837	.817
1.60	.917	.580	.635	.795	.918	.812	.812	.843	.852	.884	.856	.841
1.80	.916	.627	.675	.808	.915	.818	.820	.849	.847	.882	.852	.834
2.00	.908	.620	.668	.800	.909	.822	.817	.846	.847	.884	.852	.835
2.20	.900	.630	.675	.802	.900	.823	.810	.840	.805	.865	.829	.798
2.40	.885	.651	.690	.800	.885	.825	.811	.842	.787	.853	.817	.777
2.60	.860	.650	.690	.785	.858	.815	.800	.840	.725	.818	.757	.718
2.80	.550	.345	.390	.325	.670	.360	.340	.440	.050	.060	.110	.060
3.00	.379	.320	.360	.318	.348	.348	.322	.423	.080	.088	.190	.088
3.20	.315	.240	.290	.268	.332	.324	.290	.395	.150	.158	.270	.145
3.40	.250	.151	.190	.218	.289	.288	.255	.353	.120	.132	.140	.090
3.60	.231	.130	.150	.217	.266	.280	.249	.351	.000	.000	.000	.000
3.80	.258	.140	.160	.228	.270	.298	.267	.376	.000	.000	.000	.000
4.00	.283	.140	.175	.220	.280	.290	.260	.365	.000	.000	.000	.000
4.20	.200	.090	.125	.143	.210	.210	.178	.288	.000	.000	.000	.000
4.40	.100	.020	.030	.025	.080	.070	.040	.115	.000	.000	.000	.000
4.60	.008	.008	.010	.007	.002	.003	.000	.009	.000	.000	.000	.000
4.80	.000	.005	.009	.000	.000	.000	.000	.000	.000	.000	.000	.000
5.00	.000	.001	.008	.000	.000	.000	.000	.000	.000	.000	.000	.000

Transmittance

λ(μ)	Corning Glass Number 3718	3750	3780	3850	3961	3962	3965	3966	4010	4015	4060	4084
.22	.000	.000	.000	.000	.000	.000	.000	.000	.000	.000	.000	.000
.24	.000	.000	.000	.000	.000	.000	.000	.000	.000	.000	.000	.000
.26	.000	.000	.000	.000	.000	.000	.000	.000	.000	.000	.000	.000
.28	.000	.000	.000	.000	.000	.000	.000	.000	.000	.000	.000	.000
.30	.000	.000	.000	.000	.000	.000	.000	.000	.000	.000	.000	.000
.32	.004	.000	.000	.000	.000	.000	.018	.055	.000	.000	.000	.000
.34	.030	.000	.000	.000	.000	.018	.192	.375	.000	.000	.001	.018
.36	.550	.215	.000	.005	.020	.125	.430	.630	.000	.000	.021	.128
.38	.665	.327	.000	.350	.085	.270	.558	.710	.000	.000	.080	.248
.40	.480	.113	.000	.675	.185	.395	.636	.781	.000	.000	.178	.216
.41	.443	.088	.000	.749	.218	.426	.651	.788	.000	.000	.228	.180
.42	.465	.088	.000	.788	.248	.453	.666	.795	.000	.000	.281	.151
.43	.560	.135	.000	.812	.269	.474	.678	.800	.000	.000	.335	.136
.44	.675	.255	.006	.828	.290	.494	.693	.806	.000	.000	.388	.140
.45	.748	.410	.028	.841	.313	.519	.709	.816	.000	.005	.434	.163
.46	.780	.472	.058	.850	.331	.538	.724	.824	.006	.025	.473	.200
.47	.803	.570	.092	.858	.346	.556	.737	.831	.021	.073	.506	.247
.48	.800	.555	.088	.865	.361	.570	.748	.836	.050	.145	.527	.303
.49	.802	.550	.095	.870	.370	.582	.756	.840	.100	.245	.535	.370
.50	.824	.597	.152	.874	.376	.590	.762	.842	.160	.350	.528	.430
.51	.862	.720	.325	.878	.377	.594	.765	.843	.220	.455	.503	.465
.52	.894	.825	.595	.881	.373	.593	.765	.841	.252	.537	.460	.467
.53	.904	.853	.717	.883	.364	.588	.761	.838	.247	.582	.400	.438
.54	.905	.860	.763	.884	.354	.579	.764	.833	.207	.594	.325	.376
.55	.906	.864	.783	.884	.342	.569	.746	.827	.153	.572	.252	.303
.56	.907	.867	.795	.883	.331	.559	.736	.821	.096	.525	.183	.225
.57	.907	.869	.799	.882	.317	.547	.724	.813	.054	.457	.125	.157
.58	.907	.870	.804	.880	.298	.529	.706	.802	.026	.385	.083	.107
.59	.907	.870	.811	.877	.276	.508	.685	.790	.011	.311	.053	.073
.60	.908	.876	.826	.876	.251	.481	.662	.775	.004	.245	.033	.048
.61	.908	.880	.835	.875	.225	.452	.636	.757	.000	.190	.020	.034
.62	.909	.881	.842	.874	.299	.423	.610	.739	.000	.145	.012	.024
.63	.909	.884	.848	.875	.217	.392	.577	.719	.000	.115	.007	.018
.64	.910	.885	.854	.876	.147	.359	.546	.698	.000	.098	.004	.015
.65	.910	.887	.859	.877	.125	.326	.515	.675	.000	.084	.001	.012
.66	.911	.891	.864	.880	.104	.297	.482	.652	.000	.075	.000	.010
.67	.913	.896	.869	.883	.086	.265	.450	.630	.000	.075	.000	.009
.68	.914	.900	.873	.885	.063	.235	.418	.603	.000	.071	.000	.008
.69	.915	.901	.877	.887	.055	.206	.385	.576	.000	.065	.000	.007
.70	.915	.904	.880	.888	.042	.279	.352	.550	.000	.067	.000	.007
.71	.915	.905	.883	.888	.032	.155	.322	.524	.000	.070	.000	.007
.72	.915	.906	.885	.889	.025	.133	.294	.496	.000	.075	.000	.007
.73	.915	.907	.885	.888	.018	.114	.266	.472	.000	.080	.000	.007
.74	.915	.907	.882	.887	.014	.097	.242	.448	.000	.084	.000	.008
.75	.914	.906	.882	.886	.010	.084	.220	.424	.000	.086	.000	.008
.80	.902	.875	.855	.863	.000	.033	.120	.310	.000	.109	.000	.013
1.00	.899	.860	.882	.820	.000	.002	.038	.158	.000	.215	.018	.100
1.20	.898	.882	.898	.850	.000	.002	.040	.161	.007	.393	.158	.303
1.40	.880	.810	.855	.894	.002	.018	.100	.270	.058	.549	.404	.548
1.60	.882	.805	.855	.905	.007	.050	.190	.390	.162	.663	.612	.710
1.80	.900	.844	.907	.904	.008	.057	.201	.408	.299	.740	.740	.791
2.00	.897	.819	.909	.895	.011	.070	.228	.435	.422	.783	.817	.830
2.20	.888	.720	.900	.870	.021	.105	.277	.475	.518	.803	.840	.792
2.40	.865	.668	.890	.850	.037	.140	.320	.512	.597	.817	.862	.808
2.60	.840	.570	.860	.800	.050	.005	.339	.515	.634	.813	.870	.740
2.80	.460	.075	.620	.200	.022	.092	.135	.085	.270	.460	.520	.010
3.00	.282	.028	.465	.165	.033	.133	.200	.230	.260	.418	.620	.033
3.20	.252	.016	.420	.120	.054	.150	.247	.294	.204	.345	.642	.128
3.40	.175	.007	.370	.070	.048	.112	.165	.200	.131	.235	.631	.130
3.60	.150	.003	.351	.045	.003	.008	.016	.018	.121	.192	.634	.006
3.80	.168	.000	.368	.067	.007	.015	.035	.048	.125	.200	.600	.016
4.00	.182	.000	.370	.080	.006	.010	.020	.022	.123	.200	.557	.006
4.20	.120	.000	.300	.040	.005	.013	.022	.021	.070	.130	.422	.001
4.40	.020	.000	.140	.005	.001	.001	.000	.001	.002	.015	.135	.000
4.60	.002	.000	.010	.000	.000	.000	.000	.000	.000	.000	.012	.000
4.80	.000	.000	.000	.000	.000	.000	.000	.000	.000	.000	.002	.000
5.00	.000	.000	.000	.000	.000	.000	.000	.000	.000	.000	.000	.000

Transmittance

λ(μ)	4303	4305	4308	4309	4445	Corning Glass Number 4602	4784	5030	5031	5070	5071	5073
.22	.000	.000	.000	.000	.000	.000	.000	.000	.000	.000	.000	.000
.24	.000	.000	.000	.000	.000	.000	.000	.000	.000	.000	.000	.000
.26	.000	.000	.000	.000	.000	.000	.000	.000	.000	.000	.000	.000
.28	.000	.000	.000	.000	.000	.000	.000	.000	.000	.000	.000	.000
.30	.000	.000	.000	.000	.000	.001	.000	.000	.016	.000	.000	.000
.32	.000	.000	.000	.022	.001	.106	.000	.000	.145	.012	.045	.090
.34	.011	.060	.190	.394	.018	.505	.009	.038	.420	.310	.330	.540
.36	.188	.380	.580	.740	.114	.755	.200	.285	.685	.628	.340	.757
.38	.390	.590	.740	.831	.248	.827	.450	.595	.820	.745	.420	.810
.40	.545	.723	.826	.884	.400	.835	.596	.770	.884	.712	.665	.830
.41	.588	.750	.840	.887	.454	.845	.627	.799	.894	.600	712	.786
.42	.624	.770	.850	.890	.505	.846	.648	.808	.895	.430	.716	.705
.43	.654	.786	.857	.892	.550	.851	.666	.797	.890	.290	.694	.620
.44	.680	.798	.862	.894	.593	.856	.680	.767	.872	.170	.655	.525
.45	.698	.809	.867	.897	.631	.857	.697	.738	.865	.097	.612	.436
.46	.712	.815	.869	.898	.659	.856	.717	.702	.864	.055	.568	.365
.47	.715	.814	.866	.897	.678	.861	.735	.628	.845	.035	.531	.313
.48	.705	.802	.856	.894	.689	.866	.750	.522	.805	.023	.501	.275
.49	.678	.780	.838	.885	.687	.869	.763	.406	.750	.017	.482	.252
.50	.636	.740	.810	.872	.673	.870	.768	.288	.684	.015	.469	.237
.51	.570	.685	.770	.850	.641	.869	.767	.186	.601	.013	.463	.231
.52	.480	.610	.714	.817	.586	.866	.753	.105	.495	.013	.461	.230
.53	.387	.525	.650	.781	.520	.863	.725	.053	.388	.014	.464	.235
.54	.288	.430	.576	.736	.437	.865	.676	.022	.295	.016	.473	.245
.55	.205	.340	.502	.683	.355	.869	.615	.007	.198	.018	.486	.260
.56	.132	.255	.422	.627	.275	.868	.525	.000	.113	.023	.502	.278
.57	.082	.184	.345	.565	.202	.863	.427	.000	.057	.030	.522	.300
.58	.047	.127	.277	.505	.144	.856	.328	.000	.025	.038	.540	.324
.59	.026	.087	.218	.447	.102	.848	.235	.000	.008	.048	.555	.348
.60	.013	.057	.170	.393	.068	.838	.157	.000	.000	.058	.571	.373
.61	.006	.036	.131	.341	.046	.824	.102	.000	.000	.070	.587	.395
.62	.001	.022	.100	.296	.031	.806	.058	.000	.000	.083	.600	.415
.63	.000	.013	.074	.256	.020	.787	.032	.000	.000	.096	.612	.433
.64	.000	.007	.056	.221	.013	.767	.017	.000	.000	.108	.622	.450
.65	.000	.004	.042	.191	.008	.745	.007	.000	.000	.120	.633	.466
.66	.000	.002	.033	.167	.006	.722	.002	.000	.000	.135	.644	.482
.67	.000	.001	.025	.146	.003	.695	.000	.000	.000	.140	.638	.498
.68	.000	.000	.020	.128	.001	.665	.000	.000	.000	.165	.674	.515
.69	.000	.000	.016	.116	.000	.634	.000	.000	.000	.182	.686	.531
.70	.000	.000	.013	.104	.000	.600	.000	.000	.000	.200	.700	.548
.71	.000	.000	.010	.096	.000	.565	.000	.000	.000	.220	.712	.566
.72	.000	.000	.009	.088	.000	.531	.000	.004	.024	.245	.725	.586
.73	.000	.000	.007	.083	.000	.496	.000	.047	.119	.268	.736	.606
.74	.000	.000	.006	.079	.000	.463	.000	.190	.330	.295	.749	.675
.75	.000	.000	.005	.075	.000	.430	.000	.440	.580	.323	.759	.642
.80	.000	.000	.008	080	.003	.258	.000	.890	.917	.505	.815	.750
1.00	.000	.005	.045	.188	.056	.019	.000	.753	.868	.860	.885	.872
1.20	.013	.060	.166	.375	.269	.009	.000	.455	.720	.890	.897	.890
1.40	.080	.180	.342	.542	.527	.016	.003	.100	.285	.892	.902	.892
1.60	.210	.342	.500	.658	.701	.035	.038	.056	.162	.890	.902	.892
1.80	.350	.482	.617	.732	.792	.059	.142	.052	.140	.878	.890	.877
2.00	.475	.590	.690	.772	.839	.048	.275	.075	.175	.860	.865	.860
2.20	.560	.653	.720	.778	.850	.038	.345	.172	.295	.840	.830	.840
2.40	.635	.704	.752	.791	.870	.045	.404	.330	.483	.812	.795	.805
2.60	.663	.710	.748	.770	.861	.066	.340	.382	.530	.804	.752	.775
2.80	.260	.370	.370	.250	.280	.018	.045	.030	.030	.550	.500	.500
3.00	.249	.250	.203	.212	.395	.000	.000	.010	.001	.390	.308	.340
3.20	.202	.212	.145	.168	.451	.000	.000	.065	.026	.222	.145	.180
3.40	.135	.136	.084	.107	.449	.000	.000	.002	.006	.120	.070	.090
3.60	.124	.120	.078	.077	.470	.000	.000	.000	.000	.078	.032	.063
3.80	.132	.125	.082	.083	.470	.000	.000	.000	.000	.078	.029	.060
4.00	.132	.131	.094	.090	.448	.000	.000	.000	.000	.093	.037	.075
4.20	.078	.073	.050	.042	.320	.000	.000	.000	.000	.079	.020	.048
4.40	.008	.009	.008	.002	.100	.000	.000	.000	.000	.020	.004	.014
4.60	.000	.000	.000	.000	.007	.000	.000	.000	.000	.002	.000	.000
4.80	.000	.000	.000	.000	.000	.000	.000	.000	.000	.000	.000	.000
5.00	.000	.000	.000	.000	.000	.000	.000	.000	.000	.000	.000	.000

Transmittance

						Corning Glass Number						
λ(μ)	5113	5120	5300	5330	5433	5543	5562	5572	5840	5850	5860	5874
.22	.000	.000	.000	.000	.000	.000	.000	.000	.000	.000	.000	.000
.24	.000	.000	.000	.000	.000	.000	.000	.000	.000	.000	.000	.000
.26	.000	.000	.000	.000	.000	.000	.000	.000	.000	.000	.000	.000
.28	.000	.000	.000	.000	.000	.000	.000	.000	.000	.000	.000	.000
.30	.000	.000	.000	.002	.000	.000	.000	.000	.001	.039	.000	.000
.32	.000	.000	.000	.250	.000	.000	.000	.045	.242	.490	.008	.031
.34	.000	.000	.000	.622	.000	.000	.012	.325	.600	.790	.179	.228
.36	.035	.018	.000	.796	.100	.120	.205	.660	.682	.858	.340	.447
.38	.200	.540	.000	.835	.350	.380	.495	.805	.392	.850	.085	.378
.40	.371	.670	.000	.865	.585	.600	.717	.874	.000	.788	.000	.032
.41	.371	.790	.000	.850	.636	.635	.748	.873	.000	.720	.000	.004
.42	.337	.805	.000	.823	.665	.646	.761	.865	.000	.630	.000	.000
.43	.272	.560	.000	.783	.674	.635	.759	.857	.000	.522	.000	.000
.44	.198	.386	.000	.725	.665	.602	.742	.845	.000	.410	.000	.000
.45	.118	.485	.000	.650	.635	.550	.713	.832	.000	.290	.000	.000
.46	.055	.475	.000	.555	.577	.465	.662	.808	.000	.175	.000	.000
.47	.013	.370	.008	.455	.467	.335	.565	.765	.000	.125	.000	.000
.48	.000	.385	.026	.355	.327	.190	.435	.693	.000	.022	.000	.000
.49	.000	.685	.085	.270	.205	.090	.300	.605	.000	.005	.000	.000
.50	.000	.660	.125	.197	.120	.040	.197	.523	.000	.000	.000	.000
.51	.000	.390	.106	.145	.060	.013	.110	.430	.000	.000	.000	.000
.52	.000	.305	.094	.110	.024	.002	.051	.328	.000	.000	.000	.000
.53	.000	.175	.064	.085	.008	.000	.022	.247	.000	.000	.000	.000
.54	.000	.610	.149	.068	.005	.000	.012	.216	.000	.000	.000	.000
.55	.000	.817	.147	.055	.006	.000	.015	.246	.000	.000	.000	.000
.56	.000	.230	.087	.043	.007	.000	.018	.285	.000	.000	.000	.000
.57	.000	.125	.013	.032	.000	.000	.011	.258	.000	.000	.000	.000
.58	.000	.000	.000	.022	.000	.000	002	.175	.000	.000	.000	.000
.59	.000	.006	.000	.016	.000	.000	.000	.116	.000	.000	.000	.000
.60	.000	.180	.000	.012	.000	.000	.000	.113	.000	.000	.000	.000
.61	.000	.545	.000	.009	.000	.000	.000	.123	.000	.000	.000	.000
.62	.000	.825	.000	.010	.000	.000	.000	.126	.000	.000	.000	.000
.63	.000	.838	.000	.013	.000	.000	.000	.120	.000	.000	.000	.000
.64	.000	.878	.000	.015	.000	.000	.000	.111	.000	.000	.000	.000
.65	.000	.893	.000	.015	.000	.000	.000	.120	.000	.000	.000	.000
.66	.000	.883	.000	.014	.000	.000	.000	.165	.000	.000	.000	.000
.67	.000	.820	.000	.013	.000	.000	.000	.265	.000	.000	.000	.000
.68	.000	.705	.000	.015	.000	.000	.000	.425	.000	.000	.000	.000
.69	.000	.743	.000	.022	.000	.000	.001	.615	.000	.029	.000	.000
.70	.000	.860	.000	.041	.000	.000	.004	.756	.007	.160	.000	.017
.71	.000	.876	.000	.085	.000	.001	.004	.837	.020	.385	.000	.075
.72	.000	.815	.000	.165	.000	.002	.004	.874	.037	.615	.000	.168
.73	.000	.435	.000	.285	.000	.002	.002	.889	.060	.760	.000	.257
.74	.000	.045	.000	.460	.000	.002	.002	.895	.086	.843	.000	.320
.75	.000	.055	.000	.645	.000	.001	.001	.898	.080	.878	.000	.335
.80	.000	.030	.000	.900	.005	.002	.003	.895	.009	.890	.000	.218
1.00	.000	.860	.003	.800	.010	.015	.020	.880	.000	.716	.000	.050
1.20	.000	.770	.026	.570	.030	.020	.047	.690	.000	.169	.000	.011
1.40	.000	.550	.072	.410	.060	.040	.095	.640	.004	.042	.004	.012
1.60	.000	.620	.200	.405	.116	.060	.154	.625	.002	.036	.002	.010
1.80	.000	.580	.265	.425	.172	.090	.223	.635	.000	.048	.000	.010
2.00	.010	.580	.374	.476	.343	.247	.410	.750	.000	.168	.000	.018
2.20	.071	.747	.533	.535	.475	.400	.528	.780	.000	.338	.000	.036
2.40	.190	.400	.370	.552	.575	.520	.600	.780	.000	.492	.000	.074
2.60	.203	.530	.523	.540	.580	.532	.603	.745	.000	.510	.000	.088
2.80	.100	.250	.265	.007	.162	.132	.360	.470	.000	.170	.000	.003
3.00	.080	.080	.202	.012	.195	.151	.280	.315	.000	.128	.000	.020
3.20	.068	.072	.180	.082	.131	.105	.172	.200	.000	.084	.000	.053
3.40	.030	.029	.131	.003	.079	.065	.100	.100	.000	.065	.000	.050
3.60	.021	.017	.103	.000	.061	.030	.053	.049	.002	.070	.000	.000
3.80	.023	.021	.093	.000	.068	.032	.040	.037	.001	.082	.000	.000
4.00	.040	.025	.112	.000	.071	.039	.050	.042	.005	.090	.000	.000
4.20	.019	.010	.069	.000	.020	.015	.017	.020	.002	.055	.000	.000
4.40	.001	.001	.007	.000	.000	.000	.000	.002	.000	.002	.000	.000
4.60	.000	.000	.000	.000	.000	.000	.000	.000	.000	.000	.000	.000
4.80	.000	.000	.000	.000	.000	.000	.000	.000	.000	.000	.000	.000
5.00	.000	.000	.000	.000	.000	.000	.000	.000	.000	.000	.000	.000

Transmittance

λ (μ)	5900	5970	7380	7740	7905	7910	8364	9780	9782	9788	9830	9863
						Corning Glass Number						
.22	.000	.000	.000	.000	.000	.012	.000	.000	.000	.000	.000	.000
.24	.000	.000	.000	.000	.360	.505	.000	.000	.000	.000	.000	.054
.26	.000	.000	.000	.000	.495	.780	.000	.000	.000	.000	.000	.482
.28	.000	.000	.000	.004	.590	.855	.000	.000	.000	.000	.000	.731
.30	.000	.000	.000	.321	.720	.877	.000	.000	.000	.000	.000	.831
.32	.000	.138	.000	.722	.825	.900	.000	.000	.000	.000	.000	.862
.34	.008	.600	.000	.851	.880	.903	.002	.015	.000	.060	.001	.854
.36	.150	.799	.440	.889	.910	.905	.083	.290	.060	.470	.059	.816
.38	.445	.742	.795	.900	.915	.906	.136	.590	.445	.770	.160	.620
.40	.678	.190	.892	.916	.920	.920	.296	.725	.747	.885	.130	.090
.41	.688	.029	.904	.915	.920	.920	.273	.705	.790	.891	.044	.018
.42	.635	.000	.910	.915	.920	.921	.232	.770	.818	.902	.004	.003
.43	.386	.000	.913	.914	.920	.923	.191	.778	.836	.905	.000	.000
.44	.522	.000	.915	.913	.920	.924	.157	.801	.847	.906	.000	.000
.45	.458	.000	.916	.913	.922	.925	.144	.814	.855	.906	.000	.000
.46	.400	.000	.917	.914	.922	.925	.140	.823	.860	.906	.000	.000
.47	.350	.000	.917	.915	.922	.925	.141	.832	.863	.906	.000	.000
.48	.306	.000	.917	.915	.923	.925	.146	.839	.863	.905	.000	.000
.49	.275	.000	.918	.915	.930	.926	.152	.843	.859	.904	.000	.000
.50	.246	.000	.918	.915	.925	.926	.166	.843	.848	.900	.014	.000
.51	.223	.000	.919	.915	.925	.927	.178	.838	.825	.893	.180	.000
.52	.196	.000	.919	.916	.925	.928	.190	.824	.784	.880	.175	.000
.53	.172	.000	.919	.916	.923	.928	.196	.798	.720	.862	.018	.000
.54	.154	.000	.919	.916	.926	.929	.198	.756	.627	.831	.000	.000
.55	.148	.000	.920	.917	.923	.929	.197	.697	.515	.787	.050	.000
.56	.151	.000	.920	.917	.925	.930	.199	.615	.380	.728	.265	.000
.57	.146	.000	.919	.918	.925	.930	.206	.518	.255	.655	.165	.000
.58	.125	.000	.918	.919	.925	.930	.217	.414	.150	.570	.035	.000
.59	.102	.000	.918	.920	.925	.930	.222	.302	.075	.475	.004	.000
.60	.093	.000	.920	.920	.925	.930	.215	.215	.032	.380	.000	.000
.61	.087	.000	.920	.920	.925	.930	.196	.135	.010	.290	.000	.000
.62	.081	.000	.920	.920	.925	.930	.175	.080	.003	.210	.000	.000
.63	.070	.000	.920	.919	.926	.930	.161	.042	.000	.145	.000	.000
.64	.061	.000	.920	.919	.927	.931	.156	.021	.000	.094	.000	.000
.65	.055	.000	.920	.918	.927	.931	.162	.008	.000	.059	.000	.000
.66	.055	.000	.920	.917	.927	.932	.176	.003	.000	.035	.000	.000
.67	.059	.000	.920	.916	.928	.932	.200	.000	.000	.020	.000	.000
.68	.065	.000	.920	.916	.927	.932	.228	.000	.000	.010	.075	.022
.69	.068	.007	.921	.915	.927	.932	.248	.000	.000	.005	.380	.106
.70	.068	.036	.921	.915	.928	.932	.251	.000	.000	.001	.642	.234
.71	.066	.085	.922	.914	.926	.933	.237	.000	.000	.000	.694	.332
.72	.064	.145	.922	.912	.926	.933	.223	.000	.000	.000	.666	.383
.73	.060	.222	.922	.910	.926	.933	.210	.000	.000	.000	.607	.384
.74	.057	.323	.921	.909	.927	.934	.197	.000	.000	.000	.531	.358
.75	.055	.385	.921	.907	.928	.934	.180	.000	.000	.000	.445	.322
.80	.050	.287	.918	.890	.930	.932	.110	.000	.000	.000	.045	.175
1.00	.085	.032	.910	.860	.930	.928	.032	.000	.000	.000	.000	.119
1.20	.180	.021	.910	.860	.925	.928	.032	.000	.000	.005	.007	.016
1.40	.295	.109	.906	.870	.925	.930	.062	.018	.000	.080	.000	.005
1.60	.405	.088	.910	.892	.931	.930	.137	.131	.011	.266	.000	.007
1.80	.495	.040	.903	.896	.931	.930	.182	.317	.085	.430	.157	.011
2.00	.590	.008	.900	.897	.934	.929	.171	.440	.216	.512	.041	.029
2.20	.628	.002	.898	.875	.934	.835	.184	.440	.278	.455	.000	.048
2.40	.640	.009	.890	.850	.930	.890	.225	.440	.325	.433	.000	.060
2.60	.630	.018	.860	.820	.920	.780	.258	.280	.212	.252	.057	.051
2.80	.470	.030	.375	.140	.908	.180	.145	.060	.040	.060	.020	.000
3.00	.248	.030	.425	.360	.880	.695	.130	.000	.000	.000	.000	.000
3.20	.121	.030	.380	.490	.861	.760	.125	.000	.000	.000	.000	.000
3.40	.032	.020	.310	.270	.670	.620	.099	.000	.000	.000	.000	.000
3.60	.010	.015	.270	.010	.111	.080	.115	.000	.000	.000	.000	.000
3.80	.010	.020	.275	.040	.270	.240	.142	.000	.000	.000	.000	.000
4.00	.010	.030	.260	.013	.170	.150	.172	.000	.000	.000	.000	.000
4.20	.006	.017	.180	.026	.250	.230	.158	.000	.000	.000	.000	.000
4.40	.001	.002	.040	.004	.085	.080	.078	.000	.000	.000	.000	.000
4.60	.000	.000	.000	.000	.050	.020	.010	.000	.000	.000	.000	.000
4.80	.000	.000	.000	.000	.000	.000	.003	.000	.000	.000	.000	.000
5.00	.000	.000	.000	.000	.000	.000	.000	.000	.000	.000	.000	.000

TRANSMISSION OF WRATTEN FILTERS

Compiled by Allie C. Peed, Jr. for The Eastman Kodak Company

Data condensed from Kodak Wratten Filters for Scientific and Technical Use published by the Eastman Kodak Company, manufacturers of the filters.

The following pages give (1) percentage luminous transmittance at wave lengths from 400 to 700μ at intervals of 10μ for the standard illuminant "C" adopted by the International Commission of Illumination, (2) dominant wavelength in millimicrons, and (3) percentage of excitation purity. Values of wave length followed by "c" indicate the complementary wave lengths of purple filters which do not have a dominant wave length.

All colorimetric specifications are based on the 1931 standard ICI colorimetric and luminosity data.

The transmittance data are given as representing standard samples of the filters. They are intended only for the information of users in choosing filters which will meet their requirements. Values taken from the tables of data should not be used by research workers as representing precisely the absorption characteristics of a particular filter. If such precise data are needed, they should be determined for the particular filter being used.

Where the spectra extend into the ultraviolet this fact is indicated by an asterisk (*) in the transmission tables immediately beneath the filter number, and quantitative data are not given. The manufacturer should be consulted for this information. Transmission in the ultraviolet of wave lengths less than 330μ will be eliminated in the case of cemented filters, as glass absorbs ultraviolet radiation of wave lengths shorter than about 330μ.

Stability ratings are given as three letter combinations following the filter description in the table below. In establishing the stability classifications each filter is exposed to a selected light source for a specific time interval. The following grading system is used to describe the result:

Class A — stable
Class B — relatively stable
Class C — somewhat unstable
Class D — unstable

The classification letters, for example, AAA, describe the stability to the following three exposure tests in this order:
1. Two weeks' exposure to daylight in a south window
2. Twenty-four hours' exposure to a "Fade-Ometer"
3. Two weeks' exposure at two feet from a 1000-watt tungsten lamp.

Filters are supplied in two forms: as lacquered gelatin film, or as a gelatin film cemented between pieces of optical glass. Filters in glass are cemented between sheets of plane-parallel glass, which is surfaced in quantities and is of sufficient accuracy for general photographic work, and for most scientific purposes.

Most Wratten Gelatin Filters are stocked in 2- or 3-inch squares. Stocks of 2- or 3-inch square filters cemented in glass are maintained only in filters usually used for general photographic work.

The booklet "Kodak Filters and Lens Attachments" gives more valuable information on this subject.

FILTER DATA

No.	Description, use, and stability
	Colorless
0	For compensating thickness of other gelatin filters in optical systems, AAA.
1	Absorbs ultraviolet below 360 mμ, DDD.
1A	Kodak Skylight Filter — Reduces excess bluishness in outdoor color photographs in open shade under a clear, blue sky, ACA.
	Yellow
2B	Absorbs ultraviolet below 410 mμ, ACA.
3	Light yellow, CCD.
3N5	No. 3 plus 0.5 neutral density, AAA.
4	Light yellow — Approximate correction on panchromatic materials for outdoor scenes, including sky, CCC.
6	K1 — Light yellow — Partial correction outdoors, BBA.
8	K2 — Yellow — Full correction outdoors on Type B panchromatic materials. Widely used for proper sky, cloud, and foliage rendering. Green separation for Fluorescence Process, AAA.
8N5	No. 8 plus 0.5 neutral density, AAA.
9	K3 — Deep yellow. Moderate contrast in outdoor photography (with black-and-white films), AAA.
11	X1 — Greenish yellow. Correction for tungsten light on Type B panchromatic materials; also for daylight correction with Type C panchromatic materials in making outdoor portraits, darkening skies, or lightening foliage, AAA.
12	Minus blue. Haze cutting in aerial photography, AAA.
13	X2 — Yellow green. Correction for Type C panchromatic materials in tungsten light, ABA.
15	G — Deep yellow. Overcorrection in landscape photography. Contrast control in copying and in aerial infrared photography, AAA.
16	Blue absorption, AAB.
18A	Transmits ultraviolet and infrared only (glass), AAA.
	Oranges and Reds
21	Blue and blue-green absorption, CBB.
22	Yellow-orange. For increasing contrast in blue preparations in microscopy. Mercury yellow, BAC.
23A	Light red. Two-color projection — contrast effects, BAB.
24	Red for two-color photography (daylight or tungsten). White-flame-arc tricolor projection, AAB.

No.	Description, use, and stability
25	A — Tricolor red for direct color separation. Contrast effects in commercial photography and in outdoor scenes. Two-color general viewing. Aerial infrared photography and haze cutting, AAA.
26	Stereo red, AAA.
29	Red color separation from transparencies and for the Kodak Fluorescence Process. Strong contrast effects. Copying blueprints. Tungsten tricolor projection, AAA.
	Magentas and Violets
30	Green absorption, BBC.
31	Green absorption, CCA.
32	Minus green, CCD.
33	Strong green absorption, CCB.
34	Violet, CDD.
34A	Blue separation — Kodak Fluorescence Process, DCC.
35	Contrast in microscopy, CDD.
36	Dark violet, CCC.
	Blues and Blue-greens
38	Red absorption, BCA.
38A	Red absorption. Increasing contrast in visual microscopy, BBB.
39	Contrast control in printing motion-picture duplicates (glass) AAA.
40	Green for two-color photography (tungsten), CBC.
44	Minus red — Two-color general viewing, DDD.
44A	Minus red, DDD.
45	Contrast in microscopy, DDD.
45A	Highest resolving power in visual microscopy, CDC.
46	Blue projection (experimental), DDD.
47	Tricolor blue for direct color separation and from Kodak Ektacolor Film for Dye Transfer. Contrast effects in commercial photography. Tungsten and white-flame-arc tricolor projection, BBC.
47B	Tricolor blue for color separation from transparencies and from Kodak Ektacolor Film for Graphic Arts, BBB.
48	Green and red absorption, CBC.
48A	Green and red absorption, AAB.
49	Dark blue, BCB.
49B	Very dark blue, BBB.
50	Very dark blue. Mercury violet, CCC.
	Greens
52	Light green, AAB.

TRANSMISSION OF WRATTEN FILTERS (Continued)
FILTER DATA

No.	Description, use, and stability
53	Medium green, CCB Very dark green, AAA.
55	Stereo green, BBC.
56	Very light green, CBC.
57	Green for two-color photography (daylight), CBC.
57A	Light green, BBC.
58	Tricolor green for direct color separation. Contrast effects in commercial photography and microscopy, BBC.
59	Green for tricolor projection (white-flame-arc), BBB.
59A	Very light green, BBB.
60	Green for two-color photography (tungsten), BDC.
61	Green color separation from transparencies and Kodak Ektacolor Film. Tricolor projection (tungsten), ABC.
64	Red absorption (light), CDB.
65	Red absorption, ADB.
65A	Red absorption, CCD.
66	Contrast effects in microscopy and medical photography, DDC.
67A	Red absorption (light). Two-color projection, CDC.
Narrow-band	
70	Dark red. Infrared photography. Color separation for Kodak Ektacolor Film (with tungsten), ABC.
72B	Dark orange-yellow, CCC.
73	Dark yellow-green, ABB.
74	Dark green. Mercury green, BBC.
75	Dark blue-green, ACC.
76	Dark violet (compound filter), DDD.
77	Transmits 546 mμ mercury line (glass plus gelatin), AAA.
77A	Transmits 546 mμ mercury line (glass plus gelatin), AAA.
Photometrics	
78	Bluish. Photometric filter (visual), BAB.
78AA	Bluish. Photometric filter (visual), BAA.
78A	Bluish. Photometric filter (visual), AAA.
78B	Bluish. Photometric filter (visual), AAA.
78C	Bluish. Photometric filter (visual), BAA.
86	Yellowish. Photometric filter (visual), BBA.
86A	Yellowish. Photometric filter (visual), AAA.

No.	Description, use, and stability
†86B	Yellowish. Photometric filter (visual), BCA.
†86C	Yellowish. Photometric filter (visual), AAA.
Light Balancing	
80A	For Kodachrome Film, Daylight Type, and photographic flood lamps, ABA.
81	Yellowish. For warmer color rendering.
81A	Yellowish. For Kodak Ektachrome Film, Type B, with photographic flood lamps.
81B	Yellowish. For warmer color rendering.
81C	Yellowish. For Kodachrome Film, Type A, with flash lamps.
81D	Yellowish. For Kodachrome Film, Type A, with flash lamps.
81EF	Yellowish. For Kodak Ektachrome Film, Type B, with flash lamps.
82	Bluish. For cooler color rendering.
82A	Bluish. For Kodachrome Film, Type A, with 3200 K lamps.
82B	Bluish. For cooler color rendering.
82C	Bluish. For cooler color rendering.
83	Yellowish. For 16 mm Commercial Kodachrome Film and daylight exposure, BBB.
85	Orange. For Type A Kodak color films and daylight exposure, BAA.
85B	Orange. For Kodak Ektachrome Film, Type B, and daylight exposure, BAB.
Miscellaneous	
79	Photographic sensitometry. Corrects 2360 K to 5500 K, AAA.
87	For infrared photography. Absorbs visual.
87C	Absorbs visual, transmits infrared.
88A	For infrared photography. Absorbs visual.
89B	For infrared photography, AAA.
90	Narrow-band viewing filter for judging brightness scale of scenes, CCD.
96	Neutral filters for controlling luminance, AAB.
97	Dichroic absorption, AAA.
102	Correction filter for Barrier-layer photocell, ABA.
106	Correction filter for S-4 type photocell, AAA.

Percent transmittance

Wave length	No. 0 *	No. 1	No. 1A *	No. 2B *	No. 3	No. 3N5	No. 4	No. 6 *	No. 8	No. 8N5	No. 9	No. 11 *	No. 12 *
400	88.0	85.0	59.0	19.0	—	—	—	7.40	—	—	—	—	—
10	88.5	85.5	76.0	48.0	—	—	—	8.32	—	—	—	—	—
20	88.9	86.0	82.0	67.0	—	—	—	10.4	—	—	—	0.16	—
30	89.3	86.5	84.6	75.3	0.36	—	—	13.5	—	—	—	0.29	—
40	89.6	87.0	86.0	80.0	1.78	—	—	18.9	—	—	—	0.56	—
50	89.8	87.4	86.8	83.0	11.5	1.59	—	27.6	—	—	—	1.32	—
60	89.9	87.8	87.2	85.2	38.0	9.40	6.9	39.0	0.25	0.16	—	4.00	—
70	90.1	88.2	87.5	86.7	68.0	18.5	42.0	52.3	5.50	2.0	1.78	12.0	—
80	90.3	88.5	87.3	88.1	80.8	23.5	74.0	65.8	19.0	6.3	8.31	26.0	—
90	90.4	88.7	86.8	88.8	85.2	25.5	84.7	76.8	41.0	13.2	20.7	43.7	—
500	90.5	88.9	86.3	89.5	86.9	26.3	87.5	83.5	63.5	20.0	34.5	55.0	1.50
10	90.6	89.1	85.5	89.9	87.8	26.7	88.5	87.0	78.0	24.3	48.8	60.0	17.3
20	90.7	89.3	84.8	90.3	88.4	27.0	89.1	88.4	84.1	26.7	62.0	60.2	55.0
30	90.7	89.5	84.3	90.5	89.0	27.2	89.4	89.0	86.5	28.0	76.0	57.8	77.8
40	90.8	89.7	84.0	90.6	89.5	27.5	89.6	89.4	87.7	28.6	83.8	54.2	86.0
50	90.8	89.9	83.9	90.7	89.8	27.8	89.8	89.7	88.4	29.0	87.0	50.0	88.4
60	90.9	90.1	84.1	90.8	90.1	27.9	90.0	89.9	88.8	29.3	88.3	44.8	89.4
70	90.9	90.2	84.8	90.9	90.4	28.0	90.2	90.1	89.2	29.5	88.8	38.9	89.7
80	90.9	90.3	86.0	90.9	90.6	28.4	90.4	90.3	89.5	29.6	89.1	33.1	90.1
90	91.0	90.4	87.4	91.0	90.7	29.0	90.6	90.5	89.8	29.8	89.3	27.6	90.3
600	91.0	90.5	88.5	91.1	90.8	29.5	90.8	90.6	90.1	29.9	89.5	22.7	90.4
10	91.0	90.5	89.5	91.2	90.9	29.5	90.9	90.7	90.3	29.6	89.7	19.0	90.5
20	91.0	90.6	90.2	91.3	91.0	29.3	91.0	90.8	90.5	29.4	89.8	14.9	90.7
30	91.0	90.6	90.6	91.3	91.0	29.1	91.1	90.9	90.7	29.1	89.9	11.4	90.8
40	91.1	90.7	90.8	91.4	91.1	29.0	91.2	91.0	90.9	28.8	90.0	9.10	90.9
50	91.1	90.7	91.0	91.4	91.2	20.4	91.3	91.1	91.0	28.9	90.1	8.05	91.0
60	91.1	90.8	91.1	91.5	91.3	29.6	91.4	91.2	91.1	29.2	90.1	7.50	91.1
70	91.1	90.8	91.1	91.5	91.4	29.8	91.5	91.2	91.2	29.4	90.2	7.05	91.2
80	91.1	90.9	91.1	91.6	91.5	30.0	91.5	91.3	91.3	29.5	90.2	6.50	91.2
90	91.1	90.9	91.1	91.7	91.6	30.2	91.6	91.4	91.4	29.7	90.3	6.10	91.2
700	91.1	91.0	91.1	91.8	91.7	31.0	91.6	91.3	91.5	30.2	90.3	6.20	91.3
Luminous transmit.	90.8	89.9	85.9	90.5	88.3	27.4	87.8	87.5	82.7	27.0	76.6	40.2	73.8
Dominant wave lgth.	571.0	575.0	498.0	570.0	569.5	570.5	569.5	570.3	571.8	572.0	574.4	550.3	576.1
Excitation purity.	0.8	1.5	1.2	5.7	50.0	56.3	64.0	44.7	85.2	84.0	91.4	60.7	97.8

* Some transmission below 400 mμ. Consult the manufacturer.

Percent transmittance

Wave length	No. 13 •	No. 15 •	No. 16 •	No. 18A •	No. 21	No. 22	No. 23A •	No. 24 •	No. 25	No. 26	No. 29	No. 30 •	No. 31 •
400	—	—	—	—	—	—	—	—	—	—	—	48.6	13.8
10	—	—	—	—	—	—	—	—	—	—	—	47.4	14.5
20	—	—	—	—	—	—	—	—	—	—	—	48.5	16.4
30	0.18	—	—	—	—	—	—	—	—	—	—	50.1	25.5
40	0.50	—	—	—	—	—	—	—	—	—	—	49.4	42.7
50	1.35	—	—	—	—	—	—	—	—	—	—	43.0	50.2
60	4.08	—	—	—	—	—	—	—	—	—	—	26.5	40.4
70	11.0	—	—	—	—	—	—	—	—	—	—	13.8	22.6
80	23.5	—	—	—	—	—	—	—	—	—	—	5.00	8.20
90	39.0	—	—	—	—	—	—	—	—	—	—	0.63	1.85
500	50.8	—	—	—	—	—	—	—	—	—	—	—	0.12
10	55.2	1.00	—	—	—	—	—	—	—	—	—	—	—
20	56.5	16.0	3.00	—	—	—	—	—	—	—	—	—	—
30	55.0	52.1	22.0	—	—	—	—	—	—	—	—	—	—
40	51.0	70.7	48.0	—	2.50	—	—	—	—	—	—	—	—
50	46.0	84.3	69.5	—	29.0	0.25	—	—	—	—	—	—	—
60	39.2	87.5	79.5	—	65.0	19.0	—	—	—	—	—	0.10	—
70	32.0	88.7	84.0	—	80.6	60.0	11.0	—	—	—	—	10.0	—
80	25.1	89.3	86.3	—	85.4	81.0	47.0	4.55	—	—	—	45.0	—
90	18.2	89.7	87.8	—	87.3	87.0	69.6	37.3	12.6	2.90	—	76.0	0.63
600	13.5	90.0	89.0	—	88.1	88.5	82.7	72.3	50.0	30.0	—	87.4	26.0
10	9.60	90.1	89.6	—	88.7	89.0	85.8	82.9	75.0	63.2	10.0	89.5	67.2
20	6.40	90.2	90.0	—	89.0	89.5	87.2	86.4	82.6	78.9	45.3	90.2	84.0
30	3.66	90.3	90.2	—	89.5	89.8	87.9	87.8	85.5	84.0	71.4	90.5	88.1
40	2.20	90.4	90.3	—	89.9	90.0	88.5	88.5	86.7	86.1	82.7	90.7	89.8
50	1.58	90.5	90.4	—	90.2	90.1	89.0	89.0	87.6	87.2	86.6	90.8	90.2
60	1.74	90.6	90.5	—	90.4	90.2	89.4	89.3	88.2	88.1	88.4	90.9	90.4
70	2.62	90.6	90.6	—	90.5	90.3	89.6	89.7	88.5	88.5	89.4	91.0	90.5
80	3.55	90.7	90.7	—	90.5	90.4	89.8	89.9	89.0	88.9	90.0	91.1	90.7
90	4.48	90.7	90.8	0.25	90.6	90.5	90.0	90.2	89.3	89.2	90.3	91.1	90.8
700	5.25	90.8	90.8	1.20	90.6	90.6	90.2	90.3	89.5	89.5	90.4	91.1	91.0
Luminous transmit.	34.5	66.2	57.7	0.0014	45.6	35.8	25.0	17.8	14.0	11.7	6.3	26.6	12.9
Dominant wave lgth.	542.0	579.3	582.7	700.0	588.9	595.1	602.7	610.6	615.1	619.0	631.6	498.6c	513.1c
Excitation purity.	57.5	99.0	99.3	100.0	99.9	99.9	100.0	100.0	100.0	100.0	100.0	62.4	81.9

Percent transmittance

Wave length	No. 32 •	No. 33 •	No. 34 •	No. 34A •	No. 35 •	No. 36 •	No. 38 •	No. 38A •	No. 39 •	No. 40 •	No. 44 •	No. 44A •	No. 45 •
400	38.0	0.85	64.0	—	48.0	36.5	60.5	33.4	85.2	—	0.44	2.52	—
10	37.9	0.71	70.1	0.1	57.0	45.5	66.5	41.2	78.2	—	0.36	3.39	—
20	40.0	1.17	72.0	40.0	57.6	45.5	72.5	53.0	70.5	—	0.63	6.30	—
30	43.0	1.69	68.4	69.7	47.5	32.7	75.3	58.0	63.3	—	3.63	17.4	—
40	55.5	5.36	58.2	68.7	29.5	15.2	76.2	58.8	53.6	—	13.1	32.7	5.00
50	66.0	14.3	42.3	56.2	12.3	3.7	75.9	57.6	42.5	—	25.4	41.8	19.0
60	66.0	12.4	25.2	40.5	3.5	0.35	74.8	55.2	28.5	3.16	36.5	48.1	29.5
70	57.0	5.00	12.1	23.8	0.25	—	73.4	51.9	17.3	21.6	46.5	51.7	34.4
80	40.0	0.50	2.7	9.2	—	—	71.6	48.5	10.2	44.7	53.6	52.9	35.7
90	21.0	—	0.2	2.3	—	—	69.5	44.6	4.00	61.4	56.8	52.2	34.5
500	9.56	—	—	0.33	—	—	66.7	40.2	1.33	70.2	55.8	49.8	29.7
10	2.51	—	—	—	—	—	63.9	35.8	0.35	72.4	50.9	44.8	21.5
20	0.13	—	—	—	—	—	60.8	31.7	—	70.5	42.1	36.8	11.5
30	—	—	—	—	—	—	57.0	27.2	—	64.8	30.5	26.8	3.80
40	—	—	—	—	—	—	52.6	22.3	—	55.5	18.6	16.8	0.85
50	—	—	—	—	—	—	48.0	17.6	—	44.2	8.99	8.20	—
60	—	—	—	—	—	—	42.8	12.9	—	32.5	3.59	2.95	—
70	—	—	—	—	—	—	37.0	8.78	—	20.3	0.80	0.91	—
80	—	—	—	—	—	—	30.6	5.65	—	9.56	—	0.10	—
90	—	—	—	—	—	—	25.5	3.48	—	3.20	—	—	—
600	6.04	—	—	—	—	—	20.9	2.09	—	1.10	—	—	—
10	41.0	0.80	—	0.13	—	—	16.8	1.15	—	0.32	—	—	—
20	75.0	24.9	—	1.0	—	—	12.9	0.59	—	—	—	—	—
30	86.1	60.8	—	6.3	—	—	10.0	0.28	—	—	—	—	—
40	89.0	78.0	0.4	22.0	—	—	7.79	0.13	—	—	—	—	—
50	90.0	85.0	4.0	45.0	0.1	—	6.68	—	—	—	—	—	—
60	90.6	87.5	20.7	65.0	3.0	0.21	6.20	—	—	—	—	—	—
70	90.7	88.7	45.2	77.3	19.0	7.5	5.91	—	—	—	—	—	—
80	90.8	89.4	66.5	85.0	43.5	29.0	5.41	—	—	0.50	0.80	—	—
90	90.9	89.8	78.8	88.2	66.0	55.0	4.90	—	—	4.06	6.99	0.18	—
700	91.0	90.0	85.0	89.8	77.7	71.3	5.00	—	17.8	23.5	1.60	—	1.00
Luminous transmit.	12.5	5.2	1.3	2.9	0.45	0.25	42.5	17.3	1.2	33.6	15.6	14.4	5.2
Dominant wave lgth.	551.7c	498.0c	424.0	564.8c	566.8	566.4c	483.5	478.9	450.6	516.2	589.1	483.4	481.5
Excitation purity.	79.6	88.3	94.4	91.4	96.3	97.8	41.8	69.8	98.9	48.5	72.9	77.2	88.4

* Some transmission below 400 mμ. Consult the manufacturer.

Percent transmittance

Wave length	No. 45A	No. 46	No. 47	No. 47B	No. 48	No. 48A	No. 49	No. 49B	No. 50	No. 52	No. 53	No. 54	No. 55
400	—	1.20	7.80	16.0	0.96	5.65	3.30	1.70	0.45	2.18	—	—	—
10	—	0.60	17.4	29.5	3.16	10.0	4.28	2.00	0.39	1.51	—	—	—
20	—	0.80	34.0	43.6	8.25	16.0	6.93	3.55	0.59	0.80	—	—	—
30	1.00	5.98	47.0	50.0	15.0	21.0	11.2	7.00	2.63	0.44	—	—	—
40	8.81	19.0	50.3	47.2	22.6	25.0	18.9	13.0	8.90	0.41	—	—	—
50	17.4	30.1	48.3	36.0	30.3	26.2	25.6	17.4	14.0	0.69	—	—	—
60	20.9	33.8	43.4	25.0	33.2	22.9	24.0	14.8	12.3	1.45	—	—	0.20
70	21.6	32.1	36.2	13.2	29.6	16.5	15.7	7.60	5.36	2.70	0.10	—	2.90
80	20.5	27.0	28.5	4.5	22.4	9.55	6.93	2.76	1.55	4.90	0.7	—	13.1
90	18.0	20.2	19.6	1.3	14.1	4.27	2.14	0.40	0.10	8.50	2.14	—	34.2
500	14.4	11.1	0.36	0.17	7.30	1.58	0.46	—	—	13.3	4.47	—	53.4
10	10.1	4.39	—	—	2.64	0.48	—	—	—	18.2	7.24	0.10	67.0
20	5.60	1.66	—	—	0.50	—	—	—	—	23.7	10.7	0.31	69.3
30	2.52	0.35	—	—	—	—	—	—	—	28.5	14.0	0.64	65.1
40	0.4	—	—	—	—	—	—	—	—	32.1	16.6	0.89	56.7
50	0.10	—	—	—	—	—	—	—	—	33.1	17.3	0.93	45.0
60	—	—	—	—	—	—	—	—	—	31.0	15.4	0.62	33.1
70	—	—	—	—	—	—	—	—	—	25.6	11.4	0.21	20.7
80	—	—	—	—	—	—	—	—	—	19.1	6.90	—	9.00
90	—	—	—	—	—	—	—	—	—	12.6	3.60	—	2.70
600	—	—	—	—	—	—	—	—	—	7.78	1.41	—	0.40
10	—	—	—	—	—	—	—	—	—	4.17	0.40	—	—
20	—	—	—	—	—	—	—	—	—	2.34	0.15	—	—
30	—	—	—	—	—	—	—	—	—	1.38	—	—	—
40	—	—	—	—	—	—	—	—	—	0.80	—	—	—
50	—	—	—	—	—	—	—	—	—	0.54	—	—	—
60	—	—	—	—	—	—	—	—	—	0.36	—	—	—
70	—	—	—	—	—	—	—	—	—	0.27	—	—	—
80	—	—	—	—	—	—	—	—	—	0.23	—	—	0.66
90	0.20	0.25	—	—	—	—	—	—	—	0.19	—	—	6.90
700	2.24	0.85	—	—	—	—	—	—	—	0.17	—	—	27.8
Luminous transmit.	2.8	2.4	2.8	0.78	1.86	0.88	0.69	0.36	0.26	20.1	9.0	0.032	31.4
Dominant wave lgth.	477.6	470.4	463.7	479.8	466.5	458.0	457.9	455.5	455.9	553.3	551.1	546.1	530.2
Excitation purity.	89.7	94.9	95.8	69.1	96.1	98.3	98.9	99.3	99.4	77.3	89.7	97.0	68.4

Percent transmittance

Wave length	No. 56	No. 57	No. 57A	No. 58	No. 59	No. 59A	No. 60	No. 61	No. 64	No. 65	No. 65A	No. 66	No. 67A
400	—	—	—	—	—	—	—	—	9.00	—	—	12.3	1.10
10	—	—	—	—	—	—	—	—	9.20	—	—	13.0	0.93
20	—	—	—	—	—	—	—	—	8.75	0.23	—	15.0	1.28
30	—	—	0.19	—	—	0.16	—	—	9.20	0.61	0.16	18.4	3.16
40	—	—	0.87	—	—	0.37	—	—	11.3	1.58	1.32	23.2	6.40
50	—	—	—	—	0.40	1.26	0.19	—	15.5	4.10	5.50	31.2	10.5
60	0.16	0.44	2.56	—	1.90	4.57	1.38	—	23.3	9.00	13.0	42.2	17.7
70	3.12	3.10	7.80	0.23	7.70	13.2	5.38	—	34.4	16.8	24.9	55.5	28.5
80	13.0	13.1	21.6	1.38	21.0	30.0	15.0	0.33	46.8	24.9	36.6	68.4	41.4
90	34.5	31.9	41.7	4.90	41.5	50.8	32.0	4.00	56.6	31.3	45.1	77.6	52.1
500	59.0	50.5	58.8	17.7	59.0	66.0	48.4	16.6	62.1	33.7	45.8	82.7	57.9
10	73.0	60.6	67.9	38.8	67.7	73.0	57.2	32.3	62.9	32.4	39.7	84.6	58.8
20	79.0	63.3	70.1	52.2	69.8	75.1	59.2	40.0	59.1	27.5	29.7	84.0	55.4
30	79.9	61.0	67.6	53.6	67.2	73.2	55.5	39.6	51.6	20.7	17.8	82.6	47.5
40	77.5	55.0	61.8	47.6	61.5	68.5	47.5	34.5	41.3	13.7	7.90	79.1	36.6
50	72.6	47.1	53.5	38.4	54.0	62.0	36.8	26.3	28.0	6.50	2.40	73.7	25.0
60	66.1	37.3	43.3	27.8	45.0	54.4	25.2	17.3	16.2	1.66	0.32	67.1	14.2
70	58.0	26.5	31.6	17.4	35.0	44.5	14.4	9.70	7.95	0.40	—	58.8	5.50
80	46.1	16.6	19.4	9.0	24.0	33.0	6.3	4.40	3.10	—	—	47.2	1.40
90	33.8	8.69	9.70	3.50	14.0	22.0	1.82	1.66	0.80	—	—	34.5	0.28
600	24.0	3.70	4.50	1.50	7.95	14.6	0.48	0.38	—	—	—	24.4	—
10	18.7	1.60	2.00	0.41	4.90	10.5	0.10	—	—	—	—	18.5	—
20	13.2	0.49	0.87	—	2.70	6.92	—	—	—	—	—	13.7	—
30	7.22	—	0.22	—	1.00	3.16	—	—	—	—	—	7.70	—
40	3.02	—	—	—	0.17	1.07	—	—	—	—	—	3.00	—
50	1.48	—	—	—	—	0.50	—	—	—	—	—	1.46	—
60	1.91	—	—	—	—	0.91	—	—	—	—	—	1.91	—
70	7.95	—	—	—	0.63	3.00	—	—	—	—	—	6.17	—
80	23.0	—	0.16	—	4.00	10.0	—	—	—	—	—	19.9	—
90	44.1	—	1.15	—	12.0	20.0	2.10	—	—	0.10	—	0.20	42.6
700	64.8	—	3.17	0.53	22.6	30.0	8.70	—	4.50	—	2.18	63.1	0.40
Luminous transmit.	52.8	32.5	37.2	23.7	38.7	45.8	26.1	16.8	25.0	9.6	9.8	58.3	22.4
Dominant wave lgth.	552.3	536.4	534.0	540.2	538.3	541.4	525.7	536.8	497.3	496.6	492.7	512.3	499.8
Excitation purity.	78.2	69.2	62.1	88.1	66.0	59.3	62.2	85.4	55.0	67.8	77.4	21.5	55.8

* Some transmission below 400 mμ. Consult the manufacturer.

Percent transmittance

Wave length	No. 70	No. 72B	No. 73	No. 74	No. 75	No. 76 •	No. 77	No. 77A	No. 78 •	No. 78AA •	No. 78A •	No. 78B •
400	—	—	—	—	—	0.22	—	—	37.2	43.0	56.0	64.1
10	—	—	—	—	—	0.18	—	—	41.7	46.0	58.6	66.5
20	—	—	—	—	—	0.29	—	—	44.2	48.7	61.0	68.4
30	—	—	—	—	—	1.38	—	—	44.6	49.8	61.8	69.5
40	—	—	—	—	—	3.50	—	—	44.2	49.7	61.8	70.0
50	—	—	—	—	—	3.50	—	—	41.7	48.0	61.0	69.4
60	—	—	—	—	1.97	1.92	—	—	38.0	44.9	58.7	67.5
70	—	—	—	—	10.0	0.51	—	—	33.8	40.3	55.0	65.4
80	—	—	—	—	17.4	—	—	—	27.5	35.6	51.0	62.9
90	—	—	—	—	18.0	—	—	—	23.5	30.9	47.1	59.8
500	—	—	—	—	13.0	—	—	—	19.5	26.5	43.5	57.0
10	—	—	—	0.96	7.35	—	0.30	0.10	15.8	23.4	40.0	54.2
20	—	—	—	7.95	3.20	—	9.10	5.35	13.8	20.3	36.9	51.4
30	—	—	—	14.6	0.83	—	13.5	1.90	11.8	17.8	34.4	49.3
40	—	—	—	12.9	0.14	—	46.1	35.0	10.5	16.6	32.7	48.1
50	—	—	—	7.60	—	—	78.0	71.8	9.56	14.9	31.2	46.7
60	—	—	2.24	3.06	—	—	75.8	63.1	8.53	13.2	29.4	45.0
70	—	—	5.97	0.83	—	—	8.00	—	7.77	12.1	28.0	43.6
80	—	1.26	4.56	0.12	—	—	1.00	—	7.41	11.6	27.5	43.1
90	—	1.26	2.00	—	—	—	0.32	—	6.93	11.1	27.0	42.9
600	—	5.89	0.56	—	—	—	16.2	1.60	6.45	10.40	26.0	41.8
10	—	5.25	0.10	—	—	—	52.1	32.1	5.50	9.20	24.1	40.0
20	—	2.88	—	—	—	—	83.0	78.0	4.80	7.70	21.8	37.6
30	—	1.26	—	—	—	—	84.9	79.5	3.94	6.50	19.7	35.5
40	—	0.48	—	—	—	—	88.1	86.5	3.46	5.60	18.6	34.2
50	0.63	0.14	—	—	—	—	89.8	89.2	3.24	5.50	18.4	33.6
60	10.5	—	—	—	—	—	89.8	89.0	3.16	5.60	18.5	34.0
70	35.0	—	—	—	—	—	85.5	79.5	3.39	5.80	18.7	34.1
80	55.2	—	—	—	—	—	76.1	62.5	3.45	6.10	19.0	34.5
90	70.0	—	—	—	—	0.13	75.0	62.4	3.51	6.10	19.3	34.8
700	79.0	—	—	—	0.14	1.24	86.5	83.0	3.90	6.50	20.2	36.0
Luminous transmit.	0.31	0.74	1.3	4.0	1.9	0.046	32.3	25.5	10.7	15.8	31.6	46.7
Dominant wave lgth.	675.6	604.9	574.9	538.6	487.7	449.2	579.9	581.5	471.1	473.4	475.7	477.2
Excitation purity.	100.0	100.0	100.0	96.7	90.4	99.7	99.0	99.1	63.0	54.5	33.7	20.7

Percent transmittance

Wave length	No. 78C •	No. 79 •	No. 80A •	No. 81 •	No. 81A •	No. 81B •	No. 81C •	No. 81D •	No. 81EF •	No. 82 •	No. 82A •	No. 82B •
400	74.9	24.0	67.6	77.7	65.1	55.1	46.1	38.2	30.7	83.0	80.1	76.7
10	76.6	26.0	73.1	78.1	65.9	55.8	46.6	38.4	31.5	83.7	80.8	78.0
20	77.9	29.0	76.8	79.0	67.6	57.7	49.0	41.0	34.3	84.6	81.6	79.2
30	78.9	31.0	7.77	80.5	70.2	61.0	52.5	45.5	38.6	85.1	82.2	79.7
40	79.4	32.2	76.5	81.9	72.8	64.5	57.2	50.0	43.2	85.4	82.4	79.7
50	79.5	32.7	73.0	83.0	74.8	67.2	60.5	53.9	47.4	85.4	82.4	79.2
60	79.3	31.4	69.0	83.7	76.0	69.1	63.0	56.5	50.2	85.0	81.7	78.0
70	78.6	28.8	63.6	84.3	77.1	70.6	64.2	58.1	52.0	84.6	80.7	76.3
80	77.8	25.6	57.6	84.6	77.8	71.3	65.0	59.0	53.0	84.0	79.3	74.4
90	76.7	22.2	51.3	84.9	78.3	71.8	65.7	60.0	54.0	83.3	78.0	72.1
500	75.5	19.3	45.2	85.3	78.6	72.6	66.4	60.8	55.4	82.6	76.6	70.2
10	74.2	16.8	39.4	85.4	79.0	72.9	66.5	61.1	56.2	82.0	75.3	68.3
20	73.0	14.2	34.2	85.5	79.5	73.2	67.0	61.6	57.0	81.4	74.0	66.5
30	72.1	12.7	30.0	86.0	80.4	74.5	68.8	62.5	59.5	81.0	73.1	65.5
40	71.5	11.8	27.1	86.5	81.5	76.0	71.0	66.1	62.7	80.8	72.7	65.0
50	70.7	11.0	24.8	86.8	82.3	77.0	72.0	67.3	64.5	80.6	72.4	64.5
60	69.8	9.76	23.5	87.0	82.6	77.6	72.5	68.0	65.3	80.4	71.8	63.8
70	69.0	8.81	22.6	87.1	82.7	77.8	72.7	68.3	65.8	80.2	71.5	63.2
80	68.8	8.50	22.6	87.1	82.8	78.0	73.0	68.5	66.0	80.2	71.5	63.2
90	68.6	8.29	23.2	87.4	83.1	78.2	74.0	69.5	66.5	80.3	71.7	63.4
600	68.0	7.56	23.7	87.6	84.0	79.1	75.6	72.0	68.1	80.2	71.5	63.0
10	66.7	6.45	23.2	88.1	85.0	81.0	78.5	75.0	71.6	79.3	70.3	61.5
20	65.0	5.13	21.0	88.8	86.1	83.1	80.8	78.0	74.7	78.4	68.5	59.0
30	63.8	4.17	18.2	89.2	87.0	84.2	82.1	79.8	77.0	77.5	66.9	56.9
40	63.0	3.47	15.8	89.4	87.4	85.1	83.0	80.8	78.4	76.8	65.5	55.0
50	62.7	3.16	14.5	89.5	87.7	85.6	83.5	81.5	79.2	76.5	64.8	54.1
60	63.0	3.09	13.8	89.8	88.0	86.0	84.1	82.1	80.1	76.2	64.6	53.7
70	63.3	3.16	13.4	90.0	88.2	86.5	84.8	83.0	80.9	76.1	64.5	53.7
80	63.4	3.16	12.7	90.1	88.5	87.0	85.5	83.7	81.8	76.1	64.4	53.5
90	63.6	3.16	11.7	90.3	89.0	87.5	86.1	84.6	82.9	76.2	64.2	53.4
700	65.0	3.31	11.5	90.5	89.2	88.0	86.8	85.5	84.0	77.1	64.6	54.1
Luminous transmit.	70.4	11.3	28.4	86.8	82.0	76.9	72.0	67.4	64.0	80.7	72.5	64.6
Dominant wave lgth.	479.8	474.8	471.7	576.7	577.5	577.8	577.4	579.5	579.0	477.5	476.6	475.6
Excitation purity.	6.8	52.8	45.9	2.9	6.0	8.7	10.7	14.7	19.0	3.0	6.3	10.2

* Some transmission below 400 mµ. Consult the manufacturer.

TRANSMISSION OF WRATTEN FILTERS (Continued)

	Percent transmittance												
Wave length	No. 82C	No. 83	No. 85	No. 85B	No. 86	No. 86A	No. 86B	No. 86C	No. 89B	No. 90	No. 96	No. 97	No. 102
400	73.4	13.5	6.0	1.59	0.50	8.00	20.0	44.0	—	—	4.28	—	1.12
10	75.0	13.1	18.0	9.32	0.81	12.2	26.1	55.0	—	—	4.91	—	0.96
20	76.4	13.5	28.4	15.5	1.55	16.7	31.6	62.0	—	—	5.50	—	0.89
30	77.2	14.1	33.4	19.0	2.88	21.5	37.5	66.6	—	—	6.17	—	0.96
40	77.2	15.6	36.2	20.8	5.50	27.8	44.0	70.8	—	—	6.92	—	1.23
50	76.6	17.8	38.1	22.1	9.10	34.2	50.1	74.3	—	—	7.50	—	1.86
60	75.2	21.0	40.4	24.3	13.5	40.4	55.4	76.8	—	—	7.81	—	3.23
70	73.2	25.5	43.0	27.5	17.8	45.0	59.5	78.7	—	—	8.15	0.22	6.45
80	70.7	30.2	45.3	30.9	21.3	48.7	62.5	80.2	—	—	8.47	0.43	14.0
90	68.1	35.8	47.2	34.3	24.5	51.2	64.6	81.2	—	—	8.60	0.39	21.6
500	65.7	43.5	48.9	38.3	26.8	52.8	66.0	81.7	—	—	8.73	0.15	30.7
10	63.5	46.3	49.2	40.7	27.9	53.4	66.4	81.9	—	—	8.85	—	41.4
20	61.5	47.2	48.2	40.6	28.6	53.7	66.6	82.0	—	—	8.90	—	51.3
30	59.9	48.3	48.3	40.7	30.4	55.0	67.6	82.4	—	—	9.01	—	59.4
40	59.1	49.6	49.2	41.6	32.5	56.5	69.0	83.0	—	—	9.07	—	64.2
50	58.3	51.8	51.0	43.2	35.0	58.5	70.2	83.5	—	—	9.20	—	66.7
60	57.2	56.5	55.8	47.1	41.2	63.0	73.0	84.6	—	9.00	9.30	—	66.3
70	56.2	65.0	64.5	53.0	53.0	70.9	78.1	86.8	—	30.5	9.20	—	63.0
80	56.1	75.5	75.0	68.1	67.5	79.0	84.0	88.9	—	34.3	9.19	—	58.0
90	56.0	83.0	83.0	78.1	76.5	85.2	87.5	89.9	—	25.2	9.54	—	51.9
600	55.0	87.3	87.2	85.0	85.0	88.1	89.3	90.6	—	16.1	9.64	—	45.2
10	53.0	89.3	88.9	88.0	88.1	89.8	90.3	91.0	—	11.3	9.73	—	37.8
20	50.2	90.4	90.0	89.6	89.6	90.5	90.7	91.1	—	7.40	9.56	—	30.5
30	47.4	90.8	90.5	90.3	90.4	90.8	90.9	91.2	—	2.91	9.27	—	25.0
40	45.2	91.0	90.7	90.7	90.7	91.1	91.1	91.3	—	0.76	9.10	—	20.6
50	44.1	91.1	90.9	90.9	91.0	91.2	91.2	91.4	—	0.29	9.07	—	17.5
60	43.6	91.3	91.0	91.0	91.1	91.3	91.3	91.5	—	0.41	9.00	—	15.2
70	43.5	91.5	91.0	91.2	91.2	91.4	91.4	91.6	—	2.30	9.13	—	13.7
80	43.1	91.5	91.0	91.3	91.3	91.4	91.5	91.6	0.10	9.52	9.08	0.44	12.8
90	42.8	91.5	91.0	91.3	91.3	91.5	91.6	91.6	1.58	28.5	9.21	5.02	12.1
700	43.5	91.5	91.0	91.3	91.3	91.5	91.6	91.6	11.2	51.9	9.52	18.7	12.0
Luminous transmit.	58.1	61.4	62.5	55.5	49.7	67.1	75.5	85.4	0.017	9.8	9.1	0.041	50.8
Dominant wave lgth.	477.2	581.5	587.7	585.7	585.7	581.7	579.6	577.6	700	583.1	572.4	555.0c	564.9
Excitation purity.	14.5	55.4	30.3	48.0	69.7	37.1	24.1	9.0	100	100.0	12.1	48.0	80.0

	Percent transmittance												
Wave length	No. 106	CC-05R	CC-10R	CC-20R	CC-30R	CC-40R	CC-50R	CC-05B	CC-10B	CC-20B	CC-30B	CC-40B	CC-50B
400	—	81.0	73.0	61.5	51.6	42.5	36.4	87.0	85.5	82.2	80.2	77.0	74.1
10	—	81.0	72.4	60.0	50.0	40.0	33.9	87.5	86.4	84.0	82.5	80.3	78.4
20	0.10	81.0	72.0	58.6	48.2	38.2	31.9	87.7	87.2	85.0	84.0	82.2	80.7
30	0.20	81.1	71.6	57.7	47.0	36.8	30.5	88.0	87.5	85.3	84.3	82.5	81.1
40	0.35	81.2	71.5	57.2	46.4	36.0	29.7	88.1	87.5	85.0	83.5	81.3	79.8
50	0.58	81.4	71.6	57.2	46.4	36.1	29.6	88.1	87.2	83.9	81.9	78.7	76.6
60	0.98	81.7	72.4	58.5	47.5	37.5	31.0	87.9	86.4	82.5	80.3	75.9	72.9
70	1.5	82.3	73.7	60.6	49.9	40.0	33.6	87.5	85.3	80.3	76.2	72.0	67.9
80	2.3	82.8	74.9	62.0	52.0	42.5	35.9	87.0	84.0	77.8	72.5	67.5	62.7
90	3.5	83.2	75.8	63.5	53.9	44.8	37.9	86.2	82.4	74.2	68.3	62.3	56.6
500	5.2	83.3	76.6	64.6	55.2	46.1	39.4	85.2	80.5	71.2	63.8	56.7	50.1
10	7.7	83.0	76.1	64.0	54.5	46.0	38.5	84.4	78.6	67.7	58.7	51.0	44.5
20	10.7	82.4	74.9	61.5	51.6	42.5	35.0	83.5	77.0	64.0	54.4	46.0	38.6
30	15.1	81.6	73.5	59.4	48.5	38.5	31.5	82.6	75.2	61.5	50.7	41.6	34.1
40		81.2	72.5	57.8	46.5	36.6	29.4	82.1	73.9	59.5	48.3	39.9	31.3
50		81.1	72.4	57.1	45.6	35.6	28.7	81.5	73.0	57.5	46.3	35.9	29.5
60		81.4	72.8	58.0	46.9	36.4	29.7	81.4	72.7	57.9		36.1	28.6
70		82.5	74.5	60.6	49.2	39.2	32.7		73.0	59.3	50.7	40.8	30.5
80			77.3	65.0	54.6	45.0	38.6		73.9	61.6	53.1	43.5	33.4
90			80.7	71.0	61.8	53.5	47.6		75.1	63.5		44.7	36.0
600				77.0	70.5	64.0	58.7		76.3	64.5		44.3	37.8
10				82.0	77.8		69.8		76.7	64.3		42.5	
20				86.2	83.5				76.5	63.1		40.2	
30					87				76.6	61.6		39.0	
40									74.5	60.6		38.4	
50									74.0			37.7	
60									73.8			36.5	
70									73.3			35.4	

TRANSMISSION OF WRATTEN FILTERS (Continued)

Percent transmittance

Wave length	CC-05G	CC-10G	CC-20G	CC-30G	CC-40G	CC-50G	CC-05Y	CC-10Y	CC-20Y	CC-30Y	CC-40Y	CC-50Y
400	80.0	73.1	58.8	48.0	39.7	32.0	81.0	74.5	61.3	50.5	43.0	34.5
10	80.7	72.9	57.8	46.5	38.1	30.3	80.6	73.2	59.0	47.4	39.5	30.5
20	81.0	72.8	57.3	45.8	37.3	29.5	80.4	72.6	57.8	46.0	37.5	29.0
30	81.4	72.7	57.0	45.5	36.5	29.0	80.4	72.5	57.5	45.6	36.5	28.7
40	81.6	73.0	57.3	45.8	36.6	29.1	80.6	72.8	57.8	46.5	36.8	29.5
50	82.1	73.9	58.4	46.9	38.1	30.6	81.2	74.0	59.5	48.5	38.5	31.5
60	83.0	75.5	61.4	50.3	41.5	34.3	82.5	76.0	63.0	52.5	42.5	36.2
70	84.4	78.0	65.4	55.8	47.0	40.5	83.9	78.5	67.5	48.2	48.8	43.5
80	85.6	80.4	70.0	61.8	53.5	47.8	85.3	81.2	72.3	64.9	56.2	54.0
90	86.8	83.0	75.2	68.9	61.3	57.0	87.0	84.4	78.0	72.4	66.0	64.0
500	87.9	85.9	80.3	76.4	70.7	68.0	88.4	87.2	84.0	81.0	77.0	75.5
10	88.7	87.5	83.8	80.9	77.8	75.3	89.5	89.0	88.0	86.6	85.5	84.2
20	89.0	88.1	84.9	82.3	79.5	77.5	90.0	90.0	89.6	89.1	89.0	88.5
30	89.0	88.0	84.6	81.7	79.4	77.0	90.4	90.4	90.0	89.7	89.9	89.6
40	89.0	87.6	83.7	80.5	77.8	74.8	90.7	90.7	90.6	90.4	90.2	90.0
50	88.6	87.1	82.4	78.6	75.8	72.2	90.9	90.9	90.8	90.6	90.4	90.3
60	88.1	86.3	80.9	76.2	72.9	68.8	91.0	91.0	90.9	90.8	90.7	90.6
70	87.5	85.3	79.0	73.5	69.3	64.8	91.3	91.3	91.0	90.9	90.8	90.7
80	87.0	84.1	77.0	70.4	65.3	60.3	91.4	91.4	91.1	91.0	90.8	90.7
90	86.4	82.8	74.5	67.2	61.9	55.9	91.4	91.4	91.2	91.1	90.9	90.8
600	85.7	81.5	72.0	64.1	57.7	51.7	91.4	91.4	91.3	91.2	90.9	90.8
10	85.0	80.0	69.5	60.7	53.7	47.3	91.4	91.4	91.3	91.2	90.9	90.9
20	84.0	78.5	66.5	57.2	49.8	42.5	91.4	91.4	91.3	91.2	91.0	90.9
30	83.0	76.9	63.8	53.7	45.5	38.0	91.5	91.5	91.4	91.3	91.0	91.0
40	82.2	75.6	61.5	50.8	42.0	34.6	91.5	91.5	91.4	91.3	91.0	91.0
50	81.9	74.9	60.1	49.1	40.2	32.5	91.5	91.5	91.4	91.3	91.1	91.1
60	81.5	74.4	59.4	48.1	39.5	31.5	91.5	91.5	91.4	91.3	91.1	91.1
70	81.4	74.0	58.8	47.5	38.5	31.0	91.5	91.5	91.4	91.4	91.2	91.2
80	81.1	73.5	58.1	46.6	37.6	29.9	91.5	91.5	91.4	91.4	91.2	91.2
90	81.1	73.2	57.6	46.0	36.6	28.9	91.5	91.5	91.4	91.4	91.3	91.3
700	81.5	73.5	58.0	46.4	36.5	28.7	91.5	91.5	91.4	91.4	91.3	91.3
Luminous transmit.	87.2	84.5	77.8	72.2	67.7	63.3	90.4	90.1	89.1	88.2	87.4	86.9
Dominant wave lgth.	553.0	555.5	555.0	554.0	554.3	553.4	572.0	571.3	571.4	571.3	571.3	571.2
Excitation purity.	2.3	5.2	10.9	15.8	21.1	25.9	5.3	9.6	18.8	28.3	35.7	42.0

Percent transmittance

Wave length	CC-05M	CC-10M	CC-20M	CC-30M	CC-40M	CC-50M	CC-05C	CC-10C	CC-20C	CC-30C	CC-40C	CC-50C
400	87.6	86.6	85.6	84.2	82.3	80.9	87.3	86.0	83.9	82.3	80.4	78.8
10	88.2	87.7	86.6	85.7	84.6	83.6	88.2	87.5	85.2	84.5	83.4	82.7
20	88.6	88.0	87.0	85.9	85.2	84.4	88.7	88.1	86.5	86.0	85.3	84.8
30	88.7	88.0	86.9	85.6	84.4	83.6	89.0	88.6	87.5	87.0	86.3	85.9
40	88.7	87.9	86.0	84.7	82.5	81.4	89.3	89.0	87.7	87.3	86.6	86.1
50	88.6	87.5	84.9	82.8	80.0	78.1	89.5	89.1	87.8	87.5	86.6	86.0
60	88.4	86.5	83.1	80.0	76.1	73.7	89.6	89.1	87.7	87.3	86.4	85.7
70	87.8	85.2	80.8	76.4	71.3	68.0	89.7	89.0	87.5	87.0	85.8	85.2
80	87.0	83.6	77.9	72.1	65.8	61.7	89.7	89.0	87.2	86.5	85.3	84.3
90	86.0	81.8	74.4	67.0	60.0	55.0	89.7	89.0	87.0	86.0	84.4	83.4
500	85.0	79.7	70.5	61.7	53.7	48.1	89.6	89.0	86.5	85.2	83.5	82.3
10	83.8	77.5	66.7	56.5	47.7	41.6	89.6	88.7	86.0	84.4	82.4	80.8
20	82.7	75.3	63.4	52.0	42.8	36.3	89.5	88.5	85.2	83.5	81.1	
30	81.8	73.7	60.5	48.6	39.0	31.9	89.4	88.0	84.3	82.4	79.6	
40	81.3	72.5	58.6	46.6	36.7	29.8	89.2	87.5	83.4	81.0	79.6	
50	81.2	72.2	58.0	46.0	36.0	29.1	88.9	87.0	82.3	79.0		
60	82.2	72.8	58.3	46.5	36.7	29.7	88.5	86.1	80.5	76.7		
70	84.4	72.6	60.5	49.8	40.2	32.3	88.0	85.0	78.5	74.		
80	85.8	77.2	64.9	55.6	46.2	39.0	87.5	83.8	76.1			
90	88.0	80.8	70.6	63.3	54.9	48.7	87.0	82.5	73.9			
600	89.3	84.5	77.1	71.6	64.9	59.9	86.4	81.0	71.			
10	90.2	87.0	82.2	79.2	74.9	70.	85.5	79.5				
20	90.6	88.9	86.1	84.1	81.		77.9					
30	90.8	90.0	88.7	87.4	89.3							
40	91.0		90.0	89.3	90.							
50	91.1		90.5	90.								
60	91.2		90.8									
70	91.3		91.0									
80	91.4	91.	91.2									
90	91.5	91.	91.4									
700			91.5									
Luminous transmit.			77.5									
Dominant wave lgth.	84.2											

Percent transmittance

Wave length	No. 82C	No. 83	No. 85	No. 85B	No. 86	No. 86A	No. 86B	No. 86C	No. 89B	No. 90	No. 96	No. 97	No. 102
400	73.4	13.5	6.0	1.59	0.50	8.00	20.0	44.0	—	—	4.28	—	1.12
10	75.0	13.1	18.0	9.32	0.81	12.2	26.1	55.0	—	—	4.91	—	0.96
20	76.4	13.5	28.4	15.5	1.55	16.7	31.6	62.0	—	—	5.50	—	0.89
30	77.2	14.1	33.4	19.0	2.88	21.5	37.5	66.6	—	—	6.17	—	0.96
40	77.2	15.6	36.2	20.8	5.50	27.8	44.0	70.8	—	—	6.92	—	1.23
50	76.6	17.8	38.1	22.1	9.10	34.2	50.1	74.3	—	—	7.50	—	1.86
60	75.2	21.0	40.4	24.3	13.5	40.4	55.4	76.8	—	—	7.81	0.22	3.23
70	73.2	25.5	43.0	27.5	17.8	45.0	59.5	78.7	—	—	8.15	0.22	6.45
80	70.7	30.2	45.3	30.9	21.3	48.7	62.5	80.2	—	—	8.47	0.43	14.0
90	68.1	35.8	47.2	34.3	24.5	51.2	64.6	81.2	—	—	8.60	0.39	21.6
500	65.7	43.5	48.9	38.3	26.8	52.8	66.0	81.7	—	—	8.73	0.15	30.7
10	63.5	46.3	49.2	40.7	27.9	53.4	66.4	81.9	—	—	8.85	—	41.4
20	61.5	47.2	48.2	40.6	28.6	53.7	66.6	82.0	—	—	8.90	—	51.3
30	59.9	48.3	48.3	40.7	30.4	55.0	67.6	82.4	—	—	9.01	—	59.4
40	59.1	49.6	49.2	41.6	32.5	56.5	69.0	83.0	—	—	9.07	—	64.2
50	58.3	51.8	51.0	43.2	35.0	58.5	70.2	83.5	—	—	9.20	—	66.7
60	57.2	56.5	55.8	47.1	41.2	63.0	73.0	84.6	—	9.00	9.30	—	66.3
70	56.2	65.0	64.5	56.0	53.0	70.9	78.1	86.0	—	30.5	9.20	—	63.0
80	56.1	75.5	75.0	68.1	67.5	79.0	84.0	88.9	—	34.3	9.19	—	58.0
90	56.0	83.0	83.0	78.1	76.5	85.2	87.5	89.9	—	25.2	9.54	—	51.9
600	55.0	87.3	87.2	85.0	85.0	88.1	89.3	90.6	—	16.1	9.64	—	45.2
10	53.0	89.3	88.9	88.0	88.1	89.8	90.3	91.0	—	11.3	9.73	—	37.8
20	50.2	90.4	90.0	89.6	89.6	90.5	90.7	91.1	—	7.40	9.56	—	30.5
30	47.4	90.8	90.5	90.3	90.4	90.8	90.9	91.2	—	2.91	9.27	—	25.0
40	45.2	91.0	90.7	90.7	90.7	91.1	91.1	91.3	—	0.76	9.10	—	20.6
50	44.1	91.1	90.9	90.9	91.0	91.2	91.2	91.4	—	0.29	9.07	—	17.5
60	43.6	91.3	91.0	91.0	91.1	91.3	91.3	91.5	—	0.41	9.00	—	15.2
70	43.5	91.5	91.0	91.2	91.2	91.4	91.4	91.6	—	2.30	9.13	—	13.7
80	43.1	91.5	91.0	91.3	91.3	91.4	91.5	91.6	0.10	9.52	9.08	0.44	12.8
90	42.8	91.5	91.0	91.3	91.3	91.5	91.6	91.6	1.58	28.5	9.21	5.02	12.1
700	43.5	91.5	91.0	91.3	91.3	91.5	91.6	91.6	11.2	51.9	9.52	18.7	12.0
Luminous transmit.	58.1	61.4	62.5	55.5	49.7	67.1	75.5	85.4	0.017	9.8	9.1	0.041	50.8
Dominant wave lgth.	477.2	581.5	587.7	585.7	585.7	581.7	579.6	577.6	700	583.1	572.4	555.0	564.9
Excitation purity.	14.5	55.4	30.3	48.0	69.7	37.1	24.1	9.0	100	100.0	12.1	48.0	80.0

Percent transmittance

Wave length	No. 106	CC-05R	CC-10R	CC-20R	CC-30R	CC-40R	CC-50R	CC-05B	CC-10B	CC-20B	CC-30B	CC-40B	CC-50B
400	—	81.0	75.0	61.5	51.0	42.0	35.3	87.3	86.1	84.0	83.0	77.0	74.1
10	—	81.0	72.4	60.0	50.0	40.0	33.9	87.5	86.4	84.0	82.5	80.3	78.4
20	0.10	81.0	72.0	58.6	48.2	38.2	31.9	87.7	87.2	85.0	84.0	82.2	80.7
30	0.20	81.1	71.6	57.7	47.0	36.8	30.5	88.0	87.5	85.3	84.3	82.5	81.1
40	0.35	81.2	71.5	57.2	46.4	36.0	29.7	88.1	87.5	85.0	83.5	81.3	79.8
50	0.58	81.4	71.6	57.2	46.4	36.1	29.6	88.1	87.2	83.9	81.9	78.7	76.6
60	0.98	81.7	72.4	58.5	47.5	37.5	31.0	87.9	86.4	82.5	79.5	75.9	72.9
70	1.5	82.3	73.7	60.6	49.9	40.0	33.4	87.5	85.3	80.3	76.2	72.0	67.9
80	2.3	82.8	74.9	62.0	52.0	42.5	35.9	87.0	84.0	77.8	72.5	67.5	62.7
90	3.5	83.2	75.8	63.5	53.9	44.8	37.9	86.2	82.4	74.2	68.3	62.3	56.6
500	5.2	83.3	76.6	64.6	55.2	46.1	39.4	85.2	80.5	71.2	63.8	56.7	50.1
10	7.7	83.0	76.1	64.0	54.5	46.0	38.5	84.4	78.6	67.7	58.7	51.0	44.5
20	10.7	82.4	74.9	61.5	51.6	42.5	35.0	83.5	77.0	64.0	54.4	46.0	38.6
30	15.1	81.6	73.5	59.4	48.5	38.5	31.5	82.6	75.2	61.5	50.7	41.6	34.1
40	20.2	81.2	72.5	57.8	46.5	36.6	29.4	82.1	73.9	59.5	48.3	39.0	31.3
50	25.7	81.1	72.4	57.1	45.6	35.6	28.7	81.5	73.0	58.0	46.6	36.9	29.5
60	31.0	81.4	72.8	58.0	46.9	36.4	29.7	81.4	72.7	57.5	45.9	35.9	28.6
70	35.6	82.5	74.5	60.6	49.2	39.2	32.7	81.4	73.0	57.9	46.3	36.1	28.7
80	43.2	83.9	77.3	65.0	54.6	45.0	38.6	81.9	73.9	59.3	47.9	37.8	30.5
90	53.8	85.7	80.7	71.0	61.8	53.5	47.6	82.7	75.1	61.6	50.3	40.8	33.4
600	65.6	87.6	84.0	77.0	70.5	64.0	58.7	83.4	76.3	63.5	53.0	43.5	36.0
10	77.0	89.0	86.5	82.0	77.8	73.0	69.8	83.6	76.7	64.5	54.6	44.7	37.8
20	82.8	90.0	88.9	86.2	83.5	80.8	78.3	83.5	76.5	64.3	54.4	44.3	37.5
30	86.0	90.6	89.9	88.1	87.2	85.1	84.2	83.2	76.6	63.1	53.2	42.5	35.6
40	87.6	91.1	90.5	89.8	89.2	88.0	87.5	82.8	74.5	61.6	51.5	40.2	33.7
50	88.7	91.2	90.8	90.5	90.3	89.5	89.2	82.5	74.0	60.6	50.3	39.0	32.4
60	89.5	91.3	91.1	90.8	90.6	90.4	90.1	82.5	73.8	60.1	49.5	38.4	31.8
70	90.0	91.4	91.3	91.0	90.9	90.8	90.7	82.3	73.3	59.6	49.0	37.7	31.0
80	90.5	91.6	91.5	91.2	91.1	91.0	91.1	82.0	72.8	58.6	48.2	36.5	30.0
90	90.8	91.7	91.7	91.4	91.4	91.3	91.1	81.9	72.5	58.1	47.2	35.4	29.0
700	91.0	91.9	91.9	91.5	91.5	91.4	91.2	82.2	73.0	58.5	47.5	35.6	29.0
Luminous transmit.	34.6	83.7	77.0	65.3	55.9	47.3	41.3	82.8	75.5	62.3	52.0	42.8	35.7
Dominant wave lgth.	589.4	605.0	597.8	604.2	605.8	605.5	608.5	459.0	462.0	460.0	461.0	463.2	462.5
Excitation purity.	95.2	2.0	4.7	8.5	12.3	17.3	21.4	2.8	6.3	13.2	20.2	27.7	34.2

* Some transmission below 400 mμ. Consult the manufacturer.

Percent transmittance

Wave length	CC-05G *	CC-10G *	CC-20G *	CC-30G *	CC-40G *	CC-50G *	CC-05Y *	CC-10Y *	CC-20Y *	CC-30Y *	CC-40Y *	CC-50Y *
400	80.0	73.1	58.8	48.0	39.7	32.0	81.0	74.5	61.3	50.5	43.0	34.5
10	80.7	72.9	57.8	46.5	38.1	30.3	80.6	73.2	59.0	47.4	39.5	30.5
20	81.0	72.8	57.3	45.8	37.3	29.5	80.4	72.6	57.8	46.0	37.5	29.0
30	81.4	72.7	57.0	45.5	36.5	29.0	80.4	72.5	57.5	45.6	36.5	28.7
40	81.6	73.0	57.3	45.8	36.6	29.1	80.6	72.8	57.8	46.5	36.8	29.5
50	82.1	73.9	58.4	46.9	38.1	30.6	81.2	74.0	59.5	48.5	38.5	31.5
60	83.0	75.5	61.4	50.3	41.5	34.3	82.5	76.0	63.0	52.5	42.5	36.2
70	84.4	78.0	65.4	55.8	47.0	40.5	83.9	78.5	67.5	48.2	48.8	43.5
80	85.6	80.4	70.0	61.8	53.5	47.8	85.3	81.2	72.3	64.9	56.2	54.0
90	86.8	83.0	75.2	68.9	61.3	57.0	87.0	84.4	78.0	72.4	66.0	64.0
500	87.9	85.9	80.3	76.4	70.7	68.0	88.4	87.2	84.0	81.0	77.0	75.5
10	88.7	87.5	83.8	80.9	77.8	75.3	89.5	89.0	88.0	86.6	85.5	84.2
20	89.0	88.1	84.9	82.3	79.5	77.5	90.0	90.0	89.6	89.1	89.0	88.5
30	89.0	88.0	84.6	81.7	79.4	77.0	90.4	90.4	90.0	89.7	89.9	89.6
40	89.0	87.6	83.7	80.5	77.8	74.8	90.7	90.7	90.6	90.4	90.2	90.0
50	88.6	87.1	82.4	78.6	75.8	72.2	90.9	90.9	90.8	90.6	90.4	90.3
60	88.1	86.3	80.9	76.2	72.9	68.8	91.0	91.0	90.9	90.8	90.7	90.6
70	87.5	85.3	79.0	73.5	69.3	64.8	91.3	91.3	91.0	90.9	90.8	90.7
80	87.0	84.1	77.0	70.4	65.3	60.3	91.4	91.4	91.1	91.0	90.8	90.7
90	86.4	82.8	74.5	67.2	61.9	55.9	91.4	91.4	91.2	91.1	90.9	90.8
600	85.7	81.5	72.0	64.1	57.7	51.7	91.4	91.4	91.3	91.2	90.9	90.8
10	85.0	80.0	69.5	60.7	53.7	47.3	91.4	91.4	91.3	91.2	90.9	90.9
20	84.0	78.5	66.5	57.2	49.8	42.5	91.4	91.4	91.3	91.2	91.0	90.9
30	83.0	76.9	63.8	53.7	45.5	38.0	91.5	91.5	91.4	91.3	91.0	91.0
40	82.2	75.6	61.5	50.8	42.0	34.6	91.5	91.5	91.4	91.3	91.0	91.0
50	81.9	74.9	60.1	49.1	40.2	32.5	91.5	91.5	91.4	91.3	91.1	91.1
60	81.5	74.4	59.4	48.1	39.5	31.5	91.5	91.5	91.4	91.3	91.1	91.1
70	81.4	74.0	58.8	47.5	38.5	31.0	91.5	91.5	91.4	91.4	91.2	91.2
80	81.1	73.5	58.1	46.6	37.6	29.9	91.5	91.5	91.4	91.4	91.2	91.2
90	81.1	73.2	57.6	46.0	36.6	28.9	91.5	91.5	91.4	91.4	91.3	91.3
700	81.5	73.5	58.0	46.4	36.5	28.7	91.5	91.5	91.4	91.4	91.3	91.3
Luminous transmit.	87.2	84.5	77.8	72.2	67.7	63.3	90.4	90.1	89.1	88.2	87.4	86.9
Dominant wave lgth.	553.0	555.5	555.0	554.0	554.3	553.4	572.0	571.3	571.4	571.3	571.3	571.2
Excitation purity.	2.3	5.2	10.9	15.8	21.1	25.9	5.3	9.6	18.8	28.3	35.7	42.0

Percent transmittance

Wave length	CC-05M *	CC-10M *	CC-20M *	CC-30M *	CC-40M *	CC-50M *	CC-05C *	CC-10C *	CC-20C *	CC-30C *	CC-40C *	CC-50C *
400	87.6	86.6	85.6	84.2	82.3	80.9	87.3	86.0	83.9	82.3	80.4	78.8
10	88.2	87.7	86.6	85.7	84.6	83.6	88.2	87.5	85.2	84.5	83.4	82.7
20	88.6	88.0	87.0	85.9	85.2	84.4	88.7	88.1	86.5	86.0	85.3	84.8
30	88.7	88.0	86.9	85.6	84.4	83.6	89.0	88.6	87.5	87.0	86.3	85.9
40	88.7	87.9	86.0	84.7	82.5	81.4	89.3	89.0	87.7	87.3	86.6	86.1
50	88.6	87.5	84.9	82.8	80.0	78.1	89.5	89.1	87.8	87.5	86.6	86.0
60	88.4	86.5	83.1	80.0	76.1	73.7	89.6	89.1	87.7	87.3	86.4	85.7
70	87.8	85.2	80.8	76.4	71.3	68.0	89.7	89.0	87.5	87.0	85.8	85.2
80	87.0	83.6	77.9	72.1	65.8	61.7	89.7	89.0	87.2	86.5	85.3	84.3
90	86.0	81.8	74.4	67.0	60.0	55.0	89.7	89.0	87.0	86.0	84.4	83.4
500	85.0	79.7	70.5	61.7	53.7	48.1	89.6	89.0	86.5	85.2	83.5	82.3
10	83.8	77.5	66.7	56.5	47.7	41.6	89.6	88.7	86.0	84.4	82.4	80.8
20	82.7	75.3	63.4	52.0	42.8	36.3	89.5	88.5	85.2	83.5	81.1	79.2
30	81.8	73.7	60.5	48.6	39.0	31.9	89.4	88.0	84.3	82.4	79.6	77.3
40	81.3	72.5	58.6	46.6	36.7	29.8	89.2	87.5	83.4	81.0	77.7	75.0
50	81.2	72.2	58.0	46.0	36.0	29.1	88.9	87.0	82.3	79.0	75.3	72.2
60	81.5	72.8	58.3	46.5	36.7	29.7	88.5	86.1	80.5	76.7	72.7	69.0
70	82.5	74.6	60.5	49.8	40.2	32.3	88.0	85.0	78.5	74.0	69.3	65.0
80	84.0	77.3	64.9	55.6	46.2	39.0	87.5	83.8	76.1	70.9	65.4	60.5
90	85.8	80.8	70.6	63.3	54.9	48.7	87.0	82.5	73.9	67.5	61.6	55.8
600	88.0	84.5	77.1	71.6	64.9	59.9	86.4	81.0	71.2	64.1	57.6	51.3
10	89.3	87.0	82.2	79.2	74.9	70.7	85.5	79.5	68.5	60.4	53.4	36.2
20	90.2	88.9	86.1	84.1	81.4	79.2	84.5	77.9	65.5	56.7	49.2	42.3
30	90.6	90.0	88.7	87.4	86.0	84.5	83.8	76.3	62.7	53.1	45.0	38.0
40	90.8	90.5	90.0	89.3	88.7	87.6	83.3	75.1	60.8	50.4	42.0	34.9
50	91.0	90.7	90.5	90.2	90.0	89.7	82.8	74.4	59.5	48.8	40.2	32.9
60	91.1	91.0	90.8	90.8	90.4	90.4	82.5	74.0	58.5	48.0	39.4	32.0
70	91.2	91.2	91.0	91.0	90.7	90.7	82.4	73.6	57.9	47.2	38.6	31.0
80	91.3	91.3	91.2	91.1	91.0	91.0	82.0	73.0	57.5	46.0	37.5	29.9
90	91.4	91.4	91.4	91.3	91.3	91.3	82.0	72.8	57.4	45.5	36.7	29.1
700	91.5	91.5	91.5	91.5	91.5	91.5	82.5	74.0	58.5	46.4	37.3	29.8
Luminous transmit.	84.2	77.9	67.1	58.1	50.0	44.0	88.0	85.1	78.9	74.8	70.5	66.7
Dominant wave lgth.	541.0	547.5	551.2	550.0	550.3	551.2	489.2	487.5	486.5	486.2	486.1	485.5
Excitation purity.	3.5	7.4	14.4	21.5	28.3	34.0	1.6	4.1	8.9	12.8	17.5	20.2

* Some transmission below 400 mμ. Consult the manufacturer.

TRANSMISSION OF WRATTEN FILTERS (Continued)

Wave length	Percent transmittance				Wave length	Percent transmittance			
	No. 87	No. 87C	No. 88A	No. 89B		No. 87	No. 87C	No. 88A	No. 89B
700	—	—	—	11.2	30	74.1	17.8	84.7	88.8
10	—	—	—	32.4	40	77.7	28.2	85.5	89.0
20	—	—	—	57.6	50	81.4	41.0	86.1	89.2
30	—	—	7.4	69.1	60	84.0	53.8	86.6	89.4
40	0.10	—	32.8	77.6	70	85.4	61.6	87.2	89.6
50	2.19	—	56.3	83.1	80	86.8	69.2	87.5	89.8
60	7.95	—	69.2	85.0	90	87.8	74.1	87.8	89.9
70	17.4	—	74.2	86.1	900	88.4	78.5	88.0	90.0
80	31.6	—	77.6	87.0	10	88.8	81.5	88.2	90.1
90	43.7	—	79.7	87.7	20	89.1	83.6	88.4	90.2
800	53.8	0.32	81.4	88.1	30	89.1	85.1	88.6	90.3
10	61.7	3.20	82.6	88.4	40	89.1	86.0	88.8	90.4
20	69.2	8.90	83.7	88.6	50	89.1	87.0	89.0	90.5

* Some transmission below 400 mμ. Consult the manufacturer.

TRANSMISSIBILITY FOR RADIATIONS

Ratio of the transmitted light to the incident light for a definite thickness of the substance, usually 1 cm.

Glass

Glass in general is opaque to the ultra-violet and infrared. Uviol glass is transparent to the longer radiations of the ultra-violet.

Coefficient of transparency of glass for visible and ultra-violet radiations.

Normal incidence, thickness 1 cm.

Wave length microns	0.309	0.330	0.347	0.357	0.361	0.375	0.384	0.388	0.396
Crown, ordinary	—	—	—	—	—	.947	—	—	—
Crown, borosilicate	0.08	0.65	0.88	—	0.95	—	0.972	0.975	0.986
Flint, ordinary	—	—	—	0.72	—	—	—	0.904	—
Flint, heavy	—	—	0.01	—	0.16	—	0.58	—	—

Normal incidence, thickness 1 cm.

Wave length microns	0.400	0.415	0.419	0.425	0.434	0.455	0.500	0.580	0.677
Crown, ordinary	0.964	—	0.952	—	0.960	0.981	—	0.986	0.990
Crown, borosilicate	—	0.985	—	0.993	—	—	0.993	—	—
Flint, ordinary	—	0.939	—	—	—	—	1.00	—	—
Flint, heavy	—	—	—	0.905	—	—	—	—	—

Quartz

Quartz is very transparent to the ultra-violet and to the visible spectrum, but opaque for the infrared beyond 7.0μ.

(Pflüger.)

Wave length, microns	0.19	0.20	0.21	0.22
Transmission for 1 mm	.67	.84	.92	.94

Fluorite

Fluorite is very transparent to the ultra-violet, nearly to 0.10μ. Coefficient of transparency at $\lambda = 186$ is found by Pflüger to be 0.80.

For the infrared the values are given in a table below.

Rock Salt and Sylvine and Fluorite
Transparency for the Infrared.
Thickness 1 cm.

Wave length microns	Rock salt	Sylvine KCl	Fluorite
8.	—	—	.844
9.	0.995	1.000	.543
10.	.995	.988	.164
12.	.993	.995	.010
14.	.931	.975	.000
16.	.661	.936	—
18.	.275	.862	—
19.	.096	.758	—
20.7	.006	.585	—
23.7	.000	.155	—

COLORIMETRY

Selected from Judd, Jour, Opt. Soc. Amer. 23, 359 (1933)
Recommendations of the International Commission on Illumination

Standard Illuminants

A. Gas-filled tungsten incandescent lamp of color temperature 2848°K.

B. Noon Sunlight. Lamp as above in combination with the Davis-Gibson filter for converting color temperature 2848° to 4800°K.

   The filter is to be composed of a layer one centimeter thick of each of two separate solutions B_1 and B_2, contained in a double cell of colorless optical glass.

Solution B_1
Copper sulfate ($CuSO_4 \cdot 5H_2O$)	2.452 g
Mannite ($C_6H_8(OH)_6$)	2.452 g
Pyridine (C_5H_5N)	30.00 cc
Distilled water to make	1000 cc

Solution B_2
Cobalt ammonium sulfate ($CoSO_4 \cdot (NH_4)_2SO_4 \cdot 6H_2O$)	21.71 g
Copper sulfate ($CuSO_4 \cdot 5H_2O$)	16.11 g
Sulfuric acid (density 1.835)	10.0 cc
Distilled water to make	1000 cc

C. Average Daylight. Lamp as in A in combination with Davis-Gibson filter for converting color temperature 2848° to 6500°K.

   The filter is composed of a layer one centimeter thick of each of two separate solutions C_1 and C_2, contained in a double cell made of colorless optical glass.

Solution C_1
Copper sulfate ($CuSO_4 \cdot 5H_2O$)	3.412 g
Mannite ($C_6H_8(OH)_6$)	3.412 g
Pyridine (C_5H_5N)	30.0 cc
Distilled water to make	1000 cc

Solution C_2
Cobalt ammonium sulfate ($CoSO_4 \cdot (NH_4)_2SO_4 \cdot 6H_2O$)	30.580 g
Copper sulfate ($CuSO_4 \cdot 5H_2O$)	22.520 g
Sulfuric acid (density 1.835)	10.0 cc
Distilled water to make	1000 cc

See R. Davis and K. S. Gibson Bur. Stds. Misc. Pub. No. 114, Jan. 1931 or Bur. Stds. Jour. Research 7,796 (1931).

Standard Coordinate System

The tristimulus system of color specification is based on four chosen stimuli consisting of homogeneous radiant energy of wave lengths

700.0	546.1	435.8

mµ and of standard illuminant B (see above).

To establish the system of specification coordinates are assigned as follows:

Stimulus	x	v	z
700.0 mµ	0.73467	0.26533	0.00000
546.1 mµ	0.27376	0.71741	0.00883
435.8 mµ	0.16658	0.00886	0.82456
Standard illuminant B:	0.34842	0.35161	0.29997

The Standard Observer

The "standard observer" is determined below by the specification for the equal energy spectrum both in fractions, x, y, z of the total amount for each wave length interval of 5 mµ and directly $\bar{x}, \bar{y}, \bar{z}$. The fractional values are known as the trillnear coordinates or trichromatic coefficients of the spectrum; the direct values as the distribution functions or coefficients.

The sum of the trichromatic coefficients is unity, that is x + y + z = 1. Therefore the value of z may be and often is omitted from a specification.

Relative Visibility

The value of y given in the table is the standard visibility function or relative visibility.

Wave length mµ	Trichromatic coefficients			Distribution coefficients for equal energy			Wave length mµ	Trichromatic coefficients			Distribution coefficients for equal energy		
	x	y	z	x	y (Rel. Vis.)	z		x	y	z	x	y (Rel. Vis.)	z
380	0.1741	0.0050	0.8209	0.0014	0.0000	0.0065	495	0.0235	0.4127	0.5638	0.0147	0.2586	0.3533
385	0.1740	0.0050	0.8210	0.0022	0.0001	0.0105	500	0.0082	0.5384	0.4534	0.0049	0.3230	0.2720
390	0.1738	0.0049	0.8213	0.0042	0.0001	0.0201	505	0.0039	0.6548	0.3413	0.0024	0.4073	0.2123
395	0.1736	0.0049	0.8215	0.0076	0.0002	0.0362	510	-0.0139	0.7502	0.2359	0.0093	0.5030	0.1582
400	0.1733	0.0048	0.8219	0.0143	0.0004	0.0679	515	-0.0389	0.8120	0.1491	0.0291	0.6082	0.1117
405	0.1730	0.0048	0.8222	0.0232	0.0006	0.1102	520	0.0743	0.8338	0.0919	0.0633	0.7100	0.0782
410	0.1726	0.0048	0.8226	0.0435	0.0012	0.2074	525	0.1142	0.8262	0.0596	0.1096	0.7932	0.0573
415	0.1721	0.0048	0.8231	0.0776	0.0022	0.3713	530	0.1547	0.8059	0.0394	0.1655	0.8620	0.0422
420	0.1714	0.0051	0.8235	0.1344	0.0040	0.6456	535	0.1929	0.7816	0.0255	0.2257	0.9149	0.0298
425	0.1703	0.0058	0.8239	0.2148	0.0073	1.0391	540	0.2296	0.7543	0.0161	0.2904	0.9540	0.0203
430	0.1689	0.0069	0.8242	0.2839	0.0116	1.3856	545	0.2658	0.7243	0.0099	0.3597	0.9803	0.0134
435	0.1669	0.0086	0.8245	0.3285	0.0168	1.6230	550	0.3016	0.6923	0.0061	0.4334	0.9950	0.0087
440	0.1644	0.0109	0.8247	0.3483	0.0230	1.7471	555	0.3373	0.6589	0.0038	0.5121	1.0002	0.0057
445	0.1611	0.0138	0.8251	0.3481	0.0298	1.7826	560	0.3731	0.6245	0.0024	0.5945	0.9950	0.0039
450	0.1566	0.0177	0.8257	0.3362	0.0380	1.7721	565	0.4087	0.5896	0.0017	0.6784	0.9786	0.0027
455	0.1510	0.0227	0.8263	0.3187	0.0480	1.7441	570	0.4441	0.5547	0.0012	0.7621	0.9520	0.0021
460	0.1440	0.0297	0.8263	0.2908	0.0600	1.6692	575	0.4788	0.5202	0.0010	0.8425	0.9154	0.0018
465	0.1355	0.0399	0.8246	0.2511	0.0739	1.5281	580	0.5125	0.4866	0.0009	0.9163	0.8700	0.0017
470	0.1241	0.0578	0.8181	0.1954	0.0910	1.2876	585	0.5448	0.4544	0.0008	0.9786	0.8163	0.0014
475	0.1096	0.0868	0.8036	0.1421	0.1126	1.0419	590	0.5752	0.4242	0.0006	1.0263	0.7570	0.0011
480	0.0913	0.1327	0.7760	0.0956	0.1390	0.8130	595	0.6029	0.3965	0.0006	1.0567	0.6949	0.0010
485	0.0687	0.2007	0.7306	0.0580	0.1693	0.6162	600	0.6270	0.3725	0.0005	1.0622	0.6310	0.0008
490	0.0454	0.2950	0.6596	0.0320	0.2080	0.4652	605	0.6482	0.3514	0.0004	1.0456	0.5668	0.0006

Wave length mμ	Trichromatic coefficients			Distribution coefficients for equal energy		
	x	y	z	x	y (Rel. Vis.)	z
610	0.6658	0.3340	0.0002	1.0026	0.5030	0.0003
615	0.6801	0.3197	0.0002	0.9384	0.4412	0.0002
620	0.6915	0.3083	0.0002	0.8544	0.3810	0.0002
625	0.7006	0.2993	0.0001	0.7514	0.3210	0.0001
630	0.7079	0.2920	0.0001	0.6424	0.2650	0.0000
635	0.7140	0.2859	0.0001	0.5419	0.2170	0.0000
640	0.7190	0.2809	0.0001	0.4479	0.1750	0.0000
645	0.7230	0.2770	0.0000	0.3608	0.1382	0.0000
650	0.7260	0.2740	0.0000	0.2835	0.1070	0.0000
655	0.7283	0.2717	0.0000	0.2187	0.0816	0.0000
660	0.7300	0.2700	0.0000	0.1649	0.0610	0.0000
665	0.7311	0.2689	0.0000	0.1212	0.0446	0.0000
670	0.7320	0.2680	0.0000	0.0874	0.0320	0.0000
675	0.7327	0.2673	0.0000	0.0636	0.0232	0.0000
680	0.7334	0.2666	0.0000	0.0468	0.0170	0.0000
685	0.7340	0.2660	0.0000	0.0329	0.0119	0.0000
690	0.7344	0.2656	0.0000	0.0227	0.0082	0.0000
695	0.7346	0.2654	0.0000	0.0158	0.0057	0.0000
700	0.7347	0.2653	0.0000	0.0114	0.0041	0.0000
705	0.7347	0.2653	0.0000	0.0081	0.0029	0.0000
710	0.7347	0.2653	0.0000	0.0058	0.0021	0.0000
715	0.7347	0.2653	0.0000	0.0041	0.0015	0.0000
720	0.7347	0.2653	0.0000	0.0029	0.0010	0.0000
725	0.7347	0.2653	0.0000	0.0020	0.0007	0.0000
730	0.7347	0.2653	0.0000	0.0014	0.0005	0.0000
735	0.7347	0.2653	0.0000	0.0010	0.0004	0.0000
740	0.7347	0.2653	0.0000	0.0007	0.0003	0.0000
745	0.7347	0.2653	0.0000	0.0005	0.0002	0.0000
750	0.7347	0.2653	0.0000	0.0003	0.0001	0.0000
755	0.7347	0.2653	0.0000	0.0002	0.0001	0.0000
760	0.7347	0.2653	0.0000	0.0002	0.0001	0.0000
765	0.7347	0.2653	0.0000	0.0001	0.0000	0.0000
770	0.7347	0.2653	0.0000	0.0001	0.0000	0.0000
775	0.7347	0.2653	0.0000	0.0000	0.0000	0.0000
780	0.7347	0.2653	0.0000	0.0000	0.0000	0.0000
			Totals	21.3713	21.3714	21.3715

SPECIFIC ROTATION

Specific rotation or rotatory power is given in degrees per decimeter for liquids and solutions and in degrees per millimeter for solids; + signifies right handed rotation, — left. Specific rotation varies with the wave length of light used, with temperature and, in the case of solutions, with the concentration. When sodium light is used, indicated by D in the wavelength column, a value of $\lambda = 0.5893$ may be assumed.

Optical rotatory power for a large number of organic compounds will be found in the International Critical Tables, Vol. VII; for sugars, Vol. II.

Solids

Substance	Wave length μ	Rotation deg./mm	Substance	Wave length μ	Rotation deg./mm
Cinnabar (HgS)	D	+32.5	Quartz (continued)	0.3609	+63.628
Lead hyposulfate	D	5.5		0.3582	64.459
Potassium hyposulfate	D	8.4		0.3433	69.454
Quartz	0.7604	12.668		0.3441	70.587
	0.7184	14.304		0.3402	72.448
	0.6867	15.746		0.3360	74.571
	0.6562	17.318		0.3296	76.375
	0.5905	21.684		0.3247	80.459
	0.5889	21.727		0.3180	84.972
	0.5269	27.543		0.2747	121.052
	0.4861	32.773		0.2571	143.266
	0.4307	42.604		0.2313	190.426
	0.4101	47.481		0.2265	201.824
	0.3968	51.193		0.2194	220.731
	0.3933	52.155		0.2143	235.972
	0.3820	55.625	Sodium bromate	D	2.8
	0.3726	58.894	Sodium chlorate	D	3.13

Liquid

Liquid	Temp. °C	Wave length μ	Specific rotation deg./dm
Amyl alcohol	—	D	−5.7
Camphor	204	D	+70.33
Cedar oil	15	D	−30 to −40
Citrol oil	15	D	+62
Ethyl malate $(C_2H_5)_2C_4H_4O_5$	11	D	−10.3 to −12.4
Menthol	35.2	D	−49.7
Nicotine $C_{10}H_{14}N_2$	10—30	D	−162
	20	0.6563	−126
	20	0.5351	−207.5
	20	0.4861	−253.5
Turpentine $C_{10}H_6$	20	D	−37
	20	0.6563	−29.5
	20	0.5351	−45
	20	0.4861	−54.5

Solutions

Corrections for values of the specific rotation for concentration are given in the lat column. c indicates concentration in grams per 100 milliliters of solution; d indicates the concentration in grams per 100 grams of solution.

Substance	Solvent	Temp. °C	Wave length μ	Specific rotation deg./ dm	Correction for concentration or temperature
Albumen	water	—	D	− 25 to −38	
Arabinose	water	20	D	− 105.0	
Camphor	alcohol	20	D	+ 54.4 −.135d for d = 45—91	
	benzene	20	D	+ 56 −.166d for d = 47—90	
	ether	—	D	+ 57	
Dextrose d-glucose $C_6H_{12}O_6$	water	20	D	+ 52.5 + .025d for d = 1—18	
			.5461	+ 62.03 + .04257c for c = 6—32	
Galactose	water	—	D	+ 83.9 + .078d − .21t for d = 4—36 and t = 10—30°C	
l-Glucose (β)	water	20	D	− 51.4	
Invert sugar $C_6H_{12}O_6$	water	20	D	− 19.7 −.036c for c = 9—35 $\alpha t = \alpha_{20} + .304(t − 20) + .00165 (t − 20)^2$ for t = 3—30°C	
		25	.5461	− 21.5	
Lactose	water	20	D	+ 52.4 + .072 (20° − t) for c = 5	
			.5461	+ 61.9 + .085(20° − t) for c = 5	
Levulose fruit sugar	water	25	D	− 88.5 −.145d for d = 2.6—18.6	
		25	.5461	− 105.30	
Maltose	water	20	D	+ 138.48 −.01837d for d = 5—35	
		25	.5461	+ 153.75	
Mannose	water	20	D	+ 14.1 c = 10.2	
Nicotine	water	20	D	− 77 for d = 1—16	
	benzene	20	D	− 164 for d = 8—100	
Potassium tartrate	water	20	D	+ 27.14 + .0992c − .00094c^2 for c = 8—50	
Quinine sulfate	water	17	D	− 214	
Santonin	alcohol	20	D	− 161.0 c = 1.78	
		20	D	+ 693 c = 4.05	
	chloroform	20	D	− 202.7 + 309d for d = 75—96.5	
	alcohol	20	.6867	+ 442 c = 4.05	
			.5269	+ 991 c = 4.05	
			.4861	+ 1323 c = 4.05	
Sodium potassium tartrate (Rochelle salt)	water	20	D	+ 29.75 −.0078c	
Sucrose (cane sugar) $C_{12}H_{22}O_{11}$	water	20	D	+ 66.412 + .01267d − .00376d^2 for d = 0—50 $\alpha t = \alpha_{20}[1 − .00037 (t − 20)]$ for t = 14—30°C	

Sucrose dissolved in water, 20°C.

μ	Spec. rot.	μ	Spec. rot.	μ	Spec. rot.
670.8(Li)	+ 50.51	510.6 (Cu)	+ 90.46	435.3(Fe)	+ 128.5
643.8(Cd)	55.04	508.6(Cd)	91.16	433.7(Fe)	129.8
636.2(Zn)	56.51	481.1(Zn)	103.07	431.5(Fe)	130.7
589.3(Na)	66.45	480.0(Cd)	103.62	428.2(Fe)	133.6
578.2(Cu)	69.10	472.2(Zn)	107.38	427.2(Fe)	134.2
578.0(Hg)	69.22	468.0(Zn)	109.49	426.1(Fe)	134.9
570.0(Cu)	71.24	467.8(Cd)	109.69	419.1(Fe)	140.0
546.1(Hg)	78.16	438.4(Fe)	126.5	414.4(Fe)	144.2
521.8(Cu)	86.21	437.6(Fe)	127.2	388.9(Fe)	166.7
515.3(Cu)	88.68	435.8(Hg)	128.49	383.3(Fe)	171.8
				382.6(Fe)	173.1

Substance	Solvent	°C	μ	Spec. rot.	Correct.
Tartaric acid (ord)	water	20	D	+ 15.06 −.131c	
		20	.6563	7.75	
		20	D	8.86	
		20	.5351	9.65 for d = 41	
		20	.4861	9.37	
Turpentine	alcohol	20	D	−37 −.00482d −.00013d for d = 0—90	
	benzene	20	D	−37 −.0265d for d = 0—91	
Xylose	water	20	D	+ 19.13 d = 2.7	

OPTICAL ROTATION OF ACIDS AND BASES

Optical rotation of acids and bases commonly used in the resolution of racemic substances. Compiled by F. E. Ray.

Name	Formula	Solvent	Cone. %	α_D
Bromocamphor-sulfonic acid. K salt	$C_{10}H_{15}O_4BrS$	H_2O	—	72.1
Camphorsulfonic acid	$C_{10}H_{16}O_4S$	H_2O	—	23.9
Chlorocamphor-sulfonic acid	$C_{10}H_{15}ClO_4S$	H_2O	—	49.6
Codeinesulfonic acid	$C_{18}H_{21}NO_6S$	H_2O	3	−190.1
Hydroxybutyric acid	$C_4H_8O_3$	H_2O	3.3	−24.8
Lactic acid	$C_3H_6O_3$	H_2O	10.5	3.8
Malic acid	$C_3H_6O_5$	H_2O	—	2.4
Mandelic acid	$C_8H_8O_3$	H_2O	2.01	155.5
Methylene-camphor	$C_{11}H_{16}O$	C_2H_5OH	—	127
Phenylsuccinic acid	$C_{10}H_{10}O_4$	C_2H_5OH	1.5	148
Tartaric acid	$C_4H_6O_6$	C_2H_5OH and H_2O	—	3 to 25*
Brucine	$C_{23}H_{26}N_2O_4$	C_2H_5OH	5.4	−85
Cinchonidne	$C_{19}H_{22}N_2O$	C_2H_5OH	1.0	−111.0
Cinchonine	$C_{19}H_{22}N_2O$	$CHCl_3$	0.6	+ 209.6
Cocaine	$C_{17}H_{21}NO_4$	50% C_2H_5OH	1.1	−35.4
Coniine	$C_8H_{17}N$	$CHCl_3$	4	8.0
Codeine	$C_{18}H_{21}NO_3$	C_2H_5OH	5	−135.8
Hydrastine	$C_{21}H_{21}NO_6$	50% C_2H_5OH	0.2	115
Menthol	$C_{10}H_{20}O$	C_2H_5OH	9.6	−50.6
Menthylamine	$C_{10}H_{21}N$	C_2H_5OH	11.3	−31.9
Narcotine	$C_{22}H_{23}NO_7$	$CHCl_3$	2.6	±200.0
Quinidine	$C_{20}H_{23}N_2O_4$	C_2H_5OH	1.0	+ 233.6
Quinine	$C_{20}H_{23}N_2O_2$	C_6H_6	0.6	−136
Thebaine	$C_{19}H_{21}NO_2$	$CHCl_3$	5	−229.5
Strychnine	$C_{21}H_{22}N_2O_2$	C_2H_5OH	0.9	−128

* Varies greatly with temperature, solvent, and cone.

MAGNETIC ROTATORY POWER

The rotation of the plane of polarization of light by transparent substances subjected to a magnetic field may be applied to problems of molecular structure. This rotatory effect is known as the Faraday effect. Investigations of this effect by E. Verdet showed the angle of rotation (α) to depend on the nature of the substance and to be proportional to the length (l) of the column of the substance which the light traverses and to the strength (H) of the magnetic field. Thus

$$\alpha = \Lambda l H$$

where Λ is the Verdet constant for the experimental material.

Other symbols in the table have the following significance:

λ = Wavelength in μ

t = Degrees centigrade

ϱ_M = Molecular magnetic rotation of the substance under consideration compared to that of water determined in the same apparatus in the same magnetic field. Thus

$$\varrho_M = M'\alpha'\varrho'/M'\alpha'\varrho$$

where M is the molecular weight, α the angle of rotation and ϱ the density of the given substance, and M′, α′, and ϱ′ are the same quantities of water.

$[\Lambda]_M^{\lambda,T}$ = Molecular rotatory value of the substance at the given temperature and at the wavelength λ. Values are in (radians)(gauss⁻¹)(cm⁻¹).

Values in this table are reproduced by permission from Volume 3, Tables de Constantes et Donnees Nuimeriques, "Pouvoir Rotatorie Magnetique, Effet Magneto-Optique de Kerr."

A much greater listing of values for both organic and inorganic compounds is in the above publication.

Formula	Name	λ	t	ϱ_M	$10^5[\Lambda]_M^{\lambda,T}$	$10^5\Lambda^{\lambda,T}$ Verdet
CH_4	Methane	578	0	—	—	*17.4
CCl_4	Tetrachloromethane	589	25.1	6.58	45.3	1.60
$CHCl_3$	Trichloromethane	589	20.0	5.535	37.98	1.60
$CHBr_3$	Tribromomethane	589	17.9	11.63	80.00	3.13
CH_2O_2	Formic acid	589	20.8	1.671	11.50	1.046
CH_2Cl_2	Dichloromethane	589	11.9	4.31	29.7	1.60
CH_2Br_2	Dibromomethane	589	15.9	8.11	55.8	2.74
CH_2I_2	Diiodomethane	589	15.0	10.83	129.5	1.51
CH_3Cl	Monochloromethane	589	23.0	2.99	20.5	1.37
CH_3Br	Monobromomethane	589	1.5	4.64	31.9	2.04
CH_3I	Monoiodomethane	589	19.5	9.01	63.2	3.35
CH_4O	Methylalcohol	589	18.7	1.640	11.28	0.958
CH_5N	Methylamine	578	—	—	—	*22.7
CF_2Cl_2	Difluorodichloromethane (Freon)	—	—	—	—	*32.7
CH_3O_2N	Mononitromethane	589	9.9	1.86	12.8	0.826
CH_4ON_2	Urea (carbamide) 40% aqueous solution	578	20.0	2.38	22.7	—
C_2H_2	Acetylene (ethyne)	578	—	—	—	*33.0
C_2H_4	Ethylene (ethene)	578	—	—	—	*34.5
C_2H_6	Ethane	578	—	—	—	*23.5
C_2N_2	Cyanogen	—	—	—	—	—
$C_2H_2O_4$	Oxalic acid (aqueous sol. 8.3%)	578	20	2.88	20.6	—
$C_2H_2O_4 \cdot 2H_2O$	Oxalic acid (alcohol sol. 16.5%)	578	24	2.82	20.2	—
C_2H_3Br	Vinylbromide	589	7.8	6.22	42.8	2.10
C_2H_3N	Acetonitrile (ethanonitrile)	589	25.0	2.32	16.0	*21.0

* Verdet constant factor = 10^6.

Formula	Name	λ	t	ϱ_M	$10^5[\Lambda]_M^{\lambda,t}$	$10^5\Lambda^{\lambda,t}$ Verdet
C_2H_4O	Ethyleneoxide (1,2-epoxyethane)	589	8.0	1.935	13.3	0.92
C_2H_4O	Acetaldehyde (ethanal)	589	16.3	2.38	16.4	1.00
$C_2H_4O_2$	Acetic acid	589	21.0	2.525	17.37	1.044
$C_2H_4O_2$	Formic acid methylester (methylformate)	589	16.5	2.49	17.1	0.96
$C_2H_4Cl_2$	Ethylidenechloride (1,1-dichloroethane)	589	14.4	5.33	36.7	1.51
$C_2H_4Cl_2$	Ethylenechloride (1,2-dichloroethane)	589	14.4	5.49	37.7	1.65
$C_2H_4Br_2$	Ethylenebromide (1,2-dibromoethane)	589	15.2	9.70	66.7	2.66
C_2H_5Cl	Monochloroethane	589	5.0	4.04	27.8	1.36
C_2H_5Br	Monobromoethane	589	19.7	5.85	40.2	1.82
C_2H_5I	Monoiodoethane	589	18.1	10.07	69.3	2.95
C_2H_6O	Ethylalcohol (ethanol)	589	16.8	2.780	19.13	1.131
$C_2H_6O_2$	Glycol (1,2-ethanediol)	589	15.1	2.94	20.2	1.25
C_2H_6S	Ethylmercaptan	578	16.0	5.52	39.5	1.85
C_2H_7N	Ethylamine (aminoethane)	589	5.8	3.61	24.8	*34.5
$C_2H_2O_2Cl_2$	Dichloroacetic acid (dichloroethanoic acid)	589	13.5	5.30	3.65	1.52
$C_2H_3O_2Cl$	Chloroacetic acid (chloroethanoic acid)	589	64.5	3.89	26.7	1.33
$C_2H_3O_2Cl_3$	Chloralhydrate (2-2-2-trichloro-1,1-ethanediol)	589	54.6	7.10	48.8	1.65
$C_2H_5O_2N$	Mononitroethane	589	10.2	2.84	19.5	0.946
C_3H_8	Propane	578	—	—	—	*34.0
C_3H_4O	Acrolein (propenal)	578	20.0	4.74	34.0	1.76
$C_3H_4O_3$	Pyruvic acid (2-oxopropanoic acid)	589	14.5	3.56	24.2	1.21
$C_3H_4O_4$	Malonic acid (propandioic acid) 2 n-aqueous sol.	589	23.0	3.47	—	—
C_3H_6O	Allyl alcohol	589	18.3	4.68	32.2	1.60
C_3H_6O	Propyl alcohol (1-propanol)	589	13.6	3.33	22.9	1.09
C_3H_6O	Acetone (2-propanone)	589	20.0	3.472	23.89	1.1136
$C_3H_6O_2$	Propionic acid (propanoic acid)	589	20.3	3.462	23.82	1.10
$C_3H_6O_2$	Formic acid ethylester (ethylmethanoate)	589	18.8	3.56	24.5	1.05
$C_3H_6O_2$	Acetic acid methylester (methylacetate)	589	20.0	3.42	23.5	1.03
C_3H_7Cl	Propylchloride (1-chloropropane)	589	16.1	5.04	34.7	1.34
C_3H_7Cl	Isopropylchloride (2-chloropropane)	589	17.2	5.16	35.5	1.34
C_3H_7Br	Propylbromide (1-bromopropane)	589	19.2	6.88	47.8	1.79
C_3H_7Br	Isopropylbromide (2-bromopropane)	589	17.1	7.00	48.2	1.77
C_3H_7I	Propyliodide (1-iodopropane)	589	18.1	11.08	76.2	2.69
C_3H_7I	Isopropyliodide (2-iodopropane)	589	26.3	11.18	76.9	2.63
C_3H_8O	n-Propyl alcohol (1-propanol)	589	15.6	3.77	25.9	1.20
C_3H_8O	Isopropyl alcohol (2-propanol)	589	20.0	3.90	26.8	1.23
$C_3H_8O_3$	Glycerine (1,2,3-propanetriol)	589	16.0	4.11	28.3	1.33
C_3H_9N	n-Propylamine	589	9.6	4.56	31.4	1.33
$C_3H_5O_9N_3$	Nitroglycerine	589	13.5	5.405	37.2	0.900
$C_3H_7O_2N$	1-Nitropropane	589	18.9	3.82	26.3	1.018
C_4H_6	1,3-Butadiene (erythrene)	589	15.0	7.94	54.6	2.16
C_4H_8	1-Butene (α-butylene)	589	15.0	5.53	38.0	1.39
C_4H_8	cis-2-Butene (β-butylene)	589	15.0	5.27	36.3	1.38
C_4H_8	trans-2-Butene	589	15.0	5.07	34.9	1.29
C_4H_{10}	Butane	589	15.0	4.59	31.6	1.09
C_4H_{10}	Isobutane (2-methylpropane)	589	15.0	4.87	33.5	1.11
$C_4H_2O_3$	Maleic anhydride (cis-butenedioic anhydride)	589	25.0	4.5	31.0	—
C_4H_4O	Furan (furfuran)	589	20.0	5.48	37.7	1.78
$C_4H_4O_4$	Maleic acid (cis-butendioic acid) 2 n aqueous solution	589	25.0	5.63	38.7	—
C_4H_4S	Thiophene (thiofuran)	589	20.0	9.40	64.7	2.83
$C_4H_6O_3$	Acetic anhydride (ethanoic anhydride)	589	20.0	4.28	29.5	1.115
$C_4H_6O_4$	Succinic acid (butanedioic acid) 5.9% aqueous sol.	578	20.0	4.68	33.5	—
$C_4H_6O_6$	Tartaric acid (47.8% aqueous sol.)	578	20.0	4.79	34.3	—
$C_4H_8O_2$	n-Butyric acid (butanoic acid)	589	18.8	4.47	30.8	1.15
$C_4H_8O_2$	Ethylacetate (ethyl ethanoate)	589	20.0	4.47	30.8	1.08
$C_4H_8O_2$	Propionic acid methylester (methylpropanoate)	589	20.0	4.37	30.1	1.07
$C_4H_8O_3$	Lactic acid methylester (methyl 2-hydroxypropanoate)	589	—	4.66	32.1	—
$C_4H_{10}O$	Ethyl ether (ethoxyethane)	589	20.0	4.78	32.9	1.09
$C_4H_{10}O$	n-Butyl alcohol (1-butanol)	589	20.0	4.60	31.6	1.23
$C_4H_{10}O$	Isobutyl alcohol (2-methyl-1-propanol)	589	17.7	4.94	34.0	1.27
$C_4H_{10}O$	sec-Butyl alcohol (methylethylcarbinol)	589	20.0	4.91	33.8	1.27
C_5H_6	Cyclopentadiene	589	15.0	7.03	48.8	2.02
C_5H_8	1,3-Pentadiene	589	15.0	8.80	60.5	2.08
C_5H_8	Isoprene (2-methyl-1,3-butadiene)	589	15.0	8.80	60.5	2.08
C_5H_8	Cyclopentene	589	15.0	5.69	39.1	1.52
C_5H_{10}	1-Pentene	589	15.0	6.45	44.4	1.39
C_5H_{10}	Isopentane (2-methyl-1-butane)	589	15.0	6.36	43.8	1.39
C_5H_{10}	Cyclopentane	589	20.0	4.89	33.6	1.23
C_5H_{12}	Pentane	589	15.0	5.60	38.5	1.15
C_5H_{12}	Isopentane (2-methylbutane)	589	15.0	5.75	39.6	1.17
$C_5H_4O_2$	Furfural (2-furancarbonal)	578	20.0	7.01	50.2	2.06
C_5H_5N	Pyridine	589	11.9	8.76	60.3	2.58
$C_5H_8O_4$	Glutaric acid (pentanedioic acid) 2 N aqueous sol.	589	16.0	5.48	37.7	—
$C_5H_{10}O_2$	Propionic acid ethylester	589	20.0	5.46	37.6	1.13
$C_5H_{10}O_2$	Acetic acid propylester	589	15.7	5.45	37.5	1.13
$C_5H_{14}N_2$	Cadaverine (1,5-pentanediamine)	589	14.7	7.49	51.6	1.53
C_6H_6	Benzene	589	15.0	11.27	77.5	3.00
C_6H_{12}	Cyclohexane	589	20.0	5.66	39.0	1.24
C_6H_{14}	Hexane	589	15.0	6.62	45.5	1.20
$C_6H_4Cl_2$	1,4-Dichlorobenzene (p-dichlorobenzene)	589	64.5	13.55	93.2	2.69
C_6H_5F	Fluorobenzene (phenylfluoride)	589	19.0	9.96	68.5	2.51
C_6H_5Cl	Chlorobenzene (phenylchloride)	589	15.0	12.51	86.1	2.92
C_6H_5Br	Bromobenzene (phenylbromide)	589	15.0	14.51	99.8	3.26
C_6H_5I	Iodobenzene (phenyliodide)	589	15.0	19.11	131.4	4.06
C_6H_6O	Phenol (hydroxybenzene)	589	39.0	12.07	83.5	3.21
C_6H_7N	Aniline (aminobenzene)	589	15.0	16.08	110.6	4.18
$C_6H_{11}Cl$	Chlorocyclohexane (cyclohexylchloride)	589	13.0	7.50	51.6	1.46
$C_6H_{12}O_3$	Paraldehyde (paraacetaldehyde)	589	17.3	6.66	45.8	1.19
$C_6H_{12}O_6$	Glucose $11H_2O$ (dextrose) (1 M aqueous sol.)	589	15.0	6.72	46.2	—
$C_6H_{12}O_6$	Galactose $10H_2O$ (1 M aqueous sol.)	589	15.0	6.89	47.4	—
$C_6H_{12}O_6$	Fructose $10H_2O$ (levulose) (1 M aqueous sol.)	589	15.0	6.73	46.3	—
$C_6H_{14}O$	2-Hexanol (butylmethylcarbinol)	589	20.0	6.89	47.4	1.31
$C_6H_{14}O$	3-Hexanol (ethylpropylcarbinol)	589	20.0	6.85	47.1	1.30
$C_6H_{14}O$	2-Methyl-3-Pentanol (ethylisopropylcarbinol)	589	20.0	6.90	47.5	1.32
$C_6H_4O_4N_2$	1,3-Dinitrobenzene (m-Dinitrobenzene)	589	17.1	9.65	66.4	2.17
$C_6H_5O_2N$	Nitrobenzene	589	15.0	9.36	64.4	2.17
C_7H_8	Toluene (methylbenzene)	589	15.0	12.16	83.7	2.71
C_7H_{14}	1-Heptene (α-heptylene)	589	18.0	8.48	58.3	1.43
C_7H_{16}	Heptane	589	15.0	7.61	52.7	1.23

MAGNETIC ROTATORY POWER (Continued)

Formula	Name	λ	t	ϱ_M	$10^8[\Lambda]_M^{\lambda,t}$	$10^5\Lambda^{\lambda,t}$ Verdet
C_7H_5N	Benzonitrile (benzenecarbonitrile)	589	15.7	11.85	81.5	2.74
$C_7H_6O_2$	Benzoic acid (20% alcohol sol.)	578	20.0	11.8	84.7	—
C_7H_7Cl	o-Chlorotoluene (2-chloro-1-methylbenzene)	589	15.4	13.72	94.3	2.95
C_7H_7Cl	p-Chlorotoluene (4-chloro-1-methylbenzene)	589	15.2	13.25	90.8	2.65
C_7H_7Br	o-Bromotoluene (2-bromo-1-methylbenzene)	589	16.7	15.67	107.3	3.08
C_7H_7Br	p-Bromotoluene (4-bromo-1-methylbenzene)	589	39.0	15.09	103.6	2.88
C_7H_8O	o-Cresol (o-methylphenol)	589	16.0	13.38	92.1	3.07
C_7H_8O	m-Cresol (m-methylphenol)	589	17.9	12.77	87.6	2.89
C_7H_8O	p-Cresol (p-methylphenol)	589	17.0	12.86	88.5	2.91
C_7H_9N	o-Toluidine (o-methylaniline)	589	17.3	17.18	118.2	3.79
C_7H_9N	m-Toluidine (m-methylaniline)	589	15.0	16.21	111.5	3.56
C_7H_9N	p-Toluidine (p-methylaniline)	589	50.0	15.92	109.5	3.37
$C_7H_{14}O$	Enanthaldehyde (heptanal)	589	16.2	7.42	51.1	1.26
$C_7H_{16}O$	1-Heptanol (n-heptylalcohol)	589	12.6	7.85	54.0	1.33
$C_7H_{16}O$	2-Heptanol (amylmethylcarbinol)	589	20.0	7.94	54.6	1.32
$C_7H_{16}O$	3-Heptanol (butylethylcarbinol)	589	20.0	7.86	54.1	1.37
$C_7H_7O_2N$	o-Nitrotoluene	589	18.0	10.80	74.3	2.16
$C_7H_7O_2N$	p-Nitrotoluene	589	54.3	10.20	70.2	1.97
C_8H_{10}	Ethylbenzene (phenylethane)	589	15.0	13.41	92.3	2.80
C_8H_{10}	o-Xylene (1,2-dimethylbenzene)	589	15.0	13.36	91.9	2.62
C_8H_{10}	m-Xylene (1,3-dimethylbenzene)	589	15.0	12.82	88.2	2.47
C_8H_{10}	p-Xylene (1,4-dimethylbenzene)	589	15.0	12.80	88.1	2.46
C_8H_{16}	1-Octene (α-octylene)	589	15.0	9.00	65.5	1.44
C_8H_{16}	2-Octene (β-octylene)	589	15.0	9.33	64.2	1.43
C_8H_{18}	Octane	589	15.0	8.65	59.5	1.26
$C_8H_{18}O$	1-Octanol (n-octylalcohol)	589	20.0	8.88	61.1	1.33
$C_8H_{18}O$	2-Octanol (methylhexylcarbinol)	589	20.0	9.00	61.8	1.34
$C_8H_{18}O$	3-Octanol (ethylamylcarbinol)	589	20.0	8.90	61.2	1.33
C_9H_{12}	o-Ethyltoluene (1-ethyl-2-ethylbenzene)	589	15.0	14.56	100.2	2.32
C_9H_{12}	m-Ethyltoluene (1-ethyl-3-ethylbenzene)	589	15.0	14.18	97.6	2.91
C_9H_{12}	p-Ethylbenzene (1-ethyl-4-ethylbenzene)	589	15.0	13.98	96.2	2.37
C_9H_{12}	Mesitylene (1-3-5-trimethylbenzene)	589	15.0	13.36	91.9	2.28
C_9H_{20}	Nonane	589	15.0	9.70	66.7	1.28
$C_{10}H_8$	Naphthalene	589	89.5	24.98	171.8	4.47
$C_{10}H_{20}$	1-Decene (n-decylene)	589	21.0	11.65	80.1	1.45
$C_{10}H_{22}$	Decane	589	15.0	10.70	73.6	1.30
$C_{10}H_7Cl$	1-Chloronaphthalene (α-chloronaphthalene)	578	18.0	28.15	201.5	4.91
$C_{10}H_7Br$	1-Bromonaphthalene (α-bromonaphthalene)	578	20.0	31.05	222.0	5.19
$C_{10}H_8O$	β-Naphthol (2-hydroxynaphthalene)	578	136.0	27.1	194.0	4.80
$C_{10}H_9N$	1-Naphthylamine (α-naphthylamine)	589	32.6	37.23	256.1	6.84
$C_{10}H_{12}O_2$	Isoeugenol (4-propenylguaiacol)	589	19.3	21.44	147.5	3.55
$C_{10}H_{12}O_2$	Eugenol (4-allylguaiacol)	589	15.4	18.72	128.8	2.88
$C_{10}H_{12}O_2$	Benzoic acid propylester (n-propylbenzoate)	589	15.4	14.87	102.3	2.20
$C_{10}H_{12}O_2$	o-Toluic acid ethylester	589	15.2	15.06	103.6	2.25
$C_{10}H_{12}O_2$	p-Toluic acid ethylester	589	15.0	14.74	101.4	2.18
$C_{10}H_{12}O_2$	α-Toluic acid ethylester (ethylphenylacetate)	589	14.0	14.09	103.1	2.25
$C_{10}H_{12}O_3$	Methylisalicyclic acid ethylester	589	18.6	17.14	117.9	2.50
$C_{10}H_{15}N$	N,N-Diethylaniline (N-phenyldiethylamine)	589	15.3	25.16	173.1	3.74
$C_{10}H_{16}O$	Camphor	589	14.0	9.26	63.7	—
$C_{10}H_{18}O$	α-Terpineol	589	16.0	10.84	76.6	1.56
$C_{10}H_{18}O$	Citronellal	589	14.5	11.48	79.0	1.51
$C_{10}H_{18}O_4$	Dipropylsuccinate	589	11.4	10.36	71.3	1.22
$C_{10}H_{18}O_6$	Tartaric acid dipropylester (propyltartrate)	589	15.4	10.83	74.5	1.24
$C_{10}H_{20}O$	Menthol	589	45.2	10.51	72.3	1.40
$C_{11}H_{24}$	Undecane	589	20.5	11.65	80.1	1.31
$C_{12}H_{26}$	Dodecane	589	21.5	12.71	87.4	1.32
$C_{12}H_{22}O_{11}$	Saccharose 19H₂O (1 M aqueous sol.)	589	15.0	12.59	86.6	—
$C_{12}H_{22}O_{11}$	Maltose 20H₂O (1 M aqueous sol.)	589	15.0	12.69	87.3	—
$C_{12}H_{22}O_{11}$	Lactose 41H₂O (1 M aqueous sol.)	589	18.3	12.71	87.5	—
$C_{14}H_{10}$	Phenanthrene	578	100.0	39.7	284.0	5.84
$C_{16}H_{34}$	Hexadecane	589	15.0	16.8	115.6	1.35
$C_{18}H_{14}$	1,2-Diphenylbenzene	589	15.0	40.2	276.3	4.70
$C_{18}H_{14}$	1,3-Diphenylbenzene (m-phenyldiphenyl)	589	15.0	41.0	282.4	—
$C_{18}H_{22}$	1,6-Diphenylhexane	589	20.0	—	—	2.75

TRANSPARENCY TO OPTICAL DENSITY CONVERSION TABLE

Transparency of a layer of material is defined as the ratio of the intensity of the tra-smitted light to that of the incident light. Opacity is the reciprocal of the transparency. Optical density is the common logarithm of the opacity.

Thus,

$$\text{Transparency} = \frac{I_t}{I_i}$$

$$\text{Opacity} = \frac{1}{\text{Transparency}} = \frac{I_i}{I_t}$$

$$\text{Optical density} = \log_{10}\left(\frac{I_i}{I_t}\right)$$

where I = Intensity of incident light
I_t = Intensity of transmitted light.

Trans.	Density	Trans.	Density	Trans.	Density	Trans.	Density	Trans.	Density	Trans.	Density	Trans.	Density	Trans.	Density
0.000	—	.005	2.301	.010	2.000	.015	1.824	.020	1.699	.025	1.602	.030	1.523	.035	1.456
.001	3.000	.006	2.222	.011	1.959	.016	1.796	.021	1.678	.026	1.585	.031	1.509	.036	1.444
.002	2.699	.007	2.155	.012	1.921	.017	1.770	.022	1.658	.027	1.569	.032	1.495	.037	1.432
.003	2.523	.008	2.097	.013	1.886	.018	1.745	.023	1.638	.028	1.553	.033	1.482	.038	1.420
.004	2.398	.009	2.046	.014	1.854	.019	1.721	.024	1.620	.029	1.538	.034	1.469	.039	1.409

TRANSPARENCY TO OPTICAL DENSITY CONVERSION TABLE (Continued)

Trans.	Density	Trans.	Density	Trans.	Density	Trans.	Density	Trans.	Density	Trans.	Density	Trans.	Density	Trans.	Density
.040	1.398	.126	.8996	.212	.6737	.299	.5243	.386	.4134	.473	.3251	.560	.2518	.647	.1891
.041	1.387	.127	.8962	.213	.6716	.300	.5229	.387	.4123	.474	.3242	.561	.2510	.648	.1884
.042	1.377	.128	.8928	.214	.6696	.301	.5215	.388	.4112	.475	.3233	.562	.2503	.649	.1877
.043	1.367	.129	.8894	.215	.6676	.302	.5200	.389	.4101	.476	.3224	.564	.2487	.650	.1871
.044	1.357	.130	.8861	.216	.6655	.303	.5186	.390	.4089	.477	.3215	.565	.2479	.651	.1864
.045	1.347	.131	.8827	.217	.6635	.304	.5171	.391	.4078	.478	.3206	.566	.2472	.652	.1857
.046	1.337	.132	.8794	.218	.6615	.305	.5157	.392	.4067	.479	.3197	.567	.2464	.653	.1851
.047	1.328	.133	.8761	.219	.6596	.306	.5143	.393	.4056	.480	.3188	.568	.2457	.654	.1844
.048	1.319	.134	.8729	.220	.6576	.307	.5128	.394	.4045	.481	.3179	.569	.2449	.655	.1838
.049	1.310	.135	.8697	.221	.6556	.308	.5114	.395	.4034	.482	.3170	.570	.2441	.656	.1831
.050	1.301	.136	.8665	.222	.6536	.309	.5100	.396	.4023	.483	.3161	.571	.2434	.657	.1824
.051	1.292	.137	.8633	.223	.6517	.310	.5086	.397	.4012	.484	.3152	.572	.2426	.658	.1818
.052	1.284	.138	.8601	.224	.6498	.311	.5072	.398	.4001	.485	.3143	.573	.2418	.659	.1811
.053	1.276	.139	.8570	.225	.6478	.312	.5058	.399	.3990	.486	.3134	.574	.2411	.660	.1805
.054	1.268	.140	.8539	.226	.6459	.313	.5045	.400	.3979	.487	.3125	.575	.2403	.661	.1798
.055	1.260	.141	.8508	.227	.6440	.314	.5031	.401	.3969	.489	.3107	.576	.2396	.662	.1791
.056	1.252	.142	.8477	.228	.6421	.315	.5017	.402	.3958	.490	.3098	.577	.2388	.663	.1785
.057	1.244	.143	.8447	.229	.6402	.316	.5003	.403	.3947	.491	.3089	.578	.2381	.664	.1778
.058	1.237	.144	.8416	.230	.6383	.317	.4989	.404	.3936	.492	.3080	.579	.2373	.665	.1772
.059	1.229	.145	.8386	.231	.6364	.318	.4976	.405	.3925	.493	.3072	.580	.2366	.666	.1765
.060	1.222	.146	.8356	.232	.6345	.319	.4962	.406	.3915	.494	.3063	.581	.2358	.667	.1759
.061	1.215	.147	.8327	.233	.6326	.320	.4949	.407	.3904	.495	.3054	.582	.2351	.668	.1752
.062	1.208	.148	.8297	.234	.6308	.321	.4935	.408	.3893	.496	.3045	.583	.2343	.669	.1746
.063	1.201	.149	.8268	.235	.6289	.322	.4921	.409	.3883	.497	.3036	.584	.2336	.670	.1739
.064	1.194	.150	.8239	.236	.6271	.323	.4908	.410	.3872	.498	.3028	.585	.2328	.671	.1733
.065	1.187	.151	.8210	.237	.6253	.324	.4895	.411	.3862	.499	.3019	.586	.2321	.672	.1726
.066	1.180	.152	.8182	.238	.6234	.325	.4881	.412	.3851	.500	.3010	.587	.2314	.673	.1720
.067	1.174	.153	.8153	.239	.6216	.326	.4868	.414	.3830	.501	.3002	.588	.2306	.674	.1713
.068	1.168	.154	.8125	.240	.6198	.327	.4855	.415	.3819	.502	.2993	.589	.2299	.675	.1707
.069	1.161	.155	.8097	.241	.6180	.328	.4841	.416	.3809	.503	.2984	.590	.2291	.676	.1701
.070	1.155	.156	.8069	.242	.6162	.329	.4828	.417	.3799	.504	.2975	.591	.2284	.677	.1694
.071	1.149	.157	.8041	.243	.6144	.330	.4815	.418	.3788	.505	.2967	.592	.2277	.678	.1688
.072	1.143	.158	.8013	.244	.6126	.331	.4802	.419	.3778	.506	.2959	.593	.2269	.679	.1681
.073	1.137	.159	.7986	.245	.6108	.332	.4789	.420	.3768	.507	.2950	.594	.2262	.680	.1675
.074	1.131	.160	.7959	.246	.6091	.333	.4776	.421	.3757	.508	.2941	.595	.2255	.681	.1668
.075	1.125	.161	.7932	.247	.6073	.334	.4763	.422	.3747	.509	.2933	.596	.2248	.682	.1662
.076	1.119	.162	.7905	.248	.6056	.335	.4750	.423	.3737	.510	.2924	.597	.2240	.683	.1655
.077	1.114	.163	.7878	.249	.6038	.336	.4737	.424	.3726	.511	.2916	.598	.2233	.684	.1649
.078	1.108	.164	.7852	.250	.6021	.337	.4724	.425	.3716	.512	.2907	.599	.2226	.685	.1643
.079	1.102	.165	.7825	.251	.6003	.339	.4698	.426	.3706	.513	.2899	.600	.2219	.686	.1637
.080	1.097	.166	.7799	.252	.5986	.340	.4685	.427	.3696	.514	.2890	.601	.2211	.687	.1630
.081	1.092	.167	.7773	.253	.5969	.341	.4673	.428	.3685	.515	.2882	.602	.2204	.688	.1624
.082	1.086	.168	.7747	.254	.5952	.342	.4660	.429	.3675	.516	.2873	.603	.2197	.689	.1618
.083	1.081	.169	.7721	.255	.5935	.343	.4647	.430	.3665	.517	.2865	.604	.2190	.690	.1612
.084	1.076	.170	.7696	.256	.5918	.344	.4634	.431	.3655	.518	.2857	.605	.2182	.691	.1605
.085	1.071	.171	.7670	.257	.5901	.345	.4622	.432	.3645	.519	.2848	.606	.2175	.692	.1599
.086	1.066	.172	.7645	.258	.5884	.346	.4609	.433	.3635	.520	.2840	.607	.2168	.693	.1593
.087	1.060	.173	.7620	.259	.5867	.347	.4597	.434	.3625	.521	.2831	.608	.3161	.694	.1586
.088	1.055	.174	.7594	.260	.5850	.348	.4584	.435	.3615	.522	.2823	.609	.2154	.695	.1580
.089	1.051	.175	.7570	.261	.5834	.349	.4572	.436	.3605	.523	.2815	.610	.2147	.696	.1574
.090	1.046	.176	.7545	.263	.5800	.350	.4559	.437	.3595	.524	.2807	.611	.2140	.697	.1568
.091	1.041	.177	.7520	.264	.5784	.351	.4547	.438	.3585	.525	.2798	.612	.2132	.698	.1562
.092	1.036	.178	.7496	.265	.5768	.352	.4535	.439	.3575	.526	.2790	.613	.2125	.699	.1555
.093	1.032	.179	.7471	.266	.5751	.353	.4522	.440	.3565	.527	.2782	.614	.2118	.700	.1549
.094	1.027	.180	.7447	.267	.5735	.354	.4510	.441	.3556	.528	.2774	.615	.2111	.701	.1543
.095	1.022	.181	.7423	.268	.5719	.355	.4498	.442	.3546	.529	.2766	.616	.2104	.702	.1537
.096	1.018	.182	.7399	.269	.5702	.356	.4486	.443	.3536	.530	.2757	.617	.2097	.703	.1531
.097	1.013	.183	.7375	.270	.5686	.357	.4473	.444	.3526	.531	.2749	.618	.2090	.704	.1524
.098	1.009	.184	.7352	.271	.5670	.358	.4461	.445	.3516	.532	.2741	.619	.2083	.705	.1518
.099	1.004	.185	.7328	.272	.5654	.359	.4449	.446	.3507	.533	.2733	.620	.2076	.706	.1512
.100	1.000	.186	.7305	.273	.5638	.360	.4437	.447	.3497	.534	.2725	.621	.2069	.707	.1506
.101	.9957	.187	.7282	.274	.5622	.361	.4425	.448	.3487	.535	.2717	.622	.2062	.708	.1500
.102	.9914	.188	.7258	.275	.5607	.362	.4413	.449	.3478	.536	.2708	.623	.2055	.709	.1493
.103	.9872	.189	.7235	.276	.5591	.363	.4401	.450	.3468	.537	.2700	.624	.2048	.710	.1487
.104	.9830	.190	.7212	.277	.5575	.364	.4389	.451	.3458	.538	.2692	.625	.2041	.711	.1481
.105	.9788	.191	.7190	.278	.5560	.365	.4377	.452	.3449	.539	.2684	.626	.2034	.712	.1475
.106	.9747	.192	.7167	.279	.5544	.366	.4365	.453	.3439	.540	.2676	.627	.2027	.713	.1469
.107	.9706	.193	.7144	.280	.5528	.367	.4353	.454	.3429	.541	.2668	.628	.2020	.714	.1463
.108	.9666	.194	.7122	.281	.5513	.368	.4342	.455	.3420	.542	.2660	.629	.2013	.715	.1457
.109	.9626	.195	.7100	.282	.5498	.369	.4330	.456	.3410	.543	.2652	.630	.2007	.716	.1451
.110	.9586	.196	.7077	.283	.5482	.370	.4318	.457	.3401	.544	.2644	.631	.2000	.717	.1445
.111	.9547	.197	.7055	.284	.5467	.371	.4306	.458	.3391	.545	.2636	.632	.1993	.718	.1439
.112	.9508	.198	.7033	.285	.5452	.372	.4295	.459	.3382	.546	.2628	.633	.1986	.719	.1433
.113	.9469	.199	.7011	.286	.5436	.373	.4283	.460	.3372	.547	.2620	.634	.1979	.720	.1427
.114	.9431	.200	.6990	.287	.5421	.374	.4271	.461	.3363	.548	.2612	.635	.1972	.721	.1421
.115	.9393	.201	.6968	.288	.5406	.375	.4260	.462	.3354	.549	.2604	.636	.1965	.722	.1415
.116	.9356	.202	.6946	.289	.5391	.376	.4248	.463	.3344	.550	.2596	.637	.1959	.723	.1409
.117	.9318	.203	.6925	.290	.5376	.377	.4237	.464	.3335	.551	.2589	.638	.1952	.724	.1403
.118	.9281	.204	.6904	.291	.5361	.378	.4225	.465	.3325	.552	.2581	.639	.1945	.725	.1397
.119	.9244	.205	.6882	.292	.5346	.379	.4214	.466	.3316	.553	.2573	.640	.1938	.726	.1391
.120	.9208	.206	.0861	.293	.5331	.380	.4202	.467	.3307	.554	.2565	.641	.1932	.727	.1385
.121	.9172	.207	.6840	.294	.5317	.381	.4191	.468	.3298	.555	.2557	.642	.1925	.728	.1379
.122	.9137	.208	.6819	.295	.5302	.382	.4179	.460	.3288	.556	.2549	.643	.1918	.729	.1373
.123	.9101	.209	.6799	.296	.5287	.383	.4168	.470	.3279	.557	.2541	.644	.1911	.730	.1367
.124	.9066	.210	.6778	.297	.5272	.384	.4157	.471	.3270	.558	.2534	.645	.1904	.731	.1361
.125	.9031	.211	.6757	.298	.5258	.385	.4145	.472	.3260	.559	.2526	.646	.1898	.732	.1355

Trans.	Density	Trans.	Density	Trans.	Density	Trans.	Density	Trans.	Density
.733	.1349	.791	.1018	.849	.0711	.907	.0424	.965	.0155
.734	.1343	.792	.1013	.850	.0706	.908	.0419	.966	.0150
.735	.1337	.793	.1007	.851	.0701	.909	.0414	.967	.0146
.736	.1331	.794	.1002	.852	.0696	.910	.0410	.968	.0141
.737	.1325	.795	.0996	.853	.0690	.911	.0405	.969	.0137
.738	.1319	.796	.0991	.854	.0685	.912	.0400	.970	.0132
.739	.1314	.797	.0985	.855	.0680	.913	.0395	.971	.0128
.740	.1308	.798	.0980	.856	.0675	.914	.0391	.972	.0123
.741	.1302	.799	.0975	.857	.0670	.915	.0386	.973	.0119
.742	.1296	.800	.0969	.858	.0665	.916	.0381	.974	.0114
.743	.1290	.801	.0964	.859	.0660	.917	.0376	.975	.0110
.744	.1284	.802	.0958	.860	.0655	.918	.0371	.967	.0106
.745	.1278	.803	.0953	.861	.0650	.919	.0367	.977	.0101
.746	.1273	.804	.0948	.862	.0645	.920	.0362	.978	.0097
.747	.1267	.805	.0942	.863	.0640	.921	.0357	.979	.0092
.748	.1261	.806	.0937	.804	.0635	.922	.0353	.980	.0088
.749	.1255	.807	.0931	.865	.0630	.923	.0348	.981	.0083
.750	.1249	.808	.0926	.866	.0625	.924	.0343	.982	.0079
.751	.1244	.809	.0921	.867	.0620	.925	.0339	.983	.0074
.752	.1238	.810	.0915	.808	.0615	.926	.0334	.984	.0070
.753	.1232	.811	.0910	.869	.0610	.927	.0329	.985	.0066
.754	.1226	.812	.0904	.870	.0605	.928	.0325	.986	.0061
.755	.1221	.813	.0899	.871	.0600	.929	.0320	.987	.0057
.756	.1215	.814	.0894	.872	.0595	.930	.0315	.988	.0052
.757	.1209	.815	.0888	.873	.0590	.931	.0310	.989	.0048
.758	.1203	.816	.0883	.874	.0585	.932	.0306	.990	.0044
.759	.1198	.817	.0878	.975	.0580	.933	.0301	.991	.0039
.760	.1192	.818	.0872	.876	.0575	.934	.0296	.992	.0035
.761	.1186	.819	.0867	.877	.0570	.935	.0292	.993	.0030
.762	.1180	.820	.0862	.878	.0565	.936	.0287	.994	.0026
.763	.1175	.821	.0856	.879	.0560	.937	.0282	.995	.0022
.764	.1169	.822	.0851	.880	.0555	.938	.0278	.996	.0017
.765	.1163	.823	.0846	.881	.0550	.939	.0273	.997	.0013
.766	.1158	.824	.0841	.882	.0545	.940	.0269	.998	.0009
.767	.1152	.825	.0835	.883	.0540	.941	.0264	.999	.0004
.768	.1146	.826	.0830	.884	.0535	.942	.0260	1.000	.0000
.769	.1141	.827	.0825	.885	.0530	.943	.0255		
.770	.1135	.828	.0820	.886	.0526	.944	.0250		
.771	.1129	.829	.0815	.887	.0521	.945	.0246		
.772	.1124	.830	.0809	.888	.0516	.946	.0241		
.773	.1118	.831	.0804	.889	.0511	.947	.0237		
.774	.1113	.832	.0799	.890	.0506	.948	.2032		
.775	.1107	.833	.0794	.891	.0301	.949	.0227		
.776	.1102	.834	.0788	.892	.0496	.950	.0223		
.777	.1096	.835	.0783	.893	.0491	.951	.0218		
.778	.1090	.836	.0778	.894	.0487	.952	.0214		
.779	.1085	.837	.0773	.895	.0482	.953	.0209		
.780	.1079	.838	.0767	.896	.0477	.954	.0204		
.781	.1073	.839	.0762	.897	.0472	.955	.0200		
.782	.1068	.840	.0757	.898	.0467	.956	.0195		
.783	.1062	.841	.0752	.899	.0462	.957	.0191		
.784	.1057	.842	.0747	.900	.0458	.958	.0186		
.785	.1051	.843	.0742	.901	.0453	.959	.0182		
.786	.1046	.844	.0736	.902	.0448	.960	.0177		
.787	.1040	.845	.0731	.903	.0443	.961	.0173		
.788	.1035	.846	.0726	.904	.0438	.962	.0168		
.789	.1029	.847	.0721	.905	.0434	.963	.0164		
.790	.1024	.848	.0716	.906	.0429	.964	.0159		

DENSITY OF VARIOUS SOLIDS

The approximate density of various solids at ordinary atmospheric temperature.
In the case of substances with voids such as paper or leather the bulk density is indicated rather than the density of the solid portion.

(Selected principally from the Smithsonian Tables.)

Substance	Grams per cu. cm	Pounds per cu. ft.	Substance	Grams per cu. cm	Pound per cu. ft.	Substance	Grams per cu. cm	Pounds per cu. ft.
Agate	2.5–2.7	156–168	Glass, common	2.4–2.8	150–175	Tallow, beef	0.94	59
Alabaster, carbon-			flint	2.9–5.9	180–370	mutton	0.94	59
ate	2.69–2.78	168–173	Glue	1.27	79	Tar	1.02	66
sulfate	2.26–2.32	141–145	Granite	2.64–2.76	165–172	Topaz	3.5–3.6	219–223
Albite	2.62–2.65	163–165	Graphite*	2.30–2.72	144–170	Tourmaline	3.0–3.2	190–200
Amber	1.06–1.11	66–69	Gum arabic	1.3–1.4	81–87	Wax, sealing	1.8	112
Amphiboles	2.9–3.2	180–200	Gypsum	2.31–2.33	144–145	Wood (seasoned)		
Anorthite	2.74–2.76	171–172	Hematite	4.9–5.3	306–330	alder	0.42–0.68	26–42
Asbestos	2.0–2.8	125–175	Hornblende	3.0	187	apple	0.66–0.84	41–52
Asbestos slate	1.8	112	Ice	0.917	57.2	ash	0.65–0.85	40–53
Asphalt	1.1–1.5	69–94	Ivory	1.83–1.92	114–120	balsa	0.11–0.14	7–9
Basalt	2.4–3.1	150–190	Leather, dry	0.86	54	bamboo	0.31–0.40	19–25
Beeswax	0.96–0.97	60–61	Lime, slaked	1.3–1.4	81–87	basswood	0.32–0.59	20–37
Beryl	2.69–2.7	168–169	Limestone	2.68–2.76	167–171	beech	0.70–0.90	43–56
Biotite	2.7–3.1	170–190	Linoleum	1.18	74	birch	0.51–0.77	32–48
Bone	1.7–2.0	106–125	Magnetite	4.9–5.2	306–324	blue gum	1.00	62
Brick	1.4–2.2	87–137	Malachite	3.7–4.1	231–256	box	0.95–1.16	59–72
Butter	0.86–0.87	53–54	Marble	2.6–2.84	160–177	butternut	0.38	24
Calamine	4.1–4.5	255–280	Meerschaum	0.99–1.28	62–80	cedar	0.49–0.57	30–35
Calcspar	2.6–2.8	162–175	Mica	2.6–3.2	165–200	cherry	0.70–0.90	43–56
Camphor	0.99	62	Muscovite	2.76–3.00	172–187	dogwood	0.76	47
Caoutchouc	0.92 0.99	57 62	Ochre	3.5	218	ebony	1.11–1.33	69–83
Cardboard	0.69	43	Opal	2.2	137	elm	0.54–0.60	34–37
Celluloid	1.4	87	Paper	0.7–1.15	44–72	hickory	0.60–0.93	37–58
Cement set	2.7–3.0	170 190	Paraffin	0.87–0.91	54 57	holly	0.76	47
Chalk	1.9–2.8	118–175	Peat blocks	0.84	52	juniper	0.56	35
Charcoal, oak	0.57	35	Pitch	1.07	67	larch	0.50 0.56	31 35
pine	0.28–0.44	18–28	Porcelain	2.3–2.5	143–156	lignum vitae	1.17–1.33	73–83
Cinnabar	8.12	507	Porphyry	2.6–2.9	162–181	locust	0.67–0.71	42–44
Clay	1.8–2.6	112–162	Pressed wood			logwood	0.91	57
Coal, anthracite	1.4–1.8	87–112	pulp board	0.19	12	mahogany		
bituminous	1.2–1.5	75–94	Pyrite	4.95–5.1	309–318	Honduras	0.66	41
Cocoa butter	0.89–0.91	56–57	Quartz	2.65	165	Spanish	0.85	53
Coke	1.0–1.7	62–105	Resin	1.07	67	maple	0.62–0.75	39–47
Copal	1.04–1.14	65–71	Rock salt	2.18	136	oak	0.60–0.90	37–56
Cork	0.22–0.26	14–16	Rubber, hard	1.19	74	pear	0.61–0.73	38–45
Cork linoleum	0.54	34	Rubber, soft			pine, pitch	0.83–0.85	52–53
Corundum	3.9–4.0	245–250	commercial	1.1	69	white	0.35–0.50	22–31
Diamond	3.01–3.52	188–220	pure gum	0.91–0.93	57–58	yellow	0.37–0.60	23–37
Dolomite	2.84	177	Sandstone	2.14–2.36	134–147	plum	0.66–0.78	41–49
Ebonite	1.15	72	Serpentine	2.50–2.65	156–165	poplar	0.35–0.5	22–31
Emery	4.0	250	Silica, fused trans-			satinwood	0.95	59
Epidote	3.25–3.50	203–218	parent	2.21	138	spruce	0.48–0.70	30–44
Feldspar	2.55–2.75	159–172	translucent	2.07	129	sycamore	0.40–0.60	24–37
Flint	2.63	164	Slag	2.0–3.9	125–240	teak, Indian	0.66–0.88	41–55
Fluorite	3.18	198	Slate	2.6–3.3	162–205	African	0.98	61
Galena	7.3–7.6	460–470	Soapstone	2.6–2.8	162–175	walnut	0.64–0.70	40–43
Gamboge	1.2	75	Spermacéti	0.95	59	water gum	1.00	62
Garnet	3.15–4.3	197–268	Starch	1.53	95	willow	0.40–0.60	24–37
Gas carbon	1.88	117	Sugar	1.59	99			
Gelatin	1.27	79	Talc	2.7–2.8	168–174			

* Some values reported as low as 1.6.

WEIGHT OF ONE GALLON OF WATER (U.S. GALLONS)

The weights are for dry air at the same temperature as the water up to 40°C and at a barometric pressure corrected to 760 mm and against brass weights of 8.4 density at 0°C. Above 40°C the temperature of the air is assumed to be 20°C, i.e., the water is allowed to cool to 20°C prior to the weighings being made. The volumetric computations are based upon the relations that one liter = 1 dm³ and that 1 dm³ = 61.023744 in.³

Temperature (°C)	Weight in vacuo (g)	Weight in vacuo (lb)	Weight in air (g)	Weight in air (lb)	Temperature (°C)	Weight in vacuo (g)	Weight in vacuo (lb)	Weight in air (g)	Weight in air (lb)
0	3784.856	8.34417	3780.543	8.33467	25	3774.291	8.32088	3770.340	8.31217
1	3785.078	8.34466	3780.781	8.33518	26	3773.320	8.31870	3769.364	8.31001
2	3785.233	8.34500	3780.953	8.33556	27	3772.277	8.31644	3768.352	8.30778
3	3785.326	8.34520	3781.060	8.33580	28	3771.218	8.31410	3767.306	8.30548
4	3785.355	8.34527	3781.105	8.33590	29	3770.123	8.31169	3766.224	8.30309
5	3785.325	8.34520	3781.090	8.33587	30	3768.995	8.30920	3765.109	8.30063
6	3785.235	8.34500	3781.015	8.33570	31	3768.995	8.30664	3763.961	8.29810
7	3785.089	8.34468	3780.884	8.33541	32	3766.641	8.30401	3762.780	8.29550
8	3784.887	8.34424	3780.698	8.33500	33	3765.416	8.30131	3761.568	8.29283
9	3784.633	8.34368	3780.358	8.33447	34	3764.160	8.29854	3760.324	8.29008
10	3784.326	8.34300	3780.167	8.33383	35	3762.874	8.29571	3759.050	8.28728
11	3783.966	8.34221	3779.821	8.33307	40	3756.018	8.28059	3752.255	8.27230
12	3783.557	8.34130	3779.426	8.33220	45	3748.41	8.2638	3744.42	8.2550
13	3783.099	8.34030	3778.983	8.33122	50	3740.19	8.2457	3736.22	8.2369
14	3782.597	8.33919	3778.495	8.33014	55	3731.34	8.2261	3727.37	8.2174
15	3782.049	8.33798	3777.962	8.32897	60	3721.91	8.2054	3717.95	8.1966
16	3781.458	8.33668	3777.415	8.32770	65	3711.88	8.1832	3707.93	8.1745
17	3780.824	8.33528	3776.764	8.32633	70	3701.35	8.1600	3697.42	8.1514
18	3780.148	8.33379	3776.103	8.32487	75	3690.30	8.1357	3686.38	8.1270
19	3779.430	8.33221	3775.398	8.32332	80	3678.72	8.1101	3674.81	8.1015
20	3778.672	8.33054	3774.653	8.32167	85	3666.68	8.0836	3662.78	8.0750
21	3777.873	8.32877	3773.868	8.31994	90	3654.15	8.0560	3650.27	8.0474
22	3777.035	8.32693	3773.044	8.31813	95	3641.21	8.0274	3637.34	8.0189
23	3776.158	8.32499	3772.180	8.31622	100	3627.81	7.9979	3623.95	7.9894
24	3775.243	8.32298	3771.279	8.31424					

TEMPERATURE CORRECTION FOR VOLUMETRIC SOLUTIONS

This table gives the correction to various observed volumes of water, measured at the designated temperatures to give the volume at the standard temperature, 20°C. Conversely, by subtracting the corrections from the volume desired at 20°C, the volume that must be measured out at the designated temperatures in order to give the desired volume at 20°C., will be obtained. It is assumed that the volumes are measured in glass apparatus having a coefficient of cubical expansion of 0.000025 per degree centigrade. The table is applicable to dilute aqueous solutions having the same coefficient of expansion as water.

Temperature of measurement, °C.	Capacity of apparatus in milliliters at 20°C. 2,000	1,000	500	400	300	250	150	Temperature of measurement, °C	Capacity of apparatus in milliliters at 20°C. 2,000	1,000	500	400	300	250	150
	Correction in milliliters to give volume of water at 20°C.								Correction in milliliters to give volume of water at 20°C.						
15	+1.54	+0.77	+0.38	+0.31	+0.23	+0.19	+0.12	24	−1.61	−.81	−.40	−.32	−.24	−.20	−.12
16	+1.28	+.64	+.32	+.26	+.19	+.16	+.10	25	−2.07	−1.03	−.52	−.41	−.31	−.26	−.15
17	+.99	+.50	+.25	+.20	+.15	+.12	+.07	26	−2.54	−1.27	−.64	−.51	−.38	−.32	−.19
18	+.68	+.34	+.17	+.14	+.10	+.08	+.05	27	−3.03	−1.52	−.76	−.61	−.46	−.38	−.23
19	+.35	+.18	+.09	+.07	+.05	+.04	+.03	28	−3.55	−1.77	−.89	−.71	−.53	−.44	−.27
21	−.37	−.18	−.09	−.07	−.06	−.05	−.03	29	−4.08	−2.04	−1.02	−.82	−.61	−.51	−.31
22	−.77	−.38	−.19	−.15	−.12	−.10	−.06	30	−4.62	−2.31	−1.16	−.92	−.69	−.58	−.35
23	−1.18	−.59	−.30	−.24	−.18	−.15	−.09								

In using the above table to correct the volume of certain standard solutions to 20°C. more accurate results will be obtained if the numerical values of the corrections are increased by the percentages given below:

Solution	Normality N	N/2	N/10
HNO₃	50	25	6
H₂SO₄	45	25	5
NaOH	40	25	5
KOH	40	20	4

TEMPERATURE CORRECTION FOR GLASS VOLUMETRIC APPARATUS

This table gives the correction to be added to actual capacity (determined at certain temperatures) to give the capactiy at the standard temperature, 20°C. Conversely, by subtracting the corrections from the indicated capacity of an instrument standard at 20°C. the corresponding capacity at other temperatures is obtained. The table assumes for the cubical coefficient of expansion of glass 0.000025 per degree centigrade. The coefficients of expansion of glasses used for volumetric instruments vary from 0.000023 to 0.000028.

Temperature in degrees C.	2,000 ml	1,000 ml	500 ml	400 ml	300 ml	250 ml	Temperature in degrees C.	2,000 ml	1,000 ml	500 ml	400 ml	300 ml	250 ml
15	+ 0.25	+ 1.12	+ 0.06	+ 0.05	+ 0.04	+ 0.031	23	− .15	− .08	− .04	− .03	− .02	− .019
16	+ .20	+ .10	+ .05	+ 0.04	+ .03	+ .025	24	− .20	− .10	− .05	− .04	− .03	− .025
17	+ .15	+ .08	+ .04	+ .03	+ .02	+ .019	25	− .25	− .12	− .06	− .05	− .04	− .031
18	+ .10	+ .05	+ .02	+ .02	+ .02	+ .012	26	− .30	− .15	− .08	− .06	− .04	− .038
19	+ .05	+ .02	+ .01	+ .01	+ .01	+ .006	27	− .35	− .18	− .09	− .07	− .05	− .044
21	− .05	− .02	− .01	− .01	− .01	− .006	28	− .40	− .20	− .10	− .08	− .06	− .050
22	.10	− .05	− .02	− .02	− .02	− .012	29	.45	.22	.11	.09	.07	.056
							30	− .50	− .25	− .12	− .10	− .08	− .062

DENSITY OF VARIOUS LIQUIDS

(Selected from Smithsonian Tables.)

Liquid	Grams per cu. cm	Pounds per cu. ft.	Temp °C	Liquid	Grams per cu. cm	Pounds per cu. ft.	Temp °C
Acetone	0.792	49.4	20°	Naphtha, petroleum ether	0.665	41.5	15
Alcohol, ethyl	0.791	49.4	20				
methyl	0.810	50.5	0	wood	0.848—0.810	52.9—50.5	0
Benzene	0.899	56.1	0	Oils:			
Carbolic acid	0.950—0.965	59.2—60.2	15	castor	0.969	60.5	15
Carbon disulfide	1.293	80.7	0	cocoanut	0.925	57.7	15
tetrachloride	1.595	99.6	20	cotton seed	0.926	57.8	16
Chloroform	1.489	93.0	20	creosote	1.040—1.100	64.9—68.6	15
Ether	0.736	45.9	0	linseed, boiled	0.924	58.8	15
Gasoline	0.66—0.69	41.0—43.0		olive	0.918	57.3	15
Glycerin	1.260	78.6	0	Sea water	1.025	63.99	15
Kerosene	0.82	51.2		Turpentine (spirits)	0.87	54.3	
Mercury	13.6	849.0		Water	1.00	62.43	4
Milk	1.028—1.035	64.2—64.6					

DENSITY OF ALCOHOL

Density of Ethyl Alcohol in Grams Per Cubic Centimeter, Computed from Mendeleeff's Formula

(Selected from Smithsonian Tables.)

Temp. °C	0	1	2	3	4	5	6	7	8	9
0	.80625	.80541	.80457	.80374	.80290	.80207	.80123	.80039	.79956	.79872
10	.79788	.79704	.79620	.79535	.79451	.79367	.79283	.79198	.79114	.79029
20	.78945	.78860	.78775	.78691	.78606	.78522	.78437	.78352	.78267	.78182
30	.78097	.78012	.77927	.77841	.77756	.77671	.77585	.77500	.77414	.77329

HYDROMETERS AND DENSITY UNITS

Alcoholometer. — For testing alcoholic solutions; the scale shows the per cent of alcohol by volume; 0°—100° is the per cent.

Ammoniameter. — For testing ammonia solutions; scale 0°—40°; to convert to sp. gr. multiply by 3 and deduct from 1000.

Barktrometer or *Barkometer.* — For testing tanning liquor; scale 0°—80° Bk; the number to the right of the sp. gr. is the degree Bk; thus, 1.025 sp. gr. is 25° Bk.

Baumé. — There are two kinds in use; heavy Bé, for liquids heavier than water and light Bé for liquids lighter than water. In the former, 0° corresponds to a sp. gr. 1.000 (water at 4°C.) and 66° corresponds to a sp. gr. 1.842; in the lighter than water scale, 0° Bé is equivalent to the gravity of a 10% solution of sodium chloride and 60° Bé corresponds to a sp. gr. of 0.745. For Baumé degrees on the scale of densities greater than unity, the following equation gives the means of conversion:

$$Sp.gr. = \frac{m}{m-d}$$

where m = 145 (in the United States)
m = 144 (old scale used in Holland)
m = 146.78 (New scale or Gerlach scale)
d = Baumé reading

Beck's Hydrometer has 0° corresponding to sp. gr. 1.000 and 30° to sp. gr. 0.850; equal divisions on the scale are continued as far as required in both directions.

Brix Saccharometer or *Balling Saccharometer* shows directly the per cent of sugar (sucrose) by weight at the temperature indicated on the instrument, usually 17.5°C.; i.e., degrees Brix is the per cent sugar.

Cartier's Hydrometer floats in water at the 10° scale division and 30° corresponds to 32° Be.

Oleometer. — For vegetable and sperm oils; scale 50°—0° corresponds to sp. gr. 0.870—0.970.

Soxhlet's Lactometer, for determining the density of milk, has a scale from 25° (sp. gr. 1.025) to 35° (sp. gr. 1.035) divided into suitable scale divisions.

Twaddell Hydrometers have the scale so arranged that the reading multiplied by 5 and added to 1000 gives the sp. gr. with reference to water as 1000; it is always used for densties greater than water.

HYDROMETER CONVERSION TABLES

Showing the Relation between Density (C. G. S.)
and Degrees Baumé for Densities less than Unity

Density	Degrees Baumé				
	.00	.01	.02	.03	.04
0.60	103.33	99.51	95.81	92.22	88.75
.70	70.00	67.18	64.44	61.78	59.19
.80	45.00	42.84	40.73	38.68	36.67
.90	25.56	23.85	22.17	20.54	18.94
1.00	10.00				

Density	Degrees Baumé				
	.05	.06	.07	.08	.09
0.60	85.38	82.12	78.95	75.88	72.90
.70	56.67	54.21	51.82	49.49	47.22
.80	34.71	32.79	30.92	29.09	27.30
.90	17.37	15.83	14.33	12.86	11.41
1.00					

Showing the Relation between Density (C. G. S.) and
Baumé and Twaddell Scales for Densities above Unity

Density	Degrees Baumé	Degrees Twaddell	Density	Degrees Baumé	Degrees Twaddell	Density	Degrees Baumé	Degrees Twaddell	Denisty	Degrees Baumé	Degrees Twaddell
1.00	0.00	0	1.20	24.17	40	1.41	42.16	82	1.61	54.94	122
1.01	1.44	2	1.21	25.16	42	1.42	42.89	84	1.62	55.49	124
1.02	2.84	4	1.22	26.15	44	1.43	43.60	86	1.63	56.04	126
1.03	4.22	6	1.23	27.11	46	1.44	44.31	88	1.64	56.58	128
1.04	5.58	8	1.24	28.06	48	1.45	45.00	90	1.65	57.12	130
1.05	6.91	10	1.25	29.00	50	1.46	45.68	92	1.66	57.65	132
1.06	8.21	12	1.26	29.92	52	1.47	46.36	94	1.67	58.17	134
1.07	9.49	14	1.27	30.83	54	1.48	47.03	96	1.68	58.69	136
1.08	10.74	16	1.28	31.72	56	1.49	47.68	98	1.69	59.20	138
1.09	11.97	18	1.29	32.60	58	1.50	48.33	100	1.70	59.71	140
1.10	13.18	20	1.30	33.46	60	1.51	48.97	102	1.71	60.20	142
1.11	14.37	22	1.31	34.31	62	1.52	49.60	104	1.72	60.70	144
1.12	15.54	24	1.32	35.15	64	1.53	50.23	106	1.73	61.18	146
1.13	16.68	26	1.33	35.98	66	1.54	50.84	108	1.74	61.67	148
1.14	17.81	28	1.34	36.79	68	1.55	51.45	110	1.75	62.14	150
1.15	18.91	30	1.35	37.59	70	1.56	52.05	112	1.76	62.61	152
1.16	20.00	32	1.36	38.38	72	1.57	52.64	114	1.77	63.08	154
1.17	21.07	34	1.37	39.16	74	1.58	53.23	116	1.78	63.54	156
1.18	22.12	36	1.38	39.93	76	1.59	53.80	118	1.79	63.99	158
1.19	23.15	38	1.39	40.68	78	1.60	54.38	120	1.80	64.44	160
			1.40	41.43	80						

DENSITY OF D_2O

G. S. Kell

t, °C.	ϱ, G./Cc.	t, °C.	ϱ, G./Cc.	t, °C.	ϱ, G./Cc.	t, °C.	ϱ, G./Cc.
0	1.10469	20	1.10534	50	1.09570	80	1.07824
3.813	1.10546	25	1.10445	55	1.09325	85	1.07475
5	1.10562	30	1.10323	60	1.09060	90	1.07112
10	1.10599	35	1.10173	65	1.08777	95	1.06736
11.185	1.10600	40	1.09996	70	1.08475	100	1.06346
15	1.10587	45	1.09794	75	1.08158	101.431	1.06232

VOLUME PROPERTIES OF WATER AT 1 atm*

	ρ, kg m^{-3},	$10^6\,\alpha$, K^{-1},	$10^6\,\kappa T$/bar^{-1}		ρ, kg m^{-3},	$10^6\,\alpha$, K^{-1},	$10^6\,\kappa T$/bar^{-1}
	Equation 1	Equation 1	Equation 2		Equation 1	Equation 1	Equation 2
-30	983.857	-1400.0	80.79	38	992.9683	369.79	44.3051
-25	989.588	-955.9	70.94	39	992.5973	377.59	44.2697
-20	993.550	-660.6	64.25	40	992.2187	385.30	44.2391
-15	996.286	-450.3	59.44	41	991.8327	392.91	44.2131
-10	998.120	-292.4	55.83	42	991.4394	400.43	44.1917
-9	998.398	-265.3	55.22	43	991.0388	407.85	44.1747
-8	998.650	-239.5	54.64	44	990.6310	415.19	44.1620
-7	998.877	-214.8	54.08	45	990.2162	422.45	44.1536
-6	999.080	-191.2	53.56	46	989.7944	429.63	44.1494
-5	999.259	-168.6	53.06	47	989.3657	436.73	44.1494
-4	999.417	-146.9	52.58	48	988.9303	443.75	44.1533
-3	999.553	-126.0	52.12	49	988.4881	450.71	44.1613
-2	999.669	-106.0	51.69	50	988.0393	457.59	44.1732
-1	999.765	-86.7	51.28	51	987.5839	464.40	44.189
0	999.8425	-68.05	50.8850	52	987.1220	471.15	44.209
1	999.9015	-50.09	50.5091	53	986.6537	477.84	44.232
2	999.9429	-32.74	50.1505	54	986.1791	484.47	44.259
3	999.9672	-15.97	49.8081	55	985.6982	491.04	44.290
4	999.9750	0.27	49.4812	56	985.2111	497.55	44.324
5	999.9668	16.00	49.1692	57	984.7178	504.01	44.362
6	999.9432	31.24	48.8712	58	984.2185	510.41	44.403
7	999.9043	46.04	48.5868	59	983.7132	516.76	44.448
8	999.8512	60.41	48.3152	60	983.2018	523.07	44.496
9	999.7838	74.38	48.0560	61	982.6846	529.32	44.548
10	999.7026	87.97	47.8086	62	982.1615	535.53	44.603
11	999.6081	101.20	47.5726	63	981.6327	541.70	44.662
12	999.5004	114.08	47.3474	64	981.0981	547.82	44.723
13	999.3801	126.65	47.1327	65	980.5578	553.90	44.788
14	999.2474	138.90	46.9280	66	980.0118	559.94	44.857
15	999.1026	150.87	46.7331	67	979.4603	565.95	44.928
16	998.9460	162.55	46.5475	68	978.9032	571.91	45.003
17	998.7779	173.98	46.3708	69	978.3406	577.84	45.081
18	998.5986	185.15	46.2029	70	977.7726	583.74	45.162
19	998.4082	196.08	46.0433	71	977.1991	589.60	45.246
20	998.2071	206.78	45.8918	72	976.6203	595.43	45.333
21	997.9955	217.26	45.7482	73	976.0361	601.23	45.424
22	997.7735	227.54	45.6122	74	975.4466	607.00	45.517
23	997.5415	237.62	45.4835	75	974.8519	612.75	45.614
24	997.2995	247.50	45.3619	76	974.2520	618.46	45.714
25	997.0479	257.21	45.2472	77	973.6468	624.15	45.817
26	996.7867	266.73	45.1392	78	973.0366	629.82	45.922
27	996.5162	276.10	45.0378	79	972.4212	635.46	46.031
28	996.2365	285.30	44.9427	80	971.8007	641.08	46.143
29	995.9478	294.34	44.8537	81	971.1752	646.67	46.258
30	995.6502	303.24	44.7707	82	970.5446	652.25	46.376
31	995.3440	312.00	44.6935	83	969.9091	657.81	46.497
32	995.0292	320.63	44.6221	84	969.2686	663.34	46.621
33	994.7060	329.12	44.5561	85	968.6232	668.86	46.748
34	994.3745	337.48	44.4956	86	967.9729	674.37	46.878
35	994.0349	345.73	44.4404	87	967.3177	679.85	47.011
36	993.6872	353.86	44.3903	88	966.6576	685.33	47.148
37	993.3316	361.88	44.3452	89	965.9927	690.78	47.287

VOLUME PROPERTIES OF WATER AT 1 atm* (continued)

	ρ, kg m^{-3},	$10^6 \; \alpha$, K^{-1},	$10^6 \; \kappa T$/bar^{-1}	
	Equation 1	Equation 1	Equation 2	Equation 3
90	965.3230	696.23	47.429	47.428
91	964.6486	701.66	47.574	47.574
92	963.9693	707.08	47.722	47.722
93	963.2854	712.49	47.874	47.873
94	962.5967	717.89	48.028	48.028
95	961.9033	723.28	48.185	48.185
96	961.2052	728.67	48.346	48.346
97	960.5025	734.04	48.509	48.510
98	959.7951	739.41	48.676	48.677
99	959.0831	744.78	48.846	48.847
100	958.3665	750.14	49.019	49.020
101	957.645	755.5		49.20
102	956.920	760.8		49.38
103	956.189	766.2		49.56
104	955.454	771.5		49.74
105	954.715	776.9		49.93
106	953.971	782.2		50.13
107	953.222	787.6		50.32
108	952.469	792.9		50.52
109	951.712	798.3		50.72
110	950.950	803.6		50.93
115	947.073	830.4		52.01
120	943.085	857.4		53.17
125	938.987	884.7		54.43
130	934.778	912.3		55.79
135	930.459	940.3		57.24
140	926.029	968.9		58.80
145	921.487	998.0		60.47
150	916.832	1027.8		62.25

Equations:

$$\rho/\text{kg m}^{-3} = (999.83952 + 16.945176 \, t - 7.9870401 \times 10^{-3} \, t^2 - 46.170461 \times 10^{-6} \, t^3 + 105.56302 \times 10^{-9} \, t^4 - 280.54253 \times 10^{-12} \, t^5)/(1 + 16.879850 \times 10^{-3} \, t) \qquad (1)$$

$$10^6 \; \kappa_T/\text{bar}^{-1} = (50.88496 + 0.6163813 \, t + 1.459187 \times 10^{-3} \, t^2 + 20.08438 \times 10^{-6} \, t^3 - 58.47727 \times 10^{-9} \, t^4 + 410.4110 \times 10^{-12} \, t^5)/(1 + 19.67348 \times 10^{-3} \, t) \qquad (2)$$

$$10^6 \; \kappa_T/\text{bar}^{-1} = (50.884917 + 0.62590623 \, t + 1.3848668 \times 10^{-3} \, t^2 + 21.603427 \times 10^{-6} \, t^3 - 72.087667 \times 10^{-9} \, t^4 + 465.45054 \times 10^{-12} \, t^5)/(1 + 19.859983 \times 10^{-3} \, t) \qquad (3)$$

$$\kappa_S = (\partial \ln \rho / \partial P)_S = \frac{1}{\rho U^2} \qquad (4)$$

* Density ρ, thermal expansivity $\alpha = - (\partial \ln \rho / \partial T)_p$, and isothermal compressibility $\kappa T = (\partial \ln \rho / \partial p)T$. For purposes of this table, ordinary water is that with a maximum density of 999.972 kg m^{-3}. Equation 4 for the compressibility should be used for temperatures $0 < t < 100°$C, and Equation 3 for 100 $< + < 150°$C. The liquid is metastable below 0°C and above 100°C. Values below 0°C were obtained by extrapolation, and no claim is made for their accuracy.

Reprinted with permission from Kell, G. S., *J. Chem. Eng. Data*, 20(1), 97, 1975. Copyright by the American Chemical Society.

DENSITY AND VOLUME OF MERCURY
Based on the Density of Mercury at 0° C. by Thiesen and Scheel
(Selected from Smithsonian Tables)

Temp. °C.	Mass in gr. per ml.	Vol. of 1 gr. in ml.	Temp. °C.	Mass in gr. per ml.	Vol. of 1 gr. in ml.	Temp. °C.	Mass in gr. per ml.	Vol. of 1gr. in ml.
−10	13.6202	0.0734205	17	5536	7813	90	13.3762	0.0747594
−9	6177	4338	18	5512	7947	100	3522	8939
−8	6152	4472	19	5487	8081	110	3283	50285
−7	6128	4606	20	13.5462	0.0738215	120	3044	1633
−6	6103	4739	21	5438	8348	130	2805	2982
−5	13.6078	0.0734873	22	5413	8482	140	13.2567	0.0754334
−4	6053	5006	23	5389	8616	150	2330	5688
−3	6029	5140	24	5364	8750	160	2093	7044
−2	6004	5273	25	13.5340	0.0738883	170	1856	8402
−1	5979	5407	26	5315	9017	180	1620	9764
0	13.5955	0.0735540	27	5291	9151	190	13.1384	0.0761128
1	5930	5674	28	5266	9285	200	1148	2495
2	5906	5808	29	5242	9419	210	0913	3865
3	5881	5941	30	13.5217	0.0739552	220	0678	5239
4	5856	6075	31	5193	9686	230	0443	6616
5	13.5832	0.0736209	32	5168	9820	240	13.0209	0.0767996
6	5807	6342	33	5144	9953	250	12.9975	9381
7	5782	6476	34	5119	40087	260	9741	70769
8	5758	6610	35	13.5095	0.0740221	270	9507	2161
9	5733	6744	36	5070	0354	280	9273	3558
10	13.5708	0.0736877	37	5046	0488	290	12.9039	0.0774958
11	5684	7011	38	5021	0622	300	8806	6364
12	5659	7145	39	4997	0756	310	8572	7774
13	5634	7278	40	13.4973	0.0740891	320	8339	9189
14	5610	7412	50	4729	2229	330	8105	80609
15	13.5585	0.0737546	60	4486	3569	340	12.7872	0.0782033
16	5561	7680	70	4244	4910	350	7638	3161
			80	4003	6252	360	7405	4900

SULFURIC ACID
SPECIFIC GRAVITY OF AQUEOUS SULFURIC ACID SOLUTIONS
AT $\frac{20°}{4°}$ C.

Be.	Sp. gr.	Per cent H₂SO₄	G. per liter	Lbs. per cu. ft.	Lbs. per gal.	Be.	Sp. gr.	Per cent H₂SO₄	G. per liter	Lbs. per cu. ft.	Lbs. per gal.
0.7	1.0051	1	10.05	0.6275	0.0839	41.8	1.4049	51	716.5	44.73	5.979
1.7	1.0118	2	20.24	1.263	0.1689	42.5	1.4148	52	735.7	45.93	6.140
2.6	1.0184	3	30.55	1.907	0.2550	43.2	1.4248	53	755.1	47.14	6.302
3.5	1.0250	4	41.00	2.560	0.3422	44.0	1.4350	54	774.9	48.37	6.467
4.5	1.0317	5	51.59	3.220	0.4305	44.7	1.4453	55	794.9	49.62	6.634
5.4	1.0385	6	62.31	3.890	0.5200	45.4	1.4557	56	815.2	50.89	6.803
6.3	1.0453	7	73.17	4.568	0.6106	46.1	1.4662	57	835.7	52.17	6.974
7.2	1.0522	8	84.18	5.255	0.7025	46.8	1.4768	58	856.5	53.47	7.148
8.1	1.0591	9	95.32	5.950	0.7955	47.5	1.4875	59	877.6	54.79	7.324
9.0	1.0661	10	106.6	6.655	0.8897	48.2	1.4983	60	899.0	56.12	7.502
9.9	1.0731	11	118.0	7.369	0.9851	48.9	1.5091	61	920.6	57.47	7.682
10.8	1.0802	12	129.6	8.092	1.082	49.6	1.5200	62	942.4	58.83	7.865
11.7	1.0874	13	141.4	8.825	1.180	50.3	1.5310	63	964.5	60.21	8.049
12.5	1.0947	14	153.3	9.567	1.279	51.0	1.5421	64	986.9	61.61	8.236
13.4	1.1020	15	165.3	10.32	1.379	51.7	1.5533	65	1010	63.03	8.426
14.3	1.1094	16	177.5	11.08	1.481	52.3	1.5646	66	1033	64.46	8.618
15.2	1.1168	17	189.9	11.85	1.584	53.0	1.5760	67	1056	65.92	8.812
16.0	1.1243	18	202.4	12.63	1.689	53.7	1.5874	68	1079	67.39	9.008
16.9	1.1318	19	215.0	13.42	1.795	54.3	1.5989	69	1103	68.87	9.207
17.7	1.1394	20	227.9	14.23	1.902	55.0	1.6105	70	1127	70.38	9.408
18.6	1.1471	21	240.9	15.04	2.010	55.6	1.6221	71	1152	71.90	9.611
19.4	1.1548	22	254.1	15.86	2.120	56.3	1.6338	72	1176	73.44	9.817
20.3	1.1626	23	267.4	16.69	2.231	56.9	1.6456	73	1201	74.99	10.02
21.1	1.1704	24	280.9	17.54	2.344	57.5	1.6574	74	1226	76.57	10.24
21.9	1.1783	25	294.6	18.39	2.458	58.1	1.6692	75	1252	78.15	10.45
22.8	1.1862	26	308.4	19.25	2.574	58.7	1.6810	76	1278	79.75	10.66
23.6	1.1942	27	322.4	20.13	2.691	59.3	1.6927	77	1303	81.37	10.88
24.4	1.2023	28	336.6	21.02	2.809	59.9	1.7043	78	1329	82.99	11.09
25.2	1.2104	29	351.0	21.91	2.929	60.5	1.7158	79	1355	84.62	11.31
26.0	1.2185	30	365.6	22.82	3.051	61.1	1.7272	80	1382	86.26	11.53
26.8	1.2267	31	380.3	23.74	3.173	61.6	1.7383	81	1408	87.90	11.75
27.6	1.2349	32	395.2	24.67	3.298	62.1	1.7491	82	1434	89.54	11.97
28.4	1.2432	33	410.3	25.61	3.424	62.6	1.7594	83	1460	91.16	12.19
29.1	1.2515	34	425.5	26.56	3.551	63.0	1.7693	84	1486	92.78	12.40
29.9	1.2599	35	441.0	27.53	3.680	63.5	1.7786	85	1512	94.38	12.62
30.7	1.2684	36	456.6	28.51	3.811	63.9	1.7872	86	1537	95.95	12.83
31.4	1.2769	37	472.5	29.49	3.943	64.2	1.7951	87	1562	97.49	13.03
32.2	1.2855	38	488.5	30.49	4.077	64.5	1.8022	88	1586	99.01	13.23
33.0	1.2941	39	504.7	31.51	4.212	64.8	1.8087	89	1610	100.5	13.42
33.7	1.3028	40	521.1	32.53	4.349	65.1	1.8144	90	1633	101.9	13.63
34.5	1.3116	41	537.8	33.57	4.488	65.3	1.8195	91	1656	103.4	13.82
35.2	1.3205	42	554.6	34.62	4.628	65.5	1.8240	92	1678	104.8	14.00
35.9	1.3294	43	571.6	35.69	4.770	65.7	1.8279	93	1700	106.1	14.19
36.7	1.3384	44	588.9	36.76	4.914	65.8	1.8312	94	1721	107.5	14.36
37.4	1.3476	45	606.4	37.86	5.061	65.9	1.8337	95	1742	108.7	14.54
38.1	1.3569	46	624.2	38.97	5.209	66.0	1.8355	96	1762	110.0	14.70
38.9	1.3663	47	642.2	40.09	5.359	66.0	1.8364	97	1781	111.2	14.87
39.6	1.3758	48	660.4	41.23	5.511	66.0	1.8361	98	1799	112.3	15.02
40.3	1.3854	49	678.8	42.38	5.665	65.9	1.8342	99	1816	113.4	15.15
41.1	1.3951	50	697.6	43.55	5.821	65.8	1.8305	100	1831	114.3	15.28

DENSITY AND COMPOSITION OF FUMING SULFURIC ACID

Actual H_2SO_4, %	Specific gravity	Equiv. H_2SO_4, %	Weight, lb./cu. ft.	Weight, lb. per U.S. gal.	Comb. H_2O, %	Free SO_3, %	Total SO_3, %	SO_3, lb./cu. ft.
100	1.839	100.00	114.70	15.33	18.37	0	81.63	93.63
99	1.845	100.22	115.07	15.38	18.19	1	81.81	94.14
98	1.851	100.45	115.33	15.41	18.00	2	82.00	94.57
97	1.855	100.67	115.70	15.46	17.82	3	82.18	95.08
96	1.858	100.89	115.88	15.49	17.64	4	82.36	95.44
95	1.862	101.13	116.13	15.52	17.45	5	82.55	95.87
94	1.865	101.35	116.32	15.55	17.27	6	82.73	96.23
93	1.869	101.58	116.57	15.58	17.08	7	82.92	96.66
92	1.873	101.80	116.82	15.61	16.90	8	83.10	97.12
91	1.877	102.02	117.07	15.64	16.72	9	83.28	97.50
90	1.880	102.25	117.26	15.67	16.57	10	83.47	97.88
89	1.884	102.47	117.51	15.70	16.35	11	83.65	98.30
88	1.887	102.71	117.69	15.73	16.17	12	83.83	98.66
87	1.891	102.92	117.94	15.76	15.98	13	84.02	99.09
86	1.895	103.15	118.19	15.79	15.80	14	84.20	99.52
85	1.899	103.38	118.44	15.82	15.61	15	84.39	99.95
84	1.902	103.60	118.63	15.86	15.43	16	84.57	100.33
83	1.905	103.82	118.81	15.89	15.25	17	84.75	100.69
82	1.909	104.05	119.06	15.92	15.06	18	84.94	101.13
81	1.911	104.28	119.28	15.95	14.88	19	85.12	101.45
80	1.915	104.50	119.50	15.98	14.70	20	85.30	101.93
79	1.920	104.73	119.75	16.01	14.51	21	85.49	102.37
78	1.923	104.95	119.94	16.04	14.33	22	85.67	102.75
77	1.927	105.18	120.19	16.07	14.14	23	85.86	103.20
76	1.931	105.40	120.44	16.10	13.96	24	86.04	103.63
75	1.934	105.62	120.62	16.12	13.78	25	86.22	104.00
74	1.939	105.85	120.94	16.16	13.59	26	86.41	104.50
73	1.943	106.08	121.18	16.19	13.41	27	86.59	104.93
72	1.946	106.29	121.37	16.22	13.28	28	86.72	105.31
71	1.949	106.53	121.56	16.25	13.04	29	86.96	105.71
70	1.952	106.75	121.75	16.28	12.86	30	87.14	106.09
69	1.955	106.97	121.93	16.30	12.68	31	87.32	106.47
68	1.958	107.20	122.12	16.33	12.49	32	87.51	106.87
67	1.961	107.42	122.31	16.35	12.31	33	87.69	107.25
66	1.965	107.65	122.56	16.38	12.12	34	87.88	107.71
65	1.968	107.87	122.74	16.40	11.94	35	88.06	108.08
64	1.972	108.10	122.99	16.43	11.76	36	88.24	108.53
63	1.976	108.33	123.24	16.46	11.57	37	88.43	108.98
62	1.979	108.55	123.43	16.50	11.39	38	88.61	109.37
61	1.981	108.77	123.55	16.52	11.21	39	88.79	109.70
60	1.983	109.00	123.74	16.54	11.02	40	88.98	110.10
59	1.985	109.22	123.80	16.55	10.84	41	89.16	110.38
58	1.987	109.45	123.93	16.56	10.65	42	89.35	110.83
57	1.989	109.68	124.05	16.58	10.47	43	89.53	111.06
56	1.991	109.90	124.18	16.60	10.29	44	89.71	111.40
55	1.993	110.13	124.30	16.62	10.10	45	89.90	111.75
50	2.001	111.25	124.80	16.68	9.18	50	90.72	113.34
40	2.102	113.50	131.10	17.53	7.35	60	92.65	121.46
30	1.982	115.75	123.62	16.50	5.51	70	94.49	116.81
20	1.949	118.00	121.56	16.25	3.67	80	96.33	117.10
10	1.911	120.25	119.19	15.92	1.84	90	98.16	117.00
0	1.857	122.50	115.83	15.50	0.00	100	100.00	115.83

* By permission from the 7th edition of Chemical Plant Control Data, Chemical Construction Corporation (1957).

DENSITY OF MOIST AIR

The density of dry air may be determined by computation from the general relation $D = D_0(T_0/T)(P/P_0)$ where D_0 represents a known density at absolute temperature T_0 and pressure P_0 and D, the density at absolute temperature T and pressure P.

The density of moist air may be determined by a similar relation:

$D = 1.2929 (273.13/T) [B - 0.3783 e)/760]$ where T is the absolute temperature; B, the barometric pressure in mm, and e the vapor pressure of the moisture in the air in millimeters. The density will then be the product of two terms, each of which may be found by use of the tables which follow.

The first factor, $1.2929 (273.13/T)$, may be found directly in Table I for various temperatures. For convenience, temperatures are given in the table in °C although the values of the factor have been computed with absolute temperatures. The tabular values actually represent the density of dry air at various temperatures and 760 mm pressure.

The second factor, $[(B - 0.3783 e/760]$, must be obtained in two steps: First — the numerator of the expression is obtained by subtracting $0.3783 e$ from the barometric pressure. The quantity $0.3783 e$ may be found directly from the dew point in Table II. If the wet and dry bulb thermometer readings are known e may be found in the table Reduction of Psychrometric Observations given in the section Hygrometric and Barometric Tables. $0.3783 e$ may then be found by calculation or read from the table. Second — the value of the whole factor for any value of $B - 0.3783 e$ may be obtained from Table III.

The product of the above two factors will give the required density in a g/l.

To facilitate obtaining approximate values of the density for ordinary pressures and temperatures, a table of products is given which may be entered with the temperatures in °C and the corrected (for moisture) value of the barometric pressure in mm to obtain density.

As an illustration of the use of the tables, let it be desired to find the density of air for a barometric pressure of 750 mm, a dew point of 10°C, and air temperature of 20°C.

From the dew point, the value of $0.3783 e$ is found in Table II to be 3.48 mm. $750 - 3.48 = 746.52$, the corrected pressure. The pressure factor for this value found in Table III by interpolation is 0.98226.

The temperature factor from Table I is 1.2047.

$$1.2047 \times 0.98224 = 1.1833 \text{ g/}l.$$

To obtain the value directly from Table IV, enter it for 20°C and 746.5 mm which gives by interpolation 1.183 g/l.

TABLE I
$(1.2929 \times 273.13/T)$

(Besides being a necessary part of the determination of the density of moist air, the values in this table are actually the density of dry air in g/l at 760 mm pressure for various temperatures.)

Temp. °C		0	1	2	3	4	5	6	7	8	9
−50	1.5	826	897	969	*042	*115	*189	*264	*339	*415	*491
−40	1.5	147	213	278	345	412	479	547	616	686	756
−30	1.4	524	584	645	706	767	829	892	955	*019	*083
−20	1.3	951	*006	*062	*118	*175	*232	*289	*347	*406	*465
−10	1.3	420	472	523	575	628	680	734	787	841	896
−0	1.2	929	977	*024	*073	*121	*170	*219	*269	*319	*370
+0	1.2	929	882	835	789	742	697	651	606	561	517
10	1.2	472	428	385	342	299	256	214	171	130	088
20	1.2	047	006	*965	*925	*885	*845	*805	*766	*727	*688
30	1.1	649	611	573	535	498	460	423	387	350	314
40	1.1	277	242	206	170	135	100	065	031	*996	*962
50	1.0	928	895	861	828	795	762	729	697	664	632
60	1.0	600	569	537	506	475	444	413	382	352	322

TABLE II
Vapor Pressure — Value of $0.3783 e$

Dew point °C	Vapor press. e mm (ice)	0.3783 e	Dew point °C	Vapor press. e mm (water)	0.3783 e	Dew point °C	Vapor press. e mm (water)	0.3783 e	Dew point °C	Vapor press. e mm (ice)	0.3783 e	Dew point °C	Vapor press. e mm (water)	0.3783 e	Dew point °C	Vapor press. e mm (water)	0.3783 e
−50	0.029	0.01	−15	1.252	0.47	0	4.58	1.73	15	12.79	4.84	30	31.86	12.05	45	71.97	27.23
−45	0.054	0.02	−14	1.373	0.52	1	4.92	1.86	16	13.64	5.16	31	33.74	12.76	46	75.75	28.66
−40	0.096	0.04	−13	1.503	0.57	2	5.29	2.00	17	14.54	5.50	32	35.70	13.51	47	79.70	30.15
−35	0.169	0.06	−12	1.644	0.62	3	5.68	2.15	18	15.49	5.86	33	37.78	14.29	48	83.83	31.71
−30	0.288	0.11	−11	1.798	0.68	4	6.10	2.31	19	16.49	6.24	34	39.95	15.11	49	88.14	33.34
−25	0.480	0.18	−10	1.964	0.74	5	6.54	2.47	20	17.55	6.64	35	42.23	15.98	50	92.6	35.03
−24	0.530	0.20	− 9	2.144	0.81	6	7.01	2.65	21	18.66	7.06	36	44.62	16.88	51	97.3	36.81
−23	0.585	0.22	− 8	2.340	0.89	7	7.51	2.84	22	19.84	7.51	37	47.13	17.83	52	102.2	38.66
−22	0.646	0.24	− 7	2.550	0.96	8	8.04	3.04	23	21.09	7.98	38	49.76	18.82	53	107.3	40.59
−21	0.712	0.27	− 6	2.778	1.05	9	8.47	3.26	24	22.40	8.47	39	52.51	19.86	54	112.7	42.63
−20	0.783	0.30	− 5	3.025	1.14	10	9.21	3.48	25	23.78	9.00	40	55.40	20.96	55	118.2	44.72
−19	0.862	0.33	− 4	3.291	1.24	11	9.85	3.73	26	25.24	9.55	41	58.42	22.10	56	124.0	46.91
−18	0.947	0.36	− 3	3.578	1.35	12	10.52	3.98	27	26.77	10.13	42	61.58	23.30	57	130.0	49.18
−17	1.041	0.39	− 2	3.887	1.47	13	11.24	4.25	28	28.38	10.74	43	64.89	24.55	58	136.3	51.56
−16	1.142	0.43	− 1	4.220	1.60	14	11.99	4.54	29	30.08	11.38	44	68.35	25.86	59	142.8	54.02
															60	149.6	56.59

TABLE III

Pressure Factor — $[(B - 0.3783\,e)/760]$

The figures in the body of the table give values of the whole term $[(B - 0.3783\,e)/760]$ for various values of the numerator $(B - 0.3783\,e)$ expressed at the left and top.

Press. mm corr.	0	1	2	3	4	5	6	7	8	9
80	.10526	.10658	.10789	.10921	.11053	.11184	.11316	.11447	.11579	.11711
90	.11842	.11974	.12105	.12237	.12368	.12500	.12632	.12763	.12895	.13026
100	.13158	.13289	.13421	.13553	.13684	.13816	.13947	.14079	.14211	.14342
110	.14474	.14605	.14737	.14868	.15000	.15132	.15263	.15395	.15526	.15658
120	.15789	.15921	.16053	.16184	.16316	.16447	.16579	.16711	.16842	.16974
130	.17105	.17237	.17368	.17500	.17632	.17763	.17895	.18026	.18158	.18289
140	.18421	.18553	.18684	.18816	.18947	.19079	.19211	.19342	.19474	.19605
150	.19737	.19868	.20000	.20132	.20263	.20395	.20526	.20658	.20789	.20921
160	.21053	.21184	.21316	.21447	.21579	.21711	.21842	.21974	.22105	.22237
170	.22368	.22500	.22632	.22763	.22895	.23026	.23158	.23289	.23421	.23553
180	.23684	.23816	.23947	.24079	.24211	.24342	.24474	.24605	.24737	.24868
190	.25000	.25132	.25263	.25395	.25526	.25658	.25789	.25921	.26053	.26184
200	.26316	.26447	.26579	.26711	.26842	.26974	.27105	.27237	.27368	.27500
210	.27632	.27763	.27895	.28026	.28158	.28289	.28421	.28553	.28684	.28816
220	.28947	.29079	.29211	.29342	.29474	.29605	.29737	.29868	.30000	.30132
230	.30263	.30395	.30526	.30658	.30789	.30921	.31053	.31184	.31316	.31447
240	.31579	.31711	.31842	.31974	.32105	.32237	.32368	.32500	.32632	.32763
250	.32895	.33026	.33158	.33289	.33421	.33553	.33684	.33816	.33947	.34079
260	.34211	.34342	.34474	.34605	.34737	.34868	.35000	.35132	.35263	.35395
270	.35526	.35658	.35789	.35921	.36053	.36184	.36316	.36447	.36579	.36711
280	.36842	.36974	.37105	.37237	.37368	.37500	.37632	.37763	.37895	.38026
290	.38158	.38289	.38421	.38553	.38684	.38816	.38947	.39079	.39211	.39342
300	.39474	.39605	.39737	.39868	.40000	.40132	.40263	.40395	.40526	.40658
310	.40789	.40921	.41053	.41184	.41316	.41447	.41579	.41711	.41842	.41974
320	.42105	.42237	.42368	.42500	.42632	.42763	.42895	.43026	.43158	.43289
330	.43421	.43553	.43684	.43816	.43947	.44079	.44211	.44342	.44474	.44605
340	.44737	.44868	.45000	.45132	.45263	.45395	.45526	.45658	.45789	.45921
350	.46053	.46184	.46316	.46447	.46579	.46711	.46842	.46974	.47105	.47237
360	.47368	.47500	.47632	.47763	.47895	.48026	.48158	.48289	.48421	.48553
370	.48684	.48816	.48947	.49079	.49211	.49342	.49474	.49605	.49737	.49868
380	.50000	.50132	.50263	.50395	.50526	.50658	.50789	.50921	.51053	.51184
390	.51316	.51447	.51579	.51711	.51842	.51974	.52105	.52237	.52368	.52500
400	.52632	.52763	.52895	.53026	.53158	.53289	.53421	.53553	.53684	.53816
410	.53947	.54079	.54211	.54342	.54474	.54605	.54737	.54868	.55000	.55132
420	.55263	.55395	.55526	.55658	.55789	.55921	.56053	.56184	.56316	.56447
430	.56579	.56711	.56842	.56974	.57105	.57237	.57368	.57500	.57632	.57763

Press. mm corr.	0	1	2	3	4	5	6	7	8	9
440	.57895	.58026	.58158	.58289	.58421	.58553	.58684	.58816	.58947	.59079
450	.59211	.59342	.59474	.59605	.59737	.59868	.60000	.60132	.60263	.60395
460	.60526	.60658	.60789	.60921	.61053	.61184	.61316	.61447	.61579	.61711
470	.61842	.61974	.62105	.62237	.62368	.62500	.62632	.62763	.62895	.63026
480	.63158	.63289	.63421	.63553	.63684	.63816	.63947	.64079	.64211	.64342
490	.64474	.64605	.64737	.64868	.65000	.65132	.65263	.65395	.65526	.65658
500	.65790	.65921	.66053	.66184	.66316	.66447	.66579	.66711	.66842	.66974
510	.67105	.67237	.67368	.67500	.67632	.67763	.67895	.68026	.68158	.68290
520	.68421	.68553	.68684	.68816	.68947	.69079	.69211	.69342	.69474	.69605
530	.69737	.69868	.70000	.70132	.70263	.70395	.70526	.70658	.70790	.70921
540	.71053	.71184	.71316	.71447	.71579	.71711	.71842	.71974	.72105	.72237
550	.72368	.72500	.72632	.72763	.72895	.73026	.73158	.73290	.73421	.73553
560	.73684	.73816	.73947	.74079	.74211	.74342	.74474	.74605	.74737	.74868
570	.75000	.75132	.75263	.75395	.75526	.75658	.75790	.75921	.76053	.76184
580	.76316	.76447	.76579	.76711	.76842	.76974	.77105	.77237	.77368	.77500
590	.77632	.77763	.77895	.78026	.78158	.78290	.78421	.78553	.78684	.78816
600	.78947	.79079	.79211	.79342	.79474	.79605	.79737	.79688	.80000	.80132
610	.80263	.80395	.80526	.80658	.80790	.80921	.81053	.81184	.81316	.81447
620	.81579	.81711	.81842	.81974	.82105	.82237	.82368	.82500	.82632	.82763
630	.82895	.83026	.83158	.83290	.83421	.83553	.83684	.83816	.83947	.84079
640	.84211	.84342	.84474	.84605	.84737	.84868	.85000	.85132	.85263	.85395
650	.85526	.85658	.85790	.85921	.86053	.86184	.86316	.86447	.86579	.86711
660	.86842	.86974	.87105	.87237	.87368	.87500	.87632	.87763	.87895	.88026
670	.88158	.88290	.88421	.88553	.88684	.88816	.88947	.89079	.89211	.89342
680	.89474	.89605	.89737	.89868	.90000	.90132	.90263	.90395	.90526	.90658
690	.90790	.90921	.91053	.91184	.91316	.91447	.91579	.91711	.91842	.91974
700	.92105	.92237	.92368	.92500	.92632	.92763	.92895	.93026	.93158	.93290
710	.93421	.93553	.93684	.93816	.93947	.94079	.94211	.94342	.94474	.94605
720	.94737	.94868	.95000	.95132	.95263	.95395	.95526	.95658	.95790	.95921
730	.96053	.96184	.96316	.96447	.96579	.96711	.96842	.96974	.97105	.97237
740	.97368	.97500	.97632	.97763	.97895	.98026	.98158	.98290	.98421	.98553
750	.98684	.98816	.98947	.99079	.99211	.99342	.99474	.99605	.99737	.99868
760	1.0000	1.0013	1.0026	1.0039	1.0053	1.0066	1.0079	1.0092	1.0105	1.0118
770	1.0132	1.0145	1.0158	1.0171	1.0184	1.0197	1.0211	1.0224	1.0237	1.0250
780	1.0263	1.0276	1.0289	1.0303	1.0316	1.0329	1.0342	1.0355	1.0368	1.0382
790	1.0395	1.0408	1.0421	1.0434	1.0447	1.0461	1.0474	1.0487	1.0500	1.0513

TABLE IV

Density of Moist Air

Values in the body of the table give the density of moist air in g/ℓ for a limited range of temperatures and corrected pressure values $(B - 0.3783\,e)$. The latter may be obtained by use of Table II.

°C	600	610	620	630	640	650	660	670	680	690
5	1.0024	1.0191	1.0358	1.0525	1.0692	1.0859	1.1026	1.1193	1.1361	1.1528
6	.99876	1.0154	1.0321	1.0487	1.0654	1.0820	1.0986	1.1153	1.1319	1.1486
7	.99521	1.0118	1.0284	1.0450	1.0616	1.0781	1.0947	1.1113	1.1279	1.1445
8	.99165	1.0082	1.0247	1.0412	1.0578	1.0743	1.0908	1.1074	1.1239	1.1404
9	.98818	1.0047	1.0211	1.0376	1.0541	1.0705	1.0870	1.1035	1.1199	1.1364
10	.98463	1.0010	1.0175	1.0339	1.0503	1.0667	1.0831	1.0995	1.1159	1.1323
11	.98115	.99751	1.0139	1.0302	1.0466	1.0629	1.0793	1.0956	1.1120	1.1283
12	.97776	.99406	1.0104	1.0267	1.0430	1.0592	1.0755	1.0918	1.1081	1.1244
13	.97436	.99061	1.0068	1.0231	1.0393	1.0556	1.0718	1.0880	1.1043	1.1205
14	.97097	.98715	1.0033	1.0195	1.0357	1.0519	1.0681	1.0843	1.1004	1.1166
15	.96757	.98370	.99983	1.0160	1.0321	1.0482	1.0643	1.0805	1.0966	1.1127
16	.96426	.98033	.99641	1.0125	1.0286	1.0446	1.0607	1.0768	1.0928	1.1089
17	.96086	.97688	.99290	1.0089	1.0247	1.0409	1.0570	1.0730	1.0890	1.1050
18	.95763	.97359	.98955	1.0055	1.0215	1.0374	1.0534	1.0694	1.0853	1.1013
19	.95431	.97022	.98613	1.0020	1.0179	1.0338	1.0497	1.0656	1.0816	1.0975
20	.95107	.96693	.98278	.99864	1.0145	1.0303	1.0462	1.0620	1.0779	1.0937
21	.94784	.96364	.97944	.99524	1.0110	1.0294	1.0426	1.0584	1.0742	1.0900
22	.94460	.96035	.97609	.99184	1.0076	1.0233	1.0391	1.0548	1.0706	1.0863
23	.94144	.95714	.97283	.98852	1.0042	1.0199	1.0356	1.0513	1.0670	1.0827
24	.93829	.95393	.96957	.98521	1.0008	1.0165	1.0321	1.0478	1.0634	1.0790
25	.93513	.95072	.96630	.98189	.99748	1.0131	1.0286	1.0442	1.0598	1.0754
26	.93197	.94750	.96304	.97858	.99411	1.0096	1.0252	1.0407	1.0562	1.0718
27	.92889	.94437	.95986	.97534	.99083	1.0063	1.0218	1.0373	1.0528	1.0682
28	.92581	.94124	.95668	.97211	.98754	1.0030	1.0184	1.0338	1.0493	1.0647
29	.92273	.93811	.95350	.96888	.98426	.99969	1.0150	1.0304	1.0458	1.0612
30	.91965	.93498	.95031	.96564	.98097	.99629	1.0116	1.0270	1.0423	1.0576
31	.91665	.93193	.94721	.96249	.97777	.99304	1.0083	1.0236	1.0389	1.0542
32	.91365	.92888	.94411	.95934	.97457	.98979	1.0050	1.0203	1.0355	1.0507
33	.91065	.92583	.94101	.95619	.97137	.98654	1.0017	1.0169	1.0321	1.0473
34	.90773	.92286	.93800	.95313	.96826	.98338	.99851	1.0136	1.0288	1.0439
35	.90473	.91981	.93490	.94998	.96506	.98013	.99521	1.0103	1.0254	1.0405

°C	700	710	720	730	740	750	760	770	780	790
5	1.1695	1.1862	1.2029	1.2196	1.2363	1.2530	1.2697	1.2864	1.3031	1.3198
6	1.1652	1.1819	1.1985	1.2152	1.2318	1.2485	1.2651	1.2817	1.2984	1.3150
7	1.1611	1.1777	1.1943	1.2108	1.2274	1.2440	1.2606	1.2772	1.2938	1.3104
8	1.1569	1.1735	1.1900	1.2065	1.2230	1.2396	1.2561	1.2726	1.2892	1.3057
9	1.1529	1.1694	1.1858	1.2023	1.2188	1.2352	1.2517	1.2682	1.2846	1.3011
10	1.1487	1.1651	1.1816	1.1980	1.2144	1.2308	1.2472	1.2636	1.2800	1.2964
11	1.1447	1.1610	1.1774	1.1937	1.2101	1.2264	1.2428	1.2592	1.2755	1.2919
12	1.1407	1.1570	1.1733	1.1896	1.2059	1.2222	1.2385	1.2548	1.2711	1.2874
13	1.1368	1.1530	1.1692	1.1855	1.2017	1.2180	1.2342	1.2504	1.2667	1.2829
14	1.1328	1.1490	1.1652	1.1814	1.1975	1.2137	1.2299	1.2461	1.2623	1.2784
15	1.1288	1.1450	1.1611	1.1772	1.1933	1.2095	1.2256	1.2417	1.2579	1.2740
16	1.1250	1.1410	1.1571	1.1732	1.1893	1.2053	1.2214	1.2375	1.2535	1.2696
17	1.1210	1.1370	1.1530	1.1691	1.1851	1.2011	1.2171	1.2331	1.2491	1.2651
18	1.1172	1.1332	1.1492	1.1651	1.1811	1.1970	1.2130	1.2290	1.2449	1.2609
19	1.1134	1.1293	1.1452	1.1611	1.1770	1.1929	1.2088	1.2247	1.2406	1.2565
20	1.1096	1.1254	1.1413	1.1572	1.1730	1.1888	1.2047	1.2206	1.2364	1.2522
21	1.1058	1.1216	1.1374	1.1532	1.1690	1.1848	1.2006	1.2164	1.2322	1.2480
22	1.1020	1.1178	1.1335	1.1493	1.1650	1.1808	1.1965	1.2122	1.2280	1.2437
23	1.0984	1.1140	1.1294	1.1454	1.1611	1.1768	1.1925	1.2082	1.2239	1.2396
24	1.0947	1.1103	1.1259	1.1416	1.1572	1.1729	1.1885	1.2041	1.2198	1.2354
25	1.0910	1.1066	1.1222	1.1377	1.1533	1.1689	1.1845	1.2001	1.2157	1.2313
26	1.0873	1.1028	1.1184	1.1339	1.1494	1.1650	1.1805	1.1960	1.2116	1.2271
27	1.0837	1.0992	1.1147	1.1302	1.1456	1.1611	1.1766	1.1921	1.2076	1.2230
28	1.0801	1.0955	1.1110	1.1264	1.1418	1.1573	1.1727	1.1881	1.2036	1.2190
29	1.0765	1.0919	1.1073	1.1227	1.1380	1.1534	1.1688	1.1842	1.1996	1.2149
30	1.0729	1.0883	1.1036	1.1189	1.1342	1.1496	1.1649	1.1802	1.1956	1.2109
31	1.0694	1.0847	1.1000	1.1153	1.1305	1.1458	1.1611	1.1764	1.1917	1.2069
32	1.0659	1.0812	1.0964	1.1116	1.1268	1.1421	1.1573	1.1725	1.1878	1.2030
33	1.0624	1.0776	1.0928	1.1080	1.1231	1.1383	1.1535	1.1687	1.1839	1.1990
34	1.0590	1.0742	1.0893	1.1044	1.1195	1.1347	1.1498	1.1649	1.1801	1.1952
35	1.0555	1.0706	1.0857	1.1008	1.1158	1.1309	1.1460	1.1611	1.1762	1.1912

DENSITY OF DRY AIR

At the Temperature t, and under the Pressure H cm of Mercury the Density of Air
$$= \frac{0.001293}{1+0.00367\,t}\frac{H}{76}.$$
Units of this table are grams per milliliter
(From Miller's Laboratory Physics, Ginn & Co., publishers, by permission.)

t	Pressure H in Centimeters						Proportional Parts		
	72.0	73.0	74.0	75.0	76.0	77.0			
°									17
10	0.001182	0.001198	0.001215	0.001231	0.001247	0.001264	cm		
11	178	193	210	227	243	259	0.1	2	
12	173	190	206	222	239	255	0.2	3	
13	169	186	202	218	234	251	0.3	5	
14	165	181	198	214	230	246	0.4	7	
							0.5	8	
							0.6	10	
							0.7	12	
15	0.001161	0.001177	0.001193	0.001210	0.001226	0.001242	0.8	14	
16	157	173	189	205	221	238	0.9	15	
17	153	169	185	201	217	233			16
18	149	165	181	197	213	229	cm		
19	145	161	177	193	209	225	0.1	2	
							0.2	3	
							0.3	5	
							0.4	6	
20	0.001141	0.001157	0.001173	0.001189	0.001205	0.001221	0.5	8	
21	137	153	169	185	201	216	0.6	10	
22	134	149	165	181	197	212	0.7	11	
23	130	145	161	177	193	208	0.8	13	
24	126	142	157	173	189	204	0.9	14	
									15
							cm		
25	0.001122	0.001138	0.001153	0.001169	0.001185	0.001200	0.1	1	
26	118	134	149	165	181	196	0.2	3	
27	115	130	146	161	177	192	0.3	4	
28	111	126	142	157	173	188	0.4	6	
29	107	123	138	153	169	184	0.5	7	
							0.6	9	
							0.7	10	
							0.8	12	
30	0.001104	0.001119	0.001134	0.001150	0.001165	0.001180	0.9	13	

Density of dry air at 20C and 760mm Hg = 1.204 mg/cm^3. (*Rev. Mod. Phys.*, 52, Part II, S33, 1980.)

DENSITY OF WATER

The temperature of maximum density for pure water, free from air = 3.98C (277.13K)

t, °C	d, gm/ml
0	0.99987
3.98	1.00000
5	0.99999
10	0.99973
15	0.99913
18	0.99862
20	0.99823
25	0.99707
30	0.99567
35	0.99406
38	0.99299
40	0.99224
45	0.99025
50	0.98807
55	0.98573
60	0.98324
65	0.98059
70	0.97781
75	0.97489
80	0.97183
85	0.96865
90	0.96534
95	0.96192
100	0.95838

THERMODYNAMIC AND TRANSPORT PROPERTIES OF AIR

From NASA Technical Note D-7488 by David J. Poferl and Roger Svehla (1973). The following three tables list the thermodynamic and transport properties of air over the temperature range of 300-2800K at pressures of 20, 30, and 40 atm. Factors for converting viscosity, specific heat at constant pressure, thermal conductivity, and enthalpy from cgs units to SI and English units are

Viscosity:

$$1 \frac{g}{(cm)(sec)} = 0.1 \frac{(N)(sec)}{m^2}$$

$$= 6.72\times10^{-2} \frac{lbm}{(ft)(sec)}$$

$$= 241.9 \frac{lbm}{(ft)(hr)}$$

$$= 2.089\times10^{-3} \frac{(lbf)(sec)}{ft^2}$$

Thermal conductivity:

$$1 \frac{cal}{(cm)(sec)(K)} = 418.4 \frac{W}{(m)(K)}$$

$$= 0.8064 \frac{Btu}{(ft)^2(sec)(°F/in.)}$$

$$= 6.72\times10^{-2} \frac{Btu}{(ft^2)(sec)(°F/ft)}$$

$$= 241.9 \frac{Btu}{(ft)^2(hr)(°F/ft)}$$

Specific heat at constant pressure:

$$1 \frac{cal}{(g)(K)} = 4.184 \frac{J}{(g)(K)}$$

$$= 1 \frac{Btu}{(lbm)(°F)}$$

Enthalpy:

$$1 \frac{cal}{g} = 4.184 \frac{J}{g}$$

$$= 1.8 \frac{Btu}{lbm}$$

PROPERTIES AT 20 ATM

Temperature, T, K	Isentropic exponent, γ	Molecular weight, m	Viscosity, μ g/(cm)(sec)	Specific heat at constant pressure, c_p, cal/(g)(K)	Thermal conductivity, k, cal/(cm)(sec)(K)	Prandtl number, Pr	Enthalpy, h, cal/g
2800	1.2309	28.890	821×10^{-6}	0.3850	473×10^{-6}	0.669	747.0
2700	1.2374	28.915	800	.3707	439	.676	709.2
2600	1.2437	28.933	779	.3588	409	.683	672.7
2500	1.2498	28.945	758	.3488	384	.688	637.4
2400	1.2556	28.953	736	.3404	362	.692	602.9
2300	1.2612	28.958	715	.3332	343	.696	569.3
2200	1.2666	28.961	694	.3270	325	.698	536.2
2100	1.2719	28.963	672	.3214	309	.699	503.8
2000	1.2772	28.964	651	.3163	294	.700	471.9
1900	1.2825	28.965	629	.3115	280	.701	440.6
1800	1.2879		607	.3070	266	.702	409.6
1700	1.2933		585	.3025	252	.702	379.2
1600	1.2989		563	.2981	239	.703	349.1
1500	1.3045		540	.2939	226	.703	319.5
1400	1.3103		517	.2897	213	.704	290.3
1300	1.3162		494	.2855	200	.704	261.6
1200	1.3224		470	.2814	188	.705	233.2
1100	1.3288		445	.2773	175	.705	205.3
1000	1.3356		419	.2730	162	.705	177.8
900	1.3439	28.964	391	.2681	148	.706	150.7
800	1.3537		362	.2626	135		124.2
700	1.3646		331	.2568	121		98.2
600	1.3759		299	.2511	106		72.8
500	1.3865		265	.2461	92		48.0
400	1.3951		227	.2422	78		23.6
300	1.4000		184	.2401	63		−.5

PROPERTIES AT 30 ATM

Temperature, T, K	Isentropic exponent, γ	Molecular weight, m	Viscosity, μ, g/(cm)(sec)	Specific heat at constant pressure, c_p, cal/(g)(K)	Thermal conductivity, k, cal/(cm)(sec)(K)	Prandtl number, Pr	Enthalpy, h, cal/g
2800	1.2340	28.904	821×10^{-6}	0.3773	460×10^{-6}	0.674	745.0
2700	1.2399	28.924	800	.3652	430	.680	707.9
2600	1.2456	28.939	779	.3548	403	.686	671.9
2500	1.2512	28.949	758	.3462	380	.690	636.8
2400	1.2566	28.956	736	.3387	359	.694	602.6
2300	1.2619	28.960	715	.3321	341	.696	569.1
2200	1.2671	28.962	694	.3263	324	.698	536.1
2100	1.2722	28.964	672	.3211	309	.699	503.8
2000	1.2773	28.965	651	.3161	294	.700	471.9
1900	1.2826		629	.3115	280	.701	440.5
1800	1.2879		607	.3069	266	.702	409.6
1700	1.2933		585	.3025	252	.702	379.2
1600	1.2988		563	.2981	239	.703	349.1
1500	1.3045		540	.2939	226	.703	319.5
1400	1.3103		517	.2897	213	.704	290.3
1300	1.3162		494	.2855	200	.704	261.6
1200	1.3223		470	.2814	188	.705	233.2
1100	1.3288		445	.2773	175	.705	205.3
1000	1.3356		419	.2730	162	.705	177.8
900	1.3439	28.964	391	.2681	148	.706	150.7
800	1.3537		362	.2626	135		124.2
700	1.3646		331	.2568	121		98.2
600	1.3759		299	.2511	106		72.8
500	1.3865		265	.2461	92		48.0
400	1.3951		227	.2422	78		23.6
300	1.4000		184	.2401	63		−.5

PROPERTIES AT 40 ATM

Temperature, T, K	Isentropic exponent, γ	Molecular weight, m	Viscosity, μ, g/(cm)(sec)	Specific heat at constant pressure, c_p, cal/(g)(K)	Thermal conductivity, k, cal/(cm)(sec)(K)	Prandtl number, Pr	Enthalpy, h, cal/g
2800	1.2360	28.913	821×10^{-6}	0.3727	452×10^{-6}	0.677	743.8
2700	1.2414	28.930	800	.3619	424	.683	707.1
2600	1.2468	28.943	779	.3526	399	.688	671.4
2500	1.2521	28.951	758	.3446	377	.692	636.5
2400	1.2573	28.957	736	.3376	358	.695	602.4
2300	1.2623	28.961	715	.3315	340	.697	569.0
2200	1.2673	28.963	694	.3260	324	.699	536.1
2100	1.2724	28.964	672	.3209	308	.700	503.8
2000	1.2774	28.965	651	.3160	294	.700	471.9
1900	1.2826		629	.3114	279	.701	440.5
1800	1.2879		607	.3069	266	.702	409.6
1700	1.2933		585	.3025	252	.702	379.2
1600	1.2988		563	.2981	239	.703	349.1
1500	1.3045		540	.2939	226	.703	319.5
1400	1.3103		517	.2897	213	.704	290.4
1300	1.3162		494	.2855	200	.704	261.6
1200	1.3223		470	.2814	188	.705	233.2
1100	1.3288		445	.2773	175	.705	205.3
1000	1.3356		419	.2730	162	.705	177.8
900	1.3439		391	.2681	148	.706	150.7
800	1.3537	28.964	362	.2626	135		124.2
700	1.3646		331	.2568	121		98.2
600	1.3759		299	.2511	106		72.8
500	1.3865		265	.2461	92		48.0
400	1.3951		227	.2422	78		23.6
300	1.4000		184	.2401	63		−.5

ISOTHERMAL COMPRESSIBILITY OF LIQUIDS

J. C. McGowan

The figures in this table are for isothermal compressibilities in cgs units. The compiler suggests that, provided the pressure is not too high, the reciprocal of the isothermal compressibility varies linearly with the pressure. This suggestion also appears in papers by J. R. Macdonald, *Rev. Mod. Phys.,* 38, 669, 1966, and O. L. Anderson, *J. Phys. Chem. Solids,* 27, 547, 1966. The papers vary somewhat as to how far this linearity persists.

Liquid	Temp., °C[a]	Isothermal compressibility $\times 10^{10}$ m² N⁻¹ At 1 atm (or 1.013×10^5 Nm²)	At 1000 atm (or 1.013×10^8 Nm³)	Ref.	Liquid	Temp., °C[a]	Isothermal compressibility $\times 10^{10}$ m² N⁻¹ At 1 atm (or 1.013×10^5 Nm²)	At 1000 atm (or 1.013×10^8 Nm³)	Ref.
Acetic acid	15	8.75	—	1		50	5.33	—	1
	20	9.08	—	1		60	5.64	—	1
	30	9.72	—	1		65	5.84	3.76	5
	40	10.37	—	1		70	5.97	—	1
	50	11.11	—	1		80	6.32	—	1
	60	11.91	—	1		85	6.56	4.04	5
	70	12.77	—	1		90	6.70	—	1
	80	13.68	—	1	Anisole	21	6.67	—	2
Acetic acid, ethyl ester	0	9.78	—	1		30	7.04	—	2
	10	10.36	—	1		45	7.72	—	2
	20	11.32	—	1		60	8.50	—	2
	30	12.37	—	1		81	9.79	—	2
	40	13.52	—	1		100	11.25	—	2
	50	14.78	—	1		120	13.07	—	2
	60	16.21	—	1		140	15.45	—	2
	70	17.90	—	1	Benzene	0	8.09	—	1
Acetone	20	12.75	—	3		10	8.73	—	1
	20	12.29	—	10		10	8.64	—	3
	25	12.39	6.02	4		20	9.44	—	1
	30	13.34	—	10		20	9.37	—	3
	40	15.61	—	3		20	9.54	—	10
	40	14.64	—	10		25	9.67	5.07	6
	50	16.03	—	10		25	9.7	—	15
Aniline	0	4.08	—	1		30	10.27	—	10
	10	4.30	—	1		30	10.18	—	1
	20	4.53	—	1		30	10.12	—	3
	25	4.67	3.23	5		35	10.43	5.28	6
	40	5.04	—	1		39.5	10.91	—	15
	45	5.22	3.48	5		40	11.05	—	10
						40	11.00	—	1

ISOTHERMAL COMPRESSIBILITY OF LIQUIDS (continued)

Liquid	Temp., °C[a]	Isothermal compressibility × 10^10 m^2 N^{-1} At 1 atm (or 1.013 × 10^5 Nm^2)	At 1000 atm (or 1.013 × 10^8 Nm^3)	Ref.	Liquid	Temp., °C[a]	Isothermal compressibility × 10^10 m^2 N^{-1} At 1 atm (or 1.013 × 10^5 Nm^2)	At 1000 atm (or 1.013 × 10^8 Nm^3)	Ref.
	40	10.96	–	3		26.9	11.15	–	12
	45	11.32	5.50	6		30	11.28	–	10
	50	11.90	–	10		30	11.29	–	1
	50	11.89	–	1		30	11.18	–	3
	50	11.83	–	3		35	11.95	5.52	7
	50.1	11.91	–	15		37.5	11.99	–	15
	55	12.29	5.73	6		40	12.20	–	10
	60	12.78	–	10		40	12.23	–	1
	60	12.83	–	1		40	12.16	–	3
	60	12.96	–	15		40	11.96	–	16
	65	13.39	5.98	6		45	12.54	5.75	7
	70	13.72	–	10		50	13.20	–	10
	70	14.13	–	1		50	13.32	–	1
	75.9	14.95	–	15		50	13.26	–	3
	80	15.44	–	1		50.3	13.28	–	15
Benzene, bromo-	25	6.68	4.09	5		55	13.63	5.97	7
	45	7.52	4.39	5		55	13.51	–	15
	65	8.50	4.72	5		60	14.26	–	10
	85	9.65	5.06	5		60	14.52	–	1
Benezene, chloro-	0	6.61	–	1		62.6	14.84	–	15
	10	7.02	–	1		65	14.87	6.22	7
	20	7.45	–	1		70	15.43	–	10
	20	7.38	–	3		70	15.77	–	1
	25	7.51	4.39	5		75	16.70	–	15
	30	7.89	–	1	Chloroform	−33.1	7.13	–	12
	30	7.84	–	3		−23.1	7.61	–	12
	40	8.39	–	1		−13.1	8.12	–	12
	40	8.32	–	3		−3.1	8.66	–	12
	45	8.55	4.73	5		0	8.48	–	1
	50	8.92	–	1		0	8.55	–	3
	50	8.83	–	3		6.9	9.30	–	12
	60	9.50	–	1		10	9.17	–	1
	65	9.76	5.10	5		10	9.19	–	3
	70	10.13	–	1		16.9	10.35	–	12
	80	10.79	–	1		20	9.98	–	1
	85	11.23	5.49	5		20	9.94	–	3
Benzene, nitro-	0	4.41	–	1		20	10.15	–	10
	10	4.67	–	1		25	9.74	5.34	4
	20	4.93	–	1		26.9	11.15	–	12
	25	5.03	3.39	5		30	10.86	–	1
	30	5.23	–	1		30	10.81	–	3
	40	5.49	–	1		30	10.99	–	10
	45	5.59	3.64	5		40	11.84	–	1
	65	6.24	3.91	5		40	11.79	–	3
	85	6.99	4.20	5		40	11.94	–	10
n-Butyl alcohol	0	8.10	–	11		50	12.90	–	1
Carbon disulphide	0	8.04	–	1		50	12.99	–	10
	0	7.95	–	3		60	14.06	–	1
	10	8.64	–	1	Cyclohexane	25	11.20	–	16
	10	8.54	–	3		25	11.40	–	15
	20	9.13	–	10		35	12.19	–	16
	20	9.26	–	1		37.6	12.67	–	15
	20	9.19	–	3		40	12.56	–	16
	30	9.72	–	10		45	13.31	–	16
	30	9.92	–	1		50.1	14.15	–	15
	30	9.96	–	3		55	14.35	–	16
	40	10.65	–	1		60	15.20	–	16
	40	10.57	–	10		62.4	15.76	–	15
	40	10.89	–	3		75	17.84	–	15
	50	11.48	–	1	n-Decane	25	10.95	–	16
	50	11.95	–	3		35	11.76	–	16
Carbon tetrachloride	−22.9	7.84	–	12		45	12.65	–	16
	−13.1	8.31	–	12		60	14.11	–	16
	−3.1	8.87	–	12	Dodecane	25	9.87	–	16
	0	8.98	–	1		35	10.52	–	16
	0	8.85	–	3		37.8	9.9	–	16
	6.9	9.59	–	12		45	11.27	–	16
	10	9.70	–	1		60	12.51	–	16
	10	9.57	–	3		79.4	12.8	–	16
	10	9.45	–	16		98.9	14.4	–	16
	16.9	10.35	–	12		115.0	16.1	–	16
	20	10.46	–	1		135.0	18.3	–	16
	20	10.34	–	3	Ethane 1,1,2,2-tetrachloro-	25	6.17	3.88	4
	20	10.40	–	10	Ethyl alcohol	0	9.87	–	1
	25	10.67	5.30	7		0	9.87	–	11
	25	10.77	–	15		0	9.63	–	3
	25	10.58	–	16					

ISOTHERMAL COMPRESSIBILITY OF LIQUIDS (continued)

Liquid	Temp., °C[a]	Isothermal compressibility × 10¹⁰ m²N⁻¹ At 1 atm (or 1.013 × 10⁵ Nm²)	At 1000 atm (or 1.013 × 10⁸ Nm³)	Ref.
	10	10.49	–	1
	10	10.30	–	3
	20	11.19	–	1
	20	10.98	–	3
	30	11.91	–	1
	30	11.80	–	3
	40	12.74	–	1
	40	12.61	–	3
	50	13.70	–	1
	50	13.60	–	3
	60	14.74	–	1
	70	15.93	–	1
	75	16.67	–	1
Ethyl bromide	0	10.76	–	1
	10	11.78	–	1
	20	12.94	–	1
	30	14.23	–	1
	40	15.52	–	1
Ethylene, 1,2-dichloro-(trans)	25	11.19	5.62	4
Ethylene, tetrachloro-	25	7.56	4.45	4
Ethylene, trichloro-	25	8.57	4.99	4
Ethylene, chloride	0	6.91	–	1
	10	7.42	–	1
	20	7.97	–	1
	20	7.82	–	10
	25	7.78	4.54	4
	30	8.41	–	10
	30	8.58	–	1
	40	9.09	–	10
	40	9.25	–	1
	50	9.86	–	10
	50	9.99	–	1
	60	10.66	–	10
	60	10.83	–	1
	70	11.54	–	10
	70	11.76	–	1
	80	12.79	–	1
Ethyl ether	0	15.10	–	1
	0	15.07	–	3
	10	16.81	–	1
	10	16.52	–	3
	20	18.65	–	1
	20	18.44	–	3
	30	20.90	–	1
	30	20.80	–	3
	35	24.15	–	1
Ethyl iodide	0	8.45	–	1
	10	9.12	–	1
	20	9.82	–	1
	30	10.59	–	1
	40	11.44	–	1
	50	12.38	–	1
	60	13.40	–	1
	70	14.49	–	1
Glycol	25	3.72	2.73	7
	45	4.00	2.89	7
	65	4.32	3.05	7
	85	4.70	3.24	7
	105	5.14	3.44	7
n-Hendecane	25	10.81	–	17
	35	11.50	–	17
	50	12.47	–	17
n-Heptane	25	14.40	6.18	8
	35	15.69	–	17
	45	17.12	–	8
	60	19.62	–	8
1-Heptanol	0	7.05	–	11
n-Hexadecane	25	8.67	–	17
	35	9.43	–	17
	50	10.29	–	17
n-Hexane	25	16.72	6.51	8
	35	18.44	–	8
	45	20.33	–	8
	60	23.84	–	8
1-Hexanol	0	7.47	–	11

Liquid	Temp., °C[a]	At 1 atm (or 1.013 × 10⁵ Nm²)	At 1,000 atm (or 1.013 × 10⁸ Nm³)	Ref.
Mercury	0	0.40	–	14
	20	0.40	0.39	14
	40	0.41	–	14
	80	0.42	–	14
	120	0.44	–	14
	160	0.46	–	14
Methanol	0	10.62	–	1
	0	10.68	–	11
	0	10.78	–	3
	10	11.34	–	1
	10	11.45	–	3
	20	12.11	–	1
	20	12.18	–	3
	30	12.93	–	1
	30	12.98	–	3
	40	13.85	–	1
	40	13.82	–	3
	50	14.76	–	3
Methylene bromide	-33.1	4.95	–	12
	-23.1	5.11	–	12
	-13.1	5.50	–	12
	-3.1	5.81	–	12
	6.9	6.13	–	12
	16.9	6.47	–	12
	26.9	6.85	–	12
Methylene chloride	25	9.74	5.31	4
Methyl iodide	-33.1	6.86	–	12
	-23.1	7.41	–	12
	-13.1	7.97	–	12
	-3.1	8.33	–	12
	6.9	9.13	–	12
	16.9	9.71	–	12
	26.9	10.33	–	12
n-Nonane	25	11.77	–	8
	35	12.68	–	8
	45	13.66	–	8
	60	15.33	–	8
nOctadecane	60	9.4	5.1	13
	79.4	10.4	5.5	13
	98.9	11.6	5.8	13
	115.0	12.8	6.1	13
	135	14.4	6.4	13
n-Octane	25	12.80	–	8
	35	13.86	–	8
	45	15.04	–	8
	60	15.33	–	8
1-Octanol	0	6.82	–	11
n-Pentadecane	37.8	9.1	–	13
	60	10.2	5.2	13
	79.4	11.7	5.5	13
	98.9	13.2	5.8	13
	115	14.7	6.1	13
	135	16.8	6.4	13
Pentanol	0	7.71	–	11
Phenol	46	5.61	–	2
	60	6.05	–	2
	80	6.78	–	2
	110	8.12	–	2
	125	8.88	–	2
	150	10.30	–	2
	175	12.35	–	2
n-Propyl alcohol	0	8.43	–	11
Toluene	-59.3	5.27	–	9
	-41.1	5.93	–	9
	-19.2	6.87	–	9
	0	7.83	–	1
	0	7.97	–	3
	0	7.84	–	9
	10	8.38	–	1
	10	8.44	–	3
	20	8.96	–	1
	20	8.94	–	3
	30	9.60	–	1
	30	9.49	–	3
	40	10.33	–	1
	40	10.14	–	3
	50	11.13	–	1

ISOTHERMAL COMPRESSIBILITY OF LIQUIDS (continued)

Liquid	Temp., °C[a]	Isothermal compressibility × 10^10 m²N⁻¹			Liquid	Temp. °C[a]	Isothermal compressibility × 10^10 m²N⁻¹		
		At 1 atm (or 1.013 × 10⁵ Nm²)	At 1000 atm (or 1.013 × 10⁸ Nm³)	Ref.			At 1 atm (or 1.013 × 10⁵Nm²)	At 1,000 atm (or 1.013 × 10⁸Nm³)	Ref.
	50	10.90	–	3		65	4.48	3.42	7
	60	11.99	–	1		70	4.49	–	1
	70	12.95	–	1		75	4.55	3.47	7
Water	0	5.01	–	1		80	4.57	–	1
	0	5.04	–	9		85	4.65	3.53	7
	10	4.78	–	1		90	4.68	–	1
	20	4.58	–	1		100	4.80	–	1
	25	4.57	3.48	7	m-Xylene	0	7.44	–	1
	30	4.46	–	1		10	7.94	–	1
	34.8	4.44	–	9		20	8.46	–	1
	35	4.48	3.42	7		30	9.03	–	1
	40	4.41	–	1		40	9.63	–	1
	45	4.41	3.40	7		50	10.25	–	1
	50	4.40	–	1		60	11.01	–	1
	55	4.44	3.40	7		70	11.77	–	1
	60	4.43	–	1		80	12.56	–	1

REFERENCES

1. Tyrer, D., *J. Chem. Soc.,* 105, 2534, 1914.
2. Lutskii, A. E. and Solonko, V. N., *Russ. J. Phys. Chem.,* 38, 602, 1964.
3. Fryer, E. B., Hubbard, J. C., and Andrews, D. H., *J. Am. Chem. Soc.,* 51, 759, 1929.
4. Newitt, D. M. and Weale, K. E., *J. Chem. Soc.,* p. 3092, 1951.
5. Gibson, R. E. and Loeffler, O. H., *J. Am. Chem. Soc.,* 61, 2515, 1939.
6. Gibson, R. E. and Kincaid, J. F., *J. Am. Chem. Soc.,* 60, 511, 1938.
7. Gibson, R. E. and Loeffler, O. H., *J. Am. Chem. Soc.,* 63, 898, 1941.
8. Aicart, E., Tardajos, G., and Diaz Pena, M., *J. Chem. Eng. Data,* 26, 22, 1981.
9. Marshall, J. G., Staveley, L. A. K., and Hart, K. R., *Trans. Faraday Soc.,* 52, 23, 1956.
10. Staveley, L. A. K., Tupman, W. I., and Hart, K. R., *Trans. Faraday Soc.,* 51, 323, 1955.
11. McKinney, W. P., Skinner, G. F., and Staveley, L. A. K., *J. Chem. Soc.,* p. 2415, 1959.
12. Harrison, D. and Moelwyn-Hughes, E. A., *Proc. R. Soc. London Ser. A,* 239, 230, 1957.
13. Cutler, W. G., McMickle, R. H., Webb, W., and Schiessler, R. W., *J. Chem. Phys.,* 29, 727, 1958.
14. Moelwyn-Hughes, E. A., *J. Phys. Colloid Chem.,* 55, 1246, 1951.
15. Holder, G. A. and Walley, E., *Trans. Faraday Soc.,* 58, 2095, 1962.
16. Aicart, E., Tardajos, G., and Diaz Pena, M., *J. Chem. Eng. Data,* 25, 140, 1980.
17. Blinowska, A. and Brostow, W., *J. Chem. Thermodyn.,* 7, 787, 1975.

COEFFICIENT OF FRICTION

Compiled by Harold Minshall

The coefficient of friction between two surfaces is the ratio of the force required to move one over the other to the total force pressing the two together. If F is the force required to move one surface over another and W, the force pressing the surfaces together, the coefficient of friction,

$$\mu = \frac{F}{W}$$

Materials	Condition	Temperature °C	μ (Static)
A. STATIC FRICTION			
Non Metals			
Glass on glass	clean	—	0.9–1.0
" " "	lubricated with paraffin oil	—	0.5–0.6
" " "	" " liquid fatty acids	—	0.3–0.6
" " "	" " solid hydrocarbons, alcohols or fatty acids	—	0.1
" " metal	clean	—	0.5–0.7
" " "	lubricated	—	0.2–0.3
Diamond on diamond	clean	—	0.1
" " "	lubricated	—	0.05–0.1
" " metal	clean	—	0.1–0.15
" " "	lubricated	—	0.1
Sapphire on sapphire	clean or lubricated	—	0.2
" " steel	" " "	—	0.15
Hard carbon on carbon	clean	—	0.16
" " "	lubricated	—	0.12–0.14
Graphite on graphite	clean or lubricated	—	0.1
" " "	outgassed	—	0.5–0.8
" " steel	clean or lubricated	—	0.1
Mica on mica	freshly cleaved	—	1.0
" " "	contaminated	—	0.2–0.4
Crystals of $NaNO_3$, KNO_3, NH_4Cl on self	clean	—	0.5
" " " " " " "	lubricated with long chain polar compounds	—	0.12
Tungsten carbide on tungsten carbide	clean	room	0.17
Tungsten carbide on tungsten carbide	outgassed	room	0.58
" " " " "	clean	820	0.35
" " " " "	"	970	0.40
" " " " "	"	1010	0.45
" " " " "	"	1160	0.5
" " " " "	"	1220	0.7
" " " " "	"	1440	1.2
" " " " "	"	1000	1.8
" " graphite	outgassed	room	0.62
" " " "	clean	"	0.15
" " " "	"	800	0.32
" " " "	"	910	0.30
" " " "	"	1000	0.25
" " " "	"	1120	0.29
" " " "	"	1220	0.26
" " " "	"	1300	0.25
" " " "	"	1410	0.25
" " " "	"	1800	0.24
" " " "	"	2030	0.25
" " steel	"	—	0.4–0.6
" " " "	lubricated	—	0.1–0.2
Polymethyl methacrylate on self	clean	—	0.8
" " steel	"	—	0.4–0.5
Polystyrene on self	"	—	0.5
" " steel	"	—	0.3–0.35
Polyethylene on self	"	—	0.2
" " steel	"	—	0.2
Polytetrafluoroethylene on self	"	—	0.04
" " steel	"	—	0.04
Nylon on nylon[1]	"	—	0.15–0.25
Silk on silk	commercially clean	—	0.2–0.3
Cotton on cotton (thread)	" "	—	0.3
" " (from cotton wool)	" "	—	0.6
Rubber on solids	" "	—	1–4
Wood on wood	" " and dry	—	0.25–0.5
" " "	" " " wet	—	0.2
" " metals	" " " dry	—	0.2–0.6
" " "	" " " wet	—	0.2
" " brick	" "	—	0.6
" " leather	" "	—	0.3–0.4
Leather on metal	" "	—	0.6
" " " "	" " and wet	—	0.4
" " " "	greasy	—	0.2
Brake material on cast iron	commercially clean	—	0.4
" " " " "	" " and wet	—	0.2
" " " " "	lubricated with mineral oil	—	0.1
Wool fiber on horn	clean (against scales)	—	0.8–1.0
" " " "	" (with scales)	—	0.4–0.6
" " " "	greasy (against scales)	—	0.5–0.8
" " " "	" (with scales)	—	0.3–0.4
Metals			
Steel on steel	clean	20	0.58
" " "	vegetable oil lubricant (a) castor oil	20	0.095
		100	0.105
" " "	(b) rape	20	0.105
		100	0.105
" " "	(c) olive	20	0.105
		100	0.105
" " "	(d) coconut	20	0.08
		100	0.08

[1] Registered trade name.

Materials	Condition	Temperature °C	μ (Static)
A. STATIC FRICTION (Cont.)			
Metals (Cont.)			
Steel on steel	Animal oil lubricant		
" " "	(a) sperm	20	0.10
		100	0.10
" " "	(b) pale whale	20	0.095
		100	0.095
" " "	(c) neatsfoot	20	0.095
		100	0.095
" " "	(d) lard	20	0.085
		100	0.085
	Mineral oil lubricant		
" " "	(a) light machine	20	0.16
		100	0.19
" " "	(b) thick gear	20	0.125
		100	0.15
" " "	(c) solvent refined	20	0.15
		100	0.20
" " "	(d) heavy motor	20	0.195
		100	0.205
" " "	(e) extreme pressure	20	0.09–0.1
		100	0.09–0.1
" " "	(f) graphited oil	20	0.13
		100	0.15
" " "	(g) B.P. Paraffin	20	0.18
		100	0.22
" " "	lubricated with trichloroethylene	20	0.33
" " "	" " benzene	20	0.48
" " "	" " glycerol	20	0.2
" " "	" " ethyl alcohol	20	0.43
" " "	" " butyl alcohol	room	0.3
" " "	" " octyl	"	0.23
" " "	" " decyl	"	0.16
" " "	" " cetyl	"	0.10
" " "	lubricated with nonane	room	0.26
" " "	" " decane	"	0.23
" " "	" " acetic acid	"	0.5
" " "	" " proprionic acid	"	0.4
" " "	" " valeric acid	"	0.17
" " "	" " caproic acid	"	0.12
" " "	" " pelargonic acid	"	0.11
" " "	" " capric acid	"	0.11
" " "	" " lauric acid	"	0.11
" " "	" " myristic acid	"	0.11
" " "	" " oleic acid	20–100	0.08
" " "	" " palmitic acid	room	0.11
" " "	" " stearic acid	"	0.10
" " hard steel	" " rape oil	—	0.14
" " "	" " castor oil	—	0.12
" " "	" " mineral oil	—	0.16
" " "	" " long chain fatty acid	—	0.09
" " cast iron	" " rape oil	—	0.11
" " "	" " castor oil	—	0.15
" " "	" " mineral oil	—	0.21
" " "	clean	—	0.4
" " gun metal	lubricated with rape oil	—	0.15
" " "	" " castor oil	—	0.16
" " "	" " mineral oil	—	0.21
" " bronze	" " rape oil	—	0.12
" " "	" " caster oil	—	0.12
" " "	" " mineral oil	—	0.16
" " lead	" " "	—	0.5
" " "	" " long chain fatty acid	—	0.22
" " base white metal	" " mineral oil	—	0.1
" " " " "	" " long chain fatty acid	—	0.08
" " "	clean	—	0.55
" " tin	lubricated with mineral oil	—	0.6
" " "	" " long chain fatty acid	—	0.21
" " white metal, tin base	" " mineral oil	—	0.1
" " " " " "	" " long chain fatty acid	—	0.07
" " "	clean	—	0.8
" " sintered bronze	lubricated with mineral oil	—	0.13
" " brass	" " "	—	0.19
" " "	" " castor oil	—	0.11
" " "	" " long chain fatty acid	—	0.13
" " "	clean	—	0.35
" " copper–lead alloy	"	—	0.22
" " Wood's alloy	"	—	0.7
" " phosphor bronze	"	—	0.35
" " aluminum bronze	"	—	0.45
" " constantan	"	—	0.4
[2] " " indium film deposited on steel	4 kg load, clean	—	0.08
[2] " "	8 kg " " "	—	0.04
[2] " " " " " " " silver	4 kg " " "	—	0.1
[2] " "	8 kg " " "	—	0.07
[2] " " lead film deposited on copper	4 kg " " "	—	0.18
[2] " "	8 kg " " "	—	0.12
[2] " " copper film deposited on steel	4 kg " " "	—	0.3
[2] " " " " " " "	8 kg " " "	—	0.2
[2] Al on Al	in air or O_2	—	1.9
" "	" H_2O vapor	—	1.1
[3] Cu on Cu	" H_2 or N_2	—	4.0
" "	" air or O_2	—	1.6

[2] Hemispherical steel slider having 0.6 cm. diameter. The thin, 10^{-3} to 10^{-4} cm., thin metallic films were deposited on various substrates as indicated. Amonton's Law is not obeyed in this case.

[3] The metals which were spectroscopically pure were outgassed in a vacuum prior to other gases being admitted. When clean and in vacuum there is gross seizure.

Materials	Condition	Temperature °C	μ (Static)
A. STATIC FRICTION (Cont.)			
Metals (Cont.)			
[3]Au on Au	in H_2 or N_2	—	4.0
,, ,, ,,	,, air or O_2	—	2.8
,, ,, ,,	,, H_2O vapor	—	2.5
[3]Fe on Fe	,, in air or O_2	—	1.2
,, ,, ,,	,, H_2O vapor	—	1.2
[3]Mo on Mo	,, air or O_2	—	0.8
,, ,, ,,	,, H_2O vapor	—	0.8
[3]Ni on Ni	,, H_2 or N_2	—	5.0
,, ,, ,,	,, air or O_2	—	3.0
,, ,, ,,	,, H_2O vapor	—	1.6
[3]Pt on Pt	,, air or O_2	—	3.0
,, ,, ,,	,, H_2O vapor	—	3.0
[3]Ag on Ag	,, air or O_2	—	1.5
,, ,, ,,	,, H_2O vapor	—	1.5
Various Materials on Snow and Ice			
Ice on ice	clean	0	0.05–0.15
,, ,, ,,	,,	−12	0.3
,, ,, ,,	,,	−71	0.5
,, ,, ,,	,,	−82	0.5
,, ,, ,,	,,	−110	0.5
Polymethylmethylacrylate	on wet snow	0	0.5
,,	,, dry ,,	0	0.3
,,	,, ,, ,,	−10	0.34
,,	,, ,, ,,	−32	0.4
Polyester of teraphthalic acid and ethylene glycol	,, wet ,,	0	0.5
Polyester of teraphthalic acid and ethylene glycol	,, dry ,,	0	0.35
Polyester of teraphthalic acid and ethylene glycol	,, ,, ,,	−10	0.38
[1]Nylon	on wet snow	0	0.4
,,	,, dry ,,	0	0.3
,,	,, ,, ,,	−10	0.3
Polytetrafluoroethylene	,, wet ,,	0	0.05
,,	,, dry ,,	0	0.02
,,	,, ,, ,,	−10	0.08
,,	,, ,, ,,	−32	0.1
Paraffin wax	,, wet ,,	0	0.06
,, ,,	,, dry ,,	0	0.06
,, ,,	,, ,, ,,	−10	0.35
,, ,,	,, ,, ,,	−32	0.4
Swiss wax	,, wet ,,	0	0.05
,, ,,	,, dry ,,	0	0.03
,, ,,	,, ,, ,,	−10	0.2
,, ,,	,, ,, ,,	−32	0.2
Ski wax	,, wet ,,	0	0.1
,, ,,	,, dry ,,	0	0.04
,, ,,	,, ,, ,,	−10	0.2
,, ,,	,, ,, ,,	−32	0.2
,, laquer	,, wet ,,	0	0.2
,, ,,	,, dry ,,	0	0.1
,, ,,	,, ,, ,,	−10	0.4
,, ,,	,, ,, ,,	−32	0.4
Aluminum	,, wet ,,	0	0.4
,,	,, dry ,,	0	0.35
,,	,, ,, ,,	−10	0.38

Materials	Condition	Temperature °C	μ (Kinetic)
B. KINETIC FRICTION Various Materials			
Unwaxed hickory	4 m/sec on dry snow	−3	0.08
Waxed ,,	0.1 m/sec ,, wet ,,	0	0.14
,, ,,	0.1 m/sec ,, dry ,,	0	0.04
,, ,,	0.1 m/sec ,, ,, ,,	−3	0.09
,, ,,	4 m/sec ,, ,, ,,	−3	0.03
Waxed hickory	0.1 m/sec on dry snow	−10	0.18
,, ,,	0.1 m/sec ,, ,, ,,	−40	0.4
Ice on ice	4m/sec, clean	0	0.02
,, ,, ,,	,, ,, ,,	−10	0.035
,, ,, ,,	,, ,, ,,	−20	0.050
,, ,, ,,	,, ,, ,,	−40	0.075
,, ,, ,,	,, ,, ,,	−60	0.085
,, ,, ,,	,, ,, ,,	−80	0.09
Ebonite	4m/sec on ice	0	0.02
,,	,, ,, ,,	−10	0.05
,,	,, ,, ,,	−20	0.065
,,	,, ,, ,,	−40	0.085
,,	,, ,, ,,	−60	0.10
,,	,, ,, ,,	−80	0.11
Brass	4m/sec on ice	0	0.02
,,	,, ,, ,,	−10	0.075
,,	,, ,, ,,	−20	0.085
,,	,, ,, ,,	−40	0.115
,,	,, ,, ,,	−60	0.14
,,	,, ,, ,,	−80	0.15
Natural rubber, vulcanized	100m/min on ground glass, clean	—	1.07
,, ,, ,,	100m/min ,, ,, ,, , wetted with water	—	0.94
,, ,, ,,	100m/min on concrete, clean	—	1.02
,, ,, ,,	100m/min ,, ,, , wetted with water	—	0.97
,, ,, ,,	100m/min on bitumen, clean	—	1.07
,, ,, ,,	100m/min ,, ,, , wetted with water	—	0.95
,, ,, ,,	100m/min on rubber flooring or rubber tread vulcanisate, clean	—	1.16
,, ,, ,,	100m/min on bitumen containing rubber powder, clean	—	1.15 (Varies with quantity of powder)
,, ,, ,,	100m/min on bitumen containing rubber powder, wetted with water	—	1.03

[1] Registered trade name. [3] The metals which were spectroscopically pure were outgassed in a vacuum prior to other gases being admitted. When clean and in vacuum there is gross seizure.

LOW MELTING POINT ALLOYS

Melting point °C	Name	Composition, wt %				
−48	Binary Eutectic	Cs 77.0	K 23.0			
−40	Binary Eutectic	Cs 87.0	Rb 13.0			
−30	Binary Eutectic	Cs 95.0	Na 5.0			
−11	Binary Eutectic	K 78.0	Na 22.0			
−8	Binary Eutectic	Rb 92.0	Na 8.0			
10.7	Ternary Eutectic	Ga 62.5	In 21.5	Sn 16.0		
10.8	Ternary Eutectic	Ga 69.8	In 17.6	Sn 12.5		
17	Ternary Eutectic	Ga 82.0	Sn 12.0	Zn 6.0		
33	Binary Eutectic	Rb 68.0	K 32.0			
46.5	Quinternary Eutectic	Sn 10.65	Bi 40.63	Pb 22.11	In 18.1	Cd 8.2
47	Quinternary Eutectic	Bi 44.7	Pb 22.6	Sn 8.3	Cd 5.3	In 19.1
58.2	Quaternary Eutectic	Bi 49.5	Pb 17.6	Sn 11.6	In 21.3	
60.5	Ternary Eutectic	In 51.0	Bi 32.5	Sn 16.5		
70	Wood's Metal	Bi 50.0	Pb 25.0	Sn 12.5	Cd 12.5	
70	Lipowitz's Metal	Bi 50.0	Pb 26.7	Sn 13.3	Cd 10.0	
70	Binary Eutectic	In 67.0	Bi 33.0			
91.5	Ternary Eutectic	Bi 51.6	Pb 40.2	Cd 8.2		
95	Ternary Eutectic	Bi 52.5	Pb 32.0	Sn 15.5		
97	Newton's Metal	Bi 50.0	Sn 18.8	Pb 31.2		
98	D'Arcet's Metal	Bi 50.0	Sn 25.0	Pb 25.0		
100	Onion's or Lichtenberg's Metal	Bi 50.0	Sn 20.0	Pb 30.0		
102.5	Ternary Eutectic	Bi 54.0	Sn 26.0	Cd 20.0		
109	Rose's Metal	Bi 50.0	Pb 28.0	Sn 22.0		
117	Binary Eutectic	In 52.0	Sn 48.0			
120	Binary Eutectic	In 75.0	Cd 25.0			
123	Malotte's Metal	Bi 46.1	Sn 34.2	Pb 19.7		
124	Binary Eutectic	Bi 55.5	Pb 44.5			
130	Ternary Eutectic	Bi 56.0	Sn 40.0	Zn 4.0		
140	Binary Eutectic	Bi 58.0	Sn 42.0			
140	Binary Eutectic	Bi 60.0	Cd 40.0			
183	Eutectic solder	Sn 63.0	Pb 37.0			
185	Binary Eutectic	Tl 52.0	Bi 48.0			
192	Soft solder	Sn 70.0	Pb 30.0			
198	Binary Eutectic	Sn 91.0	Zn 9.0			
199	Tin foil	Sn 92.0	Zn 8.0			
199	White Metal	Sn 92.0	Sb 8.0			
221	Binary Eutectic	Sn 96.5	Ag 3.5			
226	Matrix	Bi 48.0	Pb 28.5	Sn 14.5	Sb 9.0	
227	Binary Eutectic	Sn 99.25	Cu 0.75			
240	Antimonial Tin solder	Sn 95.0	Sb 5.0			
245	Tin-silver solder	Sn 95.0	Ag 5.0			

HARDNESS OF MATERIALS

Material	Value	Material	Value
Agate	6–7	Indium	1.2
Alabaster	1.7	Iridium	6–6.5
Alum	2–2.5	Iridosmium	7
Aluminum	2–2.9	Iron	4–5
Alundum	9+	Kaolinite	2.0–2.5
Amber	2–2.5	Lead	1.5
Andalusite	7.5	Lithium	0.6
Anthracite	2.2	Loess (0°)	0.3
Antimony	3.0–3.3	Magnesium	2.0
Apatite	5	Magnetite	6
Aragonite	3.5	Manganese	5.0
Arsenic	3.5	Marble	3–4
Asbestos	5	Meerschaum	2–3
Asphalt	1–2	Mica	2.8
Augite	6	Opal	4–6
Barite	3.3	Orthoclase	6
Bell-metal	4	Osmium	7.0
Beryl	7.8	Palladium	4.8
Bismuth	2.5	Phosphorus	0.5
Boric acid	3	Phosphorbronze	4
Boron	9.5	Platinum	4.3
Brass	3–4	Plat-iridium	6.5
Cadmium	2.0	Potassium	0.5
Calamine	5	Pumice	6
Calcite	3	Pyrite	6.3
Calcium	1.5	Quartz	7
Carbon	10.0	Rock salt (halite)	2
Carborundum	9–10	Ross' metal	2.5–3.0
Cesium	0.2	Rubidium	0.3
Chromium	9.0	Ruthenium	6.5
Copper	2.5–3	Selenium	2.0
Corundum	9	Serpentine	3–4
Diamond	10	Silicon	7.0
Diatomaceous earth	1–1.5	Silver	2.5–4
Dolomite	3.5–4	Silver chloride	1.3
Emery	7–9	Sodium	0.4
Feldspar	6	Steel	5–8.5
Flint	7	Stibnite	2
Fluorite	4	Strontium	1.8
Galena	2.5	Sulfur	1.5–2.5
Gallium	1.5	Talc	1
Garnet	6.5–7	Tellurium	2.3
Glass	4.5–6.5	Tin	1.5–1.8
Gold	2.5–3	Topaz	8
Graphite	0.5–1	Tourmaline	7.3
Gypsum	1.6–2	Wax (0°)	0.2
Hematite	6	Wood's metal	3
Hornblende	5.5	Zinc	2.5

COMPARISON OF HARDNESS VALUES OF VARIOUS MATERIALS ON MOHS AND KNOOP SCALES*

Compiled by Laurence S. Foster

Substance	Formula	Mohs value	Knoop value
Talc	$3MgO \cdot 4SiO_2 \cdot H_2O$	1	
Gypsum	$CaSO_4 \cdot 2H_2O$	2	32
Cadmium	Cd	..	37
Silver	Ag	..	60
Zinc	Zn	..	119
Calcite	$CaCO_3$	3	135
Fluorite	CaF_2	4	163
Copper	Cu	..	163
Magnesia	MgO	..	370
Apatite	$CaF_2 \cdot 3Ca_3(PO_4)_2$	5	430
Nickel	Ni	..	557
Glass (soda lime)		..	530
Feldspar (orthoclase)	$K_2O \cdot Al_2O_3 \cdot 6SiO_2$	6	560
Quartz	SiO_2	7	820
Chromium	Cr	..	935
Zirconia	ZrO_2	..	1160
Beryllia	BeO	..	1250
Topaz	$(AlF)_2SiO_4$	8	1340
Garnet	$Al_2O_3 \cdot 3FeO \cdot 3SiO_2$	..	1360
Tungsten carbide alloy	WC, Co	..	1400–1800
Zirconium boride	ZrB_2	..	1550
Titanium nitride	TiN	9	1800
Tungsten carbide	WC	..	1880
Tantalum carbide	TaC	..	2000
Zirconium carbide	ZrC	..	2100
Alumina	Al_2O_3	..	2100
Beryllium carbide	Be_2C	..	2410
Titanium carbide	TiC	..	2470
Silicon carbide	SiC	..	2480
Aluminum boride	AlB	..	2500
Boron carbide	B_4C	..	2750
Diamond	C	10	7000

MOHS HARDNESS SCALE

Hardness number	Original scale	Modified scale
1	Talc	Talc
2	Gypsum	Gypsum
3	Calcite	Calcite
4	Fluorite	Fluorite
5	Apatite	Apatite
6	Orthoclase	Orthoclase
7	Quartz	Vitreous silica
8	Topaz	Quartz or Stellite
9	Corundum	Topaz
10	Diamond	Garnet
11	....	Fused Zirconia
12	....	Fused Alumina
13	....	Silicon Carbide
14	....	Boron Carbide
15	....	Diamond

* Acknowledgment is made to N. W. Thibault, Norton Company, Worcester, Massachusetts, for many of Knoop hardness values. Cf. R. F. Geller, "A Study of Ceramics for Nuclear Reactors," Nucleonics, Vol. 7, No. 4, Table 1, pp. 8–9 (Oct. 1950). V. E. Lysaght, Indentation Hardness Testing, Reinhold 1949.

SURFACE TENSION OF LIQUID ELEMENTS

Gernot Lang

The following data were collected from many sources. As a result their accuracy varies. Users of data in this table are advised that:

1. As a rule, results from the "sessile drop" and "maximum bubble pressure" as well as from the "pendant drop" methods are preferable to results obtained from other methods for metals with very high melting points.
2. Values of single measurements are usually not as well supported by experiments as those of serial measurements at various temperatures.
3. Values in parentheses can be considered improbable.

Element	Purity (wt.%)	σ_{mp} (dyn/cm)	Atm.	t_1 °C	σ_{t1} (dyn/cm)	t_2 °C	σ_{t2} (dyn/cm)	t_3 °C	σ_{t3} (dyn/cm)	Method	Ref.
Ag	99.99		H_2	1000	916					Bubble pressure	134
	–	(785)	vac.							Pendant drop	182
	99.96		H_2	1000	893	1150	862	1250	849	Bubble pressure	127
	–		vac.	1000	908					Sessile drop	51
	99.995		H_2	1000	907	1100	894	1200	876	Bubble pressure	128
	99.999	(828)	vac.							Pendant drop	65
	99.99		Ar, H_2	1000	890					Bubble pressure	169
	spect. pure	921		$\sigma = 1136 - 0.174\,T\,(°K)\,(1300-2200°K)$						Bubble pressure	26
		918	–	$\sigma = 918 - 0.149\,(t - t_{mp})\,(t°C)$						Bubble pressure	204
	99.999		Ar	980	905±10	1108	890±10			Sessile drop	Z3
Al	99.99	860±20	Ar							Bubble pressure	102
	99.72		vac.	950	840					Bubble pressure	46
	99.7	863±25	Ar	$\sigma_t = (863±25) - 0.33\,(t - t_{mp})\,(t°C)$						Bubble pressure	113
	99.99	865	vac.	$\sigma_t = 865 - 0.14\,(t - t_{mp})\,(t°C)$						Sessile drop	137, 56
	99.99	(825)	Ar	$(\sigma_t = 825 - 0.05\,(T - 993))\,(T°K)$						Bubble pressure	38
	99.99	866	He	$\sigma_t = 866 - 0.15\,(t - t_{mp})\,(t°C)$						Sessile drop	10
	99.999	873	He	1600	725 $\sigma_t = 873 - 0.15\,(t - t_{mp})\,(t°C)$					Sessile drop	199
	99.99	(760)	vac.	$(\sigma_t = 910 - 0.202\,T)\,(T°K\ 930-1090°K)$						Sessile drop	24
	99.998	(915)	Ar	$(\sigma_t = 915 - 0.51\,(t - t_{mp}))\,(t°C)\ (660-800°C)$						Sessile drop	Z6
	99.996	855±6	Ar	$\sigma_t = 855 - 0.104\,(t - t_{mp})\,(t°C)\ (660-911°C)$						Bubble pressure	Z13
Au		(754)	vac.							Pendant drop	182
	99.999	1130	He	1200	1070	1300	1020			Sessile drop	86
	99.999	(731)	vac.							Pendant drop	65
	99.999		Ar	1108	1130±10					Sessile drop	Z3
B	99.8	1060±50	vac.							Sessile & pendant drop	187
Ba	–		Ar	720	224					Bubble pressure	3
	99.5	276		$\sigma = 351 - 0.075\,T\,(°K)\,(1410-1880°K)$						Bubble pressure	25
Be	99.98		vac.	1500	1100					Sessile drop	58
Bi		376	vac.							Drop pressure	155
	99.99	376	vac.							Drop pressure	156
	99.90		H_2	800	343	1000	328			Bubble pressure	134
			vac.	450	(382)					Electro-capillarity	118
	99.9	380±10	Ar							Bubble pressure	103
			–	350	362					Drop pressure	76
			vac.	700	350					Sessile drop	29
	99.98	380±10	Ar							Bubble pressure	105
			–	450	380					Drop pressure	120
			vac.	300	379					Sessile drop	64
		378	vac., Ar, H_2							Drop weight	4
	99.99995	375		$\sigma = 423 - 0.088\,T\,(T°K)\,(1352-1555°K)$						Bubble pressure	26
	99.999	380±3	Ar	$\sigma_t = 380 - 0.142\,(t - t_{mp})\,(MP-555°C)\,(t°C)$						Bubble pressure	Z13a
Ca	–		Ar	850	337					Bubble pressure	3
	p.a.	360		$\sigma = 472 - 0.100\,T\,(T°K)\,(1445-1655°K)$						Bubble pressure	25
Cd			–	450	600					Drop pressure	119
			–	400	600					Bubble pressure	13
			–	350	586					Drop pressure	69
	99.9	(550±10)	Ar							Bubble pressure	103
				390	604					Bubble pressure	1
		(525±30)	H_2							Solid state curvature	Z8
	99.9999	590±5	–			(non linear)				Sessile drop	Z12

Element	Purity (wt.-%)	σ_{mp} (dyn/cm)	Atm.	σ_{t1} t_1,°C	σ_{t1} σ(dyn/cm)	σ_{t2} t_2,°C	σ_{t2} σ(dyn/cm)	σ_{t3} t_3,°C	σ_{t3} σ(dyn/cm)	Method	Ref.
Co			Ar	1550	1836					Sessile drop	108
	99.99		vac., Al$_2$O$_3$	1520	1800					Sessile drop	46
	99.99		He, Al$_2$O$_3$	1520	(1630)					Sessile drop	46
	99.99		He, BeO	1520	(1640)					Sessile drop	46
	99.99		He, MgO	1520	(1560)					Sessile drop	46
	99.99		H$_2$, Al$_2$O$_3$	1520	1780					Sessile drop	46
	99.99		He	1520	(1620)					Bubble pressure	46
	99.99		H$_2$	1520	(1590)					Bubble pressure	46
			vac.	1500	1870					Sessile drop	7
				1600	(1640)					Sessile drop	135
				1600	(1600)					Sessile drop	135
			vac.	1600	1815					Sessile drop	53
	99.99		vac., Al$_2$O$_3$	1600	1812					Sessile drop	54
		(1520)	H$_2$, He							Bubble pressure	61
	99.99		H$_2$, He	1550	1845					Bubble pressure	63
	99.9983	1880	vac.							Pendant drop	5
	99.99			1550	1780					Bubble pressure	59
Cr	–		vac.	1950	1590±50					Sessile drop	47
	99.9997	1700±50	Ar							Dynam. drop weight	6
Cs			Ar	62	68.4					Pendant drop	195
			Ar	62	67.5	146	62.9			Bubble pressure	193
	99.95		Ar	39	69.5	494	42.8	642	34.6	Bubble pressure	24
	99.99	73.74	Ar	\multicolumn{6}{l}{$\sigma = 73.74-1.791 \cdot 10^{-2}\,(t-t_{mp})-9.610 \cdot 10^{-5}\,(t-t_{mp})^2 + 6.629 \cdot 10^{-8}\,(t-t_{mp})^3\ (t°C)\ (71-1011°C)$}			Bubble pressure	26
	99.995	68.6	He	\multicolumn{6}{l}{$\sigma = 68.6-0.047\,(t-t_{mp})\ (t°C)\ (52-1100°C)$}			Bubble pressure	121
Cu			Ar	1120	1269±20					Sessile drop	11
	–	(1150)	vac.							Pendant drop	35
			Ar	1120	1285±10					Sessile drop	108
	99.98		H$_2$	1100	1301	1165	1295	1255	1287	Bubble pressure	134
		1270	vac.							Sessile drop	7
			vac.	1120	1285					Sessile drop	198
				1440	1298					Bubble pressure	129
		(1085)	vac.							Pendant drop	182
	99.99		He	1250	1290					Bubble pressure	62
	99.99		H$_2$	1250	1300					Bubble pressure	62
	99.9	(1180±40)	Ar							Bubble pressure	103
	–			1100	1220					Sessile drop	136
	–			1183	(1130)					Sessile drop	200
	–		vac.	1150	1370					Sessile drop	52
	99.997	1355	He, H$_2$							Sessile drop + Bubble pressure	138
	99.997	1352	vac.	\multicolumn{6}{l}{$\sigma_t = 1352-0.17\,(t-t_{mp})\ (t°C)$}			Sessile drop	137
	99.997	1358	Ar	\multicolumn{6}{l}{$\sigma_t = 1358-0.20\,(t-t_{mp})\ (t°C)$}			Bubble pressure	137
	–		Ar, He	1120	1285±10					Sessile drop	110
	99.99		H$_2$, He	1550	1265					Bubble pressure	63
	99.99999	1300	vac.							Pendant drop	5
			vac.	1130	1268±60					Sessile drop	15
	99.98		Ar	1600	1230					Bubble pressure	205
	99.999		N$_2$	1100	1341	1150	1338	1200	1335	Bubble pressure	128
	99.9	(1127)	vac.							Pendant drop	65
	99.99		Ar, H$_2$	1100	1320					Bubble pressure	169
Fe				1570	(1731)					Sessile drop	95
	–			1550	1860					Sessile drop	74
		1720	He							Sessile drop	74
				1580–1760	(880)					Bubble pressure	100
				1570	(1632)					Sessile drop	93
	99.99		He	1650	(1610)					Sessile drop	46
	99.99		He	1650	(1430)					Sessile drop	46
	99.99		H$_2$	1650	(1400)					Sessile drop	46
	–	(1384)	vac.							Pendant drop	182
		(1700)	vac.							Sessile drop	7
			vac., He	1550	1865					Sessile drop	209
	99.99		He	1650	(1430)					Bubble pressure	62
	99.99		H$_2$	1650	(1400)					Bubble pressure	62
				1650	(1640)					Sessile drop	136
		(1650)	He, H$_2$							Bubble pressure	61
	99.985		Ar	1550	1788					Sessile drop	110
	–	(1560)								Sessile drop	68

Element	Purity (wt.-%)	σ_{mp} (dyn/cm)	Atm.	σ_{t1} t, °C	σ_{t1} σ (dyn/cm)	σ_{t2} t, °C	σ_{t2} σ (dyn/cm)	σ_{t3} t, °C	σ_{t3} σ (dyn/cm)	Method	Ref.
	99.94		vac., Al$_2$O$_3$	1560	(1710)					Sessile drop	207
	99.9998	1880	vac.							Pendant drop	5
	99.93	1860±40	He							Sessile drop	116
		(1510)	vac.							Drop weight	139
	99.97		vac., BeO	1550	1830±6					Sessile drop	45
	Armco		Ar, N$_2$	1550	1795					Bubble pressure	164
			vac.	1550	1754					Sessile drop	42
	99.987		vac.	1550	(1730)					Sessile drop	159
	99.85	(1619)	vac.							Pendant drop	65
	99.69		He, Al$_2$O$_3$	1550	(1727)					Sessile drop	158
	99.69		H$_2$, Al$_2$O$_3$	1550	(1734)					Sessile drop	158
	–	1760±20	He, H$_2$	$\sigma = 1760-0.35\ (t-t_{mp})\ (t°C)$						Sessile drop	157
	99.9992	1773	He, H$_2$	$\sigma = 773+0.65\ t\ (t°C)\ (1550-1780°C)$						Oscillating drop	Z2
			–	1550	1780					Sessile drop	Z11
Fr	–		–	100	58.4					calculated	145
Ga		704	Ar, He							Bubble pressure	197
	–	725±10	Ar							Bubble pressure	104
			vac.	350	718					Sessile drop	64
	–		He, Al$_2$O$_3$	1500	559					Sessile drop	57
	99.9998	718	vac., Al$_2$O$_3$	$\sigma = 718-0.101\ (t-t_{mp})\ (t°C)$						Sessile drop	85
Ge		621.4								Drop weight	110
		650	vac.	1200	530					Sessile drop	99
			vac.	1000	630					Sessile drop	Z5
		632±5	N$_2$, He							Solid state curvature	Z8
Hf		(1460)	vac.							Pendant drop	131
	97.5+2.5 Zr	1630	vac.							Pendant drop	5
Hg			H$_2$	20	(542)					Pendant drop	165
			vac.	15	487.6 ± 1.1					Sessile drop	Z14
			air	20	(435.5)					Oscillating jet	175
			vac.	20	472					Oscillating jet	73
			vac.	20	(402)					Drop pressure	146
				25	476					Drop weight	75
			vac.	20	486.4 ± 1.1					Sessile drop	Z14
			vac.	20	(432)					Sessile drop	170
			H$_2$	25	476					Drop pressure	78
				25	472					Drop weight	81
				25	(464)					Sessile drop	82
			H$_2$	19	473					Bubble pressure	173
				20	(437)					Capillary depression	144
			vac.	20	480					Drop weight	21
				25	485.3 ± 1.1					Sessile drop	Z14
				25	(435)					Sessile drop	91
			vac.	25	473					Drop weight	31
				25	488					Sessile drop	32
				25	(498)					Sessile drop	30
			vac.	20	(420)					Sessile drop	174
				25	476					Sessile drop	90
			vac.	20	(410)					Drop pressure	178
			vac.	20	(455)					Drop pressure	39
				25	484±1.5					Sessile drop	89
			vac.	22	(468)					Drop pressure	162
			vac.	20	(465.2)	103	449.7	350	387.1	Drop pressure	162
				25	484.9±1.8					Sessile drop	Z7
			vac.	20	485.5±1.0	$\sigma_t = 489.5-0.20\ (t°C)$				Drop pressure	17
			Ar	20	(454.7)					Bubble pressure	188
				21	(350.5)					Bubble pressure	23
	99.99		He, H$_2$	20	475					Bubble pressure	62
	99.9		Ar	20	(500±15)					Bubble pressure	103
			vac.	-10	487					Drop pressure	69
			vac.	25	483.5±1.0					Sessile drop	140
				22	(465)					Bubble pressure	194
				25	485.1					Sessile drop	142
				16.5	487.3	25	485.4±1.2			Sessile drop	171
				23–25	482.8±9.7					Contact angle	18
				$\sigma_t = 468.7-1.61 \cdot 10^{-1}\,t-1.815 \cdot 10^{-4}\,t^2\ (t°C)$						–	206
			vac.	20	484.6±1.3					Pendant drop	132

Element	Purity (wt.-%)	σ_{mp} (dyn/cm)	Atm.	t_1, °C	σ_{t1} (dyn/cm)	t_2, °C	σ_{t2} (dyn/cm)	t_3, °C	σ_{t3} (dyn/cm)	Method	Ref.
			Ar	25	480					Bubble pressure	172
			vac.	20	482.5±3.0					Bubble pressure	177
					$\sigma_t = 485.5-0.149t-2.84 \cdot 10^{-4} t^2$ (t°C)						
			Ar	21.5	484.9±0.3					Bubble pressure	Z13a
			vac.	50	480.0 ± 1.1					Sessile drop	Z14
In	99.95	559	H_2	600	515					Capillary method	133
				623	540					Capillary method	77
	99.995	556.0	Ar, He							Bubble pressure	196
			vac.	185	592					Drop pressure	69
			H_2	600	514						16
				300	541					Sessile drop	83
	99.999		Ar	200	556	400	535	550	527.8	Bubble pressure	123
	99.9994		vac.	350	539					Drop pressure	106
	99.9999	560±5	–		$\sigma = 568.0-0.04t-7.08 \cdot 10^{-5} t^2$ (t°C)					Sessile drop	Z12
Ir	99.9980	2250	vac.							Pendant drop	5
K	99.895	101	Ar							Bubble pressure	189
		110.3±1	–								84
		117	vac.		$\sigma = 117-0.66\,(t-t_{mp})$					Drop weight	185
	–		Ar	87	112	457	80	677	64.8	Bubble pressure	24
	99.986	116.95	Ar		$\sigma = 116.95-6.742 \cdot 10^{-2}\,(t-t_{mp})-3.836 \cdot 10^{-5}\,(t-t_{mp})^2$ $+3.707 \cdot 10^{-8}\,(t-t_{mp})^3$ (t°C) (77-983°C)					Bubble pressure	172
	99.936	(79.2)	He		$(\sigma = 76.8-70.3 \cdot 10^{-4}\,(t-400))$ (t°C) (600-1126°C)					Bubble pressure	121
		95±9.5	–							Drop weight	Z9
	99.97	111.35 ±0.64	He		$\sigma = 115.51-0.0653 \cdot$ (t°C) (70-713°C)					Bubble pressure	Z10
Li	99.95		Ar	180	397.5	300	380	500	351.5	Bubble pressure	189
	99.98		Ar	287	386	922	275	1077	253	Bubble pressure	24
Mg			Ar	681	563	789	532	894	502	Bubble pressure	208
	99.8		N_2	670	552	700	542	740	528	Bubble pressure	149
	99.9		Ar	700	550±15					Bubble pressure	114
	99.91	(525±10)	Ar							Bubble pressure	105
	99.5	583			$\sigma = 721-0.149\,T$ (T°K) (1125-1326°K)					Bubble pressure	25
Mn	99.9985	1100 ± 50	Ar							Dynam. drop weight	6
	99.94		vac.	1550	1030					Sessile drop	159
	–		–	1550	1010					Sessile drop	Z11
Mo	99.7	(1915)	vac.							Pendant drop	148
		2080	vac.							Drop weight	139
	99.9996	2250	vac.							Pendant drop	5
	99.98	2049	vac.							Pendant drop	65
	–	2130	vac.							Pendant drop	Z1
Na			Ar	110	205.7	263	198.2			Bubble pressure	208
	99.995	191	Ar							Bubble pressure	188
			vac.	123	198	129	198.5			Drop volume	2
				140	190					Sessile drop	28
		200.2±0.6								–	84
		202	vac.		$\sigma = 202-0.092\,(t-t_{mp})$; 100-1000C					Drop weight	185
	p.a.		Ar	617	144	764	130	855	120.4	Bubble pressure	24
	99.96	210.12	Ar		$\sigma = 210.12-8.105 \cdot 10^{-2}\,(t-t_{mp})-8.064 \cdot 10^{-5}\,(t-t_{mp})^2$ $+3.380 \cdot 10^{-8}\,(t-t_{mp})^3$ (t°C) (141-992°C)					Bubble pressure	172
	99.982	187.4	He		$\sigma = 144-0.108\,(t-500)$ (t°C) (400-1125°C)					Bubble pressure	121
Nb, Cb	99.9986	1900	vac.							Pendant drop	5
	99.99	(1827)	vac.							Pendant drop	65
	–	2020	vac.							Pendant drop	Z1
Nd		688	Ar	1186	674					Bubble pressure	124
Ni	99.7		He	1470	(1615)					Sessile drop	94
	99.7		H_2	1470	(1570)					Sessile drop	94
	99.7		vac.	1470	1735					Sessile drop	94
		1725	vac.							Sessile drop	141
			vac.	1475	1725					Sessile drop	115
	–		Ar	1550	(1934)					Sessile drop	108
	99.99		vac., Al_2O_3	1520	1740					Sessile drop	46
	99.99		He, Ar, Al_2O_3	1520	1770					Sessile drop	46
	99.99		H_2, Al_2O_3	1520	(1600)					Sessile drop	46
	99.99		He, MgO	1470	(1530)					Sessile drop	46
	99.99		He, BeO	1470	(1500)					Sessile drop	46
	99.99		H_2	1530	(1650)					Bubble pressure	46

SURFACE TENSION OF LIQUID ELEMENTS (*Continued*)

Element	Purity (wt.-%)	σ_{mp} (dyn/cm)	Atm.	σ_{t1} t_1 °C	σ (dyn/cm)	σ_{t2} t_2 °C	σ (dyn/cm)	σ_{t3} t_3 °C	σ (dyn/cm)	Method	Ref.
	99.99		H_2	1470	(1530)					Bubble pressure	46
	99.99		He	1470	(1490)					Bubble pressure	46
		1725	vac.							Sessile drop	7
	99.99			1600	(1600)					Sessile drop	135
			vac.	1500	1720					Sessile drop	50
	99.99		vac., Al_2O_3	1550	1780					Sessile drop	54
				1550	1735					Bubble pressure	60
			H_2, He	1470	1700					Bubble pressure	61
	99.99		vac.	1640	1705					Sessile drop	56
	–		vac., Al_2O_3	1560	1810					Sessile drop	207
	99.9991	1770±13	vac.							Drop weight	5
	99.9991	1728±10	vac.							Drop weight	5
	99.9991	1822±8	vac.							Pendant drop	5
		(1670)	vac.							Drop weight	139
		1760	vac.							Sessile drop	55
		(1687)	vac.							Pendant drop	65
	–		He	1500	1745					Sessile drop	57
	–	1809±20	H_2, He Al_2O_3	$\sigma = 1770-0.39\,(t-1550)\,(t°C)$						Sessile drop	157
	99.99975	(1977)	He	$\sigma = 1665+0.215t\,(t°C)\,(1475-1650°C)$						Oscillating drop	Z2
Os	99.9998	2500	vac.							Pendant drop	5
P(white)				50	69.7	68.7	64.95			Bubble pressure	80
Pb	99.98		H_2, N_2	340	448	390	442	440	439	Bubble pressure	71
			air	360	452					Ring removal	97
		451	vac.	425	440					Drop pressure	101
		450	He							Pendant drop	87
				350–450	450					Ring removal	13
	99.998	480	H_2							Capillary method	133
				623	474					Capillary method	77
			vac.	362	455					Drop pressure	69
				700	428					Sessile drop	27
	99.9		H_2	1000	388					Bubble pressure	134
	99.9	(410±5)	Ar							Bubble pressure	104
				350	445					Drop pressure	153
	99.98		vac.	340	442	400	435			Drop pressure	70
	99.9995	470		$\sigma = 538-0.114\,T\,(T°K)\,(1440-1970°K)$						Bubble pressure	26
	99.9994		vac.	450	438					Drop pressure	107
	99.999		He	1600	310					Sessile drop	199
				390	456					Bubble pressure	1
		424±10	Air							Solid state curvature	Z8
	99.999	470	Ar	$\sigma_t = 470-0.164\,(t-t_{mp})\,(MP-535°C)\,(t°C)$						Bubble pressure	Z13a
Pd		1470	vac.							Sessile drop	50
	99.998	1500	vac.							Pendant drop	5
	99.998	1460	He							Sessile drop	199
Pt		1869	CO_2							Drop weight	167
	99.84	(1740±20)	vac.							Drop volume	48
	99.999		Ar	1800	(1699±20)					Sessile drop	109
	99.9980	1865	vac.							Pendant drop	5
Pu		550±55								–	186
Rb		(77±5)	vac.							Drop diffusion in quartz tube	201
	99.8		Ar	52	84	477	55	632	46.8	Bubble pressure	24
	99.92	91.17	Ar	$\sigma = 91.17-9.189\cdot10^{-2}\,(t-t_{mp})+7.228\cdot10^{-5}\,(t-t_{mp})^2$ $-3.830\cdot10^{-8}\,(t-t_{mp})^3\,(t°C)\,(104-1006°C)$						Bubble pressure	172
	99.997	85.7	He	$\sigma = 85.7-0.054\,(t-t_{mp})\,(t°C)\,(53-1115°C)$						Bubble pressure	121
Re	99.4	2610	vac.							Pendant drop	148
	99.9999	2700	vac.							Pendant drop	5
Ru	99.9980	2250	vac.							Pendant drop	5
Rh		1940	vac.							Sessile drop	50
	99.9975	2000	vac.							Pendant drop	5
S	–	60.9	vac.	250	51.1					Pendant drop	143

Element	Purity (wt.-%)	σ_{mp} (dyn/cm)	Atm.	σ_{t1} t_1°C	(dyn/cm)	σ_{t2} t_2°C	(dyn/cm)	σ_{t3} t_3°C	(dyn/cm)	Method	Ref.
Sb			H_2	640	349	700	349	974	342	Drop weight	20
			H_2	750	368	900	361	1100	348	Bubble pressure	40
			vac.	640	367.9	762	364.9			Drop weight	131
	99.5	383	H_2, N_2	675	384	800	380			Bubble pressure	71
	99.99	395±20	Ar							Bubble pressure	104
	99.15	395±20	Ar							Bubble pressure	105
	99.999		N_2	800	359	1000	351	1100	345	Bubble pressure	128
	99.995		Ar	650	350.2	700	347.6	800	345.0	Bubble pressure	123
	99.999		He	1600	320					Sessile drop	199
Se	—		Ar	230–250	88.0±5					Bubble pressure	105
Si			He	1450	725					Pendant drop	87
			vac.	1550	720					Sessile drop	68
	99.99		vac.	1550	750					Sessile drop	41
	99.9999		Ar	1500	825					Pendant drop	43
Sn	—		N_2	275	612	500	572	800	520	Bubble pressure	149
			air	280	523	340	520			Ring removal	97
	99.99	537	vac.	500	524	600	508			Drop pressure	154
		530	He							Pendant drop	87
			H_2	489	543	572	528	692	503	Conical capillaries	9
			—	250	536					Drop pressure	179
			—	450	530					Drop pressure	119
			—	250	545					Drop pressure	112
	99.93		vac.	250	549	400	539	600	526	Drop pressure	156
	99.998	566	H_2							Capillary method	133
			—	623	559					Capillary method	77
		610	vac.							Pendant drop	182
			—	800	500					Sessile drop	7
			—	300	538					Drop pressure	190
			—	300	(527)					Drop pressure	76
			—	290	546					Sessile drop	190
	99.99		H_2, He	600	530					Bubble pressure	62
	99.9	(526±10)	Ar							Bubble pressure	104
			vac.	290	600					Sessile drop	147
	99.965		H_2	740	508	950	489.5	1115	479.5	Bubble pressure	127
	99.89	543.7								Bubble pressure	33
		562	vac.							Sessile drop	55
			vac.	300	554					Sessile drop	64
	99.999	590	vac.							Sessile drop	86
			H_2	290	(520)	290	(524)	(vac.)		Sessile drop	67
	99.9999		H_2	246	552.7					Sessile drop	203
	99.9994		vac.	350	537					Drop pressure	106
	99.999	555.8±1.9		$\sigma = 566.84 - 4.76 \cdot 10^{-1} t$ (t°C)						Bubble pressure	176
	99.96	552	vac.	1000	470					Sessile drop	Z5
	99.96	552	Ar	$\sigma_t = 552 - 0.167$ (t-t$_{mp}$) (MP-500°C) (t°C)						Bubble pressure	Z13a
Sr			Ar	775	288	830	282	893	282	Bubble pressure	125
	99.5	303	Ar	$\sigma = 392 - 0.085 T$ (T°K) (1152–1602°K)						Bubble pressure	25
Ta		2360	vac.							Pendant drop	88
		2030	vac.							Pendant drop	88
		1910	vac.							Drop weight	139
	99.9983	2150	vac.							Pendant drop	5
	99.9	(1884)	vac.							Pendant drop	65
Te	99.4	186±2	Ar							Bubble pressure	105
			vac.	460	178±1.5					Capillary method	184
			vac.	475	(162)					Electro-capillarity	117
		178		$\sigma = 178 - 0.024$ (t-t$_{mp}$) (t°C)						Bubble pressure	204
Ti	98.7	1510	vac.							Capillary method	44
	99.92	1390	Ar							Pendant drop	151
		1460	vac.							Drop weight	139
	99.9991	1650	vac.							Pendant drop	5
	99.0		vac.	1680	1576					Drop weight	191
	99.99999		vac.	1680	1588					Drop weight	191
	99.85	(1880)	vac.							Pendant drop	65
	99.69	1402	vac.							Pendant drop	65
Tl		464.5	Ar							Bubble pressure	192
			—	450	452					Drop pressure	119
	—		vac.	450	450					Electro-capillarity	117

Element	Purity (wt.-%)	σ_{mp} (dyn/cm)	Atm.	σ_{t1}		σ_{t2}		σ_{t3}		Method	Ref.
				t_1°C	σ (dyn/cm)	t_2°C	σ (dyn/cm)	t_3°C	σ (dyn/cm)		
	99.999	467		$\sigma = 536 - 0.119T$ (T°K) (1270–1695°K)						Bubble pressure	26
	99.999		vac.	450	450					Drop pressure	107
U		1500±75								–	186
		1550	Ar							Bubble pressure	34
	99.94	(1294)	vac.							Pendant drop	65
V	99.9977	1950	vac.							Pendant drop	5
	–	(1760)	vac.							Pendant drop	Z1
W	–	2310	vac.							Pendant drop	35
	99.9999	2500	vac.							Pendant drop	5
	99.8	2220	vac.							Pendant drop	148
	99.9	(2000)	vac.							Pendant drop	65
Zn	99.9	750 ± 20	Ar							Bubble pressure	104
	99.99	757.0 ± 5	vac.							Sessile drop	202
	99.999	761.0	vac.							Sessile drop	202
	99.9999	767.5	vac.							Sessile drop	202
Zr		1400	Ar							Drop weight	151
	99.5	1411 ± 70	vac.							Drop weight	180
	99.9998	1480	vac.							Pendant drop	5
	99.7	(1533)	vac.							Pendant drop	65

SURFACE TENSION OF LIQUID HALOGENS

John Fredrickson

Element	Purity (wt.-%)	σ_{mp} (dyn/cm)	Atm.	σ_{t1}		σ_{t2}		σ_{t3}		Method	Ref.
				t_1°C	σ (dyn/cm)	t_2°C	σ (dyn/cm)	t_3°C	σ (dyn/cm)		
F			Vac.	-203.16	17.9	-197.86	16.2	-192.16	14.6	Capillary method	210
		22.33	Vac.	$\sigma = -39.10 - 0.2797(t°C)$						Calculated	211
Cl			Air	-61.5	31.61	-44.5	28.38	-28.7	25.23	Capillary method	212
		39.19	Air	$\sigma = 19.736 - 0.1927(t°C)$						Calculated	211
Br	99.999		Air	0	45.5	25	40.9	50	36.4	Capillary method	213
		46.8	Air	$\sigma = 45.5 - 0.182(t°C)$						Calculated	213
I	99.98	55.7	N_2	130	53.1	150	50.1	170	47.7	Bubble pressure	211
				$\sigma = 70.74 - 0.135(t°C)$						Calculated	211

SURFACE TENSION OF LIQUID ELEMENTS (*Continued*)

References

1. Abdel-Aziz Abol Hassan, Neue Hütte, **15**, 304 (1970).
2. Addison, Addison, Kerridge and Lewis, J. Chem. Soc., 2262 (1955).
3. Addison, Coldrey and Pulham, J. Chem. Soc., 1227 (1963).
4. Addison and Raynor, J. Chem. Soc., 965 (1966).
5. Allen, Trans. AIME, **227**, 1175 (1963).
6. Allen, Trans. AIME, **230**, 1357 (1964).
7. Allen and Kingery, Trans. AIME, **215**, 30 (1959).
8. Astakhov, Penin and Dobkina, Zh. Fiz. Khim., **20**, 403 (1946).
9. Atterton and Hoar, J. Inst. Met., **81**, 541 (1953).
10. Ayushina, Levin and Geld, Zh. Fiz. Khim., **42**, 2799 (1968).
11. Baes and Kellogg, J. Metals, **15**, 643 (1953).
12. Baker and Gilbert, J. Amer. Chem. Soc., **62**, 2479 (1940).
13. Bakradse and Pines, Zh. Tekhn. Fiz., **23**, 1548 (1953).
14. Becker, Harders and Kornfeld, Arch. f. Eisenh., **20**, 363 (1949).
15. Belforti and Lepie, Trans. AIME, **227**, 80 (1963).
16. Bergh, J. Electrochem. S., **109**, 1199 (1962).
17. Bering and Ioileva, Doklady A.N., **93**, 85 (1953).
18. Biery and Oblak, Ind. Eng. Chem., Fund., **5**, 121 (1966).
19. Bircumshaw, Phil. Mag., **2**, 341 (1926).
20. Bircumshaw, Phil. Mag., **3**, 1286 (1927).
21. Bircumshaw, Phil. Mag., **6**, 510 (1928).
22. Bircumshaw, Phil. Mag., **12**, 596 (1931).
23. Bobyk, Przemysl Chem., **39**, 423 (1960).
24. Bohdanski and Schins, J. Inorg. Nucl. Chem., **29**, 2173 (1967).
25. Bohdanski and Schins, J. Inorg. Nucl. Chem., **30**, 2331 (1968).
26. Bohdanski and Schins, J. Inorg. Nucl. Chem., **30**, 3362 (1968).
27. Bradhurst and Buchanan, J. Phys. Chem., **63**, 1486 (1959).
28. Bradhurst and Buchanan, Austral. J. Chem., **14**, 397 (1961).
29. Bradhurst and Buchanan, Austral. J. Chem., **14**, 409 (1961).
30. Bradley, J. Phys. Chem., **38**, 234 (1934).
31. Brown, Phil. Mag., **13**, 578 (1932).
32. Burdon, Trans. Farad. Soc., **28**, 866 (1932).
33. Cahill and Kirshenbaum, J. Inorg. Nucl. Chem., **26**, 206 (1964).
34. Cahill and Kirshenbaum, J. Inorg. Nucl. Chem., **27**, 73 (1965).
35. Calverley, Proc. Phys. Soc., **70**, 1040 (1957).
36. Coffman and Parr, Ind. Eng. Chem., **19**, 1308 (1927).
37. Cook, Phys. Review, **34**, 513 (1929).
38. de L. Davies and West, J. Inst. Met., **92**, 208 (1964).
39. Didenko and Pokrovski, Doklady A.N., **31**, 233 (1941).
40. Drath and Sauerwald, Z. allg. anorg. Chem., **162**, 301 (1927).
41. Dshemilev, Popel and Zarevski, Fiz. Met.i Met., **18**, 83 (1964).
42. Dyson, Trans. AIME, **227**, 1098 (1963).
43. Eljutin, Kostikov and Levin, Izv. Vys. Uch. Sav., Tsvetn. Met., **2**, 131 (1970).
44. Eljutin and Maurakh, Izv. A.N., OTN, **4**, 129 (1956).
45. Eremenko, Ivashchenko and Bogatyrenko, The Role of Surface Phenomena in Metallurgy, 37 (1963).
46. Eremenko, Ivashchenko, Fessenko and Nishenko, Izv. A.N., OTN, **7**, 144 (1958).
47. Eremenko and Naidich, Izv. A.N., OTN, **2**, 111 (1959).
48. Eremenko and Naidich, Izv. A.N., OTN, **6**, 129 (1959).
49. Eremenko and Naidich, Izv. A.N., OTN, **6**, 100 (1961).
50. Eremenko and Naidich, Izv. A.N., OTN, **2**, 53 (1960).
51. Eremenko and Naidich, The Role of Surface Phenomena in Metallurgy, 65 (1963).
52. Eremenko, Naidich and Nossonovich, Zh. Fiz. Khim., **34**, 1018 (1960).
53. Eremenko and Nishenko, Ukr. Khim. Zh., **26**, 423 (1960).
54. Eremenko and Nishenko, Zh. Fiz. Khim., **35**, 1301 (1961).
55. Eremenko and Nishenko, Ukr. Khim. Zh., **30**, 125 (1964).
56. Eremenko, Nishenko and Naidich, Izv. A.N., OTN, **3**, 150 (1961).
57. Eremenko, Nishenko and Skljarenko, Izv. A. N., OTN, **2**, 188 (1966).

References (Continued)

58. Eremenko, Nishenko and Taj-Shou-Vej., Izv. A.N., OTN, **3**, 116 (1960).
59. Eremenko and Vassiliu, Ukr. Khim. Zh., **31**, 557 (1965).
60. Eremenko, Vassiliu and Fessenko, Zh. Fiz. Khim., **35**, 1750 (1961).
61. Fessenko, Zh. Fiz. Khim., **35**, 707 (1961).
62. Fessenko and Eremenko, Ukr. Khim. Zh., **26**, 198 (1960).
63. Fessenko, Vassiliu and Eremenko, Zh. Fiz. Khim., **36**, 518 (1962).
64. Flechsig, Thesis, Techn. Univ. Berlin, (1964).
65. Flint, J. Nucl. Mat., **16**, 260 (1965).
66. Gans, Pawlek and Roepenack, Z. Metallkunde, **54**, 147 (1963).
67. Gans and Parthey, Z. Metallkunde, **57**, 19 (1966).
68. Geld and Petrushevski, Izv. A.N., OTN, **3**, 160 (1961).
69. Gratzianski and Rjabov, Zh. Fiz. Khim., **33**, 487, 1253 (1959).
70. Gratzianski, Rjabov and Tobolich, Ukr. Khim. Zh., **29**, 1219 (1963).
71. Greenaway, J. Inst. Met., **74**, 133 (1947).
72. Grunmach, Ann.d. Physik, **3**, 660 (1900).
73. Hagemann, Thesis, Univ. of Freiburg/Br., (1914).
74. Halden and Kingery, J. Phys. Chem., **59**, 557 (1955).
75. Harkins and Ewing, J. Amer. Chem., Soc., **42**, 2539 (1920).
76. Herczynska, Z. Phys. Chem., **214**, 355 (1960).
77. Hoar and Melford, Trans. Farad. Soc., **53**, 315 (1957).
78. Hogness, J. Amer. Chem. Soc., **43**, 1621 (1921).
79. Humenik and Kingery, J. Amer. Ceram. Soc., **37**, 18 (1954).
80. Hutchinson, Trans. Farad. Soc., **39**, 229 (1943).
81. Iredale, Phil. Mag., **45**, 1088 (1923).
82. Iredale, Phil. Mag., **48**, 177 (1924).
83. Jacobj, Thesis, Univ. of Braunschweig, (1962).
84. Jordan and Lane, Austral. J. Chem., **18**, 1711 (1965).
85. Karasaev, Sadumkin and Kukhno, Zh. Fiz. Khim., **41**, 654 (1967).
86. Kaufman and Whalen, Acta Metallurg., **13**, 797 (1965).
87. Keck and van Horn, Phys. Review, **91**, 512 (1953).
88. Kelly and Calverley, SERL-Report, **80**, 53 (1959).
89. Kemball, Trans. Farad. Soc., **42**, 526 (1946).
90. Kernaghan, Phys. Review, **37**, 990 (1931).
91. Kernaghan, Phys. Review, **49**, 414 (1936).
92. Kingery, J. Amer. Ceram. Soc., **37**, 42 (1954).
93. Kingery, Kolloid-Z., **161**, 95 (1958).
94. Kingery and Humenik, J. Phys. Chem., **57**, 359 (1953).
95. Kingery and Norton, AEC-Progr. Rep., **NYO-6296**, (1954).
96. Klyachko, Zavodskaja Labor., **6**, 1376 (1937).
97. Klyachko and Kunin, Zavodskaja Labor., **14**, 66 (1948).
98. Klyachko and Kunin, Doklady A.N., **64**, 64 (1949).
99. Kolesnikova, Izv. Vys. Uch. Sav., Tsvetn. Met., **9** (1960).
100. Kolesnikova and Samarin, Izv. A.N., OTN, **5**, 63 (1956).
101. Konstantinov, Thesis, State Univ. Moscow, (1950).
102. Korolkov, Izv. A.N., OTN, **2**, 35 (1956).
103. Korolkov: Litein'e Svojstva Metallov i Splavov, Isdatelstvo, A.N., SSSR, 37 (1960).
104. Korolkov and Bychkova, Issled. Splav. Tsvetn. Met., **2**, 122 (1960).
105. Korolkov and Igumnova, Izv. A.N., OTN, **6**, 95 (1961).
106. Kovalchuk, Kusnezov and Kotlovanova, Zh. Fiz. Khim., **42**, 1754 (1968).
107. Kovalchuk, Kusnezov and Butuzova, Zh. Fiz. Khim., **42**, 2265 (1968).
108. Kozakevitch and Urbain, J. Iron-& Steel-Inst., **186**, 167 (1957).
109. Kozakevitch and Urbain, C.R., Paris, **253**, 2229 (1961).
110. Kozakevitch and Urbain, Mém. Sci. Rev. Met., **58**, 401, 517, 931 (1961).
111. Krause, Sauerwald and Michalke, Z. anorg. allg. Chem., **181**, 353 (1929).
112. Kristian, Thesis, State Univ. Moscow, (1954).
113. Kubichek, Izv. A.N., OTN, **2**, 96 (1959).
114. Kubichek and Malzev, Izv. A.N., OTN, **3**, 144 (1959).
115. Kurkjian and Kingery, J. Phys. Chem., **60**, 961 (1956).
116. Kurochkin, Baum and Borodulin, Fiz. Met.i Met., **15**, 461 (1963).

117. Kusnezov, The Role of Surface Phenomena in Metallurgy, 72 (1963).
118. Kusnezov, Djakova and Malzeva, Zh. Fiz. Khim., 33, 1551 (1959).
119. Kusnezov, Kochergin, Tishchenko and Posdynsheva, Doklady A.N., 92, 1197 (1953).
120. Kusnezov, Popova and Duplina, Zh. Fiz. Khim., 36, 880 (1962).
121. Kyrianenko and Solovev, Teplofiz. Vysok. Temp., 8, 537 (1970).
122. Lasarev, Zh. Fiz. Khim., 36, 405 (1962).
123. Lasarev, Zh. Fiz. Khim., 38, 325 (1964).
124. Lasarev and Pershikov, Doklady A.N., 146, 143 (1962).
125. Lasarev and Pershikov, Zh. Fiz. Khim., 37, 907 (1963).
126. Lasarev and Pugachevich, Doklady A.N., 134, 132 (1960).
127. Lauermann, Metzger and Sauerwald, Z. Phys. Chem., 216, 42 (1961).
128. Lauermann and Sauerwald, Z. Metallkunde, 55, 605 (1964).
129. Lucas, C.R., Paris, 248, 2336 (1959).
130. Mack, Davis and Bartell, J. Phys. Chem., 45, 846 (1941).
131. Matuyama, Sci. Rep. RITU, 16, 555 (1927).
132. Melik-Gajkazan, Woronchikhina and Sakharova, Elektrokhim., 4, 1420 (1968).
133. Melford and Hoar, J. Inst. Met., 85, 197 (1957).
134. Metzger, Z. Phys. Chem., 211, 1 (1959).
135. Monma and Suto, J. Jap. Inst. Met., 24, 167 (1960).
136. Monma and Suto, J. Inst. Met., 1, 69 (1960).
137. Naidich and Eremenko, Fiz. Met.i Met., 11, 883 (1961).
138. Naidich, Eremenko, Fessenko, Vassiliu and Kirichenko, Zh. Fiz. Khim., 35, 694 (1961).
139. Namba and Isobe, Sci. Pap. Inst. Phys. Chem. Res., Tokyo, 57, 5154 (1963).
140. Nicholas, Joyner, Tessem and Olson, J. Phys. Chem., 65, 1375 (1961).
141. Norton and Kingery, AEC-Progr. Rep., NYO 4632, (1955).
142. Olson and Johnson, J. Phys. Chem., 67, 2529 (1963).
143. Ono and Matsushima, Sci. Rep. RITU, 9, 309 (1957).
144. Oppenheimer, Z. anorg. allg. Chem., 171, 98 (1928).
145. Osminin, Zh. Fiz. Khim., 43, 2610 (1969).
146. Palacios, An. Soc. Esp. Fis., 18, 294 (1920).
147. Parthey, Thesis, Techn. Univ. Berlin, (1961).
148. Pekarev, Izv. Vys. Uch. Sav., Tsvetn. Met., 6, 111 (1963).
149. Pelzel, Berg-u. Hütt. Mon. Hefte, Leoben, 93, 248 (1948).
150. Pelzel, Berg-u. Hütt. Mon. Hefte, Leoben, 94, 10 (1949).
151. Peterson, Kedesdy, Keck and Schwarz, J. Appl. Phys., 29, 213 (1958).
152. Poindexter, Phys. Review, 27, 820 (1926).
153. Pokrovski, Ukr. Khim. Zh., 7, 845 (1962).
154. Pokrovski and Galanina, Zh. Fiz. Khim., 23, 324 (1949).
155. Pokrovski and Kristian, Zh. Fiz. Khim., 28, (1954).
156. Pokrovski and Saidov, Fiz. Met.i Met., 2, 546 (1956).
157. Popel, Shergin and Zarevski, Zh. Fiz. Khim., 43, 2365 (1969).
158. Popel, Smirnov, Zarevski, Dshemilev and Pastukhov, Izv. A.N., 1, 62 (1965).
159. Popel, Zarevski and Dshemilev, Fiz. Met.i Met., 18, 468 (1964).
160. Popesco, C.R., Paris, 172, 1474 (1921).
161. Portevin and Bastien, C.R., Paris, 202, 1072 (1936).
162. Pugachevich, Zh. Fiz. Khim., 25, 1365 (1951).
163. Pugachevich and Altynov, Doklady A.N., 86, 117 (1952).
164. Pugachevich and Yashkichev, The Role of Surface Phenomena in Metallurgy, 46 (1963).
165. Quincke, Ann. d. Physik, 134, 356 (1868).
166. Quincke, Ann. d. Physik, 135, 621 (1868).
167. Quincke, Ann. d. Physik, 138, 141 (1869).
168. Quincke, Ann. d. Physik, 61, 267 (1897).
169. Raue, Metzger and Sauerwald, Metall, 20, 1040 (1966).
170. Richards and Boyer, J. Amer. Chem. Soc., 43, 290 (1921).
171. Roberts, J. Chem. Soc., 1907 (1964).
172. Roehlich jun., Tepper and Rankin, J. Chem. Eng. Data, 13, 518 (1968).
173. Sauerwald and Drath, Z. anorg. allg. Chem., 154, 79 (1926).
174. Sauerwald, Schmidt and Pelka. Z. anorg. allg. Chem., 223, 84 (1935).

175. Schmidt, Ann. d. Physik, **39**, 1108 (1912).
176. Schwaneke and Falke, US-Bur. Min., Inv. Rep. No. 7372, (1970).
177. Schwaneke, Falke and Miller, US-Bur. Min., Inv. Rep. No. 7340, (1970).
178. Semenchenko and Pokrovski, Uspekhii Khim., **6**, 945 (1937).
179. Semenchenko, Pokrovski and Lasarev, Doklady A.N., **89**, 1021 (1953).
180. Shunk and Burr, Trans. ASM, **55**, 786 (1962).
181. Siedentopf, Ann. d. Physik, **61**, 235 (1897).
182. Smirnova and Ormont, Zh. Fiz. Khim., **33**, 771 (1959).
183. Smith, J. Inst. Met., **12**, 168, 20 (1914).
184. Smith and Spitzer, J. Phys. Chem., **66**, 946 (1962).
185. Solovev and Makarova, Teplofiz. Vysok. Temp., **4**, 189 (1966).
186. Spriet, Mém. Sci. Rev. Mét., **60**, 531 (1963).
187. Tavadse, Bairamashvili, Khantadse and Zagareishvili, Doklady A.N., **150**, 544 (1963).
188. Taylor, J. Inst. Met., **83**, 143 (1954).
189. Taylor, Phil. Mag., **46**, 867 (1955).
190. Thyssen, Thesis, Humboldt-Univ. Berlin, (1960).
191. Tille and Kelly, Brit. J. Appl. Phys., **146**, 717 (1963).
192. Timofeyevicheva and Lasarev, Doklady A.N., **138**, 412 (1961).
193. Timofeyevicheva and Lasarev, Doklady A.N., 358 (1962).
194. Timofeyevicheva and Lasarev, Kolloidnyi-Zh., **24**, 227 (1962).
195. Timofeyevicheva, Lasarev and Pershikov, Doklady A.N., **143**, 618 (1962).
196. Timofeyevicheva and Pugachevich, Doklady A.N., **124**, 1093 (1959).
197. Timofeyevicheva and Pugachevich, Doklady A.N., **134**, 840 (1960).
198. Urbain and Lucas, Proc. N.P.L., Teddington, **9**, Pap. 4E (1959).
199. Watolin, Esin, Ukhov and Dubinin, Trudy Inst. Met. Sverdlovsk, **10**, 73 (1962).
200. Whalen and Humenik, Trans. AIME, **218**, 952 (1960).
201. Wegener, Z. Physik, **143**, 548 (1956).
202. White, Trans. AIME, **236**, 796 (1966).
203. White, Met. Review, **124**, 73 (1968).
204. Wobst and Rentzsch, Z. Phys. Chem., **240**, 36 (1969).
205. Yashkichevich and Lasarev, Izv. A.N., OTN, **1**, 170 (1964).
206. Yung Lee, Ind. Eng. Chem., Prod. Res. Dev., **7**, 66 (1968).
207. Zarevski and Popel, Fiz. Met.i Met., **13**, 451 (1962).
208. Zhivov, Trudy WAMI, SSSR, **14**, 99 (1937).
209. Zsin-Tan Wan, Karassev and Samarin, Izv. A.N., OTN, **1**, 30, 49 (1960).
210. Elverum and Doescher, J. Chem. Phys., **20**, 1834 (1952).
211. Fredrickson, J. Colloid Interfac. Sci., **48**, 506 (1974).
212. Johnson and McIntosh, J. Am. Chem. Soc., **31**, 1138 (1909).
213. Chao and Stenger, Talanta, **11**, 271 (1963).

Addendum

Z1. Eljutin, Kostikow and Penkow, Poroshk. Met., **9**, 46 (1970).
Z2. Fraser, Lu, Hamielec and Murarka, Met. Trans., **2**, 817 (1971).
Z3. Bernard and Lupis, Met. Trans., **2**, 555 (1971).
Z4. Rhee, J. Am. Ceram. Soc., **53**, 386 (1970).
Z5. Naidich, Perevertailo and Shuravlev, Zh. Fiz. Chim., **45**, 991 (1971).
Z6. Körber and Löhberg, Gießereiforschung, **23**, 173 (1971).
Z7. Ziesing, Austral. J. Phys., **6**, 86 (1953).
Z8. Sangster and Carman, J. Chem. Phys., **23**, 1142 (1955).
Z9. Primak and Quarterman, J. Phys. Chem., **58**, 1051 (1954).
Z10. Cooke, HTLMHTTM, Oak Ridge, Vol. I, p. 66, (Nov. 1964).
Z11. Ofizerow, Izv. A.N., Metally, **4**, 91 (1971).
Z12. White, Met. Trans., **3**, 1933 (1972).
Z13. Lang, Aluminum (Germany), **49**, 231 (1972).
Z13a. Lang, J. Inst. Met. (to be published).
Z14. **Perry and Roberts**, *J. Chem. Eng. Data*, **26**, 266 (1981).

SURFACE TENSION

SURFACE TENSION OF INORGANIC SOLUTES IN WATER

% = Weight % of solute
γ = Surface tension in dynes/cm.

Solute	T°C								
HCl	20	%	1.78	3.52	6.78	12.81	16.97	23.74	35.29
		γ	72.55	72.45	72.25	71.85	71.75	70.55	65.75
HNO₃	20	%	4.21	8.64	14.99	34.87			
		γ	72.15	71.65	70.95	68.75			
H₂SO₄	25	%	4.11	8.26	12.18	17.66	21.88	29.07	33.63
		γ	72.21	72.55	72.80	73.36	73.91	74.80	75.29
HClO₄	25	%	4.86	10.01	20.38	30.36	53.74	63.47	72.25
		γ	71.18	70.34	69.21	68.57	69.02	69.73	69.01
KOH	18	%	2.73	5.31	10.08	17.57			
		γ	73.95	74.85	76.55	79.75			
NaOH	18	%	2.72	5.66	16.66	30.56	35.90		
		γ	74.35	75.85	83.05	96.05	101.05		
NH₄OH	18	%	1.72	3.39	4.99	9.51	17.37	34.47	54.37
		γ	71.65	70.65	69.95	67.85	65.25	61.05	57.05
KCl	20	%	0.74	3.60	6.93	13.88	18.77	22.97	24.70
		γ	72.99	73.45	74.15	75.55	76.95	78.25	78.75
LiCl	25	%	5.46	7.37	10.17	13.95			
		γ	74.23	75.10	76.30	78.10			

Solute	T°C								
NaCl	20	%	0.58	2.84	5.43	10.46	14.92	22.62	25.92
		γ	72.92	73.75	74.39	76.05	77.65	80.95	82.55
PbCl	25	%	21.57	28.52	37.74				
		γ	75.20	76.80	79.20				
BaCl₂	25	%	9.26	16.73	25.58				
		γ	73.50	74.93	76.38				
MgCl	20	%	0.94	4.55	8.69	16.00	22.30	25.44	
		γ	73.07	74.00	75.75	79.15	82.95	85.75	
NaBr	20	%	4.89	9.33	13.37	23.00			
		γ	73.45	74.05	74.75	76.55			
Al₂(SO₄)₃	25	%	2.54	4.06	9.40	14.60	19.32	23.54	25.50
		γ	72.32	72.92	73.51	74.71	76.06	78.30	79.73
MgSO₄	20	%	1.19	5.68	10.75	19.41	24.53		
		γ	73.01	73.78	74.85	77.35	79.25		
Na₂SO₄	20	%	2.76	6.63	12.44				
		γ	73.25	74.15	75.45				
Na₂CO₃	20	%	2.58	5.03	9.59	13.72			
		γ	73.45	74.05	75.45	76.75			
NaNO₃	20	%	0.85	4.08	7.84	14.53	29.82	37.30	47.06
		γ	72.87	73.75	73.95	75.15	78.35	80.25	87.05

SURFACE TENSION OF INORGANIC SOLUTES IN ORGANIC SOLVENTS

% = Mol %
γ = Surface tension in dynes/cm.

Solute	Solvent	T°C							
LiCl	Ethyl alcohol	14	%	0.72	2.30	4.62			
			γ	22.90	23.17	23.26			
LiBr	Ethyl alcohol	14	%	0.95	2.60				
			γ	23.08	23.35				
LiI	Ethyl alcohol	14	%	1.43	2.87	5.08	10.21	19.47	26.92
			γ	23.11	23.56	24.39	26.03	28.87	31.95
KI	Methyl alcohol	14	%	0.81	1.52	2.68			
			γ	23.76	24.11	24.71			
CaCl₂	Ethyl alcohol	24	%	0.94	1.98	3.79	7.47		
			γ	22.62	22.58	23.23	23.97		
NaI	Methyl alcohol	14	%	0.76	1.48	4.33	8.55	12.53	
			γ	22.83	23.29	24.85	27.41	29.75	

Solute	Solvent	T°C							
NaI	Ethyl alcohol	24	%	0.45	1.80	3.63	4.54	6.02	10.46
			γ	22.47	22.82	23.41	23.52	24.00	25.07
NaI	Acetone	14	%	0.93	2.08	5.07	6.53		
			γ	24.22	24.40	25.04	25.12		
ZnI₂	Methyl alcohol	22	%	0.90	2.79	5.07			
			γ	22.97	24.23	25.84			
ZnI₂	Ethyl alcohol	24	%	0.41	1.72	3.42	6.90		
			γ	22.70	22.90	23.71	25.49		
H₂SO₄	Ethyl ether	17	%	3	10	40	75	90	
			γ	17.30	19.55	32.50	46.30	46.83	
H₂SO₄	Nitrobenzene	17	%	3	10	40	75	90	
			γ	42.73	43.96	46.05	47.52	48.25	

SURFACE TENSION OF ORGANIC COMPOUNDS IN WATER

% = Weight % of solute
γ = Surface tension in dynes/cm.

Solute	T°C								
Acetic acid	30	%	1.00	2.475	5.001	10.01	30.09	49.96	69.91
		γ	68.00	64.40	60.10	54.60	43.60	38.40	34.30
Acetone	25	%	5.00	10.00	20.00	50.00	75.00	100.00	
		γ	55.50	48.90	41.10	30.40	26.80	24.20	23.00
Acetonitrile	20	%	1.13	3.35	11.77	20.20	37.58	61.33	81.22
		γ	69.02	63.03	47.61	39.06	31.84	30.02	29.02
o-Aminobenzoic acid	25	%	12.35	22.36	30.45	37.44			
		γ	71.96	73.23	74.54	75.79			
m-Aminobenzoic acid	25	%	12.35	22.36	30.45	37.44			
		γ	73.30	74.59	76.16	77.89			
p-Aminobenzoic acid	25	%	12.35	22.36	30.45	37.44			
		γ	73.38	74.79	76.32	78.20			
Aminobutyric acid	25	%	4.96	9.34	13.43				
		γ	71.91	71.67	71.40				
Ammonium lactate	29	%	30.00	50.00	60.00	70.00	80.00	90.00	
		γ	35.40	34.40	35.40	35.60	38.20	44.50	
n-Butanol	30	%	0.04	0.41	9.53	80.44	86.05	94.20	97.40
		γ	69.33	60.38	26.97	23.69	23.47	23.29	22.25
n-Butyric acid	25	%	0.14	0.31	1.05	8.60	25.00	79.00	100.00
		γ	69.00	65.00	56.00	33.00	28.00	27.00	26.00
Dioxan	26	%	0.44	2.20	4.70	11.14	20.17	35.20	55.00
		γ	69.83	65.64	62.45	56.90	51.57	45.30	39.27
Dioxan	26	%	67.68	76.45	83.02	91.90	95.60	97.77	
		γ	36.95	35.80	35.00	33.95	33.60	33.10	

Solute	T°C								
Formic acid	30	%	1.00	5.00	10.00	25.00	50.00	75.00	100.00
		γ	70.07	66.20	62.78	56.29	49.50	43.40	36.51
Glycerol	18	%	5.00	10.00	20.00	30.00	50.00	85.00	100.00
		γ	72.90	72.90	72.40	72.00	70.00	66.00	63.00
Glycine	25	%	3.62	6.98	10.12	13.10			
		γ	72.54	73.11	73.74	74.18			
Hydrocinnamic acid	21.5	%	7.02	12.62	18.39	26.09	31.06	38.25	47.93
		γ	69.08	66.49	63.63	59.25	56.14	52.96	47.24
Methyl acetate	25	%	0.66	1.29	2.29	3.56			
		γ	66.33	62.92	58.22	55.08			
Morpholine	20	%	8.56	19.39	30.41	50.45	69.93	80.14	92.00
		γ	67.80	62.62	59.15	52.85	47.05	43.62	41.60
Potassium lactate	29	%	40.00	50.00	60.00	70.00			
		γ	66.40	66.40	65.40	63.40			
Phenol	20	%	0.024	0.047	0.118	0.417	0.941	3.76	5.62
		γ	72.60	72.20	71.30	66.50	61.10	46.00	42.30
n-Propanol	25	%	0.1	0.5	1.0	50.0	60.0	80.0	90.0
		γ	67.10	56.18	49.30	24.34	24.15	23.66	23.41
Propionic acid	25	%	1.91	5.84	9.80	21.70	49.80	73.90	100.00
		γ	60.00	49.00	44.00	36.00	32.00	30.00	26.00
Sodium lactate	29	%	1.00	10.00	30.00	40.00	50.00	60.00	70.00
		γ	70.40	69.60	68.50	64.80	45.40	56.70	60.70
Sucrose	25	%	10.00	20.00	30.00	40.00	55.00		
		γ	72.50	73.00	73.40	74.10	75.70		

SURFACE TENSION OF ORGANIC COMPOUNDS
IN ORGANIC SOLVENTS

% = Weight % of solvent
γ = Surface tension dynes/cm.

Solute	Solvent	T°C	% γ					
Acetic acid	Benzene	35	%	10.45	25.53	34.28	43.93	68.77
			γ	25.40	25.21	25.32	25.43	25.99
Acetic acid	Acetone	25	%	25.63	50.83	75.62		
			γ	27.50	26.61	24.90		
Acetone	Ethyl ether	30	%	21.83	40.98	61.29	75.69	89.03
			γ	16.75	17.49	19.15	19.80	21.00
Acetonitrile	Ethyl alcohol	20	%	10.83	32.42	49.90	69.78	80.19
			γ	22.92	23.92	24.36	25.08	26.51
Aniline	Cyclohexane	32	%	14.35	37.65	50.67	72.28	96.46
			γ	24.21	24.51	24.50	25.61	37.45
Carbontetrachloride	Benzene	50	%	30.40	51.69	62.38	78.29	88.62
			γ	24.39	24.09	23.78	23.47	23.21
Cyclohexane	Nitrobenzene	15	%	5.00	10.00			
			γ	37.59	33.03			
Naphthalene	Benzene	79.5	%	20.00	40.00	50.00	60.00	80.00
			γ	23.42	26.70	27.80	29.20	31.70
Naphthalene	p-Nitrophenol	121	%	10.21	29.74	45.94	58.01	67.57
			γ	29.90	31.80	33.70	34.80	41.80

SURFACE TENSION OF METHYL ALCOHOL
IN WATER

% = Volume % of alcohol
γ = Surface tension dynes/cm.

T°C	%	7.5	10.00	25.0	50.0	60.0	80.0	90.0	100.0
20	γ	60.90	59.04	46.38	35.31	32.95	27.26	25.36	22.65
30	γ	59.33	57.27	45.30	34.52	32.26	26.48	24.42	21.58
50	γ	56.19	55.01	43.24	32.95	30.79	25.01	22.55	19.52

SURFACE TENSION OF ETHYL ALCOHOL
IN WATER

% = Volume % of alcohol
γ = Surface tension dynes/cm.

T°C	%	5.00	10.00	24.00	34.00	48.00	60.00	72.00	80.00	96.00
20	γ	—	—	—	33.24	30.10	27.56	26.28	24.91	23.04
40	γ	54.92	48.25	35.50	31.58	28.93	26.18	24.91	23.43	21.38
50	γ	53.35	46.77	34.32	30.70	28.24	25.50	24.12	22.56	20.40

WATER AGAINST AIR

Temperature °C	Surface tension dynes/cm.	Temperature °C	Surface tension dynes/cm.
−8	77.0	25	71.97
−5	76.4	30	71.18
0	75.6	40	69.56
5	74.9	50	67.91
10	74.22	60	66.18
15	73.49	70	64.4
18	73.05	80	62.6
20	72.75	100	58.9

INTERFACIAL TENSION

Surface Tension at the Interface Between Two Liquids
(Each liquid saturated with the other)

Liquids	Temperature °C	γ
Benzene — Mercury	20	357
Ethyl ether — Mercury	20	379
Water — Benzene	20	35.00
Water — Carbon tetrachloride	20	45.
Water — Ethyl ether	20	10.7
Water — Heptylic acid	20	7.0
Water — n-Hexane	20	51.1
Water — Mercury	20	375.
Water-n-Octane	20	50.8
Water-n-Octyl alcohol	20	8.5

SURFACE TENSION OF VARIOUS LIQUIDS

Name	Formula	In contact with	Temperature °C	Surface tension dynes/cm
Acetaldehyde	C_2H_4O	- -vapor	20	21.2
Acetaldoxime	C_2H_5NO	- -vapor	35	30.1
Acetamide	C_2H_5NO	- -vapor	85	39.3
Acetanilide	C_2H_5NO	- -vapor	120	35.6
Acetic acid	$C_2H_4O_2$	- -vapor	10	28.8

Substance Name	Formula	In contact with	Temperature °C	Surface tension dynes/cm
	C_2H_4O	--vapor	20	27.8
	C_2H_4O	--vapor	50	24.8
Acetic anhydride	$C_4H_6O_3$	--vapor	20	32.7
Acetone	C_3H_6O	--air or vapor	0	26.21
	C_3H_6O	--air or vapor	20	23.70
	C_3H_6O	--air or vapor	40	21.16
Acetonitrile	C_2H_3N	--vapor	20	29.30
Acetophenone	C_8H_8O	--vapor	20	39.8
Acetyl chloride	C_2H_3ClO	--vapor	14.8	26.7
Acetylene	C_2H_2	--vapor	−70.5	16.4
Acetylsalicylic acid (in aq. sol.)	$C_9H_8O_4$	--vapor	25.9	60.06
Allyl alcohol	C_3H_6O	--air or vapor	20	25.8
Allyl isothiocyanate	C_4H_5NS	--air or vapor	20	34.5
Ammonia	NH_3	--vapor	11.1	23.4
	NH_3	--vapor	34.1	18.1
Aniline	C_6H_7N	--air	10	44.10
	C_6H_7N	--vapor	20	42.9
	C_6H_7N	--air	50	39.4
Argon	A	--vapor	−188	13.2
Azoxybenzene	$C_{12}H_{10}N_2O$	--vapor	51	43.34
Benzaldehyde	C_7H_6O	--air	20	40.04
Benzene	C_6H_6	--air	10	30.22
	C_6H_6	--air	20	28.85
	C_6H_6	--saturated with vapor	20	28.89
	C_6H_6	--air	30	27.56
Benzonitrile	C_7H_5N	--air	20	39.05
Benzophenone	$C_{13}H_{10}O$	--air or vapor	20	45.1
Benzylamine	C_7H_9N	--vapor	20	39.5
Benzyl alcohol	C_7H_8O	--air or vapor	20	39.0
Bromine	Br_2	--air or vapor	20	41.5
Bromobenzene	C_6H_5Br	--air	20	36.5
Bromoform	$CHBr_3$	--vapor	20	41.53
p-Bromophenol	C_6H_5BrO	--vapor	74.4	42.36
d-sec-Butyl alcohol	$C_4H_{10}O$	--vapor	10	23.5
n-Butyl alcohol	$C_4H_{10}O$	--air or vapor	0	26.2
	$C_4H_{10}O$	--air or vapor	20	24.6
	$C_4H_{10}O$	--air or vapor	50	22.1
tert-Butyl alcohol	$C_4H_{10}O$	--air or vapor	20	20.7
n-Butylamine	$C_4H_{11}N$	--nitrogen	41	19.7
n-Butyric acid	$C_4H_8O_2$	--air	20	26.8
Carbon bisulfide	CS_2	--vapor	20	32.33
Carbon dioxide	CO_2	--vapor	20	1.16
	CO_2	--vapor	−25	9.13
Carbon tetrachloride	CCl_4	--vapor	20	26.95
	CCl_4	--vapor	100	17.26
	CCl_4	--vapor	200	6.53
Carbon monoxide	CO	--vapor	−193	9.8
	CO	--vapor	−203	12.1
Chloral	C_2HCl_3O	--vapor	19.4	25.34
Chlorine	Cl_2	--vapor	20	18.4
	Cl_2	--vapor	−30	25.4
	Cl_2	--vapor	−40	27.3
	Cl_2	--vapor	−50	29.2
	Cl_2	--vapor	−60	31.2
Chloroacetic acid	$C_2H_2Cl_2O_2$	--nitrogen	25.7	35.4
Chlorobenzene	C_4H_5Cl	--vapor	20	33.56
Chloroform	$CHCl_3$	--air	20	27.14
o-Chlorophenol	C_6H_5ClO	--vapor	12.7	42.25
Cyclohexane	C_6H_{12}	--air	20	25.5
Dichloroacetic acid	$C_2H_2Cl_2O_2$	--nitrogen	25.7	35.4
Dichloroethane	$C_2H_4Cl_2$	--air	35.0	23.4
Diethylamine	$C_4H_{11}N$	--air	56	16.4
Diethylaniline	$C_{10}H_{15}N$	--vapor	20	34.2
Diethyl carbonate	$C_5H_{10}O$	--air	20	26.31
Diethyl oxalate	$C_6H_{10}O_4$	--vapor	20	32.0
Diethyl phthalate	$C_{12}H_{14}O_4$	--vapor	20	37.5
Diethyl sulfate	$C_4H_{12}O_4S$	--air	13	34.61
Dimethylamine	C_2H_7N	--nitrogen	0	18.1
	C_2H_7N	--nitrogen	5	17.7
Dimethylaniline	C_8H_{11}	--air or vapor	20	36.6
1,5-Dimethyl-2-phenyl-3-pyrazolone	$C_{11}H_{12}N_2O$	--vapor	25.9	63.63
Dimethyl sulfate	$C_2H_6O_4S$	--air	18	40.12
Diphenylamine	$C_{12}H_{11}N$	--air or vapor	80	37.7
Ethyl acetate	$C_4H_8O_2$	--air	0	26.5
	$C_4H_8O_2$	--air	20	23.9
	$C_4H_8O_2$	--air	50	20.2
Ethyl acetoacetate	$C_6H_{10}O_3$	--air or vapor	20	32.51
Ethyl alcohol	C_2H_6O	--air	0	24.05
	C_2H_6O	--vapor	10	23.61
	C_2H_6O	--vapor	20	22.75
	C_2H_6O	--vapor	30	21.89
Ethylamine	C_2H_7N	--nitrogen	0	21.3
	C_2H_7N	--nitrogen	9.9	20.4

Substance		In contact with	Temperature °C	Surface tension dynes/cm
Name	Formula			
Ethylaniline	$C_8H_{11}N$	--air or vapor	20	36.6
Ethylbenzene	C_8H_{10}	--vapor	20	29.20
Ethylbenzoate	$C_9H_{10}O_2$	--vapor	20	35.5
Ethyl bromide	C_2H_5Br	--vapor	20	24.15
Ethyl chloroformate	$C_3H_5ClO_2$	--vapor	15.1	27.5
Ethyl Cinnamate	$C_{11}H_{12}O_2$	--air	20	38.37
Ethylene bromide	$C_2H_4Br_2$	--vapor	20	38.37
Ethylene chloride	$C_2H_4Cl_2$	--air	20	24.15
Ethylene oxide	C_2H_4O	--vapor	−20	30.8
	C_2H_4O	--vapor	0.0	27.6
	C_2H_4O	--vapor	20	24.3
Ethyl ether	$C_4H_{10}O$	--vapor	20	17.01
	$C_4H_{10}O$	--vapor	50	13.47
Ethyl format	$C_3H_6O_2$	--air or vapor	20	23.6
Ethyl iodide	C_2H_5I	--vapor	20	29.4
Ethyl nitrate	$C_2H_5NO_3$	--air or vapr	20	28.7
dl-Ethyl lactate	$C_5H_{10}O_3$	--air	20	29.9
Ethyl mercaptan	C_2H_6S	--air or vapor	20	22.5
Ethyl salicylate	$C_9H_{10}O_3$	--vapor	20.5	38.33
Formamide	CH_3NO	--vapor	20	58.2
Formic acid	CH_2O_2	--air	20	37.6
Furfural	$C_5H_4O_2$	--air or vapor	20	43.5
Gelatin solution (1%)		--water	2.85	8.3
Glycerol	$C_3H_8O_3$	--air	20	63.4
	$C_3H_8O_3$	--air	90	58.6
	$C_3H_8O_3$	--air	150	51.9
Glycol	$C_2H_6O_2$	--air or vapor	20	47.7
Helium	He	--vapor	−269	.12
	He	--vapor	−270	.239
	He	--vapor	−271.5	.353
n-Hexane	C_6H_{14}	--air	20	18.43
Hydrazine	N_2H_4	--vapor	25	91.5
Hydrogen	H_2	--vapor	−255	2.31
Hydrogen cyanide	HCN	--vapor	17	10.2
Hydrogen peroxide	H_2O_2	--vapor	18.2	76.1
Isobutyl alcohol	$C_4H_{10}O$	--vapor	20	23.0
Isobutylamine	$C_4H_{11}N$	--air	68	17.6
Isobutyl chloride	C_4H_9Cl	--air	20	21.94
Isobutyric acid	$C_4H_8O_2$	--air or vapor	20	25.2
Isopentane	C_5H_{12}	--air	20	13.72
Isopropyl alcohol	C_3H_8O	air or vapor	20	21.7
Methyl acetate	$C_3H_6O_2$	--air or vapor	20	24.6
Methyl alcohol	CH_4O	--air	0	24.49
	CH_4O	--air	20	22.61
	CH_4O	--vapor	50	20.14
Methylamine	CH_3NH_2	--nitrogen	−12	22.2
	CH_3NH_2	--vapor	−20	23.0
	CH_3NH_2	--nitrogen	−70	29.2
N-Methylaniline	C_7H_9N	--air or vapor	20	39.6
Methyl benzoate	$C_8H_8O_2$	--air or vapor	20	37.6
Methyl chloride	CH_3Cl	--air	20	16.2
Methyl ether	C_2H_6O	--vapor	−10	16.4
	C_2H_6O	--vapor	−40	21
Methylene chloride	CH_2Cl_2	--air	20	26.52
Methylene iodide	CH_2I_2	--air	20	50.76
Methyl ethyl ketone	C_4H_8O	--air or vapor	20	24.6
Methyl formate	$C_2H_4O_2$	--vapor	20	25.08
Methyl iodide	CH_3I	--air	43.5	25.8
Methyl propionate	$C_4H_8O_2$	--air or vapor	20	24.9
Methyl salicylate	$C_8H_8O_3$	--nitrogen	94	31.9
Methyl sulfide	C_2H_6S	--vapor	11.1	26.50
Naphthalene	$C_{10}H_8$	--air or vapor	127	28.8
Neon	Ne	--vapor	−248	5.50
Nitric acid (98.8%)	HNO_3	--air	11.6	42.7
Nitrobenzene	$C_6H_5NO_2$	--air or vapor	20	43.9
Nitroethane	$C_2H_5NO_2$	--air or vapor	20	32.2
Nitrogen	N_2	--vapor	−183	6.6
	N_2	--vapor	−193	8.27
	N_2	--vapor	−203	10.53
Nitrogen tetra oxide	N_2O_4	--vapor	19.8	27.5
Nitromethane	CH_3NO_2	--vapor	20	36.82
Nitrous oxide	N_2O	--vapor	20	1.75
n-Octane	C_8H_{18}	--air	20	21.80
n-Octyl alcohol	$C_8H_{18}O$	--air	20	27.53
Oleic acid	$C_{18}H_{34}O_2$	--air	20	32.50
Oxygen	O_2	--vapor	−183	13.2
Oxygen (65%)	O_2	--air	−190.5	12.2
	O_2	--vapor	−193	15.7
	O_2	--vapor	−203	18.3
Paraldehyde	$C_6H_{12}O_3$	--air	20	25.9
Phenetole	$C_8H_{10}O$	--vapor	20	32.74
Phenol	C_6H_6O	--air or vapor	20	40.9
	C_6H_6O	--air or vapor	30	39.88

Substance		In contact with	Temperature °C	Surface tension dynes/cm
Name	Formula			
Phenylhydrazine	$C_6H_8N_2$	- -vapor	20	46.1
Phosphorus tribromide	PBr_3	- -air	24	45.8
Phosphorus trichloride	PCl_3	- -vapor	20	29.1
Phosphorus triiodide	PI_3	- -vapor	75.3	56.5
Propionic acid	$C_3H_6O_2$	- -vapor	20	26.7
n-Propyl acetate	$C_5H_{10}O_2$	- -air or vapor	20	24.3
n-Propyl alcohol	C_3H_8O	- -vapor	20	23.78
n-Propylamine	C_3H_9N	- -air	20	22.4
n-Propyl bromide	C_3H_7Br	- -vapor	71	19.65
n-Propyl chloride	C_3H_7Cl	- -air	47	18.2
n-Propyl formate	$C_4H_8O_2$	- -vapor	20	24.5
Pyridine	C_8H_5N	- -air	20	38.0
Quinoline	C_9H_7N	- -air	20	45.0
Ricinoleic acid	$C_{18}H_{34}O_3$	- -air	16	35.81
Selenium	Se	- -air	217	92.4
Styrene	C_8H_8	- -air	19	32.14
Sulfuric acid (98.5%)	H_2SO_4	- -air or vapor	20	55.1
Tetrabromoethane 1,1,2,2-	$C_2H_2Br_4$	- -air	20	49.67
Tetrachloroethane 1,1,2,2-	$C_2H_2Cl_4$	- -air	22.5	36.03
Tetrachloroethylene	C_2Cl_4	- -vapor	20	31.74
Toluene	C_7H_8	- -vapor	10	27.7
	C_7H_8	- -vapor	20	28.5
	C_7H_8	- -vapor	30	27.4
m-Toluidine	C_7H_9N	- -air	20	36.9
o-Toluidine	C_7H_9N	- -air or vapor	20	40.0
p-Toluidine	C_7H_9N	- -air	50	34.6
Trichloroacetic acid	$C_2HCl_3O_2$	- -nitrogen	80.2	27.8
Trichloroethane 1,1,2-	$C_2H_3Cl_3$	- -air	114	22.0
Triethyl phosphate	$C_6H_{15}O_4P$	- -air	15.5	30.61
Trimethylamine	C_3H_9N	- -nitrogen	−4	17.3
Triphenylcarbinol	$C_{19}H_{16}O$	- -vapor	165.8	30.38
Vinyl acetate	$C_4H_6O_2$	- -vapor	20	23.95
	$C_4H_6O_2$	- -vapor	25	23.16
	$C_4H_6O_2$	- -vapor	30	22.54
Water	H_2O	- -air	18	73.05
m-Xylene	C_8H_{10}	- -vapor	20	28.9
o-Xylene	C_8H_{10}	- -air	20	30.10
p-Xylene	C_8H_{10}	- -vapor	20	28.37

VISCOSITY

Viscosity. — All fluids possess a definite resistance to change of form and many solids show a gradual yielding to forces tending to change their form. This property, a sort of internal friction, is called viscosity; it is expressed in dyne-seconds per cm^2 or poises. Dimensions, $—[m\,l^{-1}\,t^{-1}]$. If the tangential force per unit area, exerted by a layer of fluid upon one adjacent is one dyne for a space rate of variation of the tangential velocity of unity, the viscosity is one poise.

Kinematic viscosity is the ratio of viscosity to density. The c.g.s. unit of kinematic viscosity is the **stoke**.

Flow of liquids through a tube; where l is the length of the tube, r its radius, p the difference of pressure at the ends, η the coefficient of viscosity, the volume escaping per second,

$$v = \frac{\pi p r^4}{8l\eta} \text{ (Poiseuille).}$$

The volume will be given in cm^3 per second if l and r are in cm, p in dynes per cm^2 and η in poises or dyne-seconds per cm^2.

VISCOSITY OF WATER BELOW 0°C

White-Twining 1914

Temperature	Viscosity centipoises	Temperature	Viscosity centipoises
0°C	1.798	−7.23	2.341
−2.10	1.930	−8.48	2.458
−4.70	2.121	−9.30	2.549
−6.20	2.250		

ABSOLUTE VISCOSITY OF WATER AT 20°C

Swindells, J. R. Coe, Jr., and T. B. Godfrey, Journal of Research, National Bureau of Standards **48**, 1, 1952.
The value found for the viscosity of water at 20°C was 0.010019 ± 0.000003 poise.
The value **0.01002** poise is to be used as the absolute value of the viscosity of water for calibration purposes.

VISCOSITY CONVERSION TABLE

Poise $= $ c.g.s. unit of absolute viscosity $= \dfrac{gm}{sec \times cm}$

Stoke $= $ c.g.s. unit of kinematic viscosity $= \dfrac{gm}{sec \times cm \times density\ (t°F)}$

Centipoise $= 0.01$ poise
Centistoke $= 0.01$ stoke
Centipoises $= $ Centistokes $\times$ density (at given temperature)

To convert poises to $\dfrac{lb}{sec \times ft}$ or $\dfrac{lb}{hr \times ft}$ multiply by 0.0672 or 242 respectively.

Centi-stokes	Saybolt Seconds at			Redwood Seconds at			Engler Degrees at all temps.	Centi-stokes	Saybolt Seconds at			Redwood Seconds at			Engler Degrees at all temps.
	100°F	130°F	210°F	70°F	140°F	200°F			100°F	130°F	210°F	70°F	140°F	200°F	
2.0	32.6	32.7	32.8	30.2	31.0	31.2	1.14	28.0	132.1	132.4	133.0	115.3	116.5	118.0	3.82
3.0	36.0	36.1	36.3	32.7	33.5	33.7	1.22	30.0	140.9	141.2	141.9	123.1	124.4	126.0	4.07
4.0	39.1	39.2	39.4	35.3	36.0	36.3	1.31	32.0	149.7	150.0	150.8	131.0	132.3	134.1	4.32
5.0	42.3	42.4	42.6	37.9	38.5	38.9	1.40	34.0	158.7	159.0	159.8	138.9	140.2	142.2	4.57
6.0	45.5	45.6	45.8	40.5	41.0	41.5	1.48	36.0	167.7	168.0	168.9	146.9	148.2	150.3	4.83
7.0	48.7	48.8	49.0	43.2	43.7	44.2	1.56	38.0	176.7	177.0	177.9	155.0	156.2	158.3	5.08
8.0	52.0	52.1	52.4	46.0	46.4	46.9	1.65	40.0	185.7	186.0	187.0	163.0	164.3	166.7	5.34
9.0	55.4	55.5	55.8	48.9	49.1	49.7	1.75	42.0	194.7	195.1	196.1	171.0	172.3	175.0	5.59
10.0	58.8	58.9	59.2	51.7	52.0	52.6	1.84	44.0	203.8	204.2	205.2	179.1	180.4	183.3	5.85
11.0	62.3	62.4	62.7	54.8	55.0	55.6	1.93	46.0	213.0	213.4	214.5	187.1	188.5	191.7	6.11
12.0	65.9	66.0	66.4	57.9	58.1	58.8	2.02	48.0	222.2	222.6	223.8	195.2	196.6	200.0	6.37
14.0	73.4	73.5	73.9	64.4	64.6	65.3	2.22	50.0	231.4	231.8	233.0	203.3	204.7	208.3	6.63
16.0	81.1	81.3	81.7	71.0	71.4	72.2	2.43	60.0	277.4	277.9	279.3	243.5	245.3	250.0	7.90
18.0	89.2	89.4	89.8	77.9	78.5	79.4	2.64	70.0	323.4	324.0	325.7	283.9	286.0	291.7	9.21
20.0	97.5	97.7	98.2	85.0	85.8	86.9	2.87	80.0	369.6	370.3	372.2	323.9	326.6	333.4	10.53
22.0	106.0	106.2	106.7	92.4	93.3	94.5	3.10	90.0	415.8	416.6	418.7	364.4	367.4	375.0	11.84
24.0	114.6	114.8	115.4	99.9	100.9	102.2	3.34	*100.0	462.0	462.9	465.2	404.9	408.2	416.7	13.16
26.0	123.3	123.5	124.2	107.5	108.6	110.0	3.58								

* At higher values use the same ratio as above for 100 centistokes; e.g., 110 centistokes = 110 × 4.620 Saybolt seconds at 100°F.

To obtain the Saybolt Universal viscosity equivalent to a kinematic viscosity determined at $t°F$, multiply the equivalent Saybolt Universal viscosity at 100°F by $1 + (t-100)0.000064$; e.g., 10 centistokes at 210°F are equivalent to 58.8 × 1.0070, or 59.2 Saybolt Universal seconds at 210°F.

VISCOSITY CONVERSION
Kinematic

To convert from	To	Multiply by	To convert from	To	Multiply by
cm²/sec (Stokes)	Centistokes	10^2	ft²/sec	cm²/sec. (Stokes)	9.29×10^2
	ft²/hr.	3.875		cm²/sec.×10^2 (centistokes)	9.29×10^4
	ft²/sec.	1.076×10^{-3}		ft²/hr	3.60×10^3
	in.²/sec.	1.550×10^{-1}		in.²/sec.	1.44×10^2
	m²/hr.	3.600×10^{-1}		m²/hr.	3.345×10^2
cm²/sec × 10^2 (Centistokes)	cm²/sec. (Stokes)	1×10^{-2}	in.²/sec	cm²/sec. (Stokes)	6.452
	ft²/hr.	3.875×10^{-2}		cm²/sec.×10^2 (centistokes)	6.452×10^2
	ft²/sec.	1.076×10^{-5}		ft²/hr.	2.50×10
	in.²/sec.	1.550×10^{-3}		ft²/sec.	6.944×10^{-3}
	m²/hr.	3.600×10^{-3}		m²/hr.	2.323
ft²/hr	cm²/sec (Stokes)	2.581×10^{-1}	m²/hr	cm²/sec. (Stokes)	2.778
	cm/sec × 10^2 (centistokes)	2.581×10		cm²/sec.×10^2 (centistokes)	2.778×10^2
	ft²/sec.	2.778×10^{-4}		ft²/hr.	1.076×10
	in.²/sec.	4.00×10^{-2}		ft²/sec.	2.990×10^{-3}
	m²/hr.	9.290×10^{-2}		in.²/sec.	4.306×10^{-1}

VISCOSITY CONVERSION
Absolute

Absolute viscosity = kinematic viscosity × density; lb = mass pounds; lb$_F$ = force pounds

To convert from	To	Multiply by	To convert from	To	Multiply by
gm/(cm)(sec) [Poise]	gm/(cm)(sec)(10^2) [Centipoise]	10^2		lb/(in.)(sec)	5.60×10^{-5}
	kg/(m)(hr)	3.6×10^2		(gm$_F$)(sec)/cm²	1.02×10^{-5}
	lb/(ft)(sec)	6.72×10^{-2}		(lb$_F$)(sec)/in.² [Reyn]	1.45×10^{-7}
	lb/(ft)(hr)	2.419×10^2		(lb$_F$)(sec)/ft²	2.089×10^{-5}
	lb/(in.)(sec)	5.6×10^{-3}	kg/(m)(hr)	gm/(cm)(sec)	2.778×10^{-3}
	(gm$_F$)(sec)/cm²	1.02×10^{-3}		gm/(cm)(sec)(10^2) [Centipoise]	2.778×10^{-1}
	(lb$_F$)(sec)/in.² [Reyn]	1.45×10^{-5}		lb/(ft)(sec)	1.867×10^{-4}
	(lb$_F$)(sec)/ft²	2.089×10^{-3}		lb/(ft)(hr)	6.720×10^{-1}
gm/(cm)(sec)(10^2) [Centipoise]	gm/(cm)(sec) [Poise]	10^{-2}		lb/(in.)(sec)	1.555×10^{-5}
	kg/(m)(hr)	3.6		(gm$_F$)(sec)/cm²	2.833×10^{-4}
	lb/(ft)(sec)	6.72×10^{-4}		(lb$_F$)(sec)/in.² [Reyn]	4.029×10^{-8}
	lb/(ft)(hr)	2.419		(lb$_F$)(sec)/ft²	5.801×10^{-6}

To convert from	To	Multiply by	To convert from	To	Multiply by
lb/(ft)(sec)	gm/(cm)(sec) [Poise]	1.488×10^1		$(lb_F)(sec)/ft^2$	3.73×10^{-1}
	gm/(cm)(sec)(10^2) [Centipoise]	1.488×10^3	$(gm_F)(sec)/cm^2$	gm/(cm)(sec) [Poise]	9.807×10^2
	kg/(m)(hr)	5.357×10^3		gm/(cm)(sec)(10^2) [Centipoise]	9.807×10^4
	lb/(ft)(hr)	3.60×10^3		kg/(m)(hr)	3.530×10^5
	lb/(in.)(sec)	8.333×10^{-2}		lb/(ft)(sec)	6.590×10
	$(gm_F)(sec)/cm^2$	1.518×10^{-1}		lb/(ft)(hr)	2.372×10^5
	$(lb_F)(sec)/in.^2$ [Reyn]	2.158×10^{-4}		lb/(in.)(sec)	5.492
				$(lb_F)(sec)/in.^2$ [Reyn]	1.422×10^{-2}
	$(lb_F)(sec)/ft^2$	3.108×10^{-2}		$(lb_F)(sec)/ft^2$	2.048
lb/(ft)(hr)	gm/(cm)(sec) [Poise]	4.134×10^{-3}	$(lb_F)(sec)/in.^2$ [Reyn]	gm/(cm)(sec) [Poise]	6.895×10^4
	gm/(cm)(sec)(10^2) [Centipoise]	4.134×10^{-1}		gm/(cm)(sec)(10^2) [Centipoise]	6.895×10^6
	kg/(m)(hr)	1.488		kg/(m)(hr)	2.482×10^7
	lb/(ft)(sec)	2.778×10^{-4}		lb/(ft)(sec)	4.633×10^3
	lb/(in.)(sec)	2.315×10^{-5}		lb/(ft)(hr)	1.668×10^7
	$(gm_F)(sec)/cm^2$	4.215×10^{-6}		lb/(in.)(sec)	3.861×10^5
	$(lb_F)(sec)/in.^2$ [Reyn]	5.996×10^{-8}		$(gm_F)(sec)/cm^2$	7.031×10
	$(lb_F)(sec)/ft^2$	8.634×10^{-6}		$(lb_F)(sec)/ft^2$	1.440×10^2
lb/(in.)(sec)	gm/(cm)(sec) [Poise]	1.786×10^2	$(lb_F)(sec)/ft^2$	gm/(cm)(sec) [Poise]	4.788×10^2
	gm/(cm)(sec)(10^2) [Centipoise]	1.786×10^4		gm/(cm)(sec)(10^2) [Centipoise]	4.788×10^4
	kg/(m)(hr)	6.429×10^4		kg/(m)(hr)	1.724×10^5
	lb/(ft)(sec)	1.20×10		lb/(ft)(sec)	3.217×10
	lb/(ft)(hr)	4.32×10^4		lb/(ft)(hr)	1.158×10^5
	$(gm_F)(sec)/cm^2$	1.821×10^{-1}		lb/(in.)(sec)	2.681
	$(lb_F)(sec)/in.^2$ [Reyn]	2.590×10^{-3}		$(gm_F)(sec)/cm^2$	4.882×10^{-1}
				$(lb_F)(sec)/in.^2$ [Reyn]	6.944×10^{-3}

THE VISCOSITY OF WATER 0°C TO 100°C

Contribution from the National Bureau of Standards, not subject to copyright.

°C	η(cp)	°C	η(cp)	°C	η(cp)	°C	η(cp)
0	1.787	26	0.8705	52	0.5290	78	0.3638
1	1.728	27	.8513	53	.5204	79	.3592
2	1.671	28	.8327	54	.5121	80	.3547
3	1.618	29	.8148	55	.5040	81	.3503
4	1.567	30	.7975	56	.4961	82	.3460
5	1.519	31	.7808	57	.4884	83	.3418
6	1.472	32	.7647	58	.4809	84	.3377
7	1.428	33	.7491	59	.4736	85	.3337
8	1.386	34	.7340	60	.4665	86	.3297
9	1.346	35	.7194	61	.4596	87	.3259
10	1.307	36	.7052	62	.4528	88	.3221
11	1.271	37	.6915	63	.4462	89	.3184
12	1.235	38	.6783	64	.4398	90	.3147
13	1.202	39	.6654	65	.4335	91	.3111
14	1.169	40	.6529	66	.4273	92	.3076
15	1.139	41	.6408	67	.4213	93	.3042
16	1.109	42	.6291	68	.4155	94	.3008
17	1.081	43	.6178	69	.4098	95	.2975
18	1.053	44	.6067	70	.4042	96	.2942
19	1.027	45	.5960	71	.3987	97	.2911
20	1.002	46	.5856	72	.3934	98	.2879
21	0.9779	47	.5755	73	.3882	99	.2848
22	.9548	48	.5656	74	.3831	100	.2818
23	.9325	49	.5561	75	.3781		
24	.9111	50	.5468	76	.3732		
25	.8904	51	.5378	77	.3684		

The above table was calculated from the following empirical relationships derived from measurements in viscometers calibrated with water at 20°C (and one atmosphere), modified to agree with the currently accepted value for the viscosity at 20° of 1.002 cp:

$$0° \text{ to } 20°C: \log_{10} \eta_T = \frac{1301}{998.333 + 8.1855(T-20) + 0.00585(T-20)^2} - 1.30233$$

(R. C. Hardy and R. L. Cottington, J.Res.NBS 42, 573 (1949).)

$$20° \text{ to } 100°C: \log_{10} \frac{\eta_T}{\eta_{20}} = \frac{1.3272(20-T) - 0.001053(T-20)^2}{T + 105}$$

(J. F. Swindells, NBS, unpublished results.)

VISCOSITY OF LIQUIDS

Viscosity of liquids in centipoises (cp) including elements, inorganic and organic compounds and mixtures.

Liquid	Temp. °C	Viscosity cp	Liquid	Temp. °C	Viscosity cp
Acetaldehyde	0	.2797		700	1.26
	10	.2557		800	1.08
	20	.22		850	1.05
Acetanilide	120	2.22	Benzaldehyde	25	1.39
	130	1.90	Benzene	0	.912
Acetic acid	15	1.31		10	.758
	18	1.30		20	.652
	25.2	1.155		30	.564
	30	1.04		40	.503
	41	1.00		50	.442
	59	.70		60	.392
	70	.60		70	.358
	100	.43		80	.329
anhydride	0	1.24	Benzonitrile	25	1.24
	15	.971	Benzophenone	55	4.79
	18	.90		120	1.38
	30	.783	Benzyl alcohol	20	5.8
	100	.49	Benzylamine	25	1.59
Acetone	−92.5	2.148	Benzylaniline	33	2.18
	−80.0	1.487		130	1.20
	−59.6	.932	Benzyl ether	0	10.5
	−42.5	.695		20	5.33
	−30.0	.575		40	3.21
	−20.9	.510	Bismuth	285	1.61
	−13.0	.470		304	1.662
	−10.0	.450		365	1.46
	0	.399		451	1.280
	15	.337		600	.998
	25	.316	Bromine, liq	−4.3	1.31
	30	.295		0	1.241
	41	.000		12.6	1.07
Acetonitrile	0	.442		10	1.0
	15	.375		19.5	.995
	25	.345		28.9	.911
Acetophenone	11.9	2.28	o-Bromoaniline	40	0.13
	23.5	1.59	m-Bromoaniline	20	6.81
	25.0	1.617		40	3.70
	50.0	1.246		80	1.70
	80.0	.734	p-Bromoaniline	80	1.81
Air, liq	−192.3	.172	Bromobenzene	15	1.196
Alcohol. See *Ethyl, Methyl,* etc.				30	.985
			Bromoform	15	2.152
Allyl alcohol	0	2.145		25	1.89
	15	1.49		30	1.741
	20	1.363	Butyl acetate	0	1.004
	30	1.07		20	.732
	40	.914		40	.563
	70	.553	n-Butyl alcohol	−50.9	36.1
Allylamine	130	.506		−30.1	14.7
Allyl chloride	15	.347		−22.4	11.1
	30	.300		−14.1	8.38
Ammonia	−69	.475		0	5.186
	−50	.317		15	3.379
	−40	.276		20	2.948
	−33.5	.255		30	2.30
n-Amyl acetate	11	1.58		40	1.782
	45	.805		50	1.411
alcohol	15	4.65		70	.930
	30	2.99		100	.540
ether	15	1.188	sec-Butyl alcohol	15	4.21
Aniline	−6	13.8	n-Butyl bromide	15	.626
	0	10.2	n-Butyl chloride	15	.469
	5	8.06	Butyl chloride, tertiary	15	.543
	10	6.50	n-Butyl formate	0	.940
	15	5.31		20	.689
	20	4.40	Butyric acid	0	2.286
	25	3.71		15	1.81
	30	3.16		20	1.540
	35	2.71		40	1.120
	40	2.37		50	.975
	50	1.85		70	.760
	60	1.51		100	.551
	70	1.27	Cadmium, liq	349	1.44
	80	1.09		506	1.18
	90	.935		603	1.10
	100	.825	Carbolic acid. See *Phenol.*		
Anisol	0	1.78	Carbon dioxide, liq., pressure that of saturated vapor	0	.099
	20	1.32		10	.085
	40	1.12		20	.071
Antimony, liq	645	1.55		30	.053

Liquid	Temp. °C	Viscosity cp	Liquid	Temp. °C	Viscosity cp
disulfide................	−13	.514		−40	.461
	−10	.495		−20	.362
	0	.436		0	.2842
	5	.380		17	.240
	20	.363		20	.2332
	40	.330		25	.222
Carbon tetrachloride......	0	1.329		40	.197
	15	1.038		60	.166
	20	.969		80	.140
	30	.843		100	.118
	40	.739	Ethyl acetate............	0	.582
	50	.651		8.96	.516
	60	.585		10	.512
	70	.524		15	.473
	80	.468		20	.455
	90	.426		25	.441
	100	.384		30	.400
Cetyl alcohol............	50	13.4		50	.345
Chlorine, liq............	−76.5	.729		75	.283
	−70.5	.680	Ethyl alcohol...........	−98.11	44.0
	−60.2	.616		−89.8	28.4
	−52.4	.566		−71.5	13.2
	−35.4	.494		−59.42	8.41
	0	.385		−52.58	6.87
Chlorobenzene..........	15	.900		−32.01	3.84
	20	.799		−17.59	2.68
	40	.631		−.30	1.80
	80	.431		0	1.773
	100	.367		10	1.466
Chloroform.............	−13	.855		20	1.200
	0	.700		30	1.003
	8.1	.643		40	.834
	15	.596		50	.702
	20	.58		60	.592
	25	.542		70	.504
	30	.514	Ethyl alcohol, anh.	−148	8,470
	39	.500		−146	5,990
o-Chlorophenol..........	25	4.11		−130	467
	50	2.015	Ethyl aniline............	25	2.04
m-Chlorophenol.........	25	11.55	Ethylbenzene	17	.691
p-Chlorophenol..........	50	4.99	Ethyl benzoate.........	20	2.24
Copper, liq.............	1,085	3.36	Ethyl bromide..........	−120	5.6
	1,100	3.33		−100	2.89
	1,150	3.22		−80	1.81
	1,200	3.12		0	.487
o-Cresol...............	40	4.49		10	.441
m-Cresol...............	10	43.9		15	.418
	20	20.8		20	.402
	40	6.18		30	.348
p-Cresol...............	40	7.00	n-Ethyl butyrate..........	15	.711
Creosote...............	20	12.0	Ethyl carbonate..........	15	.868
Cycloheptane...........	13.5	1.64	Ethylene bromide.........	0	2.438
Cyclohexane...........	17	1.02		17	1.95
Cyclohexanol..........	20	68		20	1.721
Cyclohexene............	13.5	.696		40	1.286
	20	.66		67.3	.922
Cyclooctane............	13.5	2.35		70	.903
Cyclopentane...........	13.5	.493		82.2	.750
nDecane...............	20	.92		99.0	.648
Diethylamine...........	25	.346	chloride...............	0	1.077
	25	.367		15	.887
Diethylaniline...........	.5	3.84		19.4	.800
	20.0	2.18		40	.652
	25.0	1.95		50	.565
Diethylcarbinol..........	15.0	7.34		70	.479
Diethylketone...........	15	.493	glycol.................	20	19.9
Dimethylaniline..........	10	1.69		40	9.13
	20	1.41		60	4.95
	25	1.285		80	3.02
	30	1.17		100	1.99
	40	1.04	oxide.................	−49.8	.577
	50	.91		−38.2	.488
Dimethyl-α-naphthylamine.	130	.868		−21.0	.394
Dimethyl-β-naphthylamine.	130	.952		0	.320
Diphenyl...............	70	1.49	Ethyl formate............	20	.402
	100	.97	iodide................	0	.727
Diphenylamine..........	130	1.04		15	.617
Dodecane..............	25	1.35		20	.592
Ether (diethyl-)..........	−100	1.69		40	.495
	−80	.958		70	.391
	−60	.637	malate................	24.7	3.016
			oxalate...............	15	2.31

Liquid	Temp. °C	Viscosity cp	Liquid	Temp. °C	Viscosity cp
propionate	15	.564	Isopropyl alcohol	15	2.86
Eugenol	0	29.9		30	1.77
	20	9.22	Isoquinoline	25	3.57
	40	4.22	Isosafrol	25	3.981
Fluorobenzene	20	.598	Lead, liq	350	2.58
	40	.478		400	2.33
	60	.389		441	2.116
	80	.329		500	1.84
	100	.275		551	1.70
Formamide	0	7.55		600	1.38
	25	3.30		703	1.349
Formic acid	7.59	2.3868		844	1.185
	10	2.262	Menthol, liq	55.6	6.29
	20	1.804		74.6	2.47
	30	1.465		99.0	1.04
	40	1.219	Mercury	−20	1.855
	70	.780		−10	1.764
	100	.549		0	1.685
Furfural	0	2.48		10	1.615
	25	1.49		19.02	1.56
Glucose	22	9.1×10^{15}		20	1.554
	30	6.6×10^{13}		20.2	1.55
	40	2.8×10^{11}		30	1.499
	60	9.3×10^{7}		40	1.450
	80	6.6×10^{5}		40.8	1.45
	100	2.5×10^{4}		41.86	1.44
Glycerin	−42	6.71×10^{8}		50	1.407
	−36	2.05×10^{6}		60	1.367
	−25	2.62×10^{5}		70	1.331
	−20	1.34×10^{5}		80	1.298
	−15.4	6.65×10^{4}		90	1.268
	−10.8	3.55×10^{4}		100	1.240
	−4.2	1.49×10^{4}		150	1.130
	0	12,110		200	1.052
	5	8,280		250	.995
	15	2,330		300	.950
	20	1,490		340	.921
	25	954	Methyl acetate	0	.484
	30	629		20	.381
Glycerin trinitrate	10	69.2		40	.320
	20	36.0	Methyl alcohol	−98.30	13.9
	30	21.0	(Methanol)	−84.23	6.8
	40	13.6		−72.55	4.36
	60	6.8		−44.53	1.98
Heptane	0	.524		−22.29	1.22
	17	.461		0	.82
	20	.409		15	.623
	25	.386		20	.597
	40	.341		25	.547
	70	.262		30	.510
n-Heptyl alcohol	15	8.53		40	.456
Hexadecane	20	3.34		50	.403
Hexane	0	.401	Methyl amine	0	.236
	17	.374	aniline	25	2.02
	20	.326		30	1.55
	25	.294	chloride	20	.1834
	40	.271	Methylene bromide	15	1.09
	50	.248		30	0.92
Hydrazine	1	1.29	chloride	15	.449
	10	1.12		30	.393
	20	.97	Methyl iodide	0	.606
Hydrogen, liq		.011		15	.518
Iodine, liq	116	2.27		20	.500
Iodobenzene	15	1.74		30	.460
Iron, 2.5% carbon, liq	1,400	2.25		40	.424
Isoamyl acetate	8.97	1.030	Naphthalene	80	.967
	19.91	.872		100	.776
alcohol	10	6.20	Nitric acid	0	2.275
amine	25	.724		10	1.770
Isobutyl alcohol	15	4.703	Nitrobenzene	2.95	2.91
amine	25	.553		5.69	2.71
Isobutyric acid	15	1.44		5.94	2.71
	30	1.13		9.92	2.48
Isoeugenol	25	26.72		14.94	2.24
Isoheptane	0	.481		20.00	2.03
	20	.384	Nitromethane	0	.853
	40	.315		25	.620
Isohexane	0	.376	o-Nitrotoluene	0	3.83
	20	.306		20	2.37
	40	.254		40	1.63
Isopentane	0	.273		60	1.21
	20	.223	m-Nitrotoluene	20	2.33

Liquid	Temp. °C	Viscosity cp	Liquid	Temp. °C	Viscosity cp
	40	1.60	n-Propyl alcohol..........	0	3.883
	60	1.18		15	2.52
p-Nitrotoluene...........	60	1.20		20	2.256
n-Nonane................	20	.711		30	1.72
n-Octane................	0	.706		40	1.405
	16	.574	n-Propyl alcohol..........	50	1.130
	20	.542		70	.760
	40	.433	Propyl aldehyde..........	10	.47
Octodecane.............	40	2.86		20	.41
n-Octylalcohol...........	15	10.6		40	.33
Oil, castor..............	10	2,420	bromide.............	0	.651
	20	986		20	.524
	30	451		40	.433
	40	231	chloride............	0	.436
	100	16.9		20	.352
cottonseed............	20	70.4		40	.291
cylinder, filtered........	37.8	240.6	n-Propyl ether...........	15	.448
	100	18.7	Pyridine................	20	.974
cylinder, dark..........	37.8	422.4	Salicylic acid............	10	3.20
	100	24.0		20	2.71
linseed...............	30	33.1		40	1.81
	50	17.6	Salol..................	45	.746
	90	7.1	Sodium bromide..........	762	1.42
machine, light........	15.6	113.8		780	1.28
	37.8	34.2	chloride, liq........	841	1.30
	100	4.9		896	1.01
machine, heavy	15.6	660.6		924	.97
	37.8	127.4	nitrate, liq............	308	2.919
Oil, olive..............	10	138.0		348	2.439
	20	84.0		398	1.977
	40	36.3		418	1.828
	70	12.4	Stearic acid.............	70	11.6
rape..................	0	2,530	Sucrose (cane sugar).......	109	2.8×10^6
	10	385		124.6	1.9×10^5
	20	163	Sulfur (gas free)..........	123.0	10.94
	30	96		135.5	8.66
soya bean..............	20	69.3		149.5	7.09
	30	40.6		156.3	7.19
	50	20.6		158.2	7.59
	90	7.8		159.2	9.48
sperm.................	15.6	42.0		159.5	14.45
	37.8	18.5		160.0	22.83
	100.0	4.6		160.3	77.32
Oleic acid................	30	25.6		165.0	500.0
Pentadecane.............	22	2.81		171.0	4,500.0
Pentane.................	0	.289		184.0	16,000.00
	20	.240		190.5	19,700.0
o-Phenetidine............	0	16.5		197.5	21,300.0
	20	6.08		200.0	21,500.0
	30	4.22		210.0	20,500.0
m-Phenetidine...........	30	12.9		217.0	19,100.0
p-Phenetidine...........	20	12.9		220.0	18,600.0
	30	8.3	Sulfur dioxide, liq..........	-33.5	.5508
Phenol.................	18.3	12.7		-10.5	.4285
	50	3.49		0.1	.3936
	60	2.61	Sulfuric acid..............	0	48.4
	70	2.03		15	32.8
	90	1.26		20	25.4
Phenylcyanide...........	.28	1.96		30	15.7
	20.0	1.33		40	11.5
Phosphorus, liq...........	21.5	2.34	Sulfuric acid..............	50	8.82
	31.2	2.01		60	7.22
	43.2	1.73		70	6.09
	50.5	1.60		80	5.19
	60.2	1.45	Tetrachloroethane........	15	1.844
	69.7	1.32	Tetradecane..............	20	2.18
	79.9	1.21	Tin, liq.................	240	2.12
Potassium bromide, liq.....	745	1.48		280	1.678
	775	1.34		300	1.73
	805	1.19		301	1.680
nitrate, liq............	334	2.1		400	1.43
	358	1.7		450	1.270
	333	2.97		500	1.20
	418	2.00		600	1.08
Propionic acid...........	10	1.289		604	1.045
	15	1.18		750	.905
	20	1.102	Toluene.................	0	.772
	40	.845		17	.61
Propyl acetate...........	10	.66		20	.590
	20	.59		30	.526
	40	.44		40	.471

VISCOSITY OF LIQUIDS (Continued)

Liquid	Temp. °C	Viscosity cp	Liquid	Temp. °C	Viscosity cp
	70	.354		70	.728
o-Toluidine	20	4.39	Turpentine, Venice	17.3	1.3×10^5
m-Toluidine	20	3.81	n-Undecane	20	1.17
p-Toluidine	50	1.80	o-Xylene (xylol)	0	1.105
Triacetin	17	28.0		16	.876
Tributyrin	20	11.6		20	.810
Trichlorethane	20	1.2		40	.627
Tridecane	23.3	1.55	m-Xylene (xylol)	0	.806
Triethylcarbinol	20	6.75		15	.650
Tripalmitin	70	16.8		20	.620
Tristearin	75	18.5		40	.497
Turpentine	0	2.248	p-Xylene (xylol)	16	.696
	10	1.783		20	.648
	20	1.487		40	.513
	30	1.272	Zinc, liq	280	1.68
	40	1.071		357	1.42
				389	1.31

VISCOSITY OF GASES

Gas or vapor	Temp. °C	Viscosity micropoises	Gas or vapor	Temp. °C	Viscosity micropoises
Acetic acid, vap	119.1	107.0	Benzene, vap	14.2	73.8
Acetone, vap	100	93.1		131.2	103.1
	119.0	99.1		194.6	119.8
	190.4	118.6		252.5	134.3
	247.7	133.4		312.8	148.4
	306.4	148.1	Bromine, vap	12.8	151
Acetylene	0	93.5		65.7	170
Air	−194.2	55.1		99.7	188
	100.1	66.7		100.7	200
	−104.0	113.0		179.7	227
	−69.4	133.3		220.3	248
	−31.6	153.9	Bromoform, vap	151.2	253.0
	0	170.8	Butyl alcohol, n, vap	116.0	147
	18	182.7	tert, vap	82.9	160
	40	190.4	chloride, n, vap	78	149.5
	54	195.8	iodide, vap	130	202
	74	210.2	β-Butylene	18.8	74.4
	229	263.8		100.4	94.5
	334	312.3		200	119.2
	357	317.5	Butyric acid, vap	161.7	130.0
	409	341.3	Carbon dioxide	−97.8	89.6
	466	350.1		−78.2	97.2
	481	358.3		−60.0	106.1
	537	368.6		−40.2	115.5
	565	375.0		−21	129.4
	620	391.6		−19.4	126.0
	638	401.4		0	139.0
	750	426.3		15	145.7
	810	441.9		19	149.9
	923	464.3		20	148.0
	1034	490.6		30	153
	1134	520.6		32	155
				35	156
Alcohol. See *Ethyl, Methyl*, etc.				40	157
Ammonia	−78.5	67.2		99.1	186.1
	0	91.8		104	188.9
	20	98.2		182.4	222.1
	50	109.2		235	241.5
	100	127.9		302.0	268.2
	132.9	139.9		490	330.0
	150	146.3		685	380.0
	200	164.6		850	435.8
	250	181.4		1052	478.6
	300	198.7	disulfide, vap	0	91.1
Argon	0	209.6		14.2	96.4
	20	221.7		114.3	130.3
	100	269.5		190.2	156.1
	200	322.3		309.8	196.6
	302	368.5	monoxide	−191.5	56.1
	401	411.5		−78.5	127
	493	448.4		0	166
	584	481.5		15	172
	714	525.7		21.7	175.3
	827	563.2		126.7	218.3
Arsenic hydride (Arsine)	0	145.8		227.0	254.8
	15	114.0		276.9	271.4
	100	198.1	tetrachloride, vap	76.7	195.0

Gas or vapor	Temp. °C	Viscosity micropoises
Carbon tetrachloride, vap	127.9	133.4
	200.2	156.2
	314.9	190.2
Chlorine	12.7	129.7
	20	132.7
	50	146.9
	100	167.9
	150	187.5
	200	208.5
	250	227.6
Chloroform, vap	0	93.6
	14.2	98.9
	100	129
	121.3	135.7
	189.1	157.9
	250.0	177.6
	307.5	194.7
Cyanogen	0	92.8
	17	98.7
	100	127.1
Ethane	−78.5	63.4
	0	84.8
	17.2	90.1
	50.8	100.1
	100.4	114.3
	200.3	140.9
Ether (diethyl), vap	0	67.8
	14.2	71.6
	100	95.5
	121.8	98.3
	159.4	107.9
	189.9	115.2
	251.0	130.0
	277.8	135.8
Ethyl acetate, vap	0	68.4
	100	94.3
	128.1	101.8
	158.6	109.8
	192.9	119.5
	212.5	126
alcohol, vap	100	108
	130.2	117.3
	170.7	129.3
	191.8	135.5
	212.5	140
	251.7	151.9
	308.7	167.0
bromide, vap	38.4	186.5
butyrate, vap	119.8	160.0
chloride, vap	0	93.7
Ethylene	−75.7	69.9
	−44.1	76.9
	−38.6	78.5
	0	90.7
	13.8	95.4
	20	100.8
Ethylene	50	110.3
	100	125.7
	150	140.3
	200	154.1
	250	166.6
bromide, vap	131.6	221.0
chloride, vap	83.5	168.0
Ethyl formate, vap	99.8	92
iodide, vap	72.3	216.0
Helium	−257.4	27.0
	−252.6	35.0
	−191.6	87.1
	0	186.0
	20	194.1
	100	228.1
	200	267.2
	250	285.3
	282	299.2
	407	343.6
	486	370.6
	606	408.7
	676	430.3
	817	471.3
Hydrogen	−257.7	5.7

Gas or vapor	Temp. °C	Viscosity micropoises
	−252.5	8.5
	−198.4	33.6
	−183.4	38.8
	−113.5	57.2
	−97.5	61.5
	−31.6	76.7
	0	83.5
	20.7	87.6
	28.1	89.2
	129.4	108.6
	229.1	126.0
	299	138.1
	412	155.4
	490	167.2
	601	182.9
	713	198.2
	825	213.7
bromide	18.7	181.9
	100.2	234.4
chloride	12.5	138.5
	16.5	140.7
	18	142.6
	100.3	182.2
iodide	20	165.5
	50	201.8
	100	231.6
	150	262.7
	200	292.4
	250	318.9
phosphide	0	106.1
	15	112.0
	100	143.8
sulfide	0	116.6
Hydrogen sulfide	17	124.1
	100	158.7
Iodine, vap	124.0	184
	170.0	204
	205.4	220
	247.1	240
Isobutyl acetate, vap	16.1	76.4
	116.4	155.0
alcohol, vap	108.4	144.5
bromide, vap	92.3	179.5
butyrate, vap	156.9	167.0
chloride, vap	68.5	150.0
iodide, vap	120	204.7
Isopentane, vap	25	69.5
	100	86.0
Isopropyl alcohol, vap	99.8	109
	120.3	103.1
	198.4	124.8
	293.1	148.8
bromide, vap	60	176.0
chloride, vap	37.0	148.5
iodide, vap	89.3	201.5
Krypton	0	232.7
	15	246
Mercury, vap	273	494
	313	551
	369	641
	380	654
Methane	−181.6	34.8
	−78.5	76.0
	0	102.6
	20	108.7
	100.0	133.1
	200.5	160.5
	284	181.3
	380	202.6
	499	226.4
Methyl acetate, vap	99.8	98
	100	100
	143.3	113.9
	218.5	134.8
alcohol, vap	66.8	135.0
	111.3	125.9
	217.5	102.0
	311.5	192.1
chloride	−15.3	92
	0	96.9
	15.0	104

Gas or vapor	Temp. °C	Viscosity micropoises	Gas or vapor	Temp. °C	Viscosity micropoises
	99.1	137		829	501.2
	182.4	168	n-Pentane, vap............	25	67.6
	302.0	211		100	84.1
iodide, vap.............	44	232	Propane..................	17.9	79.5
Neon....................	0	297.3		100.4	100.9
	20	311.1		199.3	125.1
	100	364.6	n-Propyl alcohol, vap.......	99.9	93
	200	424.8		121.7	102.5
	250	453.2		209.7	126.7
Neon....................	285	470.8		273.0	143.4
	429	545.4	bromide, vap............	99.8	119
	502	580.2	Propylene................	16.7	83.4
	594	623.0		49.9	93.5
	686	662.6		100.1	107.6
	827	721.0		199.4	133.8
Nitric oxide (NO).........	0	178	Propyl iodide, vap.........	102	210.0
	20	187.6	Sulfur dioxide.............	−75.0	85.8
	100	227.2		−20.0	107.8
	200	268.2		0	115.8
Nitrogen.................	−21.5	156.3		0	117
	10.9	170.7		18	124.2
	27.4	178.1		20.5	125.4
	127.2	219.1		100.4	161.2
	226.7	255.9		199.4	203.8
	299	279.7		293	244.7
	490	337.4		490	311.5
	825	419.2	Trimethylbutane. (2,2,3-),		
Nitrosyl chloride..........	15	113.9	vap..................	70.3	73.4
	100	150.4		132.2	82.7
	200	192.0		262.1	104.8
Nitrous oxide (N₂O).......	0	135	Trimethylethylene, vap.....	25	70.1
	26.9	148.8		100	86.9
	126.9	194.3	Water, vap...............	100	125.3
n-Nonane, vap............	100.3	63.3		150	144.5
	202.1	78.1		200	163.5
n-Octane, vap............	100.4	67.5		250	182.7
	202.2	84.0		300	300.1
Oxygen..................	0	189		350	221.8
	19.1	201.8		400	241.2
	127.7	256.8	Xenon...................	0	210.1
	227.0	301.7		16.5	223.5
	283	323.3		20	226.0
	402	369.3		127	300.9
	496	401.3		177	335.1
	608	437.0		227	365.2
	690	461.2		277	395.4

DIFFUSIVITIES OF GASES IN LIQUIDS

Solute	Solvent	Temp., °C	Diffusivity x 10^5, cm²/sec
H_2	n-Hexane	25.4	16.36
H_2	Cyclohexane	25.4	7.08
H_2	Ethylene glycol	25.4	0.75
H_2	Carbon tetrachloride	0	6.28
O_2	Cyclohexane	29.6	5.31
O_2	Carbon tetrachloride	25.4	3.71
O_2	Ethanol	29.6	2.64
N_2	Carbon tetrachloride	0	2.44
CH_4	Glycerol	25.4	0.95
C_2H_6	n-Hexane	30	6.00
C_2H_6	n-Heptane	30	5.60
C_2H_2	Water	0	1.10
H_2S	Water	16	1.77
CO_2	Amyl alcohol	25	1.91
CO_2	Isobutyl alcohol	25	2.20
SO_2	n-Heptane	20	2.70
SO_2	n-Nonane	20	2.50
SO_2	n-Decane	20	2.40

DIFFUSION COEFFICIENTS IN AQUEOUS SOLUTIONS AT 25°

The diffusion coefficient D may be defined by either of the equations

$$J = -D \frac{\partial c}{\partial x}$$

or

$$\frac{\partial c}{\partial t} = D \frac{\partial^2 c}{\partial x^2}$$

when diffusion occurs in the x-direction only. Here J is the diffusion-flux across unit area normal to the x-direction, $\frac{\partial c}{\partial x}$ is the concentration-gradient at a fixed time, $\frac{\partial c}{\partial t}$ is the rate of change of concentration with time at a fixed distance. If J is expressed in mole cm^{-2} sec^{-1} and c in mole cm^{-3}, x in cm, and t in sec, D will be given in units of cm^2 sec^{-1}. In general D varies somewhat with concentration. The values below are a selection from measurements by modern high-precision methods, mainly by H. S. Harned and collaborators, R. H. Stokes and collaborators, L. J. Gosting and collaborators, and L. G. Longsworth.

For strong electrolytes at infinite dilution, limiting diffusion coefficients may be calculated by the Nernst relation:

$$D = \frac{RT}{F^2} \left[\frac{(\nu_1 + \nu_2)(\lambda_1^0 \lambda_2^0)}{\nu_1 |Z_1| (\lambda_1^0 + \lambda_2^0)} \right]$$

where R = gas constant, F = Faraday, T = absolute temperature, λ_1^0 and λ_2^0 are cation and anion limiting equivalent conductances, ν_1 and ν_2 are the numbers of cations and anions formed from one "molecule" of electrolyte, and Z_1 is the cation valency. Concentrations, unless expressed otherwise are as molarities and the diffusion coefficients are expressed as $10^5 D/cm^2$ sec^{-1} at 25°C.

DIFFUSION COEFFICIENTS OF STRONG ELECTROLYTES
Molarity

Solute	0.01	0.1	1.0
HCL	....	3.050	3.436
HBr	....	3.156	3.87
LiCl	1.312	1.269	1.302
LiBr	....	1.279	1.404
LiNO₃	1.276	1.240	1.293
NaCl	1.545	1.483	1.484
NaBr	....	1.517	1.596
NaI	....	1.520	1.662
KCl	1.917	1.844	1.892
KBr	....	1.874	1.975
KI	....	1.865	2.065
KNO₃	1.846	....	....
KClO₄	1.790	....	....
CaCl₂	1.188	1.110	1.203

C = 0.005M

Solute	0.01	0.1	1.0
BaCl₂	1.265	1.159	1.179
Na₂SO₄	1.123	....	....
MgSO₄	0.710	....	....
LaCl₃	1.105	....	....
K₄Fe(CN)₆	1.183	....	....

DIFFUSION COEFFICIENTS OF WEAK AND NON-ELECTROLYTES
Concentration

Solute	Concentration	Coefficient
Glucose	0.39%	0.673
Sucrose	0.38%	0.521
Raffinose	0.38%	0.434
Sucrose	Zero	0.5226
Mannitol	Zero	0.682
Penta-erythritol	Zero	0.761
Glycolamide	Zero	1.142
Glycine	Zero	1.064
α-alanine	0.32%	0.910
β-alanine	0.31%	0.933
Amino-benzoic acid ortho	0.24%	0.840
Amino-benzoic acid meta	0.24%	0.774
Amino-benzoic acid para	0.23%	0.843
Citric acid	0.1 M	0.661

DIFFUSION OF GASES INTO AIR

Gas or vapor	Temp. °C	Coefficient of diffusion, sq. cm/sec	Observer
Alcohol, vapor.......	40.4	0.137	Winkelmann
Carbon dioxide......	0.0	0.139	Mean of various
Carbon disulfide......	19.9	0.102	Winkelmann
Ether, vapor.........	19.9	0.089	Winkelmann
Hydrogen.............	0.0	0.634	Obermayer
Oxygen..............	0.0	0.178	Obermayer
Water, vapor.........	8.0	0.239	Guglielmo

RADIOACTIVE TRACER DIFFUSION DATA FOR PURE METALS

John Askill

The data in these tables are the most reliable set of radioactive tracer diffusion data for pure metals published in the literature from 1938 through December, 1970. For a complete listing of all published data on this subject up to December 1968 see "Tracer Diffusion Data for Metals, Alloys and Simple Oxides" by John Askill, published by Plenum Press, New York, 1970.

The diffusion coefficient D_T at a temperature $T(^\circ K)$ is given by the following relation:

$$D_T = D_0 e^{-Q/RT}$$

Abbreviations used in the tables are:

A.R.G. = Autoradiography	S. = Single Crystal
R.A. = Residual Activity	$\perp$c. = Perpendicular to c Direction
S.D. = Surface Decrease	‖c. = Parallel to c Direction
S.S. = Serial Sectioning	99.95 = 99.95%
P. = Polycrystalline	

Solute (tracer)	Material (metal, crystalline form and purity)		Temperature range, °C	Form of analysis	Activation energy, Q, Kcal/mole	Frequency factor, D_0, cm²/sec	Reference
Aluminum							
Ag¹¹⁰	S	99.999	371–655	S.S.	27.83	0.118	1
Al²⁷	S		450–650	S.S.	34.0	1.71	2
Au¹⁹⁸	S	99.999	423–609	S.S.	27.0	0.077	3
Cd¹¹⁵	S	99.999	441–631	S.S.	29.7	1.04	3
Ce¹⁴¹	P	99.995	450–630	R.A.	26.60	1.9×10^{-6}	5
Co⁶⁰	S	99.999	369–655	S.S.	27.79	0.131	1
Cr⁵¹	S	99.999	422–654	S.S.	41.74	464	1
Cu⁶⁴	S	99.999	433–652	S.S.	32.27	0.647	1
Fe⁵⁹	S	99.99	550–636	S.S.	46.0	135	3
Ga⁷²	S	99.999	406–652	S.S.	29.24	0.49	1
Ge⁷¹	S	99.999	401–653	S.S.	28.98	0.481	1
In¹¹⁴	P	99.99	400–600	S.S., R.A.	27.6	0.123	4
La¹⁴⁰	P	99.995	500–630	R.A.	27.0	1.4×10^{-6}	5
Mn⁵⁴	P	99.99	450–650	S.S.	28.8	0.22	2
Mo⁹⁹	P	99.995	400–630	R.A.	13.1	1.04×10^{-9}	6
Nb⁹⁵	P	99.95	350–480	R.A.	19.65	1.66×10^{-7}	7
Nd¹⁴⁷	P	99.995	450–630	R.A.	25.0	4.8×10^{-7}	5
Ni⁶³	P	99.99	360–630	R.A.	15.7	2.9×10^{-8}	8
Pd¹⁰³	P	99.995	400–630	R.A.	20.2	1.92×10^{-7}	9
Pr¹⁴²	P	99.995	520–630	R.A.	22.07	3.38×10^{-7}	5
Qu¹¹⁴	P		448–620	R.A.	29.1	0.09	10
Sm¹⁵³	P	99.995	450–630	R.A.	22.88	3.45×10^{-7}	5
Sn¹¹³	P		400–600	S.S., R.A.	28.5	0.245	4
V⁴⁸	P	99.999	400–630	R.A.	19.6	6.05×10^{-8}	11
Zn⁶⁵	S	99.999	357–653	S.S.	28.86	0.259	1
Beryllium							
Ag¹¹⁰	S⊥c	99.75	650–900	R.A.	43.2	1.76	12
Ag¹¹⁰	S‖c	99.75	650–900	R.A.	39.3	0.43	12
Be⁷	S⊥c	99.75	565–1065	R.A.	37.6	0.52	13
Be⁷	S‖c	99.75	565–1065	R.A.	39.4	0.62	13
Fe⁵⁹	S	99.75	700–1076	R.A.	51.6	0.67	12
Ni⁶³	P		800–1250	R.A.	58.0	0.2	14
Cadmium							
Ag¹¹⁰	S	99.99	180–300		25.4	2.21	15
Cd¹¹⁵	S	99.95	110–283	R.A.	19.3	0.14	16
Zn⁶⁵	S	99.99	180–300	—	19.0	0.0016	15
Calcium							
C¹⁴		99.95	550–800	R.A.	29.8	3.2×10^{-5}	17
Ca⁴⁵		99.95	500–800	R.A.	38.5	8.3	17
Fe⁵⁹		99.95	500–800	R.A.	23.3	2.7×10^{-3}	17
Ni⁶³		99.95	550–800	—	28.9	1.0×10^{-5}	17
U²³⁵		99.95	500–700	R.A.	34.8	1.1×10^{-5}	17
Carbon							
Ag¹¹⁰	⊥c		750–1050	R.A.	64.3	9280	18
C¹⁴			2000–2200	—	163	5	19
Ni⁶³	⊥c		540–920	R.A.	47.2	102	18
Ni⁶³	‖c		750–1060	R.A.	53.3	2.2	18
Th²²⁸	⊥c		1400–2200	R.A.	145.4	1.33×10^{-5}	18
Th²²⁸	‖c		1800–2200	R.A.	114.7	2.48	18
U²³²	⊥c		1400–2200	R.A.	115.0	6760	18
U²³²	‖c		1400–1820	R.A.	129.5	385	18
Chromium							
C¹⁴	P		1200–1500	R.A.	26.5	9.0×10^{-3}	20
Cr⁵¹	P	99.98	1030–1545	S.S.	73.7	0.2	21
Fe⁵⁹	P	99.8	980–1420	R.A.	79.3	0.47	22
Mo⁹⁹	P		1100–1420	R.A.	58.0	2.7×10^{-3}	20
Cobalt							
C¹⁴	P	99.82	600–1400	R.A.	34.0	0.21	23
Co⁶⁰	P	99.9	1100–1405	S.S.	67.7	0.83	24
Fe⁵⁹	P	99.9	1104–1303	S.S.	62.7	0.21	24
Ni⁶³	P		1192–1297	R.A.	60.2	0.10	25
S³⁵	P	99.99	1150–1250	R.A.	5.4	1.3	26
Copper							
Ag¹¹⁰	S, P		580–980	R.A.	46.5	0.61	27
As⁷⁶	P		810–1075	R.A.	42.13	0.20	28
Au¹⁹⁸	S, P		400–1050	S.S.	42.6	0.03	29
Cd¹¹⁵	S	99.98	725–950	S.S.	45.7	0.935	30

RADIOACTIVE TRACER DIFFUSION DATA FOR PURE METALS (Continued)

Solute (tracer)	Material (metal, crystalline form and purity)		Temperature range, °C	Form of analysis	Activation energy, Q, Kcal/mole	Frequency factor, D_0, cm² sec	Reference
Ce^{141}	P	99.999	766–947	R.A.	27.6	2.17×10^{-8}	31
Cr^{51}	S, P		800–1070	R.A.	53.5	1.02	32
Co^{60}	S	99.998	701–1077	S.S.	54.1	1.93	33
Cu^{67}	S	99.999	698–1061	S.S.	50.5	0.78	34
Eu^{152}	P	99.999	750–970	S.S., R.A.	26.85	1.17×10^{-7}	31
Fe^{59}	S, P		460–1070	R.A.	52.0	1.36	32
Ga^{72}			—	—	45.90	0.55	35
Ge^{68}	S	99.998	653–1015	S.S.	44.76	0.397	36
Hg^{203}	P		—	—	44.0	0.35	35
Lu^{177}	P	99.999	857–1010	R.A.	26.15	4.3×10^{-9}	31
Mn^{54}	S	99.99	754–950	S.S.	91.4	10^7	37
Nb^{95}	P	99.999	807–906	R.A.	60.06	2.04	38
Ni^{63}	P		620–1080	R.A.	53.8	1.1	39
Pd^{102}	S	99.999	807–1056	S.S.	54.37	1.71	40
Pm^{147}	P	99.999	720–955	R.A.	27.5	3.62×10^{-8}	31
Pt^{195}	P		843–997	S.S.	37.5	4.8×10^{-4}	41
S^{35}	S	99.999	800–1000	R.A.	49.2	23	42
Sb^{124}	S	99.999	600–1000	S.S.	42.0	0.34	43
Sn^{113}	P		680–910	—	45.0	0.11	44
Tb^{160}	P	99.999	770–980	R.A.	27.45	8.96×10^{-9}	31
Tl^{204}	S	99.999	785–996	S.S.	43.3	0.71	45
Tm^{170}	P	99.999	705–950	R.A.	24.15	7.28×10^{-9}	31
Zn^{65}	P	99.999	890–1000	S.S.	47.50	0.73	46
Germanium							
Cd^{115}	S		750–950	R.A.	102.0	1.75×10^9	47
Fe^{59}	S		775–930	R.A.	24.8	0.13	48
Ge^{71}	S		766–928	S.S.	68.5	7.8	49
In^{114}	S		600–920	—	39.9	2.9×10^{-4}	50
Sb^{124}	S		720–900	—	50.2	0.22	51
Te^{125}	S		770–900	S.S.	56.0	2.0	52
Tl^{204}	S		800–930	S.S.	78.4	1700	53
Gold							
Ag^{110}	S	99.99	699–1007	S.S.	40.2	0.072	54
Au^{198}	S	99.97	850–1050	S.S.	42.26	0.107	224
Co^{60}	P	99.93	702–948	R.A.	41.6	0.068	55
Fe^{59}	P	99.93	701–948	R.A.	41.6	0.082	55
Hg^{203}	S	99.994	600–1027	—	37.38	0.116	56
Ni^{63}	P	99.96	880–940	S.S.	46.0	0.30	57
Pt^{195}	P, S	99.98	800–1060	S.S.	60.9	7.6	58
β-Hafnium							
Hf^{181}	P	97.9	1795–1995	S.S.	38.7	1.2×10^{-3}	59
Indium							
Ag^{110}	S⊥c	99.99	25–140	S.S.	12.8	0.52	60
Ag^{110}	S‖c	99.99	25–140	S.S.	11.5	0.11	60
Au^{198}	S	99.99	25–140	S.S.	6.7	9×10^{-3}	60
In^{114}	S⊥c	99.99	44–144	S.S.	18.7	3.7	61
In^{114}	S‖c	99.99	44–144	S.S.	18.7	2.7	61
Tl^{204}	S	99.99	49–157	S.S.	15.5	0.049	62
α-Iron							
Ag^{110}	P		748–888	S.S.	69.0	1950	63
Au^{198}	P	99.999	800–900	R.A.	62.4	31	64
C^{14}	P	99.98	616–844	R.A.	29.3	2.2	65
Co^{60}	P	99.995	638–768	R.A.	62.2	7.19	62
Cr^{51}	P	99.95	775–875	R.A.	57.5	2.53	66
Cu^{64}	P	99.9	800–1050	R.A.	57.0	0.57	67
Fe^{55}	P	99.92	809–889	—	60.3	5.4	68
K^{42}	P	99.92	500–800	R.A.	42.3	0.036	69
Mn^{54}	P	99.97	800–900	R.A.	52.5	0.35	70
Mo^{99}	P		750–875	R.A.	73.0	7800	71
Ni^{63}	P	99.97	680–800	R.A.	56.0	1.3	72
P^{32}	P		860–900	R.A.	55.0	2.9	73
Sb^{124}	P		800–900	R.A.	66.6	1100	74
V^{48}	P		755–875	R.A.	55.4	1.43	75
W^{185}	P		755–875	R.A.	55.1	0.29	75
γ-Iron							
Be^7	P	99.9	1100–1350	R.A.	57.6	0.1	76
C^{14}	P	99.34	800–1400	—	34.0	0.15	23
Co^{60}	P	99.98	1138–1340	S.S.	72.9	1.25	77
Cr^{51}	P	99.99	950–1400	R.A.	69.7	10.8	78
Fe^{59}	P	99.98	1171–1361	S.S.	67.86	0.49	79
Hf^{181}	P	99.99	1110–1360	R.A.	97.3	3600	78
Mn^{54}	P	99.97	920–1280	R.A.	62.5	0.16	70
Ni^{63}	P	99.97	930–2050	R.A.	67.0	0.77	72
P^{32}	P	99.99	950–1200	R.A.	43.7	0.01	80
S^{35}	P		900–1250	R.A.	53.0	1.7	81
V^{48}	P	99.99	1120–1380	R.A.	69.3	0.28	78
W^{185}	P	99.5	1050–1250	R.A.	90.0	1000	82
δ-Iron							
Co^{60}	P	99.995	1428–1521	R.A.	61.4	6.38	83
Fe^{59}	P	99.95	1428–1492	S.S.	57.5	2.01	83
P^{32}	P	99.99	1370–1460	R.A.	55.0	2.9	73
Lanthanum							
Au^{198}	P	99.97	600–800	S.S.	45.1	1.5	84
La^{140}	P	99.97	690–850	S.S.	18.1	2.2×10^{-2}	84
Lead							
Ag^{110}	P	99.9	200–310	R.A.	14.4	0.064	85
Au^{198}	S	99.999	190–320	S.S.	10.0	8.7×10^{-3}	86

Solute (tracer)	Material (metal, crystalline form and purity)		Temperature range, °C	Form of analysis	Activation energy, Q, Kcal/mole	Frequency factor, D_0, cm²/sec	Reference
Cd^{115}	S	99.999	150–320	S.S.	21.23	0.409	87
Cu^{64}	S		150–320	S.S.	14.44	0.046	88
Pb^{204}	S	99.999	150–320	S.S.	25.52	0.887	87
Tl^{205}	P	99.999	207–322	S.S.	24.33	0.511	89
Lithium							
Ag^{110}	P	92.5	65–161	S.S.	12.83	0.37	90
Au^{195}	P	92.5	47–153	S.S.	10.49	0.21	90
Bi	P	99.95	141–177	S.S.	47.3	5.3×10^{13}	91
Cd^{115}	P	92.5	80–174	S.S.	16.05	2.35	90
Cu^{64}	P	99.98	51–120	S.S.	9.22	0.47	93
Ga^{72}	P	99.98	58–173	S.S.	12.9	0.21	93
Hg^{203}	P	99.98	58–173	S.S.	14.18	1.04	93
In^{114}	P	92.5	80–175	S.S.	15.87	0.39	90
Li^{6}	P	99.98	35–178	S.S.	12.60	0.14	94
Na^{22}	P	92.5	52–176	S.S.	12.61	0.41	90
Pb^{204}	P	99.95	129–169	S.S.	25.2	160	91
Sb^{124}	P	99.95	141–176	S.S.	41.5	1.6×10^{10}	91
Sn^{113}	P	99.95	108–174	S.S.	15.0	0.62	91
Zn^{65}	P	92.5	60–175	S.S.	12.98	0.57	92
Magnesium							
Ag^{110}	P	99.9	476–621	S.S.	28.50	0.34	95
Fe^{59}	P	99.95	400–600	R.A.	21.2	4×10^{-6}	96
In^{114}	P	99.9	472–610	S.S.	28.4	5.2×10^{-2}	95
Mg^{28}	S⊥c		467–635	S.S.	32.5	1.5	97
Mg^{28}	S∥c		467–635	S.S.	32.2	1.0	97
Ni^{63}	P	99.95	400–600	R.A.	22.9	1.2×10^{-5}	96
U^{235}	P	99.95	500–620	R.A.	27.4	1.6×10^{-5}	96
Zn^{65}	P	99.9	467–620	S.S.	28.6	0.41	95
Molybdenum							
C^{14}	P	99.98	1200–1600	R.A.	41.0	2.04×10^{-2}	99
Co^{60}	P	99.98	1850–2350	S.S.	106.7	18	100
Cr^{51}	P		1000–1500	R.A.	54.0	2.5×10^{-4}	20
Cs^{134}	S	99.99	1000–1470	R.A., A.R.G.	28.0	8.7×10^{-11}	101
K^{42}	S		800–1100	R.A.	25.04	5.5×10^{-9}	102
Mo^{99}	P		1850–2350	S.S.	96.9	0.5	103
Na^{24}	S		800–1100	R.A.	21.25	2.95×10^{-9}	102
Nb^{95}	P	99.98	1850–2350	S.S.	108.1	14	100
P^{32}	P	99.97	2000–2200	S.S.	80.5	0.19	104
Re^{186}	P		1700–2100	A.R.G.	94.7	0.097	105
S^{35}	S	99.97	2220–2470	S.S.	101.0	220	106
Ta^{182}	P		1700–2150	R.A.	83.0	3.5×10^{-4}	20
U^{235}	P	99.98	1500–2000	R.A.	76.4	7.6×10^{-3}	107
W^{185}	P	99.98	1700–2260	S.S.	110.4	1.7	108
Nickel							
Au^{198}	S, P	99.999	700–1075	S.S.	55.0	0.02	109
Be^{7}	P	99.9	1020–1400	R.A.	46.2	0.019	76
C^{14}	P	99.86	600–1400	—	34.0	0.012	23
Co^{60}	P	99.97	1149–1390	R.A.	65.9	1.39	110
Cr^{51}	P	99.95	1100–1270	S.S.	65.1	1.1	111
Cu^{64}	P	99.95	1050–1360	S.S.	61.7	0.57	111
Fe^{59}	P		1020–1263	S.S.	58.6	0.074	112
Mo^{99}	P		900–1200	R.A.	51.0	1.6×10^{-3}	20
Ni^{63}	P	99.95	1042–1404	S.S.	68.0	1.9	111
Pu^{238}	P		1025–1125	A.R.G.	51.0	0.5	113
Sb^{124}	P	99.97	1020–1220	—	27.0	1.8×10^{-5}	114
Sn^{113}	P	99.8	700–1350	A.R.G.	58.0	0.83	115
V^{48}	P	99.99	800–1300	R.A.	66.5	0.87	11
W^{185}	P	99.95	1100–1300	S.S.	71.5	2.0	116
Niobium							
C^{14}	P		800–1250	R.A.	32.0	1.09×10^{-5}	117
Co^{60}	P	99.85	1500–2100	A.R.G.	70.5	0.74	118
Cr^{51}	S		943–1435	S.S.	83.5	0.30	119
Fe^{55}	P	99.85	1400–2100	A.R.G.	77.7	1.5	118
K^{42}	S		900–1100	R.A.	22.10	2.38×10^{-7}	102
Nb^{95}	P, S	99.99	878–2395	S.S.	96.0	1.1	120
P^{32}	P	99.0	1300–1800	S.S.	51.5	5.1×10^{-2}	104
S^{35}	S	99.9	1100–1500	R.A.	73.1	2600	121
Sn^{113}	P	99.85	1850–2400	S.S.	78.9	0.14	122
Ta^{182}	P, S	99.997	878–2395	S.S.	99.3	1.0	120
Ti^{44}	S		994–1492	S.S.	86.9	0.099	123
U^{235}	P	99.55	1500–2000	R.A.	76.8	8.9×10^{-3}	107
V^{48}	S	99.99	1000–1400	R.A.	85.0	2.21	124
W^{185}	P	99.8	1800–2200	R.A.	91.7	5×10^{-4}	125
Palladium							
Pd^{103}	S	99.999	1060–1500	S.S.	63.6	0.205	126
Phosphorus							
P^{32}	P		0–44	S.S.	9.4	1.07×10^{-3}	127
Platinum							
Co^{60}	P	99.99	900–1050	—	74.2	19.6	129
Cu^{64}	P		1098–1375	S.S.	59.5	0.074	41
Pt^{195}	P	99.99	1325–1600	S.S.	68.2	0.33	130
Potassium							
Au^{198}	P	99.95	5.6–52.5	S.S.	3.23	1.29×10^{-3}	131
K^{42}	S	99.7	−52–61	S.S.	9.36	0.16	132
Na^{22}	P	99.7	0–62	S.S.	7.45	0.058	133
Rb^{86}	P	99.95	0.1–59.9	S.S.	8.78	0.090	134

Solute (tracer)	Material (metal, crystalline form and purity)		Temperature range, °C	Form of analysis	Activation energy, Q, Kcal/mole	Frequency factor, D_0, cm²/sec	Reference
γ-Plutonium							
Pu^{238}	P		190–310	S.S.	16.7	2.1×10^{-5}	135
δ-Plutonium							
Pu^{238}	P		350–440	S.S.	23.8	4.5×10^{-3}	136
ε-Plutonium							
Pu^{238}	P		500–612	R.A.	18.5	2.0×10^{-2}	137
α-Praseodymium							
Ag^{110}	P	99.93	610–730	S.S.	25.4	0.14	138
Au^{195}	P	99.93	650–780	S.S.	19.7	4.3×10^{-2}	138
Co^{60}	P	99.93	660–780	S.S.	16.4	4.7×10^{-2}	138
Zn^{65}	P	99.96	766–603	S.S.	24.8	0.18	139
β-Praseodymium							
Ag^{110}	P	99.93	800–900	S.S.	21.5	3.2×10^{-2}	138
Au^{195}	P	99.93	800–910	S.S.	20.1	3.3×10^{-2}	138
Ho^{166}	P	99.96	800–930	S.S.	26.3	9.5	140
In^{114}	P	99.96	800–930	S.S.	28.9	9.6	140
La^{140}	P	99.96	800–930	S.S.	25.7	1.8	140
Pr^{142}	P	99.93	800–900	S.S.	29.4	8.7	140
Zn^{65}	P	99.96	822–921	S.S.	27.0	0.63	139
Selenium							
Fe^{59}	P		40–100	R.A.	8.88	—	141
Hg^{203}	P	99.996	25–100	R.A.	1.2	—	141
S^{35}	S⊥c		60–90	S.D.	29.9	1700	142
S^{35}	S∥c		60–90	S.D.	15.6	1100	142
Se^{75}	P		35–140	—	11.7	1.4×10^{-4}	143
Silicon							
Au^{198}	S		700–1300	S.S.	47.0	2.75×10^{-3}	145
C^{14}	P		1070–1400	R.A.	67.2	0.33	146
Cu^{64}	P		800–1100	R.A.	23.0	4×10^{-2}	147
Fe^{59}	S		1000–1200	R.A.	20.0	6.2×10^{-3}	148
Ni^{63}	P		450–800	—	97.5	1000	149
P^{32}	S		1100–1250	R.A.	41.5	—	150
Sb^{124}	S		1190–1398	R.A.	91.7	12.9	151
Si^{31}	S	99.99999	1225–1400	S.S.	110.0	1800	146
Silver							
Au^{198}	P	99.99	718–942	S.S.	48.28	0.85	54
Ag^{110}	S	99.999	640–955	S.S.	45.2	0.67	152
Cd^{115}	S	99.99	592–937	S.S.	41.69	0.44	153
Co^{60}	S	99.999	700–940	—	48.75	1.9	154
Cu^{64}	P	99.99	717–945	S.S.	46.1	1.23	155
Fe^{59}	S	99.99	720–930	S.S.	49.04	2.42	156
Ge^{77}	P		640–870	S.S.	36.5	0.084	157
Hg^{203}	P	99.99	653–948	S.S.	38.1	0.079	155
In^{114}	S	99.99	592–937	S.S.	40.80	0.41	153
Ni^{63}	S	99.99	749–950	S.S.	54.8	21.9	158
Pb^{210}	P		700–865	S.S.	38.1	0.22	159
Pd^{102}	S	99.999	736–939	S.S.	56.75	9.56	140
Ru^{103}	S	99.99	793–945	S.S.	65.8	180	160
S^{35}	S	99.999	600–900	R.A.	40.0	1.65	161
Sb^{124}	P	99.999	780–950	S.S., R.A.	39.07	0.234	162
Sn^{113}	S	99.99	592–937	S.S.	39.30	0.255	153
Te^{125}	P		770–940	R.A.	38.90	0.47	163
Tl^{204}	P		640–870	S.S.	37.9	0.15	157
Zn^{65}	S	99.99	640–925	S.S.	41.7	0.54	164
Sodium							
Au^{198}	P	99.99	1.0–77	S.S.	2.21	3.34×10^{-4}	165
K^{42}	P	99.99	0–91	S.S.	8.43	0.08	133
Na^{22}	P	99.99	0–98	S.S.	10.09	0.145	166
Rb^{86}	P	99.99	0–85	S.S.	8.49	0.15	133
Tantalum							
C^{14}	P		1450–2200	S.S.	40.3	1.2×10^{-2}	167
Fe^{59}	P		930–1240	—	71.4	0.505	168
Mo^{99}	P		1750–2220	R.A.	81.0	1.8×10^{-3}	20
Nb^{95}	P, S	99.996	921–2484	S.S.	98.7	0.23	169
S^{35}	P	99.0	1970–2110	R.A.	70.0	100	170
Ta^{182}	P, S	99.996	1250–2200	S.S.	98.7	1.24	226
Tellurium							
Hg^{203}	P		270–440	—	18.7	3.14×10^{-5}	171
Se^{75}	P		320–440	—	28.6	2.6×10^{-2}	171
Tl^{204}	P		360–430	—	41.0	320	172
Te^{127}	S⊥c	99.9999	300–400	S.S.	46.7	3.91×10^{4}	173
Te^{127}	S∥c	99.9999	300–400	S.S.	35.5	130	173
α-Thallium							
Ag^{110}	P⊥c	99.999	80–250	S.S.	11.8	3.8×10^{-2}	174
Ag^{110}	P∥c	99.999	80–250	S.S.	11.2	2.7×10^{-2}	174
Au^{198}	P⊥c	99.999	110–260	S.S.	2.8	2.0×10^{-5}	174
Au^{198}	P∥c	99.999	110–260	S.S.	5.2	5.3×10^{-4}	174
Tl^{204}	S⊥c	99.9	135–230	S.S.	22.6	0.4	175
Tl^{204}	S∥c	99.9	135–230	S.S.	22.9	0.4	175
β-Thallium							
Ag^{110}	P	99.999	230–310	S.S.	11.9	4.2×10^{-2}	174
Au^{198}	P	99.999	230–310	S.S.	6.0	5.2×10^{-4}	174
Tl^{204}	S	99.9	230–280	S.S.	20.7	0.7	175

Solute (tracer)	Material (metal, crystalline form and purity)		Temperature range, °C	Form of analysis	Activation energy, Q, Kcal/mole	Frequency factor, D_0, cm²/sec	Reference
α-Thorium							
Pa²³¹	P	99.85	770–910	—	74.7	126	176
Th²²⁸	P	99.85	720–880	—	71.6	395	176
U²³³	P	99.85	700–880	—	79.3	2210	176
Tin							
Ag¹¹⁰	S⊥c		135–225	S.S.	18.4	0.18	177
Ag¹¹⁰	S‖c		135–225	S.S.	12.3	7.1×10^{-3}	177
Au¹⁹⁸	S⊥c		135–225	S.S.	17.7	0.16	177
Au¹⁹⁸	S‖c		135–225	S.S.	11.0	5.8×10^{-3}	177
Co⁶⁰	S, P		140–217	R.A.	22.0	5.5	178
In¹¹⁴	S⊥c	99.998	181–221	S.S.	25.8	34.1	179
In¹¹⁴	S‖c	99.998	181–221	S.S.	25.6	12.2	179
Sn¹¹³	S⊥c	99.999	160–226	S.S.	25.1	10.7	180
Sn¹¹³	S‖c	99.999	160–226	S.S.	25.6	7.7	180
Tl²⁰⁴	P	99.999	137–216	S.S.	14.7	1.2×10^{-3}	181
α-Titanium							
Ti⁴⁴	P	99.99	700–850	R.A.	35.9	8.6×10^{-6}	182
β-Titanium							
Ag¹¹⁰	P	99.95	940–1570	S.S.	43.2	3×10^{-3}	183
Be⁷	P	99.96	915–1300	R.A.	40.2	0.8	184
C¹⁴	P	99.62	1100–1600	R.A.	20.0	3.02×10^{-3}	185
Cr⁵¹	P	99.7	950–1600	A.R.G.	35.1	5×10^{-3}	186
					61.0	4.9	
Co⁶⁰	P	99.7	900–1600	S.S.	30.6	1.2×10^{-2}	186
					52.5	2.0	
Fe⁵⁹	P	99.7	900–1600	A.R.G.	31.6	7.8×10^{-3}	186
					55.0	2.7	
Mo⁹⁹	P	99.7	900–1600	S.S.	43.0	8.0×10^{-3}	186
					73.0	20	
Mn⁵⁴	P	99.7	900–1600	S.S.	33.7	6.1×10^{-3}	186
					58.0	20	
Nb⁹⁵	P	99.7	1000–1600	A.R.G.	39.3	5.0×10^{-3}	186
					73.0	20	
Ni⁶³	P	99.7	925–1600	A.R.G.	29.6	9.2×10^{-3}	186
					51.5	2.0	
P³²	P	99.7	950–1600	S.S.	74.1	3.62×10^{-3}	187
					56.5	5	
Sc⁴⁶	P	99.95	940–1590	S.S.	32.4	4.0×10^{-3}	183
Sn¹¹³	P	99.7	950–1600	S.S.	31.6	3.8×10^{-4}	187
					69.2	10	
Ti⁴⁴	P	99.95	900–1540	S.S.	31.2	3.58×10^{-4}	188
					60.0	1.09	
U²³⁵	P	99.9	900–1400	R.A.	29.3	5.1×10^{-4}	189
V⁴⁸	P	99.95	900–1545	S.S.	32.2	3.1×10^{-4}	190
					57.2	1.4	
W¹⁸⁵	P	99.94	900–1250	R.A.	43.9	3.6×10^{-3}	191
Zr⁹⁵	P	98.94	920–1500	R.A.	35.4	4.7×10^{-3}	191
Tungsten							
C¹⁴	P	99.51	1200–1600	R.A.	53.5	8.91×10^{-3}	99
Fe⁵⁹	P		940–1240	—	66.0	1.4×10^{-2}	168
Mo⁹⁹	P		1700–2100	R.A.	101.0	0.3	20
Nb⁹⁵	P	99.99	1305–2367	S.S.	137.6	3.01	192
Re¹⁸⁶	S		2100–2400	R.A.	141.0	19.5	193
Ta¹⁸²	P	99.99	1305–2375	S.S.	139.9	3.05	192
W¹⁸⁵	P	99.99	1800–2403	S.S.	140.3	1.88	192
α-Uranium							
U²³⁴	P		580–650	—	40.0	2×10^{-3}	194
β-Uranium							
Co⁶⁰	P	99.999	692–763	S.S.	27.45	1.5×10^{-2}	195
U²³⁵	P		690–750	R.A.	44.2	2.8×10^{-3}	196
γ-Uranium							
Au¹⁹⁵	P	99.99	785–1007	S.S.	30.4	4.86×10^{-3}	197
Co⁶⁰	P	99.99	783–989	S.S.	12.57	3.51×10^{-4}	198
Cr⁵¹	P	99.99	797–1037	S.S.	24.46	5.37×10^{-3}	198
Cu⁶⁴	P	99.99	787–1039	S.S.	24.06	1.96×10^{-3}	198
Fe⁵⁵	P	99.99	787–990	S.S.	12.0	2.69×10^{-4}	198
Mn⁵⁴	P	99.99	787–939	S.S.	13.88	1.81×10^{-4}	198
Nb⁹⁵	P	99.99	791–1102	S.S.	39.65	4.87×10^{-2}	198
Ni⁶³	P	99.99	787–1039	S.S.	15.66	5.36×10^{-4}	198
U²³³	P	99.99	800–1070	S.S.	28.5	2.33×10^{-3}	227
Zr⁹⁵	P		800–1000	R.A.	16.5	3.9×10^{-4}	228
Vanadium							
C¹⁴	P	99.7	845–1130	S.S.	27.3	4.9×10^{-3}	199
Cr⁵¹	P	99.8	960–1200	R.A.	64.6	9.54×10^{-3}	200
Fe⁵⁹	P		960–1350	S.S.	71.0	0.373	201
P³²	P	99.8	1200–1450	R.A.	49.8	2.45×10^{-2}	202
S³⁵	P	99.8	1320–1520	R.A.	34.0	3.1×10^{-2}	184
V⁴⁸	S, P	99.99	880–1360	S.S.	73.65	0.36	203
V⁴⁸	S, P	99.99	1360–1830	S.S.	94.14	214.0	203
Yttrium							
Y⁹⁰	S⊥c		900–1300	R.A.	67.1	5.2	204
Y⁹⁰	S‖c		900–1300	R.A.	60.3	0.82	204
Zinc							
Ag¹¹⁰	S⊥c	99.999	271–413	S.S.	27.6	0.45	205
Ag¹¹⁰	S‖c	99.999	271–413	S.S.	26.0	0.32	205

Solute (tracer)	Material (metal, crystalline form and purity)		Temperature range, °C	Form of analysis	Activation energy, Q, Kcal/mole	Frequency factor, D_0, cm²/sec	Reference
Au[198]	S⊥c	99.999	315–415	S.S.	29.72	0.29	206
Au[198]	S‖c	99.999	315–415	S.S.	29.73	0.97	206
Cd[115]	S⊥c	99.999	225–416	S.S.	20.12	0.117	206
Cd[115]	S‖c	99.999	225–416	S.S.	20.54	0.114	206
Cu[64]	S⊥c	99.999	338–415	S.S.	29.92	2.0	206
Cu[64]	S‖c	99.999	338–415	S.S.	29.53	2.22	207
Ga[72]	S⊥c		240–403	S.S.	18.15	0.018	207
Ga[72]	S‖c		240–403	S.S.	18.4	0.016	207
Hg[203]	S⊥c		260–413	S.S.	20.18	0.073	208
Hg[203]	S‖c		260–413	S.S.	19.70	0.056	208
In[114]	S⊥c		271–413	S.S.	19.60	0.14	205
In[114]	S‖c		271–413	S.S.	19.10	0.062	205
Sn[113]	S⊥c		298–400	S.S.	18.4	0.13	209
Sn[113]	S‖c		298–400	S.S.	19.4	0.15	209
Zn[65]	S⊥c	99.999	240–418	S.S.	23.0	0.18	210
Zn[65]	S‖c	99.999	240–418	S.S.	21.9	0.13	210
α-Zirconium							
Cr[51]	P	99.9	700–850	R.A.	18.0	1.19×10^{-8}	211
Fe[55]	P		750–840	—	48.0	2.5×10^{-2}	212
Mo[99]	P		600–850	R.A.	24.76	6.22×10^{-8}	213
Nb[95]	P	99.99	740–857	R.A.	31.5	6.6×10^{-6}	182
Sn[113]	P		300–700	A.R.G.	22.0	1.0×10^{-8}	214
Ta[182]	P	99.6	700–800	R.A.	70.0	100	215
V[48]	P	99.99	600–850	R.A.	22.9	1.12×10^{-8}	124
Zr[95]	P	99.95	750–850	S.S.	45.5	5.6×10^{-4}	216
β-Zirconium							
Be[7]	P	99.7	915–1300	R.A.	31.1	8.33×10^{-2}	184
C[14]	P	96.6	1100–1600	R.A.	34.2	3.57×10^{-2}	217
Ce[141]	P		880–1600	R.A.	41.4	3.16	218
					74.1	42.17	
Co[60]	P	99.99	920–1600	S.S.	21.82	3.26×10^{-3}	219
Cr[51]	P	99.9	700–850	R.A.	18.0	1.19×10^{-8}	211
Fe[55]	P		750–840	—	48.0	2.5×10^{-2}	212
Mo[99]	P		900–1635	R.A.	35.2	1.99×10^{-6}	218
					68.55	2.63	
Nb[95]	P		1230–1635	R.A.	36.6	7.8×10^{-4}	220
P[32]	P	99.94	950–1200	R.A.	33.3	0.33	221
Sn[113]	P		300–700	A.R.G.	22.0	1×10^{-8}	214
Ta[182]	P	99.6	900–1200	R.A.	27.0	5.5×10^{-5}	215
U[235]	P		900–1065	S.S.	30.5	5.7×10^{-4}	222
V[48]	P	99.99	870–1200	R.A.	45.8	7.59×10^{-3}	223
V[48]	P	99.99	1200–1400	R.A.	57.7	0.32	223
W[185]	P	99.7	900–1250	R.A.	55.8	0.41	223
Zr[95]	P		1100–1500	S.S.	30.1	2.4×10^{-4}	225

REFERENCES

1. N. L. Peterson and S. J. Rothman, *Phys. Rev.* **B1** (8), 3264 (1970).
2. T. S. Lundy and J. F. Murdock, *J. Appl. Phys.* **33** (5), 1671 (1962).
3. W. B. Alexander and L. M. Slifkin, *Phys. Rev.* **B1** (8), 3274 (1970).
4. M. S. Anand and R. P. Agarwala *Phys. Stat. Solidi* **A1** (1), 41K (1970).
5. S. P. Murarka and R. P. Agarwala, *Indian At. Energy Comm.* **Report BARC-368** (1968).
6. A. R. Paul and R. P. Agarwala, *J. Appl. Phys.* **38** (9), 3790 (1967).
7. G. P. Tiwari and B. D. Sharma, *Trans. Indian Inst. Metals* **20**, 83 (1967).
8. K. Hirano, R. P. Agarwala, and M. Cohen, *Acta Met.* **10** (9), 857 (1962).
9. M. S. Anand and R. P. Agarwala, *Trans. AIME* **239** (11), 1848 (1967).
10. S. Badrinarayanan and H. B. Mathers, *Int. J. Appl. Radiat. Isotopes* **19** (4), 353 (1968).
11. S. P. Murarka, M. S. Anand, and R. P. Agarwala, *Acta Met.* **16** (1), 69 (1968).
12. M. C. Naik, J. M. Dupony, and Y. Adda, *Mem. Sci. Rev. Met.* **63**, 488 (1966).
13. J. M. Dupony, J. Mathie, and Y. Adda, *Mem. Sci. Rev. Met.* **63**, 481 (1966).
14. V. M. Ananyn, V. P. Gladkov, V. S. Zotov, and D. M. Skorov, *At. Energ:* **29** (3), 220 (1970).
15. W. Hirschwald and W. Schroedter, *Z. Physik. Chem. N.F.* **53**, 392 (1967).
16. W. Chomka, *Zess. Nauk Politech Gdansk Fizyka* **1**, 39 (1967).
17. L. V. Pavlinov, A. M. Gladyshev, and V. N. Bikov, *Fiz. Metal Metalloved.* **26** (5), 823 (1968).
18. J. R. Wolfe, D. R. McKenzie, and R. J. Borg, *J. Appl. Phys.* **36** (6), 1906 (1965).
19. M. A. Kanter, *Phys. Rev.* **107**, 655 (1957).
20. E. V. Borisov, P. L. Gruzin, and S. V. Zemskii, *Zashch Pokryt Metal* **2**, 104 (1968).
21. J. Askill and D. H. Tomlin, *Phil. Mag.* **11** (111), 467 (1965).
22. H. W. Paxton and R. A. Wolfe, *Trans. AIME* **230**, 1426 (1964).
23. I. I. Kovenski, *Fiz. Metal. i Metalloved.* **16**, 613 (1963).
24. H. W. Mead and C. E. Birchenall, *Trans. Met. Soc. AIME* **203** (9), 994 (1955).
25. K. Hirano, R. P. Agarwala, B. L. Averbach, and M. Cohen, *J. Appl. Phys.* **33** (10), 3049 (1962).
26. M. M. Pavlyuchenko and I. F. Kononyuk, *Dokl. Akad. Nauk Belorusskoi SSR* **8**, 157 (1964).
27. G. Barreau, G. Brunel, G. Azeron, P. Lacombe, *C. R. Acad. Sci. Paris* Ser C **270** (6), 514 (1970).
28. S. M. Klotsman, A. Ya Rabovskii, V. K. Talinskii, and A. N. Timofeeu, *Fiz. Met. Metalloved* **29** (4), 803 (1970).
29. A. Chatterjee and D. J. Fabian, *Acta Met.* **17** (9), 1141 (1969).
30. T. Hirone, N. Kunitomi, M. Sakamoto, and H. Yamaki, *J. Phys. Soc. Japan* **13** (8), 838 (1958).
31. S. Badrinarayanan and H. B. Mathur, *Indian J. Pure Appl. Phys.* **8** (6), 324 (1970).
32. G. Barreau, G. Brunel, and G. Cizeron, *C. R. Acad. Sci.* Ser. C **272** (7), 618 (1971).
33. C. A. Machiet, *Phys. Rev.* **109** (6), 1964 (1958).
34. S. J. Rothman and N. L. Peterson, *Phys. Stat. Solidi* **35**, 305 (1969).
35. C. T. Tomizuka, cited by D. Lazarus, *Solid State Phys.* **Vol. 10** (1960).
36. F. D. Reinke and C. E. Dahlstrom, *Phil. Mag.* **22** (175), 57 (1970).
37. A. Ikushima, *J. Phys. Soc. Japan* **14**, 111 (1959).
38. M. C. Saxena and B. D. Sharma, *Trans. Indian Inst. Metals* **23** (3), 16 (1970).
39. G. Brunel, G. Cizeron, P. Lacombe, *C. R. Acad. Sci. Paris* Ser. C **270** (4), 393 (1970).
40. N. L. Peterson, *Phys. Rev.* **132** (6), 2471 (1963).
41. R. D. Johnson and B. H. Faulkenberry, *ASD-TDR*-63-625 (July 1963).
42. F. Moya, G. E. Moya-Gontier and F. Cabane-Brouty, *Phys. Stat. Solidi* **35** (2), 893 (1969).
43. M. C. Inman and L. W. Barr, *Acta Met.* **8** (2), 112 (1960).
44. P. P. Kuzmenko, L. F. Ostrovskii, and V. S. Kovalchuk, *Fiz. Tverd. Tela* **4**, 490 (1962).
45. S. Komura and N. Kunitomi, *J. Phys. Soc. Japan* **18** (Suppl. 2), 208 (1963).
46. S. M. Klotsman, Ya A. Rabovskii, V. K. Talinskii, and A. N. Timofeev, *Fiz. Metal. Metalloved.* **28** (6), 1025 (1969).
47. V. E. Kosenko, *Fiz. Tverd. Tela* **1**:1622 (1959); *Soviet Phys.—Solid State (English transl.)* **1**, 1481 (1959).
48. A. A. Bugai, V. E. Kosenko, and E. G. Miselynuk, *Sh. Tekhn. Fiz.* **27** (1), 207 (1957); *NP-tr*-448, P. 219 (1960).
49. H. Letaw, W. M. Portnoy, and L. Slifkin, *Phys. Rev.* **102**, 636 (1956).
50. A. V. Sandulova, M. I. Droniuk, and V. M. P'dak, *Fiz. Tverd. Tela* **3**, 2913 (1961).

51. B. I. Boltaks, V. P. Grabtchak, and T. D. Dzafarov, *Fiz. Tverd. Tela* **6**, 3181 (1964).

52. V. D. Ignatkov and V. E. Kosenko, *Fiz. Tverd. Tela* **4** (6), 1627 (1962): *Soviet Phys.—Solid State (Eng. transl.)* **4** (6), 1627 (1962).

53. V. I. Tagirov and A. A. Kuliev, *Fiz. Tverd. Tela* **4** (1), 272 (1962) *Soviet Phys.—Solid State (Eng. transl.)* **4** (1), 196 (1962).

54. W. C. Mallard, A. B. Gardner, R. F. Bass, and L. M. Slifkin, *Phys. Rev.* **129** (2), 617 (1963).

55. D. Duhl, K. Hirano, and M. Cohen, *Acta Met.* **11** (1), 1 (1963).

56. A. J. Mortlock and A. H. Rowe, *Phil. Mag.* **11** (114), 115 (1965).

57. J. E. Reynolds, B. L. Averbach, and M. Cohen, *Acta Met.* **5**, 29 (1957).

58. A. J. Mortlock, A. H. Rowe, and A. D. LeClaire, *Phil. Mag.* **5**, 803 (1960).

59. F. R. Winslow and T. S. Lundy, *Trans. Met. Soc. AIME* **233**, 1790 (1965).

60. T. R. Anthony and D. Turnbull, *Phys. Rev.* **151**, 495 (1966).

61. V. M. Amonenko, A. M. Blinkin, and I. G. Ivantsov, *Fiz. Metal. i Metalloved.* **17** (1), 56 (1964); *Phys. Metals Metallog. (USSR) (Eng. transl.)* **17** (1), 54 (1964).

62. D. W. James and G. M. Leak, *Phil. Mag.* **14**, 701 (1966).

63. A. Bondy and V. Levy, *C. R. Acad. Sci. Ser C* **272** (1), 19 (1971).

64. R. J. Borg and D. Y. F. Lai, *Acta Met.* **11** (8), 861 (1963).

65. C. G. Homan, *Acta Met.* **12**, 1071 (1964).

66. A. M. Huntz, M. Aucouturier, and P. Lacombe, *C. R. Acad. Sci. Paris Ser. C* **265** (10), 554 (1967).

67. M. S. Anand and R. P. Agarwala, *J. Appl. Phys.* **37**, 4248 (1966).

68. R. Angers and F. Claisse, *Can. Met. Quart.* **7** (2), 73 (1968).

69. A. V. Tomilov and G. V. Shcherbedinskii, *Fiz. Khim. Met. Mater.* **3** (3), 261 (1967).

70. K. Nohara and K. Hirano, *Proc. Intl. Conf. Sci. Technol. Iron Steel Tokyo*, Sept 1970.

71. V. T. Borisov, V. M. Golikov, and G. V. Shcherbedinskii, *Fiz. Metal. i Metalloved.* **22** (1), 159 (1966).

72. K. Hirano, M. Cohen, and B. L. Averbach, *Acta Met.* **9** (5), 440 (1961).

73. G. Seibel, *Compt. Rend.* **256** (22), 4661 (1963).

74. G. Bruggeman and J. Roberts, *J. Met.* **20** (8), 54 (1968).

75. V. D. Lyubimov, *Izv. Akad. Nauk. SSSR. Met.* **3** 201 (1969); V. D. Lyubimov, V. A. Tskhai, and G. B. Bogomolov, *Trudy Inst. Khim. Akad. Nauk. SSSR Ural Filial* **17**, 44, 48 (1970).

76. G. V. Grigorev and L. V. Pavlinov, *Fiz. Metal. i Metalloved.* **25** (5), 836 (1968).

77. T. Suzuoka, *Japan. Inst. Metals* **2**, 176 (1961).

78. A. W. Bowen and G. M. Leak, *Met. Trans.* **1** (6), 1695 (1970).

79. Th. Heumann and R. Imm, *J. Phys. Chem. Solids* **29** (9), 1613 (1968).

80. P. L. Gruzin and V. V. Mural, *Fiz. Metal. i Metalloved.* **16** (4), 551 (1963); *Phys. Metals Metallog. (USSR) (Eng. transl.)* **16** (4), 50 (1963).

81. A. Hoshino and R. Ataki, *Tetsu to Hagane* **56** (2), 252 (1970).

82. K. Sato, *Trans. Japan Inst. Metals* **5**, 91 (1964).

83. D. W. James and G. M. Leak, *Phil. Mag.* **14**, 701 (1966).

84. M. P. Dariel, G. Erez, and G. M. J. Schmidt, *Phil. Mag.* **19** (161), 1053 (1969).

85. P. P. Kuzmenko, G. P. Grinevich and B. A. Danilchenko, *Fiz. Met. Metalloved* **29** (2), 318 (1970).

86. G. V. Kidson, *Phil. Mag.* **13**, 247 (1966).

87. J. W. Miller, *Phys. Rev.* **181** (3), 1095 (1969).

88. B. F. Dyson, T. Anthony, and D. Turnbull, *J. Appl. Phys.* **37**, 2370 (1966).

89. H. A. Resing and N. H. Nachtrieb, *Phys. Chem. Solids* **21** (1/2), 40 (1961).

90. A. Ott and A. Lodding, *Int. Conf. Vac. and Interstitials in metals. Julich* **1**, 43 (1968); *Z. Naturforsch* **23A**, 1683 (1968), 2126 (1968).

91. A. Ott, A. Lodding, and D. Lazarus, *Phys. Rev.* **188** (3), 1088 (1969).

92. A. Ott, *J. Appl. Phys.* **40** (6), 2395 (1969); J. N. Mundy, A. Ott, L. Lowenberg, and A. Lodding, *Phys. Stat. Solidi.* **35** (1), 359 (1969).

93. A. Ott, *Z. Naturforschg* **25a** (10), 1477 (1970).

94. A. Lodding, J. N. Mundy, and A. Ott, *Phys. Stat. Solidi* **38** (2), 559 (1970).

95. K. Lal, *Commis. Energ. At. Report.* **CEA-R** 3136 (1967).

96. L. V. Pavlinov, A. M. Gladyshev, and V. N. Bikov, *Fiz. Metal Metalloved* **26** (5), 823 (1968).

97. P. G. Shewmon, *Trans. Met. Soc. AIME* **206**, 918 (1956).

98. Y. V. Borisov, P. L. Gruzin, and L. V. Pavlinov, *Met. i Metalloved. Chistykh Metal.* **1** 213 (1959), translated in *JPRS-5195.*

99. A. Y. Nakonechnikov, L. V. Pavlinov, and V. N. Bikov, *Fiz. Metal i Metalloved.* **22**, 234 (1966).

100. J. Askill, *Phys. Status Solidi* **9** (2), K113 (1965).

101. L. V. Pavlinov and A. A. Kordev, *Fiz. Metal. Metalloved.* **29** (6), 1326 (1970).

102. G. N. Dubinin, G. P. Benediktova, M. G. Karpman, and G. V. Shcherbedinskii, *Khim—Term Obrab Stali Splavov* **6**, 129 (1969).

103. J. Askill and D. H. Tomlin, *Phil. Mag.* **8** (90), 997 (1963).

104. B. A. Vandyshev and A. S. Panov, *Fiz. Met. Metalloved.* **26** (3), 517 (1968).

105. S. Z. Benediktova, G. N. Dubinin, M. G. Kapman, and G. V. Shcherbedinskii, *Metalloved. i Term. Obrabotka Metal.* **5**, 5 (1966).

106. B. A. Vandyshev and A. S. Panov, *Fiz. Metal. i Metalloved.* **25** (2), 321 (1968).

107. L. V. Pavlinov, A. Y. Nakonechnikov, and V. N. Bikov, *At. Energ. (USSR)* **19** 521 (1965).

108. J. Askill, *Phys. Status Solidi* **23**, K21 (1967).

109. A. Chatterjee and D. J. Fabian, *J. Inst. Metals* **96** (6), 186 (1968).

110. A. Hassner and W. Lange, *Phys. Status Solidi* **8**, 77 (1965).

111. K. Monma, H. Suto, and H. Oikawa, *J. Japan Inst. Metals* **28**, 188 (1964).

112. A. Ya. Shinyaev, *Izv. Akad. Nauk. S.S.S.R. Metl.* **4**, 182 (1969).

113. J. J. Blechet, A. VanGryeynest, and D. Calais, *J. Nucl. Mater.* **28** (2), 177 (1968).

114. P. P. Kuzmenko and G. P. Grinevich, *Fiz. Tverd. Tela.* **4** (11), 3266 (1962); *Soviet Physics—Solid State* **4** (11), 2390 (1962).

115. S. Z. Bokshtein, S. T. Kishkin, and L. M. Moroz, *Investigation of the Structure of Metals by Radioactive Isotope Methods; State Publishing House of the Ministry of Defense Industry, Moscow* (1959); *AEC-tr-4505* (1961).

116. M. S. Anand, S. P. Murarka, and R. P. Agarwala, *J. Appl. Phys.* **36** (12), 3860 (1965).

117. V. D. Lyubimov, *Izv. Akad. Nauk. S.S.S.R. Metal* **3**, 201 (1969); V. D. Lyubimov, V. A. Tskhai, and G. B. Bogmolov, *Toudy Inst. Khim. Akad. Nauk. S.S.S.R. Ural Filial* **17**, 44, 48 (1970).

118. R. F. Peart, D. Graham, and D. H. Tomlin, *Acta Met.* **10**, 519 (1962).

119. J. Pelleg, *J. Metals* **20** (8), 54 (1968); *J. Less Common Metals* **17**, 319 (1969); *Phil. Mag.* **19** (157), 25 (1969).

120. T. S. Lundy, F. E. Winslow, R. E. Pawel, and C. J. McHargue. *Trans. Met. Soc. AIME* **223**, 1533 (1965).

121. B. A. Vandyshev and A. S. Panov, *Izv. Akad. Nauk. S.S.S.R.* **1**, 206 (1968).

122. J. Askill, *Phys. Status Solidi* **9** (3), K167 (1965).

123. J. Pelleg, *Phil. Mag.* **21** (172), 735 (1970).

124. R. P. Agarwala, S. P. Murarka, and M. S. Anand, *Acta Met.* **16** (1), 61 (1968).

125. G. B. Fedorov, F. I. Zhomov, and E. A. Smirnov, *Met. Metalloved. Chist. Metal* **8**, 145 (1969).

126. N. L. Peterson, *Phys. Rev.* **136** (2A), A568 (1964).

127. N. H. Nachtrieb and G. S. Handler, *J. Chem. Phys.* **23**, 1187 (1955).

129. J. Kucera and T. Zemcik, *Can. Met. Quart.* **7** (2), 83 (1968).

130. G. V. Kidson and R. Ross, *Proc. 1st UNESCO Int. Conf. Radioisotopes in Scientific Res.*, p. 185 (1958).

131. F. A. Smith and L. W. Barr, *Phil. Mag.* **21** (171), 633 (1970).

132. J. N. Mundy, T. E. Miller, and R. J. Porte, *Phys. Rev.* **B3** (8), 2445 (1971).

133. L. W. Barr, J. N. Mundy, and F. A. Smith, *Phil. Mag.* **16**, 1139 (1967).

134. F. A. Smith and L. W. Barr, *Phil. Mag.* **20** (163), 205 (1969).

135. R. E. Tate and G. R. Edwards, *Symposium on Thermodynamics with Emphasis on Nuclear Materials and Atomic Solids*, **pp. 105–113** in **IAEA Symposium Vol. II**, International Atomic Energy Agcy., Vienna (1966).

136. R. E. Tate and E. M. Cramer, *Trans. Met. Soc. AIME* **230**, 639 (1964).

137. M. Dupuy and D. Calais, *Trans. AIME* **242**, 1679 (1968).

138. M. P. Dariel, G. Erez, and G. M. J. Schmidt, *J. Appl. Phys.* **40** (7), 2746 (1969).

139. M. P. Dariel, *Phil. Mag.* **22** (177), 653 (1970).

140. M. P. Dariel, G. Erez, and G. M. J. Schmidt, *Phil. Mag.* **18** (161), 1045 (1969).

141. A. A. Kuliev and D. N. Nasledov, *Zh. Tekhn. Fiz.* **28** (2), 259 (1958); *Soviet Phys.—Tech. Phys. (Eng. transl.)* **3** (2), 235 (1958).

142. P. Braetter and H. Gobrecht, *Phys. Stat. Sol* **41** (2), 631 (1970).

143. B. I. Boltaks, and B. T. Plachenov, *Soviet Phys.—Tech. Phys. (Eng. transl.)* **27** (10), 2071 (1957).

144. R. F. Peart, *Phys. Status Solidi* **15**, K119 (1966).

145. W. R. Wilcox and T. J. LaChapelle, *J. Appl. Phys.* **35** (1) 240 (1964).

146. R. C. Newman and J. Wakefield, *Phys. Chem. Solids* **19** (3), 230 (1961).

147. B. I. Boltaks and I. I. Sozinov, *Zh. Tekhn. Fiz.* **28** (3), 679 (1958); *Soviet Phys.—Tech. Phys. (Eng. transl.)* **3** (3), 636 (1958).

148. J. D. Struthers, *J. Appl. Phys.* **27** (12), 1560 (1956); errata **28** (4), 516 (1957).

149. H. P. Bonzel, *Phys. Status Solidi* **20**, 493 (1967).

150. S. Mackawa, *J. Phys. Soc. Japan* **17** (10), 1592 (1962).

151. J. J. Rahan, M. E. Pickering, and J. Kennedy, *J. Electrochem. Soc.* **106** (8), 705 (1959).

152. S. J. Rothman, N. L. Peterson, and J. T. Robinson, *Phys. Stat. Sol.* **39** (2), 635 (1970).

153. C. T. Tomizuka and L. M. Slifkin, *Phys. Rev.* **96**, 610 (1954).

154. J. Bernardini, A. Combe-Brun, and J. Cabane, *Ser. Met.* **4** (12), 985 (1970).

155. A. Sawatskii and F. E. Jaumot, *Trans. AIME* **209**, 1207 (1957).

156. J. G. Mullen, *Phys. Rev.* **121**, 1649 (1961).

157. R. E. Hoffman, *Acta Met.* **6**, 95 (1958).

158. T. Hirone, S. Miura, and T. Suzuoka, *J. Phys. Soc. Japan* **16** (12), 2456 (1961).

159. R. E. Hoffman, D. Turnbull, and E. W. Hart, *Acta Met.* **3**, 417 (1955).

160. C. B. Pierce and D. Lazarus, *Phys. Rev.* **114**, 686 (1959).

161. N. Barbouth, J. Ouder, and J. Cabane, *C. R. Acad. Sci. Paris Ser. C* **264** (12), 1029 (1967).

162. V. N. Kaigorodov, Ya. A. Rabovskii, and V. K. Talinskii, *Fiz. Metal. i Metalloved.* **24** (4), 661 (1967).

163. V. N. Kaigorodov, S. M. Klotsman, A. N. Timofeev, and I. Sh. Traktenberg, *Fiz. Metal. Metalloved.* **28** (1), 120 (1969).

164. A. Sawatskii and F. E. Jaumot, *Phys. Rev.* **100**, 1627 (1955).

165. L. W. Barr, J. N. Mundy, and F. A. Smith, *Phil. Mag.* **20** (164), 389 (1969).
166. J. N. Mundy, L. W. Barr, and F. A. Smith, *Phil. Mag.* **14**, 785 (1966).
167. P. Son, S. Ihara, M. Miyake, and T. Sano, *J. Japan Inst. Met.* **33** (1), 1 (1969).
168. V. P. Vasilev, I. F. Kamardin, V. I. Skatskill, S. G. Chernomorchenko, and G. N. Shuppe, *Trudy Sred. Gos. Uni. im V.I. Lenina* **65**, 47 (1955); *Translation AEC-tr-4272.*
169. R. E. Pawel and T. S. Lundy, *J. Phys. Chem. Solids* **26**, 937 (1965).
170. B. A. Vandyshev and A. S. Panov, *Izv. Akad. Nauk. SSSR Metal* **1**, 244 (1969).
171. Sh. Movalanov and A. A. Kuliev, *Fiz. Tverd. Tela* **4** (2), 542 (1962).
172. N. I. Ibraginov, M. G. Shachtachtinskii, and A. A. Kuliev, *Fiz. Tverd. Tela* **4**, 3321 (1962).
173. R. N. Ghoshtagore, *Phys. Rev.* **155** (3), 698 (1967).
174. T. R. Anthony, B. F. Dyson, and D. Turnbull, *J. Appl. Phys.* **39** (3), 1391 (1968).
175. G. A. Shirn, *Acta Met.* **3**, 87 (1955).
176. F. Schmitz and M. Fock, *J. Nucl. Materials* **21**, 317 (1967).
177. B. F. Dyson, *J. Appl. Phys.* **37**, 2375 (1966).
178. W. Chomka and J. Andruszkiewicz, *Nukleonika* **5** (10), 611 (1960).
179. A. Sawatskii, *J. Appl. Phys.* **29** (9), 1303 (1958).
180. D. Yolokoff, S. May, and Y. Adda, *Compt. Rend.* **251** (3), 2341 (1960).
181. L. Bartha and T. Szalay, *Int. J. Appl. Radiat. Isotopes* **20** (2), 825 (1969).
182. F. Dyment and C. M. Libanati, *J. Mager. Sci.* **3** (4), 349 (1968).
183. J. Askill, *Phys. Stat Sol.* **B43** (1), K1 (1971).
184. L. V. Pavlinov, G. V. Grigorev, and G. O. Gromyko, *Izv. Akad. Nauk. SSSR Metal* **3**, 207 (1969).
185. A. Y. Nakonechnikov, L. V. Pavlinov, and V. N. Bikov, *Fiz. Metal i Metalloved* **22**, 234 (1966).
186. G. B. Gibbs, D. Graham, and D. H. Tomlin, *Phil. Mag.* **8** (92), 1269 (1963).
187. J. Askill and G. B. Gibbs, *Phys. Status Solidi* **11**, 557 (1965).
188. J. F. Murdock, T. S. Lundy, and E. E. Stansbury, *Acta Met.* **12** (9), 1033 (1964).
189. L. V. Pavlinov, *Fiz. Metal Metalloved* **30** (4), 800 (1970).
190. J. F. Murdock, T. S. Lundy, and E. E. Stansbury, *Acta Met.* **12** (9), 1033 (1964).
191. L. V. Pavlinov, *Fiz. Metal. i Metalloved.* **24** (2), 272 (1967).
192. R. E. Pawel and T. S. Lundy, *Acta Met.* **17** (8), 979 (1969).
193. L. N. Larikov, V. M. Tyshkevich, and L. F. Chorna, *Ukr. Fiz. Zh.* **12** (6), 983 (1967).
194. Y. Adda, A. Kirianenko, and C. Mairy, *Compt. Rend.* **253**, 445 (1961); *J. Nucl. Mater.* **6** (1), 130 (1962).
195. M. P. Dariel, M. Blumenfeld, and G. Kimmel, *J. Appl. Phys.* **41** (4), 1480 (1970).
196. G. B. Federov, E. A. Smirnov, and S. S. Moiseenko, *Met. Metalloved. Chist. Metal.* **7**, 124 (1968).

197. S. J. Rothman, *J. Nucl. Mater.* **3** (1), 77 (1961).
198. N. L. Peterson and S. J. Rothman, *Phys. Rev.* **136** (3A), A842 (1964).
199. P. Son, S. Ihara, M. Miyake, and T. Sano, *J. Japan Inst. Met.* **33** (1), 1 (1969).
200. H. W. Paxton and R. A. Wolfe, *Trans. AIME* **230**, 1426 (1964).
201. M. G. Coleman, *Ph.D. Thesis Univ. of Illinois* (1967); *Univ. Microf.* **68–1725** (1968).
202. B. A. Vandychev and A. S. Panov, *Izv. Akad. Nauk SSSR Metal* **2**, 231 (1970).
203. R. F. Peart, *J. Phys. Chem. Solids* **26**, 1853 (1965).
204. D. S. Gorney and R. M. Altovskii, *Fiz. Metal Metalloved* **30** (1), 85 (1970).
205. J. H. Rosolowski, *Phys. Rev.* **124** (6), 1828 (1961).
206. P. B. Ghate, *Phys. Rev.* **130** (1), 174 (1963).
207. A. P. Batra and H. B. Huntington, *Phys. Rev.* **145**, 542 (1966).
208. A. P. Batra and H. B. Huntington, *Phys. Rev.* **154** (3), 569 (1967).
209. J. S. Warford and M. B. Huntington, *Phys. Rev.* **B1** (4), 1867 (1970).
210. N. L. Peterson and S. J. Rothman, *Phys. Rev.* **163** (3), 645 (1967).
211. R. P. Agarwala, S. P. Murarka, and M. S. Anand, *Trans. Met. Soc. AIME* **233**, 986 (1965).
212. A. M. Blinkin and V. V. Vorobiov, *Ukr. Fiz. Zh.* **9** (1), 91 (1964).
213. R. P. Agarwala and A. R. Paul, *Proc. Nucl. Rad. Chem. Symp. Poona, India* p. **542** (1967).
214. P. L. Gruzin, V. S. Emelyanov, G. G. Ryabova, and G. B. Federov, *Proc. 2nd Intern. Conf. Peaceful Uses At. Energy,* **19**:187 (1958).
215. Y. V. Borisov, Y. G. Godin, P. L. Gruzin, A. I. Evstyukhin, and V. S. Yemelyanov, *Met. i Met. Izdatel. Akad. Nauk. SSSR Moscow* 1958. *Translation NP-TR-448* p. **196** (1960).
216. P. Flubacher, *E. I. R. Bericht* **49**, May (1963).
217. R. A. Andriyevskii, V. N. Zagraykin, and G. Ya. Meshcheryakov, *Phys. Met. Metalloved.* **19** (3), 146 (1966).
218. R. P. Agarwala and A. R. Paul, Report CONF-670335, *Int. Conf. Vacancies and Interstitials in Metals,* Julich (1968) **1**, 105 (1968), *Report BARC-377* (1968).
219. G. V. Kidson and G. J. Young, *Phil. Mag.* **20** (167), 1047 (1969).
220. G. B. Federov, E. A. Smirnov, and S. M. Novikov, *Met. Metalloved Chist. Metal* **8**, 41 (1969).
221. B. A. Vandychev and A. S. Panov, *Izv. Akad. Nauk. SSSR Metal* **2**, 231 (1970).
222. G. B. Federov, E. A. Smirnov, and F. I. Zhomov, *Met. Metalloved. Chist. Metal.* **7**, 116 (1968).
223. L. V. Pavlinov, *Fiz. Metal i Metalloved* **24** (2), 272 (1967).
224. H. M. Gilder and D. Lazarus, *J. Phys. Chem. Solids* **26**, 2081 (1965).
225. G. V. Kidson and J. McGurn, *Can. J. Phys.* **39** (8), 1146 (1961).
226. R. E. Pawel and T. S. Lundy, *J. Phys. Chem. Solids* **26**, 937 (1965).
227. S. J. Rothman, L. T. Lloyd, and A. L. Harkness, *Trans. Met. Soc. AIME* **218** (4), 605 (1960).
228. G. B. Federov, E. A. Smirnov, and V. N. Gusev, *At. Energ.* **27** (2), 149 (1969).

VISCOSITY AND THERMAL CONDUCTIVITY OF OXYGEN AS A FUNCTION OF TEMPERATURE

Conversion Factors

T, °K	→ T, °F: multiply by (9/5) then subtract 459.67
T, °K	→ T, °C: subtract 273.15
T, °K	→ T, °R: multiply by (9/5)
P, atm	→ P, psia: multiply by 14.69595

P, atm	→ p, N/m^2: multiply by 1.01325×10^5
η, g/cm-s	→ η, N-s/m^2
η, g/cm-s	→ η, lb$_m$/ft-s
λ, W/m-°K	→ λ, cal/cm-s-°K
λ, W/m-°K	→ λ, Btu ft-hr-°R

Temperature, °K	Viscosity, g/cm-s $10^3 \eta_0$	Thermal conductivity, W/m-K $10^3 \lambda_0$	Temperature, °K	Viscosity, g/cm-s $10^3 \eta_0$	Thermal conductivity, W/m-K $10^3 \lambda_0$
80	0.0585	6.94	280	0.1955	24.86
90	0.0663	7.95	290	0.2012	25.62
100	0.0740	8.96	300	0.2068	26.38
110	0.0818	9.96	310	0.2122	27.14
120	0.0894	10.94	320	0.2176	27.89
130	0.0970	11.92	330	0.2230	28.62
140	0.1045	12.89	340	0.2282	29.36
150	0.1118	13.85	350	0.2334	30.10
160	0.1191	14.78	360	0.2385	30.85
170	0.1261	15.70	370	0.2435	31.58
180	0.1331	16.60	380	0.2485	32.32
190	0.1399	17.48	390	0.2534	33.06
200	0.1465	18.35	400	0.2583	33.79
210	0.1530	19.21	410	0.2631	34.54
220	0.1595	20.05	420	0.2678	35.28
230	0.1658	20.88	430	0.2725	36.02
240	0.1719	21.69	450	0.2818	37.49
250	0.1780	22.49	470	0.2908	38.95
260	0.1840	23.29	490	0.2997	40.41
270	0.1898	24.08	510	0.3085	41.86

Temperature, °K	Viscosity, g/cm-s $10^3\,\eta_0$	Thermal conductivity, W/m-K $10^3\,\lambda_0$	Temperature, °K	Viscosity, g/cm-s $10^3\,\eta_0$	Thermal conductivity, W/m-K $10^3\,\lambda_0$
530	0.3170	43.29	1,210	0.5571	86.32
550	0.3255	44.72	1,240	0.5662	89.97
570	0.3338	46.14	1,270	0.5753	89.61
590	0.3420	47.56	1,300	0.5843	91.24
610	0.3501	48.97	1,330	0.5932	92.85
630	0.3580	50.37	1,360	0.6021	94.45
650	0.3659	51.77	1,390	0.6109	96.06
670	0.3736	53.16	1,420	0.6196	97.64
690	0.3813	54.54	1,450	0.6282	99.22
710	0.3888	55.89	1,480	0.6368	100.78
730	0.3963	57.25	1,510	0.6454	102.35
750	0.4037	58.59	1,540	0.6538	103.90
770	0.4110	59.90	1,570	0.6622	105.45
790	0.4183	61.22	1,600	0.6706	106.98
820	0.4290	63.16	1,630	0.6789	108.52
850	0.4396	65.09	1,660	0.6871	110.05
880	0.4500	66.98	1,690	0.6953	111.58
910	0.4603	68.89	1,720	0.7034	113.09
940	0.4705	70.72	1,750	0.7116	114.61
970	0.4805	72.52	1,780	0.7196	116.12
1,000	0.4905	74.32	1,810	0.7276	117.63
1,030[a]	0.5002	76.07	1,840	0.7355	119.12
1,060	0.5100	77.84	1,870	0.7435	120.63
1,090	0.5196	79.56	1,900	0.7514	122.13
1,120	0.5291	81.27	1,930	0.7592	123.62
1,150	0.5384	82.97	1,960	0.7670	125.11
1,180	0.5478	84.66	1,980	0.7721	126.11
			2,000	0.7773	127.10

[a]Data for all temperatures in excess of 1,000 K are extrapolated.

From Hanley, H. J. M., McCarty, R. D., and Sengers, J. V. *Viscosity and Thermal Conductivity Coefficients of Gaseous and Liquid Oxygen*, NASA CR 2440, National Aeronautics and Space Administration, Washington, D.C., August 1974 (available from National Technical Information Service, Springfield, Va. 22151).

THERMAL CONDUCTIVITY OF COMPRESSED OXYGEN

In this table, thermal conductivity, λ, is in the unit milliwatt/m-K.

Conversion Factors

T, °K → T, °F: multiply by (9/5) then subtract 459.67
T, °K → T, °C: subtract 273.15
T, °K → T, °R: multiply by (9/5)
P, atm → P, psia: multiply by 14.69595
P, atm → P, N/m²: multiply by 1.01325 × 10⁵

η, g/cm-s → η, Ns/m²: multiply by 10^{-1}
η, g/cm-s → η, lb_m/ft-s: multiply by 0.0671969
λ, W/m-K → λ, cal/cm-s-K: multiply by (1/418.4)
λ, W/m-K → λ, Btu/ft-hr-°R: multiply by 0.578176

T, °K \ P, atm	1	5	10	15	20	25	30	35	40	45	50	55
80	164.7	164.9	165.2	165.4	165.7	165.9	166.2	166.4	166.7	166.9	167.1	167.4
90	151.7	152.0	152.3	152.6	152.9	153.3	153.6	153.9	154.2	154.5	154.8	155.1
100	9.5	138.4	138.8	139.2	139.6	140.0	140.4	140.8	141.1	141.5	141.9	142.3
110	10.4	11.4	124.9	125.3	125.8	126.3	126.8	127.3	127.7	128.2	128.6	129.1
120	11.4	12.2	13.5	110.9	111.5	112.1	112.8	113.3	113.9	114.5	115.1	115.6
130	12.3	13.1	14.2	15.6	96.3	97.1	98.0	98.8	99.6	100.3	101.1	101.8
140	13.2	14.0	15.0	16.1	17.5	19.3	81.6	82.9	84.1	85.2	86.3	87.3
150	14.1	14.8	15.7	16.7	17.9	19.2	20.9	23.3	27.8	68.4	70.4	72.1
160	15.8	15.7	16.5	17.4	18.4	19.5	20.7	22.3	24.2	26.7	30.5	37.3
170	15.9	16.5	17.3	18.1	19.0	19.9	20.9	22.1	23.4	24.9	26.8	29.0
180	16.8	17.4	18.1	18.8	19.6	20.4	21.3	22.3	23.3	24.4	25.7	27.1
190	17.7	18.2	18.9	19.5	20.3	21.0	21.8	22.6	23.5	24.4	25.4	26.4
200	18.5	19.0	19.6	20.3	20.9	21.6	22.3	23.0	23.8	24.6	25.4	26.3
210	19.3	19.8	20.4	21.0	21.6	22.3	22.9	23.6	24.2	24.9	25.7	26.4
220	20.2	20.6	21.2	21.7	22.3	22.9	23.5	24.1	24.7	25.4	26.0	26.7
230	21.0	21.4	21.9	22.5	23.0	23.6	24.1	24.7	25.3	25.9	26.5	27.1
240	21.7	22.2	22.7	23.2	23.7	24.2	24.8	25.3	25.8	26.4	26.9	27.5
250	22.5	22.9	23.4	23.9	24.4	24.9	25.4	25.9	26.4	26.9	27.5	28.0
260	23.3	23.7	24.1	24.6	25.1	25.6	26.0	26.5	27.0	27.5	28.0	28.5
270	24.0	24.4	24.9	25.3	25.8	26.2	26.7	27.1	27.6	28.1	28.5	29.0
280	24.8	25.1	25.6	26.0	26.4	26.9	27.3	27.8	28.2	28.6	29.1	29.5
290	25.5	25.9	26.3	26.7	27.1	27.5	28.0	28.4	28.8	29.2	29.6	30.1
300	26.2	26.6	27.0	27.4	27.8	28.2	28.6	29.0	29.4	29.8	30.2	30.6
310	27.0	27.3	27.7	28.1	28.5	28.8	29.2	29.6	30.0	30.4	30.8	31.2
320	27.7	28.0	28.4	28.7	29.1	29.5	29.9	30.2	30.6	31.0	31.4	31.7
330	28.4	28.7	29.0	29.4	29.8	30.1	30.5	30.9	31.2	31.6	31.9	32.3
340	29.1	29.3	29.7	30.1	30.4	30.8	31.1	31.5	31.8	32.2	32.5	32.9
350	29.7	30.0	30.4	30.7	31.1	31.4	31.7	32.1	32.4	32.8	33.1	33.4
360	30.4	30.7	31.0	31.4	31.7	32.0	32.4	32.7	33.0	33.3	33.7	34.0
370	31.1	31.4	31.7	32.0	32.3	32.7	33.0	33.3	33.6	33.9	34.3	34.6
380	31.8	32.0	32.3	32.7	33.0	33.3	33.6	33.9	34.2	34.5	34.8	35.1
390	32.4	32.7	33.0	33.3	33.6	33.9	34.2	34.5	34.8	35.1	35.4	35.7
400	33.1	33.3	33.6	33.9	34.2	34.5	34.8	35.1	35.4	35.7	36.0	36.3

THERMAL CONDUCTIVITY OF COMPRESSED OXYGEN (Continued)

T,°K \ P,atm	60	65	70	80	90	100	110	120	130	150	175
80	167.6	167.9	168.1	168.6	169.1	169.5	170.0	170.4	170.9	171.8	172.8
90	155.4	155.7	156.0	156.6	157.1	157.7	158.3	158.8	159.4	160.4	161.8
100	142.6	143.0	143.3	144.0	144.7	145.4	146.1	146.8	147.4	148.7	150.3
110	129.5	130.0	130.4	131.2	132.1	132.9	133.7	134.5	135.3	136.8	138.6
120	116.2	115.7	117.2	118.3	119.3	120.3	121.2	122.2	123.1	124.9	127.0
130	102.5	103.2	103.9	105.2	106.4	107.6	108.8	109.9	111.0	113.1	115.6
140	88.3	89.2	90.1	91.8	93.4	95.0	96.4	97.8	99.1	101.6	104.5
150	73.5	74.9	76.2	78.5	80.5	82.5	84.2	85.9	87.5	90.5	93.8
160	50.2	57.8	61.3	65.6	68.6	71.0	73.1	75.1	76.9	80.2	83.9
170	31.7	35.1	39.2	47.9	54.7	59.3	62.7	65.3	67.6	71.3	75.3
180	26.7	38.4	32.4	36.9	42.0	47.0	51.3	55.0	58.1	63.0	67.7
190	27.8	28.8	30.1	33.0	36.2	39.6	43.2	46.5	49.7	55.1	60.5
200	27.2	28.1	29.1	31.3	33.6	36.1	38.7	41.4	44.0	48.9	54.3
210	27.2	28.0	28.8	30.5	32.4	34.4	36.4	38.5	40.6	44.9	49.8
220	27.4	28.1	28.8	30.3	31.9	33.5	35.2	36.9	38.7	42.3	46.6
230	27.7	28.3	29.0	30.3	31.7	33.1	34.6	36.1	37.6	40.7	44.5
240	28.1	28.7	29.2	30.5	31.7	33.0	34.3	35.6	36.9	39.7	43.1
250	28.5	29.1	29.6	30.7	31.8	33.0	34.2	35.4	36.6	39.0	42.1
260	29.0	29.5	30.0	31.0	32.1	33.1	34.2	35.3	36.4	38.7	41.5
270	29.5	29.9	30.4	31.4	32.4	33.3	34.4	35.4	36.4	38.4	41.0
280	30.0	30.4	30.9	31.8	32.7	33.6	34.6	35.5	36.5	38.4	40.8
290	30.5	30.9	31.3	32.2	33.1	33.9	34.8	35.7	36.6	38.4	40.6
300	31.0	31.4	31.8	32.7	33.5	34.3	35.1	36.0	36.8	38.5	40.6
310	31.6	31.9	32.3	33.1	33.9	34.7	35.5	36.3	37.1	38.7	40.6
320	32.1	32.5	32.8	33.6	34.3	35.1	35.8	36.6	37.4	38.9	40.8
330	32.7	33.0	33.4	34.1	34.8	35.5	36.2	37.0	37.7	39.1	40.9
340	33.2	33.5	33.9	34.6	35.3	36.0	36.7	37.3	38.0	39.4	41.1
350	33.8	34.1	34.4	35.1	35.8	36.4	37.1	37.7	38.4	39.7	41.4
360	34.3	34.6	35.0	35.6	36.2	36.9	37.5	38.2	38.6	40.1	41.7
370	34.9	35.2	35.5	36.1	36.7	37.9	38.0	38.6	39.2	40.4	42.0
380	35.4	35.7	36.0	36.6	37.2	37.8	38.4	39.0	39.6	40.8	42.3
390	36.0	36.3	36.6	37.2	37.8	38.3	38.9	39.5	40.1	41.2	42.6
400	36.6	36.9	37.1	37.7	38.3	38.8	39.4	40.0	40.5	41.6	43.0

From Hanley, H. J. M., McCarty, R. D., and Sengers, J. V., *Viscosity and Thermal Conductivity Coefficients of Gaseous and Liquid Oxygen,* NASA CR-2440, National Aeronautics and Space Administration, Washington, D.C., August 1974 (available from National Technical Information Service, Springfield, Va. 22151).

TRANSPORT PROPERTIES OF OXYGEN AT SATURATION

Conversion Factors

$T, {}^\circ K \rightarrow T, {}^\circ F$: multiply by (9/5) then subtract 459.67
$T, {}^\circ K \rightarrow T, {}^\circ C$: subtract 273.15
$T, {}^\circ K \rightarrow T, {}^\circ R$: multiply by (9/5)
$P, atm \rightarrow P, psia$: multiply by 14.69595
$P, atm \rightarrow P, N/m^2$: multiply by 1.01325×10^5
$\eta, g/cm\text{-}s \rightarrow \eta, Ns/m^2$: multiply by 10^{-1}
$\eta, g/cm\text{-}s \rightarrow \eta, lb_m/ft\text{-}s$: multiply by 0.0671969
$\lambda, W/m\text{-}K \rightarrow \lambda, cal/cm\text{-}s\text{-}K$: multiply by (1/418.4)
$\lambda, W/m\text{-}K \rightarrow \lambda, Btu/ft\text{-}hr\text{-}{}^\circ R$: multiply by 0.578176

Temperature, °K	Pressure, atm	Viscosity, mg/cm-s		Thermal conductivity, mW/m-K	
		Vapor	Liquid	Vapor	Liquid
80	0.30	0.059	2.652	7.4	164.7
90	0.98	0.068	1.971	8.5	151.8
100	2.51	0.079	1.504	9.9	138.2
110	5.36	0.092	1.188	11.5	124.3
120	10.09	0.108	0.971	13.6	110.3
130	17.26	0.129	0.824	16.3	95.9
140	27.52	0.158	0.696	20.1	80.6
142	30.00	0.166	0.668	21.1	77.4
144	32.64	0.175	0.638	22.3	74.0
146	35.45	0.185	0.607	23.6	70.6
148	38.44	0.197	0.574	25.2	67.0
150	41.61	0.211	0.537	27.2	63.2
152	45.00	0.231	0.494	29.9	58.7
154	48.65	0.269	0.424	35.2	52.0

From Hanley, H. J. M., McCarty, R. D., and Sengers, J. V., *Viscosity and Thermal Conductivity Coefficients of Gaseous and Liquid Oxygen,* NASA CR-2440, National Aeronautics and Space Administration, Washington, D.C., August 1974 (available from National Technical Information Service, Springfield, Va. 22151).

PHYSICAL CONSTANTS OF OZONE AND OXYGEN

Physical Constant	Ozone (O_3)	Oxygen (O_2)
Molecular Weight	47.9982 g/g-mol	31.9988 g/g-mol
Boiling Point (760 mm)	$-111.9 \pm 0.3°C$	$-182.97°C$
Melting Point	$-192.7 \pm 0.2°C$	$-218.4°C$
Critical Temperature	$-12.1 \pm 0.1°C$	$-118.574°C$
Critical Pressure	54.6 atm	49.77 atm
Critical Density	0.437 g/cc	0.436 g/cc
Critical Volume	147.1 cc/mol	73.37 cc/mol
Gas Density (0°C) (760 mm pressure)	2.144 g/liter	1.429 g/liter
Liquid Density		
$-112°C$	1.358 g/cc	
$-183°C$	1.571 g/cc	1.14 g/cc
$-195.4°C$	1.614 g/cc	1.201 g/cc
Surface Tension		
$-195°C$	43.8 ± 0.1 dyne/cm	
$-182.7°C$	38.1 ± 0.2 dyne/cm	
$-183.0°C$	38.4 ± 0.7 dyne/cm	13.2 dyne/cm
Heat Capacity of Liquid		
-183 to $-145°C$	0.45 cal/g°C	
Heat Capacity of Gas		
$-173°C$	7.95 cal/g mol°C	
0°C	9.10 cal/g mol°C	
25°C	9.37 cal/g mol°C	
100°C		6.979 cal/g mol°C
127°C	10.44 cal/g mol°C	
Viscosity of Liquid		
$-195.6°C$	4.14 ± 0.05 cp	
$-183.0°C$	1.57 ± 0.02 cp	0.1958 cp
Heat of Vaporization		
$-112°C$	75.6 cal/g	
$-182.9°C$		50.9 cal/g
Heat of Formation		
25°C	-34.4 Kg cal/mol	
Free Energy		
25°C	32.4 Kg cal/mol	
Van der Waals Constant (a)	3.545 atm liter²/mol²	1.36 atm liter²/mol²
Van der Waals Constant (b)	0.04903 liter/mol	0.03803 liter/mol
Magnetic Susceptibility		
gas ($\times 10^{-6}$)	0.002 cgs units	10.6.2 cgs units
liq ($\times 10^{-6}$)	0.150 cgs units	000.0 cgs units
Thermal Conductivity of Liquid		
$-195.8°C$	5.21 cal/sec cm°C $\times 10^4$	
$-183.0°C$	5.31 cal/sec cm°C $\times 10^4$	
$-105.0°C$	5.42 cal/sec cm°C $\times 10^4$	
$-128.0°C$	5.52 cal/sec cm°C $\times 10^4$	

Phase Boundaries—Ozone-Oxygen System
$-183°C$ 29.8 and 72.4 wt % O_3
$-195.4°C$ 9   and 90.8 wt % O_3
Consolute Temperature—Ozone-Oxygen System
$-180 \pm 0.5°C$
Coefficient of Thermal Expansion for Liquid Ozone

Temp. °C	α
-195.6	1.62
-183.0	1.58
-148.0	1.47
-123.0	1.41
-112.0	1.35
-98.8	1.31

PHYSICAL CONSTANTS OF CLEAR FUSED QUARTZ

Based on information contained in Fused Quartz Catalogue Q-7A General Electric Company.

Property	Clear fused quartz	Property	Clear fused quartz
Density	2.2 g./c.c.	Annealing Point	(approx.) 1140°C
Hardness	4.9 (Mohs')	Strain Point	1070°C
Tensile Strength	7,000 p.s.i.	Electrical Resistance	9.5 $\log_{10}$ R for cm.³ at 350°C
Compressive Strength	>160,000 p.s.i.	Dielectric Constant	3.75 at 20°C. 1 Mc.
Bulk Modulus	(approx.) 5.3×10^6 p.s.i.	Dielectric Loss Factor	less than .0004 at 20°C. 1 Mc.
Rigidity Modulus	4.5×10^6 p.s.i.	Dissipation Factor	less than .0001 at 20°C. 1 Mc.
Young's Modulus	10.4×10^6 p.s.i.	Index of Refraction	1.4585
Poisson's Ratio	.16		
Coefficient of Thermal Expansion	(av.) 5.5×10^{-7} cm./cm./°C $\begin{cases} 20°C \\ 320°C \end{cases}$	Velocity of Sound—Shear Wave	3.75×10^5 cm./sec.
Thermal Conductivity	.0033 g. cal./cm.²/sec./°C/cm.	Velocity of Sound—Compressional Wave	5.90×10^5 cm./sec.
Specific Heat	.18 g. cal./gm.		
Softening Point	(approx.) 1665°C	Sonic Attenuation	less than .033 db/ft./mc.

FIXED POINT PROPERTIES OF OXYGEN

From NASA SP-3071, "ASRDI Oxygen Technology Survey, Thermophysical Properties", Volume I (1972), edited by Hans M. Roder and Lloyd A. Weber. This NASA publication contains an extensive bibliography and a discussion of basis for selection of these data for oxygen. The publication is available from the National Technical Information Service, Springfield, Virginia 22151.

PROPERTIES ↓ CONDITIONS →	Triple Point			Normal Boiling Point		Critical Point ††	Standard Conditions	
	Solid	Liquid	Vapor	Liquid	Vapor		STP (0°C)	NTP (20°C)
Temperature (K)		54.351		90.180		154.576	273.15	293.15
Pressure (mmHg)		1.138		760		37,823	760	760
Density (mole/cm³) x 10³	42.46	40.83	0.000336	35.65	0.1399	13.63	0.04466	0.04160
Specific Volume (cm³/mole) x 10⁻³	0.02355	0.02449	2975	0.028047	7.1501	0.07337	22.392	24.038
Compressibility Factor, $Z = \frac{PV}{RT}$	–	0.0000082	0.9986	0.00379	0.9662	0.2879	0.9990	0.9992
Heats of Fusion & Vaporization (J/mole)	444.8	7761.4		6812.3		0		
Specific Heat C_s @saturation (J/mole-K)	46.07	53.313	-108.7	54.14	-53.2	(very large)	–	–
C_p, @ constant pressure	–	53.27	29.13	54.28	30.77	(very large)	29.33	29.40
C_v, @ constant volume	–	35.65	20.81	29.64	21.28	(38.7)	20.96	21.04
Specific Heat Ratio, $\gamma = C_p/C_v$	–	1.494	1.400	1.832	1.446	(large)	1.40	1.40
Enthalpy (J/mole)	-6634.4	-6189.6	1571.8	-4270.3	2542.0	1032.2	7937.8	8525.1
Internal Energy (J/mole)	-6634.4	-6189.6	1120.0	-4273.1	1817.5	662.3	5668.9	6089.5
Entropy (J/mole-K)	58.92	67.11	209.54	94.17	169.68	134.42	202.4	204.5
Velocity of Sound (m/sec)		1159	141	903	178	164	315	326
Viscosity, μ, (N-sec/m³) x 10³		0.6194	0.003914	0.1958	0.00685	(0.031)	0.01924	0.02036
(centipoise)‡‡		0.6194	0.003914	0.1958	0.00685	(0.031)	0.01924	0.02036
Thermal Conductivity (mW/cm-K), k		1.929	0.04826	1.515	0.08544	(*)	0.2428	0.2575
Prandtl Number, $N_{pr} = \mu C_p/k$		5.344	0.7392	2.193	0.7714		0.7259	0.7265
Dielectric Constant, ϵ	(1.614)	1.5687	1.000004	1.4870	1.00166	1.17082	1.00053	1.00049
Index of Refraction, $n = \sqrt{\epsilon}$ †	(1.271)	1.2525	1.000002	1.219	1.00083	1.0820	1.00027	1.00025
Surface Tension (N/m) x 10³		22.65		13.20		0		
Equiv. Vol./Vol. Liquid at NBT	0.8397	0.8732	106,068	1	254.9	2.616	798.4	857.1

† Long Wavelengths
* Anomalously Large

Gas Constant: R = 62,365.4 cm³-mm Hg/mole-K
†† Values in parenthesis are estimates

Molecular Weight = 31.9988
"mole" = gram mole
‡‡ Units for poise are: g/cm-sec

FIXED POINTS AND PHASE EQUILIBRIUM BOUNDARIES FOR PARAHYDROGEN

a. Triple point

P_t = 0.0704 bar
T_t = 13.803 K
ρ_t (liquid) = 38.21 mol/ℓ

b. Normal boiling point

P_b = 1.01325 bar
T_b = 20.268 K
ρ_b (liquid) = 35.11 mol/ℓ
ρ_b (gas) = 0.6636 mol/ℓ

c. Critical point

P_c = 12.928 bar
T_c = 32.976 K
ρ_c = 15.59 mol/ℓ

Note: Some data indicate that the true critical temperature is probably closer to 32.93 K. However, that value is pending further verfication.

d. Melting pressures: in atmospheres

$$P = P_t + (T - T_t) [A_1 e^{-\alpha/T} + A_2 T]$$

A_1 = 30.3312
A_2 = 0.6667
α = 5.693

e. Liquid vapor coexistence densities

Liquid, density in mol/cm³ :

$$\rho \text{ sat } \ell = \rho_c + A_1 (\Delta T)^{0.380} + A_2 (\Delta T) + A_3 (\Delta T)^{4/3} + A_4 (\Delta T)^{5/3} + A_5 (\Delta T)^2$$

A_1 = 7.323 4603 × 10^{-3}
A_2 = −4.407 4261 × 10^{-4}
A_3 = 6.620 7946 × 10^{-4}
A_4 = −2.922 6363 × 10^{-4}
A_5 = 4.008 4907 × 10^{-5}
$\Delta T = T_c - T$

Vapor $T_b \leqslant T \leqslant T_c$, density in mol/cm³ :

$$\rho \text{ sat } G = \rho_c + A_1 (\Delta T)^{0.370} + A_2 (\Delta T) + A_3 (\Delta T)^{0.7} + A_4 (\Delta T)^{0.8}$$

A_1 = −7.196 7724 × 10^{-3}
A_2 = 1.449 5527 × 10^{-3}
A_3 = 3.240 3120 × 10^{-3}
A_4 = −4.464 0177 × 10^{-3}

f. Vapor pressure: in atmospheres

For T ⩽ 29 K:

$$\log_{10} P_a = A_1 + \frac{A_2}{T + A_3} + A_4 T$$

A_1 = 2.000 620
A_2 = −50.09 708
A_3 = 1.0044
A_4 = 1.748 495 × 10^{-2}

For T > 29 K:

$$P = P_a + A_5 (T - 29)^3 + A_6 (T - 29)^5 + A_7 (T - 29)^7$$

A_5 = 1.317 × 10^{-3}
A_6 = −5.926 × 10^{-5}
A_7 = 3.913 × 10^{-6}

From Weber, L. A., Thermodynamic and Related Properties of Parahydrogen from the Triple Point to 300 K at Pressures to 1000 Bar, NASA SP-3008, NBSIR 74-374, 1975.

Temperature °C	Density (g/cm³)								Viscosity (centipoise)		
	Na			K	Alloys (wt % K)				Na	Alloys (wt % K)	
					43.4		78.6			43.3	66.0
	(a)	(b)	(c)		Experimental	(d) Calculated	Experimental	(d) Calculated			
100	.927	.927	.9265	.819	.887	.890	.847	.850	.705	.540	.529
200	.904	.904	.9037	.795	.862	.867	.823	.827	.450	.379	.354
300	.882	.880	.8805	.771	.838	.843	.799	.802	.345	.299	.276
400	.859	.856	.8570	.747	.814	.818	.775	.778	.284	.245	.229
500	.834	.831	.8331	.723	.789	.794	.751	.754	.234	.207	.195[e]
600	.809	.808	.8089	.701	.765	.771	.727	.732	.210	.178[e]	.168[e]
700	.783	.784	.....	.676	.740	.745	.703	.705	.186[e]	.257[e]	.146[e]
800	.757	.760	.....	....	....	....	....	....	.165[e]	.....	....
900	....	....	.....	....	....	....	....	....	.150[e]	.....	....

Temperature, °C	Thermal conductivity (watts/cm²-°C/cm)				Electrical resistivity (microohms)			Heat capacity[g] (cal/°C-g)		
	Na		Alloy (wt % K)		Na[f]	Alloy (wt % K)		K	Alloy (wt % K)	
	Experimental	Calculated	56.5	77.0		56.5	78.0		44.8	78.26
100	......	....	....	.238[h]	8.99	41.61	45.63	.1940	.2690	.2248
200	.815	.808	.249	.247	13.52	47.23	51.33	.1887	.2612	.2169
300	.757	.755	.262	.259	17.52	54.33	58.58	.1894	.2553	.2122
400	.712	.710	.269	.262	21.93	62.21	65.65	.126	.2512	.2097
500	.668	.672	.271	.259	26.96	69.37	73.48	.1818	.2498	.2088
600	.627[e]	.639	....	.255	32.65	78.29	82.61	.1825	.2484	.2092
700	.590[e]	.610	....	......	39.05	88.23	91.76	.1846	.2497	.2108
800	.547[e]	.583	....	......	46.15	99.68	104.51	.1883	.2529	.2133
900	......	....	......	......	......	......	......	.....	.....	.....

Vapor Pressure, mm Hg

Temperature °C	Na	K	Alloys (wt % K)	
			56	78
127	2.23×10^{-6}	. . . .	. . . .	. . . .
227	1.15×10^{-3}	2.88×10^{-2}	1.57×10^{-3}	1.81×10^{-3}
327	5.03×10^{-2}	9.27×10^{-1}	5.73×10^{-2}	6.14×10^{-2}
427	.881	9.26	3.53	5.06
527	7.53	52.22	23.0	31.87
627	39.98	201.25	101.35	136.50
727	148.5	588.62	328.7	431.85
827	453.7	1421.0	864.2	1099.52
927	1127.8	. . . .	. . . .	. . . .
1027	2522.4	. . . .	. . . .	. . . .
1127	4696.8	. . . .	. . . .	. . . .

NOTES: (a) From plotted data.

(b) Epstein equation: $d_t = 0.9514 - 2.392 \times 10^{-4} \, t°C$.

(c) Thomson and Garelis: $d_t = 0.9490 - 22.3 \times 10^{-5} t°C - 1.75 \times 10^{-8} t^2 °C$.

(d) Formula to calculate density: $V = M_K \cdot V_K + M_{Na} \cdot V_{Na}$ (where V, V_K and V_{Na} are the specific volumes (reciprocal of density) of the alloy, K and Na respectively, M_K and M_{Na} the mole fraction of the elements).

(e) Extrapolated by calculation, Epstein equation:

$$K = \frac{2.433 \times 10^{-2}(t + 273.16)}{6.8393 + 3.3873 \times 10^{-2}t + 1.7235 \times 10^{-5}t^2}$$

(f) Epstein equation: $\eta_t = 10.892 + 0.015272t + 3.6746 \times 10^{-5}t^2 - \dfrac{379.26}{t}$.

(g) Formula to calculate heat capacity: $C = W_{Na} \cdot C_{Na} + W_K \cdot C_K$ (where C, C_{Na} and C_K are the heat capacity of the alloy, Na and K respectively, W_{Na} and W_{Ka} the weight fractions of Na and K respectively in the alloy).

(h) 150°C.

MECHANICAL AND PHYSICAL PROPERTIES OF WHISKERS

From NASA SP-5055

This table lists some nominal values for physical and mechanical properties of some whiskers. The strength of whiskers is influenced by temperature, time, surface conditions, surface films or corrosion, crystallographic orientation, impurities and testing techniques.

Material	Tensile Strength (σ_t) lb/in.$^2 \times 10^{-6}$	Young's Modulus (E) lb/in.$^2 \times 10^{-6}$	Specific Gravity (S) lb/in.$^2 \times 10^{-6}$	$\dfrac{(\sigma_t)}{(S)}$ inch	$\dfrac{(E)}{(S)}$ inch
Graphite	3.0	. .	137	21,700	. . . .
Al$_2$O$_3$	3.0	76	250	12,000	300,000
Iron	1.8	28	485	3,700	58,000
Si$_3$N$_4$	2.0	55	193	10,000	285,000
SiC	3.0	70	187	16,000	380,000
Si	1.1	26	143	7,000	182,000

ABSOLUTE VISCOSITY OF LIQUID SODIUM AND POTASSIUM

Sodium

T (°K)	η (10^{-2} poise)	ν (cm^3/g)	$\eta\nu^{1/3}$ (10^3 poise cm/g$^{-1/3}$)	$1/T\nu$ [10^3g/ (cm$^3 \cdot$°K)]
		Experimental range		
371.00	0.690	1.078,75	7.0766	2.4987
473	.450	1.106,56	4.6544	1.9106
573	.340	1.135,72	3.5482	1.5366
673	.278	1.166,86	2.9268	1.2734
773	.239	1.200,34	2.54057	1.0776
873	.212	1.236,25	2.2754	0.9266
973	.193	1.274,37	2.0925	.8065
1073	.179	1.315,79	1.9615	.7083
1173	.167	1.360,54	1.8505	.6266
1203	.164	1.373,62	1.8230	.6052

Potassium

T (°K)	η (10^{-2} poise)	ν (cm^3/g)	$\eta\nu^{1/3}$ (10^3 poise cm/g$^{-1/3}$)	$1/T\nu$ [10^3g/ (cm$^3 \cdot$°K)]
		Experimental range		
336.9	0.560	1.2062$_7$	5.961$_2$	2.4606
400	.384	1.22911	4.113$_4$	2.0340
500	.276	1.26711	2.986$_6$	1.5784
600	.221	1.3075$_3$	2.416$_6$	1.2747
700	.185	1.3506$_2$	2.045$_0$	1.0577
800	.162	1.3966$_6$	1.810$_8$	0.8949$_7$
900	.147	1.4459$_2$	1.6623	.7684$_6$
1000	.132	1.4988$_0$	1.5106	.6672$_0$
1100	.121	1.5556$_9$	1.4020	.5843$_5$
1200	.113	1.6170$_7$	1.3264	.5153$_3$
1300	.106	1.6835$_0$	1.2610	.4569$_1$
1400	.100	1.7556$_1$	1.2064	.4068$_7$

PHYSICAL PROPERTIES OF PIGMENTS

From Rutherford J. Gettens and George L. Stout. *Painting Materials*, Dover Publications, Inc., New York, 1966. Reprinted by permission of the Publisher.

OPAQUE WHITE PIGMENTS

Pigment Name and Chemical Composition[1]	Specific Gravity[2]	Particle Characteristics[3]	Refractive Index[4]
Titanium calcium white, TiO_2 (25%) + $CaSO_4$ (75%)	3.10	prism. or ragged gr.	mostly 1.8–2.0 (irr.) (bi.) [M*]
Titanium dioxide (rutile) + $CaSO_4$	3.25	...	Av. 1.98 [TPC]
Titanium barium white, TiO_2 (25%) + $BaSO_4$ (75%)	4.30	min. round. gr.	$n_\Sigma c$ 1.7–2.5 [M*]
White lead (basic sulfate) $PbSO_4 \cdot PbO$	6.46	...	Av. 1.93 [TPC]
Lithopone	4.3	...	Av. 1.84 [TPC]
Lithopone (regular), ZnS (28–30%), $BaSO_4$ (72–70%)	4.30	fine comp. gr.	2.3 (ZnS)–1.64 ($BaSO_4$) [M]
Zinc white (ordinary), ZnO	5.65	v. fine cryst. gr.	ϵ2.02, ω2.00 [M]
(acicular), ZnO	...	spicules, fourlets	ϵ2.02, ω2.00 [M*]
White lead (basic carbonate), $2PbCO_3 \cdot Pb(OH)_2$	6.70	v. fine cryst.	ϵ1.94, ω2.09 [M]
Antimony oxide, Sb_2O_3	5.75	v. fine cryst.	valentinite, α2.18, γ and β2.35 [LB, M*] senarmonite, 2.09 (isot.)
Zirconium oxide (baddeleyite), ZrO_2	5.69	...	χ2.13, γ2.20, β2.19 [LB], Av. 2.40 [TPC]
Titanium dioxide (anatase), TiO_2	3.9	min. round. gr.	ϵ and ω 2.5 (w. bi.) [M*]
(rutile), TiO_2	4.2	round. or prism. gr.	ϵ2.9, ω2.6 [M*]

TRANSPARENT WHITE PIGMENTS†

Diatomaceous earth, SiO_2	2.31	min. fossil forms	n mostly 1.435, some 1.40 [M*]
Aluminum stearate, $Al(C_{18}H_{35}O_2)_3$	0.99	agg. of spher. gr.	1.49 (w. bi.) [W]
Pumice (volcanic glass), Na, K, Al, silicate	...	vesicular vitr. frag.	c 1.50 (isot.) [M*]
Aluminum hydrate, $Al(OH)_3$	2.45	v. fine amorph. part.	$n_\Sigma c$ 1.50–1.56 [M*]
Gypsum, $CaSO_4 \cdot 2H_2O$	2.36	fine cryst. gr.	α1.520, γ1.530, β1.523 [LB]
Silica quartz), SiO_2	2.66	cryst. frag.	ϵ1.553, ω1.544 [LB]
(chalcedony), SiO_2	2.6	crypt. agg.	ϵ, ω1.54 [LB, M*]
China clay (kaolinite), $Al_2O_3 \cdot 2SiO_2 \cdot 2H_2O$	2.60	fine, vermicular cryst.	α1.558, γ1.565, β1.564 (all ± .005) [LB, M*]
Talc, $3MgO \cdot 4SiO_2 \cdot H_2O$	2.77	platy frag.	α1.539, γ1.589, β1.589 [LB]
Mica (muscovite), $H_2KAl_3(SiO_4)_3$	2.89	platy frag.	α1.563, γ1.604, β1.599 [LB]
Anhydrite, $CaSO_4$	2.93	cryst. frag.	α1.570, γ1.614, β1.575 [LB]
Chalk (whiting), $CaCO_3$	2.70	hollow spherulites	$\epsilon_\Sigma c$ 1.510, $\omega_\Sigma c$ 1.645 [M*]
Barytes (barite, nat.), $BaSO_4$	4.45	cryst. frag.	α1.636, γ1.648, β1.637 [LB]
(blanc fixe, art.), $BaSO_4$	4.36	v. fine cryst. agg.	1.62–1.64 [M*]
Barium carbonate, $BaCO_3$ (witherite)	4.3	...	α1.529, γ1.677, β1.676 [LB]

†Also called "Extender" or "Inert" White Pigments.

IRON OXIDE PIGMENTS

Ochre, yellow (goethite), $Fe_2O_3 \cdot H_2O$, clay, etc.	2.9–4.0	irr. spherulites	n_Σ2.0 (isot. part); $(\alpha, \beta)_\Sigma$ 2.05–2.31; γ_Σ 2.08–2.40 (bi. part) [M*]
Sienna, raw (goethite), $Fe_2O_3 \cdot H_2O$, clay, etc.	3.14	uneven spherulites	1.87–2.17 (mostly 2.06) (isot.) [M*]
Sienna, burnt, Fe_2O_3, clay, etc.	3.56	uneven, round. part.	c1.85 (var.) (isot.) [M]
Umber, raw, $Fe_2O_3 + MnO_2 + H_2O$, clay, etc.	3.20	uneven, round. gr.	mostly 1.87–2.17 [M*]
Umber, burnt, $Fe_2O_3 + MnO_2$, clay, etc.	3.64	uneven, round. gr.	mostly 2.2–2.3 [M*]
Iron oxide red (haematite), Fe_2O_3	5.2	min. cryst.	ϵ_{Li} 2.78, ω_{Li} 3.01 [M]

RED AND ORANGE PIGMENTS

Pigment Name and Chemical Composition[1]	Specific Gravity[2]	Particle Characteristics[3]	Refractive Index[4]
Red lead, Pb_3O_4 (c95%)	8.73	crypt. agg.	2.42_{Li} (w. bi.; pleo.) [M]
**Realgar, As_2S_2	3.56	cryst. frag.	a_{Li} 2.46, γ_{Li} 2.61, β_{Li} 2.59 [LB]
Molybdate orange, $Pb(Mo,S,Cr,P)O_4$	...	min. round. gr.	β_{Li} 2.55 (s. bi.) [M*]
Chrome orange, $PbCrO_4 \cdot Pb(OH)_2$	6.7	tabular cryst.	a2.42, γ2.7 $+$, β2.7 [M*]
Cadmium red lithopone, $CdS(Se) + BaSO_4$	4.30	min. round. gr.	2.50–2.76 (for CdS(Se)part) (isot.) [M*]
Cadmium red, $CdS(Se)$	4.5	min. round. gr.	2.64 (bright red)–2.77 (deep red) (isot.) [M*]
Antimony vermilion, Sb_2S_3	...	v. fine red glob.	n_{Li} 2.65 (isot.) [M*]
Vermilion (art.), HgS	8.09	hexagonal gr. and prisms	ϵ_{Li} 3.14, ω_{Li} 2.81 [M]
**(nat., cinnabar), HgS	8.1	cryst. frag.	ϵ_{Li} 3.146, ω_{Li} 2.819 [LB]
Quinacridone red, $C_{20}H_{12}O_2N_2$ (gamma)	1.5	thin plates	n_{Na} Av. 2.04 [Du P]

YELLOW PIGMENTS

Pigment Name and Chemical Composition[1]	Specific Gravity[2]	Particle Characteristics[3]	Refractive Index[4]
**Gamboge, organic resin	...	irr. amorph. part.	1.582–1.586 [W]
**Indian yellow, $C_{19}H_{18}O_{11}Mg \cdot 5H_2O$	...	prisms, plates	1.67 (w. bi.) [M*]
Cobalt yellow, $CoK_3(NO_2)_6 \cdot H_2O$	...	fine dendritic cryst.	1.72 1.70 (isot.) [W]
Zinc yellow, $4ZnO \cdot 4CrO_3 \cdot K_2O \cdot 3H_2O$	3.46	min. spher. gr.	1.84–1.9 (irr.; bi.) [M*]
Strontium yellow, $SrCrO_4$	...	small needles	a, β (or ω) 1.92, γ (or ϵ) 2.01 (‖ext.) [M*]
Barium yellow, $BaCrO_4$	4.49	v. fine cryst. gr.	1.94–1.98 (bi.) [M]
**Naples yellow, $Pb_3(SbO_4)_2$	...	round. gr.	2.01–2.28 (isot.) [M*]
Chrome yellow (med.), $PbCrO_4$	5.96	fine prism. gr.	$a_{620m\mu}$ < 2.31, $\gamma_{650m\mu}$ 2.49 [M]
Cadmium yellow lithopone, $CdS + BaSO_4$	4.25	fine comp. gr.	2.39–2.40 (for CdS part) [M*]
Cadmium yellow, CdS	4.35	min. round. gr.	2.35–2.48 (isot.) [M*]
Massicot (litharge), PbO	9.40	min. flakes	a_{Li} 2.51, γ_{Li} 2.71, β_{Li} 2.61 [M]
**Orpiment, As_2S_3	3.4	min. flakes	a_{Li} 2.4 $\pm$, γ_{Li} 3.02, β_{Li} 2.81 [LB]

GREEN PIGMENTS

Pigment Name and Chemical Composition[1]	Specific Gravity[2]	Particle Characteristics[3]	Refractive Index[4]
Phthalocyanine green, chloro-copper phthalocyanine	2.1	laths	$n_{580m\mu}$ 1.40 [ACC]
**Verdigris (copper basic acetate), $Cu(C_2H_3O_2)_2 \cdot 2Cu(OH)_2$	...	cryst. frag.	a1.53, γ1.56 [M]
**Chrysocolla, $CuSiO_3 \cdot \eta H_2O$	2.4	crypt. agg.	a1.575, γ1.598, β1.597 [LB]
Green earth (celadonite and glauconite), Fe, Mg, Al, K, hydrosilicate	2.5–2.7	round. irr. gr.	n var. c 1.62, (porous) [M*]
Emerald green (Paris green), $Cu(C_2H_3O_2)_2 \cdot 3Cu(AsO_2)_2$	3.27	spherulites and disks	a_{Σ}1.71, γ_{Σ}1.78 (w. pleo.) [M*]
**Malachite, $CuCO_3 \cdot Cu(OH)_2$	4.0	cryst. frag.	a1.655, γ1.909, β1.875 [LB]
Cobalt green, $CoO \cdot \eta ZnO$	...	spher. gr.	1.94–2.0 (w. bi.) [M*]
Viridian (chromium oxide, transparent), $Cr_2O_3 \cdot 2H_2O$	3.32	spherul. gr.	a, β_{Σ} 1.82, γ_{Σ} 2.12 [M*]
Chrome green (med.), $Fe_4[Fe(CN)_6]_3 + PbCrO_4$	4.06	fine green agg.	c 2.4 (cf. Prussian blue and chrome yellow)
Chromium oxide green, opaque, Cr_2O_3	5.10	fine cryst. agg.	n_{Li} 2.5 [M]

BLUE PIGMENTS

Pigment Name and Chemical Composition[1]	Specific Gravity[2]	Particle Characteristics[3]	Refractive Index[4]
Phthalocyanine blue, copper phthalocyanine	1.6	laths	Av. 1.38 [DuP]
Ultramarine blue (art.), $Na_{8-10}Al_6Si_6O_{24}S_{2-4}$	2.34	uniform small round. gr.	n 1.51 green, 1.63 red (isot.) [M]
**(nat., lazurite), $3Na_2O \cdot 3Al_2O_3 \cdot 6SiO_2 \cdot 2Na_2S$	2.4	angular, broken frag.	1.50± (isot.) [LB]
**Maya blue, Fe, Mg, Ca, Al, silicate (?)	...	porous irr. agg.	$\beta\Sigma$1.54 (irr.; bi. and pleo.) [M*]
Smalt, K, Co(Al), silicate (glass)	...	splintery, vitr. frag.	1.49–1.52 [M*]
Prussian blue, $Fe_4[Fe(CN)_6]_3$	1.83	colloidal agg.	$1.56_{460m\mu}$ [M*]
**Egyptian blue, $CaO \cdot CuO \cdot 4SiO_2$	...	cryst. frag.	ϵ1.605, ω1.635 [APL]
Manganese blue, $BaMnO_4 + BaSO_4$	...	gr. and stubby prisms	c 1.65 [W]
**Blue verditer, $2CuCO_3 \cdot Cu(OH)_2$	...	fibrous agg.	$a\Sigma$1.72, $\gamma\Sigma$slightly $>$ 1.74 [M*]
Cobalt blue, $CoO \cdot Al_2O_3$	3.83	round. gr.	n var.; max. c 1.74blue (isot.) [M]
**Azurite, $2CuCO_3 \cdot Cu(OH)_2$	3.80	cryst. frag.	a1.730, γ1.838, β1.758 [LB]
Cerulean blue, $CoO \cdot \eta SnO_2$	...	round gr.	1.84 (isot.) [M*]

VIOLET PIGMENTS

Ultramarine violet	...	round. gr. (blue, rose and violet)	c 1.56 (isot.) [M*]
Cobalt violet, $Co_3(PO_4)_2$	...	round. gr.	ϵ1.65–1.79 (dull violet), ω1.68–1.81 (salmon) (s. bi.) [M*]
Manganese violet, $(NH_4)_2Mn_2(P_2O_7)_2$	...	fine cryst. gr.	a1.67, γ1.75, β1.72 (for violet) [M]
Quinacridone violet, $C_{20}H_{12}O_2N_2$ (beta)	1.5	thin plates	Av. 2.02 [DuP]

BROWN PIGMENTS

Sepia (organic)	...	angular frag.	(opaque) [M*]
Asphaltum (bitumen)	...	irr. amorph. part.	1.64–1.66 [M*]
Van Dyke brown (bituminous earth)	1.66	irr. amorph. part.	1.62–1.69 [M*]

BLACK PIGMENTS

Bone black, $C + Ca_3(PO_4)_2$	2.29	irr. coarse grains	1.65–1.70 (for larger translucent gr.) [M]
Lamp black, C	1.77	min. round. part.	(opaque)
Charcoal black, C	...	irr. splintery part.	(opaque)
Graphite, C	2.36	irr. plates	(opaque) [M]

[1]Abbreviations: art. = artificial; med. = medium; nat. = natural. The chemical formulas are those commonly accepted in chemical and mineralogical literature, but they may not compare exactly with structural formulas based on x-ray diffraction data or even on critical chemical analysis.

[2]The figures for specific gravity of the artificial pigments are mainly from H. A. Gardner, pp. 710–712, and those on the mineral pigments are chiefly from E. S. Larsen and H. Berman.

[3]Symmetry terms (monoclinic, orthorhombic, etc.) are omitted because pigments are so finely divided that it is rare when observations on crystal symmetry can be made. The term "spherulitic," as used here means aggregates that tend toward radial structure and spherical shape. "Amorphous" describes materials that are microscopically formless but may be truly crystalline on the basis of x-ray diffraction data. Abbreviations: agg. = aggregate(s); amorph. = amorphous; comp. = composite; crypt. = cryptocrystalline; cryst. = crystal(s); frag. = fragment(s); glob. = globule(s); gr. = grain(s); irr. = irregular; min. = minute; part. = particle(s); prism. = prismatic; round. = rounded; spher. = spheroidal; spherul. = spherulitic; var. = variable; v. = very; vitr. = vitreous.

[4]Unless otherwise indicated, all refractive index measurements are by sodium light. Σ is the symbol used by H. E. Merwin to indicate greater or less indefiniteness or irregularity in the case of aggregates, especially in respect to refractive index. Abbreviations: bi. = birefringent; c = circa; ext. = extinction; isot. = isotropic; || = parallel; pleo. = pleochroic; s. = strongly; w. = weakly. The letters in brackets refer to the authorities for the refractive index data: M = H. E. Merwin; M* = H. E. Merwin, data by private communication, hitherto unpublished; W = C. D. West data by private communication, hitherto unpublished; LB = E. S. Larsen and H. Berman; APL = A. P. Laurie and co-authors; ACC = A. C. Cooper, "The refractive index of organic pigments. Its determination and significance," *Journal Oil & Colour Chemists Association*, Vol. 31 (1948), pp. 343–357; TPC = Titanium Pigment Corporation; Du P = E. I. Du Pont de Nemours & Co.
**Chiefly of historical interest.

CRITICAL TEMPERATURES AND CRITICAL PRESSURES OF THE ELEMENTS[a]

Critical temperatures are listed as degrees kelvin and critical pressures as megapascals. Some conversion factors which may be useful are:

$$
\begin{array}{lll}
\text{Megapascals} & \times \quad 9.8692 & = \quad \text{atmospheres (760mm Hg)} \\
\text{Megapascals} & \times \quad 14.504 & = \quad \text{pounds per square inch} \\
\text{Megapascals} & \times \quad 10^{-6} & = \quad \text{newtons per square meter (pascals)} \\
\text{Megapascals} & \times \quad 10.1972 & = \quad \text{kilograms per square centimeter} \\
\text{Megapascals} & \times \quad 10 & = \quad \text{bars} \\
\text{Megapascals} & \times \quad 7.501 & \times \quad 10^3 = \text{millimeters Hg at } 0°C \\
\text{Megapascals} & \times \quad 4.014 & = \quad \text{inches of } H_2O \text{ at } 4°C
\end{array}
$$

Element	Symbol	T_c, K	P_c, MPa
Argon	Ar	150.8	4.87
Arsenic	As	1673	22.3
Bromine	Br_2	588	10.3
Chlorine	Cl_2	416.9	7.977
Deuterium (equilibium)	D_2	38.2	1.65
Deuterium (normal)	D_2	38.4	1.66
Fluorine	F_2	144.3	5.215
Helium-3	He^3	3.31	0.114
Helium-4	He^4	5.19	0.227
Hydrogen (equilibium)	H_2	32.98	1.293
Hydrogen (normal)	H_2	33.2	1.297
Hydrogen deuteride	HD	36.0	1.48
Iodine	I_2	819	—
Mercury	Hg	1765	151.0
Krypton	Kr	209.4	5.50
Neon	Ne	44.40	2.76
Nitrogen	N_2	126.2	3.39
Oxygen	O_2	154.58	5.043
Ozone	O_3	261.1	5.57
Phosphorus	P	994	—
Radon	Rn	377	6.28
Rubidium	Rb	2105	—
Selenium	Se	1766	27.2
Sulfur	S	1314	20.7
Tritium	T_2	40.0	—
Xenon	Xe	289.73	5.840

The data are from "Vapor-Liquid Critical Properties", March, 1978, D. Ambrose and R. Townsend, National Physical Laboratory, Teddington, Middlesex TW11 OLW, UK. These data are used with permission of the copyright owner.

PHYSICAL CONSTANTS OF LIQUID ALKALI METALS

Element	mp, °K	bp, °K	T_c, °K	P_c, MPa	Heat of fusion, KJ mol^{-1}	Heat of vaporization KJ mol^{-1}
Cesium	301.55	942.4	2051 ± 10	11.73	2.14	66
Lithium	453.69	1615	3223 ± 600	68.9 ± 20%	2.99	134.7
Potassium	336.4	1033	2220 ± 25	16.39 ± 0.02	2.33	77.08
Rubidium	312.04	959.2	2083 ± 20	14.54 ± 0.5	2.34	69
Sodium	370.96	1156	2508.7 ± 12.5	25.64 ± 0.02	2.6	89.6 ± 0.5%

Table I (Continued)

Formula	Name	Critical temp. T_c °C	Critical press. P_c atm.	Formula	Name	Critical temp. T_c °C	Critical press. P_c atm.
C₇H₈	Toluene	320.8	41.6	C₈H₁₈	Butane, 2,2,3,3-tetramethyl-	270.8	24.5
C₇H₈O	Anisole, (see Benzene, methoxy-)	368.5	41.3	C₈H₁₈	Heptane, 4-methyl-	290	25.6
C₇H₈O	Benzene, methoxy-	368.5	41.3	C₈H₁₈	Hexane, 2,4-dimethyl-	282	25.8
C₇H₈O	o-Cresol	424.4	49.4	C₈H₁₈	n-Octane	296	24.8
C₇H₈O	m-Cresol	432	45	C₈H₁₈	Pentane, 2,2,3-trimethyl-	294	28.2
C₇H₈O	p-Cresol	431.4	50.8	C₈H₁₈O	1-Octyl alcohol	385.5	26.5
C₇H₉N	Aniline, methyl-	428.4	51.3	C₈H₁₉N	Dibutyl amine	322.6	—
C₇H₉N	2,3-Lutidine	382.3	—	C₉H₇N	Quinoline	508.8	—
C₇H₉N	2,4-Lutidine	374	—	C₉H₁₂	Benzene, n-propyl-	365	31.2
C₇H₉N	2,6-Lutidine	350.6	—	C₉H₁₂	Benzene, 1,2,3-trimethyl-	391.3	31
C₇H₉N	3,4-Lutidine	410.6	—	C₉H₁₂	Cumene (see Benzene, isopropyl-)	362.7	31.2
C₇H₉N	3,5-Lutidine	394.1	—	C₉H₁₂	Benzene, isopropyl-	362.7	31.2
C₇H₁₄	Cyclohexane, methyl-	299.1	34.3	C₉H₂₀	n-Nonane	321	22.5
C₇H₁₄	Cyclopentane, ethyl-	296.3	33.5	C₉H₂₁	Tripropyl amine	304.3	—
C₇H₁₄	1-Heptene	264.1	—	C₁₀H₈	Naphthalene	474.8	40.6
C₇H₁₄O₂	Acetic acid, isoamyl- (Isoamyl acetate)	326.1	28	C₁₀H₁₂O	Ether, diphenyl-	494	30.9
C₇H₁₆	Butane, 2,2,3-trimethyl-	258.3	29.8	C₁₀H₁₄	Benzene, 1-isopropyl-4-methyl-	385.5	27.7
C₇H₁₆	n-Heptane	267.1	27	C₁₀H₁₄	Benzene, 1,2,3,5-tetramethyl-	402.8	28.6
C₇H₁₆	Hexane, 2-methyl-	257.9	27.2	C₁₀H₁₄	p-Cymene, (see Benzene, 1-isopropyl-4-methyl-)	—	—
C₇H₁₆	Pentane, 3-ethyl-	267.6	28.6		Isodurene, (see Benzene, 1,2,3,5-tetramethyl-)	—	—
C₇H₁₆	Pentane, 2,4-dimethyl-	247.1	27.4	C₁₀H₁₄O	3-p-Cymenol	425.1	33
C₇H₁₆O	1-Heptyl alcohol	365.3	29.4	C₁₀H₁₄O	Thymol, (see 3-p-Cymenol)	425.1	33
C₇F₁₆	n-Heptane, perfluoro-	201.7	16	C₁₀H₁₈	Decalin, cis-	418	28.7
C₈H₈	Styrene	374.4	39.4	C₁₀H₁₈	Decalin, trans-	408	28.7
C₈H₁₀	Benzene, ethyl-	343.9	36.9	C₁₀H₂₂	n-Decane	344.4	20.8
C₈H₁₀	o-Xylene	359	35.7	C₁₁H₁₀	Naphthalene, 1-methyl-	498.8	32.1
C₈H₁₀	m-Xylene	346	34.7	C₁₂H₁₀	Biphenyl	495	31.8
C₈H₁₀	p-Xylene	345	33.9	C₁₂H₁₁N	Diphenylamine	615.5	—
C₈H₁₀O	Xylenol	449.7	56.4	C₁₂H₁₈	Benzene, hexamethyl-	494	23.5
C₈H₁₀O	Phenetol (see Benzene, ethoxy-)	374	33.8	C₁₂H₁₈	Mellitine, (see Benzene, hexamethyl-)	—	—
C₈H₁₀O	Benzene, ethoxy-	374	33.8	C₁₂H₂₆	n-Dodecane	386	17.9
C₈H₁₁N	Aniline, N,N-dimethyl-	414.4	35.8	C₁₂H₂₇N	Tributyl amine	365.2	—
C₈H₁₁N	Aniline, N-ethyl-	425.4	—				
C₈H₁₆	n-Octene	305	25.5				

Table II

Name	Formula	Critical temp. T_c °C	Critical press. P_c atm.	Name	Formula	Critical temp. T_c °C	Critical press. P_c atm.
Ammonia	NH₃	132.5	112.5	Hydrogen iodide	HI	150	81.9
Argon	Ar	−122.3	48	Hydrogen sulfide	H₂S	100.4	88.9
Boron tribromide	BBr₃	300	—	Hydrazine	N₂H₂	380	—
Boron trichloride	BCl₃	178.8	38.2	Iodine	I₂	512	116
Boron trifluoride	BF₃	−12.26	49.2	Krypton	Kr	−63.8	54.3
Carbon dioxide	CO₂	31	72.9	Neon	Ne	−228.7	26.9
Carbon disulfide	CS₂	279	78	Nitric oxide	NO	−93	64
Carbon monoxide	CO	−140	34.5	Nitrogen dioxide	NO₂	157.8	100
Carbonyl sulfide	COS	104.8	65	Nitrogen	N₂	−147	33.5
Chlorine	Cl₂	144	76.1	Nitrous oxide	N₂O	36.5	71.7
Cyanogen	C₂N₂	126.6	—	Oxygen	O₂	−118.4	50.1
Deuterium	D₂	−234.8	16.4	Ozone	O₃	−5.16	67
Fluorine	F₂	−129	55	Phosphine	PH₃	51.3	64.5
Germanium tetrachloride	GeCl₄	276.9	38	Radon	Rn	104.04	62
Helium	He	−267.9	2.26	Silane, chlorotrifluoro-	SiClF₃	34.5	34.2
Hydrogen	H₂	−239.9	12.8	Silane	SiH₄	−3.46	47.8
Hydrogen bromide	HBr	90	84.5	Silicon tetrachloride	SiCl₄	232.8	—
Hydrogen chloride	HCl	51.4	82.1	Silicon tetrafluoride	SiF₄	−14.06	36.7
Hydrogen deuteride	HD	237.3	14.6	Sulfur dioxide	SO₂	157.8	77.7
Hydrogen cyanide	HCN	183.5	48.9	Stannic chloride	SnCl₄	318.7	37
Hydrogen fluoride	HF	188	64	Water	H₂O	374.1	218.3
				Xenon	Xe	16.6	58

DISSOCIATION PRESSURE OF CALCIUM CARBONATE

Temp. °C	mm/Hg	Temp. °C	mm/Hg	Temp. °C	mm/Hg	Temp. °C	mm/Hg
550	0.41	727	44	819	235	894	716
587	1.0	736	54	830	255	898	760 atm.
605	2.3	743	60	840	311	906.5	1.151
671	13.5	748	70	852	381	937	1.770
680	15.8	749	72	857	420	1082.5	8.892
691	19.0	777	105	871	537	1157.7	18.687
701	23.0	786	134	881	603	1226.3	34.333
703	25.5	795	150	891	684	1241	39.094
711	32.7	800	183				

CRITICAL TEMPERATURES AND PRESSURES

Compiled by Rudolf Loebel

Table I —Organic compounds
Table II—Inorganic compounds

Table I

Formula	Name	Critical temp. T_c °C	Critical press. P_c atm.	Formula	Name	Critical temp. T_c °C	Critical press. P_c atm.
$CHClF_2$	Methane, monochlorodifluoro-	96	48.5	C_4H_4O	Furan	213.8	52.5
$CHCl_2F$	Methane, dichloromonofluoro-	178.5	51	C_4H_4S	Thiophene	307	56.2
$CHCl_3$	Methane, trichloro- (Chloroform)	263	54	C_4H_6	1,2-Butadiene	171	44.4
CHF_3	Methane, trifluoro- (Fluoroform)	25.9	46.9	C_4H_6	1,3-Butadiene	152	42.7
CH_2Cl_2	Methylene chloride	237	60	C_4H_8	n-Butene	146	39.7
CH_3NO_2	Methane, nitro-	314.8	62.3	C_4H_8	2-Butene, cis-	160	40.5
CH_3Br	Methane, monobromide-	194	83.4	C_4H_8	2-Butene, trans-	155	41.5
CH_3Cl	Methane, monochloro-	143.8	65.9	C_4H_8O	1,2-Butylene oxide	243	—
CH_3F	Methane, monofluoro-	44.6	58	C_4H_8O	Ketone, ethyl methyl- (2-Butanone)	262	41
CH_3I	Methane, monoiodo-	254.8	72.7	$C_4H_8O_2$	Butanoic acid	355	52
CH_4	Methane	−82.1	45.8	$C_4H_8O_2$	p-Dioxane	314.8	51.4
CH_4O	Methanol (Methyl alcohol)	240	78.5	$C_4H_8O_2$	Acetic acid, ethyl- (Ethyl acetate)	250.4	37.8
CH_4S	Methylmercaptan	196.8	71.4	$C_4H_8O_2$	Propanoic acid, methyl- (Methyl propionate)	257.4	39.3
CH_5N	Methylamine	156.9	40.2	$C_4H_8O_2$	Formic acid, propyl-(Propyl formate)	264.9	40.1
$CBrF_3$	Methane, monobromotrifluoro-	67	50.3	C_4H_9Cl	n-Butane, monochloride-	269	—
$CClF_3$	Methane, monochlorotrifluoro-	28.85	38.2	C_4H_{10}	n-Butane	152	37.5
CCl_2F_2	Methane, dichlorodifluoro-	111.5	39.6	C_4H_{10}	Isobutane	134.7	35.9
CCl_3F	Methane, trichloromonofluoro-	198	43.2	$C_4H_{10}O$	Butanol (n-Butyl alcohol)	289.8	43.6
CCl_4	Methane, tetrachloro- (Carbon tetrachloride)	283.1	45	$C_4H_{10}O$	sec-Butyl alcohol	263	41.4
CF_4	Methane, tetrafluoro-	−45.7	41.4	$C_4H_{10}O$	tert-Butyl alcohol	235	39.2
C_2H_2	Acetylene	35.5	61.6	$C_4H_{10}O$	Ether, diethyl-	192.6	35.6
C_2H_2	Ethyne (see Acetylene)	35.5	61.6	$C_4H_{10}O$	Isobutyl alcohol	277	42.4
$C_2H_2F_2$	Ethylene, 1,1-difluoro-	30.1	—	$C_4H_{10}O_2$	Glycol, diethylene-	407	—
C_2H_3N	Acetonitrile	274.7	47.7	$C_4H_{10}S$	Diethyl sulfide	283.8	39.1
$C_2H_3F_3$	Ethane, 1,1,1-trifluoro-	73.1	—	$C_4H_{11}N$	Butyl amine	287.9	—
C_2H_4	Ethene	9.9	50.5	$C_4H_{11}N$	Diethyl amine	223.3	36.6
C_2H_4	Ethylene (see Ethene)	9.9	50.5	$C_4H_{11}N$	Isobutyl amine	266.7	—
C_2H_4O	Acetaldehyde	187.8	54.7	C_4F_{10}	Butane, perfluoro-	113.2	23
C_2H_4O	Ethylene oxide	195.8	71	C_5H_5N	Pyridine	346.8	—
$C_2H_4O_2$	Acetic acid	321.6	57.1	C_5H_8	Cyclopentene	232.94	47.2
$C_2H_4O_2$	Formic acid, methyl- (Methyl formate)	214	59.2	C_5H_{10}	Cyclopentane	238.6	44.6
$C_2H_4Cl_2$	Ethane, 1,1-dichloro-	249.8	50	C_5H_{10}	1-Pentene	191	35.5
$C_2H_4F_2$	Ethane, 1,1-difluoro-	386.7	—	$C_5H_{10}O$	Ketone, diethyl-	287.8	36.9
C_2H_5Br	Ethane, monobromo-	230.8	61.5	$C_5H_{10}O$	Propanoic acid, ethyl- (Ethyl propionate)	272.9	33
C_2H_5Cl	Ethane, monochloro-	187.2	52	$C_5H_{10}O_2$	n-Butanoic acid, methyl- (n-Methyl butyrate)	281.3	34.3
C_2H_5F	Ethane, monofluoro-	102.16	49.6	$C_5H_{10}O_2$	Acetic acid, n-propyl (n-Propyl acetate)	276	32.9
C_2H_6	Ethane	32.2	48.2	$C_5H_{10}O_2$	n-Valeric acid	378	37.6
C_2H_6O	Ether, dimethyl-	127	52.6	$C_5H_{11}N$	Piperidine	320.8	44.1
C_2H_6O	Ethanol (Ethyl alcohol)	243	63	C_5H_{12}	Butane, 2-methyl- (See Isopentane)	187.8	32.9
$C_2H_6O_2$	Glycol, ethylene-	(374)*	—	C_5H_{12}	Isopentane	187.8	32.9
C_2H_6S	Dimethylsulfide	229.9	54.5	C_5H_{12}	Neopentane	160.6	31.6
C_2H_6S	Ethylmercaptan	225.5	54.2	C_5H_{12}	n-Pentane	196.6	33.3
C_2H_7N	Dimethylamine	164.6	52.4	C_5H_{12}	Propane, 2,2-dimethyl (see Neopentane)	160.6	31.6
C_2H_7N	Ethylamine	183.2	55.5	$C_5H_{12}O$	Isoamyl alcohol	309.77	—
$C_2Br_2F_4$	Ethane, dibromotetrafluoro-	214.5	—	C_6H_5Br	Benzene, bromo-	397	44.6
C_2ClF_5	Ethane, chloropentafluoro-	80	—	C_6H_5Cl	Benzene, chloro-	359.2	44.6
$C_2Cl_2F_4$	Ethane, 1,2-dichlorotetrafluoro-	145.7	—	C_6H_5F	Benzene, fluoro-	286.95	44.6
$C_2Cl_3F_3$	Ethane, trichlorotrifluoro-	214.2	33.7	C_6H_6	Benzene	288.9	48.6
$C_2Cl_4F_2$	Ethane, tetrachlorodifluoro-	278	32.9	C_6H_6O	Phenol	421.1	60.5
C_2F_6	Ethane, hexafluoro-	24.3	—	C_6H_7N	Aniline	425.6	52.3
C_3H_4	Propadiene	120	43.6	C_6H_7N	α-Picoline	348	—
C_3H_4	Allene (see Propadiene)	120	43.6	C_6H_7N	β-Picoline	371.7	—
C_3H_4	Acetylene, methyl-	127.8	52.8	C_6H_7N	γ-Picoline	372.5	—
C_3H_4	Propyne (see Acetylene, methyl-)	127.8	52.8	C_6H_{10}	Cyclohexene	287.3	—
C_3H_5N	Ethyl cyanide	290.8	41.3	C_6H_{10}	1,5-Hexadiene	234.4	32.6
C_3H_5N	Propionitrile (see Ethyl cyanide)	290.8	41.3	C_6H_{12}	Cyclohexane	280.4	40
C_3H_5OCl	Epichlorohydrin	(323)	—	C_6H_{12}	Cyclopentane, methyl-	259.5	37.4
C_3H_5Cl	Propene, 3-chloro- (allyl-chloride)	240.3	46.5	C_6H_{12}	1-Hexene	231	31.1
C_3H_6	Propylene	91.9	45.4	$C_6H_{12}O_2$	Formic acid, n-amyl (n-Amyl formate)	302.6	34.1
C_3H_6	Cyclopropane	124.7	—	$C_6H_{12}O_2$	Acetic acid, n-butyl- (n-Butyl acetate)	305.9	30.7
C_3H_6O	Acetone	235.5	47	$C_6H_{12}O_2$	Butanoic acid, ethyl- (Ethyl butyrate)	293	30.2
C_3H_6O	Allylalcohol	272	55.5	$C_6H_{12}O_2$	Propanoic acid, propyl- (Propyl propionate)	304.8	30.7
C_3H_6O	Propylene oxide	209	48.6	$C_6H_{12}O_3$	Paraldehyde	290	—
$C_3H_6O_2$	Formic acid, ethyl- (Ethyl formate)	235.3	46.3	C_6H_{14}	Butane, 2,3-dimethyl-	226.8	30.9
$C_3H_6O_2$	Acetic acid, methyl- (Methyl acetate)	233.7	46.3	C_6H_{14}	n-Hexane	234.2	29.9
$C_3H_6O_2$	Propanoic acid	337.6	53	C_6H_{14}	Pentane, 2-methyl-	224.3	30
C_3H_7Cl	n-Propane, monochloro-	230	45.2	$C_6H_{14}O$	Ether, isopropyl-	226.9	28.4
C_3H_8	Propane	96.8	42	$C_6H_{14}O$	1-Hexyl alcohol	313.5	—
C_3H_8O	Ether, ethyl methyl- (Methoxyethane)	164.7	43.4	$C_6H_{14}O_4$	Glycol, triethylene-	(437)	—
C_3H_8O	Glycol, 1,2-propylene-	351	—	$C_6H_{15}N$	Dipropyl amine	277	31
C_3H_8O	Isopropyl alcohol	235	47	$C_6H_{15}N$	Hexyl amine	318.8	—
C_3H_8O	n-Propyl alcohol	263.6	51	$C_6H_{15}N$	Triethyl amine	258.9	30
$C_3H_8O_3$	Glycerol	(452)	—	C_7H_5N	Benzonitrile	426.2	41.6
C_3H_9N	Isopropylamine	209.7	—	C_7H_8	Benzene, methyl- (see Toluene)	320.8	41.6
C_3H_9N	n-Propylamine	223.8	46.8				

*( ) uncertain

DEFINITIONS

AB — A prefix attached to the names of the practical electric units to indicate the corresponding unit in the cgs electromagnetic system (emu), e.g. abampere, abvolt.

Abcoulomb — The abcoulomb, the emu of charge, is defined as the charge which passes a given surface in 1 sec if a steady current of one abampere flows across the surface. Its dimensions are, therefore, $cm^{1/2}\ g^{1/2}$, which differ from the dimensions of the statcoulomb by a factor which has the dimensions of a speed. This relationship is connected with the fact that the ratio $2K_e/K_m$ must have the value of the square of the speed of light in any consistent system of units. It follows further that

$$1\ abcoulomb = 2.99793 \times 10^{10}\ statcoulomb,$$

the speed of light in vacuo being $(2.99793 \pm 0.000003) \times 10^{10}$ cm/sec.

Aberration — 1. In astronomy, the apparent angular displacement of the position of a celestial body in the direction of motion of the observer, caused by the combination of the velocity of the observer and the velocity of light. See constant of aberration, planetary aberration. Compare parallax. 2. In optics, a specific deviation from perfect imagery, as, for example: spherical aberration, coma, astigmatism, curvature of field, and distortion.

Aberrations (of image) — Distortions in shape, color, focus, and density of images caused by imperfect optical elements (i.e., lens, prism, mirror, screen, etc.) Types such as coma, astigmatism, field curvature, distortion, and chromatic and spherical aberrations.

Absolute humidity — See Humidity.

Absolute pressure — See Pressure.

Absolute temperature — Temperature reckoned from the absolute zero. See Temperature.

Absolute units — A system of units based on the smallest possible number of independent units. Specifically, units of force, work, energy, and power not derived from or dependent on gravitation.

Absolute zero — The theoretical temperature at which molecular motion vanishes and a body would have no heat energy; the zero point of the Kelvin and Rankine temperature scales.

Absolute zero may be interpreted as the temperature at which the volume of a perfect gas vanishes or, more generally, as the temperature of the cold source which would render a Carnot cycle 100% efficient. The value of absoute zero is now estimated to be $-273.15°C$, $-459.67°F$, 0 K, and 0° Rankine.

Absorption — 1. Penetration of a substance into the body of another. 2. Transformation into other forms suffered by radiant energy passing through a material substances.

Absorption coefficient (a) — 1. A measure of the amount of normally incident radiant energy absorbed through a unit distance or by a unit mass of absorbing medium. Compare transmission coefficient.

The absorption coefficiet is frequently identified as follows:

$$I_{\lambda x} = I_{\lambda o}\ e^{-k^\lambda x}$$

where $I_{\lambda x}$ is the flux density of radiation of wavelength λ, initially of flux density $I_{\lambda o}$, after traversing a distance x in some absorbing medium.

2. In acoustics, the ratio of the sound energy absorbed by a surface of a medium (or material) exposed to a sound field or sound radiation to the sound energy incident on the surface. The stated values of this ratio are to hold for an infinite area of the surface. The conditions under which measurements of absorption coefficients are made are to be stated explicitly.

Three types of absorption coefficients associated with three methods of measurement are: chamber absorption coefficient, obtained in a certain reverberation chamber; free-wave absorption coefficient, obtained when a plane, progressive, sound wave is incident on the surface of the medium; sabine absorption coefficient, obtained when the sound is incident from all directions on the sample. See Absorption Factor.

Absorption cross sections — In radar, the ratio of the amount of power removed from a beam by absorption of radio energy by a target to the power in the beam incident upon the target. Compare scattering cross section. See Cross section.

Absorption factor — The ratio of the intensity loss by absorpotion to the total original intensity of radiation. If I_o represents the original intensity, I_r, the intensity of reflected radiation, I_t, the intensity of the transmitted radiation, the absorption factor is given by the expression

$$\frac{I_o - (I_r + I_t)}{I_o}$$

Also called coefficient of absorption.

Absorption, Lambert's law — If I_o is the original intensity, I the intensity after passing through a thickness x of a material whose absorption coefficient is k,

$$I = I_o e^{-kx}$$

The index of absorption k' is given by the relation $k = (4\pi k'n)/\lambda$ where n is the index of refraction and λ the wave length in vacuo. The mass absorption is given by k/d when d is the density. The transmission factor is given by I/I_o.

Absorption spectrum — The array of absorption lines and absorption bands which results from the passage of radiant energy from a continuous source through a selectively absorbing medium cooler than the source. See electromagnetic spectrum.

The absorption spectrum is a characteristic of the absorbing medium, just as an emission spectrum is a characteristic of a radiator.

An absorption spectrum formed by a monatomic gas exhibits discrete dark lines, whereas that formed by a polyatomic gas exhibits ordered arrays (bands) of dark lines, which appear to overlap. This type of absorption is often referred to as line absorption. The spectrum formed by a selectively absorbing liquid or solid is typically continuous in nature (continuous absorption).

The spectrum obtained by the examination of light from a source, itself giving a continuous spectrum after this light has passed through an abosrbing medium in the gaseous state. The absorption spectrum will consist of dark lines or bands, being the reverse of the emission spectrum of the absorbing substance.

When the absorbing medium is in the solid or liquid state the spectrum of the transmitted light shows broad dark regions which are not resolvable into lines and have no sharp or distinct edges.

Absorptive index — The imaginary part of the complex index of refraction of a medium. It represents the energy loss by absorption and has a nonzero value for all media which are not dielectrics. Also called index of absorption. Compare absorption coefficient.

Absorptive power or absorptivity for any body is measured by the fraction of the radiant energy falling upon the body which is absorbed or transformed into heat. This ratio varies with the character of the surface and the wave length of the incident energy. It is the ratio of the radiation absorbed by any substance to that absorbed under the same conditions by a black body.

Abvolt — The cgs electromagnetic unit of potential difference and electromotive force. It is the potential difference that must exist between two points in order that one erg of work be done when one abcoulomb of charge is moved from one point to the other. One abvolt is 10^{-8} V.

Acceleration — The time rate of change of velocity in either speed or direction. Cgs unit, 1 cm/sec/sec. Dimensions, — $[l\ t^{-2}]$ — See Angular acceleration.

Acceleration due to gravity — The acceleration of a body freely falling in a vacuum. The International Committee on Weights and Measures has adopted as a standard or accepted value, 980.665 cm/sec² or 32.174 ft/sec².

Acceleration due to gravity at any latitude and elevation — If ϕ is the latitude and H the elevation in centimeters the acceleration in cgs units is, $g = 980.616 - 2.5928 \cos 2\phi + 0.0069 \cos^2 2\phi - 3.086 \times 10^{-6}$ H. (Helmert's equation)

Accelerators — Machines for speeding up subatomic particles to energies running into millions of electron volts. See Betatron, Cyclotron, etc.

Acceptor — In transistors, the P-type semiconductor, the electrode containing trivalent impurities (boron, gallium, or indium) to increase the number of holes which can accept electrons. Contrast with donor.

Achromatic — A term applied to lenses signifying their more or less complete correction for chromatic aberration.

Acid — For many purposes it is sufficient to say that an acid is a hydrogen-containing substance which dissociates on solution in water to produce one or more hydrogen ions. More generally, however, acids are defined according to other concepts. The Bronsted concept states that an acid is any compound which can furnish a proton. Thus NH_4^+ is an acid since it can give up a proton:

$$NH_4^+ \rightleftharpoons NH_3 + H^+$$

and NH_3 is a base since it accepts a proton.

A still more general concept is that of G. N. Lewis which defines an acid as anything which can attach itself to something with an unshared pair of electrons. Thus in the reaction

$$H^+ + :N{\overset{H}{\underset{H}{-}}}H \rightleftharpoons NH_4^+$$

the NH_3 is a base because it possesses an unshared pair of electrons. This latter concept explains many phenomena, such as the effect of certain substances other than hydrogen ions in the changing of the color of indicators. It also explains acids and bases in nonaqueous systems as liquid NH_3 and SO_2.

Acoustic velocity (symbol a) = speed of sound.

Actinic — Pertaining to electromagneic radiation capable of initiating photochemical reactions, as in photography or the fading of pigments.

Because of the particularly strong action of ultra violet radiation on photochemical processes, the term has come to be almost synonymous with ultraviolet, as in actinic rays.

Actinide Series — Elements of atomic numbers 89 to103 analogous to the lanthanide series of the so-called rare earths.

Actinometer — The general name for any instrument used to measure the intensity of radiant energy, particularly that of the sun. See Actinometry. See also Bolometer, Dosimeter, Photometer, Radiometer.

Actinometers may be classified, according to the quantities which they measure, in the following manner: (a) pyrheliometer, which measures the intensity of direct solar radiation; (b) pyranometer, which measures global radiation (the combined intensity of direct solar radiation and diffuse sky radiation); and (c) pyrgeometer, which measures the effective terrestrial radiation.

Action is measured by the product of work by time. Cgs units of action are the erg-second and the joule-second. Dimensions, — $[m \, l_2 \, t^{-1}]$. Planck's quantum or constant of action is $(6.62517 - 0.00023) \times 10^{-27}$ erg-sec.

Active mass of a substance is the number of gram molecular weights per liter in solution, or in gaseous form.

Activity coefficient — A factor which, when multiplied by the molecular concentration yields the active mass. The activity coefficient is evaluated by thermodynamic calculations, usually from data on the emf of certain cells, or the lowering of the freezing point of certain solutions. It is a correction factor which makes the thermodynamic calculations correct.

Adiabatic — A body is said to undergo an adiabatic change when its condition is altered without gain or loss of heat. The line on the pressure volume diagram representing the above change is called an adiabatic line.

Adiabatic atmosphere — A model atmosphere in which the pressure decreases with height according to:

$$p = p_o \, [1 - (- \, gz/c_{p,d}T_o)] \, c_{p,d}R_d$$

where p_o and T_o are the pressure and temperature (°K) at sea level or other datum; z is the geometric height; R_d is the gas constant for dry gas; $c_{p,d}$ is the specific heat for dry gas at constant pressure; and g is the acceleration of gravity. Also called dry-adiabatic atmosphre, convective atmosphere, homogeneous atmosphere. See Homogeneous atmosphere, Barotropy.

Adiabatic process — A thermodynamic change of state of a system in which there is no transfer of heat or mass across the boundaries of the system. In an adiabatic process, compression always results in warming, expansion in cooling. See Diabatic process.

Adsorption — The condensation of gases, liquids, or dissolved substances on the surfaces of solids is called adsorption.

Air columns, frequency of vibration in — See Organ pipes.

Albedo — The ratio of the amoung of electromagnetic radiation reflected by a body to the amount incident upon it, often expressed as a percentage, as, the albedo of the earth is 34%.

The concept defined above is identical with reflectance. However, albedo is more commonly used in astronomy and meteorology and reflectance in physics.

Albedo is sometimes used to mean the flux of the reflected radiation as, the earth albedo is 0.64 cal/cm². This usage should be discouraged.

The albedo is to be distinguished from the spectral reflectance, which refers to one specific wavelength (monochromatic radiation).

Usage varies somewhat with regard to the esact wavelength interval implied in albedo figures; sometimes just the visible portion of the spectrum is considered, sometimes the totality of wavelengths in the solar spectrum.

Alfvén speed — The speed at which Alfvén waves are propagated along the magnetic field.

For a perfectly conducting fluid with a mass density of 1 kg/m^3 in a magnetic field of 10,000 gauss, the Alfvén speed is about 1000 m/sec while the speed of sound in air is about 300 m/sec.

Alfvén wave — A transverse wave in a magneto-hydrodynamic field in which the driving force is the tension introduced by he magnetic field along the lines of force. Also called magneto-hydrodynamic wave.

The dynamics of such waves are analogous to those in a vibrating string, the phase speed C being given by

$$C^2 = \mu H^2 / 4\pi\rho$$

where μ is the permeability; H is the magnitude of the magnetic field; and ϱ is the fluid density. Dissipative effects due to fluid viscosity and electrical resistance may also be present.

Allobar — A form of an element differing in isotopic composition from the naturally occurring form.

Allotropy — The property shown by certain elements or being capable of existence in more than one form, due to differences in the arrangement of atoms or molecules. See Monotropic and Enantiotropic.

Alpha (α)-particle, or alpha-ray — One of the particles emitted in radioactive decay. It is identical with the nucleus of the helium atom and consists, therefore, of two protons plus two neutrons bound together. A moving alpha particle is strongly ionizing and so loses energy rapidly in traversing through matter. Natural alpha particles will traverse only a few centimeters of air before coming to rest.

Alternating current, (A-C) — Current in which the charge-flow periodically reverses, as opposed to direct current, and whose average value is zero. Alternating current usually implies a sinusoidal variation of current and voltage. This behavior is represented mathematically in various ways:

$$I = I_o \cos(2\pi ft + \phi)$$
$$I = I_o \angle \phi$$
$$I = I_1^{i\omega t}$$

where fi is the frequency; $\omega \equiv 2\pi f$, the pulsatance, or radian frequency; ϕ the phase angle; I_o the amplitude; and I_1 the complex amplitude. In the complex rotation, it is understood that the actual current is the real part of I. For circuits involving also a capacitance C in farads and L in henrys, the impedance becomes,

$$\sqrt{R^2 + \left(2\pi fL - \frac{1}{2\pi fC}\right)^2}$$

Altitudes with the barometer — If b_1 and b_2 denote the corrected barometer readings at two stations, t the mean of the temperatures, t_1 and t_2 of the air at the two stations, e_1 and e_2 the tension of water vapor at the two stations, h the mean height above sea level, ϕ the latitude; then the difference in elevation in centimeters is H = 1,843,000 (log b_1 − log b_2) (1 + 0.00367t) (1 + 0.0026 cos 2ϕ + 0.00002h + 3/8k), where

$$k = \frac{1}{2}\left(\frac{e_1}{b_1} + \frac{e_2}{b_2}\right)$$

An approximate formula, sufficient for differences not over 1000 m is

$$H = 1,600,000 \, \frac{b_1 - b_2}{b_1 + b_2} \, (1 + 0.004t).$$

DEFINITIONS (Continued)

Amorphous — Without definite form, not crystallized.

Ampere (unit of electric current) is the constant current which, if maintained in two straight parallel conductors of infinite length, of negligible circular sections, and placed 1 m apart in a vacuum, will produce between these conductors a force equal to 2×10^{-7} newton per meter of length.

Ampère's law — A vector equation relating current flow and magnetic fields. The law states that if a current (I) flows through an infinitesimal distance (dl) along a line, at some point (P) a distance (r) away there is produced an infinitesimal element of a magnetic field (dH) such that

$$|dH| = I \sin \theta \, dl / |r^2|.$$

where θ is the angle between the direction of current flow and the line joing P and dl. More compactly,

$$d H = I \, d l \times r/r^3$$

Ampere's rule — A positive charge moving horizontally is deflected by a force to the right if it is moving in a region where the magnetic field is vertically upward. This may be generalized to currents in wires by recallig that a current in a certain direction is equivalent to the motion of positive charges in that direction. The force felt by a negative charge is opposite to that felt by a positive charge.

Ampere-turn — A measure of magnetomotive force, especially as developed by an electric current, defined as the magnetomotive force developed by a coil of one turn through which a current of one ampere flows, that is, 1.26 Gb.

Amplitude — The maximum value of the displacement in an oscillatory motion.

Amplitude modulation — The variation of the amplitude of a wave in a way that corresponds to another wave. In general, the positive or negative envelope of the modulated wave (carrier wave) contains enough information to allow the modulating wave to be recovered, provided that the carrier wave has at least twice the frequency and twice the peak amplitude of the modulating signal. The modulated carrier C(t) is given by

$$C(t) = A_o \, [1 - k \, M(t)] \, \cos \omega_o t,$$

Where M(t) is the modulation.

Amplitude (of wave) — A measure of the maximum displacement of the wave crest from its undisturbed position (or the maximum electric or magnetic field strength of an electromagnetic wave).

AMU — The atomic mass unit (AMU), a unit of mass equal to 1/12 the mass of the carbon atom of mass number 12. On the atomic mass scale $^{12}C \equiv 12$.

$$
\begin{aligned}
\text{I AMU} \quad &= 931.4812(52) \text{ MeV} \\
&= 1.6605655(86) \times 10^{-27} \text{ (SI units)} \\
&= 1.6605655(86) \times 10^{-24} \text{ (cgs units)}
\end{aligned}
$$

The numbers in parentheses are the standard deviation uncertainties in the last digits of the quoted value, computed on the basis of internal consistency.

Angle — The ratio between the arc and the radius of the arc. Units of angle, the radian, the angle subtended by an arc equal to the radius; the degree, 1/360 part of the total angle about a point. Dimensions, a numeric.

Angstrom — A unit of length, used especially in expressing the length of light waves, equal to one ten-thousandth of a micron, or one hundred-millionth of a centimerer (1×10^{-8} cm).

Angular acceleration — The time rate of change of angular velocity either in angular speed or in direction of the axis of rotation (precession). Cgs unit, 1 radian/sec/sec. Dimensions [t^{-2}].

If the initial angular velocity is ω_o, and the velocity after time t is ω_t, the angular acceleration,

$$\alpha = \frac{\omega_t - \omega_o}{t}$$

The angular velocity after time t,

$$\omega_t = \omega_o + \alpha t$$

The angle swept out in time t,

$$\theta = \omega_o t + \tfrac{1}{2}\alpha t^2$$

The angular velocity after movement through the arc θ,

$$\omega = \sqrt{\omega_o{}^2 + 2\alpha\theta}$$

In the above equations, for angular displacement in radians, angular velocity will be in radians per second and angular acceleration in radians per second per second.

Angular aperture of an objective is the largest angular extent of wave surface which it can transmit.

Angular harmonic motion or harmonic motion of rotation — Periodic, oscillatory angular motion in which the restoring torque is proportional to the angular displacement. Torsional vibration.

Angular momentum or moment of momentum — Quantity of angular motion measured by the product of the angular velocity and the moment of inertia. Cgs unit, unnamed, its nature is expressed by g-cm²/sec. Dimensions, [$ml^2 \, t^{-1}$].

The angular momentum of a mass whose moment of inertia is I, rotating with angular velocity w, is Iw.

Angular velocity — Time rate of angular motion about an axis. Cgs unit, radian/sec. Dimensions, [t^{-1}].

If the angle described in time t is θ, the angular velocity,

$$\omega = \frac{\theta}{t}$$

θ in radians and t in seconds gives ω in radians per second.

Anhydride (of acid or base) — An oxide which when combined with water gives an acid or base.

Anion — A negatively charged ion.

Anisotropic — Exhibiting different properties when tested along aes in different directions.

Anode — The electrode at which oxidation occurs in a cell. It is also the electrode toward which anions travel due to the electrical potential. In spontaneous cells the anode is considered negative. In nonspontaneous or electrolytic cells the anode is considered positive.

Antiferromagnetic materials — Those in which the magnetic moments of atoms or ions tend to assume an ordered arrangement in zero applied field, such that the vector sum of the moments is zero, below a characteristic temperature called the Neel Point. The permeability of antiferromagnetic materials is comparable to that of paramagnetic materials. Above the Neel Point, these materials become paramagnetic.

Anti-matter — Matter consisting of antiparticles.

Anti-particle — Any particle with a charge of opposite sign to the same particle in normal matters.

Thus, the proton has a positive charge; the anti-proton, a negative charge. When a particle and its anti-particle collide, both may disappear with the creation of lighter particles; this process is called annihilation.

Aperture ratio — The ratio of the useful diameter of a lens to its focal length. It is the reciprocal of the f-number.

In application to an optical instrument, rather than to a lens, numerical aperture is more commonly used. The aperture ratio is then twice the tangent of the angle whose sine is the numerical aperture.

Apochromat — A term applied to photographic and microscope objectives indicating the highest degree of color correction.

Archimedes principle — A body wholly or partly immersed in a fluid is buoyed up by a force equal to the weight of the fluid displaced. A body of volume V cm³ immersed in a fluid of density ϱ grams per cm³ is buoyed up by a force in dynes,

$$F = \rho g V$$

where g is the accleration due to gravity.

A floating body displaces its own weight of liquid.

Area, unit of — The square centimeter. The area of a square whose sides are one centimeter in length. Other units of area are similarly derived. Dimensions, [L²].

Arrhenius theory of electrolytic dissociation states that the molecule of an electrolyte can give rise to two or more electrically charged atoms or ions.

Astigmatism is an error of spherical lenses peculiar to the formation of images by oblique pencils. The image of a point when astigmatism is present will consist of two focal lines at right angles to each other and separated by a measurable distance along the axis of the pencil. The error is not eliminated by reduction of aperture as is spherical aberration.

Astronomical unit (abbr. AU) — 1. A unit of length, usually defined as the distance from the earth to the sun, 149,599,000 km. This value for the AU was derived from radar observations of the distance of Venus. The value given in astronomical ephemerides, 149,599,000 km, was derived from observations of the minor planet Eros.

2. The unit of distance in terms of which, in the Kepler Third Law, $n^2a^3 = k^2(1 + m)$, the semimajor axis a of an elliptical orbit must be expressed in order that the numerical value of the Gaussian constant k may be exactly 0.01720209895 when the unit of time is the ephemeris day.

In astronomical units, the mean distance of the earth from the sun, calculated by the Kepler law from the observed mean motion n and adopted mass m, is 1.00000003.

Atmosphere, standard — 1 standard atmosphere = 1,013,250 dyn/cm² i.e., 101,325 N/m².

Atmospheric radiation — Infrared radiation emitted by or being propagated through the atmosphere. See Insolation.

Atmospheric radiation, lying almost entirely within the wavelength interval of from 3 to 80 μm, provides one of the most important mechanisms by which the heat balance of the earth-atmosphere system is maintained. Infrared radiation emitted by the earth's surface (terrestrial radiation) is partially absorbed by the water vapor of the atmosphere which in turn reemits it, partly upward, partly downward. This secondarily emitted radiation is then, in general, repeatedly absorbed and reemitted, as the radiant energy progresses through the atmosphere. The downward flux, or counterradiation, is of basic importance in the greenhouse effect; the upward flux is essential to the radiative balance of the planet.

Atom — The smallest particle of an element which can enter into a chemical combination. All chemical compounds are formed of atoms, the difference between compounds being attributable to the nature, number, and arrangement of their constituent atoms. See Isotopes, Nuclear atom.

Atomic bomb — An explosive that derives its energy from the fission or fusion of atomic nuclei.

Atomic energy — 1. The constitutive internal energy of the atom which was absorbed when it was formed. 2. Energy derived from the mass converted into energy in nuclear transformations. See Einstein's formula.

Atomic mass (atomic weight) — The mass of a neutral atom of a nuclide. It is usually expressed in terms of the physical scale of atomic masses, that is, in atomic mass units (AMU). See AMU.

Atomic mass unit (AMU) — A measure of atomic mass, defined as equal to 1/12 the mass of a carbon atom of mass 12.

Atomic number (symbol Z) — An interger that expresses the positive charge of the nucleus in multiples of the electronic charge e. It is the number of electrons outside the nucleus of a neutral (un-ionized) atom and, according to widely accepted theory, the number of protons in the nucleus.

An element of atomic number Z occupies the Zth place in the periodic table of the elements. Its atom has a nucleus with a charge + Ze, which is normally surrounded by Z electrons, each of charge −e.

For example, the carbon isotope $_6C^{14}$ has an atomic number of 6 and an atomic mass of 14.

Atomic structure — According to the currently accepted view, the atom consists of a central part, called nucleus, and a number of electrons (called orbital or planetary electrons) circling about the latter, like planets about the sun. The nucleus is of a high specific weight; it contains most of the mass of the entire atom (its mass is considered equal to the atomic mass) and is composed of

positively charged particles, called protons (the number of which always equals the atomic number, (Z), and particles of 0 charge, called neutrons (the number of which equals the difference between the atomic weight and the atomic number, A − Z). The diameter of the nucleus is between 10^{-13} and 10^{-12} cm, and the relatively vast distance in which the orbital electrons circle about it is illustrated by the fact that this nuclear diameter is only 10^{-4} to 10^{-5} of the entire atomic diameter. While the nucleus carries an integral number of positive charges (an integral number of protons) each of 1.6×10^{-19} coulomb, each electron carries one negative charge of 1.6×10^{-19} coulomb, and the number of orbital electrons is equal to the number of protons in the nucleus (i.e. to the atomic number, Z), so that the atom as a whole has a net charge of 0. The electrons are arranged in successive shells (q.v.) around the nucleus; the maximum number of electrons in each shell is determined by natural laws, and the extranuclear electronic structure of the atom is characteristic of the element. The electrons in the inner shells are tightly bound to the nucleus; this inner structure can be altered by high-energy particles, γ-rays of radium, of X-rays. The electrons in the outer shells are responsible for the chemical properties of the element. See Bohr's atomic theory, Heisenberg's theory, Shell, and Subshell.

Atomic theory — All elementary forms of matter are composed of very small unit quantities called atoms. The atoms of a given element all have the same size and weight. The atoms of different elements have different sizes and weights. Atoms of the same or different elements unite with each other to form very small unit quantities of compound substances called molecules.

Atomic weight — Atomic weight is the relative weight of the atom on the basis of $^{12}C \equiv 12$. For a pure isotope, the atomic weight rounded off to the nearest integer gives the total number of nucleons (neutrons and protons) making up the atomic nucleus. If these weights are expressed in grams they are called gram atomic weights. See Isotopes and Atomic mass.

Atomic weight (abbr. at. wt.) — The weight of an atom according to a scale of atomic weight units, awu, valued as one-twelfth the mass of the carbon atom ($C^{12} = 12.00000$).

Thus expressed, the atomic weight to the nearest integer is identical with the mass number.

Attenuation coefficient (symbol α) — A measure of the space rate of attenuation of any transmitted electromagnetic radiation. The attenuation coefficient is defined by

$$dI = -\alpha I_0 dx$$

or

$$I = I_0 e^{-\alpha x}$$

where I is the flux density at the selected point in space; I_o is the flux density at the source; and α is the attenuation coefficient.

ATTO — A prefix meaning one quintillionth or 10^{-18}. Symbol is a.

Avogadro's law — Equal volumes of different gases at the same pressure and temperature contain the same number of molecules.

Avogadro's number — The number of molecules in one mole or gram-molecular weight of a substance. A number of values of the Avogadro number, which is usually denoted by N, have been found by various methods, generally lying withina range of 1% about the value 6.022045×10^{23}/gm.

Avogadro's principle (or theory) — The numbers of molecules present in equal volumes of gases at the same temperature and pressure are equal.

Azimuth — 1. Horizontal direction or bearing. Compare azimuth angle. 2. In navigation, the horizontal direction of a celestial point from a terrestrial point, expressed as the angular distance from a reference direction, usually measured from 0° at the reference direction clockwise through 360°.

An azimuth is often designated as true, magnetic, compass, grid, or relative as the reference direction is true, magnetic, compass, grid north, or heading, respectively. Unless otherwise specified, the term is generally understood to apply to true azimuth, which may be further defined as the arc of the horizon, or the angle at the zenith, between the north part of the celestial meridian or principal vertical circle and a vertical circle, measured from 0° at the north part of the principal vertical circle clockwise through 360°.

3. In astronomy, the direction of a celestial point from a terrestrial point measured clockwise from the north or the south point of the meridian plane. 4. In

surveying, the horizontal direction of an object measured clockwise from the south point of the meridian plane.

In surveying, an aximuth of a celestial body is called an astronomic azimuth.

Azimuth angle — 1. Azimuth measured from 0° at the north or south reference direction clockwise or counterclockwise through 90° or 130°.

Azimuth angle is labeled with the reference direction as a prefix and the direction of measurement from the reference direction as a suffix. Thus, azimuth angle S 144° W is 144° west of south, or azimuth 324°. When azimuth angle is measured through 180°, it is labeled N or S to agree with the latitude and E or W to agree with the meridian angle.

2. In surveying, an angle in triangulation or in traverse through which the computation of azimuth is carried.

Babo's law — The addition of a nonvolatile solid to a liquid in which it is soluble lowers the vapor pressure of the solvent in proportion to the amount of substance dissolved.

Balmer series of spectral lines. The wave lengths of a series of lines in the spectrum of hydrogen were given in angstroms by the equation

$$\lambda = 3646 \ \frac{N^2}{N^2 - 4}$$

where N is an integer having values greater than 2.

Bar — International unit of pressure 10^6 dyn/cm². Unfortunately some writers have used this term for 1 dyn/cm². 1 bar = 0.987 atm = 1000 mbars = 29.53 in of mercury. See Torr.

Barn — Unit for measuring capture cross sections (q.v.) of elements. One barn = 10^{-24} cm²/nucleus.

Barotropy — The state of a fluid in which surfaces of constant density (or temperature) are coincident with surfaces of constant pressure; it is the state of zero baroclinity. Mathematically, the equation of batortopy states that the gradients of the density and pressure fields are proportional:

$$\Delta\rho = B\Delta p$$

where ϱ is the density; p is the pressure; and B is a function of thermodynamic variables, called the coefficient of barotropy.

With the equation of state, this relation determines the spatial distribution of all state parameters once these are specified on any surface. For a homogeneous atmosphere, B = 0; for an adiabatic atmosphere,

$$B = c_v/c_p \ RT$$

where c_v and c_p are the specific heats at constant volume and pressure, respectively; R is the gas constant; and T is the Kelvin temperature; for an isothermal atmosphere, B = 1/RT.

Barye — The pressure unit of the centimeter-gram-second system of physical units; equal to one dyne per square centimeter (0.001 mbar). Unfortunately some writers have used this term for the bar which is equal to 10^6 dyn/cm².

Bases — For many purposes it is sufficient to say that a base is a substance which dissociates on solution in water to produce one or more hydroxyl ions. More generally, however, bases are defined according to other concepts. The Bronsted concept states that a base is any compound which can accept a proton. Thus NH_3 is a base since it can accept a proton to form ammonium ions.

$$NH_3 + H^+ \rightleftharpoons NH_4^+$$

A still more general concept is that of G. N. Lewis which defines a base as anything which has an unshared pair of electrons. Thus in the reaction

$$H^+ + :N{-}H \rightleftharpoons NH_4^+$$

the NH_3 is a base because it possesses an unshared pair of electrons. This latter concept explains many phenomena, such as the effect of certain substances other than hydrogen ions in the changing of the color of indicators. It also explains acids and bases in nonaqueous systems as liquid NH_3 and SO_2.

Beam (of energy) — The locus of all series of wave-fronts projected from the source and directed toward given objects or positions in space.

Beam splitter — A device to produce two separate beams from one incident beam. This can be done with prisms or halfsilvered mirrors.

Beat(s) — Two vibrations of slightly different frequencies f_1 and f_2 when added together, produce in a detector sensitive to both these frequencies, a regularly varying response which rises and falls at the "beat" frequency $f_b = |f_1 - f_2|$. It is important to note that a resonator which is sharply tuned to f_b alone will not resound at all in the presence of these two beating frequencies. See Combination Frequencies.

Beat frequencies — The beat of two different frequencies of signals on a nonlinear circuit when they combine or beat together. It has a frequency equal to the difference of the two applied frequencies.

Beating — A wave phenomenon in which two or more periodic quantities of different frequencies produce a resultant having pulsations of amplitude.

This process may be contolled to produce a desired beat frequency. See Heterodyne.

Beer's law — If two solutions of the same colored compound be made in the same solvent, one of which is, say, twice the concentration of the other, the absorption due to a given thickness of the first solution should be equal to that of twice the thickness of the second.

Mathematically this may be expressed $l_1c_1 = l_2c_2$ when the intensity of light passing through the two solutions is a constant and if the intensity and wave length of light incident upon each solution are the same.

Bel — The fundamental division of a lagarithmic scale for expressing the ratio of two amounts of power, the number of bels denoting such a ratio being the logarithm to the base 10 of this ratio.

With P_1 and $Pli2$ designating two amounts of power and N the number of bels denoting their ratio, N = $\log^{10}(P_1/P_2)$ bels.

Bernoulli law or Bernoulli theorem — (After Daniel Bernoulli, 1700—1782, Swiss scientist.) 1. In aeronautics, a law or theorem stating that in a flow of incompressible fluid the sum of the static pressure and the hynamic pressure along a streamline is constant if gravity and frictional effects are disregarded.

From this law is follows that where there is a velocity increase in a fluid flow there must be a corresponding pressure decrease. Thus an airfoil, by increasing the velocity of the flow over its upper surface, derives lift from the decreased pressure.

2. As originally formulated, a statement of the conservation of energy (per unit mass) for a nonviscous fluid in steady motion. The specific energy is composed of the kinetic energy $u^2/2$, where u is the speed of the fluid; the potential energy gz, where g is the acceleration of gravity and z is the height above an arbitrary reference level; and the work done by the pressure forces of a compressible fluid $\int v \, dp$, where p is the pressure, v is the specific volume, and the integration is always with respect to values of p and v on the same parcel. Thus, the relationship

$$\frac{v^2}{2} + gz + \int v \, dp = \text{Constant along a streamline}$$

is valid for a compressible fluid in steady motion, since the streamline is also the path. If the motion is also irrotational, the same constant holds for the entire fluid.

Berthelot principle of maximum work — Of all possible chemical processes which can proceed without the aid of external energy, that process always takes place which is accompanied by the greatest evolution of heat. This law holds good for low temperatures only and does not account for endothermic reations.

Beta (β)-particle, (Beta ray) — One of the particles which can be emitted by a radioactive atomic nucleus. It has a mass about 1/1837. that of the proton. The negatively charged beta particle is identical with the ordinary electron, while the positively charged type (positron) differs from the electron in having equal but opposite electrical properties. The emission of an electron entails the change of a neutron into a proton inside the nucleus. The emission of a positron is similarly associated with the change of a proton into a neutron. Beta particles have no independent existance inside the nucleus, but are created at the instant of emission. See Neutrino.

Betatron — An accelerator used to impart high velocities to electrons (beta particles). Propellant is an electromagnetic field. A five to six Mev betatron can produce X-rays equivalent to the gamma radiation of 10 to 20 g of radium.

Bevatron — A six or more billion electron volt accelerator of protons and other atomic particles. Makes use of a Cockcroft-Walton transformer cascade accelerator and a linear (q.v.) as well as an electromagnetic field in the build-up.

DEFINITIONS (Continued)

Binary notation — A system of positional notation in which the digits are coefficients of powers of the base 2 in same way as the digits in the conventional decimal system are coefficients of powers of the base 10.

Binary notation employs only two digits, 1 and 0, therefore is used extensively in computers where the on and off positions of a switch or storage device can represent the two digits.

In decimal notation $111 = (1\times10^2)+(1\times10^1)+(1\times10^0) = 100+10+1 =$ one hundred and eleven.

In binary notation $111 = (1\times2^2)+(1\times2^1)+(1\times2^0) = 4+2+1 =$ seven.

Balck body — If, for all values of the wave length of the incident radiant energy, all of the energy is absorbed the body is called a black body.

Bohr radius (symbol a^o) — The smallest possible radius of an electron orbit in the Bohr model of the atom, 5.29167×10^9 cm.

Bohr's atomic theory — The theory that atoms can exist for a duration solely in certain states, characterized by definite electronic orbits, i.e., by definite energy levels of their extra-nuclear electrons, and in these stationary states they do not emit radiation; the jump of an electron from an orbit to another of a smaller radius is accompanied by monochromatic radiation.

Bolometer — An instrument which measures the intensity of radiant energy by employing a thermally sensitive electrical resistor; a type of actinometer. Also called actinic balance. Compare radiometer.

Two identical, blackened, thermally sensitive electrical resistors are used in a Wheatstone bridge circuit. Radiation is allowed to fall on one of the elements, causing a change in its resistance. The change is a measure of the intensity of the radiation.

Boltzmann constant (symbol k) — The ratio of the universal gas constant to Avogadro number; equal to 1.38054×10^{-16} erg/°K. Sometimes called gas constant per molecule, Boltzmann universal conversion factor.

Bouguer law — A relationship describing the rate of decrease of flux density of a plane parallel beam of monochromatic radiation as it penetrates a medium which both scatters and absorbs at that wavelength. This law may be expressed

$$dI_\lambda = -\alpha_\lambda I_\lambda\, dx$$

or

$$I_\lambda = I_{\lambda 0} e^{-\alpha_\lambda x}$$

where I is the flux density of the radiation; α_λ is the attenuation coefficient (or extinction coefficient) of the medium at wavelength λ; $I_{\lambda 0}$ is the flux density at the source; and x is the distance from the source. Sometimes called Beer law, Lambert law of absorption. See Absorption coefficient, Scattering coefficient.

Boussinesq approximation — The assumption (frequently used in the theory of convection) that the fluid is incompressible except insofar as the thermal expansion produces a buoyancy, represented by a term $g\alpha T$, where g is the acceleration of gravity; α is the coefficient of thermal expansion; and T is the perturbation temperature.

Boyle's law for gases — At a constant temperature the volume of a given quantity of any gas varies inversely as the pressure to which the gas is subjected. For a perfect gas, changing from pressure p and volume v to pressure p' and volume v' without change of temperature,

$$pv = p'v'$$

Boyle-Mariotte law — The empirical generalization that for many so-called perfect gases, the product of pressure p and volume V is constnat in an isothermal process:

$$pV = F(T)$$

where the function F of the temperature T cannot be specified without reference to other laws (e.g., Charles-Gay-Lussac law). Also called Boyle law, Mariotte Law.

Brayton cycle — (After George B. Brayton, American engineer.) Same as Joule cycle.

Breakdown potential = dielectric strength.

Breeder, Reactor (Breeder pile) — A nuclear chain reactor in which transmutation produces a greater number of fissionable atoms than the number of parent atoms consumed.

Bremsstrahlung (German, braking radiation) — Electromagnetic radiation produced by the rapid change in the velocity of an electron or another fast, charged particle as it approaches an atomic nucleus and is deflected by it. See Bremsstrahlung effect.

Bremsstrahlung effect — The emission of electromagnetic radiation as a consequence of the acceleration of charged elementary particles, such as electrons, under the influence of the attractive or repulsive force fields of atomic nuclei near which the charged particle moves.

In cosmic-ray shower production, bremsstrahlung effects givve rise to emission of gamma rays as electrons encounter atmospheric nuclei. The emission of radiation in the bremsstrahlung effect is merely one instance of the general rule that electromagnetic radiation is emitted only when electric charges undergo acceleration.

Brewster window — An aperture through which light can enter into a new medium at an angle to the interface such that

$$\tan\theta_B = \frac{n_b}{n_a}$$

where θ_B is the Brewster angle. n_b and n_a are the indices of refraction of the media a and b, and light enters b from a.

Brewster's law — The tangent of the polarizing angle for a substance is equal to the index of refraction. The polarizing angle is that angle of incidence for which the reflected polarized ray is at right angles to the refracted ray. If n is the index of refraction and θ the polarizing angle, $n = \tan\theta$.

Brightness is measured by the flux emitted per unit emissive area as projected on a plane normal to the line of sight. The unit of brightness is that of a perfectly diffusing surface giving out one lumen per square centimeter of projected surface and is called the lambert. The millilambert (0.001 lambert) is a more convenient unit. Candle per square centimeter is the brightness of a surface which has, in the direction considered, a luminous intensity of 1 candle/cm².

British thermal unit — The quantity of heat required to raise the temperature of one pound of water 1°F at, or near, its point of maximum density (39.1°F). The Btu is equivalent to 0.252 kilogram-calorie or 1055 joules.

Brownian movement — A continuous agitation of particles in a colloidal solution caused by unbalanced impacts with molecules of the surrounding medium. The motion may be observed with a microscope when a strong beam of light is caused to traverse the solution across the line of sight.

Bulk modulus — The modulus of volume elasticity,

$$M_B = \frac{p_2 - p_1}{\frac{v_1 - v_2}{v_1}}$$

where p_1, p_2; v_1, v_2 are the initial and final pressure and volume respectively. It is the reciprocal of the coefficient of compressibility.

Calorie — The amount of heat necessary to raise 1 g of water at 15°C, 1°C. There are various calories depending upon the interval chosen. Sometimes the unit is written as the gram-calorie or the kilogram calorie, the meaning of which is evident. The calorie may be defined in terms of its mechanical equivalent. The National Bureau of Standards defines the calorie as 4.18400 joules. At the International Steam Table Conference held in London in 1929 the international calorie was defined at 1/860 of the international watt hour, which makes it equal to 4.1860 international joules.

With the adoption of the absolute system of electrical units, this becomes 1/859.858 watt hours or 4.18674 joules. The Btu was defined at the same time as 251.996 international calories.

Calutron — An apparatus operating on the principle of the mass spectrograph and used for separating U^{235} from U^{238}.

Candela — The candela is the luminous intensity, in the direction of the normal, of a black body surface 1/600,000 m² in area, at the temperature of solidification of platinum under a pressure of 101,325 N/m².

DEFINITIONS (Continued)

Candle (new unit) — 1/60 of the intensity of 1 cm² of a blackbody radiator at the temperature of solidification of platinum (2045K).

Capacitance is measured by the charge which must be communicated to a body to raise its potential one unit. Electrostatic unit capacitance is that which requires one electrostatic unit of charge to raise the potential one electrostatic unit. The farad = 9×10^{11} electrostatic units. A capacitance of one farad requires one coulomb of electricity to raise its potential one volt. Dimensions, $[\epsilon l]$; $[\mu^{-1}l^{-1}t^2]$.

A conductor charged with a quantity Q to a potential V has a capacitance,

$$C = \frac{Q}{V}$$

Capacitance of a spherical conductor of radius r,

$$C = Kr$$

Capacitance of two concentric spheres of radii r and r′

$$C = K \frac{rr'}{r - r'}$$

Capacitance of a parallel plate condenser, the area of whose plates is A and the distance between them d,

$$C = \frac{KA}{4\pi d}$$

Capacitances will be given in electrostatic units if the dimensions of condensers are substituted in centimeters. K is the dielectric constant of the medium.

Capillary constant or specific cohesion,

$$a^2 = \frac{2T}{(d_1 - d_2)g} = hr$$

where T is surface tension, d_1 and d_2, the densities of the two fluids, g the acceleration due to gravity, h the height of rise in a capillary tube of radius r. See Surface tension.

Carbon cycle — A sequence of atomic nuclear reactions and spontaneous radioactive decay which serves to convert matter into energy in the form of radiation and high-speed particles, and which is regarded as one of the principal sources of the energy of the sun and other similar stars.

This cycle, first suggested by Bethe in 1938, gets its name from the fact that carbon plays the role of a kind of catalyst in that it is both used by and produced by the reaction, but is not consumed itself. Four protons are, in net, converted into an alpha particle and two positrons (with accompanying neutrinos); and three gamma-ray emissions are emitted directly in addition to the two gamma emissions that ensue from annihilation of the positrons by ambient electrons. This cycle sets in at stellar interior temperatures of the order of 5 million degrees Kelvin.

An even simpler reaction, the proton-proton reaction, is also believed to occur within the sun and may be of equal or greater importance.

Carnot cycle — An idealized reversible thermodynamic cycle. The Carnot cycle consists of four stages: (a) an isothermal expansion of the gas at temperature T_1; (b) an adiabatic expansion to temperature T_2; (c) an isothermal compression at temperature T_2; (d) an adiabatic compression to the original state of the gas to complete the cycle. See Carnot engine, Thermodynamic efficiency.

In a Carnot cycle, the net work done is the difference between the heat input Q_1 at higher temperature T_1 and the heat extracted Q_2 at the lower temperature T_2.

Carnot engine — An idealized reversible heat engine working in a Carnot cycle. It is the most efficient engine that can operate between two specified temperatures; its efficiency is equivalent to the thermodynamic efficiency. The Carnot engine is capable of being run either as a conventional engine or as a refrigerator.

Cassegrain telescope — A reflecting telescope in which a small hyperboloidal mirror reflects the convergent beam from the paraboloidal primary mirror through a hole in the primary mirror to an eyepiece in back of the primary mirror. Also called Cassegrainian telescope, Cassegrain. See Newtonian Telescope.

Catalytic agent — A substance which by its mere presence alters the velocity of a reaction, and may be recovered unaltered in nature or amount at the end of the reaction.

Cathode — The electrode at which reduction occurs. It is the negative electrode in a cell through which current is being forced, but it is the positive pole of a battery. In a vacuum tube, the cathode is the electrode from which electrons are liberated. See Anode.

Cation — A positively charged ion.

Cauchy's dispersion formula —

$$n = A + \frac{B}{\lambda^2} + \frac{C}{\lambda^4} + \dots$$

An empirical expression giving an approximate relation between the refractive index n of a medium and the wavelength λ of the light; A, B, and C being constants for a given medium.

Celsius — See Temperature, Celsius, in this section.

CENTI — A prefix meaning 1/100 or 10^{-2}. Symbol is c.

Centigrade temperature scale (abbr. C) — A temperature scale with the ice point at 0° and the boiling point of water at 100°. Now called Celsius temperature scale.

Conversion to the Fahrenheit temperature scale is according to the formula

$$°C = 5/9 \ (°F - 32)$$

Centipoise — A standard unit of viscosity, equal to 0.01 poise, the c.g.s. unit of viscosity. Water at 20°C has a viscosity of 1.002 centipoise or 0.01002 poise.

Centripetal force — The force required to keep a moving mass in a circular path. Centrifugal force is the name given to the reaction against centripetal force.

Chain reaction — In general, any self-sustaining process, whether molecular or nuclear, the products of which are instrumental in, and directly contribute to the propagation of the process. Specifically, a fission chain reaction, where the energy liberated or particles produced (fission products) by the fission of an atom cause the fission of other atomic nuclei, which in turn propagate the fission reaction in the same manner.

Charles-Gay-Lussac law — An empirical generalization that in a gaseous system at constant pressure, the temperature increase and the relative volume increase stand in approximately the same proportion for all so-called perfect gases. Mathematically,

$$t - t_0 = (1/c \ [(v - v_0)/v_0]$$

where t is temperature; v is volume; and c is a coefficient of thermal expansion independent of the particular gas. If the centigrade temperature scale is used and v_o is the volume at 0°C, then the value of the constant c is approximately 1/273. Also called Charles law, Gay-Lussac law.

Charles law = Charles-Gay-Lussac law.

Chemiluminescence — Emission of light during a chemical reaction.

Christiansen effect — When finely powdered substances, such as glass or quartz, are immersed in a liquid of the same index of refraction complete transparency can only be obtained for monochromatic light. If white light is employed the transmitted color corresponds to the particular wave-length for which the two substances, solid and liquid have exactly the same index of refraction. Due to differences in dispersion the indices of refraction will match for only a narrow band of the spectrum.

Chromatic aberration — Due to the difference in the index of refraction for different wave lengths, light of various wave lengths from the same source cannot be focused at a point by a simple lens. This is called chromatic aberration.

Circularly polarized wave — An electromagnetic wave for which the electric or the magnetic field vector, or both, at a point describe a circle.

This term is usually applied to transverse waves.

Circular mil — The area of a circle with a diameter of 0.001 in, a unit used for the measurement of small circular areas, such as the cross section of a wire. One circular mil = 7.85×10^{-7} in².

Circular polarization — The polarization of a wave radiated by a constant electric vector rotating in a plane so as to describe a circle. See Elliptical polarization.

Circulation — 1. The flow or motion of a fluid in or through a given area or volume. 2. A precise measure of the average flow of fluid along a given closed curve. Mathematically, circulation is the line integral.

$$\oint v \cdot dr$$

about the closed curve, where v is the fluid velocity, and dr is a vector element of the curve.

By Stokes theorem, the circulation about a plane curve is equal to the total vorticity of the fluid enclosed by the curve.

The given curve may be fixed in space or may be defined by moving fluid parcels.

Circulation integral — The line integral of an arbitrary vector taken around a closed curve. Thus,

$$\oint a \cdot dr$$

is the circulation integral of the vector a around the closed curve; dr is an infinitesimal vector element of the curve. If the vector is the velocity, this integral is called the circulation.

Clapeyron-Clausius equation — The differential equation relating pressure to temperature in a system in which two phases of a substance are in equilibrium.

$$dp/dT = L/(T\Delta V)$$

where p is pressure; T is temperature; L is the latent heat of the phase change; and ΔV is the difference in volume of the phases. Also called Clapeyron equation, Clausius-Clapeyron equation.

Clausius-Clapeyron equation = Clapeyron-Clausius equation.

Cloud chamber — An apparatus containing moist air or other gas which on sudden expansion condenses moisture to droplets on dust particles or other nuclei. Thus charged particles or ions in the space become nuclei and their numbers and behavior, when properly illuminated, may be studied.

Coefficient of compressibility — The relative decrease of the volume of a gaseous system with increasing pressure in an isothermal process. This coefficient is

$$-(1/V) (\partial V/\partial p)_T$$

where V is the volume; p is the pressure; and T is the temperature. The reciprocal of this quantity is the bulk modulus. Also called compressibility. Compare coefficient of thermal expansion, coefficient of tension.

Coefficient of tension — The relative increase of pressure of a system with increasing temperature in an isochoric process. In symbols this quantity is

$$(1/p) (\partial p/\partial T)_V$$

where p is pressure; T is temperature; and V is volume. Compare coefficient of compressibility, coefficient of thermal expansion.

Coefficient of thermal expansion — The ratio of the change of length per unit length (linear), or change of volume per unit volume (volumninal), to the change of temperature.

Coherence length — The maximum tolerable optical path length difference between two energy beams which are forming an interference pattern. This will vary with the degree of spectral purity of the source producing the beams. For example a perfect monochromatic source would have an infinite coherence length.

Coherent (additon) — The vector addition of both the amplitude and phases of different waves of the same frequency at a given time or at a given position.

Coherent oscillator (abbr Coho) — An oscillator which provides a reference

by which the radio frequency phase difference of successive received pulses may be recognized. See Coherent reference.

Coherent reference — The reference signal, usually of stable frequency, to which other signals are phase-locked to establish coherence throughout a system.

Coherent (source) — A source radiating coherent waves.

Coherent (waves) — Waves whose frequencies are equal and whose phases are related to each other at a given time or at a given place in space. Coherence can be of two types, temporal and spatial.

Cold working — Deforming metal plastically at a temperature lower than the recrystallization temperature.

Colligative property — A property numerically the same for a group of substances, independent of their chemical nature.

Colloid — A phase dispersed to such a degree that the surface forces become an important factor in determining its properties.

In general particles of colloidal dimensions are approximately 10Å to 1 µm in size. Colloidal particles are often best distinguished from ordinary molecules due to the fact that colloidal particles cannot diffuse through membranes which do allow ordinary molecules and ions to pass freely.

Colloidal system — An intimate mixture of two substances one of which, called the dispersed phase (or colloid) is uniformly distributed in a finely divided state through the second substance, called the dispersion medium (or dispersing medium). The dispersion medium may be a gas, a liquid, or a solid, and the dispersed phase may also be any of these, with the exception that one does not speak of a colloidal system of one gas in another. Also called colloidal dispersion, colloidal suspension.

A system of liquid or solid particles colloidally dispersed in a gas is called an aerosol. A system of solid substance or water insoluble liquid colloidally dispersed in liquid water is called a hydrosol. There is no sharp line of demarcation between true solutions and colloidal systems on the one hand, or between mere suspensions and colloidal systems on the other. When the particles of the dispersed phase are smaller than about 1 millimicron in diameter, the system begins to assume the properties of a true solution; when the particles dispersed are much greater than 1 µm, separation of the dispersed phase from the dispersing medium becomes so rapid that the system is best regarded as a suspension.

Coma — An aberration of spherical lenses, occurring in the case of oblique incidence, when the bundle of rays forming the image is unsymmetrical. The image of a point is comet shaped, hence the name.

Combination frequencies — Two vibrations of arbitrary frequencies f_1 and f_2 when applied simultaneously to a nonlinear (distorting) device will excite it to a motion containing not only the original frequencies, but also members of a set of "combination" set of frequencies given by $f_c = mf_1 + nf_2$ where m and n are integers. A resonator sharply tuned to any one of these frequencies which may be produced in the nonlinear device will resound to it with an amplitude depending on the type of nonlinearity. The superheterodyne radio receiver depends on this phenomenon.

Combining volumes — Under comparable conditions of pressure and temperature the volume ratios of gases involved in chemical reactions are simple whole numbers.

Combining weight of an element or radical is its atomic weight divided by its valence.

Combining weights, law of — If the weights of elements which combine with each other be called their "combining weights," then elements always combine either in the ratio of their combining weights or of simple multiples of these weights.

Compensation Point — The temperature (below the Néel Point) at which, in some ferrimagnetic compounds, the saturation magnetization becomes zero.

Complementary color — Either of a pair of spectrum colors that when combined give a white or nearly white light.

Component substances, law of — Every material consists of one substance, or is a mixture of two or more substances, each of which exhibits a specific set of properties, independent of the other substances.

DEFINITIONS (Continued)

Compounds are substances containing more than one constituent element and having properties, on the whole, different from those which their constituents had as elementary substances. The composition of a given pure compound is perfectly definite, and is always the same no matter how that compound may have been formed.

Compressibility — Reciprocal of the bulk modulus.

Compton effect, (Compton recoil effect) — Elastic scattering of photons by electrons results in decrease in frequency and increase of wave length of X-rays and gamma-rays when scattered by free electrons.

Compton electron — An orbital electron of an atom which has been ejected from its orbit as a result of an impact by a high-energy quantum of radiation (X-ray or gamma-ray). Also called Compton recoil electron.

Compton wavelength (symbol λ_c) — Of a particle, the distance h/mc, where h is the Planck constant, m is the mass of the particle, and c is the velocity of light.
The Compton wavelength of the electron (symbol λ_c) is 2.4261×10^{-10} centimeter; of the proton (symbol $\lambda_{c,p}$) is 1.32140×10^{-13} centimeter.

Computer-generated holograms — A hologram made synthetically and based on computer calculations of amplitude and/or phase.

Concentration — The amount of a substance in weight, moles, or equivalents contained in unit volume.

Condensers in parallel and series — If c_1, c_2, c_3, etc. represent the capacitances of a series of condensers and C their combined capacitance,

when in parallel,

$$C = c_1 + c_2 + c_3 \ldots$$

when in series,

$$\frac{1}{C} = \frac{1}{c_1} + \frac{1}{c_2} + \frac{1}{c_3} \cdots$$

Conductance — The reciprocal of resistance, is measured by the ratio of the current flowing through a conductor to the difference of potential between its ends. The practical unit of conductance, the mho, the conductance of a body through which one ampere of current flows when the potential difference is one volt. The conductance of a body in mho is the reciprocal of the value of its resistance in ohms. Dimensions, $[\varepsilon\, l\, t^{-1}]$. $[\mu^{-1}\, l^{-1}\, t]$.

Conductivity, electrical, is measured by the quantity of electricity transferred across unit area, per unit potential gradient per unit time. Reciprocal of resistivity. Volume conductivity or specific conductance, $k = 1/\varrho$ where ϱ is the volume resistivity. Mass conductivity = k/d where d is density. Equivalent conductivity $A = k/c$ where c is the number of equivalents per unit volume of solution. Molecular conductivity $\mu = k/m$ where m is the number of moles per unit volume of solution. Dimensions: volume conductivity, $[\varepsilon\, t^{-1}]$; $[\mu^{-1}\, l^{-2}\, t]$, mass conductivity, $[\varepsilon\, m^{-1}\, l^3\, t^{-1}]$; $[\mu^{-1}\, m^{-1}\, lt]$.

Conductivity, thermal — Time rate of transfer of heat by conduction, through unit thickness, across unit area for unit difference of temperature. It is measured as calories per second per square centimeter for a thickness of one centimeter and a difference of temperature of 1°C.
If the two opposite faces of a rectangular solid are maintained at temperatures t_1 and t_2 the heat conducted across the solid of section a and thickness d in a time T will be,

$$Q = \frac{K\,(t_2 - t_1)aT}{d}$$

K is a constant depending on the nature of the substance, designated as the specific heat conductivity. K is usually given for Q in calories, t_1 and t_2 in °C, a in cem², T in sec, and d in cm. See Heat conductivity.

Conductors — A class of bodies which are incapable of supporting electric strain. A charge given to a conductor spread to all parts of the body.

Conjugate foci — Under proper conditions light divergent from a point on or near the axis of a lens or spherical mirror is focused at another point. The point of convergence and the position of the source are interchangeable and are called conjugate foci.

Conservation of energy — The principle that the total energy of an isolated system remains constant if no interconversion of mass and energy takes place.
This principle takes into account all forms of energy in the system; it therefore provides a constraint on the conversions from one form to another.

Conservation of energy, law of — Energy can neither be created nor destroyed and therefore the total amount of energy in the universe remains constant.

Conservation of mass — In all ordinary chemical changes, the total of the reactants is always equal to the total mass of the products.

Conservation of momentum, law of — For any collision, the vector sum of the momenta of the colliding bodies after collison equal the vector sum of their momenta before collision. If two bodies of masses m_1 and m_2 have, before impact velocities v_1 and v_2 and after impact velocities u_1 and u_2

$$m_1 u_1 + m_2 u_2 = m_1 v_1 + m_2 v_2$$

Constant of aberration — The maximum aberration of a star observed from the surface of the earth, 20.49 sec of arc.
The maximum occurs at the time the direction of motion of the earth in its orbit is at right angles to a line from the earth to the star.

Constitutive property — A property which depends on the constitution or structure of the molecule.

Contrast of fringes — The relative difference between the brightness or density of successive bright and dark fringes on a hologram or interferogram.

Cooling — Processing highly radioactive materials to attain lesser radioactivity for subsequent use or handling.

Coriolis acceleration — An acceleration of a particle moving in a relative coordinate system. The total acceleration of the particle, as measured in an inertial coordinate system, may be expressed as the sum of the acceleration within the relative system, the acceleration of the relative system itself, and the coriolis acceleration.
Physically, coriolis acceleration may be considered as coming from the conservation of momentum in a body moving in a direction not parallel to the axis of rotation of the relative system.
Mathematically, coriolis acceleration comes from the differentiation of terms containing the angular velocity ω in the expression for the absolute velocity of the particle.
In the case of the earth, moving with angular velocity ω, a particle moving relative to the earth with velocity v has the coriolis acceleration $2\omega \times v$. If Newton laws are to be applied in the relative system, the coriolis acceleration and the acceleration of the relative system must be treated as forces.

Cosmic rays — Highly penetrating radiations which strike the earth, assumed to originate in interstellar space. They are classed as: primary, coming from the assumed source, and secondary, those induced in upper atmospheric nuclei by collisions with primary cosmic rays.

Cosmotron — A particle accelerator capable of giving them energies to billions of electron volts.

Couette flow — The shearing flow of a fluid between two parallel surfaces in relative motion. A two-dimensional steady flow without pressure gradient in the direction of flow and caused by the tangential movement of the bounding surfaces. The only practical type is the flow between concentric rotating cylinders (as of the oil in a cylindrical bearing).

Counterradiation — The downward flux of atmospheric radiation passing through a given level surface, usually taken as the earth's surface. Also called back radiation.
This result of infrared (long-wave) absorption and reemission by the atmosphere is the principal factor in the greenhouse effect.

Coulomb (unit of quantity of electricity) — the quantity of electricity transported in 1 sec by a current of 1 A. A unit quantity of electricity. It is the quantity of electricity which must pass through a circuit to deposit 0.0011180 g of silver from a solution of silver nitrate. An ampere is 1 coulomb/sec. A coulomb is also the quantity of electricity on the positive plate of a condenser of one-farad capacity when the electromotive force is 1 v.

Couple — Two equal and oppositely directed parallel but not colinear forces acting upon a body form a couple. The moment of the couple or torque is given by the product of one of the forces by the perpendicular distance between them. Dimension, $[m \, l^2 \, t^{-2}]$.

Couple acting on a magnet of magnetic moment ml in a field of strength H. If the magnet is perpendicular to the direction of the field

$$C = Hml = HM$$

If the angle between the magnet and the field is θ

$$C = Hml \sin \theta$$

The couple will be in dyn-cm for cgs electromagnetic units of H, m and l.

Critical mass — The minimum mass the fissile material must have in order to maintain a spontaneous fission chain reaction. For pure U^{235} it is computed to be about 20 lb.

Critical point — The thermodynamic state in which liquid and gas phases of a substance coexist in equilibrium at the highest possible temperature. At higher temperatures than the critical no liquid phase can exist. For water substance the critical point is

$$P_* = 2.21 \times 10^5 \text{ mbar}$$
$$T = 647°K$$
$$v = 3.10 \text{ g/cm}^3$$

where P_* is the saturation vapor pressure of the water vapor; T is the Kelvin temperature; and v is the specific volume.

Critical temperature — 1. The temperature above which a substance cannot exist in the liquid state, regardless of the pressure. 2. As applied to materials, the temperature at which a change in phase takes place causing an appreciable change in the properties of the material.

Cross section (Nuclear cross section) — A measure of the probability of a particular process. The nuclear cross section is expressed by a/bc where a is the number of processes occurring, b the number of incident particles, and c the number of target nuclei per cm². There are nuclear cross sections for fission, for slow neutron capture, for Compton collision, and for ionization by electron impact.

Crossed polarizer — A dual polarization filter and transducer which transforms varying orientations of polarized waves into an amplitude output.

Cryohydrate — The solid which separates when a saturated solution freezes. It contains the solvent and the solute in the same proportions as they were in the saturated solution.

Cryopumping — The process of removing gas from a system by condensing it on a surface maintained at very low temperatures.

Cryotron — A device based upon the principle that superconductivity established at temperatures near absolute zero is destroyed by the application of a magnetic field.

Crystal — The "ideal crystal" is a homogeneous portion of crystalline matter, (q.v.) whether bounded by faces or not.
Crystalline matter is matter that possesses a triperiodic structure on the atomic scale. It is characterized by discontinuous vectorial properties that give rise to "crystal planes" [(1) crystal growth (faces); (2) cohesion (cleavage planes); (3) twinning (twin planes); (4) gliding (gliding planes); (5) X-ray, electron, or neutron diffraction ("reflecting" planes); all of which are parallel to lattice planes.]

Curie — The curie is the SI unit rate of radioactive decay; the quantity of any radioactive nuclide which undergoes 3.7×10^{10} disintegrations/sec. The symbol for this unit is Ci. 1 Ci = 3.7×10^{10} Bq.

Curie's law — The intensity of magnetization,

$$I = \frac{AH}{T}$$

where H, is the magnetic field strength, T the absolute temperature and A Curie's constant. Used for paramagnetic substances.

Curie point — All ferro-magnetic substances have a definite temperature of transition at which the phenomena of ferro-magnetism disappear and the sub-

stances become merely paramagnetic. This temperature is called the "Curie Point" and is usually lower than the melting point.

Curie-Weiss law — The Curie law was modified by Weiss to state that the susceptibility of a paramagnetic substance above the Curie point varies inversely as the excess of the temperature above that point.
This law is not valid at or below the Curie point.

Current (electric) — The rate of transfer of electricity. The transfer at the rate of one electrostatic unit of electricity in one second is the electrostatic unit of current. The electromagnetic unit of current is a current of such strength that one centimeter of the wire in which it flows is pushed sideways with a force of 1 dyn when the wire is at right angles to a magnetic field of unit intensity. The practical unit of current is the ampere, a transfer of one coulomb per second, which is one tenth the electromagnetic unit. The international ampere is the unvarying electric current which, when passed through a solution of silver nitrate in accordance with certain specifications, deposits silver at the rate of 0.00111800 g/sec. The international ampere is equivalent to 0.999835 absolute ampere. The ampere-turn is the magnetic potential produced between the two faces of a coil of one turn carrying one ampere. Dimensions, $[\varepsilon^{1/2} \, m^{1/2} \, l^{1/2} \, t^{-2}]$; $[\mu^{-1/2} \, m^{1/2} \, l^{1/2} \, t^{-1}]$.

Current in a simple circuit — The current in a circuit including an external resistance R and a cell of electromotive force E and internal resistance,

$$I = \frac{E}{R + r}$$

If E is in volts and r and R in ohms the current will be in amperes.
For two cells in parallel,

$$I = \frac{E}{R + \dfrac{r}{2}}$$

For two cells in series,

$$I = \frac{2E}{R + 2r}$$

CW laser — Continuous wave laser — a laser that radiates its energy in an uninterpreted beam.

Cyclotron — The magnetic resonance accelerator for imparting very great velocities to heavier nuclear particles without the use of excessive voltages.

Dalton = atomic mass unit.

Dalton's law of partial pressures — The pressure exerted by a mixture of gases is equal to the sum of the separate pressures which each gas would exert if it alone occupied the whole volume. This fact is expressed in the following formula:p

$$PV = V(p_1 + p_2 + p_3, \text{etc.})$$

Day — 1. The duration of one rotation of the earth, or another celestial body, on its axis.
A day is measured by successive transits of a reference point on the celestial sphere over the meridian, and each type takes its name from the reference used. Thus, for a solar day the reference is the sun; a mean solar day if the mean sun; and an apparent solar day if the apparent sun. For a lunar day the reference is the moon for a sidereal day the vernal equinox; for a constituent day an astre fictif or fictitious star. The expression lunar day refers also to the duration of one rotation of the moon with respect to the sun. A Julian day is the consecutive number of each day, beginning with January 1, 4713 BC.
2. A period of 24 hr beginning at a specified time, as the civil day beginning at midnight, or the astronomical day beginning at noon.

DeBroglie wavelength — In quantum mechanics, a wavelength (λ) attributed to a particle by virtue of its momentum. In general

$$\lambda = \frac{h}{mv} = \frac{h}{m_0 v} \left(1 - \frac{v^2}{c^2}\right)^{1/2}$$

where m is the observed mass of the particle, m_* is its rest mass, v is its velocity, c is the velocity of light, and h is Planck's constant.

Debye-Falkenhagen effect — The increase in the conductance of an electrolytic solution produced by alternating currents of sufficiently high frequencies over that observed with low frequencies or with direct current.

DEFINITIONS (Continued)

Debye length — A theoretical length which describes the maximum separation at which a given electron will be influenced by the electric field of a given positive ion. Sometimes referred to as the Debye shielding distance or plasma length.

It is well known that charged particles interact through their own electric fields. In addition, Debye has shown that the attractive force between an electron and ion which would otherwise exist for very large separations is indeed cut off for a critical separation due to the presence of other positive and negative charges in between. This critical separation or Debye length decreases for increased plasma density.

Decay — Diminution of a radioactive substance due to nuclear emission of alpha or beta particles, gamma rays or positrons.

Decay constant — 1. = attenuation constant. 2. (symbol λ) A constant relating the instant rate of radioactive decay of a radioactive species to the number of atoms N present at a given time t. Thus,

$$-(\partial N/\partial t) = \lambda N$$

If N_o is the number of atoms present at time zero then

$$N = N_0 e^{-\lambda t}$$

Decay product — A nuclide resulting from the radioactive disintegration of a radionuclide, being formed either directly or as the result of succssive transformations in a radioactive series. Also called daughter, daughter element.

A decay product may be either radioactive or stable.

Deci — A prefix meaning one tenth or 10^{-1}. Symbol is d.

Decibel (abbr. db) — 1. A dimensionless measure of the ratio of two powers, equal to 10 times the logarithm to the base 10 of the ratio of two powers P_1/P_2. One tenth of a bel.

The power P_2 may be some reference power; in electricity, the reference power is sometimes taken as 1 milliwatt (abbr. dbm); in acoustics, the decibel is often taken as 20 times the common logarithm of the sound pressure ratio, with the reference pressure as $0.0002 \, dyn/cm^2$.

Declination — 1. (symbol δ) Angular distance north or south of the celestial equator; the arc of an hour circle between the celestial equator and a point on the celestial sphere, measured northward or southward from the celestial equator through $90°$, and labeled N or S to indicate the direction of measurement. 2. (symbol D) Magnetic declination. See Equatorial system.

Decomposition is the chemical separation of a substance into two or more substances, which may differ from each other and from the original substances.

Definite proportions, law of — In every sample of each compound substance the proportions by weight of the constituent elements are always the same.

Degree — Angle subtended at the center by a circular arc which is 1/360 of the circumference.

Degree of association $(1 - \alpha)$ — The degree of association of an electrolytic solution is the percentage of ions associated into nonconducting species, such as ion-pairs.

Degree of dissociation (or ionization) in general, α — The degree of dissociation (or ionization) of an electrolytic solution is the percentage of solute (or electrolyte) in the dissociated (or ionized) state in solution. Classically this degree is obtained from conductance measurements from the ratio, Λ/Λ_i where Λ_i is the equivalent conductance an electrolytic solution would have at some finite concentration if it were completely dissociated into ions at that concentration if it were completely dissociated into ions at that concentration. (See ionogens). This symbol is also used to denote the fraction of free ions in a solution when simple ions, ion pairs, and clusters higher than ion pairs are present. (See ionophores).

Degree of freedom — The number of the variables determining the state of a system (usually pressure, temperature, and concentrations of the components) to which arbitrary values can be assigned.

Deka — A prefix meaning ten or 10^1. Symbol is da.

Delayed neutrons — Neutrons emitted by excited nuclei in a radioactive process, so called because they are emitted an appreciable time after the fission. Compare prompt neutrons.

Delta ray — 1. An electron ejected by recoil when a rapidly moving alpha particle or other charged particle passes through matter. 2. By extension any secondary ionizing particle ejected by recoil when a primary particle passes through matter.

Density — Concentration of matter, measured by the mass per unit volume. Dimensions, $[m \, l^{-3}]$.

Density (of film) — The logarithm of the reciprocal of the optical transmission of the film.

Depletion layer — In a semiconductor, a region in which the mobile carrier charge density is insufficient to neutralize the net fixed charge density of donors and acceptors. Also called barrier.

Detenation wave — A shock wave in a combustible mixture, which originates as a combustion wave.

Deuterium (symbol D, d) — A heavy isotope of hydrogen having one proton and one neutron in the nucleus.

The symbol D is often used to designate deuterium in compounds, as HDO for molecules of that composition. Official chemical nomenclature uses the designation d with a number which designates the carbon atom to which the deuterium is bound; e.g., 2-d propane designates CH_3CHDCH_3.

Deuteron — Nucleus of the deuterium atom or the ion of deuterium. Its structure — one neutron and one proton.

Dewpoint — The temperature to which a given parcel of air must be cooled at constant pressure and constant water-vapor content in order for saturation to occur; the temperature at which the saturation vapor pressure of the parcel is equal to the actual vapor pressure of the contained water vapor. Any further cooling usually results in the formation of dew or frost. Also called dewpoint temperature.

When this temperatue is below $0°C$, it is sometimes called the frost point.

Diabatic process — A process in a thermodynamic system in which there is a transfer of heat across the boundaries of the system.

Diabatic process is preferred to nonadiabatic process.

Diamagnetic materials — Are those within which an externally applied magnetic field is slightly reduced because of an alteration of the atomic electron orbits produced by the field. Diamagnetism is an atomic-scale consequence of the Lenz law of induction. The permeability of diagmagnetic materials is slightly less than that of empty space.

Dielectric — A material having a relatively low electrical conductivity; an insulator; a substance that contains few or no electrons and which can support electrostatic stresses. The principal properties of a dielectric are its dielectric constant (the factor by which the electric field strength in a vacuum exceeds that in the dielectric for the same distribution of charge), its dielectric loss (the amount of energy it dissipates as heat when placed in a varying electric field), and its dielectric strength (the maximum potential gradient it can stand without breaking down).

In an electromagnetic field, the centers of the nonpolar molecules of a dielectric are displaced, and the polar molecules become oriented close to the field. The net effect is the appearance of charges at the boundaries of the dielectric. The frictional work done in orientation absorbs energy from the field which appears as heat. When the field is removed the orientation is lost by thermal agitation and so the energy is not regained. If free-charge carriers are present they too can absorb energy.

A good dielectric is one in which the absorption is a minimum. A vacuum is the only perfect dielectric. The quality of an imperfect dielectric is its dielectric strength; and the accumulation of charges within an imperfect dielectric is termed dielectric absorption.

Dielectric constant (symbol ϵ) — for a given substance, the ratio of the capacity of a condenser with that substance as dielectric to the capacity of the same condenser with a vacuum for dielectric. It is a measure, therefore, of the amount of electrical charge a given substance can withstand at a given electric field strength; it should not be confused with dielectric strength.

The dielectric constant ϵ is a function of temperature and frequency and is wirtten as a complex quantity

$$\epsilon = \epsilon' - i\epsilon''$$

where ϵ' is the part that determines the displacement current and ϵ'' the dielectric absorption (see dielectric). For a nonabsorbing, nonmagnetic material ϵ' is equal to the square of the index of refraction and the relation holds only at the particular frequency where these conditions apply.

$$F = \frac{QQ'}{\epsilon r'}$$

where F is the force of attraction between two charges Q and Q' separated by a distance r in a uniform medium.

Dielectrics or insulators or nonconductors — A class of bodies supporting an electric strain. A charge on one part of a nonconductor is not communicated to any other part.

Dielectric strength — A measure of the resistance of a dielectric to electrical breakdown under the influence of strong electric fields; usually expressed in volts per centimeter. Sometimes called breakdown potential.

Diffraction — That phenomena produced by the spreading of waves around and past obstacles which are comparable in size to their wavelength.

Diffraction efficiency — Ratio of energy projected into the reconstructed image to the energy illuminating the hologram.

Diffraction fanning — The fanning out of a light or energy beam as it pours through a very narrow aperture (opening).

Diffraction grating — If s is the distance between the rulings, d the angle of diffraction, then the wave length where the angle of incidence is 90% is (for the nth order spectrum),

$$\lambda = \frac{s \sin d}{n}$$

If I is the angle of incidence, d the angle of diffraction, s the distance between the rulings, n the order of the spectrum, the wave length is,

$$\lambda = \frac{s}{n}(\sin I + \sin d).$$

A mask or special aperture used to break up a white light beam or composite energy beam into its various spectral components through the mechanism of diffraction.

Diffuse reflection — Scattering at all angles from the point of reflection.

Diffuse sky radiation — Solar radiation reaching the earth's surface after having been scattered from the direct solar beam by molecules or suspensoids in the atmosphere. Also called the skylight, diffuse skylight, sky radiation.

Of the total light removed from the direct solar beam by scattering in the atmosphere (approximately 25% of the incident radiation), about two-thirds ultimately reaches the earth as diffuse sky radiation.

Diffusion — If the concentration (mass of solid per unit volume of solution) at one surface of a layer of liquid is d_1 and at the other surface d_2, the thickness of the layer h and the area under consideration A, then the mass of the substance which diffuses through the cross-section A in time t is,

$$m = \Delta A \frac{(d_2 - d_1)t}{h}$$

where Δ is the coefficient of diffusion.

Diffusion Coefficient — If the concentration (mass of solid per unit volume of solution) at one surface of a layer of liquid is d_1, and at the other surface d_2, the thickness of the layer is h, the area under consideration is A, and the mass of a given substance which diffuses through the cross section A in time t is m, then the diffusion coefficient is defined as

$$D = \frac{mh}{A(d_2 - d_1)t}$$

Diffusivity — A measure of the rate of diffusion of a substance, expressed as the diffusivity coefficient K. When K is constant, the diffusion equation is

$$\frac{\partial q}{\partial t} = K \nabla^2 q$$

where q is the substance diffused; ∇^2 is the Laplacian operator; and t is time. The diffusivity has dimensions of a length times a velocity; it varies with the

property diffused, and for any given property it may be considered a constant or a function of temperature, space, etc., depending on the context. Also called coefficient of diffusion. See conductivity, kinematic viscosity, exchange coefficients.

In the case of molecular diffusion the length dimension is the mean free path of the molecules. By analogy, in eddy diffusion, length becomes the mixing length. The coefficient is then called the eddy diffusivity, and is in general several orders of magnitude larger than the molecular diffusivity.

Diffusivity of heat — is given by Δ in the equation

$$\frac{dH}{dt} = -\Delta sd \frac{dT}{dx} \, dy \, dz$$

where dH is the quantity of heat passing through the area dy dz in the direction of x in a time dt. The rate of variation of temperature along x is given by dT/dx, s is specific heat and d, density. Dimensions, $[l^2 t^{-1}]$.

Dimensional formulae — If mass, length, and time are considered fundamental quantities, the relation of other physical quantities and their units to these three may be expressed by a formula involving the symbols l-m and t respectively, with appropriate exponents. For example; the dimensional formula for volume would be expressed, $[l^3]$; velocity, $[t^{-1}]$; force $[mlt^{-2}]$. Other fundamental quantities used in dimensional formulae may be indicated as follows: θ, temperature, ϵ the dielectric constant of a vacuum; μ, the magnetic permeability of a vacuum.

Diminution of pressure at the side of a moving stream — If a fluid of density d moves with a velocity v, the dimunution of pressure due to the motion is (neglecting viscosity),

$$p = \frac{1}{2} dv^2$$

Dip — The angle measured in a vertical plane between the direction of the earth's magnetic field and the horizontal.

Dipole — (1) A combination of two electrically or magnetically charged particles of opposite sign which are separated by a very small distance. (2) Any system of charges, such as a circulating current, which has the properties that: (a) no forces act on it in a uniform field; (b) a torque proportional to sin θ, where θ is the angle between the dipole axis and a uniform field, does act on it; (c) it produces a potential which is proportional to the inverse square of the distance from it.

Dipole moment — A mathematical entity; the product of one of the charges of a dipole unit by the distance separating the two dipolar charges. In terms of the definition of a dipole (2), the dipole moment p is related to the torque T, and the field strength E (or B) through the equation:

$$T = p \times E$$

Dipole moment, molecular — It is found from measurements of dielectric constant (i.e. by its temperature dependence, as in the Debye equation for total polarization) that certain molecules have permanent dipole moments. These moments are associated with transfer of charge within the molecule and provide valuable information as to the molecular structure.

Directly ionizing particles — are charged particles (electrons, protons, alpha particles, etc.) having sufficient kinetic energy to produce ionization by collision.

Direct solar radiation — In actinometry, that portion of the radiant energy received at the instrument direct from the sun, as distinguished from diffuse sky radiation, effective terrestrial radiation, or radiation from any other source. See global radiation.

Direct solar radiation is measured by pyrheliometers.

Dispersion — The difference between the index of refraction of any substance for any two wave lengths is a measure of the dispersion for these wave lengths, called the coefficient of dispersion.

Dispersion forces — The force of attraction between molecules possessing no permanent dipole. The interaction energy is given by

$$U_D = -\frac{3}{4} h \frac{V_0 \alpha^2}{r^6}$$

where h is Planck's constant, V_0 a characteristic frequency of the molecule, r the distance between the molecules, and α the polarizability.

DEFINITIONS (Continued)

Dispersive power — If n_1 and n_2 are the indices of refraction for wave lengths λ_1 and λ_2 and n the mean index or that for sodium light, the dispersive power for the specified wave length is,

$$\omega = \frac{n_2 - n_1}{n - 1}$$

Displacement is a reaction in which an elementary substance displaces and sets free a constituent element from a compound.

Displacement or elongation at any instant. The distance of a vibrating or oscillating particle from its position of equilibrium.

Dissociation-field effect — The increased dissociation (or ionization) of the molecules of weak electrolytes under the influence of high electrical fields (potential gradients).

Distribution law — A substance distributes itself between two immiscible solvents so that the ratio of its concentrations in the two solvents is approximately a constant (and equal to the ratio of the solubilities of the substance in each solvent). The above statement requires modification if more than one molecular species is formed.

Donor — In transistors, the N-type semiconductor, the electrode containing impurities which increase the number of available electrons. Contrast acceptor.

Doppler broadening — The broadening of either an emission line or an absorption line due to random motions of molecules of the gas that is emitting or absorbing the radiant energy. See pressure broadening.

In the case of an emitting gas, for example, those molecules which are approaching the observer as they emit quanta of radiant energy will, because of the Doppler effect, appear to send out a train of waves of slightly shorter wavelength than that characteristic of a stationary molecule, while receding molecules will appear to emit slightly longer wavelengths. The net effect, averaged over many molecules, is to superimpose, on the natural line width, a bell-shaped broadening that is proportional to the square root of the absolute temperature of the gas.

Doppler effect — The change in frequency with which energy reaches a receiver when the receiver and the energy source are in motion relative to each other. Also called Doppler shift.

In the case of sound, or any other wave motion where a real medium of propagation exists (excepting, therefore, light and other electromagnetic radiations) one must distinguish two principal cases: If the source is in motion with speed v relative to a medium which propagates the waves in question at speed c, then the resting observer receives waves emitted with actual frequency f as if they had a frequency f′ given by the Doppler equation

$$f' = f/[1 \pm (v/c)]$$

where the positive sign refers to the case of the source receding from the observer, and vice versa for the negative sign. If, on the other hand, the source is at rest relative to the propagating medium while the observer moves with speed v relative to the source,

$$f' = f''[1 \pm (v/c)]$$

where the positive sign now refers to the case of observer approaching the source.

For electromagnetic radiation,

$$f/f' = [1 \pm (v/c)]/[1 \pm (v/c)]$$

where the top signs represent the source receding from the observer and the bottom signs, approaching the observer.

Dosimeter — 1. An instrument for measuring the ultraviolet in solar and sky radiation. Compare actinometer. 2. A device, worn by persons working around radioactive material, which indicates the dose of radiation to which they have been exposed.

Double decomposition consists of a simple exchange of the parts of two substances to form two new substances.

Double pass transmittance hologram — A hologram whose object wave was transmitted through the transparent object media to a mirror, reflected back through again, and recorded on the plate.

Dulong and Petit, law of — The specific heats of the several elements are inversely proportional to their atomic weights. The atomic heats of solid elements are constant and approximately equal to 6.3. Certain elements of low atomic weight and high melting point have, however, much lower atomic heats at ordinary temperatures.

Dynamic height — The height of a point in the atmosphere expressed in a unit proportional to the geopotential at that point. Since the geopotential at altitude z is numerically equal to the work done when a particle of unit mass is lifted from sea level up to this height, the dimensions of dynamic height are those of potential energy per unit mass. Also called geodynamic height.

The standard unit of dynamic height H_d is the dynamic meter (or geodynamic meter), defined as $10/\sec^2$; it is related to the geopotential φ, the geometric height z in meters, and the geopotential height Z in geopotential meters by

$$d\varphi = 10 dH_d = 9.8 dZ = g dz$$

where g is the acceleration of gravity in meters per second squared. (Some sources prefer to give the constants 10 and 9.8 the units of meters per second squared so that the units of φ and Z would be the same as those of the geometric height.) The dynamic meter is about 2% longer than the geometric meter and the geopotential meter. One of the practical advantages of the dynamic height over the geometric height is that when the former is introduced into the hydrostatic equation the variable acceleration of gravity is eliminated. In meteorological height calculations, geopotential height is more often used than dynamic height.

Dynamic pressure (symbol q) = The pressure of a fluid resulting from its motion, equal to one-half the fluid density times the fluid velocity squared ($1/2 \varrho V^2$). In incompressible flow, dynamic pressure is the difference between total pressure and static pressure. Also called kinetic pressure. Compare impact pressure.

Dynamic Viscosity — Of a fluid, the ratio of the shearing stress to the shear of the motion. It is independent of the velocity distribution, the dimensions of the system, etc. and for a gas it is independent of pressure except at very low pressures. Also called coefficient of molecular viscosity, coefficient of viscosity.

For the dynamic viscosity μ of a perfect gas, the kinetic theory of gases gives

$$\mu = 1/3 \, (\rho c l)$$

where ϱ is the gas density, c is the average speed of the random heat motion of the gas molecules and is proportional to the square root of the temperature, and L is the mean free path. For dry air at 0° C, the dynamic viscosity is about 1.7×10^{-4} g/cm/sec.

Whereas the dynamic viscosity of most gases increases with increasing temperature, that of most liquids, including water, decreases rapidly with increasing temperature.

Dyne — The force necessary to give acceleration of one centimeter per second per second to one gram of mass.

Earth Current — A large-scale surge of electric charge within the earth's crust, associated with a disturbance of the ionosphere.

Current patterns of quasi-circular form and extending over areas the size of whole continents have been identified and are known to be closely related to solar-induced variations in the extreme upper atmosphere.

Eddy current — A current induced in a mass of conducting material by a varying magnetic field. Also called Foucault current.

Eddy viscosity — The turbulent transfer of momentum by eddies giving rise to an internal fluid friction, in a manner analogous to the action of molecular viscosity in laminar flow, but taking place on a much larger scale.

The value of the coefficient of eddy viscosity (an exchange coefficient) is of the order of 10^4 cm²/sec, or 100,000 times the molecular kinematic viscosity.

Effective neutron cycle time — The lifetime of an average neutron within a reactor from the time it is produced to the time it is fission captured.

This average takes into account delayed as well as prompt neutrons.

Effective radius of the earth — A fictitious value for the radius of the earth, used in place of the geometrical radius to correct for atmospheric refraction when the index of refraction in the atmosphere changes linearly with height. See modified index of refraction.

Under conditions of standard refraction the effective radius of the earth is

8.5×10^8 m, or four thirds the geometrical radius. If the effective radius is used in ray tracing diagrams, the rays may be drawn as though they were traveling in straight lines.

Effective terrestrial radiation — The amount by which outgoing infrared terrestrial radiation of the earth's surface exceeds downcoming infrared counter-radiation from the atmosphere. Also called nocturnal radiation, effective radiation. See Actinometer.

It is to be emphasized that this amount is a positive quantity, of the order of several tenths of a langley per minute, at all times of day (except under conditions of low overcast clouds). It typically attains its diurnal maximum during the midday hours when high soil temperatures create high rates of outgoing terrestrial radiation. (For this reason the synonym nocturnal radiation is apt to lead to slight confusion.) However, in daylight hours the effective terrestrial radiation is generally much smaller than the insolation, while at night it typically dominates the energy budget of the earth's surface.

Einstein theory for mass-energy equivalence — The equivalence of a quantity of mass m and a quantity of energy E by the formula $E = mc^2$. The conversion factor c^2 is the square of the velocity of light.

Elastic collision — A collision between two particles in which no change occurs in the internal energy of the particles, or in the sum of their kinetic energies. Commonly referred to as a billiard-ball collision.

Elasticity — The property by virtue of which a body resists and recovers from deformation produced by force.

Elastic limit — The smallest value of the stress producing permanent alteration.

Elastic moduli —
Young's modulus by stretching — If an elongation s is produced by the weight of the mass m, in a wire of length l, and radius r, the modulus,

$$M = \frac{mgl}{\pi r^2 s}$$

Young's modulus by bending, bar supported at both ends. If a flexure s is produced by the weight of mass m, added midway between the supports speared by a distance l for a rectangular bar with vertical dimensions of cross-section a and horizontal dimension b, the modulus is,

$$M = \frac{mgl^3}{4sa^3 b}$$

For a cylindrical bar of radius r,

$$M = \frac{mgl^3}{12\pi r^4 s}$$

For a bar supported at one end. In the case of a rectangular bar as described above,

$$M = \frac{4mgl^3}{sa^3 b}$$

For a round bar supported at one end,

$$M = \frac{4mgl^3}{3\pi r^4 s}$$

Modulus of rigidity — If a couple C (= mgx) produces a twist of θ radians in a bar of length l and radius r, the modulus is

$$M = \frac{2Cl}{\pi r^4 \theta}$$

The substitution in the above formulae for the elastic coefficients of m in grams, g in cm/sec^2, l, a, b, and r in cm, s in cm, and C in dyne-cm will give moduli in dyn/cm^2. The dimensions of elastic moduli are the same as of stress, $[ml^{-1} t^{-2}]$.

Coefficient of restitution — Two bodies moving in the same straight line, with velocities v_1 and v_2 respectively, collide and after impact move with velocities v_3 and v_4. The coefficient of restitution is

$$C = \frac{v_4 - v_3}{v_2 - v_1}$$

Electret n — A piece of dielectric material that has a permanent electric polarity; the electrostatic analog of a permanent magnet.

Electric dipole — A pair of equal and opposite charges an infinitesimal distance apart.

Electric field intensity is measured by the force exerted on unit charge. Unit field intensity is the field which exerts the force of one dyne on unit positive charge. Dimensions, $[\epsilon^{-1/2} m^{1/2} l^{-1/2} t^{-1}]$; . $[\mu^{1/2} m^{1/2} l^{1/2} t^{-2}]$.

The field intensity or force exerted on unit charge at a point distant r from a charge q in a vacuum

$$H = \frac{q}{r^2}$$

If the dielectric in the above cases is not a vacuum the dielectric constant ϵ must be introduced. The formula becomes

$$H = \frac{q}{\epsilon r^2}$$

The value of ϵ is frequently considered unity for air. If the dielectric constant of a vacuum is considered unity the value for air at 0°C and 760 mm pressure is 1.000576.

Electric potential — The work per unit charge spent in moving a charged body in an electric field from a reference point to a point of interest (P). Commonly, the reference point is chosen as infinity. The potential (V) is positive if work is done on the charge and negative if work is required of the charge to move in the existing field. Analytically, assuming an electric field of intensity E,

$$V = \int_\infty^P E \cdot ds$$

where ds is a vector element of the path from ∞ to P.

Electrochemical equivalent of an ion is the mass liberated by the passage of unit quantity of electricity.

Electrolysis — If a current i flows for a time t and deposits a metal whose electrochemical equivalent is e, the mass deposited is

$$m = eit$$

The value of e is usually given for mass in grams, i in amperes and t in seconds.

Electrolytic cell constant, J_c — The cell constant of an electrolytic cell is the resistance in ohms of that cell when filled with a liquid of unit resistance.

Electrolytic dissociation or ionization theory — When an acid, base or salt is dissolved in water or any other dissociating solvent, a part or all of the molecules of the dissolved substance are broken up into parts called ions, some of which are charged with positive electricity and are called cations, and an equivalent number of whch are charged with negative electricity and are called anions.

Electrolytic solution tension theory (or the Helmholtz double layer theory) — When a metal, or any other substance capable of existing in solution as ions, is placed in water or any other dissociating solvent, a part of the metal or other substances passes into solution in the form of ions, thus leaving the remainder of the metal or substances charged with an equivalent amount of electricity of opposite sign from that carried by the ions. This establishes a difference in potential between the metal and the solvent in which it is immersed.

Electromagnetic radiation — Energy, propagated through space or through material media in the form of an advancing disturbance in electric and magnetic fields existing in space or in the media. The term radiation, alone, is used commonly for this type of energy, although it actually has a broader meaning. Also called electromagnetic energy or simply radiation. See Electromagnetic spectrum.

Electromagnetic spectrum — The ordered array of known electromagnetic radiations, extending from the shortest cosmic rays, through gamma rays, X-rays, ultraviolet radiation, visible radiation, infrared radiation, and including microwave and all other wavelengths of radio energy See Ebsorption spectrum.

The division of this continuum of wavelengths (or frequencies) into a number of named subportions is rather arbitrary and, with one or two exceptions, the boundaries of the several subportions are only vaguely defined. Nevertheless,

DEFINITIONS (Continued)

to each of the commonly identified subportions there correspond characteristic types of physical systems capable of emitting radiation of those wavelengths. Thus, gamma rays are emitted from the nuclei of atoms as they undergo any of several types of nuclear rearrangements; visible light is emitted, for the most part, by atoms whose planetary electrons are undergoing transitions to lower energy states; infrared radiations are associated with characteristic molecular vibrations and rotations; and radio waves, broadly speaking, are emitted by virtue of the accelerations of free electrons as, for example, the moving electrons in a radio antenna wire.

Electromotive force is defined as that which causes a flow of current. The electromotive force of a cell is measured by the maximum difference of potential between its plates. The electromagnetic unit of potential difference is that against which one erg of work is done in the transfer of electromagnetic unit quantity. The volt is that potential difference against which one joule of work is done in the transfer of one coulomb. One volt is equivalent to .0⁸ electromagnetic units of potential. The international volt is the electrical potential which when steadily applied to a conductor whose resistance is one international ohm will cause a current of one international ampere to flow. The international volt = 1.00033 absolute volts. The electromotive force of a Weston standard cell is 1.0183 int. volts at 20°C. Dimensions, $[\epsilon^{-1/2} m^{1/2} t^{-1}]$, $[\mu^{1/2} m^{1/2} l^{3/2} t^{-2}]$.

Electromotive series is a list of the metals arranged in the decreasing order of their tendencies to pass into ionic form by losing electrons.

Electron — The electron is a small particle having a unit negative electrical charge, a small mass, and a small diameter. Its charge is $(4.80294 \pm 0.00008) \times 10^{-10}$ absolute electrostatic units, its mass. $\frac{1}{1837}$ that of the hydrogen nucleus, and its diameter about 10^{-12} cm. The electron has a rest mass of 0.9109534 10^{-30}kg and a magnetic moment of 9.284832×10^{-24}J T^{-1}. Every atom consists of one nucleus and one or more electrons. Cathode rays and Beta rays are electrons.

Elecron-volt (ev) — Energy acquired by any charged particle carrying unit electronic charge when it falls through a potential difference of one volt. 1 electron-volt = $(1.60207 \pm 0.00007 \times 10^{-12}$ erg or 1.6020×10^{-19} joule. Multiples of this unit are also in common use: the kilo-, million-, and billion electron-volt. 1 kev = 10^3ev; 1 mev = 10^6ev; and 1 bev = 10^9 ev.

An ev is associated through the Planck constant with a photon of wave length 1.2395 μm.

Electrophoretic effect — The slowing down owing to interionic attraction and repulsion, of the movement of an ion with its solvent molecules in the forward direction by ions of opposite charge with their solvent molecules moving in the reverse direction under an applied electrical field (potential gradient).

Electrostatic unit — 1. In the cgs system, the measure of electrostatic charge, defined as a charge which, if concentrated at one point in a vacuum, would repel, with a force of 1 dyn, an equal and like charge placed 1 cm away. 2. (pl.) A system of electrical units based on the electrostatic unit.

Elements are substances which cannot be decomposed by the ordinary types of chemical change, or made by chemical union.

Elliptical polarization — The polarization of a wave radiated by an electric vector rotating in a plane and simultaneously varying in amplitude so as to describe an ellipse.

Elongation — In tensile testing the elongation of a specimen is the increase in gage length, after rupture, referred to the original gage length. It is reported as percentage elongation.

Emission spectrum — The array of wavelengths and relative intensities of electromagnetic radiation emitted by a given radiator.

Each radiating substance has a unique, characteristic emission spectrum, just as every medium of transmission has its individual absorption spectrum.

Emissive power or emissivity is measured by the energy radiated from unit area of a surface in unit time for unit difference of temperature between the surface in question and surrounding bodies. For the cgs system the emissive power is given in ergs per second per square centimeter with the radiating surface at 1° absolute and the surroundings at absolute zero. See Radiation formula.

Emissivity (symbol E∞) — A property of a material, measured as the emittance of a specimen of the material that is thick enough to be completely opaque and has an optically smooth surface.

Emittance (symbol E, ϵ) — 1. The radiant flux per unit area emitted by a body. 2. The ratio of the emitted radiant flux per unit area of a sample to that of a black body radiator at the same temperature and under the same conditions.

Spectral emittance refers to emittance measured at a specified wavelength.

Because of the two common meanings of emittance, it should be defined when used unless the context allows no misinterpretation.

Emulsion — The coating on a film or plate which is sensitive to the light illuminating it.

Enantiotropic — Crystal capable of existing in reversible equilibrium with each other.

Energy — The capability of doing work. Potential energy is energy due to position of one body with respect to another or to the relative parts of the same body. Kinetic energy is energy due to motion. Cgs units, the erg, the energy expended when a force of 1 dyn acts through a distance of 1 cm; the joule is 1 × 10⁷ ergs. Dimensions, $[ml^2 t^{-2}]$.

The potential energy of a mass m, raised through a distance h, where g is the acceleration due to gravity is

$$E = mgh.$$

The kinetic energy of mass m, moving with a velocity v, is

$$E = \tfrac{1}{2}mv^2$$

Energy will be given in ergs if m is in grams, g in centimeters per second square, h in centimeters and v in centimeters per second.

Energy of a charge in ergs where Q is the charge and V the potential in electrostatic units.

$$E = \tfrac{1}{2}QV$$

Energy of the electric field — If H is the electric field intensity in electrostatic units and K the specific inductive capacity, the energy of the field in ergs per cm³ is

$$E = \frac{KH^2}{8\pi}$$

Energy of rotation — If a mass whose moment of inertia about an axis is I, rotates with angular velocity ω about this axis, the kinetic energy of rotation will be,

$$E = \tfrac{1}{2}I\omega^2$$

Energy will be given in ergs if I is in g-cm² and ω in radians per second.

Enthalpy, or heat content, is a thermodynamic quantity. It is equal to the sum of the internal energy of a system plus the product of the pressure-volume work done on the system. Thus

$$H = E + pv$$

where

H	=	enthalpy or heat content
E	=	internal energy of the system
p	=	pressure
v	=	volume.

Entropy — 1. A measure of the extent to which the energy of a system is unavailable. A mathematically defined thermodynamic function of state, the increase in which gives a measure of the energy of a system which has ceased to be available for work during a certain process

$$ds = (du + pdv)/T \geqslant dq/T$$

where s is specific entropy; us is specific internal energy; p is pressure; v is specific volume; T is Kelvin temperature; and q is heat per unit mass. For reversible processes,

$$ds = dq/T$$

In terms of potential temperature 0,

$$ds = c_p (d\theta/\theta)$$

F-82

where cp is the specific heat at constant pressure. See third law of thermodynamics.

In an adiabatic process, the entropy increases if the process is irreversible and remains unchanged if the process is reversible. Thus, since all natural processes are irreversible. and remains unchanged if the process is reversible. Thus, since all natural processes are irreversible, it is said that in an isolated system the entropy is always increasing as the system tends toward equilibrium, a statement which may be considered a form of the second law of thermodynamics.

2. In communication theory, average information content.

Ephemeris day — 86,400 ephemeris seconds. See Ephemeris time.

Ephemeris second (unit of time) — is exactly 1/31 556 925.974 7 of the tropical year of 1900, January, 0 days, and 12 hr ephemeris time.

Ephemeris time (abbr. E.T.) — The uniform measure of time defined by the laws of dynamics and determined in principle from the orbital motions of the planets, specifically the orbital motion of the earth as represented by Newcomb's Tables of the Sun. Compare universal time.

Beginning with the volume for 1960 the American Ephemeris and Nautical Almanac uses ephemeris time as the tabular argument in the fundamental ephemerides of the sun, moon, and planets.

A gravitational ephemeris expresses the position of a celestial body as a function of ephemeris time; and, at any instant, the measure of ephemeris time is the value of the argument at which the ephemeris position is the same as the actual position at the instant. The ephemeris time at any instant is obtained from observation by directly comparing observed position of the sun, moon, and planets with graviatational ephemerides of their coordinates; observations of the moon are the most effective and expeditious for this purpose. An accurate determination, however, requires observations over a more or less extended period; in practice, it takes the form of determining the time correction ΔT that must be applied to universal time (U.T.) to obtain ephemeris time:

$$E.T. = U.T. + \Delta T$$

The universal time at any instant may be obtained with little delay from observations of the diurnal motions.

The fundamental epoch from which ephemeris time is reckoned is the epoch that Newcomb designated as 1900 January 0, Greenwich mean noon, but which actually is 1900 January 0 day 12 hr E.T.; the instant to which this designation is assigned is the instant near the beginning of the calendar year A.D. 1900 when the geometric mean longitude of the Sun referred to the mean equinox of date was 279 degrees 41 minutes 48.04 seconds. Ephemeris time is the measure of time in which Newcomb's Tables of the Sun agree with observation.

The primary unit of ephemeris time is the tropical year, defined by the mean motion of the sun in longitude at the epoch 1900 January 0 day 12 hr E.T.; its length in ephemeris days is determined by the coefficient of T in Newcomb's expression for the geometric mean longitude of the sun L referred to the mean equinox of date, given among the elements of the sun.

Equatorial system — A set of celestial coordinates based on the celestial equator as the primary great circle; usually declination and hour angle or sidereal hour angle. Also called equinoctial system of coordinates, celestial equato system of coordinates.

Equilibrium, chemical — A state of affairs in which a chemical reaction and its reverse reaction are taking place at equal velocities, so that the concentrations of reacting substances remain constant.

Equilibrium constant — The product of the concentrations (or activities) of the substances produced at equilibrium in a chemical reaction divided by the product of concentrations of the reacting substances, each concentration raised to that power which is the coef-icient of the substance in the chemical equation.

Equivalent conductance of an electrolyte is defined as the conductance of a volume of solution containing one equivalent weight of dissolved substance when placed between two parallel electrodes 1 em apart, and large enough to contain between them all of the solution. ′ is never determined directly, but is calculated from the specific conductance. If C is the concentration of a solution in gram equivalents per liter, then the concentration per cubic centimeter is $C/1000$, and the volume containing one equivalent of the solute is, therefore, $1000/C$. Since L_3 is the conductance of a centimeter cube of the solution, the conductance of $1000/C$ cc, and hence ′ will be

$$\Delta = \frac{1000 L_s}{C}$$

Equivalent temperature — 1. Isobaric equivalent temperature; the temperature that an air parcel would have if all water vapor were condensed out at constant pressure, the latent heat released being used to heat the air,

$$T_{i,e} = T[1 + (Lw/c_p T)]$$

where $T_{i,e}$ is the isobaric equivalent temperature; T is the temperature; w is the mixing ratio; L is the latent heat: and c_p is the specific heat of air at constant pressure. 2. Adiabatic equivalent temperature; The temperature that an air parcel would have after undergoing the following (physically unrealizable) process: dry-adiabatic expansion until saturated; pseudoadiabatic expansion until all moisture is precipitated out; dry-adiabatic expansion until saturated; pseudoadiabatic expansion until saturated; pseudoadiabatic expansion until all moisture is precipitated out; dry-adiabatic compression to the initial pressure. This is the equivalent temperature as read from a thermodynamic chart and is always greater than the isobaric equivalent temperature:

$$T_{a,e} = T \exp(Lw/c_p T)$$

where $T_{a,e}$ is the adiabatic equivalent temperature. Also called pseudoequivalent temperature.

Equivalent weight or combining weight of an element or ion is its atomic or formula weight divided by its valence. Elements entering into combination always do so in quantities proportional to their equivalent weights.

In oxidation-reduction reactions the equivalent weight of the reacting substances is dependent upon the change in oxidation number of the particular substance.

erg — The unit of energy or work in the centimeter-gram-second system; the work performed by a force of 1 dyne acting through a distance of 1 cm.

Escape velocity — The radial speed which a particle or larger body must attain in order to escape from the gravitational field of a planet or star. When friction is neglected, the escape velocity is

$$\sqrt{2Gm/r}$$

where G is the universal gravitational constant m is the mass of the planet or star; and r is the radial distance from the center of the planet or star. Also called escape speed.

Ettinghausen's effect (Von Ettinghausen's) — When an electric current flows across the lines of force of a magnetic field an electromotive force is observed which is at right angles to both the primary current and the magnetic field: a temperature gradient is observed which has the opposite direction to the Hall electromotive force.

Eutectic — A term applied to the mixture of two or more substances which has the lowest melting point.

Exchange coefficients — Coefficients of eddy flux (e.g., of momentum, heat, water vapor, etc) in turbulent flow, defined in analogy to those of the kinetic theory of gases (see eddy). Also called austausch coefficients, eddy coefficients, interchange coefficients.

The exchange-coefficient hypothesis states that the mean eddy flux per unit area of a conservative quantity (suitably expressed) is proportional to the gradient of the mean value of this quantity, that is,

$$\text{Mean flux per unit area} = -C_e (d\bar{E}/dN)$$

where G_e is the exchange coefficient; E is the mean value of the quantity; and N is the direction normal to the surface. In strict analogy to molecular properties C_e would be constant, for turbulent flow C_e turns out to depend on time and location. See eddy viscosity diffusivity.

Expansion of gases — *Charles' law or Gay-Lussac's law* — The volume of a gas at constant pressure increases proportionately to the absolute temperature. If V_1 and V_2 are volumes of the same mass of gas at absolute temperatures, T_1 and T_2,

$$\frac{V_1}{V_2} = \frac{T_1}{T_2}$$

For an original volume V_o at 0°C the volume at t° C (at constant pressure) is

$$V_t = V_0 (1 + 0.00367t).$$

DEFINITIONS (Continued)

General law for gases —

$$p_t v_t = p_0 v_0 \left(1 + \frac{t}{273}\right)$$

where p_0, v_0, p_t, v_t represent the pressure and volume at $0°$ and $t°C$ or

$$\frac{p_1 v_1}{T_1} = \frac{p_2 v_2}{T_2}$$

where p_1, v_1 and T_1 represent pressure, volume and absolute temperature in one case and p_2, v_2 and T_2 the same quantities for the same mass of gas in another.

The law may also be expressed:

$$pv = RmT$$

where m is the mass of gas at absolute temperature T. R is the gas constant which depends on the units used. Boltzmann's molecular gas constant is obtained by expressing m in terms of the number of molecules.

For volume in cm^3, pressure in dynes per cm^2 and temperature in Centigrade degrees on the absolute scale $R = 8.3136 \times 10^7$.

Reduction of a gas volume to $0°C$, 760 mm pressure: – If V is the original volume at $0°C$ and 760 mm pressure will be,

$$V_0 = \frac{V}{(1 + \alpha t)} \frac{H}{760}$$

If d is the original density the density at $0°C$ and 760 mm pressure will be

$$d_0 = d(1 + \alpha t) \frac{760}{H}$$

$\alpha = 0.00367$ approximately.

Extinction coefficient — The extinction coefficient ε is identified as

$$dI = -\varepsilon I \, dx$$

or

$$I = I_0 e^{-\varepsilon x}$$

where I is the illuminance (luminous flux density) at the selected point in space, I_0 is the illuminance at the light source, and x is the distance from the source.

When so used, the extinction coefficient equals the sum of the medium's absorption coefficient and scattering coefficient, each computed as a weighted average over all wavelengths in the visible spectrum. As long as scattering effects are primary, a in the lower atmosphere, the value of the extinction coefficient is a function of the particle size of atmospheric suspensoids. It varies in order of magnitude from 10/km with very low visibility to 0.01/km in very clear air.

Extinction cross section = scattering cross section.

Extraterrestrial radiation — In general, solar radiation received just outside the earth's atmosphere.

Fahrenheit temperature scale (abbr. F) — A temperature scale with the ice point at $32°$ and the boiling point of water at $212°$.
Conversion with the Celsius (centigrade) temperature scale (abbr. C) is by the formula

$$F = 9/5 \, C + 32$$

Falling bodies — For bodies falling from rest conditions are as for uniformly accelerated motion except that $v_0 = O$ and g is the acceleration due to gravity. The formulae become, air resistance neglected,

$$v_t = gt, \ s = \tfrac{1}{2}gt^2 \ v_3 = \sqrt{2gs}.$$

For bodies projected vertically upward, if v is the velocity in projection, the time to reach greatest height, neglecting the resistance of the air,

$$t = \frac{v}{g}$$

Greatest height,

$$h = \frac{v^2}{2g}$$

See Projectiles,

Farad (unit of electric capacitance) — the capacitance of a capacitor between the plates of which there appears a difference of potential of 1 when it is charged by a quantity of electricity equal to 1 coulomb.

Faraday, determination of (1960) By the National Bureau of Standards which uses an electrochemical method that dissolves, rather than deposits, silver from a solution. The new value, $96,516 \pm 2$ coulombs (physical scale) or $96,489 \pm 2$ coulombs (chemical scale). NBS used its mass, time, and elecrical standards in measuring the faraday, and have found that its value agreed within 22 ppm. with the one obtained by an independent physical method using the omegration.

Faraday constant (symbol F) — The product of the Avogadro constant N_A and the elementary charge e, $F = N_A e = 96,489 \pm 2$ coulombs/mol.

Faraday effect — The rotation of the plane of polarization produced when plane-polarized light is passed through a substance in a magnetic field, the light traveling in a direction parallel to the lines of force. For a given substance, the rotation is proportional to the thickness traversed by the light and to the magnetic field strength.

Faraday's laws — In the process of electrolytic changes equal quantities of electricity charge or discharge equivalent quantities of ions at each electrode.
One gram equivalent weight of matter is chemically altered at each electrode for 96,489 int. coulombs, or one faraday, of electricity passed through the electrolyte.

Far field (diffraction pattern) — Diffraction pattern produced at a large range from an object which is identical to that which would be produced at an infinite range from the object. This is also called a Fraunhofer diffraction pattern.

Fast neutron — A neutron of 100,000 electron-volts or greater energy.

Fast reactor — A reactor containing no moderator, so that all the fissions take place at energies on the order of 100,000 electron-volts or higher.

Femto — A prefix meaning one quadrillionth or 10^{-15}. Symbol is f.

Fermat's principle — The path followed by light (or other waves) passing through any collection of media from one specified point to another, is that path for which the time of travel is least.

Fermi (abbr. f) — A unit of length equal to 10^{-13} cm.

Ferrimagnetic Materials — Those in which the magnetic moments of atoms or ions tend to assume an ordered but nonparallel arrangement in zero applied field, below a characteristic temperature called the Néel Point. In the usual case, within a magnetic domain, a substantial net magnetization results from the antiparallel alignment of neighboring nonequivalent sublattices. The macroscopic behavior is similar to that in ferromagnetism. Above the Neel Point, these materials become paramagnetic.

Ferrogmatic Materials — Those in which the magnetic moments of atoms or ions in a magnetic domain tend to be aligned parallel to one another in zero applied field, below a characteristic temperature called the Curie Point. Complete ordering is achieved only at the absolute zero of temperature. Within a magnetic domain, at absolute zero, the magnetization is equal to the sum of the magnetic moments of the atoms or ions per unit volume. Bulk matter, consisting of many small magnetic domains, has a net magnetization which depends upon the magnetic history of the specimen (hysteresis effect). The permeability depends on the magnetic field, and can reach values of the order of 10^6 times that of free space. Above the Currie Point, these materials become paramagnetic.

First law of thermodynamics — A statement of the conservtion of energy for thermodynamic systems (not necessarily in equilibrium). The fundamental form requires that the heat absorbed by the system serve either to raise the internal energy of the system or to do work on the environment:

$$dq = du + dw$$

where dq is the heat added per unit mass; du is the increment of specific internal energy; and dw is the specific work done by the system on the environment. Although dq and dw are not perfect differentials, their difference, dv, is always a perfect differential. Example of the application of this equation: in an adiabatic free expansion of gas into a vacuum, all three terms are zero.

For reversible processes the mechanical work is equal to the expansion against the pressure forces, i.e.,

$$dw = pdv$$

where p is the pressure and v is the specific volume. For a perfect gas, the internal energy change is proportional to the temperature change,

$$du = c_v dT$$

where c_v is the specific heat at constant volume and T is the Kelvin temperature. Therefore, the form of the first law usually used in meteorological applications is

$$dq = c_v dT + pdv$$

Use of the equation of state yields an alternative form,

$$dq = c_p dT - dp$$

where c_p is the specific heat at constant pressure.

For open systems the variation of total rather than specific quantities is important:

$$dQ = dU + pdV - hdm$$

where Q is the total heat; U is the total internal energy; V is the volume; m is the mass of the system; and h is the specific enthalpy.

If a system contains the possibility of nonmechanical work, such as work done against an electric field, this work must be included in the first law.

See Second law of thermodynamics, Third law of thermodynamics, Energy equations.

Fission — The splitting of an atomic nucleus into two more-or-less equal fragments.

Fission may occur spontaneously or may be induced by capture of bombarding particles. In addition to the fission fragments, neutrons and gamma rays are usually produced during fission.

Fleming's rule — A simple rule for relating the directions of the flux, motion, and e.m.f. in an electric machine. The forefinger, second finger and thumb, placed at right-angles to each other, represent respectively the directions of flux, e.m.f., and motion or torque. If the right hand is used the conditions are those obtaining in a generator and if the left hand is used the conditions are those obtaining in a motor.

Fluidity — The reciprocal of viscosity. The cgs unit is the rhe, the reciprocal of the poise. Dimensions, $[m^{-1} lt]$.

Fluorescence — The property of emitting radiation as the result of absorption of radiation from some other source. The emitted radiation persists only as long as the exposure is subjected to radiation which may be either electrified particles or waves. The fluorescent radiation generally has a longer wave length than that of the absorbed radiation. If the fluorescent radiation includes waves of the same length as that of the absorbed radiation it is termed resonance radiation.

f/n or f number (of optics) — The ratio of effective focal length to lens diameter.

Focal length — The distance between the optical center of a lens, or the surface of a mirror, and its focus.

Focal plane — A plane parallel to the plane of a lens or mirror and passing through the focus.

Focal point = focus, in optics.

Focus (plural focuses) — 1. That point at which parallel rays of light meet after being refracted by a lens or reflected by a mirror. Also called focal point. 2. A point having specific significance relative to a geometrical figure. See Ellipse, Hyperbola, Parabola.

Footcandle (abbr. fc) — A unit of illuminance, incident light, or illumination equal to 1 lm/ft². This is the illuminance provided by a light source of one candle at a distance of 1 ft, hence the name. Compare lux, phot.

Full sunlight with zenith sun produces an illuminance of the order of 10,000 footcandles on a horizontal surface at the earth's surface. Full moonlight provides an illuminance of only about 0.02 footcandle also at earth's surface. Adequate illumination for steady reading is taken to be about 10 footcandles; that for close machine work is about 30 to 40 footcandles.

Foot-lambert (abbr. ft-1) — A unit of luminance (or brightness) equal to $1/\pi$ candle per square foot, or 1 lm/ ft².

In Great Britain this is also called the equivalent footcandle.

Force — That which changes the state of rest or motion in matter, measured by the rate of change of momentum. Absolute unit, the dyne, the force which will produce an acceleration of one centimeter per second per second in a gram mass. The gram weight or weight of a gram mass is the cgs gravitational unit. The poundal is that force which will give an acceleration of one foot per second per second to a pound mass. Dimensions, $[mlt^{-2}]$.

The force F required to produce an acceleration a in a mass m is given by

$$F = ma$$

If m is substituted in grams and a in cm per sec², F will be given in dynes.

Force between two charges, Coulomb's law — If two charges q and q′ are at a distance r in a vacuum, the force between them is,

$$F = \frac{qq'}{r^2}$$

The force will be given in dynes if q and q′ are in electrostatic units and r in centimeters.

Force between two magnetic poles — If two poles of strength m and m′ are separated by a distance r in a medium whose permeability is μ (unity for a vacuum), the force between them is,

$$F = \frac{mm'}{\mu r^2}$$

Force will be given in dynes if r is in cm and m and m′ are in cgs units of pole strength.

The strength of a magnetic field at a point distance r from an isolated pole of strength m is

$$H = \frac{m}{\mu r^2}$$

The field will be given in gauss if m and r are in cgs units.

Formula, chemical — A combination of symbols with their subscripts representing the constituents of a substance and their proportions by weight.

Fourier transform plane — Same as spatial frequency plane.

Fraunhofer's lines — When sunlight is examined through a spectroscope it is found that the spectrum is traversed by an enormous number of dark lines parallel to the length of the slit. These dark lines are known as Fraunhofer's lines. Kirchoff conceived the idea that the sun is surrounded by layers of vapors which act as filters of the white light arising from incandescent solids within and which abstract those rays which correspond in their periods of vibration to those of the components of the vapors. Thus reversed or dark lines are obtained due to the absorption by the vapor envelop, in place of the bright lines found in the emission spectrum.

Frequency — Rate of oscillation; units: 1 cycle sec⁻¹ = 1 Hertz = 1 Hz. 1 megacycle = 1 megahertz = 10^6 Hz. One gigahertz = 10^9 Hz.

Frequency modulation — A form of angle modulation in which the frequency of the carrier is made to vary in accordance with the information to be transmitted, that is, given a carrier

$$C(t) = A \cos(\omega_0 t + \phi),$$

and an information function g(t), the signal transmitted is

$$C_m(t) = A \cos[(\omega_0 + g(t) - \Delta\omega)(t) + \phi].$$

DEFINITIONS (Continued)

When used for radio communications this system has the advantages of greater immunity from noise and other interference, at the cost of increased bandwidth.

Frequency of vibrating strings — The fundamental frequency of a stretched string is given by

$$n = \frac{1}{2l} \sqrt{\frac{T}{m}}$$

where l is the length, T, the tension and m the mass per unit length.

For a string or wire of circular section of length l, tension T, density d, and radius r, the frequency of the fundamental is

$$n = \frac{1}{2rl} \sqrt{\frac{T}{\pi d}}$$

The frequency in vibrations per second will be given if T is in dynes, r and l in cm and d in g per cm³.

Frequency (of waves) — Number of like phase (peaks, troughs) wavefronts passing a given point in a unit of time.

Frequency standard (atomic or molecular frequency standard) — The standard is the transition between the hyperfine levels $F = 4, M = 0$ and $F = 3, M = 0$ of the ground state $^2S_{1/2}$ of the cesium-133 atom, unperturbed by external fields. This frequency of transition is assigned the value 9 192 631 770 hertz.

Fresnel — A measure of frequency, defined as equal to 10^{12} cycles/sec.

Friction, coefficient of — The coefficient of friction between two surfaces is the ratio of the force required to move one over the other to the total force pressing the two together.

If F is the force required to move one surface over another and W, the force pressing the surfaces together, the coefficient of friction,

$$k = \frac{F}{W}$$

Fringe — The locus of maximum constructive interference (light fringe) or destructive interference regions in a space where two or more coherent waves intersect. Fringes can be in two or three dimensions.

Fringe control — Methods of adjusting the position and/or characteristics of the fringe pattern of a holographic interferogram.

Fugacity (symbol f) — In thermodynamics, a measure of the tendency of a substance to escape by some chemical process from the phase in which it exists.

Fundamental units — See Mass, Length, and Time.

Fusion (atomic) — A nuclear reaction involving the combination of smaller atomic nuclei or particles into larger ones with the release of energy from mass transformation. This is also called a thermo-nuclear reaction by reason of the extremely high temperature required to initiate it.

Gal — 1 gal = cm/sec/sec. Therefore, where the value of gravity is 980 this is the same as 980 gal. The milligal is now quite commonly used since it is approximately one part in a thousand of the normal gravity of the earth.

Gamma (γ) rays (nuclear X-rays) — May be emitted from radioactive substances. They are quanta of electromagnetic wave energy similar to but of much higher energy than ordinary X-rays. The energy of a quantum is equal to hv ergs, where h is Planck's constant (6.6254×10^{-27} erg sec) and v is the frequency of the radiation. Gamma rays are highly penetrating, an appreciable fraction being able to traverse several centimeters of lead.

Gas — A state of matter in which the molecules are practically unrestricted by cohesive forces. A gas has neither definite shape nor volume.

Gas constant (symbol R, R*, R₀) — The constant factor in the equation of state for perfect gases. The universal gas constant is

$$R_0 = 8.3143 \text{ joules}/^\circ\text{K-mol}$$
$$= 1.9872 \text{ cal}/^\circ\text{K-mol}$$

The gas constant for a particular gas, specific gas constant,

$$r = R/m$$

where m is the molecular weight of the gas. See Boltzmann constant.

Gas thermometer — where Pi0, P_s and P_x represent the total pressure with the bulb at 0°C, at the boiling-point of water and at the unknown temperature respectively, t, the temperature of steam and t_x the unknown temperature,

$$t_x = t_s \frac{P_x - P_0}{P_s - P_0}$$

(approximately). The total pressure on the gas in the bulb is the algebraic sum of barometric pressure at the time and that measured by the manometer.

Gauss — The cgs emu of magnetic induction (flux density). It is equal to 1 maxwell per cm². It has such a value that magnetic field at a velocity of 1 cm, in an induction mutually perpendicular, the induced emf is one abvolt.

Gay-Lussac's law — See Charles' law.

Gay-Lussac's law of combining volumes — If gases interact and form a gaseous product, the volumes of the reacting gases and the volumes of the gaseous products are to each other in simple proportions, which can be expressed by small whole numbers.

Gee — A suffix meaning earth, as in perigee, apogee. See Perigee, note.

Geiger counter — Detector for radioactivity depending upon ionized particles that affect its mechanism. As its name indicates, it both detects and makes a count of them possible.

Geomagnetic pole — Either of two antipodal points marking the intersection of the earth's surface with the extended axis of a dipole assumed to be located at the center of the earth and approximating the source of the actual magnetic field of the earth.

That pole in the Northern Hemisphere (latitude, 78½° N; longitude, 69° W) is designated north geomagnetic pole, and that pole in the Southern Hemisphere (latitude, 78½° S, longitude, 111°E) is designated south geomagnetic pole. The great circle midway between these poles is called geomagnetic equator. The expression geomagnetic pole should not be confused with magnetic pole, which relates to the actual magnetic field of the earth.

Geometric mean — A measure of central position. The geometric mean of n quantities equals the nth root of the product of the quantities.

Geopotential — The geopotential Φ of a point at a height z above mean sea level is the work which must be accomplished against gravity in elevating a unit mass from sea level to height z.

$$\Phi = \int_0^z g \, dz$$

where g is the local acceleration of gravity at height z. For most metrological work geopotential is given in the units geopotential meter (gpm). By definition, 1 gpm = 9.8×10^4 cm² sec⁻². For most purposes the geopotential can be assumed to equal the geometric height.

Geopotential height — The height of a given point in the atmosphere in units proportional to the potential energy of unit mass (geopotential) at this height, relative to sea level.

The relation, in the cgs system, between the geopotential height H and the geometric height Z is

$$H = \frac{1}{980} \int_0^Z g \, dZ$$

where g is the acceleration of gravity, so that the two heights are numerically interchangeable for most meteorological purposes. Also, 1 geopotential meter is equal to 0.98 dynamic meter. See dynamic height.

At the present time, the geopotential height unit is used for all aerological reports, by convention of the World Meteorological Organization.

Geopotential meter — A unit of length used in measuring geopotential height; 1 geopotential meter is equal to 0.98 dynamic meter. See Dynamic height.

Gibbs free energy = Gibbs function.

Gibbs function — a mathematically defined thermodynamic function of state, which is constant during a reversible isobaric-isothermal process. Also called Gibbs free energy, thermodynamic potential. Compare Helmholtz function.

In symbols the specific Gibbs function g is

DEFINITIONS (Continued)

$$g = h - Ts$$

where h is specific enthalpy; T is Kelvin temperature; and s is specific entropy. By use of the first law of thermodynamics for reversible processes,

$$dg = -s\,dT + dp$$

Gibbs' phase rule — $F = C + 2 - P$, F, the number of degrees of freedom of a system, is the number of variable factors (temperature, pressure and concentration) of the components, which must be arbitrarily fixed in order that the condition of the system m ay be perfectly defined. C, the number of the components of the system, is chosen equal to the smallest number of independently variable constituents by means of which the composition of each phase participating in the state of equilibrium can be expressed in the form of a chemical equation; the components must be chosen from among the constituents which are present when the system is in a state of true equilibrium and which take part in that equilibrium; as components are chosen the smallest number of such constituents necessary to express the composition of each phase participating in the equilibrium, zero and negative quantities of components being permissible; in any system the number of components is definite, but may alter with changes in conditions of experiment; a qualitative but not quantitative freedom of selection of components is allowed, the choice being influenced by suitability and simplicity of application. P, the number of phases of the system, are the homogeneous, mechanically separable and physically distinct portions of a heterogeneous system; the number of phases capable of existence varies greatly in different systems; there can never be more than one gas or vapor phase since all gases are miscible in all proportions, a heterogeneous mixture of solid substances forms as many phases as there are substances present.

Giga — A prefix meaning billion or 10^9. Symbol is G.

Gilbert (abbr. Gb) — The cgs emu of magnetomotive force. 1 Gb = $10/4\pi$ ampere-turns.

Global radiation — The total of direct solar radiation and diffuse sky radiation received by a unit horizontal surface.

Graham's law — The relative rates of diffusion of gases under the same conditions are inversely proportional to the square roots of the densities of those gases.

Gram atom or gram atomic weight — The mass in grams numerically equal to the atomic weight.

Gram equivalent of a substance is the weight of a substance displacing or otherwise reacting with 1.008 g of hydrogen or combining with one-half of a gram atomic weight (8.00 g) of oxygen.

Gram molecular weight or gram molecule — A mass in grams of a substance numerically equal to its molecular weight. Gram mole.

Gram mole, gram formula weight, gram equivalent — Mass in grams numerically equal to the molecular weight, formula weight or chemical equivalent, respectively.

Gravitation — The acceleration produced by the mutual attraction of two masses, directed along the line joining their centers of masses, and of magnitude inversely proportional to the square of the distance between the two centers of mass.

This acceleration on a unit mass has the magnitude $G(m/r^2)$, where m is the mass of the attracting body, r is the distance between the centers of mass, and G is the gravitational constant equal to $6.670 \pm 0.005 \times 10^{-11}\,\mathrm{N}\,\mathrm{n}^2\,\mathrm{kg}^{-2}$.

In the case of masses in the earth's gravitational field, m is the mass of the earth, equal to 5.975×10^{27} g. However, the rotation of the earth and atmosphere modifies this field to produce the field of gravity.

Gravitational constant (symbol G) — The coefficient of proportionality in Newton law of gravitation

$$G = 6.670 \pm 0.005 \times 10^{-11}\,\mathrm{Nm}^2\,\mathrm{kg}^{-2}$$

Also called constant of gravitation, Newtonian universal constant of gravitation.

Gravity — 1. Viewed from a frame of reference fixed in the earth, force imparted by the earth to a mass which is at rest relative to the earth. Since the

earth is rotating, the force observed as gravity is the resultant of the force of gravitation and the centrifugal force arising from this rotation and the use of an earthbound rotating frame of reference. It is directed normal to sea level and to its geopotential surfaces. See Virtual gravity, Geopotential height.

The magnitude of the force of gravity at sea level decreases from the poles, where the centrifugal force is zero, to the equator, where the centrifugal force is a maximum but directed opposite to the force of gravitation. This difference is accentuated by the shape of the earth, which is nearly that of an oblate spheroid of revolution slightly depressed at the poles. Also, because of the asymmetric distribution of the mass of the earth, the force of gravity is not directed precisely toward the earth's center.

The magnitude of the force of gravity is usually called either gravity, acceleration of gravity, or apparent gravity.

Half life — The average time required for one half the atoms in a sample of radioactive element to decay.

The half life t½ is given by

$$t_{1/2} = (\ln 2)/\lambda$$

where λ is the decay constant.

Half-wave plate — An electro-optical material used to rotate the plane of polarization of a light beam.

Hall constant — In an electrical conductor, the constant of proportionality R in the relation

$$E_h = RJ \times H$$

where E_h is transverse electric field (Hall field); J is current density; and H is magnetic field.

The sign of the majority carrier can be inferred from the sign of the Hall constant.

Hall effect — When a steady current is flowing in a steady magnetic field, electromotive forces are developed which are at right angles both to the magnetic force and to the current and are proportional to the product of the intensity of the current, the magnetic force and the sine of the angle between the directions of these quantities.

Hall mobility — A measure of the flow of charged particles perpendicular to both a magnetic and an electric field.

Hardness — Property of substances determined by their ability to abrade or indent one another. An arbitrary scale of hardness is based upon ten selected minerals. For metals the diameter of the indentation made by a hardened steel sphere (Brinnell) or the height of rebound of a small drop hammer (Shore Scleroscope) serve to measure hardness.

Harmonic — A sinusoidal quantity whose frequency is an integral multiple of some fundamental frequency, that is, given a quantity, x(t), where

$$x(t) = A\cos(\omega t + \phi),$$

its harmonics, $h_n(x)$ are of the form

$$h_n(x) = A_n\cos(n\,\omega t + \phi_n),$$

where n is an integer larger than 1.

Harmonic motion — See Simple harmonic motion and Angular harmonic motion.

Heat — Energy transferred by a thermal process.

Heat can be measured in terms of the dynamical units of energy, as the erg, joule, etc., or in terms of the amount of energy required to produce a definite thermal change in some substance, as, for example, the energy required per degree to raise the temperature of a unit mass of water at some temperature (calorie, Btu).

Heat capacity — That quantity of heat required to increase the temperature of a system or substance one degree of temperature. It is usually expressed in calories per degree centigrade or joules per degree centigrade.

Molar heat capacity is the quantity of heat necessary to raise the temperature of one molecular weight of the substance one degree.

Heat effect — The heat in calories developed in a circuit by an electric current of I amperes flowing through a resistance of R ohms, with a difference of potential E volts for a time t seconds.

$$H = \frac{RI^2 t}{4.18} = \frac{Eit}{4.18}$$

Heat equivalent, or latent heat, or fusion — The quantity of heat necessary to change one gram of solid to a liquid with no temperature change. Dimensions, $[l^2 t^{-2}]$.

Heat of combustion of a substance is the amount of heat evolved by the combustion of 1 g mol wt of the substance.

Heat quantity — The cgs unit of heat is the calorie, the quantity of heat necessary to change the temperature of one gram of water from 3.5°C to 4.5°C (called a small calorie). If the temperature change involved is from 14.5 to 15.5°C, the unit is the normal calorie. The mean calorie is $\frac{1}{100}$ the quantity of heat necessary to raise one gram of water from 0°C to 100°C. The large calorie is equal to 1000 small calories. The British thermal unit is the heat required to raise the temperature of one pound of water at its maximum density, 1°F. It is equal to about 252 calories. Dimensions of energy, $[ml^2t^{-2}]$.

Hecto — A prefix meaning hundred or 10^2. Symbol is h.

Hehner number (value) — A number expressing the percentage (i.e. grams per hundred grams) of water-insoluble fatty acids in an oil or fat.

Heisenberg's theory of atomic structure — The currently accepted view of the structure of atom, formulated by Heisenberg in 1934, according to which the atomic nuclei are built of nucleons, which may be protons or neutrons, while the extranuclear shells consist of electrons only. The nucleons are held together by nuclear forces of attraction, with exchange forces operating between them. The number of protons is equal to the atomic number (Z) of the element, the number of neutrons is equal to the difference between the mass number and the atomic number (A — Z). The number of excess neutrons, i. e. the excess of neutrons over protons, is of paramount importance for the radioactive properties or stability of the element.

Helmholtz free energy = Helmholtz function.

Helmholtz function (symbol a) — A mathematically defined thermodynamic function of state, the decrease in which during a reversible isothermal process is equal to the work done by the system. The Helmholtz function is

$$a = u - Ts$$

where u is specific internal energy; T is Kelvin temperature; and s is specific entropy. By use of the first law of thermodynamics for reversible processes,

$$da = -s\, dT - dw$$

where dw is the work done per unit mass by the system. Also called Helmholtz free energy, work function. Compare Gibbs function.

Henry (unit of electric inductance) — The inductance of a closed circuit in which an electromotive force of 1 V is produced when the electric current in the circuit varies uniformly at a rate of 1 A/sec.

Henry's law — The mass of a slightly soluble gas that dissolves in a definite mass of a liquid at a given temperature is very nearly directly proportional to the partial pressure of that gas. This holds for gases which do not unite chemically with the solvent.

Hertz — The measure of frequency, defined as equal to 1 cycle/sec.

Hess' law of constant heat summation — The amount of heat generated by a chemical reaction is the same whether reaction takes place in one step or in several steps, or all chemical reactions which start with the same original substances and end with the same final substances liberate the same amounts of heat, irrespective of the process by which the final state is reached.

Heterodyne — To mix two radio signals of different frequencies to produce a third signal which is of lower frequency; i.e., to produce beating.

Radar receivers are of the heterodyne type (as contrasted to the superregenerative type) because the very high radio frequencies used in radar are difficult to amplify. A target signal is heterdyned with a current of lower frequency produced by a klystron oscillator and the resulting intermediate-frequency signal can then be highly amplified for subsequent presentation or analysis.

Hobbmann orbit — A minimum energy transfer orbit.

Hole — A mobile vacancy in the electronic valence of structure of a semiconductor which acts like an electron with a positive charge.

Holocamera — A device for recording or forming a hologram of an object or subject.

Hologram — A recording or picture of a three-dimensional wavefront.

Holographic — Pertaining to or using the principles of holography. For example, holographic equipment uses the principles of holography for its operation.

Holographic matched filter — A particular type of hologram which when illuminated by the type wave it is matched to will transmit a pure plane wave. This plane wave is usually focused into a correlation spot.

Holography — A recording and viewing process which allows reconstruction of three-dimensional images of diffuse objects.

Homogeneous atmosphere — 1. A hypothetical atmosphere in which the density is constant with height.

The lapse rate of temperature in such an atmosphere is known as the autoconvective lapse rate and is equal to g/R (or approximately 3.4°C/100 m) where g is the acceleration of gravity and R is the gas constant for air. A homogeneous atmosphere has a finite total thickness which is given by R_dT_v/g, where R_d is the gas constant for dry air and T_v is the virtual temperature (°K) at the surface. For a surface temperature of 273°K, the vertical extent of the homogeneous atmosphere on the earth is approxiamtely 8000 m. At the top of such an atmosphere both the pressure and absolute temperature vanish.

2. With respect to radio propagation, an atmosphere which has a constant index of refraction, or one in which radio waves travel in straight lines at constant speed. Free space is the ideal homogeneous atmosphere in this sense. 3. Same as adiabatic atmosphere. See Barotrophy.

Hooke's law — Within the elastic limit of any body the ratio of the stress to the strain produced is constant.

Horsepower — A measure of power, defined as equal to 33,000 foot pounds/min, or 746 watts.

Humidity, absolute — Mass of water vapor present in unit volume of the atmosphere, usually measured as grams per cubic meter. It may also be expressed in terms of the actual pressure of the water vapor present.

Huygens' principle — A very general principle applying to all forms of wave motion which states that every point on the instantaneous position of an advancing phase front (wave front) may be regarded as a source of secondary spherical wavelets. The position of the phase front a moment later is then determined as the envelope of all the secondary wavelets (ad infinitum).

Huygens' theory of light — This theory states that light is a disturbance traveling through some medium, such as the ether. Thus light is due to wave motion in ether.

Every vibrating point on the wave-front is regarded as the center of a new disturbance. These secondary disturbances traveling with equal velocity, are enveloped by a surface identical in its properties with the surface from which the secondary disturbances start and this surface forms the new wave-front.

Hydrogen equivalent of a substance is the number of replaceable hydrogen atoms in 1 mol or the number of atoms of hydrogen with which 1 mol could react.

Hydrogen ion concentration — The concentration of hydrogen ions in solution when the concentration is expressed as gram-ionic weights per liter. A convenient form of expressing hydrogen ion concentration is in terms of the negative logarithm of this concentration. The negative logarithm of the hydrogen ion concentration is called pH.

Water at 25°C has a concentration of H ion of 10^{-7} and of OH ion of 10^{-7} moles per liter. Thus the pH of water is 7 at 24°C. A greater accuracy is obtained if one substitutes the thermodynamic activity of the ion for its concentration.

Hydrolysis is a double decomposition reaction involving the splitting of water into its ions and the formation of a weak acid or base or both.

Hydrostatic equation — In numerical equations, the form assumed by the vertical component of the vector equation of motion when all coriolis, earth-curvature, frictional, and vertical-acceleration terms are considered negligible

DEFINITIONS (Continued)

compared with those involving the vertical pressure force and the force of gravity. Thus,

$$dP = -\rho g dZ$$

where P is the atmospheric pressure; ρ is the density; g is the acceleration of gravity; and Z is the geometric height.

Hydrostatic pressure at a distance h from the surface of a liquid of density d,

$$P = hdg$$

The total force on an area A due to hydrostatic pressure,

$$F = P A = Ahdg$$

Force in dynes and pressure in dynes per cm² will be given if h is in cm, d in g per cm³ and g in cm per sec.²

Hyperon — Any article with mass intermediate between that of the neutron and the deuteron. See Meson.

Hypersonic — 1. Pertaining to hypersonic flow. 2. Pertaining to speeds of Mach 5 or greater.

Hypersonic flow In aerodynamics, flow of a fluid over a body at speeds much greater than the speed of sound and in which the shock waves start at a finite distance from the surface of the body. Compare supersonic flow.

Hysteresis — The magnetization of a sample of iron or steel due to a magnetic field which is made to vary through a cycle of values, lags behind the field. This phenomenon is called hysteresis.
Steinmetz' equation for hysteresis gives the loss of energy in ergs per cycle per cm³,

$$W = \eta B^{1.6}$$

where B is the maximum induction in maxwells per cm² and η the coefficient of hysteresis.

Ice point — The temperature at which a mixture of air-saturated pure water and pure ice may exist in equilibrium at a pressure of one standard atmosphere.
By decision of the Tenth General Conference on Weights and Measures, Paris, October 1954 the ice point was established as 273.15°K.

Ideal gas — A gas which conforms to Boyle's law and has zero heat of free ezpansion (or also obeys Charles' law). Also called perfect gas.

Illuminance — The total luminous flux received on a unit area of a given real or imaginary surface, expressed in such units as the footcandle, lux, or phot. Illuminance is analogous to irradiance, but is to be distinguished from the latter in that illuminance refers only to light and contains the luminous efficiency weighting factor necessitated by the nonlinear wavelength-response of the human eye. Compare luminous intensity.
The only difference between illuminance and illumination is that the latter always refers to light incident upon a material surface.
A distinction should be drawn, as well, between illuminance and luminance. The latter is a measure of the light coming from a surface; thus, for a surface which is not self-luminous, luminance is entirely dependent upon the illuminance upon that surface and its reflection properties.

Illumination on any surface is measured by the luminous flux incident on unit area. The units in use are: the lux, (abbreviation lx) one lumen per square meter; the phot, (abbreviation ph) one lumen per square centimeter; the footcandle, (abbreviation fc) one lumen per square foot.

Image redundancy — Multiple storage of the same image.

Impact pressure — 1. That pressure of a moving fluid brought to rest which is in excess of the pressure the fluid has when it does not flow, i.e., total pressure less static pressure.
Impact pressure is equal to dynamic pressure in incompressible flow, but in compressible flow impact pressure includes the pressure change owing to the compressibility effect.
2. A measured quantity obtained by placing an open-ended tube, known as an impact tube or pitot tube, in a gas stream and noting the pressure in the tube on a suitable manometer.

Since the pressure is exerted at a stagnation point, the impact pressure is sometimes referred to as the stagnation pressure or total pressure.

Incandescence — Emission of light due to high temperature of the emitting material. Any other emission of light is called luminescence.

Incoherent holography — Holograms produced initially from conventional photographs or incoherent optical equipment.

Indeterminacy principle (Uncertainty principle) — The postulate that it is impossible to determine simultaneously both the exact position and the exact momentum of an electron. So this aspect of electronics can only be expressed as a probability.

Index of refraction (symbol n) — 1. A measure of the amount of refraction (a property of a dielectric substance). It is the ratio of the wavelength or phase velocity of an electromagnetic wave in a vacuum to that in the substance. Also called refractive index, absolute index of refraction, absolute refractive index, refractivity. See modified index of refraction, N-unit, potential index of refraction.
It can be a function of wavelength, temperature, and pressure. If the substance is nonabsorbing and nonmagnetic at any wavelength, then n^2 is equal to the dielectric constant at that wavelength.
The complex index of refraction is obtained when the attenuation of the wave per radian, called the absorptive index k, is paired with the index of refraction. It is written

$$n^* = n(1 - ik)$$

When the wave passes from one medium n_1 to another n_2, the angle of incidence ϕ and the angle of refraction ϕ', both measured with respect to the normal to the interface, are related by

$$\sin\phi/\sin\phi' = n_1{}^*/n_2{}^* = \text{constant}$$

which becomes, for a nonabsorbing medium, the ratios of the (noncomplex) indices of refraction. In the particular case that medium 2 is a vacuum, this ratio is the index of refraction of medium 1. This is known as Snell law, named after Willebrord Snell who discovered it about 1621.
2. A measure of the amount of refraction experienced by a ray as it passes through a refraction interface, i.e., a surface separating two media of different densities. It is the ratio of the absolute indices of refraction of the two media (see sense 1 above). Also called refractive index, relative index of refraction.

Indicators are substances which change from one color to another when the hydrogen ion concentration reaches a certain value, different for each indicator.

Indirectly ionizing particles are uncharged particles (neutrons, photons, etc.) which can liberate directly ionizing particles or can initiate nuclear transformations.

Induced electromotive force in a circuit is proportional to the rate of change of magnetic flux through the circuit.

$$E = -\frac{d\phi}{dt}$$

where $d\phi$ is the change of magnetic flux in a time dt. The induced current will be given by

$$I = \frac{d\phi}{Rdt}$$

where R is the resistance of the circuit.

Inductance — The change in magnetic field due to the variation of a current in a conducting circuit causes an induced counter electromotive force in the circuit itself. This phenomenon is known as self-induction. If an electromotive force is induced in a neighboring circuit the term mutual induction is used. Inductance may thus be distinguished as self- or mutual and is measured by the electromotive force produced in a conductor by unit rate of variation of the current. Units of inductance are the centimeter (absolute electromagnetic) and the henry, which is equal to 10^9 centimeters of inductance. The henry is that inductance in which an induced electromotive force of one volt is produced when the inducing current is changed at the rate of one ampere per second. Dimensions, $[\epsilon^{-1} l^{-1} t^2]$; $[\mu l]$.

DEFINITIONS (Continued)

Induction — Any change in the intensity or direction of a magnetic field causes an electromotive force in any conductor in the field. The induced electromotive force generates an induced current if the conductor forms a closed circuit.

Inertia — The resistance offered by a body to a change of its state of rest or motion, a fundamental property of matter. Dimension, $[m]$.

Information content — Containing or transmitting data involving new knowledge. When applied to waves or wavefronts, it includes both amplitude and phase of all parts of a wavefront at a given instant of time.

Infrared radiation (abbr. IR) — Electromagnetic radiation lying in the wavelength interval from about 75 μm to an indefinite upper boundary sometimes arbitrarily set at 1000 μm (0.01 cm). Also called long-wave radiation.

Inline holography — Hologram produced by single reference beam interferences with waves diffracted or scattered from a small object.

Insolation — (Contracted from incoming solar radiation.) 1. In general, solar radiation received at the earth's surface. See terrestrial radiation, extraterrestrial radiation, direct solar radiation, global radiation, effective terrestrial radiation, diffuse sky radiation, atmospheric radiation. 2. The rate at which direct solar radiation is incident upon a unit horizontal surface at any point on or above the surface of the earth. Compare solar constant.

Intensity of Illumination (properly called Illumination)—Illumination in lux of a screen by a source of illuminating power P at a distance r meters, for normal incidence,

$$I = \frac{P}{r^2}$$

If two sources of illuminating power P_1 and P_2 produce equal illumination on a screen when at distances r_1 and r_2 respectively,

$$\frac{P_1}{r_1{}^2} = \frac{P_2}{r_2{}^2} \quad \text{or} \quad \frac{P_1}{P_2} = \frac{r_1{}^2}{r_2{}^2}$$

If I_o is the illumination when the screen is normal to the incident light, then I is the illumination when the screen is at an angle θ. Thus,

$$I = I_0 \cos \theta$$

Intensity of magnetization is given by the quotient of the magnetic moment of a magnet by its volume. Unit intensity of magnetization is the intensity of a magnet which has unit magnetic moment per cubic centimeter. Dimensions, $[\epsilon^{-1/2} \ m^{-1/2} \ l^{-3/2}]$; $[\mu^{1/2} \ m^{1/2} \ l^{-1/3} \ t^{-1}]$.

Intensity of radiation is the rate of transfer of energy across unit areas by the radiation. In all forms of energy transfer by waves (radiation) the intensity I is given by $I = U_v$ where U is the energy density of the wave in the medium, and v is the velocity of propagation of the wave. The energy density U is always proportional to the square of the wave amplitude.

Intensity of sound depends upon the energy of the wave motion. The intensity is measured by the energy in ergs transmitted per second through 1 cm³ of surface. The energy in ergs per cm³ in a sound wave is given by

$$E = 2\pi^2 \, dn^2 \, a^2$$

where d is density in g per cm³, n is frequency in vib. per sec and a is amplitude in cm. The energy reaching the ear in unit time will also be proportional to the velocity of propagation.

Interference hologram — A holographic interferogram produced by the superposition of two or more hologram exposures.

Interference (of waves) — The coherent addition or subtraction of two different wavefronts, which usually forms a third wave different from the first two.

Interference pattern — The pattern of light and dark fringes produced when two or more coherent waves interfere or intersect.

Interferogram — The record of an interference pattern produced by holography or by conventional optical interference techniques.

Interferometry (holographic) — The process of measuring very small movements or deformations by recording or observing interference wave patterns (either light, electronic, or acoustic).

Interionic attraction — The electrostatic attraction between ions of unlike charge (sign).

Interionic repulsion — The electrostatic repulsion between ions of like charge (sign).

Internal energy — A mathematically defined thermodynamic function of state, interpretable through statistical mechanics as a measure of the molecular activity of the system. It appears in the first law of thermodynamics as

$$du - dq - dw$$

where du is the increment of specific internal energy, dq the increment of heat, and dw the increment of work done by the system per unit mass. The differential du is a perfect differential. Its integral therefore introduces a constant of integration, the zero-point internal energy, so that care must be taken when absolute values of the internal energy are employed.

International Practical Kelvin Temperature Scale of 1960 and the **International Practical Celsius Temperature Scale of 1960** are defined by a set of interpolation equations based on the following reference temperatures:

	°K	°C
Oxygen, liquid-gas equilibrium	90.18	−182.97
Water, solid-liquid equilibrium	273.15	0.00
Water, solid-liquid-gas equilibrium	273.16	0.01
Water, liquid-gas equilibrium	373.15	100.00
Zinc, solid-liquid equilibrium	692.655	419.505
Sulphur, liquid-gas equilibrium	717.75	444.6
Silver, solid-liquid equilibrium	1233.95	960.8
Gold, solid-liquid equilibrium	1336.15	1063.0

International System of Units (abbr. SI) — The metric system of units based on the meter, kilogram, second, ampere, Kelvin degree, and candela. Also called MSKA system.

Iodine number (value) — A number expressing the percentage (i.e. grams per 100 grams) of iodine absorbed by a substance. It is a measure of the proportion of unsaturated linkages present and is usually determined in the analysis of oils and fats.

Ion — An ion is an atom or group of atoms that is not electrically neutral but instead carries a positive or negative electric charge. Positive ions are formed when neutral atoms or molecules lose valence electrons; negative ions are those which have gained electrons.

Ion atmosphere (or continuus charge distribution) — In the electrostatic effects between ions the term ion atmosphere denotes a continuous charge distribution, or charge density, $\varrho(r)$, which is a continuous function of r, the distance from the reference ion, rather than a discrete or discontinuous charge distribution. The ion atmosphere extends from $r = a$ to $r = O(V^{1/3}) \approx \infty$, where V is the volume of the system, and acts electrostatically somewhat like a sphere of charge $-e$ at a distance, k^{-1}, from the reference ion of charge $+e$ (see below for definition of k^{-1}).

Ionic equivalent conductance, λ — The ionic equivalent conductance is the equivalent conductance of an individual ion constituent of the solute (or electrolyte) of an electrolytic solution. This symbol is also used to designate the equivalent conductance of complex ions, ion pairs, ion clusters, etc., in combination with simple ions.

Ionic mobility, μ — The mobility of an ion at any finite equivalent concentration is the velocity with which the ion moves under unit potential gradient. Its unit is $cm^2 \ sec^{-1} \ volt^{-1} \ equiv^{-1}$ or $cm^2 \ ohm^{-1} \ F^{-1}$ where F is the Faraday expressed in coulombs (or ampere seconds) $equiv^{-1}$.

Ionization — The process by which neutral atoms or groups of atoms become electrically charged, either positively or negatively, by the loss or gain of electrons; or the state of a substance whose atoms or groups of atoms have become thus charged.

Ionization potential — The work (expressed in electron volts) required to remove a given electron from its atomic orbit and place it at rest at an infinite

DEFINITIONS (Continued)

distance. It is customary to list values in eectron volts (ev.) 1 ev. = 23,053 calories per mole.

Ionizing radiation is any radiation consisting of directly or indirectly ionizing particles or a mixture of both.

Ionogens — Substances, like acetic acid (HAc), which, although in the pure state are nonelectrolytic neutral molecules, can react with certain solvents to form products which rearrange to ion pairs which then dissociate to give conducting solutions. As an example:

$$HAc + H_2O \rightleftharpoons HAc \cdot H_2O \rightleftharpoons H_3O^+ \cdot Ac^- \rightleftharpoons H_3O^+ + Ac^-$$

Ionophores — Substances, like sodium chloride, which exist only as ionic lattices in the pure crystalline form, and which when dissolved in an appropriate solvent give conductances which change according to some fractional power of the concentration. Such solutions possess no neutral molecules which can dissociate, but may contain associated ions.

Ionosphere — The atmospheric shell characterized by a high ion density. Its base is at about 70 or 80 km and it extends to an indefinite height.

Ion size or "ion-size" parameter, a (or a_i) — The ion size is formally considered to be the sum of the ionic radii of the oppositely charged ions in contact. The ion size is also called the "distance of closest approach" of the ions, or the "ion-size" parameter. Generally the ion size is greater than the sum of the crystal radii, and the "ion-size" parameter may include several factors which contribute to its numerical value.

Irradiance — A measure of the rate of energy falling on a given area.

Isentropic — Of equal or constant entropy with respect to either space or time.

Isobaric — Of equal or constant pressure, with respect to either space or time.

Isobars — For chemistry, elements of the same atomic mass but of different atomic numbers. The sum of their nucleons is the same but there are more protons in one than in the other.

Isochoric — Of equal or constant volume, usually applied to a thermodynamic process during which the volume of the system remains unchanged. Compare isosteric.

Isoclinic line — A line through points on the earth's surface having the same magnetic dip.

Isomer — 1. One of two or more nuclides having the same mass number A and atomic number Z, but existing for measurable times in different quantum states with different energies and radioactive properties. 2. One of two or more molecules having the same atomic composition and molecular weight, but differing in geometrical configuration.

Isomerism — Existence of molecules having the same number and kinds of atoms but in different configurations.

Isotherm — A line of equal or constant temperature. A distinction is made, infrequently, between a line representing equal temperature in space, choroisotherm, and one representing constant temperature in time, chronoisotherm.

Isothermal — When a gas passes through a series of pressure and volume variations without change of temperature the changes are called isothermal. A line on a pressure-volume diagram representing these changes is called an isothermal line.

Isotope — 1. One of several nuclides having the same number of protons in their nuclei, and hence belonging to the same element, but differing in the number of neutrons and therefore in mass number A, or in energy content (isomers). For example, $_6C^{12}$, $_6C^{13}$, and $_6C^{14}$ are carbon isotopes. Small quantitative differences in chemical properties exist between isotopes.

Isotropic — In general, pertaining to a state in which a quantity or spatial derivatives thereof are independent of direction. Also called isotropous.

Joule constant — The ratio between heat and work units from experiments based on the first law of thermodynamics: 4.1858×10^7 ergs per 15° calorie. Also called mechanical equivalent of heat.

Joule — A unit of energy or work in the MKS system; the work done when the point of application of 1 newton is displaced a distance of 1 meter in the direction of the force.

$$1 \text{ joule} = 10^7 \text{ ergs} = 1 \text{ watt second}$$

Joule cycle — (After James Prescott Joule, 1818—1889, English physicist.) An ideal cycle for engines consisting of isentropic compression of the working substance, addition of heat at constant pressure, isentropic expansion to ambient pressure, and exhaust at constant pressure. Also called Brayton cycle.

Joule's law — A law stating that the amount of heat H produced in a conductor by the flow of a steady current l is given by

$$H = Kl^2 Rt.$$

where R is the resistance of the conductor, t is the time for which the current flows, and K, the constant of proportionality, has the value 0.2390 calories per joule when R is in ohms and l in amperes.

Joule-Thomson effect — The decrease in temperature which takes place when a gas expands through a throttling device as a nozzle. Also called Joule-Kelvin effect.
The rate of change of temperature T with pressure p in the Joule-Thomson effect is called the Joule-Thomson coefficient (symbol μ):

$$\mu = \frac{dT}{dp}\bigg]_h$$

where h denotes constant enthalpy.
For the Joule-Thomson effect to take place the gas must initially be below its inversion temperature; if above the inversion temperature, the gas will gain heat on expansion. The inversion temperature of hydrogen, for example, is approximately −183°C

Junction — In a semiconductor device, a region of transition between semiconducting regions of different electrical properties.

Kelvin — The kelvin, the unit of thermodynamic temperature, is the fraction 1/273.16 of the thermodynamic temperature of the triple point of water. The decision was made at the 13th General Conference on Weights and Measures on October 13, 1967 that the name of the unit of thermodynamic temperature would be changed from degree Kelvin (symbol: °K) to kelvin (symbol: K). The name (kelvin) and symbol (K) are to be used for expressing temperature intervals. The former convention which expressed a temperature interval in degrees Kelvin or, abbreviated, deg. K is dropped. However, the old designations are acceptable temporarily as alternatives to the new ones. One may also express temperature intervals in degrees Celsius.

Kelvin temperature scale (abbr. K)—An absolute temperature scale independent of the thermometric properties of the working substance. On this scale, the difference between two temperatures T_1 and T_2 is proportional to the heat converted into mechanical work by a Carnot engine operating between the isotherms and adiabats through T_1 and T_2. Also called absolute temperature scale, thermodynamic temperature scale.
For convenience the Kelvin degree is identified with the Celsius degree. The ice point in the Kelvin scale is 273.15°K. The triple point of water, the fundamental reference point, is 273.16°K. See Absolute zero, Rankine temperature scale.

Kepler's laws — I. The planets move about the sun in ellipses, at one focus of which the sun is situated. II. The radius vector joining each planet with the sun describes equal areas in equal times. III. The cubes of the mean distances of the planets from the sun are proportional to the squares of their times of revolution about the sun.

Kerr cell — A cell that contains electrodes immersed in nitrobenzene or other liquid. It shows double refraction in high degree and with short time lag. Used in devices in which light intensity is changed rapidly according to the voltage applied to the electrons.

Kerr effect — When plane polarized light is incident on the pole of an electromagnet, polished so as to act like a mirror, the plane of polarization of the reflected light is not the same when the magnet is "on" as when it is "off." It was found that the direction of rotation was opposite to that of the currents exciting the pole from which the light was reflected.

Kilo — A prefix meaning thousand or 10^3. Symbol is k.

DEFINITIONS (Continued)

Kilogram (unit of mass)—is the mass of a particular cylinder of platinum-iridium alloy, called the International Prototype Kilogram, which is preserved in a vault at Sèvres, France, by the International Bureau of Weights and Measures.

Kilometer (abbr. km) — A unit of distance in the metric system.
1 kilometer = 3280.8 feet = 1093.6 yards = 1000 meters = 0.62137 statute miles = 0.53996 nautical miles.

Kinematic viscosity (symbol v) — A coefficient defined as the ratio of the dynamic viscosity of a fluid to its density.
The kinematic viscosity of most gases increases with increasing temperature and decreasing pressure. For dry air at 0°C, the kinematic viscosity is about 0.13 square centimeter per second. In the theory of atmospheric turbulence the kinematic viscosity is usually replaced by the kinematic eddy viscosity to account for the increased internal friction due to turbulence.

Kinetic energy (symbol E_k) — The energy which a body possesses as a consequence of its motion, defined as one-half the product of its mass m and the square of its speed v, $\frac{1}{2}mv^2$. The kinetic energy per unit volume of a fluid parcel is thus $\frac{1}{2}\varrho v^2$, where ϱ is the density and v the speed of the parcel. See Potential energy.
For relativistic speeds the kinetic energy is given by

$$E_k = mc^2 - m_o c^2$$

where c is the velocity of light in a vacuum, m_o is the rest mass, and m is the moving mass.

Kinetic theory, expression for pressure —

$$P = \frac{1}{3} N\, mv^2$$

where N is the number of molecules in unit volume, m the mass of each molecule and v^2 the mean square of the velocity of the molecules.

Kinetic theory of gases — Gases are considered to be made up of minute, perfectly elastic particles which are ceaselessly moving about with high velocities, colliding with each other and with the walls of the containing vessel. The pressure exerted by a gas is due to the combined effect of the impacts of the moving molecules upon the walls of the containing vessel, the magnitude of the pressure being dependent upon the kinetic energy of the molecules and their number.

Kirchhoff law—The radiation law which states that at a given temperature the ratio of the emissivity to the absorptivity for a given wavelength is the same for all bodies and is equal to the emissivity of an ideal black body at that temperature and wavelength.

Kirchhoff's laws —
I. The algebraic sum of the currents which meet at any point is zero.
II. In any closed circuit the algebraic sum of the products of the current and the resistance in each conductor in the circuit is equal to the electromotive force in the circuit.

Kirchhoff's laws of radiation — The relation between the powers of emission and the powers of absorption for rays of the same wave-length is constant for all bodies at the same temperature. First, a substance when excited by some means or other possesses a certain power of emission; it tends to emit definite rays, whose wave-lengths depend upon the nature of the substance and upon the temperature. Second, the substance exerts a definite absorptive power, which is a maximum for the rays it tends to emit. Third, at a given temperature the ratio between the emissive and the absorptive power for a given wave-length is the same for all bodies, and is equal to the emissive power of a perfectly black body.

Knot — A nautical mile per hour, 1.1508 statute miles per hour.

Kohlrausch law of independent migration of ions — The value of the equivalent conductance, as the concentration approaches zero, is equal to the sum of the limiting ionic equivalent conductances of the ions constituting the solute of the electrolytic solution.

Kundt's law — On approaching an absorption band from the red side of the spectrum the refractive index is abnormally increased by the presence of the band, while the approach is from the blue side and the index is abnormally decreased.

Lambert (abbr. L or l) — A unit of luminance (or brightness) equal to $1/\pi$ candle per square centimeter. Physically, the lambert is the luminance of a perfectly diffusing white surface receiving an illuminance of 1 lumen/cm².

Lambert law — A law of physics which states that the radiant intensity (flux per unit solid angle) emitted in any direction from a unit radiating surface varies as the cosine of the angle between the normal to the surface and the direction of the radiation. The radiance (or luminance) of a radiating surface is, therefore, independent of direction. Also called Lambert cosine law.

Lambert law of absorption = Bouguer law.

Lambert's law of absorption — Each layer of equal thickness absorbs an equal fraction of the light which traverses it.

Lambert's law of illumination — The illumination of a surface on which the light falls normally from a point source is inversely proportional to the square of the distance of the surface from the source. If the normal to the surface makes an angle with the direction of the rays, the illumination is proportional to the cosine of that angle.

Landau damping — The damping of a space charge wave by electrons which move at the phase velocity of the wave and gain energy transferred from the wave.

Langley — A unit of energy per unit area, equal to 1 gram-calorie/cm² commonly employed in radiation theory.

Langmuir probe — A small metallic conductor or pair of conductors inserted within a plasma in order to sample the plasma current.

Lanthanide series (Lanthanides) — Rare earth elements of atomic numbers 57 through 71, which have chemical properties similar to lanthanum (#57).

Laplacian speed of sound — The phase speed of a sound wave in a compressible fluid if the expansions and compressions are assumed to be adiabatic. This speed a is given by the formula

$$a^2 = (c_p/c_v)\, RT$$

where c_p and c_v are the specific heats at constant pressure and volume, respectively; R is the gas content; and T is the Kelvin temperature. The value of this speed under standard conditions in dry air is 331 m/sec. Compare Newtonian speed of sound. See Acoustic velocity.

Laser — A device in which the excitation energy of resonant atomic or molecular systems is used to coherently amplify or generate light. Consider a system of particles with energy levels E_1, E_2 where $E_1 < E_2$. A particle passing between these levels absorbs or emits a quantum of radiation of frequency

$$v = (E_2 - E_1)/h$$

where h is Planck's constant. If N_1 is the number of particles in E_1 and N_2 the number in E_2, it follows from the laws of thermodynamics that at a positive absolute temperature $N_2 < N_1$. As a result of this, a wave of the appropriate frequency would be attenuated. If by some means N_2 can be made to exceed N_1 by a reasonably large margin, an incoming wave of frequency v will stimulate transitions from E_2 to E_1 and in the process become coherently amplified. For sustained oscillation to occur feedback must be provided; this is often done by covering the ends of the chamber in which the reaction occurs with reflective material and arranging the chamber to be resonant at the frequency of oscillation.

Latent heat of vaporization — The quantity of heat necessary to change one gram of liquid to vapor without change of temperature, measured as calories per gram. Dimensions, $[l^2 t^{-2}]$.

Lattice energy — The energy required to separate the ions of a crystal to an infinite distance from each other.

LeChatelier's principle — If some stress is brought to bear upon a system in equilibrium, a change occurs, such that the equilibrium is displaced in a direction which tends to undo the effect of the stress.

Length — The name "micron", for a unit of length equal to 10^{-6} meter, and the symbol "μ" which has been used for it were dropped by action of the 13th General Conference on Weights and Measures on October 13, 1967. The symbol "μ" is to be used solely as an abbreviation for the prefix "micro-", standing for the multiplication by 10^{-6}. Thus the length previously designated as 1 micron, should be designated 1 μm.

Length, units of — Meter. 1. (Abbr. m) The basic unit of length of the metric system, defined as 1,650,763.73 wavelengths in vacuo of the unperturbed transition $2p^{10}$-$5d^5$ in krypton.[66]

Effective 1 July 1959 in the U.S. system of measures, 1 yard = 0.9144 meter, exactly, or 1 meter = 1.094 yards = 39.37 inches. The standard inch is exactly 25.4 millimeters.

Lenses — For a single thin lens whose surfaces have radii of curvature r_1 and r_2 whose principal focus is F, the index of refraction n, and conjugate focal distances f_1 and f_2,

$$\frac{1}{F} = \frac{1}{f_1} + \frac{1}{f_2} = (n-1)\left(\frac{1}{r_1} + \frac{1}{r_2}\right)$$

For a thick lens, of thickness t,

$$F = \frac{n r_1 r_2}{(n-1)[n(r_1 + r_2) - t(n-1)]}$$

Combinations of lenses—If f_1 and f_2 are the focal lengths of two thin lenses separated by a distance d the focal length of the system,

$$F = \frac{f_1 f_2}{f_1 + f - d}$$

Lenz's law — When an electromotive force is induced in a conductor by any change in the relation between the conductor and the magnetic field, the direction of the electromotive force is such as to produce a current whose magnetic field will oppose the change.

Limiting equivalent conductance, Λ_o — The limiting equivalent conductance of an electrolytic solution, Λ_o, is expressed by $\Lambda_m \equiv \lim (\sigma_{corr}/c)$ where σ_{corr} is solution conductance corrected for solvent conductance and c is the equivalent concentration. Λ_o is the value which Λ approaches as the solution is diluted so far that the effects of interionic forces become negligible (and dissociation, in the case of ionogens, is essentially complete).

Limiting ionic equivalent conductance, λ_o — The limiting ionic equivalent conductance of an individual ion constituent of the solute (or electrolyte) of an electrolytic solution is given by $\lambda_o \equiv \lim (\lambda/c)$. This symbol is also used to designate the limiting equivalent conductances of complexions, ion pairs, ion clusters, etc., in combination with simple ions.

Limiting ionic mobility, u^o — The limiting mobility of an individual ion of a solute (or electrolyte) is given by $u \equiv \lim u$.

Limiting molar conductance, $\Lambda^m{}_o$ — The limiting molar conductance of an electrolytic solution, $\Lambda^m{}_o$, is expressed by $\Lambda^m{}_o \equiv \lim (\sigma_{corr}/m)$ where σ_{corr} is solution conductance corrected for solvent conductance and m is the molar concentration. $\Lambda^m{}_o$ is the value which Λ^m approaches as the solution is diluted so far that the effects of interionic forces become negligible. Seldomly used.

Line of force — A term employed in the description of an electric or magnetic field. A line such that its direction at every point is the same as the direction of the force which would act on a small positive charge (or pole) placed at that point. A line of force is defined as starting from a positive charge (or pole) and ending on a negative charge (or pole).

The line (of force) is also used as a unit of magnetic flux, equivalent to the maxwell.

Lissajous figures — The path described by a particle which is simultaneously displaced by two simple harmonic motions at right angles, when the periods of the two motions are in the ratio of two small whole numbers, shows a variety of characteristic curves called Lissajous figures.

Liquid — A state of matter in which the molecules are relatively free to change their positions with respect to each other, but restricted by cohesive forces so as to maintain a relatively fixed volume.

Logarithm — A number m related to another number p by

$$B^m = p$$

where B is an arbitrarily chosen number larger than 1. Usually m is written $\log_B p$. Logarithms have the basic properties that

$$\log pq = \log p + \log q$$

and

$$\log p^q = q \log p$$

In the systems most often used B = 10 (common logarithms) or B = e = 2.71828...(natural logarithms). The common logarithm of p is written $\log_{10} p$ or simply log p; the natural logarithm of p is written $\log_e p$ or ln p. Logarithmic adi.

Longitudinal wave energy — Waves whose amplitude displacement is in the same or opposite direction as the motion.

Loschmidt's number — The number of molecules per unit volume of an ideal gas at 0°C and normal atmospheric pressure.

$$n_0 = (2.68719 \pm 0.0001) \times 10^{19} \text{ cm}^{-3}$$

Loudness — The psychological response of the ear which is related to the physical quantity intensity. The loudness of a sound depends on frequency also since the ear responds more strongly to some frequency bands than to others. The loudness is roughly related to the cube root of the intensity, and for many purposes it is convenient to represent loudness as proportional to the logarithm of the intensity.

Lumen — The lumen is the unit of luminous flux. It is equal to the luminous flux through a unit-solid angle (steradian) from a uniform point source of one candle, or to the flux on a unit surface all points of which are at unit distance from a uniform point source of one candle.

Luminescence — Light emission by a process in which kinetic heat energy is not essential for the mechanism of excitation.

Electroluminescence is luminescence from electrical discharges—such as sparks or arcs. Excitation in these cases results mostly from electron or ion collision by which the kinetic energy of electrons or ions, accelerated in an electric field, is given up to the atoms or molecules of the gas present and causes light emission. Chemiluminescence results when energy, set free in a chemical reaction, is converted to light energy. The light from many chemical reactions and from many flames is of this type. Photoluminescence, or fluorescence, results from excitation by absorption of light. The term phosphorescence is usually applied to luminescence which continues after excitation by one of the above methods has ceased. Compare incandescence.

Luminous (energy) — Energy whose wavelength is such that the eye is sensitive to it.

Luminous flux — The total visible energy emitted by a source per unit time is called the total luminous flux from the source. The unit of flux, the lumen, is the flux emitted in unit solid angle (steradian) by a point source of one candle luminous intensity. A uniform point source of one candle intensity thus emits 4π lumens.

Luminous intensity or candlepower is the property of a source of emitting luminous flux and may be measured by the luminous flux emitted per unit solid angle. The accepted unit of luminous intensity is the international candle. The Hefner unit, which is equivalent to 0.9 international candles, is the intensity of a lamp of specified design burning amyl acetate, called the Hefner lamp.

The mean horizontal candlepower is the average intensity measured in a horizontal plane passing through the source. The mean spherical candlepower is the average candlepower measured in all directions and is equal to the total luminous flux in lumens divided by 4π.

Lux — A photometric unit of illuminance or illumination equal to 1 lumen/m². Compare footcandle, phot.

Magnetic Anisotropy — In ferro- or ferrimagnetic crystals. it is found that the magnetization prefers to lie along certain crystal directions. These are termed easy directions of magnetization. Work must be expended to turn the magnetization away from these easy directions. That work as a function of crystal direction defines the anisotropy energy surface. Directions associated with a maximum of the anisotropy energy are termed hard directions of magnetization. In general, the energy difference between easy and hard directions

DEFINITIONS (Continued)

decreases as the temperature is increased, and vanishes at the Curie or Néel point.

Magnetic Domains — The magnetization of a ferromagnetic or a ferrimagnetic material tends to break up into regions called domains separated by thin transition regions called domain walls. Within the volume of a domain, the magnetization has its saturation value, and is directed along a single direction. The magnetizations of other domains are directed along different directions in such a way that the net magnetization of the whole sample may be zero. The application of an external magnetic field first causes some domains to grow by the motion of their walls. At higher fields, the magnetizations of the resulting domains rotate toward parallelism with the field.

Magnetic field due to a current — The intensity of the magnetic field in oersted at the center of a circular conductor of radius r in which a current I in absolute electromagnetic units is flowing,

$$H = \frac{2\pi I}{r}$$

If the circular coil has n turns the magnetic intensity at the center is,

$$H = \frac{2\pi n I}{r}$$

The magnetic field in a long solenoid of n turns per centimeter carrying a current I in absolute electromagnetic units

$$H = 4\pi n I$$

If I is given in amperes the above formulae become,

$$H = \frac{2\pi I}{10r}, \qquad H = \frac{2\pi n I}{10r}, \qquad H = \frac{4\pi n I}{10}$$

Magnetic field due to a magnet — At a point on the magnetic axis prolonged, at a distance r cm from the center of the magnet of length 2l whose poles are + m and −m and magnetic moment M, the field strength in oersted is,

$$H = \frac{4mlr}{(r^2 - l^2)^2}$$

If r is large compared with l,

$$H = \frac{2M}{r^3}$$

At a point on a line bisecting the magnet at right angles, with corresponding symbols,

$$H = \frac{2ml}{(r^2 + l^2)^3}$$

For large value of r,

$$H = \frac{M}{r^3}$$

Magnetic field intensity or magnetizing force — Is measured by the force acting on unit pole. Unit field intensity, the oersted, is that field which exerts a force of one dyne on unit magnetic pole. The field intensity is also specified by the number of lines of force intersecting unit area normal to the field, equal numerically to the field strength in oersted. Magnetizing force is measured by the space rate of variation of magnetic potential and as such its unit may be the gilbert per centimeter. The gamma (γ) is equivalent to 0.00001 oersted. Dimensions, $[\epsilon^{1/2} m^{1/2} l^{1/2} t^{-2}]$; $[\mu^{-1/2} m^{1/2} l^{-1/2} t^{-1}]$.

Magnetic flux through any area perpendicular to a magnetic field is measured as the product of the area by the field strength. The units of magnetic flux, the maxwell is the flux through a square centimeter normal to a field of one gauss. The line is also a unit of flux. It is equivalent to the maxwell. Dimensions, $[\epsilon^{-1/2} m^{1/2} l^{1/2}]$; $[\mu^{1/2} m^{1/2} t^{-1}]$.

Magnetic induction resulting when any substance is subjected to a magnetic field is measured as the magnetic flux per unit area taken perpendicular to the direction of the flux. The unit is the maxwell per square centimeter or its equivalent, the gauss. Dimensions, $[\epsilon^{-1/2} m^{1/2} l^{-3/2}]$; $[\mu^{1/2} m^{1/2} l^{-1/2} t^{-1}]$.

If a substance of permeability μ is placed in a magnetic field H the magnetic induction in the substance,

$$M = \mu H$$

If I is the magnetic moment for unit volume, or intensity of magnetization,

$$M = H + 4\pi I$$

The susceptibility,

$$k = \frac{I}{H} \qquad \mu = 1 + 4\pi k$$

Magnetic moment of a magnet is measured by the torque experienced when it is at right angles to a uniform field of unit intensity. The value of the magnetic moment is given by the product of the magnetic pole strength by the distance between the poles. Unit magnetic moment is that possessed by a magnet formed by two poles of opposite sign and of unit strength, 1 cm apart. Dimensions, $[\mu^{1/2} m^{1/2} l^{5/2} t^{-1}]$; $[\epsilon^{-1/2} m^{1/2} l^{3/2}]$.

If the poles are separated by a distance which is great compared with the dimensions of the magnet, the magnetic moment of a magnet of length l whose poles have values of + m and −m is,

$$m = ml$$

Magnetic permeability is a property of materials modifying the action of magnetic poles placed therein and modifying the magnetic induction resulting when the material is subjected to a magnetic field or magnetizing force. The permeability of a substance may be defined as the ratio of the magnetic induction in the substance to the magnetizing field to which it is subjected. The permeability of a vacuum is unity. Dimensions, $[\epsilon^{-1} l^{-2} t^2]$; $[\mu]$.

Magnetic pole or quantity of magnetism — Two unit quantities of magnetism concentrated at points unit distance apart in a vacuum repel each other with unit force. If the distance involved is 1 cm and the force 1 dyn, the quantity of magnetism at each point is one cgs unit of magnetism. Dimensions, $[\epsilon^{-1/2} m^{1/2} l^{1/2}]$; $[\mu^{1/2} m^{1/2} l^{3/2} t^{-1}]$.

Magnetic potential or magnetomotive force at a point measured by the work required to bring unit positive pole from an infinite distance (zero potential) to the point. The unit is the gilbert, that magnetic potential against which an erg of work is done when unit magnetic pole is transferred. Dimensions, $[\epsilon^{1/2}, m^{1/2} l^{3/2} t^{-2}]$; $[\mu^{-1/2} m^{1/2} l^{1/2} t^{-1}]$.

Magnetostriction — Change in sample dimensions as the magnitude or the direction of the magnetization in a crystal is changed.

Magnifying power of an optical instrument is the ratio of the angle subtended by the image of the object seen through the instrument to the angle subtended by the object when seen by the unaided eye. In the case of the microscope or simple magnifier the object as viewed by the unaided eye is supposed to be a distance of 25 cm (10 in.).

Maser — An abbreviation based on the English description of the function of this device, microwave amplification by stimulated emission of radiation.

Mass — Quantity of matter. **Units of mass**—the gram is $\frac{1}{1000}$ the quantity of matter in the International Prototype Kilogram; one of the three fundamental units of the cgs system. The British standard of mass is the pound, of which a standard is preserved by the government. The United States standard mass is the avoirdupois pound defined as 1/2.20462 kg.

The primary standard of the unit of mass is the international prototype of the kilogram kept at the BIPM. The mass of 1-kg secondary standards of platinum-iridium or of stainless steel is compared with the mass of the prototype by means of balances whose precision can reach 1 in 10^8 or better.

By an easy operation a series of masses can be standardized to obtain multiples and submultiples of the kilogram.

Mass — The inertial resistance of a body to acceleration, considered, in classical physics, to be a conserved quantity independent of speed. It can be shown that the parameters m_1, m_2 appearing in the gravitation equation are equivalent to the masses of the bodies in the sense given above. In relativistic physics it is shown that when the speed of a body becomes an appreciable fraction of c, the speed of light mass is given by

$$m = m_0 / [1 - (v/c)^2]^{1/2}$$

where m_o is the rest mass of the body and v is its speed relative to the observer who finds its mass to be m. Further, it has been shown that mass and energy are interconvertible as given by Einstein's equation $E = mc^2$.

Mass action, law of — At a constant temperature the product of the active masses on one side of a chemical equation when divided by the product of the active masses on the other side of the chemical equation is a constant, regardless of the amounts of each substance present at the beginning of the action.

At constant temperature the rate of the reaction is proportional to the concentration of each kind of substance taking part in the reaction.

Mass and Weight — "1 The kilogram is the unit of mass; it is equal to the mass of the international prototype of the kilogram;

"2 The word weight* denotes a quantity of the same nature as a force; the weight of a body is the product of its mass and the acceleration due to gravity; in particular, the standard weight of a body is the product of its mass and the standard acceleration due to gravity;

"3 The value adopted in the international Service of Weights and Measures for the standard acceleration due to gravity is 980.665 cm/s^2, value already stated in the laws of some countries."

Mass by weighing on a balance with unequal arms — If W_1 is the value for one side, W_2 the value for the other, the true mass,

$$W = \sqrt{W_1 W_2}$$

Mass defect — Difference between atomic mass and mass number of a nuclide. See Packing fraction.

Mass-energy equivalence — The equivalence of a quantity of mass and a quantity of energy when the two quantities are related by the equation $E = mc^2$. The conversion factor c^2 is the square of the velocity of light. The relationship was developed from relativity theory, but has been experimentally confirmed.

Mass number — The whole number nearest the value of the atomic mass of an element as expressed in atomic mass units.

The mass number is assumed to represent the total number of protons and neutrons in the atomic nucleus of the element and is therefore equal to the atomic number plus the number of the neutrons. The mass number of an atom is usually written as a superscript to the element symbol, as in O^{18}, an isotope of oxygen with mass number 18.

Maxwell — The cgs emu magnetic flux is the flux through a cm^2 normal to a field at 1 cm from a unit magnetic pole.

Maxwell's rule — A law stating that every part of an electric circuit is acted upon by a force tending to move it in such a direction as to enclose the maximum amount of magnetic flux.

Mechanical equivalent of heat is the quantity of energy which, when transformed into heat, is equivalent to unit quantity of heat; $4.18 \times 10^7 \text{ ergs} = 1$ calorie (20°C).

Mega — A prefix meaning million or 10^6. Symbol is M.

Meson — Two types of particles of mass intermediate between that of the electron and proton have been discovered in cosmic radiation and in the laboratory. The one particle with mass about $215m_o$ is called μ-meson, the other with about $280m_o$, π-meson. Mesons of both positive and negative charge have been found and there is now reasonably good evidence for neutral mesons. Both types of mesons decay spontaneously. Some evidence exists for a meson of mass about $1000m_o$.

Metallic elements in general are distinguished from the non-metallic elements by their luster, malleability, conductivity and usual ability to form positive ions. Nonmetallic elements are not malleable, have low conductivity and never form positive ions.

Metamagnetic Materials — Those which are antiferromagnetic in weak fields, but which become ferromagnetically ordered in strong applied fields.

Meter — The meter is the SI base unit of length and is the length equal to 1 650 763.73 wavelengths in vacuum of the radiation corresponding to the transition between the levels $2p_{10}$ and $5d_5$ of the krypton-86 atom.

Mev — Mev = million electron volts.

Micro — A prefix meaning one millionth or 10^{-6}. Symbol is μ.

Mie scattering — That which is produced by spherical particles without special regard to comparative size of radiation wavelength ad particle diameter.

Milli — A prefix meaning 1/1000 or 10^{-3}. Symbol is m.

Minimum deviation — The deviation or change of direction of light passing through a prism is a minimum when the angle of incidence is equal to the angle of emergence. If D is the angle of minimum deviation and A the angle of the prism, the index of refraction of the prism for the wave length used is,

$$n = \frac{\sin \frac{1}{2}(A + D)}{\sin \frac{1}{2}A}$$

Minute of arc — $\frac{1}{60}$ of a degree.

Mixtures consist of two or more substances intermingled with no constant percentage composition, and with each component retaining its essential original properties.

Moderator — A material used for slowing down neutrons in an atomic pile or reactor. Usually graphite or "heavy water" (deuterium oxide).

Modified index of refraction — An atmospheric index of refraction mathematically modified so that when its gradient is applied to energy propagation over a hypothetical flat earth it is substantially equivalent to propagation over the true curved earth with the actual index of refraction. Also called refractive modulus, modified refractive index. Compare potential index of refraction.

The modified index of refraction is usually expressed in M-units; mathematically

$$M = \left(n - 1 + \frac{h}{a}\right)10^5 = N + \left(\frac{h}{a}\right)10^6$$

where n is the index of refraction at a point in the atmosphere; h is the height above mean sea level of that point; a is the radius of the earth; and N is the index of refraction in N-units.

In ray tracing problems, the vertical gradient dM/dh can be used directly to obtain a ray path curvature that is relative to the curvature of the earth, i.e.,

$$\frac{dM}{dh} = \frac{dN}{dh} + \frac{10^5}{a} = \frac{10^6}{ka}$$

where k is a value by which the earth's radius is multiplied to get the radius of curvature of the ray path; ka is called the effective earth radius.

Modulus of elasticity — The stress required to produce unit strain, which may be a change of length (Young's modulus): a twist or shear (modulus of rigidity or modulus of torsion) or a change of volume (bulk modulus), expressed in dynes per square centimeter. Dimensions, the same as of stress, $[m \, l^{-1} \, t^{-2}]$.

Moire patterns — Pattern resulting from interference beats between two sets of periodic structures in an image.

Molal solution contains 1 mol/1000 g of solvent.

Molar conductance, Λ^m — The molar conductance of an electrolytic solution is the conductance of a solution containing 1 g mol of the solute (or electrolyte) when measured in a like manner to equivalent conductance. Seldomly used.

Molar solution contains 1 mol or g mol wt of the solute in 1 ℓ of solution.

Mole — the amount of substance of a system which contains as many elementary entities as there are atoms in 0.012 kg of carbon 12.

Note. When the mole is used, the elementary entities must be specified and may be atoms, molecules, ions, electrons, other particles, or specified groups of such particles.

Molecular volume — Volume occupied by 1 mol. Numerically equal to the molecular weight divided by the density.

---

* USA Editors' note: In the USA weight is the commonly used term for mass. Because of the dual use of the term weight as a quantity, this term should be avoided in technical practice. (American National Standard Z210.1)

DEFINITIONS (Continued)

Molecular weight — The sum of the atomic weights of all the atoms in a molecule.

Molecule — The smallest unit quantity of matter which can exist by itself and retain all the properties of the original substance.

Mol volume — The volume occupied by a mol or a gram molecular weight of any gas measured at standard conditions is 22.414 l.

Moment of force or torque — The effectiveness of a force to produce rotation about an axis, measured by the product of the force and the perpendicular distance from the line of action of the force to the axis. Cgs unit—the dyne-centimeter. Dimensions, $[m\, l^2\, t^{-2}]$. If a force F acts to produce rotation about a center at a distance d from the line in which the force acts, the force has a torque,

$$L = Fd.$$

Moment of inertia — A measure of the effectiveness of mass in rotation. In the rotation of a rigid body not only the body's mass, but the distribution of the mass about the axis of rotation determines the change in the angular velocity resulting from the action of a given torque for a given time. Moment of inertia in rotation is analogous to mass (inertia) in simple translation. The cgs unit is g-cm². Dimensions, $[m\, l^2]$.

If m_1, m_2, m_3, etc. represent the masses of infinitely small particles of a body; r_1, r_2, r_3, etc. their respective distances from an axis of rotation, the moment of inertia about this axis will be

$$I = (m_1 r_1{}^2 + m_2 r_2{}^2 + m_3 r_3{}^2 + \ldots)$$

or

$$I = \Sigma\,(mr^2)$$

Momentum — Quantity of motion measured by the product of mass and velocity. Cgs unit, one gram-centimeter per second. Dimensions, $[m\, l\, t^{-1}]$.
A mass m moving with velocity v has a momentum,

$$M = mv$$

If a mass m has its velocity changed from v_1 to v_2 by the action of a force F for a time t,

$$mv_2 - mv_1 = Ft.$$

Monochromatic emissive power is the ratio of the energy of certain defined wave lengths radiated at definite temperatures to the energy of the same wave lengths radiated by a black body at the same temperature and under the same conditions.

Monochromatic (source) — All source radiation is exactly of the same wavelength. This is never achieved in practice even with a laser, so the term "quasi-monochromatic" is used to mean nearly of the same wavelength for all practical purposes.

Monotropic — Crystal forms one of which is always metastable with respect to the other.

Mosley's law — The frequencies of the characteristic X-rays of the elements show a strict linear relationship with the square of the atomic number.

Motion, laws of — See Newton's law of motion.

Multiple proportions, law of — If two elements form more than one compound, the weights of the first element which combine with a fixed weight of the second element are in the ratio of integers to each other.

Nano — A prefix meaning one billionth or 10^{-9}. Symbol is n.

Néel Point — The temperature at which ferrimagnetic and antiferromagnetic materials become paramagnetic.

Negatron — 1. A term used for electron when it is necessary to distinguish between (negative) electrons and positrons.
2. A four element vacuum tube which displays a negative resistance characteristic.

Nernst effect — When heat flows across the lines of magnetic force, there is observed an electromotive force in the mutually perpendicular direction.

Neutralization is a reaction in which the hydrogen ion of an acid and the hydroxyl ion of a base unite to form water, the other product being a salt.

Neutrino — An electrically neutral particle of very small (probably zero) rest mass and of spin quantum number ½. When the spin is oriented parallel to the linear momentum the particle is the antineutrino. When the spin is oriented antiparallel to the linear momentum the particle is the neutrino. Postulated by Pauli in explaining the beta decay process.
Whenever a beta (positron) particle is created in a radioactive decay so is an antineutrino (neutrino). The two particles and the parent nucleus share between them the available energy and momentum. Neutrinos and antineutrinos can penetrate amounts of matter measured in light years without appreciable attenuation. Detected by Reines and Cowan using antineutrinos from fission reactors and large scintillation detectors.

Neutron — A neutral elementary particle of mass number 1. It is believed to be a constituent particle of all nuclei of mass number greater than 1. It is unstable with respect to beta-decay, with a half life of about 12 min. It produces no detectable primary ionization in its passage through matter, but interacts with matter predominantly by collisions and to a lesser extent, magnetically. Some properties of the neutron are: rest mass, 1.00894 atomic mass unit; charge, 0; spin quantum number, ½; magnetic moment, −1.9125 nuclear Bohr magnetrons.

Neutron cross section — See cross section.

Newton (unit of force) — that force which gives to a mass of 1 kg an acceleration of 1 m/sec².

Newtonian speed of sound — An approximation to the speed of sound a in a perfect gas given by the relation

$$a^2 = p/\rho$$

where p is pressure and ϱ is density. Compare Laplacian speed of sound. See Acoustic velocity.

Newtonian telescope — A reflecting telescope in which a small plane mirror reflects the convergent beam from the objective to an eyepiece at one side of the telescope. After the second reflection the rays travel approximately perpendicular to the longitudinal axis of the telescope. See Cassegrain telescope.

Newton laws of motion — A set of three fundamental postulates forming the basis of the mechanics of rigid bodies, formulated by Newton in 1687.
The first law is concerned with the principle of inertia and states that if a body in motion is not acted upon by an external force, its momentum remains constant (law of conservation of momentum). The second law asserts that the rate of change of momentum of a body is proportional to the force acting upon the body and is in the direction of the applied force. A familiar statement of this is the equation

$$F = ma$$

where F is vector sum of the applied forces, m is the mass, and a is the vector acceleration of the body. The third law is the principle of action and reaction, stating that for every force acting upon a body there exists a corresponding force of the same magnitude exerted by the body in the opposite direction.

Newton's law of cooling — The rate of cooling of a body under given conditions is proportional to the temperature difference between the body and its surroundings.

Nodal points — Two points on the axis of a lens such that a ray entering the lens in the direction of one, leaves as if from the other and parallel to the original direction.

Noncondensable gas — A gas whose temperature is above its critical temperature, so that it cannot be liquefied by increase of pressure alone.

Normal atmosphere — This is defined as the pressure exerted by a vertical column of 76 cm of mercury of density 13.5951 g/cm³ at a place where the gravitational acceleration is g = 980.665 cm/sec².

$$1\ \text{atm} = 1.01325 \times 10^6\ \text{dyn/cm}^2\ \text{(exactly)}$$

$$= 1.01325\ \text{Pa (exactly)}$$

$$= 14.696\ \text{psi}$$

$$= 29.921\ \text{in of Hg at } 32°\ \text{F}$$

$$1\text{mm of Hg} = 1\ \text{Torr} = 1333.22\ \text{dyn/cm}^2$$

Normal salt — An ionic compound containing neither replaceable hydrogen nor hydroxyl ions.

Normal solution — contains one gram molecular weight of the dissolved substance divided by the hydrogen equivalent of the substance (that is, one gram equivalent) per liter of solution.

Nuclear atom — The atom of each element consists of a small dense nucleus which includes most of the mass of the atom. The nucleus is made up of roughly equal numbers of neutrons and protons. The positive charges of the protons enables the nucleus to surround itself with a set of negatively charged electrons which move around the nucleus in complicated orbits with well defined energies. The outermost electrons which are least tightly bound to the nucleus play the dominant part in determining the physical and chemical properties of the atom. There are as many electrons in orbits as there are protons in the nucleus.

Nuclear cross section (symbol σ) — A measure of the probability that the reaction will take place which is defined by

$$dl = In\ \sigma\ dx$$

where I is the intensity of the particle beam; n is the number of target nuclei per cubic centimeter of target; σ is the cross section for the specified process, expressed in square centimeters; and x is the target thickness in centimeters. See barn.

Nuclear fusion — See fusion.

Nuclear isomers — Isotopes of elements having the same mass number and atomic number but differing in radioactive properties such as half-life period.

Nuclear magneton symbol M_N — A unit of magnetic moment of the proton equal to 5.0505×10^{-24} erg per gauss.

Nucleon — Any particle found in the structure of an atom's nucleus. The most plentiful ones are neutrons and protons.

Nucleus — The dense central core of the atom, in which most of the mass and all of the positive charge is concentrated. The charge on the nucleus, an integral multiple of Z of the electronic charge, is the essential factor which distinguishes one element from another. Z is called the atomic number and gives the number of protons in the nucleus, which includes a roughly equal number of neutrons. The mass number A gives the total number of neutrons plus protons. See Isotopes and Nuclear Atom.

Nuclide — A species of atom distinguished by the constitution of its nucleus. The nuclear constitution is specified by the number of protons, Z; number of neutrons, N; and energy content. (Or, by the atomic number, Z; mass number A (= N + Z) and atomic mass).

Numerical aperture is the sine of half the angular aperture, used as a measure of the optical power of an objective.

Nusselt number (symbol N_{Nu}) — (After Wilhelm Nusselt, 1882- , German engineer.) A number expressing the ratio of convective to conductive heat transfer between a solid boundary and a moving fluid, defined as hl/k where h is the heat-transfer coefficient, l is the characteristic length, and k is the thermal conductivity of the fluid.

Objective — The lens or combination of lenses which receives light rays from an object and refracts them to form an image in the focal plane of the eyepiece of an optical instrument, such as a telescope. Also called object glass.

Object wave — The scattered or reflected wave from the object which it is desired to image or "reconstruct".

Oersted — The cgs emu of magnetic intensity exists at a point where a force of 1 dyn acts upon a unit magnetic pole at that point, i.e., the intensity 1 cm from a unit magnetic pole.

Ohm (unit of electric resistance) — The electric resistance between two points of a conductor when a constant difference of potential of 1 volt, applied between these two points, produces in this conductor a current of 1 ampere, this conductor not being the source of any electromotive force.

Ohm's law — Current in terms of electromotive force E and resistance R.

$$I = \frac{E}{R}$$

The current is given in amperes when E is in volts and R in ohms.

Optical correlation — Process of determinating the similarity of an optical signal or wave form to a reference-stored signal or wave form. The reference is usually stored as a matched filter.

Optical path length — The total phase change between the source and a given position in the energy beam as measured along the direction of travel of the beam or wavefronts.

Optical pyrometer — A device for measuring the temperature of an incandescent radiating body by comparing its brightness for a selected wavelength interval within the visible spectrum with that of a standard source; a monochromatic radiation pyrometer.

Optical thickness — Specifically, in calculations of the transfer of radiant energy, the mass of a given absorbing or emitting material lying in a vertical column of unit cross-sectional area and extending between two specific levels. Also called optical depth.

If z_1 and z_2 are the lower and upper limits, respectively, of a layer in which the variation of a density ρ of some absorbing or emitting substance is given as a function of height z, then the quantity

$$\int_{z_1}^{z_2} \rho(z)\ dz$$

is called the optical thickness of that substance within that particular layer.

Order (of a bright fringe) — Proportional to the path difference between the two wave components producing a fringe (measured in integral numbers of wavelengths). No path difference produces zero order fringes.

Order (of a dark fringe) — Proportional to the path difference between the two wave components producing a fringe (measured as one-half the quantity of integral half wavelengths less one; i.e., path difference of 3.2 a wavelength produces a 1st order dark fringe).

Organ pipes — The frequency of vibration of a closed pipe or other air column of length l, where V is the velocity of sound in air, for the fundamental and first three overtones respectively is,

$$n_0 = \frac{V}{4l}, \qquad n_1 = \frac{3V}{4l}, \qquad n_2 = \frac{5V}{4l}, \qquad n_3 = \frac{7V}{4l}$$

For an open pipe,

$$n_0 = \frac{V}{2l}, \qquad n_1 = \frac{2V}{2l}, \qquad n_2 = \frac{3V}{2l}, \qquad n_3 = \frac{4V}{2l}$$

Osmotic-pressure effect — An enhancement in the velocity of the central ion, in the direction of the applied external field, as a result of more collisions on the central ion from ions behind the central ion than from ions in front of it.

Overall heat-transfer coefficient — The value U, in British thermal units per hour per square foot per °F in the equation

$$Q = UA(t_1 - t_2)$$

where Q is heat flow per unit time; A is area; and t is temperature.

Oxidation is any process which increases the proportion of oxygen or acid-forming or radical in a compound.

Packing fraction — Packing fraction is the difference between the actual mass of the isotope and the nearest whole number divided by the mass number. Thus the packing fraction is equal to (M − A)/A where M is the actual mass and A is the mass number. For example, one of the chlorine isotopes has a mass of 32.9860. The packing fraction for this isotope is

$$\frac{32.9860 - 33.0000}{33.0000} = -0.00042$$

Packing fractions are usually expressed as parts per 10,000 and so the packing fraction for this isotope of chlorine is written as −4.2. Since oxygen is taken as the standard, elements with positive packing fractions are less stable than oxygen, those with negative packing fractions are more stable.

It is positive for nuclides with mass number less than 16 or greater than 180, and negative for most others.

Parallax — The difference in the apparent direction or position of an object when viewed from different points expressed as an angle.

For bodies of the solar system, parallax is measured from the surface of the earth and its center and is called geocentric parallax, varying with the body's altitude and distance from the earth. The geocentric parallax when a body is in the horizon is called horizontal parallax and is the angular semidiameter of the earth as seen from the body. Parallax of the moon is called lunar parallax. For stars, parallax is measured from the earth and the sun, and is called annual, heliocentric, or stellar parallax. Compare aberration

Paramagnetic materials — Those within which an applied magnetic field is slightly increased by the alignment of electron orbits. The slight diamagnetic effect in materials having magnetic dipole moments is overshadowed by this paramagnetic alignment. As the temperature increases this paramagnetism disappears leaving only diamagnetism. The permeability of paramagnetic materials is slightly greater than that of empty space.

Parsec — A unit of length equal to the distance from the sun to a point having a heliocentric parallax of 1 second (1˝), used as a measure of stellar distance.

The name parsec is derived from the words parallax second. 1 parsec = pc
= 3.084×10^{13} kilometers
= 206,265 astronomical units
= 3.262 light years.

Partial pressure — The pressure exerted by a designated component or components of a gaseous mixture.

Pascal — The unit of pressure in the SI system; 1 pascal equals 1 newton per square meter.

Pascal's law — Pressure exerted at any point upon a confined liquid is transmitted undiminished in all directions.

Pauli exclusion principle — The principle that no pair of identical particles can simultaneously occupy the same quantum state. The principle applies to electrons, protons, and neutrons and accounts for the shell configurations of extranuclear electrons and for the shell structure of nuclei themselves.

Peltier effect — When a current flows across the junction of two unlike metals it gives rise to an absorption or liberation of heat. If the current flows in the same direction as the current at the hot junction in a thermoelectric circuit of the two metals, heat is absorbed; if it flows in the same direction as the current at the cold junction of the thermoelectric circuit heat is liberated.

Pendulum — For a simple pendulum of length l, for a small amplitude, the complete period,

$$T = 2\pi \sqrt{\frac{l}{g}} \quad \text{or } g = 4\pi^2 \frac{l}{T^2}$$

T will be given in seconds if l is in cm and g in cm per sec². For a sphere suspended by a wire of negligible mass where d is the distance from the knife edge to the center of the sphere whose radius is r, the length of the equivalent simple pendulum,

$$l = d + \frac{2r^2}{5d}$$

If the period is P for an arc θ, the time of vibration in an infinitely small arc is approximately

$$T = \frac{P}{1 + \frac{1}{4} \sin^2 \frac{\theta}{4}}$$

For a compound pendulum, if a body of mass m be suspended from a point about which its moment of inertia is I with its center of gravity a distance h below the point of suspension, the period

$$T = 2\pi \sqrt{\frac{I}{mgh}}$$

Penning effect — An increase in the effective ionization rate of a gas due to the pressure of a small number of foreign metastable atoms.

For instance, a neon atom has a metastable level at 16.6 V and if there are a few neon atoms in a gas of argon which has an ionization potential of 15.7 V, a collision between the neon metastable atom with an argon atom may lead to ionization of the argon. Thus, the energy which is stored in the metastable atom can be used to increase the ionization rate. Other gases where this effect is used are helium, with a metastable level at 19.8 V, and mercury, with an ionization level at 10.4 V.

Perfect fluid — In simplifying assumptions, a fluid chiefly characterized by lack of viscosity and, usually, by incompressibility. Also called an ideal fluid, inviscid fluid. See Perfect gas.

A perfect fluid is sometimes further characterized as homogeneous and continuous.

Perfect gas — A gas which has the following characteristics: (a) it obeys the Boyle-Mariotte law and the Charles-Gay-Lussac law, thus satisfying the equation of state for perfect gases; (b) it has internal energy as a function of temperature alone; and (c) it has specific heats with values independent of temperature. Also called ideal gas. Compare perfect fluid.

The normal volume of a perfect gas is 2.24136×10^4 cm³/mol.

Perigee — That orbital point nearest the earth when the earth is the center of attraction.

Perihelion — That point in a solar orbit which is nearest the sun.

Period in uniform circular motion is the time of one complete revolution. In any oscillatory motion it is the time of a complete oscillation. Dimension, [t].

Periodic law — Elements when arranged in the order of their atomic weights or atomic numbers show regular variations in most of their physical and chemical properties.

Permeance, the reciprocal of reluctance. Unit permeance is the permeance of a cylinder one square centimeter crosssection and one centimeter length taken in a vacuum. Dimensions $[\varepsilon^{-1} l^{-1} t^2]$; $[\mu l]$.

Phase angle — The phase difference of two periodically recurring phenomena of the same frequency, expressed in angular measure.

Phase of oscillatory motion — The fraction of a whole period which has elapsed since the moving particle last passed through its middle position in a positive direction.

Phase (of wave) — The distance between the position of an amplitude crest of a wave train and a reference position measured in units of wavelength, degrees, or radians (one wavelength equals 360°, or 2π radians).

Phon — The unit of loudness level. The loudness level of a sound in phons is equal to the sound pressure level in decibels re 0.0002 microbar of a pure tone of 1000 Hz which a group of listeners judge to be equally loud.

Phosphorescence — Emission of light which continues after the exciting mechanism has ceased. See luminescence. Compare fluorescence. An example of phosphorescence is the glowing of an oscilloscope screen after the exciting beam of electrons has moved to another part of the screen.

Phot — A photometric unit of illuminance or illumination equal to 1 lumen per square centimeter. Compare footcandle, lux.

Photoelectric effect — The emission of an electron from a surface as the surface absorbs a photon of electromagnetic radiation. Electrons so emitted are termed photoelectrons.

Photoelectron — An electron which has been ejected from its parent atom by interaction between that atom and a high-energy photon.

Photographic density — The density D of silver deposit on a photographic plate or film is defined by the relation

$$D = \log O$$

where O is the opacity. If I_o and I are the incident and transmitted intensities respectively the opacity is given by I_o/I. The transparency is the reciprocal of the opacity or I/I_o.

Photometer — An instrument for measuring the intensity of light or the relative intensity of a pair of lights. Also called illuminometer.

Photon — According to the quantum theory of radiation, the elementary

DEFINITIONS (Continued)

quantity, or quantum, of radiant energy. It is regarded as a discrete quantity having a momentum equal to hv/c, where h is Planck constance, v is the frequency of the radiation, and c is the speed of light in a vacuum. The photon is never at rest, has no electric charge and no magnetic moment, but does have a spin moment. The energy of a photon (the unit quantum of energy) is equal to hv. Photons are generated in collisions between nuclei or electrons and in any other process in which an electrically charged particle changes its momentum. Conversely photons can be absorbed (i.e., annihilated) by any charged particle.

Pico — A prefix meaning one trillionth or 10^{-12}. Symbol is p.

Piezo-electric effect — The phenomemon exhibited by certain crystals of expansion along one axis and contraction along another when subjected to an electric field. Conversely compression of certain crystals generate an electro static voltage across the crystal. Piezoelectricity is only possible in crystal classes which do not possess a center of symmetry.

Pinch effect — When an electric current, either direct or alternating, passes through a liquid conductor, that conductor tends to contract in cross-section, due to electromagnetic forces.

Pitch — Psychological response of the ear, primarily dependent upon the frequency of viration of the air. The intensity of the sound also has a certain effect on the pitch. Pitch of a screw is the axial distance between adjacent turns of a single thread on the screw.

Planck's constance (h) — A universal constant of nature which relates the energy of a quantum of radiation to the frequency of the oscillator which emitted it. It has the dimensions of action (energy × time). Expressed by E = hv where E is the energy of the quantum and v is its frequency. Its numerical value is $6.626176 (36) \times 10^{-27}$ erg sec.

Planck law — An expression for the variation of monochromatic radiant flux per unit area of source as a function of wavelength of black body radiation at a given temperature, it is the most fundamental of the radiation laws. Mathematically, Planck law is

$$dw = [c_1 \lambda^{-5} / (e^{c_2/T\lambda} - 1)] \, d\lambda$$

where dw is the radiant flux from a black body in the wavelength interval $d\lambda$, centered around wavelength λ, per unit area of black-body surface at temperature T; c_1 and c_2 are radiation constants.

Planetary aberration — A displacement in the apparent position of a planet in the celestial sphere due to the relative movement of the observer and the planet. See Aberration.

Plane wave — A wave in which the wave fronts are everywhere parallel planes normal to the direction of propagation.

Plasma — An assembly of ions, electrons, neutral atoms and molecules in which the motion of the particles is dominated by electromagnetic interaction. The temperature of the collection of these particles is sufficiently high for the ionization to be above 5%. The plasma taken as a whole is electrically neutral. A plasma is further characterized by relatively large intermolecular distances, large amounts of energy stored in the internal energy of the particles and the presence of a plasma sheath at all boundaries of the plasma. Plasmas are sometimes referred to as the fourth state of matter.

Plutonium — A fissile element, artificially produced in the pile by neutron bombardment of U^{238}.

Poise — A unit of coefficient of viscosity, defined as the tangential force per unit area (dyn/cm^2) required to maintain unit difference in velocity (1 cm/sec) between two parallel planes separated by 1 cm of fluid;

$$1 \text{ poise} = 1 \text{ dyne sec}/cm^2 = 1 \text{ g/cm sec.}$$

Poiseuille flow — The steady laminar flow of a fluid through a narrow horizontal circular cylinder according to the relation

$$u = (1/4\mu) (\partial p/\partial x) (a^2 - r^2)$$

where u is the fluid velocity along the cylinder's axis at a distance r from the cylinder's axis; μ is the dynamic viscosity of the fluid: a is the cylinder radius; and $\partial p/\partial x$ is the pressure gradient along the axis of the cylinder. The velocity profile across the cylinder is seen to be parabolic, and this relation affords a convenient experimental means of determining a fluid's viscosity. Also called Hagen-Poiseuille flow. Compare Couette flow.

Poisson constant (symbol μ) — The ratio of the gas constant to the specific heat of a gas at constant pressure.

Poisson distribution — A one-parameter discrete frequency distribution giving the probability that n points (or events) will be (or occur) in an interval (or time) x, provided that these points are individually independent and that the number occurring in a subinterval does not influence the number occurring in any other nonoverlapping subinterval. It has the form

$$f(n,x) = e^{-\sigma x} (\sigma x)^n / n!$$

The mean and variance are both σx, and σ is the average density (or rate) with which the events occur. When σx is large, the Poisson distribution approaches the normal distribution. The binomial distribution approaches the Poisson when the number of events n becomes large and the probability of success P becomes small in such a way that $nP \rightarrow \sigma x$.

Poisson's ratio is the ratio of the transverse contraction per unit dimension of a bar of uniform cross-section to its elongation per unit length, when subjected to a tensile stress.

Polarized light — Light which exhibits different directions at right angles to the line of propagation is said to be polarized. Specific rotation is the power of liquids to rotate the plane of polarization. It is stated in terms of specific rotation or the rotation in degrees per decimeter per unit density.

Polymorphism — The ability to exist in two or more crystalline forms.

Positron — A particle of the same mass M_e, as an ordinary electron. It has a positive electrical charge of exactly the same amount as that of an ordinary electron (which is sometimes called negatron). Positrons are created either by the radioactive decay of certain unstable nuclei or, together with a negatron in a collision between an energetic (more than one Mev) photon and an electrically charged particle (or another photon). A positron does not decay spontaneously but on passing through matter it sooner or later collides with an ordinary electron and in this collision the positron-negatron pair is annihilated. The rest energy of the two particles, which is given by Einstein's relation E = mc^2 and amounts to 1.0216 mev altogether, is converted into electromagnetic radiation in the form of one or more photons.

Potential (electric) at any point is measured by the work necessary to bring unit positive charge from an infinite distance. Difference of potential between two points is measured by the work necessary to carry unit positive charge from one to the other. If the work involved is one erg we have the electrostatic unit of potential. Dimensions, $[\varepsilon^{-1/2} m^{1/2} l^{1/2} t^{-1}]$, $[\mu^{1/2} m^{1/2} l^{2/3} t^{-1/2}]$. The potential at a point due to a charge q at a distance r in a medium whose dielectric constant is ε is,

$$V = \frac{q}{\varepsilon r}$$

Potential energy — Energy possessed by a body by virtue of its position in a gravity field in contrast with kinetic energy, that possessed by virtue of its motion.

Potential index of refraction — An atmospheric index of refraction so formulated that it would have no height variation in an adiabatic atmosphere. Also called potential refractive index. Compare modified index of refraction. The potential index of refraction is usually expressed in terms of B-units.

Pound (abbr. lb) — 1. A unit of mass equal in the U.S. to 0.45359237 kg, exactly. 2. Specifically, a unit of measurement of the thrust or force of a reaction engine representing the weight the engine can move, as an engine with 100,000 pounds of thrust. See Poundal, Pound mass. 3. The force exerted on 1 pound mass by the standard acceleration of gravity. See Gravity, sense 2.

Poundal — Force necessary to give 1 lb mass acceleration of 1 ft/sec².

Pound mass — 1. A mass equal to 0.45359237 kg. 2. A unit of measure of the inertial property equal to the mass of a body weighing 1 lb at the standard acceleration of gravity (980.665 cm/sec²).

Pound weight — A force equal to the earth's attraction for a mass of 1 lb. This force, acting on a 1 lb mass, will produce an acceleration of 32.1747 ft/sec².

DEFINITIONS (Continued)

Power — The time rate at which work is done. Units of power, the watt, one joule (ten million ergs) per second; the kilowatt is equal to 1000 watts; the horse-power, 33,000 foot-pounds per minute, is equal to 746 watts. Dimensions, $[m\,l^2\,t^{-3}]$. If an amount of work W is done in time t the power or rate of doing work is

$$P = \frac{W}{t}$$

Power will be obtained in watts if W is expressed in joules (10^7 ergs) and t in sec.

Power in watts for alternating current —

$$P = EI \cos \phi$$

where E and I are the effective values of the electromotive force and current in volts and amperes respectively and ϕ the phase angle between the current and the impressed electromotive force. The ratio,

$$\frac{P}{EI} = \cos \phi$$

is called the power factor.

Power developed by a direct current — The power in watts developed by an electric current flowing in a conductor, where E is the difference of potential at its terminals in volts, R its resistance in ohms, and I the current in amperes,

$$P = EI = RI^2$$

The work done in joules in a time t sec is,

$$W = EIt : RI^2\, t.$$

Power ratios in telephone engineering are measured in decibels. The gain or loss of power expressed in decibels is ten times the logarithm of the power ratio. By reference to an arbitrarily chosen "power level" the actual power may be expressed in decibels. The numerical values thus used will not be proportional to the actual power level but roughly to the sensation on the ear produced when the electrical power is converted into sound. A difference of 1 decibel in the power supply to a telephone receiver produces approximately the smallest change in volume of sound which a normal ear can detect.

Pressure — Force applied to, or distributed, over a surface; measured as force per unit area. Cgs unit, the barye, one dyne per square centimeter. The megabarye is equal to 10^6 dynes per square centimeter. Pressure is also measured by the height of the column of mercury or water which it supports. Dimensions, $[m\,l^{-1}\,t^{-2}]$.

The pressure due to a force F distributed over an area A,

$$P = \frac{F}{A}$$

Absolute pressure — Pressure measured with respect to zero pressure.
Gauge pressure — Pressure measured with respect to that of the atmosphere.

Primary colors — Any three colors that when mixed in suitable proportions produce any color. Primary colors may be subtractive, where the primaries absorb colors from white light (e.g., magenta, cyan, yellow; red, blue, yellow) used in developing color photography; or additive, where the primaries form a color by the addition of their light (e.g., red, green, blue).

Principal focus of a lens or spherical mirror is the point of convergence of light coming from a source at an infinite distance.

Projectiles — For bodies projected with velocity v at an angle a above the horizontal, the time to highest point of flight,

$$t = \frac{v \sin a}{g}$$

Total time of flight to reach the original horizontal plane,

$$T = \frac{2v \sin a}{g}$$

Maximum height,

$$h = \frac{v^2 \sin^2 a}{2g}$$

Horizontal range,

$$R = \frac{v^2 \sin 2a}{g}$$

In the above equations the resistance of the air is neglected, g is the acceleration due to gravity.

Prompt neutrons — In nuclear fission, those neutrons released coincident with the fission process, as opposed to the neutrons subsequently released.

Proton — A positively charged subatomic particle having a mass of 1.67252×10^{-24} g, slightly less than that of a neutron but about 1836 times greater than that of an electron.

Proton-proton reaction — A thermonuclear reaction in which two protons collide at very high velocities and combine to form a deuteron. The resultant deuteron may capture another proton to form tritium and the latter may undergo proton capture to form helium.

Proton storm — The flux of protons sent into space by a solar flare.

Pulsed laser — A laser that radiates its energy during short bursts of times (pulses) and then is inactive until the next burst or pulse. The frequency of these pulses is called the pulse repetition frequency (PRF) of the laser.

Purkinje effect — A phenomenon associated with the human eye, making it more sensitive to blue light when the illumination is poor (less than about 0.1 lumen/ft²) and to yellow light when the illumination is good.

Pyron — A unit of radiant intensity of electromagnetic radiation equal to 1 calorie/cm²/min.

Quality or timbre of sound depends on the coexistence with the fundamental of other vibrations of various frequencies and amplitudes.

Quantity of electricity or charge — The electrostatic unit of charge, the quantity which when concentrated at a point and placed at unit distance from an equal and similarly concentrated quantity, is repelled with unit force. If the distance is one centimeter and force of repulsion one dyne and the surrounding medium a vacuum, we have the electrostatic unity of quantity. The electrostatic unit of quantity may be defined as that transferred by electrostatic unit current in unit time. The quantity transferred by one ampere in one second is the coulomb, the practical unit. The faraday is the electrical charge carried by one gram equivalent. The coulomb = 3×10^9 electrostatic units. Dimensions, $[\varepsilon^{1/2} m^{1/2} l^{1/2} t^{-1}]$; $[\mu^{-1/2} m^{1/2} l^{1/2}]$.

Quantum — Unit quantity of energy postulated in the quantum theory. The photon is a quantum of the electromagnetic field, and in nuclear field theories, the meson is considered to be the quantum of the nuclear field.

Quantum theory — The theory first stated by Max Planck (before the Physical Society of Berlin on December 14, 1900) that all electromagnetic radiation is emitted and absorbed in quanta, each of magnitude hv, h being the Planck constant and v the frequency of the radiation.

Quasi-monochromatic — Nearly of the same wavelength (see monochromatic).

Rad — An ionizing radiation unit corresponding to an absorption of energy in any medium of 100 ergs/g (1 rad in tissue = 100/93 rep).

Radar — A system of radio detection and ranging which detects objects by beaming rf pulses that are reflected back by the object and measures its distance by the time elapsed between transmission and reception. The strength of the echo signal is determined by the radar equation

$$W_R = W_T \frac{G^2 \lambda^2 \sigma}{(4\pi)^3 R^4}$$

where W_R is received echo power. W_T is transmitted power, G is antenna gain, λ is radar carrier wavelength, σ is target cross section, and R is range.

DEFINITIONS (Continued)

Radar nautical mile — The time that a radar signal takes to hit a target one nautical mile away and return; 12.261 microseconds.

Radian — The angle subtended at the center of a circle by an arc equal in length to a radius of the circle. It is equal to $360°/2\pi$ or approximately 57 degrees 17 minutes 44.8 seconds. The radian is the SI unit of plane angle.

Radiation — The emission and propagation of energy through space or through a material medium in the form of waves.

The term may be extended to include streams of sub-atomic particles as alpha-rays, or beta-rays, and cosmic rays as well as electromagnetic radiation. Often used to designate the energy alone without reference to its character. In the case of light this energy is transmitted in bundles (photons).

Radiation formula, Planck's — The emission power of a black body at wave length λ may be written

$$E\lambda = \frac{c_1 \lambda^{-5}}{e^{c_2/\lambda T} - 1}$$

where c_1 and c_2 are constants with c_1 being 3.7403×10^{10} microwatts microns⁴ per cm² or 3.7403×10^{-12} watt cm², c_2 being 14384 micron degrees and T the absolute temperature.

Radiation laws — 1. The four physical laws which, together, fundamentally describe the behavior of black-body radiation: (a) the Kirchoff law is essentially a thermodynamic relationship between emission and absorption of any given wavelength at a given temperature; (b) the Planck law describes the variation of intensity of black-body radiation at a given temperature, as a function of wavelength; (c) the Stefan-Boltzmann law relates the time rate of radiant energy emission from a black body to its absolute temperature; (d) the Wien law relates the wavelength of maximum intensity emitted by a black body to its absolute temperature. 2. All the more inclusive assemblage of empirical and theoretical laws describing all manifestations of radiative phenomena; e.g., Bouguer law and Lambert law.

Radioactive nuclides — Atoms that disintegrate by emission of corpuscular or electromagnetic radiations. The rays most commonly emitted are alpha or beta or gamma rays. The three classes are:

1. Primary, which have half-life times exceeding 10^8 years. These may be alpha-emitters or beta-emitters.
2. Secondary, which are formed in radioactive transformations starting with U^{238}, U^{235}, or Th^{232}.
3. Induced, having geologically short lifetimes and formed by induced nuclear reactions occurring in nature. All these reactions result in transmutation.

Radioactivity — 1. Spontaneous disintegration of atomic nuclei with emission of corpuscular or electromagnetic radiations.
2. The number of spontaneous disintegrations per unit mass and per unit time of a given unstable (radioactive) element, usually measured in curies.

Radiometer — An instrument for detecting and, usually, measuring radiant energy. Compare bolometer. See Actinometer, Photometer.

Radius of gyration may be defined as the distance from the axis of rotation at which the total mass of a body might be concentrated without changing its moment of inertia. The product of total mass and the square of the radius of gyration will give (the) moment of inertia.

Raman scattering — Raman scattering of light from a gas, liquid, or solid is that scattering in which a shift in wavelength from that of the usually monochromatic radiation occurs. The amount of shift is a function of the scattering particles and wavelengths.

Rankine cycle — An idealized thermodynamic cycle consisting of two constant-pressure processes and two isentropic processes.

Rankine scale of temperature — The Rankine scale, with its size of degree equal to that of the Fahrenheit scale, also has its zero at absolute zero of temperature. Thus, $T°R = t°F + 459.67$. Therefore, $0°R = -459.67°F$ and the normal boiling point of water is $671.67°R$.

Raoult's law — Molar weights of non-volatile nonelectrolytes when dissolved in a definite weight of a given solvent under the same conditions lower the solvent's freezing point, elevate its boiling point and reduces its vapor pressure equally for all such solutes.

Rayleigh number — The nondimensional ratio between the product of buoyancy forces and heat advection and the product of viscous forces and heat conduction in a fluid. It is written as

$$N_{Ra} = \frac{g\,\Delta_z T\,\alpha d^3}{\nu k}$$

where g is the acceleration of gravity; $\Delta_z T$ is a characteristic vertical temperature difference in the characteristic depth d; α is the coefficient of expansion, ν is the kinematic viscosity; and k is the thermometric conductivity.

Rayleigh scattering — This is a coherent scattering in which the intensity of the light of wavelength λ, scattered in any direction making an angle θ with the incident light, is directly proportional to $1 + \cos^2 \theta$ and inversely proportional to λ^4. The latter point is noteworthy in that it shows how much greater the scattering of the short wavelengths is. These relations apply when the scattering particles are much smaller than the wavelength of the radiation. Thus the sky is blue because blue light is scattered more than red. The unscattered light is, of course, complementary to blue, i.e., orange or yellow, which explains the "warm" hues of the sunset.

Real image — Image formed by converging rays which form a focused image in space.

Reduction is any process which increases the proportion of hydrogen or base-forming elements or radicals in a compound. Reduction is also the gaining of electrons by an atom, an ion, or an element thereby reducing the positive valence of that which gained the electron.

Reflecting telescope — A telescope which collects light by means of a concave mirror.

**Reflection coefficient or reflectivity is the ratio of the light reflected from a surface to the total incident light. The coefficient may refer to diffuse or to specular reflection. In general it varies with the angle of incidence and with the wavelength of the light.

Reflection of light by a transparent medium in air (Fresnel's formulae) — If i is the angle of incidence, r the angle of refraction, n_1 the index of refraction for air (nearly equal to unity), n_2 index of refraction for a medium, then the ratio of the reflected light to the incident light is,

$$R = \tfrac{1}{2}\left(\frac{\sin^2(i-r)}{\sin^2(i+r)} + \frac{\tan^2(i-r)}{\tan^2(i+r)}\right)$$

If $i = 0$ (normal incidence), and $n_1 = 1$ (approximate for air),

$$R = \left(\frac{n_2 - 1}{n_2 + 1}\right)^2$$

Refracting telescope — A telescope which collects light by means of a lens or system of lenses. Also called refractor.

Refraction — See Index of refraction, Snell's law.

Refraction — The bending of a light or energy wave as it passes through material with varying wave velocities or indexes of refraction.

Refraction at a spherical surface — If u be the distance of a point source, v the distance of the point image or the intersection of the refracted ray with the axis, n_1 and n_2 the indices of refraction of the first and second medium, and r the radius of curvature of the separating surface,

$$\frac{n_2}{v} + \frac{n_1}{u} = \frac{n_2 - n_1}{r}$$

If the first medium is air the equation becomes,

$$\frac{n}{v} + \frac{1}{u} = \frac{n - 1}{r}$$

Refractivity — 1. The algebraic difference between an index of refraction and unity.

For the atmosphere, refractivity may be more conveniently expressed in N-units:

$$N = (n - 1)\,10^5$$

DEFINITIONS (Continued)

The deviation of the refractivity at any altitude from the usual standard profile is expressed in B-units (for radio frequencies up to 20 kilomegacycles):

$$B = N + 0.012h$$

where h is altitude in feet.

The deviation of the refractivity at any altitude from the gradient at which the refraction curvature of a tangential ray will match the curvature of the earth may be expressed in M-units:

$$M = N + 0.048h$$

where 0.048 is 10^6 divided by the radius of the earth in feet.

Relative biological effectiveness (RBE) — The biological effectiveness of any type of energy of ionizing radiation in producing a specific damage (e.g., leukemia, anemia, sterility, carcinogenesis, catasacts, shortening of life span, etc.).

Relative humidity — The ratio of the quantity of water vapor present in the atmosphere to the quantity which would saturate at the existing temperature. It is also the ratio of the pressure of water vapor present to the pressure of saturated water vapor at the same temperature.

Relativistic mass equation — The equation

$$m = m_0 \left[1 - (v^2/c^2) \right]^{-1/2} = m_0/(1 - \beta^2)^{1/2}$$

where $\beta \equiv v/c$ for the relativistic mass m of a particle or body of rest mass m_0 when its velocity is v. See Relativistic velocity.

Relativistic particle — A particle with a velocity so large that its relativistic mass exceeds its rest mass by an amount which is significant for the computation or other considerations at hand. See Relativistic velocity.

Relativistic velocity — A velocity sufficiently high that some properties of a particle of this velocity have values significantly different from those obtaining when the particle is at rest. See Rest mass.

The property of most interest is the mass. For many purposes, the velocity is relativistic when it exceeds about one tenth the velocity of light.

Relaxation-field effect — The delay in the ion atmosphere in maintaining its symmetry around a central ion as the central ion moves in the forward direction under an applied electrical field (potential gradient).

Reluctance is that property of a magnetic circuit which determines the total magnetic flux in the circuit when a given magnetomotive force is applied. Unit, the reluctance of one centimeter length and one square centimeter cross-section of space taken in a vacuum. Dimensions, $[\epsilon l\ t^{-2}]$; $[\mu^{-1}\ l^{-1}]$.

Reluctivity or specific reluctance is the reciprocal of magnetic permeability. The reluctivity of empty space is taken as unity. Dimensions, $[\epsilon l^2\ t^{-2}]$; $[\mu^{-1}]$.

Rem — Abbreviation for roentgen-equivalent-man.

Resistance is a property of conductors depending on their dimensions, material and temperature which determines the current produced by a given difference of potential. The practical unit of resistance, the ohm is that resistance through which a difference of potential of one volt will produce a current of one ampere. The international ohm is the resistance offered to an unvarying current by a column of mercury at 0°C; 14,4521 in mass, of constant cross-sectional area and 106.300 cm in length, sometimes called the legal ohm. Dimensions, $[\epsilon^{-1}\ l^{-1}\ t]$; $[\mu l\ t^{-1}]$.

Resistance of a conductor at 0°C, of length l, cross-section s and specific resistance ϱ

$$R_0 = \rho\,\frac{l}{s}$$

The resistivity may be expressed as ohm-cm when R is in ohms, l in cm and s in cm².

Resistance of a conductor at a temperature t whose resistance at 0°C is R_0 and whose temperature resistance coefficient is α

$$R_t = R_0\,(1 + \alpha t)$$

Resistance of conductors in series and parallel — The total resistance of any number of resistances joined in series is the sum of the separate resistances. The total resistance of conductors in parallel whose separate resistances are r_1, r_2, $r_3 \cdots r_n$ is given by the formula

$$\frac{1}{R} = \frac{1}{r_1} + \frac{1}{r_2} + \frac{1}{r_3} \cdots + \frac{1}{r_n}$$

Where R is the total resistance. For two terms this becomes,

$$R = \frac{r_1\,r_2}{r_1 + r_2}$$

Resistance, specific (Resistivity) — A proportionality factor characteristic of different substances equal to the resistance that a centimeter cube of the substance offers to the passage of electricity, the current being perpendicular to two parallel faces. It is defined by the expression:

$$R = \rho\,\frac{l}{A}$$

where R is the resistance of a uniform conductor, l is its length, A is its cross sectional area, and ϱ is its resistivity. Resistivity is usually expressed in ohm-centimeters.

Resistivity (symbol ϱ) — In electricity, a characteristic proportionality factor equal to the resistance of a centimeter cube of a substance to the passage of an electric current perpendicular to two parallel faces. Also called specific resistance.

$$R = \rho(l/A)$$

where R is the resistance of a uniform conductor, l is its length, A is its cross-sectional area, and ϱ is its resistivity.

Resolving power of a telescope or microscope is indicated by the minimum separation of two objects for which they appear distinct and separate when viewed through the instrument.

Resonance — 1. The phenomenon of amplification of a free wave or oscillation of a system by a forced wave or oscillation of exactly equal period. The forced wave may arise from an impressed force upon the system or from a boundary condition. The growth of the resonant amplitude is characteristically linear in time. 2. Of a system in forced oscillation, the condition which exists when any change, however small, in the frequency of excitation causes a decrease in the response of the system.

Resonance (chemical) — The moving of electrons from one atom of a molecule or ion to another atom of that molecule or ion. It is simply the oriented movement of the bonds between atoms.

Restitution, coefficient of, for two bodies on impact. The ratio of the difference in velocity, after impact to the difference before impact.

Rest mass — According to relativistic theory, the mass which a body has when it is at absolute rest. Mass increases when the body is in motion according to

$$m = m_0 / \sqrt{1 - (v^2/c^2)}$$

where m is its mass in motion, m_0 is its rest mass; v is the body's speed of motion, and c is the speed of light.

Newtonian physics, in contrast with relativistic physics, makes no distinction between rest mass and mass in general.

Reverberation — 1. The persistence of sound in an enclosed space, as a result of multiple reflections after the sound source has stopped. 2. The sound that persists in an enclosed space, as a result of repeated reflection or scattering, after the source of the sound has stopped.

Reversible reaction — One which can be caused to proceed in either direction by suitable variation in the conditions of temperature, volume, pressure or of the quantities of reacting substances.

Rochon prism — A birefringent electro-optical crystal which divides incident unpolarized optical beam into two polarized components.

DEFINITIONS (Continued)

Roentgen (R) — That quantity of X or gamma radiation such that the associated corpuscular emission per 0.001293 g of dry air (equals 1 cc at 0°C and 760 mm Hg) produces, in air, ions carrying 1 esu of quantity of electricity of either sign.

Roentgen equivalent man (REM) — That amount of ionizing radiation of any type which produces the same damage to man as 1 roentgen of about 200-kv X radiation (1 rem = 1 rad in tissue/RBE). It should be noted that, when the physical dose is measured in rep units, the approximate definition is used: 1 rem ≈ 1 rep/RBE.

Roentgen equivalent physical (REP) — That amount of ionizing radiation of any type which results in the absorpion of energy at the point in question in soft tissue to the extent of 93 ergs/g. It is approximately equal to 1 roentgen of about 200-kv X radiation in soft tissue.

Root-mean-square error (symbol σ). In statistics, the square root of the arithmetic mean of the squares of the deviations of the various items from the arithmetic mean of the whole. Also termed standard deviation.

Rotatory power is the power of rotating the plane of polarized light, given in general by θ/l where θ is the total rotation which occurs in a distance l.
The molecular or atomic rotatory power is the product of the specific rotatory power by the molecular or atomic weight. Magnetic rotatory power is given by

$$\theta/e\, H\cos\alpha$$

where H the intensity of the magnetic field, and α the angle between the field and the direction of the light.

Rydberg formula — A formula, similar to that of Balmer, for expressing the wave-numbers (ν) of the lines in a spectral series:

$$V = R\left[\frac{1}{(n+a)^2} - \frac{1}{(m+b)^2}\right]$$

where n and m are integers and $m > n$, a and b are constants for a particular series, and R is the Rydberg constant, 109737.3 cm^{-1} for hydrogen.

Salt — Any substance which yields ions, other than hydrogen or hydroxyl ions. A salt is obtained by displacing the hydrogen of an acid by a metal.

Scale height (symbol h, h_s) — A measure of the relationship between density and temperature at any point in an atmosphere; the thickness of a homogeneous atmosphere which would give the observed temperature:

$$h = kT/mg = R^*T/Mg$$

where k is the Boltzmann constant; T is the absolute temperature; m and M are the mean molecular mass and weight, respectively, of the layer; g is the acceleration of gravity; and R* is the universal gas constant. Compare virtual height.

Scattered (light) — Reflection of light from a surface in all directions in a nonuniform manner.

Scattering Coefficient — A measure of the attenuation due to scattering of radiation as it traverses a medium containing scattering particles. Also called total scattering coefficient.

Scattering cross section — The hypothetical area normal to the incident radiation that would geometrically intercept the total amount of radiation actually scattered by a scattering particle. It is also defined, equivalently, as the cross-section area of an isotropic scatterer (a sphere) which would scatter the same amount of radiation as the actual amount. Also called extinction cross section, effective area.

Schlieren (photography) — A picture or image in which density gradients in a volume of flow are made visible. The image is produced by refraction and scattering from regions of changing refractive index.

Scintillation counter — The combination of phosphor, photomultiplier tube, and associated circuits for counting scintillations. Also called scintillating counter.

Scintillometer — An instrument which detects radiation by emitting flashes of light.

Second — The second is the unit of time of the International System of Units. The definition adopted at the October 13, 1967 meeting of the 13th General Conference on Weights and Measures is: "The second is the duration of 9 192 631 770 periods of the radiation corresponding to the transition between the two hyperfine levels of the ground state of the atom of cesium 133." The frequency 9 192 631 770 (Hz) which the definition assigns to the cesium radiation was carefully chosen to make it impossible, by any existing experimental evidence, to distinguish the new second from the "ephemeris second" based on the earth's motion. Therefore no changes need to be made in data stated in terms of the old standard in order to convert them to the new one. The atomic definition has two important advantages over the previous definition: (1) it can be realized (i.e., generated by a suitable clock) with sufficient precision, ± 1 part per hundred billion (10^{11}) or better, to meet the most exacting demands of modern metrology; and (2) it is available to anyone who has access to or who can build an atomic clock controlled by the specified cesium radiation. (A description of such clocks is given in "Atomic Frequency Standards," NBS Tech. News Bull. 45, 8—11 (Jan., 1961). For more recent developments and technical details, see R. E. Beehler, R. C. Mockler, and J. M. Richardson, "Cesium Beam Atomic Time and Frequency Standards," Metrologia 1, 114—131 (July, 1965)). In addition one can compare other high-precision clocks directly with such a standard in a relatively short time — an hour or so compared against years with the astronomical standard. Laboratory-type atomic clocks are complex and expensive, so that most clocks and frequency generators will continue to be calibrated against a standard such as the NBS Frequency Standard, controlled by a cesium atomic beam, at the Radio Standards Laboratory in Boulder, Colorado. In most cases the comparison will be by way of the standard-frequency and time-interval signals broadcast by NBS radio stations WWV, WWVH, WWVB, and WWVL.

Second law of thermodynamics — An inequality asserting that it is impossible to transfer heat from a colder to a warmer system without the occurrence of other simultaneous changes in the two systems or in the environment.
It follows from this law that during an adiabatic process, entropy cannot decrease. For reversible adiabatic processes entropy remains constant, and for irreversible adiabatic processes it increases.
Another equivalent formulation of the law is that it is impossible to convert the heat of a system into work without the occurrence of other simultaneous changes in the system or its environment. This version, which requires an engine to have a cold source as well as a heat source, is particularly useful in engineering applications. See first law of thermodynamics.

Second of arc — 1/60 of a minute of arc.

Seebeck effect — If a circuit consists of two metals, one junction hotter than the other, a current flows in the circuit. The direction of the flow depends on the metals and the temperature of the junctions.

Semiconductor — An electronic conductor, with resistivity in the range between metals and insulators, in which the electrical charge carrier concentration increases with increasing temperature over some temperature range. Certain semiconductors possess two types of carriers, namely, negative electrons and positive holes.

Sensitiveness of a balance — Assuming the three knife edges of a balance to lie on a straight line; if M is the weight of the beam, h the distance of the center of gravity below the knife edge, L the length of the balance arms and m a small mass added to one pan, the deflection θ produced is given by

$$\tan\theta = \frac{mL}{Mh}$$

Shear strength — "The stress, usually expressed in pounds per square inch, required to produce fracture when impressed perpendicularly upon the cross-section of a material."

Shell — According to Pauli's exclusion principle (q.v.), the extranuclear electrons do not circle around the nucleus all in orbits of the same radius, but are arranged in orbits at various distances from the nucleus. The extranuclear orbital electrons are thus assumed to be arranged in a series of concentric spheres, called shells, which are designated, in the order of increasing distance from the nucleus, as K, L, M, N, O, P, and Q shells. The number of the electrons which each of these shells can contain is limited. All electrons arranged in the same shell have the same principal quantum number. The electrons in the same shell are grouped into various subshells (q.v.), and all the electrons in the same subshell have the same orbital angular momentum. See Subshell.

Sidereal day. The duration of one rotation of the earth on its axis, with respect to the vernal equinox. It is measured by successive transits of the vernal equinox over the upper branch of a meridian.

Because of the precession of the equinoxes, the sidereal day thus defined is slightly less than the period of rotation with respect to the stars, but the difference is less than 0.01 sec. The length of the mean sidereal day is 24 hr of sidereal time or 23 hr 56 min 4.09054 sec of mean solar time.

Sidereal month — The average period of revolution of the moon with respect to the stars, a period of 27 days 7 hr 43 min 11.5 sec, or approximately $27\frac{1}{3}$ days.

Sidereal time — Time based upon the rotation of the earth relative to the vernal equinox.

Sidereal time may be designated as local or Greenwich as the local or Greenwich meridian is used as the reference. When adjusted for nutation, to eliminate slight irregularities in the rate, it is called mean sidereal time.

Sidereal year — The period of one apparent revolution of the earth around the sun, with respect to the stars, averaging 365 days 6 hr 9 min 9.55 sec in 1955, and increasing at the rate of 0.000095 sec annually.

Because of the precession of the equinoxes this is about 20 min longer than a tropical year.

Simple harmonic motion — Periodic oscillatory motion in a straight line in which the restoring force is proportional to the displacement. If a point moves uniformly in a circle, the motion of its projection on the diameter (or any straight line in the same plane) is simple harmonic motion.

If r is the radius of the reference circle, ω the angular velocity of the point in the circle, θ the angular displacement at the time t after the particle passes the midpoint of its path, the linear displacement,

$$x = r \sin \theta = r \sin \omega t$$

The velocity at the same instant,

$$v = r\omega \cos \theta = \omega\sqrt{r^2 - x^2}$$

The acceleration,

$$a = -\omega^2 x.$$

The force for a mass m,

$$F = m\omega^2 x = -\frac{4\pi^2 m x}{T^2}$$

The period,

$$T = 2\pi\sqrt{\frac{x}{a}}$$

In the above equations the cgs system calls for x and r in centimeters, v in centimeters per second, a in centimeters per second squared, T in seconds, M in grams, F in dynes, θ in radians and ω in radians per second.

Simple machine — A contrivance for the transfer of energy and for increased convenience in the performance of work.

Mechanical advantage is the ratio of the resistance overcome to the force applied. Velocity ratio is the ratio of the distance through which force is applied to the distance through which resistance is overcome.

Efficiency is the ratio of the work done by a machine to the work done upon it.

If a force f applied to a machine through a distance S results in a force F exerted by the machine through a distance s, neglecting friction,

$$fS = Fs$$

The theoretical mechanical advantage or velocity ratio in the above case is,

$$\frac{S}{s}$$

Actually, the force obtained from the machine will have a smaller value than will satisfy the equation above. If F′ be the actual force obtained, the practical mechanical advantage will be,

$$\frac{F'}{f}$$

The efficiency of the machine,

$$E = \frac{F's}{fS}$$

Snell's law of refraction — If i is the angle of incidence, r the angle of refraction, v the velocity of light in the first medium, v′ the velocity in the second medium, the index of refraction n,

$$n = \frac{\sin i}{\sin r} = \frac{v}{v'}$$

Solar constant — The rate at which solar radiation is received outside the earth's atmosphere on a surface normal to the incident radiation and at the earth's mean distance from the sun.

Measurements of solar radiation at the earth's surface by the Smithsonian Institution for several decades give a best value for the solar constant of 1.934 calories per square centimeter per minute. Measurements from rockets of the intensity of the ultraviolet end of the spectrum have corrected this value to 2.00 calories per square centimeter per minute with a probable error of ± 2%.

Solar wind — Streams of plasma flowing approximately radially outward from the sun.

Solid — A state of matter in which the relative motion of the molecules is restricted and they tend to retain a definite fixed position relative to each other, giving rise to crystal structure. A solid may be said to have a definite shape and volume.

Solid angle — Measured by the ratio of the surface of the portion of a sphere enclosed by the conical surface forming the angle, to the square of the radius of the sphere. Unit of solid angle, the steradian, the solid angle which encloses a surface on the sphere equivalent to the square of the radius. Dimensions, unity.

Solubility of one liquid or solid in another is the mass of a substance contained in a solution which is in equilibrium with an excess of the substance. Under these conditions the solution is said to be saturated. Solubility of a gas is the ratio of concentration of gas in the solution to the concentration of gas above the solution.

Solubility product or precipitation value is the product of the concentration of the ions of a substance in a saturated solution of the substance. These concentrations are frequently expressed as moles of solute per liter of solution.

Solute — That constituent of a solution which is considered to be dissolved in the other, the solvent. The solvent is usually present in larger amount than the solute.

A solution is saturated if it contains at given temperature as much of a solute as it can retain in the presence of an excess of that solute.

A true solution is a mixture, liquid, solid or gaseous, in which the components are uniformly distributed throughout the mixture. The proportion of the constituents may be varied within certain limits.

Solvent — That constituent of a solution which is present in larger amount; or, the constituent which is liquid in the pure state, in the case of solutions or solids or gases in liquids.

Sone — A unit of loudness. A simple tone of frequency 1000 cycles/sec, 40 decibels above a listener's threshold, produces a loudness of 1 sone.

Spatial frequency plane — The focusing plane of an optical lens or system where the image represents the spatial Fourier transform of the object spatial function.

Specific conductance, σ_{sp} — The specific conductance, or conductivity, of a conductor of electricity is the conductance of the material between opposite sides of a cube, 1 cm in each direction. The unit of specific conductance is ohm^{-1} cm^{-1} or mho cm^{-1}.

Specific gravity — The ratio of the mass of a body to the mass of an equal volume of water at 4°C or other specified temperature. Dimensions, unity.

Specific heat of a substance is the ratio of its thermal capacity to that of water at 15°C. Dimensions, unity.

If a quantity of heat H calories is necessary to raise the temperature of m grams of a substance from t_1 to t_2°C, the specific heat, or more properly, thermal capacity of the substance,

$$s = \frac{H}{m(t_2 - t_1)}$$

Specific heat by the method of mixtures — Where a mass m_1 of the substance is heated to a temperature t_1, then placed in a mass of water m_2 t a temperature t_2 contained in a calorimeter with stirrer (of same material) of mass m_3, specific heat of the calorimeter c, t_3 the final temperature

$$m_1 s(t_1 - t_3) = (m_3 c + m_2)(t_3 - t_2)$$

Black's ice calorimeter — If a body of mass m and temperature t melts a mass m' of ice, its temperature being reduced to 0°C, the specific heat of the substance is,

$$s = \frac{80.1\, m'}{mt}$$

Bunsen's ice calorimeter — A body of mass m at temperature t causes a motion of the mercury column of l centimeters in a tube whose volume per unit length is v. The specific heat is

$$s = \frac{884 lv}{mt}$$

Specific inductive capacity — The ratio of the capacitance of a condenser with a given substance as dielectric to the capacitance of the same condenser with air or a vacuum as dielectric is called the specific inductive capacity. The ratio of the dielectric constant of a substance to that of a vacuum.

Specific rotation — If there are n grams of active substance in v cubic centimeters of solution and the light passes through l decimeters, r being the observed rotation in degrees, the specific rotation (for 1 centimeter),

$$[\alpha] = \frac{rv}{nl}$$

Specific volume is the reciprocal of density. Dimensions $[m^{-1} l^3]$.

Spectral series are spectral lines or groups of lines which occur in an orderly sequence.

Specular reflection — Mirror scattering at one angle from the point of reflection.

Speed — Time rate of motion measured by the distance moved over in unit time. Cgs unit, 1 cm/sec. Dimension $[l\, t^{-1}]$.

Speed of sound (symbol c_s) — The speed of propagation of sound waves. In the atmosphere

$$c_t = [\gamma\, (R^*/M_0)\, T_M]^{1/2}$$

where γ is the ratio of specific heat of air at constant pressure to that a constant volume, R^* is the universal gas constant, M_o is the mean molecular weight of air at sea level, and T_M is the molecular scale temperature.

At sea level in the standard atmosphere, the speed of sound is 340.294 meters per second (1116.45 ft/sec).

The concept of the speed of sound in the atmosphere loses its applicability at about 90 km where the mean free path of air molecules approaches the wavelengths of sound waves.

Spherical aberration — When large surfaces of spherical mirrors or lenses are used the light divergent from a point source is not exactly focused at a point. The phenomenon is known as spherical aberration. For axial pencils the error is known as axial spherical aberration; for oblique pencils, coma.

Spherical mirrors — If R is the radius of curvature, F principal focus, and f_1 and f_2 any two conjugate focal distances,

$$\frac{1}{f_1} + \frac{1}{f_2} = \frac{1}{F} = \frac{2}{R}$$

If the linear dimensions of the object and image be O and I respectively and u and v their distances from the mirror,

$$\frac{O}{I} = \frac{u}{v}$$

Spin — In nuclear physics, used to describe the angular momentum of elementary particles or of nuclei.

Spontaneous-ignition temperature — In testing fuels, the lowest temperature of a plate or other solid surface adequate to cause ignition in air of a fuel upon the surface.

Stagnation pressure — 1. The pressure at a stagnation point. 2. In compressible flow, the pressure exhibited by a moving gas or liquid brought to zero velocity by an isentropic process. 3. Equals total pressure. 4. Equals impact pressure.

Because of the lack of a standard meaning, stagnation pressure should be defined when it is used.

Standard conditions for gases — Measured volumes of gases are quite generally recalculated to 0°C temperature and 760 mm pressure, which have been arbitrarily chosen as standard conditions.

Stark effect — The splitting of a single spectrum line into multiple lines which occurs when the emitting material is placed in a strong electric field. The observed effect depends on the angle between the direction of the field and the direction of observation. The effect is due to the shifting of the energy states of certain orbits which all have the same energy in zero field.

Statcoulomb — The unit of electric charge in the metric system. 3×10^9 statcoulombs = 1 coulomb.

Static pressure (symbol p) — 1. The pressure with respect to a stationary surface tangent to the mass-flow velocity vector. 2. The pressure with respect to a surface at rest in relation to the surrounding fluid.

Stationary or standing waves are produced in a medium by the simultaneous transmission, in opposite directions of two similar wave motions. Fixed points of minimum amplitude are called nodes. A segment extends from one node to the next. An antinode or loop is the point of maximum amplitude between two nodes.

Statute mile — 5280 feet = 1.6093 kilometers = 0.869 nautical mile. Also called land mile.

Stefan-Boltzmann constant (symbol σ) — A universal constant of proportionality between the radiant emittance of a black body and the fourth power of the body's absolute temperature; 5.6697×10^{-5} erg per centimeter squared second °K⁴.

Stefan-Boltzmann law — A law stating that the total radiation E from a black body is given by

$$E = \sigma\, T^4$$

where T is the absolute temperature of the body and σ is the Stefan-Boltzmann constant, equal to 5.672×10^{-5} erg/sec cm² deg⁴.

Steradian — The steradian is the SI unit of solid angle.

The steradian is the solid angle which, having its vertex in the center of a sphere, cuts off an area of the surface of the sphere equal to that of a square with sides of length equal to the radius of the sphere. There are 4π steradians in a sphere.

Stereoscopic image — An image which appears as a three-dimensional object located in space.

Stoichiometric — Pertaining to weight relations in chemical reactions.

Stoke — See under Viscosity.

Stokes' law — 1. Gives the rate of fall of a small sphere in a viscous fluid. When a small sphere falls under the action of gravity through a viscous medium it ultimately acquires a constant velocity,

$$V = \frac{2ga^2(d_1 - d_2)}{9\eta}$$

where a is the radius of the sphere, d_1 and d_2 the densities of the sphere and the medium respectively, and η the coefficient of viscosity. V will be in cm per sec if g is in cm per sec², a in cm, d_1 and d_2 in g per cm³ and η in dyne-sec per cm² or poises.

2. The empirical law stating that the wavelength of light emitted by a flu-

DEFINITIONS (Continued)

orescent material is longer than that of the radiation used to excite the fluorescence. In modern language the emitted photons carry off less energy than is brought in by the exciting photons; the details accord with the energy conservation principle.

Strain — The deformation resulting from a stress measured by the ratio of the change to the total value of the dimension in which the change occurred. Dimensions, unity.

Stress — The force producing or tending to produce deformation in a body measured by the force applied per unit area Cgs units, one dyne per square centimeter. Dimensions, $[m\ l^{-1}\ t^{-2}]$.

Subshell — The electrons within the same shell (energy level) of the atom are characterized by the same principal quantum number (n), and are further divided into groups according to the value of their azimuthal quantum numbers (l); the electrons which possess the same azimuthal quantum number for the same principal quantum number are considered to occupy the same subshell (or sublevel). The individual subshells are designated with the letters s, p, d, f, g, and h, as follows:

l value	designation of subshell
0	s
1	p
2	d
3	f
4	g
5	h

An electron assigned to the s-subshell is called an s-electron, one assigned to the p-subshell is referred to as a p-electron, etc. In formulae of electron structure, the value of the principal quantum number (n) is prefixed to the letter indicating the azimuthal quantum number (l) of the electron; thus, e.g., a 4f-electron is an electron which has the principal quantum number 4 (i.e., assigned to the N-shell) and the orbital angular momentum 3 (f-subshell).

Substance, amount of — See Mole.

Supersonic flow — In aerodynamics, flow of a fluid over a body at speeds greater than the acoustic velocity and in which the shock waves start at the surface of the body. Compare hypersonic flow.

Surface density of electricity — Quantity of electricity per unit area. Dimensions, $[\epsilon^{1/2}\ m^{1/2}\ l^{-1/2}\ t^{-1}]$; $[\mu^{-1/2}\ m^{1/2}\ l^{-1/2}]$.

Surface density of magnetism — Quantity of magnetism per unit area. Dimensions, $[\epsilon^{-1/2}\ m^{1/2}\ l^{-1/2}]$, $[\mu^{1/2}\ m^{1/2}\ l^{-1/2}\ t^{-1}]$.

Surface tension — Two fluids in contact exhibit phenomena, due to molecular attractions which appear to arise from a tension in he surface of separation. It may be expressed as dynes per centimeter or as ergs per square centimeter. Dimensions, $[m\ t^{-2}]$.
The total force along a line of length l on the surface of a liquid whose surface tension is T,

$$F = lT.$$

Capillary tubes — If a liquid of density d rises a height h in a tube of internal radius r the surface tension is,

$$T = \frac{rhdg}{2}$$

The tension will be in dynes per cm if r and h are in cm, d in g per cm³ and g in cm per sec².
Drops and bubbles — Pressure in dynes per cm² due to surface tension on a drop of radius r centimeter for a liquid whose surface tension is T dynes per centimeter,

$$P = \frac{2T}{r}$$

For a bubble of mean radius r cm,

$$P = \frac{4T}{r}$$

Susceptibility (magnetic) is measured by the ratio of the intensity of magnetization produced in a substance to the magnetizing force or intensity of field to which it is subjected. The susceptibility of a substance will be unity when unit intensity of magnetization is produced by a field of one gauss. Dimensions, $[\epsilon^{-1}\ l^{-2}\ t^2]$, $[\mu]$.

"Synthetic" aperture" sidelooking radar — A sidelooking radar that generates very high resolution data by integrating its return signals during the time that the physical aircraft antenna or aperture is traveling through a large distance (making up the synthetic aperture).

Tangent galvanometer — A tangent galvanometer with n turns, of radius r, in the earth's field H, has a deflection θ. The current flowing is,

$$i = \frac{Hr}{2\pi n}\ \tan\theta$$

If $2\pi n = G$ (the galvanometer constant).

$$i = \frac{H}{G}\ \tan\theta$$

Tektite — Small glassy bodies containing no crystals, composed of at least 65% silicon dioxide, bearing no relation to the geological formations in which they occur, and believed to be of extraterrestrial origin.
Tektites are found in certain large areas called strewn fields. They are named, as are minerals, with the suffix ite, as australite, found in Australia, billitonite, indochinite, and rizalite, found in Southeast Asia, bediasite from Texas, and moldavite from Bohemia and Moravia.

Temperature may be defined as the condition of a body which determines the transfer of heat to or from other bodies. Particularly it is a manifestation of the average translational kinetic energy of the molecules of a substance due to heat agitation. The customary unit of temperature is the Centigrade degree, 1/100 the difference between the temperature of melting ice and that of water boiling under standard atmospheric pressure. Celsius scale — the Celsius temperature scale is a designation of the scale also known as the centigrade scale. The degree Fahrenheit is 1/180, and the degree Reaumur 1/80 the same difference of temperature.
The fundamental temperature cale is the absolute, thermodynamic or Kelvin scale in which the temperature measure is based on the average kinetic energy per molecule of a perfect gas. The zero of the Kelvin scale is −273.15°C. The temperature scale adopted by the International Bureau of Weights and Measures is that of the constant volume hydrogen gas thermometer. The magnitude of the degree in both these scales is defined as 1/100 the difference between the temperature of melting ice and that of boiling water at 760 mm pressure. Frequently the Kelvin scale is defined as degrees C + 273.15 and the Rankine scale as degrees F + 459.67°F.
The fundamental temperature scale is now defined by means of the equation

$$\theta(X) = 273.15°K\ \frac{X}{X_3}$$

where θ denotes the temperature; X the thermometric property (P, V, ...); the subscript 3 refers to the triple point of water; and 273.16°K is the arbitrary fixed point for the temperature associated with the triple point of water.
The ideal gas temperature θ, (numerically equal to the Kelvin temperature), in particular, is defined by either of the two equations:

$$\theta = \begin{cases} 273.15° \lim_{P_3 \to 0} \dfrac{P}{P_3}, \text{const. V} \\[2ex] 273.15° \lim_{P \to 0} \dfrac{V}{V_3}. \text{const. P} \end{cases}$$

Temperature resistance coefficient — The ratio of the change of resistance in a wire due to a change of temperature of 1°C to its resistance at 0°C. Dimension, $[\theta-1]$.

Tera — A prefix meaning trillion or 10^{12}. Symbol is T.

Terrestrial radiation — 1. The total infrared radiation emitted from the earth's surface; to be carefully distinguished from effective terrestrial radiation, atmospheric radiation (which is sometimes erroneously used as a synonym for terrestrial radiation), and insolation. Also called earth radiation, eradiation.

Thermal capacity of a substance is the quantity of heat necessary to produce unit change of temperature in unit mass. It is ordinarily expressed as calories per gram per degree Centigrade. Numerically equivalent to specific heat.

F-106

DEFINITIONS (Continued)

Thermal capacity or water equivalent — The total quantity of heat necessary to raise any body or system unit temperature, measured as calories per degree centigrade in the cgs system. Dimension, [m].

Thermal expansion — The coefficient of linear expansion or expansivity is the ratio of the change in length per degree C to the length at 0°C. The coefficient of volume expansion (for solids) is approximately three times the linear coefficient. The coefficient of volume expansion for liquids is the ratio of the change in volume per degree to the volume at 0°C. The value of the coefficient varies with temperature. The coefficient of volume expansion for a gas under constant pressure is nearly the same for all gases and temperatures and is equal to 0.00367 for 1°C. Dimension, $[\theta^{-1}]$.

If l_0 is the length at 0°C, α the coefficient of linear expansion, the length at t°C is,

$$l_t = l_0 (1 + \alpha t)$$

General formula for thermal expansion — The rate of thermal expansion varies with the temperature. The genral equation giving the magnitude m, (length or volume) at a temperature t, where m_0 is the magnitude at 0° C, is

$$m_t = m_0 (1 + \alpha t + \beta t^2 + \gamma t^3 \ldots)$$

where α, β, γ, etc. are empirically determined coefficients.

Volume expansion — If V represents volume and β the coefficient of expansion,

$$V_t = V_0 (1 + \beta t).$$

For solids,

$$\beta = 3\alpha \text{ (approximately)}.$$

Thermal neutrons — Neutrons slowed down by a moderator to an energy of a fraction of an electron volt — about 0.025 ev. at 15°

Thermionic emission — The emission of ions or electrons from a metal as the result of heating. The thermionic current density, J, in amperes/cm² is equal to $AT_2 e^{-\phi/kT}$ where A is a constant, T is degrees Kelvin, k is Boltzmann's constant, and ϕ is the work function of the emitter.

Thermodynamic efficiency — In thermodynamics, the ratio of the work done by a heat engine to the total heat supplied by the heat source. Also called thermal efficiency, Carnot efficiency.

Thermodynamic probability — Under specified conditions, the number of equally likely states in which a substance may exist. The thermodynamic probability P is related to the entropy S by

$$S = k \ln P$$

where k is Boltzmann constant. See Third law of thermodynamics.

Thermodynamic temperature scale — The Kelvin thermodynamic scale is recognized as the fundamental scale to which all temperatures should ultimately be referable. The kelvin, unit of thermodynamic temperature, is the fraction 1/273.16 of the thermodynamic temperature of the triple point of water. The unit kelvin and its symbol K are used to express an interval or a difference of temperature. In addition to the thermodynamic temperature (symbol T), expressed in kelvins, use is also made of Celsius temperature (symbol t) defined by the equation

$$t = T - T_0$$

where T_0 = 273.15 K by definition. The unit "degree Celsius" is equal to the unit "kelvin", but "degree Celsius" is a special name in place of "kelvin" for expressing Celsius temperature. A temperature interval or a Celsius temperature difference can be expressed in degrees Celsius as well as in kelvins.

Thermodynamics, law of —
I. When mechanical work is transformed into heat or heat into work, the amount of work is always equivalent to the quantity of heat.
II. It is impossible by any continuous self-sustaining process for heat to be transferred from a colder to a hotter body.

Thermoelectric power is measured by the electromotive force produced by a thermocouple for unit difference of temperature between the two junctions. It

varies with the average temperature and is usually expressed in microvolts per degree C. It is customary to list the thermoelectric power of the various metals with respect to lead. Dimensions, $[\epsilon^{-1/2} m^{1/2} t^{-1} \theta^{-1}]$; $[\mu^{1/2} m^{1/2} l^{1/2} t^{-1} \theta^{-1}]$.

Thickness or average radius of ion atmosphere, κ^{-1} — The average distance of the ion atmosphere from the reference ion in angstrom units. This average distance decreases in magnitude with the square root of the ionic concentration. Mathematically, K^{-1} is the distance at which the average charge, dq, in a spherical shell of volume $4\pi r^2 dr$ reaches a maximum using the continuous density, ϱ (r), approximation.

Third law of thermodynamics — The statement that every substance has a finite positive entropy, and that the entropy of a crystalline substance is zero at the temperature of absolute zero. See Thermodynamic probability.

Modern quantum theory has shown that the entropy of crystals at 0° absolute is not necessarily zero. If the crystal has any asymmetry, it may exist in more than one state; and there is, in addition, an entropy residue deriving from nuclear spin.

Thomson thermoelectric effect is the designation of the potential gradient along a conductor which accompanies a temperature gradient. The magnitude and direction of the potential varies with the substance.

The coefficient of the Thomson effect or specific heat of electricity is expressed in joules per coulomb per degree Centigrade. Dimensions, $[\epsilon^{-1/2} m^{1/2} l^{1/2} t^{-1} \theta^{-1}]$, $[\mu^{1/2} m^{1/2} l^{1/2} t^{-2} \theta^{-1}]$.

Time, unit of — The fundamental invariable unit of time is the ephemeris second, which is defined as 1/31,556,925.9747 of the tropical year for 1900 January $0^d 12^h$ ephemeris time. The ephemeris day is 86,400 ephemeris seconds.

The former unit of time was the mean solar second, defined as 1/86,400 of the mean solar day.

Torque prouced by the action of one magnet on another — The turning moment experienced by a magnet of pole strength m′ and length 2l′ placed at a distance r from another magnet of length 2l and pole strength m, where the center of the first magnet is on the axis (extended) of the second and the axis of the first is perpendicular to the axis of the second,

$$C = 8 \; \frac{mm'll'}{r^3} = \frac{2MM'}{r^3}$$

If the first magnet is deflected through an angle θ, the expression becomes,

$$C = \frac{2MM'}{r^3} \cos \theta$$

Torr — Provisional international standard term to replace the English term millimeter of mercury and its abbreviation mm of Hg (or the French mm de Hg).

The torr is defined as 1/760 of a standard atmosphere or 1,013,250/760 dynes per square centimeter. This is equivalent to defining the torr as 1333.22 microbars and differs by only one part in 7 million from the International Standard millimeter of mercury. The prefixes milli and micro are attached without hyphenation.

Torsional vibration — See Angular harmonic motion.

Total pressure — 1. Equals stagnation pressure. 2. Equals impact pressure. 3. The pressure a moving fluid would have if it were brought to rest without losses. 4. The pressure determined by all the molecular species crossing the imaginary surface.

Total reflection — When light passes from any medium to one in which the velocity is greater, refraction ceases and total reflection begins at a certain critical angle of incidence θ such that

$$\sin \theta = \frac{1}{n}$$

where n is the index of the first medium with respect to the second. If the second medium is air n has the ordinary value for the first medium. For any other second medium,

$$n = \frac{n_1}{n_2}$$

where n_1 and n_2 are the ordinary indices of refraction for the first and second medium respectively.

Tractive force of a magnet — If a magnet with induction B has a pole face of area A the force is,

$$F = \frac{B^2 A}{8\pi}$$

If B and A are in cgs units, F will be in dynes.

Transducer (acoustic) — A device to convert electrical oscillatory input energy into mechanical acoustical wave energy.

Transference (or transport) number — The transference number of each ion of a solute (or electrolyte) in an electrolytic solution is the fraction of the total current carried by that ion, and is given by the ratio of the mobility of the ion to the sum of the mobilities of the ions of the solute constituting the electrolytic solution.

Transmutation — A nuclear change producing a new element from an old one.

Transuranic elements — Elements of atomic numbers above 92. All of them are radioactive and are products of artificial nuclear changes. All are members of the actinide group.

Transverse electromagnetic energy — Wave whose electric field and magnetic field displacements are at right angles to each other and to the direction of propagation (motion) of the wave.

Triangle or polygon of forces — If three or more forces acting on the same point are in equilibrium, the vectors representing them form, when added, a closed figure.

Triple point — The thermodynamic state at which three phases of a substance exist in equilibrium.

The triple point of water occurs at a saturation vapor pressure of 6.11 millibar and at a temperature of 273.16° K.

Tritium — An isotope of hydrogen with a mass of three, structure, two neutrons, and one proton in its nucleus.

Ultrasound camera — A camera that converts a sound pressure "acoustic image" into an electrical TV-like image by means of the piezo-electric effect.

Uncertainty principle — See Indeterminancy.

Uniform circular motion — If r is the radius of a circle, v the linear speed in the arc, ω the angular velocity and T the period or time of one revolution,

$$\omega = \frac{v}{r} = \frac{2\pi}{T}$$

The acceleration toward the center is

$$a = \frac{v^2}{r} = \omega^2 r = \frac{4\pi^2 r}{T^2}$$

The centrifugal force for a mass m,

$$F = \frac{mv^2}{r} = m\omega^2 r = \frac{4\pi^2 mr}{T^2}$$

In the above equations ω will be in radians per second and a in cm per sec² if r is in cm, v in cm per sec and T in sec. F will be in dynes if mass is in grams and other units as above.

Application to the solar system — If M is the mass of the sun, G the constant of gravitation, P the period of the planet and r the distance of the planet from the sun, then the mass of the sun

$$M = \frac{4\pi^2 r^3}{GP^2}, \ (G = 6.670 \times 10^{-8} \text{ for cgs units})$$

If P is the period and r the distance of a satellite revolving around the planet, the above expression for M gives the mass of the planet. The formula is written on the assumption that the orbit of the planet or satellite is circular, which is only approximately true.

Uniformly accelerated rectilinear motion — If v_o is the initial velocity, v, the velocity after time t, the acceleration,

$$a = \frac{v_t - v_0}{t}$$

The velocity after time t,

$$v_t = v_0 + at$$

Space passed over in time t,

$$s = v_0 t + \tfrac{1}{2}at^2$$

Velocity after passing over space s,

$$v = \sqrt{v_0^2 + 2as}$$

Space passed over in the nth second

$$s = v_0 + \tfrac{1}{2}a(2n - 1)$$

In the above and following similar equations the values of the space, velocity, and acceleration must be substituted in the same system. For space in cm, velocity will be in centimeter per second and acceleration in centimeter per second per second.

Unit — Specific magnitude of a quantity, set apart by appropriate definition, which is to serve as a basis of comparison or measurement for other quantities of the same nature.

Universal time — Time defined by the rotational motion of the earth and determined from the apparent diurnal motions which reflect this rotation; because of variations in the rate of rotation, universal time is not rigorously uniform. Also called Greenwich mean time. Compare ephemeris time.

In the years preceding 1960 the arguments of the ephemerides in the American Ephemeris and Nautical Almanac were designated as universal time.

Valence of an atom of an element is that property which is measured by the number of atoms of hydrogen (or its equivalent) one atom of that element can hold in combination if negative, or can displace in a reaction if it is positive.

Valence electrons of the atom are electrons which are gained, lost or shared in chemical reactions.

Van Allen belt, Van Allen radiation belt — The zone of high-intensity particulate radiation surrounding the earth beginning at altitudes of approximately 1000 km.

The radiation of the Van Allen belt is composed of protons and electrons temporarily trapped in the earth's magnetic field. The intensity of radiation varies with the distance from the earth.

Van der Waals' equation of state — This equation is expressed by:

$$\left(p + \frac{a}{v^2}\right)(v - b) = RT$$

It makes allowance both for the volume occupied by the molecules and for the attractive force between the molecules. b is the effective volume of molecules in one mole of gas. a is a measure of the attractive force between the molecules. For values of R, a, and b see index for table of Van der Waals' constants for gases.

Van't Hoff's principle — If the temperature of interacting substances in equilibrium is raised, the equilibrium concentrations of the reaction are changed so that the products of that reaction which absorb heat are increased in quantity, or if the temperature for such an equilibrium is lowered, the products which evolve heat in their formation are increased in amounts.

Vapor — The words vapor and gas are often used interchangeably. Vapor is more frequently used for a substance which, though present in the gaseous phase, generally exists as a liquid or solid at room temperature. Gas is more frequently used for a substance that generally exists in the gaseous phase at room temperature. Thus one would speak of iodine or carbon tetrachloride vapors and of oxygen gas.

Vapor pressure — The pressure exerted when a solid or liquid is in equilibrium with its own vapor. The vapor pressure is a function of the substance and of the temperature.

DEFINITIONS (Continued)

Vectors, composition of — If the angle between two vectors is A, and their magnitude a and b, their resultant,

$$C = \sqrt{a^2 + b^2 - 2ab \cos A}.$$

Velocity — Time rate of motion in a fixed direction. Cgs units, one centimeter per second. Dimensions, $[l\, t^{-1}]$.

If s is space passed over in time t, the velocity,

$$\bar{v} = \frac{s}{t}$$

Velocity of a compressional wave — The velocity of a compressional wave in an elastic medium, in terms of elasticity E (bulk modulus) and density d,

$$V = \sqrt{\frac{E}{d}}$$

For the velocity of sound in air, where p is the pressure and d the density,

$$V = \sqrt{\frac{1.4 p}{d}}$$

Velocity of efflux of a liquid — If h is the distance from the opening to the free surface of the liquid, the velocity of efflux is

$$V = \sqrt{2gh}$$

The above is the theoretical discharge velocity disregarding friction and the shape of orifice. For water issuing through a circular opening with sharp edges of area, A, the volume discharged per second is given approximately by,

$$Q = 0.624 \sqrt{2gh}$$

Velocity of sound, variation with temperature — The velocity in meters per sec at any temperature t in °C is given approximately by

$$V = V_0 \sqrt{1 + \frac{t}{273}}$$

$$V = 331.5 + 0.607t$$

The variation with humidity is given by the equation

$$V_d = V_h \sqrt{1 - \frac{e}{p} \cdot \frac{\gamma \omega}{\gamma a} - \frac{5}{8}}$$

where V_d is the velocity in dry air, V_h that in air at barometric pressure p in which the pressure of water vapor is e. γ_ω and γ_a are the specific heat ratios for water vapor and for air respectively.

Velocity of a transverse wave is a stretched cord. If T is the tension of the cord and m the mass per unit length,

$$V = \sqrt{\frac{T}{m}}$$

Velocity of water waves — If the depth h is small compared with the wave length, the velocity,

$$V = \sqrt{gh}$$

In deep water for a wave length λ,

$$V = \sqrt{\frac{g\lambda}{2\pi}}$$

If the wavelength is very small, less than about 1.6 cm, the velocity increases as the wave length decreases and is expressed by the following,

$$V = \sqrt{\frac{2\pi T}{\lambda d} + \frac{g\lambda}{2\pi}}$$

where T is the surface tension and d the density of the liquid V will be given in cm per sec if h and λ are in cm, g in cm per sec², T in dynes per cm and d in g per cm³.

Velocity of a wave — The velocity of propagation in terms of wavelength λ and the period T or frequency n is,

$$V = \frac{\lambda}{T} = n\lambda$$

Virtual gravity — The force of gravity on an atmospheric parcel, reduced by centrifugal force due to the motion of the parcel relative to the earth. The virtual gravity g* is

$$g^* = g = V^2/a - 2\Omega_n V$$

where g is the magnitude of the acceleration of gravity; V is the parcel speed; a is the earth's radius; and Ω_n is the component of the earth's angular velocity vector normal to the motion of the parcel.

For reasonable atmospheric values, the correction terms are of the order of 0.01% of the magnitude of gravity. The identity of g* and g is implied by the assumption of hydrostatic equilibrium.

Virtual height — The apparent height of an ionized atmospheric layer determined from the time interval between the transmitted signal and the ionospheric echo at vertical incidence, assuming that the velocity of propagation is the velocity of light in a vacuum over the entire path. See Ionospheric recorder. Compare scale height.

Virtual image — An image that cannot be shown on a surface but is visible, as in a mirror.

Virtual mass — The actual mass of a body, plus its apparent additional mass.

Viscosity — All fluids possess a definite resistance to change of form and many solids show a gradual yielding to forces tending to change their form. This property, a sort of internal friction, is called viscosity; it is expressed in dyne-seconds per cm² or poises. Dimensions, $[m\, l^{-1}\, t^{-1}]$. If the tangential force per unit area, exerted by a layer of fluid upon one adjacent is one dyne for a space rate of variation of the tangential velocity of unity, the viscosity is one poise.

Kinematic viscosity is the ratio of viscosity to density. The c. g. s. unit of kinematic viscosity is the stoke.

Flow of liquids through a tube; where l is the length of the tube, r its radius, p the difference of pressure at the ends, η the coefficient of viscosity, the volume escaping per second,

$$v = \frac{\pi p r^4}{8 l \eta} \text{ (Poiseuille).}$$

The volume will be given in cm³ per second if l and r are in cm, p in dynes per cm² and η in poises or dyne-seconds per cm².

Viscosity effect — An alteration in the velocity of a given ion as a result of the contribution to the bulk viscosity owing to the ions of opposite charge. This effect applies to ions of large size.

Visibility is measured by the ratio of the luminous flux in lumens to the total radiant energy in ergs per second or in watts.

Volt — The unit of electric potential difference and electromotive force, equal to the difference of electric potential between two points of a conductor carrying a constant current of 1 ampere when the power dissipated between these points equals 1 watt.

Volume, unit of — The cubic centimeter, the volume of a cube whose edges are one centimeter in length. Other units of volume are derived in a similar manner. Dimension, $[l^3]$.

Volume velocity — Volume velocity is the rate of alternating flow of the medium through a specified surface due to a sound wave.

Walden's rule, $\Lambda_o \eta_o$ — Walden's rule states that the product of the limiting equivalent conductance of an electrolytic solution, Λ_o, and the viscosity of the solvent, η_o, in which the solute (or electrolyte) is dissolved is a constant at a particular temperature. Walden's rule is an approximation which would be valid only for ions which behave hydrodynamically like Stokes spheres in a continuum.

DEFINITIONS (Continued)

Watt — Rate of doing work or expending power. A watt is that power which gives rise to the production of energy at a rate of 1 joule per second or 10^7 ergs per second.

Wave — A wave is a disturbance which is propagated in a medium in such a manner that at any point in the medium the quantity serving as measure of disturbance is a function of the time while at any instant the displacement at a point is a function of the position of the point.

Wavefront — Surface whose points in an energy beam are all of equal phase or optical path length.

Wavelength — Distance between successive wavefronts of like phase (i.e., from peak to peak, or trough to trough).

Wave motion — A progressive disturbance propagated in a medium by the periodic vibration of the particles of the medium. Transverse wave motion is that in which the vibration of the particles is perpendicular to the direction of propagation. Longitudinal wave motion is that in which the vibration of the particles is parallel to the direction of propagation.

Wave number (symbol ν) — The reciprocal of wavelength; the number of waves per unit distance in the direction of propagation; or, sometimes 2π times this quantity.

In spectroscopy, wave number is usually expressed in reciprocal centimeters, as 100,000 cm^{-1} (100,000 per centimeter).

Weber (unit of magnetic flux) — The magnetic flux which, linking a circuit of one turn, produces in it an electromotive force of 1 volt as it is reduced to zero at a uniform rate in 1 second.

Weight — The force with which a body is attracted toward the earth. Cgs unit, — the dyne. Dimensions, $[m\, l\, t^{-2}]$.

Although the weight of a body varies with its location, the weights of various standards of mass are often used as units of force as, pound weight, or pound force, gram weight, etc. The weight of mass m, where g is the acceleration due to gravity,

$$W = mg$$

The weight will be given in dynes when m is in grams and g in centimeter per second squared.

Wheatstone's bridge — If the resistances r_1, r_2, r_3, and r_4 form the arms of a Wheatstone's bridge in order as the circuit (omitting cell and galvanometer connections) is traced, when the bridge is balanced,

$$\frac{r_1}{r_2} = \frac{r_4}{r_3} \qquad \text{or} \qquad \frac{r_1}{r_4} = \frac{r_2}{r_3}$$

Wien distribution law — A relation, derived on purely thermodynamic reasoning by Wien, between the monochromatic emittance of an ideal black body and that body's temperature.

$$J\lambda/T^5 = f(\lambda, T)$$

where J_λ is the monochromatic emittance (emissive power) of a black body at wavelength λ and absolute temperature T, and $f(\lambda, T)$ is a function which cannot be determined purely on classical thermodynamic grounds. Conpare Wien law.

Wien effect — The increase in the conductance of an electrolytic solution produced by high electrical fields (potential gradients).

Wien's displacement law — When the temperature of a radiating black body increases, the wave length corresponding to maximum energy decreases in such a way that the product of the absolute temperature and wave length is constant.

$$\lambda_{max}\, T = w$$

w is known as Wien's displacement constant.

Work — When a force acts against resistance to produce motion in a body the force is said to do work. Work is measured by the product of the force acting and the distance moved through against the resistance. Cgs units of work, the erg, a force of one dyne acting through a distance of one centimeter. The joule is 1×10^7 ergs. Dimensions, $[m\, l^2\, t^{-2}]$. The foot-pound is the work required to raise a mass of one pound a vertical distance of one foot where g = 32.174 ft/sec². The foot-poundal is the work done by a force of one poundal acting through a distance of one foot. The International joule, a unit of electrical energy, is the work expended per second by a current of one International ampere flowing through one International ohm. The kilowatt-hour is the total amount of energy developed in one hour by a power of one kilowatt.

If a force F act through a space s, the work done is

$$W = Fs$$

Work will be given in ergs if F is in dynes and s in centimeters.

Work done in rotation. If a torque L dyne-cm acts through an angle θ radians, the work done in ergs is

$$W = L\theta$$

Work function — The energy required for an electron to escape a solid surface. See Helmholtz function.

X-rays — A type of radiation of higher frequency than visible light but lower than gamma rays. Usually produced by high energy electrons impinging upon a metal target.

X units — X-ray wavelengths have been measured in two kinds of units. The older measurements are given in X units (XU) which are based on the effective lattice constant of rock salt being 2,814.00 XU. More recently X-ray wavelengths have been directly connected, through measurements with ruled gratings, to the wavelengths in the optical region and through them to the standard meter. It turned out that the XU which was originally intended as 10^{-11} cm was 0.202% larger than this value. It has become customary to give X-ray wavelengths in Angstrom units (A) when the absolute scale is used (1 A = 10^{-8} cm). The two are related by

$$1000\ XU = 1.00202 \pm 0.00003\ A$$

and wavelengths given in XU must be multiplied by 1.00202 and then divided by 1000 in order to convert them into Angstrom units.

Yield point — The stress at which a marked increase in deformation takes place without increase in the load.

Yield strength — "The stress at which a material exhibits a specified permanent set."

Zeeman effect — The splitting of a spectrum line into several symmetrically disposed components, which occurs when the source of light is placed in a strong magnetic field. The components are polarized, the directions of polarization and the appearance of the effect depending on the direction from which the source is viewed relative to the lines of force.

SI DERIVED UNITS WITH SPECIAL NAMES

Name	Quantity	Symbol	Expression in terms of other units	Expression in terms of SI base units
Becquerel	Activity (of a radio-nuclide)	Bq		s^{-1}
Coulomb	Quantity of electric-ity, electric charge	C	$A \cdot s$	$s \cdot A$
Degrees Celsius	Celsius temperature	°C		K
Gray	Absorbed dose, spe-cific energy im-parted, kerma, ab-sorbed dose index	Gy	J/kg	$m^2 \cdot s^{-2}$
Farad	Capacitance	F	C/V	$m^{-2} \cdot kg^{-1} \cdot s^4 \cdot A^2$
Henry	Inductance	H	Wb/A	$m^2 \cdot kg \cdot s^{-2} \cdot A^{-2}$
Hertz	Frequency	Hz		s^{-1}
Joule	Energy, work, quan-tity of heat	J	$N \cdot m$	$m^2 \cdot kg \cdot s^{-2}$
Lumen	Luminous flux	lm		$cd \cdot sr^a$
Lux	Illuminance	lx	lm/m^2	$m^{-2} \cdot cd \cdot sr^a$
Newton	Force	N		$m \cdot kg \cdot s^{-2}$
Ohm	Electric resistance		V/A	$m^2 \cdot kg \cdot s^{-3} \cdot A^{-2}$
Pascal	Pressure, stress	Pa	N/m^2	$m^{-1} \cdot kg \cdot s^{-2}$
Siemens	Conductance	S	A/V	$m^{-2} \cdot kg^{-1} \cdot s^3 \cdot A^2$
Tesla	Magnetic flux density	T	Wb/m^2	$kg \cdot s^{-2} \cdot A^{-1}$
Volt	Electric potential, po-tential difference, electromotive force	V	W/A	$m^2 \cdot kg \cdot s^{-3} \cdot A^{-1}$
Watt	Power, radiant flux	W	J/s	$m^2 \cdot kg \cdot s^{-3}$
Weber	Magnetic flux	Wb	$V \cdot s$	$m^2 \cdot kg \cdot s^{-2} \cdot A^{-1}$

a In this expression the steradian is treated as a base unit.

PRIMARY FIXED POINTS OF INTERNATIONAL PRACTICAL TEMPERATURE SCALE-68 (IPTS-68)

Primary fixed points of IPTS-68 are given in degrees Celsius and kelvin. Also shown are the comparative values from previous international temperature scales.

Fixed point	ITS-27, °C	ITS-48, °C	IPTS-48, °C	IPTS-68		
				°C	K	Uncertainty (K)
Equilibrium between the liquid and vapor phases of water (boiling point of water)	100.000	100	100	100	373.15	0.005
Equilibrium between the solid and liquid phases of zinc (freezing point of zinc)				419.58	692.73	0.03
Equilibrium between the liquid and vapor phases of sulfur (boiling point of sulfur)	444.60	444.600	444.6			
Equilibrium between the solid and liquid phases of silver (freezing point of silver)	960.5	960.8	960.8	961.93	1,235.08	0.2
Equilibrium between the solid and liquid phases of gold (freezing point of gold)	1,063	1,063.0	1,063	1,064.43	1,337.58	0.2

From Sparks, L. L., *ASRDI Oxygen Technology Survey,* Vol. 4, Scientific and Technical Information Office, National Aeronautics and Space Administration, Washington, D. C., 1974.

STANDARD TYPES OF STAINLESS AND HEAT RESISTING STEELS
Chemical Ranges and Limits
Subject to Tolerances for Check Analyses
By permission of American Iron and Steel Institute

Chemical Composition, per cent

Type Number	C	Mn Max.	P Max.	S Max.	Si Max.	Cr	Ni	Mo	Zr	Se	Cb-Ta	Ta	Al	N
⁻201	0.15 Max.	5.50/ 7.50	0.060	0.030	1.00	16.00/ 18.00	3.50/ 5.50							0.25 Max.
⁻202	0.15 Max.	7.50/ 10.00	0.060	0.030	1.00	17.00/ 19.00	4.00/ 6.00							0.25 Max.
⁻301	0.15 Max.	2.00	0.045	0.030	1.00	16.00/ 18.00	6.00/ 8.00							
⁻302	0.15 Max.	2.00	0.045	0.030	1.00	17.00/ 19.00	8.00/ 10.00							
⁻302B	0.15 Max.	2.00	0.045	0.030	2.00/ 3.00	17.00/ 19.00	8.00/ 10.00							
⁻303	0.15 Max.	2.00	0.20	0.15 Min.	1.00	17.00/ 19.00	8.00/ 10.00	0.60* Max.	0.60* Max.					
⁻303 Se	0.15 Max.	2.00	0.20	0.06	1.00	17.00/ 19.00	8.00/ 10.00			0.15 Min.				
⁻304	0.08 Max	2.00	0.045	0.030	1.00	18.00/ 20.00	8.00/ 12.00							
⁻304L	0.03 Max.	2.00	0.045	0.030	1.00	18.00/ 20.00	8.00/ 12.00							
⁻305	0.12 Max.	2.00	0.045	0.030	1.00	17.00/ 19.00	10.00/ 13.00							
⁻308	0.08 Max.	2.00	0.045	0.030	1.00	19.00/ 21.00	10.00/ 12.00							
⁻309	0.20 Max.	2.00	0.045	0.030	1.00	22.00/ 24.00	12.00/ 15.00							
⁻309S	0.08 Max.	2.00	0.045	0.030	1.00	22.00/ 24.00	12.00/ 15.00							
⁻310	0.25 Max.	2.00	0.045	0.030	1.50	24.00/ 26.00	19.00/ 22.00							
⁻310S	0.08 Max.	2.00	0.045	0.030	1.50	24.00/ 26.00	19.00/ 22.00							
⁻314	0.25 Max.	2.00	0.045	0.030	1.50/ 3.00	23.00/ 26.00	19.00/ 22.00							
⁻316	0.08 Max.	2.00	0.045	0.030	1.00	16.00/ 18.00	10.00/ 14.00	2.00/ 3.00						
⁻316L	0.03 Max.	2.00	0.045	0.030	1.00	16.00/ 18.00	10.00/ 14.00	2.00 3.00						
⁻317	0.08 Max.	2.00	0.045	0.030	1.00	18.00/ 20.00	11.00/ 15.00	3.00/ 4.00						
⁻321	0.08 Max.	2.00	0.045	0.030	1.00	17.00/ 19.00	9.00/ 12.00				5 × C Min.			
⁻347	0.08 Max.	2.00	0.045	0.030	1.00	17.00/ 19.00	9.00/ 13.00					10 × C Min.		
⁻348	0.08 Max.	2.00	0.045	0.030	1.00	17.00/ 19.00	9.00/ 13.00				10 × C Min.	0.10 Max.		
**403	0.15 Max.	1.00	0.040	0.030	0.50	11.50/ 13.00								
°405	0.08 Max.	1.00	0.040	0.030	1.00	11.50/ 14.50							0.10/ 0.30	
**410	0.15 Max.	1.00	0.040	0.030	1.00	11.50/ 13.50								
**414	0.15 Max.	1.00	0.040	0.030	1.00	11.50/ 13.50	1.25/ 2.50							
**416	0.15 Max.	1.25	0.06	0.15 Min.	1.00	12.00/ 14.00		0.60* Max.	0.60* Max.					
**416 Se	0.15 Max.	1.25	0.06	0.06	1.00	12.00/ 14.00				0.15 Min.				
**420	Over 0.15	1.00	0.040	0.030	1.00	12.00/ 14.00								

* At producer's option; reported only when intentionally added.
** Heat treatable.
⁻ Not heat treatable.
° Essentially not heat treatable.

F-112

Chemical Composition, per cent

Type Number	C	Mn Max.	P Max.	S Max.	Si Max.	Cr	Ni	Mo	Zr	Se	Cb-Ta	Ta	Al	N
°430	0.12 Max.	1.00	0.040	0.030	1.00	14.00/ 18.00								
°430F	0.12 Max.	1.25	0.06	0.15 Min.	1.00	14.00/ 18.00		0.60* Max.	0.60* Max.					
°430F Se	0.12 Max.	1.25	0.06	0.06	1.00	14.00/ 18.00				0.15 Min.				
**431	0.20 Max.	1.00	0.040	0.030	1.00	15.00/ 17.00	1.25/ 2.50							
**440A	0.60/ 0.75	1.00	0.040	0.030	1.00	16.00/ 18.00		0.75 Max.						
**400B	0.75/ 0.95	1.00	0.040	0.030	1.00	16.00/ 18.00		0.75 Max.						
**440C	0.95/ 1.20	1.00	0.040	0.030	1.00	16.00/ 18.00		0.75 Max.						
°446	0.20 Max.	1.50	0.040	0.030	1.00	23.00/ 27.00								0.25 Max.
**501	Over 0.10	1.00	0.040	0.030	1.00	4.00/ 6.00		0.40/ 0.65						
**502	0.10 Max.	1.00	0.040	0.030	1.00	4.00/ 6.00		0.40/ 0.65						

* At producer's option; reported only when intentionally added.
** Heat treatable.
⁻ Not heat treatable.
° Essentially not heat treatable.

STANDARD TEST SIEVES (WIRE CLOTH)

Sieve Designation		Nominal Sieve Opening in	Permissible Variation of Average Opening from the Standard Sieve Designation	Maximum Opening Size for Not More than 5 percent of Openings	Maximum Individual Opening	Nominal Wire Diameter, mm[a]
Standard	Alternative					
(1)	(2)	(3)	(4)	(5)	(6)	(7)
125 mm	5 in.	5	±3.7 mm	130.0 mm	130.9 mm	8.0
106 mm	4.24 in.	4.24	±3.2 mm	110.2 mm	111.1 mm	6.40
100 mm	4 in.	4	±3.0 mm	104.0 mm	104.8 mm	6.30
90 mm	$3\frac{1}{2}$ in.	3.5	±2.7 mm	93.6 mm	94.4 mm	6.08
75 mm	3 in.	3	±2.2 mm	78.1 mm	78.7 mm	5.80
63 mm	$2\frac{1}{2}$ in.	2.5	±1.9 mm	65.6 mm	66.2 mm	5.50
53 mm	2.12 in.	2.12	±1.6 mm	55.2 mm	55.7 mm	5.15
50 mm	2 in.	2	±1.5 mm	52.1 mm	52.6 mm	5.05
45 mm	$1\frac{3}{4}$ in.	1.75	±1.4 mm	46.9 mm	47.4 mm	4.85
37.5 mm	$1\frac{1}{2}$ in.	1.5	±1.1 mm	39.1 mm	39.5 mm	4.59
31.5 mm	$1\frac{1}{4}$ in.	1.25	±1.0 mm	32.9 mm	33.2 mm	4.23
26.5 mm	1.06 in.	1.06	±0.8 mm	27.7 mm	28.0 mm	3.90
25.0 mm	1 in.	1	±0.8 mm	26.1 mm	26.4 mm	3.80
22.4 mm	$\frac{7}{8}$ in.	0.875	±0.7 mm	23.4 mm	23.7 mm	3.50
19.0 mm	$\frac{3}{4}$ in.	0.750	±0.6 mm	19.9 mm	20.1 mm	3.30
16.0 mm	$\frac{5}{8}$ in.	0.625	±0.5 mm	16.7 mm	17.0 mm	3.00
13.2 mm	0.530 in.	0.530	±0.41 mm	13.83 mm	14.05 mm	2.75
12.5 mm	$\frac{1}{2}$ in.	0.500	±0.39 mm	13.10 mm	13.31 mm	2.67
11.2 mm	$\frac{7}{16}$ in.	0.438	±0.35 mm	11.75 mm	11.94 mm	2.45
9.5 mm	$\frac{3}{8}$ in.	0.375	±0.30 mm	9.97 mm	10.16 mm	2.27
8.0 mm	$\frac{5}{16}$ in.	0.312	±0.25 mm	8.41 mm	8.58 mm	2.07
6.7 mm	0.265 in.	0.265	±0.21 mm	7.05 mm	7.20 mm	1.87
6.3 mm	$\frac{1}{4}$ in.	0.250	±0.20 mm	6.64 mm	6.78 mm	1.82
5.6 mm	No. $3\frac{1}{2}$	0.223	±0.18 mm	5.90 mm	6.04 mm	1.68
4.75 mm	No. 4	0.187	±0.15 mm	5.02 mm	5.14 mm	1.54
4.00 mm	No. 5	0.157	±0.13 mm	4.23 mm	4.35 mm	1.37
3.35 mm	No. 6	0.132	±0.11 mm	3.55 mm	3.66 mm	1.23
2.80 mm	No. 7	0.111	±0.095 mm	2.975 mm	3.070 mm	1.10
2.36 mm	No. 8	0.0937	±0.080 mm	2.515 mm	2.600 mm	1.00
2.00 mm	No. 10	0.0787	±0.070 mm	2.135 mm	2.215 mm	0.900
1.70 mm	No. 12	0.0661	±0.060 mm	1.820 mm	1.890 mm	0.810
1.40 mm	No. 14	0.0555	±0.050 mm	1.505 mm	1.565 mm	0.725
1.18 mm	No. 16	0.0469	±0.045 mm	1.270 mm	1.330 mm	0.650
1.00 mm	No. 18	0.0394	±0.040 mm	1.080 mm	1.135 mm	0.580
850 μm	No. 20	0.0331	±35 μm	925 μm	970 μm	0.510
710 μm	No. 25	0.0278	±30 μm	775 μm	815 μm	0.450
600 μm	No. 30	0.0234	±25 μm	660 um	695 μm	0.390
500 μm	No. 35	0.0197	±20 μm	550 μm	585 μm	0.340
425 μm	No. 40	0.0165	±19 μm	471 μm	502 μm	0.290
355 μm	No. 45	0.0139	±16 μm	396 μm	425 μm	0.247
300 μm	No. 50	0.0117	±14 μm	337 μm	363 μm	0.215
250 μm	No. 60	0.0098	±12 μm	283 μm	306 μm	0.180
212 μm	No. 70	0.0083	±10 μm	242 μm	263 μm	0.152
180 μm	No. 80	0.0070	±9 μm	207 μm	227 μm	0.131
150 μm	No. 100	0.0059	±8 μm	174 μm	192 μm	0.110
125 μm	No. 120	0.0049	±7 μm	147 μm	163 μm	0.091
106 μm	No. 140	0.0041	±6 μm	126 μm	141 μm	0.076
90 μm	No. 170	0.0035	±5 μm	108 μm	122 μm	0.064
75 μm	No. 200	0.0029	±5 μm	91 μm	103 μm	0.053
63 μm	No. 230	0.0025	±4 μm	77 μm	89 μm	0.044
53 μm	No. 270	0.0021	±4 μm	66 μm	76 μm	0.037
45 μm	No. 325	0.0017	±3 μm	57 μm	66 μm	0.030
38 μm	No. 400	0.0015	±3 μm	48 μm	57 μm	0.025

[a] The average diameter of the warp and of the shoot wires, taken separately, of the cloth of any sieve shall not deviate from the nominal values by more than the following:

Sieves coarser than 600 μm	5 percent
Sieves 600 to 125 μm	$7\frac{1}{2}$ percent
Sieves finer than 125 μm	10 percent

PHYSICAL PROPERTIES OF GLASS SEALING AND LEAD WIRE MATERIALS

From General Electric Company

	Copper OFC	Copper 0.02P	Dumet	42 Ni	Gas Free 42% Ni	42-6 Ni Cr.	46 Ni	52 Ni.	27 Cr.	® Kovar	W	Mo	Nickel 200
Analysis: Carbon			Copper Clad 42% Nickel See GE Spec. DS 8311-01	0.10	0.05	0.10	0.10	0.10	0.15	0.02			0.06
Manganese				0.50	0.50	0.50	0.50	0.50	0.60	0.30			0.25
Silicon				0.25	0.25	0.25	0.25	0.25	0.40	0.20			0.05
Chromium				—	—	5.75	—	—	28.00	—			
Nickel				42	42	42.5	46	51	0.50	29			99.5
Copper	99.95	99.90 P 0.02		—	—		—	—		—			0.05
Other				Bal.Fe	Ti0.4 Bal.Fe	Bal.Fe	Ti0.4 Bal.Fe	Bal.Fe	Bal.Fe	Co 17 Bal.Fe			
Density: grams/cc	8.94	8.94	8.26 to 8.32	8.12	8.12	8.12	8.17	8.30	7.60	8.36	19.3	10.2	8.89
lbs. per cu. in.	0.323	0.323	0.298—0.301	0.293	0.293	0.293	0.295	0.300	0.274	0.302	0.697	0.369	0.321
Thermal Conductivity 20-100°/C Cal/cm/sec/cm²/°C	0.948	0.8	0.2—0.3	0.025	0.025	0.029	0.028	0.032	0.054	0.04	0.31	0.34	0.15
Btu/in/hr/sq. ft/°F	2750	2320	580—870	74	74	84	81	93	158	116	900	1000	435
Electrical Resistivity (20°C) Microhm—cm	1.71	2.03	7.3 to 12.0	72	72	95	46	43	63	49	5.5	5.2	9.5
Ohm per cir. mil ft.	10.3	12.2	44 to 72	430	430	570	275	258	380	294	33	31	57
Elec. Cond. % IACS	101	85	23 to 14	2.3	2.3	1.8	3.6	3.9	2.8	3.4	31	33	18
Curie Temperature °C	—	—	380	380	380	295	460	530	610	435	—	—	360
Melting Temperature °C	1083	1083	— —	1425	1425	1425	1425	1425	1425	1450	3410	2610	1455
°F	1981	1981	— —	2597	2597	2597	2597	2597	2597	2642	6170	4730	2651
Specific Heat cal/gr.	0.092	0.092	0.11	0.12	0.12	0.12	0.12	0.12	0.14	0.11	0.033	0.066	0.13
Thermal Expansion in/in/°Cx10⁷			Radial										
25-100°C	168	168	60 to 80	50.1	43.4	65.5	71.0	99.5	94.6	58.6	45	51	133
25-200°C	172	172		47.1	44.1	70.0	70.7	101.6	100.5	52.0	46		139
25-300°C	177	177	60 to 80	47.6	46.1	82.6	75.0	101.0	105.3	51.3	46		144
25-350°C	178	178	65 to 85	50.5		90.4	74.4	100.0	107.0	48.9	46		146
25-400°C	181	181	80 to 100	62.5	64.1	100.0	74.3	100.0	107.8	50.6	46		148
25-500°C	183	183	100 to 140	83.2	85.6	115.0	86.8	102.1	111.2	61.5	46	57	172
25-600°C	188	188		99.0	100.1	125.8	100.2	110.0	112.6	78.0			
Mechanical Properties (Annealed) Ultimate Str. (1000 psi)	35	35	74	82	80	80	82	80	85	75	490	120	70
Yield Str. (1000 psi)			50	34	34	40	34	40	55	50	360	110	25
% Elong. (2")			30	30	30	30	30	35	25	30	8	30	45
Rockwell Hardness				B76	B76	B80	B76	B83	B85	B68	C25	B88	B62
Elastic Modulus (10⁶ psi)	16	16		21.5	21.0	23	23	24	30	20	50	47	30

PHYSICAL PROPERTIES OF GLASS SEALING AND LEAD WIRE MATERIALS (Continued)

Column groups: **Metal Alloy Lead Wires** (Nickel 211 – 70-30 Brass); **Clad Materials** (Also see Dumet) (Kulgrid® – 30% CCFe); **Pure Metals** (Also see W and Mo glass sealing materials) (Au – Ti)

Nickel 211	AISI Type 302	AISI Type 316	AISI Type 430	400 Monel	600 Inconel	Advance®	CDA 752	70-30 Brass	Kulgrid®	40% CCFe	30% CCFe	Au	Ag	Al	Pt	Ta	Fe	Ti
0.10	0.15 Max	0.08 Max	0.12 Max	0.12	0.04		0.50 Max											
4.75	2.00 Max	2.00 Max	1.00 Max	0.90	0.20													
0.05	1.00 Max	1.00 Max	1.00 Max	0.15														
	17–19	16–18	14–18		16													
95.0	8–10	10–14	0.50 Max	66	76	43	18	70	27									
		—		31.5		Bal.	65			37.5	26							
		Mo 2–3			Fe 7.20		Bal. Zn	Bal. Zn	Cu Core	Fe Core	Fe Core							
8.72	7.9	7.9	7.7	8.84	8.41	8.9	8.73	8.53	8.89	8.15	8.15	19.3	10.5	2.69	21.45	16.6	7.87	4.51
0.315	0.285	0.285	0.278	0.319	0.304	0.322	0.316	0.308	0.321	0.294	0.294	0.698	0.379	0.097	0.775	0.600	0.284	0.163
0.12	0.04	0.04	0.05	0.062	0.036	0.051	0.08	0.29		0.46	0.38	0.71	1.00	0.57	0.165	0.13	0.18	0.43
350	116	116	145	180	104	148	232	845		1330	1100	2050	2900	1650	480	380	520	1250
18.3	72	74	60	48.2	98.1	49	28.7	6.16	2.3	4.4	5.9	2.19	1.629	2.65	9.83	12.45	9.71	7.0
110	433	445	361	290	590	294	173	37	14	26.4	35.3	13	9.8	16	59	75	58	42
9	2.3	2.3	2.9	3.6	1.7	3.5	6	28	70	40	30	78	105	65	16	14	18	25
352			610	43/60	−125	—	—	—		770	770	—	—	—	—	—	770	—
1427	1421	1399	1510	1349	1427	1210	1110	954				1063	960	660	1769	3000	1536	1668
2600	2590	2550	2750	2460	2600	2210	2030	1750				1945	1760	1220	3217	5425	2797	3035
0.13	0.12	0.12	0.11	0.102	0.109	0.094	0.09	0.09		0.10	0.10	0.031	0.056	0.225	0.031	0.034	0.11	0.124
133	166	166	101	140	115	149						142	196	239	91	65	122	88
139				145										243			129	91
144	171	175	110	150			160	199						253				
146																		
148				155													138	94
153	180	180	120	160		163						152	206		96	66		
				165										287			145	97
	90	85	75	85	100	60	60	50				20	22	13	40	55	40	
	37	35	45	40	45		30	—					8	5	12		20	
	55	55	30	40	40		40	60				45	48	40	30	30	45	
	B82	B80	B82	B68	B74		—	—										
30	29	29	29	26	31	18	18	16		24	24	11.6	11	9	21.3	27	30	16.8

WIRE TABLES COMPARISON OF WIRE GAUGES

Diameter of Wire in Inches

Gauge No.	Brown & Sharpe	Birmingham or Stubs'	Washburn & Moen	Imperial or Brit. Std.	Stubs' Steel	U.S. Std. plate
00000000						
0000000				.500		
000000				.464		.46875
00000				.432		.4375
0000	.4600	.454	.3938	.400		.40625
000	.4096	.425	.3625	.372		.375
00	.3648	.380	.3310	.348		.34375
0	.3249	.340	.3065	.324		.3125
1	.2893	.300	.2830	.300	.227	.28125
2	.2576	.284	.2625	.276	.219	.26525
3	.2294	.259	.2437	.252	.212	.25
4	.2043	.238	.2253	.232	.207	.234375
5	.1819	.220	.2070	.212	.204	.21875
6	.1620	.203	.1920	.192	.201	.203125
7	.1443	.180	.1770	.176	.199	.1875
8	.1285	.165	.1620	.160	.197	.171875
9	.1144	.148	.1483	.144	.194	.15625
10	.1019	.134	.1350	.128	.191	.140625
11	.09074	.120	.1205	.116	.188	.125
12	.08081	.109	.1055	.104	.185	.109375
13	.07196	.095	.0915	.092	.182	.09375
14	.06408	.083	.0800	.080	.180	.078125
15	.05707	.072	.0720	.072	.178	.0703125
16	.05082	.065	.0625	.064	.175	.0625
17	.04526	.058	.0540	.056	.172	.05625
18	.04030	.049	.0475	.048	.168	.05
19	.03589	.042	.0410	.040	.164	.04375
20	.03196	.035	.0348	.036	.161	.0375
21	.02846	.032	.0318	.032	.157	.034375
22	.02535	.028	.0286	.028	.155	.03125
23	.02257	.025	.0258	.024	.153	.028125
24	.02010	.022	.0230	.022	.151	.025
25	.01790	.020	.0204	.020	.148	.021875
26	.01594	.018	.0181	.018	.146	.01875
27	.01419	.016	.0173	.0164	.143	.0171875
28	.01264	.014	.0162	.0149	.139	.015625
29	.01126	.013	.0150	.0136	.134	.0140625
30	.01003	.012	.0140	.0124	.127	.0125
31	.008928	.010	.0132	.0116	.120	.0109375
32	.007950	.009	.0128	.0108	.115	.01015625
33	.007080	.008	.0118	.0100	.112	.009375
34	.006304	.007	.0104	.0092	.110	.00859375
35	.005614	.005	.0095	.0084	.108	.0078125
36	.005000	.004	.0090	.0076	.106	.00703125
37	.004453		.0085	.0068	.103	.006640625
38	.003965		.0080	.0060	.101	.00625
39	.003531		.0075	.0052	.099	
40	.003145		.0070	.0048	.097	
41			.0066	.0044	.095	
42			.0062	.0040	.092	
43			.0060	.0036	.088	
44			.0058	.0032	.085	
45			.0055	.0028	.081	
46			.0052	.0024	.079	
47			.0050	.0020	.077	
48			.0048	.0016	.075	
49			.0046	.0012	.072	
50			.0044	.0010	.069	

Diameter of Wire in Centimeters

Gauge No.	Brown & Sharpe	Birmingham or Stubs'	Washburn & Moen	Imperial or Brit. Std.	Stubs' Steel	U.S. Std. plate
00000000						
0000000			1.245	1.27		1.27
000000			1.172	1.18		1.191
00000			1.093	1.10		1.111
0000	1.168	1.15	1.000	1.02		1.032
000	1.040	1.08	0.9208	0.945		0.9525
00	0.9266	0.965	0.8407	0.884		0.8731
0	0.8252	0.864	0.7785	0.823		0.7938
1	0.7348	0.762	0.7188	0.762	0.577	0.7144
2	0.6543	0.721	0.6668	0.701	0.556	0.6747
3	0.5827	0.658	0.6190	0.640	0.538	0.6350
4	0.5189	0.605	0.5723	0.589	0.526	0.5953
5	0.4620	0.559	0.5258	0.538	0.518	0.5556
6	0.4115	0.516	0.4877	0.488	0.511	0.5159
7	0.3665	0.457	0.4496	0.447	0.505	0.4763
8	0.3264	0.419	0.4115	0.406	0.500	0.4366
9	0.2906	0.376	0.3767	0.366	0.493	0.3969
10	0.2588	0.304	0.3429	0.325	0.485	0.3572
11	0.2305	0.305	0.3061	0.295	0.478	0.3175
12	0.2053	0.277	0.2680	0.264	0.470	0.2778
13	0.1828	0.241	0.232	0.234	0.462	0.2381
14	0.1628	0.211	0.203	0.203	0.457	0.1984
15	0.1450	0.183	0.183	0.183	0.452	0.1786
16	0.1291	0.165	0.159	0.163	0.445	0.1588
17	0.1150	0.147	0.137	0.142	0.437	0.1429
18	0.1024	0.124	0.121	0.122	0.427	0.1270
19	0.09116	0.107	0.104	0.102	0.417	0.1111
20	0.08118	0.089	0.0884	0.0914	0.409	0.09525
21	0.07229	0.081	0.0808	0.0813	0.399	0.08731
22	0.06439	0.071	0.0726	0.0711	0.394	0.07938
23	0.05733	0.064	0.0655	0.0610	0.389	0.07144
24	0.05105	0.056	0.0584	0.0559	0.384	0.06350
25	0.04547	0.051	0.0518	0.0508	0.376	0.05556
26	0.04049	0.046	0.0460	0.0457	0.371	0.04763
27	0.03604	0.041	0.0439	0.0417	0.363	0.04366
28	0.03211	0.036	0.0411	0.0378	0.353	0.03969
29	0.02860	0.033	0.0381	0.0345	0.340	0.03572
30	0.02548	0.030	0.0356	0.0315	0.323	0.03175
31	0.02268	0.025	0.0335	0.0295	0.305	0.02778
32	0.02019	0.023	0.0325	0.0274	0.292	0.02580
33	0.01798	0.020	0.0300	0.0254	0.284	0.02381
34	0.01601	0.018	0.0264	0.0234	0.279	0.02183
35	0.01426	0.013	0.024	0.0213	0.274	0.01984
36	0.01270	0.010	0.023	0.0193	0.269	0.01786
37	0.01131		0.022	0.0173	0.262	0.01687
38	0.01007		0.020	0.0152	0.257	0.01588
39	0.008969		0.019	0.0132	0.251	
40	0.007988		0.018	0.0122	0.246	
41			0.017	0.0112	0.241	
42			0.016	0.0102	0.234	
43			0.015	0.0091	0.224	
44			0.015	0.0081	0.216	
45			0.014	0.0071	0.206	
46			0.013	0.0061	0.201	
47			0.013	0.0051	0.196	
48			0.012	0.0041	0.191	
49			0.012	0.0030	0.183	
50			0.011	0.0025	0.175	

TWIST DRILL AND STEEL WIRE GAUGE

Inches

No.	Size	No.	Size	No.	Size	No.	Size	No.	Size
1	0.2280	17	0.1730	33	0.1130	49	0.0730	65	0.0350
2	0.2210	18	0.1695	34	0.1110	50	0.0700	66	0.0330
3	0.2130	19	0.1660	35	0.1100	51	0.0670	67	0.0320
4	0.2090	20	0.1610	36	0.1065	52	0.0635	68	0.0310
5	0.2055	21	0.1590	37	0.1040	53	0.0595	69	0.02925
6	0.2040	22	0.1570	38	0.1015	54	0.0550	70	0.0280
7	0.2010	23	0.1540	39	0.0995	55	0.0520	71	0.0260
8	0.1990	24	0.1520	40	0.0980	56	0.0465	72	0.0250
9	0.1960	25	0.1495	41	0.0960	57	0.0430	73	0.0240
10	0.1935	26	0.1470	42	0.0935	58	0.0420	74	0.0225
11	0.1910	27	0.1440	43	0.0890	59	0.0410	75	0.0210
12	0.1890	28	0.1405	44	0.0860	60	0.0400	76	0.0200
13	0.1850	29	0.1360	45	0.0820	61	0.0390	77	0.0180
14	0.1820	30	0.1285	46	0.0810	62	0.0380	78	0.0160
15	0.1800	31	0.1200	47	0.0785	63	0.0370	79	0.0145
16	0.1770	32	0.1160	48	0.0760	64	0.0360	80	0.0135

Centimeters

No.	Size	No.	Size	No.	Size	No.	Size	No.	Size
1	0.5791	17	0.4394	33	0.2870	49	0.1854	65	0.0889
2	0.5613	18	0.4305	34	0.2819	50	0.1778	66	0.0838
3	0.5410	19	0.4216	35	0.2794	51	0.1702	67	0.0813
4	0.5309	20	0.4089	36	0.2705	52	0.1613	68	0.0787
5	0.5220	21	0.4039	37	0.2642	53	0.1511	69	0.0743
6	0.5182	22	0.3988	38	0.2578	54	0.1397	70	0.0711
7	0.5105	23	0.3912	39	0.2527	55	0.1321	71	0.0660
8	0.5055	24	0.3861	40	0.2489	56	0.1181	72	0.0635
9	0.4978	25	0.3797	41	0.2438	57	0.1092	73	0.0610
10	0.4915	26	0.3734	42	0.2375	58	0.1067	74	0.0572
11	0.4851	27	0.3658	43	0.2261	59	0.1041	75	0.0533
12	0.4801	28	0.3569	44	0.2184	60	0.1016	76	0.0508
13	0.4699	29	0.3454	45	0.2083	61	0.0991	77	0.0457
14	0.4623	30	0.3264	46	0.2057	62	0.0965	78	0.0406
15	0.4572	31	0.3048	47	0.1994	63	0.0940	79	0.0368
16	0.4496	32	0.2946	48	0.1930	64	0.0914	80	0.0343

DIMENSIONS OF WIRE

Stubs' Gauge

Giving the diameter and cross-section in English and metric system for the Birmingham or Stubs' gauge.

Gauge No.	Diameter in in.	Section in sq. in.	Diameter in cm	Section in sq. cm	Gauge No.	Diameter in in.	Section in sq. in.	Diameter in cm	Section in sq. cm
0000	.454	.16188	1.1532	1.0444	17	.058	.0026421	.14732	.017046
000	.425	.14186	1.0795	.9152	18	.049	.0018857	.12446	.012166
00	.380	.11341	.9652	.7317	19	.042	.0013854	.10668	.008938
0	.340	.09079	.8636	.5858	20	.035	.0009621	.08890	.006207
1	.300	.07069	.7620	.4560	21	.032	.0008042	.08128	.005189
2	.284	.06335	.7214	.4087	22	.028	.0006158	.07112	.003973
3	.259	.05269	.6579	.3399	23	.025	.0004909	.06350	.003167
4	.238	.04449	.6045	.2870	24	.022	.0003801	.05588	.002452
5	.220	.03801	.5588	.2452	25	.020	.0003142	.05080	.002027
6	.203	.03237	.5156	.20881	26	.018	.0002545	.04572	.0016417
7	.180	.02545	.4572	.16147	27	.016	.0002011	.04064	.0012972
8	.165	.02138	.4191	.13795	28	.014	.0001539	.03556	.0009932
9	.148	.01720	.3759	.11099	29	.013	.0001327	.03302	.0008563
10	.134	.01410	.3404	.09098	30	.012	.0001181	.03048	.0007297
11	.120	.011310	.3048	.07297	31	.010	.00007854	.02540	.0005067
12	.109	.009331	.2769	.06160	32	.009	.00006362	.02286	.0004104
13	.095	.007088	.2413	.04573	33	.008	.00005027	.02032	.0003243
14	.083	.005411	.2108	.03491	34	.007	.00003848	.01778	.0002483
15	.072	.004072	.1829	.02627	35	.005	.00001963	.01270	.0001267
16	.065	.0033183	.15610	.021409	36	.004	.00001257	.01016	.0000811

British Standard Gauge

Giving the diameter and cross-section in English and metric system for the British Standard Gauge.

Gauge No.	Diameter in in.	Section in sq. in.	Diameter in cm	Section in sq. cm	Gauge No.	Diameter in in.	Section in sq. in.	Diameter in cm	Section in sq. cm
0000000	.500	.1963	1.2700	1.267	23	.024	.0004524	.06096	.002922
000000	.464	.1691	1.1786	1.091	24	.022	.0003801	.05588	.002452
00000	.432	.1466	1.0973	.9456	25	.020	.0003142	.05080	.002027
0000	.400	.1257	1.0160	.8107	26	.0180	.0002545	.04572	.0016417
000	.372	.1087	.9449	.7012	27	.0164	.0002112	.04166	.0013628
00	.348	.0951	.8839	.6136	28	.0148	.0001728	.03759	.0011099
0	.324	.0825	.8230	.5319	29	.0136	.0001453	.03454	.0009363
1	.300	.07069	.7620	.4560	30	.0124	.0001208	.03150	.0007791
2	.276	.05983	.7010	.3858	31	.0116	.00010568	.02946	.0006818
3	.252	.04988	.6401	.3218	32	.0108	.00009161	.02743	.0005910
4	.232	.04227	.5893	.2727	33	.0100	.00007854	.02540	.0005067
5	.212	.03530	.5385	.2277	34	.0092	.00006648	.02337	.0004289
6	.192	.02895	.4877	.28679	35	.0084	.00005542	.02134	.0003575
7	.176	.02433	.4470	.15696	36	.0076	.00004536	.01930	.0002927
8	.160	.02010	.4064	.12973	37	.0068	.00003632	.01727	.0002343
9	.144	.01629	.3658	.10507	38	.0060	.00002827	.01524	.0001824
10	.128	.01287	.3251	.08302	39	.0052	.00002124	.01321	.0001370
11	.116	.010568	.2946	.06818	40	.0048	.00001810	.01219	.0001167
12	.104	.008495	.2642	.05480	41	.0044	.00001512	.01118	.0000982
13	.092	.006648	.2337	.04289	42	.0040	.00001257	.01016	.0000811
14	.080	.005027	.2032	.03243	43	.0036	.00001018	.00914	.0000656
15	.072	.004071	.1829	.02627	44	.0032	.00000804	.00813	.0000519
16	.064	.003217	.16256	.020755	45	.0028	.00000616	.00711	.0000397
17	.056	.002463	.14224	.015890	46	.0024	.00000452	.00610	.0000212
18	.048	.001810	.12192	.011675	47	.0020	.00000314	.00508	.0000203
19	.040	.001257	.10160	.008107	48	.0016	.00000201	.00406	.0000129
20	.036	.001018	.09144	.006567	49	.0012	.00000113	.00305	.0000073
21	.032	.0008042	.08128	.005189	50	.0010	.00000079	.00254	.0000051
22	.028	.0006158	.07112	.003973					

PLATINUM WIRE

Mass in Grams per Foot

B. & S. Gauge	Diameter, inches	Mass, g per ft.	B. & S. Gauge	Diameter, inches	Mass, g per ft.
10	.1019	37.5	23	.02257	1.8
11	.09074	28.0	24	.02010	1.4
12	.08081	22.0	25	.01790	1.1
13	.07196	17.5	26	.01594	0.9
14	.06408	14.0	27	.01420	0.7
15	.05707	11.0	28	.01264	0.6
16	.05082	9.0	29	.01126	0.45
17	.04526	7.0	30	.01003	0.35
18	.04030	5.7	31	.008928	0.28
19	.03589	4.4	32	.007950	0.22
20	.03196	3.4	33	.007080	0.17
21	.02846	2.9	34	.006305	0.15
22	.02535	2.3	35	.005615	0.11

WIRE TABLE, STANDARD ANNEALED COPPER

American Wire Gauge (B. & S.) English Units

Gauge No.	Diameter in mils at 20°C	Circular mils	Sq. inches	Ohms per 1000 feet* 0°C (32°F)	20°C (68°F)	50°C (122°F)	75°C (167°F)	Gauge No.	Diameter in mils at 20°C	Circular mils	Sq. inches	Ohms per 1000 feet* 0°C (32°F)	20°C (68°F)	50°C (122°F)	75°C (167°F)
0000	460.0	211600	0.1662	.04516	.04901	.05479	.05961	19	35.89	1288	.001012	7.418	8.051	9.001	9.792
000	409.6	167800	.1318	.05695	.06180	.06909	.07516	20	31.96	1022	.0008023	9.355	10.15	11.35	12.35
00	364.8	133100	.1045	.07181	.07793	.08712	.09478	21	28.45	810.1	.0006363	11.80	12.80	14.31	15.57
0	324.9	105500	.08289	.09055	.09827	.1099	.1195	22	25.35	642.4	.0005046	14.87	16.14	18.05	19.63
1	289.3	83690	.06573	.1142	.1239	.1385	.1507	23	22.57	509.5	.0004002	18.76	20.36	22.76	24.76
2	257.6	66370	.05213	.1440	.1563	.1747	.1900	24	20.10	404.0	.0003173	23.65	25.67	28.70	31.22
3	229.4	52640	.04134	.1816	.1970	.2203	.2396	25	17.90	320.4	.0002517	29.82	32.37	36.18	39.36
4	204.3	41740	.03278	.2289	.2485	.2778	.3022	26	15.94	254.1	.0001996	37.61	40.81	45.63	49.64
5	181.9	33100	.02600	.2887	.3133	.3502	.3810	27	14.20	201.5	.0001583	47.42	51.47	57.53	62.59
6	162.0	26250	.02062	.3640	.3951	.4416	.4805	28	12.64	159.8	.0001255	59.80	64.90	72.55	78.93
7	144.3	20820	.01635	.4590	.4982	.5569	.6059	29	11.26	126.7	.00009953	75.40	81.83	91.48	99.52
8	128.5	16510	.01297	.5788	.6282	.7023	.7640	30	10.03	100.5	.00007894	95.08	103.2	115.4	125.5
9	114.4	13090	.01028	.7299	.7921	.8855	.9633	31	8.928	79.70	.00006260	119.9	130.1	145.5	158.2
10	101.9	10380	.008155	.9203	.9989	1.117	1.215	32	7.950	63.21	.00004964	151.2	164.1	183.4	199.5
11	90.74	8234	.006467	1.161	1.260	1.408	1.532	33	7.080	50.13	.00003937	190.6	206.9	231.3	251.6
12	80.81	6530	.005129	1.463	1.588	1.775	1.931	34	6.305	39.75	.00003122	240.4	260.9	291.7	317.3
13	71.96	5178	.004067	1.845	2.003	2.239	2.436	35	5.615	31.52	.00002476	303.1	329.0	367.8	400.1
14	64.08	4107	.003225	2.327	2.525	2.823	3.071	36	5.000	25.00	.00001964	382.2	414.8	463.7	504.5
15	57.07	3257	.002558	2.934	3.184	3.560	3.873	37	4.453	19.83	.00001557	482.0	523.1	584.8	636.2
16	50.82	2583	.002028	3.700	4.016	4.489	4.884	38	3.965	15.72	.00001235	607.8	659.6	737.4	802.2
17	45.26	2048	.001609	4.666	5.064	5.660	6.158	39	3.531	12.47	.000009793	766.4	831.8	929.8	1012
18	40.30	1624	.001276	5.883	6.385	7.138	7.765	40	3.145	9.888	.000007766	966.5	1049	1173	1276

Resistance at the stated temperatures of a wire whose length is 1000 feet at 20°C.

ALLOWABLE CARRYING CAPACITIES OF CONDUCTORS

(National Electrical Code)

The ratings in the following tabulation are those permitted by the National Electrical Code for flexible cords and for interior wiring of houses, hotels, office buildings, industrial plants, and other buildings.

The values are for copper wire. For aluminum wire the allowable carrying capacities shall be taken as 84% of those given in the table for the respective sizes of copper wire with the same kind of covering.

Size A.W.G.	Area Circular Mils	Diameter of Solid Wires Mils	Rubber Insulation Amperes	Varnished Cambric Insulation Amperes	Other Insulations and Bare Conductors Amperes	Size A.W.G.	Area Circular Mils	Diameter of Solid Wires Mils	Rubber Insulation Amperes	Varnished Cambric Insulation Amperes	Other Insulations and Bare Conductors Amperes
18	1,624.	40.3	3*		6†	4	41,740.	204.3	70	85	90
16	2,583.	50.8	6*		10†	3	52,630.	229.4	80	95	100
14	4,107.	64.1	15	18	20	2	66,370.	257.6	90	110	125
12	6,530.	80.8	20	25	30	1	83,690.	289.3	100	120	150
10	10,380.	101.9	25	30	35	0	105,500.	325.0	125	150	200
8	16,510.	128.5	35	40	50	00	135,100.	364.8	150	180	225
6	26,250.	162.0	50	60	70	000	167,800.	409.6	175	210	275
5	33,100.	181.9	55	65	80	0000	211,600.	460	225	270	325

* The allowable carrying capacities of No. 18 and 16 are 5 and 7 amperes respectively, when in flexible cords.
† The allowable carrying capacities of No. 18 and 16 are 10 and 15 amperes respectively, when in cords for portable heaters. Types AFS, AFSJ, HC, HPD, and HSJ.

WIRE TABLE, STANDARD ANNEALED COPPER

American Wire Gauge (B. & S.) English Units (Continued)

Gauge No.	Pounds per 1000 feet	Feet per pound	Feet per ohm* 0°C (32°F)	20°C (68°F)	50°C (122°F)	75°C (167°F)	Gauge No.	Pounds per 1000 feet	Feet per pound	Feet per ohm* 0°C (32°F)	20°C (68°F)	50°C (122°F)	75°C (167°F)
0000	640.5	1.561	22140	20400	18250	16780	19	3.899	256.5	134.8	124.2	111.1	102.1
000	507.9	1.968	17560	16180	14470	13300	20	3.092	323.4	106.9	98.50	88.11	80.99
00	402.8	2.482	13930	12830	11480	10550	21	2.452	407.8	84.78	78.11	69.87	64.23
0	319.5	3.130	11040	10180	9103	8367	22	1.945	514.2	67.23	61.95	55.41	50.94
1	253.3	3.947	8758	8070	7219	6636	23	1.542	648.4	53.32	49.13	43.94	40.39
2	200.9	4.977	6946	6400	5725	5262	24	1.223	817.7	42.28	38.96	34.85	32.03
3	159.3	6.276	5508	5075	4540	4173	25	0.9699	1031	33.53	30.90	27.64	25.40
4	126.4	7.914	4368	4025	3600	3309	26	.7692	1300	26.59	24.50	21.92	20.15
5	100.2	9.980	3464	3192	2855	2625	27	.6100	1639	21.09	19.43	17.38	15.98
6	79.46	12.58	2747	2531	2264	2081	28	.4837	2067	16.72	15.41	13.78	12.67
7	63.02	15.87	2179	2007	1796	1651	29	.3836	2607	13.26	12.22	10.93	10.05
8	49.98	20.01	1728	1592	1424	1309	30	.3042	3287	10.52	9.691	8.669	7.968
9	39.63	25.23	1370	1262	1129	1038	31	.2413	4145	8.341	7.685	6.875	6.319
10	31.43	31.82	1087	1001	895.6	823.2	32	.1913	5227	6.614	6.095	5.452	5.011
11	24.92	40.12	861.7	794.0	710.2	652.8	33	.1517	6591	5.245	4.833	4.323	3.974
12	19.77	50.59	683.3	629.6	563.2	517.7	34	.1203	8310	4.160	3.833	3.429	3.152
13	15.68	63.80	541.9	499.3	446.7	410.6	35	.09542	10480	3.299	3.040	2.719	2.499
14	12.43	80.44	429.8	396.0	354.2	325.6	36	.07568	13210	2.616	2.411	2.156	1.982
15	9.858	101.4	340.8	314.0	280.9	258.2	37	.06001	16660	2.075	1.912	1.710	1.572
16	7.818	127.9	270.3	249.0	222.8	204.8	38	.04759	21010	1.645	1.516	1.356	1.247
17	6.200	161.3	214.3	197.5	176.7	162.4	39	.03774	26500	1.305	1.202	1.075	.9886
18	4.917	203.4	170.0	156.6	140.1	128.8	40	.02993	33410	1.035	.9534	.8529	.7840

* Length at 20°C of a wire whose resistance is 1 ohm at the stated temperatures.

WIRE TABLE, STANDARD ANNEALED COPPER (Continued)

American Wire Gauge (B. & S.) English Units

Gauge No.	Diameter in mils at 20°C	Ohms per pound 0°C (32°F)	Ohms per pound 20°C (68°F)	Ohms per pound 50°C (122°F)	Lbs. per ohm 20°C (68°F)
0000	460.0	0.00007051	0.00007652	0.00008554	13070
000	409.6	.0001121	.0001217	.0001360	8219
00	364.8	.0001783	.0001935	.0002163	5169
0	324.9	.0002835	.0003076	.0003439	3251
1	289.3	.0004507	.0004891	.0005468	2044
2	257.6	.0007166	.0007778	.0008695	1286
3	229.4	.001140	.001237	.001383	808.6
4	204.3	.001812	.001966	.002198	508.5
5	181.9	.002881	.003127	.003495	319.8
6	162.0	.004581	.004972	.005558	201.1
7	144.3	.007284	.007905	.008838	126.5
8	128.5	.01158	.01257	.01405	79.55
9	114.4	.01842	.01999	.02234	50.03
10	101.9	.02928	.03178	.03553	31.47
11	90.74	.04656	.05053	.05649	9.79
12	80.81	.07404	.08035	.08983	12.45
13	71.96	.1177	.1278	.1428	7.827
14	64.08	.1872	.2032	.2271	4.922
15	57.07	.2976	.3230	.3611	3.096
16	50.82	.4733	.5136	.5742	1.947
17	45.26	.7525	.8167	.9130	1.224
18	40.30	1.197	1.299	1.452	.7700
19	35.89	1.903	2.065	2.308	.4843
20	31.96	3.025	3.283	3.670	.3046
21	28.46	4.810	5.221	5.836	.1915
22	25.35	7.649	8.301	9.280	.1205
23	22.57	12.16	13.20	14.76	.07576
24	20.10	19.34	20.99	23.46	.04765
25	17.90	30.75	33.37	37.31	.02997
26	15.94	48.89	53.06	59.32	.01885
27	14.20	77.74	84.37	94.32	.01185
28	12.64	123.6	134.2	150.0	.007454
29	11.26	196.6	213.3	238.5	.004688
30	10.03	312.5	339.2	379.2	.002948
31	8.928	497.0	539.3	602.9	.001854
32	7.950	790.2	857.6	958.7	.001166
33	7.080	1256	1364	1524	.0007333
34	6.305	1998	2168	2424	.0004612
35	5.615	3177	3448	3854	.0002901
36	5.000	5051	5482	6128	.0001824
37	4.453	8032	8717	9744	.0001147
38	3.965	12770	13860	15490	.00007215
39	3.531	20310	22040	24640	.00004538
40	3.145	32290	35040	39170	.00002854

Gauge No.	Diameter in mm at 20°C	Cross section in mm² at 20°C	Ohms per kilometer* 0°C	Ohms per kilometer* 20°C	Ohms per kilometer* 50°C	Ohms per kilometer* 75°C
0000	11.68	107.2	0.1482	0.1608	0.1798	0.1956
000	10.40	85.03	.1868	.2028	.2267	.2466
00	9.266	67.43	.2356	.2557	.2858	.3110
0	8.252	53.48	.2971	.3224	.3604	.3921
1	7.348	42.41	.3746	.4066	.4545	.4944
2	6.544	33.63	.4724	.5127	.5731	.6235
3	5.827	26.67	.5956	.6465	.7227	.7862
4	5.189	21.15	.7511	.8152	.9113	.9914
5	4.621	16.77	.9471	1.028	1.149	1.250
6	4.115	13.30	1.194	1.296	1.449	1.576
7	3.665	10.55	1.506	1.634	1.827	1.988
8	3.264	8.366	1.899	2.061	2.304	2.506
9	2.906	6.634	2.395	2.599	2.905	3.161
10	2.588	5.261	3.020	3.277	3.663	3.985
11	2.305	4.172	3.807	4.132	4.619	5.025
12	2.053	3.309	4.801	5.211	5.825	6.337
13	1.828	2.624	6.054	6.571	7.345	7.991
14	1.628	2.081	7.634	8.285	9.262	10.08
15	1.450	1.650	9.627	10.45	11.68	12.71
16	1.291	1.309	12.14	13.17	14.73	16.02
17	1.150	1.038	15.31	16.61	18.57	20.20
18	1.024	.8231	19.30	20.95	23.42	25.48
19	.9116	.6527	24.34	26.42	29.53	32.12
20	.8118	.5176	30.69	33.31	37.24	40.51
21	.7230	.4105	38.70	42.00	46.95	51.08
22	.6438	.3255	48.80	52.96	59.21	64.41
23	.5733	.2582	61.54	66.79	74.66	81.22
24	.5106	.2047	77.60	84.21	94.14	102.4
25	.4547	.1624	97.85	106.2	118.7	129.1
26	.4049	.1288	123.4	133.9	149.7	162.9
27	.3606	.1021	155.6	168.9	188.8	205.4
28	.3211	.08098	196.2	212.9	238.0	258.9
29	.2859	.06422	247.4	268.5	300.1	326.5
30	.2546	.05093	311.9	338.6	378.5	411.7
31	.2268	.04039	393.4	426.9	477.2	519.2
32	.2019	.03203	496.0	538.3	601.8	654.7
33	.1798	.02540	625.5	678.8	758.8	825.5
34	.1601	.02014	788.7	856.0	956.9	1041
35	.1426	.01597	994.5	1079	1207	1313
36	.1270	.01267	1254	1361	1522	1655
37	.1131	.01005	1581	1716	1919	2087
38	.1007	.007967	1994	2164	2419	2632
39	.08969	.006318	2514	2729	3051	3319
40	.07987	.005010	3171	3441	3847	4185

* Resistance at the stated temperatures of a wire whose length is 1 kilometer at 20°C.

Gauge No.	Diameter in mm at 20°C	Kilograms per kilometer	Meters per gram	Meters per ohm* 0°C	Meters per ohm* 20°C	Meters per ohm* 50°C	Meters per ohm* 75°C
0000	11.68	953.2	0.001049	6749	6219	5563	5113
000	10.40	755.9	.001323	5352	4932	4412	4055
00	9.266	599.5	.001668	4245	3911	3499	3216
0	8.252	475.4	.002103	3366	3102	2774	2550
1	7.348	377.0	.002652	2669	2460	2200	2022
2	6.544	299.0	.003345	2117	1951	1745	1604
3	5.827	237.1	.004217	1679	1547	1384	1272
4	5.189	188.0	.005318	1331	1227	1097	1009
5	4.621	149.1	.006706	1056	972.9	870.2	799.9
6	4.115	118.2	.008457	837.3	771.5	690.1	634.4
7	3.665	93.78	.01066	664.0	611.8	547.3	503.1
8	3.264	74.37	.01345	526.6	485.2	434.0	399.0
9	2.906	58.98	.01696	417.6	384.8	344.2	316.4
10	2.588	46.77	.02138	331.2	305.1	273.0	250.9
11	2.305	37.09	.02696	262.6	242.0	216.5	199.0
12	2.053	29.42	.03400	208.3	191.9	171.7	157.8
13	1.828	23.33	.04287	165.2	152.2	136.1	125.1
14	1.628	18.50	.05406	131.0	120.7	108.0	99.24
15	1.450	14.67	.06816	103.9	95.71	85.62	78.70
16	1.291	11.63	.08595	82.38	75.90	67.90	62.41
17	1.150	9.226	.1084	65.33	60.20	53.85	49.50
18	1.024	7.317	.1367	51.81	47.74	42.70	39.25
19	0.9116	5.803	.1723	41.09	37.86	33.86	31.13
20	.8118	4.602	.2173	32.58	30.02	26.86	24.69
21	.7230	3.649	.2740	25.84	23.81	21.30	19.58
22	.6438	2.894	.3455	20.49	18.88	16.89	15.53
23	.5733	2.295	.4357	16.25	14.97	13.39	12.31
24	.5106	1.820	.5494	12.89	11.87	10.62	9.764
25	.4547	1.443	.6928	10.22	9.417	8.424	7.743
26	.4049	1.145	.8736	8.105	7.468	6.680	6.141
27	.3606	0.9078	1.102	6.428	5.922	5.298	4.870
28	.3211	.7199	1.389	5.097	4.697	4.201	3.862
29	.2859	.5709	1.752	4.042	3.725	3.332	3.063
30	.2546	.4527	2.209	3.206	2.954	2.642	2.429
31	.2268	.3590	2.785	2.542	2.342	2.095	1.926
32	.2019	.2847	3.512	2.016	1.858	1.662	1.527
33	.1798	.2258	4.429	1.599	1.473	1.318	1.211
34	.1601	.1791	5.584	1.268	1.168	1.045	.9606
35	.1426	.1420	7.042	1.006	0.9265	0.8288	.7618
36	.1270	.1126	8.879	0.7974	.7347	.6572	.6041
37	.1131	.08931	11.20	.6324	.5827	.5212	.4791
38	.1007	.07083	14.12	.5015	.4621	.4133	.3799
39	.08969	.05617	17.80	.3977	.3664	.3278	.3013
40	.07987	.04454	22.45	.3154	.2906	.2600	.2390

Gauge No.	Ohms per kilogram 0°C	Ohms per kilogram 20°C	Ohms per kilogram 50°C	Grams per ohm 20°C
0000	0.0001554	0.0001687	0.0001886	5928000
000	.0002472	.0002682	.0002999	3728000
00	.0003930	.0004265	.0004768	2344000
0	.0006249	.0006782	.0007582	1474000
1	.0009936	.001078	.001206	927300
2	.001580	.001715	.001917	583200
3	.002512	.002726	.003048	366800
4	.003995	.004335	.004846	230700
5	.006352	.006893	.007706	145100
6	.01010	.01096	.01225	91230
7	.01606	.01743	.01948	57380
8	.02553	.02771	.03098	36080
9	.04060	.04406	.04926	22690
10	.06456	.07007	.07833	14270
11	.1026	.1114	.1245	8976
12	.1632	.1771	.1980	5645
13	.2595	.2817	.3149	3550
14	.4127	.4479	.5007	2233
15	.6562	.7122	.7961	1404
16	1.043	1.132	1.266	883.1
17	1.659	1.801	2.013	555.4
18	2.638	2.863	3.201	349.3
19	4.194	4.552	5.089	219.7
20	6.670	7.238	8.092	138.2
21	10.60	11.51	12.87	86.88
22	16.86	18.30	20.46	54.64
23	26.81	29.10	32.53	34.36
24	42.63	46.27	51.73	21.61
25	67.79	73.57	82.25	13.59
26	107.8	117.0	130.8	8.548
27	171.4	186.0	207.9	5.376
28	272.5	295.8	330.6	3.381
29	433.3	470.3	525.7	2.126
30	689.0	747.8	836.0	1.337
31	1096	1189	1329	0.8410
32	1742	1891	2114	.5289
33	2770	3006	3361	.3326
34	4404	4780	5344	.2092
35	7003	7601	8497	.1316
36	11140	12090	13510	.08274
37	17710	19220	21480	.05204
38	28150	305360	34160	.03273
39	44770	48590	54310	.02058
40	71180	77260	88360	.01294

* Length at 20°C of a wire whose resistance is 1 ohm at the stated temperatures.

ALUMINUM WIRE TABLE
Hard-Drawn Aluminum Wire at 20°C (or, 68°F) American Wire Gauge (B. & S.) English Units

Gauge No.	Diameter in mils	Cross section Circular mils	Cross section Square inches	Ohms per 1000 ft.	Pounds per 1000 ft.	Pounds per ohm	Feet per ohm	Gauge No.	Diameter in mm	Cross section in mm²	Ohms per kilometer	Kilograms per kilometer	Grams per ohm	Meters per ohm
0000	460	212000	0.166	0.0804	195	2420	12400	0000	11.7	107	.0264	289	1100000	3790
000	410	168000	.132	.101	154	1520	9860	000	10.4	85.0	.333	230	690000	3010
00	365	133000	.105	.128	122	957	7820	00	9.3	67.4	.419	182	434000	2380
0	325	106000	.0829	.161	97.0	602	6200	0	8.3	53.5	.529	144	273000	1890
1	289	83700	.0657	.203	76.9	379	4920	1	7.3	42.4	.667	114.	172000	1500
2	258	66400	.0521	.256	61.0	238	3900	2	6.5	33.6	.841	90.8	108000	1190
3	229	52600	.0413	.323	48.4	150	3090	3	5.8	26.7	1.06	72.0	67900	943
4	204	41700	.0328	.408	38.4	94.2	2450	4	5.2	21.2	1.34	57.1	42700	748
5	182	33100	.0260	.514	30.4	59.2	1950	5	4.6	16.8	1.69	45.3	26900	593
6	162	26300	.0206	.648	24.1	37.2	1540	6	4.1	13.3	2.13	35.9	16900	470
7	144	20800	.0164	.817	19.1	23.4	1220	7	3.7	10.5	2.68	28.5	10600	373
8	128	16500	.0130	1.03	15.2	14.7	970	8	3.3	8.37	3.38	22.6	6680	296
9	114	13100	.0103	1.30	12.0	9.26	770	9	2.91	6.63	4.26	17.9	4200	235
10	102	10400	.00815	1.64	9.55	5.83	610	10	2.59	5.26	5.38	14.2	2640	186
11	91	8230	.00647	2.07	7.57	3.66	484	11	2.30	4.17	6.78	11.3	1660	148
12	81	6530	.00513	2.61	6.00	2.30	384	12	2.05	3.31	8.55	8.93	1050	117
13	72	5180	.00407	3.29	4.76	1.45	304	13	1.83	2.62	10.8	7.08	657	92.8
14	64	4110	.00323	4.14	3.78	.911	241	14	1.63	2.08	13.6	5.62	413	73.6
15	57	3260	.00256	5.22	2.99	.573	191	15	1.45	1.65	17.1	4.46	260	58.4
16	51	2580	.00203	6.59	2.37	.360	152	16	1.29	1.31	21.6	3.53	164	46.3
17	45	2050	.00161	8.31	1.88	.227	120	17	1.15	1.04	27.3	2.80	103	36.7
18	40	1620	.00128	10.5	1.49	.143	95.5	18	1.02	.823	34.4	2.22	64.7	29.1
19	36	1290	.00101	13.2	1.18	.0897	75.7	19	0.91	.653	43.3	1.76	40.7	23.1
20	32	1020	.000802	16.7	.939	.0564	60.0	20	.81	.518	54.6	1.40	25.6	18.3
21	28.5	810	.000636	21.0	.745	.0355	47.6	21	.72	.411	68.9	1.11	16.1	14.5
22	25.3	642	.000505	26.5	.591	.0223	37.8	22	.64	.326	86.9	0.879	10.1	11.5
23	22.6	509	.000400	33.4	.468	.0140	29.9	23	.57	.258	110	.697	6.36	9.13
24	20.1	404	.000317	42.1	.371	.00882	23.7	24	.51	.205	138	.553	4.00	7.24
25	17.9	320	.000252	53.1	.295	.00555	18.8	25	.45	.162	174	.438	2.52	5.74
26	15.9	254	.000200	67.0	.234	.00349	14.9	26	.40	.129	220	.348	1.58	4.55
27	14.2	202	.000158	84.4	.185	.00219	11.8	27	.36	.102	277	.276	0.995	3.61
28	12.6	160	.000126	106.	.147	.00138	9.39	28	.32	.0810	349	.219	.626	2.86
29	11.3	127	.0000995	134.	.117	.000868	7.45	29	.29	.0642	440	.173	.394	2.27
30	10.0	101	.0000789	169.	.0924	.000546	5.91	30	.25	.0509	555	.138	.248	1.80
31	8.9	79.7	.0000626	213.	.0733	.000343	4.68	31	.227	.0404	700	.109	.156	1.43
32	8.0	63.2	.0000496	269.	.0581	.000216	3.72	32	.202	.0320	883	.0865	.0979	1.13
33	7.1	50.1	.0000394	339.	.0461	.000136	2.95	33	.180	.0254	1110	.0686	.0616	0.899
34	6.3	39.8	.0000312	428.	.0365	.0000854	2.34	34	.160	.0201	1400	.0544	.0387	.712
35	5.6	31.5	.0000248	540.	.0290	.0000537	1.85	35	.143	.0160	1770	.0431	.0244	.565
36	5.0	25.0	.0000196	681.	.0230	.0000338	1.47	36	.127	.0127	2230	.0313	.0193	.448
37	4.5	19.8	.0000156	860.	.0182	.0000212	1.17	37	.113	.0100	2820	.0271	.00963	.355
38	4.0	15.7	.0000123	1080.	.0145	.0000134	0.924	38	.101	.00810	3550	.0215	.00608	.282
39	3.5	12.5	.00000979	1360.	.0113	.00000840	.733	39	.090	.0063	4480	.0171	.00381	.223
40	3.1	9.9	.0000077	1720.	.0091	.00000528	.581	40	.080	.0050	5640	.0135	.00240	.177

CROSS-SECTION AND MASS OF WIRES
U. S. Measure

Diameters are given in mils (1 mil = .001 in.), and area in square mils (1 sq. mil = .000001 sq. in.). For sections and masses for one-tenth the diameters given, divide by 100 and for sections and masses for ten times the diameter multiply by 100.

Diam. in mils	Cross-sec. in sq. mils	Pounds per foot Copper, density 8.90	Iron, density 7.80	Brass, density 8.56	Aluminum, density 2.67	Diam. in mils	Cross-sec. in sq. mils	Pounds per foot Copper, density 8.90	Iron, density 7.80	Brass, density 8.56	Aluminum, density 2.67
10	78.54	0.000303	0.0002656	0.0002915	0.0000909	40	1256.64	0.004849	0.004249	0.004664	0.001455
11	95.03	0367	03214	03527	01100	41	1320.25	5094	4465	4900	1528
12	113.10	0436	03825	04197	01309	42	1385.44	5346	4685	5141	1604
13	132.73	0512	04488	04926	01536	43	1452.20	5603	4911	5389	1681
14	153.94	0594	05206	05713	01782	44	1520.53	5867	5142	5643	1760
15	176.71	0682	05976	06558	02045	45	1590.43	6137	5378	5902	1841
16	201.06	0776	06799	07461	02327	46	1661.90	6412	5620	6167	1924
17	226.98	0876	07675	08423	02627	47	1734.94	6694	5867	6438	2008
18	254.47	0982	08605	09443	02946	48	1809.56	6982	6119	6715	2095
19	283.53	1094	09588	10522	03282	49	1885.74	7276	6377	6998	2183
20	314.16	1212	1062	1166	03636	50	1963.50	7576	6640	7287	2273
21	346.36	1336	1171	1285	04009	51	2042.82	7882	6908	7581	2365
22	380.13	1467	1286	1411	04400	52	2123.72	8194	7181	7881	2458
23	415.48	1603	1405	1542	04809	53	2206.18	8512	7460	8187	2554
24	452.39	1746	1530	1679	05237	54	2290.22	8837	7744	8499	2651
25	490.87	1894	1660	1822	05682	55	2375.83	9167	8034	8817	2750
26	530.93	2046	1795	1970	06147	56	2463.01	9504	8329	9140	2851
27	572.56	2209	1936	2125	06628	57	2551.76	9846	8629	9470	2954
28	615.75	2376	2082	2285	07127	58	2642.08	10195	8934	9805	3058
29	660.52	2549	2234	2451	07646	59	2733.97	10549	9245	10146	3165
30	706.86	2727	2390	2623	08182	60	2827.43	1091	956	1049	3273
31	754.77	2912	2552	2801	08737	61	2922.47	1128	988	1085	3383
32	804.25	3103	2720	2985	09309	62	3019.07	1165	1021	1120	3495
33	855.30	3300	2892	3174	09900	63	3117.25	1203	1054	1157	3608
34	907.92	3503	3070	3369	10509	64	3216.99	1241	1088	1194	3724
35	962.11	3712	3253	3570	1114	65	3318.31	1280	1122	1231	3841
36	1017.88	3927	3442	3777	1178	66	3421.19	1320	1157	1270	3960
37	1075.21	4149	3636	3990	1245	67	3525.65	1360	1192	1308	4081
38	1134.11	4376	3844	4218	1316	68	3631.68	1401	1228	1348	4204
39	1194.59	4609	4040	4433	1383	69	3739.28	1443	1264	1388	4328

U. S. Measure

Diam. in mils	Cross-sec. in sq. mils	Pounds per foot				Diam. in mils	Cross-sec. in sq. mils	Pounds per foot			
		Copper, density 8.90	Iron, density 7.80	Brass, density 8.56	Aluminum, density 2.67			Copper, density 8.90	Iron, density 7.80	Brass, density 8.56	Aluminum, density 2.67
70	3848.45	0.01485	0.01302	0.01429	0.004456	86	5808.80	0.02241	0.01964	0.02156	0.006724
71	3959.19	1528	1339	1469	4583	87	5944.68	2294	2010	2206	6881
72	4071.50	1571	1377	1511	4713	88	6082.12	2347	2057	2257	7040
73	4185.39	1615	1415	1553	4845	89	6221.14	2400	2104	2309	7201
74	4300.84	1660	1454	1596	4978	90	6361.73	2455	2151	2360	7364
75	4417.86	1705	1494	1639	5114	91	6503.88	2509	2199	2414	7528
76	4536.46	1751	1534	1684	5251	92	6647.61	2565	2248	2467	7695
77	4656.63	1797	1575	1728	5390	93	6792.91	2621	2297	2521	7863
78	4778.36	1844	1616	1773	5531	94	6939.78	2678	2347	2575	8033
79	4901.67	1892	1658	1819	5674	95	7088.22	2735	2397	2630	8205
80	5026.55	1939	1700	1865	5818	96	7238.23	2793	2448	2686	8378
81	5153.00	1988	1743	1912	5965	97	7389.81	2851	2499	2742	8554
82	5281.02	2038	1786	1960	6113	98	7542.96	2910	2551	2799	8731
83	5410.61	2088	1830	2008	6263	99	7697.69	2970	2603	2857	8910
84	5541.77	2138	1874	2057	6415	100	7853.98	3030	2656	2915	9091
85	5674.50	2189	1919	2106	6568						

Metric Measure

Diameters are given in thousandths of a centimeter and area of section in square thousandths of a centimeter. 1 $(cm/1000)^2$ = .000001 sq. cm. For sections and masses for diameters 1/10 or 10 times those of the table, divide or multiply by 100.

Diam. in thousandths of a cm	Cross-section in square thousandths of a cm	Grams per meter				Diam. in thousandths of a cm	Cross-section in square thousandths of a cm	Grams per meter			
		Copper, density 8.90	Iron, density 7.80	Brass, density 8.56	Aluminum, density 2.67			Copper, density 8.90	Iron, density 7.80	Brass, density 8.56	Aluminum, density 2.67
10	78.54	0.06990	0.06126	0.06723	0.02097	55	2375.83	2.114	1.853	2.034	0.6343
11	95.03	.08458	.07412	.08135	.02537	56	2463.01	.192	.921	.108	.6576
12	113.10	.10065	.08822	.09681	.03020	57	2551.76	.271	.990	.184	.6813
13	132.73	.11813	.10353	.11362	.03544	58	2642.08	.351	2.061	.262	.7054
14	153.94	.13701	.12008	.13177	.04110	59	2733.97	.433	.132	.340	.7300
15	176.71	.1573	.1378	.1513	.04718	60	2827.43	2.516	2.205	2.420	.7549
16	201.06	.1789	.1568	.1721	.05368	61	2922.47	.601	.280	.502	.7803
17	226.98	.2020	.1770	.1943	.06060	62	3019.07	.687	.355	.584	.8061
18	254.47	.2265	.1985	.2178	.06794	63	3117.25	.774	.431	.668	.8323
19	283.53	.2523	.2212	.2427	.07570	64	3216.99	.863	.509	.760	.8589
20	314.16	.2796	.2450	.2689	.08388	65	3318.31	2.953	2.588	2.840	.8860
21	346.36	.3083	.2702	.2965	.09248	66	3421.19	3.045	.669	.929	.9135
22	380.13	.3383	.2965	.3254	.10149	67	3525.65	.138	.750	3.018	.9413
23	415.48	.3698	.3241	.3557	.11093	68	3631.68	.232	.833	.109	.9697
24	452.39	.4026	.3529	.3872	.12079	69	3739.28	.328	.917	.201	.9984
25	490.87	.4369	.3829	.4202	.1311	70	3848.45	3.426	3.003	3.295	1.028
26	530.93	.4725	.4141	.4545	.1418	71	3959.19	.524	.088	.389	.057
27	572.56	.5096	.4466	.4901	.1529	72	4071.50	.624	.176	.485	.087
28	615.75	.5480	.4803	.5271	.1644	73	4185.39	.725	.265	.583	.117
29	660.52	.5879	.5152	.5654	.1764	74	4300.84	.828	.355	.682	.148
30	706.86	.6291	.5514	.6051	.1887	75	4417.86	3.932	3.446	3.782	1.180
31	754.77	.6717	.5887	.6461	.2015	76	4536.46	4.037	.538	.883	.211
32	804.25	.7158	.6273	.6884	.2147	77	4656.63	.144	.632	.986	.243
33	855.30	.7612	.6671	.7321	.2284	78	4778.36	.253	.727	4.090	.276
34	907.92	.8081	.7082	.7772	.2424	79	4901.67	.362	.823	.177	.309
35	962.11	.856	.7504	.8236	.2569	80	5026.55	4.474	3.921	4.303	1.342
36	1017.88	.906	.7939	.8713	.2718	81	5153.00	.586	4.019	.411	.376
37	1075.21	.957	.8387	.9204	.2871	82	5281.02	.700	.119	.521	.410
38	1134.11	1.012	.8866	.9730	.3035	83	5410.61	.815	.220	.631	.445
39	1194.59	.063	.9318	1.0230	.3190	84	5541.77	.932	.323	.744	.480
40	1256.64	1.118	.980	1.076	.3355	85	5674.50	5.050	4.426	4.857	1.515
41	1320.25	.175	1.030	.130	.3525	86	5808.80	.170	.531	.972	.551
42	1385.44	.233	.081	.186	.3699	87	5944.68	.291	.637	5.089	.587
43	1452.20	.292	.133	.243	.3877	88	6082.12	.413	.744	.206	.624
44	1520.53	.353	.186	.302	.4060	89	6221.14	.537	.852	.325	.661
45	1590.43	1.415	1.241	1.361	.4246	90	6361.73	5.662	4.962	5.446	1.699
46	1661.90	.479	.296	.423	.4437	91	6503.88	.788	5.073	.567	.737
47	1734.94	.544	.353	.485	.4632	92	6647.61	.916	.185	.690	.775
48	1809.56	.611	.411	.549	.4832	93	6792.91	6.046	.298	.815	.814
49	1885.74	.678	.471	.614	.5035	94	6939.78	.176	.413	.940	.853
50	1963.50	1.748	1.532	1.681	.5243	95	7088.22	6.309	5.529	6.068	1.893
51	2042.82	.818	.593	.753	.5454	96	7238.23	.442	.646	.196	.933
52	2123.72	.890	.657	.818	.5670	97	7389.81	.577	.764	.326	.973
53	2206.18	.964	.721	.888	.5891	98	7542.96	.713	.884	.457	2.014
54	2290.22	2.038	.786	.960	.6115	99	7697.69	.851	6.004	.589	.055
						100	7853.98	6.990	6.126	6.723	2.097

RESISTANCE OF WIRES

The following table gives the approximate resistance of various metallic conductors. The values have been computed from the resistivities at 20°C, except as otherwise stated, and for the dimensions of wire indicated. Owing to differences in purity in the case of elements and of composition in alloys, the values can be considered only as approximations.

The following dimensions have been adopted in the computations.

B. & S. gauge	Diameter		B. & S. gauge	Diameter	
	mm	mils 1 mil = .001 in.		mm	mils 1 mil = .001 in.
10	2.588	101.9	26	0.4049	15.94
12	2.053	80.81	27	0.3606	14.20
14	1.628	64.08	28	0.3211	12.64
16	1.291	50.82	30	0.2546	10.03
18	1.024	40.30	32	0.2019	7.950
20	0.8118	31.96	34	0.1601	6.305
22	0.6438	25.35	36	0.1270	5.000
24	0.5106	20.10	40	0.07987	3.145

B. & S. No.	Ohms per cm	Ohms per ft.	B. & S. No.	Ohms per cm	Ohms per ft.	B. & S. No.	Ohms per cm	Ohms per ft.	B. & S. No.	Ohms per cm	Ohms per ft.
*Advance (0°C) $\varrho = 48. \times 10^{-6}$ ohm cm			Aluminum $\varrho = 2.828 \times 10^{-6}$ ohm cm			Eureka (0°C) $\varrho = 47. \times 10^{-6}$ ohm cm			Excello $\varrho = 92. \times 10^{-6}$ ohm cm		
10	.000912	.0278	10	.0000538	.00164	10	.000893	.0272	10	.00175	.0533
12	.00145	.0442	12	.0000855	.00260	12	.00142	.0433	12	.00278	.0847
14	.00231	.0703	14	.000136	.00414	14	.00226	.0688	14	.00442	.135
16	.00367	.112	16	.000216	.00658	16	.00359	.109	16	.00703	.214
18	.00583	.178	18	.000344	.0105	18	.00571	.174	18	.0112	.341
20	.00927	.283	20	.000546	.0167	20	.00908	.277	20	.0178	.542
22	.0147	.449	22	.000869	.0265	22	.0144	.440	22	.0283	.861
24	.0234	.715	24	.00138	.0421	24	.0230	.700	24	.0449	1.37
26	.0373	1.14	26	.00220	.0669	26	.0365	1.11	26	.0714	2.18
27	.0470	1.43	27	.00277	.0844	27	.0460	1.40	27	.0901	2.75
28	.0593	1.81	28	.00349	.106	28	.0580	1.77	28	.114	3.46
30	.0942	2.87	30	.00555	.169	30	.0923	2.81	30	.181	5.51
32	.150	4.57	32	.00883	.269	32	.147	4.47	32	.287	8.75
34	.238	7.26	34	.0140	.428	34	.233	7.11	34	.457	13.9
36	.379	11.5	36	.0223	.680	36	.371	11.3	36	.728	22.1
40	.958	29.2	40	.0564	1.72	40	.938	28.6	40	1.84	56.0
Brass $\varrho = 7.00 \times 10^{-6}$ ohm cm			Climax $\varrho = 87. \times 10^{-6}$ ohm cm			German silver $\varrho = 33. \times 10^{-6}$ ohm cm			Gold $\varrho = 2.44 \times 10^{-6}$ ohm cm		
10	.000133	.00406	10	.00165	.0504				10	.0000464	.00141
12	.000212	.00645	12	.00263	.0801	10	.000627	.0191	12	.0000737	.00225
14	.000336	.0103	14	.00418	.127	12	.000997	.0304	14	.000117	.00357
16	.000535	.0163	16	.00665	.203	14	.00159	.0483	16	.000186	.00568
18	.000850	.0259	18	.0106	.322	16	.00252	.0768	18	.000296	.00904
20	.00135	.0412	20	.0168	.512	18	.00401	.122	20	.000471	.0144
22	.00215	.0655	22	.0267	.815	20	.00638	.194	22	.000750	.0228
24	.00342	.104	24	.0425	1.30	22	.0101	.309	24	.00119	.0363
26	.00543	.166	26	.0675	2.06	24	.0161	.491	26	.00189	.0577
27	.00686	.209	27	.0852	2.60	26	.0256	.781	27	.00239	.0728
28	.00864	.263	28	.107	3.27	27	.0323	.985	28	.00301	.0918
30	.0137	.419	30	.171	5.21	28	.0408	1.24	30	.00479	.146
32	.0219	.666	32	.272	8.28	30	.0648	1.97	32	.00762	.232
34	.0348	1.06	34	.432	13.2	32	.103	3.14	34	.0121	.369
36	.0552	1.68	36	.687	20.9	34	.164	4.99	36	.0193	.587
40	.140	4.26	40	1.74	52.9	36	.260	.794	40	.0487	1.48
						40	.659	20.1			
Constantan (0°C) $\varrho = 44.1 \times 10^{-6}$ ohm cm			Copper, annealed $\varrho = 1.724 \times 10^{-6}$ ohm cm			Iron $\varrho = 10. \times 10^{-6}$ ohm cm			Lead $\varrho = 22. \times 10^{-6}$ ohm cm		
10	.000838	.0255	10	.0000328	.000999	10	.000190	.00579	10	.000418	.0127
12	.00133	.0406	12	.0000521	.00159	12	.000302	.00921	12	.000665	.0203
14	.00212	.0646	14	.0000828	.00253	14	.000481	.0146	14	.00106	.0322
16	.00337	.103	16	.000132	.00401	16	.000764	.0233	16	.00168	.0512
18	.00536	.163	18	.000209	.00638	18	.00121	.0370	18	.00267	.0815
20	.00852	.260	20	.000333	.0102	20	.00193	.0589	20	.00425	.130
22	.0135	.413	22	.000530	.0161	22	.00307	.0936	22	.00676	.206
24	.0215	.657	24	.000842	.0257	24	.00489	.149	24	.0107	.328
26	.0342	1.04	26	.00134	.0408	26	.00776	.237	26	.0171	.521
27	.0432	1.32	27	.00169	.0515	27	.00979	.299	27	.0215	.657
28	.0545	1.66	28	.00213	.0649	28	.0123	.376	28	.0272	.828
30	.0866	2.64	30	.00339	.103	30	.0196	.598	30	.0432	1.32
32	.138	4.20	32	.00538	.164	32	.0312	.952	32	.0687	2.09
34	.219	6.67	34	.00856	.261	34	.0497	1.51	34	.109	3.33
36	.348	10.6	36	.0136	.415	36	.789	2.41	36	.174	5.29
40	.880	26.8	40	.0344	1.05	40	.200	6.08	40	.439	13.4

* Trade mark.

B. & S. No.	Ohms per cm	Ohms per ft.	B. & S. No.	Ohms per cm	Ohms per ft.	B. & S. No.	Ohms per cm	Ohms per ft.	B. & S. No.	Ohms per cm	Ohms per ft.
Magnesium $\varrho = 4.6 \times 10^{-6}$ ohm cm			Manganin $\varrho = 44. \times 10^{-6}$ ohm cm			Platinum $\varrho = 10. \times 10^{-6}$ ohm cm			Silver (18°C) $\varrho = 1.629 \times 10^{-6}$ ohm cm		
10	.0000874	.00267	10	.000836	.0255	10	.000190	.00579	10	.0000310	.000944
12	.000139	.00424	12	.00133	.0405	12	.000302	.00921	12	.0000492	.00150
14	.000221	.00674	14	.00211	.0644	14	.000481	.0146	14	.0000783	.00239
16	.000351	.0107	16	.00336	.102	16	.000764	.0233	16	.000124	.00379
18	.000559	.0170	18	.00535	.163	18	.00121	.0370	18	.000198	.00603
20	.000889	.0271	20	.00850	.259	20	.00193	.0589	20	.000315	.00959
22	.00141	.0431	22	.0135	.412	22	.00307	.0936	22	.000500	.0153
24	.00225	.0685	24	.0215	.655	24	.00489	.149	24	.000796	.0243
26	.00357	.109	26	.0342	1.04	26	.00776	.237	26	.00126	.0386
27	.00451	.137	27	.0431	1.31	27	.00979	.299	27	.00160	.0486
28	.00568	.173	28	.0543	1.66	28	.0123	.376	28	.00201	.0613
30	.00903	.275	30	.0864	2.63	30	.0196	.598	30	.00320	.0975
32	.0144	.438	32	.137	4.19	32	.0312	.952	32	.00509	.155
34	.0228	.696	34	.218	6.66	34	.0497	1.51	34	.00809	.247
36	.0363	1.11	36	.347	10.6	36	.0789	2.41	36	.0129	.392
40	.0918	2.80	40	.878	26.8	40	.200	6.08	40	.0325	.991
Molybdenum $\varrho = 5.7 \times 10^{-6}$ ohm cm			Monel Metal $\varrho = 42. \times 10^{-6}$ ohm cm			Steel, piano wire (0°C) $\varrho = 11.8 \times 10^{-6}$ ohm cm			Steel, invar (35% Ni) $\varrho = 81. \times 10^{-6}$ ohm cm		
10	.000108	.00330	10	.000798	.0243	10	.000224	.00684	10	.00154	.0469
12	.000172	.00525	12	.00127	.0387	12	.000357	.0109	12	.00245	.0746
14	.000274	.00835	14	.00202	.0615	14	.000567	.0173	14	.00389	.119
16	.000435	.0133	16	.00321	.0978	16	.000901	.0275	16	.00619	.189
18	.000693	.0211	18	.00510	.156	18	.00143	.0437	18	.00984	.300
20	.00110	.0336	20	.00811	.247	20	.00228	.0695	20	.0156	.477
22	.00175	.0534	22	.0129	.393	22	.00363	.110	22	.0249	.758
24	.00278	.0849	24	.0205	.625	24	.00576	.176	24	.0396	1.21
26	.00443	.135	26	.0326	.994	26	.00916	.279	26	.0629	1.92
27	.00558	.170	27	.0411	1.25	27	.0116	.352	27	.0793	2.42
28	.00704	.215	28	.0519	1.58	28	.0146	.444	28	.100	3.05
30	.0112	.341	30	.0825	2.51	30	.0232	.706	30	.159	4.85
32	.0178	.542	32	.131	4.00	32	.0368	1.12	32	.253	7.71
34	.0283	.863	34	.209	6.36	34	.0586	1.79	34	.402	12.3
36	.0450	1.37	36	.331	10.1	36	.0931	2.84	36	.639	19.5
40	.114	3.47	40	.838	25.6	40	.236	7.18	40	1.62	49.3
*Nichrome $\varrho = 150. \times 10^{-6}$ ohm cm			Nickel $\varrho = 7.8 \times 10^{-6}$ ohm cm			Tantalum $\varrho = 15.5 \times 10^{-6}$ ohm cm			Tin $\varrho = 11.5 \times 10^{-6}$ ohm cm		
10	.0021281	.06488	10	.000148	.00452	10	.000295	.00898	10	.000219	.00666
12	.0033751	.1029	12	.000236	.00718	12	.000468	.0143	12	.000348	.0106
14	.0054054	.1648	14	.000375	.0114	14	.000745	.0227	14	.000553	.0168
16	.0085116	.2595	16	.000596	.0182	16	.00118	.0361	16	.000879	.0268
18	.0138383	.4219	18	.000948	.0289	18	.00188	.0574	18	.00140	.0426
20	.0216218	.6592	20	.00151	.0459	20	.00299	.0913	20	.00222	.0677
22	.0346040	1.055	22	.00240	.0730	22	.00476	.145	22	.00353	.108
24	.0548088	1.671	24	.00381	.116	24	.00757	.231	24	.00562	.171
26	.0875760	2.670	26	.00606	.185	26	.0120	.367	26	.00893	.272
28	.1394328	4.251	27	.00764	.233	27	.0152	.463	27	.0113	.343
30	.2214000	6.750	28	.00963	.294	28	.0191	.583	28	.0142	.433
32	.346040	10.55	30	.0153	.467	30	.0304	.928	30	.0226	.688
34	.557600	17.00	32	.0244	.742	32	.0484	1.47	32	.0359	1.09
36	.885600	27.00	34	.0387	1.18	34	.0770	2.35	34	.0571	1.74
38	1.383832	42.19	36	.0616	1.88	36	.122	3.73	36	.0908	2.77
40	2.303872	70.24	40	.156	4.75	40	.309	9.43	40	.230	7.00

B. & S. No.	Ohms per cm	Ohms per ft.	B. & S. No.	Ohms per cm	Ohms per ft.
Tungsten $\varrho = 5.51 \times 10^{-6}$ ohm cm			Zinc (0°C) $\varrho = 5.75 \times 10^{-6}$ ohm cm		
10	.000105	.00319	10	.000109	.00333
12	.000167	.00508	12	.000174	.00530
14	.000265	.00807	14	.000276	.00842
16	.000421	.0128	16	.000439	.0134
18	.000669	.0204	18	.000699	.0213
20	.00106	.0324	20	.00111	.0339
22	.00169	.0516	22	.00177	.0538
24	.00269	.0820	24	.00281	.0856
26	.00428	.130	26	.00446	.136
27	.00540	.164	27	.00563	.172
28	.00680	.207	28	.00710	.216
30	.0108	.330	30	.0113	.344
32	.0172	.524	32	.0180	.547
34	.0274	.834	34	.0286	.870
36	.0435	1.33	36	.0454	1.38
40	.110	3.35	40	.115	3.50

ELECTRICAL RESISTIVITY AND TEMPERATURE COEFFICIENTS OF ELEMENTS

Element	Temperature °C	Microhm-Cm	Temperature Coefficient per °C	Element	Temperature °C	Microhm-Cm	Temperature Coefficient per °C
Aluminum, 99.996%	20	2.6548	0.00429[20 i]	Nickel	20	6.84	0.0069[0-100]
Antimony	0	39.0		Niobium (Columbium)[g]	0	12.5	
Arsenic	20	33.3		Osmium	20	9.5	0.0042[0-100]
Beryllium[a]	20	4.0	0.025[20 i]	Palladium	20	10.54	0.00374[0-60 g]
Bismuth	0	106.8		Phosphorus, white	11	1×10^{17}	
Boron	0	1.8×10^{12}		Platinum, 99.85%	20	10.6	0.003927[0-100]
Cadmium	0	6.83	0.0042[0 i]	Plutonium	107	141.4	
Calcium	0	3.91	0.00416[0 i]	Potassium	0	6.15	
Carbon[b]	0	1375.0		Praseodymium	25	68	0.00171[0-25]
Cerium	25	75.0	0.00087[0-25]	Rhenium	20	19.3	0.00395[0-100]
Cesium	20	20		Rhodium	20	4.51	0.0042[0-100]
Chromium	0	12.9	0.003[0 i]	Rubidium	20	12.5	
Cobalt	20	6.24	0.00604[0-100]	Ruthenium	0	7.6	
Copper	20	1.678	0.0068[0-500 g]	Samarium	25	88.0	0.00184[0-25]
Dysprosium[c]	25	57.0	0.00119[0-25]	Scandium[h]	22	61.0	0.00282[0-25]
Erbium	25	107.0	0.00201[0-25]	Selenium[k]	0	10^6	
Europium	25	90.0		Silicon	0	$3\text{-}4 \times 10^{6j}$	
Gadolinium	25	140.5	0.00176[0-25]	Silver	20	1.586	0.0061[0-100 g]
Gallium[d]	20	17.4		Sodium	0	4.2	
Germanium[e]	22	46×10^6		Strontium	20	23.0	
Gold	20	2.24	0.0083[0-100 g]	Sulfur, yellow	20	2×10^{23}	
Hafnium	25	35.1	0.0038[25 i]	Tantalum	25	12.45	0.00383[0-100]
Holmium	25	87.0	0.00171[0-25]	Tellurium	25	4.36×10^6	
Indium	20	8.37		Thallium	0	18.0	
Iodine	20	1.3×10^{15}		Thorium	0	13.0	0.0038[0-100]
Iridium	20	5.3	0.003925[0-100]	Thulium	25	79.0	0.00195[0-25]
Iron, 99.99%	20	9.71	0.00651[20 i]	Tin	0	11.0	0.0047[0-100]
Lanthanum	25	5.70	0.00218[0-25]	Titanium	20	42.0	
Lead	20	20.648	0.00336[20-40]	Tungsten	27	5.65	
Lithium	0	8.55		Uranium		30.0	
Lutetium	25	79.0	0.00240[0-25]	Vanadium	20	24.8-26.0	
Magnesium[f]	20	4.45	0.0165[20 i]	Ytterbium	25	29.0	0.0013[0-25]
Manganese α	23-100	185.0		Yttrium	25	57.0	0.0027[0-25]
Mercury	50	98.4		Zinc	20	5.916	0.00419[0-100]
Molybdenum	0	5.2		Zirconium	20	40.0	0.0044[20 i]
Neodymium	25	64.0	0.00164[0-25]				

a   Annealed, comm. pure.
b   Graphite.
c   Polycrystalline.
d   Hard Wire.
e   Intrinsic Ge.
f   Polycrystalline.
g   High Purity.
h   Zone refined bar.
i   Data not available to indicate range over which coefficient is valid.
j   Very sensitive to purity.
k   Crystalline.

ELECTRICAL RESISTIVITY OF THE ALKALI METALS

Metal	Atomic no.	Atomic wt.	Density at 293.15 K	M.P. (°K)	B.P. (°K)	Resistivity (μ-ohm)	at °K
Li	3	6.941	0.534	453.7	1617	8.53	273.15
						9.28	293.15
						11.45	350.
						15.59(s)	453.7
						24.80(l)	453.7
Na	11	22.989	0.971	371.0	1157	4.33	273.15
						4.77	293.15
						6.23	350.
						6.86(s)	371.
						9.43(l)	371.
K	19	39.098	0.871	336.35	1032	6.49	273.15
						7.20	293.15
						9.22(s)	336.35
						13.95(l)	336.35
						14.64	350.
Rb	37	85.4678	1.53	312.64	961	11.54	273.15
						12.84	293.15
						14.21(s)	312.64
						22.52(l)	312.64
						25.42	350.
Cs	55	132.9054	1.873	301.55	944	18.75	273.15
						20.46	293.15
						21.16(s)	301.55
						36.93(l)	301.55
						42.11	350.

LOW MELTING POINT SOLDERS

(From N.B.S. Circular 492)

Pb	Sn	Bi	Liquidus Temperature, °F
25	25	50	266
50	37.5	12.5	374
25	50	25	336

Nominal Composition, Weight Percent

AWS-ASTM Classification	Solidus, °F	Liquidus, °F	Brazing Temperature Range, °F	Ag	Al	As	Au	B	Be	Bi	C	Cd	Cr	Cu	Fe	Li	Mg	Mn	Ni	P	Pb	Sb	Si	Sn	Ti	Zn	Other Each	Other Total	AWS-ASTM Classification
ALUMINUM-SILICON																													
BAlSi-2	1070	1135	1110-1150		Bal.									0.25	0.8			0.10					6.8-8.2			0.20	0.05	0.15	BAlSi-2
BAlSi-3	970	1085	1060-1120		Bal.							0.15		3.3-4.7	0.8		0.15	0.15					9.3-10.7			0.20	0.05	0.15	BAlSi-3
BAlSi-4	1070	1080	1080-1120		Bal.									0.30	0.8		0.10	0.15					11.0-13.0			0.20	0.05	0.15	BAlSi-4
BAlSi-5	1070	1095	1090-1120		Bal.									0.30	0.8		0.05	0.05					9.0-11.0		0.20	0.10	0.05	0.15	BAlSi-5
COPPER-PHOSPHORUS																													
BCuP-1	1310	1650	1450-1700											Bal.						4.75-5.25								0.15	BCuP-1
BCuP-2	1310	1460	1350-1550											Bal.						7.00-7.25								0.15	BCuP-2
BCuP-3	1190	1485	1300-1500	4.75-5.25										Bal.						5.75-6.25								0.15	BCuP-3
BCuP-4	1190	1335	1300-1450	5.75-6.25										Bal.						7.00-7.50								0.15	BCuP-4
BCuP-5	1190	1475	1300-1500	14.50-15.50										Bal.						4.75-5.25								0.15	BCuP-5
SILVER																													
BAg-1	1125	1145	1145-1400	44-46								23-25		14-16												14-18		0.15	BAg-1
BAg-1a	1160	1175	1175-1400	49-51								17-19		14.5-16.5												14.5-18.5		0.15	BAg-1a
BAg-2	1125	1295	1295-1550	34-36								17-19		25-27												19-23		0.15	BAg-2
BAg-3	1170	1270	1270-1500	49-51								15-17		14.5-16.5					2.5-3.5							13.5-17.5		0.15	BAg-3
BAg-4	1240	1435	1435-1650	39-41										29-31					1.5-2.5							26-30		0.15	BAg-4
BAg-5	1250	1370	1370-1550	44-46										29-31												23-27		0.15	BAg-5
BAg-6	1270	1425	1425-1600	49-51										33-35												14-18		0.15	BAg-6
BAg-7	1145	1205	1205-1400	55-57										21-23											4.5-5.5		15-19		0.15	BAg-7
BAg-8	1435	1435	1435-1650	71-73										Bal.														0.15	BAg-8
BAg-8a	1410	1410	1410-1600	71-73										Bal.		0.15-0.3											0.15	BAg-8a
BAg-13	1325	1575	1575-1775	53-55										Bal.												4.0-6.0		0.15	BAg-13
BAg-18	1115	1125	1325-1550	59-61										Bal.						0.025				9.5-10.5				0.15	BAg-18
BAg-19	1435	1635	1610-1800	92-93										Bal.		0.15-0.3											0.15	BAg-19
PRECIOUS METALS																													
BAu-1	1815	1860	1860-2000				37.0 +1 −0							Bal.														0.15	BAu-1
BAu-2	1635	1635	1635-1850				79.5 +1 −0							Bal.														0.15	BAu-2
BAu-3	1785	1885	1885-1995				34.5 +1 −0							Bal.					2.5-3.5								0.15	BAu-3
BAu-4	1740	1740	1740-1840				81.5 +1 −0												Bal.								0.15	BAu-4
COPPER AND COPPER-ZINC																													
BCu-1	1980	1980	2000-2100	0.01										99.90 min.						0.075	0.02							0.10	BCu-1
BCu-1a	1980	1980	2000-2100											99.0 min.														0.30[a]	BCu-1a
BCu-2	1980	1980	2000-2100											86.5 min.														0.50[c]	BCu-2[b]
RBCuZn-A	1630	1650	1670-1750	0.01*										57-61	*				0.05*				*	0.25-1.00		Bal.		0.50[d]	RBCuZn-A[e]
RBCuZn-D	1690	1715	1720-1800	0.01*										46-50					9.0-11.0	0.25	0.05*		0.04-0.25			Bal.		0.50[d]	RBCuZn-D[e]
MAGNESIUM																													
BMg-1	830	1110	1120-1160		8.3-9.7				0.0002-0.0008					0.05	0.005		Bal.	0.15 min	0.005				0.05			1.7-2.3		0.30	BMg-1
BMg-2	770	1050	1080-1130		11.0-13.0												Bal.									4.5-5.5		0.30	BMg-2
BMg-2a	770	1050	1080-1130		11.0-13.0				0.0002-0.0008								Bal.									4.5-5.5		0.30	BMg-2a

[a] Total other elements requirement pertains only to the metallic elements for this filler metal.

[b] These chemical requirements pertain only to the copper oxide and do not include requirements for the organic vehicle in which the copper oxide is suspended.

[c] Total other elements requirement pertains only to metallic elements for this filler metal. The following limitations are placed on the nonmetallic elements:

Constituent	per cent (max)
Chlorides	0.4
Sulfates	0.1
Oxygen	remainder
Nitric acid insoluble	0.3
Acetone soluble matter	0.5

[d] Total other elements, including the elements marked with an asterisk (*), shall not exceed the value specified.

[e] This AWS-ASTM classification is intended to be identical with the same classification that appears in the Specification for Copper and Copper-Alloy Welding Rods (AWS Designation A5.7; ASTM Designation B 259).

AWS-ASTM Classification	Solidus, °F	Liquidus, °F	Brazing Temperature Range, °F	Ag	Al	As	Au	B	Be	Bi	C	Cd	Cr	Cu	Fe	Li	Mg	Mn	Ni	P	Pb	Sb	Si	Sn	Ti	Zn	Other Each	Other Total	AWS-ASTM Classification
NICKEL																													
BNi-1	1790	1900	1950-2200					2.75-4.00			0.6-0.9		13.0-15.0		4.0-5.0				Bal.				3.0-5.0					0.50	BNi-1
BNi-2	1780	1830	1850-2150					2.75-3.5			0.15		6.0-8.0		2.0-4.0				Bal.				4.0-5.0					0.50	BNi-2
BNi-3	1800	1900	1850-2150					2.75-3.5			0.06				1.5				Bal.				4.0-5.0					0.50	BNi-3
BNi-4	1800	1950	1850-2150					1.0-2.2			0.06				1.5				Bal.				3.0-4.0					0.50	BNi-4
BNi-5	1975	2075	2100-2200								0.15		18.0-20.0						Bal.				9.75-10.5					0.50	BNi-5
BNi-6	1610	1610	1700-1875								0.15								Bal.	10.0-12.0								0.50	BNi-6
BNi-7	1630	1630	1700-1900										11.0-15.0						Bal.	9.0-11.0								0.50	BNi-7
SOLDERS—ASTM DESIGNATION B-32-60T, REVISED 1966																													
70A	361	378			0.005 max	0.03 max								0.08 max	0.02 max						30			70		0.005 max			70A
70B					0.005 max	0.03 max								0.08 max	0.02 max						30	0.2-0.5		70		0.005 max			70B
63A	361	361			0.005 max	0.03 max								0.08 max	0.02 max						37	0.12 max		63		0.005 max			63A
63B					0.005 max	0.03 max								0.08 max	0.02 max						37	0.2-0.5		63		0.005 max			63B
60A	361	374			0.005 max	0.03 max								0.08 max	0.02 max						40	0.12 max		60		0.005 max			60A
60B					0.005 max	0.03 max								0.08 max	0.02 max						40	0.2-0.5		60		0.005 max			60B
50A	361	421			0.005 max	0.03 max								0.08 max	0.02 max						50	0.12 max		50		0.005 max			50A
50B	360	420			0.005 max	0.03 max								0.08 max	0.02 max						50	0.2-0.5		50		0.005 max			50B
45A	361	441			0.005 max	0.03 max								0.08 max	0.02 max						55	0.12 max		45		0.005 max			45A
45B					0.005 max	0.03 max								0.08 max	0.02 max						55	0.2-0.5		45		0.005 max			45B
40A	361	460			0.005 max	0.03 max								0.08 max	0.02 max						60	0.12 max		40		0.005 max			40A
40B	360	460			0.005 max	0.03 max								0.08 max	0.02 max						60	0.2-0.5		40		0.005 max			40B
40C	365	448			0.005 max	0.03 max								0.08 max	0.02 max						58	1.8-2.4		40		0.005 max			40C
35A	361	477			0.005 max	0.03 max								0.08 max	0.02 max						65	0.25 max		35		0.005 max			35A
35B					0.005 max	0.03 max								0.08 max	0.02 max						65	0.2-0.5		35		0.005 max			35B
35C	365	470			0.005 max	0.03 max								0.08 max	0.02 max						63.2	1.6-2.0		35		0.005 max			35C
30A	361	491			0.005 max	0.03 max								0.08 max	0.02 max						70	0.25 max		30		0.005 max			30A
30B					0.005 max	0.03 max								0.08 max	0.02 max						70	0.2-0.5		30		0.005 max			30B
30C	364	482			0.005 max	0.03 max								0.08 max	0.02 max						68.4	1.4-1.8		30		0.005 max			30C
25A	361	511			0.005 max	0.03 max								0.08 max	0.02 max						75	0.25 max		25		0.005 max			25A
25B	360	510			0.005 max	0.03 max								0.08 max	0.02 max						75	0.2-0.5		25		0.005 max			25B
25C					0.005 max	0.03 max								0.08 max	0.02 max						73.7	1.1-1.5		25		0.005 max			25C
20B	361	531			0.005 max	0.03 max								0.08 max	0.02 max						80	0.2-0.5		20		0.005 max			20B
20C	363	517			0.005 max	0.03 max								0.08 max	0.02 max						79	0.8-1.2		20		0.005 max			20C
15B	440	550			0.005 max	0.03 max								0.08 max	0.02 max						85	0.2-0.5		15		0.005 max			15B
10B					0.005 max	0.03 max								0.08 max	0.02 max						90	0.2-0.5		10		0.005 max			10B
5A	518	594			0.005 max	0.03 max								0.08 max	0.02 max						95	0.12 max		5*		0.005 max			5A
5B					0.005 max	0.03 max								0.08 max	0.02 max						95	0.2-0.5		5*		0.005 max			5B
2A					0.005 max	0.03 max								0.08 max	0.02 max						98	0.12 max		2**		0.005 max			2A
2B					0.005 max	0.03 max								0.08 max	0.02 max						98	0.2-0.5		2**		0.005 max			2B
2.5S	579	579			0.005 max	0.03 max								0.08 max	0.02 max						97.5	0.4		0***		0.005 max			2.5S
1.5S	588	588	2.3-2.7		0.005 max	0.03 max								0.08 max	0.02 max						97.5	0.4		1****		0.005 max			1.5S
95TA	452	464	1.3-1.7		0.005 max	0.03 max								0.08 max	0.02 max						0.2	4.5-5.5		95		0.005 max			95TA
96.5TS			3.3-3.7		0.005 max	0.03 max								0.08 max	0.02 max						0.2	0.2-0.5		96.5		0.005 max			96.5TS

* Permissible tin range, 4.5-5.5%.
** Permissible tin range, 1.5-2.5%.
*** Tin maximum, 0.25%.
**** Permissible tin range, 0.75-1.25%.

PHYSICAL DATA FOR THE PLANETS, THEIR SATELLITES AND SOME ASTEROIDS

Compiled by W. Joseph Armento, 1970

Data was gathered for inclusion in this table from some of the latest available references. Calculations were performed to verify the internal consistency of the interrelated quantities. The deviations are based on the spread of the data as gathered from the selected references and their respective uncertainties.

Certain data which were calculated include the length of the solar day, rotational velocity, centrifical force, isosynchronous satellite data, escape velocity, orbital velocity, synodic period, equilibrium temperature, solar constant, gravitational constant, and all conversions to "earth" units. In the last calculations, a standard deviation is carried in the data to indicate the error arising from the uncertainties in the Earth measurements.

Sidereal time is the time required for the body to return to the equivalent position as defined by the stars. An isosynchronous satellite placed over a point on an equator with no perturbations in the orbit, no eccentricity, and no deflection from the equator will appear to remain stationary in the sky (or remain over the same position on the body); the angular velocity of the body equals the angular velocity of the satellite. The synodic period is the time from conjunction to conjunction. The letter "R" indicates a retrograde or opposite motion to that of the Sun/Earth system.

The exponents for large numbers are expressed in the following form:

$$N \times 10^m = N\,EM$$
$$\text{i.e. } 1.991 \pm 0.002\,E^{30} \text{ is equivalent}$$
$$\text{to } 1.991 \times 10^{30} \pm 0.002 \times 10^{30}$$

Body	Radius of Body (km)			Ellipticity (Oblateness)	Mass (kg)	Average Density (g/cc.)
	Average	Equatorial	Polar			
Sun (Sol)	695950.0 ± 810.0	695950.0 ± 810.0	695950.0 ± 810.0	≤ 0.0017	1.991 ± 0.002 E30	1.410 ± 0.002
Mercury	2433.0 ± 35.0	2433.0 ± 35.0	Unknown	≤ 0.0292	3.181 ± 0.110 E23	5.431 ± 0.300
Venus	6053.0 ± 7.8	6053.0 ± 7.8	6053.0 ± 7.8	0.0000 ± 0.0018	4.883 ± 0.034 E24	5.256 ± 0.043
Earth (Terra)	6371.315 ± 0.437	6378.533 ± 0.437	6356.912 ± 0.437	0.003393 ± 0.000097	5.979 ± 0.004 E24	5.519 ± 0.004
Moon (Luna)	1738.3 ± 1.1	1738.7 ± 1.1	1737.6 ± 1.1	0.006327 ± 0.008947	7.354 ± 0.066 E22	3.342 ± 0.031
Mars	3380.0 ± 20.0	3386.0 ± 20.0	3369.0 ± 20.0	0.0050 ± 0.0084	6.418 ± 0.024 E23	3.907 ± 0.079
Phobos	10.4 ± 5.0	11.3 ± 5.0	8.8 ± 5.0	0.22 ± 0.63	2.72 E16†	3.5 est.
Deimos	5.0 ± 4.0	5.0 ± 4.0	Unknown	—	1.83 E15†	3.5 est.
Jupiter	69758.0 ± 139.0	71370.0 ± 254	66644.0 ± 42.0	0.0662 ± 0.0036	1.901 ± 0.003 E27	1.337 ± 0.015
Io	1726.0 ± 109.0	1726.0 ± 109.0	Unknown	—	7.87 ± 0.15 E22	3.654 ± 0.696
Europa	1488.0 ± 70.0	1488.0 ± 70.0	Unknown	—	4.78 ± 0.15 E22	3.464 ± 0.501
Ganymede	2529.0 ± 54.0	2529.0 ± 54.0	Unknown	—	1.54 ± 0.03 E23	2.273 ± 0.152
Callisto	2416.0 ± 137.0	2416.0 ± 137.0	Unknown	—	7.35 ± 2.65 E22	1.244 ± 0.496
V (Amalthea)	87.0 ± 23.0	87.0 ± 23.0	Unknown	—	8.3 E18†	3.0 est.
VI (Hestia)	67.0 ± 16.0	67.0 ± 16.0	Unknown	—	3.8 E18†	3.0 est.
VII (Hera)	20.0 ± 9.0	20.0 ± 9.0	Unknown	—	1.0 E17†	3.0 est.
VIII (Poseidon)	18.0 ± 8.0	18.0 ± 8.0	Unknown	—	7.3 E16†	3.0 est.
IX (Hades)	12.0 ± 3.0	12.0 ± 3.0	Unknown	—	2.2 E16†	3.0 est.
X (Demeter)	9.0 ± 3.0	9.0 ± 3.0	Unknown	—	9.2 E15†	3.0 est.
XI (Pan)	12.0 ± 4.0	12.0 ± 4.0	Unknown	—	2.2 E16†	3.0 est.
XII (Adrastea)	10.0 ± 2.0	10.0 ± 2.0	Unknown	—	1.3 E16†	3.0 est.
Saturn	58219.0 ± 262.0	60369.0 ± 218.0	54148.0 ± 379.0	0.1030 ± 0.0073	5.684 ± 0.022 E26	0.688 ± 0.010
Mimas	272.0 ± 23.0	272.0 ± 23.0	Unknown	—	3.7 ± 2.0 E19	0.439 ± 0.262
Enceladus	299.0 ± 21.0	299.0 ± 21.0	Unknown	—	7.4 ± 3.0 E19	0.661 ± 0.302
Tethys	581.0 ± 74.0	581.0 ± 74.0	Unknown	—	4.9 E20†	0.6 est.
Dione	598.0 ± 86.0	598.0 ± 86.0	Unknown	—	5.4 E20†	0.6 est.
Rhea	890.0 ± 37.0	890.0 ± 37.0	Unknown	—	1.8 E21†	0.6 est.
Titan	2379.0 ± 194.0	2379.0 ± 194.0	Unknown	—	1.19 ± 0.55 E23	2.110 ± 1.103
Hyperion	201.0 ± 17.0	201.0 ± 17.0	Unknown	—	6.8 E19†	2.0 est.
Iapetus	647.0 ± 166.0	647.0 ± 166.0	Unknown	—	2.3 E21†	2.0 est.
Phoebe	132.0 ± 23.0	132.0 ± 23.0	Unknown	—	1.9 E19†	2.0 est.
X (Themis/Janus)	241.0 ± 16.0	241.0 ± 16.0	Unknown	—	1.2 E20†	2.0 est.
Uranus	23470.0 ± 732.0	24043.0 ± 339.0	22362.0 ± 1254.0	0.0700 ± 0.0572	8.682 ± 0.031 E25	1.603 ± 0.149
Ariel	311.0 ± 25.0	311.0 ± 25.0	Unknown	—	5.0 E20†	4.0 est.
Umbriel	201.0 ± 20.0	201.0 ± 20.0	Unknown	—	1.4 E20†	4.0 est.
Titania	300.0 ± 32.0	300.0 ± 32.0	Unknown	—	2.1 E21†	4.0 est.
Oberon	401.0 ± 28.0	401.0 ± 28.0	Unknown	—	1.1 E21†	4.0 est.
Miranda	121.0 ± 11.0	121.0 ± 11.0	Unknown	—	3.0 E19†	4.0 est.
Neptune	22716.0 ± 869.0	22716.0 ± 869.0	Unknown	≤ 0.0796	1.027 ± 0.014 E26	2.272 ± 0.263
Triton	2008.0 ± 55.0	2008.0 ± 55.0	Unknown	—	1.46 ± 0.05 E23	4.305 ± 0.383
Nereid	136.0 ± 34.0	136.0 ± 34.0	Unknown	—	5.0 ± 3.0 E19	4.7 ± 4.6
Pluto	5700.0 ± 413.0	5700.0 ± 413.0	Unknown	≤ 0.1562	1.08 ± 1.00 E24	1.65 ± 1.57
Typical Asteroid	20.0 ± 10.0	20.0 ± 10.0	Unknown	—	1.17 E17†	3.5 est.
1—Ceres	372.0 ± 22.0	372.0 ± 22.0	Unknown	—	6.0 ± 3.0 E20	2.8 ± 1.5
2—Pallas	238.0 ± 22.0	238.0 ± 22.0	Unknown	—	1.8 ± ? E20	3.2 ± ?
3—Juno	103.0 ± 14.0	103.0 ± 14.0	Unknown	—	2.0 ± ? E19	4.4 ± ?
4—Vesta	195.0 ± 100.0	195.0 ± 100.0	Unknown	—	1.0 ± ? E20	3.2 ± ?
5—Astraea	Unknown	Unknown	Unknown	—	Unknown	3.5 est.
6—Hebe	110.0 ± 10.0	110.0 ± 10.0	Unknown	—	2.0 ± ? E19	3.6 ± ?
7—Iris	100.0 ± 10.0	100.0 ± 10.0	Unknown	—	1.5 ± ? E19	3.6 ± ?
10—Hygiea	160.0 ± 13.0	160.0 ± 13.0	Unknown	—	6.0 ± ? E19	3.5 ± ?
15—Eunomia	140.0 ± 12.0	140.0 ± 12.0	Unknown	—	4.0 ± ? E19	3.5 ± ?
16—Psyche	140.0 ± 12.0	140.0 ± 12.0	Unknown	—	4.0 ± ? E19	3.5 ± ?
51—Nemansa	40.0 ± 6.0	40.0 ± 6.0	Unknown	—	9.0 ± ? E17	3.4 ± ?
323—Brucia	Unknown	Unknown	Unknown	—	Unknown	3.5 est.
433—Eros	12.0 ± 5.0	12.0 ± 5.0	Unknown	—	5.0 ± ? E15	0.7 ± ?
511—Davida	130.0 ± 11.0	130.0 ± 11.0	Unknown	—	3.0 ± ? E19	3.3 ± ?
588—Achilles	28.0 ± 5.0	28.0 ± 5.0	Unknown	—	3.22 E14†	3.5 est
617—Patrochus	Unknown	Unknown	Unknown	—	Unknown	3.5 est.
744—Hidalgo	18.0 ± 8.0	18.0 ± 8.0	Unknown	—	3.96 E14†	3.5 est.
1048—Feodosia	Unknown	Unknown	Unknown	—	Unknown	3.5 est.
1221—Amor	3.0 ± 1.5	3.0 ± 1.5	Unknown	—	3.96 E14†	3.5 est.
1566—Icarus	0.53 ± 0.13	0.7 ± 0.1	0.3 ± 0.2	0.57 ± 0.36	5.0 ± ? E12	8.1 ± ?
1620—Geographos	1.5 ± 0.8	1.5 ± 0.8	Unknown	—	5.0 ± ? E13	3.5 ± ?
—Adonis	1.0 ± 0.5	1.0 ± 0.5	Unknown	—	1.47 E13†	3.5 est.
—Apollo	2.0 ± 1.0	2.0 ± 1.0	Unknown	—	1.17 E14†	3.5 est.
—Hermes	1.0 ± 0.5	1.0 ± 0.5	Unknown	—	1.47 E13†	3.5 est.

† Value of mass calculated from an estimated density.

Body	Semimajor axis of Revolution (km) a	Eccentricity (x/a)	Aphelion (km) (a + x)	Perhelion (km) (a − x)	Longitude of Perhelion (Deg.)	Longitude of Ascending Node (Deg.)
Sun (Sol)	—	—				
Mercury	5.795 ± 0.013 E7	0.2056 ± 0.0002	6.986 ± 0.016 E7	4.604 ± 0.010 E7	76.583	47.667
Venus	1.0811 ± 0.0010 E8	0.0068 ± 0.0001	1.0885 ± 0.0010 E8	1.0737 ± 0.0010 E8	130.783	76.183
Earth (Terra)	1.4957 ± 0.0007 E8	0.0167 ± 0.0001	1.5207 ± 0.0007 E8	1.4707 ± 0.0007 E8	101.983	
Moon (Luna)	3.84403 ± 0.00006 E5	0.0549 ± 0.0000	4.05506 ± 0.00006 E5	3.63299 ± 0.00006 E5	Not tabulated	Unknown
Mars	2.2784 ± 0.0022 E8	0.0934 ± 0.0002	2.4912 ± 0.0024 E8	2.0656 ± 0.0020 E8	335.033	49.133
Phobos	9.408 ± 0.034 E3	Unknown	—	—	Not tabulated	None*
Deimos	2.3457 ± 0.0265 E4	Unknown	—	—	Not tabulated	None*
Jupiter	7.7814 ± 0.0017 E8	0.0484 ± 0.0002	8.1580 ± 0.0024 E8	7.4048 ± 0.0022 E8	13.417	99.883
Io	4.219 ± 0.003 E5	Unknown	—	—	Not tabulated	None*
Europa	6.712 ± 0.002 E5	Unknown	—	—	Not tabulated	None*
Ganymede	1.071 ± 0.001 E6	Unknown	—	—	Not tabulated	None*
Callisto	1.8830 ± 0.005 E6	Unknown	—	—	Not tabulated	None*
V (Amalthea)	1.814 ± 0.005 E5	Unknown	—	—	Not tabulated	None*
VI (Hestia)	1.14 ± 0.02 E7	Unknown	—	—	Not tabulated	Unknown
VII (Hera)	1.16 ± 0.04 E7	Unknown	—	—	Not tabulated	Unknown
VIII (Poseidon)	2.35 ± 0.01 E7	Unknown	—	—	Not tabulated	Unknown
IX (Hades)	2.37 ± 0.02 E7	Unknown	—	—	Not tabulated	Unknown
X (Demeter)	1.178 ± 0.005 E7	Unknown	—	—	Not tabulated	Unknown
XI (Pan)	2.26 ± 0.02 E7	Unknown	—	—	Not tabulated	Unknown
XII (Adrastea)	2.11 ± 0.01 E7	Unknown	—	—	Not tabulated	Unknown
Saturn	1.4270 ± 0.0009 E9	0.0543 ± 0.0021	1.5045 ± 0.0031 E9	1.3495 ± 0.0031 E9	91.950	113.167
Mimas	1.816 ± 0.007 E5	Unknown	—	—	Not tabulated	Unknown
Enceladus	2.394 ± 0.016 E5	Unknown	—	—	Not tabulated	Unknown
Tethys	2.952 ± 0.004 E5	Unknown	—	—	Not tabulated	Unknown
Dione	3.781 ± 0.015 E5	Unknown	—	—	Not tabulated	Unknown
Rhea	5.284 ± 0.019 E5	Unknown	—	—	Not tabulated	Unknown
Titan	1.227 ± 0.007 E6	Unknown	—	—	Not tabulated	Unknown
Hyperion	1.4873 ± 0.0066 E6	Unknown	—	—	Not tabulated	Unknown
Iapetus	3.571 ± 0.019 E6	Unknown	—	—	Not tabulated	Unknown
Phoebe	1.2983 ± 0.0072 E7	0.17 ± 0.01	1.5190 ± 0.0155 E7	1.0776 ± 0.0143 E9	Not tabulated	Unknown
X (Themis/Janus)	1.58 ± 0.15 E5	Unknown	—	—	Not tabulated	Unknown
Uranus	2.8703 ± 0.0015 E9	0.0460 ± 0.0015	3.0023 ± 0.0046 E9	2.7383 ± 0.0045 E9	169.75	73.717
Ariel	1.918 ± 0.003 E5	Unknown	—	—	Not tabulated	None*
Umbriel	2.672 ± 0.002 E5	Unknown	—	—	Not tabulated	None*
Titania	4.386 ± 0.009 E5	Unknown	—	—	Not tabulated	None*
Oberon	5.861 ± 0.006 E5	Unknown	—	—	Not tabulated	None*
Miranda	1.294 ± 0.014 E5	Unknown	—	—	Not tabulated	None*
Neptune	4.4999 ± 0.0038 E9	0.0082 ± 0.0007	4.5368 ± 0.0050 E9	4.4630 ± 0.0049 E9	44.117	131.167
Triton	3.531 ± 0.004 E5	Unknown	—	—	Not tabulated	Unknown
Nereid	5.9 ± 0.2 E6	Unknown	—	—	Not tabulated	Unknown
Pluto	5.909 ± 0.019 E9	0.2481 ± 0.0006	7.375 ± 0.024 E9	4.443 ± 0.015 E9	270.059	131.278
Typical Asteroid	2.9 ± 1.5 E8	0.10 ± 0.05	3.2 ± 1.7 E8	2.6 ± 1.4 E8	Unknown	Unknown
1—Ceres	4.141 ± 0.003 E8	0.079 ± 0.002	4.468 ± 0.009 E8	3.814 ± 0.009 E8	Unknown	Unknown
2—Pallas	4.141 ± 0.003 E8	0.236 ± 0.003	5.118 ± 0.013 E8	3.164 ± 0.013 E8	Unknown	Unknown
3—Juno	3.994 ± 0.001 E8	0.259 ± 0.001	5.028 ± 0.004 E8	2.920 ± 0.004 E8	Unknown	Unknown
4—Vesta	3.531 ± 0.001 E8	0.089 ± 0.001	3.845 ± 0.004 E8	3.217 ± 0.004 E8	Unknown	Unknown
5—Astraea	3.848 ± ? E8	0.190 ± ?	4.580 ± ? E8	3.117 ± ? E8	Unknown	Unknown
6—Hebe	3.629 ± ? E8	0.20 ± ?	4.35 ± ? E8	2.90 ± ? E8	Unknown	Unknown
7—Iris	3.567 ± ? E8	0.23 ± ?	4.39 ± ? E8	2.75 ± ? E8	Unknown	Unknown
10—Hygiea	4.173 ± ? E8	0.10 ± ?	4.59 ± ? E8	3.76 ± ? E8	Unknown	Unknown
15—Eunomia	3.956 ± ? E8	0.18 ± ?	4.67 ± ? E8	3.24 ± ? E8	Unknown	Unknown
16—Psyche	4.392 ± ? E8	0.14 ± ?	4.98 ± ? E8	3.76 ± ? E8	Unknown	Unknown
51—Nemansa	3.539 ± ? E8	0.06 ± ?	3.75 ± ? E8	3.33 ± ? E8	Unknown	Unknown
323—Brucia	3.564 ± ? E8	0.301 ± ?	4.637 ± ? E8	2.491 ± ? E8	Unknown	Unknown
433—Eros	2.179 ± 0.004 E8	0.222 ± 0.002	2.663 ± 0.007 E8	1.695 ± 0.005 E8	Unknown	Unknown
511—Davida	4.759 ± ? E8	0.18 ± ?	5.616 ± ? E8	3.902 ± ? E8	Unknown	Unknown
588—Achilles	7.793 ± 0.005 E8	0.148 ± 0.001	8.946 ± 0.010 E8	6.640 ± 0.009 E8	Unknown	Unknown
617—Patrochus	7.785 ± 0.005 E8	0.141 ± 0.002	8.883 ± 0.017 E8	6.687 ± 0.016 E8	Unknown	Unknown
744—Hidalgo	8.660 ± 0.004 E8	0.656 ± 0.001	1.4341 ± 0.0011 E9	2.979 ± 0.009 E8	Unknown	Unknown
1048—Feodosia	4.088 ± ? E8	0.180 ± ?	4.824 ± ? E8	3.352 ± ? E8	Unknown	Unknown
1221—Amor	2.878 ± ? E8	0.437 ± ?	4.135 ± ? E8	1.620 ± ? E8	Unknown	Unknown
1566—Icarus	1.613 ± 0.003 E8	0.828 ± 0.002	2.949 ± 0.006 E8	2.77 ± 0.03 E7	Unknown	Unknown
1620—Geographos	1.861 ± ? E8	0.34 ± ?	2.494 ± ? E8	1.228 ± ? E8	Unknown	Unknown
—Adonis	2.942 ± ? E8	0.779 ± ?	5.234 ± ? E8	6.50 ± ? E7	Unknown	Unknown
—Apollo	2.221 ± 0.005 E8	0.566 ± 0.001	3.473 ± 0.008 E8	9.64 ± 0.03 E7	Unknown	Unknown
—Hermes	1.934 ± ? E8	0.630 ± ?	3.152 ± ? E8	7.15 ± ? E7	Unknown	Unknown

* To equator of primary.

PHYSICAL DATA FOR THE PLANETS, THEIR SATELLITES AND SOME ASTEROIDS (Continued)

Body	Sidereal Period of Revolution (Solar Seconds)	Equatorial Rotation Period (Solar Seconds)	Average Solar Day (Solar Seconds)	Rotational Velocity at Equator (km/sec)	Centrifugal Force at Equator (cm/sec²)	Altitude of an Isosynchronous Satellite (km)
Sun (Sol)	—	2.125 ± ? E6†	—	2.0578 ± 0.0024	0.6085 ± 0.0016	24054600.0 ± 1100.0
Mercury	7.60234 ± 0.00173 E6	5.06775 ± 0.00020 E6	1.52004 ± 0.00110 E7	0.0030 ± 0.0000	0.00037 ± 0.00001	237298.0 ± 49.0
Venus	1.94141 ± 0.00017 E7	R 2.10747 ± 0.09243 E7	R 1.01052 ± 0.04996 E7	0.0018 ± 0.0000	0.0000538 ± 0.0000001	1534693.0 ± 11.0
Earth (Terra)	3.1558150 ± 0.0000000 E7	8.616406 ± 0.000001 E4	8.639988 ± 0.000010 E4	0.4651 ± 0.0000	3.39177 ± 0.00052	35767.0 ± 0.6
Moon (Luna)	2.36055 ± 0.00009 E6	2.36055 ± 0.00009 E6	2.55139 ± 0.00009 E6	0.0046 ± 0.0000	0.0012319 ± 0.0000017	86673.1 ± 1.6
Mars	5.93553 ± 0.00016 E7	8.86427 ± 0.00001 E4	8.87753 ± 0.00034 E4	0.2400 ± 0.0014	1.70126 ± 0.02247	17049.4 ± 6.4
Phobos	2.75468 ± 0.00097 E4	*	2.75596 ± 0.00098 E4	0.0026 ± 0.0011	0.05879 ± 0.05817	232.0 ± ?
Deimos	1.09077 ± 0.00009 E5	*	1.09278 ± 0.00010 E5	0.00029 ± 0.00023	0.00166 ± 0.00297	283.0 ± ?
Jupiter	3.74320 ± 0.00037 E8	3.54300 ± 0.00000 E4‡	3.54334 ± 0.00050 E4	12.6568 ± 0.0450	224.46 ± 1.79	87688.0 ± 359.0
Io	1.52859 ± 0.00009 E5	*	1.52922 ± 0.00023 E5	0.0709 ± 0.0045	0.2916 ± 0.0412	12860.0 ± 1718.0
Europa	3.06824 ± 0.00017 E5	*	3.07076 ± 0.00046 E5	0.0305 ± 0.0014	0.0624 ± 0.0066	18175.0 ± 2226.0
Ganymede	6.18175 ± 0.00009 E5	*	6.19198 ± 0.00088 E5	0.02570 ± 0.00055	0.0262 ± 0.0012	43740.0 ± 4155.0
Callisto	1.44193 ± 0.00000 E6	*	1.44751 ± 0.00020 E6	0.01053 ± 0.00060	0.00459 ± 0.00058	61263.0 ± 19485.0
V (Amalthea)	4.30358 ± 0.00000 E4	*	4.30408 ± 0.00060 E4	0.01270 ± 0.00336	0.1854 ± 0.1096	2087.0 ± ?
VI (Hestia)	2.16639 ± 0.00199 E7	Unknown	Unknown	Unknown	Unknown	Unknown
VII (Hera)	2.24467 ± 0.00173 E7	Unknown	Unknown	Unknown	Unknown	Unknown
VIII (Poseidon)	R 6.38496 ± 0.00086 E7	Unknown	Unknown	Unknown	Unknown	Unknown
IX (Hades)	R 6.54048 ± 0.01469 E7	Unknown	Unknown	Unknown	Unknown	Unknown
X (Pan)	2.23085 ± 0.03110 E7	Unknown	Unknown	Unknown	Unknown	Unknown
XI (Pan)	R 6.03677 ± 0.01987 E7	Unknown	Unknown	Unknown	Unknown	Unknown
XII (Adrastea)	R 5.4000 ± 0.01296 E7	Unknown	Unknown	Unknown	Unknown	Unknown
Saturn	9.29604 ± 0.00035 E8	3.68400 ± 0.00000 E4§	3.6414 ± 0.00020 E4	10.2961 ± 0.0372	175.60 ± 1.42	48797.0 ± 308.0
Mimas	8.17603 ± 0.06912 E4	*	8.17674 ± 0.00913 E4	0.0209 ± 0.0018	0.1606 ± 0.0305	475.0 ± 297.0
Enceladus	1.18515 ± 0.00242 E5	*	1.18530 ± 0.00242 E5	0.0159 ± 0.0011	0.0840 ± 0.0132	907.0 ± 397.0
Tethys	1.62743 ± 0.00639 E5	*	1.62772 ± 0.00539 E5	0.0224 ± 0.0029	0.0866 ± 0.0241	2222.0 ± ?
Dione	2.36848 ± 0.00648 E5	*	2.36905 ± 0.00548 E5	0.0158 ± 0.0023	0.0421 ± 0.0135	3107.0 ± ?
Rhea	3.89820 ± 0.00890 E5	*	3.89983 ± 0.00531 E5	0.0143 ± 0.0006	0.0231 ± 0.0022	6796.0 ± ?
Titan	1.37912 ± 0.00225 E6	*	1.38117 ± 0.00225 E6	0.0108 ± 0.0009	0.00494 ± 0.00090	70161.0 ± 22891.0
Hyperion	1.83876 ± 0.00078 E6	*	1.84240 ± 0.00079 E6	0.000687 ± 0.000058	0.000235 ± 0.000044	7092.0 ± ?
Iapetus	6.84720 ± 0.01469 E6	Unknown	Unknown	Unknown	Unknown	Unknown
Phoebe	R 4.74509 ± 0.1814 E7	Unknown	Unknown	Unknown	Unknown	Unknown
X (Themis/Janus)	6.4145 ± ? E4	Unknown	Unknown	Unknown	Unknown	Unknown
Uranus	2.65114 ± 0.00026 E9	R 3.88080 ± 0.01920 E4	R 3.88074 ± 0.01921 E4	3.8930 ± 0.0901	63.03 ± 3.26	36370.0 ± 791.0
Ariel	R 2.17745 ± 0.00017 E5	*	R 2.17727 ± 0.00035 E5	0.00897 ± 0.00072	0.0259 ± 0.0047	3117.0 ± ?
Umbriel	R 3.58050 ± 0.00000 E5	*	R 3.58002 ± 0.00050 E5	0.0062 ± 0.0014	0.0062 ± 0.0014	2886.0 ± ?
Titania	R 7.52189 ± 0.00001 E6	*	R 7.51976 ± 0.00004 E5	0.00353 ± 0.00035	0.00349 ± 0.00050	?
Oberon	R 1.16321 ± 0.00001 E6	*	R 1.16270 ± 0.00016 E6	0.00418 ± 0.00027	0.00217 ± 0.00018	?
Miranda	R 1.22169 ± 0.00120 E5	*	R 1.22163 ± 0.00021 E5	0.00622 ± 0.00057	0.03201 ± 0.00651	?
Neptune	5.20027 ± 0.00037 E9	5.66520 ± 0.02410 E4	5.66520 ± 0.00241 E4	2.5194 ± 0.0968	27.94 ± 2.40	59506.0 ± 1229.0
Triton	R 5.07712 ± 0.00095 E5	Unknown	R 5.07663 ± 0.00108 E5	Unknown	Unknown	Unknown
Nereid	3.91390 ± 0.69980 E7	Unknown	3.94358 ± 0.70013 E7	Unknown	Unknown	Unknown
Pluto	7.83735 ± 0.01058 E9	5.52320 ± 0.00430 E5	5.52349 ± 0.01139 E5	0.0648 ± 0.0047	0.0738 ± 0.0120	76510.0 ± 584.0
Typical Asteroid	8.5224 ± ? E7					
1—Ceres	1.4515 ± 0.0009 E8	3.96 ± 3.60 E4	4.0 ± 3.8 E4	0.0032 ± 0.0030	0.0503 ± 0.0994	11259.0 ± 11019.0
2—Pallas	1.4558 ± 0.0017 E8	Unknown	Unknown	Unknown	Unknown	Unknown
3—Juno	1.3772 ± 0.0009 E8	Unknown	Unknown	Unknown	Unknown	Unknown
4—Vesta	1.1457 ± 0.0009 E8	1.92 ± 0.06 E4	1.92 ± 0.06 E4	0.0635 ± 0.0337	2.088 ± 2.455	130109.0 ± 119463.0
5—Astraea	1.3029 ± 0.0035 E8	Unknown	Unknown	Unknown	Unknown	Unknown
6—Hebe	1.1923 ± 0.0009 E8	Unknown	Unknown	Unknown	Unknown	Unknown
7—Iris	1.1612 ± 0.0009 E8	Unknown	Unknown	Unknown	Unknown	Unknown
10—Hygiea	1.7643 ± 0.0009 E8	Unknown	Unknown	Unknown	Unknown	Unknown
15—Eunomia	1.3556 ± 0.0009 E8	Unknown	Unknown	Unknown	Unknown	Unknown
16—Psyche	1.5777 ± 0.0009 E8	Unknown	Unknown	Unknown	Unknown	Unknown

PHYSICAL DATA FOR THE PLANETS, THEIR SATELLITES AND SOME ASTEROIDS (Continued)

Body	Sidereal Period of Revolution (Solar Seconds)	Equatorial Rotation Period (Solar Seconds)	Average Solar Day (Solar Seconds)	Rotational Velocity at Equator (km/sec)	Centrifugal Force at Equator (cm/sec²)	Altitude of an Isosynchronous Satellite (km)
51 — Nemansa	1.1491 ± 0.0009 E8	Unknown	—	Unknown	Unknown	Unknown
323 — Brucia	1.1612 ± 0.0035 E8	Unknown	—	Unknown	Unknown	Unknown
433 — Eros	5.5556 ± 0.0086 E7	Unknown	—	Unknown	Unknown	Unknown
511 — Davida	1.7902 ± 0.0009 E8	Unknown	—	Unknown	Unknown	Unknown
588 — Achilles	3.7688 ± 0.0190 E8	Unknown	—	Unknown	Unknown	Unknown
617 — Patrochus	3.7489 ± 0.0035 E8	Unknown	—	Unknown	Unknown	Unknown
744 — Hidalgo	4.3943 ± 0.0112 E8	Unknown	—	Unknown	Unknown	Unknown
1048 — Feodosia	1.4265 ± 0.0035 E8	Unknown	—	Unknown	Unknown	Unknown
1221 — Amor	8.4240 ± 0.0346 E7	Unknown	—	Unknown	Unknown	Unknown
1566 — Icarus	3.5538 ± 0.0086 E7	9.72 ± 3.51 E4	9.72 ± 3.51 E4	0.0000453 ± 0.0000088	0.000287 ± 0.000144	264.0 ± 186.0
1620 — Geographos	4.3805 ± 0.0086 E7	Unknown	—	Unknown	Unknown	Unknown
— Adonis	8.7091 ± 0.0346 E7	Unknown	—	Unknown	Unknown	Unknown
— Apollo	5.7110 ± 0.0346 E7	Unknown	—	Unknown	Unknown	Unknown
— Hermes	4.6397 ± 0.0346 E7	Unknown	—	Unknown	Unknown	Unknown

* Estimated to be equal to the period of revolution (shows same face to primary).

† At latitude 30°—2.290 E6; 60°—2.678 ± 6; 90°—2.851 E6.

‡ Polar—3.5741 E4.

§ Polar—3.8280 E4.

PHYSICAL DATA FOR THE PLANETS, THEIR SATELLITES AND SOME ASTEROIDS (Continued)

Body	Gravity (cm/sec²)				Escape Velocity (km/sec)	Orbital Velocity (km/sec)		
	Mean	Polar	Equatorial	Equatorial— Uncorrected		Semimajor	Perihelion	Aphelion
Sun (Sol)	27372.0 ± 7.0	27370.0 ± 11.0	27369.0 ± 116.0	27369.0 ± 116.0	617.23 ± 0.44	—	—	—
Mercury	357.8 ± 16.1	357.8 ± 19.1	357.8 ± 21.7	357.8 ± 21.7	4.1725 ± 0.0493	47.828 ± 0.566	58.921 ± 0.697	38.824 ± 0.460
Venus	887.4 ± 6.6	887.4 ± 7.4	887.4 ± 7.4	887.4 ± 7.4	10.365 ± 0.020	35.017 ± 0.066	35.256 ± 0.067	34.780 ± 0.066
Earth (Terra)	980.7 ± 0.9	985.1 ± 1.0	975.0 ± 1.0	978.4 ± 1.0	11.179 ± 0.003	29.771 ± 0.007	30.272 ± 0.008	29.278 ± 0.007
Moon (Luna)	162.0 ± 1.5	162.2 ± 1.5	162.0 ± 1.5	162.0 ± 1.5	2.3735 ± 0.0054	1.0176 ± 0.0023	1.0751 ± 0.0024	0.9632 ± 0.0022
Mars	374.0 ± 4.6	376.5 ± 7.8	355.7 ± 7.8	372.7 ± 7.8	5.0282 ± 0.0173	24.121 ± 0.083	26.490 ± 0.091	21.964 ± 0.076
Phobos	1.67 ± ?	2.33 ± ?	Unknown	1.41 ± ?	0.01865 ± 0.00501	0.6739 ± 0.1818	Unknown	Unknown
Deimos	0.49 ± ?	Unknown	Unknown	Unknown	0.00699 ± 0.00312	0.4268 ± 0.2156	Unknown	Unknown
Jupiter	2601.0 ± 11.0	2850.0 ± 17.0	2260.0 ± 23.0	2485.0 ± 23.0	60.238 ± 0.007	13.052 ± 0.016	13.700 ± 0.017	12.435 ± 0.015
Io	176.0 ± 22.0	Unknown	Unknown	Unknown	2.464 ± 0.088	17.320 ± 0.617	Unknown	Unknown
Europa	144.0 ± 14.0	Unknown	Unknown	Unknown	2.068 ± 0.057	13.732 ± 0.377	Unknown	Unknown
Ganymede	160.0 ± 8.0	Unknown	Unknown	Unknown	0.900 ± 0.012	10.871 ± 0.140	Unknown	Unknown
Callisto	84.0 ± 32.0	Unknown	Unknown	Unknown	2.013 ± 0.192	8.194 ± 0.783	Unknown	Unknown
V (Amalthea)	7.28 ± ?	Unknown	Unknown	Unknown	0.1125 ± ?	26.414 ± 3.914	Unknown	Unknown
VI (Hestia)	5.61 ± ?	Unknown	Unknown	Unknown	0.08666 ± ?	3.332 ± 0.453	Unknown	Unknown
VII (Hera)	1.67 ± ?	Unknown	Unknown	Unknown	0.02587 ± ?	3.303 ± 0.860	Unknown	Unknown
VIII (Poseidon)	1.51 ± ?	Unknown	Unknown	Unknown	0.02329 ± ?	2.321 ± 0.579	Unknown	Unknown
IX (Hades)	1.00 ± ?	Unknown	Unknown	Unknown	0.01552 ± ?	2.311 ± 0.326	Unknown	Unknown
X (Demeter)	0.73 ± ?	Unknown	Unknown	Unknown	0.01164 ± ?	3.278 ± 0.613	Unknown	Unknown
XI (Pan)	1.00 ± ?	Unknown	Unknown	Unknown	0.01552 ± ?	2.366 ± 0.445	Unknown	Unknown
XII (Adrastea)	0.84 ± ?	Unknown	Unknown	Unknown	0.01294 ± ?	2.449 ± 0.275	Unknown	Unknown
Saturn	1117.0 ± 11.0	1291.0 ± 25.0	863.0 ± 16.0	1038.0 ± 16.0	36.056 ± 0.097	9.6383 ± 0.0261	10.177 ± 0.031	9.1284 ± 0.0282
Mimas	3.33 ± 1.89	Unknown	Unknown	Unknown	0.1346 ± 0.0193	14.436 ± 2.075	Unknown	Unknown
Enceladus	5.51 ± 2.36	Unknown	Unknown	Unknown	0.1815 ± 0.0197	12.573 ± 1.376	Unknown	Unknown
Tethys	9.72 ± ?	Unknown	Unknown	Unknown	0.3361 ± ?	11.322 ± 0.807	Unknown	Unknown
Dione	10.01 ± ?	Unknown	Unknown	Unknown	0.3460 ± ?	10.004 ± 0.807	Unknown	Unknown
Rhea	14.89 ± ?	Unknown	Unknown	Unknown	0.5149 ± ?	8.4628 ± 0.1974	Unknown	Unknown
Titan	140.0 ± 69.0	Unknown	Unknown	Unknown	0.8161 ± 0.1014	5.5536 ± 0.6938	Unknown	Unknown
Hyperion	11.21 ± ?	Unknown	Unknown	Unknown	0.2123 ± ?	5.0442 ± 0.2396	Unknown	Unknown
Iapetus	36.09 ± ?	Unknown	Unknown	Unknown	0.6834 ± ?	3.2554 ± 0.4694	Unknown	Unknown
Phoebe	7.36 ± ?	Unknown	Unknown	Unknown	0.1304 ± ?	1.7073 ± 0.1677	2.0270 ± 0.1991	1.4300 ± 0.1412
X (Themis/Janus)	13.85 ± ?	Unknown	Unknown	Unknown	0.2546 ± ?	1.5476 ± 0.0575	Unknown	Unknown
Uranus	1049.0 ± 66.0	1156.0 ± 164.0	937.0 ± 101.0	1000.0 ± 101.0	22.194 ± 0.387	6.7951 ± 0.1187	7.1161 ± 0.1215	6.4902 ± 0.1133
Ariel	14.69 ± ?	Unknown	Unknown	Unknown	0.4645 ± ?	5.4898 ± 0.2471	Unknown	Unknown
Umbriel	22.43 ± ?	Unknown	Unknown	Unknown	0.3003 ± ?	4.6512 ± 0.2589	Unknown	Unknown
Titania	55.77 ± ?	Unknown	Unknown	Unknown	0.7468 ± ?	3.6304 ± 0.1302	Unknown	Unknown
Oberon	44.72 ± ?	Unknown	Unknown	Unknown	0.5989 ± ?	3.1405 ± 0.1227	Unknown	Unknown
Miranda	13.50 ± ?	Unknown	Unknown	Unknown	0.1807 ± ?	6.6837 ± 0.3433	Unknown	Unknown
Neptune	1325.0 ± 103.0	1325.0 ± 145.0	1297.0 ± 177.0	1325.0 ± 177.0	24.536 ± 0.531	5.4276 ± 0.1176	5.4723 ± 0.1186	5.3833 ± 0.1167
Triton	241.0 ± 16.0	Unknown	Unknown	Unknown	3.1116 ± 0.0546	4.4006 ± 0.0773	Unknown	Unknown
Nereid	18.0 ± 14.0	Unknown	Unknown	Unknown	0.2213 ± 0.0454	1.0765 ± 0.2282	Unknown	Unknown
Pluto	221.0 ± 207.0	221.0 ± 210.0	221.0 ± 212.0	221.0 ± 212.0	5.0230 ± 1.1804	4.7365 ± 1.1166	6.1024 ± 1.4387	3.6763 ± 0.86670
Typical Asteroid	1.95 ± 1.95	1.95 ± 2.76	1.95 ± 3.38	1.95 ± 3.38	0.02795 ± 0.00781	21.380 ± 9.067	23.637 ± 10.060	19.339 ± 8.231
1—Ceres	28.9 ± 14.8	Unknown	Unknown	Unknown	0.4634 ± 0.0599	17.892 ± 2.315	19.366 ± 2.506	16.530 ± 2.139
2—Pallas	21.2 ± ?	Unknown	Unknown	Unknown	0.3174 ± ?	17.892 ± 0.925	22.757 ± 1.178	14.067 ± 0.728
3—Juno	12.6 ± ?	Unknown	Unknown	Unknown	0.1608 ± ?	18.218 ± 1.385	23.747 ± 1.805	13.977 ± 1.062
4—Vesta	17.5 ± ?	17.5 ± ?	15.4 ± ?	17.5 ± ?	0.2613 ± ?	19.376 ± 5.556	21.184 ± 6.075	17.722 ± 5.082
5—Astraea	Unknown	Unknown	Unknown	Unknown	Unknown	Unknown	Unknown	Unknown
6—Hebe	11.0 ± ?	Unknown	Unknown	Unknown	0.1556 ± ?	19.113 ± 0.971	23.408 ± 1.190	15.605 ± 0.793
7—Iris	10.0 ± ?	Unknown	Unknown	Unknown	0.1413 ± ?	19.278 ± 1.078	24.365 ± 1.362	15.233 ± 0.853
10—Hygiea	15.6 ± ?	Unknown	Unknown	Unknown	0.2235 ± ?	17.823 ± 0.810	19.704 ± 0.895	16.122 ± 0.732
15—Eunomia	13.6 ± ?	Unknown	Unknown	Unknown	0.1951 ± ?	18.306 ± 0.877	21.959 ± 1.052	15.260 ± 0.731
16—Psyche	13.6 ± ?	Unknown	Unknown	Unknown	0.1951 ± ?	17.413 ± 0.834	20.048 ± 0.961	15.124 ± 0.725
51—Nemansa	37.5 ± ?	Unknown	Unknown	Unknown	0.1731 ± ?	19.354 ± 1.623	20.552 ± 1.723	18.226 ± 1.528
323—Brucia	Unknown	Unknown	Unknown	Unknown	Unknown	Unknown	Unknown	Unknown
433—Eros	0.0172 ± ?	Unknown	Unknown	Unknown	0.00203 ± ?	24.665 ± 5.756	30.912 ± 7.214	19.681 ± 4.593
511—Davida	0.880 ± ?	Unknown	Unknown	Unknown	0.0478 ± ?	16.690 ± 0.789	20.021 ± 0.947	13.913 ± 0.658
588—Achilles	0.204 ± ?	Unknown	Unknown	Unknown	0.0107 ± ?	13.042 ± 1.303	15.139 ± 1.512	11.236 ± 1.122
617—Patrochus	Unknown	Unknown	Unknown	Unknown	Unknown	12.372 ± 3.075	27.146 ± 6.748	5.639 ± 1.402
744—Hidalgo	0.606 ± ?	Unknown	Unknown	Unknown	0.0148 ± ?	Unknown	Unknown	Unknown
1048—Feodosia	Unknown	Unknown	Unknown	Unknown	Unknown	21.46 ± ?	34.29 ± ?	13.43 ± ?
1221—Amor	0.29 ± ?	Unknown	Unknown	Unknown	0.1325 ± ?	28.668 ± 3.833	93.458 ± 12.507	8.794 ± 1.177
1566—Icarus	0.120 ± ?	0.357 ± ?	0.069 ± ?	0.069 ± ?	0.00112 ± ?	26.689 ± 7.957	38.029 ± 11.337	18.731 ± 5.584
1620—Geographos	0.015 ± ?	Unknown	Unknown	Unknown	0.00067 ± ?	21.23 ± ?	60.23 ± ?	7.483 ± ?
—Adonis	0.098 ± ?	Unknown	Unknown	Unknown	0.0441 ± ?	24.431 ± 6.844	46.408 ± 13.001	12.861 ± 3.603
—Apollo	0.29 ± ?	Unknown	Unknown	Unknown	0.1073 ± ?	26.18 ± ?	54.95 ± ?	12.47 ± ?
—Hermes	0.098 ± ?	Unknown	Unknown	Unknown	0.0441 ± ?			

PHYSICAL DATA FOR THE PLANETS, THEIR SATELLITES AND SOME ASTEROIDS (Continued)

Body	Inclination of Equator to Orbit (Deg.)	Inclination of Orbit to Ecliptic (Deg.)	Inclination of Equator to Ecliptic (Deg.)	Inclination of Orbit to Equator of Primary (Deg.)	Inclination of Orbit to Orbit of Primary (Deg.)	Average Synodic Period (Solar Seconds)	Albedo	Avg Temp (°K) Night	Avg Temp (°K) Day	Avg Temp (°K) Equilibrium	Solar Constant (cal/cm²/min)	Surface Pressure (atmospheres)	Atmospheric Composition (in cm-atm) (1 cm-atm = 2.69 E19 molecules per cm²)
Sun (Sol)	0.0	—	7.0	—	—	—	—	—	—	5776	88682.0 (est.)	Unknown	H₂/He major
Mercury	R 179.0 ± 1.0	7.0028	Unknown	—	—	1.00149 E7	0.076 ± 0.022	13	683	633	12.79	1 E-4 (est.)	150–3550 CO₂ (surface outgassings) SO₂?; Ar?
Venus	Unknown	3.40	Unknown	—	—	5.04505 E7	0.76	233	720	463	3.675	10–25‡	90% CO₂, 0.4% H₂O, 0.4% O₂, balance N₂ + inerts
Earth (Terra)	23.45	—	23.45	—	—	—	0.36 ± 0.06	275	295	394	1.920	1.0	78% N₂, 21% O₂, 1% Ar (see supplementary tables)
Moon (Luna)	6.683	5.12	1.533	Unknown	—	6.73861 E7	0.086 ± 0.025	100	365	394	1.920	~0	small amounts of outgassing. SO₂?, Ar?, Xe?
Mars	25.20	1.850	Unknown	—	—	—	0.152 ± 0.003	170	300	319	0.827	0.006–0.009	8300 ± 800 CO₂, 3900 CO, 4.35 ± 1.87 H₂O, ≤20. O₂, bal N₂
Phobos	0.0*	Unknown	Unknown	2.0	Unknown	Not tabulated	0.065	Unknown	Unknown	319	0.827	None	—
Deimos	0.0*	Unknown	Unknown	2.0	Unknown	Not tabulated	Unknown	Unknown	Unknown	319	0.827	None	—
Jupiter	3.117	1.309	Unknown	—	—	3.44637 E7	0.54 ± 0.04	123	313	173	0.0709	10¹–10⁵	1.20 E7 H₂, 2.35 E6 He, 1.5 E4 CH₄, 7 E2 NH₃, 0.198 H₂O, no O₂/N₂? (CH₄ likely)
Io	0.0*	Unknown	Unknown	0.0	3.1	Not tabulated	0.37	Unknown	Unknown	173	0.0709	Trace	None detected
Europa	0.0*	Unknown	Unknown	0.0	3.1	Not tabulated	0.39	Unknown	Unknown	173	0.0709	None?	None detected
Ganymede	0.0*	Unknown	Unknown	0.0	3.1	Not tabulated	0.20	Unknown	Unknown	173	0.0709	None?	None detected
Callisto	0.0*	Unknown	Unknown	0.0	3.1	Not tabulated	0.03	Unknown	Unknown	173	0.0709	None?	None detected
V (Amalthea)	0.0*	Unknown	Unknown	0.0	3.1	Not tabulated	Unknown	Unknown	Unknown	173	0.0709	None	
VI (Hestia)	Unknown	Unknown	Unknown	28.5	Unknown	Not tabulated	Unknown	Unknown	Unknown	173	0.0709	None	
VII (Hera)	Unknown	Unknown	Unknown	28.0	Unknown	Not tabulated	Unknown	Unknown	Unknown	173	0.0709	None	
VIII (Poseidon)	Unknown	Unknown	Unknown	R 147.5 ± 1.0	Unknown	Not tabulated	Unknown	Unknown	Unknown	173	0.0709	None	
IX (Hades)	Unknown	Unknown	Unknown	R 156.5 ± 1.0	Unknown	Not tabulated	Unknown	Unknown	Unknown	173	0.0709	None	
X (Demeter)	Unknown	Unknown	Unknown	28.3	Unknown	Not tabulated	Unknown	Unknown	Unknown	173	0.0709	None	
XI (Pan)	Unknown	Unknown	Unknown	R 163.4	Unknown	Not tabulated	Unknown	Unknown	Unknown	173	0.0709	?	
XII (Adrastea)	Unknown	Unknown	Unknown	R ?	Unknown	Not tabulated	Unknown	Unknown	Unknown	173	0.0709	None	
Saturn	26.75	2.493	Unknown	—	—	3.26671 E7	0.57 ± 0.09	103	223	128	0.0211	10²–10⁴	H₂/He major, 2.5 E4 CH₄, 2.5 E2 NH₃
Mimas	Unknown	Unknown	Unknown	1.5	27 ± 2	Not tabulated	Unknown	Unknown	Unknown	128	0.0211	None	
Enceladus	Unknown	Unknown	Unknown	0.0	26.8	Not tabulated	Unknown	Unknown	Unknown	128	0.0211	None	
Tethys	Unknown	Unknown	Unknown	1.1	27 ± 1	Not tabulated	Unknown	Unknown	Unknown	128	0.0211	None	
Dione	Unknown	Unknown	Unknown	0.0	26.8	Not tabulated	Unknown	Unknown	Unknown	128	0.0211	None	
Rhea	Unknown	Unknown	Unknown	0.3	26.8 ± 0.3	Not tabulated	Unknown	Unknown	Unknown	128	0.0211	None	
Titan	Unknown	Unknown	Unknown	0.3	26.8 ± 0.6	Not tabulated	Unknown	Unknown	Unknown	128	0.0211	?	traces NH₃, CH₄ (est. at 2 E4)
Hyperion	Unknown	Unknown	Unknown	0.6	27.0 ± 1.0	Not tabulated	Unknown	Unknown	Unknown	128	0.0211	?	
Iapetus	Unknown	Unknown	Unknown	14.7	Unknown	Not tabulated	Unknown	Unknown	Unknown	128	0.0211	None	
Phoebe	Unknown	Unknown	Unknown	R 174.7	R 150.0	Not tabulated	Unknown	Unknown	Unknown	128	0.0211	None	
X (Themis/Janus)	Unknown	Unknown	R ?	Unknown	Unknown	Not tabulated	Unknown	Unknown	Unknown	128	0.0211	None	
Uranus	R 97.983	0.773	R ?	—	—	3.19383 E7	0.65 ± 0.03	103	123	90	0.0052	2–3	H₂/He major, 1.5 E5 CH₄, trace NH₃ (?)
Ariel	0.0*	R ?	R ?	0.0	R 98.0	Not tabulated	Unknown	Unknown	Unknown	90	0.0052	None	
Umbriel	0.0*	R ?	R ?	0.0	R 98.0	Not tabulated	Unknown	Unknown	Unknown	90	0.0052	None	
Titania	0.0*	R ?	R ?	0.0	R 98.0	Not tabulated	Unknown	Unknown	Unknown	90	0.0052	None	
Oberon	0.0*	R ?	R ?	0.0	R 98.0	Not tabulated	Unknown	Unknown	Unknown	90	0.0052	None	
Miranda	0.0*	R ?	R ?	0.0	R 98.0	Not tabulated	Unknown	Unknown	Unknown	90	0.0052	None	
Neptune	29.0	1.779	Unknown	—	—	3.17508 E7	0.68 ± 0.08	103	123	72	0.0021	5–10 (est.)	H₂/He major, 2.5 E5 CH₄, trace NH₃ (?)
Triton	Unknown	Unknown	Unknown	R 150.0 ± 10.0	Unknown	Not tabulated	Unknown	Unknown	Unknown	72	0.0021	Trace?	trace CH₄ (at < 8 E2)
Nereid	Unknown	Unknown	Unknown	Unknown	Unknown	Not tabulated	Unknown	Unknown	Unknown	72	0.0021	None	
Pluto	90.0 (?)	17.146	Unknown	—	—	3.16857 E7	0.13 ± 0.05	43	63	63	0.0012	Unknown	None detected

PHYSICAL DATA FOR THE PLANETS, THEIR SATELLITES AND SOME ASTEROIDS (Continued)

Body	Symbol	Escape Velocity — Solar Escape from Position of (km/sec)	(in Earth Units)	Escape Velocity — Solar Escape from Surface of (km/sec)	(in Earth Units)	Mean Escape Velocity km/sec	Mean Orbital Velocity km/sec	Mean Rotational Velocity km/sec	Isosynchronous Satellite Data — Altitude (km)	Orbital Velocity (km/sec)	Escape Velocity from (km/sec)	Solar Escape from (km/sec)
Sun	☉	—	—	617.2 ± 0.9	14.17 ± 0.02	617.2 ± 0.9	—	2.058 ± 0.002	24055600.0 ± 29000.0	73.19 ± 0.43	103.5 ± 0.6	—
Mercury	☿	67.64 ± 0.10	1.607 ± 0.002	67.77 ± 0.14	1.556 ± 0.003	4.173 ± 0.070	47.83 ± 0.10	0.0030 ± 0.0000	237301.0 ± 4268.0	0.297 ± 0.003	0.420 ± 0.004	67.64 ± 0.14
Venus	♀	49.52 ± 0.06	1.176 ± 0.001	50.60 ± 0.06	1.161 ± 0.001	10.36 ± 0.03	35.02 ± 0.04	0.0018 ± 0.0000	1534720.0 ± 45230.0	0.459 ± 0.017	0.650 ± 0.024	49.53 ± 0.31
Earth	⊕ or ♁	42.10 ± 0.05	1.000 ± 0.001	43.56 ± 0.04	1.000 ± 0.001	11.18 ± 0.00	29.77 ± 0.01	0.4651 ± 0.0000	35775.0 ± 13.0	3.073 ± 0.002	4.346 ± 0.003	42.33 ± 0.05
Moon*	☽	1.439 ± 0.000	0.0342 ± 0.0000	2.775 ± 0.001	0.0637 ± 0.0000	2.373 ± 0.011	1.018 ± 0.002	0.0046 ± 0.0000	86667.0 ± 275.0	0.235 ± 0.002	0.333 ± 0.003	1.477 ± 0.002*
Mars	♂	34.11 ± 0.04	0.8102 ± 0.0009	34.48 ± 0.05	0.7915 ± 0.0011	5.028 ± 0.024	24.13 ± 0.044	0.2400 ± 0.0014	17032.0 ± 118.0	1.447 ± 0.012	2.046 ± 0.017	34.18 ± 0.08
Jupiter	♃	18.46 ± 0.02	0.4384 ± 0.0004	63.00 ± 0.17	1.446 ± 0.004	60.24 ± 0.10	13.06 ± 0.03	12.66 ± 0.05	89304.0 ± 338.0	28.21 ± 0.26	39.89 ± 0.37	43.96 ± 2.92
Saturn	♄	13.63 ± 0.02	0.3238 ± 0.0004	38.55 ± 0.14	0.8850 ± 0.0032	36.06 ± 0.14	9.639 ± 0.017	10.30 ± 0.04	50964.0 ± 552.0	18.62 ± 0.34	26.33 ± 0.49	29.65 ± 2.56
Uranus	⛢	9.611 ± 0.012	0.2283 ± 0.0002	24.18 ± 0.35	0.5551 ± 0.0080	22.19 ± 0.55	6.796 ± 0.012	3.893 ± 0.091	36938.0 ± 1939.0	9.780 ± 0.825	13.83 ± 1.17	16.84 ± 3.23
Neptune	♆	7.676 ± 0.010	0.1823 ± 0.0002	25.71 ± 0.52	0.5901 ± 0.0119	24.54 ± 0.75	5.428 ± 0.010	2.519 ± 0.096	59504.0 ± 3122.0	9.119 ± 1.031	12.82 ± 1.46	15.01 ± 3.76
Pluto	♇	6.699 ± 0.011	0.1591 ± 0.0002	8.371 ± 0.237	0.1921 ± 0.0054	5.019 ± 1.57	4.737 ± 0.011	0.0648 ± 0.0047	76470.0 ± 25960.0	0.935 ± 0.847	1.322 ± 1.198	6.828 ± 0.317

*Escape from Earth

PHYSICAL DATA FOR THE PLANETS, THEIR SATELLITES AND SOME ASTEROIDS (Continued)

Body	Radius of Body (Earth Radii)			Period of Rotation (Sidereal Days)	Mass (in Earth Masses)	Density (in Earth Units)	Gravitational Constant (gR²)	
	Mean	Equatorial	Polar				km³/sec²	in Earth Units
Sun (Sol)	109.227 ± 0.128	109.232 ± 0.127	109.232 ± 0.129	24.66225 ± 0.00003	332999.0 ± 402.0	0.2555 ± 0.0009	1.3256 ± 0.0047 E + 11	3.330 ± 0.012 E + 5
Mercury	0.3819 ± 0.0055	0.3819 ± 0.0055	Unknown	58.8151 ± 0.0023	0.0532 ± 0.0018	0.9554 ± 0.0528	2.118 ± 0.113 E + 4	5.32 ± 0.28 E - 2
Venus	0.9500 ± 0.00010	0.9500 ± 0.0012	0.9500 ± 0.0012	R 224.588 ± 10.727	0.8167 ± 0.0057	0.9524 ± 0.0076	3.251 ± 0.026 E + 5	0.817 ± 0.006
Earth (Terra)	1.00000 ± 0.00010	1.00114	0.99774 ± 0.00010	1.000000 ± 0.000002	1.0000	1.0000	3.981 ± 0.004 E + 5	1.000 ± 0.001
Moon (Luna)	0.27283 ± 0.00017	0.27290 ± 0.00017	0.27272 ± 0.00017	27.3960 ± 0.0010	0.01230 ± 0.00011	0.6056 ± 0.0056	4.896 ± 0.045 E + 3	0.0123 ± 0.0001
Mars	0.5306 ± 0.0031	0.5314 ± 0.031	0.5288 ± 0.0031	1.028766 ± 0.000002	0.10734 ± 0.00041	0.7188 ± 0.0130	4.273 ± 0.073 E + 4	0.1073 ± 0.0018
Phobos	0.0016 ± 0.0008	0.0018 ± 0.0008	0.0014 ± 0.0008	0.319702 ± 0.000113	4.5426 ± ? E - 9	0.63 ± ?	1.81 ± ? E - 3	4.54 ± ? E - 9
Deimos	0.0008 ± 0.0006	0.0008 ± 0.0006	Unknown	1.26592 ± 0.00005	3.0657 ± ? E - 10	0.63 ± ?	0.12 ± ? E - 4	3.07 ± ? E - 10
Jupiter	10.949 ± 0.022	11.202 ± 0.040	10.460 ± 0.007	0.411192 ± 0.000000	317.929 ± 0.545	0.2422 ± 0.0015	1.266 ± 0.007 E + 8	317.9 ± 1.9
Io	0.2709 ± 0.0171	0.2709 ± 0.0171	Unknown	1.77405 ± 0.00010	0.01316 ± 0.00025	0.662 ± 0.126	5.240 ± 0.941 E + 3	0.0132 ± 0.0024
Europa	0.2335 ± 0.0110	0.2335 ± 0.0110	Unknown	3.56093 ± 0.00020	0.00799 ± 0.00025	0.628 ± 0.091	3.183 ± 0.435 E + 3	0.0080 ± 0.0011
Ganymede	0.3969 ± 0.0085	0.3969 ± 0.0085	Unknown	7.17440 ± 0.00020	0.02580 ± 0.00052	0.412 ± 0.003	1.025 ± 0.065 E + 3	0.0026 ± 0.0002
Callisto	0.3792 ± 0.0215	0.3792 ± 0.0215	Unknown	16.7347 ± 0.0000	0.01239 ± 0.00443	0.225 ± 0.090	4.894 ± 1.931 E + 3	0.0123 ± 0.0049
V (Amalthea)	0.0137 ± 0.0036	0.0137 ± 0.0036	Unknown	0.499463 ± 0.000000	1.3840 ± ? E - 6	0.5 ± ?	0.551 ± ?	1.38 ± ? E - 6
VI (Hestia)	0.0105 ± 0.0025	0.0105 ± 0.0025	Unknown	Unknown	6.3205 ± ? E - 7	0.5 ± ?	0.252 ± ?	6.32 ± ? E - 7
VII (Hera)	0.0031 ± 0.0014	0.0031 ± 0.0014	Unknown	Unknown	1.6809 ± ? E - 8	0.5 ± ?	6.69 ± ? E - 3	1.68 ± ? E - 8
VIII (Poseidon)	0.0028 ± 0.0013	0.0028 ± 0.0013	Unknown	R ?	1.2258 ± ? E - 8	0.5 ± ?	4.88 ± ? E - 3	1.23 ± ? E - 8
IX (Hades)	0.0019 ± 0.0005	0.0019 ± 0.0005	Unknown	R ?	3.6310 ± ? E - 9	0.5 ± ?	1.45 ± ?	3.63 ± ? E - 9
X (Demeter)	0.0014 ± 0.0005	0.0014 ± 0.0005	Unknown	Unknown	1.5322 ± ? E - 9	0.5 ± ?	6.10 ± ? E - 3	1.53 ± ? E - 9
XI (Pan)	0.0019 ± 0.0006	0.0019 ± 0.0006	Unknown	R ?	3.6310 ± ? E - 9	0.5 ± ?	1.45 ± ?	3.63 ± ? E - 9
XII (Adrastea)	0.0016 ± 0.0003	0.0016 ± 0.0003	Unknown	R ?	2.1024 ± ? E - 9	0.5 ± ?	8.37 ± ?	2.10 ± ? E - 9
Saturn	9.1377 ± 0.0411	9.4751 ± 0.0342	8.4987 ± 0.0595	0.427556 ± 0.000000	95.0661 ± 0.3734	0.1246 ± 0.0018	3.784 ± 0.050 E + 7	95.06 ± 1.27
Mimas	0.0427 ± 0.0036	0.0427 ± 0.0036	Unknown	0.948890 ± 0.008022	6.188 ± 3.345 E - 6	0.080 ± 0.047	2.464 ± 1.456	6.188 ± 3.658 E - 6
Enceladus	0.0469 ± 0.0033	0.0469 ± 0.0033	Unknown	1.37546 ± 0.00281	1.238 ± 0.502 E - 5	0.120 ± 0.055	4.927 ± 2.224	1.238 ± 0.559 E - 5
Tethys	0.0912 ± 0.0116	0.0912 ± 0.0116	Unknown	1.88876 ± 0.00742	8.244 ± ? E - 5	0.11 ± ?	3.282 ± ? E + 1	8.24 ± ? E - 5
Dione	0.0939 ± 0.0135	0.0939 ± 0.0135	Unknown	2.74880 ± 0.00752	8.990 ± ? E - 5	0.11 ± ?	3.579 ± ? E + 1	8.99 ± ? E - 5
Rhea	0.1397 ± 0.0058	0.1397 ± 0.0058	Unknown	4.52416 ± 0.00453	2.964 ± ? E - 4	0.11 ± ?	1.180 ± ? E + 2	2.96 ± ? E - 4
Titan	0.3734 ± 0.0304	0.3734 ± 0.0304	Unknown	16.0057 ± 0.0261	0.01991 ± 0.00092	0.382 ± 0.020	7.923 ± 4.093 E + 3	1.990 ± 1.028 E - 2
Hyperion	0.0315 ± 0.0027	0.0315 ± 0.0027	Unknown	21.3402 ± 0.0091	1.138 ± ? E - 5	0.36 ± ?	4.53 ± ?	1.14 ± ? E - 5
Iapetus	0.1015 ± 0.0261	0.1015 ± 0.0261	Unknown	Unknown	3.795 ± ? E - 4	0.36 ± ?	1.51 ± ? E + 2	3.79 ± ? E - 4
Phoebe	0.0207 ± 0.0036	0.0207 ± 0.0036	Unknown	R ?	3.223 ± ? E - 6	0.36 ± ?	1.78 ± ?	3.22 ± ? E - 6
X (Themis/Janus)	0.0378 ± 0.0025	0.0378 ± 0.0025	Unknown	Unknown	1.962 ± ? E - 5	0.36 ± ?	7.81 ± ?	1.96 ± ? E - 5
Uranus	3.6837 ± 0.1149	3.7740 ± 0.0877	3.5098 ± 0.1968	R 0.450397 ± 0.002228	14.521 ± 0.053	0.2905 ± 0.0272	5.781 ± 0.510 E + 6	14.52 ± 1.28
Ariel	0.0488 ± 0.0039	0.0488 ± 0.0039	Unknown	R 25.2710 ± 0.0002	8.430 ± ? E - 5	0.72 ± ?	3.36 ± ? E + 1	8.43 ± ? E - 5
Umbriel	0.0315 ± 0.0031	0.0315 ± 0.0031	Unknown	R 4.15544 ± 0.00000	2.276 ± ? E - 5	0.72 ± ?	9.06 ± ?	2.28 ± ? E - 5
Titania	0.0785 ± 0.0050	0.0785 ± 0.0050	Unknown	R 8.72973 ± 0.0001	3.502 ± ? E - 4	0.72 ± ?	1.39 ± ? E + 2	3.50 ± ? E - 4
Oberon	0.0629 ± 0.0044	0.0629 ± 0.0044	Unknown	R 13.4999 ± 0.0001	1.806 ± ? E - 4	0.72 ± ?	7.19 ± ? E + 1	1.81 ± ? E - 4
Miranda	0.0190 ± 0.0017	0.0190 ± 0.0017	Unknown	R 1.41787 ± 0.00139	4.964 ± ? E - 6	0.72 ± ?	1.98 ± ?	4.96 ± ? E - 6
Neptune	3.5654 ± 0.1364	3.5654 ± 0.1364	Unknown	0.657490 ± 0.002797	17.177 ± 0.234	0.3790 ± 0.0438	6.838 ± 0.746 E + 6	17.18 ± 1.87
Triton	0.3152 ± 0.0086	0.3152 ± 0.0086	Unknown	Unknown	0.02442 ± 0.00084	0.780 ± 0.069	9.721 ± 0.823 E + 3	2.442 ± 0.207 E - 2
Nereid	0.0213 ± 0.0053	0.0213 ± 0.0053	Unknown	Unknown	8.363 ± 5.018 E - 6	0.860 ± 0.826	3.329 ± 3.087	8.363 ± 7.755 E - 6
Pluto	0.8946 ± 0.0648	0.8946 ± 0.0648	Unknown	6.4101 ± 0.0050	0.1806 ± 0.1673	0.2523 ± 0.2399	7.191 ± 6.819 E + 4	0.1806 ± 0.1713
Typical Asteroid								
1— Ceres	0.0031 ± 0.0016	0.0031 ± 0.0016	Unknown	0.4596 ± 0.4178	1.962 ± ? E - 8	0.63 ± ?	7.81 ± ? E - 3	1.96 ± ? E - 8
2— Pallas	0.0584 ± 0.0035	0.0584 ± 0.0035	Unknown	Unknown	1.004 ± 0.502 E - 5	0.504 ± 0.267	3.995 ± 2.106	1.004 ± 0.529 E - 5
3— Juno	0.0374 ± 0.0035	0.0374 ± 0.0035	Unknown	Unknown	3.011 ± ? E - 6	0.58 ± ?	1.20 ± ?	3.01 ± ? E - 6
4— Vesta	0.0162 ± 0.0022	0.0162 ± 0.0022	Unknown	Unknown	3.345 ± ? E - 6	0.79 ± ?	1.33 ± ?	3.35 ± ? E - 6
5— Astraea	0.0306 ± 0.0157	0.0306 ± 0.0157	Unknown	0.2228 ± 0.0070	1.673 ± ? E - 5	0.58 ± ?	6.66 ± ?	1.67 ± ? E - 5
6— Hebe	0.0173 ± ?	0.0173 ± ?	Unknown	Unknown	Unknown	Unknown	Unknown	Unknown
7— Iris	0.0153 ± ?	0.0153 ± ?	Unknown	Unknown	3.345 ± ? E - 6	0.65 ± ?	1.33 ± ?	3.35 ± ? E - 6
10— Hygiea	0.0251 ± ?	0.0251 ± ?	Unknown	Unknown	2.509 ± ? E - 6	0.65 ± ?	1.00 ± ?	2.51 ± ? E - 6
15— Eunomia	0.0220 ± ?	0.0220 ± ?	Unknown	Unknown	1.004 ± ? E - 5	0.63 ± ?	3.99 ± ?	1.00 ± ? E - 5
					6.690 ± ? E - 6	0.63 ± ?	2.66 ± ?	6.69 ± ? E - 6

PHYSICAL DATA FOR THE PLANETS, THEIR SATELLITES AND SOME ASTEROIDS (Continued)

Body	Radius of Body (Earth Radii)			Period of Rotation (Sidereal Days)	Mass (in Earth Masses)	Density (in Earth Units)	Gravitational Constant (gR^2)	
	Mean	Equatorial	Polar				km^3/sec^2	in Earth Units
16—Psyche	0.0220 ± ?	0.0220 ± ?	Unknown	Unknown	6.690 ± ? E − 6	0.63 ± ?	2.66 ±	6.69 ± ? E − 6
51—Nemansa	0.0063 ±	0.0063 ± ?	Unknown	Unknown	1.505 ± ? E − 7	0.61 ± ?	0.060 ± ?	1.51 ± ? E − 7
323 — Brucia	Unknown	Unknown	Unknown	Unknown	Unknown	Unknown	Unknown	Unknown
433 — Eros	0.0019 ± 0.0008	0.0019 ± 0.0008	Unknown	Unknown	8.363 ± ? E − 10	0.13 ± ?	3.33 ± ? E − 4	8.36 ± ? E − 10
511 — Davida	0.0204 ± ?	0.0204 ± ?	Unknown	Unknown	5.018 ± ? E − 6	0.59 ± ?	2.00 ± ?	5.02 ± ? E − 6
588 — Achilles	0.0044 ± 0.0008	0.0044 ± 0.0008	Unknown	Unknown	5.382 ± ? E − 11	0.63 ± ?	2.14 ± ? E − 5	5.38 ± ? E − 11
617—Patrochus	Unknown	Unknown	Unknown	Unknown	Unknown	Unknown	Unknown	Unknown
744—Hidalgo	0.0028 ± 0.0013	0.0028 ± 0.0013	Unknown	Unknown	6.620 ± ? E − 11	0.63 ± ?	2.64 ± ? E − 5	6.62 ± ? E − 11
1048—Feodosia	Unknown	Unknown	Unknown	Unknown	Unknown	Unknown	Unknown	Unknown
1221—Amor	0.00047 ± ?	0.00047 ± ?	Unknown	Unknown	6.620 ± ? E − 11	0.63 ± ?	2.64 ± ? E − 5	6.67 ± ? E − 11
1566—Icarus	0.00008 ± 0.00002	0.00011 ± 0.00002	0.00005 ± 0.00003	1.128 ± 0.407	8.363 ± ? E − 13	1.47 ± ?	3.33 ± ? E − 6	8.36 ± ? E − 13
1620—Geographos	0.00024 ± ?	0.00024 ± ?	Unknown	Unknown	8.363 ± ? E − 12	0.64 ± ?	3.33 ± ? E − 7	8.36 ± ? E − 12
—Adonis	0.00016 ± ?	0.00016 ± ?	Unknown	Unknown	2.452 ± ? E − 12	0.63 ± ?	9.76 ± ? E − 7	2.45 ± ? E − 12
—Apollo	0.00031 ± 0.00016	0.00031 ± 0.00016	Unknown	Unknown	2.892 ± ? E − 11	0.63 ± ?	1.15 ± ? E − 5	2.89 ± ? E − 11
—Hermes	0.00016 ± ?	0.00016 ± ?	Unknown	Unknown	2.452 ± ? E − 12	0.63 ± ?	9.76 ± ? E − 7	2.45 ± ? E − 12

PHYSICAL DATA FOR THE PLANETS, THEIR SATELLITES AND SOME ASTEROIDS (Continued)

Body	Radius of Orbit (Au)			Period of Revolution (Sidereal Years)	Velocity (in Earth Units)			Gravity (in g-s)		Volume (in Earth Units)
	Mean	Aphelion	Perihelion		Rotational	Orbital	Escape	Polar	Equatorial	
Sun (Sol)	—	—	—	—	4.4241	—	55.089	27.915	27.908	1303150.0 ± 4580.0
Mercury	0.38744 ± 0.00049	0.46710 ± 0.00107	0.30779 ± 0.00071	0.240899 ± 0.000055	0.00649	1.6040	0.3723	Unknown	0.3648	0.0557 ± 0.0024
Venus	0.72281 ± 0.00075	0.72772 ± 0.00076	0.71789 ± 0.00075	0.615185 ± 0.000005	R 0.00388	1.1745	0.9286	0.9049	0.9049	**0.8575** ± 0.0033
Earth (Terra)	1.00000	1.01670 ± 0.00068	0.98330 ± 0.00066	1.000000 ± 0.000000	1.0000	1.0000	1.0000	1.005	0.9942	1.0000
Moon (Luna)	0.0025701 ± 0.0000012	0.0027111 ± 0.0000013	0.0024290 ± 0.0000011	0.074800 ± 0.000003	0.00995	0.03423	0.2116	0.1654	0.1652	0.02031 ± 0.00004
Mars	1.5233 ± 0.0016	1.6656 ± 0.0018	1.3810 ± 0.0015	1.88082 ± 0.00005	0.5160	0.8087	0.4489	0.3839	0.3627	0.1493 ± 0.0027
Phobos	0.0000629 ± 0.0000002	Unknown	Unknown	0.0008729 ± 0.0000003	0.00554	0.02262	0.001666	Unknown	Unknown	4.344 ± 6.268 E − 9
Deimos	0.000157 ± 0.000001	Unknown	Unknown	0.0034564 ± 0.0000001	0.00062	0.01433	0.000624	Unknown	Unknown	4.833 ± 11.599 E − 10
Jupiter	5.2025 ± 0.0027	5.4543 ± 0.0030	4.9507 ± 0.0028	11.8613 ± 0.0117	27.212	0.4396	5.375	2.906	2.305	1312.5 ± 7.9
Io	0.002821 ± 0.000002	Unknown	Unknown	0.0048437 ± 0.0000003	0.1525	0.5805	0.2196	Unknown	Unknown	0.01988 ± 0.00377
Europa	0.004488 ± 0.000002	Unknown	Unknown	0.0097225 ± 0.0000005	0.0655	0.4597	0.1848	Unknown	Unknown	0.0127 ± 0.0018
Ganymede	0.007161 ± 0.000007	Unknown	Unknown	0.0195884 ± 0.0000005	0.0553	0.3658	0.0804	Unknown	Unknown	0.0625 ± 0.0040
Callisto	0.012589 ± 0.000007	Unknown	Unknown	0.0456912 ± 0.0000007	0.0226	0.2752	0.1795	Unknown	Unknown	0.0545 ± 0.0093
V (Amalthea)	0.001213 ± 0.000003	Unknown	Unknown	0.0013637 ± 0.0000005	0.0273	0.8860	0.01005	Unknown	Unknown	2.546 ± 2.019 E − 6
VI (Hestia)	0.07622 ± 0.00134	Unknown	Unknown	0.686475 ± 0.000631	Unknown	0.1117	0.00774	Unknown	Unknown	1.163 ± 0.833 E − 6
VII (Hera)	0.07756 ± 0.00267	Unknown	Unknown	0.711280 ± 0.000548	Unknown	0.1107	0.00231	Unknown	Unknown	3.093 ± 4.176 E − 8
VIII (Poseidon)	0.15712 ± 0.00067	Unknown	Unknown	R 2.02324 ± 0.00027	R ?	R 0.0779	0.00208	Unknown	Unknown	2.255 ± 3.007 E − 8
IX (Hades)	0.15845 ± 0.00134	Unknown	Unknown	R 2.07252 ± 0.00462	R ?	R 0.0775	0.00139	Unknown	Unknown	**6.681** ± 5.011 E − 9
X (Demeter)	0.07876 ± 0.00034	Unknown	Unknown	0.706901 ± 0.009855	Unknown	0.1101	0.00139	Unknown	Unknown	2.819 ± 2.819 E − 9
XI (Pan)	0.1151 ± 0.0013	Unknown	Unknown	R 1.91290 ± 0.00630	R ?	R 0.0795	0.00139	Unknown	Unknown	6.681 ± 6.681 E − 9
XII (Adrastea)	1.14107 ± 0.00067	Unknown	Unknown	R 1.71113 ± 0.00411	R ?	R 0.0822	0.00115	Unknown	Unknown	3.866 ± 2.320 E − 9
Saturn	9.5407 ± 0.0075	10.0587 ± 0.0215	9.0226 ± 0.0213	29.4568 ± 0.0111	22.136	0.3235	3.2232	1.316	0.8800	763.0 ± 10.3
Mimas	0.001214 ± 0.000005	Unknown	Unknown	0.0025908 ± 0.0000219	0.0449	0.4832	0.0121	Unknown	Unknown	7.781 ± 1.974 E − 5
Enceladus	0.001601 ± 0.000011	Unknown	Unknown	0.0037554 ± 0.0000077	0.0341	0.4228	0.0162	Unknown	Unknown	1.034 ± 0.218 E − 4
Tethys	0.001874 ± 0.000003	Unknown	Unknown	0.0051569 ± 0.0000202	0.0482	0.3792	0.02873	Unknown	Unknown	7.584 ± 2.897 E − 4
Dione	0.002528 ± 0.000010	Unknown	Unknown	0.0075051 ± 0.0000205	0.0341	0.3365	0.03078	Unknown	Unknown	8.268 ± 3.567 E − 4
Rhea	0.003533 ± 0.000013	Unknown	Unknown	0.0123524 ± 0.0000124	0.0308	0.2839	0.04602	Unknown	Unknown	2.726 ± 0.340 E − 3
Titan	0.008204 ± 0.000047	Unknown	Unknown	0.0437009 ± 0.0000713	0.0233	0.1862	0.0729	Unknown	Unknown	5.206 ± 1.274 E − 2
Hyperion	0.009944 ± 0.000044	Unknown	Unknown	0.0582658 ± 0.0000247	0.00148	0.1691	0.01895	Unknown	Unknown	3.140 ± 0.797 E − 5
Iapetus	0.02388 ± 0.00013	Unknown	Unknown	0.216971 ± 0.000466	Unknown	0.1094	0.06099	Unknown	Unknown	1.047 ± 0.806 E − 3
Phoebe	0.08680 ± 0.00048	0.10156 ± 0.00104	0.07205 ± 0.00096	R 1.50360 ± 0.00575	R ?	R 0.0574	0.01245	Unknown	Unknown	8.893 ± 4.648 E − 6
X (Themis/Janus)	0.001056 ± 0.000001	Unknown	Unknown	0.002033 ± ?	Unknown	0.5201	0.02273	Unknown	Unknown	5.412 ± 1.078 E − 5
Uranus	19.190 ± 0.013	20.073 ± 0.032	18.308 ± 0.032	84.0081 ± 0.0824	R 8.3697	0.2282	1.9821	1.179 ± 0.168	0.9554 ± 0.168	49.987 ± 4.676
Ariel	0.001282 ± 0.000002	Unknown	Unknown	R 0.0068998 ± 0.0000005	R 0.01929	R 0.1842	0.04150	Unknown	Unknown	1.163 ± 0.280 E − 4
Umbriel	0.001786 ± 0.000002	Unknown	Unknown	R 0.0113457 ± 0.0000005	R 0.0075	R 0.1560	0.02682	Unknown	Unknown	3.140 ± 0.937 E − 5
Titania	0.002932 ± 0.000006	Unknown	Unknown	R 0.0238350 ± 0.0000205	R 0.00898	R 0.1218	0.06663	Unknown	Unknown	4.833 ± 0.928 E − 4
Oberon	0.003919 ± 0.000004	Unknown	Unknown	R 0.0366593 ± 0.0000003	R 0.00466	R 0.1054	0.05336	Unknown	Unknown	2.493 ± 0.522 E − 4
Miranda	0.0008651 ± 0.0000094	Unknown	Unknown	R 0.0038712 ± 0.0000038	R 0.01338	R 0.2242	0.01615	Unknown	Unknown	6.850 ± 1.868 E − 6
Neptune	30.086 ± 0.029	30.332 ± 0.036	29.839 ± 0.036	164.784 ± 0.012	5.4166	0.1822	2.1875	Unknown	1.323 ± 0.210	45.322 ± 5.201
Triton	0.002361 ± 0.000003	Unknown	Unknown	R 0.0160881 ± 0.0000030	Unknown	R 0.1477	0.2777	Unknown	Unknown	3.130 ± 0.257 E − 2
Nereid	0.03945 ± 0.00134	Unknown	Unknown	1.24022 ± 0.22175	Unknown	0.0362	0.0197	Unknown	Unknown	9.726 ± 7.294 E − 6
Pluto	39.507 ± 0.128	49.308 ± 0.162	29.705 ± 0.099	248.35 ± 0.34	0.1394	0.1591	0.4482	Unknown	0.225 ± 0.217	0.7160 ± 0.1556
Typical Asteroid	1.94 ± 1.00	2.13 ± 1.11	1.75 ± 0.91	2.70 ± ?	0.00682	0.7181	0.00250	Unknown	Unknown	3.093 ± 4.640 E − 8
1—Ceres	2.7686 ± 0.0024	2.9873 ± 0.0061	2.5499 ± 0.0060	4.5994 ± 0.0029	Unknown	0.6007	0.0413	Unknown	Unknown	1.990 ± 0.353 E − 5
2—Pallas	2.7686 ± 0.0024	3.4220 ± 0.0088	2.1152 ± 0.0085	4.6131 ± 0.0054	Unknown	0.6007	0.0283	Unknown	Unknown	5.213 ± 1.445 E − 6
3—Juno	2.6703 ± 0.0014	3.3619 ± 0.0032	1.9787 ± 0.0029	4.3640 ± 0.0054	Unknown	0.6107	0.0144	Unknown	Unknown	4.225 ± 1.723 E − 6
4—Vesta	2.3608 ± 0.0013	2.5709 ± 0.0027	2.1507 ± 0.0026	3.6304 ± 0.0029	0.1372	0.6510	0.0233	Unknown	Unknown	2.867 ± 4.411 E − 5
5—Astraea	2.57 ± ?	3.06 ± ?	2.08 ± ?	3.5816 ± 0.0111	Unknown	0.6229	Unknown	Unknown	Unknown	Unknown
6—Hebe	2.43 ± ?	2.91 ± ?	1.94 ± ?	3.7781 ± 0.0029	Unknown	0.6409	0.0139	Unknown	Unknown	5.146 ± 1.404 E − 6
7—Iris	2.38 ± ?	2.93 ± ?	1.84 ± ?	3.6796 ± 0.0029	Unknown	0.6477	0.0126	Unknown	Unknown	3.866 ± 1.160 E − 6
10—Hygiea	2.79 ± ?	3.07 ± ?	2.51 ± ?	5.5906 ± 0.0029	Unknown	0.5973	0.0199	Unknown	Unknown	1.584 ± 0.386 E − 5
15—Eunomia	2.64 ± ?	3.12 ± ?	2.17 ± ?	4.2956 ± 0.0029	Unknown	0.6141	0.0174	Unknown	Unknown	1.061 ± 0.273 E − 5

PHYSICAL DATA FOR THE PLANETS, THEIR SATELLITES AND SOME ASTEROIDS (Continued)

Body	Radius of Orbit (Au) Mean	Aphelion	Perhelion	Period of Revolution (Sidereal Years)	Velocity Rotational	Velocity Orbital	Velocity Escape	Gravity Polar	Gravity Equatorial	Volume (in Earth Units)
16—Psyche	2.92 ± ?	3.33 ± ?	2.51 ± ?	4.9993 ± 0.0029	Unknown	0.5839	0.0174	Unknown	Unknown	1.061 ± 0.273 E − 5
51—Nemansa	2.37 ± ?	2.51 ± ?	2.22 ± ?	3.6412 ± 0.0029	Unknown	0.6510	0.0154	Unknown	Unknown	2.475 ± 1.114 E − 7
323—Brucia	2.38 ± ?	3.10 ± ?	1.67 ± ?	3.6796 ± 0.0111	Unknown	0.6473	Unknown	Unknown	Unknown	Unknown
433—Eros	1.4568 ± 0.0028	1.7803 ± 0.0045	1.1334 ± 0.0036	1.7604 ± 0.0027	Unknown	0.8289	0.000210	Unknown	Unknown	6.681 ± 8.351 E − 9
511—Davida	3.18 ± ?	3.75 ± ?	2.61 ± ?	5.6727 ± 0.0029	Unknown	0.5604	0.00495	Unknown	Unknown	8.495 ± 2.156 E − 6
588—Achilles	5.2103 ± 0.0041	5.9814 ± 0.0071	4.4391 ± 0.0063	11.942 ± 0.060	Unknown	0.4362	0.00111	Unknown	Unknown	8.488 ± 4.547 E − 8
617—Patrochus	5.2049 ± 0.0041	5.9388 ± 0.0114	4.4710 ± 0.0110	11.879 ± 0.11	Unknown	0.4362	Unknown	Unknown	Unknown	Unknown
744—Hidalgo	5.7899 ± 0.0038	9.5881 ± 0.0086	1.9917 ± 0.0059	13.924 ± 0.35	Unknown	0.4161	0.0015	Unknown	Unknown	2.255 ± 3.007 E − 8
1048—Feodosia	2.73 ± ?	3.23 ± ?	2.24 ± ?	4.5202 ± 0.0111	Unknown	0.6038	Unknown	Unknown	Unknown	Unknown
1221—Amor	1.92 ± ?	2.77 ± ?	1.08 ± ?	2.6694 ± 0.0111	Unknown	0.7205	0.0119	Unknown	Unknown	1.044 ± 1.566 E − 10
1566—Icarus	1.0784 ± 0.0021	1.9714 ± 0.0044	0.1855 ± 0.0022	1.1198 ± 0.3027	0.000098	0.9631	0.0001	Unknown	3.772 ± 5.637 E − 4	5.684 ± 4.070 E − 13
1620—Geographos	1.24 ± ?	1.67 ± ?	0.82 ± ?	1.3881 ± 0.3027	Unknown	0.8960	0.000059	Unknown	Unknown	1.305 ± 2.088 E − 11
—Adonis	1.97 ± ?	3.50 ± ?	0.43 ± ?	2.7597 ± 0.0110	Unknown	0.7120	0.0039	Unknown	Unknown	3.866 ± 5.800 E − 12
—Apollo	1.4849 ± 0.0034	2.3254 ± 0.0055	0.6445 ± 0.0071	1.3097 ± 0.0110	Unknown	0.8188	0.0096	Unknown	Unknown	3.093 ± 4.640 E − 11
—Hermes	1.29 ± ?	2.11 ± ?	0.48 ± ?	1.4702 ± 0.0110	Unknown	0.7205	0.0039	Unknown	Unknown	3.866 ± 5.800 E − 12

Body	Inclination of Equator to Orbit (Deg.)	Inclination of Equator to Ecliptic (Deg.)	Inclination of Orbit to Ecliptic (Deg.)	Inclination of Orbit to Equator of Primary (Deg.)	Inclination of Orbit to Orbit of Primary (Deg.)	Average Synodic Period (Solar Seconds)	Albedo	Night Temp (°K)	Day Temp (°K)	Equilibrium Temp (°K)	Solar Constant (cal/cm²/min)	Surface Pressure (atmospheres)	Atmospheric Composition (in cm-atm) (1 cm-atm = 2.69 E19 molecules per cm²)
Typical Asteroid	Unknown	Unknown	10.0	—	—	5.01157 E7	Unknown	Unknown	Unknown	283	0.275	None	—
1—Ceres	Unknown	Unknown	10.4 ± 0.3	—	—	4.03256 E7	Unknown	Unknown	Unknown	237	0.250	None	—
2—Pallas	Unknown	Unknown	34.8 ± 0.1	—	—	4.02926 E7	Unknown	Unknown	Unknown	237	0.250	None	—
3—Juno	Unknown	Unknown	13.0 ± 0.1	—	—	4.09392 E7	Unknown	Unknown	Unknown	241	0.269	None	—
4—Vesta	Unknown	Unknown	7.11 ± 0.03	—	—	4.35554 E7	0.20 ± 0.11	Unknown	Unknown	256	0.345	None	—
5—Astraea	Unknown	Unknown	5.33 ± ?	—	—	4.16452 E7	Unknown	Unknown	Unknown	246	0.290	None	—
6—Hebe	Unknown	Unknown	14.8 ± ?	—	—	4.29177 E7	Unknown	Unknown	Unknown	253	0.326	None	—
7—Iris	Unknown	Unknown	5.5 ± ?	—	—	4.33355 E7	Unknown	Unknown	Unknown	255	0.338	None	—
10—Hygiea	Unknown	Unknown	3.8 ± ?	—	—	3.84326 E7	Unknown	Unknown	Unknown	236	0.247	None	—
15—Eunomia	Unknown	Unknown	11.8 ± ?	—	—	4.11341 E7	Unknown	Unknown	Unknown	242	0.274	None	—
16—Psyche	Unknown	Unknown	3.1 ± ?	—	—	3.94490 E7	Unknown	Unknown	Unknown	230	0.225	None	—
51—Nemansa	Unknown	Unknown	9.9 ± ?	—	—	4.35065 E7	Unknown	Unknown	Unknown	256	0.343	None	—
323—Brucia	Unknown	Unknown	24.2 ± ?	—	—	4.33355 E7	Unknown	Unknown	Unknown	255	0.338	None	—
433—Eros	Unknown	Unknown	10.8 ± 0.1	—	—	7.30584 E7	Unknown	Unknown	Unknown	326	0.905	None	—
511—Davida	Unknown	Unknown	15.7 ± ?	—	—	3.83119 E7	Unknown	Unknown	Unknown	221	0.190	None	—
588—Achilles	Unknown	Unknown	10.3 ± 0.1	—	—	3.44422 E7	Unknown	Unknown	Unknown	173	0.071	None	—
617—Patrochus	Unknown	Unknown	22.1 ± ?	—	—	3.44589 E7	Unknown	Unknown	Unknown	173	0.071	None	—
744—Hidalgo	Unknown	Unknown	42.5 ± 0.1	—	—	3.39999 E7	Unknown	Unknown	Unknown	164	0.058	None	—
1048—Feodosia	Unknown	Unknown	53.8 ± ?	—	—	4.05229 E7	Unknown	Unknown	Unknown	238	0.257	None	—
1221—Amor	Unknown	Unknown	11.9 ± ?	—	—	5.04625 E7	Unknown	Unknown	Unknown	284	0.519	None	—
1566—Icarus	Unknown	Unknown	23.0 ± 0.1	—	—	2.95038 E8	Unknown	Unknown	Unknown	377	1.651	None	—
1620—Geographos	Unknown	Unknown	13.3 ± ?	—	—	1.12878 E8	Unknown	Unknown	Unknown	353	1.240	None	—
—Adonis	Unknown	Unknown	1.5 ± ?	—	—	4.94920 E7	Unknown	Unknown	Unknown	281	0.496	None	—
—Apollo	Unknown	Unknown	6.4 ± 0.1	—	—	7.05344 E7	Unknown	Unknown	Unknown	323	0.871	None	—
—Hermes	Unknown	Unknown	6.2 ± ?	—	—	9.86736 E7	Unknown	Unknown	Unknown	346	1.140	None	—

*Assumed value.
†Conflicting satellite data.

HOHMANN ELLIPSE TRANSFER DATA (MINIMUM ENERGY OF TRANSFER)

W. Joseph Armento

The "from" body is the body of departure and the "to" body is the body of arrival. The three lines of data for each planetary entry are:

    top—from perhelion of inner planet to aphelion of outer planet
    middle—from mean of inner planet to mean of outer planet
    bottom—from aphelion of inner planet to perhelion of outer planet

The orbital parameters and velocities are given for the transfer orbit. The orbital velocity, change in velocity necessary to reach the new Hohmann orbital velocity, and the maximum change in velocity to reach the Hohmann orbital velocity are given for departure. For arrival data, look in the table listing the arrival planet as "from" and the departure planet as "to". The maximum Hohmann velocity change is assumed to be a single impulse to overcome the necessary energy for change in velocity and escape velocity from the surface of the departure planet.

Body	Mean orbital Semimajor axis (km × 10⁻⁶)	Eccentricity ε	Aphelion (km × 10⁻⁶)	Perhelion (km × 10⁻⁶)	Orbital Velocity (km/sec) Mean	Aphelion	Perhelion	Orbital Velocity of Departure Body (km/sec)	Time of one way trip (½ orbit) (Solar Seconds)	Departure Δv orbit (km/sec)	Maximum Δv (km/sec)
From Mercury to:											
Venus	77.440	0.4055	103.85	46.035	41.38	26.91	63.62	58.92	5.880 E6	4.699	6.284
	83.030	0.3021	108.11	57.950	39.96	29.26	54.58	47.83	6.528 E6	6.748	7.934
	88.620	0.2116	107.37	69.864	38.68	31.20	47.95	38.83	7.198 E6	9.124	10.03
Earth	99.052	0.5352	152.07	46.035	36.58	20.13	66.49	58.92	8.506 E6	7.569	8.643
	103.76	0.4415	149.57	57.950	35.75	22.25	57.43	47.83	9.119 E6	9.596	10.46
	108.47	0.3559	147.07	69.864	34.96	24.10	50.72	38.83	9.747 E6	11.90	12.61
Mars	147.58	0.6881	249.12	46.035	29.97	12.88	69.72	58.92	1.547 E7	10.80	11.58
	142.89	0.5945	227.84	57.950	30.46	15.36	60.40	47.83	1.474 E7	12.57	13.24
	138.21	0.4945	206.56	69.864	30.97	18.01	53.25	38.83	1.402 E7	14.43	15.02
Jupiter	430.92	0.8932	815.80	46.035	17.54	4.167	73.84	58.92	7.718 E7	14.91	15.49
	418.04	0.8614	778.14	57.950	17.81	4.860	65.26	47.83	7.375 E7	17.43	17.92
	405.17	0.8276	740.48	69.864	18.09	5.556	58.89	38.83	7.037 E7	20.06	20.49
Saturn	775.26	0.9406	1504.5	46.035	13.08	2.287	74.76	58.92	1.862 E8	15.83	16.37
	742.47	0.9220	1427.0	57.950	13.36	2.693	66.31	47.83	1.746 E8	18.48	18.94
	709.69	0.9016	1349.5	69.864	13.67	3.110	60.07	38.83	1.631 E8	21.24	21.65
Uranus	1524.2	0.9698	3002.3	46.035	9.326	1.155	75.32	58.92	5.134 E8	16.39	16.92
	1464.1	0.9604	2870.3	57.950	9.515	1.352	66.97	47.83	4.834 E8	19.14	19.59
	1404.1	0.9502	2738.3	69.864	9.717	1.552	60.83	38.83	4.539 E8	22.01	22.40
Neptune	2291.4	0.9799	4536.8	46.035	7.606	0.7662	75.51	58.92	9.464 E8	16.59	17.10
	2278.9	0.9746	4499.9	57.950	7.627	0.8655	67.21	47.83	9.337 E8	19.38	19.83
	2266.4	0.9692	4463.0	69.864	7.648	0.9569	61.13	38.83	9.310 E8	22.30	22.69
Pluto	3710.5	0.9876	7375.0	46.035	5.977	0.4723	75.66	58.92	1.950 E9	16.73	17.24
	2983.5	0.9806	5909.0	57.950	6.666	0.6601	67.31	47.83	1.406 E9	19.48	19.93
	2256.4	0.9690	4443.0	69.864	7.665	0.9612	61.13	38.83	9.248 E8	22.30	22.69
From Venus to:											
Mercury	77.440	0.4055	108.85	46.035	41.38	26.91	63.62	34.78	5.880 E6	7.873	13.02
	83.030	0.3021	108.11	57.950	39.96	29.26	54.59	35.02	6.528 E6	5.763	11.86
	88.620	0.2116	107.37	69.864	38.68	31.20	47.95	35.26	7.198 E6	4.058	11.13
Earth	129.72	0.1723	152.07	107.37	31.97	26.86	38.04	35.26	1.275 E7	2.787	10.73
	128.84	0.1609	149.57	108.11	32.08	27.27	37.73	35.02	1.262 E7	2.712	10.71
	127.96	0.1494	147.07	108.85	32.19	27.69	37.42	34.78	1.249 E7	2.635	10.69
Mars	178.25	0.3976	249.12	107.37	27.27	17.90	41.54	35.26	2.053 E7	6.283	12.12
	167.97	0.3564	227.84	108.11	28.09	19.35	40.78	35.02	1.878 E7	5.766	11.86
	157.70	0.3098	206.56	108.85	28.99	21.05	39.94	34.78	1.709 E7	5.161	11.58
Jupiter	461.59	0.7674	815.80	107.37	16.95	6.148	46.71	35.26	8.557 E7	11.46	15.45
	443.12	0.7560	778.14	108.11	17.30	6.447	46.40	35.02	8.048 E7	11.39	15.40
	424.66	0.7437	740.48	108.85	17.67	6.774	46.09	34.78	7.551 E7	11.30	15.34
Saturn	805.93	0.8668	1504.5	107.37	12.83	3.426	48.01	35.26	1.974 E8	12.75	16.43
	767.55	0.8592	1427.0	108.11	13.14	3.617	47.75	35.02	1.835 E8	12.73	16.42
	729.18	0.8507	1349.5	108.85	13.48	3.829	47.48	34.78	1.699 E8	12.70	16.39
Uranus	1554.9	0.9309	3002.3	107.37	9.234	1.746	48.83	35.26	5.290 E8	13.57	17.08
	1489.2	0.9274	2870.3	108.11	9.435	1.831	48.62	35.02	4.958 E8	13.60	17.10
	1423.6	0.9235	2738.3	108.85	9.650	1.924	48.40	34.78	4.634 E8	13.62	17.12
Neptune	2322.6	0.9538	4536.8	107.37	7.556	1.162	49.12	35.26	9.655 E8	13.86	17.31
	2304.0	0.9531	4499.9	108.11	7.586	1.176	48.94	35.02	9.542 E8	13.92	17.36
	2285.9	0.9524	4463.0	108.85	7.616	1.189	48.77	34.78	9.430 E8	13.98	17.41
Pluto	3741.2	0.9713	7375.0	107.37	5.953	0.7183	49.34	35.26	1.974 E9	14.08	17.48
	3008.6	0.9641	5909.0	108.11	6.638	0.8979	49.08	35.02	1.424 E9	14.06	17.47
	2275.9	0.9522	4443.0	108.85	7.632	1.195	48.76	34.78	9.368 E8	13.98	17.40

Body	Mean orbital Semimajor axis (km × 10⁻⁶)	Eccentricity ε	Aphelion (km × 10⁻⁶)	Perhelion (km × 10⁻⁶)	Orbital Velocity (km/sec) Mean	Aphelion	Perhelion	Orbital Velocity of Departure Body (km/sec)	Time of one way trip (½ orbit) (Solar Seconds)	Departure Δᵥ orbit (km/sec)	Maximum Δᵥ (km/sec)
From Earth to:											
	99.052	0.5352	152.07	46.035	36.58	20.13	66.49	29.28	8.506 E6	9.150	14.45
Mercury	103.76	0.4415	149.57	57.950	35.75	22.25	57.43	29.77	9.119 E6	7.523	13.47
	108.47	0.3559	147.07	69.864	34.96	24.10	50.72	30.27	9.747 E6	6.178	12.77
	129.72	0.1723	152.07	107.37	31.97	26.86	38.04	29.28	1.275 E7	2.416	11.44
Venus	128.84	0.1609	149.57	108.11	32.08	27.27	37.73	29.77	1.262 E7	2.500	11.46
	127.96	0.1494	147.07	108.85	32.19	27.69	37.42	30.27	1.249 E7	2.583	11.47
	198.10	0.2576	249.12	147.07	25.87	19.88	33.67	30.27	2.406 E7	3.396	11.68
Mars	188.70	0.2074	227.84	149.57	26.51	21.48	32.71	29.77	2.237 E7	2.942	11.56
	179.31	0.1519	206.56	152.07	27.19	23.33	31.69	29.28	2.072 E7	2.411	11.44
	481.44	0.6945	815.80	147.07	16.59	7.046	39.08	30.27	9.114 E7	8.810	14.23
Jupiter	463.85	0.6776	778.14	149.57	16.91	7.412	38.56	29.77	8.620 E7	8.789	14.22
	446.27	0.6592	740.48	152.07	17.24	7.811	38.03	29.28	8.134 E7	8.755	14.20
	825.78	0.8219	1504.5	147.07	12.67	3.962	40.53	30.27	2.047 E8	10.25	15.17
Saturn	788.28	0.8103	1427.0	149.57	12.97	4.199	40.06	29.77	1.910 E8	10.29	15.19
	750.79	0.7975	1349.5	152.07	13.29	4.461	39.59	29.28	1.775 E8	10.31	15.21
	1574.7	0.9066	3002.3	147.07	9.176	2.031	41.46	30.27	5.392 E8	11.18	15.81
Uranus	1509.9	0.9009	2870.3	149.57	9.370	2.139	41.05	29.77	5.062 E8	11.28	15.88
	1445.2	0.8948	2738.3	152.07	9.578	2.257	40.64	29.28	4.740 E8	11.36	15.94
	2341.9	0.9372	4536.8	147.07	7.524	1.355	41.79	30.27	9.779 E8	11.51	16.05
Neptune	2324.7	0.9357	4499.9	149.57	7.552	1.377	41.42	29.77	9.671 E8	11.65	16.15
	2307.5	0.9341	4463.0	152.07	7.580	1.399	41.06	29.28	9.564 E8	11.78	16.24
	3761.0	0.9609	7375.0	147.07	5.937	0.8384	42.04	30.27	1.990 E9	11.77	16.23
Pluto	3029.3	0.9506	5909.0	149.57	6.615	1.053	41.58	29.77	1.439 E9	11.81	16.26
	2297.5	0.9338	4443.0	152.07	7.596	1.405	41.06	29.28	9.502 E8	11.78	16.24
From Mars to:											
	147.58	0.6881	249.12	46.035	29.97	12.88	69.72	21.97	1.547 E7	9.081	10.38
Mercury	142.89	0.5945	227.84	57.950	30.46	15.36	60.40	24.12	1.474 E7	8.761	10.10
	138.21	0.4945	206.56	69.864	30.97	18.01	53.25	26.49	1.402 E7	8.479	9.858
	178.25	0.3976	249.12	107.37	27.27	17.90	41.54	21.97	2.053 E7	4.060	6.463
Venus	167.97	0.3564	227.84	108.11	28.09	19.35	40.78	24.12	1.878 E7	4.770	6.931
	157.70	0.3098	206.56	108.85	28.99	21.05	39.94	26.49	1.709 E7	5.444	7.411
	198.10	0.2576	249.12	147.07	25.87	19.88	33.67	21.97	2.406 E7	2.088	5.444
Earth	188.70	0.2074	227.84	149.57	26.51	21.48	32.71	24.12	2.237 E7	2.646	5.682
	179.31	0.1519	206.56	152.07	27.19	23.33	31.69	26.49	2.072 E7	3.161	5.939
	511.18	0.5959	815.80	206.56	16.10	8.104	32.00	26.49	9.972 E7	5.514	7.462
Jupiter	502.99	0.5470	778.14	227.84	16.23	8.785	30.00	24.12	9.733 E7	5.881	7.737
	494.80	0.4965	740.48	249.12	16.37	9.494	28.22	21.97	9.496 E7	6.256	8.026
	855.52	0.7586	1504.5	206.56	12.45	4.613	33.60	26.49	2.159 E8	7.105	8.704
Saturn	827.42	0.7246	1427.0	227.84	12.66	5.058	31.68	24.12	2.054 E8	7.556	9.076
	799.32	0.6883	1349.5	249.12	12.38	5.533	29.97	21.97	1.950 E8	8.010	9.457
	1604.4	0.8713	3002.3	206.56	9.090	2.384	34.66	26.49	5.545 E8	8.165	9.589
Uranus	1549.1	0.8529	2870.3	227.84	9.251	2.606	32.84	24.12	5.260 E8	8.713	10.06
	1493.7	0.8332	2738.3	249.12	9.421	2.842	31.23	21.97	4.981 E8	9.269	10.55
	2371.7	0.9129	4536.8	206.56	7.477	1.595	35.04	26.49	9.966 E8	8.548	9.917
Neptune	2363.9	0.9036	4499.9	227.84	7.489	1.685	33.28	24.12	9.916 E8	9.160	10.45
	2356.1	0.8943	4463.0	249.12	7.501	1.772	31.75	21.97	9.867 E8	9.785	11.00
	3790.8	0.9455	7375.0	206.56	5.914	0.9897	35.34	26.49	2.014 E9	8.846	10.17
Pluto	3068.4	0.9257	5909.0	227.84	6.573	1.291	33.47	24.12	1.467 E9	9.352	10.62
	2346.0	0.8938	4443.0	249.12	7.517	1.780	31.75	21.97	9.804 E8	9.781	11.00
From Jupiter to:											
	430.92	0.8932	815.80	46.036	17.54	4.167	73.84	12.44	7.718 E7	8.269	60.80
Mercury	418.04	0.8614	778.14	57.950	17.81	4.860	65.26	13.05	7.375 E7	8.193	60.79
	405.17	0.8276	740.48	69.865	18.09	5.556	58.89	13.70	7.037 E7	8.144	60.79
	461.59	0.7674	815.80	107.37	16.95	6.148	46.71	12.44	8.557 E7	6.287	60.57
Venus	443.12	0.7560	778.14	108.11	17.30	6.447	46.40	13.05	8.048 E7	6.606	60.60
	424.66	0.7437	740.48	108.85	17.67	6.774	46.09	13.70	7.551 E7	6.926	60.64
	481.44	0.6945	815.80	147.07	16.59	7.046	39.08	12.44	9.114 E7	5.390	60.48
Earth	463.85	0.6776	778.14	149.57	16.91	7.412	38.56	13.05	8.620 E7	5.641	60.50
	446.27	0.6592	740.48	152.07	17.24	7.811	38.03	13.70	8.134 E7	5.890	60.53
	511.18	0.5959	815.80	206.56	16.10	8.104	32.00	12.44	9.972 E7	4.332	60.40
Mars	502.99	0.5470	778.14	227.84	16.23	8.785	30.00	13.05	9.733 E7	4.268	60.39
	494.80	0.4965	740.48	249.12	16.37	9.494	28.22	13.70	9.496 E7	4.206	60.39
	1122.5	0.3403	1504.5	740.48	10.87	7.624	15.49	13.70	3.245 E8	1.790	60.27
Saturn	1102.6	0.2942	1427.0	778.14	10.97	8.097	14.85	13.05	3.159 E8	1.797	60.27
	1082.7	0.2465	1349.5	815.80	11.07	8.604	14.23	12.44	3.074 E8	1.797	60.27
	1871.4	0.6043	3007.3	740.48	8.417	4.180	16.95	13.70	6.985 E8	3.248	60.33
Uranus	1824.2	0.5743	2870.3	778.14	8.525	4.439	16.37	13.05	6.723 E8	3.320	60.33
	1777.0	0.5409	2738.3	815.80	8.637	4.715	15.82	12.44	6.463 E8	3.389	60.33
	2638.6	0.7194	4536.8	740.48	7.088	2.864	17.55	13.70	1.169 E9	3.845	60.36
Neptune	2639.0	0.7051	4499.9	778.14	7.088	2.947	17.04	13.05	1.170 E9	3.992	60.37
	2639.4	0.6909	4463.0	815.80	7.087	3.030	16.58	12.44	1.170 E9	4.141	60.38

Body	Mean orbital Semimajor axis (km × 10⁻⁶)	Eccentricity ε	Aphelion (km × 10⁻⁶)	Perhelion (km × 10⁻⁶)	Orbital Velocity (km/sec) Mean	Aphelion	Perhelion	Orbital Velocity of Departure Body (km/sec)	Time of one way trip (½ orbit) (Solar Seconds)	Departure Δv orbit (km/sec)	Maximum Δv (km/sec)
Pluto	4057.7	0.8175	7375.0	740.48	5.716	1.811	18.04	13.70	2.230 E9	4.339	60.40
	3343.6	0.7673	5909.0	778.14	6.297	2.285	17.35	13.05	1.668 E9	4.299	60.39
	2629.4	0.6897	4443.0	815.80	7.101	3.043	16.57	12.44	1.163 E9	4.135	60.38
From Saturn to:											
Mercury	775.26	0.9406	1504.5	46.036	13.08	2.287	74.76	9.129	1.862 E8	6.841	36.71
	742.47	0.9220	1427.0	57.950	13.36	2.693	66.31	9.639	1.746 E8	6.946	36.73
	709.69	0.9016	1349.5	69.860	13.67	3.110	60.07	10.18	1.631 E8	7.067	36.75
Venus	805.93	0.8668	1504.5	107.37	12.83	3.426	45.01	9.129	1.974 E8	5.702	36.51
	767.55	0.8592	1427.0	108.11	13.14	3.617	47.75	9.639	1.835 E8	6.021	36.56
	729.18	0.8507	1349.5	108.85	13.48	3.829	47.48	10.18	1.699 E8	6.348	36.62
Earth	825.78	0.8219	1504.5	147.07	12.67	3.962	40.53	9.129	2.047 E8	5.167	36.43
	788.28	0.8103	1427.0	149.57	12.97	4.199	40.06	9.639	1.910 E8	5.440	36.47
	750.79	0.7975	1349.5	152.07	13.29	4.461	39.59	10.18	1.775 E8	5.716	36.51
Mars	855.52	0.7586	1504.5	206.56	12.45	4.613	33.60	9.129	2.159 E8	4.516	36.35
	827.42	0.7246	1427.0	227.84	12.66	5.058	31.68	9.639	2.054 E8	4.581	36.35
	799.32	0.6883	1349.5	249.12	12.88	5.533	29.97	10.18	1.950 E8	4.644	36.36
Jupiter	1122.5	0.3403	1504.5	740.48	10.87	7.624	15.49	9.129	3.245 E8	1.504	36.10
	1102.6	0.2942	1427.0	778.14	10.97	8.097	14.85	9.639	3.159 E8	1.541	36.10
	1082.7	0.2465	1349.5	815.80	11.07	8.604	14.23	10.18	3.074 E8	1.573	36.10
Uranus	2175.9	0.3798	3002.3	1349.5	7.806	5.233	11.64	10.18	8.758 E8	1.465	36.09
	2148.6	0.3359	2870.3	1427.0	7.855	5.539	11.14	9.639	8.593 E8	1.502	36.10
	2121.4	0.2908	2738.3	1504.5	7.905	5.860	10.67	9.129	8.430 E8	1.536	36.10
Neptune	2943.2	0.5415	4536.8	1349.5	6.712	3.660	12.31	10.18	1.378 E9	2.129	36.13
	2963.4	0.5185	4499.9	1427.0	6.689	3.767	11.88	9.639	1.392 E9	2.239	36.13
	2983.7	0.4958	4463.0	1504.5	6.666	3.870	11.48	9.129	1.406 E9	2.352	36.14
Pluto	4362.3	0.6906	7375.0	1349.5	5.513	2.358	12.89	10.18	2.486 E9	2.710	36.17
	3668.0	0.6110	5909.0	1427.0	6.012	2.954	12.23	9.639	1.917 E9	2.595	36.16
	2973.7	0.4941	4443.0	1504.5	6.677	3.885	11.47	9.129	1.399 E9	2.345	36.14
From Uranus to:											
Mercury	1524.2	0.9698	3002.3	46.036	9.326	1.155	75.32	6.490	5.134 E8	5.336	22.82
	1464.1	0.9604	2870.3	57.950	9.516	1.352	66.97	6.796	4.834 E8	5.444	22.85
	1404.1	0.9502	2738.3	69.865	9.717	1.552	60.83	7.116	4.539 E8	5.564	22.88
Venus	1554.9	0.9309	3002.3	107.37	9.234	1.746	48.83	6.490	5.290 E8	4.744	22.69
	1489.2	0.9274	2870.3	108.11	9.435	1.831	48.62	6.796	4.958 E8	4.965	22.74
	1423.6	0.9235	2738.3	108.85	9.650	1.924	48.40	7.116	4.634 E8	5.192	22.69
Earth	1574.7	0.9066	3002.3	147.07	9.176	2.031	41.46	6.490	5.392 E8	4.460	22.63
	1509.9	0.9009	2870.3	149.57	9.370	2.139	41.05	6.796	5.062 E8	4.657	22.67
	1445.2	0.8948	2738.3	152.07	9.578	2.257	40.64	7.116	4.740 E8	4.859	22.72
Mars	1604.4	0.8813	3002.3	206.56	9.090	2.384	34.66	6.490	5.545 E8	4.106	22.57
	1549.1	0.8529	2870.3	227.84	9.251	2.606	32.84	6.796	5.260 E8	4.190	22.58
	1493.7	0.8332	2738.3	249.12	9.421	2.842	31.23	7.116	4.981 E8	4.275	22.60
Jupiter	1817.4	0.6043	3002.3	740.48	8.417	4.180	16.95	6.490	6.985 E8	2.310	22.31
	1824.2	0.5743	2870.3	778.14	8.525	4.439	16.37	6.796	6.723 E8	2.358	22.31
	1777.0	0.5409	2738.3	815.80	8.637	4.715	15.82	7.116	6.463 E8	2.402	22.32
Saturn	2175.9	0.3798	3002.3	1349.5	7.806	5.233	11.64	6.490	8.758 E8	1.257	22.23
	2148.6	0.3359	2870.3	1427.0	7.855	5.539	11.14	6.796	8.593 E8	1.258	22.23
	2121.4	0.2908	2738.3	1504.5	7.905	5.860	10.67	7.116	8.430 E8	1.257	22.23
Neptune	3637.5	0.2472	4536.8	2738.3	6.037	4.690	7.771	7.116	1.893 E9	0.6544	22.20
	3685.1	0.2211	4499.9	2870.3	5.998	4.790	7.510	6.796	1.930 E9	0.7139	22.20
	3732.7	0.1957	4463.0	3002.3	5.960	4.888	7.266	6.490	1.968 E9	0.7757	22.20
Pluto	5056.6	0.4585	7375.0	2738.3	5.120	3.120	8.403	7.116	3.102 E9	1.287	22.23
	4389.6	0.3461	5909.0	2870.3	5.496	3.830	7.885	6.796	2.509 E9	1.089	22.22
	3722.7	0.1935	4443.0	3002.3	5.968	4.906	7.260	6.490	1.960 E9	0.7691	22.20
From Neptune to:											
Mercury	2291.4	0.9799	4536.8	46.037	7.606	0.7662	75.51	5.384	9.464 E8	4.617	24.97
	2278.9	0.9746	4499.9	57.948	7.627	0.8655	67.21	5.428	9.387 E8	4.562	24.96
	2266.4	0.9692	4463.0	69.864	7.648	0.9569	61.13	5.473	9.310 E8	4.516	24.95
Venus	2322.1	0.9538	4536.8	107.37	7.556	1.162	49.12	5.384	9.655 E8	4.221	24.90
	2304.0	0.9531	4499.9	108.11	7.586	1.176	48.94	5.428	9.542 E8	4.252	24.90
	2285.9	0.9524	4463.0	108.85	7.616	1.189	48.77	5.473	9.430 E8	4.283	24.91
Earth	2341.9	0.9372	4536.8	147.07	7.524	1.355	41.79	5.384	9.779 E8	4.029	24.86
	2324.7	0.9357	4499.9	149.57	7.552	1.377	41.42	5.428	9.667 E8	4.051	24.87
	2307.5	0.9341	4463.0	152.07	7.580	1.399	41.06	5.473	9.564 E8	4.073	24.87
Mars	2371.7	0.9129	4536.8	206.56	7.477	1.595	35.04	5.384	9.966 E8	3.788	24.83
	2363.9	0.9036	4499.9	227.84	7.489	1.685	33.28	5.428	9.916 E8	3.743	24.82
	2356.1	0.8943	4463.0	249.12	7.501	1.772	31.75	5.473	9.867 E8	3.700	24.81
Jupiter	2638.6	0.7194	4536.8	740.48	7.088	2.864	17.55	5.384	1.169 E9	2.520	24.66
	2639.0	0.7051	4499.9	778.14	7.088	2.947	17.04	5.428	1.170 E9	2.480	24.66
	2639.4	0.6909	4463.0	815.80	7.088	3.030	16.58	5.473	1.170 E9	2.442	24.66

HOHMANN ELLIPSE TRANSFER DATA (Continued)

Body	Mean orbital Semimajor axis (km × 10⁻⁶)	Eccentricity ε	Aphelion (km × 10⁻⁶)	Perhelion (km × 10⁻⁶)	Orbital Velocity (km/sec) Mean	Aphelion	Perhelion	Orbital Velocity of Departure Body (km/sec)	Time of one way trip (½ orbit) (Solar Seconds)	Departure Δ_v orbit (km/sec)	Maximum Δ_v (km/sec)
Saturn	2943.2	0.5415	4536.8	1349.5	6.712	3.660	12.31	5.384	1.378 E9	1.723	24.60
	2963.4	0.5185	4499.9	1427.0	6.689	3.767	11.88	5.428	1.392 E9	1.661	24.59
	2983.7	0.4958	4463.0	1504.5	6.666	3.870	11.48	5.473	1.406 E9	1.602	24.59
Uranus	3637.5	0.2472	4536.8	2738.3	6.037	4.690	7.771	5.384	1.893 E9	0.6933	24.54
	3685.1	0.2211	4499.9	2870.3	5.998	4.790	7.510	5.428	1.930 E9	0.6375	24.54
	3732.7	0.1957	4463.0	3002.3	5.960	4.888	7.266	5.473	1.968 E9	0.5845	24.54
Pluto	5919.0	0.2460	7375.0	4463.0	4.733	3.682	6.084	5.473	3.929 E9	0.6112	24.54
	5204.4	0.1354	5909.0	4499.9	5.047	4.404	5.784	5.428	3.240 E9	0.3557	24.54
	4489.9	−0.1045	4443.0	4536.8	5.434	5.491	5.377	5.384	2.596 E9	−0.0061	24.54
From Pluto to:											
Mercury	3710.5	0.9876	7375.0	46.037	5.977	0.4723	75.66	3.676	1.950 E9	3.204	5.955
	2983.5	0.9806	5909.0	57.948	6.666	0.6601	67.31	4.737	1.406 E9	4.077	6.466
	2256.4	0.9690	4443.0	69.864	7.665	0.9612	67.13	6.103	9.248 E8	5.141	7.185
Venus	3741.2	0.9713	7375.0	107.37	5.953	0.7183	49.34	3.676	1.974 E9	2.958	5.826
	3008.6	0.9641	5909.0	108.11	6.638	0.8979	49.08	4.737	1.424 E9	3.839	6.319
	2275.9	0.9522	4443.0	108.85	7.632	1.195	48.76	6.103	9.368 E8	4.908	7.020
Earth	3761.0	0.9609	7375.0	147.07	5.937	0.8384	42.04	3.676	1.990 E9	2.838	5.766
	3029.3	0.9506	5909.0	149.57	6.615	1.053	41.58	4.737	1.439 E9	3.684	6.226
	2297.5	0.9338	4443.0	152.07	7.596	1.405	41.06	6.103	9.502 E8	4.697	6.874
Mars	3790.8	0.9455	7375.0	206.56	5.914	0.9897	35.34	3.676	2.014 E9	2.687	5.693
	3068.4	0.9257	5909.0	227.84	6.573	1.291	33.47	4.737	1.467 E9	3.446	6.088
	2346.0	0.8938	4443.0	249.12	7.517	1.780	31.75	6.103	9.804 E8	4.323	6.624
Jupiter	4057.7	0.8175	7375.0	740.48	5.716	1.811	18.04	3.676	2.230 E9	1.865	5.355
	3343.6	0.7673	5909.0	778.14	6.297	2.285	17.35	4.737	1.668 E9	2.452	5.586
	2629.4	0.6897	4443.0	815.80	7.101	3.043	16.57	6.103	1.163 E9	3.060	5.879
Saturn	4362.3	0.6906	7375.0	1349.5	5.513	2.358	12.89	3.676	2.486 E9	1.318	5.190
	3668.0	0.6110	5909.0	1427.0	6.012	2.954	12.23	4.737	1.917 E9	1.782	5.326
	2973.7	0.4941	4443.0	1504.5	6.677	3.885	11.47	6.103	1.399 E9	2.217	5.487
Uranus	5056.6	0.4585	7375.0	2738.3	5.120	3.120	8.103	3.676	3.102 E9	0.5564	5.050
	4389.6	0.3461	5909.0	2870.3	5.496	3.830	7.885	4.737	2.500 E9	0.9065	5.101
	3722.7	0.1935	4443.0	3002.3	3.968	4.906	7.260	6.103	1.960 E9	1.197	5.160
Neptune	5919.0	0.2460	7375.0	4463.0	4.733	3.682	6.084	3.676	3.929 E9	5.164	5.019
	5204.4	0.1354	5909.0	4499.9	5.047	4.404	5.784	4.737	3.240 E9	0.3323	5.030
	4489.9	−0.1045	4443.0	4536.8	5.434	5.491	5.377	6.103	2.596 E9	0.6117	5.056

CONSTANTS FOR SATELLITE GEODESY

Defining Constants

1. Number of ephemeris seconds in 1 tropical year (1900) .... s = 31 556 925.9747
2. Gaussian gravitational constant, defining the a. u. .... k = 0.017 202 09895

Primary Constants

3. Velocity of light in meters per second (in vacuum) .... c = $2.99792458(1.2) \times 10^8$
4. Dynamical form-factor for Earth .... J_2 = 0.001 082 7
5. Sidereal mean motion of Moon in radians per second (1900)................. $n_{(}*$ = $2.661 699 489 \times 10^{-6}$
6. General precession in longitude per tropical century (1900)................... p = 5025".64
7. Constant of nutation (1900) .... N = 9".210

Auxiliary Constants and Factors

$k/86400$, for use when the unit of time is 1 second .... k' = $1.990 983 675 \times 10^{-7}$

Number of seconds of arc in 1 radian... = 206 264.806

Factor for constant of aberration (note 10) .... F_1 = 1.000 142

Factor for mean distance of Moon (note 12) .... F_2 = 0.999 093 142

Factor for parallactic inequality (note 15) .... F_3 = 49853".2

Derived Constants

8. Solar parallax............................ $\arcsin(\alpha_e/A) = \pi.$ = 8".79405 (8".794)
9. Light-time for unit distance .... $A/c' = \tau_A$ = 499^{s}012
   = 1^s/0.002 003 96
10. Constant of aberration .... $F_1 k' \tau_A = \kappa$ = 20".4958 (20".496)
11. Ratio of masses of Sun and Earth + Moon .... $S/E(1 + \mu)$ = 328 912
12. Perturbed mean distance of Moon, in meters .... $F_2(GE(1 + \mu)/n_{(}*{}^2)^{\frac{1}{3}} = \alpha_{(}$ = $384 400 \times 10^3$
13. Constant of sine parallax for Moon............................ $\alpha_c/\alpha_{(} = \sin \pi_{(}$ = 3422".451

14. Constant of lunar inequality .... $\dfrac{\mu}{1 + \mu} \dfrac{\alpha_{(}}{A} = L$ = 6".43987 (6".440)

15. Constant of parallactic inequality.................... $F_3 \dfrac{1 - \mu}{1 + \mu} \dfrac{\alpha_{(}}{A} = P_{(}$ = 124".986

THE EARTH: ITS MASS, DIMENSIONS AND OTHER RELATED QUANTITIES

From NASA Technical Translation NASA TT-F-533

TABLE A

Quantity	Unit of Measurement	Symbol	Numerical Value	Sources; Remarks[1]
Mass	Proportion of the mass of the sun	M	1/331950	I. D. Zhongolovich
	gram		$5.9763 \cdot 10^{27}$	
Major Orbital semi-axis	Astronomical unit	a_{orb}	1.000000	1961 data of Soviet radar determinations
	km		149,457,000	
Distance from sun at perihelion	a.u.	r_π	0.983298	for 1962
Distance from sun at aphelion	a.u.	r_α	1.016744	for 1962
Moment of perihelion passage		T_π	Jan. 2, $4^{hr}52^m$	for 1962 USSR Astron. Yearbook for 1962
Moment of aphelion passage		T_α	Jul 4 $5^{hr}05^m$	for 1962
Siderial rotation period around sun	sec	P_{orb}	$31.558 \cdot 10^6$	
Mean rotational velocity	km/sec	U_{orb}	29.8	
Mean equatorial radius	km	$\bar{a}$	6,378.245 6,378.077	A. A. Izotov, 1950 I. D. Zhongolovich, 1956
Mean polar compression		α	1/298.3 1/296.6 1/298.2	A. A. Izotov, 1950 I. D. Zhongolovich, 1952 D. G. King-Hele, R. Merson, 1959. On observations on the movements of artificial earth satellites.
Difference in equatorial and polar semi-axes	km	$a - c$	21.382 21.500	A. A. Izotov, 1950 I. D. Zhongolovich, 1956
Compression of meridian of major equatorial axis		α_a	1/295.2	I. D. Zhongolovich, 1952
Compression of meridian of minor equatorial axis		α_b	1/298.0	I. D. Zhongolovich, 1952
Equatorial compression		ϵ	1/30 000 1/32 000	A. A. Izotov, 1950 I. D. Zhongolovich, 1952

[1] The source does not indicate whether the value given is generally accepted or has merely been calculated by the author, G. N. Katterfel'd.

TABLE A Continued

Quantity	Unit of Measurement	Symbol	Numerical Value	Sources; Remarks[1]
Difference in equatorial semi-axes	m	$a - b$	213 199	A. A. Izotov, 1950 I. D. Zhongolovich, 1952
Meridian of longitude of minor equatorial semi-axis		λ_a	15°E - 6°W	A. A. Izotov, 1950 I. D. Zhongolovich, 1952
Meridian of longitude of minor equatorial axis		λ_b	105°E - 75° W; 84°E - 96°W	A. A. Izotov, 1950 I. D. Zhongolovich, 1952
Difference in polar semi-axes	m	$C_N - C_S$	~ 70 <100	I. D. Zhongolovich, 1952 Based on observations of artificial satellites[1]
Polar asymmetry		η	$\sim 1.10^{-5}$	
Mean acceleration of gravity at equator		g_e	978,057.3	I. D. Zhongolovich, 1952
Mean acceleration of gravity at poles	milligals (mgl)	g_ρ	983,225.1	
Difference in acceleration of gravity at pole and at equator		$g_p - g_e$	+5,167.8	
Difference in acceleration of gravity at equator	mgl	$g_a - g_b$	+30.2	I. D. Zhongolovich, 1952
Difference in acceleration of gravity at poles		$g_N - g_S$	+30	I. D. Zhongolovich, 1952
Mean acceleration of gravity for entire surface of terrestrial ellipsoid	mgl	g	979,783.0	I. D. Zhongolovich, 1952
Mean radius	km	R	6,370.949	
Area of surface	km²	S	$510.0501 \cdot 10^6$	
Volume	km³	V	$1,083.1579 \cdot 10^9$	
Mean density	gr/cm³	δ	5.5170	I. D. Zhongolovich, 1952
Siderial rotational period	sec	P	86,164.09	
Angular rotational velocity	rad/sec	ω	$7.292116 \cdot 10^{-5}$	
Mean equatorial rotational velocity	km/sec	ν	0.465	

TABLE A Continued

Quantity	Unit of Measurement	Symbol	Numerical Value	Sources; Remarks[1]
Ratio of centrifugal force to attractive force at equator		q	$\dfrac{1}{289}$	
Ratio of centrifugal force to force of gravity at equator		q_c	$0.0034677 = \dfrac{1}{288}$	I. D. Zhongolovich, 1952
Coefficients characterizing the radial distribution of densities within the earth		κ_1 κ	0.966 0.331	Based on observations of artificial satellites; several larger values were given earlier[1]
Radius of inertia	km; proportion of mean radius	R_i	3,674.735 0.5768	
Geocentric latitude of inertial parallel		ϕ_i	$54°47'$	
Moment of inertia	$gr \cdot cm^2$	I	$8.070 \cdot 10^{44}$	
Moment of rotation	$gr \cdot cm^2/sec$	L	$5.885 \cdot 10^{40}$	
Relative true secular braking of earth's rotation due to tidal friction		$\dfrac{\Delta\omega_e}{\omega}$	$-4.2 \cdot 10^{-8}$ per century	
Relative proper secular acceleration of earth's rotation		$\dfrac{\Delta\omega_i}{\omega}$	$+1.4 \cdot 10^{-8}$ per century	N. N. Pariyskiy, 1955
Relative observed secular braking of earth's rotation		$\dfrac{\Delta\omega}{\omega}$	$-2.8 \cdot 10^{-8}$ per century	
Mean rotational velocity of terrestrial radius due to abyssal compression	cm/century	$\dfrac{\Delta R}{\Delta t}$	~5 Assumed invariability of mass (M) 4.5 and distribution of masses (κ)	B. Meyermann, 1928, 1928a N. N. Pariyskiy, 1955
Secular variation in potential gravitational energy of earth accompanying reduction of terrestrial radius by 5 cm and corresponding increase in earth's kinetic energy	erg/century	ΔE	$\sim 17 \cdot 10^{30}$	Assumed uniformity of compression of entire planet. P. N. Kropotkin, 1948 and A. T. Aslanyan, 1955

[1] On the basis of materials published in the Astron. J., Vol. 64, 1272, 1959.

TABLE A Continued

Quantity	Unit of Measurement	Symbol	Numerical Value	Sources; Remarks[1]
Probable value of total energy of tectonic deformation of earth	erg/century	E_t	$\sim 1 \cdot 10^{30}$	With allowance for earthquakes, volcanic eruptions and other forms of tectonic activity P. N. Kropotkin, 1948
Secular loss of heat of earth through radiation into space	erg/century cal/century	$\Delta' E_k$	$1 \cdot 10^{30}$ $2.4 \cdot 10^{22}$	P. N. Kropotkin, 1948
Portion of earth's kinetic energy transformed into heat as a result of lunar and solar tides in the hydrosphere	erg/century cal/century	$\Delta'' E_k$	$0.11 \cdot 10^{30}$ $0.26 \cdot 10^{22}$	Heiskanen, 1922 and de Sitter, 1927
Difference in duration of days in March and August	sec	ΔP	0.0025 (March-Aug.)	N. N. Pariyskiy, 1955
Corresponding relative annual variation in earth's rotational velocity		$\dfrac{\Delta^* \omega}{\omega}$	$2.9 \cdot 10^{-8}$ (Aug.-March)	N. N. Pariyskiy, 1955
Presumed variation in earth's radius between August and March	cm	$\Delta^* R$	-9.2 (Aug.-March)	
Annual variation in level of world ocean	cm	Δh_0	~ 10 (Sept.-March)	N. N. Pariyskiy, 1955

The Earth's Lithosphere, Hydrosphere, Atmosphere and Biosphere

TABLE B

Quantity	Unit of Measurement	Symbol	Numerical Value	Source
Area of continents	km²; in % of area of surface of earth	S_C	$149 \cdot 10^6$ 29.2	E. Kossina, 1933
Area of world ocean	km²; in % of area of surface of earth	S_o	$361 \cdot 10^6$ 70.8	E. Kossina, 1933
Mean height of continents above sea level	m	h_C	875	E. Kossina, 1933
Mean depth of world ocean	m	h_o	$=3794$	Morskoy Atlas, Vol. II, 1953
Mean position of earth's surface with respect to sea level	m	h_m	$=2430$	E. Kossina, 1933

The Earth's Lithosphere, Hydrosphere, Atmosphere and Biosphere (*Continued*)

TABLE B Continued

Quantity	Unit of Measurement	Symbol	Numerical Value	Source
Mean thickness of lithosphere within the limits of the continents	km	$h_{c.l.}$	35	M. Yuing and F. Press, 1955
Mean thickness of lithosphere within the limits of the ocean	km	$h_{o.l.}$	4.7	Kh. Khess, 1955
Mean rate of thickening of continental lithosphere	m/10^6 yr	$\frac{\Delta h}{\Delta t}$	10 – 40	V. I. Popov, 1955
Mean rate of horizontal extension of continental lithosphere	km/10^6 yr	$\frac{\Delta l}{\Delta t}$	0.75 – 20	V. I. Popov, 1955
Mass of lithosphere	gr	m_l	$2.367 \cdot 10^{25}$	A. Poldervart, 1955
Amount of water released from the mantle and core in the course of geological time	gr		$3.400 \cdot 10^{24}$	Kalp, 1951
Total reserve of water in the mantle	gr		$2 \cdot 10^{26}$	A. P. Vinogradov, 1959
Present day content of free and bound water in the earth's lithosphere	gr		$2.2 - 2.6 \cdot 10^{24}$ $1.8 - 2.7 \cdot 10^{24}$	Kalp, 1951 A. Poldervart, 1955
Mass of hydrosphere	gr	m_h	$1.664 \cdot 10^{24}$	A. Poldervart, 1955
Amount of oxygen bound in the earth's crust	gr		$1.300 \cdot 10^{24}$	A. Poldervart, 1955
Amount of free oxygen	gr		$1.5 \cdot 10^{21}$	A. Poldervart, 1955
Mass of atmosphere	gr	m_a	$5.136 \cdot 10^{21}$	A. Poldervart, 1955
Mass of biosphere	gr	m_b	$1.148 \cdot 10^{19}$	A. Poldervart, 1955
Mass of living matter in the biosphere	gr		$3.6 \cdot 10^{17}$	A. Poldervart, 1955
Density of living matter on dry land	gr/cm^2		0.1	A. Poldervart, 1955
Density of living matter in ocean	gr/cm^3		$15 \cdot 10^{-8}$	A. Poldervart, 1955

ESTIMATED AGE OF EARTH

TABLE C

Object of Study	Age in billions of years	Source
Fossils of the most ancient organisms	2.7	L. Arens, 1955
Most ancient known terrestrial rocks:		
Mica (biotite) found within migmatites on the Kola Peninsula and at the Great Rapids of the Voron'ya River in 1958;	3.6	A. A. Polkanov and E. K. Gerling, 1961
Rock found in South Africa	4	A. L. Heils, 1960[1]
Lithosphere	~ 4	V. I. Baranov, 1958
Earth	~ 5	A. P. Vinogradov, 1959 and others

[1] See *Priroda*, No. 11, 1960, p. 113.

Approximate Scale of Geologic Time,[1]
Based on the Data of Soviet Research, 1960.

TABLE D

Era	Period or Epoch	Beginning and end, in millions of years	Approximate Duration, in millions of years
Cenozoic	Quaternian Contemporary	0 – 10,000 yrs ± 2,000 yrs	8 – 12,000 years
	Pleistocene	10,000 – 1,000,000 yrs ± 50,000 yrs	1
	Tertiary Pliocene	1 – 10	9
	Miocene	10 – 25	15
	Oligocene	25 – 40	15 } 69
	Eocene	40 – 60	20
	Paleocene	60 – 70	10 } 570
Mesozoic	Cretaceous	70 – 140	70
	Jurassic	140 – 185	45
	Triassic	185 – 225	40
Paleozoic	Permian	225 – 270	45
	Carboniferous	270 – 320	50
	Devonian	320 – 400	80
	Silurian	400 – 420	20
	Ordovician	420 – 480	60
	Cambrian	480 – 570	90
	Pre-Cambrian IV (Riphean)[2]	570 – 1,200	630
	Pre-Cambrian III (Proterozoic)[3]	1,200 – 1,900	700
	Pre-Cambrian II (Archean)	1,900 – 2,700	800
	Pre-Cambrian I (Catarchean)	2,700 – 3,500	800
	Pregeological era	3,500 – 5,000	1,500

[1] Based on the geochronological scale of the Commission for Determining the Absolute Age of Geological Formations, published in *Izv. AN SSSR*, Geological series, No. 10, 1960.

[2] Proterozoic II.

[3] Proterozoic I.

ACCELERATION DUE TO GRAVITY AND LENGTH OF THE SECONDS PENDULUM

Based on the formula of the U. S. Coast and Geodetic Survey. The length of the simple pendulum whose period is two seconds, that is which beats seconds, is computed in each case from the corresponding value of the acceleration.

Latitude	Acceleration due to gravity		Length of seconds pendulum	
°	cm/sec.^{2}	ft./sec.^{2}	cm	in.
0	978.039	32.0878	99.0961	39.0141
5	978.078	32.0891	99.1000	39.0157
10	978.195	32.0929	99.1119	39.0204
15	978.384	32.0991	99.1310	39.0279
20	978.641	32.1076	99.1571	39.0382
25	978.960	32.1180	99.1894	39.0509
30	979.329	32.1302	99.2268	39.0656
31	979.407	32.1327		
32	979.487	32.1353		
33	979.569	32.1380		
34	979.652	32.1407		
35	979.737	32.1435	99.2681	39.0819
36	979.822	32.1463		
37	979.908	32.1491		
38	979.995	32.1520		
39	980.083	32.1549		
40	980.171	32.1578	99.3121	39.0992
41	980.261	32.1607		
42	980.350	32.1636		
43	980.440	32.1666		
44	980.531	32.1696		
45	980.621	32.1725	99.3577	39.1171
46	980.711	32.1755		
47	980.802	32.1785		
48	980.892	32.1814		
49	980.981	32.1844		
50	981.071	32.1873	99.4033	39.1351
51	981.159	32.1902		
52	981.247	32.1931		
53	981.336	32.1960		
54	981.422	32.1988		
55	981.507	32.2016	99.4475	39.1525
56	981.592	32.2044		
57	981.675	32.2071		
58	981.757	32.2098		
59	981.839	32.2125		
60	981.918	32.2151	99.4891	39.1689
65	982.288	32.2272	99.5266	39.1836
70	982.608	32.2377	99.5590	39.1964
75	982.868	32.2463	99.5854	39.2068
80	983.059	32.2525	99.6047	39.2144
85	983.178	32.2564	99.6168	39.2191
90	983.217	32.2577	99.6207	39.2207

Free Air Correction for Altitude

−0.0003086 cm/sec.^{2}/m for altitude in meters.
−0.000003086 ft./sec.^{2}/ft. for altitude in feet.

Altitude meters	Correction cm/sec.^{2}	Altitude feet	Correction ft./sec.^{2}
200	−0.0617	200	−0.000617
300	0.0926	300	0.000926
400	0.1234	400	0.001234
500	0.1543	500	0.001543
600	0.1852	600	0.001852
700	0.2160	700	0.002160
800	0.2469	800	0.002469
900	0.2777	900	0.002777

DATA IN REGARD TO THE EARTH

Quadrant of the equator, 10,019,150 meters, 6,225.60 miles.
Quadrant of the meridian, 10,002,290 meters, 6,215.12 miles.
1° latitude at the equator = 69.41 miles.
1° latitude at the pole = 68.70 miles.
Mean surface density of the continents, 2.67 g/cm^3, 166.7 lb./ft.^{3}
Land area, 148.847 × 10^6 km^2, 57.470 × 10^6 sq. mi.
Ocean area, 361.254 × 10^6 km^2, 139.480 × 10^6 sq. mi.
Highest mountain, Everest, 8,840 meters, 29,003 ft.
Greatest sea depth, 10,430 meters, 34,219 ft.
Thermal gradient of the earth, higher at increasing depths, 30°C per km, 48°C per mi. (uncertain).

ELEMENTS IN THE EARTH'S CRUST

A. Demayo

The elements in this table are listed in decreasing average concentration as they occur in the earth's crust. They are average concentrations and there will be variations in composition from point to point throughout the earth's crust.

Element	Concentration (mg/kg)	Element	Concentration (mg/kg)
Oxygen	4.64 × 10^5	Ytterbium	3.0 × 10^0
Silicon	2.82 × 10^5	Beryllium	2.8 × 10^0
Aluminum	8.32 × 10^4	Erbium	2.8 × 10^0
Iron	5.63 × 10^4	Uranium	2.7 × 10^0
Calcium	4.15 × 10^4	Bromine	2.5 × 10^0
Sodium	2.36 × 10^4	Tantalum	2.0 × 10^0
Magnesium	2.33 × 10^4	Tin	2.0 × 10^0
Potassium	2.09 × 10^4	Arsenic	1.8 × 10^0
Titanium	5.70 × 10^3	Molybdenum	1.5 × 10^0
Hydrogen	1.40 × 10^3	Tungsten	1.5 × 10^0
Phosphorus	1.05 × 10^3	(Wolfram)	
Manganese	9.50 × 10^2	Europium	1.2 × 10^0
Fluorine	6.25 × 10^2	Holmium	1.2 × 10^0
Barium	4.25 × 10^2	Cesium	1.0 × 10^0
Strontium	3.75 × 10^2	Terbium	9 × 10^{-1}
Sulfur	2.60 × 10^2	Iodine	5 × 10^{-1}
Carbon	2.00 × 10^2	Lutetium	5 × 10^{-1}
Zirconium	1.65 × 10^2	Thulium	4.8 × 10^{-1}
Vanadium	1.35 × 10^2	Thallium	4.5 × 10^{-1}
Chlorine	1.30 × 10^2	Antimony	2 × 10^{-1}
Chromium	1.00 × 10^2	Cadmium	2 × 10^{-1}
Rubidium	9.0 × 10^1	Bismuth	1.7 × 10^{-1}
Nickel	7.5 × 10^1	Indium	1 × 10^{-1}
Zinc	7.0 × 10^1	Mercury	8 × 10^{-2}
Cerium	6.0 × 10^1	Silver	7 × 10^{-2}
Copper	5.5 × 10^1	Selenium	5 × 10^{-2}
Yttrium	3.3 × 10^1	Palladium	1 × 10^{-2}
Lanthanum	3.0 × 10^1	Helium	8 × 10^{-3}
Neodymium	2.8 × 10^1	Neon	3 × 10^{-3}
Cobalt	2.5 × 10^1	Platinum	5 × 10^{-3}
Scandium	2.2 × 10^1	Rhenium	5 × 10^{-3}
Lithium	2.0 × 10^1	Gold	4 × 10^{-3}
Niobium	2.0 × 10^1	Osmium	1.5 × 10^{-3}
(Columbium)		Iridium	1 × 10^{-3}
Nitrogen	2.0 × 10^1	Rhodium	1 × 10^{-3}
Gallium	1.5 × 10^1	Ruthenium	1 × 10^{-3}
Lead	1.25 × 10^1	Tellurium	1 × 10^{-3}
Boron	1.0 × 10^1	Krypton	1 × 10^{-4}
Thallium	9.6 × 10^0	Xenon	3 × 10^{-5}
Praeseodymium	8.2 × 10^0	Protactinium	1.4 × 10^{-6}
Samarium	6.0 × 10^0	Radium	9 × 10^{-7}
Gadolinium	5.4 × 10^0	Actinium	5.5 × 10^{-10}
Germanium	5.4 × 10^0	Polonium	2 × 10^{-10}
Argon	3.5 × 10^0	Radon	4 × 10^{-13}
Dysprosium	3.0 × 10^0		

CHEMICAL COMPOSITION OF ROCKS

Reprinted from "Sedimentary Rocks" (1948) with the permission of F. J. Pettijohn, author, and Harper Brothers, publishers.

Element	Average igneous rock	Average shale	Average sandstone	Average limestone	Average sediment
SiO_2	59.14	58.10	78.33	5.19	57.95
TiO_2	1.05	0.65	0.25	0.06	0.57
Al_2O_3	15.34	15.40	4.77	0.81	13.39
Fe_2O_3	3.08	4.02	1.07	0.54	3.47
FeO	3.80	2.45	0.30		2.08
MgO	3.49	2.44	1.16	7.89	2.65
CaO	5.08	3.11	5.50	42.57	5.89
Na_2O	3.84	1.30	0.45	0.05	1.13
K_2O	3.13	3.24	1.31	0.33	2.86
H_2O	1.15	5.00	1.63	0.77	3.23
P_2O_5	0.30	0.17	0.08	0.04	0.13
CO_2	0.10	2.63	5.03	41.54	5.38
SO_3		0.64	0.07	0.05	0.54
BaO	0.06	0.05	0.05		
C		0.80			0.66
	99.56	100.00	100.00	99.84	99.93

SOLAR SPECTRAL IRRADIANCE

From NASA Technical Report R-351, "The Solar Constant and the Solar Spectrum Measured from a Research Aircraft". Report edited by Matthew P. Thekaekara, Goddard Space Flight Center, Greenbelt, Maryland 20771. Discussion of previously reported values and of measurements and calculations leading to the following table are contained in the NASA Technical Report R-351.

SOLAR SPECTRAL IRRADIANCE—PROPOSED STANDARD CURVE

λ —Wavelength in microns
P_λ —Solar spectral irradiance averaged over small bandwidth centered at λ, in Watts cm$^{-2}\mu^{-1}$ *
A_λ —Area under the solar spectral irradiance curve in the wavelength range 0 to λ, mW cm^{-2}
D_λ —Percentage of the solar constant associated with wavelengths shorter than λ
Solar Constant—135.30 mW cm^{-2}, or 1.940 cal min^{-1} cm^{-2}

λ	P_λ	A_λ	D_λ	λ	P_λ	A_λ	D_λ	λ	P_λ	A_λ	D_λ
0.120	0.000010	0.00059992	0.00044	0.475	0.2044	25.6001	18.921	2.4	0.0064	129.695	95.858
0.140	0.000003	0.00072999	0.00053	0.480	0.2074	26.6296	19.681	2.5	0.0054	130.285	96.294
0.150	0.000007	0.00077999	0.00057	0.485	0.1976	27.6421	20.430	2.6	0.0048	130.795	96.671
0.160	0.000023	0.00092999	0.00068	0.490	0.1950	28.6236	21.155	2.7	0.0043	131.250	97.007
0.170	0.000063	0.00135999	0.00100	0.495	0.1960	29.6011	21.878	2.8	0.00390	131.660	97.3103
0.180	0.000125	0.00229999	0.00169	0.500	0.1942	30.5766	22.599	2.9	0.00350	132.030	97.5838
0.190	0.000271	0.00427999	0.00316	0.505	0.1920	31.5421	23.312	3.0	0.00310	132.360	97.8277
0.200	0.00107	0.010984	0.0081	0.510	0.1882	32.4926	24.015	3.1	0.00260	132.645	98.0383
0.210	0.00229	0.027784	0.0205	0.515	0.1833	33.4214	24.701	3.2	0.00226	132.888	98.2179
0.220	0.00575	0.067984	0.0502	0.520	0.1833	34.3379	25.379	3.3	0.00192	133.097	98.3724
0.225	0.00649	0.098584	0.0728	0.525	0.1852	35.2591	26.059	3.4	0.00166	133.276	98.5047
0.230	0.00667	0.131484	0.0971	0.530	0.1842	36.1826	26.742	3.5	0.00146	133.432	98.6200
0.235	0.00593	0.162984	0.1204	0.535	0.1818	37.0976	27.418	3.6	0.00135	133.573	98.7238
0.240	0.00630	0.193559	0.1430	0.540	0.1783	37.9979	28.084	3.7	0.00123	133.702	98.8192
0.245	0.00723	0.227384	0.1680	0.545	0.1754	38.8821	28.737	3.8	0.00111	133.819	98.9056
0.250	0.00704	0.263059	0.1944	0.550	0.1725	39.7519	29.380	3.9	0.00103	133.926	98.9847
0.255	0.0104	0.306659	0.226	0.555	0.1720	40.6131	30.017	4.0	0.00095	134.025	99.0579
0.260	0.0130	0.365159	0.269	0.560	0.1695	41.4669	30.648	4.1	0.00087	134.116	99.1252
0.265	0.0185	0.443909	0.328	0.565	0.1705	42.3169	31.276	4.2	0.00078	134.198	99.1861
0.270	0.0232	0.548159	0.405	0.570	0.1712	43.1711	31.907	4.3	0.00071	134.273	99.2412
0.275	0.0204	0.657159	0.485	0.575	0.1719	44.0289	32.541	4.4	0.00065	134.341	99.2915
0.280	0.0222	0.763659	0.564	0.580	0.1715	44.8874	33.176	4.5	0.00059	134.403	99.3373
0.285	0.0315	0.897909	0.663	0.585	0.1712	45.7441	33.809	4.6	0.00053	134.459	99.3787
0.290	0.0482	1.09715	0.810	0.590	0.1700	46.5971	34.439	4.7	0.00048	134.509	99.4160
0.295	0.0584	1.36365	1.007	0.595	0.1682	47.4426	35.064	4.8	0.00045	134.556	99.4504
0.300	0.0514	1.63815	1.210	0.600	0.1666	48.2796	35.683	4.9	0.00041	134.599	99.482195
0.305	0.0603	1.91740	1.417	0.605	0.1647	49.1079	36.295	5.0	0.0003830	134.63905	99.511500
0.310	0.0689	2.24040	1.655	0.610	0.1635	49.9284	36.902	6.0	0.0001750	134.91805	99.717708
0.315	0.0764	2.60365	1.924	0.620	0.1602	51.5469	38.098	7.0	0.0000990	135.05505	99.818965
0.320	0.0830	3.00215	2.218	0.630	0.1570	53.1329	39.270	8.0	0.0000600	135.13455	99.877723
0.325	0.0975	3.45340	2.552	0.640	0.1544	54.6899	40.421	9.0	0.0000380	135.18355	99.913939
0.330	0.1059	3.96190	2.928	0.650	0.1511	56.2174	41.550	10.0	0.0000250	135.21505	99.937220
0.335	0.1081	4.49690	3.323	0.660	0.1486	57.7159	42.657	11.0	0.0000170	135.23605	99.952742
0.340	0.1074	5.03565	3.721	0.670	0.1456	59.1869	43.744	12.0	0.0000120	135.25055	99.963458
0.345	0.1069	5.57140	4.117	0.680	0.1427	60.6284	44.810	13.0	0.0000087	135.26090	99.971108
0.350	0.1093	6.11190	4.517	0.690	0.1402	62.0429	45.855	14.0	0.0000055	135.26800	99.976356
0.355	0.1083	6.65590	4.919	0.700	0.1369	63.4284	46.879	15.0	0.0000049	135.27320	99.980199
0.360	0.1068	7.19365	5.316	0.710	0.1344	64.7849	47.882	16.0	0.0000038	135.27755	99.983414
0.365	0.1132	7.74365	5.723	0.720	0.1314	66.1139	48.864	17.0	0.0000031	135.28100	99.985964
0.370	0.1181	8.32190	6.150	0.730	0.1290	67.4159	49.826	18.0	0.0000024	135.28375	99.987997
0.375	0.1157	8.90640	6.582	0.740	0.1260	68.6909	50.769	19.0	0.0000020	135.28595	99.989623
0.380	0.1120	9.47565	7.003	0.750	0.1235	69.9384	51.691	20.0	0.0000016	135.28775	99.990953
0.385	0.1098	10.0301	7.413	0.800	0.1107	75.7934	56.018	25.0	0.000000610	135.29328	99.995036
0.390	0.1098	10.5791	7.819	0.850	0.0988	81.0309	59.889	30.0	0.000000300	135.29555	99.996718
0.395	0.1189	11.1509	8.241	0.900	0.0889	85.7234	63.358	35.0	0.000000160	135.29670	99.997568
0.400	0.1429	11.8054	8.725	0.950	0.0835	90.0334	66.543	40.0	0.000000094	135.29734	99.998037
0.405	0.1644	12.5736	9.293	1.000	0.0746	93.9859	69.464	50.0	0.000000038	135.29800	99.998525
0.410	0.1751	13.4224	9.920	1.100	0.0592	100.675	74.409	60.0	0.000000019	135.29828	99.998736
0.415	0.1774	14.3036	10.571	1.200	0.0484	106.055	78.385	80.0	0.000000007	135.29854	99.998928
0.420	0.1747	15.1839	11.222	1.300	0.0396	110.455	81.637	100.0	0.000000003	135.29864	99.999002
0.425	0.1693	16.0439	11.858	1.400	0.0336	114.115	84.342	1000.0	0	135.30000	100.000000
0.430	0.1639	16.8769	12.473	1.500	0.0287	117.230	86.645				
0.435	0.1663	17.7024	13.083	1.600	0.0244	119.885	88.607				
0.440	0.1810	18.5706	13.725	1.700	0.0202	122.115	90.255				
0.445	0.1922	19.5036	14.415	1.800	0.0159	123.920	91.589				
0.450	0.2006	20.4856	15.140	1.900	0.0126	125.345	92.642				
0.455	0.2057	21.5014	15.891	2.000	0.0103	126.490	93.489				
0.460	0.2066	22.5321	16.653	2.100	0.0090	127.455	94.202				
0.465	0.2048	23.5606	17.413	2.200	0.0079	128.300	94.826				
0.470	0.2033	24.5809	18.167	2.300	0.0068	129.035	95.370				

*The variable $P_\lambda(\lambda)$, expressed in the units Watts cm$^{-2}\mu^{-1}$, is related to $P_\nu(\nu)$, expressed in the units Watts cm^{-2} Hz^{-1}, through the relation $P_\nu(\nu) = (d\lambda/d\nu)P_\lambda(\lambda) = (c/\nu^2)P_\lambda(\lambda)$ where c is the speed of light $(2.998 \times 10^{14}\,\mu/\text{sec})$ and ν is the frequency. More explicitly, $P_\nu(\nu) = (2.998 \times 10^{14})/\nu^2\ P_\lambda(c/\nu)$.

The conversion of $P_\nu(\nu)$ or $P_\lambda(\lambda)$ to *Spectral Energy Density* is given by $\mu(\nu)$ [joule cm^{-3} Hz^{-1}] $= (1/c)P_\nu(\nu)$ and $\mu(\lambda)$ [joule cm^{-3} μ^{-1}] $= (1/c)P_\lambda(\lambda)$ where c now has the value, 2.998×10^{10} cm sec^{-1}.

TOTAL MONTHLY SOLAR RADIATION IN A CLOUDLESS SKY

Total radiation is the sum of the direct and scattered radiation that strikes the earth's surface. It is influenced by the degree of cloudiness, atmospheric transparency, duration of sunshine, and height of elevation at which the measurements are taken. Deviations from values in this table are bound to occur; however, it is believed that the deviations do not exceed ±10% for the summer months and ±15% for the winter months. The radiation units in this table are kcal/cm^2.

$\varphi°$	January	February	March	April	May	June	July	August	September	October	November	December
90 N	0	0	0.1	10.0	21.9	26.0	23.8	12.9	2.4	0	0	0
85	0	0	0.7	10.2	21.8	25.8	23.4	13.1	3.0	0	0	0
80	0	0	2.4	10.8	21.4	25.2	23.0	13.4	4.3	0.5	0	0
75	0	0.5	4.0	11.7	21.0	24.5	22.2	13.8	5.8	1.3	0	0
70	0	1.6	6.0	13.1	20.5	23.6	21.2	14.6	7.5	2.7	0.5	0
65	0.7	2.8	8.0	14.5	20.1	22.8	21.0	15.6	9.5	4.3	1.4	0.2
60	1.8	4.3	9.9	16.0	20.8	22.9	21.4	16.7	11.3	6.1	2.6	1.1
55	3.1	6.2	11.7	17.3	21.4	23.4	21.9	17.9	12.9	7.8	4.0	2.3
50	4.8	8.2	13.3	18.5	22.2	23.7	22.6	19.1	14.4	9.7	5.8	3.9
45	6.7	10.3	14.8	19.5	22.6	23.9	23.2	20.1	15.8	11.5	7.8	5.9
40	8.8	12.2	16.4	20.3	23.0	24.0	23.4	20.9	17.0	13.2	9.7	7.7
35	10.7	14.0	17.6	21.0	23.0	24.0	23.6	21.6	18.1	14.7	11.4	9.7
30	12.5	15.5	18.6	21.4	23.0	23.8	23.4	21.8	19.1	16.1	13.1	11.5
25	14.1	16.8	19.5	21.6	23.0	23.4	23.1	21.8	19.8	17.4	14.6	13.1
20	15.5	17.9	20.2	21.6	22.5	22.8	22.6	21.6	20.4	18.5	16.1	14.7
15	16.9	19.0	20.8	21.4	21.9	22.0	21.9	21.2	20.9	19.3	17.4	16.1
10	18.1	19.8	21.1	21.2	21.2	21.0	21.1	20.6	21.2	20.1	18.5	17.5
5	19.3	20.4	21.4	21.0	20.2	19.9	20.0	20.0	21.3	20.6	19.5	18.8
0	20.2	20.9	21.5	20.4	19.3	18.8	19.1	19.3	21.2	21.2	20.4	19.2
5S	20.1	21.4	21.4	19.9	18.3	17.6	17.9	18.3	20.8	21.4	21.8	21.0
10	22.0	21.8	21.1	19.2	17.7	16.3	16.3	17.3	20.4	21.4	21.8	22.0
15	22.6	22.0	20.6	18.3	16.0	14.9	15.4	16.1	19.7	21.3	22.4	22.9
20	23.2	22.0	20.0	17.2	14.7	13.4	14.9	14.9	18.9	21.0	22.0	23.6
25	23.6	22.0	19.4	16.0	13.0	12.0	12.6	13.6	18.0	20.6	22.7	24.1
30	23.9	21.8	18.6	14.9	11.9	10.8	11.1	12.1	17.0	20.1	23.0	24.6
35	24.0	21.3	17.6	13.6	10.4	9.0	9.6	10.6	15.9	19.5	23.0	25.0
40	24.0	20.6	16.4	12.2	8.7	7.3	8.1	9.0	14.3	18.7	22.8	25.2
45	24.0	19.9	15.2	10.7	7.1	5.5	6.3	7.3	13.4	17.7	22.4	25.2
50	23.6	18.9	13.8	9.2	5.4	3.8	4.6	5.5	12.0	16.6	21.8	25.0
55	23.2	17.8	12.3	7.5	3.8	2.3	3.0	3.8	10.3	15.4	21.2	24.6
60	22.6	16.6	10.8	5.6	2.4	1.0	1.6	2.3	8.5	14.1	21.0	24.4
65	22.4	15.3	9.1	3.9	1.1	0.1	0.4	1.0	6.7	12.6	20.8	24.5
70	22.6	14.2	7.3	2.2	0.1	0	0	0	5.0	11.4	21.0	24.9
75	23.2	13.4	5.7	0.9	0	0	0	0	3.5	10.4	21.2	25.4
80	24.0	12.8	4.3	0	0	0	0	0	2.1	9.7	21.9	26.0
85	24.6	12.4	2.9	0	0	0	0	0	0.9	9.2	22.4	26.6
90	24.9	12.3	1.7	0	0	0	0	0	0	9.0	22.6	27.0

ELEMENTS IN SEA WATER

A. Demayo

The elements in sea water are listed in decreasing average concentration. These are average concentrations and there will be variations in composition as a function of the location where the sample was collected.

Element	Concentration (mg/l)	Element	Concentration (mg/l)	Element	Concentration (mg/l)
Oxygen	8.57×10^5	Iron	1×10^{-2}	Selenium	9×10^{-5}
Hydrogen	1.08×10^5	Indium	$<2 \times 10^{-2}$	Germanium	7×10^{-5}
Chlorine	1.90×10^4	Molybdenum	1×10^{-2}	Xeon	5.2×10^{-5}
Sodium	1.05×10^4	Zinc	1×10^{-2}	Chromium	5×10^{-5}
Magnesium	1.35×10^3	Nickel	5.4×10^{-3}	Thorium	5×10^{-5}
Sulfur	8.85×10^2	Arsenic	3×10^{-3}	Gallium	3×10^{-5}
Calcium	4.00×10^2	Copper	3×10^{-3}	Mercury	3×10^{-5}
Potassium	3.80×10^2	Tin	3×10^{-3}	Lead	3×10^{-5}
Bromine	6.5×10^1	Uranium	3×10^{-3}	Zirconium	2.2×10^{-5}
Carbon	2.8×10^1	Krypton	2.5×10^{-3}	Bismuth	1.7×10^{-5}
Strontium	8.1×10^0	Manganese	2×10^{-3}	Lanthanum	1.2×10^{-5}
Boron	4.6×10^0	Vanadium	2×10^{-3}	Gold	1.1×10^{-5}
Silicon	3×10^0	Titanium	1×10^{-3}	Niobium	1×10^{-5}
Fluorine	1.3×10^0	Cesium	5×10^{-4}	Thallium	$<1 \times 10^{-5}$
Argon	6×10^{-1}	Cerium	4×10^{-4}	Hafnium	$<8 \times 10^{-6}$
Nitrogen	5×10^{-1}	Antimony	3.3×10^{-4}	Helium	6.9×10^{-6}
Lithium	1.8×10^{-1}	Silver	3×10^{-4}	Selenium	$<4 \times 10^{-6}$
Rubidium	1.2×10^{-1}	Yttrium	3×10^{-4}	Tantalum	$<2.5 \times 10^{-6}$
Phosphorus	7×10^{-2}	Cobalt	2.7×10^{-4}	Beryllium	6×10^{-7}
Iodine	6×10^{-2}	Neon	1.4×10^{-4}	Protoactinium	2×10^{-9}
Barium	3×10^{-2}	Cadmium	1.1×10^{-4}	Radium	6×10^{-11}
Aluminum	1×10^{-2}	Tungsten	1×10^{-4}	Radon	6×10^{-16}

THE pH OF NATURAL MEDIA AND ITS RELATION TO THE PRECIPITATION OF HYDROXIDES

Reprinted from "Principles of Geochemistry" (1952) with the permission of Brian Mason, author, and John Wiley and Sons, publishers.

pH	Precipitation of hydroxides	Natural media	pH
11 }	Magnesium		11
10 }		Alkali soils	10
9 }	Bivalent manganese		{ 9
8 }		Seawater	{ 8
7	Bivalent iron	River water	7
6	Zinc copper	Rain water	6
5	Aluminum		5
4		Peat water	4
3 }	Trivalent iron	Mine waters	3
2 }		Acid thermal springs	{ 2
1			{ 1

PROPERTIES OF THE EARTH'S ATMOSPHERE AT ELEVATIONS UP TO 160 KILOMETERS

The average atmosphere up to 160 km based on pressure and density data obtained on rocket flights above White Sands, New Mexico.
Havens, Koll, and LaGow, Journal of Geophysical Research, March, 1952.

Altitude km above sea level	Pressure mm Hg	Density gm/meters^3	Temperatures °K (N_2, O_2) M = 29	Temperatures °K (N_2, O) M = 24	Velocity of Sound m. sec	Mean Free Path cm (N_2)
0	760	1220	290		345	6.5×10^{-6}
10	210	425	230		310	1.9×10^{-5}
20	42	92	210		295	8.6×10^{-5}
30	9.5	19	235		315	4.2×10^{-4}
40	2.4	4.3	260		325	1.8×10^{-3}
50	7.5×10^{-1}	1.3	270		330	6.1×10^{-3}
60	2.1×10^{-1}	3.8×10^{-1}	260		325	2.1×10^{-2}
70	5.4×10^{-2}	1.2×10^{-1}	210		295	6.6×10^{-1}
80	1.0×10^{-2}	2.5×10^{-2}	190		280	3.2×10^{-1}
90	1.9×10^{-3}	4.0×10^{-3}	210		295	2.0
100	4.2×10^{-4}	8.0×10^{-4}	240		315	10.0
110	1.2×10^{-4}	2.0×10^{-4}	270	220	330	40.0
120	3.5×10^{-5}	5.0×10^{-5}	330	270	370	1.5×10^2
130	1.5×10^{-5}	2.0×10^{-5}	390	320	400	4.0×10^2
140	7×10^{-6}	7.0×10^{-6}	450	370	430	1.0×10^3
150	3×10^{-6}	3.0×10^{-6}	510	420	460	2.5×10^3
160	2×10^{-6}	1.5×10^{-6}	570	470	480	5.0×10^3

VELOCITY OF SEISMIC WAVES

Depth km	Longitudinal or condensational km/sec.	Transverse or distortional km/sec.
0–20	5.4 –5.6	3.2
20–45	6.25–6.75	3.5
1300	12.5	6.9
2400	13.5	7.5

ATMOSPHERIC AND METEOROLOGICAL DATA

Total mass of the atmosphere, estimated by Ekholm, 5.2×10^{21} g, 11.4×10^{18} pounds, 5.70×10^{16} tons.
Evidence of extent: twilight, 63 km, 39 mi.: meteors, 200 km, 124 mi.: aurora 44–360 km, 27–224 mi.

*Distance to Earth.

U.S. STANDARD ATMOSPHERE (1976)

From, "U.S. Standard Atmosphere, 1976", National Oceanic and Atmospheric Administration, National Aeronautics and Space Administration and the United States Air Force, 1976. The above referenced book is available from the Superintendent of Documents, U.S. Government Printing Office, Washington, D.C. 20402. The book contains considerably more extensive tables than those presented below plus the development of the equations used in the calculations of the tables as well as a discussion of the bases for selection of constants used in the equations.

The U.S. Standard Atmosphere, 1976 is an idealized, steady-state representation of the earth's atmosphere from the surface to 1000 km, as it is assumed to exist in a period of moderate solar activity. The air is assumed to be dry, and at heights sufficiently below 86 km, the atmosphere is assumed to be homogeneously mixed with a relative-volume composition leading to a mean molecular weight M. The molecular weights and assumed fractional-volume composition of sea-level dry air were

Gas species	Molecular weight M_i (kg/kmol)	Fractional volume F_i (dimensionless)
N_2	28.0134	0.78084
O_2	31.9988	0.209476
Ar	39.948	0.00934
CO_2	44.00995	0.000314
Ne	20.183	0.00001818
He	4.0026	0.00000524
Kr	83.80	0.00000114
Xe	131.30	0.000000087
CH_4	16.04303	0.000002
H_2	2.01594	0.0000005

SYMBOLS, ABBREVIATIONS, AND UNITS RELATING TO FOLLOWING TABLE

C_s Speed of sound expressed as meters per second.

g   Acceleration due to gravity expressed as meters per second.

H   Geopotential altitude expressed in meters. The unit of measurement of geopotential is the standard geopotential meter (m′) which represents the work done by lifting a unit mass 1 geometric meter, through a region in hich the acceleration of gravity is uniformly 9.80665 m/s². The geopotential at any point with respect to sea level (assumed zero potential), expressed in geopotential meters, is called geopotential altitude.

H_p Pressure scale height. The quantity, $H_pRT/(g·m)$, which has the dimensions of length, is the quantity commonly associated with the concept of scale height, and is the defining form of pressure scale height used in this model.

K   Kelvin, the thermodynamic temperature

L   Mean free path. This is the mean value of the distances traveled by each of the neutral particles in a selected volume between successive collisions with other particles in the volume.

m   Meter, the unit of length.

M   Molecular weight expressed as kg/kmol or lb/lbmol.

n   The number of neutral atmospheric gas particles per cubic meter of the atmosphere at the geometric height Z or the geopotential height H.

N   Newton, a force having the units kg·m/s².

P   Pressure, expressed as millibars (100N/m²).

R   The gas constant, $R = 8.31432 \times 10^3$ N·m/(kmol·K).

s   Seconds, time.

T   Temperature, degrees Kelvin.

V   Mean air particle speed, expressed as meters per second, is the arithmetic average of the speeds of all air particles in the volume element being considered.

Z   Geometric altitude expressed in kilometers.

η   Kinematic viscosity in the units m²/s.

x   Thermal conductivity in the units J/m·s·K.

μ   Dynamic viscosity in the units N·s/m².

ν   Mean collision frequency. Average speed of air particles within a selected volume divided by the mean free path of the particles in that volume. Thus $ν = V/L$.

ϱ   Density expressed as kg/m².

Altitude Z (m)	Altitude H (m)	Acceleration due to gravity g (m/s²)	Pressure scale height H_p (m)	Number density n (m⁻³)	Particle speed V (m/s)	Collision frequency ν (s⁻¹)	Mean free path L (m)	Molecular weight M (kg/kmol)	Temperature T (K)	Pressure P (mb)	Density ρ (kg/m³)	Sound speed C_s (m/s)	Dynamic viscosity μ (N·s/m²)	Kinematic viscosity η (m²/s)	Thermal conductivity (J/m·s·K)
-5000	-5004	9.8221	9371.8	4.0151 (+25)	484.15	1.1506 (+10)	4.2078 (-8)	28.964	320.676	1.7776 (+3)	1.9311 (+0)	358.99	1.9422 (-5)	1.0058 (-5)	2.7882 (-5)
-4500	-4503	9.8205	9278.2	3.8445 (+25)	481.69	1.0961 (+10)	4.3945 (-8)	28.964	317.421	1.6848 (+3)	1.8491 (+0)	357.16	1.9273 (-5)	1.0423 (-5)	2.7634 (-5)
-4000	-4003	9.8190	9184.5	3.6795 (+25)	479.22	1.0437 (+10)	4.5915 (-8)	28.964	314.166	1.5959 (+3)	1.7697 (+0)	355.32	1.9123 (-5)	1.0806 (-5)	2.7384 (-5)
-3500	-3502	9.8175	9090.8	3.5201 (+25)	476.73	9.9328 (+9)	4.7995 (-8)	28.964	310.913	1.5109 (+3)	1.6930 (+0)	353.48	1.8972 (-5)	1.1206 (-5)	2.7134 (-5)
-3000	-3001	9.8159	8997.1	3.3660 (+25)	474.23	9.4481 (+9)	5.0193 (-8)	28.964	307.659	1.4297 (+3)	1.6189 (+0)	351.63	1.8820 (-5)	1.1625 (-5)	2.6884 (-5)
-2500	-2501	9.8144	8903.4	3.2171 (+25)	471.71	8.9824 (+9)	5.2515 (-8)	28.964	304.406	1.3520 (+3)	1.5473 (+0)	349.76	1.8668 (-5)	1.2065 (-5)	2.6632 (-5)
-2000	-2001	9.8128	8809.6	3.0733 (+25)	469.19	8.5357 (+9)	5.4973 (-8)	28.964	301.154	1.2778 (+3)	1.4782 (+0)	347.89	1.8515 (-5)	1.2525 (-5)	2.6380 (-5)
-1500	-1500	9.8113	8715.9	2.9346 (+25)	466.65	8.1056 (+9)	5.7571 (-8)	28.964	297.902	1.2069 (+3)	1.4114 (+0)	346.00	1.8361 (-5)	1.3009 (-5)	2.6126 (-5)
-1000	-1000	9.8097	8622.1	2.8007 (+25)	464.09	7.6934 (+9)	6.0324 (-8)	28.964	294.651	1.1393 (+3)	1.3470 (+0)	344.11	1.8206 (-5)	1.3516 (-5)	2.5872 (-5)
-500	-500	9.8082	8528.3	2.6715 (+25)	461.53	7.2980 (+9)	6.3240 (-8)	28.964	291.400	1.0747 (+3)	1.2849 (+0)	342.21	1.8050 (-5)	1.4048 (-5)	2.5618 (-5)
0	0	9.8066	8434.5	2.5470 (+25)	458.94	6.9189 (+9)	6.6332 (-8)	28.964	288.150	1.01325 (+3)	1.2250 (+0)	340.29	1.7894 (-5)	1.4607 (-5)	2.5326 (-5)
500	500	9.8051	8340.7	2.4269 (+25)	456.35	6.5555 (+9)	6.9613 (-8)	28.964	284.900	9.5461 (+2)	1.1673 (+0)	338.37	1.7737 (-5)	1.5195 (-5)	2.5106 (-5)
1000	1000	9.8036	8246.9	2.3113 (+25)	453.74	6.2075 (+9)	7.3095 (-8)	28.964	281.651	8.9876 (+2)	1.1117 (+0)	336.43	1.7579 (-5)	1.5813 (-5)	2.4849 (-5)
1500	1500	9.8020	8153.0	2.2000 (+25)	451.12	5.8743 (+9)	7.6795 (-8)	28.964	278.402	8.4559 (+2)	1.0581 (+0)	334.49	1.7420 (-5)	1.6463 (-5)	2.4591 (-5)
2000	1999	9.8005	8059.2	2.0928 (+25)	448.48	5.5554 (+9)	8.0728 (-8)	28.964	275.154	7.9501 (+2)	1.0066 (+0)	332.53	1.7260 (-5)	1.7147 (-5)	2.4333 (-5)
2500	2499	9.7989	7965.3	1.9897 (+25)	445.82	5.2504 (+9)	8.4912 (-8)	28.964	271.906	7.4691 (+2)	9.5695 (-1)	330.56	1.7099 (-5)	1.7868 (-5)	2.4073 (-5)
3000	2999	9.7974	7871.4	1.8905 (+25)	443.15	4.9588 (+9)	8.9367 (-8)	28.964	268.659	7.0121 (+2)	9.0925 (-1)	328.58	1.6938 (-5)	1.8628 (-5)	2.3813 (-5)
3500	3498	9.7959	7777.5	1.7952 (+25)	440.47	4.6802 (+9)	9.4113 (-8)	28.964	265.413	6.5780 (+2)	8.6340 (-1)	326.59	1.6775 (-5)	1.9429 (-5)	2.3552 (-5)
4000	3997	9.7943	7683.6	1.7036 (+25)	437.76	4.4141 (+9)	9.9173 (-8)	28.964	262.166	6.1660 (+2)	8.1935 (-1)	324.59	1.6612 (-5)	2.0275 (-5)	2.3290 (-5)
4500	4497	9.7928	7589.7	1.6156 (+25)	435.05	4.1602 (+9)	1.0457 (-7)	28.964	258.921	5.7752 (+2)	7.7704 (-1)	322.57	1.6448 (-5)	2.1167 (-5)	2.3028 (-5)
5000	4996	9.7912	7495.7	1.5312 (+25)	432.31	3.9180 (+9)	1.1034 (-7)	28.964	255.676	5.4048 (+2)	7.3643 (-1)	320.55	1.6282 (-5)	2.2110 (-5)	2.2765 (-5)
5500	5495	9.7897	7401.8	1.4502 (+25)	429.56	3.6871 (+9)	1.1650 (-7)	28.964	252.431	5.0539 (+2)	6.9747 (-1)	318.50	1.6116 (-5)	2.3107 (-5)	2.2500 (-5)
6000	5994	9.7882	7307.8	1.3725 (+25)	426.79	3.4671 (+9)	1.2310 (-7)	28.964	249.187	4.7217 (+2)	6.6011 (-1)	316.45	1.5949 (-5)	2.4161 (-5)	2.2236 (-5)
6500	6493	9.7866	7213.8	1.2980 (+25)	424.00	3.2577 (+9)	1.3016 (-7)	28.964	245.943	4.4075 (+2)	6.2431 (-1)	314.39	1.5781 (-5)	2.5278 (-5)	2.1970 (-5)
7000	6992	9.7851	7119.8	1.2267 (+25)	421.20	3.0584 (+9)	1.3772 (-7)	28.964	242.700	4.1105 (+2)	5.9002 (-1)	312.31	1.5612 (-5)	2.6461 (-5)	2.1703 (-5)
7500	7491	9.7836	7025.8	1.1585 (+25)	418.37	2.8689 (+9)	1.4583 (-7)	28.964	239.457	3.8299 (+2)	5.5719 (-1)	310.21	1.5442 (-5)	2.7714 (-5)	2.1436 (-5)
8000	7990	9.7820	6931.7	1.0932 (+25)	415.53	2.6888 (+9)	1.5454 (-7)	28.964	236.215	3.5651 (+2)	5.2579 (-1)	308.11	1.5271 (-5)	2.9044 (-5)	2.1168 (-5)
8500	8489	9.7805	6837.7	1.0308 (+25)	412.67	2.5178 (+9)	1.6390 (-7)	28.964	232.974	3.3154 (+2)	4.9576 (-1)	305.98	1.5099 (-5)	3.0457 (-5)	2.0899 (-5)
9000	8987	9.7789	6743.6	9.7110 (+24)	409.79	2.3555 (+9)	1.7397 (-7)	28.964	229.733	3.0800 (+2)	4.6706 (-1)	303.85	1.4926 (-5)	3.1957 (-5)	2.0630 (-5)
9500	9486	9.7774	6649.5	9.1413 (+24)	406.89	2.2016 (+9)	1.8482 (-7)	28.964	226.492	2.8584 (+2)	4.3966 (-1)	301.70	1.4752 (-5)	3.3553 (-5)	2.0359 (-5)
10000	9984	9.7759	6555.4	8.5976 (+24)	403.97	2.0558 (+9)	1.9651 (-7)	28.964	223.252	2.6499 (+2)	4.1351 (-1)	299.53	1.4577 (-5)	3.5251 (-5)	2.0088 (-5)
10500	10483	9.7743	6461.3	8.0790 (+24)	401.03	1.9177 (+9)	2.0912 (-7)	28.964	220.013	2.4540 (+2)	3.8857 (-1)	297.35	1.4400 (-5)	3.7060 (-5)	1.9816 (-5)
11000	10981	9.7728	6367.2	7.5854 (+24)	398.07	1.7871 (+9)	2.2274 (-7)	28.964	216.774	2.2699 (+2)	3.6480 (-1)	295.15	1.4223 (-5)	3.8988 (-5)	1.9543 (-5)
11500	11479	9.7713	6364.6	7.0157 (+24)	397.95	1.6525 (+9)	2.4081 (-7)	28.964	216.650	2.0984 (+2)	3.3743 (-1)	295.07	1.4216 (-5)	4.2131 (-5)	1.9533 (-5)
12000	11977	9.7697	6365.6	6.4857 (+24)	397.95	1.5277 (+9)	2.6049 (-7)	28.964	216.650	1.9399 (+2)	3.1194 (-1)	295.07	1.4216 (-5)	4.5574 (-5)	1.9533 (-5)
12500	12475	9.7682	6366.6	5.9958 (+24)	397.95	1.4123 (+9)	2.8178 (-7)	28.964	216.650	1.7934 (+2)	2.8838 (-1)	295.07	1.4216 (-5)	4.9297 (-5)	1.9533 (-5)
13000	12973	9.7667	6367.6	5.5430 (+24)	397.95	1.3056 (+9)	3.0479 (-7)	28.964	216.650	1.6579 (+2)	2.6660 (-1)	295.07	1.4216 (-5)	5.3325 (-5)	1.9533 (-5)
13500	13471	9.7651	6368.6	5.1244 (+24)	397.95	1.2070 (+9)	3.2969 (-7)	28.964	216.650	1.5327 (+2)	2.4646 (-1)	295.07	1.4216 (-5)	5.7680 (-5)	1.9533 (-5)
14000	13969	9.7636	6369.6	4.7375 (+24)	397.95	1.1159 (+9)	3.5662 (-7)	28.964	216.650	1.4170 (+2)	2.2786 (-1)	295.07	1.4216 (-5)	6.2391 (-5)	1.9533 (-5)
14500	14467	9.7621	6370.6	4.3799 (+24)	397.95	1.0317 (+9)	3.8574 (-7)	28.964	216.650	1.3100 (+2)	2.1066 (-1)	295.07	1.4216 (-5)	6.7485 (-5)	1.9533 (-5)
15000	14965	9.7605	6371.6	4.0493 (+24)	397.95	9.5380 (+8)	4.1723 (-7)	28.964	216.650	1.2111 (+2)	1.9476 (-1)	295.07	1.4216 (-5)	7.2995 (-5)	1.9533 (-5)
16000	15960	9.7575	6373.6	3.4612 (+24)	397.95	8.1528 (+8)	4.8812 (-7)	28.964	216.650	1.0352 (+2)	1.6647 (-1)	295.07	1.4216 (-5)	8.5397 (-5)	1.9533 (-5)
17000	16955	9.7544	6375.6	2.9587 (+24)	397.95	6.9691 (+8)	5.7102 (-7)	28.964	216.650	8.8497 (+1)	1.4230 (-1)	295.07	1.4216 (-5)	9.9901 (-5)	1.9533 (-5)
18000	17949	9.7513	6377.6	2.5292 (+24)	397.95	5.9576 (+8)	6.6797 (-7)	28.964	216.650	7.5652 (+1)	1.2165 (-1)	295.07	1.4216 (-5)	1.1686 (-4)	1.9533 (-5)
19000	18943	9.7483	6379.6	2.1622 (+24)	397.95	5.0931 (+8)	7.8135 (-7)	28.964	216.650	6.4674 (+1)	1.0400 (-1)	295.07	1.4216 (-5)	1.3670 (-4)	1.9533 (-5)
20000	19937	9.7452	6381.6	1.8486 (+24)	397.95	4.3543 (+8)	9.1393 (-7)	28.964	216.650	5.5293 (+1)	8.8910 (-2)	295.07	1.4216 (-5)	1.5989 (-4)	1.9533 (-5)
21000	20931	9.7422	6411.0	1.5742 (+24)	398.81	3.7161 (+8)	1.0732 (-6)	28.964	217.581	4.7289 (+1)	7.5715 (-2)	296.38	1.4267 (-5)	1.8843 (-4)	1.9611 (-5)
22000	21924	9.7391	6442.3	1.3413 (+24)	399.72	3.1733 (+8)	1.2596 (-6)	28.964	218.574	4.0475 (+1)	6.4510 (-2)	297.05	1.4322 (-5)	2.2201 (-4)	1.9695 (-5)
23000	22917	9.7361	6473.6	1.1437 (+24)	400.62	2.7119 (+8)	1.4772 (-6)	28.964	219.567	3.4668 (+1)	5.5006 (-2)	297.72	1.4376 (-5)	2.6135 (-4)	1.9778 (-5)
24000	23910	9.7330	6504.9	9.7591 (+23)	401.53	2.3194 (+8)	1.7312 (-6)	28.964	220.560	2.9717 (+1)	4.6938 (-2)	298.39	1.4430 (-5)	3.0743 (-4)	1.9862 (-5)
25000	24902	9.7300	6536.2	8.3341 (+23)	402.43	1.9852 (+8)	2.0272 (-6)	28.964	221.552	2.5492 (+1)	4.0084 (-2)	299.06	1.4484 (-5)	3.6135 (-4)	1.9945 (-5)
26000	25894	9.7269	6567.5	7.1225 (+23)	403.33	1.7004 (+8)	2.3720 (-6)	28.964	222.544	2.1883 (+1)	3.4257 (-2)	299.72	1.4538 (-5)	4.2439 (-4)	2.0029 (-5)
27000	26886	9.7239	6598.9	6.0916 (+23)	404.23	1.4575 (+8)	2.7734 (-6)	28.964	223.536	1.8799 (+1)	2.9298 (-2)	300.39	1.4592 (-5)	4.9805 (-4)	2.0112 (-5)
28000	27877	9.7208	6630.2	5.2138 (+23)	405.12	1.2502 (+8)	3.2404 (-6)	28.964	224.527	1.6161 (+1)	2.5076 (-2)	301.05	1.4646 (-5)	5.8405 (-4)	2.0195 (-5)
29000	28868	9.7178	6661.6	4.4657 (+23)	406.01	1.0732 (+8)	3.7832 (-6)	28.964	225.518	1.3904 (+1)	2.1478 (-2)	301.71	1.4699 (-5)	6.8437 (-4)	2.0278 (-5)
30000	29859	9.7147	6692.0	3.8278 (+23)	406.91	9.2192 (+7)	4.4137 (-6)	28.964	226.509	1.1970 (+1)	1.8410 (-2)	302.37	1.4753 (-5)	8.0134 (-4)	2.0361 (-5)
31000	30850	9.7117	6724.3	3.2833 (+23)	407.79	7.9251 (+7)	5.1456 (-6)	28.964	227.500	1.0312 (+1)	1.5792 (-2)	303.02	1.4806 (-5)	9.3759 (-4)	2.0443 (-5)
32000	31840	9.7087	6755.7	2.8133 (+23)	408.68	6.8175 (+7)	5.9945 (-6)	28.964	228.490	8.8906 (+0)	1.3555 (-2)	303.67	1.4859 (-5)	1.0962 (-3)	2.0526 (-5)
33000	32830	9.7056	6831.2	2.4062 (+23)	410.90	5.8522 (+7)	7.0212 (-6)	28.964	230.973	7.6730 (+0)	1.1573 (-2)	304.67	1.4992 (-5)	1.2955 (-3)	2.0733 (-5)
34000	33819	9.7026	6915.4	2.0558 (+23)	413.35	5.0297 (+7)	8.2182 (-6)	28.964	233.743	6.6341 (+0)	9.8874 (-3)	306.49	1.5140 (-5)	1.5312 (-3)	2.0963 (-5)
35000	34808	9.6995	6999.5	1.7597 (+23)	415.79	4.3307 (+7)	9.6010 (-6)	28.964	236.513	5.7459 (+0)	8.4634 (-3)	308.30	1.5287 (-5)	1.8062 (-3)	2.1193 (-5)
36000	35797	9.6965	7083.7	1.5090 (+23)	418.22	3.7356 (+7)	1.1196 (-5)	28.964	239.282	4.9852 (+0)	7.2579 (-3)	310.10	1.5433 (-5)	2.1264 (-3)	2.1422 (-5)

Altitude		Acceleration due to gravity	Pressure scale height	Number density	Particle speed	Collision frequency	Mean free path	Molecular weight	Temperature	Pressure	Density	Sound speed	Dynamic viscosity	Kinematic viscosity	Thermal conductivity
Z(m)	H(m)	g(m/s²)	H_p (m)	n (m⁻³)	V (m/s)	ν (s⁻¹)	L (m)	M (kg/kmol)	T(K)	P(mb)	ϱ(kg/m³)	C_A(m/s)	μ (N·s/m²)	η (m²/s)	(J/m·s·K)
38000	37774	9.6904	7252.1	1.1158(+23)	423.03	2.7939(+7)	1.5141(−5)	28.964	244.818	3.7713(+0)	5.3666(−3)	313.67	1.5723(−5)	2.9297(−3)	2.1878(−5)
40000	39750	9.6844	7420.6	8.3077(+22)	427.78	2.1036(+7)	2.0336(−5)	28.964	250.350	2.8714(+0)	3.9957(−3)	317.19	1.6009(−5)	4.0066(−3)	2.2331(−5)
42000	41724	9.6783	7589.2	6.2266(+22)	432.48	1.5939(+7)	2.7133(−5)	28.964	255.878	2.1996(+0)	2.9948(−3)	320.67	1.6293(−5)	5.4404(−3)	2.2781(−5)
44000	43698	9.6723	7757.9	4.6965(+22)	437.13	1.2152(+7)	3.5973(−5)	28.964	261.403	1.6949(+0)	2.2948(−3)	324.12	1.6573(−5)	7.3371(−3)	2.3229(−5)
46000	45669	9.6662	7926.7	3.5640(+22)	441.72	9.3182(+6)	4.7404(−5)	28.964	266.925	1.3134(+0)	1.7142(−3)	327.52	1.6851(−5)	9.8305(−3)	2.3764(−5)
48000	47640	9.6602	8042.4	2.7376(+22)	444.79	7.2075(+6)	6.1713(−5)	28.964	270.650	1.0229(+0)	1.3167(−3)	329.80	1.7037(−5)	1.2939(−2)	2.3973(−5)
50000	49610	9.6542	8047.4	2.1351(+22)	444.79	5.6201(+6)	7.9130(−5)	28.964	270.650	7.9779(−1)	1.0269(−3)	329.80	1.7037(−5)	1.6591(−2)	2.3973(−5)
52000	51578	9.6482	8004.3	1.6750(+22)	443.46	4.3966(+6)	1.0086(−4)	28.964	269.031	6.2214(−1)	8.0562(−4)	328.81	1.6956(−5)	2.1047(−2)	2.3843(−5)
54000	53545	9.6421	7845.3	1.3286(+22)	438.90	3.4515(+6)	1.2716(−4)	28.964	263.524	4.8337(−1)	6.3901(−4)	325.43	1.6600(−5)	2.6104(−2)	2.3400(−5)
56000	55511	9.6361	7686.2	1.0488(+22)	434.29	2.6961(+6)	1.6108(−4)	28.964	258.019	3.7362(−1)	5.0445(−4)	322.01	1.6402(−5)	3.2514(−2)	2.2955(−5)
58000	57476	9.6316	7527.0	8.2390(+21)	429.63	2.0952(+6)	2.0506(−4)	28.964	252.518	2.8723(−1)	3.9627(−4)	318.56	1.6121(−5)	4.0682(−2)	2.2508(−5)
60000	59439	9.6241	7367.8	6.4387(+21)	424.93	1.6195(+6)	2.6239(−4)	28.964	247.021	2.1958(−1)	3.0968(−4)	315.07	1.5837(−5)	5.1141(−2)	2.2058(−5)
65000	64342	9.6091	6969.1	3.3934(+21)	412.95	8.2945(+5)	4.9787(−4)	28.954	233.292	1.0929(−1)	1.6321(−4)	306.19	1.5116(−5)	9.2617(−2)	2.0926(−5)
70000	69238	9.5942	6569.9	1.7222(+21)	400.64	4.0839(+5)	9.8104(−4)	28.954	219.585	5.2209(−2)	8.2829(−5)	297.06	1.4377(−5)	1.7357(−1)	1.9780(−5)
75000	74125	9.5793	6244.9	8.3003(+20)	390.30	1.9175(+5)	2.0354(−3)	28.954	208.399	2.3881(−2)	3.8458(−5)	289.40	1.3759(−5)	3.4465(−1)	1.8834(−5)
80000	79006	9.5644	5961.7	3.8378(+20)	381.05	8.6559(+4)	4.4022(−3)	28.954	198.639	1.0524(−2)	1.8458(−5)	282.54	1.3208(−5)	7.1557(−1)	1.8001(−5)
85000	83878	9.5496	5678.0	1.7090(+20)	371.59	3.8378(+4)	9.8858(−3)	28.954	188.893	4.4568(−3)	8.2196(−6)	275.52	1.2647(−5)	1.5386(+0)	1.7162(−5)
85500	84365	9.5481	5649.6	1.5727(+20)	370.63	3.4501(+4)	1.0743(−2)	28.954	187.920	4.0802(−3)	7.5641(−6)	274.81	1.2590(−5)	1.6645(+0)	1.7078(−5)
86000	84852	9.5466	5621	1.447(+20)	369.7	3.17(+4)	1.17(−2)	28.95	186.87	3.7338(−3)	6.958(−6)	—	—	—	—
90000	88744	9.5348	5636	7.116(+19)	369.9	1.56(+4)	2.37(−2)	28.9	186.87	1.8359(−3)	3.416(−6)	—	—	—	—
95000	93610	9.5200	5727	2.920(+19)	372.6	6.44(+4)	5.79(−2)	28.78	188.42	7.5966(−4)	1.393(−6)	—	—	—	—
100000	98451	9.5052	6009	1.189(+19)	381.4	2.68(+3)	1.42(−1)	28.56	195.08	3.2011(−4)	5.604(−7)	—	—	—	—
110000	108129	9.4759	7723	2.144(+18)	431.7	5.48(+2)	7.88(−1)	27.2	240.00	7.1042(−5)	9.708(−8)	—	—	—	—
120000	117777	9.4466	12091	1.930(+17)	539.3	1.63(+2)	3.31(+0)	25.2	360.00	2.5382(−5)	2.222(−8)	—	—	—	—
130000	127395	9.4175	16288	9.322(+16)	625.0	7.1(+1)	8.8(+0)	24.7	469.27	1.2505(−5)	8.152(−9)	—	—	—	—
140000	136983	9.3886	20025	5.186(+16)	691.9	3.8(+1)	1.8(+1)	24.1	559.63	7.2028(−6)	3.831(−9)	—	—	—	—
150000	146542	9.3597	23380	3.162(+16)	746.5	2.3(+1)	3.3(+1)	23.10	634.39	4.5422(−6)	2.076(−9)	—	—	—	—
160000	156072	9.3310	26414	2.055(+16)	792.2	1.5(+1)	5.3(+1)	22.45	696.29	2.1210(−6)	1.233(−9)	—	—	—	—
170000	165572	9.3024	29175	1.400(+16)	831.3	1.0(+1)	8.2(+1)	21.90	747.57	1.5271(−6)	7.815(−10)	—	—	—	—
180000	175043	9.2740	31703	9.887(+15)	865.3	7.2(+0)	1.2(+2)	21.45	790.07	1.1266(−6)	3.581(−10)	—	—	—	—
190000	184486	9.2457	34030	7.182(+15)	895.1	5.2(+0)	1.7(+2)	21.34	825.16	8.4736(−7)	2.541(−10)	—	—	—	—
200000	193899	9.2175	36183	4.040(+15)	921.6	3.9(+0)	2.4(+2)	21.30	854.56	6.4756(−7)	1.846(−10)	—	—	—	—
210000	—	—	—	2.420(+15)	966.5	2.3(+0)	—	21.37	878.84	5.0149(−7)	1.367(−10)	—	—	—	—
220000	212641	9.1615	40043	1.515(+15)	1003.0	1.4(+0)	4.2(+2)	18.56	899.01	3.1059(−7)	7.858(−11)	—	—	—	—
240000	231268	9.1061	43405	6.509(+14)	1033.5	9.3(−1)	7.0(+2)	18.85	929.73	1.9894(−7)	4.742(−11)	—	—	—	—
260000	249784	9.0511	46346	4.405(+14)	1058.7	6.1(−1)	1.1(+3)	18.24	950.99	1.3076(−7)	2.971(−11)	—	—	—	—
280000	268187	8.9966	48925	3.029(+14)	1079.7	4.2(−1)	1.7(+3)	17.73	965.75	8.7704(−8)	1.916(−11)	—	—	—	—
300000	286480	8.9427	51193	2.109(+14)	1097.4	2.9(−1)	2.6(+3)	17.29	976.01	5.9796(−8)	1.264(−11)	—	—	—	—
320000	304663	8.8892	53199	1.485(+14)	1112.4	2.0(−1)	3.8(+3)	16.91	983.16	4.1320(−8)	8.503(−12)	—	—	—	—
340000	322738	8.8361	54996	1.056(+14)	1125.5	1.4(−1)	5.6(+3)	15.57	988.15	2.0384(−8)	5.805(−12)	—	—	—	—
360000	340705	8.7836	56637	4.678(+13)	1137.4	1.0(−1)	8.0(+3)	15.27	991.65	1.4518(−8)	4.013(−12)	—	—	—	—
380000	358565	8.7315	58178	2.192(+13)	1148.5	7.2(−2)	1.1(+4)	15.98	994.10	6.4468(−9)	2.803(−12)	—	—	—	—
400000	376320	8.6799	59678	1.097(+13)	1177.4	3.3(−2)	1.6(+4)	15.25	995.83	3.0236(−9)	1.184(−12)	—	—	—	—
450000	420250	8.5529	63644	5.950(+12)	1215.0	1.6(−2)	3.6(+4)	14.33	998.22	1.5137(−9)	5.215(−13)	—	—	—	—
500000	463540	8.4286	68785	3.540(+12)	1271.5	8.4(−3)	7.7(+4)	13.09	999.24	8.2130(−10)	2.384(−13)	—	—	—	—
550000	506202	8.3070	76427	2.311(+12)	1356.4	4.8(−3)	1.5(+5)	12.51	999.67	4.8865(−10)	1.137(−13)	—	—	—	—
600000	548252	8.1880	88244	1.637(+12)	1476.0	3.1(−3)	2.8(+5)	12.00	999.85	3.1908(−10)	5.712(−14)	—	—	—	—
650000	589701	8.0716	105992	1.234(+12)	1627.0	2.2(−3)	4.8(+5)	11.58	999.93	2.2599(−10)	3.070(−14)	—	—	—	—
700000	630563	7.9576	130630	9.717(+11)	1793.9	1.7(−3)	7.3(+5)	11.54	999.97	1.7036(−10)	1.788(−14)	—	—	—	—
750000	670850	7.8460	161074	7.876(+11)	1954.3	1.4(−3)	1.0(+6)	10.40	999.99	1.3415(−10)	1.136(−14)	—	—	—	—
800000	710574	7.7368	193862	6.505(+11)	2089.6	1.2(−3)	1.4(+6)	4.12	1000.00	1.0873(−10)	7.824(−15)	—	—	—	—
850000	749747	7.6298	224737	5.442(+11)	2192.6	1.0(−3)	1.7(+6)	3.34	1000.00	8.9816(−11)	5.759(−15)	—	—	—	—
900000	788380	7.5250	250894		2266.4	8.7(−4)	2.1(+6)		1000.00	7.5138(−11)	4.453(−15)	—	—	—	—
950000	826484	7.4224	271754		2318.1	7.5(−4)	2.6(+6)		1000.00		3.561(−15)	—	—	—	—
1000000	864071	7.3218	288203				3.1(+6)		1000.00			—	—	—	—

FIGURE 1.   Molecular-diffusion and eddy-diffusion coefficients as a function of geometric altitude.

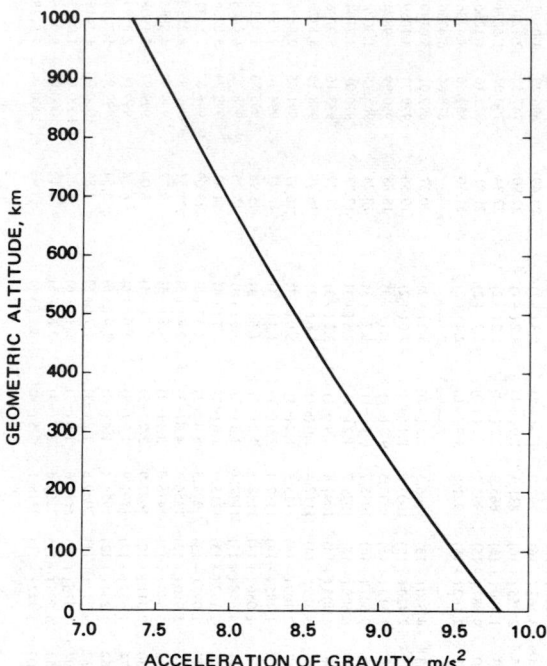

FIGURE 2.   Acceleration of gravity as a function of geometric altitude.

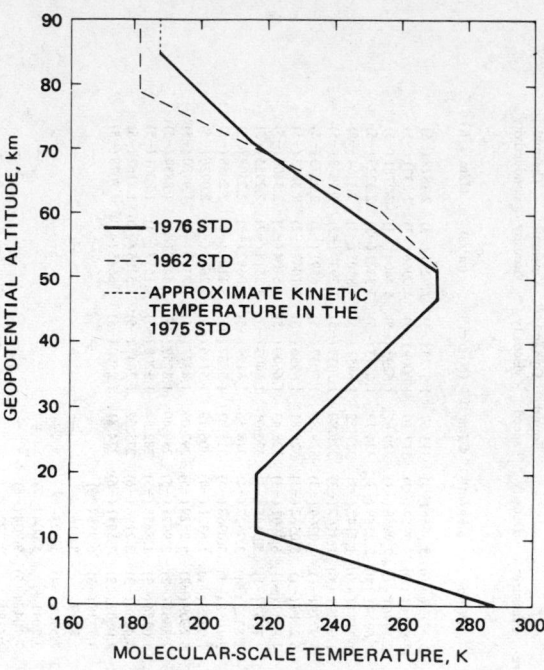

FIGURE 3.   Molecular-scale temperature as a function of geopotential altitude.

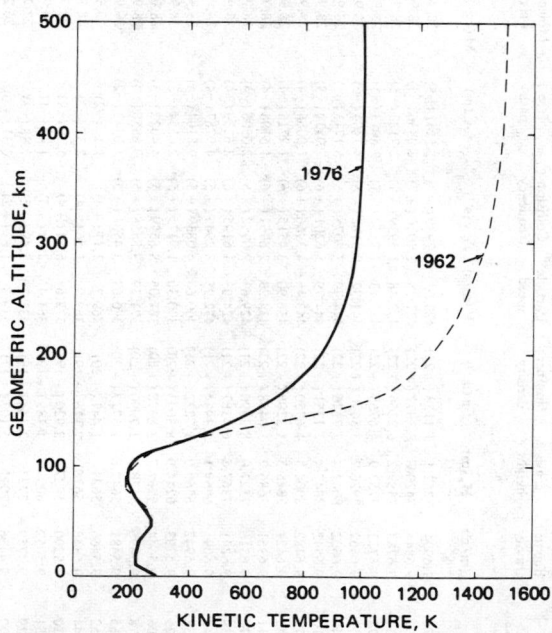

FIGURE 4.   Kinetic temperature as a function of geometric altitude.

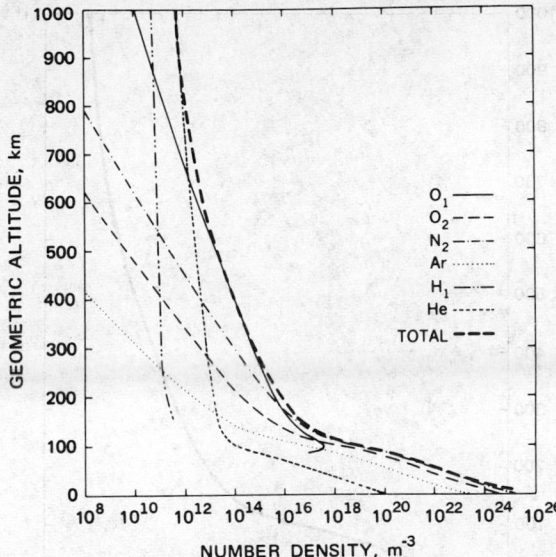

FIGURE 5.   Number density of individual species and total number density as a function of geometric altitude.

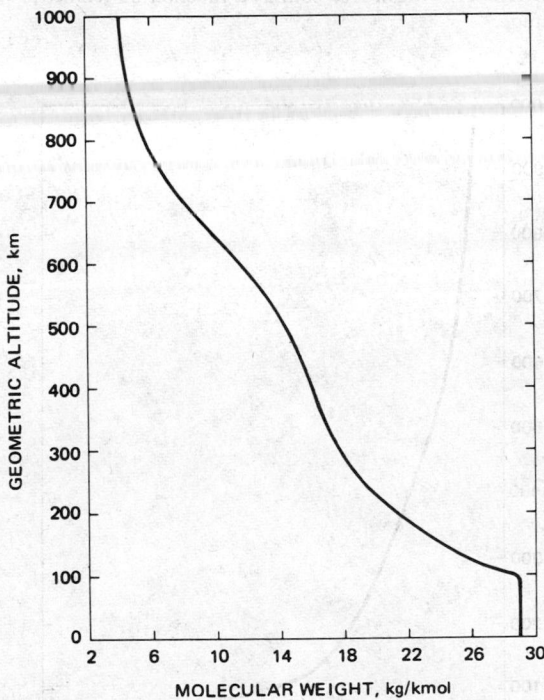

FIGURE 6.   Mean molecular weight as a function of geometric altitude.

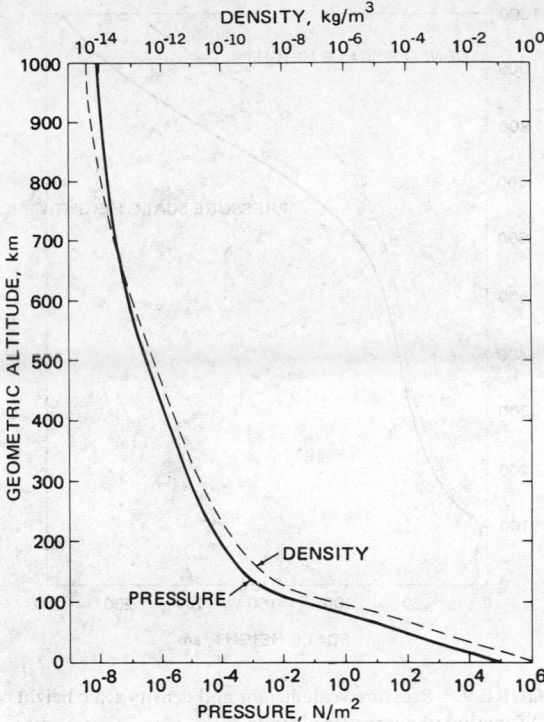

FIGURE 7.   Total pressure and mass density as a function of geometric altitude.

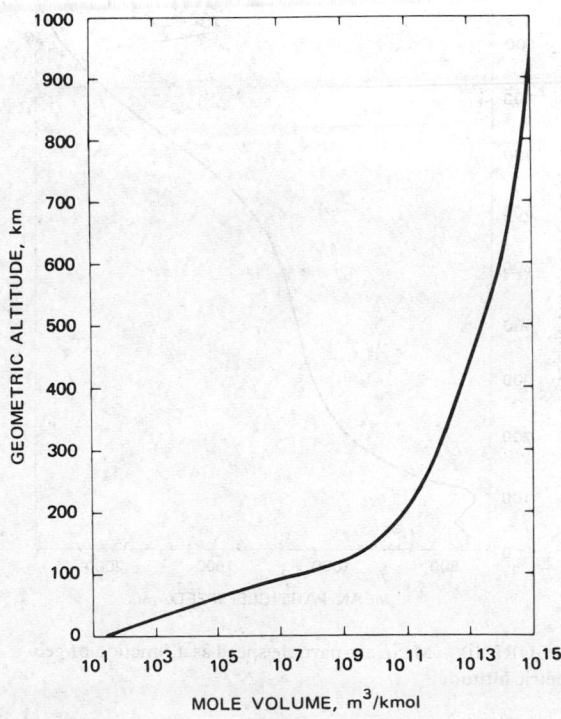

FIGURE 8.   Mole volume as a function of geometric altitude.

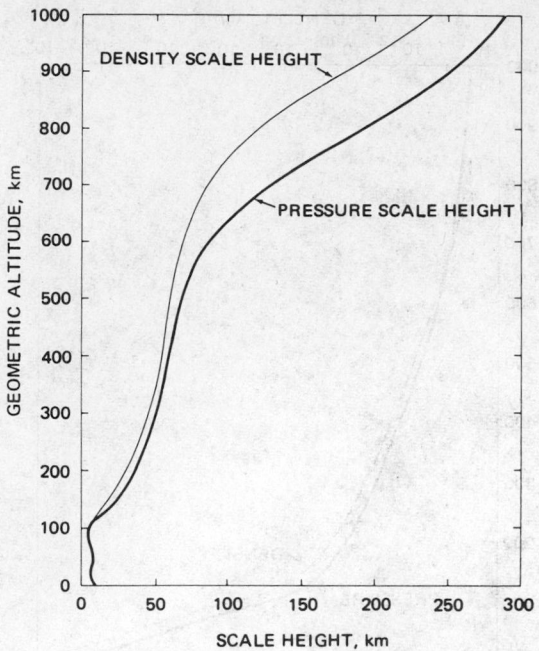

FIGURE 9.   Pressure scale height and density scale height as a function of geometric altitude.

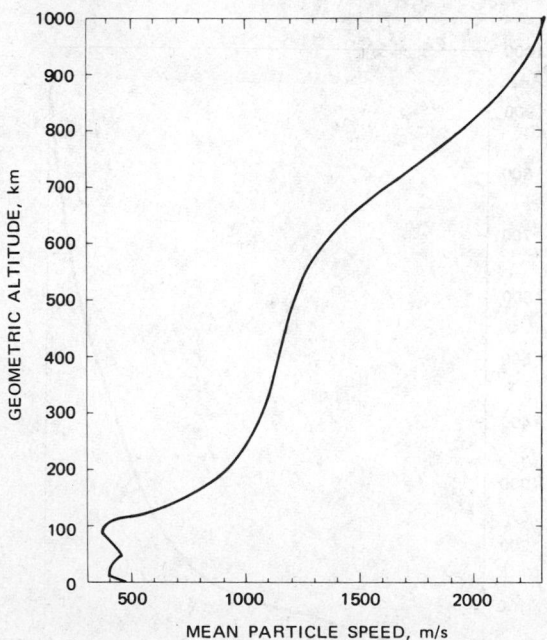

FIGURE 10.   Mean air-particle speed as a function of geometric altitude.

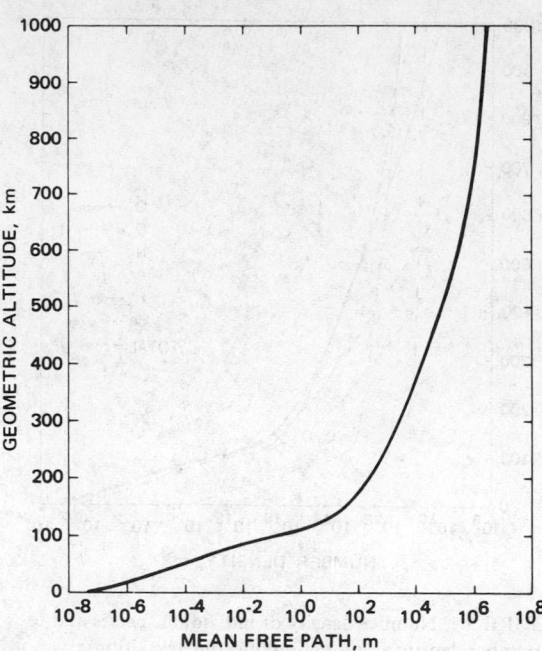

FIGURE 11.   Mean free path as a function of geometric altitude.

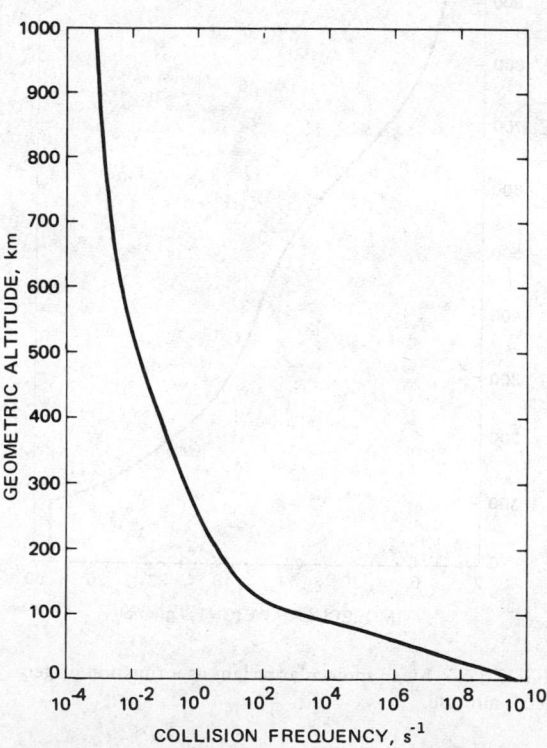

FIGURE 12.   Collision frequency as a function of geometric altitude.

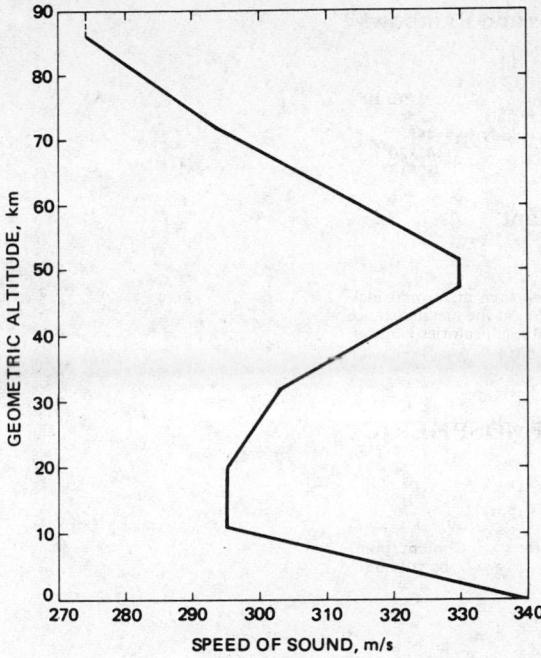

FIGURE 13. Speed of sound as a function of geometric altitude.

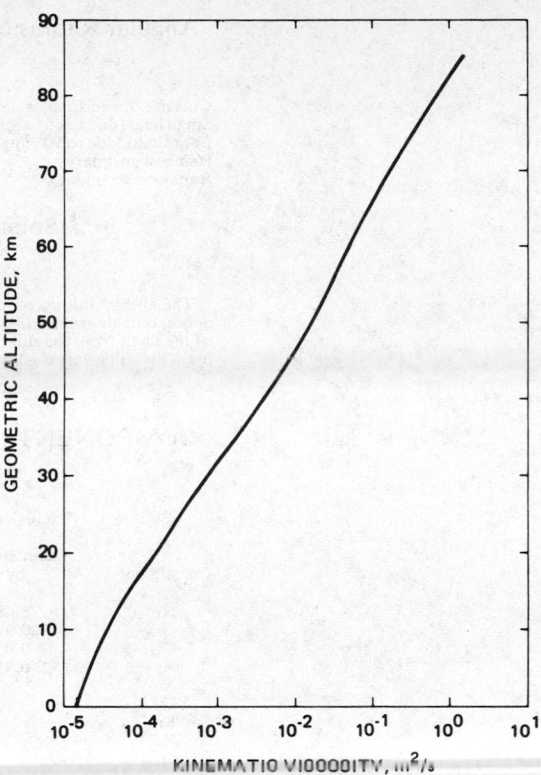

FIGURE 15. Kinematic viscosity as a function of geometric altitude.

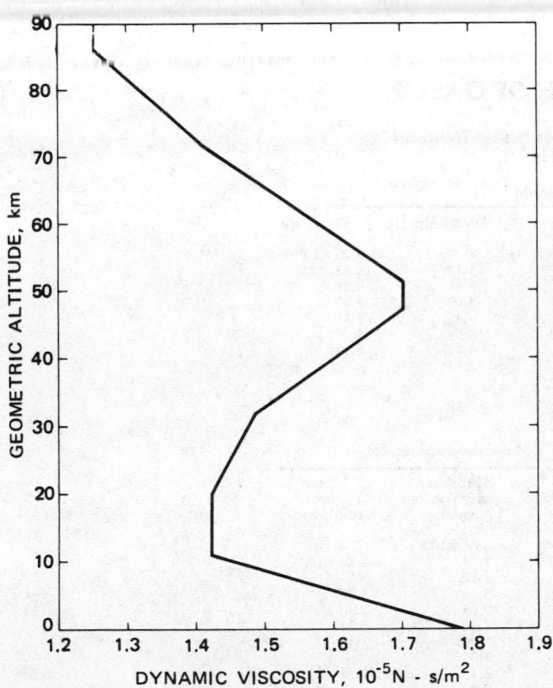

FIGURE 14. Dynamic viscosity as a function of geometric altitude.

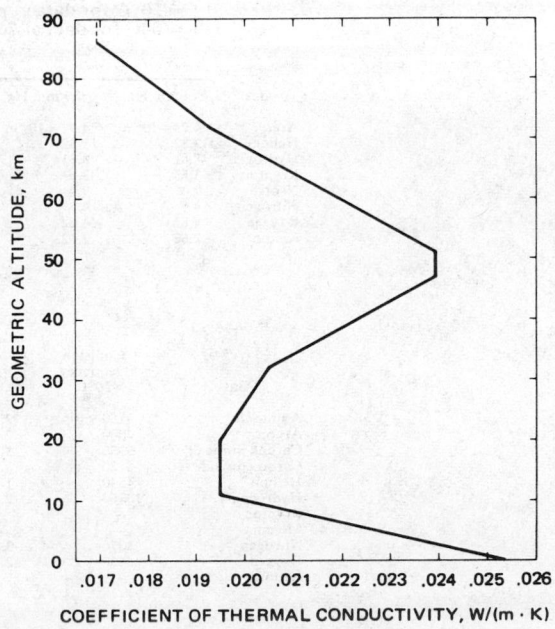

FIGURE 16. Coefficient of thermal conductivity as a function of geometric altitude.

F-161

Angular Radius of Halos and Rainbows

Coronae due to small water drops	1° to 10°
Small halo, due to 60° angles of ice crystals	22°
Large halo, due to 90° angles of ice crystals	46°
Rainbow, primary	41°20′
Rainbow, secondary	52°15′

Solar Constant

The energy falling on one sq. cm. area at normal incidence, outside the earth's atmosphere, at the mean distance of the earth from the sun equals 2.00 small calories per minute. This value varies ±2%.

COMPONENTS OF ATMOSPHERIC AIR

(Exclusive of water vapor)

Constituent	Content (per cent) by volume	Content (ppm) by volume
N_2	78.084±0.004	—
O_2	20.946±0.002	—
CO_2	0.033±0.001	—
Ar	0.934±0.001	—
Ne	—	18.18 ±0.04
He	—	5.24 ±0.004
Kr	—	1.14 ±0.01
Xe	—	0.087±0.001
H_2	—	0.5
CH_4	—	2
N_2O	—	0.5 ±0.1

MEAN FREE PATH OF GASES

t = 20°C for data at pressures below 760mm Hg
t = 0°C for data at 760mm Hg

Gas	Pressure				
	1 mm Hg	0.1 mm Hg	0.01 mm Hg	0.001 mm Hg	760 mm Hg
Argon	4.73×10^{-5}m	4.73×10^{-4}m	4.73×10^{-3}m	4.73×10^{-2}	6.20×10^{-8}m
Helium	13.32	13.32	13.32	13.32	13.32
Hydrogen	8.81	8.81	8.81	8.81	8.81
Krypton	3.63	3.63	3.63	3.63	3.63
Neon	9.4	9.4	9.4	9.4	9.4
Nitrogen	4.5	4.5	4.5	4.5	
Oxygen	4.82	4.82	4.82	4.82	6.33
Xenon	2.62	2.62	2.62	2.62	

Gas	Collision frequency 20°C	Molecular diameter, cm		
		From viscosity	From van der Waal's equation	From heat conductivity
Ammonia	9150×10^6	2.97×10^{-8}	3.08×10^{-8}	—
Argon	4000	2.88	2.94	2.86×10^{-6}
Carbon monoxide	5100	3.19	3.12	—
Carbon dioxide	6120	3.34	3.23	3.40
Helium	4540	1.90	2.65	2.30
Hydrogen	10060	2.40	2.34	2.32
Krypton	—	—	(3.69)	3.14
Mercury	—	—	3.01	
Nitrogen	5070	3.15	3.15	3.53
Oxygen	4430	2.98	2.92	
Xenon	—	—	4.02	3.42

ATMOSPHERIC ELECTRICITY

HANS DOLEZALEK

Fair-weather electricity. Electrically, a fair-weather situation is given when the activity of local generators is low. Usually, this is supposed to be the case when there are no hydrometeors (rain, snow, hail, fog, droplets) and no highly charged dust in the air, cloudiness not more than 3/10 and not concentrated overhead, wind speed below 5 m/s (no white caps on sea). Under these conditions the atmospheric electric parameters over land have the average or typical values as shown in the table.

The "Classical Picture" is a group of hypotheses to explain the electrification of the atmosphere, the main hypothesis assuming that the ionosphere is charged by the thunderstorms to a high positive potential everywhere the same but varying in time. The classical picture is probably fundamentally correct but it has never been proven and probably requires supplements and corrections. In particular, extraterrestrial influences probably have to be admitted.

ELECTRICAL PARAMETERS OF THE CLEAR (FAIR-WEATHER) ATMOSPHERE, PERTINENT TO THE CLASSICAL PICTURE OF ATM. ELECTRICITY

All currents and fields listed are part of the global circuit, i.e., circuits of local generators are not included. Values are subject to variations due to latitude and altitude of the point of observation above sea level, locality with respect to sources of disturbances, meteorological and climatological factors, and man-made changes.

Part of atmosphere for which the values are calculated (Elements are in free, cloudless atmosphere):	Currents, I, in A; and current densities, i, in A/m^2	Potential Differences, U, in V; field strength E in V/m; U = 0 at sea level	Resistances, R, in Ω; Columnar resistances, R_c, in Ωm^2; and resistivities, ρ, in Ω m	Conductances, G, in Ω^{-1}; Columnar conductances G_c, in $\Omega^{-1} m^{-2}$; total conductivities, γ, in $\Omega^{-1} m^{-1}$	Capacitances, C, in F; Columnar capacitances, C_c, in $F m^{-2}$ and capacitivities, ε, in $F m^{-1}$	Time constants τ, in seconds
Volume element at about sea level, one cubic meter	$i = 3 \times 10^{-12}$*	$E_0 = 1.2 \times 10^2$*	$\rho_0 = 4 \times 10^{13}$	$\gamma_0 = 2.5 \times 10^{-14}$	$\varepsilon_0 = 8.9 \times 10^{-12}$*	$\tau_0 = 3.6 \times 10^2$
Lower column of 1 m^2 cross section from sea level to 2 km height	same as above	at upper end: $U_1 = 1.8 \times 10^5$	$R_{cl} = 6 \times 10^{16}$	$G_{cl} = 1.7 \times 10^{-17}$	$C_{cl} = 4.4 \times 10^{-15}$	$\tau_{cl} = 2.6 \times 10^2$
Volume element at about 2 km height, 1 m^3	same as above	$E_2 = 6.6 \times 10^1$	$\rho_2 = 2.2 \times 10^{13}$(*)	$\gamma_2 = 4.5 \times 10^{-14}$	$\varepsilon_2 = 8.9 \times 10^{-12}$	$\tau_2 = 2 \times 10^2$
Center column of 1 m^2 cross section from 2 to 12 km	same as above	at upper end: $U_m = 3.15 \times 10^5$	$R_{cm} = 4.5 \times 10^{16}$	$G_{cm} = 5 \times 10^{-17}$	$C_{cm} = 8.8 \times 10^{-16}$	$\tau_{cm} = 1.8 \times 10^1$
Volume element at about 12 km height, 1 m^3	same as above	$E_{12} = 4.2 \times 10^0$	$\rho_{12} = 1.3 \times 10^{13}$(*)	$\gamma_{12} = 4.0 \times 10^{-14}$	$\varepsilon_{12} = 8.9 \times 10^{-12}$	$\tau_{12} = 1.2 \times 10^1$
Upper column of 1 m^2 cross section from 12 to 65 km height	same as above	at upper end: $U_u = 3.5 \times 10^5$	$R_{cu} = 1.5 \times 10^{16}$	$G_{cu} = 2.5 \times 10^{-17}$	$C_{cu} = 1.67 \times 10^{-16}$	$\tau_{cu} = 6.7 \times 10^0$
Whole column of 1 m^2 cross section from 0 to 65 km height	same as above	at upper end: $U = 3.5 \times 10^5$	$R_c = 1.2 \times 10^{17}$	$G_c = 8.3 \times 10^{-18}$	$C_c = 1.36 \times 10^{-16}$	$\tau_c = 1.64 \times 10^1$
Total spherical capacitor area: $5 \times 10^{14} m^2$	$1 = 1.5 \times 10^3$	$U = 3.5 \times 10^5$*	$R = 2.4 \times 10^2$	$G = 4.2 \times 10^{-3}$	$C = 6.8 \times 10^{-2}$	$\tau = 1.64 \times 10^1$

Values with a star, *, are rough average values from measurement. A star in parentheses, (*), points to a typical value from one or a few measurements. All other values have been calculated from starred values, under the assumption, that at 2 km 50% and at 12 km 90% of the columnar resistance is reached. Voltage drop along one of the partial columns can be calculated by subtracting the value for the lower column from that of the upper one. Columnar resistances, conductances, and capacitances are valid for that particular part of the column which is indicated at left. Capacitances are calculated with the formula for plate capacitors, and this fact must be considered also for the time constants for columns.

According to measurements, U, the potential difference between 0 m and 65 km may vary by a factor of approx. 2 (two). The total columnar resistance, R_c, is estimated to vary up to a factor of 3 (three), the variation being due to either reduction of conductivity in the exchange layer (about lowest 2 km of this table) or to the presence of high mountains, in both cases the variation is caused in the troposphere. Smaller variations in the stratosphere and mesosphere discussed because of aerosol there. The air-earth current density in fair-weather varies by a factor of 3 to 6 accordingly. Conductivity near the ground varies by a factor of about 3 (three) but only decreasing, increase of conductivity due to extraordinary radioactivity is a singular event. The field strength near the ground varies as a consequence of variations of air-earth current density and conductivity from about 1/3 to 10 of the value quoted in the table. Conductivity near the ground shows a diurnal and an annual variation which depends strongly on the locality; air-earth current density shows a diurnal and annual variation because the earth-ionosphere potential difference undergoes such variations, and because also the columnar resistance is supposed to have a diurnal and probably an annual variation.

Conductivities and air-earth current densities on high mountains are by factors of up to 10 greater than at sea level. Conductivity decreases when atmospheric humidity increases. Values for space charges not quoted because measurements are too few to allow calculation of average values. Values of parameters over the oceans still rather uncertain.

Ions. Ionization rate at ground level over land: from radioactive substances in the ground: 4.0; from radioactive substances in the air: 4.6; from cosmic radiation 1.5–1.8, total (10.1–10.4) $\times 10^6$ ion pairs per m^3 and sec. Over ocean far from shore: cosmic radiation only, but in spite of smaller ionization over the ocean conductivity is nearly equal to that over land because of longer ion life time (a consequence of smaller aerosol density).

Contribution from radioactive material over land decreases with altitude in the free air, from cosmic radiation increases (up to a certain level) and depends on latitude. Number densities of fast ions over land and ocean average at $(4-5) \times 10^8$ m^{-3}, varying with height by only about one order of magnitude up to ionospheric C-layer. Their life time varies from about 10 s in aerosol-rich air to 300 s in pure air. All values quoted are approximate averages.

Ion classes in lower atmosphere	Designation	Mobilities in m^2 (Vs)	Size ranges, radii in m
Primary atmospheric ions:	Fast ions (also called small or light ions)	$k \geq 10^{-4}$	$r < 6.6 \times 10^{-10}$
Secondary atmospheric ions: slow ions (also called large or heavy ions	Fast intermediate ions	$10^{-4} > k \geq 10^{-2}$	$6.6 \times 10^{-10} < r < 7.8 \times 10^{-9}$
	Slow intermediate ions	$10^{-2} > k \geq 10^{-1}$	$7.8 \times 10^{-9} < r < 2.5 \times 10^{-8}$
	Langevin ions	$10^{-1} > k \geq 2.5$	$2.5 \times 10^{-8} < r < 5.7 \times 10^{-8}$
	Ultralarge ions	$k < 2.5$	$5.7 \times 10^{-8} < r < 10^{-7}$

Electric conduction current in atmosphere carried nearly exclusively by the fast ions because number densities of the intermediate ions generally one order of magnitude smaller.

Nature of ions in pure air ("Standard Atmosphere"): initially all molecules are ionized with equal probability according to their abundance, but within nanoseconds these ions transform into the "fast ions" with the following composition:

positive ions: $(H_3O)^+(H_2O)_n$

negative ions: $O_2^-(H_2O)_n$, or $CO_4^-(H_2O)_n$;

the n depending on atmospheric temperature and water vapor pressure, varies very often in the life time of an ion. Most probable values of n for positive ions: troposphere 6 or 7, stratosphere 4, mesosphere 3 or 2.

Thunderstorm electricity. Clouds are generally more or less electrified. In thunderstorms, mechanical, thermodynamical and maybe chemical energy is transformed into electrical energy (generator) with separate poles. Most thunderstorms have during most of their life time (phases: development 10–20, mature 15–30, decay 30 min) a positive charge center above and a negative one below, probably another positive one near the base. Electrification occurs (according to processes not yet fully understood) in periods of tens of minutes and spaces of about 1 km^3, and there may be several such centers in a cloud. Thunderclouds mostly reach beyond freezing level, in moderate latitudes from about 3 to 10 km, in the tropics often up to 16–20 km, seldom more. 1000 to 2000 thunderstorms active at any time. Updrafts in storms up to 30 m/s. Water content 10^8–10^9 kg. Total energy about 10^{15} joule, electric energy 10^{12}–10^{13} J. Electric potential vs. ground at negative and positive charge centers: minus and plus 10–100 MV, resp.; field strength not yet measured reliably. Electric resistances of an atmospheric vertical column of 50 km^2 cross section: several 100 $M\Omega$ each between ground and negative center, negative and positive center, and positive center and ionosphere. Total current flowing upwards towards ionosphere derived from (still unsufficient) measurements of current densities: 0.5–2 A per thunderstorm cell. Current below cloud carried by conduction by fair-weather ions, conduction by corona ions, convection by vertical winds, precipitation, and lightning. Field strength at ground under storm: varying in strength and polarity, often 10 kV/m and more.

Atmospheric Electricity by H. Israël. Tabular material reproduced from this publication. Permission received from Israel Program for Scientific Translations.

Lightning, during its life time of up to a second, undergoes numerous variations and all its parameters change by orders of magnitude in these. Also, there are many different forms of lightning. Even disregarding particular forms such as ball lightning, we find at least four classification criteria which, taken together, give a great number of differences: intercloud, intracloud, and cloud-to-ground discharges and combinations of these: short high-current and long low-current flashes; lightnings beginning in the cloud and moving towards ground, and lightnings moving upwards; lightnings lowering positive and lightnings lowering negative charges, and those which do first the one and then the other. The usual cloud-to-ground lightning begins with a stepped leader of low luminosity, probably a meter or so in diameter, followed by a return stroke with a diameter in the order of centimeters, temperatures of about 30,000 K, and pressures of up to a MN/m^2. After this, a dart leader may again move downwards causing a second return stroke. This may repeat several times (multiple stroke flash).

Typical voltage drop in ground or other conductors after lightning impact in the neighborhood: 10 kV/m (dangerous!).

Intracloud lightnings observed with up to 100 km length.

There are probably about 100 lightnings occurring on earth at any time. Lightning frequency over oceans only about 1/10 from that over continents. Diurnal variations on continents show maximum in late afternoon and early evening, over oceans late evening until after midnight.

Energy delivered to a stroke about 100 kJ/m.

Upward lightnings initiated mostly from high masts and towers and mountains. Over flat terrain mostly downward flashes, probably generally met by short (decameters) upwards moving darts.

Height from which a lightning points at a target probably several decameters, depending on conductivity distribution in ground.

Long-lasting low-current (hundreds of amperes) flashes more dangerous to man and more damaging to objects (forest fires) than short high-current flashes.

Sferics, etc. Lightning generates a sudden variation of electric field (decreasing with r^{-3}), a quickly varying magnetic field (r^{-2}) and an electromagnetic radiation (r^{-1}) from a few hertz (in case of excitation of ionosphere-earth resonance range around the globe) through VLF (maximum intensity, range up to 10 Mm and more depending on direction and time of day) into VHF. Main frequency of thunder about 200 Hz.

Data for a normal cloud-to-ground lightning discharge bringing negative charge to earth. The values listed are intended to convey a rough feeling for the various physical parameters of lightning. No great accuracy is claimed since the results of different investigators are often not in good agreement. These values may in fact, depend on the particular environment in which the lightning discharge is generated. The choice of some of the entries in the table is arbitrary.

	Minimum*	Representative	Maximum*
Stepped leader			
Length of step, *m*	3	50	200
Time interval between steps, μsec	30	50	125
Average velocity of propagation of stepped leader, m/sec†	1.0×10^5	1.5×10^5	2.6×10^6
Charge deposited on stepped-leader channel, coulombs	3	5	20
Dart leader			
Velocity of propagation, m/sec†	1.0×10^6	2.0×10^6	2.1×10^7
Charge deposited on dart-leader channel, coulombs	0.2	1	6
Return stroke‡			
Velocity of propagation, m/sec†	2.0×10^7	5.0×10^7	1.4×10^8
Current rate of increase, kA/μsec §	<1	10	>80
Time to peak current, μsec §	<1	2	30
Peak current, kA §		10–20	110
Time to half of peak current μsec	10	40	250
Charge transferred excluding continuing current, coulombs	0.2	2.5	20
Channel length, km	2	5	14
Lightning flash			
Number of strokes per flash	1	3–4	26
Time interval between strokes in absence of continuing current, msec	3	40	100
Time duration of flash, sec	10^{-2}	0.2	2¶
Charge transferred including continuing current, coulombs	3	25	90

* The words maximum and minimum are used in the sense that most measured values fall between these limits.

† Velocities of propagation are generally determined from photographic data and thus represent "two-dimensional" velocities. Since many lightning flashes are not vertical, values stated are probably slight underestimates of actual values.

‡ First return strokes have slower average velocities of propagation, slower current rates of increase, longer times to current peak, and generally larger charge transfer than subsequent return strokes in a flash.

§ Current measurements are made at the ground.

¶ A lightning flash lasting 15 to 20 sec has been reported by Godlonton (1896).

Lightning by M. A. Uman, Table 1.1, page 4. Tabular material reproduced from this publication. Permission received from McGraw-Hill Book Company.

OUR SOLAR SYSTEM

Object	Average distance from sun (astronomical units)[a]	Radius (Earth = 1)[b]	Mass (Earth = 1)[b]	Density (g/cc) (water = 1)	Atmospheric constituents	Surface (or interior) pressure (bars)	Surface (or cloud top) temperature (K)	Number of satellites
Mercury	0.38	0.38	0.05	5.4	Helium, argon, neon, hydrogen	$>10^{-12}$	600—700 (day); ~95 (night)	0
Venus	0.72	0.96	0.82	5.2	Carbon dioxide (96%), nitrogen (~3%), helium, sulfur dioxide, water vapor, argon, neon, carbon monoxide, oxygen, sulfuric acid, hydrogen chloride, hydrogen fluoride, hydrogen sulfide	90	~730	0
Earth	1.00	1.00	1.00	5.5	Nitrogen (78%), oxygen (21%), argon (0.9%), water vapor, carbon dioxide, neon, methane, krypton, helium, xenon, hydrogen, nitrous oxide, carbon monoxide, nitrogen dioxide, sulfur dioxide, ozone	1.0	250—310	1
Mars	1.52	0.53	0.11	4.0	Carbon dioxide (95%), nitrogen (3%), argon (~2%), water vapor, oxygen, neon, carbon monoxide, krypton, xenon, ozone	0.005 to 0.008	~200—245	2
Jupiter	5.20	10.8	318	1.3	Hydrogen (89%), helium (11%), methane, ammonia, water, ethane, acetylene, phosphine, germanium tetrahydride, hydrogen cyanide?, ammonium hydrosulfide	($<10^6$)	(~140)	16
Saturn	9.5	9.0	95	0.7	Hydrogen (94%), helium (6%), methane, ammonia, ethane, phosphine, acetylene, propane?, methylacetylene?	($>10^6$)	(~100?)	17
Uranus	19.2	4.1	15	1.2	Hydrogen (<90%), methane, helium?	($>10^6$)	(~50—60)	5
Neptune	30.0	3.85	17	1.7	Hydrogen (<90%), methane, helium?	($>10^6$)	(~50—60)	2
Pluto	39.5	~0.22 to 0.25?	~0.002?	<2	Methane, argon?, neon?	$\sim10^{-3}$ to 10^{-4}?	~50—60	1
Moon	—	0.27	0.0123	3.3	Neon, argon, helium	$\sim2 \times 10^{-14}$	370 (day); 120 (night)	—

[a] Astronomical unit defined as average distance between the centers of the sun and Earth (1.496×10^8 km).
[b] Compared to Earth as unity: Earth radius is 6371 km; mass is 5.975×10^{24} kg.

Reprinted with permission from Chemical and Engineering News, page 30, August 10, 1981. Copyright 1981 by the American Chemical Society.

COSMIC RADIATION

A. Gregory and R. W. Clay

THE NATURE OF COSMIC RAYS

Primary cosmic radiation, in the form of high-energy nuclear particles, electrons, and photons, from outside the solar system and from the Sun, continually bombards our atmosphere.

Secondary radiation resulting from the interaction of primary cosmic rays with atmospheric gas, is present at sea-level and throughout the atmosphere. The secondary radiation is collimated by absorption and scattering in the atmosphere and consists of a number of components associated with different particle species. High-energy primary particles can produce a large number of secondary particles forming an extensive air shower. Thus, a number of particles may be detected in coincidence at sea level. Primary particle energies range from $\sim 10^8$ eV to $\sim 10^{20}$ eV.[1.15]

PRIMARY PARTICLE ENERGY SPECTRUM

Figure 1 shows the spectrum of primary particle energies. In differential form it is roughly a power law (with an index of -3). There appears to be a knee (a steepening) at about 10^{15} eV and an ankle (a flattening) above $\sim 10^{18}$ eV.

At energies below $\sim 10^{13}$ eV, solar system magnetic fields and plasma can modulate the primary component and Figure 2 shows the extent of this modulation between solar maximum and minimum.[10.12]

PRIMARY PARTICLE ENERGY DENSITY

If the above spectrum is corrected for solar effects the energy density above a particle energy of 10^9 eV outside the solar system is found to be $\sim 5 \times 10^5$ eV m^{-3}. As the threshold energy is increased, the energy density decreases rapidly, being 2×10^4 eV m^{-3} above 10^{12} eV and 10^2 eV m^{-3} above 10^{15} eV.[16]

PRIMARY PARTICLE ISOTROPY

This is measured as an anisotropy ($I_{max} - I_{min}/I_{max} + I_{min}$) $\times 100\%$, where I, the intensity (m^{-2}s^{-1}sr^{-1}), is usually measured with an angular resolution of a few degrees.

The anisotropy is small and energy dependent. It is roughly constant at between 0.05 and 0.1% for energies between 10^{11} eV and 10^{14} eV and increases at higher energies roughly as $0.4 \times$ (energy (eV)/10^{16})$^{1/2}$% up to $\sim 10^{18}$ eV.[16]

PRIMARY PARTICLE COMPOSITION

The composition of low energy cosmic rays (it is uncertain at high energies) is close to universal abundances except where propagation effects are present, e.g. Li, Be, B which are spallation products, are over-abundant by ~ 6 orders of magnitude.

Composition at 10^{11} eV/Nucleus

Charge	1	2	(3—5)	(6—8)	(10—14)	(16—24)	(26—28)	$\geqslant 30$
% Composition	50	25	1	12	7	4	4	0.1

($\sim 10\%$ uncertainty)

Cosmic ray composition at low energies is often quoted at a fixed *energy per nucleon*. When presented in this way, protons constitute roughly 90% of the flux, helium nuclei $\sim 10\%$ and the remainder sum to a total of $\sim 1\%$.

Certain radioactive isotopic ratios show lifetime effects. The ratio of Be^{10}/Be^9 abundances is used to measure an 'age' of cosmic rays since Be^{10} is unstable (half life = 1.6×10^6 years). A ratio of 0.6 would be expected in the absence of Be^{10} decay and a ratio of ~ 0.2 is found experimentally.[10.13]

PRIMARY ELECTRONS

Primary electrons constitute $\sim 1\%$ of the cosmic ray beam. The positron to electron ratio is about 10%.[14]

ANTIMATTER IN THE PRIMARY BEAM

The ratio of antiprotons to protons in the primary cosmic ray beam (at ~ 100 GeV) is $\sim 5 \times 10^{-4}$.[11]

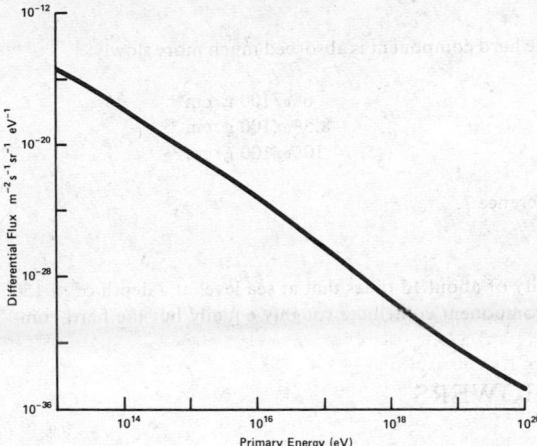

FIGURE 1. Energy Spectrum of Cosmic Ray Particles. This spectrum is of a differential form and can be converted to an integral spectrum by integrating over all energies above a required threshold (E). Insofar as this spectrum approximates to a power law of index −3, a simple conversion to the integral at an energy E/1.8 is obtained by multiplying the differential flux by the energy and dividing by 0.62.

FIGURE 2. Energy Spectrum of Cosmic Ray Particles at Lower Energies. (a) Solar minimum proton energy spectrum, (b) solar maximum proton energy spectrum, (c) γ-ray energy spectrum, (d) local interstellar electron spectrum.

PRIMARY GAMMA RAYS

The flux of primary gamma rays is low at high energies. At 1 GeV the ratio of gamma rays to protons is about 10^{-6}. The arrival directions of these gamma rays are strongly concentrated in the plane of the Milky Way although there is also a diffuse near-isotropic background flux.

Since the absorption cross section for gamma rays above 100 MeV is approximately 20 mbarn/electron, less than 10% of gamma rays reach mountain altitudes.[16]

SEA LEVEL COSMIC RADIATION

The sea level cosmic ray dose is 30 millirad. year^{-1} and the sea level ionisation is 2.22×10^{6} ion pairs m^{-}s^{-1}. The sea level flux has a soft component, which can be absorbed in ∼10cm of lead (∼100 g·cm^{-2} of absorber) and a more penetrating (largely muon) hard component. The sea level radiation is a secondary component produced in the atmosphere and its flux is dependent somewhat on the solar cycle and the geomagnetic latitude of the observer.

Absolute Flux of Hard Component[2]

Vertical integral intensity	(I) ∼ 100 m^{-2}s^{-1}sr^{-1}
Angular dependence	I(θ) ∼ I(O) cos²θ
Integrated intensity	∼ 200 m^{-2}s^{-1}

Flux of Soft Component
In free air the soft component comprises about one third of the total cosmic ray flux.

Latitude Effect
The geomagnetic field influences the trajectories of lower energy cosmic rays approaching the Earth. As a result, the background flux is reduced by about 7% at the geomagnetic equator. The effect decreases towards the poles and is negligible at latitudes above about 40°.

Flux of Protons
The proton component is strongly attenuated by the atmosphere with an attenuation length (reduction by a factor of e) of ∼ 120 g·cm^{-2}, it constitutes about 1% of the total vertical sea level flux.

Absorption

The soft component is absorbed in about 100 g·cm⁻² of matter. The hard component is absorbed much more slowly:

Absorption in lead	$6\%/100$ g·cm⁻²
Absorption in rock	$8.5\%/100$ g·cm⁻²
Absorption in water	$10\%/100$ g·cm⁻²

Absorption for depths less than 100 g·cm⁻² is given by Greisen in Reference 7.

Altitude Dependence

Cosmic ray background in the atmosphere has a maximum intensity of about 15 times that at sea level at a depth of $\sim$ 150 g·cm⁻² (15 km altitude). At maximum intensity the soft and hard component contribute roughly equally but the hard component is then attenuated more slowly.[8]

COSMIC RAY SHOWERS

High energy cosmic rays ($< 10^{13}$eV) produce measurable cascades of secondary particles in the atmosphere. The primary particle progressively loses energy which is transferred through the production of successive generations of secondary particles to a cascade of hadrons, an electromagnetic shower component (electrons and gamma rays) and muons. The secondary particles are relativistic and all travel effectively at the speed of light. As a result, they reach sea level at approximately the same time, but, due to Coulomb scattering and production angles, are spread laterally into a disk-like shower front with a typical width of $\sim$ 100 m and thickness $\sim$ 2 to 3 m. The number of particles at sea level is roughly proportional to the primary energy.

Number of particles at sea level $\sim 10^{-10} \times$ energy (eV). At altitudes below a few kilometres, the number of particles in a shower attenuates with an *attenuation length* $\sim$ 200 g·cm⁻², i.e., $N = N_0 \times \exp(-\text{depth}/200)$. The rate of observation of showers of a given size at different depths of absorber attenuates with an *absorption length* of $\sim$ 100 g·cm⁻².[15]

ATMOSPHERIC BACKGROUND LIGHT FROM COSMIC RAYS

Cosmic ray particles produce Cerenkov light in the atmosphere and produce fluorescent light through the excitation of atmospheric molecules.

Cerenkov Light

High energy charged particles will emit Cerenkov light in air if their energies are above $\sim$30 MeV (electrons). This threshold is pressure (and hence altitude) dependent. A typical Cerenkov light pulse (at sea level, 100 m from core, 10^{16} eV primary energy, in the wavelength band 430 to 530 nm). has a width of a few nanoseconds and in this time has a flux of $\sim 10^{14}$ photons m⁻²s⁻¹. For comparison, the general night sky background flux is $\sim 6 \times 10^{11}$ photons m⁻²sec⁻¹sr⁻¹ in the same wavelength band.

Fluorescent Light

Cosmic ray particles in the atmosphere excite atmospheric molecules which emit fluorescent light. This is weak compared to the Cerenkov component when viewed in the direction of the incident cosmic ray particle but is emitted isotropically. Typical pulse widths are expected to be longer than 50 ns and may be up to a few microseconds for distant large showers.[5,9]

CERENKOV EFFECTS IN TRANSPARENT MEDIA

Background cosmic ray particles will produce Cerenkov light in transparent material with a photon yield

$$\sim \frac{2\pi}{137} \sin^2 \theta_c \int_{\lambda_1}^{\lambda_2} \frac{d\lambda}{\lambda^2} \text{ photons (unit length)}^{-1}$$

where θ_c, the Cerenkov angle, $= \cos^{-1}(1/\text{refractive index})$. This background light is known to affect sensitive light detectors, e.g. photomultipliers, and can be a major source of background noise.[6]

EFFECTS ON ELECTRONIC COMPONENTS

If background cosmic rays pass through electronic components, they may deposit sufficient energy to affect the state of, e.g. a transistor flip-flop. This effect may be significant where reliability is of great importance and the background flux is high. For instance, it has been estimated that in communication satellite operation an error rate of $\sim 2 \times 10^{-3}$ per transistor per year may be found. Permanent damage may also result. A significant error rate may be found even at sea level in large electronic memories. This error rate is dependent on the sensitivity of the component devices to the deposition of electrons in their sensitive volumes.[17]

BIOPHYSICAL SIGNIFICANCE

When cosmic rays interact with living tissue they produce radiation damage, the amount of damage depending on the total dose of radiation. Radiation doses are commonly measured in rads (radiation absorbed dose $\equiv$ 100 ergs g^{-1}) or rems (radiation equivalent-man $\equiv$ Quality factor $\times$ rad). The quality factor of radiation depends on the type of particle and its energy for most cosmic ray applications it will be $\sim$1.

At sea level the cosmic radiation dose rate is small compared with doses from other sources, but both the quantity and quality of the radiation change rapidly with altitude. Approximate dose rates under various conditions

Conditions	Sea level	Cosmic rays		
		10 km (subsonic jets)	18 km (supersonic transport)	Mean total dose rate at sea level
Dose (mrem year^{-1})	30	2000	10,000	300

Astronauts would be subject to radiation from galactic (0.05 rads d^{-1}) and solar ($\sim$ few hundred rads per solar flare) cosmic rays as well as large fluxes of low energy radiation when passing through the Van Allen belts ($\sim$0.3 rads per traverse).

Both astronauts and S.S.T. travellers would be subject to a small flux of low energy heavy nuclei stopping in the body. Such particles are capable of destroying cell nuclei and could be particularly harmful in the early stages of development of an embryo. The rates of heavy nuclei stopping in tissue in supersonic transports and spacecraft are approximately as follows (from Reference 1 and 2)

Conditions (altitude)	16 km SST	20 km SST	Spacecraft
Stopping nuclei (cm^3 tissue)$^{-1}$ hr^{-1}	5.10^{-4}	5.10^{-3}	0.15

CARBON DATING

Radiocarbon is produced in the atmosphere due to the action of cosmic ray slow neutrons. Solar cycle modulation of the low energy cosmic rays causes an anticorrelation of the atmospheric ^{14}C activity with sunspot number with a mean amplitude of $\sim$0.5%. In the long term, modulation of cosmic rays by a varying geomagnetic field may be important.[4]

PRACTICAL USES OF COSMIC RAYS

There are few direct practical uses of cosmic rays. Their attenuation in water and snow have however enabled automatic monitors of water and snow depth to be constructed, and a search for hidden cavities in pyramids has been carried out using a muon 'telescope'.

OTHER EFFECTS

Stellar X-rays have been observed to affect the transmission times of radio signals between distant stations by altering the depth of the ionospheric reflecting layer. It has also been suggested that variations in ionization of the atmosphere due to solar modulation may have observable effects on climatic conditions.

REFERENCES

1. Allkofer, O. C., *Introduction to Cosmic Radiation,* Verlag Karl Thiemig, Munchen, Germany, 1975.
2. Allkofer, O. C., *J. Phys.,* Sect. G, 1, L51, 1975.
3. Allkofer, O. C. and Heinrich, W., *Health Phys.,* 27, 543, 1974.
4. Burchuladze, A. A., Pagava, S. V., Povinec, P., Togonidze, G. I., and Usacev, S., Proc. 16th Int. Cosmic Ray Conf. Kyoto, 3, 201, Univ. of Tokyo, 3, 201, 1979.
5. Cassiday, G. L., et al., Proc. 15th Int Cosmic Ray Conf., Plovdiv, Bulgarian Academy of Sciences, 8, 258, 1977.
6. Clay, R. W. and Gregory, A. G., *J. Phys.,* Sect. A, 10, 135, 1977.
7. Greisen, K., *Physical Rev.,* 63, 323, 1943.
8. Hayakawa, S., *Cosmic Ray Physics,* Wiley-Interscience, New York, 1969.
9. Jelley, J. V., *Prog. in Elementary Particle and Cosmic Ray Physics,* 9, 41, 1967.
10. Juliusson, E., Proc. 14th Int. Cos Ray Conf. Munich, 8, 2689, Max-Planck Institute for Extraterrestriche Physik, Munchen, Germany, 1975.

11. Kiraly, P., Szabelski, J., Wdowczyk, J., and Wolfendale, A. W., *Nature,* 293, 120, 1981.
12. Linsley, J., *Origin of Cosmic Rays,* I.A.U. Symposium 94, 53, D Reidel Publishing, Dordrecht, Holland, 1981.
13. Meyer, P., *Origin of Cosmic Rays,* I.A.U. Symposium 94, 7, D. Reidel Publishing, Dordrecht, Holland, 1981.
14. Tan, L. C., and Ng, L. K., *J. Phys.,* Sect. G, 7, 1135, 1981.
15. Wilson, J. G., *Cosmic Rays,* Wykeham Publishing, London, 1976.
16. Wolfendale, A. W., *Pramana,* 12, 631, 1979.
17. Ziegler, J. F., *IEEE Trans. Electron Devices,* ED-28, 560, 1981.

CRYSTAL IONIC RADII OF THE ELEMENTS

Numerical values of the radii of the ions may vary depending on how they were measured. They may have been calculated from wavefunctions and determined from the lattice spacings or crystal structure of various salts. Different values are obtained depending on the kind of salt used or the method of calculating. Data for many of the rare-earth ions were furnished by F. H. Spedding and K. Gschneidner.

Element	Charge	Atomic number	Radius in A	Element	Charge	Atomic number	Radius in A	Element	Charge	Atomic number	Radius in A
Ac	+3	89	1.18	Hf	+4	72	0.78	Ra	+2	88	1.43
Ag	+1	47	1.26	Hg	+1	80	1.27	Rb	+1	37	1.47
	+2		0.89		+2		1.10	Re	+4	75	0.72
Al	+3	13	0.51	Ho	+3	67	0.894		+7		0.56
Am	+3	95	1.07	I	−1	53	2.20	Rh	+3	45	0.68
	+4		0.92		+5		0.62	Ru	+4	44	0.67
Ar	+1	18	1.54		+7		0.50	S	−2	16	1.84
As	−3	33	2.22	In	+3	49	0.81		+2		2.19
	+3		0.58	Ir	+4	77	0.68		+4		0.37
	+5		0.46	K	+1	19	1.33		+6		0.30
At	+7	85	0.62	La	+1	57	1.39	Sb	−3	51	2.45
Au	+1	79	1.37		+3		1.061		+3		0.76
	+3		0.85	Li	+1	3	0.68		+5		0.62
B	+1	5	0.35	Lu	+3	71	0.85	Sc	+3	21	0.732
	+3		0.23	Mg	+1	12	0.82	Se	−2	34	1.91
Ba	+1	56	1.53		+2		0.66		−1		2.32
	+2		1.34	Mn	+2	25	0.80		+1		0.66
Be	+1	4	0.44		+3		0.66		+4		0.50
	+2		0.35		+4		0.60		+6		0.42
Bi	+1	83	0.98		+7		0.46	Si	−4	14	2.71
	+3		0.96	Mo	+1	42	0.93		−1		3.84
	+5		0.74		+4		0.70		+1		0.65
Br	−1	35	1.96		+6		0.62		+4		0.42
	+5		0.47	N	−3	7	1.71	Sm	+3	62	0.964
	+7		0.39		+1		0.25	Sn	−4	50	2.94
C	−4	6	2.60		+3		0.16		−1		3.70
	+4		0.16		+5		0.13		+2		0.93
Ca	+1	20	1.18	NH₄	+1		1.43		+4		0.71
	+2		0.99	Na	+1	11	0.97	Sr	+2	38	1.12
Cd	+1	48	1.14	Nb	+1	41	1.00	Ta	+5	73	0.68
	+2		0.97		+4		0.74	Tb	+3	65	0.923
Ce	+1	58	1.27		+5		0.69		+4		0.84
	+3		1.034	Nd	+3	60	0.995	Tc	+7	43	0.979
	+4		0.92	Ne	+1	10	1.12	Te	−2	52	2.11
Cl	−1	17	1.81	Ni	+2	28	0.69		−1		2.50
	+5		0.34	Np	+3	93	1.10		+1		0.82
	+7		0.27		+4		0.95		+4		0.70
Co	+2	27	0.72		+7		0.71		+6		0.56
	+3		0.63	O	−2	8	1.32	Th	+4	90	1.02
Cr	+1	24	0.81		−1		1.76	Ti	+1	22	0.96
	+2		0.89		+1		0.22		+2		0.94
	+3		0.63		+6		0.09		+3		0.76
	+6		0.52	Os	+4	76	0.88		+4		0.68
Cs	+1	55	1.67		+6		0.69	Tl	+1	81	1.47
Cu	+1	29	0.96	P	−3	15	2.12		+3		0.95
	+2		0.72		+3		0.44	Tm	+3	69	0.87
Dy	+3	66	0.908		+5		0.35	U	+4	92	0.97
Er	+3	68	0.881	Pa	+3	91	1.13		+6		0.80
Eu	+3	63	0.950		+4		0.98	V	+2	23	0.88
	+2		1.09		+5		0.89		+3		0.74
F	−1	9	1.33	Pb	+2	82	1.20		+4		0.63
	+7		0.08		+4		0.84		+5		0.59
Fe	+2	26	0.74	Pd	+2	46	0.80	W	+4	74	0.70
	+3		0.64		+4		0.65		+6		0.62
Fr	+1	87	1.80	Pm	+3	61	0.979	Y	+3	39	0.893
Ga	+1	31	0.81	Po	+6	84	0.67	Yb	+2	70	0.93
	+3		0.62	Pr	+3	59	1.013		+3		0.858
Gd	+3	64	0.938		+4		0.90	Zn	+1	30	0.88
Ge	−4	32	2.72	Pt	+2	78	0.80		+2		0.74
	+2		0.73		+4		0.65	Zr	+1	40	1.09
	+4		0.53	Pu	+3	94	1.08		+4		0.79
H	−1	1	1.54		+4		0.93				

BOND LENGTHS BETWEEN CARBON AND OTHER ELEMENTS

Prepared by Olga Kennard.

The tables are based on bond distance determinations, by experimental methods, mainly X-ray and electron diffraction, and include values published up to January 1, 1956. In the present tables, for the sake of completeness individual values of bond distances of lower accuracy are quoted with limits of error indicated where possible. Values for tungsten and bismuth should be treated with particular caution.

According to the statistical theory of errors if an average quantity $\bar{\mu}$ and a standard deviation σ can be evaluated there is a 95% probability that the true value lies within the interval $\bar{\mu} \pm 2\sigma$. Too much reliance should, however, not be placed on σ values in bond distance determinations since the derivation of these certain sources of error may have been neglected.

Values of the bond lengths and the limits of error are each given in Ångstrom units.

Reproduced by permission from International Tables for X-ray Crystallography.

BOND LENGTHS BETWEEN CARBON AND OTHER ELEMENTS

Reference: HCP and "Tables of interatomic distances" Chem. Soc. of London, 1958

Group	Bond type	Element						
I	All types	H** 1.056 − 1.115						
II		Be 1.93	Hg 2.07 ± 0.01					
III		B 1.56 ± 0.01	Al 2.24 ± 0.04	In 2.16 ± 0.04				
IV	All types	C**	Ge	Si	Sn	Pb		
	Alkyls (CH₃XH₃)	1.54 − 1.20	1.98 ± 0.03	1.865 ± 0.008	2.143 ± 0.008	2.29 ± 0.05		
	Aryl (C₆H₅XH₃)			1.84 ± 0.01	2.18 ± 0.02			
	Neg. Subst. (CH₃XCl₃)			1.88 ± 0.01				
V	All types	N**	P	As	Sb	Bi		
	Paraffinic (CH₃)₃X	1.47 − 1.1	1.87 ± 0.02	1.98 ± 0.02	2.202 ± 0.016	2.30*		
VI		O** 1.43 − 1.15	S** 1.81 − 1.55	Cr 1.92 ± 0.04	Se 1.98 − 1.71	Te 2.05 ± 0.14	Mo 2.08 ± 0.04	W 2.06 ± 0.01*
VII	Paraffinic (monosubstituted) (CH₃X)	F 1.831 ± 0.005	Cl 1.767 ± 0.002	Br 1.937 ± 0.003	I 2.13₆ ± 0.01			
	Paraffinic (disubstituted) (CH₂X₂)	1.334 ± 0.004	1.767 ± 0.002	1.937 ± 0.003	2.13₆ ± 0.1			
	Olefinic (CH₂:CHY)	1.32₅ ± 0.1	1.72 ± 0.01	1.89 ± 0.01	2.003 ± 0.003			
	Aromatic (C₆H₅X)	1.70 ± 0.01	1.70 ± 0.01	1.83 ± 0.01	2.05 ± 0.02			
	Acetylenic (HC:CX)		1.635 ± 0.004	1.79₅ ± 0.01	1.99 ± 0.02			
VIII		Fe 1.84 ± 0.02	Co 1.83 ± 0.02	Ni 1.82 ± 0.03	Pd 2.27 ± 0.04			

* Error uncertain.
** See following individual tables.

CARBON-CARBON

Single Bond

Paraffinic	1.541 ± 0.003
In diamond (18°C)	1.54452 ± 0.00014

Partial Double Bond

(1) Shortening of single bond in presence of carbon carbon double bond, e.g. (CH₃)₂C:CH₂; or of aromatic ring e.g. C₆H₅. CH₃	1.53 ± 0.01
(2) Shortening in presence of a carbon oxygen double bond e.g. CH₃CHO	1.516 ± 0.005
(3) Shortening in presence of two carbon-oxygen double bonds, e.g. (CO₂H)₂	1.49 ± 0.01
(4) Shortening in presence of one carbon-carbon triple bond, e.g. CH₃C:CH	1.460 ± 0.003
(5) In compounds with tendency to dipole formation, e.g. C:C.C:N	1.44 ± 0.01
(6) In graphite (at 15°C)	1.4210 ± 0.0001
(7) In aromatic compounds	1.395 ± 0.003
(8) In presence of two carbon carbon triple bonds, e.g. HC:C.C:CH	1.373 ± 0.004

Double Bond

(1) Simple	1.337 ± 0.006
(2) Partial triple bond, e.g. CH₂:C:CH₂	1.309 ± 0.005

Triple Bond

(1) Simple, e.g. C₂H₂	1.204 ± 0.002
(2) Conjugated, e.g. CH₃.(C:C)₂.H	1.206 ± 0.004

CARBON-HYDROGEN

(1) Paraffinic (a) in methane	1.091
(b) in monosubstitured carbon	1.101 ± 0.003
(c) in disubstituted carbon	1.073 ± 0.004
(d) in trisubstituted carbon	1.070 ± 0.007
(2) Olefinic, e.g. CH₂:CH₂	1.07 ± 0.01
(3) Aromatic in C₆H₆	1.084 ± 0.006
(4) Acetylenic, e.g. CH:C.X	1.056 ± 0.003
(5) Shortening in presence of a carbon triple bond, e.g. CH₃CN	1.115 ± 0.004
(6) In small rings, e.g. (CH₂)₃S	1.081 ± 0.007

CARBON-NITROGEN

Single Bond

(1) Paraffinic (a) 4 co-valent nitrogen	1.479 ± 0.005
(b) 3 co-valent nitrogen	1.472 ± 0.005
(2) In C−N= e.g. CH₃NO₂	1.475 ± 0.010
(3) Aromatic in C₆H₅NHCOCH₃	1.426 ± 0.012
(4) Shortened (partial double bond) in heterocyclic systems, e.g. C₅H₅N	1.352 ± 0.005
(5) Shortened (partial double bond) in N−C=O e.g. HCONH₂	1.322 ± 0.003

Triple Bond

(1) in R.C:N	1.158 ± 0.002

CARBON-OXYGEN

Single Bond
(1) Paraffinic 1.43 ± 0.01
(2) Strained e.g. epoxides 1.47 ± 0.01
(3) Shortened (partial double bond) as in carboxylic acids or 1.36 ± 0.01
through influence of aromatic ring, e.g. salicylic acid

Double Bond
(1) In aldehydes, ketones, caboxylic acids, esters 1.23 ± 0.01
(2) In zwitterion forms, e.g. DL serine 1.26 ± 0.01
(3) Shortened (partial triple bond) as in conjugated systems 1.207 ± 0.006
(4) Partial triple bond as in acyl halides or isocyanates 1.17 ± 0.01

CARBON-SULPHUR

Single Bond
(1) Paraffinic, e.g. CH_3SH $1.81(5) \pm 0.01$
(2) Lengthened in presence of fluorine, e.g. $(CF_3)_2S$ $1.83(5) \pm 0.01$
(3) Shortened (partial double bond) as in heterocyclic systems, 1.73 ± 0.01
e.g. C_4H_4S

Double Bond
(1) In ethylene thiourea 1.71 ± 0.02
(2) Shortened (partial triple bond) in presence of second caron 1.558 ± 0.003
double bond, e.g. COS

BOND LENGTHS OF ELEMENTS

Element	Bond	Å	Element	Bond	Å
Ac	Ac–Ac	3.756	Np (α-form, 20° C)	Np–Np	2.60 (orthorhombic)
Ag (25°C)	Ag–Ag	2.8894	(β-form, 313°C)		2.76 (tetragon)
Al (25°C)	Al–Al	2.863	(γ-form, 600°C)		3.05 (b.c.c.)
As	As–As	2.49	O_2	O–O	1.208
As₄	As–As	2.44 ± 0.03	O_3 angle $116.8 \pm 0.5°$		1.278 ± 0.003
Au (25°C)	Au–Au	2.8841	Os (20°C)	Os–Os	2.6754
B₂	B–B	1.589	P black	P–P	2.18
Ba (room temp.)	Ba–Ba	4.347	P₄	P–P	2.21 ± 0.02
Be (α-form, 20°C)	Be–Be	2.2260	Pa	Pa–Pa	3.212
Bi (25°C)	Bi–Bi	3.09	Pb (25°C)	Pb–Pb	3.5003
Br₂	Br–Br	2.290	Pd (25°C)	Pd–Pd	2.7511
Ca (α-form, 18°C)	Ca–Ca	3.947 (f.c.c.)	Po (α-form, 10°C)	Po–Po	3.345 (cubic)
(β-form, 500°C		3.877 (b.c.c.)	(β-form, 75°C)		3.359 (rh. hedr.)
Cd (21°C)	Cd–Cd	2.9788	Pr (α-form)	Pr–Pr	3.640 (tetrag.)
Cl₂	Cl–Cl	1.988	(β-form)		3.649 (f.c.c.)
Ce	Ce–Ce	3.650	Pt (20°C)	Pt–Pt	2.746
Co (18°C)	Co–Co	2.5061	Pu (γ-form, 235°C)	Pu–Pu	3.026 (f.c.c.)
Cr (α-form, 20°C)	Cr–Cr	2.4980	(δ-form, 313°C)		3.279 (f.c.c.)
(β-form, >1850°C)		2.61	(ε-form, 500°C)		3.150 (b.c.c.)
Cs (−10°C)	Cs–Cs	5.309	Rb (20°C)	Rb–Rb	4.95
Cu (20°C)	Cu–Cu	2.5560	Re (room temp.)	Re–Re	2.741
Dy	Dy–Dy	3.503	Rh (20°C)	Rh–Rh	2.6901
Er	Er–Er	3.468	Ru (25°C)	Ru–Ru	2.6502
Eu	Eu–Eu	3.989	S₂	S–S	1.887
F₂	F–F	1.417 ± 0.001	S₈	S–S	2.07 ± 0.02
Fe (α-form, 20°C)	Fe–Fe	2.4823 (b.c.c.)	Sb (25°C)	Sb–Sb	2.90
(γ-form, 916°C)		2.578 (f.c.c.)	Sc (room temp.)	Sc–Sc	3.212
(δ-form, 1394°C)		2.539 (b.c.c.)	Se (20°C)	Se–Se	2.321
Ga (20°C)	Ga–Ga	2.442	Se₂	Se–Se	2.152 ± 0.003
Gd (20°C)	Gd–Gd	3.573	Se₈	Se–Se	2.32 ± 0.003
Ge (20°C)	Ge–Ge	2.4498	Si (20°C)	Si–Si	2.3517
H₂	H–H in H₂	0.74611	Sn (α-form, 20°C)	Sn–Sn diamond	2.8099
	H–D in HD	0.74136	(β-form, 25°C)	type lattice	3.022 (tetrag.)
	D–D in D₂	0.74164	Sr (α-form, 25°C)	Sr–Sr	4.302 (f.c.c.)
He	He–He in [He₂]*	1.08₀	(β-form, 248°C)		4.32 (h.c.p.)
Hf (α-form, 24°C)	Hf–Hf	3.1273 (h.c.p.)	(γ-form, 614°C)		4.20 (b.c.c.)
Hg (−46°C)	Hg–Hg	3.005	Ta (20°C)	Ta–Ta	2.86
Ho	Ho–Ho	3.486	Tb	Tb–Tb	3.525
I₂	I–I	2.662	Tc (room temp.)	Tc–Tc	2.703
In (20°C)	In–In	3.2511	Te (25°C)	Te–Te	2.864
Ir (room temp.)	Ir–Ir	2.714	Th (α-form, 25°C)	Th–Th	3.595 (f.c.c.)
K (78°K)	K–K	4.544	(β-form, 1450°C)		3.56 (b.c.c.)
La (α-form)	La–La	3.739 (h.c.p.)	Ti (α-form, 25°C)	Ti–Ti	2.8956 (h.c.p.)
(β-form)		3.745 (f.c.c.)	(β-form, 900°C)		2.8636 (b.c.c.)
Li (20°C)	Li–Li	3.0390	Tl (α-form, 18°C)	Tl–Tl	3.4076 (h.c.p.)
Lu	Lu–Lu	3.435	(β-form, 262°C)		3.362 (b.c.c.)
Mg (25°C)	Mg–Mg	3.1971	Tm	Tm–Tm	3.447
Mn (γ-form, 1095°C)	Mn–Mn	2.7311 (f.c.c.)	U (α-form)	U–U	2.77
(δ-form, 1134°C)		2.6679 (b.c.c.)	(β-form, 805°C)		3.058 (b.c.c.)
Mo (20°C)	Mo–Mo	2.7251	V (30°C)	V–V	2.6224
N₂	N–N	$1.0975_8 \pm 0.0001$	W (25°C)	W–W	2.7409
Na (20°C)	Na–Na	3.7157	Y	Y–Y	3.551
Nb (20°C)	Nb–Nb	2.8584	Yb	Yb–Yb	3.880
Nd	Nd–Nd	3.628	Zn (25°C)	Zn–Zn	2.6694
Ni (18°C)	Ni–Ni	2.4916	Zr	Zr–Zr	3.179

BOND LENGTH AND ANGLE VALUES BETWEEN ELEMENTS

Elements	In	Bond length (Å)	Bond angle (°)
Boron			
B–B	B_2H_6	1.770 ± 0.013	H–B–H 121.5 ± 7.5
B–Br	BBr	1.88$_7$	–
	BBr_3	1.87 ± 0.02	Br–B–Br 120 ± 6
B–Cl	BCl	1.715$_7$	–
	BCl_3	1.72 ± 0.01	Cl–B–Cl 120 ± 3
B–F	BF	1.262	–
	BF_3	1.29$_5$ ± 0.01	F–B–F 120
B–H	Hydrides	1.21 ± 0.02	–
B–H Bridge	Hydrides	1.39 ± 0.02	–
B–N	$(BClNH)_3$	1.42 ± 0.01	B–N–B 121
B–O	BO	1.2049	–
	$B(OH)_3$	1.362 ± 0.005 (av.)	O–B–O 119.7
Nitrogen			
N–Cl	NO_2Cl	1.79 ± 0.02	–
N–F	NF_3	1.36 ± 0.02	F–N–F 102.5 ± 1.5
N–H	$[NH_4]^+$	1.034 ± 0.003	–
	NH	1.038	–
	ND	1.041	–
	HNCS	1.013 ± 0.005	H–N–C 130.25 ± 0.25
N–N	N_3H	1.02 ± 0.01	H–N–N′ 112.65 ± 0.5
	N_2O	1.126 ± 0.002	–
	$[N_2]^+$	1.116$_2$	–
N–O	NO_2Cl	1.24 ± 0.01	O–N–O 126 ± 2
	NO_2	1.188 ± 0.005	O–N–O 134.1 ± 0.25
N=O	N_2O	1.186 ± 0.002	–
	$[NO]^+$	1.0619	–
N–Si	SiN	1.572	–
Oxygen			
O–H	$[OH]^+$	1.0289	–
	OD	0.9699	–
	H_2O_2	0.960 ± 0.005	O–O–H 100 ± 2
O–O	H_2O_2	1.48 ± 0.01	–
	$[O_2]^+$	1.227	–
	$[O_2]^-$	1.26 ± 0.02	–
	$[O_2]^{--}$	1.49 ± 0.02	–
Phosphorus			
P–D	PD	1.429	–
P–H	$[PH_4]^+$	1.42 ± 0.02	–
P–N	PN	1.4910	–
P–S	$PSBr_x(Cl_a,F_a)$	1.86 ± 0.02	–
Sulfur			
S–Br	$SOBr_2$	2.27 ± 0.02	Br–S–Br 96 ± 2
S–F	SOF_2	1.585 ± 0.005	F–S–F 92.8 ± 1
S–D	SD	1.3473	–
	SD_2	1.345	–
S–O	SO_2	1.4321	O–S–O 119.54
	$SOCl_2$	1.45 ± 0.02	–
S–S	S_2Cl_2	2.04 ± 0.01	–
Silicon			
Si–Br	$SiBr_4$	2.17 ± 1.01	–
Si–Cl	$SiCl_4$	2.03 ± 1.01 (av.)	–
Si–F	SiF_4	1.561 ± 0.003 (av.)	–
SiH	SiH_4	1.480 ± 0.005	–
Si–O	$[SiO]^+$	1.504	–
Si–Si	Si_2Cl_2	2.30 ± 0.02	–

BOND LENGTHS AND ANGLES OF CHEMICAL COMPOUNDS

A. Inorganic Compounds

Compound	Formula	Bond lengths in Å		Bond angles (°)	
Ammonia	NH_3	N–H	1.008 ± 0.004	H–N–H	1.07.3 ± 0.2
Antimony tribromide	$SbBr_3$	Sb–Br	2.51 ± 0.02	Br–Sb–Br	97 ± 2
Antimony trichloride	$SbCl_3$	Sb–Cl	2.352 ± 0.005	Cl–Sb–Cl	99.5 ± 1.5
Antimony triiodide	SbI_3	Sb–I	2.67 ± 0.03	I–Sb–I	99.0 ± 1
Arsenic tribromide	$AsBr_3$	As–Br	2.33 ± 0.02	Br–As–Br	100.5 ± 1.5
Arsenic trichloride	$AsCl_3$	As–Cl	2.161 ± 0.004	Cl–As–Cl	98.4 ± 0.5
Arsenic trifluoride	AsF_3	As–F	1.712 ± 0.005	F–As–F	102.0 ± 2
Arsenic triiodide	AsI_3	As–I	2.55 ± 0.03	I–As–I	101.0 ± 1.5
Arsenic trioxide	As_4O_6	As–O	1.78 ± 0.02	O–As–O	99.0 ± 2
				As–O–As	128.0 ± 2
Arsine	AsH_3	As–H	1.5192 ± 0.002	H–As–H	91.83 ± 0.33
Bismuthum tribromide	$BiBr_3$	Bi–Br	2.63 ± 0.02	Br–Bi–Br	100.0 ± 4
Bismuthum trichloride	$BiCl_3$	Bi–Cl	2.48 ± 0.02	Cl–Bi–Cl	100.0 ± 6
Bromosilane	SiH_3Br	Si–H	1.57 ± 0.03	H–Si–H	111.3 ± 1
Chlorine dioxide	ClO_2	Cl–O	1.49	O–Cl–O	118.5
Chlorogermane	GeH_3Cl	Ge–H	1.52 ± 0.03	H–Ge–H	110.9 ± 1.5
Chlorosilane	SiH_3Cl	Si–H	1.483 ± 0.001	H–Si–H	110 ± 0.03
		Si–Cl	2.0479 ± 0.007		

BOND LENGTHS AND ANGLES OF CHEMICAL COMPOUNDS
(CONTINUED)

A. Inorganic Compounds (Continued)

Compound	Formula	Bond lengths in Å		Bond angles (°)	
Chromium oxychloride	Cr(OCl)$_2$	Cr–O	1.57 ± 0.03	O–Cr–O	105 ± 4
		Cr–Cl	2.12 ± 0.02	Cl–Cr–Cl	113 ± 3
				Cl–Cr–O	109 ± 3
Cyanuric triazide	C$_3$N$_{12}$	C–N	1.38	C–N=C	113.0
Dichlorosilane	SiH$_2$Cl$_2$	Si–H	1.46		
		SiCl	2.02 ± 0.03	Cl–Si–Cl	110 ± 1
Difluorodiazine	N$_2$F$_2$	N–F	1.44 ± 0.04		
		N–N	1.25 ± 0.02	F–N=N	115 ± 5
Difluoromethylsilane	Ch$_3$SiHf$_2$			F–Si–F	106 ± 0.5
				H–Si–C	116.2 ± 1
				C–Si–F	109.8 ± 0.5
Disilicon hexachloride (hexachlorisilane)	Si$_2$Cl$_6$	Si–Cl	2.02 ± 0.02	Cl–Si–Cl	109.5 ± 1
		Si–Si	2.34 ± 0.06		
Fluorosilane	SiH$_3$F	Si–H	1.460 ± 0.01	H–Si–H	109.3 ± 0.3
		Si–F	1.595 ± 0.002		
Hydrogen phosphide	PH$_3$	P–H	1.415 ± 0.003	H–P–H	93.3 ± 0.2
Hydrogen selenide	SeH$_2$	Se–H	1.47	H–Se–H	91.0
Hydrogen sulfide	SH$_2$	S–H	1.3455	H–S–H	93.3
Hydrogen telluride	Te–H$_2$			H–Te–H	89.5 ± 1
Iodo silane	Si–H$_3$I	Si–H	1.48 ± 0.01	H–Si–H	109.9 ± 0.4
		Si–I	2.4 ± 0.09		
Methylgermane	Ch$_3$GeH$_3$			H–C–H	108.2 ± 0.5
				H–Ge–H	108.6 ± 0.5
Nitrosyl bromide	NOBr	O–N	1.15 ± 0.04	Br–N=O	117 ± 3
		N–Br	2.14 ± 0.02		
Nitrosyl chloride	NOCl	N–O	1.14 ± 0.02	Cl–N=O	113.0 ± 2
		N–Cl	1.97 ± 0.01		
Nitrosyl fluoride	NOF	N–O	1.13	F–N=O	110.0
		N–F	1.52		
cis-Nitrous acid	NO(OH)	H–O	0.98	O–N=O	114 ± 2
		O–N	1.46		
		N–O'	1.20		
trans-Nitrous acid	NO(OH)	H–O	0.98	O–N=O	118 ± 2
		O–N	1.46		
		NO'	1.20		
Oxygen chloride	OCl$_2$	O–Cl	1.70$_1$ ± 0.02	Cl–O–Cl	110.8 ± 1
Oxygen fluoride	OF$_2$	O–F	1.418	F–O–F	103.2
Phosphorus oxychloride	POCl$_3$	P–Cl	1.99$_9$ ± 0.02	Cl–P–Cl	103.5 ± 1
Phosphorus oxysulfide	P$_4$O$_6$S$_4$	P–O	1.61 ± 0.02	P–O–P	128.5 ± 1.5
				O–P–O	101.5 ± 1
		P–S	1.85 ± 0.02	O–P–S	116.5 ± 1
Phosphorus pentoxide	P$_4$O$_{10}$	P–O	1.62 ± 0.02	O–P–O	101.5 ± 1
		P–O'	1.38 ± 0.02	O–P–O'	116.5 ± 1
				P–O–P	123.5 ± 1
Phosphorus tribromide	PBr$_3$	P–Br	2.18 ± 0.03	Br–P–Br	101.5 ± 1.5
Phosphorus trichloride	PCl$_3$	P–Cl	2.043 ± 0.003	Cl–P–Cl	100.1 ± 0.3
Phosphorus trifluoride	PF$_3$	P–F	1.535	F–P–F	100.0
Phosphorus trioxide	P$_4$O$_6$	P–O	1.65 ± 0.02	O–P–O	99.0 ± 1
				P–O–P	127.5 ± 1
Stibine	SbH$_3$	Sb–H	1.7073 ± 0.0025	H–Sb–H	91.3 ± 0.33
Sulfur dichloride	SCl$_2$	S–Cl	1.99 ± 0.03	Cl–S–Cl	101.0 ± 4
Sulfur dioxide	SO$_2$	S–O	1.4321	O–S–O	119.536
Sulfur monochloride	S$_2$Cl$_2$	S–Cl	2.01 ± 0.07	Cl–S–S	104.5 ± 0.25
Sulfurylchloride	SO$_2$Cl$_2$	S–O	1.43 ± 0.02	O–S–O	119.75 ± 5
		S–Cl	1.99 ± 0.02	Cl–S–Cl	111.20 ± 2
				Cl–O–O	106.5 ± 2
Tellurium bromide	TeBr$_2$	Te–Br	2.51 ± 0.02	Br–Te–Br	98.0 ± 3
Tribromo silane	SiHBr$_3$	Si–Br	2.16 ± 0.03	Br–Si–Br	110.5 ± 1.5
Trichloro germane	GeHCl$_3$	Ge–H	1.55 ± 0.04	Cl–Ge–Cl	108.3 ± 0.2
Trichloro silane	Si–HCl$_3$	Si–H	1.47	Cl–Si–Cl	109.4 ± 0.3
Trifluorochlorosilane	SiClF$_3$	Si–F	1.560 ± 0.005	F–Si–F	108.5 ± 1
		Si–Cl	1.989 ± 0.018		
Trifluorochlorogermane	GeClF$_3$	Ge–F	1.688 ± 0.0017	F–Ge–F	107.7 ± 1.5
		Gr–Cl	2.067 ± 0.005		
Trifluorosilane	SiHF$_3$	Si–H	1.455 ± 0.01		
		Si–F	1.565 ± 0.005	F–Si–F	108.3 ± 0.5
Vanadium oxytrichloride	VOCl$_3$	V–O	1.56 ± 0.04	Cl–V–Cl	111.2 ± 2
		V–Cl	2.12 ± 0.03	Cl–V–O	108.2 ± 2
Water	H$_2$O	O–H	0.958$_4$	H–O–H	104.45

B. Organic Compounds

Compound	Formula	Bond lengths in Å		Bond angles (°)	
Acetaldehyde	CH$_3$COH	C–H	1.09	C–C=O	121 ± 2
		C–C	1.50 ± 0.02		
		C–O	1.22 ± 0.02		
Bromomethane	Ch$_3$Br	C–H	1.11 ± 0.01	H–C–H	111.2 ± 0.5
		C–Br	1.929		
Carbon tetrachloride	CCl$_4$	C–Cl	1.766 ± 0.003	Cl–C–Cl	109.5
		Cl–Cl	2.887 ± 0.004		
Chloromethane	CH$_3$Cl	C–H	1.11 ± 0.01	H–C–H	110 ± 2
		C–Cl	1.784 ± 0.003		
Dichloromethane	CH$_2$Cl$_2$	C–H	1.068 ± 0.005	H–C–H	112 ± 0.3
		C–Cl	1.7724 ± 0.0005	Cl–C–Cl	111.8

Compound	Formula	Bond lengths in Å		Bond angles (°)	
Difluorochloromethane	$CHClF_2$	C–H	1.06	F–C–F	110.5 ± 1
		C–Cl	1.73 ± 0.03	Cl–C–Cl	110.5 ± 1
		C–F	1.36 ± 0.03		
1,1-Difluoroethylene	$C_2H_2F_2$	C–H	1.07 ± 0.02	F–C=C	125.2 ± 0.2
		C–F	1.321 ± 0.015	H–C–H	117 ± 7
		C–C	1.311 ± 0.035		
Difluoromethane	CH_2F_2	C–F	1.360 ± 0.005	F–C–F	108.2 ± 0.8
		C–H	1.09 ± 0.03	H–C–H	112.5 ± 6
p-Dinitrobenzene	$C_6H_4(NO_2)_2$	C–C	1.38	O–N=O	124.0
		C–N	1.48		
		N–O	1.21		
Dithio oxamide	$NH_2CSCSNH_2$	C–C	1.54	N–C=S	124.8
		C–N	1.30	S–C–C	124.87
		C–S	1.66	N–C–C	155.25
Ethane	C_2H_6	C–H	1.107	H–C–H	109.3
		C–C	1.536		
Ethylidene fluoride	CH_3CHF_2	C–F	1.345 ± 0.001	F–C–F	109.15 ± 0.001
		C–C	1.540	C–C–F	109.4
		C–H	1.100	H–C–C	110.2
Fluorochloromethane	CH_2ClF	C–H	1.078 ± 0.005	Cl–C–F	100.0 ± 0.1
		C–Cl	1.759 ± 0.003		
		C–F	1.378 ± 0.006		
Fluorotrichloromethane	$CFCl_3$	C–Cl	1.76 ± 0.02	Cl–C–Cl	113 ± 3
		C–F	1.44 ± 0.04		
Formaldehyde	CH_2O	C–H	1.060 ± 0.038	H–C–H	125.8 ± 7
		C–O	1.230 ± 0.017		
Formamide	$HCONH_2$	C–O	1.25₆	N–C=O	121.5
		C–N	1.300		
Formic acid	$HCOOH$	C–O'	1.245	O=C–O'	124.3
		C–O	1.312	H–C–O'	117.8
		C–H	1.085	C–O–H	107.8
		O–H	0.95		
Glycine	NH_2CH_2COOH	C–C	1.52	C–C–O	119.0
		C–O	1.27	O–C–O	122.0
		C–N	1.39	C–C–N	112.0
Hexachloroethane	C_2Cl_6	C–Cl	1.74 ± 0.01	Cl–C–Cl	109.3 ± 0.01
		C–C	1.57 ± 0.06		
Iodomethane	CH_3I	C–H	1.11 ± 0.01	H–C–H	111.4 ± 0.1
		C–I	2.139		
Methane	CH_4	C–H	1.091		
Methanethiol	CH_3SH	C–H	1.1040 ± 0.002		110.3 ± 0.2
		C–S	1.8177 ± 0.0002	H–S–C	100.3 ± 0.2
		S–H	1.329 ± 0.004		
Methanol	CH_3OH	C–H	1.096 ± 0.01	H–C–H	109.3 ± 0.75
		C–O	1.427 ± 0.007	C–O–H	108.9 ± 2
		O–H	0.956 ± 0.015		
Methylamine	CH_3NH_2	C–H	1.093	H–C–H	109.5 ± 1
		C–N	1.474 ± 0.005	H–N–H	105.8 ± 1
		N–H	1.014		
Methylether	$(CH_2)_2O$	C–O	1.43 ± 0.03	C–O–C	110.0 ± 3
Methylnitrite	CH_3NO_2	C–H	1.00	O–N=O	127 ± 4
		C–N	1.49 ± 0.02		
		N–O	1.22 ± 0.1		
Methylsulfide	$(CH_3)_2S$	C–S	1.82 ± 0.01	C–S–C	105 ± 3
		C–H	1.06	H–C–H	109.5
Oxamide	$NH_2COCONH_2$	C–C	1.54	N–C=O	125.7 ± 0.3
		C–O	1.24		
		C–N	1.32		
Phosgene	CCl_2O	C–Cl	4.745 ± 0.004	Cl–C=O	124.3 ± 0.1
		C–O	1.166 ± 0.002	Cl–C–Cl	111.3 ± 0.1
Propylene	C_3H_6			C–C=C	124.75 ± 0.3
Propynal	CHC.COH	C_1–C_2	1.204	C–C=O	123.0
		C_2–C_3	1.46	C–C–H	120.0
		C–O	1.21		
		C_1–H	1.06		
		C_3–H	1.08		
Tribromomethane	$CHBr_2$	C–H	1.068 ± 0.01	Br–C–Br	110.8 ± 0.3
		C–Br	1.930 ± 0.003		
Trichlorobromomethane	$CBrCl_3$	C–Br	1.936	Cl–C–Cl	111.2 ± 1
		C–Cl	1.764		
Trifluorochloromethane	$CClF_3$	C–Cl	1.751 ± 0.004	F–C–F	108.6 ± 0.4
		C–F	1.328 ± 0.02		
Trifluoromethane	CHF_3	C–H	1.098	F–C–F	108 ± 0.75
		C–F	1.332 ± 0.008		
Triiodomethane	CHI_3	C–I	2.12 ± 0.04	I–C–I	113.0 ± 1
Trimethylamine	$(CH_3)_3N$	C–N	1.47 ± 0.01	C–N–C	108.0 ± 4
		C–H	1.06	H–C–H	109.5
Trimethylarsine	$(CH_3)_3As$	C–H	1.09	C–As–C	96 ± 5
		C–As	1.98 ± 0.02		
Trimethylphosphine	$(CH_3)_3P$	C–H	1.09	C–P–C	100.0 ± 4
		C–P	1.87 ± 0.02		

STRENGTHS OF CHEMICAL BONDS*

J. A. Kerr

The strength of a chemical bond, $D(R - X)$, often known as the bond dissociation energy, is defined as the heat of the reaction: $RX \rightarrow R + X$. It is given by $D(R - X) = \Delta Hf°(R) + \Delta Hf°(X) - \Delta Hf°(RX)$. Some authors list bond strengths for O K but here the values for 298 K are given because more thermodynamic data are available for this temperature. Bond strengths, or bond dissociation energies, are not equal to, and may differ considerably from, mean bond energies derived solely from thermochemical data on molecules and atoms.

BOND STRENGTHS IN DIATOMIC MOLECULES

These have usually been measured spectroscopically or by mass spectrometric analysis of hot gases effusing from a Knudsen cell. Excellent accounts of these and other methods are given in Reference 109. The errors quoted are those given in the original paper or review article. The references have been chosen primarily as a key to the literature. It should not be assumed that the author referred to was responsible for the determination quoted, as the reference may be only to a review article.

$D°_0$ values have been converted to $D°_{298}$ by use of the simple relation

$$D°_{280} = D°_0 + (3/2)RT$$

The table has been arranged in alphabetical order with no regard to electronegativity or any other property of the atoms.

Table 1
BOND STRENGTHS IN DIATOMIC MOLECULES

Molecule	$D°_{298}$/kcal mol^{-1}	$D°_{298}$/kJ mol^{-1}	Ref.	Molecule	$D°_{298}$/kcal mol^{-1}	$D°_{298}$/kJ mol^{-1}	Ref.
Ag–Ag	39 ± 2	163 ± 8	66	Al–U	78 ± 7	326 ± 29	127
Ag–Al	43.9 ± 2.2	183.7 ± 9.2	64	Ar–Ar	1.13 ± 0.01	4.73 ± 0.04	52,201
Ag–Au	48.5 ± 2.2	202.9 ± 9.2	5	Ar–He	0.93	3.89	192
Ag–Bi	46 ± 10	193 ± 42	192	Ar–Hg	1.47	6.15	192
Ag–Br	70 ± 7	293 ± 29	109	Ar–I	2.4	10.0	31
Ag–Cl	81.6	341.4	141	Ar–K	1.0	4.2	172
Ag–Cu	41.6 ± 2.2	174.1 ± 9.2	114	As–As	91.3 ± 2.5	382.0 ± 10.5	193
Ag–D	54.2	226.8	172	As–Cl	107	448	66
Ag–Eu	30.3 ± 3	126.8 ± 12.6	50	As–D	64.6	270.3	172
Ag–F	84.7 ± 3.9	354.4 ± 16.3	109	As–F	98	410	172
Ag–Ga	43 ± 4	180 ± 15	32	As–Ga	50.1 ± 0.3	209.6 ± 1.2	71
Ag–Ge	41.7 ± 5.0	174.9 ± 20.9	216	As–H	84	352	75
Ag–H	54 ± 2	226 ± 8	109	As–N	139 ± 30	582 ± 126	66
Ag–Ho	29.5 ± 4.0	123.4 ± 16.7	45	As–O	115 ± 2	481 ± 8	66
Ag–I	56 ± 7	234 ± 29	109	As–P	103.6 ± 3.0	433.5 ± 12.6	131
Ag–In	42 ± 4	176 ± 17	25	As–S	~114	~478	103,224
Ag–Li	42.4 ± 1.5	177.4 ± 6.3	218	As–Sb	79.0 ± 1.3	330.5 ± 5.4	85a
Ag–Mn	24 ± 5	100 ± 21	192	As–Se	23	96	227
Ag–Na	33.1 ± 2.5	138.5 ± 10.5	236	As–Tl	47.4 ± 3.5	198 .3 ± 14.6	238
Ag–Nd	<50	<209	179	At–At	~19	~80	80
Ag–O	53.7	224.7	258	Au–Au	53.8 ± 0.5	221.3 ± 2.1	196
Ag–S	51.9	217.2	258	Au–B	87.9 ± 2.5	367.8 ± 10.5	121
Ag–Se	48.4	202.5	258	Au–Ba	38 ± 14	159 ± 59	109
Ag–Sn	32.5 ± 5.0	136.0 ± 20.9	2	Au–Be	68 ± 2	285 ± 8	109
Ag–Te	46.8	195.8	258	Au–Bi	70 ± 20	293 ± 84	192
Al–Al	44.5 ± 2.2	186.2 ± 9.2	37,142	Au–Ca	46 ± 23	193 ± 96	109
Al–As	42.9	179.5	230	Au–Ce	77.9 ± 3.5	325.9 ± 14.6	133
Al–Au	77.9 ± 1.5	325.9 ± 6.3	128	Au–Cl	82 ± 2.3	343 ± 9.6	109
Al–Br	106 ± 2	444 ± 8	66	Au–Co	51.3 ± 3.0	214.6 ± 12.6	177,258
Al–Cl	118 ± 3	494 ± 13	66	Au–Cr	51.4 ± 1.5	215.1 ± 6.3	66
Al–Cu	51.8 ± 2.5	216 ± 10.5	228	Au–Cu	56.3 ± 2.2	235.6 ± 9.2	5
Al–D	69.5	290.8	192	Au–D	76.1	318.1	172
Al–F	158.6 ± 1.5	663.6 ± 6.3	66	Au–Eu	57.6 ± 2.5	241.0 ± 10.5	50
Al–H	68.1 ± 1.5	284.9 ± 6.3	66	Au–Fe	44.7 ± 4.0	187.0 ± 16.7	177
Al–I	88.4 ± 0.5	369.9 ± 2.1	66	Au–Ga	70.2 ± 3.6	293.7 ± 15.1	32
Al–Li	42.0 ± 3.5	175.7 ± 14.6	142	Au–Ge	66.2 ± 3.5	277.0 ± 14.6	216
Al–N	71 ± 23	297 ± 96	109	Au–H	75 ± 2.3	314 ± 9.6	109
Al–O	121.3 ± 2.5	507.5 ± 10.5	107,225	Au–Ho	63.9 ± 4.0	267.4 ± 16.7	45,196
Al–P	51.8 ± 3.0	216.7 ± 12.6	66	Au–La	80.4 ± 5.0	336.4 ± 20.9	133
Al–Pd	60.8 ± 2.9	254.4 ± 12.0	48	Au–Li	68.0 ± 1.6	284.5 ± 6.5	218
Al–S	89.3 ± 1.9	373.6 ± 8.0	277	Au–Lu	79.4 ± 4.0	332.2 ± 16.7	111
Al–Sb	51.7 ± 1.4	216 ± 6.0	232	Au–Mg	58 ± 10	243 ± 42	192
Al–Se	80.7 ± 2.4	337.7 ± 10.1	277	Au–Mn	44.3 ± 3.0	185.4 ± 12.6	253
Al–Si	60 ± 3	251 ± 12	37	Au–Na	51.4 ± 3.0	215.1 ± 12.6	142
Al–Te	64.0 ± 2.4	267.8 ± 10.1	277	Au–Nd	71.5 ± 5.0	299.2 ± 20.9	133

* Revised to 30 June 1981.

Table 1 (continued)
BOND STRENGTHS IN DIATOMIC MOLECULES

Molecule	D°_{298}/kcal mol^{-1}	D°_{298}/kJ mol^{-1}	Ref.	Molecule	D°_{298}/kcal mol^{-1}	D°_{298}/kJ mol^{-1}	Ref.
Au–Ni	59 ± 5	247 ± 21	177	Bi–Se	67.0 ± 1.4	280.3 ± 5.9	276
Au–O	54.6	228.5	258	Bi–Te	55.5 ± 2.7	232.2 ± 11.3	276
Au–Pb	31 ± 10	130 ± 42	192	Bi–Tl	29 ± 3	121 ± 13	72
Au–Pd	34.2 ± 5.0	143.1 ± 20.9	6	Br–Br	46.333 ± 0.001	193.857 ± 0.004	14,172,202
Au–Pr	72.9 ± 5.0	305.0 ± 20.9	133	Br–C	67 ± 5	280 ± 21	109
Au–Rh	55.2 ± 7	230.9 ± 29	50	Br–Ca	74.3 ± 2.2	310.9 ± 9.2	165,187
Au–S	100 ± 6	418 ± 25	110	Br–Cd	38 ± 23	159 ± 96	109
Au–Sc	67.0 ± 4.0	280.3 ± 16.7	134	Br–Cl	52.3 ± 0.2	218.8 ± 0.8	109
Au–Se	58.1	243.1	258	Br–Co	79 ± 10	331 ± 42	192
Au–Si	74.5 ± 2.9	311.7 ± 12.1	281	Br–Cr	78.4 ± 5.8	328.0 ± 24.3	109
Au–Sn	58.4 ± 4.0	244.3 ± 16.7	2	Br–Cs	95.0 ± 1.0	397.5 ± 4.2	192
Au–Sr	63 ± 10	264 ± 42	192	Br–Cu	79 ± 6	331 ± 25	109
Au–Tb	70.1	293.3	114,196	Br–D	88.61	370.75	172
Au–Te	75.9	317.6	258	Br–F	59.8 ± 0.2	250.2 ± 0.6	41,43
Au–U	76 ± 7	318 ± 29	127	Br–Fe	59 ± 23	247 ± 96	109
Au–V	57.5 ± 2.9	240.6 ± 12.1	147	Br–Ga	106 ± 4	444 ± 17	66
Au–Y	72.0 ± 4.0	301.3 ± 16.7	134	Br–Ge	61 ± 7	255 ± 29	109
B–B	71 ± 5	297 ± 21	66	Br–H	87.55	366.31	172
B–Br	101 ± 5	423 ± 21	109	Br–Hg	17.4 ± 1	72.8 ± 4.2	66
B–C	107 ± 7	448 ± 29	192	Br–I	42.8 ± 0.1	179.1 ± 0.4	109
B–Ce	73 ± 5	305 ± 21	192	Br–In	99 ± 5	414 ± 21	66
B–Cl	128	536	12	Br–K	91.5 ± 2.0	382.8 ± 8.4	66
B–D	81.5 ± 1.5	341.0 ± 6.3	192	Br–Li	101 ± 5	423 ± 21	66
B–F	183 ± 3	766 ± 13	66	Br–Mg	⩽78.2	⩽327.2	172
B–H	79.8	333.9	174	Br–Mn	75.1 ± 2.3	314.2 ± 9.6	109
B–I	91 ± 5	381 ± 21	192	Br–N	66 ± 5	276 ± 21	109
B–Ir	122.9 ± 4.1	514.2 ± 17.2	281	Br–Na	88.5 ± 3.0	370.3 ± 12.6	66
B–La	81 ± 15	339 ± 63	192	Br–Ni	86 ± 3	360 ± 13	109
B–N	93 ± 5	389 ± 21	66	Br–O	56.2 ± 0.1	235.1 ± 0.4	66
B–O	192.7 ± 1.2	806.3 ± 5.0	275	Br–Pb	59 ± 9	247 ± 38	100
B–P	82.9 ± 4.0	346.9 ± 16.7	123	Br–Rb	93 ± 3	389 ± 13	109
B–Pd	78.7 ± 5.0	329.3 ± 20.9	281	Br–Sb	75 ± 14	314 ± 59	109
B–Pt	114.2 ± 4.0	477.8 ± 16.7	208	Br–Se	106 ± 15	444 ± 63	192
B–Rh	113.7 ± 5.0	475.7 ± 20.9	281	Br–Se	71 ± 20	297 ± 84	192
B–Ru	106.8 ± 5.0	446.9 ± 20.9	281	Br–Si	82 ± 12	343 ± 50	109
B–S	138.8 ± 2.2	580.7 ± 9.2	275	Br–Sn	⩾132	⩾552	223a
B–Sc	66 ± 15	276 ± 63	192	Br–Sr	79.6 ± 2.2	333.1 ± 9.2	165
B–Se	110.4 ± 3.5	461.9 ± 14.6	275	Br–Ti	105	439	192
B–Si	68.9	288.3	285	Br–Tl	79.8 ± 0.4	333.9 ± 1.7	22
B–Te	84.7 ± 4.8	354.4 ± 20.1	275	Br–V	105 ± 10	439 ± 42	192
B–Th	71	297	115	Br–W	78.7	329.3	176a
B–Ti	66 ± 15	276 ± 63	192	Br–Y	116 ± 20	485 ± 84	192
B–U	77 ± 8	322 ± 34	192	Br–Zn	34 ± 7	142 ± 29	192
B–Y	70 ± 15	293 ± 63	192	C–C	145 ± 5	607 ± 21	66
Ba–Br	86.4 ± 2.2	361.5 ± 9.2	166,187	C–Ce	109 ± 7	456 ± 29	120
Ba–Cl	104.2 ± 2.0	436.0 ± 8.4	162,166	C–Cl	95 ± 7	397 ± 29	222
Ba–D	⩽46.3	⩽193.7	172	C–D	81.6	341.4	172
Ba–F	140.3 ± 1.6	587.0 ± 6.7	161	C–F	132	552	156
Ba–H	42 ± 3.5	176 ± 15	109	C–Ge	110 ± 5	460 ± 21	109
Ba–I	72.3 ± 1.0	302.5 ± 4.2	73,189	C–H	80.8	338.1	172
Ba–O	130.4 ± 1.0	545.6 ± 4	92,95,261	C–Hf	129 ± 6	540 ± 25	265
Ba–Pd	53.0 ± 1.2	221.8 ± 5.0	129	C–I	50 ± 5	209 ± 21	109
Ba–Rh	62.0 ± 6.0	259.4 ± 25.1	129	C–Ir	149.3 ± 3.0	624.7 ± 12.6	208
Ba–S	95.6 ± 4.5	400.0 ± 18.8	56	C–La	121 ± 15	506 ± 63	192
Be–Be	14	59	78	C–N	184 ± 1	770 ± 4	67
Be–Br	91 ± 20	381 ± 84	192	C–O	257.3 ± 0.1	1076.5 ± 0.4	66
Be–Cl	92.8 ± 2.2	388.3 ± 9.2	99,171	C–Os	⩾142	⩾594	130
Be–D	48.53	203.06	172	C–P	122.7 ± 2	513.4 ± 8	259
Be–F	138 ± 10	577 ± 42	63,99	C–Pt	146.3 ± 1.5	612.1 ± 6.3	279
Be–H	47.8 ± 0.3	200.0 ± 1.3	53	C–Rh	139.5 ± 1.5	583.7 ± 6.3	279
Be–O	105.4 ± 3	441.0 ± 13	261	C–Ru	154.9 ± 3	648.1 ± 13	113
Be–S	89 ± 14	372 ± 59	109	C–S	167.0 ± 2.0	698.7 ± 8.4	84,154
Bi–Bi	47.9 ± 1.8	200.4 ± 7.5	240,243	C–Sc	94 ± 15	393 ± 63	192
Bi–Br	63.9 ± 1.0	267.4 ± 4.2	61	C–Se	141.1 ± 1.4	590.4 ± 5.9	254
Bi–Cl	72 ± 1	301 ± 8	63	C–Si	107.9	451.5	77,285
Bi–D	67.8	283.7	206	C–Th	116 ± 6	485 ± 25	265
Bi–F	62 ± 7	259 ± 29	109	C–Ti	104 ± 6	435 ± 25	265
Bi–Ga	38 ± 4	159 ± 17	235	C–U	111 ± 7	464 ± 29	120
Bi–H	⩽67.7	⩽283.3	206	C–V	112 ± 15	469 ± 63	120
Bi–I	52.1 ± 1.1	218.0 ± 4.6	62	C–Y	100 ± 15	418 ± 63	120
Bi–O	81.9 ± 1.4	342.7 ± 5.9	276	C–Zr	134 ± 6	561 ± 25	265
Bi–P	67 ± 3	280 ± 13	131	Ca–Ca	3.58 ± 0.11	14.98 ± 0.46	11
Bi–Pb	33.9 ± 3.5	141.8 ± 14.6	243	Ca–Cl	95 ± 3	398 ± 13	162
Bi–S	75.4 ± 1.1	315.5 ± 4.6	276	Ca–D	⩽40.6	⩽169.9	172
Bi–Sb	60 ± 1	251 ± 4	190	Ca–F	126 ± 5	527 ± 21	26,168

Table 1 (continued)

BOND STRENGTHS IN DIATOMIC MOLECULES

Molecule	D_{298}°/kcal mol^{-1}	D_{298}°/kJ mol^{-1}	Ref.	Molecule	D_{298}°/kcal mol^{-1}	D_{298}°/kJ mol^{-1}	Ref.
Ca–H	40.1	167.8	109	Co–S	79	331	258
Ca–I	63.0 ± 2.5	263.6 ± 10.5	189	Co–Si	66 ± 4	276 ± 17	280
Ca–O	92 ± 2	385 ± 8	92,96,261	Cr–Cr	37 ± 5	155 ± 21	181
Ca–S	80.7 ± 4.5	337.7 ± 18.8	56,172	Cr–Cu	37 ± 5	155 ± 21	183
Cd–Cd	2.7 ± 0.2	11.3 ± 0.8	109	Cr–F	106.3 ± 4.7	444.8 ± 19.7	186
Cd–Cl	49.8	208.4	172	Cr–Ge	40.6 ± 7	169.9 ± 29	182
Cd–F	73 ± 5	305 ± 21	23	Cr–H	67 ± 12	280 ± 50	109
Cd–H	16.5 ± 0.1	69.0 ± 0.4	109	Cr–I	68.6 ± 5.8	287.0 ± 24.3	109
Cd–I	33 ± 5	138 ± 21	109	Cr–N	90.3 ± 4.5	377.8 ± 18.8	262
Cd–In	33	138	192	Cr–O	102 ± 7	427 ± 29	66
Cd–O	90.1	377.0	139	Cr–S	79	331	83
Cd–S	48	201	209	Cs–Cs	9.97 ± 0.22	41.75 ± 0.93	172
Cd–Se	46.8	195.8	136	Cs–F	122.9 ± 2	514.2 ± 8	192
Ce–Ce	58.6	245.2	133	Cs–H	42.6 ± 0.9	178.2 ± 3.8	242,271
Ce–F	139 ± 10	582 ± 42	192	Cs–Hg	2	5	172
Ce–Ir	140	586	126	Cs–I	81 ± 1	339 ± 4	192
Ce–N	124 ± 5	519 ± 21	122	Cs–O	71 ± 6	297 ± 25	192
Ce–O	190 ± 3	795 ± 13	4	Cu–Cu	48.2 ± 2	201.7 ± 4	228
Ce–Os	121 ± 8	507 ± 33	130	Cu–D	64.6	270.3	172
Ce–Pd	77.0	322.2	46	Cu–F	98.8 ± 3	413.4 ± 13	91
Ce–Pt	133	557	126	Cu–Ga	51.6 ± 3.6	215.9 ± 15.1	32
Ce–Rh	131	548	126	Cu–Ge	49.9 ± 5	208.8 ± 21	216
Ce–Ru	127 ± 6	531 ± 25	130	Cu–H	64	268	29,172
Ce–S	136	569	19,57	Cu–I	47 ± 5	197 ± 21	109
Ce–Se	118.2 ± 3.5	494.4 ± 14.6	215	Cu–Li	46.1 ± 2.1	192.9 ± 8.8	218
Ce–Te	93 ± 10	389 ± 42	192	Cu–Mn	37.9 ± 4	158.6 ± 17	183
Cl–Cl	58.062 ± 0.001	242.933 ± 0.004	172,202	Cu–Na	42.1 ± 4.0	176.2 ± 16.7	237
Cl–Co	93	389	199	Cu–Ni	49.2 ± 4	205.9 ± 17	183
Cl–Cr	87.5 ± 5.8	366.1 ± 24.3	109	Cu–O	65.2	272.8	17,258
Cl–Cs	105 ± 5	439 ± 21	66	Cu–S	66	276	258
Cl–Cu	91.5 ± 1.1	382.8 ± 21	144	Cu–Se	60	251	258
Cl–D	104.32	436.47	172	Cu–Sn	42.3 ± 4	177.0 ± 17	2
Cl–Eu	~78	~326	105	Cu–Te	66.6	278.7	1
Cl–F	61.24	256.3	172,219	D–D	105.957	443.324	172
Cl–Fe	~84	~352	109	D–F	137.8	576.6	172
Cl–Ga	115 ± 3	481 ± 13	66	D–Ga	<65.2	<272.8	197
Cl–Ge	~103	~431	172	D–Ge	≤77	≤322	172
Cl–H	103.13	431.49	172	D–H	104.978 ± 0.001	439.228 ± 0.004	172
Cl–Hg	24 ± 2	100 ± 8	109	D–Hg	10.05	42.08	172
Cl–I	50.5 ± 0.1	211.3 ± 0.4	109	D–In	58.8	246.0	172
Cl–In	105 ± 2	439 ± 8	66	D–Li	57.4066 ± 0.0011	240.1892 ± 0.0046	173,269
Cl–K	102 ± 2	427 ± 8	66	D–Mg	32.3	135.1	172
Cl–Li	112 ± 3	469 ± 13	66	D–Ni	≤72.4	≤302.9	172
Cl–Mg	78.3 ± 0.5	327.6 ± 2.1	94,162	D–Pt	≤83.7	≤350.2	172
Cl–Mn	86.2 ± 2.3	360.7 ± 9.6	109	D–S	84	352	172
Cl–N	79.8 ± 2.3	333.9 ± 9.6	39	D–Si	72.3	302.5	172
Cl–Na	98 ± 2	410 ± 8	66	D–Sr	≥65.9	≥275.7	172
Cl–Ni	89 ± 5	372 ± 21	109	D–Zn	21.2	88.7	172
Cl–O	65 ± 1	272 ± 4	66	Dy–F	127	531	296
Cl–P	69 ± 10	289 ± 42	192	Dy–O	146 ± 10	611 ± 42	27
Cl–Pb	72 ± 7	301 ± 29	109	Dy–S	99 ± 10	414 ± 42	192
Cl–Ra	82 ± 18	343 ± 75	109	Dy–Se	77 ± 10	322 ± 42	192
Cl–Rb	107 ± 5	448 ± 21	66	Dy–Te	56 ± 10	234 ± 42	192
Cl–Sb	86 ± 12	360 ± 50	109	Er–F	135 ± 4	565 ± 17	296
Cl–Sc	79	331	284	Er–O	146 ± 3	611 ± 13	251
Cl–Se	77	322	192	Er–S	100 ± 10	418 ± 42	192
Cl–Si	109 ± 10	456 ± 42	192	Er–Se	78 ± 10	326 ± 42	192
Cl–Sm	≥101 ± 3	≥423 ± 13	292	Er–Te	57 ± 10	239 ± 42	192
Cl–Sn	99 ± 4	414 ± 17	192	Eu–Eu	8.0 ± 4.0	33.5 ± 16.7	50
Cl–Sr	97 ± 3	406 ± 13	162,166	Eu–F	126.1 ± 4.4	527.6 ± 18.4	74,294
Cl–Ti	118	494	192	Eu–O	112.2 ± 2.4	469.5 ± 10.0	8,74,214
Cl–Tl	89.1 ± 0.5	372.8 ± 2.1	22	Eu–Rh	55.9 ± 8.0	233.9 ± 33.5	50
Cl–V	114 ± 15	477 ± 63	192	Eu–S	86.6 ± 3.1	362.3 ± 13.0	215,256
Cl–W	101 ± 10	423 ± 42	192	Eu–Se	72 ± 3.5	301 ± 14.6	18,215
Cl–Xe	1.6	6.7	172	Eu–Te	58 ± 3.5	243 ± 14.6	18,215
Cl–Y	126 ± 20	527 ± 42	192	F–F	37.83	158.3	68,172
Cl–Yb	~77	~322	105	F–Ga	138 ± 3.5	577 ± 14.6	212
Cl–Zn	54.7 ± 4.7	228.9 ± 19.7	58	F–Gd	141.1 ± 6.5	590.4 ± 27.2	294
Cm–O	176	736	250	F–Ge	116 ± 5	485 ± 21	90
Co–Co	40 ± 6	167 ± 25	180	F–H	136.2	569.9	172
Co–Cu	38.7 ± 4.0	161.9 ± 16.7	183	F–Hg	~43	~180	172
Co–F	104 ± 15	435 ± 63	192	F–Ho	129	540	296
Co–Ge	57 ± 6	239 ± 25	182	F–I	≤64.9	≤271.6	7,42,60
Co–I	68 ± 5	285 ± 21	192	F–In	121 ± 3.5	506 ± 14.6	212
Co–O	88 ± 5	368 ± 21	27	F–K	118.9 ± 0.6	497.5 ± 2.5	13

Table 1 (continued)
BOND STRENGTHS IN DIATOMIC MOLECULES

Molecule	D°_{298}/kcal mol⁻¹	D°_{298}/kJ mol⁻¹	Ref.	Molecule	D°_{298}/kcal mol⁻¹	D°_{298}/kJ mol⁻¹	Ref.
F–La	143 ± 10	598 ± 42	192	H–P	71	297	172
F–Li	138 ± 5	577 ± 21	66	H–Pb	42 ± 5	176 ± 21	172
F–Lu	136 ± 10	569 ± 42	192	H–Pt	≤80	≤335	172
F–Mg	110.4 ± 1.2	461.9 ± 5.0	161	H–Rb	40 ± 5	167 ± 21	109
F–Mn	101.2 ± 3.5	423.4 ± 14.6	185	H–S	82.3 ± 2.9	344.3 ± 12.1	173
F–Mo	111.1	464.8	164	H–Sc	~43	~180	246
F–N	82	342	172	H–Se	73 ± 0.5	305 ± 2.1	145
F–Na	124	519	172	H–Si	≤71.5	≤299.2	172
F–Nd	130.3 ± 3.0	545.2 ± 12.6	293	H–Sn	63 ± 4	264 ± 17	109
F–Ni	104	435	192	H–Sr	39 ± 2	163 ± 8	109
F–O	53 ± 4	222 ± 17	44	H–Te	64 ± 0.5	268 ± 2.1	145
F–P	105 ± 23	439 ± 96	109	H–Ti	~38	~159	245
F–Pb	85 ± 2	356 ± 8	292	H–Tl	45 ± 2	188 ± 8	109
F–Pm	129 ± 10	540 ± 42	192	H–Yb	38 ± 9	159 ± 38	109
F–Pr	139 ± 11	582 ± 46	192	H–Zn	20.5 ± 0.5	85.8 ± 2.1	109
F–Pu	128.7 ± 7	538.5 ± 29	184	He–He	0.9	3.8	172
F–Rb	118 ± 5	493.7 ± 21	66	He–Hg	1.58	6.61	192
F–S	81.9 ± 1.2	342.7 ± 5.0	167	Hf–C	131 ± 15	548 ± 63	192
F–Sb	105 ± 23	439 ± 96	109	Hf–N	128 ± 7	534 ± 29	191
F–Sc	140.8 ± 3	589.1 ± 13	295	Hf–O	189.8 ± 2.0	794.1 ± 8.4	4
F–Se	81 ± 10	339 ± 42	192	Hg–Hg	4.1 ± 0.5	17.2 ± 2.1	109
F–Si	132.1 ± 0.5	552.7 ± 2.1	100	Hg–I	9	38	289
F–Sm	126.9 ± 4.4	531.0 ± 18.4	74,291,294	Hg–K	1.97 ± 0.05	8.24 ± 0.21	192
F–Sn	111.5 ± 3	466.5 ± 13	292	Hg–Li	3.3	13.9	172
F–Sr	129.5 ± 1.6	541.8 ± 6.7	161	Hg–Na	2.2	9.2	172,300
F–Tb	134 ± 10	561 ± 42	192	Hg–Rb	2.0	8.4	172
F–Ti	136 ± 8	569 ± 34	297	Hg–S	51	213	209
F–Tl	106.4 ± 4.6	445.2 ± 19.3	22	Hg–Se	≤40	≤167	209
F–Tm	136 ± 10	569 ± 42	192	Hg–Te	≤34	≤142	192
F–V	141 ± 15	590 ± 63	192	Hg–Tl	1	4	151
F–W	131 ± 15	548 ± 63	192	Ho–Ho	20 ± 4	84 ± 17	45
F–Xe	4.0 ± 0.1	16.8 ± 0.4	273	Ho–O	148	619	251
F–Y	144.6 ± 5.0	605.0 ± 20.9	295	Ho–S	102.4 ± 3.5	428.4 ± 14.6	251
F–Yb	≥124.6 ± 2.3	≥521.3 ± 9.6	14,105,290	Ho–Se	80 ± 4	335 ± 17	18
F–Zn	88 ± 15	368 ± 63	192	Ho–Te	62 ± 4	259 ± 17	18
F–Zr	149 ± 15	623 ± 63	192	I–I	36.456	152.532	172,202
Fe–Fe	24 ± 5	100 ± 21	204	I–In	80	335	12
Fe–Ge	50.4 ± 7	210.9 ± 29	182	I–K	79 ± 3	331 ± 13	66
Fe–O	97.7 ± 3	408.8 ± 13	157	I–Li	84 ± 3	352 ± 13	66
Fe–S	77	322	83	I–Mg	~68	~285	20
Fe–Si	71 ± 6	297 ± 25	280	I–Mn	67.6 ± 2.3	282.8 ± 9.6	109
Ga–Ga	33 ± 5	138 ± 21	192	I–N	38 ± 4	159 ± 17	192
Ga–H	<65.5	<274.1	197	I–Na	72 ± 2	301 ± 8	66
Ga–I	81 ± 2.3	339 ± 9.6	109	I–Ni	70 ± 5	293 ± 21	109
Ga–Li	31.8 ± 3.5	133.1 ± 14.6	142	I–O	43	180	172
Ga–O	91	381	30	I–Pb	47 ± 9	197 ± 38	109
Ga–P	54.9 ± 3.0	229.7 ± 12.6	137	I–Rb	80 ± 3	335 ± 13	66
Ga–Sb	45.9 ± 3.0	208.8 ± 12.6	231	I–Si	70	293	172
Ga–Te	60 ± 6	251 ± 25	278	I–Sn	56 ± 10	234 ± 42	192
Gd–O	171 ± 4	716 ± 17	251	I–Sr	64.5 ± 1.4	269.9 ± 5.9	189
Gd–S	125.9 ± 2.5	526.8 ± 10.5	106,251	I–Te	46 ± 10	193 ± 42	192
Gd–Se	103 ± 3.5	431 ± 14.6	18	I–Ti	74 ± 10	310 ± 42	192
Gd–Te	82 ± 3.5	343 ± 14.6	18	I–Tl	65 ± 2	272 ± 8	20
Ge–Ge	65.4 ± 5	273.6 ± 21	216	I–Zn	33 ± 7	138 ± 29	109
Ge–H	≤76.9	≤321.8	16,189a	I–Zr	73	305	188
Ge–Ni	67 ± 3	280 ± 13	183	In–In	24 ± 2	100 ± 8	192
Ge–O	157.3	658.1	172	In–Li	22.1 ± 3.5	92.5 ± 14.6	142
Ge–Pd	63.2	264.4	226	In–O	86 ± 5	360 ± 21	66
Ge–S	131.7 ± 0.6	551.0 ± 2.5	57,79,84	In–P	47.3 ± 2.0	197.9 ± 8.5	230
Ge–Se	115.7	484.1	172	In–S	69 ± 4	289 ± 17	55
Ge–Si	72 ± 5	301 ± 21	109	In–Sb	36.3 ± 2.5	151.9 ± 10.5	69
Ge–Te	99	414	53a	In–Se	59 ± 4	247 ± 17	55
H–H	104.155	435.783	172	In–Te	52 ± 4	218 ± 17	55
H–Hg	9.523	39.843	172	Ir–O	84 ± 5	352 ± 21	192
H–I	71.32	298.39	172	Ir–Si	110.6 ± 5.0	462.8 ± 20.9	281
H–In	58.1	243.0	172	Ir–Th	137	573	112
H–K	43.8 ± 3.5	183.3 ± 14.6	172	K–K	13.7 ± 1.0	57.3 ± 4.2	192
H–Li	56.895 ± 0.001	238.049 ± 0.004	286	K–Kr	1.1	4.6	172
H–Mg	30.2 ± 0.7	126.4 ± 5.9	9,10,89	K–Li	19.6 ± 1.0	82.0 ± 4.2	299
H–Mn	56 ± 7	234 ± 29	109	K–Xe	1.2	5.1	172
H–N	≤81	≤339	172	K–Na	15.5 ± 1.0	64.9 ± 4.2	299
H–Na	44	184	172	K–O	66 ± 3	276 ± 13	88
H–Ni	≤72	≤301	172	Kr–Kr	1.25	5.23	35,172
H–O	102.2	427.5	172	Kr–O	<2	<8	192

Table 1 (continued)
BOND STRENGTHS IN DIATOMIC MOLECULES

Molecule	D°_{298}/kcal mol⁻¹	D°_{298}/kJ mol⁻¹	Ref.	Molecule	D°_{298}/kcal mol⁻¹	D°_{298}/kJ mol⁻¹	Ref.
La–La	59 ± 5	247 ± 21	284	O–Si	193.5 ± 2.6	809.6 ± 10.9	200
La–N	124 ± 10	519 ± 42	192	O–Sm	136.9 ± 2.0	572.8 ± 8.4	74,159,251
La–O	191 ± 3	799 ± 13	138,251	O–Sn	131 ± 5	548 ± 21	66
La–Rh	126 ± 4	527 ± 17	49	O–Sr	99.4 ± 2	415.9 ± 8	92,96,261
La–S	137.0 ± 0.4	573.4 ± 1.8	57,176,268	O–Ta	192.4 ± 3	805.0 ± 13	255
La–Se	114 ± 4	477 ± 17	18,215	O–Tb	169 ± 3	707 ± 13	251
La–Te	91 ± 4	381 ± 17	18	O–Te	93.4 ± 2.0	390.8 ± 8.4	211,298
La–Y	48.3	202.1	284	O–Th	204 ± 3	854 ± 13	170,217
Li–Li	26.34 ± 1	106.48 ± 4.2	282,290	O–Ti	159.3 ± 1.5	666.5 ± 6.3	86,158,207
Li–Na	21.6 ± 1.5	90.4 ± 6.3	299	O–Tm	133 ± 3	557 ± 13	251
Li–O	81.4 ± 1.5	340.6 ± 6.3	163,198	O–U	181.9 ± 4.0	761.1 ± 16.7	267
Lu–Lu	34 ± 8	142 ± 34	192	O–V	149	623	101
Lu–O	166 ± 3	695 ± 13	251	O–W	156 ± 6	653 ± 25	27
Lu–Pt	96 ± 8	402 ± 34	116	O–Xe	8.7	36.4	192
Lu–S	121.2 ± 3.5	507.1 ± 14.6	251	O–Y	170.9 ± 3.0	715.1 ± 3.0	34,267
Lu–Se	100 ± 4	418 ± 17	18	O–Yb	99.8 ± 1.2	417.6 ± 5.0	59,291
Lu–Te	78 ± 4	326 ± 17	18	O–Zn	67.9	284.1	140
Mg–Mg	2.044 ± 0.001	8.552 ± 0.004	203	O–Zr	181.6 ± 2	759.8 ± 8.4	4,213
Mg–O	86.6 ± 0.5	362.3 ± 2.1	93,97	P–P	117.0 ± 2.5	489.5 ± 10.5	125
Mg–S	56	234	56	P–Pt	≤99.6 ± 4	≤416.7 ± 17	257
Mn–Mn	6.2	25.9	179	P–Rh	84.4 ± 4	353.1 ± 17	257
Mn–O	86	360	57	P–S	106 ± 2	444 ± 8	81
Mn–S	72 ± 4	301 ± 17	287	P–Sb	85.3	356.9	195
Mn–Se	57.2 ± 2.2	239.3 ± 9.2	260,288	P–Se	86.9 ± 2.4	363.6 ± 10	81
Mo–Mo	97 ± 5	406 ± 20	145a	P–Si	86.9	363.6	252
Mo–Nb	109 ± 6	454 ± 25	145a	P–Te	71.2 ± 2.4	297.9 ± 10.0	81
Mo–O	145.1 ± 8	607.1 ± 34	38	P–Th	90	377	115
N–N	225.94 ± 0.14	945.33 ± 0.59	172	P–Tl	50 ± 3	209 ± 13	232
N–O	150.71 ± 0.03	630.57 ± 0.13	172	P–U	71 ± 5	297 ± 21	192
N–P	147.5 ± 5.0	617.1 ± 20.9	125	P–W	73 ± 1	305 ± 4	124
N–Pu	113 ± 15	473 ± 63	192	Pb–Pb	81 ± 6	339 ± 25	132
N–S	111 ± 5	464 ± 21	192	Pb–S	82.7 ± 0.4	346.0 ± 1.7	276
N–Sb	72 ± 12	301 ± 50	109	Pb–Se	72.4 ± 1	302.9 ± 4	276
N–Sc	112 ± 20	469 ± 84	192	Pb–Te	60 ± 3	251 ± 13	276
N–Se	91 ± 15	381 ± 63	192	Pd–Pd	17	75	205
N–Si	105 ± 9	439 ± 38	109	Pd–Si	74.9 ± 3.3	313.4 ± 13.8	281
N–Ta	146 ± 20	611 ± 84	192	Pm–S	101 ± 15	423 ± 63	192
N–Th	138.0 ± 0.5	577.4 ± 2.1	119	Pm–Se	81 ± 15	339 ± 63	192
N–Ti	111	464	263	Pm–Te	61 ± 15	255 ± 63	192
N–U	127.0 ± 0.5	531.4 ± 2.1	117	Po–Po	44.7	187.0	172
N–V	114.1 ± 2	477.4 ± 8	98	Pr–S	117.7 ± 1.1	492.5 ± 4.6	104
N–Xe	5.5	23.0	150	Pr–Se	106.7 ± 5.5	446.4 ± 23.0	215
N–Y	115 ± 15	481 ± 63	192	Pr–Te	78 ± 10	326 ± 42	192
N–Zr	135 ± 6	565 ± 25	119	Pt–Si	119.8 ± 4.3	501.2 ± 18.0	281
Na–Na	18.4	77.0	249	Pt–Th	132	552	112
Na–O	61.3 ± 4	256.5 ± 17	169	Rb–Rb	10.9 ± 0.5	45.6 ± 2.1	234
Na–Rb	14 ± 0.9	59 ± 3.8	109	Rh–Rh	68.2 ± 5.0	285.4 ± 20.9	47,233
Nb–O	180 ± 3	753 ± 13	192	Rh–Si	94.4 ± 4.3	395.0 ± 18.0	281
Nd–Nd	<39	<163	192	Rh–Th	123 ± 5	513 ± 21	135
Nd–O	168 ± 8	703 ± 34	251	Rh–Ti	93.4 ± 3.5	390.8 ± 14.6	47
Nd–S	112.7	471.5	18	Rh–U	124 ± 4	519 ± 17	135
Nd–Se	92 ± 4	385 ± 17	18,215	Ru–Si	94.9 ± 5.0	397.1 ± 20.9	281
Nd–Te	73 ± 4	305 ± 17	18	Ru–Th	141.4 ± 10	591.6 ± 42	113
Ne–Ne	0.94	3.93	272	S–S	101.65	425.28	172
Ni–Ni	62.6 ± 0.6	261.9 ± 2.5	244	S–Sb	90.5	378.7	102
Ni–O	93.6 ± 0.9	391.6 ± 3.8	244	S–Sc	114.3 ± 3.0	478.2 ± 12.6	268,274
Ni–S	82.3	344.3	83	S–Se	88.7 ± 1.6	371.1 ± 6.7	85
Ni–Si	76 ± 4	318 ± 17	280	S–Si	149	623	172
Np–O	172 ± 7	720 ± 29	3	S–Sm	93	389	104
O–O	119.016 ± 0.048	498.340 ± 0.200	28,51	S–Sn	111 ± 0.8	464 ± 3.2	79
O–Os	<142	<594	27	S–Sr	81	339	33
O–P	142.6	596.6	82	S–Tb	123 ± 10	515 ± 42	192
O–Pb	90.3 ± 1.0	377.8 ± 4.2	221,276	S–Te	81 ± 5	339 ± 21	79
O–Pd	56 ± 7	234 ± 29	27	S–Ti	101.8 ± 1.8	425.9 ± 7.5	87
O–Pm	161 ± 15	674 ± 63	192	S–Tm	88 ± 10	368 ± 42	192
O–Pr	180 ± 4	753 ± 17	251	S–U	124.9 ± 2.3	522.6 ± 9.6	267
O–Pt	83 ± 8	347 ± 34	27	S–V	117 ± 4	490 ± 16	223
O–Pu	163 ± 15	682 ± 63	27	S–Y	126.3 ± 2.5	528.4 ± 10.5	57,266
O–Rb	61 ± 20	255 ± 84	27	S–Yb	40	167	192
O–Rh	90 ± 15	377 ± 63	27	S–Zn	49 ± 3	205 ± 13	70
O–Ru	115 ± 15	481 ± 63	27	S–Zr	137.5 ± 4.0	575.3 ± 16.7	267
O–S	124.69 ± 0.03	521.70 ± 0.13	27	Sb–Sb	71.5 ± 1.5	299.2 ± 6.3	69,194
O–Sb	89 ± 20	372 ± 84	27	Sb–Te	66.3 ± 0.9	277.4 ± 3.8	239,270
O–Sc	164.4 ± 2.3	687.9 ± 9.6	34	Sb–Tl	26.1	109.2	232
O–Se	≤102.6	≤429.3	172	Sc–Sc	38.9 ± 5	162.8 ± 21	114

Table 1 (continued)
BOND STRENGTHS IN DIATOMIC MOLECULES

Molecule	D_{298}°/kcal mol^{-1}	D_{298}°/kJ mol^{-1}	Ref.	Molecule	D_{298}°/kcal mol^{-1}	D_{298}°/kJ mol^{-1}	Ref.
Sc—Se	92 ± 4	385 ± 17	192	Sn—Te	86.0	359.8	172
Sc—Te	69 ± 4	289 ± 17	192	Tb—Tb	31.4 ± 6.0	131.4 ± 25.1	196
Se—Se	79.5 ± 0.1	332.6 ± 0.4	85,276	Tb—Te	81 ± 10	339 ± 42	192
Se—Si	131	548	172	Te—Te	62.6	261.9	172,276
Se—Sm	79.1 ± 3.5	331.0 ± 14.6	215	Te—Ti	69 ± 4	289 ± 17	192
Se—Sn	95.9 ± 1.4	401.3 ± 5.9	54	Te—Tm	66 ± 10	276 ± 42	192
Se—Sr	~68	~285	21	Te—Y	81 ± 3	339 ± 13	192
Se—Tb	101 ± 10	423 ± 42	192	Te—Zn	23	96	172
Se—Te	71.0 ± 1.2	297.1 ± 5.0	79,85	Th—Th	≤69	≤289	115
Se—Ti	91 ± 10	381 ± 42	192	Ti—Ti	33.8 ± 5	141.4 + 21	178
Se—Tm	66 ± 10	276 ± 42	192	Tl—Tl	<22	<92	172
Se—V	83 ± 5	347 ± 21	192	U—U	53 ± 5	222 ± 21	192
Se—Y	104 ± 3	435 ± 13	192	V—V	57.9 ± 5	242.3 ± 21	178
Se—Zn	32.6 ± 3.0	136.4 ± 12.6	70	Xe—Xe	1.56 ± 0.07	6.53 ± 0.30	36
Si—Si	78.1 ± 2.4	326.8 ± 10.0	37,172	Y—Y	38 ± 5	159 ± 21	192
Si—Te	108	452	172	Yb—Yb	4.9 ± 4	20.5 ± 17	143
Sm—Te	65.1 ± 3.5	272.4 ± 14.6	215	Zn—Zn	7	29	249
Sn—Sn	46.7 ± 4	195.4 ± 17	2				

REFERENCES

1. Abbasov, A. S., Azizov, T. Kh., Alleva, N. A., Aliev, I. Ya., Mustafaev, F. M., and Mamedov, A. N., *Zh. Fiz. Khim.*, 50, 2172, 1976.
2. Ackerman, M., Drowart, J., Stafford, F. E., and Verhaegen, G., *J. Chem. Phys.*, 36, 1557, 1962.
3. Ackermann, R. J., Faircloth, R. F., Rauh, E. G., and Thorn, R. J., *J. Inorg. Nucl. Chem.*, 28, 111, 1966.
4. Ackermann, R. J. and Rauh, E. G., *J. Chem. Phys.*, 60, 2266, 1974.
5. Ackerman, M., Stafford, F. E., and Drowart, J., *J. Chem. Phys.*, 33, 1784, 1960.
6. Ackerman, M., Stafford, F. E., and Verhaegen, G., *J. Chem. Phys.*, 36, 1560, 1962.
7. Appelman, E. H. and Clyne, M. A. A., *J. Chem. Soc. Faraday Trans. 1*, 71, 2072, 1975.
8. Balducci, G., Gigli, G., and Guido, M., *J. Chem. Phys.*, 67, 147, 1977.
9. Balfour, W. J. and Cartwright, H. M., *Astron. Astrophys. Suppl. Ser.*, 26, 389, 1976.
10. Balfour, W. J. and Lingren, B., *Can. J. Chem.*, 56, 767, 1978.
11. Balfour, W. J. and Whitlock, R. E., *J. Chem. Soc. D*, p. 1231, 1971.
12. Barrow, R. F., *Trans. Faraday Soc.*, 56, 952, 1960.
13. Barrow, R. F. and Caunt, A. D., *Proc. R. Soc. London Ser. A*, 219, 120, 1953.
14. Barrow, R. F., Clark, T. C., Coxon, J., and Yee, K. K., *J. Mol. Spectrosc.*, 51, 428, 1974.
15. Barrow, R. F. and Chojnicki, A. H., *J. Chem. Soc. Faraday Trans. 2*, 71, 728, 1975.
16. Barrow, R. F. and Deutsch, E. W., *Proc. Chem. Soc.*, p. 122, 1960.
17. Belyaev, V. N., Lebedeva, N. L., Krasnov, K. S., and Gurvich, L. V., *Uchebn. Zaved. Khim. Khim. Fekhnol.*, 21, 1698, 1978.
18. Bergman, C., Coppens, P., Drowart, J., and Smoes, S., *Trans. Faraday Soc.*, 66, 800, 1970.
19. Bergman, C. and Gingerich, K. A., *J. Phys. Chem.*, 76, 2332, 1972.
20. Berkowitz, J. and Chupka, W. A., *J. Chem. Phys.*, 45, 1287, 1966.
21. Berkowitz, J. and Chupka, W. A., *J. Chem. Phys.*, 45, 4289, 1966.
22. Berkowitz, J. and Walter, T., *J. Chem. Phys.*, 49, 1184, 1968.
23. Besenbruch, G., Kana'an, A. S., and Margrave, J. L., *J. Phys. Chem.*, 69, 3174, 1965.
24. Birks, J. W., Gabelnick, S. D., and Johnston, H. S., *J. Mol. Spectrosc.*, 57, 23, 1975.
25. Biron, M., *C. R. Acad. Ser. B*, 265, 1026, 1427, 1967.
26. Blue, G. D., Green, J. W., Bautista, R. G., and Margrave, J. L., *J. Phys. Chem.*, 67, 877, 1963,
27. Brewer, L. and Rosenblatt, G. M., *Adv. High Temp. Sci.*, 2, 1, 1969.
28. Brix, P. and Herzberg, G., *J. Chem. Phys.*, 21, 2240, 1953.
29. Bulewicz, E. M. and Sugden, T. M., *Trans. Faraday Soc.*, 52, 1475, 1956.
30. Burns, R. P., *J. Chem. Phys.*, 44, 3307, 1966.
31. Burns, G., Le Roy, L. J., Morris, D. J., and Blake, J. A., *Proc. R. Soc. London Ser. A*, 316, 81, 1970.
32. Carbonel, M., Bergman, C., and Laffite, M., *Colloq. Int. Cent. Nat. Rech. Sci.*, 210, 311, 1972.
33. Cater, E. D. and Johnson, E. W., *J. Chem. Phys.*, 47, 5353, 1967.
34. Chalek, C. L. and Gole, J. L., *Chem. Phys.*, 19, 59, 1977.
35. Chashchina, G. I. and Shreider, E. Ya., *Zh. Prikl. Spektrosk.*, 21, 696, 1974.
36. Chashchina, G. I. and Shreider, E. Ya., *Zh. Prikl. Spektrosk.*, 25, 163, 1976.
37. Chatillon, C., Allibert, M., and Pattoret, A., *C. R. Acad. Sci. Ser. C*, 280, 1505, 1975.
38. Choudary, U. V., Gingerich, K. A., and Kingcade, J. E., *J. Less Common Met.*, 42, 111, 1975.
39. Clarke, T. C. and Clyne, M. A. A., *Trans. Faraday Soc.*, 66, 877, 1970.
40. Clements, R. M. and Barrow, R. F., *Trans. Faraday Soc.*, 64, 2893, 1968.
41. Clyne, M. A. A., Curran, A. H., and Coxon, J. A., *J. Mol. Spectrosc.*, 63, 43, 1976.
42. Clyne, M. A. A. and McDermid, I. S., *J. Chem. Soc. Faraday Trans. 2*, 72, 2252, 1976.

43. Clyne, M. A. A. and McDermid, I. S., *J. Chem. Soc. Faraday Trans. 2*, 74, 644, 1978.
44. Clyne, M. A. A. and Watson, R. T., *Chem. Phys. Lett.*, 12, 344, 1971.
45. Cocke, D. L. and Gingerich, K. A., *J. Phys. Chem.*, 75, 3264, 1971.
46. Cocke, D. L. and Gingerich, K. A., *J. Phys. Chem.*, 76, 2332, 1972.
47. Cocke, D. L. and Gingerich, K. A., *J. Chem. Phys.*, 60, 1958, 1974.
48. Cocke, D. L., Gingerich, K. A., and Chang, C.-A., *J. Chem. Soc. Faraday Trans. 1*, 72, 268, 1976.
49. Cocke, D. L., Gingerich, K. A., and Kordis, J., *High Temp. Sci.*, 5, 474, 1973.
50. Cocke, D. L., Gingerich, K. A., and Kordis, J., *High Temp. Sci.*, 7, 61, 1975.
51. CODATA recommended key values for thermodynamics, 1973, *J. Chem. Thermodyn.*, 7, 1, 1975.
52. Colbourn, E. A. and Douglas, A. E., *J. Chem. Phys.*, 65, 1741, 1976.
53. Colin, R. and De Greef, D., *Can. J. Phys.*, 53, 2142, 1975.
53a. Colin, R. and Drowart, J., *J. Phys. Chem.*, 68, 428, 1964.
54. Colin, R. and Drowart, J., *Trans. Faraday Soc.*, 60, 673, 1964.
55. Colin, R. and Drowart, J., *Trans. Faraday Soc.*, 64, 2611, 1968.
56. Colin, R., Goldfinger, P., and Jeunehomme, M., *Trans. Faraday Soc.*, 60, 306, 1964.
57. Coppens, P., Smoes, S., and Drowart, J., *Trans. Faraday Soc.*, 63, 2140, 1967.
58. Corbett, J. D. and Lynde, R. A., *Inorg. Chem.*, 6, 2199, 1967.
59. Cosmovici, C. B., D'Anna, E., D'Innocenzo, A., Leggieri, G., Perrone, A., and Dirscherl, R., *Chem. Phys. Lett.*, 47, 241, 1977.
60. Coxon, J. A., *Chem. Phys. Lett.*, 33, 136, 1975.
61. Cubicciotti, D., *Inorg. Chem.*, 7, 208, 1968.
62. Cubicciotti, D., *Inorg. Chem.*, 7, 211, 1968.
63. Cubicciotti, D., *J. Phys. Chem.*, 71, 3066, 1967.
64. Cuthill, A. M., Fabian, D. J., and Shu-Shou-Shen, S., *J. Phys. Chem.*, 77, 2008, 1973.
65. Dagdigian, P. J., Cruze, H. W., and Zare, R. N., *J. Chem. Phys.*, 62, 1824, 1975.
66. Darwent, B. de B., *Bond Dissociation Energies in Simple Molecules*, NSRDS-NBS 31, National Bureau of Standards, Washington, D.C., 1970.
67. Davis, D. D. and Okabe, H., *J. Chem. Phys.*, 49, 5526, 1968.
68. De Corpo, J. J., Steiger, R. P., Franklin, J. L., and Margrave, J. L., *J. Chem. Phys.*, 53, 936, 1970.
69. De Maria, G., Drowart, J., and Inghram, M. G., *J. Chem. Phys.*, 31, 1076, 1959.
70. De Maria, G., Goldfinger, P., Malaspina, L., and Piacente, V., *Trans. Faraday Soc.*, 61, 2146, 1965.
71. De Maria, G., Malaspina, L., and Piacente, V., *J. Chem. Phys.*, 52, 1019, 1970.
72. De Maria, G., Malaspina, L., and Piacente, V., *J. Chem. Phys.*, 56, 1978, 1972.
73. Dickson, C. R., Kinney, J. B., and Zare, R. N., *Chem. Phys.*, 15, 243, 1976.
74. Dickson, C. R. and Zare, R. N., *Chem. Phys.*, 7, 361, 1975.
75. Dixon, R. N. and Lambertson, H. M., *J. Mol. Spectrosc.*, 25, 12, 1968.
76. Drowart, J., in *Phase Stability in Metals and Alloys*, Rudman, P. S., Ed., McGraw-Hill, New York, 1967, 305.
77. Drowart, J., De Maria, G., and Inghram, M. G., *J. Chem. Phys.*, 29, 1015, 1958.
78. Drowart, J. and Goldfinger, P., *Angew. Chem.*, 6, 581, 1967.
79. Drowart, J. and Goldfinger, P., *Q. Rev. (London)*, 20, 545, 1966.
80. Drowart, J. and Honig, R. E., *J. Phys. Chem.*, 61, 980, 1957.
81. Drowart, J., Myers, C. E., Szwarc, R., Vander Auwera-Mahieu, A., and Uy, O. M., *High Temp. Sci.*, 5, 482, 1973.
82. Drowart, J., Myers, C. E., Szwarc, R., Vander Auwera-Mahieu, A., and Uy, O. M., *J. Chem. Soc. Faraday Trans. 2*, 68, 1749, 1972.
83. Drowart, J., Pattoret, A., and Smoes, S., *Proc. Br. Ceramic Soc.*, No. 8, 67, 1967.
84. Drowart, J., Smets, J., Reynaert, J. C., and Coppens, P., *Adv. Mass Spectrom.*, 7A, 647, 1978.
85. Drowart, J. and Smoes, S., *J. Chem. Soc. Faraday Trans. 2*, 73, 1755, 1977.
85a. Drowart, J., Smoes, S., and Vander Auwera-Mahieu, A., *J. Chem. Thermodyn.*, 10, 453, 1978.
86. Dubois, L. H. and Gole, J. L., *J. Chem. Phys.*, 66, 779, 1977.
87. Edwards, J. G., Franklin, H. F., and Gilles, P. W., *J. Chem. Phys.*, 54, 545, 1971.
88. Ehlert, T. C., *High Temp. Sci.*, 9, 237, 1977.
89. Ehlert, T. C., Hilmer, R. M., and Beauchamp, E. A., *J. Inorg. Nucl. Chem.*, 30, 3112, 1968.
90. Ehlert, T. C. and Margrave, J. L., *J. Chem. Phys.*, 41, 1066, 1964.
91. Ehlert, T. C. and Wang, J. S., *J. Phys. Chem.*, 81, 2069, 1977.
92. Engelke, F., Sander, R. K., and Zare, R. N., *J. Chem. Phys.*, 65, 1146, 1976.
93. Evans, P. J. and Mackie, J. C., *Chem. Phys.*, 5, 277, 1974.
94. Farber, M. and Srivastava, R. D., *Chem. Phys. Lett.*, 42, 567, 1976.
95. Farber, M. and Srivastava, R. O., *High Temp. Sci.*, 7, 74, 1975.
96. Farber, M. and Srivastava, R. O., *High Temp. Sci.*, 8, 73, 1976.
97. Farber, M. and Srivastava, R. D., *High Temp. Sci.*, 8, 195, 1976.
98. Farber, M. and Srivastava, R. D., *J. Chem. Soc. Faraday Trans. 1*, 69, 390, 1973.
99. Farber, M. and Srivastava, R. D., *J. Chem. Soc. Faraday Trans. 1*, 70, 1581, 1974.
100. Farber, M. and Srivastava, R. D., *J. Chem. Soc. Faraday Trans. 1*, 74, 1089, 1978.
101. Farber, M., Uy, O. M., and Srivastava, R. D., *J. Chem. Phys.*, 56, 5312, 1972.
102. Faure, F. M., Mitchell, M. J., and Bartlett, R. W., *High Temp. Sci.*, 4, 181, 1972.

103. Faure, F. M., Mitchell, M. J., and Bartlett, R. W., *High Temp. Sci.*, 5, 128, 1973.
104. Fenochka, B. V. and Gorkienko, S. P., *Zh. Fiz, Khim.*, 47, 2445, 1973.
105. Filippenko, N. V., Motozov, E. V., Giricheva, N. I., and Krasnev, K. S., *Izv. Vyssh. Ucheb. Zaved Khim. Technol.*, 15, 1416, 1972.
106. Fries, J. A. and David, C. E., *J. Chem. Phys.*, 68, 3978, 1978.
107. Fu, C. M. and Burns, R. P., *High Temp. Sci.*, 8, 353, 1976.
108. Fujishiro, S., *Trans. Jpn. Inst. Met.*, 1, 125, 1960.
109. Gaydon, A. G., *Dissociation Energies and Spectra of Diatomic Molecules*, 3rd ed., Chapman and Hall, London, 1968.
110. Gingerich, K. A., *Chem. Commun.*, 580, 1970.
111. Gingerich, K. A., *Chem. Phys. Lett.*, 13, 262, 1972.
112. Gingerich, K. A., *Chem. Phys. Lett.*, 23, 270, 1973.
113. Gingerich, K. A., *Chem. Phys. Lett.*, 25, 523, 1974.
114. Gingerich, K. A., *Chimia*, 26, 619, 1972.
115. Gingerich, K. A., *High Temp. Sci.*, 1, 258, 1969.
116. Gingerich, K. A., *High Temp. Sci.*, 3, 415, 1971.
117. Gingerich, K. A., *J. Chem. Phys.*, 47, 2192, 1967.
118. Gingerich, K. A., *J. Chem. Phys.*, 49, 14, 1968.
119. Gingerich, K. A., *J. Chem. Phys.*, 49, 19, 1968.
120. Gingerich, K. A., *J. Chem. Phys.*, 50, 2255, 1969.
121. Gingerich, K. A., *J. Chem. Phys.*, 54, 2646, 1971.
122. Gingerich, K. A., *J. Chem. Phys.*, 54, 3720, 1971.
123. Gingerich, K. A., *J. Chem. Phys.*, 56, 4239, 1972.
124. Gingerich, K. A., *J. Phys. Chem.*, 68, 768, 1964.
125. Gingerich, K. A., *J. Phys. Chem.*, 73, 2734, 1969.
126. Gingerich, K. A., *J. Chem. Soc. Faraday Trans. 2*, 70, 471, 1974.
127. Gingerich, K. A. and Blue, G. D., *J. Chem. Phys.*, 47, 5447, 1967.
128. Gingerich, K. A. and Blue, G. D., *J. Chem. Phys.*, 59, 186, 1973.
129. Gingerich, K. A. and Choudary, U. V., *J. Chem. Phys.*, 68, 3265, 1978.
130. Gingerich, K. A. and Cocke, D. L., *Inorg. Chim. Acta*, 28, L171, 1970.
131. Gingerich, K. A., Cocke, D. L., and Kordis, J., *J. Phys. Chem.*, 78, 603, 1974.
132. Gingerich, K. A., Cocke, D. L., and Miller, F., *J. Chem. Phys.*, 64, 1027, 1976.
133. Gingerich, K. A. and Finkbeiner, H. C., *J. Chem. Phys.*, 54, 2621, 1971.
134. Gingerich, K. A. and Finkbeiner, H. C., Proc. 9th Rare Earth Res. Conf. 2, 795, 1971.
135. Gingerich, K. A. and Gupta, S. K., *J. Chem. Phys.*, 69, 505, 1978.
136. Goldfinger, P. and Jeunehomme, M., *Trans. Faraday Soc.*, 59, 2851, 1963.
137. Gingerich, K. A. and Piacente, V., *J. Chem. Phys.*, 54, 2498, 1971.
138. Gole, J. L. and Chalek, C. L., *J. Chem. Phys.*, 65, 4384, 1976.
139. Grade, M., Hirschwald, W., and Stolze, F., *Ber. Bunsenges. Phys. Chem.*, 82, 152, 1978.
140. Grade, M., Hirschwald, W., and Stolze, F., *Z. Phys. Chem. Frankfurt am Main*, 100, 165, 1976.
141. Graeber, P. and Weil, K. G., *Ber. Bunsenges. Phys. Chem.*, 76, 417, 1972.
142. Guggi, D. J., Neubert, A., and Zmbov, K. F., Conf. Int. Thermodyn. Chim. [C.R.] 4th, 3, 124, 1975.
143. Guido, M. and Balducci, G., *J. Chem. Phys.*, 57, 5611, 1972.
144. Guido, M., Gigli, G., and Balducci, G., *J. Chem. Phys.*, 57, 3731, 1972.
145. Gunn, S. R., *J. Phys. Chem.*, 68, 949, 1964.
145a. Gupta, S. K., Atkins, R. M., and Gingerich, K. A., *Inorg. Chem.*, 17, 3211, 1978.
146. Gupta, S. K. and Gingerich, K. A., *J. Chem. Phys.*, 69, 4318, 1978.
147. Gupta, S. K., Pelino, M., and Gingerich, K. A., *J. Chem. Phys.*, 70, 2044, 1979.
148. Hariharan, A. V. and Eick, H. A., *J. Chem. Thermodyn.*, 6, 373, 1974.
149. Hastie, J. W., *J. Chem. Phys.*, 57, 4556, 1972.
150. Herman, R. and Herman, L., *J. Phys. Radium*, 24, 73, 1963.
151. Herzberg, G., *Molecular Spectra and Molecular Structure. I. Spectra of Diatomic Molecules*, 2nd ed., Van Nostrand, New York, 1950.
152. Herzberg, G., *J. Mol. Spectrosc.*, 33, 147, 1970.
153. Herzberg, G. J. and Johns, J. W. G., *Astrophys. J.*, 158, 399, 1969.
154. Hildenbrand, D. L., *Chem. Phys. Lett.*, 15, 379, 1972.
155. Hildenbrand, D. L., *Chem. Phys. Lett.*, 20, 127, 1973.
156. Hildenbrand, D. L., *Chem. Phys. Lett.*, 32, 523, 1975.
157. Hildenbrand, D. L., *Chem. Phys. Lett.*, 34, 352, 1975.
158. Hildenbrand, D. L., *Chem. Phys. Lett.*, 44, 281, 1976.
159. Hildenbrand, D. L., *Chem. Phys. Lett.*, 48, 340, 1977.
160. Hildenbrand, D., *J. Chem. Phys.*, 48, 2457, 1968.
161. Hildenbrand, D. L., *J. Chem. Phys.*, 48, 3657, 1968.
162. Hildenbrand, D. L., *J. Chem. Phys.*, 52, 5751, 1970.
163. Hildenbrand, D. L., *J. Chem. Phys.*, 57, 4556, 1972.
164. Hildenbrand, D. L., *J. Chem. Phys.*, 65, 614, 1976.

165. Hildenbrand, D. L., *J. Chem. Phys.*, 66, 3526, 1977.
166. Hildenbrand, D. L., *J. Chem. Phys.*, 66, 3526, 1977.
167. Hildenbrand, D. L., *J. Chem. Phys.*, 77, 897, 1973.
168. Hildenbrand, D. L. and Murad, E., *J. Chem. Phys.*, 44, 1524, 1966.
169. Hildenbrand, D. L. and Murad, E., *J. Chem. Phys.*, 53, 3403, 1970.
170. Hildenbrand, D. L. and Murad, E., *J. Chem. Phys.*, 61, 1232, 1974.
171. Hildenbrand, D. L. and Theard, L. P., *J. Chem. Phys.*, 50, 5350, 1969.
172. Huber, K. P. and Herzberg, G., *Molecular Spectra and Molecular Structure Constants of Diatomic Molecules*, Van Nostrand, New York, 1979.
173. Ihle, H. R. and Wu, C. H., *J. Chem. Phys.*, 63, 1605, 1975.
174. Johns, J. W. C., Grimm, F. A., and Porter, R. F., *J. Mol. Spectrosc.*, 22, 435, 1967.
175. Johns, J. W. C. and Ramsey, D. A., *Can. J. Phys.*, 39, 210, 1961.
176. Jones, R. W. and Gole, J. L., *Chem. Phys.*, 20, 311, 1977.
176a. Kaposi, O., *Magy. Kem. Foly.*, 83, 356, 1977.
177. Kant, A., *J. Chem. Phys.*, 49, 5144, 1968.
178. Kant, A. and Lin, S-S., *J. Chem. Phys.*, 51, 1644, 1969.
179. Kant, A., Lin, S-S., and Strauss, B., *J. Chem. Phys.*, 49, 1983, 1968.
180. Kant, A., and Strauss, B. H., *J. Chem. Phys.*, 41, 3806, 1964.
181. Kant, A. and Strauss, B., *J. Chem. Phys.*, 45, 3161, 1966.
182. Kant, A. and Strauss, B., *J. Chem. Phys.*, 49, 3579, 1968.
183. Kant, A., Strauss, B., and Lin, S-S., *J. Chem. Phys.*, 52, 2384, 1970.
184. Kent, R. A., *J. Am. Chem. Soc.*, 90, 5657, 1968.
185. Kent, R. A., Ehlert, T. C., and Margrave, J. L., *J. Am. Chem. Soc.*, 86, 5090, 1964.
186. Kent, R. A. and Margrave, J. L., *J. Am. Chem. Soc.*, 87, 3582, 1965.
187. Khitrov, A. N., Ryabova, V. G., and Gurvich, L. V., *Teplofiz Vys. Tempo.*, 11, 1126, 1973.
188. Kleinschmidt, D., Cubicciotti, D., and Hildenbrand, D. L., *J. Electrochem. Soc.*, 125, 1543, 1978; *Proc. Electrochem. Soc.*, 78, 217, 1978.
189. Kleinschmidt, P. D. and Hildenbrand, D. L., *J. Chem. Phys.*, 68, 2819, 1978.
189a. Klynning, L. and Lindgren, B., *Arkiv. Fysik.*, 32, 575, 1966.
190. Kohl, F. J. and Carlson, K. D., *J. Am. Chem. Soc.*, 90, 4814, 1968.
191. Kohl, F. J. and Stearns, C. A., *J. Phys. Chem.*, 78, 273, 1974.
192. Kondratiev, V. N., *Bond Dissociation Energies, Ionization Potentials and Electron Affinities*, Mauka Publishing House, Moscow, 1974.
193. Kordis, J. and Gingerich, K. A., *J. Chem. Eng. Data*, 18, 135, 1973.
194. Kordis, J. and Gingerich, K. A., *J. Chem. Phys.*, 58, 5141, 1973.
195. Kordis, J. and Gingerich, K. A., *J. Phys. Chem.*, 76, 2336, 1972.
196. Kordis, J., Gingerich, K. A., and Seyse, R. J., *J. Chem. Phys.*, 61, 5114, 1974.
197. Kronekvist, M., Lagerqvist, A., and Neuhaus, H., *J. Mol. Spectrosc.*, 39, 516, 1971.
198. Kudo, H., Wu, C. H., and Ihle, H. R., *J. Nucl. Mater.*, 78, 380, 1978.
199. Kulkarni, M. P. and Dadape, V. V., *High Temp. Sci.*, 3, 277, 1971.
200. Kvande, H. and Wahlbeck, P. G., *High Temp. High Pressures*, 8, 45, 1976.
201. LeRoy, R. J., *J. Chem. Phys.*, 57, 573, 1972.
202. LeRoy, R. J. and Bernstein, R. B., *Chem. Phys. Lett.*, 5, 42, 1970.
203. Li, K. C. and Stwalley, W. C., *J. Chem. Phys.*, 59, 4423, 1973.
204. Lin, S-S., and Kant, A., *J. Phys. Chem.*, 73, 2450, 1969.
205. Lin, S-S., Strauss, B., and Kant, A., *J. Chem. Phys.*, 51, 2282, 1969.
206. Lindgren, B. and Nilsson, Ch., *J. Mol. Spectrosc.*, 55, 407, 1975.
207. Liu, M. B. and Wahlbeck, P. G., *J. Chem. Phys.*, 63, 1694, 1975.
208. McIntyre, N. S., Vander Auwera-Mahieu, A., and Drowart, J., *Trans. Faraday Soc.*, 64, 3006, 1968.
209. Marquart, J. R. and Berkowitz, J., *J. Chem. Phys.*, 39, 283, 1963.
210. Martin, E. and Barrow, R. F., *Phys. Scr.*, 17, 501, 1978.
211. Muenow, D. W., Hastie, J. W., Hauge, R., Bautista, R., and Margrave, J. L., *Trans Faraday Soc.*, 65, 3210, 1969.
212. Murad, E., Hildenbrand, D. L., and Main, R. P., *J. Chem. Phys.*, 45, 263, 1966.
213. Murad, E. and Hildenbrand, D. L., *J. Chem. Phys.*, 63, 1139, 1975.
214. Murad, E. and Hildenbrand, D. L., *J. Chem. Phys.*, 65, 3250, 1976.
215. Nagai, S., Shinmei, M., and Yokokawa, T., *J. Inorg. Nucl. Chem.*, 36, 1904, 1974.
216. Neckel, A. and Sodeck, G., *Monatsh. Chem.*, 103, 367, 1972.
217. Neubert, A. and Zmbov, K. F., *High Temp. Sci.*, 6, 303, 1974.
218. Neubert, A. and Zmbov, K. F., *J. Chem. Soc. Faraday Trans. 1*, 70, 2219, 1974.
219. Nordine, P. C., *J. Chem. Phys.*, 61, 224, 1974.
220. O'Hare, P. A. G., *J. Chem. Phys.*, 52, 2992, 1970.
221. Oldenberg, R. C., Dickson, C. R., and Zare, R. N., *J. Mol. Spectrosc.*, 58, 283, 1975.
222. Ovcharenko, I. E., Ya, Kuzyankov, Y., and Tatevaskii, V. M., *Opt. Spektrosk.*, 19, 528, 1965.
223. Owzarski, T. P. and Franzen, H. F., *J. Chem. Phys.*, 60, 1113, 1974.
223a. Parr, T. P., Behrens, R., Freedman, A., and Heron, R. R., *Chem. Phys. Lett.*, 56, 71, 1978.

224. Pashinkin, A. S., Molodyk, A. D., Belousov, V. I., Strel'chenko, S. S., and Fedorova, V. A., *Izv. Akad. Nauk. USSR Neorg. Mater.,* 10, 1600, 1974.

225. Pasternack, L. and Dagdigian, P. J., *J. Chem. Phys.,* 67, 3854, 1977.

226. Peeters, R., Vander Auwera-Mahieu, A., and Drowart, J., *Z. Naturforsch. Teil A,* 26, 327, 1971.

227. Pelevin, O. V., Mil'vidskii, M. G., Belyaev, A. I., and Khotin, B. A., *Izv. Akad. Nauk SSSR Neorg. Mater.,* 2, 924, 1966.

228. Perakis, J., Chatillon, C., and Pattoret, A., *C. R. Acad. Sci. Ser. C,* 276, 1357, 1973.

229. Petzel, T., *High Temp. Sci.,* 6, 246, 1974.

230. Piacente, V. and Balducci, G., *Dyn. Mass Spectrom.,* 4, 295, 1976.

231. Piacente, V. and Balducci, G., *High Temp. Sci.,* 6, 254, 1974.

232. Piacente, V-.V. and Balducci, G., *Adv. Mass Spectrom.,* 7A, 626, 1978.

233. Piacente, V., Balducci, G., and Bardi, G., *J. Less-Common Met.,* 37, 123, 1974.

234. Piacente, V., Bardi, G., and Malaspina, L., *J. Chem. Thermodyn.,* 5, 219, 1973.

235. Piacente, V. and Desideri, A., *J. Chem. Phys.,* 57, 2213, 1972.

236. Piacente, V. and Gingerich, K. A., *High Temp. Sci.,* 9, 189, 1977.

237. Piacente, V. and Gingerich, K. A., *Z. Naturforsch, Teil A,* 28, 316, 1973.

238. Piacente, V. and Malaspina, L., *J. Chem. Phys.,* 56, 1780, 1972.

239. Porter, R. F. and Spencer, C. W. J., *J. Chem. Phys.,* 32, 943, 1960.

240. Prasad, R., Venugopal, V., and Sood, D. D., *J. Chem. Thermodyn.,* 9, 593, 1977.

241. Ringstrom, U., *Ark, Fys.,* 27, 227, 1964.

242. Ringstrom, U., *J. Mol. Spectrosc.,* 36, 232, 1970.

243. Rovner, L., Drowart, A., and Drowart, J., *Trans. Faraday Soc.,* 63, 2910, 1967.

244. Rutner, E. and Haury, G. L., *J. Chem. Eng. Data,* 19, 19, 1974.

245. Scott, P. R. and Richards, W. G., *J. Phys. B,* 7, 500, 1974.

246. Scott, P. R. and Richards, W. G., *J. Phys. B,* 7, 1679, 1974.

247. Shardanand, A., *Phys. Rev.,* 160, 67, 1967.

248. Shenyavskaya, E. A., Mal'tsev, A. A., Kataev, D. I., and Gurvich, L. V., *Opt. Spektrosk.,* 26, 937, 1969.

249. Siegel, B., *Q. Rev. (London),* 19, 77, 1965.

250. Smith, P. K. and Peterson, D. E., *J. Chem. Phys.,* 52, 4963, 1970.

251. Smoes, S., Coppens, P., Bergman, C., and Drowart, J., *Trans. Faraday Soc.,* 65, 682, 1969.

252. Smoes, S., Depiere, D., and Drowart, J., *Rev. Int. Hautes Temp. Refractaires Paris,* 9, 171, 1972.

253. Smoes, S. and Drowart, J., *Chem. Commun.,* p. 534, 1968.

254. Smoes, S. and Drowart, J., *J. Chem. Soc. Faraday Trans. 2,* 73, 1746, 1977.

255. Smoes, S., Drowart, J., and Myers, C. E., *J. Chem. Thermodyn.,* 8, 225, 1976.

256. Smoes, S., Drowart, J., and Welter, J. M., *J. Chem. Thermodyn.,* 9, 275, 1977; *Adv. Mass Spectrom.,* 7A, 622, 1978.

257. Smoes, S., Huguet, R., and Drowart, J., *Z. Naturforsch. Teil A,* 26, 1934, 1971.

258. Smoes, S., Mandy, F., Vander Auwera-Mahieu, A., and Drowart, J., *Bull. Soc. Chim. Belg.,* 81, 45, 1972.

259. Smoes, S., Myers, C. E., and Drowart, J., *Chem. Phys. Lett.,* 8, 10, 1971.

260. Smoes, S., Pattje, W. R., and Drowart, J., *High Temp. Sci.,* 10, 109, 1978.

261. Srivastava, R. D., *High Temp Sci.,* 8, 225, 1976.

262. Srivastava, R. D. and Farber, M., *High Temp. Sci.,* 5, 489, 1973.

263. Stearns, C. A. and Kohl, F. J., *High Temp. Sci.,* 2, 146, 1970.

264. Stearns, C. A. and Kohl, F. J., *High Temp. Sci.,* 5, 113, 1973.

265. Stearns, C. A. and Kohl, F., *High Temp. Sci.,* 6, 284, 1974.

266. Steiger, R. A. and Cater, E. D., *High Temp. Sci.,* 7, 204, 1975.

267. Steiger, R. A. and Cater, E. D., *High Temp. Sci.,* 7, 288, 1975.

268. Stwalley, W. C., *J. Chem. Phys.,* 65, 2038, 1970.

269. Stwalley, W. C., Way, K. R., and Velasco, R., *J. Chem. Phys.,* 60, 3611, 1974.

270. Sullivan, C. L., Zehe, M. J., and Carlson, K. D., *High Temp. Sci.,* 6, 80, 1974.

271. Tam, A. C. and Happer, W., *J. Chem. Phys.,* 64, 2456, 1976.

272. Tanaka, Y., Yushina, K., and Freeman, D. E., *J. Chem. Phys.,* 59, 564, 1973.

273. Tellinghuisen, J., Tisone, G. C., Hoffman, J. M., and Hays, A. K., *J. Chem. Phys.,* 64, 4796, 1976.

274. Tuenge, R. T., Laabs, F., and Franzen, H. F., *J. Chem. Phys.,* 65, 2400, 1976.

275. Uy, O. M. and Drowart, J., *High Temp. Sci.,* 2, 293, 1970.

276. Uy, O. M. and Drowart, J., *Trans. Faraday Soc.,* 65, 3221, 1969.

277. Uy, O. M. and Drowart, J., *Trans. Faraday Soc.,* 67, 1293, 1971.

278. Uy, O. M., Muenow, D. W., Ficalora, P. J., and Margrave, J. L., *Trans. Faraday Soc.,* 64, 2998, 1968.

279. Vander Auwera-Mahieu, A. and Drowart, J., *Chem. Phys. Lett.,* 1, 311, 1967.

280. Vander Auwera-Mahieu, A., McIntyre, N. S., and Drowart, J., *Chem. Phys. Lett.,* 4, 198, 1969.

281. Vander Auwera-Mahieu, A., Peeters, R., McIntyre, N. S., and Drowart, J., *Trans. Faraday Soc.,* 66, 809, 1970.

282. Velasco, R., Ottinger, C., and Zare, R. N., *J. Chem. Phys.,* 51, 5522, 1969.

283. Verhaegen, G., Ph.D. Thesis, University of Brussels, 1965.

284. Verhaegen, G., Smoes, S., and Drowart, J., *J. Chem. Phys.,* 40, 239, 1964.

285. Verhaegen, G., Stafford, F. E., and Drowart, J., *J. Chem. Phys.,* 40, 1622, 1964.

286. Way, K. R. and Stwalley, W. C., *J. Chem. Phys.*, 59, 5298, 1973.
287. Wiedemeier, H. and Gilles, P. W., *J. Chem. Phys.*, 42, 2765, 1965.
288. Wiedemeier, H. and Goyette, W. J., *J. Chem. Phys.*, 48, 2936, 1968.
289. Wieland, Von K., *Z. Elektrochem.*, 64, 761, 1960.
290. Wu, C. H., *J. Chem. Phys.*, 65, 3181, 1976; 65, 2040, 1976.
291. Yokozeki, A. and Menzinger, M., *Chem. Phys.*, 14, 427, 1976.
292. Zmbov, K. F., Hastie, J. W., and Margrave, J. L., *Trans. Faraday Soc.*, 64, 861, 1968.
293. Zmbov, K. F. and Margrave, J. L., *J. Chem. Phys.*, 45, 3167, 1966.
294. Zmbov, K. F. and Margrave, J. L., *J. Inorg. Nucl. Chem.*, 29, 59, 1976.
295. Zmbov, K. F. and Margrave, J. L., *J. Chem. Phys.*, 47, 3122, 1967.
296. Zmbov, K. F. and Margrave, J. L., *J. Phys. Chem.*, 70, 3379, 1966.
297. Zmbov, K. F. and Margrave, J. L., *J. Phys. Chem.*, 71, 2893, 1967.
298. Zmbov, K. F. and Miletic, M. M., *Glas. Hem. Drus. Beograd*, 43, 521, 1978.
299. Zmbov, K. F., Wu, C. H., and Ihle, H. R., *J. Chem. Phys.*, 67, 4603, 1977.
300. Zollweg, R. J., *Contrib. Pap. Int. Conf. Phenomena of Inoization Gases. 11th*, 402, 1973.

HEATS OF FORMATION OF GASEOUS ATOMS FROM ELEMENTS IN THEIR STANDARD STATES

For elements that are diatomic gases in their standard states these are readily obtained from the bond strength. For elements that are crystalline in their standard states they are derived from vapor pressure data.

Table 2
HEATS OF FORMATION OF GASEOUS ATOMS FROM ELEMENTS IN THEIR STANDARD STATES

Atom	$\Delta H^\circ_{f(298)}$/kcal mol^{-1}	$\Delta H^\circ_{f(298)}$/kJ mol^{-1}	Ref.	Atom	$\Delta H^\circ_{f(298)}$/kcal mol^{-1}	$\Delta H^\circ_{f(298)}$/kJ mol^{-1}	Ref.
Ag	68.1 ± 0.2	284.9 ± 0.8	1	Na	25.85 ± 0.15	108.16 ± 0.63	2
Al	78.8 ± 1.0	329.7 ± 4.0	1	Nb	172.4 ± 1	721.3 ± 4	2
As	72.3 ± 3	302.5 ± 13	2	Ni	102.8 ± 0.5	430.1 ± 2.1	2
Au	88.0 ± 0.5	368.2 ± 2.1	2	O	59.553 ± 0.024	249.17 ± 0.10	1
B	139 ± 3	560 ± 12	1	Os	188 ± 1.5	787 ± 6.3	2
Ba	42.5 ± 1	177.8 ± 4	2	P	79.4 ± 1.0	332.2 ± 4.2	2
Be	77.5 ± 1.5	324.3 ± 6.3	2	Pb	46.62 ± 0.3	195.06 ± 1.3	2
Bi	50.1 ± 0.5	209.6 ± 2.1	2	Pd	90.0 ± 0.5	376.6 ± 2.1	2
Br	26.86 ± 0.03	112.38	3	Pt	135.2 ± 0.3	565.7 ± 1.3	2
C	171.29 ± 0.11	716.67 ± 0.44	1	Pu	87.1 ± 4	364.4 ± 17	2
Ca	42.6 ± 0.4	178.2 ± 1.7	2	Rb	19.6 ± 0.1	82.0 ± 0.4	2
Cd	26.72 ± 0.15	111.80 ± 0.63	2	Re	185 ± 1.5	774 ± 6.3	2
Ce	101 ± 3	423 ± 13	2	Rh	133 ± 1	557 ± 4	2
Cl	29.031 ± 0.002	121.466 ± 0.008	3	Ru	155.5 ± 1.5	648.5 ± 6.3	2
Co	102.4 ± 1	428.4 ± 4	2	S	66.20 ± 0.06	276.98 ± 0.25	1
Cr	95 ± 1	398 ± 4	2	Sb	63.2 ± 0.6	264.4 ± 2.5	2
Cs	18.7 ± 0.1	78.2 ± 0.4	2	Sc	90.3 ± 1	377.8 ± 4	2
Cu	80.7 ± 0.3	337.6 ± 1.2	1	Se	54.3 ± 1	227.2 ± 4	2
Er	75.8 ± 1	317.1 ± 4	2	Si	108 ± 2	450 ± 8	1
F	18.92	79.14	3	Sm	49.4 ± 0.5	206.7 ± 2.1	2
Fe	99.3 ± 0.3	415.5 ± 1.3	2	Sn	72.2 ± 0.5	302.1 ± 2.1	2
Ga	65.4 ± 0.5	273.6 ± 2.1	2	Sr	39.1 ± 0.5	163.6 ± 2.1	2
Ge	89.5 ± 0.5	374.5 ± 2.1	2	Ta	186.9 ± 0.6	782.0 ± 2.5	2
H	52.077 ± 0.001	217.890 ± 0.005	3	Te	47.0 ± 0.5	196.7 ± 2.1	2
Hf	148 ± 1	619 ± 4	2	Th	137.5 ± 0.5	575.3 ± 2.1	2
Hg	14.69 ± 0.03	61.46 ± 0.13	2	Ti	112.3 ± 0.5	469.9 ± 2.1	2
I	25.688 ± 0.010	107.477 ± 0.040	3	Tl	43.55 ± 0.1	182.21 ± 0.4	2
In	58 ± 1	243 ± 4	2	U	126 ± 3	527 ± 13	2
Ir	160 ± 1	669 ± 4	2	V	122.9 ± 0.3	514.2 ± 1.3	2
K	21.42 ± 0.05	89.62 ± 0.21	2	W	203.1 ± 1	849.8 ± 4	2
Li	38.6 ± 0.4	161.5 ± 1.7	2	Y	101.5 ± 0.5	424.7 ± 2.1	2
Mg	35.0 ± 0.3	146.4 ± 1.3	2	Yb	36.35 ± 0.2	152.09 ± 0.8	2
Mn	67.7 ± 1	283.3 ± 4	2	Zn	31.17 ± 0.05	130.42 ± 0.20	1
Mo	157.3 ± 0.5	658.1 ± 2.1	2	Zr	145.5 ± 1	608.8 ± 4	2
N	112.97 ± 0.10	472.68 ± 0.40	1				

REFERENCES

1. CODATA recommended Key values for thermodynamics, 1975, *J. Chem. Thermodyn.*, 8, 603, 1976.
2. Brewer, L. and Rosenblatt, G. M., *Adv. High Temp. Chem.*, 2, 1, 1969.
3. Huber, K. P. and Herzberg, G., *Molecular Spectra and Molecular Structure Constants of Diatomic Molecules*, Van Nostrand, New York, 1979.

BOND STRENGTHS IN POLYATOMIC MOLECULES

The values below refer to a temperature of 298 K and have mostly been determined by kinetic methods (see References 18 and 71 following Table 3 for a full discussion of the methods).

Some have been calculated from the heats of formation of the species involved according to the equations:

$$D(R - X) = \Delta H_f^o(\dot{R}) + \Delta H_f^o(\dot{X}) - \Delta H_f^o(RX)$$

$$D(R - R) = 2\Delta H_f^o(\dot{R}) - \Delta H_f^o(RR)$$

The sources of the data on the heats of formation are given in the references following Table 3.

An attempt has been made to list all the important values obtained by methods that are considered to be valid. The references are intended to serve as a guide to the literature.

Table 3
BOND STRENGTHS IN POLYATOMIC MOLECULES

Bond	D_{298}^o/kcal mol^{-1}	D_{298}^o/kJ mol^{-1}	Ref.
H−CH	102 ± 2	427 ± 8	1,28
H−CH$_2$	110 ± 2	460 ± 8	1,28
H−CH$_3$	105.1 ± 0.15	439.7 ± 0.6	8
H−ethynyl	125 ± 1	523 ± 4	94
H−vinyl	≥108 ± 2	≥452 ± 8	63
H−C$_2$H$_5$	98 ± 1	410 ± 4	63
H−propargyl	90 ± 1	377 ± 4	80,127
H−allyl	86.6 ± 1.5	362.3 ± 6.3	115
H−cyclopropyl	106.3 ± 0.3	444.8 ± 1.1	9
H−n-C$_3$H$_7$	98 ± 1	410 ± 4	63
H−i-C$_3$H$_7$	95 ± 1	398 ± 4	63
H−CH$_2$C CCH$_3$	90.7	379.5	127
H−cyclobutyl	96.5 ± 1	403.8 ± 4	50,92
H−cyclopropylcarbinyl	97.4 ± 1.6	407.5 ± 6.7	91
H−CH(CH$_3$)CH$_2$	83 ± 1	347 ± 4	63
H−CH$_2$C(CH$_3$)$_2$	85.6 ± 1.1	358.2 ± 4.6	125,126
H−c-C$_4$H$_7$	93 ± 1	398 ± 4	63
H−t-C$_4$H$_9$	92 ± 1	385 ± 4	63
H−cyclopentadien-1,3-yl-5	81.2 ± 1.2	339.7 ± 5.0	61
H−pentadien-1,4-yl-3	80 ± 1	335 ± 4	63
H−cyclopentenyl-3	82.3 ± 1	344.3 ± 4	60
H−spiropentyl	98.8 ± 1	413.4 ± 4	50
H−C(CH$_3$)$_2$C CH	79	331	79,116
H−cyclopentyl	94.5 ± 1	395. ± 4	50,59
H−C(CH$_3$)$_2$CH$_2$	77.2 ± 1.5	323.0 ± 6.3	124
H−neo−C$_5$H$_{11}$	100.3 ± 1	419.7 ± 4	88
H−C$_6$H$_5$	110.2 ± 2.0	461.1 ± 8.4	25
H−cyclohexadien-1,3-yl-5	70 ± 5	293 ± 21	70
H−cyclohexyl	95.5 ± 1	400 ± 4	50
H−CH$_2$C(CH$_3$):C(CH$_3$)$_2$	78.0 ± 1.1	326.4 ± 4.6	113
H−C(CH$_3$)$_2$C(CH$_3$)$_2$	76.3 ± 1.1	319.2 ± 4.6	113
H−CH$_2$C$_6$H$_5$	87.9 ± 1.5	367.8 ± 6.3	115
H−cycloheptatrien-1,3-yl-7	73.4	307.1	131
H−norbornyl	96.7 ± 2.5	404.6 ± 10.5	97
H−cycloheptyl	92.5 ± 1	387.0 ± 4	50
H−CN	120 ± 1	502 ± 4	42
H−CH$_2$CN	~93	~389	75
H−CH(CH$_3$)CN	89.9 ± 2.3	376.1 ± 9.6	74
H−C(CH$_3$)$_2$CN	86.5 ± 2.0	361.9 ± 8.4	76
H−CH$_2$NH$_2$	94.6 ± 2.0	395.8 ± 8.4	31,116
H−CHO	87 ± 1	364 ± 4	63
H−CH$_2$OH	94 ± 2	393 ± 4	63
H−COCH$_3$	86.0 ± 0.8	359.8 ± 3.4	44
H−CH$_2$CHO	94.8 ± 2.0	396.6 ± 8.4	114
H−CH$_2$OCH$_3$	93 ± 1	389 ± 4	63
H−CH(CH$_3$)OH	93.0 ± 1.0	389.1 ± 4.2	2
H−COCH$_2$	87.1 ± 1.0	364.4 ± 4.2	3
H−CH$_2$COCH$_3$	98.3 ± 1.8	411.3 ± 7.5	77,122
H−COC$_2$H$_5$	87.4 ± 1.0	365.7 ± 4.2	136
H−CH(OH)CH$_2$	81.6 ± 1.8	341.4 ± 7.5	5
H−C(CH$_3$)$_2$OH	91 ± 1	381 ± 4	63
H−tetrahydrofuran-2-yl	92 ± 1	385 ± 4	63
H−CH(CH$_3$)COCH$_3$	92.3 ± 1.4	386.2 ± 5.9	123
H−COC$_6$H$_5$	86.9 ± 1	363.6 ± 4	121
H−COOCH$_3$	92.7 ± 1	387.9 ± 4	119
H−CH$_2$OCOC$_6$H$_5$	100.2 ± 1.3	419.2 ± 5.4	120
H−COCF$_3$	91.0 ± 2	380.7 ± 8	6
H−CH$_2$F	101 ± 2	423 ± 8	72
H−CHF$_2$	101 ± 2	423 ± 8	72

Table 3 (continued)
BOND STRENGTHS IN POLYATOMIC MOLECULES

Bond	D^o_{298}/kcal mol^{-1}	D^o_{298}/kJ mol^{-1}	Ref.
H–CF$_3$	106 ± 1	444 ± 4	63
H–CF$_2$Cl	104 ± 1	435 ± 4	90
H–CH$_2$Cl	100.9	422.2	58
H–CHCl$_2$	99.0	414.2	58
H–CCl$_3$	95.8 ± 1	400.8 ± 4	93
H–CH$_2$Br	102.0	426.8	58
H–CHBr$_2$	103.7	433.9	58
H–CBr$_3$	96.0 ± 1.6	401.7 ± 6.7	78
H–CH$_2$I	103 ± 2	431 ± 8	63
H–CHI$_2$	103 ± 2	431 ± 8	63
H–CH$_2$CF$_3$	106.7 ± 1.1	446.4 ± 4.6	139
H–CF$_2$CH$_3$	99.5 ± 1	416.3 ± 4.2	104
H–C$_2$F$_5$	103.1 ± 1.5	431.4 ± 6.3	10
H–CCl$_2$CHCl$_2$	94 ± 2	393 ± 8	37,57
H–C$_2$Cl$_5$	95 ± 2	398 ± 8	37,56
H–n-C$_3$F$_7$	104 ± 2	435 ± 8	71
H–CHClCH$_2$	88.6 ± 1.4	370.7 ± 5.9	4
H–CH$_2$Si(CH$_3$)$_3$	99.2 ± 1	415.1 ± 4.2	46
H–NH$_2$	110 ± 2	460 ± 8	21,64
H–NHCH$_3$	103 ± 2	431 ± 8	64
H–N(CH$_3$)$_2$	95 ± 2	398 ± 8	64
H–NHC$_6$H$_5$	86.4 ± 2	361.5 ± 8	34
H–N(CH$_3$)C$_6$H$_5$	84.9 ± 2	355.2 ± 8	34
H–NO	≤49.5	≤207.1	29
H–NF$_2$	75.7 ± 2.5	316.7 ± 10.5	99
H–N$_3$	85	356	66
H–OH	119 ± 1	498 ± 4	71
H–OCH$_3$	104.4 ± 1	436.8 ± 4	11,37
H–OC$_2$H$_5$	104.2 ± 1	436.0 ± 4	11,37
H–OC(CH$_3$)$_3$	105.1 ± 1	439.7 ± 4	11,37,110
H–OCH$_2$C(CH$_3$)$_3$	102.3 ± 1.5	428.0 ± 6.3	110
H–OC$_6$H$_5$	88 ± 5	368 ± 21	24
H–O$_2$H	87.2 ± 1.0	364.8 ± 4.2	1,66a
H–O$_2$CCH$_3$	106	443	71,143
H–O$_2$CC$_2$H$_5$	110 ± 4	460 ± 8	71
H–O$_2$Cn-C$_3$H$_7$	103 ± 4	431 ± 8	71
H–ONO	78.3 ± 0.5	327.6 ± 2.1	1,19
H–ONO$_2$	101.2 ± 0.5	423.4 ± 2.1	1,19
H–SH	90.5 ± 1.1	378.7 ± 4.6	67
H–SCH$_3$	≥88	≥368	71
H–SO	41.3	172.8	137
H–SiH$_3$	90 ± 2	376 ± 8	48,111
H–Si(CH$_3$)$_3$	90 ± 2.6	376 ± 11	134
H–SiF$_3$	100 ± 1	419 ± 4	47
H–SiCl$_3$	91.3 ± 1.4	382 ± 6	135
CH≡CH	230 ± 2	962 ± 8	1,28
CH$_2$=CH$_2$	172 ± 2	720 ± 8	1,28
CH$_3$–CH$_3$	88 ± 2	368 ± 8	71
CH$_3$–CH$_2$C CH	76.0	318.0	80
CH$_3$–CH$_2$C(CH$_3$)$_2$	72.0 ± 0.8	301.2 ± 3.4	125
CH$_3$–C(CH$_3$)$_2$C CH	70.7 ± 1.5	295.8 ± 6.3	79
CH$_3$–C(CH$_3$)$_2$CH$_2$	69.4	290.4	20
n-C$_3$H$_7$–CH$_2$CH$_3$	70.7	295.8	128
C$_6$H$_5$CH$_2$–C$_2$H$_5$	69 ± 2	289 ± 8	71
C$_6$H$_5$CH(CH$_3$)–CH$_3$	71	297	20
C$_6$H$_5$CH$_2$–n-C$_3$H$_7$	67 ± 2	280 ± 8	71
CH$_3$–CN	123.9 ± 0.7	518.4 ± 2.9	94
C$_2$H–CN	144 ± 1	603 ± 4	94
CH$_3$–CH$_2$CN	72.7 ± 2	304.2 ± 8	20,71
CH$_3$–CH(CH$_3$)CN	78.8 ± 2	329.7 ± 8	74
C$_2$H$_5$–CH$_2$CN	76.9 ± 1.7	321.8 ± 7.1	75
CH$_3$–C(CH$_3$)$_2$CN	74.7 ± 1.6	312.6 ± 6.7	20,71,76
C$_6$H$_5$C(CH$_3$)(CN)–CH$_3$	59.9	250.6	20
CN–CN	128 ± 1	536 ± 4	42
C$_6$H$_5$CH$_2$–CH$_2$NH$_2$	65.7	274.9	31
C$_6$H$_5$CH$_2$CO–CH$_2$C$_6$H$_5$	65.4	273.6	20
C$_6$H$_5$CO–CF$_3$	73.8	308.8	20
CH$_3$CO–COCH$_3$	67.4 ± 2.3	282.0 ± 9.6	81
C$_6$H$_5$CH$_2$–COOH	68.1	284.9	20
C$_6$H$_5$CH$_2$–O$_2$CCH$_3$	67	280	20
C$_6$H$_5$CO–COC$_6$H$_5$	66.4	277.8	20
C$_6$H$_5$CH$_2$–O$_2$CC$_6$H$_5$	69	289	20
(C$_6$H$_5$CH$_2$)$_2$CH–COOH	59.4	248.5	20
CH$_2$F–CH$_2$F	88 ± 2	368 ± 8	72
CH$_3$–CF$_3$	101.2 ± 1.1	423.4 ± 4.6	112

Table 3 (continued)
BOND STRENGTHS IN POLYATOMIC MOLECULES

Bond	D°_{298}/kcal mol^{-1}	D°_{298}/kJ mol^{-1}	Ref.
$CF_2=CF_2$	76.3 ± 3	319.2 ± 13	142
CF_3-CF_3	96.9 ± 2	405.4 ± 8	35
$C_6F_5-C_6F_5$	116.6 ± 5.9	488.4 ± 24.5	109
$C_6H_5CH_2-NH_2$	71.9 ± 1	300.8 ± 4	64
$C_6H_5NH-CH_3$	67.7	283.3	20
$C_6H_5CH_2-NHCH_3$	68.7 ± 1	287.4 ± 4	64
$C_6H_5N(CH_3)-CH_3$	65.2	272.8	20
$C_6H_5CH_2-N(CH_3)_2$	60.9 ± 1	254.8 ± 4	64
CF_3-NF_2	65 ± 2.5	272 ± 10.5	132
$CH_2=N_2$	$\leqslant 41.7 \pm 1$	$\leqslant 174.5 \pm 4$	89
$CH_3N:N-CH_3$	52.5	219.7	20
$C_2H_5N:N-C_2H_5$	50.0	209.2	20
$i\text{-}C_3H_7N:N-i\text{-}C_3H_7$	47.5	198.7	20
$n\text{-}C_4H_9N:N-n\text{-}C_4H_9$	50.0	209.2	20
$i\text{-}C_4H_9N:N-i\text{-}C_4H_9$	49.0	205.0	20
$s\text{-}C_4H_9N:N-s\text{-}C_4H_9$	46.7	195.4	20
$t\text{-}C_4H_9N:N-t\text{-}C_4H_9$	43.5	182.0	20
$C_6H_5CH_2N:N-CH_2C_6H_5$	37.6	157.3	20
$CF_3N:N-CF_3$	55.2	231.0	20
C_6H_5-NO	51.5 ± 1	215.5 ± 4	26
CF_3-NO	31	130	23
C_6F_5-NO	50.5 ± 1	211.3 ± 4	23,26
CCl_3-NO	32	134	23
$CN-NO$	28.8 ± 2.5	120.5 ± 10.5	65
$t\text{-}C_4H_9-NO\ t\text{-}C_4H_9$	29	121	23
$C_2H_5-NO_2$	62	259	1,117
$O=CO$	127.2 ± 0.1	532.2 ± 0.4	40
$CH_3-OC(CH_3)_2$	66.3	277.4	141
$CH_3-OC_6H_5$	57 ± 2	239 ± 8	100
$CH_3-OCH_2C_6H_5$	67.0	280.3	33
$C_2H_5-OC_2H_5$	64.0	267.8	33
$CH_3CH_2-OC_2H_5$	50.6	211.7	33
$CH_3-O_2SCH_3$	66.8	279.5	20,71
$\text{allyl}-O_2SCH_3$	49.6	207.5	20,71
$C_6H_5CH_2-O_2SCH_3$	52.9	221.3	20,71
$C_6H_5S-CH_3$	67.5 ± 2.0	282.4 ± 8.4	31
$C_6H_5CH_2-SCH_3$	59.4 ± 2.0	248.5 ± 8.4	31
$F-CH_3$	108 ± 3	452 ± 13	45,71
$F-CF_2Cl$	117 ± 6	490 ± 25	55
$F-CFCl_2$	110 ± 6	460 ± 25	55
$Cl-CN$	97 ± 1	406 ± 4	42
$Cl-COC_6H_5$	74 ± 3	310 ± 13	71
$Cl-CF_3$	86.1 ± 0.8	360.2 ± 3.3	36
$Cl-CF_2Cl$	76 ± 2	318 ± 8	55
$Cl-CCl_2F$	73 ± 2	305 ± 8	54
$Cl-CCl_3$	70.4 ± 1	294.6 ± 4	93
$Cl-C_2F_5$	82.7 ± 1.7	346.0 ± 7.1	36
$Cl-CF_2CF_2Cl$	78 ± 2	326 ± 8	55
$Cl-NF_2$	$\leqslant 65$	$\leqslant 272$	30
$Cl-SiCl_3$	111	466	135
$Br-CH_3$	70.0 ± 1.2	292.9 ± 5.0	49
$Br-C_6H_5$	74.8 ± 1.3	313 ± 5.4	84
$Br-CN$	83 ± 1	347 ± 4	42
$Br-CH_2COCH_3$	62.5	261.5	141
$Br-COC_6H_5$	64.2	268.6	20
$Br-CHF_2$	69 ± 2	289 ± 8	95
$Br-CF_3$	70.6 ± 1.0	295.4 ± 4.2	51,53
$Br-CCl_3$	55.7 ± 1	233.1 ± 4.2	93
$Br-CBr_3$	56.2 ± 1.8	235.1 ± 7.5	78
$Br-CF_2CH_3$	68.6 ± 1.3	287.0 ± 5.4	105
$Br-C_2F_5$	68.7 ± 1.5	287.4 ± 6.3	36,52,53
$Br-n\text{-}C_3F_7$	66.5 ± 2.5	278.2 ± 10.5	36
$Br-NF_2$	$\leqslant 53$	$\leqslant 222$	30
$I-CH_3$	56.3 ± 1	235.6 ± 4	71
$I-\text{norbornyl}$	62.5 ± 2.5	261.5 ± 10.5	97
$I-CN$	73 ± 1	305 ± 4	42
$I-CF_3$	53.2 ± 1.0	222.6 ± 4.2	96
$I-CF_2CH_3$	52.1 ± 1.0	218.0 ± 4.2	103
$I-CH_2CF_3$	56.3 ± 1	235.6 ± 4	138
$I-C_2F_5$	51.2 ± 1.0	214.2 ± 4.2	96,140
$I-n\text{-}C_3F_7$	49.8 ± 1.0	208.4 ± 4.2	96
$I-i\text{-}C_3F_7$	49.8 ± 1.0	208.4 ± 4.2	96
$I-n\text{-}C_4F_9$	49.0 ± 1.0	205.0 ± 4.2	96
$I-C_6H_5$	63.7 ± 1.2	266.3 ± 4.8	82
$I-C_6F_5$	~ 66.2	~ 277	86
$I-SH$	49.4 ± 2	206.7 ± 8	68

Table 3 (continued)
BOND STRENGTHS IN POLYATOMIC MOLECULES

Bond	D°_{298}/kcal mol^{-1}	D°_{298}/kJ mol^{-1}	Ref.
$C_2H_5-ZnC_2H_5$	~47.5	~198.7	85
$CH_3-Ga(CH_3)_2$	59.5	249.0	20
CH_3-CdCH_3	54.4	227.6	20
$C_2H_5-Sn(C_2H_5)_3$	~57	~239	39
CH_3-HgCH_3	57.5	240.6	83
$C_2H_5-HgC_2H_5$	43.7 ± 1	182.8 ± 4	87
$n\text{-}C_3H_7-Hg\,n\text{-}C_3H_7$	47.1	197.1	20
$i\text{-}C_3H_7-Hg\,i\text{-}C_3H_7$	40.7	170.3	20
$C_6H_5-HgC_6H_5$	68	285	20
$CH_3-Tl(CH_3)_2$	36.4 ± 0.6	152.3 ± 2.5	108
$CH_3-Pb(CH_3)_3$	49.4 ± 1	206.7 ± 4	62
BH_3-BH_3	35	146	20
NH_2-NH_2	70.8 ± 2	296.2 ± 8	20
NH_2-NHCH_3	64.8	271.1	20
$NH_2-N(CH_3)_2$	62.7	262.3	20
$NH_2-NHC_6H_5$	51.1	213.8	20
$NO-NO_2$	9.5 ± 0.5	39.8 ± 2.1	71
NO_2-NO_2	12.9 ± 0.5	54.0 ± 2.1	71
NF_2-NF_2	21 ± 1	88 ± 4	40
$O-N_2$	40	167	1,19
$O-NO$	73	305	1,19
CH_3O-NO	41.8 ± 0.9	174.9 ± 3.8	11,15
C_2H_5O-NO	42.0 ± 1.3	175.7 ± 5.5	11,16
$n\text{-}C_3H_7O-NO$	40.1 ± 1.8	167.8 ± 7.5	11
$i\text{-}C_3H_7O-NO$	41.0 ± 1.3	171.5 ± 5.5	11,17
$n\text{-}C_4H_9O-NO$	42.5 ± 1.5	177.8 ± 6.3	11
$i\text{-}C_4H_9O-NO$	42.0 ± 1.5	175.7 ± 6.3	11
$s\text{-}C_4H_9O-NO$	41.5 ± 0.8	173.6 ± 3.4	11,13
$t\text{-}C_4H_9O-NO$	40.9 ± 0.8	171.1 ± 3.4	11,14,27
$HO-NCH_3$	49.7	208.0	20
$Cl-NF_2$	~32	~134	1,102
$I-NO$	17.3 ± 1	72.4 ± 4	129
$I-NO_2$	18.3 ± 1	76.6 ± 4	129
$HO-OH$	51 ± 1	213 ± 4	71
CH_3O-OCH_3	37.6 ± 0.2	157.3 ± 0.8	7,11,12
$HO-OC(CH_3)_3$	46.3 ± 1.9	193.7 ± 8.0	1,11,37
$C_2H_5O-OC_2H_5$	37.9	158.6	11
$n\text{-}C_3H_7O-O\,n\text{-}C_3H_7$	37.1	155.2	11
$i\text{-}C_3H_7O-O\,i\text{-}C_3H_7$	37.7	157.7	11
$s\text{-}C_4H_9O-O\,s\text{-}C_4H_9$	36.4 ± 1	152.3 ± 4	133
$t\text{-}C_4H_9O-O\,t\text{-}C_4H_9$	38.0	159.0	11
$neo\text{-}C_5H_{11}O-O\,neo\text{-}C_5H_{11}$	36.4 ± 1	152.3 ± 4	101
CF_3O-OCF_3	46.2	193.3	43
$(CF_3)_3CO-OC(CF_3)_3$	35.5 ± 1.1	148.7 ± 4.4	69
$O-O_2ClF$	58.4	244.3	20
$CH_3CO_2-O_2CCH_3$	30.4 ± 2	127.2 ± 8	20,71
$C_2H_5CO_2-O_2CC_2H_5$	30.4 ± 2	127.2 ± 8	20,71
$n\text{-}C_3H_7CO_2-O_2C\,n\text{-}C_3H_7$	30.4 ± 2	127.2 ± 8	20,71
$O-SO$	132 ± 2	552 ± 8	40
$F-OCF_3$	43.5 ± 0.5	182.0 ± 2.1	38
$HO-Cl$	60 ± 3	251 ± 13	71
$O-ClO$	59 ± 3	247 ± 13	40
$HO-Br$	56 ± 3	234 ± 13	71
$HO-I$	56 ± 3	234 ± 13	71
ClO_3-ClO_4	58.4	244.3	20
$O=PF_3$	130 ± 5	543.9 ± 21	71
$O=PCl_3$	122 ± 5	511 ± 21	71
$O=PBr_3$	119 ± 5	498 ± 21	71
$F-SF_5$	89.9 ± 3.4	376.1 ± 14.2	73
SiH_3-SiH_3	81 ± 4	339 ± 17	111
$(CH_3)_3Si-Si(CH_3)_3$	80.5	336.8	41
$(C_6H_5)_3Si-Si(C_6H_5)_3$	88 ± 7	368 ± 29	22

REFERENCES

1. A value calculated from one of the above thermochemical equations taking data from the references quoted.
2. Alfassi, Z. B. and Golden, D. M., *J. Phys. Chem.*, 76, 3314, 1972.
3. Alfassi, Z. B. and Golden, D. M., *J. Am. Chem. Soc.*, 95, 319, 1973.
4. Alfassi, Z. B., Golden, D. M., and Benson, S. W., *Int. J. Chem. Kinet.*, 5, 155, 1973: Trenwith, A. B., *Int. J. Chem. Kinet.*, 5, 67, 1973.
5. Alfassi, Z. B. and Golden, D. M., *Int. J. Chem. Kinet.*, 5, 295, 1973: Trenwith, A. B., *J. Chem. Soc. Faraday Trans. 1*, 69, 1737, 1973.
6. Amphlett, J. C. and Whittle, E., *Trans. Faraday Soc.*, 66, 2016, 1970.

7. Barker, J. R., Benson, S. W., and Golden, D. M., *Int. J. Chem. Kinet.*, 9, 31, 1977.
8. Baghal-Vayjooee, M. H., Colussi, A. J., and Benson, S. W., *J. Am. Chem. Soc.*, 100, 3214, 1978; *Int. J. Chem. Kinet.*, 11, 147, 1979.
9. Baghal-Vayjooee, M. H. and Benson, S. W., *J. Am. Chem. Soc.*, 101, 2838, 1979.
10. Bassett, J. E. and Whittle, E., *J. Chem. Soc. Faraday Trans. 1*, 68, 492, 1972.
11. Batt, L., Christie, K., Milne, R. T., and Summers, A. J., *Int. J. Chem. Kinet.*, 6, 877, 1974.
12. Batt, L. and McCulloch, R. D., *Int. J. Chem. Kinet.*, 8, 491, 1976.
13. Batt, L. and McCulloch, R. D., *Int. J. Chem. Kinet.*, 8, 911, 1976.
14. Batt, L. and Milne, R. T., *Int. J. Chem. Kinet.*, 8, 59, 1976.
15. Batt, L., Milne, R. T., and McCulloch, R. D., *Int. J. Chem. Kinet.*, 9, 567, 1977.
16. Batt, L. and Milne, R. T., *Int. J. Chem. Kinet.*, 9, 549, 1977.
17. Batt, L. and Milne, R. T., *Int. J. Chem. Kinet.*, 9, 141, 1977.
18. Benson, S. W., *J. Chem. Educ.*, 42, 502, 1965.
19. Benson, S. W., *Thermochemical Kinetics*, 2nd ed., John Wiley & Sons, New York, 1976.
20. Benson, S. W. and O'Neal, H. E., *Kinetic Data on Gas Phase Unimolecular Reactions*, National Bureau of Standards, NSRDS-NBS, Washington, D.C., 21, 1970.
21. Bohme, D. K., Hemsworth, R. S., and Rundle, H. W., *J. Chem. Phys.*, 59, 77, 1973.
22. Calle, L. M. and Kana'an, A. S., *J. Chem. Thermodyn.*, 6, 935, 1974.
23. Carmichael, P. J., Gowenlock, B. G., and Johnson, C. A. F., *J. Chem. Soc. Perkin Trans. 2*, 1853, 1973.
24. Carson, A. S., Fine, D. H., Gray, P., and Laye, P. G., *J. Chem. Soc. B*, p. 1611, 1971.
25. Chamberlain, G. A. and Whittle, E., *Trans. Faraday Soc.*, 67, 2077, 1971.
26. Choo, K. Y., Golden, D. M., and Benson, S. W., *Int. J. Chem. Kinet.*, 7, 713, 1975.
27. Choo, K. Y., Mendenhall, G. D., Golden, D. M., and Benson, S. W., *Int. J. Chem. Kinet.*, 6, 813, 1974.
28. Chupka, W. A. and Lifshitz, C., *J. Chem. Phys.*, 48, 1109, 1968.
29. Clement, M. J. Y. and Ramsay, D. A., *Can. J. Phys.*, 39, 205, 1961.
30. Clyne, M. A. A. and Connor, J., *J. Chem. Soc. Faraday Trans. 2*, 68, 1220, 1972.
31. Colussi, A. J. and Benson, S. W., *Int. J. Chem. Kinet.*, 9, 295, 1977.
32. Colussi, A. J. and Benson, S. W., *Int. J. Chem. Kinet.*, 9, 307, 1977.
33. Colussi, A. J., Zabel, F., and Benson, S. W., *Int. J. Chem. Kinet.*, 9, 161, 1977.
34. Colussi, A. J. and Benson, S. W., *Int. J. Chem. Kinet.*, 10, 1139, 1978.
35. Coomber, J. W. and Whittle, E., *Trans. Faraday Soc.*, 63, 1394, 1967.
36. Coomber, J. W. and Whittle, E., *Trans. Faraday Soc.*, 63, 2656, 1967.
37. Cox, J. D. and Pilcher, G., *Thermochemistry of Organic and Organometallic Compounds*, Academic Press, New York, 1970.
38. Czarnarski, J., Castellano, E., and Shumacher, H. J., *Chem. Commun.*, p. 1255, 1968.
39. Daly, M. and Price, S. J., *Can. J. Chem.*, 54, 1814, 1976.
40. Darwent, D. de B., *Bond Dissociation Energies in Simple Molecules*, National Bureau of Standards. NSRDS-NBS, Washington, D.C., 31, 1970.
41. Davidson, I. M. T. and Howard, A. B., *J. Chem. Soc. Faraday Trans. 1*, 71, 69, 1975.
42. Davis, D. D. and Okabe, H., *J. Chem. Phys.*, 49, 5526, 1968.
43. Descamps, B. and Forst, W., *J. Phys. Chem.*, 80, 933, 1976.
44. Devore, J. A. and O'Neal, H. E., *J. Phys. Chem.*, 73, 2644, 1969.
45. Dibeler, V. H. and Reese, R. M., *J. Res. Natl. Bur. Stand.*, 54, 127, 1955.
46. Doncaster, A. M. and Walsh, R., *J. Chem. Soc. Faraday Trans. 1*, 72, 2908, 1976.
47. Doncaster, A. M. and Walsh, R., *Int. J. Chem. Kinet.*, 10, 101, 1978.
48. Doncaster, A. M. and Walsh, R., *J. Chem. Soc. Chem. Commun.*, 904, 1979.
49. Ferguson, K. C., Okafo, E. N., and Whittle, E., *J. Chem. Soc. Faraday Trans. 1*, 69, 295, 1973.
50. Ferguson, K. C. and Whittle, E., *Trans. Faraday Soc.*, 67, 2618, 1971; Jones, S. H. and Whittle, E., *Int. J. Chem. Kinet.*, 2, 479, 1970.
51. Ferguson, K. C. and Whittle, E., *J. Chem. Soc. Faraday Trans. 1*, 68, 295, 1972.
52. Ferguson, K. C. and Whittle, E., *J. Chem. Soc. Faraday Trans. 1*, 68, 306, 1972.
53. Ferguson, K. C. and Whittle, E., *J. Chem. Soc. Faraday Trans. 1*, 68, 641, 1972.
54. Foon, R. and Tait, K. B., *J. Chem. Soc. Faraday Trans. 1*, 68, 104, 1972.
55. Foon, R. and Tait, K. B., *J. Chem. Soc. Faraday Trans. 1*, 68, 1121, 1972.
56. Franklin, J. A., Huybrechts, G. H., and Cillien, C., *Trans. Faraday Soc.*, 65, 2094, 1969.
57. Franklin, J. A., and Huybrechts, G. H., *Int. J. Chem. Kinet.*, 1, 1, 1969.
58. Furuyama, S., Golden, D. M., and Benson, S. W., *J. Am. Chem. Soc.*, 91, 7564, 1969.
59. Furuyama, S., Golden, D. M., and Benson, S. W., *Int. J. Chem. Kinet.*, 2, 83, 1970.
60. Furuyama, S., Golden, D. M., and Benson, S. W., *Int. J. Chem. Kinet.*, 2, 93, 1970.
61. Furuyama, S., Golden, D. M., and Benson, S. W., *Int. J. Chem. Kinet.*, 3, 237, 1971.
62. Gilroy, K. M., Price, S. J., and Webster, N. J., *Can. J. Chem.*, 50, 2639, 1972.
63. Golden, D. M., and Benson, S. W., *Chem. Rev.*, 69, 125, 1969.
64. Golden, D. M., Solly, R. K., Gac, N. A., and Benson, S. W., *J. Am. Chem. Soc.*, 94, 363, 1972.
65. Gowenlock, B. G., Johnson, C. A. F., Keary, C. M., and Pfab, J., *J. Chem. Soc. Perkin Trans. 2*, 351, 1975.
66. Gray, P., *Q. Rev. (London)*, 17, 441, 1963.

66a. Howard, C. J. *J. Am. Chem. Soc.,* 102, 6937, 1980.
67. Hwang, R. J. and Benson, S. W., *Int. J. Chem. Kinet.,* 11, 579, 1979.
68. Hwang, R. J. and Benson, S. W., *J. Am. Chem. Soc.,* 101, 2615, 1979.
69. Ireton, R., Gordon, A. S., and Tardy, D. C., *Int. J. Chem. Kinet.,* 9, 769, 1977.
70. James, D. G. L. and Suart, R. D., *Trans. Faraday Soc.,* 64, 2752, 1968.
71. Kerr, J. A., *Chem. Rev.,* 66, 465, 1966.
72. Kerr, J. A. and Timlin, D. M., *Int. J. Chem. Kinet.,* 3, 427, 1971.
73. Kiang, T., Estler, R. C., and Zare, R. N., *J. Chem. Phys.,* 70, 5925, 1979.
74. King, K. D. and Goddard, R. D., *J. Am. Chem. Soc.,* 97, 4504, 1975.
75. King, K. D. and Goddard, R. D., *Int. J. Chem. Kinet.,* 7, 837, 1975.
76. King, K. D. and Goddard, R. D., *J. Phys. Chem.,* 80, 546, 1976.
77. King, K. D., Golden, D. M., and Benson, S. W., *J. Am. Chem. Soc.,* 92, 5541, 1970.
78. King, K. D., Golden, D. M., and Benson, S. W., *J. Phys. Chem.,* 75, 987, 1971.
79. King, K. D., *Int. J. Chem. Kinet.,* 9, 907, 1977.
80. King, K. D., *Int. J. Chem. Kinet.,* 10, 545, 1978.
81. Knoll, H., Scherker, K., and Geiseler, G., *Int. J. Chem. Kinet.,* 5, 271, 1973.
82. Kominar, R. J., Krech, M. J., and Price, S. J. W., *Can. J. Chem.,* 54, 2981, 1976.
83. Kominar, R. J. and Price, S. J., *Can. J. Chem.,* 47, 991, 1969.
84. Kominar, R. J., Krech, M. J., and Price, S. J. W., *Can. J. Chem.,* 56, 1589, 1978.
85. Koski, A. A., Price, S. J. W., and Trudell, B. C., *Can. J. Chem.,* 54, 482, 1976.
86. Krech, M. J., Price, S. J. W., and Yared, W. F., *Int. J. Chem. Kinet.,* 6, 257, 1974.
87. LaLonde, A. C. and Price, S. J. W., *Can. J. Chem.,* 49, 3367, 1971.
88. Larson, C. W., Hardwidge, E. A., and Rabinovitch, B. S., *J. Chem. Phys.,* 50, 2769, 1969.
89. Laufer, A. H. and Okabe, H., *J. Am. Chem. Soc.,* 93, 4137, 1971.
90. Leyland, L. M., Majer, J. R., and Robb, J. C., *Trans. Faraday Soc.,* 66, 898, 1970.
91. McMillen, D. F., Golden, D. M., and Benson, S. W., *Int. J. Chem. Kinet.,* 3, 359, 1971.
92. McMillen, D. F., Golden, D. M., and Benson, S. W., *Int. J. Chem. Kinet.,* 4, 487, 1972.
93. Mendenhall, G. D., Golden, D. M., and Benson, S. W., *J. Phys. Chem.,* 77, 2707, 1973.
94. Okabe, H. and Dibeler, V. H., *J. Chem. Phys.,* 59, 2430, 1973.
95. Okafo, E. N. and Whittle, E., *J. Chem. Soc. Faraday Trans. 1,* 70, 1366, 1974.
96. Okafo, E. N. and Whittle, E., *Int. J. Chem. Kinet.,* 7, 287, 1975.
97. O'Neal, H. E., Bagg, J. W., and Richardson, W. H., *Int. J. Chem. Kinet.,* 2, 493, 1970.
98. O'Neal, H. E. and Benson, S. W., in *Free Radicals,* Kochi, J. K., Ed., John Wiley & Sons, New York, 1973, 275.
99. Pankratov, A. V., Zercheninov, A. N., Chesnokov, V. I., and Zhdanova, N. N., *Zh. Fiz. Khim.,* 43, 394, 1969.
100. Paul, S. and Back, M. H., *Can. J. Chem.,* 53, 3330, 1975.
101. Perona, M. J. and Golden, D. M., *Int. J. Chem. Kinet.,* 5, 55, 1973.
102. Petry, R. C., *J. Am. Chem. Soc.,* 89, 4600, 1967.
103. Pickard, J. M. and Rodgers, A. S., *Int. J. Chem. Kinet.,* 8, 809, 1976.
104. Pickard, J. M. and Rodgers, A. S., *J. Am. Chem. Soc.,* 99, 695, 1977.
105. Pickard, J. M. and Rodgers, A. S., *Int. J. Chem. Kinet.,* 9, 759, 1977.
106. Piper, L. G., *J. Chem. Phys.,* 70, 3417, 1979.
107. Potzinger, P. and Lampe, F. W., *J. Phys. Chem.,* 74, 719, 1970.
108. Price, S. J., Richard, J. P., Rufeldt, R.C., and Jacko, M. G., *Can. J. Chem.,* 51, 1397, 1973.
109. Price, S. J. W. and Sapiano, H. J., *Can. J. Chem.,* 57, 1468, 1979.
110. Reed, K. J. and Brauman, J. L., *J. Am. Chem. Soc.,* 97, 1625, 1975.
111. Ring, M. A., Puentes, M. J., and O'Neal, H. E., *J. Am. Chem. Soc.,* 92, 4845, 1970; Steele, W. C. and Stone, F. G. A., *J. Am. Chem. Soc.,* 84, 3599, 1962; Steele, W. C., Nichols, L. D., and Stone, F. G. A., *J. Am. Chem. Soc.,* 84, 441, 1962.
112. Rodgers, A. S. and Ford, W. G. F., *Int. J. Chem. Kinet.,* 5, 965, 1973.
113. Rodgers, A. S. and Wu, M. C. R., *J. Am. Chem. Soc.,* 95, 6913, 1973.
114. Rossi, M. and Golden, D. M., *Int. J. Chem. Kinet.,* 11, 715, 1979.
115. Rossi, M. and Golden, D. M., *J. Am. Chem. Soc.,* 101, 1230, 1979.
116. Sen Sharma, D. K. and Franklin, J. L., *J. Am. Chem. Soc.,* 95, 6562, 1973.
117. Shaw, R., *Int. J. Chem. Kinet.,* 5, 261, 1973.
118. Simmons, J. W., Hase, W. L., Phillips, R. J., Porter, E. J., and Growcock, F. B., *Int. J. Chem. Kinet.,* 7, 879, 1975.
119. Solly, R. K. and Benson, S. W., *Int. J. Chem. Kinet.,* 1, 427, 1969.
120. Solly, R. K. and Benson, S. W., *Int. J. Chem. Kinet.,* 3, 509, 1971.
121. Solly, R. K. and Benson, S. W., *J. Am. Chem. Soc.,* 93, 1592, 1971.
122. Solly, R. K., Golden, D. M., and Benson, S. W., *Int. J. Chem. Kinet.,* 2, 11, 1970.
123. Solly, R. K., Golden, D. M., and Benson, S. W., *Int. J. Chem. Kinet.,* 2, 381, 1970.
124. Trenwith, A. B., *Trans. Faraday Soc.,* 66, 2805, 1970.
125. Trenwith, A. B. and Wrigley, S. P., *J. Chem. Soc. Faraday Trans. 1,* 73, 817, 1977.
126. Tsang, W., *Int. J. Chem. Kinet.,* 5, 929, 1973.
127. Tsang, W., *Int. J. Chem. Kinet.,* 10, 687, 1978.

128. Tsang, W., *Int. J. Chem. Kinet.*, 10, 1119, 1978.
129. van der Bergh, H. and Troe, J., *J. Chem. Phys.*, 64, 736, 1976.
130. Vanderwielen, A. J., Ring, M. A., and O'Neal, H. E., *J. Am. Chem. Soc.*, 97, 993, 1975.
131. Vincow, G., Dauben, H. J., Hunter, F. R., and Volland, W. V., *J. Am. Chem. Soc.*, 91, 2823, 1969.
132. Walker, L. C., *J. Chem. Thermodyn.*, 4, 219, 1972.
133. Walker, R. F. and Phillips, L., *J. Chem. Soc. A*, 2103, 1968.
134. Walsh, R. and Wells, J. M., *J. Chem. Soc. Faraday Trans. 1*, 72, 100, 1976.
135. Walsh, R. and Wells, J. M., *J. Chem. Soc. Faraday Trans. 1*, 72, 1212, 1976.
136. Watkins, K. W. and Thompson, W. W., *Int. J. Chem. Kinet.*, 5, 791, 1973.
137. White, J. N. and Gardiner, W. C., *Chem. Phys. Lett.*, 58, 470, 1978.
138. Wu, E-C. and Rodgers, A. S., *Int. J. Chem. Kinet.*, 5, 1001, 1973.
139. Wu, E-C. and Rodgers, A. S., *J. Phys. Chem.*, 78, 2315, 1974.
140. Wu, E-C. and Rodgers, A. S., *J. Am. Chem. Soc.*, 98, 6112, 1976.
141. Zabel, F., Benson, S. W., and Golden, D. M., *Int. J. Chem. Kinet.*, 10, 295, 1978.
142. Zmbov, K. F., Uy, O. M., and Margrave, J. L., *J. Am. Chem. Soc.*, 90, 5090, 1968.
143. Cook, K. D. and Taylor, J. W., *Int. J. Mass Spectrom. Ion Phys.*, 30, 93, 1979.

HEATS OF FORMATION OF FREE RADICALS

The heats of formation of the free radicals are related to the corresponding bond strengths by the equations

$$D(R - X) = \Delta H_f^\circ(\dot{R}) + \Delta H_f^\circ(\dot{X}) - \Delta H_f^\circ(RX)$$

or

$$D(R - R) = 2\Delta H_f^\circ(\dot{R}) - \Delta H_f^\circ(RR)$$

For an excellent review of the methods of determining the heats of formation of free radicals the reader is referred to "Thermochemistry of Free Radicals" by H. E. O'Neal and S. W. Benson in *Free Radicals,* Kochi, J. K., Ed., John Wiley & Sons, New York, 1973, 275.

The references are the same as given for Table 3.

Table 4
HEATS OF FORMATION OF FREE RADICALS

Radical	$\Delta H^\circ_{f(298)}$/kcal mol^{-1}	$\Delta H^\circ_{f(298)}$/kJ mol^{-1}	Ref.
CH	142 ± 1	594 ± 4	1,28
CH_2	92 ± 1	385 ± 4	1,28
$^1CH_2$	101 ± 3	423 ± 13	118
CH_3	35.1 ± 0.15	146.9 ± 0.63	8
Ethynyl	128 ± 1	536 ± 4	94
Vinyl	≥68	≥285	63
C_2H_5	25.9 ± 1.3	108.4 ± 5.4	63
Propargyl	82.1 ± 1.0	343.5 ± 4.2	80,127
Allyl	39.4 ± 1.5	164.9 ± 6.3	115
Cyclopropyl	66.9 ± 0.25	279.9 ± 1.1	9
n-C_3H_7	22.6 ± 1.8	94.6 ± 7.5	63
i-C_3H_7	18.2 ± 1.5	76.2 ± 6.3	63
$CH_3C{\equiv}CCH_2$	73.6	308	127
Methallyl	30.4 ± 1.3	127.2 ± 5.4	63
Isobutenyl	30.0 ± 1.0	125.5 ± 4.2	125,126
Cyclopropylcarbinyl	51.5 ± 1.6	215.5 ± 6.7	91
Cyclobutyl	51.2 ± 1.0	214.2 ± 4.2	1,37,50,92
s-C_4H_9	13.0 ± 2	54.4 ± 8	63
t-C_4H_9	7.6 ± 1.2	31.8 ± 5.0	63
Cyclopentadien-1,3-yl-5	60.9 ± 1.2	254.8 ± 5.0	61
Pentadienyl	52.9 ± 1.2	221.3 ± 5.0	63
Cyclopentenyl-3	38.4 ± 1	160.7 ± 4	60
Spiropentyl	91.0 ± 1	380.7 ± 4	1,37,50
$(CH_3)_2CC{\equiv}CH$	61.5 ± 2.0	257.3 ± 8.4	79,116
Dimethylallyl	18.3 ± 1.5	76.6 ± 6.3	124
Cyclopentyl	24.3 ± 1	101.7 ± 4	1,37,50,59
Neo-C_5H_{11}	7.6 ± 1	31.8 ± 4	1,37,88
Phenyl	77.7 ± 2.0	325.1 ± 8.4	25
Cyclohexadien-1,3-yl-5	44 ± 5	184 ± 21	70
Cyclohexyl	13.9 ± 1	58.2 ± 4	1,37,50
$(CH_3)_2C{=}C(CH_3)CH_2$	9.6 ± 1.1	40.2 ± 4.6	113
Benzyl	47.8 ± 1.5	200.0 ± 6.3	115
Cycloheptatrien-1,3-yl-7	64.8	271.1	131
Norbornyl	32.6 ± 2.5	136.4 ± 10.5	97
Cycloheptyl	12.2 ± 1	51.1 ± 4	1,37,50
$C_6H_5CHCH_3$	38	159	20,98
CN	101 ± 1	423 ± 4	42
CH_2NH_2	37.0 ± 2.0	154.8 ± 8.4	32,116
CH_3NH	45.4 ± 1	190.0 ± 4	64
CH_2CN	58.5 ± 2.2	209.6 ± 9.6	74

Table 4 (continued)
HEATS OF FORMATION OF FREE RADICALS

Radical	$\Delta H^\circ_{f(298)}$/kcal mol⁻¹	$\Delta H^\circ_{f(298)}$/kJ mol⁻¹	Ref.
$(CH_3)_2N$	38.5 ± 1	161.1 ± 4	64
CH_3CHCN	50.1 ± 2.3	209.6 ± 9.6	74
$(CH_3)_2CCN$	39.8 ± 2.0	166.5 ± 8.4	76
C_6H_5NH	55.1	230.5	34
$C_6H_5NCH_3$	53.2	222.6	34
$C_6H_5C(CH_3)CN$	54.4	227.6	20,98
CHO	7.7 ± 1.2	32.2 ± 5.0	63,98
CH_2OH	-6.2 ± 1.5	-25.9 ± 6.3	63,98
CH_3O	3.8 ± 0.2	15.9 ± 0.8	11,13
CH_3CO	-5.8 ± 0.4	-24.3 ± 1.7	63,98
CH_2CHO	3.0 ± 2.0	12.6 ± 8.4	114
CH_3CHOH	-15.2 ± 1.0	-63.6 ± 4.2	2
CH_3OCH_2	-2.8 ± 1.2	-11.7 ± 5.0	63,98
C_2H_5O	-4.1	-17.2	11
$CH_2=CHCO$	17.3	72.4	3
C_2H_5CO	-10.2 ± 1.0	-42.7 ± 4.2	136
CH_3COCH_2	-5.7 ± 2.6	-23.9 ± 10.9	77
$CH_2=CHCHOH$	0.0	0.0	5
$(CH_3)_2COH$	-26.6 ± 1.1	-111.3 ± 4.6	63,98
$CH_3COCHCH_3$	-16.8 ± 1.7	-70.3 ± 7.1	123
$n\text{-}C_3H_7O$	-9.9	-41.4	11
$i\text{-}C_3H_7O$	-12.5	-52.3	11
Tetrahydrofuran-2-yl	-4.3 ± 1.5	-18.0 ± 6.3	20,98
$n\text{-}C_4H_9O$	-14.7	-61.5	98
$s\text{-}C_4H_9O$	-16.6 ± 0.8	-69.5 ± 3.3	13
$(CH_3)_3CO$	-21.7	-90.8	11,13
C_6H_5O	11.4 ± 2.0	47.7 ± 8.4	24,33,10D
C_6H_5CO	26.1 ± 2	109.2 ± 8	98,121
COOH	-53.3	-223.0	20,121
$COOCH_3$	-40.4 ± 1	-169.0 ± 4	98,119
CH_3COO	-49.6 ± 1	-207.5 ± 4	98
C_2H_5COO	-54.6 ± 1	-228.5 ± 4	98
$n\text{-}C_3H_7COO$	-59.6 ± 1	-249.4 ± 4	98
$C_6H_5COOCH_2$	-16.7 ± 2.0	-69.9 ± 8.4	98,120
CFO	$\leqslant -44$	$\leqslant -184$	98
CH_3S	34.2 ± 2.0	143.1 ± 8.4	31
C_6H_5S	56.8 ± 2.0	237.7 ± 8.4	31
CH_3SO_2	-57.2	-239.3	98
CH_2F	-7.8 ± 2	-32.6 ± 8	72
CHF_2	-59.2 ± 2	-247.7 ± 8	72
CF_2	-46.4 ± 2.2	-194.1 ± 9.2	95,142
CF_3	-112.0 ± 3.6	-468.6 ± 15.1	63,98
C_6F_5	-92.6 ± 2.9	-387.4 ± 12.0	109
CCl_2	57	239	98
$CHCl_2$	24.1	100.8	58
CF_2Cl	-64.3 ± 2	-269.0 ± 8	55,90
$CFCl_2$	-23	-96	54
CCl_3	19.0 ± 1	79.5 ± 4	93
CH_2Br	41.5	173.6	58
$CHBr_2$	54.3	227.2	58
CH_2I	55.0 ± 1.6	230.1 ± 6.7	63,98
CHI_2	79.8 ± 2.2	333.9 ± 9.2	63,98
CH_3CF_2	-72.3 ± 2	-302.5 ± 8.4	104
CF_3CH_2	-123.6 ± 1.2	-517.1 ± 5.0	139
C_2F_5	-217.3 ± 1	-909.2 ± 4	98,140
$CHCl_2CCl_2$	5.6 ± 2	23.4 ± 8	1,37,57
CF_2ClCF_2	-164 ± 4	-686 ± 17	55
C_2Cl_5	8.4 ± 2	35.2 ± 8	1,37,56
C_6F_5	-130.9 ± 2	-547.7 ± 8	26
NH	84.2 ± 2.3	352.3 ± 9.6	106
NH_2	47.2 ± 1	197.5 ± 4	21,64,98
NF_2	8 ± 1	34 ± 4	1,19,40
PH_2	33 ± 2	138 ± 8	98
HO	9.2 ± 1	38.5 ± 4	71
HO_2	2.5 ± 0.6	10.5 ± 2.5	66a
HS	33.6 ± 1.1	140.6 ± 4.6	67
SiH_2	58.6 ± 3.5	245.2 ± 14.6	130
SiH_3	49.1 ± 3	205.4 ± 13	107,111
CH_3SiH	50.9 ± 3.5	213.0 ± 14.6	130
$(CH_3)_3Si$	-2.6	-10.9	41
$C_6H_5Si(CH_3)_2$	39	163	98
$(C_6H_5)_2SiCH_3$	78	326	98
$(C_6H_5)_3Si$	116.2	486.2	22
SiH_3SiH	64.5 ± 3.5	269.9 ± 14.6	130
SiF_3	-235 ± 14	-983 ± 60	47
GeH_3	57	239	98

BOND STRENGTHS OF SOME ORGANIC MOLECULES

Bond strengths at 298 K expressed in kcal/mol for some organic molecules of the general formula R–X are presented below. Some are experimental values taken from the preceding tables; the remainder are calculated from the heats of formation of the radicals, listed above, and the heats of formation of the parent compounds from sources indicated by the references below. The table also includes the bond strengths for ammonia, water, and the hydrogen halides.

Table 5
BOND STRENGTHS OF SOME ORGANIC MOLECULES

R \ X	H	F	Cl	Br	I	OH	NH$_2$	CH$_3$O	CH$_3$	CH$_3$CO	NO	CF$_3$
H	104.207	135.9	103.2	87.4	71.4	119	110	104	104	86	50	106
CH$_3$	105	109[2]	84[1]	70[1]	56[1]	91[1]	87[1]	81[1]	88[1]	80[1]	42	101
C$_2$H$_5$	98	—	81[1]	68[1]	53[1]	91[1]	84[1]	81[1]	85[1]	77[1]	42	—
i-C$_3$H$_7$	95	106[1]	81[1]	68[1]	53[1]	92[1]	85[1]	81[1]	84[1]	75[1]	41	—
t-C$_4$H$_9$	92	—	81[1]	67[1]	50[1]	91[1]	84[1]	81[1]	82[1]	71[1]	41	—
C$_6$H$_5$	110	125[1]	95[1]	80	64[1]	110[1]	104[1]	98[1]	100[1]	93[1]	52	—
C$_6$H$_5$CH$_2$	85	—	69[1]	55[1]	45[1]	78[1]	—	—	72[1]	63[1]	—	—
CCl$_3$	96	102[1]	70	56	—	—	—	—	88[4]	—	32	—
CF$_3$	106	130[1]	86	71	53	—	—	—	99[1]	—	31	97
C$_2$F$_5$	103[6]	127[6]	85[6]	68[6]	53[6]	—	—	—	—	—	—	—
CH$_3$CO	86	119[1]	81[1]	67[1]	50[1]	106[1]	99[3]	97	80[1]	67	—	—
CN	120	—	—	83	73	—	—	—	—	—	29	—
C$_6$F$_5$	114	114[5]	92[5]	—	66[5]	107[5]	—	—	105[5]	—	50[5]	60[5]

REFERENCES

1. Cox, J. D. and Pilcher, G., *Thermochemistry of Organic and Organometallic Compounds*, Academic Press, London, 1970.
2. Lacher, J. R. and Skinner, H. A., *J. Chem. Soc. A*, 1034, 1968.
3. Benson, S. W. et al., *Chem. Rev.*, 69, 279, 1969.
4. Hu, A. T., Sinke, G. C., and Mintz, M. J., *J. Chem. Thermodyn.*, 4, 239, 1972.
5. Choo, K. Y., Mendenhall, G. D., Golden, D. M., and Benson, S. W., *Int. J. Chem. Kinet.*, 6, 813, 1974.
6. Wu, E.-C. and Rodgers, A. S., *J. Am. Chem. Soc.*, 98, 6112, 1976.

THE MADELUNG CONSTANT AND CRYSTAL LATTICE ENERGY

Donald F. Swinehart

If U is the crystal lattice energy and M is the Madelung constant, then

$$U = \frac{N\,M\,z_i z_j\,e^2}{r}\,(1 - 1/n)^a$$

Substance	Ion type	Crystal form[b]	M
Sodium chloride, NaCl	M^+, X^-	FCC	1.74756
Cesium chloride, CsCl	M^+, X^-	BCC	1.76267
Calcium chloride, $CaCl_2$	$M^{++}, 2X^-$	Cubic	2.365
Calcium fluoride, fluorite, CaF_2	$M^{++}, 2X^-$	Cubic	2.51939
Cadmium chloride, $CdCl_2$	$M^{++}, 2X^-$	Hexagonal	2.244[c]
Cadmium iodide, $CdI_2(\alpha)$	$M^{++}, 2X^-$	Hexagonal	2.355[c]
Magnesium fluoride, MgF_2	$M^{++}, 2X^-$	Tetragonal	2.381[c]
Cuprous oxide, cuprite, Cu_2O	$2M^+, X^=$	Cubic	2.22124
Zinc oxide, ZnO	$M^{++}, X^=$	Hexagonal	1.4985[c]
Sphalerite, zinc blende, ZnS	$M^{++}, X^=$	FCC	1.63806
Wurtzite, ZnS	$M^{++}, X^=$	Hexagonal	1.64132[c]
Titanium dioxide, anatase, TiO_2	$M^{4+}, 2X^=$	Tetragonal	2.400[c]
Titanium dioxide, rutile, TiO_2	$M^{4+}, 2X^=$	Tetragonal	(2.408) − (2.055) $(0.721 - c/a)^{2\,d}$
β-Quartz, SiO_2	$M^{4+}, 2X^=$	Hexagonal	2.2197[c]
Corundum, Al_2O_3	$2M^{3+}, 3X^=$	Rhombohedral	4.1719

[a] N is Avogadro's number, z_i and z_j are the integral charges on the ions (i.e., in units of the electronic charge), and e is the charge on the electron in electrostatic units (e = 4.803 × 10^{-10} ESU). r is the shortest distance between cation-anion pairs in centimeters. Then U is in ergs.

[b] FCC = face centered cubic, BCC is body centered cubic.

[c] For tetragonal and hexagonal crystals the value of M depends on the details of the lattice parameters.

[d] For rutile-type structures, M is given to a good approximation by inserting the proper c/a ratio. c/a for rutile itself is 0.721 and $M = 2.408$.

Note: Several variations of the equation for U appear in the literature with corresponding variations in M. In general, using literature values for M requires caution. The Born exponent, n, may be obtained from the following table:

THE BORN EXPONENT, n

Ion type	n
He, Li$^+$	5
Ne, Na$^+$, F$^-$	7
Ar, K$^+$, Cu$^+$, Cl$^-$	9
Kr, Rb$^+$, Ag$^+$, Br$^-$	10
Xe, Cs$^+$, Au$^+$, I$^-$	12

For a crystal with a mixed-ion type, an average of the values of n in this table is to be used (6 for LiF, for example).

VALUES OF THE GAS CONSTANT, R

Coleman J. Major

The numerical value of the gas constant, R, defined by the equation $PV = nRT$, depends upon the units of P, V, n, and T. A large number of values of the constant may be calculated. The accompanying table gives 84 values of R in a convenient form using the most common units of pressure and volume. It also incorporates both the pound and gram mole and both Rankine and Kelvin temperature scales. Various combinations of metric and English units may, therefore, be used without the necessity of converting each variable to a common system of units. Conversion factors and constants used for computing the values of R are listed at the bottom of the table.

The following example illustrates the use of the table:

Calculate: The volume in ft^3 occupied by 2 lb. moles of a gas at 15°C at a pressure of 32.2 ft. of water, assuming the ideal gas law.

Solution: $15°C + 273.2 = 288.2°K$

Enter the top of the table under the column headed "ft H_2O" and proceed downward to the value of 44.6 for R. (Note that this lines up horizontally with the desired units of ft^3, °K, and lb. moles shown on the left side of the table)

$$V = \frac{nRT}{P} = \frac{2 \times 44.6 \times 288.2}{32.2} = 798 \ ft^3$$

Values of Gas Constant, $R = \dfrac{PV}{nT}$

Absolute Pressure

Volume	Temp.	moles	Atm	psia	mm Hg	cm Hg	in Hg	in H₂O	ft H₂O
ft³	°K	gm	0.00290	0.0426	2.20	0.220	0.0867	1.18	0.0982
		lb	1.31	19.31	999	99.9	39.3	535	44.6
	°R	gm	0.00161	0.02366	1.22	0.122	0.0482	0.655	0.0546
		lb	0.730	10.73	555	55.5	21.8	297	24.8
cm³	°K	gm	82.05	1206	62,400	6240	2450	33,400	2780
		lb	37,200	547,000	2.83×10^7	2.83×10^6	1.11×10^6	1.51×10^7	1.26×10^6
	°R	gm	45.6	670	34,600	3460	1360	18,500	1550
		lb	20,700	304,000	1.57×10^7	1.57×10^6	619,000	8.41×10^6	701,000
liters	°K	gm	0.08205	1.206	62.4	6.24	2.45	33.4	2.78
		lb	37.2	547	28,300	2830	1113	15,140	1262
	°R	gm	0.0456	0.670	34.6	3.46	1.36	18.5	1.55
		lb	20.7	304	15,700	1570	619	8410	701

Conversion Factors and Constants

1 lb. = 453.59 gm
1 atm = 14.696 psia
1 atm = 760 mm Hg
1 atm = 76 cm Hg
1 atm = 29.921 in Hg
1 atm = 406.79 in H_2O
1 atm = 33.90 ft H_2O
1 atm = 1.01325 Pa

359.0 ft^3/lb mole
22,414 cm^3/gm mole
1 inch = 2.54 cm
Std. temp. = 273.15°K or 491.67°R
28.31605 liters = 1 ft^3
$R = 8.31441 (26) \times 10^7$ erg °K^{-1} mol^{-1}
    = 1.9872 cal °K^{-1} mol^{-1}

RECOMMENDED CONSISTENT VALUES OF THE FUNDAMENTAL PHYSICAL CONSTANTS

The numbers in parentheses are the standard deviation uncertainties in the last digits of the quoted value, computed on the basis of internal consistency.

Quantity	Symbol	Value	Uncertainty, ppm
1. Permeability of Vacuum	μ_0	$4\pi \times 10^{-7}$ H m^{-1}	
		$= 12.5663706144 \times 10^{-7}$ H m^{-1}	
2. Speed of Light in Vacuum	c	$2.99792458(1.2) \times 10^8$ m s^{-1}	0.004
3. Permittivity of Vacuum	$\epsilon_0 = (\mu_0 c^2)^{-1}$	$8.85418782(7) \times 10^{-12}$ F m^{-1}	0.008
4. Fine Structure Constant, $\mu_0 c e^2 /2h$	α	$0.0072973506(60)$	0.82
	α^{-1}	$137.03604(11)$	0.82
5. Elementary Charge	e	$1.6021892(46) \times 10^{-19}$ C	2.9
6. Planck Constant	h	$6.626176(36) \times 10^{-34}$ J Hz^{-1}	5.4
	$h = h/2\pi$	$1.0545887(57) \times 10^{-34}$ J s	5.4
7. Avogadro Constant	N_A	$6.022045(31) \times 10^{23}$ mol^{-1}	5.1
8. Atomic Mass Unit	$1u = (10^{-3}$ kg mol$^{-1})/N_A$	$1.6605655(86) \times 10^{-27}$ kg	5.1
9. Electron Rest Mass	m_e	$0.9109534(47) \times 10^{-30}$ kg	5.1
		$5.4858026(21) \times 10^{-4}$ u	0.38
10. Muon Rest Mass	m_μ	$1.883566(11) \times 10^{-28}$ kg	5.6
		$0.11342920(26)$ u	2.3
11. Proton Rest Mass	m_p	$1.6726485(86) \times 10^{-27}$ kg	5.1
		$1.007276470(11)$ u	0.011
12. Neutron Rest Mass	m_n	$1.6749543(86) \times 10^{-27}$ kg	5.1
		$1.008665012(37)$ u	0.037
13. Ratio, Proton Mass to Electron Mass	m_p/m_e	$1,836.15152(70)$	0.38
14. Ratio, Muon Mass to Electron Mass	m_μ/m_e	$206.76865(47)$	2.3
15. Specific Electron Charge	e/m_e	$1.7588047(49) \times 10^{11}$ C kg^{-1}	2.8
16. Faraday Constant	$\mathscr{F} = N_A e$	$9.648456(27) \times 10^4$ C mol^{-1}	2.8
17. Magnetic Flux Quantum	$\Phi_0 = h/2e$	$2.0678506(54) \times 10^{-15}$ Wb	2.6
	h/e	$4.135701(11) \times 10^{-15}$ J Hz^{-1} C^{-1}	2.6
18. Josephson Frequency-Voltage Ratio	$2e/h$	$483.5939(13)$ THz V^{-1}	2.6
19. Quantum of Circulation	$h/2m_e$	$3.6369455(60) \times 10^{-4}$ J Hz^{-1} kg^{-1}	1.6
	h/m_e	$7.273891(12) \times 10^{-4}$ J Hz^{-1} kg^{-1}	1.6
20. Rydberg Constant	R_∞	$1.097373177(83) \times 10^7$ m^{-1}	0.075
21. Bohr Radius	$a_0 = \alpha/4\pi R_\infty$	$0.52917706(44) \times 10^{-10}$ m	0.82
22. Electron Compton Wavelength	$\lambda_c = \alpha^2/2R_\infty$	$2.4263089(40) \times 10^{-12}$ m	1.6
	$\lambda_c = \lambda_c/2\pi = \alpha a_0$	$3.8615905(64) \times 10^{-13}$ m	1.6
23. Classical Electron Radius	$r_e = \mu_0 e^2/4\pi m_e = \alpha \lambda_c$	$2.8179380(70) \times 10^{-15}$ m	2.5
24. Electron g-Factor	$\frac{1}{2}g_e = \mu_e/\mu_B$	$1.0011596567(35)$	0.0035
25. Muon g-Factor	$\frac{1}{2}g_\mu$	$1.00116616(31)$	0.31
26. Proton Moment in Nuclear Magnetons	μ_p/μ_N	$2.7928456(11)$	0.38
27. Bohr Magneton	$\mu_B = e\hbar/2m_e$	$9.274078(36) \times 10^{-24}$ J T^{-1}	3.9
28. Nuclear Magneton	$\mu_N = e\hbar/2m_p$	$5.050824(20) \times 10^{-27}$ J T^{-1}	3.9
29. Electron Magnetic Moment	μ_e	$9.284832(36) \times 10^{-24}$ J T^{-1}	3.9
30. Proton Magnetic Moment	μ_p	$1.4106171(55) \times 10^{-26}$ J T^{-1}	3.9
31. Proton Magnetic Moment in Bohr Magnetons	μ_p/μ_B	$1.521032209(16) \times 10^{-3}$	0.011
32. Ratio, Electron to Proton Magnetic Moments	μ_e/μ_p	$658.2106880(66)$	0.010
33. Ratio, Muon Moment to Proton Moment	μ_μ/μ_p	$3.1833402(72)$	2.3
34. Muon Magnetic Moment	μ_μ	$4.490474(18) \times 10^{-26}$ J T^{-1}	3.9
35. Proton Gyromagnetic Ratio	γ_p	$2.6751987(75) \times 10^8$ s^{-1} T^{-1}	2.8
36. Diamagnetic Shielding Factor, Spherical H$_2$O Sample	$1 + \sigma(H_2O)$	$1.000025637(67)$	0.067
37. Proton Gyromagnetic Ratio (uncorrected)	γ'_p	$2.6751301(75) \times 10^8$ s^{-1} T^{-1}	2.8
	$\gamma'_p/2\pi$	$42.57602(12)$ MHz T^{-1}	2.8
38. Proton Moment in Nuclear Magnetons (uncorrected)	μ'_p/μ_N	$2.7927740(11)$	0.38
39. Proton Compton Wavelength	$\lambda_{C,p} = h/m_p c$	$1.3214099(22) \times 10^{-15}$ m	1.7
	$\lambda_{C,p} = \lambda_{C,p}/2\pi$	$2.1030892(36) \times 10^{-16}$ m	1.7
40. Neutron Compton Wavelength	$\lambda_{C,n} = h/m_n c$	$1.3195909(22) \times 10^{-15}$ m	1.7
	$\lambda_{C,n} = \lambda_{C,n}/2\pi$	$2.1001941(35) \times 10^{-16}$ m	1.7
41. Molar Gas Constant	R	$8.31441(26)$ J mol^{-1} K^{-1}	31
42. Molar Volume, Ideal Gas ($T_0 = 273.15$ K, $p_0 = 1$ atm)	$V_m = RT_0/p_0$	$0.02241383(70)$ m^3 mol^{-1}	31
43. Boltzmann Constant	$k = R/N_A$	$1.380662(44) \times 10^{-23}$ J K^{-1}	32
44. Stefan-Boltzmann Constant	$\sigma = (\pi^2/60)k^4/\hbar^3 c^2$	$5.67032(71) \times 10^{-8}$ W m^{-2} K^{-4}	125
45. First Radiation Constant	$c_1 = 2\pi\hbar c^2$	$3.741832(20) \times 10^{-16}$ W m^2	5.4
46. Second Radiation Constant	$c_2 = \hbar c/k$	$0.01438786(45)$ m K	31
47. Gravitational Constant	G	$6.6720(41) \times 10^{-11}$ N m^2 kg^{-2}	615

Data from CODATA Bulletin No. 11, ICSU CODATA Central Office, CODATA Secretariat: 51 Boulevard de Montmorency, 75016 Paris, France (copies of this bulletin are available at no cost from this office).

ENERGY CONVERSION FACTORS

Quantity	Value	Unit	Error (ppm)
1 kg	5.609538(24)	10^{29} MeV	4.4
1 amu	931.5016(26)	MeV	2.8
Electron mass	0.5110041(16)	MeV	3.1
Proton mass	938.2592(52)	MeV	5.5
Neutron mass	939.5527(52)	MeV	5.5
1 electron volt	1.6021917(70)	10^{-19} J	4.4
		10^{-12} erg	
	2.4179659(81)	10^{14} Hz	3.3
	8.065465(27)	10^{5} m^{-1}	3.3
		10^{3} cm^{-1}	
	1.160485(49)	10^{4} K	42
Energy-wavelength conversion	1.2398541(41)	10^{-6} eV·m	3.3
		10^{-4} eV·cm	
Rydberg constant, R_∞	2.179914(17)	10^{-18} J	7.6
		10^{-11} erg	
	13.605826(45)	eV	3.3
	3.2898423(11)	10^{15} Hz	0.35
	1.578936(67)	10^{5} K	43
Bohr magneton, μ_B	5.788381(18)	10^{-5} eV T^{-1}	3.1
	1.3996108(43)	10^{10} Hz T^{-1}	3.1
	46.68598(14)	m^{-1}·T^{-1}	3.1
		10^{-2} cm^{-1}·T^{-1}	
	0.671733(29)	K T^{-1}	43
Nuclear magneton, μ_n	3.152526(21)	10^{-8} eV T^{-1}	6.8
	7.622700(42)	10^{6} Hz T^{-1}	5.5
	2.542659(14)	10^{-2} m^{-1}·T^{-1}	5.5
		10^{-4} cm^{-1}·T^{-1}	
	3.65846(16)	10^{-4} K T^{-1}	44
Gas constant, R_0	8.20562(35)	10^{-2} m^3·atm kmole^{-1} K^{-1}	42
Standard volume of ideal gas, V_0	22.4136	m^3 kmole^{-1}	

INFRARED CORRELATION CHART No. 1

Prepared from information supplied by Beckman Instruments

WAVENUMBERS IN KAYSERS

WAVELENGTH IN MICRONS

LEGEND
S = Strong   SP = Sharp
M = Medium   V = Vary
               W = Weak

Based on work done by Colthup.

ALKANE GROUPS
ALKENE
ALKYNE
AROMATIC
ETHERS
ALCOHOLS
ACIDS

F-200

INFRARED CORRELATION CHART No. 1 (Con't.)

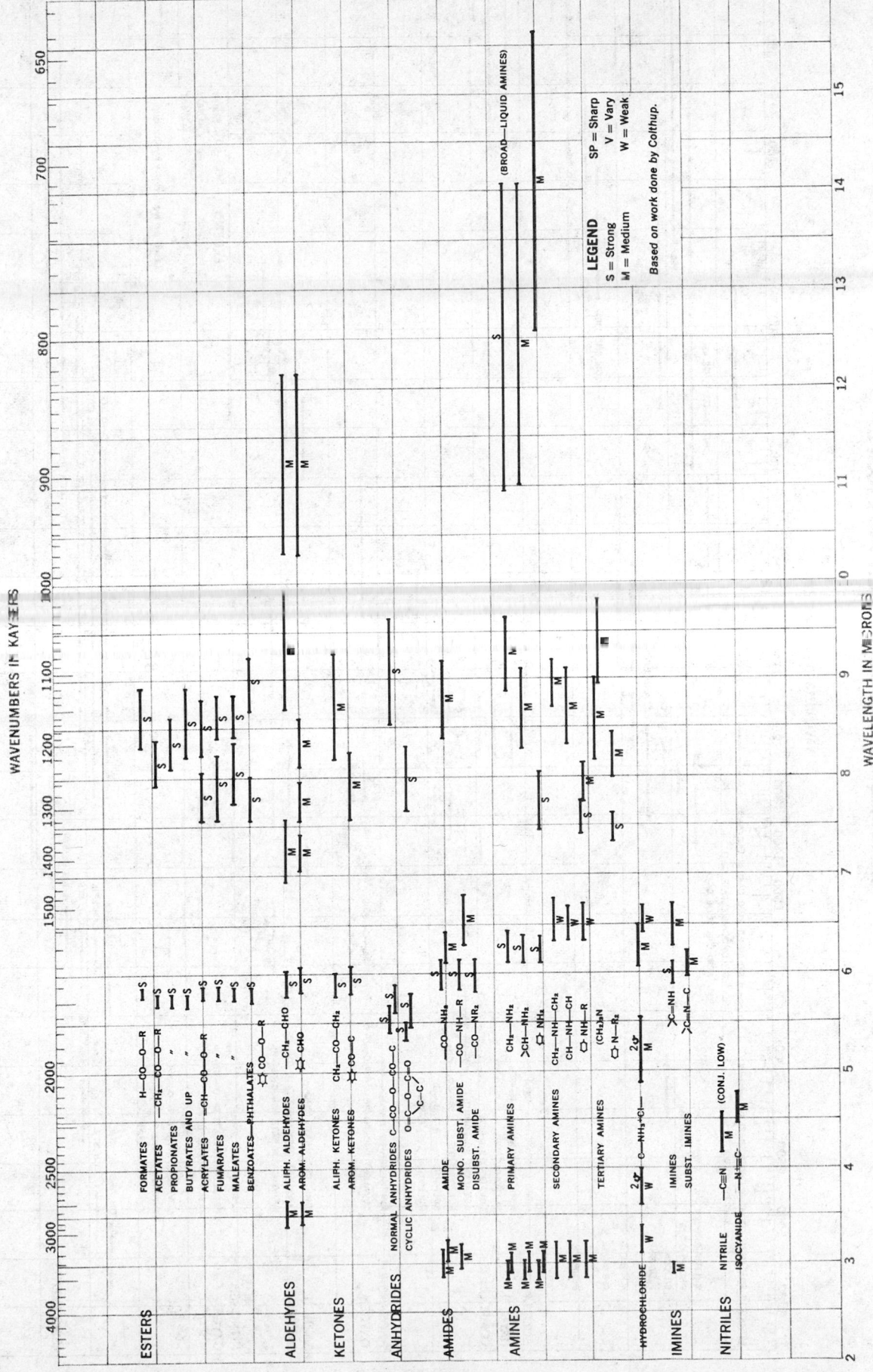

INFRARED CORRELATION CHART No. 1 (Con't.)

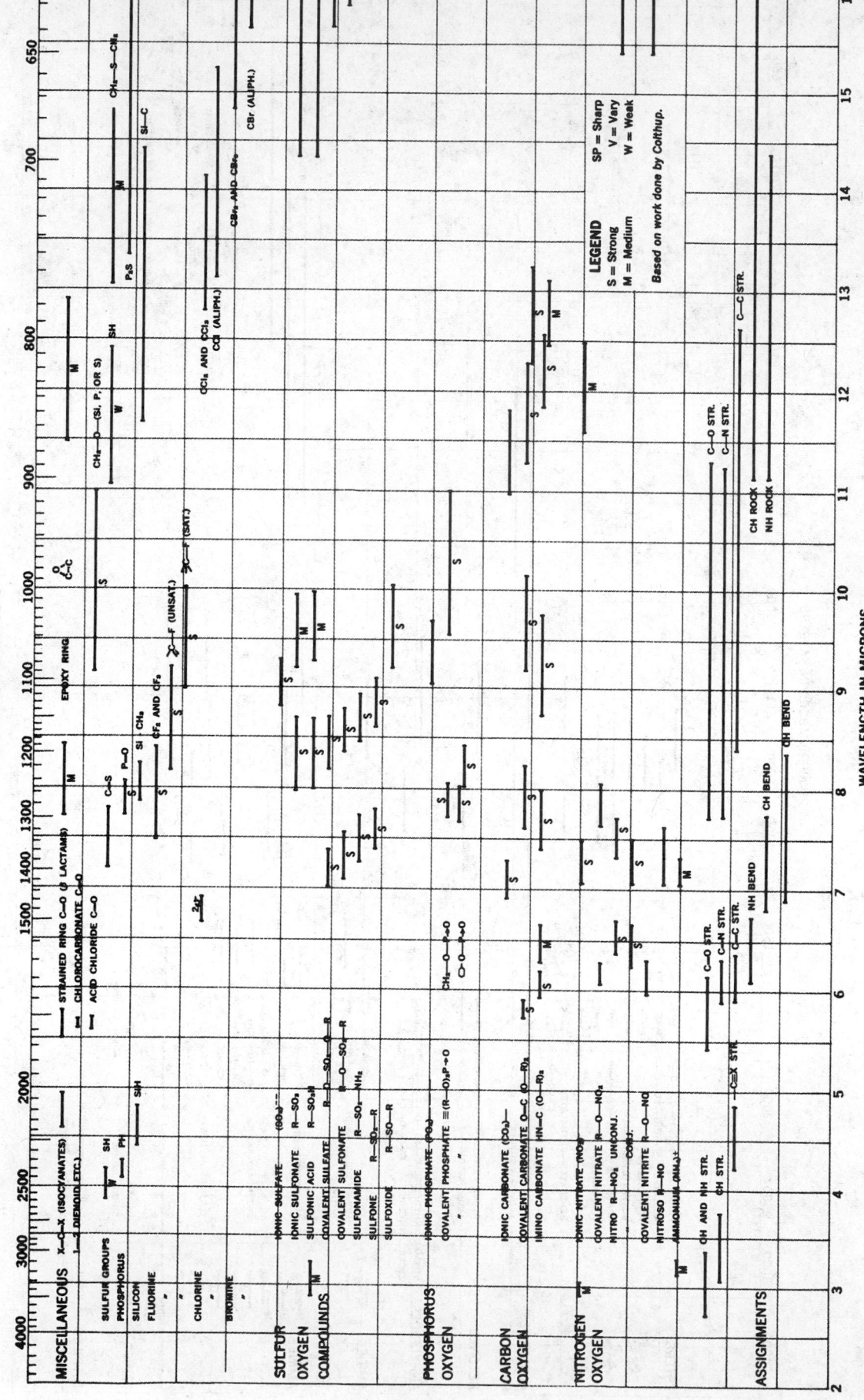

WAVENUMBERS IN KAYSERS

WAVELENGTH IN MICRONS

LEGEND

S = Strong
M = Medium
W = Weak

SP = Sharp
V = Vary

Based on work done by Colthup.

F-202

INFRARED CORRELATION CHART No. 2

Prepared from information supplied by Beckman Instruments

This chart presents some information regarding structure, double-bond vibrations, hydrogen stretching and triple-bond vibrations.

HYDROGEN STRETCHING AND TRIPLE-BOND VIBRATIONS, 3750-2000 CM.⁻¹

DOUBLE-BOND VIBRATIONS, ETC. 2000-1500 CM.⁻¹

WAVENUMBERS IN KAYSERS

WAVELENGTH IN MICRONS

Left-side group labels:

- CH₃
- CH₂
- CH
- CH=CH
- CH=CH₂
- >C=CH₂
- CH≡CR
- RC≡CR'
- CHO
- AROMATIC CH
- COOH
- OH
- C=O
- NH₂
- NH
- =NH
- CONH₂
- CONH (open-chain)
- CONH (lactams)
- NH₃⁺RCOOH
- NH₄⁺ClRCOOH
- NH₂ RCOO—X⁺
- PYRIDINE, Etc.
- AZIDES
- C≡N
- αβ — UNSATURATED NITRILES
- C≡N—
- P—H
- NH₄⁺
- CN⁻, OCN⁻, Etc.
- P—OH
- SH
- Si H

Right-side group labels (DOUBLE-BOND):

- ALLENES
- —CH=CH₂
- >C=CH₂
- C=C
- C=C (conjugated)
- AROMATIC (all types of substitution except as below)
- AROMATIC (para- and unsymmetrical trisubstitution)
- AROMATIC (vicinal trisubstitution)
- AROMATIC (conjugated)
- AROMATIC (all types)
- C=N
- C=N (conjugated or cyclic)
- N=N
- CONH₂ (see chart 4 for CO)
- CONH (open-chains only)
- NH₂
- NH
- NH₃⁺
- NH₄⁺CLR
- COO⁻
- C—NO₂
- O—NO₂
- N—NO₂
- —O—N=O
- PYRIDINE, Etc.
- PYRIMIDINE, Etc.
- TROPOLONES

Annotations within chart:

- (MONOMER) M
- FREE (V)
- POLYMERS (S)
- SINGLE BRIDGE (V)
- CHELATES (W)
- FREE 2 BANDS (M)
- BONDED (S)
- FREE BAND (M)
- BONDED (S)
- FREE (M)
- BONDED (M)
- FREE (M)
- BONDED (M)
- (W) MOST
- CONTINUOUS BAND SERIES
- 2 BANDS(S)
- W;BROAD DIMER
- CHARACTERISTIC BAND PATTERNS FOR VARIOUS TYPES OF SUBSTITUTION
- BONDED (S)
- FREE (S)
- BONDED (S)
- FREE (S)

LEGEND

S = Strong	SP = Sharp
M = Medium	V = Vary
W = Weak	

Based on work done by Bellamy.

INFRARED CORRELATION CHART No. 3

Prepared from information supplied by Beckman Instruments

This chart presents some correlations between structure and the carbonyl vibrations of some classes of organic compounds. In all cases the absorption bands are strong and fall within the range of 1900-1500 cm^{-1}.

WAVENUMBERS IN KAYSERS

4000	3000	2500	2000	1500	1400

SATURATED KETONES AND ACIDS
$\alpha\beta$—UNSATURATED KETONES
ARYL KETONES
$\alpha\beta$—,$\alpha'\beta'$—UNSATURATED AND DIARYL KETONES
α—HALOGEN KETONES
$\alpha\alpha'$—HALOGEN KETONES
CHELATED KETONES
6-MEMBERED RING KETONES
5-MEMBERED RING KETONES
4-MEMBERED RING KETONES
SATURATED ALDEHYDES
$\alpha\beta$—UNSATURATED ALDEHYDES
$\alpha\beta$—,$\alpha'\beta'$—UNSATURATED ALDEHYDES
CHELATED ALDEHYDES
$\alpha\beta$—UNSATURATED ACIDS
α—HALOGEN ACIDS
ARYL ACIDS
INTRAMOLECULARLY BONDED ACIDS
IONISED ACIDS
SATURATED ESTERS 6- AND 7-RING LACTONES
$\alpha\beta$—UNSATURATED AND ARYL ESTERS
VINYL ESTERS, α—HALOGEN ESTERS
SALICYLATES AND ANTHRANILATES
CHELATED ESTERS
5-RING LACTONES
$\alpha\beta$—UNSATURATED 5-RING LACTONES
THIOL ESTERS
ACID HALIDES
CHLOROCARBONATES
ANHYDRIDES (open-chain)   SEPARATION 60 CM^{-1}
ANHYDRIDES (cyclic)   SEPARATION 60 CM^{-1}
ALKYL PEROXIDES   SEPARATION 25 CM^{-1}
ARYL PEROXIDES   SEPARATION 25 CM^{-1}
PRIMARY AMIDES (CO)   FREE BONDED
SECONDARY AMIDES AND —δ LACTAMS (CO)   FREE BONDED
TERTIARY AMIDES (CO)
γ—LACTAMS   FUSED RINGS UNFUSED
β—LACTAMS   FUSED UNFUSED RINGS

3	4	5	6	7

WAVELENGTH IN MICRONS

F-204

INFRARED CORRELATION CHART No. 4

Prepared from information supplied by Beckman Instruments

This chart presents some correlations between structure and single-bond vibrations for a number of classes of compounds having absorption between 1500-650 cm⁻¹

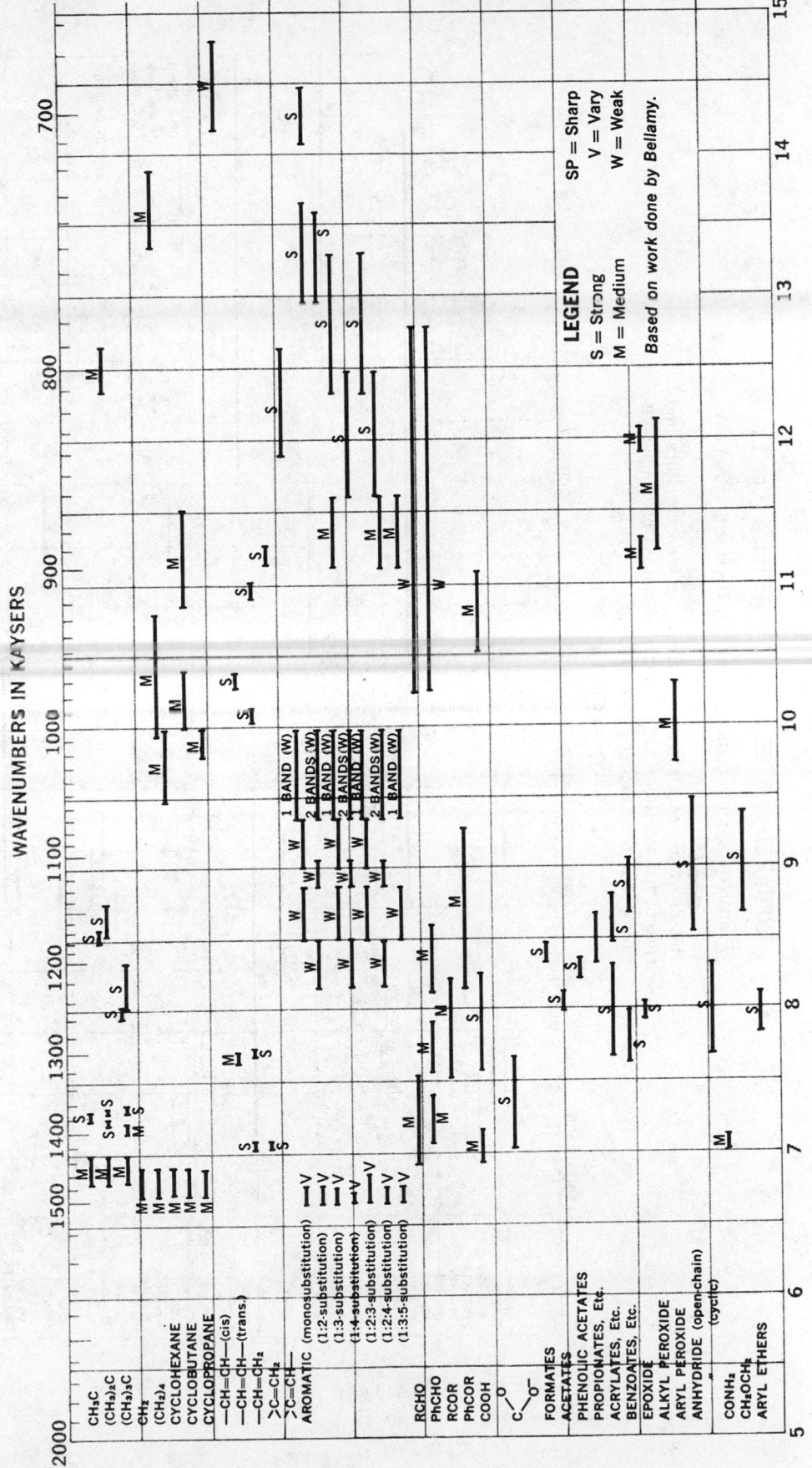

LEGEND

SP = Sharp   V = Vary

S = Strong   W = Weak

M = Medium

Based on work done by Bellamy.

INFRARED CORRELATION CHART No. 4 (Con't.)

Prepared from information supplied by Beckman Instruments

This chart presents some correlations between structure and single-bond vibrations for a number of classes of compounds having absorption between 1500-650 cm⁻¹.

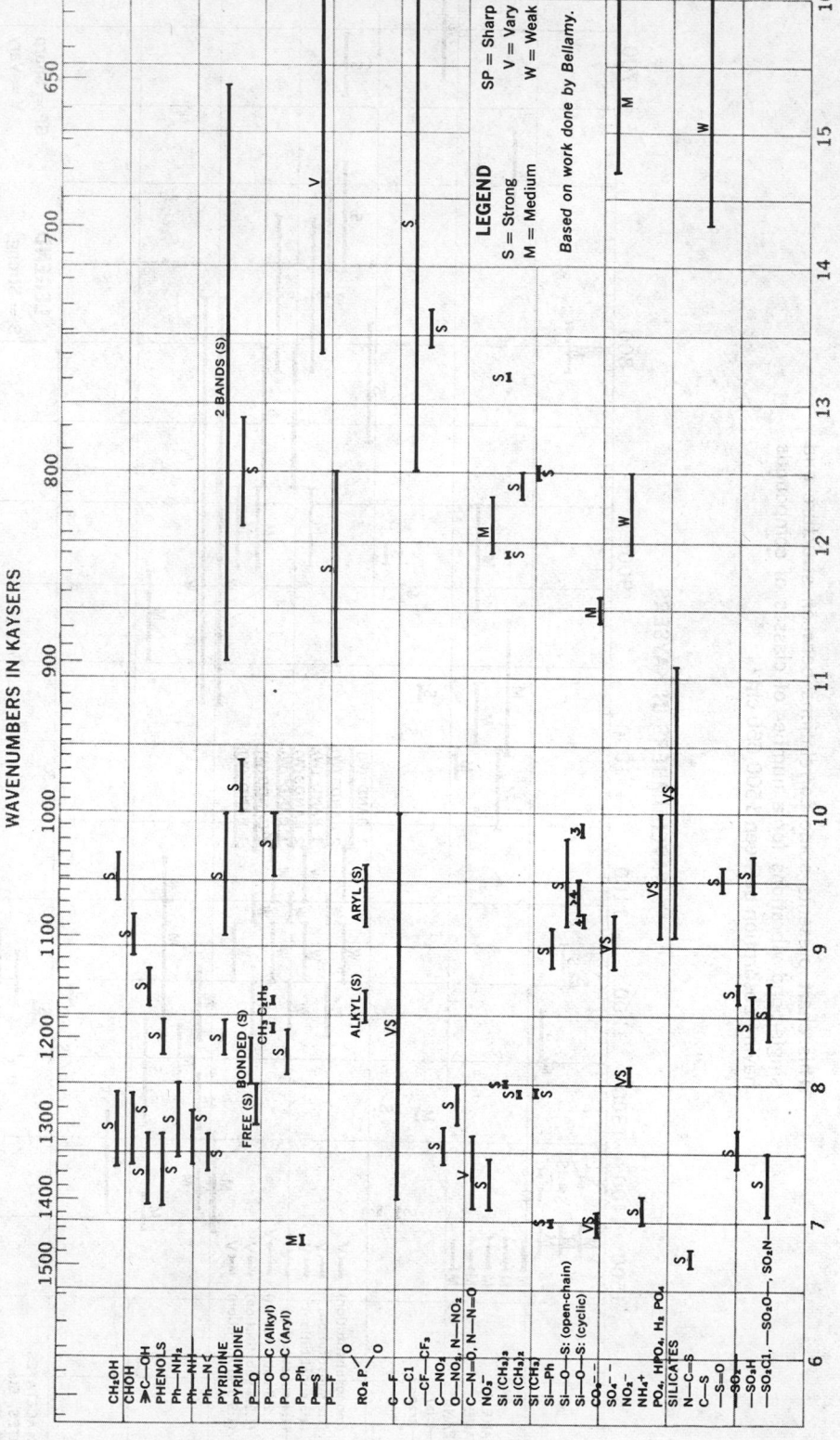

FAR INFRARED VIBRATIONAL FREQUENCY CORRELATION CHART

Based on evidence compiled by James E. Stewart of Beckman Instruments.
This chart shows the vibrational frequency correlation in the far infrared region.
Because research is continuing in the far infrared region, this chart is not all-inclusive.

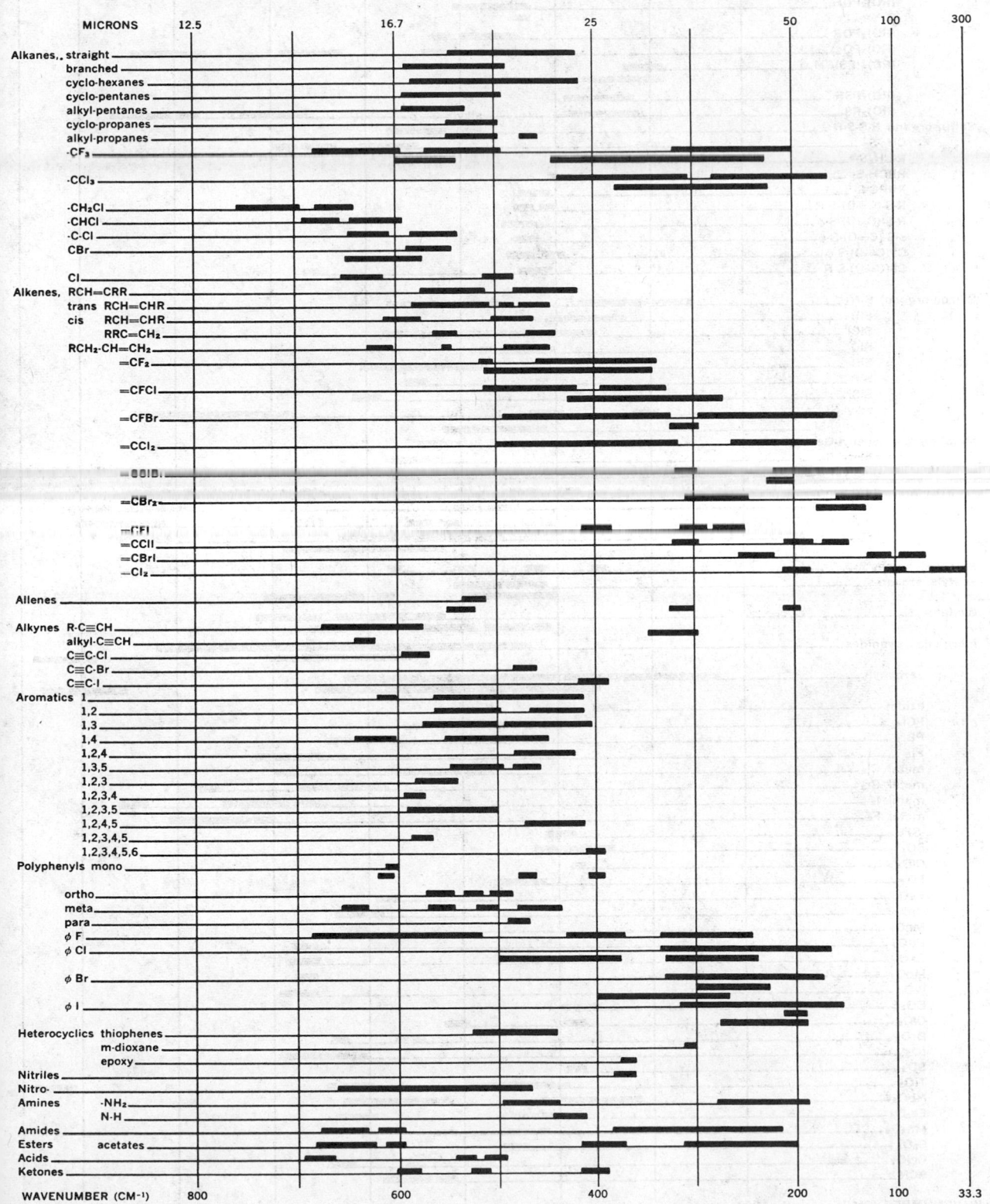

FAR INFRARED VIBRATIONAL FREQUENCY CORRELATION CHART (Con't.)

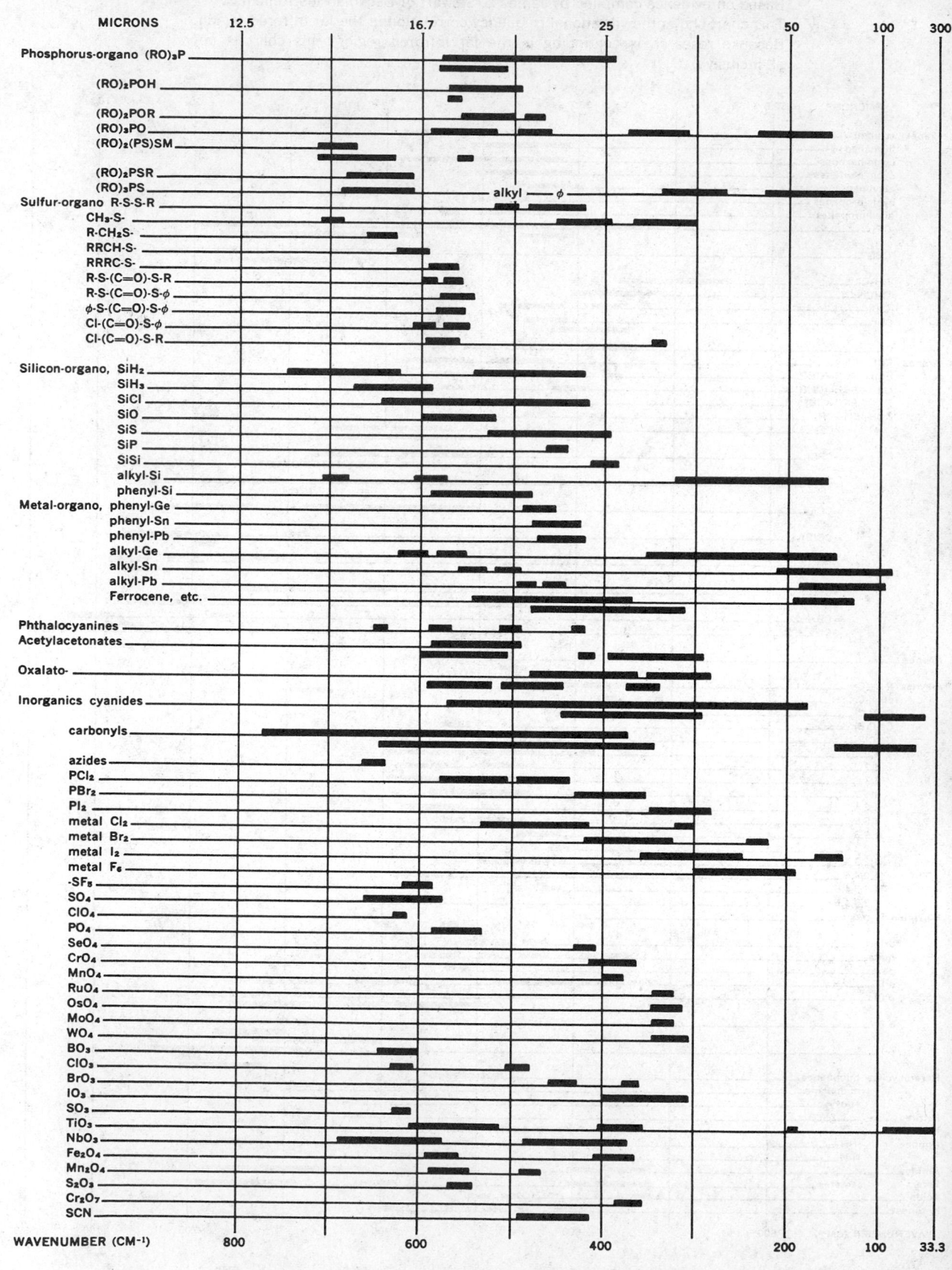

CHART OF CHARACTERISTIC FREQUENCIES BETWEEN $\sim 700\text{--}300$ cm^{-1}

Freeman F. Bentley, Lee D. Smithson, and Adele L. Rozek

This chart summarizes the characteristic frequencies known to occur between approximately 700–300^{-1}. Those who anticipate using this region of the spectrum should consult "Infrared Spectra and Characteristic Frequencies between $\sim$700–300 cm^{-1}" by Interscience Publishers, a division of John Wiley and Sons, Inc. for a complete discussion of the characteristic frequencies summarized in this chart, a large collection of infrared spectra (700–300 cm^{-1}) of most of the common organic and inorganic compounds, and an extensive bibliography of references to infrared data below $\sim$700 cm^{-1}.

In this chart the black horizontal bars indicate the range of the spectrum in which the characteristic frequencies have been observed to occur in the compounds investigated. The number of compounds investigated is given immediately to the right of the names or structures of the compounds. Obviously those characteristic frequency ranges based upon a limited number of compounds should be used with caution.

The letters above the bars indicate the relative intensities of the absorption bands. These intensities are based upon the strongest band in the spectra (700–300 cm^{-1}) of specific classes of compounds investigated, and they cannot be compared accurately with the intensities given for other classes.

When known, the specific vibration giving rise to the characteristic frequency is printed in abbreviated form immediately to the right of the bar indicating the frequency range except when lack of space prevents this. When there can be no ambiguity, this information may be printed other than to the right. In doubtful cases, arrows are used for clarification.

Naturally, the characteristic frequencies vary in their specificity and analytical value. The user is, therefore, cautioned to use this chart with some reserve. After reviewing this chart, the reader should be aware that there are many characteristic frequencies in the 700–300 cm^{-1} region. Used cautiously, this chart can be of considerable value in the elucidation of structures of unknown compounds.

It is important to emphasize that the region of the infrared between $\sim$700–300 cm^{-1} should be used in conjunction with the more conventional 5000–700 cm^{-1} region. Much of the value of the 700–300 cm^{-1} region can only be realized after interpreting the spectrum between 5000–700 cm^{-1}.

The following symbols and abbreviations are used:

Symbol or Abbreviation	Definition
αCCC	In-plane bending of benzene ring
Antisym.	Antisymmetrical (Asymmetrical)
$\sim$	Approximately
β	In-plane bending of ring substituent bond
δ	In-plane bending
γ	Out-of-plane bending
i.p.	In-plane
m	Medium
ν	Stretching
ν_s	Symmetrical stretching
ν_{as}	Antisymmetrical stretching
o.p.	Out-of-plane
$\parallel$	Parallel
$\perp$	Perpendicular
ϕ	Phenyl
ϕCC	Out-of-plane bending of aromatic ring
r	Rocking
s	Strong
sh	Shoulder
Sym.	Symmetrical
v	Variable
w	Weak
"X" Sensitive	An aromatic vibrational mode whose frequency position is greatly dependent on the nature of the substituent.

CHART OF CHARACTERISTIC FREQUENCIES BETWEEN ～ 700–300 cm⁻¹

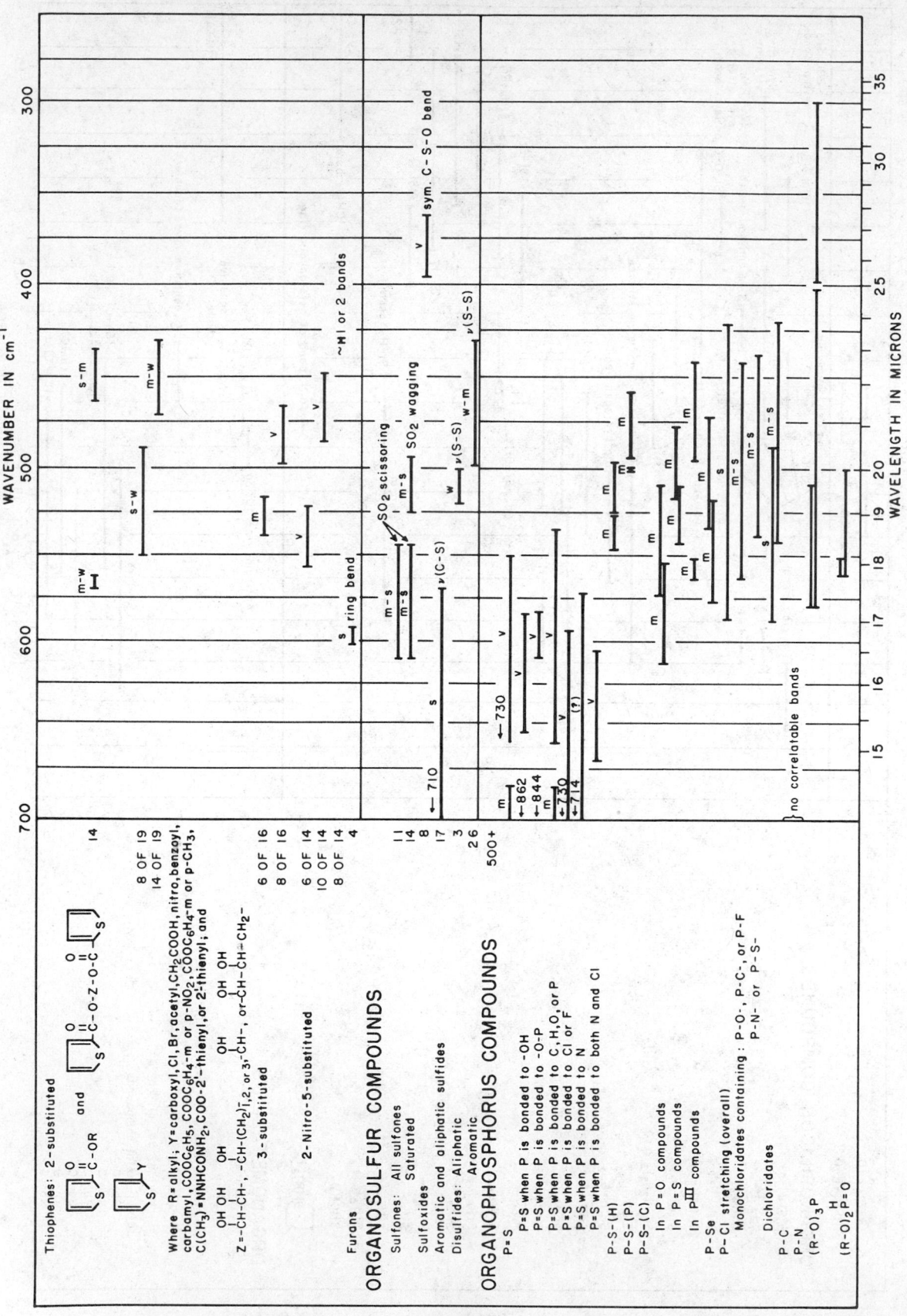

WAVENUMBER IN cm^{-1}

300   400   500   600   700

WAVELENGTH IN MICRONS

35   30   25   20   19   18   17   16   15

Mono- and disubstituted benzenes with the following substituents:

- $-COOH$ 7
- $-COOR(R \neq H)$ 10
- $O=C-CH_3$ 5
- $-CHO$ 8 OF 9
- $O=C-NH_2$ 2
- $O=C-O$ 17
- $-NO_2$ 23
- $-OCH_3$ 11
- $-CN$ 10
- $-OH$ 15 OF 22
- $-NH_2$ 22
- $-F$ 9
- $-Cl$ 8
- $-Br$ 14 OF 24
- $-CF_3$ 12
- $-CH_3$ 4
- $-C_2H_5$ 33
- $-C_3H_7$, $-C_4H_9$, $-C_{10}H_{21}$ 7
- $-iC_3H_7$ 4

37 OF 43

BIPHENYLS   9

NAPHTHALENES   18

HETEROCYCLICS

Pyridines: 2-substituted   14
           3-substituted   13

Pyrimidines: 2-substituted   11
            4-substituted   8

Imidazoles: 4 (or 5)-substituted   6
           4,5-disubstituted   9
           1,4,5-trisubstituted   11

Purines: 2-substituted   7
        6-substituted   32
        8-substituted   8
        2,6-disubstituted   21

Band labels:

- m-s,b $\delta(\underset{O}{C}-C-O-)$?
- $\beta(CCN)$ m
- $\beta(C-OH)$ w
- $\beta(C-NH_2)$ w
- $\beta(C-F)$ mw-s
- $\beta(C-Cl)$ m-s
- $\beta(C-Br)$ and $\alpha(CCC)$ s
- $\beta(CX)$ m
- $r(CO_2)$ o.p   s
- $r(CC_2$ or $\delta(OCO)$?   s
- v
- $\delta-CCC$/i.p.   s
- $\beta(CCC)$ i.p.   s
- $N-C=C$ bend   s
- $-(OH)$ s,b $\rightarrow$720
- $\delta(NO_2)$ w-m
- $\delta(COC)$ m-s
- $\phi(CC) + \delta(CN)$
- $r(CF_3)$ s
- $\delta(CCC)$ m-s
- $\delta(CCC)$ m
- m-s H
- CC o.p. ring   m-s
- CC o.p. ring   m-s
- o.p. skeletal bend   v
- skeletal distortion i.p.   m-s
- $\phi(CC)$ s
- $\phi(CC)$ s
- m
- m-s
- s
- m
- w-m
- v
- A'' o.p ring bend   v
- $\alpha(CCC)$ m-s
- $\alpha(CCC)$ m-s
- o.p. ring bend   m-s
- γ, N-H   s
- m-s
- m-s
- m-s
- $\gamma(CH)$ m-s
- v
- s
- v
- s

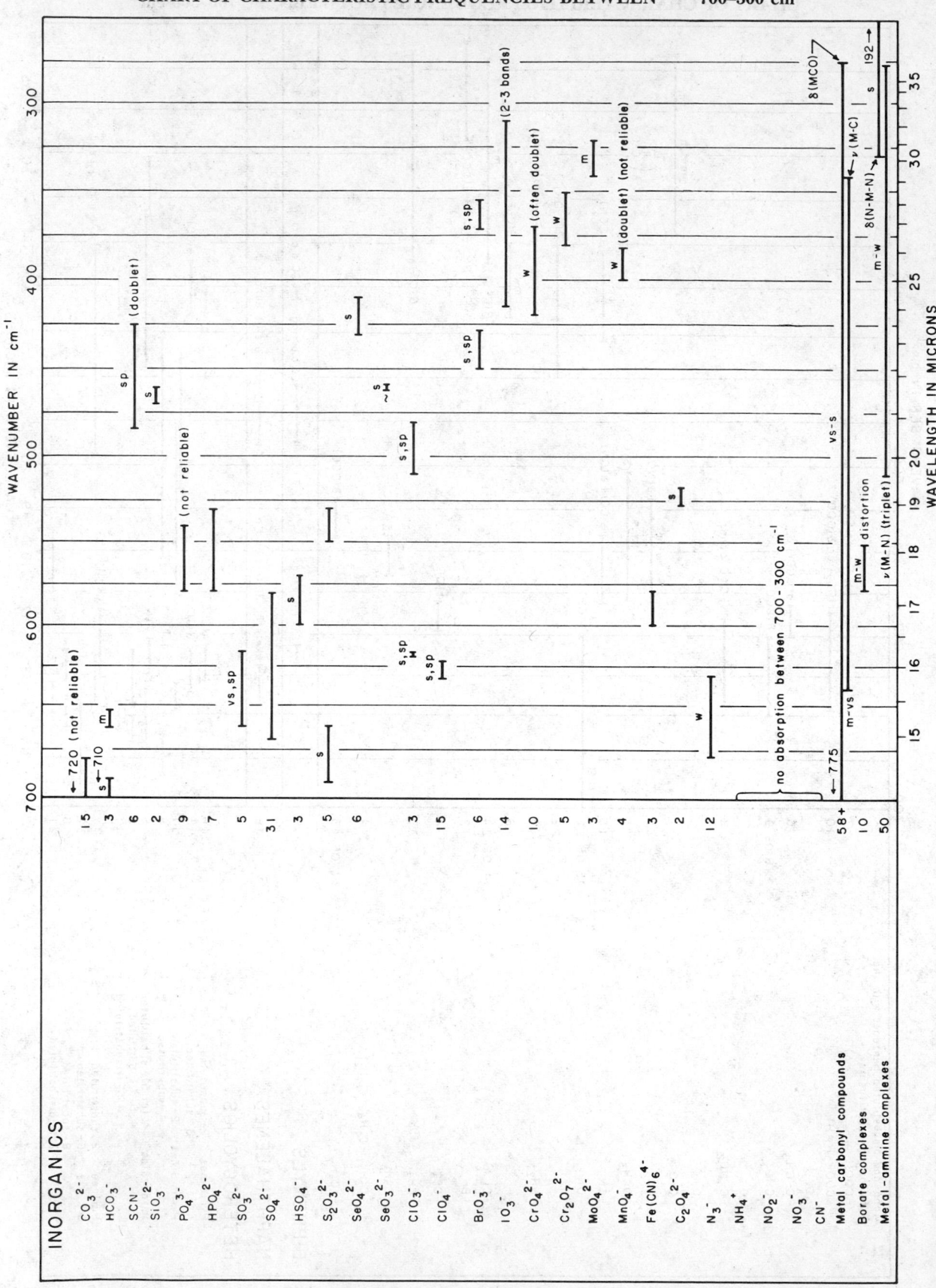

CHARACTERISTIC NMR SPECTRAL POSITIONS
FOR HYDROGEN IN ORGANIC STRUCTURES

By permission from Erno Mohacsi, J. of Chemical Education, 41, 38 (1964)

This table is useful for quick qualitative determination of proton spectrum lines by providing a tabulation of line positions obtained using tetramethylsilane as an internal reference. The listing has been kept as simple as possible for this purpose. The proton spectrum lines are arranged according to the chemical shift relative to tetramethylsilane and are given in values of τ and δ. The purpose of this table is to supplement tables available in standard references and to summarize information available in the literature.

TABLES OF PARTICLE PROPERTIES

April 1982

M. Aguilar-Benitez, R.L. Crawford, R. Frosch, G.P. Gopal, R.E. Hendrick, R.L. Kelly, M.J. Losty, L. Montanet,

F.C. Porter, A. Rittenberg, M. Roos, L.D. Roper, T. Shimada, R.E. Shrock, T.G. Trippe, Ch. Walck, C.G. Wohl, G.P. Yost

(Closing date for data: Jan. 1, 1982)

From 'Review of Particle Properties'
Physics Letters
Vol. 111B
April 1982

Stable Particle Table

For additional parameters, see Addendum to this table.

Quantities in italics have changed by more than one (old) standard deviation since April 1980.

Particle	$I^G(J^P)C_n{}^a$	Mass^b (MeV) Mass^2 (GeV^2)	Mean life^b (sec) $c\tau$ (cm)	Partial decay mode		
				Mode	Fraction^b	p or $p_{max}{}^c$ (MeV/c)
PHOTON						
γ	$0,1(1^-)-$	$(<6\times10^{-22})$	-------	stable		
LEPTONS						
ν_e	$J=\frac{1}{2}$	$(<0.000046)^d$	stable $(>3\times10^8 m_{\nu_e}$ (MeV))	stable		
e	$J=\frac{1}{2}$	0.5110034 ±0.0000014	stable $(>2\times10^{22}y)$	stable		
ν_μ	$J=\frac{1}{2}$	$0(<0.52)$	stable $(>1.1\times10^5 m_{\nu_\mu}$ (MeV))	stable		
				$\mu^- \rceil$ (or $\mu^+ \to$CC)		
μ	$J=\frac{1}{2}$	105.65943 ±0.00018 $m^2=0.01116392$	2.19714×10^{-6} ±0.00007 $c\tau=6.5868\times10^4$	$e^-\bar{\nu}\nu$	$(98.6\pm0.4)\%$	53
				$e^-\bar{\nu}\nu\gamma$	$^e(1.4\quad)\%$	53
				$\dagger[e^-\nu_e\bar{\nu}_\mu$	$(<9\quad)\%]$	53
				$e^-\gamma$	$(<1.9\quad)\times10^{-10}$	53
				$e^-e^+e^-$	$(<1.9\quad)\times10^{-9}$	53
				$e^-\gamma\gamma$	$(<5\quad)\times10^{-8}$	53
ν_τ	$J=\frac{1}{2}$	<250				
				$\tau^- \rceil$ (or $\tau^+ \to$CC)		
τ	$J=\frac{1}{2}$	1784.2 ±3.2 $m^2=3.18$	$(4.6\pm1.9)\times10^{-13}$ $c\tau=0.014$	$\mu^-\bar{\nu}\nu$	$(18.5\pm1.2)\%$	889
				$e^-\bar{\nu}\nu$	$(16.2\pm1.0)\%$	892
				hadron$^-$ neutrals	$(37.0\pm3.2)\%$	
				$3(\text{hadron}^\pm)$ neutrals	$(28.4\pm3.0)\%$	
				$5(\text{hadron}^\pm)$ neutrals	$(<6\quad)\%$	
				$\dagger[3(\text{hadron}^\pm)\nu$	$(13\pm8)\%$	
				$3(\text{hadron}^\pm)\nu(\geq1\gamma)$	$(15\pm7)\%]$	
				$\dagger[\pi^-\nu$	$(10.7\pm1.6)\%$	887
				$\rho^-\nu$	$(21.6\pm3.6)\%$	726
				K^- neutrals	(small)	
				$\pi^-\pi^-\pi^+\nu$	$(7\pm5)\%$	864
				$\pi^-\pi^-\pi^+(\geq0\pi^0)\nu$	$(18\pm7)\%]$	864
				$\dagger[K^{*-}(892)\nu$	$(1.7\pm0.7)\%$	669
				$K^{*-}(1430)\nu$	$(<0.9\quad)\%$	316
				$\pi^-\rho^0\nu$	$(5.4\pm1.7)\%]$	718

(continued next page)

Particle	$I^G(J^P)C_n$ [a]	Mass [b] (MeV) Mass^2 (GeV^2)	Mean life [b] (sec) $c\tau$ (cm)	Mode	Fraction [b]	p or P_{max} [c] (MeV/c)
				Partial decay mode		

τ (continued)

$\tau^- \to\!\!\rceil$ (or $\tau^+ \to CC$)

Mode	Fraction	p or P_{max} (MeV/c)
e^- chgd. parts. $+\ \mu^-$ chgd. parts.	(<4)%	
$\mu^-\gamma$	(<5.5)$\times 10^{-4}$	889
$e^-\gamma$	(<6.4)$\times 10^{-4}$	892
$\mu^-\mu^+\mu^-$	(<4.9)$\times 10^{-4}$	876
$e^-\mu^+\mu^-$	(<3.3)$\times 10^{-4}$	886
$\mu^-e^+e^-$	(<4.4)$\times 10^{-4}$	889
$e^-e^+e^-$	(<4.0)$\times 10^{-4}$	892
$\mu^-\pi^0$	(<8.2)$\times 10^{-4}$	884
$e^-\pi^0$	(<2.1)$\times 10^{-3}$	887
μ^-K^0	(<1.0)$\times 10^{-3}$	819
e^-K^0	(<1.3)$\times 10^{-3}$	823
$\mu^-\rho^0$	(<4.4)$\times 10^{-4}$	722
$e^-\rho^0$	(<3.7)$\times 10^{-4}$	726

NONSTRANGE MESONS [a]

$\pi^+ \to\!\!\rceil$ (or $\pi^- \to CC$)

$\pi^\pm$ $\quad 1^-(0^-)$ $\quad$ 139.5673 ± 0.0007 $\quad m^2 = 0.0194790$ $\quad m_{\pi\pm} - m_{\mu\pm} = 33.9079 \pm 0.0007$ $\quad$ 2.6030×10^{-8} ± 0.0023 $\quad c\tau = 780.4$ $\quad (\tau^+ - \tau^-)/\bar{\tau} = (0.05\pm 0.07)\%$ (test of CP?)

Mode	Fraction	p or P_{max} (MeV/c)
$\mu^+\nu$	100%	30
$e^+\nu$	(1.267$\pm$0.023)$\times 10^{-4}$	70
$\mu^+\nu\gamma$	[e](1.24$\pm$ 0.25)$\times 10^{-4}$	30
$e^+\nu\gamma$	[e](5.6 $\pm$ 0.7)$\times 10^{-8}$	70
$e^+\nu\pi^0$	(1.02$\pm$ 0.07)$\times 10^{-8}$	5
$e^+\nu e^+e$	(<3)$\times 10^{-9}$	70

π^0 $\quad 1^-(0^-)$ $\quad$ 134.9630 ± 0.0038 $\quad m^2 = 0.018215$ $\quad m_{\pi\pm} - m_{\pi 0} = 4.6043 \pm 0.0037$ $\quad$ 0.83×10^{-16} ± 0.06 $S=1.8^*$ $\quad c\tau = 2.5\times 10^{-6}$

Mode	Fraction	p or P_{max} (MeV/c)
$\gamma\gamma$	(98.797 $\pm$ 0.030)%	67
γe^+e^-	(1.213)%	67
$\gamma\gamma\gamma$	(<3.8)$\times 10^{-7}$	67
$e^+e^-e^+e^-$	[f](3.32)$\times 10^{-5}$	67
$\gamma\gamma\gamma\gamma$	(<4)$\times 10^{-6}$	67
e^+e^-	(2.2 $^{+2.4}_{-1.1}$)$\times 10^{-7}$	67
$\nu\nu$	(<2.4)$\times 10^{-5}$	67

η $\quad 0^+(0^-)+$ $\quad$ 548.8 ± 0.6 $S=1.4^*$ $\quad m^2 = 0.3012$ $\quad \Gamma = (0.83\pm 0.12)$keV

Neutral decays $(70.9\pm 0.7)\%$

Mode	Fraction	p or P_{max} (MeV/c)
$\gamma\gamma$	(39.1 $\pm$ 0.8)%	274
$3\pi^0$	(31.8 $\pm$ 0.8)% $\quad S=1.1^*$	180
$\pi^0\gamma\gamma$	[g](<0.3)%	258

Charged decays $(29.1\pm 0.7)\%$

Mode	Fraction	p or P_{max} (MeV/c)
$\pi^+\pi^-\pi^0$	(23.7 $\pm$ 0.5)%	175
$\pi^+\pi^-\gamma$	(4.91$\pm$ 0.13)%	236
$e^+e^-\gamma$	(0.50$\pm$ 0.12)%	274
$\mu^+\mu^-\gamma$	(3.1 $\pm$ 0.4)$\times 10^{-4}$	253
e^+e^-	(<3)$\times 10^{-4}$	274
$\mu^+\mu^-$	(6.5 $\pm$ 2.1)$\times 10^{-6}$	253
$\pi^+\pi^-e^+e^-$	(0.13$\pm$ 0.13)%	236
$\pi^+\pi^-\gamma\gamma$	(<0.21)%	236
$\pi^+\pi^-\pi^0\gamma$	(<6)$\times 10^{-4}$	175
$\pi^+\pi^-$	(<0.15)%	236
$\pi^0e^+e^-$	(<5)$\times 10^{-5}$	258
$\pi^0\mu^+\mu^-$	(<5)$\times 10^{-6}$	211
$\pi^0\mu^+\mu^-\gamma$	(<3)$\times 10^{-6}$	211

Stable Particle Table *(cont'd)*

Particle	$I^G(J^P)C_n$ [a]	Mass [b] (MeV) / Mass^2 (GeV^2)	Mean life [b] (sec) / $c\tau$ (cm)	Mode	Fraction [b]	p or p_{max} [c] (MeV/c)
				Partial decay mode		

STRANGE MESONS [a]

$K^+ \rceil$ (or $K^- \to CC$)

Particle	$I^G(J^P)C_n$ [a]	Mass (MeV) / Mass^2 (GeV^2)	Mean life (sec) / $c\tau$ (cm)	Mode	Fraction [b]	p or p_{max} (MeV/c)
$K^{\pm}$	$\frac{1}{2}(0^-)$	493.667 ±0.015 $m^2 = 0.2437$	1.2371×10^{-8} ±0.0026 S=1.9[*] $c\tau=370.9$ $(\tau^+ - \tau^-)/\bar{\tau}=$ $(0.11\pm0.09)\%$ (test of CPT) S=1.2[*]	$\mu^+\nu$	$(\ 63.50\pm\ 0.16)\%$	236
				$\pi^+\pi^0$	$(\ 21.16\pm\ 0.15)\%$	205
				$\pi^+\pi^+\pi^-$	$(\ 5.59\pm\ 0.03)\%$ S=1.1[*]	125
				$\pi^+\pi^0\pi^0$	$(\ 1.73\pm\ 0.05)\%$ S=1.4[*]	133
				$\pi^0\mu^+\nu$	$(\ 3.20\pm\ 0.09)\%$ S=1.7[*]	215
				$\pi^0 e^+\nu$	$(\ 4.82\pm\ 0.05)\%$ S=1.1[*]	228
		$m_{K^\pm}-m_{K^0}=-4.01$ ±0.13 S=1.1[*]		$\mu^+\nu\gamma$	$^e(\ 5.8\ \pm\ 3.5\ )\times10^{-3}$	236
				$\pi^0\pi^0 e^+\nu$	$(\ 1.8\ ^{+2.4}_{-0.6}\ )\times10^{-5}$	207
				$\pi^+\pi^- e^+\nu$	$(\ 3.90\pm\ 0.15)\times10^{-5}$	203
				$\pi^+\pi^+ e^-\bar{\nu}$	$(\ <1.2\ )\times10^{-8}$	203
				$\pi^+\pi^-\mu^+\nu$	$(\ 1.4\ \pm\ 0.9\ )\times10^{-5}$	151
				$\pi^+\pi^+\mu^-\bar{\nu}$	$(\ <3.0\ )\times10^{-6}$	151
				$e^+\nu$	$(\ 1.54\pm\ 0.07)\times10^{-5}$	247
				$e^+\nu\gamma$ (SD+)[h]	$(\ 1.52\pm\ 0.23)\times10^{-5}$	247
				$e^+\nu\gamma$ (SD−)[h]	$(\ <1.6\ )\times10^{-4}$	247
				$\pi^+\pi^0\gamma$	$^{i,e}(\ 2.75\pm\ 0.16)\times10^{-4}$	205
				$\pi^+\pi^+\pi^-\gamma$	$^e(\ 1.0\ \pm\ 0.4\ )\times10^{-4}$	125
				$\pi^0\mu^+\nu\gamma$	$^e(\ <6\ )\times10^{-5}$	215
				$\pi^0 e^+\nu\gamma$	$^e(\ 3.7\ \pm\ 1.4\ )\times10^{-4}$	228
				$\pi^+ e^+ e^-$	$(\ 2.7\ \pm\ 0.5\ )\times10^{-7}$	227
				$\pi^- e^+ e^+$	$(\ <1\ )\times10^{-8}$	227
				$\pi^+\mu^+\mu^-$	$(\ <2.4\ )\times10^{-6}$	172
				$\pi^+\gamma\gamma$	$^e(\ <3.5\ )\times10^{-5}$	227
				$\pi^+\gamma\gamma\gamma$	$^e(\ <3.0\ )\times10^{-4}$	227
				$\pi^+\nu\bar{\nu}$	$(\ <1.4\ )\times10^{-7}$	227
				$\pi^+\gamma$	$(\ <4\ )\times10^{-6}$	227
				$\pi^+ e^+\mu^\pm$	$(\ <7\ )\times10^{-9}$	214
				$\pi^+ e^-\mu^+$	$(\ <5\ )\times10^{-9}$	214
				$e^+\nu\nu\bar{\nu}$	$(\ <6\ )\times10^{-5}$	247
				$\mu^+\nu\nu\bar{\nu}$	$(\ <6\ )\times10^{-6}$	236
				$\mu^+\nu e^+ e^-$	$(\ 11\ \pm\ 3\ )\times10^{-7}$	236
				$\mu^-\nu e^+ e^+$	$(\ <2.0\ )\times10^{-8}$	236
				$e^+\nu e^+ e^-$	$(\ 2\ ^{+2}_{-1}\ )\times10^{-7}$	247
				$\mu^+\nu_e$	$(\ <4\ )\times10^{-3}$	236
$\dfrac{K^0}{\bar{K}^0}$	$\frac{1}{2}(0^-)$	497.67 ±0.13 S=1.1[*] $m^2=0.24768$		50 % K_{Short}, 50% K_{Long}		
K_S^0	$\frac{1}{2}(0^-)$		0.8923×10^{-10} ±0.0022 $c\tau=2.675$	$\pi^+\pi^-$	$(\ 68.61\pm\ 0.24)\%$ S=1.1[*]	206
				$\pi^0\pi^0$	$(\ 31.39\ )\%$	209
				$\pi^+\pi^-\gamma$	$^e(\ 1.85\pm\ 0.10)\times10^{-3}$	206
				$\mu^+\mu^-$	$(\ <3.2\ )\times10^{-7}$	225
				$e^+ e^-$	$(\ <3.4\ )\times10^{-4}$	249
				$\gamma\gamma$	$(\ <0.4\ )\times10^{-3}$	249

Stable Particle Table *(cont'd)*

Particle	$I^G(J^P)C_n$ [a]	Mass [b] (MeV) Mass^2 (GeV^2)	Mean life [b] (sec) $c\tau$ (cm)	Mode	Fraction [b]	p or p_{max} [c] (MeV/c)
K^0_L	$\frac{1}{2}(0^-)$	5.183×10^{-8} ± 0.040 $c\tau=1554$		$\pi^0\pi^0\pi^0$	$(\ 21.5\ \pm\ 1.0\ )\%$ S=1.7*	139
				$\pi^+\pi^-\pi^0$	$(\ 12.39\pm\ 0.20)\%$ S=1.3*	133
				$\pi^\pm\mu^\mp\nu$	$(\ 27.1\ \pm\ 0.4\ )\%$ S=1.4*	216
				$\pi^\pm e^\mp\nu$	$(\ 38.7\ \pm\ 0.5\ )\%$ S=1.5*	229
				†$[\pi e\nu\gamma$	[e]$(\ 1.3\ \pm\ 0.8\ )\%]$	229
	$m_{K_L}-m_{K_S} = 0.5349\times10^{10}\ \hbar\ \sec^{-1}$ ± 0.0022			$\pi^+\pi^-$	[j]$(\ 0.203\pm\ 0.005)\%$ S=1.1*	206
				$\pi^0\pi^0$	[j]$(\ 0.094\pm\ 0.018)\%$ S=1.5*	209
				$\pi^+\pi^-\gamma$	[e]$(\ 4.41\pm\ 0.32)\times10^{-5}$	206
				$\pi^0\gamma\gamma$	$(\ <2.4\ )\times10^{-4}$	231
				$\gamma\gamma$	$(\ 4.9\ \pm\ 0.4\ )\times10^{-4}$	249
				$e\mu$	[k]$(\ <6\ )\times10^{-6}$	238
				$\mu^+\mu^-$	$(\ 9.1\ \pm\ 1.9\ )\times10^{-9}$	225
				$\mu^+\mu^-\gamma$	$(\ 2.8\ \pm\ 2.8\ )\times10^{-7}$	225
				$\pi^0\mu^+\mu^-$	$(\ <1.2\ )\times10^{-6}$	177
				e^+e^-	[k]$(\ <2.0\ )\times10^{-7}$	249
				$e^+e^-\gamma$	$(\ 1.7\ \pm\ 0.9\ )\times10^{-5}$	249
				$\pi^0 e^+e^-$	$(\ <2.3\ )\times10^{-6}$	231
				$\pi^+\pi^- e^+e^-$	$(\ <8.8\ )\times10^{-6}$	206
				$\pi^0\pi^\pm e^\mp\nu$	$(\ 6.2\ \pm\ 2.0\ )\times10^{-5}$	207

CHARMED NONSTRANGE MESONS [a]

$D^+\ \rceil$ (or $D^-\rightarrow$CC)

Particle	$I^G(J^P)C_n$ [a]	Mass (MeV) Mass^2 (GeV^2)	Mean life (sec) $c\tau$ (cm)	Mode	Fraction	p or p_{max} (MeV/c)
$D^\pm$	$\frac{1}{2}(0^-)$	1869.4 ± 0.6 $m^2=3.495$	$(9.1^{+2.2}_{-1.9})\times10^{-13}$ $c\tau=0.027$	$e^\pm$ anything	$(\ 19\ ^{+4}_{-3}\ )\%$	
				K^- anything	$(\ 16\ \pm\ 4\ )\%$	
				$\bar{K}^0$ any + K^0 any	$(\ 48\ \pm\ 15\ )\%$	
				K^+ anything	$(\ 6.0\ \pm\ 3.3\ )\%$	
	$m_{D^\pm}-m_{D^0}= 4.7$ ± 0.3			η anything	[l]$(\ <13\ )\%$	
				†$[K^-\pi^+\pi^+$	$(\ 4.6\ \pm\ 1.1\ )\%$	845
				$K^-\pi^+\pi^+\pi^+\pi^-$	$(\ <4\ )\%$	772
				$\bar{K}^0\pi^+$	$(\ 1.8\ \pm\ 0.5\ )\%$	862
				$\bar{K}^0\pi^+\pi^0$	$(\ 13\ \pm\ 8\ )\%$	845
				$\bar{K}^0\pi^+\pi^+\pi^-$	$(\ 8.4\ \pm\ 3.5\ )\%$	814
				$\bar{K}^0 K^+$	$(\ 0.45\pm\ 0.30)\%$	792
				$K^+K^-\pi^+$	$(\ <0.6\ )\%$	744
				$K^+\pi^+\pi^-$	$(\ <0.23\ )\%$	845
				$\pi^+\pi^0$	$(\ <0.5\ )\%$	925
				$\pi^+\pi^+\pi^-$	$(\ <0.37\ )\%]$	908
				†$[\bar{K}^{*0}\pi^+$	$(\ <3.7\ )\%]$	714

$D^0\ \rceil$ (or $\bar{D}^0\rightarrow$CC)

Particle	$I^G(J^P)C_n$ [a]	Mass (MeV) Mass^2 (GeV^2)	Mean life (sec) $c\tau$ (cm)	Mode	Fraction	p or p_{max} (MeV/c)
D^0 $\bar{D}^0$	$\frac{1}{2}(0^-)$	1864.7 ± 0.6 $m^2=3.477$	$(4.8^{+2.4}_{-1.5})\times10^{-13}$ $c\tau=0.014$	$e^\pm$ anything	$(\ <6\ )\%$	
				K^- anything	$(\ 44\ \pm\ 10\ )\%$ S=1.3*	
				$\bar{K}^0$ any + K^0 any	$(\ 33\ \pm\ 10\ )\%$	
				K^+ anything	$(\ 8\ \pm\ 3\ )\%$	
				η anything	[l]$(\ <13\ )\%$	
				†$[K^-\pi^+$	$(\ 2.4\ \pm\ 0.4\ )\%$	861
				$K^-\pi^+\pi^0$	$(\ 9.3\ \pm\ 2.8\ )\%$	844
				$K^-\pi^+\pi^+\pi^-$	$(\ 4.5\ \pm\ 1.3\ )\%$ S=1.4*	812
				$K^-\pi^+\pi^0\pi^0$	(\ seen\ )	815
				$\bar{K}^0\pi^0$	$(\ 2.2\ \pm\ 1.1\ )\%$	860
				$\bar{K}^0\pi^+\pi^-$	$(\ 4.2\ \pm\ 0.8\ )\%$	842
	$\dfrac{\Gamma(D^0\rightarrow\bar{D}^0\rightarrow K^+\pi^-)}{\Gamma(D^0\rightarrow K\pi)} < 0.16$			$\pi^+\pi^-$	$(\ 7.9\ \pm\ 3.8\ )\times10^{-4}$	922
				$\pi^+\pi^+\pi^-\pi^-$	$(\ <9\ )\times10^{-4}$	768
				K^+K^-	$(\ 2.7\ \pm\ 0.8\ )\times10^{-3}]$	791
				†$[K^{*-}\pi^+$	$(\ 3.4\ \pm\ 1.4\ )\%$	711
				$\bar{K}^{*0}\pi^0$	$(\ 1.4\ ^{+2.3}_{-1.4}\ )\%$	711
				$K^-\rho^+$	$(\ 7.2\ ^{+3.0}_{-3.1}\ )\%$	679
				$\bar{K}^0\rho^0$	$(\ 0.1\ ^{+0.6}_{-0.1}\ )\%]$	677

Stable Particle Table *(cont'd)*

Particle	$I^G(J^P)C_n$ [a]	Mass [b] (MeV) Mass^2 (GeV^2)	Mean life [b] (sec) $c\tau$ (cm)	Partial decay mode		p or p_{max} [c] (MeV/c)
				Mode	Fraction [b]	

CHARMED STRANGE MESON [a]

				$F^+\rceil$ (or $F^- \to CC$)		
$F^\pm$	$0(0^-)$ [m]	2021 ±15	$(2.2^{+2.8}_{-1.1})\times10^{-13}$	$\eta\pi^+$	(seen)	930
				$\eta\pi^+\pi^+\pi^-$	(seen)	885
				$\eta'\varkappa^+\pi^+\pi^-$	(seen)	713
				$\rho^+\phi$	(seen)	467

$\to B$

NONSTRANGE BARYONS [a]

p	$\frac{1}{2}(\frac{1}{2}^+)$	938.2796 ±0.0027 $m^2=0.880369$	stable ($\geq 8\times10^{30}$y)	stable		
				$\lvert q_p\rvert - \lvert q_e\rvert < 10^{-21}\lvert q_e\rvert$ [n]		
n	$\frac{1}{2}(\frac{1}{2}^+)$	939.5731 ±0.0027 $m^2=0.882798$	925 ± 11 $c\tau=2.77\times10^{13}$	$pe^-\bar\nu$	100%	1
				$p\nu\bar\nu$ (chg. noncons.)	(<9)$\times10^{-24}$	1
		$m_p-m_n=-1.29343$ ±0.00004		$\lvert q_n\rvert < 10^{-21}\lvert q_e\rvert$ [n]		

STRANGENESS −1 BARYONS [a]

Λ	$0(\frac{1}{2}^+)$	1115.60 ±0.05 $S=1.2^*$ $m^2=1.2446$ $m_\Lambda-m_{\Sigma^0}=-76.86$ ±0.08	2.632×10^{-10} ±0.020 S=1.6* $c\tau=7.89$	$p\pi^-$ $n\pi^0$ $pe^-\nu$ $p\mu^-\nu$ $p\pi^-\gamma$	(64.2 $\pm$ 0.5)% (35.8 $\pm$ 0.5)% (8.35$\pm$ 0.15)$\times10^{-4}$ (1.57$\pm$ 0.35)$\times10^{-4}$ [e](8.5 $\pm$ 1.4)$\times10^{-4}$	100 104 163 131 100
Σ^+	$1(\frac{1}{2}^+)$	1189.36 ±0.06 $S=1.8^*$ $m^2=1.4146$	0.800×10^{-10} ±0.004 $c\tau=2.40$	$p\pi^0$ $n\pi^+$ $p\gamma$ $n\pi^+\gamma$ $\Lambda e^+\nu$ $n\mu^+\nu$ $ne^+\nu$ pe^+e^-	(51.64 $\pm$ 0.30)% (48.36)% (1.20$\pm$ 0.13)$\times10^{-3}$ S=1.4* [e](4.5 $\pm$ 0.5)$\times10^{-4}$ (2.0 $\pm$ 0.5)$\times10^{-5}$ (<3.0)$\times10^{-5}$ (<5)$\times10^{-6}$ (<7)$\times10^{-6}$	189 185 225 185 71 202 224 225
		$m_{\Sigma^+}-m_{\Sigma^-}=-7.97$ ±0.07 $S=1.3^*$	$\dfrac{\Gamma(\Sigma^+\to \ell^+n\nu)}{\Gamma(\Sigma^-\to \ell^-n\nu)} < .04$			
Σ^0	$1(\frac{1}{2}^+)$ [p]	1192.46 ±0.08 $m^2=1.4220$	5.8×10^{-20} ±1.3 $c\tau=1.7\times10^{-9}$	$\Lambda\gamma$ Λe^+e^- $\Lambda\gamma\gamma$	100% [g](5.45)$\times10^{-3}$ (<3)%	74 74 74
Σ^-	$1(\frac{1}{2}^+)$	1197.34 ±0.05 $m^2=1.4336$ $m_{\Sigma^0}-m_{\Sigma^-}=-4.88$ ±0.06	1.482×10^{-10} ±0.011 S=1.3* $c\tau=4.44$	$n\pi^-$ $ne^-\nu$ $n\mu^-\nu$ $\Lambda e^-\nu$ $n\pi^-\gamma$	100% (1.08$\pm$ 0.04)$\times10^{-3}$ (0.45$\pm$ 0.04)$\times10^{-3}$ (0.61$\pm$ 0.05)$\times10^{-4}$ [e](4.6 $\pm$ 0.6)$\times10^{-4}$	193 230 210 79 193

Stable Particle Table *(cont'd)*

Particle	$I^G(J^P)C_n$ [a]	Mass [b] (MeV) Mass² (GeV²)	Mean life [b] (sec) cτ (cm)	Partial decay mode		
				Mode	Fraction [b]	p or p_{max} [c] (MeV/c)

STRANGENESS −2 BARYONS [a]

Particle	$I^G(J^P)C_n$	Mass	Mean life	Mode	Fraction	p or p_{max}
Ξ^0	$\frac{1}{2}(\frac{1}{2}^+)$ [q]	1314.9 ±0.6 m²=1.729	2.90×10⁻¹⁰ ±0.10 cτ =8.69	$\Lambda\pi^0$	100%	135
				$\Lambda\gamma$	(0.5 ± 0.5)%	184
				$\Sigma^0\gamma$	(<7)%	117
				$p\pi^-$	(<3.6)×10⁻⁵	299
				$pe^-\nu$	(~1.3)×10⁻³	323
		$m_{\Xi^0}-m_{\Xi^-} = -6.4$ ±0.6		$\Sigma^+e^-\nu$	(<1.1)×10⁻³	120
				$\Sigma^-e^+\nu$	(<0.9)×10⁻³	112
				$\Sigma^+\mu^-\nu$	(<1.1)×10⁻³	65
				$\Sigma^-\mu^+\nu$	(<0.9)×10⁻³	49
				$p\mu^-\nu$	(<1.3)×10⁻³	309
Ξ^-	$\frac{1}{2}(\frac{1}{2}^+)$ [q]	1321.32 ±0.13 m²=1.7459	1.641×10⁻¹⁰ ±0.016 cτ =4.92	$\Lambda\pi^-$	100%	139
				$\Lambda e^-\nu$	(2.9 ± 1.1)×10⁻⁴	190
				$\Sigma^0 e^-\nu$	(<1.4)×10⁻⁴	123
				$\Lambda\mu^-\nu$	(3.5 ± 3.5)×10⁻⁴	163
				$\Sigma^0\mu^-\nu$	(<8)×10⁻⁴	70
				$n\pi^-$	(<1.1)×10⁻³	303
				$ne^-\nu$	(<3.2)×10⁻³	327
				$n\mu^-\nu$	(<1.5)%	313
				$\Sigma^-\gamma$	(<1.2)×10⁻³	118
				$p\pi^-\pi^-$	(<4)×10⁻⁴	223
				$p\pi^-e^-\nu$	(<4)×10⁻⁴	304
				$p\mu^-\mu^-\nu$	(<4)×10⁻⁴	250
				$\Xi^0 e^-\nu$	(<2.3)×10⁻³	6

STRANGENESS −3 BARYON [a]

Particle	$I^G(J^P)C_n$	Mass	Mean life	Mode	Fraction	p or p_{max}
Ω^-	$0(\frac{3}{2}^+)$ [q]	1672.45 ±0.32 m²=2.7971	0.819×10⁻¹⁰ ±0.027 cτ =2.46	ΛK^-	(68.6 ± 1.3)%	211
				$\Xi^0\pi^-$	(23.4 ± 1.3)%	294
				$\Xi^-\pi^0$	(8.0 ± 0.8)%	290
				$\Xi^0 e^-\nu$	(~1)%	319
				$\Xi^0(1530)\pi^-$	(~2)×10⁻³	
				$\Lambda\pi^-$	(<1.3)×10⁻³	449
				$\Xi^-\gamma$	(<3.1)×10⁻³	314

NONSTRANGE CHARMED BARYON [a]

Particle	$I^G(J^P)C_n$	Mass	Mean life	Mode	Fraction	p or p_{max}
Λ_c^+	$0(\frac{1}{2}^+)$ [r]	2282.2 ±3.1 S=1.8* m²=5.21	$(1.1^{+0.9}_{-0.4})\times10^{-13}$ cτ =0.003	$pK^-\pi^+$	(2.2 ± 1.0)%	820
				$p\bar{K}^0$	(1.1 ± 0.7)%	870
				$p\bar{K}^0\pi^+\pi^-$	(<4)%	751
				$\Lambda\pi^+$	(0.6 ± 0.5)%	861
				$\Lambda\pi^+\pi^+\pi^-$	(< 3.1, seen)%	804
				$\Sigma^0\pi^+$	(seen)	822
				†[pK^{*0}	(0.48± 0.30)%	682
				$\Delta^{++}K^-$	(0.45± 0.27)%	707
				$pK^{*-}\pi^+$	(seen)]	576
				e^+ anything	(4.5 ± 1.7)%	
				†[pe^+ anything	(1.8 ± 0.9)%	
				Λe^+ anything	(1.1 ± 0.8)%]	

$\rightarrow \Lambda_b^0$
→ searches for massive neutrinos and lepton mixing
→ ν bounds from astrophysics and cosmology
→ heavy lepton searches
→ weak gauge boson searches
→ free quark searches
→ magnetic monopole searches
→ charm searches and evidence
→ bottom hadron searches
→ top hadron searches
→ other stable particle searches

Stable Particle Table

e [w] Magnetic Moment
$1.001\ 159\ 652\ 209 \pm .000\ 000\ 000\ 031 \quad \dfrac{e\hbar}{2m_e c}$

μ [w] $1.001\ 165\ 924 \pm .000\ 000\ 009 \quad \dfrac{e\hbar}{2m_\mu c}$

μ Decay parameters[s]

$\rho = 0.752 \pm 0.003 \qquad \eta = -0.12 \pm 0.21$
$\xi \cdot P_\mu = 0.972 \pm 0.014 \qquad \delta = 0.755 \pm 0.009 \qquad h = 1.01 \pm 0.06$
$|g_A/g_V| = 0.86^{+0.33}_{-0.11} \qquad \phi = 180° \pm 15°$

η

Mode	Left-right asymmetry	Sextant asymmetry	Quadrant asymmetry
$\pi^+\pi^-\pi^0$	$(0.12 \pm .17)\%$	$(0.19 \pm 0.16)\%$	$(-0.17 \pm 0.17)\%$
$\pi^+\pi^-\gamma$	$(0.88 \pm .40)\%$		

$\beta = 0.047 \pm 0.062$

K[±]

Mode	Partial rate (sec^{-1})	
$\mu\nu$	$(51.33 \pm 0.17) \times 10^6$	S=1.2*
$\pi\pi^0$	$(17.10 \pm 0.13) \times 10^6$	S=1.1*
$\pi\pi^+\pi^-$	$(4.52 \pm 0.02) \times 10^6$	S=1.1*
$\pi\pi^0\pi^0$	$(1.40 \pm 0.04) \times 10^6$	S=1.4*
$\mu\pi^0\nu$	$(2.58 \pm 0.07) \times 10^6$	S=1.7*
$e\pi^0\nu$	$(3.90 \pm 0.04) \times 10^6$	S=1.1*

Slope parameters for $K \to 3\pi$ [f]

	g	
$K^+ \to \pi^+\pi^+\pi^-$	$g = -0.215 \pm .004$	S=1.4*
$K^- \to \pi^-\pi^-\pi^+$	$g = -0.217 \pm .007$	S=2.5*
$K^\pm \to \pi^0\pi^0\pi^\pm$	$g = 0.607 \pm .030$	S=1.3*
$K_L^0 \to \pi^+\pi^-\pi^0$	$g = 0.670 \pm .014$	S=1.6*

See Data Card Listings for quadratic coefficients.

K$_S^0$

$\pi^+\pi^-$	$^j (0.7689 \pm .0033) \times 10^{10}$	
$\pi^0\pi^0$	$^j (0.3517 \pm .0029) \times 10^{10}$	S=1.1*

K$_L^0$

$\pi^0\pi^0\pi^0$	$(4.14 \pm 0.20) \times 10^6$	S=1.7*
$\pi^+\pi^-\pi^0$	$(2.39 \pm 0.04) \times 10^6$	S=1.2*
$\pi\mu\nu$	$(5.23 \pm 0.09) \times 10^6$	S=1.3*
$\pi e\nu$	$(7.47 \pm 0.11) \times 10^6$	S=1.3*
$\pi^+\pi^-$	$^j (3.91 \pm 0.10) \times 10^4$	S=1.1*
$\pi^0\pi^0$	$^j (1.81 \pm 0.36) \times 10^4$	S=1.5*

Form factors for $K_{\ell 3}$ decays

$K_{\ell 3}^+ \begin{cases} \lambda_+^e = 0.029 \pm .004 \\ \lambda_+^\mu = 0.026 \pm .008 \ \text{S=1.5*} \\ \lambda_0^\mu = -0.003 \pm .007 \ \text{S=1.5*} \end{cases}$
$K_{\ell 3}^0 \begin{cases} \lambda_+^e = 0.0300 \pm .0016 \ \text{S=1.2*} \\ \lambda_+^\mu = 0.034 \pm .006 \ \text{S=2.5*} \\ \lambda_0^\mu = 0.020 \pm .007 \ \text{S=2.5*} \end{cases}$

See Data Card Listings for ξ, f_s, and f_t.

CP violation parameters[u,j]

$|\eta_{+-}| = (2.274 \pm .022) \times 10^{-3} \qquad |\eta_{00}| = (2.33 \pm .08) \times 10^{-3} \ \text{S=1.1*}$
$\phi_{+-} = (44.6 \pm 1.2)° \qquad \phi_{00} = (54 \pm 5)°$
$|\eta_{+-0}|^2 < 0.12 \qquad |\eta_{000}|^2 < 0.28 \qquad \delta = (0.330 \pm .012)\%$

$\Delta S = -\Delta Q$
$\text{Re } x = 0.009 \pm .020 \ \text{S=1.4*} \qquad \text{Im } x = -0.004 \pm .026 \ \text{S=1.1*}$

	Magnetic moment ($e\hbar/2m_p c$)	Decay parameters[v] Measured α	φ(degree)	γ	Derived Δ(degree)	Coupling Constant Ratios
p	$2.7928456 \pm .0000011$					
n [w]	$-1.91304184 \pm .00000088$	$pe^-\nu$				$g_A/g_V = -1.255 \pm 0.006$ $\phi_{AV} = (180.11 \pm 0.17)°$
Λ [w]	$-0.613 \pm .004$	$p\pi^-$ 0.642 ± 0.013	$(-6.5 \pm 3.5)°$	0.76	$(7.7 \pm 4.1)°$	
		$n\pi^0$ 0.646 ± 0.044				
		$pe\nu$				$g_A/g_V = -0.690 \pm 0.034$ S=1.4*
Σ[+]	$2.33 \pm .13$	$p\pi^0$ -0.979 ± 0.016	$(36 \pm 34)°$	0.17	$(187 \pm 6)°$	
		$n\pi^+$ $+0.068 \pm 0.013$	$(167 \pm 20)°$	-0.97	$(-73^{+134}_{-10})°$	
		$p\gamma$ -0.72 ± 0.29	S=1.1*			
Σ[−]	$-1.41 \pm .25$	$n\pi^-$ -0.068 ± 0.008	$(10 \pm 15)°$	0.98	$(249^{+12}_{-116})°$	
		$ne^-\nu$				$g_A/g_V = \pm(0.385 \pm 0.070)$ S=2.3*
		$\Lambda e^-\nu$				$g_V/g_A = 0.14 \pm 0.24$ S=1.6* $g_{WM}/g_A = 2.4 \pm 1.7$
Ξ[0]	$-1.250 \pm .014$	$\Lambda\pi^0$ -0.413 ± 0.022 S=2.0*	$(21 \pm 12)°$	0.85	$(218^{+12}_{-18})°$	
Ξ[−]	$-1.85 \pm .75$	$\Lambda\pi^-$ -0.434 ± 0.015 S=1.4*	$(2 \pm 6)°$ S=1.1*	0.90	$(184 \pm 12)°$	
Ω[−]		ΛK^- -0.10 ± 0.38 S=1.2*				

→ Indicates an entry in the Stable Particle Data Card Listings not entered in the Stable Particle Table.

* S = Scale factor = $\sqrt{\chi^2/(N-1)}$, where $N \approx$ number of experiments. S should be ≈ 1. If $S > 1$, we have enlarged the error of the mean, $\delta\bar{x}$; i.e., $\delta\bar{x} \to S\delta\bar{x}$. This convention is still inadequate, since if $S \gg 1$ the experiments are probably inconsistent, and therefore the real uncertainty is probably even greater than $S\delta x$. See text, and ideograms in Stable Particle Data Card Listings.

† Square brackets indicate subreactions of some previous unbracketed decay mode(s). Reactions in one set of brackets may overlap with reactions in another set of brackets.

a. The baryon number B, strangeness S, and charm C of the hadrons which appear in the tables are as follows:

Mesons (B=0)	S	C	Baryons (B=1)	S	C
π, η	0	0	p,n	0	0
K^+, K^0	+1	0	Λ, Σ	1	0
$K^-, \bar{K}^0$	−1	0	Ξ	−2	0
D^+, D^0	0	+1	Ω^-	−3	0
$D^-, \bar{D}^0$	0	−1	Λ_c^+	0	+1
F^+	+1	+1			
F^-	−1	−1			

b. Quoted upper limits correspond to a 90% confidence level.

c. In decays with more than two bodies, p_{max} is the maximum momentum that any particle can have.

d. 99% confidence level. Lower limit from same experiment, > 14 eV, not yet confirmed. See Stable Particle Data Card Listings.

e. See Stable Particle Data Card Listings for energy limits used in this measurement.

f. Theoretical value; see also Stable Particle Data Card Listings.

g. See note in Stable Particle Data Card Listings.

h. Structure-dependent part with positive (SD+) and negative (SD−) photon helicity.

i. The direct emission branching fraction is $(1.56 \pm .35)\times 10^{-5}$.

j. The $K_S^0 \to \pi\pi$ and $K_L^0 \to \pi\pi$ rates (and branching fractions) are from our branching fraction and rate fits and do not include results of K_L, K_S interference experiments. The $|\eta_{+-}|$ and $|\eta_{00}|$ values given in the addendum are these rates combined with the $|\eta_{+-}|$ and $|\eta_{00}|$ results from interference experiments.

k. The stronger limit $<2\times10^{-9}$ of Clark et al., Phys. Rev. Lett. **26**, 1667 (1971) is not listed because of possible (but unknown) systematic errors. See Stable Particle Data Card Listings.

ℓ. This is a weighted average of $D^\pm$ (44%) and D^0 (56%) branching fractions.

m. Quantum numbers shown are favored but not yet established. See Stable Particle Data Card Listings.

n. Limit from neutrality-of-matter experiments. Assumes $|q_n| = |q_p| - |q_e|$.

p. J^P not measured for Σ^0. Assumed same as $\Sigma^\pm$ to allow isotriplet association.

q. P for Ξ and J^P for Ω^- not yet measured. Values shown are SU(3) predictions.

r. J^P for Λ_c^+ not yet measured. Values shown are SU(4) predictions.

s. $|g_A/g_V|$ defined by $g_A^2 = |C_A|^2+|C'_A|^2$, $g_V^2 = |C_V|^2+|C'_V|^2$, and $\Sigma<\bar{e}|\Gamma_i|\mu><\bar{\nu}|\Gamma_i(C_i+C'_i\gamma_5)|\nu>$; ϕ defined by $\cos\phi = -Re(C_A^*C'_V+C'_AC_V^*)/g_Ag_V$, P_μ is muon longitudinal polarization [for more details, see text Section VI A].

t. The definition of the slope parameter of the Dalitz plot is as follows [see also text Section VI B.1]: $|M|^2 = 1 + g\left[\dfrac{s_3-s_0}{m_{\pi^+}^2}\right]$

u. The definition for the CP violation parameters is as follows [see also text Section VI B.3]:

$$\eta_{+-} = |\eta_{+-}|e^{i\phi_{+-}} = \frac{A(K_L^0 \to \pi^+\pi^-)}{A(K_S^0\to\pi^+\pi^-)} \qquad \eta_{00} = |\eta_{00}|e^{i\phi_{00}} = \frac{A(K_L^0\to\pi^0\pi^0)}{A(K_S^0\to\pi^0\pi^0)}$$

$$\delta = \frac{\Gamma(K_L^0\to\ell^+)-\Gamma(K_L^0\to\ell^-)}{\Gamma(K_L^0\to\ell^+)+\Gamma(K_L^0\to\ell^-)}, \qquad |\eta_{+-0}|^2 = \frac{\Gamma(K_S^0\to\pi^+\pi^-\pi^0)^{CP\ viol.}}{\Gamma(K_L^0\to\pi^+\pi^-\pi^0)}, \qquad |\eta_{000}|^2 = \frac{\Gamma(K_S^0\to\pi^0\pi^0\pi^0)^{CP\ viol.}}{\Gamma(K_L^0\to\pi^0\pi^0\pi^0)}.$$

v. The definition of these quantities is as follows [for more details on sign convention, see text Section VI B]:

$$\alpha = \frac{2|s||p|\cos\Delta}{|s|^2+|p|^2} \qquad \beta = \sqrt{1-\alpha^2}\sin\phi \qquad g_A, g_V, g_{WM}\ \text{defined by}\ <B_f|\gamma_\lambda(g_V-g_A\gamma_5)+(g_{WM}/m_{B_i})\sigma^{\lambda\nu}q_\nu|B_i>$$

$$\beta = \frac{-2|s||p|\sin\Delta}{|s|^2+|p|^2} \qquad \gamma = \sqrt{1-\alpha^2}\cos\phi \qquad \phi_{AV}\ \text{defined by}\ g_A/g_V = |g_A/g_V|e^{i\phi_{AV}}$$

w. For limits on electric dipole moment, see Data Card Listings. Forbidden by P and T invariance.

Meson Table

April 1982

In addition to the entries in the Meson Table, the Meson Data Card Listings contain all substantial claims for meson resonances. See contents of Meson Data Card Listings below.

Quantities in italics are new or have changed by more than one (old) standard deviation since April 1980.

J^P	G\I	0	1	½						Partial decay mode		
N	+	ϵ	ρ	K^*	$I^G(J^P)C_n$	Mass M (MeV)	Full Width Γ (MeV)	$M^2 \pm \Gamma M^a$ (GeV^2)		Mode	Fraction(%) [Upper limits (%) are 90% CL]	p or $p_{max}^{\ b}$ (MeV/c)
	−	ω/ϕ	δ									
A	+−	η	π	K	⊢— *estab.*							

				NONSTRANGE MESONS				
$\pi^\pm$	$1^-(0^-)+$	139.57	0.0	0.019479		See Stable Particle Table		
π^0		134.96	7.95 eV ±0.55 eV	0.018215				
η	$0^+(0^-)+$	548.8 ±0.6	0.83 keV ±0.12 keV	0.301 ±0.000	Neutral Charged	70.9 29.1	See Stable Particle Table	
$\rho(770)$	$1^+(1^-)-$	769‡ ±3§	154‡ ±5§	0.591 ±0.118	$\pi\pi$ $\pi\gamma$ $\mu^+\mu^-$ e^+e^- $\eta\gamma$	≈ 100 *0.044 ±0.005* *0.0067 ±0.0012^d* *0.0043 ±0.0005^d* seen‡	358 372 370 384 189
	M and Γ from neutral mode.				For upper limits, see footnote e		
$\omega(783)$	$0^-(1^-)-$	782.6 ±0.2 S=1.1*	9.9 ±0.3	0.612 ±0.008	$\pi^+\pi^-\pi^0$ $\pi^0\gamma$ $\pi^+\pi^-$ $\pi^0\mu^+\mu^-$ e^+e^- $\eta\gamma$	89.9 ±0.5 8.7 ±0.5 1.4 ±0.2 *0.010 ±0.002* 0.0072 ±0.0007 S=1.3* seen‡	327 380 366 349 391 199
					For upper limits, see footnote f		
$\eta'(958)$	$0^+(0^-)+^\ddagger$	957.57 ±0.25	0.28 ±0.10	0.917 ±0.0003	$\eta\pi\pi$ $\rho^0\gamma$ $\omega\gamma$ $\gamma\gamma$ $\mu^+\mu^-\gamma$	65.3 ±1.6 30.0 ±1.6 2.8 ±0.5 1.9 ±0.2 *0.009 ±0.002*	231 170 159 479 467
					For upper limits, see footnote g		
$S^*(975)$	$0^+(0^+)+$	975^c ±4 S=1.4*	33^c ±6	0.951 ±0.032	$\pi\pi$ $K\bar{K}$	*78 ±3* *22 ±3*	467
	See note on $\pi\pi$ and $K\bar{K}$ S wave.‡						
$\delta(980)^\ddagger$	$1^-(0^+)+$	983^h ±2	54^h ±7	0.966 ±0.053	$\eta\pi$ $K\bar{K}$	seen seen	320
$\phi(1020)$	$0^-(1^-)-$	1019.61 ±0.07	4.21 ±0.13	1.040 ±0.004	K^+K^- $K_L K_S$ $\pi^+\pi^-\pi^0$ (incl. $\rho\pi$) $\eta\gamma$ $\pi^0\gamma$ e^+e^- $\mu^+\mu^-$ $\pi^+\pi^-$	49.1 ±1.0 S=1.3* 34.6 ±1.0 S=1.3* 14.8 ±0.7 S=1.2* 1.5 ±0.2 0.14 ±0.05 0.031 ±0.001 0.025 ±0.003 *0.02 ±0.01*	127 111 462 362 501 510 499 490
					For upper limits, see footnote i		
$H(1190)$	$0^-(1^+)-$	*1190* *±60*	*320* *±50*	1.416 ±0.381	$\rho\pi$	*seen*	327
	Seen in one experiment only.						
$B(1235)$	$1^+(1^+)-$	1233 ±10§	137 ±10§	1.52 ±0.17	$\omega\pi$ [D/S amplitude ratio = 0.29 ±0.05]	only mode seen	349
					For upper limits, see footnote j		

Meson Table *(cont'd)*

J^P	G	I	0	1	½
N	+		ε	ρ	K*
	−		ω/φ	δ	
A	+−		η	π	K

$\longmapsto$ *estab.*

Particle	$I^G(J^P)C_n$	Mass M (MeV)	Full Width Γ (MeV)	M^2 $\pm \Gamma M^a$ (GeV²)	Mode	Fraction(%) [Upper limits (%) are 90% CL]		p or $p_{max}^{\,b}$ (MeV/c)
f(1270)	$0^+(2^+)+$	1273 ±5§	179 ±20§	1.62 ±0.23	$\pi\pi$	83.1 ±1.9	S=1.4*	621
					$2\pi^+2\pi^-$	2.8 ±0.4	S=1.2*	558
					$K\bar{K}$	2.9 ±0.2	S=1.2*	397
					$\gamma\gamma$	0.0016 ±0.0003		637
					$\pi^+\pi^-2\pi^0$	seen		561
					For upper limits, see footnote k			
A₁(1270)	$1^-(1^+)+$	1275‡ ±30	315‡ ±45	1.63 ±0.40	$\rho\pi$	dominant		389
					$\pi(\pi\pi)_{S-wave}$	seen		599
D(1285)	$0^+(1^+)+$	1283 ±5§	26 ±5§	1.65 ±0.03	$K\bar{K}\pi$	11 ±3		302
					$\eta\pi\pi$	49 ±6		482
					†[$\delta\pi$	36 ±7]		236
					4π (prob. $\rho\pi\pi$)‡	40 ±7		564
ε(1300)	$0^+(0^+)+$	~1300	200–600		$\pi\pi$	~90		635
					$K\bar{K}$	~10		418
					$\eta\eta$			348
	See note on $\pi\pi$ and $K\bar{K}$ S wave. ‡							
π(1300)	$1^-(0^-)+$	1300§ ±100§	200–600		$\rho\pi$	seen		407
					$\pi(\pi\pi)_{S-wave}$	seen		612
	Not a well-established resonance.							
A₂(1320)	$1^-(2^+)+$	1318 ±5§	110 ±5§	1.74 ±0.14	$\rho\pi$	70.1 ±2.2		419
					$\eta\pi$	14.5 ±1.2		534
					$\omega\pi\pi$	10.6 ±2.5		361
					$K\bar{K}$	4.8 ±0.5		434
					$\eta'\pi$	< 2 (CL=97%)		286
					$\pi\gamma$	0.45 ±0.11		652
					$\gamma\gamma$	0.0007 ±0.0004		659
E(1420)‡	$0^+(1^+)+$	1418 ±10§	52 ±10§	2.01 ±0.07	$K\bar{K}\pi$ (prob. $K^*\bar{K}+K\bar{K}^*$)	seen		423
					$\eta\pi\pi$	possibly seen		565
					†[$\delta\pi$	possibly seen]		348
f′(1515)	$0^+(2^+)+$	1520 ±10§	75 ±10§	2.31 ±0.11	$K\bar{K}$	dominant		574
					$\pi\pi$	possibly seen		747
ρ′(1600)	$1^+(1^-)-$	1600‡ ±20§	300‡ ±100§	2.56 ±0.48	4π (incl. $\rho\pi^+\pi^-$,A₁π)	large		738
					$\pi\pi$	< 30‡		788
					$K^*\bar{K} + \bar{K}^*K$	~15		388
					$\eta\pi\pi$	~13		675
					$K\bar{K}$	~1		630
					e^+e^-	seen		800
ω(1670)	$0^-(3^-)-$	1668 ±5	166 ±15§ S=1.1*	2.78 ±0.28	3π	seen		806
					†[$\rho\pi$	seen]		648
					5π	seen		740
					†[$\omega\pi\pi$ (prob. Bπ)	seen]		616
A₃(1680)‡	$1^-(2^-)+$	1680§ ±30§	250§ ±50§	2.82 ±0.42	fπ	55 ±5		337
					$\rho\pi$	36 ±6		656
					$\pi(\pi\pi)_{S-wave}$	9 ±5		813
					For upper limits, see footnote ℓ			
φ′(1680)	$0^-(1^-)-$	1684 ±15§	126 ±22	2.84 ±0.21	$K^*\bar{K} + \bar{K}^*K$	dominant		541
					$\omega\pi\pi$	seen		624
					$K\bar{K}$	seen		682
g(1690)‡	$1^+(3^-)-$	1691 ±5§	200§ ±20§	2.86 ±0.34	2π	23.8 ±1.3		834
					4π (incl. $\pi\pi\rho,\rho\rho,$A₂$\pi,\omega\pi$)	70.9 ±1.9		787
					$K\bar{K}\pi$ (incl. $K^*\bar{K}$)	3.8 ±1.2		625
					$K\bar{K}$	1.5 ±0.3	S=1.3*	684

$\rightarrow$ J^P, M, and Γ from the 2π and $K\bar{K}$ modes.
$\rightarrow$
$\rightarrow$
$\rightarrow$

Meson Table *(cont'd)*

J^P	G	I 0 1 ½						Partial decay mode		
N	+	ϵ ρ K^*	$I^G(J^P)C_n$	Mass M	Full Width Γ	M^2		Mode	Fraction(%)	p or p_{max}
	−	ω/ϕ δ				$\pm\Gamma M^a$			[Upper limits (%) are 90% CL]	b
A	+−	η π K	⊢ estab.	(MeV)	(MeV)	(GeV²)				(MeV/c)

h(2040)	$0^+(4^+)+$	2040§ ±20§	150§ ±50§	4.16 ±0.31		$\pi\pi$	seen	1010
						$K\bar{K}$	seen	890

$\eta_c(2980)$	$0^+(\ )+$	2981 ±6	< 20	8.89		$\eta\pi^+\pi^-$	seen	1426
						$2(\pi^+\pi^-)$	seen	1458
						$K^+K^-\pi^+\pi^-$	seen	1343
						$p\bar{p}$	seen	1158

$J/\psi(3100)$	$0^-(1^-)-$	3096.9 ±0.1	0.063 ±0.009	9.591 ±0.000		e^+e^-	7.4 ± 1.2	1548
						$\mu^+\mu^-$	7.4 ± 1.2	1545
						hadrons + radiative	85 ± 2	

Decay modes into stable hadrons

Mode	Fraction	p
†[$2(\pi^+\pi^-)\pi^0$	3.7 ± 0.5	1496
$3(\pi^+\pi^-)\pi^0$	2.9 ± 0.7	1433
$\pi^+\pi^-\pi^0K^+K^-$	1.2 ± 0.3	1368
$4(\pi^+\pi^-)\pi^0$	0.9 ± 0.3	1345
$\pi^+\pi^-K^+K^-$	0.72 ± 0.23	1407
$p\bar{p}\pi^+\pi^-$	0.53 ± 0.06	1107
$2(\pi^+\pi^-)$	0.4 ± 0.1	1517
$3(\pi^+\pi^-)$	0.4 ± 0.2	1466
$n\bar{n}\pi^+\pi^-$	0.38 ± 0.36	1106
$\Xi\bar{\Xi}$	0.32 ± 0.08	818
$2(\pi^+\pi^-)K^+K^-$	0.31 ± 0.13	1320
$K^0_S K^\pm\pi^\mp$	0.26 ± 0.07	1440
$\Sigma^+\Sigma^-$	0.24 ± 0.26	988
$p\bar{p}\eta$	0.23 ± 0.04	948
$p\bar{p}$	0.22 ± 0.02	1232
$p\bar{n}\pi^-$ or $\bar{p}n\pi^+$	0.21 ± 0.02	1174
$n\bar{n}$	0.18 ± 0.09	1231
$p\bar{p}\pi^+\pi^-\pi^0$	0.16 ± 0.06^m	1033
$\Sigma^0\bar{\Sigma}^0$	0.13 ± 0.04	988
$\Lambda\bar{\Lambda}$	0.11 ± 0.02	1074
$p\bar{p}\pi^0$	0.11 ± 0.01	1176
$2(K^+K^-)$	0.07 ± 0.03	1131
K^+K^-	0.022 ± 0.008	1468
$\pi^+\pi^-$	0.011 ± 0.005	1542
$\Lambda\bar{\Sigma}$	< 0.015	1032
$K^0_S K^0_L$	< 0.009]	1466

Decay modes into hadronic resonances

Mode	Fraction	p
†[$\rho\pi$	1.22 ± 0.12	1449
$\omega 2\pi^+2\pi^-$	0.85 ± 0.34	1392
ρA_2	0.84 ± 0.45	1126
$\omega\pi\pi$	0.68 ± 0.19	1435
$K^{*0}(892)\bar{K}^{*0}(1430)+$c.c.	0.67 ± 0.26	1007
$K^\pm K^{*\mp}(892)$	0.34 ± 0.05	1373
$B^\pm(1235)\pi^\mp$	0.29 ± 0.07	1299
$K^0\bar{K}^{*0}(892)+$c.c.	0.27 ± 0.06	1373
ωf	0.23 ± 0.08 S=1.2*	1144
$\phi\pi^+\pi^-$	0.21 ± 0.09	1365
$\eta'p\bar{p}$	0.18 ± 0.06	596
$\phi K\bar{K}$	0.18 ± 0.08	1176
$\omega p\bar{p}$	0.16 ± 0.03	768
$\omega K\bar{K}$	0.16 ± 0.10	1265
$\phi\eta$	0.10 ± 0.06	1320
$\phi f'(1515)$	0.08 ± 0.05	874
$\pi^\pm A^\mp$	< 0.43	1263
$K^{*0}(1430)\bar{K}^{*0}(1430)$	< 0.29	584
$K^0\bar{K}^{*0}(1430)+$c.c.	< 0.2	1154
$K^\pm K^{*\mp}(1430)$	< 0.2	1154
$\phi 2\pi^+2\pi^-$	< 0.15	1318
$\phi\eta'$	< 0.13	1192
$K^{*0}(892)\bar{K}^{*0}(892)$	< 0.05	1266
ϕf	< 0.037	1037
$\omega f'$	< 0.016]	1006

Radiative decay modes

Mode	Fraction	p
†[$\gamma\iota(1440)$	0.55 ± 0.22^n	1224
$\gamma\eta'$	0.36 ± 0.05	1400
γf	0.15 ± 0.04	1287
$\gamma\eta$	0.086 ± 0.009	1500
$\gamma\pi^0$	0.007 ± 0.005	1546
$\gamma D(1285)$	< 0.6	1283
2γ	< 0.05	1548
$\gamma f'(1515)$	< 0.03	1175
$\gamma p\bar{p}$	< 0.01	1232
3γ	< 0.006]	1548

$\chi(3415)$	$0^+(0^+)+$	3415.0 ±1.0		11.662		$2(\pi^+\pi^-)$ (incl. $\pi\pi\rho$)	4.3 ± 0.9	1679
						$\pi^+\pi^-K^+K^-$ (incl. $\pi K\bar{K}^*$)	3.4 ± 0.9	1580
						$3(\pi^+\pi^-)$	1.7 ± 0.6	1633
						$\pi^+\pi^-$	0.9 ± 0.2	1702
						$\gamma J/\psi(3100)$	0.8 ± 0.2	303
						K^+K^-	0.8 ± 0.2	1635
						$p\bar{p}\pi^+\pi^-$	0.6 ± 0.2	1320
						For upper limits, see footnote o		

Meson Table *(cont'd)*

J^P	G	I	0	1	½
N	+	ϵ	ρ	K*	
	−	ω/ϕ	δ		
A	+−	η	π	K	⊢──┘ *estab.*

	$I^G(J^P)C_n$	Mass M (MeV)	Full Width Γ (MeV)	M^2 $\pm\Gamma M^a$ (GeV^2)	Mode	Fraction(%) [Upper limits (%) are 90% CL]	p or P_{max} ^{b} (MeV/c)
p_c or $\chi(3510)$	$0^+(1^+)+$	3510.0 ±0.6		12.320	$\gamma J/\psi(3100)$	28 ±3	389
					$3(\pi^+\pi^-)$	2.4 ±0.9	1683
					$2(\pi^+\pi^-)$ (incl. $\pi\pi\rho$)	1.8 ±0.5	1727
					$\pi^+\pi^-K^+K^-$ (incl. $\pi K\bar{K}^*$)	1.0 ±0.4	1632
					$\pi^+\pi^- p\bar{p}$	0.15 ±0.10	1381
					For upper limits, see footnote p		
$\chi(3555)$	$0^+(2^+)+$	3555.8 ±0.6		12.644	$\gamma J/\psi(3100)$	15.7 ±1.7	429
					$2(\pi^+\pi^-)$ (incl. $\pi\pi\rho$)	2.3 ±0.5	1750
					$\pi^+\pi^-K^+K^-$ (incl. $\pi K\bar{K}^*$)	2.0 ±0.5	1656
					$3(\pi^+\pi^-)$	1.2 ±0.8	1706
					$\pi^+\pi^- p\bar{p}$	0.35 ±0.14	1410
					$\pi^+\pi^-$	0.20 ±0.11	1772
					K^+K^-	0.16 ±0.12	1708
					For upper limits, see footnote q		
→ $\psi(3685)$	$0^-(1^-)-$	3686.0 ±0.1	0.215 ±0.040	13.587 ±0.001	e^+e^-	0.9 ±0.1	1843
					$\mu^+\mu^-$	0.8 ±0.2	1840
					hadrons + radiative	98.1 ±0.3	

$$m_{\psi(3685)} - m_{\psi(3100)} = 589.06 \pm 0.13$$

Radiative decay modes				Decay modes into hadrons		
†[$\gamma\chi(3415)$	8.2 ±1.4	261		†[$J/\psi\pi^+\pi^-$	33 ±2	477
$\gamma\chi(3510)$	8.0 ±1.3	172		$J/\psi\pi^0\pi^0$	17 ±2	481
$\gamma\chi(3555)$	7.4 ±1.3	128		$J/\psi\eta$	2.0 ±0.4 §	196
$\gamma\eta_c(2980)$	0.43 +0.26	638		$2(\pi^+\pi^-)\pi^0$	0.35 ±0.15	1799
$\gamma\eta'_c(3590)$	0.2 to 1.3	91		$\pi^+\pi^-K^+K^-$	0.16 ±0.04	1726
$\gamma\gamma$ $^\delta$	<0.5 (CL=95%)	1841		$J/\psi\pi^0$	0.10 ±0.03	528
$\gamma\eta$	<0.02	1802		$p\bar{p}\pi^+\pi^-$	0.08 ±0.02	1491
$\gamma\eta'$	<0.02	1719		$K^{*0}(892)K^\pm\pi^\mp$	0.067 ±0.025	1674
$\gamma\iota(1440)$	<0.018^n]	1570		$2(\pi^+\pi^-)$	0.05 ±0.01	1817
				$\rho^0\pi^+\pi^-$	0.042 ±0.015	1751
				$p\bar{p}$	0.019 ±0.005	1586
				$3(\pi^+\pi^-)$	0.015 ±0.010	1774
				K^+K^-	0.010 ±0.007	1776
				$\pi^+\pi^-$	0.008 ±0.005	1838
				$\rho\pi$	<0.1	1760
				$\Lambda\bar{\Lambda}$	<0.04]	1467

	$I^G(J^P)C_n$	Mass M (MeV)	Full Width Γ (MeV)	M^2 $\pm\Gamma M^a$ (GeV^2)	Mode	Fraction(%)	p or P_{max} (MeV/c)
$\psi(3770)$	$(1^-)-$	3770 ±3	25 ±3	14.213 ±0.094	e^+e^-	0.0011 ±0.0002	1885
					$D\bar{D}$	dominant	242

$$m_{\psi(3770)} - m_{\psi(3685)} = 83.9 \pm 2.4 \quad S=1.8^*$$

	$I^G(J^P)C_n$	Mass M (MeV)	Full Width Γ (MeV)	M^2 $\pm\Gamma M^a$ (GeV^2)	Mode	Fraction(%)	p or P_{max} (MeV/c)
$\psi(4030)$	$(1^-)-$	4030§ ±5§	52 ±10	16.241 ±0.210	e^+e^-	0.0014 ±0.0004	2015
					hadrons	dominant	
$\psi(4160)$	$(1^-)-$	4159 ±20	78 ±20	17.297 ±0.324	e^+e^-	0.0010 ±0.0004	2079
					hadrons	dominant	
$\psi(4415)$	$(1^-)-$	4415 ±6	43 ±20§	19.492 ±0.190	e^+e^-	0.0010 ±0.0003 S=1.4*	2208
					hadrons	dominant	
$\Upsilon(9460)$	$(1^-)-$	9456 ±10	0.042 ±0.015	89.416 ±0.0004	$\mu^+\mu^-$	3.2 ±0.7	4727
					e^+e^-	2.8 ±1.1	4728
$\Upsilon(10020)$	$(1^-)-$	10016 ±10	0.030 ±0.010	100.320 ±0.0003	$\mu^+\mu^-$	seen	5007
					e^+e^-	1.7 ±0.6	5008
					$\Upsilon(9460)\pi\pi$	30 ±6	472

$$m_{\Upsilon(10020)} - m_{\Upsilon(9460)} = 559 \pm 3$$

	$I^G(J^P)C_n$	Mass M (MeV)	Full Width Γ (MeV)	M^2 $\pm\Gamma M^a$ (GeV^2)	Mode	Fraction(%)	p or P_{max} (MeV/c)
$\Upsilon(10350)$	$(1^-)-$	10347 ±10		107.060	e^+e^-	seen	5174

$$m_{\Upsilon(10350)} - m_{\Upsilon(9460)} = 891 \pm 4$$

J^P G I | 0 | 1 | ½ N + | ϵ | ρ | K* − | ω/ϕ | δ A +− | η | π | K	$I^G(J^P)C_n$ → *estab.*	Mass M (MeV)	Full Width Γ (MeV)	M^2 $\pm\Gamma M^a$ (GeV^2)	Partial decay mode		
					Mode	Fraction(%) [Upper limits (%) are 90% CL]	p or p_{max} b (MeV/c)
$\Upsilon(10570)$	$(1^-)-$	10569 ±10	14 ±5	111.704 ±0.15	e^+e^-	0.0019 ± 0.0008	5285

$m_{\Upsilon(10570)} - m_{\Upsilon(9460)} = 1113 \pm 4$

STRANGE MESONS

	$I^G(J^P)C_n$	Mass M (MeV)	Full Width Γ (MeV)	$M^2 \pm \Gamma M^a$ (GeV^2)	Mode	Fraction(%)	p or p_{max} (MeV/c)
K^+	$1/2(0^-)$	493.67		0.244	See Stable Particle Table		
K^0		497.67		0.248			
$K^*(892)$	$1/2(1^-)$	891.8 ±0.4 S=1.2*	50.8 ±0.9	0.795 ±0.045	$K\pi$ $K\gamma$ $K\pi\pi$	≈ 100 0.15 ± 0.07 < 0.07 (CL=95%)	288 309 216

M and Γ from charged mode; $m^0 - m^\pm = 6.7 \pm 1.2$ MeV.

$Q_1(1280)$	$1/2(1^+)$	$1270^\S$ $\pm10^\S$	$90^\S$ $\pm20^\S$	1.61 ±0.11	$K\rho$ $\kappa\pi$ $K^*\pi$ $K\omega$ $K\epsilon$	42 ± 6 28 ± 4 16 ± 5 11 ± 2 3 ± 2	45 299
$\kappa(1350)$	$1/2(0^+)$	~1350	~250	1.82 ±0.34	$K\pi$	seen	574

See note on $K\pi$ S wave.[‡]

$Q_2(1400)$	$1/2(1^+)$	1414 ±13	180 ±10	2.00 ±0.25	$K^*\pi$ $K\rho$ $K\epsilon$ $K\omega$	94 ± 6 3 ± 3 2 ± 2 1 ± 1	410 308 294
$K^*(1430)$	$1/2(2^+)$	$1434^\S$ $\pm5^\S$	$100^\S$ $\pm10^\S$	2.06 ±0.14	$K\pi$ $K^*\pi$ $K^*\pi\pi$ $K\rho$ $K\omega$ $K\eta$	44.8 ± 2.3 S=2.7* 24.6 ± 2.0 S=1.1* 13.0 ± 2.6 S=1.1* 8.8 ± 1.1 S=1.3* 4.2 ± 1.5 $5 \pm 5^\S$	623 424 374 334 320 492
$L(1770)$[‡]	$1/2(2^-)$	$\sim1770^\S$	$\sim200^\S$	3.13 ±0.35	$K^*(1430)\pi$ $K^*(892)\pi$ Kf	dominant seen seen	278 652

See note on $L(1770)$.[‡]

$K^*(1780)$[‡]	$1/2(3^-)$	$1775^\S$ $\pm10^\S$	$140^\S$ $\pm20^\S$	3.15 ±0.25	$K\pi\pi$ †[$K\rho$ †[$K^*\pi$ $K\pi$	large large] large] $17 \pm 5^\S$	793 616 654 812

See note on $K^*(1780)$.[‡]

CHARMED, NONSTRANGE MESONS

D^+	$1/2(0^-)$	1869.4		3.495	See Stable Particle Table		
D^0		1864.7		3.477			
$D^{*+}(2010)$	$1/2(1^-)$	2010.1 ±0.7	< 2.0	4.041	$D^0\pi^+$ $D^+\pi^0$ $D^+\gamma$	64 ± 11 28 ± 9 8 ± 7	39 38 136

$m_{D^{*+}} - m_{D^0} = 145.4 \pm 0.2$ MeV

$D^{*0}(2010)$	$1/2(1^-)$	2007.2 ±2.1	< 5	4.029	$D^0\pi^0$ $D^0\gamma$	55 ± 15 45 ± 15	44 137

CHARMED, STRANGE MESON

F^+	$0(0^-)$	2021		4.084	See Stable Particle Table		

Meson Table *(cont'd)*

Contents of Meson Data Card Listings

entry	$I^G(J^P)C_n$	entry	$I^G(J^P)C_n$	entry	$I^G(J^P)C_n$	entry	$I(J^P)$
	Non-strange (S = 0; C, B = 0)					**Strange ($\|S\| = 1$; C, B = 0)**	
π	$1^-(0^-)+$	f' (1515)	$0^+(2^+)+$	$\to \delta$ (2450)	$1^-(6^+)+$	K	$1/2(0^-)$
η	$0^+(0^-)+$	ρ' (1600)	$1^+(1^-)-$	$\to e^+e^-$ (1100−2200)		K^* (892)	$1/2(1^-)$
ρ (770)	$1^+(1^-)-$	$\to \theta$ (1640)	$0^+(2^+)+$	$\to \bar{N}N$ (1400−3600)		Q_1 (1280)	$1/2(1^+)$
ω (783)	$0^-(1^-)-$	ω (1670)	$0^-(3^-)-$	$\to X$ (1900−3600)		κ (1350)	$1/2(0^+)$
η' (958)	$0^+(0^-)+$	A_3 (1680)	$1^-(2^-)+$	η_c (2980)	$+$	Q_2 (1400)	$1/2(1^+)$
S^* (975)	$0^+(0^+)+$	ϕ' (1680)	$0^-(1^-)-$	J/ψ (3100)	$0^-(1^-)-$	$\to K'$ (1400)	$1/2(0^-)$
δ (980)	$1^-(0^+)+$	g (1690)	$1^+(3^-)-$	χ (3415)	$0^+(0^+)+$	K^* (1430)	$1/2(2^+)$
ϕ (1020)	$0^-(1^-)-$	$\to \phi$ (1850)	0	P_c or χ(3510)	$0^+(1^+)+$	$\to L$ (1580)	$1/2(2^-)$
H (1190)	$0^-(1^+)-$	$\to X$ (1850)	(2^+)	χ (3555)	$0^+(2^+)+$	$\to K^*$ (1650)	$1/2(1^-)$
B (1235)	$1^+(1^+)-$	$\to S$ (1935)		$\to \eta_c'$ (3590)	$+$	L (1770)	$1/2(2^-)$
$\to \rho'$ (1250)	$1^+(1^-)-$	$\to \delta$ (2030)	$1^-(4^+)+$	ψ (3685)	$0^-(1^-)-$	K^* (1780)	$1/2(3^-)$
f (1270)	$0^+(2^+)+$	h (2040)	$0^+(4^+)+$	ψ (3770)	$(1^-)-$	$\to K^*$ (2060)	$1/2(4^+)$
A_1 (1270)	$1^-(1^+)+$	$\to \pi$ (2050)	$1^-(3^+)+$	ψ (4030)	$(1^-)-$	$\to K^*$ (2200)	
$\to \eta$ (1275)	$0^+(0^-)+$	$\to \pi$ (2100)	$1^-(2^-)+$	ψ (4160)	$(1^-)-$		**Charmed ($\|C\| = 1$)**
D (1285)	$0^+(1^+)+$	$\to \rho$ (2150)	$1^+(1^-)-$	ψ (4415)	$(1^-)-$	D (1870)	$1/2(0^-)$
ϵ (1300)	$0^+(0^+)+$	$\to \epsilon$ (2150)	$0^+(2^+)+$	Υ (9460)	$(1^-)-$	D^* (2010)	$1/2(1^-)$
π (1300)	$1^-(0^-)+$	$\to \rho$ (2250)	$1^+(3^-)-$	Υ (10020)	$(1^-)-$	F (2020)	$0 \; (0^-)$
A_2 (1320)	$1^-(2^+)+$	$\to \epsilon$ (2300)	$0^+(4^+)+$	Υ (10330)	$(1^-)-$	$\to F^*$ (2140)	
E (1420)	$0^+(1^+)+$	$\to \rho$ (2350)	$1^+(5^-)-$	Υ (10570)	$(1^-)-$		**Bottom (Beauty) ($\|B\| = 1$)**
						$\to B$ (5200)	
						$\to$ Exotics	

$\to$ Indicates an entry in the Meson Data Card Listings not entered in the Meson Table. We do not regard these as established resonances. All the entries in the Listings can be found in the Table of Contents of the Meson Data Card Listings immediately preceding these footnotes.

‡   See Meson Data Card Listings.

*   Quoted error includes scale factor $S = \sqrt{\chi^2/(N-1)}$. See footnote to Stable Particle Table.

†   Square brackets indicate a subreaction of the previous (unbracketed) decay mode(s).

§   This is only an educated guess; the error given is larger than the error on the average of the published values. (See the Meson Data Card Listings for the latter.)

a.   ΓM is approximately the half-width of the resonance when plotted against M^2.

b.   For decay modes into ≥ 3 particles, p_{max} is the maximum momentum that any of the particles in the final state can have. The momenta have been calculated by using the averaged central mass values, without taking into account the widths of the resonances.

c.   From pole position $(M - i\Gamma/2)$.

d.   The e^+e^- branching fraction is from $e^+e^- \to \pi^+\pi^-$ experiments only. The $\omega\rho$ interference is then due to $\omega\rho$ mixing only, and is expected to be small. See note in the Meson Data Card Listings. The $\mu^+\mu^-$ branching fraction is compiled from 3 experiments, each possibly with substantial $\omega\rho$ interference. The error reflects this uncertainty; see notes in the Meson Data Card Listings. If $e\mu$ universality holds, $\Gamma(\rho^0 \to \mu^+\mu^-) = \Gamma(\rho^0 \to e^+e^-) \times 0.99785$.

e. Empirical limits on fractions for other decay modes of $\rho(770)$ are $\pi^{\pm}\eta < 0.8\%$ (CL=84%), $\pi^{+}\pi^{+}\pi^{-}\pi^{-} < 0.15\%$, $\pi^{\pm}\pi^{+}\pi^{-}\pi^{0} < 0.2\%$ (CL=84%).

f. Empirical limits on fractions for other decay modes of $\omega(783)$ are $\pi^{+}\pi^{-}\gamma < 5\%$, $\pi^{0}\pi^{0}\gamma < 1\%$, η + neutral(s) $< 1.5\%$, $\mu^{+}\mu^{-} < 0.02\%$.

g. Empirical limits on fractions for other decay modes of $\eta'(958)$ are $\pi^{+}\pi^{-} < 2\%$ (CL=84%), $\pi^{+}\pi^{-}\pi^{0} < 5\%$ (CL=84%), $\pi^{+}\pi^{+}\pi^{-}\pi^{-} < 1\%$ (CL=95%), $\pi^{+}\pi^{+}\pi^{-}\pi^{-}\pi^{0} < 1\%$ (CL=84%), $6\pi < 1\%$, $\pi^{+}\pi^{-}e^{+}e^{-} < 0.6\%$, $\pi^{0}e^{+}e^{-} < 1.3\%$ (CL=84%), $\eta e^{+}e^{-} < 1.1\%$, $\pi^{0}\rho^{0} < 4\%$, $\eta\mu^{+}\mu^{-} < 1.5 \times 10^{-5}$, $\pi^{0}\mu^{+}\mu^{-} < 6 \times 10^{-5}$.

h. The mass and width are from the $\eta\pi$ mode only. If the $K\bar{K}$ channel is strongly coupled, the width may be larger.

i. Empirical limits on fractions for other decay modes of $\phi(1020)$ are $\pi^{+}\pi^{-}\gamma < 0.7\%$, $\omega\gamma < 5\%$ (CL=84%), $\rho\gamma < 2\%$ (CL=84%), $2\pi^{+}2\pi^{-}\pi^{0} < 1\%$ (CL=95%), $2\pi^{+}2\pi^{-} < 0.1\%$.

j. Empirical limits on fractions for other decay modes of B(1235) are $\pi\pi < 15\%$, $K\bar{K} < 2\%$ (CL=84%), $4\pi < 50\%$ (CL=84%), $\phi\pi < 1.5\%$ (CL=84%), $\eta\pi < 25\%$, $(\bar{K}K)^{\pm}\pi^{0} < 8\%$, $K_{S}K_{S}\pi^{\pm} < 2\%$, $K_{S}K_{L}\pi^{\pm} < 6\%$.

k. Empirical limits (CL=95%) on fractions for other decay modes of f(1270) are $\eta\pi\pi < 1\%$, $K^{0}K^{-}\pi^{+}$ + c.c. $< 0.5\%$, $\eta\eta < 2\%$.

ℓ. Empirical limits on fractions for other decay modes of $A_{3}(1680)$ are $\eta\pi < 10\%$, $5\pi < 10\%$.

m. Includes $p\bar{p}\pi^{+}\pi^{-}\gamma$ and excludes $p\bar{p}\eta$, $p\bar{p}\omega$, $p\bar{p}\eta'$.

n. The $\iota(1440)$ evidence is listed under E(1420); see E(1420) mini-review.

o. Empirical limits on fractions for other decay modes of $\chi(3415)$ are $2\gamma < 0.17\%$, $p\bar{p} < 0.011\%$.

p. Empirical limits on fractions for other decay modes of $\chi(3510)$ are $(\pi^{+}\pi^{-}$ and $K^{+}K^{-}) < 0.2\%$, $\gamma\gamma < 0.16\%$, $p\bar{p} < 0.13\%$.

q. Empirical limits on fractions for other decay modes of $\chi(3555)$ are $2\gamma < 0.06\%$, $p\bar{p} < 0.10\%$, $J/\psi\pi^{+}\pi^{-}\pi^{0} < 1.5\%$.

Established Nonets, and octet-singlet mixing angles θ obtained from the Gell-Mann-Okubo mass formula [Appendix II, Eq. (3)]. Of the two isosinglets, the "mainly octet" one is written first, followed by a semicolon. The angle $\delta = \theta - 35.3°$ measures the deviation from ideal mixing.

$(J^{P})C_{n}$	Nonet members	$\theta_{\text{lin.}}$	$\theta_{\text{quadr.}}$	$\delta_{\text{lin.}}$	$\delta_{\text{quadr.}}$
$(0^{-})+$	π, K, η; η'	$-24.4 \pm 0.1°$	$-11.1 \pm 0.2°$	$-59.7 \pm 0.1°$	$-46.4 \pm 0.2°$
$(1^{-})-$	ρ, K^{*}, ϕ; ω	$35.9 \pm 0.5°$	$38.6 \pm 0.4°$	$0.6 \pm 0.5°$	$3.3 \pm 0.4°$
$(2^{+})+$	A_{2}, $K^{*}(1430)$, f'; f	$26 \pm 3°$	$28 \pm 3°$	$-9 \pm 3°$	$-7 \pm 3°$
$(1^{+})+^{\dagger}$	A_{1}, Q_{A}, E; D	$52 \pm 13°$	$51 \pm 12°$	$16 \pm 13°$	$15 \pm 12°$

† $m(Q_{A})$ is assumed to be the average of $m(Q_{1})$ and $m(Q_{2})$.

More generally, because of unitarity, the mixing angles are energy dependent and complex above the first threshold (see Appendix II C), which is important especially for the scalar and the axial mesons. Note also that the two axial strange mesons (Q_{1} and Q_{2}) are mixtures of the exact SU(3) states: $Q_{1} = \cos\phi\, Q_{A} + \sin\phi\, Q_{B}$, $Q_{2} = -\sin\phi\, Q_{A} + \cos\phi\, Q_{B}$. Below we give the mixing angles δ and ϕ obtained in a unitary mixing scheme using both masses and widths as input data:

$(J^{P})C_{n}$	Nonet members	Mixing angles
$(1^{+})+$	A_{1}, Q_{A}, E; D	$\delta_{DE}(1283) = 14° + i1°$
		$\delta_{DE}(1418) = 25° + i8°$
$(1^{+})-$	B, Q_{B}, H'†; H	$\delta_{HH'}(1190) = -6° + i4°$
		$\delta_{HH'}(1400) = -15° + i10°$
		$\phi_{Q_{1}Q_{2}}(1270) = 50° + i3°$
		$\phi_{Q_{1}Q_{2}}(1414) = 61° - i3°$
$(0^{+})+$	δ, κ, S^{*}; ϵ	$\delta_{S^{*}\epsilon}(975) = +4° + i29°$
		$\delta_{S^{*}\epsilon}(1300) = -33° + i7°$

† as yet, not seen experimentally

Baryon Table

April 1982

The following short list gives the name, the nominal mass, the quantum numbers (where known), and the status of each of the Baryon States in the Data Card Listings. States with 3- or 4-star status are included in the Baryon Table below; the others are omitted because the evidence for the existence of the effect and/or for its interpretation as a resonance is open to question.

N(939)	P11	****	Δ(1232)	P33	****	Λ(1115)	P01	****	Σ(1193) P11 ****
N(1440)	P11	****	Δ(1550)	P31	**	Λ(1405)	S01	****	Σ(1385) P13 ****

N(939) P11 ****   Δ(1232) P33 ****   Λ(1115) P01 ****   Σ(1193) P11 ****   Ξ(1317) P11 ****
N(1440) P11 ****   Δ(1550) P31 **   Λ(1405) S01 ****   Σ(1385) P13 ****   Ξ(1530) P13 ****
N(1520) D13 ****   Δ(1600) P33 ***   Λ(1520) D03 ****   Σ(1480) *   Ξ(1630) **
N(1535) S11 ****   Δ(1620) S31 ****   Λ(1600) P01 ***   Σ(1560) **   Ξ(1680) S11 **
N(1540) P13 *   Δ(1700) D33 ****   Λ(1670) S01 ****   Σ(1580) D13 **   Ξ(1820) 13 ***
N(1650) S11 ****   Δ(1900) S31 ***   Λ(1690) D03 ****   Σ(1620) S11 **   Ξ(1940) **
N(1675) D15 ****   Δ(1905) F35 ****   Λ(1800) S01 ***   Σ(1660) P11 ***   Ξ(2030) 1 ***
N(1680) F15 ****   Δ(1910) P31 ****   Λ(1800) P01 ***   Σ(1670) D13 ***   Ξ(2120) *
N(1700) D13 ****   Δ(1920) P33 ***   Λ(1800) G09 Dead   Σ(1670) **   Ξ(2250) *
N(1710) P11 ****   Δ(1930) D35 ****   Λ(1800) *   Σ(1690) **   Ξ(2370) 1 **
N(1720) P13 ****   Δ(1940) D33 *   Λ(1820) F05 ****   Σ(1750) S11 ***   Ξ(2500) **
N(1990) F17 **   Δ(1950) F37 ****   Λ(1830) D05 ****   Σ(1770) P11 Dead
N(2000) F15 **   Δ(2150) S31 *   Λ(1890) P03 ****   Σ(1775) D15 ****   Ω(1672) P03 ****
N(2080) D13 ***   Δ(2160) *   Λ(2000) *   Σ(1840) P13 *
N(2100) S11 *   Δ(2200) G37 **   Λ(2020) F07 *   Σ(1880) P11 **   Λc(2282) ****
N(2100) P11 *   Δ(2300) H39 **   Λ(2100) G07 ****   Σ(1915) F15 ****
N(2190) G17 ****   Δ(2350) D35 *   Λ(2110) F05 ***   Σ(1940) D13 ***   Σc(2450) **
N(2200) D15 ***   Δ(2400) F37 *   Λ(2325) D03 *   Σ(2000) S11 *
N(2220) H19 ****   Δ(2400) G39 ***   Λ(2350) ****   Σ(2030) F17 ****   Λb(5500) *
N(2250) G19 ****   Δ(2420) H311 ***   Λ(2585) ***   Σ(2070) F15 *
N(2600) I111 ***   Δ(2500) *   Σ(2080) P13 **   Dibaryons
N(2700) K113 *   Δ(2750) I313 *   Σ(2100) G17 *   NN(2170) 1D2 ***
N(2800) G19 *   Δ(2850) ***   Σ(2250) ****   NN(2250) 3F3 ***
N(3000) *   Δ(2950) K315 *   Σ(2455) ***   NN(?) *
N(3030) ***   Δ(3230) ***   Σ(2620) ***   ΔN(2130) 3S1 ***
N(3245) *   Σ(3000) **   ΞN(?) *
N(3690) *   Z0(1780) P01 *   Σ(3170) *
N(3755) *   Z0(1865) D03 *
  Z1(1900) P13 *
  Z1(2150) *
  Z1(2500) *

****   Good, clear, and unmistakable.
***   Good, but in need of clarification or not absolutely certain.
**   Needs confirmation.
*   Weak.

Particle[a]	$I(J^P)L_{2I \cdot 2J}^{b}$	P_{beam}^{c} (GeV/c) $\sigma = 4\pi\lambda^2$ (mb)	Mass[d] M (MeV)	Full[e] width Γ (MeV)	M^2 [f] ±ΓM (GeV²)	Mode[g]	Fraction[h] (%)	p[i] (MeV/c)
			S=0 I=1/2 NUCLEON RESONANCES (N)					
p	$1/2(1/2^+)$		938.3		0.880	See Stable Particle Table		
n			939.6		0.883			
N(1440)	$1/2(1/2^+)P'_{11}$	p = 0.61 $\sigma = 31.0$	1400 to 1480 (200)	120 to 350	2.07 ±0.29	Nπ	50–70	397
						Nη	8–18	†
						Nππ	~30	342
						Δπ	12–28 ⎤*	143
						Nρ	~ 7 ⎟	†
						Nε	~ 5 ⎦	†
N(1520)	$1/2(3/2^-)D'_{13}$	p = 0.74 $\sigma = 23.5$	1510 to 1530 (125)	100 to 140	2.31 ±0.19	Nπ	50–60	456
						Nη	< 1	149
						Nππ	35–50	410
						Δπ	15–25 ⎤*	228
						Nρ	15–25 ⎟	†
						Nε	< 5 ⎦	†
N(1535)	$1/2(1/2^-)S'_{11}$	p = 0.76 $\sigma = 22.4$	1520 to 1560 (150)	100 to 250	2.36 ±0.23	Nπ	35–50	467
						Nη	40–65	182
						Nππ	~ 5	422
						Δπ	~ 1 ⎤*	242
						Nρ	~ 3 ⎟	†
						Nε	~ 2 ⎦	†

Particle[a]	$I(J^P)L_{2I \cdot 2J}$[b]	P_{beam}^c (GeV/c) $\sigma = 4\pi\lambda^2$ (mb)	Mass[d] M (MeV)	Full[e] width Γ (MeV)	M^2 [f] ±ΓM (GeV²)	Mode[g]	Fraction[h] (%)	p[i] (MeV/c)
N(1650)	$1/2(1/2^-)S''_{11}$	p = 0.96 σ = 16.4	1620 to 1680	100 to 200 (150)	2.72 ±0.25	Nπ	55–65	547
						Nη	~ 1	346
						ΛK	5–10	161
						ΣK	3–10	†
						Nππ	~30	511
						Δπ	4–15 *	344
						Nρ	~20	†
						Nε	< 5	†
N(1675)	$1/2(5/2^-)D'_{15}$	p = 1.01 σ = 15.4	1660 to 1690	120 to 180 (155)	2.81 ±0.26	Nπ	30–40	563
						Nη	< 2	374
						ΛK	< 1	209
						Nππ	55–70	529
						Δπ	50–65 *	364
						Nρ	~ 5	†
N(1680)	$1/2(5/2^+)F'_{15}$	p = 1.01 σ = 15.2	1670 to 1690	110 to 140 (125)	2.82 ±0.21	Nπ	55–65	567
						Nη	< 1	379
						Nππ	~40	532
						Δπ	~12 *	369
						Nρ	~10	†
						Nε	~20	†
N(1700)	$1/2(3/2^-)D''_{13}$	p = 1.05 σ = 14.5	1670 to 1730	70 to 120 (100)	2.89 ±0.17	Nπ	8–12	580
						Nη	~ 4	400
						ΛK	~ 1	250
						Nππ	~85	547
						Δπ	15–40 *	385
						Nρ	~ 5	†
						Nε	<40	†
N(1710)	$1/2(1/2^+)P''_{11}$	p = 1.07 σ = 14.2	1680 to 1740	90 to 130 (110)	2.92 ±0.19	Nπ	10–20	587
						Nη	5–35	410
						ΛK	5–15	264
						ΣK	2–10	138
						Nππ	>50	554
						Δπ	10–25 *	393
						Nρ	25–65	48
						Nε	15–40	†
N(1720)	$1/2(3/2^+)P''_{13}$	p = 1.09 σ = 13.9	1690 to 1800	125 to 250 (200)	2.96 ±0.34	Nπ	10–20	594
						Nη	3–6	420
						ΛK	2–12	278
						ΣK	2–5	162
						Nππ	~70	561
						Δπ	~20 *	401
						Nρ	45–70	104
						Nε	~20	†
N(1990)	$1/2(7/2^+)F_{17}$	p = 1.62 σ = 8.34	1950 to 2050	120 to 400 (350)	3.96 ±0.70	Nπ	~ 5	766
						Nη	~ 3	648
						ΛK	seen	554
						ΣK	seen	497
→ N(2080)	$1/2(3/2^-)D'''_{13}$	p = 1.82 σ = 7.26	2030 to 2100	115 to 300 (275)	4.33 ±0.57	Nπ	~10	821
						ΛK	seen	627
→						ΣK	seen	576
→ N(2190)	$1/2(7/2^-)G_{17}$	p = 2.07* σ = 6.21	2120 to 2230	200 to 500 (350)	4.80 ±0.77	Nπ	~14	888
						Nη	~ 2	790
						ΛK	< 1	712
N(2200)	$1/2(5/2^-)D''_{15}$	p = 2.10 σ = 6.12	1900 to 2230	150 to 400 (300)	4.84 ±0.66	Nπ	~ 8	894
						Nη	seen	797
						ΛK	seen	718
N(2220)	$1/2(9/2^+)H_{19}$	p = 2.14 σ = 5.97	2150 to 2300	300 to 500 (400)	4.93 ±0.89	Nπ	~18	905
						Nη	~ 1	811

Baryon Table *(cont'd)*

Particle[a]	$I(J^P)L_{2I \cdot 2J}$[b]	P_{beam}^c (GeV/c) $\sigma = 4\pi\lambdabar^2$ (mb)	Mass[d] M (MeV)	Full[e] width Γ (MeV)	M^2[f] ±ΓM (GeV²)	Mode[g]	Fraction[h] (%)	p[i] (MeV/c)
N(2250)	$1/2(9/2^-)G'_{19}$	p = 2.21 σ = 5.74	2130 to 2270	200 to 500 (300)	5.06 ±0.68	Nπ Nη	~10 ~2	923 831
N(2600)	$1/2(11/2^-)I_{111}$	p = 3.12 σ = 3.86	2580 to 2700	>300 (400)	6.76 ±1.04	Nπ	~5	1126
N(3030)	1/2(?)	p = 4.41 σ = 2.62	~3030	~400 (400)	9.18 ±1.21	Nπ	(J+1/2)x <0.1[j]	1366

S=0 I=3/2 DELTA RESONANCES (Δ)

Particle[a]	$I(J^P)L_{2I \cdot 2J}$[b]	P_{beam}^c (GeV/c) $\sigma = 4\pi\lambdabar^2$ (mb)	Mass[d] M (MeV)	Full[e] width Γ (MeV)	M^2[f] ±ΓM (GeV²)	Mode[g]	Fraction[h] (%)	p[i] (MeV/c)
Δ(1232)	$3/2(3/2^+)P'_{33}$	p = 0.30 σ = 95.0	1230 to 1234	110 to 120 (115)	1.52 ±0.14	Nπ Nγ	99.4 0.6	227 259
	Δ(++) pole position:[k] M−iΓ/2 = (1210.6±0.5) − i(49.7±0.3)[ℓ]							
	Δ(0) pole position:[k] M−iΓ/2 = (1210.3±1.0) − i(53.0±1.0)[ℓ]							
Δ(1600)	$3/2(3/2^+)P''_{33}$	p = 0.87 σ = 18.7	1500 to 1900	150 to 350 (250)	2.56 ±0.40	Nπ Nππ ⌈Δπ ⌊Nρ	15–25 ~80 20–65⌉• <10⌋	512 473 301 †
Δ(1620)	$3/2(1/2^-)S'_{31}$	p = 0.91 σ = 17.7	1600 to 1650	120 to 160 (140)	2.62 ±0.23	Nπ Nππ ⌈Δπ ⌊Nρ	25–35 ~70 35–50⌉• <40⌋	526 488 318 †
Δ(1700)	$3/2(3/2^-)D'_{33}$	p = 1.05 σ = 14.5	1630 to 1740	190 to 300 (250)	2.89 ±0.43	Nπ Nππ ⌈Δπ ⌊Nρ	10–20 ~85 <50⌉• ~40⌋	580 547 385 †
Δ(1900)	$3/2(1/2^-)S''_{31}$	p = 1.44 σ = 9.71	1850 to 2000	130 to 300 (150)	3.61 ±0.29	Nπ ΣK	6–12 ~10	710 410
Δ(1905)	$3/2(5/2^+)F_{35}$	p = 1.16 σ = 9.63	1890 to 1920	250 to 400 (300)	3.63 ±0.57	Nπ ΣK Nππ ⌈Δπ ⌊Nρ	8–15 < 3 ~80 10–30⌉• ~60⌋	713 415 687 542 421
Δ(1910)	$3/2(1/2^+)P''_{31}$	p = 1.46 σ = 9.54	1850 to 1950	200 to 330 (220)	3.65 ±0.42	Nπ ΣK Nππ ⌈Δπ ⌊Nρ	20–25 2–20 >40 small⌉• <40⌋	716 421 691 545 426
Δ(1920)	$3/2(3/2^+)P'''_{33}$	p = 1.48 σ = 9.39	1860 to 2160	190 to 300 (250)	3.69 ±0.48	Nπ ΣK	14–20 ~5	722 431
Δ(1930)	$3/2(5/2^-)D'_{35}$	p = 1.50 σ = 9.21	1890 to 1960	150 to 350 (250)	3.72 ±0.48	Nπ ΣK	4–14 <10	729 441
Δ(1950)	$3/2(7/2^+)F'_{37}$	p = 1.54 σ = 8.91	1910 to 1960	200 to 340 (240)	3.80 ±0.47	Nπ ΣK Nππ ⌈Δπ ⌊Nρ	35–45 < 1 ~60 ~40⌉• ~20⌋	741 460 716 574 469
Δ(2420)	$3/2(11/2^+)H_{311}$	p = 2.64 σ = 4.68	2380 to 2450	300 to 500 (300)	5.86 ±0.73	Nπ	5–15	1023
Δ(2850)	$3/2(?^+)$	p = 3.85 σ = 3.05	2800 to 2900	~400 (400)	8.12 ±1.14	Nπ	(J+1/2)x ~0.25[j]	1266
Δ(3230)	3/2(?)	p = 5.08 σ = 2.25	3200 to 3350	~440 (440)	10.43 ±1.42	Nπ	(J+1/2)x ~0.05[j]	1475

Baryon Table *(cont'd)*

Particle [a]	$I(J^P)L_{1 \cdot 2J}$ [b]	P_{beam}^c (GeV/c) $\sigma = 4\pi\lambda^2$ (mb)	Mass [d] M (MeV)	Full [e] width Γ (MeV)	M^2 [f] $\pm\Gamma M$ (GeV²)	Partial decay mode Mode	Fraction [h] (%)	p [i] (MeV/c)
			S=−1 I=0 LAMBDA RESONANCES (Λ)					
Λ	$0(1/2^+)$		1115.6		1.245	See Stable Particle Table		
$\Lambda(1405)$	$0(1/2^-)S'_{01}$	Below K^-p threshold	1405 $\pm 5^{\,\ell}$	$40 \pm 10^{\,\ell}$	1.97 ± 0.06	$\Sigma\pi$	100	152
$\Lambda(1520)$	$0(3/2^-)D'_{03}$	p = 0.395 $\sigma = 82.2$	1519.4 $\pm 1.0^{\,\ell}$	$15.6 \pm 1.0^{\,\ell}$	2.31 ± 0.02	$N\overline{K}$	45 ±1	244
						$\Sigma\pi$	42 ±1	267
						$\Lambda\pi\pi$	10 ±1	252
						$\Sigma\pi\pi$	0.9 ±0.1	152
$\Lambda(1600)$	$0(1/2^+)P'_{01}$	p = 0.58 $\sigma = 41.6$	1560 to 1700 (150)	50 to 250	2.56 ± 0.24	$N\overline{K}$	15–30	343
						$\Sigma\pi$	10–60	336
$\Lambda(1670)$	$0(1/2^-)S''_{01}$	p = 0.74 $\sigma = 28.5$	1660 to 1680 (35)	25 to 50	2.79 ± 0.06	$N\overline{K}$	15–25	414
						$\Sigma\pi$	20–60	393
						$\Lambda\eta$	15–35	64
$\Lambda(1690)$	$0(3/2^-)D''_{03}$	p = 0.78 $\sigma = 26.1$	1685 to 1695 (60)	50 to 70	2.86 ± 0.10	$N\overline{K}$	20–30	433
						$\Sigma\pi$	20–40	409
						$\Lambda\pi\pi$	~25	415
						$\Sigma\pi\pi$	~20	350
$\Lambda(1800)$	$0(1/2^-)S'''_{01}$	p = 1.01 $\sigma = 17.6$	1720 to 1850 (300)	200 to 400	3.24 ± 0.54	$N\overline{K}$	25–40	528
						$\Sigma\pi$	seen	493
						$\Sigma(1385)\pi$	seen	345
						$N\overline{K}{}^*(892)$	seen	†
$\Lambda(1800)$	$0(1/2^+)P''_{01}$	p = 1.01 $\sigma = 17.6$	1750 to 1850 (150)	50 to 250	3.24 ± 0.27	$N\overline{K}$	20–50	528
						$\Sigma\pi$	10–40	493
						$\Sigma(1385)\pi$	seen	345
						$N\overline{K}{}^*(892)$	30–60	†
$\Lambda(1820)$	$0(5/2^+)F_{05}$	p = 1.06 $\sigma = 16.5$	1815 to 1825 (80)	70 to 90	3.29 ± 0.15	$N\overline{K}$	55–65	545
						$\Sigma\pi$	8–14	508
						$\Sigma(1385)\pi$	5–10	362
$\Lambda(1830)$	$0(5/2^-)D_{05}$	p = 1.08 $\sigma = 16.0$	1810 to 1830 (95)	60 to 110	3.35 ± 0.17	$N\overline{K}$	3–10	553
						$\Sigma\pi$	35–75	515
						$\Sigma(1385)\pi$	>15	371
$\Lambda(1890)$	$0(3/2^+)P'_{03}$	p = 1.21 $\sigma = 13.6$	1850 to 1910 (100)	60 to 200	3.57 ± 0.19	$N\overline{K}$	20–35	599
						$\Sigma\pi$	3–10	559
						$\Sigma(1385)\pi$	seen	420
						$N\overline{K}{}^*(892)$	seen	233
$\Lambda(2100)$	$0(7/2^-)G_{07}$	p = 1.68 $\sigma = 8.68$	2090 to 2110 (200)	100 to 250	4.41 ± 0.42	$N\overline{K}$	25–35	751
						$\Sigma\pi$	~ 5	704
						$\Lambda\eta$	< 3	617
						ΞK	< 3	483
						$\Lambda\omega$	< 8	443
						$N\overline{K}{}^*(892)$	10–20	514
$\Lambda(2110)$	$0(5/2^+)F''_{05}$	p = 1.70 $\sigma = 8.54$	2090 to 2140 (200)	150 to 250	4.45 ± 0.42	$N\overline{K}$	5–25	757
						$\Sigma\pi$	10–40	711
						$\Lambda\omega$	seen	455
						$\Sigma(1385)\pi$	seen	589
						$N\overline{K}{}^*(892)$	10–60	524
$\Lambda(2350)$	$0(9/2^+)$	p = 2.29 $\sigma = 5.84$	2340 to 2370 (150)	100 to 250	5.52 ± 0.35	$N\overline{K}$	~12	915
						$\Sigma\pi$	~10	867
$\Lambda(2585)$	$0(\,?\,)$	p = 2.92 $\sigma = 4.35$	~2585	~300 (300)	6.68 ± 0.78	$N\overline{K}$	$(J+1/2)x$ ~1.0 [j]	1060

Particle[a]	$I(J^P)L_{1 \cdot 2J}$[b]	P_{beam}[c] (GeV/c) $\sigma = 4\pi\lambda^2$ (mb)	Mass[d] M (MeV)	Full[e] width Γ (MeV)	M^2[f] ±ΓM (GeV²)	Mode	Fraction[h] (%)	p[i] (MeV/c)
							Partial decay mode	

Particle[a]	$I(J^P)L_{1 \cdot 2J}$[b]	P_{beam}[c] (GeV/c) $\sigma = 4\pi\lambda^2$ (mb)	Mass[d] M (MeV)	Full[e] width Γ (MeV)	M^2[f] ±ΓM (GeV²)	Mode	Fraction[h] (%)	p[i] (MeV/c)
S=−1 I=1 SIGMA RESONANCES (Σ)								
Σ	$1(1/2^+)$		(+)1189.4		1.415	See Stable Particle Table		
			(0)1192.5		1.422			
			(−)1197.3		1.434			
Σ(1385)	$1(3/2^+)P'_{13}$	Below K^-p threshold	(+)1382.3±0.4 S=1.6[m] (0)1382.0±2.5 S=1.6[m] (−)1387.4±0.6 S=2.2[m]	35±1 S=1.0[m] ~35 40±2 S=1.9[m]	1.92 ±0.05	Λπ Σπ	88±2 12±2	208 127
Σ(1660)	$1(1/2^+)P'_{11}$	p = 0.72 σ = 29.8	1630 to 1690	40 to 200 (100)	2.76 ±0.17	N$\overline{K}$ Λπ Σπ	10–30 seen seen	405 439 385
Σ(1670)	$1(3/2^-)D''_{13}$	p = 0.74 σ = 28.5	1665 to 1685	40 to 80 (60)	2.79 ±0.10	N$\overline{K}$ Λπ Σπ	7–13 5–15 30–60	414 447 393
Σ(1750)	$1(1/2^-)S''_{11}$	p = 0.91 σ = 20.7	1730 to 1800	60 to 160 (90)	3.06 ±0.16	N$\overline{K}$ Λπ Σπ Ση	10–40 seen < 8 15–55	486 507 455 81
Σ(1775)	$1(5/2^-)D_{15}$	p = 0.96 σ = 19.0	1770 to 1780	105 to 135 (120)	3.15 ±0.21	N$\overline{K}$ Λπ Σπ Σ(1385)π Λ(1520)π	37–43 14–20 2–5 8–12 17–23	508 525 474 324 198
Σ(1915)	$1(5/2^+)F'_{15}$	p = 1.26 σ = 12.8	1900 to 1935	80 to 160 (120)	3.67 ±0.23	N$\overline{K}$ Λπ Σπ Σ(1385)π	5–15 seen seen < 5	618 622 577 440
Σ(1940)	$1(3/2^-)D'''_{13}$	p = 1.32 σ = 12.1	1900 to 1950	130 to 300 (220)	3.76 ±0.43	N$\overline{K}$ Λπ Σπ Σ(1385)π Λ(1520)π Δ(1232)$\overline{K}$ N$\overline{K}$*(892)	<20 seen seen seen seen seen seen	637 639 594 460 354 410 320
Σ(2030)	$1(7/2^+)F_{17}$	p = 1.52 σ = 9.93	2025 to 2040	150 to 200 (180)	4.12 ±0.37	N$\overline{K}$ Λπ Σπ ΞK Σ(1385)π Λ(1520)π Δ(1232)$\overline{K}$ N$\overline{K}$*(892)	17–23 17–23 5–10 < 2 5–15 10–20 10–20 < 5	702 700 657 412 529 430 498 438
Σ(2250)	1(?)	p = 2.04 σ = 6.76	2210 to 2280	60 to 150 (100)	5.06 ±0.23	N$\overline{K}$ Λπ Σπ	<10 seen seen	851 842 803
Σ(2455)	1(?)	p = 2.57 σ = 5.08	~2455	~120 (120)	6.03 ±0.29	N$\overline{K}$	(J+1/2)x ~0.2 [j]	981
Σ(2620)	1(?)	p = 3.02 σ = 4.19	~2600	~200 (200)	6.86 ±0.52	N$\overline{K}$	(J+1/2)x ~0.3 [j]	1081

Baryon Table *(cont'd)*

Particle[a]	$I(J^P)L_{2I \cdot 2J}$[b]	P_{beam}^c (GeV/c) $\sigma = 4\pi\lambda^2$ (mb)	Mass[d] M (MeV)	Full[e] width Γ (MeV)	M^2[f] $\pm\Gamma M$ (GeV^2)	Partial decay mode Mode	Fraction (%)	p^i (MeV/c)
			S=−2 I=1/2 CASCADE RESONANCES (Ξ)					
Ξ	$1/2(1/2^+)$		(0)1314.9 (−)1321.3		1.729 1.746	See Stable Particle Table		
$\Xi(1530)$	$1/2(3/2^+)P_{13}$		(0)1531.8±0.3 S = 1.3^m (−)1535.0±0.6	9.1±0.5 10.1±1.9	2.34 ±0.02	$\Xi\pi$	100	148
$\Xi(1820)$	$1/2(3/2\ )$		1823 ±6$^\ell$	20^{+15}_{-10} $^\ell$	3.31 ±0.04	$\Lambda\bar{K}$ $\Sigma\bar{K}$ $\Xi\pi$ $\Xi(1530)\pi$	~45 ~10 small ~45	396 306 413 231
$\Xi(2030)$	$1/2(\ ?\ )$		2024 ±6$^\ell$	16^{+15}_{-5} $^\ell$	4.12 ±0.03	$\Lambda\bar{K}$ $\Sigma\bar{K}$ $\Xi\pi$ $\Xi(1530)\pi$	~20 ~80 small small	587 524 573 418
Ω^-	$0(3/2^+)$		1672.4±0.3		2.797	See Stable Particle Table		
Λ_c^+	$0(1/2^+)$		2282±3		5.21	See Stable Particle Table		

Particle[a]	$I(J^P)2S\,{}^{+1}L_J$[b]	P_{beam}^c (GeV/c) $\sigma = 4\pi\lambda^2$ (mb)	Mass[d] M (MeV)	Full[e] width Γ (MeV)	M^2[f] $\pm\Gamma M$ (GeV^2)	Partial decay mode Mode	Fraction[h] (%)	p^i (MeV/c)
			S=0 DIBARYONS					
d	$0(1^+)$		1875.6		3.518			
NN(2170)	$1(2^+){}^1D_2$	p = 1.26 σ = 16.5	2140 to 2190	50 to 125 (90)	4.71 ±0.20	NN πd	10–20 seen	545 241
NN(2250)	$1(3^-){}^3F_3$	p = 1.49 σ = 12.7	2200 to 2400	75 to 225 (150)	5.06 ±0.34	NN πd	20–30 seen	621 318
			S=−1 DIBARYONS					
ΛN(2130)	$1/2(1^+){}^3S_1$	p = 0.64 σ = 61.5	2100 to 2200	5 to 25 (15)	4.54 ±0.03	ΛN	seen	282

Baryon Table (cont'd)

→ Each arrow in the left-hand margin indicates there is an entry in the Data Card Listings for a baryon that is not well enough established (status less than 3 stars) to be included here. There is a short list of *all* the baryons in the Listings, whatever their status, at the front of this Table.

†. This mode is energetically forbidden when the nominal mass of the decaying resonance (and of any resonance in the final state) is used, but is in fact allowed due to the finite width of the resonance(s).

*. The modes in brackets are sub-reactions of the first preceding unbracketed mode.

a. The nominal mass here (in MeV) is used for identification. See column 4 for the actual mass.

b. When there is more than one baryon with the same quantum numbers, one prime is attached to the spectroscopic symbol for the first of them (e.g., S_{11}'), two primes to the second, etc.

c. The quantities here are calculated using the nominal mass of column 1.

d. Usually a conservatively large range of masses rather than a statistical average of various determinations of the mass is given. In these cases, the mass determinations are nearly entirely from various phase-shift analyses of more or less the same data. It is thus not appropriate to treat the determinations as independent measurements or to average them together. The masses, widths, and branching fractions in this Table are Breit-Wigner parameters. The Data Card Listings also include pole parameters where they are available.

e. Usually a conservatively large range of widths rather than a statistical average of various determinations of the width is given (see note d for the reason). The nominal value in parentheses is then simply a best guess.

f. The quantities here are calculated using the nominal mass of column 1 and the nominal width of column 5.

g. For information on the $N\gamma$ decay modes of the N and Δ baryons, see the mini-review on these states in the Listings.

h. Most of the inelastic branching fractions come from partial-wave analyses, and these determine $\sqrt{xx'}$, where x and x' are the elastic and inelastic branching fractions, not x' directly. Thus any uncertainty (and it is often considerable) in x carries over into x'. When x' so determined is really poorly known, we here simply note that the mode is seen. The values of $\sqrt{xx'}$ are given in the Data Card Listings.

i. For a 2-body decay mode, this is the momentum of the decay products in the rest frame of the decaying particle. For a mode with more than two decay products, this is the maximum momentum any of the products can have in this frame. The nominal mass of column 1 is used, as is the nominal mass of any resonance in the final state.

j. The size of the bump in the total cross section gives $(J + 1/2)x$, where x is the elastic branching fraction, but the value of J is not known.

k. These pole positions are from fits to phase shifts (without Coulomb corrections). The Data Card Listings now include pole positions and residues for most of the N and Δ resonances. See Sect. J of the N and Λ mini-review in the Listings for a brief discussion of the advantages of pole parameters over the usual Breit-Wigner parameters.

ℓ. The error given here is only an educated guess. It is larger than the error on the weighted average of the published values (the error on this average is given in the Listings).

m. The error given here has been scaled up by the "S factor" (see the * footnote to the Stable Particle Table for how S is defined) because the various measurements disagree more seriously than one would expect from statistics.

CHARACTERISTICS OF PARTICLES AND PARTICLE DISPERSOIDS

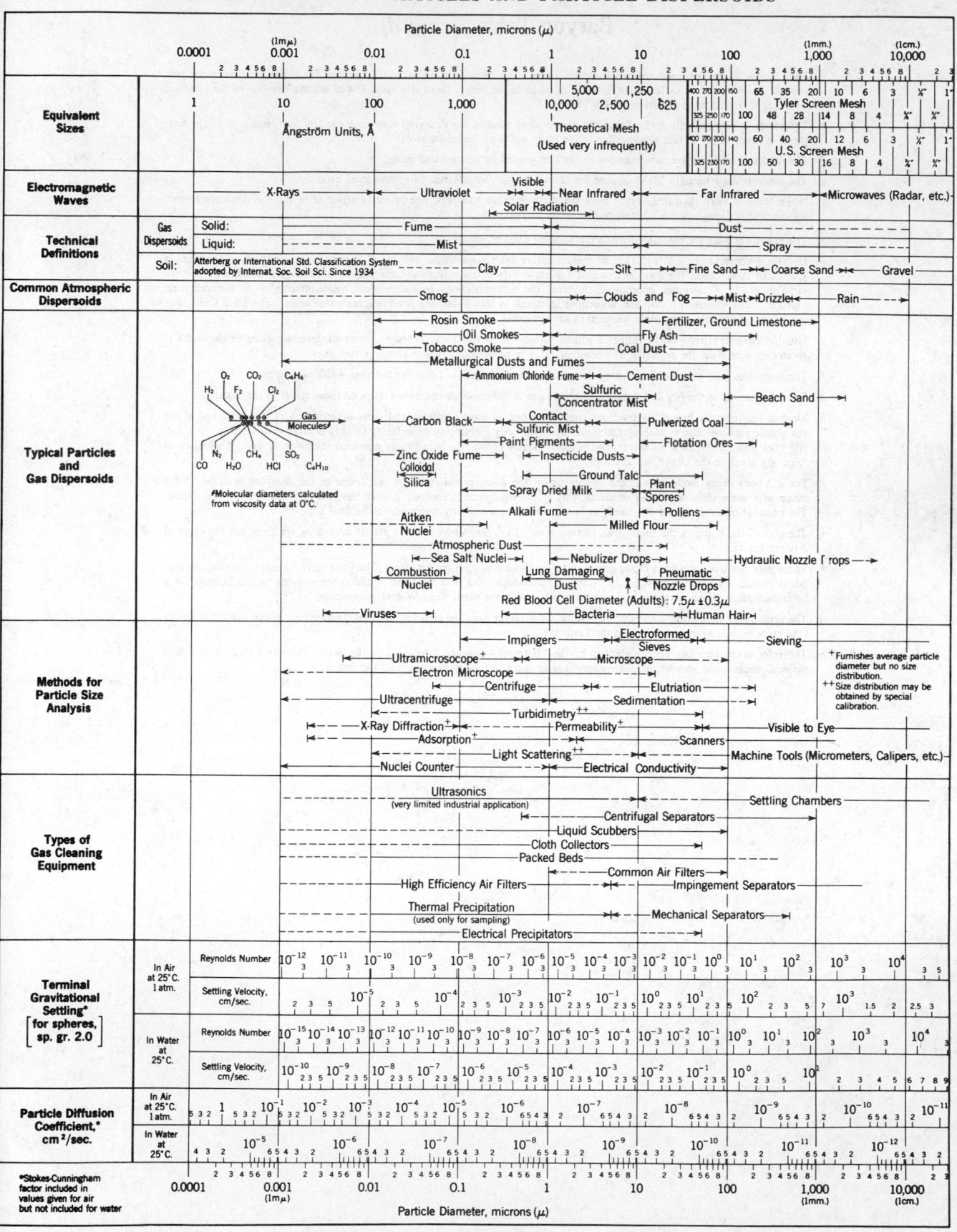

C. E. Lapple, Stanford Research
Institute Journal, Vol. 5, p.95
(Third Quarter, 1961)

The following list of abbreviations is intended to cover those in common use in chemistry and physics.   Symbols are presented in a separate list following the abbreviations.

Abbreviation	Meaning
A.	Acre
Å	Ångström unit
a.	Are
a.	Acid
abs.	Absolute
abt.	About
a.c.	Alternating current
acet.	Acetone
acet. a.	Acetic acid
al.	Alcohol
alk.	Alkali
alt.	Altitude
amal.	Amalgam; amalgamated
amor. or amorph.	Amorphous
amp.	Ampere
anh.	Anhydrous
antilog	Antilogarithm
ap.	Apothecaries'
appr.	Approximately
aq.	Aqua; aqueous; water
aq. reg.	Aqua regia
asym.	Asymmetrical
atm. or atmos.	Atmosphere (atmospheric)
At. No.	Atomic number
At. Wt.	Atomic weight
aux.	Auxiliary
Av. or av. or avoir.	Average / Avoirdupois
bar.	Barometer
bbl.	Barrel
bd.	Board
Bé	Beaumé (degrees)
B.G.	Birmingham gauge (hoop and sheet)
b.h.p.	Brake horse power
bl.	Blue
blk.	Black
B.M.	Board measure
b.p.	Boiling point
br.	Brown
BTU	British thermal unit
bu.	Bushel
B.W.G.	Birmingham wire gauge
bz.	Benzene
C	Centigrade
c	Carat; centi-
c.	Cold
ca	Candle
ca.	Circa, about; approximately
cal.	Calorie (gram)
cc. or c.c.	Cubic centimeter
cd.	Cord
c. cm	Cubic centimeter
Cent.	Centigrade
centi-	Prefix meaning 1/100
cf.	Confer, compare
c.f.m.	Cubic foot per minute
cgs	Centimeter-gram-second system of units
cgse	Cgs electrostatic system
cgsm	Cgs electromagnetic system
ch.	Chain
chl.	Chloroform
cir.	Circular
circum.	Circumference
cl	Centiliter
cm	Centimeter
cm²	Square centimeter
cm³	Cubic centimeter
c.m.	Circular mil
coef.	Coefficient
colog	Cologarithm
colorl.	Colorless
comm'l	Commercial
conc.	Concentrated
cond.	Condensing
const.	Constant
cos	Cosine
$\cos^{-1}$	Arc or angle whose cosine is...; anti-cosine of; inverse cosine of
cosec	Cosecant
cosh	Hyperbolic cosine
$\cosh^{-1}$	Inverse hyperbolic cosine
cot	Cotangent
$\cot^{-1}$	Arc or angle whose cotangent is...
coth	Hyperbolic cotangent
$\coth^{-1}$	Inverse hyperbolic cotangent
covers	Coversed sine
c.p.	Candle power; circular pitch; center of pressure
cry. or cryst.	Crystalline; crystals
csc	Cosecant
$\csc^{-1}$	Arc or angle whose cosecant is...
csch	Hyperbolic cosecant
csch^{-1}	Inverse hyperbolic cosecant
CTU	Centigrade thermal unit
cu.	Cubic
cu. cm	Cubic centimeter
cu. ft.	Cubic foot
cu. in.	Cubic inch
cu. m	Cubic meter
cu. yd.	Cubic yard
cwt.	Hundredweight
cyl.	Cylinder
d.	Derivative; deci-; Decomposes; day
d.	Dextrorotary
d.c.	Direct current
dec.	Decomposes
deci-	Prefix meaning 1/10
def.	Definition(s)
deg	Thermometric degree; absolute C unless contrary is indicated
deka-	Prefix meaning 10
deliq.	Deliquescent
den. or dens.	Density
dg	Decigram
diam.	Diameter
dil.	Dilute
dissd.	Dissolved
Deka-	Deka-
dk.	Dark
dkg.	Dekagram
dkl	Dekaliter
dkm	Dekameter
dkm²	Square dekameter
dkm³	Cubic dekameter
dks	Dekastere
dl	Deciliter
dm	Decimeter
dm²	Square decimeter
dm³	Cubic decimeter
d.p.	Diametral pitch; double pole
dr.	Dram
dr. ap. or ℥ ap.	Dram, apothecaries'
dr. av. or ℥ av.	Dram, avoirdupois
dr. fl. or ℥ fl.	Dram, fluid
dr. t. or ℥ t.	Dram, troy
ds	Decistere
dwt.	Pennyweight
efflor.	Efflorescent
e.g.	Exempli gratia, for example
e.h.p.	Effective horse power
E.L.	Elastic limit
em	Cgsm unit of quantity of electricity
emf or e.m.f.	Electromotive force
es	Electrostatic or cgse unit of quantity of electricity
etc.	Et cetera, and so forth
eth.	Ether
eth. acet.	Ethyl acetate
et. seq.	Et sequentes, and the following
evap.	Evaporation
ex.	Excess
exp	Exponential function
exp.	Explodes
exsec	Exterior secant
F	Fahrenheit
F.	From
fahr.	Fahrenheit
fath.	Fathom
feath.	Feathery
f.h.p.	Friction horse power
fir.	Firkin
fl.	Fluid
fl. dr.	Dram, fluid
fl. oz.	Ounce, fluid
fluores.	Fluorescent
fps	Foot-pound-second system of units
fpse	Foot-pound-second electrostatic system
fpsm	Foot-pound-second electromagnetic system
F.S.	Factor of safety
ft.	Foot
ft.²	Square foot
ft.³	Cubic foot
ft.-lb.	Foot-pound
fur.	Furlong
G	Gravitation constant
g	Gram
g-cal. or g.-cal.	Gram calorie
gal.	Gallon
gel.	Gelatinous
gi.	Gill
glac.	Glacial
glit.	Glittering
glyc.	Glycerine
gm.	Gram
gr.	Gray; grain
grn.	Green
gyr.	Gyration
h	Hecto-
h.	Hot; hour
ha	Hectare
hecto-	Prefix meaning 100
hex.	Hexagonal
hg	Hectogram
hhd.	Hogshead
hl	Hectoliter
hm	Hectometer
hm²	Square hectometer
hm³	Cubic hectometer
hor. or horiz.	Horizontal
h.-p.	High-Pressure
HP or h.p.	Horse power
h.p.-hr.	Horse power-hour
hr.	Hour
hyg.	Hygroscopic
i.	Insoluble
ibid.	Ibidem, in the same place
i.e.	Id est, that is
ign.	Ignites
i.h.p.	Indicated horse power
in.	Indigo; inch
in.²	Square inch
in.³	Cubic inch
inc.	Inclusive
in.-lb.	Inch-pound
insol.	Insoluble
Int.	International
iso.	Isotropic
isom.	Isometric
isoth.	Isothermal
k	Kilo-
kg	Kilogram
kg-cal.	Kilogram-calorie
kg-m	Kilogram-meter
kilo-	Prefix meaning 1,000
kl	Kiloliter
km	Kilometer
km²	Square kilometer
km³	Cubic kilometer
kva.	Kilovolt-ampere
kw.	Kilowatt
kw.-hr.	Kilowatt-hour
l	Liter
l.	Long
l	Laevorotary
lat.	Latitude
lb.	Pound
lb. ap.	Pound, apothecaries'
lb. av.	Pound, avoirdupois
lb. t.	Pound, troy
leaf.	Leaflets
lgr.	Ligroin
li.	Link
lin.	Linear
liq.	Liquid
lim.	Limit
ln	Natural hyperbolic or Napierian logarithm
log or log.	Logarithm
$\log_e$	Logarithm to the base e; natural, hyperbolic or Napierian logarithm
$\log_{10}$	Common logarithm; logarithm to the base 10
long.	Longitude
lng.	Long
l.-p.	Low-pressure
lt.	Light
lust.	Lustrous
♏	Minim or drop
m	Meter; milli-
m²	Square meter
m³	Cubic meter
m.	Minute
m.	Meta-
max.	Maximum
med.	Medium
meth.	Methyl
meth. al.	Methyl alcohol
m.e.p.	Mean effective pressure
met.	Metallic
mg	Milligram
m.h.c.p.	Mean horizontal candle power
mi.	Mile
mic.	Microscopic
micro-	Prefix meaning 1/1,000,000 or 10^{-6}
micro-micro	Prefix meaning 10^{-12}
milli-	Prefix meaning 1/1,000
milli-micro	Prefix meaning 10^{-9}
min or min.	Minute
min.	Minim; minimum; mineral
ml	Milliliter
m.l.h.c.p.	Mean lower hemispherical candle power
mm	Millimeter
mm²	Square millimeter
mm³	Cubic millimeter
mmf or m.m.f.	Magnetomotive force
mol.	Molecule
Mol. Wt.	Molecular weight
monocl.	Monoclinic
m.p.	Melting point
m.s.c.p.	Mean spherical candle power
myria-	Prefix meaning 10,000 or 10^4
mμ	Millimicron; millimicro-
N	Numeric; number (in mathematical tables)
n.	Normal
n	Refractive index
need.	Needles
o	Ortho-
Obs.	Observer
octahdr.	Octahedral
oil turp.	Oil of turpentine
or.	Orange
oz.	Ounce
oz. ap. or ℥ ap.	Ounce, apothecaries'
oz. av. or ℥ av.	Ounce, avoirdupois
oz. fl. or ℥ fl.	Ounce, fluid
oz. t. or ℥ t.	Ounce, troy
p	Para-
p. ct.	Per cent
perp.	Perpendicular
p.f.	Power factor
pk.	Peck
pl.	Plates
powd.	Powder
pr.	Prisms
precip. or p'p't'd	Precipitated
p. sol.	Partly soluble
pt.	Point; pint
purp.	Purple
pyr.	Pyridine
Q	Quantity
q	Quintal
qt.	Quart
q.v.	Quod vide, which see
R	Réaumur; radioactive mineral
rac	Racemic
rad	Radian, measure of angle
rad.	Radius
rd.	Rod
reg.	Regular
rev.	Revolution
rhbdr.	Rhombohedral
rhomb.	Rhombic or orthorhombic
R.M.S.	Square root of mean square
r.p.m.	Revolutions per minute
s	Stere
s.	Scruple; soluble; second
s. ap. or ℈	Scruple, apothecaries'
sat. or sat'd	Saturated
sc.	Scales
S.E.	Siemens unit
sec or sec.	Second (mean solar unless contrary is stated)
sec	Secant
$\sec^{-1}$	Arc or angle whose secant is...
sech	Hyperbolic secant
sech^{-1}	Inverse hyperbolic secant
segm.	Segment
sh.	Short
sin	Sine
$\sin^{-1}$	Arc or angle whose sine is...
sinh	Hyperbolic sine
$\sinh^{-1}$	Inverse hyperbolic sine
sl.	Slightly
sm.	Small
sol.	Solution; soluble
soln.	Solution
sp.	Specific
specit.	Specification
sp. gr.	Specific gravity
sq.	Square
sq. ch.	Square chain
sq. ft.	Square foot
sq. in.	Square inch
sq. m.	Square mile
sq. rd.	Square rod
sq. yd.	Square yard
std.	Standard
subl.	Sublimes
sym.	Symmetrical
t.	Metric ton
t.	Troy
tab. or tabl.	Tablets
tan	Tangent
$\tan^{-1}$	Arc or angle whose tangent is...
tanh	Hyperbolic tangent
$\tanh^{-1}$	Inverse hyperbolic tangent
temp.	Temperature
tetr. or tetrag.	Tetragonal
tn.	Ton
tr.	Transition
tricl.	Triclinic
trig.	Trigonal
trim.	Trimetric
T.S.	Tensile strength
turp.	Turpentine
Tw.°	Degrees Twaddell, hydrometer scale
ult.	Ultimate
uns.	Unsymmetrical
U.S.	United States of America; universal system of lens apertures
v.	Very
v.	Vide, see
vel. or veloc.	Velocity
vers	Versed sine
vert	Vertical
visc.	Viscous
vol.	Volume
volt.	Volatilizes
w.	Water
wh.	White
wt.	Weight
yd.	Yard
yel.	Yellow
yr.	Year
μ	Micron; micro-
μμ	Micromicron; micromicro-

SPELLING AND SYMBOLS FOR UNITS

From "Units of Weight and Measure"
L. B. Chisholm, National Bureau of Standards
Miscellaneous Publication 286 (May, 1967)

The spelling of the names of units as adopted by the National Bureau of Standards is that given in the list below. The spelling of the metric units is in accordance with that given in the law of July 28, 1866, legalizing the Metric System in the United States.

Following the name of each unit in the list below is given the symbol that the Bureau has adopted. Attention is particularly called to the following principles:

1. No period is used with symbols for units. Whenever "in" for inch might be confused with the preposition "in", "inch" should be spelled out.

2. The exponents "2" and "3" are used to signify "square" and "cubic," respectively, instead of the symbols "sq" or "cu," which are, however, frequently used in technical literature for the U. S. Customary units.

3. The same symbol is used for both singular and plural.

Some Units and Their Symbols

Unit	Symbol	Unit	Symbol	Unit	Symbol
acre	acre	fathom	fath	millimeter	mm
are	a	foot	ft	minim	minim
barrel	bbl	furlong	furlong	ounce	oz
board foot	fbm	gallon	gal	ounce, avoirdupois	oz avdp
bushel	bu	grain	grain	ounce, liquid	liq oz
carat	c	gram	g	ounce, troy	oz tr
Celsius, degree	°C	hectare	ha	peck	peck
centare	ca	hectogram	hg	pennyweight	dwt
centigram	cg	hectoliter	hl	pint, liquid	liq pt
centiliter	cl	hectometer	hm	pound	lb
centimeter	cm	hogshead	hhd	pound, avoirdupois	lb avdp
chain	ch	hundredweight	cwt	pound, troy	lb tr
cubic centimeter	cm³	inch	in	quart, liquid	liq qt
cubic decimeter	dm³	International		rod	rod
cubic dekameter	dam³	Nautical Mile	INM	second	s
cubic foot	ft³	Kelvin, degree	°K	square centimeter	cm²
cubic hectometer	hm³	kilogram	kg	square decimeter	dm²
cubic inch	in³	kiloliter	kl	square dekameter	dam²
cubic kilometer	km³	kilometer	km	square foot	ft²
cubic meter	m³	link	link	square hectometer	hm²
cubic mile	mi³	liquid	liq	square inch	in²
cubic millimeter	mm³	liter	liter	square kilometer	km²
cubic yard	yd³	meter	m	square meter	m²
decigram	dg	microgram	μg	square mile	mi²
deciliter	dl	microinch	μin	square millimeter	mm²
decimeter	dm	microliter	μl	square yard	yd²
dekagram	dag	micron	μm	stere	stere
dekaliter	dal	mile	mi	ton, long	long ton
dekameter	dam	milligram	mg	ton, metric	t
dram, avoirdupois	dr avdp	milliliter	ml	ton, short	short ton
				yard	yd

SYMBOLS AND TERMINOLOGY FOR PHYSIOCHEMICAL QUANTITIES AND UNITS

Reproduced by permission of the Assistant Secretary, Publications, of IUPAC from the *Manual of Symbols and Terminology for Physiochemical Quantities and Units (1979),* International Union of Pure and Applied Chemistry, Division of Physical Chemistry, Commission on Symbols, Terminology and Units. The above referenced Manual is published by and is available from Pergamon Press, Inc., Maxwell House, Fairview Park, Elmsford, New York 10523.

This Handbook, the *Handbook of Chemistry and Physics,* also contains a section on symbols, terminology, etc. for Physics which immediately follows this section on symbols, terminology, etc. for Chemistry. Rules and guidance which apply to both Chemistry and Physics and which do not require separate interpretation are reproduced only in the section, "Symbols, Units and Nomenclature in Physics" and are not reproduced in the section, "Symbols and Terminology for Physiochemical Quantities and Units". Paragraphs in the following are numbered the same as those in the complete IUPAC Manual.

1. PHYSICAL QUANTITIES AND SYMBOLS FOR PHYSICAL QUANTITIES

1.2 Base Physical Quantities

Physical quantities are generally organized in a dimensional system built upon seven base quantities. These base quantities, each of which has its own dimension, and the symbols used to denote them, are as follows:

Base physical quantity	Symbol for quantity
Length	l
Mass	m
Time	t
Electric current	I
Thermodynamic temperature	T
Amount of substance	n
Luminous intensity	I_v

Luminous intensity is seldom if ever needed in physical chemistry.

One of these independent base quantities is of special importance to chemists but until recently had no generally accepted name, although units such as the mole have been used for it. The name 'amount of substance' is now reserved for this quantity.

The definition of amount of substance, as of all other physical quantities (see Section 5), has nothing to do with any choice of *unit*, and in particular has nothing to do with the particular unit of amount of substance called the mole (see Section 3.6). It is now as inconsistent to call n the 'number of moles' as it is to call m the 'number of kilograms' or l the 'number of meters', since n, m, and l are symbols for quantities not for numbers.

The amount of a substance is proportional to the number of *specified* elementary entities of that substance. The proportionality factor is the same for all substances; its reciprocal is the Avogadro constant. The specified elementary entity may be an atom, a molecule, an ion, a radical, an electron, etc., or any *specified* group of such particles.

1.3 Derived Physical Quantities

All other physical quantities are regarded as being derived from, and as having dimensions derived from, the seven independent base physical quantities by definitions involving only multiplication, division, differentiation, and/or integration. Examples of derived physical quantities are given, often with brief definitions, in Section 2.

1.4 Use of the Words 'Specific' and 'Molar' in the Names of Physical Quantities

The word 'specific' before the name of an extensive physical quantity is restricted to the meaning 'divided by mass'. For example specific volume is the volume divided by the mass. When the extensive quantity is represented by a capital letter, the corresponding specific quantity may be represented by the corresponding lower case letter. *Examples;* volume: V; specific volume: $v=V/m$; heat capacity at constant pressure: C_p; specific heat capacity at constant pressure: $c_p=C_p/m$.

The word 'molar' before the name of an extensive quantity is restricted to the meaning 'divided by amount of substance'. For example molar volume is the volume divided by the amount of substance. The subscript m attached to the symbol for the extensive quantity denotes the corresponding molar quantity. *Examples;* volume: V; Gibbs energy: G; molar volume: $V_m=V/n$; molar Gibbs energy: $G_m=G/n$.

The subscript m may be omitted when there is no risk of ambiguity. Lower case letters may be used to denote molar quantities when there is no risk of misinterpretation.

The symbol X_B, where X denotes an extensive quantity and B is the chemical symbol for a substance, denotes the partial molar quantity of the substance B defined by the relation:

$$X_B = (\partial X/\partial n_B)T,p,n_c,\ldots$$

For a pure substance B the partial molar quantity X_B and the molar quantity X_m are identical. The partial molar quantity X_B of pure substance B, which is identical with the molar quantity X_m of pure substance B, may be denoted by $X_{B,}*$ where the superscript * denotes 'pure', so as to distinguish it from the partial molar quantity X_B of substance B in a mixture.

1.6 Printing of Subscripts and Superscripts

Subscripts or superscripts which are themselves symbols for physical quantities or numbers should be printed in italic (sloping) type and all others in roman (upright) type. *Examples*; C_p for heat capacity at constant pressure, but C_B for heat capacity of substance B.

2. RECOMMENDED NAMES AND SYMBOLS FOR QUANTITIES IN CHEMISTRY AND PHYSICS

The following list contains the recommended symbols for the most important quantities likely to be used by chemists. Whenever possible the symbol used for a physical quantity should be that recommended. In a few cases where conflicts were foreseen alternative recommendations have been made. Bold-faced italic (sloping) as well as ordinary italic (sloping) type can also sometimes be used to resolve conflicts. Further flexibility can be obtained by the use of capital letters as variants for lower-case letters, and *vice versa*, when no ambiguity is thereby introduced.

For example, d and D may be used instead of d_i and d_e for internal and external diameter in a context in which no quantity appears, such as diffusion coefficient, for which the recommended symbol is D. Again, the recommended symbol for power is P and for pressure is p or P, but P and p may be used for two powers or for two pressures; if power and pressure appear together, however, P should be used only for power and p only for pressure, and necessary distinctions between different powers or between different pressures should be made by the use of subscripts or other modifying signs.

When the above recommendations are insufficient to resolve a conflict or where a need arises for other reasons, an author is of course free to choose an *ad hoc* symbol. Any *ad hoc* symbol should be particularly carefully defined.

In the following list, where two or more symbols are indicated for a given quantity and are separated only by commas (without parentheses), they are on an equal footing; symbols within parentheses are reserve symbols.

Any description given after the name of a physical quantity is merely for identification and is not intended to be a complete definition.

Vector notation (bold-faced italic or sloping type) is used where appropriate in Section 2.6; it may be used when convenient also for appropriate quantities in other Sections.

2.1 Space, time, and related quantities

2.1.01	length	l
2.1.02	height	h
2.1.03	radius	r
2.1.04	diameter	d
2.1.05	path, length of arc	s
2.1.06	wavelength	λ
2.1.07	wavenumber: $1/\lambda$	$\sigma^a, \tilde{\nu}^b$
2.1.08	plane angle	$\alpha, \beta, \gamma, \theta, \phi$
2.1.09	solid angle	ω, Ω
2.1.10	area	$A, S, A_s{}^c$
2.1.11	volume	V
2.1.12	time	t
2.1.13	frequency	ν, f
2.1.14	circular frequency: $2\pi\nu$	ω
2.1.15	period: $1/\nu$	T
2.1.16	characteristic time interval, relaxation time, time constant	τ
2.1.17	velocity	v, u, w, c
2.1.18	angular velocity: $d\phi/dt$	ω
2.1.19	acceleration	a
2.1.20	acceleration of free fall	g

2.2 Mechanical and related quantities

2.2.01	mass	m
2.2.02	reduced mass	μ
2.2.03	specific volume (volume divided by mass)	v
2.2.04	density (mass divided by volume)	ρ
2.2.05	relative density (ratio of the density to that of a reference substance)	d
2.2.06	moment of inertia	I
2.2.07	momentum	p
2.2.08	force	F
2.2.09	weight	$G, (W)$
2.2.10	moment of force	M
2.2.11	angular momentum	L
2.2.12	work (force times path)	w, W
2.2.13	energy	E
2.2.14	potential energy	E_p, V, Φ
2.2.15	kinetic energy	E_k, T, K
2.2.16	Hamiltonian function	H
2.2.17	Lagrangian function	L
2.2.18	power (energy divided by time)	P
2.2.19	pressure	$p, (P)$
2.2.20	normal stress	σ

---

a   In solid-state studies, wavevector k is used ($|k| = 2\pi/\lambda$).

b   For electromagnetic radiation referred to a vacuum $\tilde{\nu} = \nu/c = 1/\lambda_{vac}$ is preferred.

c   The symbol A_s may be used when necessary to avoid confusion with the symbol A for Helmholtz energy.

2.2.21	shear stress	τ
2.2.22	linear strain (relative elongation): $\Delta l/l_0$	ϵ, e
2.2.23	volume strain (bulk strain): $\Delta V/V_0$	θ
2.2.24	modulus of elasticity (normal stress divided by linear strain, Young's modulus)	E
2.2.25	shear modulus (shear stress divided by shear angle)	G
2.2.26	compressibility: $-V^{-1}(dV/dp)$	κ
2.2.27	compression (bulk) modulus: $-V_0(\Delta p/\Delta V)$	K
2.2.28	velocity of sound	c
2.2.29	viscosity	$\eta, (\mu)$
2.2.30	fluidity: $1/\eta$	ϕ
2.2.31	kinematic viscosity: η/ρ	ν
2.2.32	friction coefficient (frictional force divided by normal force)	$\mu, (f)$
2.2.33	surface tension	γ, σ
2.2.34	angle of contact	θ
2.2.35	diffusion coefficient	D
2.2.36	mass transfer coefficient (dimension of length divided by time)	k_d

2.3 Molecular and related quantities

2.3.01	relative atomic mass of an element formerly called 'atomic weight')[a]	A_r
2.3.02	relative molecular mass of a substance (formerly called 'molecular weight')[b]	M_r
2.3.03	molar mass (mass divided by amount of substance)	M
2.3.04	Avogadro constant	L, N_A
2.3.05	number of molecules or other entries	N
2.3.06	amount of substance[c]	$n, (\nu)$
2.3.07	mole fraction of substance B: $n_B/\Sigma_i n_i$	x_B, y_B
2.3.08	mass fraction of substance B	w_B
2.3.09	volume fraction of substance B	ϕ_B
2.3.10	molality of solute substance B (amount of B divided by mass of solvent)[d]	m_B
2.3.11	amount-of-substance concentration of substance B (amount of B divided by the volume of the solution)[e]	$c_B, [B]$
2.3.12	mass concentration of substance B (mass of B divided by the volume of the solution)	ρ_B
2.3.13	surface concentration, surface excess	Γ
2.3.14	collision diameter of a molecule	d, σ
2.3.15	mean free path	l, λ
2.3.16	collision number (number of collisions divided by volume and by time)	Z
2.3.17	grand partition function (system)	Ξ
2.3.18	partition function (system)	Q, Z
2.3.19	partition function (particle)	q, z
2.3.20	statistical weight	g
2.3.21	symmetry number	σ, s
2.3.22	characteristic temperature	Θ

2.4 Thermodynamic and related quantities

2.4.01	thermodynamic temperature, absolute temperature	T
2.4.02	Celsius temperature	t, θ [f]
2.4.03	(molar) gas constant	R
2.4.04	Boltzmann constant	k
2.4.05	heat	q, Q [g]
2.4.06	work	w, W [g]
2.4.07	internal energy	$U, (E)$
2.4.08	enthalpy: $U + pV$	H
2.4.09	entropy	S
2.4.10	Helmholtz energy: $U - TS$	A
2.4.11	Massieu function: $-A/T$	J
2.4.12	Gibbs energy: $H - TS$	G
2.4.13	Planck function: $-G/T$	Y
2.4.14	compression factor: pV_m/RT	Z
2.4.15	heat capacity	C
2.4.16	specific heat capacity (heat capacity divided by mass; the name 'specific heat' is not recommended)	c
2.4.17	ratio C_p/C_V	$\gamma, (\kappa)$
2.4.18	Joule–Thompson coefficient	μ
2.4.19	thermal conductivity	λ, k
2.4.20	thermal diffusivity: $\lambda/\rho c_p$	a
2.4.21	coefficient of heat transfer (density of heat flow rate divided by temperature difference)	h
2.4.22	cubic expansion coefficient: $V^{-1}(\partial V/\partial T)_p$	a
2.4.23	isothermal compressibility: $-V^{-1}(\partial V/\partial p)_T$	κ
2.4.24	pressure coefficient: $(\partial p/\partial T)_v$	β
2.4.25	chemical potential of substance B	μ_B
2.4.26	absolute activity of substance B: $\exp(\mu_B/RT)$	λ_B
2.4.27	fugacity	$f, \tilde{p}$
2.4.28	osmotic pressure	Π
2.4.29	ionic strength: $(I_m = \frac{1}{2}\Sigma_i m_i z_i^2$ or $I_c = \frac{1}{2}\Sigma_i c_i z_i^2)$	I
2.4.30	activity, relative activity of substance B	a_B
2.4.31	activity coefficient, mole fraction basis	f_B
2.4.32	activity coefficient, molality basis	γ_B
2.4.33	activity coefficient, concentration basis	y_B
2.4.34	osmotic coefficient	ϕ

2.5 Chemical reactions

[a] The ratio of the average mass per atom of an element to 1/12 of the mass of an atom of nuclide ^{12}C. *Example:* A_r(Cl) = 35.453.

[b] The ratio of the average mass per formula unit of a substance to 1/12 of the mass of an atom of nuclide ^{12}C. *Example:* M_r(KCl) = 74.555.

[c] See Section 1.2.

[d] A solution having a molality equal to 0.1 mol kg^{-1} is sometimes called a 0.1 molal solution or a 0.1 m solution.

[e] This quantity may be simply called 'concentration' when there is no risk of ambiguity. A solution with an amount-of-substance concentration of 0.1 mol dm^{-3} is often called a 0.1 molar solution or a 0.1 M solution.

[f] Where symbols are needed to represent both time and Celsius temperature, t is the preferred symbol for time and θ for Celsius temperature.

[g] It is recommended that $q > 0$ and $w > 0$ both indicate *increase* of energy of the system under discussion. Thus $\Delta U = q + w$.

2.5.01	stoichiometric coefficient of substance B (negative for reactants, positive for products)	ν_B
2.5.02	general equation for a chemical reaction	$0 = \Sigma_B \nu_B B$
2.5.03	extent of reaction: $(d\xi = dn_B/\nu_B)$	ξ
2.5.04	rate of reaction: $d\xi/dt$ (see Section 11)	$\dot{\xi}, J$
2.5.05	rate of increase of concentration of substance B: dc_B/dt	ν_B, r_B
2.5.06	rate constant	k
2.5.07	affinity of a reaction: $-\Sigma_B \nu_B \mu_B$	$A, (\ )$
2.5.08	equilibrium constant	K
2.5.09	degree of dissociation	a

2.6 Electricity and magnetism

2.6.01	elementary charge (of a proton)	e
2.6.02	quantity of electricity	Q
2.6.03	charge density	ρ
2.6.04	surface charge density	σ
2.6.05	electric current	I
2.6.06	electric current density	j
2.6.07	electric potential	V, ϕ
2.6.08	electric potential difference: IR	$U, \Delta V, \Delta \phi$
2.6.09	electric field strength	E
2.6.10	electric displacement	D
2.6.11	capacitance	C
2.6.12	permittivity: $(D = \epsilon E)$	ϵ
2.6.13	permittivity of vacuum	ϵ_0
2.6.14	relative permittivity[a]: ϵ/ϵ_0	$\epsilon_r, (\epsilon)$
2.6.15	dielectric polarization: $D - \epsilon_0 E$	P
2.6.16	electric susceptibility: $\epsilon_r - 1$	χ_e
2.6.17	electric dipole moment	p, p_e
2.6.18	permanent dipole moment of a molecule	p, μ
2.6.19	induced dipole moment of a molecule	p, p_i
2.6.20	electric polarizability of a molecule	a
2.6.21	magnetic flux	Φ
2.6.22	magnetic flux density, magnetic induction	B
2.6.23	magnetic field strength	H
2.6.24	permeability: $(B = \mu H)$	μ
2.6.25	permeability of vacuum	μ_0
2.6.26	relative permeability: μ/μ_0	μ_r
2.6.27	magnetization: $(B/\mu_0) - H$	M
2.6.28	magnetic susceptibility: $\mu_r - 1$	$\chi, (\chi_m)$
2.6.29.1	Bohr magneton	μ_B
2.6.29.2	nuclear magneton	μ_N
2.6.29.3	g-factor	g
2.6.29.4	gyromagnetic ratio, magnetogyric ratio	γ
2.6.30	electromagnetic moment: $(E_p = -m \cdot B)$	m, μ

2.6.31	resistance	R
2.6.32	resistivity (formerly called specific resistance): $(E = \rho j)$	ρ
2.6.33	conductivity (formerly called specific conductance): $(j = \kappa E)$	$\kappa, (\sigma)$
2.6.34	self-inductance	L
2.6.35	mutual inductance	M, L_{12}
2.6.36	reactance	X
2.6.37	impedance (complex impedance): $R + iX$	Z
2.6.38	loss angle	δ
2.6.39	admittance (complex admittance): $1/Z$	Y
2.6.40	conductance: $(Y = G + iB)$	G
2.6.41	susceptance: $(Y = G + iB)$	B

2.7 Electrochemistry

2.7.01	Faraday constant	F
2.7.02	charge number of an ion B (positive for cations, negative for anions)	z_B
2.7.03	charge number of a cell reaction	$n, (z)$
2.7.04	electromotive force	E, E_{MF}
2.7.05	electrochemical potential of ionic component B: $\mu_B + z_B F\phi$	$\tilde{\mu}_B$
2.7.06	electric mobility (velocity divided by electric field strength)	u, μ
2.7.07	electrolytic conductivity (formerly called specific conductance)	$\kappa, (\sigma)$
2.7.08	molar conductivity of electrolyte or ion[b]: κ/c	Λ, λ [c]
2.7.09	transport number (transference number or migration number)	t
2.7.10	overpotential	η
2.7.11	exchange current density	j_0
2.7.12	electrochemical transfer coefficient	α
2.7.13	electrokinetic potential (zeta potential)	ζ
2.7.14	thickness of diffusion layer	δ
2.7.15	inner electric potential	ϕ
2.7.16	outer electric potential	ψ
2.7.17	surface electric potential difference: $\phi - \psi$	χ

2.8 Light and related electromagnetic radiation[d]

2.8.01	Planck constant	h
2.8.02	Planck constant divided by 2π	$\hbar$
2.8.03	radiant energy	Q [e]
2.8.04	radiant flux, radiant power	Φ [e], P
2.8.05	radiant intensity: $d\Phi/d\omega$	I [e]
2.8.06	radiance: $(dI/dS)/\cos\theta$	L [e]
2.8.07	radiant excitance: $d\Phi/dS$	M [e]
2.8.08	irradiance: $d\Phi/dS$	E [e]

[a] Also called dielectric constant, and sometimes denoted by D, when it is independent of E.

[b] The word molar, contrary to the general rule given in Section 1.4, here means 'divided by amount-of-substance concentration'.

[c] The formula unit whose concentration is c must be specified.
Example: $\lambda(Mg^{2+}) = 2\lambda(\frac{1}{2}Mg^{2+})$. Λ is used for an electrolyte and λ for individual ions.

[d] References to the symbols used in defining the quantities in 2.8 are as follows:

l	2.1.01	$\tilde{\nu}$	2.1.07	θ	2.1.08	ω	2.1.09
S	2.1.10	V	2.1.11	t	2.1.12	c_B	2.3.11
ρ_B	2.3.12	Φ	2.8.04	I	2.8.05	E	2.8.08
T	2.8.12	A	2.8.13.1	B	2.8.13.2	α	2.8.14.2
k	2.8.15	n	2.8.21.1				

[e] The same symbol is often used also for the corresponding luminous quantity. Subscripts e for energetic and v for visible may be added whenever confusion between these quantities might otherwise occur.

2.9 Transport properties

2.10 Symbols for particular cases of physical quantities

It is much more difficult to make detailed recommendations on symbols for physical quantities in particular cases than in general cases. The reason is the incompatibility between the need for specifying numerous details and the need for keeping the printing reasonably simple. Among the most awkward things to print are superscripts to subscripts and subscripts to subscripts. Examples of symbols to be avoided are:

$$\lambda_{NO_3^-} \qquad \Delta H_{25°C} \qquad (pV)_{0\,°C}^{p=0}$$

The problem is vastly reduced if it is recognized that two different kinds of notation are required for two different purposes. In the formation of general fundamental relations the most important requirement is a notation that is easy

a   These names and symbols are in agreement with those adopted jointly by the International Commission of Illumination (CIE) and the International Electrotechnical Commission (IEC).
b   The terms extinction (for 2.8.13.1) and extinction coefficient (for 2.8.14.1) are unsuitable because extinction is reserved for diffusion of radiation rather than absorption. Molar absorptivity (for 2.8.17.1) should be avoided because the meaning, absorptance per unit length, has been accepted internationally for the term absorptivity.
c   The word specific, contrary to the general rule given in Section 1.4, here means 'divided by mass concentration'.
d   The word molar, contrary to the general rule given in Section 1.4, here means 'divided by amount-of-substance concentration'.
e   For measurement on solutions, $1/T$ is ordinarily replaced by T_0/T where T_0 is the internal transmittance of the solvent medium and T is the internal transmittance of the solution. If a double-beam spectrometer is used in solution spectrometry, T_0/T is given directly, provided the boundary and container influences have been equalized between the two cells; in addition to the physical matching of the sample and reference cells this requires that there be no significant difference between $n_{solvent}$ and $n_{solution}$.
f   For measurements on solutions, it is tacitly assumed that the solution obeys the Beer-Lambert law unless the solute concentration is specified. The temperature should be specified.
g   The name Sherwood number, symbol Sh has been widely used.

to understand and easy to remember. In applications to particular cases, in quoting numerical values, and in tabulation, the most important requirement is complete elimination of any possible ambiguity even at the cost of an elaborate notation.

The advantage of a dual notation is already to some extent accepted in the case of concentration. The recommended notation for the formulation of the equilibrium constant K_c for the general reaction:

$$0 = \Sigma_B \nu_B B$$

is

$$K_c = \Pi_B (c_B)^{\nu_B}$$

but when we turn to a particular example it is better to use a notation such as:

$$Br_2 + H_2O = HOBr + H^+ + Br^-$$

$$\frac{[HOBr][H^+][Br^-]}{[Br_2]} = K_c$$

$$K_c(25\,°C) = 6 \times 10^{-9}\,mol^2\,dm^{-6}$$

Once the principle of dual notation is accepted, its adaptability and usefulness become manifest in all fields of physical chemistry. It will here be illustrated by just a few examples.

The general relation between the molar conductivity of an electrolyte and the molar conductivities of the two ions is written most simply and most clearly as:

$$\Lambda = \lambda^+ + \lambda^-$$

but when it comes to giving values in particular cases a much more appropriate notation is:

$$\lambda\tfrac{1}{2}Mg^{2+} = 53\,S\,cm^2\,mol^{-1}\ at\ 25°C$$
$$\lambda(Cl^-) = 76\,S\,cm^2\,mol^{-1}\ at\ 25°C$$
$$\Lambda\tfrac{1}{2}MgCl_2) = 129\,S\,cm^2\,mol^{-1}\ at\ 25°C$$
$$\Lambda(MgCl_2) = 258\,S\,cm^2\,mol^{-1}\ at\ 25°C$$

The general relation between the partial molar volumes of the two components A and B of a binary mixture is written most simply:

$$n_A dV_A + n_B dV_B = 0 \qquad (T,p\ const.)$$

But when it comes to specifying values, a completely different notation is called for, such as:

$$V(K_2SO_4, 0.1\,mol\,dm^{-3}\ in\ H_2O, 25\,°C) = 48\,cm^3\,mol^{-1}$$

Each kind of notation is appropriate to its purpose.

A last example will be given relating to optical rotation. The relations between the angle α of rotation of the plane of polarization and the amount n, or the number N of molecules, of the optically active substance in the path of a light beam of cross-section A can be clearly expressed in the form:

$$\alpha = n\alpha_n/A = N\alpha_N/A$$

where α_n is the molar optical rotatory power and α_N the molecular optical rotatory power. When on the other hand it is desired to record an experimental measurement, an appropriate notation would be:

$$\alpha(589.3\,nm, 20\,°C, sucrose, 10\,g\,dm^{-3}\ in\ H_2O, 10\,cm) = +66.470°$$

2.11 Recommended superscripts

The following superscripts are recommended:

° or *pure substance
∞ infinite dilution

3. UNITS AND SYMBOLS FOR UNITS

3.5 The International System of Units

The name International System of Units has been adopted by the Conférence Générale des Poids et Mesures for the system of units based on a selected set of dimensionally independent *SI Base Units*.

The SI Base Units are the meter, kilogram, second, ampere, kelvin, candela, and mole. In the International System of Units there is one and only one *SI Unit* for each physical quantity. This is either the appropriate SI Base Unit itself (see Section 3.7) or the appropriate *SI Derived Unit* formed by multiplication and/or division of two or more SI Base Units (see Section 3.10). A few such SI Derived Units have been given special names and symbols (see Section 3.9). There are also two *SI Supplementary Units* for which it is not decided whether they are SI Base Units or SI Derived Units (see Section 3.8).

Any of the approved decimal prefixes, called *SI Prefixes*, may be used to construct decimal multiples or submultiples of SI Units (see Section 3.11).

It is recommended that only units composed of SI Units and SI Prefixes be used in science and technology.

3.6 Definitions of the SI Base Units

Meter — The meter is the length equal to 1650763.73 wavelengths in vacuum of the radiation corresponding to the transition between the levels $2p_{10}$ and $5d_5$ of the krypton-86 atom.

Kilogram — The kilogram is the unit of mass; it is equal to the mass of the international prototype of the kilogram.

Second — The second is the duration of 9192631770 periods of the radiation corresponding to the transition between the two hyperfine levels of the ground state of the caesium-133 atom.

Ampere — The ampere is that constant current which, if maintained in two straight parallel conductors of infinite length, of negligible cross-section, and placed 1 meter apart in vacuum, would produce between these conductors a force equal to 2×10^{-7} newton per meter of length.

Kelvin — The kelvin, unit of thermodynamic temperature, is the fraction 1/273.16 of the thermodynamic temperature of the triple point of water.[a]

Candela — The candela is the luminous intensity, in the perpendicular direction, of a surface of 1/600000 square meter of a black body at the temperature of freezing platinum under a pressure of 101325 newtons per square meter.

Mole — The mole is the amount of substance of a system which contains as many elementary entities as there are atoms in 0.012 kilogram of carbon-12. When the mole is used, the elementary entities must be specified and may be atoms, molecules, ions, electrons, other particles, or specified groups of such particles. *Some examples of the use of the mole:*

- 1 mole of $HgCl$ has a mass of 236.04 grams
- 1 mole of Hg_2Cl_2 has a mass of 472.08 grams
- 1 mole of Hg_2^{2+} has a mass of 401.18 grams and a charge of 192.97 kilocoulombs
- 1 mole of $\frac{1}{2}Ca^{2+}$ has a mass of 20.04 grams and a charge of 96.49 kilocoulombs
- 1 mole of $Cu_{0.5}Zn_{0.5}$ has a mass of 64.46 grams
- 1 mole of $Fe_{0.91}S$ has a mass of 82.88 grams
- 1 mole of e^- has a mass of 548.60 micrograms, a charge of -96.49 kilocoulombs, and contains 6.02×10^{23} electrons
- 1 mole of a mixture containing the mole fractions $x(N_2) = 0.7809$, $x(O_2) = 0.2905$, $x(Ar) = 0.0093$, and $x(CO_2) = 0.0003$ has a mass of 28.964 grams
- 1 mole of photons whose frequency is 10^{14} Hz has energy 39.90 kilojoules

(The numerical values in these examples are approximate.)

3.7 Names and symbols for SI Base Units

Physical quantity	Name of SI Unit	Symbol for SI Unit
Length	meter	m
Mass	kilogram	kg
Time	second	s
Electric current	ampere	A
Thermodynamic temperature	kelvin	K
Amount of substance[b]	mole	mol
Luminous intensity	candela	cd

a   In October 1967 the thirteenth Conférence Générale des Poids et Mesures recommended that the kelvin, symbol K, be used for thermodynamic temperature and for thermodynamic temperature interval, and that the unit-symbols °K and deg be abandoned.

b   See Section 1.2.

3.8 Names and symbols for SI Supplementary Units

Physical quantity	Name of SI Unit	Symbol for SI Unit
Plane angle	radian	rad
Solid angle	steradian	sr

3.9 Special names and symbols for certain SI Derived Units

Physical quantity	Name of SI Unit	Symbol for SI Unit	Definition of SI Unit
Force	newton	N	$m\ kg\ s^{-2}$
Pressure, stress	pascal	Pa	$m^{-1}\ kg\ s^{-2}\ (=N\ m^{-2})$
Energy	joule	J	$m^2\ kg\ s^{-2}$
Power	watt	W	$m^2\ kg\ s^{-3}\ (=J\ s^{-1})$
Electric charge	coulomb	C	$s\ A$
Electric potential difference	volt	V	$m^2\ kg\ s^{-3}\ A^{-1}\ (=J\ A^{-1}\ s^{-1})$
Electric resistance	ohm	Ω	$m^2\ kg\ s^{-3}\ A^{-2}\ (=V\ A^{-1})$
Electric conductance	siemens	S	$m^{-2}\ kg^{-1}\ s^3\ A^2\ (=A\ V^{-1} = \Omega^{-1})$
Electric capacitance	farad	F	$m^{-2}\ kg^{-1}\ s^4\ A^2\ (=A\ s\ V^{-1})$
Magnetic flux	weber	Wb	$m^2\ kg\ s^{-2}\ A^{-1}\ (=V\ s)$
Inductance	henry	H	$m^2\ kg\ s^{-2}\ A^{-2}\ (=V\ A^{-1}\ s)$
Magnetic flux density	tesla	T	$kg\ s^{-2}\ A^{-1}\ (=V\ s\ m^{-2})$
Luminous flux	lumen	lm	$cd\ sr$
Illuminance	lux	lx	$m^{-2}\ cd\ sr$
Frequency	hertz	Hz	s^{-1}
Activity (of radioactive source)	becquerel	Bq	s^{-1}
Absorbed dose (of radiation)	gray	Gy	$m^2\ s^{-2}\ (=J\ kg^{-1})$

3.10 SI Derived Units and Unit-symbols for other quantities
(This list is not exhaustive)

Physical quantity	SI Unit	Symbol for SI Unit
Area	square meter	m^2
Volume	cubic meter	m^3
Density	kilogram per cubic meter	$kg\ m^{-3}$
Velocity	meter per second	$m\ s^{-1}$
Angular velocity	radian per second	$rad\ s^{-1}$
Acceleration	meter per second squared	$m\ s^{-2}$
Kinematic viscosity, diffusion coefficient	square meter per second	$m^2\ s^{-1}$
Dynamic viscosity	newton-second per square meter	$N\ s\ m^{-2}$
Molar entropy, molar heat capacity	joule per kelvin mole	$J\ K^{-1}\ mol^{-1}$
Concentration	mole per cubic meter	$mol\ m^{-3}$
Electric field strength	volt per meter	$V\ m^{-1}$
Magnetic field strength	ampere per meter	$A\ m^{-1}$
Luminance	candela per square meter	$cd\ m^{-2}$

3.12 The degree Celsius

Physical quantity	Name of unit	Symbol for unit	Definition of unit
Celsius temperature	degree Celsius	°C[a]	°C = K

The Celsius temperature t, is defined by $t = T - T_0$ where T_0 = 273.15 K. This leads to $t/°C = T/K - 273.15$.

3.13 Decimal fractions and multiples of SI Units having special names

The following units do not belong to the International System of Units, but in view of existing practice the Comité

[a] The ° sign and the letter following form one symbol and there should be no space between them. *Example:* 25 °C not 25° C.

International des Poids et Mesures has considered (1969) that it was preferable to keep them for the time being (along with several other specified units not particularly relevant to chemistry) for use with those of the International System.

Physical quantity	Name of unit	Symbol for unit	Definition of unit
Length	ångström	Å	10^{-10} m
Cross section	barn	b	10^{-28} m^2
Volume	liter[a]	l, L	10^{-3} m^3
Mass	tonne	t	10^3 kg
Pressure	bar	bar	10^5 Pa

Other units with special names based on the c.g.s. system and the electromagnetic c.g.s. system[b] are preferably not to be used; among these are the erg (10^{-7} J), the dyne (10^{-5} N), the poise (0.1 Pa s), the stokes (10^{-4} m^2 s^{-1}), the gauss (corresponding to 10^{-4} T),[b] the oersted (corresponding to $1000/4\pi$ A m^{-1}),[b] and the maxwell (corresponding to 10^{-8} WB).[b] The name micron and symbol μ should not be used for the unit of length, 10^{-6} m, which has the SI name micrometer and symbol μm.

3.14 Some other units now exactly defined in terms of the SI units

The CIPM (1969) recognized that users of the SI will wish to employ with it certain units not part of it but which are important and are widely used. These units are given in the following table. The combination of units of this table with SI units to form compound units should, however, be authorized only in limited cases.

Units in use with the International System

Name of unit	Symbol	Definition of unit
minute	min	60 s
hour	h	3 600 s
day	d	86 400 s
degree	°	$(\pi/180)$ rad
minute	'	$(\pi/10\ 800)$ rad
second	"	$(\pi/648\ 000)$ rad

In view of existing practice, as in the case of those units listed in Section 3.13, the CIPM (1969) has considered it preferable to retain the following units for the time being, for use with those of the SI. The definitions given in the fourth column of this table are exact.

Units to be used with the International System for a limited time

Physical quantity	Name of unit	Symbol for unit	Definition of unit
Radioactivity	Curie	Ci	3.7×10^{10} Bq
Exposure to X or γ radiation	röntgen	R	2.58×10^{-4} C kg^{-1}
Ionizing radiation absorbed	rad	rad[c]	10^{-2} Gy

The use of the following units is to be progressively discouraged and eventually abandoned. In the meantime it is recommended that any author who uses these units will define them in terms of SI units once in each publication in which he uses them. The definitions given here are exact. This list is not exhaustive.

Other units generally deprecated

Physical quantity	Name of unit	Symbol for unit	Definition of unit
Length	inch	in	2.54×10^{-2} m
Mass	pound (avoirdupois)	lb	0.453 592 37 kg
Force	kilogram-force	kgf	9.806 65 N
Pressure	standard atmosphere[d]	atm	101 325 Pa

[a] By decision of the Twelfth Conférence Générale des Poids et Mesures in October 1964, the old definition of the liter (1.000 028 dm^3) was rescinded. The word liter is now regarded as a special name for the cubic decimeter. Neither the word liter nor its symbol should be used to express results of high precision. The alternative symbol L was recommended by CIPM to CGPM in 1978.

[b] The electromagnetic c.g.s. system is a three-dimensional system of units in which the unit of electric current and units for other electric and magnetic quantities are considered to be derived from the centimeter, gram, and second as base units. The electric and magnetic units of this system cannot strictly speaking be compared to the corresponding units of the SI, which has four dimensions when only units derived from the meter, kilogram, second, and ampere are considered.

[c] Where there is a risk of confusion with the symbol for radian, rd may be used as the symbol for the unit, rad.

[d] The phrase 'standard atmosphere' remains admissable for the reference pressure 101 325 Pa.

Pressure	torr	Torr	$\dfrac{101\ 325}{760}$ Pa
Pressure	conventional millimeter of mercury [a]	mmHg	13.5951 × 980.665 × 10^{-2} Pa
Energy	kilowatt-hour	kW h	3.6×10^6 J
Energy	thermochemical calorie	cal_{th}	4.184 J
Energy	British thermal unit	Btu	1055.055 852 62 J
Thermodynamic temperature	degree Rankine	°R	(5/9) K

3.15 Units defined in terms of the best available experimental values of certain physical constants

It is necessary to recognize outside the International System some units, useful in specialized fields, the values of which expressed in SI units can be obtained only by experiment and are therefore not known exactly. Among such units recognized by the CIPM (1969) that are relevant to chemistry are the following:

Physical quantity	Name of unit	Symbol for unit	Conversion factor
Energy	electronvolt	eV	1 eV ≈ 1.6021892 × 10^{-19} J
Mass	(unified) atomic mass unit	μ	1 u ≈ 1.6605655 × 10^{-27} kg

3.16 'International' electrical units

These units are obsolete having been replaced by the 'absolute' (SI) units in 1948. The conversion factors which should be used with electrical measurements quoted in 'international' units depend on where and when the instruments used to make the measurements were calibrated. The following two sets of conversion factors refer respectively to the 'mean international' units estimated by the ninth Conférence Générale des Poids et Mesures in 1948, and to the 'US international' units estimated by the National Bureau of Standards (USA) as applying to published measurements made with instruments calibrated by them prior to 1948.

- 1 'mean international ohm'   = 1.0049 Ω
- 1 'mean international volt'   = 1.00034 V
- 1 'US international ohm'   = 1.000495 Ω
- 1 'US international volt'   = 1.000330 V

3.17 Electrical and magnetic units belonging to unit-systems other than the International System of Units

Definitions of units used in the obsolescent 'electrostatic CGS' and 'electromagnetic CGS' unit-systems can be found in References 13.1.05 and 13.2.

Another 'electrostatic CGS' unit used in chemistry for electric dipole moment is the debye, symbol D. 1D = $(10^{-21}/c)$ A m² ≈ 3.3356×10^{-30} C m.

5. PHYSICAL QUANTITIES, UNITS, AND NUMERICAL VALUES

The value of a *physical quantity* is equal to the product of a *numerical value* and a *unit*:

$$\text{physical quantity} = \text{numerical value} \times \text{unit.}$$

Neither any physical quantity, nor the symbol used to denote it, should imply a particular choice of unit.

Operations on equations involving physical quantities, units, and numerical values, should follow the ordinary rules of algebra.

Thus the physical quantity called the critical pressure and denoted by p_c has the value for water:

$$p_c = 221.2 \text{ bar} \quad \text{or better} \quad p_c = 22.12 \text{ MPa.}$$

These equations may equally well be written in the forms:

$$p_c/\text{bar} = 221.1 \quad \text{or better} \quad p_c/\text{MPa} = 22.12,$$

which are especially useful for the headings in tables and as labels on the axes of graphs.

[a] The conventional millimeter of mercury, symbol mmHg (not mm Hg), is the pressure exerted by a column exactly 1 mm high of a fluid of density exactly 13.5951 g cm⁻³ in a place where the acceleration of free fall is exactly 980.665 cm s⁻². The mmHg differs from the Torr by less than 2×10^{-7} Torr.

6. RECOMMENDED MATHEMATICAL SYMBOLS[a]

Mathematical operators (for example d and Δ) and mathematical constants (for example e and π should always be printed in roman (upright) type. Letter symbols for numbers other than mathematical constants should be printed in italic type.

equal to	$=$
not equal to	$\neq$
identically equal to	$\equiv$
corresponds to	$\triangleq$
approximately equal to	$\approx$
approaches	$\rightarrow$
asymptotically equal to	$\simeq$
proportional to	$\propto \quad \sim$
infinity	∞
less than	$<$
greater than	$>$
less than or equal to	$\leqslant$
greater than or equal to	$\geqslant$
much less than	$\ll$
much greater than	$\gg$
plus	$+$
minus	$-$
multiplied by	$\times \quad \cdot$
a divided by b	$\frac{a}{b} \quad a/b \quad ab^{-1}$
magnitide of a	$\lvert a \rvert$
a raised to the power n	a^n
square root of a	$a^{1/2} \quad a^{\frac{1}{2}} \quad \sqrt{a} \quad \sqrt{a}$
nth root of a	$a^{1/n} \quad a^{\frac{1}{n}} \quad \sqrt[n]{a} \quad \sqrt[n]{a}$
mean value of a	$\langle a \rangle \quad \bar{a}$
natural logarithm of a	$\ln a \quad \log_e a$
decadic logarithm of a	$\lg a \quad \log_{10} a \quad \log a$
binary logarithm of a	$\operatorname{lb} a \quad \log_2 a$
exponential of a	$\exp a \quad e^a$

7. SYMBOLS FOR CHEMICAL ELEMENTS, NUCLIDES, AND PARTICLES

7.1 Definitions

A nuclide is a species of atoms of which each atom has identical atomic number (proton number) and identical mass number (nucleon number). Different nuclides having the same value of the atomic number are named isotopes or isotopic nuclides. Different nuclides having the same mass number are named isobars or isobaric nuclides.

7.2 Elements and nuclides

Symbols for chemical elements should be written in roman (upright) type. The symbol is not followed by a full stop except when it occurs at the end of a sentence in text. *Examples:* Ca C H He.

The nuclide may be specified by attaching numbers. The mass number should be placed in the left superscript position; the atomic number, if desired, may be placed as a left subscript. The number of atoms per molecule is indicated as a right subscript. Ionic charge, or state of excitation, or oxidation number[b] may be indicated in the right superscript space. *Examples;* Mass number: $^{14}N_2$, $^{35}Cl^-$; Ionic charge: Cl^-, Ca^{2+}, PO_4^{3-} or $PO_4{}^{3-}$; Excited electronic state: He^*, NO^*; Oxidation number: $Pb_2^{II}Pb^{IV}O_4$, $K_6 M^{IV}Mo_9O_{32}$ (where M denotes a metal).

7.4 Abbreviated notation for nuclear reactions

The meaning of the symbolic expression indicating a nuclear reaction should be the following:

$$\text{initial nuclide} \left(\begin{array}{cc} \text{incoming particles(s)} & \text{outgoing particles(s)} \\ \text{or quanta} & \text{, or quanta} \end{array} \right) \text{final nuclide}$$

Examples: $^{14}N(\alpha,p)^{17}O$; $^{23}Na(\gamma,3n)^{20}Na$; $^{59}Co(n,\gamma)^{60}Co$; $^{31}P(\gamma,pn)^{29}Si$.

[a] Taken from Reference 13.1.11 where a more comprehensive list can be found.
[b] For a more detailed discussion see Reference 13.4.

8. SYMBOLS FOR SPECTROSCOPY[a]

8.1 General rules

A letter-symbol indicating the quantum state of *a system* should be printed in capital upright type. A letter-symbol indicating the quantum state of *a single particle* should be printed in lower case upright type.

9. CONVENTIONS CONCERNING THE SIGNS OF ELECTRIC POTENTIAL DIFFERENCES, ELECTROMOTIVE FORCES, AND ELECTRODE POTENTIALS [b]

9.1 The electric potential difference for a galvanic cell

The cell should be represented by a diagram, for example:

$$Zn \mid Zn^{2+} \mid Cu^{2+} \mid Cu$$

The electric potential difference ΔV is equal in sign and magnitude to the electric potential of a metallic conducting lead on the right minus that of an identical lead on the left.

When the reaction of the cell is written as:

$$\tfrac{1}{2}Zn + \tfrac{1}{2}Cu^{2+} \to \tfrac{1}{2}Zn^{2+} + \tfrac{1}{2}Cu$$

this implies a diagram so drawn that this reaction takes place when positive electricity flows through the cell from left to right. If this is the direction of the current when the cell is short-circuited, as it will be in the present example (unless the ratio $[Cu^{2+}]/[Zn^{2+}]$ is extremely small), the electric potential difference will be positive.

If, however, the reaction is written as:

$$\tfrac{1}{2}Cu + \tfrac{1}{2}Zn^{2+} \to \tfrac{1}{2}Cu^{2+} + \tfrac{1}{2}Zn$$

this implies the diagram:

$$Cu \mid Cu^{2+} \mid Zn^{2+} \mid Zn$$

and the electric potential difference of the cell so specified will be negative (unless the ratio $[Cu^{2+}]/[Zn^{2+}]$ is extremely small).

The limiting value of the electric potential difference for zero current through the cell is called the electromotive force and denoted by E_{MF} or E.

9.2 Electrode potential

The so-called electrode potential of an electrode (half-cell) is defined as the electromotive force of a cell in which the electrode on the left is a *standard hydrogen electrode* and that on the right is the electrode in question. For example, for the zinc electrode (written as $Zn^{2+} \mid Zn$) the cell in question is:

$$Pt \mid H_2 \mid H^+ \mid Zn^{2+} \mid Zn$$

The reaction taking place at the zinc electrode is:

$$Zn^{2+} + 2e^- \to Zn$$

The latter is to be regarded as an abbreviation for the reaction in the mentioned cell:

$$Zn^{2+} + H_2 \to Zn + 2H^+$$

In the standard state the electromotive force of this cell has a negative sign and a value of -0.763 V. The standard electrode potential of the zinc electrode is therefore -0.763 V.

The symbol $Zn \mid Zn^{2+}$ on the other hand implies the cell:

$$Zn \mid Zn^{2+} \mid H^+ \mid H_2 \mid Pt$$

---

[a] Taken from Reference 13.2. For further details see Reference 13.10.
[b] The conventions given here are in accordance with the 'Stockholm Convention' of 1953.

in which the reaction is:

$$Zn + 2H^+ \rightarrow Zn^{2+} + H_2$$

The electromotive force of this cell should *not* be called an electrode potential.

10. THE QUANTITY pH[a]

10.1 Operational definition

In all existing national standards the definition of pH is an operational one. The electromotive force E_x of the cell:

reference electrode | concentrated KCl solution :: solution X | H$_2$ | Pt

is measured and likewise the electromotive force E_s of the cell:

reference electrode | concentrated KCl solution :: solution S | H$_2$ | Pt

both cells being at the same temperature throughout and the reference electrodes and bridge solutions being identical in the two cells. The pH of the solution X, denoted by pH(X), is then related to the pH of the solution S, denoted by pH(S), by the definition:

$$pH(X) = pH(S) + \frac{(E_s - E_x)F}{RT \ln 10}$$

where R denotes the gas constant, T the thermodynamic temperature, and F the Faraday constant. Thus defined the quantity pH is a number.

To a good approximation, the hydrogen electrodes in both cells may be replaced by other hydrogen-ion-responsive electrodes, e.g. glass or quinhydrone. The two bridge solutions may be any molality not less than 3.5 mol kg^{-1}, provided they are the same (see Reference 13.5).

10.2 Standards

The difference between the pH of two solutions having been defined as above, the definition of pH can be completed by assigning a value of pH at each temperature to one or more chosen solutions designated as standards.

If the definition of pH given above is adhered to strictly, then the pH of a solution may be slightly dependent on which standard solution is used. These unavoidable deviations are caused not only by imperfections in the response of the hydrogen-ion electrodes but also by variations in the liquid junctions resulting from the different ionic compositions and mobilities of the several standards and from differences in the geometry of the liquid-liquid boundary. In fact such variations in measured pH are usually too small to be of practical significance. Moreover, the acceptance of several standards allows the use of the following alternative definition of pH.

The electromotive force E_x is measured, and likewise the electromotive forces E_1 and E_2 of two similar cells with the solution X replaced by the standard solutions S_1 and S_2 such that the E_1 and E_2 values are on either side of, and as near as possible to, E_x. The pH of solution X is then obtained by assuming linearity between pH and E, that is to say:

$$\frac{pH(X) - pH(S_1)}{pH(S_2) - pH(S_1)} = \frac{E_x - E_1}{E_2 - E_1}$$

This procedure is especially recommended when the hydrogen-ion-responsive electrode is a glass electrode.

11. DEFINITION OF RATE OF REACTION AND RELATED QUANTITIES

11.1 Rate of reaction

For the reaction

$$0 = \Sigma_B \nu_B B$$

the extent of reaction ξ is defined according to 2.5.03 by

---

[a] The symbol pH is an exception to the general rules given in Section 1.5.

$$d\xi = v_B^{-1} dn_B$$

where n_B is the amount, and v_B is the stoichiometric number, of the substance B.

It is recommended that the *rate of reaction* be defined as the rate of increase of the extent of reaction, namely

$$\dot{\xi} = d\xi/dt = v_B^{-1} dn_B/dt$$

This definition is independent of the choice of B and is valid regardless of the conditions under which a reaction is carried out, e.g. it is valid for a reaction in which the volume varies with time, or for a reaction involving two or more phases, or for a reaction carried out in a flow reactor.

If both sides of this equation are divided by any specified volume V, not necessarily independent of time, and not necessarily that of a single phase in which the reaction is taking place, then

$$V^{-1} d\xi/dt = V^{-1} v_B^{-1} {}^{-1} dn_B/dt$$

If the specified volume V is independent of time, then

$$V^{-1} d\xi/dt = v_B^{-1} d(n_B/V)/dt$$

If this specified volume V is such that

$$n_B/V = c_B \text{ or } [B]$$

where c_B or $[B]$ is the amount-of-substance concentration of B, then

$$V^{-1} d\xi/dt = v_B^{-1} dc_B/dt \text{ or } v_B^{-1} d[B]/dt$$

The quantity

$$dn_B/dt(= v_B d\xi/dt)$$

may be called the rate of formation of B, and the quantity

$$V^{-1} v_B^{-1} dn_B/dt(= V^{-1} d\xi/dt)$$

may be called the rate of reaction divided by volume, and the quantity

$$v_B = dc_B/dt \text{ or } d[B]/dt$$

which has often been called the rate of reaction, may be called the rate of increase of the concentration of B, but none of these three quantities should be called the rate of reaction.

11.2 Order of reaction

If it is found *experimentally* that the rate of increase of the concentration of B is given by

$$v_B \propto [C]^c [D]^d \dots$$

then the reaction is described as of order c with respect to C, or order d with respect to D, ..., and of overall order $(c + d + \dots)$.

11.3 Labelling of elementary processes

Elementary processes should be labelled in such a manner that reverse processes are immediately recognizable. *Example:*

Elementary process	Label	Rate of constant
$Br_2 + M \rightarrow 2Br + M$	1	k_1
$Br + H_2 \rightarrow HBr + H$	2	k_2
$H + Br_2 \rightarrow HBr + Br$	3	k_3
$H + HBr \rightarrow H_2 + Br$	-2	k_{-2}
$2Br + M \rightarrow Br_2 + M$	-1	k_{-1}

11.4 Collision number

The collision number defined as the number of collisions per unit time and per unit volume and having dimensions $(\text{time})^{-1} \times (\text{volume})^{-1}$ should be denoted by Z.

The collision number divided by the product of two relevant concentrations (or by the square of the relevant concentration) and by the Avogadro constant is a second-order rate constant having dimensions $(\text{time})^{-1} \times (\text{volume}) \times (\text{amount of substance})^{-1}$ and should be denoted by z. Thus $z = Z/Lc_A c_B$.

13. REFERENCES

13.1   The ISO 31 International Standard Series will, when complete, form a comprehensive publication dealing with quantities and units in various fields of science and technology. The following parts have so far been published and can be purchased in any country belonging to ISO from the 'Member Body', usually the national standardizing organization of the country.

13.1.00   'Part 0: General principles concerning quantities, units and symbols', 1st edition, April 1974.

13.1.01   'Part I: Quantities and units of space and time', 1st edition, March 1978.

13.1.02   'Part II: Quantities and units of periodic and related phenomena', 1st edition, March 1978.

13.1.03   'Part III: Quantities and units of mechanics', 1st edition, March 1978.

13.1.04   'Part IV: Quantities and units of heat', 1st edition, March 1978.

13.1.05   'Part V: Quantities and units of electricity and magnetism', 1st edition due late 1978.

13.1.06   'Part VI: Quantities and units of light and related electromagnetic radiations', 1st edition, April 1973.

13.1.07   'Part VII: Quantities and units of acoustics', 1st edition, March 1978.

13.1.08   'Part VIII: Quantities and units of physical chemistry and molecular physics', 1st edition, April 1973.

13.1.09   'Part IX: Quantities and units of atomic and nuclear physics', 1st edition, April 1973.

13.1.10   'Part X: Quantities and units of nuclear reactions and ionizing radiations', 1st edition, July 1973.

13.1.11   'Part XI: Mathematical signs and symbols for use in the physical sciences and technology', 1st edition, March 1978.

13.1.12   'Part XII: Dimensionless parameters', 1st edition, April 1975.

13.1.13   'Part XIII: Quantities and units of solid state physics', 1st edition, December 1975.

13.2   'Symbols, Units and Nomenclature in Physics', Document UIP 20 (SUN 65−3), published by IUPAP, 1978. This document supersedes Document UIP 9 (SUN 61−44) with the same title, which was published by IUPAP in 1961. Also published in *Physica*, (1978), 93A, 1.

13.3   'Manual of Physicochemical Symbols and Terminology', published for IUPAC by Butterworths Scientific Publications, London, 1959. This document was reprinted in the *J. Am. Chem. Soc.*, (1960), 82, 5517.

13.4   'Nomenclature of Inorganic Chemistry, 1970', 2nd ed., *Pure and Applied Chemistry*, (1971), 28, 1. This document has been issued for IUPAC also in book form by Butterworths, Borough Green, Sevenoaks, Kent TN158PH, UK.

13.5   *Pure and Applied Chemistry*, (1960), 1, 163.

13.6   *Pure and Applied Chemistry*, (1964), 9, 453.

13.7   *Pure and Applied Chemistry*, (1970), 21, 1.

13.8   'Le Système International d'Unités (SI)', Bureau International des Poids et Mesures, 2e Ed., 1973, OFFILIB, 48, rue Gay-Lussac, F 75 Paris 5.

13.8.01   'The International System of Units (SI)', National Bureau of Standards Special Publication 330, 1977 Edition, SD Catalog No. C 13.10: 330/4, US Government Printing Office, Washington, DC 20402.

13.8.02   'SI The International System of Units', National Physical Laboratory, Third edition, 1977, Her Majesty's Stationery Office, London.

13.9.01   'Definitions, Terminology and Symbols in Colloid and Surface Chemistry', Part I, *Pure and Applied Chemistry*, (1972), 31, 577 and Part II, 'Heterogeneous Catalysis', *Pure and Applied Chemistry*, (1976), 46, 71.

13.9.02   'Electrochemical Nomenclature', *Pure and Applied Chemistry*, (1974), 37, 503.

13.10   'Report on Notation for the Spectra of Polyatomic Molecules', *J. Chem. Phys.*, (1955) 23, 1997. For notation for diatomic molecules, see F. A. Jenkins, *J. Opt. Soc. Am.*, (1953), 43, 425.

13.11   'Recommended Consistent Values of the Fundamental Physical Constants, 1973', *CODATA Bulletin No. 11*, (December 1973), Committee on Data for Science and Technology of the International Council of Scientific Unions.

13.12   'The 1973 Least-Squares Adjustment of the Fundamental Constants', E. R. Cohen and B. N. Taylor, *J. Phys. Chem. Ref. Data*, (1973), 2, 663.

APPENDIX I

DEFINITION OF ACTIVITIES AND RELATED QUANTITIES

A.I.1 *Chemical potential and absolute activity*

The chemical potential μ_B of a substance B in a mixture of substances B, C, . . . , is defined by

$$\mu_B = (\partial G / \partial n_B)_{T,p,n_c} \cdots$$

where G is the Gibbs energy of the mixture, T is the thermodynamic temperature, p is the pressure, and n_B, n_c, . . . , are the amounts of the substances B, C, . . . , in the mixture.

(In molecular theory the symbol μ_B is sometimes used for the quantity μ_B/L where L is the Avogadro constant, but this usage is not recommended.)

The absolute activity λ_B of the substance B in the mixture is a number defined by

$$\lambda_B = \exp(\mu_B/RT) \quad \text{or} \quad \mu_B = RT \ln \lambda_B$$

where R is the gas constant.

The definitions given below often take simpler, though perhaps less familiar, forms when they are expressed in terms of absolute activity rather than in terms of chemical potential. Each of the definitions given below is expressed in both of these ways.

1. Pure substances

A.I.2 *Properties of pure substances*

The superscript * attached to the symbol for a property of a substance denotes the property of the *pure* substance. It is sometimes convenient to treat a mixture of constant composition as a pure substance.

A.I.3 *Fugacity of a pure gaseous substance*

The fugacity $f_B{}^*$ of a pure gaseous substance B is a quantity with the same dimensions as pressure, defined in terms of the absolute activity $\lambda_B{}^*$ of the pure gaseous substance B by

$$f_B{}^* = \lambda_B{}^* \lim_{p \to 0} (p/\lambda_B{}^*) \qquad (T \text{ const.})$$

or in terms of the chemical potential μ_B by

$$RT \ln f_B{}^* = \mu_B{}^* + \lim_{p \to 0} (RT \ln p - \mu_B{}^*) \qquad (T \text{ const.})$$

where p is the pressure of the gas and T is its thermodynamic temperature. It follows from this definition that

$$\lim_{p \to 0} (f_B{}^*/p) = 1 \qquad (T \text{ const.})$$

and that

$$RT \ln(f_B{}^*/p) = \int_0^p (V_B{}^* - RT/p)\, dp \qquad (T \text{ const.})$$

where $V_B{}^*$ is the molar volume of the pure gaseous substance B.

A pure gaseous substance B is treated as an *ideal gas* when the approximation $f_B{}^* = p$ is used. The ratio $(f_B{}^*/p)$ may be called the fugacity coefficient.

The name activity coefficient has sometimes been used for this ratio but is not recommended.

2. Mixtures

A.I.4 *Definition of a mixture*

The word *mixture* is used to describe a gaseous or liquid or solid phase containing more than one substance, when the substances are all treated in the same way (contrast the use of the word *solution* in Section A.I.9).

A.I.5 *Partial pressure*

The partial pressure p_B of a substance B in a *gaseous* mixture is a quantity with the same dimensions as pressure defined by

$$p_B = y_B p$$

where y_B is the mole fraction of the substance B in the gaseous mixture and p is the pressure.

A.I.6 *Fugacity of a substance in a gaseous mixture*

The fugacity f_B of the substance B in a gaseous mixture containing mole fractions $y_B, y_C, \dots$, of the substances B, C, $\dots$, is a quantity with the same dimensions as pressure, defined in terms of the absolute activity λ_B of the substance B in the gaseous mixture by

$$f_B = \lambda_B \lim_{p \to 0} (y_B p / \lambda_B) \qquad (T \text{ const.})$$

or in terms of the chemical potential μ_B by

$$RT \ln f_B = \mu_B + \lim_{p \to 0} \{ RT \ln(y_B p) - \mu_B \} \qquad (T \text{ const.})$$

It follows from this definition that

$$\lim_{p \to 0} (f_B / y_B p) = 1 \qquad (T \text{ const.})$$

and that

$$RT \ln(f_B / y_B p) = \int_0^p (V_B - RT/p) \, dp \qquad (T \text{ const.})$$

where V_B is the partial molar volume (see Section 1.4) of the substance B in the gaseous mixture.

A gaseous mixture of B, C, $\dots$, is treated as an *ideal gaseous mixture* when the approximations $f_B = y_B p, f_C = y_C p$, $\dots$, are used. It follows that $pV = (n_B + n_C + \dots) RT$ for an ideal gaseous mixture of B, C, $\dots$

The ratio $(f_B / y_B p)$ may be called the fugacity coefficient of the substance B. The name activity coefficient has sometimes been used for this ratio but is not recommended.

When $y_B = 1$ the definitions given in this Section for the fugacity of a substance in a gaseous mixture reduce to those given in Section A.I.3 for the fugacity of a pure gaseous substance.

A.I.7 *Activity coefficient of a substance in a liquid or solid mixture*

The activity coefficient f_B of a substance B in a liquid or solid mixture containing mole fractions $y_B, y_C, \dots$, of the substances B, C, $\dots$, is a number defined in terms of the absolute activity λ_B of the substance B in the mixture by

$$f_B = \lambda_B / \lambda_B^* x_B$$

where λ_B^* is the absolute activity of the pure substance B at the same temperature and pressure, or in terms of the chemical potential μ_B by

$$RT \ln(x_B f_B) = \mu_B - \mu_B^*$$

where μ_B^* is the chemical potential of the pure substance B at the same temperature and pressure.

It follows from this definition that

$$\lim_{x_B \to 1} f_B = 1 \qquad (T, p \text{ const.})$$

A.I.8 *Relative activity of a substance in a liquid or solid mixture*

The relative activity a_B of a substance B in a liquid or solid mixture is a number defined by

$$a_B = \lambda_B / \lambda_B^*$$

or by

$$RT \ln a_B = \mu_B - \mu_B^*$$

where the other symbols are as defined in Section A.I.7.

It follows from this definition that

$$\lim_{x_B \to 1} a_B = 1 \qquad (T, p \text{ const.})$$

A mixture of substances B, C, $\dots$, is treated as an *ideal mixture* when the approximations $a_B = x_B, a_C = x_C, \dots$, and consequently $f_B = 1, f_C = 1, \dots$, are used.

3. Solutions

A.I.9 *Definition of solution*

The word *solution* is used to describe a liquid or solid phase containing more than one substance, when for convenience one of the substances, which is called the *solvent* and may itself be a mixture, is treated differently from the other substances, which are called *solutes*. When, as is often but not necessarily the case, the sum of the mole fractions of the solutes is small compared with unity, the solution is called a *dilute solution*. In the following definitions the solvent substance is denoted by A and the solute substances by B, C, ...

A.I.10 *Properties of infinitely dilute solutions*

The superscript ∞ attached to the symbol for a property of a solution denotes the property of an *infinitely dilute solution*.

For example if V_B denotes the partial molar volume (see Section 1.4) of the solute substance B in a solution containing molalities $m_B, m_C, \ldots$, or mole fractions $x_B, x_C, \ldots$, of solute substances B, C, . . . , in a solvent substance A, then

$$V_B^{\infty} = \lim_{\Sigma_i m_i \to 0} V_B = \lim_{\Sigma_i x_i \to 0} V_B \qquad\qquad (T, p \text{ const.})$$

where i = B, C, ...

Similarly if V_A denotes the partial molar volume of the *solvent* substance A, then

$$V_A^{\infty} = \lim_{\Sigma_i m_i \to 0} V_A = \lim_{\Sigma_i x_i \to 0} V_A = V_A{}^* \qquad\qquad (T, p \text{ const.})$$

where $V_A{}^*$ is the molar volume of the pure solvent substance A.

A.I.11 *Activity coefficient of a solute substance in a solution*

The activity coefficient γ_B of a *solute* substance B in a solution (especially in a dilute liquid solution) containing molalities $m_B, m_C, \ldots$, of solute substances B, C, . . . , in a solvent substance A, is a number defined in terms of the absolute activity λ_B of the solute substance B in the solution by

$$\gamma_B = (\lambda_B/m_B)/(\lambda_B/m_B)^{\infty} \qquad\qquad (T, p \text{ const.})$$

or in terms of the chemical potential μ_B by

$$RT \ln(m_B\gamma_B) = \mu_B - RT \ln m_B)^{\infty} \qquad\qquad (T, p \text{ const.})$$

It follows from this definition that

$$\gamma_B{}^{\infty} = 1 \qquad\qquad (T, p \text{ const.})$$

The name activity coefficient with the symbol y_B may be used for the quantity similarly defined but with amount-of-substance concentration c_B (see Section 2.3) in place of molality m_B.

Another activity coefficient, called the *rational activity coefficient* of a solute substance B and denoted by $f_{x,B}$ is sometimes used. It is defined in terms of the absolute activity λ_B by

$$f_{x,B} = (\lambda_B/x_B)/(\lambda_B/x_B)^{\infty} \qquad\qquad (T, p \text{ const.)})$$

or in terms of the chemical potential μ_B by

$$RT \ln(x_B f_{x,B}) = \mu_B - (\mu_B - RT \ln x_B)^{\infty} \qquad\qquad (T, p \text{ const.})$$

where x_B is the mole fraction of the solute substance B in the solution. The rational activity coefficient $f_{x,B}$ is related to the (practical) activity coefficient γ_B by the formula

$$f_{x,B} = \gamma_B(1 + M_A\Sigma_i m_i) = \gamma_B/(1 - \Sigma_i x_i)$$

A solution of solute substances B, C, . . . , in a solvent substance A is treated as an *ideal dilute solution* when the activity coefficients are approximated to unity, for example $\gamma_B = 1, \gamma_C = 1, \ldots$.

A.I.12 *Relative activity of a solute substance in a solution*

The relative activity a_B of a *solute* substance B in a solution (especially in a dilute liquid solution) containing molalities $m_B, m_C, \ldots$, of solute substances B, C, . . . , in a solvent substance A, is a number defined in terms of the absolute activity λ_B by

$$a_B = (\lambda_B/m^\ominus)/(\lambda_B/m_B)^\infty = m_B\gamma_B/m^\ominus \qquad\qquad (T, p \text{ const.})$$

or in terms of the chemical potential μ_B by

$$RT \ln a_B = \mu_B - RT \ln m^\ominus - (\mu_B - RT \ln m_B)^\infty$$
$$= RT \ln(m_B\gamma_B/m^\ominus)$$

where $m^\ominus$ is a standard value of molality (usually chosen to be 1 mol kg^{-1}) and where the other symbols are defined in Section A.I.11.

It follows from this definition of a_B (compare Section A.I.8) that

$$(a_B m^\ominus/m_B)^\infty = 1 \qquad\qquad (T, p \text{ const.})$$

The name activity is often used instead of the name relative activity for this quantity.

The name relative activity with the symbol $a_{c,B}$ may be used for the quantity similarly defined but with concentration c_B (see Section 2.3) in place of molality m_B, and a standard value $c^\ominus$ of concentration (usually chosen to be 1 mol dm^{-3}) in place of the standard value $m^\ominus$ of molality.

Another relative activity, called the *rational relative activity* of the solute substance B and denoted by $a_{x,B}$, is sometimes used. It is defined in terms of the absolute activity λ_B by

$$a_{x,B} = \lambda_B/(\lambda_B/x_B)^\infty = x_B f_{x,B} \qquad\qquad (T, p \text{ const.})$$

or in terms of the chemical potential μ_B by

$$RT \ln a_{x,B} = \mu_B - (\mu_B - RT \ln x_B)^\infty$$
$$= RT \ln(x_B f_{x,B}) \qquad\qquad (T, p \text{ const.})$$

where x_B is the mole fraction of the substance B in the solution. The rational relative activity $a_{x,B}$ is related to the (practical) relative activity a_B by the formula

$$a_{x,B} = a_B m^\ominus M_A$$

A.I.13 *Osmotic coefficient of the solvent substance in a solution*

The osmotic coefficient ϕ of the *solvent* substance A in a solution (especially in a dilute liquid solution) containing molalities m_B, $m_C, \ldots$, of solute substances B, C, $\ldots$, is a number defined in terms of the absolute activity λ_A of the solvent substance A in the solution by

$$\phi = (M_A \Sigma_i m_i)^{-1} \ln(\lambda_A*/\lambda_A)$$

where λ_A* is the absolute activity of the pure solvent substance A at the same temperature and pressure, and M_A is the molar mass of the solvent substance A, or in terms of the chemical potential μ_A* by

$$\phi = (\mu_A* - \mu_A)/RTM_A \Sigma_i m_i$$

where μ_A* is the chemical potential of the pure solvent substance A at the same temperature and pressure.

For an *ideal dilute solution* as defined in Section A.I.11 or A.I.12 it can be shown that $\phi = 1$.

Another osmotic coefficient, called the *rational osmotic coefficient* of the solvent substance A and denoted by ϕ_x, is sometimes used. It is defined in terms of the absolute activity λ_A by

$$\phi_x = \ln(\lambda_A/\lambda_A*)/\ln x_A = \ln(\lambda_A/\lambda_A*)/\ln(1 - \Sigma_i x_i)$$

or in terms of the chemical potential μ_A by

$$\phi_x = (\mu_A - \mu_A*)/RT \ln x_A = (\mu_A - \mu_A*)/RT \ln(1 - \Sigma_i x_i)$$

where x_A is the mole fraction of the solvent substance A in the solution. The rational osmotic coefficient ϕ_x is related to the (practical) osmotic coefficient ϕ by the formula

$$\phi_x = \phi M_A \Sigma_i m_i/\ln)1 + M_A \Sigma_i m_i) = -\phi M_A \Sigma_i m_i/\ln(1 - \Sigma_i x_i)$$

A.I.14 *Relative activity of the solvent substance in a solution*

The relative activity a_A of the *solvent* substance A in a solution (especially in a dilute liquid solution) containing molalities $m_B, m_C, \ldots$, or more fractions $x_B, x_C, \ldots$, of solute substances B, C, $\ldots$, is a number defined in terms of the absolute activity λ_A of the solvent substance A in the solution by

$$a_A = \lambda_A/\lambda_A{}^* = \exp(-\phi M_A \Sigma_i m_i) = (1 - \Sigma_i x_i)^{\phi_x}$$

or in terms of the chemical potential μ_A by

$$RT \ln a_A = \mu_A - \mu_A{}^* = -RT\phi M_A \Sigma_i m_i = \phi_x RT \ln(1 - \Sigma_i x_i)$$

where the other quantities are as defined in Section A.I.13.

Note: The definition in this Section of the relative activity of the *solvent* in a *solution*, is identical with the definition in Section A.I.8 of the relative activity of any substance in a *mixture*. See also Section A.I.12.

SYMBOLS, UNITS AND NOMENCLATURE IN PHYSICS

From Document U.I.P. 20 (1978)
International Union of Pure and Applied Physics
Used by permission of the Secretary

1 PHYSICAL QUANTITIES–GENERAL RECOMMENDATIONS †)

Note: The German, Italian, Russian, and Spanish translations of this term are 'physikalische Größe', 'grandezza fizika', 'ФИЗИЧЕСКАЯ ВЕЛИЧИНА', and 'magnitud fisica', respectively.

1.1 Physical quantities

A physical quantity (French: 'grandeur physique') is equivalent to the product of the *numerical value*, i.e. a pure number, and a *unit*:

$$\text{physical quantity} = \text{numerical value} \times \text{unit}.$$

For a physical quantity with symbol a this relation is usually represented in the form $a = \{a\} \cdot [a]$, where $\{a\}$ stands for the numerical value and $[a]$ stands for the symbol for the unit. For dimensionless physical quantities the unit often has no name or symbol and is not explicitly indicated (see section 9.1 "unit systems").

Examples:

$$E = 200\,\text{J} \qquad n = 1{,}55 \quad \text{(for quartz)}$$
$$F = 27\,\text{N} \qquad \nu = 3 \times 10^8\,\text{Hz}.$$

1.2 Symbols for physical quantities–General rules

1. *Symbols for physical quantities* should be *single letters* of the Latin or Greek alphabet with or without modifying signs: subscripts, superscripts, primes, etc.

   Remarks:
   (a) An exception to this rule is given by the two-letter symbols which are used to represent dimensionless combinations of physical quantities (see section 7.14 "dimensionless parameters"). If such a symbol, composed of two letters, appears as a factor in a product, it is recommended to separate this symbol from the other symbols by a dot or by a space or by brackets. It can be raised to a positive or negative power without using brackets.
   (b) Abbreviations, i.e. shortened forms of names or expressions, such as p.f. for partition function, should not be used in physical equations. These abbreviations in the text should be written in ordinary roman type.

2. *Symbols for physical quantities* should be printed in *italic* (or *sloping*) *type*.

   Remark: It is recommended to consider as a guiding principle for the printing of indices the criterion: only indices which are symbols for physical quantities should be printed in italic (sloping) type.

---

†) For further details see International Standard I.S.O. 31/0: *General principles concerning quantities, units and symbols.*

Examples:

upright indices	sloping indices
C_g (g = gas)	p in C_p
g_n (n = normal)	n in $\sum_n a_n \psi_n$
μ_r (r = relative)	x in $\sum_x a_x b_x$
E_k (k = kinetic)	i, k in g_{ik}
χ_e (e = electric)	x in p_x

3. Symbols for vectors and tensors

To avoid the usage of subscripts it is often convenient to indicate vectors and tensors of the second rank by letters of a special type. The following choice is recommended:

(*a*) Vectors should be printed in bold italic (sloping) type, e.g. **A**, **a**.

(*b*) Tensors of the second rank should be printed in bold sans serif italic (sloping) type, e.g. **S**, **T**.

Remark: When such type is not available, a vector may be indicated by an arrow and a tensor by a double arrow above the symbol; e.g. $\vec{A}$, $\vec{\vec{S}}$.

1.3 Simple mathematical operations

1. Addition and subtraction of two physical quantities are indicated by:

$$a + b \quad \text{and} \quad a - b.$$

2. Multiplication of two physical quantities may be indicated in one of the following ways:

$$ab \quad a \cdot b \quad a \times b.$$

3. Division of one quantity by another quantity may be indicated in one of the following ways:

$$\frac{a}{b} \quad a/b \quad ab^{-1}$$

or in any other way of writing the product of a and b^{-1}.

These procedures can be extended to cases where one of the quantities or both are themselves products, quotients, sums or differences of other quantities.

If necessary, brackets have to be used in accordance with the rules of mathematics. If the solidus is used to separate the numerator from the denominator and if there is any doubt where the numerator starts or where the denominator ends, brackets should be used.

Examples:

expressions with a horizontal bar	same expressions with a solidus
$\dfrac{a}{bcd}$	a/bcd
$\dfrac{2}{9}\sin kx$	$(2/9)\sin kx$

expressions with a horizontal bar	same expressions with a solidus
$\dfrac{a}{b} - c$	$a/b - c$
$\dfrac{a}{b - c}$	$a/(b - c)$
$\dfrac{a - b}{c - d}$	$(a - b)/(c - d)$
$\dfrac{a}{c} - \dfrac{b}{d}$	$a/c - b/d$

Remark: It is recommended that in expressions like:

$$\sin\{2\pi(x - x_0)/\lambda\} \qquad \exp\{(r - r_0)/\sigma\}$$
$$\exp\{-V(r)/kT\} \qquad \sqrt{(\varepsilon/c^2)}$$

the argument should always be placed between brackets, except when the argument is a simple product of two quantities: e.g. $\sin kx$. When the horizontal bar above the square root is used no brackets are needed.

2 UNITS–GENERAL RECOMMENDATIONS

2.1 Symbols for units–General rules

1. *Symbols for units* of physical quantities should be printed in *roman (upright)* type.

2. *Symbols for units* should not contain a full stop (period) and should remain unaltered in the plural, e.g.: 7 cm and *not* 7 cms.

3. *Symbols for units* should be printed in *lower case* roman (upright) type. However, the symbol for a unit, derived from a proper name, should start with a capital roman letter, e.g.: m (metre); A (ampere); Wb (weber); Hz (hertz).

2.2 Prefixes–General rules

1. The *prefixes* which should be used to indicate decimal multiples or sub-multiples of a unit are given in table 1.

Table 1

deci; *déci*	$(= 10^{-1})$	d	deca; *déca*	$(= 10^{1})$	da
centi; *centi*	$(= 10^{-2})$	c	hecto; *hecto*	$(= 10^{2})$	h
milli; *milli*	$(= 10^{-3})$	m	kilo; *kilo*	$(= 10^{3})$	k
micro; *micro*	$(= 10^{-6})$	μ	mega; *méga*	$(= 10^{6})$	M
nano; *nano*	$(= 10^{-9})$	n	giga; *giga*	$(= 10^{9})$	G
pico; *pico*	$(= 10^{-12})$	p	tera; *téra*	$(= 10^{12})$	T
femto; *femto*	$(= 10^{-15})$	f	peta; *peta*	$(= 10^{15})$	P
atto; *atto*	$(= 10^{-18})$	a	exa; *exa*	$(= 10^{18})$	E

2. *Compound prefixes*, formed by juxtaposition of two or more prefixes, are not to be used.

Not: mμs,		*but:* ns	(nanosecond)
Not: kMW,		*but:* GW	(gigawatt)
Not: μμF,		*but:* pF	(picofarad)

3. When the symbol of a prefix is placed before the symbol of a unit, the *combination of the two symbols* should be considered as *one new symbol*, which can be raised to a positive or negative power without using brackets.

Examples: cm^3 mA^2 μs^{-1}.

Remark:

cm^3 *means always* $(0,01 \text{ m})^3$ *but never* $0,01 \text{ m}^3$
μs^{-1} *means always* $(10^{-6} s)^{-1}$ *but never* $10^{-6} s^{-1}$.

2.3 Mathematical operations

1. Multiplication of two units may be indicated in one of the following ways:

$$N \, m \qquad N \cdot m.$$

2. Division of one unit by another unit may be indicated in one of the following ways:

$$\frac{m}{s} \qquad m/s \qquad m \, s^{-1}$$

or by any other way of writing the product of m and s^{-1}.
Not more than one solidus should be used.

Examples:

Not: cm/s/s,	*but:* $cm/s^2 = cm \, s^{-2}$
Not: J/K/mol,	*but:* $J/K \, mol = J \, K^{-1} \, mol^{-1}$

3 NUMBERS

1. Numbers should be printed in *upright type*.

2. The *decimal* sign is a comma on the line (,). In documents in the English language a comma or a dot on the line (.) may be used.
If the magnitude of the number is less than unity, the decimal sign should be preceded by a zero.

3. The sign for *multiplication* of numbers is a cross (×) or a dot half-high (·). If a dot is used as a decimal sign, a dot should not be used as the multiplication sign.

Example: 2.3×3.4 or $2,3 \cdot 3,4$.

4. Division of one number by another number may be indicated in the following ways:

or by writing it as the product of numerator and the inverse first power of the denominator. In such cases the number under the inverse power should always be placed between brackets.

Remark: When the solidus is used and when there is any doubt where the numerator starts or the denominator ends, brackets should be used, as in the case of quantities (see section 1.3).

5. To facilitate the reading of *large numbers*, the digits may be grouped in *groups of three*, but *no* comma or point should be used except for the decimal sign.

  Example: 2 573,421 736.

4 SYMBOLS FOR CHEMICAL ELEMENTS, NUCLIDES, AND PARTICLES

1. *Symbols for chemical elements* should be written in *roman (upright) type*. The symbol is not followed by a full stop.

  Examples: Ca   C   H   He.

2. The *nucleon number* (*mass number*) of a *nuclide* is shown as a left superscript (e.g. ^{14}N).

3. The right subscript position is used to indicate the *number of atoms* of a nuclide in a molecule (e.g. $^{14}N_2$).

4. The right superscript position should be used, if required, for indicating a *state of ionization* (e.g. Ca^{2+}, PO_4^{3-}) or an *excited state* (e.g. $^{110}Ag^m$, He^*).

5. The attached numeral specifying the *spectrum of a z-fold ionized atom* is the roman number $z + 1$.

  Examples: CaII, AlIII, HI (spectrum of the *neutral* hydrogen atom).

  Remark: Roman numbers in right superscript position may indicate the *oxidation number* (e.g. $Pb_2^{II}Pb^{IV}O_4$; $K_6M^{IV}Mo_9O_{32}$, where M denotes a metal).

6. *Symbols for particles and quanta*

  It is recommended to use the notation listed in table 2.

Table 2

nucleon	N	pion	π
proton	p	K-meson	K
neutron	n		
		electron	e
Λ-particle	Λ	muon	μ
Σ-particle	Σ	neutrino	ν
Ξ-particle	Ξ		
Ω-particle	Ω	photon	γ
deuteron	d		
triton	t		
helion ($^3\text{He}^{2+}$)	h		
α-particle	α		

The charge of a particle may be indicated by adding the superscript +, − or 0.

Examples: π^+ π^- π^0, p^+ p^-, e^+ e^-.

If in connection with the symbols p and e no charge is indicated, these symbols should refer to the positive proton and the negative electron, respectively.

The bar ¯, or sometimes the tilde ˜, above the symbol of a particle is used to indicate the corresponding anti-particle.

5 QUANTUM STATES

5.1 General rules

A letter symbol indicating the quantum state of *a system* should be printed in capital upright type.

A letter symbol indicating the quantum state of *a single particle* should be printed in lower case upright type.

5.2 Atomic spectroscopy

The letter symbols indicating atomic quantum states are:

$L, l = 0$: S, s	$L, l = 4$: G, g	$L, l = 8$: L, l
$= 1$: P, p	$= 5$: H, h	$= 9$: M, m
$= 2$: D, d	$= 6$: I, i	$= 10$: N, n
$= 3$: F, f	$= 7$: K, k	$= 11$: O, o.

A right subscript indicates the total angular momentum quantum number J or j. A left superscript indicates the spin multiplicity $2S + 1$.

Example: $^2\text{P}_{3/2}$-state   ($J = 3/2$, spin multiplicity 2)
    $\text{p}_{3/2}$-electron ($j = 3/2$).

An atomic electron configuration is indicated symbolically by:

$$(nl)^\kappa \quad (n'l')^{\kappa'} \ldots$$

Instead of $l = 0, 1, 2, 3 \ldots$ one uses the quantum state symbol s, p, d, f, ...

Example: the atomic electron configuration: $(1s)^2 (2s)^2 (2p)^3$.

5.3 Molecular spectroscopy

The letter symbols, indicating molecular electronic quantum states are in the case of *linear molecules:*

$$\Lambda, \lambda = 0: \quad \Sigma, \sigma$$
$$= 1: \quad \Pi, \pi$$
$$= 2: \quad \Delta, \delta$$

and for *non-linear molecules*

$$A, a; \quad B, b; \quad E, e; \quad \text{etc.}$$

Remarks: A left superscript indicates the spin multiplicity. For molecules having a symmetry centre the parity symbol g or u, indicating respectively symmetric or antisymmetric behaviour on inversion, is attached as a right subscript. A + or − sign attached as a right superscript indicates the symmetry as regards reflection in any plane through the symmetry axis of the molecules.

Examples: Σ_g^+, Π_u, $^2\Sigma$, $^3\Pi$, etc.

The letter symbols indicating the vibrational angular momentum states in the case of *linear molecules* are

$$l = 0: \quad \Sigma$$
$$= 1. \quad \Pi$$
$$= 2: \quad \Delta.$$

5.4 Nuclear spectroscopy

The spin and parity assignment of a nuclear state is

$$J^\pi$$

where the parity symbol π is + for even and − for odd parity.

Examples: 3^+, 2^-, etc.

A shell model configuration is indicated symbolically by:

$$(nlj)^\kappa \quad (n'l'j')^{\kappa'}$$

where the first bracket refers to the proton shell and the second to the neutron shell. Negative values of κ or κ' indicate holes in a completed shell. Instead of $l = 0, 1, 2, 3, \ldots$ one uses the quantum state symbol s, p, d, f, ...

Example: the nuclear configuration: $(1\,d\,3/2)^3 (1\,f\,7/2)^2$.

5.5 Spectroscopic transitions

1. The upper level and the lower level are indicated by ′ and ″ respectively.

   Examples: $h\nu = E' - E''$ $\sigma = T' - T''$.

2. A spectroscopic transition should be indicated by writing the upper state first and the lower state second, connected by a dash.

   Examples: $^2P_{1/2} - {}^2S_{1/2}$ for an electronic transition
   $(J', K') - (J'', K'')$ for a rotational transition
   $v' - v''$ for a vibrational transition.

3. Absorption transition and emission transition may be indicated by arrows ← and → respectively.

Examples: $^2P_{1/2} \to {}^2S_{1/2}$ emission from $^2P_{1/2}$ to $^2S_{1/2}$
      $(J', K') \leftarrow (J'', K'')$ absorption from (J'', K'') to (J', K').

4. The difference between two quantum numbers should be that of the upper state minus that of the lower state.

Example: $\Delta J = J' - J''$.

5. The indications of the branches of the rotation band should be as follows:

$$\Delta J = J' - J'' = -2: \quad \text{O-branch}$$
$$= -1: \quad \text{P-branch}$$
$$= \ \ 0: \quad \text{Q-branch}$$
$$= +1: \quad \text{R-branch}$$
$$= +2: \quad \text{S-branch.}$$

6 NOMENCLATURE

1. Use of the words 'specific' and 'molar'

The word 'specific' in the English name for an extensive physical quantity should be restricted to the meaning 'divided by mass (mass of the system, if this consists of more than one component or of more than one phase)'.

Examples:

specific volume	volume/mass
specific energy	energy/mass
specific heat capacity	heat capacity/mass.

The word 'molar' in the English name for an extensive physical quantity should be restricted to the meaning 'divided by amount of substance (amount of substance of the system, if this consists of more than one component or of more than one phase)'.

Examples:

molar volume	volume/amount of substance
molar energy	energy/amount of substance
molar heat capacity	heat capacity/amount of substance.

The symbol X_B, where X denotes an extensive quantity and B is the chemical symbol for a substance, denotes the partial molar quantity of the substance B defined by the relation:

$$X_B = (\partial X / \partial n_B)_{T,p,n_C,\dots}$$

For a pure substance B the partial molar quantity X_B and the molar quantity X_m are identical. The partial molar quantity X_B of pure substance B, which is identical with the molar quantity X_m of pure substance B, may be denoted by X_B^* where the superscript * denotes 'pure', so as to distinguish it from the partial molar quantity X_B of substance B in a mixture.

2. Notation for covariant character of coupling

S	Scalar coupling	A	Axial vector coupling
V	Vector coupling	P	Pseudoscalar coupling
T	Tensor coupling.		

3. *Abbreviated notation for a nuclear reaction*

The meaning of the symbolic expression indicating a nuclear reaction should be the following:

initial ⎛ incoming particle(s)   outgoing particle(s) ⎞ final

nuclide ⎝ or quanta   ,   or quanta ⎠ nuclide.

Examples: $^{14}N(\alpha, p)^{17}O$ $^{59}Co(n, \gamma)^{60}Co$

            $^{23}Na(\gamma, 3n)^{20}Na$ $^{31}P(\gamma, pn)^{29}Si$.

4. *Character of transitions*

    Multipolarity of transition:

              electric or magnetic monopole   E0 or M0
              ,,   ,,   ,,   dipole   E1 or M1
              ,,   ,,   ,,   quadrupole E2 or M2
              electric or magnetic octupole   E3 or M3
              ,,   ,,   ,, 2^n-pole   En or Mn.

    Parity change in transition:

              transition *with* parity change: yes
              transition *without* parity change: no.

5. *Nuclide*

A species of *atoms*, identical as regards atomic number (proton number) and mass number (nucleon number) should be indicated by the word *nuclide*, not by the word isotope.

Different nuclides having the same atomic number should be indicated as *isotopes* or *isotopic nuclides*.

Different nuclides having the same mass number should be indicated as *isobars* or *isobaric nuclides*.

6. *Sign of polarization vector (Basel Convention)*

In nuclear interactions the positive polarization of particles with spin $\frac{1}{2}$ is taken in the direction of the vector product

$$k_i \times k_o$$

where k_i and k_o are the circular wave vectors of the incoming and outgoing particles, respectively.

7. *Description of polarization effects (Madison Convention)*

In the symbolic expression for a nuclear reaction A(b, c)D an arrow placed over a symbol denotes a particle which is initially in a polarized state or whose state of polarization is measured.

Examples:

    A($\vec{b}$, c)D   polarized incident beam
    A($\vec{b}$, $\vec{c}$)D   polarized incident beam; polarization of the outgoing particle c measured (polarization transfer)
    A(b, $\vec{c}$)D   unpolarized incident beam; polarization of the outgoing particle c measured
    $\vec{A}$(b, c)D   unpolarized beam incident on a polarized target
    $\vec{A}$(b, $\vec{c}$)D   unpolarized beam incident on a polarized target; polarization of the outgoing particle c measured
    A($\vec{b}$, c)$\vec{D}$ polarized incident beam; measurement of the polarization of the target.

7 RECOMMENDED SYMBOLS FOR PHYSICAL QUANTITIES

Remarks:

(1) Where several symbols are given for one quantity, and no special indication is made, they are on equal footing.

(2) In general no special attention is paid to the name of the quantity.

(3) Where there is more than one form for a greek letter ($\varepsilon, \epsilon; \vartheta, \theta; \kappa, \varkappa; \varphi, \phi$) either form may be used. The form ϖ of the letter pi may be used as though it were a different letter.

7.1 Space and time

space coordinates; *coordonnées d'espace*	(x, y, z)
position vector; *vecteur de position*	r
length; *longueur*	l
breadth; *largeur*	b
height; *hauteur*	h
radius; *rayon*	r
thickness; *épaisseur*	d, δ
diameter; *diamètre:* $d = 2r$	d
element of path; *élément de parcours*	ds
area; *aire, superficie*	A, S
volume; *volume*	$V, (v)$
plane angle; *angle plan*	$\alpha, \beta, \gamma, \theta, \vartheta, \varphi$
solid angle; *angle solide*	ω, Ω
wave length; *longueur d'onde*	λ
wave number; *nombre d'onde:* $\sigma = 1/\lambda$	σ †)
wave vector; *vecteur d'onde*	$\boldsymbol{\sigma}$
circular wave number; *nombre d'onde circulaire:* $k = 2\pi/\lambda$	K
circular wave vector; *vecteur d'onde circulaire*	$\boldsymbol{K}$
attenuation coefficient; *constante d'affaiblissement*:	
$\quad F(x) = \exp(-\alpha x) \cos \beta x$	α
phase coefficient; *constante de phase*	β
propagation coefficient; *constante de propagation:* $\gamma = \alpha + i\beta$	γ
time; *temps*	t
period, periodic time; *période, durée d'une période*	T
frequency; *fréquence:* $\nu = 1/T$	ν, f
circular frequency, pulsatance; *pulsation:* $\omega = 2\pi\nu$	ω
relaxation time; *temps de relaxation:* $F(t) = \exp(-t/\tau)$	τ
damping coefficient; *coefficient d'amortissement*:	
$\quad F(t) = \exp(-\delta t) \sin \omega t$	δ
logarithmic decrement; *décrément logarithmique*:	
$\quad \Lambda = T\delta = T/\tau$	Λ
velocity; *vitesse:* $v = ds/dt$	u, v
angular velocity; *vitesse angulaire:* $\omega = d\varphi/dt$	ω
acceleration; *accélération:* $a = dv/dt$	a
angular acceleration; *accélération angulaire:* $\alpha = d\omega/dt$	α
acceleration of free fall; *accélération de la pesanteur*	g
standard --; -- *normale*	g_n
speed of light in empty space; *vitesse de la lumière dans le vide*	c
v/c	β
relativistic coordinates; *coordonnées relativistes*:	
$\quad x_0 = ct, x_1 = x, x_2 = y, x_3 = z, x_4 = ict$	$(x_0 x_1 x_2 x_3)$ $(x_1 x_2 x_3 x_4)$

†) In molecular spectroscopy $\bar{\nu}$ is also used.

7.2 Mechanics

mass; *masse* — m

(mass) density; *masse volumique:* $\rho = m/V$ — ρ

relative density; *densité relative:* $d = \rho/\rho(H_2O)$ — d

specific volume; *volume massique:* $v = V/m = 1/\rho$ — v

reduced mass; *masse réduite:* $\mu = m_1 m_2/(m_1 + m_2)$ — μ

momentum; *quantité de mouvement:* $p = mv$ — p

angular momentum; *moment cinétique:* $L = r \times p$ — L

second moment of plane area; *moment quadratique d'une aire plane:* $I_{a,y} = \int x^2 dx dy$ — I_a

second polar moment of plane area; *moment quadratique polaire d'une aire plane:* $I_p = \int (x^2 + y^2) dx dy$ — I_p

moment of inertia; *moment d'inertie:* $I_z = \int (x^2 + y^2) dm$ — I, J

force; *force* — F

torque, moment of a couple; *torque, moment d'un couple* — T

weight; *poids* — $G, (W, P)$

moment of force; *moment d'une force* — M

pressure; *pression* — p

normal stress; *contrainte normale* — σ

shear stress; *contrainte tangentielle, cission* — τ

gravitational constant; *constante de gravitation:* $F(r) = G m_1 m_2/r^2$ — G

linear strain, relative elongation; *dilatation linéique relative:* $\varepsilon = \Delta l/l_0$ — ε

modulus of elasticity, Young's modulus; *module d'élasticité longitudinale:* $\sigma = E \varepsilon$ — E

shear strain; *glissement unitaire* — γ

shear modulus; *module d'élasticité de glissement:* $\tau = G\gamma$ — G

volume strain, bulk strain; *dilatation volumique relative:* $\theta = \Delta V/V_0$ — θ

bulk modulus; *module de compression:* $p = -K\theta$ — K

Poisson ratio, *rapport de Poisson* — μ, ν

viscosity; *viscosité* — $\eta, (\mu)$

kinematic viscosity; *viscosité cinématique:* $\nu = \eta/\rho$ — ν

friction coefficient; *coefficient de frottement* — $\mu, (f)$

surface tension; *tension superficielle* — γ, σ

energy; *énergie* — E, W

potential energy; *énergie potentielle* — E_p, V, Φ

kinetic energy; *énergie cinétique* — E_k, T, K

work; *travail* — W, A

power; *puissance* — P

efficiency; *rendement* — η

Hamiltonian function; *fonction de Hamilton* — H

Lagrangian function; *fonction de Lagrange* — L

principal function of Hamilton; *fonction principale de Hamilton:* $W = \int L dt$ — W, S_p

characteristic function of Hamilton; *fonction caractéristique de Hamilton:* $S = 2 \int T dt$ — S

generalized coordinate; *coordonnée généralisée* — q, q_i

generalized momentum; *moment généralisé* — p, p_i

action integral; *intégrale d'action:* $J = \oint p dq$ — J

7.3 Molecular physics

number of molecules; *nombre de molécules.* — N

number density of molecules; *nombre volumique de molécules*
  $n = N/V$ n

Avogadro constant; *constante d'Avogadro* L, N_A

molecular mass; *masse moléculaire* m

molecular velocity vector with (magnitudes of) components; $c, (c_x, c_y, c_z)$
  vecteur vitesse moléculaire et ses coordonnées $\boldsymbol{u}, (u_x, u_y, u_z)$

molecular position vector with coordinates; *vecteur position*
  moléculaire et ses coordonnées $\boldsymbol{r}, (x, y, z)$

molecular momentum vector with (magnitudes of) components;
  vecteur quantité de mouvement moléculaire et ses
  coordonnées $\boldsymbol{p}, (p_x, p_y, p_z)$

average velocity; *vitesse moyenne* $c_0, \boldsymbol{u}_0, \langle c \rangle, \langle \boldsymbol{u} \rangle$

average speed; *vitesse moyenne* $\bar{c}, \bar{u}, \langle c \rangle, \langle u \rangle$

most probable speed; *vitesse la plus probable* $\hat{c}, \hat{v}$

mean free path; *libre parcours moyen* l

molecular attraction energy parameter; *paramètre d'énergie*
  d'attraction moléculaire ε

interaction energy between molecules i and j; *énergie*
  d'interaction entre les molécules i et j φ_{ij}, V_{ij}

velocity distribution function; *fonction de distribution des*
  vitesses: $n = \int f \, dc_x dc_y dc_z$ $f(c)$

Boltzmann function; *fonction de Boltzmann* H

generalized coordinate; *coordonnée généralisée* q

generalized momentum; *moment généralisé* p

volume in γ phase space; *volume dans l'espace γ* Ω

thermodynamic temperature; *température thermodynamique* T

Boltzmann constant; *constante de Boltzmann* k

$1/kT$ (in exponential functions; *dans les fonctions*
  exponentielles) β

molar gas constant; *constante molaire des gaz* R

partition function; *fonction de partitions* Q, Z

symmetry number; *facteur de symétrie* s

diffusion coefficient; *coefficient de diffusion* D

thermal diffusion coefficient; *coefficient de thermodiffusion* D_T

thermal diffusion ratio; *rapport de thermodiffusion* k_T

thermal diffusion factor; *facteur de thermodiffusion* α_T

characteristic temperature; *température caractéristique* Θ

Debye temperature; *température de Debye:* $\Theta_D = h\nu_D/k$ Θ_D

Einstein temperature; *température d'Einstein:* $\Theta_E = h\nu_E/k$ Θ_E

rotational temperature; *température de rotation:* $\Theta_r = h^2/8\pi^2 Ik$ Θ_r

vibrational temperature; *température de vibration:* $\Theta_v = h\nu/k$ Θ_v

7.4 *Thermodynamics* †)

quantity of heat; *quantité de chaleur* Q

work; *travail* W, A

thermodynamic temperature; *température thermodynamique* T

Celsius temperature; *température Celsius* t, ϑ ††)

entropy; *entropie* S

internal energy; *énergie interne* U

Helmholtz function, *fonction de Helmholtz, énergie*
  libre: $F = U - TS$ F

enthalpy; *enthalpie:* $H = U + pV$ H

---

†) The index m is added in the case of molar quantities, if needed, to distinguish them from quantities referring to the whole system. For specific quantities (see section 6.1) lower case letters are used.

††) When Celsius temperature and time come together, t is reserved for time.

Gibbs function; *fonction de Gibbs, enthalpie libre:*
  $G = H - TS$ G

Massieu function; *fonction de Massieu:* $J = -F/T$ J

Planck function; *fonction de Planck:* $Y = -G/T$ Y

pressure coefficient; *coefficient de pression:* $\beta = (\partial p/\partial T)_V$ β

relative pressure coefficient; *coefficient de pression*
  relative: $\alpha_p = (1/p)(\partial p/\partial T)_V$ α_p

compressibility; *compressibilité:* $\kappa = -(1/V)(\partial V/\partial p)_T$ κ

linear expansion coefficient; *dilatabilité linéique* α_l

cubic expansion coefficient; *dilatabilité volumique:*
  $\alpha = (1/V)(\partial V/\partial T)_p$ α, γ

thermal conductivity; *conductivité thermique* λ

specific heat capacity; *chaleur massique:* $c = C/m$ c_p, c_V

heat capacity; *capacité thermique* C_p, C_V

Joule–Thomson coefficient; *coefficient de Joule–Thomson* μ

isentropic exponent; *exposant isentropique* *):
  $\kappa = -(V/p)(\partial p/\partial V)_S$ κ

ratio of specific heat capacities; *rapport des chaleurs*
  massiques $\gamma, (\kappa)$

heat flow rate; *flux thermique* $\Phi, (q)$

heat current density; *densité de flux thermique* $q, (\varphi)$

thermal diffusivity; *diffusivité thermique:* $a = \lambda/\rho c_p$ a

7.5 Electricity and magnetism **)

quantity of electricity; *quantité d'électricité*	Q
charge density; *charge volumique*	ρ
surface charge density; *charge surfacique*	σ
electric potential; *potentiel électrique*	V, φ
potential difference, tension; *différence de potentiel, tension*	U, V
electromotive force; *force électromotrice*	E
electric field strength; *champ électrique*	$\boldsymbol{E}$
electric flux, *flux électrique*	Ψ
electric displacement; *déplacement électrique*	$\boldsymbol{D}$
capacitance; *capacité*	C
permittivity; *permittivité:* $\boldsymbol{D} = \varepsilon \boldsymbol{E}$	ε
permittivity of vacuum, electric constant; *permittivité du vide,*	
constante électrique	ε_0
relative permittivity; *permittivité relative:* $\varepsilon_r = \varepsilon/\varepsilon_0$	ε_r
dielectric polarization; *polarisation diélectrique* : $\boldsymbol{D} = \varepsilon_0 \boldsymbol{E} + \boldsymbol{P}$	$\boldsymbol{P}$
electric susceptibility; *susceptibilité électrique*	χ_e
polarizability; *polarisabilité*	α, γ
electric dipole moment; *moment dipolaire électrique*	$\boldsymbol{p}$
electric current; *courant électrique*	I
electric current density; *densité de courant électrique*	j, J
magnetic field strength; *champ magnétique*	$\boldsymbol{H}$
magnetic potential difference; *différence de*	
potentiel magnétique	U_m
magnetomotive force; *force magnétomotrice:* $F_m = \oint H_s \, ds$	F_m
magnetic induction, magnetic flux density; *induction*	
magnétique, densité de flux magnétique	$\boldsymbol{B}$
magnetic flux; *flux magnétique*	Φ
permeability; *perméabilité:* $\boldsymbol{B} = \mu \boldsymbol{H}$	μ

 *) For an ideal gas the isentropic exponent, κ, is equal to the ratio of specific heat
    capacities, γ.

**) Written according to the rationalized, 4-dimensional system of quantities, see Appendix
    I, section 2.

permeability of vacuum, magnetic constant; *perméabilité du vide, constante magnétique*	μ_0
relative permeability; *perméabilité relative:* $\mu_r = \mu/\mu_0$	μ_r
magnetization; *aimantation:* $B = \mu_0(H + M)$	M
magnetic susceptibility; *susceptibilité magnétique*	χ_m
electromagnetic moment; *moment électromagnétique* $E_p = -m \cdot B$	μ, m
resistance; *résistance*	R
reactance; *réactance*	X
quality factor; *facteur de qualité:* $Q = |X|/R$	Q
impedance; *impédance:* $Z = R + iX$	Z
admittance; *admittance:* $Y = 1/Z = G + iB$	Y
conductance; *conductance*	G
susceptance; *susceptance*	B
resistivity; *résistivité*	ρ
conductivity; *conductivité* $\gamma = 1/\rho$	γ, σ
self inductance; *inductance propre*	L
mutual inductance; *inductance mutuelle*	M, L_{12}
coupling coefficient; *coefficient de couplage:* $k = L_{12}/(L_1 L_2)^{1/2}$	k
phase number; *nombre de phases*	m
loss angle; *angle de pertes*	δ
number of turns; *nombre de tours*	N
power; *puissance*	P
electromagnetic energy density; *énergie électromagnetique volumique*	w
Poynting vector; *vecteur de Poynting*	S
magnetic vector potential; *potentiel vecteur magnétique*	A

7.6 Radiation, light †)

radiant energy; *énergie rayonnante*	$Q, (Q_e), W$
radiant energy density; *énergie rayonnante volumique*	w
spectral concentration of radiant energy density (in terms of wavelength); *énergie rayonnante volumique spectrique (en longueur d'onde):* $w = \int w_\lambda \, d\lambda$	w_λ
radiant flux, radiant power; *flux énergétique, puissance rayonnante:* $\Phi = \int \Phi_\lambda \, d\lambda$	$\Phi, (\Phi_e), P$
radiant flux density; *densité de flux énergétique*	φ
radiant intensity; *intensité énergétique:* $\Phi = \int I \, d\Omega$	$I, (I_e)$
spectral concentration of radiant intensity (in terms of frequency); *intensité énergétique spectrique (en fréquence):* $I = \int I_\nu \, d\nu$	$I_\nu, (I_{e\nu})$
irradiance; *éclairement énergétique:* $\Phi = \int E \, dS$	$E, (E_e)$
radiance; *luminance énergétique:* $I = \int L \cos \vartheta \, dS$	$L, (L_e)$
radiant excitance; *excitance énergétique:* $\Phi = \int M \, dS$	$M, (M_e)$
Stefan–Boltzmann constant; *constante de Stefan–Boltzmann:* $\sigma = 2\pi^5 k^4/15 h^3 c^2$	σ
first radiation constant; *première constante de rayonnement:* $c_1 = 2\pi h c^2$	c_1
second radiation constant; *deuxième constante de rayonnement:* $c_2 = hc/k$	c_2

†) In several cases, the same symbol is used for a pair of corresponding radiant and luminous quantities with the understanding that subscripts e for energetic and v for visible will be added, whenever confusion between these quantities might otherwise occur.

emissivity; *émissivité:* $\varepsilon = M/M_B$ ε
   (M_B: radiant excitance of a black body radiator)

luminous efficacy; *efficacité lumineuse:* $K = \Phi_v/\Phi_e$ K

spectral luminous efficacy; *efficacité lumineuse spectrale:*
   $K(\lambda) = \Phi_{v\lambda}/\Phi_{e\lambda}$ $K(\lambda)$

maximum spectral luminous efficacy; *efficacité lumineuse*
   spectrale maximale K_m

luminous efficiency; *efficacité lumineuse relative:* $V = K/K_m$ V

spectral luminous efficiency; *efficacité lumineuse relative*
   spectrale: $V(\lambda) = K(\lambda)/K_m$ $V(\lambda)$

quantity of light; *quantité de lumière* $Q, (Q_v)$

luminous flux; *flux lumineux* $\Phi, (\Phi_v)$

luminous intensity; *intensité lumineuse:* $\Phi = \int I \, d\Omega$ $I, (I_v)$

spectral concentration of luminous intensity (in terms of wave
   number); *intensité lumineuse spectrique (en nombre*
   d'onde): $I = \int I_\sigma \, d\sigma$ $I_\sigma, (I_{v\sigma})$

illuminance, illumination; *éclairement lumineux:* $\Phi = \int E \, dS$ $E, (E_v)$

luminance; *luminance:* $I = \int L \cos \vartheta \, dS$ $L, (L_v)$

luminous excitance; *excitance lumineuse:* $\Phi = \int M \, dS$ $M, (M_v)$

absorptance; *facteur d'absorption:* Φ_a/Φ_0 α *)

reflectance; *facteur de réflexion:* Φ_r/Φ_0 ρ *)

transmittance; *facteur de transmission:* Φ_{tr}/Φ_0 τ *)

linear attenuation coefficient; *coefficient d'atténuation linéique* μ

linear absorption coefficient; *coefficient d'absorption linéique* a

speed of light in empty space; *vitesse de la lumière dans le vide* c

refractive index; *indice de réfraction:* $n = c/c_n$ n

7.7 Acoustics

velocity of sound; *vitesse du son* c

velocity of longitudinal waves; *vitesse longitudinale* c_l

velocity of transversal waves; *vitesse transversale* c_t

group velocity; *vitesse de groupe* c_g

sound energy flux; *flux d'énergie acoustique* P, P_a

reflexion factor; *facteur de réflexion:* P_r/P_0 ρ

acoustic absorption factor; *facteur d'absorption*
   acoustique: $1 - \rho$ $\alpha_a, (\alpha)$

transmission factor; *facteur de transmission:* P_{tr}/P_c τ

dissipation factor; *facteur de dissipation:* $\alpha_a - \tau$ δ

loudness level; *niveau d'isosonie* L_N

sound power level; *niveau de puissance acoustique* L_P

sound pressure level; *niveau de pression acoustique* L_p

7.8 Quantum mechanics

complexe conjugate of Ψ; *complexe conjugué de Ψ* Ψ^*

probability density; *densité de probabilité:* $P = \Psi^*\Psi$ P

probability current density; *densité de courant de probabilité:*
   $S = (\hbar/2im)(\Psi^*\nabla\Psi - \Psi\nabla\Psi^*)$ S

charge density of electrons; *charge volumique d'électrons:*
   $\rho = -eP$ ρ

electric current density of electrons; *densité de courant*
   électrique d'électrons: $j = -eS$ j

---

*) $\alpha(\lambda)$, $\rho(\lambda)$, and $\tau(\lambda)$ designate spectral absorptance, $\Phi_{a\lambda}/\Phi_{0\lambda}$, spectral reflectance,
   $\Phi_{r\lambda}/\Phi_{0\lambda}$, and spectral transmittance, $\Phi_{tr\lambda}/\Phi_{0\lambda}$, respectively.

Dirac bra vector; *vecteur bra de Dirac*	$\langle\ \vert$
Dirac ket vector; *vecteur ket de Dirac*	$\vert\ \rangle$
expectation value of A; *valeur moyenne de A*	$\langle A\rangle$, $\bar{A}$
commutator of A and B; *commutateur de A et B:*	
$[A,B] = AB - BA$	$[A,B]$, $[A,B]_-$
anticommutator of A and B; *anticommutateur de A et B:*	
$[A,B]_+ = AB + BA$	$[A,B]_+$
matrix element; *élément de matrice:* $A_{ij} = \int \varphi_i^*(A\varphi_j)\,d\tau$	A_{ij}
Hermitian conjugate of operator A; *conjugué Hermitien de l'opérateur A*	$A^\dagger$
momentum operator in coordinate representation; *opérateur de quantité de mouvement*	$+(\hbar/i)\nabla$
annihilation operators; *opérateurs d'annihilation*	a, b, α, β
creation operators; *opérateurs de création*	$a^\dagger, b^\dagger, \alpha^\dagger, \beta^\dagger$

7.9 Atomic and nuclear physics

nucleon number, mass number; *nombre de nucléons, nombre de masse*	A
proton number, atomic number; *nombre de protons, nombre atomique*	Z
neutron number; *nombre de neutrons:* $N = A - Z$	N
elementary charge (equal to charge of proton); *charge élémentaire (égale à la charge du proton)*	e
electron mass; *masse de l'électron*	m, m_e
proton mass; *masse du proton*	m_p
neutron mass; *masse du neutron*	m_n
meson mass; *masse du méson*	m_π,
nuclear mass; *masse nucléaire* (of nucleus: ^{A}X)	m_N, m_N (^{A}X)
atomic mass; *masse atomique* (of nuclide: ^{A}X)	m_a, m_a (^{A}X)
(unified) atomic mass constant; *constante (unifiée) de masse atomique:* $m_u = m_a(^{12}\mathrm{C})/12$	m_u
relative atomic mass; *masse atomique relative:* m_a/m_u	A_r
Planck constant; *constante de Planck* ($\hbar = h/2\pi$)	h
principal quantum number (qu.n.); *nombre quantique (n.qu.) principal*	n, n_i
orbital angular momentum qu.n.; *n.qu. de moment angulaire orbital*	L, l_i
spin qu.n.; *n. qu. de spin*	S, s_i
total angular momentum qu.n.; *n.qu. de moment angulaire total*	J, j_i
magnetic qu.n.; *n.qu. magnétique*	M, m_i
nuclear spin qu.n.; *n.qu. de spin nucléaire*	I, J †)
hyperfine qu.n.; *n.qu. hyperfin*	F
rotational qu.n.; *n.qu. de rotation*	J, K
vibrational qu.n.; *n.qu. de vibration*	v
quadrupole moment; *moment quadripolaire*	Q
Rydberg constant; *constante de Rydberg* ††)	R_∞
Bohr radius; *rayon de Bohr* ††)	a_0
fine structure constant; *constante de structure fine* ††)	α
mass excess; *excès de masse:* $m_a - Am_u$	Δ
packing fraction; *packing fraction:* Δ/Am_u	f
nuclear radius; *rayon nucléaire:* $R = r_0 A^{1/3}$	R

†) I is used in atomic physics, J in nuclear physics.
††) See for definition: Appendix I, section 3.

magnetic moment of particle; *moment magnétique d'une particule*	μ
magnetic moment of proton; *moment magnétique du proton*	μ_p
magnetic moment of neutron; *moment magnétique du neutron*	μ_n
magnetic moment of electron; *moment magnétique électronique*	μ_e
Bohr magneton; *magnéton de Bohr* †)	μ_B
nuclear magneton; *magnéton nucléaire*	μ_N
g-factor; *facteur g:* e.g. $g = \mu/I\mu_N$	g
gyromagnetic ratio, gyromagnetic coefficient; *rapport gyromagnétique, coefficient gyromagnétique* †)	γ
Larmor circular frequency; *pulsation de Larmor* †)	ω_L
level width; *largeur d'un niveau*	Γ
mean life; *vie moyenne*	τ
reaction energy; *énergie de réaction*	Q
cross section; *section efficace*	σ
macroscopic cross section; *section efficace macroscopique:* $\Sigma = n\sigma$	Σ
impact parameter; *paramètre de collision*	b
scattering angle; *angle de diffusion*	$\vartheta, \theta, \varphi$
internal conversion coefficient; *coefficient de conversion interne*	α
disintegration energy; *énergie de désintégration*	Q
half life; *demi-vie*	$T_{1/2}$
reduced half life; *demi-vie reduite*	$fT_{1/2}$
decay constant, disintegration constant; *constante de désintégration*	λ
activity; *activité*	A
Compton wavelength; *longueur d'onde de Compton:* $\lambda_1 = h/mc$	λ_1
electron radius; *rayon de l'electron* †)	r_e
linear attenuation coefficient; *coefficient d'atténuation linéique*	μ, μ_1
atomic attenuation coefficient; *coefficient d'atténuation atomique*	μ_a
mass attenuation coefficient; *coefficient d'atténuation massique*	μ_m
linear stopping power; *pouvoir d'arrêt linéaire*	S, S_1
atomic stopping power; *pouvoir d'arrêt atomique*	S_a
linear range; *distance de pénétration linéaire*	R, R_1
recombination coefficient; *coefficient de recombinaison*	α

7.10 Solid state physics

lattice vector: a translation vector which maps the crystal lattice on itself; *vecteur du réseau: vecteur qui reproduit par translation le réseau cristallin sur lui-même*	R, R_0
fundamental translation vectors for the crystal lattice; *vecteurs de base de la maille cristalline:* $R = n_1 a_1 + n_2 a_2 + n_3 a_3$, where n_1, n_2, n_3 are integers	a_1, a_2, a_3 a, b, c
(circular) reciprocal lattice vector; *vecteur du réseau réciproque:* $G \cdot R = m \cdot 2\pi$, where m is an integer	G
(circular) fundamental translation vectors for the reciprocal lattice; *vecteur de base de la maille du réseau réciproque:* $a_i \cdot b_k = 2\pi\delta_{ik}$ ††), where δ_{ik} is Kronecker delta symbol	b_1, b_2, b_3 a^*, b^*, c^*
lattice plane spacing; *espacement entre plans réticulaires*	d

†) See for definition: Appendix I, section 3.
††) In crystallography however $a_i \cdot b_k = \delta_{ik}$.

Bragg angle; *angle de Bragg*	ϑ
order of reflexion; *ordre de réflexion*	n
short range order parameter; *paramètre d'ordre à courte distance (local)*	σ
long range order parameter; *paramètre d'ordre à grande distance*	s
Burgers vector; *vecteur de Burgers*	b
particle position vector; *vecteur de position d'une particule* *)	r, R
equilibrium position vector of ion; *vecteur de position d'équilibre d'un ion*	R_0
displacement vector of ion; *vecteur de déplacement d'un ion*	u
normal coordinates; *coordonnées normales*	Q_i
polarization vector; *vecteur de polarisation*	e
Debye–Waller factor; *facteur de Debye–Waller*	D
Debye circular wave number; *nombre d'onde circulaire de Debye*	q_D
Debye circular frequency; *pulsation de Debye*	ω_D
Grueneisen parameter; *paramètre de Grüneisen:* $\gamma = \alpha V / \kappa C_v$ (α: cubic expansion coefficient; κ: compressibility)	γ, Γ
Madelung constant; *constante de Madelung*	α
mean free path of electrons; *libre parcours moyen des électrons*	l, l_e
mean free path of phonons; *libre parcours moyen des phonons*	Λ, l_{ph}
drift velocity; *vitesse de mouvement*	v_{dr}
mobility; *mobilité*	μ
one electron wave function; *fonction d'onde monoélectronique*	$\psi(r)$
Bloch wave function; *fonction d'onde de Bloch:* $\psi_k(r) = u_k(r) \exp(ik \cdot r)$	$u_k(r)$
density of states; *densité (électronique) d'état:* $dN(E)/dE = N_E(E) = \rho(E)$	N_E, ρ
(spectral) density of vibrational modes; *densité spectrale de modes de vibration*	g, N_ω
exchange integral; *intégrale d'échange*	J
resistivity tensor; *tenseur de résistivité*	ρ_{ik}
electric conductivity tensor; *tenseur de conductibilité électrique*	σ_{ik}
thermal conductivity tensor; *tenseur de conductibilité thermique*	λ_{ik}
residual resistivity; *résistivité résiduelle*	ρ_R
relaxation time; *temps de relaxation*	τ
Lorenz coefficient; *coefficient de Lorenz:* $L = \lambda/\sigma T$	L
Hall coefficient; *coefficient de Hall*	A_H, R_H
Ettinghausen coefficient; *coefficient d'Ettinghausen*	A_E
first Ettinghausen–Nernst coefficient; *premier coefficient d'Ettinghausen–Nernst*	A_N
first Righi–Leduc coefficient; *premier coefficient de Righi–Leduc*	A_{RL}
thermoelectromotive force between substances a and b; *force thermoélectromotrice entre deux substances a et b*	E_{ab}
Seebeck coefficient for substances a and b; *coefficient de Seebeck pour deux substances a et b:* $S_{ab} = dE_{ab}/dT$	S_{ab}, ε_{ab}
Peltier coefficient for substances a and b; *coefficient de Peltier pour deux substances a et b*	Π_{ab}
Thomson coefficient; *coefficient de Thomson*	$\mu, (\tau)$

*) To distinguish between electron and ion position vectors, lower case and capital letters are used, respectively.

work function; *travail d'extraction* *): $\Phi = e\varphi$ φ

Richardson constant; *constante de Richardson:*
    $j = AT^2 \exp(-\Phi/kT)$ A

electron number density; *nombre volumique électronique*
    (*densité électronique*) **) n, n_n, n_-

hole number density; *nombre volumique de trous*
    (*densité de trous*) **) p, n_p, n_+

donor number density; *nombre volumique de donneurs*
    (*densité de donneurs*) n_d

acceptor number density; *nombre volumique d'accepteurs*
    (*densité d'accepteurs*) n_a

intrinsic number density; *nombre volumique intrinsèque*
    (*densité intrinsèque*): $n_i = (n \cdot p)^{1/2}$ n_i

energy gap: *bande d'énergie interdite (énergie gap)* E_g

donor ionization energy; *énergie d'ionisation de donneur* E_d

acceptor ionization energy; *énergie d'ionisation d'accepteur* E_a

Fermi energy; *énergie de Fermi* E_F, ε_F

circular wave vector, propagation vector (of particles); *vecteur*
    d'onde, vecteur de propagation (de particules) k

circular wave vector, propagation vector (of phonons); *vecteur*
    d'onde, vecteur de propagation (de phonons) q

Fermi circular wave vector; *vecteur de Fermi* k_F

electron annihilation operator; *opérateur d'annihilation*
    d'électron a

electron creation operator; *opérateur de création d'électron* $a^\dagger$

phonon annihilation operator; *opérateur d'annihilation de*
    phonon b

phonon creation operator; *opérateur de création de phonon* $b^\dagger$

effective mass; *masse effective* **) m_n^*, m_p^*

mobility; *mobilité* **) μ_n, μ_p

mobility ratio; *rapport de mobilité:* $b = \mu_n/\mu_p$ b

diffusion coefficient; *coefficient de diffusion* **) D_n, D_p

diffusion length; *longueur de diffusion* **) L_n, L_p

carrier life time; *durée de vie de porteur* **) τ_n, τ_p

characteristic (Weiss) temperature; *température*
    caractéristique (de Weiss) Θ, Θ_W

Curie temperature; *température de Curie* T_C

Néel temperature; *température de Néel* T_N

superconductor critical transition temperature; *température*
    critique de transition supraconductrice T_c

superconductor (thermodynamic) critical field strength; *champ*
    critique (thermodynamique) d'un supraconducteur H_c

superconductor critical field strengths (type II); *champ critique*
    d'un supraconducteur (type II)† H_{c1}, H_{c2}, H_{c3}

superconductor energy gap; *bande interdite du supraconducteur* Δ

London penetration depth; *profondeur de pénétration de*
    London λ_L

coherence length; *longueur de cohérence* ξ

Landau–Ginzburg parameter; *paramètre de Landau–*
    Ginzburg: $\kappa = \lambda_L/\xi\sqrt{2}$ κ

---

*) The symbol W is used for the quantity $W = \Phi + \mu$, where μ is the electron chemical potential which, at $T = 0$, *is equal to the Fermi energy* E_F.

**) In general subscripts n and p or $-$ and $+$ may be used to denote electrons and holes, respectively.

†) H_{c1}: for magnetic flux entering the superconductor, H_{c2}: for disappearing of bulk superconductivity, H_{c3}: for disappearing of surface superconductivity.

flux (or fluxoid) quantum; *quantum de flux:* $\Phi_0 = h/2e$ **) $\qquad$ Φ_0

Miller indices; *indices de Miller* $\qquad$ h_1, h_2, h_3
$\qquad\qquad\qquad\qquad\qquad\qquad\qquad\qquad\qquad\qquad\qquad$ h, k, l

single plane or set of parallel planes in lattice; *plan simple ou*
famille de plans réticulaires parallèles dans un réseau $\qquad$ (h_1, h_2, h_3)
$\qquad\qquad\qquad\qquad\qquad\qquad\qquad\qquad\qquad\qquad\qquad$ (h, k, l)

full set of planes in lattice equivalent by symmetry; *famille de*
plans réticulaires équivalents par symétrie $\qquad$ $\{h_1, h_2; h_3\}$
$\qquad\qquad\qquad\qquad\qquad\qquad\qquad\qquad\qquad\qquad\qquad$ $\{h, k, l\}$

direction in lattice; *rangée réticulaire* $\qquad$ $[u, v, w]$
full set of directions in lattice equivalent by symmetry; *famille*
de rangées réticulaires équivalentes par symétrie $\qquad$ $\langle u, v, w \rangle$

Note: If the letter symbols are replaced by numbers in the bracketed expressions, it is customary to omit the commas. A negative numerical value is commonly indicated by a bar above the number, e.g. $(\bar{1}10)$.

7.11 Molecular spectroscopy

quantum number (qu.n.) of component of electronic orbital
angular momentum vector along symmetry axis; *nombre*
$\quad$ *quantique (n.qu.) de la composante du moment angulaire*
$\quad$ *orbital électronique suivant l'axe de symétrie* $\qquad$ Λ, λ_i
qu.n. of component of electronic spin along symmetry axis;
$\quad$ *n.qu. de la composante du spin électronique suivant l'axe de*
$\quad$ *symétrie* $\qquad$ Σ, σ_i
qu.n. of total electronic angular momentum vector along
$\quad$ symmetry axis; *n.qu. du moment angulaire total électronique*
$\quad$ *suivant l'axe de symétrie:* $\quad \Omega = |\Lambda + \Sigma|$ $\qquad$ Ω, ω_i
qu.n. of electronic spin; *n.qu. du spin électronique* $\qquad$ S
qu.n. of nuclear spin; *n.qu. du spin nucléaire* $\qquad$ I
qu.n. of vibrational mode; *n.qu. d'une mode de vibration* $\qquad$ v
degeneracy of vibrational mode; *degré de dégénérescance*
$\quad$ *d'une mode de vibration* $\qquad$ d
qu.n. of vibrational angular momentum; *n.qu. du moment*
$\quad$ *angulaire vibrationnel* (L.M.) $\qquad$ l
qu.n. of total angular momentum; *n.qu. du moment angulaire*
$\quad$ *total* (excluding nuclear spin) $\qquad$ J
qu.n. of component of J in direction of external field; *n.qu. de*
$\quad$ *la composante de J dans la direction du champ extérieur* $\qquad$ M, M_J
qu.n. of component of S in direction of external field; *n.qu. de*
$\quad$ *la composante de S dans la direction du champ extérieur* $\qquad$ M_S
qu.n. of total angular momentum; *n.qu. du moment angulaire*
$\quad$ *total* (including nuclear spin) $F = J + I$ $\qquad$ F
qu.n. of component of F in direction of external field; *n.qu. de*
$\quad$ *la composante de F dans la direction du champ extérieur* $\qquad$ M_F
qu.n. of component of I in direction of external field; *n.qu. de*
$\quad$ *la composante de I dans la direction du champ extérieur* $\qquad$ M_I
qu.n. of component of angular momentum along axis; *n.qu. de*
$\quad$ *la composante du moment angulaire suivant l'axe* (L.M. and
$\quad$ S.T.M.; excluding electron and nuclear spin; for
$\quad$ L.M.: $\quad (K = |\Lambda + l|)$ $\qquad$ K

**) $2e/h = 1/\Phi_0$ is also called characteristic constant for macroscopic coherence in super-
conductors.

qu.n. of total angular momentum; *n.qu. du moment angulaire*
total (L.M. and S.T.M.; excluding electron and nuclear
spin: $J = N + S$ *) N

qu.n. of component of angular momentum along symmetry P
axis; *n.qu. de la composante du moment angulaire suivant*
l'axe de symétrie (L.M. and S.T.M., excluding nuclear spin;
for L.M.: $P = |K + \Sigma|$ **)

electronic term; *terme électronique*: $T_e = E_e/hc$ * * *) T_e

vibrational term; *terme de vibration*: $G = E_{vibr}/hc$ G

coefficients in expression for vibrational term (for D.M.);
     coefficients de l'expression d'un terme de vibration:
     $G - v_e(v + \frac{1}{2}) - xv_e(v + \frac{1}{2})^2$ $\sigma_e, x\sigma_e$

coefficients in expression for vibrational term (for P.M.);
     coefficients de l'expression d'un terme de vibration:
     $G = \Sigma\sigma_j(v_j + \frac{1}{2}d_j) + \frac{1}{2}\underset{j}{\Sigma}\underset{k}{\Sigma}\, x_{jk}(v_j + \frac{1}{2}d_j)(v_k + \frac{1}{2}d_k)$ σ_j, x_{jk}

rotational term; *terme de rotation*: $F = E_{rot}/hc$ F

principal moments of inertia, *moments principaux d'inertie*
     $I_A \leqslant I_B \leqslant I_C$ †) I_A, I_B, I_C

rotational constants; *constantes de rotation:*
     $A = h/8\pi^2 cI_A$, etc. †) A, B, C

total term; *terme total*: $T = T_e + G + F$ T

Remark: L.M. = linear molecules. S.T.M. = symmetric top molecules.
D.M. = diatomic molecules. P.M. = polyatomic molecules.
See for further details: Report on Notation for the Spectra of Polyatomic
Molecules (Joint Commission for Spectroscopy of I.U.P.A.P. and I.A.U. 1954)
J. Chem. Phys. **23** (1955) 1997.

7.12 Chemical physics

relative atomic mass; *masse atomique relative* A_r

relative molecular mass; *masse moléculaire relative* M_r

amount of substance; *quantité de matière* n, ν ††)

molar mass of substance B; *masse molaire de la substance* B M_B

concentration of substance B; *concentration de la substance*
     B: $c_B = n_B/V$ c_B

mole fraction of substance B; *fraction molaire de la*
     substance B x_B

mass fraction of substance B; *titre en masse de la substance* B w_B

volume fraction of substance B; *titre en volume de la substance*
     B φ_B

mole ratio of solution; *rapport molaire d'une solution* r

molality of solution; *molalité d'une solution* m

chemical potential of substance B; *potentiel chimique de la*
     substance B †††) μ_B

absolute activity of substance B (dimensionless); *activité*
     absolue de la substance B (*sans dimension*):
     $\lambda_B = \exp(\mu_B/kT)$ λ_B

relative activity; *activité relative* a_B

---

  *) Case of loosely coupled electron spin.
  **) Case of tightly coupled electron spin.
  ***) All energies are taken here with respect to the ground state as reference level.
  †) For diatomic molecules use I and $A = h/8\pi^2 cI$.
  ††) ν may be used as an alternative to n, when n is used for number density of particles.
†††) Referred to one particle.

activity coefficients; *coefficients d'activité*	γ_B, f_B
osmotic pressure; *pression osmotique*	Π
osmotic coefficient; *coefficient osmotique*	g, φ
stoichiometric number of substance B; *nombre stœchiométrique de la substance* B	ν_B
affinity; *affinité*	A
extent of reaction; *état d'avancement d'une réaction*	ξ
equilibrium constant; *constante d'équilibre*	K
charge number of ion; *nombre de charge d'un ion, électrovalence*	z
Faraday constant; *constante de Faraday*	F
ionic strength; *force ionique*	I
reduced activity of substance B; *activité réduite de la substance* B: $z_B = 2\pi mkT/h^2)^{3/2}\lambda_B$	z_B

7.13 Plasma physics

energy of particle; *énergie d'une particule*	ε
dissociation energy (e.g. of molecule X); *énergie de dissociation (par ex., d'une molécule* X)	$E_d, E_d(X)$
electron affinity; *affinité électronique*	E_{ea}
ionization energy; *énergie d'ionisation*	E_i
degree of ionization; *degré d'ionisation*	x
charge number of ion (positive or negative); *charge ionique (positif ou négatif)*	z
number density of ions of charge number z; *densité ionique des ions de charge z* †)	n_z
degree of ionization for charge number $z \geqslant 1$; *degré d'ionisation pour un nombre de charge $z \geqslant 1$:* $x_z = n_z/(n_z + n_{z-1})$	x_z
neutral particle temperature; *température des neutres*	T_n
ion temperature; *température ionique*	T_i
electron temperature; *température électronique*	T_e
electron number density; *densité électronique*	n_e
electron plasma circular frequency; *fréquence de plasma:* $\omega_{pe}^2 = n_e e^2/\varepsilon_0 m_e$	ω_{pe}
Debye length; *longueur de Debye*	λ_D
charge of particle; *charge d'une particule*	q
electron cyclotron circular frequency; *fréquence cyclotron électronique:* $\omega_{ce} = (e/m_e)B$	ω_{ce}
ion cyclotron circular frequency; *fréquence cyclotron ionique:* $\omega_{ci} = (ze/m_i)B$	ω_{ci}
reduced mass; *masse réduite:* $\mu = m_1 m_2/(m_1 + m_2)$	μ, m_r
impact parameter; *paramètre d'impact*	b
mean free path; *libre parcours moyen*	l, λ
collision frequency; *fréquence de collision*	ν_{coll}, ν_c
mean time interval between collisions; *intervalle de temps moyenne entre collisions:* $\tau_{coll} = 1/\nu_{coll}$	τ_{coll}, τ_c
cross section; *section efficace:* $\sigma = 1/ln$	σ
(electron) ionization efficiency; *efficacité d'ionisation (électronique):* $s_e = (\rho_0/\rho)dN/dx$	s_e

(dN: number of ion pairs formed by an ionizing electron travelling through dx in the plasma at gas density ρ; ρ_0: gas density at $p_0 = 1$ Torr, $T_0 = 273,15$ K)

†) When only singly charged ions need to be considered, n_{-1} and n_{+1} may be represented by n_- and n_+.

rate coefficient; *taux de réaction*	k
one-body rate coefficient; *taux de réaction unimoléculaire:* $-dn_A/dt = k_m n_A$	k_m
relaxation time; *temps de relaxation:* (e.g. $\tau = 1/k_m$)	τ
two-body rate coefficient, binary rate coefficient; *taux de réaction binaire:* (e.g. $X + Y \rightarrow XY + h\nu$) $dn_{XY}/dt = k_b n_X n_Y$	k_b
three-body rate coefficient, ternary rate coefficient; *taux de réaction ternaire:* (e.g. $X + Y + M \rightarrow XY + M^*$) $dn_{XY}/dt = k_t n_M n_X n_Y$	k_t
Townsend (electron) ionization coefficient; *coefficient de Townsend* *)	α
Townsend (ion) ionization coefficient; *coefficient ionique de Townsend*	β
secondary electron emission coefficient; *taux d'émission secondaire*	γ
drift velocity; *vitesse de mouvement*	v_{dr}
mobility; *mobilité:* $\mu = v_{dr}/E$	μ
positive or negative ion diffusion coefficient: *coefficient de diffusion des ions*	D_+, D_-
electron diffusion coefficient; *coefficient de diffusion des électrons*	D_e
ambipolar (ion–electron) diffusion coefficient; *coefficient de diffusion ambipolaire:* $D_a = (D_+\mu_e + D_e\mu_+)/(\mu_+ + \mu_e)$	D_a, D_{amb}
characteristic diffusion length; *longueur caractéristique de diffusion*	L_D, Λ
ionization frequency; *fréquence d'ionisation*	ν_i
ion–ion recombination coefficient; *coefficient de recombinaison ion–ion:* $dn_-/dt = -\alpha_i n_- n_+$	α_i
electron–ion recombination coefficient; *coefficient de recombinaison électron–ion:* $dn_e/dt = -\alpha_e n_e n_+$	α_e
plasma pressure; *pression cinétique du plasma*	p
magnetic pressure; *pression magnétique:* $p_m = B^2/2\mu$ (μ: permeability)	p_m
magnetic pressure ratio; *coefficient β:* $\beta = p/p_m$ (p_m: magnetic pressure outside the plasma)	β
magnetic diffusivity; *diffusivité magnétique:* $\nu_m = 1/\mu\sigma$ (σ: electric conductivity; μ: permeability)	ν_m, η_m
Alfvén speed; *vitesse d'Alfvén:* $v_A = B/(\mu\rho)^{1/2}$ (ρ: (mass) density; μ: permeability)	v_A

7.14 Dimensionless parameters †)

1. Momentum transport

Reynolds number; *nombre de Reynolds:* $Re = vl/\nu$	Re
Euler number; *nombre d'Euler:* $Eu = \Delta p/\rho v^2$	Eu
Froude number; *nombre de Froude:* $Fr = v(lg)^{-1/2}$	Fr
Grashof number; *nombre de Grashof:* $Gr = l^3 g\gamma\Delta T/\nu^2$	Gr
Weber number; *nombre de Weber:* $We = \rho v^2 l/\sigma$	We
Mach number; *nombre de Mach:* $Ma = v/c$	Ma
Knudsen number; *nombre de Knudsen:* $Kn = \lambda/l$	Kn
Strouhal number; *nombre de Strouhal:* $Sr = lf/v$	Sr

where the symbols used in the definitions denote, respectively, l: a characteristic length; v: a characteristic velocity; ΔT: a characteristic temperature difference; Δp: pressure difference; ρ: (mass) density; η: viscosity; ν:

*) The same name is also used for $\eta = \alpha/E$, where E denotes electric field strength.

†) These symbols are those recommended in the International Standard I.S.O. 31 Part XII.

kinematic viscosity (η/ρ); σ: surface tension; g: acceleration of free fall; γ: cubic expansion coefficient $(-\rho^{-1}(\partial\rho/\partial T)_p)$; λ: mean free path; f: a characteristic frequency; c: velocity of sound. $\Delta\rho/\rho$ equals $\gamma\Delta T$.

2. Transport of heat

Fourier number; *nombre de Fourier*: $Fo = a\Delta t/l^2$ *Fo*

Péclet number; *nombre de Péclet*: $Pe = vl/a = Re \cdot Pr$ *Pe*

Rayleigh number; *nombre de Rayleigh*:

$Ra = l^3 g\gamma\Delta T/va = Gr \cdot Pr$ *Ra*

Nusselt number; *nombre de Nusselt*: $Nu = hl/\lambda$ *Nu*

Stanton number; *nombre de Stanton*: $St = h/\rho v c_p = Nu \cdot Pe^{-1}$ *St*

where the symbols used in the definitions denote, respectively, l: a characteristic length; v: a characteristic velocity; Δt: a characteristic time interval; ΔT: a characteristic temperature difference; g: acceleration of free fall; ρ: (mass) density; η: viscosity; ν: kinematic viscosity (η/ρ); c_p: specific heat capacity at constant pressure; γ: cubic expansion coefficient $(-\rho^{-1}(\partial\rho/\partial T)_p)$; λ: thermal conductivity; a: thermal diffusivity $(\lambda/\rho c_p)$; h: coefficient of heat transfer (heat/(time × cross sectional area × temperature difference)).

3. Transport of matter in a binary mixture

Fourier number for mass transfer; *nombre de Fourier pour*
 transfert de masse: $Fo^* = D\Delta t/l^2 = Fo/Le$ *Fo**

Péclet number for mass transfer; *nombre de Péclet pour*
 transfert de masse: $Pe^* \times vl/D = Re \cdot Sc = Pe \cdot Le$ *Pe**

Grashof number for mass transfer; *nombre de Grashof pour*
 transfert de masse: $Gr^* = l^3 g\beta'\Delta x/\nu^2$ *Gr**

Nusselt number for mass transfer; *nombre de Nusselt pour*
 transfert de masse: $Nu^* = kl/\rho D$ *Nu**

Stanton number for mass transfer; *nombre de Stanton pour*
 transfert de masse: $St^* = k/\rho v = Nu^*/Pe^*$ *St**

where the symbols used in the definitions denote, respectively, l: a characteristic length; v: a characteristic velocity; Δt: a characteristic time interval; Δx: a characteristic difference of mole fraction; g: acceleration of free fall; ρ: (mass) density; ν: kinematic viscosity (η/ρ); β': $\beta' = -\rho^{-1}(\partial\rho/\partial x)_{T,p}$; D: diffusion coefficient; k: mass transfer coefficient (mass/(time × cross sectional area × mole fraction difference)); γ: cubic expansion coefficient $(-\rho^{-1}(\partial\rho/\partial T)_p)$. The quantity $-\Delta\rho/\rho$ equals $\gamma\Delta T + \beta'\Delta x$.

4. Dimensionless constants of matter

Prandtl number; *nombre de Prandtl*: $Pr = \nu/a$ *Pr*

Schmidt number; *nombre de Schmidt*: $Sc = \nu/D$ *Sc*

Lewis number; *nombre de Lewis*: $Le = a/D = Sc/Pr$ *Le*

where the symbols used in the definitions denote, respectively, ρ: (mass) density; η: viscosity; ν: kinematic viscosity (η/ρ); D: diffusion coefficient; c_p: specific heat capacity at constant pressure; λ: thermal conductivity; a: thermal diffusivity $(\lambda/\rho c_p)$.

5. Magnetohydrodynamics

Magnetic Reynolds number; *nombre de Reynolds*
 magnétique: $Rm = v\mu\sigma l$ *Rm*

Alfvén number; *nombre d'Alfvén*: $Al = v/v_A$ *Al*

Hartmann number; *nombre de Hartmann*: $Ha = Bl(\sigma/\rho\nu)^{1/2}$ *Ha*

Cowling number (second Cowling number); *nombre de Cowling*
 (deuxième nombre de Cowling): $Co = B^2/\mu\rho v^2 = Al^{-2}$ *Co, Co₂*

first Cowling number; *premier nombre de Cowling*:

$Co_1 = B^2 l\sigma/\rho v = Rm \cdot Co_2 = Ha^2/Re$ *Co₁* where the symbols used in the definitions denote, respectively, ρ: (mass) density; l: a characteristic length; υ: a characteristic velocity; η: viscosity; ν: kinematic viscosity (η/ρ); μ: magnetic permeability; B: magnetic flux density; σ: electric conductivity; υ_A: Alfvén speed $(B(\rho\mu)^{-1/2})$.

8 RECOMMENDED MATHEMATICAL SYMBOLS

8.1 General symbols

equal to; *égal à*	$=$
not equal to; *différent de*	$\neq$
identically equal to; *égal identiquement à*	$\equiv$
by definition equal to; *égal par définition à*	$\overset{\mathrm{def}}{=}$
corresponds to; *correspond à*	$\triangleq$
approximately equal to; *égal environ à*	$\approx$
asymptotically equal to, *asymptotiquement égal à*	$\simeq$
proportional to; *proportionnel à*	$\sim, \propto$
approaches; *tend vers*	$\rightarrow$
greater than; *supérieur à*	$>$
less than; *inférieur à*	$<$
much greater than; *tres supérieur à*	$\gg$
much less than; *tres inférieur à*	$\ll$
greater than or equal to; *superiéur ou egál a*	$\geqslant, \geqq, \geq$
less than or equal to; *inférieur ou égal à*	$\leqslant, \leqq, \leq$
plus; *plus*	$+$
minus; *moins*	$-$
plus or minus; *plus ou moins*	$\pm$
a multiplied by b; *a multiplié par b*	$ab, a \cdot b, a \times b$
a divided by b; *a divisé par b*	$a/b, \dfrac{a}{b}, ab^{-1}$
ratio of the circumference of a circle to its diameter; *rapport de la circonférence d'un cercle à son diamètre*	π
a raised to the power n; *a puissance n*	a^n
magnitude of a; *valeur absolue de a*	$|a|$
square root of a; *racine carrée de a*	$\sqrt{a}, \surd a, a^{1/2}$
mean value of a; *valeur moyenne de a*	$\bar{a}, \langle a \rangle$
factorial p; *factorielle p*	$p!$
binomial coefficient; *coefficient binomial: n!/p!(n − p)!*	$\binom{n}{p}$
infinity; *infini*	∞

8.2 Letter symbols

Letter symbols and letter expressions for *mathematical operations* should be written in *roman* (i.e. *upright*) type.

exponential of x; *exponentielle de x*	$\exp x, \mathrm{e}^x$
base of natural logarithms; *base des logarithmes népériens*	e
logarithm to the base a of x; *logarithme de base a de x*	$\log_a x$
natural logarithm of x; *logarithme népérien de x*	$\ln x, \log_\mathrm{e} x$
common logarithm of x; *logarithme décimal de x*	$\lg x, \log_{10} x$
binary logarithm of x; *logarithme binaire de x*	$\mathrm{lb}\, x, \log_2 x$
summation; *somme*	Σ
product; *produit*	Π
finite increase of x; *accroissement fini de x*	Δx †)

†) Greek capital delta, not triangle.

variation of x; *variation de x*	δx
total differential of x; *différentielle totale de x*	dx
function of x; *fonction de x*	$f(x), \mathrm{f}(x)$
composite function of f and g; *fonction composée de f et g*:	
$\quad (g \circ f)(x) = g(f,x)$	$g \circ f$
limit of $f(x)$; *limite de $f(x)$*	$\displaystyle\lim_{x \to a} f(x), \lim_{x \to a} f(x)$
derivative of f; *dérivée de f*	$\dfrac{df}{dx}, df/dx, f'$
partial derivative of f; *dérivée partielle de f*	$\dfrac{\partial f}{\partial x}, \partial f/\partial x, \partial_x f, f_x$
total differential of f; *différentielle totale de f*:	
$\quad df(x,y) = (\partial f/\partial x)_y\, dx + (\partial f/\partial y)_x\, dy$	df
variation of f; *variation de f*	δf
Dirac delta function; *fonction delta de Dirac*	
$\quad \delta(r) = \delta(x)\delta(y)\delta(z)$	$\delta(x), \delta(r)$
Kronecker delta symbol; *symbole delta de Kronecker*	δ_{ij}
Unit step function; *fonction unité*:	
$\quad \varepsilon(t) = 1$ for $t > 0$, $\varepsilon(t) = 0$ for $t < 0$	$\varepsilon(t)$
signum a; *signum a*: sgn $a = a/|a|$ for $a \neq 0$	sgn a
greatest integer $\leq a$; *le plus grand entier $\leq a$*	ent a

8.3 Circular and hyperbolic functions

sine of x, *sinus x*	sin x
cosine of x, *cosinus x*	cos x
tangent of x, *tangente x*	tan x, tg x
cotangent of x, *cotangente x*	cot x, ctg x
secant of x, *sécante x*	sec x
cosecant of x, *cosécante x*	cosec x

It is recommended to use for the *inverse circular functions* the symbolic expressions for the corresponding circular function preceded by the letters: arc

Examples: arcsin x, arccos x, arctan x, etc.

It is recommended to use for the *hyperbolic functions* the symbolic expressions for the corresponding circular function, followed by the letter: h

Examples; sinh x, cosh x, tanh x, etc.

It is recommended to use for the *inverse hyperbolic functions* the symbolic expression for the corresponding hyperbolic function preceded by the letters: ar

Examples: arsinh x, arcosh x, etc.

8.4 Complex quantities

imaginary unit; *unité imaginaire* ($i^2 = -1$)	i, j
real part of z; *partie réelle de z*	Re z, z'
imaginary part of z; *partie imaginaire de z*	Im z, z''
modulus of z; *module de z*	$|z|$
phase, argument of z; *phase, argument de z*: $z = |z| \exp i\varphi$	arg z, φ
complex conjugate of z, conjugate of z; *complexe conjugué de z, conjugué de z*	z^*

Remark: Sometimes the notation $\bar{z}$ is used for the complex conjugate of z.

8.5 Symbols for special values of periodic quantities

Symbols for special values of periodic quantities are given in table 3.

Table 3

	case A*)	case B*)
instantaneous value; *valeur instantanée*	x	x
r.m.s. value**); *valeur efficace*	$\tilde{x}$	X
maximum value†) *valeur maximale*	$\hat{x}$	$\hat{x}, \hat{X}$
mean value††) *valeur moyenne*	$\bar{x}, \langle x \rangle$	$\bar{x}, \langle x \rangle$

*) A is the case in which only a lower case or only a capital letter may be used for the quantity. B is the case in which both, lower case and capital letters, may be used for the same quantity.

**) The r.m.s. value is here defined as

$$\tilde{x} = \left(T^{-1} \int_0^T [x(t)]^2 dt \right)^{1/2};$$

the symbols x_{rms} and x_{eff} are also used.

†) The minimum value of x may be indicated by x_{min} or $\check{x}$.

††) The mean value is here defined as

$$\bar{x} = T^{-1} \int_0^T x(t) dt.$$

8.6 Vector calculus*)

vector; *vecteur*	a, A
absolute value; *valeur absolue*	$|A|, A$
unit vector; *vecteur unitaire:* $a/|a|$	e_a
unit vectors; *vecteurs unitaires*	e_x, e_y, e_z, i, j, k †)
scalar product of a and b; *produit scalaire de a et b*	$a \cdot b$
vector product of a and b; *produit vectoriel de a et b*	$a \times b \quad a \wedge b$
dyadic product of a and b; *produit dyadique de a et b*	ab
differential vector operator, nabla; *opérateur vectoriel, nabla*	$\partial/\partial r, \nabla$
gradient; *gradient*	$\mathrm{grad}\,\varphi, \nabla\varphi$
divergence; *divergence*	$\mathrm{div}\,A, \nabla \cdot A$
curl; *rotationnel*	$\mathrm{curl}\,A, \ \mathrm{rot}\,A, \ \nabla \times A$
Laplacian; *Laplacien*	$\triangle\varphi, \nabla^2\varphi$
Dalembertian; *Dalembertien*	$\Box\varphi$
second order tensor; *tenseur du second ordre*	$\mathbf{A}$

*) See also sections 1.2.3 and 1.3.2.
†) $1_x, 1_y, 1_z$ are also used.

scalar product of tensors **S** and **T**; *produit scalaire des tenseurs **S** et **T**:* $(\Sigma_{i,k} S_{ik} T_{ki})$ **S** : **T**

tensor product of tensors **S** and **T**; *produit tensoriel des tenseurs **S** et **T**:* $(\Sigma_k S_{ik} T_{kl})$ **S** · **T**

product of tensor **S** and vector **A**; *produit du tenseur **S** et vecteur **A**:* $(\Sigma_k S_{ik} A_k)$ **S** · **A**

8.7 Matrix calculus

matrix; *matrice* A

$$\begin{pmatrix} a_{11} \ldots a_{1n} \\ \vdots \qquad \vdots \\ a_{m1} \ldots a_{mn} \end{pmatrix}$$

product of A and B: *produit de A et B* AB

inverse of A: *inverse de A* A^{-1}

unit matrix; *matrice unité* E, I

transpose of matrix A: *matrice transposée de A:* $\tilde{A}_{ij} = A_{ji}$ $\tilde{A}$

complex conjugate of A: *matrice complexe conjugueé de A:* $A_{ij}^* = (A_{ij})^*$ A^*

Hermitian conjugate of A; *matrice conjuguée Hermitienne de A:* $A_{ij}^\dagger = A_{ji}^*$ $A^\dagger$

determinant of A; *déterminant de A* $\det A$

trace of A; *trace de A* $\mathrm{Tr}\, A$

Pauli matrices; *matrices de Pauli:* $\boldsymbol{\sigma},$

$$\sigma_x = \begin{pmatrix} 0 & 1 \\ 1 & 0 \end{pmatrix} \qquad \sigma_y = \begin{pmatrix} 0 & -i \\ i & 0 \end{pmatrix} \qquad \sigma_z = \begin{pmatrix} 1 & 0 \\ 0 & -1 \end{pmatrix}$$ $\sigma_x, \sigma_y, \sigma_z$
  $\sigma_1, \sigma_2, \sigma_3$

unit matrix; *matrice unité:* $I = \begin{pmatrix} 1 & 0 \\ 0 & 1 \end{pmatrix}$ I

Dirac (4×4) matrices; (4×4) *matrices de Dirac†):*

$$\alpha_x = \begin{pmatrix} 0 & \sigma_x \\ \sigma_x & 0 \end{pmatrix} \quad \alpha_y = \begin{pmatrix} 0 & \sigma_y \\ \sigma_y & 0 \end{pmatrix} \quad \alpha_z = \begin{pmatrix} 0 & \sigma_z \\ \sigma_z & 0 \end{pmatrix}$$ $\boldsymbol{\alpha}$
  $\alpha_x, \alpha_y, \alpha_z$

$$\beta = \begin{pmatrix} I & 0 \\ 0 & -I \end{pmatrix}$$ β

8.8 Theory of sets

is an element of; *est un élément de:* $x \in A$ $\in$

is not an element of, *n'est pas un élément de:* $x \notin A$ $\notin$

contains as element; *contient comme élément:* $A \ni x$ $\ni$

set of elements; *ensemble des éléments* $\{a_1, a_2, \ldots\}$

the set of integers; *ensemble des nombres entiers* $\mathbf{Z}$

the set of rational numbers; *ensemble des nombres rationnels* $\mathbf{Q}$

the set of real numbers; *ensemble des nombres réels* $\mathbf{R}$

the set of complex numbers; *ensemble des nombres complexes* $\mathbf{C}$

set of elements of A for which $p(x)$ is true; *ensemble des éléments de A pour lequels p(x) est vrai* $\{x \in A | p(x)\}$

is contained as subset in; *est contenu comme sous-ensemble dans:* $B \subseteq A$ $\subseteq (\subset)$

contains; *contient:* $A \supseteq B$ $\supseteq (\supset)$

is properly contained in; *est strictement contenu dans* $\subset (\subsetneq)$

†) Sometimes a different representation is used.

contains properly; *contient strictement*   ⊃ (⊃)

union; *réunion:* $A \cup B = \{x \mid x \in A \quad \text{or} \quad x \in B\}$ ∪
intersection; *intersection:*
   $A \cap B = \{x \mid x \in A \quad \text{and} \quad x \in B\}$ ∩
difference; *différence:* $A \backslash B = \{x \mid x \in A \quad \text{and} \quad x \notin B\}$ \
complement of; *complément de:* $\complement A = \{x \mid x \notin A\}$ ∁

8.9 Symbolic logic

conjunction: $p \wedge q$ means "*p* and *q*"; *conjonction: $p \wedge q$*
   signifie "p et q"   ∧
disjunction: $p \vee q$ means "*p* or *q* or both"; *disjonction:*
   $p \vee q$ signifie "p ou q ou les deux"   ∨
negation; *négation*   ¬
implication; *implication*   ⇒
   equivalence, bi-implication; *équivalence, bi-implication*   ⇔
universal quantifier; *quantificateur universel*   ∀
existential quantifier; *quantificateur existential*   ∃

9. INTERNATIONAL SYMBOLS FOR UNITS

9.1 Unit systems

In a system of physical quantities and equations between them, a certain number of quantities, which *by convention* are regarded as dimensionally independent, form a set of *base quantities* for the whole system. All other quantities can be defined as *derived quantities* in terms of the base quantities and can be expressed in terms of powers of the base quantities by algebraic relations.

A *coherent system of units* is a system, based on a certain set of base units well defined in terms of actual physical phenomena, in which all *derived units* are expressed as products of powers of the base units by algebraic relations, analogous to the corresponding quantities, dropping numerical factors.

The expression of a quantity as a product of powers of the base quantities (neglecting their vectorial or tensorial character and all numerical factors including their sign) is called the *dimensional product* or the *dimension* of the quantity with respect to the chosen set of base quantities or base dimensions. The exponents of the powers, to which the various base quantities or base dimensions are raised, are called the *dimensional exponents*.

Physical quantities, which have as their dimension a product of powers of the base quantities or base dimensions with all dimensional exponents equal to zero, are called *dimensionless quantities* or *quantities of the dimension 1*.

The number of the base units of the unit system, coherent with respect to the system of quantities and equations in question, is equal to that of the corresponding set of base quantities. The base units themselves are well defined samples of the base quantities: base quantity and corresponding base unit are of the same dimension.

A derived unit, i.e. the unit for a derived quantity of the coherent unit system, is formed as a product of powers of the base units, with the same exponents as those in the product of powers of the base quantities in the algebraic expression for the derived quantity in question, and without introducing numerical factors. Derived units and their symbols are expressed algebraically in terms of base units and their symbols, respectively, by means of the mathematical signs for multiplication and division. Several of these algebraic expressions for derived units in terms of base units can be replaced by special names and their symbols which can themselves be used to form other derived units and their symbols, respectively (see sections 9.2 and 9.3).

The values of dimensionless quantities, e.g. relative density, relative permeability or refractive index, are expressed by pure numbers. The corresponding unit, which is the ratio of a unit to itself, is the *dimensionless unit* of the coherent unit system and may be expressed by the number 1.

9.2 The International System of Units (SI)

The name *International System of Units* with the international abbreviation SI has been adopted by the Conférence Générale des Poids et Mesures (CGPM) in 1960. It is based on the seven base units (CGPM 1960 and 1971) listed in table 4.

Table 4. SI base units

base quantity; *grandeur de base*	SI base unit; *unité de base SI*	
	name; *nom*	symbol; *symbole*
length; *longueur*	metre; *mètre*	m
mass; *masse*	kilogram; *kilogramme*	kg
time; *temps*	second; *seconde*	s
electric current; *courant électrique*	ampere; *ampère*	A
thermodynamic temperature; *température thermodynamique*	kelvin; *kelvin*	K
amount of substance; *quantité de matière*	mole; *mole*	mol
luminous intensity; *intensité lumineuse*	candela; *candela*	cd

The SI base units have been defined by the CGPM in the following way:

(1) metre; *mètre*

The metre is the length equal to 1 650 763,73 wavelengths in vacuum of the radiation corresponding to the transition between the levels $2p_{10}$ and $5d_5$ of the krypton-86 atom. (11th CGPM (1960), Resolution 6.)

(2) kilogram; *kilogramme*

The kilogram is the unit of mass; it is equal to the mass of the international prototype of the kilogram. (1st CGPM (1889) and 3rd CGPM (1901), Declarations.)

(3) second; *seconde*

The second is the duration of 9 192 631 770 periods of the radiation corresponding to the transition between the two hyperfine levels of the ground state of the caesium-133 atom. (13th CGPM (1967), Resolution 1.)

(4) ampere; *ampère*

The ampere is that constant current which, if maintained in two straight parallel conductors of infinite length, of negligible circular cross section, and placed 1 metre apart in vacuum, would produce between these conductors a force equal to 2×10^{-7} newton per metre of length. (9th CGPM (1948), Resolutions 2 and 7.)

(5) kelvin; *kelvin*

The kelvin, unit of thermodynamic temperature, is the fraction 1/273,16 of the thermodynamic temperature of the triple point of water. (13th CGPM (1967), Resolution 4.)

The 13th CGPM (1967, Resolution 3) also decided that the unit kelvin and its symbol K should be used to express an interval or a difference of temperature. *Note:* In addition to the thermodynamic temperature (symbol T), expressed in kelvins, use is also made of Celsius temperature (symbol t) defined by the equation

$$t = T - T_0$$

where $T_0 = 273,15$ K by definition. Celsius temperature is expressed in degree Celsius; *degré Celsius* (symbol °C). The unit "degree Celsius" is equal to the unit "kelvin" and an interval or a difference of Celsius temperature may also be expressed in degrees Celsius.

(6) mole; *mole*

The mole is the amount of substance of a system which contains as many elementary entities as there are atoms in 0.012 kilogram of carbon 12.
When the mole is used, the elementary entities must be specified and may be atoms, molecules, ions, electrons, other particles, or specified groups of such particles.
The mole is a base unit of the International System of Units. (14th CGPM (1971), Resolution 3.)

(7) candela; *candela*

The candela is the luminous intensity, in the perpendicular direction, of a surface of 1/600 000 square metre of a black body at the temperature of freezing platinum under a pressure of 101 325 newtons per square metre. (13th CGPM (1967), Resolution 5.)

The coherent units of the system based on these base units are called: *SI units.* *Derived SI units having special names and symbols* are listed in table 5.

Table 5. Derived SI units with special names

quantity; *grandeur*	derived SI unit; *unité SI dérivée*			
	name; *nom*	symbol; *symbole*	expression in terms of base units; *expression en unités de base*	expression in terms of other SI units; *expression en d'autres unités SI*
plane angle; *angle plan*	radian; *radian*	rad	$m \cdot m^{-1}$	
solid angle; *angle solide*	steradian; *stéradian*	sr	$m^2 \cdot m^{-2}$	
frequency; *fréquence*	hertz; *hertz*	Hz	s^{-1}	
force; *force*	newton; *newton*	N	$m \cdot kg \cdot s^{-2}$	J/m
pressure; *pression*	pascal; *pascal*	Pa	$m^{-1} \cdot kg \cdot s^{-2}$	N/m^2
energy, work, quantity of heat; *énergie, travail, quantité de chaleur*	joule; *joule*	J	$m^2 \cdot kg \cdot s^{-2}$	$N \cdot m$
power, radiant flux; *puissance, flux énergétique*	watt; *watt*	W	$m^2 \cdot kg \cdot s^{-3}$	J/s
quantity of electricity, electric charge; *quantité d'électricité, charge électrique*	coulomb; *coulomb*	C	$s \cdot A$	$A \cdot s$
electric potential, potential difference, electromotive force; *tension électrique, différence de potentiel, force électromotrice*	volt; *volt*	V	$m^2 \cdot kg \cdot s^{-3} \cdot A^{-1}$	W/A
capacitance; *capacité électrique*	farad; *farad*	F	$m^{-2} \cdot kg^{-1} \cdot s^4 \cdot A^2$	C/V
electric resistance; *résistance électrique*	ohm; *ohm*	Ω	$m^2 \cdot kg \cdot s^{-3} \cdot A^{-2}$	V/A
conductance; *conductance*	siemens; *siemens*	S	$m^{-2} \cdot kg^{-1} \cdot s^3 \cdot A^2$	A/V
magnetic flux; *flux d'induction magnétique*	weber; *weber*	Wb	$m^2 \cdot kg \cdot s^{-2} \cdot A^{-1}$	$V \cdot s$
magnetic flux density; *induction magnétique*	tesla; *tesla*	T	$kg \cdot s^{-2} \cdot A^{-1}$	Wb/m^2
inductance; *inductance*	henry; *henry*	H	$m^2 \cdot kg \cdot s^{-2} \cdot A^{-2}$	Wb/A
luminous flux; *flux lumineux*	lumen; *lumen*	lm		$cd \cdot sr$
illuminance; *éclairement lumineux*	lux; *lux*	lx		$m^{-2} \cdot cd \cdot sr$

Table 5 (Continued)

| quantity; *grandeur* | derived SI unit; *unité SI dérivée* | | | |
	name; *nom*	symbol; *symbole*	expression in terms of base units; *expression en unités de base*	expression in terms of other SI units; *expression en d'autres unités SI*
activity; *activité*	becquerel; *becquerel*	Bq	s^{-1}	
absorbed dose; *dose absorbée*	gray; *gray*	Gy	$m^2 \cdot s^{-2}$	J/kg

Although it might be thought that SI units can only be base units or derived units, the 11th Conférence Générale des Poids et Mesures (1960) admitted a third class of SI units called *supplementary units*, for which it declined to state, whether they were base units or derived units. To the class of supplementary units belong only the radian as SI unit for plane angle and the steradian as SI unit for solid angle; they may be regarded either as base units or as derived units.

In physics, radian and steradian are, in general, regarded as derived units (see table 5). In some special fields, e.g. in electromagnetic radiation and light or in particle scattering problems, the steradian is sometimes *formally* treated as a base unit; then the symbol sr must *not* be replaced by the number 1.

The usage of the SI units and their decimal multiples and sub-multiples formed by using the prefixes given in table 1 (see section 2.2.1) has been especially recommended.

Remark: Among the SI base units the unit of mass is the only whose name, for historical reasons, contains a prefix. Names of decimal multiples and sub-multiples of the SI unit of mass are formed by attaching prefixes to the word "gram" (Comité International des Poids et Mesures (CIPM) 1967, Recommendation 2).

Several *sub-systems* of the SI are used in different fields of science and technology. The two most important are:

(a) *The metre-kilogram-second system (MKS system)*
The MKS system is a coherent unit system for mechanics, based on the three base units metre, kilogram and second for the three base quantities length, mass and time.

(b) *The metre-kilogram-second-ampere system (MKSA system)*
The MKSA system is a coherent unit system for mechanics, electricity and magnetism, based on the four base units metre, kilogram, second and ampere for the four base quantities length, mass, time and electric current. This system has been given the name *"Giorgi system"* by the International Electrotechnical Commission in 1935.

9.3 The centimetre-gram-second system (CGS system)

The CGS system is a coherent system of units based on the three base units listed in table 6.

Table 6. CGS base units

base quantity; *grandeur de base*	base unit; *unité de base*	
	name; *nom*	symbol; *symbole*
length; *longueur*	centimetre; *centimètre*	cm
mass; *masse*	gram; *gramme*	g
time; *temps*	second; *seconde*	s

Derived CGS units having special names and symbols are listed in table 7.

Table 7. Derived CGS units with special name.

quantity; *grandeur*	derived CGS unit; *unité CGS dérivée*		
	name; *nom*	symbol; *symbole*	expression in terms of base units; *expression en unités de base*
force; *force*	dyne	dyn	$cm \cdot g \cdot s^{-2}$
energy; *énergie*	erg	erg	$cm^2 \cdot g \cdot s^{-2}$
viscosity; *viscosité*	poise	P	$cm^{-1} \cdot g \cdot s^{-1}$
kinematic viscosity; *viscosité cinématique*	stokes	St	$cm^2 \cdot s^{-1}$
acceleration of free fall; *accélération de la pesanteur*	gal	Gal	$cm \cdot s^{-2}$

The CGS system enlarged by the kelvin (K) as unit of thermodynamic temperature (see section 9.2) and the mole (mol) as unit of amount of substance (see section 9.2) or by the candela (cd) as unit of luminous intensity (see section 9.2) is also used in *mechanics including thermodynamics* or *photometry*, respectively.

Derived units (from cm, g, s and cd, including sr, see section 9.2), in the field of *photometry, having special names and symbols* are listed in table 8.

Table 8. CGS units in photometry with special name.

quantity; *grandeur*	derived unit; *unité dérivée*		
	name; *nom*	symbol; *symbole*	expression; *expression*
luminance; *luminance*	stilb	sb	$cm^{-2} \cdot cd$
illuminance; *éclairement lumineux*	phot	ph	$cm^{-2} \cdot cd \cdot sr$

ALPHABET TABLE

Greek letter	Greek name	English equivalent	RUSSIAN letter	English equivalent
A α	Alpha	(ä)	А а	(ä)
B β	Beta	(b)	Б б	(b)
Γ γ	Gamma	(g)	В в	(v)
Δ δ	Delta	(d)	Г г	(g)
E ε	Epsilon	(a)	Д д	(d)
Z ζ	Zeta	(z)	Е е	(ye)
H η	Eta	(a)	Ж ж	(zh)
Θ θ	Theta	(th)	З з	(z)
I ι	Iota	(ē)	И и	(i, ē)
K κ	Kappa	(k)	Й й	(ē) 7
Λ λ	Lambda	(l)	К к	(k)
M μ	Mu	(m)	Л л	(l)
N ν	Nu	(n)	М м	(m)
Ξ ξ	Xi	(ks)	Н н	(n)
O o	Omicron	(a)	О о	(ô, o)
Π π	Pi	(p)	П п	(p)
P ρ	Rho	(r)	Р р	(r)
Σ σ s	Sigma	(s)	С с	(s)
T τ	Tau	(t)	Т т	(t)
Υ υ	Upsilon	(ü, ōō)	У у	(ōō)
Φ φ	Phi	(f)	Ф ф	(f)
X χ	Chi	(H)	Х х	(kh)
Ψ ψ	Psi	(ps)	Ц ц	(ts)
Ω ω	Omega	(ō)	Ч ч	(ch)
			Ш ш	(sh)
			Щ щ	(shch)
			Ъ ъ	8
			Ы ы	(ē)
			Ь ь	9
			Э э	(e)
			Ю ю	(ū)
			Я я	(yä)

AMERICAN STANDARD ABBREVIATIONS FOR SCIENTIFIC AND ENGINEERING TERMS

Reproduced by permission of The American Society of Mechanical Engineers

Introductory Notes

Scope and Purpose

1. The Executive Committee of the Sectional Committee on Scientific and Engineering Symbols and Abbreviations has made the following distinction between symbols and abbreviations: Letter symbols are letters used to represent magnitudes or physical quantities in equations and mathematical formulas. Abbreviations are shortened forms of names or expressions employed in texts and tabulations, and should not be used in equations.

Fundamental Rules

2. Abbreviations should be used sparingly in text and with due regard to the context and to the training of the reader. Terms denoting units of measurement should be abbreviated in the text only when preceded by the amounts indicated in numerals; thus "several inches," "one inch," "12 in." In tabular matter, specifications, maps, drawings, and texts for special purposes, the use of abbreviations should be governed only by the desirability of conserving space.
3. Short words such as ton, day, and mile should be spelled out.
4. Abbreviations should not be used where the meaning will not be clear. In case of doubt, spell out.
5. The same abbreviation is used for both singular and plural, as "bbl" for barrel and barrels.
6. The use of conventional signs for abbreviations in text is not recommended; thus "per," not /; "lb," not #; "in," not ". Such signs may be used sparingly in tables and similar places for conserving space.
7. The period should be omitted except in cases where the omission would result in confusion.
8. The letters of such abbreviations as ASA should not be spaced (not A S A).
9. The use in text of exponents for the abbreviations of square and cube and of the negative exponents for terms involving "per" is not recommended. The superior figures are usually not available on the keyboards of typesetting and linotype machines and composition is therefore delayed. There is also the likelihood of confusion with footnote reference numbers. These shorter forms are permissible in tables and are sometimes difficult to avoid in text.
10. A sentence should not begin with a numeral followed by an abbreviation. Abbreviations for names of units are to be used only after numerical values, such as 25 ft or 110 v.

Abbreviations*

absolute	abs
acre	spell out
acre-foot	acre-ft
air horsepower	air hp
alternating-current (as adjective)	a-c
ampere	amp
ampere-hour	amp-hr
amplitude, an elliptic function	am.
Angstrom unit	A
antilogarithm	antilog
atmosphere	atm
atomic weight	at. wt
average	avg
avoirdupois	avdp
azimuth	az or α
barometer	bar
barrel	bbl
Baumé	Bé
board feet (feet board measure)	fbm
boiler pressure	spell out
boiling point	bp
brake horsepower	bhp
brake horsepower-hour	bhp-hr
Brinell hardness number	Bhn
British thermal unit[1]	Btu or B
bushel	bu
calorie	cal
candle	c
candle-hour	c-hr
candlepower	cp
cent	c or ¢
center to center	c to c
centigram	cg
centiliter	cl
centimeter	cm
centimeter-gram-second (system)	cgs
chemical	chem
chemically pure	cp
circular	cir
circular mils	cir mils
coefficient	coef
cologarithm	colog
concentrate	conc

conductivity	cond
constant	const
continental horsepower	cont hp
cord	cd
cosecant	csc
cosine	cos
cosine of the amplitude, an elliptic function	cn
cost, insurance, and freight	cif
contangent	cot
coulomb	spell out
counter electromotive force	cemf
cubic	cu
cubic centimeter cu cm, cm³ (liquid, meaning milliliter. ml)	
cubic foot	cu ft
cubic feet per minute	cfm
cubic feet per second	cfs
cubic inch	cu in.
cubic meter	cu m or m³
cubic micron	cu μ or cu mu or μ^3
cubic millimeter	cu mm or mm³
cubic yard	cu yd
current density	spell out
cycles per second	spell out or c
cylinder	cyl
day	spell out
decibel	db
degree[1]	deg or °
degree centigrade	C
degree Fahrenheit	F
degree Kelvin	K
degree Réaumur	R
delta amplitude, an elliptic function	dn
diameter	diam
direct-current (as adjective)	d-c
dollar	$
dozen	doz
dram	dr
efficiency	eff
electric	elec
electromotive force	emf
elevation	el
equation	eq

---

* These forms are recommended for readers whose familiarity with the terms used makes possible a maximum of abbreviations. For other classes of readers editors may wish to use less contracted combinations made up from this list. For example, the list gives the abbreviation of the term "feet per second" as "fps". To some readers ft per sec will be more easily understood. ¹Abbreviation recommended by the A.S.M.E. Power Test Codes Committee. B = 1 Btu, kB = 1000 Btu, mB = 1,000,000 Btu. The A.S.H. & V.E. recommends the use of Mb = 1000 Btu and Mbh = 1000 Btu per hr.

¹ There are circumstances under which one or the other of these forms is preferred. In general the sign ° is used where space conditions make it necessary, as in tabular matter, and when abbreviations are cumbersome, as in some angular measurements, i.e., 59° 23' 42". In the interest of simplicity and clarity the Committee has recommended that the abbreviation for the temperature scale, F, C, K, etc., always be included in expressions for numerical temperatures, but, wherever feasible, the abbreviation for "degree" be omitted; as 69 F.

external	ext
farad	spell out or f
feet board measure (board feet)	fbm
feet per minute	fpm
feet per second	fps
fluid	fl
foot	ft
foot-candle	ft-c
foot-Lambert	ft-L
foot-pound	ft-lb
food-pound-second (system)	fps
foot-second (see cubic feet per second)	
franc	fr
free aboard ship	spell out
free alongside ship	spell out
free on board	fob
freezing point	fp
frequency	spell out
fusion point	fnp
gallon	gal
gallons per minute	gpm
gallons per second	gps
grain	spell out
gram	g
gram-calorie	g-cal
greatest common divisor	gcd
haversine	hav
hectare	ha
henry	H
high-pressure (adjective)	h-p
hogshead	hhd
horsepower	hp
horsepower-hour	hp-hr
hour	hr
hour (in astronomical tables)	h
hundred	C
hundredweight (112 lb)	cwt
hyperbolic cosine	cosh
hyperbolic sine	sinh
hyperbolic tangent	tanh
inch	in
inch-pound	in-lb
inches per second	ips
indicated horsepower	ihp
indicated horsepower-hour	ihp-hr
inside diameter	ID
intermediate-pressure (adjective)	i-p
internal	int
joule	J
kilocalorie	kcal
kilocycles per second	kc
kilogram	kg
kilogram-calorie	kg-cal
kilogram-meter	kg-m
kilograms per cubic meter or kg/m³	kg per cu m
kilograms per second	kgps
kiloliter	kl
kilometer	km
kilometers per second	kmps
kilovolt	kv
kilovolt-ampere	kva
kilowatt	kw
kilowatthour	kwhr
lambert	L
latitude	lat or ϕ
least common multiple	lcm
linear foot	lin ft
liquid	liq
lira	spell out
liter	l
logarithm (common)	log
logarithm (natural)	log or ln
longitude	long. or λ
low-pressure (as adjective)	l-p
lumen	l*
lumen-hour	l-hr*
lumens per watt	lpw
mass	spell out
mathematics (ical)	math
maximum	max
mean effective pressure	mep
mean horizontal candlepower	mhcp
megacycle	spell out
megohm	spell out

melting point	mp
meter	m
meter-kilogram	m-kg
mho	spell out
microampere	μa or mu a
microfarad	μf
microinch	μin.
micromicrofarad	$\mu\mu$f
micromicron	$\mu\mu$ or mu mu
micron	μ or mu
microvolt	μv
microwatt	μw or mu w
mile	spell out
miles per hour	mph
miles per hour per second	mphps
milliampere	ma
milligram	mg
millihenry	mh
millilambert	mL
milliliter	ml
millimeter	mm
millimicron	mμ or m mu
million	spell out
million gallons per day	mgd
millivolt	mv
minimum	min
minute	min
minute (angular measure)	
minute (time) (in astronomical tables)	m
mole	spell out
molecular weight	mol. wt
month	spell out
National Electrical Code	NEC
ohm	spell out or Ω
ohm-centimeter	ohm-cm
ounce	oz
ounce-foot	oz-ft
ounce-inch	oz-in
outside diameter	OD
parts per million	ppm
peck	pk
penny (pence — New British)	p.
pennyweight	dwt
per	(see Fundamental Rules)
peso	spell out
pint	pt
potential	spell out
potential difference	spell out
pound	lb
pound-foot	lb-ft
pound-inch	lb-in.
pound sterling	£
pounds per brake horsepower-hour	lb per bhp-hr
pounds per cubic foot	lb per cu ft
pounds per square foot	psf
pounds per square inch	psi
pounds per square inch absolute	psia
power factor	spell out or pf
quart	qt
radian	spell out
reactive kilovolt-ampere	kvar
reactive volt-ampere	var
revolutions per minute	rpm
revolutions per second	rps
rod	spell out
root mean square	rms
secant	sec
second	sec
second (angular measure)	"
second-foot (see cubic feet per second)	
second (time) (in astronomical tables)	s
shaft horsepower	shp
shilling	s
sine	sin
sine of the amplitude, an elliptic function	sn
specific gravity	sp gr
specific heat	sp ht
spherical candle power	scp
square	sq
square centimeter	sq cm or cm²
square foot	sq ft
square inch	sq in.
square kilometer	sq km or km²
square meter	sq m or m²
square micron	sq μ or sq mu or μ²

* The International Commission on Illumination has changed the symbol for lumen to lm, and the symbol for lumen-hour to lm-hr. This nomenclature is used in American Standard for Illuminating Engineering Nomenclature and Photometric Standards (ASA Z7.1-1942).

AMERICAN STANDARD ABBREVIATIONS (Continued)

square millimeter	sq mm or mm²	versed sine	vers
square root of mean square	rms	volt	v
standard	std	volt-ampere	va
stere	s	volt-coulomb	spell out
tangent	tan	watt	w
temperature	temp	watthour	whr
tensile strength	ts	watts per candle	wpc
thousand	M	week	spell out
thousand foot-pounds	kip-ft	weight	wt
thousand pound	kip	yard	yd
ton	spell out	year	yr
ton-mile	spell out		

ABBREVIATIONS OF COMMON UNITS OF WEIGHT AND MEASURE

From NBS Miscellaneous Publication No. 233.
Spelling and Abbreviations of Units

The spelling of the names of units as adopted by the National Bureau of Standards is that given in the list below. The spelling of the metric units is in accordance with that given in the law of July 28, 1866, legalizing the metric system in the United States.

Following the name of each unit in the list below is given the abbreviation which the Bureau has adopted. Attention is particularly called to the following principles:

1. The period is omitted after all abbreviations of units, except where the abbreviation forms an English word.
2. The exponents "²" and "³" are used to signify "square" and "cubic", respectively, instead of the abbreviations "sq" or "cu," which are, however, frequently used in technical literature for the United States Customary units. In conformity with this principle the abbreviation for cubic centimeter is "cm³" (instead of "cc" or "c cm"). The term "cubic centimeter," as used in chemical work, is, in fact, a misnomer, since the unit actually used is the "milliliter" of which "ml" is the correct abbreviation.
3. The use of the same abbreviation for both singular and plural is recommended. This practice is already established in expressing metric units and is in accordance with the spirit and chief purpose of abbreviations.
4. It is also suggested that, unless all the text is printed in capital letters, only small letters be used for abbreviations, except in such case as, A for angstrom, etc., where the use of capital letters is general.

LIST OF THE MOST COMMON UNITS OF WEIGHT AND MEASURE AND THEIR ABBREVIATIONS

Unit	Abbreviation	Unit	Abbreviation	Unit	Abbreviation
acre	acre	dram, avoirdupois	dr avdp	ounce, apothecaries	oz ap or ʒ
angstrom	A	dram, fluid	fl dr	ounce, avoirdupois	oz avdp
are	a	fathom	fath	ounce, fluid	fl oz
avoirdupois	avdp	foot	ft	ounce, troy	oz t
barrel	bbl	furlong	fur.	peck	pk
board foot	fbm	gallon	gal	pennyweight	dwt
bushel	bu	grain	grain	pint	pt
carat	c	gram	g	pound	lb
centare	ca	hectare	ha	pound, apothecaries	lb ap
centigram	cg	hectogram	hg	pound, avoirdupois	lb avdp
centiliter	cl	hectoliter	hl	pound, troy	lb t
centimeter	cm	hectometer	hm	quart	qt
chain	ch	hogshead	hhd	rod	rd
cubic centimeter	cm³	hundredweight	cwt	scruple, apothecaries	s ap or ℈
cubic decimeter	dm³	inch	in.	square centimeter	cm²
cubic dekameter	dkm³	kilogram	kg	square chain	ch²
cubic foot	ft³	kiloliter	kl	square decimeter	dm²
cubic hectometer	hm³	kilometer	km	square dekameter	dkm²
cubic inch	in.³	link	li	square foot	ft²
cubic kilometer	km³	liquid	liq	square hectometer	hm²
cubic meter	m³	liter	liter	square inch	in.²
cubic mile	mi³	meter	m	square kilometer	km²
cubic millimeter	mm³	metric ton	t	square link	li²
cubic yard	yd³	microgram*	μg	square meter	m²
decigram	dg	microinch	μin.	square mile	mi²
deciliter	dl	microliter*	μl	square millimeter	mm²
decimeter	dm	micron	μ	square rod	rd²
decistere	ds	mile	mi	square yard	yd²
dekagram	dkg	milligram	mg	stere	s
dekaliter	dkl	milliliter	ml	ton	ton
dekameter	dkm	millimeter	mm	ton, metric	t
dekastere	dks	millimicron	mμ	troy	t
dram	dr	minim	min or ℔	yard	yd
dram, apothecaries	dr ap or ʒ	ounce	oz		

* The abbreviations γ and λ for microgram and microliter, respectively have been advocated by some authorities.

CONVERSION FACTORS

(From various U. S. Government and IUPAC publications and from calculations based on values given in these publications)

To convert from	To	Multiply by	To convert from	To	Multiply by
Abamperes	Amperes	10	Acre-inches	Gallons (U.S.)	27154.286
"	E.M. cgs. units of current	1	Amperes	Abamperes	0.1
"	E.S. cgs. units	2.997930×10^{10}	"	Amperes (Int.)	1.000165
"	Faradays (chem.)/sec	1.036377×10^{-4}	"	Cgs. units of current	1
"	Faradays (phys.)/sec	1.036086×10^{-4}	"	Mks. units of current	1
"	Statamperes	2.997930×10^{10}	"	Coulombs/sec	1
Abamperes/cm	E.M. cgs. units of surface charge density	1	"	Coulombs (Int.)/sec	1.000165
"	E.S. cgs. units	2.997930×10^{10}	"	Faradays (chem.)/sec	1.036377×10^{-5}
Abamperes/sq. cm.	Amperes/circ. mil	5.0670748×10^{-5}	"	Faradays (phys.)/sec	1.036086×10^{-5}
"	Amperes/sq. cm	10	"	Statamperes	2.997930×10^{9}
"	Amperes/sq. inch	64.516	Amperes (Int.)	Amperes	0.999835
Abampere-turns	Ampere-turns	10	"	Coulombs/sec	0.999835
Abampere-turns/cm	Ampere-turns/cm	10	"	Coulombs (Int.)/sec	1
Abcoulombs	Ampere-hours	0.0027777	"	Faradays (chem.)/sec*	1.03623×10^{-5}
"	Coulombs	10	"	Faradays (phys.)/sec*	1.03592×10^{-5}
"	Electronic charges	6.24196×10^{19}	Amperes/meter	Cgs. units of surface current density	0.01
"	E.M. cgs. units of charge	1	"	E.M. cgs. units	0.001
"	E.S. cgs. units	2.997930×10^{10}	"	E.S. cgs. units	2.997930×10^{7}
"	Faradays (chem.)	1.036377×10^{-4}	"	Mks. units	1
"	Faradays (phys.)	1.036086×10^{-4}	Amperes/sq. meter	Cgs. units of volume current density	0.0001
"	Statcoulombs	2.997930×10^{10}	"	E.M. cgs. units	1×10^{-5}
Abfarads	E.M. cgs. units of capacitance	1	"	E.S. cgs. units	299793.0
"	E.S. cgs. units	8.987584×10^{20}	"	Mks. units	1
"	Farads	1×10^{9}	Amperes/sq. mil	Abamperes/sq. cm	15500.031
"	Microfarads	1×10^{15}	"	Amperes/sq. cm	1.5500031×10^{5}
"	Statfarads	8.987584×10^{20}	Ampere-hours	Abcoulombs	360
Abhenries	E.M. cgs. units of induction	1	"	Coulombs	3600
"	E.S. cgs. units	1.112646×10^{-21}	"	Faradays (chem.)*	0.0373096
"	Henries	1×10^{-9}	"	Faradays (phys.)*	0.0372551
Abmhos	E.M. cgs. units of conductance	1	Ampere-turns	Cgs. units of magnetomotive force	1.2566371
"	E.S. cgs. units	8.987584×10^{20}	"	E.M. cgs. units	1.2566371
"	Megamhos	1000	"	E.S. cgs. units	3.767310×10^{10}
"	Mhos	1×10^{9}	"	Gilberts	1.2566371
"	Statmhos	8.987584×10^{20}	Ampere-turns/weber	Cgs. units of reluctance	1.256637×10^{-8}
Abohms	E.M. cgs. units of resistance	1	"	E.M. cgs. units	1.256637×10^{-8}
"	Megohms	1×10^{-15}	"	E.S. cgs. units	1.129413×10^{13}
"	Microhms	0.001	"	Gilberts/maxwell	1.256637×10^{-8}
"	Ohms	1×10^{-9}	Ångström units	Centimeters	1×10^{-8}
"	Statohms	1.112646×10^{-21}	"	Inches	3.9370079×10^{-9}
Abohm-cm	Circ. mil-ohms/ft	0.0060153049	"	Microns	0.0001
"	E.M. cgs. units of resistivity	1	"	Millimicrons	0.1
"	Microhm-inches	0.00039370079	"	Wave length of orange-red line of krypton 86	0.000165076373
"	Ohm-cm	1×10^{-9}	"	Wave length of red line of cadmium	0.000155316413
Abvolts	Microvolts	0.01	Ares	Acres	0.024710538
"	Millivolts	1×10^{-5}	"	Sq. dekameters	1
"	Volts	1×10^{-8}	"	Sq. feet	1076.3910
"	Volts (Int.)	9.99670×10^{-9}	"	Sq. ft. (U.S. Survey)	1076.3867
Abvolts/cm	E.M. cgs. units of electric field intensity	1	"	Sq. meters	100
"	E.S. cgs. units	3.335635×10^{-11}	"	Sq. miles	3.8610216×10^{-5}
"	Volts/cm	1×10^{-8}	Atmospheres	Bars	1.01325
"	Volts/inch	2.54×10^{-8}	"	Cm. of Hg (0°C.)	76
"	Volts/meter	1×10^{-6}	"	Cm. of H_2O (4°C.)	1033.26
Acres	Sq. cm	40468564	"	Dynes/sq. cm	1.01325×10^{6}
"	Sq. ft.	43560	"	Ft. of H_2O (39.2°F.)	33.8995
"	Sq. ft. (U.S. Survey)	43559.826	"	Grams/sq. cm	1033.23
"	Sq. inches	6272640	"	In. of Hg (32°F.)	29.9213
"	Sq. kilometers	0.0040468564	"	Kg./sq. cm	1.03323
"	Sq. links (Gunter's)	1×10^{5}	"	Mm. of Hg (0°C.)	760
"	Sq. meters	4046.8564	"	Pounds/sq. inch	14.6960
"	Sq. miles (statute)	0.0015625	"	Tons (short)/sq. ft	1.05811
"	Sq. perches	160	"	Torrs	760
"	Sq. rods	160	Atomic mass units (chem.)*	Electron volts	9.31395×10^{8}
"	Sq. yards	4840	"	Grams*	1.66024×10^{-24}
Acre-feet	Cu. feet	43560	Atomic mass units (phys.)*	Electron volts	9.31141×10^{8}
"	Cu. meters	1233.4818	"	Grams*	1.65979×10^{-24}
"	Cu. yards	1613.333			
Acre-inches	Cu. feet	3630			
"	Cu. meters	102.79033			

* For factors for C = 12 scale or those which can be derived from same see table on Value for General Physical Constants.

To convert from	To	Multiply by	To convert from	To	Multiply by
Bags (Brit.)	Bushels (Brit.)	3	B.t.u.	Kw.-hours (Int.)	0.000292827
Barns	Sq. cm.	1×10^{-24}	"	Liter-atm	10.4053
Barrels (Brit.)*	Bags (Brit.)	1.5	"	Tons of refrig. (U.S. std.)	3.46995×10^{-6}
"	Barrels (U.S., dry)	1.415404	"	Watt-seconds	1054.35
"	Barrels (U.S., liq.)	1.372513	"	Watt-seconds (Int.)	1054.18
"	Bushels (Brit.)	4.5	B.t.u. (IST.)	B.t.u.	1.00065
"	Bushels (U.S.)	4.644253	B.t.u. (mean)	B.t.u.	1.00144
"	Cu. feet	5.779568	"	B.t.u. (IST.)	1.00078
"	Cu. meters	0.1636591	"	B.t.u. (39°F.)	0.996415
"	Gallons (Brit.)	36	"	B.t.u. (60°F.)	1.00113
"	Liters	163.6546	"	Hp.-hours	0.000393317
Barrels (petroleum, U.S.)	Cu. feet	5.614583	"	Joules	1055.87
"	Gallons (U.S.)	42	"	Kg.-meters	107.669
"	Liters	158.98284	"	Kw.-hours	0.000293297
Barrels (U.S., dry)	Barrels (U.S. liq.)	0.969696	"	Kw.-hours (Int.)	0.000293248
"	Bushels (U.S.)	3.2812195	"	Liter-atm	10.4203
"	Cu. feet	4.083333	"	Watt-hours	0.293297
"	Cu. inches	7056	"	Watt-hours (Int.)	0.293248
"	Cu. meters	0.11562712	B.t.u. (39°F.)	B.t.u.	1.00504
"	Quarts (U.S., dry)	105	"	B.t.u. (IST.)	1.00439
Barrels (U.S., liq.)	Barrels (U.S., dry)	1.03125	"	B.t.u. (mean)	1.00360
"	Barrels (wine)	1	"	B.t.u. (60°F)	1.00473
"	Cu. feet	4.2109375	"	Joules	1059.67
"	Cu. inches	7276.5	B.t.u. (60°F.)	B.t.u.	1.00031
"	Cu. meters	0.11924047	"	B.t.u. (IST.)	0.999657
"	Gallons (Brit.)	26.22925	"	B.t.u. (mean)	0.998873
"	Gallons (U.S., liq.)	31.5	"	B.t.u. (39°F.)	0.995291
"	Liters	119.23713	B.t.u./hr	Cal., kg./hr	0.251996
Bars	Atmospheres	0.986923	"	Ergs/sec	2.928751×10^{6}
"	Baryes	1×10^{6}	"	Foot-pounds/hr	777.649
"	Cm. of Hg (0°C.)	75.0062	"	Horsepower	0.000392752
"	Dynes/sq. cm.	1×10^{6}	"	Horsepower (boiler)	2.98563×10^{-5}
"	Ft. of H$_2$O (60°F.)	33.4883	"	Horsepower (electric)	0.000392594
"	Grams/sq. cm.	1019.716	"	Horsepower (metric)	0.000398199
"	In. of Hg (32°F.)	29.5300	"	Kilowatts	0.000292875
"	Kg./sq. cm.	1.019716	"	Lb. ice melted/hr	0.0069714
"	Millibars	1000	"	Tons of refrig. (U.S. comm.)	8.32789×10^{-5}
"	Pounds/sq. inch	14.5038	"	Watts	0.292875
Baryes	Atmospheres	9.86923×10^{-7}	B.t.u./min	Cal., kg./min	0.251996
"	Bars	1×10^{-6}	"	Ergs/sec	1.75725×10^{8}
"	Dynes/sq. cm.	1	"	Foot-pounds/min	777.649
"	Grams/sq. cm.	0.001019716	"	Horsepower	0.0235651
"	Millibars	0.001	"	Horsepower (boiler)	0.00179138
Bels	Decibels	10	"	Horsepower (electric)	0.0235556
Board feet	Cu. cm.	2359.7372	"	Horsepower (metric)	0.0238920
"	Cu. feet	0.083333	"	Joules/sec	17.5725
"	Cu. inches	144	"	Kg.-meters/min	107.514
Bolts of cloth	Linear feet	120	"	Kilowatts	0.0175725
"	Meters	36.576	"	Lb. ice melted/hr	0.41828
Bougie decimales	Candles (Int.)	1.00	"	Tons of refrig. (U.S. comm.)	0.00499673
B.t.u.	B.t.u. (IST.)**	0.999346	"	Watts	17.5725
"	B.t.u. (mean)	0.998563	B.t.u. (mean)/min	B.t.u. (mean)/hr	60
"	B.t.u. (39°F.)	0.994982	"	Cal., kg. (mean)/hr	15.1197
"	B.t.u. (60°F.)	0.999689	"	Cal., kg. (mean)/min	0.251996
"	Cal. gm.	251.99576	"	Ergs/sec	1.75978×10^{8}
"	Cal., gm. (IST.)	251.831	"	Foot-pounds/min	778.768
"	Cal., gm. (mean)	251.634	"	Horsepower	0.0235990
"	Cal., gm. (20°C.)	252.122	"	Horsepower (boiler)	0.00179396
"	Cu. cm.-atm	10405.6	"	Horsepower (electric)	0.0235895
"	Ergs	1.05435×10^{10}	"	Horsepower (metric)	0.0239264
"	Foot-poundals	25020.1	"	Joules/sec	17.5978
"	Foot-pounds	777.649	"	Kg.-meters/min	107.669
"	Gram-cm	1.07514×10^{7}	"	Kilowatts	0.0175978
"	Hp.-hours	0.000392752	"	Lb. ice-melted/hr	0.41828
"	Hp.-years	4.48347×10^{-8}	B.t.u./lb	Cal., gm./gram	0.555555
"	Joules	1054.35	"	Cu. cm.-atm./gram	22.9405
"	Joules (Int.)	1054.18	"	Cu. ft.-atm./lb	0.367471
"	Kg.-meters	107.514	"	Cu. ft.-(lb./sq. in.)/lb	5.40034
"	Kw.-hours	0.000292875	"	Foot-pounds/lb	777.649
			"	Hp.-hr./lb	0.000392752

* Barrel (Brit., liq.) = Barrel (Brit., dry)
** International Steam Table.

To convert from	To	Multiply by	To convert from	To	Multiply by
B.t.u./lb	Joules/gram	2.32444	Calories, gm.**	B.t.u.	0.0039683207
B.t.u. (mean)/lb	Cal., gm. (mean)/gram	0.555555	"	B.t.u. (IST.)	0.00396573
"	Cu. cm.-atm./gram	22.9735	"	B.t.u. (mean)	0.00396262
"	Foot-pounds/lb	778.768	"	B.t.u. (39°F.)	0.00394841
"	Hp.-hr./lb	0.000393317	"	B.t.u. (60°F.)	0.00396709
"	Joules/gram	2.32779	"	Cal., gm. (IST.)	0.999346
B.t.u./sec	B.t.u./hr	3600	"	Cal., gm. (mean)	0.998563
"	B.t.u./min	60	"	Cal., gm. (15°C.)	0.999570
"	Cal., kg./hr.	907.185	"	Cal., gm. (20°C.)	1.00050
"	Cal., kg./min	15.1197	"	Cal., kg.	0.001
"	Cheval-vapeur	1.43352	"	Cal., kg. (IST.)	0.000999346
"	Ergs/sec.	1.05435 × 10¹⁰	"	Cal., kg. (mean)	0.000998563
"	Foot-pounds/sec.	777.649	"	Cal., kg. (15°C.)	0.000999570
"	Horsepower	1.41391	"	Cal., kg. (20°C.)	0.00100050
"	Horsepower (boiler)	0.107483	"	Cu. cm.-atm.	41.2929
"	Horsepower (electric)	1.41334	"	Cu. ft.-atm.	0.00145824
"	Horsepower (metric)	1.43352	"	Ergs	4.184 × 10⁷
"	Kg.-meters/sec.	107.514	"	Foot-poundals	99.2878
"	Kilowatts	1.05435	"	Foot-pounds	3.08596
"	Kilowatts (Int.)	1.05418	"	Gram-cm	42664.9
"	Watts	1054.35	"	Hp.-hours	1.55857 × 10⁻⁶
"	Watts (Int.)	1054.18	"	Joules	4.184
B.t.u. (mean)/sec.	Ergs/sec.	1.05587 × 10¹⁰	"	Joules (Int.)	4.18331
"	Foot-pounds/sec.	778.768	"	Kg.-meters	0.426649
"	Horsepower	1.41594	"	Kw.-hours	1.162222 × 10⁻⁶
"	Horsepower (boiler)	0.107637	"	Liter-atm.	0.0412917
"	Horsepower (electric)	1.41537	"	Watt-hours	0.001162222
"	Horsepower (metric)	1.43558	"	Watt-hours (Int.)	0.00116203
"	Watts	1055.87	"	Watt-seconds	4.184
B.t.u./sq. ft.	Cal., gm./sq. cm.	0.271246	Calories, gm. (mean)	B.t.u.	0.00397403
B.t.u./sq.ft. × min)			"	Cal., gm	1.00144
"	Hp./sq. ft.	0.0235651	"	Cal., gm. (IST.)	1.00078
"	Kw./sq. ft.	0.0175725	"	Cal., gm. (20°C.)	1.00194
"	Watts/sq. in	0.122031	"	Cal., kg. (mean)	0.001
Buckets (Brit.)	Cu. cm.	18184.35	"	Cu. cm.-atm.	41.3523
"	Gallons (Brit.)	4	"	Cu. ft.-atm.	0.00146034
Bushels (Brit.)	Bags (Brit.)	0.333333	"	Ergs	4.19002 × 10⁷
"	Bushels (U.S.)	1.032056	"	Foot-poundals	99.4308
"	Cu. cm	36368.70	"	Foot-pounds	3.09040
"	Cu. feet	1.284348	"	Hp.-hours	1.56081 × 10⁻⁶
"	Cu. inches	2219.354	"	Joules	4.19002
"	Dekaliters	3.636768	"	Joules (Int.)	4.18933
"	Gallons (Brit.)	8	"	Kg.-meters	0.427263
"	Hectoliters	0.3636768	"	Kw.-hours	1.16390 × 10⁻⁶
"	Liters	36.36768	"	Liter-atm.	0.0413511
Bushels (U.S.)*	Barrels (U.S.), dry	0.3047647	"	Watt-seconds	4.19002
"	Bushels (Brit.)	0.9689395	Calories, gm. (15°C.)	B.t.u.	0.00397003
"	Cu. cm	35239.07	"	Cal., gm	1.00043
"	Cu. feet	1.244456	"	Cal., gm. (IST.)	0.999776
"	Cu. inches	2150.42	"	Cal., gm. (mean)	0.998992
"	Cu. meters	0.03523907	"	Cal., gm. (20°C.)	1.00093
"	Cu. yards	0.04609096	"	Joules	4.18580
"	Gallons (U.S., dry)	8	"	Joules (Int.)	4.18511
"	Gallons (U.S., liq.)	9.309177	Calories, gm. (20°C.)	B.t.u.	0.00396633
"	Liters	35.23907	"	Cal., gm	0.999498
"	Ounces (U.S., fluid)	1191.575	"	Cal., gm. (IST.)	0.998845
"	Pecks (U.S.)	4	"	Cal., gm. (mean)	0.998061
"	Pints (U.S., dry)	64	"	Cal., gm. (15°C.)	0.999068
"	Quarts (U.S., dry)	32	"	Joules	4.18190
"	Quarts (U.S., liq.)	37.23671	"	Joules (Int.)	4.18121
Butts (Brit.)	Bushels (U.S.)	13.53503	Calories, kg	B.t.u.	3.9683207
"	Cu. feet	16.84375	"	B.t.u. (IST.)	3.96573
"	Cu. meters	0.4769619	"	B.t.u. (mean)	3.96262
"	Gallons (U.S.)	126	"	B.t.u. (60°F.)	3.96709
			"	Cal., gm	1000
			"	Cal., kg. (mean)	0.998563
Cable lengths	Fathoms	120	"	Cal., kg. (15°C.)	0.999570
"	Feet	720	"	Cal., kg. (20°C.)	1.00050
"	Meters	219.456	"	Cu. cm.-atm.	41292.86

* Stricken or struck bushel. A heaped bushel for apples of 2747.715 cu. inches was established by the U.S. Court of Customs Appeals on Feb. 15, 1912. A heaped bushel equal to 1¼ stricken bushels is also known.

** This is the calorie as defined by the U.S. National Bureau of Standards and is equal to 4.18400 joules.

To convert from	To	Multiply by	To convert from	To	Multiply by
Calories, *kg.*	Ergs	4.184×10^{10}	Cal., *gm.*-cm.	B.t.u.-ft.	0.0671969
"	Foot-poundals	99287.8	$\overline{\text{(hr.} \times \text{sq. cm.} \times \text{°C.)}}$	$\overline{\text{(hr.} \times \text{sq. ft.} \times \text{°F.)}}$	
"	Foot-pounds	3085.96	"	B.t.u.-inch	0.806363
"	Gram-cm	4.26649×10^{7}		$\overline{\text{(hr.} \times \text{sq. ft.} \times \text{°F.)}}$	
"	Hp.-hours	0.00155857	Cal., *gm.*-cm./sq. cm	B.t.u.-inch/sq. ft	1.4514530
"	Joules	4184	Cal., *gm.*-sec.	Planck's constant	6.31531×10^{33}
"	Kw.-hours	0.001162222	Cal., *gm.*-sec./Avog.		
"	Liter-atm	41.2917	No. (chem.)*	Planck's constant	1.04849×10^{10}
"	Watt-hours	1.162222	Cal., *gm.*-sec./Avog.		
Calories, *kg.* (mean)	B.t.u.	3.97403	No. (phys.)*	Planck's constant	1.04821×10^{10}
"	B.t.u. (IST.)	3.97144	Candles (English)	Candles (Int.)	1.04
"	B.t.u. (mean)	3.9683207	"	Hefner units	1.16
"	B.t.u. (60°F.)	3.97280	Candles (German)	Candles (English)	1.01
"	Cal., *gm.*	1001.44	"	Candles (Int.)	1.05
"	Cal., *gm.* (IST.)	1000.78	"	Hefner units	1.17
"	Cal., *gm.* (mean)	1000	Candles (Int.)	Candles (English)	0.96
"	Cal., *gm.* (15°C.)	1000.10	"	Candles (German)	0.95
"	Cal., *gm.* (20°C.)	1001.94	"	Candles (pentane)	1.00
"	Ergs	4.19002×10^{10}	"	Hefner units	1.11
"	Foot-poundals	99430.8	"	Lumens (Int.)/steradian	1
"	Foot-pounds	3090.40	Candles (pentane)	Candles (Int.)	1.00
"	Gram-cm	4.27263×10^{7}	Candles/sq. cm	Candles/sq. inch	6.4516
"	Hp.-hours	0.00156081	"	Candles/sq. meter	10000
"	Joules	4190.02	"	Foot-lamberts	2918.6351
"	Kg.-meters	427.263	"	Lamberts	3.1415927
"	Kw.-hours (Int.)	0.00116370	Candles/sq. ft	Candles/sq. inch	0.0069444
"	Liter-atm	41.3511	"	Candles/sq. meter	10.763910
"	Watt-hours	1.16390	"	Foot-lamberts	3.1415927
Cal., *gm.*/°C	B.t.u./°F	0.00220462	"	Lamberts	0.0033815822
"	Joules/°F	2.324444	Candles/sq. inch	Candles/sq. cm	0.15500031
"	Joules (Int.)/°F	2.32406	"	Candles/sq. foot	144
Cal., *gm.*/gram	B.t.u./lb	1.8	"	Foot-lamberts	452.38934
"	Foot-pounds/lb	1399.77	"	Lamberts	0.48694784
"	Joules/gram	4.184	Candle power (spher.)	Lumens	12.566370
"	Watt-hours/gram	0.001162222	Carats (parts of gold		
Cal., *gm.*/(gram × °C)	B.t.u./(lb. × °C.)	1.8	per 24 of mixture)	Milligrams/gram	41.6666
"	B.t.u./(lb. × °F.)	1	Carats (1877)	Grains	3.168
"	Cal., *kg.*/(kg. × °C.)	1	"	Milligrams	205.3
"	Joules/(gram × °C.)	4.184	Carats (metric)	Grains	3.08647
"	Joules/(lb. × °F.)	1054.35	"	Grams	0.2
Cal., *gm.*/hr	B.t.u./hr	0.0039683207	"	Milligrams	200
"	Ergs/sec	11622.222	Carcel units	Candles (Int.)	9.61
"	Watts	0.001162222	Centals	Kilograms	45.359237
Cal., *gm.* (mean)/hr	B.t.u. (mean)/hr	0.0039683207	"	Pounds	100
"	Ergs/sec	11639.0	Centares	Ares	0.01
"	Watts	0.00116390	"	Sq. feet	10.763910
Cal., *kg.*/hr	Watts	1.162222	"	Sq. inches	1550.0031
Cal., *gm.*/min	B.t.u./min	0.0039683207	"	Sq. meters	1
"	Ergs/sec	697333.3	"	Sq. yards	1.1959900
"	Watts	0.069733	Centigrams	Grains	0.15432358
Cal., *gm.* (mean)/min	B.t.u. (mean)/min	0.0039683207	"	Grams	0.01
"	Ergs/sec	698337	Centiliters	Cu. cm	10
"	Joules/sec	0.0698337	"	Cu. inches	0.6102545
"	Watts	0.0698337	"	Liters	0.01
Cal., *kg.*/min	Kg. ice melted/min	0.012548	"	Ounces (U.S., fluid)	0.3381497
"	Lb. ice melted/min	0.027665	Centimeters	Ångström units	1×10^{8}
"	Watts	69.7333	"	Feet	0.032808399
Cal., *gm.*/sec	B.t.u./sec	0.0039683207	"	Feet (U.S. Survey)	0.032808333
"	Ergs/sec	4.184×10^{7}	"	Hands	0.098425197
"	Foot-pounds/sec	3.08596	"	Inches	0.39370079
"	Horsepower	0.00561084	"	Links (Gunter's)	0.049709695
"	Watts	4.184	"	Links (Ramden's)	0.032808399
Cal., *gm.* (mean)/sec	Ergs/sec	4.19002×10^{7}	"	Meters	0.01
"	Watts	4.19002	"	Microns	10000
Cal., *gm.*/(sec. ×			"	Miles (naut., Int.)	5.3995680×10^{-6}
sq. cm.)	B.t.u./(hr. × sq. ft.)	13272.1	"	Miles (statute)	6.2137119×10^{-6}
"	Cal., *gm.*/(hr. × sq. cm.)	3600	"	Millimeters	10
"	Watts/sq. cm	4.184	"	Millimicrons	1×10^{7}
Cal., *gm.*/(sec. ×			"	Mils	393.70079
sq. cm. × °C.)	B.t.u./(hr. × sq. ft. × °F.)	7373.38	"	Picas (printer's)	2.3710630
Cal., *gm.*/sq. cm	B.t.u./sq. ft	3.68669	"	Points (printer's)	28.452756

* For factors for C = 12 scale or those which can be derived from same table on Values for General Physical Constants.

To convert from	To	Multiply by	To convert from	To	Multiply by
Centimeters	Rods	0.0019883878	Circumferences	Minutes	21600
"	Wave length of orange-red line of krypton 86	16507.6373	"	Radians	6.2831853
			"	Seconds	1296000
"	Wave length of red line of cadmium	15531.6413	Cords	Cord-feet	8
"	Yards	0.010936133	"	Cu. feet	128
Cm. of Hg (0°C.)	Atmospheres	0.013157895	"	Cu. meters	3.6245734
"	Bars	0.0133322	Cord-feet	Cords	0.125
"	Dynes/sq. cm	13332.2	"	Cu. feet	16
"	Ft. of H_2O (4°C.)	0.446050	Coulombs	Abcoulombs	0.1
"	Ft. of H_2O (60°F.)	0.446474	"	Ampere-hours	0.0002777
"	In. of Hg (0°C.)	0.39370070	"	Ampere-seconds	1
"	Kg./sq. meter	135.951	"	Coulombs (Int.)	1.000165
"	Pounds/sq. ft	27.8450	"	Electronic charge	6.24196×10^{18}
"	Pounds/sq. inch	0.193368	"	E.M. cgs. units of electric charge	0.1
"	Torrs	10	"	E.S. cgs. units of electric charge	2.997930×10^9
Cm. of H_2O (4°C.)	Atmospheres	0.000967814	"	Faradays (chem.)	1.036377×10^{-5}
"	Dynes/sq. cm	980.638	"	Faradays (phys.)	1.036086×10^{-5}
"	Pounds/sq. inch	0.0142229	"	Mks. units of electric charge	1
Centimeters/sec	Feet/min	1.9685039	"	Statcoulombs	2.997930×10^9
"	Feet/sec	0.032808399	Coulombs/cu. meter	E.M. cgs. units of volume charge density	1×10^{-7}
"	Kilometers/hr	0.036	"	E.S. cgs. units	2997.930
"	Kilometers/min	0.0006	Coulombs/sq. cm	Abcoulombs/sq. cm	0.1
"	Knots (Int.)	0.019438445	"	Cgs. units of polarization, and surface charge density	1
"	Meters/min	0.6	Cubic centimeters	Board feet	0.00042377600
"	Miles/hr	0.022369363	"	Bushels (Brit.)	2.749617×10^{-5}
"	Miles/min	0.00037282272	"	Bushels (U.S.)	2.837759×10^{-5}
Cm./(sec. × sec.)	Kilometers/(hr. × sec.)	0.036	"	Cu. feet	3.5314667×10^{-5}
"	Miles/(hr. × sec.)	0.022369363	"	Cu. inches	0.061023744
Centimeters/year	Inches/year	0.39370079	"	Cu. meters	1×10^{-6}
Centipoises*	Grams/(cm × sec.)	0.01	"	Cu. yards	1.3079500×10^{-6}
"	Poises	0.01	"	Drachms (Brit., fluid)	0.28156080
"	Pound/(ft. × hr.)	2.4190883	"	Drams (U.S., fluid)	0.27051218
"	Pounds/(ft. × sec.)	0.00067196898	"	Gallons (Brit.)	0.0002199694
Centistokes*	Stokes	0.01	"	Gallons (U.S., dry)	0.00022702075
Chains (Gunter's)	Centimeters	2011.68	"	Gallons (U.S., liq.)	0.00026417205
"	Chains (Ramden's)	0.66	"	Gills (Brit.)	0.007039020
"	Feet	66	"	Gills (U.S.)	0.0084535058
"	Feet (U.S. Survey)	65.999868	"	Liters	0.001
"	Furlongs	0.1	"	Ounces (Brit., fluid)	0.03519510
"	Inches	792	"	Ounces (U.S., fluid)	0.033814023
"	Links (Gunter's)	100	"	Pints (U.S., dry)	0.0018161660
"	Links (Ramden's)	66	"	Pints (U.S., liq.)	0.0021133764
"	Meters	20.1168	"	Quarts (Brit.)	0.0008798775
"	Miles (statute)	0.0125	"	Quarts (U.S., dry)	0.00090808298
"	Rods	4	"	Quarts (U.S., liq.)	0.0010566882
"	Yards	22	Cu. cm./gram	Cu. ft./lb	0.016018463
Chains (Ramden's)	Centimeters	3048	Cu. cm./sec	Cu. ft./min	0.0021188800
"	Chains (Gunter's)	1.515151	"	Gal. (U.S.)/min	0.015850323
"	Feet	100	"	Gal. (U.S.)/sec	0.00026417205
"	Feet (U.S. Survey)	99.999800	Cu. cm.-atm	B.t.u.	9.61019×10^{-5}
Cheval-vapeur	Horsepower (metric)	1	"	B.t.u. (mean)	9.59637×10^{-5}
Cheval-vapeur-heures	Joules	2647795	"	Cal., gm	0.0242173
Circles	Degrees	360	"	Cal., gm. (mean)	0.0241824
"	Grades	400	"	Cu. ft.-atm	3.5314667×10^{-5}
"	Minutes	21600	"	Joules	0.101325
"	Radians	6.2831853	"	Watt-hours	2.81458×10^{-5}
"	Signs	12	Cu. cm.-atm./gram	B.t.u./lb	0.0435911
Circular inches	Circular mm	645.16	"	Cal., gm./gram	0.0242173
"	Sq. cm	5.0670748	"	Cu. ft.-(lb./sq. in.)/lb	0.235406
"	Sq. inches	0.78539816	"	Ft.-lb./lb	33.8985
Circular mm	Sq. cm	0.0078539816	"	Joules/gram	0.101325
"	Sq. inches	0.0012173696	"	Kg.-meters/gram	0.0103323
"	Sq. mm	0.78539816	"	Kw.-hr./gram	2.81458×10^{-8}
Circular mils	Circular inches	1×10^{-6}	Cubic decimeters	Cu. cm	1000
"	Sq. cm	5.0670748×10^{-6}	"	Cu. feet	0.035316667
"	Sq. inches	7.8539816×10^{-7}	"	Cu. inches	61.023744
"	Sq. mm	0.00050670748	"	Cu. meters	0.001
"	Sq. mils	0.78539816			
Circumferences	Degrees	360			
"	Grades	400			

* See also special table on viscosity.

To convert from	To	Multiply by	To convert from	To	Multiply by
Cubic decimeters.....	Cu. yards...............	0.0013079506	Cubic inches...........	Ounces (U.S., fluid).......	0.55411255
"	Liters...................	1	"	Pecks (U.S.).............	0.0018601017
Cubic dekameters....	Cu. decimeters...........	1×10^6	"	Pints (U.S., dry).........	0.029761628
"	Cu. feet.................	35314.667	"	Pints (U.S., liq.).........	0.034632035
"	Cu. inches..............	6.1023744×10^7	"	Quarts (U.S., dry)........	0.014880814
"	Cu. meters..............	1000	"	Quarts (U.S., liq.)........	0.017316017
"	Liters...................	999972	Cu. in. of H_2O (4°C.).	Pounds of H_2O........	0.0361263
Cubic feet...........	Acre-feet................	2.2956841×10^{-5}	Cu. in. of H_2O (60°F.).	Pounds of H_2O.......	0.0360916
"	Board feet...............	12	Cubic meters...........	Acre-feet................	0.00081071319
"	Bushels (Brit.)..........	0.7786049	"	Barrels (Brit.)..........	6.110261
"	Bushels (U.S.)..........	0.80356395	"	Barrels (U.S., dry).......	8.648490
"	Cords (wood)............	0.0078125	"	Barrels (U.S., liq.).......	8.3864145
"	Cord-feet...............	0.0625	"	Bushels (Brit.)..........	27.49617
"	Cu. centimeters..........	28316.847	"	Bushels (U.S.)..........	28.377593
"	Cu. meters..............	0.028316847	"	Cu. cm.................	1×10^6
"	Gallons (U.S., dry).......	6.4285116	"	Cu. feet................	35.314667
"	Gallons (U.S., liq.).......	7.4805195	"	Cu. inches..............	61023.74
"	Liters...................	28.316847	"	Cu. yards...............	1.3079506
"	Ounces (Brit., fluid)......	996.6143	"	Gallons (Brit.)..........	219.9694
"	Ounces (U.S., fluid)......	957.50649	"	Gallons (U.S., liq.).......	264.17205
"	Pints (U.S., liq.).........	59.844156	"	Hogshead...............	4.1932072
"	Quarts (U.S., dry).......	25.714047	"	Liters...................	1000
"	Quarts (U.S., liq.).......	29.922078	"	Pints (U.S., liq.).........	2113.3764
Cu. ft. of H_2O (39.2°F.)	Pounds of H_2O........	62.4262	"	Quarts (U.S., liq.).......	1056.6882
Cu. ft. of H_2O (60°F.)	Pounds of H_2O........	62.3663	"	Steres..................	1
Cu. ft./hr	Acre-feet/hr.............	2.2956841×10^{-5}	Cu. meters/min	Gal. (Brit.)/min.........	219.9694
"	Cu. cm./sec.............	7.8657907	"	Gal. (U.S.)/min.........	264.1721
"	Cu. ft./day.............	24	"	Liters/min..............	999.972
"	Gal. (U.S.)/hr...........	7.4805195	Cu. millimeters	Cu. cm.................	0.001
"	Liters/hr................	28.31605	"	Cu. inches..............	6.1023744×10^{-5}
Cu. ft./min	Acre-feet/hr.............	0.0013774105	"	Cu. meters..............	1×10^{-9}
"	Acre-feet/min............	2.2956841×10^{-5}	"	Minims (Brit.)..........	0.01689365
"	Cu. cm./sec.............	471.94744	"	Minims (U.S.)...........	0.016230731
"	Cu. ft./hr...............	60	Cu. yards...........	Bushels (Brit.)..........	21.02233
"	Gal. (U.S.)/min..........	7.4805195	"	Bushels (U.S.)..........	21.696227
"	Liters/sec...............	0.4719342	"	Cu. cm.................	764554.86
Cu. ft./lb	Cu. cm./gram............	62.427961	"	Cu. feet................	27
"	Millimeters/gram.........	62.42621	"	Cu. inches..............	46656
Cu. ft./sec...........	Acre-inches/hr...........	0.99173553	"	Cu. meters..............	0.76455486
"	Cu. cm./sec.............	28316.847	"	Gallons (Brit.)..........	168.1787
"	Cu. yards/min...........	2.222222	"	Gallons (U.S., dry).......	173.56981
"	Gal. (U.S.)/min..........	448.83117	"	Gallons (U.S., liq.).......	201.97403
"	Liters/min..............	1698.963	"	Liters..................	764.55486
"	Liters/sec...............	28.31605	"	Quarts (Brit.)..........	672.7146
Cu. ft. of H_2O (60°F.)/sec	Lb. of H_2O/min........	3741.98	"	Quarts (U.S., dry).......	694.27926
Cu. ft.-atm	B.t.u...................	2.72130	"	Quarts (U.S., liq.).......	807.89610
"	Cal., *gm*...............	685.756	Cu. yd./min........	Cu. ft./sec.............	0.45
"	Cu. cm.-atm............	28316.847	"	Gal. (U.S.)/sec..........	3.3662338
"	Cu. ft.-(lb/sq. in.).....	14.6960	"	Liters/sec..............	12.74222
"	Foot-pounds............	2116.22	Cubits.............	Centimeters............	45.72
"	Hp.-hours..............	0.00106880	"	Feet...................	1.5
"	Joules..................	2869.20	"	Inches.................	18
"	Kg.-meters.............	292.577			
"	Kw.-hours..............	0.000797001	Daltons (chem.)......	Grams.................	1.66024×10^{-24}
Cubic inches...........	Barrels (Brit.)..........	0.0001001292	Daltons (phys.).....	Grams.................	1.65979×10^{-24}
"	Barrels (U.S., dry).......	0.00014172336	Days (mean solar)....	Days (sidereal).........	1.00273791
"	Board feet..............	0.0069444	"	Hours (mean solar)......	24
"	Bushels (Brit.)..........	0.0004505815	"	Hours (sidereal)........	24.065710
"	Bushels (U.S.)..........	0.00046502544	"	Years (calendar)........	0.0027397260
"	Cu. cm.................	16.387064	"	Years (sidereal)........	0.0027378031
"	Cu. feet................	0.00057870370	"	Years (tropical)........	0.0027379093
"	Cu. meters..............	1.6387064×10^{-5}	Days (sidereal).......	Days (mean solar).......	0.99726957
"	Cu. yards...............	2.1433470×10^{-5}	"	Hours (mean solar)......	23.934470
"	Drams (U.S., fluid)......	4.4329004	"	Hours (sidereal)........	24
"	Gallons (Brit.)..........	0.003604652	"	Minutes (mean solar)...	1436.0682
"	Gallons (U.S., dry).......	0.0037202035	"	Minute (sidereal).......	1440
"	Gallons (U.S., liq.).......	0.0043290043	"	Second (sidereal).......	86400
"	Liters..................	0.016387064	"	Years (calendar)........	0.0027322454
"	Milliliters..............	16.387064	"	Years (sidereal)........	0.0027303277
"	Ounces (Brit., fluid)......	0.5767444	"	Years (tropical)........	0.0027304336
			Decibels.............	Bels...................	0.1
			Decimeters............	Centimeters............	10

To convert from	To	Multiply by
Decimeters	Feet	0.32808399
"	Feet (U.S. Survey)	0.328083333
"	Inches	3.9370079
"	Meters	0.1
Decisteres	Cu. meters	0.1
Degrees	Circles	0.0027777
"	Minutes	60
"	Quadrants	0.0111111
"	Radians	0.017453293
"	Seconds	3600
Degrees/cm	Radians/cm	0.017453293
Degrees/foot	Radians/cm	0.00057261458
Degrees/inch	Radian/cm	0.0068713750
Degrees/min	Degrees/sec	0.0166666
"	Radians/sec	0.00029088821
"	Revolutions/sec	4.629629×10^{-5}
Degrees/sec	Radians/sec	0.017453293
"	Revolutions/min	0.166666
"	Revolutions/sec	0.0027777
Dekaliters	Pecks (U.S.)	1.135136
"	Pints (U.S., dry)	18.16217
Dekameters	Centimeters	1000
"	Feet	32.808399
"	Feet (U.S. Survey)	32.808333
"	Inches	393.70079
"	Kilometers	0.01
"	Meters	10
"	Yards	10.93613
Demals	Gram-equiv./cu. decimeter	1
Drachms (Brit., fluid)	Cu. cm	3.551631
"	Cu. inches	0.2167338
"	Drams (U.S., fluid)	0.9607594
"	Milliliters	3.551531
Drams (apoth. or troy)	Drams (avdp.)	2.1942857
"	Grains	60
"	Grams	3.8879346
"	Ounces (apoth. or troy)	0.125
"	Ounces (avdp.)	0.13714286
"	Scruples (apoth.)	3
Drams (avdp.)	Drams (apoth. or troy)	0.455729166
"	Grains	27.34375
"	Grams	1.7718452
"	Ounces (apoth. or troy)	0.056966146
"	Ounces (avdp.)	0.0625
"	Pennyweights	1.1393229
"	Pounds (apoth. or troy)	0.0047471788
"	Pounds (avdp.)	0.00390625
"	Scruples (apoth.)	1.3671875
Drams (U.S., fluid)	Cu. cm	3.6967162
"	Cu. inches	0.22558594
"	Drachms (Brit., fluid)	1.040843
"	Gills (U.S.)	0.03125
"	Milliliters	3.696588
"	Minims (U.S.)	60
"	Ounces (U.S., fluid)	0.125
"	Pints (U.S., liq.)	0.0078125
Dynes	Grains	0.01573663
"	Grams	0.001019716
"	Newtons	0.00001
"	Poundals	7.2330138×10^{-5}
"	Pounds	2.248089×10^{-6}
Dynes/cm	Ergs/sq. cm	1
"	Ergs/sq. mm	0.01
"	Grams/cm	0.001019716
"	Poundals/inch	0.00018371855
Dynes/cu. cm	Grams/cu. cm	0.001019716
"	Poundals/cu. inch	0.0011852786
Dynes/sq. cm	Atmospheres	9.86923×10^{-7}
"	Bars	1×10^{-6}
"	Baryes	1
"	Cm. of Hg (0°C.)	7.50062×10^{-5}

To convert from	To	Multiply by
Dynes/sq. cm	Cm. of H_2O (4°C.)	0.001019745
"	Grams/sq. cm	0.001019716
"	In. of Hg (32°F.)	2.95300×10^{-5}
"	In. of H_2O (4°C.)	0.000401474
"	Kg./sq. meter	0.01019716
"	Poundals/sq. in.	0.00046664510
"	Pounds/sq. in.	1.450377×10^{-5}
Dyne-centimeters	Ergs	1
"	Foot-poundals	2.3730360×10^{-6}
"	Foot-pounds	7.37562×10^{-8}
"	Gram-cm	0.001019716
"	Inch-pounds	8.85075×10^{-7}
"	Kg.-meters	1.019716×10^{-8}
"	Newton-meters	1×10^{-7}
Electron volts	Ergs	1.60219×10^{-12}
"	Grams	1.78253×10^{-33}
Electronic charges	Abcoulombs	1.60209×10^{-20}
"	Coulombs	1.60209×10^{-19}
"	Statcoulombs	4.80296×10^{-10}
Electronic charges/kg	Statcoulombs/dyne	4.89766×10^{-16}
E.S. cgs. units of induction flux	E.M. cgs. units	2.997930×10^{10}
E.S. cgs. units of magnetic charge	E.M. cgs. units	2.997930×10^{10}
E.S. cgs. units of magnetic field intensity	E.M. cgs. units	3.335635×10^{-11}
Ells	Centimeters	114.3
"	Inches	45
Ergs	B.t.u.	9.48451×10^{-11}
"	Cal., gm.	2.39006×10^{-8}
"	Cal., kg	2.39006×10^{-11}
"	Cal., kg (20°C.)	2.39006×10^{-11}
"	Cu. cm.-atm	9.86923×10^{-7}
"	Cu. ft.-atm	3.48529×10^{-11}
"	Cu. ft.-(lb./sq. in.)	5.12196×10^{-10}
"	Dyne-cm	1
"	Electron volts	6.24145×10^{11}
"	Foot-poundals	2.3730360×10^{-6}
"	Foot-pounds	7.37562×10^{-8}
"	Gram-cm	0.001019716
"	Joules	1×10^{-7}
"	Joules (Int.)	9.99835×10^{-8}
"	Kw.-hours	2.777777×10^{-14}
"	Kg.-meters	1.019716×10^{-8}
"	Liter-atm	9.86895×10^{-10}
"	Watt-sec	1×10^{-7}
Ergs/(gram-mol. × °C.)	Foot-pounds/(lb.-mol. × °F.)	1.85863×10^{-5}
Ergs/sec	B.t.u./min	5.69071×10^{-9}
"	Cal., gm./min	1.43403×10^{-6}
"	Dyne-cm./sec	1
"	Foot-pounds/min	4.42537×10^{-6}
"	Gram-cm./sec	0.001019716
"	Horsepower	1.34102×10^{-10}
"	Joules/sec	1×10^{-7}
"	Kilowatts	1×10^{-10}
"	Watts	1×10^{-7}
Ergs/sq. cm	Dynes/cm	1
"	Ergs/sq. mm	0.01
Ergs/sq. mm	Dynes/cm	100
"	Ergs/sq. cm	100
Erg-sec	Planck's constant	1.50932×10^{26}
Farads	Abfarads	1×10^{-9}
"	E.M. cgs. units	1×10^{-9}
"	E.S. cgs. units	8.987584×10^{11}
"	Farads (Int.)	1.000495
"	Microfarads	1×10^{6}
"	Statfarads	8.98758×10^{11}
Farads (Int.)	Farads	0.999505
Fathoms	Centimeters	182.88

To convert from	To	Multiply by	To convert from	To	Multiply by
Fathoms	Feet	6	Feet/(sec. × sec.)	Meters/(sec. × sec.)	0.3048
"	Inches	72	"	Miles/(hr. × sec.)	0.68181818
"	Meters	1.8288	Firkins (Brit.)	Bushels (Brit.)	1.125
"	Miles (naut., Int.)	0.00098747300	"	Cu. cm.	40914.79
"	Miles (statute)	0.001136363	"	Cu. feet	1.444892
"	Yards	2	"	Firkins (U.S.)	1.200949
Feet	Centimeters	30.48	"	Gallons (Brit.)	9
"	Chains (Gunter's)	0.01515151	"	Liters	40.91364
"	Fathoms	0.166666	"	Pints (Brit.)	72
"	Feet (U.S. Survey)	0.99999800	Firkins (U.S.)	Barrels (U.S., dry)	0.29464286
"	Furlongs	0.00151515	"	Barrels (U.S., liq.)	0.28571429
"	Inches	12	"	Bushels (U.S.)	0.96678788
"	Meters	0.3048	"	Cu. feet	1.203125
"	Microns	304800	"	Firkins (Brit.)	0.8326747
"	Miles (naut., Int.)	0.00016457883	"	Liters	34.06775
"	Miles (statute)	0.000189393	"	Pints (U.S., liq.)	72
"	Rods	0.060606	Foot-candles	Lumens/sq. ft.	1
"	Ropes (Brit.)	0.05	"	Lumens/sq. meter	10.763910
"	Yards	0.333333	"	Lux	10.763910
Feet (U.S. Survey)	Centimeters	30.480061	"	Milliphots	1.0763910
"	Chains (Gunter's)	0.015151545	Foot-lamberts	Candles/sq. cm.	0.00034262591
"	Chains (Ramden's)	0.010000020	"	Candles/sq. ft.	0.31830989
"	Feet	1.0000020	"	Millilamberts	1.0763910
"	Inches	12.000024	"	Lamberts	0.0010763910
"	Links (Gunter's)	1.5151545	"	Lumens/sq. ft	1
"	Links (Ramden's)	1.0000020	Foot-poundals	B.t.u.	3.99678 × 10⁻⁵
"	Meters	0.30480061	"	B.t.u. (IST.)	3.99417 × 10⁻⁵
"	Miles (statute)	0.00018939432	"	B.t.u. (mean)	3.99104 × 10⁻⁵
"	Rods	0.060606182	"	Cal., gm.	0.0100717
"	Yards	0.33333400	"	Cal., gm. (IST.)	0.0100651
Feet of air (1 atm., 60°F.)	Atmospheres	3.6083 × 10⁻⁵	"	Cal., gm. (mean)	0.0100573
"	Ft. of Hg (32°F.)	0.00089970	"	Cu. cm.-atm	0.415890
"	Ft. of H₂O (60°F.)	0.0012244	"	Cu. ft.-atm	1.46870 × 10⁻⁵
"	In. of Hg (32°F.)	0.0010796	"	Dyne-cm	4.2140110 × 10⁵
"	Pounds/sq. inch	0.00053027	"	Ergs	4.2140110 × 10⁵
Feet of Hg (32°F.)	Cm. of Hg (0°C.)	30.48	"	Foot-pounds	0.0310810
"	Ft. of H₂O (60°F.)	13.6085	"	Hp.-hours	1.56974 × 10⁻⁸
"	In. of H₂O (60°F.)	163.302	"	Joules	0.042140110
"	Ounces/sq. inch	94.3016	"	Joules (Int.)	0.0421332
"	Pounds/sq. inch	5.89385	"	Kg.-meters	0.00429710
Feet of H₂O (4°C.)	Atmospheres	0.0294990	"	Kw.-hours	1.17056 × 10⁻⁸
"	Cm. of Hg (0°C.)	2.24192	"	Liter-atm	0.000415879
"	Dynes/sq. cm.	29889.8	Foot-pounds	B.t.u.	0.00128593
"	Grams/sq. cm.	30.4791	"	B.t.u. (IST.)	0.00128509
"	In. of Hg (32°F.)	0.882646	"	B.t.u. (mean)	0.00128408
"	Kg./sq. meter	304.791	"	Cal., gm.	0.324048
"	Pounds/sq. inch	0.433515	"	Cal., gm. (IST.)	0.323836
Feet/hour	Cm./hr	30.48	"	Cal., gm. (mean)	0.323582
"	Cm./min	0.508	"	Cal., gm. (20°C.)	0.324211
"	Cm./sec	0.0084666	"	Cal., kg.	0.000324048
"	Feet/min	0.0166666	"	Cal., kg. (IST.)	0.000323836
"	Inches/hr	12	"	Cal., kg. (mean)	0.000323582
"	Kilometers/hr	0.0003048	"	Cu. ft.-atm	0.000472541
"	Kilometers/min	5.08 × 10⁻⁶	"	Dyne-cm	1.35582 × 10⁷
"	Knots (Int.)	0.0001645788	"	Ergs	1.35582 × 10⁷
"	Miles/hr	0.000189393	"	Foot-poundals	32.1740
"	Miles/min	3.156565 × 10⁻⁶	"	Gram-cm	13825.5
"	Miles/sec	5.2609428 × 10⁻⁸	"	Hp.-hours	5.05050 × 10⁻⁷
Feet/minute	Cm./sec	0.508	"	Joules	1.35582
"	Feet/sec	0.0166666	"	Kg.-meters	0.138255
"	Kilometers/hr	0.018288	"	Kw.-hours	3.76616 × 10⁻⁷
"	Meters/min	0.3048	"	Kw.-hours (Int.)	3.76554 × 10⁻⁷
"	Meters/sec	0.00508	"	Liter-atm	0.0133805
"	Miles/hr	0.01136363	"	Newton-meters	1.3558180
Feet/second	Cm./sec	30.48	"	Lb. H₂O evap. from and at 212°F	1.3245 × 10⁻⁶
"	Kilometers/hr	1.09728	"	Watt-hours	0.000376616
"	Kilometers/min	0.018288	Foot-pounds/hr	B.t.u./min	2.14321 × 10⁻⁵
"	Meters/min	18.288	"	B.t.u. (mean)/min	2.14013 × 10⁻⁵
"	Miles/hr	0.68181818	"	Cal., gm./min	0.00540080
"	Miles/min	0.01136363	"	Cal., gm. (mean)/min	0.00539304
Feet/(sec. × sec.)	Kilometers/(hr. × sec.)	1.09728	"	Ergs/min	2.25970 × 10⁵

To convert from	To	Multiply by	To convert from	To	Multiply by
Foot-pounds/hr	Foot-pounds/min	0.0166666	Gallons (U.S., dry)	Cu. inches	268.8025
"	Horsepower	5.050505×10^{-7}	"	Gallons (U.S., liq.)	1.16364719
"	Horsepower (metric)	5.12055×10^{-7}	"	Liters	4.404760
"	Kilowatts	3.76616×10^{-7}	Gallons (U.S., liq.)	Acre-feet	3.0688833×10^{-6}
"	Watts	0.000376616	"	Barrels (U.S., liq.)	0.031746032
"	Watts (Int.)	0.000376554	"	Barrels (petroleum, U.S.)	0.023809524
Foot-pounds/min	B.t.u./sec	2.14321×10^{-5}	"	Bushels (U.S.)	0.10742088
"	B.t.u. (mean)/sec	2.14013×10^{-5}	"	Cu. centimeters	3785.4118
"	Cal., gm./sec	0.00540080	"	Cu. feet	0.133680555
"	Cal., gm. (mean)/sec	0.00539304	"	Cu. inches	231
"	Ergs/sec	2.25970×10^{5}	"	Cu. meters	0.0037854118
"	Foot-pounds/sec	0.0166666	"	Cu. yards	0.0049511317
"	Horsepower	3.030303×10^{-5}	"	Gallons (Brit.)	0.8326747
"	Horsepower (metric)	3.07233×10^{-5}	"	Gallons (U.S., dry)	0.85936701
"	Joules/sec	0.0225970	"	Gallons (wine)	1
"	Joules (Int.)/sec	0.0225932	"	Gills (U.S.)	32
"	Kilowatts	2.25970×10^{-5}	"	Liters	3.7854118
"	Watts	0.0225970	"	Minims (U.S.)	61440
Foot-pounds/lb	B.t.u./lb	0.00128593	"	Ounces (U.S., fluid)	128
"	B.t.u. (IST.)/lb	0.00128509	"	Pints (U.S., liq.)	8
"	B.t.u. (mean)/lb	0.00128408	"	Quarts (U.S., liq.)	4
"	Cal., gm./gm	0.000714404	Gallons (U.S.) of H₂O (4°C.) in air	Lb. of H₂O	8.33585
"	Cal., gm. (IST.)/gram	0.000713937			
"	Cal., gm. (mean)/gram	0.000713377	Gallons (U.S.) of H₂O (60°F.) in air	Lb. of H₂O	8.32823
"	Hp.-hr./lb	5.05050×10^{-7}			
"	Joules/gram	0.00298907	Gallons (U.S.)/day	Cu. ft./hr	0.0055700231
"	Kg.-meters/gram	0.000304800	Gallons (Brit.)/hr	Cu. meters/min	7.576812×10^{-5}
"	Kw.-hr./gram	8.30296×10^{-10}	Gallons (U.S.)/hr	Acre-feet/hr	3.0688833×10^{-6}
Foot-pounds/sec	B.t.u./min	0.0771556	"	Cu. ft./hr	0.1336805
"	B.t.u. (mean)/min	0.0770447	"	Cu. meters/min	6.3090197×10^{-5}
"	B.t.u./sec	0.00128593	"	Cu. yd./min	8.2518861×10^{-5}
"	B.t.u. (mean)/sec	0.00128408	"	Liters/hr	3.7854118
"	Cal., gm./sec	0.324048	Gal. (Brit.)/sec	Cu. cm./sec	4546.087
"	Cal., gm. (mean)/sec	0.323582	Gal. (U.S.)/sec	Cu. cm./sec	3785.4118
"	Ergs/sec	1.35582×10^{7}	"	Cu. ft./min	8.020833
"	Gram-cm./sec	13825.5	"	Cu. yd./min	0.29706790
"	Horsepower	0.00181818	"	Liters/min	227.1183
"	Joules/sec	1.35582	Gammas	Grams	1×10^{-6}
"	Kilowatts	0.00135582	"	Micrograms	1
"	Watts	1.35582	Gausses	E.M. cgs. units of magnetic flux density	1
"	Watts (Int.)	1.35559			
Furlongs	Centimeters	20116.8	"	E.S. cgs. units	3.335635×10^{-11}
"	Chains (Gunter's)	10	"	Gausses (Int.)	0.999670
"	Chains (Ramden's)	6.6	"	Maxwells/sq. cm	1
"	Feet	660	"	Lines/sq. cm	1
"	Inches	7920	"	Lines/sq. inch	6.4516
"	Meters	201.168	Gausses (Int.)	Gausses	1.000330
"	Miles (naut., Int.)	0.10862203	Gausses/oersted	E.M. cgs. units of permeability	1
"	Miles (statute)	0.125			
"	Rods	40	"	E.S. cgs. units	1.112646×10^{-21}
"	Yards	220	Geepounds	Slugs	1
			"	Kilograms	14.5939
Gallons (Brit.)	Barrels (Brit.)	0.027777	Gigameters	Meters	1×10^{9}
"	Bushels (Brit.)	0.125	Gilberts	Abampere-turns	0.079577472
"	Cu. centimeters	4546.087	"	Ampere-turns	0.79577472
"	Cu. feet	0.1605436	"	E.M. cgs. units of mmf., or magnetic potential	1
"	Cu. inches	277.4193			
"	Drachms (Brit. fluid)	1280	"	E.S. cgs. units	2.997930×10^{10}
"	Firkins (Brit.)	0.111111	"	Gilberts (Int.)	1.000165
"	Gallons (U.S., liq.)	1.200949	Gilberts (Int.)	Gilberts	0.999835
"	Gills (Brit.)	32	Gilberts/cm	Ampere-turns/cm	0.79577472
"	Liters	4.545960	"	Ampere-turns/in	2.0212678
"	Minims (Brit.)	76800	"	Oersteds	1
"	Ounces (Brit., fluid)	160	Gilberts/maxwell	Ampere-turns/weber	7.957747×10^{7}
"	Ounces (U.S., fluid)	153.7215	"	E.M. cgs. units of reluctance	1
"	Pecks (Brit.)	0.5			
"	Lb. of H₂O (62°F.)	10	"	E.S. cgs. units	8.987584×10^{20}
Gallons (U.S., dry)	Barrels (U.S., dry)	0.038095592	Gills (Brit.)	Cu. cm	142.0652
"	Barrels (U.S., liq.)	0.036941181	"	Gallons (Brit.)	0.03125
"	Bushels (U.S.)	0.125	"	Gills (U.S.)	1.200949
"	Cu. centimeters	4404.8828	"	Liters	0.1420613
"	Cu. feet	0.15555700	"	Ounces (Brit., fluid)	5

To convert from	To	Multiply by	To convert from	To	Multiply by
Gills (Brit.)	Ounces (U.S., fluid)	4.803764	Grams/cu. cm	Pounds/gal. (U.S., dry)	9.7111064
"	Pints (Brit.)	0.25	"	Pounds/gal. (U.S., liq.)	8.3454044
Gills (U.S.)	Cu. cm	118.29412	Grams/cu. meter	Grains/cu. ft.	0.43699572
"	Cu. inches	7.21875	Grams/liter	Parts/million*	1000
"	Drams (U.S., fluid)	32	"	Lb./cu. ft.	0.06242621
"	Gallons (U.S., liq.)	0.03125	"	Lb./gal. (U.S.)	8.345171×10^{-3}
"	Gills (Brit.)	0.8326747	Grams/milliliter	Grams/cu. cm	1
"	Liters	0.1182908	"	Pounds/cu. ft.	62.42621
"	Minims (U.S.)	1920	"	Pounds/gallon (U.S.)	8.345171
"	Ounces (U.S., fluid)	4	Grams/sq. cm	Atmospheres	0.000967841
"	Pints (U.S., liq.)	0.25	"	Bars	0.000980665
"	Quarts (U.S., liq.)	0.125	"	Cm. of Hg. (0°C.)	0.0735559
Grades	Circles	0.0025	"	Dynes/sq. cm	980.665
"	Circumferences	0.0025	"	In. of Hg (32°F.)	0.0289590
"	Degrees	0.9	"	Kg./sq. meter	10
"	Minutes	54	"	Mm. of Hg (0°C.)	0.735559
"	Radians	0.015707963	"	Poundals/sq. inch	0.457623
"	Revolutions	0.0025	"	Pounds/sq. inch	0.014223343
"	Seconds	3240	Grams/ton (long)	Milligrams/kg.	0.98420653
Grains	Carats (metric)	0.32399455	Grams/ton (short)	Milligrams/kg.	1.1023113
"	Drams (apoth. or troy)	0.016666	Grams-cm	B.t.u.	9.30113×10^{-8}
"	Drams (avdp.)	0.036571429	"	B.t.u. (IST.)	9.29505×10^{-8}
"	Dynes	63.5460	"	B.t.u. (mean)	9.28776×10^{-8}
"	Grams	0.06479891	"	Cal., *gm*	2.34385×10^{-5}
"	Milligrams	64.79891	"	Cal., *gm*. (IST.)	2.34231×10^{-5}
"	Ounces (apoth. or troy)	0.0020833	"	Cal., *gm*. (mean)	2.34048×10^{-5}
"	Ounces (avdp.)	0.0022857143	"	Cal., *gm*. (15°C.)	2.34284×10^{-5}
"	Pennyweights	0.041666	"	Cal., *gm*, (20°C.)	2.34502×10^{-5}
"	Pounds (apoth. or troy)	0.000173611	"	Cal., *kg*	2.34385×10^{-8}
"	Pounds (avdp.)	0.00014285714	"	Cal., *kg*. (IST.)	2.34231×10^{-8}
"	Scruples (apoth.)	0.05	"	Cal., *kg*. (mean)	2.34048×10^{-8}
"	Tons (metric)	6.479891×10^{-8}	"	Dyne-cm	980.665
Grains/cu. ft.	Grams/cu. meter	2.2883519	"	Ergs	980.665
Grains/gal. (U.S.)	Parts/million*	17.11854	"	Foot-poundals	0.00232715
"	Pounds/million gal	142.8571	"	Foot-pounds	7.2330138×10^{-5}
Grams	Carats (metric)	5	"	Hp.-hours	3.65303×10^{-11}
"	Decigrams	10	"	Joules	9.80665×10^{-5}
"	Dekagrams	0.1	"	Kw.-hours	2.72407×10^{-11}
"	Drams (apoth. or troy)	0.25720597	"	Kw.-hours (Int.)	2.72362×10^{-11}
"	Drams (avdp.)	0.56438339	"	Newton-meters	9.80665×10^{-5}
"	Dynes	980.665	"	Watt-hours	2.72407×10^{-8}
"	Grains	15.432358	Gram-cm./sec.	B.t.u./sec.	9.30113×10^{-8}
"	Kilograms	0.001	"	Cal., *gm*./sec.	2.34385×10^{-5}
"	Micrograms	1×10^6	"	Ergs-sec.	980.665
"	Myriagrams	0.0001	"	Foot-pounds/sec.	7.2330138×10^{-5}
"	Ounces (apoth. or troy)	0.032150737	"	Horsepower	1.31509×10^{-7}
"	Ounces (avdp.)	0.035273962	"	Joules/sec.	9.80665×10^{-5}
"	Pennyweights	0.64301493	"	Kilowatts	9.80665×10^{-8}
"	Poundals	0.0709316	"	Kilowatts (Int.)	9.80503×10^{-8}
"	Pounds (apoth. or troy)	0.0026792289	"	Watts	9.80665×10^{-5}
"	Pounds (avdp.)	0.0022046226	Gram/sq. cm	Pounds/sq. inch	0.000341717
"	Scruples (apoth.)	0.77161792	Gram wt.-sec./sq. cm	Poises	980.665
"	Tons (metric)	1×10^{-6}	Gravitational con-		
Grams/cm	Dynes/cm	980.665	stants	Cm./(sec. × sec.)	980.621
"	Grams/inch	2.54	"	Ft./(sec. × sec.)	32.1725
"	Kg./km	100			
"	Kg./meter	0.1	Hands	Centimeters	10.16
"	Poundals/inch	0.180166	"	Inches	4
"	Pounds/ft.	0.067196898	Hectares	Acres	2.4710538
"	Pounds/inch	0.0055997415	"	Ares	100
"	Tons (metric)/km	0.1	"	Sq. cm	1×10^8
Grams/(cm. × sec.)	Poises	1	"	Sq. feet	107639.10
"	Lb./(ft. × sec.)	0.06719690	"	Sq. meters	10000
Grams/cu. cm	Dynes/cu. cm	980.665	"	Sq. miles	0.0038610216
"	Grains/milliliter	15.43279	"	Sq. rods	395.36861
"	Grams/milliliter	1	Hectograms	Grams	100
"	Poundals/cu. inch	1.16236	"	Poundals	7.09316
"	Pounds/circ. mil-ft.	3.4049170×10^{-7}	"	Pounds (apoth or troy)	0.26792289
"	Pounds/cu. ft.	62.427961	"	Pounds (avdp.)	0.22046226
"	Pounds/cu. inch	0.036127292	Hectoliters	Bushels (Brit.)	2.749694
"	Pounds/gal. (Brit.)	10.02241	"	Bushels (U.S.)	2.837839

* Based on density of 1 gram/ml.

To convert from	To	Multiply by
Hectoliters	Cu. cm	1.00028×10^5
"	Cu. feet	3.531566
"	Gallons (U.S., liq.)	26.41794
"	Liters	100
"	Ounces (U.S.) fluid	3381.497
"	Pecks (U.S.)	11.35136
Hectometers	Centimeters	10000
"	Decimeters	1000
"	Dekameters	10
"	Feet	328.08399
"	Meters	100
"	Rods	19.883878
"	Yards	109.3613
Hectowatts	Watts	100
Hefner units	Candles (English)	0.86
"	Candles (German)	0.85
"	Candles (Int.)	0.90
"	10-cp. pentane candles	0.090
Henries	Abhenries	1×10^9
"	E.M. cgs. units	1×10^9
"	E.S. cgs. units	1.112646×10^{-12}
"	Henries (Int.)	0.999505
"	Millihenries	1000
"	Mks. (r or nr) units	1
"	Stathenries	1.112646×10^{-12}
Henries (Int.)	Henries	1.000495
Henries/meter	Cgs. units of permeability	795774.72
"	E.M. cgs. units	795774.72
"	E.S. cgs. units	8.854156×10^{-16}
"	Gausses/oersted	795774.72
"	Mks. (nr) units	0.079577472
"	Mks. (r) units	1
Hogsheads	Butts (Brit.)	0.5
"	Cu. feet	8.421875
"	Cu. inches	14553
"	Cu. meters	0.2384809
"	Gallons (Brit.)	52.458505
"	Gallons (U.S.)	63
"	Gallons (wine)	63
"	Liters	238.47427
Horsepower*	B.t.u. (mean)/hr	2542.48
"	B.t.u./min	42.4356
"	B.t.u. (mean)/sec	0.706243
"	Cal., *gm.*/hr	6.41616×10^5
"	Cal., *gm.* (IST.)/hr	6.41196×10^5
"	Cal., *gm.* (mean)/hr	6.40693×10^5
"	Cal., *gm.*/min	10693.6
"	Cal., *gm.* (IST.)/min	10686.6
"	Cal., *gm.* (mean)/min	10678.2
"	Ergs/sec	7.45700×10^9
"	Foot-pounds/hr	1980000
"	Foot-pounds/min	33000
"	Foot-pounds/sec	550
"	Horsepower (boiler)	0.0760181
"	Horsepower (electric)	0.999598
"	Horsepower (metric)	1.01387
"	Joules/sec	745.700
"	Kilowatts	0.745700
"	Kilowatts (Int.)	0.745577
"	Tons of refrig. (U.S., comm.)	0.21204
"	Watts	745.700
Horsepower (boiler)	B.t.u. (mean)/hr	33445.7
"	Cal., *gm.*/min	140671.6
"	Cal., *gm.* (mean)/min	140469.4
"	Cal., *gm.* (15°C.)/min	140611.1
"	Cal., *gm.*, (20°C.)/min	140742.2
"	Ergs/sec	9.80950×10^{10}
"	Foot-pounds/min	434107
"	Horsepower	13.1548
"	Horsepower (electric)	13.1495

To convert from	To	Multiply by
Horsepower (boiler)	Horsepower (metric)	13.3372
"	Horsepower (water)	13.1487
"	Joules/sec	9809.50
"	Kilowatts	9.80950
"	Lb. H$_2$O evap. per hr. from and at 212°F	34.5
Horsepower (electric)	B.t.u./hr	2547.16
"	B.t.u. (IST.)/hr	2545.50
"	B.t.u. (mean)/hr	2543.50
"	Cal., *gm.*/sec	178.298
"	Cal., *kg.*/hr	641.874
"	Ergs/sec	7.46×10^9
"	Foot-pounds/min	33013.3
"	Foot-pounds/sec	550.221
"	Horsepower	1.00040
"	Horsepower (boiler)	0.0760487
"	Horsepower (metric)	1.0142777
"	Horsepower (water)	0.999942
"	Joules/sec	746
"	Kilowatts	0.746
"	Watts	746
Horsepower (metric)	B.t.u./hr	2511.31
"	B.t.u. (IST.)/hr	2509.66
"	B.t.u. (mean)/hr	2507.70
"	Cal., *gm.*/hr	6.32838×10^5
"	Cal., *gm.* (IST.)/hr	6.32425×10^5
"	Cal., *gm.* (mean)/hr	6.31929×10^5
"	Ergs/sec	7.35499×10^9
"	Foot-pounds/min	32548.6
"	Foot-pounds/sec	542.476
"	Horsepower	0.986320
"	Horsepower (boiler)	0.0740782
"	Horsepower (electric)	0.985923
"	Horsepower (water)	0.985866
"	Kg.-meters/sec	75
"	Kilowatts	0.735499
"	Watts	735.499
Horsepower (water)	Foot-pounds/min	33015.2
"	Horsepower	1.00046
"	Horsepower (boiler)	0.0760531
"	Horsepower (electric)	1.00006
"	Horsepower (metric)	1.01434
"	Kilowatts	0.746043
Horsepower-hours	B.t.u.	2546.14
"	B.t.u. (IST.)	2544.47
"	B.t.u. (mean)	2542.48
"	Cal., *gm.*	641616
"	Cal., *gm.* (IST.)	641196
"	Cal., *gm.* (mean)	640693
"	Foot-pounds	1.98×10^6
"	Joules	2.68452×10^6
"	Kg.-meters	273745
"	Kw.-hours	0.745700
"	Watt-hours	745.700
Hp.-hr./lb.	B.t.u./lb	2546.14
"	Cal., *gm.*/gram	1414.52
"	Cu. ft.-(lb./sq. in.)/lb	13750
"	Foot-pounds/lb	1980000
"	Joules/gram	5918.35
Hours (mean solar)	Days (mean solar)	0.0416666
"	Days (sidereal)	0.041780746
"	Hours (sidereal)	1.00273791
"	Minutes (mean solar)	60
"	Minutes (sidereal)	60.164275
"	Seconds (mean solar)	3600
"	Seconds (sidereal)	3609.8565
"	Weeks (mean calendar)	0.0059523809
Hours (sidereal)	Days (mean solar)	0.41552899
"	Days (sidereal)	0.0416666
"	Hours (mean solar)	0.99726957
"	Minutes (mean solar)	59.836174

* Mechanical horsepower, equal to 550 ft.-lb./sec.

To convert from	To	Multiply by	To convert from	To	Multiply by
Hours (sidereal)	Minutes (sidereal)	60	Joules (abs.)	Cal., *kg.* (mean)	0.000238662
Hundredweights (long)	Kilograms	50.802345	" "	Cu. ft.-atm	0.000348529
"	Pounds	112	" "	Ergs	1×10^7
"	Quarters (Brit., long)	4	" "	Foot-poundals	23.730360
"	Quarters (U.S., long)	0.2	" "	Foot-pounds	0.737562
"	Tons (long)	0.05	" "	Gram-cm	10197.16
Hundredweights (short)	Kilograms	45.359237	" "	Hp.-hours	3.72506×10^{-7}
"	Pounds (advp.)	100	" "	Joules (Int.)	0.999835
"	Quarters (Brit., short)	4	" "	Kg.-meters	0.1019716
"	Quarters (U.S., short)	0.2	" "	Kw.-hours	2.7777×10^{-7}
"	Tons (long)	0.044642857	" "	Liter-atm	0.00986895
"	Tons (metric)	0.045359237	" "	Volt-coulombs (Int.)	0.999835
"	Tons (short)	0.05	" "	Watt-hours (abs.)	0.0002777777
			" "	Watt-hours (Int.)	0.000277732
Inches	Ångström units	2.54×10^8	" "	Watt-sec.	1
"	Centimeters	2.54	" "	Watt-sec. (Int.)	0.999835
"	Chains (Gunter's)	0.00126262	Joules (Int.)	B.t.u.	0.000948608
"	Cubits	0.055555	"	B.t.u. (IST.)	0.000947988
"	Fathoms	0.013888	"	B.t.u. (mean)	0.000947244
"	Feet	0.083333	"	Cal. *gm.*	0.239045
"	Feet (U.S. Survey)	0.083333167	"	Cal., *gm.* (IST.)	0.238888
"	Links (Gunter's)	0.126262	"	Cal., *gm.* (mean)	0.238702
"	Links (Ramden's)	0.083333	"	C.h.u.	0.000527004
"	Meters	0.0254	"	C.h.u. (IST.)	0.000526660
"	Mils	1000	"	C.h.u. (mean)	0.000526247
"	Picas (printer's)	6.0225	"	Cu. cm.-atm	9.87086
"	Points (printer's)	72.27000	"	Cu. ft.-atm	0.000348586
"	Wave length of orange-red line of krypton 86	41929.399	"	Dyne-cm	1.000165×10^7
"	Wave length of the red line of cadmium	39450.369	"	Ergs	1.000165×10^7
"	Yards	0.027777	"	Foot-poundals	23.73428
Inches of Hg (32°F.)	Atmospheres	0.0334211	"	Foot-pounds	0.737684
"	Bars	0.0338639	"	Gram-cm	10198.8
"	Dynes/sq. cm.	33863.9	"	Joules (abs.)	1.000165
"	Ft. of air (1 atm., 60°F.)	926.24	"	Kw.-hours	2.77824×10^{-7}
"	Ft. of H_2O (39.2°F.)	1.132957	"	Liter-atm	0.00987058
"	Grams/sq. cm.	34.5316	"	Volt-coulombs	1.000165
"	Kg./sq. meter	345.316	"	Volt-coulombs (Int.)	1
"	Mm. of Hg (60°C.)	25.4	"	Watt-sec.	1.000165
"	Ounces/sq. inch	7.85847	"	Watt-sec. (Int.)	1
Inches of Hg (32°F.)	Pounds/sq. ft.	70.7262	Joules/(abcoulomb × °F.)	Joules/(coulomb × °C.)	0.18
Inches of Hg (60°F.)	Atmospheres	0.0333269	Joules/amp.-hr.	Joules/abcoulomb	0.002777
"	Dynes/sq. cm.	39768.5	"	Joules/statcoulomb	9.265653×10^{-14}
"	Grams/sq. cm.	34.4343	Joules/coulomb	Joules/abcoulomb	10
"	Mm. of Hg (60°F.)	25.4	"	Volts	1
"	Ounces/sq. inch	7.83633	Joules/(coulomb × °F.)	Joules/(coulomb × °C.)	1.8
"	Pounds/sq. ft.	70.5269	Joules/°C	B.t.u./°F	0.000526917
Inches of H_2O(4°C.)	Atmospheres	0.0024582	"	Cal., *gm.*/°C	0.239006
"	Dynes/sq. cm.	2490.82	"	Cal., *gm.* (mean)/°C	0.238662
"	In. of Hg (32°F.)	0.0735539	Joules/electronic charge	Joules/abcoulomb	6.24196×10^{19}
"	Kg./sq. meter	25.3993	Joules/(electronic charge × °C.)	Joules/(coulomb × °C.)	6.24196×10^{18}
"	Ounces/sq. ft.	83.2350	Joules/(gram × °C.)	B.t.u./(lb. × °F.)	0.239006
"	Ounces/sq. inch	0.578020	"	Cal., *gm.*/(gram × °C.)	0.239006
"	Pounds/sq. ft.	5.20218	Joules (Int.)/(gram °C.)	B.t.u./(lb. × °F.)	0.239045
"	Pounds/sq. inch	0.03612628	"	Cal., *gm.* (mean)/(gram × °C.)	0.238702
Inches/hr.	Cm./hr.	2.54	Joules/sec. (abs.)	B.t.u./min	0.0569071
"	Feet/hr.	0.0833333	"	Cal., *gm.*/min	14.3403
"	Miles/hr.	1.578282×10^{-5}	"	Cal., *kg.*/min	0.0143403
Inches/min	Cm./hr.	152.4	"	Cal., *kg.* (mean)/min	0.0143197
"	Feet/hr.	5	"	Dyne-cm./sec	1×10^7
"	Miles/hr.	0.000946969	"	Ergs/sec	1×10^7
			"	Foot-pounds/sec	0.737562
Joules (abs.)	B.t.u.	0.000948451	"	Gram-cm./sec	10197.16
" "	B.t.u. (IST.)	0.000947831	"	Horsepower	0.00134102
" "	B.t.u. (mean)	0.000947088	"	Watts	1
" "	Cal., *gm.*	0.239006	"	Watts (Int.)	0.999835
" "	Cal., *gm.* (IST.)	0.238849	Joules (Int.)/sec	B.t.u./min	0.0569165
" "	Cal., *gm.* (mean)	0.238662			
" "	Cal., *gm.* (15°C.)	0.238903			
" "	Cal., *gm.* (20°C.)	0.239126			

To convert from	To	Multiply by	To convert from	To	Multiply by
Joules (Int.)/sec	B.t.u. (mean)/min	0.0568347	Kilogram-meters	Hp.-hours	3.65304×10^{-6}
"	Cal., gm./min	14.3427	"	Joules	9.80665
"	Cal., kg./min	0.0143427	"	Joules (Int.)	9.80503
"	Dyne-cm./sec	1.000165×10^7	"	Kw.-hours	2.72407×10^{-6}
"	Ergs/sec	1.000165×10^7	"	Liter-atm	0.0967814
"	Foot-pounds/min	44.2610	"	Newton-meters	9.80665
"	Foot-pounds/sec	0.737684	"	Watt-hours	0.00272407
"	Gram-cm./sec	10198.8	"	Watt-hours (Int.)	0.00272362
"	Horsepower	0.00134124	Kilogram-meters/sec	Watts	9.80665
"	Watts	1.000165	Kilolines	Maxwells	1000
"	Watts (Int.)	1	"	Webers	1×10^{-5}
			Kiloliters	Cu. centimeters	1×10^6
Kilderkins (Brit.)	Cu. cm	81829.57	"	Cu. feet	35.31566
"	Cu. feet	2.889784	"	Cu. inches	61025.45
"	Cu. inches	4993.55	"	Cu. meters	1.000028
"	Cu. meters	0.08182957	"	Cu. yards	1.307987
"	Gallons (Brit.)	18	"	Gallons (Brit.)	219.9755
Kilograms	Drams (apoth. or troy)	257.20597	"	Gallons (U.S., dry)	227.0271
"	Drams (avdp.)	564.38339	"	Gallons (U.S., liq.)	264.1794
"	Dynes	980665	"	Liters	1000
"	Grains	15432.358	Kilometers	Astronomical units	6.68878×10^{-9}
"	Hundredweights (long)	0.019684131	"	Centimeters	100000
"	Hundredweights (short)	0.022046226	"	Feet	3280.8399
"	Ounces (apoth. or troy)	32.150737	"	Feet (U.S. Survey)	3280.833
"	Ounces (avdp.)	35.273962	"	Light years	1.05702×10^{-13}
"	Pennyweights	643.01493	"	Meters	1000
"	Poundals	70.931635	"	Miles (naut., Int.)	0.53995680
"	Pounds (apoth. or troy)	2.6792289	"	Miles (statute)	0.62137119
"	Pounds (avdp.)	2.2046226	"	Myriameters	0.1
"	Quarters (Brit., long)	0.078736522	"	Rods	198.83878
"	Quarters (U.S. long)	0.0039368261	"	Yards	1093.6133
"	Scruples (apoth.)	771.61702	Kilometers/hr	Cm./sec	27.7777
"	Slugs	0.068521177	"	Feet/hr	3280.8399
"	Tons (long)	0.00098420653	"	Feet/min	54.680665
"	Tons (metric)	0.001	"	Knots (Int.)	0.53995680
"	Tons (short)	0.0011020115	"	Meters/sec	0.277777
Kilograms/cu. meter	Grams/cu. cm	0.001	"	Miles (statute)/hr	0.62137119
"	Lb./cu. ft	0.062427961	Kilometers/(hr. × sec.)	Cm./(sec. × sec.)	27.7777
"	Lb./cu. inch	3.6127292×10^{-5}	"	Ft./(sec. × sec.)	0.91134442
Kg. of ice melted/hr	Tons of refrig. (U.S., comm.)	0.026336	"	Meters/(sec. × sec.)	0.277777
Kilograms/sq. cm	Atmospheres	0.967841	Kilometers/min	Cm./sec	1666.666
"	Bars	0.980665	"	Feet/min	3280.8399
"	Cm. of Hg (0°C.)	73.5559	"	Kilometers/hr	60
"	Dynes/sq. cm	980665	"	Knots (Int.)	32.397408
"	Ft. of H$_2$O (39.2°F.)	32.8093	"	Miles/hr	37.282272
"	In. of Hg (32°F.)	28.9590	"	Miles/min	0.62137119
"	Pounds/sq. inch	14.223343	Kilovolts/cm	Abvolts/cm	1×10^{11}
Kilograms/sq. meter	Atmospheres	9.67841×10^{-5}	"	Microvolts/meter	1×10^{11}
"	Bars	9.80665×10^{-5}	"	Millivolts/meter	1×10^8
"	Dynes/sq. cm	98.0665	"	Statvolts/cm	3.335635
"	Ft. of H$_2$O (39.2°F.)	0.00328093	"	Volts/inch	2540
"	Grams/sq. cm	0.1	Kilowatts	B.t.u./hr	3414.43
"	In. of Hg. (32°F.)	0.00289590	"	B.t.u. (IST.)/hr	3412.19
"	Mm. of Hg (0°C.)	0.0735559	"	B.t.u. (mean)/hr	3409.52
"	Pounds/sq. ft	0.20481614	"	B.t.u. (mean)/min	56.8253
"	Pounds/sq. in	0.0014223343	"	B.t.u. (mean)/sec	0.947088
Kilograms/sq. mm	Pounds/sq. ft	204816.14	"	Cal., gm. (mean)/hr	859184
"	Pounds/sq. in	1422.3343	"	Cal., gm. (mean)/min	14319.7
"	Tons (short)/sq. in	0.71116716	"	Cal., gm. (mean)/sec	238.662
Kilogram sq. cm	Pounds sq. ft	0.0023730360	"	Cal., kg. (mean)/hr	859.184
"	Pounds sq. in	0.34171719	"	Cal., kg. (mean)/min	14.3197
Kilogram-meters	B.t.u. (mean)	0.00928776	"	Cal., kg. (mean)/sec	0.238662
"	Cal., gm. (mean)	2.34048	"	Cu. ft.-atm./hr	1254.70
"	Cal., kg. (mean)	0.00234048	"	Ergs/sec	1×10^{10}
"	Cu. ft.-atm	0.00341790	"	Foot-poundals/min	1.42382×10^6
"	Dynes-cm	9.80665×10^7	"	Foot-pounds/hr	2.65522×10^6
"	Ergs	9.80665×10^7	"	Foot-pounds/min	44253.7
"	Foot-poundals	232.715	"	Foot-pounds/sec	737.562
"	Foot-pounds	7.23301	"	Gram-cm./sec	1.019716×10^7
"	Gram-cm	100000	"	Horsepower	1.34102

To convert from	To	Multiply by	To convert from	To	Multiply by
Kilowatts	Horsepower (boiler)	0.101942	Lamberts	Candles/sq. cm.	0.31830989
"	Horsepower (electric)	1.34048	"	Candles/sq. ft.	295.71956
"	Horsepower (metric)	1.35962	"	Candles/sq. inch	2.0536081
"	Joules/hr.	3.6×10^6	"	Foot-lamberts	929.0304
"	Joules (IST.)/hr.	3.59941×10^6	"	Lumens/sq. cm.	1
"	Joules/sec.	1000	Lasts (Brit.)	Liters	2909.414
"	Kg.-meters/hr.	3.67098×10^5	Leagues (naut., Brit.)	Feet	18240
"	Kilowatts (Int.)	0.999835	"	Kilometers	5.559552
"	Watts (Int.)	999.835	"	Leagues (naut., Int.)	1.0006393
Kilowatts (Int.)	B.t.u./hr.	3414.99	"	Leagues (statute)	1.151515
"	B.t.u. (IST.)/hr.	3412.76	"	Miles (statute)	3.454545
"	B.t.u. (mean)/hr.	3410.08	Leagues (naut., Int.)	Fathoms	3038.0577
"	B.t.u. (mean)/min.	56.8347	"	Feet	18228.346
"	B.t.u. (mean)/sec.	0.947244	"	Kilometers	5.556
"	Cal., gm. (mean)/hr.	859326	"	Leagues (statute)	1.1507794
"	Cal., gm. (mean)/min.	14322.1	"	Miles (statute)	3.4523383
"	Cal., kg./hr.	860.563	Leagues (statute)	Fathoms	2640
"	Cal., kg. (IST.)/hr.	860	"	Feet	15840
"	Cal., kg. (mean)/hr.	859.326	"	Kilometers	4.828032
"	Cu. cm.-atm./hr.	3.55351×10^7	"	Leagues (naut., Int.)	0.86897625
"	Cu. ft.-atm./hr.	1254.91	"	Miles (naut., Int.)	2.6069287
"	Ergs/sec.	1.000165×10^{10}	"	Miles (statute)	3
"	Foot-poundals/min.	1.42406×10^6	Light years	Astronomical units	63279.5
"	Foot-pounds/min.	44261.0	"	Kilometers	9.46055×10^{12}
"	Foot-pounds/sec.	737.684	"	Miles (statute)	5.87851×10^{12}
"	Gram-cm./sec.	1.01988×10^7	Lines	Maxwells	1
"	Horsepower	1.34124	Lines (Brit.)	Centimeters	0.211666
"	Horsepower (boiler)	0.101959	"	Inches	0.083333
"	Horsepower (electric)	1.34070	Lines/sq. cm.	Gausses	1
"	Horsepower (metric)	1.35985	Lines/sq. inch	Gausses	0.15500031
"	Joules/hr.	3.60059×10^6	"	Webers/sq. inch	1×10^{-8}
"	Joules (Int.)/hr.	3.6×10^6	Links (Gunter's)	Chains (Gunter's)	0.01
"	Kg.-meters/hr.	367158	"	Feet	0.66
"	Kilowatts	1.000165	"	Feet (U.S. Survey)	0.65999868
Kilowatt-hours	B.t.u. (mean)	3409.52	"	Inches	7.92
"	Cal., gm. (mean)	859184	"	Meters	0.201168
"	Foot-pounds	2.65522×10^6	"	Miles (statute)	0.000125
"	Hp.-hours	1.34102	"	Rods	0.04
"	Joules	3.6×10^6	Links (Ramden's)	Centimeters	30.48
"	Kg.-meters	367098	"	Chains (Ramdens)	0.01
"	Lb. H$_2$O evap. from and at 212°F.	3.5168	"	Feet	1
"	Watt-hours	1000	"	Inches	12
"	Watt-hours (Int.)	999.835	Liters	Bushels (Brit.)	0.02749617
Kilowatt-hours (Int.)	B.t.u. (mean)	3410.08	"	Bushels (U.S.)	0.02837759
"	Cal., gm. (IST.)	860000	"	Cu. centimeters	1000
"	Cal., gm. (mean)	859326	"	Cu. feet	0.035314667
"	Cu. cm-atm.	3.55351×10^7	"	Cu. inches	61.023744
"	Cu. ft.-atm.	1254.91	"	Cu. meters	0.001
"	Foot-pounds	2.65566×10^6	"	Cu. yards	0.0013079506
"	Hp.-hours	1.34124	"	Drams (U.S., fluid)	270.51218
"	Joules	3.60059×10^6	"	Gallons (Brit.)	0.2199694
"	Joules (Int.)	3.6×10^6	"	Gallons (U.S., dry)	0.22702075
"	Kg.-meters	367158	"	Gallons (U.S., liq.)	0.26417205
Kw.-hr./gram	B.t.u./lb.	1.54876×10^6	"	Gills (Brit.)	7.039020
"	B.t.u. (IST.)/lb.	1.54774×10^6	"	Gills (U.S.)	8.4535058
"	B.t.u. (mean)/lb.	1.54653×10^6	"	Hogsheads	0.004193212
"	Cal., gm./gram	860421	"	Minims (U.S.)	16320.75
"	Cal., gm. (mean)/gram	859184	"	Ounces (Brit., fluid)	35.19510
"	Cu. cm.-atm./gram	3.55292×10^7	"	Ounces (U.S., fluid)	33.814023
"	Cu. ft.-atm./lb.	569124	"	Pecks (Brit.)	0.1099848
"	Hp.-hr./lb.	608.277	"	Pecks (U.S.)	0.1135105
"	Joules/gram	3.6×10^6	"	Pints (Brit.)	1.795756
Knots (Int.)	Cm./sec.	51.4444	"	Pints (U.S., dry)	1.8161660
"	Feet/hr.	6076.1155	"	Pints (U.S., liq.)	2.1133764
"	Feet/min.	101.26859	"	Quarts (Brit.)	0.8798775
"	Feet/sec.	1.6878099	"	Quarts (U.S., dry)	0.90808298
"	Kilometers/hr.	1.852	"	Quarts (U.S., liq.)	1.0566882
"	Meters/min.	30.8666	Liters/min.	Cu. ft./min.	0.0353147
"	Meters/sec.	0.514444	"	Cu. ft./sec.	0.000588578
"	Miles (naut., Int.)/hr.	1	"	Gal. (U.S., liq.)/min.	0.26417226
"	Miles (statute)/hr.	1.1507794	Liters/sec.	Cu. ft./min.	2.118880
			"	Cu. ft./sec.	0.0353147

To convert from	To	Multiply by	To convert from	To	Multiply by
Liters/sec	Cu. yards/min	0.0784771	Meters	Links (Ramden's)	3.2808399
"	Gal. (U.S., liq.)/min	15.850342	"	Megameters	1×10^{-6}
"	Gal. (U.S., liq.)/sec	0.2641723	"	Miles (naut., Brit.)	0.00053961182
Liter-atm	B.t.u	0.0961045	"	Miles (naut., Int.)	0.00053995680
"	B.t.u. (IST.)	0.0960417	"	Miles (statute)	0.00062137119
"	B.t.u. (mean)	0.0959664	"	Millimeters	1000
"	Cal., gm	24.2179	"	Millimicrons	1×10^{9}
"	Cal., gm. (IST.)	24.2021	"	Mils	39370.079
"	Cal., gm. (mean)	24.1831	"	Rods	0.19883878
"	Cu. ft.-atm	0.0353157	"	Yards	1.0936133
"	Foot-poundals	2404.55	Meters of Hg (0°C.)	Atmospheres	1.3157895
"	Foot-pounds	74.7356	"	Ft. of H₂O (60°F.)	44.6474
"	Hp.-hours	3.77452×10^{-5}	"	In. of Hg (32°F.)	39.370079
"	Joules	101.328	"	Kg./sq. cm	1.35951
"	Joules (Int.)	101.311	"	Pounds/sq. inch	19.3368
"	Kg.-meters	10.3326	Meters/hr	Feet/hr	3.2808399
"	Kw.-hours	2.81466×10^{-5}	"	Feet/min	0.054680665
Liter-atm. (lat. 45°)	Joules	101.323	"	Knots (Int.)	0.00053995680
Lumens	Candle power (spher.)	0.079577472	"	Miles (statute)/hr	0.00062137119
Lumens (at 5550 Å)	Watts	0.0014705882	Meters/min	Cm./sec	1.666666
Lumens/sq. cm	Lamberts	1	"	Feet/min	3.2808399
"	Phots	1	"	Feet/sec	0.054680665
Lumens/(sq. cm. × steradian)	Lamberts	3.1415927	"	Kilometers/hr	0.06
Lumens/sq. ft	Foot-candles	1	"	Knots (Int.)	0.032397408
"	Foot-lamberts	1	"	Miles (statute)/hr	0.037282272
"	Lumens/sq. meter	10.763910	Meters/sec	Feet/min	196.85039
Lumens/(sq. ft. × steradian)	Millilamberts	3.3815822	"	Feet/sec	3.2808399
Lumens/sq. meter	Foot-candles	0.09290304	"	Kilometers/hr	3.6
"	Lumens/sq. ft	0.09290304	"	Kilometers/min	0.06
"	Phots	0.0001	"	Miles (statute)/hr	2.2369363
Lux	Foot-candles	0.09290304	Meters/(sec. × sec.)	Kilometers/(hr.)(sec.)	3.6
"	Lumens/sq. meter	1	"	Miles/(hr. × sec.)	2.2369363
"	Phots	0.0001	Meter-candles	Lumens/sq. meter	1
Maxwells	E.M. cgs. units of induction flux	1	Mhos	Abmhos	1×10^{-9}
"	E.S. cgs. units	3.335635×10^{-11}	"	Cgs. units of conductance	1
"	Gauss-sq. cm	1	"	E.M. cgs. units	1×10^{-9}
"	Lines	1	"	E.S. cgs. units	8.987584×10^{11}
"	Maxwells (Int.)	0.999670	"	Mhos (Int.)	1.000495
"	Volt-seconds	1×10^{-8}	"	Mks. (r or nr) units	1
"	Webers	1×10^{-8}	"	Ohms⁻¹	1
Maxwells (Int.)	Maxwells	1.000330	"	Siemen's units	1
Maxwells/sq. cm	Maxwells/sq. in	6.4516	"	Statmhos	8.987584×10^{11}
"	Maxwells (Int.)/sq. cm	0.999670	Mhos (Int.)	Abmhos	9.99505×10^{-10}
Maxwells (Int.)/sq. cm	Maxwells/sq. cm	1.000330	"	Mhos	0.999505
Maxwells/sq. inch	Maxwells/sq. cm	0.15500031	Mhos/meter	Abmhos/cm	1×10^{-11}
Megalines	Maxwells	1×10^{6}	"	Mhos (Int.)/meter	1.000495
Megmhos/cm	Abmhos/cm	0.001	Mho-ft./circ. mil	Mhos/cm	6.0153049×10^{6}
"	Megmhos/inch cube	2.54	Microfarads	Abfarads	1×10^{-15}
"	(Microhm-cm.)⁻¹	1	"	Farads	1×10^{-6}
Megmhos/inch	Megmhos/cm	0.39370079	"	Statfarads	8.987584×10^{5}
"	(Microhm-inches)⁻¹	1	Micrograms	Grams	1×10^{-6}
Megohms	Microhms	1×10^{12}	"	Milligrams	0.001
"	Ohms	1×10^{6}	Microhenries	Henries	1×10^{-6}
"	Statohms	1.112646×10^{-6}	"	Stathenries	1.112646×10^{-18}
Megohms⁻¹	Micromhos	1	Microhms	Abohms	1000
Meters	Ångström units	1×10^{10}	"	Megohms	1×10^{-12}
"	Centimeters	100	"	Ohms	1×10^{-6}
"	Chains (Gunter's)	0.049709695	"	Statohms	1.112646×10^{-18}
"	Chains (Ramden's)	0.032808399	Microhm-cm	Abohm-cm	1000
"	Fathoms	0.54680665	"	Circ. mil-ohms/ft	6.0153049
"	Feet	3.2808399	"	Microhm-inches	0.39370079
"	Feet (U.S. Survey)	3.280833	"	Ohm-cm	1×10^{-6}
"	Furlongs	0.0049709695	Microhm-inches	Circ. mil-ohms/ft	15.278875
"	Inches	39.370079	"	Michrom-cm	2.54
"	Kilometers	0.001	Micromicrofarads	Farads	1×10^{-12}
"	Links (Gunter's)	4.9709695	Micromicrons	Ångström units	0.01
			"	Centimeters	1×10^{-10}
			"	Inches	$3.9370079 \times 10^{-11}$
			"	Meters	1×10^{-12}
			"	Microns	1×10^{-6}
			Microns	Ångström units	10000

To convert from	To	Multiply by	To convert from	To	Multiply by
Microns	Centimeters	0.0001	Milligrams	Grains	0.015432358
"	Feet	3.2808399×10^{-6}	"	Grams	0.001
"	Inches	3.9370079×10^{-5}	"	Ounces (apoth. or troy)	3.2150737×10^{-5}
"	Meters	1×10^{-6}	"	Ounces (avdp.)	3.5273962×10^{-5}
"	Millimeters	0.001	"	Pennyweights	0.00064301493
"	Millimicrons	1000	"	Pounds (apoth. or troy)	2.6792289×10^{-6}
Miles (naut., Brit.)	Cable lengths (Brit.)	8.4444	"	Pounds (avdp.)	2.2046226×10^{-6}
"	Fathoms	1013.333	"	Scruples (apoth.)	0.00077161792
"	Feet	6080	Milligrams/assay ton	Milligrams/kg	34.285714
"	Meters	1853.184	"	Ounces (troy)/ton (avdp.)	1
"	Miles (Adm., Brit.)	1	Milligrams/gm	Dynes/cm	0.980665
"	Miles (naut., Int.)	1.0006393	"	Pounds/inch	5.5997415×10^{-6}
"	Miles (statute)	1.151515	Milligrams/gram	Carats (parts gold per 24 of mixture)	0.024
Miles (naut., Int.)	Cable lengths	8.4390493	"	Grams/ton (short)	907.18474
"	Fathoms	1012.6859	"	Milligrams/assay ton	29.166666
"	Feet	6076.1155	"	Ounces (avdp.)/ton (long)	35.84
"	Feet (U.S. Survey)	6076.1033	"	Ounces (avdp.)/ton (short)	32
"	Kilometers	1.852	"	Ounces (troy)/ton (long)	32.6666
"	Leagues (naut., Int.)	0.333333	"	Ounces (troy)/ton (short)	29.1666
"	Meters	1852	Milligrams/inch	Dynes/cm	0.386089
"	Miles (geographical)	1	"	Dynes/inch	0.980665
"	Miles (naut. Brit.)	0.99936110	"	Grams/cm	0.00039370079
"	Miles (statute)	1.1507794	"	Grams/inch	0.0001
Miles (statute)	Centimeters	160934.4	Milligrams/kg	Pounds (avdp.)/ton (short)	0.002
"	Chains (Gunter's)	80	Milligrams/liter	Grains/gal. (U.S.)	0.05841620
"	Chains (Ramden's)	52.8	"	Grams/liter	0.001
"	Feet	5280	"	Parts/million*	1
"	Feet (U.S. Survey)	5279.9894	"	Lb./cu. ft.	6.242621×10^{-5}
"	Furlongs	8	Milligrams/mm	Dynes/cm	9.80665
"	Inches	63360	Millihenries	Abhenries	1×10^{6}
"	Kilometers	1.609344	"	Henries	0.001
"	Light years	1.70111×10^{-13}	"	Stathenries	1.112646×10^{-15}
"	Links (Gunter's)	8000	Millilamberts	Candles/sq. cm	0.00031830989
"	Meters	1609.344	"	Candles/sq. inch	0.0020536081
"	Miles (naut., Brit.)	0.86842105	"	Foot-lamberts	0.9290304
"	Miles (naut., Int.)	0.86897624	"	Lamberts	0.001
"	Myriameters	0.1609344	"	Lumens/sq. cm	0.001
"	Rods	320	"	Lumens/sq. ft	0.9290304
"	Yards	1760	Milliliters	Cu. cm	1
Miles/hr	Cm./sec	44.704	"	Cu. inches	0.06102545
"	Feet/hr	5280	"	Drams (U.S., fluid)	0.2705198
"	Feet/min	88	"	Gills (U.S.)	0.008453742
"	Feet/sec	1.466666	"	Liters	0.001
"	Kilometers/hr	1.609344	"	Minims (U.S.)	16.23119
"	Knots (Int.)	0.86897624	"	Ounces (Brit., fluid)	0.03519609
"	Meters/min	26.8224	"	Ounces (U.S., fluid)	0.03381497
"	Miles/min	0.0166666	"	Pints (Brit.)	0.001759804
Miles/(hr. × min.)	Cm./(sec. × sec.)	0.7450666	"	Pints (U.S., liq.)	0.002113436
Miles/(hr. × sec.)	Cm./(sec. × sec.)	44.704	Millimeters	Ångström units	1×10^{7}
"	Ft./(sec. × sec.)	1.466666	"	Centimeters	0.1
"	Kilometers/(hr. × sec.)	1.609344	"	Decimeters	0.01
"	Meters/(sec. × sec.)	0.44704	"	Dekameters	0.0001
Miles/min	Cm./sec	2682.24	"	Feet	0.0032808399
"	Feet/hr	316800	"	Inches	0.039370079
"	Feet/sec	88	"	Meters	0.001
"	Kilometers/min	1.609344	"	Microns	1000
"	Knots (Int.)	52.138574	"	Mils	39.370079
"	Meters/min	1609.344	"	Wave length of orange-red line of krypton 86	1650.76373
"	Miles/hr	60	"	Wave length of red line of cadmium	1553.16413
Millibars	Atmospheres	0.000986923	Millimeters of Hg (0°C.)	Atmospheres	0.0013157895
"	Bars	0.001	"	Bars	0.00133322
"	Baryes	1000	"	Dynes/sq. cm	1333.224
"	Dynes/sq. cm	1000	"	Grams/sq. cm	1.35951
"	Grams/sq. cm	1.019716	"	Kg./sq. meter	13.5951
"	In. of Hg (32°F.)	0.0295300	"	Pounds/sq. ft.	2.78450
"	Pounds/sq. ft.	2.088543	"	Pounds/sq. inch	0.0193368
"	Pounds/sq. inch	0.0145038	"	Torrs	1
Milligrams	Carats (1877)	0.004871			
"	Carats (metric)	0.005			
"	Drams (apoth. or troy)	0.00025720597			
"	Drams (advp.)	0.00056438339			

* Density of 1 gram per milliliter of solvent.

To convert from	To	Multiply by	To convert from	To	Multiply by
Millimicrons	Ångström units	10	Oersteds	Gilberts/cm	1
"	Centimeters	1×10^{-7}	"	Oersteds (Int.)	1.000165
"	Inches	3.9370079×10^{-8}	Oersteds (Int.)	Oersteds	0.999835
"	Microns	0.001	Ohms	Abohms	1×10^{9}
"	Millimeters	1×10^{-6}	"	Cgs. units of resistance	1
Milliphots	Foot-candles	0.9290304	"	Megohms	1×10^{-6}
"	Lumens/sq. ft	0.9290304	"	Microhms	1×10^{6}
"	Lumens/sq. meter	10	"	Ohms (Int.)	0.999505
"	Lux	10	"	Statohms	1.112646×10^{-12}
"	Phots	0.001	Ohms (Int.)	Ohms	1.000495
Millivolts	Statvolts	3.335635×10^{-6}	Ohms (mil, foot)	Circ. mil-ohms/ft	1
"	Volts	0.001	"	Ohm-cm	1.6624261×10^{-7}
Minims (Brit.)	Cu. cm	0.05919385	Ohm-cm	Circ. mil-ohms/ft	6.0153049×10^{6}
"	Cu. inches	0.003612230	"	Microhm-cm	1×10^{6}
"	Milliliters	0.05919219	"	Ohm-inches	0.39370079
"	Ounces (Brit., fluid)	0.0020833333	Ohm-inches	Ohm-cm	2.54
"	Scruples (Brit., fluid)	0.05	Ohm-meters	Abohm-cm	1×10^{11}
Minims (U.S.)	Cu. cm	0.061611520	"	E.M. cgs. units	1×10^{11}
"	Cu. inches	0.0037597656	"	E.S. cgs. units	1.112646×10^{-10}
"	Drams (U.S., fluid)	0.0166666	"	Mks. units	1
"	Gallons (U.S., liq.)	1.6276042×10^{-5}	"	Statohm-cm	1.112646×10^{-10}
"	Gills (U.S.)	0.0005208333	Ounces (apoth. or troy)	Dekagrams	3.1103486
"	Liters	6.160979×10^{-5}	"	Drams (apoth. or troy)	8
"	Milliliters	0.06160979	"	Drams (avdp.)	17.554286
"	Ounces (U.S., fluid)	0.002083333	"	Grains	480
"	Pints (U.S., liq.)	0.0001302083	"	Grams	31.103486
Minutes (angular)	Degrees	0.0166666	"	Milligrams	31103.486
"	Quadrants	0.000185185	"	Ounces (avdp.)	1.0971429
"	Radians	0.00029088821	"	Pennyweights	20
"	Seconds (angular)	60	"	Pounds (apoth. or troy)	0.0833333
Minutes (mean solar)	Days (mean solar)	0.0006944444	"	Pounds (avdp.)	0.068571429
"	Days (sidereal)	0.00069634577	"	Scruples (apoth.)	24
"	Hours (mean solar)	0.0166666	"	Tons (short)	3.4285714×10^{-5}
"	Hours (sidereal)	0.016712298	Ounces (avdp.)	Drams (apoth. or troy)	7.291666
"	Minutes (sidereal)	1.00273791	"	Drams (avdp.)	16
Minutes (sidereal)	Days (mean solar)	0.00069254831	"	Grains	437.5
"	Minutes (mean solar)	0.99726957	"	Grams	28.349523
"	Months (mean calendar)	2.2768712×10^{-5}	"	Hundredweights (long)	0.00055803571
"	Seconds (sidereal)	60	"	Hundredweights (short)	0.000625
Minutes/cm	Radians/cm	0.00029088821	"	Ounces (apoth. or troy)	0.9114583
Months (lunar)	Days (mean solar)	29.530588	"	Pennyweights	18.229166
"	Hours (mean solar)	708.73411	"	Pounds (apoth. or troy)	0.075954861
"	Minutes (mean solar)	42524.047	"	Pounds (avdp.)	0.0625
"	Second (mean solar)	2.5514428×10^{6}	"	Scruples (apoth.)	21.875
"	Weeks (mean calendar)	4.2186554	"	Tons (long)	2.7901786×10^{-5}
Months (mean calendar)	Days (mean solar)	30.416666	"	Tons (metric)	2.8349527×10^{-5}
"	Hours (mean solar)	730	"	Tons (short)	3.125×10^{-5}
"	Months (lunar)	1.0300055	Ounces (Brit., fluid)	Cu. cm	28.41305
"	Weeks (mean calendar)	4.3452381	"	Cu. inches	1.733870
"	Years (calendar)	0.08333333	"	Drachms (Brit., fluid)	8
"	Years (sidereal)	0.083274845	"	Drams (U.S., fluid)	7.686075
"	Years (tropical)	0.083278075	"	Gallons (Brit.)	0.00625
Myriagrams	Grams	10000	"	Milliliters	28.41225
"	Kilograms	10	"	Minims (Brit.)	480
"	Pounds (avdp.)	22.046226	"	Ounces (U.S., fluid)	0.9607594
			Ounces (U.S., fluid)	Cu. cm	29.573730
Newtons	Dynes	100000	"	Cu. inches	1.8046875
"	Pounds	0.22480894	"	Cu. meters	2.9573730×10^{-5}
Newton-meters	Dyne-cm	1×10^{7}	"	Drams (U.S., fluid)	8
"	Gram-cm	10197.162	"	Gallons (U.S., dry)	0.0067138047
"	Kg.-meters	0.10197162	"	Gallons (U.S., liq.)	0.0078125
"	Pound-feet	0.73756215	"	Gills (U.S.)	0.25
Noggins (Brit.)	Cu. cm	142.0652	"	Liters	0.029572702
"	Gallons (Brit.)	0.03125	"	Minims (U.S.)	480
"	Gills (Brit.)	1	"	Ounces (Brit., fluid)	**1.040843**
			"	Pints (U.S., liq.)	**0.0625**
Oersteds	Ampere-turns/inch	2.0212678	"	Quarts (U.S., liq.)	**0.03125**
"	Ampere-turns/meter	79.577472	Ounces/sq. inch	Dynes/sq. cm	4309.22
"	E.M. cgs. units of magnetic field intensity	1	"	Grams/sq. cm	4.3941849
"	E.S. cgs. units	2.997930×10^{10}	"	In. of H_2O (39.2°F.)	1.73004
			"	In. of H_2O (60°F.)	1.73166

CONVERSION FACTORS (Continued)

To convert from	To	Multiply by	To convert from	To	Multiply by
Ounces/sq. inch	Pounds/sq. ft	9	Pints (Brit.)	Minims (Brit.)	9600
"	Pounds/sq. inch	0.0625	"	Ounces (Brit., fluid)	20
Ounces (avdp.)/ton (long)	Milligrams/kg	27.901786	"	Pints (U.S., dry)	1.032056
Ounces (avdp.)/ton (short)	Milligrams/kg	31.25	"	Pints (U.S., liq.)	1.200949
			"	Quarts (Brit.)	0.5
			"	Scruples (Brit., fluid)	480
Paces	Centimeters	76.2	Pints (U.S., dry)	Bushels (U.S.)	0.015625
"	Chains (Gunter's)	0.0378788	"	Cu. cm	550.61047
"	Chains (Ramden's)	0.025	"	Cu. inches	33.6003125
"	Feet	2.5	"	Gallons, (U.S., dry)	0.125
"	Hands	7.5	"	Gallons (U.S., liq.)	0.14545590
"	Inches	30	"	Liters	0.5505951
"	Ropes (Brit.)	0.125	"	Pecks (U.S.)	0.0625
Palms	Centimeters	7.62	"	Quarts (U.S., dry)	0.5
"	Chains (Ramden's)	0.0025	Pints (U.S., liq.)	Cu. cm	473.17647
"	Cubits	0.1666666	"	Cu. feet	0.016710069
"	Feet	0.25	"	Cu. inches	28.875
"	Hands	0.75	"	Cu. yards	0.00061889146
"	Inches	3	"	Drams (U.S., fluid)	128
Parsecs	Kilometers	3.08572×10^{13}	"	Gallons (U.S., liq.)	0.125
"	Miles (statute	1.91738×10^{13}	"	Gills (U.S.)	4
Parts/million*	Grains/gal. (Brit.)	0.07015488	"	Liters	0.4731632
"	Grains/gal. (U.S.)	0.05841620	"	Milliliters	473.1632
"	Grams/liter	0.001	"	Minims (U.S.)	7680
"	Milligrams/liter	1	"	Ounces (U.S., fluid)	16
Pecks (Brit.)	Bushels (Brit.)	0.25	"	Pints (Brit.)	0.8326747
"	Coombs (Brit.)	0.0625	"	Quarts (U.S., liq.)	0.5
"	Cu. cm	9092.175	Planck's constant	Erg-seconds	6.6255×10^{-27}
"	Cu. inches	554.8385	"	Joule-seconds	6.6255×10^{-34}
"	Gallons (Brit.)	2	"	Joule-sec./Avog. No. (chem.)	3.9905×10^{-10}
"	Gills (Brit.)	64	Points (printer's)	Centimeters	0.03514598
"	Hogsheads	0.03812537	"	Inches	0.013837
"	Kilderkins (Brit.)	0.111111	"	Picas	0.0833333
"	Liters	9.091920	Poises**	Cgs. units of absolute viscosity	1
"	Pints (Brit.)	16	"	Grams/(cm. × sec.)	1
"	Quarterns (Brit., dry)	4	Poise-cu. cm./gram	Sq. cm./sec	1
"	Quarters (Brit., dry)	0.03125	Poise-cu. ft./lb.	Sq. cm./sec	62.427960
"	Quarts (Brit.)	8	Poise-cu. in./gram	Sq. cm./sec	16.387064
"	Quarts (U.S., dry)	8.256449	Poles/sq. cm	E.M. cgs. units of magnetization	1
Pecks (U.S.)	Barrels (U.S., dry)	0.076191185	Pottles (Brit.)	Gallons (Brit.)	0.5
"	Bushels (U.S.)	0.25	"	Liters	2.272980
"	Cu. cm	8809.7675	Poundals	Dynes	13825.50
"	Cu. feet	0.311114005	"	Grams	14.09808
"	Cu. inches	537.605	"	Pounds (avdp.)	0.0310810
"	Gallons (U.S., dry)	2	Pounds (apoth. or troy)	Drams (apoth. or troy)	96
"	Gallons (U.S., liq.)	2.3272944	"	Drams (avdp.)	210.65143
"	Liters	8.809521	"	Grains	5760
"	Pints (U.S., dry)	16	"	Grams	373.24172
"	Quarts (U.S., dry)	8	"	Kilograms	0.37324172
Pennyweights	Drams (apoth. or troy)	0.4	"	Ounces (apoth. or troy)	12
"	Drams (avdp.)	0.87771429	"	Ounces (avdp.)	13.165714
"	Grains	24	"	Pennyweights	240
"	Grams	1.55517384	"	Pounds (avdp.)	0.8228571
"	Ounces (apoth. or troy)	0.05	"	Scruples (apoth.)	288
"	Ounces (avdp.)	0.054857143	"	Tons (long)	0.00036734694
"	Pounds (apoth. or troy)	0.0041666	"	Tons (metric)	0.00037324172
"	Pounds (avdp.)	0.0034285714	"	Tons (short)	0.00041142857
Perches (masonry)	Cu. feet	24.75	Pounds (avdp.)	Drams (apoth. or troy)	116.6666
Phots	Foot-candles	929.0304	"	Drams (avdp.)	256
"	Lumens/sq. cm	1	"	Grains	7000
"	Lumens/sq. meter	10000	"	Grams	453.59237
"	Lux	10000	"	Hundredweights (long)	0.00892857
Picas (printer's)	Centimeters	0.42175176	"	Hundredweights (short)	0.01
"	Inches	0.166044	"	Kilograms	0.45359237
Pints (Brit.)	Cu. cm	568.26092	"	Ounces (apoth. or troy)	14.583333
"	Gallons (Brit.)	0.125	"	Ounces (avdp.)	16
"	Gills (Brit.)	4			
"	Gills (U.S.)	4.803797			
"	Liters	0.5682450			

* Based on density of 1 gram/ml. for the solvent.
** See separate table of "viscosity conversion factors".

To convert from	To	Multiply by	To convert from	To	Multiply by
Pounds (avdp.)	Pennyweights	291.6666	Quarterns (Brit., liq.)	Gallons (Brit.)	0.03125
"	Poundals	32.1740	"	Liters	0.1420613
"	Pounds (apoth. or troy)	1.215277	Quarters (U.S., long)	Kilograms	254.0117272
"	Scruples (apoth.)	350	"	Pounds (avdp.)	560
"	Slugs	0.0310810	Quarters (U.S., short)	Kilograms	226.796185
"	Tons (long)	0.00044642857	"	Pounds	500
"	Tons (metric)	0.00045359237	Quarts (Brit.)	Cu. cm	1136.522
"	Tons (short)	0.0005	"	Cu. inches	69.35482
Pounds of H₂O evap. from and at 212°F.	B.t.u.	970.9	"	Gallons (Brit.)	0.25
"	B.t.u. (IST.)	970.2	"	Gallons (U.S., liq.)	0.3002373
"	B.t.u. (mean)	960.4	"	Liters	1.136490
"	Joules	1.0237×10^6	"	Quarts (U.S., dry)	1.032056
"	Joules (Int.)	1.0234×10^6	"	Quarts (U.S., liq.)	1.200949
Pounds/cu. ft.	Grams/cu. cm	0.016018463	Quarts (U.S., dry)	Bushels (U.S.)	0.03125
"	Kg./cu. meter	16.018463	"	Cu. cm	1101.2209
Pounds/cu. inch	Grams/cu. cm	27.679905	"	Cu. feet	0.038889251
"	Grams/liter	27.68068	"	Cu. inches	67.200625
"	Kg./cu. meter	27679.905	"	Gallons (U.S., dry)	0.25
Pounds/gal. (Brit.)	Pounds/cu. ft.	6.228839	"	Gallons (U.S., liq.)	0.29091180
Pounds/gal. (U.S., liq.)	Grams/cu. cm	0.11982643	"	Liters	1.1011901
"	Pounds/cu. ft.	7.4805195	"	Pecks (U.S.)	0.125
Pounds/inch	Grams/cm	178.57967	"	Pints (U.S., dry)	2
"	Grams/ft.	5443.1084	Quarts (U.S., liq.)	Cu. cm	946.35295
"	Grams/inch	453.59237	"	Cu. feet	0.033420136
"	Ounces/cm	6.2992	"	Cu. inches	57.75
"	Ounces/inch	16	"	Drams (U.S., fluid)	256
"	Pounds/meter	39.370079	"	Gallons (U.S., dry)	0.21484175
Pounds/minute	Kilograms/hr	27.2155422	"	Gallons (U.S., liq.)	0.25
"	Kilograms/min	0.45359237	"	Gills (U.S.)	8
Pounds of H₂O (39.2°F.)/min	Cu. ft./min	0.01601891	"	Liters	0.9463264
"	Gal. (U.S.)/min	0.1198298	"	Ounces (U.S., fluid)	32
"	Liters/min	0.45359237	"	Pints (U.S., liq.)	2
Pounds/sq. ft.	Atmospheres	0.000472541	"	Quarts (Brit.)	0.8326747
"	Bars	0.000478803	"	Quarts (U.S., dry)	0.8593670
"	Cm. of Hg (0°C.)	0.0359131	Quintals (metric)	Grams	100000
"	Dynes/sq. cm	478.803	"	Hundredweights (long)	1.9684131
"	Ft. of air (1 atm., 60°F.)	13.096	"	Kilograms	100
"	Grams/sq. cm	0.48824276	"	Pounds (avdp.)	220.46226
"	In. of Hg (32°F.)	0.0141390			
"	In. of H₂O (39.2°F.)	0.192227	Radians	Circumferences	0.15915494
"	Kg./sq. meter	4.8824276	"	Degrees	57.295779
"	Mm. of Hg (0°C.)	0.359131	"	Minutes	3437.7468
Pounds/sq. inch	Atmospheres	0.0680460	"	Quadrants	0.63661977
"	Bars	0.0689476	"	Revolutions	0.15915494
"	Cm. of Hg (0°C.)	5.17149	"	Seconds	206264.81
"	Cm. of H₂O (4°C.)	70.3089	Radians/cm	Degrees/cm	57.295779
"	Dynes/sq. cm	68947.6	"	Degrees/ft.	1746.3754
"	Grams/sq. cm	70.306958	"	Degrees/inch	145.53128
"	In. of Hg (32°F.)	2.03602	"	Minutes/cm	3437.7468
"	In. of H₂O (39.2°F.)	27.6807	Radians/sec.	Degrees/sec.	57.295779
"	Kg./sq. cm	0.070306958	"	Revolutions/min	9.5492966
"	Mm. of Hg (0°C.)	51.7149	"	Revolutions/sec.	0.15915494
Pound wt.-sec./sq. ft.	Poises	478.803	Radians/(sec. × sec.)	Revolutions/(min. × min.)	572.95779
Pound wt.-sec./sq. in.	Poises	68947.6	"	Revolutions/(min. × sec.)	9.5492966
Puncheons (Brit.)	Cu. meters	0.31797510	"	Revolutions/(sec. × sec.)	0.15915494
"	Gallons (Brit.)	69.94467	Register tons	Cu. feet	100
"	Gallons (U.S.)	84	"	Cu. meters	2.8316847
Quadrants	Minutes	5400	Revolutions	Degrees	360
"	Radians	1.5707963	"	Grades	400
Quarterns (Brit., dry)	Buckets (Brit.)	0.125	"	Quadrants	4
"	Bushels (Brit.)	0.0625	"	Radians	6.2831853
"	Cu. cm	2273.044	Reyns*	Centipoises	6.89476×10^6
"	Gallons (Brit.)	0.5	Rhes	Poises⁻¹	1
"	Liters	2.272980	Rods	Centimeters	502.92
"	Pecks (Brit.)	0.25	"	Chains (Gunter's)	0.25
Quarterns (Brit., liq.)	Cu. cm	142.0652	"	Chains (Ramden's)	0.165
			"	Feet	16.5
			"	Feet (U.S. Survey)	16.499967
			"	Furlongs	0.025
			"	Inches	198
			"	Links (Gunter's)	25

* See also separate table of "viscosity conversion factors".

To convert from	To	Multiply by	To convert from	To	Multiply by
Rods.............	Links (Ramden's)........	16.5	Sq. centimeters.......	Sq. mm...............	100
"	Meters...............	5.0292	"	Sq. mils.............	155000.31
"	Miles (statute)..........	0.003125	"	Sq. rods.............	3.9536861×10^{-6}
"	Perches...............	1	"	Sq. yards.............	0.00011959900
"	Yards...............	5.5	Sq. chains (Gunter's)	Acres...............	0.1
Rods (Brit., volume)..	Cu. feet...............	1000	"	Sq. feet.............	4356
"	Cu. meters.............	28.316847	"	Sq. ft. (U.S. Survey).....	4355.9826
Roods (Brit.).......	Acres...............	0.25	"	Sq. inches.............	627264
"	Ares...............	10.117141	"	Sq. links (Gunter's)....	10000
"	Sq. perches.............	40	"	Sq. meters.............	404.68564
"	Sq. yards.............	1210	"	Sq. miles.............	0.00015625
Ropes (Brit.).......	Feet...............	20	"	Sq. rods.............	16
"	Meters...............	6.096	"	Sq. yards.............	484
"	Yards...............	6.6666666	Sq. chains (Ramden's)	Acres...............	0.22956841
			"	Sq. feet.............	10000
Scruples (apoth.).....	Drams (apoth. or troy)..	0.333333	"	Sq. ft. (U.S. Survey)....	9999.9600
"	Drams (avdp.)............	0.73142857	"	Sq. inches.............	1.44×10^{6}
"	Grains...............	20	"	Sq. links (Ramden's)....	10000
"	Grams...............	1.2959782	"	Sq. meters.............	929.0304
"	Ounces (apoth. or troy)....	0.041666	"	Sq. miles.............	0.00035870064
"	Ounces (avdp.)............	0.045714286	"	Sq. rods.............	36.730946
"	Pennyweights.............	0.833333	"	Sq. yards.............	1111.111
"	Pounds (apoth. or troy)....	0.003472222	Sq. decimeters.......	Sq. cm...............	100
"	Pounds (avdp.)............	0.0028571429	"	Sq. inches.............	15.500031
Scruples (Brit., fluid)	Minims (Brit.)............	20	Square degrees.......	Steradians.............	0.00030461742
Seams (Brit.).......	Bushels (Brit.)............	8	Sq. dekameters....	Acres...............	0.024710538
"	Cu. feet...............	10.27479	"	Ares...............	1
"	Liters...............	290.9414	"	Sq. meters.............	100
Seconds (angular)....	Degrees...............	0.000277777	"	Sq. yards.............	119.59900
"	Minutes...............	0.0166666	Sq. feet..........	Acres...............	2.295684×10^{-5}
"	Radians...............	4.8481368×10^{-6}	"	Ares...............	0.0009290304
Seconds (mean solar)..	Days (mean solar)...	1.1574074×10^{-5}	"	Sq. cm...............	929.0304
"	Days (sidereal)............	1.1605763×10^{-5}	"	Sq. chains (Gunter's).....	0.00022956841
"	Hours (mean solar)......	0.0002777777	"	Sq. ft. (U.S. Survey).....	0.99999600
"	Hours (sidereal)............	0.00027853831	"	Sq. inches.............	144
"	Minutes (mean solar).....	0.0166666	"	Sq. links (Gunter's)....	2.2956841
"	Minutes (sidereal)........	0.016712298	"	Sq. meters.............	0.09290304
"	Seconds (sidereal)........	1.00273791	"	Sq. miles.............	3.5870064×10^{-8}
Seconds (sidereal)....	Days (mean solar)...	1.1542472×10^{-5}	"	Sq. rods.............	0.0036730946
"	Days (sidereal)............	1.1574074×10^{-5}	"	Sq. yards.............	0.111111
"	Hours (mean solar)......	0.00027701932	Sq. feet (U.S. Survey).	Acres...............	$2.29569330 \times 10^{-5}$
"	Hours (sidereal)............	0.000277777	"	Sq. centimeters.......	929.03412
"	Minutes (mean solar).....	0.016621159	"	Sq. chains (Ramden's)....	0.00010000040
"	Minutes (sidereal)........	0.0166666	"	Sq. feet.............	1.0000040
"	Seconds (mean solar).....	0.99726957	Sq. hectometers.......	Sq. meters.............	10000
Siemen's units.......	Same as Mhos		Sq. inches..........	Circ. mils.............	1273239.5
Skeins.............	Feet...............	360	"	Sq. cm...............	6.4516
"	Meters...............	109.728	"	Sq. chains (Gunter's).....	1.5942251×10^{-6}
Slugs.............	Geepounds.............	1	"	Sq. decimeters.........	0.064516
"	Kilograms.............	14.5939	"	Sq. feet.............	0.0069444
"	Pounds (avdp.)............	32.1740	"	Sq. ft. (U.S. Survey).....	0.0069444167
Slugs/cu. ft.......	Grams/cu. cm.........	0.515379	"	Sq. links (Gunter's)....	0.01594225
Space (entire).......	Hemispheres.............	2	"	Sq. meters.............	0.00064516
"	Steradians.............	12.566371	"	Sq. miles.............	$2.4909767 \times 10^{-10}$
Spans.............	Centimeters.............	22.86	"	Sq. mm...............	645.16
"	Fathoms.............	0.125	"	Sq. mils.............	1×10^{6}
"	Feet...............	0.75	Sq. inches/sec.......	Sq. cm./hr.............	23225.76
"	Inches...............	9	"	Sq. cm./sec.............	6.4516
"	Quarters (Brit. linear).....	1	"	Sq. ft./min.............	0.416666
Spherical right angles.	Hemispheres.............	0.25	Sq. kilometers.......	Acres...............	247.10538
"	Spheres.............	0.125	"	Sq. feet.............	1.0763910×10^{7}
"	Steradians.............	1.5707963	"	Sq. ft. (U.S. Survey).....	1.0763867×10^{7}
Sq. centimeters.......	Ares...............	1×10^{-6}	"	Sq. inches.............	1.5500031×10^{9}
"	Circ. mm.............	127.32395	"	Sq. meters.............	1×10^{6}
"	Circ. mils.............	197352.52	"	Sq. miles.............	0.38610216
"	Sq. chains (Gunter's).....	2.4710538×10^{-7}	"	Sq. yards.............	1.1959900×10^{6}
"	Sq. chains (Ramden's).....	1.0763910×10^{-7}	Sq. links (Gunter's)...	Acres...............	1×10^{-5}
"	Sq. decimeters............	0.01	"	Sq. cm...............	404.68564
"	Sq. feet.............	0.0010763910	"	Sq. chains (Gunter's).....	0.0001
"	Sq. ft. (U.S. Survey).....	0.0010763867	"	Sq. feet.............	0.4356
"	Sq. inches.............	0.15500031	"	Sq. ft. (U.S. Survey).....	0.43559826
"	Sq. meters.............	0.0001	"	Sq. Inches.............	62.7264

To convert from	To	Multiply by	To convert from	To	Multiply by
Sq. links (Ramden's)..	Acres................	2.2956841×10^{-5}	Statfarads..........	Farads...............	1.112646×10^{-12}
" ..	Sq. feet.............	1	"	Microfarads..........	1.112646×10^{-6}
Sq. meters..........	Acres................	0.00024710538	Stathenries........	Abhenries............	8.987584×10^{20}
"	Ares.................	0.01	"	E.M. cgs. units of induct-	
"	Hectares.............	0.0001		ance................	8.987584×10^{20}
"	Sq. cm..............	10000	"	E.S. cgs. units.........	1
"	Sq. feet.............	10.763910	"	Henries..............	8.987584×10^{11}
"	Sq. inches...........	1550.0031	"	Millihenries..........	8.987584×10^{14}
"	Sq. kilometers........	1×10^{-6}	Statohms..........	Abohms..............	8.987584×10^{20}
"	Sq. links (Gunter's).....	24.710538	"	E.S. cgs. units..........	1
"	Sq. links (Ramden's).....	10.763910	"	Ohms................	8.987584×10^{11}
"	Sq. miles............	3.8610216×10^{-7}	Statvolts..........	Abvolts..............	2.997930×10^{10}
"	Sq. mm.............	1×10^{6}	"	Volts................	299.7930
"	Sq. rods............	0.039536861	Statvolts/cm......	Volts/cm.............	299.7930
"	Sq. yards............	1.1959900	"	Volts/inch...........	761.4742
Sq. miles..........	Acres................	640	Statvolts/inch......	Volts/cm.............	118.0287
"	Hectares.............	258.99881	Steradians.........	Hemispheres..........	0.15915494
"	Sq. chains (Gunter's).....	6400	"	Solid angles..........	0.079577472
"	Sq. feet.............	2.7878288×10^{7}	"	Spheres..............	0.079577472
"	Sq. ft. (U.S. Survey)....	2.78288×10^{7}	"	Spher. right angles.....	0.63661977
"	Sq. kilometers........	2.5899881	"	Square degrees........	3282.8063
"	Sq. meters...........	2589988.1	Steres............	Cubic meters.........	1
"	Sq. rods............	102400	"	Decisteres...........	10
"	Sq. yards............	3.0976×10^{6}	"	Dekasteres...........	0.1
Sq. millimeters......	Circ. mm............	1.2732395	"	Liters...............	999.972
"	Circ. mils............	1973.5252	Stilbs............	Candles/sq. cm.......	1
"	Sq. cm..............	0.01	"	Candles/sq. inch......	6.4516
"	Sq. inches...........	0.0015500031	"	Lamberts............	3.1415927
"	Sq. meters...........	1×10^{-6}	Stokes*............	Cgs. units of kinematic	
Sq. mils...........	Circ. mils............	1.2732395		viscosity.............	1
"	Sq. cm..............	6.4516×10^{-6}	"	Sq. cm./sec..........	1
"	Sq. inches...........	1×10^{-6}	"	Sq. inches/sec........	0.15500031
"	Sq. mm.............	0.00000410	"	Poise sq. cm./gram....	1
Sq. rods..........	Acres................	0.00625	Stones (Brit., legal)...	Centals (Brit.)........	0.14
"	Ares.................	0.2529285264			
"	Hectares.............	0.002529285264	Tons (long)........	Dynes...............	9.96402×10^{8}
"	Sq. cm..............	252928.5264	"	Hundredweights (long)....	20
"	Sq. feet.............	272.25	"	Hundredweights (short)...	22.4
"	Sq. ft. (U.S. Survey).....	272.24891	"	Kilograms............	1016.0469
"	Sq. inches...........	39204	"	Ounces (avdp.)........	35840
"	Sq. links (Gunter's).....	625	"	Pounds (apoth. or troy)....	2722.22
"	Sq. links (Ramden's)....	272.25	"	Pounds (avdp.)........	2240
"	Sq. meters...........	25.29285264	"	Tons (metric).........	1.0160469
"	Sq. miles............	9.765625×10^{-6}	"	Tons (short)..........	1.12
"	Sq. yards............	30.25	Tons (metric).......	Dynes...............	9.80665×10^{8}
Sq. yards.........	Acres................	0.00020661157	"	Grams...............	1×10^{6}
"	Ares.................	0.0083612736	"	Hundredweights (short)...	22.046226
"	Hectares.............	8.3612736×10^{-5}	"	Kilograms............	1000
"	Sq. cm..............	8361.2736	"	Ounces (avdp.)........	35273.962
"	Sq. chains (Gunter's).....	0.0020661157	"	Pounds (apoth. or troy)....	2679.2289
"	Sq. chains (Ramden's).....	0.0009	"	Pounds (avdp.)........	2204.6226
"	Sq. feet.............	9	"	Tons (long)..........	0.98420653
"	Sq. ft. (U.S. Survey).....	8.9999640	"	Tons (short)..........	1.1023113
"	Sq. inches...........	1296	Tons (short)........	Dynes...............	8.89644×10^{8}
"	Sq. links (Gunter's).....	20.661157	"	Hundredweights (short)...	20
"	Sq. links (Ramden's).....	9	"	Kilograms............	907.18474
"	Sq. meters...........	0.83612736	"	Ounces (avdp.)........	32000
"	Sq. miles............	$3.228305785 \times 10^{-7}$	"	Pounds (apoth. or troy)....	2430.555
"	Sq. perches (Brit.).....	0.033057851	"	Pounds (avdp.)........	2000
"	Sq. rods............	0.033057851	"	Tons (long)..........	0.89285714
Statamperes.......	Abamperes...........	3.335635×10^{-11}	"	Tons (metric).........	0.90718474
"	Amperes.............	3.335635×10^{-10}	Tons of refrig. (U.S.,		
"	E.M. cgs. units of current.	3.335635×10^{-11}	comm.)............	B.t.u. (IST.)/hr.......	12000
"	E.S. cgs. units........	1	"	B.t.u. (IST.)/min........	200
Statcoulombs......	Ampere-hours........	9.265653×10^{-14}	"	Cal., kg. (IST.)/hr......	3023.949
"	Coulombs............	3.335635×10^{-10}	"	Horsepower..........	4.71611
"	Electronic charges.....	2.082093×10^{9}	"	Kg. of ice melted/hr....	37.971
"	E.M. cgs. units of electric		"	Lb. of ice melted/hr......	83.711
	charge..............	3.335635×10^{-11}	Tons of refrig. (U.S.,		
Statfarads........	E.M. cgs. units of capaci-		std.).............	B.t.u. (IST.).........	288000
	tance..............	1.112646×10^{-21}	"	B.t.u. (mean).........	287774
"	E.S. cgs. units........	1	"	Cal., kg. (IST.).......	72574.8

* See also separate table of "viscosity conversion factors".

To convert from	To	Multiply by
Tons of refrig. (U.S., std.)	Cal., *kg.* (mean)	72517.9
"	Lb. of ice melted	2009.1
Tons (long)/sq. ft	Atmospheres	1.05849
"	Dynes/sq. cm	1.07252×10^6
"	Grams/sq. cm	1093.6638
"	Pounds/sq. ft	2240
Tons (short)/sq. ft	Atmospheres	0.945082
"	Dynes/sq. cm	957.605
"	Grams/sq. cm	976.486
"	Pounds/sq. inch	13.8888
Tons (long)/sq. in	Atmospheres	152.423
"	Dynes/sq. cm	1.54443×10^8
"	Grams/sq. cm	157487.59
Tons (short)/sq. in	Dynes/sq. cm	1.37895×10^8
"	Kg./sq. mm	1406.139
"	Pounds/sq. inch	2000
Torrs (*or* Tors)	Millimeters of Hg (0°C.)	1
Townships (U.S.)	Acres	23040
"	Sections	36
"	Sq. miles	36
Tuns	Gallons (U.S.)	252
"	Hogsheads	4
Volts	Abvolts	1×10^8
"	Mks. (r *or* nr) units	1
"	Statvolts	0.003335635
"	Volts (Int.)	0.999670
Volts (Int.)	Volts	1.000330
Volts/°C	Joules/(coulomb × °C.)	1
Volt-coulombs	Joules (Int.)	0.999835
Volt-coulombs (Int.)	Joules	1.000165
Volt-electronic charge-seconds	Planck's constant*	2.41814×10^{14}
Volt-faraday (chem.)-seconds	Planck's constant*	1.45650×10^{38}
Volt-faraday (phys.)-seconds	Planck's constant	1.45690×10^{38}
Volt-seconds	Maxwells	1×10^8
Watts	B.t.u./hr	3.41443
"	B.t.u. (mean)/hr	3.40952
"	B.t.u. (mean)/min	0.0568253
"	B.t.u./sec	0.000948451
"	B.t.u. (mean)/sec	0.000947088
"	Cal., *gm.*/hr	860.421
"	Cal., *gm.* (mean)/hr	859.184
"	Cal., *gm.* (20°C.)/hr	860.853
"	Cal., *gm.*/min	14.3403
"	Cal., *gm.* (IST.)/min	14.3310
"	Cal., *gm.* (mean)/min	14.3197
"	Cal., *kh.*/min	0.0143403
"	Cal., *kg.* (IST.)/min	0.0143310
"	Cal., *kg.* (mean)/min	0.0143197
"	Ergs/sec	1×10^7
"	Foot-pounds/min	44.2537
"	Horsepower	0.00134102
"	Horsepower (boiler)	0.000101942
"	Horsepower (elec.)	0.00134048
"	Horsepower (metric)	0.00135962
"	Joules/sec	1
"	Kilowatts	0.001
"	Liter-atm./hr	35.5282
Watts (Int.)	B.t.u./hr	3.41499
"	B.t.u. (mean)/hr	3.41008
"	B.t.u./min	0.569165
"	B.t.u. (mean)/min	0.0568347
"	Cal., *gm.*/hr	860.563
"	Cal., *gm.* (mean)/hr	859.326
"	Cal., *kg.*/min	0.0143427
"	Cal., *kg.* (IST.)/min	0.0143333
"	Cal., *kg.* (mean)/min	0.0143221

To convert from	To	Multiply by
Watts (Int.)	Ergs/sec	1.000165×10^7
"	Joules (Int.)/sec	1
"	Watts	1.000165
Watts/sq. cm	B.t.u./(hr. × sq. ft.)	3172.10
"	Cal., *gm.*/(hr. × sq. cm.)	860.421
"	Ft.-lb./(min. × sq. ft.)	41113.1
Watts/sq. in	B.t.u./(hr. × sq. ft.)	491.677
"	Cal., *gm.*/(hr. × sq. cm.)	133.365
"	Ft.-lb./(min. × sq. ft.)	6372.54
Watt-hours	B.t.u.	3.41443
"	B.t.u. (mean)	3.40952
"	Cal., *gm*	860.421
Watt-hours	Cal., *kg.* (mean)	0.859184
"	Cal., *gm.* (mean)	859.184
"	Foot-pounds	2655.22
"	Hp.-hours	0.00134102
"	Joules	3600
"	Joules (Int.)	3599.41
"	Kg.-meters	367.098
"	Kw.-hours	0.001
"	Watt-hours (Int.)	0.999835
Watt-sec	Foot-pounds	0.737562
"	Gram-cm	10197.16
"	Joules	1
"	Liter-atm	0.00986895
"	Volt-coulombs	1
Wave length of orange-red line of krypton 86	Ångström units	6057.80211
"	Millimeters	0.000605780211
Wave length of red line of cadmium	Ångström units	6438.4696
"	Millimeters	0.00064384696
Webers	Cgs. units of induction flux	1×10^8
"	E.M. cgs. units of induction flux	1×10^8
"	Lines	1×10^8
"	Maxwells	1×10^8
"	Mks. units of induction flux	1
"	**Mks. nr** units of magnetic charge	0.079577472
"	**Mks. r** units of magnetic charge	1
"	Volt-seconds	1
Webers/sq. cm	Gausses	1×10^8
"	Lines/sq. cm	1×10^8
"	Lines/sq. inch	6.4516×10^8
Webers/sq. in	Gausses	1.5500031×10^7
Weeks (mean calendar)	Days (mean solar)	7
"	Days (sidereal)	7.0191654
"	Hours (mean solar)	168
"	Hours (sidereal)	168.45997
"	Minutes (mean solar)	10080
"	Minutes (sidereal)	10107.598
"	Months (lunar)	0.23704235
"	Months (mean calendar)	0.23013699
"	Years (calendar)	0.019178082
"	Years (sidereal)	0.019164622
"	Years (tropical)	0.019165365
Weys (Brit., mass.)	Pounds (avdp.)	252
Yards	Centimeters	91.44
"	Chains (Gunter's)	0.4545454
"	Chains (Ramden's)	0.03
"	Cubits	2
"	Fathoms	0.5
"	Feet	3
"	Feet (U.S. Survey)	2.9999940
"	Furlongs	0.00454545
"	Inches	36
"	Meters	0.9144
"	Poles (Brit.)	0.181818

* For factors for C = 12 scale or those which can be derived from same see table on Values for General Physical Constants.

CONVERSION FACTORS (Continued)

To convert from	To	Multiply by	To convert from	To	Multiply by
Yards..............	Quarters (Brit., linear)....	4	Years (sidereal)......	Days (sidereal)........	366.25640
"	Rods....................	0.181818	"	Years (calendar)........	1.0007024
"	Spans...................	4	"	Years (tropical)..........	1.0000388
Years (calendar).....	Days (mean solar).......	365	Years (tropical).....	Days (mean solar)......	365.24219
"	Hours (mean solar)......	8760	"	Days (sidereal)........	366.24219
"	Minutes (mean solar).....	525600	"	Hours (mean solar)......	8765.8126
"	Months (lunar)..........	12.360065	"	Hours (sidereal)........	8789.8126
"	Months (mean calendar)...	12	"	Months (mean calendar)...	12.007963
"	Seconds (mean solar).....	3.1536×10^7	"	Seconds (mean solar).....	3.1556926×10^7
"	Weeks (mean calendar)....	52.142857	"	Seconds (sidereal).......	3.1643326×10^7
"	Years (sidereal)..........	0.99929814	"	Weeks (mean calendar)....	52.177456
"	Years (tropical)..........	0.99933690	"	Years (calendar)..........	1.0006635
Years (leap)........	Days (mean solar).......	366	"	Years (sidereal)..........	0.99996121
Years (sidereal)......	Days (mean solar).......	365.25636			

DEFINED VALUES AND EQUIVALENTS

Meter.. **(m)**   1 650 763.73 wave lengths in vacuo of the unperturbed transition $2p_{10} - 5d_5$ in ^{86}Kr

Kilogram.. **(kg)**   mass of the international kilogram at Sèvres, France

Second... **(s)**   1/31 556 925.974 7 of the tropical year at 12^h ET, 0 January 1900

Degree Kelvin... **(°K)**   defined in the thermodynamic scale by assigning 273.16 °K to the triple point of water (freezing point, 273.15 °K = 0 °C)

Unified atomic mass unit.. **(u)**   1/12 the mass of an atom of the ^{12}C nuclide

Mole... **(mol)**   amount of substance containing the same number of atoms as 12 g of pure ^{12}C

Standard acceleration of free fall.................................. **(g$_n$)**   9.806 65 m s^{-2}, 980.665 cm s^{-2}

Normal atmospheric pressure...................................... **(atm)**   101 325 N m^{-2}, 1 013 250 dyn cm^{-2}

Thermochemical calorie.. **(cal$_{th}$)**   4.1840 J, 4.1840 $\times 10^7$ erg

International Steam Table calorie.................................. **(cal$_{IT}$)**   4.1868 J, 4.1868 $\times 10^7$ erg

Liter... **(l)**   0.001 m^3, 1000 cm^3 (recommended by GCWM, 1964)

Inch... **(in)**   0.0254 m. 2.54 cm

Pound (avdp)... **(lb)**   0.453 592 37 kg, 453.592 37 g

FACTORS FOR THE CONVERSION OF (LOG$_{10}$X) TO (RT LOG$_e$X)

Units are in Calories

t°C	0	1	2	3	4	5	6	7	8	9
0	1249.4	1254.0	1258.6	1263.2	1267.7	1272.3	1276.9	1281.45	1286.0	1290.6
10	1295.2	1299.8	1304.3	1308.9	1313.5	1318.0	1322.6	1327.2	1331.8	1336.3
20	1340.9	1345.5	1350.1	1354.6	1359.2	1363.8	1368.4	1372.9	1377.5	1382.1
30	1386.7	1391.2	1395.8	1400.4	1405.0	1409.5	1414.1	1418.7	1423.2	1427.8
40	1432.4	1437.0	1441.5	1446.1	1450.7	1455.3	1459.8	1464.4	1469.0	1473.6
50	1478.1	1482.7	1487.3	1491.9	1496.4	1501.0	1505.6	1510.2	1514.7	1519.3
60	1523.9	1528.5	1533.0	1537.6	1542.2	1546.7	1551.3	1555.9	1560.5	1565.0
70	1569.6	1574.2	1578.8	1583.3	1587.9	1592.5	1597.1	1601.6	1606.2	1610.8
80	1615.4	1619.9	1624.5	1629.1	1633.7	1638.2	1642.8	1647.4	1651.9	1656.5
90	1661.1	1665.7	1670.2	1674.8	1679.4	1684.0	1688.5	1693.1	1697.7	1702.3
100	1706.8	1711.4	1716.0	1720.6	1725.1	1729.7	1734.3	1738.9	1743.4	1748.0
110	1752.6	1757.2	1761.7	1766.3	1770.9	1775.4	1780.0	1784.6	1789.2	1793.7
120	1798.3	1802.9	1807.5	1812.0	1816.6	1821.2	1825.8	1830.3	1834.9	1839.5
130	1844.1	1848.6	1853.2	1857.8	1862.4	1866.9	1871.5	1876.1	1880.6	1885.2
140	1889.8	1894.4	1898.9	1903.5	1908.1	1912.7	1917.2	1921.8	1926.4	1931.0
150	1935.5	1940.1	1944.7	1949.3	1953.8	1958.4	1963.0	1967.6	1972.1	1976.7
160	1981.3	1985.9	1990.4	1995.0	1999.6	2004.1	2008.7	2013.3	2017.9	2022.4
170	2027.0	2031.6	2036.2	2040.7	2045.3	2049.9	2054.5	2059.0	2063.6	2068.2
180	2072.8	2077.3	2081.9	2086.5	2091.1	2095.6	2100.2	2104.8	2109.4	2113.9
190	2118.5	2123.1	2127.6	2132.2	2136.8	2141.4	2145.9	2150.5	2155.1	2159.7
200	2164.2	2168.8	2173.4	2178.0	2182.5	2187.1	2191.7	2196.3	2200.8	2205.4

Differences

	Tenths	Units	Tens
1	.5	4.6	45.7
2	.9	9.1	91.5
3	1.4	13.7	137.2
4	1.8	18.3	183.0
5	2.3	22.9	228.7
6	2.7	27.4	274.4
7	3.2	32.0	320.2
8	3.7	36.6	365.9
9	4.1	41.2	411.7

Conversion Tables

Equivalent second order rate constants

A \ B	cm^3 mol^{-1} s^{-1}	dm^3 mol^{-1} s^{-1}	m^3 mol^{-1} s^{-1}	cm^3 $molecule^{-1}$ s^{-1}	$(mm\ Hg)^{-1}$ s^{-1}	atm^{-1} s^{-1}	ppm^{-1} min^{-1}	m^2 kN^{-1} s^{-1}
$1\ cm^3\ mol^{-1}\ s^{-1} =$	1	10^{-3}	10^{-6}	1.66×10^{-24}	$1.604 \times 10^{-5}\ T^{-1}$	$1.219 \times 10^{-2}\ T^{-1}$	2.453×10^{-9}	$1.203 \times 10^{-4}\ T^{-1}$
$1\ dm^3\ mol^{-1}\ s^{-1} =$	10^3	1	10^{-3}	1.66×10^{-21}	$1.604 \times 10^{-2}\ T^{-1}$	$12.19\ T^{-1}$	2.453×10^{-6}	$1.203 \times 10^{-1}\ T^{-1}$
$1\ m^3\ mol^{-1}\ s^{-1} =$	10^6	10^3	1	1.66×10^{-18}	$16.04\ T^{-1}$	$1.219 \times 10^4\ T^{-1}$	2.453×10^{-3}	$120.3\ T^{-1}$
$1\ cm^3\ molecule^{-1}\ s^{-1} =$	6.023×10^{23}	6.023×10^{20}	6.023×10^{17}	1	$9.658 \times 10^{18}\ T^{-1}$	$7.34 \times 10^{21}\ T^{-1}$	1.478×10^{15}	$7.244 \times 10^{19}\ T^{-1}$
$1\ (mm\ Hg)^{-1}\ s^{-1} =$	$6.236 \times 10^4\ T$	$62.36\ T$	$6.236 \times 10^{-2}\ T$	$1.035 \times 10^{-19}\ T$	1	760	4.56×10^{-2}	7.500
$1\ atm^{-1}\ s^{-1}$	$82.06\ T$	$8.206 \times 10^{-2}\ T$	$8.206 \times 10^{-5}\ T$	$1.362 \times 10^{-22}\ T$	1.316×10^{-3}	1	6×10^{-5}	9.869×10^{-3}
$1\ ppm^{-1}\ min^{-1} =$ at 298 K, 1 atm total pressure	4.077×10^8	4.077×10^5	407.7	6.76×10^{-16}	21.93	1.667×10^4	1	164.5
$1\ m^2\ kN^{-1}\ s^{-1} =$	$8314\ T$	$8.314\ T$	$8.314 \times 10^{-3}\ T$	$1.38 \times 10^{-20}\ T$	0.1333	101.325	6.079×10^{-3}	1

To convert a rate constant from one set of units A to a new set B find the conversion factor for the row A under column B and multiply the old value by it, e.g. to convert $cm^3\ molecule^{-1}\ s^{-1}$ to $m^3\ mol^{-1}\ s^{-1}$ multiply by 6.023×10^{17}.

Table adapted from High Temperature Reaction Rate Data No. 5, The University, Leeds (1970).

Equivalent third order rate constants

A \ B	cm^6 mol^{-2} s^{-1}	dm^6 mol^{-2} s^{-1}	m^6 mol^{-2} s^{-1}	cm^6 $molecule^{-2}$ s^{-1}	$(mm\ Hg)^{-2}$ s^{-1}	atm^{-2} s^{-1}	ppm^{-2} min^{-1}	m^4 kN^{-2} s^{-1}
$1\ cm^6\ mol^{-2}\ s^{-1} =$	1	10^{-6}	10^{-12}	2.76×10^{-48}	$2.57 \times 10^{-10}\ T^{-2}$	$1.48 \times 10^{-4}\ T^{-2}$	1.003×10^{-19}	$1.447 \times 10^{-8}\ T^{-2}$
$1\ dm^6\ mol^{-2}\ s^{-1} =$	10^6	1	10^{-6}	2.76×10^{-42}	$2.57 \times 10^{-4}\ T^{-2}$	$148\ T^{-2}$	1.003×10^{-13}	$1.447 \times 10^{-2}\ T^{-2}$
$1\ m^6\ mol^{-2}\ s^{-1} =$	10^{12}	10^6	1	2.76×10^{-36}	$257\ T^{-2}$	$1.48 \times 10^8\ T^{-2}$	1.003×10^{-7}	$1.447 \times 10^4\ T^{-2}$
$1\ cm^6\ molecule^{-2}\ s^{-1} =$	3.628×10^{47}	3.628×10^{41}	3.628×10^{35}	1	$9.328 \times 10^{37}\ T^{-2}$	$5.388 \times 10^{43}\ T^{-2}$	3.64×10^{28}	$5.248 \times 10^{39}\ T^{-2}$
$1\ (mm\ Hg)^{-2}\ s^{-1} =$	$3.89 \times 10^9\ T^2$	$3.89 \times 10^3\ T^2$	$3.89 \times 10^{-3}\ T^2$	$1.07 \times 10^{-38}\ T^2$	1	5.776×10^5	3.46×10^{-5}	56.25
$1\ atm^{-2}\ s^{-1} =$	$6.733 \times 10^3\ T^2$	$6.733 \times 10^{-3}\ T^2$	$6.733 \times 10^{-9}\ T^2$	$1.86 \times 10^{-44}\ T^2$	1.73×10^{-6}	1	6×10^{-11}	9.74×10^{-5}
$1\ ppm^{-2}\ min^{-1} =$ at 298 K, 1 atm total pressure	9.97×10^{18}	9.97×10^{12}	9.97×10^6	2.75×10^{-29}	2.89×10^4	1.667×10^{10}	1	1.623×10^6
$1\ m^4\ kN^{-2}\ s^{-1} =$	$6.91 \times 10^7\ T^2$	$69.1\ T^2$	$6.91 \times 10^{-5}\ T^2$	$1.904 \times 10^{-40}\ T^2$	0.0178	1.027×10^4	6.16×10^{-7}	1

From *J. Phys. Chem. Ref. Data*, 9, 470, 1980, by permission of the authors and the copyright owner, the American Institute of Physics.

DIMENSIONLESS GROUPS

John P. Catchpole and George Fulford

Reprinted from Industrial & Engineering Chemistry, Vol. 58, No. 3, pp. 47 to 60. Copyright 1966 by the American Chemical Society and reprinted by permission of the copyright owners and the authors. See also the Supplementary Table which follows this table.

Dimensionless groups are frequently generated in the analysis of a complex engineering problem. The more common groups thus generated are easily recognized, while the less common ones are not. Unless the less common existing groups are recognized, an already named group could unknowingly be renamed. Table A provides a tool that may be used to avoid this occurrence, by listing the groups by the variables of which they consist. These variables—i.e., length, density, diffusivity, viscosity, etc.—are further subdivided into their exponents to which they are raised in the groups in question. Thus, Reynolds number is listed under the exponent $+1$ for the variables, length, fluid velocity, and density, and the exponent -1 for viscosity.

To illustrate the use of the tables in the analysis of a problem, the group $(kE/\eta\sigma T^3)$ might be generated in the solution of a complex heat transfer problem. From Table A the groups containing the constituent variables are checked and the groups are listed:

Thermal conductivity	(k^{+1})	**F11, L7, R1**
Modulus of elasticity	(E^{+1})	**C1, E13, R1**
Stefan-Boltzmann coefficient	(η^{-1})	**R1, T6**
Surface tension	(σ^{-1})	**B9, C3, E8, L6, R1, W1**
Temperature	(T^{-3})	**R1, T6**

It is immediately apparent that the only group common to all the categories listed is the Radiation number, **R1**, which is equivalent to the previously unidentified group.

The symbol assigned to a dimensionless group is usually the first two letters of its names. Several groups, however, have nonstandard symbols, particularly in the groups which are named after persons. These symbols are listed in the nomenclature.

NOMENCLATURE

a	=	annulus or clearance width, L
A	=	area, L^2
A^*	=	cooling area/unit volume, $1/L$
b	=	bearing breadth, L
B	=	groups **B6, B11**
c	=	specific heat, $L^2/\theta^2 T$
c_A	=	concentration, M/L^3
c_b	=	specific vapor capacity (mass/unit mass/unit pressure change), $L\theta^2/M$
$c_d \atop c_D$ }	=	group **D13**
c_D	=	group **R9**
c_f	=	group **H4**
c_H	=	mass capacity, L^3/M
c_m	=	specific heats at constant pressure and volume, $L^2/\theta^2 T$
$c_p \atop c_v$ }	=	heat capacity, $L^2/\theta^2 T$
c_q	=	group **F9**
c_Q	=	group **S11**
c_S	=	group **C10**, dimensional concentration, M/L^3
C	=	groups **C11, C4**
C_a	=	diameter, L
d	=	equivalent diameter (of particles, etc.), L
d_e	=	hydraulic diameter, L
d_h	=	diffusivity (molecular, unless noted otherwise), L^2/θ
D	=	binary bulk diffusion coefficient, L^2/θ
D_{AB}	=	Knudsen diffusion coefficient, L^2/θ
D_{KA}	=	voidage; porosity ($^-$)
e	=	surface emissivity ($^-$)
e^*	=	modulus of elasticity, $M/L\theta^2$
E	=	activation energy, L^2/θ^2
E_a	=	bulk modulus, $M/L\theta^2$
E_b	=	frequency, $1/\theta$, or Group **F1**
f	=	group **F11**
$f(M)$	=	force, ML/θ^2
F	=	force per unit length of bearing, M/θ^2
F_b	=	group **F6**
$F(M)$	=	resistance force in flow, ML/θ^2
F_R	=	acceleration due to gravity, L/θ^2
g	=	mass velocity (mass flux density; mass transfer coefficient), $M/\theta L^2$
G	=	heat transfer coefficient, $M/T\theta^3$
h	=	convective heat transfer coefficient, $M/T\theta^3$
h_c	=	energy change per unit mass ($= g \times$ head), L^2/θ^2
H	=	fluid head, L
H'	=	field strength, $Q/L\theta$
H_e	=	homochronicity number
H_0	=	group **F8**
$I(M)$	=	heat liberated per unit volume per unit time, $M/L\theta^3$
j	=	groups **J2, J3**
j_H, j_M		

J	=	average free path/average velocity, θ, or group **L6**
k	=	thermal conductivity, $ML/T\theta^3$
k_c	=	mass transfer coefficient, L/θ
K	=	groups **K2, K10, N5**
K_1	=	group **A4**
$\bar{K}_E$	=	group **E4**
K_F	=	group **C2**
K_P	=	group **P13**
K_Q	=	group **H5**
K_r	=	group **E11**
$\bar{K}_g$	=	group **A1**
$\bar{K}_E$	=	group **E4**
$\bar{K}_r$	=	group **E12**
$\bar{K}_{rE}$	=	group **E13**
$\bar{K}_s$	=	group **R1**
$\bar{K}_\sigma$	=	group **C1**
$\bar{K}_{\sigma g}$	=	group **C2**
L	=	characteristic dimension (except as noted), L
L_m	=	distance from midpoint to surface, L
m_T	=	group **T3**
M	=	group **M10**
M_H	=	group **H2**
n	=	concentration, wt./wt. ($^-$)
n^*	=	specific mass content, mass/mass ($^-$)
n_m	=	moisture content, wt./wt. bone dry gas ($^-$)
N	=	rate of rotation, $1/\theta$, and groups **M3, N4**
N_{Ba}	=	groups **B7, B8**
N_c	=	group **C5**
N_{cv}	=	group **C10**
N_D	=	groups **D7, D14**
N_E	=	group **E1**
N_F	=	group **F4**
N_H	=	groups **H1, H9**
N_K	=	group **K1**
N_{KnA}	=	Knudsen number for diffusion (see Addendum)
N_1	=	group **N2**
N_P	=	group **P7**
N_{ef}	=	group **R7**
N_{s1}	=	group **S8**
N_{s2}	=	group **S9**
N'_T	=	group **N9**
P	=	pressure, $M/L\theta^2$
P	=	plasticity number (see Addendum)
p_b	=	bearing pressure, $M/L\theta^2$
p_s	=	static pressure, $M/L\theta^2$
p_v	=	vapor pressure, $M/L\theta^2$
p_σ	=	capillary pressure, $M/L\theta^2$
Δp_F	=	frictional pressure drop, $M/L\theta^2$
q	=	heat flux (heat flow/unit time), ML^2/θ^3
q^*	=	heat flux density (heat flux/unit area), M/θ^3
Q	=	heat liberated/unit mass, L^2/θ^2
r	=	latent heat of phase change, L^2/θ^2
r_v	=	heat of vaporization, L^2/θ^2
R	=	radius, L
R_H	=	hydraulic radius, L
R_2'	=	group **R5**
R_M	=	group **M7**
R_V	=	group **V3**
$\mathcal{R}$	=	gas constant, $L^2/\theta^2 T$
s	=	humid heat, $L^2/\theta^2 T$
S	=	particle area/particle volume, L^2/L^3, and group **M6**
St	=	group **S14**
t	=	temperature, T
T	=	absolute temperature, T
$\Delta t, \Delta T$	=	temperature difference, T
U^+	=	group **P10**
U	=	reaction rate, $M/L^3\theta$
v_s	=	velocity of surface (solid), L/θ
V	=	fluid velocity, L/θ, and group **V1**
V_A	=	velocity of Alfven magnetic waves, L/θ
V_f	=	volumetric flow rate, L^3/θ
V_l	=	velocity of light, L/θ
V_m	=	mass flow rate, M/θ
V_s	=	velocity of sound, L/θ
w	=	circumferential velocity, L/θ
W	=	volume of system, L^3
W^*	=	gross volume, L^3
x	=	entry length; distance from entrance, L
y^+	=	group **P11**
Z	=	group **O2**
α	=	thermal diffusivity (temperature conductivity), L^2/θ

β = coefficient of bulk expansion, $1/T$, and group **D12**
β^* = Dufour coefficient, T
γ = specific gravity $(^-)$ and group **R3**
$\dot{\gamma}$ = rate of shear, $1/\theta$
Γ = rate of change of temperature of medium, T/θ
δ = Soret or thermogradient coefficient, $1/T$, and group **D11**
Δ = difference in quantity
ε = height of roughness, L and group **A3**
ε_D = eddy mass diffusivity, L^2/θ
ζ = diffusion tortuosity $(^-)$
η = radiation coefficient (Stefan-Boltzmann coefficient), $M/T^4\theta^3$
θ = time, θ
θ_r = relaxation time, θ
λ = mean free path, L
μ = dynamic viscosity, $M/L\theta$
μ_e = magnetic permeability, ML/Q^2
μ_p = rigidity coefficient, $M/L\theta$
ξ = permeability, L^2

π = 3.1416. . . .
Π = power to agitator or impeller, ML^2/θ^3
ρ = density, M/L^3
ρ_a = group **P3**
σ = surface tension, M/θ^2 and group **S12**
σ_c = group **C6**
σ_e = electrical conductivity, $Q^2\theta/L^3M$
σ_t = group **T4**
τ = group **T8**
τ_w = wall shear stress, $M/L\theta^2$
τ_y = yield stress, $M/L\theta^2$
φ = group **D9**
ψ = groups **N8, P14, R10**
ω = angular velocity (of fluid, unless noted otherwise), $1/\theta$
Ω = mass transfer potential (concn.), M/L^3
$-$ (bar over) = mean value
N.B.: $(F) = \left(\dfrac{ML}{\theta^2}\right)$; $(H) = \left(\dfrac{ML^2}{\theta^2}\right)$

TABLES FOR IDENTIFYING DIMENSIONLESS GROUPS
TABLE A

PHYSICAL PROPERTIES
General Physical Properties

Parameter	Symbol	Dimensions	Exponent	Group
Coefficient of bulk expansion	β	$1/T$	-1	E4, 12, G2
			$+1$	G5, R5, K9, 6
Density	ρ	M/L^3	-2	M1
			-1	A1, B1, 10, C1, 2, 6, D3, 13, E3, 9, 10, 13, F1, 11, H5, J1, K4, K10, L7, M3, 6, N2, 3, 4, P7, 9, R10, 13, S4, 13, 15
			$-\frac{2}{3}$	C9, J3
			$-\frac{1}{2}$	E2, L11, O2, P13
			$+\frac{1}{3}$	J3, K5, 53
			$+\frac{1}{2}$	D10, G3, P10, 11, R4, W2
			$+\frac{2}{3}$	F2, N8, S2
			$+1$	A5, B1, B6, 9, C5, 7, D5, 7, 13, E4, 6, 7, 8, H6, 11, J1, 3, 4, K1, L9, P1, R5, 11, 13, S17, T1, T6, V1, W1, W3
			$+2$	C10, G1, 5, K9, R5, 6, T2
Density gradient	$d\rho/dL$	M/L^4	$+1$	R13
Diffusivity (molecular unless noted otherwise)	D, a_m, ε_D	L^2/θ	-1	B5, B7[4], D2, K4[1], K7[2], L2, 7, 10, N7, P2, P9, S4, 13
			$-\frac{2}{3}$	J3
			$-\frac{1}{2}$	T3
			$+1$	D12, F12[1], K7[3], L2[4], L9[1]
Diffusivity (surface)	D_s	ML/θ^2	$+1$	S18
Diffusion tortuosity	ζ	$-$	-1	K7
Molecular weight	M	$-$	-1	D14, K9, S6
			$+1$	D14
Permeability (packed bed)	ξ	L^2	$+\frac{1}{2}$	L6
Porosity (voidage)	e		-1	B4
			$-\frac{1}{2}$	L6
			$+1$	K7
Specific weight	γ	$-$	$\pm\frac{1}{3}$	F2
Surface tension	σ	M/θ^2	-3	C2
			-2	C1
			-1	B9, C3, 13, E8, L6, R1, W1, 3
			$-\frac{1}{2}$	D10, G3, O2, P13, R4, W2
			$+1$	M9, S17, 18

[1] Coefficient of potential diffusion in mass transfer.
[2] Knudsen diffusion coefficient.
[3] Binary bulk diffusion coefficient.
[4] Effective diffusivity $(D + \varepsilon_D)$ (molecular + eddy transfer).

PHYSICAL PROPERTIES
Electrical and Magnetic Properties

Parameter	Symbol	Dimensions	Exponent	Group
Current density	I	$Q/L^2\theta$	$+1$	K10
Electrical conductivity	σ_e	$Q^2\theta/L^3M$	-1	E6
			$+\frac{1}{2}$	H2
			$+1$	L11, M3, 7
Field strength	H_e	$Q/L\theta$	-2	J4
			$+1$	H2, L11
			$+2$	K9, M3, 6, S6
Magnetic permeability	μ_e	ML/Q^2	-1	E5, J4
			$+1$	H2, M6, 7
			$+\frac{3}{2}$	L11
			$+2$	M3
Voltage	E	$ML^2/Q\theta^2$	$+1$	K10

Thermal Properties

Parameter	Symbol	Dimensions	Exponent	Group
Humid heat	S	L^2/θ^2T	-1	P15
Latent heats of phase change	λ, r	L^2/θ^2	-1	A3, E11, 12, 13, J1, K10, M1
			$+1$	B13, C9, K8, 11, N5
Ratio of specific heats	γ	$-$	$\pm\frac{1}{2}$	L5
Specific heat	C, c	L^2/θ^2T	-1	B2, 13, D3, 15, E1, F11, J2, K8, 11, L7, M10, N5, R3, S13
			$-\frac{1}{2}$	J2
			$+\frac{1}{2}$	F6, 7, 8
			$+\frac{2}{3}$	J2
			$+1$	A3, E4, 12, G4, J1, 4, L9, P1, 8, R3, 5, 6, 7
Surface emissivity	e^*	$-$	-1	T6
Temperature conductivity (thermal diffusivity)	α	L^2/θ	-1	L9, P1, 12, R5
			$+1$	C13, L7, 10
Thermal conductivity	k or λ	$ML/T\theta^3$	-3	M1
			-2	R6
			-1	B4, 12, C7, 9, 10, D4, G4, K3, L9, N6, P1, 5, 8, R5, S14

Rheological and Elastic Behavior

Parameter	Symbol	Dimensions	Exponent	Group
Modulus of elasticity	E	$M/L\theta^2$	-1	C5, E4, H11
			$+1$	C1, E13, R1
			$+3$	A1
Rate of shear	$\dot{\gamma}$	$1/\theta$	$+1$	T8
Rigidity coefficient	μ_p	$M/L\theta$	-2	H6
			-1	B3
Shear stress	τ	$M/L\theta^2$	-1	E5
			$-\frac{1}{2}$	P10
			$+\frac{1}{2}$	P11
Viscosity (in all cases kinematic viscosity has been written as μ/ρ)	μ	$M/L\theta$	-2	A1, 5, G1, 5, K1, 9, S17, T2
			-1	B6, C10, D5, 7, E6, 7, H8, L1, 3, 9, O1, P4, 11, R5, 6, 11, S9, 19, T1, V1

Rheological and Elastic Behavior

Parameter	Symbol	Dimensions	Exponent	Group
			$-\tfrac{2}{3}$	F2, K5, N8
			$-\tfrac{1}{2}$	H2
			$-\tfrac{1}{3}$	S2, 3
			$+\tfrac{1}{2}$	E2
			$+\tfrac{2}{3}$	C9, J2, 3
			$+1$	B12, C3, 13, E3, 5, M1, O2, P8, 9, S4, 8, 15, T8
			$+2$	C1
			$+4$	C2
Viscosity (surface)	μ_s	M/θ	$+1$	S19
Yield stress	τ_y	$M/L\theta^2$	$+1$	B3, H6

LENGTHS, AREAS AND VOLUMES
Characteristic Linear Dimensions
(In all cases kinematic viscosity has been written as μ/ρ)

Parameter	Symbol	Dimensions	Exponent	Group
General characteristic linear dimension	Various	L	-5	P7
			-3	F9, L1
			-2	D9, 12, E3, F11, 12, H4, 5, N2, 3, 4, O1, R9, S8, 9, 15
			-1	B1, 10, E2, 5, 10, F1, 13, G4, H10, K6, L4, R6, 10, 15, 16, S6, 19, T5, W4
			$-\tfrac{1}{2}$	B11, D7, F14, O2
			$+\tfrac{1}{2}$	R4, T1, W2
			$+1$	B3, 4, 5, 7, 10, C7, D1, 3, 5, 11, 13, E7, F1, 2, 5, G3, H2, K3, 4, 6, 10, L3, 4, 9, 11, M1, 3, 7, N6, 7, 8, O1, P1, 2, 11, R8, 10, 11, 16, S2, 7, 14, 16, 17, 18, T3, W1
			$\pm\tfrac{1}{2}$	D7, H7, T1
			$+2$	B9, D2, 4, E8, H6, K9, O1, P4, 5, 12, S8, 9, V1
			$+3$	A6, C10, G1, 5, K1, R5
			$+4$	T2
			$+5$	R6
Dimension of agitator, impeller, etc.	Various	L	-5	P7
			-3	F9
			-2	D7, H4
			$+1$	D11, F15
			$+3$	W3
Film thickness	L_f	L	$+1$	N8
Furnace half-width	L	L	-1	T7
Larmor radius	L_L	L	$+1$	L4
Mean free path	λ	L	$+1$	K6
Particle dimension	d_e	L	-2	S15
			$+1$	F2
			$+3$	A5, G1
Pore or nozzle radius	R	L	$+1$	M12, T7
Reactor length	L	L	$+1$	B7
			-1	C3, L9
Thickness of liquid layer	L	L	$+1$	H2, L11
			$+2$	L9
			$+3$	R5
			$+4$	T2

Areas

Parameter	Symbol	Dimensions	Exponent	Group
Area	A	L^2	-1	D9, F6, 7, 8
			$+1$	B2
Area/unit volume	S, A^*	$1/L$	-1	B6
			$+1$	M11, 12

Volumes

Parameter	Symbol	Dimensions	Exponent	Group
Volume	—	L^3	$+$	A5, H9, M11

TIMES AND FREQUENCIES

Parameter	Symbol	Dimensions	Exponent	Group
Time	θ	θ	-1	D8
			$+1$	D8, 12, E3, F11, 12, H1, 10, S15, T5
Frequency	f'	$1/\theta$	$+1$	H1, 10, S16, V1

TEMPERATURES AND CONCENTRATIONS (DRIVING FORCES)
Concentrations and Related Quantities

Parameter	Symbol	Dimensions	Exponent	Group
Dimensional concentration	C, c_A	M/L^3	-1	D1, 2, R2
			$-\tfrac{1}{2}$	T3
			$+1$	B10
Dimensionless concentration—e.g., wt./wt. inert material, etc.	n	—	-1	P6
			$+1$	A4, K8
Mass capacity	C_m	L^3/M	-1	R2
Mole fraction	Y, Z	—	±1	D14
Specific mass content, mass/unit mass	n^*	—	-1	K4
Surface concentration	Γ'	M/L^2	±1	S18
Vapor capacity (porous body)	C_b	$L\theta^2/M$	$+1$	B13, R2

Temperatures, Temperature Differences

Parameter	Symbol	Dimensions	Exponent	Group
Temperature, temperature difference	$T, \Delta T$	T	-3	R1*, T6*
			-1	A4, 6*, B12, 13, C4*, 7, 10, D3*, 4*, 15, E1, G2, 6*, K3, 8, 9, 11, L9, N5, P5*, 12*, R14*, S6
			$-\tfrac{1}{2}$	L5
			$+\tfrac{1}{2}$	F6*, 7*, 8*
			$+1$	C4, G5, 6, J1, 4, L9, M1, P6, R5, 7, 14*
			$+3$	S14*
Rate of temperature change	—	T/θ	$+1$	P12

* Absolute temperature; others—temperature differences.

VELOCITIES, RATES, FLUXES, TRANSFER COEFFICIENTS
Velocities

Parameter	Symbol	Dimensions	Exponent	Group
Angular velocity (rate of rotation)	N	$1/\theta$	-3	P7
			-2	H4, L1
			-1	F9, R15
			$-\tfrac{1}{2}$	E2
			$+1$	S8, 11, 12, T1
			$+2$	F15, T2, W3
Fluid velocity	V	L/θ	-3	H5
			-2	C6, 11, D13, E9, 10, F1, M6, N2, 3, 4, R7, 8, 9, 10
			-1	A2, B3, C12, 14, D1, 3, F10, H7, 8, K10, L3, M3, P4, S13, 16
			$+1$	A2, B6, 7, 11, C3, 12, 14, D5, 7, E5, 7, F10, 14, H10, K2, L5, 8, M2, 4, 7, P1, 2, 3, 10, R4, 11, 15, S1, 2, 3, T5, 6, V2, W2, 4,
			$+2$	B1, 12, C5, 11, D15, E1, 11, F13, H11, W1
			$+3$	C7
Impeller or agitator circumference	U_s	L/θ	-2	P14
			-1	D9
Light	V_l	L/θ	-1	L8
Sound	V_s	L/θ	-1	M2, N1, S1
Waves	V_w	L/θ	-1	V1
			-1	P3
Velocity gradient	dV/dL	$1/\theta$	-2	R12
Velocity of Alfven waves	V_A	L/θ	-1	A2, K2, M4
			$+1$	A2, N1
			$+2$	C11
Velocity of bearing surface	v_s	L/θ	$+1$	H8, O1, S9

Flow Rates (Mass Fluxes)

Parameter	Symbol	Dimensions	Exponent	Group
Mass flow rate (mass flux)	V_m	M/θ	-1	B2, M11, T7
			-1	F6, 7, 8, G4, T7
Mass flux density (mass flux/unit area)	G	$M/L^2\theta$	-1	J2, 3, P15, S13
			$+1$	K4, M11

Flow Rates (Mass Fluxes)

Parameter	Symbol	Dimensions	Exponent	Group
Mass flux/unit volume (reaction rate)	U	$M/L^3\theta$	$+\frac{1}{2}$	T3
Reaction rate constant	K	L/θ	$+1$	D1, 2, 3, 4
Volumetric flow rate	V_f	L^3/θ	-1	S5
			-1	H9
			$-\frac{1}{2}$	D11
			$+\frac{1}{2}$	S11, 12
			$+1$	D9, F9

Heat Fluxes

Parameter	Symbol	Dimensions	Exponent	Group
Heat flux (heat flow/unit time)	q	ML^2/θ^3	$+1$	H5
Heat flux/unit area	q^*	M/θ^3	$+1$	K3, R6
Heat liberated/unit mass	Q	L^2/θ^2 (H/M)	$+1$	D3, 4
Rate of heat liberation/unit volume (heat source power)	j	$M/L\theta^3$	$+1$	P5

Transfer Coefficients

Parameter	Symbol	Dimensions	Exponent	Group
Heat transfer coefficient	h	$M/T\theta^3$	$+1$	B2, 4, C9, J2, N6, P15, S13
		$(H/L^2T\theta)$	$+4$	M1
Mass transfer coefficient	k_c	L/θ	$+1$	B5, J3, L9, N7, S5, 7

FORCE, HEAD, POWER, PRESSURE
Forces

Parameter	Symbol	Dimensions	Exponent	Group
Force (resistance)	F, F_R	ML/θ^2	$-\frac{1}{3}$	S2
			$+\frac{1}{3}$	K5
			$+1$	N2, 3, 4, R9
Force/unit length	F_b	M/θ^2	$+1$	H8, O1, S9

Heads, Power

Parameter	Symbol	Dimensions	Exponent	Group
Fluid head	H'	L	$-\frac{3}{4}$	S11
			$+1$	H4

Heads, Power

Parameter	Symbol	Dimensions	Exponent	Group
Head (energy per unit mass of fluid $= gH'$)	H	L^2/θ^2	-1	P3, T4
			$-\frac{3}{4}$	S12
			$+\frac{1}{4}$	D11
			$+1$	A6, P14, T4
Power	π	ML^2/θ^3 (LF/θ)	$+1$	L1, P7

Pressures

Parameter	Symbol	Dimensions	Exponent	Group
Pressure	P	$M/L\theta^2$	-1	F6, 7, 8, H9, S8, T8
			$+1$	B13, C6, L6, P13, R2
Pressure drop	$\Delta P, dP$	$M/L\theta$	$+1$	E9, 10, F1, H9, L3, R10
Pressure gradient	$\Delta P/L$	$M/L^2\theta^2$	$+1$	E10, F1, K1, P4, R10

CONSTANTS AND MISCELLANEOUS QUANTITIES
Gravity Acceleration

Parameter	Symbol	Dimensions	Exponent	Group
Gravity acceleration	—	L/θ^2	-2	A1
			-1	B1, F13, 15, M1, S6
			$-\frac{3}{4}$	S11
			$-\frac{1}{2}$	B11, F14, P13
			$-\frac{1}{3}$	C9, S3
			$+\frac{1}{3}$	F2, N8
			$+1$	A5, B9, C2, 10, D13, E8, G1, 5, H4, R5, 6, 8, 13

Other Quantities

Parameter	Symbol	Dimensions	Exponent	Group
Avogadro's number	N	$1/M$	$+1$	K9, S6
Boltzmann's constant	k		-1	K9, S6
Dufour coefficient	β	T	$+1$	A3, 4, F3
Energy of activation	E_a	L^2/θ^2	$+1$	A6
Gas constant	$\mathscr{R}$	L^2/θ^2T	-1	A6
			$-\frac{1}{2}$	L5
Shape factor	ξ^*	—	$+1$	V2
Soret coefficient	δ	$1/T$	$+1$	F3, P6
Stefan constant	η	$M/T^4\theta^3$	-1	R1, T6
			$+1$	S14

TABLE B
ALPHABETICAL LIST OF NAMED GROUPS

Serial No.	Name	Symbol	Definition	Significance	Field of Use	Reference
A1	Acceleration number	K_g	$E^3/\rho g^2 \mu^2 = (N_{Re}N_{Fr_1})^2/(H_0)^3$	Group dependent only on physical properties	Accelerated flow	25
A2	Alfven number	N_{Al}	V_A/V (or V/V_A) [cf. Cowling No. Kármán No. (2), magnetic mach number]	Ratio of Alfven wave velocity/fluid velocity	Magneto-fluid dynamics	5, 6
A3	Anonymous group (1)	ε	$\beta^* c/r$ [see also Fedorov No. (2)]		Transfer processes	39
A4	Anonymous group (2)	K_1	$\beta^* \Delta n/\Delta t$; Δt = temp. diff. $[T]$; Δn = conc. diff. [-]		Transfer processes	67
A5	Archimedes number	N_{Ar}	$\frac{gL^3\rho_o}{\mu^2}(\rho-\rho_o)$; ρ_o = fluid density; ρ = particle density (cf. N_{Ga1})	N_{Re}, gravitational force/viscous force	Fluidization, motion of liquids due to density differences	6, 13
A6	Arrhenius group	—	$E_a/\mathcal{R}T$	Activation energy/potential energy of fluid	Reaction rates	5
B1	Bagnold number	B	$3c\alpha\rho_s V^2/4d\rho_p g$, ρ_s = gas density, ρ_p = particle density	Drag force/gravitational force	Saltation studies	12
B2	Bansen number	N_{Ba}	$h_r A_w/V_{mc}$; h_r = radiant heat transfer coefficient; A_w = wall area of channel; (cf. N_{St})	Heat transferred by radiation/thermal capacity of fluid	Radiation	1
B3	Bingham number	N_{Bm}	$\tau_y L/\mu_p V$ (L = channel width)	Ratio of yield stress/viscous stress	Flow of Bingham plastics	5
B4	Biot number (heat transfer)	N_{Bih}	hL_m/k (in French literature, "Biot No." $= N_{Nu}$)	Midplane thermal internal resistance/surface film resistance	Unsteady state heat transfer	5, 6, 13
B5	Biot number (mass transfer)	N_{Bim}	$k_c L/D_{int}$; L = thickness of layer, D_{int} = diffusivity at interface	Mass transfer rate at interface/mass transfer rate in interior of solid wall thickness L	Mass transfer between fluid and solid	13
B6	Blake number	B	$V\rho/[\mu(1-e)S]$	Inertial force/viscous force	Beds of particles	6
B7	Bodenstein number	N_{Bo}	$VL/D_a = N_{Pe_m}$; L = reactor length, D_a = axial diffusivity (effective) (L^2/θ)		Diffusion in reactors	6
B8	Boltzmann number	N_{Bo}	$\equiv$ Thring radiation group			1
B9	Bond number	N_{Bo}	$(\rho-\rho')L^2 g/\sigma = N_{We_1}/N_{Fr_1}$ if $\rho-\rho' \cong \rho$ (gas in liq.); ρ = drop or bubble density; ρ' = medium density	Gravitational force/surface tension force	Atomization, motion of bubbles and drops	5
B10	Bouguer number	N_{Bu}, B	$3C_D/4\rho_D D\lambda$, C_D = wt. dust/unit bed volume (M/L^3), λ_r = mean path for radiation (L), ρ_D = dust density, R = mean particle radius. Also $N_{Bu} = kL$; L = characteristic dimension, k = absorption coefficient of medium		Radiant heat transfer to dust—gas streams	66
B11	Boussinesq number	B	$V/(2gR_H)^{1/2}$ (cf. N_{Fr2})	(Inertia force/gravitational force)$^{1/2}$	Wave behavior in open channels	6
B12	Brinkman number	N_{Br}	$\mu V^2/k\Delta t$; Δt = temp. diff.	Heat generation/heat transferred	Viscous flow	6
B13	Bulygin number	N_{Bu}	$r \cdot \frac{C_b}{C_q} \cdot \frac{P}{t_m-t_o} t_m$ = temp. of medium, medium, t_o = init. temp. of body	Heat for vaporization/heat to bring liquid to boiling point	Heat transfer during evaporation	6, 13, 14
C1	Capillary number	$K\sigma$	$\mu^2 E/\rho\sigma^2 = (N/_{We_1})^2/H_0 \cdot (N_{Re})^2$	Depends only on physical properties	Action of surface tension in flowing media	25
C2	Capillarity-buoyancy number (physical properties group) (film No.)[a]	$K\sigma_g, K_F$	$g\mu^4/\rho\sigma^3 \cdot = \sqrt{K\sigma/Kg} = (N_{We_1})^3/(N_{Fr_1})(N_{Re})^4$	Depends only on physical properties and g	Effects of surface tension and acceleration in flowing media (two-phase flow)	21, 25
C3	Capillary number	Ca	$\mu V/\sigma = N_{We}/N_{Re}$	Viscous force/surface-tension force	Atomization, two-phase flow	6
C4	Carnot number	Ca, N_{Ca}	$(T_2-T_1)/(T_2)$; T_1, T_2 = abs. temp. of two heat sources or sinks	Theoretical efficiency of Carnot cycle operating between T_1 and T_2		21
C5	Cauchy number	N_c	$\rho V^2/E_b = (N_{Ma})^2$ = Hooke No.	Inertia force/compressibility force	Compressible flow	5, 25
C6	Cavitation number	σ_c	$[(p-P_v)/\rho]/(V^2/2)p$ = local static pressure (abs.); P_v = vapor pressure	Excess of local static head over vapor pressure head/velocity head	Cavitation	5
C7	Clausius number	Cl, N_{Cl}	$V^3 L\rho/k\Delta T$; ΔT = temp. diff.		Heat conduction in forced flows	25
C8	Colburn number	—	Same as Schmidt number			5
C9	Condensation number (1)	N_{Co}	$(h/k)(\mu^2/\rho^2 g)^{1/3}$	$N_{Co} = N_{Nu}\left[\frac{(\text{viscous force})}{(\text{gravity force})} \times \frac{1}{Re}\right]^{1/3}$	Condensation	5
C10	Condensation number (2)	N_{Cv}	$L^3\rho^2 gr/k\mu\Delta t$; r = latent heat of condensation		Condensation on vertical walls	5
C11	Cowling number	C	$(V_A/V)^2 \equiv$ (Alfven number)^2		Magneto-fluid dynamics	6
C12	Craya-Curtet number	C_t	$V_k/(V_d^2-V_k^2/2)^{1/2}$; V_k = kinematic mean velocity, V_d = dynamic mean velocity		Radiant heat transfer	3, 27

[a] Very similar to Hu and Kintner's pH factor for drops and bubbles [*A.I.Ch.E. J.* 1, 42 (1955)].

Serial No.	Name	Symbol	Definition	Significance	Field of Use	Reference
C13	Crispation group	N_{Cr}	$\mu\alpha/\sigma^*L$; σ^* = undisturbed surface tension; L = layer thickness		Convection currents	55
C14	Crocco number	N_{Cr}	$V/V_{max} = \left[1 + \dfrac{2}{(\gamma-1)(N_{Ma})^2}\right]^{-1/2}$ V_{max} = maximum velocity of gas expanding adiabatically	Velocity/maximum velocity	Compressible flow	5
D1	Damköhler group I	$Da\text{I}$	UL/V_{cA}	Chemical reaction rate/bulk mass flow rate	Chemical reaction, momentum, and heat transfer	5
D2	Damköhler group II	$Da\text{II}$	UL^2/D_{cA}	Chemical reaction rate/molecular diffusion rate	Chemical reaction, momentum, and heat transfer	5
D3	Damköhler group III	$Da\text{III}$	$QUL/C_p\rho Vt$	Heat liberated/bulk transport of heat	Chemical reaction, momentum, and heat transfer	5
D4	Damköhler group IV	$Da\text{IV}$	QUL^2/kt	Heat liberated/conductive heat transfer	Chemical reaction, momentum, and heat transfer	5
D5	Damköhler group V	$Da\text{V}$	$= (N_{Re})$			5
D6	Darcy number		$4f$; see Fanning friction factor			5
D7	Dean number	N_D	$(VL\rho/\mu)(L/2R)^{1/2}$; L = pipe diam.; R = radius of curvature of bend	N_{Re} (centrifugal force/inertial force)	Flow in curved channels	5, 6
D8	Deborah number	D	θ_r/θ_o; θ_o = observation time	Relaxation time/observation time	Rheology	49
D9	Delivery number	ϕ	V_f/Aw; A = impeller area = $\pi d^2/4 \equiv$ [Diameter No.]$^{-3}$ [Speed No.]$^{-1}$		Flow machines	25
D10	Deryagin number	De	$L(\rho g/2\sigma)^{1/2}$; L = film thickness	Film thickness/capillary length	Coating	58
D11	Diameter group	δ	$(\pi/4)^{1/2}(2H)^{1/4}d/(V_f)^{1/2}$, d = impeller diam. = [pressure No.]$^{1/4}$ × [delivery No.]$^{-1/2}$		Flow machines	25
D12	Diffusion group	β	$D\theta/L_s^2$; D = diffusivity of solute through stationary solution contained in solid; cf. N_{Fom}		Mass transfer	37
D13	Drag coefficient	$C_d = C_D$	$(\rho - \rho')L_s/\rho V^2$; ρ = density of object; ρ' = density of medium; cf. f, ψ, N_e	Gravity force/inertial force	Free settling velocities, etc.	5
D14	Drew number	N_D	$\dfrac{Z_A(M_A - M_B) + M_B}{(Z_A - Y_{AW})(M_B - M_A)}\ln\dfrac{M_V}{M_W}$; M_A, M_B = mol. wt. of components A and B; M_V, M_W = mol. wt. of mixture in vapor and at wall; Y_{AW} = mole fraction of A at wall; Z_A = mole fraction of A in diffusing stream		Boundary layer mass transfer rates; velocity profile distortion; drag coefficients for binary system	22
D15	Dulong number	Du, N_{Du}	$V^2/C_p\Delta T$ = Eckert No.			25
E1	Eckert number	N_E	$V_\infty^2/C_p\Delta T$, V_∞ = velocity of fluid far from body (=2/recovery factor, q.v. $\equiv$ Dulong No.)		Compressible flow	5, 24
E2	Ekman number		$(\mu/2\rho\omega L^2)^{1/2} = (N_{Ro}/N_{Re})^{1/2}$	(Viscous force/Coriolis force)$^{1/2}$	Magneto-fluid dynamics	6
E3	Elasticity number (1)	N_{El_1}	$\theta_r\mu/\rho L^2$ = pipe radius	Elastic force/inertial force	Viscoelastic flow	5
E4	Elasticity number (2)	$\overline{K}_E$	$\rho C_p/\beta E \equiv$ [Gay Lussac No.] × [Hooke No.] ÷ [Dulong No.]	Depends on physical properties only	Effect of elasticity in flow processes	25
E5	Ellis number	N_{El}	$\mu_0 V/2\tau_{1/2}R$; μ_0 = zero shear viscosity, $\tau_{1/2}$ = shear stress when $\mu = \mu_0/2[M/L\theta^2]$; R = tube radius		Flow of non-Newtonian liquids	38
E6	Elsasser number	N_{El}	$\rho/\mu\sigma_e\mu_e \equiv N_{Re}/[$magnetic Reynolds No.$]$		Magneto-fluid dynamics	6
E7	Entry Reynolds Number	K_E	$\chi/d_h \cdot N_{Re} = \dfrac{\chi V\rho}{\mu}$; χ = entry length	As N_{Re}	Entry or inlet processes	25
E8	Eötvös number	N_{Eo}	$(\rho - \rho')L^2 g/\sigma \equiv$ Bond No., q.v.			6
E9	Euler number (1)	N_{Eu_1}	$\Delta P_F/\rho V^2$; ΔP_F = pressure drop due to friction	Friction head/2 × velocity head	Fluid friction in conduits	5, 13, 25
E10	Euler number (2)	N_{Eu_2}	$d(-dp/dL)\rho V^2$; d = pipe diam., dp/dL = pressure gradient $\equiv 2 \times$ Fanning friction factor		Fluid friction in conduits	
E11	Evaporation number	K_r	V^2/r [r = heat of vaporization (L^2/θ^2)]		Evaporation processes	25
E12	Evaporation number (2)	$\overline{K}_r$	$C_p/r\beta$ (r as in E11) $\equiv$ (Gay Lussac No.) × (E11)/(Dulong No.)		Evaporation processes	25
E13	Evaporation-elasticity number	$\overline{K}_{rE}$	$E/r\rho = K_r/$Hooke number		Evaporation processes	25
F1	Fanning friction factor	f	$d\Delta p_F/2\rho V^2 L$, d = dimension of cross section; L = length (cf. resistance coeff., Ne)	Shear stress at wall expressed as number of velocity heads	Fluid friction in conduits	5
F2	Fedorov number (1)	F_{e_1}, N_{Fe_1}	$d\sqrt[3]{\dfrac{4g\rho^2}{3\mu^2}\left[\dfrac{\gamma M}{\gamma g} - 1\right]}$ d_e = equiv. particle diam.; γM = sp. gr. of particles; γg = sp. gr. of gas (cf. N_{Ar})		Fluidized beds	6, 13

Serial No.	Name	Symbol	Definition	Significance	Field of Use	Reference
F3	Fedorov number (2)	F_{e_2}, N_{Fe_2}	$\delta\beta^* = K_1 Pn = \varepsilon \times K_a Pn$	Mass transfer analogy of Posnov number	Transport processes	39, 67
F4	Fenske number	N_F		Number of stages in separation process		6
F5	Fineness coefficient	ψ	$L/W_D^{1/3}$; W_D = volume displacement $[L^3]$		Ship modeling	26, 43
F6 F7 F8	Fliegner numbers	$F(M_a)$ $f(M_a)$ $I(M_a)$	$\{$ Functions of ratio of specific heats and mach number $V_{m}(cT)^{1/2}/A(p_s + \rho V^2) = [\gamma M_a/(\gamma-1)^{1/2}]$ $\left[1 + \dfrac{(\gamma-1)Ma^2}{2}\right]^{1/2}$ = impulse Fliegner number; γ = ratio of specific heats, Ma = mach number, A = flow area			43
F9	Flow coefficient	C_Q	V_f/Nd, d = impeller diam.		Power required by fans, etc.	32
F10	Fluidization number		V/V_{init}, V_{init} = velocity for initial fluidization	Fluid velocity in fluidized bed/that at start of fluidization	Fluidization	59
F11	Fourier number (heat transfer)	N_{Foh}	$k\theta/\rho C_p L_m^2$		Unsteady state heat transfer	5, 13, 14, 25
F12	Fourier number (mass transfer)	N_{Fom}	$D\theta/L^2 = k_c\theta/L$ (cf. D12)		Unsteady state mass transfer	13
F13	Froude number (1)	N_{Fr_1}	$V^2/gL \equiv (N_{Fr_2})^2$ (cf. Reech No., Boussinesq No., Vedernikov No.)	Inertial force/gravitational force	Wave and surface behavior	5, 13, 25
F14	Froude number (2)	N_{Fr_2}	$V/\sqrt{gL} \equiv (N_{Fr_1})^{1/2}$ (cf. Boussinesq No.)	Velocity of open channel flow/speed of very small gravity wave	Open channel flow; free surfaces	7, 50
F15	Froude No. (rotating)	Fr	DN^2/g; D = impeller diam.		Agitation	64
G1	Galileo number	N_{Ga_1}	$L^3 g\rho^2/\mu^2$ (cf. N_{Ar}, Nusselt thickness group)	$N_{Ga_1} = N_{Re} \times$ gravity force/viscous force	Circulation of viscous liquid, thermal expansion	5, 13
G2	Gay Lussac number	Ga, N_{Ga_2}	$1/\beta\Delta T$		Thermal expansion processes	25
G3	Goucher number	N_{Go}	$R(\rho g/2\sigma)^{1/2}$; R = wall or wire radius	Gravitational force/surface tension force$^{1/2}$	Coating	24
G4	Graetz (Grätz) number	N_{Gz}	$V_m C_p/kL$	Thermal capacity fluid/convective heat transfer	Streamline flow	5, 25
G5	Grashof number	N_{Gr}	$L^3\rho^2 g\beta\Delta t/\mu^2 \equiv N_{Re}/(N_{Ga_1}) = (N_{Re})^2/(N_{Ga_1})(N_{Ga_2})$	$N_{Gr} = N_{Re} \times$ (buoyancy force/viscous force)	Free convection	5, 13, 25
G6	Gukhman number	Gu, N_{Gu}	$(t_a - t_m)/T_a$; t_a, T_a = temp. (°C., °K.) of hot gas stream, t_m = temp. of moist surface (wet bulb temp.)	Thermodynamic criterion of evaporation under isobaric adiabatic conditions	Convective heat transfer in evaporation	6, 20, 17, 33
G7	Guldberg Waage group	N_{Gw}	Given by equation relating volumes of reacting gases and reaction products		Chemical reaction in blast furnaces	5
H1	Hall coefficient	N_H	$f_c J$ (f_c = cyclotron frequency, J = av. free path/av. veloc.)		Magneto-fluid dynamics	7
H2	Hartmann number	M_H	$(\mu_e^2 H_e^2\sigma_e L^2/\mu)^{1/2} \equiv (SR_M N_{Re})^{1/2} \equiv (N_{Re}N)^{1/2}$	Magnetically induced stress/hydrodynamic shear stress (magnetic body force/viscous force)$^{1/2}$	Magneto-fluid dynamics	5, 6
H3	Hatta number	β	$\gamma/\tanh\gamma$; $\gamma = (rCD)^{1/2}/k_c$, r = reaction rate constant $[L^3/M\theta]$ [a modified Hatta number has also been defined 35]		Gas absorption with chemical reaction	41
H4	Head coefficient	C_H	$gH'/N^2 d^2$ (d = impeller diam.)		Flow in pumps and fans	32
H5	Heat transfer number	K_Q	$q/V^3 L^2 \rho$		Heat transfer in stream	25
H6	Hedstrom number	N_{He}	$\tau_y L^2\rho/\mu^2 \equiv (N_{Re}) \times (N_{Bm})$		Flow of Bingham plastics	5
H7	Helmholtz resonator group		$(d^3/W)^{1/2}/Ma$	Proportional to frequency × residence time	Pulsating combustion	28
H8	Hersey number		$F_b/\mu v_s$ (cf. truncation number)	Load force/viscous force	Lubrication	6
H9	Hodgson number	N_H	$Wf\Delta\rho_F/\overline{V}_{f,ps}$	Time constant of system/period of pulsation	Pulsating gas flow	5
H10	Homochronous number	H_{o_1}	$V\theta/L$ (θ = time for liquid to move characteristic distance L)	Duration of process/time for liquid to move through L	Choice of time scales	13
H11	Hooke number	H_{o_2}	$\rho V^2/E \equiv$ Vauchy No., q.v.		Elasticity of flowing media	25
J1	Jakob modulus	Ja	$C_P\rho_L\Delta t/r\rho_v$ (ρ_L, ρ_v = densities of liquid and vapor; Δt = liquid superheat temperature diff.)	Maximum bubble radius/thickness of superheated film	Boiling	6
J2	J-factor (heat transfer)	j_H	$(h/c_p G)(c_p\mu/k)^{2/3} \equiv (N_{Nu})/(N_{Re})(N_{Pr})^{1/3}$		Heat, mass and momentum transfer theory	5
J3	J-factor (mass transfer)	j_M	$(k_c\rho/G)(\mu/\rho D)^{2/3} \equiv (k_c\rho/G)(N_{Sc})^{2/3}$			5
J4	Joule number	J	$2\rho C_p\Delta t/\mu_e H_e^2 \equiv 2(N_{Re})(R_M)/(M_H)^2(N_E)$	Joule heating energy/magnetic field energy	Magneto-fluid dynamics	6
K1	Kármán number (1)	N_K	$\rho d^3(-dp/dL)/\mu^2$ (d = pipe diam., dp/dL = pressure gradient) $\equiv 2(N_{Re})^2 f^{1/2}$		Fluid friction in conduits	5

Serial No.	Name	Symbol	Definition	Significance	Field of Use	Reference
K2	Kármán number (2)	K	V/V_A (see Alfven No.)		Magneto-fluid dynamics	6
K3	Kirpichev number for heat transfer	Ki_q, N_{Ki_q}	$q^* L/k \Delta t$ (cf. N_{Bih}, N_{Nu})	Intensity external heat transfer/internal heat transfer intensity	Heat transfer	6, 14
K4	Kirpichev number for mass transfer	Ki_m	$GL/D\rho n^*$ (cf. N_{Pem}, N_{Bim})	Intensity external mass transfer/internal mass transfer intensity	Mass transfer	13, 14
K5	Kirpitcheff number		$(\rho F_R/\mu^2)^{1/3} = [(N_{Re})^2 cf]^{1/3}$		Flow around obstacles	34
K6	Knudsen number (1)	N_{Kn}	λ/L	Length of mean free path/characteristic dimension	Low pressure gas flow	5
K7	Knudsen number (2)	N_{KnA}	$\rho D/_{AB} D_{KA} \zeta$	Bulk diffusion/Knudsen diffusion	Gaseous diffusion in packed beds	6
K8	Kossovich number	K_o, N_{Ko}	$r_v \Delta n_m/c \Delta t$	Heat used for evaporation/heat used in raising temperature of body	Convective heat transfer during evaporation	6, 14
K9	Kronig number	Kr	$4L^2 \beta \rho^2 \Delta t E_S^2 N[\alpha + 2/3(p_e^2/kT)]/u^2 M$ E_S = electric field at surface, N = Avogadro's Number, α = polarization coefficient, p_e = molecular dipole moment, k = Boltzmann's constant, M = molecular weight	(N_{Re}) (electrostatic force/viscous force)	Convective heat transfer	6, 44
K10	Kutateladze number (1)	Ku	$IEL/\rho V u'$; I = current density $[Q/L^2\theta]$, E = voltage $[ML/Q\theta^2]$, u' = enthalpy $[L^2/\theta^2]$		Electric arcs in gas streams	65
K11	Kutateladze number (2)	K	$r_v/c_p(t_o - t_w)$, (t_o, t_w = stream, wall temp.)		Combined heat and mass transfer in evaporation	17
L1	Lagrange group (1)	La_1	$\Pi/\mu L^3 N^2$; L = characteristic dimension of agitator $= N_{Re} N_p$		Agitation	6
L2	Lagrange number (2)	La_2	$(D + \varepsilon_D)/D$	Combined molecular and eddy mass transfer rate/molecular mass transfer rate	Mass transfer in turbulent systems	31
L3	Lagrange number (3)	La_3	$\Delta PR/\mu V$		Magneto-fluid dynamics	6
L4	Larmor number	R_{Lo}	L_L/L; (L_L = Larmor radius)		Magneto-fluid dynamics	23
L5	Laval number	La	$V/\left(\dfrac{2\gamma}{\gamma+1} RT\right)^{1/2}$; γ = ratio of specific heats	Linear velocity/critical velocity of sound	Compressible flow	56
L6	Leverett function	J	$(\xi/e)^{1/2}(P\sigma/\sigma)$	Characteristic dimension of surface curvature/characteristic dimension of pores	Two-phase flow in porous media	6
L7	Lewis No.	N_{Le}	$k/\rho c_p D = \alpha/D \equiv N_{Sc}/N_{Pr}$, (N.B.: Lewis number is sometimes defined as reciprocal of this quantity)		Combined heat and mass transfer	5, 25
L8	Lorentz number	N_{Lo}	V/V_1; (V_1 = velocity of light)	Fluid velocity/velocity of light	Magneto-fluid dynamics	6
L9	Luikov (Lÿkov) number	Lu	$k_c L/\alpha = k_c L_\rho C_p/k$	Mass diffusivity/thermal diffusivity; rate of extension of mass transfer field/rate of extension of heat transfer field	Combined heat and transfer	6, 13
L10	Lukomskii number	Lu	α/a_m; a_m = potential conductivity of mass transfer $[L^2/\theta]$		Combined heat and mass transfer	36
L11	Lundquist number	N_{Lu}	$\sigma_e H_o \mu_e^{3/2} L/\rho^{1/2} = M_H(R_M/N_{Re})^{1/2}$ (L = thickness of fluid layer)		Magneto-fluid dynamics	6
L12	Lyashchenko number	Ly	$\equiv N_{Re}^3/N_{Ar}$	Magnetic body force/square root of product of the inertia and buoying force.	Fluidization	19
L13	Lykoudis number	N_{Ly}	$(\mu_e H_o)^2 \dfrac{\sigma_e}{\rho}\left[\dfrac{L}{g\beta\Delta t}\right]^{1/2} \equiv (M_H)^2/(N_{Gr})^{1/2}$		Magneto-fluid dynamics	6
M1	McAdams group		$h^4 L\mu\Delta t/k^3 \rho^2 gr$	Constant for given surface orientation	Condensation	5
M2	Mach number	N_{Ma}, Ma	V/V_s; (V_s = velocity of sound in fluid) $\equiv$ $v/\sqrt{E_b/\rho}$; (E_b = bulk modulus of fluid) (cf. Sarrau number)	Linear velocity/sonic velocity	Compressible flow	5–7, 25, 50
M3	Magnetic force parameter	N	$\mu_e^2 H_o^2 \sigma_e L/\rho V$	Magnetic body force/inertia force; resistance time of fluid in field/relaxation time of lines force	Magnetic-fluid dynamics	6
M4	Magnetic mach number	M_{Ma}	V/V_e (see Alfven number)		Magneto-fluid dynamics	6
M5	Magnetic Oseen number	k	$\frac{1}{2}(1 - N_{Al}^2)R_M$	Magnetic force/inertia force	Magneto-fluid dynamics	2
M6	Magnetic pressure number	S	$\mu_e H_o^2/\rho V^2$	Magnetic pressure/2 × dynamic pressure	Magneto-fluid dynamics	5
M7	Magnetic Reynolds number	R_M	$\sigma_c \mu_e LV$ (cf. velocity number)	Mass transport diffusivity/magnetic diffusivity	Magneto-fluid dynamics	5
M8	Maievskii number		$\equiv Ma$		Compressible flow	12
M9	Marangoni number	N_{Ma}	$\dfrac{\Delta\sigma}{\Delta t}\dfrac{\Delta t}{\Delta L} L^2/\mu\alpha$; L = layer thickness		Cellular convection	46

Serial No.	Name	Symbol	Definition	Significance	Field of Use	Reference
M10	Margoulis number	M	$\equiv N_{st}$		Forced convection	6, 7
M11	Merkel number	N_{Me}	$GA^* W^*/(V_m)$ gas	Mass of water transferred in cooling per unit humidity difference/ mass of dry gas	Cooling towers, liquid-gas contact	5
M12	Miniovich number	Mn	SR/e; R = pore radius		Drying	36
M13	Mondt number	N_{Mo}		Convective/conductive heat transfer	Heat transfer	6
N1	Naze number	N_a	$V_A/V_s \equiv (N_{Ma}.N_{A1})$	Velocity Alfven wave/ velocity of sound	Magneto-fluid dynamics	6
N2	Newton inertial force group	N_I	$F/\rho V^2 L^2$	Imposed force/inertial group	Agitation	5
N3	Newton number	N_e	$F_R/\rho V^2 L^2$; $(cf. f, \psi)$	Resistance force/inertia force	Friction in fluid flow	25
N4	Number of velocity heads	N	$(F/\rho L^2)/(V^2/2)$	Imposed head/velocity head	Friction in conduits	6
N5	Number for similarity of phys. and chem. changes	K	$r/C_p \Delta t$	Heat flow for phase change/superheat (supercooling) of one of the phases	Changes of phase	14
N6	Nusselt number	N_{Nu}	$hL/k = (N_{Re}N_{St})$ (cf. N_{Bih}, Ki_g)	Total heat transfer/conductive heat transfer	Forced convection	5, 13, 25
N7	Nusselt number for mass transfer	Nu_m, N_{Nu_m}	$k_c L/D = N_{Sh}$	Intensity of mass flux at interface/specific flux by pure molecular diffusion in layer of thickness, L	Mass transfer	13, 20
N8	Nusselt film thickness group	ψ, N_T	$L_f(\rho^2 g/\mu^2)^{1/3} \equiv (N_{Ga})^{1/3}$; ($L_f$ = film thickness)	$= (N_{Re})^{1/3}$ (gravitational force/viscous force)$^{1/3}$	Falling films	6
O1	Ocvirk number		$(F_b/\mu V_s)(a/R)^2(D/b)^2$; ($v_s$ = shaft surface velocity; R = shaft radius; D = shaft diam.) (cf. N_S)	Load force/viscous force	Lubrication	6
O2	Ohnesorge number	Z	$\mu/(\rho L\sigma)^{1/2} = (N_{We_1})^{1/2}/(N_{Re})$	Viscous force/(inertia force × surface tension force)$^{1/2}$	Atomization	5
P1	Péclet number (heat)	Pe, N_{Pe}	$LV\rho C_p/\lambda = LV/\alpha = (N_{Re}N_{Pr})$	Bulk heat transfer/conductive heat transfer	Forced convection	5, 13, 25
P2	Péclet number (mass)	N_{Pe_m}	$LV/D = (N_{Re}N_{Sc})$	Bulk mass transfer/diffusive mass transfer	Mass transfer	5
P3	Pipeline parameter	ρn	$V_w V_o/2H'_s$ (V_w = velocity waterhammer wave, V_o = initial velocity H'_s = static head $\times g[L^2/\theta^2]$	Maximum pressure rise in water hammer/2 × static pressure	Water hammer	5
P4	Poiseuille number		$D^2(-dp/dL)/\mu V(D$ = pipe diam., dp/dL = pressure gradient)	$= 32$ for laminar flow in round pipe	Laminar fluid friction	5
P5	Pomerantsev number	P_o	$jL^2/k(t_m-t_o)(t_m, t_o$ = temp. of medium, initial temp. of body) cf. Damköhler Group IV)		Heat transfer with heat sources in medium	6, 14, 60
P6	Posnov number	Pn	$\delta\Delta t/(\Delta n_m)$ (cf. Fe_2)		Combined heat and mass transfer	13, 14, 67
P7	Power number	N_P	$\Pi/L^5\rho N^3$	Drag on (agitator impeller) or inertial force	Power consumption by agitators, fans, pumps, etc	5, 32
P8	Prandtl number	N_{Pr}	$C_p\mu/k$ = Da IV/Da III × Da V	Momentum diffusivity/ thermal diffusivity	Forced and free convection	5, 13, 25
P9	Prandtl number (mass transfer)	Pr_m	$\mu/\rho D = N_{Sc}$, v (used in Russian, German literature)	See Schmidt number		13
P10	Prandtl velocity ratio	u^+	$V/(\tau_w/\rho)^{1/2}$ (V = local fluid velocity)	Inertial force/wall shear force$^{1/2}$	Turbulence studies	5
P11	Prandtl dimensionless	y^+	$L(\rho\tau_w)^{1/2}/\mu$ (L = distance from wall, etc.)		Turbulence studies	
P12	Predvoditelev number	Pd	$\Gamma L^2/\alpha t_o = \left(\dfrac{dt^*}{\alpha(N_{Fo})}\right)_{max}$ where t_o = init. temp. of body, t^* = temp. of medium relative to its initial temp.	Rate of change of temp. of medium/rate of change of temp. of body	Heat transfer	13, 14, 60
P13	Pressure number (1)	K_P	$P/\{g\sigma(\rho'-\rho'')\}^{1/2}(\rho', \rho''$ = density of liquid gas)	Absolute pressure in system (pressure jump on interface)		14
P14	Pressure number (2)		$H/\frac{1}{2}2U_s^2(U_s$ = circumferential velocity) $\equiv$ [diameter No.]$^{-2}$ × [Speed No.]$^{-2}$		Flow machines (turbines, pumps, etc.)	25
P15	Psychrometric ratio		h_c/Gs	Heat transfer by convection/heat transfer by mass transfer	Wet and dry bulb thermometry	5
R1	Radiation number	$\bar{K}_s$	$kE/\eta\sigma T^3 = (N_{We_1})/$(Hooke No.) × (Stefan No.)		Radiant transfer	25
R2	Ramzin number	Ra	$\dfrac{C_p P}{C_m(\Delta\Omega)} = \dfrac{\text{(Bulygin No.)}}{\text{(Kosovich No.)}}$		Molar mass transfer	40
R3	Ratio of specific heats	γ	C_p/C_v (specific heats at constant pressure, volume)		Compressible flow	5
R4	Rayleigh number (1)	N_{Ra_1}	$V(\rho L/\sigma)^{1/2} = N_{We_1}$ (q.v.)	See N_{We}	Breakup of liquid jets	5
R5	Rayleigh number (2)	R'_2	$L^3\rho^2 g\beta c_p\Delta t/\mu k = L^3\rho g\beta\Delta t/\mu\alpha = (N_{Gr}) \cdot (N_{Pr})$		Free convection	5, 25

Serial No.	Name	Symbol	Definition	Significance	Field of Use	Reference
R6	Rayleigh number (3)	Ra_3	$q^* L^5 \rho^2 g \beta C_p / \mu k^2 x = (N_{Gr})(N_{Pr})$ $(N_{Nu})(L/x)$; $(L = $ pipe diam.$)$		Combined free and forced convection in vertical tubes	6
R7	Recovery factor	N_{rf}	$C_p(t_{aw} - t_m)/V^2$; t_{aw} = attained adiabatic wall temp. t_m = temp. of moving medium (cf. Eckert No.)	Actual temp. recovery/ theoretical temp. recovery	Convective heat transfer in compressible flow	5
R8	Reech number		$= 1/(N_{Fr})$ q.v.		Wave and surface behavior	6
R9	Resistance coefficient (1)	C_f	$F_R/\frac{1}{2}\rho V^2 L^2$ (cf. drag coeff., Newton number, Fanning factor)		Flow resistance	25
R10	Resistance coefficient (2)	ψ	Δp. $D_H/\frac{1}{2}\rho V^2 L$ (Δp = pressure drop over length, L) (cf. R9)		Fluid friction in conduits	25
R11	Reynolds number	N_{Re}	$LV\rho/\mu$	Inertia force/viscous force	Dynamic similarity	5, 25
R12	Reynolds number (rotating)	Re	$L^2 N \rho/\mu$; L = impeller diam.		Agitation	45
R13	Richardson number	N_{Ri}	$-(g/\rho)(d\rho/dL)/(dV/dL)w^2$ [L = height of liquid layer, $(dV/dL)_w$ = velocity gradient at wall]	Gravity force/inertial force	Stratified flow of multi-layer systems	54
R14	Romankov number	R'_o	T_D/T_{PROD}	Dry bulb temperature (abs.)/(product temperature (abs.)	Drying	6, 52
R15	Rossby number	N_{Ro}	$V/2\omega_e L \sin \Lambda$ (ω_e = angular velocity of earth's rotation [$1/\theta$]; Λ = angle between axis of earth's rotation and direction of fluid motion [$^-$])	Inertia force/Coriolis force	Effect of earth's rotation on flow in pipes	5
R16	Roughness factor		ε/L		Fluid friction	5
S1	Sarrau number		$\equiv$ mach number, q.v.		Compressible flow	6
S2	Schiller number (1)		$LV(\rho^2/\mu F_R)^{1/3}$		Flow around obstacles	34
S3	Schiller number (2)	Sch	$V\left[\frac{3}{4} \cdot \frac{\rho\gamma_m}{g\mu(\gamma_M - \gamma_m)}\right]^{1/3}$; V = velocity in fluidized bed; γ_m, γ_M = specific gravity of medium and material in bed		Fluidization	66
S4	Schmidt number	N_{Sc}	$\mu/\rho D$ (cf. N_{Pr_m}) ($=$ Da II/Da I Da V)	Kinetic viscosity/molecular diffusivity	Diffusion in flowing	5, 25
S5	Semenov number	Sm	k_c/K; K = reaction rate constant [L/θ]		Reaction kinetics	11
S6	Senftleben number	S_e	$NE_s^2[\alpha + 2/3(\rho_o^2/kT)] \cdot [1/4LM_n]$ Kronig number, q.v.		Convective heat transfer	6
S7	Sherwood number	N_{Sh}	$k_c L/D$ (also termed Taylor number)	Mass diffusivity/molecular diffusivity	Mass transfer	5
S8	Sommerfeld number (1)	N_{S_1}	$(\mu N/P_b)(D/a)^2$ (D = shaft diam., (cf.) Ocvirk number)	Viscous force/load force	Lubrication	5
S9	Sommerfeld number (2)	N_{S_2}	$(F_b/\mu V_s)(a/R)^2$ (V_s = veloc. of shaft surface; R = shaft radius) ($N_{S_2} = 4/\pi N_{S_1}$)	Viscous force/load force	Lubrication	6
S10	Spalding function	Sp	$-\left(\frac{\delta\theta}{\partial u^+}\right)_{u^+=0}$; $\theta = (T - T_\infty)/(T_w - T_\infty)$, T_w = wall temperature, T_∞ = free stream temp., u^+ = Prandtl velocity ratio	Dimensionless temp. gradient at wall	Convection	18, 33
S11	Specific speed	C_s	$N(V_f)^{1/2}/(gH')^{3/4}$ (H' = head of liquid produced by one stage) (cf. speed number)		Pumps and compressors	8, 32
S12	Speed number	σ	$(4\pi)^{1/2}(V_f)^{1/2} N/(2H)^{3/4}$ = (delivery number)$^{1/2}$ × pressure number)$^{-3/4}$ (cf. specific speed)		Flow machines	25
S13	Stanton number	N_{St}	$h/C_p\rho V = h/C_p G = (N_{Nu})/(N_{Re})(N_{Pr})$	Heat transferred/thermal capacity of fluid	Forced convection	5, 6, 13, 25
S14	Stefan number	St	$\eta L T^3/k$		Heat radiation	25
S15	Stokes number	St	$\mu\theta_v/\rho L^2$ (θ_v = vibration time) $\equiv$ $(N_{S_1})^{-1}(N_{Re})^{-1}$		Particle dynamics	6
S16	Strouhal number	N_{S_1}, Sr	fL/V (cf. N_{Th})		Vortex streets; unsteady-state flow	6, 25
S17	Suratman number	Su	$\rho L\sigma/\mu^2 = (N_{Re})^2/(N_{We}) = (Z)^{-2}$		Particle dynamics	6
S18	Surface elasticity number	N_{El}	$-\frac{\Gamma'}{D_S} L \frac{(\partial\sigma)}{(\partial\Gamma')}$; Γ' = surface concentration of surfactant in undisturbed state, D_S = surface diffusivity, L = film thickness		Convection cells	10
S19	Surface viscosity number	N_{Vi}	$\mu_S/\mu L$; μ_S = surface viscosity, [M/θ], L = film thickness		Convection cells	10
T1	Taylor number (1)	N_{Ta_1}	$\omega_c(R_a)^{1/2}a^{3/2}\rho/\mu$; ($\omega_c$ = angular velocity of cylinder; R_a = mean radius of annulus)		Stability of flow pattern in annulus with rotating cylinder	5
T2	Taylor number (2)	N_{Ta_2}	$(2\omega L^2\rho/\mu)^2[\omega$ = rate of spin $(1/\theta)$; L = height of fluid layer]	α(Coriolis force/viscous force)^2	Effect of rotation on free convection	6
T3	Thiele modulus	m_T	$Q^{1/2}U^{1/2}L/k^{1/2}t^{1/2} = (Da IV)^{1/2}$		Diffusion in porous catalysts	5
T4	Thoma number	σ_T	$(H_a - H_s - H_v)/H$ (H = total head; H_a = atm. pressure head; H_s = suction head; H_v = vapor pressure head)	Net positive suction head/total head	Cavitation in pumps	5
T5	Thomson number	N_{Th}	$\theta V/L$; θ = characteristic time (cf. N_{S_1})		Fluid flow	6
T6	Thring radiation group		$\rho C_p V/e^*\eta T^3$ (cf. Boltzmann number)	Bulk heat transport/heat transport by radiation	Radiation	5

TABLE B (Continued)

Serial No.	Name	Symbol	Definition	Significance	Field of Use	Reference
T7	Thring-Newby criterion	θ	$[(V_{m_i} + V_{mo})/V_{mo}](R/L)$; V_{mo}, V_{m_i} = mass flow rates of nozzle fluid and surrounding fluid $[M/\theta]$; R = equivalent nozzle radius; L = furnace half width		Combustion of fuels	*4*
T8	Truncation number	r	$\mu\gamma/P$ (*cf.* Hersey number)	Shear stress/normal stress	Viscous flow	*6*
V1	Valensi number	V	$\omega L^2 \rho/\mu$; ω = circular oscillation frequency when $\mu = 0$ $[1/\theta]$		Oscillations of drops and bubble-	*57*
V2	Vedernikov number	V	$\zeta^*\xi^*\bar{V}/(V_w - \bar{V}) \equiv \zeta^*\xi^*(N_{Fr_2})$ (ζ^* = exponent of hydraulic radius in formula $[^-]$; ζ^* = shape factor of channel section; V_w = absolute velocity of disturbance wave)	Generalized Froude number	Instability of open-channel flow	*6, 61*
V3	Velocity number	R_v	$\equiv$ Magnetic Reynolds number, *q.v.*			*5, 6*
W1	Weber number (1)	N_{We_1}	$V^2\rho L/\sigma = (N_{We_2})^2$	Inertia force/surface tension force	Bubble formation, etc.	*2, 25*
W2	Weber number (2)	N_{We_2}	$V(\rho L/\sigma)^{1/2} = (N_{We_1})^{1/2}$			
W3	Weber number (rotating)	W_e	$L^3N^2\rho/\sigma$; L = impeller diameter		Agitation	*64*
W4	Weissenberg number	N_{We}	ω_3V/ω_1L; $\omega_3 = \int_0^\infty sG(s)\,ds$, $\omega_1 = \int_0^\infty G(s)\,ds$, G = relaxation modulus of linear viscoelasticity, s = recoverable elastic strain	Viscoelastic force/viscous force	Viscoelastic flow	*63*

REFERENCES

1. Adrianov, V. N., Shorin, S. N., *AIAA J.* **1**, 1729 (1963).
2. Ahlstrom, H. G., *J. Fluid Mech.* **15**, 205 (1963).
3. Becker, H. A., Hottel, H. C., Williams, G. C., "Ninth Symposium (International) on Combustion," p. 7 Academic Press, New York, 1963.
4. Baer, J. M., Chigier, N. A., Lee, K. B., *Ibid.* p 803.
5. Boucher, D. F., Alves, G. E., *Chem. Eng. Progr.* **55** (9), 55 (1959).
6. Ibid., **59** (8), 75 (1963).
7. British Standard 1991, "Recommendations for Letter Symbols, Signs and Abbreviations. Part 1. Chemical Engineering, Nuclear Science, and Applied Chemistry," British Standards Institution, London, 1961.
8. Brown, G. G., *et al.*, "Unit Operations," Wiley, New York, 1950.
9. Buckingham, E., *Phys. Rev.* **4**, 345 (1914).
10. Berg, J. C., Acrivos, A., *Chem. Eng. Sci.* **20**, 737 (1965).
11. Chukhanov, Z. F., *Intern. J. Heat Mass Transfer* **6**, 691 (1963).
12. Dallavalle, J. M., "Micromeritics," 2nd ed., Pitman, New York, 1948.
13. El'perin, I. T., *Inzh. Fiz. Zh. Akad. Nauk Belorussk. SSR* **4** (1), 131 (1963).
14. El'perin, I. T., *Intern. J. Heat Mass Transfer* **5**, 349 (1962).
15. Engel, F. V. A., *Z.V.D.I.* **107**, 671, 793 (1965).
16. Faller, A. J., *J. Fluid Mech.* **15**, 560 (1963).
17. Fedorov, B. I., *Inzh. Fiz. Zh. Akad. Nauk Belorussk. SSR* **7** (1), 21 (1964).
18. Gardner, G. O., Kestin, J., *Intern. J. Heat Mass Transfer* **6**, 289 (1963).
19. Gel'perin, I. T., Ainshtein, V. G., Goĭkhman, I. D., *Inzh. Fiz. Zh. Akad. Nauk Belorussk. SSR* **7** (7), 15 (1964).
20. Grassmann, P., *Chem. Ing.-Tech.* **31**, 148 (1959).
21. Grassman, P., Lemaire, L. H., *Ibid.*, **30**, 450 (1958).
22. Greene, D. F., Ph.D. Thesis, Columbia Univ., 1961 [*Dissertation Abstr.* **24** (8), 3248 (1964)].
23. Gukhman, A. A., "Introduction to the Theory of Similarity," Academic Press, New York, 1965.
24. Gutfinger, C., Tallmadge, J. A., *A.I.Ch.E. J.* **10**, 774 (1965).
25. Hahnemann, H. W., "Die Umstellung auf das internationale Einheitensystem in Mechanik und Wärmetechnik," VDI-Verlag, Düsseldorf, 1959.
26. Holt, M., "Dimensional Analysis" in "Handbook of Fluid Dynamics," V. L. Streeter, ed., McGraw-Hill, New York, 1961.
27. Hottel, H. C., Sarofim, A. F., *Intern. J. Heat Mass Transfer* **8**, 1153 (1965).
28. Hottel, H. C., Williams, G. C., Jensen, W. P., Tobey, A. C., Burrage, P. M. R., p. 923 in "Ninth Symposium (International) on Combustion," Academic Press, New York, 1963.
29. Huntley, H. E., "Dimensional Analysis," MacDonald & Co., London, 1952.
30. Johnson, S. P., "Survey of Flow Calculation Methods," p. 98, Pre-printed Papers & Program, Aeronautic & Hydraulic Divisions, A.S.M.E. Summer Meeting, June 19–21, Univ. of Calif. and Stanford Univ., 1934.
31. Kafarov, V. V., *Zh. Prikl. Khim.* **29**, 40 (1956).
32. Kay, J. M., "An Introduction to Fluid Mechanics & Heat Transfer," Cambridge Univ. Press, 1957.
33. Kestin, J., Persen, L. N., *Intern. J. Heat Mass Transfer* **5**, 143 (1962).

34. Klinkenberg, A., Mooy, H. H., *Chem. Eng. Progr.* **44**, 17 (1948).
35. Koide, K., Kubota, H., Shindo, M., *Chem. Eng.* (Japan), **28** (8), 657 (1964).
36. Lykov, A. V., Mikhaĭlov, Yu. A., "Theory of Energy & Mass Transfer," Prentice-Hall, Englewood Cliffs, N.J., 1961.
37. McCabe, W. L., Smith, J. C., "Unit Operations of Chemical Engineering," McGraw-Hill, New York, 1956.
38. Matsuhisa, S., Bird, R. B., *A.I.Ch.E. J.* **11**, 588 (1965).
39. Mikhaĭlov, Yu. A., Bornikova, R. M. *Inzh. Fiz. Zh. Akad. Nauk Belorussk. SSR* **6** (10), 43 (1963).
40. Mikhaĭlov, Yu. A., Romanina, I. V., *Ibid.*, **7** (1), 49 (1964).
41. Miyauchi, T., Nakano, K., Obata, K., Kimura, S., *Chem. Eng.* (Japan) **26** (9), 999 (1962).
42. Mkhitaryan, A. M., "Hydraulics & Fundamentals of Gas Dynamics," Israel Program for Scientific Translations, Jerusalem, 1964.
43. Mordell, D. L., Wu, J. H. T., *Can. Aeronaut. Space J.* **9** (4), 117 (1963).
44. Motulevich, V. P., Eroshenko, V. M., Petrov, Yu. P., in "Physics of Heat Exchange & Gas Dynamics," A. S. Predvoditelev, ed., Consultants Bureau, New York, 1963.
45. Nagata, S., *Chem. Eng.* (Japan) **27** (8), 592 (1962).
46. Nield, D. A., *J. Fluid Mech.* **19**, 341 (1964).
47. Potter, J. M. F., B.Sc. Thesis, Dept. of Chem. Engrg., Univ. of Birmingham, England, 1959.
48. Rayleigh, Lord, *Phil. Mag.* **48**, 321 (1899).
49. Reiner, M., *Phys. Today* **17** (1), 62 (1964).
50. Rouse, H. (ed.), "Engineering Hydraulics," Wiley, New York, 1950.
51. Rouse, H., Ince, S., "History of Hydraulics," Iowa Institute of Hydraulic Research, State University of Iowa, 1957.
52. Sazhin, B. S., *Inzh. Fiz. Zh. Akad. Nauk Belorussk. SSR* **5** (6), 13 (1962).
53. Sazhin, B. S., Miklin, Yu. A., *Ibid.*, **6** (10), 57 (1963).
54. Schlichting, H., "Boundary Layer Theory," 4th ed., McGraw-Hill, New York, 1960.
55. Scriven, L. E., Sternling, C. V., *J. Fluid Mech.* **19**, 321 (1964).
56. Sillem, H., *Z.V.D.I.* **106**, 398 (1964).
57. Szebehely, V. G., p. 771 in "Proc. 2nd U.S. Nat. Congress of Appl. Mech.," Ann Arbor, Mich., June 1954; A.S.M.E., New York, 1955.
58. Tallmadge, J. A., Labine, R. A., Wood, B. H., *Ind. Eng. Chem. Fundamentals* **4**, 400 (1965).
59. Tamarin, A. I., *Inzh. Fiz. Zh. Akad. Nauk Belorussk. SSR* **6** (7), 19 (1963).
60. Tartakovskiĭ, D. F., *Ibid.*, **7** (1), 71 (1964).
61. Vedernikov, V. V., *Compt. Rend. Acad. Sci. U.R.S.S.* **48**, 239 (1945); **52**, 207 (1946).
62. Weber, M., *Jahrb. Schiffbautechn. Ges.* **20**, 355 (1919).
63. White, J. L., *J. Appl. Polymer. Sci.* **8**, 2339 (1964).
64. Yamaguchi, I., Yabuta, S., Nagata, S., *Chem. Eng.* (Japan) **27** (8), 576 (1963).
65. Yas'ko, O. I., *Inzh.-Fiz. Zh. Akad. Nauk Belorussk. SSR* **7** (12), 112 (1964).
66. Zabrodskiĭ, S. S., "Flow & Heat Transfer in Fluidized Beds," to be published shortly by M.I.T. Press.
67. Zhuravleva, V. P., *Inzh.-Fiz. Zh. Akad. Nauk Belorussk. SSR* **6** (9), 73 (1963).

DIMENSIONLESS GROUPS (Supplementary Table)

George D. Fulford and John P. Catchpole

Reprinted from Industrial and Engineering Chemistry, Vol. 60, No. 3, pp. 71 to 78. Copyright 1968 by the American Chemical Society and reprinted by permission of the copyright owners and the authors.

TABLE II. ALPHABETICAL
LIST OF NEW GROUPS

Serial No.	Name	Symbol	Definition	Significance	Field of Use	Reference
A0	Absorption No.	Ab	$kc_L\sqrt{\dfrac{xL_f}{DV_f'}}$ kc_L = liquid side mass transfer coefficient; x = length of wetted surface; L_f = film thickness; V_f' = volume flow rate per wetted perimeter $[L^2/\theta]$	Dimensionless mass transfer coefficient	Gas absorption in wetted wall column	40
A1a	Advance ratio	J	V/ND V = forward speed; D = propeller diameter	Special form of Strouhal No.	Propeller studies	27
A1b	Aeroelasticity parameter	—	$\equiv$ Cauchy No., q.v.	Inertia force/compressibility force	Compressible flow	27
A4a	Anonymous group 3	ε	$Dx/V_f'L_f$ (symbols as in Absorption No.). $\varepsilon = (Ab)^2/(N_{Sh})^2$	Dimensionless diffusivity	Gas absorption in wetted wall column	40
A4b	Anonymous group 4	$1/\alpha$ $(1/\beta)$	$\tau_w R/V_\infty\mu$; R = cylinder radius; V_∞ = velocity outside boundary layer	Frictional force/viscous force dimensionless skin friction)	Laminar boundary layer flow	12, 28, 35
B1a	Bairstow No.	—	V/V_{sw} V_{sw} = velocity of sound at wall (cf. Mach. No.)	Previously used for Mach No., now largely obsolete	—	27
B2a	Batchelor No.	—	$VL\sigma_e/V_t^2\varepsilon_e$ ε_e = electrical permittivity $[Q^2\theta^2/L^3M]$		Magnetofluid dynamics	27
B13a	Buoyancy parameter	—	$\dfrac{\Delta T}{T}\dfrac{gL}{V^2} = (N_{Gr})/(N_{Re})^2 = \dfrac{\Delta T}{T}\left(\dfrac{1}{N_{Fr}}\right)$	Buoyancy force/inertia force	Free convection	27
D6a	Darcy No. (2)	Da_2	VL/D'; D' = permeability coefficient of porous medium $[L^2/\theta]$	Inertia force/permeation force	Flow in porous media	19
D8a	Generalized Deborah No.	N_2	$\sqrt{I_e - I_w} \cdot \theta_n$ I_e = invariant of rate of strain tensor (sec.$^{-2}$); I_w = invariant of vorticity tensor (sec.$^{-2}$); θ_n = natural time (sec.)	Generalization of group D8	Rheology	2
D14a	Dufour No.	Du_2	$\mathcal{R}\Theta n_{10}'/c_p$ Θ = thermodiffusion constant = $(D_T/D)/n_{10}n_{20}$ [—]; D_T = thermal diffusion coefficient $[L^2/\theta]$; n_{10}', $n_{20}' = n_1'/n'$, n_2'/n'; n' = total No. of molecules = $n_1' + n_2'$; n_1', n_2' = No. of molecules of components 1, 2, in binary mixture; also $Du_2 = (D_T/D)p/\rho c_p Tn_{20}$	Heat of isothermal mass transfer/enthalpy of unit mass of mixture	Thermodiffusion	22
E1a	Einstein No.	—	V/V_l (V_l = speed of light) (cf. Lorentz No.)	Fluid velocity/velocity of light	Magnetofluid dynamics	27
E4a	Electric field parameter	R_E	$E/V\mu_e H_e$		Magnetohydrodynamics	26
E4b	Electrical characteristic No.	El	$\rho(d\chi/dT)L^2 \cdot \Delta T \cdot E_1^2/\mu^2$ E_1 = electrical field strength $[ML/Q\theta^2]$; χ = dielectric susceptibility $[Q^2\theta^2/ML^3]$		Electrical effects on transfer processes	6
E4c	Electrical Nusselt No.	N_u	VL/D^*; $D^* = \frac{1}{2}(D^+ + D^-)$ $[L^2/\theta]$ D^+, D^- = diffusion coefficients of ions (cf. group P2)	Convection current/diffusion current	Electrochemistry	10
E4d	Electrical Reynolds No. (1)	—	$\varepsilon_e V/Q'Lb'$ ε_e = electrical permittivity $[Q^2\theta^2/L^3M]$; Q' = space charge density $[Q/L^3]$; b' = carrier mobility $[Q\theta/M]$		Electrical effects in flow	27
E4e	Electrical Reynolds No. (2)	—	Alternate name for group E4c, q.v.			10
E10a	Modified Euler No.	Eu'	$H_L' \rho_L g/V_G^2\rho_G$ H_L' = head of liquid on tray $[L]$; ρ_L, ρ_G = densities of liquid, vapor; V_G = vapor velocity based on free area, $[L/\theta]$	Friction head/velocity head	Flow of vapor across mass transfer trays	13
E13a	Expansion No.	Ex	$\left(\dfrac{gd}{V^2}\right)\left(\dfrac{\rho_L - \rho_G}{\rho_L}\right)$ d = bubble diam., V = bubble veloc., ρ_L, ρ_G = densities of liquid, gas	$1/N_{Fr} \times$ density ratio	Rise of bubbles	13
F12a	Fourier No. (flow)	Fo	$\mu\theta/\rho L^2$		Unsteady state flow problems	17
F12b	Frank-Kamenetskii No.	δ	$\dfrac{Q''}{k}\dfrac{E_a}{\mathcal{R}T_o^2}L^2k_o\exp(-E_a/\mathcal{R}T)$ Q'' = heat liberated per unit mass of material/unit volume $[1/L\theta^2]$; k = thermal conductivity of reacting mixture $[ML/T\theta^3]$; k_o = preexponential constant in Arrhenius equation $[M/\theta]$	Dimensionless heat effect of reaction	Heat transfer in reacting systems	24
F12c	Frequency parameter	—	$\omega'L/V = 2\pi \times$ Strouhal No.; $2\pi/\omega'$ = period of motion $[\theta]$	Special form of Strouhal No. (cf. also T5)	Unsteady state flow, etc.	27
F12d	Frequency No. (2)	N_f	$\omega_r L/V$; L = packing element diameter $[L]$; V = interstitial fluid velocity $[L/\theta]$; ω_r = radial frequency (radians/sec.) $[1/\theta]$ (cf. groups H10, S16, T5)	Special form of group F12c	Flow in packed or fluidized beds	23, 29

TABLE II. ALPHABETICAL LIST OF NEW GROUPS (Continued)

Serial No.	Name	Symbol	Definition	Significance	Field of Use	Reference
F12e	Frössling No. (heat transfer)	Fs_h	$(N_{Nu} - 2)/(N_{Re}^{1/2} N_{Pr}^{1/3})$	Special dimensionless heat transfer coefficient	Heat transfer to spheres in turbulent streams	8a
F12f	Frössling No. (mass transfer)	Fs_m	$(N_{Sh} - 2)/(N_{Re}^{1/2} N_{Sc}^{1/3})$	Special dimensionless mass transfer coefficient	Mass transfer to spheres in turbulent streams	8a
G2a	Geometric No.	Ge	h^*/H^*; $h^* = $ surface area of packing element/perimeter $[L]$; $H = $ height of packing $[L]$.	Dimensionless packed bed height	Mass transfer in packed beds	5a
G2b	Goertler parameter		$\dfrac{VL_b\rho}{\mu}\left(\dfrac{L_b}{R_c}\right)^{1/2}$ $L_b = $ boundary layer momentum thickness, $[L]$; $R_c = $ radius of curvature $[L]$	Modified Reynolds No.	Boundary layer flow on curved surfaces	27
G5a	Diffusional Grashof No.	Gr_{AB}	$L^3 \rho^2 g \beta'_A \Delta n'_A/\mu^2$ $n'_A = $ mass fraction of species A, $[-]$; $\beta'_A = $ coefficient of density change with n'_A, $[-]$	Buoyant forces $\times$ inertia forces/(viscous forces)^2	Interphase transfer by free convection (density changes caused by concentration differences)	37
H8a	Hess No.	Ge	$(KL^2/a_m)(C_o)^{n-1}$ $n = $ order of reaction $[-]$; $C_o = $ initial concn. $[M/L^3]$; $a_m = $ mass transfer conductivity of reaction products $[L^2/\theta]$; $K = $ reaction rate constant, $\left[\dfrac{1}{\theta}\dfrac{L^{3n-3}}{M^{n-1}}\right]$		Heat and mass transfer with chemical and phase changes	22
H9a	Homochronicity No.	Ho_3	$N\theta$ ($N = $ mixer r.p.m., $\theta = $ mixing time)		Mixing, agitation	31
H11a	Hydraulic resistance group	Γc	$\Delta p_p/\rho_L g L = We_4 \times$ Laplace No. $= N_{Eu_1} \times N_{Fr_1}$ $\Delta p_p = $ pressure drop across liquid on tray $[M/L\theta^2]$; $\rho_L = $ liquid density; $L = $ depth of liquid on tray	Characterizes development of interfacial area per unit area of tray	Pressure drop in distillation columns	15
I1	Ilyushin No.	I	$(Vd\rho/\mu) \cdot 4\tau_D/3V^2\rho = 4\tau_D/3V^2\rho \cdot N_{Re}$; $\tau_D = $ max. dynamic slip resistance $[M/L\theta^2]$		Flow of viscoplastic fluids	20a
K7a	Knudsen No. for diffusion	$N_{Kn_{A2}}$	$3eD_{AB}/4\zeta K_{oA}\bar{u}_A$ $K_{oA} = $ Knudsen flow permeability constant; $\bar{u}_A = $ equilibrium mean molecular speed of species A	Differs from K7 by numerical constant	Gaseous diffusion in packed beds	34
K7b	Kondrat'ev No.	Kn	ΨN_{Bih}; $N_{Bih} = $ heat transfer Biot No.; $\Psi = $ temp. field nonuniformity parameter $(t_s - t_a)/(\bar{t} - t_n)$; $t_a = $ temp. of surrounding medium; $t_s = $ body surface temp.; $\bar{t} = $ body mean temp.		Heat transfer between fluid and body	21
L3a	Laplace No.	La_4	$\Delta p_p \cdot L/\sigma = \Gamma\zeta/We_1$ $\Delta p_p = $ pressure drop across liquid on tray $[M/L\theta^2]$; $L = $ depth of liquid on tray		Interfacial behavior on distillation trays	32
L5a	Lebedev No.	Le_2	$eb_T(t_n - t_a)/c_b\rho\rho_s$ $b_T = $ intensity of vapor expansion in capillaries of body on heating $[M/L^3T]$; $t_a = $ temp. of surrounding medium; $t_o = $ initial temp.; $\rho_s = $ density of solid $[M/L^3]$	Molar expansion flux/molar vapor transfer flux	Drying of porous materials	22
L5b	Leroux No.	—	$\equiv$ cavitation No. C6			27
L7a	Turbulent Lewis No.	Le_T	$c_p\rho\varepsilon_D/k_T = l_D/l_T = \varepsilon_D/\varepsilon_T$ $k_T = $ eddy thermal conductivity $[LM/T\theta^3]$; $l_D, l_T = $ mixing lengths for mass, heat transfer $[L]$; $\varepsilon_T = $ eddy thermal diffusivity $[L^2/\theta]$		Combined turbulent heat and mass transfer	11
L7b	Lewis-Semenov No.	—	$\equiv 1/N_{Le}$; $N_{Le} = $ group L7			27
L7c	Lock No.	—	$\rho R^4 i a'/I$ $\rho = $ fluid density; $a' = $ rotor lift curve slope $[L^2/M]$; $i = $ blade chord $[L]$; $R = $ rotor radius $[L]$; $I = $ moment of inertia of blade about hinge $[L^4]$		Rotor blade dynamics	27
M3a	Magnetic Grashof No.	—	$4\pi\sigma_e\mu_e(\mu/\rho) \cdot N_{Gr}$ $N_{Gr} = $ group G5		Magnetofluid dynamics	7
M4a	Magnetic No.	R_m	$\mu_e H_e(\sigma_e L/\rho V)^{1/2} = $ (magnetic force parameter)$^{1/2}$	See group M3	Magnetohydrodynamics	26
M5a	Magnetic Prandtl No.	—	$\sigma_e\mu_e\dfrac{\mu}{\rho}$ cf. Magnetic Grashof No., group M3a	Magnetic Reynolds No./Reynolds No. (properties of fluid)	Magnetofluid dynamics	27
P3a	Plasticity No.	P	$\equiv$ Bingham No.			41
P7a	Modified Power No.	N'_P	$\dfrac{\Pi}{L^5\rho N^3}\left(\dfrac{D'_e}{L'}\right)(\Delta\omega)^{-1/2}\dfrac{1}{(N_bN_s)^{0.67}}$ $D'_e = $ effective agitator diameter $[L]$; $L'_e = $ effective agitator height; $\Delta\omega = $ wall proximity factor; $N_b = $ No. of blades on agitator; $N_s = $ effective No. of blade edges		Agitation	36
P8a	Total Prandtl No.	Pr	$\dfrac{\varepsilon_M + (\mu/\rho)}{\varepsilon_T + \alpha}$ $\varepsilon_M, \varepsilon_T = $ eddy transfer coefficient for momentum, heat $[L^2/\theta]$	Total momentum diffusivity/total thermal diffusivity	Heat transfer in combined turbulent and laminar flows	25
P8b	Turbulent Prandtl No.	Pr_T	$\varepsilon_M/\varepsilon_T = l/l_T$ $\varepsilon_M, \varepsilon_T = $ eddy viscosity, eddy thermal diffusivity $[L^2/\theta]$; $l, l_T = $ mixing lengths for momentum, heat transfer	Eddy momentum diffusivity/eddy thermal diffusivity	Heat transfer in turbulent flow	9, 25

TABLE II. ALPHABETICAL LIST OF NEW GROUPS (Continued)

Serial No.	Name	Symbol	Definition	Significance	Field of Use	Reference
P12a	Predvoditelev No. (mass transfer)	Pd_m	$(\Gamma_m L^2/a_m)\,(N_{Fo_m})$ Γ_m = rate of change of mass transfer potential of medium, (mass/unit mass)/time $[1/\theta]$; a_m = mass conductivity of material $[L^2/\theta]$; N_{Fo_m} = group F12	Rate of change of concn. of medium/ rate of change of concn. of body	Mass transfer	22
P15a	Pulsation No.	N_{Pu}	$fd_e\rho/G$ d_e = equiv. diam. of channel		Transfer to pulsed fluid	20
R0a	Radial frequency parameter (1)	—	$\omega_r D/V^2,\ \omega_r \alpha/V^2$ D = diffusivity or dispersion coefficient of packed bed $[L^2/\theta]$, ω_r = radial frequency (radians/sec.) $[1/\theta]$		Packed and fluidized beds	18, 33, 39
R0b	Radial frequency parameter (2)	—	$\omega_r L^2/\alpha$ ω_r as in group R0a; L = tube radius $[L]$		Packed and fluidized beds	29, 39
R0c	Radial frequency parameter (3)	—	$\omega_r^2 DL/V^3$ (Quantities as in groups R0a and R0b)	$\dfrac{(\text{Group R0a})^2}{(\text{Group P2})}$	Packed and fluidized beds	18, 33, 39
R0d	Radial frequency parameter (4)	—	$L(\omega_r/2D)^{1/2} \equiv [\frac{1}{2}(\text{group R0b})]^{1/2}$ (quantities as in groups R0a and R0b)	Analog of Wave No.	Packed and fluidized beds	39
R1a	Radiation parameter	Φ	$e^+\eta T_w^3 d_h/k$ e^+ = function of mean surface emissivity of walls, $[-]$; T_w = wall temp. (abs.) $[T]$		Effect of radiation on convective mass transfer	14
R6a	Reaction enthalpy No.	N_H	$(\Delta u)_A (\Delta n_A)/C_p(\Delta T)$ $(\Delta u)_A$ = enthalpy of reaction per unit mass of A produced $[L^2/\theta^2]$, n_A = mass fraction of species A $[-]$	Change in reaction energy/change in sensible energy	Interphase transfer with chemical reaction	37
R15a	Rossby No.	—	$V/\omega L$	More general form of group R15		27
S4a	Schmidt No. (2)	—	$\equiv$ Semenov No. (2)	No longer used		27
S4b	Schmidt No. (3)	$(Sc)_3$	$\mu\chi/\rho\sigma_e L^2$ χ as in group E4b	Diffusivity of vorticity/mass diffusivity of ions	Electrochemistry	10
S4c	Total Schmidt No.	Sc	$\dfrac{\varepsilon_M+(\mu/\rho)}{\varepsilon_D+D}$ ε_M = eddy viscosity $[L^2/\theta]$	Total momentum diffusivity/total mass diffusivity	Mass transfer in combined laminar and turbulent flows	25
S4d	Turbulent Schmidt No.	Sc_T	$\varepsilon_M/\varepsilon D = l/l_D$ ε_M = eddy viscosity, $[L^2/\theta]$, l, l_D = mixing lengths for momentum, mass transfer $[L]$	Eddy momentum diffusivity/eddy mass diffusivity	Mass transfer in turbulent flow	25
S5a	Semenov No. (2)	—	$\equiv 1/N_{Le}$ (see group L7)			27
S7a	Smoluchowski No.	—	$L/\lambda = 1/N_{Kn}$ N_{Kn} = group K6	See group K6		27
S9a	Soret No.	S_o	$\Theta n'_{20}$ (definitions as group D14a)	Dimensionless thermodiffusion coefficient	Coupled heat and mass transfer	22
S10a	Spalding No.	B'	$c_p\Delta T/(r_v - q_r/V_m)$; q_r = radiant heat flux $[ML^2/\theta^3]$; V_m = rate of mass transfer $[M/\theta]$	Ratio (sensible heat/ latent heat) for evaporated material	Droplet evaporation	33a
S13a	Stark No.	Sk	$\eta T^3 L/k$ L = thickness of layer $[L]$; ($\equiv$ Stefan No.)		Radiant heat transfer	1, 30
S14a	Stewart No.	—	$\mu_e^2 H_e^2 \sigma_e L/V\rho$	$(\text{Hartmann No.})^2/$ Reynolds No.	Magnetofluid dynamics	38
S15a	Stokes No. (2)	St_2	$1.042\,m_f g\rho(1-\rho/\rho_f)R^{*3}/\mu^2$; ρ, μ = density, viscosity of fluid; m_f, ρ_f = mass, density of float; R^* = tube radius/float radius $[-]$		Calibration of rotameters	10a
S18a	Surface tension No.	T_S	$\mu^2/h^*\sigma\rho$; h^* = surface area of packing element/perimeter $[L]$		Mass transfer in packed columns	5a
T5a	Thompson No.	N_{Th_2}	$\dfrac{-\Delta t}{\Delta L}L^2\left(\dfrac{d\sigma}{dT}\right)/\mu\alpha$ (cf. Marangoni No.)		Cellular convection	3
T7a	Thrust coefficient	T_c	$F_T/\rho V^2 d^2$ F_T = thrust force $[ML/\theta^2]$; V = forward speed $[L/\theta]$; d = tip diameter $[L]$		Propeller studies	27
T7b	Torque coefficient	Q_c	$F'/\rho V^2 d^3$ F' = propeller torque $[ML^2/\theta^2]$		Propeller studies	27
T7c	Transiency groups	K_p K_Q	$\dfrac{1}{\left[\dfrac{\partial p}{\partial L}\right]}\cdot\dfrac{\partial\left[\dfrac{\partial p}{\partial L}\right]}{\partial(Fo_f)}$ $\dfrac{1}{(N_{Re})}\cdot\dfrac{\partial(N_{Re})}{\partial(Fo_f)}$ $\partial p/\partial L$ = pressure gradient in flow direction $[M/L^2\theta^2]$; Fo_f = group F12a		Transient flow behavior	17
W0a	Wave No.	k	$L(\omega_r/2\alpha)^{1/2}$ w_r = radial frequency (radians/sec.) $[1/\theta]$	Heat transfer analog group R0d	Cyclic heat transfer	4
W3a	Weber No. (3)	We_3	$\sigma/\rho_L g L^2$ ρ_L = liquid density $[M/L^3]$; L = depth of liquid on tray $[L]$	Surface tension force/ gravity force	Interfacial area determination in distillation equipment	32
W4a	Generalized Weissenberg No.	N_1	$\sqrt{I_e}\theta_n$ (definitions as group D8a)	Generalization of group W4	Rheology	2

TABLE III. TABLES FOR IDENTIFYING DIMENSIONLESS GROUPS

These tables incorporate those from the earlier publication (5)

Basic dimensions are taken to be: Length [L]
Mass [M]
Electrical charge [Q]
Temperature [T]
Time [θ]

PHYSICAL PROPERTIES
General Physical Properties

Parameter	Symbol	Dimensions	Exponent	Groups
Coefficient of thermal bulk expansion	β	$1/T$	-1	E4, E12, G2
			$+1$	G5, K9, M3a, R5, R6
Coefficient of density change with concn. n_A	β_A	—	$+1$	G5a
Density	ρ	M/L^3	-2	M1
			-1	A1, B1, B10, C1, C2, C6, D3, D13, D14a, E3, E9, E10, E10a, E13, E13a, F1, F11, F12a, H5, H11a, I1, J1, K4, K10, L5a, L5b, L7, M3, M3a, M5a, M6, N2, N3, N4, P7, P8a, P9, R0a, R9, R10, R13, S4, S4b, S4c, S13, S14a, S15, S15a, S18a, T7a, T7b, T7c, W3a
			$-\frac{2}{3}$	C9, J3
			$-\frac{1}{2}$	E2, L11, M4a, O2, P13
			$+\frac{1}{3}$	J3, K5, S3
			$+\frac{1}{2}$	D10, G3, P10, P11, R4, W0a, W2
			$+\frac{2}{3}$	F2, N8, S2
			$+1$	A1b, B1, B6, B9, C5, C7, D5, D7, D13, E4, E4b, E6, E7, E8, E10a, E13a, G2b, H6, H11, I1, J1, J3, J4, K1, L7a, L7b, L7c, L9, M3a, P1, P8a, P15a, R0b, R5, R11, R13, S4a, S5a, S15a, S17, T1, T5a, T6, T7c, V1, W1, W3
			$+2$	A5, C10, G1, G5, G5a, K9, R5, R6, S15a, T2
Density gradient	$d\rho/dL$	M/L^4	$+1$	R13
Diffusivity (molecular unless noted otherwise)	D, a_m	L^2/θ	-1	B5, B7[4], D2, D14a, E4c, E4e, H8a[1], K4[1], K7[2], L2, L7, L10, N7, P2, P9, P12a[1], S4, S4c, S9a
			$-\frac{2}{3}$	J3
			$-\frac{1}{2}$	A0, R0d
			$+1$	A4a, F12[1], K7[3], K7a, L2[4], L7b, L9[1], R0a, R0c, S4a, S5a
Diffusivity (eddy)	E_D, ε_D	L^2/θ	-1	S4c, S4d
			$+1$	L2, L7a
Diffusivity (surface)	D_S	ML/θ^2	-1	S18
Diffusion tortuosity	ζ	—	-1	K7, K7a
Dispersion (permeability) coefficient	D'	L^2/θ	-1	D6a
			$+1$	R0a, R0c
Molecular weight	M	M	-1	D14, K9, S6
			$+1$	D14
Permeability (packed bed)	ξ	L^2	$+\frac{1}{2}$	L6
Porosity (voidage)	e	—	-1	B6, M12
			$-\frac{1}{2}$	L6
			$+1$	K7, K7a, L5a
Specific gravity	γ	—	$\pm\frac{1}{2}$	F2
Surface tension	σ	M/θ^2	-3	C2
			-2	C1
			-1	B9, C3, C13, E8, L3a, L6, R1, S18a, W1, W3
			$-\frac{1}{2}$	D10, G3, O2, P13, R4, W2
			$+1$	M9, S17, S18, T5a, W3a

PHYSICAL PROPERTIES
General Physical Properties

Parameter	Symbol	Dimensions	Exponent	Groups
Thermal diffusion coefficient	D_T	L^2/θ	$+1$	D14a, S9a

[1] Coefficient of potential diffusion in mass transfer.
[2] Knudsen diffusion coefficient.
[3] Binary bulk diffusion coefficient.
[4] Effective diffusivity (molecular and eddy) = $D + \varepsilon_D$.

Electrical and Magnetic Properties

Parameter	Symbol	Dimensions	Exponent	Groups
Carrier suspectibility	b'	$Q\theta/M$	-1	E4d
Current density	I	$Q/L^2\theta$	$+1$	K10
Dielectric susceptibility	χ	$Q^2\theta^2/L^3M$	$+1$	E4b, S4b
Electrical conductivity	σ_e	$Q^2\theta/L^3M$	-1	E6, S4b
			$+\frac{1}{2}$	H2, M4a
			$+1$	B2a, L11, M3, M3a, M5a, M7, S14a
Electrical permittivity	ε_e	$Q^2\theta^2/L^3\theta$	-1	B2a
			$+1$	E4d
Field strength	H_e	$Q/L\theta$	-2	J4
			-1	D4a
			$+1$	H7, L11, M4a
			$+2$	K9, M3, M6, S6, S14a
Magnetic permeability	μ_e	ML/Q^2	-1	E4a, E6, J4
			$+1$	H7, M3a, M4a, M5a, M6, M7
			$+\frac{1}{2}$	L11
			$+2$	M3, S14a
Space charge density	Q	Q/L^3	-1	E4d
Voltage	E	ML/θ^2Q	$+1$	E4a, K10
			$+2$	E4b

Thermal Properties

Parameter	Symbol	Dimensions	Exponent	Groups
Eddy thermal conductivity	k_T	$ML/T\theta^3$	-1	L7a
Eddy thermal diffusivity	ε_T	L^2/θ	-1	L7a, P8a, P8b
Humid heat	s	L^2/θ^2T	-1	P15
Latent heats of phase change	λ, r	L^2/θ^2	-1	A3, E11, E12, E13, J1, M1, S10a
			$+1$	B13, C10, K8, K11, N5
Ratio of specific heats	γ	—	$\pm\frac{1}{2}$	L5
Specific heat	C, c_p	L^2/θ^2T	-1	B2, B13, D3, D14a, D15, E1, F11, J2, K8, K11, L7, M10, N5, R0a, R3, R6a, S13
			$-\frac{1}{3}$	J2
			$+\frac{1}{2}$	F6, F7, F8, W0a
			$+\frac{2}{3}$	J2
			$+1$	A3, E4, E12, G4, J1, J4, L7a, L7b, L9, P1, P8, P8a, R0b, R3, R5, R6, R7, S5a, S10a, T5a
Surface emissivity	e^*	—	-1	T6
			$+1$	R1a
Temperature conductivity (thermal diffusivity)	$\alpha = \dfrac{k}{\rho c_p}$	L^2/θ	-1	L7b, L9, P1, P8a, P12, R0b, R5, S5a, T5a
			$-\frac{1}{2}$	W0a
			$+1$	C13, L7, L10, R0a
Thermal conductivity	k, λ	$ML/T\theta^3$	-3	M1
			-2	R6
			-1	B4, B12, C7, C9, C10, D4, F12b, G4, K3, K7b, L7b, L9, N6, P1, P5,

TABLE III. TABLES FOR IDENTIFYING DIMENSIONLESS GROUPS (Continued)

Thermal Properties

Parameter	Symbol	Dimensions	Exponent	Groups
				P8, R0b, R1a, R5, S4a, S5a, S13a, S14, T5a
			$-\frac{2}{3}$	J2
			$-\frac{1}{2}$	T3, W0a
			$+1$	F11, L7, R0a, R1

Rheological and Elastic Behavior

Parameter	Symbol	Dimensions	Exponent	Groups
Eddy viscosity	e_M	L^2/θ	$+1$	P8a, P8b, S4c, S4d
Invariants of rate-of-strain and vorticity tensors	I_e, I_w	θ^{-2}	$+\frac{1}{2}$	D8a, W4a
Maximum dynamic slip resistance	τ_D	$M/L\theta^2$	$+1$	I1
Modulus of elasticity	E	$M/L\theta^2$	-1	A1b, C5, E4, H11
			$+1$	C1, E13, R1
			$+3$	A1
Rate of shear	$\dot\gamma$	$1/\theta$	$+1$	T8
Rigidity coefficient	μ_p	$M/L\theta$	-2	H6
			-1	B3, P3a
Shear stress	τ	$M/L\theta^2$	-1	E5
			$-\frac{1}{2}$	P10
			$+\frac{1}{2}$	P11
			$+1$	A4b
Viscosity (in all cases, kinematic viscosity has been written as μ/ρ)	μ	$M/L\theta$	-2	A1, A5, E4b, G1, G5, G5a, K1, K9, S15a, S17, T2
			-1	A4b, B6, C10, D5, D7, E6, E7, G2b, H8, I1, L1, L3, M3a, O1, P4, P11, R5, R6, R11, S9, S19, T1, T5a, T7c, V1
			$-\frac{2}{3}$	F2, K5, N8
			$-\frac{1}{2}$	H2
			$-\frac{1}{3}$	S2, S3
			$+\frac{1}{2}$	E2
			$+\frac{2}{3}$	C9, J2, J3
			$+1$	B12, C3, C13, E3, E5, F12a, M1, M3a, M5a, O2, P8, P8a, P9, S4, S4b, S4c, S8, S15, T7c, T8
			$+2$	C1, S18a
			$+4$	C2
Viscosity (surface)	μ_s	M/θ	$+1$	S19
Yield stress	τ_y	$M/L\theta^2$	$+1$	B3, H6, P3a

LENGTHS, AREAS, AND VOLUMES

Characteristic Lengths

Parameter	Symbol	Dimensions	Exponent	Groups
General length dimension	Various	L	-5	P7, P7a
			-3	F9, L1, S15a, T7b
			-2	D9, D12, E3, F11, F12, F12a, H4, H5, N2, N3, N4, O1, R9, S4b, S8, S9, S15, T7a, W3a
			-1	A1a, B1, B10, E2, E4d, E5, E10, F1, F13, G2a, G4, H10, H11a, K6, L4, P7a, R6, R10, R15, R15a, R16, S6, S7a, S18a, S19, T5, T5a, T7, T7c, W4
			$-\frac{1}{2}$	B11, D7, F14, G2b, O2
			$+\frac{1}{2}$	A0, G2b, M4a, R4, T1, W2
			$+1$	A4a, A4b, B2a, B3, B4, B5, B7, B10, B13a, C7, D1, D3, D5, D6a, D11, D13, E4c, E4e, E7, E10, E13a, F1, F2, F5, F12c, F12d, F15, G2a, G2b, G3, H2, I1, K3, K4, K6, K7b, K10, L3, L3a, L4, L7c, L9, L11, M1, M3, M7, N6, N7, N8, P1, P2, P3a, P7a, P11, P15a, R0c, R0d, R1a, R8, R10,

Characteristic Lengths

Parameter	Symbol	Dimensions	Exponent	Groups
				R11, R16, S2, S7, S7a, S13a, S14, S14a, S16, S17, S18, T3, T7c, W0a, W1
			$+\frac{3}{2}$	D7, H7, T1
			$+2$	B9, D2, D4, E4b, E8, F12b, H6, H8a, K9, O1, P4, P5, P12, P12a, R0b, S8, S9, T5a, T7c, V1
			$+3$	A5, C10, G1, G5, G5a, K1, M3a, R5, S15a, W3
			$+4$	L7c, T2
			$+5$	R6
Dimension of agitator, propeller, etc.	Various	L	-5	P7, P7a
			-3	F9, T7b
			-2	H4, T7a
			-1	A1a, P7a
			$+1$	D11, F15, L7c, P7a
			$+3$	W3
			$+4$	L7c
Film thickness	L_f	L	-1	A4a
			$+1$	N8
Furnace half-width	L	L	-1	T7
Larmor radius	L_L	L	$+1$	L4
Mean free path	λ	L	-1	S7a
			$+1$	K6
Mixing length-mass	l_D	L	-1	S4d
			$+1$	L7a
Mixing length-heat	l_T	L	-1	L7a, P8b
Mixing length-momentum	l	L	$+1$	P8b, S4d
Particle dimension	d_e	L	-2	S15
			$+1$	F2, F12d
			$+3$	A5, G1
Pore or nozzle radius	R	L	$+1$	B7
Thickness of liquid layer	L	L	-1	C13, M9, T5a
			$+1$	H2, L11
			$+2$	M9, T5a
			$+3$	R5
			$+4$	T2

Areas

Parameter	Symbol	Dimensions	Exponent	Groups
Area	A	L^2	-1	D9, F6, F7, F8
			$+1$	B2
Area/unit volume	S, A^*	L^{-1}	-1	B6
			$+1$	M11, M12

Volumes

Parameter	Symbol	Dimensions	Exponent	Groups
Volume	W, W^*	L^3	$+1$	A5, H9, M11

TIMES AND FREQUENCIES

Parameter	Symbol	Dimensions	Exponent	Groups
Time	θ	θ	-1	D8, T7c
			$+1$	D8, D8a, D12, E3, F11, F12, F12a, H1, H9a, H10, P12a, S15, T5, W4a
Frequency	f', ω'	$1/\theta$	$+1$	F12c, H1, P15a, S16, V1
Radial frequency[1] ($= 2\pi f'$)	ω, (radians/time)	$1/\theta$	$+\frac{1}{2}$	R0d, W0a
			$+1$	F12d, R0a, R0b
			$+2$	R0c

[1] Also see angular velocity.

TEMPERATURES AND CONCENTRATIONS (DRIVING FORCES)

Concentrations and Related Quantities

Parameter	Symbol	Dimensions	Exponent	Groups
Dimensional concentration	C, c_A	M/L^3	$(n-1)$	H8a (nth order reaction)
			-1	D1, D2, R2
			$+1$	B10
Dimensionless concentration (e.g., wt./wt., mass fraction)	n	—	-1	D14a, P6, S9a
			$+1$	A4, D14a, G5a, K8, R6a, S9a

TABLE III. TABLES FOR IDENTIFYING DIMENSIONLESS GROUPS (Continued)

Concentrations and Related Quantities

Parameter	Symbol	Dimensions	Exponent	Groups
Mole fraction	Y, Z	—	± 1	D14
Rate of change of mass transfer potential (mass per unit mass/time)	Γ_m	$1/\theta$	$+1$	P12a
Specific mass content (mass/unit mass)	n^*	—	-1	K4
Surface concentration	Γ'	M/L^2	± 1	S18
{ Vapor capacity } { Mass capacity } (porous body)	c_b	$L\theta^2/M$	-1 $+1$	L5a B13, R2
Vapor expansion intensity on heating	h_T	$M/L^3 T$	$+1$	L5a

Temperatures, Temperature Differences

Parameter	Symbol	Dimensions	Exponent	Groups
Dimensional temperature, temperature difference	$T, \Delta T$	T	-3	R1*, T6*
			-2	F12b*
			-1	A4, A6*, B12, B13, B13a*, C4*, C7, C10, D3*, D4*, D14a*, D15, E1, E4b*, F12b*, G2, G6*, K3, K7b, K8, K9, K11, N5, P5*, P12*, R6a, R14*, S6, T5a
			$-\frac{1}{2}$	L5
			$+\frac{1}{2}$	F6*, F7*, F8*
			$+1$	B13a, C4, E4b, G5, G6, J1, J4, K7b, L5a, M1, M3a, P6, R5, R7, R14*, S10a, T5a
			$+3$	R1a*, S13a*, S14*
Dimensionless temperature difference	Ψ	—	$+1$	K7b
Rate of temperature change	Γ	T/θ	$+1$	P12

* Absolute temperature.

VELOCITIES, RATES, FLUXES, TRANSFER COEFFICIENTS

Velocities

Parameter	Symbol	Dimensions	Exponent	Groups
Angular velocity (rate of rotation) (see also frequency)	N, ω	$1/\theta$	-3	P7, P7a
			-2	H4, L1
			-1	A1a, F9, R15, R15a
			$-\frac{1}{2}$	E2
			$+1$	H9a, S8, S11, S12, T1
			$+2$	F15, T2, W3
Equilibrium molecular speed of A	u_A	L/θ	-1	K7a
General velocity (usually of fluid)	V	L/θ	-3	H5, R0c
			-2	B13a, C6, C11, D13, E9, E10, E13a, F1, I1, L5b, M6, N2, N3, N4, R0a, R7, R8, R9, R10, T7a, T7b
			-1	A2, A4b, B3, C12, C14, D1, D3, E4a, F10, F12c, F12d, H7, H8, I1, K10, L3, M3, P3a, P4, S13, S14a, S16, T7c, V2
			$-\frac{1}{2}$	M4a
			$+1$	A1a, A2, B1a, B2a, B6, B7, B11, C3, C12, C14, D5, D6a, D7, E1a, E4c, E4d, E4e, E5, E7, F10, F14, G2b, H7, H10, I1, K2, L5, L8, M2, M4, M7, P1, P2, P3, P10, R4, R11, R15, R15a, S1,

Velocities

Parameter	Symbol	Dimensions	Exponent	Groups
				S2, S3, T5, T6, T7c, V2, W2, W4
			$+2$	A1b, B1, B12, C5, C11, D15, E1, E11, F13, H11, P3, W1
Velocity of impeller or agitator periphery	U_s	L/θ	-2	P14
			-1	D9
Velocity of light	V_l	L/θ	-2	B2a
			-1	E1a, L8
Velocity of sound	V_s	L/θ	-1	B1a, M2, N1, S1
Velocity of waves	V_w	L/θ	-1	V2
			$+1$	P3
Velocity of Alfven waves	V_A	L/θ	-1	A2, K2, M4
			$+1$	A2, N1
			$+2$	C11
Velocity of bearing surface	v_s	L/θ	-1	H8, O1, S9
Velocity gradient	dV/dL	$1/\theta$	-2	R13

Flow Rates, Mass Fluxes, etc.

Parameter	Symbol	Dimensions	Exponent	Groups
Mass flow rate (mass flux)	V_m	M/θ	-1	B2, M11, T7
			$+1$	F6, F7, F8, G4, S10a, T7
Mass flow rate/unit area (mass flux density)	G	$M/L^2\theta$	-1	J2, J3, P15, P15a, S13
			$+1$	K4, M11
Mass flux/unit volume (reaction rate)	U	$M/L^2\theta$	$+\frac{1}{2}$	T3
			$+1$	D1, D2, D3, D4
Reaction rate constant (first order)	K	L/θ	-1	S5
Reaction rate constant (nth order)	K_n	$\dfrac{1}{\theta}\dfrac{L^{3n-3}}{M^{n-1}}$	$+1$	I10a
Volumetric flow rate	V_f	L^3/θ	-1	H9
			$-\frac{1}{2}$	D11
			$+\frac{1}{2}$	S11, S12
			$+1$	D9, F9
Volumetric flow rate/wetted perimeter	$V_{f'}$	L^2/θ	-1	A4a
			$-\frac{1}{2}$	A0

Heat Fluxes and Related Quantities

Parameter	Symbol	Dimensions	Exponent	Groups
Heat flux (heat flow/unit time)	q	ML^2/θ^3	-1	S10a
			$+1$	H5
Heat flux/unit area	q^*	M/θ^2	$+1$	K3, R6
Heat of reaction/unit mass of material reacting	Δu	L^2/θ^2	$+1$	R6a
Heat liberated/unit mass	Q	L^2/θ^2	$+1$	D3, D4
Heat liberated per unit mass of material reacting/unit volume	Q''	$1/L\theta^2$	$+1$	F12b
Heat liberated/unit volume (heat source power)	j	$M/L\theta^3$	$+1$	P5

Transfer Coefficients

Parameter	Symbol	Dimensions	Exponent	Groups
Heat transfer coefficient	h	$M/T\theta^3$	$+1$	B2, B4, C9, J2, K7b, N6, P15, S13
			$+4$	M1
Mass transfer coefficient	k_c	L/θ	$+1$	A0, B5, J3, L9, N7, S5, S7

FORCE, TORQUE, POWER, PRESSURE, HEAD

Force, Torque, Power

Parameter	Symbol	Dimensions	Exponent	Groups
Force	F, F_R	ML/θ^2	$-\frac{1}{2}$	S2
			$+\frac{1}{2}$	K5
			$+1$	N2, N3, N4, R9, T7a
Force/unit length	F_b	M/θ^2	$+1$	H8, O1, S9
Torque	F	ML^2/θ^2	$+1$	T7b
Power	Π	ML^2/θ^3	$+1$	L1, P7, P7a

TABLE III. TABLES FOR IDENTIFYING DIMENSIONLESS GROUPS (Continued)

Head, Pressure

Parameter	Symbol	Dimensions	Exponent	Groups
Fluid head (ft. fluid)	H'	L	$-\frac{3}{2}$	S12
			$+1$	E10a, H4
Head (energy/unit mass of fluid $= gH'$)	H	L^2/θ^2	-1	P3, T4
			$-\frac{3}{2}$	S11
			$+\frac{1}{2}$	D11
			$+1$	P14, T4
Pressure	P, p	$M/L\theta^2$	-1	F6, F7, F8, H9, L5a, S8, T8
			$+1$	B13, C6, D14a, L5b, L6, P13, R2
Pressure drop	$\Delta P, dP$	$M/L\theta^2$	$+1$	E9, E10, F1, H9, H11a, K1, L3, L3a, R10
Pressure gradient	$\Delta P/L$	$M/L^2\theta^2$	-1	T7c
			$+1$	E10, F1, H11a, K1, P4, R10, T7c

Other Quantities

Parameter	Symbol	Dimensions	Exponent	Groups
Arrhenius pre-exponential factor	k_0	M/θ	$+1$	F12b
Avogadro's No.	N	$1/M$	$+1$	K9, S6
Boltzmann's constant	k	$ML^2/\theta^2 T$	-1	K9, S6
Dufour coefficient	β^*	T	$+1$	A3, A4, F3
Energy of activation/unit mass	E_a	L^2/θ^2	$+1$	A6, F12b
Gas constant/unit mass	$\mathscr{R}$	$L^2/\theta^2 T$	-1	A6, F12b
			$-\frac{1}{2}$	L5
			$+1$	D14a
Knudsen flow permeability	K_{oA}	L	-1	K7a
Mass of float	m_f	M	$+1$	S15a
Moment of inertia	I	L^4	-1	L7c
Rotor lift curve slope	a'	L^2/M	$+1$	L7c
Shape factor	ξ^*	—	$+1$	V2
Soret coefficient	δ	$1/T$	$+1$	F3, P6
Stefan-Boltzmann coefficient (radiation coefficient)	η	$M/T^4\theta^3$	$+1$	R1, T6
		—	$+1$	R1a, S13a, S14
Thermodiffusion constant	Θ	—	$+1$	D14a, S9a

CONSTANTS AND MISCELLANEOUS QUANTITIES

Gravity Acceleration

Parameter	Symbol	Dimensions	Exponent	Groups
Gravity acceleration	g	L/θ^2	-2	A1
			-1	B1, F13, F15, H11a, M1, S6, W3a
			$-\frac{3}{2}$	S11
			$-\frac{1}{2}$	B11, F14, P13
			$-\frac{1}{3}$	C9, S3
			$+\frac{1}{3}$	F2, N8
			$+1$	A5, B9, B13a, C2, C10, D13, E8, E10a, E13a, G1, G5, G5a, H4, M3a, R5, R6, R8, R13, S15a

REFERENCES

1. Adrianov, V. N., "Radiation-Conductive and Radiation-Convective Heat Transfer," in "Teplo- i Massoperenos," A. V. Lȳkov and B. M. Smol'skii, Eds., Vol. II, pp. 92–102, Nauka i Tekhnika, Minsk, 1965.
2. Astarita, G., *Ind. Eng. Chem. Fundamentals* 6, 257 (1967).
3. Berg, J. C., Acrivos, A., Boudart, M., "Evaporative Convection," in "Advances in Chemical Engineering," T. B. Drew et al., Eds., Vol. 6, pp. 61–123, Academic Press, New York, 1966.
4. Carslaw, H. S., Jaeger, J. C., "Conduction of Heat in Solids," 2nd ed., Oxford University Press, 1959.
5. Catchpole, J. P., Fulford, G., *Ind. Eng. Chem.* 58 (3), 46 (1966).
5a. Costa Novella, E., Costa Lopez, J., *An. Real. Soc. Espan. Fis. Quim.*, Ser. B, 63 (2), 145 (1967).
6. Coulson, J. M., Porter, J. E., *Trans. Inst. Chem. Engrs.* 44, T388 (1966).
7. Ede, A. J., *Intern. J. Heat Mass Transfer* 9, 837 (1966).
8. Gall, C. E., Hudgins, R. R., *J. Chem. Educ.* 42, 611 (1965).
8a. Galloway, T. R., Sage, B. H., *Int. J. Heat Mass Transfer* 10, 1195 (1967).
9. Gardner, G. O., Kestin, J., *Ibid.*, 6, 289 (1963).
10. Gibbins, J. C., Hignett, E. T., *Electrochim. Acta* 11, 815 (1966).
10a. Gilmont, R., Maurer, P. W., *Instr. Control Systems* 34, 2070 (1961).
11. Ginzburg, I. P., "Heat and Mass Transfer from Bodies Interacting with Gas and Liquid Streams," in "Teplo- i Massoperenos," A. V. Lȳkov and B. M. Smol'skii, Eds., Vol. II, pp. 3–26, Nauka i Tekhnika, Minsk, 1965.
12. Glauert, M. B., Lighthill, M. J., *Proc. Roy. Soc.* 230A, 188 (1955).
13. Hobler, T., "Mass Transfer and Absorbers," Pergamon Press, New York, 1966.
14. Irvine, T. F., Stein, R. P., Simon, G. A., "Effect of Radiation on Convection in a Flat Channel," in "Teplo- i Massoperenos," A. V. Lȳkov and B. M. Smol'skii, Eds., Vol. II, pp. 78–91, Nauka i Tekhnika, Minsk, 1965.
15. Kasatkin, A. G., Dȳtnerskii, Yu. N., Kochergin, N. V., "Mass and Heat Transfer in Tray Columns," in "Teplo- i Massoperenos," A. V. Lȳkov and B. M. Smol'skii, Eds., Vol. IV, pp. 12–17, Nauka i Tekhnika, Minsk, 1966.
16. Kerber, R., *Chem.-Ing.-Technik* 38, 1133 (1966).
17. Kochenov, I. S., Kuznetsov, Yu. N., "Unsteady Flow in Pipes," in "Teplo-Massoperenos," A. V. Lȳkov and B. M. Smol'skii, Eds., Vol. II, pp. 306–314, Nauka i Tekhnika, Minsk, 1965.
18. Kramers, H., Alberda, G., *Chem. Eng. Sci.* 2, 173 (1953).
19. Laslo, A., *Inzh.-Fiz. Zh.* 10 (1), 60 (1966).
20. Lemlich, R., Armour, J. C., *Chem. Engr. Progr. Symp. Ser.* 61 (57), 83 (1965).
20a. Leshchii, N. P., Mochernyuk, D. Yu, *Izv. Vysshikh Uchebn. Zavedenii, Nefti Gaz* 10 (2), 77 (1967).
21. Luikov (Lȳkov), A. V., "Heat and Mass Transfer in Capillary-Porous Bodies," Pergamon Press, New York, 1966.
22. Luikov (Lȳkov), A. V. Mikhailov, Yu. A., "Theory of Energy and Mass Transfer," revised Engl. ed., Pergamon Press, New York, 1965.
23. McHenry, K. W., Wilhelm, R. H., *A.I.Ch.E. J.* 3, 83 (1957).
24. Merzhanov, A. G., "Problems of Heat Transfer in the Theory of Thermal Detonation," in "Teplo- i Massoperenos," A. V. Lȳkov and B. M. Smol'skii, Eds., Vol. IV, pp. 259–72, Nauka i Tekhnika, Minsk, 1966.
25. Opfell, J. B., Sage, B. H., "Turbulence in Thermal and Material Transport," in "Advances in Chemical Engineering," T. B. Drew et al., Ed., Vol. I, pp. 241–88, Academic Press, New York, 1956.
26. Pai, Shih-I., "Magnetohydrodynamics of Channel Flow," in "Advances in Hydroscience," Ven Te Chow, Ed., Vol. III, pp. 63–110, Academic Press, New York, 1966.
27. Pankhurst, R. C., "Dimensional Analysis and Scale Factors," Chapman and Hall, London, 1964.
28. Pechoč, V., Research Institute of Chemical Fibers, Svit, Czechoslovakia, private communication, 1966.
29. Philip, J. R., *Austral. J. Phys.* 16 (4), 454 (1963).
30. Postolnik, Yu. S., *Inzh.-Fiz. Zh.* 8 (1), 64 (1965).
31. Prochazka, J., Landau, J., *Collection Czech. Chem. Commun.* 26, 2961 (1961).
32. Rodionov, A. I., Kashnikov, A. M., Radikovskii, V. M., "Determination of Interfacial Contact Areas and Heat and Mass Transfer Coefficients on Sieve Trays," in "Teplo- i Massoperenos," A. V. Lȳkov, B. M. Smol'skii, Eds., Vol. IV, pp. 28–37, Nauka i Tekhnika, Minsk, 1966.
33. Rosen, J. B., Winsche, W. E., *J. Chem. Phys.* 18, 1587 (1950).
33a. Ross, L. L., Hoffman, T. W., Proc. 3rd Internat. Heat Transfer Conference, Chicago, August 1966, 5, 50 (A.I.Ch.E., New York, 1966).
34. Rothfield, L. B., *A.I.Ch.E. J.* 9, 19 (1963).
35. Sakiadis, B. S., *Ibid.*, 54, 467 (1961).
36. Skelland, A. H. P., "Non-Newtonian Flow and Heat Transfer," Wiley, New York, 1967.
37. Stewart, W. E., *Chem. Eng. Progr. Symp. Ser.* 61 (58), 16 (1965).
38. Tsinober, A. B., *Appl. Mech. Rev.* 18, 3751 (1965).
39. Turner, G. A., Dept. of Chemical Engineering, University of Waterloo, Waterloo, Ontario, private communication, 1967.
40. Vyazovov, V. V., *Zh. Tekhn. Fiz.* 10, 1519 (1940).
41. Wilkinson, W. L., "Non-Newtonian Fluids," Pergamon Press, London, 1960.

NATIONAL STANDARD REFERENCE DATA SYSTEM PUBLICATIONS

The National Standard Reference Data System was established in 1963 as a means of coordinating on a national scale the production and dissemination of critically evaluated reference data in the physical sciences. Under the Standard Reference Data Act (Public Law 90-396) the National Bureau of Standards of the U.S. Department of Commerce has the primary responsibility in the Federal Government for providing reliable scientific and technical reference data. The Office of Standard Reference Data of NBS coordinates a complex of data evaluation centers, located in university, industrial, and other Government laboratories as well as within the National Bureau of Standards, which are engaged in the compilation and critical evaluation of numerical data on physical and chemical properties retrieved from the world scientific literature.

The primary focus of the NSRDS is on well-defined physical and chemical properties of well-characterized materials or systems. An effort is made to assess the accuracy of data reported in the primary research literature and to prepare compilations of critically evaluated data which will serve as reliable and convenient reference sources for the scientific and technical community.

This Publication List includes NSRDS data compilations, critical reviews, and related publications which are available from various sources. Further information may be obtained from:

Office of Standard Reference Data
National Bureau of Standards
Washington, D. C. 20234
Telephone: (301) 921-2467

JOURNAL OF PHYSICAL AND CHEMICAL REFERENCE DATA (REPRINTS TO VOLUMES 1 TO 8 NO. 2)

Reprints from Volume 1 (1972)

1. *Gaseous Diffusion Coefficients,* T. R. Marrero and E. A. Mason, Vol. 1, No. 1, 1—118, 1972. $7.00.

2. *Selected Values of Critical Supersaturation for Nucleation of Liquids from the Vapor,* G. M. Pound, Vol. 1, No. 1, 119—134, 1972. $3.00.

3. *Selected Values of Evaporation and Condensation Coefficients of Simple Substances,* G. M. Pound, Vol. 1, No. 1, 135—146, 1972. $3.00.

4. *Atlas of the Observed Absorption Spectrum of Carbon Monoxide between 1060 and 1900 Å,* S. G. Tilford and J. D. Simmons, Vol. 1, No. 1, 147—188, 1972. $4.50.

*5. *Tables of Molecular Vibrational Frequencies, Part 5,* T. Shimanouchi, Vol. 1, No. 1, 189—216, 1972. $4.00.

6. *Selected Values of Heats of Combustion and Heats of Formation of Organic Compounds Containing the Elements C, H, N, O, P, and S,* Eugene S. Dolmalski, Vol. 1, No. 2, 221—278, 1972. $5.00.

7. *Thermal Conductivity of the Elements,* C. Y. Ho, R. W. Powell, and P. E. Liley, Vol. 1, No. 2, 279—422, 1972. $7.50.

8. *The Spectrum of Molecular Oxygen,* Paul H. Krupenie, Vol. 1, No. 2, 423—534, 1972. $6.50.

9. *A Critical Review of the Gas-Phase Reaction Kinetics of the Hydroxyl Radical,* W. E. Wilson, Jr., Vol. 1, No. 2, 535—574, 1972. $4.50.

10. *Molten Salts: Volume 3, Nitrates, Nitrites, and Mixtures, Electrical Conductance, Density, Viscosity, and Surface Tension Data,* G. J. Jenz, Ursula Krebs, H. F. Siegenthaler, and R. P. T. Tomkins, Vol. 1, No. 3, 581—746, 1972. $8.50.

11. *High Temperature Properties and Decomposition of Inorganic Salts—Part 3, Nitrates and Nitrites,* Kurt H. Stern, Vol. 1, No. 3, 747—772, 1972. $4.00.

12. *High-Pressure Calibration: A Critical Review,* D. L. Decker, W. A. Bassett, L. Merrill, H. T. Hall, and J. D. Barnett, Vol. 1, No. 3, 773—830, 1972. $5.00.

13. *The Surface Tension of Pure Liquid Compounds,* Joseph J. Jasper, Vol. 1, No. 4, 841—1009, 1972. $8.50.

14. *Microwave Spectra of Molecules of Astrophysical Interest, I. Formaldehyde, Formamide, and Thioformaldehyde,* Donald R. Johnson, Frank J. Lovas, and William H. Kirchhoff, Vol. 1, No. 4, 1011—1045, 1972. $4.50.

15. *Osmotic Coefficients and Mean Activity Coefficients of Uni-univalent Electrolytes in Water at 25°C,* Walter J. Hamer and Yung Chi Wu, Vol. 1, No. 4, 1047—1099, 1972. $5.00.

16. *The Viscosity and Thermal Conductivity Coefficients of Gaseous and Liquid Fluorine,* H. J. M. Hanley and R. Prydz, Vol. 1, No. 4, 1101—1113, 1972. $3.00.

Reprints from Volume 2 (1973)

17. *Microwave Spectra of Molecules of Astrophysical Interest. II. Methylenimine,* William H. Kirchhoff, Donald R. Johnson, and Frank J. Lovas, Vol. 2, No. 1, 1—10, 1973. $3.00.

18. *Analysis of Specific Heat Data in the Critical Region of Magnetic Solids,* F. J. Cook, Vol. 2, No. 1, 11—24, 1973. $3.00.

19. *Evaluated Chemical Kinetic Rate Constants for Various Gas Phase Reactions,* Keith Schofield, Vol. 2, No. 1, 25—84, 1973. $5.00.

20. *Atomic Transition Probabilities for Forbidden Lines of the Iron Group Elements. (A Critical Data Compilation for Selected Lines),* M. W. Smith and W. L. Wiese, Vol. 2, No. 1, 85—120, 1973. $4.50.

* Superseded by Reprint No. 103.

*21. *Tables of Molecular Vibrational Frequencies, Part 6,* T. Shimanouchi, Vol. 2, No. 1, 121—162, 1973. $4.50.

22. *Compilation of Energy Band Gaps in Elemental and Binary Compound Semiconductors and Insulators,* W. H. Strehlow and E. L. Cook, Vol. 2, No. 1, 163—200, 1973. $4.50.

23. *Microwave Spectra of Molecules of Astrophysical Interest. III. Methanol,* R. M. Lees, F. J. Lovas, W. H. Kirchhoff, and D. R. Johnson, Vol. 2, No. 2, 205—214, 1973. $3.00.

24. *Microwave Spectra of Molecules of Astrophysical Interest, IV. Hydrogen Sulfide,* Paul Helminger, Frank C. De Lucia, and William H. Kirchhoff, Vol. 2, No. 2, 215—224, 1973. $3.00.

*25. *Tables of Molecular Vibrational Frequencies, Part 7,* T. Shimanouchi, Vol. 2, No. 2, 225—256, 1973. $4.00.

26. *Energy Levels of Neutral Helium (^{4}He I),* W. C. Martin, Vol. 2, No. 2, 257—266, 1973. $3.00.

27. *Survey of Photochemical and Rate Data for Twenty-eight Reactions of Interest in Atmosphere Chemistry,* R. F. Hampson, Ed., W. Braun, R. L. Brown, D. Garvin, J. T. Herron, R. E. Huie, M. J. Kurylo, A. H. Laufer, J. D. McKinley, H. Okabe, M. D. Sheer, W. Tsang, and D. H. Stedman, Vol. 2, No. 2, 267—312, 1973. $4.50.

28. *Compilation of the Static Dielectric Constant of Inorganic Solids,* K. F. Young and H. P. R. Frederikse, Vol. 2, No. 2, 313—410, 1973. $6.50.

29. *Soft X-Ray Emission Spectra of Metallic Solids: Critical Review of Selected Systems,* A. J. McAlister, R. C. Dobbyn, J. R. Cuthill, and M. L. Williams, Vol. 2, No. 2, 411—426, 1973. $3.00.

30. *Ideal Gas Thermodynamic Properties of Ethane and Propane,* J. Chao, R. C. Wilhoit, and B. J. Zwolinski, Vol. 2, No. 2, 427—438, 1973. $3.00.

31. *An Analysis of Coexistence Curve Data for Several Binary Liquid Mixtures Near Their Critical Points,* A. Stein and G. F. Allen, Vol. 2, No. 3, 443—466, 1973. $4.00.

32. *Rate Constants for the Reactions of Atomic Oxygen (O^3P) with Organic Compounds in the Gas Phase,* John T. Herron and Robert E. Huie, Vol. 2, No. 3, 467—518, 1973. $5.00.

33. *First Spectra of Neon, Argon, and Xenon 136 in the 1.2—4.0 μm Region,* Curtis J. Humphreys, Vol. 2, No. 3, 519—530, 1973. $3.00.

34. *Elastic Properties of Metals and Alloys, I. Iron, Nickel, and Iron-Nickel Alloys,* H. M. Ledbetter and R. P. Reed, Vol. 2, No. 3, 531—618, 1973. $6.00.

35. *The Viscosity and Thermal Conductivity Coefficients of Dilute Argon, Krypton, and Xenon,* H. J. M. Hanley, Vol. 2, No. 3, 619—642, 1973. $4.00.

36. *Diffusion in Copper and Copper Alloys. Part I. Volume and Surface Self-Diffusion in Copper,* Daniel B. Butrymowicz, John R. Manning, and Michael E. Read, Vol. 2, No. 3, 643—656, 1973. $3.00.

37. *The 1973 Least-Squares Adjustment of the Fundamental Constants,* E. Richard Cohen and B. N. Taylor, Vol. 2, No. 4, 663—734, 1973. $5.50.

38. *The Viscosity and Thermal Conductivity Coefficients of Dilute Nitrogen and Oxygen,* H. J. M. Hanley and James F. Ely, Vol. 2, No. 4, 735—756, 1973. $4.00.

39. *Thermodynamic Properties of Nitrogen Including Liquid and Vapor Phases from 63K to 2000K with Pressures to 10,000 Bar,* Richard T. Jacobsen and Richard B. Stewart, Vol. 2, No. 4, 757—922, 1973. $8.50.

40. *Thermodynamic Properties of Helium 4 from 2 to 1500K at Pressures to 10^8Pa,* Robert D. McCarty, Vol. 2, No. 4, 923—1042, 1973. $7.00.

Reprints from Volume 3 (1974)

41. *Molten Salts: Volume 4, Part 1, Fluorides and Mixtures, Electrical Conductance, Density, Viscosity, and Surface Tension Data,* G. J. Janz, G. L. Gardner, Ursula Krebs, and R. P. T. Tomkins, Vol. 3, No. 1, 1-115, 1974. $7.00.

42. *Ideal Gas Thermodynamic Properties of Eight Chloro- and Fluoromethanes,* A. S. Rodgers, J. Chao, R. C. Wilhoit, and B. J. Zwolinski, Vol. 3, No. 1, 117—140, 1974. $4.00.

43. *Ideal Gas Thermodynamic Properties of Six Chloroethanes,* J. Chao, A. S. Rodgers, R. C. Wilhoit, and B. J. Zwolinski, Vol. 3, No. 1, 141—162, 1974. $4.00.

44. *Critical Analysis of Heat-Capacity Data and Evaluation of Thermodynamic Properties of Ruthenium, Rhodium, Palladium, Iridium, and Platinum from 0 to 300K. A Survey of the Literature Data on Osmium,* George T. Furukawa, Martin L. Reilly, and John S. Gallagher, Vol. 3, No. 1, 163—209, 1974. $4.50.

45. *Microwave Spectra of Molecules of Astrophysical Interest, V. Water Vapor,* Frank C. De Lucia, Paul Helminger, and William H. Kirchhoff, Vol. 3, No. 1, 211—219, 1974. $3.00.

46. *Microwave Spectra of Molecules of Astrophysical Interest, VI. Carbonyl Sulfide and Hydrogen Cyanide,* Arthur G. Maki, Vol. 3, No. 1, 221—244, 1974. $4.00.

47. *Microwave Spectra of Molecules of Astrophysical Interest, VII. Carbon Monoxide, Carbon Monosulfide, and Silicon Monoxide,* Frank J. Lovas and Paul H. Krupenie, Vol. 3, No. 1, 245—257, 1974. $3.00.

48. *Microwave Spectra of Molecules of Astrophysical Interest, VIII. Sulfur Monoxide,* Eberhard Tiemann, Vol. 3, No. 1, 259—268, 1974. $3.00.

*49. *Tables of Molecular Vibrational Frequencies, Part 8,* T. Shimanouchi, Vol. 3, No. 1, 269—308, 1974. $4.50.

50. *JANAF Thermochemical Tables, 1974 Supplement,* M. W. Chase, J. L. Curnutt, A. T. Hu, H. Prophet, A. N. Syverud, and L. C. Walker, Vol. 3, No. 2, 311—480, 1974. $8.50.

51. *High Temperature Properties and Decomposition of Inorganic Salts, Part 4. Oxy-Salts of the Halogens,* Kurt H. Stern, Vol. 3, No. 2, 481—526, 1974. $4.50.

52. *Diffusion in Copper and Copper Alloys, Part II. Copper-Silver and Copper-Gold Systems,* Daniel B. Butrymowicz, John R. Manning, and Michael E. Read, Vol. 3, No. 2, 527—602, 1974. $5.50.

53. *Microwave Spectral Tables, I. Diatomic Molecules,* Frank J. Lovas and Eberhard Tiemann, Vol. 3, No. 3, 609—770, 1974. $8.50.

54. *Ground Levels and Ionization Potentials for Lanthanide and Actinide Atoms and Ions,* W. C. Martin, Lucy Hagan, Joseph Reader, and Jack Sugar, Vol. 3, No. 3, 771—780, 1974. $3.00.

55. *Behavior of the Elements at High Pressures,* John Francis Cannon, Vol. 3, No. 3, 781—824, 1974. $4.50.

56. *Reference Wavelengths from Atomic Spectra in the Range 15 Å to 25000 Å,* Victor Kaufman and Bengt Edlen, Vol. 3, No. 4, 825—895, 1974. $5.50.

57. *Elastic Properties of Metals and Alloys, II. Copper,* H. M. Ledbetter and E. R. Naimon, Vol. 3, No. 4, 897—935, 1974. $4.50.

58. *A Critical Review of H-Atom Transfer in the Liquid Phase: Chlorine Atom, Alkyl Trichloromethyl, Alkoxy, and Alkylperoxy Radicals,* D. G. Hendry, T. Mill, L. Piszkiewicz, J. A. Howard, and H. K. Eigenmann, Vol. 3, No. 4, 937—978, 1974. $4.50.

59. *The Viscosity and Thermal Conductivity Coefficients for Dense Gaseous and Liquid Argon, Krypton, Xenon, Nitrogen, and Oxygen,* H. J. M. Hanley, R. D. McCarty, and W. M. Haynes, Vol. 3, No. 4, 979—1017, 1974. $4.50.

Reprints from Volume 4 (1975)

60. *JANAF Thermochemical Tables, 1975 Supplement,* M. W. Chase, J. L. Curnutt, H. Prophet, R. A. McDonald, and A. N. Syverud, Vol. 4, No. 1, 1—175, 1975. $8.50.

61. *Diffusion in Copper and Copper Alloys, Part III. Diffusion in Systems Involving Elements of the Groups IA, IIA, IIIB, IVB, VB, VIB, and VIIB,* Daniel B. Butrymowicz, John R. Manning, and Michael E. Read, Vol. 4, No. 1, 177—249, 1975. $6.00.

62. *Ideal Gas Thermodynamic Properties of Ethylene and Propylene,* Jing Chao and Bruno J. Zwolinski, Vol. 4, No. 1, 251—261, 1975. $3.00.

63. *Atomic Transition Probabilities for Scandium and Titanium (A Critical Data Compilation of Allowed Lines),* W. L. Wiese and J. R. Fuhr, Vol. 4, No. 2, 263—352, 1975. $6.00.

64. *Energy Levels of Iron, Fe I through Fe XXVI,* Joseph Reader and Jack Sugar, Vol. 4, No. 2, 353—440, 1975. $6.00.

65. *Ideal Gas Thermodynamic Properties of Six Fluoroethanes,* S. S. Chen, A. S. Rodgers, J. Chao, R. C. Wilhoit, and B. J. Zwolinski, Vol. 4, No. 2, 441—456, 1975. $3.00.

66. *Ideal Gas Thermodynamic Properties of the Eight Bromo- and Iodomethanes,* S. A. Kudchadker and A. P. Kudchadker, Vol. 4, No. 2, 457—470, 1975. $3.00.

67. *Atomic Form Factors, Incoherent Scattering Functions, and Photon Scattering Cross Sections,* J. H. Hubbell, Wm. J. Veigele, E. A. Briggs, R. T. Brown, D. T. Cromer, and R. J. Howerton, Vol. 4, No. 3, 471—538, 1975. $5.50.

68. *Binding Energies in Atomic Negative Ions,* H. Hotop and W. C. Lineberger, Vol. 4, No. 3, 539—576, 1975. $4.50.

69. *A Survey of Electron Swarm Data,* J. Dutton, Vol. 4, No. 3, 577—856, 1975. $12.00.

70. *Ideal Gas Thermodynamic Properties and Isomerization of n-Butane and Isobutane,* S. S. Chen, R. C. Wilhoit, and B. J. Zwolinski, Vol. 4, No. 4, 859—870, 1975. $3.00.

71. *Molten Salts: Volume 4, Part 2, Chlorides and Mixtures, Electrical Conductance, Density, Viscosity, and Surface Tension Data,* G. J. Janz, R. P. T. Tomkins, C. B. Allen, J. R. Downey, Jr., G. L. Gardner, U. Krebs, and S. K. Singer, Vol. 4, No. 4, 871—1178, 1975. $13.00.

72. *Property Index to Volumes 1—4 (1972—1975),* Vol. 4, No. 4, 1179—1192, 1975. $3.00.

Reprints from Volume 5 (1976)

73. *Scaled Equation of State Parameters for Gases in the Critical Region,* J. M. L. Levelt-Sengers, W. L. Greer, and J. V. Sengers, Vol. 5, No. 1, 1—51, 1976. $5.00.

74. *Microwave Spectra of Molecules of Astrophysical Interest, IX. Acetaldehyde,* A. Bauder, F. J. Lovas, and D. R. Johnson, Vol. 5, No. 1, 53—77, 1976. $4.00.

75. *Microwave Spectra of Molecules of Astrophysical Interest, X. Isocyanic Acid,* G. Winnewisser, W. H. Hocking, and M. C. L. Gerry, Vol. 5, No. 1, 79—101, 1976. $4.00.

76. *Diffusion in Copper and Copper Alloys, Part IV. Diffusion in Systems Involving Elements of Group VIII,* Daniel B. Butrymowicz, John R. Manning, and Michael E. Read, Vol. 5, No. 1, 103—200, 1976. $6.50.

77. *A Critical Review of the Stark Widths and Shifts of Spectral Lines from Non-Hydrogenic Atoms,* N. Konjevic and J. R. Roberts, Vol. 5, No. 2, 209—257, 1976. $5.00.

78. *Experimental Stark Widths and Shifts for Non-Hydrogenic Spectral Lines of Ionized Atoms,* N. Konjevic and W. L. Wiese, Vol. 5, No. 2, 259—308, 1976. $5.00.

79. *Atlas of the Absorption Spectrum of Nitric Oxide (NO) between 1420 and 1250 Å,* E. Miescner and F. Alberti, Vol. 5, No. 2, 309—317, 1976. $3.00.

80. *Ideal Gas Thermodynamic Properties of Propanone and 2-Butanone,* J. Chao and B. J. Zwolinski, Vol. 5, No. 2, 319—328, 1976. $3.00.

81. *Refractive Index of Alkali Halides and Its Wavelength and Temperature Derivatives,* H. H. Li, Vol. 5, No. 2, 329—528, 1976. $9.50.

82. *Tables of Critically Evaluated Oscillator Strengths for the Lithium Isoelectronic Sequence,* G. A. Martin and W. L. Wiese, Vol. 5, No. 3, 537—570, 1976. $4.50.

83. *Ideal Gas Thermodynamic Properties of Six Chlorofluoromethanes,* S. S. Chen, R. C. Wilhoit, and B. J. Zwolinski, Vol. 5, No. 3, 571—580, 1976. $3.00.

84. *Survey of Superconductive Materials and Critical Evaluation of Selected Properties,* B. W. Roberts, Vol. 5, No. 3, 531—821, 1976. $12.50.

85. *Nuclear Spins and Moments,* Gladys H. Fuller, Vol. 5, No. 4, 835—1092, 1976. $11.50.

86. *Nuclear Moments and Moment Ratios as Determined by Mössbauer Spectroscopy,* J. G. Stevens and B. D. Dunlop, Vol. 5, No. 4, 1093—1121, 1976. $4.00.

87. *Rate Coefficients for Ion-Molecule Reactions, I. Ions Containing C and H,* L. Wayne Sieck and Sharon G. Lias, Vol. 5, No. 4, 1123—1146, 1976. $4.00.

88. *Microwave Spectra of Molecules of Astrophysical Interest, XI. Silicon Sulfide,* Eberhard Tiemann, Vol. 5, No. 4, 1147—1156, 1976. $3.00.

89. *Property Index and Author Index to Volume 1—5 (1972—1976),* Vol. 5, No. 4, 1161—1183, 1976. $4.00.

Reprints from Volume 6 (1977)

90. *Diffusion in Copper and Copper Alloys, Part V. Diffusion in Systems Involving Elements of Group VA,* Daniel B. Butrymowicz, John R. Manning, and Michael E. Read, Vol. 6, No. 1, 1—50, 1977. $5.00.

91. *The Calculated Thermodynamic Properties of Superfluid Helium-4,* James S. Brooks and Russell J. Donnelly, Vol. 6, No. 1, 51—104, 1977. $5.00.

92. *Thermodynamic Properties of Normal and Deuterated Methanols,* S. S. Chen, R. C. Wilhoit, and B. J. Zwolinski, Vol. 6, No. 1, 105—112, 1977. $3.00.

93. *The Spectrum of Molecular Nitrogen,* Alf Lofthus and Paul H. Krupenie, Vol. 6, No. 1, 113—307, 1977. $9.50.

94. *Energy Levels of Chromium, Cr I through Cr XXIV,* Jack Sugar and Charles Corliss, Vol. 6, No. 2, 317—383, 1977. $5.50.

95. *The Activity and Osmotic Coefficients of Aqueous Calcium Chloride at 298.15K,* Bert R. Staples and Ralph L. Nuttall, Vol. 6, No. 2, 385—407, 1977. $4.00.

96. *Molten Salts: Volume 4, Part 3, Bromides and Mixtures; Iodides and Mixtures — Electrical Conductance, Density, Viscosity, and Surface Tension Data,* G. J. Janz, R. P. T. Tomkins, C. B. Allen, J. R. Downey, Jr., and S. K. Singer, Vol. 6, No. 2, 409—596, 1977. $9.00.

97. *The Viscosity and Thermal Conductivity for Dense Gaseous and Liquid Methane,* H. J. M. Hanley, W. M. Haynes, and R. D. McCarty, Vol. 6, No. 2, 597—609, 1977. $3.00.

98. *Phase Diagrams and Thermodynamic Properties of Ternary Copper-Silver Systems,* Y. Austin Chang, Daniel Goldberg, and Joachim P. Neumann, Vol. 6, No. 3, 621—673, 1977. $5.00.

99. *Crystal Data Space-Group Tables,* Alan D. Mighell, Helen M. Ondik, and Bettijoyce Breen Molino, Vol. 6, No. 3, 675—829, 1977. $8.00.

100. *Energy Levels of One-Electron Atoms,* Glen W. Erickson, Vol. 6, No. 3, 831—869, 1977. $4.50.

101. *Rate Constants for Reactions of ClO_x of Atmospheric Interest,* R. T. Watson, Vol. 6, No. 3, 871—917, 1977. $4.50.

102. *NMR Spectral Data: A Compilation of Aromatic Proton Chemical Shifts in Mono- and Di-Substituted Benzenes,* B. L. Shapiro and L. E. Mohrmann, Vol. 6, No. 3, 919—991, 1977.$5.50.

103. *Tables of Molecular Vibrational Frequencies. Consolidated Volume II,* T. Shimanouchi, Vol. 6, No. 3, 993—1102, 1977. $6.50.

104. *Effects of Isotopic Composition, Temperature, Pressure, and Dissolved Gases on the Density of Liquid Water,* George S. Kell, Vol. 6, No. 4, 1109—1132, 1977. $4.00.

105. *Viscosity of Water Substance — New International Formulation and Its Background,* A. Nagashima, Vol. 6, No. 4, 1133—1166, 1977. $4.50.

106. *A Correlation of the Existing Viscosity and Thermal Conductivity Data of Gaseous and Liquid Ethane,* H. J. M. Hanley, K. E. Gubbins, and S. Murad, Vol. 6, No. 4, 1167—1180, 1977. $3.00.

107. *Elastic Properties of Zinc: A Compilation and Review,* H. M. Ledbetter, Vol. 6, No. 4, 1181—1204, 1977. $4.00.

108. *Behavior of the AB-Type Compounds at High Pressures and High Temperatures,* Leo Merrill, Vol. 6, No. 4, 1205—1252, 1977. $4.50.

109. *Energy Levels of Manganese, Mn I through Mn XXV,* Charles Corliss and Jack Sugar, Vol. 6, No. 4, 1253—1329, 1977. $5.50.

Reprints from Volume 7 (1978)

110. *Tables of Atomic Spectral Lines for the 10,000 Å to 40,000 Å Region,* Michael Outred, Vol. 7, No. 1, 1—262, 1978. $11.50.

111. *Evaluated Activity and Osmotic Coefficients for Aqueous Solutions: The Alkaline Earth Metal Halides,* R. N. Goldberg and R. L. Nuttall, Vol. 7, No. 1, 263—310, 1978. $4.50.

112. *Microwave Spectra of Molecules of Astrophysical Interest, XII. Hydroxyl Radical,* Robert A. Beaudet and Robert L. Poynter, Vol. 7, No. 1, 311—362, 1978. $5.00.

113. *Ideal Gas Thermodynamic Properties of Methanoic and Ethanoic Acids,* Jing Chao and Bruno J. Zwolinski, Vol. 7, No. 1, 363, 1978. $3.00.

114. *Critical Review of Hydrolysis of Organic Compounds in Water Under Environmental Conditions,* W. Mabey and T. Mill, Vol. 7, No. 2, 383—415, 1978. $4.50.

115. *Ideal Gas Thermodynamic Properties of Phenol and Cresols,* S. A. Kudchadker, A. P. Kudchadker, R. C. Wilhoit, and B. J. Zwolinski, Vol. 7, No. 2, 417—423, 1978. $3.00.

116. *Densities of Liquid $CH_{4-a}X_a$ (X = Br,I) and $CH_{4-(a+b+c+d)}F_aCl_bBr_cI_d$ Halomethanes,* A. P. Kudchadker, S. A. Kudchadker, P. R. Patnaik, and P. P. Mishra, Vol. 7, No. 2, 425—439, 1978. $3.00.

117. *Microwave Spectra of Molecules of Astro-physical Interest, XIII. Cyanoacetylene,* W. J. Lafferty and F. J. Lovas, Vol. 7, No. 2, 441—493, 1978. $5.00.

118. *Atomic Transition Probabilities for Vanadium, Chromium, and Manganese (A Critical Data Compilation of Allowed Lines),* S. M. Younger, J. R. Fuhr, G. A. Martin, and W. L. Wiese, Vol. 7, No. 2, 495—629, 1978. $7.50.

119. *Thermodynamic Properties of Ammonia,* Lester Haar and John S. Gallagher, Vol. 7, No. 3, 635—792, 1978. $8.00.

120. *JANAF Thermochemical Tables, 1978 Supplement,* M. W. Chase, Jr., J. L. Curnutt, R. A. McDonald, and A. N. Syverud, Vol. 7, No. 3, 793—940, 1978. $8.00.

121. *Viscosity of Liquid Water in the Range −8 °C to 150 °C,* Joseph Kestin, Mordechai Sokolov, and William A. Wakeham, Vol. 7, No. 3, 941—948, 1978. $3.00.

122. *The Molar Volume (Density) of Solid Oxygen in Equilibrium with Vapor,* H. M. Roder, Vol. 7, No. 3, 949—957, 1978. $3.00.

123. *Thermal Conductivity of Ten Selected Binary Alloy Systems,* C. Y. Ho, M. W. Ackerman, K. Y. Wu, S. G. Oh, and T. N. Havill, Vol. 7, No. 3, 959—1177, 1978. $10.00.

124. *Semi-Empirical Extrapolation and Estimation of Rate Constants for Abstraction of H from Methane by H, O, HO, and O_2,* Robert Shaw, Vol. 7, No. 3, 1179—1190, 1978. $3.00.

125. *Energy Levels of Vanadium, V I through V XXIII,* Jack Sugar and Charles Corliss, Vol. 7, No. 3, 1191—1262, 1978. $5.50.

126. *Recommended Atomic Electron Binding Energies, 1s to $6p_{3/2}$, for the Heavy Elements, Z = 84 to 103,* F. T. Porter and M. S. Freedman, Vol. 7, No. 4, 1267—1284, 1978. $4.00.

127. *Ideal Gas Thermodynamic Properties of $CH_{4-(a+b+c+d)}F_aCl_bBr_cI_d$ Halomethanes,* Shanti A. Kudchadker and Arvind P. Kudchadker, Vol. 7, No. 4, 1285—1307, 1978. $4.00.

128. *Critical Review of Vibrational Data and Force Field Constants for Polyethylene,* John Barnes and Bruno Fanconi, Vol. 7, No. 4, 1309—1321, 1978. $3.00.

129. *Tables of Molecular Vibrational Frequencies,* Part 9, Takehiko Shimanouchi, Hiroatsu Matsuura, Yoshiki Ogawa, and Issei Harada, Vol. 7, No. 4, 1323—1443, 1978. $7.00.

130. *Microwave Spectral Tables, II. Triatomic Molecules,* Frank J. Lovas, Vol. 7, No. 4, 1445—1750, 1978. $13.00.

Reprints from Volume 8 (1979)

131. *Energy Levels of Titanium, Ti I through Ti XXII,* Charles Corliss and Jack Sugar, Vol. 8, No. 1, 1—62, 1979. $5.00.

132. *The Spectrum and Energy Levels of the Neutral Atom of Boron (B I),* G. A. Odintzova and A. R. Striganov, Vol. 8, No. 1, 63—67, 1979. $3.00.

133. *Relativistic Atomic Form Factors and Photon Coherent Scattering Cross Sections,* J. H. Hubbell and I. Øverbø, Vol. 8, No. 1, 69—105, 1979. $4.50.

134. *Microwave Spectra of Molecules of Astrophysical Interest, XIV. Vinyl Cyanide (Acrylonitrile),* M. C. L. Gerry, K. Tamada, and G. Winnewisser, Vol. 8, No. 1, 107—123, 1979. $4.00.

135. *Molten Salts: Volume 4, Part 4, Mixed Halide Melts. Electrical Conductance, Density, Viscosity, and Surface Tension Data,* G. J. Janz, R. P. T. Tomkins, and C. B. Allen, Vol. 8, No. 1, 125—302, 1979. $9.00.

136. *Atomic Radiative and Radiationless Yields for K and L Shells,* M. O. Krause, Vol. 8, No. 2, 307—327, 1979. $4.00.

137. *Natural Widths of Atomic K and L Levels, $K\alpha$ X-Ray Lines and Several KLL Auger Lines,* M. O. Krause and J. H. Oliver, Vol. 8, No. 2, 329—338, 1979. $3.00.

138. *Electrical Resistivity of Alkali Elements,* T. C. Chi, Vol. 8, No. 2, 339—438, 1979. $6.50.

* Do not include supplements.

139. *Electrical Resistivity of Alkaline Earth Elements,* T. C. Chi, Vol. 8, No. 2, 439—497, 1979. $5.00.

140. *Vapor Pressures and Boiling Points of Selected Halomethanes,* A. P. Kudchadker, S. A. Kudchadker, P. R. Shukla, and P. R. Patnaik, Vol. 8, No. 2, 499—517, 1979. $4.00.

141. *Ideal Gas Thermodynamic Properties of Selected Bromoethanes and Iodoethanes,* S. A. Kudchadker and A. P. Kudchadker, Vol. 8, No. 2, 519—526, 1979. $3.00.

142. *Thermodynamic Properties of Normal and Deuterated Naphthalenes,* S. S. Chen, S. A. Kudchadker, and R. C. Wilhoit, Vol. 8, No. 2, 527—535, 1979. $3.00.

143. *Microwave Spectra of Molecules of Astrophysical Interest, XV. Propyne,* A. Bauer, D. Boucher, J. Burie, J. Demaison, and A. Dubrulle, Vol. 8, No. 2, 537—558, 1979. $4.00.

144. *A Correlation of the Viscosity and Thermal Conductivity Data of Gaseous and Liquid Propane,* P. M. Holland, H. J. M. Hanley, K. E. Gubbins, and J. M. Haile, Vol. 8, No. 2, 559—575, 1979. $4.00.

Supplements

Physical and Thermodynamic Properties of Aliphatic Alcohols, R. C. Wilhoit and B. J. Zwolinski, Vol. 2, Supplement 1, 420 pages, 1973. Hardcover — $33.00; Softcover — $30.00.

Thermal Conductivity of the Elements. A Comprehensive Review, C. Y. Ho, R. W. Powell, and P. E. Liley, Vol. 3, Supplement 1, 796 pages, 1974. Hardcover — $60.00; Softcover — $55.00.

Energetics of Gaseous Ions, H. M. Rosenstock, K. Draxl, B. W. Steiner, and J. T. Herron, Vol. 6, Supplement 1, 783 pages, 1977. Hardcover — $70.00; Softcover — $65.00.

Evaluated Kinetic Data for High Temperature Reactions, Volume 4. Homogeneous Gas Phase Reactions of Halogen- and Cyanide-Containing Species, D. L. Baulch, J. Duxbury, S. J. Grant, and D. C. Montague, Vol. 9, Supplement 1, 700 pages est., 1980, in press.

Journal of Physical and Chemical Reference Data Subscription Prices* for Volume 8 (1979)

	Member (ACS or AIP member or affiliated society)	Regular rate
USA, Canada, Mexico	$25.00	$100.00
Foreign (surface mail)	29.00	104.00
Air freight Europe	32.00	106.00
Air freight Asia	39.00	114.00

Subscriptions to be sent to:
American Chemical Society
Office of the Comptroller
1155 Sixteenth Street, N. W.
Washington, D. C. 20036

Inquiries on microfilm should be sent to:
American Institute of Physics
Department S/F
335 East 45th Street
New York, New York 10017

Special Reprint Packages

H. *Thermodynamic Properties of Industrially Important Compounds*, (14 parts) consisting of reprint nos: 6, 30, 42, 43, 62, 65, 66, 70, 80, 83, 92, 113, 115, 127. If purchased individually — $52.00; package price — $35.00.

I. *Molecular Vibrational Frequencies*, (3 parts) consisting of consolidated Volume I, NSRDS-NBS 39; reprint nos: 103, 129. If purchased individually — $18.60; package price — $12.00.

J. *Chemical Data Relevant to Air Pollution*, (7 parts) consisting of reprint nos: 1, 9, 19, 27, 32, 87, 101. If purchased individually — $34.50; package price — $23.00.

K. *Metal Properties Data*, (12 parts) consisting of reprint nos: 29, 34, 36, 44, 52, 57, 61, 76, 90, 98, 107, 123. If purchased individually — $63.00; package price — $42.00.

L. *JANAF Thermochemical Tables*, (4 parts) consisting of Second Edition of JANAF Thermochemical Tables 1971, NSRDS-NBS 37; reprint nos: 50, 60, 120. If purchased individually — $40.60; package price — $32.00.

M. *Thermophysical Properties of Fluids*, (9 parts) consisting of reprint nos: 16, 35, 38, 39, 40, 59, 97, 106, 119. If purchased individually — $45.00; package price — $28.00.

N. *Atomic Energy Levels and Spectra*, (14 parts) consisting of reprint nos: 26, 33, 54, 56, 64, 68, 77, 78, 94, 100, 109, 110, 125, 126. If purchased individually — $72.00; package price — $50.00.

O. *Molecules in Space*, (13 parts) consisting of reprint nos: 14, 17, 23, 24, 45, 46, 47, 48, 74, 75, 88, 112, 117. If purchased individually — $47.50; package price — $34.00.

P. *Molecular Spectra*, (6 parts) consisting of reprint nos: 4, 8, 53, 79, 93, 130. If purchased individually — $45.00; package price — $33.00.

Q. *Atomic Transition Probabilities*, (4 parts) consisting of reprint nos: 20, 63, 82, 118. If purchased individually — $22.50; package price — $15.00.

NSRDS-NBS SERIES

NSRDS-NBS 1, *National Standard Reference Data System Plan of Operation*, E. L. Brady and M. B. Wallenstein, 1964.

NSRDS-NBS 2, *Thermal Properties of Aqueous Uniunivalent Electrolytes*, V. B. Parker, 1965.

NSRDS-NBS 3, Sec. 1, *Selected Tables of Atomic Spectra, Atomic Energy Levels and Multiplet Tables, Si II, Si III, Si IV*, C. E. Moore, 1965.

NSRDS-NBS 3, Sec. 2, *Selected Tables of Atomic Spectra, Atomic Energy Levels and Multiplet Tables, Si I*, C. E. Moore, 1967.

NSRDS-NBS 3, Sec. 3, *Selected Tables of Atomic Spectra, Atomic Energy Levels and Multiplet Tables, CI, CII, CIII, CIV, CV, CVI*, C. E. Moore, 1970.

NSRDS-NBS 3, Sec. 4, *Selected Tables of Atomic Spectra, Atomic Energy Levels and Multiplet Tables, N IV, N V, N VI, N VII*, C. E. Moore, 1971.

NSRDS-NBS 3, Sec. 5, *Selected Tables of Atomic Spectra, Atomic Energy Levels and Multiplet Tables, N I, N II, N III*, C. E. Moore, 1975.

NSRDS-NBS 3, Sec. 6, *Selected Tables of Atomic Spectra, Atomic Energy Levels and Multiplet Tables, H I, D, T*, C. E. Moore, 1972.

NSRDS-NBS 3, Sec. 7, *Selected Tables of Atomic Spectra, Atomic Energy Levels and Multiplet Tables, O I*, C. E. Moore, 1976.

NSRDS-NBS 3, Sec. 8, *Selected Tables of Atomic Spectra, Atomic Energy Levels and Multiplet Tables, O VI, O VII, O VIII*, C. E. Moore, 1979.

NSRDS-NBS 3, Sec. 9, *Selected Tables of Atomic Spectra, Atomic Energy Levels and Multiplet Tables, O IV, O V*, C. E. Moore, 1979, in press.

NSRDS-NBS 3, Sec. 10, *Selected Tables of Atomic Spectra, Atomic Energy Levels and Multiplet Tables, O III*, C. E. Moore, 1980, in press.

NSRDS-NBS 3, Sec. 11, *Selected Tables of Atomic Spectra, Atomic Energy Levels and Multiplet Tables, O II*, C. E. Moore, 1980, in press.

NSRDS-NBS 4, *Atomic Transition Probabilities: Vol. I. Hydrogen Through Neon*, W. L. Wiese, M. W. Smith, and B. M. Glennon, 1966.

NSRDS-NBS 5, *The Band Spectrum of Carbon Monoxide*, P. H. Krupenie, 1966.

NSRDS-NBS 6, *Tables of Molecular Vibrational Frequencies*, Part 1, T. Shimanouchi, 1967; superseded by NSRDS-NBS 39.

NSRDS-NBS 7, *High Temperature Properties and Decomposition of Inorganic Salts*, Part 1, Sulfates, K. H. Stern and E. L. Weise, 1966.

NSRDS-NBS 8, *Thermal Conductivity of Selected Materials*, R. W. Powell, C. Y. Ho, and P. E. Liley, 1966.

NSRDS-NBS 9, *Tables of Bimolecular Gas Reactions*, A. F. Trotman-Dickenson and G. S. Milne, 1967.

NSRDS-NBS 10, *Selected Values of Electric Dipole Moments for Molecules in the Gas Phase*, R. D. Nelson, Jr., D. R. Lide, Jr., and A. A. Maryott, 1967.

NSRDS-NBS 11, *Tables of Molecular Vibrational Frequencies*, Part 2, T. Shimanouchi, 1967; superseded by NSRDS-NBS 39.

NSRDS-NBS 12, *Tables for the Rigid Asymmetric Rotor: Transformation Coefficients from Symmetric to Asymmetric Bases and Expectation Values of P^2_z, P^4_z, P^6_z*, R. H. Schwendeman, 1968.

NSRDS-NBS 13, *Hydrogenation of Ethylene on Metallic Catalysts*, J. Horiuti and K. Miyahara, 1968.

NSRDS-NBS 14, *X-Ray Wavelengths and X-Ray Atomic Energy Levels*, J. A. Bearden, 1967.

NSRDS-NBS 15, *Molten Salts: Vol. 1. Electrical Conductance, Density, and Viscosity Data*, G. J. Janz, F. W. Dampier, G. R. Lakshminarayanan, P. K. Lorenz, and R. P. T. Tomkins, 1968.

NSRDS-NBS 16, *Thermal Conductivity of Selected Materials*, Part 2, C. Y. Ho, R. W. Powell, and P. E. Liley, 1968.

NSRDS-NBS 17, *Tables of Molecular Vibrational Frequencies*, Part 3, T. Shimanouchi, 1968; superseded by NSRDS-NBS 39.

NSRDS-NBS 18, *Critical Analysis of the Heat-Capacity Data of the Literature and Evaluation of Thermodynamic Properties of Copper, Silver, and Gold from 0 to 300 K*, G. T. Furukawa, W. G. Saba, and M. L. Reilly, 1968.

NSRDS-NBS 19, *Thermodynamic Properties of Ammonia as an Ideal Gas*, L. Haar, 1968.

NSRDS-NBS 20, *Gas Phase Reaction Kinetics of Neutral Oxygen Species*, H. S. Johnston, 1968.

NSRDS-NBS 21, *Kinetic Data on Gas Phase Unimolecular Reactions*, S. W. Benson and H. E. O'Neil, 1970.

NSRDS-NBS 22, *Atomic Transition Probabilities, Vol. II. Sodium Through Calcium, A Critical Data Compilation*, W. L. Wiese, M. W. Smith, and B. M. Miles, 1969.

NSRDS-NBS 23, *Partial Grotrian Diagrams of Astrophysical Interest*, C. E. Moore and P. W. Merrill, 1968.

NSRDS-NBS 24, *Theoretical Mean Activity Coefficients of Strong Electrolytes in Aqueous Solutions from 0 to 100 °C*, Walter J. Hamer, 1968.

NSRDS-NBS 25, *Electron Impact Excitation of Atoms*, B. L. Moiseiwitsch and S. J. Smith, 1968.

NSRDS-NBS 26, *Ionization Potentials, Appearance Potentials and Heats of Formation of Gaseous Positive Ions*, J. L. Franklin et al., 1969; superseded by *J. Phys. Chem. Reference Data*, 6, Supplement 1, 1977.

NSRDS-NBS 27, *Thermodynamic Properties of Argon from the Triple Point to 300 K at Pressures to 1000 Atmospheres*, A. L. Gosman, R. D. McCarty, and J. G. Hust, 1969.

NSRDS-NBS 28, *Molten Salts: Vol. 2, Section 1. Electrochemistry of Molten Salts: Gibbs Free Energies and Excess Free Energies from Equilibrium-Type Cells*, G. J. Janz and C. G. M. Dijkhuis; Section 2, Surface Tension Data, G. J. Janz, G. R. Lakshminarayanan, R. P. T. Tomkins and J. Wong, 1969.

NSRDS-NBS 29, *Photon Cross Sections, Attenuation Coefficients, and Energy Absorption Coefficients from 10 keV to 100 GeV*, J. H. Hubbell, 1969.

NSRDS-NBS 30, *High Temperature Properties and Decomposition of Inorganic Salts*, Part 2. Carbonates, K. H. Stern and E. L. Weise, 1969.

NSRDS-NBS 31, *Bond Dissociation Energies in Simple Molecules*, B. deB. Darwent, 1970.

NSRDS-NBS 32, *Phase Behavior in Binary and Multicomponent Systems at Elevated Pressures: n-Pentane and Methane-n-Pentane*, V. M. Berry and B. H. Sage, 1970.

NSRDS-NBS 33, *Electrolytic Conductance and Conductances of the Halogen Acids in Water*, W. J. Hamer and H. J. DeWane, 1970.

NSRDS-NBS 34, *Ionization Potentials and Ionization Limits Derived from the Analyses of Optical Spectra*, C. F. Moore, 1970.

NSRDS-NBS 35, *Atomic Energy Levels as Derived from the Analyses of Optical Spectra, Vol. I, ^1H to ^{23}V; Vol. II, ^{24}Cr to ^{41}Nb, Vol. III, ^{42}Mo to ^{57}La, ^{72}Hf to ^{89}Ac*, C. E. Moore, 1971.

NSRDS-NBS 36, *Critical Micelle Concentrations of Aqueous Surfactant Systems*, P. Mukerjee and K. J. Mysels, 1971.

NSRDS-NBS 37, *JANAF Thermochemical Tables*, Second Edition, D. R. Stull, H. Prophet, 1971.

NSRDS-NBS 38, *Critical Review of Ultraviolet Photoabsorption Cross Sections for Molecules of Astrophysical and Aeronomic Interest*, R. D. Hudson, 1971.

NSRDS-NBS 39, *Tables of Molecular Vibrational Frequencies, Consolidated Volume I*, T. Shimanouchi, 1972.

NSRDS-NBS 40, *A Multiplet Table of Astrophysical Interest, Revised Edition, Part I. Table of Multiplets; Part II. Finding List of All Lines in the Table of Multiplets*, C. E. Moore, 1972.

NSRDS-NBS 41, *Crystal Structure Transformations in Binary Halides*, C. N. R. Rao and M. Natarajan, 1972.

NSRDS-NBS 42, *Selected Specific Rates of Reactions of the Solvated Electron in Alcohols*, E. Watson, Jr. and S. Roy, 1972.

NSRDS-NBS 43, *Selected Specific Rates of Reactions of Transients from Water in Aqueous Solution, I. Hydrated Electron*, M. Anbar, M. Bambenek, and A. B. Ross, 1972.

NSRDS-NBS 43, Supplement, *Selected Specific Rates of Reactions of Transients from Water in Aqueous Solution. Hydrated Electron, Supplemental Data,* A. B. Ross, 1975.

NSRDS-NBS 44, *The Radiation Chemistry of Gaseous Ammonia,* D. B. Peterson, 1974.

NSRDS-NBS 45, *Radiation Chemistry of Nitrous Oxide Gas. Primary Processes, Elementary Reactions, and Yields,* G. R. A. Johnson, 1973.

NSRDS-NBS 46, *Reactivity of the Hydroxyl Radical in Aqueous Solutions,* Leon M. Dorfman and Gerald E. Adams, 1973.

NSRDS-NBS 47, *Tables of Collision Integrals and Second Virial Coefficients for the (m,6,8) Intermolecular Potential Function,* Max Klein, H. J. M. Hanley, Francis J. Smith, and Paul Holland, 1974.

NSRDS-NBS 48, *Radiation Chemistry of Ethanol: A Review of Data on Yields, Reaction Rate Parameters, and Spectral Properties of Transients,* Gordon A. Freeman, 1974.

NSRDS-NBS 49, *Transition Metal Oxides, Crystal Chemistry, Phase Transition, and Related Aspects,* C. N. R. Rao and G. V. Subba Rao, 1974.

NSRDS-NBS 50, *Resonances in Electron Impact on Atoms and Diatomic Molecules,* George J. Schulz, 1973.

NSRDS-NBS 51, *Selected Specific Rates of Reactions of Transients from Water in Aqueous Solution, II. Hydrogen Atom,* Michael Anbar, Farhataziz, and A. B. Ross, 1975.

NSRDS-NBS 52, *Electronic Absorption and Internal and External Vibrational Data of Atomic and Molecular Ions Doped in Alkali Halide Crystals,* J. C. Jain, A. V. R. Warrier, and S. K. Agarwal, 1974.

NSRDS-NBS 53, *Crystal Structure Transformations in Inorganic Nitrites, Nitrates and Carbonates,* C. N. R. Rao, B. Prakash, and M. Natarajan, 1975.

NSRDS-NBS 54, *Radiolysis of Methanol: Product Yields, Rate Constants and Spectroscopic Parameters of Intermediates,* J. H. Baxendale and Peter Wardman, 1975.

NSRDS-NBS 55, *Property Index to NSRDS Data Compilations, 1964—1972,* David R. Lide, Jr., Gertrude B. Sherwood, Charles H. Douglass, Jr., and Herman M. Weisman, 1975.

NSRDS-NBS 56, *Crystal Structure Transformations in Inorganic Sulfates, Phosphates, Perchlorates, and Chromates,* B. Prakash and C. N. R. Rao, 1975.

NSRDS-NBS 57, *Yields of Free Ions Formed in Liquids by Radiation,* A. O. Allen, 1976.

NSRDS-NBS 58, *Drift Mobilities and Conduction Band Energies of Excess Electrons in Dielectric Liquids,* A. O. Allen, 1976.

NSRDS-NBS 59, *Selected Specific Rates of Reactions of Transients from Water in Aqueous Solution, III. Hydroxyl Radical and Perhydroxyl Radical and Their Radical Ions,* Farhataziz and Alberta B. Ross, 1977.

NSRDS-NBS 60, *Atomic Energy Levels — The Rare Earth Elements, The Spectra of Lanthanum Cerium, Praseodymium, Neodymium, Promethium, Samarium, Europium, Gadolinium, Terbium, Dysprosium, Holmium, Erbium, Thulium, Ytterbium, and Lutetium,* W. C. Martin, Romuald Zalubas, and Lucy Hagan, 1978.

NSRDS-NBS 61, Part I, *Physical Properties Data Compilations Relevant to Energy Storage I. Molten Salts: Eutectic Data,* George J. Janz, Carolyn B. Allen, Joseph R. Downey, Jr., and R. P. T. Tomkins, 1978.

NSRDS-NBS 61, Part II, *Physical Properties Data Compilations Relevant to Energy Storage II. Molten Salts: Data on Single and Multi-Component Systems,* G. J. Janz, C. B. Allen, N. P. Bansal, R. M. Murphy, and R. P. T. Tomkins, 1979.

NSRDS-NBS 61, Part III, *Physical Properties Data Compilations Relevant to Energy Storage, III. Batteries: Data on Sodium Beta and Beta´ Aluminas,* G. Miller, 1979.

NSRDS-NBS 61, Part IV, *Physical Properties Data Compilations Relevant to Energy Storage. IV. Flywheels: Data on Metals and Metal Alloys,* H. M. Ledbetter, 1979, in press.

NSRDS-NBS 61, Part V, *Physical Properties Data Compilations Relevant to Energy Storage. V. Flywheels: Data on Selected Non-Metallic Composite Materials,* H. E. Pebly, 1979, in press.

NSRDS-NBS 61, Part VI, *Physical Properties Data Compilations Relevant to Energy Storage. VI. Hydrogen: Data on Selected Metal Hydrides,* G. C. Carter, 1979, in press.

NSRDS-NBS 62, *Compilation of Rate Constants for the Reaction of Metal Ions in Unusual Valency States,* George V. Buxton and Robin M. Sellers, 1978.

NSRDS-NBS 63, *EPA/NIH Mass Spectral Data Base, Volumes I to IV. Molecular Weights 30—1674, and 1978 Index,* S. R. Heller and G. W. A. Milne, 1978.

NSRDS-NBS 64, *Rate Coefficients for Ion-Molecule Reactions, II. Organic Ions Other Than Those Containing Only C and H,* L. W. Sieck, 1979.

NSRDS-NBS 65, *Rate Constants for Reactions of Inorganic Radicals in Aqueous Solution,* A. B. Ross and P. Neta, 1979.

NSRDS DATA PUBLICATIONS FROM OTHER PUBLISHERS

Coblentz Society Evaluated Infrared Reference Spectra, edited and published by the Coblentz Society, sponsored by the Joint Committee on Atomic and Molecular Physical Data (1969—1974). Available from Sadtler Research Laboratories, Philadelphia, PA. 19104. (Microfilm available from the Coblentz Society.) Vol. 6, 1000 Spectra — $295.00; Vol. 7, 1000 Spectra — $295.00; Vol. 8, 1000 Spectra — $295.00; Vol. 9, 1000 Spectra — $295.00; Vol. 10, 1000 Spectra — $295.00; Cumulative Coblentz Indexes — $50.00.

Combustion Fundamentals for Waste Incineration, sponsored by the American Society of Mechanical Engineers Research Committee on Industrial and Municipal Wastes, 1974 (H00087). $30.00. Published by the American Society of Mechanical Engineers, New York.

Compilation of Electron Collision Cross Section Data for Modeling Gas Discharge Lasers, L. J. Kieffer, JILA Information Center Report No. 13, 1973. COM-74-11661.*

A Compilation of YN Reactions, Particle Data Group, 1970. UCRL-20000 YN.*

Compilation of Low Energy Electron Collision Cross Section Data. Part I, L. J. Kieffer, JILA Information Center Report No. 6, 1969. PB189127.*

Compilation of Low Energy Electron Collision Cross Section Data. Part II, L. J. Kieffer, JILA Information Center Report No. 7, 1969. AD696467.*

Contributions to the Data on Theoretical Metallurgy: XVI. Thermodynamic Properties of Nickel and Its Inorganic Compounds, Alla D, Mah and L. B. Pankratz, 1976. U. S. Bureau of Mines Bulletin 668.*

Crystal Data, Determinative Tables, 3rd ed., Vol. 1. Organic Compounds, J. D. H. Donnay and Helen M. Ondik, editors, 1972. Published jointly by the U. S. Department of Commerce, National Bureau of Standards and the Joint Committee on Powder Diffraction Standards. $30.00.

Crystal Data, Determinative Tables, 3rd ed., Vol. 2. Inorganic Compounds, J. D. H. Donnay and Helen M. Ondik, editors, 1973. Published jointly by the U. S. Department of Commerce, National Bureau of Standards and the Joint Committee on Powder Diffraction Standards. $50.00.

Crystal Data, Determinative Tables, 3rd ed., Vol. 3. Organic Compounds 1967—1974, Olga Kennard, David G. Watson, John R. Rogers, editors; J. D. H. Donnay, honorary editor, 1978. Published jointly by the U. S. Department of Commerce, National Bureau of Standards and the JCPDS-International Centre for Diffraction Data. $65.00.

Crystal Data, Determinative Tables, 3rd ed., Vol. 4. Inorganic Compounds 1967—1969, Helen M. Ondik and Alan D. Mighell, editors; J. D. H. Donnay, honorary editor, 1978. Published jointly by the U. S. Department of Commerce, National Bureau of Standards and the JCPDS-International Centre for Diffraction Data. $65.00.

Special price for 4-volume set of *Crystal Data*. $185.00. Order the volumes of *Crystal Data* from: JCPDS-International Centre for Diffraction Data, 1601 Park Lane, Swarthmore, PA 19081.

Derived Thermodynamic Properties of Ethylene, Roland H. Harrison and Donald R. Douslin, *Journal of Chemical and Engineering Data*, 22, 1, 24—30, January 1977.

The Desk Book of Infrared Spectra, Clara D. Craver, Editor, 1977. $90.00. The Coblentz Society, P. O. Box 9952, Kirkwood, MO 63122.

Diffusion Rate Data and Mass Transport Phenomena in Copper Systems, D. B. Butrymowicz, J. R. Manning, and M. E. Read, 1977. Published by the International Copper Research Association, Inc., New York (INCRA Monograph V). $20.00.

Dissociation in Heavy Particle Collisions, G. W. McClure and J. M. Peek, 1972. Published by Wiley-Interscience Series in Atomic and Molecular Collisional Processes. ISBN 0-471-58165-8.

Electron Impact Ionization Cross-Section Data for Atoms, Atomic Ions, and Diatomic Molecules: I. Experimental Data, L. J. Kieffer and G. H. Dunn, *Rev. Mod. Phys.*, 38, 1, 1966.

Excitation in Heavy Particle Collisions, E. W. Thomas, 1972. Published by Wiley-Interscience Series in Atomic and Molecular Collisional Processes. ISBN 0-471-85890-0. $31.00.

Infrared Spectra of Halogenated Hydrocarbons, Clara D. Craver, Editor, 1977. $40.00. The Coblentz Society, P. O. Box 9952, Kirkwood, MO 63122.

Infrared Spectra of Plasticizers and Other Additives, Clara D. Craver, Editor, 1977. $60.00. The Coblentz Society, P. O. Box 9952, Kirkwood, MO 63122.

Ion Molecule Reactions, E. W. McDaniel, V. Cermak, A. Dalgarno, E. E. Ferguson, and L. Friedman, 1970. Published in the Wiley-Interscience Series in Atomic and Molecular Collisional Processes, Wiley-Interscience, New York. ISBN 0-471-57386-3. $29.50.

K°₁N Interactions — A Compilation, Particle Data Group, 1972. LBL-55.*

LNG Materials & Fluids, A User's Manual of Property Data in Graphic Format, First Edition, Douglas Mann, General Editor, Cryogenics Division, National Bureau of Standards, 1977. Loose leaf $35.00. Make check or money order payable to "National Bureau of Standards, Department of Commerce" and send to: LNG Materials and Fluids Users' Manual, Cryogenic Division, National Bureau of Standards, Boulder, CO 80302. For additional information, contact Douglas Mann at the same address, or telephone (303) 499-1000, ext. 3652.

Low Energy Electron-Collision Cross Section Data. Part III: Total Scattering; Differential Elastic Scattering, L. J. Kieffer, *Atomic Data*, 2, 4, 293—391, 1971.

Metallic Shifts in NMR: A Review of the Theory and Comprehensive Critical Data Compilation of Metallic Materials, G. C. Carter, L. H. Bennett, and D. J. Kahan, in 4 volumes, 2350 pp., 1977. Volume 20 in *Progress in Materials Science*, Pergamon Press, Oxford, New York. ISBN 0-08-021143-7. $300.00.

NN and ND Interactions (above 0.5 GeV/c) — A Compilation, Particle Data Group, 1970. UCRL-20000 NN.*

NN and ND Interactions — A Compilation, Particle Data Group, 1972. LBL-58.*

* Available from the National Technical Information Service.

Nuclear Wallet Cards, V. S. Shirley and C. M. Lederer, editors, 1979. Copies may be obtained from the National Nuclear Data Center, Brookhaven National Laboratory.

πN Partial-Wave Amplitudes, Particle Data Group, 1970. UCRL-20030 πN.*

πN Two-Body Scattering Data I. A User's Guide to the Lovelace-Almehed Data Tape, Particle Data Group, 1973. LBL-63.*

π + p, π + n, and π + d Interactions — A Compilation: Parts I and II, Particle Data Group, 1973. LBL-53.*

Pressure, Volume, Temperature Relations of Ethylene, D. R. Douslin and R. H. Harrison, *J. Chem. Thermodyn.,* 76, 8, 301—330, Aug. 1976.

Review of Particle Properties, Particle Data Group, *Phys. Letters 75B,* No. 1, 1978.

Selected Thermodynamic Values and Phase Diagrams for Copper and Some of Its Binary Alloys, R. Hultgren and P. D. Desai, 1973. Published by the International Copper Research Association, Inc., New York; Monograph I. $20.00.

Selected Values of the Thermodynamic Properties of Binary Alloys, R. Hultgren, P. D. Desai, D. T. Hawkins, M. Gleiser, and K. K. Kelley, 1973. Published by the American Society for Metals, Metals Park, Ohio. $30.00.

Selected Values of the Thermodynamic Properties of the Elements, R. Hultgren, P. D. Desai, D. T. Hawkins, M. Gleiser, K. K. Kelley, and D. D. Wagman, 1973. Published by the American Society for Metals, Metals Park, Ohio. $20.00.

Table of Isotopes, 7th edition, C. Michael Lederer and Virginia S. Shirley, editors; Edgardo Browne, Janis M. Dairiki, and Raymond E. Doebler, 1978. John Wiley & Sons, New York. Clothbound, ISBN 0-471-04179-3 — $40.00; paperbound, ISBN 0-471-04180-7 — $26.25.

Thermodynamic Data for Waste Incineration, sponsored by the American Society of Mechanical Engineers Research Committee on Industrial and Municipal Wastes, 1979. Order from the ASME Order Department as H00141 — $30.00. Members — $15.00.

Thermodynamic Properties of Aqueous Copper Systems, Paul Duby, 1977. Published by the International Copper Research Association, Inc., New York. (Monograph IV). $20.00.

Thermodynamic Properties of Copper and Its Inorganic Compounds, E. G. King, Alla D. Mah, and L. B. Pankratz, 1973. Published by the International Copper Research Association, Inc., New York. (Monograph II). $20.00.

Thermodynamic Properties of Copper-Slag Systems, Paul Duby, 1976. Published by the International Copper Research Association, Inc., New York. (Monograph III). $20.00.

Theory of Charge Exchange, R. A. Mapleton, 1971. Published by Wiley-Interscience Series in Atomic and Molecular Collisional Processes. ISBN 0-471-56781-7.

Theory of the Ionization of Atoms by Electron Impact, M. R. H. Rudge, *Rev. Mod. Phys.,* 40 (3), 564, 1968.

Total Electron-Atom Collision Cross Sections at Low Energies — A Critical Review, Benjamin Bederson and L. J. Kieffer, *Rev. Mod. Phys.,* 43 (4), 601, 1971.

OTHER NBS COMPILATIONS OF DATA

NBS Tech. Note 270-3, *Selected Values of Chemical Thermodynamic Properties, Tables for the First Thirty-Four Elements in the Standard Order of Arrangement,* D. D. Wagman, W. H. Evans, V. B. Parker, I. Halow, S. M. Bailey, and R. H. Schumm, 1968.

NBS Tech. Note 270-4, *Selected Values of Chemical Thermodynamic Properties, Tables for Elements 35 through 53 in the Standard Order of Arrangement,* D. D. Wagman, W. H. Evans, V. B. Parker, I. Halow, S. M. Bailey, and R. H. Schumm, 1969.

NBS Tech. Note 270-5, *Selected Values of Chemical Thermodynamic Properties, Tables for Elements 54 through 61 in the Standard Order of Arrangement,* D. D. Wagman, W. H. Evans, et al., 1971.

NBS Tech. Note 270-6, *Selected Values of Chemical Thermodynamic Properties, Tables for the Alkaline Earth Elements (Elements 92 through 97 in the Standard Order of Arrangement),* V. B. Parker, D. D. Wagman, and W. H. Evans, 1971.

NBS Tech. Note 270-7, *Selected Values of Chemical Thermodynamic Properties, Tables for the Lanthanide (Rare Earth) Elements (Elements 62 through 76 in the Standard Order of Arrangement),* R. H. Schumm, D. D. Wagman, S. Bailey, W. H. Evans, and V. B. Parker, 1973.

NBS Tech. Note 270-8, *Selected Values of Chemical Thermodynamic Properties, Tables for the Alkali Metals (Elements 98 through 103 in the Standard Order of Arrangement),* D. D. Wagman, et al., 1980, in press.

NBS Tech. Note 270-9, *Selected Values of Chemical Thermodynamic Properties, Tables for the Actinide Elements (Elements 77 through 90 in the Standard Order of Arrangement),* D. D. Wagman, et al., 1980, in process.

NBS Tech. Note 438, *Compendium of ab-initio Calculations of Molecular Energies and Properties,* M. Krauss, 1967.

NBS Tech. Note 474, *Critically Evaluated Transition Probabilities for Ba I and II,* B. M. Miles and W. L. Wiese, 1969.

NBS Tech. Note 484, *A Review of Rate Constants of Selected Reactions of Interest in Re-Entry Flow Fields in the Atmosphere,* M. H. Bortner, 1969.

NBS Tech. Note 968, *Total Photon Absorption Cross Section Measurements, Theoretical Analysis and Evaluation for Energies above 10 MeV,* H. A. Gimm and John H. Hubbell, 1978.

NBS Tech. Note 983, *Properties of Selected Superconductive Materials, 1978 Supplement,* B. W. Roberts, 1978. Supplements *J. Phys. Chem. Reference Data,* 5 (3), 581—821, 1976, Reprint No. 84.

NBS Spec. Publ. 513, *Reaction Rate and Photochemical Data for Atmospheric Chemistry — 1977,* Robert F. Hampson, Jr. and David Garvin, editors, 1978.

NBS Monograph 70, Vol. I. *Microwave Spectral Tables, Diatomic Molecules,* P. F. Wacker, M. Mizushima, J. D. Petersen, and J. R. Ballard, 1964.

NBS Monograph 70, Vol. II. *Microwave Spectral Tables, Line Strengths of Asymmetric Rotors,* P. F. Wacker and M. R. Pratto, 1964.

NBS Monograph 70, Vol. III. *Microwave Spectral Tables, Polyatomic Molecules with Internal Rotation,* P. F. Wacker, M. S. Cord, D. G. Burkhard, J. D. Petersen, and R. F. Kukol, 1969.

NBS Monograph 70, Vol. IV. *Microwave Spectral Tables, Polyatomic Molecules without Internal Rotation,* M. S. Cord, J. D. Petersen, M. S. Lojko, and R. H. Haas, 1968.

NBS Monograph 70, Vol. V. *Microwave Spectral Tables, Spectral Line Listing,* M. S. Cord, M. S. Lojko, and J. D. Petersen, 1968.

NBS Monograph 94, *Thermodynamic and Related Properties of Parahydrogen from the Triple Point to 100 K at Pressures to 340 Atmospheres,* H. M. Roder, L. A. Weber, and R. D. Goodwin, 1965.

NBS Monograph 115, *The Calculation of Rotational Energy Levels and Rotational Line Intensities in Diatomic Molecules,* Jon T. Hougen, 1970.

NBS Monograph 134, *Space Groups and Lattice Complexes,* Werner Fisher, Hans Burzlaff, Erwin Hellner, and J. D. H. Donnay, 1973.

NBS Monograph 153, *The First Spectrum of Hafnium (Hf I),* William F. Meggers and Charlotte E. Moore, 1976.

NBS Monograph 158, *Handbook of Rock Salt Properties Data,* L. H. Gevantman, editor, 1979, in press.

NBSIR 75-770, *Equation of State for Ammonia,* L. Haar and J. Gallagher, 1975.

NBSIR 75-968, *Selected Thermochemical Data Compatible with the CODATA Recommendations,* V. B. Parker, D. D. Wagman, and D. Garvin, 1976.

NBSIR 76-1034, *Chemical Thermodynamic Properties of Compounds of Sodium, Potassium, and Rubidium: An Interim Tabulation of Selected Material,* D. D. Wagman, W. H. Evans, V. B. Parker, and R. H. Schumm, 1976.

NBSIR 76-1061, *Annotated Bibliography on Proton Affinities,* K. Hartman, S. Lias, P. J. Ausloos, and H. M. Rosenstock, 1976.

NBSIR 77-860, *Provisional Thermodynamic Functions of Propane from 85 to 700 K at Pressures to 700 Bar,* R. D. Goodwin, 1977.

NBSIR 77-865, *Thermodynamic and Related Properties of Oxygen from the Triple Point to 300 K at Pressures to 1000 Bar,* L. A. Weber, 1977.

NBSIR 77-1300, *A Computer Assisted Evaluation of the Thermochemical Data of the Compounds of Thorium,* D. D. Wagman, R. H. Schumm, and V. B. Parker, 1977.

NBSIR 78-1479, *Thermodynamic Properties of Miscellaneous Materials,* Eugene S. Domalski, William H. Evans, and Thomas L. Jobe, Jr., 1978.

TRANSLATIONS FROM THE RUSSIAN

A number of products of the Russian standard reference data system have been translated with the support of the National Science Foundation at the request of the National Bureau of Standards. These Russian translations are available from NTIS.

Electric Conductivity of Ferroelectrics, V. M. Gurevich, 1969. Original published by the Committee on Standards, Measures and Measuring Instruments of the USSR Council of Ministers — Government Standards Service. TT70-50180, 1971, vi + 362 pp.

Handbook of Hardness Data, A. A. Ivan'ko; G. V. Samsonov, editor, 1968. Original published by the Academy of Sciences of the Ukrainian SSR, Institute of Materials Sciences. TT70-50177, 1971, iii + 66 pp.

Handbook of Phase Diagrams of Silicate Systems, Vol. I. Binary Systems, Second Revised Edition, N. A. Toropov, V. P. Barzakovskii, V. V. Lapin, and N. N. Kurtseva; N. A. Toropov, editor, 1969. Original published by the Academy of Sciences of the USSR Grebenshchikov Institute of Silicate Chemistry. TT71-50040, 1972, viii + 723 pp.

Handbook of Phase Diagrams of Silicate Systems. Vol. II. Metal-Oxygen Compounds in Silicate Systems, Second Revised Edition, N. A. Toropov, V. P. Barzakovskii, I. A. Bondar, and Yu, P. Udalov; N. A. Toropov, editor, 1969. Original published by the Academy of Sciences of the USSR Grebenshchikov Institute of Silicate Chemistry. TT71-50041, 1972, iv + 325 pp.

Handbook of Thermodynamic Data, G. B. Naumov, B. N. Ryzhenko, and I. L. Khodakovsky; A. I. Tugarinov, editor, 1971. Original published by the USSR Academy of Sciences, Atomizdat. PB226-722, xi + 328 pp.

Heavy Water, Thermophysical Properties, Ya. Z. Kazavchinskii et al.; V. A. Kirillin, editor, 1963. TT70-50094, 1971, vii + 263 pp., 2 charts.

Isochoric Heat Capacity of Water and Steam, Kh. I. Amirkhanov, G. V. Stepanov, B. G. Alibekov; M. P. Vukalovich, editor, 1969. Original published by the Dagestan Branch of the USSR Academy of Sciences. TT72-52002, xii + 203 pp.

Properties of Liquid and Solid Hydrogen, B. N. Esel'son, Yu. P. Blagoi, V. N. Grigor'ev, V. G. Manzhelii, S. A. Mikhailenko, and N. P. Neklyudov, 1969. Original published by the Committee on Standards, Measures and Measuring Instruments of the USSR Council of Ministers — Government Standards Service. TT70-50179, 1974, iii + 123 pp.

Rate Constants of Gas Phase Reactions, Reference Book, V. N. Kondratiev, 1970. Original published by the Academy of Sciences of the USSR Order-of-Lenin Institute of Chemical Physics. Translated by L. Holtschlag and R. Fristrom. COM-72-10014, 1972, vi + 428 pp.

Surface Ionization, E. Ya. Zandberg and N. L. Ionov, 1969. Original published by the Science Publishing House. Translated by E. Harnik. TT70-50148, 1971, xii + 355 pp.

Thermodynamic and Thermophysical Properties of Combustion Products, Volume I. Computation Methods, V. E. Alemasov, A. F. Dregalin, A. P. Tishin, and V. A. Khudyakov; V. P. Glushko, editor, 1971. Original published by the Academy of Sciences of the USSR. TT74-50019, 1974, xx + 433 pp.

Thermodynamic and Thermophysical Properties of Combustion Products, Volume II. Oxygen Based Propellants, V. E. Alemasov, A. F. Dregalin, V. A. Khudyakov, and V. N. Kostin; V. P. Glushko, editor, 1972. Original published by the Academy of Sciences of the USSR. TT74-50032, x + 495 pp.

Thermodynamic and Thermophysical Properties of Combustion Products, Volume III. Oxygen- and Air-Based Propellants, V. E. Alemasov, A. F. Dregalin, A. P. Tishin, V. A. Khudyakov, and V. N. Kostin; V. P. Glushko, editor, 1973. Original published by the Academy of Sciences of the USSR. TT75-50007, 1975, x + 622 pp.

Thermodynamic and Thermophysical Properties of Combustion Products, Volume IV. Nitrogen Tetroxide-Based Propellants, V. E. Alemasov, A. F. Dregalin, A. P. Tishin, V. A. Khudyakov, and V. N. Kostin; V. P. Glushko, editor, 1973. Original published by the Academy of Sciences of the USSR. TT76-50007, 1976, x + 530 pp.

Thermodynamic and Thermophysical Properties of Helium, N. V. Tsederberg, V. N. Popov, and N. A. Morozova; A. F. Alyab'ev, editor, 1969. Original published by Atomizdat. TT70-50096, 1971, v + 255 pp.

Thermodynamic and Transport Properties of Ethylene and Propylene, I. A. Neduzhii et al., 1971. Original published by the State Committee of Standards of the Soviet Ministry, USSR, State Office of Standards and Reference Data Series: Monograph No. 8. Translation published June 1972. NBSIR 75-763.

Thermophysical Properties of Air and Air Components, A. A. Vasserman, Ya. A. Kazavchinskii, and V. A. Rabinovich; A. M. Zhuravlev, editor, 1966. Original published by the Academy of Sciences of the USSR. TT70-50095, 1971, x + 383 pp., 8 charts.

Thermophysical Properties of Freon-22, A. V. Kletskii, 1970. Original published by the Committee of Standards, Measures and Measuring Instruments of the USSR Council of Ministers — Government Standards Service. TT70-50178, 1971, iv + 67 pp., 1 chart.

Thermophysical Properties of Gases and Liquids, No. 1, V. A. Rabinovich, 1968. Original published by the Committee of Standards, Measures and Measuring Instruments of the USSR Council of Ministers — Government Standards Service, Series: Physical Constants and Properties of Substances. TT69-55091, 1970, viii + 207 pp.

The translation of volumes 2, 3, and 4 of this series have the English title Thermophysical Properties of Matter and Substances.

Thermophysical Properties of Gaseous and Liquid Methane, V. A. Zagoruchenko and A. M. Zhuravlev, 1969. Original published by the Committee of Standards, Measures and Measuring Instruments of the USSR Council of Ministers — Government Standards Service. TT70-50097, 1970, viii + 243 pp.

Thermophysical Properties of Liquid Air and Its Components, A. A. Vasserman and V. A. Rabinovich, 1968. Original published by the Committee of Standards, Measures and Measuring Instruments of the USSR Council of Ministers — Government Standards Service. TT69-55092, 1970, viii + 235 pp., 2 charts.

Thermophysical Properties of Matter and Substances, Volume 2, V. A. Rabinovich, 1970. Original published by the Committee of Standards, Measures and Measuring Instruments of the USSR Council of Ministers — Government Standards Service, Series: Physical Constants and Properties of Matter. TT72-52001, 1974, xi + 384 pp.

Thermophysical Properties of Matter and Substances, Volume 3, V. A. Rabinovich, 1971. Original published by the Committee of Standards, Measures and Measuring Instruments of the USSR Council of Ministers — Government Standards Service, Series: Physical Constants and Properties of Matter. TT73-52009, 1974, xiii + 343 pp.

Thermophysical Properties of Matter and Substances, Volume 4, V. A. Rabinovich, 1971. Original published by the Committee of Standards, Measures and Measuring Instruments of the USSR Council of Ministers — Government Standards Service, Series: Physical Constants and Properties of Matter. TT73-52029, vii + 173 pp.

The translation of Volume 1 of this series has the English title Thermophysical Properties of Gases and Liquids, No. 1.

CRITICAL BIBLIOGRAPHIES AND INDEXES FROM OTHER PUBLISHERS

Bibliography of High Pressure Research, Leo Merrill, compiler, Volume XII, 1979. Bimonthly — $9.00; overseas airmail — $13.00. Subscriptions and information requests should be sent to: High Pressure Data Center, 5093 HBLL, Brigham Young University, Provo, UT 84602 USA.

Binary Fluorides, Free Molecular Structures and Force Fields, A Bibliography (1957—1975), D. T. Hawkins, L. S. Bernstein, W. E. Falconer, and W. Klemperer, IFI/Plenum, New York, 1976.

Biweekly List of Papers on Radiation Chemistry, Annual Cumulation with Keyword and Author Indexes, Vol. VII, 1974, 1975. COM-75-11475.*

Biweekly List of Papers on Radiation Chemistry, Annual Cumulation with Keyword and Author Indexes, Vol. VIII, 1975, 1976. PB 257 025.*

Biweekly List of Papers on Radiation Chemistry, Annual Cumulation with Keyword and Author Indexes, Vol. IX, 1976, 1977. PB 268 699.*

Biweekly List of Papers on Radiation Chemistry, Annual Cumulation with Keyword and Author Indexes, Vol. X, 1977, 1978. PB 291 876.*

Biweekly List of Papers on Radiation Chemistry and Photochemistry, Annual Cumulation with Keyword and Author Indexes, Vol. XI, 1978, 1979. Available from the Radiation Chemistry Data Center.

* Available from the National Technical Information Service.

Subscriptions to the *Biweekly List of Papers on Radiation Chemistry and Photochemistry (Vol. XII, 1979)* and the *Annual Cumulations* for the previous year (Vol. XI, 1978) are available on a subscription basis from the: Radiation Chemistry Data Center, Radiation Laboratory, University of Notre Dame, Notre Dame, IN 46556 USA.

Please write to the Data Center for further information.

Bulletin of Chemical Thermodynamics, IUPAC Commission on Thermodynamics and Thermochemistry, Robert D. Freeman, editor; former title: Bulletin of Thermodynamics and Thermochemistry. Volume 21, 1978 — $25.00. Subscriptions and orders for single copies and back issues should be addressed to: Bulletin of Chemical Thermodynamics, Department of Chemistry, Oklahoma State University, Stillwater, OK 74074 USA.

Checks should be made payable, in U.S. funds, to: Bulletin of Chemical Thermodynamics.

A Compilation of Volumes I—IV of Bibliography on High Pressure Research with Author and Subject Indexes, Volumes I and II, J. F. Cannon and L. Merrill, 1972. Published by and available from the High Pressure Data Center, 5093 HBLL, Brigham Young University, Provo, UT 84602 USA. In microfiche — $14.00.

Compilation of Volumes V—IX of the Bibliography on High Pressure Research with Author and Subject Indexes, L. Merrill, 1979, in press. Will be available from the High Pressure Data Center.

Equilibrium Properties of Fluid Mixtures, A Bibliography on Fluids of Cryogenic Interest, M. J. Hiza, A. J. Kidnay, and R. C. Miller, IFI/Plenum, New York, 1975. ISBN 0-306-66001-6 — $37.50.

"A Guide to Sources of Information on Materials," R. S. Marvin and G. B. Sherwood, Chapter 10, pp. 603—627, in *Handbook of Materials Science, Volume III, Nonmetallic Materials and Publications,* C. T. Lynch, editor, CRC Press, Cleveland, Ohio, 1975.

Index and Cumulative List of Papers on Radiation Chemistry, Volume IV, Numbers 27—52, July through December 1971, January 1972. COM-72-10266.*

Index and Cumulative List of Papers on Radiation Chemistry, Volume V, Numbers 1—26, January through June 1972, July 1972. COM-72-10621.*

Index and Cumulative List of Papers on Radiation Chemistry, Volume V, Numbers 27—52, July through December 1972, January 1973. COM-73-10281.*

Index and Cumulative List of Papers on Radiation Chemistry, Volume VI, Numbers 1—26, January through June 1973, July 1973. COM-73-11878/8AS.*

Index and Cumulative List of Papers on Radiation Chemistry, Volume VI, Numbers 27—52, July through December 1973, January 1974. COM-74-10899.*

An Indexed Compilation of Experimental High Energy Physics Literature, Particle Data Group, LBL90, 1978.*

Liquid Vapor Equilibria on Systems of Interest in Cryogenics, A Survey, A. J. Kidnay, M. J. Hiza, and R. C. Miller, *Cryogenics,* 13, 10, 575—599, 1973.

Mossbauer Effect Data Index, 1958—1965, A. H. Muir, K. J. Ando and H. M. Coogan, Interscience Publishers, New York, 1967.

Mossbauer Effect Data Index, Covering the 1966—1968 Literature, J. G. Stevens and V. E. Stevens, IFI/Plenum, New York, 1975. ISBN 0-306-65162-9 — $49.50.

Mossbauer Effect Data Index, Covering the 1969 Literature, 1971. ISBN 0-306-65140-8 — $37.50.

Mossbauer Effect Data Index, Covering the 1970 Literature, 1972. ISBN 0-306-65141-6 — $37.50.

Mossbauer Effect Data Index, Covering the 1971 Literature, 1973. ISBN 0-306-65142- — $37.50.

Mossbauer Effect Data Index, Covering the 1972 Literature, 1973. ISBN 0-306-65143-2 — $37.50.

Mossbauer Effect Data Index, Covering the 1973 Literature, 1975. ISBN 0-306-65144-0 — $49.50.

Mossbauer Effect Data Index, Covering the 1974 Literature, 1975. ISBN 0-306-65145-9 — $49.50.

Mossbauer Effect Data Index, Covering the 1975 Literature, 1977. ISBN 0-306-65146- — $49.50.

Mossbauer Effect Data Index, Covering the 1976 Literature, 1978. ISBN 0-306-65149-1 — $49.50.

Mossbauer Effect Reference and Data Journal, Volume 1 (1978); 10 issues per year; J. G. Stevens, V. E. Stevens, and W. L. Gettys, editors. For further information write to: Mossbauer Effect Data Center, University of North Carolina — Asheville, Asheville, NC 28804 USA. Telephone (704) 258-0200.

Thesaurus for Radiation Chemistry, Alberta B. Ross, 1977. UND-RCDC-1.*

Weekly List of Papers on Radiation Chemistry, Index and Cumulation, Volume IV, Numbers 1—26, January through June 1971, August 1971. COM-71-01103.*

CRITICAL BIBLIOGRAPHIES AND INDEXES IN NBS SERIES

NBS Misc. Publ. 281, *Bibliography on Flames Spectroscopy, Analytical Applications, 1800 to 1966,* R. Mavrodineanu, 1967.

NBS Spec. Publ. 306-1, *Bibliography on the Analyses of Optical Atomic Spectra, Section 1,* ^1H-^{23}V, C. E. Moore, 1968.

NBS Spec. Publ. 306-2, *Bibliography on the Analyses of Optical Atomic Spectra, Section 2,* ^{24}Cr-^{41}Nb, C. E. Moore, 1969.

NBS Spec. Publ. 306-3, *Bibliography on the Analyses of Optical Atomic Spectra, Section 3,* ^{42}Mo-^{57}La, ^{72}Hf-^{89}Ac, C. E. Moore, 1969.

NBS Spec. Publ. 306-4, *Bibliography on the Analyses of Optical Atomic Spectra, Section 4,* ^{57}La-^{71}Lu, ^{89}Ac-^{99}Es, C. E. Moore, 1969.

NBS Spec. Publ. 324, *The NBS Alloy Data Center: Permuted Materials Index,* G. C. Carter, D. J. Kahan, L. H. Bennett, J. R. Cuthill, and R. C. Dobbyn, 1971.

NBS Spec. Publ. 349, *Heavy-Atom Kinetic Isotope Effects, An Indexed Bibliography,* M. J. Stern and M. Wolfsberg, 1972.

NBS Spec. Publ. 362, *Chemical, Kinetics in the C-O-S and H-N-O-S Systems: A Bibliography — 1899 through June 1971,* F. Westley, 1972.

NBS Spec. Publ. 363, *Bibliography on Atomic Energy Levels and Spectra, July 1968 through June 1971*, L. Hagan and W. C. Martin, 1972.

NBS Spec. Publ. 363, Supplement 1, *Bibliography on Atomic Energy Levels and Spectra, July 1971 through June 1975*, Lucy Hagan, 1977.

NBS Spec. Publ. 363, Supplement 2, *Bibliography on Atomic Energy Levels and Spectra, July 1975 through December 1978*, R. Zalubas and A. Albright, 1979, in press.

NBS Spec. Publ. 366, *Bibliography on Atomic Line Shapes and Shifts (1899 through March 1972)*, J. R. Fuhr, W. L. Wiese, and L. J. Roszman, 1972.

NBS Spec. Publ. 366, Supplement 1, *Bibliography on Atomic Line Shapes and Shifts (April 1972 through June 1973)*, J. R. Fuhr, L. J. Roszman, and W. L. Wiese, 1974.

NBS Spec. Publ. 366, Supplement 2, *Bibliography on Atomic Line Shapes and Shifts (July 1973 through May 1975)*, J. R. Fuhr, G. A. Martin, and B. Specht, 1975.

NBS Spec. Publ. 366, Supplement 3, *Bibliography on Atomic Line Shapes and Shifts (June 1975 through June 1978)*, J. R. Fuhr, B. J. Miller, and G. A. Martin, 1978.

NBS Spec. Publ. 369, *Soft X-Ray Emission Spectra of Metallic Solids: Critical Review of Selected Systems and Annotated Spectral Index*, A. J. McAlister, R. C. Dobbyn, J. R. Cuthill, and M. L. Williams, 1974.

NBS Spec. Publ. 371, *A Supplementary Bibliography of Kinetic Data on Gas Phase Reactions of Nitrogen, Oxygen, and Nitrogen Oxides*, F. Westley, 1973.

NBS Spec. Publ. 371-1, *Supplementary Bibliography of Kinetic Data on Gas Phase Reactions of Nitrogen, Oxygen, and Nitrogen Oxides (1972—1973)*, Francis Westley, 1975.

NBS Spec. Publ. 380, *Photonuclear Reaction Data, 1973*, E. G. Fuller, H. M. Gerstenberg, H. Vander Molen, and T. C. Dunn, 1973.

NBS Spec. Publ. 380, Supplement 1, *Photonuclear Reactions Data, July 1973 to December 1977*, E. G. Fuller and H. M. Gerstenberg, 1978.

NBS Spec. Publ. 381, *Bibliography of Ion-Molecule Reaction Rate Data (January 1950 through October 1971)*, George A. Sinnott, 1973.

NBS Spec. Publ. 392, *Vibrationally Excited Hydrogen Halides: A Bibliography on Chemical Kinetics of Chemiexcitation and Energy Transfer Processes (1958 through 1973)*, Francis Westley, 1974.

NBS Spec. Publ. 396-1, *Critical Surveys of Data Sources: Mechanical Properties of Metals*, R. B. Gavert, R. L. Moore, and J. H. Westbrook, 1974.

NBS Spec. Publ. 396-2, *Critical Surveys of Data Sources: Ceramics*, Dorothea M. Johnson and James F. Lynch, 1975.

NBS Spec. Publ. 396-3, *Critical Surveys of Data Sources: Corrosion of Metals*, Ronald B. Diegle and Walter K. Boyd, 1976.

NBS Spec. Publ. 396-4, *Critical Surveys of Data Sources: Electrical and Magnetic Properties of Metals*, M. J. Carr, R. B. Cavert, R. L. Moore, H. W. Wawrousek, and J. H. Westbrook, 1976.

NBS Spec. Publ. 426, *Bibliography of Low Energy Electron and Photon Cross Section Data (through December 1974)*, Lee J. Kieffer, 1976.

NBS Spec. Publ. 426, Supplement 1, *Bibliography of Low Energy Electron and Photon Cross Section Data (January 1975 through December 1977)*, J. W. Gallagher, J. R. Rumble, E. C. Beatty, 1979.

NBS Spec. Publ. 428, *Bibliography of Infrared Spectroscopy through 1960, Parts 1, 2, and 3*, C. N. R. Rao, S. K. Dikshit, S. A. Kudchadker, D. S. Gupta, V. A. Narayan, and J. J. Comeford, 1976.

NBS Spec. Publ. 449, *Chemical Kinetics of the Gas Phase Combustion of Fuels (A Bibliography on the Rates and Mechanisms of the Oxidation of Aliphatic C_1—C_{10} Hydrocarbons and of Their Oxygenated Derivatives)*, Francis Westley, 1976.

NBS Spec. Publ. 454, *An Annotated Bibliography of Sources of Compiled Thermodynamic Data Relevant to Biochemical and Aqueous Systems (1930—1975) Equilibrium, Enthalpy, Heat Capacity, and Entropy Data*, George T. Armstrong and Robert N. Goldberg, 1976.

NBS Spec. Publ. 478, *Nitrogen Oxychlorides: A Bibliography on Data for Physical and Chemical Properties of $ClNO$, $ClNO_2$ and $ClNO_3$*, F. Westley, 1977.

NBS Spec. Publ. 485, *A Bibliography of Sources of Experimental Data Leading to Activity or Osmotic Coefficients for Polyvalent Electrolytes in Aqueous Solution*, R. N. Goldberg, B. R. Staples, R. L. Nuttail, R. Arbuckle, 1977.

NBS Spec. Publ. 505, *Bibliography on Atomic Transition Probabilities (1914 through October 1977)*, J. R. Fuhr, B. J. Miller, and G. A. Martin, 1978.

NBS Spec. Publ. 537, *A Bibliography of Sources of Experimental Data Leading to Thermal Properties of Binary Aqueous Electrolyte Solutions*, D. Smith-Magowan and R. N. Goldberg, 1979.

NBS Tech. Note 291, *A Bibliography on Ion-Molecule Reactions, January 1900 to March 1966*, F. N. Harllee, H. M. Rosenstock, and J. T. Herron, 1966.

NBS Tech. Note 464, *The NBS Alloy Data Center: Function, Bibliographic System, Related Data Centers, and Reference Books*, G. C. Carter, L. H. Bennett, J. R. Cuthill, and D. J. Kahan, 1968.

NBS Tech. Note 554, *Annotated Accession List of Data Compilations of the NBS Office of Standard Reference Data*, H. M. Weisman and G. B. Sherwood, 1970.

NBS Tech. Note 848, *A Bibliography of the Russian Reference Data Holdings of the Library of the Office of Standard Reference Data*, Gertrude B. Sherwood and Howard J. White, Jr., 1974.

NBS-OSRDB-70-1, Vol. 1, *High Pressure Bibliography 1900—1968. Volume I. Section I — Bibliography, Section II. Author Index*, Leo Merrill, 1970.

NBS-OSRDB-70-1, Vol. 2, *High Pressure Bibliography 1900—1968. Volume II. Subject Index*, Leo Merrill, 1970.

NBS-OSRDB-70-2, *The NBS Alloy Data Center: Author Index*, G. C. Carter, D. J. Kahn, L. H. Bennett, J. R. Cuthill, and R. C. Dobbyn, 1970.

NBS-OSRDB-70-3, *Semiempirical and Approximate Methods for Molecular Calculations — Bibliography and KWIC Index*, George A. Henderson and Sandra Frattali, 1969.

NBS-OSRDB-70-4, *Bibliography of Photoabsorption Cross Section Data*, Robert D. Hudson and Lee J. Kieffer, 1970.

NBS-OSRDB-71-1, *Bibliography on Properties of Defect Centers in Alkali Halides*, S. C. Jain, S. A. Khan, H. K. Sehgal, V. K. Garg, and R. K. Jain, 1971.

NBS-OSRDB-71-2, *A Bibliography of Kinetic Data on Gas Phase Reactions of Nitrogen, Oxygen, and Nitrous Oxides*, Francis Westley, 1970.

The following lists the name and prices of certain publications of
   Fachinformationszentrum
      Energie, Physik, Mathematik Gmbh
   Karlsruhe Nuclear Research Centre
   Kernforschungzentrum
   D-75014 Eggenstein
   Leopoldshafen 2
   West Germany

PHYSIK DATEN/PHYSICS DATA

Price/plus postage

1—2 (1977): *Survey Index of Pion-Nucleon Scattering Data,* March 1977 Edition, by K. H Augenstein, G. Hohler, E. Pietarinen and H. M. Staudenmaier, 66 pages. This issue supersedes the previous edition published as No. 1—1 (1975). — out of print

2—1 (1976): *Stopping Cross Sections of Elements with Z = 2 to 87 for Li ions with Energies between 80 keV and 840 keV,* by W. Neuwirth, W. Pietsch and U. Hauser, 10 pages. A compilation of experimental and calculated stopping cross sections. — DM 3.50

3—1 (1976): *Datensammlungen in der Physik./Data Compilations in Physics,* by H. Behrens and G. Ebel, 226 pages. A bibliography of 1457 existing tables and compilations from all fields of physics. — DM 14.—

3—2 (1977): *Datensammlungen in der Physik./ Data Compilations in Physics,* by H. Behrens and G. Ebel. 116 pages. Supplement to No. 3—1 containing 561 further references to tables and compilations. — DM 7.—

3—3 (1978): *Datensammlungen in der Physik./Data Compilations in Physics,* by H. Behrens and G. Ebel. 78 pages. Supplement to No. 3—1 and 3—2 containing 348 further references to tables and compilations. — DM 8.—

3—4 (1979): *Datensammlungen in der Physik./Data Compilations in Physics,* by H. Behrens and G. Ebel. 86 pages. Supplement to No. 3—1, 3—2, and 3—3 containing 418 further references to tables and compilations. — DM 11.50

4—2 (1979): *Compilation of Coupling Constants and Low-Energy Parameters,* 1978 Edition, reprinted from *Nuclear Physics B,* 147, 189—276, 1979, by M. M. Nagels, T. A., Rijken, J. J. de Swart, G. C. Oades, J. L. Petersen, A. C. Irving, C. Jarlskog, W. Pfeil, H. Pilkuhn, and H. P. Jakob. 92 pages. A compilation of coupling constants and other parameters which characterize the interactions of elementary particles at low energies. This issue supersedes the previous edition published as No. 4—1 (1976). — DM 7.50

5—1 (1976): *Gases and Carbon in Metals (Thermodynamics, Kinetics, and Properties), Part I: Alkali Metals, Alkaline Earth Metals, Light Metals (Li, Na, K, Rb, Cs; Ca, Sr, Ba; Be, Mg, Al),* by E. Fromm, H. Jehn, and G. Hörz. 30 pages. — DM 9.50

5—2 (1977): *Gases and Carbon in Metals (Thermodynamics, Kinetics, and Properties), Part II: Group II B to V B Metals (Zn, Cd, Ga, In, Tl, Ci, Ge, Sn, Pg; Bi),* by H. Jehn, E. Fromm, and G. Hörz. 40 pages. — DM 9.50

5—3 (1978): *Gases and Carbon in Metals (Thermodynamics, Kinetics, and Properties),* Part III: Group III A Metals, Rare Earth Metals (Sc, Y; La, Ce, Pr, Nd, Sm, Eu, Gd, Tb, Dy, Ho, Er, Tm, Yb, Lu), by E. Fromm, H. Jehn and G. Hörz. 32 pages. — DM 14.—

5—4 (1978): *Gases and Carbon in Metals (Thermodynamics, Kinetics, and Properties),* Part IV: Actinides (U, Th, Pu, Pa, Np, Am, Cm, Bk, Cf), by H. Jehn, E. Fromm, and G. Hörz. 34 pages. — DM 14.50

5—5 (1979): *Gases and Carbon in Metals (Thermodynamics, Kinetics, and Properties),* Part V: Group IV A Metals (1): Titanium (Ti), by H. Jehn, E. Fromm, and G. Hörz. 84 pages. — DM 9.—

5—6 (1979): *Gases and Carbon in Metals (Thermodynamics, Kinetics, and Properties),* Part VI: Group IV A Metals (2); Zirconium, Hafnium (Zr, Hf), by H. Jehn, H. Speck, E. Fromm, and G. Hörz. 80 pages. — DM 12.—

5—7 (1981): *Gases and Carbon in Metals (Thermodynamics, Kinetics, and Properties),* Part VII: Group V A Metals (1): Vanadium (V), by G. Hörz, H. Speck, E. Fromm and H. Jehn. 70 pages. — DM 9.50

5—8 (1981): *Gases and Carbon in Metals (Thermodynamics, Kinetics, and Properties),* Part VIII: Group V A Metals (2): Niobium (Nb), by G. Hörz, H. Speck, E. Fromm, and H. Jehn, 120 pages. — DM 16.—

5—9 (1981): *Gases and Carbon in Metals (Thermodynamics, Kinetics, and Properties),* Part IX: Group V A Metals (3): Tantalum (Ta), by G. Hörz, H. Speck, E. Fromm, and H. Jehn. 78 pages. — DM 14.—

5—10 (1980): *Gases and Carbon in Metals (Thermodynamics, Kinetics, and Properties),* Part X: Group VI A Metals (1): Chromium, Tungsten (Cr,W), by H. Jehn, H. Speck, E. Fromm, and G. Hörz. 66 pages. — DM 9.—

5—11 (1980): *Gases and Carbon in Metals (Thermodynamics, Kinetics, and Properties),* Part XI: Group VI A Metals (2): Molybdenum (Mo), by H. Jehn, H. Speck, E. Fromm, and G. Hörz. 56 pages. — DM 9.—

5—12 (1980): *Gases and Carbon in Metals (Thermodynamics, Kinetics, and Properties),* Part XII: Group VII A Metals: Manganese, Technetium, Rhenium (Mn, Tc, Re), by H. Jehn, H. Speck, E. Fromm, and G. Hörz. 32 pages. — DM 7.50

5—13 (1981): *Gases and Carbon in Metals (Thermodynamics, Kinetics, and Properties),* Part XIII: Ferrous Metals (1): Iron-Hydrogen (Fe-H), by E. Fromm, H. Speck, H. Jehn, and G. Hörz. 108 pages. — DM 16.50

5—14 (1981): *Gases and Carbon in Metals (Thermodynamics, Kinetics, and Properties),* Part XIV: Ferrous Metals (2): Iron-Carbon (Fe-C), by E. Fromm, H. Speck, H. Jehn, and G. Hörz. 122 pages. — DM 23.50

5—15 (1982): *Gases and Carbon in Metals (Thermodynamics, Kinetics, and Properties),* Part XV: Ferrous Metals (3) Iron-Nitrogen (Fe-N), by E. Fromm, H. Jehn, W. Hehn, and H. Speck. 94 pages. — DM 17.50

6—1 (1976): *Shapes of Beta Spectra,* by H. Behrens and L. Szybisz. 52 pages. A compilation of all measured beta spectrum shape factors. — DM 4.50

7—1 (1977): *Compilation of Pion Photoproduction Data,* by D. Menze, W. Pfeil, and R. Wilcke. 318 pages. Tables of differential and integrated cross sections, recoil nucleon polarizations, and polarized photon and polarized target asymmetries for photon energies up to 18 GeV. — DM 19.—

8—1 (1977): *Optical Properties of Some Insulators in the Vacuum Ultraviolet Region,* by. R.-P. Haelbich, M. Iwan, and E. E. Koch. XII, 186 pages. Tables of available optical properties and curves of absorption spectra for alkali halides and rare gas solids in the wavelength range 2000 to 2 Å. — DM 16.—

9—1 (1977):   *Bibliography of Microwave Spectroscopy. 1945—1975,* by B. Starck, R. Mutter, C. Spreter, K. Kettemann, A. Boggs, M. Botskor, and M. Jones. X, 664 pages.   DM 18.—

10—1 (1978):   *Graphs of Neutron Capture Cross Sections of Fission Product Isotopes from FPLIB-65/ENDF/B-IV,* by M. Mattes. XXII, 276 pages.   DM 26.50

11—1 (1978):   *Nucleon-Nucleon Scattering Data,* by J. Bystricky and F. Lehar. In two parts, 782 pages.   DM 62.—
—1 (Pt. I):   Part I, 202 pages.
—1 (Pt. II):   Part II, 580 pages.

11—2 (1981):   *Nucleon-Nucleon Scattering Data. Summary Tables,* 1981 Edition, by J. Bystricky and F. Lehar. 232 pages. This issue supersedes the previous edition published as No. 11—1 (Pt. I) (1978).   DM 34.40

11—3 (1981):   *Nucleon-Nucleon Scattering Data. Detailed Tables,* (Supplement 1), by J. Bystricky and F. Lehar. 132 pages. Supplement to No. 11—1 (Pt. II) (1978).   DM 29.—

12—1 (1979):   *Handbook of Pion-Nucleon Scattering,* by G. Höhler, F. Kaiser, R. Koch, and E. Pietarinen. 450 pages.   DM 35.—

13—1 (1979):   *Evaluation of the Cross-Sections for the Reactions $^{24}Mg(n,p)\,^{24}Na$, $^{64}Zn(n,p)\,^{64}Cu$, $^{63}Cu(n,2n)\,^{62}Cu$ and $^{90}Zr(n,2n)\,^{89}Zr$,* by S. Tagesen, H. Vonach, and B. Strohmaier. 154 pages.   DM 16.—

13—2 (1980):   *Evaluation of the Cross-Sections for the Reactions $^{19}F(n,2n)^{18}F$, $^{31}P(n,p)^{31}Si$, $^{93}Nb(n,n')^{93m}Nb$ and $^{103}Rh(n,n')^{103m}Rh$,* by B. Strohmaier, S. Tagesen, and H. Vonach. 134 pages.   DM 16.—

13—3 (1981):   *Evaluation of the Cross-Sections for the Reaction $^{27}Al(n,\alpha)^{24}Na$,* by S. Tagesen and H. Vonach. 76 pages.   DM 14.—

14—1 (1979):   *Compilation of Data from Hadronic Atoms,* by H. Poth. 138 pages.   DM 8.50

15 (1979):   *Karlsruhe Charged Particle Reaction Data Compilation,* by H. Munzel, H. Klewe-Nebenius, J. Lange, G. Pfennig, and K. Hemberle. Loosleaf collection, XLVI, 1652 pages.   DM 149.—
—1   Entry 1—35. XIV, 762 pages.
—2   Entry 36—80. XIV, 818 pages.
—Index   Index to Entries 1—35 and 36—80. XVIII, 72 pages.

16—1 (1979):   *The Oxygen Framework in Garnet and its Occurrence in the Structures of $Na_3Al_2Li_3F_{12}$, $Ca_3Al_2(OH)_{12}$, $RhBi_4$ and Hg_3TeO_6,* by E. Hellner, R. Gerlich, E. Koch, and W. Fischer. 40 pages.   DM 7.50

16—2 (1981):   *The Homogeneous Frameworks of the Cubic Crystal Structures,* by E. Hellner, E. Koch, and A. Reinhardt. 78 pages.   DM 45.—

17—1 (1981):   *Compilation of Experimental Values of Internal Conversion Coefficients and Ratios for Nuclei with $Z \lesssim 60$,* by H. H. Hansen. 274 pages.   DM28.—

18—1 (1981):   *Optical Properties of Metals,* Part I: The Transition Met Metals (0.1 ⩽ hv ⩽ 500 eV) Ti, V, Cr, Mn, Fe, Co, Ni, Zr, Nb, Mo, Ru, Rh, Pd, Hf, Ta, W, Re, Os, Ir, Pt), by J. H. Weaver, C. Krafka, D. W. Lynch, and E. E. Koch. XII, 344 pages.   DM 33.—

18—2 (1981):   *Optical Properties of Metals,* Part II: Noble Metals, Aluminium, Scandium, Yttrium, the Lanthanides and the Actinides (0.1 ⩽ hv ⩽ 500 eV) (Cu, Ag, Au; Al; Sc; Y; La, Ce, Pr, Nd, Sm, Eu, Gd, Tb, Dy, Ho, Er, Tm, Yb, Lu; Th, U, Am), by J. H. Weaver, C. Krafka, D. W. Lynch, and E. E. Koch. XIV, 320 pages.   DM 35.—

21—1 (1982):   *Bibliography of Gas Phase Electron Diffraction 1930—1979,* by J. Buck, M. Maier, R. Mutter, U. Seiter, C. Spreter, B. Starck, J. Hargittai, O. Kennard, D. G. Watson, A. Lohr, T. Pirzadeh, H. G. Schirdewahn, and Z. Maier. 714 pages.   DM 65.—

22—1 (1981):   *Phenomenological Analyses of Nucleon-Nucleon Scattering,* by P. Kroll. 280 pages.   DM 32.50

23—1 (1982):   *Temperature Dependence of Thermal Neutron Scattering Cross Sections for Hydrogen Bound in Moderators,* by Jürgen Keinert, 135 pages.   DM 28.00

INDEX

wires, resistance of, tables for several metals, F-123—124
Alpha (α)-particle, definition, F-69
Alpha-ray, see Alpha particle
Alternating current, definition and equations, F-69
Altitudes with the barometer, formulae, F-69
Aluminon (solution), preparation of, D-136
Aluminum (Al 13)
  atomic number, B-1, 4
  atomic weight, B-4
  boiling point, B-4
  electronic configuration, B-1
  gravimetric factors, logs of, B-159
  history, B-4
  line spectra of, E-192—193
  melting point, B-4
  optical properties, E—369
  production, general methods, B-4
  properties, in general, B-4
  properties, specific, see under Elements, Inorganic compounds, Metals
  specific gravity, B-4
  symbol, B-1, 4
  transition probabilities, E-321
  uses, B-4
  valence, B-4
  wire, cross-section and mass, F-122—123
  wire table, F-121
Americium (Am 95)
  atomic number, B-1, 5
  atomic weight, B-5
  boiling point, B-5
  electronic configuration, B-1
  history, B-5
  line spectra of, E-193—194
  melting point, B-5
  price, approximate, B-5
  production, general methods, B-5
  properties in general, B-5
  properties, specific, see under Elements, Inorganic compounds
  specific gravity, B-5
  symbol, B-1, 5
  uses, B-5
  valence, B-5
Amino acids, see also Organic compounds
  density of crystallines, C-724
  ionization constants in various solvents, C-719—722
  pH values, table, C-756
  solubilities in various solvents, C-723—724
  specific rotation, table, C-720—721
Amino sugars, see Carbohydrates
Ammonia, see also Inorganic compounds
  concentrative properties of, in varying concentration, D-226
  cryogenic properties, B-362
  dissociation constants, 0° to 50°C, D-169
  thermodynamic properties, E-25
Ammoniameter, scale units, F-3
Ammonium, gravimetric factors and their logarithms, B-159—160
Ammonium chloride, see also Inorganic compounds
  concentrative properties of, in varying concentration, D-226—227
Ammonium hydroxide, see also Inorganic compounds
  concentrated, composition of, D-170
  concentrative properties of, in varying concentration, D-226
Ammonium salts
  lattice energy, D-102—103
  tetra alkyl
    lattice energy, D-112—113
Ammonium sulfate, see also Inorganic compounds
  concentrative properties of, in varying concentration, D-227
Amorphous, definition, F-70

Ampere, definition, F-70, 77
Ampere, international, definition, F-77
Ampere-turn, definition, F-70, 77
Ampere's law, definition and equations, F-70
Ampere's rule, definition, F-70
Amplitude, definition, F-70
Amplitude modulation, definition and equation, F-70
Amplitude of wave, definition, F-70
AMU, definition, F-71
Anabolic agents
  formulae, chemical, C-709—712
  properties, C-709—712
  uses, C-703
Analysing crystals, lattice spacing, E-176
Analytical reagents, organic, D-142—145
Androgens
  formulae, chemical, C-709—712
  properties, C-709—712
  uses, C-703
Anesthetic agents, local
  formulae, chemical, C-718
  properties, C-718
  uses, C-703
Angle, definition, F-70
Angle values
  bond lengths between elements, F-173
  bond lengths of chemical compounds, F-173—175
Angles in χ radians, natural trigonometric functions for, A-17—18
Angles in π radians, natural trigonometric functions, sine, tangent, cotangent, cosine, A-20
Angstrom, definition, F-70
Angular acceleration, definition and equations, F-70
Angular aperture, definition, F-70
Angular functions, relation in terms of one another, A-22
Angular harmonic motion, definition, F-70
Angular momentum, definition, F-70
Angular momentum, isotopes, table, B-233—316
Angular radius of halos and rainbows, F-162
Angular velocity, definition and equation, F-70
Anhydride, definition, F-70
Animal fats, see Fats and oils
Animal protein, see Protein, animal
Anion, definition, F-70
Anisotrophy, magnetic, definition, F-93—94
Anisotropic, definition, F-70
Anode, definition, F-70
Anonymous group, definition, F-331, 338
Antiallergic agents, see Adrenal corticosteroids
Antidiuretic agents
  formulae, chemical, C-718
  properties, C-718
  uses, C-703
Antiferromagnetic compounds, properties, E-107
Antiferromagnetic materials, definition, F-70
Antiinflammatory agents, see Adrenal corticosteroids
Anti-matter
  definition, F-70
  in primary cosmic ray beam, F-166
Antimony (Sb 51)
  atomic number, B-1, 5
  atomic weight, B-5
  boiling point, B-5
  electronic configuration, B-1
  gravimetric factors, logs of, B-160—161
  history, B-5
  line spectra of, E-194
  melting point, B-5
  production, general methods, B-5
  properties in general, B-5
  properties, specific, see under Elements, Inorganic compounds, Metals
  specific gravity, B-5
  symbol, B-1, 5
  toxicity, B-5
  uses, B-5
  valence, B-5

Antimony-sulfur compounds, electrical resistivities of some, B—218
Antineutrino, definition, F-96
Antinode, definition, F-105
Anti-particle, definition, F-70
Antirheumatic agents, see Adrenal corticosteroids
Aperture ratio, definition, F-70
Apochromat, definition, F-70
Aqua regia, preparation of, D-135
Aqueous solutions
  concentrative properties
    conversion tables, D-223
    organic and inorganic substances, tables of values, D-225—272
  condosity, table, D-273—274
  freezing point depression, table, D-273—274
  molarity, table, D-273—274
  osmolality, table, D-273—274
  osmosity, table, D-273—274
  specific conductance, table, D-273—274
Aqueous tension, inorganic compounds in saturated solution, E-42
Aqueous vapor, pressure, see Water (aqueous) vapor pressure
Archimedes number, definition, F-331
Archimedes principle, definition and formula, F-71—72
Area, unit, of definition, F-71
Argon (Ar 18)
  atomic number, B-1, 5
  atomic weight, B-5
  boiling point, B-5
  density, B-5
  eddy diffusion coefficient, as a function of geometric altitude, F-158
  electronic configuration, B-1
  history, B-5
  line spectra of, E-195—196
  melting point, B-5
  molecular diffusion coefficient, as a function of geometric altitude, F-158
  production, general methods, B-5
  properties in general, B-5
  properties, specific, see under Elements, Inorganic compounds, Gases
  symbol, B-1, 5
  transition probabilities, E-321—324
  uses, B-5
Arrhenius group, definition, F-331
Arrhenius theory of electrolytic dissociation, definition, F-71
Arsenic (As 33)
  atomic number, B-1, 5
  atomic weight, B-5
  boiling point, B-6
  electronic configuration, B-1
  gravimetric factors, logs of, B-160—161
  history, B-6
  lattice energy, D-102
  line spectra of, E-196
  melting point, B-6
  production, general methods, B-6
  properties in general, B-6
  properties, specific, see under Elements, Inorganic compounds
  specific gravity, B-6
  symbol, B-1, 5
  toxicity, B-6
  transition probabilities, E-324
  uses, B-6
  valence, B-6
Arsenic-sulfur compounds, electrical resistivities of some, B—218
Arsenides, electrical resistivities of some, B—218
Association, degree of, definition, F-78
Astatine (At 85)
  atomic number, B-1, 6
  atomic weight, B-6

MELTING AND BOILING POINTS, AND ATOMIC WEIGHTS OF THE ELEMENTS

Based on the Assigned Relative Atomic Mass of $^{12}C = 12$

The following values apply to elements as they exist in materials of terrestrial origin and to certain artificial elements. When used with the footnotes, they are reliable to ±1 in the last digit, or ±3 if that digit is in small type.

Name	Symbol	Atomic number	Atomic weight	Melting point, °C	Boiling point, °C
Actinium[k]	Ac	89	227.028	1,050	3,200 ± 300
Aluminum	Al	13	26.98154[b]	660.37	2,467
Americium	Am	95	(243)	994 ± 4	2,607
Antimony	Sb	51	121.75*	630.74	1,750
Argon[h,i]	Ar	18	39.948*[b,c,d,g]	−189.2	−185.7
Arsenic (gray)	As	33	74.9216[a]	817(28 atm)	613(sub.)
Astatine	At	85	(210)	302	337
Barium[i]	Ba	56	137.33	725	1,640
Berkelium	Bk	97	(247)	−	−
Beryllium	Be	4	9.01218[a]	1,278 ± 5	2,970(5 mm)
Bismuth	Bi	83	208.9804[a]	271.3	1,560 ± 5
Boron[h,j]	B	5	10.81[c,d,e]	2,300	2,550(sub.)
Bromine	Br	35	79.904[c]	−7.2	58.78
Cadmium[i]	Cd	48	112.41	320.9	765
Calcium[i]	Ca	20	40.08	839 ± 2	1,484
Californium	Cf	98	(251)	−	−
Carbon[h,l]	C	6	12.011[b,d]	3,652(sub)	l
Cerium[i]	Ce	58	140.12	798 ± 3	3,257
Cesium	Cs	55	132.9054[c]	28.40 ± 0.01	669.3
Chlorine	Cl	17	35.453[c]	−100.98	−34.6
Chromium	Cr	24	51.996[c]	1,857 ± 20	2,672
Cobalt	Co	27	58.9332[a]	1,495	2,870
Copper[h]	Cu	29	63.546*[c,d]	1,083.4 ± 0.2	2,567
Curium	Cm	96	(247)	1,340 ± 40	
Dysprosium	Dy	66	162.50*	1,409	2,335
Einsteinium	Es	99	(254)	−	−
Erbium	Er	68	167.26*	1,522	2,510
Europium[i]	Eu	63	151.96	822 ± 5	1,597
Fermium	Fm	100	(257)	−	−
Fluorine	F	9	18.998403[a]	−219.62	−188.14
Francium	Fr	87	(223)	(27)	(677)
Gadolinium[i]	Gd	64	157.25*	1,311 ± 1	3,233
Gallium	Ga	31	69.72	29.78	2,403
Germanium	Ge	32	72.59*	937.4	2,830
Gold	Au	79	196.9665[a]	1,064.43	3,080
Hafnium	Hf	72	178.49*	2,227 ± 20	4,602
Helium[i]	He	2	4.00260[b]	−272.2[26 atm]	−268.934
Holmium	Ho	67	164.9304[a]	1,470	2,720
Hydrogen[h]	H	1	1.0079[b,d]	−259.14	−252.87
Indium[i]	In	49	114.82	156.61	2,080
Iodine	I	53	126.9045[a]	113.5	184.35
Iridium	Ir	77	192.22*	2,410	4,130
Iron	Fe	26	55.847*	1,535	2,750
Krypton[i,j]	Kr	36	83.80	−156.6	−152.30 ± 0.10
Lanthanum[i]	La	57	138.9055*[b]	920 ± 5	3,454
Lawrencium	Lr	103	(260)	−	−
Lead[h,j]	Pb	82	207.2[d,g]	327.502	1,740
Lithium[h,i,j]	Li	3	6.941*[c,d,e]	180.54	1,342
Lutetium	Lu	71	174.967 ± 0.003	1,656 ± 5	3,315
Magnesium[i]	Mg	12	24.305[c]	648.8 ± 0.5	1,090
Manganese	Mn	25	54.9380[a]	1,244 ± 3	1,962
Mendelevium	Md	101	(257)	−	−
Mercury	Hg	80	200.59*	−38.87	356.58
Molybdenum	Mo	42	95.94	2,617	4,612
Neodymium[i]	Nd	60	144.24*	1,010	3,127
Neon[j]	Ne	10	20.179*[c]	−248.67	−246.048
Neptunium[k]	Np	93	237.0482[b]	640 ± 1	3,902
Nickel	Ni	28	58.70	1,453	2,732
Niobium (Columbium)	Nb	41	92.9064[a]	2,468 ± 10	4,742